CONCISE ENCYCLOPEDIA OF POLYMER SCIENCE AND ENGINEERING

Jacqueline I. Kroschwitz, *Executive Editor*

WILEY

A Wiley-Interscience Publication

John Wiley & Sons

New York / Chichester / Weinheim / Brisbane / Singapore / Toronto

This text is printed on acid-free paper.

Paper edition published 1998.

Library of Congress Cataloging in Publication Data:

Concise encyclopedia of polymer science and engineering / Jacqueline I. Kroschwitz, executive editor.
p. cm.
"Contains all of the subjects covered in the 17 main volumes and the supplement and index volumes of the . . . 2nd edition of the Encyclopedia of polymer science and engineering"—Pref.
"A Wiley-Interscience publication."
Includes bibliographical references.
ISBN 0-471-51253-2 (cloth); ISBN 0-471-31856-6 (paper)
1. Polymers—Dictionaries. I. Kroschwitz, Jacqueline I.
II. Encyclopedia of polymer science and engineering.
TP1087.C66 1990 89-70674
668.9′03—dc20 CIP

Printed in the United States of America

10 9 8 7 6 5 4 3 2 1

CONCISE ENCYCLOPEDIA OF POLYMER SCIENCE AND ENGINEERING

PREFACE

This compact desk reference contains all of the subjects covered in the 17 main volumes and the Supplement and Index Volumes of the world-renowned 2nd Edition of the *Encyclopedia of Polymer Science and Engineering*. The articles have been condensed by professional science writers, reviewed for accuracy by the original authors or their colleagues, and updated where necessary. This distillation, skillfully prepared to retain the key data, tables, and factual matter of the original, is a complete and self-contained encyclopedia. It is designed to serve as a ready-reference guide for students, scientists, engineers, and technologists seeking answers to questions on any aspect of polymer science and engineering.

This one million word version, like the ten million word larger work, provides both SI and common units, carefully selected key references for each article, and hundreds of tables, charts, figures, and graphs. Coverage includes every important sector, such as: polymeric materials, natural and synthetic; polymer properties, i.e. molecular, chemical, physical, electrical, mechanical, thermal and biological properties, morphology, compatibility, and stability; synthesis and reactions; characterization and analytical methods; physical processes; engineering; polymer processing; product fabrication; test methods; uses in adhesives, coatings, films, fibers, elastomers, plastics composites, and occurrence in natural materials; historical perspective; and economics.

Information from the Supplement and Index Volumes has been included alongside the main-volume articles. Although specific reference to the articles in the original work is not always made, it should be understood that further details, specific bibliographic citations, and much wider coverage of any subject may be obtained by referring to the 19-volume edition.

The editors have also carefully preserved the tradition of citing related articles in the text (*see* and *see also* citations), as well as secondary entries or cross references which cite the synonym or entry term where a subject can be located. An Index of key terms provides further access to the contents where desired.

The complete Second Edition of the *Encyclopedia* has been called a "landmark publication" and a "milestone in polymer science." This concise version presents the essence of this monumental work in a useful daily tool comparable to a handbook or dictionary from which the comprehensive, authoritative, and lucidly written data become instantly available. This new encyclopedia has been developed over five years by the writers, authors, and Wiley editorial staff. Every effort has been made by the editors to provide a work that is unsurpassed in quality and accuracy. We hope we have succeeded and that this work will serve your reference needs for years to come.

JACQUELINE I. KROSCHWITZ
Executive Editor

EDITORIAL STAFF

CONTRIBUTORS

Visvaldis Abolins, *General Electric Company, Selkirk, New York,* Poly(phenylene ether)
E. T. Adams, Jr., *Texas A&M University, College Station, Texas,* Osmometry
S. Ahuja, *Xerox Corporation, Webster, New York,* Xerographic materials
Takuzo Aida, *University of Tokyo, Tokyo, Japan,* Living polymer systems, epoxides and lactones, new developments
David Albert, *Anatrace, Inc., Maumee, Ohio,* Enzymes, immobilized
Robert L. Albright, *Rohm and Haas Company, Philadelphia, Pennsylvania,* Ion-exchange polymers
Peter Aldrich, *The Dow Chemical Company, Midland, Michigan,* Styrene polymers, dielectric properties
Harry R. Allcock, *Pennsylvania State University, University Park, Pennsylvania,* Polyphosphazenes
Richard B. Allen, *General Electric Company, Selkirk, New York,* Compatibility
Judith Allison, *Rohm & Haas, Bristol, Pennsylvania,* Processing aids
Harvey Alter, *U.S. Chamber of Commerce, Washington, D.C.,* Disposal and reuse of plastics
Gerald D. Andrews, *E.I. du Pont de Nemours & Co., Inc., Parkersburg, West Virginia,* Chlorosulfonated polyethylene
Mark P. Andrews, *AT&T Bell Laboratories, Murray Hill, New Jersey,* Metal-vapor synthesis
J. M. Antonucci, *National Bureau of Standards, Washington, D.C.,* Dental applications
David N. Armentrout, *The Dow Chemical Company, Midland, Michigan,* Styrene polymers, properties, degradation
Eugene Armour, *Eastman Kodak Company, Rochester, New York,* Photographic applications
Jett C. Arthur, Jr., *Consultant, Metairie, Louisiana,* Cellulose, graft copolymers; Cotton
T. M. Austin, *Union Carbide Corporation, South Charleston, West Virginia,* Xylylene polymers
David Aycock, *General Electric Company, Selkirk, New York,* Poly(phenylene ether)
Hartwig C. Bach, *Monsanto Company, Pensacola, Florida,* Acrylic fibers
J. K. Backus, *Mobay Corporation, Pittsburgh, Pennsylvania,* Polyurethanes
Stanley K. Baczek, *The Lubrizol Corporation, Wickliffe, Ohio,* Petroleum additives
P. Bahadur, *South Gujurat University, Surat, India,* Block copolymers
David G. Bailey, *Eastern Regional Research Center, U.S. Department of Agriculture, Philadelphia, Pennsylvania,* Leather
William J. Bailey,† *University of Maryland, College Park, Maryland,* Ladder and spiro polymers
John K. Baird, *Kelco Division of Merck Company, Inc., San Diego, California,* Xanthan
Lawrence E. Ball, *The Standard Oil Company, Cleveland, Ohio,* Methacrylonitrile polymers
Bryan Ballantyne, *Union Carbide Corporation, Charleston, West Virginia,* Toxicology
F. J. Balta-Calleja, *Institute for Structure of Matter, CSIC, Madrid, Spain,* Hardness
C. H. Bamford, *Royal Liverpool Hospital, Liverpool, United Kingdom,* Radical polymerization
Ronald Baney, *Dow Corning Corporation, Midland, Michigan,* Preceramic polymers
Eugene S. Barabas, *Consultant, Watchung, New Jersey,* N-Vinyl amide polymers
George Barany, *University of Minnesota, Minneapolis, Minnesota,* Polypeptide synthesis, solid-phase method

Pier Camillo Barbe, *Himont Italia, Centro Recerche Guilo Natta, Ferrara, Italy,* Propylene polymers
J. W. Barlow, *University of Texas, Austin, Texas,* Polymer blends
D. R. Bassett, *Union Carbide Corporation, South Charleston, West Virginia,* Xylylene polymers
O. A. Battista, *The O. A. Battista Research Institute, Fort Worth, Texas,* Cellulose, microcrystalline
William Bauer, Jr., *Rohm and Haas Company, Spring House, Pennsylvania,* Acrylic and methacrylic acid polymers
Gary A. Baum, *The Institute of Paper Chemistry, Appleton, Wisconsin,* Paper
David L. Beach, *Chevron Chemical Co., Kingwood, Texas,* Ethylene polymers, high density polyethylene; Olefin polymers
W. F. Beach, *Union Carbide Corporation, South Charleston, West Virginia,* Xylylene polymers
Ellen C. Bellantoni, *American Optical Corporation, Framingham, Massachusetts,* Contact lenses
James N. BeMiller, *Purdue University, West Lafayette, Indiana,* Glycogen; Gums, industrial
John T. Bendler, *General Electric Company, Schenectady, New York,* Transitions and relaxations
Harriet Benson, *ALZA Corporation, Palo Alto, California,* Drug applications
Jack H. Benson, *The Dow Chemical Company, Midland, Michigan,* Styrene polymers, uses
Charles A. Berglund, *The Dow Chemical Company, Midland, Michigan,* Styrene polymers, dynamic mechanical properties
L. Berkowski, *BASF Corporation, Wyandotte, Michigan,* Interpenetrating polymer networks
Guy C. Berry, *Carnegie-Mellon University, Pittsburgh, Pennsylvania,* Light scattering
T. W. Bethea, *Firestone Tire and Rubber Company, Akron, Ohio,* Butadiene polymers
L. Bevan, *North East London Polytechnic, Essex, United Kingdom,* Micromechanical measurements
R. S. Bhakuni, *The Goodyear Tire & Rubber Co., Akron, Ohio,* Tires
Joseph A. Biesenberger, *Stevens Institute of Technology, Hoboken, New Jersey,* Devolatilization
Norbert M. Bikales, *National Science Foundation, Washington, D.C.,* Nomenclature
Kurt Binder, *University of Mainz, Mainz, Federal Republic of Germany,* Diffusion, polymer–polymer
J. D. Birchall, *Imperial Chemical Industries, Runcorn, Cheshire, United Kingdom,* Cement additives
Mukul Biswas, *Indian Institute of Technology, Kharagpur, India,* Vinyl ether polymers
T. N. Bitzer, *Hexcel Corporation, Dublin, California,* Honeycombs
D. C. Blackley, *London School of Polymer Technology, London, United Kingdom,* Latices
R. R. Blanchard, *The Dow Chemical Co., Plaquemine, Louisiana,* Ethylene polymers, chlorinated polyethylene
R. W. Blevins, *Eastman Kodak Company, Rochester, New York,* Lithographic resists
Thomas P. Blomstrom, *Monsanto Chemical Company, Springfield, Massachusetts,* Vinyl acetal polymers
C. D. Blue, *Mobay Corporation, New Martinsville, West Virginia,* Polyurethanes
J. Bock, *Exxon Research, Annandale, New Jersey,* Water-soluble polymers
R. W. Body, *Hercules, Incorporated, Wilmington, Delaware,* 1,2-Epoxide polymers, polyether elastomers
Herman V. Boenig, *Research Institute of Plasma Chemistry and Technology, Carlsbad, California,* Plasma polymerization
R. T. Bogan, *Eastman Kodak Company, Kingsport, Tennessee,* Cellulose esters, inorganic; Cellulose esters, organic, survey; Cellulose esters, organic, plastics
G. Bolder, *Institut für Kunststoffverarbeitung of the RWTH Aachen, Aachen, Federal Republic of Germany,* Processing, computer control
Armen Boranian, *Consultant, Cadiz, Kentucky,* Flooring materials
John K. Borchardt, *Shell Development Company, Houston, Texas,* Oilfield applications
F. A. Bovey, *AT&T Bell Laboratories, Murray Hill, New Jersey,* Nuclear magnetic resonance
Scott D. Boyce, *The PQ Corporation, Lafayette Hill, Pennsylvania,* Silicates
P. M. Boyd, *Mobay Corporation, Pittsburgh, Pennsylvania,* Polyurethanes
Raymond F. Boyer, *Michigan Molecular Institute, Midland, Michigan,* Transitions and relaxations, $T_{\ell\ell}$ and related liquid-state transitions and relaxations
D. G. Brady, *Phillips Petroleum Company, Bartlesville, Oklahoma,* Poly(arylene sulfide)s
Robert F. Brady, Jr., *U.S. Naval Research Laboratory, Washington, D.C.,* Marine applications
Donald Brasure, *E.I. du Pont de Nemours & Co., Inc., Buffalo, New York,* Vinyl fluoride polymers
G. M. Brauer, *National Bureau of Standards, Washington, D.C.,* Dental applications
Dietrich Braun, *Deutsches Kunststoff-Institut Darmstadt, Federal Republic of Germany,* Identification, plastics
David S. Breslow, *D. S. Breslow Associates, Wilmington, Delaware,* Ethylenesulfonic acid polymers
George E. F. Brewer, *Coating Consultants, Birmingham, Michigan,* Coatings, electrodeposition
R. J. Brewer, *Eastman Kodak Company, Kingsport, Tennessee,* Cellulose esters, inorganic; Cellulose esters, organic, survey; Cellulose esters, organic, plastics
David Briggs, *ICI Petrochemicals and Plastics Division, Wilton, United Kingdom,* Surface analysis
Norman F. Brockmeier, *Amoco Chemicals Corporation, Naperville, Illinois,* Gas-phase polymerization
Aaron L. Brody, *Schotland Business Research, Inc., Princeton, New Jersey,* Packaging materials, flexible materials; Packaging materials, rigid containers
S. B. Brown, *General Electric Research and Development Center, Schenectady, New York,* Reactive extrusion
Robert A. Bubeck, *The Dow Chemical Company, Midland, Michigan,* Styrene polymers, environmental stress cracking
Timothy J. Bublick, *The DeVilbiss Company, Toledo, Ohio,* Coating methods, spray coating
Alan Buckley, *Celanese Research Center, Summit, New Jersey,* Polybenzimidazoles
H. Henning Bunge, *Dayglo Corporation, Cleveland, Ohio,* Dyeing
M. J. Bunten, *European Vinyl Corporation(UK) Ltd., Runcorn, United Kingdom,* Vinyl chloride polymers, polymerization; emulsion polymerization and plastisols
K. A. Burgess, *Consultant to the Columbian Chemicals Co., Swartz, Louisiana,* Carbon
Frank Burkett, *The Dow Chemical Company, Midland, Michigan,* Styrene polymers, processing, injection molding
W. K. Busfield, *Griffith University, Brisbane, Australia,* Polymerization, thermodynamics
George B. Butler, *University of Florida at Gainesville, Florida,* Cyclopolymerization
Walter C. Buzanowski, *The Dow Chemical Company, Midland, Michigan,* Styrene polymers, analytical methods
S. Bywater, *National Research Council of Canada, Ottawa, Canada,* Anionic polymerization
Israel Cabasso, *State University of New York-ESF at Syracuse, Syracuse, New York,* Membranes
Mukerrem Cakmak, *University of Akron, Akron, Ohio,* Orientation; Orientation processes
F. J. Cama, *Mobay Corporation, Pittsburgh, Pennsylvania,* Polyurethanes
Charles M. Campbell, *Tradewinds Technologies Inc., Phoenix, Arizona,* Product design
G. A. Campbell, *Eastman Kodak Company, Rochester, New York,* Photographic applications
Françoise Candau, *Institut Charles Sadron, Strasbourg, France,* Microemulsion polymerization

Numa J. Capiati, *PIDCOP, Bahia Blanca, Argentina,* Solid-state extrusion
D. J. Carlsson, *National Research Council of Canada, Ottawa, Canada,* Degradation
Dewey K. Carpenter, *Louisiana State University, Baton Rouge, Louisiana,* Solution properties
Robert Carpenter, *Windsor Plastics, Evansville, Indiana,* Decorating systems
Robert E. Carpenter, *Ashland Chemical Company, Bartow, Florida,* Polyesters, water-extended
Charles E. Carraher, *Florida Atlantic University, Boca Raton, Florida,* Education; Organometallic polymers; Polymers from renewable sources
Don E. Carter, *Monsanto Company, St. Louis, Missouri,* Plant design
Antonio Casale, *Snia Tecnopolimeri SpA, Milan, Italy,* Mechanochemical reactions
Richard J. Cella, *E.I. du Pont de Nemours & Co., Inc., Wilmington, Delaware,* Polyesters, elastomeric
Sebastiano Cesca, *ENICHEM SpA, Milan, Italy,* Isomerization polymerization
Gerit Challa, *Gronigen University, Gronigen, The Netherlands,* Template polymerization
William B. Chamberlin, *The Lubrizol Corporation, Wickliffe, Ohio,* Petroleum additives
Ronald R. Chance, *Allied Corporation, Morristown, New Jersey,* Diacetylene polymers; Electrically conductive polymers
Grish Chandra, *Dow Corning Corporation, Midland, Michigan,* Preceramic polymers
S. Chandrasekaran, *Ansimont USA Inc., Morristown, New Jersey,* Chlorotrifluoroethylene polymers, chlorotrifluoroethylene homopolymer
J. H. Chapman, *Mobay Corporation, Pittsburgh, Pennsylvania,* Polyurethanes
Robert A. Charton, *Norton Company, Akron, Ohio,* Tubing
Ananda M. Chatterjee, *Shell Development Company, Houston, Texas,* Butene polymers
Stephen Z. D. Cheng, *University of Akron, Akron, Ohio,* Thermodynamic properties
Herve Cheradame, *Institut National Polytechnique de Grenoble, Saint Martin d'Heres, France,* Cationic polymerization
C. I. Chiriac, *Petru Poni Institute of Macromolecular Chemistry, Jassy, Romania,* Polyureas
Lionel Choplin, *Laval University, University City, Quebec, Canada,* Pullulan
Yoshiki Chujo, *Kyoto University, Kyoto, Japan,* Ring-opening polymerization
Francesco Ciardelli, *Universita di Pisa, Pisa, Italy,* Optically active polymers
Donald C. Clagett, *General Electric Company, Pittsfield, Massachusetts,* Engineering plastics
Edward S. Clark, *University of Tennessee, Knoxsville, Tennessee,* Elasticity, hard elastic behavior
Jay B. Class, *Hercules Research Center, Wilmington, Delaware,* Resins, natural
Nye Clinton, *Union Carbide Corporation, Tarrytown, New York,* 1,2-Epoxide polymers, ethylene oxide polymers and copolymers
Roger Clough, *Sandia National Laboratories, Albuquerque, New Mexico,* Radiation-resistant polymers
A. William M. Coaker, *Consultant, at present for BFGoodrich Company, Avon Lake, Ohio,* Calendering
K. J. Coeling, *The DeVilbiss Company, Toledo, Ohio,* Coating methods, spray coating
M. L. Coleman, *E.I. du Pont de Nemours & Co., Inc., Wilmington, Delaware,* Chloroprene polymers
M. M. Coleman, *The Pennsylvania State University, University Park, Pennsylvania,* Infrared-absorption spectroscopy
Gene Collard, *Alternative Screen Printing, Ipswich, Massachusetts,* Decorating, methods
Howard Colvin, *Goodyear Chemical Research & Development Division, Akron, Ohio,* Diisopropenylbenzene polymers
J. Ross Colvin, *National Research Council of Canada, Ottawa, Canada,* Cellulose, biosynthesis
W. A. Combellick, *Goodyear Tire & Rubber Company, Akron, Ohio,* Barrier polymers
Robert P. Conger, *Consultant, Park Ridge, New Jersey,* Flooring materials
Kelsey D. Cook, *University of Tennessee, Knoxville, Tennessee,* Mass spectrometry
Anthony R. Cooper, *Lockhead Missiles & Space Company, Inc., Palo Alto, California,* Molecular weight determination
H. W. Coover, Jr., *Eastman Kodak Company, Kingsport, Tennessee,* Acrylic and methacrylic ester polymers, 2-Cyanoacrylic ester polymers
Aubert Y. Coran, *Monsanto Chemical Company, Akron, Ohio,* Vulcanization
J. J. Coughlan, *HIMONT U.S.A., Incorporated, Wilmington, Delaware,* Ethylene polymers, ultrahigh molecular weight polyethylene
J. M. G. Cowie, *Heriot-Watt University, Edinburgh, United Kingdom,* Miscibility
James W. Crary, *E.I. du Pont de Nemours & Co., Inc., Wilmington, Delaware,* Acrylic elastomers, ethylene–acrylic elastomers
Thomas E. Creighton, *Medical Research Council Laboratory of Molecular Biology, Cambridge, United Kingdom,* Proteins
Bill M. Culbertson, *Ashland Chemical Co., Columbus, Ohio,* Aminimide polymers; Maleic and fumaric polymers
Benedict S. Curatolo, *The Standard Oil Company, Cleveland, Ohio,* Methacrylonitrile polymers
Brian Dalke, *The Dow Chemical Company, Midland, Michigan,* Styrene polymers, properties, general purpose polystyrene
R. C. Daly, *Eastman Kodak Company, Rochester, New York,* Lithographic resists
William H. Daly, *Louisiana State University, Baton Rouge, Louisiana,* Vinylcycloalkane polymers; Vinyl ketone polymers
D. K. Dandge, *New Mexico Institute of Mining and Technology, Socorro, New Mexico,* Nitroso polymers
I. M. Daniel, *Illinois Institute of Technology, Chicago, Illinois,* Composites, testing
James R. Daniel, *Purdue University, West Lafayette, Indiana,* Cellulose, structure and properties
Wiley E. Daniels, *Air Products and Chemicals, Allentown, Pennsylvania,* Vinyl ester polymers
G. W. Davis, *Celanese Corporation, Charlotte, North Carolina,* Polyesters, fibers
Robert L. Dawson, *E.I. du Pont de Nemours & Co., Inc., Wilmington, Delaware,* Ethylene polymers, modified polyethylenes, chlorosulfonated polyethylene
John M. Dealy, *McGill University, Montreal, Canada,* Energy conservation
Oliver D. Deex, *Monsanto Chemical Company, Springfield, Massachusetts,* Polymerization procedures, laboratory
Phillip T. Delassus, *The Dow Chemical Company, Midland, Michigan,* Styrene polymers, diffusion; Vinylidene chloride polymers
Morton M. Denn, *University of California, Berkeley, California,* Processing, modeling
M. J. Denton, *University of Leeds, Leeds, United Kingdom,* Fibers, manufacture
Kenneth DeVries, *University of Utah, Salt Lake City, Utah,* Electron-spin resonance
E. De Witte, *Koninklijk Instituut voor het Kunspatrimonium, Brussels, Belgium,* Fine arts
Martin Dexter, *CIBA-GEIGY Corporation, Ardsley, New York,* Antioxidants
Ashok K. Dhingra, *E.I. du Pont de Nemours & Co., Inc., Wilmington, Delaware,* Fibers, engineering
Mitch G. Dibbs, *The Dow Chemical Company, Midland, Michigan,* Styrene polymers, optical properties
Richard A. Dipert, *Carolina Power and Light Company, Raleigh, North Carolina,* Flammability
Giusseppe Di Silvestro, *Universita Degli Studi de Milano, Milan, Italy,* Polymerization in clathrates
Kenneth W. Doak, *Consultant, Murrysville, Pennsylvania,* Ethylene polymers, introduction; Ethylene polymers, low density polyethylene
Philip T. Dodge, *Quantum Chemical Company, USI Division, Rolling Meadows, Illinois,* Rotational molding
Julius E. Dohany, *Pennwalt Corporation, King of Prussia, Pennsylvania,* Vinylidene fluoride polymers
Yoshiharu Doi, *Tokyo Institute of Technology, Tokyo, Japan,* Living polymer systems, Ziegler-Natta polymerization

Thomas J. Dolce, *Celanese Engineering Resins Co., Summit, New Jersey,* Acetal resins

W. Doll, *Fraunhofer-Institut für Werktoffmechanik, Freiburg, Federal Republic of Germany,* Micromechanical measurements

A. Domb, *Massachusetts Institute of Technology, Cambridge, Massachusetts,* Polyanhydrides

L. G. Donaruma, *MacroChem Corporation, Woburn, Massachusetts,* Nitroso polymers

Adam Dreiblatt, *Werner Pfleiderer Corporation, Ramsey, New Jersey,* Screw design, multiscrew

Arthur Drelich, *Chicopee, Dayton, New Jersey,* Nonwoven fabrics, survey

Michael J. Drews, *Clemson University, Clemson, South Carolina,* Flammability

W. J. Dreyer, *California Institute of Technology, Pasadena, California,* Antibodies and antigens, polymer-bound

M. P. Dreyfuss, *Michigan Molecular Institute, Midland, Michigan,* Oxetane polymers; Tetrahydrofuran polymers

P. Dreyfuss, *Michigan Molecular Institute, Midland, Michigan,* Graft copolymers; Oxetane polymers; Tetrahydrofuran polymers

Andrezej Duda, *Polish Academy of Sciences, Lodz, Poland,* Sulfur-containing polymers

J. L. Duda, *The Pennsylvania State University, University Park, Pennsylvania,* Diffusion

Hardev S. Dugal, *The Institute of Paper Chemistry, Appleton, Wisconsin,* Paper

William R. Dunnavant, *Ashland Chemical Company, Columbus, Ohio,* Foundry resins

G. G. S. Dutton, *University of British Columbia, Vancouver, British Columbia, Canada,* Polysaccharides

J. L. Eakin, *Mobay Corporation, Pittsburgh, Pennsylvania,* Polyurethanes

Dwight B. Easty, *The Institute of Paper Chemistry, Appleton, Wisconsin,* Paper

Sina Ebnesajjad, *E.I. du Pont de Nemours & Co., Inc., Buffalo, New York,* Vinyl fluoride polymers

Herbert Edelstein, *Chesebrough-Ponds Research Laboratories, Trumbull, Connecticut,* Cosmetic applications

S. Ellerstein, *Morton Thiokol, Inc. Woodstock, Illinois,* Polysulfides

C. Michael Elliott, *Colorado State University, Fort Collins, Colorado,* Electrodes, polymer-modified

B. Erman, *Bogazici University, Istanbul, Turkey,* Birefringence

Lewis Erwin, *Kraft Company, Glenview, Illinois,* Mixing

David Evans, *Creative Pultrusion, Inc., Alum Park, Pennsylvania,* Pultrusion

James S. Falcone, Jr., *The PQ Corporation, Lafayette Hill, Pennsylvania,* Silicates

Ping Fang, *Drexel University, Philadelphia, Pennsylvania,* Inorganic fibers

Mario Farina, *Universita Degli Studi de Milano, Milan, Italy,* Polymerization in clathrates

Allan H. Fawcett, *The Queen's University of Belfast, United Kingdom,* Olefin–sulfur dioxide copolymers

Lewis J. Fetters, *Exxon Research and Engineering Company, Annandale, New Jersey,* Monodisperse polymers

Max Feughelman, *University of New South Wales, New South Wales, Australia,* Keratin

James A. Finnegan, *Arthur D. Little, Inc., Cambridge, Massachusetts,* Market evaluation

Daryle Fish, *State University of New York at Syracuse, Syracuse, New York,* Polyelectrolyte complexes

John M. Fisher, *E.I. du Pont de Nemours & Co., Inc., Wilmington, Delaware,* Computer-assisted design

Robert M. Fitch, *Louis Laboratories, Johnson Wax Company, Racine, Wisconsin,* Colloids

Paul J. Flory,[†] *Stanford University, Stanford, California,* Networks

F. B. Flynn, *R. T. Vanderbilt Company, Inc., Norwalk, Connecticut,* Biocides

Joseph H. Flynn, *Sci. Thermal Research & Data Analysis, Bethesda, Maryland,* Thermal analysis

Kendall F. Foley, *Hercules, Inc., Wilmington, Delaware,* Hydrocarbon resins

Terence Foster, *American Cyanamid Company, Stamford, Connecticut,* Mining applications, mineral processing

Frederick M. Fowkes, *Lehigh University, Bethlehem, Pennsylvania,* Acid–base interactions

Daniel W. Fox, *General Electric Company, Pittsfield, Massachusetts,* Compatibility

Bruce M. Foxman, *Brandeis University, Waltham, Massachusetts,* Coordination polymers

Karl F. Freed, *University of Chicago, Chicago, Illinois,* Statistical mechanics

W. J. Freeman, *Hercules, Inc., Wilmington, Delaware,* Characterization of polymers

D. Freitag, *Farbenfabriken Bayer AG, Bayerwerk, Federal Republic of Germany,* Polycarbonates

Gino J. Frisone, *Gencorp, Akron, Ohio,* Identification, elastomers

Jane E. Frommer, *Allied-Signal Corporation, Morristown, New Jersey,* Electrically conductive polymers

Rolf E. Funer, *ICI Americas, Inc., Wilmington, Delaware,* Electronics applications, electronic compositions

B. Lionel Funt, *Simon Fraser University, Burnaby, Vancouver, Canada,* Electrochemical initiation

J. Furukawa, *Kyoto University, Kyoto, Japan,* Copolymers, alternating

Barbara Gabrys, *Imperial College, London, United Kingdom,* Scattering

Steven D. Gagnon, *BASF Corporation, Wyandotte, Michigan,* 1,2-Epoxide polymers, propylene oxide and higher 1,2-epoxide polymers

Alessandro Gandini, *Institut National Polytechnique de Grenoble, Saint Martin d'Heres, France,* Cationic polymerization; Furan polymers

Subhash V. Gangal, *E.I. du Pont de Nemours & Co., Inc., Parkersburg, West Virginia,* Tetrafluoroethylene polymers

Richard G. Gann, *United States National Bureau of Standards, Gaithersburg, Maryland,* Flammability

John A. Gannon, *CIBA-GEIGY Corporation, Ardsley, New York,* Epoxy resins

John B. Gardner, *The Dow Chemical Company, Midland, Michigan,* Styrene polymers, monomers

B. C. Gates, *University of Delaware, Newark, Delaware,* Catalysts, polymer-supported

Russell Gaudiana, *Polaroid Corporation, Cambridge, Massachusetts,* Polyesters, mainchain aromatic

Roy H. Geiss, *IBM Research, San Jose, California,* Electron diffraction

Alan N. Gent, *The University of Akron, Akron, Ohio,* Adhesion and bonding, adhesion

Steve Gerbig, *Akzo Engineering Plastics, Inc., Evansville, Indiana,* Reinforced plastics, thermoplastics

Susan W. Gersten, *Stauffer Chemical Company, Elmsford, New York,* Coordination polymers

Frank P. Gerstle, Jr., *Sandia National Laboratories, Albuquerque, New Mexico,* Composites

D. S. Gibbs, *The Dow Chemical Company, Midland, Michigan,* Vinylidene chloride polymers

Harry W. Gibson, *Signal UOP Research Center, Des Plaines, Illinois,* Acetylenic polymers

S. A. Giddings, *Formica Corporation, Cincinnati, Ohio,* Laminates

John K. Gillham, *Princeton University, Princeton, New Jersey,* Curing

Ray Giornelli, *Shell Chemical Company, Atlanta, Georgia,* Furniture

Wolfgang G. Glasser, *Virginia Polytechnic Institute and State University, Blacksburg, Virginia,* Lignin

W. Brandt Goldsworthy, *Goldsworthy Engineering Inc., Torrance, California,* Composites, fabrication

Isaac Goodman, *University of Bradford, Bradford, West Yorkshire, United Kingdom,* Polyesters

Murray Goodman, *University of California at San Diego, La Jolla, California,* Polypeptides

Bernard Gordon III, *Pennsylvania State University, University Park, Pennsylvania,* Telomerization

R. L. Gordon, *Union Camp Corporation, Savannah, Georgia,* Polyamides from fatty acids

John A. Grates, *Celanese Engineering Resins Co., Summit, New Jersey,* Acetal resins

Michael Green, *Saint Louis University School of Medicine, St. Louis, Missouri,* Proteins, biosynthesis
Gordon K. Gregory, *S.C. Johnson & Sons, Inc., Racine, Wisconsin,* Floor polishes
James R. Griffith, *Naval Research Laboratory, Washington, D.C.,* Phthalocyanine polymers
U. Grigo, *Farbenfabriken Bayer AG, Bayerwerk, Federal Republic of Germany,* Polycarbonates
Leo Gros, *Fachhochschule Frensenius, Wiesbaden, Federal Republic of Germany,* Biological activity
Gordon E. Gross, *Solar Energy Research Institute, Golden, Colorado,* Energy applications, Solar applications
Steven Grossman, *University of Lowell, Lowell, Massachusetts,* Head-to-head polymers
Walter Grot, *DuPont Experimental Station, Wilmington, Delaware,* Tetrafluoroethylene polymers, perfluorinated ionomers
Robert Grubbs, *California Institute of Technology, Pasadena, California,* Living polymer systems, olefin metathesis
Robert J. Gruber, *Xerox Corporation, Webster, New York,* Xerographic materials
Eric A. Grulke, *Michigan State University, East Lansing, Michigan,* Suspension polymerization
Wilhelm Guschlbauer, *CEN-Saclay, Gif-sur-Yvette, France,* Polynucleotides
J. T. Guthrie, *The University of Leeds, Leeds, United Kingdom,* Dyes, macromolecular
Edmund Haberstroh, *Institut für Kunststoffverarbeitung Technischen Hochschule, Aachen, Federal Republic of Germany,* Heat transfer
D. W. Hadley, *University of Reading, Reading, United Kingdom,* Mechanical properties
John R. Hagman, *E.I. du Pont de Nemours & Co., Inc., Wilmington, Delaware,* Acrylic elastomers, ethylene–acrylic elastomers
Jerry Hahnfeld, *The Dow Chemical Company, Midland, Michigan,* Styrene polymers, properties, general purpose polystyrene
Joseph D. Halford, *E.I. du Pont de Nemours & Co., Inc., Wilmington, Delaware,* Computer-assisted design
B. David Halpern, *Polysciences, Inc., Warrington, Pennsylvania,* Medical applications
Gary R. Hamed, *The University of Akron, Akron, Ohio,* Adhesion and bonding, adhesion
W. F. Hamner, *Astroturf Industries, Dalton, Georgia,* Recreational surfaces
S. J. Harasin, *Mobay Corporation, Pittsburgh, Pennsylvania,* Polyurethanes
William C. Harbison, *Swedlow, Inc., Garden Grove, California,* Casting
Bruce Hardman, *General Electric Company, Waterford, New York,* Silicones
Barbara Harley, *DNAX Research Institute, Palo Alto, California,* Drug applications
Robert J. Harper, Jr., *Agricultural Research Service, USDA, New Orleans, Louisiana,* Textile resins
J. E. Harris, *Union Carbide Corporation, Bound Brook, New Jersey,* Polysulfones
Bruce Hartmann, *U.S. Naval Surface Weapons Center, Silver Spring, Maryland,* Acoustic properties
G. Hartwig, *Kernsforschungszentrum Karlsruhe, Karlsruhe, Federal Republic of Germany,* Cryogenic properties
H. J. Harwood, *University of Akron, Akron, Ohio,* Living polymer systems, radical polymerization
Masaki Hasegawa, *University of Tokyo, Tokyo, Japan,* Polyaminotriazoles
W. Lincoln Hawkins, *Consultant, Montclair, New Jersey,* Stabilization
Raymond C. Hayes, *McClean Anderson, Inc., Milwaukee, Wisconsin,* Filament winding
Rhomie L. Heck III, *Uniroyal Chemical Group, Naugatuk, Connecticut,* Blowing agents
David M. Hedstrand, *The Dow Chemical Company, Midland, Michigan,* Dendritic polymers
Steven Heilmann, *3M Company, St. Paul, Minnesota,* Polyazlactones
Howard Heitner, *American Cyanamid Company, Stamford, Connecticut,* Mining applications, Mineral processing
Mark W. Hellums, *University of Texas at Austin, Austin, Texas,* Transport properties
J. Neil Henderson, *The Goodyear Tire & Rubber Company, Akron, Ohio,* Butadiene derivatives, polymeric; *Consultant, Hudson, Ohio,* Pentadiene polymers
Paul M. Hergenrother, *NASA Langley Research Center, Hampton, Virginia,* Acetylene-terminated prepolymers; Heat-resistant polymers; Polyquinoxalines
D. J. Herron, *Bepex Corporation, Minneapolis, Minnesota,* Drying
Richard W. Hertzberg, *Lehigh University, Bethlehem, Pennsylvania,* Fatigue
Jean E. Herz, *Institut Charles Sadron, Strasbourg, France,* Model polymers
Toshinobu Higashimura, *Kyoto University, Kyoto, Japan,* Living polymer systems, living cationic polymerization, alkenyl
H. Wayne Hill, Jr., *Hill Associates, Bartlesville, Oklahoma,* Poly(arylene sulfide)s
Philip Hodge, *University of Lancaster, Lancaster, United Kingdom,* Polymers as chemical reagents
Jonathan H. Hodgkin, *CSIRO, Melbourne, Victoria, Australia,* Chelate-forming polymers
David A. Holden, *University of Waterloo, Waterloo, Ontario, Canada,* Photochemistry
Geoffrey Holden, *Shell Development Company, Houston, Texas,* Elastomers, thermoplastic
Michael J. Hopper, *Hoechst Celanese Corp., Wiesbaden, Federal Republic of Germany,* Polyesters, films
B. A. Howell, *Central Michigan University, Mount Pleasant, Michigan,* Vinylidene chloride polymers
Jack W. Hoyt, *San Diego State University, San Diego, California,* Drag reduction
Samuel J. Huang, *The University of Connecticut, Storrs, Connecticut,* Biodegradable polymers
D. P. Hug, *HIMONT U.S.A., Incorporated, Wilmington, Delaware,* Ethylene polymers, ultrahigh molecular weight polyethylene
Kathleen A. Hughes, *Rohm & Haas, Spring House, Pennsylvania,* Solution polymerization
John L. Hull, *Hull Corporation, Hatboro, Pennsylvania,* Compression and transfer molding
Jack D. Hultman, *The Institute of Paper Chemistry, Appleton, Wisconsin,* Paper
D. Hummel, *Bepex Corporation, Minneapolis, Minnesota,* Drying
James S. Humphrey, *Pennwalt Corporation, King of Prussia, Pennsylvania,* Vinylidene fluoride polymers
G. Hurtrez, *Ecole Nationale Superieure de Chimie de Mulhouse, Mulhouse, France,* Block copolymers
Frank Iddings, *Southwest Research Institute, San Antonio, Texas,* Nondestructive testing
Yukio Imanishi, *Kyoto University, Kyoto, Japan,* Polyimidazole catalysts
Shohei Inoue, *University of Tokyo, Tokyo, Japan,* Living polymer systems, epoxides and lactones, new developments
Christopher Irwin, *Hoover Universal, Inc., Manchester, Michigan,* Blow molding
A. I. Isayev, *University of Akron, Akron, Ohio,* Thermal stress
Hatsuo Ishida, *Case Western Reserve University, Cleveland, Ohio,* Magnetic polymers
Kenneth J. Ivin, *The Queen's University of Belfast, Belfast, United Kingdom,* Methathesis polymerization; Polymerization, thermodynamics
Jalandar Y. Jadhav, *University of Massachusetts, Amherst, Massachusetts,* Polyesters, thermoplastic
Michael Jaffe, *Celanese Research Company, Summit, New Jersey,* High modulus polymers
David E. James, *Union Carbide Corporation, Bound Brook, New Jersey,* Ethylene polymers, linear low density polyethylene
Denis H. James, *The Dow Chemical Company, Midland, Michigan,* Styrene polymers, monomer
Alex M. Jamieson, *Case Western Reserve University, Cleveland, Ohio,* Solution rheology
Borek Janik, *Gelman Sciences, Inc., Ann Arbor, Michigan,* Electrophoresis
Siegfried Janocka, *Hoechst AG, Wiesbaden, Federal Republic of Germany,* Polyesters, films

Stanley Jasne, *CIBA-GEIGY Corporation, Ardsley, New York,* Polypyrroles
L. W. Jelinski, *AT&T Bell Laboratories, Murray Hill, New Jersey,* Nuclear magnetic resonance
Samson S. Jenekhe, *University of Rochester, Rochester, New York,* Ion implantation
Aubrey D. Jenkins, *The University of Sussex, Brighton, Sussex, United Kingdom,* Kinetics of polymerization
Julian F. Johnson, *University of Connecticut, Storrs, Connecticut,* Chromatography
R. N. Johnson, *Union Carbide Corporation, Bound Brook, New Jersey,* Polysulfones
R. W. Johnson, *Union Camp Corporation, Savannah, Georgia,* Polyamides from fatty acids
Martin B. Jones, *University of North Dakota, Grand Forks, North Dakota,* Oxidative polymerization
R. Sidney Jones, *Hoechst Celanese Corporation, Summit, New Jersey,* Optical data storage
E. K. Just, *Hercules, Inc., Wilmington, Delaware,* Cellulose ethers
Stuart M. Kaback, *Exxon Research and Engineering Company, Linden, New Jersey,* Patent information
R. P. Kambour, *General Electric Company, Schenectady, New York,* Crazing
Tetsua Kanamoto, *Science University of Tokyo, Tokyo, Japan,* Solid-state extrusion
Simon Kantor, *University of Massachusetts, Amherst, Massachusetts,* Polyesters, thermoplastic
Reuben H. Karol, *Rutgers University, New Brunswick, New Jersey,* Mining applications, chemical grouts
John J. Kaufman, *Witco Corporation, Petrolia, Pennsylvania,* Waxes
H. M. Kausch, *Ecole Polytechnique Federale de Lausanne, Lausanne, Switzerland,* Fracture
Yuhsuke Kawakami, *Nagoya University, Nagoya, Japan,* Macromers
Horst Kehlen, *Carl Schorlemmer Technical University at Merseburg, GDR,* Solution properties, polydispersity effects
Tominaga Keii, *Tokyo Institute of Technology, Tokyo, Japan,* Living polymer systems, Ziegler-Natta polymerization
Paul D. Kelley, *GE Plastics, Washington, West Virginia,* Acrylonitrile–butadiene–styrene polymers
Stephen S. Kelley, *Virginia Polytechnic Institute and State University, Blacksburg, Virginia,* Lignin
H. Keskkula, *University of Texas, Austin, Texas,* Polymer blends
Melvin H. Keyes, *Anatrace, Inc., Maumee, Ohio,* Enzymes, immobilized
Ishrat Khan, *Atlanta University, Atlanta, Georgia,* Vinylpyridine polymers
R. W. Kight, *Union Camp Corporation, Savannah, Georgia,* Polyamides from fatty acids
George R. Killat, *The Dow Chemical Company, Midland, Michigan,* Alkyleneimine polymers
Benjamin B. Kine, *Rohm and Haas Company, Spring House, Pennsylvania,* Acrylic and methacrylic ester polymers, survey
Yury V. Kissin, *Mobil Chemical Co., Edison, New Jersey,* Ethylene polymers, high density polyethylene; 4-Methyl-1-pentene polymers; Olefin polymers; Stereoregular styrene polymers
Jacob Klein, *The Weizmann Institute, Rehovot, Israel,* Macromolecular dynamics
D. Klempner, *University of Detroit, Detroit, Michigan,* Interpenetrating polymer networks
Peter R. Klinkowski, *Innotech Corporation, Trumbull, Connecticut,* Ultrafiltration
Nancy Kneib-Cordonier, *University of Minnesota, Minneapolis, Minnesota,* Polypeptide synthesis, solid-phase method
D. L. Knight, *The Goodyear Tire & Rubber Co., Akron, Ohio,* Tires
Raymond S. Knorr, *Monsanto Company, Pensacola, Florida,* Acrylic fibers
Frank Ko, *Drexel University, Philadelphia Pennsylvania,* Inorganic fibers
Akiro Kobayashi, *Saitama University, Urawa, Japan,* Machining
Shiro Kobayashi, *Kyoto University, Kyoto Japan,* Cyclic imino ethers, polymerization
Jeffrey T. Koberstein, *The University of Connecticut, Storrs, Connecticut,* Interfacial properties
L. Könczöl, *Fraunhofer-Institut für Werktoffmechanik, Freiburg, Federal Republic of Germany,* Micromechanical measurements
Peter W. Kopf, *Arthur D. Little, Inc., Cambridge, Massachusetts,* Phenolic resins
William J. Koros, *University of Texas at Austin, Austin, Texas,* Transport properties
Peter Kovacic, *University of Wisconsin, Milwaukee, Wisconsin,* Oxidative polymerization
Ronald Kowalski, *Exxon Chemical Company, Linden, New Jersey,* Recovery and isolation techniques
Larry Krepski, *3M Company, St. Paul, Minnesota,* Polyazlactones
Edward N. Kresge, *Exxon Chemical Company, Linden, New Jersey,* Isobutylene polymers
George A. Kruder, *Consultant, Mt. Gilead, Ohio,* Extrusion
Val Krukonis, *Phasex Corporation, Lawrence, Massachusetts,* Supercritical fluids
Gerald A. Krulik, *Morton Thiokol, Inc., Tustin, California,* Metallizing, electroless plating; Metallizing, metal-spraying
Przemyslaw Kubisa, *Polish Academy of Sciences, Lodz, Poland,* Living polymer systems, living cationic ring-opening polymerization
Joseph A. Kuczkowski, *The Goodyear Tire & Rubber Company, Akron, Ohio,* Rubber chemicals
V. Kudela, *Institute of Macromolecular Chemistry, Czechoslovak Academy of Sciences, Prague, Czechoslovakia,* Hydrogels
James E. Kuder, *Hoechst Celanese Corporation, Summit, New Jersey,* Optical data storage
Rolf G. Kuehni, *Mobay Chemical Corporation, Rockhill, South Carolina,* Dyeing
William C. Kuhlke, *Shell Chemical Company, Houston, Texas,* Furniture
Donald M. Kulich, *GE Plastics, Washington, West Virginia,* Acrylonitrile–butadiene–styrene polymers
Irving Kuntz, *Exxon Chemical Company, Linden, New Jersey,* Cyclopentadiene polymers
Albert Kveglis, *Sun Chemical Corporation, Carlstadt, New Jersey,* Printing-ink vehicles
S. L. Kwolek, *E.I. du Pont de Nemours & Co., Inc., Wilmington, Delaware,* Liquid crystalline polymers
V. L. Kyllingstad, *Hercules, Incorporated, Wilmington, Delaware,* 1,2-Epoxide polymers, polyether elastomers
S. S. Labana, *Ford Motor Company, Dearborn, Michigan,* Cross-linking
Joginder Lal, *Lal Associates, Akron, Ohio,* Elastomers, synthetic
John K. Lancaster, *Royal Aircraft Establishment, Farnborough, Hants, United Kingdom,* Abrasion and wear
J. B. Lando, *Case Western Reserve University, Cleveland, Ohio,* Epitaxial processes
L. M. Landoll, *Hercules, Inc., Oxford, Georgia,* Olefin fibers
Heimo J. Langer, *Ashland Chemical Company, Columbus, Ohio,* Foundry resins
R. Langer, *Massachusetts Institute of Technology, Cambridge, Massachusetts,* Polyanhydrides
H. J. Lanson, *LanChem Corporation, East St. Louis, Illinois,* Alkyd resins
R. P. Lattimer, *BFGoodrich Company, Brecksville, Ohio,* Antiozonants
Thomas M. Laue, *University of New Hampshire, Durham, New Hampshire,* Ultracentrifugation
Herbert G. Lauterbach, *E.I. du Pont de Nemours & Co., Inc., Wilmington, Delaware,* Fibers, engineering
R. W. Layer, *BFGoodrich Company, Brecksville, Ohio,* Antiozonants
Albert R. LeBoeuf, *Barnes-Hind Pharmaceuticals, Inc., San Diego, California,* Contact lenses
Anh LeDuy, *Laval University, University City, Quebec, Canada,* Pullulan
C. Lee, *Union Carbide Corporation, South Charleston, West Virginia,* Xylylene polymers
Julia T. Lee, *Michigan Molecular Institute, Midland, Michigan,* Literature of polymers
Stuart M. Lee, *Ford Aerospace & Communications Corp., Palo Alto, Cal-*

ifornia, Coating methods, vacuum coating; *SRI International, Menlo Park, California,* Embedding

D. G. LeGrand, *General Electric Company, Schenectady, New York,* Annealing

L. Leibler, *Universite Pierre et Marie Curie, Paris, France,* Phase transformations

Kam Leong, *Johns Hopkins University, Baltimore, Maryland,* Polyanhydrides

Alvin C. Levy, *AT&T Network Systems, Norcross, Georgia,* Fibers, optical

Richard B. Lieberman, *HIMONT Research & Development Center, Wilmington, Delaware,* Propylene polymers

J. Lieto, *University of Delaware, Newark, Delaware, Institut de Petroleochimie et de Synthese, Marsielle, France,* Catalysts, polymer-supported

Charles D. Livengood, *North Carolina State University, Raleigh, North Carolina,* Silk

James E. Loftus, *Pennsylvania State University, University Park, Pennsylvania,* Telomerization

Karl D. Loos, *Arthur D. Little, Inc., Cambridge, Massachusetts,* Market evaluation

Kyriakos Loufakis, *Universität Freiburg, Freiburg, Federal Republic of Germany,* Thermodynamic properties

J. Herbert Lowell, *Consultant, Wynnewood, Pennsylvania,* Coatings, Systems

P. J. Ludovice, *ETH-Zurich, Zurich, Switzerland,* Atomistic modeling, polymeric glasses

Robert D. Lundberg, *Exxon Chemical Company, Linden, New Jersey,* Ionic polymers

John H. T. Luong, *Biotechnology Research Institute, National Research Council of Canada, Montreal, Quebec, Canada,* Pullulan

Pierre Lutz, *Institut Charles Sadron, Strasbourg, France,* Macrocyclic polymers

Donald J. Lyman, *University of Utah, Salt Lake City, Utah,* Biomaterials

Michael L. Lynn, *3M Company, St. Paul, Minnesota,* Fluorocarbon elastomers

F. Lyon, *Columbian Chemicals Company, Swartz, Louisiana,* Carbon

K. J. Mackenzie, *Hoechst Celanese Corp., Wiesbaden, Federal Republic of Germany,* Polyesters, films; *American Hoechst Corporation, Greer, South Carolina,* Sheeting

Norbert L. Maecker, *The Dow Chemical Company, Midland, Michigan,* Styrene polymers, degradation

T. G. Majewicz, *Hercules, Inc., Wilmington, Delaware,* Cellulose ethers

Michael T. Malanga, *The Dow Chemical Company, Midland, Michigan,* Styrene polymers, polymerization

Earl W. Malcolm, *The Institute of Paper Chemistry, Appleton, Wisconsin,* Bleaching, bleaching of wood pulp; Paper

John B. Malloy, *Amoco Corporation, Chicago, Illinois,* Economic evaluation

Thomas M. Maloney, *Washington State University, Pullman, Washington,* Composition board

Michel Mandel, *University of Leiden, Leiden, the Netherlands,* Polyelectrolytes

Georg Manecke, *Institut für Organische Chemie, Freie Universitat, Berlin,* Electron-transfer polymers

John A. Manson,† *Lehigh University, Bethlehem, Pennsylvania,* Fatigue

L. T. Manzione, *AT&T Bell Laboratories, Murray Hill, New Jersey,* Reaction injection molding

T. J. Mao, *General Motors Research Laboratories, Warren, Michigan,* Sheet molding compounds

J. E. Mark, *University of Cincinnati, Ohio,* Elasticity, rubberlike

Carl D. Marotta, *Innovation Technology, Inc., Warrington, Pennsylvania,* Packaging materials, medical devices

F. Lennart Marten, *Air Products and Chemicals, Allentown, Pennsylvania,* Vinyl alcohol polymers

Kenneth N. Mathes, *Consultant, Schenectady, New York,* Electrical properties

Paul Matlock, *Union Carbide Corporation, Tarrytown, New York,* 1,2-Epoxide polymers, ethylene oxide polymers and copolymers

S. Matsuoka, *AT&T Bell Laboratories, Murray Hill, New Jersey,* Dielectric relaxation

Wayne L. Mattice, *Louisiana State University, Baton Rouge, Louisiana,* Helix-coil transitions

K. A. Mauritz, *University of Southern Mississippi, Hattiesburg, Mississippi,* Epitaxial processes

Rennie May, *ICI Americas, Inc., Wilmington, Delaware,* Polyetheretherketones

Amitava Mazumdar, *Indian Institute of Technology, Kharagpur, India,* Vinyl ether polymers

Richard A. Mazzarella, *Saint Louis University School of Medicine, St. Louis, Missouri,* Proteins, biosynthesis

Louis V. McAdams, *CIBA-GEIGY Corporation, Hawthorne, New York,* Epoxy resins

E. R. McAfee, *Mobay Corporation, Pittsburgh, Pennsylvania,* Polyurethanes

Francis B. McAndrew, *Hoechst Celanese Corporation, Summit, New Jersey,* Technical service

Robert A. McCarthy, *Springborn Group, Inc., Enfield, Connecticut,* Chemically resistant polymers

Thomas J. McCarthy, *University of Massachusetts, Amherst, Massachusetts,* Surface modification

C. G. McCarty, *Mobay Corporation, Pittsburgh, Pennsylvania,* Polyurethanes

Charles L. McCormick, *University of Southern Mississippi, Hattiesburg, Mississippi,* Agricultural applications; Water-soluble polymers

P. Y. McCormick, *E.I. du Pont de Nemours & Co., Inc., Wilmington, Delaware,* Drying

Joseph N. McDonald, *Joseph McDonald Associates, Inc., Lexington, Massachusetts,* Thermoforming

John D. McGervey, *Case Western Reserve University, Cleveland, Ohio,* Positron annihilation

Vincent D. McGinness, *Battelle Columbus Laboratories, Columbus, Ohio,* Cross-linking with radiation

John N. McGovern, *University of Wisconsin-Madison, Wisconsin,* Fibers, vegetable

Mark A. McHugh, *Johns Hopkins University, Baltimore, Maryland,* Supercritical fluids

J. M. McIntire, *Eastman Kodak Company, Kingsport, Tennessee,* Acrylic and methacrylic ester polymers, 2-cyanoacrylic ester polymers

J. E. McIntyre, *University of Leeds, Leeds, United Kingdom,* Fibers, manufacture

Samuel P. McManus, *University of Alabama in Huntsville, Huntsville, Alabama,* Space environment processing

Paul Meakin, *E.I. du Pont de Nemours & Co., Inc., Wilmington, Delaware,* Fractals

Bernard J. Meister, *The Dow Chemical Company, Midland, Michigan,* Styrene polymers, polymerization; Styrene polymers, properties, rheology

G. Menges, *Institut für Kunststoffverarbeitung, Aachen, Federal Republic of Germany,* Computer-assisted engineering

Virginia Menikheim, *Consultant, Pittsboro, North Carolina,* Polymerization procedures, industrial

Fred J. Meyer, *Ford Motor Company, Detroit, Michigan,* Adhesion and bonding, bonding

John A. Meyer, *SUNY College of Environmental Science and Forestry, Syracuse, New York,* Wood, polymer-impregnated

Ralph E. Meyer, *Products Research & Chemical Corporation, Glendale, California,* Sealants

W. Michaeli, *Carl Freudenberg, Weinheim, Federal Republic of Germany,* Die design

G. E. Miller, *Union Camp Corporation, Savannah, Georgia,* Polyamides from fatty acids

Walter J. Miller, *Pennwalt Corporation, Philadelphia, Pennsylvania,* Metallizing, vacuum metallizing

William A. Miller, *Allied Corporation, Morristown, New Jersey,* Chlorotrifluoroethylene polymers, chlorotrifluoroethylene–ethylene copolymers

Frank Millich, *University of Missouri, Kansas City, Missouri,* Polyisocyanides

N. J. Mills, *University of Birmingham, Birmingham, United Kingdom,* Optical properties

John Mitchell, Jr., *Consultant, Wilmington, Delaware,* Chemical analysis

Pradyot Mitra, *Indian Institute of Technology, Kharagpur, India,* Vinyl ether polymers

Gary N. Mock, *North Carolina State University, Raleigh, North Carolina,* Bleaching, bleaching of textiles

Eugene R. Moore, *The Dow Chemical Company, Midland, Michigan,* editor of Styrene polymers

Herbert Morawetz, *Polytechnic University, Brooklyn, New York,* History of polymer science

Paul W. Morgan, *Consultant, West Chester, Pennsylvania,* Interfacial polymerization; Liquid crystalline polymers

E. M. Mount III, *Mobil Chemical Company, Macedon, New York,* Films, manufacture

S. K. Mowdood, *The Goodyear Tire & Rubber Co., Akron, Ohio,* Tires

Stan Mruk, *Plastics Pipe Institute, New York, New York,* Pipe

Ernest C. Mueller, *The Dow Chemical Company, Midland, Michigan,* Styrene polymers, monomers

D. P. Mukherjee, *Dow Corning Wright, Arlington, Tennessee,* Sutures

Dan Mullen, *University of Minnesota, Minneapolis, Minnesota,* Polypeptide synthesis, solid-phase method

P. R. Muller, *Farbenfabriken Bayer AG, Bayerwerk, Federal Republic of Germany,* Polycarbonates

Takayuki Murayama,[†] *Monsanto Company, Springfield, Massachusetts,* Dynamic mechanical properties

L. W. Muszynski, *Pyrament-Lone Star Industries, Inc., Houston, Texas,* Polymer concrete

Riccardo A. A. Muzzarelli, *Faculty of Medicine, University of Ancona, Ancona, Italy,* Chitin

K. Nagoshi, *Kuraray Company, Ltd., Osaka, Japan,* Leatherlike materials

A. M. Nahmias, *Mobil Chemical Company, Macedon, New York,* Films

M. J. Nanjan, *Hindustan Photo Films Manufacturing Company, Ltd., Indunagar, Ootacamund, India,* Polyhydrazides and polyoxadiazoles

E. Bruce Nauman, *Rensselaer Polytechnic Institute, Troy, New York,* Bulk polymerization; Flash devolatilization

Joseph W. Nemec, *Rohm and Haas Company, Spring House, Pennsylvania,* Acrylic and methacrylic acid polymers

Nelson L. Nemerow, *University of Miami, Florida International University, Miami, Florida,* Waste treatment

M. W. Newman, *European Vinyls Corporation, Runcorn, United Kingdom,* Vinyl chloride polymers, structure and properties; Vinyl chloride polymers, vinyl chloride monomer

Seymour Newman, *Ford Motor Company, Detroit, Michigan,* Automotive applications

Tadatomi Nishikubo, *Kanagawa University, Yokohama, Japan,* Polyaminotriazoles

N. H. Nodelman, *Mobay Corporation, Pittsburgh, Pennsylvania,* Polyurethanes

W. Nouvertne, *Farbenfabriken Bayer AG, Bayerwerk, Federal Republic of Germany,* Polycarbonates

Bruce M. Novak, *California Institute of Technology, Pasadena, California,* Living polymer systems, olefin metathesis

Ronald W. Novak, *Rohm and Haas Company, Spring House, Pennsylvania,* Acrylic and methacrylic ester polymers, survey

Oskar Nuyken, *Technische Universität Munchen, Garching, Federal Republic of Germany,* Azo polymers; *Universität Bayreuth, Bayreuth, Federal Republic of Germany,* Telechelic polymers

Leroy Charles Oberholtzer, *Rockwell International, Cedar Rapids, Iowa,* Electromagnetic interference

D. F. O'Brien, *The University of Arizona, Tucson, Arizona,* Vesicles

Keith T. O'Brien, *Celanese Corporation, Summit, New Jersey,* Compounding

George Odian, *The College of Staten Island, The City University of New York, Staten Island, New York,* Chain-reaction polymerization

Eilert A. Ofstead, *Goodyear Tire & Rubber Company, Akron, Ohio,* Polyalkenamers

Richard Kafue Okine, *E.I. du Pont de Nemours & Co., Inc., Wilmington, Delaware,* Cold forming

R. Olson, *Nova Tran Corporation, Clear Lake, Wisconsin,* Xylylene polymers

Yoshitsugu Oono, *University of Illinois, Urbana, Illinois,* Statistical thermodynamics

C. M. Orlando, *General Electric Research and Development Center, Schenectady, New York,* Reactive extrusion

T. A. Orofino, *Astroturf Industries, Dalton, Georgia,* Recreational surfaces

Morris R. Ort, *Monsanto Chemical Company, Springfield, Massachusetts,* Polymerization procedures, laboratory

Robert A. Orwoll, *College of William and Mary, Williamsburg, Virginia,* Solubility of polymers

Raphael M. Ottenbrite, *Virginia Commonwealth University, Richmond, Virginia,* Controlled-release technology

Michael J. Owen, *Dow Corning Corporation, Midland, Michigan,* Antifoaming agents; Release agents

John E. Pace, *GE Plastics, Washington, West Virginia,* Acrylonitrile–butadiene–styrene polymers

P. C. Painter, *The Pennsylvania State University, University Park, Pennsylvania,* Infrared-absorption spectroscopy

Skip Palenik, *Walter C. McCrone Associates, Inc., Chicago, Illinois,* Forensic aspects

Hans P. Panzer, *American Cyanamid Company, Stamford, Connecticut,* Mining applications, mineral processing

S. Peter Pappas, *North Dakota State University, Fargo, North Dakota,* Photopolymerization

H. C. Park, *Mobil Chemical Company, Macedon, New York,* Films; Films, manufacture

Hsing-Yeh Parker, *Rohm & Haas, Bristol, Pennsylvania,* Processing aids

Stephen Pask, *Consultant, Dormagen, Federal Republic of Germany,* Telechelic polymers

Andrew J. Pasztor, Jr., *The Dow Chemical Company, Midland, Michigan,* Styrene polymers, thermal properties

Jocelyn M. Patterson, *USAF Materials Laboratory, Dayton, Ohio,* Reinforcement

D. R. Paul, *University of Texas, Austin, Texas,* Polymer blends

John Paul, *PEI Associates, Inc., Fort Worth, Texas,* Rubber reclaiming

J. M. Pearson, *Eastman Kodak Company, Rochester, New York,* Vinylcarbazole polymers

William Pearson, *The Dow Chemical Company, Midland, Michigan,* Styrene polymers, economic aspects

Warren J. Peascoe, *Uniroyal Chemical Group, Naugatuck, Connecticut,* Blowing agents

Evaristo Peggion, *University of Padua, Padua, Italy,* Polypeptides

Stanislaw Penczek, *Polish Academy of Sciences, Lodz, Poland,* Living polymer systems, living cationic ring-opening polymerization; Sulfur-containing polymers

Fred M. Peng, *Monsanto Company, Indian Orchard, Massachusetts,* Acrylonitrile polymers

Randy E. Pepper, *Georgia Institute of Technology, Atlanta, Georgia,* Refractometry

Virgil Percec, *Case Western Reserve University, Cleveland, Ohio,* Oligomers

Anton Peterlin, *Consultant, Bethesda, Maryland,* Morphology

George E. Pickering, *Arthur D. Little, Inc., Watertown, Massachusetts,* Toys

Karl A. Piez, *Collagen Corporation, Palo Alto, California,* Collagen

I. Piirma, *University of Akron, Akron, Ohio,* Colloids

Charles U. Pittman, *Mississippi State University, Mississippi State, Mississippi,* Organometallic polymers

E. P. Pleuddenmann, *Dow Corning Corporation, Midland, Michigan,* Coupling

John F. Plummer, *Emerson & Cuming, Inc., Canton, Massachusetts,* Microspheres

John W. Pochan, *Johnson Wax Corp., Racine, Wisconsin,* Acetylenic polymers

Gary W. Poehlein, *Georgia Institute of Technology, Atlanta, Georgia,* Emulsion polymerization

B. Poldermans, *Gist-brocades, Inc., Delft, The Netherlands,* Enzymes

Roger S. Porter, *University of Massachusetts, Amherst, Massachusetts,* Mechanochemical reactions; Solid-state extrusion

Rudolph Potenzone, Jr., *Polygen Corporation, Waltham, Massachusetts,* Polymer modeling

D. Powell, *Polyplan Consultants, Leicester, United Kingdom,* Construction and building applications
J. Preston, *Research Triangle Institute, Research Triangle Park, North Carolina,* Polyamides, aromatic
D. C. Prevorsek, *Allied Corporation, Morristown, New Jersey,* Dimensional stability, fibers; Ultimate properties
Gerfried Pruckmayr, *E.I. du Pont de Nemours & Co., Inc., Wilmington, Delaware,* Tetrahydrofuran polymers
Richard W. Prugh, *Hazard Reduction Engineering, Inc., Glen Mills, Pennsylvania,* Safety
Coleen Pugh, *Case Western Reserve University, Cleveland, Ohio,* Oligomers
J. R. Pyne, *Consultant, Shrewsbury, United Kingdom,* Rubber, hard
Jean-Pierre Queslel, *Manufacture Michelin, Clermont-Ferrand, France,* Birefringence; *University of Cincinnati, Cincinnati, Ohio,* Elasticity, rubberlike
R. P. Quirk, *University of Akron, Akron, Ohio,* Graft copolymers
I. S. Rai, *The Goodyear Tire & Rubber Co., Akron, Ohio,* Tires
M. Louise Rainey, *The Dow Chemical Company, Midland, Michigan,* Styrene polymers, health and safety
V. Ramaswami, *The University of Arizona, Tucson, Arizona,* Vesicles
J. C. Randall, *Exxon Chemical Company, Baytown, Texas,* Microstructure
Jerald K. Rasmussen, *3M Company, St. Paul, Minnesota,* Polyazlactones
Margit T. Ratzsch, *Carl Schorlemmer Technical University at Merseburg, GDR,* Solution properties, polydispersity effects
Chris J. Rauwendaal, *Raychem Corporation, Menlo Park, California,* Screw design, single screw
Ludwig Rebenfeld, *Textile Research Institute, Princeton, New Jersey,* Fibers
Richard A. Reck, *Akzo Chemie America, Chicago, Illinois,* Antistatic agents
H. Recker, *Institut für Kunststoffverarbeitung, Aachen, Federal Republic of Germany,* Computer-assisted engineering
Herbert Rees, *Consultant, Orangeville, Ontario, Canada,* Molds
R. W. Rees, *E.I. du Pont de Nemours & Co., Inc., Wilmington, Delaware,* Cross-linking, reversible
S. E. Reickert, *Case Western Reserve University, Cleveland, Ohio,* Epitaxial processes
A. Rembaum,[†] *California Institute of Technology, Pasadena, California,* Antibodies and antigens, polymer-bound
Paul Rempp, *Institut Charles Sadron, Strasbourg, France,* Macrocyclic polymers; Model polymers
John R. Reynolds, *University of Texas-Arlington, Arlington, Texas,* Organometallic polymers
C. K. Rhee, *BFGoodrich Company, Brecksville, Ohio,* Antiozonants
A. Rheingold, *University of Delaware, Newark, Delaware,* Inorganic polymers
William C. Rice, *University of Lowell, Lowell, Massachusetts,* Polyampholytes; Vinylpyridinium salts
Paul N. Richardson, *E.I. du Pont de Nemours & Co., Inc., Wilmington, Delaware,* Plastics processing
Douglas S. Richart, *The Polymer Corporation, Reading, Pennsylvania,* Coating methods, powder coating
J. N. Rieck, *Mobay Corporation, Pittsburgh, Pennsylvania,* Polyurethanes
Gerard Riess, *Ecole National Superieure de Chimie de Mulhouse, Mulhouse, France,* Block copolymers
John P. Riggs, *Celanese Corporation, Summit, New Jersey,* Carbon fibers
John F. Robyt, *Iowa State University, Ames, Iowa,* Dextran
T. G. Rochow, *North Carolina State University, Raleigh, North Carolina,* Light microscopy; Resinography
Ryong-Joon Roe, *University of Cincinnati, Cincinnati, Ohio,* Glass transition; X-ray diffraction
J. A. Roels, *Gist-brocades, Inc., Delft, The Netherlands,* Enzymes
E. Ron, *Massachusetts Institute of Technology, Cambridge, Massachusetts,* Polyanhydrides
Jacques Roovers, *Division of Chemistry, National Research Council of Canada, Ottawa, Canada,* Branched polymers
D. V. Rosato, *University of Lowell, Lowell, Massachusetts, and Borg-Warner Chemicals, Chicago, Illinois,* Materials, selection; Reinforced plastics, thermosets
Gerald R. Rose, *Nalco Chemical Company, Naperville, Illinois,* Flocculation
Philip I. Rose, *Eastman Kodak Co., Rochester, New York,* Gelatin
Stephen M. Rowland, *University of Utah, Salt Lake City, Utah,* Biomaterials
David Roylance, *Massachusetts Institute of Technology, Cambridge, Massachusetts,* Electron-spin resonance
Irvin I. Rubin, *Robinson Plastic Corporation, New York, New York,* Injection molding
D. R. Rueda, *Institute for Structure and Matter, CSIC, Madrid, Spain,* Hardness
George Rugman, *BFGoodrich Company, Akron, Ohio,* Belting
James P. Runt, *The Pennsylvania State University, University Park, Pennsylvania,* Crystallinity determination
Takeo Saegusa, *Kyoto University, Kyoto, Japan,* Cyclic imino ethers, polymerization; Ring-opening polymerization; Zwitterionic polymerization
David R. St. Cyr, *The Goodyear Tire & Rubber Company, Akron, Ohio,* Rubber, natural
Joseph C. Salamone, *University of Lowell, Lowell, Massachusetts,* Polyampholytes; Vinylpyridinium salts
J. Martinez Salazar, *Institute for Structure and Matter, CSIC, Madrid, Spain,* Hardness
Robert J. Samuels, *Georgia Institute of Technology, Atlanta, Georgia,* Refractometry
James H. Saunders, *Consultant, St. Louis, Missouri,* Polyamides, fibers
Hideo Sawada, *Daicel Chemical Industries Ltd., Osaka, Japan,* Depolymerization
M. Sawamoto, *Kyoto University, Kyoto, Japan,* Living polymer systems, living cationic polymerization, alkenyl
Arthur N. Scarlatti, *Hercules, Inc., Wilmington, Delaware,* Hydrocarbon resins
J. R. Schaefgan, *Consultant, Wilmington, Delaware,* Liquid crystalline polymers
R. Ted Scharenberg, *Consultant, White House Station, New Jersey,* Coating methods, roll coating
Ralph H. Schatz, *Exxon Chemical Company, Linden, New Jersey,* Isobutylene polymers
Harold A. Scheraga, *Cornell University, Ithaca, New York,* Helix-coil transitions
C. E. Schildknecht, *Gettysburg College, Gettysburg, Pennsylvania,* Diallyl and related polymers
Nicholas E. Schlotter, *Bell Communications Research, Redbank, New Jersey,* Raman spectroscopy
H. G. Schmelzer, *Mobay Corporation, Pittsburgh, Pennsylvania,* Polyurethanes
Claude J. Schmidle, *Congoleum Corporation, Trenton, New Jersey,* Flooring materials
Edward E. Schmitt, *Cooper Vision Inc., Mountain View, California,* Drug applications
Clifford K. Schoff, *PPG Industries, Inc., Allison Park, Pennsylvania,* Rheological measurements
Walter J. Schrenk, *The Dow Chemical Company, Midland, Michigan,* Films, multilayer
R. R. Schrieke, *Ballarat C.A.E., Victoria, Australia,* Casein
Herman Schroeder, *Schroeder Scientific Services, Inc., Greenville, Delaware,* Polyesters, elastomeric
Conrad Schuerch, *State University of New York, Syracuse, New York,* Polysaccharides, synthetic
Donald N. Schulz, *Exxon Research & Engineering Company, Annandale, New Jersey,* Hydrogenation; Water-soluble polymers
Rolf Christian Schulz, *University of Mainz, Mainz, Federal Republic of Germany,* Acrolein polymers
Patricia S. Schwartz, *United States Food and Drug Administration, Washington, D.C.,* Food applications
Friedrich R. Schwarzl, *Universität Erlangen-Nurnberg, Erlangen, Federal Republic of Germany,* Viscoelasticity
William H. Scouten, *Baylor University, Waco, Texas,* Chromatography, affinity
D. Seanor, *Xerox Corporation, Webster, New York,* Xerographic materials
N. D. Searle, *Consultant, Plastics and Chemicals, Bridgewater, New Jersey,* Weathering

J. K. Sears, *Consultant, St. Louis, Missouri,* Plasticizers
Jerry T. Seitz, *The Dow Chemical Company, Midland, Michigan,* Styrene polymers, properties, orientation and physical properties
Jeffrey Selley, *Consultant, Dublin, Ohio,* Polyesters, unsaturated
Michael L. Senyek, *The Goodyear Research Laboratory, Akron, Ohio,* Isoprene polymers
George A. Serad, *Celanese Fibers Operations, Charlotte, North Carolina,* Cellulose esters, organic, fibers; Polybenzimidazoles
Raymond B. Seymour, *University of Southern Mississippi, Hattiesburg, Mississippi,* Additives
C. S. Sheppard, *Pennwalt Corporation, Buffalo, New York,* Azo compounds; Peroxy compounds
David C. Sherrington, *University of Strathclyde, Glasgow, Scotland,* Reactions of polymers
Yasuhiko Shirota, *Osaka University, Osaka, Japan,* Charge-transfer complexes
F. A. Shutov, *Mendeleev Institute of Chemistry and Technology, Moscow, USSR,* Structural foams
Richard H. Siemer, *The Dow Chemical Company, Midland, Michigan,* Styrene polymers, economic aspects
Alex Silberberg, *The Weizmann Institute of Science, Rehovot, Israel,* Adsorption
Ruth Silberman-Hazony, *Consultant, Newton, Massachusetts,* Radiopaque polymers
Hans Sillescu, *University of Mainz, Mainz, Federal Republic of Germany,* Diffusion, polymer–polymer
D. H. Silvey, *BFGoodrich Company, Akron, Ohio,* Belting
Robert M. Simon, *USTEK Inc., Dublin, Ohio,* Conductive plastics
Helmut K. Singer, *American Optical Corporation, Framingham, Massachusetts,* Contact lenses
John D. Sinkey, *The Institute of Paper Chemistry, Appleton, Wisconsin,* Paper
Richard E. Skochdopole, *The Dow Chemical Company, Midland, Michigan,* Styrene polymers, processing, polystyrene foams; Styrene polymers, processing, extrusion and thermoforming
P. V. Smallwood, *ICI Chemicals & Polymers Ltd., Runcorn, United Kingdom,* Vinyl chloride polymers, polymerization, suspension and mass polymerization
Johannes Smid, *State University of New York at Syracuse, Syracuse, New York,* Polyelectrolyte complexes
Leslie E. Smith, *National Bureau of Standards, Gaithersburg, Maryland,* Polymer standards
Thomas Smith, *Xerox Corporation, Webster, New York,* Polyimidazoles
Ronald L. Smorada, *E.I. du Pont de Nemours & Co., Inc., Wilmington, Delaware,* Nonwoven fabrics, spunbonded
Arthur W. Snow, *Naval Research Laboratory, Washington, D.C.,* Phthalocyanine polymers
Mark E. Soderquist, *The Dow Chemical Company, Midland, Michigan,* Styrene polymers, high impact polystyrene
Kazuo Soga, *Tokyo Institute of Technology, Tokyo, Japan,* Living polymer systems, Ziegler-Natta polymerization
George Sollner, *Ren Plastics Inc., East Lansing, Michigan,* Tooling with plastics
John H. Southern, *Monsanto Textiles Company, Pensacola, Florida,* Solid-state extrusion
Gavin G. Spence, *Callaway Chemical Company, Columbus, Georgia,* Paper additives and resins
Leslie H. Sperling, *Lehigh University, Bethlehem, Pennsylvania,* Microphase structure; Polymers from renewable sources
G. Neil Spokes, *Consultant, Briarcliff, New York,* Test methods
E. P. Squiller, *Mobay Corporation, Pittsburgh, Pennsylvania,* Polyurethanes
M. Srinivasan, *Indian Institute of Technology, Madras, India,* Polybenzoxazinones
Neela Srinivasan, *University of North Carolina, Chapel Hill, North Carolina,* Molecular graphics
R. C. Stephenson, *European Vinyls Corporation, Runcorn, United Kingdom,* Vinyl chloride polymers, structure and properties
S. S. Sternstein, *Rensselaer Polytechnic Institute, Troy, New York,* Polymer creep
Milan Stolka, *Xerox Corporation, Webster, New York,* Photoconductive polymers
Alan W. Stratton, *Witco Corporation, Petrolia, Pennsylvania,* Waxes
Janet Strimaitis, *Zymark Corporation, Hopkinton, Massachusetts,* Robotics
L. C. E. Struik, *Plastics and Rubber Research Institute TNO, Delft, The Netherlands,* Aging, physical
D. E. Stuetz, *Celanese Research Center, Summit, New Jersey,* Polybenzimidazoles
A. Subramaniam, *Rubber Research Institute of Malaysia, Kuala Lumpur, Malaysia,* Rubber derivatives
K. W. Suh, *Dow Chemical U.S.A., Granville, Ohio,* Cellular materials
Theodore Sulzberg, *Sun Chemical Corporation, Carlstadt, New Jersey,* Printing-ink vehicles
U. W. Suter, *ETH-Zurich, Zurich, Switzerland,* Atomistic modeling, polymeric glasses
Dirk M. Sutherlin, *Consultant, Bartlesville, Oklahoma,* Polyquinolines
Masato Suzuki, *Kyoto University, Kyoto, Japan,* Zwitterionic polymerization
G. S. Swei, *Case Western Reserve University, Cleveland, Ohio,* Epitaxial processes
Graham Swift, *Rohm & Haas, Spring House, Pennsylvania,* Solution polymerization
Paul D. Sybert, *General Electric Company, Schenectady, New York,* Polybithiazoles
Hiroyuki Tadokoro, *Osaka University, Osaka, Japan,* Crystalline polymers
T. Takeshita, *E.I. du Pont de Nemours & Co., Inc., Wilmington, Delaware,* Chloroprene polymers
James R. Talbot, *Celanese Corporation, Charlotte, North Carolina,* Polyesters, fibers
Y. Y. Tan, *Gronigen University, Gronigen, The Netherlands,* Template polymerization
T. Tanaka, *Massachusetts Institute of Technology, Cambridge, Massachusetts,* Gels
Barry L. Tarmy, *Exxon Research and Engineering Company, Florham Park, New Jersey,* Reactor technology
Kohji Tashiro, *Osaka University, Osaka, Japan,* Crystalline polymers
David P. Tate, *Firestone Tire & Rubber Company, Akron, Ohio,* Butadiene polymers
M. Tatsumi, *Kansai University, Suita, Japan,* Proliferous polymerization
Carl R. Taylor, *AT&T Network Systems, Norcross, Georgia,* Fibers, optical
Charles C. Taylor, *Charles C. Taylor & Associates, Brevard, North Carolina,* Cellophane
W. S. Taylor, *R.T. Vanderbilt Company, Inc., Norwalk, Connecticut,* Biocides
Samuel C. Temin, *The Kendall Company, Lexington, Massachusetts,* Adhesive compositions; *Consultant, Needham, Massachusetts,* Pressure-sensitive adhesives and products
Fred N. Teumac, *Reeves Brothers, Inc., Spartanburg, South Carolina,* Fabrics, coated
Mrinal Thakur, *GTE Laboratories Incorporated, Waltham, Massachusetts,* Electrooptical applications; *AT&T Bell Laboratories, Murray Hill, New Jersey,* Solid-state polymerization
Curt Thies, *Washington University, St. Louis, Missouri,* Microencapsulation
Edwin L. Thomas, *University of Massachusetts, Amherst, Massachusetts,* Electron microscopy
Walter M. Thomas, *American Cyanamid Company, Stamford, Connecticut,* Acrylamide polymers
Edward V. Thompson, *University of Maine, Orono, Maine,* Thermal properties
John A. Thornton,[†] *University of Illinois, Urbana, Illinois,* Metallizing, sputtering
M. C. Throckmorton, *Consultant, Akron, Ohio,* Pentadiene polymers
James L. Throne, *Sherwood Technologies, Inc., Akron, Ohio,* Structural foams
David A. Tirrell, *University of Massachusestts, Amherst, Massachusetts,* Copolymerization

D. B. Todd, *APV Chemical Machinery Inc., South Plainfield, New Jersey,* Pelletizing

Donald A. Tomalia, *The Dow Chemical Company, Midland, Michigan,* Alkyleneimine polymers; Dendritic polymers

Paul E. Tomlins, *Imperial College, London, UK,* Scattering

Alan E. Tonelli, *AT&T Bell Laboratories, Murray Hill, New Jersey,* Conformation and configuration

Y.-C. Tong, *Tianjin Bureau of Chemical Industry, Tianjin, People's Republic of China,* Medical applications

Arnold Torkelson, *General Electric Company, Waterford, New York,* Silicones

N. W. Touchette, *Consultant, St. Louis, Missouri,* Plasticizers

Thomas D. Traugott, *The Dow Chemical Company, Midland, Michigan,* Styrene polymers, copolymers

Peter Trefonas, *Monsanto Electronic Materials Company, St. Louis, Missouri,* Polysilanes and polycarbosilanes

Sukant Tripathy, *GTE Laboratories Incorporated, Waltham, Massachusetts,* Electrooptical applications

Terry C. Troxell, *United States Food and Drug Administration, Washington, D.C.,* Food applications

Stephen Tsai, *USAF Materials Laboratory, Dayton, Ohio,* Reinforcement

Paul A. Tucker, *North Carolina State University, Raleigh, North Carolina,* Identification, fibers

L. H. Tung, *The Dow Chemical Company, Midland, Michigan,* Fractionation

Albin Turbak, *Georgia Institute of Technology, Altanta, Georgia,* Rayon

S. R. Turner, *Eastman Kodak Company, Rochester, New York,* Lithographic resists

R. P. Tye, *Dynatech Research and Development Company, Cambridge, Massachusetts,* Energy applications, thermal applications

Henri Ulrich, *The Dow Chemical Company, North Haven, Connecticut,* Isocyanate-derived polymers

A. J. Ultee, *E.I. du Pont de Nemours & Co., Inc., Waynesboro, Virginia,* Fibers, elastomeric

Ivor H. Updegraff, *Consultant, Stamford, Connecticut,* Amino resins

D. A. Upson, *Eastman Kodak Company, Rochester, New York,* Photographic applications

Sidney L. Vail, *Agricultural Research Service, USDA, New Orleans, Louisiana,* Textile resins

J. M. Valdespino, *Union Camp Corporation, Savannah, Georgia,* Polyamides from fatty acids

C. S. Van Buskirk, *Rensselaer Polytechnic Institute, Troy, New York,* Polymer creep

E. J. Vandenberg, *Arizona State University at Tempe, Tempe, Arizona,* Coordinate polymerization

Roy L. Vaz, *Tripos Associates, St. Louis, Missouri,* Molecular graphics

Earl W. Veazey, *The Dow Chemical Company, Freeport, Texas,* Films, multilayer

J. W. Verbicky, Jr., *General Electric Company, Schenectady, New York,* Polyimides

Peter H. Verdier, *National Bureau of Standards, Gaithersburg, Maryland,* Polymer standards

Gary Ver Strate, *Exxon Chemical Company, Linden, New Jersey,* Ethylene–propylene elastomers

Tyrone L. Vigo, *United States Department of Agriculture, Knoxville, Tennessee,* Cellulose derivatives

Otto Vogl, *Polytechnic University, Brooklyn, New York,* Aldehyde polymers; Head-to-head polymers

I. Voigt-Martin, *Johannes Gutenberg Universität, Mainz, Federal Republic of Germany,* Amorphous polymers

N. Vorchheimer, *Betz Laboratories, Inc., Trevose, Pennsylvania,* Polyamines and polyquaternary ammonium salts

Walter Vredenburgh, *Hercules, Inc., Clairton, Pennsylvania,* Hydrocarbon resins

J. S. Vrentas, *The Pennsylvania State University, University Park, Pennsylvania,* Diffusion

W. H. Waddell, *The Goodyear Tire & Rubber Co., Akron, Ohio,* Tires

Phillip A. Wagner, *The Dow Chemical Company, Midland, Michigan,* Styrene polymers, uses

Theodore C. Wallace, *The Dow Chemical Company, Midland, Michigan,* Styrene polymers, processing, extrusion and thermoforming

H.-C. Wang, *Exxon Chemical Company, Linden, New Jersey,* Isobutylene polymers

I. M. Ward, *University of Leeds, Leeds, United Kingdom,* Mechanical properties

William J. Ward, *University of Massachusetts, Amherst, Massachusetts,* Surface modification

F. J. Washabaugh, *Engelhard Corporation, Menlo Park, New Jersey,* Fillers

David Wei Wang, *IBM Corp., Endicott, New York,* Acrylamide polymers

D. D. Webb, *Dow Chemical U.S.A., Granville, Ohio,* Cellular materials

S. W. Webber, *The University of Texas at Austin, Austin, Texas,* Energy transfer

Thomas G. Webber, *Consultant, Vienna, West Virginia,* Colorants

O. W. Webster, *E.I. du Pont de Nemours & Co., Inc., Wilmington, Delaware,* Group-transfer polymerization

Theodore H. Wegner, *United States Department of Agriculture, Forest Products Laboratory, Madison, Wisconsin,* Wood

Edward D. Weil, *Stauffer Chemical Company, Dobbs Ferry, New York,* Phosphorus-containing polymers

Lewis B. Weisfeld, *Consultant, Jenkintown, Pennsylvania,* Regulatory considerations

R. J. Welgos, *Allied Chemical Corporation, Morristown, New Jersey,* Polyamides, plastics

Gary C. Welsh, *The Dow Chemical Company, Midland, Michigan,* Styrene polymers, processing, extrusion and thermoforming; Styrene polymers, processing, polystyrene foams

J. Wendorff, *Deutches Kunststoff Institute, Darmstadt, Federal Republic of Germany,* Amorphous polymers

Eberhard Werner, *Hoechst AG, Wiesbaden, Federal Republic of Germany,* Polyesters, films

R. A. Wessling, *The Dow Chemical Company, Midland, Michigan,* Vinylidene chloride polymers

Dwain M. White, *General Electric Company, Selkirk, New York,* Poly(phenylene ether)

James L. White, *University of Akron, Akron, Ohio,* Orientation; Orientation processes

Zeno W. Wicks, Jr., *North Dakota State University, Fargo, North Dakota,* Coatings, principles; Drying oils

Abe Widra, *University of Illinois College of Medicine, Chicago, Illinois,* Skin, synthetic

Rudolf Wiedenmann, *Siemens AG, Erlangen, Federal Republic of Germany,* Wire and cable coverings

G. D. Wignall, *Oak Ridge National Laboratories, Oak Ridge, Tennessee,* Neutron scattering

D. M. Wiles, *National Research Council of Canada, Ottawa, Canada,* Degradation

Garth L. Wilkes, *Virginia Polytechnic University, Blacksburg, Virginia,* Rheooptical properties

J. G. Williams, *Imperial College of Science and Technology, London, United Kingdom,* Fracture

Roy C. Williams, *NL Chemicals, Inc., Buffalo, New York,* Polyesters, water-soluble

Larry R. Wilson, *The Dow Chemical Company, Midland, Michigan,* Dendritic polymers

T. Lamont Wilson, *Consultant, Louisville, Kentucky,* Dielectric heating

G. Winter, *University of Melbourne, Victoria, Australia,* Casein

H. H. Winter, *University of Massachusetts, Amherst, Massachusetts,* Gel point

Robert L. Wintermyer, *The Dow Chemical Company, Midland, Michigan,* Styrene polymers, analytical methods

Fred R. Wolf, *The BFGoodrich Company, Avon Lake, Ohio,* Acrylic elastomers, survey

James F. Wolfe, *SRI International, Menlo Park, California,* Polybenzothiazoles and polybenzoxazoles

C. P. Wong, *AT&T Engineering Research Center, Princeton, New Jersey,* Electronics applications, polymeric encapsulants

Allan T. Worm, *3M Company, St. Paul, Minnesota,* Fluorocarbon elastomers

Bernhard Wunderlich, *Oak Ridge National Laboratory, Oak Ridge, Tennessee,* Thermodynamic properties

S. Yamamoto, *Kansai University, Suita, Japan,* Proliferous polymerization

Ioannis V. Yannas, *Massachusetts Institute of Technology, Cambridge, Massachusetts,* Skin, regeneration templates

Peter A. Yarnell, *Rohm and Haas Company, Spring House, Pennsylvania,* Ion-exchange polymers

Albert F. Yee, *University of Michigan, Ann Arbor, Michigan,* Impact resistance

Teh Fu Yen, *University of Southern California, Los Angeles, California,* Asphaltic materials

W. L. Young, *The Dow Chemical Co., Plaquemine, Louisiana,* Ethylene polymers, chlorinated polyethylene

Anagnostis E. Zachariades, *IBM Corporation, San Jose, California,* Solid-state extrusion

James E. Zajic,[†] *University of Texas, El Paso, Texas,* Pullulan

Guiseppe Zerbi, *Politecnico Milano, Milan, Italy,* Defects

William F. Zimmer, *Norton Company, Worcester, Massachusetts,* Abrasives

Joseph Zimmerman, *Consultant, Greenville, Delaware,* Polyamides

Paul Zoller, *E.I. du Pont de Nemours & Co., Inc., Wilmington, Delaware,* Dilatometry

[†]Deceased.

CONVERSION FACTORS, ABBREVIATIONS, AND UNIT SYMBOLS

SI UNITS (ADOPTED 1960)

A new system of measurement, the International System of Units (abbreviated SI), is being implemented throughout the world. This system is a modernized version of the MKSA (meter, kilogram, second, ampere) system, and its details are published and controlled by an international treaty organization (The International Bureau of Weights and Measures) (1).

SI units are divided into three classes:

Base Units

length	meter† (m)
mass‡	kilogram (kg)
time	second (s)
electric current	ampere (A)
thermodynamic temperature§	kelvin (K)
amount of substance	mole (mol)
luminous intensity	candela (cd)

Supplementary Units

plane angle	radian (rad)
solid angle	steradian (sr)

Derived Units and Other Acceptable Units

These units are formed by combining base units, supplementary units, and other derived units (2–4). Those derived units having special names and symbols are marked with an asterisk in the list below:

Quantity	*Unit*	*Symbol*	*Acceptable equivalent*
*absorbed dose	gray	Gy	J/kg
acceleration	meter per second squared	m/s^2	
*activity (of ionizing radiation source)	becquerel	Bq	1/s

† The spellings "metre" and "litre" are preferred by ASTM; however, "-er" is used in the *Encyclopedia*.

‡ "Weight" is the commonly used term for "mass."

§ Wide use is made of "Celsius temperature" (t) defined by

$$t = T - T_0$$

where T is the thermodynamic temperature, expressed in kelvins, and $T_0 = 273.15$ K by definition. A temperature interval may be expressed in degrees Celsius as well as in kelvins.

Quantity	*Unit*	*Symbol*	*Acceptable equivalent*
area	square kilometer	km^2	
	square hectometer	hm^2	ha (hectare)
	square meter	m^2	
*capacitance	farad	F	C/V
concentration (of amount of substance)	mole per cubic meter	mol/m^3	
*conductance	siemens	S	A/V
current density	ampere per square meter	A/m^2	
density, mass density	kilogram per cubic meter	kg/m^3	g/L; mg/cm^3
dipole moment (quantity)	coulomb meter	C·m	
*electric charge, quantity of electricity	coulomb	C	A·s
electric charge density	coulomb per cubic meter	C/m^3	
electric field strength	volt per meter	V/m	
electric flux density	coulomb per square meter	C/m^2	
*electric potential, potential difference, electromotive force	volt	V	W/A
*electric resistance	ohm	Ω	V/A
*energy, work, quantity of heat	megajoule	MJ	
	kilojoule	kJ	
	joule	J	N·m
	electronvolt†	eV†	
	kilowatt hour†	kW·h†	
energy density	joule per cubic meter	J/m^3	
*force	kilonewton	kN	
	newton	N	$kg·m/s^2$
*frequency	megahertz	MHz	
	hertz	Hz	1/s
heat capacity, entropy	joule per kelvin	J/K	
heat capacity (specific), specific entropy	joule per kilogram kelvin	J/(kg·K)	
heat transfer coefficient	watt per square meter kelvin	$W/(m^2·K)$	
*illuminance	lux	lx	lm/m^2
*inductance	henry	H	Wb/A
linear density	kilogram per meter	kg/m	
luminance	candela per square meter	cd/m^2	
*luminous flux	lumen	lm	cd·sr
magnetic field strength	ampere per meter	A/m	
*magnetic flux	weber	Wb	V·s
*magnetic flux density	tesla	T	Wb/m^2
molar energy	joule per mole	J/mol	
molar entropy, molar heat capacity	joule per mole kelvin	J/(mol·K)	
moment of force, torque	newton meter	N·m	
momentum	kilogram meter per second	kg·m/s	
permeability	henry per meter	H/m	
permittivity	farad per meter	F/m	
*power, heat flow rate, radiant flux	kilowatt	kW	
	watt	W	J/s
power density, heat flux density, irradiance	watt per square meter	W/m^2	
*pressure, stress	megapascal	MPa	
	kilopascal	kPa	
	pascal	Pa	N/m^2
sound level	decibel	dB	
specific energy	joule per kilogram	J/kg	
specific volume	cubic meter per kilogram	m^3/kg	
surface tension	newton per meter	N/m	
thermal conductivity	watt per meter kelvin	W/(m·K)	
velocity	meter per second	m/s	
	kilometer per hour	km/h	
viscosity, dynamic	pascal second	Pa·s	
	millipascal second	mPa·s	

† This non-SI unit is recognized by the CIPM as having to be retained because of practical importance or use in specialized fields (1).

Quantity	*Unit*	*Symbol*	*Acceptable equivalent*
viscosity, kinematic	square meter per second	m^2/s	
	square millimeter per second	mm^2/s	
volume	cubic meter	m^3	
	cubic decimeter	dm^3	L(liter) (5)
	cubic centimeter	cm^3	mL
wave number	1 per meter	m^{-1}	
	1 per centimeter	cm^{-1}	

In addition, there are 16 prefixes used to indicate order of magnitude, as follows:

Multiplication factor	*Prefix*	*Symbol*
10^{18}	exa	E
10^{15}	peta	P
10^{12}	tera	T
10^{9}	giga	G
10^{6}	mega	M
10^{3}	kilo	k
10^{2}	hecto	h[a]
10	deka	da[a]
10^{-1}	deci	d[a]
10^{-2}	centi	c[a]
10^{-3}	milli	m
10^{-6}	micro	μ
10^{-9}	nano	n
10^{-12}	pico	p
10^{-15}	femto	f
10^{-18}	atto	a

Note

[a]Although hecto, deka, deci, and centi are SI prefixes, their use should be avoided except for SI unit-multiples for area and volume and nontechnical use of centimeter, as for body and clothing measurement.

For a complete description of SI and its use, the reader is referred to ASTM E 380 (4).

CONVERSION FACTORS TO SI UNITS

A representative list of conversion factors from non-SI to SI units is presented herewith. Factors are given to four significant figures. Exact relationships are followed by a dagger (†). A more complete list is given in ASTM E 380–84 (4) and ANSI Z 210.1–1976 (6).

To convert from	*To*	*Multiply by*
acre	square meter (m^2)	4.047×10^{3}
angstrom	meter (m)	$1.0 \times 10^{-10\dagger}$
are	square meter (m^2)	$1.0 \times 10^{2\dagger}$
astronomical unit	meter (m)	1.496×10^{11}
atmosphere	pascal (Pa)	1.013×10^{5}
bar	pascal (Pa)	$1.0 \times 10^{5\dagger}$
barn	square meter (m^2)	$1.0 \times 10^{-28\dagger}$
barrel (42 U.S. liquid gallons)	cubic meter (m^3)	0.1590
Bohr magneton (μ_β)	J/T	9.274×10^{-24}
Btu (International Table)	joule (J)	1.055×10^{3}
Btu (mean)	joule (J)	1.056×10^{3}
Btu (thermochemical)	joule (J)	1.054×10^{3}
bushel	cubic meter (m^3)	3.524×10^{-2}
calorie (International Table)	joule (J)	4.187
calorie (mean)	joule (J)	4.1908
calorie (thermochemical)	joule (J)	$4.184^{\dagger}$
centipoise	pascal second (Pa·s)	$1.0 \times 10^{-3\dagger}$
centistokes	square millimeter per second (mm^2/s)	$1.0^{\dagger}$

† Exact.

To convert from	*To*	*Multiply by*
cfm (cubic foot per minute)	cubic meter per second (m^3/s)	4.72×10^{-4}
cubic inch	cubic meter (m^3)	1.639×10^{-5}
cubic foot	cubic meter (m^3)	2.832×10^{-2}
cubic yard	cubic meter (m^3)	0.7646
curie	becquerel (Bq)	$3.70 \times 10^{10\dagger}$
debye	coulomb·meter (C·m)	3.336×10^{-30}
degree (angle)	radian (rad)	1.745×10^{-2}
denier (international)	kilogram per meter (kg/m)	1.111×10^{-7}
	tex‡	0.1111
dram (apothecaries')	kilogram (kg)	3.888×10^{-3}
dram (avoirdupois)	kilogram (kg)	1.772×10^{-3}
dram (U.S. fluid)	cubic meter (m^3)	3.697×10^{-6}
dyne	newton (N)	$1.0 \times 10^{-5\dagger}$
dyne/cm	newton per meter (N/m)	$1.0 \times 10^{-3\dagger}$
electron volt	joule (J)	1.602×10^{-19}
erg	joule (J)	$1.0 \times 10^{-7\dagger}$
fathom	meter (m)	1.829
fluid ounce (U.S.)	cubic meter (m^3)	2.957×10^{-5}
foot	meter (m)	0.3048†
footcandle	lux (lx)	10.76
furlong	meter (m)	2.012×10^{-2}
gal	meter per second squared (m/s^2)	$1.0 \times 10^{-2\dagger}$
gallon (U.S. dry)	cubic meter (m^3)	4.405×10^{-3}
gallon (U.S. liquid)	cubic meter (m^3)	3.785×10^{-3}
gallon per minute (gpm)	cubic meter per second (m^3/s)	6.308×10^{-5}
	cubic meter per hour (m^3/h)	0.2271
gauss	tesla (T)	1.0×10^{-4}
gilbert	ampere (A)	0.7958
gill (U.S.)	cubic meter (m^3)	1.183×10^{-4}
grad	radian	1.571×10^{-2}
grain	kilogram (kg)	6.480×10^{-5}
gram-force per denier	newton per tex (N/tex)	8.826×10^{-2}
hectare	square meter (m^2)	$1.0 \times 10^{4\dagger}$
horsepower (550 ft·lbf/s)	watt (W)	7.457×10^{2}
horsepower (boiler)	watt (W)	9.810×10^{3}
horsepower (electric)	watt (W)	$7.46 \times 10^{2\dagger}$
hundredweight (long)	kilogram (kg)	50.80
hundredweight (short)	kilogram (kg)	45.36
inch	meter (m)	$2.54 \times 10^{-2\dagger}$
inch of mercury (32°F)	pascal (Pa)	3.386×10^{3}
inch of water (39.2°F)	pascal (Pa)	2.491×10^{2}
kilogram-force	newton (N)	9.807
kilowatt hour	megajoule (MJ)	3.6†
kip	newton (N)	4.48×10^{3}
knot (international)	meter per second (m/s)	0.5144
lambert	candela per square meter (cd/m^2)	3.183×10^{3}
league (British nautical)	meter (m)	5.559×10^{3}
league (statute)	meter (m)	4.828×10^{3}
light year	meter (m)	9.461×10^{15}
liter (for fluids only)	cubic meter (m^3)	$1.0 \times 10^{-3\dagger}$
maxwell	weber (Wb)	$1.0 \times 10^{-8\dagger}$
micron	meter (m)	$1.0 \times 10^{-6\dagger}$
mil	meter (m)	$2.54 \times 10^{-5\dagger}$
mile (statute)	meter (m)	1.609×10^{3}
mile (U.S. nautical)	meter (m)	$1.852 \times 10^{3\dagger}$
mile per hour	meter per second (m/s)	0.4470
millibar	pascal (Pa)	1.0×10^{2}
millimeter of mercury (0°C)	pascal (Pa)	$1.333 \times 10^{2\dagger}$

†Exact.

‡ See footnote on p. xx.

To convert from	*To*	*Multiply by*
minute (angular)	radian	2.909×10^{-4}
myriagram	kilogram (kg)	10
myriameter	kilometer (km)	10
oersted	ampere per meter (A/m)	79.58
ounce (avoirdupois)	kilogram (kg)	2.835×10^{-2}
ounce (troy)	kilogram (kg)	3.110×10^{-2}
ounce (U.S. fluid)	cubic meter (m^3)	2.957×10^{5}
ounce-force	newton (N)	0.2780
peck (U.S.)	cubic meter (m^3)	8.810×10^{-3}
pennyweight	kilogram (kg)	1.555×10^{-3}
pint (U.S. dry)	cubic meter (m^3)	5.506×10^{-4}
pint (U.S. liquid)	cubic meter (m^3)	4.732×10^{-4}
poise (absolute viscosity)	pascal second (Pa·s)	$0.10^{\dagger}$
pound (avoirdupois)	kilogram (kg)	0.4536
pound (troy)	kilogram (kg)	0.3732
poundal	newton (N)	0.1383
pound-force	newton (N)	4.448
pound-force per square inch (psi)	pascal (Pa)	6.895×10^{3}
quart (U.S. dry)	cubic meter (m^3)	1.101×10^{-3}
quart (U.S. liquid)	cubic meter (m^3)	9.464×10^{-4}
quintal	kilogram (kg)	$1.0 \times 10^{2\dagger}$
rad	gray (Gy)	$1.0 \times 10^{-2\dagger}$
rod	meter (m)	5.029
roentgen	coulomb per kilogram (C/kg)	2.58×10^{-4}
second (angle)	radian (rad)	4.848×10^{-6}
section	square meter (m^2)	2.590×10^{6}
slug	kilogram (kg)	14.59
spherical candle power	lumen (lm)	12.57
square inch	square meter (m^2)	6.452×10^{-4}
square foot	square meter (m^2)	9.290×10^{-2}
square mile	square meter (m^2)	2.590×10^{6}
square yard	square meter (m^2)	0.8361
stere	cubic meter (m^3)	$1.0^{\dagger}$
stokes (kinematic viscosity)	square meter per second (m^2/s)	$1.0 \times 10^{-4\dagger}$
tex	kilogram per meter (kg/m)	$1.0 \times 10^{-6\dagger}$
ton (long, 2240 pounds)	kilogram (kg)	1.016×10^{3}
ton (metric)	kilogram (kg)	$1.0 \times 10^{3\dagger}$
ton (short, 2000 pounds)	kilogram (kg)	9.072×10^{2}
torr	pascal (Pa)	1.333×10^{2}
unit pole	weber (Wb)	1.257×10^{-7}
yard	meter (m)	$0.9144^{\dagger}$

ABBREVIATIONS AND UNIT SYMBOLS

Following is a list of commonly used abbreviations and unit symbols appropriate for use in the *Encyclopedia*. In general they agree with those listed in *American National Standard Abbreviations for Use on Drawings and in Text (ANSI Y1.1)* (6) and *American National Standard Letter Symbols for Units in Science and Technology (ANSI Y10)* (6). Also included is a list of acronyms for a number of private and government organizations as well as common industrial solvents, polymers, and other chemicals.

Rules for Writing Unit Symbols (4):

1. Unit symbols should be printed in upright letters (roman) regardless of the type style used in the surrounding text.
2. Unit symbols are unaltered in the plural.
3. Unit symbols are not followed by a period except when used as the end of a sentence.
4. Letter unit symbols are generally written in lowercase (eg, cd for candela) unless the unit name has been derived from a proper name, in which case the first letter of the symbol is capitalized (W, Pa). Prefix and unit symbols retain their prescribed form regardless of the surrounding typography.
5. In the complete expression for a quantity, a space should be left between the numerical value and the unit symbol. For example, write 2.37 lm, *not* 2.37lm, and 35 mm, *not* 35mm. When the quantity is used in an adjectival sense, a hyphen is often

† Exact.

used, for example, 35-mm film. *Exception:* No space is left between the numerical value and the symbols for degree, minute, and second of plane angle, and degree Celsius.

6. No space is used between the prefix and unit symbols (eg, kg).
7. Symbols, not abbreviations, should be used for units. For example, use "A," not "amp," for ampere.
8. When multiplying unit symbols, use a raised dot:

N·m for newton meter

In the case of W·h, the dot may be omitted, thus:

Wh

An exception to this practice is made for computer printouts, automatic typewriter work, etc, where the raised dot is not possible, and a dot on the line may be used.

9. When dividing unit symbols use one of the following forms:

$$\mathrm{m/s} \text{ or } \mathrm{m{\cdot}s^{-1}} \text{ or } \frac{\mathrm{m}}{\mathrm{s}}$$

In no case should more than one slash be used in the same expression unless parentheses are inserted to avoid ambiguity. For example, write:

$$\mathrm{J/(mol{\cdot}K)} \text{ or } \mathrm{J{\cdot}mol^{-1}\cdot K^{-1}} \text{ or } \mathrm{(J/mol)/K}$$

but *not*

J/mol/K

10. Do not mix symbols and unit names in the same expression. Write:

joules per kilogram *or* J/kg *or* $\mathrm{J{\cdot}kg^{-1}}$

but *not*

joules/kilogram *nor* joules/kg *nor* $\mathrm{joules{\cdot}kg^{-1}}$

Abbreviations and Units

A	ampere
A	anion (eg, H*A*); mass number
a	atto (prefix for 10^{-18})
AATCC	American Association of Textile Chemists and Colorists
ABS	acrylonitrile–butadiene–styrene
abs	absolute
ac	alternating current, *n.*
a-c	alternating current, *adj.*
ac-	alicyclic
ACGIH	American Conference of Governmental Industrial Hygienists
ACS	American Chemical Society
AGA	American Gas Association
Ah	ampere hour
AIChE	American Institute of Chemical Engineers
AIME	American Institute of Mining, Metallurgical, and Petroleum Engineers
AIP	American Institute of Physics
AISI	American Iron and Steel Institute
alc	alcohol(ic)
Alk	alkyl
alk	alkaline (not alkali)
-alt-	alternating as in alternating copolymer
amt	amount
amu	atomic mass unit
ANSI	American National Standards Institute
AO	atomic orbital
AOAC	Association of Official Analytical Chemists
AOCS	American Oil Chemists' Society
APHA	American Public Health Association
API	American Petroleum Institute
aq	aqueous
Ar	aryl
ar-	aromatic
as-	asymmetric(al)
ASHRAE	American Society of Heating, Refrigerating, and Air Conditioning Engineers
ASM	American Society for Metals
ASME	American Society of Mechanical Engineers
ASTM	American Society for Testing and Materials
at no.	atomic number
at wt	atomic weight
av(g)	average
AWS	American Welding Society
b	bonding orbital
bbl	barrel
bcc	body-centered cubic
bct	body-centered tetragonal
Bé	Baumé
BET	Brunauer-Emmett-Teller (adsorption equation)
bid	twice daily
Boc	*t*-butyloxycarbonyl
BOD	biochemical (biological) oxygen demand
bp	boiling point
Bq	becquerel
C	coulomb
°C	degree Celsius
C-	denoting attachment to carbon
C_M	chain-transfer constant for monomer

C_P	chain-transfer constant for polymer
C_S	chain-transfer constant for solvent
c	centi (prefix for 10^{-2})
c	critical
ca	circa (approximately)
cd	candela; current density; circular dichroism
CFR	Code of Federal Regulations
cgs	centimeter-gram-second
CI	Color Index
cis-	isomer in which substituted groups are on same side of double bond between C atoms
cl	carload
cm	centimeter
cmil	circular mil
cmpd	compound
CNRS	Centre National de la Recherche Scientifique
CNS	central nervous system
-co-	copolymerized with
CoA	coenzyme A
COC	Cleveland open cup
COD	chemical oxygen demand
coml	commercial(ly)
conc	concentration
cp	chemically pure
cph	close-packed hexagonal
CPSC	Consumer Product Safety Commission
cryst	crystalline
cub	cubic
D	debye
D-	denoting configurational relationship
d	differential operator
d-	*dextro-*, dextrorotatory
da	deka (prefix for 10^1)
dB	decibel
dc	direct current, *n.*
d-c	direct current, *adj.*
dec	decompose
detd	determined
detn	determination
dia	diameter
dil	dilute
dl-; DL-	racemic
DMA	dimethylacetamide
DMF	dimethylformamide
DMG	dimethyl glyoxime
DMSO	dimethyl sulfoxide
DOD	Department of Defense
DOE	Department of Energy
DOT	Department of Transportation
DP	degree of polymerization
dp	dew point
DPH	diamond pyramid hardness
DS	degree of substitution
dsc	differential scanning calorimetry
dstl(d)	distill(ed)
dta	differential thermal analysis
E	Young's modulus
(*E*)-	entgegen; opposed
e	polarity factor in Alfrey-Price equation
e^-	electron

ECU	electrochemical unit
ed.	edited, edition, editor
ED	effective dose
EDTA	ethylenediaminetetraacetic acid
em	electron microscopy
emf	electromotive force
emu	electromagnetic unit
en	ethylene diamine
eng	engineering
EPA	Environmental Protection Agency
epr	electron paramagnetic resonance
ϵ	dielectric constant (unitless)
eq.	equation
esca	electron-spectroscopy for chemical analysis
esp	especially
esr	electron-spin resonance
est(d)	estimate(d)
estn	estimation
esu	electrostatic unit
η	viscosity
$[\eta]$	intrinsic viscosity
η_{inh}	inherent viscosity
η_r	relative viscosity
η_{red}	reduced viscosity
η_{sp}	specific viscosity
exp	experiment, experimental
ext(d)	extract(ed)
F	farad (capacitance)
F	faraday (96,487 C); free energy
f	femto (prefix for 10^{-15})
FAO	Food and Agriculture Organization (United Nations)
fcc	face-centered cubic
FDA	Food and Drug Administration
FEA	Federal Energy Administration
FHSA	Federal Hazardous Substances Act
fob	free on board
fp	freezing point
FPC	Federal Power Commission
FRB	Federal Reserve Board
frz	freezing
G	giga (prefix for 10^9)
G	gravitational constant = 6.67×10^{11} $N \cdot m^2/kg^2$; Gibb's free energy
g	gram
(g)	gas, only as in $H_2O(g)$
g	gravitational acceleration
-g-	graft as in graft copolymer
gc	gas chromatography
gem-	geminal
glc	gas-liquid chromatography
g-mol wt; gmw	gram-molecular weight
GNP	gross national product
gpc	gel-permeation chromatography
GRAS	Generally Recognized as Safe
grd	ground
Gy	gray
H	henry
H	enthalpy
h	hour; hecto (prefix for 10^2)
ha	hectare

HB	Brinell hardness number
Hb	hemoglobin
hcp	hexagonal close-packed
hex	hexagonal
HK	Knoop hardness number
hplc	high-pressure liquid chromatography
HRC	Rockwell hardness (C scale)
HV	Vickers hardness number
hyd	hydrated, hydrous
hyg	hygroscopic
Hz	hertz
^{i}(eg, Pr^{i})	iso (eg, isopropyl)
i-	inactive (eg, *i*-methionine)
IACS	International Annealed Copper Standard
ibp	initial boiling point
IC	inhibitory concentration
ICC	Interstate Commerce Commission
ICT	International Critical Table
ID	inside diameter; infective dose
ip	intraperitoneal
IPS	iron pipe size
IPTS	International Practical Temperature Scale (NBS)
ir	infrared
IRLG	Interagency Regulatory Liaison Group
ISO	International Organization for Standardization
IU	International Unit
IUPAC	International Union of Pure and Applied Chemistry
IV	iodine value
iv	intravenous
J	joule
K	kelvin
K	equilibrium constant
k	kilo (prefix for 10^3)
k	reaction rate constant
kg	kilogram
L	denoting configurational relationship
L	liter (for fluids only) (5)
l-	*levo*-, levorotatory
(l)	liquid, only as in NH_3(l)
LC_{50}	conc lethal to 50% of the animals tested
LCAO	linear combination of atomic orbitals
LCD	liquid crystal display
lcl	less than carload lots
LD_{50}	dose lethal to 50% of the animals tested
LED	light-emitting diode
liq	liquid
lm	lumen
ln	logarithm (natural)
LNG	liquefied natural gas
log	logarithm (common)
LPG	liquefied petroleum gas
ltl	less than truckload lots
lx	lux
M	mega (prefix for 10^6); metal (as in M*A*)
M	molar; actual mass
$\overline{M}_w$	weight-average mol wt
$\overline{M}_n$	number-average mol wt
$\overline{M}_v$	viscosity-average mol wt

m	meter; milli (prefix for 10^{-3})
m	molal
m-	meta
max	maximum
MCA	Chemical Manufacturers' Association (was Manufacturing Chemists Association)
MEK	methyl ethyl ketone
meq	milliequivalent
mfd	manufactured
mfg	manufacturing
mfr	manufacturer
MIBC	methyl isobutyl carbinol
MIBK	methyl isobutyl ketone
MIC	minimum inhibiting concentration
min	minute; minimum
mL	milliliter
MLD	minimum lethal dose
MO	molecular orbital
mo	month
mol	mole
mol wt	molecular weight
mp	melting point
MR	molar refraction
ms	mass spectrum
mxt	mixture
μ	micro (prefix for 10^{-6})
N	newton (force)
N	normal (concentration); neutron number
N-	denoting attachment to nitrogen
n (as n_D^{20})	index of refraction (for 20°C and sodium light)
^{n} (as Bu^{n}), *n*-	normal (straight-chain structure)
n	neutron
n	nano (prefix for 10^9)
na	not available
NAS	National Academy of Sciences
NASA	National Aeronautics and Space Administration
nat	natural
NBS	National Bureau of Standards
neg	negative
NEMA	National Electrical Manufacturer's Association
NF	*National Formulary*
NIH	National Institutes of Health
NIOSH	National Institute of Occupational Safety and Health
nmr	nuclear magnetic resonance
NND	New and Nonofficial Drugs (AMA)
no.	number
NOI(BN)	not otherwise indexed (by name)
NOS	not otherwise specified
nqr	nuclear quadruple resonance
NRC	Nuclear Regulatory Commission; National Research Council
NRI	New Ring Index
NSF	National Science Foundation
NTA	nitrilotriacetic acid
NTP	normal temperature and pressure (25°C and 101.3 kPa or 1 atm)
NTSB	National Transportation Safety Board

O-	denoting attachment to oxygen
o-	ortho
OD	outside diameter
ω	frequency
OPEC	Organization of Petroleum Exporting Countries
OSHA	Occupational Safety and Health Administration
owf	on weight of fiber
Ω	ohm
P	peta (prefix for 10^{15})
p	pico (prefix for 10^{-12})
p-	para
p	proton
p.	page
Pa	pascal (pressure)
PAN	polyacrylonitrile
pd	potential difference
PE	polyethylene
pH	negative logarithm of the effective hydrogen ion concentration
phr	parts per hundred of resin (rubber)
π	osmotic pressure
p-i-n	positive-intrinsic-negative
pmr	proton magnetic resonance
p-n	positive-negative
po	per os (oral)
POP	polyoxypropylene
pos	positive
PP	polypropylene
pp.	pages
ppb	parts per billion (10^9)
pph	parts per hundred
ppm	parts per million (10^6)
ppmv	parts per million by volume
ppmwt	parts per million by weight
PPO	poly(phenyl oxide)
ppt(d)	precipitate(d)
pptn	precipitation
Pr (no.)	foreign prototype (number)
PS	polystyrene
pt	point; part
PVC	poly(vinyl chloride)
pwd	powder
py	pyridine
Q	reactivity of monomer in Alfrey-Price equation
qv	quod vide (which see)
R	univalent hydrocarbon radical
(*R*)-	rectus (clockwise configuration)
r	precision of data
rad	radian; radius
rds	rate determining step
Ref.	reference
rf	radio frequency, *n.*
r-f	radio frequency, *adj.*
rh	relative humidity
ρ, d	density
RI	Ring Index
rms	root-mean square
rpm	rotations per minute
rps	revolutions per second

RT	room temperature
^{s} (eg, Bu^s); *sec*-	secondary (eg, secondary butyl)
S	siemens
(*S*)-	sinister (counterclockwise configuration)
S	entropy
S-	denoting attachment to sulfur
s-	symmetric(al)
s	second
(s)	solid, only as in $H_2O(s)$
SAE	Society of Automotive Engineers
SAMPE	Society for the Advancement of Material and Process Engineering
SAN	styrene–acrylonitrile
sat(d)	saturate(d)
satn	saturation
SBR	styrene–butadiene–rubber
sc	subcutaneous
SCF	self-consistent field; standard cubic feet
Sch	Schultz number
SFs	Saybolt Furol seconds
SI	Le Système International d'Unités (International System of Units)
sl sol	slightly soluble
SMC	sheet molding compound
sol	soluble
soln	solution
soly	solubility
sp	specific; species
SPE	Society of Plastics Engineers
sp gr	specific gravity
SPI	Society of the Plastics Industry
sr	steradian
std	standard
STP	standard temperature and pressure (0°C and 101.3 kPa)
sub	sublime(s)
SUs	Saybolt Universal seconds
syn	synthetic
^{t} (eg, Bu^t), *t*-, *tert*-	tertiary (eg, tertiary-butyl)
T	tera (prefix for 10^{12}); tesla (magnetic flux density)
t	metric ton (tonne); temperature
TAPPI	Technical Association of the Pulp and Paper Industry
τ	relaxation
TCC	Tagliabue (Tag) closed cup
tex	tex (linear density)
T_g	glass-transition temperature
tga	thermogravimetric analysis
θ	adjective of condition, solvent, temperature
THF	tetrahydrofuran
tlc	thin layer chromatography
TLV	threshold limit value
TMS	tetramethylsilane
TOC	Tagliabue (Tag) open cup
trans-	isomer in which substituted groups are on opposite sides of double bond between C atoms
TSCA	Toxic Substance Control Act

TWA	time-weighted average	vs	versus
Twad	Twaddell	v sol	very soluble
UL	Underwriters' Laboratory	W	watt
USDA	United States Department of Agriculture	Wb	weber
USP	*United States Pharmacopeia*	Wh	watt hour
uv	ultraviolet	WHO	World Health Organization (United Nations)
V	volt (emf)		
var	variable	wk	week
vic-	vicinal	yr	year
vol	volume (not volatile)	*(Z)-*	zusammen; together; atomic number

Non-SI (Unacceptable and Obsolete) Units		*Use*
Å	angstrom	nm
at	atmosphere, technical	Pa
atm	atmosphere, standard	Pa
b	barn	cm^2
bar[†]	bar	Pa
bbl	barrel	m^3
bhp	brake horsepower	W
Btu	British thermal unit	J
bu	bushel	m^3; L
cal	calorie	J
cfm	cubic foot per minute	m^3/s
Ci	curie	Bq
cSt	centistokes	mm^2/s
c/s	cycle per second	Hz
cu	cubic	exponential form
D	debye	C·m
den	denier	tex
dr	dram	kg
dyn	dyne	N
dyn/cm	dyne per centimeter	mN/m
erg	erg	J
eu	entropy unit	J/K
°F	degree Fahrenheit	°C; K
fc	footcandle	lx
fl	footlambert	lx
fl oz	fluid ounce	m^3; L
ft	foot	m
ft·lbf	foot pound-force	J
gf/den	gram-force per denier	N/tex
G	gauss	T
Gal	gal	m/s^2
gal	gallon	m^3; L
Gb	gilbert	A
gpm	gallon per minute	(m^3/s); (m^3/h)
gr	grain	kg
hp	horsepower	W
ihp	indicated horsepower	W
in.	inch	m
in. Hg	inch of mercury	Pa
in. H_2O	inch of water	Pa
in.-lbf	inch pound-force	J
kcal	kilogram-calorie	J
kgf	kilogram-force	N
kilo	for kilogram	kg

[†] Do not use bar (10^5Pa) or millibar (10^2Pa) because they are not SI units, and are accepted internationally only for a limited time in special fields because of existing usage.

Non-SI (Unacceptable and Obsolete) Units		*Use*
L	lambert	lx
lb	pound	kg
lbf	pound-force	N
mho	mho	S
mi	mile	m
MM	million	M
mm Hg	millimeter of mercury	Pa
mμ	millimicron	nm
mph	mile per hour	km/h
μ	micron	μm
Oe	oersted	A/m
oz	ounce	kg
ozf	ounce-force	N
η	poise	Pa·s
P	poise	Pa·s
ph	phot	lx
psi	pound-force per square inch	Pa
psia	pound-force per square inch absolute	Pa
psig	pound-force per square inch gauge	Pa
qt	quart	m^3; L
°R	degree Rankine	K
rd	rad	Gy
sb	stilb	lx
SCF	standard cubic foot	m^3
sq	square	exponential form
thm	therm	J
yd	yard	m

BIBLIOGRAPHY

1. The International Bureau of Weights and Measures, BIPM (Parc de Saint-Cloud, France) is described on page 22 of Ref. 4. This bureau operates under the exclusive supervision of the International Committee of Weights and Measures (CIPM).
2. *Metric Editorial Guide (ANMC-78-1)* 3rd ed., American National Metric Council, 5410 Grosvenor Lane, Bethesda, Md. 20814, 1981.
3. *SI Units and Recommendations for the Use of Their Multiples and of Certain Other Units (ISO 1000–1981),* American National Standards Institute, 1430 Broadway, New York, N.Y. 10018, 1981.
4. Based on *ASTM E 380–84 (Standard for Metric Practice)*, American Society for Testing and Materials, 1916 Race Street, Philadelphia, Pa. 19103, 1984.
5. *Fed. Regist.,* Dec. 10, 1976 (41 FR 36414).
6. For ANSI address, see Ref. 3.

R. P. Lukens
American Society for Testing and Materials

A

ABHERENTS. See RELEASE AGENTS.

ABLATIVE POLYMERS. See HEAT-RESISTANT POLYMERS.

ABRASION AND WEAR

The term abrasion is commonly used to refer to a process of wear in which there is displacement of material from a surface during relative motion against hard particles or protuberances. However, it is important to recognize that abrasion is only one among many other types of wear. In any given situation, the relative importance of all the different processes will depend on the type of material involved, eg, elastomers, plastics, plastics composites, and on the conditions of operation to which they are subjected, eg, sliding or rolling, speed, temperature, stress, etc. The subject of wear is thus one of considerable complexity and it is only within the past two or three decades that the essential scientific foundations have been laid.

Surface Contact

All surfaces prepared by conventional engineering techniques are rough and will touch only over a limited number of small, discrete areas. The real area of contact is generally very much smaller than the apparent (geometric) area. The question then arises as to whether the localized contact deformation is likely to be elastic or plastic. A simple criterion to predict this relates the mechanical properties of a material to the limiting angle of slope of an asperity beyond which plastic flow will occur during flattening into the general plane of the surface. For this condition,

$$\tan\theta \,\big|_{\text{limit}} = k\,\frac{H}{E}\,(1-\nu^2)$$

where H is the indentation hardness, E is Young's modulus, ν is Poisson's ratio, and $k = 2$ for full plasticity. Some typical values of the limiting asperity angles for full plasticity are given in Table 1 and show very clearly that polymers are much more likely to form elastic asperity contacts than are metals.

Table 1. Plasticity Criteria for Asperity Deformation

Material	θ_{crit}
polyethylene, low density	13
polytetrafluoroethylene	11.4
polyethylene, high density	8.6
nylon-6,6	6
poly(methyl methacrylate)	4.9
tool steel	3.9
low carbon steel	1.0
lead	0.25

Friction

Friction is the tangential force resisting motion when one surface is displaced relative to another under the influence of an external force and it generally arises from two types of contact. In one, the localized forces of adhesion (qv) must be overcome by shear, whereas in the other, interpenetrating asperities give rise to further deformation. The total frictional force F is thus given by the sum of two terms: $F = F_{adh} + F_{def}$.

Deformation losses are most pronounced when sliding against very rough surfaces where plastic flow may lead to microcutting and viscoelastic deformation can result in tearing, cracking, or fatigue (qv). At the other extreme, when sliding against smooth surfaces, frictional work is dissipated in a very narrow interfacial zone via the rupture of adhesive bonds. For polymers, the relevant adhesive forces are primarily of the van der Waals type, but sometimes supplemented by electrostatic interactions and hydrogen bonds. With the more rigid polymers, rupture of the adhesive junctions often leads to transfer and the formation of a layer of modified material on the polymer itself. These surface changes play a crucial role in subsequent friction and wear processes. With elastomers, however, sliding is usually restricted to the actual interface itself, leading to tearing, cracking, and fatigue or to energy dissipation via waves of detachment, Shallamach waves. Both bulk-deformation and interfacial-sliding processes lead to heat generation that may ultimately result in surface oxidation, thermal degradation (qv), or actual melting.

Wear Testing

Apparatus. Numerous devices for assessing wear are used (Fig. 1), ranging in complexity from simple laboratory apparatus for basic studies, eg, **a** and **b**, to equipment simulating specific applications, eg, **g** and **h**.

Wear Indexes. The basic parameter of interest in all wear investigations is the wear rate; a variety of units are used to specify this quantity. When the most serious consequence of wear is the loss in fit of a component, such as a bearing, it is common to use a depth wear rate $W_h = h/d$, where h is the thickness removed and d is the distance of sliding. A more fundamental parameter, which arises from several simple theories of wear, is a volumetric wear rate expressed as $W_v = V/dL$, where V is the wear volume and L is the load applied. This quantity is referred to as a specific wear rate, K-factor, wear coefficient, or abrasion factor, and has units of m^2/N. The terms wear resistance and abrasion resistance are also widely used and correspond to the reciprocal of the wear rate. In routine materials testing, wear resistance is frequently compared with that of a standard reference material, and the ratio expressed as a relative wear resistance.

Deformation Wear

Abrasion. The simplest theory of abrasive (cutting) wear can be derived by considering the penetration of hard irregularities, such as cones, into a softer material. If the base angle θ of the cone is sufficiently large to cause full plasticity, then it may readily be shown from geometrical considerations that the specific wear rate is given by

$$\frac{V}{Ld} = \frac{K\tan\theta}{\pi P_m}$$

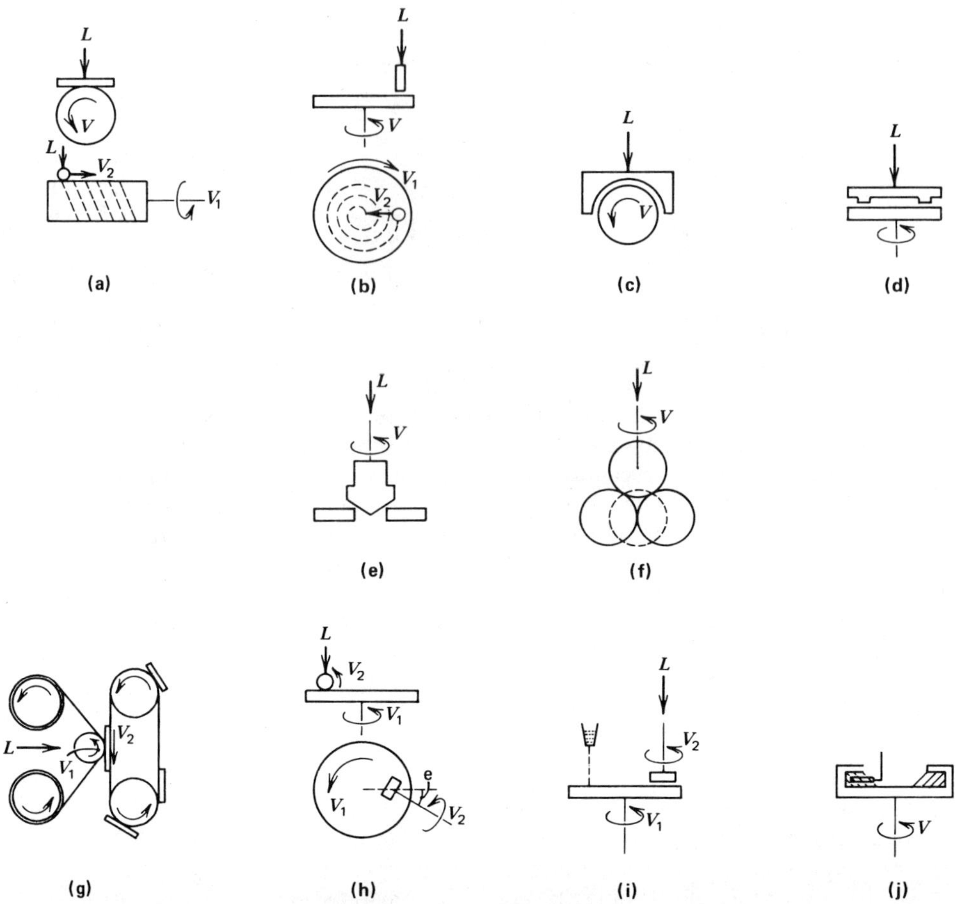

Figure 1. Schematic arrangements of various types of wear-testing apparatus: (**a**) pin on ring; (**b**) pin on disk; (**c**) block on ring; (**d**) thrust washers; (**e**) cone and washer; (**f**) four ball; (**g**) tape abrader (Armstrong); (**h**) rolling with slip (Taber); (**i**) pin and disk—loose abrasive (Olsen); (**j**) abrasive erosion (NBS). British Crown Copyright.

where L is the applied load, d is the sliding distance, P_m is the flow pressure (approximately equal to the indentation hardness), and K is the proportion of material displaced, which is removed as debris.

The wear rate is thus a product of two parameters: a material property, typified by P_m, and a geometrical factor, given in this case by $\tan \theta/\pi$. The latter will vary with the particular shape assumed for the penetrating asperities.

Erosion. When impacting abrasive particles have an appreciable velocity component parallel to the surface under attack, the mechanisms of wear and the relationships between wear and material properties are very similar to those already discussed. In general, the greater the impact strength or toughness of a material, the greater the erosion resistance.

Fatigue. Recognition of fatigue (qv) as a main mechanism of wear for polymers has arisen mainly from work in the USSR. The simplest theoretical treatment assumes indentation of a rigid hemispherical asperity of radius r into a smooth polymer plane under an applied load L. If n repeated contacts are required to produce a debris particle of dimensions comparable to the diameter of the contact spot a, then the volume per unit sliding distance V/d, $\propto a^2/n$. The value of n can be derived from the Wöhler relationship characterizing the fatigue behavior of the polymer:

$$n \propto \left(\frac{\sigma_0}{\sigma}\right)^t$$

where σ is the applied stress, σ_0 is the failure stress, and t is a material constant. From elasticity (qv) theory:

$$a \propto \left(\frac{rL}{E}\right)^{1/3} \text{ and } \sigma \propto \left(\frac{E}{r}\right)^{2/3} L^{1/3}$$

and the wear rate is thus given by

$$\frac{V}{d} \propto \sigma_0^{-t} E^{2/3(t-1)} r^{-2/3(t-1)} L^{1/3(t+2)}$$

Values of the exponent t range from 1.5–3 for elastomers to 3–10 for the more rigid plastics and their composites.

Interfacial Wear

The main characteristics of interfacial wear are energy dissipation, occurring within a narrow region adjacent to the sliding interface, and adhesion which plays a principal role. This regime is mainly associated with the more rigid plastics sliding against smooth metals; breaking of the interfacial bonds often leads to transfer to the metal counterface, localized high temperatures, and structural or chemical degradation of the plastic. An intermediate situation between deformation and interfacial wear arises when elastomers slide against smooth metals. Although the adhesive forces involved here can be large, deformation still provides the main contribution to wear.

Composites

The primary reason for adding fillers or reinforcing fibers to polymers is to improve their mechanical properties, but the effects on wear are not invariably beneficial. The most important instance of this occurs during deformation (abrasive) wear against rough surfaces where the relevant mechanical property controlling the wear rate is the product *se*. Whereas reinforcements increase the breaking strength *s*, they usually greatly decrease the elongation to break *e*, and hence the product *se* may become smaller than that of the unreinforced polymer. In consequence, reinforcement frequently increases abrasive wear rates. Similar considerations apply to the erosion of composites, either by solid particles or by liquid impact. In contrast, the effects of fillers and reinforcements on interfacial wear processes are much more favorable. Here, the fillers may support the load preferentially or modify the counterface by transfer or polishing.

Lubrication

The ideal function of a fluid lubricant is to provide complete separation between two moving surfaces so that resistance to motion only arises from viscous shear within the fluid itself. This is hydrodynamic or elastohydrodynamic lubrication. The low elastic moduli of polymers, compared with metals, facilitate the development of this lubrication regime because elastic deflections under load can become sufficient to generate the appropriate geometrical surface contours on either a macroscale or microscale. A particularly interesting feature for some polymers, eg, acetal and nylon-6,6, is that low friction can be maintained for long periods by using only traces of liquid lubricants.

Uses of polymer for wear-resistant purposes are widespread, and include bearings, gears, seals, piston rings, flooring materials, and dental restoratives. For dry-bearings, Table 2 gives a summary of the ranges of wear rates characteristic of different groups of polymeric materials. As an alternative to fabricating components from bulk materials, use is often made of thin-surface films bonded to metallic substrates so as to increase component strength and stiffness.

Table 2. Comparative Wear Rates of Various Groups of Polymers and Composites[a]

Material	Wear rate, mm^3/Nm
unfilled PTFE	ca 10^{-4}
unfilled thermoplastics	3×10^{-7}–10^{-5}
filled–reinforced thermoplastics	10^{-7}–10^{-5}
reinforced thermosets and solid lubricants	3×10^{-7}–3×10^{-6}
filled PTFE formulations	10^{-7}–10^{-6}
polyethylene, ultrahigh mol wt	10^{-7}–5×10^{-7}
woven PTFE fiber–resin thin layers	3×10^{-8}–5×10^{-7}
porous bronze–PTFE lead	3×10^{-8}–2×10^{-7}

[a] Sliding against smooth steel; frictional heating negligible.

BIBLIOGRAPHY

B. J. Briscoe, "Tribology of Polymers: State of an Art," in K. L. Mittal, ed., *Physicochemical Aspects of Polymer Surfaces,* Plenum Publishing Corp., New York, 1982, p. 387.

G. M. Bartenev and V. V. Lavrentev, *Friction and Wear of Polymers,* Elsevier, North-Holland, Amsterdam, The Netherlands, 1981.

D. I. James, ed., *Abrasion of Rubber,* MacLaren & Sons, London, 1967.

J. K. Lancaster, "Dry-Bearings: A Survey of Materials and Factors Affecting Their Performance," *Tribology* **6,** 219 (1973).

JOHN K. LANCASTER
Royal Aircraft Establishment

ABRASIVES

Abrasive products are produced by bonding particles of abrasives either to themselves or to a substrate. Naturally occurring abrasives such as corundum, garnet, silica, flint, emery, and diamond are used as well as synthetic abrasives such as aluminum oxide, alumina zirconia, silicon carbide, cubic boron nitride, and synthetic diamond. Abrasive products are divided into two main categories according to processing and usage differences: bonded abrasives and coated abrasives.

Bonded Abrasive Products

A bonded abrasive article is constructed with a three-dimensional distribution of abrasive grain, bond, and porosity, each volume fraction of which is carefully controlled. Such compositions operate sacrificially: as the abrasive grains at the surface begin to dull, they are removed by either operational wear or a dressing operation to expose other sharper grains. In this way, a relatively steady stock-removal rate can be maintained during the useful life of the article. Bond behavior depends on bond type, ie, whether it is metal, ceramic, or organic in nature. Manufacturing methods include powder technology and liquid systems.

Organic Bond Types. Organic bond types include phenolic resin, rubber, shellac, alkyd resin, epoxy resin, polyurethane, and high performance polymers.

Coated Abrasive Products

In contrast to bonded abrasives, coated abrasives are designed to have a more-or-less planar distribution of abrasive grains on a backing material that normally exhibits some flexibility. These grains are imbedded in a thin layer of adhesive called the maker coat and are usually coated over by one or more sizing coats intended to provide the necessary adhesive strength to fix the abrasive thoroughly in place. Coated products are designed to have a high degree of initial sharpness for rapid stock removal at relatively light pressures.

Backings. Backings for coated abrasives generally consist of paper (qv), vulcanized fiber, woven cloth or mesh, nonwoven fibrous materials, filament materials, or even wires or wire mesh.

Maker and Size Adhesives. A partial listing of adhesives includes hide glue; air-drying varnishes; phenol–formaldehyde (resole), urea–formaldehyde, and melamine resins; epoxies, polyurethanes, and a variety of thermoplastics (see ADHESIVE COMPOSITIONS).

Adhesives for Coated Abrasive Conversions. Adhesives used in manufacturing coated abrasive products need to be heat, shock, and moisture resistant as well as to have flexibility and high tensile strength.

Economic Considerations

The free-world market in 1982 for abrasive products was estimated at about $3,200,000,000. Of this, there were sales worth about $1,500,000,000 in the United States.

BIBLIOGRAPHY

L. Coes, Jr., *Abrasives, Applied Mineralogy Series,* Springer-Verlag, New York, 1971.

Coated Abrasives—Modern Tool of Industry, 1st ed., Coated Abrasive Manufacturers Institute, New York, 1965.

W. G. Pinkstone, *The Abrasive Ages,* A Livingston Book, Science Book Service, Sutter House, Lititz, Pa., 1974.

K. B. Lewis and W. F. Schleicher, *The Grinding Wheel,* The Grinding Wheel Institute, Cleveland, Ohio, 1976.

M. J. Collie, ed., *Industrial Abrasive Materials and Composites,* Springer-Verlag, New York, 1981.

William F. Zimmer
Norton Company

ABSORPTION. See Transport properties.

ABS POLYMERS. See Acrylonitrile–butadiene–styrene polymers.

ACCELERATORS AND ACTIVATORS. See Compounding; Rubber chemicals.

ACETALDEHYDE POLYMERS. See Aldehyde polymers.

ACETAL RESINS

Acetal resins are the family of polymers derived primarily from formaldehyde and therefore have a repeat unit $(CH_2—O)_n$. The homopolymers are completely characterized by this repeat unit and the name polyoxymethylene is more structurally descriptive. Acetal copolymers have the oxymethylene structure occasionally interrupted by the comonomer unit. The designation aldehyde polymers (qv) generally refers to polymers of aldehydes larger than formaldehyde. The term polyacetal (qv) should be reserved for polymeric products of the reactions of aldehydes and polyhydric alcohols.

Today both homopolymers and copolymers are commercially available from Celanese Engineering Resins, DuPont, Ticona Polymerwerke GmbH, Polyplastics Co., Ltd., Ultraform GmbH, Asahi Chemical Industry Co., Ltd., and Mitsubishi Gas Chemical Co. Throughout this discussion, the homopolymers and copolymers referred to are generally DuPont's Delrin and Celanese's Celcon acetal copolymer, respectively.

Physical Properties of Monomer

Acetal resins are produced from either formaldehyde or its cyclic trimer, trioxane. Formaldehyde at room temperature is a colorless, pungent gas that is highly irritating to the mucous membranes of the eyes, nose, and throat. Trioxane is the anhydrous trimer of formaldehyde. It is a white, crystalline solid with a pleasant etherlike aroma unlike that of formaldehyde. Both formaldehyde and trioxane are flammable. Table 1 lists the physical properties of each.

Table 1. Properties of Formaldehyde and Trioxane

Property	Formaldehyde	Trioxane
molecular weight	30.03	90.08
formula	CH_2O	$C_3H_6O_3$
density, g/cm^3		
−20°C	0.815	
10°C		1.17
normal boiling point, °C	−19	114.4
melting point, °C	−118	62

Chemical Properties of Monomer

Formaldehyde is extremely reactive with itself and with other chemicals. At ambient temperatures, pure formaldehyde gas polymerizes slowly, yielding a white solid. Trioxane is chemically stable. It readily depolymerizes to monomeric formaldehyde in an acid environment at elevated temperatures.

Manufacture of Monomer

Formaldehyde is manufactured from methanol. In a typical process, methanol and air are passed across a catalyst bed consisting of a silver or a metal oxide catalyst. The reactor product (consisting of formaldehyde and unconverted methanol and the by-products carbon monoxide, carbon dioxide, methyl formate, and formic acid) is cooled and then scrubbed with water in an absorption tower. The aqueous product is then purified by distillation. Acetal copolymer is commonly produced from trioxane, which is made from aqueous formaldehyde in the presence of a strong acid catalyst.

Polymerization

Mechanism. Formaldehyde is very reactive and readily polymerizes with ionic initiation. Either anionic or cationic materials such as amines, phosphines, onium salts, sulfur compounds, mineral acids, halogens, and Lewis acids, etc will initiate the polymerization reaction, as is illustrated below.

Initiation:
$$R^- + CH_2O \longrightarrow RCH_2O^- \quad \text{(anionic)}$$
$$R^+ + O{=}CH_2 \longrightarrow$$
$$[R—\overset{+}{O}{=}CH_2 \longleftrightarrow R—O—CH_2{}^+] \quad \text{(cationic)}$$

In commercial practice, it is believed that anionic initiators are used to produce acetal homopolymers. Commercial acetal copolymers are typically produced by the cationic polymerization of trioxane and 0.1 to 15 mol % of cyclic ether, such as ethylene oxide or 1,3-dioxolane.

Stabilization. Commercial acetal resin must be stabilized in order for it to be melt processed by conventional plastic-processing techniques. The end-capping reactions commonly used for the prevention of unzippering mechanisms are esterification or alkylation of the hemiacetal, ie, —O—CH_2—OH, chain ends that normally result from polymerization.

Manufacture. Although the actual manufacturing processes for acetal resins have never been revealed, the following process description is believed to be representative of those used for homopolymer and copolymer. In general, monomer is purified on site and fed to a polymerization reactor; the resulting polymer is sent through another reactor, in which end-group stabilization takes place by acetylation for the homopolymer, or hydrolysis for the copolymer. The resin is then washed and dried before a final compounding operation, in which thermal stabilizers, formaldehyde scavengers, and other additives are incorporated in the product.

Polymer Properties

Acetal resins are crystalline polymers (qv); the reported crystallinity is between 60 and 77% measured either by x ray or from density data. Some mechanical and thermal properties are given in Table 2.

Table 2. Mechanical Properties of Acetal Resin

Property	Homopolymer	Copolymer
tensile strength, yield, MPa[a], 23°C	68.9	60.6
elongation, break, %	23–75	40–75
tensile modulus, MPa[a], 23°C	3100	2825
specific gravity	1.42	1.41
melting point, °C	175	165
flow temperature, °C	184	174

[a] To convert MPa to psi, multiply by 145.

Acetal resins have excellent resistance to most chemicals except strong mineral acids. Their electrical properties are fairly typical of thermoplastic resins in general. However, they are important in combination with the other attributes of acetal resins. One significant observation is the relative constancy of dielectric constant and low losses in the useful operating range of −40 to 50°C.

Processing of Polymers

Processing techniques include injection molding, extrusion, foam molding, and rotational molding (qv).

Resin Grades

The general-purpose grades of acetal resins are characterized by differences in melt-flow properties. These grades are available with or without lubricants and in a wide variety of colors. In addition to the general-purpose resins, specialty grades have also emerged, such as Celcon GC-25A, a glass fiber-reinforced copolymer that provides higher strength and stiffness and the chemical resistance of acetals.

Economic Aspects

Total capacity in 1983 for the main producers in the United States, western Europe, and Japan was estimated at 254,000 metric tons. The recent proliferation in acetal resin suppliers is evidence of the steady growth of acetals and the acceptance of these resins as metal replacements. Reduced energy costs in the preparation of acetal resins have made them cost competitive with the metals they seek to replace in many uses. Common markets for acetal resins include consumer goods, transportation, industrial machinery, plumbing, appliances, and hardware.

Health and Safety Factors

Continued exposure to formaldehyde over a prolonged period of time is harmful. It is important, therefore, to locate processing operations in well-ventilated areas.

BIBLIOGRAPHY

M. Sittig, *Polyacetal Resins,* Gulf Publishing Co., Houston, Tex., 1963, pp. 55–66.

S. J. Baker and M. B. Price, *Polyacetals,* American Elsevier Publishers Co., New York, 1961.

R. B. Seymour and G. S. Kirshenbaum, *High Performance Polymers: Their Origin and Development,* Elsevier Science Publishing Co., Inc., New York, 1986, pp. 105–124.

THOMAS J. DOLCE
JOHN A. GRATES
Celanese Engineering Resins Co.

ACETATE FIBERS. See CELLULOSE ESTERS, ORGANIC.

ACETONE POLYMERS. See KETONE POLYMERS.

ACETYLENE-TERMINATED PREPOLYMERS

Monomers, oligomers, and polymers containing the acetylenic (ethynyl) group (—C≡C—H) and substituted acetylenic group (—C≡C—R) have received considerable attention recently. The work on thermally curable acetylene-containing materials during the last 10 years has produced significant advances: (1) a new inexpensive synthetic route to acetylene-containing compounds, (2) fundamental information on chemical structure–cure–property relationships, (3) cured resins with attractive properties, and (4) a technology providing a broad scope of resin products for various applications. The chemistry of acetylene-containing materials permits the tailoring of systems to provide cured resins with a unique combination of properties.

The acetylenic group can react under cationic, coordination, free-radical, photolytic, and thermal inducement. Only the thermally induced reaction of the acetylenic group leading to chain extension, branching, cross-linking, and rigidification is treated in this article. Oligomers and polymers containing acetylenic groups have excellent shelf lives, can be cured under moderate conditions without the evolution of volatiles, and when cured, provide resins with solvent resistance, moisture resistance, and a favorable combination of physical and mechanical properties. Cured materials from oligomers and polymers containing the acetylenic group are useful as functional [eg, films (qv) and coatings (qv)] and structural resins (eg, adhesives and composite matrices) (see also COMPOSITES; CURING).

Synthesis of Acetylenic Compounds

Ethynyl and substituted ethynyl groups have been placed on monomers, oligomers, and polymers through various synthetic routes. In general, monomers or end-capping reactants containing the acetylenic group are prepared and used to synthesize polymers or terminate oligomers. In a few cases, functional groups on oligomers and polymers are converted to acetylenic groups.

Simple Compounds Containing a Single Acetylenic Group

Phenylacetylene reacts thermally under different conditions to yield complex products composed primarily of polymeric materials. The polymers are conjugated polyenes [—(C_6H_5)C=CH—] in which three conformational isomers (trans–transoid, trans–cisoid, and cis–transoid) are possible depending on the method of synthesis and thermal history (see ACETYLENIC POLYMERS).

Simple Compounds Containing Two Acetylenic Groups

Although many diethynyl compounds have been reported, relatively little work has been disclosed regarding the nature of the product from the thermal reaction of simple diacetylenic compounds. Products from the thermal polymerization of diethynyl compounds are insoluble, which makes identification difficult.

Acetylene-Terminated Oligomers

Acetylene-terminated oligomers include phenylenes (eg, H-Resin), whose typical properties are shown in Table 1; imides (eg, HR-600), whose typical properties are shown in Table 2; acetylene-terminated phenylquinoxalines (ATPQ), whose typical properties are shown in Table 3; sulfones, whose properties are shown in Table 4; esters; phenyl-*as*-triazines; and perfluoroalkylene ether benzoxazoles.

Table 1. Typical Properties of H-Resin

density, g/cm^3	1.145
short-term use temperature in air, °C	350
flexural strength at 23°C, MPa	34.5[a]
flexural modulus at 23°C, GPa	4.82[b]

[a] To convert MPa to psi, multiply by 145.
[b] To convert GPa to psi, multiply by 145 × 10^3.

Table 2. Typical Properties of HR-600

tensile strength, MPa	96.5
tensile modulus, GPa	3.79
elongation, %	2.6
flexural strength, MPa	124
flexural modulus, GPa	4.48

[a] ASTM D790.

Table 3. Unidirectional ATPQ HT-S Graphite Fiber Composite Properties[a]

Test condition	Flexural strength[b], GPa	Flexural modulus[b], GPa	Interlaminar shear strength[c], MPa
RT	1.63	133.0	100.7
232°C after 10 min	1.42	124.8	71.0
232°C after 2000 h	1.37	126.8	51.7

[a] Cured at 316°C under 1.38 MPa (200 psi) for 4 h, calculated resin content = 32%.
[b] ASTM D790.
[c] ASTM D2344.

Table 4. Properties of Ethynyl-terminated Sulfones and UDEL

M_n of sulfone segment	η_{inh}, dL/g[a]	Gpc[b] peak retention time, min	T_g of cured resin, °C Dsc[c]	Tba[d]	Chloroform solubility of cured resin[e]
3,000	0.26	19.07	211	208	very slight swelling
9,000	0.38	17.74	202	203	swells
15,000	0.41	17.26	197	198	pronounced swelling
UDEL[f]	0.44	17.08	193	194	dissolves

[a] Inherent viscosity, 0.5% solution in chloroform at 25°C.
[b] Gel-permeation (size-exclusion) chromatography using μ-Styragel (10^5, 10^4, 10^3, 10^2 nm), chloroform as solvent.
[c] Differential scanning calorimetry at heating rate of 20°C/min, resin cure 0.5 h at 250°C in air.
[d] Torsional braid analysis at heating rate of 3°C/min, resin cure 0.5 h at 250°C in air.
[e] Determined on film cured 0.5 h at 250°C in air, chloroform soak time 1 h.
[f] Available from Union Carbide Corporation.

Table 5. Thin-film Properties[a] of PPQ[b]

Polymer, T_g [c]	Test temperature, °C	Tensile strength, MPa	Tensile modulus, GPa	Break elongation, %
void of	RT	111.1	2.44	7.1
C≡C—C_6H_5	200	50.3	1.64	12.3
containing	RT	111.1	2.57	6.2
5 mol % C≡C—C_6H_5	200	56.8	1.78	10.7

[a] ASTM D882.
[b] Cured at 200°C for 16 h and at 325°C for 1 h.
[c] Determined by thermal mechanical analysis at a heating rate of 5°C/min.

Polymers Containing Pendent Acetylenic Groups

Polymers containing pendent acetylenic groups include phenylquinoxalines (PPQ), whose typical properties are shown in Table 5, imides, sulfones, aromatic amides, esters, and phenoxy resins.

Propargyl-containing Polymers

N-Propargyl-substituted polyamides were prepared by copolymerization of 4,4′,-bis(methylaminophenyl)methane and 4,4′-bis(propargylaminophenyl)methane with isophthaloyl chloride. The thermooxidative stability for cured propargyl-substituted polyamides, as evidenced by isothermal aging at 300°C in air, was significantly poorer than for a comparable polyamide without propargyl groups.

Conclusion

Acetylene-containing materials, particularly oligomers and polymers, represent a new class of high performance functional and structural resins with excellent processibility, solvent and moisture resistance, and attractive mechanical properties.

BIBLIOGRAPHY

H. Jabloner and L. C. Cessna, *Polym. Prepr. Am. Chem. Soc. Div. Polym. Chem.* **17**(1), 169 (1976).

N. Bilow and A. L. Landis, *Natl. SAMPE Tech. Conf. Ser.* **8**, 94 (1976).

P. M. Hergenrother, *Polym. Eng. Sci.* **21**, 1072 (1981).

P. M. Hergenrother, *J. Appl. Polym. Sci.* **28**, 355 (1983).

PAUL M. HERGENROTHER
NASA Langley Research Center

ACETYLENIC POLYMERS

In the late 1950s and early 1960s, there was much interest in conducting polymers. Conjugated polymers were perceived to be desirable for this purpose. The parent member of this class of polymers, polyacetylene, has been known since 1953, when K. Ziegler patented a process employing catalysts prepared from metal alkyls and transition-metal salts for its synthesis from acetylene.

POLYACETYLENE

Properties of the Monomer (HC≡CH)

Safety Considerations. Acetylene is both flammable and explosive. Use of vacuum techniques ensures that air content never approaches the 19% volume required for combustion and eliminates the danger from flammability. Acetylene's minimum ignition temperature is 300°C. A hood should enclose all apparatus for rapid venting and dilution of any acetylene that might be liberated.

Acetylene Purification. Acetylene is commercially diluted with acetone for handling in cylinders because it cannot be safely compressed to more than 103 kPa absolute. The acetone must be removed to obtain high quality polyacetylene. Two basic purification methods have been employed: scrubbers and sublimation.

Chemical Stability of Polyacetylene

If the polymer is to be used in any practical application, the question of chemical stability of polyacetylene is extremely important. The chemical stability of polyacetylene toward organic solvents is probably good owing to its intractable nature; however, chlorinated hydrocarbons should probably be avoided because of halogen interactions with the polyene. Acids and bases will either dope the polyacetylene or react with the dopant. Care should be taken in choosing dopant–solvent combinations in any experiment.

Physical Properties of Polyacetylene

Many of the important physical properties of polyacetylene are listed in Table 1.

Table 1. Physical Properties of Polyacetylene

Property	Value
Crystal structure	
cis-polyacetylene	orthorhombic a = 7.61, b = 4.47, c = 4.39 density = 1.16 g/cm^3
trans-polyacetylene	orthorhombic a = 7.32, b = 4.24, c = 2.46 density = 1.13 g/cm^3
Morphology	
fibrillar, Shirakawa	0.02–0.08 μm diameter
amorphous, nonporous films	
swollen gels	
Molecular weight, M_n	6–10 × 10^4
Thermal behavior	
cis polymer	
1. cis→trans isomerization	145°C
2. molecular rearrangement	325°C
3. thermal decomposition	420°C
trans polymer	should exhibit events 1 and 2, provided these temperatures have not been reached previously
T_g, inferred from dynamic mechanical measurements	>150°C
Orientation effects and mechanical properties	
Young's modulus	200 MPa (cis at RT)
cis material	orientable (draw ratio 3×); produces anisotropies in conductivity
trans material	brittle
exposure to oxygen	produces brittleness

Polymerization

Acetylene is polymerized to freestanding polyacetylene film by the Shirakawa method.

Processing of Polyacetylene

Polyacetylene prepared by the Shirakawa method is a cis isomer (**1**). It can be isomerized to the thermodynamically more stable trans isomer (**2**) thermally or by electrochemical doping and undoping.

Isomerization of *Cis*- to *Trans*-polyacetylene.

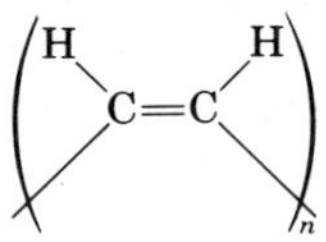

(1)

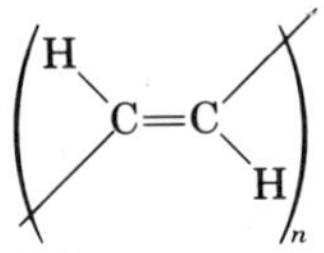

(2)

Economic Aspects

Because viable methods of processing polyacetylene have not been demonstrated, polyacetylene has not really been developed commercially.

Health and Safety Factors

Pure polyacetylene is an inocuous polymer and should present no health hazard.

Uses

The current main interests for polyacetylene are in the areas of solar cell and battery applications.

SUBSTITUTED POLYACETYLENES

Synthesis and Properties

Monosubstituted Acetylene Polymers. ***Direct Synthesis.*** Many monosubstituted acetylenes (RC≡CH) have been polymerized using a wide range of catalysts and conditions (Table 2). Polymerization conditions include homogeneous and heterogeneous Ziegler-Natta catalysts (qv); transition-metal complexes (palladium, tungsten, molybdenum, nickel, and cobalt); free-radical initiators such as benzoyl peroxide and di-*tert*-butyl peroxide; thermal polymerization; γ-radiation in solid and liquid states; cationic initiation with BF_3; and anionic initiation by butyllithium, triethylamine, and sodium amide. Only the primary and secondary alkylacetylenes can be polymerized with Ziegler-Natta catalysts. For more highly branched alkylacetylenes, arylacetylenes, and silylacetylenes, the highest molecular weights are generally obtained when they are polymerized with molybdenum- or tungsten-based catalysts such as $MoCl_5$, $Mo(CO)_6$, WCl_6, and $W(CO)_6$. Acetylenes with electron-withdrawing substituents, such as —CH_2Cl, —CH_2Br, —CF_3, —COOR, and —CN, can be polymerized anionically. Polymerization of acetylenes with strongly electron-releasing substituents has not been reported, but ferrocenylacetylene and phenylthioacetylene have been polymerized cationically with alkylboranes (BR_3).

The average of the number-average molecular weight ($\bar{M}_n$) for the soluble polymers in Table 2 is 15,000. The long wavelength absorption maxima (λ_{max}) for these polymers range from 250 to 620 nm, and the visual colors from white or colorless to black. Conductivities (σ) of the polymers range from $<10^{-19}$ to 10^{-4} S/cm for the undoped polymers.

A few terminal diacetylenes (qv) have been cyclopolymerized to conjugated polyenes. The resultant polymers are considered monosubstituted because only every other carbon of the backbone is substituted. The long wavelength absorption maxima (λ_{max}) for these cyclopolymers are 270–620 nm, similar to the values for the polymers of Table 2, but visual colors are deeper, ie, dark red to black.

Indirect Synthesis. Indirect synthesis of substituted polyacetylenes involves the elimination of some species from a preformed polymer to generate a conjugated polyene or the alteration of substituents on a preformed polyacetylene derivative by chemical reaction. All of the monosubstituted polyacetylenes prepared by indirect methods are dark in color, ie, brown or black.

Disubstituted Acetylene Polymers. At least 17 disubstituted acetylenes have now been polymerized. These include a variety of dialkyl, alkyl–silyl, and diaryl compounds, and three electron-withdrawing substituents, —CF_3, —Cl, and —CN. The dialkyl, alkyl–aryl, diaryl, and alkyl–silyl acetylene derivatives have been polymerized using transition-metal catalysts prepared from $MoCl_5$, $MoCl_6$, $Nb_2Cl_6(SC_4H_8)_3$, $TaBr_5$, and $Ta_2Cl_6(SC_4H_8)_3$, sometimes with cocatalysts such as $Sn(C_6H_5)_4$, $Al(C_2H_5)_3$, and light.

The disubstituted acetylene polymers prepared using transition-metal catalysts possess very high molecular weights, generally $>10^6$. Most of these polyenes are noteworthy for their solubility and their lack of absorption in the visible region of the spectrum. Most are colorless and possess very few free radicals.

Indirect syntheses of disubstituted acetylene polymers require either the formation of the conjugated polyene by elimination from an unconjugated polymeric precursor, or the alteration of the chemical structure of a preformed conjugated polyene.

Mechanism of Conduction

The most important mechanistic distinction is between ionic and electronic modes of conduction. In the former case, current flow is the result of macroscopic motion of ions in the solid. Electronic conduction involves the motion of electrons along the polyene chains; any motion of the counterions by relaxation is incidental to conduction.

These two types of conduction mechanism result in quite different behavior, especially as a function of time, temperature, and frequency (alternating current). It is therefore important to know which mechanism obtains in a material prior to any attempted applications. Such mechanistic studies have been carried out for only a few of the substituted polyacetylenes.

Steric Effects

The steric effects of substituents in polyacetylenes on polymer properties are very important and, in fact, are predominant in most cases, producing nonconjugated polyenes. Thus, most of the monosubstituted acetylenes in Table 2 are soluble, lack color, exhibit few free spins and broad esr line widths (ΔH), and have low intrinsic and low doped conductivities. The same properties apply to the disubstituted acetylene polymers where

Table 2. Some Monosubstituted Acetylene Polymers Prepared by Direct Polymerization $(CH{=}C(R))_n$

R	$10^{-3}\ \bar{M}_n$	Log N_s	ΔH, mT[a]	λ^b_{max}, nm	− Log $\sigma_{25°C}$, S/cm Undoped	Doped	$E_{a,cond}$, eV, undoped
CH_3		ca 0		290(o)		3[c]	
n-C_4H_9	4	16		(o)	>18		
$CH(CH_3)C_2H_5$	27	15	0.8–1	296(y)	18		
$C(CH_3)_3$	380	15		(w)	18		
$C(CH_3)_2OH$		17	0.86	(br)	>14		
$CH_2C(CH_3)_2OOCCH_3$		17	0.95	(br)	>14		
	3.4			(br)	16		0.97
CH_2Cl				(bl)	5	3[d]	
CF_3				(br)	>12	4[e]	
$CONH_2$	<0.7	18		(br)	19		1.29
CN	<0.5	18		390(bl)	17		1.49
α-$C_{10}H_7$[f]		19		(dr)	17		1.24

[a] To convert mT to gauss, multiply by 10.
[b] Color in parenthesis. Abbreviations: o = orange, br = brown, y = yellow, w = white, bl = black, dr = dark red, r = red.
[c] Composition $(C_3H_4I_{0.51})_n$.
[d] Iodine doped, composition unknown.
[e] Composition $(C_3HF_3I_{0.929})_n$.
[f] Trans cisoid.

steric effects are generally even more severe. The polymers that have substituents in the form of cyclic structures are dark in color and have spin densities of 10^{17}–10^{19}/g and narrow lines, mostly 0.3–0.5 mT (3–5G). Conductivities up to ca 1 S/cm are reported. The presence of the substituents in cyclic form appears to ameliorate their deleterious steric effects, leading to a more coplanar polyene backbone.

POLYACETYLENE ANALOGUES

Other polymers of current interest as organic conductors are the polyaryls $(Ar)_n$ analogues of polyacetylene. These include poly-p-phenylene (see OXIDATIVE POLYMERIZATION), polythiophene (see SULFUR-CONTAINING POLYMERS), polyquinolines (qv), and polypyrroles (qv).

BIBLIOGRAPHY

V. T. Skotheim, ed., *Handbook of Conducting Polymers,* Marcel Dekker Inc., New York, 1986.

J. C. W. Chien, *Polyacetylene: Chemistry, Physics and Materials Science,* Academic Press, New York, 1984.

HARRY W. GIBSON
Xerox Corporation

JOHN M. POCHAN
Eastman Kodak Company

ACID ANHYDRIDES, POLYMERIC. See POLYANHYDRIDES.

ACID–BASE INTERACTIONS

Most polymers, except for saturated hydrocarbons such as polyethylene or polypropylene, have functional sites that are electron donors (bases) or electron acceptors (acids). Even polyethylene and polypropylene may be modified by chemical or plasma treatments to provide acidic or basic surface sites. These sites provide for enhanced specific intermolecular interactions with solvents, plasticizers, other polymers (as in blends), adhering substrates, and reinforcing fillers; acid-base interactions add to the ubiquitous van der Waals interactions. Hydrogen bonds are typical acid–base bonds, for the heats of hydrogen bonding are a function of the acid strength of the hydrogen donor and the base strength of the hydrogen acceptor.

Organic Molecules

Hildebrand's solubility parameters were devised to correlate and predict the endothermic heats of mixing of liquids that interact only by van der Waals interactions; he warned against their use for hydrogen-bonding liquids. The solubility parameter δ is the square root of the cohesive energy density (the energy of vaporization per unit volume), and the partial molar heats of mixing are predicted with the equation

$$\overline{\Delta H}_1 = V_1\phi_2^2(\delta_1 - \delta_2)^2 \qquad (1)$$

where V_1 is the molar volume of component 1, ϕ_2 is the volume fraction of component 2, and δ_1 and δ_2 are the solubility parameters of components 1 and 2. Equation 1 predicts only endothermic interactions; ΔH is always positive. For ideal solutions the heat of mixing must be zero; this condition exists only when $\delta_1 = \delta_2$.

The heat of mixing of acids with bases (eg, of butanol with acetone) include the endothermic van der Waals interactions plus the exothermic acid–base interactions; this sum is generally exothermic, although it can be endothermic if the acid–base interactions are very weak. More significant results are obtained from measurements of heats of acid–base interaction in solution, for then the van der Waals interactions cancel out.

In another scheme for evaluating the strengths of organic acids and bases (electron acceptors and donors), the donor number (DN) is the actual heat of mixing of a basic liquid with antimony pentachloride in 1,2-dichloroethane, and the acceptor number (AN) is predicted from the ^{31}P nmr chemical shift of triethylphosphine oxide upon mixing with an organic acid. The nmr measurements are made without solvent, so the resulting acceptor numbers include an appreciable contribution from the van der Waals interactions.

Heats of Acid–Base Interaction of Polymers with Test Acids and Bases

Calorimetry, infrared, and nmr spectroscopic methods used for determining the heats of mixing of organic liquids can also be used for the determination of the heats of acid–base interactions of polymers. Infrared spectral shifts of the carbonyl stretching frequency have been used to determine heats of acid–base interactions of acidic and basic polymers. The heat of acid–base interaction of acids with carbonyl bases in kJ/mol was found to equal within ~1% the shift of the carbonyl peak in wave numbers (cm^{-1}); this has been confirmed.

Calorimetry may also be used for determining the heats of mixing of polymers, but since most polymers are not soluble in neutral solvents, weakly acidic or basic solvents may be used, and the heats of mixing corrected as recommended by Drago. These corrections have been found to work quite well in the case of polyvinylpyridine in benzene solution.

There have also been several efforts to determine heats of acid–base interaction of polymers with organic solvents by inverse gas chromatography. The igc method is quite effective in determining acidity or basicity of polymers by comparing the elution volumes with neutral, acidic, and basic vapors.

Heats of Mixing and Solubility

The solubility of polymers in solvents or of solvents in polymers is a direct function of the heats of mixing, the sum of the endothermic van der Waals heats of mixing, and the exothermic acid–base heats of mixing.

Acid–Base Interactions of Polymers at Interfaces

Adsorption and adhesion at polymer interfaces may be viewed as being governed by the magnitude of the surface and interfacial tensions of polymers, with hydrophilic polymer surfaces of high surface tension, and hydrophobic polymer surfaces of low surface tension. At first glance this seems reasonable, for water has a much higher surface tension than hydrocarbons, and hydrocarbons have higher surface tensions than fluorocarbons. However, remembering that surface tensions of higher liquid hydrocarbons are appreciably higher than those of very hydrophilic solvents (such as acetone, ethanol and ethyl ether), it is obvious that the hydrophilic nature of surfaces is not a simple function of surface energy, but involves the chemical nature of the surface sites. The surface and interfacial tensions of solid polymers are not measurable, but because of the great interest in these properties, it has become customary (although quite incorrect) to estimate the surface energies of polymers from contact angle measurements (see ADHESION AND BONDING; INTERFACIAL PROPERTIES). The van der Waals contribution to the surface energy of solids (γ_S^d) is indeed estimable from contact angle measurements with liquids such as methylene iodide, which has a high surface tension (γ_L) and has only van der Waals interactions with substrates. The van der Waals contribution to the surface energy of a solid is useful in predicting the van der Waals contribution to the work of adhesion W_{SL}^d to an adjacent liquid having a known value of γ_L^d:

$$W_{SL}^d = 2(\gamma_S^d \times \gamma_L^d)^{1/2} \quad (2)$$

Adsorption of polymers onto inorganic solids is an important aspect of the behavior of polymers in adhesives, surface coatings, and polymer composites; hence, it is important to characterize the acidity or basicity of the surface sites of inorganic materials used in these applications.

Flow microcalorimetry appears to be a superb tool for observing the adsorption of polymers from solution onto finely divided solids, for the rates of adsorption can be observed separately for the two important steps of the process: first, the entrapment of a polymer molecule from the flowing solution by surface sites of the powder in the adsorption bed (as evidenced by the polymer concentration flowing through the concentration detector), and secondly the gradual relaxation of polymer conformation and complexation of many of the basic sites of the polymer with the acidic surface sites (as evidenced by the generation of heat of adsorption in the calorimeter).

Adhesion of Polymers to Acidic and Basic Substrates

The mechanical behavior of adhesive joints became a science in the 1970s when it became evident that most adhesive failure mechanisms could be treated as the growth of an interfacial crack, and that the strength of an adhesive bond depends to a large extent on the thermodynamic work of debonding (the work of adhesion). Adhesive joints often fail due to interfacial penetration of water or other acidic or basic liquids between two acidic or two basic materials; this principle applies to polymer–polymer interfaces as well as polymer–inorganic interfaces.

Composites. Polymer composites have enormous interfacial areas, and their properties are very dependent on the work of adhesion across these interfaces. Composites without acid–base interfacial bonding tend to have poor mechanical properties and are subject to degradation in the presence of moisture. The role of acid–base interactions of polymer with filler begins on compounding, whether from solution or from the melt. Strong specific interfacial interactions of polymer and filler are needed to ensure that the filler particles become completely coated and strongly bonded with polymer. Complete coating assists attainment of good dispersions of the filler and eliminates filler–filler interfaces that result in clumping, air pockets, and brittleness. There are now several illustrations of such effects in the literature.

Conclusion

In two-component solutions the van der Waals interactions are endothermic and diminish solubility, whereas acid–base interactions between a solvent and solute are exothermic and enhance solubility; these principles are most important in polymer blends. In some situations the van der Waals interactions cancel out, such as on interaction of an acid and base in a solvent, or on adsorption from solution onto an interface. Acid–base interactions are of much importance in adhesion of polymers to substrates, for protection against interfacial incursion by water, and in attainment of optimum mechanical properties in composites.

BIBLIOGRAPHY

J. J. Christensen and co-workers, *Handbook of Heats of Mixing,* John Wiley & Sons, Inc., New York, 1982.

V. Gutmann, *The Donor–Acceptor Approach to Molecular Interactions,* Plenum Publishing Co., New York, 1978.

F. M. Fowkes, D. O. Tischler, J. A. Wolfe, L. A. Lannigan, C. M. Ademu-John, and M. J. Halliwell, *J. Polym. Sci. Polym. Chem. Ed.* **22,** 547 (1984); T. K. Kwei, E. Pearce, and co-workers, *J. Polym. Sci. Polym. Phys. Ed.* **24,** 1597 (1986).

K. L. Mittal, ed., *Physicochemical Aspects of Polymer Interfaces,* Vol. 2, Plenum Press, New York, 1983, pp. 739–748.

FREDERICK M. FOWKES
Lehigh University

ACID RESISTANCE. See CHEMICALLY RESISTANT POLYMERS.

ACOUSTIC EMISSION. See FATIGUE; NONDESTRUCTIVE TESTING.

ACOUSTIC PROPERTIES

Sound speed and sound absorption measurements in polymers are useful both as a probe of the molecular structure of polymers and as a source of engineering design properties. As a molecular probe, acoustic properties are related to such structural factors as the glass transition (qv), cross-link density, morphology (qv), and chemical composition. Thus, acoustic measurements can be used as a measure of any of these factors, or at least to monitor changes that may occur as a function of time, temperature, pressure, or some other variable. As a source of engineering properties, acoustic measurements are used for applications such as the absorption of unwanted sound and the construction of acoustically transparent windows.

Theory

In an unbounded isotropic solid, two types of sound wave can be propagated. In the first type, called a longitudinal wave, the polymer vibrates in the direction of wave propagation. In the second type, called a shear wave, the polymer motion is perpendicular to the direction of propagation. Longitudinal waves are sometimes called dilatational, compressional, or irrotational waves. Shear waves are sometimes called distortional, isovoluminous, or transverse waves. These two types of wave propagate independently of one another and are the only two types possible in an unbounded solid. Associated with each of the two modes of propagation is a sound speed and an absorption. Thus four parameters are required to specify the acoustic properties of a solid isotropic polymer: longitudinal speed, shear speed, longitudinal absorption, and shear absorption.

Molecular Probe. Acoustic measurements have been employed as molecular probes of polymer structure to study different aspects of molecular structure: transitions, curing (qv), density, and chemical composition in both synthetic and biological polymers.

Transitions. The acoustic properties of polymers, when plotted over broad ranges of frequency and temperature, are usually dominated by the glass transition. Typical data are shown in Figure 1.

Curing. Early measurements on polyester, phenolic, and melamine polymers made during the curing (qv) process showed a significant rise in sound speed as the materials passed from the uncured stage, through the gel stage, to the fully cured condition. Absorption increased during the gel stage and then dropped in the fully cured stage (see GELS). Increasing the cure temperature increases the final sound speed and decreases the final absorption. Later work on epoxies has emphasized the difficulty in making such measurements because of temperature changes caused by the reaction exotherm, the increase of T_g as curing proceeds, and the viscoelastic nature of the epoxies.

Density. For organic liquids, it is observed that the higher the density, the higher the sound speed. Similar behavior is found with solid polymers.

From a fundamental point of view, sound speed can be said to depend primarily on volume because of the volume dependence of the intermolecular potential. Any mechanism that changes the volume of the polymer will change the sound speed by an amount determined by the volume change, and not by the mechanism producing the volume change. This is true for such diverse mechanisms as degree of crystallinity, branching, cross-linking, and temperature change.

Sound Speed. A number of studies have been made on the effect of variations in molecular structure on sound speed. Replacing the hydrogen atoms in polyethylene with fluorine atoms lowers the sound speed, in line with the expectation that there will be a reduction in intermolecular attraction due to the larger size of the fluorine atoms. In this case, the sound speed also decreases because the polymer density increases.

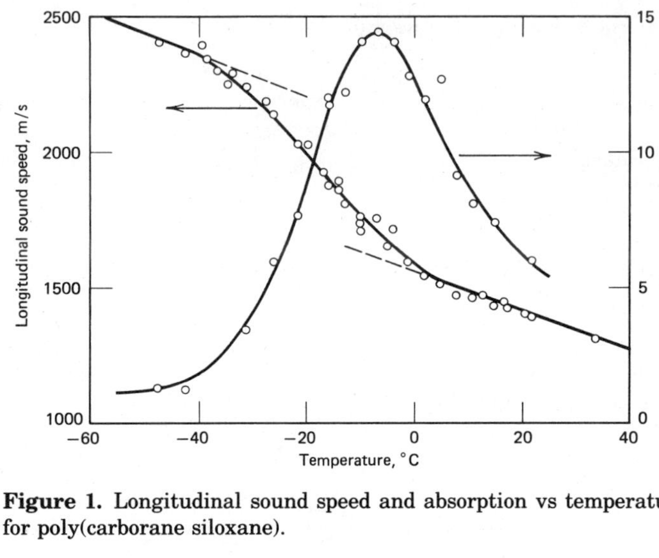

Figure 1. Longitudinal sound speed and absorption vs temperature for poly(carborane siloxane).

Similarly, in a series of poly(alkyl methacrylates), the sound speed decreases as the alkyl side-chain length increases. Again, there is a volume increase that reduces the intermolecular attraction.

Additive Properties. Many of the studies cited have considered the effect on the sound speed of substituting one component for another or an aromatic for an aliphatic, thus leading to the possibility that there is a certain sound speed associated with each chemical component. This is indeed the case, as was first demonstrated for liquids and then for solid polymers.

Biological Polymers. Perhaps the most common observation about the acoustic properties of biopolymers is that the absorption is a linear function of frequency, ie, a hysteresis absorption. Because of its appearance in very different biopolymers, hysteresis is presumed to arise whenever there is a broad spectrum of relaxation times, perhaps through the same mechanism postulated for other polymers.

Test Methods

Test methods include the immersion technique, delay rod techniques, the multiple echo technique, brillouin scattering, low frequency techniques, and surface acoustic waves.

BIBLIOGRAPHY

B. Hartmann and J. Jarzynski, *J. Appl. Phys.* **43,** 4304 (1972).

D. W. Phillips and R. A. Pethrick, *J. Macromol. Sci. Rev. Macromol. Chem.* **16,** 1 (1977).

B. Hartmann and J. Jarzynski, *J. Acoust. Soc. Am.* **56,** 1469 (1974).

J. R. Asay, D. L. Lamberson, and A. H. Guenther, *J. Appl. Phys.* **40,** 1768 (1969).

BRUCE HARTMANN
U.S. Naval Surface Weapons Center

ACROLEIN POLYMERS

Acrolein (propenal), $CH_2{=}CH{-}CH{=}O$, is the simplest unsaturated aldehyde.

Physical Properties of Monomer

Acrolein is a colorless liquid with a low viscosity and an extremely pungent smell. Hydroquinone is most frequently used as a stabilizer to prevent autooxidation and spontaneous polymerization. Some physical properties of acrolein appear in Table 1.

Table 1. Physical Properties of Monomeric Acrolein

Property	Value
formula weight, C_3H_4O	56.06
melting point, °C	−86 to −87
normal boiling point, °C	52.7
density, 20°C, g/cm³	0.8389
refractive index, n_D^{20}	1.4017
viscosity, 20°C, mPa·s	0.329
uv absorption in hexane, λ_{max}, nm	335

Chemical Properties of Monomer

Acrolein is very reactive; reactions can be carried out at the aldehyde group as well as at the C=C double bond. Furthermore, acrolein may act as a diene and dienophile in Diels-Alder reactions. Only a few of these features are being utilized for technical syntheses.

Manufacture of Monomer

The gas-phase oxidation of the propene route of acrolein synthesis is used most often today.

$$CH_2{=}CH{-}CH_3 + O_2 \longrightarrow CH_2{=}CH{-}CH{=}O + H_2O$$

Total United States acrolein production for 1984 is estimated at 25–31 kt/yr.

Polymerization of Acrolein

Acrolein can polymerize by 1,2-addition at the vinyl group, or at the carbonyl group, or it may polymerize as a conjugated diene by 1,4-addition across the carbonyl and α,β unsaturation.

Properties of Polymers

The chemical structure and the physical and chemical properties of the acrolein polymers depend on the mechanism of polymerization. Polymers formed by spontaneous or thermal polymerization (qv) are colorless to yellow powders or a horny mass of undefined structure. They are insoluble, infusible, and of no practical interest.

Most acrolein polymers that have been studied have been obtained by radical or anionic polymerization. Polymer products from radical polymerizations are colorless, light powders. They are infusible, insoluble in common solvents, and have densities of 1.32–1.37 g/cm³. The molecular weights can be 300,000 and more (determined after conversion into soluble derivatives). In contrast, polymers from anionic polymerizations are soluble in many organic solvents. They soften between 90 and 150°C. Molecular weights are much lower and can be determined by the usual techniques (see MOLECULAR-WEIGHT DETERMINATION).

Economic Aspects

By oxidative copolymerization of acrolein with acrylic acid, a product of commercial interest is obtained. This material has the Degussa trade name POC, and is a strong complexing agent that has attracted interest as a sequestering agent for water treatment.

Health and Safety Factors

Monomeric acrolein must be handled with utmost caution because of its high flammability and irritating effect on eyes and mucous membranes.

BIBLIOGRAPHY

R. C. Schulz and H. Cherdon in N. G. Gaylord, ed., *Macromolecular Syntheses*, Vol. 3, John Wiley & Sons, Inc., New York, 1969.

R. C. Schulz in G. E. Ham, ed., *Vinyl Polymerization,* Marcel Dekker, Inc., New York, 1967, Pt. 1.

R. C. Schulz, *Angew. Chem.* **76,** 357 (1964); *Int. Ed. Engl.* **3,** 416 (1964).
S. Margel and E. Wiesel, *J. Polym. Sci. Polym. Chem. Ed.* **22,** 145 (1984).

ROLF CHRISTIAN SCHULZ
University of Mainz, FRG

ACRYLAMIDE POLYMERS

In the last 20 years, the industrial use of acrylamide polymers has grown enormously. Significant commercial advances have included development of catalytic processes for the preparation of monomer, introduction of easily handled emulsion polymers, and identification of larger potential uses in enhanced oil recovery. Many other uses for polyacrylamides have been explored, but the principal outlets continue to be in water treatment, mining, and paper manufacture. Production of copolymers has settled on a few cationic and anionic comonomers and on polymer modification through controlled hydrolysis, methylolation, and the Mannich reaction. Use of acrylamide in thermosetting polymers has increased, but it is small compared to water-soluble applications (see WATER-SOLUBLE POLYMERS).

Physical Properties of Acrylamide Monomers

Acrylamide, $H_2C{=}CHCONH_2$ (2-propenamide), is a white crystalline solid. The important physical properties of acrylamide are summarized in Table 1. Acrylamide exhibits good thermal stability and long shelf life. When heated above its melting point, however, acrylamide may undergo vigorous polymerization with evolution of heat. Precautions should be taken when handling even small quantities of molten acrylamide.

Table 1. Physical Properties of Acrylamide

Property	Value
molecular weight	71.08
melting point, °C	84.5 ± 0.3
vapor pressure	
solid, Pa (mm Hg) at 25°C	0.9 (0.007)
liquid, kPa (mm Hg) at 87°C	0.27 (2.0)
density, g/mL at 30°C	1.122
equilibrium moisture content, particle size 355μm (45 mesh), at 22.8°C, 50% rh	1.7 g of water/kg of dry acrylamide
optical properties	
crystal system	monoclinic or triclinic
crystal habit	thin tabular to lamellar

In addition to the solid form, a 50% aqueous solution of acrylamide is available commercially. A large number of *N*-substituted acrylamides are known including *N*-methylolacrylamide ($CH_2{=}CHCONHCH_2OH$; mol wt 101.10; mp 75–55°C), *N,N'*-methylenebisacrylamide [$(CH_2{=}CHCONH)_2CH_2$; mol wt 154.17; mp 185–186°C (d)]; *N*-isobutoxymethylacrylamide [$(CH_2{=}CHCONHCH_2OCH_2CH(CH_3)_2$; mol wt 157.21; bp 99–100°C (4 Pa)], *N,N*-dimethylacrylamide [$(CH_2CHCON(CH_3)_2$; mol wt 99.13; bp 81°C (2.8 kPa)]; and 2-acrylamido-2-methylpropanesulfonic acid [$(CH_2{=}CHCONHCH(CH_3)_2CH_2SO_3H$; mol wt 207.25; bp 185°C (d)]. Three main routes have been used to prepare these compounds: (1) acrylonitrile–Ritter reactions, (2) acryloyl halide–amine condensations, and (3) acrylamide–formaldehyde reactions.

Chemical Properties of Acrylamide Monomers

The amide group and the electron-deficient double bond are the two reactive centers of acrylamide. The amide group is readily hydrolyzed by acid or base to yield acrylic acid. Hydrolysis of *N*-alkyl derivatives proceeds at slower rates. The presence of an electron-withdrawing group on nitrogen not only facilitates hydrolysis, but also affects the polymerization behavior of these derivatives.

Manufacture of Acrylamide Monomer

For many years, acrylamide was made by reaction of acrylonitrile with $H_2SO_4{\cdot}H_2O$, followed by separation of acrylamide from its sulfate salt by use of a base or an ion-exclusion column. This process yields satisfactory monomer, either as crystals or in solution, but also produces unwanted sulfates and waste streams. More recently, manufacturers have shifted to catalytic methods, and it is common practice to hydrate acrylonitrile over fixed-bed copper catalysts.

Polymerization

Free-radical Polymerization. ***Polymerization in Solution.*** Acrylamide polymerizes easily in the presence of free-radical initiators to form chains of the structure

$$(\text{—}CH_2\text{—}\underset{\underset{\displaystyle CONH_2}{|}}{CH}\text{—})_n,$$

where n is 20,000–300,000.

Polymerization in the Solid State. Acrylamide may be polymerized in the solid state through the use of ionizing radiation (see SOLID-STATE POLYMERIZATION).

Commercial Polymerization. The oldest, and still widely used, method of making polyacrylamide employs a dilute, eg, 9–20% aqueous solution and a conventional azo, peroxy, or redox initiator. Batch reactors are common, but continuous stirred-tank reactors may be substituted.

Other methods have been proposed to avoid the problems associated with aqueous gels. A process suitable for preparing low to moderate molecular weight polymer is to polymerize acrylamide dissolved in a mixture of water and a water-miscible liquid. Several photopolymerization (qv) methods also have been proposed.

Much attention has been given to water-in-oil emulsions, wherein a concentrated monomer solution is polymerized as the dispersed phase and a liquid hydrocarbon constitutes the continuous phase. Also, patents exist on polymerization in aqueous solution using x-ray or gamma-ray initiation.

Copolymerization. Acrylamide copolymerizes easily with many vinyl monomers. Reaction rates and molecular weights tend to be lower than those obtained in acrylamide homopolymerization, and the acrylamide content of copolymers is often higher than that of the feed. Some of the most widely used comonomers are weak acids, acrylic and methacrylic acids; strong acids, styrenesulfonic, ethylenesulfonic, and 2-acrylamido-2-methylpropanesulfonic acids; and bases, dimethylaminoethyl methacrylate, dimethylaminoethyl acrylate, diallyl dimethylammonium chloride, and their quaternary salts.

Graft and Block Copolymers. The literature contains many references to graft and block copolymers. Products so described frequently contain some acrylamide homopolymer and may not be well characterized.

Anionic Polymerization. A polymer having the principal structure $\leftarrow CH_2CH_2CONH \rightarrow_n$ is formed when acrylamide is heated in solution with a strong base. Common names for this polymer are poly(β-alanine) and nylon-3. Nylon-3 has long been of interest for textile fiber applications. Melt spinning is unsuitable because the polymer melts with decomposition, but attractive fibers have been made by wet spinning.

Polymer Properties

Properties of Solutions. Polyacrylamide is soluble in water over the entire range of concentration and has no upper or lower temperature limits. An extrapolated theta temperature in water is ca −20°C. Very dilute filtered solutions of pure polyacrylamide in water contain only individual, unassociated polymer coils. At higher concentrations, solutions contain many points of chain–chain contact and behave as gels. Large, sometimes fibrous aggregates have been observed in dilute aqueous solutions of high molecular weight polyacrylamide, even after centrifugation to remove insolubles.

Hydrolyzed polyacrylamides interact with many metal cations in solution. Aqueous solutions of these polymers also tolerate substantial amounts of organic compounds that are miscible with water. Formamide, hydrazine, ethylene glycol, and morpholine are solvents for polyacrylamide. Most other organic liquids are nonsolvents.

Flow Properties. The observed viscosity of polyacrylamide solutions in water depends not only on concentration, molecular weight, and temperature, but also on such variables as shear rate, past history of the solution, and mechanics of the measuring technique. At very low shear rates (the zero shear range) viscosity is independent of shear rate, but if the rate is increased beyond a certain point, viscosity drops markedly (shear thinning) (Fig. 1). Solution viscosity increases with molecular weight, first moderately, then much more rapidly. Solutions of polyacrylamide are viscoelastic and exhibit stresses perpendicular to the direction of flow, the Weissenberg effect. The stability of polyacrylamide solutions is adequate for many proposed applications, but changes in viscosity can occur as the result of physical stress, chemical reactions, or even subtle conformational rearrangements.

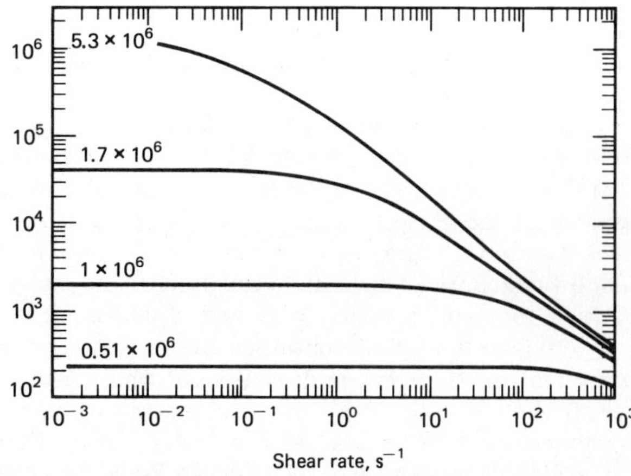

Figure 1. Effect of shear rate on viscosity; polyacrylamides of varying molecular weight at 5% concentration in water at 25°C. Courtesy of Pergamon Press, Ltd.

Covalent Gels. Aqueous solutions of high molecular weight polyacrylamide resemble gels in their appearance and physical properties. A common method of preparing gels is to copolymerize an aqueous solution of acrylamide and a diacrylate or bisacrylamide in a glass mold.

Much attention has been paid to the thermodynamics of gels. The partially hydrolyzed gels undergo sharp volume transitions from a swollen to a collapsed structure as composition of the medium, temperature, pH, membrane thickness, or electric potential is varied over very small intervals. Gels derived from acrylamide and a polyfunctional comonomer can be used in fabric coating, soil stabilizers, and water and photographic applications.

Chemical Reactivity. Polyacrylamide undergoes reactions typical of low molecular weight amides, and therefore is a useful precursor to a variety of functional derivatives.

Properties of Solid Polyacrylamide. Applications of polyacrylamide rely primarily on behavior in aqueous solution, and for this reason the properties of solid polymer have received little attention. Reported properties of the solid must be accepted with caution because it retains water strongly during ordinary drying and, when dry, rapidly picks up moisture from the environment. Physical properties of solid polyacrylamide are listed in Table 2.

Table 2. Physical Properties of Solid Polyacrylamide

Property	Reported value
density, 23°C, g/cm³	1.302
critical surface tension, mN/m[a]	30–40
approximate softening temp, °C	210
approximate weight loss by tga, °C	initial loss ca 290 98% loss ca 555
chain structure	predominantly heterotactic evidence for some head-to-head addition

[a] mN/m = dyn/cm.

Processing and Handling

At the laboratory scale, polyacrylamides can be handled safely using normal precautions to avoid ingestion or skin or eye contact. This presupposes that content of residual monomers is very low. Anyone planning to use polyacrylamides on a large scale should first consult with manufacturers for advice on suitable equipment and procedures.

Uses

Traditionally, the large markets for polyacrylamides have been in water treatment (qv), paper (qv), oil recovery, and mining (see MINING APPLICATIONS).

Economic Aspects

The principal producers of acrylamide monomer are American Cyanamid Co., The Dow Chemical Co., Nalco Chemical Co.

(internal use), Allied Colloids (internal use), Mitsubishi Chemical Industries, Mitsui Toatsu, Nitto Chemical Industries Co., and Chemische Fabrik Stockhausen & Cie. Listed selling prices on the open market in the United States in 1983 were about \$2.20/kg for crystal and \$1.65 on a 100% basis for solution. Future costs are tied to petrochemical prices and to levels of plant utilization. Total monomer capacity worldwide is on the order of 1.4×10^8 kg/yr.

There are at least two dozen suppliers of water-soluble polyacrylamides worldwide including Allied Colloids, American Cyanamid Co., Betz Laboratories, Calgon Corp., The Dow Chemical Co., Drew Chemical Corp., Hercules, Inc., and Nalco Chemical Co. in the United States. Typical 1983 prices for polyacrylamides were in the range of \$3.30–11.00/kg.

Health and Safety

Acrylamide monomer is a neurotoxin to humans. People working with acrylamide are routinely cautioned to avoid absorption through the skin and inhalation of dust or vapor. Suitable respirators and clothing, which consists of a head covering, long-sleeved coverall, impervious gloves, and rubber footwear, are recommended to avoid contact.

Several studies demonstrate that acrylamide is biodegradable. The hydrolysis of acrylamide proceeds readily both in rivers and soils.

BIBLIOGRAPHY

H. Volk and R. E. Friedrich in R. L. Davidson, ed., *Handbook of Water Soluble Gums and Resins,* McGraw-Hill Inc., New York, 1980, Chapt. 16.

D. C. MacWilliams in M. Grayson, ed., *Kirk-Othmer Encyclopedia of Chemical Technology,* 3rd ed., Vol. 1, John Wiley & Sons, Inc., New York, 1978, pp. 298–311.

D. C. MacWilliams in R. H. Yocum and E. B. Nyquist, eds., *Functional Monomers,* Vol. 1, Marcel Dekker, Inc., New York, 1973, pp. 1–197.

W. -M. Kulicke, R. Kniewske, and J. Klein, *Prog. Polym. Sci.* **8**, 373 (1982).

WALTER M. THOMAS
American Cyanamid Company

DAVID WEI WANG
IBM Corporation

ACRYLATE POLYMERS. See ACRYLIC AND METHACRYLIC ESTER POLYMERS.

ACRYLATE RUBBER. See ACRYLIC ELASTOMERS.

ACRYLIC AND METHACRYLIC ACID POLYMERS

Acrylic and methacrylic acid and their esters are primarily used to prepare polymers and copolymers. The polymeric products can be made to vary widely in physical properties through control of the nature and ratio of monomers employed in the preparation and control of the degree of cross-linking (qv) and molecular weight.

Properties of Monomers

Physical properties of acrylic and methacrylic acid are listed in Table 1. Acrylates and methacrylates undergo the usual addition reactions, eg, of halogens, hydrogen halides, and hydrogen cyanides. In general, esters, rather than the acids themselves, are employed.

Table 1. Physical Properties of Acrylic and Methacrylic Acid

Property	Value	
	Acrylic acid	Methacrylic acid
formula weight	72.06	86.10
melting point, °C	13.5	14
boiling point, 101 kPa, °C[a]	141	159–163
vapor pressure, 25°C, kPa[b]	0.57	0.13
density, 25°C, g/mL	1.045	1.015
heat of vaporization, J/g[c,d]	435	418
heat of polymerization, kJ/g[d]	1.08	0.657

[a] To convert kPa to atm, divide by 101.3.
[b] To convert kPa to mm Hg, multiply by 7.5.
[c] Latent heat of vaporization at normal boiling point.
[d] To convert J to cal, divide by 4.184.

Manufacture of Acrylic Acid Monomer

Acrylic acid monomer is manufactured either by the propylene oxidation route or the acetylene carbonylation route.

Manufacture of Methacrylic Acid Monomer

Methacrylic acid is manufactured by the acetone cyanohydrin route and the C-4 oxidation route. C-3 and C-2 routes are under development.

Handling and Storage of Acrylic and Methacrylic Acids

Acrylic acid is commercially available in better than 98% purity and usually contains 10–200 ppm of hydroquinone, monomethyl ether (MEHQ), hydroquinone (HQ), or phenothiazine as polymerization inhibitor. The product purity of methacrylic acid is generally above 99%. This is similarly stabilized to prevent polymerization during handling and storage.

In general, the monomers should not be stored for longer than one year. The acids are corrosive to many metals and should be stored only in stainless steel, glass, aluminum, or suitably lined equipment.

Health and Safety

Acrylic and methacrylic acid are moderately toxic materials when they are ingested or absorbed through the skin.

Polymerization

Linear polymers of acrylic and methacrylic acids can be prepared using the general free-radical initiation systems used for other vinyl monomers.

Copolymerization. Acrylic and methacrylic acids are readily copolymerized with many other monomers. Their versatility arises from the combination of their highly reactive double bonds and their miscibility with both water-soluble and oil-soluble monomers.

Properties of Polymers

Tacticity. The substituents of adjacent monomer units in a polymeric chain may have either the same or opposite orientations with respect to the polymer backbone. If all substituents have the same orientation with respect to the main chain, ie, ℓℓℓ or ddd, the polymer is said to be isotactic. If the configuration alternates regularly (ℓdℓdℓd), the polymer is syndiotactic. If the configurations vary at random, the polymer is atactic. In polymers prepared by free-radical polymerization, the tacticity depends on the monomer structure itself, the temperature at which polymerization is carried out, and the solvent.

Thermal Properties. When heated near 200°C in vacuum, poly(methacrylic acid) begins to lose water with formation of anhydride linkages. Poly(acrylic acid) and poly(methacrylic acid) are brittle solids that cannot be molded. Their glass-transition temperatures T_g have been reported to be 106°C and 130°C or 185°C, respectively.

Solubility. Poly(acrylic acid) and poly(methacrylic acid) are soluble to the extent of at least 1–2 pph w/v in water, dioxane, dimethylformamide, ethanol, and methanol. Poly(acrylic acid) [but not poly(methacrylic acid)] is similarly soluble in isopropyl alcohol. Acetone, diethyl ether, benzene, and aliphatic hydrocarbons are nonsolvents for both polymers.

Precipitation Temperature. The solubility of poly(acrylic acid) and poly(methacrylic acid) depends on concentration, temperature, molecular weight, and extent of neutralization. Solubility of acrylic acid polymer in water decreases as temperature is lowered. Actual precipitation does not normally take place unless salts or mineral acid are present. This use of salts to force precipitation is employed in the slurry polymerization. Poly(methacrylic acid) becomes less soluble as temperature increases. Thus, aqueous solutions that are clear gels at room temperature become cloudy at higher temperatures. In this case, the lower critical solution temperature is ca 50°C.

Viscosity. ***Dilute Solutions.*** When acrylic acid or methacrylic acid polymer is dissolved in a solvent in which the acid is not ionized, the reduced specific viscosity

$$\frac{(\eta - \eta_0)}{\eta_0 c} = \frac{\eta_{sp}}{c}$$

varies linearly with concentration and up to about 0.5% follows Huggins equation

$$\eta_{sp}/c = [\eta] + k'c\,[\eta]^2$$

where η is the intrinsic viscosity, c is concentration in grams per unit volume of solution, and k' is the Huggins constant.

Chemical Properties. ***Titration and Acid Strength.*** Acrylic acid and methacrylic acid polymers are weaker acids than the corresponding monomers and their titration curves (pH vs fraction neutralized) rise only slowly with increasing neutralization. Hence, they are good buffers in the pH range of 4–6.4.

Cation Binding. Cations behave as if bound to the polymer; this is due to the strong electrostatic forces surrounding regions of a chain bearing many ionized carbonyl groups.

Acid Strength of Copolymers. Cross-linking (qv), because it decreases the swelling and, hence, the separation of the carboxyl groups that would occur in its absence, decreases the acid strength of poly(acrylic acid) and poly(methacrylic acid). The susceptibility to modification of acid strengths by carboxylate density on the chains, and hence, their interaction with counterion, makes it possible to vary the properties of acrylic acid and methacrylic acid polymers substantially and is the basis of their use in a wide range of applications.

Economic Aspects

Manufacture of acrylates will continue to be concentrated in the United States, Europe, and Japan. Combined capacity for acrylate and methacrylate esters in 1982 (see ACRYLIC AND METHACRYLIC ESTER POLYMERS) was 2×10^6 metric tons and worth $1.30/kg. The United States' manufacturing capacity for acrylic acid and acrylate esters is over 500 kt per year, which is half of world capacity.

Uses

Acrylic acid and methacrylic acid polymers and copolymers are used in fields as varied as mining, textile manufacture, cosmetics, paper manufacture, oil-well drilling, secondary oil recovery, agricultural soil modification, and water clarification.

BIBLIOGRAPHY

J. W. Nemec and W. Bauer, Jr. in M. Grayson, ed., *Kirk-Othmer Encyclopedia of Chemical Technology,* 3rd ed., Vol. 1, Wiley-Interscience, New York, 1978, pp. 330–354.

J. W. Nemec and L. S. Kirch in M. Grayson, ed., *Kirk-Othmer Encyclopedia of Chemical Technology,* 3rd ed., Vol. 15, Wiley-Interscience, New York, 1981, pp. 346–376.

Acrylic Acid, a Summary of Safety and Handling, a joint publication of Badische Corp., Williamsburg, Va., Celanese Chemical Co., Dallas, Texas, Rohm and Haas Co., Philadelphia, Pa., and Union Carbide Corp., New York, N.Y., Feb. 1980.

H. L. Greenwald and L. S. Luskin in R. L. Davidson, ed., *Handbook of Water Soluble Gums and Resins,* McGraw-Hill Inc., New York, 1980, Chapt. 17, pp. 1–19.

JOSEPH W. NEMEC
WILLIAM BAUER, JR.
Rohm and Haas Company

ACRYLIC AND METHACRYLIC ESTER POLYMERS

Survey

Acrylate and methacrylate esters are unsymmetrically substituted ethylenes and can be represented by the generic formula

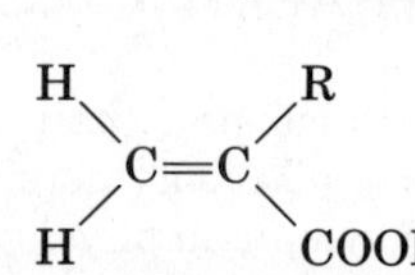

in which R = H for acrylates and R = CH_3 for methacrylates. The identity and nature of the R and R′ groups determine the properties of both the monomers and the polymers. Polymers of this class are noted for their outstanding water-clear color and the stability of their properties on aging under severe service conditions.

Physical Properties of Acrylate and Methacrylate Monomers

Acrylate Monomers. Table 1 lists the properties of commercially important acrylate esters.

Methacrylate Monomers. Properties of commercially important methacrylate monomers are listed in Table 2.

Chemical Properties of the Monomers

Acrylates and methacrylates are derivatives of ethylene. Substitution of a carbonyl group for one of the ethylene hydrogens greatly influences the reactivity of the double bond. Because of the electron-withdrawing inductive effect of the electron-

Table 1. Properties of Commercially Important Acrylate Esters

Property	Value				
	Methyl acrylate	Ethyl acrylate	*n*-Butyl acrylate	Isobutyl acrylate	2-Ethylhexyl acrylate
solubility at 23°C, parts per 100 of solvent					
in water	5	1.5	0.2	0.2	0.01
of water in ester	2.5	1.5	0.7	0.6	0.15
heat of vaporization, kJ/g[a]	0.39	0.35	0.19	0.30	0.25
specific heat, J/g°C[a]	2.00	1.97	1.93	1.93	1.93
boiling points of azeotropes					
with water, °C, 101 kPa[b]	71	81.1	94.5		
water content, %	7.2	15	40		
with methanol, °C, 101 kPa[b]	62.5	64.5			
methanol content, %	54	84.4			
with ethanol, °C, 101 kPa[b]	73.5	77.5			
ethanol content, %	42.4	72.7			
with *n*-butanol, °C, 101 kPa[b]			119		
butanol content, %			89		
heat of polymerization, kJ/mol[a]	78.7	77.8	77.4		

[a] To convert J to cal, divide by 4.184.
[b] To convert kPa to atm, divide by 101.

Table 2. Physical Properties of Commercially Available Methacrylate Monomers $CH_2{=}C(CH_3)COOR$

Compound	Molecular weight	Melting point, °C	Boiling point, °C	Refractive index, n_D^{25}	Density, d_5^{25} g/cm³	Flash point, °C COC[a]	Flash point, °C TOC[b]	Typical inhibitor ppm[c]
methacrylic acid	86.09	14	159–163[d]	1.4288	1.015	77		100 MEHQ
methyl methacrylate	100.11	−48	100–101[d]	1.4120	0.939		13	10 MEHQ
ethyl methacrylate	114.14		118–119[d]	1.4116	0.909	35	21	15 MEHQ
n-butyl methacrylate	142.19		163.5–170.5[d]	1.4220	0.889	66		10 MEHQ
isobutyl methacrylate	142.19		155[d]	1.4172	0.882		49	10 MEHQ
isodecyl methacrylate	226		120[e]	1.4410	0.878	121		10 HQ + MEHQ
lauryl methacrylate	262	−22	272–343[d]	1.444	0.868	132		100 HQ
stearyl methacrylate	332	15	310–370[d]	1.4502	0.864	>149		100 HQ
2-hydroxyethyl methacrylate	130.14	−12	95[f]	1.4505	1.064	108		1200 MEHQ
2-hydroxypropyl methacrylate	144.17	<−70	96[f]	1.4456	1.027	121		1200 MEHQ
2-dimethylaminoethyl	157.20	ca −30	68.5[f]	1.4376	0.933		74	200 MEHQ
2-*t*-butylaminoethyl methacrylate	185.25	<−70	93[f]	1.4400	0.914	11		1000 MEHQ
glycidyl methacrylate	142.1		75[f]	1.4482	0.073		84	25 MEHQ
ethylene glycol dimethacrylate	198.2		96–98[g]	1.4520	1.048	113		60 MEHQ
1,3-butylene dimethacrylate	226		110[e]	1.4502	1.011	124		200 MEHQ
trimethylolpropane trimethacrylate	338	−14	155[h]	1.471	1.06	>149		90 HQ

[a] Cleveland open cup; COC = Cleveland open cup.
[b] Tagliabue open cup; TOC = Tagliabue open cup.
[c] MEHQ = monomethylether of hydroquinone; HQ = hydroquinone.
[d] At 101 kPa (1 atm).
[e] At 0.4 kPa (3 mm Hg).
[f] At 1.3 kPa (9.8 mm Hg).
[g] At 0.53 kPa (4.0 mm Hg).
[h] At 0.13 kPa (1.0 mm Hg).

deficient carboxyl carbon atom and the electron-releasing resonance effect of the carbon–oxygen double bond, acrylates and methacrylates react readily with electrophilic, free-radical, and nucleophilic reagents.

Acrylates and methacrylates polymerize readily, rapidly, and exothermally. In order to avoid the formation of polymer when conducting other reactions of acrylic acid and its unsaturated derivatives, it is important to use mild conditions and a polymerization inhibitor. Acrylates and methacrylates undergo similar reactions, but the rates and yields of the acrylate reactions are usually higher.

Manufacture of Monomers

Raw material costs, the predicted future availability of the raw material, and plant-investment and plant-operating costs as well as the cost of waste disposal, dictate the manufacturing technology to be adopted for the competitively priced, high purity acrylate and methacrylate monomers.

Acrylate Monomers. For the past 30 years, production of acrylic has been based on acetylene. The acetylene route, however, is becoming increasingly less attractive because of the short supply of acetylene and increasing costs. Accordingly, new facilities for the manufacture of acrylates tend to be based on the oxidation of propylene.

Methacrylate Monomers. At this time in the United States, the acetone cyanohydrin process is the only one employed for the manufacture of methyl methacrylate. The most likely future route for the manufacture of methacrylates is based on C_4-oxidation technology; isobutylene or *t*-butyl alcohol are the most likely raw materials.

Handling of Acrylic and Methacrylic Monomers. Good ventilation to reduce exposure to vapors, splash-proof goggles to avoid eye contact, and protective clothing to avoid skin contact are required for safe handling of acrylic and methacrylic monomers.

Inhibition. Inhibitors such as hydroquinone (HQ), butylated hydroxytoluene (BHT), other phenolic compounds, and the monomethyl ether of hydroquinone (MEHQ) are added to acrylic monomers to stabilize them during shipment and storage.

Flammability. Methyl acrylate, ethyl acrylate, methyl methacrylate, and ethyl methacrylate have flash points below 26.7°C. They are classified by the DOT as flammable liquids, and shipping regulations require posting of ICC red labels.

Storage and Shipping. Acrylate and methacrylate monomers are shipped in bulk quantities, tank cars, or tank trucks. Mild or stainless steel are the usual materials chosen for bulk-storage facilities for these monomers.

Toxicity. Liquid methyl and ethyl acrylate are moderately toxic when taken internally, brought into contact with the eyes, or absorbed through the skin.

Properties

The nature of the alkyl group from the esterifying alcohol and the molecular weight of the polymer determine the physical and chemical properties of acrylate and methacrylate ester polymers.

Glass-transition Temperature. In methacrylate ester polymers, the T_g is influenced primarily by the nature of the alcohol group, but also to a lesser extent by the stereoregularity of the backbone chain, as can be seen in Table 3. Chemical reactivity, mechanical and dielectric relaxation (qv), viscous flow, load-bearing capacity, hardness (qv), tack, heat capacity, refractive index, thermal expansivity, creep, crystallization, and diffusion markedly differ below and above the transition region.

Table 3. Glass-transition Temperatures (T_g) of Atactic, Syndiotactic, and Isotactic Polymethacrylates, °C

	Polymer chain configuration		
Methacrylate	Atactic	Syndiotactic	Isotactic
methyl	105	160	43
ethyl	65	120	8
n-propyl	35		
isopropyl	81	139	27
n-butyl	20	81	−24
isobutyl	53	120	8
sec-butyl	60		
cyclohexyl	104	163	51

Molecular Weight. The mechanical properties of polymers increase as the molecular weight increases. However, beyond some critical molecular weight, often about 100,000 to 200,000 for amorphous polymers, the increase in property values is slight and levels off asymptotically.

Mechanical Properties Related to Polymer Structure. The first member of the acrylate series, poly(methyl acrylate), has little or no tack at room temperature; it is a tough, rubbery, and moderately hard polymer. Poly(ethyl acrylate) is more rubberlike, considerably softer, and more extensible. Poly(butyl acrylate) is softer still and much tackier.

Solution Properties. Contrary to the properties of ordinary monomeric organic molecules, if a polymer is soluble in a good solvent, it is generally soluble in all proportions. As solvent evaporates from the solution, no phase separation or precipitation occurs. The solution viscosity increases continually until a coherent film is formed.

Solubility occurs when the free energy of mixing is negative. On solution, there is more movement of polymer chains; therefore, entropy increases as a polymer dissolves. Polymer solution viscosity is dependent on the concentration of the solvent, the molecular weight of the polymer, the polymer composition, the solvent composition, and the temperature.

Optical and Other Electromagnetic Spectral Properties. Poly(methyl methacrylate) transmits light in the range of 360–1000 nm almost perfectly. Commercial grades of poly(methyl methacrylate) often contain uv radiation absorbers that block light in the 290–350 nm range, thus screening the user from sunburn and, in addition, protecting the polymer against long-term degradation from light.

Electrical Properties. The surface resistivity of poly(methyl methacrylate) is higher than that of most plastic materials. Weathering and moisture affect poly(methyl methacrylate) to only a minor degree. High resistant and nontracking characteristics have resulted in its use in high voltage applications, and its excellent weather resistance has promoted the use of poly(methyl methacrylates) for outdoor electrical applications.

Hydrolytic Stability. Methacrylate polymers have greater hydrolytic resistance to both acidic and alkaline hydrolysis than do acrylate polymers; both are far more stable than poly(vinyl acetate) and vinyl acetate copolymers. Acrylic polymers and copolymers tend to have a greater resistance both to acidic and alkaline hydrolysis than poly(vinyl acetate) and vinyl acetate copolymers.

Photochemical Stability. Acrylic polymers are not sensitive to normal uv degradation because the primary uv absorption of acrylics occurs below the solar spectrum of 290 nm.

Thermal Stability. Unless subjected to extreme heat conditions, acrylate polymers are durable, do not depolymerize, and degrade slowly. By contrast, methacrylates readily depolymerize at high conversion (95%) and at temperatures greater than 300°C.

Polymerization

Radical Polymerization. Usually, free-radical initiators such as azo compounds or peroxides are used to initiate the polymerization of acrylic monomers (see INITIATION). Photochemical and radiation-initiated polymerizations are also well known. At a constant temperature, the initial rate of the bulk or solution radical polymerization of acrylic monomers is first order with respect to monomer concentration and one-half order with respect to the initiator concentration. Rate data for polymerization of several common acrylic and methacrylic monomers initiated with 2,2-azobisisobutyronitrile (AIBN) have been determined and are shown in Table 4. The table also includes heats of polymerization and volume percent shrinkage data.

Table 4. Polymerization Data for Acrylic and Methacrylic Ester Monomers

	k_{sp}[a], 44.1°C	Heat, kJ/mol	Volume % shrinkage
acrylates			
methyl	250[b]	78.7	24.8
ethyl	313[c]	77.8	20.6
butyl	324[d]	77.4	15.7
methacrylates			
methyl	27[e]	57.7	21.0
ethyl	25[f]	57.7	18.2
butyl	41[e]	59.4	14.9

[a] The units of k_{sp} are $L^{1/2}/mol^{1/2} \cdot h^{-1}$. Initial rate of polymerization is calculated from k_{sp} and the concentration of AIBN using the equation: Initial rate of polymerization in %/h = $k_{sp}\sqrt{AIBN}$.
[b] 3 *M* soln in methyl propionate.
[c] 3 *M* soln in benzene.
[d] 1.5 *M* soln in toluene.
[e] In bulk.
[f] 2.5 *M* soln in benzene.

The vast majority of commercially prepared acrylic polymers and a substantial fraction of methacrylic polymers are copolymers. Monomeric acrylic or methacrylic esters copolymerize with one another and often several other monomers. Copolymerization greatly increases the range of available polymer properties.

Bulk Polymerization. The bulk polymerization (qv) of acrylic monomers is characterized by a rapid acceleration in the rate and the formation of a cross-linked insoluble network polymer at low conversion. The bulk polymerization of acrylic monomers is of limited commercial importance. In contrast, the bulk polymerization of methacrylic esters is the method of choice for the manufacture of poly(methyl methacrylate) sheets, rods, tubes, and molding powders.

Solution Polymerization. The solution polymerization (qv) of acrylic and methacrylic monomers to form soluble polymers or copolymers is an important commercial process for the preparation of polymers for use as coatings, adhesives, impregnates, and laminates.

Emulsion Polymerization. The principal markets for aqueous dispersion polymers made by emulsion polymerization of acrylic and methacrylic esters are the paint, paper, textile, floor polish, and leather industries where they are used as coatings or binders. Copolymers of methyl methacrylate with either ethyl acrylate or butyl acrylate are most common.

Suspension Polymerization. Suspension polymerization (qv) yields polymethacrylates in the form of tiny beads, which are primarily used as molding powders and ion-exchange resins.

Nonaqueous Dispersion Polymerization. Acrylic and methacrylic nonaqueous dispersions (NADs) are primarily utilized by the coatings industry to avoid certain difficulties associated with aqueous dispersion (emulsion) polymers.

Graft Polymerization. The most important methacrylate graft copolymers (qv) commercially are the methyl methacrylate–acrylonitrile–butadiene–styrene (MABS) and methyl methacrylate–butadiene–styrene (MBS) polymers.

Radiation-induced Polymerization. The uv cure of coatings, printing inks, and photoresists has become increasingly important in the last decade because it offers the economic advantage of requiring less energy than thermal cure and the ecological advantage of reduced solvent emission. Electron-beam curing of coatings is also becoming an area of active interest and increasing industrial importance.

Ionic Polymerization. Acrylate and methacrylate monomers do not generally polymerize by a cationic mechanism. In fact, acrylate or methacrylate functionality is often utilized as a passive pendent group for cationically polymerizable monomers.

The anionic polymerization (qv) of acrylic and methacrylic monomers to stereoregular (qv) or block copolymers (qv) is well known. These polymerizations are conducted in organic solvents, primarily using organometallic compounds as initiators.

Health and Safety Factors

In general, acrylate and methacrylate polymers are considered to be nontoxic. However, care must be exercised because additives or residual monomers present in various types of polymer can display toxicity.

During manufacture, considerable care is exercised to reduce the potential for violent polymerizations and to reduce exposure to flammable and potentially toxic monomers and solvents. Dust explosions ignited by static discharge are a recognized hazard encountered in the handling of poly(methyl methacrylate) powders or in the fabrication of poly(methyl methacrylate) plastic sheet.

Economic Aspects

Acrylates. In 1982, the worldwide production of acrylic monomers was ca 810,000 t. U.S. production is about one-half of this volume and is centered among four companies:

Producer	Capacity, t
Rohm and Haas	270,000
Celanese	171,000
Badische	158,000
Union Carbide	113,000

In 1982, the U.S. prices for ethyl and butyl acrylate were in the range of $1.10–1.30/kg.

Methacrylates. The worldwide production capacity of methyl methacrylate in 1981 was estimated to be about 1140 t. The United States accounts for about 43% of this capacity. The important U.S. producers are shown below:

Producer	1983 Capacity, t
Rohm and Haas	338,000
DuPont	162,000
CY/RO Industries	81,000

The 1982 standard price for methyl methacrylate was about $1.35/kg.

Uses

Poly(methyl methacrylate), because of its high hardness, tends to be used as a shaped object. The polyacrylates are much softer and therefore tend to find use in applications that require flexibility or extensibility.

Nonrigid applications include coatings, textiles, paper and oil additives. Rigid applications include glazing materials, biomedical appliances, and optical applications.

BIBLIOGRAPHY

E. H. Riddley, *Monomeric Acrylic Esters,* Reinhold Publishing Corp., New York, 1954.

J. W. Nemec and W. Bauer, Jr. in M. Grayson, ed., *Kirk-Othmer Encyclopedia of Chemical Technology,* 3rd ed., Vol. 1, Wiley-Interscience, New York, 1978.

Benjamin B. Kine
Ronald W. Novak
Rohm and Haas Company

2-Cyanoacrylic Ester Polymers

2-Cyanoacrylic ester polymers provide excellent adhesive bonds between a wide variety of substrates. These polymeric bonds are formed rapidly at room temperature and without the addition of a catalyst by spreading a monomeric cyanoacrylic ester, $CH_2{=}C(CN)COOR$, into a thin film that joins two adherends.

The first commercial 2-cyanoacrylic ester adhesive, Eastman 910 adhesive, was introduced in 1958. Similar adhesive compositions (qv) have been marketed under such brand names as Permabond, Super Glue, and Krazy Glue and, because of their unique performance, have been used in a broad spectrum of industrial, consumer, and medical adhesive applications.

Physical Properties

The industrially important 2-cyanoacrylic ester adhesives are usually prepared from methyl, ethyl, allyl, or butyl 2-cyanoacrylate; ethyl 2-cyanoacrylate is the most commonly used monomer. These 2-cyanoacrylic esters are formulated into a variety of adhesives having viscosities from 2 mPa·s (mPa·s = cP) to about 300 mPa·s. The set times of 2-cyanoacrylic ester adhesives are usually less than 2 min on most substrates. They generally achieve sufficient bond strength to permit handling of the bonded substrates in 10 min or less.

The physical properties of bonds of 2-cyanoacrylic ester polymers can be affected by moisture, heat, and solvents, but this does not detract from their usefulness in many applications.

Chemical Properties

The most important chemical property of 2-cyanoacrylic esters is the ability to form strong adhesive bonds rapidly. This property is derived from the highly electronegative configuration of the nitrile (—CN) and the alkoxycarbonyl (—COOR) groups.

Manufacture and Formulation

Most routes for the preparation of cyanoacrylic esters involve condensing an alkyl cyanoacetate with formaldehyde in the presence of base catalyst to yield a low molecular weight cyanoacrylic ester polymer. This polymer is then depolymerized at a high temperature to yield the 2-cyanoacrylic ester.

Economic Aspects

The world production of 2-cyanoacrylic esters for industrial, consumer, and medical adhesive applications is estimated at 500,000 kg/yr at prices ranging from about $30/kg to over $1000/kg.

Health and Safety Factors

2-Cyanoacrylic ester adhesives have pungent, unpleasant odors and may be mildly lacrimatory. If eye or skin contact occurs, the affected area should be flushed with copious quantities of water; medical attention may be needed. All 2-cyanoacrylic ester adhesives are defined as combustible liquids by the National Fire Protection Association (Std. No. 321-1973).

BIBLIOGRAPHY

H. W. Coover, Jr. and J. M. McIntire in T. Matsumoto, ed., *Tissue Adhesives in Surgery,* Medical Examination Publishing Co., Inc., New York, 1972, pp. 154–188.

H. W. Coover, Jr. and J. M. McIntire in I. Skiest, ed., *Handbook of Adhesives,* 2nd ed., Van Nostrand Reinhold Co., New York, 1977, pp. 569–580.

H. W. Coover, Jr. and J. M. McIntire in M. Grayson, ed., *Encyclopedia of Chemical Technology,* 3rd ed., Vol. 1, Wiley-Interscience, New York, 1978, pp. 408–413.

H. W. Coover, Jr.
J. M. McIntire
Eastman Kodak Company

ACRYLIC ELASTOMERS

Survey

Acrylic elastomers have the ASTM designation ACM for polymers of ethyl acrylate and other acrylates and ANM for copolymers of ethyl or other acrylates with acrylonitrile. In both cases, the M indicates a polymer having a saturated chain of the polymethylene type. The combination of a saturated backbone with polar side chains results in a class of polymers with very good resistance to heat and oil, including oils containing hypoid additives. Thus they are used for such applications as automotive transmission seals and rear axle seals. Acrylic elastomers also have good resistance to sunlight and ozone. Ethylene–acrylic elastomers are discussed in a separate section of this article.

Monomers

Monomers are divided into two types: backbone monomers, which comprise the principal proportion of the monomers and determine the physical and chemical properties of the polymer, and cure-site monomers, which are incorporated to the extent of 1–5% to introduce reactive sites for subsequent cross-linking (qv) reactions.

Backbone Monomers. The main backbone monomers used in acrylic elastomer polymerization are ethyl acrylate ($CH_2{=}CHCOOC_2H_5$), *n*-butyl acrylate ($CH_2{=}CHCOOC_4H_9$), 2-methoxyethyl acrylate ($CH_2{=}CHCOOC_2H_4OCH_3$), or 2-ethoxyethyl acrylate ($CH_2{=}CHCOOC_2H_4OC_2H_5$). Refractive index and glass-transition data for the homopolymers derived from the four most common backbone monomers are given in Table 1.

Cure-site Monomers. Current acrylic elastomers use a functional monomer containing chlorine or epoxy cure sites, either alone or in combination with a small amount of a carboxyl cure site. Of these, the chlorine type is the most common. In general, the cure-site monomers are proprietary.

Polymerization

Practically all acrylic elastomers, both commercial and experimental, are produced most often by free-radical polymerization (qv) (see EMULSION POLYMERIZATION; SUSPENSION POLYMERIZATION).

Polymer Properties

Suitably compounded and cured, acrylic elastomers are resistant to temperatures from −40 to 204°C, oils and greases at elevated temperatures, oxidation at normal and elevated temperatures, ozone at normal and elevated temperatures, aliphatic hydrocarbons, sunlight (uv) discoloration, and weathering. Typical vulcanizate properties are as follows: tensile strength (MPa), 5–16; elongation (%), 100–400; hardness (Duro A), 40–90; and compression set, 70h/150°C (%), 15–60. Table 2 provides a comparison of the acrylics (ACM) to other oil-resistant, commercially important elastomers.

Additives

Vulcanization. Because the cure-site monomer composition varies in commercial acrylic elastomers, different cure systems have been developed for specific types. The most popular cure systems currently employed for U.S. produced acrylics are soap–urea, activated thiol, lead–thiourea, soap–sulfur (or sulfur donor), diamine, and trithiocyanuric acid. Typical soaps are sodium and potassium stearate, with the latter being the more active of the two. The majority of acrylic cure systems are basic in nature; hence they are retarded by acids and accelerated by bases.

Reinforcing Agents. Acrylic elastomers are noncrystalline and amorphous and do not yield high cured gum strength. Carbon black or mineral reinforcing agents are required to develop useful properties. Of these, carbon black reinforced compounds generally yield the better overall balance of properties. Neutral to high pH alkaline pigments are specifically recommended because acidic types tend to retard the basic cure mechanism of most acrylic rubbers.

Plasticizers. Relatively low, ie, 5–10 phr, concentrations of plasticizers are sometimes utilized, primarily as process aids or to improve low temperature resistance. Type and amount are dictated by service volatility and extraction requirements.

Process Aids. Lubricating agents (ie, stearic acid) and commercial process aids are required to enhance the overall processing characteristics of acrylic compounds.

Table 1. Properties of Acrylate Polymers

	Acrylate backbone monomer			
Properties of homopolymer	Ethyl	*n*-Butyl	2-Methoxyethyl	2-Ethoxyethyl
refractive index, n_D^{20}	1.4688	1.4631	1.463	1.471
T_g, °C, from dta 10°/min	−15	−45.5	−36	−41

Table 2. Comparative Properties of Oil-resistant Specialty Elastomers

Elastomer type	Sp gr	Service temp, max, °C, 70 h/air	Service temp, continuous, °C 1000 h/air	Service temp, min, °C, flexibility	Oil resistance, 150°C, 70 h/ASTM Oil No. 3
Chloroprene	1.25	149	107	−40	fair
Nitrile rubber	1.00	149	107	−46	excellent
Chlorosulfonated PE	1.15	163	121	−46	fair
Epi Chlorohydrin	1.30	163	135	−57	excellent
Ethylene-Acrylic	0.96	204	163	−34	fair
Acrylics (ACM)	1.10	204	177	−40	excellent
Fluorosilicone	1.40	260	232	−57	excellent
Fluorocarbon	1.80	288	260	−34	excellent
Silicone	0.95	316	274	−100	fair

Antioxidants. Acrylic rubbers are inherently resistant to oxidation. However, certain nonvolatile solid diphenylamine antioxidants, eg, AgeRite Stalite S, do provide marginally improved dry heat aging resistance.

Processing

Mixing. Internal, Banbury mixing is generally preferred over open-mill mixing by the industry.

Compound Storage Stability. Acrylic compounds exhibit relatively good shelf-stability characteristics, in the range of one to four weeks, when stored under normal RT, dry conditions.

Extrusion and Calendering. In general, the use of high structure reinforcing agents, along with increased concentrations of lubricants and process aids, is usually effective for obtaining satisfactory size and surface-release characteristics for extruded and calendered finish goods applications.

Vulcanization Methods. Acrylic elastomers lend themselves to virtually all conventional cure processes. They are commonly used in compression, transfer, and injection-molded applications, but are open steam cured as well. Short cycle molded goods are usually postcured (tempered) in a circulating hot air environment to achieve optimum compression set resistance.

Bonding Characteristics. Acrylic compounds provide excellent bonds to metals and other substrates with a variety of commercially available solvent-base, curing-type adhesives.

Applications. Current commercial applications for acrylic elastomers are as follows:

Seals	Other
automatic transmission	hose, tube
valve stem	rolls
crankshaft	boots-spark plug, dust
pinion	binders-explosive, propellant
oil pan	caulk, sealants
O-ring	adhesives, coatings
grommets	plastics modification
diaphragms	fabric coating
packings	sponge-closed cell

Historically, the acrylics have been strongly associated with the automotive industry.

Health and Safety Factors

The monomers used in preparing acrylic elastomers have varying degrees of toxicity so that no polymerizations should be undertaken without first consulting the appropriate material safety data sheets.

Economic Aspects

As of July 1983, U.S. produced acrylic elastomers ranged in price from \$4.62 to 7.46/kg, fob shipping point. Worldwide producers of acrylic elastomers are as follows:

Company	Country	Trade Name
BFGoodrich Elastomers & Latex Div.	U.S.	Hycar
American Cyanamid Co.	U.S.	Cyanacryl
ENI	Italy	Elaprim
Nippon Zeon Co., Ltd. (NZ)	Japan	Nipol
TOA Paint Co., Ltd.	Japan	Acron
Nippon Oil Seal-NOK	Japan	Noxtite
Japan Synthetic Rubber Co., Ltd.	Japan	JSR AR

BIBLIOGRAPHY

C. H. Fisher, G. S. Whitby, and E. M. Beavers in G. S. Whitby, ed., *Synthetic Rubber,* John Wiley & Sons, Inc., New York, 1954, p. 900.

T. M. Vial, *Rubber Chem. Technol.* **44,** 344 (1971).

F. R. Wolf and R. D. DeMarco in R. O. Babbit, ed., *Vanderbilt Rubber Handbook,* 12th ed., R. T. Vanderbilt, Norwalk, Conn., 1978, p. 188.

R. D. DeMarco, *Rubber Chem. Technol.* **52**(1), 173 (1979).

PHILIP H. STARMER
FRED R. WOLF
The BFGoodrich Company

Ethylene–Acrylic Elastomers

Ethylene–acrylic is the term used to describe a new acrylic elastomer introduced commercially by DuPont. This elastomer, marketed under the trademark Vamac, is the result of an intensive research effort to develop an oil-resistant elastomer affording greater heat resistance than nitrile or chloroprene rubbers at a moderate cost well below that of silicone rubbers and fluoroelastomers. In addition to being heat and oil resistant, vulcanizates of ethylene–acrylic elastomer remain flexible at low temperatures and possess high resistance to degradation by weather and ozone, high strength, unique vibration-damping properties, and resistance to compression set and flex fatigue.

Polymer Composition. Ethylene–acrylic elastomer is manufactured by addition copolymerization of ethylene with methyl acrylate in the presence of a small amount of an alkenoic acid to provide sites for cross-linking by diamines.

Processing

Mixing. Compounds of ethylene–acrylic elastomers are lower in viscosity than the average rubber compounds, but can be mixed on a rubber mill or in an internal mixer using normal procedures.

Extrusion and Calendering. Most compounds of ethylene–acrylic elastomer have low nerve and yield smooth extrusions and calendered sheets.

Molding. Parts can be produced from ethylene–acrylic elastomer using compression, transfer, and injection-molding (qv) techniques in commercial operations.

Postcuring. Compression set resistance of vulcanizates that have utilized the diamine or peroxide-diamine curing system will improve substantially after the demolded vulcanizate has been heated in an oven for 1–8 h at 170–190°C.

Adhesion. Commercially available one- or two-coat adhesive systems produce rubber failure in bonds between ethylene–acrylic elastomer and metal. Adhesion to nylon, polyester, or aramid fiber cord or fabric is greatest when the cords or fabric have been treated with carboxylated nitrile rubber latex.

Shrinkage. Dimensions of ethylene–acrylic elastomer-molded parts will be smaller than those of the mold cavity. Shrinkage is made greater by an increase in amount of plasticizer, volatility of the plasticizer, mold temperature, postcure temperature, and duration of postcure.

Uses

Ethylene–acrylic elastomers are used in the automotive industry and in wire and cable jacketing. Industrial applications include pipe seals, hydraulic system seals, dampers for machinery and high speed printers, and motor lead wire insulation.

BIBLIOGRAPHY

J. F. Hagman, R. E. Fuller, W. K. Witsiepe, and R. N. Greene, *Rubber Age* **108**(5), 29 (1976).

W. R. Abell, *Ethylene–Acrylic Elastomer for Demanding Automotive Applications,* Paper No. 770864, Society of Automotive Engineers, Passenger Car Meeting, Detroit, Mich., Sept. 1977.

J. W. Crary, *Vamac Ethylene–Acrylic Elastomer—Balanced Properties for Demanding Exposures,* Contribution 372, E. I. du Pont de Nemours & Co., Inc., Wilmington, Del., May 1980.

F. V. Bailor, *RPN Technical Notebook, Rubber and Plastics News,* 24–26 (Sept. 4, 1978).

John F. Hagman
James W. Crary
E. I. du Pont de Nemours & Co., Inc.

ACRYLIC FIBERS

In 1960, the Federal Trade Commission defined an acrylic fiber as a manufactured fiber in which the fiber-forming substance is any long-chain synthetic polymer composed of at least 85% by weight of acrylonitrile units $\left(-CH_2\underset{\displaystyle CN}{\underset{|}{C}}H-\right)$. A modacrylic fiber is composed of less than 85% but at least 35% by weight of acrylonitrile units. Acrylic and modacrylic fiber world production figures for the most important geographical areas are given in Table 1.

Table 1. World Acrylic (and Modacrylic) Fiber Production, 10^3 t (1976–1982)

Area	1976	1982
western Europe	734	721
eastern Europe	172	209
U.S.	282	283
Central and South America	82	121
Japan	310	347
other	162	346
total	*1742*	*2027*

Monomer Manufacture

The discovery and development of the ammoxidation of propylene, the so-called Sohio process, appreciably reduced acrylonitrile production costs after 1960 because of the lower cost of the raw materials and the single-step nature of the process. In the Sohio process, propylene, ammonia, and oxygen react at high temperature in the presence of catalysts such as bismuth phosphomolybdate.

Structure and Morphology of Acrylic Fibers

The structure and morphology of acrylic (and to a lesser extent, modacrylic) polymers and fibers is dominated by the steric and physical interactions between neighboring nitrile groups along a given polymer chain and between adjacent polymer chains.

Polymer and Fiber Properties

Physical Properties. ***Transition Points.*** Acrylonitrile homopolymers and copolymers with a high acrylonitrile content do not have measurable melting points under normal rates of heating because of thermal instability below the melting point. On heating above ca 180°C, acrylic polymers undergo autocatalytic cyclization through the nitrile groups as well as decomposition to gaseous products such as HCN, NH_3, nitriles, etc.

Fiber Characteristics. The physical properties of acrylic and modacrylic fibers are compared with those of natural fibers and other synthetic fibers in Table 2.

Chemical Properties. One of the outstanding properties of acrylic fibers is their superior resistance to degradation (qv) by outdoor exposure, particularly their resistance to sunlight

Table 2. Physical Properties of Fibers

Property	Acrylic	Modacrylic	Nylon	Polyester	Polyolefin	Cotton	Wool
sp gr	1.14–1.19	1.28–1.37	1.14	1.38	0.90–1.0	1.54	1.28–1.32
tenacity, N/tex[a]							
dry	0.9–0.33	0.13–0.25	0.40–0.53	0.40–0.49	0.26–0.7	0.18–0.44	0.09–0.15
wet	0.14–0.24	0.11–0.23	0.35–0.68	0.40–0.49	0.26–0.7	0.21–0.53	0.07–0.14
loop/knot tenacity	0.09–0.3	0.11–0.19	0.35–0.46	0.30–0.36	0.26–0.37		
breaking elongation, %							
dry	25–45	25–45	26–32	20–30	10–30	<10	25–35
wet	29–61		30–38	20–30	10–30		25–50
initial modulus, N/tex[a]							
dry	3.5–4.9	2.6–3.5		97			
wet	3.1–4.9						
elastic recovery, %, 2% stretch	99	95–100		97		74	99
electrical resistance	high	high	very high	high	high	low	low
static buildup	moderate	moderate	very high	high	high	low	low
flammability	moderate	low	self-extinguishing	moderate	moderate	spontaneous ignition at 360°C	self-extinguishing
resistance to sunlight	excellent	excellent	poor; must be stabilized	good	poor; must be stabilized	fair–degrades	fair–degrades
resistance to chemical attack	excellent	excellent	good	good	excellent	attacked by acids	attacked by alkalies, oxidizing, and reducing agents
abrasion resistance	moderate	moderate	very good	very good	excellent	good	moderate

[a] To convert N/tex to g-f/denier, multiply by 11.3.

Table 3. Chemical Properties of Fibers

Chemical agent	Acrylic/modacrylic	Nylon	Polyester	Polyolefin	Cotton	Wool
inorganic acids						
weak	resistant	fair	resistant	resistant	hydrolyzed hot	resistant
strong	dissolves in concd HNO_3, H_2SO_4	poor	resistant, but degrades in 96% H_2SO_4	slowly oxidized	hydrolyzed, oxidized	resistant
alkalies						
dilute	resistant	good	resistant	resistant	resistant	degrades easily
concentrated	degrades hot	good	degrades hot	resistant	swells	degrades easily
oxidizing agents	resistant	resistant	resistant	moderately resistant	degrades in concd H_2SO_4	degrades
dry-cleaning solvents	insoluble	insoluble				
heat resistance	degrades slowly above 200°C	degrades slowly >200°C; melts at 250–255°C	excellent; melts at 250°C	degrades slowly and melts at 150–170°C	excellent below 150°C; burns at 390°C	degrades above 150°C; chars at 300°C
stability in sunlight	excellent	good	good	must be stabilized	fair–degrades	fair–degrades
resistance to insect attack	resistant	resistant	resistant	resistant	attacked by silverfish; not resistant to rotting	attacked by moths
resistance to biological agents, ie, mildew, fungus, bacteria, etc	resistant		resistant	resistant		resistant

and the action of microorganisms. This unique property has given acrylic fibers a special niche among fabrics used in outdoor applications, such as tenting and awning fabrics, or sandbags for riverbank stabilization. Acrylic and modacrylic fibers also are outstanding in their resistance to heat and chemical agents. The chemical properties of acrylic and modacrylic fibers are summarized and compared with other fibers in Table 3 (see also CHEMICALLY RESISTANT POLYMERS; HEAT-RESISTANT POLYMERS).

Flammability. The burning characteristics of acrylic and modacrylic fibers are compared with other important commercial fibers in Table 4.

Table 4. Burning Characteristics of Fibers

Fiber type	Reaction
cotton, rayon	chars and burns
nylon, polyester acrylic, olefin, wool	melts and burns
treated cotton	chars and self-extinguishes
modacrylic	melts and self-extinguishes
Teflon	melts and nonburning
carbon, asbestos	nonmelting and nonburning

Polymerization

Fundamental Characteristics. In the preparation of acrylic and modacrylic fibers, both polymerization and spinning help determine the ultimate properties of the fiber. The polymerization process determines the composition and molecular weight of the polymer and thus sets limits on the final properties of the fiber as well as on the spinning process.

Copolymer Composition. Most fibers made from acrylonitrile incorporate at least one other property-modifying comonomer. The comonomers most used fall into three main categories: neutral (used to improve the rate of diffusion of dyes into the fiber), ionic (used to improve dyeability and to produce water-reversible crimp bicomponent fibers), and halogen-containing monomers (used to impart flame-resistance to fibers used in home furnishings and sleepwear).

Molecular-weight Considerations. The molecular weight of the polymer must be low enough so that the polymer is readily soluble in spinning solvents, yet high enough to give a dope of moderately high viscosity.

Polymerization Methods. Commercially practiced methods of polymerization include solution, suspension, and dispersion polymerization.

Effect of Reaction Medium. Most commercial acrylic processes employ aqueous dispersion or emulsion techniques. In these reaction systems, the monomer is dispersed in two or three separate phases: a continuous aqueous phase, a monomer droplet phase, and a phase consisting of polymer particles swollen at the surface with monomer.

The heterogeneous nature of these polymerizations affects the ultimate polymer composition because the monomers are partitioned in such a way that the monomer mixture in the aqueous phase is richer in the more water-soluble monomers than the two possible organic phases. When polymerization occurs predominantly in the organic phases, the relatively more water-soluble monomers may incorporate into the copolymer in lower proportions than expected. This solubility effect is strongest for ionic monomers because this type of monomer is usually soluble in water and insoluble in the other monomers.

Commercial Polymerization Practices. Processes using a continuous stirred-tank reactor have replaced the semibatch process except when low volume specialty products are made. Typically, the reactor is charged with a certain amount of the reaction medium. The reactor feeds are then metered in at a constant rate for the entire course of the production run, which normally continues until equipment maintenance is needed. A steady state is established by taking an overflow stream at the same mass flow rate as the combined feed streams. An important advantage of this process over the semibatch process is the greatly improved control of molecular weight, dye-site level, and polymer composition.

Copolymer Composition Control. Control of copolymer composition is especially difficult if the comonomers have vapor pressures and reactivities that greatly differ from those of acrylonitrile. Heterogeneities in polymer composition and molecular weight are greatly reduced and reaction control is much more effective, however, because of the continuous stirred-tank reaction used in the acrylic fibers industry today. The reactor contents reach a natural steady state after a start-up period of three to five dwell times.

Processing

From a processing standpoint, acrylic and modacrylic polymers, as a class, have the disadvantage of melting at higher temperatures than their decomposition points. Consequently, polymers containing acrylonitrile as a principal building block must be spun to fiber from a solution or, in specialized cases, from a plasticized melt.

Acrylic polymers can be spun from solution to fiber by two fundamentally different processes: a wet- and a dry-spinning method. A hybrid method, dry jet–wet spinning, also is used.

Commercial Products

Standard Acrylic Staple and Tow. Acrylic fibers are used principally in the apparel and home furnishing markets. Much of the growth in the acrylic fibers market has come from the replacement of wool.

Filament Products. Acrylic filament yarns are produced in relatively small quantities, mainly in Europe and Japan. These fibers are valued for their high luster and silklike hand. In the silk (qv) replacement markets, continuous filament is a premium product suitable for dress fabrics, satins, and poplins.

In the general apparel markets, continuous filament acrylics find use in uniform fabrics, knitted fabrics, and fabrics with a furlike appearance. Continuous filament also is used for outdoor applications where resistance to sunlight is a critical factor.

High Bulk and Pile. High bulk acrylics are used in flat or circular knitted outerwear. The largest market for modacrylics is in high pile fabrics used for throw rugs, blankets, coats, jackets, and for lining and trimmed material.

Bicomponent Fibers. Bicomponent fibers are in demand for knit goods and craft yarns where bulk, resilience, and hand are important properties. A true bicomponent fiber consists of filaments with a compositional interface along the longitudinal axis. On opposite sides of this interface are a high shrinkage polymer and a low shrinkage composition. A helical crimp can be developed in this type of fiber and is referred to as water reversible, because the fiber loses much of its crimp when wet, but regains it on drying. Commercial examples are Monsanto's Paquel and Remember and DuPont's Wintuk.

Modacrylic and Acrylic Flame-resistant Fibers. The chief application of these modacrylic fibers is in deep-pile fabrics for blankets, apparel linings, scatter rugs, and furlike outerwear. Other applications include children's sleepwear, draperies, comforters, work clothing, and industrial uses where chemical resistance is important.

Flame-resistant acrylic fibers generally contain halogen monomers, but in amounts less than 15%. An alternative to a halogen comonomer is to blend halogen polymers into acrylic dopes.

Modifications of Acrylic Fibers. Acrylic fibers are adaptable. Their polymerization and spinning offer many opportunities for modifying the final product. Fiber properties that have received attention are improved hot-wet properties; wear comfort; softness of handle; and antipilling, antistatic, and antisoiling properties.

Acrylics for Special Uses. Acrylics have found application as asbestos substitutes and in industrial filters and battery separators, semipermeable membranes, synthetic paper, carbon and graphite fiber precursors, and moisture-absorbing products.

BIBLIOGRAPHY

B. G. Frushour and R. S. Knorr in M. Lewin and E. M. Pearce, eds., *Handbook of Fiber Science and Technology (IV): Fiber Chemistry*, Marcel Dekker, Inc., New York, 1985, Chapt. 3.

W. E. Morton and J. W. S. Hearle, *Physical Properties of Textile Fibers*, John Wiley & Sons, New York, 1975.

HARTWIG C. BACH
RAYMOND S. KNORR
Monsanto Company

ACRYLONITRILE–BUTADIENE–STYRENE POLYMERS

ABS (acrylonitrile–butadiene–styrene) polymers are elastomeric and thermoplastic composites (qv) that exhibit excellent toughness and good dimensional stability (qv). The elastomeric component, usually polybutadiene or a butadiene copolymer, exists as a discrete phase dispersed in the thermoplastic component, a copolymer of styrene and acrylonitrile (SAN). Grafting a styrene–acrylonitrile copolymer onto the elastomer permits the presence of a distinct rubber phase in a sufficiently compatible state. The flexibility in composition offered by the use of a three-monomer system combined with the option to alter molecular weight, degree of grafting, and rubber particle size and morphology allows the tailoring of properties to meet specific product needs. This versatility may be even further enhanced through the addition of other monomers to attain higher heat deflection temperatures, by matching indexes of refraction of the phases or by reducing dispersed-phase particle size to obtain transparency, and through alloying with other polymers such as polycarbonate and poly(vinyl chloride) to obtain special product features. Consequently, ABS polymers provide one of the broadest selection of properties available in an engineering thermoplastic (see ENGINEERING PLASTICS; GRAFT COPOLYMERS; POLYMER BLENDS). The range of properties typically available for ABS and its alloys is illustrated in Table 1.

Overall toughness is the prime mechanical property that prompts most users to select ABS for their applications. ABS

Table 1. Properties of ABS and Related Products

Properties	ASTM method	Units	Low impact	Medium impact	High impact	Heat resistant	Flame retardant	Plating	Clear	ABS–PVC	ABS–PC
Izod impact	D 256	J/m[a]	105–215	215–375	375–440	120–320	185–280	265–375	134	340–695	370–560
tensile strength	D 638	MPa[b]	41–52	41–47	33–44	41–52	40–50	38–44	43	30–41	45–60
tensile elongation	D 638	%	5–30	15–50	15–70	5–20	5–25	10–30	20–35	25–100	25–80
tensile modulus	D 638	GPa[c]	2.2–2.6	2.1–2.4	1.7–2.1	2.1–2.6	2.1–2.5	2.3–2.6	2.3	2.1–2.3	2.4–2.6
flexural yield strength	D 790	MPa[b]	68–87	68–80	55–68	68–90	69–86	69–80	72	54–73	79–90
flexural modulus	D 790	GPa[c]	2.3–2.6	2.2–2.5	1.8–2.2	2.1–2.8	2.3–2.8	2.3–2.7	2.4	2.1–2.3	2.2–2.6
Rockwell hardness	D 785	R	105–110	95–105	88–100	100–112	97–111	103–110	103	95–100	111–120
heat deflection temp[d]	D 648	°C									
at 1820 kPa[e]			94–100	96–100	98–104	105–121	90–107	95–100	88	70–90	114–121
at 455 kPa[e]			100–108	102–104	103–108	108–127	96–111	99–104	92	77–96	117–124
coefficient of thermal expansion	D 696	$\times 10^5$ cm/cm/°C	7.0–8.8	7.8–8.8	9.5–11.0	6.4–9.3	7.0–8.8	6.5–7.0	8.6	7.6–8.5	6.2–7.4
specific gravity	D 792		1.05–1.07	1.04–1.05	1.02–1.04	1.04–1.06	1.20–1.22	1.04–1.06	1.07	1.16–1.20	1.07–1.12

[a] To convert J/m to ft·lb/in., divide by 53.4.
[b] To convert MPa to psi, multiply by 145.
[c] To convert GPa to psi, multiply by 145,000.
[d] Annealed.
[e] To convert kPa to psi, multiply by 0.145.

is structured to effectively dissipate the energy of an impact blow. The presence of the grafted rubber domains in the SAN copolymer promotes energy-dissipative processes such as crazing and shear yielding that inhibit crack formation, which would otherwise produce catastrophic failure. ABS materials can deform in a ductile manner over a broad temperature range and at high strain rates. In comparison to most other polymers, ABS is relatively insensitive to notch radii. Also, the critical thickness effect, above which significant reductions in impact occur in some materials, is not observed in ABS.

As seen in Table 1, grades widely ranging in modulus, flow, and tensile strength can be made available, allowing considerable freedom in part design and fabrication method while maintaining adequate toughness. ABS exhibits electrical properties (qv) that are fairly constant over a wide range of frequencies and are essentially unaffected by temperature or humidity. Property values are as follows: dielectric strength, 16–18 mV/m (400–450 V/mil); arc resistance, 50–90 s; and dielectric constant, 2.4–3.3 over the frequency range of 60–10^6 Hz. ABS grades are also available with antistatic properties with surface resistivities of 10^{12}–10^{13} Ω/square.

General-purpose grades of ABS having heat deflection temperatures ranging from 82–90°C at 1.8 MPa (264 psi) are available. The high heat grades provide deflection temperatures up to 110°C at 1.8 MPa (264 psi).

ABS grades range in surface gloss from low mat finishes at <10% (60° Gardner) to very high gloss in excess of 95%. Grades are also available as color-matched pellets or in unpigmented form for customer coloring to provide a broad aesthetic choice. The chemical properties behavior of ABS may be inferred from consideration of the functional groups present within the polymer (see also CHEMICALLY RESISTANT POLYMERS).

The polar character of the nitrile group from the acrylonitrile component reduces interaction of the polymer with hydrocarbon solvents. Thus, ABS is more resistant to attack by mineral oils, waxes, and related household and commercial materials than is high impact polystyrene. Like most polymers, ABS will undergo stress cracking when brought into contact with certain chemical agents under stress. The presence of unsaturation in the polybutadiene component increases susceptibility to oxidation, halogenation, sulfonation, etc, by reactive chemical agents such as concentrated nitric acid, sulfuric acid, chlorine, etc.

Thermal oxidation of ABS can induce embrittlement of the rubber and detachment of the graft chains. Under oven-aging conditions, oxidation is diffusion limited and is restricted to the outer surface of the sample. Antioxidants have been shown to substantially improve oxidative stability.

Extended outdoor applications, especially in direct sunlight, require measures to protect against photooxidative degradation. Increased stability may be achieved by the use of stabilizing additives, pigments, and protective coatings (qv) and films (qv) (see STABILIZATION).

A broad range of flame-retardant grades is available that meets Underwriters Laboratories requirements. Most grades are obtained through the use of halogen compounds in combination with antimony oxide or are based on alloys with PVC.

Manufacture

ABS polymers are produced by the free-radical polymerization of styrene and acrylonitrile in the presence of polybutadiene or butadiene copolymers. Radical attack on the polybutadiene substrate leads to the formation of grafted styrene–acrylonitrile copolymer along with free copolymer.

Emulsion, suspension, or bulk processes are utilized for commercial production. The emulsion process involves a two-step polymerization in which a rubber substrate latex is produced and then styrene and acrylonitrile are grafted in a separate latex reaction. The suspension process is a two-part process in which a bulk polymerized prepolymer is made using a linear rubber and styrene and acrylonitrile monomers followed by polymerization in a suspension reaction. In the bulk process, all polymerization is completed in a monomer–polymer medium starting with a linear rubber dissolved in styrene and acrylonitrile monomers. Typically the bulk and suspension processes produce large rubber particles (0.5–5 μm) which are highly occluded with continuous-phase copolymer, while the emulsion process produces smaller (0.1–1 μm) less highly occluded particles.

Processing

ABS grades are available for processing by injection molding (qv), extrusion (qv), blow molding (qv), and calendering (qv). When processing ABS by any of these methods, adequate drying is required to achieve optimum part quality. Unlike some polymers, the presence of moisture alone during processing does not result in degradation of the ABS polymer. Therefore, parts exhibiting moisture splay may be ground, properly dried, blended with virgin material, and reprocessed. Secondary operations suitable for use with ABS polymers include thermoforming (qv), cold forming (qv), metallizing (qv), ultrasonic welding, hot stamping, and adhesive bonding (qv).

Applications

The estimated usage of ABS in the United States per market segment in 1986 is listed in Table 2. The largest single market was the appliance market with the major applications being refrigerator door and tank liners. Automotive applications (qv), the major area in the transportation segment, provided the second largest market. Specific uses include instrument panels, light consoles, pillar post moldings, and various other interior trim parts. Also in automobiles, ABS plating grades are used for knobs, light bezels, mirror housings, decorative trim, and grills. Pipe, fittings, and conduit represented the principal applications in the third largest market: building and construction.

Economics

The estimated worldwide production capacity and use consumption for ABS plastics in 1986 are categorized by geographic regions in Table 3. The price history of ABS in the United States is presented in Table 4 for the period from 1976 through 1986.

Table 2. Principal Markets for ABS Plastics in the United States, 1986

Market	%
appliances	18
building and construction	14
transportation	17
business machines, telephones	6
recreation	6
consumer electronics	2
modifiers	3
luggage	1
packaging	1
furniture	1
other	21
export	10

Table 3. Worldwide Capacity and Consumption of ABS Plastics, 1986

Geographic region	Capacity, 10^3 t	Consumption, 10^3 t
North America	750	550
western Europe	600	450
Far East	800	750
South America	70	50
rest of world	130	100
Total	*2350*	*1900*

Table 4. Price History of ABS Plastics in the United States

Year	$/kg
1976	1.01
1978	1.08
1980	1.41
1982	1.74
1984	1.85
1986	1.67

BIBLIOGRAPHY

Modern Plastics Encyclopedia, Vol. 59, McGraw-Hill, Inc., New York, 1982, p. 466.

Plast. Technol. **29,** 330 (1983).

Mod. Plast. **64**(1), 56 (1987).

C. B. Bucknall, *Toughened Plastics,* Applied Science Publishers Ltd., London, 1977, pp. 90–106.

W. J. Frazer, *Chem. Ind.,* 1399 (1966).

Synthetic Organic Chemicals, United States Production & Sales, USITC Publication 833, 1977; *USITC Publication 1001,* 1979; *USITC Publication 1183,* 1981; *USITC Publication 1422,* 1983; *USITC Publication 1745,* 1985.

DONALD M. KULICH
PAUL D. KELLEY
JOHN E. PACE
GE Plastics

ACRYLONITRILE POLYMERS

Physical Properties of Acrylonitrile

Acrylonitrile is a colorless liquid with the faintly pungent odor of peach pits. Its properties are summarized in Table 1. Acrylonitrile is miscible with most organic solvents.

Chemical Properties of Acrylonitrile

The presence of both a double bond and an electron-accepting nitrile group permits acrylonitrile to participate in a large number of addition reactions and polymerizations.

Reactions of the Nitrile Group. Reactions of the nitrile group include hydration, hydrolysis, and alcoholysis. The nitrile group reacts with olefins and alcohols as well as with aldehydes and methylol compounds.

Reactions of the Double Bond. Reactions of the double bond include Diels-Alder reactions, hydrogenation, halogenation, hydroformylation, and hydrodimerization. The double bond reacts with azo compounds.

Reactions of Both Functional Groups. Hydrolysis of acrylonitrile catalyzed by hydrochloric acid yields 3-chloropropionic acid. Alcoholysis and chlorination occur simultaneously in the

Table 1. Physical Properties of Acrylonitrile Monomer

Property	Value
molecular weight, g	53.06
boiling point, °C at 101.3 kPa[a]	77.3
critical pressure, kPa	3.535×10^3
critical temperature, °C	246.
cryoscopic constant, mol %/°C	2.7
density, g/L at 20°C	806.
dielectric constant at 33.5 MHz	38
dipole moment, C·m[b]	
liquid	1.171×10^{-29}
vapor	1.294×10^{-29}
entropy of polymerization,	
liquid, 25°C, J/(mol·K)	−109
explosive mixture with air at 25°C, vol %	
lower limit	3.05
upper limit	17.0 ± 0.5
flash point (tag open cup), °C	−5
freezing point, °C	−83.55 ± 0.05
Gibbs energy of formation,	
vapor at 25°C, kJ/mol	195.4
heat capacity, specific, liquid, kJ/(kg·K)	2.094
heat of combustion, liquid at 25°C, kJ/mol	-1.7615×10^3
heat of formation at 25°C, kJ/mol	
vapor	189.83
liquid	151.46
heat of fusion, kJ/mol	6.635×10^3
heat of polymerization, kJ/mol	−72.4 ± 2.1
heat of vaporization at 101.3 kPa, kJ/mol	32.65
ignition temperature, °C	481.0
molar refraction, D line	15.67
polarizability at 25°C	266.
refractive index	
n_D^{20}	1.3911–1.39142
n_D^{25}	1.3888
solubility parameter, $(J/mL)^{1/2}$	21.48
surface tension at 24°C, mN/m (dyn/cm)	27.3
surface tension of aqueous solution, c from 0 to 6 wt %, d = mN/m	$c = 0.223d - 0.0018d^2 + 0.00013d^3$

[a] To convert kPa to mm Hg, multiply by 7.5.
[b] To convert C·m to debyes, divide by 3.336×10^{-30}.

presence of sulfuric acid. Similarly, alcoholysis and hydrochlorination also occur. Addition of both ammonia and hydrogen produces trimethylenediamine and propylamine. Treatment of acrylonitrile with hydrogen peroxide at neutral to slightly alkaline pH yields glycidamide. Similarly, treatment with water, containing ammonium sulfide or a weak base, forms bis(2-carboxamidoethyl)sulfide or poly(β-alanine).

Cyanoethylation Reactions (Michael-type Additions)

Most compounds with a labile hydrogen atom can add across the double bond of acrylonitrile to form cyanoethyl groups.

Manufacture of Acrylonitrile

Today, virtually all acrylonitrile throughout the world is produced by the propylene ammoxidation process. Its development produced a drastic reduction in price and an increase in supply, paving the way for many high volume applications of acrylonitrile in fibers and plastics.

Table 2 shows the estimated raw material requirements and the by-product formation for 10^5 t/yr acrylonitrile production.

Table 2. Estimated Yields and Raw Materials for the Manufacture of 100,000 t/yr of High Purity Acrylonitrile[a]

	Amount, t/yr
Raw materials consumption[b]	
propylene	118,000
ammonia	48,000[c]
Principal by-products	
hydrocyanic acid	5,500
acetonitrile[d]	2,500

[a] Courtesy of *Hydrocarbon Processing*.
[b] Based on 100% purity.
[c] May be decreased.
[d] As purified product, recovery optional.

Properties of Polymers

Homopolymer. ***Structure and Properties.*** Polyacrylonitrile adopts the head-to-tail linkage of its monomer units with nitrile groups on alternate carbon atoms at very close proximity:

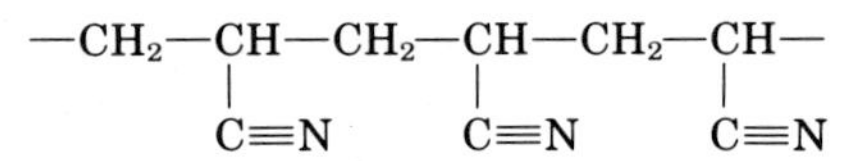

By conventional polymerization methods, polyacrylonitrile forms both isotactic and syndiotactic configurations in approximately equal proportion. However, primarily the isotactic form has been attained.

The prevailing polar nature of polyacrylonitrile provides its unique and well-known characteristics including hardness and rigidity; resistance to most chemicals and solvents, sunlight, heat, and microorganisms; slow burning and charring; reactivity toward nitrile reagents; compatibility with certain polar substances; ability to orient; and low permeability toward gases.

Morphology. The heterogeneous system of polyacrylonitrile contains crystalline, quasicrystalline, and amorphous phases.

Crystallization. Using fractionated polyacrylonitrile, crystallization has been carried out at various temperatures. Several morphological growth features have been observed, including rectangular single crystals, twinned crystals, ovals, and spherulites.

Amorphous Polyacrylonitrile. This polymer has been synthesized successfully, using bis(pentamethyleneimino)magnesium as catalyst and n-heptane as solvent.

Melting Point. Because polyacrylonitrile decomposes before reaching its melting temperature, the determination of its melting point requires rather unusual approaches. A melting point of 317°C has been obtained by dilatometry (qv).

Polarization. Polyacrylonitrile can achieve very high, persistent electrical polarization as inferred from thermally stimulated discharge analysis.

Solubility. Because of the properties of polyacrylonitrile, an active solvent capable of dissolving this polymer must be able to form hydrogen bonds with a strength comparable to that of the polymer chains and, at the same time, to separate polymer molecules with a nonpolar segment. For example, dimethylformamide is an effective solvent, but formamide, methylformamide, and diethylformamide are not. Copolymers of acrylonitrile often are soluble in dioxane, chlorobenzene, cyclohexanone, methyl ethyl ketone, acetone, dimethylformamide, butyrolactone, and tetrahydrofuran.

Barrier Properties. The remarkable barrier property of polyacrylonitrile to oxygen and carbon dioxide has been demonstrated, but high permeability toward helium is noticed.

Chemical Reactions. Polyacrylonitrile is resistant to common solvents, oils, and chemicals, but its nitrile groups and α-hydrogens do react with certain reagents. Hydration with concentrated sulfuric acid forms a solution.

Thermal Degradation. On heating, discoloration of polyacrylonitrile occurs; it first becomes yellow, progressively red, and finally black.

Thermal Decomposition. The decomposition products of pure polyacrylonitrile yarn pyrolyzed above 400°C include 16 nitriles and the predominant toxic HCN.

Copolymers. For general applications, acrylonitrile content of copolymers ranges up to ca 50% and for barrier applications, to ca 75%. Acrylonitrile copolymer properties such as rigidity, chemical resistance, melt viscosity, stability, and permeability, generally vary in proportion to the acrylonitrile content. However, the glass-transition temperature shows unusual behavior; there is a maximum or a minimum T_g in certain cases, eg, for copolymers of styrene, vinylidene chloride, and methyl methacrylate.

The principal uses of acrylonitrile are in acrylic fibers (qv), copolymers with styrene (SAN), and in combination with butadiene and styrene (ABS). Other copolymers include those of benzofuran, butadiene, carbon dioxide, 2-dimethylaminoethyl methacrylate, methyl methacrylate, poly(vinyl alcohol) with formaldehyde and hydroquinone, 4-vinylpyridine, vinylidene chloride, and the copolymer Barex 210.

Polymerization of Acrylonitrile

Homopolymerization. Pure acrylonitrile does not polymerize readily without initiators or light, but polymerization proceeds rapidly and exothermically in the presence of free radicals or anionic initiators. Polymerization can be carried out in bulk, emulsion, suspension, slurry, or solution.

Processing of Polymers

Acrylic fibers are prepared by either wet- or dry-spinning processes. Depending on the application, copolymers can be processed by various methods, including calendering (qv), casting (qv), extrusion (qv), thermoforming (qv), blow molding (qv), compression molding, and injection molding (qv).

Economic Aspects

Worldwide acrylonitrile producers and their capacities are listed in Table 3. The U.S. price and sales histories of acrylonitrile are available (Fig. 1).

Table 3. Worldwide Acrylonitrile Producers and Capacities, 1981[a]

Producer	Capacity, 1000 t
United States	
Monsanto	413
Sohio Chemical	327
DuPont	181
American Cyanamid	120
U.S. total	*1041*
Latin America total	*134*
Europe total	*1425*
Far East total	*1063*
Eastern Europe total	*505*
World total	*4168*

[a] Courtesy of R. D. Bragg, Monsanto Company.

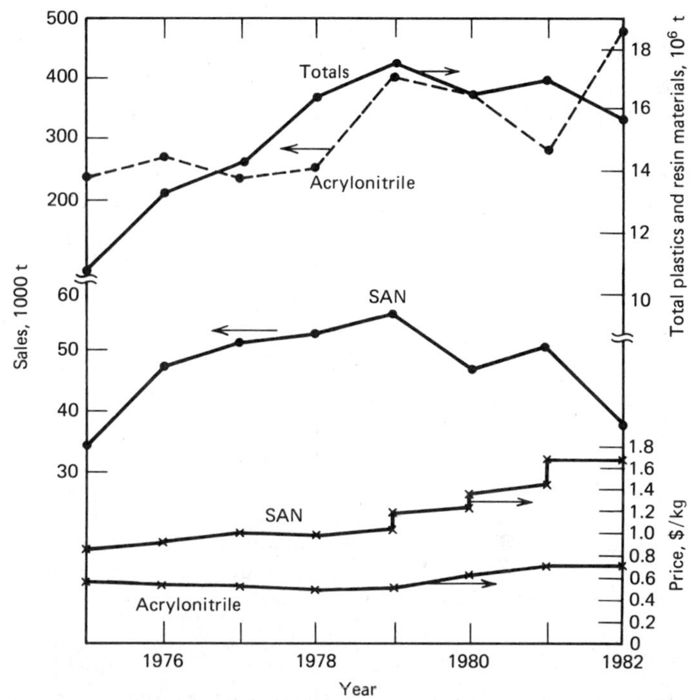

Figure 1. U.S. sales and price history of acrylonitrile and SAN compared to total plastics and resins. Courtesy of C. R. Ganote, Monsanto Co.

Health and Safety Factors

Toxicity. Acrylonitrile is a pale yellow or colorless liquid and is toxic, flammable, and explosive. Extreme care is required in handling this material.

Acrylonitrile in Polymers. The very low amount of residual acrylonitrile in finished resins or products (ca 1 ppm in acrylic and modacrylic fibers, 20–50 ppm in ABS and SAN) does not pose the threat of acrylonitrile migration or release under normal intended use and handling conditions.

Acrylonitrile Storage and Handling. Well-sealed containers of carbon or stainless steel, tin-coated metals, or brown glass bottles can be used and should be labeled "danger, contains acrylonitrile, cancer hazard." They should be properly grounded and stored in a well-ventilated area free of excessive heat, flames, sparks, or other sources of ignition. Contamination with strong acids or bases, peroxides, or other initiators should be avoided.

Uses

Historically, synthetic fibers consume more than half the acrylonitrile produced throughout the world, and ABS–SAN copolymers are the second largest users. Another large use of acrylonitrile is as the raw material for the manufacture of acrylamide, which in turn is polymerized to form products for

Table 4. Properties of SAN Copolymers[a]

Acrylonitrile,[b] wt %	Tensile strength, MPa[d]	Elongation, %	Impact, strength, J/m[e] notch	Heat distortion, °C	Solution viscosity,[c] mPa·s
5.5	42.27	1.6	26.6	72	11.1
14.0	57.37	2.2	27.1	84	13.0
27.0	72.47	3.2	27.1	88	25.7

[a] Courtesy of American Chemical Society.
[b] By Kjeldahl nitrogen analysis.
[c] 10% solution in MEK; mPa·s = cP.
[d] To convert MPa to psi, multiply by 145.
[e] To convert J/m to ft·lb/in., divide by 53.38.

water and sewage treatments, oil-well drilling applications, textile treatments, and papermaking (see ACRYLAMIDE POLYMERS).

SAN COPOLYMERS

Manufacture

SAN (styrene–acrylonitrile) copolymers are manufactured by three processes: emulsion, suspension, and continuous bulk.

Properties of SAN Copolymers

The properties of SAN copolymers are dependent on their acrylonitrile content. Table 4 illustrates some properties of the copolymers containing 5.5–27 wt % of acrylonitrile.

Polymerization of SAN Copolymers

Reactivity and Composition. The copolymer composition is related to the monomer composition by the monomer reactivity ratios. When the SAN compositions vary significantly, the copolymers become incompatible. The incompatibility of SAN causes the loss of optical clarity, mechanical strength, moldability, heat, solvent, and chemical resistance, and other important properties. Thus, the necessity of maintaining the compositional homogeneity in the manufacture of SAN copolymers is obvious.

Processing

The conventional fabrication methods, such as injection molding, blow molding, extrusion, thermoforming, and casting, are generally used.

Economic Aspects

Historically, the growth of SAN has followed the trend of total plastics sales in the United States. A primary portion (≥80%) of SAN is incorporated in ABS. Figure 1 shows U.S. sales and price histories along with total plastics sales.

Uses

SAN copolymers are used in appliances, housewares, packaging materials, automotive features, industrial applications (batteries, transmitter caps, business machines, instrument covers, tape and data reels, and medical apparatuses), and custom-molding products.

BIBLIOGRAPHY

The Chemistry of Acrylonitrile, 2nd ed., American Cyanamid Co., New York, 1959, pp. 2–10.

P. R. Pujado, B. V. Vora, and A. P. Krueding, *Hydrocarbon Process.* **56**(5), 169 (1977).

A. Noshay and J. E. McGrath, *Block Copolymers: Overview and Critical Survey,* Academic Press, Inc., New York, 1977.

Code of Federal Regulations, Title 29, Chapter XVII, Part 1910, Subpart 1045, "Occupational Safety and Health Standards for General Industry."

FRED M. PENG
Monsanto Company

ADDITION POLYMERIZATION

Addition polymerization is defined succinctly by IUPAC as polymerization by a repeated addition process. That is, monomer units are added so that the resulting chain is a perfect sum of all the atoms in the monomers. This summing distinguishes addition polymerization from condensation polymerization in which small molecules are eliminated as the monomers are assembled into polymers (see CONDENSATION POLYMERIZATION). A common type of addition polymerization is vinyl polymerization, in which molecules with carbon–carbon double bonds are the monomer units.

BIBLIOGRAPHY

"IUPAC Basic Definitions of Terms Relating to Polymers," *Pure Appl. Chem.* **40,** 477 (1974).

ADDITIVES

Additives have been an important segment of the polymer industry, and their present volume is greater than that of the entire production of plastics four decades ago. Additives used today include antiblocking agents, antifoaming agents (qv), antifogging agents, antioxidants (qv), antistatic agents (qv), biocides (qv), blowing agents (qv), colorants (qv), coupling agents (qv), curing agents, flame retardants, heat stabilizers, low profile additives, lubricants, mold-release agents, odorants, plasticizers (qv), slip agents, ultraviolet stabilizers, urethane catalysts, and viscosity control agents.

BIBLIOGRAPHY

W. O. Lundberg, ed., *Autoxidation and Antioxidants,* Wiley-Interscience, New York, 1962.

R. B. Seymour, ed., *Additives for Plastics,* Vols. 1 and 2, Academic Press, New York, 1978.

T. C. Patton, *Pigment Handbook*, John Wiley & Sons, Inc., New York, 1972.

R. Gachter and H. Muller, *Handbook—Plastic Additives*, Hanser Pub., Munich, 1983.

RAYMOND B. SEYMOUR
The University of Southern Mississippi

ADHESION AND BONDING

Adhesion

The Adhesive Bond

Mechanisms of Adhesion. Adhesion has been defined as the state in which two surfaces are held together by interfacial forces of attraction. These forces range in magnitude from the strong chemical bonds formed when two atoms share electrons, or when electric charges attract each other, to the comparatively weak attractive forces known as van der Waals interactions. Bonding may also result from mechanical interlocking or by the penetration of adhesive molecules into the substrate by diffusion.

Coupling agents are widely used to enhance chemical bonding between materials. Silanes provide a good example: one part of the molecule is able to react with SiOH groups on the surface of glass and another part is chosen to react with a resin. Virtually all glass fibers used in the manufacture of fiber glass composites are treated in this way.

Surface and Interfacial Properties

Surface Topography. Two solids will not ordinarily adhere appreciably because intimate contact between them does not occur. Their surfaces are rough on a microscale, and they may also be separated by adsorbed molecules. Because molecular forces such as van der Waals interactions drop to near zero at a distance of 0.3 nm, it is clear that close proximity is needed for strong adhesion.

Wetting and Setting. An adhesive must be able to make intimate contact with each surface and spread freely on them, ie, it should be capable of wetting the surfaces. It is applied, therefore, in the liquid state and with sufficient pressure so that it will flow into small crevices. Then, it sets to a tough solid, usually in one of three ways: by cooling, by solvent removal, or by a chemical reaction.

Thermodynamic Work of Adhesion. If a liquid is placed on a solid surface with which it has no interaction, then the reversible work W_a required to separate the two and create a unit area of new surface in each material is zero. However, in all real systems there are at least van der Waals attractions between the molecules of the liquid and the solid. In the simplest case, when the interaction energy is given by the geometric mean of the two surface energies, it is concluded that

$$W_a = (W_{cs}W_{cl})^{1/2} = 2(\gamma_s\gamma_l)^{1/2} \quad (1)$$

where W_{cs} denotes the reversible work required to create a unit area of new surface in the solid, given by 2 γ_s, and W_{cl} denotes the same property for the liquid, given by 2 γ_l, where γ_s and γ_l denote the surface free energies of the solid and liquid.

It has been proposed that surface energies comprise the sum of several types of intermolecular interaction:

$$\gamma = \gamma^d + \gamma^h + \gamma^p + \gamma^{ab} + \ldots \quad (2)$$

where the first term γ^d arises from van der Waals interactions, and the other terms arise from specific interactions: hydrogen bonding γ^h, polar γ^d, acid–base γ^{ab}, etc. Interaction between different types of molecules is postulated to occur only through the interaction of terms common to both.

When γ_s and γ_l are given values appropriate to simple nonpolar substances, about 25 mJ/m^2, the value obtained for W_a from equation 1 is only about 50 mJ/m^2. The work of detaching one adhering substance from another, denoted G_a, is found to be much larger than this, in the range of 1–10 kJ/m^2. Thus, other contributions to the mechanical strength from dissipative processes within the adherends greatly outweigh the intrinsic adhesion. Nevertheless, these other contributions depend on the existence of the intrinsic adhesion and they may well be directly proportional to its magnitude. If there is no affinity between adherends, there is certainly no mechanical strength of an adhesive bond.

Surface Treatment

In order to obtain a strong and durable adhesive joint, the surfaces of adherends are often treated before bonding. In general, these treatments alter the surface region in one or more ways: removal of a weak boundary layer, change in surface topography, or modification of the physical or chemical nature of the surface.

Polyolefins. Nearly all adhesives have surface tensions too great to allow them to wet the surfaces of solids with low surface energies, such as polyolefins and fluorocarbon polymers. Several methods have been used to modify the surface of these materials so that they can be strongly bonded with conventional adhesives.

CASING or Plasma Treatment. CASING (cross-linking by activated species or inert gas), is a process in which the surface of a solid is bombarded with ions of an inert gas at low pressure. Polyolefin surfaces treated in this manner have been bonded with typical epoxy adhesives to form strong adhesive joints.

Several mechanisms have been proposed to explain the enhanced bondability. CASING causes cross-linking (qv) reactions in the surface region of polyethylene and, in this way, can increase the joint strength. Oxidation has also been shown to occur in the surface regions. This could cause an increase in γ_s as well as in adhesion to polar adhesives, such as epoxies.

Corona Discharge. The material is exposed to a corona discharge, usually in air and at atmospheric pressure. Polyethylene treated in this way has oxygen groups (Fig. 1) and C=C double bonds introduced onto the surface. The enhanced bonding is attributed both to an increase in surface roughness and to an increase in γ.

Acid Etching. Chromic acid is commonly used to treat polyolefins before bonding. After treatment, hydroxyl, carbonyl, carboxylic acid, and —SO_3H groups have been identified in the surface regions of polyethylene and polypropylene.

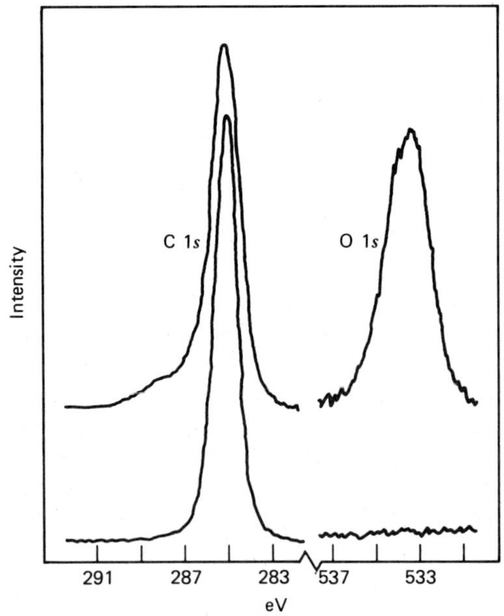

Figure 1. Esca spectra of low density polyethylene before (lower) and after (upper curve) treatment in a corona discharge.

Transcrystalline Growth. The ability to bond polyolefins is also improved by molding them against a high energy metallic substrate that induces profuse nucleation of crystallites at the surface. Columnar spherulites then grow inward and give enhanced surface strength. In addition, suitable substrates may catalyze oxidation of the polyolefin, resulting in a substantial increase in γ_s.

Surface Grafting. In one example, polyethylene was exposed to gamma irradiation in the presence of vinyl acetate monomer, which became chemically grafted to the surface. Bond strengths to an epoxy adhesive were greatly improved with this pretreatment (see GRAFT COPOLYMERS).

Fluorocarbon Polymers. Fluorocarbon polymers have very low surface energies and require treatment with aggressive etchants before they can be firmly bonded. One treatment employs metallic sodium dissolved in a mixture of naphthalene and tetrahydrofuran. Xps studies have shown the presence of unsaturation and carbonyl and carboxyl groups in treated surfaces, increasing the surface free energy and, hence, wettability.

Metals. In order to prepare a metal surface for bonding, it is etched to remove both surface contaminants and the existing oxide layer. A new stronger oxide layer is then formed under controlled conditions. The treatment of aluminum has received the most attention because of the widespread use of aluminum–epoxy bonds in aircraft structures.

On exposure to a moist environment, the bond strength may fall markedly. This deterioration is attributed to a change of oxide into hydroxide, which adheres poorly to the underlying aluminum. One method of treating aluminum before bonding is the Forest Products Laboratory (FPL) process. This consists of degreasing, alkaline cleaning, and etching in a solution containing $Na_2Cr_2O_7{\cdot}2H_2O$, H_2SO_4, and H_2O. A further enhancement in aluminum–epoxy joint durability is obtained by anodizing the aluminum after FPL treatment. One such process, phosphoric acid anodization, is thought to give joints of exceptional durability because a layer of $AlPO_4$ protects the oxide against hydrolysis.

Strength of Adhesives and Joints

Measurements of the Strength of Adhesive Bonds. Destructive mechanical tests are usually employed to measure the strength of adhesive bonds. During fracture, the adhesive and bonded members are deformed by the applied forces. All materials are imperfectly elastic to some degree, ie, they do not return all the energy expended in deforming them. This dissipation of energy causes the detachment work G_a determined from a mechanical test to be larger than that of W_a predicted from equilibrium considerations. Thus, to judge the intrinsic adhesion by the measured mechanical strength may be misleading.

Fracture Mechanics of Simple Joints. The strength of an adhesively bonded joint is generally a function of the dimensions and elastic properties of the bonded components as well as of the intrinsic strength of the interface. Fracture mechanics relates the breaking load to these factors. One method of analysis uses a simple energy criterion for fracture. An alternative method considers the stresses set up at the site of fracture. These methods are fundamentally equivalent, but energy calculations are often easier to perform.

In applying an energy criterion, an initial flaw or debond is identified at the interface. Failure is assumed to take place by growth of this initial flaw until the joint is completely broken. An energy balance is formulated for a small growth of the debond; changes in the strain energy of the joint and the potential energy of the loading device are equated to the characteristic work G_a of detachment.

Peeling Separation. The peel test is particularly simple to analyze because the elastic energy changes very little as peeling proceeds; this is because most adhering layers bend easily but do not stretch significantly under the force of peeling. Thus the work of detachment is provided directly by the loading device. For peeling at 90°, the peel force P per unit width is given by

$$P = G_a \qquad (3)$$

where G_a denotes the work of detachment per unit area of interface.

Lap Shear. When a deformable adhering layer is subjected to a force applied parallel to the bond plane, the layer becomes stretched after detachment (Fig. 2). For a linearly elastic layer, the relationship between the detachment force P per unit width and G_a is then

$$P^2 = 2tEG_a \qquad (4)$$

where E is Young's modulus of elasticity in tension for the detaching layer and t is its thickness.

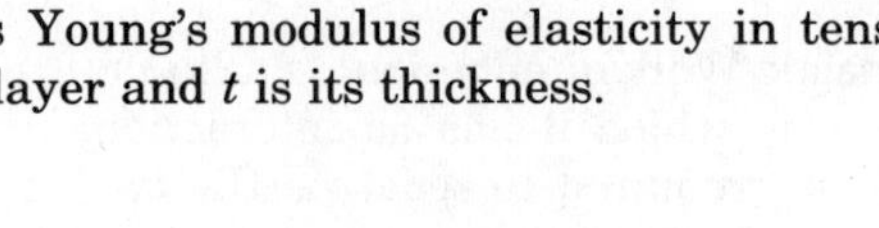

Figure 2. Detachment of an adhering layer by a force parallel to the interface; w = width.

Viscoelastic Effects in Adhesion. On peeling a pressure-sensitive adhesive from a rigid surface, higher peeling rates give higher bond strengths. However, the true adhesion is presumably a thermodynamic quantity, independent of the rate of separation. The rate dependence of the peeling force is therefore attributed to viscoelastic processes within the adhesive.

An important feature of this behavior is the opposite influence of test temperature and test rate. This is referred to as the principle of rate-temperature equivalence. For amorphous glass-forming liquids well above their glass-transition temperature (T_g), Williams, Landel, and Ferry proposed a universal relationship, ie, the WLF equation to quantify this principle:

$$\log a_T = \frac{17.4(T - T_g)}{51.6 + (T - T_g)} \qquad (5)$$

where a_T is the ratio of corresponding test rates at temperatures T and T_g. The factor a_T also represents the ratio of the rates of Brownian motion of molecular segments at T and T_g.

The utility of the WLF equation can be illustrated by an example. In Figure 3, the work of detachment is plotted as a function of peeling rate R and test temperature T for an elastomeric layer adhering to glass. The data for each temperature can be shifted along the log rate axis using displacements log a_T calculated from equation 5, to form a master curve (Fig. 4). This figure gives the work of detachment at T_g as a function of the reduced rate Ra_T. Thus, it is possible to predict the rate dependence of the work of detachment over a wide range of peel rates using data obtained at various temperatures over a narrow range of rates.

Note that the WLF equation is only applicable to simple viscoelastic substances. It does not hold for semicrystalline, rigid, or glassy adhesives.

Tack and Autohesion

Pressure-sensitive Adhesion. Some adhesives have the special property of sticking firmly to a surface after a brief contact under a light pressure. They possess little resistance to small strain deformation so that rapid wetting takes place, but they can support large strains without flowing apart easily. These features are obtained by a loose network of high molecular weight entangled molecules, diluted with special resins called tackifiers. Common tackifiers include rosin derivatives, coumarone-indene resins, terpene oligomers, aliphatic petroleum resins, and alkyl-modified phenolics.

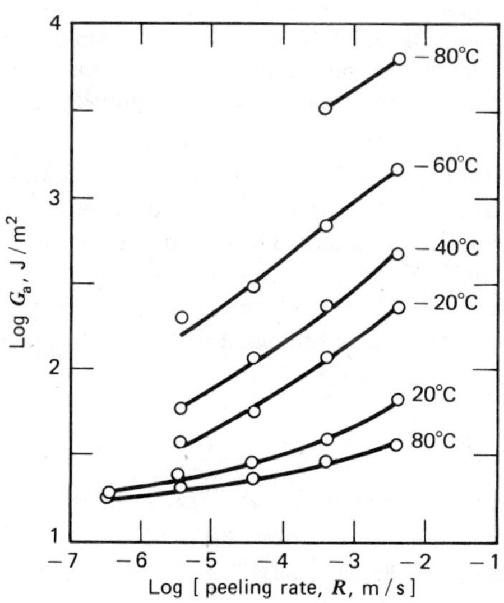

Figure 3. Detachment energy vs rate of detachment for an elastomer layer bonded to glass, tested at various temperatures.

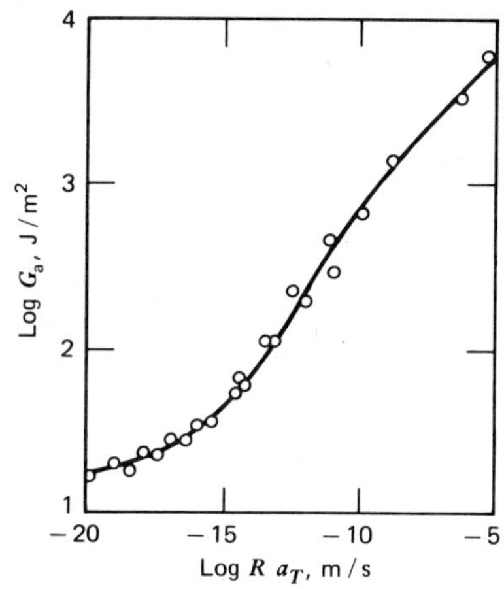

Figure 4. Master curve for detachment energy vs effective rate of detachment (Ra_T) for an elastomer bonded to glass. Data taken from Figure 3.

Rate and Temperature Effects. The peel strength of pressure-sensitive adhesives depends on the rate of peel and temperature. At low rates, the peel force increases with rate, and failure takes place within the adhesive layer, which fails by flowing apart. At a critical rate, which depends on the test temperature in accordance with the WLF rate-temperature equivalence (eq. 5), a transition takes place to interfacial failure, ie, clean separation of the adhesive from the substrate, often at much smaller peel forces. This transition occurs when the rate of deformation of the adhesive layer becomes so high that the adhesive molecules are unable to disentangle and flow apart like a liquid, but remain intertwined like a coherent elastic solid.

Above the critical rate, the peel strength depends on interfacial attractions and dissipative processes within the adherends. Below the critical rate, it is primarily a measure of the work of extending a viscous or viscoelastic liquid to the point of rupture.

Autohesion of Elastomers. For elastomeric layers to adhere together well they must also come into intimate molecular contact and resist high stress without flowing apart. This latter property, denoted green strength, is an important feature of elastomeric compounds. Amorphous elastomers, incapable of crystallizing on stretching, rely on molecular entanglements to provide cohesion and green strength. When the molecular weight of these elastomers is less than the characteristic value at which entanglements form, their green strength is low. On the other hand, when the molecular weight is much greater than the entanglement molecular weight, they cease to flow like liquids and do not readily wet on contact. A strain-crystallizing material such as natural rubber (NR) shows much better autohesion. It can be processed to a relatively low viscosity for easy wetting on contact and still exhibits green strength owing to strain-induced crystallization.

BIBLIOGRAPHY

G. R. Anderson, S. J. Bennett, and K. L. Devries, *Analysis and Testing of Adhesive Bonds,* Academic Press, Inc., New York, 1977.

S. Wu, *Polymer Interfaces and Adhesion,* Marcel Dekker, Inc., New York, 1982.

W. C. Wake, *Adhesion and the Formulation of Adhesives,* 2nd ed., Applied Science Publishers, Ltd., London, 1982.

R. L. Patrick, ed., *Treatise on Adhesion and Adhesives,* 5 Vols., Marcel Dekker, Inc., New York, 1964.

A. N. Gent
G. R. Hamed
The University of Akron

This article is based in part on a report prepared for the Office of Interdisciplinary Studies of the National Science Foundation under contract PO 82-SP-1012.

Bonding

Plastic bonds may be created in four general ways: mechanically, (Fig. 1) adhesively, through the use of fasteners (Fig. 2), and by fusion or melting techniques. Within each general method there are a variety of submethods (ie, fusion: spin welding, sonic welding, etc). While each bonding method can potentially achieve material-tearing strength, fusion methods are limited to bonding identical or melt-compatible polymers. The choice of the "best" bonding technique anticipates the potential failure modes and incorporates safeguards within the bond design. Maximum productivity is achieved through minimum fixture or clamp times, the ability to automate and the overall bonding "system cost" (not merely the bonding "prices"). Product testing and bond quality assurance can be accomplished nondestructively, if thought is given to applying NDT during the evolution of the bond design.

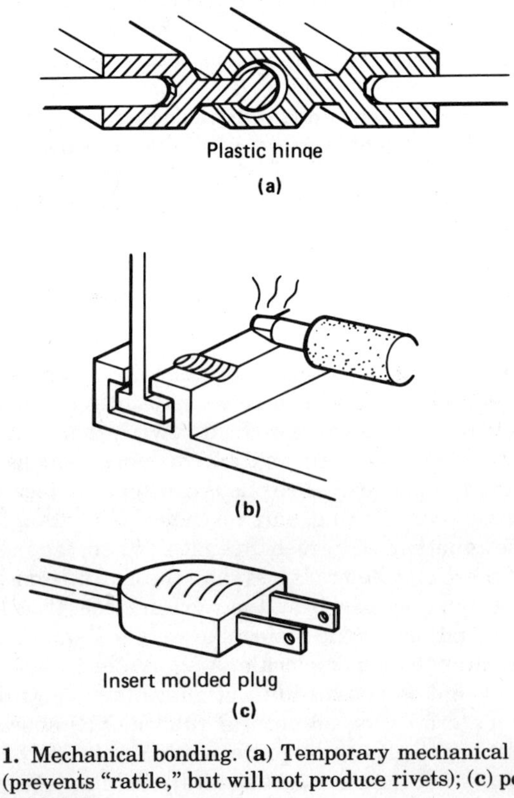

Figure 1. Mechanical bonding. (**a**) Temporary mechanical bond; (**b**) staking (prevents "rattle," but will not produce rivets); (**c**) permanent mechanical bond.

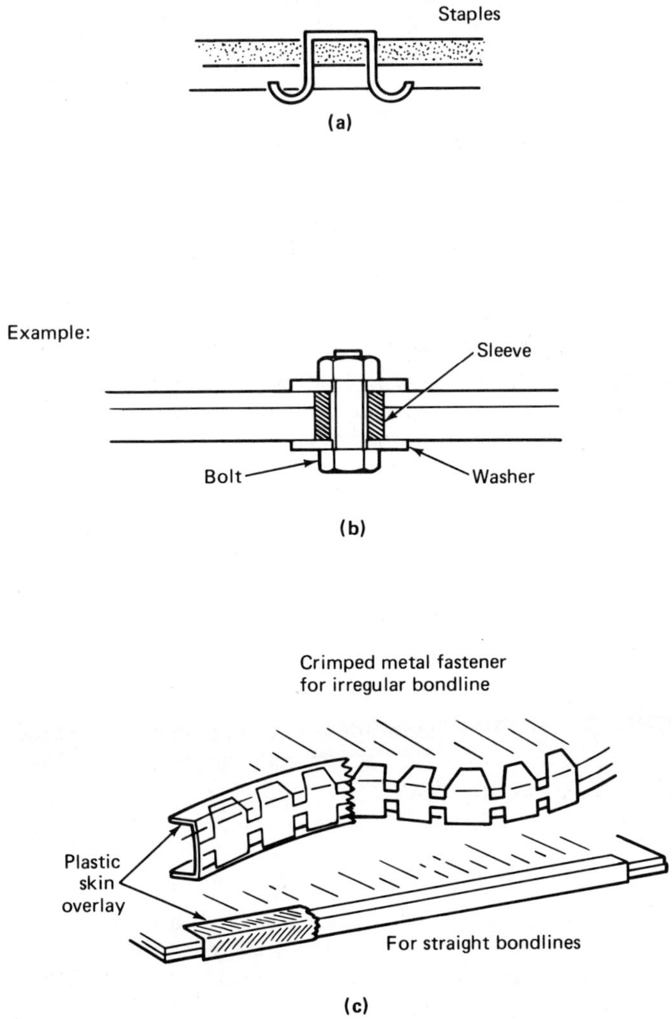

Figure 2. Fastener bonding. (**a**) Temporary fastener bond and (**b**) permanent fastener bonds must be stabilized against compression set of plastics by employing special designs; (**c**) crimped metal fasteners avoid compression set by uniform stress distribution.

Effective plastic bonding requires that the same amount of time given to the selection of a technique be given to considering other bonding aspects. These other, critically important, aspects are bond design, probable failure modes, physical bond-testing methods, NDT capability, and productivity. The success of joining plastics depends equally on all these bonding subsystems.

BIBLIOGRAPHY

F. J. Meyer and R. J. Zienert, *Adhes. Age,* 31 (Apr. 1981).

F. J. Meyer and R. J. Zienert, *Body Eng. J.,* 45 (Spring 1981).

F. J. Meyer and J. F. Smith, *Body Eng. J.,* 24 (Fall 1981).

Fred J. Meyer
Ford Motor Company

ADHESIVE COMPOSITIONS

An adhesive is a material that brings two thicker materials, ie, adherends, together; the assembly is an adhesive joint. Adhesives can be categorized both in terms of their chemical composition or polymer type and their physical forms; the latter, of course, determines preferred modes of application. Adhesive compositions frequently include a number of additives (qv) in their formulation. To enhance performance, primers and coupling agents (qv) are also employed. The adhesive joining of substrates is rapidly increasing, and competes with mechanical fastening.

Economics

Since 1960, adhesive consumption in the United States has increased almost fourfold. Table 1 estimates the current and projected market figures for important categories of adhesives in the United States.

Table 1. U.S. Adhesive Markets, in Thousand Metric Tons

Adhesives	1982	1995
Natural adhesives		
starch and dextrin	1203	1770
sodium silicate	280	382
rubber	106	150
animal	73	59
flour	50	52
other	116	107
Synthetic adhesives		
phenolic	480	615
hot melt	242	731
vinyl	351	561
rubber	361	590
urea	269	364
acrylics[a]	83	184
polyethylene[a]	54	23
epoxy	15	38
other[b]	32	308

[a] Includes sealants.
[b] Mostly new technology.

Physical Nature of Adhesives

An adhesive must be applied as a liquid, preferably of a low viscosity, both to wet the adherend surface and to flow into the crevices and asperities universally found in solid surfaces. The liquid form of the adhesive is obtained by heating to the point that facile flow occurs, dissolving or dispersing the material in a solvent, or starting with liquid monomers or oligomers that polymerize or react after application. Eventually, the adhesive must undergo a phase change, ie, by cooling, solvent evaporation, or reaction, to a solid in order for the joint to acquire the necessary strength to resist shearing forces. A notable exception is the category of pressure-sensitive adhesives, where no phase change occurs. Adhesives are produced in a number of forms including hot-melt types, solutions, aqueous dispersion, activated adhesives, film adhesives, polymerizing types, and pressure-sensitive adhesives (qv).

Adhesive Application Methods

The choice of a method, manual or mechanical, for applying adhesives depends on the geometry of the parts to be bonded, the rate of assembly desired, and the rheological characteristics of the adhesive. In addition, the application mode must accommodate the single- or two-component nature of the adhesive. Mixing of two-component systems must precede or accompany the application.

Chemical Types of Adhesives

Adhesives are classified below in terms of chemically similar types. However, few adhesives can be described adequately on that basis alone because of their complex natures. Almost every adhesive is formulated, in that it is a combination of many ingredients that are often unrelated chemically. In addition to the base polymer (or precursor), the formulated adhesive contains additives that drastically affect the properties; the same base resin may yield either rigid or elastomeric adhesives depending on the nature of the additives.

Chemical types of adhesive include carbohydrate-based adhesives (starch and dextrins), protein glues (fish glues, animal glues, casein, soybean, and blood glues), elastomer-based adhesives [natural rubber (NR), styrene–butadiene rubber (SBR), polychloroprene or neoprene, nitrile rubbers, butyl rubber, and thermoplastic elastomers], other synthetic organic adhesives [phenolic resins, amino resins, ethylene copolymers, poly(vinyl acetate), poly(vinyl acetal) adhesives, polyurethanes, epoxies, acrylic adhesives, polyamides, polyesters, high temperature adhesives, silicone adhesives, poly(vinyl chloride) copolymers, and combinations thereof called hybrids], and inorganic adhesives, the most important of which are the soluble silicates prepared by fusion of purified silica with alkali carbonate.

Surface Preparation

Some mention of the importance of surface preparation of the adherend to adhesion is warranted. Because of contamination of metallic surfaces with process oils, plastics with mold-release agents, and any surface with adventitious low energy pollutants or dirt, cleaning of the adherend is often critical to successful use of adhesives.

BIBLIOGRAPHY

C. V. Cagle, ed., *Handbook of Adhesive Bonding,* McGraw-Hill Inc., New York, 1973.

I. Skeist, ed., *Handbook of Adhesives,* 2nd ed., Van Nostrand Reinhold Co., New York, 1977.

W. C. Wake, *Adhesion and the Formulation of Adhesives,* 2nd ed., Applied Science Publishers, Ltd., London, 1982.

Adhesives, a Desk-Top Data Bank, 3rd ed., International Plastics Selector, Inc., San Diego, Calif., 1980.

Samuel C. Temin
The Kendall Company

ADSORPTION

Adsorption is the accumulation or depletion of materials at the boundary of a system with its environment, ie, at its interface. It is characterized by concentration profiles, through the surface, that may at times be very complex.

Adsorption is expressed as the amount of a particular component present per unit area of the interface in excess of the amount calculated for a reference-concentration profile. The reference-concentration profile is based on the concentration levels in the bulk phases and usually involves the assumption that the transition surface between them is geometrically sharp. The location of this transition surface must be fixed. This is done by convention and often requires that the adsorption of one of the components is zero. With the reference state defined, the adsorption can be calculated for each component of the system and can be positive, zero, or negative.

Figure 1 illustrates two cases. Figure 1**a** depicts adsorption to a preexisting surface. The preformed surface, which also is the reference surface in this case, is pictured as a wall, and concentrations C_A and C_B of the two components of the mixture are shown by the location of the line DGF (note that $C_A + C_B = 1$). Component B is positively adsorbed to the amount of the area GEF. Component A, correspondingly, is negatively adsorbed to the same extent (area GEF). The region over which the line GF deviates significantly from the line GE marks the interfacial zone. Figure 1**b** illustrates adsorption at a spontaneously formed surface. Two phases, composed of A and B in differing concentrations, are assumed to have been formed. Their concentrations are C_A, C_B and C'_A, C'_B, respectively. The concentration profile is given by the line DGFG′D′, and there is thermodynamic equilibrium. The reference surface is so sited at F that the area EFG equals the area E′FG′. For component A, the adsorption is positive to the left of F and negative to the right. Net adsorption is zero. The adsorption of component B follows a similar pattern. If the components are incompressible, the adsorption for component B will also be zero with F placed as it is. This is not the case, however, when the system is compressible, when more than two components are involved, or when different conventions are employed to locate the nominal surface.

Methods

The methods used in the study of adsorption vary according to the parameter being examined and the nature of the interface, ie, solid–liquid, liquid–liquid, or air–liquid. The properties of general interest include (1) the amount of material adsorbed to the interface. In the case of macromolecules, it is also important to know (2) the extent of contact with the interface and (3) the thickness and structure of the interfacial zone. Methods include hydrodynamic techniques, ellipsometry, neutron scattering, direct force measurement, and sedimentation volume.

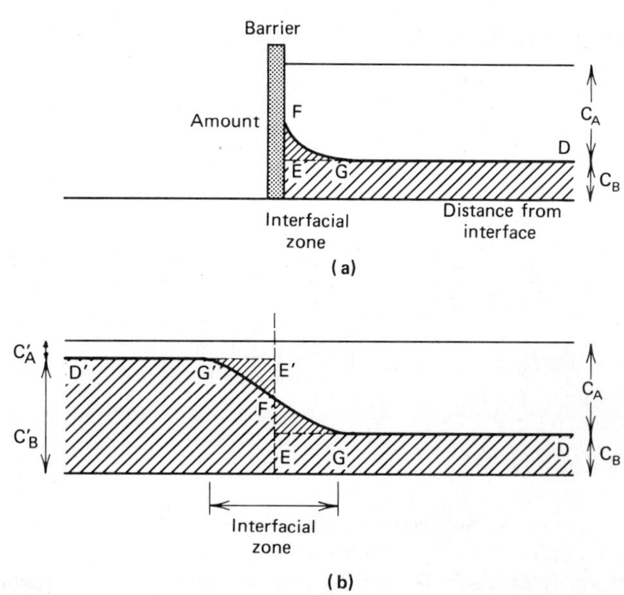

Figure 1. The interfacial zone: (**a**) adsorption to a preexisting surface; (**b**) adsorption at a spontaneously formed surface. F: location of reference surface.

Applications

Most modern applications of polymer adsorption are designed to modify the character of a surface chemically and mechanically. A macromolecular layer over the surface of particles is generally thick enough to influence the long- and short-range mechanical, electrical, and chemical interactions between them. Because macromolecules almost invariably adsorb well and resist desorption, the object of applied technology in this field has been to match surface properties so that they answer to the purpose in mind. Polymer adsorption is used in the stabilization and flocculation of particle suspensions and emulsions; in the surface modification of solids, mainly fine solid powders; in facilitating adhesion; and in biocompatibility.

BIBLIOGRAPHY

B. Vincent and S. G. Whittington, *Surf. Colloid Sci.* **12,** 1 (1982).

A. Takahashi and M. Kawaguchi, *Adv. Polym. Sci.* **46,** 1 (1982).

T. F. Tadros, ed., *The Effect of Polymers on Dispersion Properties*, Academic Press, London, 1982.

B. Vincent, *Adv. Colloid Interface Sci.* **4,** 193 (1974).

ALEX SILBERBERG
The Weizmann Institute of Science

AGING, PHYSICAL

This article is confined to the very general and important type of aging that is due to the inherent instability of the amorphous glassy state (see AMORPHOUS POLYMERS). The process is also known as volume recovery, structural relaxation, stabilization, annealing (qv), etc. The process can be characterized as follows: products of glassy polymers are usually made by shaping them in the molten (rubbery) state, followed by a fixation of the shape by rapid cooling to below the glass-transition (qv) temperature (T_g). During this cooling, the material solidifies and stiffens, but the period is too short for this process to be completed. Solidification and stiffening continue during the service life of the product; thus, many material properties change with time. For example, the elastic modulus and the yield stress increase, the ductility may decrease, the creep and stress relaxation rates decrease, the dielectric relaxation (qv) is delayed, and the propensity for crazing (qv) and shear banding is enhanced.

Typical Aspects of Aging

Generality. Aging is a consequence of the instability of the glassy state and is therefore found in all glassy materials: amorphous glassy polymers as well as low molecular weight organic glasses, substances such as bitumen, cured epoxies, inorganic glasses, and glassy metals.

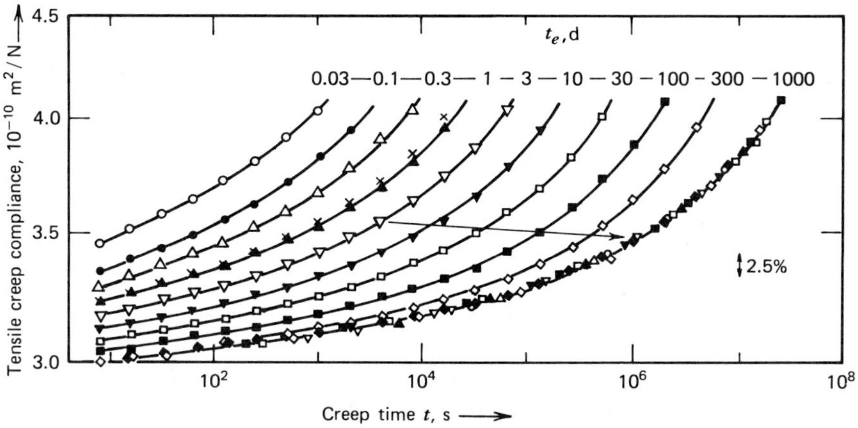

Figure 1. Small-strain tensile creep curves of rigid PVC quenched from 90°C (ie, ca 10°C above T_g) to 20°C and kept at 20 ± 0.1°C for a period of 4 yr. The different curves were measured for various values of the time t_e, in days, elapsed after the quench. The master curve at t_e = 1000 d gives the result of a superposition by shifts that were almost horizontal; the shifting direction is indicated by the arrow. The crosses (×) on the t_e = 1 d curve were found when, after 1000 d of aging, the sample was reheated to 90°C, requenched to 20°C, and remeasured for a t_e of 1 d; they show that the aging is thermoreversible. Courtesy of Elsevier.

Reversibility. In contrast to chemical aging or degradation, physical aging is a reversible process, ie, by reheating the aged material to above glass-transition temperature T_g, the original state of thermodynamic equilibrium is recovered; a renewed cooling will induce the same aging effects as before (Fig. 1).

Aging Primarily Affects Retardation Times. The prime effect of aging is to increase the relaxation–retardation times. Therefore, creep curves shift progressively along the logarithmic time scale with increasing age (Fig. 1). The shifting effect has implications for isochronal values of several properties: creep compliance decreases, the modulus increases, and the isochronal dielectric constant decreases like the creep compliance. It should, however, be realized that these changes are not the result of intrinsic changes in stiffness or polarizability (dielectric constant); the changes in the isochronal values are merely reflections of the delay in relaxation in increasing age.

Relationship of Retardation Time to Aging Time and Temperature. Retardation time τ, characterizing the position of the creep curve on the time scale, is proportional to aging time t_e and the temperature difference, $T_g - T$, according to the formula:

$$\tau \cong k\, t_e\, (T_g - T)$$

where k is a constant. This implies that the creep–relaxation rate is decreased by a factor of 10 for each 10-fold increase in aging time t_e (Fig. 1). The dependence of τ on temperature is linear rather than exponential (non-Arrhenius behavior).

Aging Occurs between T_g and T_β. The aging range runs from T_β [the first (highest) secondary transition], to T_g, eg, for PVC from ca −50 to 80°C and for polycarbonate (PC) from −120 to 150°C.

Deaging by Partial Heating

So far, the phenomena discussed pertain to isothermal aging after a quench (rapid cooling) from above to below T_g. Under such circumstances, aging always runs in the same direction (increasing retardation times). Peculiar phenomena occur, however, when after a period of aging at temperature T_1, the material is heated to a final temperature T_∞ between T_1 and $T_g (T_1 \leq T_\infty \leq T_g)$. An example for PVC ($T_g \cong 80$°C) is shown in Figure 2. Two samples are compared for which the aging times at T_1 = 20°C differ by a factor of 18. When tested at 20°C, their creep properties differ considerably, but all differences disappear when the tests are done at 50°C.

Practical Consequences of Aging

The facts about aging discussed so far have a number of practical consequences. First, data show that aging time t_e should be included in the specification of material properties, particularly for low frequency properties such as creep and stress or dielectric relaxation. The second point is that the deaging phenomena should be taken into account in experimental studies, ie, annealing studies. The third consequence of aging concerns the prediction of long-term behavior from short-time tests. In a long-term mechanical deformation process such as creep, two processes occur simultaneously: first the creep process itself and second, the aging process that continuously stiffens the material and delays the creep. Methods for predicting long-term behavior that take these simultaneous effects into con-

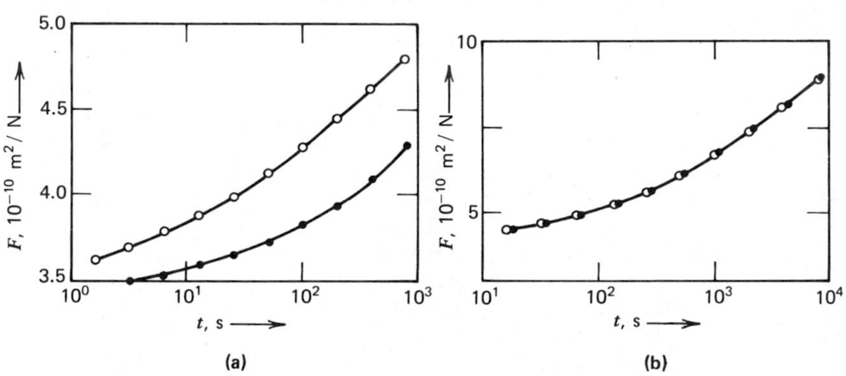

Figure 2. Small-strain tensile creep curves of rigid PVC quenched from 90 to 20°C, and aged at 20°C for ⅓ h (○, sample 1) and 6 h (●, sample 2). After these two aging histories at 20°C, the samples were tested at (**a**) 20°C and (**b**) 50°C. The equilibration time at 50°C was ½ h. Courtesy of Elsevier.

sideration have been worked out. Finally, attempts to accelerate polymer creep (qv) by performing tests at higher temperatures are generally useless for predicting the long-term creep at the working temperature.

Nonlinear Viscoelasticity, Yielding, and Fracture

Deaging effects can also be brought about by high mechanical stresses. In the extreme of yielding and necking, the deaging is complete and the material behaves as if it were actually heated to above T_g.

Aging plays a key role in the nonlinear viscoelastic behavior of plastics. When left at rest and at constant temperature, there is continuous stiffening. However, when the aged material is slightly heated or mechanically deformed, it is deaged and softened.

BIBLIOGRAPHY

A. J. Kovacs, *Fortschr. Hochpolym. Forsch.* **3,** 394 (1964).

L. C. E. Struik, *Physical Aging of Amorphous Polymers and Other Materials,* Elsevier, Amsterdam, The Netherlands, 1978.

L. C. E. Struik in W. Brostow and R. D. Corneliussen, eds., *Mechanical Failure of Plastics,* Hanser, Munich, FRG, 1983, Chapt. 15.

L. C. E. Struik
Plastics and Rubber Research Institute TNO

AGRICULTURAL APPLICATIONS

Since the introduction of plastic film in the 1930s and 1940s for greenhouse covering, fumigation, and mulching, agricultural applications of polymers have grown at an enormous rate. All principal classes of polymers, ie, plastics, coatings, elastomers, fibers, and water-soluble polymers, are presently utilized in applications that include controlled release of pesticides and nutrients, soil conditioning, seed coating, gel planting, and plant protection. Additionally, polymers are becoming increasingly important as structural components of farm buildings and machinery, for water transport and control, and in packaging of produce.

BIBLIOGRAPHY

P. Dubois, *Plastics in Agriculture,* Applied Science Publishers, Ltd., London, 1978.

A. F. Kydonieus, ed., *Controlled Release Technologies: Methods, Theory, and Applications,* Vols. I and II, CRC Press, Inc., Boca Raton, Fla., 1980.

C. L. McCormick, K. W. Anderson, and B. H. Hutchinson, *J. Macromol. Sci. Rev. Mac. Chem. Physics* **22**(1), 57 (1982).

Charles L. McCormick
University of Southern Mississippi

AIR-ASSIST FORMING. See Thermoforming.

ALBUMIN. See Proteins.

ALDEHYDE POLYMERS

Aldehyde polymerization proceeds by bond opening, chain-reaction polymerization of the carbonyl group of an aldehyde, and results in the formation of polyaldehydes (polyacetals):

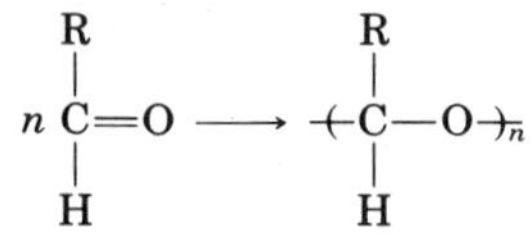

Aldehyde Monomers

Aliphatic Aldehydes. Linear or branched aliphatic aldehydes are prepared commercially; the most common aldehydes are acetaldehyde and *n*-butyraldehyde, 2-ethylhexylaldehyde, and *n*-heptaldehyde. Melting points and boiling points of aliphatic aldehydes are shown in Table 1.

Table 1. Melting Points and Boiling Points of Aliphatic Aldehydes

Aldehyde	Mp, °C	Bp, °C[a]
acetaldehyde	−123.5	20.2
propionaldehyde	−81	49
n-butyraldehyde	−97.1	74.7
2-methylpropionaldehyde	−65.9	64
n-valeraldehyde	−91.5	103.4
dl-2-methylbutyraldehyde		93
n-hexaldehyde	−56	131
n-heptaldehyde	−45	155
n-octaldehyde	−22	171
n-nonaldehyde	−17	62[b]
n-decaldehyde	−3	44[c]
n-undecaldehyde	2	55[c]
n-dodecaldehyde	13	67[c]

[a] At 101 kPa unless otherwise noted.
[b] 0.7 kPa. To convert kPa to mm Hg, multiply by 7.5.
[c] 0.07 kPa.

Functionally Substituted Higher Aldehydes. Boiling points of several functionally substituted aldehydes are listed in Table 2.

Dialdehydes. Boiling points of dialdehydes are presented in Table 3.

Table 2. Boiling Points of Functionally Substituted Aldehydes

Aldehyde	Bp, °C/kPa
3-ethoxypropionaldehyde	134/101
3-methoxypropionaldehyde	121/101
3-methoxybutyraldehyde	138–140/101
3-methylthiopropionaldehyde	166/100
3-ethylthiopropionaldehyde	185/101
3-methylthiobutyraldehyde	78–79/3.0
3-ethylthiobutyraldehyde	92–93/3.2
3-cyanopropionaldehyde	85–87/0.8
phenylacetaldehyde	91–92/2.7

Table 3. Boiling Points of Some Dialdehydes

Aldehyde	Bp, °C/kPa
glyoxal	50/101
succinaldehyde	169–170/101
glutaraldehyde	187–189/101
malealdehyde	54–61/1.3
phthalaldehyde	83–84/0.1

Haloaldehydes. Haloaldehydes are compounds in which one or more hydrogen atoms of an aliphatic aldehyde are replaced by halogen atoms, ie, fluorine, chlorine, or bromine atoms. Physical properties of the most important perhaloacetaldehydes are described in Table 4.

Table 4. Physical Properties of Haloaldehydes and Their Polymerization Ceiling Temperatures

Haloaldehyde	Bp, °C	Density, g/cm³	Ceiling temp, °C
CH_2FCHO	64		
CHF_2CHO	27		
CF_3CHO	−19	1.47	85
CH_2ClCHO	85		
$CHCl_2CHO$	89		
CCl_3CHO[a]	98	1.50	11
CBr_3CHO	174	2.73	−75
CF_2ClCHO	18	1.45	63
$CFCl_2CHO$	56	1.43	41
CF_2BrCHO	43	1.80	48
$CFBr_2CHO$	116	2.25	−7
CCl_2BrCHO	127	1.87	−15
$CClBr_2CHO$	148	2.27	−40

[a] Mp, −57°C.

Polymerization of Aldehydes

The carbonyl bond, unlike the carbon–carbon double bond, is highly polarized because of the different electronegativities of the carbon and oxygen atoms of the bond.

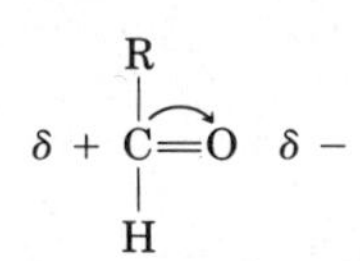

Consequently, aldehydes are susceptible to ionic, ie, both cationic and anionic polymerization. The only possible exception to this is fluoral, for which radical polymerization has been claimed.

Thermodynamics of Aldehyde Polymerization. The enthalpy of polymerization of an aldehyde carbonyl group is much smaller than the enthalpy of polymerization of a vinyl monomer with similar substitution on the double bond.

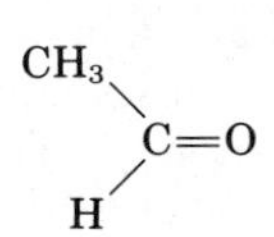

acetaldehyde, ΔH_{polym} (lc) = 0

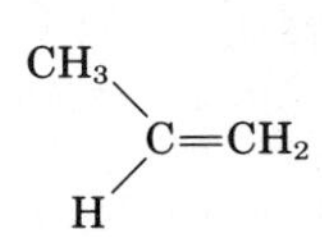

propylene, ΔH_{polym} (lc) = −84 kJ/mol

The ceiling temperature of aldehyde polymerization is relatively low, and is often below room temperature.

Acetaldehyde Polymerization. Acetaldehyde polymerization can be polymerized by cationic and anionic polymerization. In addition, acetaldehyde has been copolymerized with a number of comonomers, including formaldehyde, higher aldehydes, ketenes, and isocyanates.

Higher Aliphatic Aldehyde Polymerization. Higher aliphatic aldehydes also polymerize readily at low temperatures. Most polymerizations of higher aldehydes, up to C_{11}, have been carried out with anionic initiators, especially with the hydrocarbon-soluble lithium tertiary-butoxide. Polymers seem to be more crystalline as the length of the side chain increases or the degree of branching increases. All crystalline higher aliphatic polyaldehydes are isotactic and crystallize in a fourfold helix with an identity period ranging from 0.48 to 0.52 nm.

Cyclopolymerization of Dialdehydes. The simplest dialdehyde, glyoxal, polymerizes easily to solid polymers. Glutaraldehyde, succinaldehyde, and phthalaldehyde have been cyclopolymerized to relatively low molecular weight amorphous polymers using ionic initiators.

Polymerization of Unsaturated Aldehydes. The only unsaturated aldehyde that has been studied in detail is acrolein. Acrolein may be polymerized by radical or ionic mechanisms, but the structure of the polymer has been shown to be primarily that of a polymer that was obtained by polymerization of the carbon–carbon double bond; some polymerization seems to have also occurred by anionic polymerization through the carbonyl bond (see ACROLEIN POLYMERS).

Optically Active Polyaldehydes. Optically active polyaldehydes have been prepared from aliphatic aldehydes possessing chiral centers in the chains. The optical activity of these polyaldehydes is higher than expected from the optical activities of the corresponding monomer unit. This effect is based on the conformation rigidity or helicity of the chain near the asymmetric centers of the side chains of the polymers and does not depend on the degree of crystallinity (see also OPTICALLY ACTIVE POLYMERS).

Haloaldehyde Polymerization. Studies of the polymerization of haloaldehydes have contributed greatly to the knowledge and understanding of aldehyde polymerization. The most thoroughly studied haloaldehydes are halogenated derivatives of acetaldehyde. The polymerization of trichloroacetaldehyde (chloral) occupies a special position because of the availability of the monomer and the potential practical application of polychloral.

Haloaldehyde polymers include the polymers of all 10 possible perhalogenated (fluoro-, chloro-, bromo-) acetaldehydes.

Thermal and Mechanical Properties of Chloral Polymers

Unlike most other polymers, polychloral does not show a melting or glass transition below 220°C, at which the polymer starts to degrade by depolymerization to monomer. The polymer is of moderate crystallinity and shows properties characteristic of engineering plastics (qv).

Potential Applications

None of the higher aldehyde polymers have found commercial application, although some of the properties of the haloaldehyde polymers are attractive.

BIBLIOGRAPHY

P. Kubisa and O. Vogl, *Polymer London* **21**, 525 (1980).

O. Vogl, *Polyaldehydes,* Marcel Dekker, Inc., New York, 1967.

O. Vogl, *Polym. News* **5**, 150 (1979).

O. Vogl, H. C. Miller, and W. H. Sharkey, *Macromolecules* **5**, 656 (1972).

OTTO VOGL
Polytechnic University

ALEURITIC ACID. See RESINS, NATURAL.

ALFIN CATALYSTS. See ANIONIC POLYMERIZATION.

ALFREY-PRICE EQUATION. See COPOLYMERIZATION.

ALGIN. See POLYSACCHARIDES.

ALKALI RESISTANCE. See CHEMICALLY RESISTANT POLYMERS.

ALKENE POLYMERS. See OLEFIN POLYMERS.

ALKENYLSODIUM COMPOUNDS. See ANIONIC POLYMERIZATION.

ALKYD RESINS

Alkyd resins have been defined as the reaction product of a polybasic acid and a polyhydric alcohol. This definition includes polyester resins of which alkyds are a particular type. The specific definition that has gained general acceptance is that alkyds are polyester modified with monobasic fatty acids.

The formation of alkyd resins is a typical example of condensation polymerization. Although early investigators reported the formation of resinous products by the reaction of polybasic acids and polyhydric alcohols, it was not until 1929 that Carothers showed the broad relationship of functionality to polymer formation. His early work defined the conditions necessary for intermolecular condensation polymerization to produce linear, soluble, fusible polymers such as polyesters from glycols and dibasic acids.

At about the same time at General Electric Research Laboratory, it was shown that whereas a bifunctional polymer such as the glycol phthalate polyester is fusible and soluble, the use of a trifunctional component such as in the glyceryl phthalate polyester led to gelation long before completion of the esterification.

The conditions for the gelation of such condensation polymers has been treated mathematically by Carothers. The simplified form of the Carothers equation states that as the molecular weight becomes infinite at the gel point, $p = \frac{2}{f}$, where p is the extent of reaction, and f, the average degree of functionality or the average number of functional groups in the reacting molecules, considering only stoichiometric equivalents of interacting functional groups. The Carothers equation, when applied to bifunctional reactants, indicates that no gelation will occur even at 100% esterification.

Thus to avoid gelation the average functionality of the reactants must be approximately two. If glycerol (three hydroxyl groups) and phthalic anhydride (two carboxyl groups) are present in equimolecular proportions, the actual functionality of the glycerol is limited to two because that is the number of carboxyl groups available for reaction. If all primary hydroxyl groups of glycerol react first, a linear polymer will result. Therefore, the use of more glycerol than the stoichiometric quantity required by the available carboxyl groups reduces the functionality. An additional method for controlling functionality is the basis for the formation of monobasic acid-modified polyesters, or alkyds as they are commonly known. This involves conversion of the trifunctional polyol to a bifunctional polyol by forming a monoglyceride and then reaction with the dibasic acid to form a monobasic acid-modified linear polymer.

The monobasic acid modifies the property of the resin in two ways: first, by its capacity to control functionality and thus allow control of polymer building and second, by its physical and chemical properties. Most of the monobasic acids used in alkyd resins are derived from natural glyceride oils. The degree and nature of unsaturation in fatty acid groups affect the susceptibility to absorb oxygen and cross-link to a solid film. Nature has provided a wide array of oils with fatty acids of varying degrees of unsaturation and chain length. Variation in this component alone allows a wide gradation in film properties from soft colorless plasticizing films to hard, tough, tack-free films. In a plasticizing alkyd, fatty acids that are fully saturated or contain only one double bond are generally used.

Thus, alkyds based on nonoxidizing castor and coconut oils are used to impart flexibility and color retention. When the fatty acids present in an alkyd are derived from semidrying or drying oils, the alkyd can undergo oxidation at ambient temperature and convert to a tack-free film.

The following compounds are usually used for the polyol component of alkyd resins: glycerol, pentaerythritol, ethylene glycol, propylene glycol, diethylene glycol, sorbitol, neopentyl glycol, trimethylolpropane, and trimethylolethane. Of these, glycerol and pentaerythritol are most widely used. Pentaerythritol, which has four primary alcohol groups, forms more complex resins with phthalic anhydride than does glycerol, and its reactivity can be reduced, either by the partial replacement with various glycols (functionality of two) or by the use of larger proportions of fatty acids.

The third basic component of alkyd resins is the polybasic acid. By far, the most important is phthalic anhydride, most of which is produced by the oxidation of ortho-xylene. Increasing quantities of isophthalic acid are being used. This dibasic acid, unlike phthalic anhydride, will not form intramolecular cyclic structures, and it imparts higher molecular weights and higher viscosities. Other polybasic acids that are used to impart special properties are fumaric acid, maleic anhydride, adipic acid, azelaic acid, and dimerized fatty acids.

It becomes readily apparent that the permutations and combinations possible with only the three basic components of alkyd resins are enormous. The structure and proportions of these individual components control the ultimate properties of the polymer. Alkyds are classified according to the amount of fatty acids and phthalic anhydride present. The percentage of fatty acid modification greatly influences the properties of alkyd resins.

The components of a particular alkyd resin and the proportions of these components will vary with the intended use of the resins. By suitable variation, the polymer chemist can achieve a wide range of physical properties and performance characteristics including viscosity, color drying rate, hardness, adhesion, solubility, chemical resistance, color and gloss retention durability, and compatibility with other polymers.

Functionality and Formulation

In alkyd formulation, the polymer chemist generally seeks to prepare the resin with highest molecular weight at low acid numbers without encountering gelation. Patton has derived an "alkyd constant" based on an adaptation of Carother's classic functionality equation. His proposed constant is the ratio of $\frac{m_o}{e_A}$ and equal to unity:

$$K = \frac{m_o}{e_A} = 1$$

where m_o is the total number of acid and hydroxyl-bearing molecules in a given alkyd composition at the beginning of a condensation reaction between acid (COOH) and hydroxyl (OH) groups and e_A, the total number of acid equivalents in the reaction. Patton showed that those alkyds based on phthalic anhydride tend to have lower alkyd-constant values (1.005 ± 0.014), whereas those based on isophthalic acid tend to have higher values (1.046 ± 0.008). An alkyd constant lower than 0.05 unit from the above values indicates a probability of encountering premature gelation short of 100% reaction.

The use of alkyd constant represents a starting point in the formulation of alkyd resins. Although it provides valuable data on reactant proportions, the choice of raw materials must be carefully evaluated, and with the increasing number of new polyols, polybasic acids, and monobasic acids, this is a most important consideration in the design of alkyd resins.

In the production of alkyd resins, two basic methods are used. The most direct method of producing this polyester is the simultaneous esterification of the polyol, polybasic acid, and the fatty acid: the "fatty acid method." In the other procedure, a glyceride oil reacts with glycerol or other polyol in the presence of a catalyst at temperatures of 225–250°C: the "alcoholysis method. Interesterification occurs, and with glycerol as the polyol, the following reaction occurs.

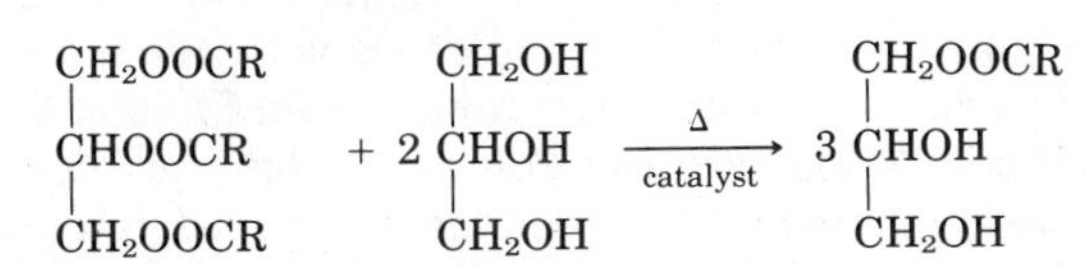

triglyceride oil glycerol monoglyceride

When a dibasic acid such as phthalic is added to the monoglyceride, a homogeneous resin will result.

Properties and Uses

Alkyds are characterized by rapid drying, good adhesion, flexibility, mar resistance, and durability. Their principal performance weakness is the ease with which the ester groups are hydrolyzed under alkaline conditions. The wide spectrum of properties of all alkyd resins is broadened by their use with many reactive chemicals and other resinous materials. Although they have been in commercial use for over 50 years, they still rank as the most widely used synthetic resins for a variety of organic coatings.

In recent years, the use of oil-free alkyds in thermosetting coatings has grown considerably. They are characterized by the absence of fatty acids and are formed by the reaction of polybasic acids with polyhydric alcohols in nonstoichiometric amounts. These products are best described as functional saturated polyesters containing unreacted OH and COOH groups. These polyesters are lighter in color than conventional alkyds because unsaturated fatty acids that contribute to color are not used. Moreover, the absence of unsaturation eliminates the post embrittlement caused by the oxidation of unsaturated fatty acids. The elimination of long chain fatty acids also improves chemical resistance. These saturated reactive polyesters display a unique hardness-to-flexibility relationship that makes it possible to attain greater hardness values at high reverse impact values than is possible with other polymer systems.

The powerful forces of the ecology lobby have resulted in antipollution legislation requiring the coatings industry to limit atmospheric emission of photochemically reactive solvents. Research on new application techniques, polymers, and crosslinking systems has led to new technologies such as powder coatings and radiation-cured coatings. Water-reducible coatings and high solids coatings are effectively lowering solid emissions. Alkyd resins and reactive saturated polyesters have been designed to meet the requirements of these new technologies and the use of these polymers is rapidly increasing in these areas.

BIBLIOGRAPHY

H. J. Lanson in R. W. Tess and G. W. Poeklin, eds., *Applied Polymer Science,* 2nd ed., American Chemical Society, Washington, D.C., 1985, Chapt. 49.

T. C. Patton, *Alkyd Resin Technology,* Wiley-Interscience, New York, 1962.

D. H. Solomon, *The Chemistry of Organic Film Formers,* John Wiley & Sons, Inc., New York, 1967.

H. J. LANSON
LanChem—an AKZO Company

ALKYLALUMINUM COMPOUNDS. See ZIEGLER-NATTA CATALYSTS.

ALKYLENEIMINE POLYMERS

In general, polyalkyleneimines are made by the nucleophilic addition of imine or amine functions to activated, strained ring

Table 1. Imine Monomer Types

I	II	III	IV
Secondary amines (unsubstituted imines)	Tertiary amines (1-substituted imines)	Tertiary amides (1-acyl substituted imines)	Iminoesters (amide isomers)
	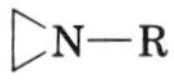	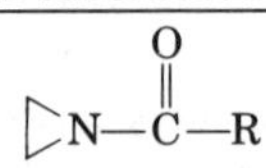	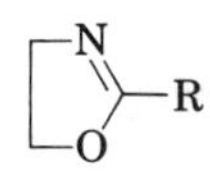 oxazolines
	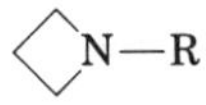	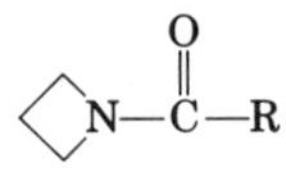	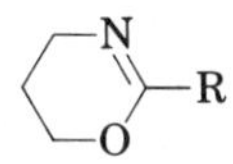 oxazines

systems or alternatively, activated double bonds. Most commercial examples involve ionic, ie, cationic–anionic, propagation processes that are driven by ring strain associated with the respective monomers. Polyalkyleneimines can be derived from the four general monomer types shown in Table 1.

Monomer Properties and Synthesis

Aziridines: Monomer Types I–III. ***Ethyleneimine.*** Ethyleneimine (aziridine) is a colorless flammable, volatile liquid that is miscible in all proportions with most organic solvents and water, but incompletely miscible with aliphatic hydrocarbons and concentrated aqueous alkali solutions. Physical properties are listed in Table 2. This monomer is very toxic and must be handled with extreme caution; it is toxic both by inhalation and skin absorption. Contact with Lewis or Brønsted acids can lead to explosive polymerization. Ethyleneimine (EI) can be synthesized by the Gabriel method, the Wenker method, or the Dow process.

Table 2. Physical Properties of Neat and Aqueous Solutions of Ethyleneimine

Property	Value
mol wt	43.07
density, g/mL	
−10°C	0.865
10°C	0.846
sp gr, % ethyleneimine solutions, at 26°C	
99.6	0.829
57.0	0.929
18	0.978
bp, °C	57
fp, °C	−78
index of refraction, n_D, at 25°C	1.4123
viscosity, mPa·s	
−10°C	0.638
10°C	0.506
surface tension, mN/m, at 25°C	32.8
flash point (closed cup), % ethyleneimine and aqueous solutions, °C	
100	−11.1
40	27
10	57
limits of flammability in air, vol %, at 101.3 kPa[a]	3.3–3.8
heat of vaporization, kJ/mol, at 20°C	33.1
dissociation constant, K_d, at 22°C	7.8×10^{-7}

[a] To convert kPa to mm Hg, multiply by 7.502.

1-Substituted Ethyleneimine. Most commercially available 1-substituted aziridines have been prepared by the addition of ethylenineimine to either cyclic oxides or activated double bonds. Properties of some *N*-substituted aziridines appear in Table 3.

1-Acyl Substituted Ethyleneimines. These can be prepared by the addition of aziridine to aliphatic or aromatic acid chlorides under Schotten-Baumann conditions (Table 3).

Oxazolines: Monomer Type IV. Iminoesters or oxazolines, specifically 2-subsituted 1,3-oxazolin-2-enes, are formed by isomerization of *N*-acyl aziridines with catalytic amounts of iodide ion, heat, or acid. Typical physical properties for 2-ethyl-1,3-oxazolin-2-ene are shown in Table 4 (see also CYCLIC IMINO ETHERS, POLYMERIZATION).

Azetidines: Monomer Types I–III. This group of compounds has received much less attention than aziridines owing to past difficulties in obtaining high yields of the monomer. Azetidine is a colorless liquid that fumes in moist air and is completely miscible with water. The boiling point is recorded as 62°C at 97.3 kPa (730 mm Hg) and 62.5°C at 99.6 kPa (747 mm Hg). There are four principal routes to azetidines: ring closure of γ-haloamines or γ-amino alkyl sulfates, formation and reduction of 1-arylsulfonazetidides, and reduction of azetidinones.

Commercial Monomer and Polymer Production

Aziridine Monomers and Polymers. Currently, these materials are produced domestically in undisclosed amounts by Cordova Chemical Co. (now a part of Virginia Chemical Co.). Polyethyleneimine is imported in small amounts from Nippon Soda in Japan and is marketed in the United States by Japan Catalytic Co., in New York, and Aceto Chemical Co., in New Jersey.

1-Substituted Aziridines. The principal producer of 1-(2-hydroxyethyl)aziridine is currently Cordova Chemical Co., which supplies it at $11–13/kg. The unsaturated derivative, 2-(1-aziridinyl)ethyl methacrylate, known commercially as iminol methacrylate, is now supplied principally by Cordova.

1-Substituted Acylaziridines. Polymers derived from this class, ie, via oxazoline monomers, are commercially available.

2-Substituted 2-Oxazolines. Dow is the principal worldwide supplier of the monomer, 2-ethyl-2-oxazoline, as well as its linear *N*-acylated (and deacylated) polyethyleneimine polymers. The price of the monomer ranges from $4.40 to $6.60/kg.

Azetidines: Monomers and Polymers. None of these imine monomers or polymers have been manufactured in commercial quantities.

Table 3. Commercial *N*-Substituted Aziridines

Name	Structure	Mol wt	Fp, °C	Bp, °C (kPa)	Density, 25°C, g/mL	n_D^{25}
N-acetylaziridine	▷N—C(=O)—CH_3	85.108	−60.24	59(6.3)	0.9838	1.4370
N-(2-hydroxyethyl)aziridine	▷NCH_2CH_2OH	87.1		167(101.3)	0.9867	1.4534
N-ethylaziridine	▷NCH_2CH_3	71.12		50.5(101.3)	0.746(20°C)	1.3964(20°C)
N-(2-cyanoethyl)aziridine	▷NCH_2CH_2CN	96.1	−66.0	187(101.3)	0.9581	1.4440
N-(2-aminoethyl)aziridine	▷$NCH_2CH_2NH_2$	86.1	−56.48	131.0(101.3)	0.899(20°C)	1.4510
1-phenethylaziridine	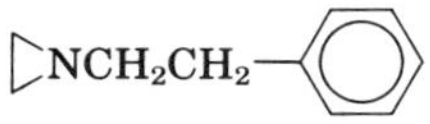	147.22	−40.97	91.7(5.3)	0.9525	1.5185
bis[2-(1-aziridinyl)ethyl] isophthalate	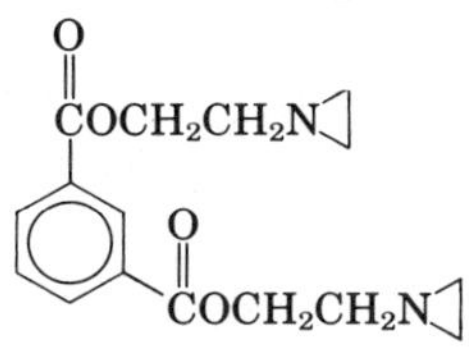	304.35	20 (est)		1.1634	1.5266

Table 4. Physical Properties of 2-Ethyl-1,3-oxazolin-2-ene

Property	Value
mol wt	99.13
bp, °C, at 101.3 kPa[a]	128.8
vapor pressure, at 25°C, kPa	1.3
fp, °C	−62
sp gr, 25°C	0.974
viscosity, at 25°C, mm^2/s	0.91
flash point (TCC), °C	34
auto ignition, °C	490
solubility	miscible with common organic solvents and water
critical temp, °C	333
critical pressure, MPa	4.044

[a] To convert kPa to mm Hg, multiply by 7.502.

N-Unsubstituted Polyalkyleneimines

Structure and Properties of Polyethyleneimines. Polyethyleneimines (PEI) consist of a large family of water-soluble, ie, hydrophilic, polyamines of varying molecular weight and degree of chemical modification. They function as weak bases and exhibit cationic character depending on the extent of protonation. In general, all PEI produced by classic cationic ring-opening polymerization techniques are believed to be highly branched, possessing primary, secondary, and tertiary amine groups in a ratio of approximately 1:2:1.

Linear Polyethyleneimines. More recently, other synthetic routes have led to PEI possessing more regular, ie, comblike rather than random, branching. Furthermore, totally linear PEI has been prepared by either modifying the aziridine polymerization conditions or using oxazoline monomers as precursors.

Formation of Branched Polyethyleneimine. ***Polymerization Mechanism.*** Polymerization of EI proceeds through a cationic polymerization (qv) chain-growth mechanism. A variety of cationic agents can be used, but proton transfer produces the same initiating species.

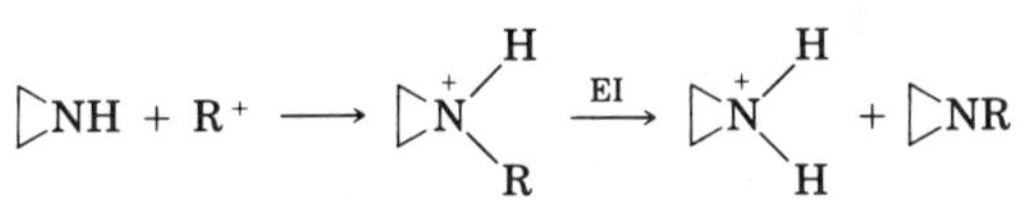

Some of the cationic agents that have been used are as follows:

- strong Brønsted acids
- carboxylic acids
- Lewis acids
- water
- organic halides
- esters of strong acids
- onium salts
- phenols
- acidic inorganic surfaces
- inorganic salts
- halogens

Storage and Handling. These polymers are readily stored neat or as aqueous solutions in stainless steel or aluminum storage tanks and present no unusual safety problems other than those associated with the corrosivity of amines. Toxicity of these products is low enough to allow federal approval for use as tie-coat adhesives for the production of certain food containers and as clarifiers for potable water.

Commercial Availability. Currently, commercial branched PEI is available from only one domestic source, Cordova Chemical Co. (now Virginia Chemical Co.).

Formation of Linear Polyethyleneimines. ***From 2-Oxazoline Monomers.*** Poly(*N*-acylethyleneimine)s are obtained by the cationic polymerization of 2-substituted 2-oxazolines.

Table 5. Physical Properties of Commercial *N*-Substituted Polyethyleneimines

	Values for		
Property	Poly[*N*-(2-hydroxyethyl)aziridine]	Poly[*N*-(2-cyanoethyl)aziridine]	Poly[*N*-(2-phenylethyl)aziridine]
structure	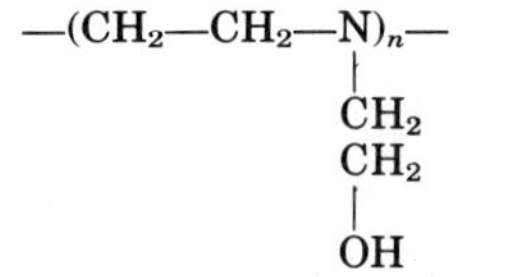	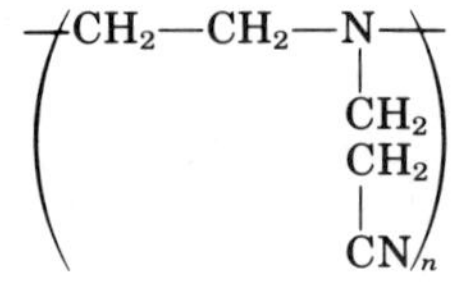	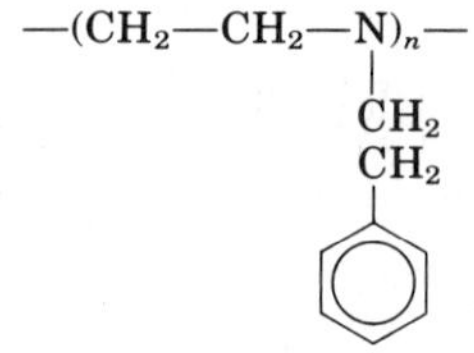
physical form	clear, colorless, viscous liquid containing <1.0% water	cream-colored powder	yellow liquid
mol wt, approx	2000	2000	2000
mp, °C		103	
density, g/cm^3			
at 25°C		1.178	1.043
at 65.5°C			1.017
flash point	none (decomposes at 271°C)		
refractive index, n_D^{25}			1.5693
Gardner color			4
pour point, °C			0
viscosity			
Pa·s			
at 25°C	800		
at 50°C	48		
at 100°C	2.6		
mm^2/s			
at 25°C			9730
at 65.6°C			320
solubility	the polymer is soluble in water, methanol, and ethanol; insoluble in benzene, toluene, and acetone	less than 1% soluble in water, methanol, acetone, benzene, carbon tetrachloride, diethyl ether, and *n*-heptane; soluble in hot DMF, DMSO, hot ethylene glycol, and dilute acids	soluble in acetone, methanol, benzene, diethyl ether, and carbon tetrachloride; insoluble in heptane and water

Polyazetidine (Polytrimethyleneimine). In general, unsubstituted azetidines parallel aziridines in their polymerization behavior.

N-Alkyl-Substituted Polyalkyleneimines

Polymers from *N*-substituted Ethyleneimines. A wide variety of alkylating agents and acids have been used to initiate the polymerization of *N*-substituted aziridine monomers. Essentially any of the catalysts shown above can be used under the appropriate conditions, and a number of lower molecular weight polymers have been produced commercially by these methods (Table 5).

Polymerization Mechanism. In contrast to *N*-unsubstituted aziridines, two different initiating species are produced as shown below with protons or alkylating agents, and this leads to two different propagation sequences.

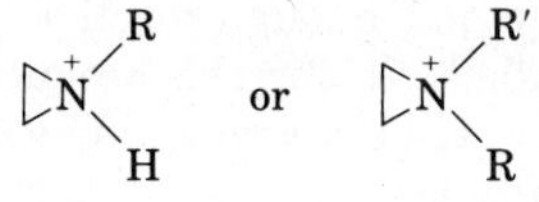

Poly(*N*-substituted azetidines). Research has shown that an azetidinium salt is the active species in the polymerization of *N*-substituted azetidines. The initiation can occur by use of proton acids or by the azetidinium salt prepared separately. The polymerization of azetidinols tends to yield low molecular weight polymers.

N-Acyl Substituted Polyalkyleneimines

Polymers from Activated Aziridines. Activated aziridines are defined as ethyleneimine compounds in which the nitrogen substituent can conjugate with the unshared pair of electrons on the aziridine nitrogen. Polymerization of these imine monomers has been demonstrated thermally and with both acid and base catalysts. Aziridine tertiary amides undergo polymerization to acylic polymer not only with the usual cationic catalysts, but also with anionic reagents.

Polymers from 2-Oxazolines. 2-Oxazolines can be readily polymerized with cationic catalysts to a wide variety of *N*-substituted polyethyleneimines with molecular weights varying from a few thousand to millions. Cationic polymerization of the commercially available 2-ethyl-2-oxazoline produces poly(*N*-propionylethyleneimine), which is offered in commercial quantities from The Dow Chemical Co. in at least three molecular weight ranges. Some physical properties are shown in Table 6.

Hydrolysis of Initial Amide Products. The preferred route to either linear partially *N*-acylated or totally unsubstituted linear polyethyleneimines is from 2-oxazolines. Both acid and basic hydrolysis have been applied to the initial poly(*N*-acylethyleneimine) products. Acid hydrolysis is preferred for most alkyl or aryl groups in the acyl moiety whereas basic hydrolysis is facile and complete for the formyl group. Physical properties depend on the degree of hydrolysis.

Table 6. Physical Properties of Poly(*N*-propionylethyleneimine)

Property	Values for 50,000 mol wt
melt viscosity, Pa·s, 1000 s^{-1} shear rate, at 190°C	53
melt flow, gm/10 min	
refractive index, n_D^{25}	1.520
specific gravity	1.14
glass-transition temp, amorphous, °C	69–71
thermal stability, tga in air, °C	380
Vicat softening temp, °C	70
solubility, 150,000 mol wt sample	
soluble, ≥50% by weight	water, methanol, ethanol, DMF, methylene chloride, acetone, THF, chloroform, acetonitrile
insoluble, <2% by weight	hexane, benzene, toluene, cyclohexane, xylene, diethyl ether, CCl_4

Linear N-Substituted PEI. Linear poly(*N*-alkylethyleneimine)s have been prepared from polymers of 2-oxazolines by either of two routes: amide reduction with metal hydrides or the Eschweiler-Clarke reaction on linear polyethyleneimine with formic acid or formaldehyde in water.

Polymers from 2-Oxazines. 2-Substituted and unsubstituted 2-oxazines have been polymerized with cationic catalysts to the corresponding linear *N*-acyl substituted poly(trimethyleneimine)s.

Polymer Modifications

Essentially all of the important modifications of polyalkyleneimines involve replacement of the hydrogen atoms on nitrogens. These transformations usually allow modification of solubility, introduction of functional groups, or irreversible control of charge density. The azetidine functionality is most commonly introduced into a polymer by the reaction of epichlorohydrin with the polymeric secondary or primary amine sites.

Copolymers

Alkyleneimine monomers have been copolymerized with a wide variety of reagents. In general, these combinations are more responsive to ionic process, but there are some instances where free-radical mechanisms are implicated.

Copolymers include random copolymers from aziridines and from activated aziridines and 2-oxazolines, graft copolymers from ethyleneimine and oxazolines, sequence alkyleneimine copolymers from aziridine and azetidines, from acyclic monomers and from oxazolines and oxazines, and block copolymers.

Applications

Commercial applications and uses for the various alkyleneimine polymers are diverse and are summarized in tabular form according to chief industries (Table 7). See also FIBERS; DYEING; ANTISTATIC AGENTS; TEXTILE RESINS; PAPER ADDITIVES AND RESINS; ION-EXCHANGE POLYMERS; LAMINATES; COATINGS; FLOCCULATION; OIL-FIELD APPLICATIONS; ENZYMES, IMMOBILIZED; MEDICAL APPLICATIONS.

Table 7. Alkyleneimine Polymer Applications

Applications–Industries	Polymer[a] type
Textile industry	
dyeing/printing	
cellulosics	B-PEI
	PAMAM
	AZET
	L-PEI
noncellulosics	B-PEI
	POXAZL
water repellency	B-PEI
	POXAZL
textile softening	B-PEI
miscellaneous	B-PEI
antistatic	B-PEI
	AZET
wood dimensional stabilizer	AZET
fibers	POXAZL
Paper industry	
wet strength	B-PEI
	PAMAM
	AZET
	PAMAM
retention aid	B-PEI
	PAMAM
sizing	B-PEI
	PAMAM
	AZET
	POXAZL
coating	B-PEI
	PAMAM
	POXAZL
	POXAZ
Plastics industry	
general	B-PEI
	POXAZL
foams	POXAZL
Separations industry	
ion exchange	B-PEI
	L-PEI
chelation-type resins	PEI
	PAMAM
	L-PEI
membranes/filters	B-PEI
	POXAZL
	L-PEI
chromatography	B-PEI
Adhesives industry	
general	B-PEI
laminating	B-PEI
	POXAZL
Coatings	
prime coating (tie coat)	B-PEI
finish coating	B-PEI
Photography	B-PEI
	POXAZL

Table 7. (*Continued*)

Applications–Industries	Polymer[a] type
Surface active agents	
surfactants	B-PEI
	POXAZL
flocculation	B-PEI
	PAMAM
de-emulsification	B-PEI
	AZET
	PAMAM
corrosion inhibitors	B-PEI
	PAMAM
emulsifiers	B-PEI
Lubricants	B-PEI
	L-PEI
Fuels	
general	B-PEI
solid propellants	B-PEI
Miscellaneous	
industrial cleaning	B-PEI
oil recovery	B-PEI
enzyme mimics	B-PEI
immobilized-enzymes application	B-PEI
medical use	B-PEI
	L-PEI
	POXAZL
electroplating	B-PEI
cosmetics	B-PEI
antimicrobial	POXAZL
flame retardant	L-PEI

[a] B-PEI = branched-polyethyleneimine
L-PEI = linear polyethyleneimine
POXAZL = polyoxazolines
POXAZ = polyoxazines
BPTMI = branched-polytrimethyleneimine
LPTMI = linear polytrimethyleneimine
PAMAM = polyamidoamines
AZET = azetidine functional polymers.

BIBLIOGRAPHY

J. A. Moore in A. Weissberger, ed., *Heterocyclic Compounds with Three- and Four-Membered Rings,* Wiley-Interscience, New York, 1964.

E. J. Goethals, ed., *Polymeric Amines and Ammonium Salts,* Pergamon Press, Inc., Elmsford, N.Y., 1980.

O. C. Dermer and G. E. Ham, *Ethyleneimine and Other Aziridines,* Academic Press, New York, 1969.

Ethyleneimine, Dow Technical Bulletin, The Dow Chemical Co., Midland, Mich., 1965.

2-Ethyl-2-Oxazoline XAS-1454 (Ethyloazoline): An Intermediate for Aminoethylation, Dow Technical Bulletin, Form #116-1030-81, The Dow Chemical Co., Midland, Mich., 1981.

PEOX (Polyethyloxazoline) Polymer, Dow Technical Bulletin, Form #171-013-85, The Dow Chemical Co., Midland, Mich., 1985.

Donald A. Tomalia
George R. Killat
The Dow Chemical Company

ALKYLENE OXIDE POLYMERS. See 1,2-Epoxide polymers; Oxetane polymers; Polyethers.

ALKYLENE SULFONE POLYMERS. See Olefin–sulfur dioxide copolymers.

ALKYLLITHIUM COMPOUNDS. See Anionic polymerization.

ALKYL METHACRYLATE POLYMERS. See Acrylic and methacrylic ester polymers.

ALKYLSODIUM COMPOUNDS. See Anionic polymerization.

ALLERGENS. See Antibodies and antigens, polymer-bound.

ALLOY RESINS. See Polymer blends.

ALLYL POLYMERS. See Diallyl and related polymers.

ALPHA-CELLULOSE. See Cellulose.

ALTERNATING COPOLYMERS. See Copolymers, alternating.

ALUMINUM ALKYLS. See Ziegler-Natta catalysts.

AMBER. See Resins, natural.

AMINIMIDE POLYMERS

The aminimide nomenclature is generally used because the ylid functional group (**1**) is isoelectronic with an amine oxide (**2**). The generalized aminimide structures (**1**) and (**2**) provide some insight into their chemical versatility. Absorption bands in the ir at 1555–1600 cm^{-1} support the resonance structures shown.

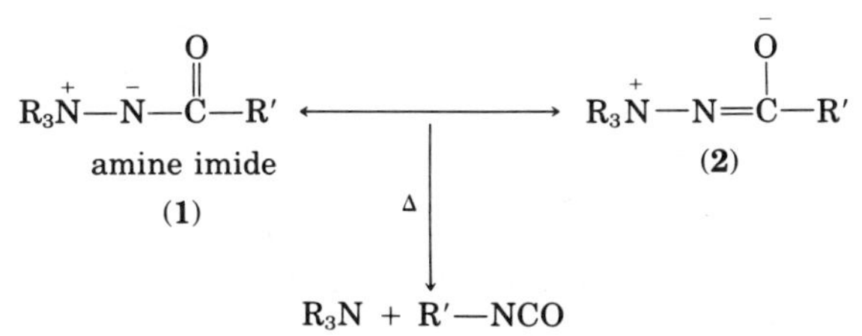

amine imide (**1**) (**2**) (**3**)

As isocyanate precursors, amine acylimides give chemists a clever method to prepare a wide variety of monomeric and polymeric, aliphatic and aromatic isocyanates (**3**) by proper selection of the R′ moiety on the aminimide. The aminimide also functions as an amine precursor, giving an easily handled source of tertiary amine.

Preparation of Monomers

Many different types of monofunctional, difunctional, etc, aminimide have been prepared at Ashland Chemical Co. and other locations. The monomers are mainly useful in radical-initiated, chain-growth polymerizations and in step-growth polymerizations or polymer derivatization reactions (see CHAIN-REACTION POLYMERIZATION; STEP-REACTION POLYMERIZATION).

Several procedures are known for preparing molecules with the amine acylimide residue. The two most attractive procedures, from a commercial view, are shown in equations 1 and 2.

$$R'COOR + NH_2\overset{+}{N}(CH_3)_3X^- + NaOR \longrightarrow R'\overset{O}{\overset{\|}{C}}\overset{-}{N}\overset{+}{N}(CH_3)_3 + 2\,ROH + NaX \quad (1)$$

$$RCOOR' + NH_2N(CH_3)_2 + R''CH\overset{O}{\frown}CH_2 \longrightarrow R{-}\overset{O}{\overset{\|}{C}}\overset{-}{N}\overset{+}{N}(CH_3)_2CH_2CH(OH)R'' + R'OH \quad (2)$$

Properties of Monomers

Aminimides are soluble in water and a number of polar organic solvents. They exhibit good shelf stability. However, when heated at or above their melting points they undergo nitrogen–nitrogen bond cleavage, rearrangement, or elimination reactions.

Polymerization

Aminimide monomers with acrylic and methacrylic residues readily homopolymerize and copolymerize, using nonoxidative, free-radical initiators.

Polymer Properties

Aminimide polymer films are hard, brittle materials. When heated at or above 125°C, the films become insoluble in water, alcohol, and other common organic solvents. Infrared spectra, taken during heating, show that the aminimide carbonyl band at 1560 cm^{-1} vanishes and a new absorption band appears at 2260 cm^{-1}, indicating isocyanate formation. Differential thermal analysis (dta) and thermogravimetric analysis (tga) studies show that trimethylamine acylimide polymers start to decompose at approximately 125°C and lose theoretical amounts of trimethylamine from 125 to 200°C.

All aminimide polymers are water soluble. Copolymers may be water soluble, depending on aminimide or hydrazinium salt content.

Copolymers with aminimide residues also suffer nitrogen–nitrogen bond-breaking reactions and elimination of tertiary amine when heated to 125°C.

Applications

With the exception of Permachem Asia company's interest in a one-package epoxy resin system, the once bright future proclaimed for aminimide monomers and polymers has never developed.

BIBLIOGRAPHY

W. J. McKillip, E. A. Sedor, B. M. Culbertson, and S. Wawzonek, *Chem. Rev.* **73,** 255 (1973).

S. Wawzonek, *Ind. Eng. Chem. Prod. Res. Dev.* **19**(3), 338 (1980).

B. M. Culbertson, *Polym. News* **5,** 104 (1978).

H. J. Langer and B. M. Culbertson, "Copolymers, Polyblends, and Composites," *Adv. Chem. Ser.* **142,** 129 (1975).

BILL M. CULBERTSON
Ashland Chemical Co.

AMINOPLASTS. See AMINO RESINS.

AMINO RESINS

Amino resins are thermosetting polymers made by combining an aldehyde with a compound containing an amino (—NH_2) group. Urea–formaldehyde accounts for over 80% of amino resins; melamine–formaldehyde accounts for most of the rest. Other aldehydes and other amino compounds are used to a very minor extent. The first commercially important amino resin appeared about 1930, or some 20 years after the introduction of phenol–formaldehyde resins and plastics (see PHENOLIC RESINS).

The principal attractions of amino resins and plastics are water solubility before curing (qv), colorlessness, excellent solvent resistance in the cured state, outstanding hardness (qv) and abrasion resistance, and good heat resistance. Limitations of these materials include release of formaldehyde during cure and, in cases such as foamed insulation, after cure. Poor outdoor weatherability for urea moldings is another limitation. Repeated cycling of wet and dry conditions will cause surface cracks. Melamine moldings have relatively good outdoor weatherability.

Amino resins are manufactured throughout the industrialized world to provide a wide variety of useful products. Adhesives represent the largest market. They are used to make plywood, chipboard, and sawdust board. Other types of amino resin adhesive are used to make laminated wood beams, parquet flooring, and furniture (see ADHESIVE COMPOSITIONS; COMPOSITION BOARD; LAMINATES).

Monomer Raw Materials

Most amino resins are based on the reaction of formaldehyde with urea or melamine.

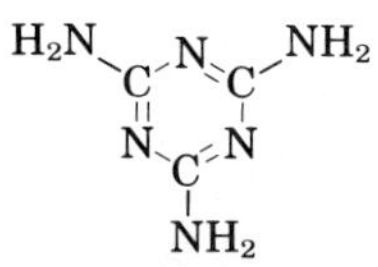

melamine

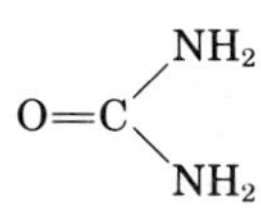

urea

Although formaldehyde will combine with many other amines, amides, and aminotriazines to form useful products, only a few are used and are of minor importance compared to products based on urea and melamine.

Polymerization Chemistry

Two main steps are recognized in the reaction of formaldehyde with amino compounds to form useful resins and plastics. The first is a simple addition of formaldehyde to introduce the hydroxymethyl group. This is known as methylolation or hydroxymethylation and is illustrated in the equation below.

$$R{-}NH_2 + HCHO \longrightarrow R{-}NH{-}CH_2OH$$

The second step is a condensation reaction that involves the linking together of monomer units with the liberation of water to form a dimer, a polymer chain, or a vast network. This is usually referred to as methylene bridge formation, polymerization, resinification, or simply cure; this is illustrated in the following equation:

$$RNH{-}CH_2OH + H_2NR \longrightarrow RNH{-}CH_2{-}NHR + H_2O$$

Formulation of Products for Specific Applications

Precise control of the course, speed, and extent of the reaction is essential for the manufacture of successful products. Factors important for achieving this control are the mole ratio of reactants, catalyst (pH of reaction mixture), and the reaction time and temperature. Amino resins are usually made by a batch process. Figure 1 illustrates the essential equipment and summarizes the processing steps needed for making amino resin syrups, cellulose-filled molding compounds, and spray-dried resins.

Coatings. Cured amino resins are far too hard and brittle to be used alone as coating materials. They are used in combination with more flexible resins to effect cure or cross-linking. Such polymers as alkyds, polyesters, epoxy resins, and certain types of acrylic and vinyl resins are used in the manufacture of surface coatings.

Textile Finishes. Most of the amino resins used commercially for finishing textile fabrics are methylolated derivatives of urea or melamine. Although these products are usually monomeric, they may contain some polymer as a by-product.

Amino Resins for Paper. Amino resins are used by the paper industry in large volume for a variety of applications (see PAPER ADDITIVES AND RESINS). The resins are divided into two classes according to the mode of application. Resins added to the fiber slurry before the sheet is formed are called wet-end additives and are used to improve wet and dry strength and stiffness. Resins applied to the surface of formed paper or board, almost invariably with other additives, are used to improve the water resistance of coatings, the sag resistance of ceiling tiles, and the scuff resistance of cartons and labels.

Other Uses. Water-soluble melamine–formaldehyde resins are used in the tanning of leather (qv) in combination with the usual tanning agents. Urea–formaldehyde resins are used in the manufacture of foams and as the binder for the sand cores used in the molds for casting hollow metal shapes. The fastest growing market for urea–formaldehyde resins is in shingles and built-up roofing. The amino compounds used in the manufacture of amino resins are nontoxic.

Economic Aspects

Representative prices of amino resins and plastics are shown in Table 1.

The largest producers of amino resins are American Cyanamid Co., Borden Chemical Co., Chembond (a subsidiary of Getty Oil), Georgia-Pacific Corp., Gulf Oil Co., Monsanto Co., and Reichhold Co. Each of these companies is believed to produce over 45 kt of resin annually (100% solids basis).

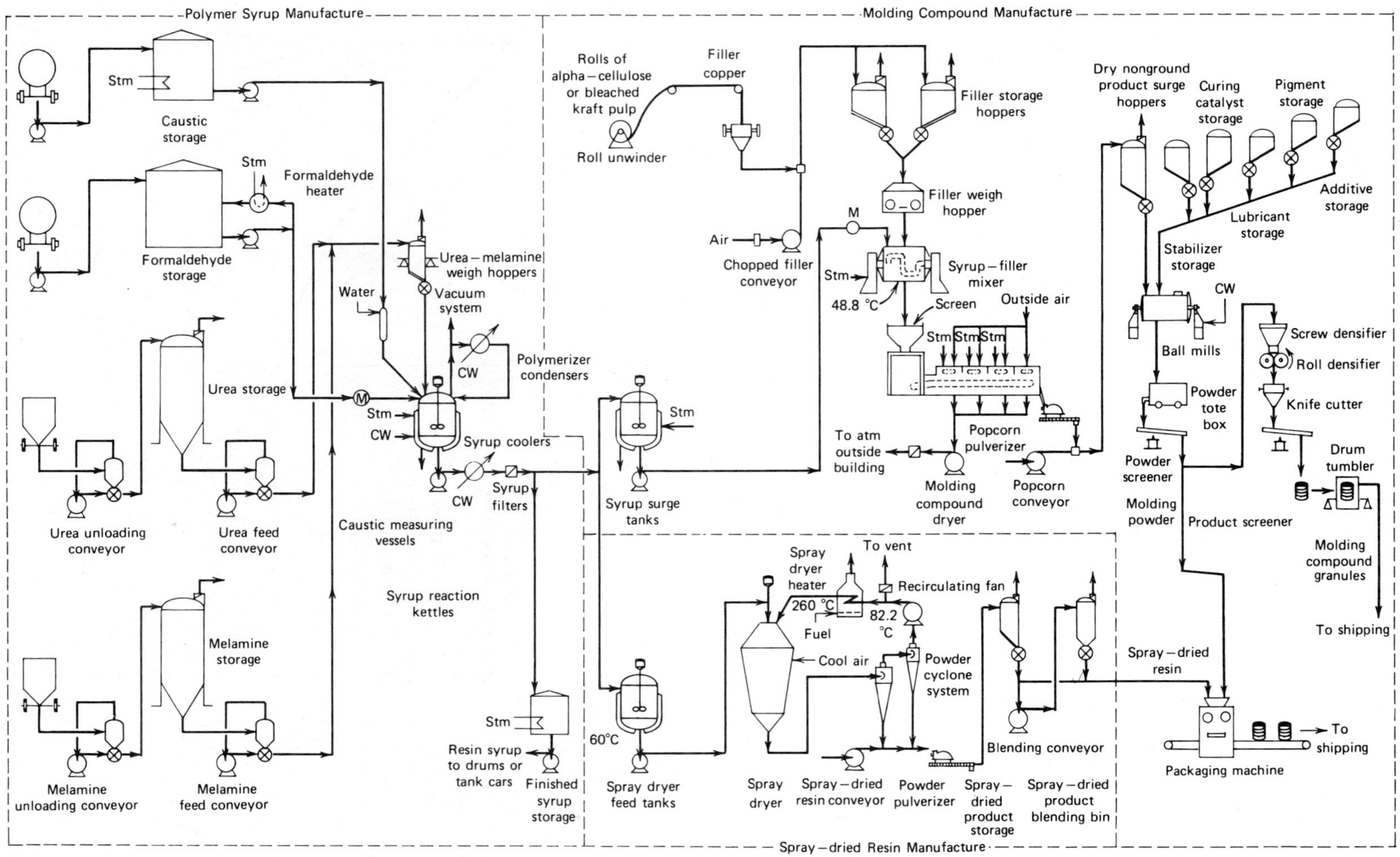

Figure 1. Urea–formaldehyde and melamine–formaldehyde resin manufacture. CW = cold water, Stm = steam. Courtesy of SRI International.

Table 1. List Prices of Representative MF and UF Resins and Plastics, 1981[a,b]

Material	Price, \$/kg[c]
melamine molding resins	
(eg, American Cyanamid's Cymel 1077 and Monsanto's Resimene 805)	1.78
urea molding resins	
(eg, Allied Chemical's Plaskon and American Cyanamid's Beetle molding compounds)	1.51
melamine laminating resins	
(eg, Monsanto's Resimene 814 and 817 and American Cyanamid's Cymel 405 and 511)	1.78
hexamethoxymethylmelamine	
(eg, American Cyanamid's Cymel 300 and 303)	1.82
butylated melamine surface-coating resin	
(eg, American Cyanamid's Cymel 248-8)	1.58
butylated urea surface-coating resin	
(eg, American Cyanamid's Beetle 227-8)	1.07
liquid urea adhesive resin	
(eg, American Cyanamid's Urac 185)	1.05
powdered urea–melamine adhesive resin	
(eg, American Cyanamid's Melurac 305)	1.43

[a] Courtesy of SRI International.
[b] Prices for melamine and urea molding compounds had risen only slightly as of June 1984.
[c] In general, the prices quoted are for carload or truckload quantities, fob the manufacturing plant where the product is made.

BIBLIOGRAPHY

J. P. Weidner, ed., *TAPPI Monogr. Ser.* **29** (1965).

T. J. Suen in C. E. Schildknecht, ed., *Polymer Processes,* Interscience Publishers, Inc., New York, 1956, Chapt. 8.

C. P. Vale and W. G. K. Taylor, *Amino Plastics,* Iliffe Books, Ltd., London, 1964.

S. A. Heap, R. E. Hunt, P. A. Rennison, R. Tattersall, W. F. Herbes, S. J. O'Brien, and R. G. Weyker in H. Mark, N. S. Wooding, and S. M. Atlas, eds., *Chemical Aftertreatment of Textiles,* Wiley-Interscience, New York, 1971, Chapt. 6.

J. C. Petropoulos, I. H. Updegraff, and L. L. Williams in J. K. Craver and R. W. Tess, eds., *Applied Polymer Science,* Organic Coatings and Plastics Chemistry Division of the American Chemical Society, Washington, D.C., 1975, Chapt. 46.

IVOR H. UPDEGRAFF
Consultant

AMINOTRIAZOLES, POLYMERIC.

See POLYAMINOTRIAZOLES.

AMORPHOUS POLYMERS

The structure of an amorphous material is characterized by the absence of a regular three-dimensional arrangement of molecules or subunits of molecules extending over distances that are large compared to atomic dimensions, ie, there is no long-range order. However, due to the close packing of the particles in the condensed state, certain regularity of the structure exists on a local scale, denoted as short-range order.

The structure, however, is not static but is subject to thermally driven fluctuations. The local structure changes continuously as a function of time due to orientational and translational molecular motions. The time scale of these motions may comprise nanoseconds up to several hundred years. The structure of the amorphous state as well as its time-dependent fluctuations can be analyzed by various scattering (qv) techniques, such as x ray, neutron, electron, and light.

General Properties of the Amorphous State

Consider a thermoplastic material in the molten state at elevated temperatures. If the melt is cooled, partial crystallization may occur if the polymer crystallizes readily. Linear polyethylene and isotactic polypropylene are examples. The resulting structure is characterized by the presence of ordered crystalline regions as well as amorphous regions. Often, however, crystallization does not take place, eg, if the chemical structure is irregular, preventing the formation of ordered regions. Examples are atactic poly(methyl methacrylate), atactic polystyrene, or random copolymers. In other cases, crystallization does not take place readily for kinetic reasons, despite a regular chemical structure of the chain molecules (see CRYSTALLIZATION KINETICS).

A transition from liquid to an amorphous solid is observed at sufficiently low temperatures. The transition is associated

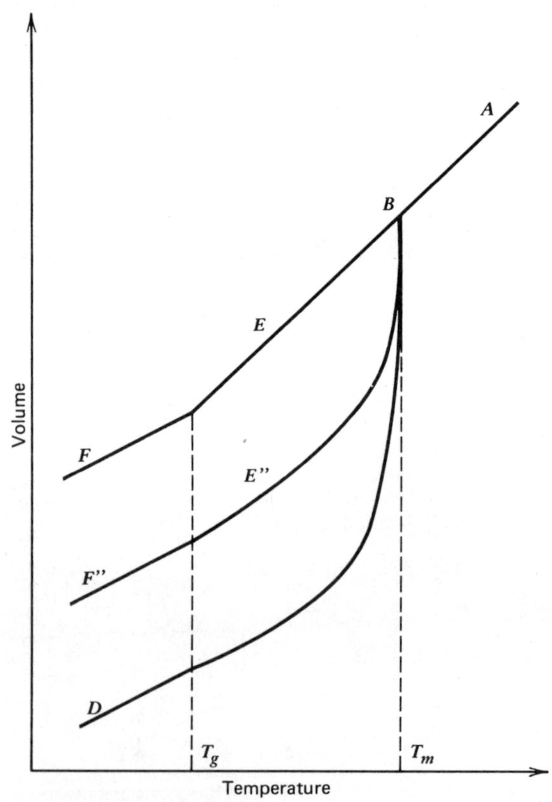

Figure 1. Volume–temperature relations for polymers. Melting and glass transition in a polymer: *EF*, no crystallization; *E''F''*, partial crystallization; and *BD*, complete crystallization. Courtesy of Taylor and Francis.

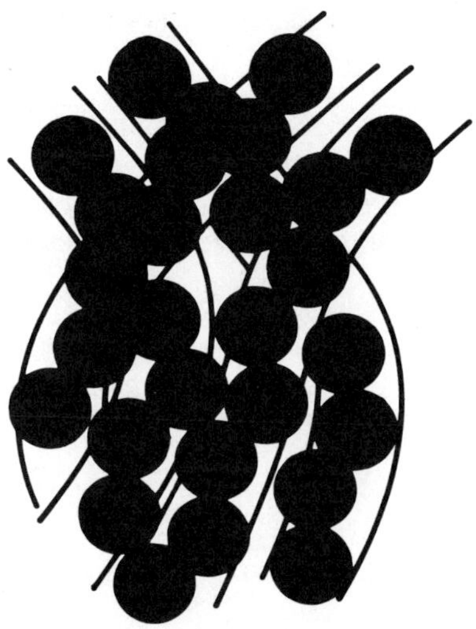

Figure 2. Short-range order in chain molecules in the condensed state.

with a bend in the volume–temperature relationship, as displayed in Figure 1. The region of the bend is termed the glass transition T_g (qv).

Whereas a melt is usually thermodynamically stable, this is not the case for the glassy state. Experimental observation indicates that the exact location of the glass transition, as well as the volume at a particular temperature in the glassy state, depend on the cooling rate. The glassy state may be considered a frozen-in liquid state. The glass transition is characterized by the restriction of certain cooperative motions, and the distribution of molecules or subunits of molecules in the glassy state does not correspond to that of the lowest free energy.

Although macroscopic data do not provide direct information on the structure of the amorphous state, some tentative conclusions can be drawn. The dense packing of molecules in the fluid or glassy amorphous state, as opposed to the dilute packing in the gaseous state, makes short-range positional order mandatory. This is illustrated schematically in Figure 2 for the particular case of subunits of chain molecules.

Experimental Techniques

Indirect experimental techniques used to study the arrangement of molecules in amorphous polymers include thermoelastic measurements, cyclization equilibrium measurements, entropy, vapor pressure, and heat of mixing determinations. Direct structure measurements all involve the interaction between particles or electromagnetic radiation and the substance in question. The necessary wavelengths and particle masses involved depend on the dimensions and type of information required. A full description of the system will require information about both its static and dynamic properties.

Chain Conformation

A large body of experimental evidence about the structure of polymers has been obtained from coherent elastic scattering experiments. For dilute solutions, the methods of coherent elastic-light scattering or small-angle x-ray scattering are well established for obtaining information about chain conformation. In the case of bulk polymers, it was first suggested that tagged molecules be used to show up the single molecule against the environment. However, contrast was poor and it was then decided that end tagging is only useful for polymers of low molecular weight; the possibility that this could change the chain conformation was recognized. Later, small-angle neutron scattering on a dilute solution of deuterium-tagged macromolecules in a protonated matrix was used in a manner analogous to that in solutions. In this case, reference is generally made to the form factor $P_{(q)}$ rather than to the structure factor because the information obtained refers to the conformation of a single molecule.

Short-Range Positional Order

Diffraction patterns in the wide angle range for the purpose of obtaining radial distribution functions have been obtained by x rays, neutrons, and electrons using coherent elastic-scattering methods. Information about atomic distributions is obtained through interferences arising from phase differences that are due to path differences.

Orientational Order

A distinguishing feature among the proposed models for the amorphous state is the amount of orientational correlations. The random coil models only require local intrachain correlations of segments, but all the bundle models or meander models are based on the assumptions of extended intersegmental correlations of various magnitude. Light scattering responds to such structural features in the case of the depolarized scattering if the chain segments themselves are sufficiently optically anisotropic.

Direct Structural Observation by Electron Microscopy

Structural information is also obtainable from direct electron microscopical observations. In order to judge which of the results obtained are reliable, details regarding the microscope transfer function must be taken into account (see also ELECTRON MICROSCOPY).

Density Fluctuations in the Amorphous State, Static Properties

Thermal density fluctuations are related to the integral over the pair correlation function or to the scattering law in the limit of $q \rightarrow 0$. Thus, the theoretical prediction for the equilibrium state is that the small-angle x-ray or neutron scattering is directly related to the isothermal compressibility. The scattering curve should be approximately constant for small values of q, allowing an extrapolation of the scattering curve to $q \rightarrow 0$. Deviations are expected to occur near a critical point, where the compressibility diverges and where the correlation length of the thermal density fluctuations also diverges.

In practice it has been observed that in the case of small-angle x-ray scattering a strong increase of the scattered intensity takes place with decreasing values of q due to a superposition of scattering components from additives, stabilizers, etc plus a component due to pure thermal density fluctuations.

Dynamic Properties of the Amorphous State as Obtained from Inelastic and Quasielastic Light Scattering

The structure of the amorphous state is subjected to time-dependent variations, owing to the existence of translational and orientational motions both of the individual segments and of the chain as a whole. The motions couple to the light because they induce variations of the local dielectric constant $\epsilon(\vec{r}, \tau)$ or the local polarizability $\alpha(\vec{r}, \tau)$. Inelastic and quasielastic scattering experiments thus allow the determination of the time laws according to which these motions occur.

Quasielastic Neutron Scattering

The scattering cross section for neutrons is given in terms of the structure factor $S(\vec{q}, \omega)$ and related to the time-dependent correlation function $G(\vec{r}, \tau)$ by:

$$\frac{d^2\sigma}{d\Omega d\omega} = \frac{1}{2\pi}\frac{\vec{k}_1}{\vec{k}_0}\int\int e^{i(\vec{q}\cdot\vec{r}-\omega\tau)}\left[\frac{\sigma_{\text{coh}}}{4\pi}G(\vec{r},\tau) + \frac{\sigma_{\text{inc}}}{4\pi}G_S(\vec{r},\tau)\right]d\vec{r}$$

$$= \frac{\vec{k}_1}{\vec{k}_0}\left[\frac{\sigma_{\text{coh}}}{4\pi}S(q,\omega) + \frac{\sigma_{\text{inc}}}{4\pi}S_S(q,\omega)\right] + \delta(q)\text{ terms}$$

As always, the structure factor for a model system must be calculated and compared with experiment. In the melt, interaction forces are expected to exist between molecules, in contrast to the situation in dilute solutions. It is therefore surprising that the viscoelastic properties of polymers below a critical molecular weight can be treated with the Rouse model, which describes the motion of an isolated chain in a heat bath. In this model, the chain segments are supposed to experience only local frictional forces. The failure of the model above a critical molecular weight is attributed to entanglements, which are expected to slow down those intramolecular motions that extend beyond the entanglement distance. In order to model the motion of polymer segments in a melt the reptation model was proposed (see MACROMOLECULAR DYNAMICS).

BIBLIOGRAPHY

P. A. Egelstaff, *An Introduction to the Liquid State,* Academic Press, London, 1967.

H. Benoit, R. Duplessix, R. Ober, M. Daoud, J. P. Cotton, B. Farnoux, and G. Jannink, *Macromolecules* **8,** 451 (1975).

F. Kohler, *The Liquid State,* Verlag Chemie, Weinheim, FRG, 1972.

P. G. de Gennes, *J. Chem. Phys.* **72,** 4756 (1980).

I. VOIGT-MARTIN
Johannes Gutenberg Universitat

J. WENDORFF
Deutsches Kunststoff-Institut

AMYLENE POLYMERS. See OLEFIN POLYMERS.

ANALYSIS OF POLYMERS. See under the various methods, eg, CHEMICAL ANALYSIS; ELECTRON DIFFRACTION; INFRARED-ABSORPTION SPECTROSCOPY; THERMAL ANALYSIS; X-RAY DIFFRACTION.

ANHYDRIDES, POLYMERIC. See POLYANHYDRIDES.

ANIDEX FIBER. See FIBERS, ELASTOMERIC.

ANILINE–ALDEHYDE RESINS. See AMINO RESINS.

ANIMAL GLUE. See ADHESIVE COMPOSITIONS; CASEIN.

ANION-EXCHANGE POLYMERS. See ION-EXCHANGE POLYMERS.

ANIONIC POLYMERIZATION

This article details polymerization that proceeds by the addition of monomers to active centers bearing a whole or partial negative charge. The active center is regenerated in each step.

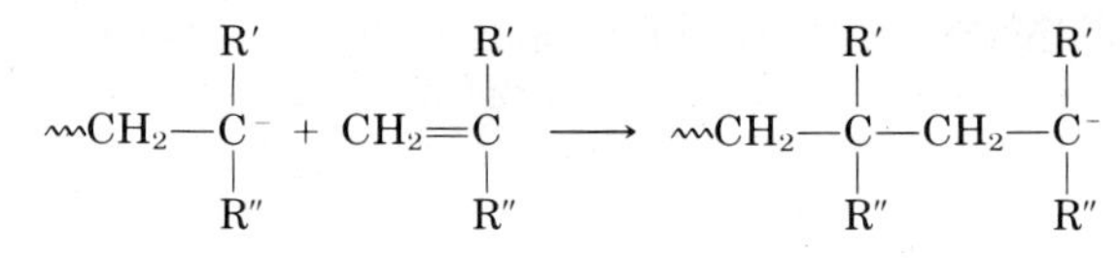
(1)

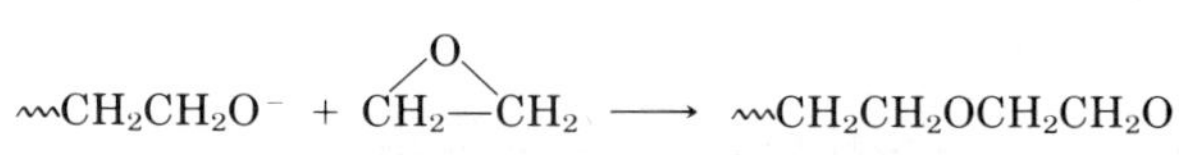
(2)

The negatively charged chain end is often associated with an alkali metal or alkaline earth metal cation, Mt^+ and Mt^{2+}. Partially covalent polarized bonds, eg, —C(R′)(R″)$^{\delta-}$—$Mt^{\delta+}$ are also included in the classification.

General Considerations

Monomers. Monomers susceptible to anionic polymerization fall into two main classes: vinyl or diene monomers, and cyclic monomers.

Initiation of Vinyl and Diene Monomers. The chief monomer-consuming process, the so-called chain-propagation step, is illustrated in equations 1 and 2. Prior to this reaction, it is necessary to form the growing chain anions by an initiation step. Anionic polymerization requires compounds of the most electropositive elements, ie, primarily the alkali metals, for chain initiation, although alkaline earth metals are also suitable. For vinyl and diene monomers (M), lithium alkyls are particularly favored initiators.

$$RLi + M \longrightarrow RM^- \ Li^+$$

Initiation of Cyclic Monomers. Cyclic monomers, such as ethylene oxide, propylene sulfide, cyclosiloxanes, lactones, and aldehydes, are polymerizable by quite weak bases. Alkali metal alkoxides have frequently been used as initiators; lactones, chloral, and formaldehyde (but not higher aldehydes) can be polymerized by tertiary amines. Lithium alkyl initiation has been described in some cases: the initiator simply serves as a generator of sulfide or alkoxy anions by reaction with the monomer.

General Mechanism. The polymerization mechanism for a number of monomers is very simple, and is composed of chain initiation and chain propagation with no termination reaction. The active centers still remain after complete polymerization, so that a new batch of monomer subsequently added will add to the existing chains and increase their degree of polymerization (qv). Hence, the term living polymers (qv) is often used to describe these systems. Under these conditions $\bar{M}_N$ is simply determined by $n\dfrac{M_0}{I_0}$ (n is the functionality of initiator) and a very sharp (Poisson) molecular-weight distribution is obtained if chain initiation is fast.

Properties of Carbanions

The simplest potential carbanions are those of alkali-metal alkyls. The properties of lithium alkyls have been extensively studied and indicate a rather complex behavior. They are soluble in hydrocarbon solvents in extensively associated forms, ie, usually tetramers or hexamers, and form reversed micelles, ie, the lithium atoms are in the aggregate center surrounded by the alkyl groups, each of which appears to be bonded to more than one lithium. Association persists in polar solvents and even in the gas phase. In crystals, ethyllithium and ethyl sodium form layer lattices; only methylpotassium crystallizes in a typical ionic lattice. This supports the conventional idea that the carbon–lithium bond in alkyls is at least partly covalent and that the ionic character increases in compounds of the heavier alkali metals.

Many of the other initiators, eg, fluorenyl, cyclopentadienyl, and benzylic derivatives of the alkali metals, and essentially all growing active centers in vinyl and diene polymerizations constitute another class of carbanionic species. The possibility of charge delocalization increases the stability of ionic forms. For example, the propagating center in styrene polymerization is a benzylic carbanion and in diene polymerization, an allylic carbanion. These also tend to be associated in hydrocarbon solvents but not in polar solvents such as tetrahydrofuran, where a small amount of dissociation to free ions is observed.

Polymerization Mechanisms

Styrene and Diene Monomers. Lithium alkyl-initiated polymerization in hydrocarbon solvents shows the simplest behavior; only chain initiation and propagation are of importance at normal temperatures (20–50°C). The initiation step is slowest with n-alkyllithiums in aliphatic solvents and fastest with *sec*-butyllithium in aromatic solvents. Slow initiation broadens the molecular-weight distribution. In polar solvents, chain initiation is always rapid, but carbanion stability is lower so that for "living" systems temperatures appreciably below 0°C are required.

Polar Monomers. n-Butyllithium has often been used as initiator for monomers such as methyl methacrylate or acrylonitrile usually at low temperatures in toluene. Only a very small fraction of the initiator produces growing chains, most is destroyed in side reactions with the monomer. The reaction products interfere with the process so that polymers of broad molecular-weight distribution are formed. With methyl methacrylate, for example, butyllithium can react in two ways with the monomer:

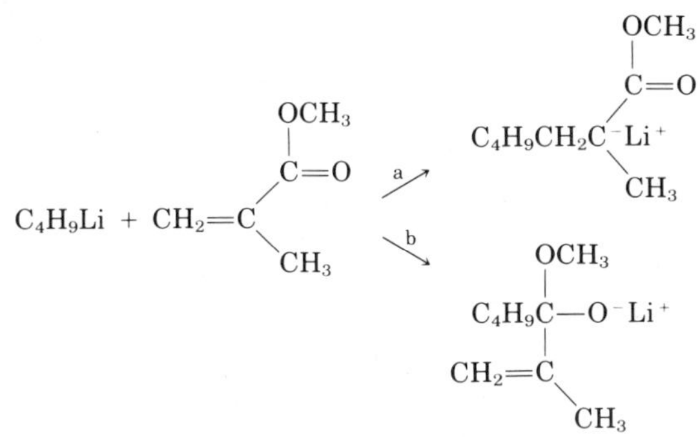

(21)

Only reaction a leads to chain growth, b leads to butyl isopropenyl ketone, which when added to a growing chain forms a very unreactive carbanion. Termination processes can also occur, particularly at temperatures approaching 0°C. In polar solvents, at low temperatures (~ −80°C) less reactive initiators can be used to produce, in many cases, termination-free living systems. Magnesium alkyls and Grignard reagents are also suitable initiators for polar monomers, but again their mechanism is complex and not well understood.

Cyclic Monomers. Polymerization of molecules with strained rings (eg, ethylene oxide, propylene sulfide, α-disubstituted propiolactones) proceed by the classic termination- and transfer-free mechanism. Polymerizations are normally carried out in polar solvents, which normally produce faster rates, but even here they are still much slower than for monomers such as styrene. Larger rings (eg, ϵ-caprolactone, octamethylcyclotetrasiloxane) have the complication of ring-chain equilibration and so produce large fractions of cyclic molecules at equilibrium. D_3 (hexamethylcyclotrisiloxane) does, however, behave almost ideally. Lactams undergo a more unusual form of anionic polymerization in which the monomer anion attacks a cyclic lactam group at the end of the polymer chain.

Anionic Copolymerization

Binary copolymerization (qv) is normally described in terms of the reactivity ratios r_1 and r_2 of the two monomers. If an individual r is greater than unity, the corresponding chain end prefers to react with its own monomer; if r is less than unity it reacts with the other monomer preferentially. Despite some difficulties in using the classic copolymer composition equation in anionic copolymerization, certain trends are clear. The general characteristic of anionic polymerization is $r_1 > 1$, $r_2 < 1$ or *vice versa,* which means that one of the two monomers is more reactive to both types of active center and is then preferentially incorporated into the polymer. The alternating copolymerization tendency common in free-radical polymerization, ie, $r_1 < 1$, $r_2 < 1$, is not found, except in cases such as styrene–1,1-diphenylethylene, where one monomer does not homopolymerize.

Stereoregulation in the Propagation Step

Microstructure of Diene Polymers. Much of the early interest in anionic polymerization was stimulated by the discovery that a highly *cis*-1,4-polyisoprene, equivalent in structure to natural rubber, is produced by lithium metal or lithium alkyl-initiated polymerization in nonpolar solvents. Anionic polymerization often offers possible control of the polymer structure under homogeneous reaction conditions, in contrast to other systems where surface effects are the controlling factor. With 1,3-dienes, for example, four modes of monomer incorporation are possible:

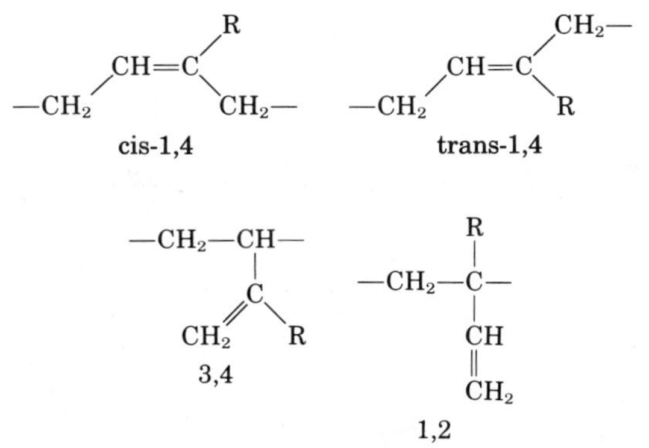
cis-1,4 trans-1,4 3,4 1,2

The 3,4- and 1,2-structures are, of course, identical for butadiene; otherwise, they are often classed together and termed vinyl unsaturation. All dienes, including butadiene, isoprene, 2,3-dimethylbutadiene, and 1,3-pentadiene, polymerize to highly 1,4-polymers in lithium–hydrocarbon solvent systems, but only for isoprene is this mostly in the cis form. The derivatives of the other alkali metals produce largely vinyl unsaturation. All alkali metals and their compounds in polar solvents such as THF, dioxane, and ether produce polymers in which the vinyl unsaturation is high. Again, as in hydrocarbon solvents, lithium compounds give the purest structure, but in this case, it is highly vinyl. High vinyl diene structures can be produced in nonpolar solvents by the presence of very small amounts of chelating agents such as TMEDA or 1,2-dipiperidinoethane.

Microstructure of Vinyl Polymers. The microstructure of vinyl polymers $-CH_2-C(R)(R')-$, can vary because of the presence of pseudoasymmetric carbons in the chain if R and R′ are not identical. Two configurations are possible: meso, $-C(R)(R')-CH_2-C(R)(R')-CH_2-$, and racemic, $-C(R)(R')-CH_2-C(R')(R)-CH_2-$. Vinyl microstructure is commonly described in terms of the configuration of three successive monomer units or triads, isotactic (mm), syndiotactic (rr) or heterotactic (mr). Poly(methyl methacrylate) prepared using lithium-based initiation in toluene is highly isotactic but prepared in tetrahydrofuran is highly syndiotactic. However, such highly ordered structures are not produced from styrene-type monomers.

Microstructure of Other Polymers. Monomers with cumulated double bonds can, in principle, be polymerized by opening either bond. With monoisocyanates, preferential opening of the N=C bond occurs to form *N*-substituted nylons. Dimethylketene appears to form a predominantly acetal structure if lithium derivatives are used but some other anionic catalysts give polymers containing some ester groups.

Block Copolymers

In principle, anionic polymerization is an excellent method of formation of copolymers containing long runs of (usually) two monomers. After butadiene homopolymers and random styrene–diene copolymers, block copolymers (qv) are the most important industrial products made by anionic polymerization. Naturally, only those monomers that, under the polymerization conditions used, have no termination or transfer reactions are suitable for the process. A living polymer of one monomer is initiated, and when it has been completely polymerized, the second monomer is added, which will then simply add to the end of the sequence of the first monomer to form a di-block copolymer. The only precaution needed is the selection of the order in which the two monomers are used, because crossover reactions may be impossible or slow in one direction. Because of this fact, tri-block polymers are commonly made by coupling together two di-block polymers with a suitable dihalide.

Functionally Terminated Polymers

α-ω-Disubstituted polymers are the most important example of such products and are particularly useful to form network polymers. The conversion of anionic active centers to —OH and —COOH groups by reaction with ethylene oxide or carbon dioxide, followed by protonation, was described in early studies of anionic copolymerization, and polymers containing such end groups are still the most common products. There are some practical problems, but these can be overcome by judicious choice of reaction conditions. An alternative approach, particularly valuable to avoid problems of preparation of difunctional initiators in some systems, is to use functionalized lithium alkyls, eg, $Li(CH_2)_nOCH(CH_3)OC_2H_5$. The protective group can be removed by hydrolysis after polymerization, α-ω Difunctionality then requires coupling of two living polymers.

Branched Polymers

The preparation of polymer molecules of well controlled branch structure is important for the assessment of the effect of branching on physical properties of polymers. Anionic polymerization is a particularly suitable method for preparing such polymers. For preparation of graft copolymers (qv), for example, two basic methods have been used. The first method involves the production of anionic centers along a polymer backbone; these are subsequently used to polymerize the same or different monomers. The second (better) method involves the production of reactive groups (eg, halides) on the polymer chain, which can then be coupled to living polymers. "Star" polymers, ie, molecules having several linear chains each linked at one end to a central molecule, are easily prepared by termination of living polymers by multifunctional silicon–chlorine compounds. "Ring" polymers, ie, long chains with no ends, can be produced by coupling α-ω difunctional living polymers with a suitable dihalide under carefully controlled conditions.

BIBLIOGRAPHY

M. Szwarc, *Adv. Polym. Sci.* **49,** 1 (1983).

M. Morton, *Anionic Polymerization, Principles and Practice,* Academic Press, Inc., New York, 1983.

S. Bywater, *Prog. Polym. Sci.* **4,** 27 (1974).

S. Bywater in C. H. Bamford and C. F. H. Tipper, eds., *Comprehensive Chemical Kinetics,* Vol. 15, Elsevier, Amsterdam, The Netherlands, 1976, Chapt. 1.

S. Bywater
National Research Council of Canada

ANNEALING

The annealing of plastics can be defined as a secondary process wherein the plastic is brought to a certain temperature, kept there for a time, and then cooled to room temperature. The primary reasons for annealing include the reduction or removal of residual stresses and strains, dimensional stabilization, reduction or elimination of defects, and improvement of physical properties.

In the plastics industry, annealing has been applied primarily to thermoplastic polymers, block copolymers, and amorphous or semicrystalline polymer blends.

Annealing becomes necessary when molded parts are excessively stressed, when maximum dimensional stability (qv) and heat resistance are required, or when certain properties must be enhanced. Improper annealing may also cause deterioration in performance.

Failure Modes

Molded thermoplastics containing residual stresses and frozen-in orientation undergo stress crazing or stress cracking when placed in the wrong environment or momentarily exposed to certain liquids; these conditions represent the primary modes of plastics failure.

Principles

Initially, annealing of plastics was of industrial interest, but it has recently become of serious concern to scientists and engineers in academic and government laboratories, and has been named physical aging (see Aging, physical). Although the work has focused on the behavior of glasses, it increases understanding of the behavior of semicrystalline polymers, block copolymers, and polymer blends because of the development of new experimental techniques to monitor the changes.

Commercial polymer blends and block copolymers typically contain other low- and intermediate-molecular-weight additives, such as plasticizers, flame retardants, and uv and thermal stabilizers. During annealing, phase and microphase separation may be enhanced and bleeding of the additives may be observed. The morphologies of polymer blends and block copolymers can be affected by processing and quenching conditions. If the melt viscosities of the polymers are not matched, compositional layering perpendicular to the direction of flow may occur. As in the case of crystalline polymers, the skin may be different both in morphology and composition. Annealing may cause more significant changes in the skin than in the interior.

Equipment

The equipment used for annealing is similar to that used for preheating thermoplastic sheets, rods, or tubes before shaping. Depending on the annealing medium, ie, liquid or air, different heating units are used.

The equipment for liquid baths, ie, water, oils, or waxes, consists of tanks or troughs that should be of suitable size. When air is the heating medium, circulating-air or air-flow ovens have proved most suitable.

Methods

Theoretically, the most desirable annealing temperatures for amorphous plastics, some polymer blends, and block copolymers is above their glass transition, where the relaxation of stress and orientation is the most rapid. However, the required temperatures may cause excessive distortion and warping. To anneal as quickly as possible, the plastic is heated to the highest possible temperature at which dimensional changes owing to strain release are within permissible ranges. This temperature can be determined by placing the plastic part in an air oven or liquid bath and gradually raising the temperature by intervals of 3–5°C until the maximum allowable change in shape or dimension occurs. This distortion temperature is dictated by thermomechanical processing history, geometry, thickness, weight, and size. The annealing temperature should then be fixed about 5°C lower.

Media

The medium affects annealing time. In a liquid medium, short times are required, because the heat transfer is much faster than in air. In addition, a more uniform distribution of temperature is achieved without additional labor cost when the molded part is immediately placed into the bath.

Effectiveness

Annealing is a secondary finishing operation that requires time, equipment, energy, and personnel. For maximum effectiveness, conditions should be determined carefully using quality-control methods. A common procedure is to monitor the process in a solvent-induced stress-crazing or stress-cracking test, in which samples are exposed to a liquid or mixture of liquids.

Economic Aspects

With the development of faster processing equipment, computer controls, and a better understanding of the effects of processing, annealing will continue to play an important role in plastics technology.

BIBLIOGRAPHY

L. C. E. Struik, *Physical Aging in Amorphous Polymers and Other Materials,* Elsevier Scientific Publishing Co., Amsterdam, 1978.

J. D. Ferry, *Viscoelastic Properties of Polymers,* John Wiley & Sons, Inc., New York, 1961.

N. E. McCrum, B. E. Read, and G. Williams, *Anelastic and Dielectric Effects in Polymeric Solids,* John Wiley & Sons, Inc., New York, 1967.

E. Martusselli, R. Palumbo, and M. Kryszewski, *Polymer Blends (Properties, Morphology and Processing),* Plenum Press, New York, 1979.

D. G. LeGrand
General Electric Company

ANTIBODIES AND ANTIGENS, POLYMER-BOUND

During the past decade, scientists have learned how to produce large amounts of homogeneous antibodies to order by altering normal antibody-producing cells so they can easily be grown in the laboratory. Each cell clone produces only one specific antibody, hence such products are termed monoclonal antibodies.

Monoclonal antibodies are already used in a wide variety of applications. They are bound to polymeric supports and used to purify molecules of many types (see CHROMATOGRAPHY AFFINITY). They also are used to test for the presence and amount of specific substances, particularly in clinical tests. They have also been used for therapy.

Immunomicrospheres

Polymer chemistry has been particularly useful for the design and synthesis of specialized latex particles, ie, microspheres (qv), to which antibodies or antigens are attached. Reagents formed in this way are referred to as immunomicropheres.

Applications

Immunomicrospheres can be structured to permit identification by radioactive labeling, fluorescence, electron opacity, and magnetic and electrophoretic separation. They have been used in drug delivery systems.

BIBLIOGRAPHY

R. H. Kennett, T. J. Mekeam, and K. B. Bechtol, eds., *Monoclonal Antibodies,* Plenum Press, New York, 1980.

A. Rembaum and W. J. Dreyer, *Science* **298,** 364 (1980).

R. S. Molday, S. P. S. Yen, and A. Rembaum, *Nature* **268,** 437 (1977).

W. P. Collins, ed., *Alternative Immunoassays,* John Wiley & Sons, Inc., New York, 1985.

W. J. Dreyer
A. Rembaum
California Institute of Technology

ANTIFOAMING AGENTS

The control or elimination of the foam that arises in many industrial processes can be the crucial factor in their operation. Polymer scientists and engineers are interested in antifoaming phenomena not only because of the use of antifoams in many polymer processes, but also because many of the components of antifoaming agents are polymeric materials.

The durability of the antifoaming effect is an important aspect of the cost factor. In most applications, subsequent additions of the agent are required to sustain performance and the required frequency of these additions has a large impact on cost effectiveness. Another factor that strongly influences the choice of any particular antifoaming agent is its ancillary surface properties. Many defoamers have wetting, dispersion, leveling, and other benefits that may determine their selection for a particular application.

Classification of Antifoaming Agents

Antifoaming agents typically contain numerous ingredients, which usually fall into one of the following categories: liquid-phase components; solid-phase components; ancillary agents, eg, emulsifiers, thickeners, and preservatives; or carriers, eg, water and solvents. Classification according to chemical type is probably the most satisfactory. Because lowering of surface tension is generally believed to be a key antifoam property, classification by the component of the antifoam primarily responsible for this property is useful. The component is usually the nonpolar portion of the liquid-phase component. The principal nonpolar entities present in the liquid-phase components of antifoaming agents are hydrocarbons, polyethers (qv), silicones (qv), and fluorocarbons.

Solid-phase antifoaming-agent components are vital ingredients in commercially important antifoam formulations. Current theory on antifoaming mechanism ascribes the active antifoaming action to this dispersed solid phase. Solid-phase components include hydrocarbons, silicones, and fluorocarbons.

Ancillary Agents. The ancillary agents can occasionally be polymeric species as well. Nonfoaming polymeric antimicrobial and preservative agents are also available to the antifoam-agent formulator.

Theory of Antifoaming

Two related mechanisms of antifoaming have been advanced. The agent dispersed in the form of fine drops enters the liquid film between bubbles and spreads as a thick duplex film, ie, a film thick enough to have two definite surfaces enclosing a bulk phase. The tensions created by the spreading duplex film lead to the rupture of the original liquid film. Dewetting of the film from the dispersed solid particles has the same effect. Alternatively, a droplet of the agent enters the liquid film between bubbles, but rather than spreading as a duplex film, produces a mixed monolayer on the surface. This monolayer, if of less coherence than the original film-stabilizing monolayer, will cause destabilization of the film.

Applications in Polymers

There are two broad categories of application in the polymer industry: foam prevention during preparation of a polymeric material and foam prevention during its use in manufacturing processes.

Economic Aspects

In 1979, the U.S. market for all antifoaming agents was probably greater than 136,000 t/yr, that is, at least equal to that in the rest of the world. Consumption has been relatively constant in recent years, and this remains a good estimate for 1985.

Health and Safety Factors

The principal health and safety concerns involve the food and drug industries where polymeric materials are used in both processing and packaging. U.S. government regulations concerning the use of additives such as antifoaming agents in food and drugs are listed in the *Code of Federal Regulations*.

BIBLIOGRAPHY

J. C. Colbert, *Foam and Emulsion Control Agents and Processes: Recent Developments,* Noyes Data Corp., Park Ridge, N.J., 1981.

I. Lichtman and T. Gammon in M. Grayson, ed., *Kirk-Othmer Encyclopedia of Chemical Technology,* 3rd ed., Vol. 7, Wiley-Interscience, New York, 1979, pp. 430–448.

S. J. Ross, *The Inhibition of Foaming, Bulletin No. 63,* Rensselaer Polytechnic Institute, Troy, N.Y., 1950.

S. Ross and G. Nishioka in P. Becher and M. N. Yudenfreund, eds., *Emulsions, Lattices and Dispersions,* Marcel Dekker, Inc., New York, 1978, pp. 237–256.

S. Ross and G. Nishioka, *J. Colloid Interface Sci.* **65,** 216 (1978).

MICHAEL J. OWEN
Dow Corning Corporation

ANTIFUNGAL AGENTS. See BIOCIDES.

ANTIGENS. See ANTIBODIES AND ANTIGENS, POLYMER-BOUND.

ANTIMONY DERIVATIVES. See ORGANOMETALLIC POLYMERS.

ANTIOXIDANTS

Antioxidants are used to protect polymers against degradation (qv) caused by reaction with atmospheric oxygen. Oxidation can occur at all stages, ie, from synthesis to final use, causing undesirable changes in mechanical, aesthetic, and electrical properties.

A large number of antioxidants designed to meet the specialized needs of the polymer industry are commercially available. In the United States, approximately 80,000 metric tons were used in 1981.

Mechanism of Uninhibited Autoxidation

The mechanism by which an organic material (RH) undergoes autoxidation involves a free-radical chain reaction:

Initiation

$$RH \longrightarrow \text{free radicals, eg, } R\cdot,\ ROO\cdot,\ RO\cdot,\ HO\cdot \qquad (1)$$

Propagation

$$R\cdot + O_2 \longrightarrow ROO\cdot \qquad (2)$$

$$ROO\cdot + RH \longrightarrow ROOH + R\cdot \qquad (3)$$

Autocatalysis

$$ROOH \longrightarrow RO\cdot + \cdot OH \qquad (4)$$

$$2\ ROOH \longrightarrow RO\cdot + ROO\cdot + H_2O \qquad (5)$$

$$ROOR \longrightarrow 2\ RO\cdot \qquad (6)$$

Termination

$$2\ R\cdot \longrightarrow R{-}R \qquad (7)$$

$$ROO\cdot + R\cdot \longrightarrow ROOR \qquad (8)$$

$$2\ ROO\cdot \longrightarrow \text{nonradical products} \qquad (9)$$

Radical Scavengers

Hydrogen-donating antioxidants (AH), such as hindered phenols and secondary aromatic amines, inhibit oxidation by competing with the polymeric substrate (RH) for peroxy radicals, thus shortening the kinetic chain length of the propagation reactions.

Hindered Amines. Hindered amines are usually classified as light stabilizers rather than antioxidants, because they are extremely effective in protecting polyolefins and other polymeric materials against photodegradation. Hindered amines inhibit photooxidation and thermal oxidation by trapping free radicals.

Peroxide Decomposers

Thermally induced homolytic decomposition of hydroperoxides to free radicals (eqs. 4 and 5) is responsible for increasing the rate of oxidation. Decomposition to nonradical species removes hydroperoxides as potential sources of oxidation initiators. Most peroxide decomposers are derived from divalent sulfur and trivalent phosphorus.

Synergism

A mixture of antioxidants that function by different mechanisms might be synergistic and provide a higher degree of protection than the sum of the stabilizing activities of each component. The most frequently used synergistic mixtures are combinations of radical scavengers and hydroperoxide decomposers.

Metal Deactivators

The ability of metal ions to catalyze oxidation can be inhibited by deactivators. These additives chelate metal ions and increase the potential difference between the oxidized and reduced states of the metal ions, decreasing the ability of the metal to produce radicals from hydroperoxides by oxidation and reduction. Examples of commercial metal deactivators used in polymers are oxalyl bis(benzylidenehydrazide), *N,N′*-bis(3,5-di-*tert*-butyl-4-hydroxyhydrocinnamoyl)hydrazine, and 2,2′-oxamidobisethyl(3,5-di-*tert*-butyl-4-hydroxyhydrocinnamate).

Ancillary Requirements

Ancillary requirements can be as important as or even more important than inherent reactivity in determining the commercial acceptability of a product. Factors requiring consideration include toxicity, cost, persistence, compatibility, color stability, physical form, and organoleptic properties.

Health and Safety Factors

Safety is assessed by subjecting the antioxidant to a series of animal toxicity tests, eg, oral, inhalation, eye, and skin tests. Mutagenicity tests also are carried out to determine possible or potential carcinogenicity.

BIBLIOGRAPHY

G. Scott, *Atmospheric Oxidation and Antioxidants,* Elsevier, Amsterdam, The Netherlands, 1965.

J. R. Shelton in W. L. Lawkins, ed., *Polymer Stabilization,* Wiley-Interscience, New York, 1972.

K. Schwarzenbach in R. Gächter and H. Müller, eds., *Plastics Additives Handbook,* Carl Hanser Verlag, Munich, 1985, pp. 1–73.

J. Voigt, *Die Stabilisierung der Kunstoffe Gegen Licht und Wärme,* Springer-Verlag, Heidelberg, 1966.

MARTIN DEXTER
CIBA-GEIGY Corporation

ANTIOZONANTS

The term antiozonant, in its broadest sense, denotes any additive that protects rubber against ozone deterioration. Most frequently, the protective effect results from a reaction with ozone, in which case the term used is chemical antiozonant. Ozone is generated by electrical discharge, but a more important source is the photolysis of oxygen on sunny days.

Structure and Properties

Rubber can be protected against ozone by the addition of waxes, inert polymers, or chemically reactive antiozonants. Waxes are generally of the paraffin or microcrystalline type. Many compounds are reported to be chemical antiozonants, and nearly all contain nitrogen. Compound classes include derivatives of 2,2,4-trimethyl-1,2-dihydroquinoline, *N*-substituted ureas or thioureas, substituted pyrroles, and nickel dithiocarbamate salts. The most effective antiozonants are *N,N'*-derivatives of *p*-phenylenediamine (*p*-PDA).

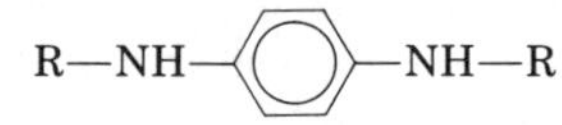

N,N'-substituted *p*-phenylenediamine

Manufacture and Production

The *N,N'*-dialkyl-*p*-PDAs are manufactured by reductively alkylating *p*-PDA with ketones. The United States annual pro-

Table 1. U.S. Production of Phenylenediamine Antiozonants, Metric Tons

	1970	1976	1981
total substituted *p*-PDAs[a]	27,353	32,559	28,091
bis(1,3-dimethylpentyl)-*p*-PDA	2,421		2,338
U.S. factory auto sales, thousands	6,547	8,498	6,253

[a] Listed in Table 2.

Table 2. U.S. Manufacturers of Commercial Antiozonants

p-Phenylenediamine	Manufacturer[a]	Trade name[a]	Price, $/kg[b]	Melting point, °C
N,N'-bis(1,4-dimethylpentyl)	Monsanto	Santoflex 77	6.13	liquid
	Universal Oil Products	UOP 788		
	Uniroyal	Flexzone 48		
N,N'-bis(1-ethyl-3-methylpentyl)	Uniroyal	Flexzone 8L	5.96	liquid
	Universal Oil Products	UOP 88		
N,N'-bis(1-methylheptyl)	Universal Oil Products	UOP 288	5.80	low melting point
N-cyclohexyl-*N'*-phenyl	Uniroyal	Flexzone 6H	6.62	110
N-(1,3-dimethylbutyl)-*N'*-phenyl	Goodyear	Wingstay 300	6.26	48–50
	Monsanto	Santoflex 13		
	Uniroyal	Flexzone 7L or 7F		
	Universal Oil Products	UOP 588		
N-isopropyl-*N'*-phenyl	Uniroyal	Flexzone 3C	6.48	78
N-(1-methylheptyl)-*N'*-phenyl	Universal Oil Products	UOP 688		
N-(1-methylpentyl)-*N'*-phenyl[c]	Uniroyal			liquid
N,N'-mixed diaryl	Goodyear	Wingstay 100	5.23	90–150
		Wingstay 200	4.96	60
N,N'-diphenyl	BFGoodrich	Agerite DPPD	8.82	144–153
	Uniroyal			
N,N'-di-2-naphthyl	BFGoodrich	Agerite White	11.73	224–230
N,N'-dicyclohexyl	Universal Oil Products	UOP 26		106

[a] Other suppliers include: R. T. Vanderbilt (Antozite), Mobay (Vulkanox), American Cyanamid (Cyzone), Pennwalt ($ANTO_3$), Vulnax (Permanax), Akron Chemical (Akrochem), Bayer (Antiozonant), Sumitomo (Antigene), ICI (Topanol and Nonox) and Rhone-Poulenc.

[b] As of Sept. 1983.

[c] Used in mixtures.

duction of substituted *p*-PDAs (Table 1) is related to automobile production. A list of commercial antiozonants is given in Table 2.

Alternatives to Chemical Antiozonants

In a number of diene rubber applications, discoloration caused by chemical antiozonants cannot be tolerated, for example, white sidewalls, gaskets, weather stripping, and sporting goods. For these applications, alternatives include ozone-resistant elastomers, blends of ozone-resistant elastomers and diene rubber, and waxes.

Health and Safety Factors

Current higher molecular-weight *N*-alkyl-*N'*-aryl or *N,N'*-dialkyl antiozonants are not primary skin irritants. The exception is *N*-isopropyl-*N'*-phenyl-*p*-PDA, which causes dermatitis.

BIBLIOGRAPHY

General References

J. C. Ambelang, R. H. Kline, O. M. Lorenz, C. R. Parks, and C. Wadelin, *Rubber Chem. Technol.* **36,** 1497 (1963).

P. M. Lewis, *NR Technol.* (1), 1 (1972).

R. E. Keller, *Rubber Chem. Technol.* **58,** 637 (1985).

R. P. Lattimer, J. Gianelos, H. E. Diem, R. W. Layer, and C. K. Rhee, *Rubber Chem. Technol.* **59,** 263 (1986).

C. K. Rhee
R. P. Lattimer
R. W. Layer
The BFGoodrich Company

ANTISEPTICS. See Biocides.

ANTISTATIC AGENTS

Static electricity is caused by the transfer of electrons across surfaces brought in contact with each other. This exchange is not necessarily symmetrical, and with two identical surfaces, one acquires an excess of electrons and the other a deficiency. During contact, the effect is not noticed. However, after separation, part of the charge is discharged and part remains on the surface. The charges are equal, but the signs opposite. The static generated is a surface phenomenon. If the substrate is conducting, the electrons are dissipated and the excess is eliminated. If the substrate is nonconductive, ie, an insulator, static electricity becomes apparent.

Function

Antistatic agents, also called antistats, promote static-charge decay of surfaces and reduce the clinging effect of fabrics, eliminate spark discharge, and prevent accumulation of dust and dirt. Applications include plastic films, containers, fibers, and many other consumer goods (see also Electromagnetic interference).

External Antistats

External antistatic agents can be applied by conventional coating methods (qv). They are usually applied from dilute aqueous or alcohol solutions.

Internal Antistats

Internal antistatic agents are incorporated in the polymer during molding or film-forming operations and tend to bloom or rise to the surface later, where the mechanism is the same as for external antistats. Ambient moisture is attracted, causing static charges to decay. Internal and external antistats for plastics are listed in Table 1.

Table 1. Surface and Internal Antistatic Agents for Plastics

Trade name	Manufacturer	Chemical type	Recommended for	Use concentration, %
surface				
Ast-1001	Merix Chemical Co.	quaternary ammonium	most plastics[a]	10–100
Kemamine Q-9702C	Humko Chemicals	quaternary ammonium	most plastics[a]	0.05–0.5
Markstat AL-22	Argus Chemical Corp.	quaternary ammonium	most plastics	0.5–2.0
Varstat 66	Sherex Chemical Co.		most plastics[a]	0.1–1.0
internal				
Armostat 375[b]	Akzo Chemie America (Noury Chemicals)	amine	PE	0.05–0.5
Armostat 450	Akzo Chemie America (Noury Chemicals)	amine	PE	2.0–4.0
Cyastat LS	American Cyanamid Corp.	quaternary ammonium	ABS, PS, PVC	0.5–2.0
Kemamine AS990	Humko Chemicals	amine	PE, PP, PS	0.05–0.5
Lubrol 12A9[b]	ICI Americas	amine	ABS, PE, PP, PS	
Markstat AS-7[b]	Argus Chemical Corp.	amine	PE, PP, EVA[c], polybutene	0.3–10
Sandin VU	Sandoz Colors & Chemicals		PVC	0.1–1.5
both				
Markstat AL-14[b]	Argus Chemical Corp.	quaternary ammonium	most plastics[a]	0.5–2.0 (surface)
			PS	3.0–7.0 (internal)
Valstat FL-163	Valchem Div.		ABS, PE, PP, PS, PVC, polyacetate, polyester	0.1–1.0

[a] Included here are ABS, acrylic, cellulose acetate and nitrate, nylon, polyacetate, polycarbonate, PE, PP, PS, and PVC (flexible and rigid).
[b] Accepted by the FDA.
[c] Ethylene–vinyl acetate copolymer.

Table 2. Antistatic Agents for Textiles

Trade name	Manufacturer	Chemical type	Remarks
Aerotex Antistatic D	American Cyanamid Corp.		durable antistat
Ampitol VAC	Dexter Chemical Co.	quaternary ammonium	cationic antistatic softener
Antistatin D	BASF		
Arkostat AC	Hoechst		
Aston 123	Refinex Onyx	polyamine resin	durable antistat
Aston MS	Refinex Onyx	quaternized fatty amine condensate	cationic antistat
Aston AP	Refinex Onyx	amine condensate	
Avitex DN	DuPont	cationic	antistatic softener
Cassastat	Cassella	cationic	
Catanac SN	American Cyanamid Corp.	cationic	
Cirrasol Z	ICI Americas		durable antistat
Elfugin UW	Sandoz Colors & Chemicals		
Igepal CO 430	GAF Corp.	nonyl phenol ethylene oxide	
Neutro-stat	Simco		antistatic spray
Nonax 1166	Henkel	polyamine resin	durable antistat
Permalose T	ICI Americas	nonionic	durable antistat for polyester
Siligen APE	BASF	quaternary ammonium	
Stanax 1166	Standard Chemical Products	polyamine resin	durable antistat
Statexan HA	Bayer		
Tinorex TC	CIBA-GEIGY		
Zelec DP	DuPont	cationic	
Zero C	Lutex Chemical Corp.	nitrogeneous polymer	

Agents for Textiles

Antistatic finishes for fibers and textiles are durable or nondurable. Nondurable finishes are usually water soluble and removed by laundering. Only a few durable antistatic finishes for textiles are available.

Most durable antistats are polyelectrolytes (qv) and variations of cross-linked polyamines with ethoxy moieties. A list of antistatic agents for textiles is given in Table 2.

Measurement of Charge Decay. Static charges on polymers decay either by conduction to the ground through the material or by leakage to the air by radiation, as represented by the following equation:

$$V_t = V_0 e^{-(K_1 + K_2)t}$$

where V_0 = initial voltage, V_t = voltage at time t in s, K_1 = conduction rate constant, and K_2 = radiation rate constant.

Conductivity and Antistatic Action. Assuming that an antistatic finish forms a continuous layer, the conductance it contributes to the fiber is proportional to its volume or weight and specific conductance of the finish; the fiber may be fine or coarse. A concentration of ca 0.1% is needed for antistatic protection of textile fabrics. The exact amount depends on the efficiency of the antistat, relative humidity, temperature, diameter of fiber, fabric structure, and on the degree of antistatic protection desired.

Internal Agents. Internal antistatic agents for fibers are usually polyethoxy compounds, eg, poly(ethylene glycol), in the 20,000 molecular-weight range. The antistat is dispersed in the fiber raw material before spinning and probably forms chemical linkages with reactive sites on the polymer.

Carbon Black and Metallic Agents. Carbon black and metals increase the electrical conductivity of resins and thereby dissipate electrical charges. Their effectiveness does not depend on increasing the moisture content of the fiber.

Agents for Resins

Antistatic agents in solid resins reduce the attraction of dust and dirt. Quaternary ammonium or similar ionic species are most commonly used as external antistats; water or alcohol solubility are essential.

Commercial Agents for Plastics

Commercial surface and internal antistatic agents used for plastics are similar to nondurable antistatics for textiles (Tables 1–3).

Carbon Black. Carbon (qv) can be used as an antistat in plastics because of its high electrical conductivity properties; antistatic activity does not depend on absorbed moisture.

Table 3. U.S. Consumption of Antistatic Agents in Plastics

Year	Metric tons
1960	182
1966	681
1972	1300
1975	1600
1980	2150
1981	2620
1982	2800
1983	2940

BIBLIOGRAPHY

General References

P. S. H. Henry, *Br. J. Appl. Phys.* **4,** Suppl. No. 2, S31 (1953).

J. Gayler, R. E. Wiggins, and J. B. Arthur, *Static Electricity, Generation, Measurement, and Its Effects on Textiles,* The School of Textiles, North Carolina State of the University of North Carolina at Raleigh, 1965.

Antistatic Additives and Surface Treatments for Fibers, Films, and Moldings, A Literature Survey 1907–1969, Science Surveys, Mountainside, N.J., 1967.

H. Marumo, *The Internal Antistatic Treatment of Thermoplastic Polymers,* paper presented at the 1st International Congress of Static Electricity, Vienna, Austria, May 3–7, 1970.

RICHARD A. RECK
Akzo Chemie America

AQUEOUS POLYMERIZATION. See SOLUTION POLYMERIZATION.

ARAMIDS. See POLYAMIDES, AROMATIC.

ARCHITECTURAL APPLICATIONS. See CONSTRUCTION AND BUILDING APPLICATIONS.

ARMS. See BRANCHED POLYMERS.

AROMATIC POLYAMIDE FIBERS. See POLYAMIDES, AROMATIC; POLYAMIDES, FIBERS.

ARSENIC-CONTAINING POLYMERS. See INORGANIC POLYMERS.

ARTIFICIAL SILK. See RAYON.

ARTIFICIAL WOOL. See PROTEINS.

ARYLENE SULFIDES, POLYMERIC. See POLY(ARYLENE SULFIDE)S.

ASPHALTIC MATERIALS

For convenience, asphalt can be classified into natural and artificial asphalts. Natural asphalts include bituminous materials laid down in natural deposits such as those in the Trinidad and Bermudez lakes and natural bitumens such as gilsonites and graphamites, which are completely soluble in carbon disulfide. Artificial asphalts include mainly petroleum-derived asphalts and, to a lesser extent, coal tar, water-gas tars, and their pitches.

A classification scheme developed in 1960 and later modified has been used for the study of asphalts and related bituminous substances (Fig. 1). Recently, well-defined solubility characteristics, such as the ternary diagram and biomarker parameters, have advanced the classification toward a more scientifically sound basis.

Industrially, asphalt can be manufactured and processed by straight reduction, propane deasphalting, and air-blowing. Specification is based on cement viscosity or residue viscosity obtained by the rolling thin-film oven test.

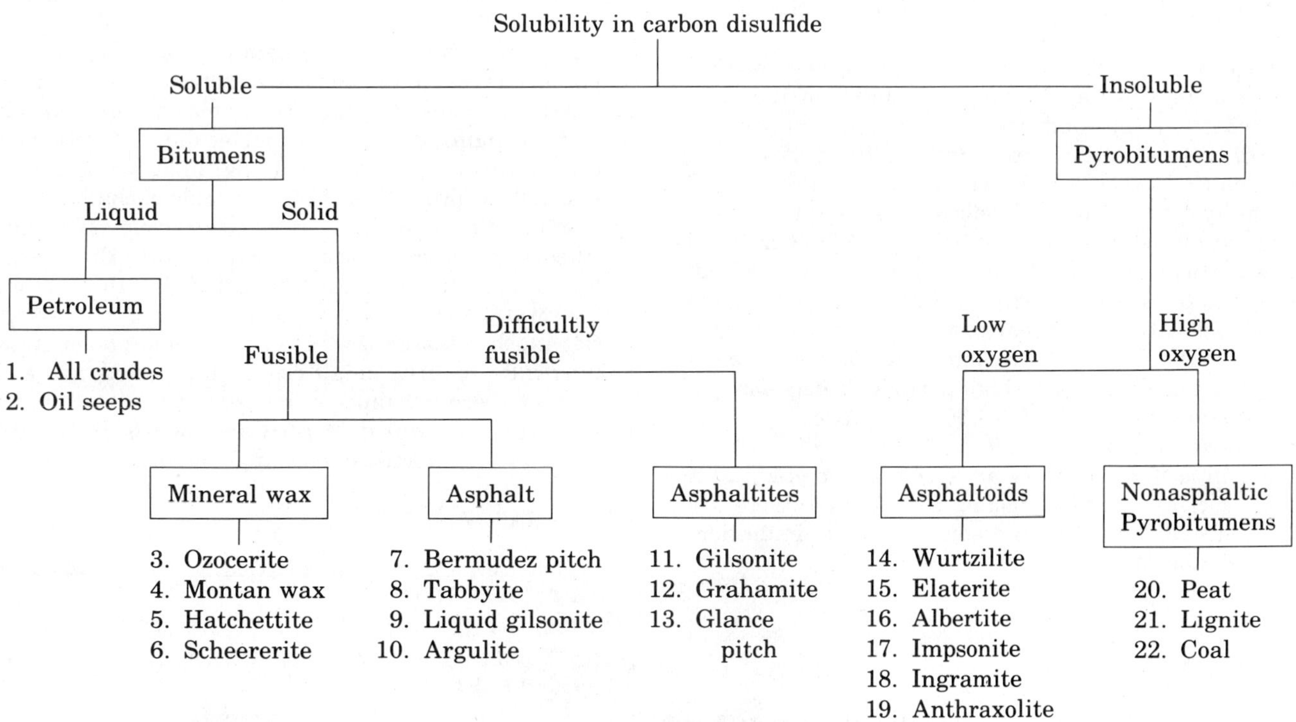

Figure 1. Designation and classification of natural asphalt.

Composition

The properties and behavior of asphalts are critically dependent on the nature of the constituents. Chemically these constituents consist of hydrocarbons and heterocyclic or nitrogen-, sulfur-, and oxygen-containing compounds. Separation of the various fractions of asphalt is usually based on their different boiling point ranges, molecular weights, and solubilities in solvents of different polarity. Techniques for obtaining narrow fractions include vacuum distillation, solvent extraction, thermal diffusion, crystallization, and others, individually or in combination, followed by chromatographic separation. Asphalt is also fractionated by solvent partitioning, selective adsorption–desorption, and chemical precipitation. The principal classes of constituents are gas oils, asphaltene, resins, carboid, carbene, and mesophase.

The Colloidal Nature of Asphalt. Asphalt can be considered a colloidal system similar to petroleum, with the lighter molecules removed. The physical properties of the asphalt are directly related to the quantity of dispersed phase (asphaltenes); the size of the micellar structures, which is dependent on the degree of adsorption of resins; and the nature of the dispersion medium, oils, and resins (maltenes).

Chemical Properties. In heavier fractions of petroleum, single molecules progressively contain C_P, C_N, and C_A portions. The aromatic ring system can participate in a number of electrophilic reactions, eg, nitration, sulfonation, halogenation, and others. Examples of the derivatization of functional groups are silylation of hydroxyls, esterification of acids, and quaternary salt formation of amines.

The polarity of asphaltics is important in hydrogen bonding, donor–acceptor charge transfer, and acid–base salt formation. Hydrogen bonding is essential to such viscosity behavior as the alteration of flow properties by antistripping agents (amines). Asphalt functionality as a result of molecules interacting with each other would affect component compatibility. The polarity affects the micellar structure and consequently the colloidal nature of the asphalt system.

Rheological Properties. Asphalt rheology is concerned with the study of physical or mechanical stability of the colloid system under varying conditions of time, temperature, pressure, load, or stress. The stress–strain patterns result from deformation or a series of deformations with time.

The ability of the asphalts to deform is intimately connected with the micelle state of the asphaltenes. Whether the asphaltene particles are in a highly dispersed or coarsely dispersed state affects the properties of these materials and their industrial applications. The coarseness of the dispersions gives rise to a variety of non-Newtonian states as shown by the following scheme from low to high particle dispersions: plasticity → pseudoplasticity → false body → thixotropy → dilatancy → Newtonian flow.

Studies in asphalt rheology are essential for practical applications, such as pavement, roofing, coating, and others. They can also be useful for testing procedures and the simulation of long-term performance.

BIBLIOGRAPHY

G. V. Chilingarian and T. F. Yen, *Bitumens, Asphalts and Tar Sands,* Vol. 7, Elsevier Scientific Publishers, Amsterdam, 1978, pp. 153–190.

M. A. Bestougeff and R. J. Byranhee in T. F. Yen and G. V. Chilingarian, eds., *Asphaltenes and Asphalt,* Vol. 1, Elsevier Scientific Publishing Co., Amsterdam, 1989.

J. W. Burger and N. C. Li, eds., *Chemistry of Asphaltenes,* Vol. 195, American Chemical Society, Washington, D.C., 1981.

T. F. YEN
University of Southern California

ATOMISTIC MODELING, POLYMERIC GLASSES

The purpose of modeling, in general, is twofold: on the one hand, the investigator tries to establish a model for a state or process to obtain a "physically correct" functional form as an aid in the interpolation and extrapolation of properties; on the other hand, the model provides the potential for new insight into the state or process modeled, if its predictions prove to be accurate. Modeling of the glassy or vitreous state of polymers is particularly important, because the peculiar condition called glass is not very well understood. This condition is often and easily realized with polymers, and polymeric glasses are technologically very relevant. Connected with the quest for structure and properties of the glassy state is the need to understand the ranges of temperature, stress (or pressure), and time (or frequency) where the conventional states, foremost among them the liquid state, undergo the so-called glass transition or glass formation and become glassy (see also GLASS TRANSITION; TRANSITIONS AND RELAXATIONS).

A glass possesses the amorphous structure of a liquid, but the density and viscosity of a solid. Analysis of the nature of the glass is complicated because its structure does not equilibrate within the time scale of conventional experiments. Thermodynamic considerations are appropriate within certain boundaries, but it is easy to commit fallacies with thermodynamic arguments when they are applied to glasses. Often, the behavior of glasses seems counterintuitive to "normal" thermodynamically trained instincts.

The chief difficulty in the detailed modeling of structure and properties of amorphous glassy polymeric solids lies in the fact that intermolecular interactions play a dominant role. For individual chains, detailed intramolecular interactions can be accounted for by the conformational analysis of chain fragments, and the properties of the ensemble of chains that make up a solution or melt can be comprehensively represented by the theory of rotational isomeric states (RIS). However, intermolecular interactions cannot be included in this method (see CONFORMATION AND CONFIGURATION).

Not much research that deals with amorphous glasses in atomistically realistic detail has been reported, and a classification of these attempts is not straightforward, but three groups may be identified: Monte Carlo models (both static and dynamic), static atomistic modeling, and molecular dynamics.

BIBLIOGRAPHY

A. M. Skvortsov, A. A. Sariban, and T. M. Birshtein, *Vysokomol. Soedin.* **5,** 1014 (1977).

F. T. Wall and W. A. Seitz, *J. Chem. Phys.* **67,** 3722 (1977).

D. N. Theodorou and U. W. Suter, *Macromolecules* **18,** 1206, 1467 (1985); **19,** 193 (1986).

U. W. SUTER
Institut für Polymere, ETH

AUTOCLAVE MOLDING.

See REINFORCED PLASTICS.

AUTOMOTIVE APPLICATIONS

Polymeric materials play an important role in vital areas of technology, such as transportation. In the 1980s, the annual consumption of polymers in the U.S. automotive industry has been on the order of 680×10^3 metric tons. The economic impact on the chemical industry is strong, particularly for materials that are utilized extensively, such as polypropylene, poly(vinyl chloride), polyurethanes, and unsaturated polyesters (for composites). About 50% of the total rubber output of the United States is used for automobile tires (qv). Table 1 covers plastic materials other than rubber used in 1983 and projects usage through 1993. The approximate plastic content of an average car in 1980 was about 64 kg (excluding elastomers, sealers, and sound deadeners).

Table 1. U.S. Plastics Usage in the Automobile Industry 1983–1993[a]

	1000 t[b]		
Material	1983	1988	1993
thermoset polyester	52.4	81.9	123.7
polyurethane	105.1	122.7	127.5
polypropylene	108.5	112.7	103.9
polyethylene	21.8	27.4	31.3
ABS	27.9	24.6	22.9
SMA[c]	17.7	19.3	7.7
modified poly(propylene oxide)	5.9	8.0	8.9
polystyrene	2.4	3.6	1.4
nylon	39.1	46.2	48.5
thermoplastic polyester	8.9	12.4	17.3
polycarbonate	5.1	11.1	22.9
acetal resins	7.5	8.6	8.8
acrylics	9.5	10.9	10.8
PVC	58.4	59.0	58.4
phenolic resins	6.4	8.3	23.9
other	15.3	18.6	23.9
Total	*491.9*	*556.7*	*636.9*

[a] Statistical estimates of polymer usage vary, depending on the yearly production of cars and whether trucks, buses, recreational vehicles, and tractors are included. Generally, elastomers for tire usage are not included. Compilations of this type are rare and are not updated in any systematic manner for public use. Wherever possible, the most recent summaries available have been used.
[b] Total compound weight, including reinforcement and filler.
[c] Styrene–maleic anhydride copolymer.

Properties of Polymeric Materials

Polymeric systems, ie, including homogeneous polymers, composites and blends, anisotropic polymeric solids, and others, afford materials with a wide range of useful properties. Blends of amorphous polymers such as ABS display toughness, flexibility, and impact strength. Composites (qv) provide a wide range of flow properties, stiffness, and strength. Crystalline or well-oriented polymers provide tensile strength needed in textile fibers and in materials of construction. In Table 2, the tensile properties of several important polymeric systems are compared to those of other materials.

Table 2. Tensile Properties of Polymers and Other Materials

	Property		Specific property[a]	
Material	E, tensile modulus, GPa[c]	TS, tensile strength, GPa[c]	$E/\rho g^b$, sp tensile modulus, km	$TS/\rho g^b$, sp tensile strength, km
Isotropic				
polystyrene	2	0.04	200	4
polycarbonate	2.5	0.06	150	6
polyethylene, high density	1.5	0.025		2.5
graphite	72	1.5		
glass	72	1.5	3,000	64
carbon steel	200	0.45	2,500	6
aluminum	70	0.4	2,600	15
Oriented				
polyethylene				
single crystal theoretical $E_{\parallel}$	180–360			
single crystal, theoretical $E_{\perp}$	2–10			
submicroscopic crystallite, exp $E_{\parallel}$	240–360			
submicroscopic crystallite, exp $E_{\perp}$	2–6			
ultraoriented	140	3	14,000	300
polyaramid[d]	130	3.8	9,000	270
flax, native	100	1	7,000	70
Composite				
fiber glass–vinyl ester				
random, 65% glass	16	0.20	850	10
unidirectional, 75% glass $E_{\parallel}$	38	0.55	1,900	30

[a] Property divided by density.
[b] g = standard acceleration of gravity; ρ = density; $g = 9.8$ m/s^2, ρ = g/cm^3.
[c] To convert GPa to psi, multiply by 145,000.
[d] Kevlar-49.

It must be emphasized that the information listed in Table 2 is incomplete for engineering design purposes, because the properties of organic polymers often greatly depend on temperature and environment and exhibit time-dependent failure modes. Similar effects with metals are usually less pronounced.

Special Design Considerations. For many applications, light, strong, durable materials are needed in place of metals. Polymers can meet a number of these needs. The range of properties that can be achieved with composites, blends, and imposed anisotropy provides materials for many applications, but an understanding of the advantages and especially the limitations of such materials is necessary for successful application.

Process Freedom and Opportunities. Improved methods of polymeric processing and fabrication include solid-state forming methods, powder sintering, reactive processing, the use of ordered (mesomorphic) fluids, and electrodeposition of coatings. Such techniques provide faster processing, often at lower energy cost with less use of troublesome solvents.

Substitutes for Metals and Other Materials. ***Technological and Economic Limitations.*** Material substitution is often impeded and even prevented by technical and economic considerations. In addition to research and development costs, new facilities and large capital investments are needed. These considerations must be carefully examined in view of anticipated benefits.

Novel Opportunities. Of immediate importance to the consumer is the substitution of lightweight composites for metals in automobiles. Recent projections suggest that the use of composites may increase from 1 to 8% of the vehicle weight during 1980–1990, representing about 70 kg of composite materials per vehicle.

Economy in Transportation

The oil embargo of 1973–1974, the passage of the Energy Policy and Conservation Act of 1975, and rapidly rising fuel costs profoundly affected the transportation industry and its use of materials. Reduction of vehicle size and the use of lightweight materials such as plastics, composites, aluminum alloys, and high strength steels are essential for meeting increasingly stringent fuel-economy standards.

Materials and Energy Consumption in Manufacture. An important study by the Owens-Corning Fiberglas Corp. in 1975 compared energy needs for the manufacture of plastic components with those of steel and aluminum parts. Because design enters into the analysis, the comparison was made on the basis of a total product rather than on a basis of energy per mass or volume. A modest energy saving was indicated for fiber-reinforced plastics (FRP) versus steel, and a severe energy penalty for aluminum.

Component Weight Savings and Cost. Substitutions in new vehicle designs are more efficient than in existing ones because of the advantage of secondary effects, such as smaller tires or engines.

Thermoplastics and Their Applications

The principal thermoplastics used in automotive applications are poly(vinyl chloride), polypropylene, and styrene and its copolymers. Others include nylons, polyesters, acrylics, and polyethylene.

Thermoplastic Blends. Polymer blending is one of the fastest growing areas for plastics, and blends account for 20% of all plastics in the United States today. ABS, a blend of poly(styrene–acrylonitrile) and a dispersed rubber, has numerous applications in vehicles, eg, interior trim and electroplated grilles.

Most blends are incompatible or two-phase systems. Blends of styrene polymers or poly(butylene terephthalate) with polycarbonate are widely used, eg, for bumpers. In rare cases, the components are thermodynamically miscible, eg, polystyrene and poly(phenylene oxide) in Noryl. For incompatible systems, morphology and composition are crucial (see COMPATIBILITY; POLYMER BLENDS).

Structural Composites

High performance plastic composites, which may be fabricated by various techniques and which utilize oriented or high concentrations of random glass fibers, offer outstanding promise in lightweight structural vehicle components. Compared with steel, oriented fiber-reinforced plastics (FRP) offer advantages in elastic strain and specific strength. However, FRP composites have a lower specific modulus than steel, even for unidirectional reinforcement, and shear strengths and ductility are generally lower. Material costs, however, are at least several times those of steel (by weight); therefore, weight must be reduced and fabricating techniques must be suitably chosen. Thermoset FRP composites in the automotive industry are dominated by polyesters, vinyl esters, and epoxy resins.

Manufacturing Techniques. At present, continuous directional fibers may be incorporated by compression molding of prepregs, filament winding (qv), and pultrusion (qv) or combinations thereof. Compression molding of rapidly curable prepregs represents the most important technique for parts of complex shape (see COMPRESSION AND TRANSFER MOLDING).

Material Comparisons. Material properties of automotive composites under fatigue and impact conditions are often unavailable. The effects of environmental conditions such as temperature, salt spray, pebble impact, and aging require further study. Process and material variation should be taken into account. Composite materials differ from metals in failure modes, which also need further study. In general, composites replace mild steel.

Graphite. Compared to other composites and aluminum, graphite-fiber composites offer the highest possible weight reduction. However, graphite is a premium material when compared to commodity glass-fiber composites.

However, the cost per kilogram saved is increased substantially when carbon fiber is substituted for glass. Nevertheless, carbon fibers (qv) will almost certainly find limited applications for such parts as drive shafts, making use of processing techniques now being developed for glass. Processing glass with carbon fibers improves handling and gives structures with superior stiffness.

Engineering Applications. It is evident that fiber glass-resin composites have a distinct advantage over steel for components requiring strength. On the other hand, they do not perform as well as steel in applications requiring stiffness, except for a simple solid beam in flexure (leaf springs) or where there is space for increased depth, as in hollow beams. In applications involving high localized loads and shear stress, such as brackets and hinges, the use of composites is limited because of their low shear strength and inability to accommodate high stress concentrations.

Exterior Body Panels

Although it is not likely that plastic-based materials will completely displace steel in exterior body applications, plastic substitution will probably continue in such areas as exterior panels, eg, hoods, in specialized versions of existing vehicles; for vehicles with sales limited to less than about 150,000/yr; special applications allowing substantial parts integration; and vehicles where plastics form the entire exterior skin. Hoods, fenders, rear-deck lids, hatchbacks, fascia, and doors represent some of the principal possibilities.

The wide variety of plastic materials and associated processes being explored for exterior body-panel applications, include compression-molded sheet-molding compound (SMC) (qv); injection-molded bulk-molding compound (BMC); reaction injection molding based on the chemistry of urethanes, caprolactam, and polyesters; resin-transfer molding; injection-molded thermoplastics, and stampable metal–plastic–metal laminates.

BIBLIOGRAPHY

Polymer Science and Engineering: Challenges, Needs, and Opportunities, National Academy Press, 1981; available from the National Research Council, Washington, D.C.

Program Planning Survey on Polymer Composite Science and Technology, prepared for National Bureau of Standards 6080.1 by Springborn Laboratories, Inc., Enfield, Conn., Oct. 1978.

Plastics in the Automotive Industry, Springborn Laboratories, Enfield, Conn., 1977.

Material Substitution in Exterior Automotive Panels, Phillip Townsend Associates, Inc., Houston, Tex., Sept. 1982.

SEYMOUR NEWMAN
Ford Motor Company

AUTOTHERMAL EXTRUSION. See EXTRUSION.

AZINES, POLYMERIC. See AZO POLYMERS.

AZIRIDINE POLYMERS. See ALKYLENEIMINE POLYMERS.

AZO COMPOUNDS

Azo compounds are organonitrogen derivatives with the characteristic —N═N— functionality and the general formula R—N═N—R′ where R and R′ can be the same or different organic groups. In the polymer industry, azo compounds are used as dyes, blowing agents, and free-radical polymerization initiators. The use of azo compounds as dyes and blowing agents is covered elsewhere in this *Encyclopedia* (see DYEING; BLOWING AGENTS). This article describes the azo compounds used as free-radical initiators. Azo compounds used commercially as free-radical initiators are listed in Table 1; all contain at least one tertiary carbon (sp^3) attached to azo nitrogen.

Table 1. Commercial Azo Initiators

Compound	Mp, °C	10-h $t_{1/2}$, °C
Symmetrical		
2,2′-azobis(4-methoxy-2,4-dimethylpentanenitrile)	55–86	40
2,2′-azobis(2,4-dimethylpentanenitrile)	46–65	52
2,2′-azobis(2-methylpropanimidamide)·2HCl	160–169	56[a]
2,2′-azobis(isobutyronitrile)	103–104	64
2,2′-azobis(2-methyl-butanenitrile)	49–51	67
4,4′-azobis(4-cyanopentanoic acid)	112–116	66[a]
1,1′-azobis(cyclohexane-carbonitrile)	114–115	88
2,2′-azobis(2-acetoxypropane)	101–103	189
Unsymmetrical		
2-(*tert*-butylazo)-4-methoxy-2,4-dimethylpentanenitrile	−11	55
2-(*tert*-butylazo)-2,4-dimethylpentanenitrile	−16	70
4-(*tert*-butylazo)-4-cyanopentanoic acid	83–85	73[a]
2-(*tert*-butylazo)isobutyronitrile	17	79
2-(*tert*-butylazo)-2-methylbutanenitrile	−25	82
1-(*tert*-amylazo)cyclohexanecarbonitrile	−12	94
1-(*tert*-butylazo)cyclohexanecarbonitrile	27–28	96
1-(*tert*-butylazo)-formamide	105	105

[a] In water.

These azo compounds decompose on heating to generate two free radicals and nitrogen:

$$R^1\text{—N═N—}R^2 \xrightarrow{\Delta} R^1\cdot + N_2 + R^2\cdot$$

The rate of decomposition is generally expressed in half-life terms, $t_{1/2}$, °C, and is related to the structure of the R groups. Such azo compounds have been used to initiate a variety of free-radical type commercial processes such as vinyl monomer copolymerizations, curing resins, preparing graft copolymers, autoxidation, halogenations, telomerizations, and anti-Markovnikov additions to terminal olefins.

Properties

The commercial symmetrical initiators given in Table 1 are solids with melting points ranging from 46 to 169°C and limited solubility in common organic solvents. The unsymmetrical azonitriles without a carboxy substituent are liquids or low melting solids with good solubility in organic solvents.

Preparation

Azo compounds have been synthesized by a variety of methods. The symmetrical azonitriles in Table 1 are produced by two general routes: (1) reaction of a ketone with hydrazine to form an azine to which hydrogen cyanide is added to produce an intermediate 1,2-disubstituted hydrazine, which is oxidized to the symmetrical azonitrile; (2) reaction of a ketone with HCN to form a cyanohydrin, which reacts with ammonia to produce an aminonitrile; this is coupled to form the symmetrical azonitrile by hypohalite oxidation.

The unsymmetrical azonitriles in Table 1 have been synthesized by the reaction of a tertiary alkylhydrazine with a ketone to form the tertiary alkylhydrazone $R_3C\text{—NHN═}CR^1R^2$ to which HCN is added to produce the unsymmetrical 1,2-disubstituted hydrazine $R_3C\text{—NHNH—}C(CN)R^1R^2$. This is oxidized to the unsymmetrical tertiary azonitrile.

A variety of other nucleophiles (such as imido, organothio, hydroperoxy, alkylperoxy, azido, cyanato, isocyanato, isothiocyanato, alkoxy, and aryloxy) react with α-chloroazo intermediates to form other symmetrical and unsymmetrical azo and bisazo compounds, some of which may have commercial value.

Economic Aspects

Total world market figures for azo initiators are not available but undoubtedly amount to several thousand metric tons annually. 2,2′-Azobis(isobutyronitrile), often referred to as AIBN, has been the most popular and widely used.

Health and Safety Factors

Azo initiators decompose with dangerous rapidity when overheated and may cause explosions in closed containers. Azonitriles are readily ignited with an open flame, and dispersion of solid azonitriles in air are explosive. Care should be taken to avoid dusty conditions; environments should always be sparkproof.

BIBLIOGRAPHY

C. S. Sheppard and V. Kamath in M. Grayson, ed., *Kirk-Othmer Encyclopedia of Chemical Technology,* 3rd ed., Vol. 13, Wiley-Interscience, New York, 1981, pp. 355–373.

S. Patai, ed., *The Chemistry of the Hydrazo, Azo, and Azoxy Groups,* John Wiley & Sons, Inc., New York, 1975, Pts. 1 and 2.

C. G. Overberger, J. P. Anselme, and J. G. Lombardino, *Organic Compounds with Nitrogen–Nitrogen Bonds,* The Ronald Press Company, New York, 1966, Chapt. 4.

C. S. SHEPPARD
Pennwalt Corporation

AZO POLYMERS

Azo-containing polymers (azo polymers) can be formally classified according to a variety of criteria. In this review, the thermal stability of the —N=N— group is the principal criterion. A further subdivision, according to the position of the —N=N— group, is given below.

Main-chain Azo Groups. Each monomer unit contains a stable or unstable azo function; in copolymers, one component contains an azo group.

$$-(R-N=N-R)_n- \qquad -(R-N=N-R)_n-(R'-)_m$$

Side-chain Azo Groups. Each monomer contains an azo group; in copolymers, one component contains an azo function.

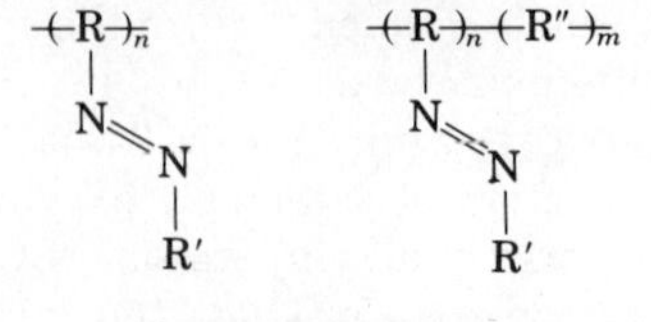

Labile Main-chain Azo Groups

Synthesis. As early as 1951, the patent literature reported polymers containing labile azo groups. The synthesis involved the conversion of aliphatic diketones having at least four carbon atoms between the carbonyl functions (**1**) to polyazines (**2**), which are converted to polyhydrazonitriles (**3**) by addition of HCN; the latter are transformed into polymeric azo compounds (**4**) by oxidation.

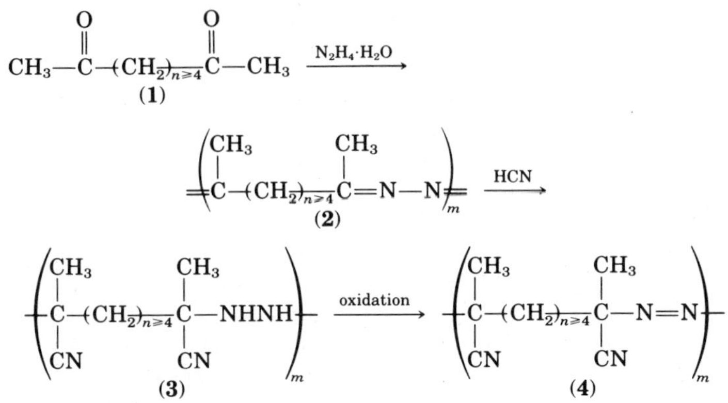

Characterization. Characteristics of azo polymers are the average number of azo groups per chain, the rate constant of thermal decomposition, and the initiator efficiency of the azo functions.

Labile Side-chain Azo Groups

Synthesis. To prepare azo polymers of the general structure

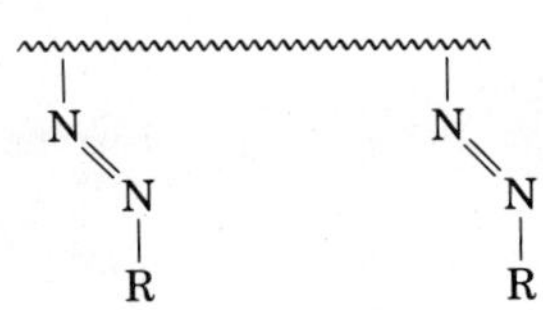

monomers containing an azo group, in addition to a polymerizable functionality, are copolymerized with other monomers:

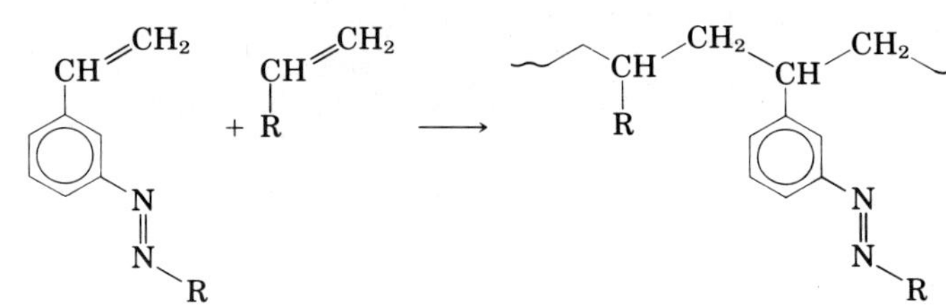

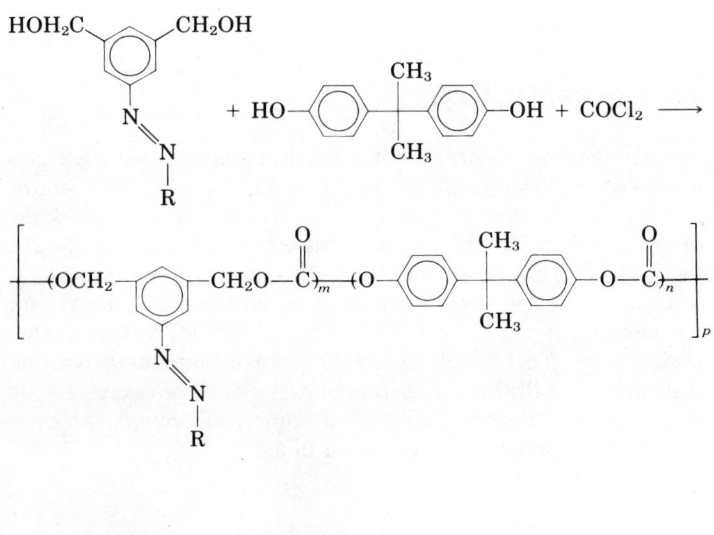

Characterization. The most important characterization parameters are molar mass, azo content, and initiator efficiency.

Stable Main-chain Azo Groups

Synthesis. Many different synthetic routes are available for the production of polymers containing stable azo functions in the backbone chain; only the most widely used are given here. Azo polyamides and polyesters can be synthesized via interfacial or solution polymerization (qv) techniques.

A different synthesis of polymers containing stable azo groups in the main chain is given below:

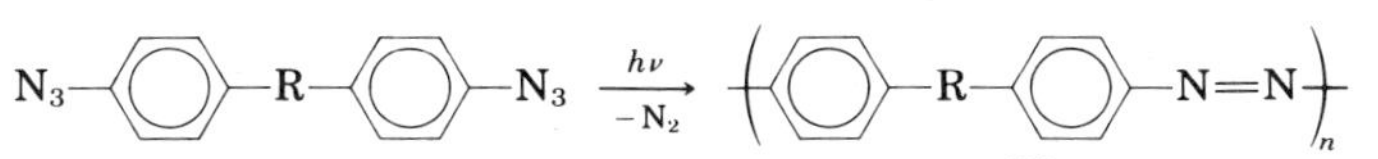

(**5**)

where R = $(CH_2)_n$, —O—$(CH_2)_n$—O—

Product (**5**) undergoes photochemical decomposition to yield brown or black solids typical of azo polymers.

The conversion of nitrosamides into esters, followed by decomposition, can also be utilized for the synthesis of polymers containing stable azo groups.

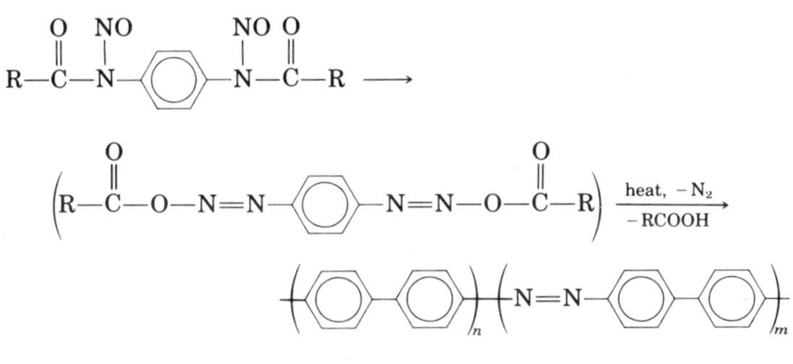

The catalyzed oxidative coupling of aromatic diamines has been intensively studied for the synthesis of polymers containing stable azo groups.

Characterization. The most important properties of azo polymers are the molar mass, azo content, solubility, color, thermal stability, film and fiber formation, cis–trans isomerization, and electrical conductivity.

Stable Side-chain Azo Groups

Synthesis. The synthesis of polymers containing side-chain azo groups has been achieved by modification of a preformed polymer, or copolymerization of a monomer having additional azo functionality with a simple monomer such as styrene or methyl methacrylate.

Characterization. Molar mass and azo content, the most important properties of polymers with side-chain azo groups, are measured by the usual methods.

Industrial Applications

Numerous patents and publications describe possible industrial uses for azo-containing polymers, such as starting material for block copolymers, starting material for graft copolymers, polymeric conductors and semiconductors, polymers with chemically bound dyes, and polymers with liquid-crystal properties. As far as is known, however, no industrial process exists.

Health and Safety Considerations

As far as the author is aware, there have been no studies on the health risks associated with handling of azo polymers. Some azo-containing polymers are explosive and should be treated accordingly.

BIBLIOGRAPHY

E. Marechál, *Prog. Org. Coat.* **10,** 251 (1982).

O. Nuyken and R. Weidner, *Advances in Polymer Science,* Vol. 73/74, Springer-Verlag, New York, 1986, pp. 145–199.

OSKAR NUYKEN
Technische Universitat München

B

BACK MOLDING. See THERMOFORMING.

BACTERICIDES. See BIOCIDES.

BAGASSE. See PAPER; POLYMERS FROM RENEWABLE SOURCES.

BAG MOLDING. See REINFORCED PLASTICS.

BAKELITE. See PHENOLIC RESINS.

BALATA. See RUBBER, NATURAL.

BARRIER POLYMERS

Barrier polymers are organic polymers displaying the ability to restrict the passage of gases, vapors, and organic liquids. Their importance as an article of commerce is recognized by the food-packaging and related industries. These polymers, when used alone or in combination with other polymers or materials, provide an efficient and economical method for the packaging and shelf-life extension of food and beverages (see also FILMS; PACKAGING MATERIALS).

Permeation

Permeation is the rate at which a gas or vapor passes through a polymeric material. The permeation of a gas or vapor through a polymer depends on the polymer, the penetrant, and the environment.

Polymer Characteristics. The nature of the pendent groups on the polymer backbone and the degree or ease of packing affect permeability. A polymer configuration leading to good packing decreases permeability. The simpler the molecular structure, the better the packing. Crystallinity is an important factor because crystallites are impermeable and reduce the permeation rate.

Penetrant. Permeation also depends on the nature of the penetrant. The rate of passage of a permeating species through a polymer matrix is governed by its solubility in the polymer and the relationship between the size of the penetrant molecule and the interstices in the polymer.

Environment. Permeation rates are affected by humidity and temperature.

Permeability

Permeability is the proportionality constant in the general equation for mass transport of a penetrant across a barrier:

$$\frac{\Delta M}{\Delta t} = P\left(\frac{A\Delta p}{L}\right)$$

where P = permeability of the barrier; $\frac{\Delta M}{\Delta t}$ = the mass of a penetrant crossing a barrier per unit time; A = area of the barrier; L = thickness of the barrier; and Δp = partial pressure difference across the barrier. Permeability, as a property of a material, is the product of permeance and thickness.

Barrier Classification

Organic polymers are classified by the degree to which they restrict the passage of gases. The categories range from high barrier (low permeability) to low barrier (high permeability). The category in which a polymer is classified may vary according to the penetrant gas. Typical low barrier polymers are given in Table 1. Polymers containing good and high barrier characteristics are given in Table 2.

Table 1. Permeability of Low Barrier Polymers

Polymer	Oxygen[a]	Carbon dioxide[a]	Moisture vapor[b]
plasticized PVC	6–400	20–600	25–188
polypropylene			
cast	30–52	100–160	1–1.75
oriented	30–32	108	0.5–1.25
polyethylene			
low density	52–96	200–540	2.5–3.75
high density	18–56		0.76–1.0
linear low density	50–140	500	3.0
polystyrene	52		17.5
EVA,[c] 12%	100–140	440–580	9.75
polybutylene	80	160	3.0

[a] mol/(m·s·Pa) × 10^{-17} at 23°C, 50% rh.
[b] mol/(m·s·Pa) 10^{-15} at 23°C, 100% rh.
[c] Poly(ethylene-*co*-vinyl acetate).

Table 2. Permeability of Good and High Barrier Polymers

Polymer	Oxygen[a]	Carbon dioxide[a]	Moisture vapor[b]
Good barrier			
polyester, PET	0.6–0.8	3–5	2.5–5.0
polyamide, nylon-6	0.4–0.6	2–2.4	40–55
poly(vinyl chloride), rigid	1.0–4	4–10	12.5–188
poly(vinyl fluoride)	0.6	2.2	10
High barrier			
cellophane, dry	0.1–0.16	0.2–1.2	1–335
poly(vinylidene chloride)[c]	0.16–0.22	0.76–1.2	0.5–1.5

[a] mol/(m·s·Pa) × 10^{-17} at 23°C, 50% rh.
[b] mol/(m·s·Pa) × 10^{-15} at 23°C, 100% rh.
[c] Saran wrap.

Processing

The principal use of barrier polymers is in packaging, especially of food and beverages, but they are becoming increasingly important in the packaging of chemicals, insecticides, and automotive supplies. These thermoplastic materials are versatile and can be fabricated into many shapes and forms. The package may be rigid such as a bottle or can, semirigid and thermoformed from a sheet, or flexible as a pouch.

No single resin is capable of producing the right balance of physical and mechanical properties, appearance, and economics for all types of packaging. The logical and practical approach is the combination or modification of existing polymers by polymer blending or alloying, chemical modification, or structuring using melt-processing techniques.

Federal Regulations

All materials used to package food and beverages must comply with governmental regulations, including the Federal Food, Drug and Cosmetics Act and all applicable food additive regulations.

BIBLIOGRAPHY

M. Salame and S. Steingiser, *Barrier Polymers,* Chemical Marketing & Economics Division of the American Chemical Society Symposia, New York, Apr. 5–8, 1976, ACS, Washington, D.C.

"Barrier Polymers," in M. Grayson, ed., *Kirk-Othmer Encyclopedia of Chemical Technology,* 3rd ed., Vol. 3, Wiley-Interscience, New York, 1978, pp. 480–502.

P. L. Heater, "Polyethylene terephthalate: PET," in R. Juran, ed., *Modern Plastics Encyclopedia,* Vol. 65, No. 11, McGraw-Hill Inc., New York, Oct. 1988, p. 46.

R. H. Foster, "Ethylene-vinyl alcohol copolymers," in R. Juran, ed., *Modern Plastics Encyclopedia,* Vol. 65, No. 11, McGraw-Hill Inc., New York, Oct. 1988, p. 76.

D. L. Clarke, "Vinylidene Chloride," in R. Juran, ed., *Modern Plastics Encyclopedia,* Vol. 65, No. 11, McGraw-Hill Inc., New York, Oct. 1988, p. 127.

W. A. Combellick
Goodyear Tire & Rubber Company

BARRIERS, VAPOR.
See Barrier polymers.

BEAD POLYMERIZATION.
See Suspension polymerization.

BEATER ADDITIVES.
See Paper additives and resins.

BELTING

A belt is defined as a continuous band of tough, flexible material used for transmitting power or conveying materials.

Conveyor Belts

Manufacture. The belt used in a conveyor system is composed of the cover and the carcass or reinforcement (Fig. 1). The carcass is the working tension component, which is protected from the environment by the cover.

The top cover is usually thicker than the bottom cover because the greatest wear is concentrated on the top or carrying side, mainly at the point of loading, where material is accelerated to belt speed. Rubber materials are designed to withstand wear and seal the carcass against water, acids, oil, and other foreign materials.

Figure 1. A typical conveyor belt, with cutaway exhibiting a top breaker ply and three plain-weave fabric plies. Courtesy of BFGoodrich Co.

The carcass is usually composed of plies or layers of fabric bonded by a coat of rubber. It may also consist of a single interwoven ply, or a single layer of cords or steel cable.

Breakers. An extra ply of open-weave fabric, called a breaker ply, is sometimes inserted between the top cover and the first ply of the carcass to give the belt added resistance against shock-loading conditions, increase the cover adhesion, and prevent tearing.

Surface. Standard belt has a smooth flat surface. Specialty belts may be ribbed or cleated.

Fasteners and Splicing. The ends of linear belt lengths are joined by vulcanized splices (or joints) or metal fasteners.

Rubber Materials

Rubber compositions contain many ingredients with a rubber content usually between 40 and 60%. The other ingredients are fine reinforcing powders (carbon black or clay), softeners to control viscosity (oils or resins), vulcanizing agents, protective chemicals, and other additives. About 10–25 such ingredients are blended under high pressure to yield a plastic, deformable mix (see Rubber chemicals). Elastomers used include natural rubber, styrene–butadiene rubber, neoprene rubber, nitrile–butadiene rubber, butyl rubber, poly(vinyl chloride), ethylene–propylene rubber, and polyurethane rubber.

Selection

The type of rubber compounds utilized in a belt depends on the service conditions, such as type, shape, and particle size of the material, belt speed, loading conditions, and exposure.

Power Transmission Belts

V-belts, V-ribbed belts, synchronous belts, and flat belts are used to transmit power from motors to other machinery such as fans, generators, grinders, crushers, lathes, presses, rollers, and pumps (Fig. 2).

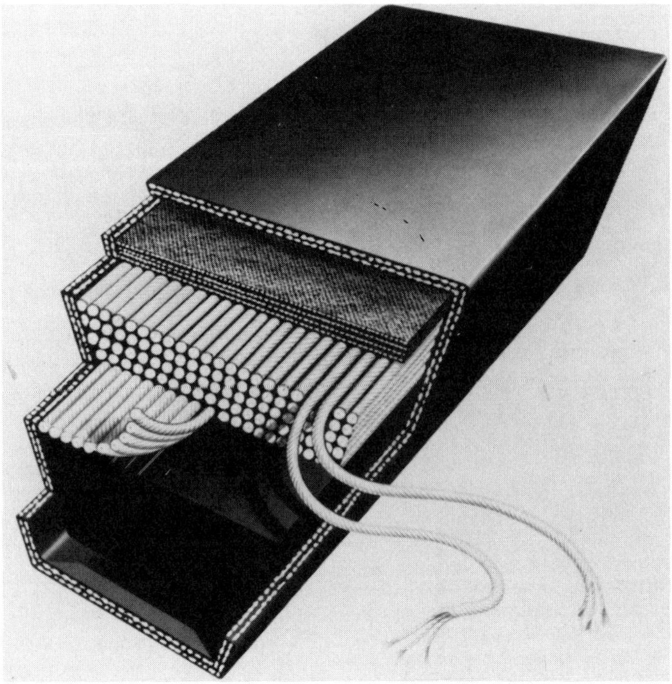

Figure 2. A premium heavy-rated industrial V-belt. Courtesy of BFGoodrich Co.

BIBLIOGRAPHY

Engineering Standards MPTA/RMA (IP20, IP22, IP24, IP25, IP26), Rubber Manufacturers Association, Washington, D.C., 1984.

D. H. SILVEY
GEORGE RUGMAN
BFGoodrich Company

BENZIMIDAZOLES, POLYMERIC. See POLYBENZIMIDAZOLES.

BENZOGUANAMINE–ALDEHYDE RESINS. See AMINO RESINS.

BENZOYL PEROXIDE. See PEROXY COMPOUNDS.

BENZYLCELLULOSE. See CELLULOSE ETHERS.

BIAXIAL ORIENTATION. See ORIENTATION PROCESSES.

BINDERS. See COATINGS; PAPER ADDITIVES AND RESINS.

BIOCIDES

The term biocide includes all chemical agents that control or destroy microorganisms. This would encompass materials variously described as bactericides, bacteriostats, mildewcides, fungicides, fungistats, microbiocides, germicides, preservatives, algicides, etc.

Synthetic polymers, such as polyolefins, polyesters, and vinyl polymers, are considered for the most part to be resistant to biological attack, except perhaps in specialized environments. Some synthetic polymers, such as the polyester from ϵ-caprolactone, have been designed to be subject to biological degradation, whereas others may be subject to attack after exposure to the weather (see BIODEGRADABLE POLYMERS; DEGRADATION).

Synthetic polymers are seldom used alone but are frequently mixed with plasticizers (qv), lubricants, stabilizers, fillers (qv), and other materials that usually change the biological resistance of the polymer and the finished composition. Such material may be vulnerable to biological attack.

Biological Environment

The biological environment includes the biological agents responsible for the deterioration of polymeric substances.

The Plant Kingdom. *Eumycetes,* or true fungi, are microorganisms of particular importance in causing the degradation of materials. The bacteria, *Schizomycetes,* which can be aerobic or anaerobic, are important in the degradation of materials such as cellulosic thickeners in water-based (latex) paints. The *Actinomycetes,* although classified as bacteria, possess many fungilike characteristics. Algae are not associated with polymer deterioration to any large extent. The chief problem with algae is their appearance and odor when growing on polymerics; furthermore, they may constitute a source of food for other organisms.

The Animal Kingdom. Organisms in the animal kingdom causing degradation of materials include insects, birds, rodents, and marine-fouling organisms.

Susceptibility to Biological Attack

Natural and synthetic polymers may be attacked by a variety of species.

Nutrients. Extensive damage is inflicted by bacteria and fungi, insects, insect larvae, rodents, and birds to natural polymers. Fermentation, putrefaction, and associated chemical changes are carried on by microorganisms by the action of enzymes specific to a substrate (ie, cellulases degrade cellulose, lipases split fats, and proteases attack proteins).

Attachments. Organisms may adhere to polymers without causing degradation. However, aesthetic or other problems, eg, loss of fuel economy and speed in ships, secondary infestations, or impairment of function may arise.

Barriers. Rodents and insects, in the attempt to destroy material blocking access to food, cause extensive damage.

Biological Degradation

Natural Polymers. ***Resins.*** Natural resins usually possess biocidal properties and are consequently resistant to biodegradation.

Lignin. Natural lignin (qv), an insoluble polymer of phenolic composition, is resistant to microbes, but is attacked by white-rot fungi and certain soil microorganisms.

Industrial Gums. Plant polysaccharides and their derivatives, known as industrial gums (qv), include plant exudates, seaweed extracts, seed extracts, starches, and others. Because they are widely used in foods and food packaging, the biocides employed must be acceptable to the FDA, consistent with the Food Additives Amendment.

Natural Rubber. Natural rubber latex is subject to microbial attack.

Proteins. Purified native proteins do not support bacterial growth and are not attacked even by strongly proteolytic organisms.

Leather. Leather is subject to degradation by bacteria and fungi during processing (tanning) and in service.

Cellulose. Cellulose (qv) and cellulose derivatives are highly susceptible to biodegradation through enzymatic hydrolysis or physical penetration.

Synthetic Polymers. Synthetic resins or polymers are inherently resistant to biological attack because of their hardness, limited moisture absorption, smooth surface, and lack of susceptibility to enzymatic systems. However, susceptibility to biodegradation varies and is affected by additives, especially plasticizers.

Additives. A product may be highly susceptible to biological attack even though the polymer on which it is based is inert or only mildly susceptible, because compounding ingredients may be of nutritive value and invite attack.

Addition of Biocides to Polymers

Biocides are usually added to the finished polymer rather than being incorporated during polymerization.

Evaluation

The effectiveness of a biocide can be determined only by field exposure under a variety of conditions. The record of performance guides the manufacturer's recommendation for dosage, areas of application, and duration of effectiveness.

Toxicology

Biocides are, of course, toxic materials. Ideally, a biocide should have a high degree of toxicity to the organisms responsible for degradation, but should be harmless to humans. In practice, all biocides must be handled with care.

Table 1. Biocides Approved by the Food and Drug Administration for Use in Food Packaging Materials

Slimicides for the manufacture of paper and paperboard in contact with food (Section 176.300)
acrolein
1,2-benzisothiazolin-3-one (27 g/t)
bis(1,4-bromoacetoxy)-2-butene
bis(trichloromethyl)sulfone
β-bromo-β-nitrostyrene (max 453 g/t of dry weight fiber)
chloroethylene bisthiocyanate
5-chloro-2-methyl-4-isothiazolin-3-one, calcium chloride (3 parts) and 2-methyl-4-isothiazolin-3-one (1 part) (1.13 kg/t)
1,2-dibromo-2,4-dicyanobutane (0.005% of dry weight fiber)
2,2-dibromo-3-nitrilopropionamide (45 g/t)
3,5-dimethyl-1,3,5-2*H*-tetrahydrothiadiazine-2-thione
dipotassium and disodium ethylenebis(dithiocarbamate)
disodium cyanodithioimidocarbonate
2-mercaptobenzothiazole
methylene(bisthiocyanate)

(Continued)

Table 1. *(Continued)*

2-nitrobutyl bromoacetate (68 g/t)
potassium 2-mercaptobenzothiazole
potassium *N*-methyldithiocarbamate
sodium dimethyldithiocarbamate
sodium 2-mercaptobenzothiazole
vinylene bisthiocyanate
For textiles (Section 177.2800)
formaldehyde
For wood (Section 178.3800)
copper 8-quinolinolate
pentachlorophenol
zinc salt of sulfonated petroleum
Preservatives
For defoaming agents used in the manufacture of coatings (Section 176.200)
formaldehyde
sodium 2-mercaptobenzothiazole
For adhesives (Section 175.105)
1,2-benzisothiazolin-3-one
benzyl bromoacetate
bis(tri-*n*-butyltin) oxide
bis(trichloromethyl)(sulfone)
1-(3-chloroallyl)-3,5,7-triaza-1-azonia-adamantane chloride
1,2-dibromo-2,4-dicyanobutane
4(diiodomethyl sulfonyl)toluene[a]
3,5-dimethyl-1,3,5(2*H*)-tetrahydrothiadiazine-2-thione
2-mercaptobenzothiazole and dimethyldithiocarbamic acid mixture, sodium salts
2-mercaptobenzothiazole, sodium or zinc salt
o-phenylphenol
oxazolidine
potassium *N*-methyldithiocarbamate
quarternary ammonium chlorides, hexadecyl and octadecyl derivatives
sodium dehydroacetate
sodium salicylate
sodium 1-hydroxy-2-(1*H*)-pyridine thione
ziram
For animal glue (Section 178.3120)
3,5-dimethyl-1,3,5(2*H*)-tetrahydrothiadiazine-2-thione
disodium cyanodithioimidocarbonate
potassium *N*-methyldithiocarbamate
sodium 2-mercaptobenzothiazole
sodium *o*-phenylphenate
zinc dimethyldithiocarbamate
zinc 2-mercaptobenzothiazole
For paper and paperboard in contact with dry foods (Section 176.180)
barium metaborate[b]
1,2-benzisothiszolin-3-one (0.0031 mg/cm^2)
bis(trichloromethyl)sulfone
oxazolidine
In contact with aqueous and fatty foods (Section 176.170)
1,2-benzisothiazolin-3-one (0.0016 mg/cm^2)
2-bromo-4′-hydroxyacetophenone (0.006%)
1,2-dibromo-2,4-dicyanobutane (0.04 wt %)
captan (*N*-trichloromethylmercapto-4-cyclohexene-1,2-dicarboximide)[c]
1-(3-chloroallyl)-3,5,7-triaza-1-azonia-adamantane chloride (0.3 wt % in latex)
copper 8-quinolinolate
sodium *o*-phenylphenate
1,3,5-triethylhexahydro-1,3,5-triazine (0.05–0.15 wt %)

[a] Against fungi.
[b] Coatings and sizing.
[c] Mold proofing.

Table 2. Some Industrial Biocides

Chemical name	Application	Suppliers	Trade names
barium metaborate (modified)	latex paint preservative	Buckman	Busan 11-M-1
1,2-benzisothiazolin-3-one	metal-working fluids, slimicide, adhesive and latex preservative	ICI United States	Proxel CRL
benzyl bromoacetate	latex, polymer preservative	Merck/Calgon	Merbac 35
bromonitrostyrene	latex, polymer preservative	Givaudan	Giv-Gard BNS
1,2-dibromo-2,4-di-cyanobutane	latex, paint, metal-working fluid preservative	Merck/Calgon	Tektamer 38
1-(3-chloroallyl)-3,5,7-triaza-1-azonia-adamantane chloride	paint preservative	Dow	Dowicil 75
diiodomethyl *p*-tolyl sulfone	paint, adhesive preservative	Abbott	Amical 48
chloroxylenol	vinyl preservative	Ferro	Ottasept
p-chloro-*m*-cresol	emulsion preservative	Cosan	PCMC
copper naphthenate	wood, cordage and textile fungicide	Harshaw Chemical Nuodex, Inc.	Uversol Nuodex Copper 6%, 8%
copper 8-quinolinolate	textile, rope, cordage, wood, plastic, paper, adhesive and paint fungicide	Napp-Lemke Ventron Nuodex, Inc.	Quinosol Cunilate Quindex
potassium *N*,*N*-dimethyldithiocarbamate	latex paint preservative	Buckman	Busan 1009
dimethyldithiocarbamic acid and 2-mercaptobenzothiazole, sodium and zinc salts of	slime control preservative for cotton fabrics, adhesives, latex	R. T. Vanderbilt	Vancide 51 and 51Z Ziram
4,4-dimethyloxazolidine 3,4,4-trimethyloxazolidine	emulsion preservative	Angus	Bioban CS-1135 Cosan 101
3,5-dimethyl-1,3,5(2*H*)-tetrahydrothiadiazine-2-thione (Mylone)	paper mill slimicide, adhesive, latex, paint, paper preservative	Stauffer Buckman Cosan Merck Troy	N521 DMTT Cosan S Metasol D3T Troysan 142
di(phenylmercuric)-dodecenyl succinate	latex paint preservative and fungicide	Nuodex, Inc.	Super Ad-It
hexahydro-1,3,5-triethyl-*s*-triazine	latex paint preservative	R. T. Vanderbilt	Vancide TH
hexahydro-1,3,4-tris-(2-hydroxyethyl)-*s*-triazine	metal-working cutting fluids preservative	Lehn & Fink Onyx	Grotan Onyxide 200
2-[(hydroxymethyl)-amino]ethanol	paint preservative	Troy Cosan	Troysan 174 Cosan 91
2-*n*-octyl-4-isothiazolin-3-one	paint mildewcide, polymer preservative	Rohm & Haas Ferro	Skane M-8 Micro-chek
oxazolidine	latex preservative	Nuodex, Inc.	Nuosept 95
10,10′-oxybisphenoxyarsine (OBPA)	vinyl, urethane polyolefin	Ventron Maag	Vinyzene BP-5 Vi-Nylate
o-phenylphenol	metal-working fluids, adhesives	Dow Chemical	Dowicide 1
pyridinethione, sodium	metal-working fluids, latex and vinyl preservative	Olin	Zinc Omadine Triadine 10
sodium *o*-phenylphenate	leather, latex paints, textile, adhesive, paper, natural and synthetic gum preservative, metal-working cutting fluids	Dow Chemical	Dowicide A
2,4,5,6-tetrachloroisophthalonitrile (chlorothalonil)	latex-paint preservative	Diamond Shamrock	Nopcocide N-96
2-(4-thiazolyl)benzimidazole (thiabendazole)	paint, adhesive, paper	Merck	Metasol TK-100
thiocyanomethyl thiobenzothiazole	paint and leather preservative	Buckman	Busan 72 Busan 1030
tributyltin oxide	latex paint, paper, pulp, textile and plastic preservative	Witco Carlisle Chem. Works M & T Chemical	Keycide X-10 TBTO TBTO
N-trichloromethylmercapto-4-cyclohexene-1,2-dicarboximide (captan)	paint, lacquer, paper, plasticizers, resin, vinyl	R. T. Vanderbilt	Vancide 89
N-(trichloromethylthio)-phthalimide (folpet)	paint and vinyl fungicide	Interstab Troy Mooney Chemical Cosan Nuodex Chemicals	Intercide TMP Troysan O Bio-Chem FPT Cosan P Fungitrol II
tris(hydroxymethyl) nitromethane	emulsion preservative	Angus	Tris-Nitro

The supplier of the biocide must provide toxicity and handling information. The EPA requires the registration of every biocide manufactured and sold in the United States. Another regulatory agency, the FDA, is responsible for the approval of all food and food packaging additives. Table 1 shows some biocides currently approved by the FDA.

Uses

The extensive use of industrial biocides is demonstrated by a 1982 world consumption of 500×10^6 dollars. The United States is the largest consumer of biocides (108×10^3 metric tons in 1978). A large portion of industrial biocides are used to prevent fungal and bacterial attack on paint, paper, wood, petroleum products, textiles, plastics, resins, adhesives, and rubber products. Table 2 shows some industrial biocides and their producers.

BIBLIOGRAPHY

A. H. Walters and J. J. Elphick, eds., *Biodeterioration of Materials,* Elsevier Publishing Company, Ltd., London, 1968.

E. T. Reese and N. Mandels, "Enzymatic Degradation," in N. M. Bikales and L. Segal, eds., *Cellulose and Cellulose Derivatives,* Wiley-Interscience, New York, 1971, pp. 1079–1094.

G. A. Greathouse and C. J. Wessel, eds., *Deterioration of Materials: Causes and Preventative Techniques,* Reinhold Publishing Corp., New York, 1954.

N. I. Sax, *Dangerous Properties of Industrial Materials,* Reinhold Book Corp., New York, 1968.

S. I. Trotz and J. J. Pitts, "Industrial Antimicrobial Agents," in M. Grayson, ed., *Kirk-Othmer Encyclopedia of Chemical Technology,* 3rd ed., Vol. 13, Wiley-Interscience, New York, 1981, pp. 223–253.

F. B. FLYNN
W. S. TAYLOR
R. T. Vanderbilt Company, Inc.

BIOCOMPATIBILITY. See BIOMATERIALS.

BIODEGRADABLE POLYMERS

One of the important current incentives for the study of biodegradable polymers is their easier disposal. Moreover, biodegradable polymers are useful for applications such as sutures, surgical implants, controlled-release formulations of drugs and agricultural chemicals, agricultural mulch, etc, and interest in them continues to increase (see also AGRICULTURAL APPLICATIONS; BIOMATERIALS; DRUG APPLICATIONS). In this article, the definition of biodegradation is restricted to degradations caused by enzymes and microorganisms *in vitro* and *in vivo.*

Testing for Biodegradation

Tests for biodegradation include soil burial, microbial degradation with cultured microorganisms and purified enzymes, degradation in blood serum and tissues, and *in vivo* degradation.

Effect of Polymer Structure on Biodegradation

Natural macromolecules, eg, protein, cellulose, and starch, are generally degraded in biological systems by hydrolysis followed by oxidation. It is not surprising, then, that most of the reported synthetic biodegradable polymers contain hydrolyzable linkages along the polymers. In order for a synthetic polymer to be degradable by enzyme catalysis, the polymer chain must be flexible enough to fit into the active site of the enzyme.

Morphology. One of the principal differences between proteins and synthetic polymers is that proteins do not have equivalent repeating units along the polypeptide chains. This irregularity results in protein chains less likely to crystallize. It is quite probable that this property contributes to the ready biodegradability of proteins. Synthetic polymers, on the other hand, generally have short repeating units, and this regularity enhances crystallization, making the hydrolyzable groups inaccessible to enzymes. It was reasoned that synthetic polymers with long repeating units would be less likely to crystallize and thus might be biodegradable; indeed, a series of poly(amide–urethane)s were found to be readily degraded by subtilisin.

$$-\!\!\left[NHCH_2CH_2O-\overset{\overset{O}{\|}}{C}NH(CH_2)_6NH\overset{\overset{O}{\|}}{C}-OCH_2CH_2NH\overset{\overset{O}{\|}}{C}-(CH_2)_x\overset{\overset{O}{\|}}{C}\right]_n$$

poly(amide–urethane)s

where $x = 2,4,8$

Selective chemical degradation of semicrystalline polymer samples shows certain characteristic changes. During degradation, the crystallinity of the sample increases rapidly at first, then levels off to a much slower rate as the crystallinity approaches 100%. This is attributed to the eventual disappearance of the amorphous portion of the sample. Scanning electron microscopy (sem) has shown that the degradation of a partially crystalline polycaprolactone film by filamentous fungi proceeds in a selective manner, with the amorphous regions being degraded prior to the degradation of the crystalline region. This is shown in Figures 1 and 2.

Molecular Weight. The dominant factor determining degradability is either polymer molecular weight or degree of crystallinity, depending on the microorganisms.

Polymers with Hydrolyzable Backbones

Polyesters. Aliphatic polyesters (qv) are perhaps the most easily biodegraded synthetic polymers known. The poly(hydroxyacid)s in particular have many important applications as biodegradable polymers.

Polyurethanes. In one study of the fungal degradation of approximately 100 polyurethanes, it was concluded that polyester-based polyurethanes are much more susceptible to degradation than polyurethanes derived from polyether diols. Poly(amide–urethane)s prepared from amino alcohols are easily degraded by the enzyme subtilisin.

Polyamides. Regular high molecular-weight polyamides (qv) such as polycaprolactam (nylon-6) and poly(hexamethylenediamine-*co*-adipic acid) (nylon-6,6) generally resist microbial and enzymatic attacks. However, oligomers of ϵ-aminohexanoic acid have been degraded by enzymes and microorganisms. Polyureas (qv), polyanhydrides (qv), poly(amide–enamine)s, and polyphosphazene (qv) also have hydrolyzable backbones.

Polymers with Carbon Backbones

Vinyl polymers, with few exceptions, are generally not susceptible to hydrolysis. Their biodegradation, if occurring at all,

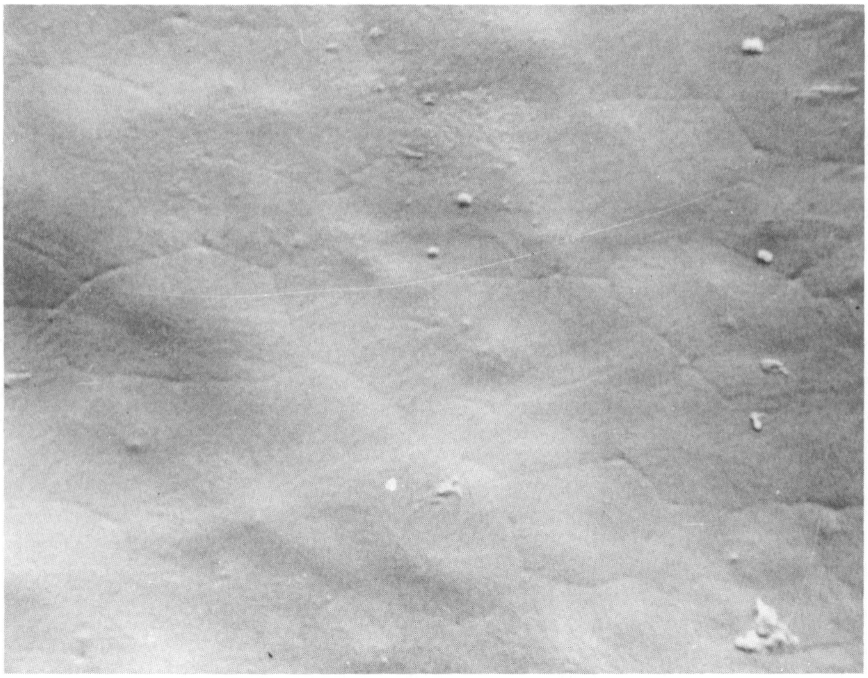

Figure 1. Undegraded polycaprolactone (PCL-700) film showing spherulite structure at film surface. Courtesy of *Organic Coatings and Applied Polymer Science Proceedings.*

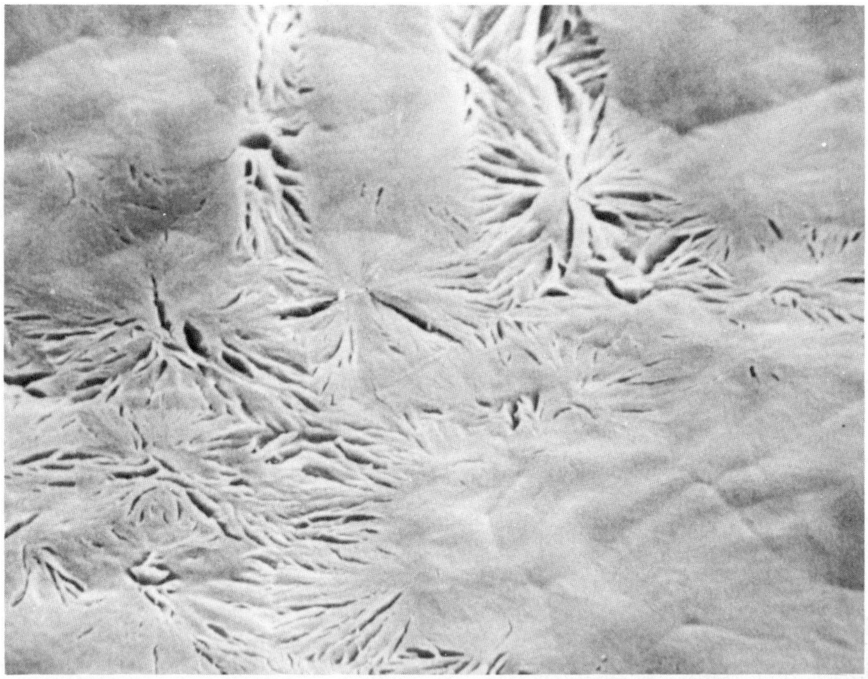

Figure 2. Partially degraded PCL-700 film showing tracks left by enzyme-secreting fungal mycelia. Courtesy of *Organic Coatings and Applied Polymer Science Proceedings.*

requires an oxidation process, and most of the biodegradable vinyl polymers contain an easily oxidizable functional group. Approaches to improve the biodegradability of vinyl polymers often include the addition of catalysts to promote their oxidation or photooxidation, or both. The incorporation of photosensitive groups, eg, ketones, into the polymers also have been attempted. Vinyl polymers that are or can be processed to be biodegradable include poly(vinyl alcohol) and poly(vinyl acetate), and polyacrylates.

BIBLIOGRAPHY

S. J. Huang, J. P. Bell, J. R. Knox, H. Atwood, D. Bansleben, M. Bitritto, W. Borghard, T. Chapin, K. W. Leong, K. Natarjan, J. Nepumuceno, M. Roby, J. Soboslai, and N. Shoemaker in J. M. Sharpley and A. M. Kaplan, eds., *Proceedings of the 3rd International Biodegradation Symposium,* Applied Science Publishers, New York, 1976, pp. 731–741.

J. E. Potts in M. Grayson, ed., *Kirk-Othmer Encyclopedia of Chemical Technology,* Suppl. Vol., Wiley-Interscience, New York, 1984, pp. 626–668.

R. D. Fields, F. Rodriguez, and R. K. Finn, *J. Appl. Polym. Sci.* **18,** 3571 (1974).

J. Kopecek and K. Ulbrich, *Prog. Polym. Sci.* **9,** 1 (1983).

Samuel J. Huang
The University of Connecticut

BIOLOGICAL ACTIVITY

During the last three decades, the study and application of biological activity of polymers have developed. There are basically three main types of polymer used in a biological environment:

1. Polymers used as biomaterials (qv), eg, in organ replacement and bone surgery.
2. Polymers serving as matrices in devices that permit controlled release of an active substance over a long period of time (see CONTROLLED RELEASE TECHNOLOGY).
3. Soluble, synthetic polymers that themselves display biological activity.

This article is confined primarily to polymers in the last category.

Features of Biological Systems

Design of polymers for use in biological systems must take into account those properties of the biological environment relevant to its interaction with polymeric materials. The interactions are conveniently considered at different levels of organization.

Molecular Level. On the molecular level, chemical reactions and intermolecular attractions are significant.

Cellular Level. Polymers cannot penetrate the cell membrane via diffusion. Depending on charge, charge distribution, molecular weight, hydrophobicity, conformation, and tacticity, polymers interact more or less strongly with cell membrane components, ie, mainly lipids, proteins, glycoproteins. First attempts at finding structure–activity relationships have been made. Cell uptake of polymers occurs by the mechanism of endocytosis, which is comparable to the ingestion of material by amoebas.

Whole Organism. As far as whole organisms are concerned, as in cells, polymers meet highly compartmentalized systems, and their access to some compartments is restricted. The parenteral administration route (subcutaneous, intraperitoneal, intramuscular, and intravascular—ways that are not at all equivalent) is the only efficient way of administering polymers. Intravenous injection leads to fast distribution to all vascularized parts of an organism.

Model Considerations for Biologically Active Polymers

The possible scope of polymers as biologically active agents was outlined by a model, which is shown in Figure 1 and has been widely discusssed and modified. Although this model describes only pharmacologically active polymers, its implications offer a helpful framework for the choice and design of biologically active polymers in general. Theoretically, the combination of different effector moieties in different domains of a polymer gives a wider scope in comparison to low molecular-weight agents, where small changes often cause the loss of desired activities. The polymer backbone itself may be biodegradable or biostable, and its molecular weight, tacticity, stereochemistry, and monomer sequence (block, random, etc) may be varied to influence the microstructure and conformation of the polymer.

Classes of Soluble Biologically Active Polymers

Polymeric Models of Enzymes and Biopolymers. Many biomolecules are polymers. Therefore, attempts have been made to synthesize model compounds that mimic, and thereby shed light on, the structure–activity relationships of their natural analogues. Table 1 lists some types of polymer that have been synthesized.

Table 1. Polymeric Models of Biopolymers

Biopolymers	Synthetic analogues
collagen	polypeptides
heparin	semisynthetic polysaccharides
enzymes	vinyl polymers containing imidazole and other nucleophiles, polyethyleneimine derivatives, polypeptides, synzymes
nucleic acids	vinyl polymers containing nucleic acid bases, polyethyleneimine derivatives, poly(4-vinylpyridine) salts

Polymers with Inherent Biological Activity. Polymers may display biological activity even if they do not contain biologically active moieties or repeating units. Table 2 gives some selected examples.

Table 2. Selected Examples of Polymers with Inherent Biological Activity

Class	Polymer	Biological activity, use
neutral	polysaccharides	plasma expanders
	copolymers containing thiazolidine	radiation protection
	poly(*N*-vinylpyrrolidinone)	plasma expanders
polycationic	poly(vinylpyridine-*N*-oxide)	antisilicotic
	polyethyleneimine	antitumor
	polyvinylimidazoline	
	polyamidoamines	heparin-complexing
	quaternized polyetheramines	antimicrobial
	polyhydroxyamines	
	polyamines	
	6,10-ionene	antitumor, antimicrobial
polyanionic	DIVEMA, ie, divinyl ether–maleic anhydride copolymer, and other polyanions	antitumor antiviral interferon-inducing immune-modulating macrophage-activating
	macromolecular sulfates	heparinlike activity

Soluble Polymers with Bound Biologically Active Moieties—Model and Reality. *The Carriers.* The ideal carrier for biologically active substances should be biodegradable or readily excreted without being metabolized (nonaccumulating), nontoxic, nonimmunogenic, and should contain functional groups for drug binding. Examples include polyvinyl derivatives, such as poly[(2-methylsulfinyl)ethyl methacrylate], poly(vinylpyrrolidinone-*co*-vinylamine), poly[*N*-(2-hydroxypropyl)methacrylamide] (PHPMA), poly(vinyl alcohol), and polyalkadienes; various biodegradable systems, such as polyphosphazenes, poly(aspartic acid), poly(β-malic acid), poly(amino acids), poly(aspartic acid amide), and polysaccharides; polyvinyl oligomers cross-linked with peptide spacers; poly(*L*-lysine); polyelectrolytes, such as (e.g. divinyl ether–maleic an-

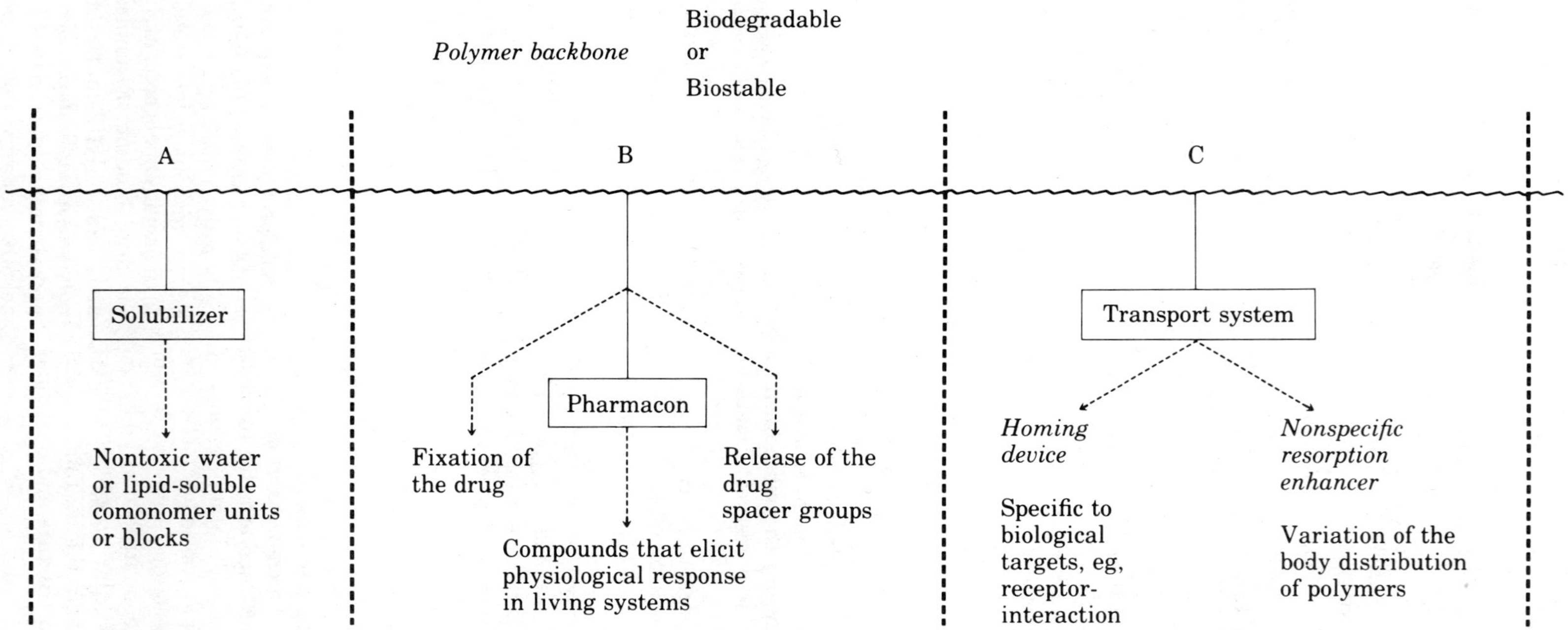

Figure 1. Model for pharmacologically active polymers. Courtesy of Gordon and Breach Science Publishers.

hydride copolymer (DIVEMA); poly(ethyleneimine); and block copolymers with hydrophobic and hydrophilic blocks.

The Biologically Active Moieties. Polymers with antitumor activity and antimicrobial activity have received special attention. One of the chief aims in fixing biologically active moieties to polymers is latentiation of biological effects.

Spacer Units. Biologically active units are usually bound to a polymer carrier through spacer units. The bound unit may be active either in a bound state or after liberation.

Targeting of Drug-carrier Systems. Animal cells contain specific surface structures (eg, receptors and surface antigens) that recognize effector molecules (eg, hormones, sugar moieties, and antibodies). Many soluble polymers do not interact strongly with cell membrane surfaces. Hormones have been attached to polymers as targeting or homing devices. Given their high specificity, antibodies have been attached to polymers as homing devices, eg, in affinity chemotherapy. Targeting to an intracellular site, the lysosomes provide a further degree of specific delivery of biologically active moieties (lysosomotropic systems). In the future, special attention must be given to mild coupling procedures leading to soluble active derivatives, and to the body distribution of these polymers.

Polymeric Cell Models and Other Particulate Drug Carriers. Among the carrier systems for biologically active systems studied, there are also cells and liposomes. "Artificial cells" produced by interfacial polymerization can serve as ultrathin capsules. These systems have been tested for enzyme deficiency therapy, as red blood cell substitutes, for perfusion by body fluids, as carriers for detoxicants, and for artificial organs, eg, kidneys.

In a different approach, lipid vesicles (liposomes) consisting of polymerizable lipids have been prepared. These simple and stable polymeric liposomes can be functionalized in order to mimic biological activities (see VESICLES).

Immune Properties of Polymers

Synthetic polymers entering an organism may elicit immune responses. It seems that, in contrast to earlier findings, all macromolecules are potentially immunogenic if given in the proper dosage and according to the proper schedule. By varying these parameters, moderate-to-low immune response could be observed with polyvinylpyrrolidinone, poly(methacrylic acid), and other vinyl polymers.

Uses

Biologically active polymers are used in biomedical research and sustained release of bioactive agents.

Test Methods

Toxicity and Immune Properties. Cell cultures provide a simple and relatively inexpensive means of preliminary tests for toxicity.

Tests of Biological Activity and Mechanisms. Again, cells in culture may be used to assess biological activities.

Testing of Interactions between Polymers and Cells. Interactions between polymers and cell membranes may be investigated by high sensitivity dsc.

Testing Biodegradation, Body Distribution, and Excretion. The fate of polymers in whole organisms is followed by labeling.

BIBLIOGRAPHY

L. Illum and S. S. Davis, *Polymers in Controlled Drug Delivery,* John Wright, Bristol, 1987.

D. F. Williams, *Critical Reviews in Biocompatibility Series,* CRC Press, Inc., Boca Raton, Fla.

Adv. Polym. Sci. **57,** 1 and 51 (1983).

H. Ringsdorf, B. Schlarb, and J. Venzmer, *Angew. Chem. Int. Ed.* **27,** 113–158 (1988).

LEO GROS
Fachhochschule Fresenius
Wiesbaden

BIOMASS.

See POLYMERS FROM RENEWABLE SOURCES.

BIOMATERIALS

Biomaterials are used to repair, restore, or replace damaged or diseased tissue or to interface with the physiological environment. Synthetic biomaterials, ie, metals, ceramics, and polymers, are often called biomedical to differentiate them from natural biomaterials. This article is confined to synthetic polymeric biomedical materials.

Recent advances in polymer science and surgery now make it possible to rebuild many parts of the human body. Polymeric materials having a vast spectrum of properties are now available. Some polymers have mechanical and surface properties that make them suitable as biomedical materials.

Biomedical polymers in medicine and surgery are currently widely used, and include intracorporeal, paracorporeal, and extracorporeal applications (inside, interfacing, and outside the body, respectively). Applications include:

Intracorporeal (implanted) materials

- temporary devices such as surgical dressings and sutures
- simple semipermanent devices, eg, tendons and reinforcing meshes
- complex devices simulating physiological processes such as the artificial kidney–blood dialysis

Paracorporeal or extracorporeal materials

- catheters and blood bags
- syringes and surgical instruments
- disposables for health-care delivery

Biomaterials science encompasses aspects of physical sciences, engineering, biology, and medicine (see also BIOLOGICAL ACTIVITY; DRUG APPLICATIONS; MEDICAL APPLICATIONS).

Biocompatibility

Traditionally, the primary focus of the search for biomedical materials has been on finding biocompatible materials. The original definition of a biocompatible biomaterial as one that is inert toward the physiological environment has been modified to include materials having minimal interaction with the environment. It is unlikely that an absolutely inert biomaterial can be found, but the physiological response to any biomaterial must be kept within acceptable bounds.

Biomedical Polymers

To choose a polymer for use in biomedical application, the scientist or clinician must specify which physical properties are required. Design parameters are based on a thorough understanding of the physiological functions and conditions under which the device must operate. Considerations include mechanical properties, purity, fabrication, stability, and tolerability.

Characterization and Evaluation. Today, any synthetic polymer system used in an implant should be fully characterized in order to establish structure–response relationships and reproducibility of results.

Sterilization

The method of sterilization of biomedical polymeric devices must not affect material properties, contaminate the device, or alter its geometry. Sterilizability is often overlooked as a property that is essential to the successful use of any biomedical polymer.

Sterilization techniques are categorized as physical or chemical. Physical methods include dry heat, steam, and ionizing radiation. Chemical sterilization methods are also widely used. Ethylene oxide is effective at relatively low temperatures, with good penetration into porous materials. Other chemical sterilants include ozone, propylene oxide, formaldehyde, glutaraldehyde, β-propiolactone, phenols, quaternary salts, and hypochlorite. All must be removed completely before use.

Physiological Defense Mechanisms

The defense mechanisms of the body must be considered in the application of implant materials. Although these processes are often discussed separately, physiological responses do not exhibit uniquely differentiable stages. The responses are continuous and often integrated among various defense mechanisms against injury or invasion of foreign materials. Defense mechanisms include inflammation and wound healing, the blood-clotting system, the kinin system, the immune system, and the complement system.

Biological Responses to Implants

Performance in an implant or extracorporeal device is the ultimate test of a biomaterial. In addition to performance, the evaluation should determine any adverse reactions and short- and long-term effects of the physiological environment on the implant. Complex relationships exist between materials properties, design and implantation parameters, and the biological response.

Blood–Materials Interactions. The correlation between biomaterial surface and blood compatibility has been much studied. Several approaches to nonthrombogenic surfaces have emerged, each emphasizing certain surface properties: surface texture; electrostatic effects; interfacial properties; preferential protein adsorption; microtextured and endothelial-seeded surfaces; "bioactive" heparinized and immobilized enzyme surfaces; and sustained drug-release surfaces.

Immunity and Infection. Foreign materials may disturb responses associated with the immune and complement systems, especially in regard to infection.

Mutagenesis and Carcinogenesis. Implant-related neoplasms seem to be rarer in human subjects than in animals. At present, the risk of implant-induced neoplasms appears very small, but long-term epidemiological studies are needed.

Inflammatory and Wound-healing Response. Acute inflammation defends the body in the case of trauma and reestablishes a physiological environment in which the wound-healing process can take place. The chronic inflammatory response to an implant is affected by its chemical, mechanical, and surface characteristics, and is usually associated with encapsulation in fibrous tissue. Soft tissue response is sensitive to surface texture.

Implant Design and Application

The proper design of an implant is critical, even if a suitable biomedical polymer is being used for its fabrication. Improper design can compromise device performance if it causes the implant to disturb the normal physiological environment.

The potential market for polymers in this area has been estimated at $\$10^9$. Achieving this will require greater knowledge of how polymers and the body interact at the molecular, cellular, and organ levels, and the long-term effects of the material. Some requirements are set forth in the Medical Device Amendments of 1976 to the Federal Food, Drug and Cosmetic Act. Trials on humans are needed. However, progress will be achieved only at the cost of risk and occasional failure.

BIBLIOGRAPHY

D. J. Lyman and W. J. Seare, *Ann. Rev. Mater. Sci.* **4,** 415 (1974).

J. Black, *Biological Performance of Materials,* Marcel Dekker, Inc., New York, 1981, p. 3.

H. E. Kambic, S. Murabayashi, and Y. Nose in M. Szycher, ed., *Biocompatible Polymers, Metals and Composites,* Technomic Publishing Co., Inc., Lancaster, Pa., 1983, pp. 179–198.

Fed. Regist. **42**(109), 22620 (June 4, 1976).

DONALD J. LYMAN
STEPHEN M. ROWLAND
The University of Utah

BIOPOLYMERS

Biopolymers include all naturally occurring, large, polymeric molecules, such as polypeptides (qv) and proteins (qv), nucleic acids or polynucleotides (qv), and polysaccharides (qv). All of these natural macromolecules are formed through condensation reactions. The term biopolymers is also applied to synthetic polymers prepared from the same or similar monomer units as the natural polymers. Each of the categories of biopolymers is covered by at least one comprehensive article in the *Encyclopedia.*

BIOSYNTHESIS. See PROTEINS, BIOSYNTHESIS.

BIOTECHNOLOGY. See ENZYMES; ENZYMES, IMMOBILIZED; POLYNUCLEOTIDES.

BIREFRINGENCE

Birefringence is the difference in the refractive indexes of two perpendicular directions in a given material. For polymeric systems, the phenomenon is commonly observed by applying an anisotropic macroscopic deformation and measuring the difference between the refractive indexes of the sample along two perpendicular directions. The birefringence observed in this manner originates essentially from the anisotropy of the individual chains constituting the medium. The necessary conditions for a chain to display birefringence are that the segments or the repeat units of the chain should be optically anisotropic, they should display anisotropic polarizabilities, and the spatial configuration of the chain should be anisotropic. The first condition is fulfilled in all polymeric species because of the inherent anisotropy of the structure of the repeat unit. The second condition is obtained by the presence of anisotropic microstructures or by the imposition of a macroscopically anisotropic deformation on the system. The chains then exhibit different polarizabilities in different directions. The sum of the polarizabilities of all chains leads to the total polarizability of the polymeric sample.

Birefringence is measured in the solid (strain birefringence) or the liquid (flow or streaming birefringence). The method has been applied to various polymeric systems for the determination of molecular parameters such as the optical anisotropy of segments of polymeric molecules, the orientation of segments under stress, and the optical response in relation to molecular morphology. The phenomenon of birefringence has been reviewed by various authors.

The technique of birefringence finds a wide spectrum of applications in the area of polymer research, ranging from the determination of structural defects in polymeric solids to more basic investigations of molecular and morphological properties. In this article, problems relating to the latter are emphasized.

The birefringence of a deformed polymeric system results from various contributions, such as the intrinsic birefringence of the chains, intermolecular correlations among chains, correlations among solvent and polymer molecules, and form birefringence. The intrinsic birefringence results from orientations of the anisotropic segments of the chains. The average intrinsic birefringence of the chain may be calculated by the application of statistical methods. Local correlations among neighboring segments of different chains are a source of birefringence observed in polymeric systems, in addition to intrinsic birefringence. Similarly, correlations among anisotropic solvent molecules and chain segments give rise to a further contribution. If the medium contains two or more phases, each having a different refractive index and at least one dimension comparable with the wavelength of light, a positive contribution to birefringence results, called the form effect. This effect is of particular importance in block copolymers, semicrystalline polymers, and in single chains subject to a flow field.

Optical Properties of the Single Chain

Polarizability of the Repeat Unit. Application of an electric field $\mathbf{E}_0$ on a polarizable medium induces a dipole moment $\mathbf{m}$, which is the difference between the dipole moment of the charges of the body before and after the application of the field $\mathbf{E}_0$. In general, the dipole moment produced is proportional to the applied field:

$$\mathbf{m} = \boldsymbol{\alpha}\,\mathbf{E}_0 \tag{1}$$

where $\boldsymbol{\alpha}$ is a second-order tensor called the polarizability tensor. Components of $\boldsymbol{\alpha}$ in terms of a coordinate system Oxyz are

$$\boldsymbol{\alpha} = \begin{bmatrix} \alpha_{xx} & \alpha_{xy} & \alpha_{xz} \\ \alpha_{yx} & \alpha_{yy} & \alpha_{yz} \\ \alpha_{zx} & \alpha_{zy} & \alpha_{zz} \end{bmatrix} \tag{2}$$

where α_{xx} is the polarizability in the direction of Ox for a field applied in the direction Ox, and α_{xy} is the polarizability in the direction Ox for a field applied in the direction Oy. The definitions of the other components of $\boldsymbol{\alpha}$ follow similarly.

The polarizability tensor for a bond is cylindrically symmetric. If the x axis is chosen along the length of the bond, $\boldsymbol{\alpha}$ may be expressed as

$$\boldsymbol{\alpha} = \begin{bmatrix} \alpha_{\parallel} & 0 & 0 \\ 0 & \alpha_{\perp} & 0 \\ 0 & 0 & \alpha_{\perp} \end{bmatrix} \tag{3}$$

where $\alpha_{\parallel}$ and $\alpha_{\perp}$ are the components of polarizability along parallel and perpendicular directions to the bond.

The polarizabilities $\alpha_{\parallel}$ and $\alpha_{\perp}$ of a bond may be determined by various experimental techniques in the gaseous, liquid, or crystalline phases. For the C—C bond, the reported values of $\Delta\alpha$ vary between 4.4×10^{-25} and 18.6×10^{-25} cm^3. The larger value was favored first, but later experimental work led to a lower value of 9.5×10^{-25} cm^3. Such large discrepancies in the values reported by different workers are often observed for other bonds as well, indicating a basic difficulty of quantitative birefringence studies.

The polarizability tensor of a molecule as a whole is obtained by tensorial addition of polarizabilities of each bond. This mode of adding the contributions from each bond of the molecule constitutes the basis of the valence-optical scheme. Although the bond additivity scheme provides a useful means of calculating the total polarizability of a chain, short-range inductive effects among bonds act as sources of polarizability commensurate with intrinsic polarizabilities of bonds. Improvements in the calculation of chain polarizability have been suggested recently. According to this scheme, the independently measured polarizabilities of larger groups of the chain are tensorially added, instead of those of the individual bonds.

The polarizability of a chain depends on its configuration. The average polarizability of a chain with two ends fixed in space is required for the formulation of the birefringence of a polymeric medium.

The Freely Jointed Chain. Figure 1 shows one end of a freely jointed chain fixed at the origin of a coordinate system Oxyz. The vector $\mathbf{r}$ denotes the vector from the origin to the other end of the chain, and Ox refers to an arbitrary fixed axis. The chain may take all possible configurations, subject to the condition that the projection of its chain vector $\mathbf{r}$ on x is fixed. The projection of $\mathbf{r}$ on x is denoted r_x. The bonds of the chain are assumed to be identical and are denoted by l. The projection of each bond on x is denoted by l_x. The number of bonds n_x having orientations such that their projections on x are in the range l_x to $l + dl_x$ is

$$dn_x = (n\beta/\sinh\beta)\exp(\beta l_x/l)(2l)^{-1}dl_x \tag{4}$$

where n is the total number of bonds of the chain and β is obtained from the solution of the expression

$$\coth\beta - 1/\beta = r_x/nl \tag{5}$$

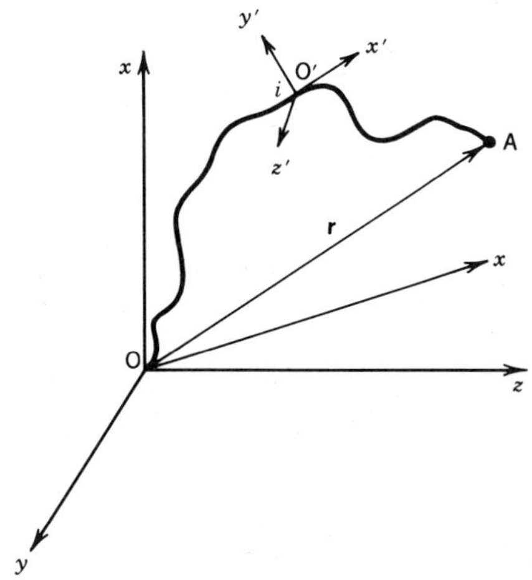

Figure 1. Coordinate systems for the chain OA; **r** denotes the end-to-end vector; Oxyz is the laboratory-fixed coordinate system; Ox is an arbitrary fixed axis. The primed system O′x′y′z′ is affixed to the ith unit.

The left side of equation 5 is called the Langevin function of β, $\mathcal{L}(\beta) \equiv \coth \beta - 1/\beta$. Equation 5 is written in terms of the Langevin function as

$$\mathcal{L}(\beta) = r_x/nl \tag{6}$$

β may be obtained formally from equation 6 as

$$\beta = \mathcal{L}^*(r_x/nl) \tag{7}$$

where $\mathcal{L}^*$ is the inverse Langevin function.

The average square of the cosine of the angle between a bond and the x axis is given as

$$\begin{aligned}\langle(l_x/l)^2\rangle &= \beta(\sinh \beta)^{-1} \int_{-1}^{1} \exp (\beta l_x/l)(l_x^2/2l^3)dl_x \\ &= 1 - 2\mathcal{L}(\beta)/\beta\end{aligned} \tag{8}$$

where the brackets denote averaging over all configurations of the chain subject to the stated conditions. Inasmuch as all bonds of the freely joined chain are equivalent, the expression given by equation 8 is valid for any bond along the chain. Using equations 7 and 8 in equation 9 leads to

$$\langle(l_x/l)^2\rangle = 1 - 2(r_x/nl)/\mathcal{L}^*(r_x/nl) \tag{9}$$

The average squared cosine of the angle between the bond and the end-to-end vector **r** is required for treatment of birefringence. For $r/nl \ll 1$, and $n \gg 1$, r_x in equation 9 may be replaced by r. If Φ is the angle between the bond and the x axis, the following expression is obtained for $(\cos^2 \Phi)$

$$\langle\cos^2 \Phi\rangle = 1 - 2(r/nl)\mathcal{L}^*(r/nl) \tag{10}$$

Series expansion of the right side of equation 10 gives

$$\langle\cos^2 \Phi\rangle \cong 1/3 + (2/5)(r/nl)^2 + (24/175)(r/nl)^4 + \cdots \tag{11}$$

For a freely jointed chain, the bond polarizability tensor is cylindrically symmetric. In a coordinate system with axes parallel and perpendicular to the bond, α is given by equation 3 for each bond. The average polarizability α_r of the chain along the fixed chain length is then given as

$$\alpha_r = n[\alpha_\parallel - 2\,\Delta\alpha(r/nl)/\mathcal{L}^*(r/nl)] \tag{12}$$

Inasmuch as the freely jointed chain is symmetric about the fixed end-to-end vector, the transverse polarizability of the chain has the same value along any transverse direction. The average polarizability of a chain in a reference system where one axis is along the fixed chain vector and the other two axes are transverse to it is

$$\alpha_r = \begin{bmatrix} \alpha_r & 0 & 0 \\ 0 & (3\overline{\alpha} - \alpha_r)/2 & 0 \\ 0 & 0 & (3\overline{\alpha} - \alpha_r)/2 \end{bmatrix} \tag{13}$$

where $3\overline{\alpha}$ is the sum of the diagonal components of α_r and $(3\overline{\alpha} - \alpha_r)/2$ is a transverse component. The difference between α_r and $(3\overline{\alpha} - \alpha_r)/2$ is then obtained as

$$\Delta\alpha_r = (3/2)(\alpha_r - \overline{\alpha}) \tag{14}$$

Substituting equation 12 into equation 14 and using $\overline{\alpha} = n(\alpha_\parallel + 2\alpha_\perp)$,

$$\begin{aligned}\Delta\alpha_r &= n\,\Delta\alpha[1 - (3r/nl)/\mathcal{L}^*(r/nl)] \\ &= n\,\Delta\alpha(3/5)(r/nl)^2[1 + (12/35)(r/nl)^2 \\ &\quad + (36/175)(r/nl)^4 + \cdots]\end{aligned} \tag{15}$$

Assuming $n \gg 1$ and $r/nl \ll 1$, equation 15 simplifies to

$$\Delta\alpha_r = (3/5)\,\Delta\alpha(r^2/nl^2) = \Gamma_2 r^2/\langle r^2\rangle_0 \tag{16}$$

where $\Gamma_2 = 3\,\Delta\alpha/5$. Equation 16 is the basic equation for the polarizability difference $\Delta\alpha_r$ for a freely jointed chain, which is used to obtain the polarizability of a polymeric medium.

The Real Chain. The expression given by equation 16 is for a hypothetical freely jointed chain. Derivation of $\Delta\alpha_r$ for a real chain has been treated, most notably by Flory. The theory of birefringence has been developed to a large extent by using the concept of the freely jointed chain. Only in a few cases has the real character of the chains been taken into consideration.

Amorphous Rubber Networks

An amorphous rubber network at a fixed deformation forms a suitable medium for the analysis of the birefringence of polymeric chains. The relationship of the measured birefringence of the macroscopic network to that of the single chain is required for this purpose. Additionally, the state of deformation at the microscopic, molecular level must be related to the externally applied macroscopic state of strain. Because of these considerations, a molecular model of the network is needed. The molecular theory of networks (qv) seems to be satisfactory in describing the qualitative aspects of birefringence. Intermolecular correlations and solvent–polymer interactions in the network are sources of difficulties in applying this technique to quantitative determination of chain polarizabilities.

Semicrystalline Polymeric Solids

Deformation mechanisms in semicrystalline solids are analyzed by determining the relative orientations of the amorphous and the crystalline phases under a macroscopically imposed strain. Determination of the overall birefringence of a sample is one of several techniques used to obtain information on the behavior of the two phases. The other techniques are wide-angle x-ray diffraction, sonic wave propagation, small-angle light and x-ray scattering, and infrared dichroism. The complex structure of the semicrystalline solid often requires a combination of these experimental techniques.

Birefringence of Amorphous Polymers in the Glassy State and the Glass–Rubber Transition Region

Unstressed glassy polymers are usually free of macroscopic birefringence. However, fast cooling after processing as in injection molding and insufficient annealing can result in residual stresses and consequently residual birefringence. It is common practice to examine moldings with polarized light to study thermal strains and orientation produced by the molding process qualitively. A quantitative analysis is generally too complicated.

In rubbers, orientational birefringence arises when optically anisotropic links (eg, chemical bonds) are physically ordered toward the direction of extension. On the basis of Gaussian chain statistics, the theories predict a relation between birefringence, stress, and yield for the orientational stress-optical coefficient B_{or}.

For uncross-linked rubbery polymers in the viscoelastic region extending from the low modulus end of the glass–rubber transition to the terminal zone, the stress-optical coefficient is independent of time during stress relaxation and creep tests and equal to the equilibrium value B_{or}. In the nonequilibrium form the orientational birefringence Δn_{or} and stress σ_{or} are written as a function of time.

$$\Delta n_{or}(t) = B_{or}\tau_{or}(t) \qquad (17)$$

In the rubbery state, orientation birefringence is related to the entropy-elastic stress.

The deformation of polymers in the glassy state is very different from that in the rubbery state; it arises from short-range processes such as valence bond twisting, twisting about single bonds, and change in intermolecular distances, in contrast to the long-range orientational processes occurring in the rubbery state. These local deformations give rise to energy-elastic distortional birefringence Δn_d and to distortional stress σ_d, which, under equilibrium conditions, determine the distortional stress-optical coefficient $B_d = \Delta n_d/\sigma_d$.

The molecular origins of distortional birefringence are not well understood, although the anisotropic local electric field can yield a significant birefringence contribution and the local tilting of anisotropic side groups toward the directions of extension can produce a net distortional birefringence of either positive or negative sign. During relaxation under nonequilibrium conditions, distortional stress and birefringence exhibit similar time dependence and the stress-optical coefficient is independent of time.

Flow Birefringence

Flow birefringence (also called streaming birefringence) is the most commonly employed method of determining the birefringence and polarizabilities of single polymeric chains. Polymeric molecules dissolved in suitable solvents are subjected to a flow field, usually in the form of shear flow. The polymeric coils undergo deformation and tend to orient along the lines of flow. Inasmuch as the chains are optically anisotropic, the distortion applied by the flow field results in birefringence. An advantage of flow birefringence over experiments in the bulk state is that relatively dilute solutions may be used and hence the contribution of intermolecular interactions can be minimized. The fundamental problem in interpreting flow birefringence results is to relate the degree of anisotropy induced in the coils to their hydrodynamic interactions with the surrounding fluid undergoing flow. Various models have been proposed to describe the hydrodynamic interactions.

In a flow birefringence experiment, the difference of birefringence along two perpendicular directions in a plane (eg, the xy plane) is measured. A suitable device for measuring flow birefringence consists of two concentric cylinders; the inner cylinder rotates while the outer cylinder remains stationary, as shown in Figure 2.

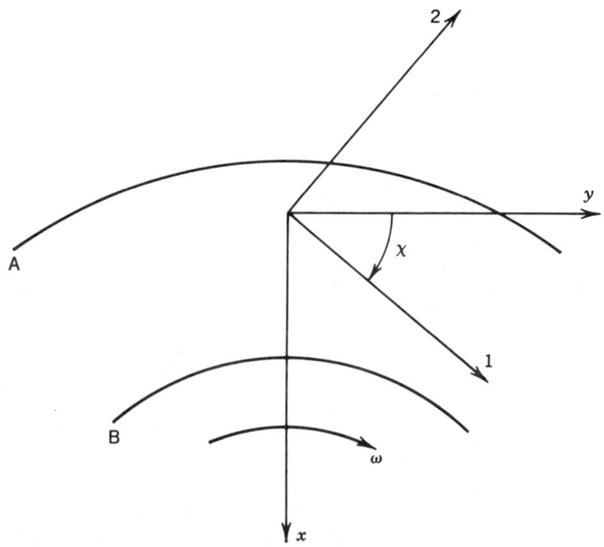

Figure 2. Schematic description of a flow birefringence experiment in shear. The fluid is confined in the region between two cylinders represented by the curves A and B. Cylinder B rotates with angular velocity ω. The y axis indicates the flow direction. Axes 1 and 2 are the principal optical directions.

Combination of flow birefringence and other techniques of physical characterization such as translational diffusion, viscosimetry, sedimentation, and light scattering can often remove ambiguities associated with any method used alone. The use of flow birefringence to characterize flexible chain macromolecules is limited by inadequacies in available theories. Qualitative information on chain conformation is obtained by comparison of birefringence data with flexible-chain theories. Chain stiffness can be determined from the theory of Kuhn for the elastic dumbbell with internal viscosity or from the theory for flexible chains.

Flow birefringence is a useful technique to study the rheology of dilute and concentrated polymer solutions and melts. Streaming birefringence has been applied to the investigation of macromolecular deformation and orientation in various flows of polymeric liquids. Obviously, only clear fluids can be investigated. Systems containing filler particles are excluded.

Birefringence is an indirect method of measuring stress distribution in a flow. Extensive studies have established proportionality between the deviatoric components of the refractive index tensor and the stress tensor. There is also no observable time lag between stress and birefringence. Rheological investigations on polymer melts are based on this linear stress-optical rule.

In most experiments relevant for practical situations in polymer processing, stresses appear to be too small for deviations from the linear stress-optical rule. This holds in particular for all experiments in shear flow. Birefringence can then be applied to obtain optical equivalents of viscosity, first normal stress difference, and equilibrium shear compliance. These data are used to test rheological equations, molecular theories, and rheological models to obtain stress distribution and orientation in flow or to investigate processing problems, such as sharkskin defects.

Apparatus for Measuring Birefringence

Birefringence experiments for polymeric solids are generally performed by applying simple elongation to a thin strip or film and measuring the difference Δn of the refractive indexes along the direction of stretch and the lateral direction. For the determination of Δn two general methods are used: the compensator and transmission methods. An experimental apparatus used in the compensator method is shown in Figure 3. The instrumental setup for the transmission technique is similar to that shown in Figure 3. The compensator *B* is removed from the setup in this method.

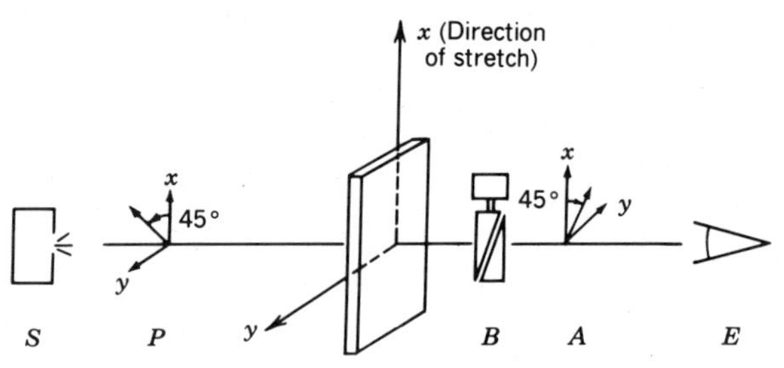

Figure 3. Apparatus for measuring birefringence.

Flow birefringence techniques for polymer melts are associated with mechanical techniques in rheometry: extensional rheometers for elongational viscosity and birefringence measurements, apparatus for stagnation flow in one-sided compression, cone-and-plate apparatus, and slit and capillary rheometers.

BIBLIOGRAPHY

G. H. Meeten, ed., *Optical Properties of Polymers,* Elsevier Applied Science Publishers, Ltd., Barking, UK, 1986.

H. Janeschitz-Kriegl, *Polymer Melt Rheology and Flow Birefringence,* Springer-Verlag, Berlin, 1983.

I. M. Ward, ed., *Development in Oriented Polymers,* Elsevier Applied Science Publishers, Ltd., Barking, UK, 1982.

P. J. Flory, *Statistical Mechanics of Chain Molecules,* Wiley-Interscience, New York, 1969.

B. Erman
Bogazici University

Jean-Pierre Queslel
Manufacture Michelin

BLEACHES, OPTICAL. See Paper.

BLEACHING

Bleaching of Wood Pulp

Bleaching refers to the manufacturing process used to improve whiteness and purity in wood pulp. Traditional bleaching processes are based on chemical modification of the raw wood pulp. The components responsible for the poor color or purity are either removed or made harmless. For unbleached pulps produced by chemical means, such as the kraft process, the reactions are degradation and removal of the offending groups. For pulps produced by mechanical means, such as the groundwood process, the principal reactions involve chemical modifications that decrease color and improve whiteness without significant removal of material.

Slightly more than 50% of the total chemical pulp produced in the United States is bleached; on a world basis the percentage is greater. Over 90% of the pulp sold on the open market, as opposed to captive use, is bleached. The effectiveness of bleaching is measured in terms of brightness, which is defined as the percent reflectance from a dry-pulp sheet on exposure to a beam of light with wavelength centered at 457 nm.

Bleaching Technology

The active chemicals used in bleaching are inorganic oxidizing or reducing agents. In general, bleaching proceeds in several stages. Each stage consists of the following steps: mixing of pulp and reactants; a dwell time at the desired temperature to allow the reactions to go to the required degree of completion; and a wash stage to remove both the exhausted chemical and the degraded wood components, eg, lignin fragments. Multistage sequences require less bleaching chemical and allow different chemicals of varying strength and cost to be used in the most efficient manner.

Multistage sequences, widely used today, are tailored to the demands of a given manufacturing process. A typical sequence for bleaching softwood pulp consists of an initial chlorine (C) stage; an extraction stage (E_1) with sodium hydroxide; a stage based on chlorine dioxide (D_1); a second caustic extraction stage (E_2); and a second chlorine dioxide stage (D_2).

Equipment

The equipment used in a classic bleach plant is outlined in Figure 1.

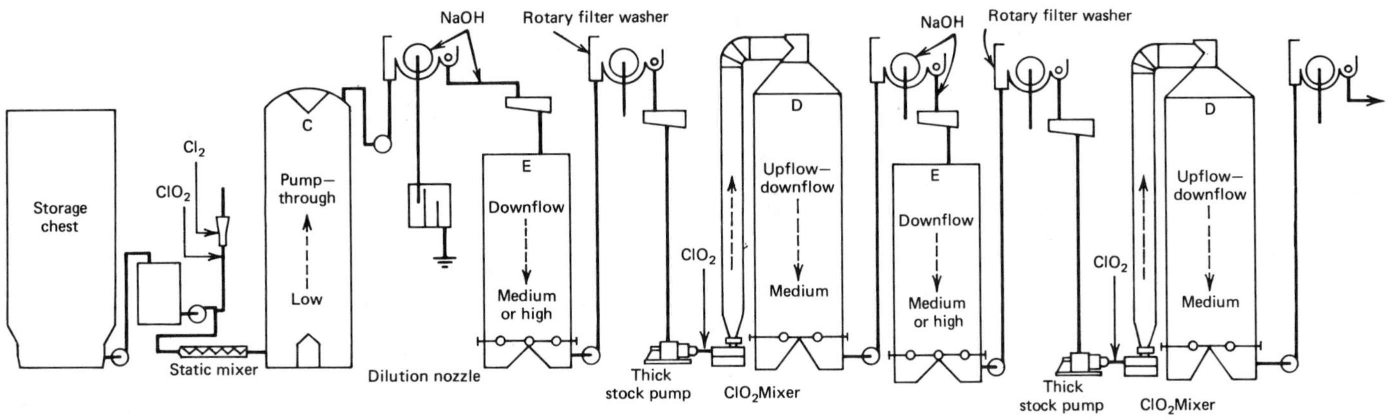

Figure 1. Diagram of a C/D EDED bleach plant.

Bleach Stages

Chemical Pulp. ***Chlorination.*** The principal purpose of chlorination is to remove lignin (qv). Like chlorination, the oxygen stage removes lignin.

Caustic Extraction. Alkaline extraction's primary function is the dissolution of lignin fragments from the previous stage; it normally follows the chlorination and first chlorine dioxide stages.

Chlorine Dioxide. The high brightness demanded of well-bleached pulps is attained by treatment with chlorine dioxide, which is selective in its reactions with lignin and therefore causes little or no loss of pulp strength as the residual lignin is oxidized.

Hypochlorite. The hypochlorite stage normally follows a CE sequence. A brightness of ca 80% can be obtained, but cellulose is severely attacked. Hypochlorite treatment reduces chlorine dioxide requirements in subsequent stages. Other bleach chemicals include ozone and hydrogen peroxide.

Visible particles or specks are undesirable in bleached pulps. Specks usually can be traced to particles that have not responded as quickly to bleaching chemicals as the rest of the pulp. Often, the only reason to attain a brightness of 90% is to ensure bleaching of slower reacting particles. Particles should be removed from the pulp before bleaching. Screening, eg, with a vortex cleaner, increased cooking and washing temperatures, and antislime agents are employed.

High-yield Mechanical Pulp. High-yield mechanical pulps are produced by the traditional stone groundwood process or on the more recent disk-refining systems. Brightness is lower than that obtained with full chemical pulps; a value of 80% is considered high. The main source of light-absorbing chromophores in mechanical pulps is in the lignin molecule. They can be reduced or oxidized to brighten the pulp. Both oxidative and reductive systems are used commercially. As in chemical pulp bleaching, multistage sequences are used, generally two stages. The oxidative stage, eg, hydrogen peroxide, is followed by a reductive stage, eg, dithionite (hydrosulfite). Other reagents have been shown to work on a laboratory scale, but only hydrogen peroxide and sodium hydrosulfite are used commercially.

Economic Aspects

The cost of operation for bleached pulp manufacture is highly site specific. Some representative cost figures are given in Table 1 for a C/D EDED and an OC/D EDED sequence.

Table 1. Economic Consideration of Bleach Plant (Pine Kraft) with a High Consistency Oxygen Stage, Capacity 750 t/d (a.d.)[a]

	Sequence	
Operating conditions	I C/D EDED	II OC/D EDED
Consumption, kg		
O O_2		17
oxidized white liquor		23
C/D Cl_2,[b]	56	28
ClO_2 as active Cl_2	14	7
E NaOH	25	17
D ClO_2 as active Cl_2	30	15
E NaOH	10	10
D ClO_2 as active Cl_2	10	10
Cost in 1982 U.S. \$/t (a.d.)		
chemicals	44.5	31.5
energy	10.5	14.5
labor, maintenance	5.5	7.0
capital (25% annuity)	19.5	27.5
Total	*80.0*	*80.5*

[a] a.d. = air-dry.
[b] Active chlorine defined as equivalent theoretical oxidizing power of Cl_2, 1 kg ClO_2 = 2.63 kg active Cl_2.

Health and Safety Factors

Many bleaching operations are potentially hazardous because of the highly reactive chemicals being used. Safe operation requires detailed knowledge of their toxic and corrosive properties. All manufacturers now provide safety data on their products, and many government and private organizations also provide safety information.

BIBLIOGRAPHY

R. P. Singh, ed., *The Bleaching of Pulp,* Tappi Press, Atlanta, Ga., 1979.

1985 Bleach Plant Operations Seminar, Tappi Press, Atlanta, Ga., 1985.

Proceedings of 1985 International Pulp Bleaching Conference, CPPA TAPPI, Tappi Press, Atlanta, Ga., 1985.

N. Irving Sax, *Dangerous Properties of Industrial Materials,* 3rd ed., Reinhold Publishing Co., New York, 1968.

EARL W. MALCOLM
The Institute of Paper Chemistry

Bleaching of Textiles

Chemistry

The most important aim of the bleaching operation is the attainment of whiteness. Another result is improved absorbency, attained to some extent by scouring. Because some of the chemical ingredients are identical, it is difficult to separate bleaching from scouring. Scouring is the saponification and emulsification of the natural oils, fats, and waxes that coat the fibers, and the synthetic oils and fats added as lubricants. Synthetic paraffin waxes are not saponifiable because they do not contain ester linkages. Removing the hydrophobic fats and waxes increases water absorbency.

Bleaching agents convert colored impurities to water-soluble or colorless particles. Color is imparted by a chromophore, ie, a moiety usually involving alternating carbon–carbon single and double bonds. Bleaching destroys these double bonds by addition (saturation) or rupture. Once the free electrons of a double bond are tied up by substitution or by rupturing the double bond, electromagnetic radiation is not absorbed and reflected in the visible region of the spectrum, and color ceases to exist. Bleaching agents used include oxygen bleaches (peroxides, perborates, and peracetic acid), chlorine bleaches (chlorine dioxide and sodium hypochlorite), and reducing bleaches.

Processing

The steps for chemical bleaching usually include wetting or retention in a machine to allow time for the chemicals to act, followed by washing to remove the active chemical and byproducts of the bleaching action. Equipment used includes saturators, J-boxes; steamers; and various types of washers.

Fibers

Although a number of chemical systems are available for bleaching, hydrogen peroxide is preferred for most fibers and is probably used in 90–95% of all bleaching in the United States. In fiber blends, the most sensitive fiber governs the choice. A guide to the selection of bleaching systems is given in Table 1.

Table 1. Guide to Fiber Bleaching

Fiber	Hydrogen peroxide[a]	Sodium chlorite[a]	Peracetic acid[a]	Persulfates[a]
cotton	1	1	1	3–
wool	1	4	1	2
silk	1	4	1	4
viscose	1–3–	1	1	3–
acetate	2–3	1	1	4
triacetate	2–3	1	1	4
nylon	2–3	1	1	4
polyester	3–3–	1	1	3–
acrylic	4	1	1	4
polyurethane	1	4	2	4

[a] 1, suitable; 2, suitable under specific conditions; 3, compatible and may be used in blends; 4, not suitable.

Energy Consumption

A textile bleachery requires chemicals, vast quantities of water, steam, electricity, labor, and expensive equipment. A study for the Department of Energy revealed that preparation accounts for 17% of the total energy consumed in textiles. In the United States, an estimated 90 PJ/yr (8.5×10^{13} Btu/yr) were consumed in the late 1970s, and the technology has not been changed.

Handling of Bleaching Chemicals

Bleaching chemicals must be handled with care, using safety goggles, rubber gloves, aprons, and boots. They should be stored in a cool, dry place away from contamination.

BIBLIOGRAPHY

B. K. Easton, *Ciba-Geigy Rev.* (3), 3 (1971).

B. L. McConnell, W. Prager, and B. K. Easton, *Text. Chem. Color.* **14**(4), 11 (1982).

K. V. Datye and A. A. Vaidya, *Chemical Processing of Synthetic Fibers and Blends,* John Wiley & Sons, Inc., New York, 1984.

M. Lewin and S. B. Sello, *Chemical Processing of Fibers and Fabrics,* Vol. 1 of *Handbook of Fiber Science and Technology,* Marcel Dekker, Inc., New York, 1984.

GARY N. MOCK
North Carolina State University

BLENDING.

See POLYMER BLENDS.

BLENDS OF POLYMERS.

See POLYMER BLENDS.

BLOCK COPOLYMERS

Block copolymers are defined as having a linear arrangement of blocks of varying monomer composition. That is, a block copolymer is a combination of two or more polymers joined end-on-end. Star-block or radial-block copolymers have branched structures and are not considered here (see BRANCHED POLYMERS). Graft copolymers (qv), although their properties might be similar to those of block copolymers if the number of grafts is limited, are also discussed in a separate article (see also DENDRITIC POLYMERS: ELASTOMERS, THERMOPLASTIC; POLYESTERS, ELASTOMERIC).

The increasing importance and interest in block copolymers arises mainly from their unique properties in solution and in the solid state, which are a consequence of their molecular structure. In particular, sequences of different chemical composition are usually incompatible and therefore have a tendency to segregate in space. Amphiphilic properties in solution and microdomain formation in the solid state are directly related to this specific molecular architecture, which can be designed by using existing monomers or polymers (see also COMPATIBILITY; MICROPHASE STRUCTURE; POLYMER BLENDS).

As an indication of interest and activity in block copolymers, more than a thousand patents were filed between 1976 and 1982, and the U.S. consumption of thermoplastic elastomers, which represents only a part of the total block copolymer production, was approximately 400,000 metric tons per year by 1985, with a growth of 9%/yr.

Synthesis Strategies

There are two general methods for the synthesis of block copolymers. In the first, active sites on a macromolecular chain are created that then initiate the polymerization of a monomer. The polymerization may be radical, anionic, or cationic, and for one active site, can be represented as follows:

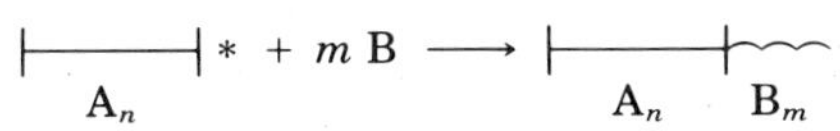

The second method, which is usually called condensation, is the reaction between chemical functional groups present at the ends of different polymers. Schematically,

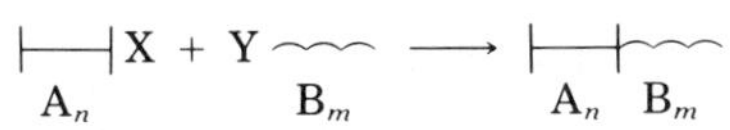

With difunctional species, the reactions lead to multiblock copolymers, also called segmented copolymers.

Radical Polymerization. Free-radical active sites can be generated using polymeric azo or peroxide derivatives as well as inifertos, eg, involving photolabile sulfur–carbon bonds. The use of polymers with end groups that can be converted into radicals by organometallic compounds is another effective approach.

Polyinitiators. Polyinitiators, eg, polyazo derivatives, in which the initiating groups are in the chain rather than the end groups, can be used for the synthesis of block copolymers. The general formula of such initiators is $(\!-R{-}N{=}N{-}R{-}N{=}N-\!)_x$, where R may be of low molecular weight, or R may be a precursor sequence of the block copolymer. A typical example of a polyinitiator where R is small is the following polyazoester:

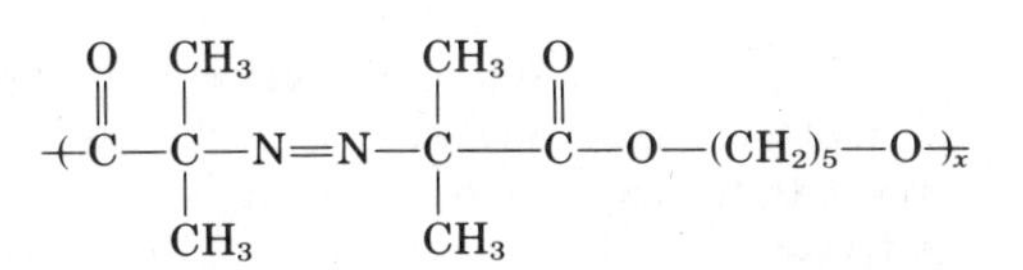

There are many other thermal, photolytic, and mechanical methods of generating radicals on the chain for the initiation of a second block.

Anionic Polymerization. Anionic polymerization (qv) is one of the preferred techniques for preparing monodisperse polymers (qv) and well-defined block copolymers because under proper conditions it lacks a termination or transfer step, ie, the system is "living." This allows the preparation of nearly pure block copolymers of desired molecular weight, composition, and structure. The most extensively studied block copolymers prepared by anionic polymerization are synthesized from monomers such as styrene, butadiene, α-methylstyrene, and isoprene with monofunctional initiators such as butyllithium, phenylisopropylpotassium, or with difunctional initiators such as alkali metal naphthalides.

Polymerization Using Coordination Catalysts. Polymerization using Ziegler-Natta catalysts (qv) has been reviewed. To prepare block copolymers, the catalyst must have a lifetime sufficient to permit sequencing of the monomers (particularly for $(AB)_n$- and ABA-type copolymers), and chain-transfer reactions must be minimal. Otherwise, the reaction product will be, at best, a mixture of homopolymers and block copolymer.

Cationic Polymerization. Whereas in the past copolymers prepared by cationic polymerization (qv) were ill-defined in structure and contaminated by a large amount of homopolymer, better control of initiation, propagation, and termination has recently been achieved. Examples of monomers copolymerized by cationic techniques are isobutylene with styrene, isobutyl vinyl ether with α-methylstyrene or *N*-vinylcarbazole, and 1,3-dioxolane with 1,3-dioxepane.

Active Centers Transformation Reactions. An alternative approach, which, in principle, should greatly extend the range of possible monomer combinations in block copolymers, is to devise processes by which the polymerization mechanism can be changed at will to suit the monomers being polymerized sequentially. The different mechanisms of polymerization already discussed can be interconverted if suitable redox reactions are developed. Such reactions, called transformation reactions, have been reviewed. All possible transformations may be represented schematically.

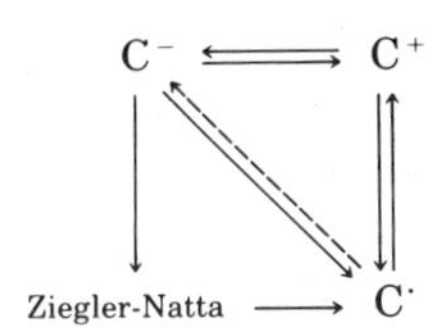

Condensation and Coupling Reactions. The principal limitation of condensation or coupling is that increasing molecular weight of the precursor polymer sequences decreases the concentration of end groups and can decrease reaction rate. Phase separation may also occur. A new interfacial polymerization (qv) method alleviates these drawbacks.

Chemical Modification of Block Copolymers. Modification of a block copolymer transforms at least one of the sequences and leads to a new copolymer. A typical example of such a modification is the hydrogenation under catalysis of the polybutadiene block in a block copolymer; polystyrene-*block*-polybutadiene-*block*-polystyrene could become polystyrene-*block*-polyethylene-*block*-polystyrene.

Fractionation and Characterization of Block Copolymers

Fractionation (qv) of a block copolymer sample is necessary to separate it into different molecular species that are homogeneous in structure, composition, and molecular weight. The different steps concerning the fractionation and characterization of copolymers with heterogeneities in structure, composition, and molecular weight can be summarized by the following scheme, showing the procedure applicable to a diblock poly A-*block*-poly B containing poly A and poly B homopolymers as an example:

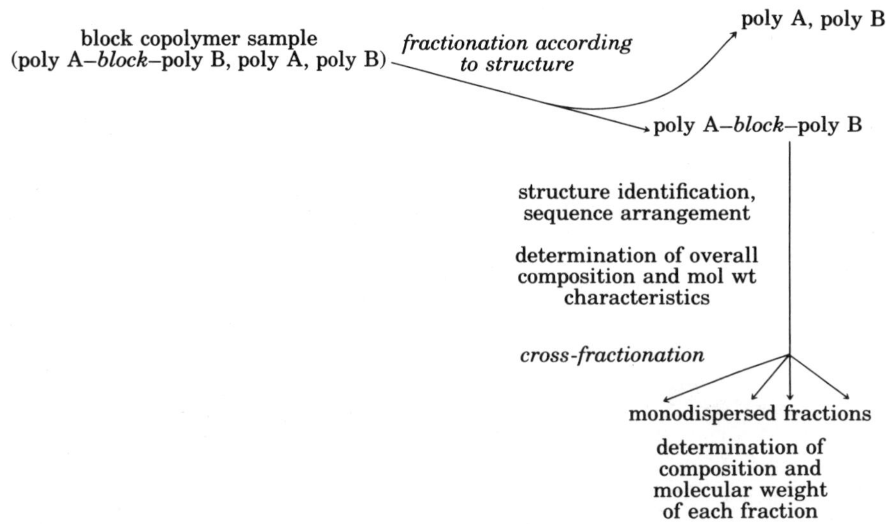

Solution and Colloidal Behavior

In general, because of the incompatible blocks, it is almost impossible to mix block copolymers at the molecular level even in dilute solutions. Block copolymers behave differently in various solvents. The dilute solution behavior, particularly conformational behavior, of these copolymers is mostly investigated in common solvents, that is, good solvents for each of the blocks. In a selective solvent, that is, a good solvent for one of the blocks, these copolymers exhibit colloidal behavior.

Dilute Solutions in Good (Common) Solvents. ***Conformation Studies.*** Dilute solutions of copolymers in good solvents exhibit characteristic behavior resulting from thermodynamic interactions of subchains with solvent molecules and each other. Two plausible conformations for block copolymers in solution have been proposed. In the segregated model, there are few heterocontacts and the molecular size of the blocks in solution is close to that expected for comparable homopolymer chains. In the pseudo-Gaussian or random structure model, the different blocks overlap to some extent, resulting in heterocontacts that expand the component coils because of the additional repulsive forces between unlike segments.

Hydrodynamic Studies. Techniques for studying hydrodynamic behavior of block copolymers in dilute solution are essentially the same as for homopolymers, if it is recognized that the incompatible blocks influence the results. The dilute solution viscosity of block copolymers depends on their molecular characteristics, interaction between the incompatible blocks, and the polymer–solvent interactions.

Dilute Solutions in Selective Solvents (Micelles). Block copolymers in a selective solvent, that is, a good solvent for one block, but a precipitant for the other, behave like typical amphiphiles. The copolymer molecules aggregate reversibly to form micelles in a manner analogous to aggregation of classical surface active agents. The micelles consist of a swollen core of the insoluble block surrounded by a flexible fringe of soluble block(s). It is generally believed that such micelles, formed through a closed association process, are spherical with narrow size distribution, but may change the shape and the distribution under certain conditions. The copolymer micelles are capable of solubilizing substances otherwise insoluble, a behavior shown by low-molecular-weight surfactant micelles. Also, the block copolymers exhibit the phenomenon of adsorption (qv) at interfaces and are therefore capable of stabilizing colloidal dispersion. The colloid-chemical aspects of block copolymers in solutions have been reviewed.

Micelles of block copolymers have been treated as monomolecular and multimolecular micelles (Fig. 1). Generally, the existence of multimolecular micelles has been verified, and such micelles have been found to form by the equilibrium association of the molecules through a closed associated process analogous to that of surfactant micelles.

Surface Activity. Two incompatible blocks make copolymers interfacially active, ie, the block copolymer molecules tend to adsorb at interfaces. The adsorption characteristics and surface activity of various water-soluble block copolymers has been studied, generally by surface tension and determining pressure-area isotherms for monolayer adsorption of the copolymers.

Stability of Colloidal Dispersions. ***Dispersion Polymerization.*** Studies on adsorption characteristics of block copolymers at interfaces suggest that ethylene oxide–propylene oxide block copolymers, poly(vinyl alcohol)-*block*-polystyrene copolymers, and polyvinylpyridine-*block*-poly(styrene sulfonate) copolymers can act as good stabilizers for colloidal dispersion. One such application lies in the stabilization of polymeric dispersed particles formed as a result of dispersion polymerization either in organic or in aqueous medium. Block copolymers sterically stabilize the dispersion by adsorbing onto the particle surface in such a way that one block acts as an anchor, whereas the second remains flexible and extends out into the dispersion medium.

Foams. The basic requirements for block copolymers used as stabilizers of rigid foams are high surface activity for nucleation and stabilization of the cells, combined with good emulsifying abilities for the raw materials, and the blowing agent. These requirements are met by designing the block copolymers according to structural characteristics (linear-branched, molecular characteristics, and presence of end groups).

Oil–Water Emulsions. Block copolymers, generally with poly(ethylene oxide) as hydrophilic blocks and poly(propylene oxide), poly(α-butylene oxide), polystyrene, etc, as the hydrophobic blocks, have been tried as emulsifiers for oil–water

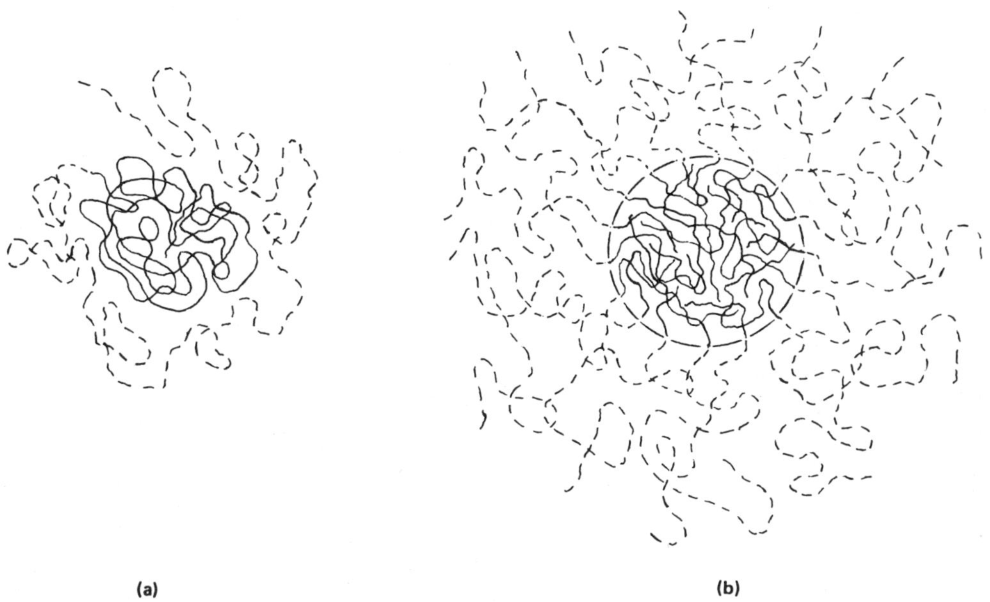

Figure 1. Block copolymer micelles in selective solvent. The insoluble block is surrounded by the solvated block. (**a**) Monomolecular; (**b**) multimolecular.

emulsions. The stability and type of emulsions formed are correlated to the nature, structure, and molecular characteristics of the block copolymer used and the other ingredients of the emulsion.

Oil–Oil Emulsions. Block copolymers act as efficient emulsifiers for stabilizing emulsions composed of two immiscible organic liquids when each of them is a selective solvent of one of the blocks of the copolymer, for example, DMF–hexane emulsions. From viscosity and particle size determinations for these oil-in-oil emulsions as a function of phase ratio and type of emulsion for block copolymers with different molecular characteristics, it was concluded that the solubility of the blocks in oil is an important parameter, and the block copolymers with the lowest degree of association, in both hexane and dimethylformamide, were generally the most efficient emulsifiers for oil-in-oil emulsions. A microphotograph of oil-in-oil emulsions is shown in Figure 2.

Microemulsions. The formation of microemulsions using styrene–ethylene oxide block copolymers in place of common surfactants has been reported. These microemulsions were formed

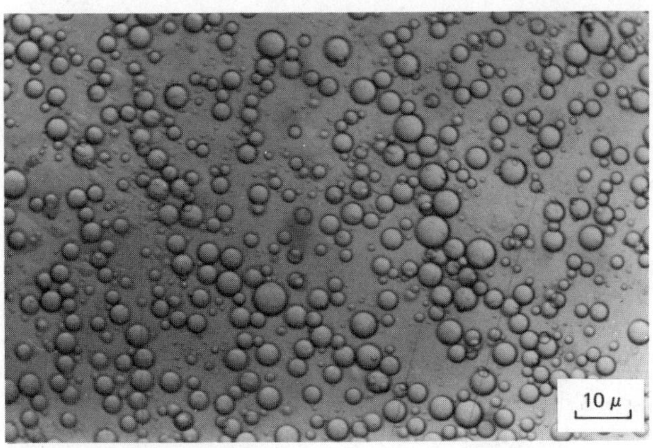

Figure 2. Microphotograph of oil-in-oil emulsion system: dimethylformamide-in-hexane emulsion stabilized by polystyrene-*block*-polyisoprene; $\bar{M}_n$ = 33,000; 44 wt % polystyrene.

by successive addition of 2-propanol to the oil–water emulsions stabilized by the block copolymer. Other block copolymers have also been used to make microemulsions (see MICROEMULSION POLYMERIZATION).

Polymeric Emulsions. It has been demonstrated that block copolymers have emulsifying properties in systems of an organic solvent and two incompatible polymers, that is, in systems in which the two phases to be emulsified consist of the same solvent, but of different polymers. These emulsions are known as polymeric oil-in-oil emulsions. A typical example is the system of polystyrene and polybutadiene in the presence of a common solvent such as toluene or styrene.

It also is possible to prepare polymeric water-in-water emulsions by taking two incompatible polymers, poly A and poly B, which are both water soluble, and emulsifying the system by addition of a water-soluble block copolymer poly A-*block*-poly B. A typical example is the system poly(ethylene oxide), poly(vinylpyridine hydrochloride), and water in the presence of poly(ethylene oxide)-*block*-poly(vinylpyridine hydrochloride).

Properties in the Solid State

Block copolymers with incompatible sequences exhibit characteristic morphological behavior and interesting properties, owing to microdomain formation in the solid state. Thermoplastic elastomers, such as styrene–butadiene block copolymers and segmented polyurethanes or polyesters, which combine the properties of a rubber and the processing characteristics of thermoplastics, have gained technical importance (see ELASTOMERS, THERMOPLASTIC; POLYURETHANES.)

Theories of Microdomain Formation. Block copolymers composed of incompatible and noncrystallizable A and B blocks form microdomain structures as a consequence of the microphase separation of the chains.

Experimental Techniques. The most suitable techniques to study the morphology of block copolymers, and especially their ordered structures, are electron microscopy, x-ray diffraction, and small-angle x-ray scattering (saxs), as well as other scattering techniques such as neutron scattering and small-angle light scattering.

Morphology of Copolymers with Amorphous Blocks. *General Aspects of Microdomains.* It has been established, especially by electron microscopy, saxs, and x-ray diffraction, that segregated microphases can be spheres, cylinders, or lamellae.

Morphology of Copolymers with Amorphous and Crystallizable Blocks. Crystallization in a block copolymer having amorphous and crystallizable blocks implies the separation of the crystallizing species in a distinct phase in which crystalline order can be reached. The existence of a crystalline phase, associated with an amorphous phase, is readily perceived by the formation of spherulites or single crystals having the characteristic features of the corresponding homopolymer in the crystalline state.

Among copolymers with amorphous and crystallizable blocks, two categories should be distinguished: diblock or triblock copolymers, such as styrene–ethylene oxide and styrene–ϵ-caprolactone copolymers, which exhibit well-organized periodic structures, and multiblock copolymers, such as segmented polyurethanes or segmented polyesters, which yield microdomain formation with no apparent periodic structure.

Morphology of Copolymers with Crystallizable Blocks. A typical example is poly(ϵ-caprolactone)-*block*-poly(ethylene oxide)-*block*-poly(ϵ-caprolactone), in which both types of block crystallize, producing a lamellar structure.

Thermal Behavior, Transitions, Dynamic Mechanical Properties. The transitions, either glass transitions (T_g) or melting

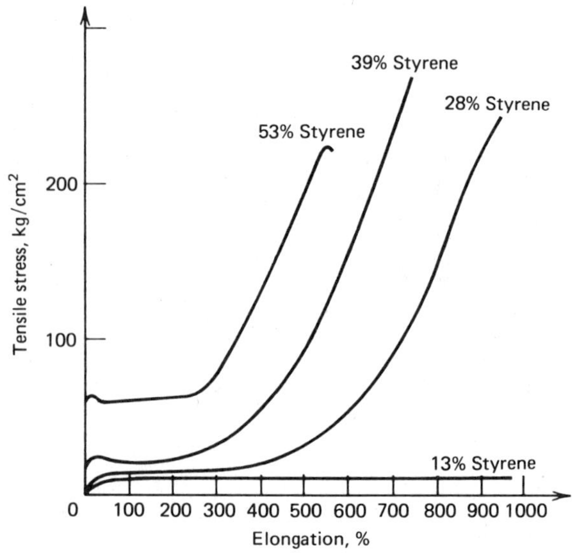

Figure 3. Stress–strain curves for SBS block copolymers of various styrene contents. To convert kg/cm² to psi, multiply by 14.

Table 1. Examples of Commercially Available Block Copolymers

Copolymer type	Trade names	Typical applications
Styrene–diene		
diblock SB; SI	Solprene (Phillips)	thermoplastic elastomers,
triblock SBS; SIS	Kraton (Shell)	molding products, plastic
star block $(SB)_x$	Cariflex (Shell)	modifiers, adhesives, films
	Tufprene (Asahi)	
	Europrene (Anic)	
	Buna BL (Bayer)	
	Eltar (Icechim-Romania)	
Hydrogenated styrene–diene		
hydrogenated SBS	Kraton G (Shell)	oxidation and weather-resistant
styrene–(ethylene–butylene)–styrene	Elaxar (Shell)	thermoplastic elastomers,
		moldings, cable insulation
Segmented polyester–polyurethane		
thermoplastic polyurethanes (TPU)	Estane (Goodrich)	thermoplastic elastomers, fibers
	Texin (Bayer, Mobay)	
	Roylar (Uniroyal)	
	Elastothane (Thiocol)	
	Pellethane (Upjohn)	
	Lycra (DuPont)	
	Spandex (DuPont)	
Segmented polyester–polyether		
	Hytrel (DuPont)	thermoplastic elastomers
	Pelprene (Toyobo)	
	Arnitel (Akzo)	
Segmented polyamide–polyether		
	Pebax (Atochem)	molding extrusion, footwear,
	Vestamid E (C.W. Huls)	automotive parts, cable insulation
Polyolefinic block copolymers		
	TPR (Uniroyal)	thermoplastic elastomers
	Somel (DuPont)	
Ethylene oxide/propylene oxide block copolymers		
	Pluronics, Tetronics (BASF Wyandotte)	surfactants

points, of microphase-separated block copolymers are commonly determined by thermal analysis (dsc) or by dynamic mechanical analysis. The time–temperature variation of the complex modulus, like the dynamic storage modulus (E') or the dynamic mechanical loss (tan δ), is influenced by the type of coupling between the microphases, which, in turn, depends on the morphology of the system.

Mechanical and Rheological Properties. As a general approach to the study of mechanical properties of block copolymers, it is useful to examine the tensile strength of a series of polystyrene-*block*-polybutadiene-*block*-polystyrene (SBS copolymers) as a function of molecular weight and composition. Figure 3 shows the characteristic features of the stress–strain curves as a function of the composition of the SBS copolymers, which, in turn, influence the morphology of the system.

From Figure 3, it appears that block copolymers having a continuous polybutadiene phase and a dispersed phase of polystyrene microdomains, generally as spheres or as cylinders, exhibit the typical behavior of a cross-linked elastomeric system. Thus, one of the most important characteristics of these elastomeric block copolymers is their ability to display rubbery properties while retaining thermoplastic processibility. The typical elastomeric properties of these thermoplastic elastomers can be attributed to the physical cross-linking resulting from domain formation in the case of triblock or multiblock copolymers having hard and soft segments (see also ELASTOMERS, THERMOPLASTIC).

The ultimate properties of block copolymer materials are strongly influenced by the microdomain morphology, and particularly by the interfacial area, which in turn governs the domain boundary phenomena.

Rheological Behavior. From various studies of styrene–butadiene block copolymers, it appears that the melt behavior of block copolymer systems is quite different from that of their corresponding homopolymers or random copolymers. Thus a block copolymer exhibits a higher melt viscosity than a random copolymer of the same molecular weight and composition, and block copolymers do not show constant viscosities at low shear rate. This complex rheological behavior is primarily attributed to the persistence of the two-phase domain structure of the block copolymer in the melt.

Transport Behavior. This characteristic property of block copolymers, involving permeation, diffusion, and solution of inert gases, has been reviewed. A generalized relationship of permeability versus composition for random, block, and graft copolymers has been established from which it appears that properties such as selective permeation or barrier properties are mainly caused by the two-phase microdomain structure of block copolymers. As a consequence, it also is possible to provide interesting information about the morphology of the block copolymers and the segmental motion in the microdomains by studies of transport and solution behavior.

Applications

Several of the block copolymers have been commercialized and are available in the marketplace. Table 1 lists examples of common trade names and some typical applications, mainly of block copolymers produced in the United States, Western Europe, and Japan. The most common uses of block copolymers are as thermoplastic elastomers, toughened thermoplastic resins, membranes, polymer blends, and surfactants.

BIBLIOGRAPHY

I. Goodman, ed., *Developments in Block Copolymers,* Vol. 1, Applied Science Publishers, London, 1982.

A. Noshay and J. E. McGrath, *Block Copolymers: Overview and Critical Survey,* Academic Press, New York, 1977.

G. Riess and P. Callot in L. H. Tung, ed., *Fractionation of Synthetic Polymers—Principles and Practices,* Marcel Dekker, Inc., New York, 1977, p. 475.

P. Gaillard, M. Ossenbach-Sauter, and G. Riess in K. Solc, ed., *Polymer Compatibility and Incompatibility,* Vol. 2, Harwood Academic Publishers, New York, 1982.

GERARD RIESS
G. HURTREZ
Ecole Nationale Supérieure de Chimie de Mulhouse
P. BAHADUR
South Gujurat University

BLOWING. See BLOWING AGENTS; BLOW MOLDING; CELLULAR MATERIALS.

BLOWING AGENTS

Blowing agents, sometimes referred to as foaming agents, produce gas and thus may be used to generate cells (gas pockets) in polymeric materials. They are classified as either physical or chemical blowing agents (CBAs). Physical blowing agents form cells by a phase change, eg, a liquid may be volatilized or a gas dissolved in a polymer under high pressure may be desorbed by decompression (see physical properties in Table 1). Chemical blowing agents produce gas by thermal decomposition and, in a few instances, via a chemical reaction with other components of the polymer system (see also CELLULAR MATERIALS). Regardless of type, a blowing agent should possess the following desirable qualities:

1. Long-term storage stability under fairly ordinary conditions.
2. Gas release over a controlled time and temperature range.
3. Low toxicity, odor, and color of both the blowing agent and its decomposition products.
4. No deleterious effects on the stability and processing characteristics of the polymer.
5. The ability to produce cells of uniform size.
6. The ability to produce a stable foam, ie, the gas must not be lost from the cell and cause it to collapse.
7. Good cost-performance relation and availability.

Expansion Process

During the expansion of the substrate by the blowing agent, cells are initiated and grow to produce the final foam. As gas is produced, an equilibrium is established between material in the gas phase and material dissolved in the solid phase. The gas dissolved in the solid phase migrates from the solution into the gas phase. The cells formed initially are under higher than ambient pressure because they must counteract the effects of

Table 1. Properties of Physical Blowing Agents

Blowing agent	Molecular weight	Boiling point, °C	Blowing efficiency[a] At boiling point	At 100°C
pentanes				
n-pentane	72.15	36.1	216	261
2,2-dimethylpropane	72.15	9.5	196	260
1-pentene	70.15	30.0	227	280
hexanes				
n-hexane	86.17	68.7	212	232
2-methylpentane	86.17	60.2	207	232
3-methylpentane	86.17	63.3	211	234
2,2-dimethylbutane	86.17	49.7	204	229
cyclohexane	84.17	80.8	266	281
heptanes				
n-heptane	100.20	98.4	206	207
2,2-dimethylpentane	100.20	79.2	193	204
2,4-dimethylpentane	100.20	80.6	193	204
3-ethylpentane	100.20	93.4	204	212
1-heptene	98.20	93.2	212	216
toluene	92.13	110.6	294	286
trichloromethane	119.39	61.2	342	382
tetrachloromethane	153.84	76.7	296	316
trichlorofluoromethane	137.38	23.8	261	329
methanol	32.04	64.6	679	752
2-propanol	60.09	82.3	378	397
isopropyl ether	102.16	67.5	198	217
methyl ethyl ketone	72.10	79.6	324	344

[a] Data computed according to the formula

$$\frac{22{,}400}{\text{mol wt}} \times (\text{density at } 25°\text{C}) \times \frac{(273 + t)}{273}$$

where t = boiling point in °C for the next to the last column and 100°C for the last column. To a first approximation vapors are treated as ideal gases.

surface tension. Because the pressure due to surface tension depends on the reciprocal of the cell radius, the pressure within the cell is reduced as the cell grows. Thus, small cells tend to disappear and large cells tend to increase in size because the gas migrates through the substrate or the cell walls break.

After the cells are formed, the foam must be stable, ie, the gas must not diffuse out of the cell too quickly, thereby causing collapse or excessive shrinkage. The stability of the foam depends on the solubility and diffusivity of the gas in the matrix.

Chemical Blowing Agents

Chemical blowing agents include sodium bicarbonate, dinitrosopentamethylenetetramine, sulfonyl hydrazides, azodicarbonamide, *p*-toluenesulfonyl semicarbazide, 5-phenyltetrazole, diisopropylhydrazodicarboxylate, 5-phenyl-3,6-dihydro-1,3,4-oxadiazine-2-one, and sodium borohydride.

Thermoset Polyester Blowing Agents

Chemical blowing agents are used to foam unsaturated polyester resins, eg, unsaturated polyester dissolved in styrene. In this system, thermal decomposition of the blowing agent cannot be applied because the heat of polymerization is not high enough to induce decomposition. Chemical reactions, however, simultaneously produce gas and free radicals. Such reactions typically involve oxidation and reduction of a hydrazine derivative and a peroxide; they are catalyzed by metals, which can be used repeatedly.

BIBLIOGRAPHY

Chemical Blowing Agents Chemistry and Decomposition Mechanisms, Bulletin ASP-4455, Uniroyal Chemical Co., Div. of Uniroyal, Inc., Naugatuck, Conn., Sept. 1977.

R. J. Bathgate and K. T. Collington, *SPE ANTEC, 39th Meeting,* May 4–7, 1981, Boston, Society of Plastics Engineers, Brookfield, Conn., 1981, p. 856.

R. L. Heck, III, "Foaming Agents," in J. Agranoff, ed., *Modern Plastics Encyclopedia,* McGraw-Hill Inc., New York, 1983, pp. 142–145.

RHOMIE L. HECK III
WARREN J. PEASCOE
Uniroyal Chemical Group

BLOW MOLDING

Although still used mainly for bottles, blow molding is increasingly used for industrial parts, such as automotive rear deck air spoilers, seat backs, toy tricycles and wheels, typewriter cases, surfboards, flexible bellows, and fuel tanks. Figure 1 shows a fuel tank being molded on an accumulator head machine.

Figure 1. Blow-molded Volkswagen fuel tank made of high-molecular-weight polyethylene. Equipment is Krupp-Kautex Model KB250-S120 with a 120-mm extruder and 15-kg accumulator head. Courtesy of Krupp-Kautex Machines, Inc.

More resins are being considered for blow molding. Although most thermoplastic resins can be blow molded, high-density polyethylene, poly(vinyl chloride), poly(ethylene terephthalate), and polypropylene are the most widely used. For industrial parts, acetal, polycarbonate, polyester elastomers, modified poly(phenylene oxide), polyurethane, and high- and ultrahigh-molecular-weight polyethylene are also considered.

General Characteristics

Blow molding produces hollow objects. Air, or occasionally nitrogen, is used to expand a hot preform or parison against a female mold cavity. A common feature of all blow-molded articles is the reentrant curve, such as a bottle opening much smaller than the body. Some articles are designed for strength. Toy tricycles and wheels utilize the strength of unit box sections. In some cases, it is important to provide very thin wall sections with relatively low stress. Another characteristic is that detail can be better controlled on the outside than on the inside, where material thickness affects the shape.

Processes

Today, three blow-molding processes are used.

1. Injection blow molding.
2. Extrusion blow molding.
3. Stretch blow molding.

In injection blow molding, melted plastic resin is injected into a parison cavity and around a core rod. This test-tube shaped parison, while still hot, is transferred on the core rod to the bottle blow-mold cavity, which, in turn, cools the part. Figure 2 illustrates a typical three-station injection blow-molding machine.

Generally, injection blow molding is used for small bottles and parts less than 500 mL in volume. The process is scrap free, with extremely accurate control of weight and neck finish. This allows the intricate shapes required by some of the special safety and tamperproof closures. However, part proportions are limited and the method is impractical for containers with handles; tooling costs are relatively high.

Extrusion blow molding, the most common process, is used for bottles or parts 250 mL in volume or larger; tanks as large as 5000 L weighing 120 kg have been blow-molded. Tooling is less expensive, and part proportions are not severely limited. Containers with handles and offset necks are easily fabricated. On the other hand, flash or scrap resin must be trimmed from each part and recycled. Operator skill is more crucial to the control of part weight and quality.

In extrusion blow molding, melted plastic resin is extruded as a tube into the air. This tube, called a parison, is captured by the two halves of the bottle blow mold. A blow pin is inserted through which air enters and expands and then cools the parison against the cold mold cavity. The action of pinching and sealing the two open ends of the parison is what creates the flash that in turn must be reclaimed.

Two variations of the process are in use. With continuous extrusion, the parison is continuously formed at the same rate as the article is molded, cooled, and removed. To avoid interference with the parison formation, the mold-clamping mechanism must move quickly to capture the parsion and return it to the blowing station where the blow pin enters. Figure 3 illustrates a typical side-shuttle, continuous-extrusion blow molder.

With intermittent extrusion, the parison is quickly extruded after the bottle is removed from the mold. The clamping mechanism does not need to move to a blowing station. Blow molding, cooling, and bottle removal all take place under the extrusion head. This permits a simple and rugged clamping system. Because of the stop–start aspect of extrusion, this process is more suitable for polyolefin and other heat-stable materials.

Stretch blow molding is used for bottles between 500 mL and 2 L in size, and occasionally as large as 25 L. The molecular biaxial orientation of certain resins enhances stiffness, impact, and barrier performance, and permits weight reductions and a lower cost material.

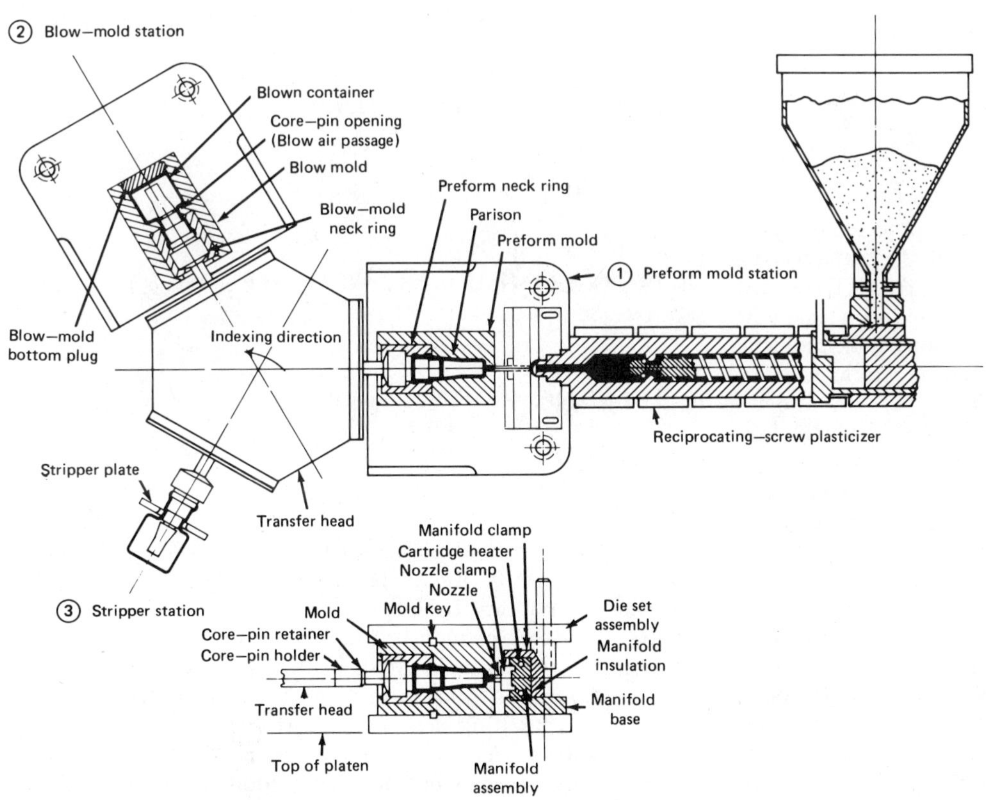

Figure 2. A typical three-station injection blow-molding machine. Courtesy of Rainville Operation, Hoover Universal, Inc.

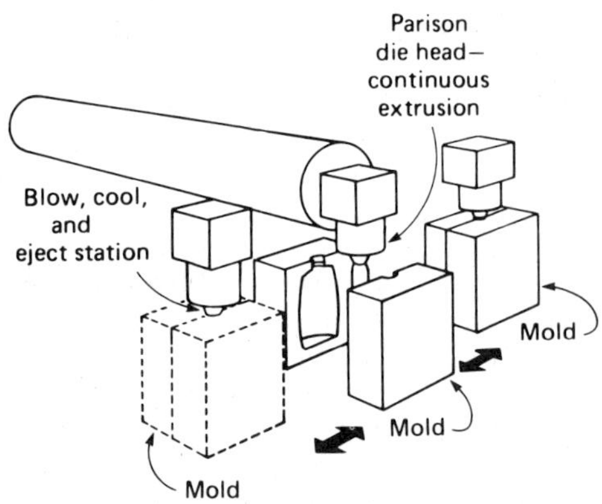

Figure 3. Shuttle continuous-extrusion, blow-molding machine. Molds on this dual-sided system move alternatingly to capture a parison.

The stretch process takes advantage of the crystallization behavior of the resin and requires the preform to be temperature-conditioned and then rapidly stretched and cooled into the bottle shape. The technique is limited to simple bottle shapes and is based on either the injection or extrusion processes. Poly(ethylene terephthalate), poly(vinyl chloride), polypropylene, and polyacryonitrile are the resins most often used.

Multilayer Blow Molding. All materials, whether metal, glass, or paper, have certain strengths and weaknesses, and advantages and disadvantages. Frequently, two or more materials can be layered and combined, eg, chromium-plated steel, laminated automobile windshields, and wax- or polyethylene-covered paperboard. In multilayer blow molding, the advantages of two or more resins are combined to produce packaging superior to that produced by a single resin.

Other Blow-molding Operations

Many related operations have been used to improve blow molding, eg, in-mold labeling, fluorination of surfaces, and internal cooling.

Designs

Extruders and Screws. All blow-molding processes require melted-plastic resin. The quality of this melt and productivity of the blow-molding equipment depend heavily on the extruder-screw design. The designs are the same as those used with other quality plastic injection and extrusion process technologies.

Injection Blow-mold Tooling. Injection blow molding requires two molds: one for molding the preform or parison, and the other for molding the bottle. The preform mold consists of the preform cavity, injection nozzle, neck-ring insert, and core rod assembly. The blow mold consists of the bottle cavity, neck-ring insert, and bottom-plug insert. All of the components, particularly with multicavity arrangements, are mounted to a die set for easier handling and setup.

Injection blow-mold tooling is designed with very precise dimensional tolerances, often held to ± 0.015 mm. The need for precision is the most crucial factor in the high cost of the tooling. However, once properly assembled, the injection blow-mold process can provide high yield, trouble-free production.

Extrusion Blow-mold Tooling. Extrusion blow molding requires only one mold; however, because the parison is formed from am extruded tube and flash is a by-product of the molding process, several tooling elements are combined to produce the finished bottle. These elements include the design of the extrusion head, head tooling, blow-pin tooling, mold-cavity tooling, and trim tooling.

Guidelines. Product design guidelines begin with a clear understanding of process and tooling limitations, such as parison blow-up ratios, ovality ratios, head-tooling sizes, etc. Second, designers must be familiar with the part performance requirements. The Society of the Plastics Industry has recommended several standard bottle practices and tests. The most important are vertical compression or top-load strength, drop-impact resistance, product compatibility and permeability, closure torque, and top-load stress-crack resistance.

Most blow-molded articles perform better with rounded, slanted, and tapered surfaces. Square or flat surfaces with sharp corners are undesirable.

BIBLIOGRAPHY

J. R. Dreps, *Plastics Eng.*, 34 (Jan./Feb. 1975).

K. Stoeckhert, *Ind. Prod. Eng.*, 462 (1980).

Publications of the Plastic Bottle Institute of the Society of The Plastics Industry, New York.

D. C. Peters, *Plastics Eng.*, 21 (Oct. 1982).

Christopher Irwin
Johnson Controls, Inc.
Plastics Machinery Division

BLOWN TUBING. See Films.

BOARD, COMPOSITION. See Composition board.

BOATS. See Marine applications.

BONDING. See Adhesion and bonding.

BORON-CONTAINING POLYMERS. See Inorganic polymers.

BOUNCING PUTTY. See Silicones.

BRANCHED POLYMERS

Long-chain branched polymers form a class of polymers between the strictly linear polymers and polymer networks. Branched polymers are characterized by the presence of branch points, ie, atoms or small groups from which more than two long chains emanate, or by the presence of more than two end groups. Almost all monomers can be converted to branched polymers. Branching reactions occur inadvertently in many polymerization processes and are independent of the polymerization mechanism. Controlled branching can also be achieved.

The industrially important conversion of linear prepolymers to networks, eg, by vulcanization or thermosetting, always passes through the initial formation of branched polymers (see INTERPENETRATING POLYMER NETWORKS; NETWORKS).

Because of their finite size, branched polymers resemble linear polymers rather than infinite networks. The properties of polymers that are directly related to their overall dimensions are strongly affected by long-chain branching. The presence of even a small fraction of long-chain branched molecules may result in properties considerably different from those of a purely linear polymer.

The functionally f of a monomer indicates the number of chemical bonds by which it can be incorporated into a macromolecular structure. Most monomers are bifunctional ($f = 2$) and lead to strictly linear chains. However, many bifunctional monomers are potentially multifunctional ($f > 2$) because of side reactions. Foremost among these is chain transfer (qv) to polymer, which creates an active center in a monomer unit of an already formed chain. Transfer to monomer and termination by disproportionation create end-standing double bonds in the polymer. When these double bonds polymerize later, they become branch points. Chain transfer to monomer is found in radical polymerization (qv) and is quite common in cationic polymerization (qv), eg, of isobutylene.

Strategies and Methods of Preparation

Branched polymers can be classified into several categories: star branched, comb, and randomly branched. Branched polymers can be prepared in a one- or two-step process. In a one-step synthesis, bifunctional and multifunctional monomers are present from the onset of polymerization. The incorporation of both monomers and the formation of branches are governed by copolymerization statistics.

In a two-step process a linear polymer is formed first. This polymer may or may not require chemical modification before being submitted to the branching reaction. Practically all graft-copolymerization techniques can be adapted to the preparation of branched polymers. Such reactions are either "branching-from" or "branching-onto" reactions. Chemical vulcanization and irradiation of preformed polymers, when kept below the gel point, belong to this type of synthesis.

Properties

Molecular Weight and Molecular Weight Distribution. The complete linking of arms into stars with constant functionality (f) leads to narrowing of the molecular-weight distribution:

$$(M_w/M_n)_{\text{star}} = 1 + [(M_w/M_n)_{\text{arm}} - 1]/f \tag{1}$$

Similar equations have been derived when backbones and branches are combined into combs.

The occurrence of a random-branching process during polymerization increases the molecular weights (M_n, M_w) and the molecular-weight distribution (M_w/M_n) relative to those of the linear polymer formed under otherwise identical conditions.

Dimensions. The radius of gyration of polymer consisting of N repeat units is given by

$$\langle S^2 \rangle = \frac{1}{2N^2} \sum_{i=1}^{N} \sum_{j=1}^{N} \overline{r}_{ij}^2 \tag{2}$$

where $\overline{r}_{ij}^2$ is the mean-square distance between the ith and jth monomer units. For example, in the unperturbed state, $\overline{r}_{ij}^2 = b^2|i - j|$, ie, $\overline{r}_{ij}^2$ is a linear function of the number of monomers separating i and j. Relative to a linear polymer with the same total number of monomers N, the sum in equation 2 for a branched polymer contains more small terms and fewer large terms. Therefore, the radius of gyration of a branched polymer is smaller than that of the linear polymer with the same N. Branched polymers are often characterized by their branching factor

$$g_{th} = \left(\frac{\langle S^2 \rangle_{o,\text{br}}}{\langle S^2 \rangle_{o,1}} \right)_M \tag{3}$$

Hydrodynamic Properties. Because the radius of gyration of a branched polymer is smaller than that of a linear polymer, it is expected that its intrinsic viscosity $[\eta]$ is also smaller and $g' = ([\eta]_{\text{br}}/[\eta]_1)_M < 1$.

Experimental Evaluation of Long-chain Branching

It is most helpful when the type of branching (structure, functionality of the branch points) can be derived from a consideration of the synthesis of the polymer. For a model branched polymer, it is possible to derive the average number of branches from its molecular weight and from that of the precursor polymers. The quantitative analysis of end groups or branch points should not be overlooked because such analysis provides an absolute number of branches per polymerized monomer. The universal calibration principle in size-exclusion chromatography is used extensively in the qualitative and quantitative evaluation of branching.

Bulk Properties

Bulk properties of polymers that depend on the local conformation of the chains are unlikely to be strongly influenced by the presence of long-chain branching. The glass-transition temperature of a branched polymer is almost always identical to that of the linear homologue. Similarly, long-chain branching has little influence on crystallinity, although the rate of crystallization may be affected. The bulk properties, governed by the overall dimensions of the polymer chains, are expected to depend on long-chain branching. Differences between the properties of branched and linear polymers are observed whenever the polymer chains are deformed and move as in flow or diffusion experiments on melts or on concentrated solutions.

Applications

The large-volume polymers, eg, polyethylene, polystyrene, poly(butadiene-*co*-styrene), poly(ethylene-*co*-vinyl acetate) are branched to some extent because they are produced radically at high conversion. Multifunctional monomers are sometimes added for the creation of a small amount of branching. In particular, anionic polymers are partially converted to multiarm stars. In these cases the aim is often at wider molecular-weight distributions and the specific properties of branched polymers.

Deformation of a polymer by continuous shearing creates nonequilibrium states that persist longer in branched polymers than in linear polymers. In these nonequilibrium states, polymers have lower viscosities, lower elasticity as measured by die swell, and produce less surface roughness on processing.

In solution, branched polymers are less sensitive to mechanical degradation than linear polymers.

BIBLIOGRAPHY

P. J. Flory, *Principles of Polymer Chemistry,* Ithaca University Press, Ithaca, N.Y., 1953.

B. H. Zimm and W. H. Stockmayer, *J. Chem. Phys.* **17,** 1301 (1949).

P. A. Small, *Adv. Polymer. Sci.* **18,** 1 (1975).

W. W. Graessley, *Acc. Chem. Res.* **10,** 332 (1977).

JACQUES ROOVERS
Division of Chemistry
National Research Council of Canada

BRANCHING. See BRANCHED POLYMERS.

BRIGHTENERS, OPTICAL. See PAPER.

BRITTLE FRACTURE. See FRACTURE.

BRITTLE TEMPERATURE. See GLASS TRANSITION.

BUILDING AND CONSTRUCTION APPLICATIONS. See CONSTRUCTION AND BUILDING APPLICATIONS.

BULK POLYMERIZATION

Bulk polymerization traditionally has been defined as the formation of polymer from pure, undiluted monomers. A common feature of most bulk polymerization and other processes not traditionally classified as such, however, is the need to process fluids of very high viscosity. The high viscosity results from the presence of dissolved polymer in a continuous liquid phase. Significant concentrations of a high-molecular-weight polymer typically increase fluid viscosities by 10^4 or more compared to the unreacted monomers. This suggests classifying a polymerization as bulk whenever a substantial concentration of polymer occurs in the continuous phase. Although this definition encompasses a wide variety of polymerization mechanisms, it leads to unifying concepts in reactor design. The design engineer must confront the polymer in its most intractable form, ie, as a high viscosity solution or polymer melt (see also REACTOR TECHNOLOGY).

In all bulk polymerizations, highly viscous polymer solutions and melts are handled. This fact tends to govern the process design and to a lesser extent, the process economics. Suitably robust equipment has been developed for the various processing steps, including stirred-tank and tubular reactors, flash devolatilizers, extruder reactors, and extruder devolatilizers. Equipment costs are high based on working volume, but the volumetric efficiency of bulk polymerizations is also high. If a polymer can be made in bulk, manufacturing economics will most likely favor this approach.

Free-radical Polymerizations

Free-radical polymerizations proceed by a complex mechanism of initiation, propagation, and termination of which the propagation and termination steps are typically very fast (see also RADICAL POLYMERIZATION). Heat removal is complicated by the rapidly increasing solution viscosity. The increasing viscosity also lowers the diffusivity of polymer chain segments, but has less effect on the diffusivity of the monomers. This leads to a common phenomenon in bulk polymerizations known as the Trommsdorff or gel effect (qv).

Polystyrene and poly(methyl methacrylate) are usually manufactured by free-radical bulk polymerizations. Poly(vinyl chloride) is usually manufactured by a free-radical process using chemical initiators at moderate temperatures, ie, below 100°C. Polyethylene and ethylene-rich copolymers are sometimes formed in a free-radical bulk process using the older, high pressure technology.

Condensation Polymerizations

Many condensation polymerizations (polycondensations) satisfy the definition of a bulk polymerization even though they differ significantly from free-radical polymerizations. High-molecular-weight polymer forms in a slow stepwise process. It requires exacting stoichiometry, a favorable reaction equilibrium, and high conversion. Changes in viscosity are relatively modest until the final stages of the reaction when substantial increases may occur. Reaction exotherms are also moderate, with most of the heat of the reaction being liberated when viscosities are low. Thus, heat removal presents little problem in bulk polycondensation.

Characteristics of Bulk Polymerizations

The common characteristic of bulk polymerization is the presence of high-molecular-weight polymer in the continuous phase.

Changes in Viscosity and Diffusivity. The most dramatic change accompanying bulk polymerization is the increase in continuous-phase viscosity, as illustrated by the viscosity data for the bulk polymerization of styrene at temperatures of commercial interest (Fig. 1).

Reaction Exotherms. Heats of polymerization range from zero upward. A high exotherm is not intrinsic to the definition of a bulk polymerization (even by the old definition), because polycaprolactam has a heat of polymerization of only 15.9 kJ/mol (3.8 kcal/mol).

Reactors

Bulk polymerization reactors include batch reactors, continuous stirred tanks, tubular reactors, and extruders.

Devolatilization

Nearly all bulk-polymerization processes require devolatilization (qv) of residual monomers, solvents, or condensation by-products.

Bulk Processes

Poly(methyl methacrylate). Poly(methyl methacrylate) (PMMA) can be made by a variety of processes; casting (qv)

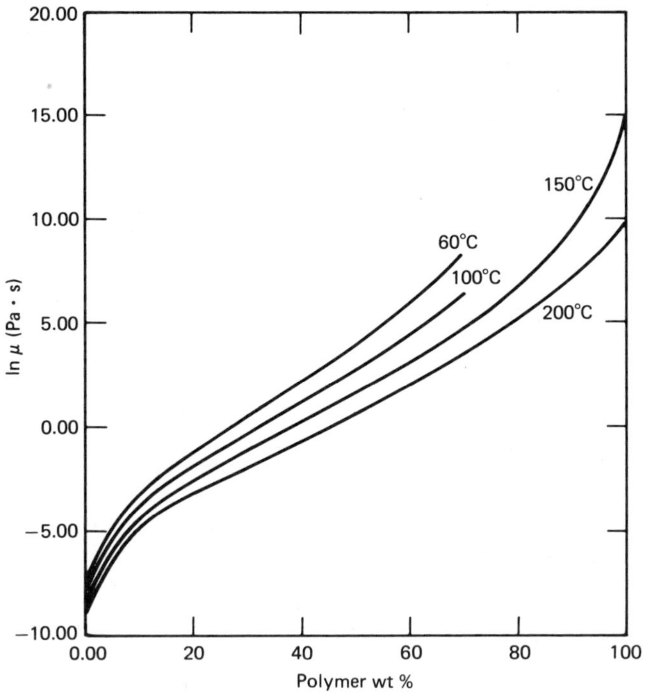

Figure 1. Viscosities in the styrene–polystyrene system; $\bar{M}_w$ = 259,000, $\bar{M}_n$ = 110,000.

accounts for perhaps half of total production (see also ACRYLIC AND METHACRYLIC ESTER POLYMERS).

Polystyrene. The first bulk continuous processes for polystyrene went into production in the early 1950s. In a typical operation, two stirred tanks in a series are followed by a tubular polymerizer and devolatilizer.

Low Density Polyethylene. The high pressure bulk process used to produce this polymer is technically obsolete, although much of the world's production is still made by this route.

Poly(ethylene terephthalate). Although a condensation polymer, the chemistry of the poly(ethylene terephthalate) reaction lends itself to continuous operation.

Polysulfone. Polysulfone is a typical condensation polymer with an exacting stoichiometry requirement. Batch reactors are therefore appropriate because the starting materials can be batch weighted with greater accuracy than they can be fed on a continuous basis.

BIBLIOGRAPHY

J. A. Biesenberger and D. H. Sebastian, *Polymerization Engineering,* John Wiley & Sons, Inc., New York, 1983.

E. B. Nauman in M. D. Baijal, ed., *Plastics Polymer Science and Technology,* John Wiley & Sons, Inc., New York, 1982, Chapt. 10.

E. BRUCE NAUMAN
Rensselaer Polytechnic Institute

BUNA RUBBER.
See BUTADIENE POLYMERS.

BUTADIENE–ACRYLONITRILE–STYRENE POLYMERS.
See ACRYLONITRILE–BUTADIENE–STYRENE POLYMERS.

BUTADIENE DERIVATIVES, POLYMERIC

This article deals with polymers obtained from derivatives of 1,3-butadiene, except those from 2-methylpentadiene (isoprene), 2-chlorobutadiene (chloroprene), 1,3-pentadiene (piperylene), and cyclopentadiene. These are discussed separately (see CHLOROPRENE POLYMERS; CYCLOPENTADIENE POLYMERS; ISOPRENE POLYMERS; PENTADIENE POLYMERS). (See also BUTADIENE POLYMERS).

The polymers discussed here are all obtained by stepwise addition of monomer units that contain the conjugated double-bond sequence C═C—C═C. There is an energy advantage of approximately 84 kJ/mol (20 kcal/mol) in the process of converting such monomers to polymers, because at least one of the two double bonds in the monomer is converted to a lower energy single bond. This energy advantage drives the polymerization and facilitates polymer production on an industrial scale. The principal difficulties are finding an economic source of monomer and establishing conditions to initiate a process without frequent termination or side reactions.

In general, polymeric butadiene derivatives are important primarily for their physical properties and secondarily for basic research. Their physical properties include elasticity, toughness, durability in various environments, and wide temperature ranges over which the properties are retained. The most important single application has been, and remains, automobile tires (qv), which consumes nearly 2 million in North America. The impetus for much of the research into other polydienes has been potential improvements in tires. The polymers considered here are currently not of industrial significance. They may be of future or past importance in tires; they may have promise as specialty polymers, ie, rubbery, plastic, or liquid; or they may be primarily of scientific interest.

Properties of Monomers

Homologues of butadiene and other monomeric butadiene derivatives are colorless, flammable liquids with odors characteristically more pungent than saturated hydrocarbons, but less objectionable than nonconjugated olefins. Polar substituents contribute to the odor. The toxicology of dienes has not been thoroughly investigated. However, because malignant tumors were produced in mice by inhalation of butadiene itself, substituted butadiene monomers should be handled carefully.

In general, the reactions of substituted butadienes are similar to those of butadiene, but the substituents affect the rates and relative importance of reactions, eg, Diels-Alder additions and all modes of polymerization.

Preparation of Monomers

The only alkylbutadienes that have been produced on large enough scale for tire rubbers appear to be isoprene, the pentadienes, and 2,3-dimethylbutadiene. Butadiene, isoprene, and the pentadienes are extracted in large quantities from petroleum fractions. A few synthetic routes suitable for industrial scale have been reported for dimethylbutadiene, ethylbutadiene, and 3-methylpentadiene.

Grignard reactions have been used frequently to prepare alkylbutadienes. A wide array of new laboratory-scale methods are appearing because of the increasing interest in pure isomers of the more complex butadiene derivatives as precursors for biologically active compounds.

Dimethylbutadiene Polymers

The monomer 2,3-dimethyl-1,3-butadiene was the first source of synthetic rubber (methyl rubber) for tires. Polymerization methods include emulsion, free radical, Ziegler-Natta, channel complex, anionic, and cationic. Dimethylbutadiene may also be used as a comonomer or activator.

Microstructure. It has been determined by 220-MHz nmr that polydimethylbutadiene homopolymer made in cyclohexane at 60°C with *n*-butyllithium has about 73% trans-1,4; 24% cis-1,4; and 3% 1,2 units. A polymer made at 60°C with a free-radical initiator has about 51 ± 3% trans, 41 ± 3% cis, and 8% 1,2.

Dimethylbutadiene Copolymers

Interesting early work in the USSR on copolymerization of butadiene and dimethylbutadiene compares initiator systems quantitatively. Although important conditions are omitted, the results illustrate striking differences. The tendency for butadiene to enter the polymer preferentially is very strong in an anionic system, weak in two Ziegler-Natta systems, and reversed in a cationic system.

2-Alkylbutadiene Polymers

Ethylbutadiene Polymers. Ethylbutadiene (2-ethyl-1,3-butadiene), the first homologue of isoprene, provides the analogous stereoregular cis-1,4 and trans-1,4 homopolymers. Some natural substances contain polymerized ethylbutadiene units, eg, juvenile hormone, although natural sources of high-molecular-weight polyethylbutadienes have not been reported. Polymerization methods include emulsion, Ziegler-Natta, and cationic.

Myrcene Polymers

Myrcene and ocimene are the most familiar of a group of natural dienes that are widespread in plants. The monomers are used commercially, for example, in perfumes, but not in polymers of significant volume, although polymerizations have been reported and patents issued.

Phenylbutadiene Polymers

During the 1960s in Italy, 2-phenylbutadiene was a serious candidate for commercialization as a cis–cis copolymer with butadiene made with an aluminum hydride–aluminum iodide–titanium tetrachloride catalyst. It improved strength, tear resistance, traction, and aging in tire compounds, but reduced rebound and abrasion resistance. Tires recapped with the copolymer showed improved traction at the expense of wear, as might be expected from the T_g increase that 2-phenylbutadiene units produce. Polymerization methods include Ziegler-Natta, anionic, and cationic.

Methylpentadiene Polymers

The methyl-1,3-pentadienes (MPD) can also be termed dimethyl-1,3-butadienes (DMB). Both nomenclatures are used, but examination of the structure dispels any confusion. Cisoid conformations are shown here; transoid conformations can be visualized.

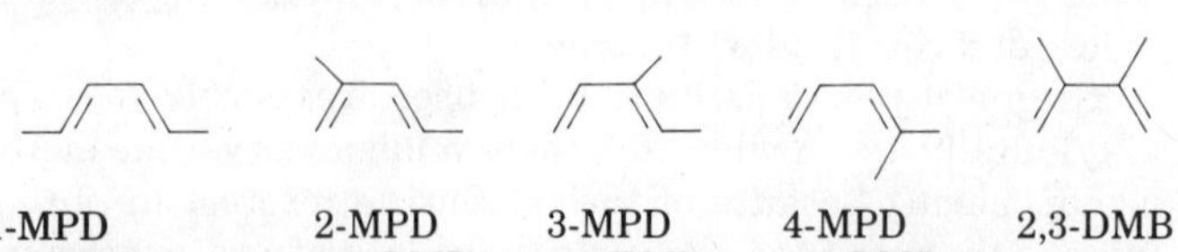

1-MPD 1,4-DMB 2,4-hexadiene; 2-MPD 1,3-DMB; 3-MPD 1,2-DMB; 4-MPD 1,1-DMB; 2,3-DMB

Polymers of 2-Methylpentadienes. 2-Methylpentadienes, along with 4-methylpentadiene, are found in petroleum refinery fractions. The cis and trans isomers,

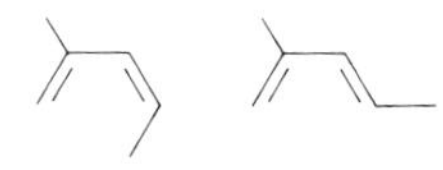

cis trans

have generally been used as a mixture in polymerization studies and treated as one monomer.

Polymers of 4-Methylpentadiene. A study of the behavior of 4-methylpentadiene in emulsion and in various media with butyllithium, lithium, sodium, and potassium showed a slow rate compared with other dienes and a different response to butyllithium in polar media. THF did not elevate the 1,2 content, although some weaker bases, such as diethyl ether, did. Similarly, termination was found in THF.

Polymers of 3-Methylpentadiene. 3-Methylpentadiene is least abundant in petroleum fractions. With $(i\text{-}C_4H_9)_3$ Al–$TiCl_4$ catalyst, only the trans isomer polymerizes; the cis isomer does not react. The product is high-molecular-weight, crystalline *cis*-1,4-poly(3-methylpentadiene), T_m 79–84°C.

Hexadiene Polymers

There are three 2,4-hexadienes:

Because of the terminal methyls, they seem poor prospects for useful polymerization. However, in 1969, a group at Osaka University showed that crystalline polymer could be obtained with Ziegler-Natta catalysts, and that cationic polymers were reasonably high in molecular weight.

1,3-Cyclohexadiene Polymers. Thermal (radical) polymerization competes weakly with dimerization, isomerization, and dismutation, giving a polymer yield of only ca 3% at 100°C.

Heterobutadiene Polymers

The numerous polymeric butadiene derivatives with one or more heteroatoms are not commercially significant, with the exception of polychloroprene (see CHLOROPRENE POLYMERS). In general, the heteroatom may be expected to promote side reactions with the organometallic components of stereoregulating catalysts, thus limiting the options for polymerization. On the other hand, free-radical and anionic polymerization should be less inhibited. Heterobutadiene polymers include halogen-substituted butadienes, cyanobutadiene polymers, and methoxybutadiene polymers.

Mechanisms for Ziegler-Natta Polymerizations

The detailed steps by which stereoregulation is accomplished continue to be a matter of intense speculation. Substituted dienes provide a severe test because of the high degree of regulation required.

BIBLIOGRAPHY

R. N. Young, R. P. Quirk, and L. J. Fetters, *Adv. Polym. Sci.* **56,** 1 (1984).

R. Ohno, Y. Tanaka, and M. Kawakami, *Polym. J.* **4**(1), 56 (1973).

T. Suzuki, Y. Tsuji, Y. Takegami, and H. J. Harwood, *Macromolecules* **12**(2), 234 (1979).

L. Porri, *Structural Order in Polymers,* International Symposium on Macromolecules, Florence, Italy, Sept. 1980, 1981, p. 51.

J. Neil Henderson
The Goodyear Tire & Rubber Company

BUTADIENE POLYMERS

Production of elastomers and plastics consumes by far the largest amount of butadiene monomer. Among the elastomers, the most important synthetic rubber and still most widely used rubber in the world, is the copolymer SBR (styrene–butadiene rubber). The butadiene–acrylonitrile copolymer (nitrile rubber or NBR) has widespread use where solvent and oil resistance are desired. Homopolymers of polybutadiene are widely used in tire tread and sidewall compounds and are important as impact modifiers in HIPS (high impact polystyrene) and ABS (acrylonitrile–butadiene–styrene) resins (see Acrylonitrile–butadiene–styrene polymers). Block copolymers (qv), such as SBS (styrene–butadiene–styrene) are important as adhesives (see Adhesive compositions). In addition, butadiene is the starting material for the preparation of hexamethylenediamine, a component of nylon-6,6.

Butadiene Monomer

Molecular Structure. 1,3-Butadiene, $CH_2{=}CH{-}CH{=}CH_2$, is the simplest conjugated diene. The cumulene isomer, 1,2-butadiene, $CH_3{-}CH{=}C{=}CH_2$, is commercially important only in molecular-weight control in the manufacture of butadiene rubbers.

The structure of 1,3-butadiene has received much theoretical attention because of its symmetry and simplicity. Heats of hydrogenation and combustion reveal that the stabilization contributed by the conjugation of the double bonds is 12.5–14.6 kJ/mol (3–3.5 kcal/mol). Electron diffraction revealed a planar molecule with bond angles C=C—C 122.4° and C=C—H 119.8°.

Physical Properties. At room temperature, 1,3-butadiene is a highly reactive, colorless gas with a mildly aromatic odor. Physical properties are given in Table 1.

Reactions. The chemistry of butadiene is prototypical of hydrocarbons containing conjugated double bonds. Butadiene is a reactive and versatile material. Reactions include self-condensations, Diels-Alder reactions, and addition reactions.

Manufacture. As world economics have changed since World War II, butadiene production has shifted from dehydrogenation of butane and butenes to by-product formation in ethylene manufacture.

Purification. Butadiene is produced from petroleum in a crude C_4 stream. For polymerization, butadiene that is at least 99 mol % pure is required. Alkynes are the most troublesome impurities. Separation of the butadiene from other C_4 products is also necessary. Simple fractional distillation is effective for removing the light (C_3) and heavy (C_5) ends from butadiene, but not for removing the various C_4s because of the closeness of the boiling points to each other and to butadiene. Also, butadiene forms azeotropes with n-butane and 2-butene. The most widely used recovery systems are extraction with aqueous cuprous ammonium acetate (CAA) and solvent extractions with furfural, acetonitrile, dimethylformamide, dimethylacetamide, or N-methylpyrrolidinone.

Table 1. Physical and Thermodynamic Properties of Butadiene

Property	Value
molecular weight	54.09
boiling point, °C at 101.3 kPa[a]	−4.413
melting point, °C	−113
freezing point, in air at 101.3 kPa[a], °C	−108.915
density, at saturation pressure at 20°C, g/mL	0.6211
pressure coefficient of boiling point at 101.3 kPa[a], °C/Pa	4.502
vapor pressure of liquid at 21°C, kPa[a]	24.5
critical temperature, °C	152
critical pressure, MPa[b]	4.33
critical density, g/mL	0.245
critical volume, mL/g[c]	4.09
flash point, °C	−76.1

[a] To convert kPa to mm Hg, multiply by 7.5.
[b] To convert MPa to atmospheres, multiply by 9.87.
[c] To convert mL/g to L/mol, multiply by 0.0540.

Health and Safety Factors. *Toxicity.* Butadiene causes tumors in male and female rats and should be considered a possible human carcinogen.

Handling and Storage. Butadiene is dangerous because of its explosiveness when mixed with air. Because of the high vapor pressure of the liquid, it must be kept under pressure.

Polymer Structure

Microstructure (qv) refers to the disposition of the double bonds present in the polymer chain and macrostructure refers to long chain features such as extent of branching, molecular weight, and molecular-weight distributions.

Microstructure. Butadiene polymerizes by addition (see Addition polymerization). Having two double bonds, it forms a variety of polymer structures. One of these, known as the vinyl or 1,2-type, results from addition across one of the double bonds:

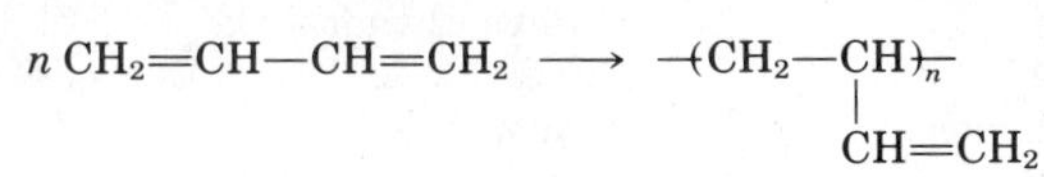

1,2-polybutadiene

Three vinyl structures are possible: the isotactic, the syndiotactic, and the heterotactic (or atactic).

A second structure arises by 1,4-addition when both double bonds participate. The remaining double bond in the polymer backbone permits two 1,4-structures, the cis-1,4 (or Z), and the trans-1,4 (or E) isomers. In the absence of side chains there is no tacticity.

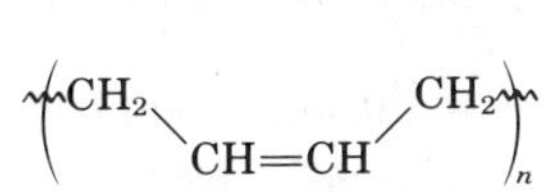

cis-1,4 isomer

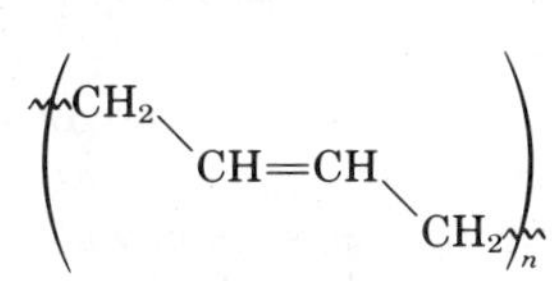

trans-1,4

In an actual polymer it is possible to have any or all of the several forms in a single chain. The structures indicated have all been synthesized individually, as well as in various mixtures, to give polymers with a preponderance of either 1,2- or 1,4-units.

Polymerization

Polymerization is either initiated or catalyzed and done in homogeneous solution, including bulk polymerization where excess monomer functions as the solvent, and in water emulsion. The propagating polymer may be an anion, cation, free-radical, or coordination complex. Ionic species and coordination species that are reactive enough to initiate polymerizations are usually too reactive to be stable in water solution, but free-radical polymerizations may be carried out in aqueous or non-aqueous media.

In contrast to homogeneous free-radical systems, emulsion polymerization (qv) affords high propagation rates and high molecular-weight products of elastomer quality. The propagating radical chains are somewhat physically isolated and thus prevented from recombining as rapidly as they would in solution or bulk media.

Styrene–Butadiene Copolymers. The synthetic rubber industry, based on the free-radical emulsion process, was created almost overnight during World War II. Styrene–butadiene rubber (SBR) created at that time gives such good tire treads that natural rubber has never regained this market. The free-radical emulsion process is still used today.

Commercial Grades. Table 2 gives trade names but covers only commercial grades of SBR and emulsion-type polybutadienes.

Table 2. Trade Names of SBR and Emulsion-type Polybutadienes

Trade name	Manufacturer
Ameripol	BFGoodrich
ASRC	American Synthetic Rubber Corp.
Buna	Buna Werke Huls G.m.b.H.
Copo	Copolymer Rubber and Chemical Corp.
Europrene, Intol, Sirel	Enoxy Chemical Ltd.
Gentro	The General Tire & Rubber Co.
JSR	Japan Synthetic Rubber Co.
Philprene	Phillips Chemical Co.
Pliofex	The Goodyear Tire & Rubber Co.
Polysar, Krylene, Krynol	Polysar, Ltd.
Sumitomo	Sumitomo Chemical Co.
Synpol	Synpol, Inc.

Acrylonitrile–Butadiene Copolymers (NBR). The generic term nitrile rubber refers to copolymers of any unsaturated nitrile and diene, but commercially are almost exclusively acrylonitrile and butadiene.

Nitrile rubbers are used primarily where resistance to swelling by oils and greases is required. A high nitrile content increases swell resistance but limits flexibility at low temperatures.

Nitrile rubbers are produced in emulsion. Manufacturers have their own trade names and code designations (Table 3).

Table 3. Trade Names of Nitrile Rubbers

Trade name	Producer
Chemigum	The Goodyear Tire & Rubber Co.
Europrene	Enoxy Chemical Ltd.
Hycar	BFGoodrich Co.
Krynac	Polymer Corporation, Ltd.
Nysyn	Copolymer Rubber Co.
Paracril	Uniroyal Inc.
JSR	Japan Synthetic Rubber Co.
Elaprim	Montepolimeri S.P.A.
Perbunan	Bayer A.G.

Other Emulsion Copolymers. Many monomers have been copolymerized with butadiene. A few have some commercial utility. A butadiene–styrene–vinylpyridine terpolymer latex is used for improving tire cord and fabric adhesion to rubber.

Latex. The first latices (qv) were polymer mixtures stabilized with antioxidant. They contained 25–30% solids, but were soon succeeded by latex of higher solids content. Latex is useful for coatings, adhesives, paper treating, foam rubber, carpet backing, and dipped goods.

Anionic Polymerization. ***Microstructure.*** Changes in the recipe for free-radical polymerization, in emulsion, solution, or bulk, cause little or no change in the microstructure of the resultant polymers, whereas changes in the recipes for alkali metal polymerization often appreciably affect the structure of the product (see also ANIONIC POLYMERIZATION).

The structures of polybutadiene obtained in hydrocarbon media with alkali metal catalysts and in an emulsion are shown in Table 4.

Table 4. Microstructure of Alkali Metal-catalyzed Butadienes

Catalyst or condition	cis-1,4, %	trans-1,4, %	1,2, %
lithium	35	52	13
sodium	10	25	65
potassium	15	40	45
rubidium	7	31	62
cesium	6	35	59
emulsion	18	64	18

Lithium-initiated Butadiene Copolymers. In lithium-initiated systems, butadiene comonomers are limited to materials such as styrene and isoprene that do not undergo terminating side reactions with strong bases. The most useful comonomer is styrene. Because of the stability and "living" nature of the allylic lithium end group, butadiene–styrene copolymers of widely different structures and properties are conveniently prepared.

Block Copolymers. The utility of lithium-based anionic initiator systems is evident in the synthesis of block copolymers (qv). The living anionic end of a polymer chain A can initiate the polymerization of another monomer B to form an A–B block copolymer.

End-group Reactions. Anionic polymerization is primarily concerned with the anionic end of the polymer chain. With monofunctional initiators, termination of the polymerizing butadiene with various reagents can give rise to many polymers

Table 5. Characteristics of Crystalline Polybutadienes

Polybutadiene	Lattice	Monomers/ unit cell	Mp, °C	Crystal density, g/cm³	Repeat distance, nm
1,4-trans, 99–100%	pseudo-hexagonal	1	145	0.97 (I) 0.93 (II)	0.485 0.465
1,4-cis, 98–99%	monoclinic	2	2	1.01	0.86
1,2-isotactic, 99%	rhombohedral	2	126	0.96	0.65
1,2-syndio-tactic, 98%	helical	3	156	0.96	0.514

containing an end-functional group. A diinitiator gives a polybutadiene with two live ends that can be terminated to give α,ω difunctional polymers. Although such polymers can be produced by oxidative degradation or free-radical polymerization by an initiator containing a functional group, this technique gives better control.

Transition-metal Catalysts. Transition-metal or coordination catalysts (ie, Ziegler-Natta catalysts (qv)) comprise mixtures of an organometallic and a transition-metal compound. These catalysts often contain a heterogeneous, insoluble phase, which may be a complex of the two components. The combination of two or more ingredients permits a wide range of activities and structures and has led to unusual, new polyolefins and polydienes including high *cis*-1,4-polyisoprene, which duplicates natural rubber, *Hevea brasiliensis* (see also ISOPRENE POLYMERS). Among polymers not found in nature, high *cis*-1,4-polybutadienes have attained commercial importance.

The transition-metal catalysts that promote cis-addition fall into three classes: titanium-based catalysts containing iodine and trialkylaluminums, cobalt-based catalysts containing alkylaluminum halides, and nickel-based catalysts containing BF_3 and trialkylaluminums. Recently, rare-earth catlysts, particularly those based on neodymium, have shown commercial potential. Other catalysts allow the preparationn of *trans*-1,4-polybutadiene and syndiotactic and isotactic 1,2-polybutadiene. Generally, π-allyl complexes of nickel, cobalt, titanium, and rhodium tend to produce 1,4-polybutadienes, whereas chromium, molybdenum, and palladium produce 1,2-polybutadienes.

Butadiene Copolymers. Using transition-metal catalysts, butadiene has been copolymerized with isoprene, 2,3-dimethylbutadiene, and styrene.

Cationic Polymerization. Cationic polymerization of butadiene is conducted at very low temperatures to reduce gel formation. Emulsion or solution polymerization of butadiene with rhodium salts gives 98 + % trans polymer and is thought to be cationic.

Polymer Properties

Uncured Polymers. Polybutadienes of the four possible structures have been prepared sufficiently pure for crystallization and structure determination (Table 5).

Cured Polymers. The microstructure of polybutadiene affects its mechanical behavior. Changes in physical properties occur as the structure approaches pure cis or pure trans; little change is evident between 25 and 80% cis content. High cis polymers have high resilience and low hysteresis. Very high cis (>96%) polymers crystallize on being stretched and, like *Hevea,* the gum stocks have high tensile strengths. High trans polymers are crystalline with high tensile strength gum stock, high hysteresis, and inferior resilience. The amorphous polymers of high vinyl content (95% 1,2) have fairly good tensile properties, low hysteresis, and poor low-temperature characteristics; T_g is approximately 0°C. Crystalline, syndiotactic 1,2-polybutadiene (70% 1,2) resembles the amorphous polymer but has higher tensile strength and elastic modulus.

The widespread use of butadiene-based elastomers in automobile tires (qv) is attributable to their favorable balance of cost, processibility and resistance to heat, abrasion, and oxidative degradation.

Although nitrile rubbers swell more in polar solvents than nonpolar, they can be used with antifreeze solutions and in similar applications. They are also used in poly(vinyl chloride) and with styrene (ABS resins). In textiles and leather, nitrile latices are used to increase abrasion resistance and as binders for pigments. In paper manufacture they are used as pigment binders, coatings for increased waterproofing and grease-proofing, and in making high strength papers.

Processing

Elastomers, unlike fibers and plastics, require fillers (qv) and other ingredients as well as curing (qv). Processing involves three steps: mastication of the raw rubber with fillers, processing aids, curatives, antioxidants, etc; molding of the rubber compound into the desired shape; and cure (vulcanization) by heat. Similar equipment is used for butadiene polymers and other rubbers. Indeed, the commercial success of a new rubber often depends on whether it can be handled in existing equipment.

Economic Aspects

Butadiene consumption is dominated by the synthetic rubber industry, which in turn is dominated by the tire market. Ap-

Table 6. U.S. Production of Butadiene-type Rubbers, 10^3 Metric Tons

Year	SBR	BR	N-Type
1952	666		13.9
1962	979		36.5
1972	1500	298	75
1977	1513	362	80
1979	1436	397	89
1981	1085	342	70
1982	876	288	45
1983	904	332	55

proximately 70% of U.S. butadiene consumption is used in rubber products, of which over 50% is for tires. Technological change and economic conditions greatly affect butadiene consumption. Production figures are shown in Table 6.

Prices. The approximate 1985 list prices of rubber are shown in Table 7.

Table 7. 1985 Rubber Prices[a]

Rubber or monomer	$/kg
butadiene monomer	0.73
SBR	
emulsion	1.23
anionic	1.25
cis-polybutadiene	1.47
nitrile rubber	2.33
natural rubber[b]	0.93

[a] Actual prices may vary considerably because of discounting.
[b] No. 1 RSS.

BIBLIOGRAPHY

F. R. Eirich, ed., *Science and Technology of Rubber,* Academic Press, New York, 1978.

M. Saltman, ed., *The Stereo Rubbers,* John Wiley & Sons, Inc., New York, 1977.

J. P. Kennedy and E. G. M. Tornqvist, eds., *Polymer Chemistry of Synthetic Elastomers, Parts I and II,* Vol. XXIII of *High Polymers: A Series of Monographs on the Chemistry, Physics, and Technology of High Polymeric Substances,* Wiley-Interscience, New York, 1968.

DAVID P. TATE
T. W. BETHEA
Firestone Tire & Rubber Company

BUTENE POLYMERS

Polybutylene (PB) refers to a series of commercial semicrystalline resins derived from high molecular-weight, predominantly isotactic poly(1-butene) homopolymer or copolymer. Poly(1-butene) can be represented by the formula $\left[CH_2CH(C_2H_5) \right]_n$, where *n* is the degree of polymerization. Unless otherwise indicated, the poly(1-butene) or PB in this article refers to isotactic homopolymer.

Properties of 1-Butene

1-Butene is a colorless gas with a boiling point of −6°C at 101.3 kPa (1 atm). 1-Butene undergoes isomerization, promoted by various catalysts, and hydrogenation. It undergoes Ziegler-Natta polymerization to form an isotactic, crystalline polymer. Cationic or free-radical initiators produce atactic poly(1-butene), which is amorphous (see also ZIEGLER-NATTA CATALYSTS).

Manufacture of 1-Butene

1-Butene is produced commercially, together with other α-olefins, by the catalytic oligomerization of ethylene.

Polymer Properties

Polymorphism. Poly(1-butene) exhibits polymorphism, including the crystal forms I (twinned hexagonal), II (tetragonal), III (orthorhombic), I′ (untwinned hexagonal), and II′.

Melt Crystallization. Poly(1-butene) crystallizes from the unstrained melt in spherulitic morphology (qv). The crystallization rate increases with increasing molecular weight. The temperature for attaining homogeneous nucleation of poly(1-butene) crystallization is below 0°C, compared to ca 81°C for linear polyethylene and 88°C for isotactic polypropylene.

Transformation in the Solid State. The metastable form-II poly(1-butene), crystallized from the melt, undergoes a spontaneous, irreversible transformation to the stable form I. This transformation is affected by temperature, pressure, mechanical stress and orientation (qv), molecular weight, tacticity, comonomers, additives, impurities, sample thickness, and γ radiation.

Mechanical Properties. Generally, PB does not exhibit necking on elongation, but rather shows uniform sample deformation and strain-hardening. The yield strength, ultimate strength, and modulus increase with increasing crystallinity and decrease with increasing temperature. The glass-transition temperature (T_g) of poly(1-butene) ranges from −17 to −25°C. The activation energy for viscous flow is 45 kJ/mol (11 kcal/mol). The properties of pipe-grade PB are listed in Table 1.

Table 1. Properties of Pipe-grade Polybutylene Resin[a] at 23°C

Property	ASTM test method	Value
melt index, g/10 min	D1238[b]	0.4
density, g/cm³	D792	0.93–0.94
tensile properties		
yield strength, MPa[c]	D638	16–18
break strength, MPa[c]	D638	32–35
elongation to break, %	D638	275–320
modulus, MPa[c]	D638	290–295
flexural modulus, MPa[c]	D790	375–380
notched Izod impact strength, J/m[d]	D256	640–800
hardness, Shore	D2240	D60
brittleness temperature, °C	D746	−18
melting temperature, °C	D3418	124–130
Vicat softening point, °C	D1525	112–114
heat deflection temperature, °C		
at 1.82 MPa[e]	D648	54–60
at 0.46 MPa[e]	D648	102–113
linear coefficient of thermal expansion, m/m/°C	D696	0.00013
thermal conductivity, W/(m·K)	C177	0.22
UL flammability	UL94	HB
dielectric constant, 10^3–10^6 Hz	D150	2.53
dissipation factor, 10^3–10^6 Hz	D150	0.0005
water absorption, in 24 h, %	D570	<0.03

[a] Compression-molded into plaques and after transformation to stable form I.
[b] Condition E.
[c] To convert MPa to psi, multiply by 145.
[d] To convert J/m to ft·lbf/in., divide by 53.38.
[e] To convert MPa to atm, divide by 0.101.

Chemical Properties. Polybutylene resins resist most acids and bases below 90°C and many chemicals, solvents, soaps, and detergents. Resistance is poor to strong oxidizing acids and to aromatic and chlorinated hydrocarbon solvents above 60°C. During thermooxidative and photooxidative degradation, PB generally undergoes chain scission rather than cross-linking. Stabilizers prevent oxidation induced by heat and ultraviolet radiation.

Solution Properties. At elevated temperatures, poly(1-butene) dissolves in benzene, toluene, decalin, tetralin, chloroform, and α-chloronaphthalene. Its solubility depends on temperature, molecular weight, and crystallinity.

Polymerization

Isotactic poly(1-butene) is manufactured by stereospecific Ziegler-Natta polymerization (qv).

Processing

PB is melt extruded to form pipe, sheet, or profile section. Conventional single-screw extrusion is used; vacuum- or pressure-sizing can be used for dimensional control in pipe manufacture. Film may be manufactured by the blown or cast process. Coextrusion and extrusion coating are also practiced. PB can also be injection molded. In melt processing of commercial PB resins, crystal forms II and I are generally encountered and the transitions must be considered in part design.

Economic Aspects

The 1986 worldwide PB market was approximately 11,200 t/yr. The bulk list price of PB resins ranges from $1.84 to $2.87/kg, depending on the grade.

Health and Safety Factors

1-Butene may be encountered when sealed containers are opened. The gas is flammable. It is not toxic at moderate concentration, but acts as an anesthetic and asphyxiant at higher concentration.

Uses

The most important commercial outlet for PB resins is in pipe, with applications that include hot- and cold-water plumbing and cold-water service. PB is also used in specialty films and film coats for packaging dry foods, meats, and medical parts. Sheet is used for abrasion-resistant lining applications. PB resins are injection molded into fusion fittings for pipe and tubing. Other uses are in adhesives and sealants formulations and as a modifying additive with other polyolefins.

BIBLIOGRAPHY

I. D. Rubin, *Poly(1-Butene)—Its Preparation and Properties,* Gordon & Breach, New York, 1968.

S. G. Kemp, *Plast. Rubber Proc. Appl.* **3**(2), 169 (1983).

A. M. Chatterjee, "Polybutylene," Chapt. 83 in I. Rubin, ed., *Handbook of Plastics Materials and Technology,* John Wiley & Sons, Inc., New York, 1990.

M. P. Schard, "Polybutylene," in *Modern Plastics Encyclopedia 1986–87,* McGraw-Hill Inc., New York, p. 38.

ANANDA M. CHATTERJEE
Shell Development Company

BUTYLENE POLYMERS.

See BUTENE POLYMERS.

BUTYL METHACRYLATE POLYMERS.

See ACRYLIC AND METHACRYLIC ESTER POLYMERS.

BUTYL RUBBER.

See ISOBUTYLENE POLYMERS.

BUTYRATE PLASTIC.

See CELLULOSE ESTERS, ORGANIC.

C

CABLE COVERINGS. See Wire and cable coverings.

CAD. See Computer-assisted design.

CAGE EFFECT. See Radical polymerization.

CALENDERING

Calendering is essentially extruding a mass of material between successive pairs of corotating, parallel rolls to form a film or sheet. Figure 1 shows the calendering process schematically. This process is widely used in the thermoplastics, rubber, textile, paper, and nonwoven fabrics industries.

With paper, textiles, and nonwoven fabrics, a preformed web is passed through a calender to improve its appearance and properties. In viscous bank forming, used for thermoplastics and rubber, the sheet is usually formed, densified, and surfaced in the calender; sometimes a preformed plastic sheet is patterned in a calender. Often plastic or rubber is laminated onto a substrate in a calender. Floor tile is made in specialized

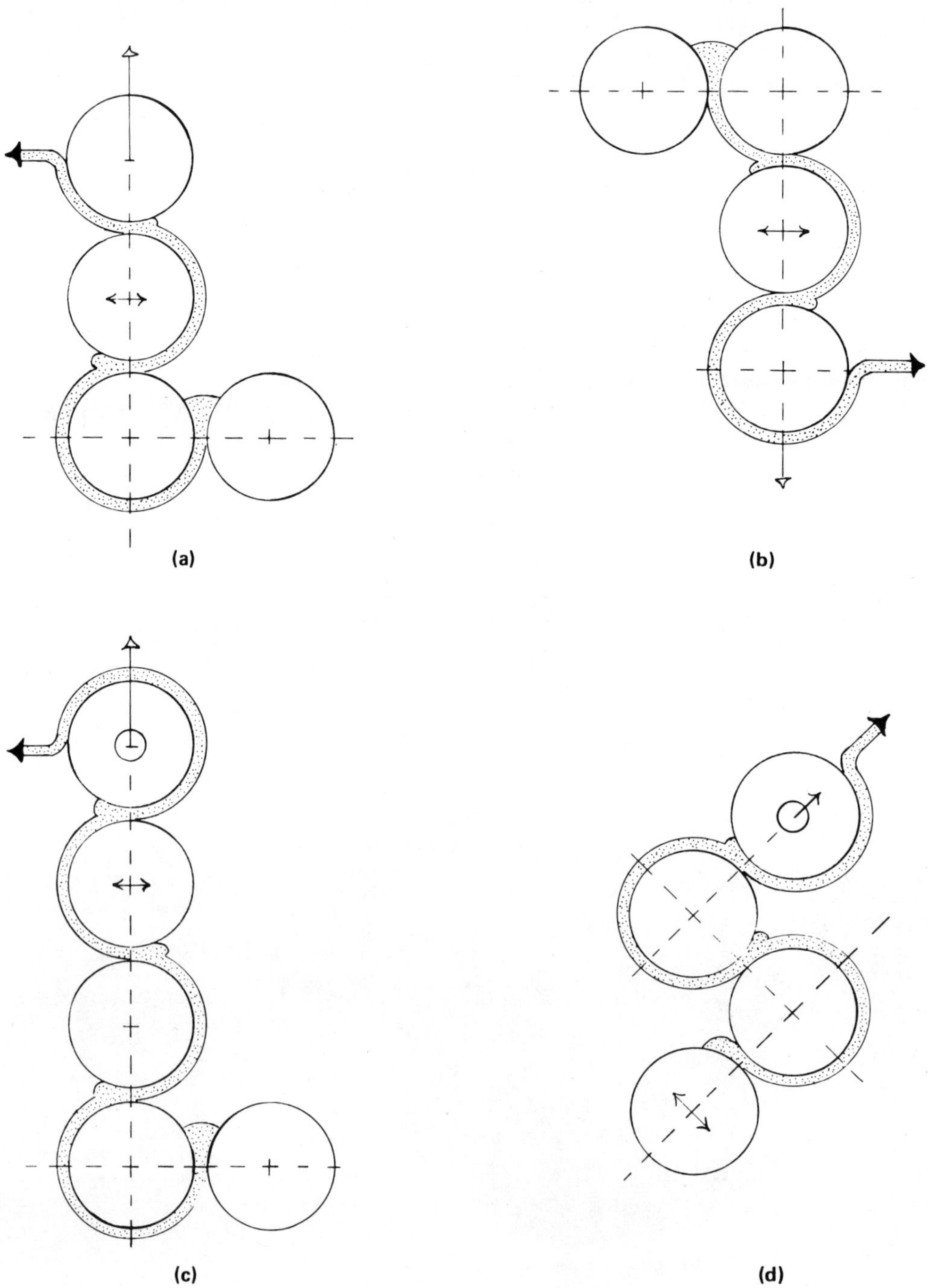

Figure 1. Calendering. (**a**) 4-Roll L Calender; (**b**) 4-roll inverted L calender; (**c**) 5-roll L calender; (**d**) 4-roll Z calender. Courtesy of Marcel Dekker, Inc.

calenders. In the metals industries, processes somewhat similar to calendering are referred to as cold rolling or hot rolling, depending on temperature conditions.

Many polymers are formed into supported and unsupported webs by calendering. These include poly(vinyl chloride) (PVC), poly(acrylonitrile–butadiene–styrene) (ABS), polyurethane (PU), styrene–butadiene rubber (SBR), natural rubber (NR), and others.

PVC and its copolymers, sometimes blended with ABS, comprise the largest volume of calendered thermoplastics and are the primary polymeric materials discussed in this article; analogous equipment and processes are used for other polymers.

Theory

A theoretical understanding of the phenomena governing calendering is necessary to address the demands on a modern calender: high speed; small thickness tolerances of the finished sheet or film, both in the calendering direction and perpendicular to it; uniform structure and surface of the film; and economical operation.

Control of Thickness across the Web. When a calendered web is wound up onto a core, close thickness control across the web is necessary. Calculations and experience in subsequent web-handling operations provide guidelines. For a web wound tightly on a core,

$$L = \frac{\pi}{1000}\left[t\left(\frac{D-d}{2t}\right)^2 + (d+t)\left(\frac{D-d}{2t}\right)\right]$$

where L = length of web on roll in m; t = thickness of web on roll (in mm); D = outside diameter (OD) of roll (in mm); and d = inside diameter (ID) of roll (in mm).

Causes of Thickness Deviations. In calendering thin polymeric webs, roll-separating forces in the final nip may be as high as 1051.1 kN/m (6000 lb/in), which may cause considerable deflection of the final calender roll. Unless this is compensated for, gauge variation across the web will occur.

Other potential causes of thickness changes across the calendered web include nonhomogeneous rheology of the stock, problems with control of stock temperature and lubricity, equipment or control system malfunction, range of roll-separating force along the length of the roll face, and temperature gradients in the rolls or use of damaged calender rolls.

Compensation for Calender-roll Deflections. In the calendering of plastics, compensation for deflection of calender rolls is normally done by combining three methods: axis crossing or skewing, roll straightening (often called roll bending), and roll crown (sometimes called roll profile). Roll crown may be ground or blocked on.

Estimating Rolls-separating Forces. Although limited by a number of simplifying assumptions, the following expression for calculating rolls-separating forces in milling and calendering plastics has been used for years:

$$F = 2\mu Vrw\left(\frac{1}{h_o} - \frac{1}{H}\right)$$

where F = total separating force between rolls in kN; μ = viscosity of stock being calendered in Pa·s; V = peripheral speed of rolls in m/s; r = radius of rolls in m; w = width of web being calendered in m; h_o = separation between rolls at the nip in mm; and H = height of stock at the nip entrance in mm.

Equipment

Information on calendering equipment and auxiliary units is updated frequently by manufacturers such as Farrel Co., Adamson, Berstorff Corp., and EKK GmbH, and is periodically reviewed. Equipment for the Luvitherm Process for making biaxially oriented, thin, rigid PVC films from emulsion PVC has been described.

Processing Considerations

Preparation of stock for calendering, conditions on the calender, the takeoff, thickness measurement, and control and windup must all be adapted to the polymer system being calendered. Other considerations include whether the polymer is laminated to a fabric on the calender; whether the web is embossed or slit in line; what finish is required, ie, glossy, semimatte, or matte; and whether the product needs special properties such as optical clarity, biaxial orientation, or very low strain recovery. General processing considerations include temperature control, calender speed, finish, gauge, orientation, and embossing.

Recent Improvements. Much progress in calendering is due to improvements in control of the overall heating and cooling processes involved in complete calender trains, as well as the speed and tension of the web at each stage.

Economic Aspects

The bulk of U.S. calendering output in plastics is PVC, with flexible sheet and film accounting for the greatest volume. The capital cost of a new calendering line is $5 $\times$ 10^6, exclusive of building costs. The fixed plus variable costs in the U.S. for operating a new line amount to $400–500/h.

Health and Safety Factors

Most of the hazards in calendering facilities are associated with the high speeds at which webs normally travel, the large forces involved in calendering, or the high temperatures.

BIBLIOGRAPHY

Nipco and Tri-Pass Rolls, Bulletin 134; Non-Woven Calenders; Calenders for Industrial Sheeting, Elastomers, Tapes, Coated Fabrics, Bulletin 223; Farrel Machinery Group, Ansonia, Conn., 1983.

R. H. Crotogino, *Temperature-Gradient Calendering;* R. H. Crotogino, S. M. Hussain, and J. D. McDonald, *Mill Application of the Calendering Equation,* Pulp and Paper Research Institute of Canada, Ponte Claire, Quebec, 1982.

A. W. M. Coaker in I. L. Gomez, ed., *Engineering with Rigid PVC: Processability and Applications,* Marcel Dekker, Inc., New York, 1984, Chapt. 8.

H. J. Bielitza and K. Marquardt, "Calendering" in R. Juran, ed., *Modern Plastics Encyclopedia,* Vol. 65, No. 11, McGraw-Hill Inc., New York, Oct. 1988, p. 224.

A. William M. Coaker
Consultant

CAM. See PROCESSING, COMPUTER CONTROL.

CAOUTCHOUC. See RUBBER, NATURAL.

CAPROLACTAM POLYMERS. See POLYAMIDES.

CARBAMATES, POLYMERIC. See POLYURETHANES.

CARBANIONIC POLYMERIZATION. See ANIONIC POLYMERIZATION.

CARBOCATIONIC POLYMERIZATION. See CATIONIC POLYMERIZATION.

CARBODIIMIDES, POLY-. See ISOCYANATE-DERIVED POLYMERS.

CARBOHYDRATES. See POLYSACCHARIDES.

CARBON

Carbon is a nonmetallic element occurring in the native state and is also obtained industrially in varying states of order and purity. Unlike the other two forms of elemental carbon, diamond and graphite, carbon black is not found in the native state; it has been synthetically produced for several millennia. Carbon black is the generic name for the family of materials consisting essentially of elemental carbon in the form of spheroidal colloidal particles and coalesced particle aggregates of colloidal size obtained by the thermal decomposition of hydrocarbons. Carbon black is the only form of carbon that is used extensively in polymers.

CARBON BLACK

Properties

Morphology and Microstructure. With the exception of coarse thermal carbon, carbon blacks rarely exist as individual spheroidal particles. Instead, particles exist in coalesced or fused aggregates forming a coherent unit. The carbon-black particle, also referred to as a domain or nodule, consists of concentric, imperfect graphitic layer planes.

Surface Area. Surface area is inversely related to particle size for smooth-surface carbon blacks and affects the performance of carbon-black–filled polymers. Surface areas are generally determined by nitrogen adsorption using the ASTM standard test method.

Surface Activity and Chemistry. The surface activity of carbon black is a function of the degree of heterogeneity of the carbon surface and the nature and number of chemical functional groups on the edges of the graphitic layer planes. The actual carbon content ranges from ca 80 to +99 wt %, depending on the manufacturing process and the aftertreatment.

The interaction between carbon black and polymer may be physical or chemical in nature, and probably both mechanisms are frequently involved. The microstructure itself is the dominant factor in the interaction with most elastomers.

Manufacture

Carbon black has been produced by five processes: lampblack, impingement, acetylene, thermal, and furnace (Table 1). The furnace process is the dominant worldwide process.

Economic Aspects

Worldwide carbon-black capacity was estimated at 5.7×10^6 t in 1983; approximately 98% was made by the furnace process. The U.S. furnace-carbon capacity represents approximately 28% of the world capacity.

Applications

In 1982, approximately 1.02×10^6 t of carbon black was consumed in the United States, 63% of which was used in rubber products for transportation (mainly tires and automotive uses), 30% in industrial rubber products (belting, hose, and molded products) and 7% in inks, coatings, plastics, and specialty applications. About 93% of carbon-black production is used in elastomers, mainly to improve strength properties. Carbon black is the principal pigment in ink, and excels in its ability to prevent damage to polymers from uv radiation.

Table 1. Properties of Carbon Blacks

Process grade	1983 U.S. production, 10^3 t	Particle diameter d, nm	Porosity, % of total surface
furnace			
rubber	954	17–110	<12
ink	38	18–70	<8
paint	6.8	10–70	<45
plastics			
color	29	18–70	<36
conductive	6.8	15–30	>35 <95
lampblack replacement	1.4	69–90	0
other	13	15–70	<75
thermal	18	150–500	0
lampblack	1.6	100–130	0
acetylene	6.8	35–50	0

Health, Safety, and Environmental Factors

The Environmental Health Association of the Carbon Black Industry, Inc. has compiled a bibliography related to the health and environmental effects of carbon black. An excellent critical review of these studies concluded that commercial carbon blacks are neither mutagenic nor carcinogenic.

ACTIVATED CARBON

Activated carbons are extremely porous carbon materials that have been carbonized and oxidized to increase their adsorptive capacity. Activated carbons are prepared from carbonaceous materials by oxidation at temperatures of 700–1000°C, usually with steam or carbon dioxide, and sometimes with the addition of inorganic chemicals to aid degradation and dehydration.

Traditional applications have included decolorization of sugar, removal of undesirable tastes and odors from drinking water, treatment of municipal and industrial wastes, and a host of other applications in the food, beverage, electroplating, and pharmaceutical industries (see also WASTE TREATMENT).

GRAPHITE

Graphite, in its natural form, consists of ordered parallel layers of planar hexagonal carbon rings. Graphite has lubricating properties that result from the weak van der Waals forces between the layer planes and the presence of an adsorbed layer of water. Graphite lubricants are particularly heat-resistant and stable.

By selection of raw materials and processing conditions, graphites can be manufactured with a wide range of properties. The graphitization step is carried out in either batch or continuous processes between 2500 and 3000°C. Manufactured graphite is a composite of filler carbon, binder carbon, and pores and has a particularly low coefficient of thermal expansion and good thermal shock resistance.

BIBLIOGRAPHY

J. B. Donnet and A. Voet, *Carbon Black,* Marcel Dekker, Inc., New York, 1976.

A Bibliography of Carbon Black Literature Concerning Occupational and Environmental Health, The Environmental Health Association of the Carbon Black Industry, Inc., May 1982.

R. W. Soffel in M. Grayson, ed., *Encyclopedia of Chemical Technology,* 3rd ed., Vol. 4, Wiley-Interscience, New York, 1978, pp. 689–709.

S. B. Seeley in M. Grayson, ed., *Encyclopedia of Chemical Technology,* 3rd ed., Vol. 4, Wiley-Interscience, New York, 1978, pp. 689–709.

F. LYON
Columbian Chemicals Company

K. A. BURGESS
Consultant

CARBONATES, POLYMERIC. See POLYCARBONATES.

CARBON FIBERS

Over the last 15 years, carbon fibers have emerged as the main reinforcement fiber for high performance composite materials. The development of strong and stiff carbon fibers for a wide range of applications and their use in lightweight structural parts are among the principal technological achievements of this period. These fibers demonstrate a unique combination of static and dynamic performance advantages in demanding uses. High strength and stiffness, light weight, improved fatigue resistance and vibration damping, corrosion resistance, good friction and wear qualities, low thermal expansion, and thermal and electrical conductivity combine to make carbon-fiber composites an attractive substitute for various metals, special alloys, and other materials. These performance characteristics, in concert with composite structure, offer a degree of versatility in design and fabrication that is unavailable in other materials (see also COMPOSITES).

The utility of carbon fiber principally depends on its use as a reinforcement in various matrices, including thermoset synthetic resins such as epoxies, bismaleimides, polyimides, vinyl esters, and to a growing extent in a number of thermoplastics; it is used to a lesser degree in carbon, metal, ceramic, and glass matrices.

The term carbon fiber generally refers to materials that have been heat treated at temperatures of 1000–3000°C and have markedly different properties and structure; they contain 92–99.99 + % carbon. In practice, the term carbon fiber sometimes refers to materials treated in the range of 1000–2000°C. The term graphite fiber refers to fibers processed at or in excess of 2500°C, although technically, these fibers lack the three-dimensional crystalline structure of graphite, except for the highest modulus, pitch-based materials. In this article, the term carbon fiber is used in its broadest sense.

Carbon fibers have been broadly categorized, as listed in Table 1, principally according to tensile or Young's modulus (or stiffness) and heat-treatment temperature.

Properties

Carbon-fiber properties are highly anisotropic; the properties of interest and advantage are in the longitudinal (fiber axis, 0°) direction and the properties transverse (or 90°) to the fiber direction are distinctly different.

Representative unidirectional properties of PAN and mesophase pitch-based carbon fibers are given in Tables 2 and 3, respectively. As tensile modulus increases (increasing degree of orientation along the fiber axis), carbon content, density, electrical conductivity (inverse of resistivity), and thermal conductivity increase, tensile strength and elongation decrease.

Composites. Structural composites are defined as products that utilize fiber reinforcements (50–70% by weight) of very high strength and stiffness in combination with polymers, metals, carbon, glass, or ceramics to form structural materials with unique property-to-weight characteristics. The matrix binds the reinforcing fibers together, forming a coherent structure, and provides a medium by which to transfer applied stresses from one filament to the next. The matrix material affects high temperature mechanical properties, transverse strength, ie, nonunidirectional, and moisture resistance, as well as other properties, and is a key factor in toughness, shear strength, and oxidation and radiation resistance. The matrix system also strongly influences the fabrication process and associated parameters for forming composite materials into intermediate and final products.

Most carbon fiber composites are based mainly on thermosetting epoxy matrices; they offer ease of fabrication, ex-

Table 1. Broad Categories of Carbon Fibers

Carbon %	Fiber type[a]	Heat-treatment temp, °C[b]	Modulus, GPa[b,c]
92–96	high strength	1200–1400	228–241
99	high modulus	1800–2500	276–380
99+	ultrahigh modulus	2800–3000+	483–690+

[a] In the UK, high strength fibers are called Type II and high modulus, Type I.
[b] Principally PAN-based carbon fibers.
[c] To convert GPa to psi, multiply by 145,000.

Table 2. Properties of PAN-based Carbon Fiber[a]

Property	High strength[b]	High modulus[c]	Ultrahigh modulus[d]
filament dia, μm	5.5–8.0	5.4–7.0	8.4
density, g/cm^3	1.75–1.80	1.78–1.81	1.96
carbon content, wt %	92–95	99–99+	99+
tensile strength, MPa[e]	3105–4555	2415–2555	1865
tensile modulus, GPa[e]	228–262	359–393	517
elongation at break, %	1.3–1.8	0.6–0.7	0.38
electrical resistivity, μΩ·m	15–18	9–10	6.5
thermal conductivity, W/(m·K)	8.1–9.3	64–70	ca 120

[a] Strand data.
[b] Thornel T-300, T-500, T-600, T-700, Union Carbide Corp.; Celion 3000, 6000, 12000, Celanese Corp.; AS2/AS4/AS6/IM6, Hercules Corp.
[c] Thornel T-50, Celion G-50, HMS, Hercules Corp.
[d] Celion GY-70.
[e] To convert MPa to psi, multiply by 145. To convert GPa to psi, multiply by 145,000.

Table 3. Properties of Mesophase Pitch-based Carbon Fiber

Property	Thornel P55S[a]	Thornel P75S[a]	Thornel P100[a]
filament dia, μm	10	10	10
density, g/cm^3	2.02	2.06	2.15
carbon content, wt %	99	99	99+
tensile strength, MPa[b]	1895	2070	2240
tensile modulus, GPa[b]	380	517	690
elongation at break, %	0.5	0.4	0.3
electrical resistivity, μΩ·m	7.5	4.6	2.5
thermal conductivity, W/(m·K)	110	185	515

[a] Thornel is a trade name of Union Carbide Corp.
[b] To convert MPa to psi, multiply by 145. To convert GPa to psi, multiply by 145,000.

cellent adhesion, low shrinkage during cure, a good balance of physical and electrical properties, and good heat resistance and stability under a wide range of environmental conditions. Other thermoset and thermoplastic matrix systems are growing in significance. Typical properties are given in Table 4.

Property Development. Tensile, or Young's modulus, an intrinsic material property, is correlated to preferred orientation, that is, the more highly oriented the layer planes in carbon fiber, the higher the tensile modulus. Young's modulus can be increased by higher temperature and higher stress (stretch) processing.

Structure of PAN Carbon Fiber and Property Development. A key characteristic of carbon fibers is high structural anisotropy, which can be achieved in a filament built up of two-dimensional, highly anisotropic graphitic crystallites arrayed in undulating sheetlike assemblies oriented along the fiber axis. The principal limitation of carbon fiber is the relatively low strain to failure, particularly in higher modulus materials. Improvements in strength-strain have reduced extrinsic defects.

Manufacture and Processing

Carbon fibers are prepared by the controlled pyrolysis of an organic precursor in fibrous form. Commercial products are based on rayon (or regenerated cellulose), pitch (petroleum and coal tar based), and PAN (polyacrylonitrile).

Precursor weight loss is a central issue in the processing and economics of carbon-fiber technology. Many materials have been evaluated as potential precursors with the objective of increasing carbonization yields and reducing fiber costs. These materials include nonheterocyclic aromatic compounds, heterocyclic aromatic compounds, and aliphatic compounds.

Although the carbon fibers based on rayon, pitch, or PAN differ in properties, uses, economics, and process chemistry and

Table 4. Properties of High Strength and High Modulus Carbon Fiber Composites in Epoxy Matrices

Property	Symbol	ASTM test method	High strength: Celion 3000[a,b] 23°C	High strength: Celion 3000[a,b] 82°C	High strength: Celion 6000[a,c] 23°C	High strength: Celion 6000[a,c] 177°C	High modulus Celion G-50[d] 23°C
Unidirectional laminate							
longitudinal, 0°							
tensile strength, MPa[e]	σ_{11}	D3039	1647	1633	1785	1757	1165[f]
modulus, GPa[e]	E_{11}		151	143	145	147	212[f]
ultimate strain, %	ϵ_{11}		1.1	1.1	1.2	1.2	0.55
compression strength, MPa[e]	σ^c_{11}	D3410	1470	1525	1221	1118	841[f]
modulus, GPa[e]	E^c_{11}		141	151	138	133	189[f]
ultimate strain, %	ϵ^c_{11}		1.7	1.1	1.1	1.4	0.45
flexure strength, 4 pt., MPa[e]	σ^f_{11}	D790	1750	2012	1997	1994	1338[f]
modulus, GPa[e]	E^f_{11}		135		133		192[f]
interlaminar shear strength, short beam, MPa[e]	σ_{ILS}	D2344	125	79.3	97.0	76.0	79
Transverse, 90°							
tensile strength, MPa[e]	σ_{22}	D3039	69.6	30.4	48.0	45.0	35.8
modulus, GPa[e]	E_{22}		9.3	9.0	9.6	9.6	7.24
ultimate strain, %	ϵ_{22}		0.78	0.33	0.52	0.56	0.49
Other							
density, g/cm³	ρ	D792	1.59	1.59	1.55	1.55	1.61
shear strength, inplane[g], MPa[e]	σ_{12}	D3518	76.6	58.0	72.0	71.0	59.3
shear modulus, inplane[g], GPa[e]	G_{12}		6.2	4.1	4.8	4.1	4.07
Poisson ratio 0° coupon	ν_{12}		0.32	0.32	0.3	0.3	0.24
coefficient thermal expansion, 10^{-6}/°C							
at 0°C	α_{11}		−0.22	−0.4	−0.31	−0.56	
at 90°C	α_{22}		15.0	27.1	19.8	35.8	
Quasi-isotropic laminate, (0° ±45°/90°)							
tensile strength, MPa[e]	σ_x	D3039	494	444	537	528	305
tensile modulus, GPa[e]	E_x		53.1	48.0	49.7	50.0	73
ultimate strain, %	ϵ_x		1.1	0.9	1.2	1.2	0.42

[a] Properties were obtained on autoclave-cured laminates using Rigidite 5208-Celion 3000 and Rigidite 5213-Celion 6000 (Celanese Corp.), respectively; prepreg manufactured by Narmco Materials, Inc. All unidirectional tension and flexural values normalized to 62 vol % fiber. Nominal fiber loading for all other properties, 60–65%.

[b] High strength carbon fiber in a high temperature epoxy matrix of the type used in aircraft and aerospace applications (5208); 3000 filaments per strand.

[c] High strength carbon fiber in a general epoxy matrix used in recreational and industrial applications (5213); 6000 filaments per strand.

[d] Properties developed with Celion G-50, 12,000 filament high modulus carbon fiber in Fiberite 934 high temperature epoxy; prepreg manufactured by Fiberite Corporation.

[e] To convert MPa to psi, multiply by 145. To convert GPa to psi, multiply by 145,000.

[f] Normalized to 62 vol % fiber; all other values in this column are for fiber loading of 60–65%.

[g] Inplane shear properties determined by tension test of ±45° balanced laminate in 0° direction.

technology, processing involves a number of common steps: extrusion of a polymer solution or melt into fine filaments; stabilization (preoxidation or thermosetting) carried out at relatively low temperatures (200–250°C), usually in air, which converts the precursor to an infusible state, stable to subsequent high temperature processing; carbonization in an inert atmosphere (usually pure nitrogen) at 1000–2000°C, with an increase in carbon content to 85–99%; graphitization in an inert environment (Ar or N_2) at temperatures in excess of 2500°C, with an increase in carbon content to greater than 99%; and surface treatment to promote adhesion to a matrix (usually a polymeric resin).

Incorporation into Composite Materials and Structures

The utility of carbon fiber, from both performance and cost considerations, is determined by its use in a composite system (combination of fiber and matrix). The variety of commercially available fiber and matrix systems provides a wide spectrum of cost and performance. By changing the orientations of the reinforcing fibers and the strength-modulus characteristics, the structural performance of a composite system can be tailored to meet specific requirements. Similarly, the choice of the matrix material influences service environment and fabricability. Different fibers can be used in the same composite system (hybrid composites) to achieve other performance advantages; notable examples include the combination of carbon fiber and glass (increased toughness and reduced cost) and carbon fiber and Kevlar (toughness, cost and enhanced specific strength).

Product Forms and Fabrication. Carbon fibers are incorporated into various resin-based composite structures by using a variety of fiber-matrix forms and fabrication procedures. These include unidirectional prepregs, fabric prepregs, filament winding, pultrusion (qv), molding compounds, and honeycomb constructions (see HONEYCOMBS).

Composite Design. The variations possible with carbon fiber and other composite systems enable design and fabrication engineers to develop a component's shape, sizing, and material configuration to attain the highest strength in highly stressed locations and directions. Unlike isotropic metals, these materials can be designed to allow stiffness in selected directions while maintaining desired flexibility in other directions. Composite systems include hybrid systems, high temperature composites, and tough composites.

New Processes—Effects on Mechanical and Electrical Properties

Based on present PAN technology, a number of new fiber variants are being developed with the objectives of improved mechanical performance, modification and control of electrical and thermal conductivity characteristics, improved thermal-oxidative resistance, and enhanced cost effectiveness.

Economic Aspects

Carbon-fiber cost is principally determined, independent of the usual volume-price relationships, by precursor cost and related mass yield in conversion and capital and energy intensity.

The carbon fiber business has developed into a complex and unusual multinational network of interlocking corporate arrangements dominated in manufacturing capacity by Japan and in consumption by the United States. Past and projected U.S. consumption are given in Table 5; worldwide consumption is approximately twice that of the United States.

Table 5. U.S. Carbon Fiber Consumption, t

Application	1981	1983	1986[a]	1990[a]
aircraft and aerospace	370	500	1250	2700
sporting goods, recreation, marine	95	250	325	410
general industrial, transportation	50	205	275	500
Total	*515*	*955*	*1850*	*3610*

[a] Projected estimates.

Health and Safety Factors

Manufacturers' materials safety information should be consulted for specifics on health and safety factors. Potential hazards include respiratory problems owing to inhalation of dust or fine fibrous particles, and electrical hazards caused by fiber, dust, and fibrous particles; explosive shorting of high voltage systems is possible.

BIBLIOGRAPHY

J. B. Donnet and R. C. Bansal, *Carbon Fibers,* Marcel Dekker, Inc., New York, 1984.

John P. Riggs
Celanese Corporation

CARBONIUM ION POLYMERIZATION. See CATIONIC POLYMERIZATION.

CARBON MONOXIDE COPOLYMERS. See OLEFIN–CARBON MONOXIDE COPOLYMERS.

N-CARBOXYANHYDRIDES. See POLYPEPTIDES.

CARBOXYMETHYLCELLULOSE. See CELLULOSE ETHERS.

CASEIN

Casein is the main protein of milk, constituting approximately 3% of cow's milk. The two types, acid and rennet, are precipitated from skim milk by acid and rennet, respectively. The term casein usually applies to the acid form; the industrial use of rennet casein is small.

Chemical Properties

Casein is a phosphoprotein, ie, it contains a significant amount of organically bound phosphorus as the phosphate ester. Its elemental composition is given in Table 1.

Table 1. Elemental Composition of Casein

Element	Percent
carbon	53.9
oxygen	22.14
hydrogen	7.13
nitrogen	15.8
sulfur	0.72
phosphorus	0.71

Physical Properties

Casein is a medium hard, granular solid that is pale cream to straw yellow in color, and has a sweet, milky odor. Acid casein is essentially insoluble but forms colloidal dispersions in aqueous solutions of acids, alkali, and some salts and organic substances. Physical properties are given in Table 2.

Table 2. Physical Properties of Casein

Property	Value
molecular weight	33,600–375,000
colloid particle diameter[a], nm	20–140
moisture at 25°C, 20% rh	8
solubility in water, g/L, 25°C	0.11
specific gravity	1.25

[a] In milk.

Micellar Structure. In milk, casein occurs as roughly spherical particles of colloidal dimensions, which gives the milk its white appearance.

Economic Aspects

The main casein-producing countries are New Zealand, France, Poland, the FRG, Ireland, and the Netherlands. The main importing countries are the United States, Japan, and the FRG. World production in the years 1974–1983 is given in Table 3.

Table 3. World Production of Casein (Thousands of Metric Tons)

year	1974–1978	1979	1980	1981	1982	1983
10^3 t	150[a]	194	202	171	166	180[b]

[a] Average.
[b] Projected.

Uses

Casein is used in food processing, adhesives, coatings, plastics and fibers (cross-linked by treatment with formaldehyde), or additives to soap and in the electrorefining of copper. It has also been suggested as an ion-exchange material for the removal of metal from solutions.

BIBLIOGRAPHY

F. Haurowitz, *The Chemistry and Functions of Proteins,* Academic Press, New York, 1963.

H. D. McKenzie, *Milk Proteins,* Academic Press, New York, 1970.

P. F. Fox, *Developments in Dairy Chemistry,* Applied Science Publications, London, 1982.

C. Ellis, *The Chemistry of Synthetic Resins,* Reinhold Publishing Corp., New York, 1935.

R. R. SCHRIEKE
Ballarat C.A.E., Victoria

G. WINTER
University of Melbourne

CASTOR OIL. See DRYING OILS.

CASTING

Casting is the process whereby a liquid is poured into a mold and allowed to react, cure, or harden to form a rigid object that reproduces the mold cavity. Many resin systems may be utilized in this process, eg, acrylic, diethylene glycol bis(allyl carbonate), epoxy, phenol–formaldehyde, polystyrene, polysulfide, and silicone (see also DIALLYL AND RELATED POLYMERS; EPOXY RESINS; PHENOLIC RESINS; POLYSULFIDES; SILICONES; STYRENE POLYMERS). Methyl methacrylate is the principal monomer used in the casting, or bulk polymerization, of acrylic plastics.

High molecular-weight cast acrylic sheet is the primary article of manufacture; however, rods, tubes, spheres, lenses, and other intricate shapes may also be produced. Acrylic sheet may be manufactured in a batch-casting process by using individual glass or metal cells or in a continuous operation usually between polished metal belts. Rods, tubes, spheres, and lenses are normally cast in metal molds using conventional or rotational casting techniques. Cast acrylic articles provide high transparency, a wide range of colors, rigidity, moderate impact strength, thermal and chemical resistance, and excellent weatherability.

In 1982, an estimated 86×10^3 metric tons of cast acrylic sheet were consumed in the United States alone. The sheet was used in a vast variety of applications such as glazing (for aircraft, architectural, automotive, mass transit, and security), signs, displays, furniture, sanitary ware, lighting fixtures, and decorative panels.

Theory; Bulk Polymerization of Methyl Methacrylate

The bulk, solvent-free, polymerization of methyl methacrylate monomer is normally conducted during casting by means of a peroxide-initiated free-radical process. During initiation, free radicals are normally generated by thermal decomposition of a peroxide catalyst added to the monomer. The free radicals attack the double bond of the vinyl monomer and generate the radical, which rapidly adds another monomer molecule, thus extending the chain by one unit and regenerating a new active end site.

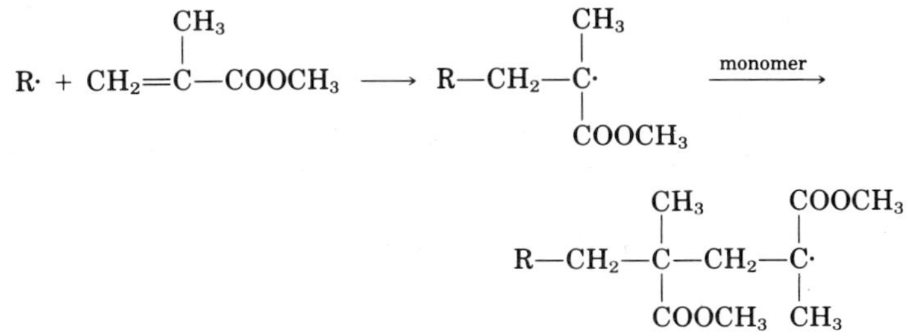

The polymerization rate is dependent on the monomer concentration, the initiator concentration, and the temperature. In the presence of dissolved molecular oxygen both initiation and subsequent chain propagation may be inhibited. The period of inhibition is proportional to the oxygen concentration in the monomer and directly related to the catalyst concentration and other inhibitors.

Equipment and Procedure

Transparent poly(methyl methacrylate) sheet is normally cast between individual plates of glass or highly polished metal, although other molds or mold surfaces can also be employed. Filled or opaque acrylic articles may be manufactured in plaster, reinforced plastic, metal, wood, or concrete molds that provide adequate heat transfer and surface-release characteristics; however, postmachining may be required to achieve the desired surface quality.

Cell Casting. Standard cell casting operations utilize two flat glass plates (slightly larger in area than the finished casting) separated by an elastomeric gasket, as shown in Figure 1.

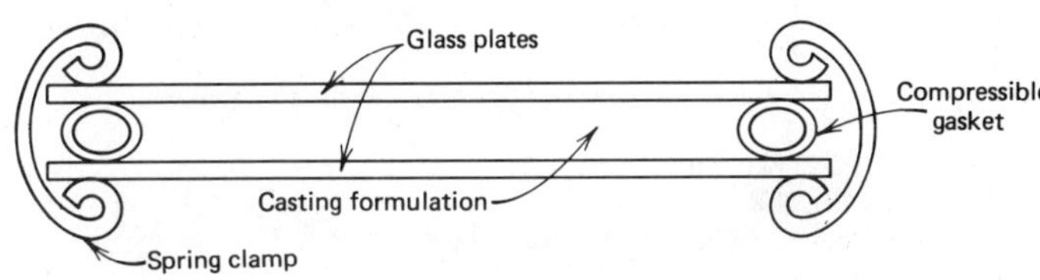

Figure 1. Conventional cell-casting mold configuration, edge view.

Continuous Casting. Continuous casting of poly(methyl methacrylate) sheeting can be accomplished in equipment utilizing two parallel, continuously conveying, endless, highly polished, stainless-steel metal belts (Fig. 2).

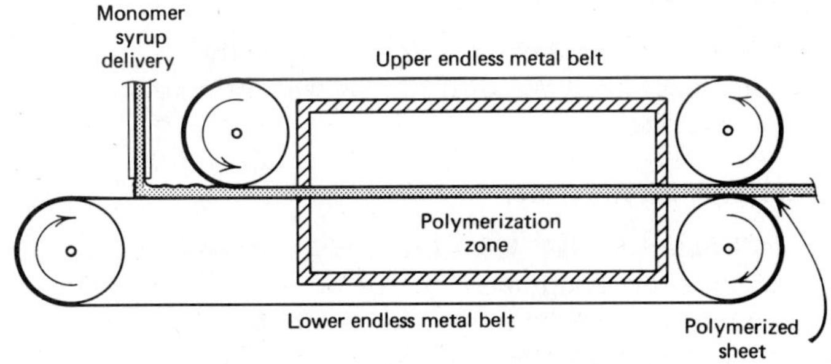

Figure 2. Continuous-casting apparatus.

Use of Acrylic Polymers in the Casting Process. Polymerization rates are increased and shrinkage is reduced by the addition of fine poly(methyl methacrylate) powder or a prepolymerized acrylic syrup to the monomer. The powder may be produced by bulk, solvent, or emulsion polymerization. The syrup may be produced in a batch or continuous process with solids limited to 15.0–40.0%.

Properties of Poly(methyl methacrylate) Sheet

Properties of cast poly(methyl methacrylate) sheet are given in Table 1.

Table 1. Properties of Cast Poly(methyl methacrylate) Sheet

Property	ASTM test method[a]	Typical performance value[b]
specific gravity	D792	1.19
refractive index	D542	1.49
light transmittance[c], %	D1003	92.0
haze, %	D1003	1.0
tensile strength max, MPa[d]	D638, 0.5 cm/min	72.4
elongation max, %		4.9
modulus of elasticity, GPa[d]		3.1
flexural strength max, MPa[d]	D790, 2.5 cm/min	110
shear strength, MPa[d]	D732	62
shear modulus, GPa	D732	1.15
maximum recommended continuous service temperature, °C		82–93
coefficient of thermal expansion, m/(m·°C)	D696	7.2×10^{-5}[e]
coefficient of thermal conductivity, W/(m·K)	Cenco-Fitch	0.19
specific heat, kJ/(kg·K)[f]		1.46
dielectric strength, V/μm	D149	20
dielectric constant 60 Hz	D257	3.7

[a] Test specimens generally conditioned per ASTM D 618, procedure B, unless otherwise designated.
[b] Test specimens 6.3 mm thick; some values may vary with thickness.
[c] Colorless.
[d] To convert MPa to psi, multiply by 145.
[e] 4.0×10^{-5} in./(in.·°F).
[f] To convert kJ/(kg·K) to Btu/(lb·°F), divide by 4.186.

Manufacture of Cast Acrylic Sheet

Considerations include choosing the optimum casting formulation, prevention of air bubbles and voids, polymerization control, chain depolymerization, and mold release.

Economic Aspects

In 1983, combined annual casting capacity in the United States, Canada, Japan, and Western Europe was estimated at 322×10^3 metric tons.

Health and Safety Factors

Methyl Methacrylate Monomer. Acrylic monomers have low to moderate toxicity properties. The most toxic of the common esters is methyl acrylate. Areas in which the monomer is used should be well ventilated and protective clothing and safety goggles worn during handling.

Methyl methacrylate monomer (boiling point = 100°C) is classified as a flammable liquid by the Department of Transportation and requires an appropriate red label during shipment. Poly(methyl methacrylate) cast acrylic articles are combustible and must be protected from flames, high heat, and ignition sources.

BIBLIOGRAPHY

U.S. Pat. 2,091,615 (Aug. 31, 1937), O. Rohm and W. Bauer (to Rohm and Haas Co.).

U.S. Pat. 2,500,728 (Mar. 14, 1950), E. G. Williams (to Imperial Chemical Industries, Ltd.); U.S. Pat. 3,371,303 (Mar. 5, 1968), O. H. Hellsund (to Swedlow, Inc.); U.S. Pat. 4,133,861 (Jan. 9, 1979), T. Kato (to Mitsubishi Rayo Co. Ltd.).

C. Schildnecht, *Polymer Processes,* Vol. X, and C. Schildnecht and I. Skeist, *Polymerization Process,* Vol. XXIX, of *High Polymers: A Series of Monographs on the Chemistry, Physics, and Technology of High Polymeric Substances;* Vol. X (Interscience Publishers, New York) 1956; Vol. XXIX, Wiley-Interscience, New York, 1977.

U.S. Pat. 3,551,541 (Dec. 29, 1970), C. Rossetti (to Rostero S.A.); U.S. Pat. 3,680,991 (Aug. 1, 1972), A. C. Case and R. C. Gordon (to American Cyanamid Co.); U.S. Pat. 3,689,022 (Sept. 5, 1972), C. Rossetti (to Rostero S.A.); U.S. Pat. 3,694,129 (Sept. 26, 1972), P. R. Daddona; U.S. Pat. 3,718,182 (Feb. 27, 1973), C. Rossetti (to Rostero S.A.).

WILLIAM C. HARBISON
Swedlow, Inc.

CATALYSTS, ALFIN. See ANIONIC POLYMERIZATION.

CATALYSTS, FRIEDEL-CRAFTS. See CATIONIC POLYMERIZATION.

CATALYSTS, POLYMER-SUPPORTED

A catalyst, by definition, increases the rate of a chemical reaction without being consumed in the reaction. In homogeneous catalysis, catalysts are used in solution. Most industrial catalysts, however, are solids, which offer the advantages of ease of separation from products and lack of corrosiveness.

Industrial catalysts include solid organic polymers, especially cross-linked polystyrene functionalized with groups such as acids, bases, and transition-metal complexes. A polymer-supported group often has catalytic properties analogous to those of the same group used in homogeneous catalysis. The comparison is not always simple, however, because the physical properties of the support can exert a strong influence on the catalyst performance.

Preparation and Characterization

Cross-linked Polystyrene. Functionalized polymers are prepared by copolymerization of functionalized monomers or by functionalization of poly(styrene-*co*-divinylbenzene).

Other Polymers. A number of organic polymers have been used as catalyst supports, including polybutadiene, poly(vinyl alcohol), poly(vinyl chloride), polyvinylpyridine, and polyamides.

Attachment of Catalytic Groups. *Ion-exchange Resins.* The polymers most widely used as catalysts are ion-exchange resins, especially those incorporating sulfonic acid groups or quaternary ammonium groups (see ION-EXCHANGE POLYMERS). Porous polymers are sulfonated with concentrated (+99%) H_2SO_4 or chlorosulfonic acid. The resin is usually in a partially swelled state with ethylene dichloride as the swelling agent.

Attachment of Transition-metal Complexes. The simplest method of incorporating a metal complex into a polymer is to treat a solid incorporating attached ligands with a solution of the complex incorporating another ligand.

Catalyst Characterization. The usual goal of the synthesis is to prepare supported species that have retained unique (molecular) identities. It is important to determine whether the supported species are indeed the desired ones.

Most identification methods are based on the stoichiometry of the synthesis reaction and on comparisons of spectra of the supported species with those of known compounds. The most commonly used method is infrared spectroscopy.

Functional-group Catalysis

The simplest polymer-supported catalysts have functional groups that act like the same groups in solution. The role of the polymer support, which is swelled by a solvent, is minimal; the polymer is considered an inert matrix holding the catalytic groups. In other words, the catalysis may be described as homogeneous catalysis in a liquidlike phase within the polymer. Examples include catalysis by water-swelled sulfonated polymers (wherein the groups are dissociated into $—SO_3^-$ groups and hydrogen ions), catalysis by polymer-supported transition-metal complexes, and triphase catalysis.

Matrix Effects

In the more interesting examples of catalysis by polymers, the support plays an important role in addition to holding the catalytic groups in a phase separate from that holding the gas or liquid reactants. There are examples of efficient catalysis by groups on supports, which in solution are inactive, and there are cooperative effects involving more than one group (bifunctional and multifunctional catalysis). The physical properties of the support, the spacing of the groups, and the affinity of the polymeric matrix for the different reactants all influence the catalyst performance.

Porous Polymers and the Importance of Surfaces. The role of the polymer catalyst in offering sites for bonding one reactant in high concentration and activating it for reaction is not always sufficient. Often the other reactants must also be accommodated in the catalyst in order to find access to the first reactant and undergo conversion. Macroporous (or macroreticular) resin beads find catalytic applications in the absence of swelling agents. Macroporous-polymer supports consist of aggregates of very small gelular microparticles interspersed with macropores (Fig. 1). Reactants and products can readily diffuse in and out of the macropores, thereby coming in contact with a large number of microparticles constituting the internal surface. Whereas gel-form beads can be efficient supports in the presence of a swelling medium, macroporous resins are efficient in both swelling and nonswelling media, including gas-phase reactants.

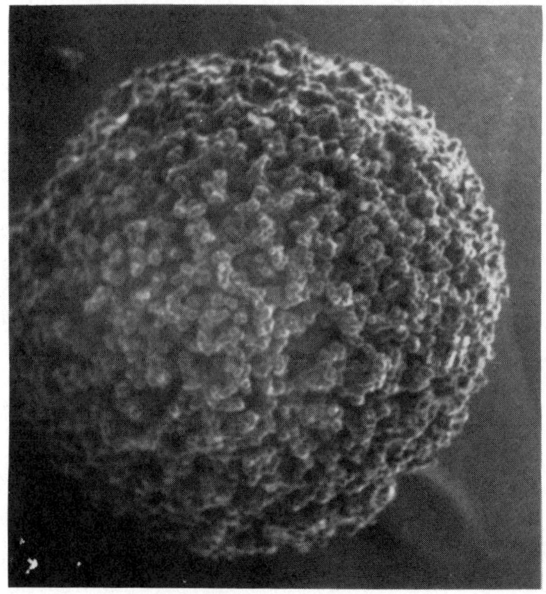

Figure 1. Scanning electron micrograph of a macroporous poly(styrene-*co*-divinylbenzene) bead (Amberlite XE-238) at 100× magnification. Courtesy of R. L. Albright, Rohm and Haas Co., Philadelphia.

Industrial Catalysts

In industrial chemical processing, the stability of a catalyst is often more important than its activity or selectivity. Because of the need for stability, most industrial catalyst supports are robust inorganic solids rather than organic polymers. The functional groups that can be attached to the polymers can similarly be attached to metal oxides such as silica. Metal complexes can also be incorporated directly onto metal-oxide surfaces without the need for intervening ligands. Such supported complexes of chromium, for example, are used industrially as olefin polymerization catalysts.

BIBLIOGRAPHY

J. Lieto, D. Milstein, R. L. Albright, J. V. Minkiewicz, and B. C. Gates, *Chemtech,* 46 (1983).

Y. Chauvin, D. Commereuc, and F. Dawans, *Prog. Polym. Sci.* **5,** 95 (1977).

C. U. Pittman in G. Wilkinson, F. G. A. Stone, and E. W. Abel, eds., *Comprehensive Organometallic Chemistry,* Vol. 8, Pergamon Press, Elmsford, N.Y., 1982.

P. Hodge and D. C. Sherrington, eds., *Polymer-Supported Reactions in Organic Synthesis,* John Wiley & Sons, Inc., New York, 1980.

B. C. GATES
University of Delaware

J. LIETO
University of Delaware and Institut de Pétroléochimie et de Synthèse Organique Industrielle, Marseille

CATALYSTS, ZIEGLER-NATTA.

See ZIEGLER-NATTA CATALYSTS.

CATION-EXCHANGE POLYMERS.

See ION-EXCHANGE POLYMERS.

CATIONIC POLYMERIZATION

The term cationic polymerization must be clearly defined. Within the context of polyaddition, cationic polymerization refers to chain-growth reactions promoted by active species possessing electrophilic character: carbenium, carbonium, oxonium, sulfonium, ammonium, and phosphonium ions, but also molecular-chain carriers, ie, mostly esters, which are sufficiently polarized to induce propagation.

This review focuses on the classic chain-growth processes. For the other processes, see, for example, ACETAL RESINS; ALDEHYDE POLYMERS; POLYAMIDES; POLYESTERS; POLY(PHENYLENE ETHER); AND STEP-REACTION POLYMERIZATION.

General Considerations

The following generalizations can be used as a rough guide to the behavior of cationic polymerizations:

The DPs of polymers obtained from alkenyl monomers are often determined by important transfer reactions, and high molecular weights are only attained by working at low or very low temperature. The ring-opening polymerizations of heterocyclic monomers are less subject to this restriction.

On the other hand, heterocyclic monomers often exhibit relatively low ceiling temperatures, whereas few alkenyl compounds are affected by this problem.

The stability of chain carriers, which determines their lifetime, obviously affects both DP and yield; it depends on several factors, but mostly on the structure of the monomer and the type of initiator. Broadly speaking, the saturated heterocyclic monomers give more stable active species than those arising from alkenyl monomers.

All polyadditions can be viewed as special cases of specific physical organic reactions in which the intermediate species manages to survive long enough to ensure the growth of a macromolecule instead of rapidly collapsing into inactive products. A cationic polymerization, then, is a special case of electrophilic addition.

Initiators

Because the active centers in cationic polymerization are electron-deficient species, they usually result from the interaction of the monomer with an acidic entity.

Brønsted Acids. A wide variety of protonic acids (HA) ranging in strength from acetic to trifluoromethanesulfonic acid has been used to initiate cationic polymerizations. Their ionization in nonnucleophilic solvents is dominated by ionic aggregations (homoconjugation) through hydrogen bonding, both for the anion and the proton:

$$3\,HA \rightleftharpoons H_2A^+ + HA_2^-$$

Iodine acts likewise and eventually gives HI in its interactions with alkenes.

Inorganic Lewis Acids. Many metal halides (MtX_n) are good initiators of cationic polymerizations, eg, BF_3, $AlBr_3$ $TiCl_4$, SbF_5, $SnCl_4$. They self-ionize in nonprotic solvents following a 2:2 equilibrium:

$$2\,MtX_n \text{ or } (MtX_n)_2 \rightleftharpoons MtX_{n+1}^- + MtX_{n-1}^+$$

In the presence of small amounts of protogenic substances (for example, water, alcohols, and carboxylic acids) they form strong complexes that play a dominant role in cocatalytic initiation. Metal salts of strong oxyacids, eg, $Hg(ClO_4)_2$ and $Al(SO_3CF_3)_3$, are also excellent initiators and behave like their halide counterparts.

Organic Lewis Acids. Among these initiators the most notable are the stable carbenium salts, eg, trityl perchlorates, tetrafluoroborates, and hexafluoroantimonates; onium salts such as $(C_2H_5)_3O^+$ BF_4^-; oxocarbenium salts such as $C_6H_5CO^+$ SbF_6^-; and double anhydrides, eg, acetyl perchlorate.

Physical Methods of Initiation. Many techniques can activate an inert system by a physical means that provides the necessary energy for initiation to occur. The monomer itself may be activated through ionization without the intervention of other substances. By far the most investigated method is the use of γ radiation or high energy electrons that kick off an electron from the monomer.

Monomers

The susceptibility of a potential monomer to undergo cationic polymerization depends primarily on the electron-donating capacity of the site to be activated (initiation) and on the survival time of the electrophilic active species (propagation). Three different nucleophilic moieties meet these requirements in principle: (*1*) carbon–carbon unsaturations which can be π-donors, as in alkenyl, alkynyl, and dienic compounds; (*2*) the lone pair of a heteroatom (oxygen, sulfur, nitrogen, and phosphorus) in certain (mostly saturated) heterocyclic compounds that are therefore n-donors; and (*3*) single bonds involving carbon atoms, as in alkanes, which can become σ-donors under special conditions.

Among π-donor monomers, isobutene, isoprene, the styrenes, and indene are the most studied compounds. *N*-Vinylcarbazole and the vinyl ethers represent the most typical examples of n, π-donor monomers, which are more nucleophilic than alkenes. Tetrahydrofuran, 1,3-dioxolane, *t*-butylaziridine, oxazoline, and cyclic siloxanes are a few structures susceptible to polymerize because of their n-donor character. These heterocycles are more nucleophilic than the π- and n,π-donor monomers.

Solvents

Following a classic subdivision, three categories of solvents are, in principle, suitable for cationic polymerization in that the various active species do not react with them: (*1*) aprotic, inert solvents, ie, aliphatic and some aromatic hydrocarbons and their haloderivatives; (*2*) aprotic, dipolar, protophobic solvents, eg, the nitroalkanes, nitrobenzene, and acetonitrile; and (*3*) amphiprotic, protogenic solvents, ie, mineral and carboxylic acids. Whereas solvents belonging to the first two classes are commonly used in cationic polymerization, those of the third class are not, although they would constitute an excellent support for the active species.

The choice of a suitable solvent for a given system depends on such factors as the polarity desired, the solvating power with respect to the different active species, possible specific interactions between solvent and initiator(s), and, of course, more practical considerations such as the melting point, ease of purification, and homogeneity of the system regarding monomer, initiator(s), and polymer.

Temperature

The choice of an optimum temperature for a given monomer–initiator–solvent system depends strongly on the nature of the system and on the specific aim of the polymerization. Temperature affects yields, degrees of polymerization, rates, side reactions, stereospecificity, and reactivity ratios in copolymerization.

Impurities

In cationic polymerization, the most obnoxious impurities are those of nucleophilic character. By far the most complex problem in this context is drying, because water can both inhibit and accelerate, depending on the nature of the system and on the residual concentration of moisture.

The Interplay among the Basic Elements of a System

A cationic polymerization system can now be defined by enumerating four basic elements: the initiator, the monomer, the solvent, and the temperature, and a fifth, less precise one, namely the concentration of residual water and other possible active impurities. The variables associated with the initiator are its electrophilicity or acidity, but also the stability and leaving properties of the counterion or the polarity of the ester formed. The variables associated with the monomer are its nucleophilicity or basicity and the reactivity and stability of the cation, or of the ester formed from it, by the initiation. The role of the solvent is determined by its polarity and possible specific solvating properties. Changes in temperature can result in purely kinetic modifications or in a more complex behavior owing to thermodynamic reasons and to possible mechanistic changes. Finally, the amount of residual moisture must be considered an important variable together with the nature and concentration of contaminants.

Experimental Techniques

Routine Syntheses. In general, the preparation of a polymer by cationic polymerization can be achieved by taking elementary precautions that can be more or less demanding according to the specific system involved. Examples of such syntheses are given in classic books of "recipes." Normally, a well-dried vessel, distilled and dried reagents and solvents, and a nitrogen blanket should ensure a satisfactory polymerization. If the appropriate temperature is chosen, molecular weights should also be satisfactory.

Fundamental Research. *Kinetic Measurements.* The rate of monomer consumption can be followed by different high vacuum techniques according to the half-life values of the systems: dilatometry, adiabatic calorimetry, stop-flow spectroscopic setups, etc.

Static Measurements. Measurements of equilibria attained during or at the end of a polymerization as well as in model systems, eg, simulating initiation, can be carried out in high vacuum devices, using uv, visible, and nmr spectroscopy; electrical conductivity; or more conventional analytical methods.

The Mechanistic Steps

Kinetics and thermodynamics of each mechanistic step of the cationic polymerization process are discussed separately.

Initiation. This discussion is organized by initiator type.

Brønsted Acids. Depending on the specific conditions of monomer, acid, possible catalyst, and polarity of the medium two distinct pathways are formulated for the formation of the ester, ie, the Ad_E2 and Ad_E3 processes.

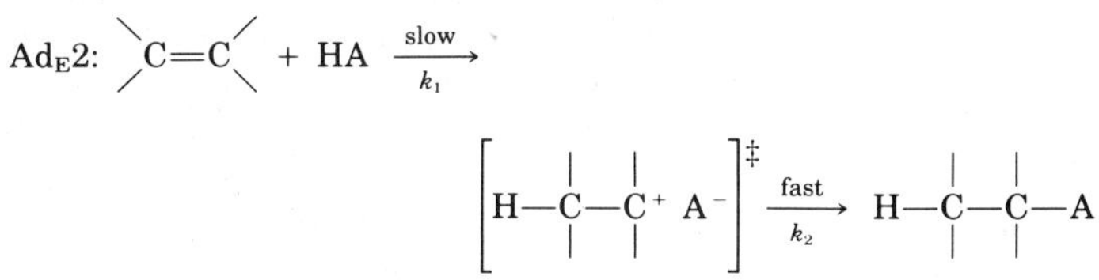

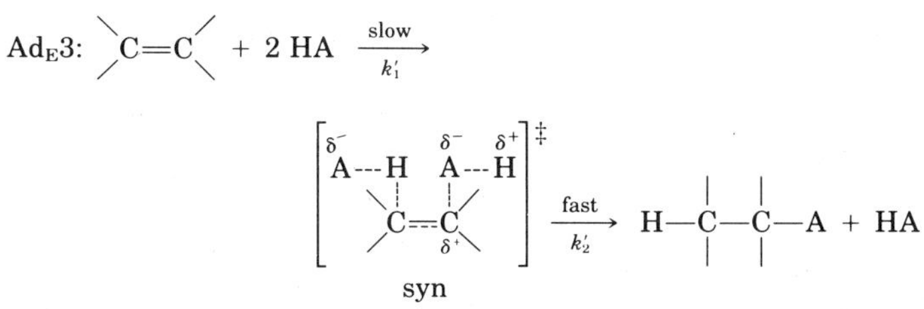

In these simplified mechanisms, the homoconjugated species H_2A^+ and HA_2^- are absent, but in specific instances, they play an important role.

Inorganic Lewis Acids. Today, two basic processes are recognized as fundamental alternative pathways of initiation when inorganic Lewis acids are used as catalysts: direct initiation and cocatalysis. Typical mechanisms of direct initiation are

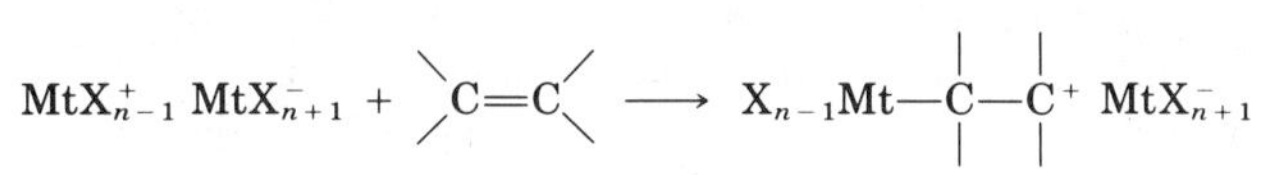

and

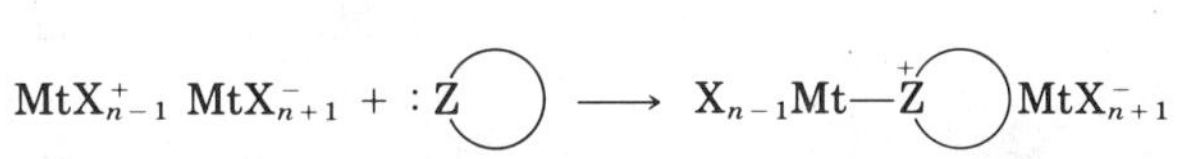

whereas cocatalysis proceeds through protonation and involves an Ad_E3-type intermediate, eg,

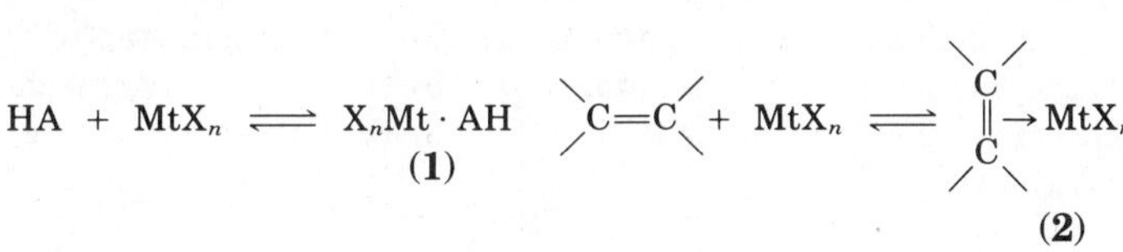

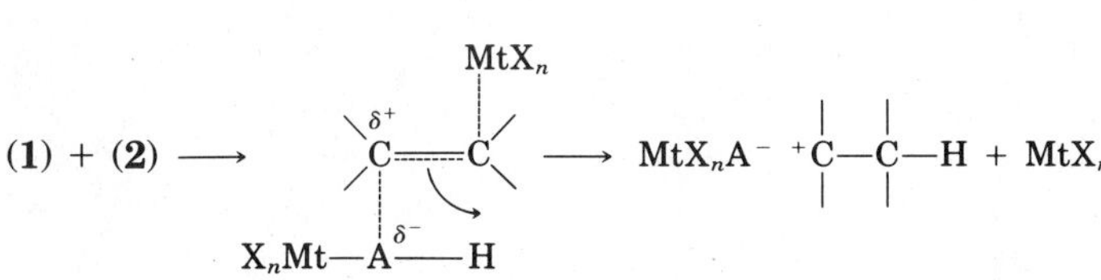

Organic Lewis Acids. With alkenyl monomers, direct electrophilic addition of $C_7H_7^+$ and $(C_6H_5)_3C^+$ to the double bond has been proven in several instances. With the heterocyclics, the situation is more complex, as shown by the behavior of different oxygen compounds. Propylene oxide gives an addition product that can then polymerize or copolymerize with THF. Addition to THF and 1,3-dioxolane is also possible, but the resulting oxonium ion is sterically hindered with respect to propagation. However, hydride-ion abstraction from the monomer also occurs and constitutes an alternative primary step leading to proper initiation. Other organic electrophiles usually add directly to the monomer.

External Energy. Most of the techniques providing external energy to initiate a cationic polymerization generate monomer cation-radicals that are then transformed into cations or dications, usually by reacting with the substrate. These processes are very rapid and not yet fully understood.

Propagation. ***Multiplicity and Structure of Active Species.*** Except for the systems in which initiation involves an electron abstraction from the monomer to give a bare cation as the only chain carrier, a set of equilibrated entities is usually formed:

polarized ester molecules ⇌ tight ion pairs ⇌ solvent-separated ion pairs ⇌ unpaired (free) ions

According to the basicity (nucleophilicity) of the monomer, the acidity of the initiator (electrophilicity), the polarity of the solvent, and the temperature, these equilibria can be shifted to give different situations, including the limiting ones characterized by the predominance of either unpaired ions or ester molecules. The two principal equilibria are those connecting esters with ion pairs and the latter with unpaired ions.

Rate Constants. Taking into account the three types of active species possibly present in a cationic polymerization, the general kinetic expression for the rate of propagation takes the more appropriate form, shown below, which illustrates the relative contributions of unpaired cations P_n^+, ion pairs $P_n^{\pm}$, and polarized ester molecules P_n^e (pseudocationic polymerization) to the overall process of chain growth.

$$-\frac{d[M]}{dt} = \Sigma R_p = [M](k_p^+ [P_n^+] + k_p^{\pm} [P_n^{\pm}] + k_p^e [P_n^e])$$

In order to arrive at individual k_p values, conventional kinetic measurements of the rate of monomer consumption must be coupled with the determination of the concentration of active species during the polymerization.

Mechanisms, Structures, and Rates. The propagation step with alkenyl monomers can be represented by two alternative mechanisms depending on whether the active species is ionic or covalent:

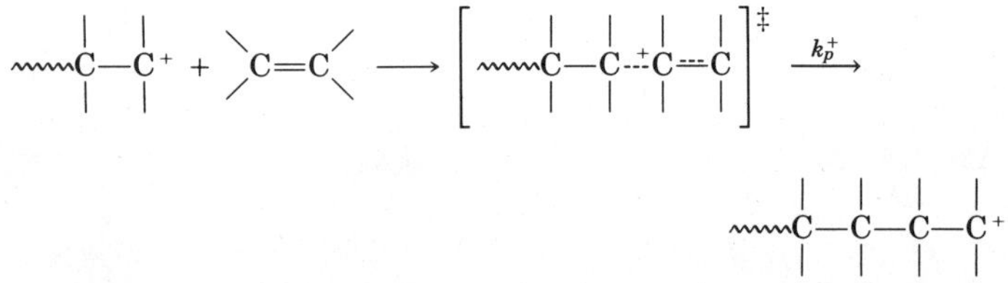

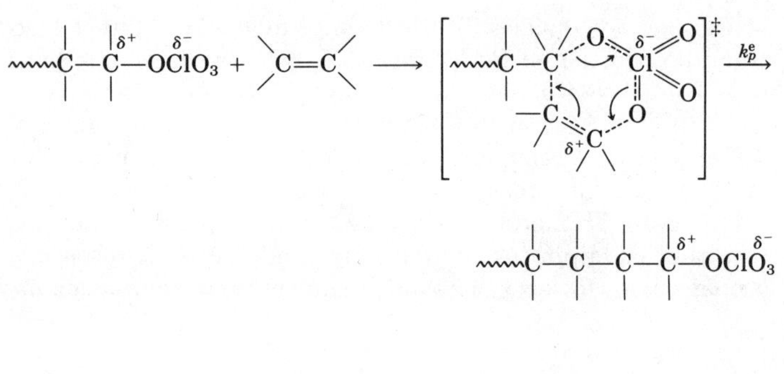

With heterocyclic monomers, the generally accepted mechanism of propagation is the equivalent of nucleophilic substitution, namely an S_N2 reaction. This is shown below for both ionic and covalent active species:

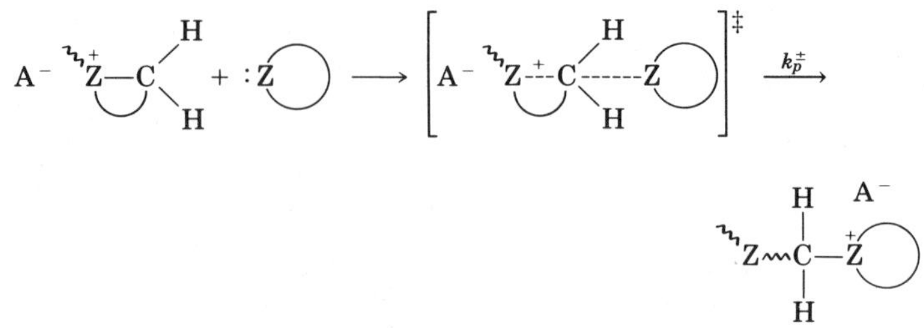

and

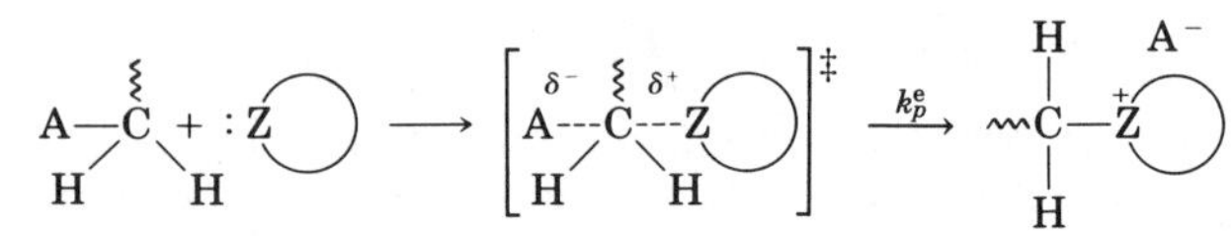

Stereochemistry and Stereospecificity. The propagation step of most heterocyclic monomers is accompanied by complete inversion of configuration, as directly proven in the polymerization of *cis*- and *trans*-dideuteroethylene oxides, *cis*- and *trans*-butene oxides, and 2-substituted 7-oxabicycloheptanes. The low temperature polymerization of butene oxides gives stereoregular polymers (qv); the cis compound yields amorphous, disyndiotactic polymer and the trans gives crystalline mesodiisotactic material.

Among π-donor monomers, only α-methylstyrene gives stereoregular polymers. Vinyl ethers have long been known to give stereoregular polymers. Few homogeneous cationic polymerizations give a notable degree of stereospecificity, but several heterogeneous systems, ie, those with insoluble initiators, can give remarkable steric control.

Isomerization Propagation. Although normal propagation predominates in the cationic polymerization of most alkenyl and heterocyclic monomers, notable exceptions have been reported with both groups of compounds. The carbenium ions derived from some olefins (such as 3-methyl-1-butene) can display a high tendency to isomerize to a more stable configuration before propagating.

Electronic rearrangements are often encountered during propagation with unconjugated diolefins having the two C=C bonds as vinyl groups or as unsaturations in bicyclic compounds (see CYCLOPOLYMERIZATION).

The other type of rearrangement with alkenyl monomer involves H^- or Cl^- shifts, ie, isomerization by transport of atoms.

Among n-donor monomers, oxazolines are an excellent example; only repeat units corresponding to ring-opening of the hypothetical monomers, N-acylaziridines, are found in the polymers, ie, isomerization occurs at every propagation step.

The ease with which 2-alkenylfurans undergo C-5 alkylation makes them prone to anomalous propagation steps with incorporation of the ring into the chain backbone. Under specific experimental conditions, the following unusual polymerization was observed:

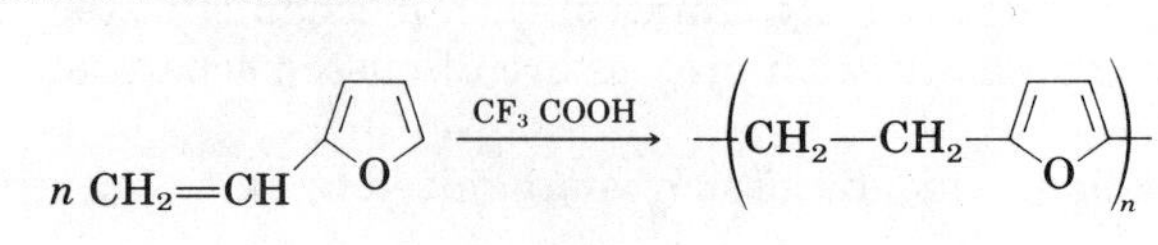

Transfer and Termination Reactions. Transfer and termination reactions are treated together because it is sometimes difficult in cationic polymerizations to classify reactions into one of the two categories unequivocally (see also CHAIN TRANSFER). Some reactions display mechanistic features typical of a retardation reaction, ie, the deactivation of a chain carrier followed by either its slow restoration or the formation of a less active one. The term chain breaking, often used to qualify these mechanisms, is not entirely correct, because when a dead polymer molecule acts as a transfer or termination agent, as frequently encountered in ring-opening polymerizations, the DP is not necessarily reduced. However, it is true that all these reactions thwart normal chain growth, ie, propagation, and constitute unwanted events in many instances.

The following discussion is organized according to the nature of the entity reacting with the growing species to cause transfer or termination.

Spontaneous Reactions. An interaction affecting active species that does not involve any other reagent is the unimolecular expulsion of acid from an ester chain carrier derived from an alkenyl monomer and a Brønsted acid, eg, in the pseudocationic polymerization of styrene by $HClO_4$.

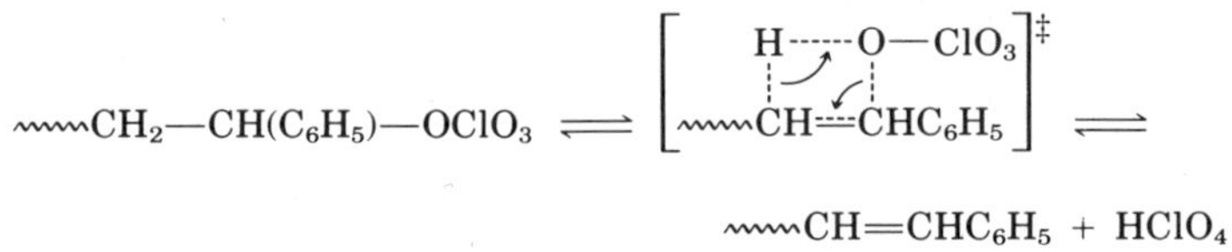

Reactions with the Counterion. In the following examples, the chain carrier is represented as an ion pair; of course, similar mechanisms arising from unpaired ions are also possible.

Brønsted Acids. Reactions with anions derived from Brønsted acids can involve spontaneous acid expulsion.

$$\sim\sim CH_2—\overset{+}{C}HR\ A^- \longrightarrow \sim\sim CH{=}CHR + HA$$

Other possible reactions are

$$\sim\sim CH_2—\overset{+}{C}HR\ A^- \longrightarrow \sim\sim CH_2—CH(R)A$$

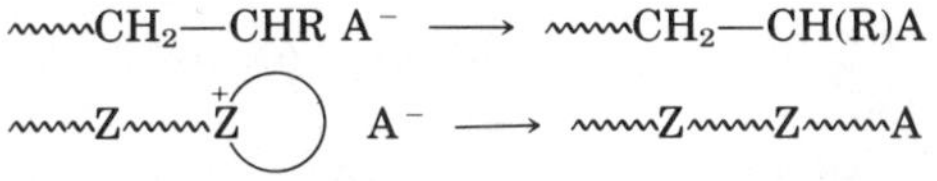

Lewis Acids. The following reactions are well known:

$$\sim\sim CH_2—\overset{+}{C}HR\ MtX_{n+1}^- \longrightarrow \sim\sim CH_2—CH(R)X + MtX_n$$

Water Cocatalysis. The following situations may arise from water-cocatalyzed initiation of alkenyl monomers.

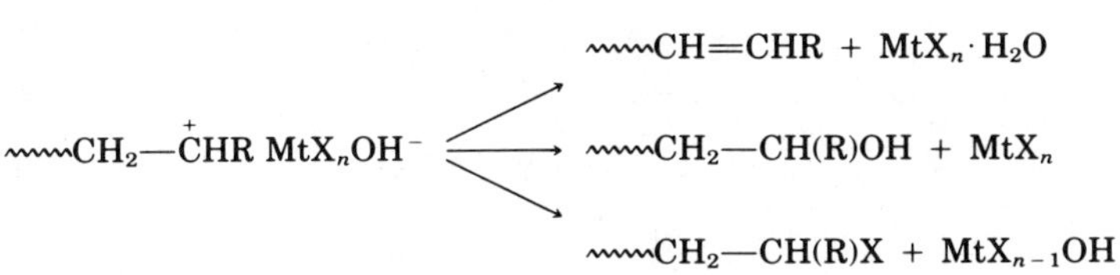

Organoaluminum Halides. Termination by alkylation or hydridation has clearly been established in systems involving organoaluminum halide initiation:

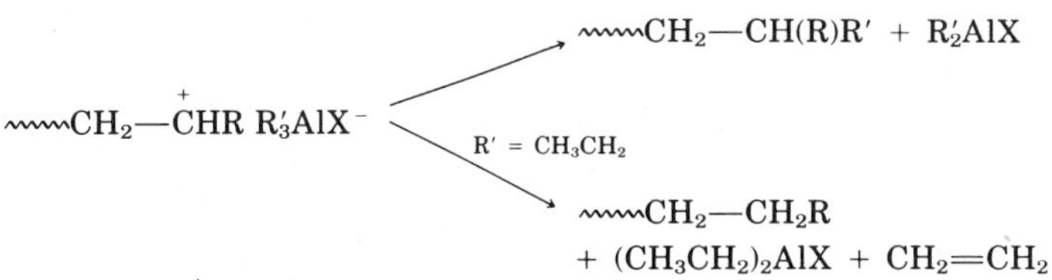

$$\sim\sim CH_2—\overset{+}{C}HR\ R'AlX_2OH^- \longrightarrow \sim\sim CH_2—CH(R)R' + AlX_2OH$$

Reactions with the Monomer. These interactions are kinetic alternatives to propagation because they involve the same two reagents. They are very rare in polymerizations of n-donor monomers, but generally important in systems comprising π- and n,π-donor monomers:

$$\sim\sim CH_2—\overset{+}{C}HR + CH_2{=}CHR \longrightarrow \sim\sim CH{=}CHR + CH_3—\overset{+}{C}HR$$

Reactions with the Polymer. Transfer and termination reactions with polymer molecules are more frequently encountered and tend to be more important with n-donor monomers than with alkenes. This is because onium ions are longer lived than carbenium ions and the polymerizations they induce are generally slower than those of alkenes. These facts, associated with the high nucleophilicity of the heteroatoms in the polymer chains, favor intermolecular and intramolecular reactions typified by the following interactions:

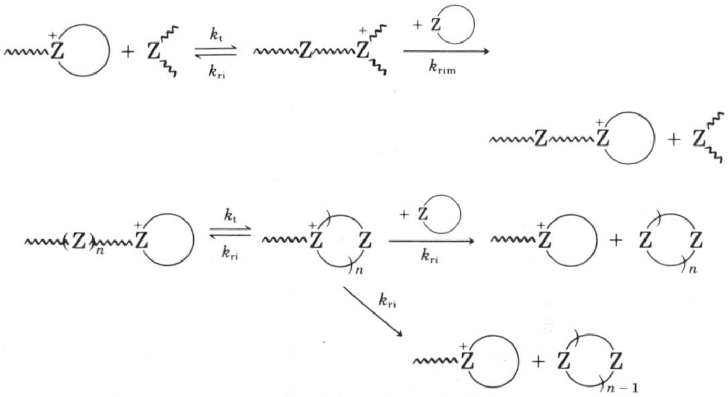

Reactions with the Solvent. Typical solvents used in cationic polymerization, ie, aliphatic hydrocarbons and CCl_4, CH_2Cl_2, C_2H_5Cl, and CH_2ClCH_2Cl, seldom participate in transfer and even less often in termination reactions, although minor and unconfirmed chain-transfer processes by alkyl chlorides have sporadically been reported. Aromatic solvents, on the other hand, definitely cause transfer by electrophilic substitution, ie, alkylation with toluene, and should therefore be avoided if high DPs are sought.

Reactions with Impurities. Strongly nucleophilic impurities kill an equivalent number of active species and this termination reaction becomes more serious as the chain carrier becomes more reactive, ie, the less basic the monomer.

Reactions with Specific Compounds. The addition of a specific transfer or terminating agent to achieve a mechanistic or structural aim is fairly common in cationic polymerization.

Molecular Weights and Distributions

In a system where both transfer and termination reactions are minimized, the average degree of polymerization approaches the value given by the ratio [*monomer*]/[*active species*]. If, moreover, initiation is fast compared with propagation and consumes all the catalyst, a narrow distribution of chain lengths will be achieved and DP $= n[M]_0/[I]_0$, where n is the number of active ends in a growing chain produced by one molecule of initiator I. This ideal situation, ie, living polymerization (qv), is not frequent in cationic processes, but examples have been found, particularly in ring-opening polymerization. More typical situations are characterized by the occurrence of various processes competing with propagation. Molecular weights are most conveniently dealt with separately according to the type of monomer involved, ie, n-donor and π-donor structures.

In the ring-opening polymerization of heterocycles, the transfer and termination reactions are mostly due to intermolecular and intramolecular reorganizations and can give rise to thermodynamically stable cyclic oligomeric structures that cause important changes in the DP and molecular weight distribution with respect to the theoretical value given above.

In the polymerization of alkenyl monomers the most important reaction affecting the DP is usually monomer transfer. A decrease in the reaction temperature reduces its importance relative to propagation, thus favoring higher molecular weights.

The Synthesis of Functionalized Oligomers and Polymers

Cationic polymerization provides some interesting synthetic routes toward oligomers and polymers with specific end groups either for direct use, such as viscostatic agents, surfactants, etc, or as reagents for various types of chain-extension process.

Copolymerization

Random Copolymers. More than 600 sets of r_1 and r_2 values have been determined in copolymerizations involving two alkenyl monomers (π- or n,π-donors). Some interesting trends emerge from a close scrutiny of this huge compilation, particularly within the contexts of the effect of the relative nucleophilic character of each monomer on the polymer composition, of the role of the copolymerization conditions (type of initiator, solvent, and temperature) on the reactivity ratios for a given monomer pair, and consequently of the nature of the active species (ionic or covalent, ie, cationic vs pseudocationic propagation) on the features of the copolymerization.

With n-donor monomers copolymerization studies are much less abundant and concern mostly oxygen heterocycles (cyclic ethers and acetals). These investigations are difficult because a fundamental approach is complicated by thermodynamic factors.

Block Copolymers. Macromolecules containing different homopolymeric sequences can be prepared through two diverse routes. The first involves the coupling of reactive homopolymers or oligomers through specific chemical interactions and is not a block copolymerization in the true sense of the term, but a condensation process that is beyond the scope of this article (see BLOCK COPOLYMERS). The second route uses either living ends on macromolecules formed in a first homopolymerization, to which a second monomer is added, or specific functions on the first homopolymer, which can be activated and which operate as starting carriers for the second polymerization.

Graft Copolymers. Three distinct processes, described below, can lead to grafting:

1. The active species are generated on the trunk polymer and branching starts from these sites (grafting from):

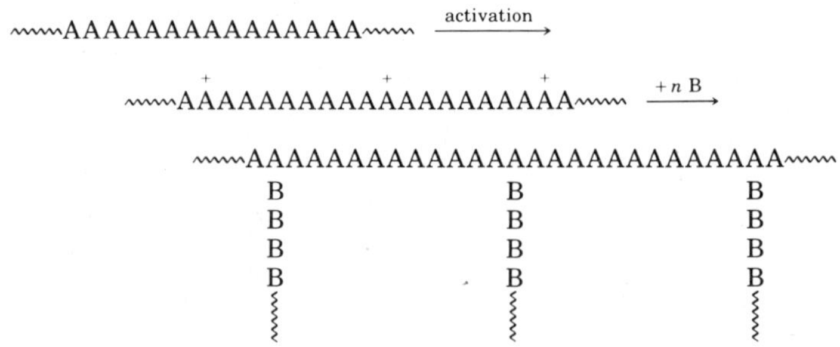

2. The monomer to be grafted is polymerized in solution and the growing chains terminate (or transfer) onto nucleophilic sites of the trunk polymer, thus giving branches (grafting onto):

3. If the growing chains react with the trunk polymer through an unsaturation, and the polymerization continues therefrom, the trunk polymer plays the role of a comonomer to the branches (grafting through):

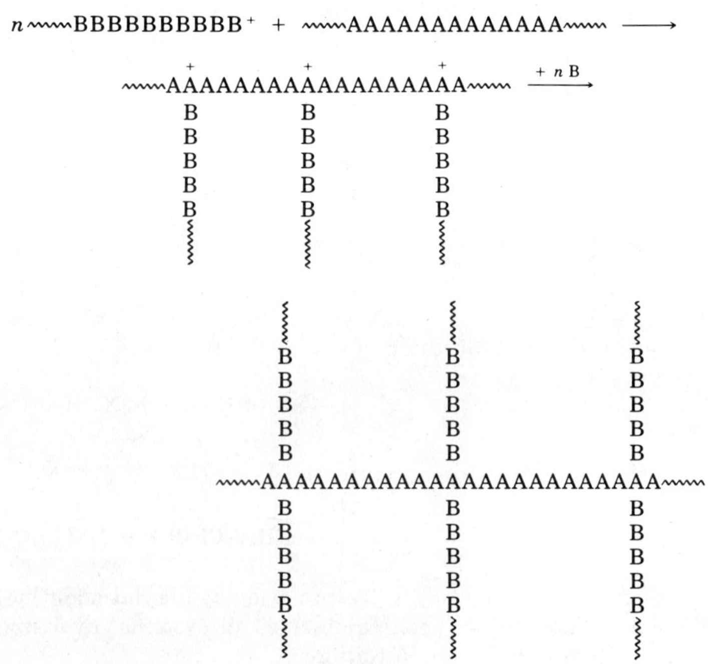

Industrial Applications

π-Donor Monomers. Some of the polymers obtained from alkenes are used as adhesives, adhesion primers (paper-polymer), caulkers, additives for motor oils, additives to sealing materials, waxes, butyl rubbers, pressure-sensitive adhesives, rubber compounding, hot-melt adhesives and coatings, tackifying agents, petroleum feedstocks, synthetic gasoline, lubricating oils, and viscosity improvers.

n, π-Donor Monomers. Some of these monomers (eg, vinyl ethers) give polymers that are exploited for their tack, medicinal properties, plasticizing effects (low DPs), and as adhesives or coating agents.

n-Donor Monomers. Polymers from heterocycles are used to replace metals in bearings and gears, as precursors for preparing polyurethanes and elastoplastic polyester–polyethers, for lining pipes, tanks, and valves, in the manufacture of water-meter parts and pump gears, and as flocculants, wet-strength additives for paper, adhesives, coatings, and surface-active agents (Polymins, BASF).

Future Development. Much of the current developmental work on adhesives and coating materials involves epoxide resins that are cured cationically. Other short-term developments should include the polymerization of various oxiranes and thiiranes for the manufacture of rubbery or crystalline thermoplastic materials, and the synthesis of functionalized oligomers of such monomers as isobutene, THF, etc, for use as specific additives according to the nature of the main chain and of the end groups, or as reagents for chain-extension processes. Cationic copolymerizations leading to block and graft copolymers seem feasible industrial propositions for the elaboration of specialty macromolecular products such as thermoplastic elastomers (see ELASTOMERS, THERMOPLASTIC).

BIBLIOGRAPHY

P. H. Plesch, ed., *The Chemistry of Cationic Polymerization,* Pergamon Press, Oxford, UK, 1963.

A. Gandini and H. Cheradame, *Adv. Polym. Sci.* **34/35** (1980).

S. Penczek, P. Kubisa, and K. Matyjaszewski, *Adv. Polym. Sci.* **37** (1980).

E. J. Goethals, ed., *Cationic Polymerization and Related Processes,* Academic Press, Inc., New York, 1984.

ALESSANDRO GANDINI
HERVÉ CHERADAME
Institut National Polytechnique de Grenoble (EFP)

CAULKING COMPOSITIONS. See SEALANTS.

CEILING TEMPERATURE. See DEPOLYMERIZATION.

CELLOPHANE

Cellophane (Table 1) is a family of transparent films that is manufactured by chemically regenerating a dissolved cellulosic compound into a continuous thin sheet. This compound, called viscose, was discovered and developed in England in 1892, but 20 years passed before a Swiss chemist, J. E. Brandenberger, known today as the father of cellophane, succeeded in developing a workable viscose-control technique and film-casting method. Late that year, viscose-based cellulose film was commercialized under the trade name Cellophane and was first produced in France by La Cellophane SA.

From that beginning, an international cellophane industry grew. At its peak, there were over 50 producers with close to 6.8×10^5 t annual capacity. With the introduction of low cost petroleum-based thermoplastic films in the early 1960s, cellophane's near-monopoly in transparent packaging markets began to erode (see FILMS; PACKAGING MATERIALS). However, even with this formidable competition, cellophane remains an exceedingly viable flexible packaging material. Today, 41 pro-

Table 1. Cellophane Properties[a,b]

Property	ASTM test	Nitrocellulose, nonheat-sealed	PVDC-solvent coated, heat-sealed	Uncoated, nonheat-sealed	Nitrocellulose, heat-sealed
specific gravity	D 1505	1.5	1.5	1.5	1.5
yield, cm²/g[c]		277	277	277	277
haze, %	D 1003	3.0	3.5	3.0	3.0
tensile strength, MPa	D 882				
MD		124	124	124	124
XD		55	55	55	55
elongation, %	D 882				
MD		20	22	16	16
XD		50	60	60	60
tensile modulus, 1% secant, MPa[d]					
MD		5170	5170	5520	5520
XD		2760	2900	2760	2760
bursting strength, MPa[a]	D 774	0.28	0.28	0.28	0.28
service-temperature range, °C		−18–177	−18–177	−18–177	−18–177
heat-seal range temperature, °C			74		77

[a] 1-mil (25.4-μm) thickness.
[b] Test conditions: 23°C and 50% rh unless otherwise stated.
[c] To convert cm²/g to in.²/lb, multiply by 70.32.
[d] To convert MPa to psi, multiply by 145.

ducers have an estimated worldwide annual capacity of 4.1 × 10^4 t. These producers represent some of the largest chemical companies in the free world, including Flexel Inc. in the United States; British Cellophane, Ltd. (BCL), a part of Courtaulds, in the UK and Canada; Sidac, a unit of Belgium's UCB, in the UK, Spain, and Belgium; Wolff, a part of Bayer AG in the FRG; and Hoechst-Celanese in Mexico and Colombia.

BIBLIOGRAPHY

C. H. Ward-Jackson, *The 'Cellophane' Story: Origins of a British Industrial Group,* British Cellophane Ltd., Bridgwater, Somerset, UK, 1977. Contains an excellent bibliography.

G. C. Inskeep and P. Van Horn, *Ind. Eng. Chem.* **44**(11), 2512–2521 (Nov. 1952). This report, a primary source for the technical description of the cellophane manufacturing process, contains a good bibliography.

CHARLES C. TAYLOR
Charles C. Taylor & Associates

CELLULAR MATERIALS

Foamed plastics, also known as cellular plastics or plastic foams, have been important since primitive man began to use wood, a cellular form of the polymer cellulose. Cellulose is the most abundant of all naturally occurring organic compounds, comprising approximately one-third of all vegetable matter in the world. The high strength : weight ratio of wood, good insulating properties of cork and balsa, and cushioning properties of cork and straw have contributed to the development of the broad range of cellular synthetic polymers in use today.

Cellular polymers have been commercially accepted for a wide variety of applications since the 1940s. The total usage of foamed plastics is expected to rise to about 2.1 × 10^6 t in 1987 and 2.8 × 10^6 t in 1995.

Nomenclature. A cellular plastic is defined as a plastic whose apparent density is decreased substantially by the presence of numerous cells disposed throughout its mass. In this article the term cellular polymer, foamed plastic, expanded plastic, and plastic foam are used interchangeably to denote all two-phase gas–solid systems in which the solid is continuous and composed of a synthetic polymer or rubber.

Theory of the Expansion Process

Foamed plastics can be prepared by various methods. The most widely used, called the dispersion process, involves dispersion of a gaseous phase throughout a fluid polymer phase and the preservation of the resultant state. Other methods of producing cellular plastics include sintering dispersed small particles and dispersing small cellular particles in the plastic. The latter processes are relatively straightforward techniques of lesser commercial importance.

The expansion process has been the subject of extensive investigation because it is the foundation of foamed plastics. In general, the expansion process consists of three steps: creation of small discontinuities or cells in a fluid or plastic phase, growth of these cells to a desired volume, and stabilization of the resultant cellular structure by physical or chemical means.

Properties

The data in Table 1 show the broad ranges of properties of rigid foamed plastics; the manufacturer should be consulted for the properties of a particular product. The properties of some commercial flexible foamed plastics are given in Table 2. These values depend on several structural variables and should be used only as general guidelines.

The properties of a foamed plastic depend on composition and geometry, often referred to as structural variables. Furthermore, they are influenced by the foam structure and the properties of the parent polymer. The polymer phase description must include the additives present.

Cells. A complete knowledge of the cell structure of a particular polymer would require the size, shape, and location of each cell. Because this is impractical, approximations are employed. Cell size has been characterized by measurements of the cell diameter and as a measurement of average cell volume. Mechanical, optical, and thermal foam properties depend on cell size.

Geometry. Cell shape is governed predominantly by final foam density and the external forces exerted on the cellular structure before its stabilization in the expanded state. In the presence of external forces, the cells may be elongated or flattened. Cell orientation can influence many properties.

Fraction Open Cells. An important characteristic of the cell structure is the extent of communication with other cells. This is expressed as fraction open cells.

Gas Composition. In closed-celled foams, the gas phase in the cells can contain blowing agent (so-called captive blowing agent), air, or other gases generated during foaming. Thermal and electrical conductivity can be profoundly influenced by the cell-gas composition.

Moisture Resistance. Plastic foams perform better than other thermal insulation when exposed to moisture, particularly where subjected to a combination of thermal and moisture gradients.

Electrical Properties. Electrical insulation takes advantage of the toughness and moisture resistance of polymers along with the decreased dielectric constant and dissipation factor of the foams. In addition, because of the low dissipation factor and rigidity, plastic foams are used in the construction of radar domes.

Environmental Aging. The response of cellular materials to light and oxygen is governed almost entirely by the composition and state of the polymer phase. Expansion into the cellular state increases the surface area; reactions of the foam with vapors and liquids are correspondingly faster than those of solid polymer. All cellular polymers deteriorate under the combined effect of light or heat and oxygen; this may be alleviated by additives (see also DEGRADATION; STABILIZATION).

Other Properties. Cellular polymers by themselves are poor materials for reducing sound transmission. They are, however, effective in absorbing sound waves of certain frequencies. Materials with open cells on the surface are particularly effective in this respect. The presence of open cells in a foam allows gases and vapors to permeate the cell structure by diffusion and convection, thus yielding high permeation rates. In closed-celled foams, the permeation of gases or vapors is governed by

Table 1. Physical Properties of Commercial Rigid Foamed Plastics

Property	ASTM test	Cellulose acetate	Epoxy	Phenolic	Polystyrene							PVC		Polyurethane				Urea–formal-dehyde
					Extruded plank		Expanded plank			Extruded sheet				Polyether		Isocyanurate		
																Bun	Laminate	
density, kg/m³[a]		96–128	32–48	32–64	35	53	16	32	80	96	160	32	64	32–48	64–128	32	32	13–19
mechanical properties																		
compressive strength, kPa[b] at 10%	D1621	862	138–172	138–620	310	862	90–124	207–276	586–896	290	469	345	1,035	138–344	482–1,896	210	117–206	34
tensile strength, kPa[b]	D1623	1,172		138–379	517		145–193	310–379	1,020–1,186	2,070–3,450	4,137–6,900	551	1,207	138–482	620–2,000	250	248–290	
flexural strength, kPa[b]	D790	1,014		172–448	1,138		193–241	379–517				586	1,620	413–689	1,380–2,400			
shear strength, kPa[b]	C273	965		103–207	241			241				241	793	138–207	413–896	180	117	
compression modulus, MPa[b]	D1621	38–90	3.9		10.3			3.4–14				13.1	35	2.0–4.1	10.3–31			
flexural modulus, MPa[b]	D790	38			41			9.0–26				10.3	36	5.5–6.2	5.5–10.3			
shear modulus, MPa[b]	C273			2.8–4.8	10.3			7.6–11.0				6.2	21	1.2–1.4	3.4–10.3		1.7	
thermal properties																		
thermal conductivity, W/(m·K)	C177	0.045–0.046	0.016–0.022	0.029–0.032	0.030		0.037	0.035	0.035	0.035	0.035	0.023		0.016–0.025	0.022–0.030	0.054	0.019	0.026–0.030
coefficient of linear expansion, 10^{-5}/°C	D696		1.5	0.9	6.3	6.3	5.4–7.2	5.4–7.2						5.4–7.2	7.2	7.2		
maximum service temperature, °C		177	205–260	132	74		74–80	74–80	74–80	77–80	80			93–121	121–149	149	149	
specific heat, kJ/(kg·K)[c]	C351				1.1									ca 0.9	ca 0.9	ca 0.9		
electrical properties																		
dielectric constant	D1673	1.12			1.19–1.20	<1.05	<1.05	1.02	1.02	1.02	1.27	1.28			1.05	1.1	1.4	
dissipation factor		20			0.028–0.031	<0.0004	<0.0004	0.0007	0.0007	0.0007	0.00011	0.00014			13	18		
moisture resistance																		
water absorption, vol %	C272	4.5		13–51	0.02	0.05	1–4	1–4	1–4									
moisture vapor transmission, g/(m·s·GPa)[b]	E96		58		35		<120	35–120	23–35	86	56	15		35–230	50–120		230	1,610–2,000

[a] To convert kg/m³ to lb/ft³, multiply by 0.0624.
[b] To convert kPa to psi, multiply by 0.145.
[c] To convert kJ/(kg·K) to Btu/(lb·°F), divide by 4.184.

Table 2. Physical Properties of Commercial Flexible Foamed Plastics

Property	ASTM test	Expanded natural rubber		Expanded SBR	Latex foam rubber		Polyethylene			Polypropylene		Polyurethane		Poly(vinyl chloride)		Silicone	
							Extruded plank	Extruded sheet	Cross-linked sheet	Modified	Sheet	Standard cushioning	High resilience type			Liquid	Sheet
density, kg/m³[a]		56	320	72	80	130	35	43	26–28	64–96	10	24	40	56	112	272	160
cell structure		closed	closed	closed	open	open	closed	closed	closed	closed		open	open	closed	closed	open	open
compression strength 25% deflection, kPa[b]	D3574, D3575						48			206	4.8	5.7	4.6				
tensile strength, kPa[b]	D3574		206	551	103		138	41		344		118	103	10.3	24	36 at 20%	
tensile elongation, %	D3574				310		60	276	276–480	1380	138–275	205	160			227	310
rebound resilience, %	D3574				73			50		75		40	62				
tear strength, (N/m)[c] × 10²	D3574						10.5	26				4.4	2.4				
maximum service temperature, °C		70	70	70			82	82	79–93	135	121					350	260
thermal conductivity, W/(m·K)	C177	0.036	0.043	0.030		0.050	0.053	0.040–0.049	0.036–0.040	0.039	0.039			0.035	0.040	0.078	0.086

[a] To convert kg/m³ to lb/ft³, multiply by 0.0624.
[b] To convert kPa to psi, multiply by 0.145.
[c] To convert N/m to lbf/in., divide by 1.75.

the composition of the polymer phrase, gas composition, density, and cellular structure (see also BARRIER POLYMERS). The penetration of visible light through foamed polystyrene has been shown to follow approximately the Beer-Lambert law of light absorption.

The resistance to rot, mildew, and fungus is related to moisture absorption. Therefore, open-celled foams support such growth better than closed-celled foams (see also BIOCIDES; BIODEGRADABLE POLYMERS).

Manufacturing Processes

Table 3 outlines production methods.

Uses

Concern over energy conservation and safety has stimulated growth in applications for insulation and cushioning in transport. A healthy economy is also expected to increase the demand for cushioning in furniture, bedding, and flooring, as well as for packaging. Structural foams (qv) are widely used as substitutes for wood, metal, or unfoamed plastics.

Cushioning is the largest single application of cellular polymers. As might be expected, the flexible foams are the main contributors to this field. The rapid growth rate of structural, packaging, and insulation applications has increased their volume over that of flexible foams during the past few years (Table 4).

Energy Considerations in Foam Insulation

The principal heat losses from buildings are through the ceilings, walls, windows, doors, floors, and foundations, and through air infiltration. A significant reduction in heat loss can be realized by insulating attics or ceilings; the improvement by

Table 3. Production of Cellular Polymers

Polymer	Extrusion	Expandable formulation	Spray	Froth foam	Compression mold	Injection mold	Sintering	Leaching
cellulose acetate	X							X
epoxy resin		X	X	X				
phenolic resin		X						
polyethylene	X	X			X	X	X	X
polystyrene	X	X				X	X	
silicones		X						
urea–formaldehyde resin				X				
urethane polymers		X	X	X		X		
latex foam rubber				X				
natural rubber	X	X			X			
synthetic elastomers	X	X			X			
poly(vinyl chloride)	X	X		X	X	X		X
ebonite					X			
polytetrafluoroethylene							X	

Table 4. U.S. Cellular-polymer Market

Applications and resins	10^3t				Annual growth, %		
	1967	1982	1987[a]	1995[a]	67–82	82–87	87–95
insulation	58	261	347	472	10.6	5.9	3.9
flooring	20	98	122	154	11.3	4.5	3.0
construction	9	136	197	288	20.2	7.8	4.8
cushioning	52	195	249	336	9.2	5.0	3.8
furniture	40	103	135	175	6.4	5.6	3.3
packaging	43	177	236	311	9.9	5.9	3.5
transportation	76	140	186	238	4.1	5.9	3.1
consumer	44	136	171	225	7.9	4.8	3.4
bedding	18	57	77	113	7.9	6.3	4.9
appliances	14	40	51	61	7.2	4.9	2.4
other	68	225	308	408	8.3	6.4	3.6
Total	*441*	*1567*	*2080*	*2781*	*8.8*	*5.8*	*3.7*
flexible urethane	181	511	658	844	7.1	5.2	3.2
rigid urethane	68	248	336	449	9.0	6.2	3.7
polystyrene	125	410	535	699	8.3	5.5	3.4
poly(vinyl chloride)	61	232	311	413	9.3	6.0	3.6
others	6	165	240	376	24.9	7.8	5.8
Total	*441*	*1567*	*2080*	*2781*	*8.8*	*5.8*	*3.7*

[a] Projected.

insulating walls is less significant. Substantial savings can be achieved by insulating basement and perimeter foundation walls.

Health and Safety Factors

Flammability. Plastic foams are organic and therefore combustible. All plastic foams should be handled, transported, and used according to manufacturers' recommendations and local and national regulations.

Virtually all plastic foams are blown with inert gas, chemical blowing agents that release inert gases, hydrocarbons containing three to five carbon atoms, chlorinated hydrocarbons, and chlorofluorocarbons such as CFC-11, CFC-12, CFC-113, and CFC-114. The hydrocarbons are flammable and pose a fire hazard during manufacturing.

Atmospheric Emissions. Certain organic compounds generate smog photochemically. Interaction with the total environment must be considered in developing environmentally acceptable blowing agents (qv). The products of combustion of plastic foams are usually carbon monoxide and carbon dioxide with smaller amounts of many other substances.

The presence of additives or unreacted monomer in certain plastic foams can limit their use where food or human contact is anticipated. The manufacturers' recommendations and existing regulations should be followed.

BIBLIOGRAPHY

K. C. Frisch and J. H. Saunders, *Plastic Foams,* Vol. 1, Parts 1 and 2, Marcel Dekker, Inc., New York, 1972 and 1973.

C. J. Benning, *Plastic Foams,* Vol. 1 and 2, Wiley-Interscience, New York, 1969.

N. C. Hilyard and co-workers, *Mechanics of Cellular Plastics,* Macmillan Publishing Co., Inc., New York, 1982.

J. L. Throne and B. C. Wendle, eds., *Engineering Guide to Structural Foams,* Technomic Publishing Co., Westport, Conn., 1976.

D. W. Suh
D. D. Webb
Dow Chemical U.S.A.

CELLULOID.

See Cellulose esters, inorganic.

CELLULOSE, BIOSYNTHESIS

Biochemistry of Cellulose Biosynthesis

Although cellulose is ubiquitous biologically, the biochemical and physical aspects of biosynthesis in the cell are just beginning to be understood. At the present time, biochemical knowledge is extensive and sure, but far from complete. Glucose, the starting material, is transformed to uridine diphosphoglucose (UDPG) by the classic pathway.

D-glucose + adenosine triphosphate (ATP) ⇌ D-glucose 6-phosphate (G6P) + adenosine diphosphate (ADP)

D-glucose 6-phosphate (G6P) ⇌ D-glucose 1-phosphate (G1P)

G1P + uridine triphosphate (UTP) ⇌ uridine diphosphoglucose (UDPG) + inorganic phosphate (P_i)

There is now circumstantial, and inconclusive, evidence that the activated glucose residues are then transferred successively from UDPG to a phosphorylated carrier lipid in both bacteria and green plants to form a transient compound that contains either cellobiose or cellotriose. Subsequently, cellobiose is split from the inferred lipid intermediate and attached to a growing, intracellular 1→4-β-glucan chain. If this mechanism proves a general one, then cellobiose may be regarded as the biosynthetic monomer of cellulose rather than glucose. In actively metabolizing bacteria, the 1→4-β-glucans are then rapidly transported to the outside of the cells through relatively large terminal pores. However, under conditions where the extracellular formation of cellulose microfibrils is repressed, as in colonies, the internal glucans tend to accumulate as electron-opaque, spherical, or ovoid globules.

Physical Chemistry

As a prelude to sketching possible mechanisms for the physical formation of the microfibril, one fact should be stressed: despite extensive research and literature on the subject, there is no consensus on how microfibrils are assembled.

At the risk of oversimplification, the mechanisms that have been suggested for the extracellular formation of cellulose microfibrils may be divided into two broad categories: one in which the microfibril is thought to be formed or shaped by a special structure (pore) or by an ordered granule (terminal complex or terminal globule), and a second in which the formation of the microfibril is thought to result spontaneously from the operation of physical–chemical forces on previously synthesized transient polymer chains. The first idea is called the ordered-granule hypothesis and the second is the association-crystallization hypothesis.

Prospects for Technical Application

By suitable manipulation of growth conditions in cellulose-producing bacteria or by the introduction of a plasmid for glucans into noncellulose-producing bacteria, it might be possible to accumulate the polymers in quantity. Probably with more difficulty, accumulation of such polymers from cultures of green plant cells might be feasible. Solutions or suspensions of the intermediate polymers may then give new properties of finish, digestibility, absorbency, or permeability to products made from a proportion of conventional cellulose.

BIBLIOGRAPHY

J. R. Colvin in J. Preiss, ed., *The Biochemistry of Plants,* Vol. 3, Academic Press, Inc., New York, 1980, p. 543.

Y. Aloni and M. Benziman in R. M. Brown, Jr., ed., *Cellulose and Other Natural Polymer Systems: Biogenesis, Structure and Degradation,* Plenum Press, New York, 1982, p. 341.

D. P. Delmer, *Adv. Carbohyd. Chem. Biochem.* **41**(sections V, 1–2), 105 (1983).

J. Ross Colvin
National Research Council of Canada

CELLULOSE, GRAFT COPOLYMERS

A cellulose graft copolymer refers to a macromolecularly modified cellulosic fiber. It is cellulose covalently linked to a polymer so that separation by solvent extraction is difficult without first degrading the cellulose, eg, by hydrolytic or oxidative depolymerization. The copolymer could be considered either as a graft copolymer, ie, with a side chain on the cellulose molecule, or as a block copolymer. In neither case can the grafted chains be proven to be solely side chains on the cellulose or to be solely at the end of the cellulose chain, ie, colinear (see also Block copolymers; Graft copolymers).

Reaction Mechanisms

Free-radical initiation is the primary method of forming cellulose graft copolymers. In some cases, macrocellulosic radicals are formed directly and initiate polymerization of vinyl monomers onto cellulose. In other cases, free radicals are formed that may abstract hydrogen atoms from cellulose to form macrocellulosic radical sites. Complexations of metallic ions with cellulose and subsequent release of the metallic ions in a lower state of oxidation also result in formation of macrocellulosic radical sites. In all cases, oxidative depolymerization of the cellulose molecule is also initiated.

Ionic-initiated grafting of vinyl monomers onto cellulose by formation of active sites on the cellulose molecule has been reported. Many cyclic monomers can be graft polymerized with cellulose by condensation reactions. Some ionic and condensation grafting reactions of monomers with cellulose are used industrially to modify the properties of cellulosic products. There is a large patent literature.

Methods

The grafting of vinyl monomers onto fibrous cellulose takes place in a heterogeneous reaction system. The cellulosic fibers are in the solid phase, whereas monomers are in vapor, liquid, or solid phases or in a solution phase as a solute or dispersed in an emulsion. During grafting, the morphological and molecular-orientation properties of the fibrous cellulose are usually altered, particularly the morphological structure. These effects, called Arthur effects, depend on the composition of the monomers and solutions and on the reaction method.

To obtain cellulosic products with contents of grafted copolymers of several percent may take from minutes to hours. Under similar experimental conditions, and with products that contain different amounts of grafted copolymers, the molecular weights of the grafted copolymers remain relatively constant. That is, at different grafted copolymer contents, each of the grafted polymeric chains has approximately the same molecular weight. These phenomena indicate that each grafted chain may polymerize very quickly. However, the accumulation of grafted polymer evidently depends on the diffusion of monomer to other radical sites and grafting of additional polymeric chains. This is in contrast to the relatively fast formation of polymeric coatings on cellulose or other types of substrate in which monomers are almost completely polymerized in a short time.

Special Methods. Free-radical–initiated grafting occurs between monomers and oligomers soluble in the same solvent as cellulose.

Grafting Reactions

Ionizing Radiation. Interactions of ionizing radiation with cellulose to form macrocellulosic radicals may be divided into those in which the energy is initially selectively absorbed by specific groups on the molecule, mainly uv radiation, and those in which the energy is randomly absorbed by the molecule, ie, high energy radiation, including both gamma and electron radiation. In each case, after absorption of the radiant energy, oxidative depolymerization of cellulose is initiated, and macrocellulosic radicals are formed.

Redox Reactions. A typical redox system used to initiate grafting reactions is ferrous ammonium sulfate and hydrogen peroxide. Cellulose is wetted with a solution of the ferrous salt and dried. A hydrogen peroxide solution of monomer is brought in contact with the cellulosic product and stirred to generate radicals and grafting.

Ceric Chelates. Ceruim (IV) ions in acidic solution form chelates with the hydroxyl groups on carbons C-2 and C-3 of the anhydroglucose unit of cellulose. Transfer of electrons from cellulose to Ce(IV) gives Ce(III), which dissociates from the chelate. The anhydroglucose ring is scissioned between carbons C-2 and C-3 and a short-lived radical is formed. In the presence of monomer, grafting reactions are initiated.

Other Oxidative Reactions. Other methods include chemical oxidation of cellulose by thermal decomposition of peroxides, ultrasonic radiation, electric-arc discharge (also called corona discharge), electrolysis, mechanical milling, and oxidation of products of chemical reactions in the presence of vinyl monomers. Chemical modification of cellulose, eg, by diazotization, thiolatation, and thiocarbonation increases the rate of oxidation and of graft copolymer formation in the presence of vinyl monomers.

Copolymer Properties

Morphological Properties. The morphological or supermolecular structure of fibrous cellulose is easily altered according to the product properties desired. Cotton fibers have an ordered morphological structure. The morphological structures of fibrous copolymers range from the initial collapsed structure of

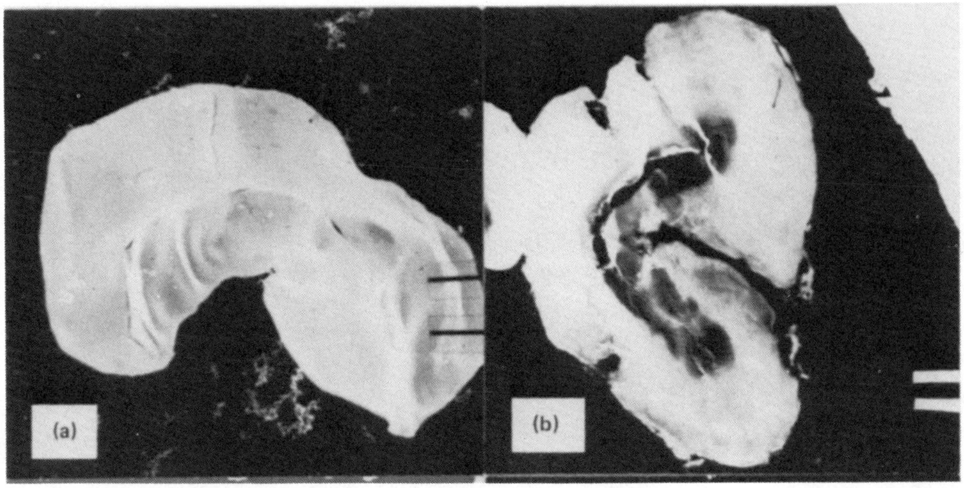

Figure 1. Electron microphotographs of morphological structures of cotton cellulose–polyacrylonitrile copolymers, as related to grafting method. (**a**) Ceric-ion initiation with acrylonitrile dissolved in water; (**b**) immersion of cellulose fiber in acrylonitrile (32%) in 80% aqueous $ZnCl_2$ after high energy irradiation.

cotton fiber to a rounded structure, as shown by electron microphotographs. The distribution of polymer within the fibrous copolymer primarily depends on the interaction of the monomer solution with the fiber and the method used to initiate copolymerization. For example, a graft copolymer of polyacrylonitrile and cellulose may have more than one morphological structure, as shown in Figure 1.

Mechanical Properties. The mechanical properties of cellulosic fibers are not significantly changed by random depolymerization during grafting under conditions that yield a satisfactory concentration of macrocellulosic radicals. Grafted fibers woven into textile-fabric constructions, or more often cellulosic fabrics grafted as is, retain many of the original mechanical properties and attain improved properties, such as wash-wear rating, wrinkle-recovery angle, and abrasion resistance.

Macromolecular Properties. In the modified cellulose copolymer product, the molecular weights of both modified cellulose and polymer are usually high. Graft copolymers of cellulose may exhibit second-order transition temperatures below the normal decomposition temperature of unmodified cellulose.

Applications

Cellulose graft copolymers are used in biomedical products, wood products, textiles, and paper products.

BIBLIOGRAPHY

J. C. Arthur, Jr. in W. M. Pasika, ed., *Advances in Macromolecular Chemistry,* Vol. 2, Academic Press, Inc., New York, 1970, pp. 1–87.

J. C. Arthur, Jr., ed., *Proceedings of the Symposium on Graft Polymerization onto Cellulose, Polymer Symposia No. 37,* Wiley-Interscience, New York, 1972.

J. C. Arthur, Jr. in J. J. Burke and V. Weiss, eds., *Block and Graft Copolymers,* Syracuse University Press, Syracuse, N.Y., 1973, pp. 295–323.

J. C. Arthur, Jr. in N. S. Allen, ed., *Developments in Polymer Photochemistry,* Vol. 1, Elsevier Applied Science Publishers Ltd., London, 1980, pp. 69–92.

JETT C. ARTHUR, JR.
Consultant

CELLULOSE, MERCERIZED.

See COTTON.

CELLULOSE, MICROCRYSTALLINE

Commercial forms of microcrystalline celluloses were first described in a patent issued in 1961. These new, stable thixotropic gel systems involve aqueous colloidal dispersions of disintegrated cellulose of constant degree of polymerization (level-off DP) at high solids concentration.

Two characteristic prerequisites for producing novel colloidal phenomena from a fibrous high polymer such as cellulose are a controlled chemical pretreatment to destroy the molecular bonds whereby microcrystals are hinged together in a network and an appropriate use of mechanical energy to disperse a sufficient amount of the unhinged microcrystals in the aqueous phase. This ensures the characteristic novel rheology and the smooth, fatlike spreadability of the resulting colloidal microcrystalline cellulose gels. This science has been expanded to include amylose, collagen, polyamide, polyester, and polyolefin aqueous microcrystal suspensoids as new articles of commerce.

Current applications of microcrystalline cellulose are given in Table 1. Microcrystalline cellulose is marketed under the trade name Avicel.

Table 1. Applications of Microcrystalline Celluloses

Application	Role of microcrystalline cellulose
tableting	inert binder, diluent, disintegrant, lubricant, sugar-coating additive
suspensions	suspending agent, stabilizer, thickening agent
emulsions	emulsifier, stabilizer
chromatography	tlc, column chromatography, and ion exchange
derivatives	controlled–particle-size MCC derivatives
ceramics	plasticity rapid prefired glaze hardening
food	nonnutritive filler reduces caloric value
cosmetics	retains volatile principles
paints	flow control, brush control, viscosity control

BIBLIOGRAPHY

O. A. Battista and P. A. Smith, *Ind. Eng. Chem.* **54**(9), 20 (1962).

O. A. Battista, *Am. Sci.* **53**(2), 151 (1965).

U.S. Pat. 2,978,446 (Apr. 4, 1961), O. A. Battista and P. A. Smith (to American Viscose Corp.).

O. A. Battista and co-workers, *Ind. Eng. Chem.* **49**(2), 333 (1956).

Fundamentals of High Polymers, Reinhold, New York, 1958, pp. 97–113.

O. A. Battista, *Microcrystal Polymer Science,* McGraw-Hill Book Company, New York, 1975, p. 20. Reprint by RSC Publishers, Fort Worth, Tex.

O. A. BATTISTA
The O. A. Battista Research Institute

CELLULOSE, REGENERATED. See RAYON.

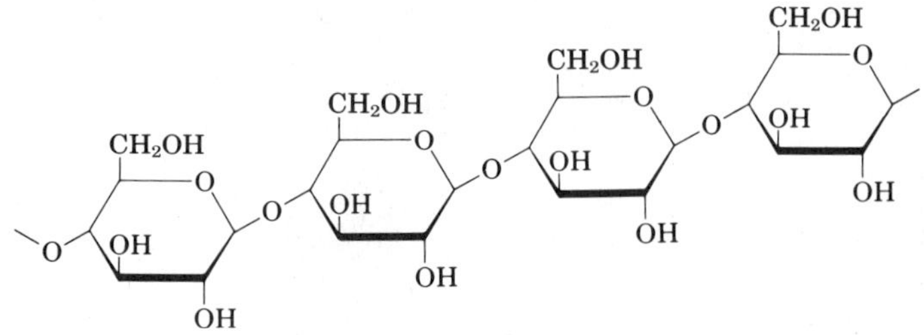

(1)

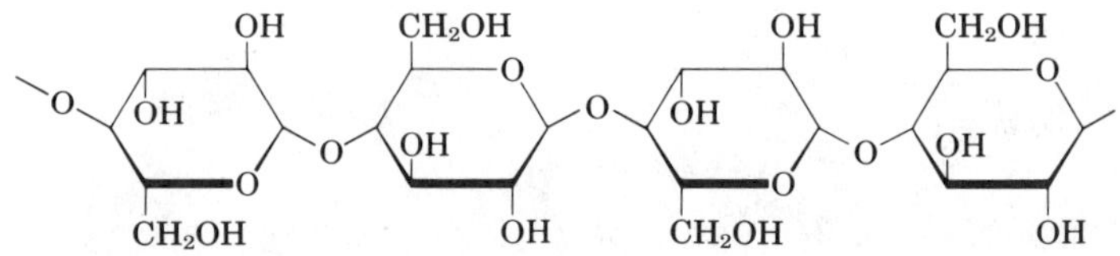

(2)

Figure 1. The cellulose structure involves D-glucose linked through β-glycoside bonds. (**1**) presents the traditional Haworth orientation for all glucose units, (**2**) shows the linear array of rings more effectively.

CELLULOSE, STRUCTURE AND PROPERTIES

Representation and Terminology

In a strict scientific sense, the term cellulose applies only to the β-(1→4)-*D*-glucan isolated from plant cell walls. This material has a structure that may be represented by (**1**) or (**2**) as shown in Figure 1 (see also POLYSACCHARIDES). Material described as cellulose typically also includes smaller amounts of hemicelluloses and lignin (qv). Ideally, material designated as cellulose should be identified by source and physicochemical treatments received.

Commonly encountered terms are *native cellulose,* cellulose as it occurs *in vivo; regenerated cellulose,* cellulose precipitated out of solution, usually alkaline; *holocellulose,* plant cell wall material minus the noncarbohydrate, lignin (holocellulose = cellulose + hemicellulose); and *hemicellulose,* noncellulose cell wall polysaccharides that are easily extractable by dilute aqueous alkali.

Physicochemical Properties

The various crystalline modifications have different physical properties and chemical reactivities. Tensile strength, relative wet–dry strength, and percent extension to break are shown in Table 1, and elastic moduli for various celluloses are given in Table 2. Other selected physical constants appear in Table 3.

Occurrence and Uses

Of the tremendous amount of cellulose produced by nature annually, only a small percentage is used by humans. Cellulose occurs naturally in food and feed crops, and is used as fuel or lumber, as purified cellulose for production of paper and textiles, and as derivatives of cellulose, which are widely employed in the manufacture of glues, varnishes, thickeners, films, foils, and plastics.

Commercial cellulose production concentrates on the naturally highly pure sources of cellulose such as cotton (qv) or easily harvested sources such as wood (qv). Of the two principal areas of cellulose consumption, ie, fibers and paper (qv), more cellulose is used in paper and paperboard manufacture.

Biosynthesis

Several organisms have been used in the study of cellulose biosynthesis, among them *Acetobacter xylinum,* which is es-

Table 1. Tensile Strength of Celluloses

Form	Tensile strength, MPa[a]		Relative wet–dry strength, %	Extension at break, %	
	Dry	Wet		Dry	Wet
ramie	900	1060	117	2.3	2.4
cotton	200–800	200–800	105	16–12	6–13
flax	824	863	105	1.8	2.2
viscose rayon	200–400	100–200	50	8–26	13–43
viscose rayon, highly oriented	610	520	86	9	9
cellulose acetate	150–200	100–120	65	21–30	29–30

[a] To convert MPa to psi, multiply by 145.

Table 2. Elastic Moduli of Some Celluloses

Form	Modulus, GPa[a]
native flax	78–108
native hemp	59–78
native ramie	48–69
mercerized ramie	80
oriented rayon	33
cellulose acetate film	4

[a] To convert GPa to psi, multiply by 145,000.

Table 3. Selected Physical Properties of Cellulose

Property	Value
specific gravity	
in benzene	1.570
in water	1.604–1.609
regenerated cellulose	1.522
x-ray density, g/mL	
crystalline portion	1.590–1.630
amorphous portion	1.482–1.489
average crystallinity, %	
native	70
regenerated	40
specific heat, J/g·°C[a]	1.22
ignition point, °C	>280
heat of combustion, kJ/g	17.43
dielectric constant	
pulp sheets	2.2–2.3
crystalline portion	5.7
insulating value, kV/cm	500

[a] To convert J to cal, divide by 4.184.

pecially useful since much is known about its carbohydrate metabolism. Some algae, such as *Valonia* and *Oocystis,* have been investigated, as have slime molds, and the preponderance of evidence suggests that UDP-glucose plays a critical role in the synthesis of β-(1→4)-glucans. The intermediacy of UDP-glucose in cellulose synthesis and other details of the process are summarized in Figure 2 (see also CELLULOSE, BIOSYNTHESIS).

Chemical Structure

The fundamental building block of cellulose is the monosaccharide D-glucose of chemical composition $C_6H_{12}O_6$.

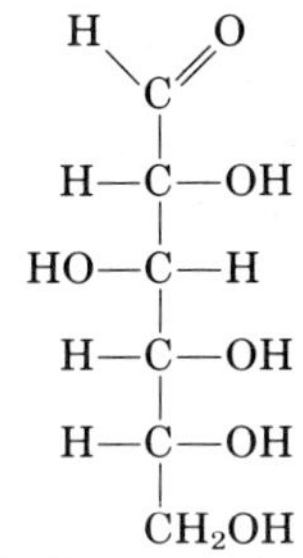

Fischer structure

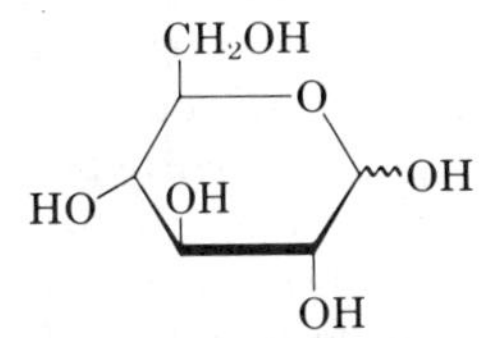

Haworth projection

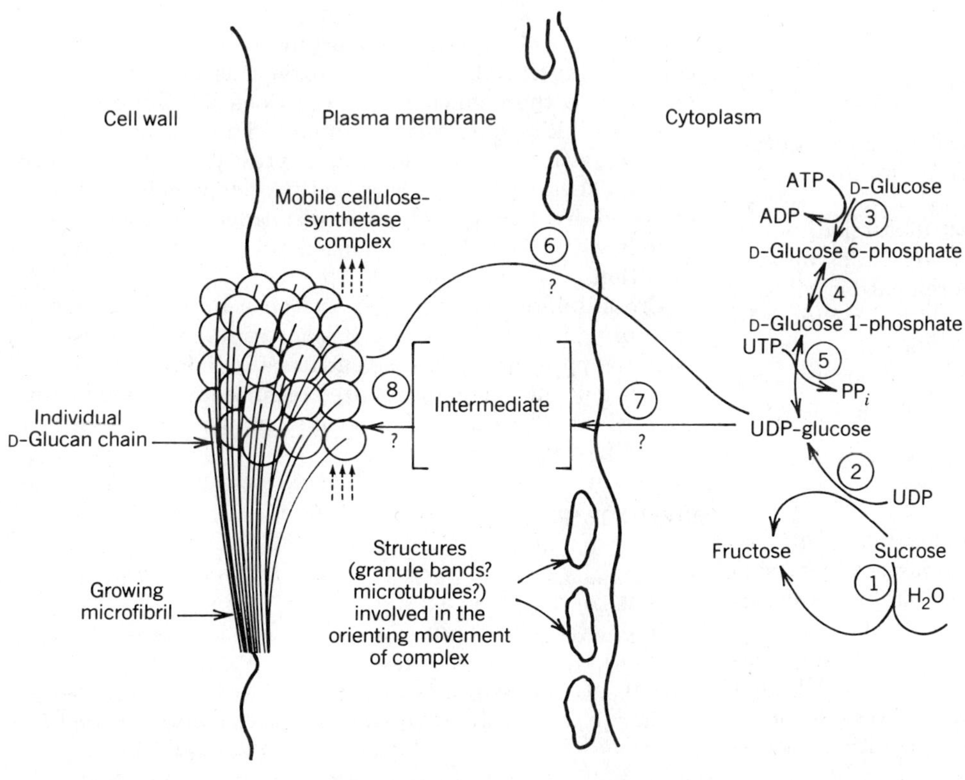

Figure 2. Biosynthetic pathways involved in cellulose synthesis. The [intermediate] is a lipid- or protein-linked monosaccharide or oligosaccharide. The designation ? indicates uncertainty about the process. Numbers ①–⑤ refer to reactions catalyzed by the following enzymes: ①, invertase; ②, sucrose synthetase; ③, hexokinase; ④, phosphoglucomutase; and ⑤, UDP-glucose pyrophosphorylase. Reactions ⑥, ⑦, and ⑧ are hypothetical pathways linking UDP-glucose and cellulose.

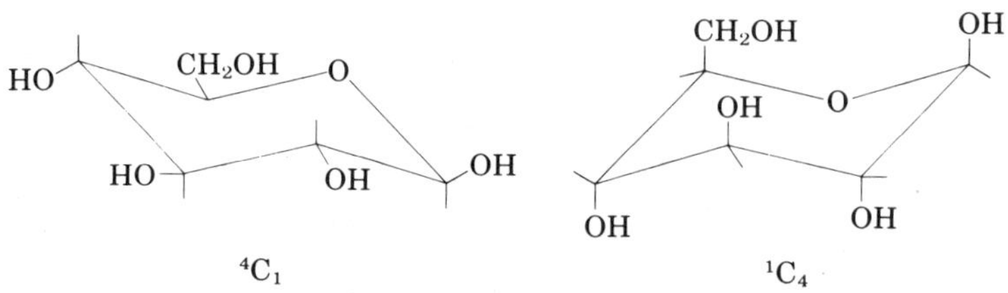

Chair conformations

Structural Physics

Crystallinity. Ir spectroscopy and x-ray diffraction studies of the organization in plant cellulose reveal that a chief portion of cellulose is arranged in crystalline micelles or crystallites, with interspersed amorphous regions of low degree of order. X-ray analysis shows that native cellulose I has a monoclinic unit cell that is disrupted by swelling in strong alkali, as in conversion to cellulose II.

Accessibility. The accessible areas of cellulose are defined as the amorphous areas plus the surfaces of the crystalline micelles.

Dispersion Properties

Adsorption. Adsorption of water by cellulose is an important process because adsorption (or desorption) is commonly involved in pulp, paper, or textile processing. Different degrees of water binding are encountered; the most firmly held is water of constitution, which is released on pyrolysis of cellulose. This water is not adsorbed, but covalently bound. Monomolecularly bound water that is hydrogen-bonded to primary and secondary hydroxyl groups in the amorphous regions and on the surfaces of the crystalline regions is less strongly bound. Least firmly bound, and hence most available, is the water adsorbed polymolecularly. This water has been referred to as capillary condensed, but is best thought of as a solid solution of water in the cellulose matrix.

Swelling. Swelling of cellulose by water and other reagents reduces crystallinity and increases accessibility. These are important factors for subsequent chemical or enzymatic modification or degradation of cellulose.

Solubility. Because cellulose is both a polar and crystalline polymer, it is difficult to find a good solvent. Alkali hydroxides are useful cellulose solvents.

Cellulose may be dissolved by indirect dispersion, which involves conversion to a cellulose derivative such as nitrate, acetate, or xanthate, and dissolution in a compatible solvent. More commonly, direct dispersion of unmodified cellulose in cuprammonium hydroxide, cupriethylene diamine, mineral acids, tetraalkylammonium hydroxides, or solutions of heavy metal complexes (Cadoxen or iron tartaric acid) produces cellulose solutions.

Degradation

Acid Hydrolysis. Like all carbohydrate glycosides, cellulose is subject to acidic hydrolysis. The degree of susceptibility varies according to the region, amorphous or crystalline, concentration and type of acid, and temperature.

Alkaline Degradation. This process has great industrial importance in both cooking and refining of wood pulp. Alkaline degradation is most prominent in the soda process, but also occurs in the kraft process. Under alkaline conditions at elevated temperature, cellulose is subject to β-elimination, ie, unzipping, to produce cellulose degradation products.

Thermal Degradation. Studies of thermal degradation are important because of the interesting small molecules produced during the process, as well as the contribution to understanding the thermal loss of desirable properties in paper and textile products.

Degradation by Electromagnetic Radiation. Cellulose may be photochemically degraded in a number of ways, including sequential degradation following photolysis of water, photosensitization by dyes or metallic oxides, or by direct photolysis.

Enzymatic Degradation of Cellulose. Degradation of cellulose by enzymes, particularly the cellulase system of *Trichoderma* species, assumes importance in wood and textile preservation, breakdown of cellulosic biomass to produce feedstock chemicals, and many other areas. Attack by cellulase causes many changes in cellulose, including loss in tensile strength, lower DP, increased moisture uptake, and fragmentation to short fibers.

Cellulose Derivatives

Cellulose derivatives include inorganic esters (cellulose nitrate), organic esters (acetates), cellulose ethers (qv) [CMC and methylcellulose (MC)] and regenerated cellulose (cellophane (qv) and rayon (qv)). See also CELLULOSE ESTERS.

Cellulose Cross-linking

Cross-linking of cellulose is important in textiles. Most of these cross-linking reactions are heterogeneous and thus proceed more slowly than homogeneous reactions of cellulose.

Cross-linking of textiles for improved dimensional stability (qv), crease resistance, and durable press properties has been accomplished with *N*-hydroxymethyl compounds, which are prepared by reaction of various nitrogen-containing compounds with formaldehyde under neutral or slightly alkaline conditions.

Cross-linking has many effects on cellulose or cellulose-containing materials, eg, textiles. Cross-linked cellulose may swell in appropriate reagents such as alkali, but is not soluble in such solvents, as underivatized cellulose is. Chemical reactivity is greatly decreased, probably because of decreased susceptibility to swelling.

BIBLIOGRAPHY

R. D. Preston, *The Physical Biology of Plant Cell Walls,* Chapman and Hall, London, 1974.

R. D. Preston in M. H. Zimmerman, ed., *The Formation of Wood in Forest Trees,* Academic Press, Inc., New York, 1975.

C. H. Haigler and M. Benziman in R. M. Brown, Jr., ed., *Cellulose and Other Natural Polymer Systems: Biogenesis, Structure, and Degradation,* Plenum Publishing Corp., New York, 1982.

D. P. Delmer in R. S. Tipson and D. Horton, eds., *Advances in Carbohydrate Chemistry,* Vol. 41, Academic Press, Inc., New York, 1983.

JAMES R. DANIEL
Purdue University

CELLULOSE DERIVATIVES

This article concerns novel and unconventional cellulose derivatives, and thus excludes conventional derivatives prepared by esterification, etherification, and graft polymerization (see CELLULOSE, GRAFT COPOLYMERS; CELLULOSE ESTERS, INORGANIC; CELLULOSE ESTERS, ORGANIC; CELLULOSE ETHERS.

Nomenclature

Deoxycelluloses denote anhydroglucose units in which the three hydroxyl groups are partially or completely replaced by other functional groups which do not contain an oxygen atom attached to the ring carbon atom, eg, R = H, halogen, NHR, SR, etc. Anhydrocelluloses are anhydroglucose units in which an intramolecular ring structure is formed by the loss of one mole of water to yield a 2,3-epoxide or 3,6-tetrahydrofuran structure.

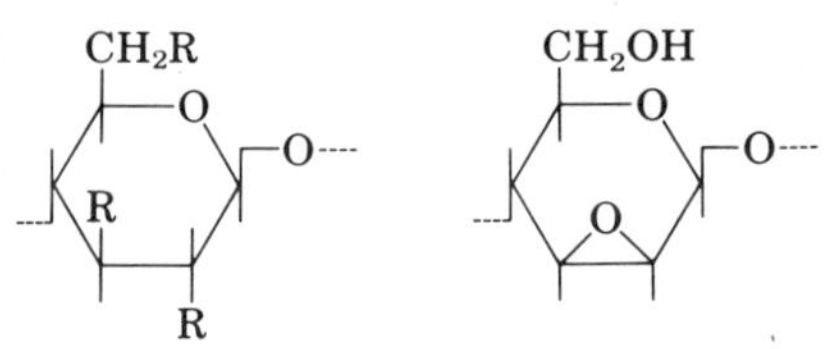

deoxycellulose 2,3-anhydrocellulose

Deoxycelluloses

A variety of reactions has been used to prepare deoxycelluloses. Many of these utilize displacement of a good leaving group, eg, tosylate or mesylate, to form the desired deoxy derivative. However, newer synthetic methods have been adapted for the preparation of these derivatives. The reactions may be classified as follows:

1. Nucleophilic substitution at saturated carbon atoms.
2. Halogenation.
3. Free-radical addition across a double bone using 5,6-cellulosene as the starting material.
4. Reactions with carbonyl-containing celluloses.
5. Other methods that cannot be classified by the preceding categories.

Halodeoxycelluloses. The most widely used method for preparing halodeoxycelluloses (cell—X, where X = F, Cl, Br, or I) is the nucleophilic displacement of good leaving groups by halides in various solvents.

Nitrogen-containing Deoxy Derivatives. Aminodeoxycelluloses and related derivatives are synthesized by nucleophilic displacement of halogen or tosylate groups and by reduction of cellulose derivatives such as hydrazones, azides, and oximes.

Thiodeoxycelluloses. The introduction of thiol groups into anhydroglucose units and subsequent oxidation to disulfide cross-links have been achieved via cellulose tosylate (cell-OTs) or chlorodeoxycellulose.

Phosphorus-containing Deoxycelluloses. Most investigations in this area have been directed toward the preparation of flame-resistant deoxycelluloses with C—P bonds that are stable to hydrolysis.

Organometallic Reactions and Derivatives. *C*-Alkyldeoxycelluloses, eg, *n*-butyldeoxycellulose, have been prepared by reaction of iododeoxycellulose with *n*-alkyllithiums in the presence of THF.

Other Cellulose Derivatives. In addition to the standard deoxycelluloses, the propargyl ether of cellulose (cell—$OCH_2C{\equiv}CH$) has been made by heating the ditosyl ester of cellulose with sodium propargylate in DMF for 4–6 h at 80–120°C or by heating sodium cellulosate with propargyl bromide in benzene for 2–24 h at 60–120°C. Transesterification of chlorodeoxycellulose with sodium salicylate in DMF afforded cellulose salicylate, which was complexed with $Hg(CH_3COO)_2$ in DMF. Reaction of cotton cellulose with pentafluorobenzoic acid and pentafluorobenzamide under basic conditions gave 4-substituted tetrafluorophenyl cellulose ethers; subsequent decarboxylation in refluxing DMSO afforded the 2,3,4,5,6-tetrafluorophenyl ether of cellulose.

Applications. Applications can be categorized into biological and biomedical uses; heat and flame resistance; wrinkle resistance and durable-press properties (see TEXTILE RESINS); and removal of water pollutants such as heavy metals.

Cellulosenes

Cellulosenes are anhydroglucose units in which one mole of water is eliminated, and are alkenes whose prefix numbers describe the position of the double bond in the ring, eg, 5,6-cellulosene. The removal of a hydroxyl group from an anhydroglucose units to prepare the corresponding alkene or cellulosene is accomplished by traditional methods, such as dehydrohalogenation in the presence of a strong base.

Anhydrocelluloses

Anhydrocelluloses are anhydroglucose units in which an intramolecular ring structure is formed by the loss of one mole of water to yield a 2,3-epoxide or 3,6-tetrahydrofuran structure. A detailed study of the ir spectra of 2,3,- and 3,6-anhydrocelluloses indicated that the band at 855 cm^{-1} could be assigned to the 2,3-anhydro ring or to a 1-CH equatorial group, whereas the bands at 800 and 840 cm^{-1} were attributed to the 3,6-anhydro ring and the 1-CH equatorial group, respectively. These anhydrocelluloses were synthesized by the reaction of 6-tosyl cellulose with aqueous alkali to prepare 3,6-anhydrocellulose and the reaction of 6-*O*-trityl-tosyl (2 or 3 position or both) cellulose with methanolic KOH.

Oxidized Celluloses and Derivatives

Oxidation products of cellulose vary markedly with the reagent, pH, temperature, and other variables. In some instances, only certain types of oxidation product are formed, eg, dialdehydes, and monocarboxyl groups; in other instances, several different types of oxidation product are formed.

Oxidized Celluloses Containing Carbonyl Groups. Aldehyde and ketone groups are the carbonyl groups most common in oxidized celluloses.

Oxidized Celluloses Containing Carboxyl Groups. The oxidation of cellulose with NO_2 and its dimer N_2O_4 has been studied with regard to solvent effects, mechanism of oxidation, and time and temperature. Although 6-carboxycellulose is the main product formed under various reaction conditions, other products have also been observed and verified.

Applications. Applications of oxidized cellulose include biological and biomedical uses; altering dye behavior; as precursors for graft polymerization; heat and flame resistance; removal of pollutants from water; and others.

BIBLIOGRAPHY

T. L. Vigo and C. M. Welch, *Textilveredlung* **8**(3), 94 (1973).

TYRONE L. VIGO
U.S. Department of Agriculture

CELLULOSE DIACETATE. See CELLULOSE ESTERS, ORGANIC.

CELLULOSE ESTERS, INORGANIC

Cellulose Nitrate

Cellulose nitrate, commonly misnomered nitrocellulose, is the oldest cellulose derivative; it is also the only cellulose inorganic ester of commercial importance. Multiton quantities of cellulose nitrate are produced annually in the United States and several foreign countries. Most recent developments concern continuous-nitration processes, lower cost raw materials, and improvement of stabilization.

Physical Properties. Physical properties of a typical cellulose nitrate are given in Table 1.

Table 1. Properties of Cellulose Nitrate[a]

Property	Value or characteristic
General	
odor and taste	none
color of film	water-white
clarity of film	excellent
specific gravity of cast film	1.58–1.65
refractive index, principal	1.51
light transmission, lower limit of substantially complete, nm	313
moisture absorption at 21°C in 24 h, 80% rh, %	1.0
water-vapor permeability at 21°C, (g·cm)/(cm²·h) × 10^{-4}	2.8
sunlight, effect on discoloration	moderate
effect on embrittlement	moderate
aging, effect of	slight
Electrical	
dielectric constant at 25–30°C, 60 Hz	7–7.5
power factor at 25–30°C, %, 60 Hz	3–5
Mechanical	
tensile strength at 23°C, 50%, rh, MPa[b]	62–110.3
elongation at 23°C and 50% rh, %	13–14
flexibility of 0.076–0.10 μm[c]	30–500
Solubility and compatibility characteristics	
solvents, principal type	esters, ketones, ether–alcohol mixtures
resins, compatible	almost all types
plasticizers, compatible	almost all types[d]
waxes and tars	incompatible
cellulose derivatives, compatible	ethylcellulose, cellulose acetate, ethyl hydroxyethylcellulose

[a] Unplasticized RS nitrocellulose, Hercules, Inc.
[b] To convert MPa to psi, multiply by 145.
[c] Double folds under 200-g tension, ASTM D 2176-69.
[d] Including vegetable oils.

Preparation. Cellulose is a polymer molecule comprised of β-1,4-linked anhydroglucose units (AGU) containing one primary and two secondary alcohol (hydroxyl) groups per AGU. Nitration of cellulose may be compared to a typical equilibrium esterification, where reaction of an alcohol and an acid (HNO_3) produces an ester and water.

Manufacture and Processing. Because of increased costs and limited availability of cotton linters and the advent of higher quality wood pulps, uniform, stable cellulose nitrate is made from wood cellulose sources. Cellulose nitrate is produced by a continuous automatic process in which the cellulose and mixed nitrating acids are fed simultaneously into the nitration vessel. The reaction slurry is passed continuously into a centrifuge which is divided into several zones. As the cellulose nitrate is advanced through each succeeding zone, the more concentrated nitrating acids are replaced with successively weaker aqueous acid washes, and finally with pure water in the last zone. The water slurry of cellulose nitrate is pumped to the purification area, and the spent nitrating acids and aqueous acid washes are recovered for reuse.

Economic Aspects. Hercules, the only U.S. producer, has a rated annual capacity of ca 41,000 t of cellulose nitrate. Demand for cellulose nitrate plastics and films has declined sharply, but cellulose nitrate continues to be an important ingredient in lacquer formulations for decorative and protective coatings and a component in various military explosives and propellants. The current price for cellulose nitrate esters ranges from ca $2.80 to $3.15/kg.

Health and Safety Factors. Cellulose nitrate wet with 30% alcohol is shipped under a "Flammable Liquid" label as defined by the ICC. In this condition, cellulose nitrate exhibits the same hazards as other flammable liquids. Cellulose nitrate should not be allowed to dry out because its rate of flammability increases as its alcohol content is lowered, and dry cellulose nitrate of high nitrogen content may explode on sudden impact. Cellulose nitrate has FDA approval for use in coatings, inks, and adhesives in contact with food.

Uses. Cellulose nitrate is used in lacquers, explosives and propellants, and plastics and films.

Cellulose Sulfate Esters

When prepared to the appropriate degree of sulfate ester substitution (DS), cellulose sulfate esters are water soluble and of interest as food thickeners or enhancers, textile sizings, viscosity modifiers for oil-drilling muds, and in pharmaceutical applications.

Cellulose Phosphate Esters

Cellulose phosphate esters are of considerable interest because of their inherent flame resistance and ion exchange capability, and consequently much research has been reported concerning their preparation and evaluation. Cellulose phosphate esters are generally prepared as the ammonium salt by the reaction of cellulose with phosphoric acid and urea at elevated temperatures (130–150°C).

Cellulose phosphate ester as the sodium salt is produced in the United States by the James River Chemical Co.; production is estimated to be less than 20,000 kg annually. This ester is used for the treatment of kidney stones and has an estimated price of $40.00/kg.

BIBLIOGRAPHY

J. Barsha in E. Ott, H. M. Spurlin, and M. W. Grafflin, eds., *Cellulose and Cellulose Derivatives,* Vol. 5, Part II of *High Polymers,* 2nd ed., Interscience Publishers, New York, 1954, pp. 713–762.

J. W. Green in R. L. Whistler and co-workers, eds., *Methods in Carbohydrate Chemistry,* Vol. III, Academic Press, New York, 1963, pp. 213–217.

G. L. Drake and W. A. Reeves in N. M. Bikales and L. Segal, eds., *Cellulose and Cellulose Derivatives,* Vol. 5, Part V of *High Polymers,* Wiley-Interscience, New York, 1971, pp. 1293–1331.

RICHARD J. BREWER
RICHARD T. BOGAN
Eastman Kodak Company

CELLULOSE ESTERS, ORGANIC

Survey

Cellulose, a polyhydroxy compound, forms esters with organic acids, anhydrides, and acid chlorides. Cellulose esters of almost any organic acid can be prepared, but because of practical limitations, esters of acids containing more than four carbon atoms have not achieved commercial significance.

Cellulose acetate is the most important organic ester because of its broad application in fibers and plastics. It is prepared in multiton quantities with degrees of substitution (DS) ranging from that needed for hydrolyzed, water-soluble monoacetate to those for fully substituted triacetate.

Although cellulose acetate remains the most widely used organic ester of cellulose, its usefulness is restricted by its moisture sensitivity, limited compatibility with other synthetic resins, and relatively high processing temperatures. Cellulose esters of higher aliphatic acids (C_3, C_4) circumvent these shortcomings with varying degrees of success. Mixed cellulose esters containing the dicarboxylate moiety, eg, cellulose acetate phthalate, have commercially useful properties such as alkaline solubility and excellent film-forming characteristics.

Properties

The properties of cellulose esters are affected by the number of acyl groups per anhydroglucose unit, acyl chain length, and the degree of polymerization (DP) (molecular weight). The properties of some cellulose triesters are given in Table 1.

The common commercial products are the primary (triacetate) and secondary (acetone-soluble, ca 39.5% acetyl, 2.45 DS) acetates; they are odorless, tasteless, and nontoxic. Their properties depend on the combined acetic acid content and molecular weight.

Liquid Crystalline Solutions. Cellulose esters, when dissolved in the appropriate solvents at the proper concentration, show liquid crystalline characteristics similar to those of other rigid-chain polymers due to an ordered arrangement of the polymer molecules in solution.

Manufacture and Processing

Simple triesters such as cellulose formate, cellulose propionate, and cellulose butyrate have been prepared and their properties studied; none of these triesters is produced in large quantities. Cellulose formate esters, prepared by reaction of cellulose with formic acid, are thermally and hydrolytically unstable. Cellulose propionate and cellulose butyrate triesters are synthe-

Table 1. Properties of Cellulose Triesters

Cellulose ester	Shrinking point, °C	Mp, °C	Char point, °C	Moisture regain, % 25% rh	Moisture regain, % 75% rh	Density, g/mL	Tensile strength, MPa[a]
cellulose[b]				5.4	15.5	1.52	
acetate		306	315	0.6	3.8	1.28	71.6
propionate	229	234	>315	0.1	1.5	1.23	48.0
butyrate	178	183	>315	0.1	0.7	1.17	30.4
valerate	119	122	>315	0	0.3	1.13	18.6
caproate	84	94	>315	0	0.2	1.10	13.7
heptylate	82	88	290	0	0.2	1.07	10.8
caprylate	82	86	315	0	0.1	1.05	8.8
caprate	87	88	310	0	0.2	1.02	6.9
laurate	89	91	>315	0	0.1	1.00	5.9
myristate	87	106	315	0	0.1	0.99	5.9
palmitate	90	105	315	0	0.1	0.99	4.9

[a] To convert MPa to psi, multiply by 145.

[b] Starting cellulose, prepared by deacetylation of commercial, medium-viscosity cellulose acetate (40.4% acetyl content).

sized with propionic or butyric anhydride in the presence of an acid catalyst by methods similar to those used in the preparation of cellulose acetate.

Cellulose esters of unsaturated acids, such as the acetate methacrylate, acetate maleate, and propionate crotonate have been prepared by treating the hydrolyzed acetate or propionate with the corresponding acyl chloride in a pyridine solvent. Cellulose esters of unsaturated acid are cross-linkable by heat or uv light; solvent-resistant films can be prepared from such esters.

Amine-containing cellulose esters, eg, the acetate *N,N*-diethylaminoacetate and propionate morpholinobutyrate, are of interest because of their unique solubility in dilute acid. Such esters are prepared by the addition of the appropriate amine to the cellulose acylate crotonate ester or by replacement of the chlorine on cellulose acylate chloroacetate esters with amines. This type of ester has been suggested for use in controlled-release, rumen-protected feed supplements for ruminants.

Mixed esters, such as cellulose acetate propionate and cellulose acetate butyrate, have desirable properties not exhibited by the acetate or the higher acyl triesters. These mixed esters are produced commercially in multiton quantities by methods similar to those for cellulose acetate; they are prepared over a wide range of acyl substitution and viscosity.

Stabilization. After hydrolysis, precipitation, and thorough washing of the cellulose esters to remove residual acids, the esters must be stabilized against thermal degradation and color development which may occur during processing, such as extrusion or injection molding (see also STABILIZATION). The thermal instability is caused by the presence of oxidizable substances and small amounts of free and combined sulfuric acid. Treatment with aqueous potassium or calcium iodide reportedly stabilizes the cellulose acetate against thermal degradation.

Cellulose Acetate. Almost all cellulose acetate, with the exception of fibrous triacetate, is prepared by a solution process employing sulfuric acid as the catalyst with acetic anhydride in an acetic acid solvent.

In the fibrous acetylation process, part or all of the acetic acid solvent is replaced with an inert diluent, such as toluene, benzene, or hexane, to maintain the fibrous structure of cellulose throughout the reaction.

Economic Aspects

Cellulose ester flake production increased to a maximum of 403,000 t in 1980, but has declined since then (Fig. 1).

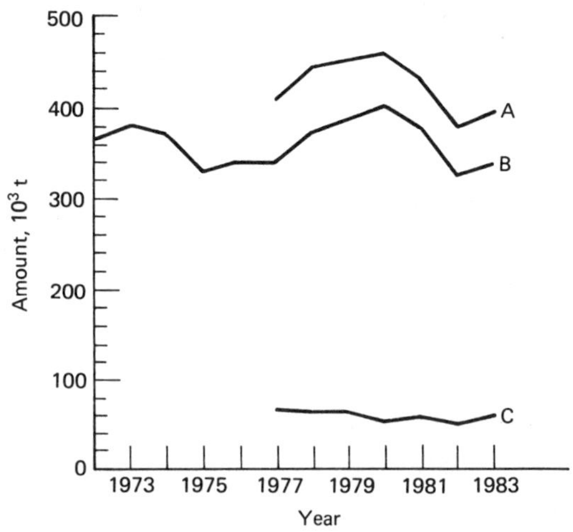

Figure 1. U.S. production of cellulose esters; **A**, total cellulose esters production; **B**, cellulose acetate flake; **C**, other cellulose esters, ie, cellulose nitrate, cellulose acetate butyrate, and cellulose acetate propionate.

Health and Safety Factors

The vapors of the organic solvents used in the preparation of cellulose ester solutions represent a potential fire, explosion, or health hazard. Cellulose esters are considered nontoxic and may be used in food-contact applications.

Uses

The cellulose esters with the largest commercial consumption are cellulose acetate, including cellulose triacetate, cellulose acetate butyrate, and cellulose acetate propionate. Cellulose acetate used for photographic film base is almost exclusively triacetate; some triacetate also is used for textile fibers because of its crystalline and heat-setting characteristics.

R. T. BOGAN
R. J. BREWER
Eastman Kodak Company

Plastics

Cellulose ester plastics have long been used for the extrusion of film and sheeting and for the injection molding of parts. Cellulose esters are thermoplastic and therefore processible by a variety of methods with readily available equipment that can be modified easily if necessary. Cellulose esters are adaptable to compression molding, injection molding, extrusion, and solution casting.

Cellulose ester plastics are noted for their toughness, surface gloss, smoothness, and clarity. Some variation in these characteristics exists among the three principal materials: acetate, acetate butyrate, and acetate propionate. Acetate is selected when the dimensional stability under the expected humidity and temperature conditions is adequate and hardness and stiffness are required. Butyrate is chosen over acetate or propionate for its weatherability, low temperature impact strength, and dimensional stability (qv). Propionate is selected over butyrate when its superior hardness, tensile strength, and stiffness are important.

Physical Properties

Physical properties of commercial cellulose acetate, acetate butyrate, and acetate propionate plastics are given in Table 1.

Additives

A fully formulated cellulosic plastic contains, in addition to the specific ester, a variety of additives (qv) including heat stabilizers, antioxidants (qv), plasticizers (qv), uv inhibitors, and coloring agents.

Table 1. Properties of Commercial Cellulose Ester Plastics

Property	ASTM test method	Cellulose acetate	Cellulose acetate butyrate	Cellulose acetate propionate
General				
molding qualities		good	good	good
compression molding temperature, °C		171–188	121–182	138–193
compression molding pressure, MPa		17.2–34.5	10.3–34.5	10.3–34.5
injection molding temperature, °C		188–227	154–210	182–221
injection molding pressure, MPa		20.7–207.0	55.2–207.0	96.6–207.0
mold shrinkage, cm/cm	D955	0.002–0.004	0.002–0.004	0.002–0.003
compression ratio, bulk factor	D1895	1.8–2.4	1.8–2.4	1.8–2.4
refractive index at 25°C	D542	1.46–1.50	1.46–1.49	1.46–1.48
Thermal				
flow temperature, °C	D569	130–180	130–175	155–190
thermal conductivity, W/(m·K)	C177	0.16–0.33	0.16–0.33	0.16–0.33
specific heat, J/(°C·g)[a]		1.2–1.6	1.2–1.6	1.2–1.6
coefficient of thermal expansion, per °C × 10^{-5}	D696	11–16	11–16	12–16
heat distortion, °C				
at 455 kPa[b]	D648	55–118	59.5–112	71–121
at 1820 kPa[b]	D648	44–113	50–99	49–110
flammability, cm/min	D635	1.27–5.08	1.27–3.81	2.54–3.81
Mechanical				
specific gravity	D792	1.27–1.34	1.15–1.22	1.19–1.23
tensile strength at fracture, MPa[b]				
at 23°C	D638	13.1–58.6	13.8–51.7	13.8–51.7
at 70°C	D638	5.5–41.4	8.3–39.3	6.7–44.8
tensile strength at yield at 73°C, MPa[b]	D638	13.8–48.3	10.3–48.3	10.3–48.3
elongation at fracture, %	D638	50–6	74–38	60–35
flexural strength at yield, MPa[b]	D790	13.8–110.3	10.3–64.1	20.7–75.9
compressive strength at yield, MPa[b]	D695	13.1–64.1	7.6–52.4	21.4–69.0
deformation under load, %				
at 13.79 MPa[b]	D621	40–1	40–1	35–1
at 6.90 MPa[b]	D621	15–1	15–1	5–1
modulus of elasticity in flexure, MPa[b]	D790	620.7–1793	482.8–1379	689.7–1931
Izod impact strength, notched J/m[c]				
at 23°C	D256	132.7–6.6	149.3–9.9	182.5–13.3
at −40°C	D758	14.3–1.9	23.8–6.6	19.0–1.9
Rockwell hardness, R scale	D785	39–120	29–117	20–120
Electrical				
dielectric strength (1/8-cm thick test specimens), V/mm	D149	250–365	250–400	300–450
volume resistivity, Ω·cm	D257	10^{10}–10^{13}	10^{10}–10^{12}	10^{12}–10^{15}
Environmental				
water absorption, 24 h, %	D570	2.0–6.5	1.0–4.0	1.0–3.0
solubles lost, %	D570	2.8–0.0	0.8–0.0	0.2–0.0
accelerated aging, 72 h at 82°C, wt lost, %	D706	12.0–0.4	4.0–0.1	2.0–0.1

[a] To convert J to cal, divide by 4.184.
[b] To convert kPa to psi, multiply by 0.145; to convert MPa to psi, multiply by 145.
[c] To convert J/m to ft·lb/in., divide by 53.38.

Compounding

Organic esters of cellulose may be processed into useful plastics by solvent casting or thermal melt processing (Fig. 1). Because of the obvious cost advantage of thermal processing over solvent casting, the former is used whenever possible.

Molding

Cellulose ester plastics are thermoplastic, ie, they are softened by heat and reharden when allowed to cool. Properly stabilized cellulose ester plastics can be injection-molded or extruded without excessive discoloration, degradation, or alteration of the physical and electrical properties. Cellulosics are hygroscopic materials and require drying before molding.

Drying

Cellulosic materials can be dried in commercial hopper-type or floor-type dryers (see also DRYING). The equipment should be large enough to allow a drying time of 2 h, although 1½ h is usually adequate. The drying air should be first dehumidified by passage through a desiccant bed and then heated to a temperature of 66–80°C. The air should flow uniformly through the pellet bed.

Economic Aspects

The sales volume of cellulosic plastics in the United States steadily declined from 1965 through 1982. Generally, the price of the mixed ester plastics (acetate butyrate and acetate pro-

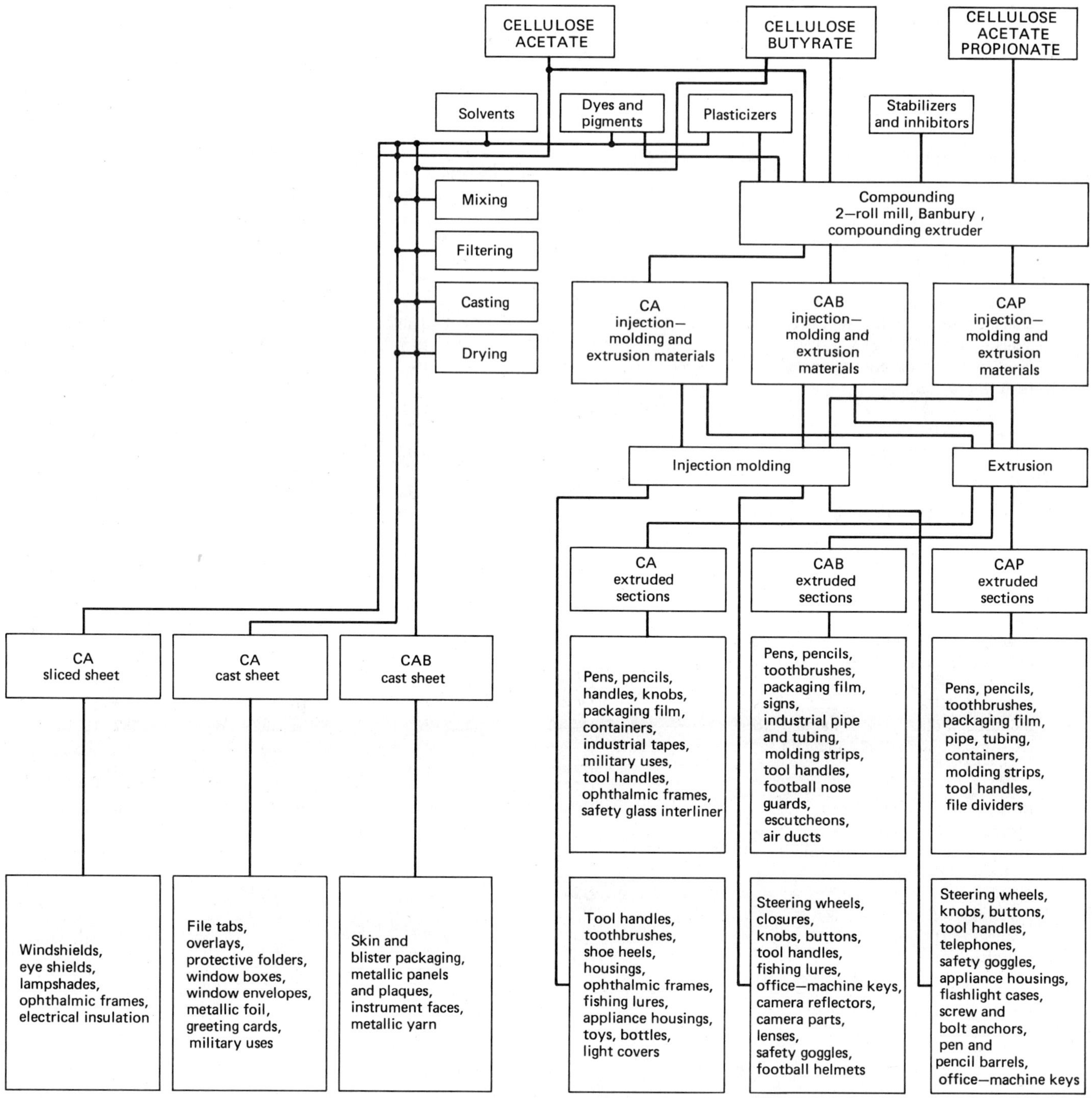

Figure 1. Processing of cellulose organic esters and products.

pionate) is slightly higher than the price of the cellulose acetate plastics, although the difference in price today is less than in earlier years.

In the United States, plastic-grade cellulose acetate is manufactured by the Eastman Chemicals Division of Eastman Kodak Co. and by the Hoechst-Celanese Chemical Co. The mixed esters for plastics, cellulose acetate butyrate and acetate propionate, are produced domestically only by the Eastman Chemicals Division of Eastman Kodak Co.

Several foreign companies in Japan and Eastern and Western Europe manufacture cellulose acetate flake for plastics. Farbenfabriken Bayer, AG, FRG, suspended production of cellulose acetate butyrate and acetate propionate during 1988.

BIBLIOGRAPHY

C. J. Malm and G. D. Hiatt in E. Ott and co-workers, eds., *Cellulose and Cellulose Derivatives*, Part II of *High Polymers*, 2nd ed., Vol. 5, John Wiley & Sons, Inc., New York, 1954, p. 766.

R. T. Bogan and co-workers in M. Grayson, ed., *Kirk-Othmer Encyclopedia of Chemical Technology*, 3rd ed., Vol. 5, Wiley-Interscience, New York, 1979, pp. 118–143.

G. D. Hiatt and W. J. Rebel in N. M. Bikales and L. Segal, eds., *Cellulose and Cellulose Derivatives,* Part V of *High Polymers,* 2nd ed., Vol. 5, Wiley-Interscience, New York, 1971.

R. T. Bogan
R. J. Brewer
Eastman Kodak Company

Fibers

Acetate fiber is the generic name for a fiber that is a partially acetylated cellulose, also known as secondary cellulose acetate. The wide use of acetate fiber in textiles is the result of attractive economics, uniform high quality, color, styling versatility, drape, hand, and other favorable aesthetic properties. Triacetate is the generic name for cellulose triacetate fiber, also known as primary cellulose acetate; this form is almost completely acetylated. Acetate and triacetate differ only moderately in the degree of acetyl substitution on cellulose, yet they have different chemical and physical properties. Triacetate fiber is hydrophobic; heat treatment develops a high degree of crystallinity that can be used to impart desirable fabric-performance characteristics.

Acetate and triacetate cause no harmful reactions when ingested. Toxic effects, skin irritations, or allergic reactions have never been reported. Cellulose acetate and triacetate are obtained by the esterification of cellulose with acetic anhydride.

Physical Properties

Cellulose triacetate develops considerable crystallinity after heat treatment, eg, at 240°C for 1 min. Many improved fiber-performance characteristics of triacetate over acetate result from the increased internal structural order.

Fibers of both cellulose acetate and triacetate have a bright, lustrous appearance. Acetate and triacetate fibers have essentially the same light-absorption characteristics in the visible spectrum; absorption in the uv region is slightly higher.

Dyeing Characteristics. Disperse dyes, high melting, crystalline compounds with low solubility in the dye bath, are most frequently used for cellulose acetate and triacetate fibers. Triacetate fibers are dyed more slowly than acetate fibers; dye carriers accelerate the rate.

Specific Gravity. The values of 1.32 for acetate and 1.30 for triacetate are accepted for fibers of combined acetic acid contents of 55 or 61.5%, respectively.

Refractive Index. The refractive index parallel to the fiber axis (ϵ) is 1.478 for acetate and 1.472 for triacetate.

Absorption and Swelling Behavior. The absorption of moisture by acetate and triacetate fibers generally depends on the relative humidity to which the fibers are exposed. However, it varies according to whether equilibrium is approached from the dry or the wet side. Absorption of water by fibers generally causes swelling roughly proportional to moisture content.

Thermal Behavior. Like most thermoplastic fibers, acetate and triacetate stick, soften, and melt.

Light Stability. Acetate and triacetate behave similarly to cotton and rayon, ie, they are somewhat more resistant than unstabilized pigmented nylon and silk and appreciably less resistant than acrylic and polyester fibers.

Electrical Behavior. Because acetate and triacetate yarns readily develop static charges, an antistatic finish is sometimes desirable.

Mechanical Properties. The mechanical and appearance properties of textile fabrics depend on the physical form of the fibers and the geometrical construction of the fabric. Properties characteristic of commercial acetate and triacetate fibers include the following: tenacity, standard is 0.10–0.12 N/tex, knot and loop is 0.09–0.10 N/tex, and wet is 0.07–0.09 N/tex; elongation, standard is 25–45% and wet is 35–50%; modulus of elasticity is 2.2–4.0 N/tex; work of rupture is 0.022 N/tex; immediate elastic recovery at 50% elongation is 58–74 N/tex and at break it is 12–14%. The inherent mechanical properties of the fiber are described by the stress–strain curve and the recovery behavior under conditions of tensile, torsional, bending, and shear loading.

Chemical Properties

Under slightly acidic or basic conditions at room temperature, acetate and triacetate fibers are resistant to chlorine bleach at the concentrations normally used in laundering. Triacetate fiber is significantly more resistant than acetate to alkalies encountered in normal textile operations. Acetate and triacetate are essentially unaffected by dilute solutions of weak acids, but strong mineral acids cause serious degradation. Acetate and triacetate fibers are not affected by the perchloroethylene dry cleaning solutions normally used in the United States and Canada. Trichloroethylene, employed to a limited extent in the UK and Europe, softens triacetate.

Resistance to Microorganisms and Insects. Resistance of triacetate to microorganisms, based on soil-burial tests, is high, approaching that of polyester, acrylic, and nylon fibers.

Manufacture

Cellulose Acetate and Triacetate Polymer. High purity chemical cellulose is esterified with acetic anhydride; a high degree of purity facilitates fiber preparation. Wood pulp is the principal source of chemical cellulose, except for special plastic-grade acetates requiring low color and high clarity, where cotton linters are used. The other raw materials, ie, acetic acid, acetic anhydride, and sulfuric acid, are commercially available in high purity grades.

Secondary Acetate Processes. Most cellulose acetate is manufactured by a solution process, ie, the cellulose acetate dissolves as it is produced. In the second most common process, the solvent process, methylene chloride replaces the acetic acid as solvent, and perchloric acid is frequently the catalyst. To obtain a secondary cellulose acetate sufficiently soluble in acetone to produce fiber, the cellulose must be completely acetylated during the dissolution step and then hydrolyzed to the required acetyl value.

Cellulose Acetate and Triacetate Fibers. ***Extrusion Processes.*** Polymer solutions are converted into fibers by extrusion (qv). The dry-extrusion process, also called dry spinning, is used primarily for acetate and triacetate. Acetone is the preferred solvent for acetate fiber and methylene chloride for triacetate fibers. Extensive solvent recovery is required for an economic process.

Anisotropic Solutions

Cellulose acetate and triacetate form an isotropic, ie, liquid crystalline, solution, but no significant commercial use exists.

Products

Fiber products include filament yarns, staples, and tows. Cellulose acetate polymer is used in plastic applications requiring toughness and clarity.

Economic Aspects

Cellulose acetate, the second-oldest synthetic fiber, is an important factor in the textile industry; 260,000 metric tons were produced worldwide in 1982. Acetate belongs to the less-expensive fibers; triacetate is slightly more expensive.

The principal textile applications of both acetate and triacetate fibers are in women's apparel and home-furnishing fabrics. Although the use of acetate fiber for textile applications has generally declined, the total worldwide production of cellulose acetate increased due to tow for cigarette filters, which rose from 154,000 to 339,000 t between 1970 and 1982. Because of its superior filtration effect on cigarette taste and its low cost, acetate is expected to supply up to 90% of the filter-cigarette market.

BIBLIOGRAPHY

L. Segal in N. M. Bikales and L. Segal, eds., *Cellulose and Cellulose Derivatives, High Polymers Series,* Vol. V, Wiley-Interscience, New York, 1971, Chapt. XVII-A.

C. J. Malm and G. D. Hiatt in E. Ott, H. M. Spurlin, and M. W. Graffin, eds., *Cellulose and Cellulose Derivatives, High Polymers Series,* 2nd ed., Vol. V, Wiley-Interscience, New York, 1954, Pt. II.

GEORGE A. SERAD
Celanese Fibers Operations

CELLULOSE ETHERS

Cellulose ethers comprise a class of cellulose derivatives, soluble in water or organic solvents, which function as thickeners, flow-control agents, suspending aids, protective colloids, water binders, liquid crystals, film formers, or thermoplastics.

Raw Materials

The basic materials for the production of cellulose ethers are cellulose and alkylating reagents. Other ingredients include alkali, usually sodium hydroxide, water, and organic solvents.

Processes

Alkylation processes are homogeneous or heterogeneous; the latter have achieved greater commercial importance and can be further subdivided into organic diluent-mediated or diluent-free processes in batch or continuous operation.

Characterization and Properties

Cellulose ethers are characterized by solution viscosity, chemical nature of the substituent, degree of substitution (DS) (qv) or molar substitution (MS), purity, rheological properties, and solubility and compatibility. Cellulose ethers, soluble in organic solvents such as ethylcellulose, have DS of 2.3–2.8. Water-soluble cellulose ethers have DS values between 0.4 and 2.0, depending on the chemical and physical nature of the substituent. Commercial hydroxyalkyl ethers, such as hydroxyethylcellulose or hydroxypropylcellulose, may have MS values of 1.0–4.0.

The ability of a cellulose ether to function as a thickener or flow-control agent (rheology control) depends largely on the degree of substitution (or molar substitution). Properties such as salt compatibility, thermal stability, colloidal stabilization, surface activity, and cross-linking depend primarily on the chemical nature of the substituent. The distribution of substituents also affects properties. Accordingly, a wide range of cellulose ethers are manufactured to meet specific needs of industrial applications.

Health and Safety Factors

Cellulose ethers generally show no evidence of toxicity to rats, dogs, guinea pigs, or humans. No damage from cellulose ethers has been observed in ecological systems. Some powders may be considered nuisance dusts; fine dusts can form explosive mixtures in air. Spilled powders become slippery when wet.

Commercial Cellulose Ethers

Sodium Carboxymethylcellulose. Sodium carboxymethylcellulose is an anionic polyelectrolyte prepared by the reaction of sodium chloroacetate with alkali cellulose. The sodium form of carboxymethylcellulose, known simply as CMC, or if food grade, as cellulose gum, is soluble in hot and cold water. Depending on purity, CMC is sold as a white to off-white powder and is available in several grades and in a variety of types by degree of substitution, viscosity, and particle size.

Properties. Typical properties of a CMC with DS 0.7 are given in Table 1.

Table 1. Typical Properties of CMC

Property	Value
Polymer	
sodium carboxymethylcellulose, dry basis, min%	99.5
moisture, max %	8.0
bulk density, g/mL	0.75
Solutions, 2%	
sp gr, at 25°C	1.0068
refractive index at 25°C	1.3355
pH	7.5
Films	
density, g/mL	1.59
refractive index	1.515

Economic Aspects. Since 1969, U.S. production has grown at an average annual rate of 1.0% to 30.8×10^3 t in 1979. Total 1984 U.S. consumption was estimated at 10^3 t; 1980 U.S. imports amounted to 2×10^3 t.

Uses. CMC is used in applications where thickening, suspending, stabilizing, binding, and film-forming properties are important.

Hydroxyethylcellulose. Hydroxyethylcellulose (HEC) is a nonionic polymer typically prepared by reaction of cellulose with ethylene oxide.

Properties. Typical properties are given in Table 2.

Table 2. Typical Properties of Hydroxyethylcellulose

Property	Value
Polymer	
ash, as Na_2SO_4, %	3.5
moisture, as shipped, max %	5.0
effect of heat, softening range, °C	135–140
weight, kg/L	1.39
Solutions, 2%	
sp gr, g/mL	1.0033
refractive index	1.336
pH	7
Films	
refractive index	1.51
sp gr at 50% rh, g/mL	1.34
moisture content (equilibrium) at 23°C, 50% rh, %	6

Economic Aspects. U.S. consumption of HEC is expected to more than double for the period between 1973 and 1984 to 28.5 $\times$ 10^3 t. Water-soluble HEC is manufactured in the United States by Hercules, Inc. as Natrosol and by Union Carbide Corp. as Cellosize.

Uses. Because of its broad compatibilities, HEC is widely used in applications and products that require thickening, water-binding, lubricating, film-forming, and protective colloid or stabilizing properties.

Carboxymethylhydroxyethylcellulose. Chemical modification of hydroxyethylcellulose with sodium chloroacetate yields the mixed ether, sodium carboxymethylhydroxyethylcellulose (CMHEC). Various grades of CMHEC, ranging in carboxymethyl DS from 0.3 to 0.5 and hydroxyethy MS from 0.7 to 2.0, are currently supplied by Hercules, Inc. at $3.85–6.05/kg.

Properties. CMHEC is a water-soluble cellulose ether with the properties of both carboxymethylcellulose (CMC) and hydroxyethycellulose (HEC). Typical properties are given in Table 3.

Table 3. Typical Properties of CMHEC

Property	Value
Polymer	
carboxymethyl, DS	0.3
hydroxyethyl, MS	0.7
Solutions, 2%	
pH	6.5–8.5
Films	
refractive index, n_D	1.530
tensile strength, 50% rh, MPa[a]	69

[a] To convert MPa to psi, multiply by 145.

Uses. CMHEC products are predominantly used in oil-recovery applications. They also are used in fracturing fluids.

Hydroxypropylcellulose. Hydroxypropylcellulose (HPC) is a thermoplastic, nonionic cellulose ether soluble in water and in a number of organic solvents. It is a surface-active agent capable of thickening and stabilizing aqueous systems. At very high concentration, HPC displays liquid crystal behavior. Current production of hydroxypropylcellulose is estimated at 2.25 $\times$ 10^3 t; HPC is sold by Hercules, Inc. under the Klucel trademark at $8.80–9.90/kg.

Properties. Properties are given in Tables 4 and 5.

Table 4. Typical Properties of HPC

Property	Value
Polymer	
appearance	off-white, odorless tasteless powder
bulk density, g/mL	0.5[a]
softening temp, °C	130
burnout temp, °C	450–500[b]
particle size through 840 μm (20 mesh), %	99
ash as Na_2SO_4, max %	0.5
moisture, max %	5.0
Solution in water	
sp gr, 2% at 30°C	1.010
refractive index	1.337
surface tension, 0.1%, mN/m (= dyn/cm)	43.6
interfacial tension, 0.1% in water vs refined mineral oil, mN/m (= dyn/cm)	12.5
bulking value, L/kg	0.33
solution pH	5.0–8.5
cloud point, °C	40–45

[a] Varies with type.
[b] In nitrogen or oxygen.

Table 5. Properties of HPC Films[a]

Property	Value
appearance	clear
specific gravity	~1.1
tensile strength, MPa[b]	13.8
modulus of elasticity, MPa[b]	414.0
elongation at break, %	~50
refractive index	1.559
moisture at 22°C, 50% rh, %	4

[a] Nonplasticized Klucel J-type films air-dried from water solution.
[b] To convert MPa to psi, multiply by 145.

Uses. HPC is used in adhesives, binders, cosmetics, encapsulation, extrusion–molding, foods, paint removers, paper, pharmaceuticals, PVC, and inks.

Methyl- and Hydroxyalkylmethylcelluloses. Methylcellulose (MC) and its alkylene oxide derivatives are nonionic polymers soluble in cold water. Hydroxyalkyl derivatives are obtained from ethylene, propylene, and butylene oxides. Depending on extent and uniformity of methyl substitution and type of hydroxyalkyl modification, these polymers constitute a series of cellulose ethers that vary in solubility, surface activity, thermal gelation, enzyme resistance, and material compatibilities.

Properties. Methylcellulose has unique solution properties. It and its congener ethers gel at specific temperatures when heated. The gel temperature is invariant with viscosity grade, but is influenced by rate of heating, shear, and additives, especially salts. Properties of methylcellulose ethers are given in Table 6.

Manufacture and Composition. Methylcellulose, first described in 1905, is prepared by the reaction of alkali cellulose with methyl chloride.

Table 6. Properties of Methylcellulose Ethers

Property	Values		
	MC	HPMC	HBMC
Powders			
physical appearance	white to off-white, odorless and tasteless powder		
apparent density, g/cm^3		0.25–0.70	
sp gr		1.39	
wt, kg/L		1.39	
relative flammability		90+[a]	
Solutions			
sp gr			
1%		1.0012	
10%		1.0245	
refractive index		1.336	
gelation temp, °C	48	54–70	49
Films[b]			
sp gr	1.39	1.29	
tensile strength at 24°C, 50% rh, MPa[c]	58.6–78.6	58.6–61	
elongation at 24°C, 50% rh, %	10–15	5–10	
refractive index	1.49		
softening point, °C		240	
melting point, °C	290–305	260	

[a] 700°C furnace.
[b] For Methocel A methylcellulose, 50 μm thickness.
[c] To convert MPa to psi, multiply by 145.

Economic Aspects. Methylcelluloses, hydroxypropylmethylcelluloses (HPMC), and hydroxybutylmethylcelluloses (HBMC) are all manufactured in the United States by The Dow Chemical Co. under the trademark Methocel. Annual U.S. capacity for these materials is estimated at 18×10^3 t. They are priced by grade: premium grades of MC and HPMC are listed at $5.94–6.27/kg and standard grades sell for $4.95–5.50/kg. Standard HBMC grades list for $4.84/kg and imported standard HEMC for $6.00/kg.

Uses. Methylcellulose is used in adhesives, agricultural chemicals, ceramics, construction products, cosmetics, foods, leather, paint, paint removers, paper products, pharmaceuticals, PVC, plywood, printing inks, resins and rubbers, textiles, tobacco.

Ethylcellulose. Ethylcellulose, the ethyl ether of cellulose, is a nonionic polymer commercially available in dry, granular form. At low degrees of substitution (DS 0.8–1.7), it is soluble in water, at high etherification (DS 2.3–2.6), in organic solvents. Only ethylcelluloses that are thermoplastic and soluble in organic solvents are commercially significant. They form tough, stable coatings and films and are employed in hot-melt applications and the casting of plastics.

Properties. Properties typical of ethylcellulose (EC) with an ethoxyl content of 48.0–49.5% (DS 2.4–2.5) are given in Table 7.

Ethylhydroxyethylcellulose (EHEC). Ethylhydroxyethylcellulose, a nonionic mixed ether known as EHEC, is available in versions soluble in water or organic solvents. An alternative to ethylcellulose, it is used in coating applications that require solubility in fast-drying, aliphatic hydrocarbons.

Properties. Properties are given in Table 8.

Table 7. Physical Properties of Ethylcellulose (EC)

Property	Value
bulking value, granule form, kg/L	0.311–0.335
electric properties	
dielectric constant, at 25°C, 1 MHz	2.8–3.9
dielectric strength, V/25 μm[a]	1500
elongation at rupture, %	7–30
flexibility, folding endurance[b]	160–2000
light transmission, 310–400 nm	practically complete
moisture absorption, %	2
softening point, °C	152–162
specific gravity	1.14
specific volume, cm^3/kg in soln	862
tensile strength, MPa[c]	46–72

[a] 250-μm film by ASTM D149-64, step-by-step.
[b] MIT double folds, 75-μm film.
[c] Dry 75-μm film; to convert MPa to psi, multiply by 145.

Table 8. Physical Properties of EHEC

Property	Value
bulk density of granular powder, g/cm^3	0.30–0.35
solution properties	
specific volume, cm^3/kg	883
refractive index	1.47
film properties[a]	
density, g/cm^3	1.12
tensile strength, MPa[b]	34–41
elongation, %	6–10
flexibility, MIT double folds	500–900

[a] 75-μm thick, unmodified film.
[b] To convert MPa to psi, multiply by 145.

Manufacture. The manufacture of either EHEC type involves stepwise reactions of alkali cellulose with ethylene oxide at ca 35°C followed by ethyl chloride at ca 110°C.

Economic Aspects. U.S. capacity of EHEC soluble in organic solvents combined with that of ethylcellulose is estimated at 3.6×10^3 t. It is more expensive than ethylcellulose. Hercules is the only U.S. supplier.

EHEC is used in inks, clear lacquers, and other coatings. Water-soluble EHEC is used in waterborne paints, building materials, pastes, polymer dispersions, detergents, tobacco, cosmetics and pharmaceuticals, and ceramics.

Cationic Cellulose Ethers. Union Carbide Corp. and National Starch and Chemical Corp. both manufacture water-soluble, cellulose-based cationics, primarily for use in cosmetics and toiletries.

Properties. Properties of Celquat resins are given in Table 9.

Preparation. Several cationic ether compositions have been patented, including diethylaminoethyl cellulose (DEAE-cellulose) and hydroxypropyltrimethylammonium chloride hydroxyethylcellulose. The latter is prepared from the caustic mediated reaction of hydroxyethylcellulose with 3-chloro-2-hydroxypropyltrimethylammonium chloride.

DEAE-cellulose can be prepared by reaction of alkali cellulose with 2-chloroethyldiethylamine hydrochloride.

Table 9. Typical Properties of Celquat

Property	Value	
	Celquat H-100	Celquat L-200
solutions		
bulk density kg/L	0.48	0.48
viscosity[a], mPa·s (= cP)	600	150
pH[b]	7	7
films		
tensile[c], MPa[d]	22	14.4
elongation[c], %	60	70
Sward Rocker hardness[e]	12	12

[a] Brookfield viscosities at 20 rpm for 2% solids in water at 21°C.
[b] 2% solids in water.
[c] Dried film equilibrated 24 h at 21°C and 50% rh.
[d] To convert MPa to psi, multiply by 145.
[e] 12.5-μm dry film equilibrated 24 h at 21°C and 50% rh.

Other Cellulose Ethers

More recent cellulose ether compositions include dihydroxypropylcelluloses, hydrophobically modified cellulose ethers, phosphorus containing ethers, and hydroxyethylhydroxypropylcellulose.

BIBLIOGRAPHY

Cellulose Gum, Chemical and Physical Properties, Hercules, Inc., Wilmington, Del., 1980.

Natrosol R Hydroxyethylcellulose, Hercules, Inc., Wilmington, Del., 1978.

Methocel R, The Dow Chemical Co., Midland, Mich., 1978.

R. L. Davidson ed., *Handbook of Water-Soluble Polymers,* McGraw-Hill Inc., New York, 1980.

E. K. Just
T. G. Majewicz
Hercules, Inc.

CELLULOSE NITRATE. See Cellulose esters, inorganic.

CELLULOSE PROPIONATE. See Cellulose esters, organic.

CELLULOSE TRIACETATE. See Cellulose esters, organic.

CELLULOSE XANTHATE. See Rayon.

CELLULOSIC FIBERS. See Cellulose esters, organic; Cotton; Fibers, vegetable; Rayon.

CELLULOSIC FILMS AND SHEETS. See Cellulose esters, inorganic; Cellulose esters, organic; Cellulose ethers; Nonwoven fabrics; Paper.

CELLULOSIC PLASTICS. See Cellulose esters, inorganic; Cellulose esters, organic; Cellulose ethers.

CEMENT ADDITIVES

Organic polymeric materials are used to improve the mechanical properties and durability of products (especially concrete) based usually, but not exclusively, on portland cement. Modification of cement and concrete products is done in three ways:

1. A preformed concrete object is impregnated with a liquid monomer that is polymerized *in situ,* giving polymer-impregnated concrete (PIC).
2. A monomer, prepolymer, or polymer latex is incorporated into the mix of cement, aggregate, and water, forming an intimate part of the hardened structure called polymer portland cement concrete (PPCC).
3. A water-soluble polymer is incorporated in a cement–water mix to give a rheology eliminating macroscopic defects owing to poor particle packing and entrained air. This recently developed technology is still experimental; the products are termed macrodefect-free (MDF) cement.

Compressive strength is enhanced and resistance to aggressive environments, such as freeze–thaw cycles or seawater, is greatly improved (Table 1). The polymeric materials form an integral part of the cement or concrete structure. Water-soluble polymeric materials influence the setting rate or the rheology of a cement–water mixture. They may contain polysaccharides to retard setting or oligomeric naphthalene or melamine sulfonate–formaldehyde condensates to increase the fluidity and reduce water requirements. Such additives influence mechanical properties by affecting the microstructure rather than by forming a separate phase.

Table 1. Properties of Concrete, Polymer-impregnated Concrete, and MDF Cement

Property	Concrete	Polymer-impregnated concrete	MDF cement
compressive strength, MPa[a]	30–40	100–140	200–300
modulus of elasticity, GPa[a]	25	40–45	45–50
direct tensile strength, MPa[a]	3–6	8–12	60–100
flexural modulus, MPa[a]	5–10	10–25	80–120
relative abrasion loss, g	12–15	3–5	
water absorption, %	5–8	0.3–1.25	
water permeability, 3×10^6 mm/yr	5–6	0.75–1.5	
coefficient of expansion, $nm^{-1} \cdot °C^{-1}$	7.0–7.5	9–10	

[a] To convert MPa to psi, multiply by 145.

So-called polymer concretes (qv) are quite different and consist of a polymerized dispersion of a particulate solid, eg, silica, in a liquid monomer, eg, a methacrylate.

BIBLIOGRAPHY

P. C. Hewlett, *Mater. Eng. Appl.* **1**(6), 335 (1979).

ACI Committee 438, *Polymers in Concrete, Document 54BR-77,* American Concrete Institute, Detroit, Mich., 1977.

Technology in the 1990's; Developments in Hydraulic Cements, The Royal Society, London, 1983.

J. D. BIRCHALL
Imperial Chemical Industries

CEMENTING AND CEMENTS.

See ADHESION AND BONDING; ADHESIVE COMPOSITIONS.

CERAMICS.

See PRECERAMIC POLYMERS.

CHAIN-LENGTH, KINETICS.

See KINETICS OF POLYMERIZATION.

CHAIN-REACTION POLYMERIZATION

Chain-reaction polymerization is one of the two processes for converting monomer to polymer; the other is step-reaction polymerization. Among the important products produced by chain polymerization are plastics such as polyethylene, polypropylene, polystyrene, poly(methyl methacrylate), poly(vinyl chloride), and polytetrafluoroethylene; elastomers such as polyisoprene, polychloroprene, polybutadiene, and polybutadiene copolymers with styrene and acrylonitrile; and fibers based on copolymers of vinylidene chloride and acrylonitrile. The utility of chain-reaction polymerization resides in the wide range of polymer structures that can be obtained through proper choice of monomer and method of initiation, ie, radical, cationic, anionic, and coordination. Further variations in polymer structure are available through copolymerization and stereochemical control.

Comparison of Step- and Chain-reaction Polymerizations

Step- and chain-reaction polymerizations differ in the length of time required for the growth of large-size, ie, polymer, molecules relative to the time required for high conversion. The typical step-reaction polymerization involves two different bifunctional reactants, referred to as monomers, each possessing a different functional group X and Y. Step-reaction polymerizations proceed by the stepwise reaction between X and Y functional groups to form dimer, trimer, tetramer, and so on until eventually polymer molecules, ie, molecules containing at least 50–100 monomer units, are formed. As each larger-size species is formed, it competes with the smaller-size species for further reaction. The average size of the molecules increases slowly with conversion, and high molecular weight polymer is not obtained until near the very end of the reaction, ie, at very high conversion.

In contrast, in chain-reaction polymerizations large-size molecules are produced almost immediately after the start of the reaction, ie, at low conversion. Chain polymerizations require an initiator, which produces an initiator species R* with a reactive center. The reactive center may be a free radical, cation, anion, or organometallic complex. Polymerization occurs by the propagation of the reactive center through successive additions of large numbers of monomer molecules in a chain reaction. This propagation process often occurs rapidly, so that hundreds or more of monomer molecules add in about a second, usually in much shorter times. However, this time interval can be considerably longer depending on the type of reactive center, monomer, temperature, and other reaction conditions.

The most common example of chain polymerization is that of monomers containing the carbon–carbon double bond. The overall reaction can be depicted as follows:

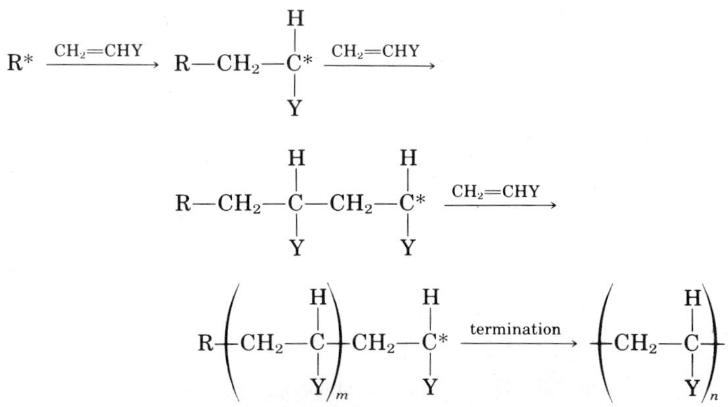

Polymerizability by Chain-reaction

Polymerization of a wide variety of unsaturated monomers is thermodynamically feasible, ie, the free energy change, ΔG for the conversion of monomer to polymer is negative. However, a negative ΔG does not ensure that polymerization will occur at a reasonable rate under a particular set of reaction conditions. Although radical, anionic, and cationic initiators are used in chain-reaction polymerizations, they cannot be used indiscriminately because all three types of initiation do not work for all monomers. The type of reactive center must be matched to the particular monomer. The carbon–carbon and carbon–oxygen double bonds are the two main types of linkage that undergo chain polymerization. Polymerization of C=C monomers is of greater industrial importance than that of C=O monomers.

Radical Chain-reaction Polymerization

Radical chain-reaction polymerization involves initiation, propagation, and termination. Thermal homolysis of an initiator such as benzoyl peroxide is the most widely used mode of initiating polymerization.

Characteristics of Radical Polymerizations. Temperature effects both polymerization rate and DP. Polymerization rate increases rapidly with increasing temperature while DP decreases. Radical polymerizations are carried out in bulk, solution, suspension, and emulsion. Many radical chain polymerizations are characterized by an autoacceleration in both R_p and DP with conversion.

Ionic Chain-reaction Polymerization

Ionic chain-reaction polymerizations require relatively low or moderate temperatures and solvating media in order for the cationic and anionic centers to propagate to polymeric size before termination. Although high polarity solvents are desirable to solvate the ions, they usually cannot be used. Ionic polymerizations are typically carried out in solvents of low or moderate polarity, eg, alkanes, chlorinated hydrocarbons, benzene, nitrobenzene, and tetrahydrofuran.

The mechanisms of ionic chain-reaction polymerizations, especially cationic chain-reaction polymerizations, are not as well understood as radical chain-reaction polymerizations because of experimental difficulties involved in their study. Compared to radical polymerization, the commercial utility of ionic polymerization is limited mostly because of its far greater selectivity. Butyl rubber, a copolymer of isobutene and isoprene, is produced commercially by cationic polymerization (qv). Anionic polymerization (qv) is used for various polymerizations of 1,3-butadiene, isoprene, styrene, and acrylates.

Copolymerization

Chain-reaction copolymerization, the polymerization of a mixture of two monomers, yields a copolymer with two different repeating units distributed along the polymer chain (see COPOLYMERIZATION). Copolymerization has practical utility for altering the properties of a homopolymer (a polymer produced from a single monomer) in a desired direction. Copolymer properties are determined by the identities of the two monomers and their relative amounts in the copolymer.

The composition of a copolymer is determined by the relative amounts of the two monomers in the reaction mixture and their relative reactivities toward copolymerization.

Stereochemistry

Stereoisomerism is possible in the polymerization of monosubstituted ethylenes. Two ordered or stereoregular polymers are possible, ie, isotactic and syndiotactic, wherein the substituent R groups on successive chiral carbons in the polymer chain have the same or opposite relative configurations, respectively. The unordered or atactic polymer structure has a random distribution of equal numbers of the two configurations.

The extent of stereospecificity in a polymerization depends on whether the active end of a propagating chain is a relatively free species or is coordinated with initiator, ie, whether or not there is free rotation about the end bond in the propagating chain. Polymer stereochemistry is of practical importance because of its effect on physical properties, especially in butadiene polymers (qv), isoprene polymers (qv), polypropylene, and many others.

BIBLIOGRAPHY

F. W. Billmeyer, Jr., *Textbook of Polymer Science,* 3rd ed., John Wiley & Sons, Inc., New York, 1984.

J. A. Brydson, *Plastics Materials,* 4th ed., Butterworth Scientific, New York, 1982.

H.-G. Elias, *Macromolecules,* Vols. 1 and 2, Plenum Press, New York, 1977.

G. Odian, *Principles of Polymerization,* 2nd ed., John Wiley & Sons, Inc., New York, 1981.

GEORGE ODIAN
The College of Staten Island
The City University of New York

CHAIN TRANSFER

Chain transfer refers to the termination of a growing polymer chain and the start of a new one. The process is mediated by a chain-transfer agent, which may be the monomer, initiator, solvent, polymer, or some species that has been added deliberately to effect chain transfer. Because chain transfer occurs in all radical polymerization (qv), it must be taken into account in any quantitative consideration of these reactions. Chain transfer always decreases the molecular weight of the product and may decrease the rate.

Applications. Chain-transfer agents are used to regulate and limit the molecular weight in a polymer reaction. They are also used in telomerization (qv), the production of very low molecular weight polymers (telomers) by chain transfer.

BIBLIOGRAPHY

J. Brandrup and E. H. Immergut, *Polymer Handbook,* 3rd ed., John Wiley & Sons, Inc., New York, 1989.

CHARACTERIZATION OF POLYMERS

Polymer characterization is an essential step in working with polymers. As a rule, such efforts are directed toward a specific purpose. In newly synthesized material, composition, molecular weight, and physical properties are determined first. Other information may be needed with respect to selection or qualification of polymers for specific properties, analysis of competitive products, or quality-control testing. Polymers may be categorized as natural or synthetic and as homopolymers or copolymers. Copolymers and terpolymers are further classified according to the method of production and the arrangement of the monomeric units. Special methods and techniques may be needed for composites and polymer blends of homopolymers and various copolymers. Thus the techniques discussed may not be uniformly applicable to all classes of polymers.

Molecular Structure

Many chemical and spectroscopic techniques are available for the characterization of primary molecular structure, ie, the arrangement or type of monomeric units in a polymer chain. Other structural information is usually sought as part of the primary molecular description regarding the sequence of monomer units and the type and amount of configurational or geometric isomerism present. That information is obtained using techniques similar to those used for determination of low molecular weight organic compounds. However, a critical problem in polymer characterization is that many of these techniques require that the analysis be performed in solution. This requirement is sometimes difficult to meet, particularly for very high molecular weight or cross-linked polymers, and a number of supplemental approaches are necessary.

Fractionation. Common fractionation techniques used to characterize molecular structure include solubility tests (precipitation, fractional dissolution (extraction), coacervate extraction, thin-layer chromatography, countercurrent distribution, and turbidimetry), and measurement of molecular size [size-exclusion (sec) or gel-permeation chromatography (gpc)]. Other fractionation techniques include thin-layer chromatography and field-flow fractionation (fff).

Spectroscopic Techniques. Spectroscopic techniques include nuclear magnetic resonance (qv), infrared and Raman spectroscopy (qv), pyrolysis-gas chromatography, mass spectrometry, uv spectroscopy, luminescence spectroscopy, and electron spectroscopy for chemical applications (esca).

Chemical Tests. For many problems, spectroscopic techniques serve to identify unknown polymers rapidly. However, depending on the information sought and the availability of equipment, simple chemical tests may serve as well or better than elaborate instrumental procedures. For example, the way in which a material burns, solubility or elemental analysis, origin, or price may serve to identify or to allow a choice to be made between one of several possible materials (see CHEMICAL ANALYSIS IDENTIFICATION).

Molecular Size

A fundamental question in polymer characterization is: How large is the polymer? The answer can be expressed as molecular weight or degree of polymerization (DP), hydrodynamic volume, radius of gyration, or some other measure of molecular dimensions.

Unlike simple compounds where the molecular weight in the sample is uniform, polymer samples are composed of polymer chains of varying length. Hence, a distribution of molecular weights is present. In the case of copolymers, chemical composition and length of the polymer chain often vary. Therefore, in expressing polymer molecular weights various average expressions are used.

Average Molecular Weights. Many techniques are available for determining average molecular weights. The choice of method is often complicated by limitations of the technique as well as by the nature of the polymer because most techniques require a sample in solution. Table 1 compares some of the techniques available.

Solution Viscosity. Solution viscosities are measured by flow rate through a capillary. Temperature and concentration are carefully controlled.

Molecular Weight Distributions. Average molecular weight information is useful; however, characterization of the breadth of the distribution of molecular weight within a polymer sample is perhaps more valuable. For example, two polymers may have exactly the same or similar average molecular weights but very different molecular weight distributions (MWD).

There are several ways to measure MWD, eg, fractionation of a polymer with a broad MWD into narrower MWD fractions and molecular weight determination (qv) of the narrow fractions. Fractionation (qv) is helpful in evaluating the true range of polydispersity of polymers with a natural narrow MWD. Size- exclusion chromatography (qv) (sec) is used for the study of MWD. However, MWD determined by sec can be influenced by a number of factors. Peak-broadening effects and incomplete resolution can give misleading information. The breadth of MWD is measured by the so-called polydispersity index, which is the ratio of the weight- to number-average molecular weight.

Molecular Organization

Polymer-chain interactions may result in an ordering of chains into regions of different supermolecular structure referred to as crystalline or amorphous in terms of the degree of ordering. Relatively small changes in the size and distribution of these ordered regions can be responsible for dramatic differences in

Table 1. Determination of Average Molecular Weight

Method	Absolute or relative	Average obtained	Mol wt range, g/mol	Characteristics
ebulliometry	A	$\overline{M}_n$	up to 10^4	low sensitivity
cryoscopy	A	$\overline{M}_n$	up to 10^4	small samples, fast
membrane osmometry	A	$\overline{M}_n$	$5 \times 10^3 \sim 10^6$	suitable membrane required
vapor-pressure osmometry	R	$\overline{M}_n$	$<3 \times 10^4$	suitable standards for accurate calibration required
end-group determination	A	$\overline{M}_n$	$10^2 \sim 3 \times 10^4$	low sensitivity at higher molecular weight, must assume two ends of same type per molecule
light scattering	A	$\overline{M}_w$	$3 \times 10^4 \sim 10^7$	low sensitivity at low molecular weight, expensive equipment, time-consuming
ultracentrifugation	A	$\overline{M}_w, \overline{M}_z$	$2 \times 10^3 \sim 10^7$	small samples, time-consuming, expensive equipment
dilute-solution viscosity	R	$\overline{M}_v$	$10^2 \sim 10^7$	fast, low cost, small samples

the physical and mechanical properties between different samples of the same polymer.

Characterization of Molecular Order. Techniques used to characterize molecular order include thermal methods, x-ray diffraction (qv), solid-state nmr, infrared and Raman spectroscopy, microscopy, inverse gas chromatography, neutron scattering (qv), and other techniques such as birefringence (qv), dichroism, and small-angle light scattering.

Characterization of Cross-linked Networks. The accurate characterization of cross-linked polymer networks is among the more difficult tasks of polymer analysis. Yet cross-links have a major effect on physical properties. Cross-linked materials are difficult to handle by solution techniques due to limited solubility. Solution techniques can be used after a degradative process has been applied. Many of the characterization tests are based on mechanical properties such as stress–strain properties, tensile properties, compression modulus, and hardness.

Swelling, Extraction, and Random Degradation. Chemical methods such as extraction of a lightly cross-linked network by solvent can be used effectively to remove noncross-linked components. Many polymer networks swell by immersion in the proper choice of solvent or by exposure to solvent vapors. Care should be taken that swelling occurs under equilibrium conditions. Alternatively, a cross-linked network may be degraded into small soluble chemical fragments.

Instrumental and Spectroscopic Methods. Thermal analysis (qv) is frequently used to characterize cross-linked polymers. Many spectroscopic methods mentioned previously may be applied to cross-linked polymers. Only limited information can be obtained, and no one technique has emerged as being completely general to determine all parameters of interest. Other techniques that have been used to characterize cross-linked systems are small-angle neutron scattering (sans), laser Raman spectroscopy, and light scattering (qv).

Mechanical, Electrical, and Physical Properties

Glass-transition Temperature and Crystalline Melting Point. Mechanical and physical properties are strongly dependent on temperature. A dramatic shift in properties is observed for linear amorphous polymers and copolymers at the glass-transition temperature T_g.

At or above the crystalline melting point T_m, polymer chains exhibit extensive translational movement or flow. The melting of semicrystalline polymers as measured by a variety of techniques occurs over a temperature range, the upper point of which is taken as T_m; this is a function of a number of properties, including molecular weight, MWD, percent crystallinity, and dimensions of the polymer crystallites and others. The most common methods used to measure T_m are dsc, dta, and hot-stage optical microscopy; nmr, x-ray, and infrared techniques are less satisfactory.

The glass-transition temperature is usually reported as a single value. However, the glass transition occurs over a temperature range and is kinetic in nature. Although many techniques are available, the measurement of T_g is more complex than is commonly appreciated. Some of the common techniques are dsc; calorimetry; dynamic mechanical methods including torsional braid, torsional pendulum, and Rheovibron; and spectroscopic techniques including nmr, infrared, and x-ray diffraction (see GLASS TRANSITION).

Mechanical Properties. The measurement of mechanical properties (qv) is concerned with load-deformation or stress–strain relationships. Forces may be applied as tension, shear, torsion, compression, and binding. Measurement techniques include stress–strain tests, creep tests, stress–relaxation tests, dynamic mechanical tests, and impact resistance (qv) tests.

Electrical Properties. Most polymers are electrical insulators or dielectrics and exhibit high resistivity (low conductance). Tests depend on the purpose to be accomplished.

Dielectric Relaxation. Dielectric-relaxation (qv) spectroscopy is the study of the frequency dependence of the electrical permittivity or dielectric constant ϵ.

Electrical Breakdown. The dielectric strength to breakdown is measured by applying increasing voltage across a sample until physical breakdown of the polymer results in catastrophic decrease in resistance.

Other Electrical Properties. Static electrical charging by contact electrification (triboelectrification) is a property of interest in commercial applications. A triboelectric series may be established that roughly follows the dielectric constants of the polymers (see ELECTRICAL PROPERTIES).

Physical Properties. ***Softening Points.*** Softening points are determined by a variety of tests, including heat-distortion temperature, Vicat softening temperature, and the shear modulus as a function of temperature.

Flow Properties and Processing. A variety of instruments and procedures are used to measure melt viscosity, including rotational and capillary techniques.

Solubility. In general, solubility is a function of molecular structure, composition, and molecular weight (see also SOLUBILITY OF POLYMERS).

Density. Density is measured by a density-gradient column or by dilatometric techniques (see DILATOMETRY).

Stability. Resistance to environmental factors determines the useful life of a polymer in an intended application. Stresses may be imposed by mechanical processes and handling; heat; radiation, including sunlight, x-rays, and nuclear radiation; gases, including oxygen, ozone, sulfur dioxide, and nitrogen dioxide; solvents and corrosive chemicals; moisture; and bacterial attack. Weatherability may be gauged in laboratory equipment such as the Atlas Type XI Weather-O-Meter.

Permeability. The permeability coefficient, diffusion coefficient, and solubility coefficient can all be measured in addition to a gas-transmission rate. Most often the gases or vapors of interest are water vapor, oxygen, carbon dioxide, and nitrogen. In addition to the inverse gas chromatographic method, permeability can be measured by many techniques. Standardized methods are given in the ASTM series for gas transmission in films and sheets and water vapor in polymers (see also BARRIER POLYMERS).

Flammability. Ignition tests determine the ease with which a material ignites. Burning tests determine the burning rate and the ability to self-extinguish on removal of the ignition source. The oxygen index test accurately determines relative flammability. Heat-contribution tests determine relative contribution to the temperature rise of combustion gases.

Smoke evolution is important in determining combustion characteristics and can be greatly affected by the addition of fire-retardant materials. Smoke development usually increases as the rate of flame propagation is decreased. Products of degradation are measured in a variety of ways, including collection, analysis, and toxicity testing of gases and pyrolyzates. Conventional spectroscopic and chromatographic methods are used for the identification of combustion degradation components. Thermal methods include tga and thermal evolution analysis (tea) to measure char yields and to develop weight-loss profiles.

BIBLIOGRAPHY

J. V. Dawkins, ed., *Developments in Polymer Characterization,* Applied Science Publishers Ltd., London, 1978. A compendium of articles on many aspects of polymer characterization.

J. E. Nielson, *Mechanical Properties of Polymers and Composites,* Vols. 1 and 2, Marcel Dekker, Inc., New York, 1974. A good introductory text dealing with mechanical property techniques.

R. Brown, ed., *Polymer Testing,* Applied Science Publishers, London, 1980. An international journal that covers all manner of testing and frequently contains articles in methodologies of analyses dealing with aging and degradative methods.

W. J. FREEMAN
Hercules, Inc.

CHARGE-TRANSFER COMPLEXES

This article discusses the role of charge-transfer interactions in polymerization reactions and the potential technological applications of polymerization reactions involving this phenomenon.

Although charge transfer or partial electron transfer to form charge-transfer complexes and complete one-electron transfer to generate ion radicals should be distinguished from each other in their quantum chemical meaning and in discussing chemical reactions, the term charge transfer is frequently used to include complete one-electron transfer.

Charge-transfer polymerizations are initiated by means of charge-transfer interactions involving monomers as one or both components of an electron donor-acceptor system. The terms charge-transfer initiation and charge-transfer propagation distinguish polymerizations involving charge-transfer phenomena in the initiation process from those involving charge-transfer complexes in the propagation process. Thermally and photochemically induced charge-transfer polymerizations involve novel types of initiation. Another class of polymerization that involves charge-transfer interactions is spontaneously initiated or free-radical–initiated 1:1-alternating radical copolymerization of an electron donor monomer with an electron acceptor monomer. Charge-transfer interactions have also been widely observed in conventional ionic and free-radical polymerizations.

Charge-transfer Interactions

Charge-transfer Complexes. Mulliken's charge-transfer theory shows that charge-transfer complexes are formed as a result of partial electron transfer from the donor to the acceptor. The degree of electron transfer varies greatly from almost zero to almost one depending on electron donor–acceptor pairs and on the electronic energy state of the charge-transfer complex. The ground and electronically excited states of the charge-transfer complex are described respectively by the following:

$$\psi_N = a\phi(\mathrm{D}\cdots\mathrm{A}) + b\phi(\mathrm{D}^+\mathrm{A}^-)$$

$$\psi_E = -b^*\phi(\mathrm{D}\cdots\mathrm{A}) + a^*\phi(\mathrm{D}^+\mathrm{A}^-)$$

$$a \approx a^* \gg b \approx b^*$$

Exciplexes. Charge-transfer complexes are formed not only in the ground state but also in the electronically excited state of either an electron donor or acceptor molecule. Such complexes that form only in the electronically excited state are termed exciplexes, which are stable in the electronically excited state but dissociative in the ground state. The formation of the exciplex is due to the enhanced donor and acceptor properties of molecules in their excited states. The electronic state of the exciplex is generally described by the following:

$$\psi(\mathrm{D}^+\mathrm{A}^-)^* = a\phi(\mathrm{D}^*\mathrm{A}) + b\phi(\mathrm{DA}^*) + c\phi(\mathrm{D}^+\mathrm{A}^-) + d\phi(\mathrm{DA})$$

Ion-radical Formation. In an extreme case of charge-transfer, complete one-electron transfer occurs thermally or photochemically to generate ion radicals.

Charge Transfer in Conventional Polymerization

Charge-transfer interactions are widely recognized in the initiation and propagation processes of polymerizations initiated by conventional ionic and free-radical initiators. Solvents and deliberately added electron donors or acceptors also affect ionic and free-radical polymerization via charge transfer with monomers or with growing chain ends.

Cationic Polymerization. Usually polymerization is initiated by the electrophilic addition to monomers of initiators or cationic species produced by the aid of cocatalysts. Charge-transfer complexation generally takes place between electron-donating monomers and initiators, eg, halogens, Lewis acids, or stable carbenium ion salts, but the role of charge-transfer interactions in initiation is not clear.

The chain propagation process in cationic polymerization (qv) is understood as involving competing charge-transfer interactions or solvation among cationic chain ends, monomers, counter anions, and solvents:

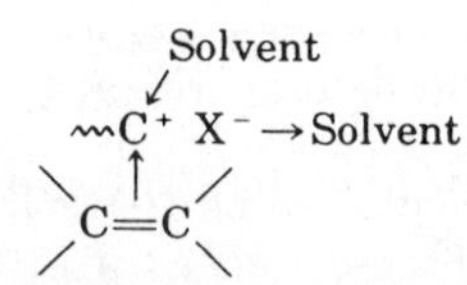

Anionic Polymerization. Initiation of anionic polymerization (qv) with initiators such as metallic lithium, sodium, or sodium naphthalide ($\mathrm{Na^+Np^{\bar{\cdot}}}$) proceeds in a two-stage process involving electron transfer. Formation of monomer anion radicals occurs in the first stage, followed by either radical coupling to generate dimer dicarbanions or the formation of dimeric anion radicals in the second stage. The dimer anion radical in some systems leads to tetramer dicarbanions; both species initiate anionic polymerization that proceeds without termination, giving rise to living polymers (qv).

Radical Polymerization. Charge-transfer interactions of growing radicals with aromatic solvents have been postulated to explain solvent effects on the rate of free-radical polymerization of vinyl monomers, eg, in the polymerization of styrene and methyl methacrylate.

Thermally Induced Charge-transfer Polymerizations

Scope and Mechanism. ***General Aspects.*** Charge-transfer polymerizations have been observed in three systems: donor monomer–electron acceptor, acceptor monomer–electron donor, and donor monomer–acceptor monomer. In the first two systems homopolymerization takes place. The reaction of the donor monomer–acceptor monomer system results in ionic homopolymerization of each monomer, free-radical copolymerization, or cycloaddition, depending on the reactivities of donor and acceptor monomers. Charge-transfer polymerizations are frequently accompanied by interrelated small-molecule reactions, which are often the key to understanding the polymerization mechanism.

Understanding the initiation mechanism of charge-transfer polymerization requires knowledge of the nature of the propagation process. The nature of propagation in spontaneous homopolymerization has been studied by standard methods, eg, effects of additives and copolymerization.

Charge-transfer complexes are common primary intermediates in thermally induced charge-transfer polymerizations, but generally are not directly responsible for initiation. Rather, they are involved in the formation of initiating species.

Photochemically Induced Charge-transfer Polymerizations

Photopolymerizations of monomers are classified into four types: (*1*) polymerization by direct irradiation, (*2*) polymerization in the presence of photoinitiators, (*3*) photoinduced charge-transfer polymerization, and (*4*) four-center polymerization in the solid state. Among these, photoinduced charge-transfer polymerizations, in which monomers are involved in charge-transfer interactions, and photopolymerizations of monomers using photoinitiator systems that involve electron transfer in the process of the generation of active species are the subject of this article (see also PHOTOPOLYMERIZATION).

Scope and Mechanism. ***General Aspects.*** Photochemical reactions of weak electron donor–acceptor systems involving vinyl monomers as components lead to polymerizations and small-molecule reactions via charge-transfer interactions or electron transfer in the electronically excited state. There have been numerous studies of the photophysics and photochemistry of electron donor–acceptor systems, and several typical features of photoinduced charge-transfer polymerizations have been noted.

Photochemically induced charge-transfer polymerizations generally proceed via monomer ion radicals that are generated by photochemical electron transfer. Transient monomer ion radicals have been detected by means of flash photolysis. Whereas photopolymerizations are usually of a free-radical character, photoinduced charge-transfer polymerizations involve ionic polymerizations as well as free-radical polymerizations with initiation by monomer ion radicals. In photoinduced charge-transfer polymerization systems, small-molecule reactions, typically cyclodimerization, also occur by means of the common intermediate, monomer ion radicals. The predominance of polymerization or cyclodimerization depends on the vinyl monomers and reaction conditions, particularly solvents. Reactivities of vinyl monomers differ between VCZ and C-vinyl monomers, eg, styrene and its derivatives.

Free-radical Copolymerizations of Electron Donor Monomers with Electron Acceptor Monomers

The previous discussion has dealt with charge-transfer phenomena in initiation. Charge-transfer complexes formed between monomer pairs may participate in the propagation process of polymerization.

General Aspects. The composition of copolymers produced by free-radical copolymerizations is greatly affected by the polarity of monomers. Free-radical copolymerizations of electron donor monomers with electron acceptor monomers tend to produce copolymers with compositions close to 1:1 over a wide range of monomer feeds, and certain polymerization systems yield 1:1-alternating copolymers regardless of the monomer-feed composition.

The following general features have been recognized for 1:1-alternating radical copolymerizations: (*1*) The formation of a weak ground-state charge-transfer complex between monomer pairs is observed spectrophotometrically in the polymerization systems. (*2*) Copolymer compositions may depend on temperature, deviating from 1:1 alternation at elevated temperatures, as reported for the α-methylstyrene–maleic anhydride (MAn) and vinyl acetate–MAn systems. There is also solvent dependence. (*3*) 1:1-Alternating radical copolymerization is initiated either spontaneously or by added free-radical initiators. (*4*) Copolymerization can be photoinitiated, as for *N*-vinylcarbazole (VCZ)–fumaronitrile (FN), VCZ–diethyl fumarate (DEF), and VCZ–MAn systems. The photoinitiation mechanism is explained in terms of the intermediacy of ion radicals generated by electron transfer in the electronically excited state. (*5*) In some systems, 1:1-alternating radical copolymerization competes with cationic polymerization of electron donor monomers, anionic polymerization of electron acceptor monomers, or cycloaddition reactions of donor monomers with acceptor monomers.

Scope of Alternating Radical Copolymerizations. Numerous 1:1-alternating radical copolymerization systems have been reported. Typical electron-accepting monomers that undergo 1:1-alternating radical copolymerization are maleic anhydride, SO_2, and di- and tri-substituted ethylenes with cyano or carboalkoxy groups.

Mechanism. The propagation mechanism for the 1:1-alternating radical copolymerization has long been a subject of controversy as to whether a charge-transfer complex formed between monomer pairs participates in the propagation process.

Potential Applications

Charge transfer and electron transfer are important concepts in electrical conduction in organic solids, photoconduction in organic solids that can be used as photoreceptors in electrophotography, conversion of solar energy into chemical and electrical energies, visible and relief image formation in some non-silver halide photographic systems, and in other applications.

BIBLIOGRAPHY

Y. Shirota and H. Mikawa, *J. Macromol. Sci. Rev. Macromol. Chem.* **16**(2), 129 (1977–1978).

H. K. Hall, Jr., *Angew. Chem. Int. Ed. Engl.* **22,** 440 (1983).

H. Hirai, *J. Polym. Sci. Macromol. Rev.* **11,** 47 (1976).

D. J. T. Hill, J. J. O'Donnell, and P. W. O'Sullivan, *Prog. Polym. Sci.* **8,** 215 (1982).

YASUHIKO SHIROTA
Osaka University

CHELATE-FORMING POLYMERS

Chelate-forming polymers are polymers that incorporate chelating (multidentate) ligands by covalent bonding. Ion-exchange resins represent the most prominent class of chelating polymer and generally contain the multifunctional ligands attached at one point (**1**). Other chelate-containing polymer systems include cross-linked polymers with ligands attached at two or more positions in the network (**2**) and linear polymers with pendent ligands forming either intramolecular chelating groups (**3**) or intermolecular chelating groups (**4**) (see also ION-EXCHANGE POLYMERS). Related materials include metallopolymers (**5**) and chemically inert polymers with adsorbed low molecular-weight ligands, (**6**) (see COORDINATION POLYMERS; ORGANOMETALLIC POLYMERS).

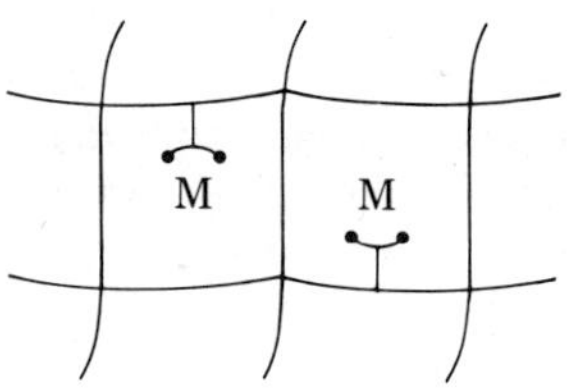

(1)

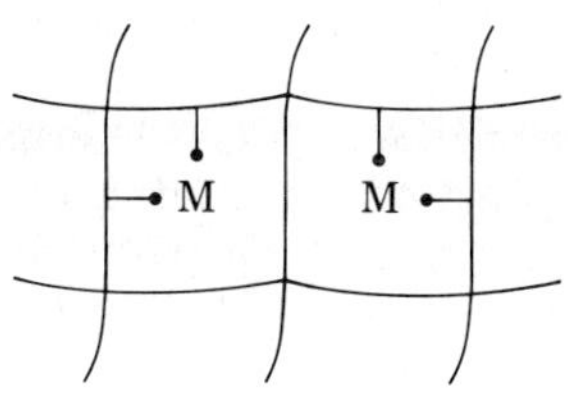

(2)

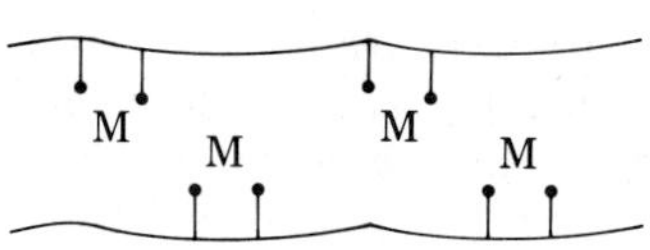

(3)

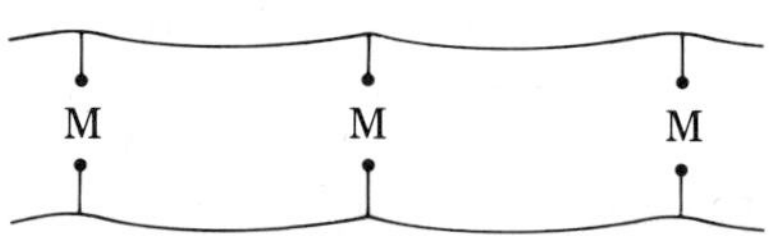

(4)

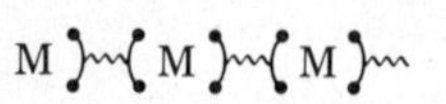

(5)

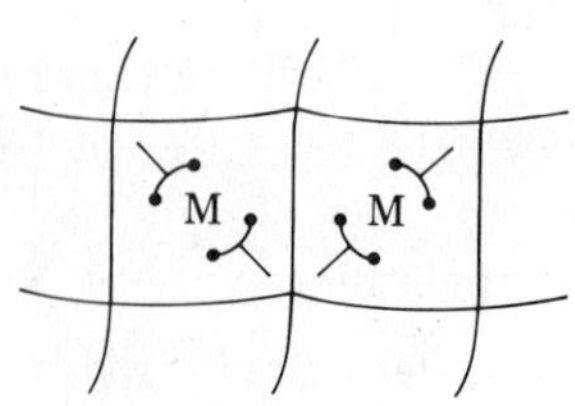

(6)

In the past, the principal commercial interest in chelate-forming polymers has been in the selective removal and recovery of multivalent metal ions from aqueous solutions. More recently, polymer chelation of monovalent metal ions (using crown complexes) and multivalent anions has been accomplished. Some of these polymer complexes are finding applications in ion-selective electrodes and ion-exchange membranes (see ELECTRODES, POLYMER-MODIFIED; MEMBRANES).

Types of Ligand

The range of monomeric chelating ligands capable of interacting with ions in solution is very wide, and many of these have been bound to polymers. Although the number of known chelating and complexing agents is very large, the donor atoms that undergo combination into the various ligand types are restricted to the nonmetallic elements of Groups V and VI; nitrogen, oxygen, and sulfur are the most common examples.

Polymer Preparation

There are various standard methods of preparing chelating ion-exchange resins, and by extension, other forms of chelating polymers such as membranes, fibers, etc. The following are the principal categories: chemical conversion of cross-linked, nonpolar vinyl polymers; chemical conversion of linear vinyl polymers, followed by cross-linking; polymerization of vinyl monomers containing chelating groups; formation of condensation polymers incorporating chelating monomers; chemical conversion of condensation polymers; and chemical conversion of natural polymers.

Theory

The intricate interactions involved in the actual complex formation of chelate-forming polymers can be very different than those for the related soluble monomeric complexes, due to the altered physical environment. The thermodynamic and kinetic effects of each step in the multistep chelate-formation process must be considered. The complexes may be cationic, neutral, or anionic, with different types of calculation for each.

Kinetics. The chelate-forming process can cover such large rate ranges that these chemical reactions can become the rate-determining step within chelate resins. The slow exchange rates of many chelating resins have caused problems with their commercial use.

Resin–Metal Equilibria. ***Stability Constants.*** There are four common differences between monomeric and polymeric metal complexes: (*1*) the stability is usually much higher than that of the complex of the corresponding monomeric analogues; (*2*) the stepwise formation of a multidentate complex is often not observed; (*3*) when a stepwise formation of a multidentate complex is observed, the stepwise stability constants usually increase with the degree of coordination, which is contrary to the behavior of the monomeric ligand; and (*4*) a distortion of polymer chains in some cases reduces the stability of its metal complex; thus, stability can strongly depend on the backbone configuration.

Polymer Selectivity. The factors that encourage ion selectivity include electroselectivity, the general preference for counterions of higher valence, which increases at low concentrations and with resins of high internal molality. Another factor increasing the selectivity with highly cross-linked resins is ionic-solvation variations and swelling-pressure changes. These factors also work with chelating polymers, but the most important consideration is comparative stability constants.

Resin Characterization

The commercial chelating resins are generally characterized by the methods used for normal ion-exchange resins, ie, by their capacity, density, porosity, moisture content, and stability. The important additional factor is selectivity, most easily determined by comparing one ion directly against another at the same concentration under the same conditions.

Applications

Ion Selectivity. The main applications for chelating polymers are based on the high selectivity of the materials for particular ions. There are many mining or pollution situations in which a precious or toxic ion is a small part of a mixture of many other ions, and if this ion can be recovered specifically, the energy and material requirements of the process can be reduced dramatically. However, large-scale commercial use of chelating resins as is common with simple ion-exchange resins has not really arrived, except in some water-softening areas where undesirable multivalent ions such as Ca^{2+} and Mg^{2+} replace monovalent ions.

Polymeric Catalysts. Chelating ion-exchange resins have been widely used as polymeric catalysts and reagents in many commercial applications (see CATALYSTS, POLYMER-SUPPORTED).

Natural Chelating Polymers. One final application of chelating polymers not usually considered, but vitally important in pollution control, is the great influence of natural chelate-forming polymers on the transport and function of metal ions in the environment.

BIBLIOGRAPHY

S. K. Sahni and J. Reedijk, *Coord. Chem. Rev.* **59,** 1 (1984).

M. Marhol, *Ion Exchangers in Analytical Chemistry: Their Properties and Use in Inorganic Chemistry,* Vol. XIV, Elsevier Scientific Publishing Co., New York, 1982.

M. Kaneko and E. Tsuchida, *J. Polym. Sci. Macromol. Rev.* **16,** 397 (1981).

H. F. Walton, *Anal. Chem.* **52,** 15 R (1980).

JONATHAN H. HODGKIN
CSIRO

CHEMICAL ANALYSIS

Chemical analysis in its broadest sense refers to the determination of chemical structure and chemically active species. It involves both direct measurement and use of specific compounds to achieve selective reactions: of a component of the substance being analyzed; to produce a readily measurable species; or to determine a reactive end product. Detection may involve simple visual or optical observation, recording of an electrical or thermal response, or use of sophisticated spectroscopic or other complex automated instrumentation with computerized controls and data-handling capabilities.

Low molecular weight materials, particularly liquids and gases, usually do not require special sample preparation because they tend to be homogeneous. High molecular weight species, such as polymers, are usually solids at normal temperatures and pressures and are likely to have a heterogeneous distribution of chemical and physical components. Hence, a special requirement in the chemical analysis of polymers is the preparation of a representative sample.

The techniques employed for chemical analyses of polymers, usually extensions of those applied to low molecular weight substances, are given below. Many of these are discussed elsewhere in the *Encyclopedia* and, therefore, will be referred to only briefly here (see CHROMATOGRAPHY; ELECTRON MICROSCOPY; ELECTRON-SPIN RESONANCE; INFRARED ABSORPTION SPECTROSCOPY; LIGHT MICROSCOPY; MASS SPECTROMETRY; MOLECULAR WEIGHT DETERMINATION; NUCLEAR MAGNETIC RESONANCE; X-RAY DIFFRACTION; THERMAL ANALYSIS).

Titrimetry
- conductometry
- coulometry
- photometry
- potentiometry
- visual

Polarography

Thermal analysis
- calorimetry
- desiccation
- differential scanning calorimetry (dsc)
- differential thermal analysis (dta)
- oven drying
- pyrolysis, combustion
- thermogravimetry (tg)

Separations
- chromatography (gc, lc)
- distillation
- evaporation

Neutron activation

Radiochemistry

Spectroscopy
- atomic absorption (aas)
- electron spectroscopy for chemical analysis (esca)
- infrared (ir)
- mass spectrometry (ms)
- nuclear magnetic resonance (nmr)
- ultraviolet–visible (uv–vis)
- x-ray fluorescence (xrf)
- x-ray spectroscopy (xrs)

Microscopy
- electron
- scanning electron (sem)
- light

Electrical measurements
- coulometry
- conductance (χ)
- hygrometry
- dielectric constant (ϵ)
- resistance (R)

Physical measurements
- density
- dynamic mechanical analysis (dma)
- melting point
- odor
- refractive index
- solubility
- vapor pressure

A polymer can be converted into another species in order to obtain information on the original substance or to provide reference material for characterization. Such reactions may involve hydrolysis, alcoholysis, esterification, transesterification, and hydrogenation in studies of branching, stereoregularity, and sequence distribution. Standard polymers (qv) are available from the National Bureau of Standards or in the form of commercial lots.

Sample Preparation

Sampling procedures depend on the information desired and the nature of the material, which may vary from fine powder to coated film to textile fiber to container coating. Most polymers are heterogeneous and cover a range of molecular weights. They usually consist of ordered and disordered areas reflecting the distribution of crystalline and amorphous regions, which may separate during solidification from the melt. A representative analytical sample can be obtained by grinding a solid or taking an aliquot of a solution. Many polyols and other condensation polymers are hygroscopic; therefore, exposure to the atmosphere should be minimized by the use of a dry box or polyethylene (PE) bag. Many physical tests require a laboratory with good temperature and humidity control; in the United States the standard conditions are 23 ± 2°C and 50 ± 5% rh.

Ashing or pyrolysis are often used to prepare samples and determine volatile and nonvolatile species as well as degradation products. Low temperature ashing in an oxygen atmosphere is a less drastic means for effective controlled decomposition.

When controlled pyrolysis is required, an inert environment, such as a nitrogen atmosphere, is used to limit the breakdown to controlled thermal degradation. This widely used technique serves as a basic method for qualitative and quantitative analyses of polymers. Often pyrolysis is coupled with spectroscopy, chromatography, mass spectrometry, etc to provide "on line" determinations of the products.

Qualitative Analysis

Many rapid tests are available for the identification of resins. They require a minimum of mobile equipment (preferably microscale) that permits field, in-plant, and laboratory testing.

The identification of a commercial resin is often complicated by the presence of stabilizers, fillers, plasticizers, antioxidants, surface active agents, and other additives. Concentrations may vary from ppm to several percent. The sample may be in the form of molding powder, film, laminates, rod, tube, or molded or machined objects. Because the components may be distributed heterogeneously in the specimen, a powdered, finely divided sample is desirable to assure uniformity.

For the detection of elements other than carbon, hydrogen, and oxygen, fusion, combustion, or pyrolysis may afford a representative sample and concentrate trace components. The most popular techniques involve sodium fusion, sodium peroxide decomposition, and oxygen-flask and Wickbold combustion microprocedures.

After combustion, elements can usually be detected colorimetrically, titrimetrically, or electrometrically.

Polymers can be simply identified by the so-called bite, bend, and burn tests. Useful information is often provided by physical appearance and behavior (see IDENTIFICATION).

Density. In the absence of significant amounts of additives, the density d indicates class identification. Density is conveniently measured by flotation tests in liquids of known densities.

Melting Point. The melting behavior of crystalline polymers serves for identification purposes. The Kofler hot-stage microscope permits observation during the softening and melting process and disappearance of crystallinity while recording the temperature.

Solubility. Behavior toward solvents aids in the purification and identification of polymers. 1,2-Dichloroethane is used for thermoplastic resins such as acrylics; toluene or xylene for polyolefins; methyl ethyl ketone for polybutadiene and copolymers; water for polymers having high concentrations of polar groups such as poly(vinyl alcohol) (PVA); formic acid for polyamides and polyurethanes; and dimethylformamide (DMF) for polyacrylonitrile (PAN) and other vinyl resins. Additives, such as fillers, must be considered.

Colorimetry. Visual observation of color may often be adapted to semiquantitative measurements. Spot tests, notably those developed by Feigl, require only a few milligrams of sample; additives may mask or reduce the sensitivity of the test. White or black porcelain spot test plates and other microequipment are available from most laboratory supply houses such as Arthur H. Thomas Co. (Philadelphia, Pa.).

Identifications. A generalized identification scheme for polymers includes (*1*) solubility determination in five liquids (THF, DMF, H_2O, HCOOH, and boiling xylene); (*2*) pyrolysis, absorption, and pH check; (*3*) identification of products from pyrolysis and hydrolysis, using chemical and colorimetric tests, and thin-layer chromatography; and (*4*) determination of density, thermal effects, and solubility of the specimen and its hydrolysis products.

Quantitative Analysis

Elemental Analysis. Instrumental detection of metals and other inorganic species in plastics is generally the most rapid method, serving for both direct and indirect analysis of ash or extract. Many instrumental techniques are based on spectroscopy, notably x-ray spectroscopy (fluorescence), atomic absorption, and emission spectrography. Each has advantages and disadvantages; most require chemical calibration.

Functional-group Analysis. Functional-group determinations are widely used for the analyses of polymers having a chemically or spectroscopically active species. Most procedures involve direct measurement of an active function or utilize a chemical reaction to provide a readily measurable group or determine the amount of an active reagent remaining after reaction is complete. The specimens are usually handled in solution.

The most widely used techniques for functional-group analysis employ titrimetry, ir and nmr spectroscopy, mass spectrometry, and some physical tests.

Other Techniques

Other analysis techniques include polarography, electron spectroscopy for chemical analysis (esca), ultraviolet-visible spectroscopy, atomic absorption spectroscopy (aas) and fluorescence spectroscopy.

BIBLIOGRAPHY

J. Haslam, H. A. Willis, and D. C. M. Squirell, *Identification and Analysis of Plastics,* 2nd ed., Butterworth, London, 1972, reprinted, Heyden & Son, Inc., Philadelphia, Pa., 1980, 1981.

J. Urbanski, *Handbook of Analysis of Synthetic Polymers and Plastics,* Harwood, Ltd., London, 1977, pp. 129–141.

J. Mitchell, Jr., ed., *Applied Polymer Analysis and Characterization,* Carl Hanser Verlag, Munich, 1987.

F. Scholl, *Additives and Processing Aids,* Vol. 3 of *Atlas of Polymer and Plastics Analysis,* 2nd rev. ed., Carl Hanser Verlag, Munich, Verlag Chemie, Weinheim, 1981.

JOHN MITCHELL, JR.
Consultant

CHEMICAL CELLULOSE.

See CELLULOSE, STRUCTURE AND PROPERTIES.

CHEMICAL COTTON.

See CELLULOSE, STRUCTURE AND PROPERTIES.

CHEMICAL GROUT.

See MINING APPLICATIONS.

CHEMICALLY RESISTANT POLYMERS

Some polymeric materials have outstanding chemical resistance. The chemical resistance of a polymer material is its ability to withstand chemical attack with minimal change in appearance, dimensions, mechanical properties, and weight over a period of time. Test conditions include the length of exposure, concentration, temperature, and internal stress. The final classification as chemically resistant depends on the application.

Mechanisms

Polymers can absorb, react chemically, become plasticized, dissolve, or be stress cracked by chemical environments. Chemical change is stimulated by heat and concentration. If the chemical exposure is limited or infrequent, the service life of the polymeric part may still be long, but some change in properties usually occurs.

Fabrication, compounding, and application affect chemical resistance. The choice of polymer is aided by the following general rules:

1. Solubility is reduced and resistance enhanced by increasing molecular weight.
2. Susceptibility to oxidation increases with unsaturation.
3. Solubility is favored by chemical similarity of solute and solvent.
4. The solubility of a polymer is reduced by chain branching and cross-linking (qv) (see also SOLUBILITY OF POLYMERS). The chemical resistance of some polymers is given in Table 1.

Table 1. Chemical Resistance of Polymers[a]

Polymers	Acids	Alkali	Oxidizing agents	Oils, greases, petroleum, and fuels	Hydrocarbons	Average value
PTFE, FEP, PFA[b]	10	10	10	10	10	10
Kalrez fluoroelastomer	10	10	10	10	10	10
ETFE, ECTFE[c]	10	10	10	10	8	9.6
PVDF[d]	10	10	10	10	7	9.4
other fluoroelastomers	9	7	10	9	8	8.8
PCTFE, PVF[e]	10	10	10	8	5	8.6
PVC—unplasticized PVC, PVDC–PVC copolymer, CPVC[f]	9	10	10	9	5	8.6
furan resins	10	10	1	10	10	8.2
poly(phenylene sulfide)	6	8	4	10	10	7.6
thermosetting polyimides	6	6	8	8	9	7.4
poly(ether sulfone), polyarylketone	8	9	6	8	6	7.4
epoxy resins	7	8	3	9	9	7.4
polysulfone	8	8	7	6	6	7.0
nitrile rubber	7	8	4	10	5	6.8
acetal resins	5	10	4	8	9	6.8
chloroprene polymers	9	8	4	8	5	6.8
unsaturated polyester, FRP[g]	7	5	5	8	8	6.6
polyolefins[h]	8	9	7	6	3	6.6
polyacrylic rubber	6	6	6	9	6	6.6
chlorosulfonated rubber	10	7	7	5	4	6.6
polysulfide rubber	4	8	4	7	9	6.4
silicone rubber	7	9	7	5	2	6.2
poly(amide-imide), polyetherimide, thermoplastic polyimides	8	5	2	8	8	6.2
polypropylene, ethylene–propylene elastomers	9	9	6	5	2	6.2
thermoplastic polyesters	6	5	4	8	8	6.2
butyl rubber	10	10	4	4	2	6.0
nylon	3	8	3	8	8	6.0

[a] Based on an arbitrary scale of increasing resistance of 0 to 10.
[b] PTFE = polytetrafluoroethylene; FEP = fluorinated ethylene–propylene copolymer; PFA = perfluoroalkoxy resins.
[c] ETFE = ethylene–tetrafluoroethylene copolymer; ECTFE = ethylene–chlorotrifluoroethylene copolymer.
[d] PVDF = poly(vinylidene fluoride).
[e] PCTFE = polychlorotrifluoroethylene; PVF = poly(vinyl fluoride).
[f] PVC = poly(vinyl chloride); PVDC = poly(vinylidene chloride); CPVC = chlorinated poly(vinyl chloride).
[g] Fiber-reinforced plastics.
[h] Except polypropylene.

BIBLIOGRAPHY

ASTM D543, Resistance of Plastics to Chemical Reagents, American Society for Testing and Materials, Philadelphia, Pa., 1984.

J. Agranoff, ed., *Modern Plastics Encyclopedia 1988,* McGraw-Hill Inc., New York, 1987 (or latest edition).

R. B. Seymour, *Plastics vs. Corrosives (Society of Plastics Engineers Monographs),* Wiley-Interscience, New York, 1982, Chapt. 14, pp. 147–262.

ROBERT A. MCCARTHY
Springborn Group, Inc.

CHITIN

Chitin is the most abundant organic skeletal component of invertebrates. It is the characteristic polysaccharide of several important phyla, eg, *Arthropoda, Annelida, Mollusca,* and *Coelenterata,* and also occurs in many fungi, eg, *Euasomycetes, Zygomycetes, Basidiomycetes,* and *Deuteromycetes.* The biosynthesis of chitin, (1→4)-linked 2-acetamido-2-deoxy-β-D-glucan, follows the pathway shown in Figure 1.

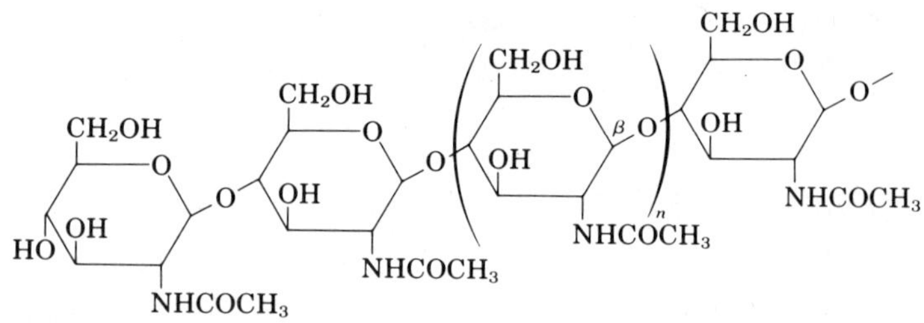

chitin

Properties

Depending on the extent of deacetylation, chitins contain from 5 to 8 wt % nitrogen, which in chitosan is mostly in the form

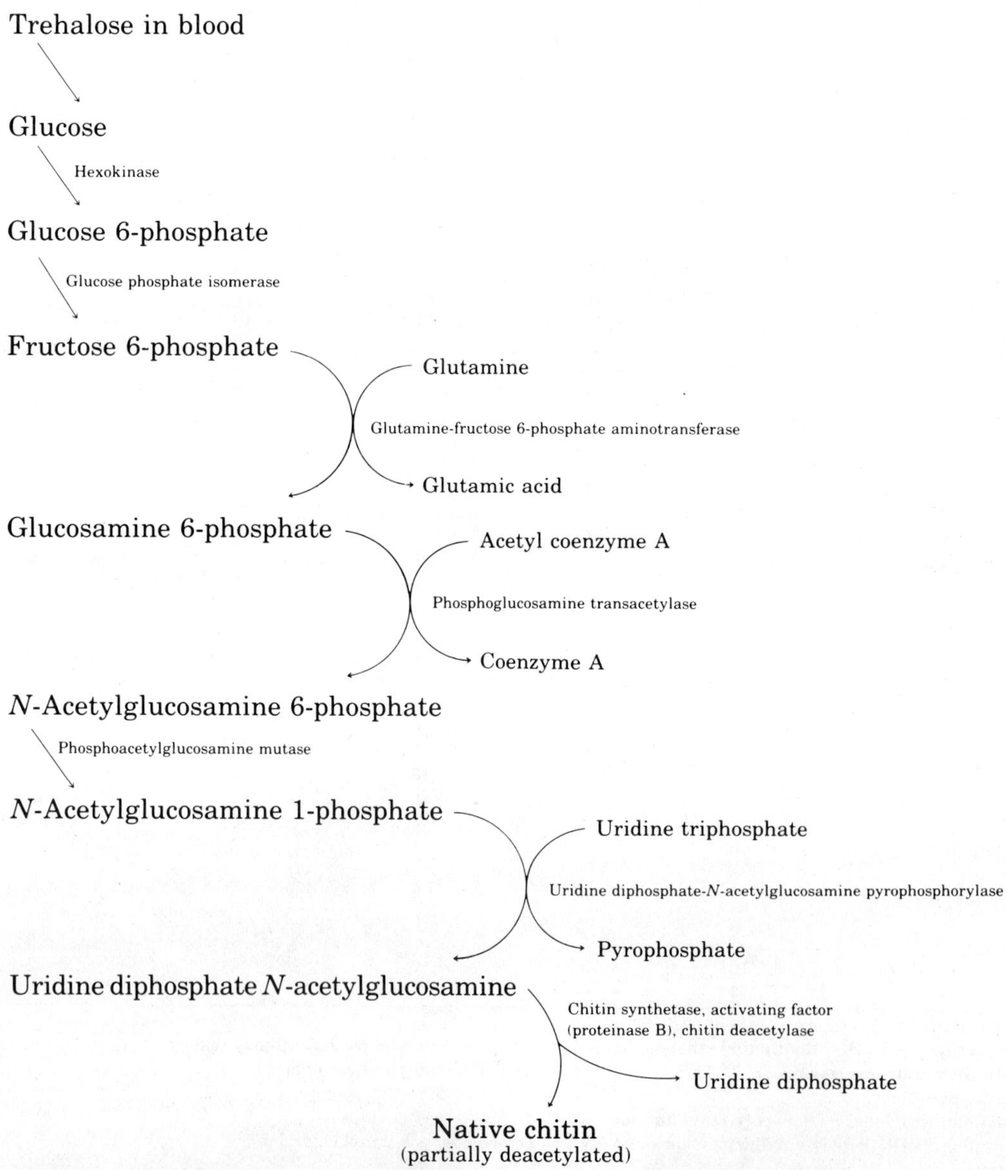

Figure 1. Pathway of chitin synthesis. Phosphoglucosamine transacetylase is also called glucosamine 6-phosphate *N*-acetyltransferase.

of primary aliphatic amino groups. Chitosan therefore undergoes the reactions typical of amines, of which *N*-acylation and the Schiff reactions are the most important. Chitosan derivatives are easily obtained under mild conditions and can be considered substituted glucans. Although it is difficult to introduce nitrogen into cellulose, C-2 substituted glucans can be obtained from chitosan.

Biodegradability. The biodegradability of chitin is important from a technological standpoint, especially in relation to its uses as a carrier for delayed release of drugs, pesticides, and other formulations. Biodegradation by chitinoclastic bacteria in soils and the oceans plays an important role in biogeochemical cycles of carbon and nitrogen. The importance of chitin from the ecochemical point of view relies on its production of natural biocenoses (or communities) and on its biodegradation and bioaccumulation in sediments. Chitin and its oligomers can be degraded by chitinases associated with chitobiases, β-*N*-acetylhexosaminidases, and lysozyme.

On an industrial scale, chitin–chitosan is currently hydrolyzed by inorganic acids to form glucosamine, for pharmaceutical uses.

Processing

Chitin is easily obtained from crab or shrimp shells and fungal mycelia. The processing of the fungal wastes from *Aspergillus niger, Mucor rouxii,* and *Streptomyces* consists of an alkali treatment that yields chitosan–glucan complexes.

The processing of crustacean shells is associated with the canning activities of the fishing industry and solves in part its waste-disposal problems, while giving origin to chitin and chitosan, whose general properties vary according to origin and processing. The latter mainly involves the removal of proteins and the dissolution of calcium carbonate that is present in crab shells in high concentrations.

Economic Aspects

The production of chitin and chitosan is currently based in Oregon, Washington, Virginia, and Japan and several countries of the third world. In 1984 the worldwide price of chitosan was ca $20.00/kg.

Health and Safety Factors

Chitin and chitosan have low toxicity, eg, the LD_{50} of chitosan in laboratory mice is 16 g/kg body weight, which is close to that of salt or sugar.

Uses

Chitosan and its derivatives, rather than chitin itself, find use in a number of fields. Examples of the use of chitosan are the coating of seeds to impart antifungal resistance and the applications in dentistry and surgery for wound healing and tissue regeneration (see SKIN, SYNTHETIC). Examples of the uses of chitosan derivatives are the recovery of trace metals from aqueous and organic media, and the formulation of cosmetics.

BIBLIOGRAPHY

R. A. A. Muzzarelli, *Chitin,* Pergamon Press, New York, 1977.

R. A. A. Muzzarelli and E. R. Pariser, eds., *Proceedings of the 1st International Conference on Chitin/Chitosan,* MIT-SG, Cambridge, Mass., 1978.

R. A. A. Muzzarelli in G. O. Aspinall, ed., *The Polysaccharides,* Vol. 3, Academic Press, New York, 1985.

R. A. A. Muzzarelli, C. Jeuniaux, and G. W. Gooday, eds., *Chitin in Nature and Technology,* Plenum Press, New York, 1986.

RICCARDO A. A. MUZZARELLI
Faculty of Medicine
University of Ancona

CHLORAL POLYMERS. See ALDEHYDE POLYMERS.

CHLORINATED POLYETHYLENE. See ETHYLENE POLYMERS.

CHLORINATED POLY(VINYL CHLORIDE). See VINYL CHLORIDE POLYMERS.

CHLORINATED RUBBER. See RUBBER DERIVATIVES.

2-CHLOROACRYLIC ESTER POLYMERS. See ACRYLIC AND METHACRYLIC ESTER POLYMERS.

2-CHLOROBUTADIENE POLYMERS. See CHLOROPRENE POLYMERS.

CHLOROBUTYL RUBBER. See ISOBUTYLENE POLYMERS.

CHLOROPRENE POLYMERS

Chloroprene Monomer

Properties. Chloroprene (mp, $-130\pm$°C; bp at 101 kPa 59.4°C) is a colorless liquid with a characteristic ethereal odor, only slightly soluble in water, but miscible with most organic solvents. Spectral data indicate that the molecule is planar and transoid in conformation.

Chloroprene is toxic and flammable and forms explosive mixtures with air. The TLV for exposure to vapor is 10 ppm by volume.

Manufacture. As the price of acetylene has increased relative to butadiene, all production has been converted to the butadiene route except where low cost acetylene is available. This route involves the chlorination of butadiene to the 3,4-dichloride and its dehydrochlorination to chloroprene; some 1,4-dichloride is always obtained in the chlorination.

Polymerization

Free-radical Polymerization. Chloroprene polymerizes readily in the presence of a free-radical initiator. Because of the 2-chlorine substituent, it is so highly reactive with itself that only a few comonomers can be used.

Emulsion Polymerization. Similar to other diene polymers, emulsion polymerization (qv) is the only practical way to accommodate high polymerization rates and achieve high molecular weight. This is an addition polymerization process in which a large number of concurrent propagating centers are isolated from one another.

Copolymerization. Except for 2,3-dichlorobutadiene, chloroprene is several times more reactive than most vinyl comonomers. A highly electron-deficient monomer such as SO_2 readily copolymerizes with chloroprene to give a 1:1-alternating copolymer as polymerization temperature drops to −78°C.

Other Methods. Although other polymerization, eg, cationic, modes are possible for chloroprene, no useful polymers have been prepared so far. A Ziegler-type coordinate polymerization provides unusual compositions.

Manufacture of Polychloroprene. Today, essentially all polychloroprenes are manufactured by emulsion polymerization, mostly batch and some continuous, and isolated by freeze rolling or drum drying.

Microstructure of Polychloroprene

Although chloroprene, present predominantly in the transoid form, can polymerize readily by several different paths, free-radical methods are used commercially to produce a relatively stereospecific 1,4-trans polymer.

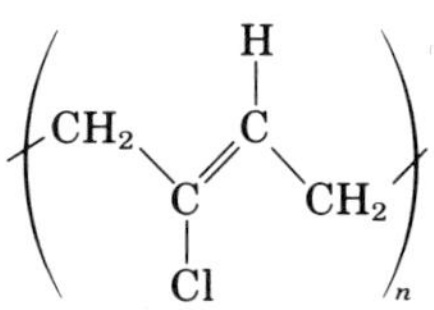

1,4-trans
(**1**)

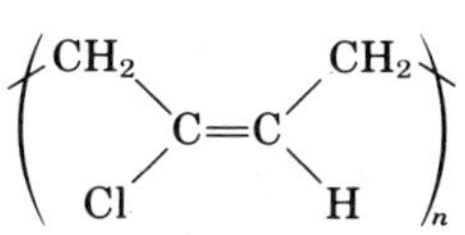

1,4-cis
(**2**)

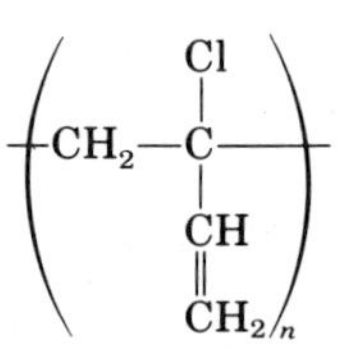

1,2
(**3**)

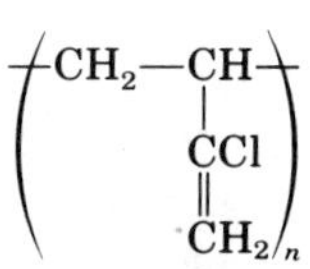

3,4
(**4**)

Crystallinity. Because polychloroprene prepared between 0 and 40°C predominantly consists of the 1,4-trans unit, both the raw and cured polymer crystallize, particularly on stretching.

Glass Transition. Because irregularities in an otherwise predominantly 1,4-trans structure can be viewed as copolymerized components, the incorporation of these groups affects the glass-transition temperature. As an extreme case, 1,4-*cis*-polychloroprene has a higher glass-transition temperature, ie, −20°C, in contrast to −49°C for all *trans*-polychloroprenes.

Aging. Because of the electronegative chlorine on the 1,4-trans double bond, the resistance of polychloroprene to air oxidation, ozone attack, or weathering is higher than that of a typical diene rubber such as SBR or natural rubber. Nevertheless, a typical polychloroprene vulcanizate eventually fails after exhibiting a severalfold increase in modulus, eg, M_{100}, and decreases in elongation and tensile properties.

Types of Neoprene

Types of neoprene include dry neoprene and neoprene latices. Dry neoprene is classified as either general-purpose (G types, W types, and T types) or special-purpose (including types used as adhesives and sealants). Latices also are classified as general-purpose or specialty types. Characteristics are given in Tables 1 and 2, respectively.

Table 1. General-purpose Neoprene Latices

Property	571	654	842A
designation	high gel	high solids low gel	fast-curing high gel
properties			
pH	12.0	12.0	12.0
solids, %	50	59	50
emulsifier	Na resinate	K resinate[a]	Na resinate
sp gr			
latex	1.11	1.13	1.11
polymer	1.23	1.23	1.23
viscosity at 30 rpm, mPa·s (=cP)	15	55	15

[a] Disproportionated.

Table 2. Special-purpose Neoprene Latices

	115	735A
main characteristic	carboxylated nonionic	sol
properties		
pH	7	12
solids, %	47	45
emulsifier	poly(vinyl alcohol)	Na resinate[a]
sp gr		
latex	1.09	1.10
polymer	1.24	1.23
viscosity at 30 rpm, mPa·s (=cP)	350	5

[a] Disproportionated.

Theory of Vulcanization of Neoprene

Vulcanization (qv) of neoprene mostly involves the highly reactive allylic chlorine from the original 1,2 and isomerized 1,2 units. It is thus possible to cure with a typical Lewis acid precursor such as ZnO in a manner similar to that suggested for chlorobutyl rubber cure by a carbonium ion mechanism. To achieve a good balance of physical properties, a useful compound contains not only metal oxides and organic accelerators, but also retarders, antioxidants, fillers, and process oils and aids.

Processing of Neoprenes

Processing Dry Neoprenes. The methods for processing neoprene reflect the principles common to the processing of most elastomers. However, some unusual features bear mention.

Mixing. With few exceptions, neoprene compounds should be mixed at as low a temperature and with as short a cycle as possible.

Calendering and Extrusion. Control of calendering is critical, particularly for G types, which are highly temperature sensitive and perform best below 70°C. The W types are generally calendered on slightly hotter rolls because they are released more easily and require a higher temperature to avoid shrinkage. Most neoprene compounds extrude best with a cooled barrel and screw, warm head, and hot die. Beyond that, neoprene can be treated like other elastomers (see also EXTRUSION).

Processing Neoprene Latices. The manufacture of articles from neoprene and natural rubber latex is similar, and the processes used with natural rubber are adaptable to neoprene in most instances; curing is an exception. Neoprene requires neither precuring nor aftercuring, but neoprene films must be cured above 100°C, at 120–140°C to attain optimum properties.

Health and Safety Factors

For all the solid neoprene polymers, routine industrial hygiene practices are recommended to avoid dirt buildup or static charges. Neoprene is free of toxicity.

Uses

The broad classification of rubber products known as mechanical and automotive goods accounts for the greatest usage of neoprene; this includes belts, hoses, and molded goods. A large percentage of the neoprene sold is used in wire and cable jackets. Other uses include building seal, concrete-highway joint seals, bridge pads, and soil-pipe gaskets. Neoprene cements are used for maintenance and roof coatings. Neoprene latices are used for wet-laminating and contact-bond adhesives, fiber binders, in coatings and dipped goods, as modifiers in elasticized bitumens and cements, and to produce cellular materials.

BIBLIOGRAPHY

P. R. Johnson, *Rubber Chem. Technol.* **49,** 650 (1976).

M. M. Coleman, D. L. Tabb, and E. G. Brame, Jr., *Rubber Chem. Technol.* **50,** 49 (1977).

J. R. Ebdon, *Polymer* **19,** 1232 (1978).

R. M. Murray and D. C. Thompson, *The Neoprenes,* E. I. du Pont de Nemours & Co., Inc., Wilmington, Del., 1963.

C. A. STEWART, JR.
T. TAKESHITA
M. L. COLEMAN
E. I. du Pont de Nemours & Co., Inc.

CHLOROSTYRENE POLYMERS.

See STYRENE POLYMERS.

CHLOROSULFONATED POLYETHYLENE.

See ETHYLENE POLYMERS.

CHLOROTRIFLUOROETHYLENE POLYMERS

Chlorotrifluoroethylene Homopolymer

High molecular weight homopolymers of chlorotrifluoroethylene and its copolymers with small amounts, ie, less than 5%, of vinylidene fluoride are thermoplastic resins that have found specialty uses based on vapor-barrier properties, superior thermal stability, and resistance to strong oxidizing agents.

Properties of Monomer

Chlorotrifluoroethylene, $CF_2{=}CFCl$, mol wt 116.47, bp −27.9°C, fp −157.5°C, is one of a series of halogenated ethylenes used as a monomer for the production of chlorofluorocarbon polymers. As is characteristic of most monomers in this class, it is a colorless gas at room temperature and pressure.

Manufacture of Monomers

The most important industrial process for the manufacture of chlorotrifluoroethylene is based on the dechlorination of 1,1,2-trichloro-1,2,2-trifluoroethane by vapor-phase dehalogenation at 500–600°C or in the liquid phase using zinc in methanol.

Structure of Polymer

Crystallinity. The crystal structure of PCTFE has been determined by x-ray diffraction. The pseudohexagonal lattice parameters are $a = 0.644$ nm and $c = 4.15$ nm. The polymer chain is helical, with an average of 16.8 monomer units in one turn of the helix (equal to c).

Microstructure. Nmr studies demonstrate that polychlorotrifluoroethylene has isotactic as well as syndiotactic sequences of monomer units, but that syndiotactic linkages predominate.

Properties of PCTFE Polymers

High Molecular Weight Plastics. The molecular weights of commercial polychlorotrifluoroethylene plastics are in the range $\bar{M}_n = 70{,}000–400{,}000$, as measured by zero strength time (ZST) and intrinsic viscosity (see MOLECULAR WEIGHT DETERMINATION).

Physical and Mechanical Properties. The unique properties of PCTFE, as compared with those of other perhalogenated resins, include rigidity, low temperature toughness, resistance to creep and cold flow, and low permeability to moisture and gases. The physical and mechanical properties of typical molding-grade resins with excellent machinability are given in Table 1, the properties of extruded films of two different degrees of crystallinity in Table 2, and mechanical properties of a low crystallinity resin at cryogenic temperatures in Table 3.

Chemical Resistance. Polychlorotrifluoroethylene is resistant to attack by most industrial chemicals, including strong acids and alkalies over a wide range of concentrations and temperatures.

Elastomers. ***Physical Properties.*** The raw-gum properties of a typical elastomeric copolymer containing 70 mol % vinylidene fluoride and 30 mol % chlorotrifluoroethylene are shown in Table 4. The raw gum can be cured with amines, dihydroxy compounds, or peroxides. The properties of typical cured vulcanizates are shown in Table 5.

The outstanding property of chlorotrifluoroethylene-based elastomer is its superior resistance to strongly oxidizing acids such as fuming nitric acid.

Table 1. Physical Properties of PCTFE Molding Resins

Property	Value
sp gr	2.075–2.185
water absorption, %	0
mold shrinkage, cm/cm	0.015–0.020
refractive index; n_D at 20°C	1.425
flammability, limiting oxygen index, %	95
Thermal properties	
crystalline mp, °C	210–212
useful temp range, °C	−200–180
specific heat, J/(mol·K)[b]	97.5
thermal conductivity, W/(m·K)	1.195×10^{-2}
Electrical properties	
dielectric constant at 10^3 Hz	2.3–2.7
dissipation factor at 10^3 Hz	0.025
dielectric strength, 1.5 mm short time, V/mm	25.4×10^3

[a] To convert MPa to psi, multiply by 145.
[b] To convert J to cal, divide by 4.184.

Table 2. Typical Properties of PCTFE Films

	Value	
Property	Low crystallinity	High crystallinity
sp gr	2.08	2.11
water absorption, %	0	0
tensile strength[a], MPa[b], MD[c]	28–42	28–48
elongation, %, MD[c]	150–250	100–200
tear strength[a], Graves, g, MD[c]	1890	2300

[a] 0.127 mm thickness.
[b] To convert MPa to psi, multiply by 145.
[c] MD = machine direction.

Table 3. Mechanical Properties of PCTFE Molding Resin at −196°C

Property	Value
tensile strength, MPa[a]	119–173
tensile elongation, %	2–4
flexural yield strength, MPa[a]	400

[a] To convert MPa to psi, multiply by 145.

Table 4. Gum Properties of Kel-F 3700 Elastomer

Property	Value
sp gr	1.85
hardness, Shore A	45
embrittlement temp, °C	−64
tensile strength, MPa[a]	2.4–4.1

[a] To convert MPa to psi, multiply by 145.

Table 5. Properties of Kel-F-3700 Vulcanizate

Property	Amine cured	Peroxide cured
sp gr	1.85	
tensile strength, MPa[a]	18.6	16
tear strength, kg/m		5360
elongation, %	400	450

[a] To convert MPa to psi, multiply by 145.

Thermal Stability. Among many thermal stabilizers claimed effective against thermal degradation at 330–350°C are 2,4-dichlorophenoxymethylbenzofuran, 2-benzylhydroquinone, 2-benzyl-4-methoxyphenol, and 2-cyclopentylhydroquinone.

Radiation Behavior. Chlorotrifluoroethylene–vinylidene fluoride copolymer elastomers are cross-linked by radiation with γ rays (see also CROSS-LINKING WITH RADIATION).

Oils and Waxes. ***Physical Properties.*** The oils range in viscosity from 5 mm^2/s (=cSt) for light oils to 1500 mm^2/s (=cSt) for heavy oils. The corresponding pour points range from −90 to 40°C. The melting points of waxes range from 55°C for soft waxes to 135°C for hard waxes, and the corresponding viscosities range from 20–70 mm^2/s (=cSt) at 100°C for the soft waxes to 80–400 mm^2/s (=cSt) at 100°C for the hard waxes. The boiling point of the waxes is ca 260°C. The greases are prepared by gelling the oils with the wax or silica gel. The pour points of the greases are above 200°C. The service temperature for the wax-thickened greases is −40 to 135°C; silica-thickened greases have an upper limit of 175°C. The densities of the various oils, greases, and waxes range from 1.8 to 2. They are soluble in most organic solvents, but solubility decreases with increasing molecular weight. The lubricity of the PCTFE oils is at least equal to that of petroleum oils.

Chemical Properties. Polychlorotrifluoroethylene oils, waxes, and greases are noted for their excellent chemical inertness, especially to oxygen and oxidizing acids. They are reactive toward amines, liquid fluorine, liquid chlorine trifluoride, and alkali metals.

Polymerization

High molecular weight homopolymers and copolymers of chlorotrifluoroethylene are prepared by free-radical–initiated polymerization in bulk, suspension, or aqueous emulsion using organic and water-soluble initiators or ionizing radiation.

Processing of Polymers

PCTFE plastics can be processed by conventional thermoplastic processing techniques such as compression and injection molding (qv) and extrusion (qv).

Economic Aspects

U.S. consumption of molding and extrusion grades of resin by application is shown below.

Market	Metric tons	10^6 $
film	225	14.5
electrical	36	3.30
chemical, mechanical	64	5.70

The U.S. market for PCTFE oils, waxes, and greases is estimated at approximately 35 t/yr. The various grades of oil range in price from $60 to 125/kg, waxes from $70 to 80/kg, and greases from $100 to 130/kg.

U.S. production of chlorotrifluoroethylene–vinylidene fluoride elastomers is approximately 45 t/yr at ca $100/kg.

Health and Safety Factors

Good industrial hygiene practice demands careful control and monitoring of trace monomer in the workplace. At molding and other melt-processing temperature, PCTFE plastics generate minute quantities of decomposition products, some of which may be toxic. Processing temperatures below 325°C and good ventilation ensure complete safety. Scrap waste should be buried, not incinerated.

PCTFE oils, waxes, and greases are subject to thermal cracking above 300°C and to rapid breakdown above 325°C; the products are toxic. Adequate ventilation is essential above 250°C.

Uses

The main uses for polychlorotrifluoroethylene plastic molded parts are in chemical process equipment, mechanical parts for cryogenic applications, and electrical parts. Extruded film and film laminates are used in high performance packaging (see also FILMS).

PCTFE oils, waxes, and greases are used as nonflammable hydraulic fluids in aerospace applications, lubricants in oxygen-manufacturing plants, pump lubricant in LOX trailers, vacuum-pump oil for handling reactive gases such as chlorine and fluorine, flotation and damping fluid in gyroscopes, stopcock grease in corrosive reactive service, and sealants.

Chlorotrifluoroethylene-based elastomers are used in O-rings, oil and shaft seals, gaskets, hose, tubing, diaphragms, protective clothing, and mechanical goods where high temperature performance, low temperature abrasion resistance, and resistance to many chemicals and oxidizing acids are the principal requirements.

S. CHANDRASEKARAN
Ansimont, USA, Inc.

Chlorotrifluoroethylene–Ethylene Copolymers

An essentially 1:1-alternating melt-processible copolymer of ethylene and chlorotrifluoroethylene (ECTFE) has been developed that retains the desirable properties of fluoropolymers, but lacks many of the deficiencies of earlier fluoropolymers, such as creep, poor abrasion resistance, and difficulty in processing. This material, although mechanically tough, is easily processed by extrusion (qv), injection molding (qv), powder coating, blow molding, and rotomolding into wire products, fibers, films, and coated and molded parts.

Preparation

Copolymerization of ethylene and chlorotrifluoroethylene has been carried out by aqueous suspension techniques, by low temperature polymerization initiated by oxygen-activated triethylboron in dichlorotetrafluoroethane, and by radiation-induced polymerization.

Morphology

The ECTFE copolymer is partially crystalline, ie, 50–55%, depending on the method of preparation. The commercial product melts at 240°C. X-ray diffraction of oriented film indicates that the unit cell of the alternating copolymer is hexagonal, with a chain repeat distance of $c = 0.502$ nm.

Properties

The outstanding characteristics of ECTFE are easy processibility by thermoplastic techniques; thermal and oxidative stability to 150–175°C; chemical inertness against numerous chemicals and solvents; mechanical toughness over a wide range of temperatures; nonflammability (UL-94 V-O classification down to 0.18-mm thickness); and good permeability resistance.

Health and Safety Factors

At molding temperatures, ECTFE copolymer can produce small quantities of hydrochloric acid and hydrofluoric acid, particularly if overheated. Adequate ventilation should be provided and other safety considerations appropriate for these gases taken.

Uses

Wire and Cable Insulation. The primary use of ECTFE copolymer is in wire and cable insulation, where its electrical properties, thermal capability at 150°C, and mechanical toughness and flame resistance in thin walls are exploited.

BIBLIOGRAPHY

Aclar R Fluorohalocarbon Film Technical Data Sheet, Allied Corp., Morristown, N.J., 1978.

Halocarbon Chlorofluorocarbon Fluids and Lubricants, Halocarbon Products Corp., Hackensack, N.J., 1978.

W. E. Ashton and C. A. Strack in R. C. Gunderson and A. W. Hart, eds., *Synthetic Lubricants,* Reinhold Publishing Corp., New York, 1962, Chapt. 6, pp. 246–259.

W. T. Miller, Jr. (Chapt. 32, pp. 567–685) and E. A. Belmore, W. M. Ewalt, and R. L. Murray (Chapt. 33, pp. 686–713) in C. Slesser and S. R. Schram, eds., *Preparation, Properties, and Technology of Fluorine and Organic Fluorocompounds,* McGraw-Hill Inc., New York, 1951.

WILLIAM A. MILLER
Allied Corporation

CHROMATOGRAPHY

Chromatography includes various experimental techniques designed to separate mixtures of compounds, making it possible to understand the composition of complex mixtures. In the last three decades, chromatography has revolutionized analytical chemistry, and the term chromatography no longer has any reference to color.

Chromatography involves a stationary and a mobile phase. Separation takes place because of partitioning between the sample components in the moving and stationary phases. The general classification of chromatographic methods is based on the nature of these phases (Table 1) (see also CHROMATOGRAPHY, AFFINITY).

Table 1. Chromatographic Methods

Mobile phase	Stationary phase	Acronym
gas	solid or liquid	gc
gas	solid	gsc
liquid	solid	lsc
gas	liquid	glc
supercritical fluid	liquid	sfc
liquid	liquid	llc[a] or lc
liquid	solid[b]	tlc

[a] Including size-exclusion or gel-permeation (sec or gpc), ion-exclusion, and high performance or high pressure chromatography (hplc).
[b] Thin-layer chromatography (tlc).

Gas–solid chromatography (gsc) and liquid–solid chromatography (lsc) are employed less than the other methods because of greater experimental difficulties. Size exclusion is a modified form of liquid–liquid chromatography that uses column packings that separate molecules on the basis of effective molecular size in solution. Because the columns operate over a molecular weight range from 10^2 to 10^7, depending on the nature of the polymer, they are invaluable in determining molecular weights and molecular weight distribution rapidly and precisely.

Polymer scientists and engineers cannot, however, be concerned only with polymers. In manufacturing, it is necessary to determine the purity of monomers and other materials, such as solvents; here, gas–liquid chromatography plays an important role. Similarly, liquid–liquid chromatography is used in the analysis of additives and other compounds in polymer formulations. Supercritical fluid–liquid chromatography is a relatively new method between gas–liquid and liquid–liquid chromatography that seems to hold much promise for the future.

The elements common to most forms of chromatography are given in Figure 1. A mobile-phase reservoir or reservoirs is needed for solvent-gradient programming. In gas chromatography, this reservoir is usually a cylinder of compressed gas. In various forms of liquid chromatography, it is a container of some kind, serving the same purpose. It should be noted that in hplc, the mobile phase contributes to the separation; this is not the case in size-exclusion chromatography.

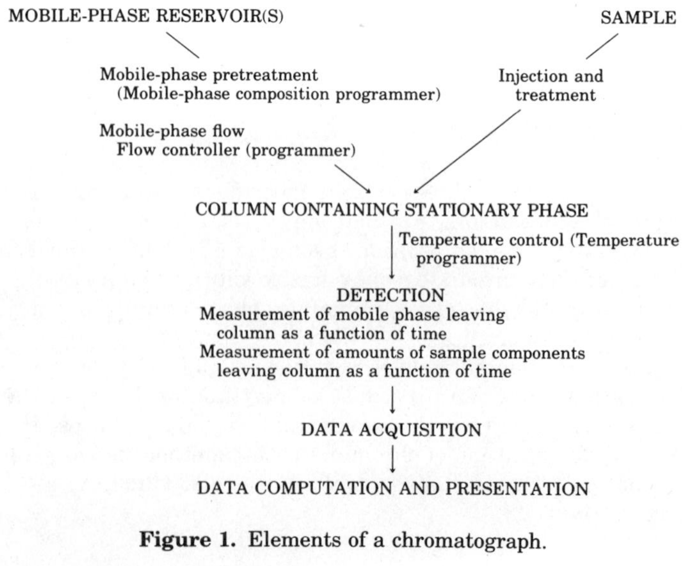

Figure 1. Elements of a chromatograph.

In most methods, some mobile-phase pretreatment is required. In gc, for example, it is frequently necessary to remove oxygen and water from the mobile-phase gas. In lc, highly purified solvents are required; they are usually available commercially. In addition, it is often necessary to degas the mobile phase for detector baseline stability.

In some methods, a mobile-phase movement must be produced. In lc, this is accomplished by pumps of various complexity. In gc, the pressure in the reservoir usually causes the flow. For both gases and liquids, some form of flow control, either built into the pump or through gas-flow regulation, is required.

Before injection, the sample may require pretreatment such as filtration. It may also need to be treated after injection. For example, in gc, the sample is transferred into an injection port heated above the column temperature to vaporize liquids rapidly. From here, it passes into a column containing the stationary phase. Separation occurs as the sample travels along this column, which is equipped with temperature control. As the mobile phase leaves the column, its progress must be monitored. The time since the injection of the sample is usually of interest. Furthermore, the amount of sample contained in the mobile phase as it leaves the column must be measured, again as a function of time. Data acquisition and computational procedures must be carried out and the data presented in usable form.

Auxiliary equipment is frequently required. In the case of solvent-gradient operation in liquid chromatography, a mobile-phase composition programmer and flow programmers are frequently used as well as temperature controls.

The measurements can be automated to any degree required; sample injection can be automated. Computer-data acquisition, computation, and presentation and computer and microprocessor control of the operating conditions are widely used. Many chromatographs operate unattended; personnel prepare and load the samples, program the computer, start the program, and are free to leave. The higher the degree of advancement and automation, the higher the costs.

Resolution and Separation Power. The concept of a theoretical plate, borrowed from distillation practice, is also useful in chromatography. A monodisperse compound is injected and the time t_R (retention time) from injection to the center of the peak is measured. The theoretical plates are calculated by various methods, two of which are illustrated in Figure 2. The first is by constructing tangents to the points of inflection of the curve and measuring the width w at the intercepts on the baseline. An equivalent method is to measure the peak height divided in two and the width at half height $w_{1/2}$.

Thin-layer Chromatography

In thin-layer chromatography (tlc), the adsorbent serves as a stationary phase. It is deposited in a thin layer, ie, from a fraction of a millimeter to 1–2-mm thick, on a flat, supporting surface, usually a piece of glass. The adsorbent is held in place by a binding agent, frequently starch. The most common adsorbent is silica gel of various grades.

The technique is very simple. The sample is dissolved in suitable solvent. A small drop or spot of the solution is placed on one end of the plate and the solvent is allowed to evaporate. The plate is placed in a developing chamber that contains enough developing solvent at the bottom to reach the end of

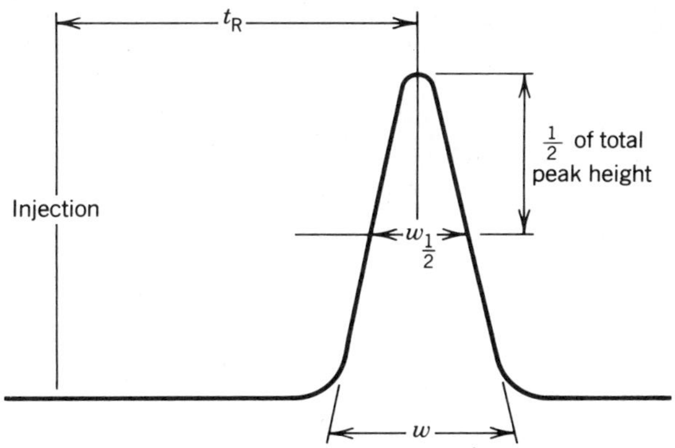

Figure 2. Parameters for the calculation of theoretical plates.

the adsorbent on the plate. The solvent flows upward through capillary action and the various components are separated as they move at different rates. The plate is removed and the location of the spots determined visually or otherwise. Dyes may help visualization, and phosphors may serve as tracers under uv light.

BIBLIOGRAPHY

W. W. Yau, J. J. Kirkland, and D. D. Bly, *Modern Size-Exclusion Liquid Chromatography,* John Wiley & Sons, Inc., New York, 1974.

L. R. Snyder and J. J. Kirkland, *Introduction to Modern Liquid Chromatography,* 2nd ed., John Wiley & Sons, Inc., New York, 1979.

F. C. Smith and R. C. Chang, *The Practice of Ion Chromatography,* John Wiley & Sons, Inc., New York, 1982.

R. P. W. Scott, ed., *Small Bore Liquid Chromatography Columns,* John Wiley & Sons, Inc., New York, 1984.

JULIAN F. JOHNSON
The University of Connecticut at Storrs

CHROMATOGRAPHY, AFFINITY

In recent years, one of the most important developments in enzymology has been the introduction of a rapid and highly efficient means of protein purification utilizing the highly selective ability of enzymes (qv) to recognize certain biological

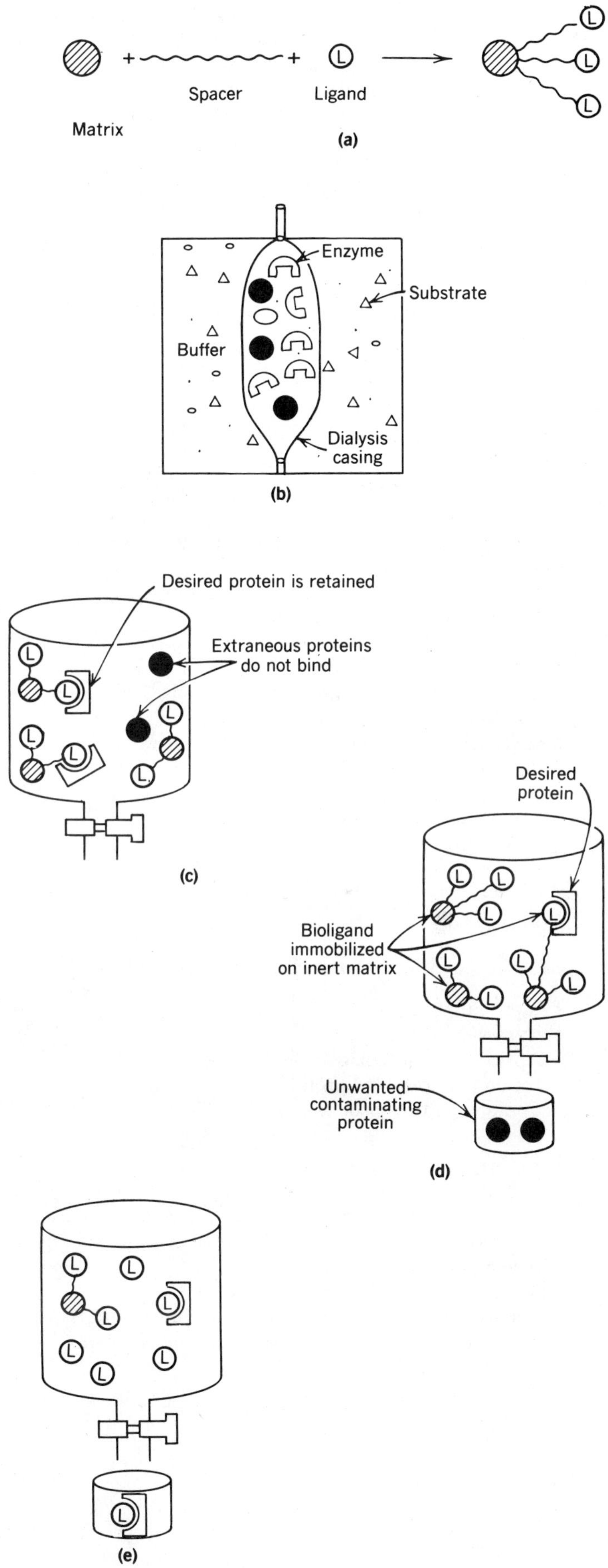

Figure 1. The steps of affinity chromatography. (**a**) A bioligand is immobolized; (**b**) a crude extract is prepared and freed from an endogenous substrate; (**c**) the substrate-free extract is applied to the chromatographic packing of the immobilized bioligand from step (**a**); (**d**) unwanted protein is removed by washing; and (**e**) the desired protein is eluted, possibly with a soluble bioligand.

compounds or their analogues. This technique, termed affinity chromatography, involves the immobilization of an appropriate ligand in such a way that the enzyme is still capable of recognizing and binding to the immobilized form of the ligand, whereas contaminating proteins (qv) have no such recognition. Virtually hundreds of proteins have been purified in this way, using a wide variety of bioligands immobilized in a matrix; enzyme inhibitors, coenzymes, antibodies, and even other enzymes may be useful bioligands in the purification of a particular protein see also ENZYMES, IMMOBILIZED. A classification of the types of affinity chromatography is given below:

Bioselective adsorption where affinity is based on biologically relevant binding:
- Group-specific ligands
  - Lectins
  - Nucleotide cofactors (NAD, AMP)
- Specific ligands
  - Rarely employed cofactors (eg, vitamin B_{12})
  - Receptor proteins
  - Immunosorbents

Chemiselective adsorption where affinity is based on chemically defined interactions:
- Without modification
  - Hydrophobic chromatography
  - Ion-exchange chromatography
- Transient protein modification
  - Covalent chromatography (active thiols, Hg^{2+}, etc)
  - Borate complexes
- Immobilized protein modifiers (IMPs)
  - Reducing agents
  - Oxidizing agents
  - Reactive-site modifiers (lysine, cysteine)

The basic application of affinity chromatography involves several steps, as illustrated in Figure 1.

Applications

Applications of affinity chromatography include:

- Protein purification
  - Enzymes
  - Antibodies and antigens
  - Binding or receptor proteins
  - Complementary proteins
  - Repressor proteins
- Separative procedures
  - Cells and viruses
  - Denatured and chemically modified proteins from native proteins
  - Nucleic acids and nucleotides
- Concentration of dilute protein solutions
- Storage of otherwise unstable proteins in immobilized form
- Investigation of kinetic sequences and mechanisms
- Medical applications
  - Extracorporeal adsorbents
  - Immunoassays

BIBLIOGRAPHY

W. H. Scouten, *Affinity Chromatography,* Wiley-Interscience, New York, 1981.

W. H. Scouten, eds. *Solid Phase Biochemistry,* Wiley-Interscience, New York, 1983.

J. Turkova, *Affinity Chromatography,* Elsevier, Amsterdam, The Netherlands, 1978.

K. Mosbach, ed. *Methods in Enzymology,* Vols. 34 and XX, Academic Press, New York, 1974 and in press.

WILLIAM H. SCOUTEN
Baylor University

CHYMOTRYPSIN.
See ENZYMES.

CIRCULAR DICHROISM.
See OPTICAL PROPERTIES.

CLASSIFICATION OF POLYMERIZATION REACTIONS

Various methods are used to classify polymerization reactions. Classification can be made by stoichiometry, mechanism of polymerization, nature of propagating species, method of initiation, medium of reaction, structure of product, and nature of reactants.

BIBLIOGRAPHY

U. W. Suter, *Basic Classification and Definitions of Polymerization Reactions,* IUPAC Committee on Macromolecular Nomenclature, IUPAC, Oxford, UK, 1989, unpublished document.

CLAYS.
See FILLERS.

COACERVATION.
See COLLOIDS.

COATED FABRICS.
See FABRICS, COATED.

COATING METHODS

Polymeric film-forming materials are applied as coatings for functional, protective purposes and for decoration. Common coating methods and process data are given in Table 1.
Most methods can be used for a wide range of base materials and coating compositions. The selection of a particular method depends on product requirements, availability of equipment, and manufacturer preference. New technologies, especially in electronics, have required the development of vacuum-coating methods (see also COATINGS; COATINGS, ELECTRODEPOSITION; DECORATING; FILMS; METALLIZING).

Table 1. Coating Methods

Coating method	Base material[a]	Coating composition[b]	Usual coating speed, m/min	Viscosity range, mPa·s (=cP)	Wet-coating thickness range, μm
air knife	b,d	R,T,X	15–600	1–500	25–60
brush	b,c,e,f,g	R,S,X,Z	30–120	100–2,000	50–200
calender	a,b,d,e	U,V,W	5–90		100–500
cast-coating	a,b,d	Q,R,S,T,V,Y	3–60	1,000–5,000	50–500
curtain	a,b,c,d,e,f	R,S,V,X,Z	20–400	100–20,000	25–250
dip	a,b,d,e,f,g	R,S,V,X,Y,Z	15–200	100–1,000	25–250
extrusion	a,b,d,e	T,U,V,W	20–900	30,000–50,000	12–50
blade	a,b	R,S,T,V,X,Y,Z	300–600	5,000–10,000	12–25
floating knife	a,b,d	R,S,T,V,X,Y,Z	3–30	500–5,000	50–250
gravure	a,b,d,e	R,S,T,U,V,Y,Z	2–450	100–1,000	12–50
kiss roll	a,b,c,d,e,f	R,S,V,X,Z	30–300	100–2,000	25–125
knife-over-blanket	a,b,d	R,S,T,V,X,Y,Z	3–30	500–5,000	50–250
knife-over-roll	a,b,c,d,e	R,S,T,U,V,X,Y,Z	3–60	1,000–10,000	50–500
offset gravure	b,d	R,S,T,Z	30–600	50–500	12–25
reverse roll	a,b,c,d,e,f	R,S,T,U,V,X,Y,Z	30–300	50–20,000	50–500
reverse-smoothing roll	a,b	R,T,X	15–300	1,000–5,000	25–75
rod	b,d	R,S,T,V,X,Y,Z	3–150	50–500	25–125
sprays					
airless spray	a,b,c,d,e,f,g	S,T,V,X,Y,Z	3–90	[c]	2–250
air spray	a,b,c,d,e,f,g	S,T,V,X,Y,Z	3–90	[c]	2–250
electrostatic	a,b,c,d,e,f,g	S,T,V,X,Y,Z	3–90	[c]	2–250
squeeze roll	a,b,c,d,e,f	R,S,T,U,V,X,Y	30–700	100–5,000	25–125
in situ polymerization	a,b,c,d,e,f,g	Y,Z	undetermined	liquid or vapor	6–2.5
powdered resin	a,b,c,e,f,g	Q	3–60		25–250[d]
electrostatic spray		Q			20–75[d]
fluidized bed	e,g	Q			200–2,000[d]

[a] Key: a, woven and nonwoven textiles; b, paper and paperboard; c, plywood and pressed fiberboards; d, plastic films and cellophane; e, metal sheet, strip, or foil; f, irregular flat items; g, irregularly shaped.

[b] Key: Q, powdered resin compositions; R, aqueous latices, emulsions, dispersions; S, organic lacquer solutions and dispersions; T, plastisol and organosol formulations; U, natural and synthetic rubber compositions; V, hot-melt compositions; W, thermoplastic masses; X, oleoresinous compositions; Y, reacting formulations, eg, epoxy and polyester; Z, plastic monomers.

[c] 80–50 s in no. 4 Ford cup.

[d] Dry thickness.

BIBLIOGRAPHY

H. L. Weiss, *Coating and Laminating Machines,* Converting Technology Co., Milwaukee, Wis., 1977, p. 158–248.

R. J. Jacobs, *Pap. Film Foil Converter* **37**(2), 47 (1963); **37**(3), 50 (1963); **37**(4), 62 (1963); **37**(5), 60 (1963); **37**(6), 60 (1963); **37**(7), 53 (1963).

R. TED SCHARENBERG
Consultant

Spray Coating

A plastic coating may be applied to articles, ie, workpieces, by spraying. This method of application is especially attractive when the articles have been previously assembled and have irregularly shaped and curved surfaces.

The material applied is frequently in the form of a paint, which is a combination of resin, solvent, diluent, additives, and pigment. The material can also be a hot thermoplastic or a polymer dissolved in a solvent.

Many different types of spray equipment are available, and they can be classified in various ways. Typical classifications are the method of atomization (airless, air, rotary, or electrostatic) and the method of deposition assist (electrostatic or nonelectrostatic). Different methods can be used in combinations and most can be used for simple one-applicator manual systems or highly complex computer-controlled automatic systems with hundreds of applicators. In an automatic installation, the applicators can be mounted on fixed stands, reciprocating or rotating machines, or even robots.

BIBLIOGRAPHY

General References

R. Tholome and G. Sorcinelli, *Ind. Finish.* (Nov. 1977).

F. A. Robinson, Jr., G. Pickering, and J. Scharfenberger, *Paint Con '84,* Hitchcock Publishing Co., Wheaton, Ill., 1984.

J. M. Lipscomb, *Surface Coating '83,* Chemical Coaters Association, Wheaton, Ill., 1983.

K. J. COELING
TIMOTHY J. BUBLICK
The DeVilbiss Company

Powder Coating

A variety of methods employing powdered plastics and resins are used commercially to apply coatings to various articles. These methods include fluidized bed, electrostatic spray, electrostatic fluidized bed, plasma spray, and hot flocking, as well as combinations and variations of these methods.

In the fluidized-bed-coating process, a finely divided plastic is placed in an open container, the bottom of which is a porous plate. Air is passed through this plate, causing the powder to be suspended in a fluidized state. The object to be coated is preheated above the melting point of the plastic powder and dipped into the fluidized bed. The powder melts and forms a coating.

In the electrostatic spray process, the coating powder is withdrawn from a reservoir in an air stream and electrostatically charged in the high voltage corona field of a spray gun. The charged particles are attracted to the grounded metal object to be coated and adhere to it by electrostatic attraction. The coated substrate is then placed in an oven and the coating is fused to form a continuous film. If the powder is sprayed on a preheated article, the powder melts and fuses directly on the hot surface; further heating to fuse or cure the coating may be required, depending on the type of coating powder and the substrate.

In the plasma-coating method, a high temperature plasma is established in an inert gas such as nitrogen, and the coating powder is introduced at the periphery of the plasma. The particles melt and are propelled at high velocity to the substrate, where they form a film.

In hot-flocking techniques, powders are usually dispersed in air and sprayed or blown onto the preheated substrate, where they melt and form a coating. In a variation of this process, small parts are preheated and dropped into a bed of powder kept in a mobile state by vibration. In this method, the parts are completely coated with an unfused layer of powder on the surface.

All powder-coating methods are generically referred to as fusion-coating processes, since at some stage in the process the material must be melted and fused to form a continuous coating.

Powder-coating processes have many advantages over conventional painting. Most significant is that they are ecologically acceptable and meet all federal and state requirements for minimum pollution. Since there are no solvents involved, there is no air pollution, and no solvent-recovery system is required.

Powder-coating Materials

Powder-coating materials are generally classified into functional and decorative grades. Functional coatings are thicker, ie, ca 250–1000 μm, and frequently applied by fluidized-bed coating; decorative coatings are thinner, ie, ca 20–100 μm, and are usually applied by electrostatic spray. A further division is made between thermoplastic and thermosetting coating powders. Thermoplastic powders are generally applied by fluidized-bed coating and account for only approximately 10% of the powder-coatings market. PVC-based products comprise the largest segment of this market, about 80%. The PVC powders offer good resistance to corrosion, abrasion, and weathering, as well as good electrical properties.

Coating powders based on nylon-11 and nylon-12 have good chemical and solvent resistance and outstanding abrasion and wear resistance. They are used as antifriction coatings on spline shafts, shifter forks, and similar articles. They are also used to coat high quality office furniture, hospital beds and other medical articles that must be frequently cleaned or sterilized, and a host of other parts where the toughness, scratch resistance, and good appearance of nylon coatings are desirable.

Functional thermosetting coating powders, generally based on epoxy resins, have a long history of applications, eg, as insulating coatings for bus bars, motor armatures and stators, and circuit boards, wire, and similar articles.

Decorative powders based on epoxy resins are made in a wide variety of grades varying in cure speed, impact resistance, flexibility, hardness, scratch and mar resistance, gloss, and special textures and effects. Epoxy powder coatings are attractive and possess a good balance of physical properties, corrosion resistance, and electrical properties; in addition, they are available in more formulations than any other decorative powder coating. However, epoxy coatings chalk rapidly on exterior exposure and tend to discolor rapidly even indoors, ie, yellow after several years of exposure to sunlight.

In the United States, the largest share of the exterior decorative market belongs to urethane polyester powders. They are based on hydroxy functional polyester resins, which are cross-linked with blocked isocyanates. Powders based on this chemistry are available in a wide variety of glosses and textures, but not to the extent of epoxy-based powders. Urethane polyesters have excellent weathering characteristics and good chemical resistance and physical properties. They are used for coating lawn furniture, pole-line transformers, garden equipment, and many other outdoor articles.

Polyester powders are also available based on acid functional polyester resins which have reacted with resins or curing agents with glycidyl functionality. When the reaction is with epoxy resins, the resulting powder coatings are referred to as hybrids and have properties similar to those associated with epoxy powder coatings. Triglycidyl isocyanurate (TGIC) is widely used as a curing agent also, yielding coatings with excellent exterior durability and mechanical properties.

Economic Aspects

After a slow start and overly optimistic growth projections, powder coating, especially the electrostatic spray process, has been fully accepted in industrial finishing. In early 1983, it was estimated that about 800 electrostatic spray installations were operating in the United States. The rate of powder growth is estimated at 10–15% over the next several years, as powder coatings gain a greater market share.

Health and Safety Factors

Any finely divided organic material can form explosive mixtures with air at certain concentrations. The most significant hazard in the manufacture and application of coating powders is the potential for a dust explosion.

With regard to the physiological hazards to personnel working with coating powders, particular attention must be paid to exposure by inhalation and skin contact. Skin and eye irritation as well as oral toxicity must also be considered. Coating powders should be treated as nuisance dusts and the concentration in air kept below the TLV of 10 mg/m^3, primarily through environmental controls.

BIBLIOGRAPHY

S. T. Harris, *The Technology of Powder Coatings,* Portcullis Press, London, UK, 1976.

D. S. Richart in M. Grayson, ed., *Kirk-Othmer Encyclopedia of Chemical Technology,* 3rd ed., Vol. 19, Wiley-Interscience, New York, 1982, pp. 1–27.

E. Miller, ed., *Users Guide to Powder Coating,* Society of Manufacturing Engineers, Dearborn, Mich., 1985.

D. S. Richart
Morton Powder Coating

Vacuum Coating

The last 30 years have witnessed an evolution in the development of polymer-coating technology in advanced applications such as microelectronics. Polymeric coatings for these high technology applications are employed to maintain certain desirable surface properties that differ from those of the bulk material or for protection of the underlying material. Good polymeric coatings can be deposited using vacuum techniques, which can be divided into physical and chemical vapor-deposition methods.

Physical vapor-deposition processes include evaporation and sputtering; chemical vapor-deposition techniques involve polymerization through pyrolysis and glow discharge (plasma). The relationship between these deposition techniques is shown in Figure 1. The starting materials may be either solid or gas, and the type of energy applied either plasma or thermal. Sputtering, for example, employs a solid target and a plasma energy source.

Electron-beam Polymerization (EB)

Electron-beam polymerization is much slower than plasma polymerization. Monomeric materials that have been irradiated to produce polymer films include silicones, butadiene, styrene, methyl methacrylate, divinyl-benzene, and epoxy resins. However, the lack of selectivity of this energy source results in contamination. Film properties can be tailored somewhat by varying the acceleration voltage.

Ion-beam Polymerization. Ion beams can be used in a manner similar to electron beams to polymerize gaseous organic monomers.

Radiation

Ulraviolet Irradiation, Photopolymerization. Over 40 years ago, it was observed that a polymer film formed on the walls of a flask containing methyl methacrylate irradiated with uv light. The mechanism postulated for this transformation involved excitation of the monomer in the gas phase, resulting in vinyl-addition polymerization.

Contact with uv light initiates film formation at rates up to several hundred micrometers per minute. The deposited polymers are generally flexible and contain little cross-linking. However, in some instances extensive side reactions have been observed to yield low molecular weight materials and other by-products.

The polymerization mechanism involves two steps: the gaseous monomers are first attached to the vessel surface by chemisorption or physical adsorption, and then activated with uv light. Substrates can be coated by uv polymerization without the necessity of a mask by using quartz optics to direct the radiation to provide a suitable pattern of uv light at the surface (see also PHOTOPOLYMERIZATION).

Gamma Radiation. Gamma radiation can be used for formation of thin polymeric films via a mechanism of free-radical polymerization similar to that occurring in uv irradiation and electron-beam polymerization.

BIBLIOGRAPHY

R. F. Bunshah and co-workers, *Deposition Technologies For Films and Coatings,* Noyes Publications, Park Ridge, N.J., 1982, Chapt. 1.

S. M. Lee in M. Grayson, ed., *Kirk-Othmer Encyclopedia of Chemical Technology,* 3rd ed., Vol. 24, Wiley-Interscience, New York, 1984, pp. 744–771.

H. K. Yasuda in M. Shen and A. T. Bell, eds., *Plasma Polymerization,* The American Chemical Society, Washington, D.C., 1979, Chapt. 2.

S. M. Lee in M. Grayson, ed., *Kirk-Othmer Encyclopedia of Chemical Technology,* 3rd ed., Vol. 10, Wiley-Interscience, New York, 1980, pp. 247–283.

Stuart M. Lee
SRI International

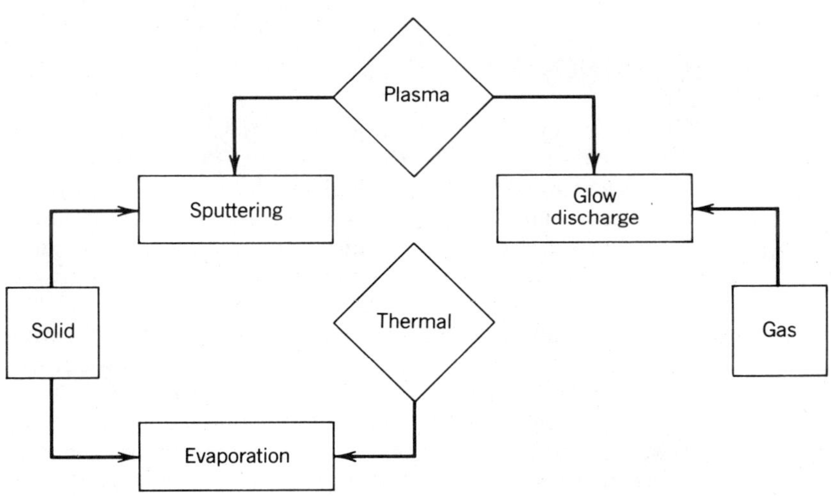

Figure 1. Relationship of vacuum deposition processes. □ starting materials; ◇ type of energy applied.

COATINGS, ELECTRODEPOSITION

Electrodeposition offers high corrosion protection, low cost, and compliance with environmental regulations. It is used for the coating of articles of various sizes including steel building trusses, car bodies, furniture, appliances, toys, and nuts and bolts. The current success of electrodeposition is due to the use of water-dispersible, synthetic, electrodepositable macroions as film formers. The process is most frequently called electrocoating, although various names, such as electropainting, elpo, e/coat, etc, are in use. Electrodeposition combines many advantages of other painting methods with new and desirable features:

1. Formation of films in highly recessed areas such as cavities, box sections, creases, and flanges results in excellent corrosion protection.
2. Use of water as a carrier virtually eliminates fire hazard and environmental pollution and reduces the cost of control equipment.
3. Low viscosity of the paint bath (approximately equal to that of water) facilitates agitation and pumping and allows fast entry and drainage of workpieces.
4. Freshly deposited coats are composed of ca 97% nonvolatile substances and therefore allow immediate gentle handling; there is no tendency to sag or wash off during cure.
5. A second coat, usually a color coat, of waterborne or solvent-borne spray paint can be applied directly over the uncured electrocoat.
6. Approximately 95% of the applied paint is utilized because the liquid paint, which adheres or fills cavities of freshly coated pieces, is rinsed back into the coating tank.
7. Overall savings, accounting for materials, labor, capital investment, energy, etc, are reported to be 20–50% when compared with spray, electrostatic-spray, or dip-coat painting (see also COATING METHODS).

Process Fundamentals

In the electrocoating process, electrically charged waterborne, film-forming organic macroions are attracted by an electrode of opposite polarity, where they are deposited (Fig. 1).

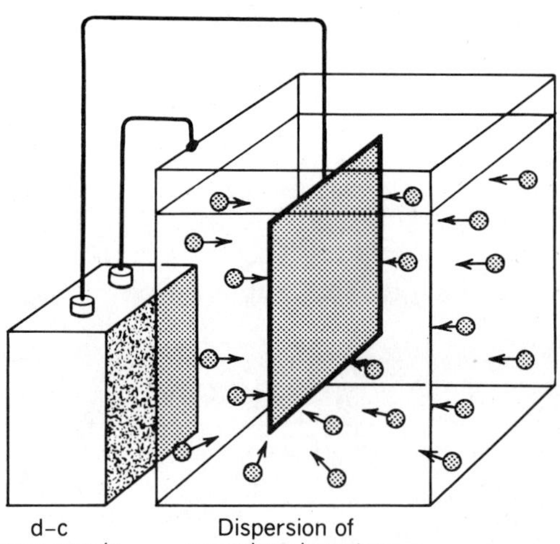

Figure 1. Diagram of the electrocoating operation; a typical time cycle is one minute.

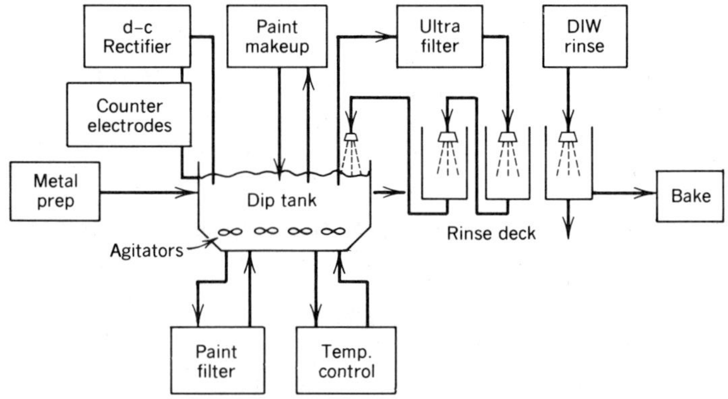

Figure 2. Components of an electrocoating installation.

Well-known paint resins with excellent properties including epoxies, acrylics, polyesters, and many others are used. Ionizable groups in resins are, for example, RCOOH or R_3N. Film-forming properties are usually associated with the chemical nature of the resin and a molecular weight range of 2,000–20,000.

Throwing Power. Throwing power is the ability to form electrodeposited films of such uniform thickness that even the most recessed areas of a workpiece are covered.

Technique and Equipment

Most electrocoating installations are composed of several distinct components as indicated in Figure 2. Some installations use both spray and dip rinses.

BIBLIOGRAPHY

R. L. Yeates, *Electropainting,* Robert Draper Ltd., Teddington, UK, 1970.

W. Machu, *Handbook of Electropainting Technology,* Electrochemical Publication, Ltd., Ayr, UK, 1978.

R. H. Chandler, *Advances in Electrophoretic Painting,* Bi- or Triannual Abstracts since 1966, R. H. Chandler, Braintree, Essex, UK.

T. Robison, Products Finishing Co., Cincinnati, Ohio, Organizer of *Electrocoat Conferences* (even years since 1982).

GEORGE E. F. BREWER
Coating Consultants

COATINGS, PRINCIPLES

This article emphasizes basic principles of coatings technology. The discussion is based primarily on practices common in the U.S. coatings industry. Although the basic principles are not affected, variations in customs, raw material costs and availability, toxic hazard regulations, and environmental controls can lead to substantial variations from country to country.

Coatings are ubiquitous in an industrialized society. The U.S. consumption in 1987 was about 3.8×10^6 m^3 with a value of about \$10 billion. Coatings are used for decorative, protective, and functional treatments of many kinds of surfaces. Coatings such as interior flat wall paints are used for their decorative value but, even in this case, provide a functional value, namely, diffuse reflectance of light to provide more uniform illumination. Some coatings, such as those on undersea pipelines, are for protective purposes. Others, such as exterior automotive coatings, fulfill both decorative and protective functions. Still others provide friction control on boat decks or car seats. Some coatings control the fouling of ship bottoms, others protect food and beverages in cans. Silicon chips, printed circuit panels, coatings on waveguide fibers for signal transmission, and magnetic coatings on videotapes and computer disks are among many so-called high-tech applications for coatings.

Each year, tens of thousands of coating types are manufactured. In general, they are composed of one or more resins or polymers, a mixture of solvents (except in powder coatings), commonly one or more pigments, and frequently several additives. Coatings can be classified into thermoplastic and thermosetting coatings. Thermoplastic coatings contain at least one polymer with sufficiently high molecular weight to provide the required mechanical strength properties without further polymerization. Thermosetting coatings contain lower molecular weight polymers and are further polymerized after application in order to achieve the appropriate properties. The term thermosetting is used not only when the coating is cured or baked in an oven after application, but also when it is dried or cured at ambient temperature.

Film Formation

Most coatings are manufactured and applied as liquids and are converted to "solid" films after application to the substrate. In the case of powder coatings, the solid powder is converted after application to a liquid, which in turn forms a solid film. The polymer systems used in coatings are amorphous materials; therefore, the term solid does not have an absolute meaning, especially in thermoplastic systems such as lacquers and most plastisols. A useful definition of a solid film is that it is a material that does not undergo detectable flow under the conditions and during the time interval of observation. Thus, a film can be defined as a solid under a set of conditions by stating the minimum viscosity at which flow is not observable in a specified time interval. A film is called dry or solid when it develops a high enough viscosity to pass a certain test. For example, it is reported that a film is "dry to touch" if its viscosity is higher than 10^3 Pa·s (10^4 P).

For practical coatings, it is not sufficient just to form a film; the film must also have a minimum level of strength depending on the end use of the product. Film strength depends on many variables, but one critical factor is molecular weight. The required molecular weight varies with the chemical composition of the polymer and with the mechanical properties required in a particular application.

Solvent Evaporation from Solutions of Thermoplastic Polymers. A solution of a copolymer of vinyl chloride, vinyl acetate, and a hydroxy-functional vinyl monomer can be used as an example. Such a copolymer, with a number-average molecular weight $\bar{M}_n = 23{,}000$ and a $T_g = 79°C$, gives coatings with good mechanical properties without cross-linking. Considering a simple system with only the resin and 2-butanone, or methyl ethyl ketone (MEK), as the sole solvent, the polymer concentration would be about 19 wt % solids (approximately 12 vol %) in order to have a viscosity of about 0.1 Pa·s (= P) for spray application. Because of its relatively high vapor pressure under application conditions, MEK evaporates rapidly. A substantial fraction of the solvent evaporates in the time interval between the coating leaving the orifice of the spray gun and its arrival on the surface being coated. As the solvent evaporates, the viscosity increases and the coating reaches the dry-to-touch state very rapidly after application and does not block under the conditions mentioned earlier. However, if the film is formed at 25°C, the dry film contains several percent retained solvent.

In the first stages of solvent evaporation from such a film, the rate of evaporation depends on the vapor pressure at the temperatures encountered during the evaporation, the ratio of surface area to volume of the film, and the rate of airflow over the surface, and is essentially independent of the presence of polymer. The rate of solvent diffusion through the film depends not only on the temperature and the T_g of the film but also on the solvent structure and solvent–polymer interactions. Film thickness is also an important factor in solvent loss and film formation.

Thermoplastic polymer-based coatings have a low solids content because the relatively high molecular weight requires large amounts of solvent to reduce the viscosity to that required for application. Air pollution regulations limiting the emission of volatile organic compounds (VOC) and the increasing cost of solvents has led to the increasing replacement of such coatings with types that require less solvent for application.

Film Formation from Solutions of Thermosetting Polymers. Substantially less solvent is required in formulating a coating from a low molecular weight resin that can be further polymerized to a higher molecular weight after application to the substrate and evaporation of the solvent. The average functionality must be at least over two per molecule, in order to ensure that the molecular weight of the final cured film is high enough for good properties.

The mechanical properties of the final film depend strongly on the T_g of the cross-linked polymer and the degree of cross-linking, that is, the cross-link density. The latter is affected by the average functionality and equivalent weight of the system and by the completeness of the reaction.

In designing resins, the ultimate T_g is controlled by the flexibility of the segments between the cross-links and the modulus above T_g is controlled by cross-link density. Another critical variable in designing thermosetting systems is the molecular weight and the functionality of the reactants. The lower the molecular weight, the higher the solids can be at application viscosity. The lower limits of molecular weight and average functionality are system-dependent.

A problem in thermosetting systems is the relationship between storage stability of the coating before application and the time and temperature required to cure the film after application. It is generally desirable to be able to store a coating for many months or even years without significant increase in viscosity that would result from cross-linking during storage. On the other hand, after application, the cross-linking reaction should proceed in a short time at as low a temperature as possible. Because reaction rates depend on the concentration of functional groups, storage life can be increased by using more dilute systems; adding more solvent increases storage life. When the solvent evaporates after application, the reaction rate increases initially. However, it is advantageous to reduce solvent concentration as much as possible, that is, to use coatings with a solids content as high as possible. As industry shifts to higher and higher solids, the problem of storage stability will become increasingly severe.

Film Formation by Coalescence of Polymer Particles. Latex paints have low solvent emissions as well as many other advantages. A latex is a dispersion of high molecular weight polymer in water. The dispersion is stabilized by charge repulsion and entropic repulsion (also called steric repulsion and osmotic repulsion). Because the latex polymer is not in solution, the rate of loss of water by evaporation is almost independent of the formula until near the end of the evaporation. When a dry film is prepared from a latex, the forces that stabilize the dispersion of latex particles must be overcome and the particles must coalesce into a continuous film.

The rate of coalescence is controlled by the free volume availability, which, in turn, depends mainly on $(T - T_g)$. The viscosity of the coalesced film is also dependent on the free volume.

Powder coatings form films by coalescence. Because the powder must not fuse or sinter during storage, the free volume at storage temperature must be sufficiently low to avoid coalescence at this stage. In general, the T_g of the powder formula would be set so that it is higher than the storage temperature. However, rapid coalescence after application requires that the $(T - T_g)$ under the baking conditions be as large as possible. Obviously, a compromise is necessary.

Resins

Alkyds. Although no longer the principal class of resins used in coatings in the United States, alkyds are still very important, especially in relatively low cost cross-linking coatings for air-dry or force-dry applications. Their principal shortcomings are embrittlement and discoloration with age and low saponification resistance. Their main advantages are low cost and relatively foolproof application characteristics, due primarily to low surface tension.

Polyesters. The term polyester is used in the coatings field almost entirely for low molecular weight hydroxy- (or occasionally carboxylic acid-) terminated, oil-free polyesters. Polyesters have been one important class of replacements for alkyd components in melamine–formaldehyde cross-linked baking enamels. Hydroxy-terminated polyesters are also used with polyfunctional isocyanates in making air-dry and force-dry coatings as well as with blocked isocyanates in coatings for higher baking temperatures and powder coatings. When direct adhesion to metal is required, polyesters are preferred over acrylic resins. On the other hand, when highest exterior durability is needed, especially over primers, acrylic resins are preferred to polyesters.

Amino Resins. Melamine–formaldehyde (MF) resins are the most widely used cross-linking agents for baking enamels. Unlike the MF resins used in plastics, textile, and paper treatment, the MF resins used in coatings are ether derivatives. The two types of resin are those made with a low ratio of formaldehyde to melamine (1,3,5-triazine-2,4,6-triamine), which contain a substantial fraction of the amine groups on the melamine substituted with only one alkoxymethyl group, and those made with excess formaldehyde, which contain a substantial fraction of the amine groups substituted with two alkoxymethyl groups.

Most amino resins used in coatings are melamine–formaldehyde resins, but other amino resins are also employed. Urea–formaldehyde resins are used in coatings for wood furniture where curing temperatures are very low and high water resistance and exterior durability are not needed. Ethers of formaldehyde derivatives of 6-phenyl-1,3,5-triazine-2,4-diamine (benzoguanamine resins) are widely used in coatings for home laundry machines because high detergent resistance is required. Ethers of formaldehyde derivatives of glycoluril–formaldehyde resins are finding increasing use, especially for applications requiring outstanding extensibility during fabrication. Glycoluril is tetrahydroimidazo (1,5-*d*) imidazole-2,5-(1*H*,3*H*)-dione. Amides react with formaldehyde to give methylol derivatives that, in turn, react with alcohols to give ethers. Thermosetting acrylics with such reactive groups derived from acrylamide were among the first thermosetting acrylic resins.

Acrylic Resins. Acrylic resins are the largest volume class of coatings resins. The term acrylic is used in a general sense to mean resins where a significant fraction of the comonomers are acrylic or methacrylic esters. Other monomers such as styrene (ethenylbenzene), vinyl acetate, and others may be included in the copolymer. High molecular weight thermoplastic acrylics, which were widely used on new automobiles, are currently restricted to repair and after-market lacquers because of the need to control solvent emissions in automobile manufacturing. Thermosetting acrylic resins continue to expand with increasing emphasis on high solids and waterborne types. Nonaqueous dispersion (NAD) acrylics, which were an intermediate step in reducing solvent emissions in automotive top coats, have been almost completely replaced by high solids thermosetting solution acrylics or "water-soluble" thermosetting acrylics. In architectural coatings, acrylic latices and styrene or vinyl acetate acrylic copolymer latices have replaced drying oil and alkyd systems almost completely except for some continuing use in gloss enamels. The consumption of acrylic latices in industrial coatings applications is small but growing. The main advantage of acrylic coatings, in most cases, is their high degree of resistance to thermal and photooxidation and to hydrolysis, giving coatings with superior color retention, resistance to embrittlement, and exterior durability.

Hydroxy-functional thermosetting acrylics are widely used in baking enamels for automobile and appliance top coats, exterior can coatings, and coil coating. Research efforts are being made to increase the solid content of such systems while maintaining their desirable properties.

Epoxy Resins. The most widely used epoxy resins in coatings are bisphenol A (BPA) epoxy resins. One use is in making drying oil fatty acid esters, so-called epoxy esters. These synthetic drying oils provide coatings with better adhesion and substantially better saponification resistance than alkyds, because the backbone linkages are stable ether groups instead of ester groups; exterior durability, however, is poor. Their principal use is in primers for steel, especially baking primers. Another important application of epoxy resins is as a component in two-package primers. The other package contains a polyfunctional amine. Epoxy resins are also used in powder coatings.

Urethane Systems. Diisocyanates such as 2,4-toluene diisocyanate (TDI) can be used to prepare polyurethanes that are analogous to alkyds by reaction with monoglycerides obtained by the transesterification of drying oils and a polyol such as glycerol. Like alkyds, the urethane oils (also called urethane alkyds and uralkyds) form dry films by autoxidation. Polyurethane coatings exhibit greater resistance to hydrolysis and abrasion than alkyds. Urethane oils have no free isocyanate groups and are considered safe for use by the public. Many coating products for the do-it-yourself market labeled as varnish are urethane oils. Their properties are superior to those of varnishes as well as those of alkyds.

Volatile Components

In most coatings, solvents are used to adjust the viscosity to the level required for the application procedure. Most methods require viscosities of 0.05–1.0 Pa·s (0.5–10 P). Many factors must be considered in the selection of the solvent or, more commonly, solvent mixtures. Except in waterborne systems, solvents are usually chosen that give true resin solutions. Although solubility parameters have been widely recommended for the selection of the appropriate resin solvent, it is clear that the concept of solubility parameters for the prediction of polymer solubility is a gross oversimplification. In time, it is to be expected that an adequate solubility theory will be developed, but for the present, the old principle that "like dissolves like" is the most useful selection criterion.

An important characteristic of solvents is their rate of evaporation. Rates of solvent loss are controlled by the vapor pressure (in turn, affected by the temperature), partial pressure of the solvent over the surface (in turn, affected by the airflow rate over the surface), and the ratio of surface area to volume. Tables of relative evaporation rates, in which *n*-butyl acetate is the standard, are widely used in selecting solvents.

The higher the ratio of surface area to volume, the more rapid is the evaporation of the solvent. When coatings are applied by spray gun, the atomized particles have a very high ratio of surface area to volume and hence solvent evaporation is much more rapid during the time when the atomized particles are traveling between the spray gun and the surface being coated than from the film of the coating. Except in the case of high solids coatings, this rapid evaporation of solvent during the atomized stage permits the formulation and spray application of coatings that level well, yet do not sag on vertical surfaces.

Toxic hazards, environmental considerations, flammability, odor, and viscosity also affect solvent selection. In high solids coatings, the effect of solvent choice on viscosity can be critical.

Pigments

Pigments in coatings provide opacity and color. Pigment content governs the gloss of the final film and can have important effects on its mechanical properties. Some pigments inhibit corrosion. Pigments affect the viscosity and hence the application properties of the coating. An important variable determining the properties of pigments is their particle size and particle-size distribution. Pigment manufacturing processes are designed to afford the particle size and particle-size distribution that provide the best compromise of properties for that pigment. In the process of drying the pigment, the particles generally aggregate. The coatings manufacturer must disperse these dry pigment aggregates in such a way as to achieve a stable dispersion where most, if not all, of the pigment particles are separated into the individual particles designed into the product by the pigment manufacturer.

The pigments used in coatings are white pigments, color pigments, and inert pigments. The ideal white pigment, when dispersed in the coating, would absorb no visible light and would result in efficient scattering of light entering the film. Color pigments selectively absorb some wavelengths of visible light more strongly than others. Inert pigments do not exhibit significant absorption or scattering of light when used in coatings. In most cases, inert pigments are used to occupy volume in the coating composition.

Pigment Dispersion. The dispersion of pigments involves wetting, separation, and stabilization. Wetting, that is, displacement of air from the surface of the pigment, requires that the surface tension of the dispersion medium be lower than the surface energy of the pigment surface. Except when the vehicle is water, wetting is rarely a problem. The pigment aggregates are separated into individual particles by a variety of dispersion equipment, which transmits shear stress of sufficient magnitude to break up the aggregates. Pigment stabilization is usually the most critical step. If the dispersion is not fully stabilized, the separated pigment particles flocculate (reagglomerate). Pigment dispersions are stabilized by charge repulsion and entropic repulsion (some authors prefer to use the terms steric repulsion and osmotic repulsion).

Pigment Volume Relationships. Most effects of pigments on film properties are related to the volume relationships rather than the weight relationships, that is, the properties vary with the volume of pigment in the dry film. Pigment volume concentration (PVC) is defined as the volume of pigment in the dry film divided by the total volume of the dry film, commonly expressed as a percentage. (Some authors use the term pigment volume content for pigment volume concentration.)

As the volume of pigment in a series of formulations is increased, properties change; at some PVC there is a fairly drastic change in a series of properties as a function of PVC. This PVC is defined as the critical pigment volume concentration (CPVC) for that system. The CPVC is the maximum PVC that can be present in a dry film, leaving sufficient solvent-free resin to adsorb on all pigment surfaces and fill all interstices between pigment particles. In other words, when the PVC is above the CPVC, there are voids in the film. Gloss decreases as PVC increases. Hiding increases as PVC increases until CPVC is reached, where the rate of increase of hiding increases more rapidly because of the presence of voids. In the unique case of rutile TiO_2, hiding passes through a maximum at about 22%. If the PVC of rutile pigmented coatings were increased above CPVC, the hiding would again start to increase rapidly. Tinting strength of white paints behaves like hiding, increasing rapidly above CPVC. Tensile strength of films increases with increasing PVC, passing through a maximum at CPVC. Stain resistance of films is poorer and removal of stains becomes more difficult with coatings with PVC above CPVC, compared to those with PVC below CPVC. Blistering is less likely to occur above CPVC. The intercoat adhesion to a primer is improved when the primer has a PVC above CPVC. The CPVC effects are due to the voids in dry films with a PVC greater than CPVC.

The CPVC depends on the pigments and pigment combinations. It is generally lower for smaller particle-size pigments, because these have a higher volume proportion of adsorbed resin than larger particle-size pigments. Particle-size distribution is very important. Because broad distribution systems pack more efficiently, they have higher CPVC values. Flocculation results in lower CPVC in a dry film as compared with the same pigment combination with a well-stabilized dispersion. Each coating application has a PVC/CPVC ratio that provides the best compromise of properties for the end use involved. By determining the appropriate PVC/CPVC ratio for some application, it can be used as a starting point in other formulations.

Color Matching. Most pigmented coatings must be color matched to a certain standard. In many cases, a significant part of the time involved in formulating a new paint is in establishing the color match and a significant part of manufacturing cost is the color-matching operation. Poor color matching is a common source of customer complaints. Problems and costs can be minimized if the customer and supplier cooperate to establish the specifications for the initial color match and for judging the acceptability of production batches.

Whenever possible, the customer should accept a color recommendation made by the coatings supplier. This effectively eliminates the time and cost involved in the initial color match and ensures that a pigment combination appropriate to the end use has been selected. If, however, the customer provides a standard for the initial color match, the coatings formulator needs the following information:

1. Possibility of a spectral (nonmetameric) match.
2. Light sources.
3. Gloss and texture.
4. Color properties required.
5. Film thickness and substrate.
6. Baking schedule.
7. Cost.
8. Tolerance for variations in closeness of match.

Flow

Viscosity of Resin Solutions. The viscosity of coatings must be adjusted to the application method to be used. It is usually between 0.05 and 1 Pa·s (0.5 and 10 P), at the shear rate involved in the application method. The viscosity of the coating is controlled by the viscosity of the resin solution, which is in turn controlled mainly by the free volume. The factors controlling free volume are temperature, resin structure, solvent structure, concentration, and solvent–resin interactions.

The temperature dependence of viscosity of resin solutions can be expressed by the WLF equation 1 where the reference temperature T_r is taken as the lowest temperature for which data are available. This relationship has been shown to hold

$$\ln \eta = \ln \eta_{T_r} - \frac{c_1(T - T_r)}{c_2 + (T - T_r)} \qquad (1)$$

for a wide variety of coatings resins and resin solutions over a wide range of concentrations.

The viscosity of high solids coatings in the range of application varies more rapidly with temperature than is the case with conventional lower solids coatings. Hence, the viscosity is reduced more by using hot-spray systems that permit a further increase in solids. The viscosity of high solids varies more with concentration than the viscosity of conventional coatings.

Viscosity of Systems with Dispersed Phases. A large proportion of coatings are pigmented and, therefore, have a dispersed phase. In latex paints, both the pigments and the principal polymer are in dispersed phases. The viscosity of a coating with dispersed phases is a function of the volume concentration of the dispersed phase and can be expressed mathematically by the Mooney equation. A convenient form is given in equation 2

$$\ln \eta = \ln \eta_e + \frac{K_E V_i}{1 - \dfrac{V_i}{\phi}} \qquad (2)$$

where η_e is the viscosity of the external phase, K_E is a shape constant (2.5 for spheres); V_i is the volume fraction of internal phase, and ϕ is the packing factor. The Mooney equation assumes rigid particles with no particle–particle interaction. It fits pigment dispersions and latices that exhibit Newtonian flow.

Flow during Application. Many methods of film application to a substrate give an uneven surface. For example, application by brush gives brush marks. These are not related to the bristles of the brush but rather to the fact that the film of wet paint is split between the substrate and the brush. Similarly, when paint is applied with a roller, the film is split with part of the paint remaining on the roller and part transferring to the substrate. Film splitting leads to a nonuniform surface. It is desirable for these nonuniformities to level out. Leveling occurs most rapidly if the viscosity of the coating on the substrate is low. In order to keep the viscosity low for a longer period, it is common to formulate brush-applied coatings with relatively slow evaporating solvents. However, low viscosity coatings applied to a vertical wall sag, that is, run down the wall unevenly. The viscosity requirements to promote leveling and minimize sagging are in conflict with each other. Solvents should be chosen that evaporate slowly enough to give a low viscosity long enough to permit reasonable leveling, but evaporate rapidly enough that the viscosity increases before serious sagging occurs.

Another type of flow defect that can occur during application is crawling. In some cases, after application of a smooth film, the liquid coating partially dewets and attempts to form balls on the substrate, leading to a highly irregular coating with very thin or even bare areas. Crawling results from applying a coating with a high surface tension over a substrate of lower surface tension. On a metal substrate like steel, it is usually the result of incomplete removal of oil from the substrate before coating. On plastic substrates, it can result from the failure to remove mold-release compounds from the surface of the plastic article before coating. In the case of low surface tension plastics such as polyethylene, it may be difficult to make a coating with a surface tension lower than that of the plastic. To avoid crawling, it may be necessary to oxidize the surface of the plastics to increase its surface tension.

Cratering is another common flow defect encountered in the application of coatings. Craters appearing on the surface are not due to the eruption of bubbles from the inside of the drying film, but are rather commonly caused by small contaminants of low surface tension being deposited on the surface of the wet, freshly applied film. Because liquids of low surface tension flow to try to cover liquids of high surface tension, the coating near the contaminating particle flows away over the surrounding area. As the coating flows, solvent evaporates, further increasing the surface tension differential and accelerating the flow. As the liquid flows away from the contaminating particle, the viscosity increases, and the flow slows to a stop, resulting in a craterlike appearance. Cratering can commonly be eliminated by the addition of a small amount of a silicone fluid to the coating.

Film Properties

The film properties required for some application can only be determined by the performance of the applied coating in practice. Because requirements and exposure conditions vary widely, devising laboratory tests to predict film performance is difficult and frequently impossible. Data banks give actual field performance as a function of environment, coatings composition, and application variables. In coatings where performance requirements are high such as for automotive exteriors, cans, marine applications, and corrosion protection, coatings suppliers and users have accumulated such data banks and use them as guides to formulate new coatings or devise new application procedures. The data on films that fail are, in a sense, even more useful than the data on successful films. As the variables in the data bank expand, it becomes increasingly possible to explain correlations between composition and performance and predict the performance of a new coating. Data banks are also very useful for checking the relationship between field performance and empirical tests.

Because following actual field performance often requires observations over many years, it is desirable to accelerate the testing process. A widely used test method is field exposure under conditions that are more drastic than commonly encountered in actual use.

Numerous paint tests are widely used. These tests can be suitable quality-control tests or useful guides along with a data bank of known behavior in considering the effect of changes in composition on properties such as impact resistance and ability to withstand deformation without cracking. In general, however, such tests are not satisfactory for predicting field performance of a new coating, although they are widely used in this way. It is unfortunately common for a specification based on quality-control tests to be used as a design criterion for a new coating.

Adhesion. In most cases, it is desirable to have a coating that is difficult to remove from the substrate to which it has been applied. An important factor controlling this property is the adhesion between the substrate and the coating. In formulating a coating, it is critical to remember that difficulty in removing a coating can also be strongly affected by how difficult it is to penetrate through the coating and how much force is required to push the coating out of the way as the coating is being removed from the substrate as well as the actual force holding the coating onto the substrate. Furthermore, the difficulty of removing the coating can be strongly affected by the roughness of the substrate. If the substrate has undercut areas that are filled with cured coating, a mechanical component makes removal of the coating even more difficult, analogous to holding two dovetailed pieces of wood together. Surface roughness affects the interfacial area between the coating and the substrate. Because the force required to remove coatings is related to the geometric surface area, whereas the forces holding the coating on to the substrate are related to the actual interfacial contact area, the difficulty of removing a coating can be increased by increasing the surface roughness. However, greater surface roughness is only of advantage if the coating penetrates completely into all irregularities, cracks, and crevices of the surface. Failure to penetrate completely can lead to less coating to interface contact than the corresponding geometric area and will leave voids between the coating and the substrate. The physical chemistry of adhesion has been reviewed (see ADHESION AND BONDING).

Exterior Durability. In many cases, an important performance requirement for coatings is exterior durability. There are many potential modes of failure. Commonly, the first indication of failure is reduction in gloss due to erosion or surface embrittlement leading to the development of roughness and cracks in the coating. In some cases, the next step is chalking, that is, the erosion of resin from the surface of the coating, leaving loose pigment particles on the surface. Chalking reduces gloss, and as it proceeds can lead to complete film erosion and also to drastic color changes caused by increased surface reflectance, which makes colors shift to lighter shades. Colors can also change if the pigment, or pigment–resin combination, undergoes photochemical degradation on exterior exposure. Photochemical oxidation and hydrolysis of the resins in the coatings are common causes of failure. On exterior exposure, films may crack or check as a result of embrittlement, reducing the elongation to break. Such films crack as they undergo expansion and contraction, for example, caused by temperature changes, especially when the substrate and the coating have different coefficients of thermal expansion. Various kinds of chemical attacks, such as acid rain, can also result in film degradation and discoloration. Retention of dirt from the atmosphere on the surface can lead to drastic color changes and blotchy appearance. Mildew can grow on the surface of many coating films, again leading to blotches of gray discoloration. In view of the large variety of exposure conditions and possible modes of failure, no laboratory test has been devised to predict field performance of coatings on exterior exposure. As noted earlier, careful accumulation of actual field-use results correlated with environmental, application, and composition variables is the most useful way of understanding the causes of failures and, hence, of being able to forecast the possible performance of a new coating material.

Although many failure mechanisms are involved, the two most common modes are hydrolysis and photochemical oxidation by free-radical chain reactions. In general, resins that do not have backbone linkages that can hydrolyze provide better exterior durability than systems with, for example, ester groups in the backbone.

Pigment selection can also be critical in formulating for high exterior durability. Some pigments are more susceptible to color change on exposure than others. Some pigments act as photosensitizers to accelerate the degradation of the resins in the presence of uv and water. The exterior durability of relatively stable coatings can be enhanced by use of additives. Ultraviolet absorbers reduce the absorption of uv by the resins (or pigments), and hence decrease the rate of photodegradation. Further improvements can result from use of free-radical trap antioxidants such as hindered phenols and especially hindered-amine light stabilizers (HALS). Photochemical and thermal degradation of coatings and their stabilization have been described.

Corrosion Control. An important function of many coatings is to protect metals, especially steel, against corrosion. Corrosion protection is required in two different situations: in one case, the steel is protected against corrosion with intact coating films, whereas in the other case, the objective is particularly to protect the steel against corrosion even though the film has been ruptured.

Protection by Intact Films. In the case of intact films, the key factors responsible for corrosion protection are adhesion of the coating to the steel in the presence of water, oxygen and water permeability, and resistance of the coating resins to saponification. If the resins in the coating are adsorbed to cover all of the steel surface and if the adsorbed groups cannot be displaced by water, oxygen and water permeating through the film cannot contact the steel and corrosion does not occur. In effect, only a monolayer is required to protect against corrosion, if the monolayer stays in place in the presence of water.

Protection by Nonintact Films. It is also possible to achieve corrosion control by coatings after the film has been ruptured. Because the methods used to achieve this control give poorer performance when the films are not ruptured, such systems should be used only when film rupture must be anticipated or when complete coverage of the steel interface cannot be achieved. There are two general techniques on a large scale: use of corrosion inhibiting pigments and use of zinc-rich primers.

The mechanism of action of corrosion-inhibiting pigments is not completely understood, but it is generally agreed that they promote oxidation at the surface of the anodic areas of the steel to form a barrier layer. For the pigments to operate, they must have a minimum solubility, but if the solubility is too high, they would be lost rapidly from the film by leaching and thus provide only short-term protection. Because the pigments are somewhat soluble, their presence in an intact film can lead to blistering and loss of adhesion. Therefore, it is undesirable to use such pigments except when needed for protection by nonintact films is anticipated.

Mechanical Properties

Coating films are viscoelastic; therefore, their mechanical properties depend on the temperature and the rate of application of stress. Their behavior is shifted toward the elastic mode with increased tensile strength to break and decreased elongation to break and with a more nearly constant modulus as a function of stress as temperature is decreased or as the rate of application of stress is increased. The shifts can be especially large if results are compared above and below the T_g. Below T_g, coatings show dominantly elastic response and, therefore, are brittle and break if the relatively low elongation to break is exceeded. Above T_g, the viscous component of deformation response is larger and films are softer (lower moduli) and less likely to break during forming. Caution is necessary in considering the relationship of T_g to formability because some materials, such as poly(methyl methacrylate) and especially bisphenol A polycarbonate, are ductile at temperatures substantially below T_g. Modulus above T_g is controlled primarily by the cross-link density. In fact, it has now been shown that the cross-link density of melamine–formaldehyde cross-linked coatings can be calculated from the storage modulus above T_g.

Abrasion resistance of floor coatings has been related to work to break of the coatings; there is no correlation with the most widely used specification test for abrasion, the Tabor Abraser test. Polyurethane coatings, in general, have superior abrasion resistance. This may be due to the presence of what might be thought of as two types of cross-links, the covalent cross-links and cross-links resulting from intermolecular hydrogen bonds between urethane groups on different molecules. At low levels of stress, the hydrogen bonds remain intact and the coating shows adequate hardness and resistance to swelling with solvent, whereas at higher levels the hydrogen bonds can be separated, permitting extension without rupture of the covalent cross-links. When the stress is removed, the hydrogen bonds re-form.

Uses

Architectural Coatings. Roughly half the volume (1.8×10^6 m^3 with a value of \$4.1 $\times 10^9$ in 1986) of all coatings sold in the United States are classified as architectural coatings. These coatings are designed to be applied to residences, offices, and for other light-duty building purposes. In contrast to most product coatings, they are designed to be applied in the field, in some cases by contractors but in large measure by "do-it-yourself" consumers. A wide range of products is involved, including flat wall paint, exterior house paints, and gloss enamels.

Product Coatings. The next largest category of coatings, about 35 vol % (in 1986, 1.28×10^6 m^3 with a value of \$3.7 $\times 10^9$), are those applied in factories to an enormous variety of products, ranging from automobiles to toys. They are called product coatings or, commonly, industrial coatings. Since these coatings are all proprietary, only limited details can be provided about any particular application. Government regulations of air quality, effluent, waste disposal, and toxic hazards are continually changing, and specific information needs immediate updating. Product coatings include metal finishing, metal primers, metal top coats, coil coatings, radiation-cure coatings, powder coatings, and wood product coatings.

Special-purpose Coatings. This heading is used in reports on the coatings industry to include coatings that do not fit under the definition of architectural or product coatings. About 15% of all U.S. coatings (5.7×10^5 m^3 with a value of \$2 billion in 1986) fall into this category. It includes heavy-duty maintenance and marine coatings, automotive refinish paints, and traffic paints.

BIBLIOGRAPHY

Monograph series published by Federation of Societies of Coatings Technology, Philadelphia, Pa. Titles issued: *Film Formation, Introduction to Polymers and Resins, Solvents, Mechanical Properties of Coatings, Corrosion Protection by Coatings, Coil Coatings, Automotive Coatings, Radiation Cured Coatings, Coating Film Defects, Application of Paints and Coatings, Organic Pigments, Inorganic Primer Pigments,* and *Marine Coatings.*

ZENO W. WICKS, JR.
North Dakota State University

COATINGS, SYSTEMS

Surface coatings represent a large segment of the chemical industry. According to the U.S. Department of Commerce, Bureau of Census, M28F, the total factory shipments of all types of coatings in the United States in 1983 totaled 3.49×10^6 m^3 (923×10^6 gal) valued at ca $\$8.55 \times 10^9$. Because of continual upgrading in performance, coatings last considerably longer in service than they did some years ago, and repainting is needed less frequently. In some cases, the film thickness of the coating has been reduced. In addition to these factors, the substitution of plastics for metal in many uses has eliminated the need for paint, so that the growth rate of the coatings industry is expected to be slow, ie, 1–2%/yr.

Surface coating compositions are usually more or less viscous liquids with three basic components: a film-forming substance or combination of substances called the binder, a pigment or combination of pigments, and a volatile liquid. The combination of binder and volatile liquid is called the vehicle; it may be a solution or a dispersion of fine binder particles in a nonsolvent. If a clear, transparent coating is desired, pigments are omitted. The volatile liquid provides a practical viscosity for packaging and application and does not normally become a part of the final film. When the liquid coating is applied to a surface, the volatile portion of the vehicle evaporates, leaving the nonvolatile binder–pigment combination as a residual film which may or may not require chemical conversion to an insoluble state. Small amounts of additives are often included to improve application, pigment settling, drying characteristics, and film properties. Most binders are either high molecular weight, nonreactive organic polymers or low-to-medium molecular weight reactive polymers capable of being further polymerized via chain-extension or cross-linking reactions to high molecular weight films.

The term organic coatings encompasses conventional paints, varnishes, enamels, lacquers, water-emulsion and solution finishes, nonaqueous dispersions (organosols), plastisols, and powder coatings. However, the term paint is often used nonspecifically to cover all of these categories as though it were synonymous with coatings. In this article, the terms are often used interchangeably (see VACUUM COATING under COATING METHODS and ELECTRONICS APPLICATIONS for discussions on less conventional coatings). The binders of conventional paints are, for the most part, raw or processed unsaturated vegetable oils, eg, linseed, dehydrated castor, soya, etc, often modified with natural or low molecular weight synthetic resins. Pigmentation provides opacity and an almost endless variety of colors. When the liquid paint is applied to a surface and the solvent evaporates, the uncured film is insolubilized by oxidative polymerization via the double bonds in the oil. For varnishes, which are unpigmented paints, the same drying oils (qv) and resins are used. The distinction between paints and enamels is not clearcut, but enamels generally contain higher molecular weight binders than paints and are therefore formulated with lower solids concentration. They are also formulated at lower pigment-binder ratios and thus exhibit superior gloss.

Lacquers are differentiated from paints and enamels in that they are formulated with thermoplastic, soluble polymers of much higher molecular weight and low chemical reactivity. Film formation occurs solely by solvent evaporation. Further polymerization in the deposited film is unnecessary because the film already possesses the requisite properties. In general, lacquer films remain permanently soluble and thermoplastic, although long exposure to sunlight induces partial or complete insolubilization in some cases. Conventional lacquers are normally solvent-borne compositions, but dispersions of polymers in water (latices) or organic liquids (organosols) that yield thermoplastic, soluble films also qualify for the lacquer designation. Polymers most frequently encountered in lacquer formulations are cellulose esters and ethers, vinyls, and acrylics.

Plastisols are dispersions of finely divided polymers in plasticizers that are nonsolvents at room temperature, but good solvents at high temperatures. Under normal storage conditions, they are stable compositions that can be coalesced to continuous films at elevated temperatures.

Coatings are applied to surfaces of all types (Table 1) to provide protection and decoration.

MATERIALS

Binders

Waterborne Systems. Vehicles in which at least 80% of the liquid is water are classified as aqueous systems. There are three categories: those in which the binder is soluble in water, those in which it is colloidally dispersed, and those in which it is emulsified to form a latex. Water-soluble binders are low molecular weight polymers (5,000–10,000). Colloidal-dispersion polymers have a molecular-weight range of 10,000–50,000, and emulsion polymers have molecular weights of 10^6 or more. The emulsion polymers have the best film properties and resistance to water. The water-soluble binders are the poorest in these respects, and the colloidally dispersed polymers are intermediate.

Nonaqueous Dispersion (Organosols). Indirect methods for preparing organosols or nonaqueous dispersions include grinding bulk- or bead-type polymers to relatively fine powders that can be dispersed in nonaqueous media; the low practical limit for particle size is ca 10 μm.

Powder Coatings. Powder coatings require no solvent for application and therefore do not present air pollution problems. Coating powders are based on both thermoplastic and thermosetting resins. The volume of thermosetting powders sold exceeds the sales of thermoplastic powders by a factor of five. For use as a powder coating, a resin should possess the following characteristics: low melt viscosity, affording a smooth, continuous film; good adhesion to the substrate; good physical properties, eg, high toughness and impact resistance; light color, permitting pigmentation in white and pastel shades; good heat and chemical resistance; and good weathering resistance. It should remain stable in storage at 25°C for at least six months.

Manufacture. Coating powders are either melt-mixed or dry-blended.

Application. Powders are applied by a fluidized-bed technique or electrostatic spray.

Powder coatings are used primarily on steel pipes. Other applications include appliances, electrical components, automotive parts, and metal furniture (see also POWDER COATING under COATING METHODS).

Plasticizers. Plasticizers (qv) are used primarily with high molecular weight linear polymers to improve flexibility, distensibility, and impact resistance; in addition, they may affect flow, leveling, and gloss.

Volatile Components

In addition to the binder, most coating vehicles contain a volatile liquid. In a solution-type coating, it is a solvent for the binder; in a dispersion-type coating, such as a latex- or orga-

Table 1. Polymer Systems Used for Industrial Coatings

Polymer systems	Coil	Metals	Appliance	Furniture	Hardboard	Lumber and plywood	Marine	Maintenance	Automotive OEM[a]	Automotive refinish	Cans	Paperboard
Natural or modified polymers												
drying oils				X	X	X	X	X			X	
cellulose esters		X		X		X			X	X		X
cellulose ethers				X						X		
Condensation systems												
alkyd resins	X	X	X	X	X	X	X	X	X	X	X	X
polyesters, high molecular weight	X	X	X		X	X						
amino resins	X	X	X	X	X	X			X	X		
phenolic resins	X	X	X				X	X	X	X	X	
polyamides		X					X	X			X	X
polyurethanes				X	X		X	X	X		X	X
epoxy resins	X	X	X	X	X		X	X	X		X	
silicones		X	X				X	X			X	
Vinyl polymers and copolymers based on:												
butadiene								X			X	
acrylic or methacrylic ester	X	X	X	X	X		X	X	X	X	X	X
vinyl acetate				X	X	X	X	X				
vinyl chloride	X	X	X	X	X	X	X	X	X		X	
vinylidene chloride							X	X				X
styrene		X	X		X			X	X	X	X	X
vinyl acetal or butyral	X	X		X				X			X	
fluorocarbons	X											
Resin combinations												
acrylic and amino	X	X	X	X	X				X		X	X
acrylic and epoxy		X	X						X		X	
acrylic and silicone	X	X						X				
alkyd and amino	X	X	X	X	X				X	X	X	X
alkyd and acrylic		X	X	X				X	X	X	X	
alkyd and epoxy		X	X					X	X			
alkyd and silicone	X	X	X									
polyester and epoxy		X	X					X	X		X	X
polyester and silicone	X	X			X							
cellulose ester and urethane				X								
alkyd, acrylic, and amino					X							
polyester and amino											X	X
phenolic and epoxy							X	X			X	
epoxy and amino											X	
phenolic and amino											X	
alkyd and vinyl chloride polymers							X	X				

[a] OEM = original equipment manufacture, paint sales to original car manufacturers as distinguished from the refinish market, ie, repair shops, etc.

nosol-based composition, it is a nonsolvent, ie, water or organic liquid.

Pigments

Pigments are finely divided, insoluble, solid particles dispersed in the coating vehicle and distributed throughout the binder in the final film. They are used in coatings primarily to obscure the surface and provide color (see also COLORANTS).

Color. The color of a pigment depends on the spectral composition of the incident light and its reflectance. Colors are compared visually and with instruments, which are usually based on the trichromaticity principle. More recently, colors have been matched by computer, resulting in substantial progress.

Light Scatter and Hiding. As an example, colored glass, when pulverized, gives powders that become progressively whiter and more opaque as pulverization proceeds owing to the scattering of light at the glass–medium (air) interface caused by the difference in the refractive indexes.

Surfactants

Surfactants are used as pigment dispersants in both nonaqueous and aqueous systems. Dispersants employed in nonaqueous systems include lecithin, zinc naphthenate, calcium naphthenate, copper oleate, and oleic acid.

Pigment dispersion is but one example of the use of surface-active agents. Surfactants play an important role in the performance of coatings; some act as emulsifying agents, others as wetting agents. Surfactants that maintain a fine state of pigment dispersion after the grinding operation are called deflocculating agents. Others, designated as antiflooding and antifloating agents, help develop uniform color. Antifoaming agents (qv) break the foam produced in the manufacture of waterborne systems. Surfactants help control the consistency and stability of paints in the package and improve application and leveling properties.

Manufacture

The manufacture of paint involves mixing, grinding, thinning, color matching and other adjustments, and packaging. In one of the older methods, pigments and vehicle were mixed into a paste of suitable consistency in a tub with rotating blades. This paste was fed into a slow-speed stone mill, where the pigments were dispersed in the liquid, eliminating unwetted agglomerates and air pockets. The paste was then fed in a continuous stream into a thinning tank equipped with rotating blades, and the remainder of the liquid was added. The product was tested for viscosity, color, and other physical properties. The approved batch was strained and filled into containers. Dispersion by old-fashioned stone mills is extremely slow and costly and is now obsolete; various types of more efficient equipment are now used, including the high-speed disperser, the sand grinder, ball mills, and roller mills.

Economic Aspects

The term trade sales is used for coatings that are marketed through retail or wholesale outlets for use by professional painters or homeowners. Price ranges for exterior architectural and automobile refinish paints are given in Tables 2 and 3.

Table 2. Approximate 1984 Paint Price Ranges[a] of Architectural Paints

Grade	$/gal
Topcoats	
lowest quality latex, ie, low acrylic level	12
intermediate acrylic-latex quality	19–20
highest acrylic-latex quality	22–23
oil or alkyd	22–24
Primers	
latex	20
oil	21

[a] From Sherwin-Williams retail-price schedule.

Table 3. Approximate 1984 Price Ranges[a] of Automobile Refinish Paints

Type	$/gal
Topcoats, matched colors	
acrylic lacquers	36–72
acrylic enamels, less hardener	34–87
alkyd enamels	30–54
polyurethane enamels, including activator	74–134
Primers and primer–surfacers	
enamel	23–26
lacquer	22–28
epoxy, including activator	42–44

[a] From DuPont price schedule.

APPLICATION PROPERTIES AND FILM FORMATION

Rheological Effects

Flow, Leveling, and Sagging. Rheology is the science of the flow response of liquids to the application of a shear stress. Flow and sagging of a paint film and pigment settling occur in response to the shearing force of gravitational pull. Leveling, influenced by gravity on vertical surfaces, is the effect of surface tension. When the shear rate of a coating is directly proportional to the applied shear stress, the system is said to exhibit Newtonian flow, which is exhibited by many paint vehicles. Likewise, pigmented systems in which the pigment particles are well dispersed may exhibit Newtonian or pseudoplastic flow. When such a paint is applied to a vertical surface, it flows until the motion is arrested by the solvent evaporation. The film levels well, but is susceptible to sagging. The conditions that promote a good sag-flow balance are just sufficient flow time to ensure good leveling followed by an immediate cessation of flow.

Film thickness has a profound effect on the sagging rate. In a Newtonian system, the sagging rate is proportional to the cube of the film thickness.

Most pigmented systems exhibit a combination of plastic flow and thixotropy. Such paints do not flow under an applied shear stress until that force exceeds a certain minimum value. The force required to overcome the resistance to flow is called the yield value. Under shear these systems exhibit lowered viscosity, but when shear ceases they return to their original state. In systems that exhibit pure plastic flow the return is immediate, whereas in thixotropic systems it is time-dependent.

Pigment Settling. Pigment particles in Newtonian or pseudoplastic systems are prone to hard settling. Even though they exhibit only a small shear stress owing to gravitational attraction, they eventually form a hard cake at the bottom of the container that is difficult to redisperse; the finer the particle size and the lower the particle density, the slower the settling process.

Filling. The ability of a coating to hide roughness in the substrate, eg, scratches and metal grain, is called filling. The same flow properties that control leveling also control filling. The other important factor is shrinkage. A coating exhibiting Newtonian flow applied to a rough surface flows from the elevations of the substrate roughness into the depressions and creates a smooth wet surface; however, as the film dries it shrinks, permitting some of the substrate roughness to reappear.

Flooding and Floating. In a coating composition containing two or more pigments differing in size, shape, or density, conditions known as flooding and floating may ensue. These conditions are characterized by an uneven distribution of pigments in the drying film.

The term flooding applies to a different concentration of one or more pigments at the surface as compared with the interior of the film. Floating is an uneven distribution of pigments at the film surface itself, observed as off-color streaks or mottled areas.

Metallic Two-tone. Coatings containing aluminum flake in conjunction with deeply colored prime pigments exhibit two-tone, depending on the types of colored pigment and vehicle. In this condition, the film is highly reflective and silvery when viewed at a 90° angle, but changes to a deeper shade approaching that of the colored pigment when viewed at other angles.

Application Methods

Coatings are applied by direct contact of the liquid coating with the substrate or deposition by atomization processes. The first category includes brushing, roller coating, dipping, flow coating, and electrodeposition; the second includes conventional spray, airless spray, hot spray, and electrostatic spray (see also COATING METHODS).

Film Formation and Curing

After a coating has been applied, solvent evaporation and rheological factors contribute to the setting of the film. Solvent evaporates from the surface of the film initially at about the same rate as it would in the absence of binder. As the film solidifies, evaporation slows down because the diffusion rate to the surface is usually slower than the evaporation rate.

Lacquer films do not cure by chemical reaction to achieve

the hardness and toughness required of a serviceable film. They simply dry by solvent evaporation and depend on the high molecular weight of the polymer to provide the required properties. Latex house paints and some nonaqueous dispersion coatings act similarly.

Electron-beam and Uv Curing. Electron-beam radiation requires a power supply and an electron-beam accelerator. The objects to be coated are sprayed electrostatically and passed behind the accelerator in an inert atmosphere. The curing rate is almost instantaneous, ca 1 s for a 125-μm film.

In uv cure, the energy source is a high intensity mercury-vapor tube that emits radiation at 315–400 nm. Although uv radiation is much weaker than that from the electron beam, curing times are between 5 s and 2 min. Coatings amenable to cure by these techniques are those that undergo free-radical polymerization.

USES

Large paint manufacturers maintain formulas and stocks of large numbers of coating materials, each designed for a specific use. These may be classified in more than one way. One convenient classification is air-drying paints that are used primarily for structures in place, ie, architectural paints, vs industrial coatings applied at the factory to mass-production items and heat-cured.

Architectural paints include exterior solvent-based paints (both oil and alkyd formulations), exterior latex paints, interior solvent-based paints (high gloss, semigloss, and flat wall paints), interior latex paints, water paints other than latex (whitewash, casein paint, cement paints, and linseed oil emulsions), floor coatings, and enamels. Industrial coatings include maintenance paints (chemically resistant coatings, traffic paints, and marine paints), automotive finishes, specialty paints (daylight fluorescent colors, fire-retardant paints, home-appliance finishes, coil coatings, wood finishes, and aerosol colors), and artists' colors (latex, oil and water-color paints, gouache, tempera, encaustic, fresco, and ethyl silicate).

Service Performance

Corrosion Mechanism. Corrosion is caused by the attack on metal by chemical or electrochemical reaction with the surrounding medium. Although all metals are subject to corrosion, the large industrial use of iron and the nonprotective character of its oxides make the corrosion protection of iron especially important.

Protection Against Corrosion. Coatings provide protection against corrosion.

Cathodic Protection. The mechanism of electrolytic corrosion demonstrates that the iron is transformed into soluble ions as a result of being anodic. By making iron the cathode, this transformation cannot occur. In many cases, protection is afforded by using an imposed electric current from an external source, which makes the iron the cathode.

Sacrificial Anodes. Without using an external source, a countercurrent for cathodic protection can be set up in the system by using a large excess of a metal with a more negative standard electrode potential than iron, eg, zinc, which is anodic with respect to the iron.

Anticorrosion Pigments. Red lead is considered to be a mixture of PbO and PbO_2. The tetravalent PbO_2 has a stronger anticorrosive effect than the divalent PbO. Calcium plumbate ($2CaO{\cdot}PbO_2$) is also used as an anticorrosive pigment for oil paints and oil-modified alkyds.

Film Requirements and Film Formation. To develop corrosion, liquid water or water that has been condensed in micropores from a relative humidity of at least 70% is required. If an organic film is applied that completely wets the surface of the metal and is in intimate contact with it, access of water to the substrate and the onset of corrosion are impeded. As a film forms from a polymeric material dissolved in a solvent, the molecules of polymer, even if they wet the substrate adequately, are never perfectly aligned. They undergo Brownian motion during the drying process and, at some point during the volatilization of solvent, the mixture gels and has only limited capability of further orientation of the polymer molecules. As more solvent evaporates, voids are left in the film. The higher the molecular weight of the polymer, the earlier the gel point and the lower the capability to orient, and hence reduced corrosion protection.

Weathering. Many materials are sensitive to outdoor weathering (qv). Organic coatings often suffer from the effects of weathering, but as long as they last, they protect the underlying substrates, particularly if they contain pigments that are opaque to uv radiation (see also DEGRADATION).

Substrate Effects. The organic coating must be chosen with careful consideration of the demands imposed on it by the substrate.

Coating Failure. The usual types of failure of organic protective coatings are chalking, cracking, checking, erosion, blistering, flaking or scaling, peeling, color change, loss of gloss and degradation due to environmental factors such as uv radiation, oxygen, heat, and moisture.

Decorative Effects. The following decorating techniques may be used: flakes of aluminum or other metals applied to give special metallic effects; special spraying techniques such as stippling; various extenders with otherwise clear coatings to give subtle dulling effects; hand-rubbing of irregular surfaces with dye-vehicle systems to give an antique effect; tipping, ie, depositing a layer of a dye coat onto an irregular surface and then wiping off the tips of the irregularities to obtain a two-tone effect.

Flexibility. Flexibility requirements often dictate the polymer system employed in coatings. To resist flexural fatigue, the polymers must be of very high molecular weight and must have a low torsional modulus.

Government Regulations

The future course of the coating industry is being seriously affected by tightening government regulations. Of chief concern is the presence of excessive amounts of toxic materials in coatings, and the pollution of water and the atmosphere by toxic effluents and volatile organic compounds (VOC). Research activity has been directed toward compliance with EPA requirements without sacrificing quality or appearance. The directions explored include reduction in solvent content by formulating at higher solids; changing to waterborne coatings; uv and electron-beam curing; and powder coatings.

BIBLIOGRAPHY

General References

H. F. Payne, ed., *Organic Coating Technology,* Vols. 1 and 2, John Wiley & Sons, Inc., New York, 1954 and 1961.

C. R. Martens in C. R. Martens, ed., *Technology of Paints, Varnishes and Lacquers,* Robert E. Krieger Publishing Co., Inc., New York, 1974.

T. C. Patton, *Paint Flow and Pigment Dispersion,* Wiley-Interscience, New York, 1964.

R. R. Myers and J. S. Long, eds., *Treatise on Coatings,* Vols. 1–4, Marcel Dekker, Inc., New York, 1967–1975.

J. HERBERT LOWELL
Consultant

COENZYMES. See ENZYMES.

COHESIVE-ENERGY DENSITY. See STATISTICAL MECHANICS.

CO-INJECTION MOLDING. See INJECTION MOLDING.

COLD FLOW. See POLYMER CREEP.

COLD FORMING

Cold forming is the process of changing the shape of a thermoplastic sheet or preform in the solid phase through plastic (permanent) deformation with the help of dies. An example is shown in Figure 1, where a preform in the shape of a disk is reshaped into an irregular cross-section by forging (plastic deformation) in closed dies. The term cold forming implies that the deformation occurs with the material at room temperature. Its use, however, has been expanded to include forming at higher temperature, ie, warm forming, but still much below the melt temperature of the material and lower than temperatures used in thermoforming (qv). Solid-phase forming (spf) encompasses cold and warm forming as well as thermoforming.

Cold-forming techniques are basically the same as those used for metal forming. In fact, most metal-forming techniques such as stamping, extrusion (qv), forging, drawing, coining, rolling, etc, are used on thermoplastics. The main differences are the time dependence of deformation and springback, or recovery, in thermoplastics. Not all thermoplastics are cold-formable, mainly because of their brittleness. Heat generation in workpieces during forming (owing to plastic working) can be significant, especially in forging. The resultant temperature rise can affect final properties.

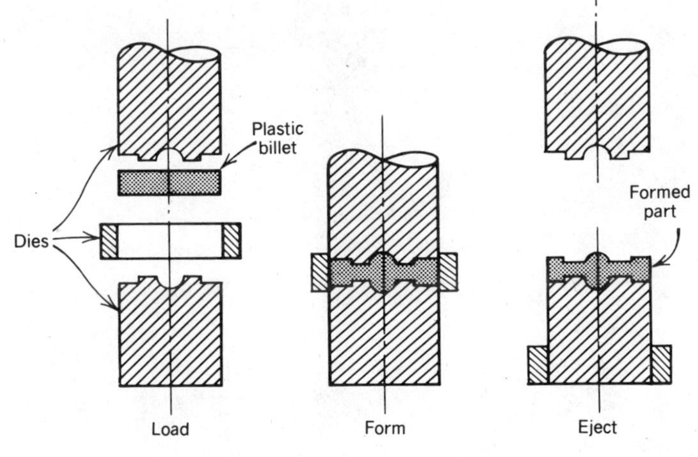

Figure 1. Forging process.

Analysis of Cold-forming Operation

In the analysis of thermoplastic-forming operations, generally the same methods are used as in metal-forming analyses. The main difference is that deformation in thermoplastics is time-dependent (viscoelastic), as is springback or strain recovery. Cold forming is currently not considered a major plastics processing technique, and its analysis has received only limited attention. However, the development of analysis techniques for metal forming has been fairly successful and they have been useful when applied to thermoplastics with appropriate modifications.

In contrast to metals and melt-processed thermoplastics, a cold-formed thermoplastic part exhibits molecular orientation along principal strain directions. Mechanical properties, such as tensile strength and modulus, are enhanced in the direction of orientation, but reduced in the normal directions.

Strain Recovery or Springback

All cold-formed materials exhibit some strain recovery or springback, ie, reversion toward initial, undeformed geometry. In thermoplastics, this process depends on temperature, deformation history, and time (as opposed to time-independent elastic springback in metals). For any given forming temperature, holding the workpiece in the deformed state for a given time, ie hold or dwell time, to allow for stress relaxation reduces the degree and rate of springback.

Materials and Equipment

The most easily cold-formed materials, depending on the forming technique, include the following:

For deep drawing, cellulose acetate butyrate, polycarbonate, polypropylene, ABS, and polyoxymethylene.

For fiber drawing and extrusion, high density polyethylene, polypropylene, and polyoxymethylene.

For forging, polypropylene, high density polyethylene, nylon, acetals, polycarbonate, ABS, and PVC.

Equipment for cold forming tends to be the same as that used for metal forming.

BIBLIOGRAPHY

R. E. Evans, *Polym. Eng. Sci.* **13**(1), 65 (1973).

K. M. Kulkarni, *Polym. Eng. Sci.* **19**(7), 474 (1979).

R. K. Okine and N. P. Suh, *Polym. Eng. Sci.* **23**(2), 61 (1983).

P. S. Hope, A. Richardson, and I. M. Ward, *J. Appl. Polym. Sci.* **26,** 2879 (1981).

G. D. M. Pushee, III, *Polym. Plast. Technol. Eng.* **6**(1), 57 (1976).

RICHARD KAFUE OKINE
E.I. du Pont de Nemours & Co., Inc.

COLD RUBBER. See BUTADIENE POLYMERS.

COLD WORKING. See COLD FORMING.

COLLAGEN

Collagen is the principal structural protein in the vertebrate body. The extracellular proteins of the main connective tissues consist of 90% or more collagen in tendon and bone, and more than 50% in skin. The different properties of these tissues are in part the result of different architectural organizations of collagen fibrils.

Collagen is, of course, a protein of a size and amino acid sequence determined by the genes of the organism and by biological processing steps that regulate its final chemistry. It has specific helical and aggregate structures, again biologically regulated.

Chemistry

Chain Composition. The type-I collagen molecule contains three polypeptide chains, termed α chains: two of one kind $\alpha1$(I), and one of another $\alpha2$(I). They are similar and homologous. Both consist of repeating triplets of sequence glycine-X-Y where X and Y can be any amino acid; proline is frequently in the X position, and hydroxyproline in the Y position.

The closely related types II and III collagen have three identical chains ($\alpha1$(II) and $\alpha1$(III), respectively), and $\alpha1$(I) forms a trimer under some conditions both *in vitro* and *in vivo*. However, the homotrimer is only partially equivalent functionally to the heterotrimer and is considered an aberrant form.

Amino Acid Composition. Qualitatively, hydroxyproline and hydroxylysine are essentially unique to collagen. Hydroxyproline is always present in mammalian collagen at about 9–10 residue %. Hydroxylysine content varies from about 0.5 to 2 residue %, depending on species, tissue, and sometimes the age of a given tissue. In other respects, the amino acid composition is unremarkable.

Carbohydrate Content. The carbohydrate content of type-I collagen is less than 1% in pure preparation.

Cross-links. Under most conditions, collagen in tissues cannot be solubilized without degradation because the molecules are covalently cross-linked. The strength of collagenous tissues is markedly reduced if cross-linking is inhibited.

Structure

Molecule. The triple-chain helix is a unique structure that defines a protein as a collagen. The stability of the helix depends on two factors: glycine (Gly) as every third amino acid in the amino acid sequence and proline and hydroxyproline as frequent occupants of the X and Y positions, respectively, of the repeating Gly-X-Y triplet.

Native Fibril. The D-periodic banded fibril seen by electron microscopy is one of the best-known characteristics of type-I collagen. Negatively stained fibrils from rat-tail tendon are shown in Figure 1.

Other Aggregates. Collagen can be precipitated in several polymorphic forms in addition to the native fibril. Although these forms have no known biological relevance, they have been useful in determining native fibrillar structure and understanding molecular properties. Polymorphic fibrillar forms include four different D-periodic symmetric fibrils, a set of D-periodic obliquely banded fibrils, and filament-long-spacing fibrils.

Tissue Architecture. Collagen used in the form of intact tissue has properties determined in part by tissue architecture, and thus the subject may be of importance to the chemist and engineer.

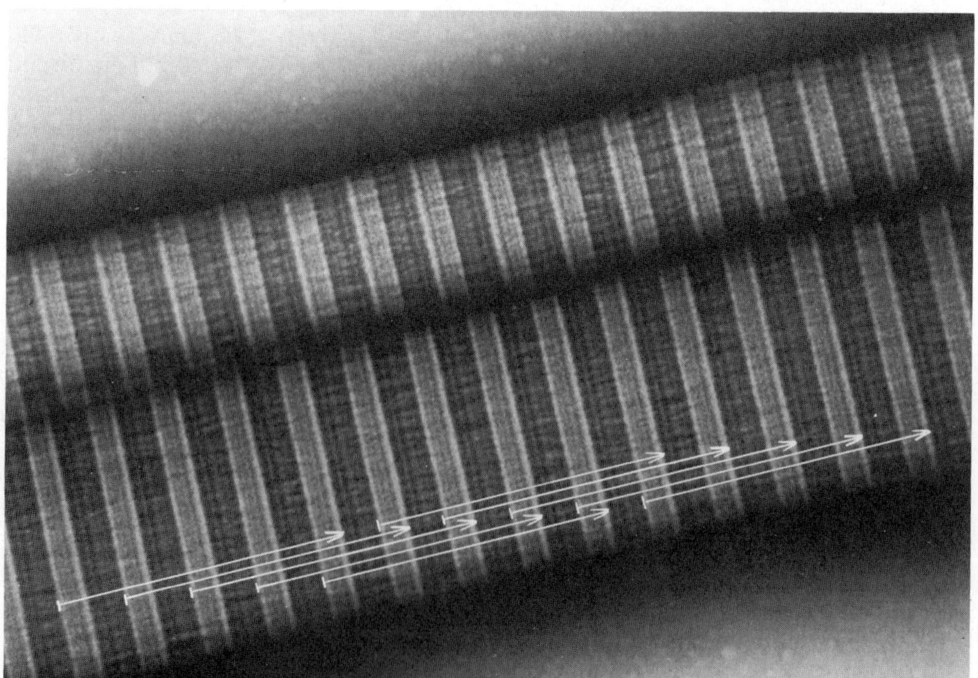

Figure 1. Electron micrograph of fibrils from rat-tail tendon negatively stained with phosphotungstic acid; the periodicity is 67 nm. The superimposed arrows represent staggered collagen molecules 300 nm long showing their axial relationship in the fibril. The arrows point in the *C*-terminal direction; the two fibrils have opposite polarities.

Biosynthesis

Type-I collagen is made as a precursor, type-I procollagen, which is longer at both the *N*- and *C*-terminal ends. The biosynthesis of type-I procollagen is similar to other proteins made to be secreted. It involves transcription of genes, processing and translation of mRNA, processing and folding of polypeptide chains, transport through the cell and secretion, extracellular processing, and assembly into the macromolecular form.

Immunology

The immunology of collagen is important for two reasons. First, because animal collagens are being used in humans, it is necessary to understand the potential immunological consequences. Second, antibodies to collagen and other proteins are powerful tools used in the laboratory for highly specific and sensitive assays (see also ANTIBODIES AND ANTIGENS, POLYMER-BOUND).

Immunogenicity. The impression gained by many investigators that collagen does not readily elicit antibodies compared to other common proteins is undoubtedly true. The reason for this behavior seems to be that species differences among type-I collagens are small.

Antigenicity. In the case of type-I collagen, although the triple helix is a poor immunogen, some denaturation may take place *in vivo,* which reveals sequential determinants. Indeed, it is common experience that polyclonal antibodies, either raised deliberately in animals or as a consequence of collagen utilization in humans, react with both native and denatured collagen.

Isolation and Purification

Soluble Collagen. Type-I collagen is not normally extractable from tissues because it is cross-linked. However, in some tissues, notably the skin of young animals, cross-linking is sufficiently slow that a small portion, at most a few percent, can be extracted in cold 0.5–1 M NaCl at neutral pH. Somewhat more can be extracted by dilute acid, eg, 0.5 M acetic acid, which cleaves aldimine cross-links. Soluble collagen is purified mainly by precipitation.

Fibrillar Collagen. Extensively cross-linked collagen from, for example, skin or tendon can be dispersed as a fine fibrillar suspension by the use of mild denaturing agents, mechanical fragmentation, and selective proteolysis of noncollagenous components.

Reconstituted Fibrillar Collagen. Collagen in solution can be reconstituted into fibrils that have a structure similar to that of native fibrils from which it was originally extracted. This self-assembly can be carried out by one of two procedures. If collagen dissolved in acid is converted to a low ionic strength solution near neutral pH, spontaneous precipitation occurs. The second procedure involves the dissolution of collagen in the cold in a buffer close to physiological ionic strength and pH.

Collagenous Tissues. Tissues consisting largely of collagen, including skin, tendon, dura, large veins, and heart valves, can be cleaved mechanically by extraction and sometimes by selective proteolysis.

Characterization

Denatured Collagen. Denatured collagen can be characterized by molecular-sieve chromatography, carboxymethyl cellulose chromatography, and sodium dodecylsulfate (SDS) gel electrophoresis.

Soluble Collagen. Native type-I collagen in solution, usually at acid pH, has been extensively characterized by physical techniques including viscosity determination, ultracentrifugation, sedimentation equilibrium, electrophoresis, intensity light scattering, dynamic light scattering, electric birefringence, nuclear magnetic resonance, optical rotary dispersion, circular dichroism, and electron microscopy of rotary-shadowed molecules (see also CHARACTERIZATION OF POLYMERS).

Fibrillar Collagen. Insoluble collagen in fibrillar form, whether never solubilized or solubilized and reconstituted, is more difficult to characterize since most physical methods require protein in solution. In addition, the form of the collagen as well as its inherent properties are usually of interest. Three useful characterization methods are readily adapted to all fibrillar (or otherwise insoluble) collagens: amino acid analysis, differential scanning calorimetry (dsc), and transmission electron microscopy (tem).

Uses

The technology used to manufacture articles of commerce such as sutures, hemostatic agents, tissue replacements, and sausage casings from fibrillar collagen is as much art as science. It is not reported in the literature but is the subject of many patents. Most process information is proprietary. Industrial uses of collagen as gelatin (qv) and leather (qv) are undoubtedly the largest. The most interesting applications of collagen are probably biomedical (Table 1).

Table 1. Biomedical Applications of Collagen

Application	Collagen form
hemostatic agents	powder, sponge, fleece
blood vessels	processed human or animal blood vessel, reconstituted fibrillar collagen
heart valves	processed porcine heart valve
tendons and ligaments	processed animal tendon, collagen–carbon fiber composite
burn dressings	processed animal skin, sponge, composites
intradermal augmentation	reconstituted fibrillar collagen
drug-delivery systems	various biodegradable forms

BIBLIOGRAPHY

M. Nimni, ed., *Collagen: Biochemistry, Biotechnology and Molecular Biology,* Vol. 3, CRC Press, Inc., Boca Raton, Fla., 1988.

"Biology, Chemistry and Pathology of Collagen," in *Ann. N.Y. Acad. Sci.,* Vol. 460, 1985, pp. 1–537.

G. N. Ramachandran and A. H. Reddi, eds., *Biochemistry of Collagen,* Plenum Press, New York, 1976.

G. N. Ramachandran, ed., *Treatise on Collagen,* Vol. 1, Academic Press, Inc., New York, 1967.

S. Gay and E. J. Miller, *Collagen in the Physiology and Pathology of Connective Tissue,* Gustav Fischer Verlag, New York, 1979.

KARL A. PIEZ
Collagen Corporation

COLLIGATIVE PROPERTIES.

See MOLECULAR WEIGHT DETERMINATION.

COLLOIDS, I

A polymer colloid is a dispersion of polymer particles in a fluid in which the size of the particles is anywhere within the colloidal range, ie, from 10 to 1000-nm equivalent diameter. The fluid is usually a liquid, aqueous or organic, although it may be a gas. This article is restricted to systems in which the fluid is a liquid and a nonsolvent for the polymer; these are often referred to as lyophobic colloids. The polymer can be organic or inorganic, natural or synthetic. The most widely known natural polymer colloid is the sap of the *hevea brasiliensis* tree; the sap is known as latex, from which natural rubber is derived.

Synthetic colloids or latices (qv) are usually made by the so-called emulsion polymerization of olefinic monomers, which involves a free-radical propagation mechanism. Millions of tons of industrial polymers are produced annually throughout the world by emulsion polymerization, including the commercially important poly(styrene-*co*-butadiene) (SBR rubber), poly(vinyl chloride) (PVC), poly(vinyl acetate) (PVAC), poly(methyl methacrylate) (PMMA), polyacrylonitrile (PAN), esters of poly(acrylic acid), and copolymers of almost any combination of these. A broad range of physical properties, ie from hard, glassy plastics to soft, rubbery elastomers, is encompassed by these materials.

Polymer colloids can also be made by dispersion techniques and condensation polymerization. In dispersion techniques, low molecular weight polymers must be used with surfactants under conditions of high shear agitation. Polyurethane and polyester dispersions made by this technique are of some commercial importance, with polyester dispersions being especially useful in electrodeposited automotive primer coatings (see also COATINGS, ELECTRODEPOSITION). Examples of condensation emulsion polymerization are the Thiokol latices, which are made by reaction of polysulfide ions in aqueous solution with an emulsion of an alkylene dihalide. The product is a polysulfide elastomer in dispersion. Such polymers are used in solid rocket propellants and as caulking compounds (see also POLYSULFIDES).

Colloidal polymer particles are generally isotropic spheres because the polymer molecules therein are in random orientations; that is, the polymers are amorphous. The distribution of particle sizes can be extremely narrow (monodisperse), broad (polydisperse), or polymodal. An attractive feature of emulsion polymerization is that it is relatively easy to make reproducible, monodisperse colloids. As with colloids in general, the specific surface areas can be large, ie, on the order of 50–100 m^2/g of polymer. The latex particles invariably carry functional groups chemically bound to the particle surface, the interface between the particle, and the surrounding medium. These groups are derived from initiator fragments, comonomers bearing such groups, or from chemical reactions such as hydrolysis of the monomer or polymer during or after polymerization. These groups often affect the stability toward coagulation of the colloid, either favorably or unfavorably.

Characterization

Particle Size. The particles in polymer colloids may be hard or soft, glassy or elastomeric. If they are hard, transmission electron microscopy (tem) is the method of choice for determining the size and distribution of particles. Very small particles, soft or hard, exhibit poor contrast in the electron microscope. Under such circumstances, negative staining may be employed.

A relatively easy method to measure weight-average particle size is by means of elastic light scattering. For larger particle sizes, the angular dependence of the intensity may be used. The equivalent hydrodynamic radius of the particles can be obtained by quasielastic light scattering (qels). Hydrodynamic chromatography (hdc) and sedimentation field flow fractionation (sfff) can be used to obtain the full-size distribution.

Surface Charge and Electrical Potential. Polymer colloids, like other lyophobic colloids, are stabilized against coagulation by electrostatic charges on the surface of the particles, or by a surface lyophilic polymer, electrostatic or steric stabilization, respectively. Surface electrical properties may be conveniently measured by dielectric spectroscopy.

Particle Morphology. Although the particles in polymer colloids are very small, they may still have internal variations in composition that may have implications for their use. A gradual change in composition with radial distance within a polymer particle may markedly affect the performance of materials derived from it. This is accomplished by the continuous addition of comonomers in such a way that the composition is gradually changed as the monomer mixture is added. Monomer-starved conditions must be maintained to keep the polymer content of the particles constantly near 100%. This ensures that the swelling of the particles is not uniform, and that the reaction zone is comprised of a shell on the outside of the particles. Such core-shell morphology can lead to dramatic differences from uniform morphology in the performance of derived films.

Surface Chemistry

Surface chemistry is important to polymer colloids for several reasons: colloidal stability depends largely on the nature of the surface groups; adsorption studies require a detailed knowledge of the interface; in the study of interactions with surfaces in connection with deposition onto such surfaces, surface chemistry must be known and reproducible; surface functional groups have been shown to be active in heterogeneous catalysis by polymer colloids; and attachment of antibodies to the surface of the particles renders them capable of immunospecific binding onto cell surfaces containing the corresponding antigens. This has been used for cell labeling, clinical diagnostics, and cancer chemotherapy, and may be used for targeted-drug delivery in the future.

Colloid Chemistry

Interparticle Interactions. The particles in lyophobic colloids tend to aggregate because there is net attraction among them at all times. Stable colloids do not aggregate, because a potential energy barrier prevents the particles from coming into contact with each other.

Electrostatic Stabilization. In the presence of ionic groups at the particle surface, an equal number of counterions of opposite charge maintain electroneutrality in the surrounding medium. A repulsion is generated by the overlap of counterion clouds between two particles during Brownian collision. Surface ionic groups are introduced by chemical reaction during formation of the polymer colloid or by adsorption of an ionic surface-active agent.

Steric Stabilization. It is possible to stabilize colloidal particles against coagulation without use of electrical charges by adsorbing or bonding lyophilic polymer molecules onto the surface of the particles. Thus, two particles coated in this manner, on a Brownian collision, repel each other strongly because of the solvation forces of the continuous medium for the adsorbed polymer.

BIBLIOGRAHY

K. E. J. Barrett, ed., *Dispersion Polymerization in Organic Media,* John Wiley & Sons, Inc., New York, 1975.

R. M. Fitch, ed., *Polymer Colloids,* Plenum Publishing Corp., New York, 1971.

I. Piirma, ed., *Emulsion Polymerization,* Academic Press, Inc., New York, 1982.

J. T. Kemshead, J. G. Treleaven, F. M. Gibson, J. Ugelstad, A. Rembaum, T. Philip, and G. D. Caine, *Lancet* **70** (Jan. 14, 1984).

ROBERT M. FITCH
Louis Laboratories
Johnson Wax Company

COLLOIDS, II

This article deals primarily with solid in liquid dispersions, where the solid is a synthetic polymer, and the colloid is made either by direct polymerization or by emulsification of a polymer solution. In order for a polymer colloid to be stable, to remain as a fine subdivision of polymer particles in a liquid (which is a nonsolvent for the polymer), it must be stabilized. The stabilization is achieved by amphipathic materials, which form a monolayer at the polymer particle–liquid interface, creating a repulsive barrier. Due to Brownian motion, particles in a dispersion are in constant motion and collide with each other and the container walls. Because there is a van der Waals attraction between particles, these collisions can cause particles to adhere. To reduce collisions leading to particle coagulation, it is necessary to provide the particles with a repulsive interaction that outweighs the attraction. Currently, stability is provided to colloidal particles by surrounding them with an electrical double layer (electrostatic stabilization) or by adsorbed or chemically attached oligomeric or polymeric materials (steric stabilization).

Preparation of Polymer Colloids

Polymer colloids can be prepared by direct polymerization or by emulsification of a polymer solution.

Properties

Whereas homopolymerizations in heterophase systems produce particles uniform in their polymer composition, interesting particle morphologies can be obtained as a result of copolymerizations. In random copolymerization, the composition of the copolymer is determined by the ratio of the monomers in the phase where the propagation takes place. If the initiation step occurs in the continuous phase (water in emulsion polymerization), the relative solubilities of the monomers, as well as that of the resulting polymers and copolymers, influence the final polymer composition in the particle.

Because particle morphology is a very important parameter in determining the performance of the polymer colloid in various applications, deliberate modifications of the morphology have been carried out. Nonuniform particle morphologies have been achieved in semibatch systems by a change of monomer feed composition during polymerization eg at first of an inner-core polymer of a certain glass-transition temperature, followed by a subsequent polymerization in the same particle of another monomer; the result is a "core-shell" polymer. Most of these industrial developments are primarily in the patent literature.

Particle Size and Distribution Determination

There are a number of ways the size of colloidal particles can be measured in a dilute dispersion. The most common method is the use of a transmission electron microscope. This type of particle analysis is especially simple for hard, high T_g polymer colloids, in which no distortion occurs under the electron beam. For soft, low T_g polymer particles, if the polymer chain contains double bonds, the addition of osmium tetroxide or irradiation with uv light can be used to harden the particles. In all cases, the size and size distribution can be obtained from micrographs with the use of a particle-size analyzer (see ELECTRON MICROSCOPY). Chromatography (qv) and light scattering (qv) are also used for analysis.

Application

Polymer colloids have found a multitude of applications. Industries involved in the production of paper, textiles, paint, adhesive, ink, and photographic materials use polymer colloids. They are also major components in water purification, waste water treatment, tertiary oil recovery, and controlled-release systems in agriculture and pharmacology. In medicine colloidal systems are used in immunology, in polymer membranes, and as drug and antibody carriers.

BIBLIOGRAPHY

I. Piirma, ed., *Emulsion Polymerization,* Academic Press, Inc., New York, 1982.

D. H. Napper, *Polymer Stabilization of Colloidal Dispersions,* Academic Press, Inc., New York, 1983.

R. M. Fitch, ed., *Polymer Colloids II,* Plenum Press, New York, 1971.

I. PIIRMA
University of Akron

COLOPHONY. See RESINOGRAPHY.

COLORANTS

Plastics can be colored by dyes and inorganic and organic pigments. Dyes soluble under the conditions of use must be completely dissolved, leaving no color streaks and little or no haze. Pigments are insoluble and consist of particles that must be dispersed by physical means.

Thermoplastics are completely polymerized, and are processed, ie, colored, at or near their melting points. Scrap may be reground and remolded, eg, polyethylene, polypropylene, poly(vinyl chloride), acetal, acrylics, ABS, nylons, cellulosics, and polystyrene. Thermoset resins are only partially polymerized when the colorants are incorporated. Polymerization is completed when the resin is molded. The result is a nonmeltable cross-linked polymer that cannot be reworked, eg, epoxies, urea–formaldehyde, melamine–formaldehyde, phenolics, and thermoset polyesters. For these, it is customary to incorporate colorants into the less reactive ingredient.

A reaction may occur between a plastic and a colorant at processing temperatures. More often, allowance must be made for thermal stability and additives (qv) such as antioxidants (qv), flame retardants, ultraviolet light absorbers, and fillers (qv). The final use of the resin often indicates the selection of colorant. Obviously, colorants must be evaluated for later processing conditions in view of the subsequent fate of the fabricated parts.

Dispersion and Stability

The process of dispersing a dry pigment into a liquid vehicle or molten plastic takes place in two steps: the pigment agglomerates are first broken up into much smaller particles; the air is then displaced from the particle surfaces to obtain a complete pigment-to-vehicle interface.

Today most plastics are colored with concentrates. After the colorants have been incorporated into the plastic, lightfastness is determined by carbon arc or xenon lamp. Tests for outdoor weathering (qv) may include exposure to the moist climate of Florida or the dryness of Arizona. Performance requirements may demand testing of tensile strength or changes in brittleness.

Dyes

Dyes should be checked for migration, sublimation, and heat stability before use. These precautions are particularly important for application to plasticized resins. Dyes include azo dyes, anthraquinone dyes, xanthene dyes, azine dyes, fluorescent dyes and others [Brilliant Sulfoflavine (Acid Yellow 7), Solvent Orange 60, basic triphenylmethane dyes, and quinoline yellows].

Pigments

Inorganic pigments include whites; blacks; iron oxides; chromium oxide greens; iron blue and chrome green; violet; ultramarine pigments; blue, green, yellow, and brown metal combinations; lead chromate and lead molybdates; cadmium pigments; titanate pigments; pearlescent pigments; and metallic pigments. Organic pigments include monoazo pigments, disazo pigments, disazo condensation pigments, quinacridone pigments, dioxazine violet, vat pigments, perylene pigments, thioindigo pigments, phthalocyanine pigments, and tetrachloroisoindolinones.

Economic Aspects

Estimates of the annual consumption of colorants for plastics for the past several years are given in Table 1.

Table 1. U.S. Consumption of Colorants for Plastics, Metric Tons

Colorant	1982	1984
Dyes		
nigrosines	1,300	1,500
oil-solubles	550	650
anthraquinones	180	210
others	190	230
Total	*2,220*	*2,590*
Inorganic pigments		
titanium dioxide	100,000	123,000
iron oxides	3,500	4,300
cadmiums	2,000	2,350
chrome yellows	2,000	2,400
molybdate oranges	1,400	1,650
others	1,200	1,400
Total	*110,100*	*135,100*
Organic pigments		
carbon black	30,000	36,500
phthalocyanine blues	1,160	1,400
phthalocyanine greens	480	600
organic reds	1,000	1,200
organic yellows	200	240
others	450	500
Total	*33,290*	*40,440*
Grand total	*145,610*	*178,130*

Health and Safety Factors

Dyes may sublime under operating conditions; dust is always a problem. The toxicity of colored plastics in food packaging, drugs, and cosmetics poses complex issues. The maximum practical ventilation should be used in handling colorants, additives, and resins.

BIBLIOGRAPHY

J. Agranoff, ed., *Modern Plastics Encyclopedia,* McGraw-Hill, Inc., New York. Presents a useful table of colorants and is a mine of information about plastics.

F. W. Billmeyer, Jr. and M. Saltzman, *Principles of Color Technology,* 2nd ed., Wiley-Interscience, New York, 1981. A basic guide to color.

K. McKlaren, *The Colour Science of Dyes and Pigments,* Adam Hilger, Ltd., Bristol, UK, 1983. An advanced treatment.

T. G. Webber, ed., *Coloring of Plastics (Society of Plastics Engineers Monographs),* John Wiley & Sons, Inc., New York, 1979. Each plastic is treated individually.

THOMAS G. WEBBER
Consultant

COMBS. See BRANCHED POLYMERS.

COMONOMERS. See COPOLYMERIZATION.

COMPATIBILITY

Polymer–polymer compatibility has received much attention recently because many polymer blends perform well in attractive new applications for which the individual polymer components are not suited. Compatible and incompatible refer to the degree of intimacy of blends, which depends on the method of measurement employed in the examination. From a practical standpoint, it is useful to refer to a polymer blend as compatible when it does not exhibit gross symptoms of polymer segregation. A blend that is heterogeneous on a macroscopic level would thus be considered incompatible. The simple observation that a blend is compatible is sufficient to establish the material as potentially useful. However, from a theoretical perspective, much can be learned about the nature of the compatibility and the expected properties of the blend by probing its morphology.

Under microscopic inspection, a miscible polymer blend consists of a single phase; on a molecular level, polymer-A molecules intermingle with polymer-B molecules as shown in Figure 1**a**. A second possible morphology for polymer blends is shown in Figure 1**b**, where polymer A forms a separate phase from polymer B. This is the morphology of an immiscible blend, which is far more common than the one-phase miscible blend. Frequently, a blend of two polymers is neither totally miscible nor totally immiscible, but falls somewhere in between. A blend of this type is partially miscible and is depicted in Figure 1**c**.

Determination of Polymer–Polymer Miscibility

Numerous techniques are used to determine if a polymer is single phase or multiphase. A quick but not totally reliable method is by transparency. Erroneous conclusions may be reached with this method if the refractive indexes of polymers A and B are similar. Microscopy can be a useful tool to determine if a blend is single phase or multiphase.

Glass-transition Temperature. The glass-transition temperature T_g of a polymer is the temperature at which the molecular chains have sufficient energy to overcome attractive forces and move vibrationally and translationally. The number and locations of the T_g values provide much insight into the nature of a polymer blend.

Other Techniques. Other techniques include the more common approaches of microscopy and calorimetry, and thermomechanical, dynamic mechanical, thermoelectrical, and dilatometric procedures.

Thermodynamics of Polymer Blend

For two polymers to be miscible, the Gibbs free energy of mixing must be negative. Thermodynamics favors misciblity (qv) in polymers that attract each other, and miscibility is more favored in blends of low molecular weight than in blends of high molecular weight polymers.

Mechanical Properties

In general, the mechanical properties of commercial polymer blends are known but the formulations are proprietary. The mechanical properties of a polymer blend may be approximated from those of the components. The typical compositional plots are step, maximum, minimum, or linear (Fig. 2). The mechanical properties of a two-phase blend are not necessarily related to those of the components, and must be evaluated in the light of the blend's morphology, chemistry, and crystallinity and the processing and testing procedures.

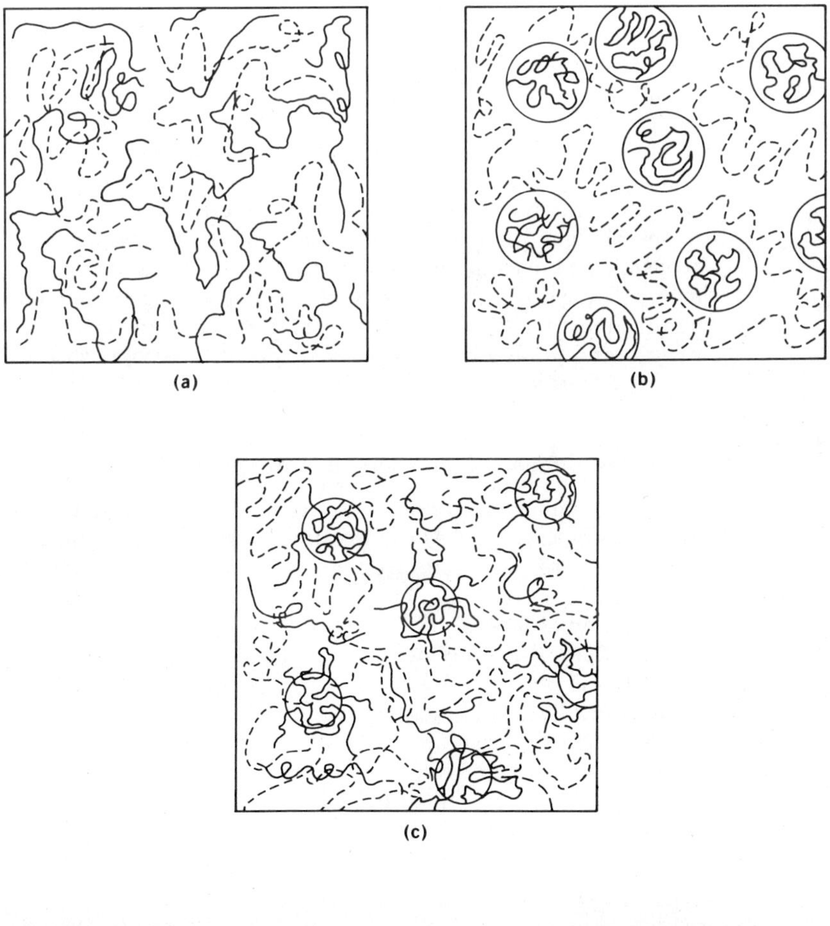

Figure 1. Morphologies of a blend of polymer A (solid line) and polymer B (dashed line): (**a**) miscible; (**b**) immiscible; (**c**) partially miscible.

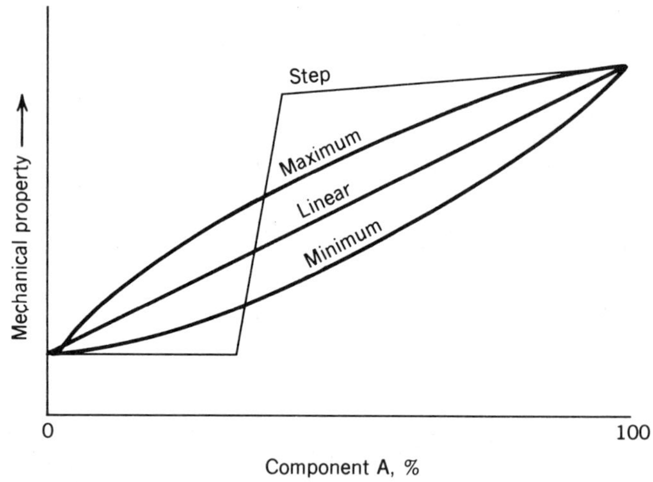

Figure 2. Possible functions of mechanical properties vs two-component compositions.

Processing

Miscible polymer blends have one phase and process much like a homopolymer or random copolymer. However, two-phase blends have unique processing characteristics. Multiphase blends can exhibit phase segregation and orientation under high shear processing conditions, which do not occur with miscible blends. Chemical grafting of one phase to another has been used commercially to minimize shear-induced agglomeration and orientation.

Polymer-Polymer Reactions

Complications arise in interpreting T_g data of blends that react with one another. Reactions between polymers may occur that appear to induce miscibility, when in reality the blend is being transformed into a copolymer. This type of reaction is most common in condensation polymers or polymers with reactive end or pendent groups.

Two-phase Polymer–Polymer Blends

Stable polymer blends can be produced from immiscible polymers by using compatibilizers that effectively prevent unmixing. For example, a block–block (A–B) copolymer may compatibilize an immiscible blend of A and B homopolymers.

Applications

The principle of compatibility provides the basis for the current explosion in custom-designed polymer blends and alloys.

BIBLIOGRAPHY

D. R. Paul and S. Newman, eds., *Polymer Blends,* Vols. I and II, Academic Press, New York, 1978.

O. Olabisi, L. M. Robeson, and M. T. Shaw, *Polymer–Polymer Miscibility,* Academic Press, New York, 1979.

D. Klempner and K. C. Frisch, eds., *Polymer Alloys I, II, III,* Plenum, New York, 1977, 1980, 1983.

Daniel W. Fox
Richard B. Allen
General Electric Company

COMPLEX CATALYSTS. See Ziegler-Natta catalysts.

COMPOSITES

The composite materials discussed in this article are those formed by aligning extremely strong and stiff continuous fibers in a polymer resin matrix or binder. The materials in this class have exceptional mechanical properties and are often termed advanced composites to distinguish them from chopped-fiber or otherwise-filled polymers. The fibers are typically 50 times stronger and 20–150 times stiffer than the matrix polymers. The role of the matrix is primarily that of a glue or binder that enables the fibers to support the applied loads. Advanced composites are more accurately described as bonded-fiber materials than as reinforced plastics.

The fibers that dominate the field of advanced composites are, in order of chronological development, S-glass, boron on a tungsten filament core, graphite or carbon, and aromatic polyamides (Kevlar). They possess the desirable properties of low density (1.44–2.7 g/cm^3) and extremely high strengths (3–4.5 GPa) and moduli (60–520 GPa).

When combined with a resin binder and laminated to support the applied loads, these fibers provide mechanical properties equal to or exceeding those of most metals. Typical fiber and unidirectional composite properties are given in Table 1.

Three factors affect the competition of advanced composites with traditional engineering materials: cost, reliability, and complexity. The cost barrier has been reduced by mass production and by fiber prices that have either fallen or remained constant. Fibers exhibit linear elastic behavior to failure at strains of only 1–3%. Engineers accustomed to designing to yield with a built-in safety margin of at least 10% strain due to plastic flow have been skeptical about using these materials as primary structures. Other reliability concerns include reported variability in properties and lack of information about response to fatigue, sustained loading, and different environments. Complexity stems from the anistropy and inhomogeneity (on the microscale) of advanced composites. In addition to the obvious directional dependence of their thermoelastic properties, composites exhibit a variety of strengths and failure modes. Thus direct substitution of a composite for an existing metal using traditional isotropic material design, manufacturing, and joining practices has led to unsuccessful designs (see also Composites, fabrication).

Principles of Composite-property Improvement

Structural materials require strength, stiffness, and toughness. Strength measures the maximum stress a material can withstand; stiffness, strain under a given stress; and toughness, resistance to stress concentrations and to the growth and propagation of flaws.

Fiber Role. Bulk polymers typically have room-temperature strengths of 30–130 MPa that are low compared with steel or aluminum alloys, and their low moduli of 2–4 GPa would result in unacceptably large deformations in structures of any size. The first principle of continuous-fiber composites is that the fibers must support all main loads and limit deformations acceptably.

Table 1. Comparative Fiber and Unidirectional Composite Properties

Fiber/composite	Elastic modulus, GPa[a]	Tensile strength, GPa[a]	Density, g/cm^3	Specific stiffness, MJ/kg	Specific strength, MJ/kg
E-glass fiber	72.4	2.4	2.54	28.5	0.95
epoxy composite	45	1.1	2.1	21.4	0.52
S-glass fiber	85.5	4.5	2.49	34.3	1.8
epoxy composite	55	2.0	2.0	27.5	1.0
boron fiber	400	3.5	2.45	163	1.43
epoxy composite	207	1.6	2.1	99	0.76
high strength graphite fiber	253	4.5	1.8	140	2.5
epoxy composite	145	2.3	1.6	90.6	1.42
high modulus graphite fiber	520	2.4	1.85	281	1.3
epoxy composite	290	1.0	1.63	178	0.61
aramid[b] fiber	124	3.6	1.44	86	2.5
epoxy composite	80	2.0	1.38	58	1.45

[a] To convert GPa to psi, multiply by 145,000.
[b] Aromatic polyamide.

Matrix Role. The matrix maintains the desired fiber orientations and spacings, transmits shear loads between layers of fibers so that they resist bending and compression, and protects the fiber from surface damage. The matrix role is more subtle and complex in the areas of strength and toughness. Glass, graphite, and boron fibers are linear elastic, brittle solids. They fail catastrophically, without plastic flow, whenever the stress on them is sufficient to cause unstable flaw growth. Flaws are typically surface notches or steps resulting from handling or processing.

Composite strength and stiffness in the direction normal to the fibers are much less than those parallel to the fibers. Advanced composites are anisotropic, with strength, stiffness, and toughness functions of direction relative to that of the fiber. Anisotropy constitutes one of the chief differences between advanced composites and isotropic structural metals.

Although both fiber and matrix are brittle, their combination produces a material that is quite tough, indeed much tougher than either of them alone. This synergism is achieved by a combination of mechanisms that tends to keep cracks small, isolated, and blunted, and that dissipates energy. These mechanisms, based in part on the heterogeneity of composites, constitute another difference from structural metals. Toughness in advanced composites is achieved without the large-scale plastic flow seen in tough metals.

Lamina Properties

The simplest representative form of advanced composites is the unidirectional ply, or lamina, consisting of fibers oriented in a single direction and bonded together by a resin. If the properties of this assemblage are known, it is possible to calculate most thermoelastic and transport properties of laminates containing arbitrary numbers and orientations of laminae. The unidirectional lamina may be considered the basic building block of more complex laminates (qv).

Micromechanical Analysis. Even with the simplification afforded by the assumption of isotropy transverse to the fiber direction, a formidable array of properties is required to fully characterize a lamina. Because of the large number of measurements needed, techniques for estimating lamina properties from properties of the constituents have been developed. This field, known as composite micromechanics, utilizes many of the principles of continuum (macro) mechanics to estimate properties and provide an understanding of composite behavior on the microscale; this enables design and fabrication of composites having the desired physical properties. Of the many micromechanics approaches in the literature, the most versatile and useful is given by the Halpin-Tsai equation.

Strength and Failure Criteria. A unidirectional lamina exhibits strengths that are a function of both fiber orientation and loading direction. If the lamina is transversely isotropic, it exhibits six distinct ultimate strengths: longitudinal tensile and compressive; transverse tensile and compressive; interfiber shear; and shear through the fiber. Predicting these from constituent properties has proven difficult. Many of the simplifying assumptions used in predicting elastic and thermophysical properties do not apply to strength analysis. Mechanical failure is usually caused by fiber defects, weak interfacial bonds, voids, or residual stresses.

Phenomenological Failure Criteria. Complexity of failure processes and the concurrent difficulty in modeling have led to the adoption of several types of empirical criteria. The most common are maximum stress, maximum strain, and various polynomials that define a failure surface. Of these, the maximum stress criterion has proved the most useful for design purposes.

Toughness. In most structures, failure originates at a stress concentration necessary for design function, ie, bolt or rivet hole, joint, cutout, etc, or at a flaw, ie, manufacturing defect, in-service damage, etc. A useful structural material must tolerate reasonable stress concentrations and flaws without failing. The material property that quantifies this tolerance is toughness. Although not so high as some metals, composite toughness is much higher than that of either resin or fiber alone. Typical fracture surface energies range from $10^3 J/m^2$ to over $10^4 J/m^2$ (see also FRACTURE AND FATIGUE, FRACTURE).

Time-dependent Deformation and Failures. A comprehensive treatment of viscoelastic deformation in composites has been presented and shows that for thermorheologically simple materials, creep and stress relaxation can be predicted in composites if the orthotropic viscoelastic constants are known. Whereas the effect of time on fiber-controlled deformation is not large, its effect on strength is pronounced. Glass fiber is especially susceptible to static fatigue, a loss of strength under sustained load. This effect is accelerated by moisture, which increases the crack-growth rate in stressed glass.

Constituent and Composite Properties

Thermosetting Resins. Epoxy resins (qv) are compatible with all fibers used in advanced composites and are the most commonly used. They adhere well to most fibers, cure with relatively little shrinkage compared with polyesters and phenolics, have no condensate on cure, are reasonably thermally stable, and are moderately priced. Disadvantages include brittleness, considerable moisture absorption, and a tedious processing cycle. Any molecule containing the oxirane ring is classified as an epoxy; only two are used extensively in advanced composites. They are the diglycidyl ether of bisphenol A (DGEBPA) and tetraglycidylmethylenedianiline (TGMDA), ie, *N,N,N′,N′*-tetraglycidyl-4,4′-diaminodiphenylmethane. For higher temperature applications (ca 350°C), a thermosetting polyimide, PMR-15, can be used. It is very brittle and requires an extended high temperature, high pressure cure cycle. Typical neat resin properties are listed in Table 2.

Table 2. Properties of Matrix Resins[a]

Property	Epoxy	Polyimides thermosetting
elastic modulus, GPa[b]	2.8–4.2	3.2
tensile strength, MPa[c]	55–130	55.8
compressive strength, MPa[c]	140	187
density, g/cm^3	1.15–1.2	1.43
thermal expansion, 10^{-6}/°C	45–65	50.4
thermal conductivity, W/(m·K)	0.17–0.21	0.36
glass-transition temperature, °C	130–250	370

[a] At RT.
[b] To convert GPa to psi, multiply by 145,000.
[c] To convert MPa to psi, multiply by 145.

Thermoplastic Resins. Polyimide can also be made in an uncross-linked thermoplastic form by a condensation reaction. This material, DuPont's NR-150, is less expensive than the corresponding thermoset and easier to process. Maximum use temperature, ie, from 300 to 325°C, is less than the polyimide thermoset. Mechanical properties are also somewhat reduced by the porosity resulting from the condensation reaction.

A new class of thermoplastics, polyetheretherketone (PEEK) (qv) was introduced several years ago by ICI Ltd. in the UK. Compatible with both glass and graphite, this material has better toughness and moisture resistance than epoxy. Its disadvantages are much higher cost ($55/kg) and lower use temperature (ca 150°C) than epoxy.

Strong Fibers. There are several features common to most fibers suitable for use in advanced composites. High strength requires an atomic structure with a closely spaced, continuous network of strong bonds. None of the fibers exhibits any ductility on the macroscale; in tension, they behave as linear elastic brittle solids. The strengths they exhibit result mainly from the reduction of stress-concentrating flaws. Most strong fibers are sensitive to surface damage caused by handling.

Glass Fibers and Composites. Glass fibers are the oldest form of strong fibers used in composite structural materials. Continuous fibers are made by a gravity extrusion process. Many types of glass fibers have been made, but only two are used in structural composites: E (for electrical) and S (for strength). E-glass is a low alkali, aluminum-borosilicate composition with excellent electrical properties and reasonable strength and modulus. In chopped form it is the staple of the fiberglass molding-compound industry. S-glass is a magnesium-aluminosilicate composition with considerably higher strength and modulus (Table 3). S-glass has been used in filament-winding applications.

Graphite Fibers and Composites. Graphite fibers have been made from a number of precursors but are now mainly produced from polyacrylonitrile (PAN), pitch, and rayon. They derive their enormous strength and modulus from the alignment of covalently bonded basal planes of carbon atoms along the axis of the fiber.

Applications for graphite composites are mainly in the aerospace and sporting goods industries. Graphite-epoxy has been used in military aircraft for both primary and secondary structural members for some years.

Aramid Fibers and Composites. Covalently bonded organics offer the potential for high strength and stiffness fibers. The ideal strength of aligned organics has been estimated between 20 and 50 GPa, and a theoretical modulus of about 200 GPa has been calculated. Although these values should be attainable in any organic in which chains are closely packed and precisely aligned, the problems in processing such materials are formidable. The only materials approaching their theoretical potential in commercial quantities are the aromatic polyamides, or aramids, marketed by DuPont under the trade name Kevlar.

Kevlar fibers and composites are tough, light, strong in tension, and reasonably priced. They also exhibit excellent ballistic resistance, which has led to their use in Army helmets and in protective vests for law enforcement officers. Kevlar 49 composites are used for filament-wound rocket motor cases and pressure vessels. Kevlar epoxy is being used increasingly in aircraft exterior facings and cabin interiors.

Boron Fibers and Composites. Boron was the first high strength, high modulus fiber to be produced. It is made by the reduction of BCl_3 in a chemical vapor-deposition process onto a hot tungsten or graphite filament. The resulting amorphous boron has excellent properties, but the process is very costly. Current prices are ca $330/kg. Properties for tungsten core material are listed in Table 3. Graphite core material is potentially cheaper, but its properties are not as consistent.

Boron-epoxy composites have excellent properties and have been used in a few military aircraft, but cannot compete with graphite composites for most applications. Boron composites have been used in some sporting goods equipment, eg, tennis rackets and fishing rods.

Table 3. Fiber Properties[a]

Property	E-glass	S-glass	HS[b] graphite	HM[c] graphite	Kevlar 29	Kevlar 49	Boron, W core
diameter, μm	3–20[d]	9	6–8	7–9	12.1	11.9	100, 140, 200
density, g/cm³	2.54	2.49	1.7–1.8	1.85	1.44	1.44	2.65, 2.45, 2.38
tensile strength, GPa[e]	2.4	4.5	3–4.5	2.4	3.5	3.6	3.5
elastic modulus, GPa[e]	72.4	85.5	234–253	345–520	59	124	386–400
thermal expansion, 10^{-6}/°C	5.0	5.6	−0.5(a)[f] 7(r)[f]	−1.2(a)[f] 12(r)[f]	−2(a)[f] 58(r)[f]	−2(a)[f] 59(r)[f]	5.4
thermal conductivity, W/(m·K)	1.86	2.55	8–25(a)[f]	105(a)[f]		3.1	2.7
cost, $/kg	1.1	22–33	66–110	220–660	11–22	22–33	330–440

[a] Data culled from many sources; should not be used for design.
[b] High strength.
[c] High modulus.
[d] Most common roving sizes are 9, 10, and 13 μm.
[e] To convert GPa to psi, multiply by 145,000.
[f] (a) = axial; (r) = radial.

Laminate Design, Analysis, and Properties

Unidirectional composites have properties suitable for one-dimensional stress states, such as might be present in a leaf spring or fishing rod. But even these apparently simple structures must withstand both bending and torsional moments. Indeed, most structures must support a variety of loads. The low values of transverse tensile and interfiber shear strengths for unidirectional composites usually require that fibers be oriented to support all principal tensile and shear loads. This is done by bonding two or more lamina together to form a laminate (see also LAMINATES).

Traditionally, structural design has focused on well-characterized, tough metals such as steel, aluminum, and titanium. Application of metal-design precepts to advanced composites can be catastrophic. Composite design requires an integrated approach involving structural, materials, and manufacturing engineering, cost and value analysis, and quality assurance and certification.

Current Status

Continuous fiber composites have become a mature technology. Improved understanding and control of processes have increased fiber mechanical properties and reduced variability. Fiber strengths have risen to the point that out-of-plane reinforcement (three-dimensional) can be employed without jeopardizing in-plane strengths. Thermoplastic resins are proving to be superior to thermosets for automated processing without sacrifice in properties. Design and production methods for composites have steadily become more sophisticated, so that they now can be considered engineering materials.

BIBLIOGRAPHY

R. M. Jones, *Mechanics of Composite Materials,* Scripta Book Co., Washington, D.C., 1975. Contains excellent treatments of anisotropic elasticity and laminated plate theory.

D. Hull, *An Introduction to Composite Materials,* Cambridge University Press, Cambridge, UK, 1981. Contains an excellent treatment of composite-strength theories for various modes of failure.

G. Lubin, ed., *Handbook of Composites,* Van Nostrand Reinhold Co., New York, 1982.

A. Kelly and N. H. Macmillan, *Strong Solids,* 3rd ed., Oxford University Press, Oxford, UK, 1986.

FRANK P. GERSTLE, JR.
Sandia National Laboratories

COMPOSITES, FABRICATION

Composite fabrication techniques have evolved from early manual labor to microprocessor-controlled systems. For convenience, terms frequently encountered in composite fabrication are defined below:

Angle of winding (wind angle)	The angle the roving band is laid with respect to the mandrel axis of rotation.
Closed-end vessels	Parts that have much smaller diameters or totally closed domes at the ends.
Debulking	A process in which air is squeezed out of prepreg laminate in order to promote adhesion.
Doff	Roving package.
Fiber payout	Transfer of rovings from delivery system to product line.
Mat	Chopped or random reinforcement material cut to the contour of a mold, usually impregnated with resin just before or during the molding process.
Overtravel	The additional carriage or eye travel beyond the ends of the part mandrel that is necessary to provide laydown of the fiber on the mandrel.
Peel ply	The outside expendable layer of a laminate, which is removed to improve bonding of additional layers.
Prepregs	Reinforcements in the form of a cloth impregnated with thermosetting resin advanced in

cure only through the B-stage. The term also covers fabrics such as jute and rayon, which have been impregnated with a thermoplastic resin, eg, vinyl or ABS.

Roving — A form of reinforcing fiber glass comprised of 8–120 (and usually 60) single strands gathered into a bundle, and treated with a coupling agent to promote adhesion of the glass to the plastic matrix.

Tape — A form of reinforcement similar to mat and broad goods, cut in 7.5–30 cm widths, and of flexible consistency for laying onto a mold. Tapes are generally prepregs and are supplied with backing paper to prevent loss of integrity prior to use.

Tow — Term used instead of roving to designate fiber strands when referring to graphite or boron reinforcements in the fiber strand form.

Wet systems — Composite fabrication processes and equipment that incorporate reinforcement impregnation as part of the process, just before raw material die entry or mold contact.

Product and Machine Design

An integrated approach is necessary to achieve the best results in product automation and economics. Cooperative effort among all disciplines is required from the start of a project.

The first step during the preliminary stage is product design (qv). Structural analysis to determine property requirements is recommended. The design factor with the greatest impact on cost and performance is fiber orientation control. Strength properties depend on fiber orientation, and departures from the optimum can result in drastic property reduction.

The most common composite-reinforcing materials are fiber glass, carbon fibers, aromatic polyamide (aramid), and boron. The term composite is used throughout to include glass fibers and other so-called advanced composite materials (see COMPOSITES).

In addition to the common roving strands, raw material is available in many other forms, such as preimpregnated and wet systems, mats (paper and nonwovens), woven specialties with knit and braided reinforcement, multiple-ply broad goods, and core materials as they apply to automation, fillers, extenders, and additives (films, coatings, plating, etc). Matrix selection is another important design consideration; resin chemistry can accommodate nearly every conceivable product application.

Methods and Processes

Common forms of processing used to produce nonstructural components are SMC [sheet-molding compound (qv)], BMC (bulk-molding compound), preform molding, injection molding (qv), and spray-up. Processes used in the manufacture of structural components, ie, those using continuous fiber reinforcements, include pultrusion (qv) (Fig. 1), filament winding (qv), tape placement (Fig. 2), and pulforming. Specialty machines have been custom designed for a wide variety of applications such as composite-faced plywood siding, graphite epoxy golf clubs, reinforced foam insulation, automotive drive shafts, artillery shells, helicopter rotor blade spars, aircraft components, space construction, and many more.

Economic Aspects

The composites industry has usually been perceived as providing specialty parts at premium prices. Today, however, com-

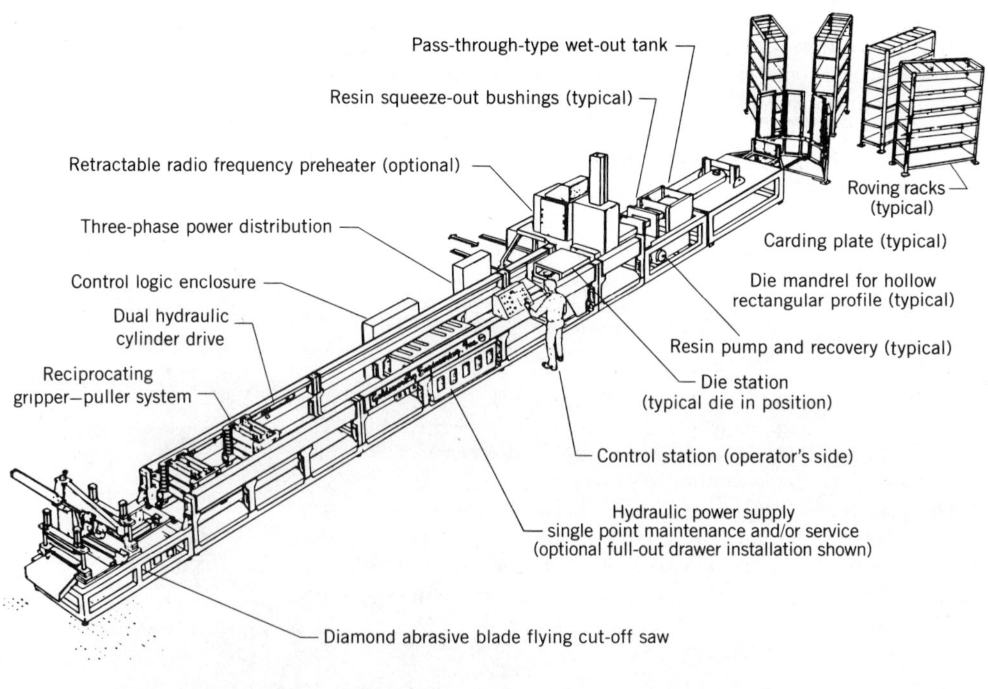

Figure 1. Typical pultrusion machine. Courtesy of Goldsworthy Engineering, Inc.

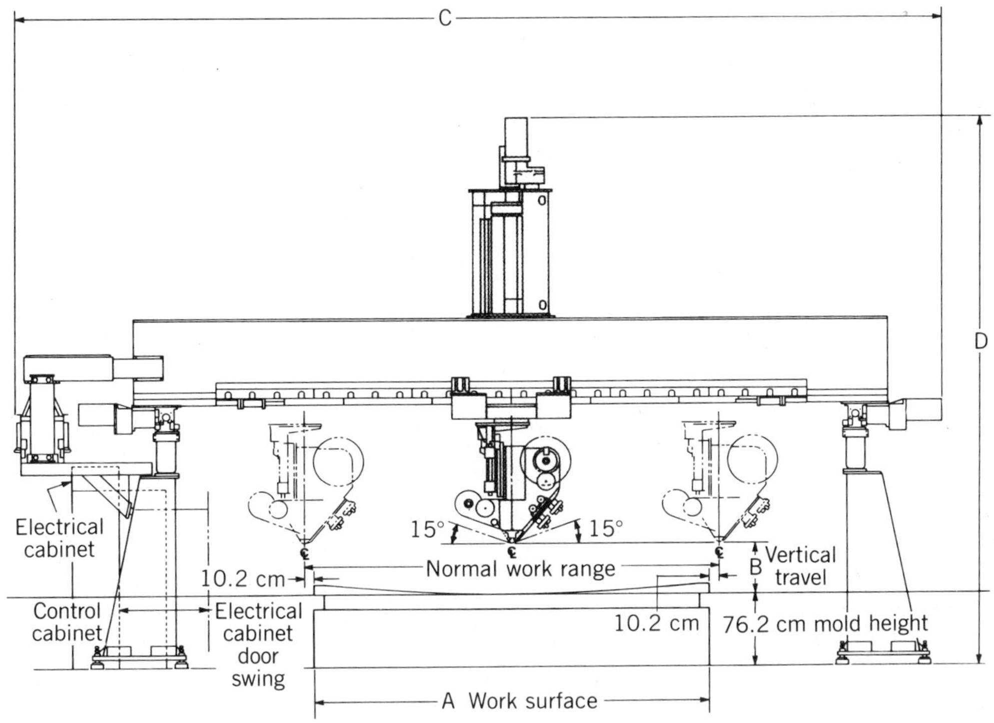

Figure 2. Line drawing of seven-axis CNC tape-placement machine. Courtesy of Cincinnati Milacron, Inc.

posites are being reclassified from "specialty" to "commodity" industry. The growth of composite product development has resulted in lower prices without loss of properties such as light weight, high stiffness, corrosion resistance, etc. Composites are now cheaper than traditional materials, and may become the economic barometer of the country, similar to steel in an earlier era.

Environmental Advantages. With respect to expensive safety and pollution-control devices, composites have a marked advantage. A steel-spring plant presents a variety of hazards and serious pollution problems, whereas composite fabrication is conducted at a console.

Quality Assurance Techniques

Quality assurance techniques for composite products fall into two categories: process control and product evaluation. Raw materials control is, of course, critical to all manufacturing processes, and is assumed in this article.

This type of manufacturing confronts problems not found elsewhere, such as process variables and determining the size and degree of a fault area in large composite pieces. Still another problem is that faults in composite materials do not propagate as in more traditional materials and are harder to locate. As a result, in-process quality assurance is of much greater importance than in traditional manufacturing. Some defects can be corrected while the process is under way; others can be detected only in the end product, and therefore corrective action must wait until another cycle (see also COMPOSITES, TESTING).

BIBLIOGRAPHY

G. Lubin, ed., *Handbook of Composites,* Van Nostrand Reinhold Co., Inc., New York, 1982.

M. M. Schwartz, *Composite Materials Handbook,* McGraw-Hill Inc., New York, 1984.

L. R. Whittington, *Whittington's Dictionary of Plastics,* Technomic Publishing Co., Inc., Westport, Conn., 1968.

W. B. Goldsworthy, *Volume Component Manufacturing,* short course presented to UCLA Extension, Composite Materials, Sept. 17–21, 1984.

W. BRANDT GOLDSWORTHY
Goldsworthy Engineering, Inc.

COMPOSITES, TESTING

The design and analysis of composite structures relies heavily on experimental data. Testing of composites serves a variety of purposes such as

1. Characterization of constituent materials, ie, fiber and matrix, for micromechanics analysis in order to predict the behavior of the lamina and hence of the laminates and structures.
2. Verification of micromechanics analysis, including effects of curing stresses, temperature, and moisture.
3. Qualification and characterization of the basic unidirectional lamina, which forms the building block of all laminated structures. This characterization, especially the determination of irreversible and ultimate properties, is essential in providing data for subsequent stress analysis using lamination theory.
4. Experimental stress analysis by measuring strain distributions and using the appropriate constitutive relations.
5. Fracture characterization, including identification of fracture mechanisms and modes, initiation, and propagation.
6. Testing under special conditions of loading, eg, fatigue, multiaxial, and high rate testing.

7. Assessment of structural integrity and reliability using nondestructive evaluation methods.
8. Life prediction through accelerated testing.

The various experimental methods include photoelastic methods, strain gauges, moiré, holographic and interferometric methods, ultrasonics, acoustic emission, x radiography, and thermography.

BIBLIOGRAPHY

J. M. Whitney, I. M. Daniel, and R. Byron Pipes, *Experimental Mechanics of Fiber Reinforced Composite Materials,* Monograph No. 4, Society for Experimental Stress Analysis, Brookfield Center, Conn., 1982.

I. M. Daniel in Z. Hashin and C. T. Herakovich, eds., *Mechanics of Composite Materials,* Pergamon Press, Elmsford, N.Y., 1982, pp. 473–496.

I. M. Daniel in G. C. Sih and A. M. Skudra, eds., *Fracture Mechanics of Composites,* North Holland Publishing Co., Amsterdam, 1985.

I. M. Daniel in A. S. Kobayashi, ed., *Handbook on Experimental Mechanics,* Prentice-Hall, Englewood Cliffs, N.J., 1987.

I. M. DANIEL
Northwestern University

COMPOSITION BOARD

The term composition board refers to panel products such as hardboard, insulating board, particleboard, medium density fiberboard, waferboard, oriented strand board, and cement-bonded boards. All of these particular boards are panels or plates rather than the small rectangular cross section associated with lumber-type board products. However, as the industry has developed, molded products with irregular shapes have also entered production. In addition, the materials used to produce panel products are now being used in the manufacture of rectangular-shaped lumber products.

Composition materials are made from comminuted raw materials, in the shape of fibers, shavings, various types of flake, and other types of particle. They are usually made from woodworking and forestry residues, low value tree species, and agricultural wastes. Some industrial and municipal refuse is also

used. Thus, previously wasted material is converted into useful products, a trend that will probably accelerate in the future.

The commercial products are available in large sheets and have smooth surfaces, uniform properties from panel to panel, and freedom from localized defects as are found in lumber and plywood products. Methods for aligning particles, particles manufactured with optimum particle geometry, new adhesives and adhesive systems, and adjustments in manufacturing parameters provide material that equal or surpass the structural capability and reliability of sawn lumber and plywood products. In some processes, composition-board products can be treated with fire retardants and preservatives, and modified to improve dimensional stability, eg, cooking to remove hemicelluloses.

Composition materials, as defined for this article, are categorized as wet-process and dry-process types. This terminology relates to the manner in which raw material is formed into the so-called mat before final compression or drying into the product in a hot press or oven. Certain fiberboard products are produced by the wet-process method. In this method, the

Figure 1. Hardboard products. Courtesy of Washington State University.

Figure 3. Insulation-board products. Courtesy of Washington State University.

fibers and appropriate chemicals are mixed in a water slurry and formed into the mat. The same types of product can also be manufactured according to the dry-process system. Here, the mat is formed by dropping the particles onto a moving belt. Most products produced are manufactured by a dry process. Dry-process products are bonded together with a synthetic adhesive, whereas wet-process products can be self-bonding. Production figures for hardboard (Fig. 1), particleboard (Fig. 2),

Figure 4. Examples of medium density fiberboard. Courtesy of Washington State University.

insulation board (Fig. 3), and medium density fiberboard (MDF) (Fig. 4) are given in Figure 5.

Material Properties

The raw material is obtained from logs, small planer shavings, fiber, or sander dust. It can come from a variety of species, with variations in density, acidity, moisture content, extractives, and extraneous matter. Density or specific gravity is an important factor in choosing the species. In general, the lower density species produce better panels.

Binders

Urea resins are the predominant binders used today, particularly in interior applications such as furniture, cabinets, underlayment, and mobile-home decking. Phenolic and isocyanate resins are favored for exterior applications, such as siding and sheathing.

Additives

Additives impart dimensional stability, fire and decay resistance, and improved strength properties. Additives include wax

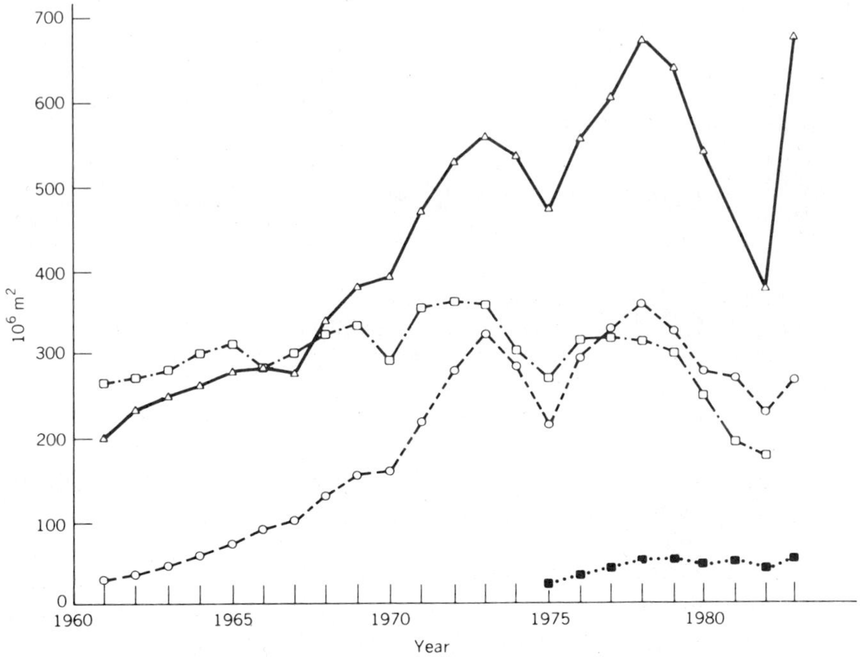

Figure 5. Production of hardboard, particleboard, insulation board, and medium density fiberboard in the United States: △-△, hardboard (3.2-mm basis); ○--○, particleboard (19-mm basis); □-·-□, insulation board (12.7-mm basis); ■···■, MDF (19-mm basis).

and oil-based additives; boric acid, mixtures of boric acid and borax, ammonium phosphates, and alumina trihydrate as fire retardants; and preservative systems based on chromated copper arsenate-type C and ammoniacal copper arsenate.

Health and Safety Factors

Problems of exposure to formaldehyde emission are being resolved by resin modification, scavengers, placing line operators in climate-controlled rooms, and good ventilation. Workers in contact with chemicals such as phenols, isocyanates, preservatives, and fire retardants take special precautions with clothing and breathing apparatus.

In wet-process plants, water pollution must be avoided by cleaning and recycling the water. In dry-process plants, dust is controlled by enclosing the offending areas and by bag houses on pneumatic air systems.

BIBLIOGRAPHY

T. M. Maloney, *Modern Particleboard and Dry-Process Fiberboard Manufacturing,* Miller Freeman Publications Inc., San Francisco, Calif., 1977, pp. 27–29.

T. M. Maloney, ed., *Proceedings of the International Particleboard Symposium,* 1–21, Washington State University, Pullman, Wash., 1967–1987.

H. J. Deppe and K. Ernst, *Taschenbuch der Spanplattentechnik* (in German), DRW-Verlag, Stuttgart, 1977.

THOMAS M. MALONEY
Washington State University

COMPOUNDING

In compounding, a polymer is modified physically and chemically by additives that change its properties. For example, the incorporation of glass fibers improves impact strength, and the incorporation of a pigment imparts color. The ways in which a polymer can be modified are manifold, and, as a result, a polymer is usually available from the compounder in a wide range of grades. These grades provide a plethora of material choices from which to make products.

Compounding also embraces the physical mixing of two or more polymers to form polymer blends (qv). Polymer blends may be miscible, eg, poly(phenylene oxide) in polystyrene, or immiscible, eg, polyurethane in polyoxymethylene.

The importance of compounding to the plastics industry cannot be overstated. A particular goal can be reached by employing various polymers, additives, compounding processes, and a variety of routes. The compounding line may be arranged in several ways; it involves many operations, each of which can be executed on more than one specific piece of equipment. This offers numerous choices, the best of which is frequently the simplest.

Ingredients

Ingredients include processing aids, modifiers, antiaging additives, flame retardants, and extenders.

Equipment

The precise quantity of additive is very important to ensure the required properties. To this end, the ingredients must be precisely weighed before compounding, using scales in a batch operation, a weight-belt feeder or continuous weighing (or metering) for direct additions to a continuous process. Possible routes to compounding are shown in Figure 1. This choice of route is not solely a technical matter, because the economics of the process, available equipment, and work loads must also be considered. Equipment includes nonintensive mixers, medium intensity mixers and intensive-fluxing mixers.

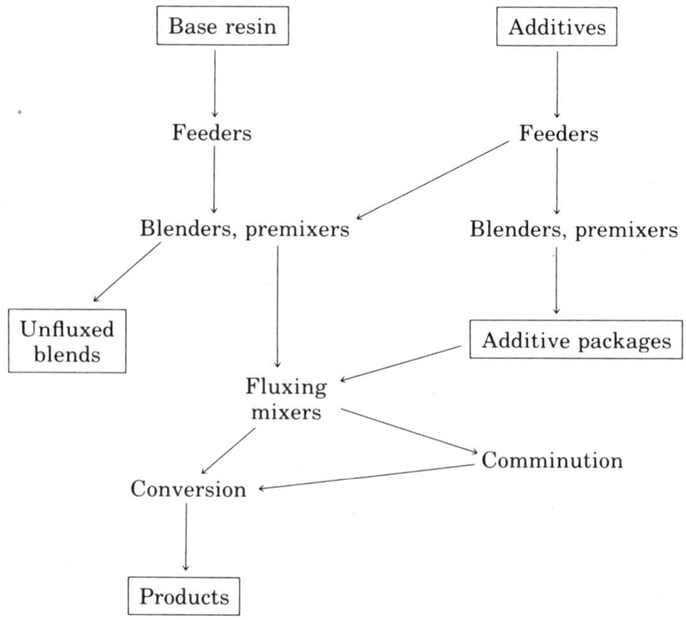

Figure 1. Compounding scheme.

BIBLIOGRAPHY

Plast. Compounding, 56 (Mar./Apr. 1981).

G. Mathews, *Polymer Mixing Technology,* Elsevier North Holland, New York; Applied Science Publishers, Barking, UK, 1982.

R. Gachter and H. Muller, eds., *Plastics Additives Handbook,* Hanser Publishers, Munich, 1985.

KEITH T. O'BRIEN
Celanese Corporation

COMPOUNDING RUBBER. See RUBBER COMPOUNDING AND PROCESSING.

COMPRESSION. See MECHANICAL PROPERTIES.

COMPRESSION AND TRANSFER MOLDING

Compression and transfer molding are the two main methods for molding parts from thermosetting plastics.

The Molding Process

In compression or transfer molding, a thermosetting molding compound is exposed to sufficient heat, generally ca 150°C, to soften or plasticize it sufficiently to flow into the mold cavity. The fluid plastic is held under pressure (>13.4 MPa = 2000 psi) for a sufficient length of time for the material to polymerize or cross-link and harden.

Compression Molding. The preferred technique has always been compression molding. The thermosetting plastic is placed in the cavity in the bottom half of an open-heated mold mounted in a press. As the press is closed, bringing the bottom half of the mold against the top half under pressure, the plastic material begins to soften and is formed by the pressure of the two mold halves. After a minute or more, depending on the material and thickness of the part, the press is opened, and the plastic has assumed the shape of the mold cavity. A simple flash mold is shown in Figure 1.

Transfer (Plunger) Molding. In transfer or plunger molding, the mold is closed before the molding compound is introduced in its fluid state through a small opening or gate leading to the mold cavity. This technique is used frequently for fragile mold sections, molded part sections > 4 mm thick, or when an insert is lightly restrained in the cavity for molding in place. In such applications, closing the mold with the granular compound, as in compression molding, might cause damage or displace the insert. When the material enters in a softened state, flow speed and pressure can be controlled to prevent damage.

In true transfer or pot-type transfer molding, the mold is closed and placed in an open press. The charge of molding compound is introduced into an open pot at the top of the mold, the plunger is placed into the pot, and the press is closed. In plunger molding, the plunger is a part of the press rather than a part of the mold. It is usually driven by a hydraulic circuit and a cylinder attached to the head of the press (Fig. 2) and can therefore be considerably smaller in protected area than the pot-type plunger. In this article, the term transfer molding refers to plunger molding, the most common type of transfer molding in use today.

Cold Molding. In cold molding, a variation of compression molding, a compound is compacted in a mold, usually at room temperature. The compound contains a high percentage of filler with a thermosetting binder. The compacted part is removed from the mold and placed in an oven. Under heating, the binder is cross-linked and together with the filler forms a rigid part. Such a process is used for making certain high temperature porcelain and ceramic parts used in the electrical industries.

Elements of the Process. Five elements make up the compression and transfer molding process: part design, material selection, mold design and construction, machines, and operators.

Processing Variables. Processing variables that need to be considered include molding compounds, (thermosetting plastics and elastomers), temperature, cure time, breathe and dwell, pressure, part size and design, and press closing and plunger speeds.

Equipment

The equipment for compression and transfer molding consists of the machine, auxiliary equipment, and molds.

Compression-molding Presses. Compression-molding presses consist of the frame, driving mechanism, and controls.

Transfer Presses. In transfer or plunger molding, the mold halves are brought together under pressure, as in compression molding. The charge of molding compound is placed into a pot (included in the mold) and driven through runners and gates into the mold cavities by means of a plunger (included with the press).

Auxiliary Equipment. Auxiliary equipment includes heating devices, preformers, ejector mechanisms, flash removal systems, postcure systems, and vacuum venting systems.

Tooling. Hardened-steel hand tools are usually selected for experimental work and low or moderate production, whereas self-contained tools are used for moderate or high production rates.

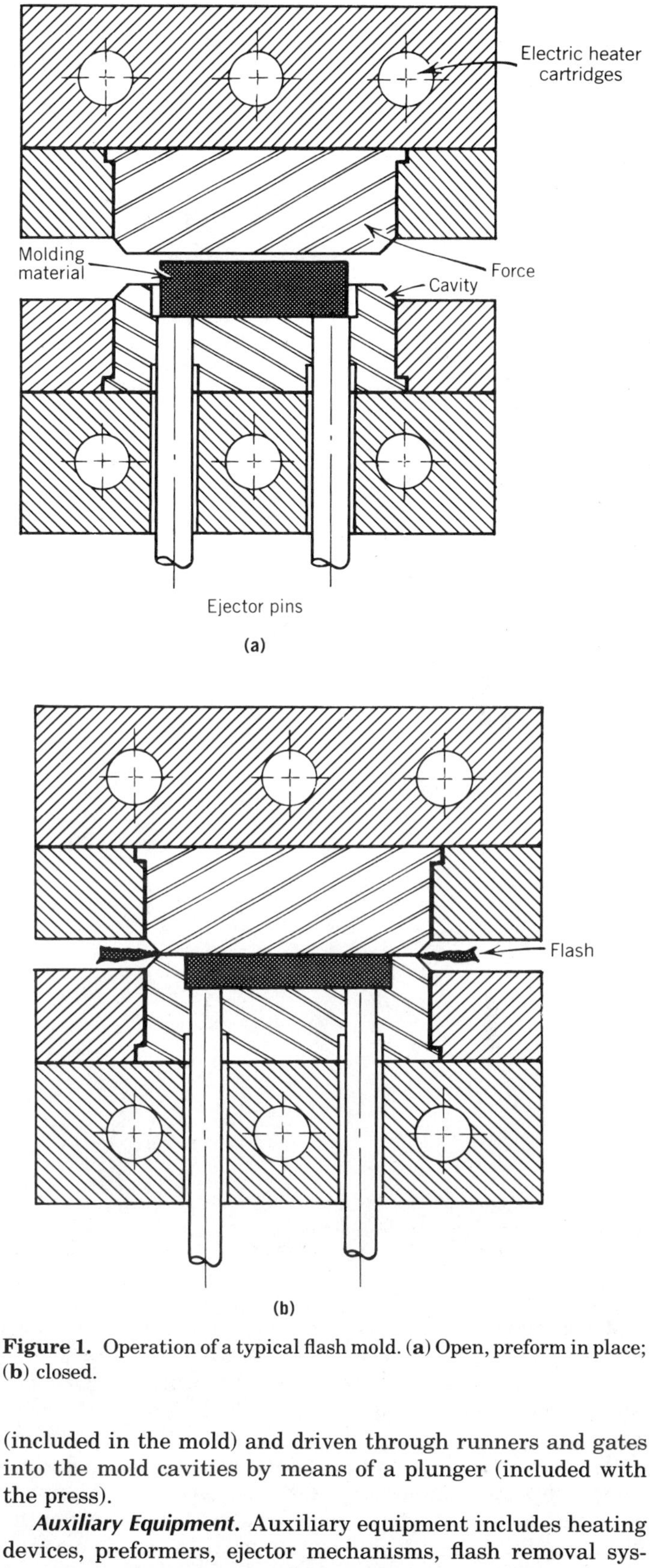

Figure 1. Operation of a typical flash mold. (**a**) Open, preform in place; (**b**) closed.

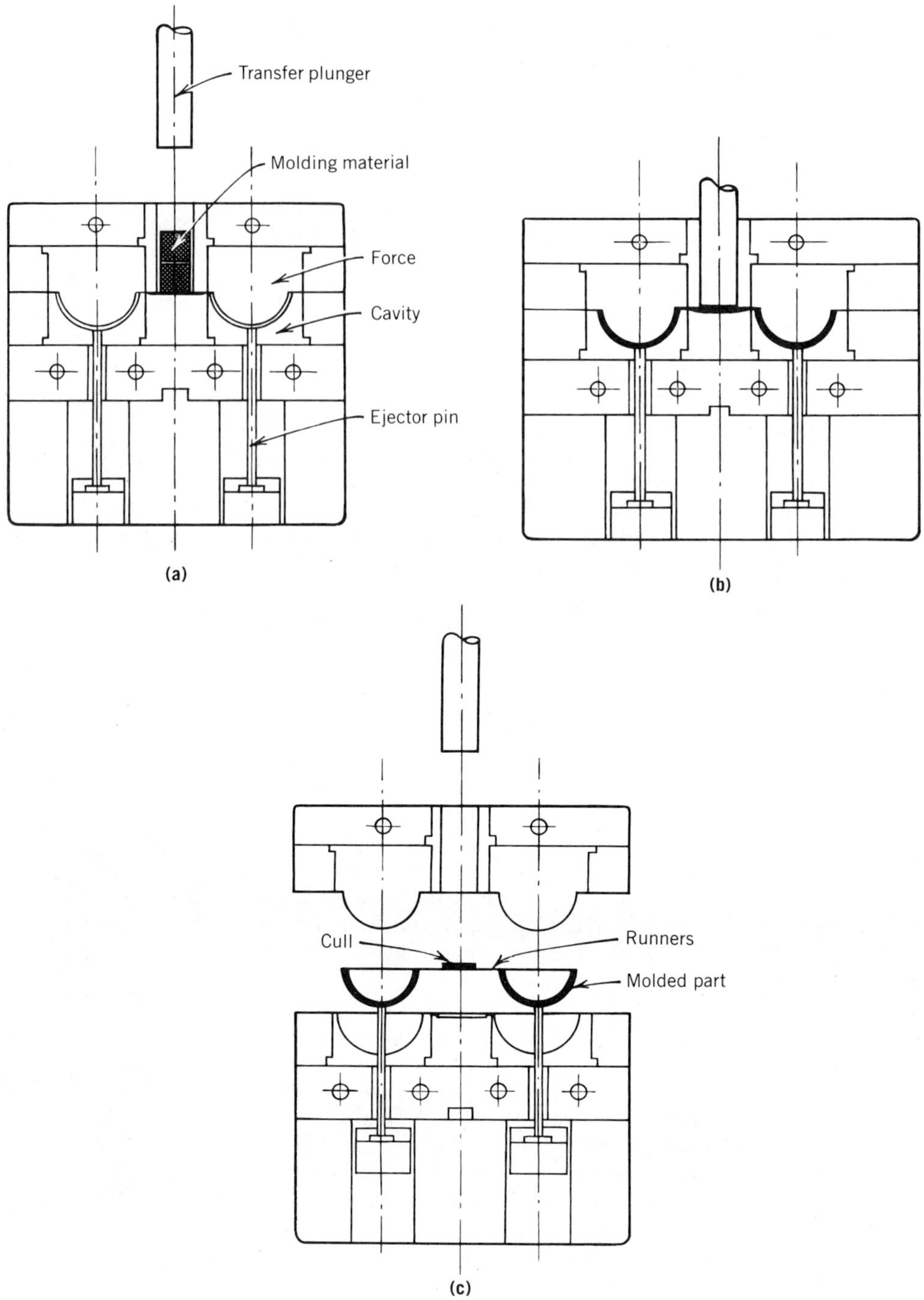

Figure 2. Plunger mold configurations. (**a**) Mold closed, pot loaded; (**b**) mold closed, plunger down; and (**c**) mold open, molded part, cull, and runners ejected.

Cavity Design. The cavity should be shaped for easy removal of the molded part. Uniform thicknesses of the plastic are desirable to minimize localized stresses from uneven curing. If an insert is to be molded in, the cavity design must provide adequate support during mold closing, plastic flow, and curing.

Maintenance. Most problems encountered with molds are due to unclean parting surfaces. Stickage can be temporarily remedied by a light application of release agent, but cavities should be highly polished to eliminate sticking. The operator should visually check the mold before every cycle.

Automatic Molding Machines

Automatic machines actuate the next molding cycle on satisfactory completion of the preceding cycle. In automatic compression molding, stripper combs or air blasts are generally used to collect the molded parts from the ejector pins, remove them from the mold area, and drop them into bins or baskets. Usually a mechanical or electronic safety device ensures that no molded parts remain in the mold area and then starts the next cycle. Should a part remain in the mold area, the safety device does not indicate an "all-clear" condition, and the machine shuts down.

In automatic compression machines, feed mechanisms meter a charge from a hopper and deliver it to the cavities in the open mold, frequently by means of a sliding plate. In automatic transfer molding, especially on smaller machines with capacities to approximately 60 g, material feed is generally effected from a hopper or preform bin to the transfer pot. After the charge has been delivered, the machine passes through the same molding cycle as the semiautomatic machine. Most large automatic machines for compression and transfer molding use auxiliary preheating of the plastic before it is fed to the mold.

Special Applications

Considerable ingenuity has been demonstrated in the application of plastics for a wide variety of uses, eg, melamine dinnerware, encapsulation of electronic devices, and screw closures.

Economic Aspects

Approximate prices of molding equipment are given in Tables 1 and 2.

Table 1. Approximate 1985 Prices of Thermosetting Molding Equipment

Characteristics	Dimension, cm	Price, $
Molds		
hand, single		
single small cavity	2.5[a]	3,000
single large cavity	10.0[a]	4,500
production		
four cavities	7.5[a]	20,000
fifty cavities	2.5[a]	50,000
Automatic preformers		
200 kN[b]	5.0	20,000
750 kN[b]	10.0	50,000
Electronic preheaters		
1 kW		4,000
10 kW		10,000

[a] Cavity diameter.
[b] 1 N = 0.101 kgf.

Table 2. Approximate 1985 Prices of Molding Presses, $

Press	300–1,500 kN[a]
manual	
compression	10,000–20,000
transfer	15,000–30,000
semiautomatic	
compression	35,000–45,000
transfer	40,000–55,000
automatic	
compression	75,000–100,000
injection	60,000–100,000

[a] 1 N = 0.101 kgf.

BIBLIOGRAPHY

J. Frados, *Plastics Engineering Handbook of the Society of the Plastics Industry, Inc.,* 4th ed., Van Nostrand Reinhold Co., Inc., New York, 1976.

Proceedings of the 42nd Annual Technical Conference of the Society of Plastics Engineers, Inc., New Orleans, La., 1984, SPE, Brookfield Center, Conn., 1982.

Proceedings of the Regional Technical Conference on Thermosets, Philadelphia, Pa., 1981, Society of Plastics Engineers, Brookfield Center, Conn., 1981.

Proceedings of the Regional Technical Conferences on Thermosets, Philadelphia, Pa., 1985, Society of Plastics Engineers, Brookfield Center, Conn., 1985.

John L. Hull
Hull Corporation

COMPUTER-ASSISTED DESIGN

The evolutionary nature of the design process offers many opportunities for cost-effective usage of computers. The computer can be used to make rapid and complex calculations, document each phase of the design, and store data. Computers are thus used in four general areas of engineering and design:

1. Engineering analysis calculations.
2. Graphical documentation (computer-aided drafting).
3. Technical data bases.
4. Digital electronic communications.

Applications

Computer assistance is expected to be integrated into the design process from initial concepts and the traditional design activities, throughout construction, and on into start-up, operation, and maintenance of the facilities, including the flow of information from the client to the design group to the constructor and back to the client. Sophisticated computer technology will merge with high speed communications technology, expedited by decreasing computer costs.

System Selection

For manual design methods, some rough measures have been devised for estimating effectiveness and progress, for example,

1. Design cost as a percentage of total plant investment, where size, nature, and complexity of the plant are considered as well as scheduling constraints (necessity for overtime, etc.).
2. Design cost based on the number of drawings by type and possibly the number of model tables to be built.
3. Duration of the design phase as a function of project size and complexity, including the determination of the point in design at which the specifications and drawings of different types can be issued to the construction organization.

Design offers a very large number of opportunities for applying computers in a cost-effective manner. Definition of the specific tasks involved as well as their relative costs is necessary if computer techniques are to be used in a rational and prudent manner. The recommended way to ensure selection of a system that meets the requirements is to prepare functional and performance specifications, solicit bids, and test the proposed hardware and software under actual operating conditions. Demonstrations staged by vendors tend to be unreliable, regardless of the honesty and integrity of the vendors.

BIBLIOGRAPHY

A-E-C Automation Newsletter, Automation Newsletter Companies, Inc., Carlsbad, Calif.

Anderson Report, Anderson Publishing Co., Simi Valley, Calif.

S. Klein Newsletter on Computer Graphics, Technology and Business Communications, Inc., Sudbury, Mass.

John M. Fisher
Joseph D. Halford
E.I. du Pont de Nemours & Company, Inc.

COMPUTER-ASSISTED ENGINEERING

Definition of Terms

In the course of only a few years, graphics-compatible computers with appropriate software have been developed from automatic drawing machines into universal tools for the design engineer. Initially they were used to support purely manual drafting and provide a convenient archive of the drawings produced. With the introduction of network control (NC) and computer numerical control (CNC) processing machines, it became necessary to convert the geometries available in the computer into control data sets and program for production machines. Computers with suitable software also provide data banks, perform calculations, and simulate cycles and events. This computer-assisted design (CAD) activity is referred to as computer-assisted engineering (CAE); the relevant programs are termed CAE software.

Hardware. Initial use of CAE programs was confined to mainframe and large minicomputers because of the capacities required. Rapid advances in hardware have produced microcomputers or personal computers (PCs) with performances adequate to run many CAE programs.

The choice of hardware installed by the user depends on two main applications of CAE software. Where CAE programs are provided on a time-sharing basis by a network operator, the user needs only to install an intelligent graphics terminal and a telephone modem in order to utilize programs accessed on a mainframe computer. Charges incurred for use are calculated from the computing capacity, storage capacity, and leased line or connection costs. With increasing use of the facility, the second option, involving purchase of a software license and operation on an in-house computer, becomes economically attractive.

Work stations with one or several user terminals are nowadays available for use with software of the type outlined above. These are based on a 32-bit computer, generally with virtual internal memory, a hard disk with some tens of megabytes (MB) (for some applications several hundred MB) and input–output (I/O) devices appropriate to the application. These include graphics screens and printers, possibly supplemented by plotters and hard-copy systems. A mouse or menu board and digitizer ensure rapid, user-friendly inputting of data and commands. A workplace often consists of a work station of this type with two screens: one for drawing, designing, and displaying simulation processes with the aid of the graphics coprocessor and a fast-screen repeat memory, the other for alphanumeric communication with the computer core (Fig. 1).

Methods of Calculation. Analytic solutions of differential equations can be performed only for solids with simple geometries (eg, plate, beam, or cylinder). The fact that computers no longer have any difficulty in performing an extremely large number of calculations in a very short time permits numerical solution of differential equations even for complex geometries of the kind usually encountered on technical solids. The real geometry of the solid to be calculated is divided into a mesh of finite elements. The elements used to describe areas are triangles, squares, rectangles, and polygons; for volumes, the corresponding three-dimensional solids such as prisms and cubes are used to occupy the space to be calculated. In the calculation, the sums of the corner points of these elements, referred to as nodes, represent the complete solid. An alternative to the finite-element method (FEM) is the finite-differences method, in which the differential equation is solved by transition to a difference calculus. A recent technique is the boundary-element method, in which the profile of the solid represents the object as a whole; the profile can be redefined at the beginning of each step in the calculation.

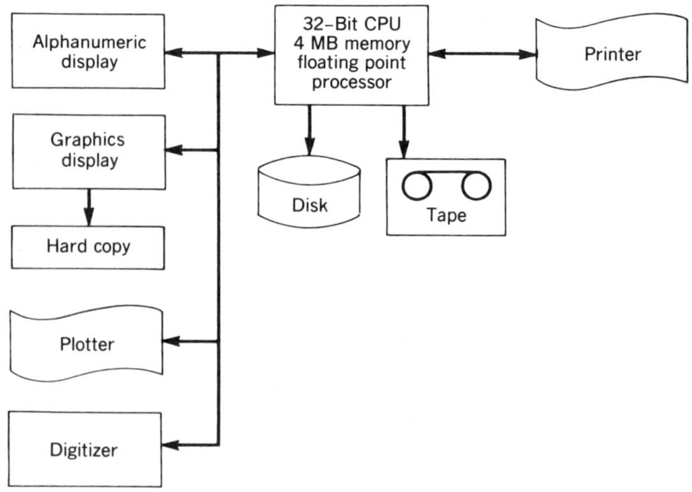

Figure 1. Configuration of a CAD/CAE work station. CPU = control processing unit.

Preprocessors. Manual creation of a finite-element (FE) mesh is exceedingly time-consuming. The computer allows automatic subdivision of the areas and solids requiring calculation into suitable elements. Programs used to structure geometries in this way are known as mesh generators. Mesh generators for two-and-a-half-dimensional meshes work interactively and semiautomatically. Figure 2 shows the FE surface mesh of a lawn-mower housing.

Postprocessors. Graphic display is nowadays invariably chosen for the presentation of numerical results. Programs that convert calculated results into graphic displays of some kind are referred to as postprocessors. They also provide facilities for the translation and rotation of the display and generally possess a zoom function. An important component of a postprocessor is its hidden-line algorithm, which enables it to calculate hidden edges for any desired perspective and suppress these during presentation.

The option of using colors, and thus filling areas, permits highly convenient displays, including shading. The visual design can be derived directly from the technical design or used to generate advertising graphics.

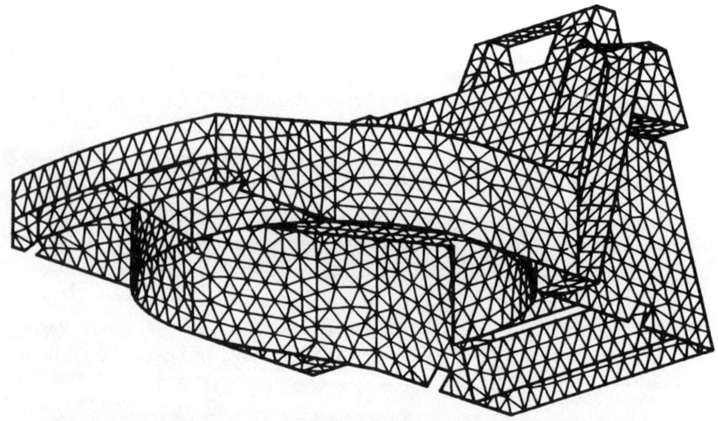

Figure 2. Finite element method (FEM) mesh of a lawn-mower housing.

Computer-assisted Engineering Programs for Plastics Technology

Use of CAE programs, already relatively widespread in mechanical engineering (design, plant planning) and electrical engineering (circuit design, layout), is now increasing in other engineering disciplines, for example in the design and simulation of chemical reactions. In chemistry, it is used for molecular calculations and reaction simulation, and in plastics technology for draft design of processing machines, design of articles produced by compression or injection molding (calculation of nonstationary flow phenomena, solidification, holding pressure, and shrinkage), part layout (including material selection, design of the article, and mechanical loading), in mold design (filling-pattern analysis, detailed rheological layout, holding-pressure analysis, thermal layout, and mechanical layout), reaction injection molding and sheet molding compound molds, layout of extrusion dies, and design and production of fiber-composite articles.

Computer-assisted Manufacturing for Injection Molding Machines

Assuming that a feasible machine–mold combination has been established by machine occupancy planning, an optimum working point for this combination can be determined. The prerequisites are the data outputs for the mold filling and holding pressure phases from the simulation program, together with the geometry data for the mold and the machine. Given these data, process variables can be transferred as machine parameters for the injection molding machine.

Further Utilization of Data

Data can be used for a great number of conceivable and useful applications. Apart from production scheduling, which becomes effective only when such data are available, the information can be exploited for the preparation of operating manuals, training documentation, and advertising material, as well as for design and production of molds, product design, and creation of networks.

BIBLIOGRAPHY

Some available Software: *CADMOULD,* Institut fur Kunststoffverarbeitung, Aachen, FRG; *MOLDFLOW,* MOLDFLOW Australia, Pty. Ltd., Boronia, Australia; *TMC,* Plastics and Computer Inc., Montclair, N.J.; *POLYCOOL,* General Electric Company, Schenectady, N.Y.

L Manzione, ed., *Computer Aided Engineering for Polymer Processing: Application of CAE in Injection Molding,* Hanser Verlag, Munich, FRG, 1989.

G. Menges, M. Effing, and E. Schmachtenberg, *Plastverarbeiter* **9,** 96 (1987).

G. Menges
H. Recker
Institut für Kunststoffverarbeitung

COMPUTER-ASSISTED MANUFACTURE.

See Processing, computer control.

COMPUTERS

Computers permeate all areas of polymer science and engineering, as they do most areas of business, technology and entertainment today. Computer technology is implied or directly discussed in most articles of the *Encyclopedia* because modern design, synthesis, characterization, and manufacture of polymers and their products are impossible without computers. This is particularly apparent in articles on analysis, eg, spectroscopic and chromatographic methods, and processing.

BIBLIOGRAPHY

T. Provder, *ACS Symp. Ser.* **197** (1982).

CONDENSATION POLYMERIZATION

Condensation polymerization is defined by IUPAC as polymerization by a repeated condensation process, ie, with elimination of simple molecules. The products of condensation polymerization reactions are termed condensation polymers. Just as esters and amides are products of simple condensation reactions of monofunctional reactants, so polyesters and polyamides are the products of condensation polymerization of difunctional acids with difunctional alcohols or amines. Water is frequently the small molecule split off during condensation; other common by-products of polycondensations are ammonia, hydrogen chloride, carbon dioxide, sodium bromide, nitrogen, methanol, etc.

BIBLIOGRAPHY

Pure Appl. Chem. **40,** 477 (1974).

CONDUCTING POLYMERS.

See Electrically conductive polymers; Photoconductive polymers.

CONDUCTIVE PLASTICS

Recent developments in conductive plastic composites using high aspect-ratio fillers have produced plastic formulations that combine the desired conductivity of metal with the processing ease and economy of plastic. These new specialty fillers take up where the traditional carbonaceous and other powder fillers leave off and increase composite conductivity from many times to many orders of magnitude higher than the case polymer. Uses for these unusual resin composites include heat exchanger, resistive and inductive heating, static dissipation, and electromagnetic interference (EMI) shielding applications.

Composite Fillers

Low Aspect-ratio Composite Fillers. Most mineral fillers and other additives to polymer composites cause only a minor reduction in the resistivity, without truly enhancing conductivity. Glass fibers added for strength and rigidity cause such reductions, as will talc and mica fillers.

Thirty years ago formulations had already been developed that significantly improved the conductivity of plastics. These formulations employed powdered fillers of metallic or carbonaceous characteristics. Along with improvements in heat conductivity, improvements in electrical conductivity were also being pursued. Plastics were compounded with various forms of carbon black to produce antistatic (resistivity, 10^{9-13} ohm·cm) and static dissipating (resistivity, 10^{5-9} ohm·cm) formulations.

High Aspect-ratio Composite Fillers. Conductive fillers of high aspect ratio (length-to-thickness) began to be tested 20 years ago. Initially, work centered around radar chaff, long, thin pieces of metal and metallized glass that were used to jam enemy radar.

Many of the processing problems encountered with the high aspect-ratio fillers are due to the key to their effectiveness, that is, their high aspect ratio. During melt processing the particles would break, distort, tear apart, and form agglomerations, thus reducing their effectiveness in the end product. The next logical step in filler development was toward a combination of greater durability and moderate aspect ratio.

Highly Effective Composite Fillers. ***Moderate Aspect-ratio Fibers.*** Aluminum fibers were first produced in the mid-1970s by a process called rapid solidification. The long ductile fibers (12–18 mm and longer) were, however, difficult to melt-process with most thermoplastics, making lay-up and casting processes the most easily employed. For melt-processing, the intermediate lengths of 3–6 mm were used.

High Strength Fibers. Fine stainless steel fibers can be produced by a repeated bundling and drawing process. Stainless steel fibers offer distinct advantages over the previously tested fillers. Being fine, typically 8 μm in diameter, a high aspect ratio is easily achieved at a relatively short fiber length. This ensures a high degree of conductivity at a low volume loading. Such products are easily processed in traditional thermoset and thermoplastic machines. A second advantage is the tough, flexible nature of the metal fiber.

Metal-flake Fillers. A variation of the rapid solidification process produces flat, rectangular metal flakes. Like many rapid-solidification processes, the range of alloys being processed is almost unlimited. The metal alloy can be modified to improve its strength and resistance to degradation during melt processing.

Metallized Mica Platelets. Nickel-plated mica combines the aspect ratio of an aluminum flake with the processibility of a fine particle and the inactivity and conductive oxide layer of nickel.

Low Aspect-ratio–Low Density Fillers. Low aspect-ratio fillers such as metallized or hollow glass spheres are also used as conductive fillers. Their spherical shape results in a very low aspect ratio, but often gives optimum performance in finely detailed parts. The most common metal coating employed is silver (Fig. 1), resulting in a very high particle-to-particle conductivity due to the inherent conductivity of the silver oxide.

Carbonaceous Fibers. Fine graphite fibers, based on polyacrylonitrile (PAN) or pitch, are also used in conductive plastic formulations. They offer some of the advantages of carbon black, eg, uniformity of conductivity even in the resin-rich skin, and reinforcement.

Compounding Techniques

Regardless of the filler type, the key to a highly conductive molded part lies in the compounding. A molding machine can be used for compounding a concentrate with a high loading of steel fibers or metallized graphite. The composite is usually precompounded to ensure product uniformity. The critical factors are filler volume loading and low shear processing.

Formulation. The choice of the resin matrix must be based on its chemical resistance, weathering, flame retardancy, and other properties, as well as on density and modulus increase.

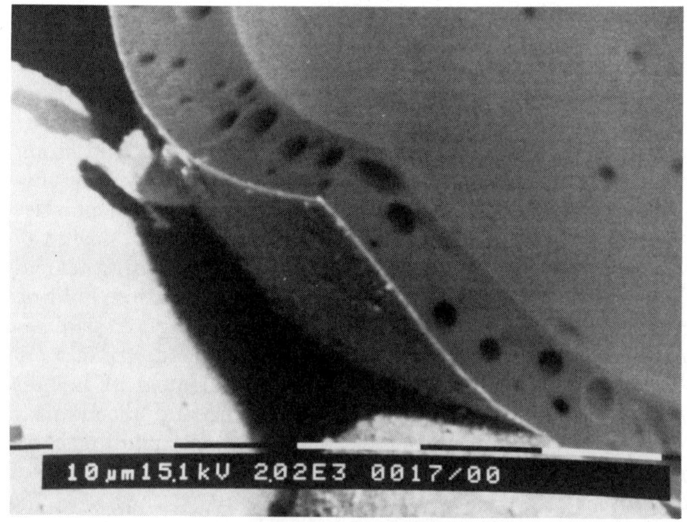

Figure 1. Silver-plated microsphere.

Extrusion. Like most filler systems, metal and metallized flakes increase the viscosity of the melt. Waxes, wetting agents, silanes, titanates, and other agents can be added to the melt or tumbled onto the filler before addition to the compounder in an effort to aid processing and reduce melt viscosity.

Equipment. Because filler aspect ratio is a key element in the conductivity of a composite, equipment and processing parameters must be chosen to minimize damage. Shear must be kept low, although good dispersion must still be maintained. Often a single-screw pelletizing extruder with worn-out elements gives better results than a more recent unit designed and maintained for ultimate color development.

Molding. Most conductive composites can be easily molded by any standard equipment. Properly formulated thermosets can be compression, transfer, or injection molded. Thermoplastics are readily injection molded under conditions typical for the base resin type, providing that the mold has been properly designed.

Measurement and Testing. Composite electrical conductivity and EMI shielding are measured by techniques developed by ASTM or related methods. The transmission-line method measures the performance of the composite under far-field conditions. A dual-chamber box or isolated room is used for near-field studies. The test results are indicative of the inherent properties of the composite, but cannot predict how the electronic device behaves in practice. The determination of flame retardancy in plastics is similar, where the laboratory tests cannot accurately predict field performance.

Thermal conductivity can be measured through the sample, perpendicular to the flat surface, or in the plane parallel to the surface. In a highly oriented molding the two directions will give greatly different values of thermal conductivity.

BIBLIOGRAPHY

A. Szczepanek, *Elektrisch leitende Kunststoffe*, 109 (1986).

L. Naturman, *Polym. Plast. Technol. Eng.*, 1 (1981).

R. M. Simon, *Designing for Thermal Conductivity with Flake Filled Composites*, 1986.

R. M. Simon, *International Symposium on Electromagnetic Compatibility,* Tokyo, October 1984.

ROBERT M. SIMON
USTEK Incorporated

CONDUCTIVITY, ELECTRICAL. See ELECTRICAL PROPERTIES.

CONDUCTIVITY, THERMAL. See THERMAL PROPERTIES.

CONDUCTORS. See CONDUCTIVE PLASTICS; ELECTRICAL PROPERTIES; ELECTRICALLY CONDUCTIVE POLYMERS; PHOTOCONDUCTIVE POLYMERS.

CONFIGURATION. See CONFORMATION AND CONFIGURATION.

CONFORMATION AND CONFIGURATION

The conformation of a polymer molecule describes the geometrical arrangement of the atoms in the chain achieved through rotations about or stretching of its chemical bonds and bending of its valence angles. Configuration denotes the stereochemical arrangements of the atoms in the polymer chain. The configuration of a polymer molecule cannot be altered without breaking and then reforming chemical bonds. In conjunction with the partition function, configuration takes on a different meaning. The configurational partition function is that portion of the partition function that depends only on the internal coordinates of the atoms and that remains after integrating out all momenta. Polymer configuration is discussed here only insofar as it affects the conformation(s) available to the chain.

A schematic representation is given in Figure 1 of a polymer carbon backbone consisting of n bonds and $n + 1$ atoms. The skeletal or backbone conformation of this polymer is determined by the set of bond vectors $\mathbf{l}_1, \mathbf{l}_2, \mathbf{l}_3, \ldots \mathbf{l}_n$. A more convenient measure of its conformation is provided by the vector $\mathbf{r}$ connecting the ends of the chain C_0 and C_n.

$$\mathbf{r} = \sum_{i=1}^{n} \mathbf{l}_i \tag{1}$$

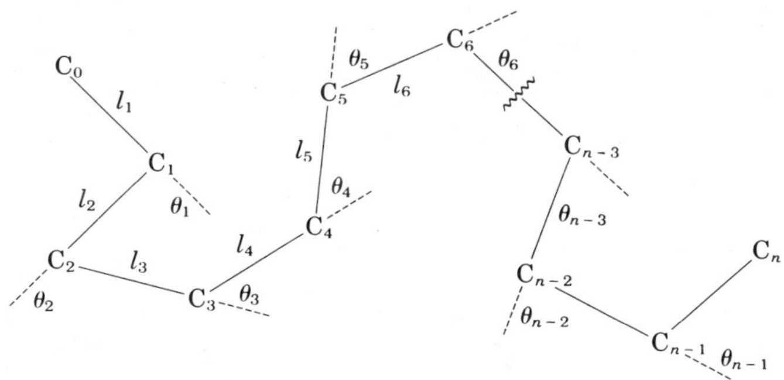

Figure 1. Schematic representation of a polymer carbon backbone.

Distributions

The ability to rotate about chemical bonds makes the number of conformations that a long polymer chain may adopt extremely large. Complete enumeration of each conformation is a hopeless task that is avoided by considering appropriate averages over the totality of available conformations.

The average conformation of a polymer chain can be described by specifying the distribution $W(\mathbf{r})$ of end-to-end chain distances $\mathbf{r}$. $W(\mathbf{r})\,d\mathbf{r}$ gives the probability that one end of a chain is within the volume element $d\mathbf{r}$ located at $\mathbf{r}$ relative to the other chain end. For long polymer chains, unperturbed by excluded volume self-intersections of long range,

$$W(\mathbf{r}) = (3/2\pi\langle r^2\rangle)^{3/2} \exp\left[-\frac{3r^2}{2\langle r^2\rangle}\right] \tag{2}$$

where $W(\mathbf{r})$ is normalized in such a way that

$$\int_0^\infty W(\mathbf{r})\, 4\pi\mathbf{r}^2\, d\mathbf{r} = 1 \tag{3}$$

and $\langle r^2\rangle$ is the mean-square magnitude of the end-to-end vector $\mathbf{r}$ averaged over all chain conformations. The spherically symmetric distribution of end-to-end vectors (eq 2) has been referred to as the Gaussian form.

The distribution of distances between segments in polymer chains of typical equilibrium flexibility departs significantly from the Gaussian form for segments separated by fewer than 20–50 backbone bonds. The resulting distribution function no longer has a simple mathematical expression, and is more conveniently expressed in graphical form.

Hypothetical Chains

Several models of hypothetical chains have been used to evaluate conformation-dependent properties of polymer molecules, such as $\langle r^2\rangle$. The simplest is the random-flight or freely joined chain, whose neighboring bonds are completely uncorrelated, ie, all directions of a given bond are equally likely and independent of the neighboring bonds. If the bonds of the polymer chain in Figure 1 are fixed in length and connected to each other at any angle with equal probability, and if rotation about each bond is unrestricted, the polymer chain is freely jointed. For the uncorrelated bonds of a freely jointed chain,

$$\langle \mathbf{l}_i \cdot \mathbf{l}_j\rangle = 0, \; i \neq j \tag{4}$$

and

$$\langle r^2\rangle = nl^2 \tag{5}$$

RIS Approximation. The discrete nature of the rotational potential has been recognized, and in the rotational isomeric state (RIS) approximation, each bond in the polymer backbone is assumed to occur in any one of a small number of discrete rotational states. These states are generally selected to coincide with the potential minima. For polymers with bonds whose rotational energy barriers do not exceed RT, the RIS model may be used to approximate the nearly continuous rotation potential by a discrete sum of rotational states.

Matrix Methods. If all bond lengths and valence angles are assumed to be fixed, the conformation of an n-bond polymer chain can be specified by assigning a rotational state to each of the $n - 2$ nonterminal bonds. Let ν be the number of rotational states, ie, usually 3, about each bond. Then there are $(\nu)^{n-2}$ possible conformations in all. The energy $E\{\phi\}$ of any conformation can be expressed in terms of the pairwise (nearest-neighbor) dependent energies $E_i(\phi_{i-1}, \phi_i)$.

$$E\{\phi\} = \sum_{i=2}^{n-1} E_i(\phi_{i-1}, \phi_i) = \sum_{i=2}^{n-1} E_{\xi\eta;i} \tag{6}$$

where ξ and η denote the rotational states of bonds $i - 1$ and i, respectively.

Statistical weights $\mu_{\xi\eta}$, or Boltzmann factors, corresponding to the energies $E_{\xi\eta}$, may be defined as

$$\mu_{\xi\eta;i} = \exp\,[-E_{\xi\eta;i}/RT] \tag{7}$$

and expressed in matrix form.

$$\mathbf{U}_i = [\mu_{\xi\eta;i}] = \begin{bmatrix} \mu_{\alpha\alpha}\mu_{\alpha\beta} & \cdots & \mu_{\alpha\nu} \\ \mu_{\beta\alpha}\mu_{\beta\beta} & \cdots & \mu_{\beta\nu} \\ \vdots & \vdots & \vdots \\ \mu_{\nu\alpha}\mu_{\nu\beta} & \cdots & \mu_{\nu\nu} \end{bmatrix} \tag{8}$$

The rows of the $\nu \times \nu$ $\mathbf{U}_i$ matrix are indexed with the states (ξ) of bond $i - 1$ and the columns with the states (η) of bond i. The statistical weight of a particular chain conformation is then simply

$$\mathbf{\Omega}_{\{\phi\}} = \prod_{i=2}^{n-1} \mu_{\xi\eta;i} \tag{9}$$

Summing equation 9 over all possible conformations leads formally to the configurational partition function

$$Z = \sum_{\{\phi\}} \mathbf{\Omega}_{\{\phi\}} = \sum_{\{\phi\}} \prod_{i=2}^{n-1} \mu_{\xi\eta;i} \tag{10}$$

Application of matrix methods, which were previously used to treat the Ising ferromagnet, leads to

$$Z = \mathbf{J}^* \left[\prod_{i=2}^{n-1} \mathbf{U}_i \right] \mathbf{J} \tag{11}$$

where $\mathbf{J}^*$ and $\mathbf{J}$ are the $1 \times \nu$ and $\nu \times 1$ row and column vectors

$$\mathbf{J}^* = [1\ 0 \cdots 0];\ \mathbf{J} = \begin{bmatrix} 1 \\ 1 \\ \vdots \\ 1 \end{bmatrix} \tag{12}$$

Equations 11 and 12 permit evaluation of the configurational partition function for a polymer chain of any length provided the RIS model, ie, the number and energies of the bond-rotational states, are known.

Long-range Interactions. Thus far, conformational characteristics of polymer chains have been treated in terms of short-range intramolecular interactions resulting in nearest-neighbor–dependent rotational states. For polymers whose range of

Table 1. U.S. Consumption of Plastics for Building and Construction Applications, 1964[a] and 1982[b], 10^3 metric tons

Resin	Building panels and siding		Fiber-particle board and plywood bonding		Other bonding and adhesives		Wire and cable covering		Decorative laminates, paneling		Wall coverings	
	1964	1982	1964	1982	1964	1982	1964	1982	1964	1982	1964	1982
ABS												
acrylic		6										
epoxy						7						
phenolic			52	526					36	14		
acetal resins												
polycarbonate												
reinforced unsaturated polyester[g]	3	60										
polyethylene												
high density								54				
low density								132				
polypropylene								5				
polystyrene												7
polyurethane												
PVC												
calendered sheet									1	10	3	2
extrusion and injection	3	97					91	146				
other					70	20						5
urea–melamine			148	372				19	12			

[a] Figures rounded to nearest 10^3 t.
[b] Polyplan Consultants.
[c] Includes roof membranes.
[d] Window frame sections, gutters and downspouts, mobile home skirt, foam sections, weather-stripping, etc.

bond-rotation interdependence goes beyond first neighbors, the statistical weight matrix (eq 8) can be revised to connect the conformation of two (or more) bonds to the conformations chosen for the preceding pair (or triad, etc).

Conformational Energetics. To calculate the conformation-dependent properties of polymer chains by using the RIS model and matrix methods, the chain geometry, ie, bond lengths and angles, and the locations and energetics of the bond rotational states must be known. Bond-length and valence-angle data are usually available from x-ray diffraction studies. Thus, it remains to select the rotational states and evaluate their relative energies, or statistical weights.

Statistical Weights. The most widely employed method of determining the RIS model for a given polymer chain utilizes rotational-state statistical weights as parameters to be determined by comparison of calculated and measured conformation-dependent properties, such as $\langle r^2 \rangle$ and its temperature coefficient.

Conformational Entropy

The conformational entropy S_{conf} of an unperturbed polymer chain can be obtained directly from the configurational partition function (eqs 11 and 12) and its temperature coefficient, according to the usual statistical thermodynamic relation

$$S_{\text{conf}} = R\left[\ln Z + \left(\frac{T}{Z}\right)\frac{dZ}{dT}\right] \qquad (13)$$

Application of the Theoretical Models

The conformational characteristics of polymer chains, which are typically deduced from calculations and measurements performed on isolated molecules, are also important to the understanding of their bulk properties. For example, three applications of the RIS model of polymer chains and the matrix methods used in its treatment would be the effects of monomer-sequence distribution on the conformational properties of a synthetic copolymer, the stereosequence-dependent ^{13}C-nmr chemical shifts of vinyl polymers, and the determination of the solution conformation(s) of cyclic polypeptides, many of which are biologically active.

BIBLIOGRAPHY

P. J. Flory, *Principles of Polymer Chemistry,* Cornell University Press, Ithaca, N.Y., 1953.

P. J. Flory, *Statistical Mechanics of Chain Molecules,* Wiley-Interscience, New York, 1969.

A. E. Tonelli, *J. Chem. Phys.* **52,** 4749 (1970); **54,** 4637 (1971); **56,** 5533 (1972).

A. E. Tonelli, *Macromolecules* **5,** 5631 (1972).

R. S. Porter and J. F. Johnson, eds., *Analytical Calorimetry,* Vol. 3, Plenum Press, New York, 1974.

J. Polym. Sci. Part B **15,** 2051 (1977).

A. E. Tonelli and F. C. Schilling, *Acc. Chem. Res.* **14,** 233 (1981).

ALAN E. TONELLI
AT&T Bell Laboratories

CONSTRUCTION AND BUILDING APPLICATIONS

Plastics have a wide range of important applications in building and engineering operations (Table 1). They are used in

Sanitary, plumbing, and hardware		Flooring		Moisture and vapor barrier[c]		Thermal insulation		Irrigation, pipe, conduit, and fittings		Building profiles[d]		Lighting fixtures		Flat glazing, skylights and corrugated sheet	
1964	1982	1964	1982	1964	1982	1964	1982	1964	1982	1964	1982	1964	1982	1964	1982
								6	60						
	14												13	22	35
			8					2	5[e]						
						49	158[f]								
2	9														
													3		30
5	59							5	95					17	89
								21[h]	215[h]						
				23	62			30[h]	32[h]						
								01	11						
	22					24	87	7	4				11		
						3	130								
		107	68		12										
								27	961	9	87		7		
			45												
	1														

[e] As pipe coating.
[f] As binder to nonplastics fibers.
[g] Includes fillers and reinforcement.
[h] Includes building profiles.

wire and cable covering (qv), as moisture and vapor barriers, in thermal insulation (as foam or binder), and in pipework and fittings. They have opened up new markets and facilitated construction methods, for example, as adhesives, sealants (qv), and siding materials; in roof lighting; and in tub and shower enclosures. In addition, paints are based on polymers, eg, acrylics, epoxies, silicones, and polyurethanes. As such, they have increased the durability and protection of traditional substrates. Other applications include polymer-bound mortars and cements, cultured marble utilizing acrylic syrup, concrete reinforced by polyester, and suspended roofs capable of enclosing large spaces. The importance of construction as a plastics materials market is illustrated by Table 2, which gives the percentages of total plastics output consumed by construction in the United States, UK, and FRG.

Table 2. Consumption of Plastics Materials by the Construction Industry and Its Percentage of National Plastics Production

	1964		1967		1980		1982	
Country	10^3 t	%	10^3 t	%	10^3 t	%	10^3 t	%
United States[a]	1000	22					3355	20.5
UK[a]	90	12	202	20			420	20
FRG[b]					1700	28		

[a] Source: Polyplan Consultants (UK).
[b] Source: Institut für das Bauen mit Kunststoffen (Darmstadt, FRG).

BIBLIOGRAPHY

Plast. Eng. **30,** 17 (1984).

L. H. Dunlap and R. Desch in M. Grayson, ed., *Kirk-Othmer Encyclopedia of Chemical Technology,* 3rd ed., Vol. 18, John Wiley & Sons, Inc., New York, pp. 87–110.

D. Powell
Polyplan Consultant

CONTACT ANGLE.

See Adhesion and bonding.

CONTACT LENSES

Contact lenses range from hard to soft. Conventional hard lenses contain no less than 95% PMMA. They are designed and fit in such a manner that the tear fluid supplies the oxygen required by the cornea. Hard and semirigid lenses permeable to oxygen have been developed; most of them are made from copolymers of siloxanes and methacrylates.

Flexible, oxygen-permeable silicone lenses are also available. Finally, soft contact lenses are prepared from polymers that absorb large quantities of water to become hydrogels; the aqueous phase of the hydrogels is oxygen-permeable.

An essential difference among the types of contact lens is the manner in which they fit in the eye. Hard contact lenses and hydrophobic flexible lenses require a relatively thick tear film between their surfaces and the cornea; this is of optical and physiological importance. Soft hydrogel lenses adhere closely to the cornea with only a tear film of capillary thickness between the lens and the cornea. These lenses have wide public acceptance. They are comfortable, easily tolerated by the eye, and offer extended-wear possibilities.

Manufacture

Starting from a polymer blank or a molded semifinished lens, most hard contact and silicone–acrylate lenses are made by standard methods of machining and polishing; CAB and PMMA lenses are also injection molded. Soft contact lenses are prepared by several methods, eg, centrifugal casting of hydrogel contact lenses. A monomer solution containing a mixture of HEMA, cross-linking agent, modifiers, and an initiator is polymerized in a mold rotating about its central axis. The procedure yields lenses in a swollen state, which are then equilibrated in physiological saline solution. A more widely used method is a standard machining technique similar to the technique used for hard lenses.

Direct casting of finished contact lenses (dry state) has been developed for several polymers. The monomer mix containing appropriate initiators is placed in disposable plastic molds and cured by either thermal- or ultraviolet-initiated polymerization. Minor finishing steps, such as edge buffing, may be used. Direct casting followed by hydration is an economical route to finished hydrogel lenses.

Cleaning and Disinfection

Cleaning is an important step in daily-wear lenses. Specially formulated surfactants remove dust, dirt, mucus, cosmetics, and microbes as well as other lipid-based contaminants before they become embedded permanently in the lens. Proteolytic enzyme-based materials are used to remove proteinaceous deposits. It is good practice to disinfect the lenses, hard or soft, whenever they are removed from the eye. Disinfecting regimens may be thermal or chemical.

Government Regulation

All contact lenses not made of PMMA are regulated by the FDA. In the United States, new contact lenses must be approved by the FDA. The FDA also has jurisdiction over lens-care products.

BIBLIOGRAPHY

U.S. Pat. 3,808,178 (Apr. 30, 1974), N. G. Gaylord (to Polycon Laboratories).

U.S. Pat. 3,220,960 (Nov. 30, 1965), O. Wichterle and D. Lim.

U.S. Pat. 3,721,657 (Mar. 20, 1973), M. Seiderman.

M. F. Refojo, *Surv. Ophthalmol.* **16,** 233 (1972).

Helmut K. Singer
Ellen C. Bellantoni
American Optical Corporation
Albert R. LeBoeuf
Barnes-Hind Pharmaceuticals, Inc.

CONTACT MOLDING.

See Reinforced plastics.

CONTAINERS.

See Packaging materials.

CONTROLLED RELEASE.

See CONTROLLED-RELEASE TECHNOLOGY; DRUG APPLICATIONS; MICROENCAPSULATION.

CONTROLLED-RELEASE TECHNOLOGY

Controlled-release technology has emerged during the 1980s as a commercially sound methodology. The achievement of predictable and reproducible release of an agent into a specific environment over an extended period of time has many significant merits. The most significant is to create a desired environment with optimal response, minimum side effects, and prolonged efficacy. This is a relatively new technology and requires an interdisciplinary scientific approach. Its most popular application is drug delivery, although it is also used in the application of agricultural agents.

Polymeric Drug Systems

Polymeric drug systems include macromolecules that elicit an efficaceous biological activity on their own and macromolecules that serve as carriers for drugs.

Polymeric Drugs. Developments in the areas of biochemistry and molecular biology have resulted in the elucidation of the structure and function of many polymer molecules essential to the homeostasis of living organisms. Consequently, there has been significant research regarding the synthesis and properties of physiologically active polymers. It is now possible to prepare macromolecules with predetermined structures, discrete molecular weights, and specific functional groups. Advantages of polymeric drugs are that the polymers exhibit delayed action, prolongation of activity, decreased rate of drug metabolism, and drug excretion.

The biological activity demonstrated by polymeric compounds may be the result of the polymer itself, as in the case of polyanions made from inactive monomers, or may be the result of polymerization or copolymerization of a known active monomeric drug. Many active drug monomers lose their activity in the polymeric form, but others exhibit enhanced activity. The polymerization of active drug monomers has in most cases decreased toxicity, but it can also have the opposite effect.

In polymer drug systems, two of the most important variables that apparently regulate biological activity are molecular structure and stereochemical configuration. These two features are manifested in most polymer systems by molecular weight, copolymer properties, chain coiling, chain folding, cross-linking, tacticity, and electrolytic character. With drug copolymers it has been shown that the nature of the nondrug comonomer is very important and that biological activity is not related to the drug comonomer alone. Certain properties characteristic of polymer systems may correlate with biological activity. Consequently, more detailed structure–reactivity studies of polymeric drugs are needed to serve as a basis for future drug design.

Polymeric drugs have been extensively applied in cancer treatment. Synthetic polyanions have shown excellent potential in many studies. Because nucleic acids and enzymes play such a large role in chromosome replication during mitosis, a considerable amount of research has been conducted in this area to control viruses. On the molecular level, analogues of nucleic acids are capable of forming complexes with adenine, cytosine, uracil, thymine, and guanine. Through complexation, these nucleic acid analogues are potential inhibitors of biosyntheses that require nucleic acids as templates.

Polymeric Drug Carriers or Conjugate Polymeric Drugs. Macromolecular drug carriers hold a great potential for drug administration. Many pharmaceutical companies have drugs that are nearing the end of their patent protection period. However, polymer-bound drugs used in new controlled-release or delivery methods can provide extended patent rights.

Most medications are micromolecular in size and are relatively free to diffuse throughout the biological system. Consequently, drugs have been inherently difficult to administer in a localized, concentrated mode within the primary target tissues and organs. Because polymers diffuse slowly and are often absorbed at interfaces, the attachment of pharmaceutical moieties produces a biopolymer with distinct pharmacological activity. These polymeric-drug carriers have desirable properties such as sustained therapy, slow drug release, prolonged activity, and drug latenation. Drug metabolism and excretion are also delayed (Table 1).

Table 1. Characteristics of Polymer Drug Carriers

depot effects	slow diffusion of drug
	slower absorption of drug
	slower elimination of drug
	nonadsorbability for topical application
pharmacokinetic	sustained-release of drug
	variable drug metabolism routes
body distribution	localization of drug
	inhibition of adsorption in some areas due to molecular weight
	cell-specific interactions and uptake of drug
	protein binding
pharmacological activity	variable activity from none to an increase
	variable toxicity
	incorporation of drug combinations on polymer backbone
	reduction of side effects such as nausea and irritation common with large doses of drugs

The method of attachment of a drug to the polymer depends on the ultimate use of the adduct. The chemical reaction conditions for the attachment of the drug to the polymer should not adversely affect the biological activity of the drug. Temporary attachment of a pharmacon is necessary if the drug is active only in the free form. A drug–polymer conjugate, where the drug is active only after being cleaved from the polymer chain is called a prodrug. This is usually the case with agents that function intracellularly. Temporary attachment usually involves a hydrolyzable bond such as an anhydride, ester, acetal, or orthoester. Permanent attachment of the drug moiety is generally used when the drug exhibits activity in the attached form. The pharmacon is usually attached away from the main polymer chain and other pendent groups by means of a spacer moiety that allows for drug–receptor interaction.

Drug targeting to a specific biological site is an enormous advantage in drug delivery because only those sites involved are affected by the drug. This precludes the transport throughout the body, which can elicit serious side effects. Ideally, a targetable drug carrier is captured by the target cell to achieve optimum drug delivery while minimizing deposition elsewhere in the host.

The ultimate fate of drugs and their metabolites is a major concern. If they are not eliminated in a reasonable time, they could promote undersirable side effects. Polymeric drug carriers are usually nonbiodegradable, and if their size is greater than 50,000 daltons, they could accumulate in the host with the potential of future deleterious effects. Consequently, polymeric systems such as *N*-(2-hydroxypropyl)methacrylamide, which was originally developed and used as a plasma extender, have been used. This polymer is well tolerated in the host and has been established to be nontoxic and nonimmunogenic. To further enhance the elimination of this macromolecular drug carrier, smaller segments of *N*-(2-hydroxylpropyl)methacrylamide have been cross-linked with biologically hydrolyzable peptides to achieve the needed macromolecular size for drug delivery.

Chemically Controlled Systems

In chemically controlled delivery systems the release of a pharmacological agent usually takes place in an aqueous environment and involves specific events. *In vitro* studies of degradation at pH 7.4 show progressive changes in molecular weight and molecular weight distribution, consistent with bulk hydrolysis *in vivo*. Results indicate that reactions by enzyme mediation in subcutaneous tissue are less important than simple hydrolysis.

Polymer Prodrug Systems. Prodrugs are compounds that are not physiologically active as constituted. The activity is usually masked by the attachment of a labile entity that is readily removed *in situ* by enzyme hydrolysis to release the active agent.

Diffusional Delivery Systems

Diffusional polymeric drug-delivery systems can be divided into either matrix or reservoir types, as shown in Figure 1. The matrix system is also known as the monolithic system; the biological agent is contained as a semisolid or dispersion within the polymer mass. In this system the active compound is dissolved or intimately mixed with the polymer, which also serves as the matrix. In some cases a continuous phase is formed and can be cast at a higher temperature into sheets, solid rods, or fibers that form a "gelled" matrix on cooling. Individual devices can then be punched from the sheets or sliced from the solid rods. The matrix method of controlled delivery was first used commercially for flea collars and dispensers such as Shell's No Pest Strip and Hercules' Hercon.

The reservoir system is composed of an inner reservoir or depot that contains the excipient which in turn is surrounded by a semipermeable membrane that controls the flow or release. The membrane reservoir system is diffusion controlled, dispensing an agent through a rate-controlling membrane. A commercial example of a reservoir device is the Ocusert developed by ALZA. This is an ocular therapeutic device that delivers pilocarpine, an ophthalmic drug for the treatment of glaucoma. The device is placed under the lower eyelid where it continuously releases the drug through a rate-controlling membrane (see DRUG APPLICATIONS).

Both systems have advantages and disadvantages. The matrix system, which is easy to make, can deliver high molecular weight drugs and can be biodegradable. However, drug release may not be constant. The membrane system gives a zero-order or constant rate of release and is easily designed, but it is expensive to make and leaks or ruptures are a potential hazard. Polymeric membranes used for drug diffusion can be macroporous, microporous, and nonporous. Equations to describe the release of drugs from these membranes take into consideration the effects of pore diameter and tortuosity on the material released.

The basic requirements for the nonbiodegradable polymers used are freedom from elutable impurities, additives, stabilizers, catalysts, and emulsifiers; stability to chemical or hydrolytic cleavage; good mechanical properties; readiness to be cast

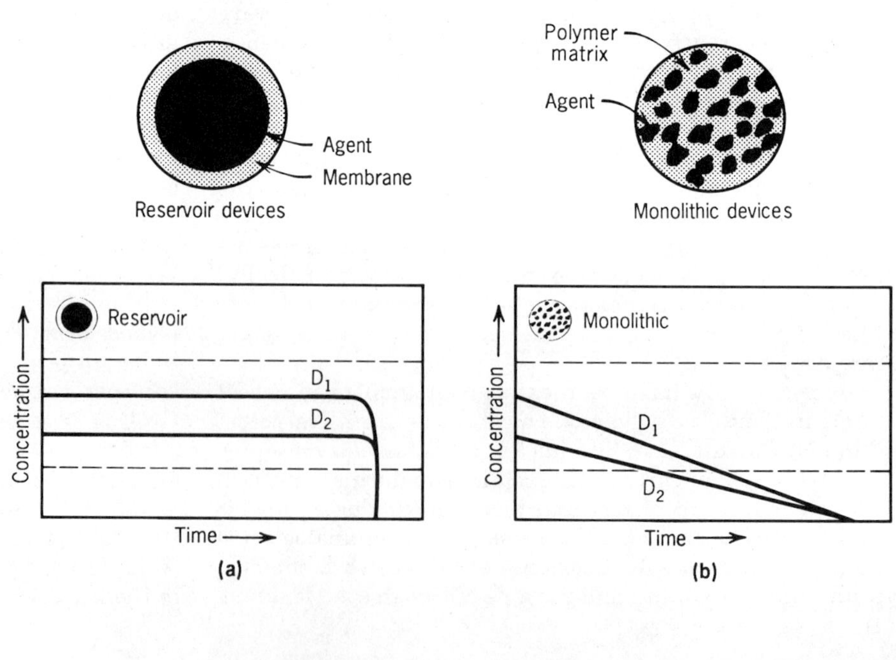

Figure 1. Diffusional drug delivery systems, schematic representations and dose-response curves. Dose 1 > dose 2 in both cases. (**a**) Reservoir device; (**b**) monolithic device.

or molded into films, rods, tubes, or other useful shapes; noncarcinogenicity or teratogenicity; nontoxicity; noninflammatory reaction with tissue; sterilizability; and no degradation of mechanical properties *in vivo*. Commonly used polymers include polyesters, polyorthoester, polyacids, hydrogels, polysaccharides, polypeptides, polyaminotriazole, and albumin beads.

Factors that affect the *in vitro* release are the concentration of drug in the receiving fluid, the composition of receiving fluid, temperature, agitation, pH, and stability of the drug. Examples of therapeutic agents delivered by diffusion from polymer matrices include narcotic antagonists, steroids, antimalarials, proteins, antibacterials, ophthalmics, anti-inflammatory agents, vitamins, and anticancer drugs.

Biodegradable Systems. The release of a substance entrapped in a polymer matrix by means of polymer biodegradation is a technology that is being utilized fruitfully. Usually the polymer matrix contains hydrolytic or enzymatic labile bonds and, as the polymer erodes due to the cleavage of these bonds, the entrapped agent is released. The terms bioerodable and biodegradable are often used interchangeably. However, the erosion process implies concurrent release of an agent with chemical breakdown, whereas biodegradation implies that the polymer disintegrates after it has completed its function. The major advantage of these systems is that surgery is not required to remove the delivery device after the administration period. The erosion–degradation products must be nontoxic and effectively eliminated from the host.

Transdermal Delivery Systems. In a broad sense, the term transdermal delivery includes all topically applied formulations intended to deliver an active ingredient into the circulatory system. Now a number of membrane-controlled, matrix-controlled, adhesive-dispersion, and microreservoir-dissolution systems are commercially available that supplement the original ointment systems. Two typical transdermal delivery devices are shown in Figure 2. One of the first was the Transderm-Scop for delivery of scopolamine. A similar system was developed for delivery of estradiol for use in postmenopausal treatment.

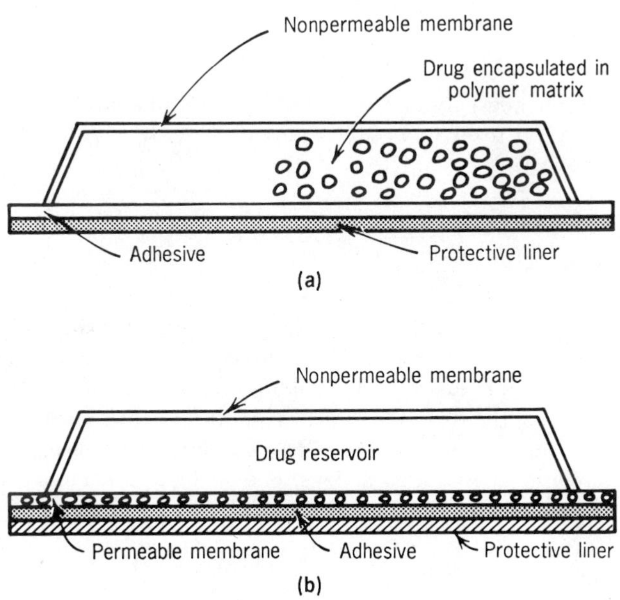

Figure 2. Two typical delivery systems. (**a**) Monolithic transdermal system; (**b**) reservoir–membrane system.

Liposomal Delivery. Liposomes were first used as models for biomembranes and artificial cells. Due to their unique structure they were investigated as potential drug-delivery devices. Because they are made primarily from biological materials, biocompatibility was not considered a problem.

Site-specific delivery of liposomes is the current research goal in this area. The ability to destabilize the lipid assembly specifically has led to the design of heat-sensitive, light-sensitive, and pH-sensitive liposomes. The chemical approach to obtaining site-specific delivery involves the inclusion or attachment of targeting moieties such as monoclonal antibodies and specific receptor-site binding carbohydrates, hormones, and peptides.

Oral Delivery. Oral delivery was among the first encapsulation-mediated drug-delivery methods used by the pharmaceutical industry. Enteric-resistant coated pills were originally introduced to resist enzyme action and gastric fluids that are acidic in the stomach (pH 1) and basic in the intestine (pH 8.4). Many oral devices are designed to disintegrate and dissolve after passage to the intestine. One principal purpose of the coatings is to prevent nausea and vomiting induced by drugs that cause local gastric irritation.

Several variables affect the rate of release of drugs from oral sustained-release preparations. The size of the tablet and concentration of drug determine the surface area for solution and the rate at which the drug will dissolve. The pore size of the inert matrix, its resistance to sloughing, the presence of water-soluble substances in the matrix, and the intrinsic solubility of the drug in an aqueous medium all affect the release rates.

Current technology involves an ion-exchange resin core. In the acidic stomach environment drugs with amine groups are positively charged due to protonation and bind to polymer matrices that have negative charges. When the delivery system passes into the intestine, the amines are neutralized and released.

More complex laminated coatings have been introduced and are referred to as sustained-release medications. The drug may be applied in a soluble form to a permeable outside polymer layer of a tablet containing a less soluble but porous core with more of the same drug trapped inside. Mucoadhesives for sustained release in the stomach for gastric retention is another method for delaying passage through the gastrointestinal tract and subsequent rapid absorption in the small intestine.

Osmotic Pumps. Miniature osmotic pumps are capable of deliverying a variety of solutions in suspension at zero-order or patterned rates for periods of hours, days, weeks, or months. The simplest osmotic pump consists of a compacted solid core of water-soluble drug that is enclosed in a selective membrane that is permeable to water but not the enclosed drug. The membrane has a small orifice from which the drug escapes at a rate consistent with the rate of water infusion through the membrane. A more sophisticated model consists of a cylindrical reservoir made from an impermeable membrane, with an orifice for the reagent to escape. This is surrounded by an osmotic material (salt), which is encased in a semipermeable membrane. The rate of delivery is determined by the water permeability of the membrane, the osmotic material, and the dimensions of the pump.

Infusion Pumps. The first infusion pump, the Infusaid Implantable Infusion Pump, was developed about 1970. This device consists of a titanium disk-shaped canister with central and peripheral refill septa for changing the drug. The peripheral septum can be used for bolus or infusion delivery of drugs or radionuclides for organ scanning and contrasting agents for angiograms. This was the first pump to obtain FDA approval for clinical use.

BIBLIOGRAPHY

R. M. Ottenbrite in M. Phillips, S. Shoemaker, R. Middlekauff, and R. Ottenbrite, eds., *The Impact of Chemistry on Biotechnology,* ACS Symposium Series 362, American Chemical Society, Washington, D.C., 1986.

P. I. Lee and W. R. Good, eds., *Controlled-Release Technology,* ACS Symposium Series 348, American Chemical Society, Washington, D.C., 1987.

P. Johnston and J. G. Lloyd-Jones, eds., *Drug Delivery Systems,* Ellis Harwood Pub., Chichester, UK, 1987.

S. Dumitriu, M. Popa, and M. Dumitriu, *J. Bioact. Compat. Polym.* **3,** Parts I and II (1988); Parts II and IV (1989).

RAPHAEL M. OTTENBRITE
Virginia Commonwealth University

CONTROL TESTS. See TEST METHODS.

CONVEYOR BELTS. See BELTING.

COORDINATE POLYMERIZATION

Coordinate or coordination polymerization was first proposed in 1956 for the unusual addition polymerization of ethylene and propylene with the transition-metal catalysts discovered by Ziegler and co-workers in 1953, and for the iron-catalyzed ring-opening polymerization of propylene oxide to crystalline polymer reported in a Dow patent. This terminology implies that the growing polymer chain is bound to a metal atom and that insertion of the monomer into the metal-bound polymer chain is preceded by, and presumably activated by, the coordination of the monomer with the metal. Many diverse monomer types have been reported to polymerize by this mechanism.

The coordination step proposed has not been unequivocally established, and other mechanisms, also unproven, have been advanced. The more general term insertion polymerization (qv) has been used for these and many varied, unusual polymerization systems to imply a hindered propagation site and to avoid implying the unproved coordination aspect.

BIBLIOGRAPHY

C. C. Price and M. Osgan, *J. Am. Chem. Soc.* **78,** 4787 (1956).

E. J. Vandenberg in C. C. Price and E. J. Vandenberg, eds., *Coordination Polymerization,* Plenum Publishing Corp., New York, 1983.

G. Natta, G. Mazzanti, P. Longi, G. Dall'Asta, and F. Bernardini, *J. Polym. Sci.* **51,** 487 (1961).

K. J. Iris, J. J. Rooney, C. D. Stewart, M. L. H. Green, and R. Mahtab, *J. Chem. Soc. Chem. Commun.* **1978,** 604 (1978).

E. J. VANDENBERG
Arizona State University at Tempe

COORDINATION POLYMERS

A coordination polymer is defined in this article as a polymeric substance containing a coordination compound or metal complex in its repeat unit. The metal complex is not necessarily part of the polymer backbone.

A general problem in the characterization of coordination polymers arises because of the extreme intractability of many of the materials and the consequent lack of available structural information. Often, where solubility exists, it leads to complete depolymerization (qv); however, molecular weight data are available for selected soluble polymers. Thus the structures depicted in the literature and here are in some cases quite speculative. Owing to the difficulty of obtaining single crystals of the materials, relatively few x-ray structure determinations have been carried out on coordination polymers, and spectroscopic techniques have provided only limited insight into structure.

Methods of Preparation

Three general techniques have been employed in the preparation of transition-metal coordination polymers.

1. Ligands capable of simultaneously coordinating to two metal atoms generate the desired polymer, ie,

$$n\mathrm{M} + n(\text{—Y—}) \longrightarrow (\text{—M—Y—})_n$$

   where M is a metal and Y a ligand.

2. Preformed polymers to which metal atoms are attached may be used to produce polymers with advantageous properties, ie,

$$n\mathrm{M} + (\text{—Y—})_n \longrightarrow \underset{\displaystyle \mathrm{M}}{\overset{(\text{—Y—})_n}{\downarrow}}$$

   Many possess catalytic activity or have potential as ion- or electron-exchanges (see also ION-EXCHANGE POLYMERS; CHELATE-FORMING POLYMERS).

3. Much of the recent interest in metal-containing polymers has focused on the use of transition-metal coordination complexes in which at least one of the ligands has functional groups that can be used in polymer formation, eg,

$$n(\text{M—Y}) \longrightarrow (\text{—M—Y—})_n$$

   Substituted metallocenes have been at the center of much of this research (see also ORGANOMETALLIC POLYMERS).

Classification of Coordination Polymers

The transition-metal coordination polymers that have been prepared may be classified into four groups according to the position of the metal in the macromolecule: (*1*) the metal is a pendent group, (*2*) the metal acts as a cross-linking agent, (*3*) the metal is enmeshed in a polymer (parquet polymers), and (*4*) the metal is an integral part of the polymer backbone.

Applications

Interest in coordination polymers developed primarily because of the potential effect of the metal on the organic group. One hope was for improved thermal stability. However, such factors as acid sensitivity, photosensitivity, and long-term stability have restricted the widespread use of these polymers. Many organic polymers also suffer from photodegradation.

Another application of coordination polymers is in the area of catalysis. Several coordination polymers have also been proposed for various uses: in the photocopying process, as ion- or electron-exchange resins, as additives to fuels for smoke reduction, or in compounds where strength is required. Coordination polymers appear to have important future application as molecular metals and as templates for the preparation of new and unusual solid materials.

BIBLIOGRAPHY

E. Tsuchida and H. Nishide, *Adv. Polym. Sci.* **24,** 1 (1977).

M. Kaneko and E. Tsuchida, *Macromol. Rev.* **16,** 397 (1981).

N. Hagihara, K. Sonogashira, and S. Takahashi, *Adv. Polym. Sci.* **41,** 149 (1981).

C. E. Carraher, Jr., *J. Chem. Ed.* **58,** 921 (1981).

D. Wohrle, *Adv. Polym. Sci.* **50,** 45 (1983).

BRUCE M. FOXMAN
Brandeis University

SUSAN W. GERSTEN
Applied Solar Energy Corporation

COPOLYMERIZATION

Copolymerization is most usefully defined as polymerization in which two or more structurally distinct monomers are incorporated into the same polymer chain. No sequence information is implied in structure **1**.

$$xA + yB \longrightarrow -\!(A)_x\!-\!(B)_y\!-$$

(**1**)

A description of copolymer structure requires the specification of composition, ie, the relative amounts of comonomers of A and B; sequence distribution, ie, the order of incorporation of A and B into the chain; and linearity. Copolymers are most often classified according to the latter two characteristics; thus the major classes of copolymers are

Statistical (random) copolymers, in which comonomers appear in irregular, unspecified sequences along the chain:

$$A + B \longrightarrow \sim\!ABBABAAABAB\!\sim$$

Alternating copolymers, in which the comonomers occur in alternation:

$$A + B \longrightarrow \sim\!ABABABABABA\!\sim$$

Block copolymers, in which long linear sequences of comonomer A are joined to long linear sequences of comonomer B:

$$A + B \longrightarrow \sim\!AAAAAABBBBBB\!\sim$$

Graft copolymers, in which chains of one comonomer are pendent from the backbone of the other:

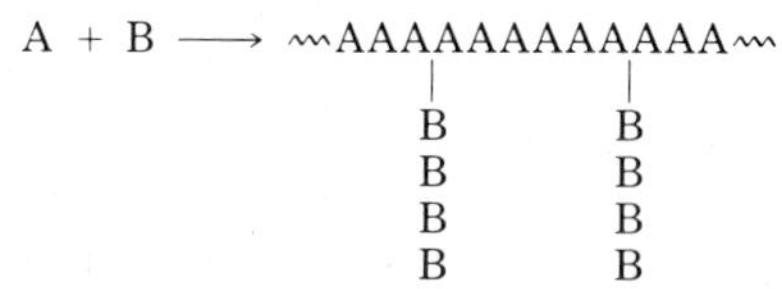

This article concerns the preparation of statistical (random) copolymers; each of the other classes of copolymers forms the subject of a separate article (see COPOLYMERS, ALTERNATING; BLOCK COPOLYMERS; GRAFT COPOLYMERS).

Uses of Copolymerization

Copolymerization is the most general and powerful method of effecting systematic changes in polymer properties, and is widely used in the production of commercial polymers and in fundamental investigations of structure–property relations. Copolymerization modifies the symmetry of the polymer chain and modulates both intramolecular and intermolecular forces, so properties such as melting point, glass temperature, crystallinity, solubility, elasticity, permeability, and chemical reactivity may be varied within wide limits. Table 1 summarizes some of the important properties of commercial ethylene copolymers and illustrates the range of properties accessible through copolymerization.

Chain-growth Copolymerization

Most theoretical and experimental studies of copolymerization have been concerned with chain-growth processes, in which the copolymer grows via successive additions of the comonomers to an active center that is radical, ionic, or organometallic (see also CHAIN-REACTION POLYMERIZATION):

$$\sim\!ABAABBA^* \begin{cases} \xrightarrow{A} \sim\!ABAABBAA^* \\ \xrightarrow{B} \sim\!ABAABBAB^* \end{cases}$$

Of primary concern are copolymer composition and sequence and copolymerization rate.

Table 1. Physical Properties of Polyethylene and Some Ethylene Copolymers

Property	HDPE[a]	LDPE[b]	EP[c]	EEA[d]	EVA[e]	Surlyn[f]
pendent group		$—C_nH_{(2n+1)}$	$—CH_3$	$—\overset{O}{\overset{\|}{C}}OC_2H_5$	$—O\overset{O}{\overset{\|}{C}}CH_3$	$—\overset{O}{\overset{\|}{C}}—O^-M^+$
density, g/cm³	0.965	0.920	0.875	0.931	0.943	0.93–0.94
tensile strength, MPa[g]	31.7	9.7	[h]	12.4	20.7	24.1–37.9
modulus, MPa[g]	1034	159	[h]	31	35.2	
yield stress, MPa[g]	29	8.3	[h]	5.9[i]	4.8[i]	13.8–17.2
melting point, °C	133	107		92	89	

[a] High density polyethylene.
[b] Low density polyethylene.
[c] Ethylene–propylene copolymer.
[d] Ethylene–ethyl acrylate copolymer.
[e] Ethylene–vinyl acetate copolymer.
[f] Zinc or sodium salt of ethylene–methacrylic acid copolymer; ionomer resin.
[g] To convert MPa to psi, multiply by 145.
[h] Too soft for measurement.
[i] Stress at 100% elongation.

Prediction of Instantaneous Copolymer Composition. The relative rates of incorporation of comonomers A and B are not, in general, equal to their relative concentrations in the initial reaction mixture. Thus, the copolymer formed at any instant differs in composition from the feed mixture, and the feed composition changes continuously from the beginning of the reaction to the end. A central problem of copolymerization theory has been the prediction of copolymer composition given a fixed, or instantaneous, feed composition.

The products of copolymerization, like those of any chemical reaction, may be dictated by the kinetics or by the thermodynamics of the reaction. With few exceptions, chain-growth copolymerization products are kinetically determined, so instantaneous copolymer composition can be predicted by a set of differential equations that describe the rates of monomer consumption. Developing these equations requires an appropriate kinetic model of the copolymerization process. Most published models (ie, the terminal model, the penultimate model, the complex-participation model, the complex-dissociation model, and the copolymerization with depropagation model) have been developed to treat copolymerization via radical intermediates, but they are potentially applicable to any chain-growth copolymerization.

The Terminal Model. In 1944, Mayo and Lewis formulated a model of the copolymerization process in which the rate constant for addition of each monomer was assumed to be dependent on the identity of the terminal unit on the growing chain. Thus, four propagation rate constants are required:

$$\sim M_1^* + M_1 \xrightarrow{k_{11}} \sim M_1M_1^*$$

$$\sim M_1^* + M_2 \xrightarrow{k_{12}} \sim M_1M_2^*$$

$$\sim M_2^* + M_1 \xrightarrow{k_{21}} \sim M_2M_1^*$$

$$\sim M_2^* + M_2 \xrightarrow{k_{22}} \sim M_2M_2^*$$

According to this model, the rates of monomer disappearance are given as

$$\frac{-d[M_1]}{dt} = k_{11}[M_1^*][M_1] + k_{21}[M_2^*][M_1] \quad (1)$$

$$\frac{-d[M_2]}{dt} = k_{12}[M_1^*][M_2] + k_{22}[M_2^*][M_2] \quad (2)$$

where $[M_1]$ and $[M_2]$ are monomer-feed concentrations, and $[M_1^*]$ and $[M_2^*]$ are the concentrations of growing chains with terminal residues derived from M_1 and M_2, respectively. The representation M^* designates an active center without specifying radical or ionic character. Dividing equation 1 by equation 2 gives

$$\frac{d[M_1]}{d[M_2]} = \frac{k_{11}[M_1^*][M_1] + k_{21}[M_2^*][M_1]}{k_{12}[M_1^*][M_2] + k_{22}[M_2^*][M_2]} \quad (3)$$

as the molar ratio of M_1 and M_2 in the copolymer. Assumption of steady-state concentrations of M_1^* and M_2^* takes the form

$$\frac{d[M_1]}{d[M_2]} = \frac{[M_1]}{[M_2]} \cdot \frac{r_1[M_1] + [M_2]}{[M_1] + r_2[M_2]} \quad (4)$$

where r_1 and r_2 are "reactivity ratios" defined as

$$r_1 = \frac{k_{11}}{k_{12}} \quad \text{and} \quad r_2 = \frac{k_{22}}{k_{21}}$$

Equation 4, widely known as the Mayo-Lewis equation, became the standard treatment of chain-growth copolymerization and has been used to correlate an enormous volume of copolymerization data. It is useful to examine the dependence of copolymer composition on monomer-feed composition for several different combinations of reactivity ratios that describe real copolymerizations:

$r_1 = r_2 = 1$ ($k_{11} = k_{12}$; $k_{21} = k_{22}$). In this case, neither active center shows any selectivity in adding monomer, and the monomers enter the copolymer in amounts determined only by their relative concentrations in the monomer feed; copolymer composition is always identical to feed composition.

$r_1r_2 = 1$ ($r_1 = 1/r_2$; $k_{11}/k_{12} = k_{21}/k_{22}$). In this case, both active centers show the same preference for addition of one of the two monomers, with the result that the copolymer is always richer in the more reactive monomer than is the feed mixture. Copolymerizations that behave in this way are often called ideal copolymerizations and may be considered special cases of the more general class for which $r_1 > 1$, $r_2 < 1$.

$r_1 = r_2 = 0$ ($k_{11} = k_{22} = 0$). In this case, each active center adds exclusively the monomer other than that which contributes the terminal residue on the growing chain. This demands the formation of an alternating copolymer of 1:1 composition regardless of feed composition.

$r_1 < 1$; $r_2 < 1$ ($k_{12} > k_{11}$; $k_{21} > k_{22}$). This case is intermediate between the first and third cases; each chain end prefers cross-propagation to homopropagation, but the preference is not absolute. Thus there is a tendency toward alternation, which grows stronger as the r_1r_2 product approaches zero. An important characteristic of such copolymerizations is the existence of a point at which monomer and polymer compositions are equal. This so-called azeotropic composition emphasizes the analogy of constant boiling mixtures in distillation. The azeotropic composition is readily calculated from equation 4, because it is defined as the point at which

$$\frac{d[M_1]}{d[M_2]} = \frac{[M_1]}{[M_2]} \tag{5}$$

$r_1 > 1$, $r_2 > 1$ ($k_{11} > k_{12}$, $k_{22} > k_{21}$). Each active center prefers homopropagation to cross-propagation. The result is a mixture of homopolymers or a copolymer with long sequences of each monomer uninterrupted by the other (a "blocky" copolymer).

Prediction of Comonomer Sequence Length. A copolymer chain of fixed length and composition can exist in an enormous number of isomeric structures that differ from one another in comonomer sequence. For several decades following the first quantitative treatments of copolymerization, sequence information was only rarely accessible through experiment. The routine use of ^{13}C nmr spectrometry and, most recently, the availability of nmr spectral editing techniques, have made the characterization of copolymer sequence a realistic objective. Each of the copolymerization models mentioned previously makes definite predictions concerning comonomer sequence lengths, and the comparison of predicted and measured sequence distributions provides a powerful test of copolymerization mechanism. In those instances where the sequence distribution is not experimentally accessible, calculation provides the only basis for inferences concerning the relation of chain microstructure and macroscopic polymer properties.

A convenient method of describing the sequence structure of a copolymer is to state the fraction of sequences of each monomer (M_1 and M_2) that are of a particular length. As an example, consider the short copolymer chain below:

$$\underbrace{M_1}_{1}M_2\underbrace{M_1M_1}_{2}M_2\underbrace{M_1}_{1}M_2M_2\underbrace{M_1M_1}_{2}$$

The chain contains four sequences of M_1 units uninterrupted by M_2. Of these, two are of length 1 and two are of length 2. Thus the number fractions of sequences of M_1 of lengths 1 (N_1^1) and 2 (N_2^1) are each 1/2 (2/4). For M_2, the corresponding fractions are $N_1^2 = 2/3$ and $N_2^2 = 1/3$. A complete description of the comonomer sequence distribution for any chain is provided by specification of N_x^1 and N_x^2 for all values of x.

The Terminal Model. The number fraction of sequences of M_1 of length x (N_x^1) is identical to the probability that any particular sequence of M_1 units is x units long. Such a sequence arises in the terminal model when a growing chain carrying a terminal M_1 unit adds $(x - 1)$ M_1s followed by an M_2. Thus,

$$N_x^1 = P_{11}^{(x-1)}P_{12} \tag{6}$$

where P_{11} is the probability that a growing chain carrying a terminal M_1 adds M_1:

$$P_{11} = \frac{k_{11}[M_1^*][M_1]}{k_{11}[M_1^*][M_1] + k_{12}[M_1^*][M_2]} = \frac{r_1}{r_1 + [M_2]/[M_1]} \tag{7}$$

and $P_{12}(= 1 - P_{11})$ is the probability that the same growing chain adds M_2. Similarly for M_2 sequences,

$$N_x^2 = P_{22}^{(x-1)}P_{21} \tag{8}$$

Equations 6 and 8 allow calculation of the comonomer sequence distribution from a knowledge of the terminal model reactivity ratios and the monomer-feed composition.

Copolymer Composition as a Function of Conversion. In general, the copolymer formed at any instant differs in composition from the feed mixture, so the feed composition itself drifts, ie, changes continuously, from the beginning of the reaction to the end. Exceptions are terminal-model copolymerizations in which $r_1 = r_2 = 1$ or in which "azeotropic" behavior is observed (ie, where the composition of the copolymer is fortuitously equal to that of the monomer feed mixture). Operationally, compositional drift can be eliminated by feeding the reactor continuously at a rate that matches the copolymerization rate, with a monomer mixture identical in composition to the copolymer that is being formed.

Prediction of Copolymerization Rate. Whereas composition and sequence are determined essentially entirely by the propagation step, a prediction of the overall rate of monomer consumption must treat initiation and termination as well. For the terminal model of radical copolymerization this requires consideration of the following reactions:

$$\left.\begin{array}{l} A \longrightarrow P_1 \\ A \longrightarrow Q_1 \end{array}\right\} \text{total rate of initiation } R_i$$

$$P_r + A \xrightarrow{k_{paa}} P_{(r+1)}$$

$$P_r + B \xrightarrow{k_{pab}} Q_{(r+1)}$$

$$Q_r + A \xrightarrow{k_{pba}} P_{(r+1)}$$

$$Q_r + B \xrightarrow{k_{pbb}} Q_{(r+1)}$$

$$\left.\begin{array}{l} P_r + P_s \xrightarrow{k_{taa}} \\ Q_r + P_s \xrightarrow{k_{tab}} \\ Q_r + Q_s \xrightarrow{k_{tbb}} \end{array}\right\} \text{dead polymer}$$

where monomers A and B yield radicals of structures P and Q, respectively.

There has been virtually no dispute concerning the handling of the initiation and propagation steps in development of equations for the overall copolymerization rate, but the termination step has been more troublesome. One model assumes that the termination reaction is chemically controlled, ie, that the three termination reactions (k_t) are kinetically distinguishable on the basis of the terminal monomer unit. The other model maintains that termination is diffusion controlled and depends on the chain composition only insofar as composition determines the diffusive properties of the growing chain.

Neither of these kinetic schemes is fully satisfactory. Chemically controlled termination is counter to observations from the chemistry of simple radicals, and problematic insofar as the cross-termination fact or ϕ varies with composition. The diffusion-control model circumvents these difficulties, but seems to provide no method of predicting or even rationalizing the observed values of $k_{t(AB)}$. Additional ambiguity arises from the fact that both treatments assume the propriety of the terminal copolymerization model, and inferences concerning the termination step rest on this assumption.

Estimation of Copolymerization Reactivity Ratios. Most experimental studies of copolymerization have assumed the validity of the terminal copolymerization model, so the critical evaluation of procedures for estimating reactivity ratios has focused on this scheme.

For many years the standard method of determining r_1 and r_2 was based on an algebraic rearrangement of the Mayo-Lewis equation (rewritten here with M_1 and M_2 as mole fractions of the two monomers in the copolymer and m_1 and m_2 as mole fractions in the feed):

$$\frac{M_1}{M_2} = \frac{m_1}{m_2}\,\frac{r_1 m_1 + m_2}{r_1 + r_2 m_2} \tag{9}$$

into the following form:

$$G = r_1 H - r_2 \tag{10}$$

where

$$G = \frac{m_1}{m_2}\left(1 - \frac{M_2}{M_1}\right) \tag{11}$$

and

$$H = \left(\frac{m_1}{m_2}\right)^2 \frac{M_2}{M_1} \tag{12}$$

Equation 10 then allows r_1 and r_2 to be determined from the slope and intercept, respectively, of a plot of G vs H. Equation 10 is linear in r_1 and r_2, whereas equation 9 is not. However, it has been pointed out that this linearization also transforms the error structure in the variables in such a way that the use of linear least-squares methods is no longer suitable. A nonlinear least-squares analysis of a composition equation that is explicit in M_1 and M_2 is preferred.

Testing of Alternative Copolymerization Models. Test methods include those using composition and sequence, stereochemistry, monomer orientation, and model reactions.

Reactivity in Copolymerization

Radical Copolymerization of Olefins. Reactivity in radical copolymerization is determined by a combination of steric, resonance, and polar factors. Although a number of theoretical and empirical treatments of reactivity are available, it is the Q,e scheme of Alfrey and Price that has achieved nearly universal acceptance. The Q,e scheme provides an empirical method of predicting copolymer composition; it is not a generally reliable basis for inferences concerning the fundamental determinants of reactivity.

Step-growth Copolymerization

Although chain-growth polymerization is the more commonly used copolymerization procedure, there are circumstances that require rapid, irreversible step-growth copolymerization. Examples include reaction-injection molding processes that combine a single diisocyanate (AA) with two different diols (BB and CC). Because one of the diols contributes to the formation of the "hard segment" of the copolymer and other forms the "soft segment," the prediction and control of comonomer sequence are vital to the optimization of material properties.

Such a reaction system has been considered using a recursive statistical approach similar to that used for the calculation of molecular weight averages in nonlinear step-growth polymerizations. When applied to copolymerization, the method yields average values of molecular weight and comonomer sequence length.

BIBLIOGRAPHY

G. E. Ham, ed., *Copolymerization,* Wiley-Interscience, New York, 1964.

G. Odian, *Principles of Polymerization,* 2nd ed., John Wiley & Sons, Inc., New York, 1981, Chapt. 6.

D. J. T. Hill, J. H. O'Donnell, and P. W. O'Sullivan, *Prog. Polym. Sci.* **8,** 215 (1982).

J. M. G. Cowie, ed., *Alternating Copolymers,* Plenum Press, New York, 1985.

DAVID A. TIRRELL
University of Massachusetts

COPOLYMERS, ALTERNATING

Alternating copolymers are a class of regular copolymers in which the constituent comonomers A and B are arranged in an essentially alternating sequence $(AB)_n$. They are a special class of multicomponent periodic polymers such as $(ABCD\ldots)_n$. The alternating copolymer is a homogeneous material, in contrast to the block copolymer, which behaves like a heterogeneous composite owing to the formation of block domains. The alternating copolymer may be regarded as a polymer of a new unit AB.

Many alternating copolymers possess glass-transition temperatures that differ from the average of those of the homopolymers and sometimes exhibit excellent mechanical properties owing to high orientation on stretching. The alternating copolymers of acrylonitrile–butadiene and propylene–butadiene have been studied as new synthetic rubbers.

A polar group can be distributed by the alternating copolym-

Table 1. Properties of Butadiene–Acrylonitrile Alternating and Random Copolymers

	Vulcanization of alternating copolymer		Vulcanization of random copolymer	
Property	60 min	90 min	60 min	90 min
At 25°C				
hardness[a]	73	74	86	85
modulus at 100% elongation, MPa[b]	4.2	4.2	10.3	11.1
tensile strength at break, MPa[b]	23.2	23.8	19.0	20.1
elongation at break, %	400	400	210	200
At 110°C				
tensile strength at break, MPa[b]	5.6		4.9	
elongation at break, %	340		210	
After immersion[c]				
swelling, %	112	112	112	112
tensile strength at break, MPa[b]	14.4	15.0	9.5	10.9
elongation at break, %	300	290	150	160

[a] Japan industrial standard.
[b] To convert MPa to psi, multiply by 145.
[c] Immersion in isooctane–toluene (70:30) mixture at 25°C for 48 h.

erization of a polar monomer, such as maleic acid or acrylic acid with styrene, isobutene, or vinylamine. These polymers form molecularly homogeneous membranes (qv).

Alternating copolymers can be prepared by two methods. One is the copolymerization (qv) through a monomer pair intermediate that is formed as a donor–acceptor charge-transfer (CT) complex, a zwitterion complex, or a cyclic compound (see CHARGE-TRANSFER COMPLEXES). The other is the highly selective cross-addition of a monomer to the growing polymer end owing to the steric hindrance or the cross-coordination of the monomer.

Properties

Butadiene–Acrylonitrile Alternating Copolymer. Butadiene–acrylonitrile (BD–AN) alternating copolymer is an excellent rubber with low glass-transition temperature and high orientation on stretching. The modulus at low extension is lower than in the random copolymer, but strength and elongation at break are higher. Physical properties are listed in Table 1.

Propylene–Butadiene Alternating Copolymer. Propylene–butadiene alternating copolymer (PBR) shows high orientation at the ultimate elongation, like natural rubber (NR) and synthetic *cis*-1,4-polyisoprene (IR). It exhibits high strength and elongation at break for carbon-loaded vulcanizates and for pure gum vulcanizate (see Table 2). Other properties are given in Table 3.

Table 2. Pure Gum Vulcanizate[a]

Property	Value
modulus at 300% elongation, MPa[b]	1.0
strength at break, MPa[b]	10.8–15.5
elongation at break, %	900–1100

[a] Recipe (parts): PBR, 100; zinc stearate, 3; stearic acid, 1; sulfur, 0.9; MBTS (bisbenzothiazolyl sulfide), 1; antioxidant, 0.1. Cured for 40–45 min at 160°C.
[b] To convert MPa to psi, multiply by 145.

Table 3. Carbon-black-loaded Rubber Vulcanizates[a]

Property	PBR		SBR-1500	BR	S–SBR	IR	NR
cross-link density	1.84	1.68	1.37	1.15	1.28	1.21	
hardness[b]	64	68	64	59	62	60	66
modulus at 300% elongation, MPa[c]	10.0	12.2	13.1	9.4	9.7	12.2	15.3
strength at break, MPa[c]	28.9	29.4	25.6	19.9	25.2	27.5	29.9
elongation at break, %	600	590	530	530	590	580	560
tear strength, kN/m[d]	52	57	49	35	47	102	
rebound, %	66	56	54	62	56	56	
temperature rise, °C	16.3	23.3	20	20.3	21.3	10.8	
permanent set, %	2.8	3.6	5.2	4.6	4.5	5.3	
Pico Index	111	162	83	147	100	75	
Akron Index	0.23	0.12	0.2	0.09	0.37	0.40	
wet skid resistance	51	52	55	35	48	52	

[a] Key: PBR = propylene–butadiene alternating copolymer; SBR = styrene–butadiene rubber; BR = butadiene rubber; S–SBR = solution-polymerized–styrene–butadiene rubber; IR = synthetic *cis*-1,4-polyisoprene; NR = natural rubber.
[b] Japan industrial standard.
[c] To convert MPa to psi, multiply by 145.
[d] To convert kN/m to ppi, multiply by 5.7.

Applications

Alternating copolymers of maleic anhydride with various olefins and vinyl ethers are commercially available, such as those with methyl vinyl ether (Gantrez-AN resin of General Aniline and Film Corp.), ethylene (EMA resin of Monsanto Chemical Co.), and styrene (SMA resin of Sinclair Petrochemicals, Inc.) (see also MALEIC AND FUMARIC POLYMERS).

A copolymer possessing acid or ester groups is useful for adhesives, pigments, and mineral binders for coated printing paper. A copolymer of maleic anhydride with vinylsuccinimide has been investigated as a membrane for amino acid dialysis; the hydrolyzed copolymer possesses acid–amine pairs in regular sequence and provides a more homogeneous molecular structure than a random copolymer. The copolymer of diethylaluminum acrylate with styrene is completely alternating.

BIBLIOGRAPHY

J. Furukawa and E. Kobayashi, *Rubber Chem. Technol.* **51,** 600 (1978).

J. Furukawa, *Vysokomol. Soedin.* **21,** 2591 (1979).

J. Furukawa, *J. Polym. Sci. Symp.* **51,** 105 (1975).

J. FURUKAWA
Kyoto University

CORROSION RESISTANCE. See CHEMICALLY RESISTANT POLYMERS; COATINGS.

COSMETIC APPLICATIONS

The cosmetics industry has made use of the technology of the chemical industry, in particular, the paint and coatings industry and surfactant and petroleum industries. Polymers used in the various technologies have been investigated by the cosmetic formulator, and the packaging industry has used new and innovative polymers in the packaging of cosmetics. These polymers must be compatible with the food and drug laws. Polymers used in formulations of cosmetics must meet stringent safety requirements because, like pharmaceuticals and food, they are in intimate contact with the consumer.

A ruling by the FDA in 1973 requires the cosmetic industry to include the ingredients on the label of a cosmetic sold in the retail market. The Cosmetic, Toiletries, and Fragrance Association (CTFA), the industry representative before the FDA, published a compendium containing the designations for most cosmetic ingredients. These names are acceptable to the FDA. The purpose of this compendium is to warn users who might have allergies. The names chosen do not, however, reflect the chemistry of all ingredients. They are designed to fit easily on a package or label.

The range of polymers used in cosmetics is surprisingly wide, from cellulosics through acrylics, from polyethylene through copolymers of styrene–maleic anhydride and others. These polymers can be classified into natural, modified natural, and synthetic substances. The first category contains gums obtained from seeds, plants, or sea or animal sources: guar, gum arabic, karaya, tragacanth, shellac, algin, chitin, proteins such as gelatin, and many other natural polymeric materials.

Natural Polymers

Natural polymers are derived from seeds, bark, or seaweed as exudates or by grinding or other means of extraction. They are usually in the form of salts of complex polysaccharides. Known in the industry as gums, they provide a great variety of uses. Most of these gums are anionic, except those of the locust bean and guar, which are nonionic. A cationic version of guar is useful in shampoo and hair rinses as conditioner and thickener.

Gums are used as viscosity builders, emulsifiers, emulsion stabilizers, and protective colloids.

Modified Natural Polymers

Nitrocellulose. Along with the water-soluble cellulose polymers, nitrocellulose is important to the cosmetics industry as nail polish. A modified natural polymer, this material is used in the formulation of hard, flexible films that, when pigmented, wear well. It transmits water vapor ("breathes") (see CELLULOSE ESTERS, INORGANIC).

Cellulose Ethers. The cellulose ethers are the most useful polymers in the cosmetics industry. They exhibit versatility and stability in almost any type of cosmetic where viscosity modification, suspension of solid particles, feel, and film forming are required. These modified natural polymers are manufactured by several companies in the United States and Europe, including The Dow Chemical Company, Union Carbide, Hercules Powder, Henkel International, and Berol Kemi Ab. Hercules Powder Co. and Henkel in a joint venture formed the Aqualon Corp. designed to market and distribute cellulosics formerly manufactured and sold by each alone.

The cellulose ethers are ethoxylated or hydroxylated, nonionic or anionic polymers drived from cellulose. Ethylcellulose is a solvent-soluble film former and viscosity improver for nonaqueous systems; cationic versions are also available (eg, Polymer JR Series).

Cationic Guar Gum. This modification of natural guar can be used both as viscosity builder and cationic conditioner for hair. It is used in shampoo, mousse, and conditioner-type formulations. It is compatible with anionic surfactants and can be used from pH 3 to 11. The material increases viscosity at low concentrations with pseudoplastic behavior.

Synthetic Polymers

Quaternary Ammonium Polymers. An early quaternary polymer, known as Living Curl, was made by Revlon, and used in a hair spray. The material is a quaternized copolymer of diethylaminoethyl methacrylate and lauryl methacrylate with good setting and holding properties for hair.

Subsequently, a large number of quaternized polymeric systems (polyquats) were synthesized. An early system still in use is Union Carbides's Polymer JR series. These are cellulose derivatives ranging in viscosity from approximately 150 mPa·s (= cP) for a 2% solution of JR-125 to 2500 mPa·s (= cP) for the 30M at the same concentration. The solutions have a soft, slippery feel and are substantive to hair and skin.

Polymer JR, LR, and Quadrisoft, as well as other cationic resins, are used as hair and skin conditioners. They modify the feel of the hair or skin by depositing a film that is orientated so that the cationic positive charge adheres to the negative charges on the surface of the substrate. The noncationic portion of the molecule has a lubricant effect.

Polyvinylpyrollidinone and Derivatives. Polyvinylpyrollidinone (PVP), produced by the GAF Corp., is used for pharmaceutical and cosmetic applications. It is nontoxic, soluble in water and alcohol, and forms films easily. With these properties, it is a good shellac substitute in hair-setting lotions and sprays. It is available in various grades (K15, 30, 60, and 90) and as a powder (see *N*-VINYL AMIDE POLYMERS).

Polyacrylates. Certain polymers based on acrylic acid are suitable for cosmetic applications. They are used as thickeners, suspending agents, and emulsion stabilizers. They have some film-forming properties, and in mixtures with PVP or PVP copolymers can be used in hair-setting products. Neutralization with alkali is necessary to achieve the required viscosity. Products vary from a viscous solution to a firm gel (see ACRYLIC AND METHACRYLIC ESTER POLYMERS).

Carbopols or Carbomers. BFGoodrich supplies Carbopols (Fig. 1) in several formulations, with varying stability to salts or pH changes. These are among the most versatile materials suitable for a wide range of cosmetic products. Presence of salts and pH control are important. Viscosity drops off rapidly as pH is lowered or salt content raised. Carbopol 1342 (Fig. 1) is especially designed for shampoos. It is less affected by salt concentration and can be used with surfactants. Carbopols have good freeze–thaw properties.

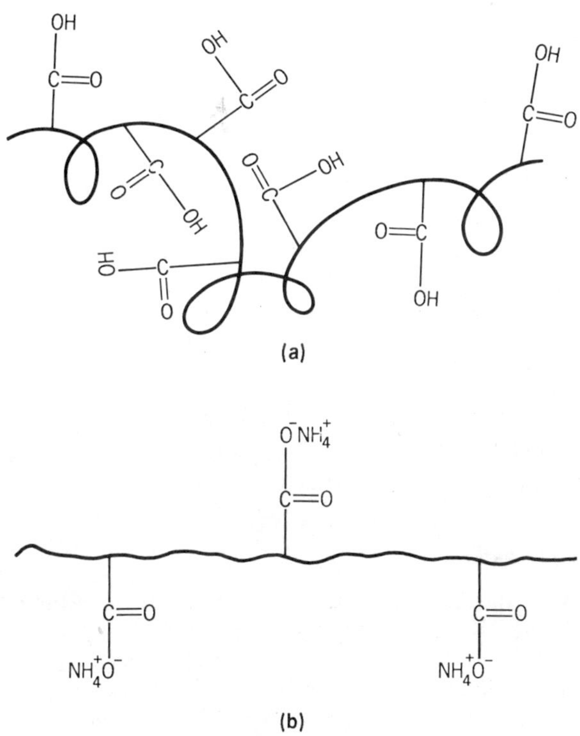

Figure 1. Carbopol molecule in (**a**) relaxed and (**b**) uncoiled state. Courtesy of BFGoodrich Co.

Polyethylenes. Polyethylenes were introduced by Allied Chemical and others as coatings, such as floor and furniture polishes; most are emulsifiable. The films are water and smear resistant. Use is made of these properties in mascaras, eyeliners, and lip products. Various types are employed, such as homopolymers, oxidized homopolymers, ethylene–acrylic acid copolymers, ethylene–vinyl acetate copolymers, and micronized versions; the Allied Chemical products are called AC polyethylene. The Epolene and Microthene series (Eastman and US Industries, respectively) are similar to the AC polyethylenes.

Silicones. Silicones, eg, polydimethylsiloxane (dimethicone), encompass a wide variety of polymers, nonionic as well as cationic species.

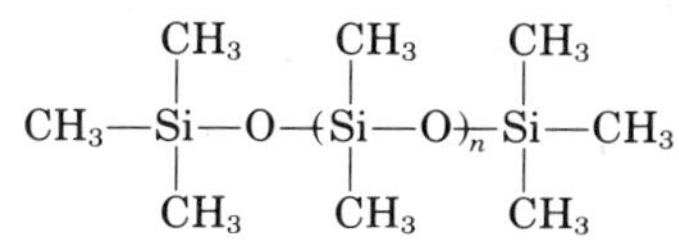

dimethicone

They are available in a range from low viscosity liquids to solids.

The silicones are strongly hydrophobic and impart this property to the formulations. They are used in deodorants and suncreens where they confer a dry, nonoily, feel. They give a shine and luster to hair. Curly hair is easily combed.

Poly(ethylene Glycol)s. Poly(ethylene glycol) systems provide important raw materials for cosmetics. Up to a molecular weight of approximately 600, they are liquids. Above molecular weight 700, they are solids ranging from soft and waxy to hard. Poly(ethylene glycol)s are water soluble and impart lubricity and solvency to creams, lotions, and shampoos. Water solutions have a cloud point when heated, depending on their molecular weight. The trade names for these materials are Carbowax (Union Carbide), Pluracol E (BASF), Jeffox (Jefferson), Polyglycol E (Dow), and others.

Other Polymers. With the exception of poly(ethylene glycol)s, the polymers discussed so far are used in large amounts in cosmetic formulations. A host of other polymers are used to some degree, eg, simple poly(ethylene glycol)s of high molecular weight, and natural protein hydrolyzates, protein isolates, and plant polymers. Ethylene oxide–propylene oxide block polymers are used as surfactants; they are called Pluronics (CTFA designation Poloxamer). Union Carbide introduced the Polyox resins ranging in molecular weight from 10^5 through 5×10^6 (see 1,2-EPOXIDE POLYMERS).

BIBLIOGRAPHY

C.T.F.A. Dictionary, 3rd ed., C.T.F.A., Washington, D.C.

R. Y. Lockhead, *Cos. Toil.* **103**(12), 23 (1988).

Products for Cosmetics, G.A.F. Corp. Wayne, N.J., 1986.

Hair Care, Dow Corning, Midland, Mich., 1982.

HERBERT EDELSTEIN
Chesebrough-Ponds Research
Laboratories

COTTON

Cotton Plant

Cotton, a plant of genus *Gossypium* (name used by Pliny) is a member of the family Malvaceae (Mallow) or the tribe *Hibisceae*. In this tribe, *Gossypium* is the most important genus from an economic point of view.

Fibers

Chemical Structure. An elemental analysis of cotton reported by Payen in 1838 and a calculated elemental composition based on structure (**1**) are shown in Table 1. The calculated composition of cellulose was based on the conclusion that cotton cellulose is a high polymer consisting of units of anhydroglucose with the empirical formula $(C_6H_{10}O_5)_n$ and structure (**1**).

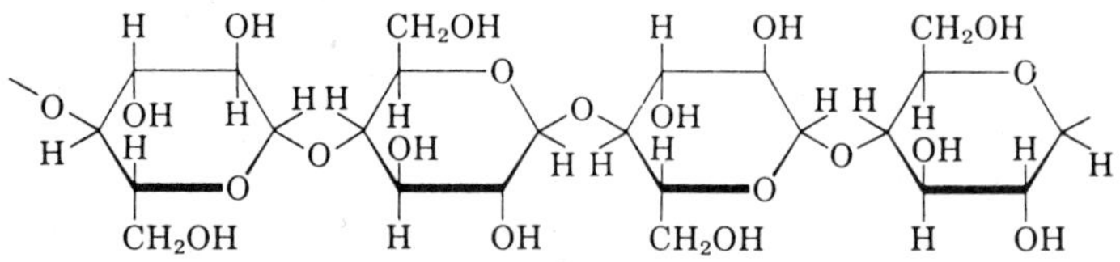

(1)

The degree of polymerization (qv) for cotton cellulose can be as high as 4,000–5,000, giving molecular weights up to 800,000.

Table 1. Elemental Analyses of Purified Cotton Cellulose, %

Element	Sample	Calculated[a]
carbon	44.35	44.45
hydrogen	6.14	6.22
oxygen	49.51	49.34

[a] Based on structure (**1**).

Cotton cellulose crystallizes in different forms. Crystalline lattice type I is the most common natural structure (see CELLULOSE, STRUCTURE AND PROPERTIES).

Composition. The compositions of a typical mature cotton fiber, as harvested and after purification, including regain moisture, are shown in Table 2.

Table 2. Typical Compositions of Cotton Fiber

	Lint, %	
Constituent	Harvested	Purified
cellulose	88.6	92–94
protein	0.9	
inorganic ash	0.5	<0.1
wax	0.7	
pectic acid	0.9	
other organic acids	0.5	
sugars	0.1	
other constituents	0.8	
water	7	6–8

Morphology. The natural physical structure of the fiber and the manufactured structure of yarns and fabrics are important factors in the chemical modification and finishing of cotton textiles. Cotton fiber, as harvested, consists of the cuticle, primary wall, secondary wall, and lumen as shown in Figure 1.

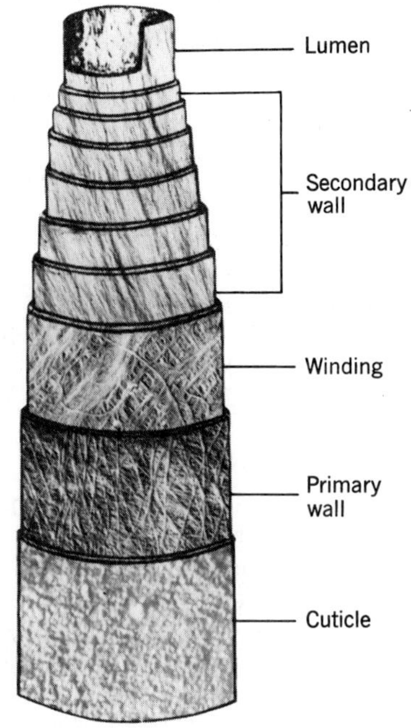

Figure 1. Principal parts of a single cotton fiber.

Cultivation

Cotton cultivation varies according to the terrain and conditions of the area, particularly climate, soil, and crop systems. The most important changes in the cultivation of cotton in the United States during the twentieth century have been increased mechanization and scientifically controlled production practices; increased production in California, Arizona, and New Mexico; and growing of the higher priced and longer-staple American–Egyptian cotton in these states and Texas. In cultivation, residual fertilizers, herbicides, insecticides, and the maintenance of nutrient balances in the soil are receiving more and more attention.

Mechanical Processing

Mechanical processing includes ginning, baling, and fiber organization.

Environmental Considerations. In the 1970s and 1980s, concern for working conditions grew, especially in the United States, because of the occurrence of respiratory problems associated with inhalation of cotton dust.

Economic Aspects

For more than 175 years, cotton has been the leading material for textile fibers. The 1980 production was estimated at 14.2×10^6 t. The 1980 world production of regenerated cellulosic fibers was estimated to be 3.2×10^6 t. In 1980, the average price for cotton fiber received by the U.S. farmers was about \$1.65/kg. In 1985, the cotton number 2 futures were traded at ca \$1.40/kg. In the United States, the cotton price is government-supported, meaning that, for eligible producers, the government will make up the difference between a set target price for fiber and the actual market price if the latter is lower than the former.

Fiber Treatments

As a textile material, cotton fibers require some modification of raw fibers before or after mechanical organization into yarns and fabrics. The modifying processes may include changes in morphological structure, molecular orientation, macromolecular structure, or organochemical structure. Treatments include sizing, bleaching, mercerization, and dyeing.

Chemical Modification

Heterogeneous alkylation and acrylation are the two most common ways to modify the properties of cotton without changing its physical form, particularly at low degrees of substitution.

Uses Other Than Fiber

Although cotton is grown primarily for fibers (lint) that can be formed into textile yarns and products, the fibrous material represents less than 40 wt % of the raw cotton harvested. Today, the residue, called cottonseed, is a source of vegetable oil and derived products.

BIBLIOGRAPHY

L. H. Bailey, *The Standard Cyclopedia of Horticulture,* The Macmillan Co., New York, 1939.

N. M. Bikales and L. Segal, eds., *Cellulose and Cellulose Derivatives,* Vol. 5 of *High Polymers: A Series of Monographs on the Chemistry, Physics, and Technology of High Polymeric Substances,* 2nd ed., Parts IV and V, John Wiley & Sons, Inc., New York, 1971.

D. Swern, ed., *Bailey's Industrial Oil and Fat Products,* John Wiley & Sons, Inc., New York, 1979.

J. G. Montalvo, Jr., ed., *Cotton Dust: Controlling an Occupational Health Hazard, American Chemical Society Symposium Series No. 189,* ACS, Washington, D.C., 1982.

JETT C. ARTHUR, JR.
Consultant

COTTON EFFECT.

See OPTICAL PROPERTIES.

COUMARONE–INDENE RESINS.

See HYDROCARBON RESINS.

COUPLING AGENTS

Coupling agents are defined primarily as materials that improve the adhesive bond of dissimilar surfaces. This must involve an increase in true adhesion, but it may also involve better wetting, rheology, and other handling properties (see also ADHESION AND BONDING). The coupling agent may also modify the interphase region to strengthen the organic and inorganic boundary layers. Only organofunctional silanes and methacrylatochrome complexes, eg, Volan, have been important commercially (Table 1).

The Nature of Adhesion

Coupling agents are compounds containing both organic and inorganic groups; they bridge the interface between resin and

Table 1. Representative Commercial Coupling Agents

Functional group	Structure no.	Chemical structure	Name
vinyl	(1)	$CH_2{=}CHSi(OCH_3)_3$	vinyltrimethoxysilane
chloropropyl	(2)	$ClCH_2CH_2CH_2Si(OCH_3)_3$	chloropropyltrimethoxysilane
epoxy	(3)	$\overbrace{CH_2CH}^{O}CH_2OCH_2CH_2CH_2Si(OCH_3)_3$	3-glycidoxypropyltrimethoxysilane
methacrylate	(4)	$CH_2{=}C(CH_3){-}COOCH_2CH_2CH_2Si(OCH_3)_3$	3-methacryloxypropyltrimethoxysilane
primary amine	(5)	$H_2NCH_2CH_2CH_2Si(OC_2H_5)_3$	3-aminopropyltriethoxysilane
diamine	(6)	$H_2NCH_2CH_2NHCH_2CH_2CH_2Si(OCH_3)_3$	N-2-aminoethyl-3-aminopropyltrimethoxysilane
mercapto	(7)	$HSCH_2CH_2CH_2Si(OCH_3)_3$	3-mercaptopropyltrimethoxysilane
cationic styryl	(8)	$CH_2{=}CHC_6H_4CH_2NHCH_2CH_2NH(CH_2)_3Si(OCH_3)_3{\cdot}HCl$	3-(N-styrylmethyl-2-aminoethylamino)propyltrimethoxysilane hydrochloride
cycloaliphatic epoxide	(9)	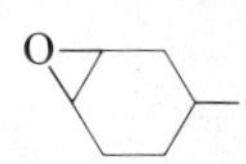	β-(3,4-epoxycyclohexyl)ethyltrimethoxysilane
chrome complex	(10)	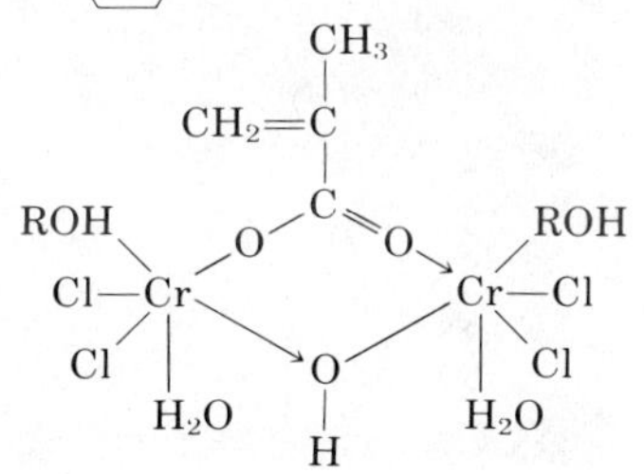	Volan
titanate	(11)	$(CH_2{=}C(CH_3){-}COO)_3TiOCH(CH_3)_2$	tris(methacryl)isopropyl titanate

reinforcement. Organofunctional silanes have R groups that form covalent bonds with the resin, whereas hydroxyl (or alkoxy) groups bond to the mineral.

Methods of Application

Fiber glass and coarse particulate fillers are generally treated with aqueous dispersions of silane coupling agents. The bath should be at the isoelectric point (IEPS) of the surface; for electrical-grade glass, ie, E-glass, the IEPS is at pH 4–6.

Economic Aspects

Silane coupling agents are relatively expensive materials used in minute amounts in reinforced composites. Various organofunctional silanes sell for $13–40/kg in commercial quantities. Others are sold as laboratory chemicals for up to $1/g or more. U.S. suppliers include Union Carbide Corp., Dow Corning Corp., Dynamit Nobel Kay-Fries Chemical Division, and SCM Specialty Chemicals; laboratory silanes are available from Petrarch Systems, (Division of Dynamit Nobel).

The principal market for silanes is in the treatment of fiber glass and particulate minerals used in polymer composites. The total market in the United States for this application in 1984 was between 1500 and 2000 metric tons.

Safety in Handling

In general, precautions in handling are determined by the nature of the organofunctional group, and by the ability to generate volatile alcohol upon hydrolysis. Of the coupling agents listed in Table 1, only silane (1) is considered a volatile material, with a flash point below 38°C. All silanes can generate methanol by hydrolysis and should be handled as potentially flammable materials.

BIBLIOGRAPHY

E. P. Plueddemann, *Silane Coupling Agents,* Plenum Press, New York, 1982, p. 132.

H. Ishida in K. L. Mittal, ed., *Adhesion Aspects of Polymeric Coatings,* Plenum Press, New York, 1983, pp. 95–106.

M. W. Ranney, S. E. Berger, and J. G. Marsden in E. P. Plueddemann, ed., *Composite Materials,* Vol. 6, Academic Press, Inc., New York, 1974, p. 134.

E. P. Plueddemann
Dow Corning Corporation

CRACKING, STRESS. See Crazing; Fracture.

CRAZING

In glassy polymers such as polystyrene (PS), the appearance of what seem to be many small cracks has long been termed crazing. Misnamed because they are not true cracks, these entities, now called crazes, usually grow normal to the direction of largest tensile stress, reflect light, and eventually fracture if stressed sufficiently. In these ways, crazes resemble cracks.

Unlike cracks, however, these entities are load bearing as a result of a web of stretched microfibrils that spans the space between the walls of what would otherwise be called cracks (Fig. 1). As a result of this ability to bear a substantial load, a craze growing near another passes by as if the other were not there. In contrast, a crack tends to veer and terminate perpendicularly in the flank of a neighboring crack, thus the appearance of so-called mud-flat cracking. This difference often provides the quickest way to judge whether the entity is a crack or a craze.

Crazes are important for several reasons. First, they are the defects in which true cracks initiate. Second, as the crack

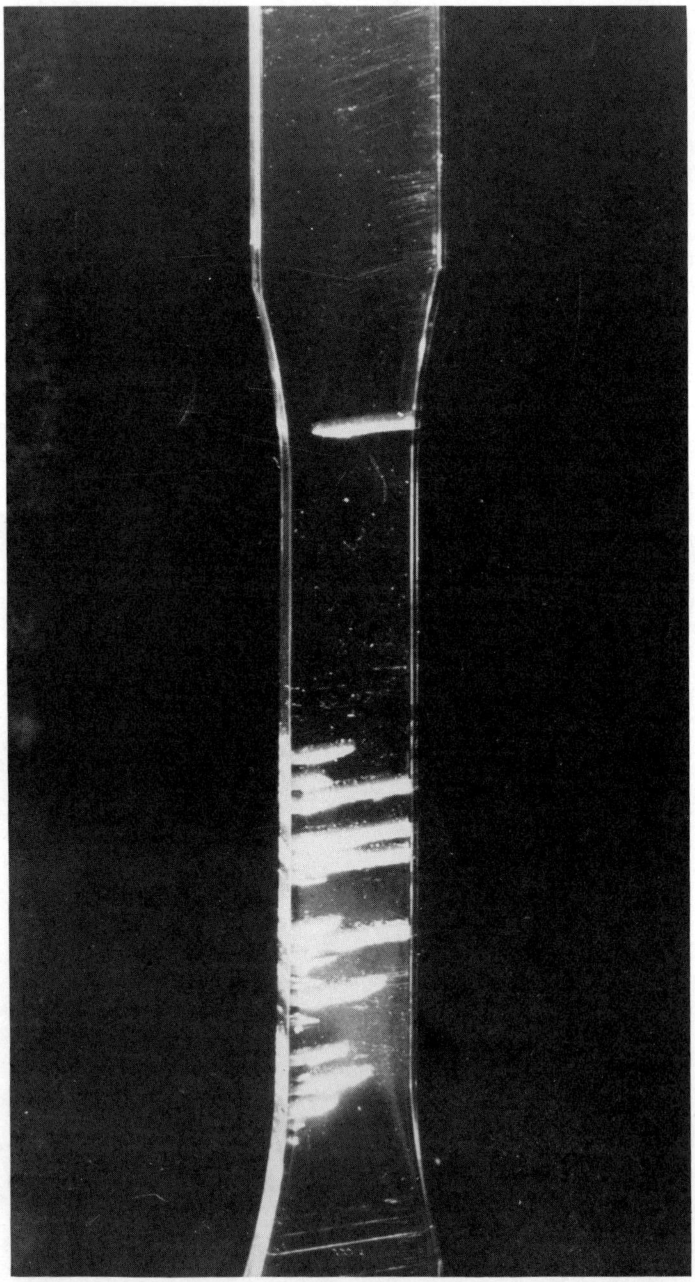

Figure 1. A bar of bisphenol A–polycarbonate containing several unusually large crazes that have run through the entire cross section without specimen failure, demonstrating that crazes are load-bearing.

propagates, more crazes are produced at the crack tip, thus constituting a plastic zone that blunts the crack. Third, their formation and deformation is the principal source of energy dissipation, ie, toughness, in many rubber-toughened plastics. Fourth, crazing, or a cavitational process akin to crazing, occurs in the shoulder of the neck formed during the cold-drawing of many semicrystalline plastics. Last, craze formation and breakdown in amorphous plastics is often markedly facilitated by small-molecule organic agents, and this susceptibility is the single greatest vulnerability of amorphous thermoplastics.

Craze Growth

After initiation, the craze tends to grow in all three dimensions. Under constant applied tensile stress, crazes, once initiated, grow in the lateral dimension either at a constant rate, eg, polystyrene, or at a rate decreasing linearly with inverse time under load, eg, poly(methyl methacrylate) containing a plasticizer and polycarbonate at high temperature.

Craze Breakdown and Crack Initiation

In glassy polymers crazes constitute the defects in which brittle cracks initiate. Moreover as the crack propagates, more crazes are produced at its tip, thus tending to resist its advance. Resistance to breakdown varies with time, stress, stress history, temperature, molecular weight, chain structure, and environment.

The craze breaks down in at least two different ways. Sometimes localized cracks initiate at certain weak spots in a craze that otherwise seems undamaged. These internal cracks grow slowly and eventually link up. This mechanism is found primarily in high molecular weight resins at temperatures well below the T_g of the craze. Closer to the T_g and in lower molecular weight resins, a craze may deteriorate by a process of fibril breakdown throughout, which results in a coarsening of the structure and general reduction in craze strength.

Environmental Effects

A wide variety of environments act as so-called stress-cracking agents for plastics. They fall into two classes: chemical corrosion agents that produce chain scission, ie, that break primary bonds; and small-molecule solvents and swelling agents, ie, agents that interfere with the cohesive forces between chains. Although the porosity of the craze and the small size of its fibrils make the craze vulnerable to both classes of agents, the second class causes the most difficulty.

Crazing in Other Resins

Crazing also occurs in crystalline plastics such as polypropylene and nylons, where it is less evident because of their lack of transparency. Cross-linking (qv) seems to bring about the total disappearance of crazing at sufficiently high cross-link densities, eg, well-cured epoxies.

Crack Propagation

Crack propagation in thermoplastics, as in metals, takes two forms. First, so-called plane-stress failure tends to occur in thin sections; in more ductile, craze-resistant materials; at higher temperatures; and at lower speeds. Shear flow in directions that result in a thinning of the region in front of the crack tip is responsible for plane-stress crack toughness. Second, so-called plane-strain fracture tends to occur in thick sections; in less ductile, more easily crazed materials; and at lower temperatures and higher speeds. Crack toughness in this case arises from the continual generation of one or more crazes at the crack tip and sometimes a degree of plane-strain shear flow.

BIBLIOGRAPHY

C. B. Bucknall, *Toughened Plastics,* Applied Science Publishers, Ltd., London, 1977.

R. P. Kambour, *Macromol. Rev.* **7,** 1 (1973).

E. J. Kramer in H. H. Kausch, ed., *Crazing in Polymers, Advances in Polymer Science,* Vol. 52/53, Springer-Verlag, Berlin, 1983, Chapt. 1.

E. J. Kramer in E. H. Andrews, ed., *Developments in Polymer Fracture,* Vol. 1, Applied Science Publishers, Ltd., London, 1979, Chapt. 3.

R. P. Kambour
General Electric Company

CREASE RESISTANCE.

See Cross-linking; Textile resins.

CREEP.

See Dimensional stability, fibers; Polymer creep.

CREPE RUBBER.

See Rubber, natural.

CRESOLS.

See Phenolic resins.

CRITICAL PHENOMENA

Phase transformations (qv) and critical phenomena influence polymer science because both focus on phenomena arising only when a large number of constituent assemblies display some cooperation.

Liquid–Gas Phase Transition

Consider the familiar substance H_2O. Two of its phases are water (liquid) and steam (gas). The phase in which a system of pure H_2O exists depends on the temperature T. Only at the boiling temperature T_b can liquid and gas phases coexist. Just below T_b, the system is entirely in the liquid phase; just above T_b, only the gas phase exists. This situation is indicated schematically in Figure 1.

The Liquid–Gas Critical Point

If $T \simeq 200°C$, the density difference $\rho_L - \rho_G$ and the latent heat ΔQ are smaller than at $T = 100°C$. In fact, as T approaches 374°C, both $\rho_L - \rho_G$ and ΔQ approach zero. The water and steam phases become indistinguishable and their coexistence curve terminates in a point called the critical point. Its coordinates are called the critical temperatures T_c, critical pressure P_c, and critical density ρ_c. Although the coordinates of the

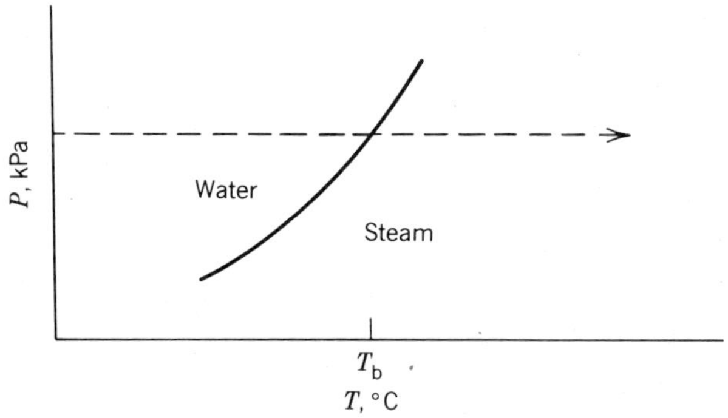

Figure 1. Constant-pressure section of the P–T phase diagram of H_2O; P = 101.3 kPa (1 atm). At the boiling or evaporation temperature, ie, T_b = 100°C, water and steam phases coexist. Just below or just above T_b, the entire system consists of a single phase. To convert kPa to atm, multiply by 0.01.

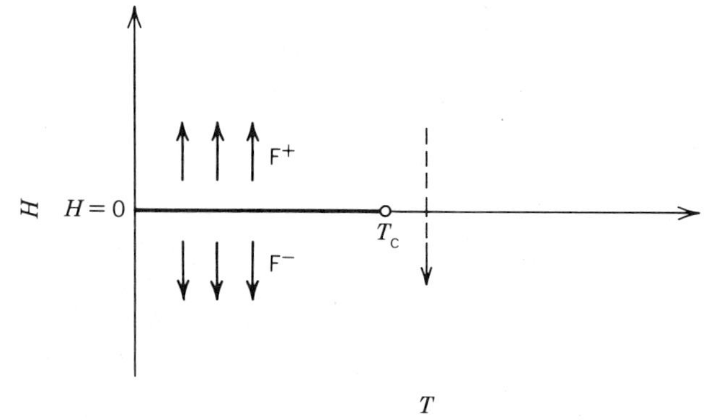

Figure 2. Projection of the H–M–T surface of a uniaxial ferromagnet in the HT plane; F^+ and F^- denote phases of predominantly up and down spins, respectively.

critical point vary widely from one fluid to another, the numerical values of the critical-point exponents and certain combinations of critical amplitude do not.

The Critical Point of a Uniaxial Ferromagnet

Iron is a ferromagnet. The magnetic moments of the constituent iron atoms tend to align with one another to produce a macroscopic magnetic moment. If, however, iron is heated above ca 700°C, a phase transition occurs and the macroscopic moment disappears.

For an idealized uniaxial ferromagnet, the behavior near the critical point is analogous to that of a simple fluid. The following correspondence of pressure P and number density $\hat{N}$,

$$-P \longleftrightarrow H$$
$$\hat{N} \longleftrightarrow M$$

is of interest; H is the external magnetic field and M is the magnetization per magnetic moment. The formal analogy is plausible: if pressure is exerted on a fluid, its density increases and if a magnetic field is applied to a uniaxial ferromagnet, its magnetization increases. Moreover, the corresponding response functions, the fluid's compressibility and the magnet's susceptibility, are analogous in their behavior near the critical point.

In the H–T projection of Figure 2, it is assumed for the sake of simplicity that the simplest uniaxial magnet is considered: each magnetic moment can orient itself in only two directions: up or down. The coexistence line separates two ferromagnetic phases of opposite polarity: the phase F^+ is characterized by a macroscopic magnetic moment in the up direction, and the phase F^- has a net moment in the down direction.

Just as in the fluid case, this coexistence curve terminates in a critical point whose coordinates are (H_c, M_c, T_c). For the simple uniaxial ferromagnet considered here, $H_c = M_c = 0$. Because the coexistence line terminates in a critical point, one can pass from the F^+ phase to the F^- phase continuously, as indicated by the dotted path shown in Figure 2.

Examples of Critical Phenomena

For both the one-component fluid and the uniaxial ferromagnet, a critical point appears at the terminus of a coexistence curve separating two different phases. Associated with these critical points are dramatic new physical phenomena.

What actually happens near a critical point is perhaps most easily exemplified by considering the uniaxial ferromagnet in the external magnetic field H at a temperature below T_c. The most striking macroscopic property of this magnet is that it is magnetized; it can, for example, pick up iron filings. Counting the number of iron fillings falling off, as a measure of magnetization, while the temperature is slowly increased at a uniform rate, it can be observed that the rate of falloff increases with the temperature. Eventually, a critical temperature T_c is reached at which this rate, $(\partial M/\partial T)_{H=0}$, diverges to infinity, in other words, above T_c no filings are left. Singularities such as this, associated with the critical points, are called critical phenomena.

Examples of additional critical phenomena are the singularities in two important response functions:

1. The constant-field specific heat C_H is the response in heat content to an infinitesimal change in temperature; C_H is singular at the same point ($H = 0$, $T = T_c$), at which the magnetization M is singular.
2. The isothermal susceptibility χ_T is the response of the magnetization of an infinitesimal increment in magnetic field. This function is also singular at the point ($H = 0$, $T = T_c$).

Interpretation in Terms of Microscopic Fluctuations

Static Fluctuations. To what microscopic features of the system do the singularities in the macroscopic thermodynamic functions M, C_H, and χ_T correspond? The specific heat in a constant magnetic field is proportional to the mean square of the fluctuations in enthalpy per particle E

$$C_H \propto \langle (E - \langle E \rangle)^2 \rangle$$

where the brackets denote a thermal average. Similarly, the susceptibility is proportional to the mean-square fluctuation in magnetization per spin M:

$$\chi_T \propto \langle (M - \langle M \rangle)^2 \rangle$$

Thus, a divergence in these response functions corresponds to a divergence in the degree to which the local values of variables such as enthalpy or magnetization fluctuate about their mean values. Indeed, because a critical system is highly susceptible, it is easy to generate such local fluctuations.

Dynamic Fluctuations. Associated with the critical point are two striking features: thermodynamic functions are singular, and the correlation length characterizing the spatial range of static correlations diverges to infinity. Recently it has become possible to learn about the time dependence of the correlations among the constituent particles and to understand the dramatic slowing-down of molecular fluctuations. This critical slowing-down can be observed by measuring the inelastic scattering of radiation from a sample near its critical point.

Scaling and Universality

The critical phenomena observed in a wide variety of systems bear striking resemblances to one another. The hypotheses of scaling and universality arose in an attempt to deal with these similarities.

Critical Exponents. In order to make quantitative comparisons among disparate systems, it is convenient to first introduce a set of parameters, the critical point exponents. Their function is to characterize quantitatively the strengths of the singularities of the thermodynamic functions, correlation lengths, and characteristic times. There is generally a different exponent for each function and each different path of approach to the critical point.

Scaling. Scaling implies that a function of two variables, $M(H,T)$, can be expressed in terms of a single scaled variable when the system is near its critical point.

Universality. Experimental *MHT* data of five magnetic materials near the critical points are given in Figure 3; the five materials are $CrBr_3$, EuO, Ni, $Y_3Fe_5O_{12}$, and the disordered phase of the ferromagnetic alloy Pd_3Fe. The fact that data near the critical point collapse onto a single scaling function supports the scaling hypotheses, although the fact that the scaling function is the same for all materials suggests that all belong to the same university class.

Renormalization Group. The renormalization-group approach to critical phenomena has provided a theoretical foundation for both the scaling and universality hypotheses. An important concept that has emerged is the idea of upper and lower marginal dimensionalities d_+ and d_-.

Polymers

Critical phenomena in polymeric systems apply to the sol–gel transition, to linear polymers, and to branched polymers.

BIBLIOGRAPHY

H. E. Stanley, *Introduction to Phase Transitions and Critical Phenomena,* 2nd ed., Oxford University Press, Inc., New York, 1985.

L. P. Kadanoff in C. Domb and M. S. Greens, eds., *Phase Transitions and Critical Phenomena,* Vol. 5a, Academic Press, Inc., New York, 1976, pp. 1–34.

H. EUGENE STANLEY
Boston University

CROSS-LINKED POLYETHYLENE.

See CROSS-LINKING; ETHYLENE POLYMERS.

CROSS-LINKING

Cross-linking reactions are those that lead to the formation of insoluble and infusible polymers in which chains are joined together to form a three-dimensional network structure. A simple cross-linking reaction is exemplified by polymer chains with several functional groups designated A that are capable of reacting among themselves to form chemical bonds A—A. If these polymer chains are exposed to conditions such that the functional groups do react, then all the chains in the reaction vessel will be tied to each other through A—A bonds. In prin-

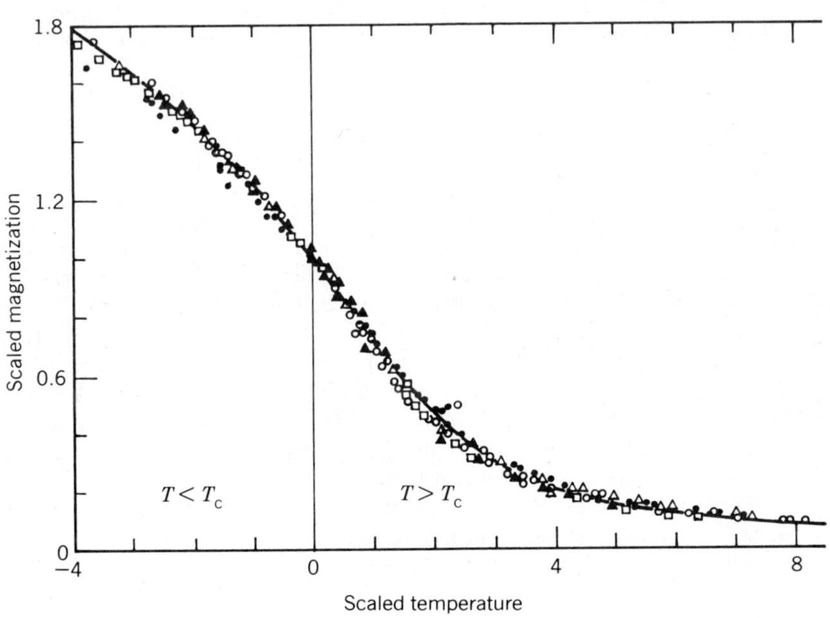

Figure 3. Experimental *MHT* data on five different magnetic materials plotted in scaled form: △ = $CrBr_3$; ○ = EuO; ▲ = Ni; ● = Pd_3Fe; □ = YIG.

ciple, the polymer molecules in the reaction vessel will have formed one giant molecule.

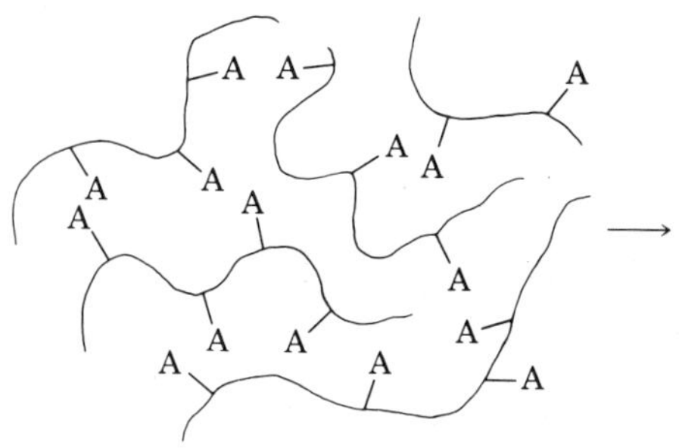

linear polymer

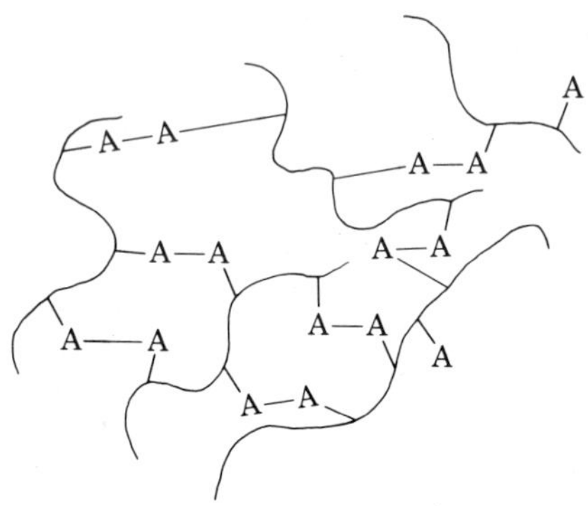

cross-linked polymer

An example of cross-linking is the vulcanization (qv) process, which makes rubber useful in applications where mechanical strength is important. Other cross-linked polymers are widely used in paints, printing inks, adhesives, sealants, encapsulants, and electrical and electronic components.

In the example above, the macromolecules were self-reacting, but this is not necessary for network formation. Cross-linking can be carried out through the use of a cross-linking agent, a molecule that has two or more groups capable of reacting with the functional groups on the polymer chain. Cross-linked polymers can also be prepared by the polymerization of monomers with an average functionality greater than two.

Reaction between trifunctional molecules and difunctional molecules initially forms low molecular weight polymers that are multifunctional and highly branched but not yet cross-linked. Cross-linking takes place only in the later part of the reaction. If a mixture of a diol and a triisocyanate is stirred, the viscosity of the mixture increases steadily as the reaction proceeds. At some extent of reaction, the viscosity undergoes a very rapid rise, the mixture becomes elastic, and begins to feel like a rubber, ie, the mixture is gelled. This extent of conversion, at which an infinite molecular weight network begins to form, is called the gel point. Chemical reaction can continue to take place in the gel state and more and more chains become linked to the network. Even when all the chains have been linked, the reaction may continue so that each chain is linked to other chains at several points. Cross-link density υ or degree of cross-linking is a measure of the total links between chains in a given mass of material. The molecular weight of the length of chain between each cross-link is denoted by M_c. In a cross-linking system a fraction of the species still have finite molecular weight. This fraction, called the sol fraction W_s, is soluble and can be extracted with suitable solvents. The cross-linked fraction is insoluble and is called the gel fraction W_g.

Gel Point Determination

Step-reaction Polymerization. A number of equations to calculate extent of reaction at the gel point (qv) have been developed. Carothers's equation (eq. 1) works well for systems undergoing step-reaction polymerization (qv) in which equivalent numbers of different reactive groups are present.

$$P_{\text{gel}} = \frac{2}{f_{\text{avg}}} \tag{1}$$

where P_{gel} is extent of reaction at the gel point, and f_{avg} is the average functionality.

$$f_{\text{avg}} = \frac{\Sigma N_i f_i}{N_i} \tag{2}$$

N_i is the number of molecules of monomer i with functionality f_i; the summation is over all monomers in the system.

Flory and others have derived expressions to predict extent of reaction at the gel point using a statistical approach. Weight-average molecular weight $\overline{M}_{w(P)}$ at an extent of reaction P can be calculated by equation 3:

$$\overline{M}_{w(P)} = \frac{\overline{M}_{w(0)}(1 + P)}{1 - P(f - 1)} \tag{3}$$

where $\overline{M}_{w(0)}$ is the weight-average molecular weight at no reaction. At the gel point, where $\overline{M}_{w(P)}$ is infinite, the necessary condition is that $1 - P(f - 1)$ be equal to zero. Therefore, conversion at the gel point is

$$P_{\text{gel}} = \frac{1}{f - 1} \tag{4}$$

The extent of reaction at the gel point for a system containing two types of mutually reactive functional groups is given by equation 5.

$$(P_{\text{gel}})^2 = \frac{1}{r(f_{\text{A}} - 1)(f_{\text{B}} - 1)} \tag{5}$$

where f_{A} is the effective average functionality (usually weight average) of molecules bearing reactive groups of type A, and f_{B} is the effective average of functionality of the molecules bearing reactive groups of type B. The ratio of A to B is r. This equation is valid only if reactive groups A and B are present in nearly stoichiometric amounts, the only reaction taking place is the reaction between A and B, all groups of type A are of equal reactivity, all groups of type B are also of equal reactivity, the reactivity of the groups does not change during the reaction (no substitution effects), and each molecule bears only one kind of functional group.

Network polymers are often formed by the reaction of functional groups on linear polymer chains. In this case the average functionality of the polymer chain is the average number of reactive groups A per chain. Average functionalities can be calculated in terms of weight average or number average. For a polymer chain in which the molecular weight between cross-linkable sites is M_c and in which the fraction of repeat units reacted is P (degree of conversion),

$$\overline{DP}_w = \frac{\bar{M}_w}{\bar{M}_c} = \frac{f(1 + P)}{1 - P(f - 1)} \tag{6}$$

where f is the average functionality of the starting polymer chains, $\overline{DP}_w$ is the weight-average degree of polymerization (qv), and $\bar{M}_w$ is the weight-average molecular weight of the polymer chains at a degree of conversion equal to P. If $\overline{DP}_{w(0)}$ is the weight-average degree of polymerization (in terms of reactive units) of the initial mixture of long chains, then

$$\overline{DP}_w = \frac{\overline{DP}_{w(0)}(1 + P)}{1 - P(\overline{DP}_{w(0)} - 1)} \tag{7}$$

At the gel point, $\overline{DP}_w$ approaches ∞; therefore

$$P_{\text{gel}} = \frac{1}{(\overline{DP}_{w(0)} - 1)} \tag{8}$$

Equation 8 is equivalent to equation 4 because $\overline{DP}_{w(0)}$ is numerically equal to the weight-average functionality of the starting polymer chains. The following relationship holds in terms of number-average degree of polymerization $\overline{DP}_n$:

$$\overline{DP}_n = \frac{\overline{DP}_{n(0)}}{1 - P\overline{DP}_{n(0)}} \tag{9}$$

and at the gel point where $\overline{DP}_n \to \infty$

$$P_{\text{gel}} = \frac{1}{\overline{DP}_{n(0)}} \tag{10}$$

Equations 7 and 9 can be used to calculate molecular-weight changes and gel points in cross-linking systems, provided the reactive groups are uniformly distributed along the polymer chains.

Chain-reaction Polymerization. The gel point of chain-type polymerization reactions (neglecting unreacted monomers present in the system) can be calculated by equation 11.

$$P_{\text{gel}} = \frac{1 - q}{q(f - 2)a_f S} \tag{11}$$

The value of q is given by

$$q = \frac{R_p}{R_p + R_{tr} + R_{td} + R_{tc}}$$

Degree of Cross-linking

Degree of cross-linking is usually expressed as number of cross-links per gram or per unit volume of polymer. If C is moles of cross-links per unit volume, n is number of network chains per unit volume, d is density of cross-linked polymer, and M_c is number-average molecular weight of the polymer segments between cross-links, then

$$C = \frac{n}{2} = \frac{d}{2\,M_c} \tag{12}$$

ie, the number of gram-moles of cross-links per gram of network polymer is equal to $1/2M_c$.

Effect of Cross-linking on Properties

Molecular Weight. Changes in the molecular weight in a cross-linking system both in the pregel and postgel regions of the cross-linking process can be predicted approximately by the foregoing relationships. In stepwise polymerization, eg, polyether network formation by homopolymerization of pentaerythritol, polydispersity increases as the reaction proceeds.

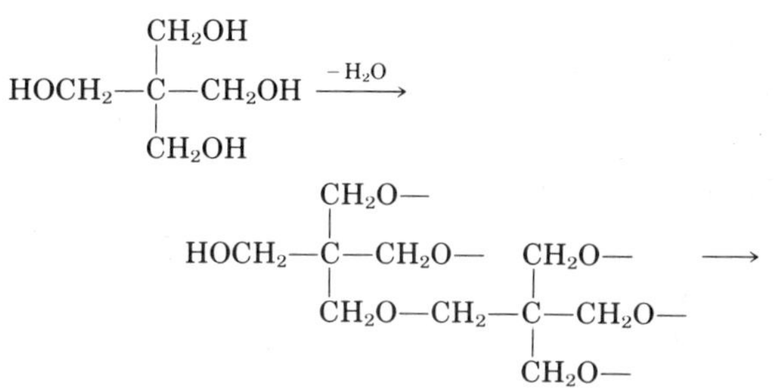

branched polymer ⟶ network

Physical Properties. Cross-linking changes and generally improves physical properties. Improvements are most significant above the glass-transition temperature. Creep, compression set, and stress relaxation are generally improved on cross-linking. Among other properties, thermal expansion and heat capacity are lowered, and heat distortion temperature, tensile strength, and refractive index are raised. Glass-transition temperature increases with increasing cross-link density.

Cross-linking Reactions

Any chemical reaction that leads to formation of a chemical bond can be used for cross-linking of polymers if the requirements of effective average functionality per molecule are met. That is, monomers, oligomers, and polymers carrying more than two reactive groups per chain can be cross-linked by reaction with cross-linking agents that carry two or more functional groups per molecule. Technologically useful cross-linking systems include

- Phenol–formaldehyde resins
  - resoles
  - novolacs
- Polyesters
  - saturated
  - alkyds
  - unsaturated
- Epoxy resins, cured with
  - amines
  - amine-terminated polyamides
  - amidoamines

- acid catalysts
- tertiary amine catalysts
- carboxylic acids
- anhydrides
- phenols

Isocyanates

Vinyl esters

Monovinyl–multivinyl systems

Urethanes
- two-component
- one-component

Amino resins
- urea–formaldehyde
- melamine–formaldehyde
- acrylic–melamine

Acrylics, with the functional groups
- hydroxyl
- glycidyl
- carboxylic
- isocyanate
- oxazolidine
- aziridine
- activated ester
- vinyl acrylics

Acetylene derivatives
- acetylene-terminated polyimide
- polyphenylene quinoxolines
- diaryl acetylenes

Polyimides

Cyanates, isocyanates, and nitriles

Halogenated hydrocarbons

Silicones

Unsaturated hydrocarbons
- sulfur cure
- peroxide cure
- urethane cure

Radiation-induced Cross-linking. Absorption of electromagnetic radiation by polymers leads to highly excited states of polymer molecules. If the excitation energy is sufficient to cause breakage of chemical bonds or create other reactive species, then the polymer molecules undergo chemical reactions including cross-linking reactions. Electron beam and ultraviolet light are the two most commonly used forms of radiation, although other types such as gamma rays, visible light, and others have also been investigated (see CROSS-LINKING WITH RADIATION; PHOTOPOLYMERIZATION.

BIBLIOGRAPHY

A. J. Chompff and S. Newman, eds., *Polymer Networks,* Plenum Press, New York, 1979.

R. W. Lenz, *Organic Chemistry of Synthetic High Polymers,* Wiley-Interscience, New York, 1967.

S. S. Labana, ed., *Chemistry and Properties of Crosslinked Networks,* Academic Press, Inc., New York, 1977.

R. S. Bauer, R. A. Dickie, and S. S. Labana, eds., *Chemistry, Properties and Applications of Crosslinking Systems,* American Chemical Society, Washington, D.C., 1988.

S. S. LABANA
Ford Motor Company

CROSS-LINKING, REVERSIBLE

Polymers are cross-linked to improve a wide variety of properties, including strength, heat resistance, and recovery from deformation. However, as a linear polymer is converted to a three-dimensional network, its thermoplastic character is lost, so reprocessing of scrap and trim material becomes impracticable. When cross-linking is combined with molding, as in rubber technology, the rate of processing is, of necessity, slower than that of thermoplastic injection molding. If the cross-links are reversible, it is possible to combine cross-linked properties under normal-use conditions with thermoplastic fabricability, leading to improved economy of operation.

The ideal reversible cross-link is thermolabile: there is strong interchain bonding at normal-use temperatures, ie, −20–100°C, but the cross-links become completely dissociated at processing temperatures of 150–250°C. It is possible to achieve this behavior by use of ionic or dispersion forces between the chains. Alternatively, the effect of temperature in shifting the equilibria of chemical reactions such as esterification can be applied to effect reversibility. A third general category of reversible cross-links consists of covalent bonds, which can be broken by a chemical reagent without affecting the main chains and subsequently reformed by a second reaction. This approach has merit in heat-sensitive polymers, provided the physical form of the polymeric article is conducive to diffusion of the reagents.

Noncovalent Cross-links

Metal-ion Cross-linking. Ionomers derived from ethylene copolymers have solid-state properties characteristic of cross-linked polymers and melt fabricability characteristic of uncross-linked thermoplastic polymers. Although it is possible to prepare ionomers from a wide variety of monomers, commercial activity has been confined to the ethylene polymer family.

Ionomers From Ethylene Copolymers. Radical copolymerization of a monocarboxylic acid, such as methacrylic or acrylic, with ethylene produces a semicrystalline product resembling low density polyethylene at low acid incorporation, ie, 1–6 mol % (see ETHYLENE POLYMERS). Partial neutralization of the carboxylic acid groups with Group I or II metal ions results in substantial solid-state property changes, but melt fabricability is retained. Neutralization can be effected by melt or solution reactions or by diffusion. Commercial products in the United States contain either sodium or zinc ions.

Elastomeric Ionomers. Amorphous, rubbery copolymers of butadiene and methacrylic acid have been commercially available for many years, and the effects of neutralizing the carboxyls by milling with metal oxides have been studied. Unusually high tensile strengths are obtained by using about twice the theoretical quantity of zinc oxide (Table 1).

Styrene-based Ionomers. Ionomers derived from styrene copolymers have been studied, but have not been commercialized. Conversion of a styrene–carboxylic acid copolymer to an ion-

Table 1. Effect of Neutralization on Tensile Strength of Elastomeric Ionomers

Ionomer	Ultimate tensile strength, MPa[a]	Ultimate elongation, %
raw polymer, ie, poly(butadiene-*co*-methacrylic acid)	<0.7	>1600
polymer neutralized with NaOH	11.7	900
polymer neutralized with NaOH and ZnO	41	400

[a] To convert MPa to psi, multiply by 145.

omer results in comparatively small property changes. The polymer retains its characteristics of high RT modulus, brittleness, and thermoplasticity. Partial neutralization reportedly raises the glass-transition temperature.

Compatibilization by Ionic Cross-linking. Ionic linking can improve the compatibility of styrene and ethylene copolymers. A milled blend of ethylene–methacrylic acid and styrene–methacrylic acid is visually inhomogeneous, and the tensile strength is only 12 MPa. Addition of zinc acetate during the milling process produces a more homogeneous blend, with a tensile strength of 21 MPa.

Halatopolymers and Crystalline Ionomers. The use of divalent ions to act as bridges between carboxyl-terminated polymers results in a family of halatopolymers with a broad range of molecular characteristics. For carboxytelechelic butadiene, viscoelastic behavior can be related to cation size, temperature, and solvent type.

Grafting of metal acrylates to crystalline polyolefins such as polypropylene has been used to obtain ionic cross-linking in polymer systems where simple copolymerization is not feasible. In general, the magnitude of property change effected by ionization is less in these grafted polymers than in random copolymers (see also GRAFT COPOLYMERS).

Amine Salt Cross-linking. Diamines are widely used in elastomer technology to form permanent, nonreversible cross-links. The chemistry usually involves splitting out of water to form an amide or imide linkage. However, diamines can also produce thermolabile, reversible linkages by amine salt formation, provided appropriate reaction conditions are selected.

Cross-linking by Secondary Valence Forces. Secondary intermolecular forces are universal in polymeric materials and can be used to provide the effect of reversible cross-linking in multiphase systems. Commercially successful thermoplastic elastomers are based on the two-phase approach, combining rubbery chains with a second hard, ie, crystalline or glassy, phase (see ELASTOMERS, THERMOPLASTIC). At temperatures above the melting point or glass transition of the hard phase, melt flow is possible. At normal use temperatures, the hard phases act as cross-linking sites, so good elastic recovery can be achieved.

Thermal Equilibrium Displacement

Thermally reversible cross-linking can be created by reversible Diels-Alder reactions, dimerization of nitroso groups, and reversible esterification reactions.

Use of Chemical Reagents

Reversible cross-linking can be produced using anhydride cross-linking, complex formation, and sulfur bridges.

Applications

Ionic cross-linking by metal ions is applied commercially to ethylene copolymer resins and specialty elastomers. These materials are used for packing of food and other products, sporting goods, automobile parts, and sealing of glass. Selective reduction and subsequent reoxidation of sulfur bridges are widely practiced in cosmetic chemistry.

Health and Safety

Surlyn ionomers and Hytrel thermoplastic elastomers have FDA clearance for food contact. The experimental polymers require testing to establish their lack of toxicity.

BIBLIOGRAPHY

A. Eisenberg and M. King, *Ion-Containing Polymers,* Academic Press, Inc., New York, 1977.

A. Noshay and J. E. McGrath, *Block Copolymers,* Academic Press, Inc., New York, 1977.

A. Y. Coran and R. P. Patel, *Rubber Chem. Technol.* **53,** 141, 781 (1980).

S. D. Gershan and co-workers in E. Sagarin, ed., *Cosmetics, Science & Technology,* Interscience Publishers, New York, 1957.

R. W. REES
E.I. du Pont de Nemours & Co., Inc.

CROSS-LINKING WITH RADIATION

The interaction of electromagnetic radiation with polymeric materials can lead to the formation of three-dimensional network structures, which generally improve the overall physical or chemical properties of the original substrate. Network structures form under a wide variety of radiation conditions (ionizing or nonionizing radiation, photochemical or thermal chemical cure systems), but this article focuses on ionizing, or high energy, radiation cross-linking processes. Both qualitative and quantitative evaluations of radiation-induced cross-linking reactions were pursued between 1944 and 1964, but in recent years commercialization of radiation processing of polymers has been emphasized.

In general, cross-linking of polymeric substrates and the related technologies involve four main variables: the type of radiation and its source; the nature of the polymer structure to be irradiated and its response characteristics; mechanisms or theories of reaction; and chemical, physical, and mechanical properties of network formation (see also CURING; PHOTOPOLYMERIZATION; RADIATION-RESISTANT POLYMERS).

Radiation

Types and Sources. Monomer, oligomer, and polymer cross-linking reactions may be induced by nonparticulate radiations, such as microwave, infrared, light energies, x rays, and gamma rays, or particulate radiations, such as alpha particles, beta particles, high energy electrons, protons, deuterons, and neutrons.

The most common radiation sources for the commercial cross-linking of organic materials are cobalt 60, low and high energy electron accelerators, light energy (ultraviolet–visible), infrared sources of energy, and plasma or glow-discharge energy sources (microwave or radio-frequency range).

Intensity. In high energy radiation–polymer reactions, similar products (cross-linked or degraded polymers) result from energy absorption by the polymer regardless of the type of radiation (gamma, pile, or electron). The concept that equal reaction products result from equal energy absorption in organic and polymeric materials irrespective of the type of high energy radiation used to cause the reaction is helpful in predicting product yields for similar materials under different conditions. The product yield or efficiency of an irradiation process can be defined as the amount of reaction (cross-linking, degradation, gas formation, formation of unsaturation, etc) produced per unit of absorbed energy. The common industrial high energy radiation sources are cobalt 60 (1.25 MeV) and high energy electrons in the range of 0.1–10 MeV.

Theories of Cross-linking

The mechanism of cross-linking by radiation is either free radical or ionic. Cross-linking by interaction with high energy (ionizing) radiation sources occurs through both mechanisms.

Cross-linking:

$$\text{polymer—CH}_2\text{CH}_2\text{—polymer} \xrightarrow[\text{direct absorption}]{\text{ionizing radiation}} \text{macroradicals} \begin{cases} \text{polymer—}\dot{\text{C}}\text{HCH}_2\text{∼∼polymer} \\ \text{polymer—}\dot{\text{C}}\text{HCH}_2\text{∼∼polymer} \end{cases} + H_2$$

$$\downarrow$$

$$\begin{array}{l} \text{polymer—CHCH}_2\text{—polymer} \\ \quad\quad\quad\quad | \longleftarrow \text{cross-link site} \\ \text{polymer—CHCH}_2\text{—polymer} \end{array}$$

Scission:

$$\underset{\text{(high molecular weight)}}{\text{polymer—polymer}} \xrightarrow{\text{ionizing radiation}} \underset{\text{(low molecular weight)}}{\text{polymer}} + \underset{\text{(low molecular weight)}}{\text{polymer}}$$

Gel formation:

$$\text{polymer} + \underset{\substack{\text{multifunctional}\\ \text{vinyl monomers}\\ \text{or oligomers}}}{(CH_2{=}CH)_n\text{—R}} \xrightarrow[\text{radiation}]{\text{ionizing}} \text{rapid gel formation}$$

Response to Irradiation

The interaction of electromagnetic radiation with polymeric materials falls into two classes: monomers, oligomers, polymers, and additives interacting with ionizing radiation or high energy electrons (0.1–10 MeV); and monomers, oligomers, polymers, catalysts, or additives undergoing photochemical or thermal reaction processes (0.01–1 eV).

Other factors to consider are the types of radiation energies and their intensities; monomer, oligomer, and polymer structures; and additives or catalysts and temperature effects on the reaction mechanism.

Monomer, Oligomer, and Polymer Structures

Monomers are defined here as singly vinyl-unsaturated organic molecules with well characterized molecular weights (100–500 g/mol).

Cross-linking oligomers contain multiple unsaturation (mol wt ca 200–1000). They are used to promote rapid free-radical-induced propagation reactions leading to highly complex polymer network (cross-linked) structures.

Thermoplastic vinyl polymers (mol wt 40×10^3–10^6) such as polyethylene, polypropylene, polystyrene, polyacrylates, and, to some degree, poly(vinyl chloride) tend to show improvements in physical or chemical properties on the formation of cross-link sites. Vinylidene polymers such as polyisobutylene, poly(α-methylstyrene), polymethacrylates, and poly(vinylidene chloride) degrade under high energy radiation. Certain other types of polymer structures (high aromatic content or thermoset) resist degradation by high energy radiation. The molecular weights of polymers (thermoplastic or reactive) used in radiation-curable coating systems are ca 1,000–40,000. The reactive polymers usually contain acrylic, methacrylic, or fumaric vinyl unsaturation along or attached to the polymer backbone.

Preformed polymer structures that cross-link directly under

Table 1. Approximate Cross-link and Scission Yields of Irradiated Polymers[a]

Polymer	G(X)[b]	G(S)[c]
polyethylene	3.0	0.88
linear low density polyethylene	2.53	0.4
polypropylene	0.6[d]	1.1[d]
natural rubber	1.1	0.22
polybutadiene	3.8	
butadiene–styrene copolymer[e]	2.8	0.39
polystyrene	0.045	<0.018
polydimethylsiloxane	2.7	<0.54
poly(alkyl acrylates)		
methyl	0.55	0.18
n-butyl	0.63	0.18
isobutyl	0.63	0.18
sec-butyl	0.47	0.18
tert-butyl	0.16	0.18
neopentyl	0.57	0.18
poly(alkyl methacrylates)		
methyl		1.22–3.5
n-butyl		2.7
isobutyl		3.9
sec-butyl		5.3
tert-butyl		9.6
poly(ethylene terephthalate)	0.035–0.14	0.07–0.17
poly(propylene oxide)	0.15	0.22
poly(vinyl chloride)	2.15	
	0.33	0.23
polycaprolactam (nylon-6)	0.5	0.6
poly(hexamethylene adipamide)(nylon-6,6)	0.5	0.6
poly(vinyl acetate)	0.1–0.3	0.06–0.17

[a] Irradiation at room temperature under vacuum or inert atmosphere (nitrogen).
[b] G(X) = cross-links per 100 eV energy absorption (average).
[c] G(S) = scissions per 100 eV.
[d] Dependent on tacticity and crystallinity.
[e] Butadiene:styrene = 77:23.

the influence of light energy (uv or visible) require special photosensitive groups to effect network formation.

Additives

Additives include cross-linking enhancers (halides, nitrous oxide, sulfur monochloride, and oxygen) and cross-linking retarders (quinones, hydroquinones, and aromatic amines).

Structural Changes

Most theoretical explanations of radiation-induced chemical or physical changes in a high molecular weight preformed polymer structure are based on changes in the molecular size distribution. Chemical changes (loss of low molecular weight fragments or changes in chemical structures) can be determined by a wide variety of spectroscopic techniques. A polymer under high energy irradiation forms a wide variety of intermediate structures (free radicals, unsaturation sites, cross-link sites, etc) related to events associated with main-chain scission of the polymer backbone. Segmentation of the polymer chains into smaller molecules changes the molecular size distribution of the system.

Cross-linking of Specific Polymers

The two main effects of high energy radiation on polymeric materials, ie, changes in the chemical and physical properties, depend on the efficiency of the cross-linking reaction, $G(X)$ = number of cross-links (consisting of two cross-linked polymer units) formed for 100 eV of radiant energy absorbed. A measure of main-chain scission is $G(S)$ = number of breaks for 100 eV of energy absorbed.

Cross-link $G(X)$ and main-chain scission $G(S)$ yields for several types of radiation-cross-linked polymer systems are given in Table 1.

Commercial Developments

The commercial development of radiation-induced modification (cross-linking, degradation, grafting) of polymeric materials applies to several manufacturing areas: tires, wire and cable insulation, heat-shrinkable film and tubing, controlled polymer modification (molecular weight reductions for ease of molding or chemical grafting of monomers onto polymer surfaces for ease of bonding, dye acceptance, etc), and curing of coatings. Examples of commercial radiation processing are given in Table 2.

BIBLIOGRAPHY

V. D. McGinniss in M. Grayson, ed., *Kirk-Othmer Encyclopedia of Chemical Technology,* 3rd ed., Vol. 19, Wiley-Interscience, New York, 1982, pp. 607–624.

S. P. Pappas, ed., *UV Curing Science and Technology,* Vol. II, Technology Marketing Corp., Norwalk, Conn., 1985.

M. Dole in M. Dole, ed., *The Radiation Chemistry of Macromolecules,* Vol. I, Academic Press, Inc., New York, 1972, pp. 265–286.

R. L. Clough and K. T. Gillen, *Radiat. Phys. Chem.* **22,** 527 (1983).

VINCENT D. MCGINNISS
Battelle Columbus Laboratories

Table 2. Commercial Uses of Radiation-processing Techniques

Substrate	Radiation process	Commercial use
polyolefins and poly(vinyl chloride)	cross-linking using Dynamitrons[a] at ca 10 kGy/s[b] in 0.4–3 MeV range or higher	wire (1.2 mm) insulation for computers and communication applications[c]
rubber materials	cross-linking with high energy electrons	radiation processing for improvement of form stability, rolling resistance, traction, and treadwear lifetimes of tires
polytetrafluoroethylene (Teflon)	degradation by high energy electron or cobalt-60 sources delivering 0.2–0.4 MGy[b]	conversion of waste Teflon materials into easily moldable powders or waxes of commercial value
wood impregnated with acrylic or methacrylic monomers	polymerization with cobalt-60 source	no-wear high performance wood floors for high traffic areas
curing of coatings and adhesives	low energy electron-processing equipment in the 100–500 keV range at 0.1–0.2 MGy[b]	ca 11 t/yr coatings and adhesives produced for the modification of wood, textiles, paper, film and metal substrates
polyolefin and poly(vinyl chloride) foams	cross-linking with high energy electrons	improved thermal stability for insulation and packaging applications[d]

[a] High power 10–100 kW.
[b] To convert Gy to rad, multiply by 100.
[c] 30% Annual growth in Japan.
[d] Japan's output of cross-linked polyethylene foam is ca 7000 t/yr.

CRYOGENIC PROPERTIES

Cryogenic applications of polymers are mainly connected with superconductivity, space technology, and liquefied gases. They include superconducting motors or pulsed magnets for fusion technology, which require materials resistant to fatigue and electrically resistive to eddy currents. Support elements for liquid–gas containers require low thermal conductivity and good mechanical strength.

The high specific strength or stiffness, thermal and electrical insulation capabilities, and fatigue-endurance limits of fiber composites are the chief reasons for their cryogenic applications. The polymeric matrix plays an important part and, for several properties, constitutes the limiting factor.

Polymer Properties

Cryogenic research on polymers offers several advantageous features:

1. Most degrees of freedom are frozen. Individual molecular motions can be studied; this is true for mechanical and dielectric activations.

Table 1. Properties of Polymers[a]

Polymers	Shear modulus G', GPa	Young's modulus E', GPa	Tensile stress, σ_{UT}, MPA	Tensile strain, ϵ_{UT}, %	Poisson's ratio, μ'	Specific heat C, J/(kg·K)	Thermal conductivity λ at 2 K, mW/(m·K)	Integral thermal expansion[b] $\Delta L/L$, %
epoxy resin[c]	2.6	8.1	179	2.1	0.37	1.6	70	−1.2
Thermoplastic polymers[d]								
HDPE[e]	3.5	9.7	175	4.0	0.30	1.0	26	−1.8 to −2.2
PTFE	2.8	8.5	81	1.2				−1.6
PS	2.1	5.0	68	2.1	0.35	4–6	25	−1.5
PSU	1.9	6.0	150	3.0	0.37			−1.1
PVC	2.6	7.9	110				20	
PC	2.2	6.0	170	3.5	0.39		25	−1.4
POM	5.0	13.0	170	1.5	0.32	1.1	30	−1.5
polyamide	2.55	7.5	190	2.2	0.30	1.6		−0.8
polyimide[f]	2.35	6.6			0.39		40	−0.8

[a] At 4.2 K.
[b] Between 4.2 and 293 K.
[c] CIBA-GEIGY Cy 221/Hy 979.
[d] Key: HDPE = high density polyethylene; PTFE = polytetrafluoroethylene; PS = polystyrene; PSU = polysulfone; PVC = poly(vinyl chloride); PC = polycarbonate; POM = polyoxymethylene.
[e] 64 vol % crystalline.
[f] Kerimid.

Table 2. Properties of Fibers and Fiber Composites

Fibers	Tensile strength σ_{UT}, GPa[a]	Tensile strain ϵ_{UT}, %	Young's modulus E', GPa[a]	Tensile stress σ_{UC}, GPa[a]	Poisson's ratio, μ'	Interlaminar shear strength, τ_{ILS}, MPa[b]	Fatigue-endurance limit[c] σ/σ_{UT}, %
carbon fiber							
T 300 (Torayca)	3.5	1.5	230				
M 40A (Torayca)	2.4	0.6	400				
A S-4 (Hercules)	2.8		210				
fiber glass, E-glass	2.0–3.4	3.1	70				
Kevlar 49	2.8	2.1	140				
In epoxy matrix[d] at 4.2 K							
Unidirectional,[e] parallel							
carbon fiber							
T 300 (Torayca)	1.6–2.0	1.2–1.5	140	0.8–1.3	0.32	120	85
M 40A (Torayca)	1.2–1.5	0.5	240		0.32	100	85
A S-4 (Hercules)	1.4–1.6		130			140	
fiber glass, E-glass	1.2–1.8	2.5–3.0	45	0.7–1.3		170	
Kevlar 49	1.3–1.6	1.5–1.8	90	0.3		38	65
Perpendicular							
carbon fiber T 300	0.05–0.06	0.4–0.5	13		0.07		
Kevlar 49			7		0.1		

[a] To convert GPa to psi, multiply by 145,000.
[b] To convert MPa to psi, multiply by 145.
[c] Of tensile threshold fatigue tests at 77 K.
[d] Epoxy resin Cy 221/Hy 979 (CIBA-GEIGY).
[e] Fiber content 60 vol %.

2. Long-wave phonons are dominant, allowing study of vibrational modes between molecular chains, ie, interchain modes.
3. Because most main-chain motions are frozen, interchain binding forces can be analyzed without regard to viscoelastic intrachain effects.
4. At very low temperatures, many properties are rather independent of chemical structure but depend on crystalline content.
5. At extremely low temperature, ie, <1 K, quantum-mechanical effects such as phonon tunneling can be detected.

At cryogenic temperatures, interchain interactions govern most polymer properties, as is evident for elasticity and most thermal properties. Dominant phonon wavelengths become large enough to establish a similar intra- and interchain phonon flux.

Relaxation processes control most polymer properties. Some mechanical properties are controlled by isothermal–adiabatic transitions at high deformation rates.

Below 200 K, dielectric and mechanical-loss tangents exhibit the same temperature dependence; their ratio is a constant that is determined by the polarity of polymers. Most polymers are brittle at low temperatures due to freezing of intrachain interactions. At very low temperatures, ie, <1 K, quantum-mechanical effects predominate in phonon interactions. Physical properties are listed in Table 1.

Polymeric Composites

The properties of composites at cryogenic temperatures are dominated by the interfacial bond or the boundary condition between filler and matrix, the brittleness of the matrix, and the residual thermal stress or strain. Cryogenic properties of fibers and composites are given in Table 2. Most fiber-composite properties are governed by anisotropic features originating in the fiber arrangement, the intrinsic fiber anisotropy, and the fiber-matrix interfacial bond (see also COMPOSITES).

At cryogenic temperatures, mechanical properties are influenced by stresses arising from differential thermal contractions, and thermal transport phenomena are influenced by the Kapitza effect arising from phonon scattering at filler-matrix boundaries (phonon mismatch).

BIBLIOGRAPHY

A. Clark, R. P. Reed, and G. Hartwig, eds., *Nonmetallic Materials and Composites at Low Temperatures,* Vol. 1, Plenum Press, New York, 1978.

A. Clark and R. P. Reed, eds., *Advances in Cryogenic Engineering,* Vol. 28, Plenum Press, New York, 1978.

G. Hartwig and D. Evans, eds., *Nonmetallic Materials and Composites at Low Temperatures,* Vols. 2 and 3, Plenum Press, New York, 1982.

A. Clark and R. P. Reed, eds., *Advances in Cryogenic Engineering,* Vol. 30, Plenum Press, New York, 1984.

G. K. White, *Experimental Techniques in Low Temperature Physics,* Clarendon Press, Oxford, UK, 1968.

G. HARTWIG
Kernforschungszentrum Karlsruhe

CRYOSCOPY.

See MOLECULAR WEIGHT DETERMINATION.

CRYSTALLINE POLYMERS

A polymer solid can be partially crystalline, such as polyethylene and polyethylene terephthalate, or noncrystalline, such as commercial poly(methyl methacrylate) (PMMA) and polystyrene. Partially crystalline polymers, called crystalline polymers, are constructed by a complicated aggregation of crystalline and amorphous regions. In the amorphous region the molecular chains are in a conformationally random state, in some cases in the theta state (see AMORPHOUS POLYMERS). In the crystalline region, polymer chains are extended in a definite characteristic structure and packed together regularly, forming a crystallite. A molecular chain is in general far longer than the crystallites, ca 10.0–20.0 nm. For example, the length of the extended planar zigzag of a polyethylene molecule of mol wt = 50,000 is about 450.0 nm. Hence, one molecular chain is considered to pass through many crystalline and noncrystalline regions successively via intermediate regions. When such a specimen is stretched under suitable conditions, the molecular chains and crystallites are oriented along the drawing direction. Figure **1a** illustrates the so-called fringed micelle model, which has been used for a long time. Figure **1b** shows the folded lamellar model, where a long chain passes in and out of the same crystallite (lamella) many times, accompanied by a folding of the chain. This chain-folding principle was developed after the discovery of the polymer single crystal grown from a dilute solution.

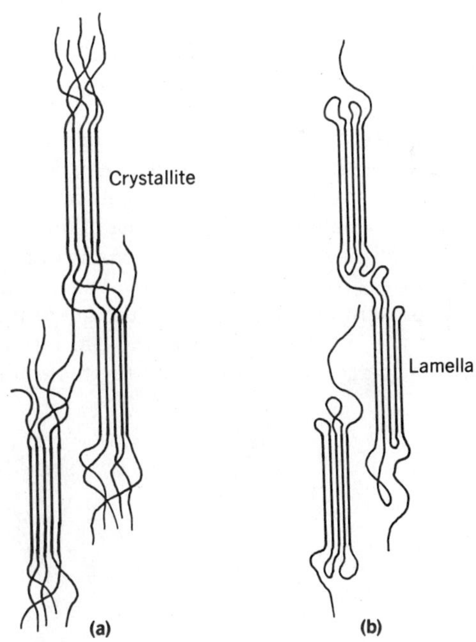

Figure 1. Molecular chains. (**a**) Fringed micelle model; (**b**) folded lamellar model.

Crystallization of polymer samples with low crystallinity is accelerated by heat treatment under suitable conditions. The weight fraction of the crystalline regions determines the degree of crystallinity, for which many methods of estimation have

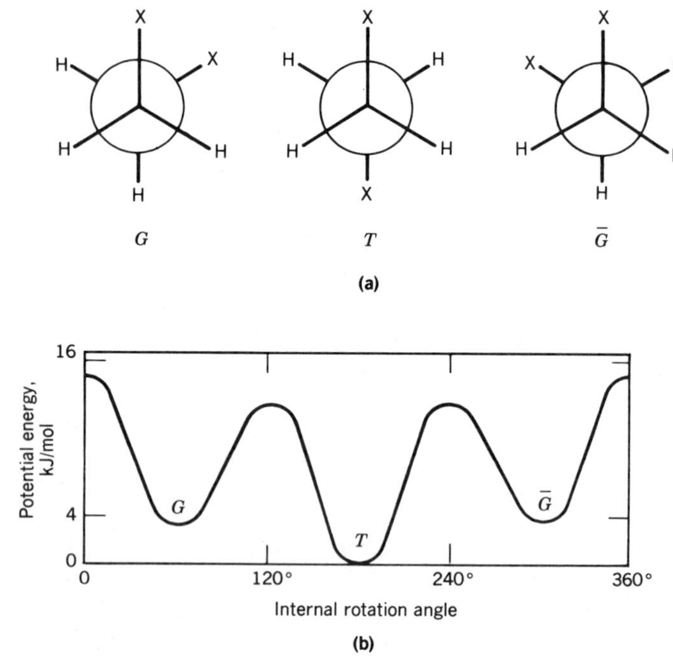

Figure 2. (**a**) The projected structures of the trans and gauche forms. (**b**) Potential-energy curve for internal rotation around the carbon-carbon bond of *n*-butane. To convert J to cal, divide by 4.184.

been developed. The term fine texture or morphology (qv) is used to indicate the state of aggregation, including the size, shape, and orientation of crystallites and their distribution, as well as the structure of the amorphous regions (Fig. 1). The recent development in processing techniques makes it possible to produce the fine texture of the highly oriented and highly crystalline state of the polymer chains throughout the sample. Physical properties of such a sample are very close to that of the crystalline region; for example, the bulk Young's modulus along the draw direction approaches that of the crystallite modulus. Therefore, the relationship between structure and properties of the crystalline regions is important, because the crystalline state is considered a limiting state of the fine texture.

The term crystalline polymers includes many kinds of structural concepts extending from microscopic to macroscopic dimensions. The present article focuses on the general aspect of the structure of the crystalline region and the intimately related physical properties as viewed from the molecular level.

Structure of Polymer Crystals. *Chain Conformation.* A stereochemically characteristic point of polymer chains is that they may take a variety of possible molecular structures or conformations, unlike the more limited possibilities for low molecular weight compounds. This is due to the infinite number of possible combinations of internal rotations about the skeletal bonds, where each bond can assume, eg, in the carbon–carbon bond, one of the three energetically stable internal rotation angles, trans (T), gauche (G), and gauche ($\bar{G}$) (Fig. 2). Some of the conformations are, however, excluded because of the steric repulsion between the nonbonded interatomic pairs. In the crystal lattice especially, the molecular conformation is uniquely determined so that the free energy of the system is minimized under conditions of a given temperature and pressure. In actual crystallization, a kinetic effect must also be considered. Molecular conformation is described in terms of the sequence of the internal rotation angles (Fig. 3).

Packing of Molecular Chains in the Crystal Lattices

The molecular chain of *it*-PP takes a (3/1) helical conformation consisting of a regular sequence of $TGTGTGTG$ . . . (Fig. 3). The one-dimensional space group contains the threefold screw axis C_3^s (or 3_1). In three-dimensional lattices, however, this threefold screw symmetry disappears and the molecular chain does not possess any symmetry element except for the identity element; that is, the symmetry of the *it*-PP molecule at a lattice point, or the site symmetry, is isomorphous to C_1 (Fig. **4a**). Thus the smallest repeating unit in the crystal lattice or the crystallographic asymmetric unit is not one monomeric unit but three monomeric units. Choosing an arbitrary point representative of such an asymmetric unit, these points are arranged in the lattice as shown in Figure **4b**. Included in the cell are four points or units, which are interrelated by the following symmetry operations:

unit 1 ⟶ unit 1	E (identity operation)
unit 1 ⟶ unit 2	i (center of symmetry)
unit 1 ⟶ unit 3	$\sigma_g(a'c)$ (glide plane)
unit 1 ⟶ unit 4	C_2^s (twofold screw axis)

The helical rotation and up-down direction are related by these symmetry operations. For example, if unit 1 is a right-hand downward helical chain, then unit 2 is a left-hand upward helix, unit 3 a left-hand downward helix, and unit 4 a right-

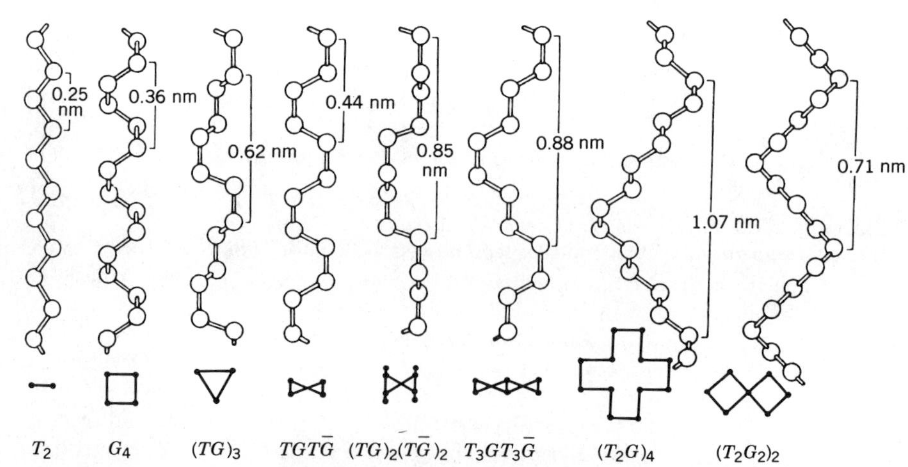

Figure 3. Possible conformations of single-bonded carbon atoms.

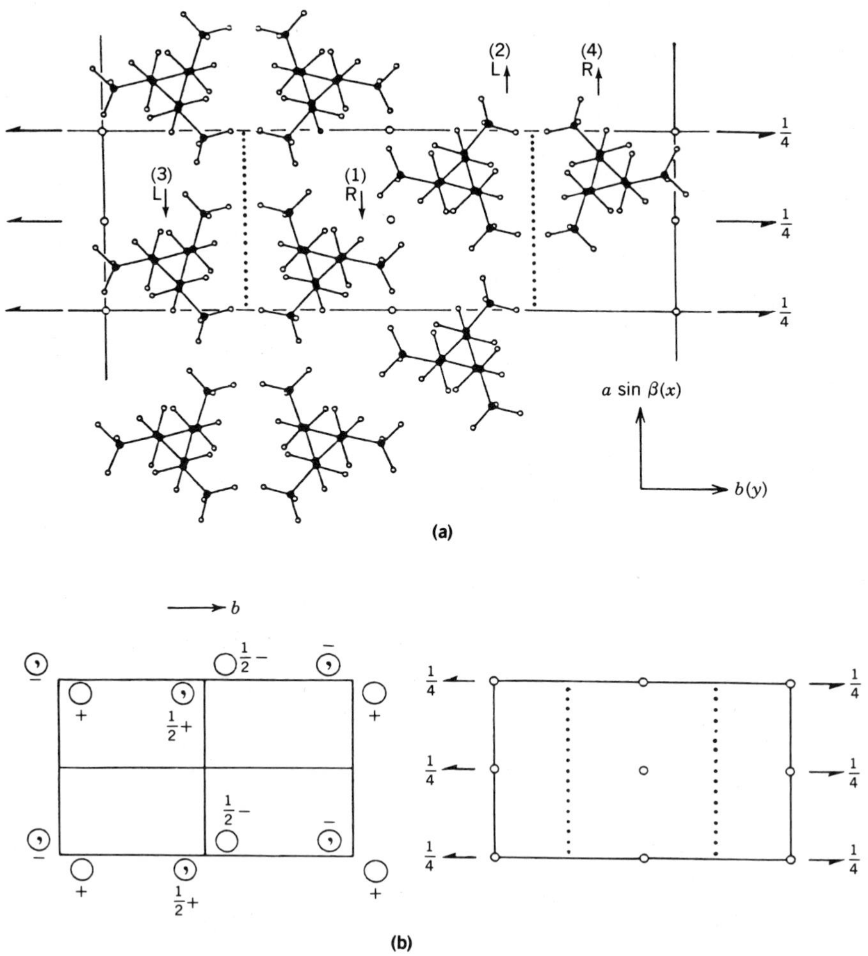

Figure 4. (**a**) Crystal structure and (**b**) space group symmetry of isotactic polypropylene; ○ and ⊙ represent the point chosen arbitrarily for the three monomeric units of each chain and its mirror image, respectively. The fractions shown in (**b**) represent the height along the z axis. $a' = a \sin \beta$.

hand upward helix (Fig. 4). This spatial symmetry corresponds to that of the space group $P1,2_1/c$, $1\text{-}C_{2h}^{5}$ as listed in the International Tables.

Figure 5 shows the crystal structure of orthorhombic polyethylene. An asymmetric unit is $(1/2)CH_2$ because this methylene group lies on the mirror plane $\sigma(ab)$. The unit cell contains two zigzag chains, four CH_2 units, and therefore eight asymmetric units. The space group is $Pnam\text{-}D_{2h}^{16}$. The correlation among the molecular group, site group, and space group is expressed as follows.

Molecular group	Site group	Space group
D_{2h} ———	C_{2h} ———	D_{2h}

Determination of Molecular and Crystal Structures

Many different methods have been utilized to determine the molecular and crystal structures of polymer crystals. X-ray diffraction and vibrational spectroscopy are most important, although neutron and electron diffractions and solid-state nmr may also be useful.

In polymer systems, the observable x-ray reflections are broad and small in number, making it difficult to construct a molecular model to be analyzed. In such a case, a combination of the x-ray diffraction and the infrared–Raman spectroscopic data may provide useful information. In particular, the existence of the bands intrinsic to the specific conformations (conformation-sensitive bands) helps to construct reasonable models. In some cases the normal-mode frequencies are calculated for the possible molecular models, and the model that gives the best fit for the observed frequencies is chosen; some examples are poly(ethylene oxide), polyallene, poly-1,3-dioxolane, and others.

Energy calculation may be a useful method of predicting the molecular conformation of a polymer chain. The calculation takes all possible intramolecular interactions into consideration: potential energy hindering the internal rotation around a single bond, repulsive force and van der Waals attraction between nonbonded atoms or groups, electrostatic interaction, and hydrogen bonding. In some cases the energy of bending of the bond angle is considered. The intermolecular interactions also include the efficient packing in the crystal lattice. These potential energy terms are summed up as functions of the internal torsional angles, for example, and the minimal energy model is searched for. Solid-state nmr spectroscopy also provides useful information for determining the molecular conformation.

Crystal Structural Analysis by X-ray Diffraction. Currently, x-ray diffraction (qv) is considered one of the most powerful methods for determining the crystal structure of the crystalline polymers. The process can be described as follows.

The lattice constants (a,b,c,α,β, and γ), necessary for describing the size and shape of the unit cell, are determined by indexing the x-ray reflections observed on the photographs or

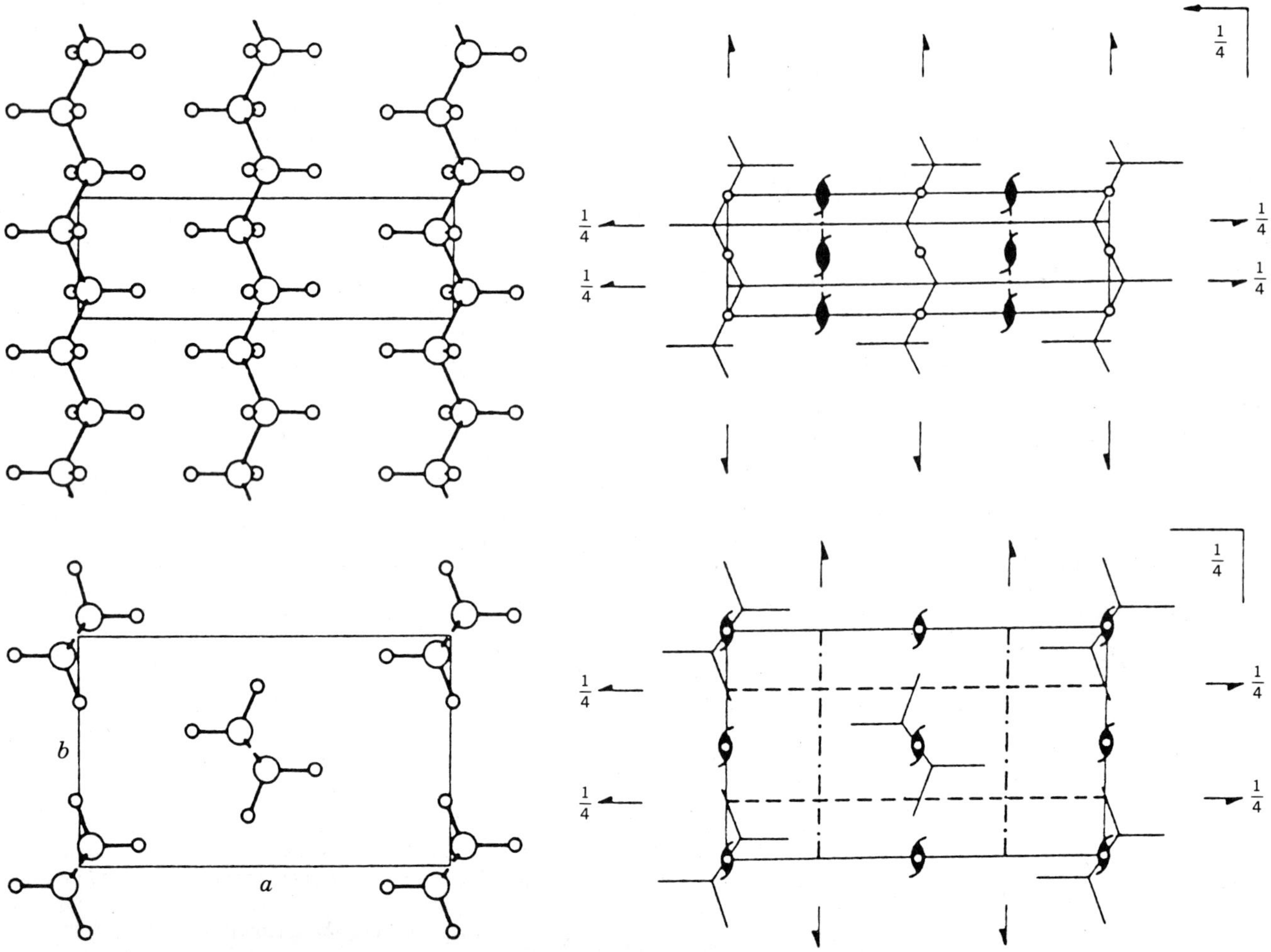

(a) (b)

Figure 5. (**a**) Crystal structure and (**b**) space group symmetry of orthorhombic polyethylene.

traced by the x-ray–counter system. The Bragg's condition of reflection is utilized:

$$2d_{hkl} \sin \theta = \lambda \qquad (1)$$

where d_{hkl} is the spacing of atomic net planes (hkl), 2θ is the Bragg's diffraction angle, and λ is the x-ray wavelength. The diffraction angle 2θ is measured for all reflections and the lattice spacing d_{hkl} in equation 1 is calculated. A set of d_{hkl} is a function of the unit cell parameters. By consistently assuming suitable indexes hkl for all observed reflections, the unit-cell parameters are determined by least-squares calculation.

The systematic absences for a specific series of reflections are investigated, to give the lattice symmetry or space group. It must be borne in mind that the determination of the unit cell and space group symmetry is difficult at the early stage of polymer crystal analysis because of the small number of reflections. This is in contrast to single crystals of low molecular weight compounds that usually have 10^3–10^4 reflections. The x-ray diffraction intensity I_{hkl} is proportional to $|F(hkl)|^2$. The systematic absence of the reflections is due to the shape of the structure factor $F(hkl)$.

$$F(hkl) = \sum_j f_j \exp\,[2\pi i(hx_j + ky_j + lz_j)] \exp\,(-B_j \sin^2 \theta/\lambda^2) \qquad (2)$$

where f_j is the atomic scattering factor of the jth atom in the unit cell, (x_j, y_j, z_j) is the fractional coordinate, and B_j is the thermal temperature factor representing the degree of thermal fluctuation of the atom. Summation covers all atoms involved in the unit cell. For example, the polymer chain with a 2_1 helical symmetry shows the systematic absence of the x-ray reflections (00l) with l = 1, 3, 5, . . . That is to say, only the reflections of the even order (002, 004, . . .) are observed. The atomic positions of the 2_1 helical chain are expressed by (x, y, z), $(-x, -y, z + ½)$, and so on in the fractional coordinate system. Substituting into equation 2 gives

$$\begin{aligned} F(00l) &\propto f \exp\,[2\pi i(0\cdot x + 0\cdot y + z\cdot l)] \\ &\quad + f \exp\,[2\pi i(0\cdot x + 0\cdot y + l\cdot(z + ½)] \\ &\propto \exp\,(2\pi izl)(1 + \exp \pi il) \\ &= \exp\,(2\pi izl)(1 + \cos \pi l) \end{aligned}$$

The last term $(1 + \cos \pi l)$ diminishes for the odd integers l. Similarly, the 3_1 helical symmetry results in the systematic absence of $00l$, $l \neq 3n$ ($n = 1, 2, 3 \ldots$). In this way, the symmetry of the crystal lattice closely relates with systematic absences of the x-ray reflections. Systematic absences are summarized in the International Tables for x-ray crystallography.

A molecular conformation suitable for a starting model is

built up by the methods described previously. These chains are packed in the unit cell so as to satisfy the space group symmetry.

The reflection intensities are observed and corrected for polarization factor, Lorentz factor, and absorption factor. The structure factor $F(hkl)$ in equation 2 is calculated for the structural model constructed in the previous section, and the parameters are refined to reproduce the observed intensities as well as possible. After determining the rough crystal structure by trial and error, the least-squares calculation is made by varying the adjustable parameters.

Combining X-ray and Energy Calculation Methods. Under constrained conditions, the least-squares calculation reduces the R factor to a satisfactorily low level. The resultant crystal structure may not necessarily be reasonable from the energetic point of view since some atomic pairs are frequently too close to be accepted. In order to have an energetic disadvantage as small as possible, the packing energy is calculated and the most stable structure is used as a starting model for the least-squares calculations of the x-ray intensity.

Vibrational Spectroscopy in Structural Investigations

Some interactions predominant in crystalline polymers can be identified by infrared–Raman vibrational spectroscopy, which can also indicate a slight change in the local structure and force field.

Crystallization-sensitive Bands. When a polymer sample is annealed at a given temperature for a certain time, several bands in the infrared and Raman spectra show a remarkable increase in intensity. These bands disappear on melting of the crystals. They are called crystallization-sensitive bands. The crystallization-sensitive band appears for the first time when monomeric units are aggregated to form a regular conformational segment or when oscillators are arranged together in a regular way. An oscillator is an abstract representation of the vibrational mode of an atomic group. The sequential length of the regular structural units necessary to detect a crystallization-sensitive band can be determined by a quantitative analysis of the relation between the band intensity and the segmental length in the copolymer samples, for example. The regular monomeric sequences in the copolymer chain are cut by introducing other types of comonomer such as deuterated monomer, slightly different chemical species, and others. However, it is important that these comonomers do not modify the type of vibrational modes of the host monomers.

Conformation-sensitive Bands. Polymer chains crystallize into various types of crystal modification of different molecular conformation. For example, the infrared spectra of the crystal modifications of poly(vinylidene fluoride) are compared in Figure 6. The spectral patterns differ considerably.

Polarized Spectra and Chain Orientation. In a drawn polymer sample, the crystrallites are usually oriented along the draw axis. The details of the crystallite orientation in the bulk sample can be obtained by x-ray diffraction. The molecular chain orientation can be estimated by utilizing the polarization character of the vibrational modes. The vibrational spectra have some advantages in detecting the local orientation of each functional group.

Vibrational Analysis. In the vibrational spectra of polymer crystals, the intramolecular modes appear in the frequency region of ca 100–4000/cm, whereas the intermolecular or external modes (lattice modes) appear in the region lower than 100/cm. Therefore, the conventionally measured infrared and Raman data (4000–400/cm) may be interpreted in a good approximation in terms of the isolated single chain, as long as the intermolecular effect (correlation splitting of bands, etc) is not taken into account. Orthorhombic polyethylene (PE) is given as an example.

Crystal Modifications and Crystalline Phase Transition

When polymers are crystallized from melt or solution, the crystalline region may exhibit various types of crystal modification, depending on the cooling rate, evaporation rate of solvent, temperature, and other conditions. These crystal modifications are different in their molecular and crystal structures as well as their physical properties. Many types of crystalline modification are reported, including a stable orthorhombic phase and metastable monoclinic phase for PE; α, β, and γ forms for *it*-PP; trigonal and orthorhombic phases for polyoxymethylene; α and γ forms for nylon-6; and others.

Crystal Modifications and Sample Preparation. Poly(vinylidene fluoride) (PVDF) (—CH_2—CF_2—) appears in at least four types of crystalline modification. Figure 7 shows the relation between the crystalline forms and sample preparation.

Polymer–Solvent Complexes. Some polymer molecules interact with solvent in the crystalline state and form a complex or an inclusion compound. For example, among the various crystalline forms of polyoxacyclobutane, form I is stable only in the presence of water, ie, form I is a hydrate. Polyethyleneimine (PEI) also gives hydrates. In the anhydrate form PEI occurs as a double helix, with the (5/1) helical chains combined by intermolecular hydrogen bonds (nitrogen–hydrogen . . . nitrogen). The hydrate is formed by absorption of water, while the molecular chains are converted into the planar-zigzag conformation; hydration depends on the water content (the ratio PEI: H_2O = 0.5, 1.5, or 2.0).

Thermally-induced Crystalline Phase Transition. Some crystalline modifications can be converted to another modification by a change in temperature. For example, polytetrafluoroethylene (PTFE) undergoes a reversible two-step phase transition at 19 and 30°C. The molecular conformation changes from a (13/6) helix to a (15/7) helix above 19°C, whereas the unit cell changes from triclinic to hexagonal. In the transition at 30°C, the conformation is disordered, although the hexagonal packing is maintained. In the crystal regions of *it*-PP and *it*-PS, the disorder-to-order phase transition occurs irreversibly by annealing. In cold-drawn *it*-PP, the (3/1) helices are packed upward and downward in a statistically disordered fashion (space group C2/c). As the annealing temperature increases, this packing disorder is gradually converted to the ordered phase where the upward and downward helices are packed in a regular fashion.

Vinylidene fluoride–trifluoroethylene (VDF–TrFE) copolymer exhibits a ferroelectric–paraelectric phase transition, the first such case for a synthetic polymer. In this transition, the electric polarization and piezoelectric constant of the film disappear above the Curie point (T_c). The temperature dependence of the dielectric constant obeys the so-called Curie-Weiss law, $1/\epsilon \propto T - T_c$. Figure 8 shows the crystal structure of the ferroelectric (low temperature) and paraelectric (high temperature) phases. The ferroelectric phase has a polar structure consisting of planar-zigzag chains with the CF_2 dipoles arranged parallel to the b axis, while in the paraelectric phase the molecular chains, which are constructed by a statistical combination of TG, $T\bar{G}$, and TT rotational isomers and so contract

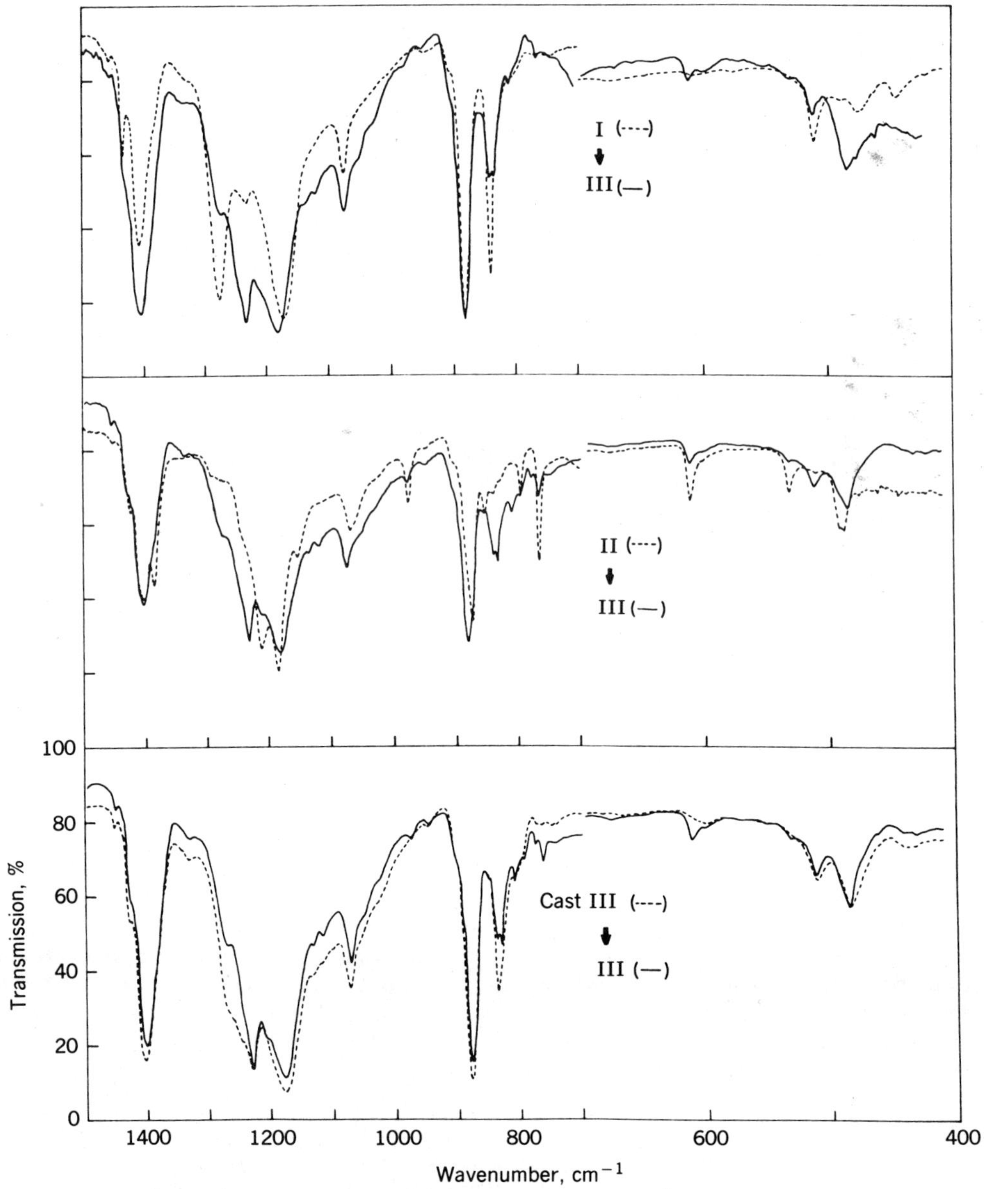

Figure 6. Comparison of the infrared spectra of poly(vinylidene fluoride) forms I, II, and as-cast III (dashed lines) with that of form III (solid line).

from the extended planar conformation, rotate violently around the molecular axis and are packed in the nonpolar unit cell of hexagonal symmetry. Such a large molecular conformational change occurring in this phase transition is accompanied by a rearrangement of the CF_2 dipoles in the unit cell, resulting in the remarkable change of the mechanical and electrical properties in the ferroelectric phase transition. This phase transitional behavior changes with the VDF content of the copolymers.

Phase Transition Induced by an External Field. *Phase Transition under Tensile Stress.* Natural rubber orients and crystallizes under tension and reverts to its original amorphous state by relaxation. Stress-induced transition is also observed in some crystalline polymers, eg, poly(tetramethylene terephthalate) (PTMT) and its block copolymers with poly(tetramethylene oxide), poly(ethylene oxide), (PEO), polyoxacyclobutane, nylon-6, poly(ethylene oxybenzoate), poly(vinylidene fluoride), polypivalolactone, keratin, and others. These stress-induced phase transitions are either reversible, where the crystal structure reverts to the original structure on relaxation, or irreversible, where the newly formed structure does not revert after relaxation. Examples of the former include PTMT, PEO, and keratin.

Phase Transitions Induced by Hydrostatic Pressure or Shearing Stress. Orthorhombic polyethylene transforms to the hexagonal phase at a temperature immediately below the melting point under a hydrostatic pressure above 300 MPa (3 kbar). The hexagonal phase is called the condis phase because the molecular chains rearrange to a conformationally disordered structure, consisting of trans sequences mixed with gauche segments. Such an orthorhombic-to-hexagonal phase transition can be observed, even at atmospheric pressure, for ultrahigh modulus PE prepared by superdrawing of dry gel. The thermal tension plays an important role because the transition occurs only under tension. By applying the shearing stress on orthorhombic polyethylene, it transforms to the monoclinic form. Under a hydrostatic pressure of 500 MPa (5 kbar) at 80–100°C the PTFE (15/7) helix transforms to the planar-zigzag conformation.

Phase Transition Induced by a High Electric Field. A typical example for such a phase transition is shown for PVDF (Fig.

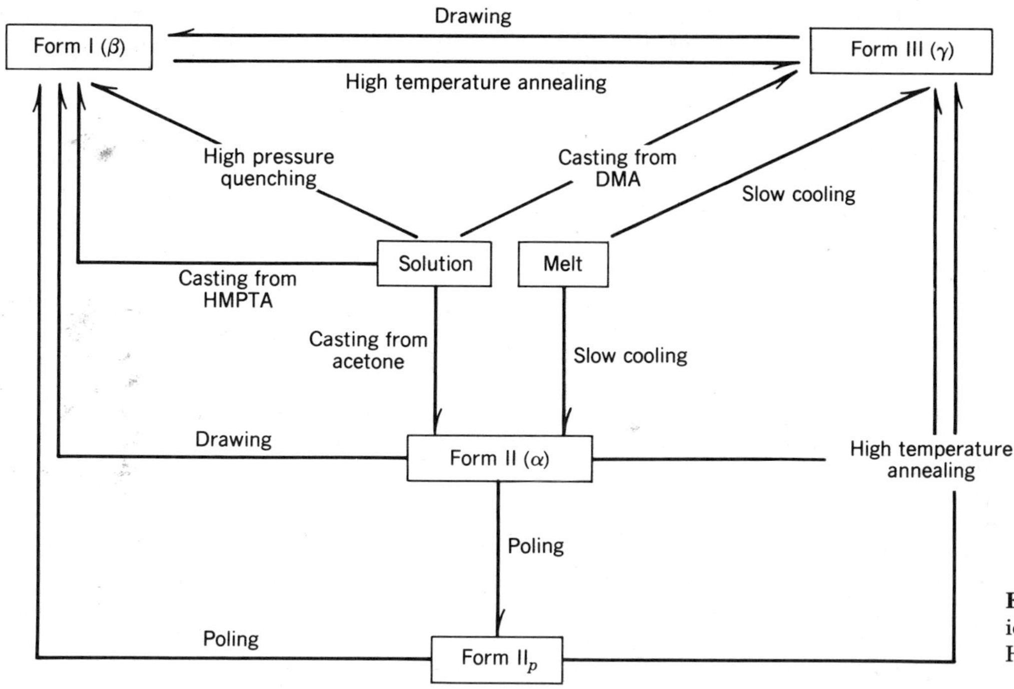

Figure 7. The transition among the various phases of poly(vinylidene fluoride). HMPTA = hexamethylphosphoric triamide.

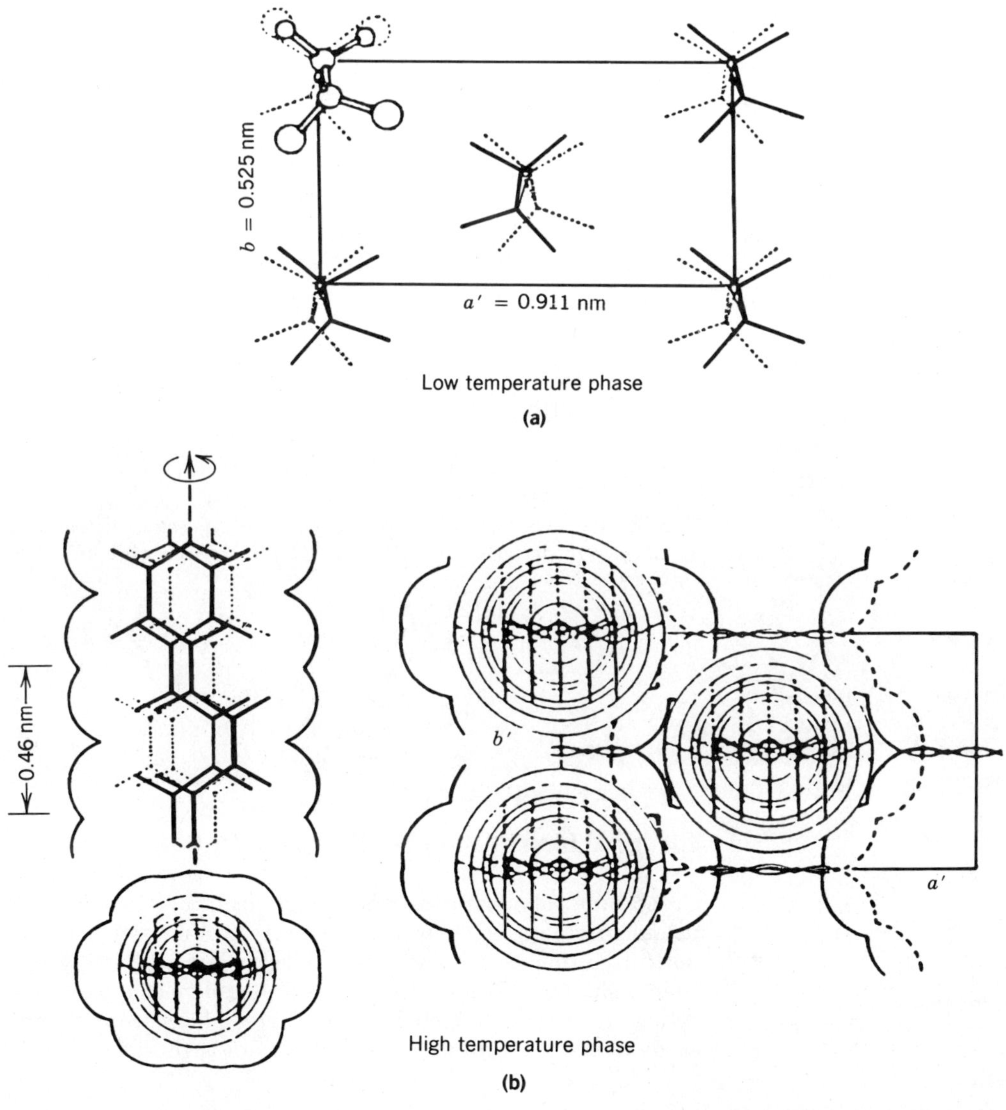

Figure 8. Crystal structure of the (**a**) ferroelectric and (**b**) paraelectric phases of vinylidene fluoride–trifluoroethylene copolymer.

7). The PVDF form II is a nonpolar crystal phase, in which the $TGT\bar{G}$ chains are related to each other by a point of symmetry. When the form II film is subjected to a high electric field, it converts to the polar crystal form II_P, where the $TGT\bar{G}$ molecular chains are arranged in parallel along the a axis and the unit cell becomes polar.

Structure and Physical Properties

In crystalline polymer samples, the physical properties of the crystalline region significantly influence the bulk properties, although the amorphous region also makes a contribution.

Crystallite Modulus and Molecular Conformation. *Experimental Evaluation of Crystallite Modulus.* It is practically impossible to treat a polymer single crystal of submicrometer-size manually. Various methods have been proposed to determine the mechanical properties of the polymer crystalline region: the change in lattice spacing induced by the externally applied tensile stress is measured by x-ray diffraction; the Young's modulus E_c is evaluated from the stress–strain curve thus obtained; the velocity of acoustic phonon propagating within the polymer crystal is estimated from the frequency-dispersion curve measured by a coherent inelastic neutron scattering; the vibrational frequency ν of the accordionlike mode, the overall stretching mode of an extended chain segment of finite length, is measured by Raman spectroscopy ($E_c \propto \nu^2$). Among these methods, x-ray diffraction may be one of the most convenient and most accurate methods.

Theoretical Interpretation. As shown in Table 1, the crystallite modulus changes from polymer to polymer. It is important to clarify these differences theoretically. Lattice dynamic treatment makes possible the calculation of the Young's modulus and the deformation mechanism of the chain subjected to tensile stress.

Experimental Proof of Molecular Deformation Mechanism. It is necessary to experimentally prove the theoretically predicted molecular deformation mechanism of polymer chains. A precise structural analysis based on x-ray diffraction measurements under tension might be the most direct method. It is, however, practically impossible to carry out such an analysis and detect minute atomic displacements of about 1 pm, because of the few broad x-ray reflections in the polymer system. On the other hand, infrared and Raman vibrational spectra taken under the application of stress are highly sensitive to slight changes in the local structure and force field, providing the possibility of monitoring the molecular deformation process.

Table 1. Crystallite Modulus along the Chain Axis, GPa[a]

Form	Polymer[b]	Observed	Calculated
planar-zigzag	PE	240	316
		(290,329)	
	PVA	250	287
	nylon-6 (α)	165 (20°C)	170
		270 (−150°C)	312
	PVDF (I)	177	237
	PDA (PDCHD)	45	40
helix, glide	*it*-PP	40	40
	POM	53 (75)	109
	PVDF (II)	59	77
	nylon-6 (γ)	27	54
aromatic polyester	PET	108	95
	PTMT (α)	13	10
	PTMT (β)	20	27
	PEOB (α)	6	2
	PEOB (β)		57
rigid-rod polymer	Kevlar	153, 200	182
	PBA	188	163 (*CTCT*)
			238 (*TTTT*)
	Nomex	88	117
	PBT	395	371
	P-*p*-P		275
	PBB		588
	PPPI		505
	cellulose (I)	127	168
	cellulose (II)	90, 110	162

[a] To convert GPa to kbar, multiply by 10^5.
[b] PDA = polydiacetylene; PET = poly(ethylene terephthalate); PBA = poly-*p*-benzamide; P-*p*-P = poly(*p*-phenylene); PBB = polybenzobisthiazole; PPPI = poly(*p*-phenylenepyromellitimide).

Three-dimensional Elastic Constants of Polymer Crystals

Theoretical calculation of three-dimensional elastic constants is desirable for further understanding of mechanical behavior of crystalline polymers. Such calculations have been carried out by several researchers. For example, the elastic and compliance constants of orthorhombic polyethylene crystal at room temperature are calculated as follows (1 GPa = 10^5 kbar).

$$\mathbf{C} = \begin{bmatrix} 7.99 & 3.28 & 1.13 & & 0 & \\ 3.28 & 9.92 & 2.14 & & & \\ 1.13 & 2.14 & 315.92 & & & \\ & & & 3.19 & & \\ & 0 & & & 1.62 & \\ & & & & & 3.62 \end{bmatrix} \text{GPa}$$

$$\mathbf{S} = \begin{bmatrix} 14.48 & -4.78 & -0.02 & & 0 & \\ -4.78 & 11.67 & -0.06 & & & \\ -0.02 & -0.06 & 0.32 & & & \\ & & & 31.31 & & \\ & 0 & & & 61.86 & \\ & & & & & 27.60 \end{bmatrix} \times 10^{-2}(\text{GPa})^{-1}$$

BIBLIOGRAPHY

H. Tadokoro, *Structure of Crystalline Polymers,* John Wiley & Sons, Inc., New York, 1979.

E. B. Wilson, Jr., J. C. Decius, and P. C. Cross, *Molecular Vibrations: The Theory of Infrared and Raman Spectra,* McGraw-Hill, Inc., New York, 1955.

T. Ito, *Polymer* **23,** 1412 (1982).

G. Wu, K. Tashiro, and M. Kobayashi, *Macromolecules* **22,** 758 (1989).

KOHJI TASHIRO
HIROYUKI TADOKORO
Osaka University

CRYSTALLINITY DETERMINATION

Crystalline polymers constitute a significant percentage of all useful polymers. As is well known, crystalline polymers are not fully ordered but are partially crystalline, typically less than about 80%. Practically all physical and mechanical properties are strongly affected by the degree of crystallinity; the study of the crystalline state in polymers is therefore of paramount importance (see also CRYSTALLIZATION KINETICS; MORPHOLOGY; CRYSTALLINE POLYMERS).

Different techniques do not necessarily measure the same degree of crystallinity. X-ray diffraction affords a direct determination of three-dimensional order, but there are still problems with absolute crystallinity measurements even for the most rigorous methods, due to the arbitrary nature of constructing the baseline delineating crystalline and amorphous scattering. Other techniques measure properties related to the degree of crystallinity. Heats of fusion are routinely determined by dsc, but are not free from uncertainties. Accurate crystallinities from density measurements are limited by the reliability of completely amorphous and crystalline density values. Estimates from infrared and Raman spectra rarely involve true crystallinity bands, but bands associated with regularity or preferred conformations. The nmr method measures the fraction of immobile protons, but protons other than those associated with the crystalline state can be immobile. The problem of proper spectral decomposition must be dealt with in the vibrational spectroscopy and nmr techniques.

Methods of crystallinity determination include volumetric methods, thermal analysis, x-ray diffraction (approximate methods, Ruland method, and small-angle x-ray scattering), vibrational spectroscopy, nuclear magnetic resonance, and other methods (inverse gas chromatography, etching techniques, and birefringence measurements).

BIBLIOGRAPHY

B. Wunderlich, *Macromolecular Physics,* Vols. 1 and 30 Academic Press, Inc., New York, 1973 and 1980.

JAMES P. RUNT
The Pennsylvania State University

CRYSTALLIZATION KINETICS

The properties of semicrystalline polymers depend on chemical structure, molecular weight and molecular weight distribution, and crystallization conditions. Crystallization conditions and the mechanisms involved in crystallization play a basic role in the characteristics of the resulting system. For this reason, control of the crystallization gives rise to a broad range of properties, among which the size of the crystallites, the degree of crystallinity, the molecular structure of the lamellar crystallites and the associated interfacial or amorphous regions, and, finally, the superstructure or morphology (qv) are the basic ones.

As a result of these parameters, the thermal and mechanodynamic properties of semicrystalline polymers vary, affecting their industrial and technological use. In fabrication processes crystallization is carried out in terms of the thermal history to which the material is subjected, because the temperature and cooling rate in the process, or the possibility of orientation in the same, varies widely, depending on the process involved and gives rise to extensive variations in the final properties of the end product.

Under appropriate conditions of temperature, pressure, and stress or under the influence of solvents, a spontaneous ordering of the macromolecules or part of them can take place. In this ordering, the chain bonds assume rotational or favored oriented states and the polymeric chains can be organized into a regular three-dimensional array. This orderly arrangement is termed the crystalline state, in which the molecular organization is complex due to its polycrystalline character and to the crystallizing entities, which are not isolated elements of structure but molecules consisting of many thousands of units. Aside from the structural conditions and thermodynamic criteria that must be met in order for crystallization to take place, it is necessary that the kinetic requirements, as far as the rate of the process is concerned, are met so that the same takes place in the time allotted for observation.

In a phase transformation such as represented by crystallization, the two basic processes are the initiation or nucleation process by which a new phase is initiated within a parent phase, and the subsequent growth of the new phase at the expense of the previous one. They are concurrent processes that have been extensively analyzed in monomeric systems. The former may occur homogeneously, by statistical fluctuations in the parent phase, or by formation of nuclei catalyzed by heterogeneity or impurities present in the melt.

Figure 1 shows a model of a crystallite and the three principal associated regions: the crystalline region with a three-dimensional ordered structure, the interfacial region, and the interzonal or amorphous regions, which consist of chain units in nonordered conformations that connect crystallites.

The crystallite size depends on the crystallization temperature and the molecular weight; consequently, a wide range of sizes may be developed. The thickness can be determined from small-angle x-ray scattering or from analysis of the Raman low-frequency longitudinal mode. The last method also gives the crystallite size distribution. Thicker crystallites are obtained when crystallization takes place under isothermal conditions with a complex thickening and a variation in size distribution. Structural differences in the lamellae-like crystallites may also be due to the molecular weight and crystallization temperature.

Higher orders of organization are observed in the crystallization of polymers. In these supermolecular structures lamella-like crystallites are assembled. Spherulites are characteristic of polymers, crystallized from the melt. Spherulites are spherically symmetric birefringent structures that display a dark Maltese cross when observed under crossed polarizers in optical microscopy. The hedrites, another superstructure, are layered or steplike structures apparently grown from a central screw dislocation. The Maltese cross is lost, and only poorly developed hedrites show some birefringence. Other crystalline morphologies vary from rods and sheetlike aggregates to a random collection of lamellae.

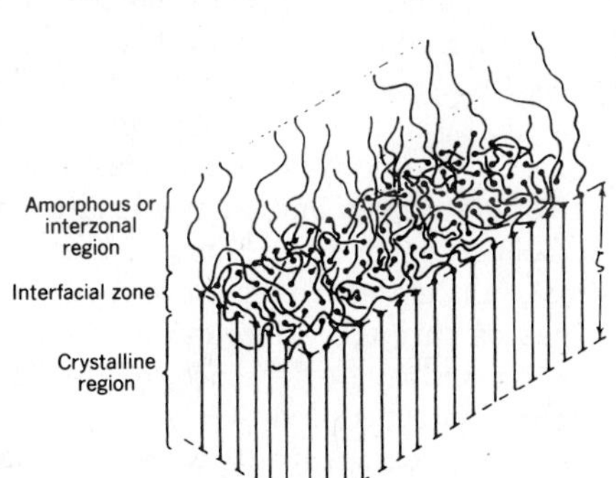

Figure 1. Schematic representation of a crystallite.

Crystallization takes place well below the melting temperature. For most polymers, high undercooling is required for crystallization. The reasons for this behavior are related to the high interfacial free energy associated with the basal plane of the crystallites and the difficulty in extracting ordered sequences of sufficient length from the disordered melt.

Polymer crystallization occurs either from a deformed melt by the application of an external force or from an undeformed melt. The latter has been carried out over a wide range of concentrations from very dilute solutions to the pure bulk polymer, but only a few investigations deal with crystallization kinetics in strained polymers.

In the most general method for crystallinity determination (qv) the extent of transformation or the degree of crystallinity may be calculated from the observed specific volumes, assuming the additive contribution of the crystalline and amorphous regions. The crystallinity at different temperatures can be calculated from the specific volume–temperature relations of the completely amorphous and completely crystalline polymers. For kinetic studies, dilatometry (qv) is the most widely used, especially for crystallization at low undercooling. High crystallization rates at high undercooling cannot be followed with this technique because crystallization starts in many cases before thermal equilibrium is reached.

Crystallization is accompanied by the evolution of latent heat. This property can be used for following the crystallization by calorimetric methods. Other techniques include wide- and small-angle x-ray diffraction, infrared spectrometry, nuclear magnetic resonance, and depolarization microscopy. More recent methods include rapid-scanning Raman spectroscopy, Fourier transformation ftir spectroscopy, and small- and wide-angle scattering measurements, which use synchroton radiation for fast isothermal crystallization.

Bulk Crystallization

Crystallization from the molten state proceeds at a finite rate at temperatures well below the melting temperature, under conditions far removed from equilibrium. The isothermal rate at which crystallinity develops in a pure homopolymer follows a universal pattern. The polymer changing from a temperature above the melting temperature to a predetermined crystallization temperature passes three stages: in the first stage, no crystallinity is observed; in the second stage, crystallization proceeds at a higher rate, which is almost autocatalytic in character and typical of phase transformation; finally, the process changes only very slowly with time and a pseudoequilibrium of crystallinity is approached.

The process is continuous, with no sudden changes observed in the isotherms. Sigmoidal-shaped isotherms, first observed in 1934, are characteristic of the crystallization of polymers. The early stages, known as primary crystallization, are distinguished from the later stages or secondary crystallization, which may proceed over many decades. The rate of the crystallization is strongly dependent on the crystallization temperature. Just below the melting temperature the crystallization rate is low, but as the temperature is lowered, the rate increases and passes through a maximum. At lower crystallization temperatures, the rate decreases again. This behavior is typical of nucleation-controlled processes.

The development of crystallinity with time has been described by different treatments. The free-growth approximation, derived by Goler and Sachs, assumes that the evolution of an individual growing center is independent of the mass already transformed as well as of the growth of other centers.

For the initial part of the transformation,

$$1 - \lambda(t) \cong Kt^n \tag{1}$$

where $1 - \lambda(t)$ is the crystallinity at time t, K is a constant, and n defines the geometry of the growth. The values of n are 2, 3, or 4 for one-, two-, or three-dimensional growth geometries, respectively.

A better description of the kinetics takes into account the mutual impingement of growing centers, with the assumption that on impingement growth ceases. The problem has been considered and the process is characterized by the Avrami relation

$$\theta = \exp(-Kt^n) \tag{2}$$

where θ is the volume fraction not transformed at time t, and K is a constant.

The degree of crystallinity is given by

$$1 - \lambda(t) = 1 - \exp(-Kt^n) \tag{3}$$

Mathematically, equation 3 reduces in form to the Goler-Sachs expression for small extents of transformation.

The exponent of time in the Avrami equation is indicative of nucleation and growth processes. Because of the simplifications introduced in the Avrami equation, details of the growth geometry and type of nucleation cannot be elucidated from the isotherms by specifying the value of the exponent n. However, the analysis of the overall crystallization data for various polymers shows good adherence to this equation for the initial part of the crystallization process.

The value of n can be obtained from the slope of the double-logarithmic plot, log log θ against log t, or from the plot of log $1 - \lambda_t$ against log t. Another method is to construct a plot of θ against log t and compare it with calculated master plots. Integral values of $n = 4$ and $n = 3$ have been found in many polymers. However, in other cases, fractional exponents from data fitting have been reported.

The influence of the melting conditions on the crystallization behavior is related to the fact that the experimental conditions do not yield a completely randomly molten system. In other cases, chemical modification or degradation processes may occur or heterogeneous nucleation catalyzed by insoluble impurities may take place. Moreover, in some polymers changes in the values of n depend on the crystallization conditions.

The influence of molecular weight on the crystallization process can be seen in Figure 2, where crystallization kinetic isotherms for molecular weight fractions of linear polyethylene have been superimposed on one another by shifting the individual isotherms along the log time axis. The influence of molecular weight on the deviations from theory and on the final level of crystallinity is clearly demonstrated. As the molecular weight increases, the limiting degree of crystallinity is sharply reduced. The crystallinity level at which the process is arrested is thus dependent on molecular weight. The structure or topology of the residual melt arrests the crystallization process,

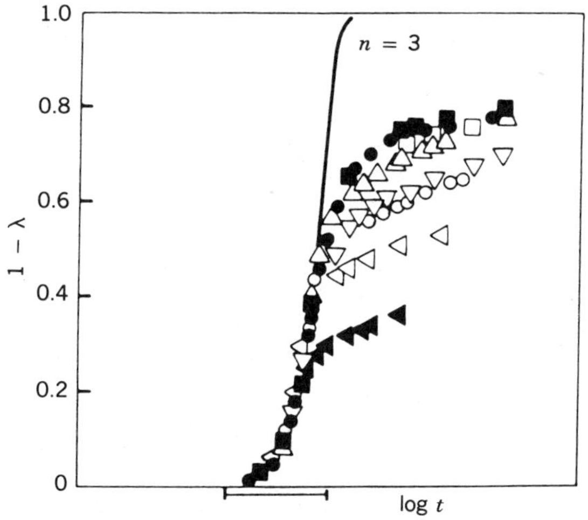

Figure 2. Plot of $1 - \gamma$ vs log t after isothermal crystallization of linear polyethylene. ——, Theoretical Avrami with $n = 3$. Mol wt: ●, 11,500; □, 14,000; ■, 20,000; △, 47,000; ▲, 76,000; ▽, 122,000; ▼, 175,000; ○, 200,000; ◁, 600,000; ◀, 1.2×10^6. Courtesy of The American Chemical Society.

which ceases at a transformation determined by molecular weight. In addition to the influence of molecular weight on the isotherm shape and on the final degree of crystallinity, the overall crystallization rate depends on the chain length.

Nucleation and Growth Processes

The crystallization rate depends mainly on temperature. Because crystallization is described by nucleation and growth mechanisms, the temperature coefficient observed in the transformation must be explained in terms of the temperature coefficients of these two processes.

Primary nucleation is the process by which a new phase is initiated within a parent phase. Nuclei can be formed either homogeneously by means of statistical fluctuations in the parent phase or heterogeneously, catalyzed by the presence of heterogeneity. In the latter process, nucleation starts on surfaces, cavities, and cracks of insoluble impurities.

After the nucleus is formed, a new layer grows after secondary nucleation, a process similar to primary nucleation. The linear growth rate corresponds to the macroscopic advance of a crystal front. The growth rate is controlled by a secondary nucleation process and the new crystallites are nucleated on the faces of existing ones.

The steady-state nucleation rate per unit volume and time can be expressed by

$$N = N_0 \exp\left(-E_D/RT - \Delta G^*/RT\right) \tag{4}$$

where N_0 is a constant, E_D is the energy of transport in the solid–liquid interface, and ΔG^* is the free energy of formation of a stable nucleus from nucleation. Because ΔG^* is a function of the temperature, the nucleation rate is also a function of temperature in accordance with equation 4.

In developing a nucleation theory appropriate to chains, the classic procedures used for monomeric systems can be adopted. The expression for the free energy of forming a nucleus is composed of two terms: one is the bulk free energy of fusion, and of opposite sign are terms that represent the excess surface free energies contributed by the interfaces present. The equations that describe the critical condition for nucleation present the same analytical form for chains of infinite molecular weight as the equations derived for a collection of monomeric molecules with a similar geometric arrangement. Theory pertinent to long chains of finite length presents significant departures from monomer theory.

For three-dimensional homogeneous nucleation, the critical conditions can be obtained by a general treatment. The free energy of fusion ΔG_f for a system composed of N molecules with x repeat units, as derived by Flory, can be written as

$$\frac{\Delta G_f}{xN} = \frac{\xi p}{xN}\,\Delta G_u + RT\left\{\frac{V_u}{V_1}\frac{(1-v_2)}{v_2} + \frac{1}{x}\right\}\ln\left(1 - \frac{\xi\rho}{xN}v_2\right) + RT\left\{\frac{\xi\rho}{xN}\cdot\frac{1}{\xi}\left[\ln v_2 D + \ln\frac{x-\xi+1}{x}\right] + \frac{\chi_1(1-v_2)^2\zeta\rho}{(1-v_2+v_2\lambda)xN}\right\} \tag{5}$$

where ΔG_u is the bulk free energy of fusion per mole of repeat unit, D is defined as $\exp\left[-2\sigma_e/RT\right]$, σ_e is the excess interfacial free energy per repeat unit as it emerges from the crystal face normal to the chain direction, ζ is the number of repeat units in the chain in the crystalline array, ρ are sequences in cross-section, V_u and V_1 are the molar volume of the polymer unit and of the diluent, v_2 is the volume fraction of the polymer, χ_1 is the polymer diluent interaction constant, and λ is the fraction amorphous content. At the beginning of the nucleation process $\lambda \simeq 1$, so that $\xi\rho/xN \to 0$ and $\ln(1 - \xi\rho/xN) \to \xi\rho/xN$.

Then the free energy of fusion is given by

$$\Delta G_f = \xi\rho\,\Delta G_u - RT\left\{\frac{V_u}{V_1}(1-v_2)\zeta\rho + \frac{1}{x}\xi\rho v_2\right\} + RT\chi_1(1-v_2)^2\xi\rho + RT\left\{\rho\ln v_2 D + \ln\left(\frac{x-\zeta+1}{x}\right)\right\} \tag{6}$$

Moreover, the free energy change in forming a cylindrically arranged nucleus is given by

$$\Delta G_d = 2\xi\sigma_u\pi^{1/2}\rho^{1/2} - \Delta G_f(\zeta\rho) \tag{7}$$

With substitution of equation 6, then

$$\Delta G_d = 2\pi^{1/2}\rho^{1/2}\sigma_u\xi - \xi\rho\left\{\Delta G'_u - \frac{RT}{x}v_2 + \frac{RT}{\xi}\left[\ln v_2 D + \ln\frac{x-\xi+1}{x}\right]\right\} \tag{8}$$

where

$$\Delta G'_u = \Delta G_u - RT\left\{\frac{V_u}{V_1}(1-v_2) - \chi_1(1-v_2)^2\right\} \tag{9}$$

The surface described by 8 contains a saddle point with coordinates defined by $(\delta\,\Delta\,G_d/\delta\,\rho)_\zeta = 0$ and $(\delta\,\Delta\,G_d/\delta\,\zeta)_\rho = 0$. From this condition the critical dimensions of such a nucleus are given by

$$\rho^{*1/2} = \frac{2\pi^{1/2}\sigma_u}{\Delta G'_u - (RT/x)v_2 - RT/(x - \xi + 1)} \tag{10}$$

$$\frac{\xi^*}{2}\left[\Delta G'_u - \frac{RT}{x}v_2 + \frac{RT}{(x - \xi^* + 1)}\right] = 2\sigma_c - RT \ln v_2 - RT \ln \left(\frac{x - \xi^* + 1}{x}\right) \tag{11}$$

$$\Delta G^* = \pi^{1/2}\sigma_u \rho^{1/2}\xi^* \tag{12}$$

These general equations reduce to some particular cases. For a pure polymer ($v_2 = 1$) and infinite molecular weight approximation, the critical values correspond to the expressions

$$\Delta G^* = \frac{8\pi\sigma_u\sigma_e}{\Delta G_u^2}; \qquad \rho^* = \frac{4\pi\sigma_u}{\Delta G_u^2}; \qquad \zeta^* = \frac{4\sigma_e}{\Delta G_u} \tag{13}$$

These expressions are identical to those for a critical-sized nucleus composed of low molecular weight species. They correspond to a three-dimensional nucleus that is formed homogeneously by statistical fluctuations, an unlikely event. However, nuclei formed heterogeneously have the same temperature dependence, although the critical free energy for nucleation can be reduced. The effects of finite molecular weight, particularly in the low molecular weight region, and of the diluents on the nucleation process are very important. Significant differences are obtained for the critical dimensions of the nuclei and for the free energy of nucleation, as indicated in equations 10–12.

Another type of nucleus of interest is that formed on a developed crystal face. The nucleation of the type described by Gibbs involves the unimolecular deposition of chain sequences. The change in free energy that accompanies the formation of a monomolecular nucleus ζ units long and ρ sequences in breadth from a collection of molten chains each of which is x units long can be expressed as

$$\Delta G_d = 2\xi\sigma_u - \xi\rho\left[\Delta G'_u - \frac{RT}{x}v_2 + \frac{RT}{\xi} \times \left(\ln v_2 D + \ln \frac{x - \xi + 1}{x}\right)\right] \tag{14}$$

The coordinates of the saddle point are given by

$$\rho^* = \frac{2\sigma_u}{\Delta G'_u - (RT/x)v_2 + RT/(x - \xi + 1)} \tag{15}$$

$$\xi^* = \frac{2\sigma_e - RT \ln v_2 - RT \ln (x - \xi + 1)/x}{\Delta G'_u - (RT/x)v_2} \tag{16}$$

$$\Delta G^* = 2\sigma_u\xi \tag{17}$$

The critical conditions for a pure polymer and infinite molecular weight are

$$\rho^* = \frac{2\sigma_u}{\Delta G_u}; \qquad \zeta^* = \frac{2\sigma_e}{\Delta G_u}; \qquad \Delta G^* = \frac{2\sigma_e\sigma_u}{\Delta G_u} \tag{18}$$

These expressions are identical to the classic results for small molecules. However, the Gibbs-type nucleus presents some unique features. It cannot become stable without growth in the chain direction, irrespective of the amount of lateral growth that takes place, in contrast with the three-dimensional nucleus, which gains stability with only lateral growth. The Gibbs-type nucleus is unstable at temperatures that are infinitesimally above the crystallization temperature, and stability is only ensured by allowing for growth along the chain direction.

If the approximation that $\Delta G_u = \Delta H_u(T/T_m)$ is used, equations 13 and 18 for the free energy of nucleation may be expressed by

$$\Delta G^* = \frac{8\pi\sigma_u\sigma_e^2T_m^2}{\Delta H_u\, \Delta T^2} \tag{19}$$

or

$$\Delta G^* = \frac{4\sigma_u\sigma_e T_m}{\Delta H_u\, \Delta T} \tag{20}$$

for the three- and two-dimensional cases, respectively. The critical dimensions may be expressed in the same terms. Thus these parameters depend on the supercooling of the system. The dimensions are infinite at the melting temperature and decrease with a reduction of temperature. At considerable supercooling they are small and insensitive to temperature. The temperature dependence of the free energy of nucleus formation is identical for homogeneous and heterogeneous nucleation, but less critical free energy is necessary to promote heterogeneous nucleation.

Studies of the temperature coefficient of the overall crystallization rate in bulk and from solution and of the spherulite and lamellar growth rate in solution have established that polymer crystallization is a nucleation-controlled process. This conclusion does not depend on the form, structure, or chain disposition within the nucleus. Therefore, the crystallization rate can be described by an equation identical in form to equation 4. The spherulite growth rate G is given by the relation

$$G = G_0 \exp(-E_D/RT) \exp(-\Delta G^*/RT) \tag{21}$$

where ΔG^* is the critical free energy for nucleation, E_D the energy of transport in the interface, and G_0 a preexponential factor. If it is assumed that growth is controlled by a Gibbs-type nucleation process and that new crystallites are nucleated on the faces of existing ones, the growth rate depends exponentially on T_m/T, according to equation 20.

On the other hand, kinetic theories of polymer crystallization are essentially surface-nucleation theories and these treatments are related to both the formation of the nucleus and its subsequent growth. However, some important limitations must be mentioned. Thus kinetic theories of regular folding assume a model of polymer crystallization with regularly folded nuclei being formed that grow into crystallites with the same structure. Moreover, a specific path has been chosen to make the Gibbs-type nucleus stable at temperatures infinitesimally above the crystallization temperature. The assumption that critical-sized nuclei are composed of regularly folded chains is not required by the theory. Although a nucleation-controlled process exists, this conclusion does not allow the specification of the molecular structure for the nucleus. Most of these theories have been formulated in terms of infinite molecular weight chains, despite the fact that the influence of molecular weight in the crystallization process is very important.

To analyze the growth rate, the kinetic theories assume proportionality of the growth rate and the total flux or nucleation rate. However, the two-dimensional growth of a platelike crystal requires successive surface nucleations, and the growth rate of a growth front G is not simply proportional to S, the surface nucleation rate.

The net growth rate, perpendicular to the crystal face, depends on the relative values of the nucleation rate i and of the lateral growth rate g and can be calculated. For the growth of low molecular weight substances on a substrate two extreme cases have been considered and adapted to polymers and named regimes I and II. This behavior is found in all crystallizing systems. In regime I, the growth step is allowed to sweep across the face of the crystal and a pause occurs before the next layer is nucleated. Thus g is very much faster than i, and the growth rate G is given by

$$G_{\mathrm{I}} = b_i L \qquad (22)$$

where b is the thickness of a growth step and L is the layer width or substrate length.

For regime II, new growth steps are allowed to nucleate before the previous layer has filled the substrate. The nucleation rate is higher than the growth rate and multiple nucleation can occur. The growth rate is thus dependent on both g and i and is given by

$$G_{\mathrm{II}} = b(ig)^{1/2} \qquad (23)$$

The overall temperature dependence of G_{I} and G_{II} are related, with the temperature dependence of i for G_{I} and of i and g for G_{II}.

The nucleation rate is given by $i \propto \exp\,(-4b\sigma_e\sigma_u/\Delta G_u\mathrm{k}T)$, and g usually has a Boltzmann-type temperature dependence.

As a consequence of different relations between the nucleation rate and the rate of lateral spreading, the different regimes of nucleation result in different temperature coefficients, which are revealed when the growth of crystallization is plotted against $1/T_c\Delta T$. Thus a transition from one regime to another as a function of undercooling is observed experimentally in several polymers.

This analysis is very formal and does not require any assumptions with regard to the structure of the nucleus. It is not concerned with regularly folded chains or with other assumptions on the interfacial structure. The kinetic theories based on coherent surface nucleation imply a model of folds with adjacent reentry, in contradiction to the evidence of a nonadjacent-type reentry. Moreover, fundamental discrepancies contradicting the model of coherent surface nucleation and subsequent sequential deposition of segments have been reported in the crystal growth of low molecular weight fractions of poly(ethylene oxide). The magnitude or the existence of the substrate or persistence length is controversial, and these theories cannot provide estimations of the length, rate of lateral spreading of a secondary nucleus, and lateral surface energy.

The last theoretical treatment to be mentioned is the adaptation of the roughening-induced theories of crystal growth, established for single crystals to polymers. The model, involving roughening of the growth surface and pinning, has been proposed to explain the effect of reentrant corners on crystal growth, eg, the formation of crystals grown in poor solvents at high temperatures, where the crystals are elongated and the faces show distinct curvatures. This is not a nucleation theory, and the annealing of the pinned sites is controlled by a high free-energy barrier to growth.

Spherulitic Growth

Spherulites are a morphological feature of bulk-crystallized polymers. They consist of a large number of lamellar crystallites, radiating in all directions from one point, with well-defined boundaries. When the specimens are observed under crossed polaroids in the polarizing microscope, large birefringent regions are seen. In most cases, they consist of circularly birefringent areas, including a dark Maltese cross with arms parallel and perpendicular to the direction of polarization. Differences in spherulitic texture depend on the chemical structure, molecular weight, and crystallization conditions; two types of spherulites exist, the loose, fibrillar nonbanded and the close-packed, banded (ring-type) spherulites. The radial symmetry of the spherulites may develop via other geometric forms. Fibrous or needlelike crystals are formed, which develop lengthwise by a branching or fanning mechanism. If the directions of branching of one crystallite relative to another are well defined, a crystallographic orientation should be developed within the aggregate. The branching itself does not effect crystallinity. Spherulites have been observed in many polymers and in a variety of nonpolymeric compounds, and were considered to be a universal mode of crystallization of macromolecules.

However, the organization of the lamellar crystallites into higher morphologies is not only spherulitic, and differently organized and even nonorganized forms have been described. The formation of the specific supermolecular structure depends on the molecular weight and molecular weight distribution, the crystallization conditions, and the structural regularity of the molecule. Structures may vary from well-developed spherulites to rods and sheetlike aggregates to random collections of lamellae. Spherulites are produced only by a limited range of molecular weights and crystallization conditions.

The kinetics of the spherulite growth is most frequently analyzed with the help of the polarizing microscope. The spherulites are usually spherical in shape during the initial stages of the crystallization, and the radii are measured before they impinge on their neighbors. The temperature range over which the growth rate can be measured depends on the number and sizes of spherulites, and these, in turn, depend on the nucleation and on the growth rates. The growth rate may be so high that the dissipation of the heat of crystallization prevents the attainment of isothermal conditions. In some cases, measurements of the growth rate have been based on the light-scattering theory of spherulites.

The spherulitic growth is typical of a nucleation-controlled process. This conclusion has been deduced from the characteristic large negative temperature coefficient of the growth rate in the vicinity of the melting temperature. Homogeneous nucleation takes place when spherulite development starts as a result of random fluctuations in the pure melt. Spherulite nucleations sporadic in space and time have been reported and have been considered evidence for homogeneous nucleation. In general, conditions appropriate to homogeneous nucleation are difficult to attain.

The critical free energy for nucleus formation is given by $\Delta G^* = K\sigma_u^2\sigma_e/\Delta G_u^2$ an expression that is equivalent to equation 13 for homogeneous nucleation (K is a constant determined by the geometry of the nucleus). Values of the product $\sigma_u^2\sigma_e$ for different polymers range from 6 to 26 aJ^3/cm^6 (6,000 to 26,000 erg^3/cm^6).

Nucleation of polymers to give spherulites has been described frequently as a heterogeneous process, which takes place at a nonconstant rate and is initiated over particles of insoluble impurities. Evidence that nucleation is heterogeneous in many systems has been obtained by the observation of the appearance of spherulites in the place where they were nucleated for the first time, even after successive meltings and crystallizations. Nucleation begins on insoluble heterogeneities which, because of the high viscosity of the melt, cannot move. Moreover, the influence of the melting conditions on the subsequent nucleation and growth processes is related to the fact that a completely molten system is not obtained under certain experimental conditions or in the presence of insoluble impurities. Remnants of polymer crystalline order are not destroyed when the melting temperature is not high enough or when the time at which the melting temperature is maintained is short and, thus, spherulites start where remnants of old spherulites remain. The number of spherulites per unit volume is reduced by heating the polymer to increasingly higher temperatures before crystallization. An increase in the crystallization or growth rates on melting at lower temperatures is probably caused by self-seeding. However, if the melting temperature is high and the residence time is long, the polymer may degrade, frequently with a reduction of molecular weight. This fact favors the appearance of bigger spherulites with growth rates that are higher than the corresponding value for intact material.

Several attempts have been made to correlate the molecular weight with the crystallization rate G. Most of these calculations are empirical and fit the data for a limited range of molecular weights and crystallization temperatures. The molecular weight dependence of G can be analyzed by considering the effect of the chain length on the free energy of nucleation.

Heterogeneous nucleation by the addition of heterogeneous nuclei is very important and is related to the reduction in spherulite sizes and an increase in the growth rate. Inorganic and organic salts have been used for this purpose in order to modify optical and mechanical properties.

The nucleation of polymers within holes, cavities, cracks, or other irregularities has been examined. The detailed nature of the substrate and the mode by which the units fill the cavities must be specified. It appears that ΔG^* and ρ^* are always reduced relative to their values for homogeneous nucleation, but there are many examples where ξ remains unaltered.

Crystallization Temperature Coefficient

The linear growth rate of spherulites can be expressed by the general equation 21, where the critical free energy for nucleation ΔG^* can take different expressions, depending on the mode of nucleation. The postulate has been made that the linear growth is controlled by monomolecular nucleation on an existing crystal face. This process involves the unimolecular deposition of chain sequences, and the growth rate depends exponentially on $T_m/T\,\Delta T$. Moreover, it has been proposed that the nucleation process involved could be three-dimensional, as the presence of an initial crystallite would influence the organization of the untransformed material. The free energy for this nucleation is a fraction of the homogeneous free energy, which is reduced for this secondary nucleation process. From the critical conditions expressed by equation 12, the growth rate depends exponentially on $T_m^2/T\,\Delta T^2$. In some instances, the experimental data fit the functional forms required for either type of nucleation and thus this critical analysis does not allow for discrimination between the two cases.

The experimental data can be treated if the equilibrium melting temperature, the heat of fusion, and the transport term are defined. The free energy of nucleation depends on the equilibrium melting temperature T_m^0 through its dependence on the free energy of fusion. In the vicinity of the melting temperature, the appropriate temperature variables are sensitive to the value assigned to T_m^0. However, the experimental determination of the equilibrium melting temperature has been elusive. Polyethylene is the only polymer for which a theoretical estimate exists, based on the thermodynamic parameters for fusion of the monomeric homologues at the equilibrium melting temperature.

Different approximations have been proposed for the transport term. The simplest approximation is to assume that it is independent of temperature with suggested values of 21–125 kJ/mol (5–30 kcal/mol), which yields an Arrhenius-type functionality. This treatment is useful only if the crystallization temperatures correspond to a narrow range just below the melting temperature. In another approximation, the Williams-Landel-Ferry (WLF) equation is used in the general form

$$E_D = \frac{C_1 T}{C_2 + T_c - T_g} \tag{24}$$

where T_g is the glass-transition temperature, and C_1 and C_2 are universal constants (C_1 = 17.2 kJ/mol or 4.12 kcal/mol and C_2 = 51.6 K). However, there is no direct proof of the validity of applying this equation to growth rate or overall rate of crystallization. In many cases, this equation has been used considering C_1 and C_2 as adjustable parameters.

Solution Crystallization

The crystallization from solution has been investigated in concentrated solutions, polymer-diluent systems, and diluted solutions. Concentrated solutions are characterized by the random distribution of polymer segments throughout the mixture, and in many systems that diluent is prevented from entering the crystal lattice. A dilute solution is characterized by a nonuniform polymer segment distribution throughout the medium, and the individual molecules are isolated from one another. These differences are important in the crystallization mechanisms. There are few studies of the kinetics of concentrated solutions, and dilute crystallization has been less extensively studied than bulk crystallization despite the numerous theoretical and experimental investigations of the resulting morphology.

Lamellar-like crystallites are the primary morphological entity observed after crystallization in dilute solutions. These crystals are about 10–30 nm thick and several micrometers in lateral dimensions. The orientation of the chain axis is preferentially perpendicular to the wide face of the crystal, and this appears to thicken through a screw dislocation mechanism that leads to a spiral structure. These crystals have been universally observed in a large variety of polymers.

The crystallite thickness tends to be remarkably uniform at a given crystallization temperature and for a given solvent. However, the strong dependence on the crystallization temperature is characteristic. The thickness is independent of molecular weight, except for low molecular weights.

A great variety of crystal habits have been observed from lamellae to dendritic forms. These habits depend on the crystallization temperature or undercooling, on the polymer–solvent interaction, and on the solution concentration.

All structures and morphological features are the result of the crystallization mechanisms. Studies of the temperature coefficient of the overall crystallization rate and lamellar growth rate in solution have shown that crystallization is a nucleation-controlled process. Therefore, the crystallite thickness is determined by the initial nucleation and the growth processes.

The crystallization of concentrated and moderately concentrated solutions displays characteristic behavior that is typified by a large negative temperature coefficient. The isotherms for a given composition at different temperatures can be superposed on one another, as is usually observed in bulk crystallization. As the polymer is diluted, the shapes of the isotherms change and the process becomes more protracted with dilution.

The analysis of the temperature coefficient of the crystallization process over the complete range in which a polymer is capable of crystallizing has aided in elucidating the role played by the nucleation process. Compositional changes must affect the free energy for nucleation and, hence, the overall rate of the process. A nucleation theory for polymer–diluent mixtures pertinent to chains of infinite molecular weight has been developed and extended to chains of finite molecular length. For a three-dimensional nucleus and finite chains, the critical dimensions and the free energy of nucleation are given in equations 10–12. These general expressions reduce to previously derived particular cases. For polymer–diluent systems, with $x \rightarrow \infty$, the critical conditions are expressed by

$$\rho^* = \frac{4\pi\sigma_u^2}{(\Delta G_u')^2} \quad (25)$$

$$\zeta^* = \frac{4\sigma_e - 2RT \ln v_2}{(\Delta G_u')^2} \quad (26)$$

$$\Delta G^* = \frac{8\pi\sigma_u^2\sigma_e}{(\Delta G_u')^2} - \frac{4\pi\sigma_u^2 RT \ln v_2}{(\Delta G_u')^2} \quad (27)$$

The change in free energy that accompanies the formation of a two-dimensional nucleus is given by equation 14; the critical conditions are

$$\rho^* = \frac{2\sigma_u}{\Delta G_u'} \quad (28)$$

$$\zeta^* = \frac{2\sigma_e - RT \ln v_2}{\Delta G_u'} \quad (29)$$

$$\Delta G^* = \frac{2\sigma_e\sigma_u - 2\sigma_u RT \ln v_2}{\Delta G_u'} \quad (30)$$

The free energy of fusion $\Delta G_u'$, given in equation 9, is a function of temperature and composition, v_2, but can be approximated by $\Delta G_u' = H_u(T_m - T)T_m$ where T_m is the melting temperature of the polymer–diluent mixture. The polymer segment concentration is uniform for concentrated solutions and v_2 can be identified with the actual concentration. However, in dilute solutions, the concentration within the swollen coil cannot be identified with the nominal composition. The theory rests heavily on the Flory-Huggins theory, which only holds for high and moderate concentrations in which the excluded volume effects are not important.

For dilute solutions ΔG^* becomes very high and, according to equation 4, nucleation only proceeds at reasonable rates at high undercooling. However, when crystallization progresses, the change in undercooling is very small. For concentrated mixtures, the equilibrium melting temperature decreases as the crystallization progresses and supercooling is lowered, with resulting attenuation in nucleation and overall rates.

The crystallization kinetics of dilute solution exhibits different characteristics as a consequence of the two extreme cases of polymer solution behavior. At sufficiently high concentration an enmeshed, intertwined, entangled assembly of chains is present. Under these conditions, an essentially uniform polymer segment concentration may be assumed throughout the solution. At high dilutions, however, the macromolecules are more or less isolated from one another, only occasionally interfacing in small clusters. At concentrations below 1%, the molecules can be considered isolated, and concentrations below this value are considered for this analysis. The kinetics of dilute solutions has been studied by dilatometry and the observation of lamellar growth. The measurements of lamellar growth rate exhibit peculiar characteristics and are more complicated than the measurements of spherulite growth rate. The growth rate along the normal to the (110) crystallographic faces is generally obtained from electron micrographs, although phase-contrast optical microscopy has also been used.

The crystallinity of solution-crystallized polymers is substantially higher than that of the corresponding bulk-crystallized samples. For the former, typical degrees of crystallinity are 80–90%. In contrast to crystallization in bulk, crystallinity depends only slightly on molecular weight. Values from 90–75% are found for polyethylene fractions between $M_w = 1.4 \times 10^4$ and 1.2×10^6 isothermally crystallized in dilute solutions of α-chloronaphthalene. Although solution-formed crystals thicken with time during annealing, thickening does not take place during isothermal crystallization from dilute solution. This fact, quite different from bulk-crystallization behavior, can be explained by the high supercooling involved in the crystallization process for dilute systems.

BIBLIOGRAPHY

J. G. Fatou, "Crystallization, Morphology and Properties of Polymers," Makromol. Chem. Suppl. **7**, 131, 1984.

L. Mandelkern "Crystallization and Melting of Polymers" in G. Allen ed. *Comprehensive Polymer Science,* Vol. 2. Pergamon Press, 1989.

R. L. Miller in R. L. Miller, ed., *Flow-induced Crystallization in Polymer Systems,* Gordon & Breach, New York, 1979.

B. Wunderlich, *Macromolecular Physics,* Vols. 1–3, Academic Press, New York, 1973, 1976, 1980.

CUMENE HYDROPEROXIDE. See PEROXY COMPOUNDS.

CUPRAMMONIUM RAYON. See RAYON.

CURING

Curing may be defined as changing the properties of resin or rubber formulations by chemical reaction. This is usually accomplished by the action of heat under varying conditions of pressure. Although the term cure is used in connection with thermosets and rubbers, this article is concerned with thermosets (see also CROSS-LINKING; GELS; NETWORKS; VULCANIZATION).

A time-temperature-transformation (TTT) cure diagram may be used to provide an intellectual framework for understanding and comparing the cure and physical properties of thermosetting systems. The main features of such a diagram, as shown in Figure 1, can be obtained by measuring the times to events that occur during isothermal cure vs temperature, ie, T_{cure}. These events include the onset of phase separation, gelation, vitrification, full cure, and devitrification.

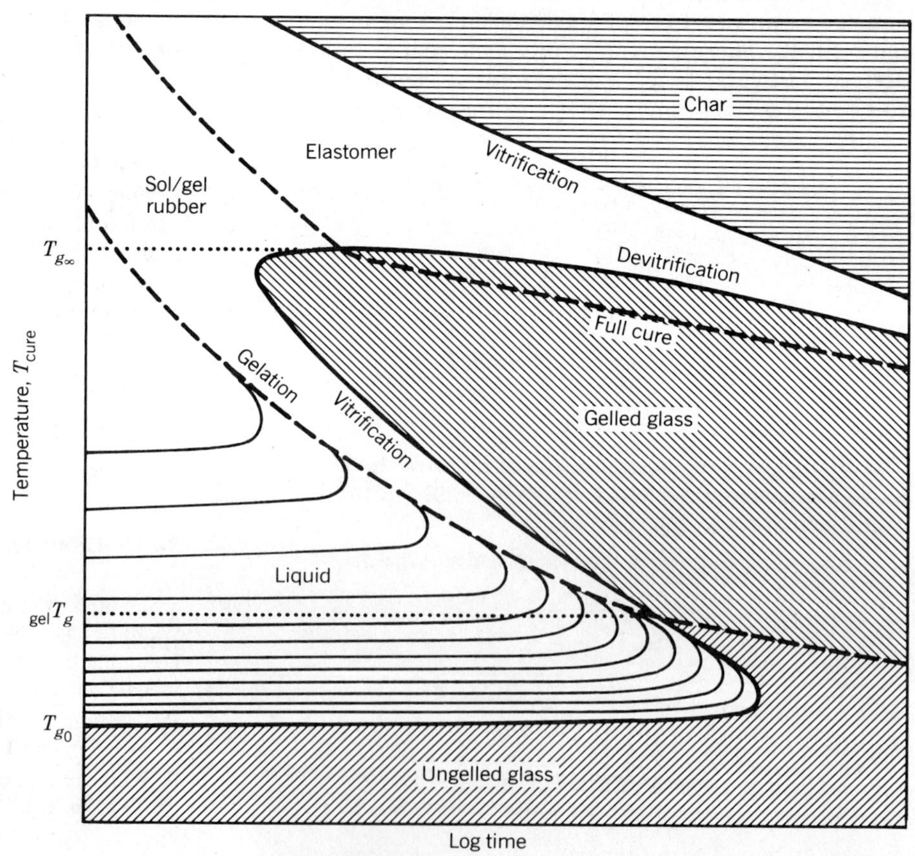

Figure 1. Time-temperature-transformation (TTT) isothermal cure diagram for a thermosetting system, showing three critical temperatures, ie, $T_{g\infty}$, ${}_{gel}T_g$, T_{g0}, and the distinct states of matter, ie, liquid, sol/gel rubber, elastomer, gelled glass, ungelled (or sol) glass, and char. The full-cure line, ie, $T_g = T_{g\infty}$, divides the gelled glass region into two parts: sol/gel glass and gel (fully cured) glass. Successive isoviscous contours shown in the liquid region differ by a factor of ten.

BIBLIOGRAPHY

J. K. Gillham, *Polym. Eng. Sci.* **26**(20), 1429 (1986).

J. K. Gillham in A. J. Kinloch, ed., *Structural Adhesives: Developments in Resins and Primers*, Elsevier Applied Science Publishers, London, 1986, Chapt. 1, pp. 1–27.

M. T. Aronhime and J. K. Gillham in K. Dusek, ed., *Epoxy Resins and Composites III, Advances in Polymer Science,* Vol. 78, Springer-Verlag, Berlin, 1986, pp. 84–113.

J. B. Enns and J. K. Gillham, *J. Appl. Polym. Sci.* **28,** 2567 (1983).

JOHN K. GILLHAM
Princeton University

CYANOACRYLIC ESTER POLYMERS. See ACRYLIC AND METHACRYLIC ESTER POLYMERS.

CYANOETHYLATION. See ACRYLONITRILE POLYMERS.

CYCLIC IMINO ETHERS, POLYMERIZATION

Polymerized or copolymerized cyclic imino ether monomers are classified into monomers having the —N=C—O— group in the ring (cyclic endoimino compounds), as shown by structures (**1**)–(**5**) and cyclic monomers having an exoimino group, as shown by (**6**)–(**8**).

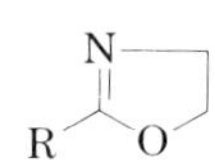
(1)

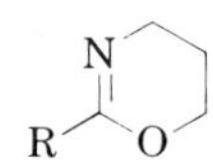
(2)

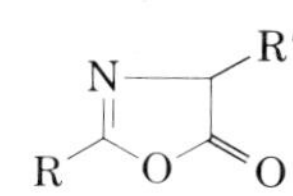
(3)

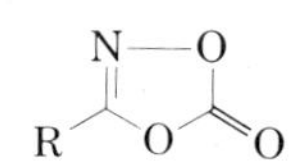
(4)

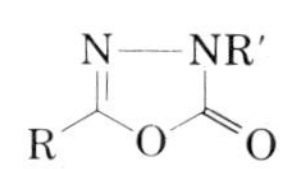
(5)

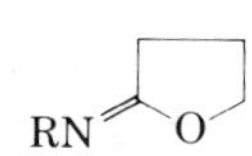
(6)

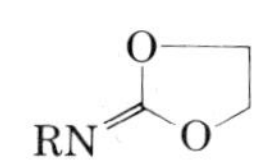
(7)

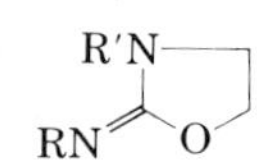
(8)

Polymerization of 2-Oxazolines

2-Oxazolines (**1**) are by far the most important monomers.

Cationic Ring-opening Polymerization. Ring-opening polymerization involving isomerization yields poly(*N*-acylethyleneimine) (**9**) (see also ALKYLENEIMINE POLYMERS).

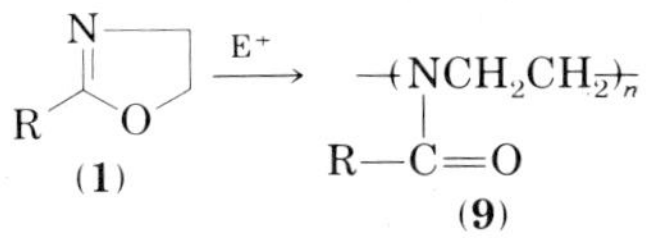

The polymerization of 2-oxazoline gives poly(*N*-formylethyleneimine) (**9**, R = H). The oxazoline ring-opening polymerization is not effected with anionic or radical initiators.

Polymerization Mechanism and Kinetics. The polymerization mechanism of 2-oxazolines is one of the best understood in polymer chemistry. Cationic propagating species are relatively stable; their reactivities depend on the substituent, counteranion derived from the initiator, solvent, and temperature.

The usual scheme of propagation for cationic ring-opening polymerization is

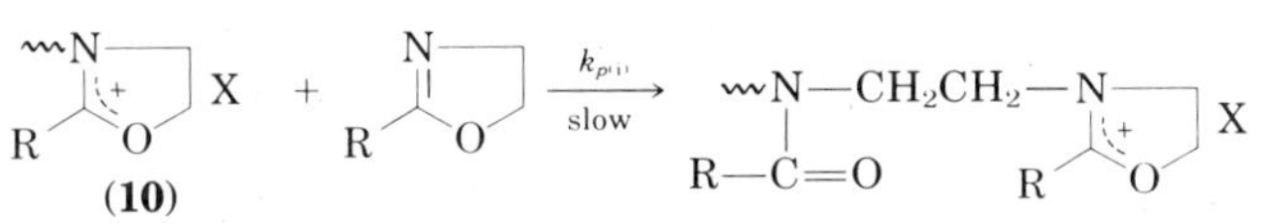

where X denotes a counteranion of low nucleophilicity.

In contrast to the above scheme, the propagation via covalent electrophilic species has been established in the system of polymerization where the nucleophilic reactivity of counter anion X^- is high. The rate-determining step is a reaction between two dipoles.

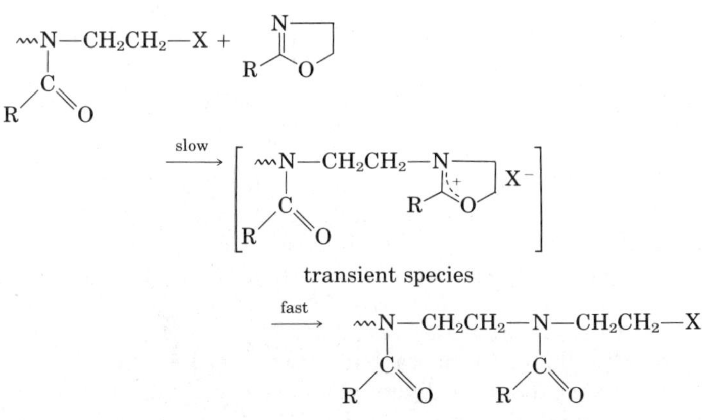

Polymerization of Other Cyclic Imino Ethers

Other cyclic imino ethers (**2–8**) have been used less for polymerization than 2-oxazolines (**1**). Substituted or unsubstituted 5,6-dihydro-4*H*-1,3-oxazines (**2**) produce poly(*N*-acyl- or *N*-formyltrimethyleneimine)s (**11**) by cationic polymerization.

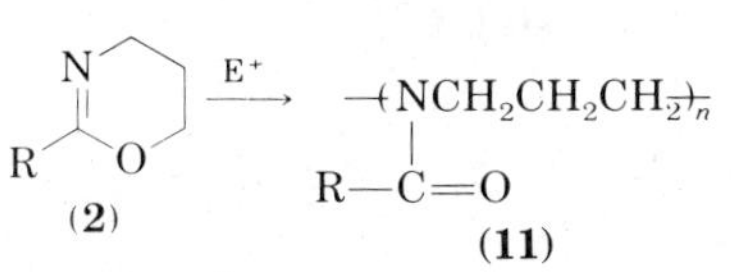

2-Unsubstituted 5-oxazolone (**3**) polymerizes with a cationic initiator to give poly(*N*-formylpeptide).

Monomers (**12**) derived from 2-imino-1,3-oxazolidine (**8**) undergo cationic ring-opening polymerization in two ways. The products are (**13a**) and (**13b**), as established by nmr, uv, ir, and mass spectra.

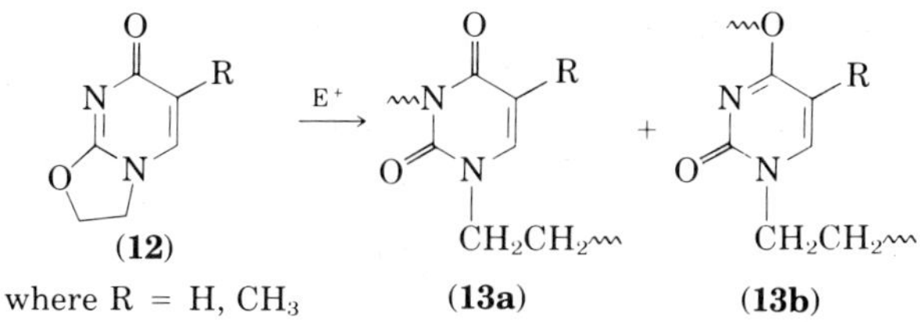

Copolymerization

Cyclic imino ethers can be copolymerized by random, block, and graft copolymerization and by the so-called "no catalyst copolymerization via zwitterion intermediates."

Uses of Poly(*N*-acylalkyleneimine)s and Polyalkyleneimines

These polymers are used in resins, gelling agents, thickening agents, adhesives, antielectrostatic agents, surface coatings, thermosetting rubber blocks, surfactants, detergents in lubricating oils, oil dispersants, suspension stabilizers for PVC, ion exchangers and coagulants, paper sizes, fire retardants, textile colorants, and complexing agents with phenolic compounds in waste water.

BIBLIOGRAPHY

S. Kobayashi and T. Saegusa, *Ring-Opening Polymerization,* Applied Science Publishers, Ltd., Essex, UK, 1984, pp. 761–807.

D. A. Tomalia, B. P. Thill, and M. J. Fazio, *Polym. J.* **12,** 661 (1980).

S. Kobayashi and T. Saegusa, *Alternating Copolymers,* Plenum Publishing Co., Ltd., London, 1985, pp. 189–238.

C. G. Overberger and Y. Morishima, *J. Polym. Sci. Polym. Chem. Ed.* **18,** 1433 (1980).

SHIRO KOBAYASHI
TAKEO SAEGUSA
Kyoto University

CYCLOPENTADIENE POLYMERS

This article describes the preparation of polymers and copolymers of 1,3-cyclopentadiene, C_5H_6 (**1**), substituted cyclopentadienes, and their dimers. The Diels-Alder dimer, dicyclopentadiene (**2**), forms rapidly from cyclopentadiene on storage and is the form handled in commerce. A mixture of 1-methyl- and 2-methylcyclopentadiene is also commercially available. Polymeric derivatives are prepared by thermal treatment and free-radical and cationic polymerization techniques. In thermal polymerization, 1,3-cyclopentadiene monomer is formed *in situ* and reacts further.

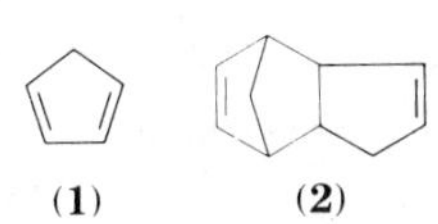

Cyclopentadiene and its dimer provide olefinic sites in terpolymers with ethylene and propylene and in blends with other olefins and diolefins that are polymerized to hydrocarbon resins. It is also used as a diluent for unsaturated polyesters and alkyds (see also ALKYD RESINS; POLYESTERS, UNSATURATED).

Manufacture of Monomer

Cyclopentadiene is prepared by the thermal cracking of naphtha, gas oils, or natural-gas liquids; steam cracking is preferred.

Polymerization

At room temperature, cyclopentadiene spontaneously forms the dimer, but heating the dimer at ca 200°C yields oligomers up to the pentamers. Oligomerization is believed to involve partial dissociation of the dimer to monomer, which reacts with dimer in a Diels-Alder reaction to form trimer, or the reaction of monomer with trimer to form tetramer, and so on.

Polymerization of cyclopentadiene initiated by $SnCl_4$, $TiCl_4$, BF_3·etherate, $AlCl(C_2H_5)_2$, and $AlCl_2C_2H_5$ at −60 to −90°C in various solvents gives polymers insoluble or partially soluble in hydrocarbons. The trichloroacetic acid-initiated polymerization of cyclopentadiene in benzene at 25°C in the presence of α-tocopherol has been studied in detail. The cationic polymerization (qv) of 1,3-dimethylcyclopentadiene can be initiated with boron trifluoride etherate, $SnCl_4$, $TiCl_4$, and triethylaluminum-titanium tetrachloride.

Copolymerization

Cyclopentadiene has been copolymerized with a wide variety of comonomers, leading to many commercial applications. Cationic copolymerization with α-methylstyrene gives high yields of copolymers relatively resistant to oxidation.

The copolymerization of 1,3-cyclopentadiene with 2-chloroethyl vinyl ether initiated with boron trifluoride etherate in toluene, methylene chloride, or nitroethane at −78°C has been reported. The preparation of isobutylene copolymers with 5–40 mol % cyclopentadiene is reported in the same study. Copolymers of 1,3-cyclopentadiene with sulfur dichloride are prepared by adding the SCl_2 to a solution of the diene in methylene dichloride; the reaction is exothermic.

Cyclopentadiene yields copolymers of regularly alternating structure with other monomers. These pairs consist of the electron-rich cyclopentadiene and an electron-acceptor monomer, often coordinated with a Lewis acid. Cyclopentadiene can be copolymerized with maleic anhydride between 80 and 205°C in the presence of peroxides.

Terpolymers of ethylene, propylene, and a small amount of a diene are termed EPDM elastomers (see ETHYLENE–PROPYLENE ELASTOMERS). Dicyclopentadiene was one of the first dienes examined for this application and is still used in commercial products.

Polycyclopentadiene prepared by cationic catalysts has been brominated. Dicyclopentadiene in the presence of a metathesis polymerization (qv) catalyst forms a thermoset polymer with high modulus and impact strength.

Health and Safety Factors

Cyclopentadiene has a recommended TLV of 75 ppm and a STEL of 150 ppm. Closed containers of cyclopentadiene monomer may rupture when storage temperatures are too high. Dicyclopentadiene has a recommended TLV of 5 ppm to prevent irritation and possible chronic effects.

Uses

Cyclopentadiene-containing polymers are used as elastomers, as diluents; in tackifier resins; as hot-melt, pressure-sensitive, and other adhesives; and to modify alkyds and drying oils.

BIBLIOGRAPHY

W. A. Thaler and D. J. Buckley, Sr., *Rubber Chem. Technol.* **49,** 967 (1976).

W. Kuran, S. Pasynkiewicz, Z. Florjanczyk, and D. Lyszkowska, *J. Polym. Sci. Polym. Chem. Ed.* **16,** 867 (1978).

F. P. Baldwin and G. Ver Strate, *Rubber Chem. Technol.* **45,** 709 (1972).

IRVING KUNTZ
Exxon Chemical Company

CYCLOPOLYMERIZATION

Cyclopolymerization is an addition polymerization that leads to introduction of cyclic structures into the main chain of the polymer. Its scope is broad, including polymerization of symmetrical dienes such as the 1,6-dienes and unsymmetrical ones as well. A variety of other multiply bonded structures, such as alkynes, carbonyls, isocyanates, epoxides, and nitriles also undergo cyclopolymerization.

Nonconjugated Symmetrical Dienes

Suitable monomers undergo cyclopolymerization by all well known initiation methods, eg, 1,2-divinylbenzene and 2,6-diphenyl-1,6-heptadiene. By suitable initiation, 1,6-heptadiene-2,6-dicarboxylic acid and its methyl and ethyl esters, and nitrile, 4-substituted 1,6-heptadienes, diacrylylmethane, diallylquaternary ammonium chlorides and bromides, *N*-alkyldiallylammonium chlorides, *N,N*-diallylmelamine, 1,6-dienes, diacrylamide, dimethacrylamide, *N*-substituted dimethacrylamides, diallylamines, piperidine polymers, *N,N*-diallylmethanesulfonamide, *N,N*-diallylethanesulfonamide, *N,N*-dimethallylmethanesulfonamide, *N,N*-dimethallylethanesulfonamide, unsaturated silanes, perfluoro-1,6-heptadiene and 4-chloroperfluoro-1,6-haptadiene, perfluorinated C_5—C_8 α,ω-dienes, 4-chloroperfluoro-1,6-heptadiene, perfluoro-1,11-dodecadiene, unsymmetrical dienes, diynes, dialdehydes, diisocyanates, diepoxides, and dinitriles also cyclize.

Nonconjugated Unsymmetrical Dienes

Methyl allyl maleate and methyl allyl fumarate were perhaps the earliest examples of unsymmetrical dienes studied. Polymer structures, ie (**1**), in which *x*, *y*, and *z* vary with monomer type and experimental conditions, were shown to be present in radical-initiated systems.

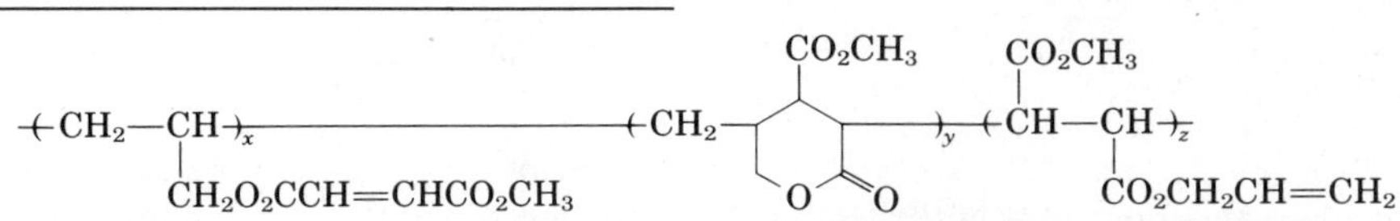

(**1**)

Although these unsymmetrical monomers generally lead to cross-linked polymers, the extent of cyclization of the soluble polymers obtained by stopping the polymerization just short of the gel time varies from 23–63%. Those polymers containing the greatest degree of cyclization possess double bonds of comparable reactivity (allyl and fumarate bonds in methyl allyl fumarate, 2-butenyl and maleate bonds in methyl-2-butenyl maleate). The structure for poly(methyl allyl fumarate), where $x{:}y{:}z = 1{:}3{:}1$, is supported by analytical and spectral data. Cyclopolymerization of the following nonconjugated unsymmetrical dienes has been studied: monoallyl esters of maleic and citraconic acids, allyl acrylate, allyl methacrylate, vinyl *trans*-cinnamate, vinyl methacrylate, allyl methacrylate, *N*-allylacrylamide, *N*-allylmethacrylamide, cinnamyl methacrylate, 3-vinylcyclopentene, methacrylate, γ-chlorocrotyl acrylate, γ-chlorocrotyl, *N*-2-phenyl-allylacrylamide, *N*-ethyl-2-phenylallylacrylamide, *N*-allyl-*N*-methylmethacrylamide, allylvinyldimethylsilane, *o*-allylstyrene, 1-allylindene, 1-allylindane, allyl allylsulfonate, *o*-isopropenylstyrene, *N*-allylmaleimide, *o*-isopropenylphenyl vinyl ether, vinyl *o*-isopropenyl benzoate, *o*-allylphenylacrylate, *o*-vinylphenyl acrylate, and 2-(*o*-allylphenoxy)ethyl acrylate.

Dienes Yielding Polycyclic Systems

Polycyclic systems have been introduced into polymers in a number of ways. For example, *cis*-1,3-divinylcyclohexane and *o*-divinylbenzene have been cyclopolymerized to polymers containing bicyclic units. In these cases, only one ring system is developed during the propagation step. Other polymerizations involve monomers that introduce at least a bicyclic unit from open chain or branched monomers, or the initial ring structure present is altered during the propagation step. Monomers that have been cyclopolymerized include triallylethyl- and tetraallylammonium bromides, triallylmethylsilane, branched triene compounds, trivinyl orthoformate, trivinyl phosphate, triallylamine hydrobromide, 2-carbethoxybicyclo[2.2.1]-2,5-heptadiene, 1-methylene-4-vinylcyclohexane, and 1,4-diene monomers.

Other Bifunctional Monomers

Other bifunctional monomers include nonconjugated diynes, dialdehydes, diisocyanates, diepoxides, dinitriles, and organometallic monomers.

Copolymerization of 1,6-Dienes with Conventional Vinyl Monomers

Copolymerization of 1,6-dienes and other cyclopolymerizable monomers with conventional vinyl monomers can proceed by two distinctly different pathways: the nonconjugated diene alone forms the cyclic structure, followed by reaction of the resulting cyclic radical (or other reactive intermediate) with the comonomer; or the comonomer contributes to formation of the cyclic structure along with the nonconjugated diene. The latter type is generally defined as cyclocopolymerization, and will be discussed later.

Cyclocopolymerization of 1,4-Dienes with Suitable Alkenes

Cyclocopolymerization incorporates both comonomers into the developing cyclic structure. Perhaps the most thoroughly explored example of this unusual type of copolymerization is the cyclocopolymer of divinyl ether and maleic anhydride (DIVEMA) (**2**).

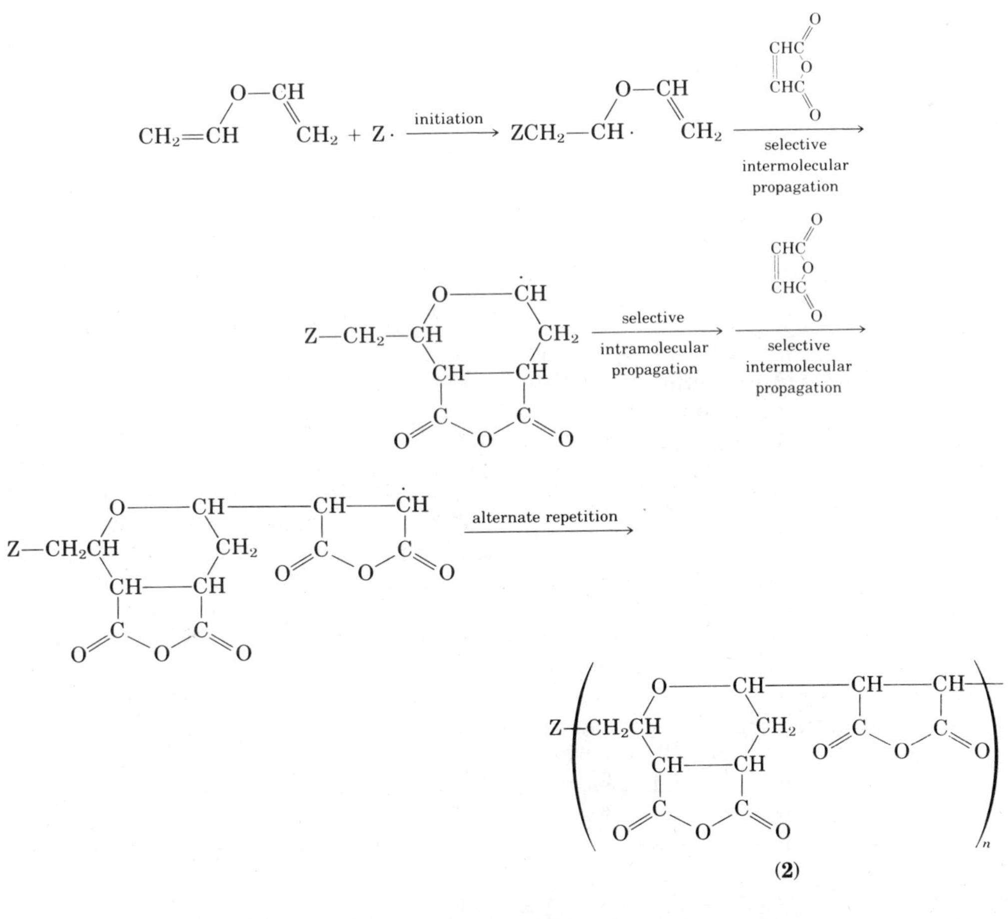

(**2**)

Cyclocopolymerization is most likely to occur according to the mechanisms shown when one monomer possesses strong electron-donor properties and the other strong electron-acceptor properties. Thus, an intermediate charge-transfer complex (or donor–acceptor complex) probably plays a significant role in controlling the regularly alternating cyclocopolymerization (see also CHARGE-TRANSFER COMPLEXES; COPOLYMERS, ALTERNATING).

Ladder Polymers

Cyclopolymerization has been used to synthesize ladder polymers from preformed polymers of general structure (**3**).

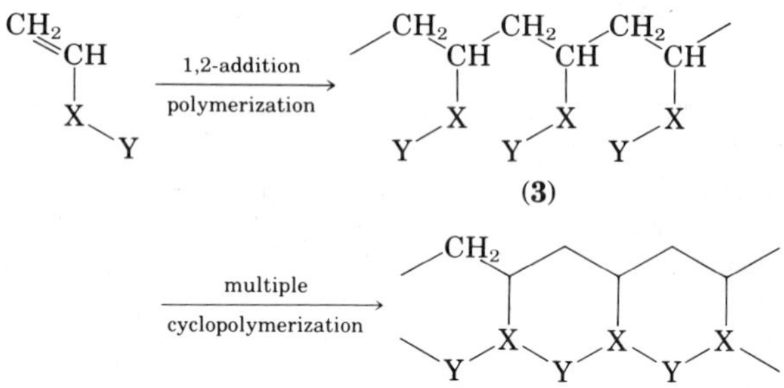

(**3**)

—X—Y = —CH=CH$_2$, —C=O, —C≡N, or —N=C=O

Perhaps the most important example of this mode of polymer modification is black orlon, a polymer obtained by heat treatment of polyacrylonitrile, and a precursor in one process for manufacture of carbon fibers (qv).

Cyclopolymerization of Monomers Containing Different Functional Groups

o-Vinylbenzaldehyde has been cyclopolymerized by cationic initiation to yield polymers containing both the predicted cyclized structures, ie, (**4**) and (**5**).

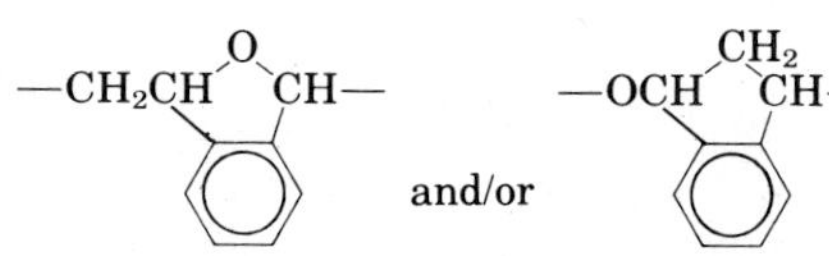

(**4**) (**5**)

The latter structure probably predominates because the aldehyde group is more reactive toward the cationic initiator than the vinyl group. The cyclopolymer also contains 20% pendent vinyl group and 5% pendent aldehyde group.

Cyclopolymerization of Diene Monomers Leading to Larger Ring Sizes

Poly(diallyl phthalate) has been commercially important for many years. The monomer is functionally capable of undergoing cyclopolymerization, and early studies led to the conclusion that soluble polymers containing 31–41% of a cyclic structure presumed to consist of 11 or 13 members could be obtained before gelation. By use of coordination catalysts, polymerization of the hydrocarbons, $CH_2{=}CH(CH_2)_nCH{=}CH_2$, in which n = 4–9, 11, 12, 14, and 18, has been studied. In all cases, both cyclic and acylic recurring units are formed; the largest extent of cyclization, ie, 25%, occurs at calculated ring size 7. However, the extent of cyclization among the larger calculated ring sizes varies only from 4 to 15%. The extents of cyclization observed correspond roughly to those that had been obtained in other cyclization reactions.

The Mechanism of Cyclopolymerization

Kinetic Studies. Important aspects of the cyclopolymerization process are embodied in a kinetic study carried out on methacrylic anhydride. The kinetic relationship between intramolecular and intermolecular propagation is derived from equations 1 and 2.

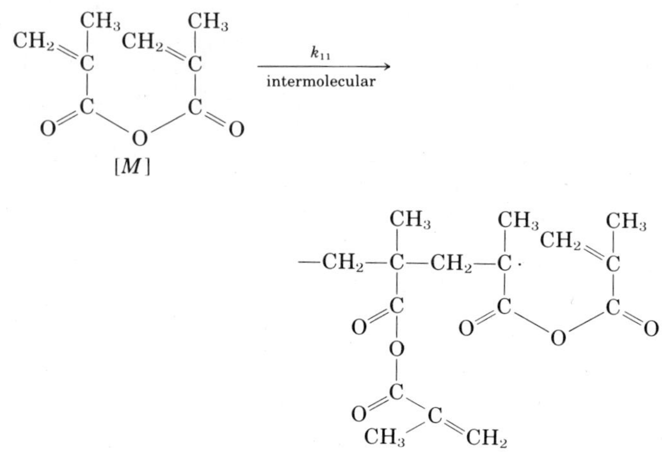

(1)

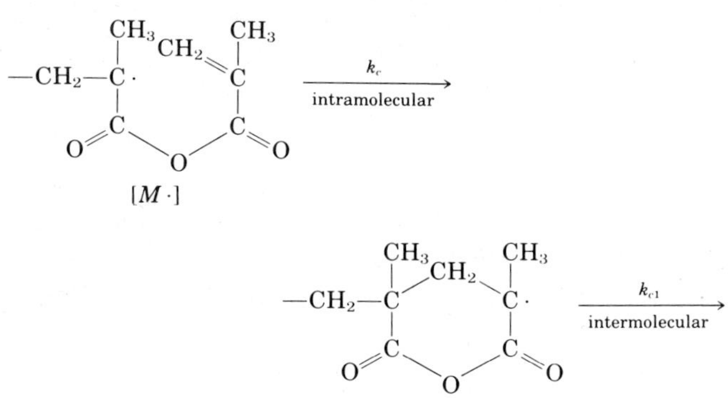

(2)

The energy of activation for the intramolecular cyclization step is higher than that for the intermolecular step by approximately 10.9 kJ/mol (2.6 kcal/mol). The rate of cyclization, however, is considerably faster than the intermolecular propagation step, suggesting a very high steric factor favoring cyclization. The value k_c/k_{11} is 2.4 mol/L, and the Arrhenius frequency factor ratio is 256 mol/L in favor of the cyclization step. Increasing steric factors, higher temperature, lower monomer concentration, poorer solvent, and higher conversion favor cyclization.

Other kinetic studies of cyclopolymerization have been conducted; the results are summarized in Table 1. They consistently indicate that the cyclic propagation step requires greater energy than the intermolecular step, but the rate of cyclization

Table 1. Cyclopolymerization Activation Parameters[a] for Typical Monomers

Monomer[b]	Solvent	Temp, °C	k_c/k_i, mol/L	E_c/E_i, kJ/mol[c]	A_c/A_i, mol/L
acrylic anhydride	cyclohexanone	45	5.9	10.0	167
o-divinylbenzene	benzene	50	2.8	7.9	10
o-divinylbenzene[d]	toluene	−78	0.1–5	22.2	2.2×10^4
o-divinylbenzene[e]	methylene chloride	−78		25.9	5×10^4
o-divinylbenzene[f]	tetrahydrofuran			9.2	15
o-divinylbenzene[g]	tetrahydrofuran			11.7	66
methacrylic anhydride	cyclohexanone	37	46		
methacrylic anhydride	dimethylformamide	0	2.4	10.9	256
methacrylic anhydride	dimethylformamide	80	6.2		
methacrylic anhydride	dimethylformamide		11.4		
divinylformal	benzene	50	200	10.9	
vinyl *trans*-cinnamate	benzene	70	13.3	~0	

[a] Calculated from the equation $t_c = k_c/k_i = A_c/A_i \exp[(-E_c + E_i)/RT]$.
[b] Radical initiation unless otherwise specified.
[c] To convert kJ to kcal, divide by 4.184.
[d] $BF_3O(C_2H_5)_2$ initiator.
[e] $(C_6H_5)_3CBF_4$ initiator.
[f] Sodium naphthalene initiation.
[g] Sodium naphthalene–$NaB(C_6H_5)_4$ initiation.

is considerably larger than that for intermolecular propagation. The values for the ratio A_c/A_i indicate a high steric factor favoring cyclization.

Ring Size in Cyclopolymerization. Formation of a primary radical in preference to the more stable secondary radical is not predicted. However, considerable evidence indicates the formation of five-membered rings rather than six-membered during cyclopolymerization, ie, the reactions are under kinetic rather than thermodynamic control.

The Mechanism of Cyclocopolymerization

Kinetic Studies. The first example of cyclocopolymerization in which both comonomers contribute to the cyclic structure was the alternating copolymerization with cyclization of divinyl ether and maleic anhydride. The mechanistic steps postulated for this copolymerization [see structure (**2**)] account for the failure of the system to cross-link in accord with the widely accepted theory of Staudinger, the essential absence of carbon–carbon double-bond content in the copolymer, a copolymer composition equivalent to a diene–olefin molar ratio of 1:2, and essentially quantitative conversion of monomers to copolymer. A wide variety of 1,4-dienes and alkenes have since been shown to undergo cyclocopolymerization in a similar manner.

A general copolymer equation has been developed for the cyclocopolymerization of 1,4-dienes (M_1) and monoolefins (M_2). The kinetic scheme is

$$
\begin{array}{ll}
m_1^{\cdot} + M_1 \xrightarrow{k_{11}} m_1^{\cdot} & m_3^{\cdot} + M_2 \xrightarrow{k_{32}} m_2^{\cdot} \\
m_1^{\cdot} + M_2 \xrightarrow{k_{12}} m_3^{\cdot} & m_c^{\cdot} + M_1 \xrightarrow{k_{c1}} m^{\cdot} \\
m_3^{\cdot} \xrightarrow{k_c} m_c^{\cdot} & m_c^{\cdot} + M_2 \xrightarrow{k_{c2}} m_2^{\cdot} \\
m_3^{\cdot} + M_1 \xrightarrow{k_{31}} m_1^{\cdot} & m_2^{\cdot} + M_1 \xrightarrow{k_{21}} m_1^{\cdot} \\
 & m_2^{\cdot} + M_2 \xrightarrow{k_{22}} m_2^{\cdot}
\end{array}
$$

where $m_1^{\cdot}$ represents the radical derived from M_1, etc.

The differential copolymer composition equation (3), which may be applied by letting $n = [M_1]/[M_2]$, the fractional ratio of monomers combined in the copolymer at low conversions, has been derived.

$$n = \frac{(1 + r_1x)\{1/[M_2] + (1/a)(1 + x/r_3)\}}{(1/a)\{(x/r_3) + (r_2/x + 2\} + (1/[M_2])\{1 + (1 + r_2/x)(1 + r_cx)^{-1}} \qquad (3)$$

A similar equation relates the relative rate of addition of diene and the rate of cyclization (eq. 4).

$$
\begin{aligned}
d[M_1]/d[M_c] &= [(K_{11} + K_{12})/K_{12}k_c](K_{31} + K_{32} + k_c) \\
&= (r_1x + 1)\{([M_1]/r_3a) + ([M_2]/a) + 1\}
\end{aligned} \qquad (4)
$$

Equation 4 applies at low conversions, where $d[M_1]/d[M_2] = [M_1]/[M_2]$, the ratio of the total fraction of diene (unsaturated and cyclic) to the fraction of diene in cyclized units in the copolymer.

If precise analytical methods are available for determining both the total fraction of diene in the copolymer and the fraction of either cyclic units or pendent vinyl groups, then by making a series of such measurements for different initial monomer-feed compositions, values for r_1, r_2, and a could be obtained from equation 4. Then the remaining two parameters, r_2 and r_c, can be obtained from equation 3.

Ring Size in Cyclocopolymerization. Early work led to the conclusion that the more thermodynamically stable six-membered ring is formed in cyclocopolymerization. By analogy to the case of cyclopolymerization of 1,6-dienes, it is apparent that the cyclocopolymerization could lead to the thermodynamically favored six-membered ring, the five-membered ring, or to a combination of both.

A satisfactory theory for cyclopolymerization must adequately account for the following: (*1*) almost all nonconjugated dienes and other similar polymerizable structures studied undergo cyclopolymerization in preference to cross-linking or other modes of propagation; (*2*) such dienes invariably undergo polymerization at a higher rate than their respective monoene counterparts; (*3*) the cyclization step has invariably been shown to be of higher activation energy than propagation without cyclization; and (*4*) the cyclization step often leads to propagation by the less thermodynamically stable reaction intermediate, leading to the less thermodynamically stable ring structure.

Although the differences in activation energies between the intramolecular and intermolecular propagation reactions suggest that the former is a slightly less favorable reaction, the high tendency for cyclization has been explained as due to a smaller decrease in entropy for such a reaction compared with intermolecular addition. Therefore, the Gibbs free energy of activation is smaller for cyclopolymerization than for intermolecular propagation.

Practical Significance of Cyclopolymerization

Cyclopolymers are used in paper-reproduction processes, water treatment, electroplating industries, cosmetics, sludge dewatering, flocculation, and as biologically active materials, binding agents, antistatic agents, electrically conductive coatings, softening agents, and electrophotographic toners.

Cyclopolymers often possess enhanced properties in comparison with their open-chain analogues. However, only in a few instances has the economics of the two competitors favored the cyclopolymer.

BIBLIOGRAPHY

G. B. Butler and R. J. Angelo, *J. Am. Chem. Soc.* **79,** 3128 (1957).

N. D. Field, *J. Org. Chem.* **25,** 1006 (1960).

G. B. Butler, G. C. Corfield, and C. Aso in A. D. Jenkins, ed., *Progress in Polymer Science,* Vol. 4, Pergamon Press, Elmsford, N.Y., 1974.

J. M. Barton, G. B. Butler, and E. C. Chapin, *J. Polym. Sci. Part A* **3,** 501 (1965).

GEORGE B. BUTLER
University of Florida at Gainesville

D

DAMMAR. See RESINS, NATURAL.

DEAD-END POLYMERIZATION. See RADICAL POLYMERIZATION.

DECORATING, METHODS

Decoration can be applied to plastics by heat-transfer labeling, offset printing, hot stamping, screen process, pad-transfer printing, painting, vacuum metallizing, electroplating, and various other coating methods (qv) (see also COATINGS; METALLIZING; PRINTING-INK VEHICLES).

Screen-process Printing

In the screen process, (Fig. 1), ink is forced through a synthetic nylon, polyester, or wire mesh in a selective manner to produce a controlled image repeatedly. Excellent deposition results in opacity, gloss, and high color intensity. Quality is paramount in three-dimensional screen-process printing.

Figure 1. Kammann K10G semiautomatic screen printer. Courtesy of Kammann Machines, Inc.

Pad-transfer Printing

Pad-transfer (Fig. 2) uses silicone rubber pads and photoengraved plates made from hard, wear-resistant tool steels. It can be used for irregular shapes, concave and convex parts, and almost all textured surfaces. In addition, up to four or five colors can be printed with little or no interim drying. This is known as the wet-on-wet process, which is employed especially for halftones.

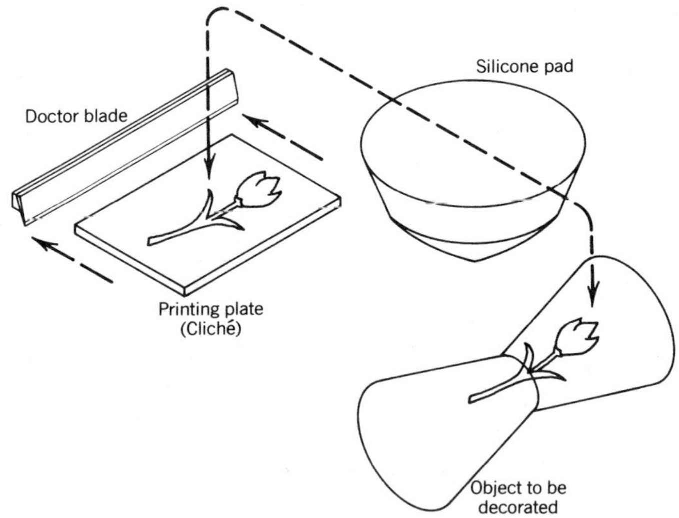

Figure 2. Pad printing or the indirect-gravure process.

BIBLIOGRAPHY

Modern Plastics Encyclopedia 1988, McGraw-Hill Inc., New York, 1988, p. 349.

GENE COLLARD
Alternative Screen Printing

DECORATING SYSTEMS

The design of a thermoplastic part that is to be decorated must be viewed as a system involving the selection of the proper molding material based on the environment in which the article must perform, design of the article and mold, and selection and application of suitable finishing materials. In many instances only one or two of these features are carefully examined and a coordination of effort is lacking. As a result, the final article may differ from that originally intended; one parameter may be downgraded to improve the acceptability of another.

Resin selection is of vital importance in determining the decorating techniques and materials utilized. Environmental impacts and physical parameters such as heat distortion, impact resistance, flammability, and flexural modulus must be considered. Decorations may be hot stamped, painted, or electroplated. Some materials are more readily plated or painted, or interface with the finishing materials more effectively.

In a design for decorative effects, the appearance of the molded article and the finishing techniques required are important. The custom processor should be consulted at the design-sketch stage to discuss the design and appearance objectives. Consultation should continue through each phase of the design process until prints are prepared for bid.

The position of a gate on a decorated product is important for high visibility articles such as nameplates, faceplates, and control panels, where sprue mark or gating points may be visible from some viewing angles. If the problem can be identified, the gate location can be changed or the process modified to mask blemishes.

Gating. Gate location and design influence the quality of the article; center gating, edge gating, and their variants are used. Each provides different trade-offs, involving the locations of the sprue or gate marks, the stress patterns, and distortion of part geometry, all of which must be evaluated separately.

Molding Control. Uniformity of the molded article is essential. Close control over dimensions assures uniformity of lettering and other decorations. Consistent dimensions are crucial to the accurate deposition of paint masks and foiling hobs. Poor tolerances can result in overspray, blushing, or bleeding at the overlap between adjacent decorated surfaces.

Mold-release materials should be avoided, since a surface so contaminated cannot be prepared for painting, screening, or electroplating. The mold design (radii and drafts) must allow for manufacture without mold-release agents.

Design of the Article. Decorating an article in a single step, such as by spray painting with one color, is a simple operation. If, however, a series of processes or steps is required, for example, in painting, foiling, and metallizing certain areas, the article must be designed to meet these requirements.

Electroplating. Electroplating is a complicated process requiring an understanding of both the plating and the molding processes. The important aspects of electroplating are related to current efficiency, the degree to which the process deposits a bright metallic film. The current efficiency therefore determines the film thickness and physical properties of the deposition, and is normally judged by current density and film thickness of the plate.

First, provisions must be made for racking. This contact must be as close to a high current density area as possible without affecting surface appearance; this is critical for proper current distribution. Another aspect related to current density is the appearance of the plate. Areas designed with a deep vertical wall are subject to insufficient coverage, usually distinguished by a brown cast. Another problem resulting in poor appearance is the problem of burning in the chrome tank, the result of excessive and overly rapid application of the plate. The selection of raw materials and incorporation of elements to improve the decorative processes should be included in the planning phase.

Nominal Wall Thickness. A completely uniform wall thickness is ideal, but rarely practical; uniform wall thickness promotes uniform cooling and minimizes warpage and shrinkage.

Ribs. Ribs are linear wall projections originating from adjacent plane surfaces. Ribs reduce the bulk or weight of articles that must meet high strength and rigidity requirements. Interconnecting ribs are frequently used to prevent bowing or warpage of large plane surfaces.

Bosses and Edges. Bosses are projections that provide for attachment and support of related components; they may be hollow or solid. Solid bosses are sometimes called studs.

It is often possible to design intricate systems of structurally interdependent bosses and ribs that provide dimensional stability, while providing for attachment of related components, maintaining proper wall thickness ratios, and reducing material usage and molding-cycle time.

The unsupported edges of a part may be stiffened or strengthened by turning the edge or changing the plane of the wall. These methods may be employed to provide edge rigidity and reduce or eliminate warpage, especially when minimum wall thickness is used.

Radius. The radius at both external and internal plane intersections should be as generous as possible to improve the flow of the molding material and minimize stress concentration when the article is under a load.

Silk Screening

Silk screening has been used for a long time to decorate plastics. The basic concepts underlying its applications are well known throughout the industry. Changes in screen mesh or paint formulation allow this technique to be used in a great variety of applications.

Silk-Screen Paints. Acrylics have been used for silk screening by the plastics industry for the last 25 years. They are easy to handle, accept most pigments readily, are stable, and give adequate to good performance for most applications. Silk-screen paints achieve adhesion through a paint-to-surface affinity rather than relying on solvent attack for bonding. As a result, many of the silk-screen formulas are identical in many respects. Formulation changes are minor and are directed at making the coating more compatible with the substrate, making the silk-screen composition more or less polar. Applications for silk-screening include the dead-front effect for back lighting requirements and halftone techniques utilized to produce woodgrains.

Hot Stamping

Hot stamping is the heat transfer of a pigmented color or metallized surface from a polyester carrier onto a thermoplastic part. A heated die or roller is used to apply pressure on a foil (Fig. 1). The pressure forces the pigmented or metallized portion of foil into contact with a nested part. The combination of heat and pressure softens the plastic substrate and activates the foil coating to form a strong chemical–physical bond. The polyester carrier is pulled away and moved forward to afford a new stamping surface. Hot stamping is an excellent procedure for applying bright beads on low abrasion areas of bezels and for applying wood grains to large surface areas such as television cabinets.

Spray Painting

Although one of the earliest decorating techniques, spray painting with a simple handgun is still widely used. For some applications reciprocating spray heads are more effective. For very large and complex articles requiring complicated and repetitive spray motions, such as automotive grilles or business machine cabinets, robotics has gained wide acceptance. Figure 2 shows a diagram of an automatic spray operation. A safe spraying operation requires ventilation or the capture of paint particulates. Local regulations regarding emissions must be observed.

The number of coatings applied is determined by the plastic substrate and the appearance requirements. Spray painting is utilized for the application of base coats for vacuum metallizing, resist coatings for electroplating, and finish coatings.

The paint is the most critical part of the system because

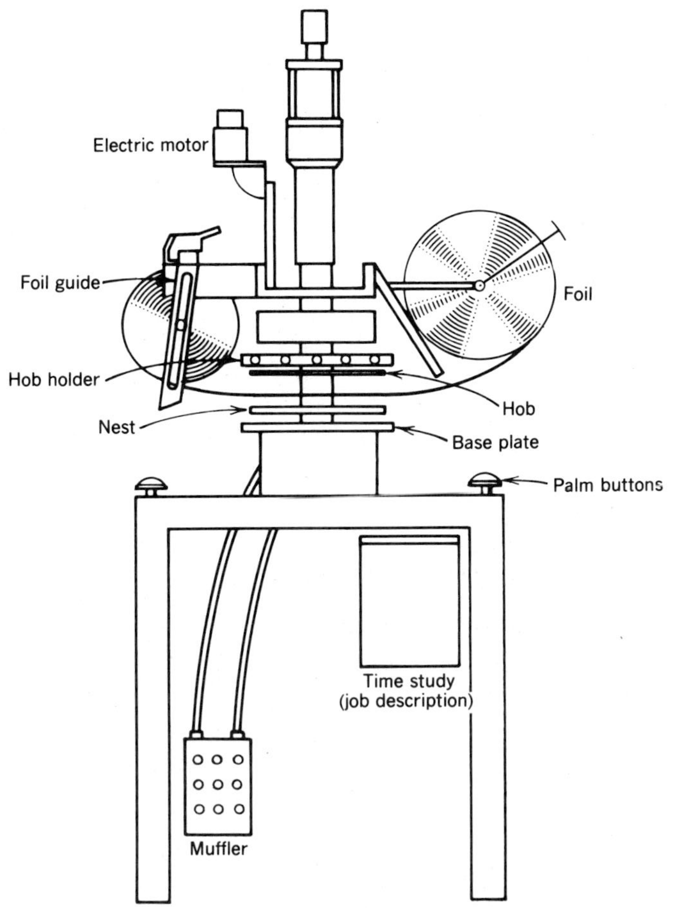

Figure 1. Hot stamp machine.

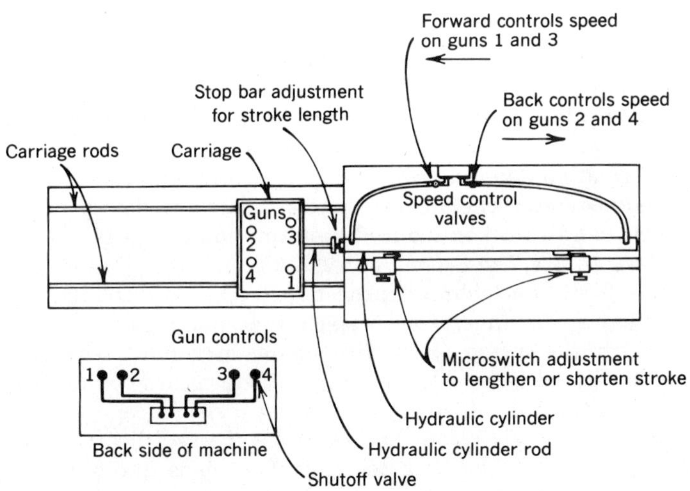

Figure 2. Automatic spray operation.

drying rate, ease of application, appearance, and physical properties of the finished coat are all a function of the paint. The choice depends on the plastic substrate and the environment in which the product will be used. Adhesion is critical and must be considered in the choice of paint.

In the past few years many manufacturers have replaced painted surfaces on plastics with molded-in textures and colors in an effort to reduce costs. However, painting remains one of the most cost-effective means of producing an attractive plastic coating that also serves as protection against excessive abrasion and ultraviolet exposure.

Vacuum Metallizing

Requirements for brightwork, specifically second surface decorative techniques on acrylics and polycarbonates, are routinely fulfilled by vacuum metallization. Formerly the dominant technique for first surface brightwork, vacuum metallization has lost a substantial portion of this market to electroplating.

Both vacuum metallization and sputtering, related yet slightly different, are available to the designer and decorator. In both processes the three basic steps are as follows: preparation and base-coat application, metallizing (qv), and backcoat or top-coat application.

Decorating Technique Combinations

Decorative finishes can be combined, for example, by resisting chromed parts and painting over chrome. This technique is used for the production of exterior parts such as grilles and headlamp bezels, but can also be employed for interior designs.

However, difficulties in the adhesion between an inert metal surface and an organic resin have threatened the growth of this technique. Improvements in coating systems, specifically two-part urethanes, and cleanliness of the final rinse systems and the chrome substrate have improved this technique, but continued efforts are required. Nevertheless, in some instances paint over chrome is an attractive choice for a particular finish.

Shielding against Electromagnetic Interference

The limitation to the growth of applications for decorative thermoplastics in the business machine market is their inability to shield against electronic noise, commonly referred to as RFI/EMI interference. In fact, the continued use of thermoplastics in some existing applications is in question because of this problem (see ELECTROMAGNETIC INTERFERENCE).

During the design stage many techniques for shielding against electromagnetic interference are available, including electrostatically shielded transformers, shielded wiring (with leads as short as possible), and good mechanical design. When these techniques have been exhausted, the level and frequency of the emission are examined. Utilizing the standardized attenuation figure, alternative shielding systems can be selected and evaluated in actual use. The ability of divergent shielding systems to attenuate the energy of an electromagnetic wave depends on several factors. Attenuation itself is actually composed of two major elements, absorption and reflection.

The conductivity and film thicknesses achievable are important factors in the selection of the shielding materials, but permeability and the ability of the medium to respond to magnetic fields also play a role. Shielding materials and techniques include copper paints, nickel paints, silver paints, graphite, arc spraying, vacuum metallizing, electroless plating, and conductive composites (see CONDUCTIVE PLASTICS).

BIBLIOGRAPHY

Standards and Guidelines for Electroplated Plastics, 3rd ed., Prentice Hall, Englewood Cliffs, N.J., 1984.

R. W. Carpenter, *Plast. Design Forum* **66** (1982).
D. Daubs, *Plast. World* **56** (1975).
H. Kurz, *Stamping on Plastics,* Kurz Brochure, 1985.

ROBERT CARPENTER
Windsor Plastics

DECORATIVE LAMINATES. See LAMINATES.

DEFECTS

In a solid polymer, different kinds of structural irregularities or defects can occur. Some defects are directly related to the molecular structure of the polymer chain; these are the subject of this article. Understanding the structural disorder in a polymeric material requires detailed knowledge of the steps followed in the formation of a macromolecular system.

The Ideal Perfect Structure

Regularity is needed in order to form a perfect structure. *Chemical regularity* of the monomer units within the chain and *stereoregularity* or *tacticity* of the chain are the results of chemical reactions. Intramolecular interactions allow the chain to take up a certain *conformational regularity,* which creates a *one-dimensional translational periodicity.* One-dimensional polymers then pack in a three-dimensional lattice, originating *crystallinity.* The lack of at least one of these properties generates defects and structural disorder.

The Real Imperfect or Disordered Structure

All polymers, even in the single-crystal phase, contain disorder of various kinds. Most physical properties of these materials strongly depend on the type and degree of structural disorder. The overall lack of regularity can be described by the term disorder, whereas a single structural irregularity can be called a structural defect.

Defects can occur as single errors in the process of polymer formation, or they may be clustered together because of a failure of the physical conditions that determine a given structural property. The formation of defects may not always be an undesired and negative process. Because defects affect the chemical and physical properties of the material, introduction of particular defects may be the only means to attain a certain property.

Chemistry. Classification of defects can start with a faulty chemical process of linking of monomer units yielding an undesired linkage. If head-to-tail (HT) linking is desired, head-to-head (HH) inversion can take place, breaking chemical regularity (see also HEAD-TO-HEAD POLYMERS).

Another type of chemical defect is based on branching, ie, the linear polymer chain branches off into two independent chains that may grow at different rates. Chain branching may result in the formation of short or long side chains (Fig. 1**a**), or in cross-linking where two chains are joined (Fig. 1**b**). Cross-linking creates special mechanical properties, eg, the elasticity of rubber is a result of cross-linking (qv) (see also BRANCHED POLYMERS; ELASTICITY; NETWORKS).

The second step in building a polymer is the stereochemical arrangement of chiral chemical units, which yields polymers with a certain stereoregularity or tacticity. Syndiotactic defects embedded in an otherwise isotactic polymer chain or vice versa are commonly found (see also STEREOREGULAR POLYMERS).

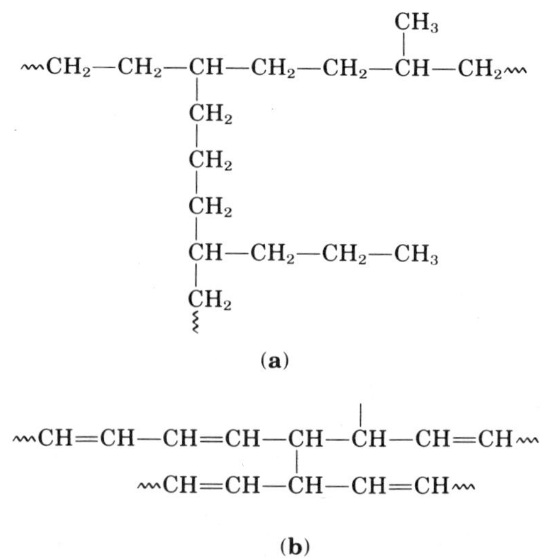

Figure 1. Chain branching or cross-linking that may occur as chemical defects during the polymerization reaction: (**a**) chain branching in polyethylene with the formation of tertiary carbon atoms at each branching point; (**b**) cross-linking in polyacetylene with the inclusion of sp^3 carbon atoms (tetrahedral structure) in a one-dimensional lattice of carbon atoms with sp^2 hybridization (trigonal, planar).

The clustering of tacticity defects can create two kinds of structures. If the tacticity defects cluster closely in certain sections of the polymer, stereoblocks are formed with a random length and distribution. However, if single-tacticity defects occur at random, regions of atactic polymers are formed. The degree of tacticity is inversely related to the concentration of stereochemical defects in a given polymeric material.

Conformation. Defects occur in the conformational coiling of polymer chains. Conformational defects occur in the process of crystallization of an otherwise chemically and stereochemically regular chain because of kinetic and energetic factors; these defects can be termed primary conformational defects. Conformational defects can also be forcefully induced because of the existence of chemical and or stereochemical defects; these defects can be termed induced conformational defects.

Defects in a Three-dimensional Lattice. In the final step of crystal formation, chains approach each other in a parallel way and find an arrangement in space that minimizes the total energy in the system. In reality, chains are finite and highly conformationally flexible; flexibility is determined by the physical conditions in which crystallization takes place.

Intrinsic Crystal Defects. In the process of crystallization from solution or melt, the long polymer chain folds like a ribbon, and the segments of conformationally regular chains then pack in a three-dimensional arrangement. Crystalline lamellae are thus formed. For a chemically and stereochemically perfect polymer chain, the lamella core can ideally be perfectly organized in a three-dimensional lattice. On the other hand, the shape and dimensions of conformational defects on the lamella surface depend on the crystallization conditions.

Crystalline and Amorphous Domains

Under normal crystallization conditions, segments of the conformationally flexible polymer chain approach each other and form bundles of parallel chains that coalesce into a three-dimensionally ordered domain. These ordered domains are called crystallites. Their interconnection is ensured by the fact that the chains are very long, and each chain may take part in the formation of several different crystallites. Inside each crystallite three-dimensional order exists; outside each crystallite the flexible chains wander and twist around in space. In the space between crystallites, three-dimensional periodicity is lacking, and amorphous surroundings can be described as a conglomerate of conformationally disordered or coiled polymer chains.

In polymer science, a material is strictly described as amorphous only if it lacks three-dimensional periodicity. The concept of amorphous is strictly the opposite of crystalline, for which three-dimensional order must exist.

Effect of Induced Conformational and Other Defects on Crystal Disorder

Around the site of an induced conformational defect, regular polymer chains can never arrange themselves in a regular and ordered three-dimensional packing.

If the concentration of induced conformational defects is not high, the chemically, stereochemically, and conformationally regular sections of the polymer chains can pack while confining the defects in the amorphous part. If the concentration of induced conformational defects is too high, the material cannot crystallize and remains solid but amorphous. Thus atactic polymers do not crystallize, as there is no possibility for the generation of highly ordered and oriented intermolecular forces to keep the molecules together in a crystalline arrangement.

Molecular Aspects of the Physical States of Polymers

Solution or Melt State. Chemical and stereochemical regularities or irregularities derive from a chemical reaction and exist in solution or melt states. Only physical treatments that break covalent bonds can change the type and number of chemical and stereoregular defects. The conformational shape of the macromolecule can, however, change according to the physical conditions, ie, the type and number of conformation defects can change. The laws of statistical thermodynamics determine the actual final shape of the molecule.

No three-dimensional order is expected to exist in the liquid or solution states. Indeed, from this viewpoint, the system is amorphous. Even if some clustering occurs in the liquid, real three-dimensional periodicity, even at very small distances, is unlikely.

Glassy State. The main difference between liquid and glassy states is in the conformational dynamics. The evolution with time, at various time scales, of the structure of the macromolecule in the glassy state is different from that in solution and in the liquid.

Smectic State. Despite different definitions of the smectic or paracrystalline state, the important structural feature is that chains generally have acquired one-dimensional periodicity or conformational regularity over long distances. In the smectic state, the system is still amorphous, as no three-dimensional periodicity exists.

Crystalline State. Sections of the chain that are chemically, stereochemically, and conformationally regular can acquire crystallinity, ie, three-dimensional order. Chemical and stereochemical defects create induced conformational defects, which together with primary conformational defects form the disordered and amorphous domains of the solid crystalline material.

Physical Methods for the Detection of Defects

Methods for detection of defects include density measurements, x-ray diffraction, calorimetric methods, and vibrational ir and Raman spectroscopy.

BIBLIOGRAPHY

"Organization of Macromolecules in the Condensed Phase," *Faraday Discuss. Chem. Soc.* **68** (1979).

G. Zerbi, R. Magni, M. Gussoni, K. Holland Moritz, A. B. Bigotto, and S. Dirlikov, *J. Chem. Phys.* **75,** 3175 (1981).

GIUSEPPE ZERBI
Politecnico Milano

DEFLASHING. See BLOW MOLDING.

DEFOAMING AGENTS. See ANTIFOAMING AGENTS.

DEFORMATION. See MECHANICAL PROPERTIES.

DEGASSING. See DEVOLATILIZATION.

DEGRADATION

In broad terms, degradation usually involves the chemical modification of the polymer by its environment, modification that is often (but not always) detrimental to the performance of the polymeric materials. Although the chemical alteration of a polymer is frequently a destructive process, for some applications degradation can be controlled and encouraged to achieve a specific property. Examples include the adjustment of the molecular weight of a thermoplastic in the extruder through controlled backbone scission and the controlled oxidative cross-linking of surface coatings. Embrittlement induced by weathering is a technologically successful way to destroy modified plastic film, in the form of either litter or agricultural mulch.

This article discusses chemical degradation of plastics and fiber-forming polymers, but excludes rubbers. Physical degradation processes [including aging (qv), annealing (qv), and crazing (qv)], deterioration attributable to microorganisms, and combustion of polymers are covered elsewhere (see also BIODEGRADABLE POLYMERS; FLAMMABILITY; RADIATION-RESISTANT POLYMERS; RUBBER, NATURAL; RUBBER RECLAIMING; STABILIZATION).

Mechanisms

From the extensive study of the degradation of many polymeric materials, a limited range of key degradative processes has now been established. From this range of reactions, it is possible to anticipate the types of processes likely to be important in the deterioration of any new polymer system.

Primary Bond-scission Reactions. All degradative processes originate from an initial bond-breaking reaction. This may represent the total extent of degradation, or it may be the prelude to a series of secondary chemical reactions leading to further bond scission, recombination, or substitution reactions. Examples of bond-cleavage processes are collected in Table 1.

Metal-catalyzed Degradation Processes. Many commercial polymers contain metallic compounds, as impurities or deliberately incorporated additives. The bulk of the latter type are pigments, whereas the former include polymerization catalyst residues or debris from processing equipment. A special case is the protective coating of metals, especially the polymer cladding of copper conductors. Polymerization catalysts such as transition metals (eg, Ti) may remain in polyolefins at 2–100 ppm, depending on work-up and catalyst efficiency. These residues have been implicated in both photo and thermal stability problems, although other work indicates that they play only a minor role.

Polymer Cross-linking and Branching. These processes may be beneficial or destructive, depending on the end use envisaged (see also CROSS-LINKING; BRANCHED POLYMERS). They are advantageous if heat deflection, creep, and softening temperatures are to be raised, eg, as in cable coating and high density polyethylene (HDPE) heat storage devices; shrinkage memory is required (shrink insulation); or molecular weights must be adjusted; but detrimental if flexibility and elongation are reduced.

Ozone Attack. The catastrophic effect of ozone (O_3) on unprotected, unsaturated rubbers is well known (see ANTIOZONANTS). Saturated polymers are also attacked by O_3, although at rates 10^4–10^6 times slower than rubbers. Table 2 provides relative ranking of O_3 susceptibility for polymer solutions.

Table 2. Ozone Sensitivity of Polymers

Polymer	Rate constant[a], L/mol·s
polybutadiene	6×10^4
polycarbonate	3
polystyrene[b]	0.3
polypropylene	8×10^{-2}, 2×10^{-2}[c]
polyethylene	5×10^{-2}

[a] Rate of O_3 attack on polymer in CCl_4 solution at 20°C unless otherwise noted.
[b] Initial product yields, based on O_3 consumption for solid PS at 20°C: peroxides, 35%; ketones, 18%; acids, 3%.
[c] PP powder.

Table 1. Common Bond-scission Reactions in Polymers

Degradative process[a]	Origin	Occurrence
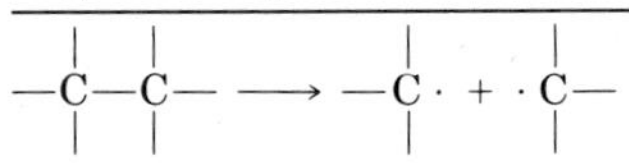	short-wave uv, γ, x-ray, electron irradiation	all polymers
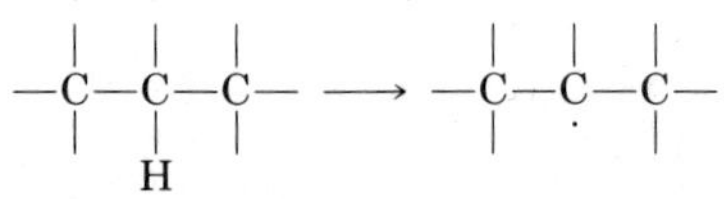	irradiation, heat, mechanical	
	uv exposure	ketone, amide, urethane, ester
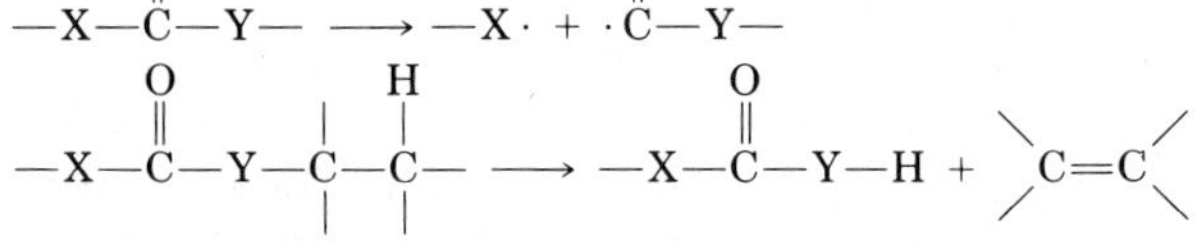	uv exposure	ketone, ester
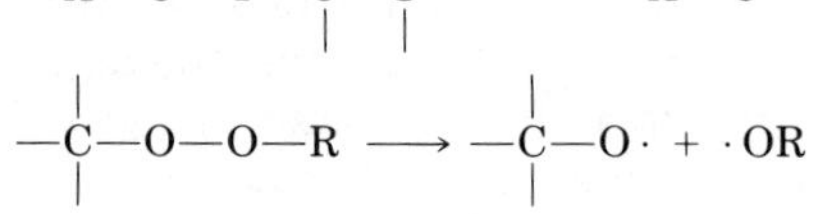	photo, thermal	any partially oxidized polymer
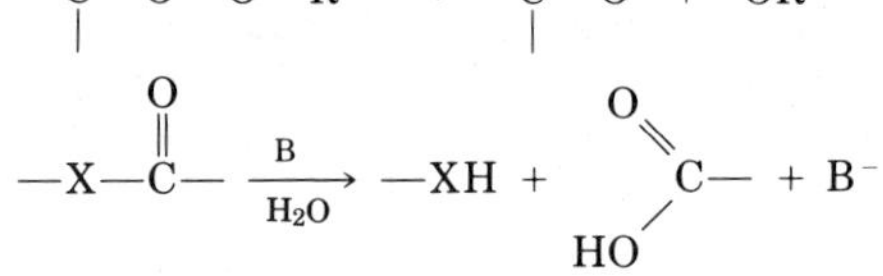	hydrolysis base-catalyzed	amides, urethanes imides, urea esters
$—X—Y \xrightarrow[H_2O]{H^+} —XH + HOY + H^+$	hydrolysis acid-catalyzed	amides, urethanes esters, siloxanes acetals
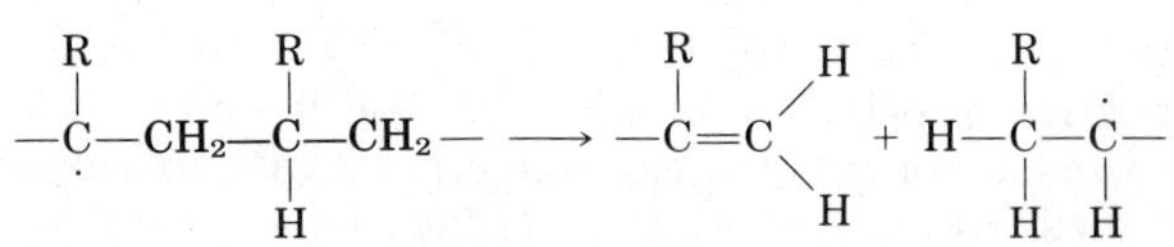	thermal scission	polyolefins
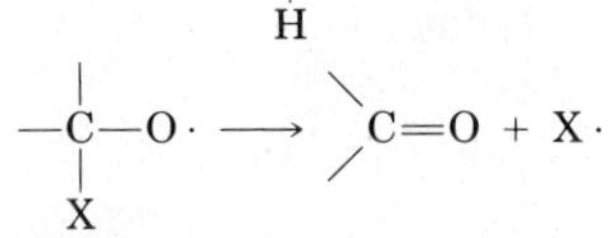	thermal scission	all oxidizing polymers
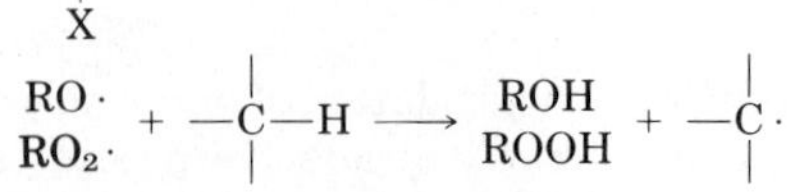	thermal attack	all oxidizing polymers
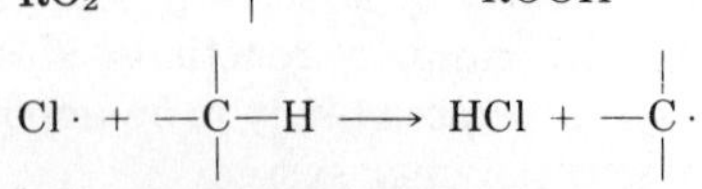	thermal attack	poly(vinyl chloride)

[a] R = alkyl or H; X,Y = hetero or carbon atoms.

Table 3. Degradation of Polymers in Air by High Energy Radiation[a]

Polymer	O_2 uptake[b], G_{O_2}	Average kinetic chain length[b]	Residual elongation at break, %	Residual elongation at break, Dose, kGy	Dose rate, kGy/h
polystyrene	1	5	50	530	0.15
			86	25	3.6
			107[c]	25	3.6×10^8
high density polyethylene	11	1.9	50	25	0.15
low density polyethylene			50	150	0.15
			77	25	3.6
			100[c]	25	3.6×10^8
poly(vinyl chloride)	39	15			
polypropylene	45	9	50	10	0.15
poly(ethylene oxide)	175	36			

[a] γ unless otherwise noted.
[b] G_{O_2} is the number of O_2 molecules consumed per 100 eV of absorbed radiation. Dose rate = 0.6 kGy/h (10 kGy = 1 Mrad).
[c] Electron-beam irradiation.

Mechanical Degradation. Electron-spin resonance spectroscopy has graphically demonstrated that stretching, grinding, milling, and any type of polymer shearing process produce free radicals as a result of main chain fracture (see also MECHANOCHEMICAL REACTIONS). In the absence of oxygen, various carbon radicals are generated from polyolefins, PMMA, etc.

Degradation by Ionizing Radiation. The effects of ionizing radiation on polymers are apparent in the deterioration of nuclear-power station components, sterilization of medical supplies, controlled cross-linking, and radiation-initiated graft polymerization (see CROSS-LINKING WITH RADIATION; RADIATION-RESISTANT POLYMERS). The interaction of high energy radiation with polymers leads to a cascade of very fast physical and physicochemical processes that culminate in the generation of free radicals and produce the detected products. Polymer sensitivity to radiation varies enormously (Table 3).

Degradation of Specific Polymers

A meaningful comparison of the intrinsic resistance of polymers from different generic classes to a particular degradative process is difficult because of the wide range of test conditions, the sensitivity of polymers to trace impurities and additives, and the diverse physical forms of the polymers, ie, thickness, shape, morphology, orientation etc. In addition, deterioration may result from exposure to two or more combined processes. For example, photodeterioration is often a composite of light attack and thermal oxidation. With these reservations, a qualitative comparison of polymers to different degradative processes is shown in Table 4. The outdoor weathering performance of polymers is compared in Table 5. Ozone and γ-irradiation sensitivities are compared in Tables 2 and 3, respectively. Resistance to thermal oxidation is a particularly difficult parameter to determine because of the key role of impurities in triggering the deterioration.

Table 4. Susceptibility of Unstabilized Polymers to Degradation

Polymer	Resistance to degradative process[a]: Thermal oxidation	Photo-oxidation, weathering	Ozone	Hydrolysis	γ-Oxidation
polyethylene	p	p	e	e	f
polypropylene	vp	vp	e	e	p
polystyrene	f	p	e	e	g
poly(methyl methacrylate)	g	e	e	g	f
poly(tetrafluoroethylene)	e	e	e	e	vp
polyamide[b]	f	f		f	f
polyacrylonitrile	p	g	e	g	f
poly(vinyl chloride)	vp	p	e	e	p
poly(ethylene terepthalate)	g	g	e	g	g
polyoxymethylene	p	p	f	f	vp
polycarbonate	f	p	g	g	g
poly(phenylene oxide)	vp	f	g	g	g
poly(ester urethane)	f	f	g	g	f
poly(ether urethane)	p	p	g	f	f
poly(*m*-phenylene isophthalamide)	e	vp	e	g	e
poly(*p*-phenylene terephthalamide)	e	vp	e	g	e

[a] Key: e = excellent, g = good, f = fair, p = poor, and vp = very poor.
[b] Nylon-6 and -6,6.

Table 5. Uv Cutoff Values and Weathering Lifetimes of Common Polymers

Polymer	Cutoff[a], nm	Outdoor lifetime[b], yr
Nonabsorbers[c]		
polyethylene	<180	0.5–1.0
polypropylene	<180	0.2
polyoxymethylene	<210	0.3
poly(vinyl chloride)	~220	0.5
poly(methyl methacrylate)	~240	>20
polyamides[d]	~240	3–4[e]
polyamide[f]	~240	~8
polystyrene	~270	~0.1[g]
Inherent absorbers[c]		
polycarbonate	~280	0.5
poly(phenylene oxide)	~280	<0.5
polyurethane[h]	~280	~2
poly(ethylene terephthalate)	~310	~3
poly(*m*-phenylene isophthalamide)	~340	~0.2
poly(*p*-phenylene terepthalamide)	~350	~0.3

[a] Wavelength at which absorbance of 10-μm film reaches 1.0. All values are approximate.
[b] Unstabilized grade, normal impurity levels. Time to ~50% loss in tensile, elongation, or impact properties.
[c] Absorption relative to solar limit of 280–285 μm depending on location.
[d] Nylons -6, -6,6, and -6,10.
[e] Probably mainly hydrolytic failure.
[f] Nylon-11.
[g] Yellowing.
[h] MDI-based; MDI = 4,4′-methylenebis(phenyl isocyanate).

Table 6. Thermal Deterioration of Polymers: Maximum Use Temperatures[a]

Polymer generic type	Film thickness, mm	Maximum use temperature, °C
poly(vinyl chloride)	2.0	50
polyethylene	1.5	50
polyoxymethylene	0.7	50
poly(methyl methacrylate)	1.5	50
polystyrene	3.1	50
poly(phenylene oxide)	1.5	50
polyamide	0.7	65
polycarbonate[b]	0.7	65
epoxy resins	3.0	90
silicones	0.7	105
poly(ethylene terephthalate)	0.2	105
alkyd resin	1.5	130
polytetrafluoroethylene	0.9	150
phenolic resin	2.0	150
polyimide	0.1	200
aramid[c]	0.2	220

[a] Recommended temperature at which 50% of the original dielectric strength, tensile, and impact properties are retained for 11,000 h under low continuous stress. In some cases, stabilizers may be present. Underwriters Laboratories data.
[b] Bisphenol A.
[c] Aromatic polyamide.

A qualitative, but nevertheless useful, comparison of heat resistance and thermal oxidation sensitivity comes from "maximum use temperature" values. These values are derived by Underwriters Laboratories and are recommended, maximum-use temperatures at which a 50% decrease in properties occurs in 11,000 h of exposure. These temperatures are listed in Table 6 and must be regarded as upper limits for completely stabilizer-free polymers because the precise composition is not given.

Polyolefins. Polyolefins are prone to degradation by thermo- and photooxidative processes, as well as the less common effects of γ-radiation and potent acidic media. In all uses, the rate of attack on PP is greater than on PE.

Poly(vinyl chloride) (PVC). If PVC were to be discovered today, its commercialization would be unlikely because of the extreme sensitivity of the polymer to both heat- and light-induced degradation. The widespread use of PVC is a tribute to the stabilization technology developed over the last 30 years.

Polytetrafluoroethylene (PTFE). PTFE has outstanding thermal, photochemical, and hydrolytic stability largely because of the strong C—F bond (ca 460 kJ/mol or 110 kcal/mol). Only ionizing radiation can cause rapid degradation of this polymer. In fact, PTFE is ca 10^3 times more sensitive to degradation by ionizing radiation than PE or PP. PTFE embrittles during irradiation in the presence of oxygen as a result of direct chain scission and the β-scission of fluoroalkyoxyl radicals.

Polyethers. Polymers with ether groups in the backbone are used as engineering plastics for molding, whereas polyethers with pendent ether groups are more often used as protective coatings or adhesives. Polyethers are prone to oxidation as well as hydrolytic attack.

Polyamides. Synthetic polyamides are widely exploited as fibers and, to a lesser but growing extent, as moldings and films. They are sensitive to heat- and light-induced oxidation, especially if a TiO_2 delustrant is present, and commercial resins are seldom available without incorporated manganese and copper stabilizers. Although the polyamides, silk (qv), and wool have good resistance to dilute alkalies, they are attacked by acids, oxidizing agents, and hypochlorite solutions. Wool and silk are both discolored by heating above ca 120°C and tenderized by light exposure.

Poly(methyl methacrylate) (PMMA). PMMA is used in extrusions as clear sheets and in moldings and protective coatings and adhesives. The resin has excellent weathering resistance and is only attacked by concentrated acids or alkalies.

Polysiloxanes. These polymers are frequently used as rubbery sealants, adhesives, or coatings where extreme resistance to weathering or high temperatures is required. They are attacked slowly under prolonged or extreme hydrolysis conditions.

Cellulose. As an abundant natural product, cellulose is used in many forms, including fibers and films. Depending on the presence of impurities or dyes, cellulose may be very sensitive to weathering, heat, and solutions of acids, alkalies, or oxidizing agents.

Poly(vinyl alcohol) (PVA). PVA forms hydrophilic fibers or films and is also used for coatings and adhesives or oil-resistant seals. The polymer is very resistant to weathering, but is degraded by strong acids (see also VINYL ALCOHOL POLYMERS).

Polacrylonitrile (PAN). PAN is used in fiber production, especially its copolymers, to produce wool-like filaments (see also ACRYLIC FIBERS); it also has valuable barrier properties for containers (see also ACRYLONITRILE POLYMERS). The fiber has high uv resistance, but discolors on heating above 150°C. The controlled degradation of PAN is an important route to carbon fibers. Although decomposed by concentrated alkalies and acids, PAN is resistant to dilute reagents.

Polystyrene (PS). Polystyrene has been widely used for many years as a general-purpose thermoplastic in the form of moldings, electrical insulation, packaging, etc. The thermal and photoinitiated degradation of PS is one of the most studied areas of polymer deterioration, both because of the practical importance of the polymer and the theoretical implications of the processes (see also STYRENE POLYMERS).

Poly(ethylene terephthalate) (PET). PET is widely used as a textile fiber and, to a lesser extent, as an engineering plastic, in moldings, and as dielectric films. The most important, practical degradative processes are hydrolysis, slow photooxidation, and cyclodepolymerization (see also POLYESTERS).

Bisphenol-A Polycarbonate (PC). Polycarbonates (qv) are widely used as engineering thermoplastics, especially where transparent, impact-resistant articles are required. PC is discolored and embrittled by light and can be sensitive to hydrolysis and heat.

Aromatic Polyamides. Fully aromatic polyamides (aramids) are used as film, paper, fiber, and coatings because of their heat resistance and the high modulus and tensile strength of para isomers. They are prone to photodegradation.

Poly(phenylene oxide) (PPO). PPO [poly(2,6-dimethyl-1,4-phenylene oxide)] is used when blended with other polymers to adjust its impact properties. The blended polymer is usually a styrene–butadiene polymer in which the rubber component is especially sensitive to oxidative degradation, which can make the blend brittle.

Poly(phenylene sulfide) (PPS). PPS displays extreme resistance to aggressive liquids (better than polyamides, PPO, PC, and polysulfones) and γ-irradiation. It has high thermal stability although it is slowly degraded by exposure to sunlight.

Polyurethane (PU). The polyurethanes are a broad class of chemically divergent polymers. The two main subgroups are poly(ether urethane)s, comprising soft (flexible) polyether segments joined by hard (rigid) urethane segments; and poly(ester urethanes), where soft aliphatic or aromatic polyester segments are joined by urethane links. In general, they undergo degradation by hydrolysis, thermal, and photoreactions as well as chain oxidation.

BIBLIOGRAPHY

B. Ranby and J. F. Rabek, *Photodegradation, Photo-oxidation and Photostabilization of Polymers: Principles and Applications,* Wiley-Interscience, New York, 1975.

D. L. Allara and W. L. Hawkins, eds., "Stabilization and Degradation of Polymers," *Adv. Chem. Ser.* **169** (1978).

A. Davis and D. Sims, *Weathering of Polymers,* Applied Science Publishers, Ltd., London, 1983.

D. J. Carlsson and D. M. Wiles, *J. Macromol. Sci. Rev. Macromol. Chem.* **14,** 65 (1976).

D. J. CARLSSON
D. M. WILES
National Research Council of Canada

DEGRADATIVE CHAIN TRANSFER.

See CHAIN TRANSFER.

DEGREE OF POLYMERIZATION

The degree of polymerization (DP) may be defined as the average number of base units per molecule if the molecules are composed of regularly repeating units, or as the average number of monomeric units (mers) per molecule. These definitions are not necessarily equivalent. It should also be noted that, except in the case of polymers that are homogeneous with respect to molecular weight, the definition is not precise unless the type of averaging is specified.

BIBLIOGRAPHY

IUPAC, *J. Polym. Sci.* **8,** 257 (1952).

DEGREE OF SUBSTITUTION

The degree of substitution (DS) of a polymer is the (average) number of substituent groups attached per base unit (in the case of condensation polymers) or per monomeric unit (in the case of addition polymers). The term has been used mainly in cellulose chemistry where each anhydroglucose (β-glucopyranose) unit has three reactive (hydroxyl) groups; degrees of substitution may therefore range from zero (cellulose itself) to three (fully substituted cellulose).

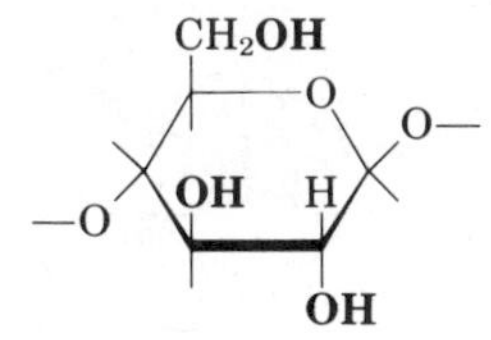

DELAMINATION.

See ADHESION AND BONDING; COMPOSITES.

DENDRITIC POLYMERS

Since the turn of the century, a substantial number of synthetic polymer systems have evolved, replacing natural materials in a variety of ways. Most notable are those involving highly branched precursors that lead to three-dimensional gels and networks. These systems are broadly recognized as thermoset types and generically include such classes as epoxy resins;

polyesters; polyurethanes; phenolic, alkyd, and melamine resins; cross-linked styrene–acrylic systems; latices; and vulcanized rubber. Their superior properties have gained wide acceptance in such diverse applications as engineering and construction materials, coatings, adhesives, optics, extraction–absorption resins, and, more recently, surgical and dental prosthetics. Annual consumption accounts for billions of dollars in commerce worldwide.

Two general categories of branching theory have evolved to explain the *in situ* formation and behavior of dendritic macromolecular species leading to infinite network systems: classic theory utilizes graphlike models not associated with dimensionality of the network space. It is a collection of classic hypotheses, including statistical and kinetic treatments with a chemical intermediate perspective. Percolation theory involves simulation of network construction according to its dimensionality. It is a more contemporary physics approach for converting monomer units to dendritic macromolecules based on lattice and percolation theories.

At least three major dendritic categories (I–III) may be visualized to account for the transformation of monomer units to infinite network systems. These categories are derived from branch cells or network substructure terms (models) as defined by various workers in their efforts to analyze simplified components of the network mathematically.

Considering analogous open-branched systems void of any looping or closure, it is easy to visualize species or components such as starbranched oligomers I(*), branch cells I(▲), branch cell assemblies II(▲), or dendritic species such as II(*) (Fig. 1). Other dendritic species might include ensembles of species II that could be of two dimeric types (eg, open dendritic species III) or looped dendritic species (III′). Higher multiples of micronetworks (eg, IV with $n = 7$) may possess both opened and looped domains. These assemblies are referred to as micronetwork clusters in much the same spirit as the formation of macroscopic solids by clustering atoms or molecules has been described.

It follows that these species represent a sampling of the local connectivity and topology patterns leading to macroscopic networks. Therefore, in the synthetic context of constructing an infinite network, these topological species may be thought of as starbranched (I) and dendritic-branched (II, III, III′, IV) multifunctional oligomers (Fig. 1). In this sense, the termini (*) may be viewed as organic moieties (functionalities) or loci from which to replicate the branch juncture multiplicity as well as connecting chains to the next tier of branch junctures. Through these terminal loci, dimensional advancement to open symmetrical micronetworks (such as A–E in Fig. 2) might be envisioned. More likely, asymmetric (defective) variations of A–E are formed, unless specific, controlled conditions are employed. Concurrently or alternatively, covalent bridging of these micronetworks might occur to produce open or looped subcritical network clusters (such as IV in Fig. 1), which may be compared to the finite clusters previously described. Growth of these clusters to a critical state ultimately leads to a gel. The collection of dendritic species I–IV is referred to as the sol

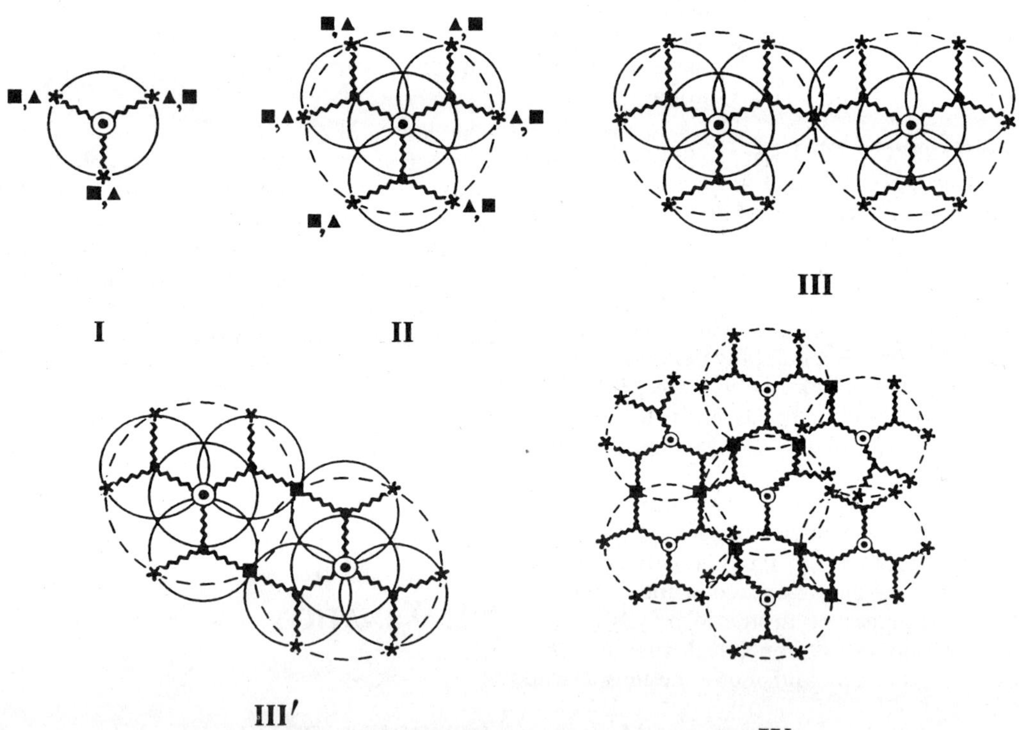

Figure 1. Two-dimensional projections of branch cells. ⊙, Initiator core (reference point). Connectivity: *, unreacted moieties (chemical functionality); ▲, attachment to a starbranched oligomer or starburst dendrimer; ■, attachment to a micronetwork or infinite network. Topology: —, branch cell or network cell; ---, starburst dendrimer or micronetwork; ∿, branch segments. I: *, starbranch oligomer, ▲, branch cell, ■, network cell; II: *, starburst dendrimer, ▲, branch cell assembly, ■, micronetwork; III: starburst dendrimer dimer; III′: looped micronetwork dimer; IV: micronetwork cluster (derived from II, partially looped).

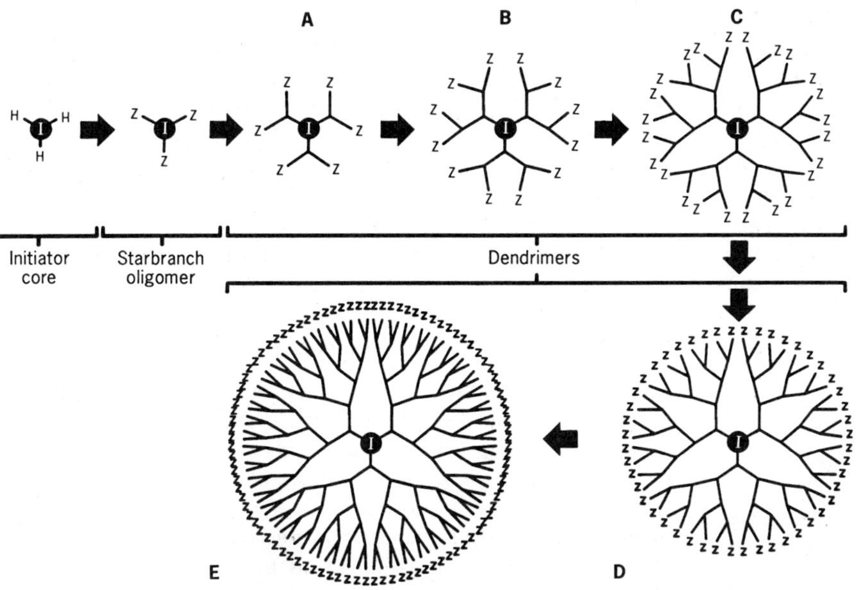

Figure 2. Two-dimensional projections of starburst dendrimers as a function of generation (A) → (E). Open micronetworks: starburst network polymers as a function of generation $[(A)_n \rightarrow (E)_n]$ (generations 1–5).

phase. The role of these dendritic species in the formation of infinite networks as a function of dimension and degree of branching may be envisioned as in Figure 3.

Attempts to monitor these transient dendritic species (I–III, Fig. 3), either *in situ* or by quenching experiments, have resulted in limited success for a variety of reasons. However, it remains of utmost importance to understand the significance and interplay of these species in the development of network properties for both theoretical and commercial purposes. This article surveys the development of a new class of dendritic macromolecules that are functional examples of the dendritic species continuum (I–III) (Fig. 3). These highly branched macromolecules are referred to as starburst dendrimers or polymers and may be thought of as ideally functionalized or branched assemblies and micronetworks (II) or clusters (IV) (Fig. 1). They are produced as stable, isolatable intermediates and may be derived from a wide variety of monomer units. Their construction involves, first, the synthesis of a branch cell (I), followed by reaction of an initiator core (I) where f (functionality) = 3 or greater with suitable monomer units to produce usually a starbranched oligomer. Additional monomer units are introduced in a controlled, concentric fashion to give discrete stages

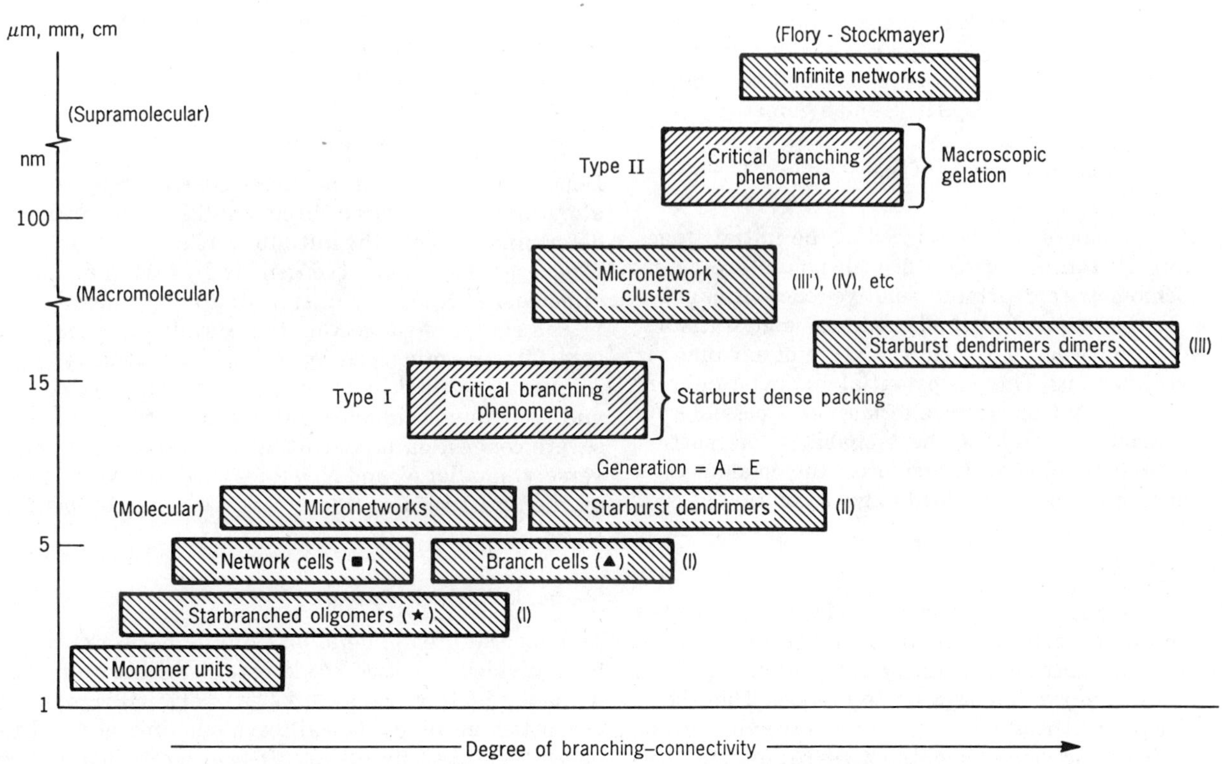

Figure 3. Comparison of molecular → macromolecular → supramolecular network sized (nm) vs degree of branching–connectivity.

(generations) of dendrimers, which are illustrated as A–E in Figure 2. Choices in initiator core (I) and monomer units can be used to control shape, size, topology, and the surfaces of these branch-cell assemblies. Advancement to higher generations can introduce precise, incremental increases in their three-dimensional size. Ideal concentric growth of generations around an initiator core may occur up to a limiting generation. As this limiting generation is approached, surface congestion forces a new (Type I) critical branching phenomenon that precedes a new phenomenon, referred to as starburst dense packing. Higher degrees of connectivity (branching) are attained by bridging either the critically branched dendrimers or their precursors to form a continuum of microclusters (IV). Proceeding to these higher degrees of branching ultimately leads to the classic Flory-Stockmayer (Type II) critical branching state that is recognized as penultimate to macroscopic gelation (infinite network formation). These dendrimers may be used as precision, macromolecular building blocks to produce a wide variety of micronetwork clusters (IV) up to and including infinite network development.

These building blocks are used in the same manner as described for the formation of macroscopic solids by clustering atoms or molecules. The dendritic species in this continuum are referred to as starburst dendrimers or clusters and are illustrated as $(A)_n \rightarrow (E)_n$ in Figure 2. The resulting infinite networks are referred to as starburst network polymers.

Starburst Branching Topology

Classic dendritic networks can be seen as the product of growth from a point of initiation. After a span of linear growth, a branching event takes place followed by continued linear growth for each branch. These new linear segments, in turn, develop branches when conditions favor the branching event, and the process repeats itself. The topology of the network formed by this process is determined by a limited number of parameters that set the distance between branch points and the characteristics of the branching, multiplicity, and symmetry.

In condensation polymerizations of the $(\mathrm{A}{-}\!\!<^{\mathrm{A}}_{\mathrm{A}} + \mathrm{B{-}B} \longrightarrow)$ type, dendritic oligomers may be formed at the initial stages of the reaction. Distances between branch points are determined by stoichiometry of reactants, relative reactivities of the acid groups in the sequential condensations, and statistics. Multiplicity of branching is 3 with a mixture of asymmetric and symmetric junctions. There is potential for deviation from dendritic growth in that macrocycle formation is possible. As the extent of reaction increases, the probability for reaction between the dendritic oligomers increases; the condensation of the oligomers may ultimately lead to gel formation.

Dendrimer Synthesis

Starburst dendrimers have been synthesized with a variety of chemical interiors, branch multiplicities, and surface groups. Generally, the construction of interior generations involves time-sequenced step-growth propagation techniques. These step-growth sequencing methods are of two types, involving either differentiated, polyfunctional reagents, excess reagents, or protection–deprotection. With these methods dendrimers are produced with various core and branch juncture multiplications. Examples include starburst polyamidoamines and amidoalcohols (arborols) by the first method and starburst polyethers, polythioethers, polyethylenimines, and polyamides by the protection–deprotection method. In all cases branch cells are introduced in a concentric, radial manner around the initiator core by either *in situ* synthesis or by coupling preformed branched reagents in a sequenced fashion. The reaction steps must be essentially quantitative to assure branching ideality and must be performed under conditions that minimize intramolecular looping or intermolecular bridging.

Dendrimer Characterization and Analysis

The structural subtleties represented in these dendrimer topologies required a variety of analytical methods for unequivocal verification. No single method could be used as conclusive proof, but rather accumulative corroboration offered by a variety of convergent analytical methods was necessary. Elemental composition was determined by C, H, N analyses and mass spectroscopy fragmentation patterns. Low angle laser light scattering, chemical ionization, fast atom bombardment mass spectroscopy, and vapor phase osmometry were used to establish predictable molecular weights vs generation. Homogeneity was assessed by size exclusion chromatography coupled with low angle laser light scattering. Interior and end group determination were done by infrared, ^{15}N, ^{13}C, and ^{1}H nmr spectroscopy, titrimetry, and stoichiometry with various reagents. ^{13}C and ^{1}H nmr (relaxation times, integral ratios vs generation), rheology studies, and electron microscopy were used to observe topological features. Electron microscopy, intrinsic viscosity measurements, and size exclusion chromatography were used to evaluate dimensions.

These methods have been used extensively for structure verification of starburst polyamidoamines, polyethers, and polyethyleneimine. In many cases, small molecule model systems were used for process optimization, defect identification, and stoichiometry studies.

Critical Physical Properties of Dendrimers

Dendrimer shapes can be nicely controlled by selecting initiator cores with desired directionalities. Dendrimer sizes are determined by both the initiator core size used and the number of generations advanced concentrically from the core. The three-dimensional character of starburst dendrimers is determined by the extent of congestion that develops around an initiator core. This is influenced by the core multiplicity (N_c), branch-segment length (l), and branch multiplicity (N_b). Higher N_c and N_b values combined with smaller (l) values can lead to severe congestion properties at even lower generations. Conversely, smaller N_c and N_b values combined with longer branch segments (l) result in less congested, more open dendrimer types.

Starburst Polymer Networks

The general area of gelation (infinite network) formation has been divided into two broad categories according to the permanency of their cross-links, namely irreversible networks characterized by (chemical) covalent linkages and reversible networks linked by (physical) van der Waals and coulombic associations. Covalent polymeric networks are completely disordered. Among the reversible networks, lamellar structures are either well-ordered gel mesophases; polymeric networks that are largely disordered but with regions of local order through physical aggregation; or particulate disordered structures.

Within these broad dimensions, Flory has described several specific network types and commented on the degree of ordering involved in their construction. The category of irreversible gels is, of course, dominated by the many classic thermoset systems involving chemical cross-links. Examples of the weak (reversible) gels are found primarily in biopolymers or colloids and have been reviewed.

Starburst dendrimers have been found to form both types of networks. They may be thought of as stoichiometric, three-dimensional, oligomeric building blocks, albeit with very specific sizes, shapes, and a variety of surface moieties. The individual dendrimers may be viewed as monomeric entities that can be covalently bridged to form various irreversible micronetwork clusters (Fig. 2) on the way to gelation. Depending on their surface moieties, dendrimers may also behave as equilibrating nanoparticles to form reversible clusters and networks with varying degrees of order. In either case, these supramolecular lattices [ie, $(A)_n - (E)_n$, Fig. 2] are referred to as starburst polymer networks.

Influence of Surface Moieties and Dendrimer Shape. Earlier discussion described the facile and varied methods by which dendrimer size and shape can be controlled. A multitude of specialized morphologies may result from the self-assembly of these microdomains to form reversible networks as a function of their surface moieties. In general, hydrophobic surface moieties have led to various micronetwork clusters (micelles), liquid crystal assemblies, or infinite network gelations. In some cases, these assemblies are shape-dependent. On the other hand, hydrophilic (ionic, saltlike) surfaces have led to a number of surface-associated micronetworks, micronetwork clusters, or microcrystalline networks, depending on the extent of surface congestion.

Influence of Surface Congestion on Network Formation. Recent work using computer-simulated molecular graphics has shown that dendrimer surface congestion in the starburst PAMAM series parallels the critical physical data described earlier. The first three generations are described as open morphology to a more closed symmetrical (spheroidal) shape at generations 3–4.

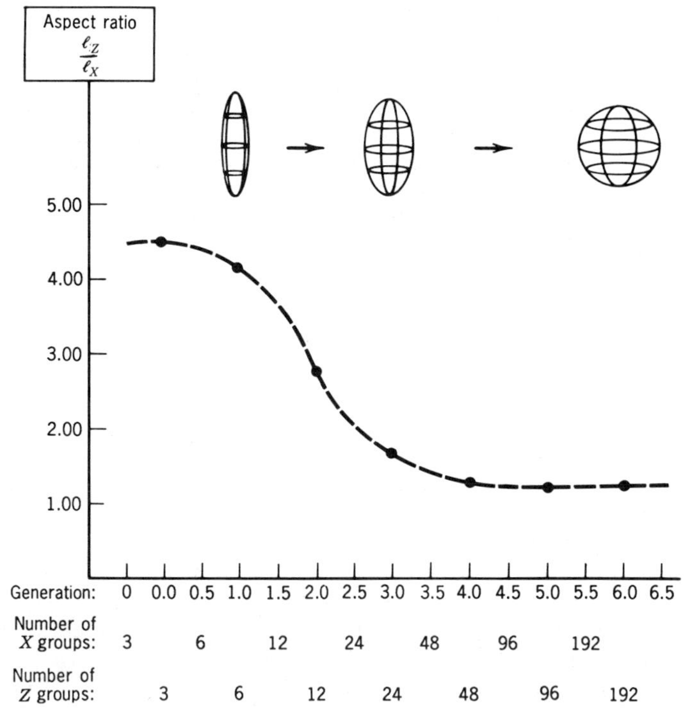

Figure 4. Aspect ratios ℓ_Z/ℓ_X derived from computer molecular modeling as a function of generation or number of surface groups designated X or Z.

Representative structures and aspect ratios associated with these morphological changes are shown in Figure 4. The extent of surface congestion, caused by the tethering of the surface groups around the initiator core, can strongly influence the molecular shape and consequently the association (crystallization) properties of these moieties. Depending on surface moieties involved, dendrimer concentrations, temperature, and other parameters, intradendrimer association or interdendrimer assembly to a variety of reversible network systems has been observed.

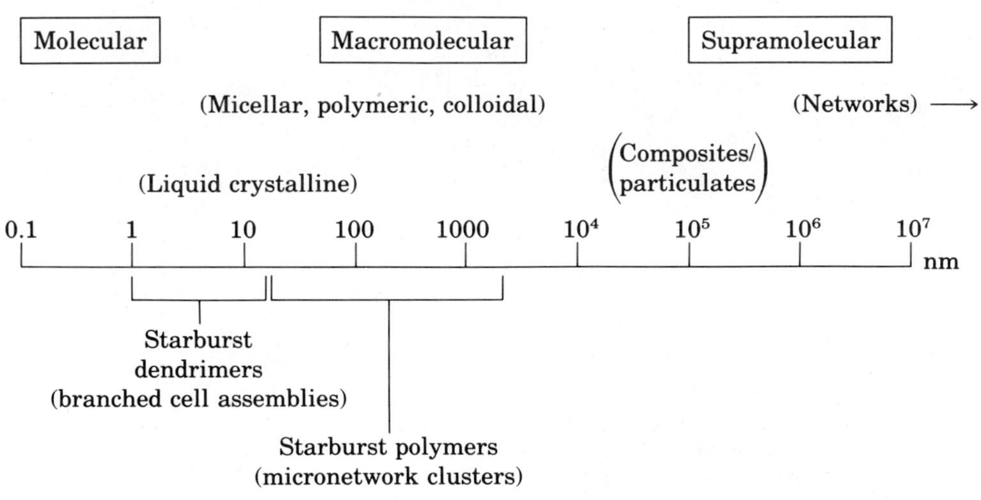

Figure 5. Dimensional hierarchy from molecular level species to infinite network, which indicates dimensional positions of starburst dendrimers and starburst polymers.

Classification of Dendritic Macromolecular Networks

The growth of dendritic macromolecules, referred to as finite clusters in the sol phase (Fig. 2), may be thought of as geometric aggregations of various branch cells or network cells as defined by Flory. Beginning as molecular species, they advance with growth through the dimensional heirarchy, shown in Figure 5, to macromolecular, supramolecular, and ultimately to infinite network systems. Traditional network-forming systems (eg, epoxy resins, urethanes, polyesters) no doubt progress through this growth process in a random fashion, thus becoming a collection of unequally segmented Gaussian chains between *f*-functional branch junctures (cross-links) that are more or less ideally branched. In contrast to the starburst dendrimers and their clusters, they are generally unordered, exhibiting primarily amorphous character. Examination of the spatial neighborhood surrounding a network cell in one of these networks shows that many other neighboring branch junctures can interpenetrate the domain defined by its first tier (generation) of branch junctures (Fig. 6). Because of these lengthy, Gaussian branch-juncture chains, as many as 25–100 branch junctures may interpenetrate this domain in a typical elastomeric network. It becomes obvious from Figure 7 that interconnection of branch junctures as a function of generations or tiers evolving from the origin of this network cell produces totally unordered zones or arrays of branch junctures. In fact, model networks that possess ideally branched, equally segmented Gaussian-chained networks do not show any order beyond the first tier by x-ray scattering. Hence, these dendritic types are referred to as branching Type I (unordered, ideal branched) (Fig. 7).

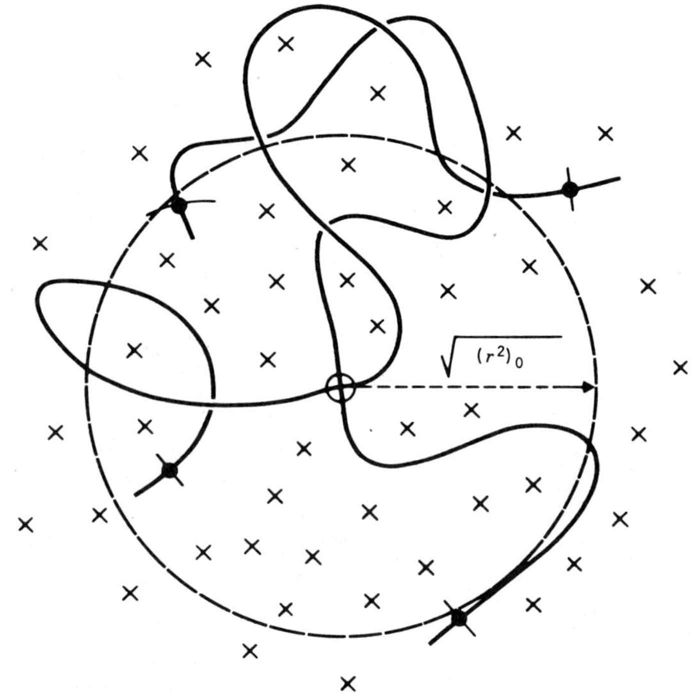

Figure 6. Neighboring network cell junctions (×) surrounding a reference network cell junction (+) together with its four topological neighbors (+) in an $f = 4$ network system. Courtesy of *J. Br. Polymer Chem.*

The PAMAM and polyether dendrimer series that are constructed from short, non-Gaussian branch segments with symmetry elements in their branch junctures may be good representatives of the branching type III, namely ordered, ideally branched dendritic macromolecules.

An intermediate example of branch-cell order is that represented by Denkewalter dendrimers. In this case, the branch segments are of unequal length. Their segment length distribution from initiator core I to surface groups should follow a

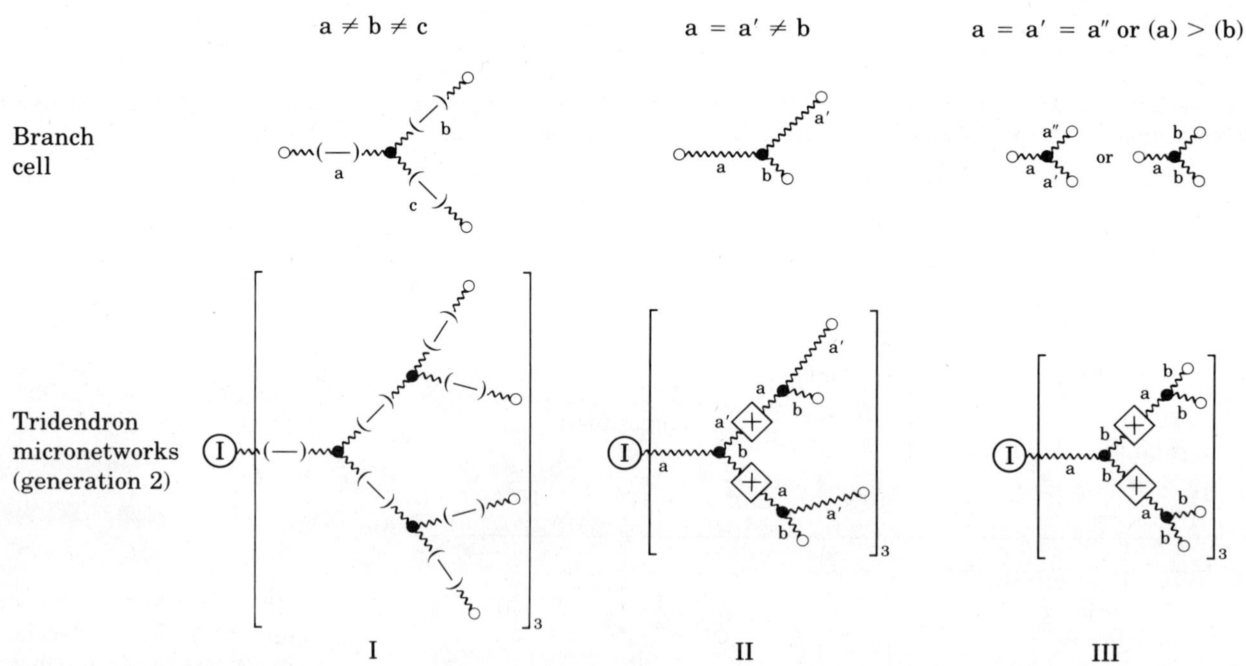

Figure 7. Branching types within the generic family of dendritic macromolecules. Type I: unordered, ideal branched; Type II: semi-ordered, ideal branched; Type III: ordered, ideal branched. (—), Gaussian branch segments; ~, non-Gaussian branch segments; ⊕, branch coupling.

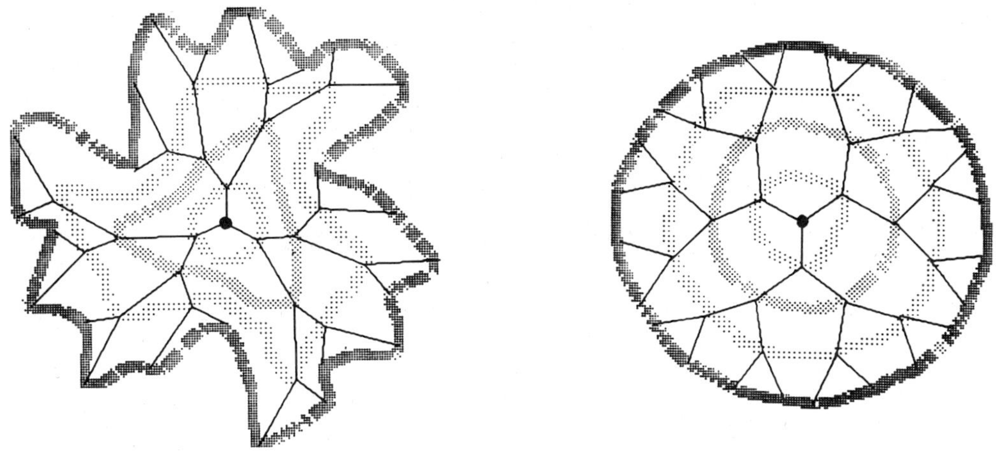

Figure 8. Comparison of concentric order surrounding a reference branch cell juncture in dendritic macromolecules of branching Type II and Type III.

binomial expansion as a function of generation. The resulting zones derived by interconnecting the branch junctures at each generation gives a mathematically driven geometric pattern that suggests semi-order compared to Types I and III dendrimers. This intermediate Type II, semi-ordered, ideally branched dendritic macromolecular class is constructed from short non-Gaussian branch segments with unequal length but possessing branch junctures with certain symmetry properties as described in Figure 8. A comparison of branching zones surrounding an initiator core I in the case of Types II and III dendrimers is shown in Figure 8.

BIBLIOGRAPHY

D. A. Tomalia, V. Berry, M. Hall, and D. M. Hedstrand, *Macromolecules* **20,** 1164 (1987).

D. A. Tomalia, M. Hall, and D. M. Hedstrand, *J. Am. Chem. Soc.* **109,** 1601 (1987).

S. M. Aharoni, C. R. Crosby III, and E. K. Walsh, *Macromolecules* **15,** 1093 (1982).

D. A. Tomalia, G. E. Kiefer, A. M. Naylor, and W. A. Goddard III, *J. Am. Chem. Soc.* **111,** 2339 (1989).

DONALD A. TOMALIA
DAVID M. HEDSTRAND
LARRY R. WILSON
The Dow Chemical Company

DENSITY. See CRYSTALLINITY DETERMINATION; DILATOMETRY.

DENSITY-GRADIENT SEDIMENTATION. See ULTRACENTRIFUGATION.

DENTAL APPLICATIONS

Dentistry has led the health-science field in taking advantage of the rapidly expanding family of polymeric materials, although the volume of dental resins used is insignificant compared with the total consumption of plastics. The search for desirable preventive and restorative dental materials has been limited to a few classes of materials because of strict requirements. Products of dental research include epoxy resins and amine–peroxide initiator systems, specially developed for rapid cures. Similar cures of polyesters and acrylics, activated chemically at ambient temperatures, have found diverse uses as cements in orthopedic and neurosurgical applications.

The largest volume of plastics is consumed in the construction and repair of prosthetic devices such as denture base, repair and relining materials, and plastic teeth. Polymers also play an important role in operative dentistry as ingredients of composite restorative materials, cements and adhesives, cavity liners, and as protective sealants for pits and fissures. Elastomeric compositions are employed to yield exact impressions of oral structures. Resilient and aesthetic devices restore soft-tissue defects of external features. Mouth protectors not only prevent injury to teeth, but also protect the head and neck of those participating in contact sports. Other devices using plastics include patterns for metal-inlay and partial denture framework castings, contoured impression trays, orthodontic and periodontic splints, temporary space maintainers and bite plates, obdurators for cleft palates, and oral implants.

Prosthetic Materials

Prosthetic materials include denture bases (Table 1) [heat-cured resins, room temperature cured resins (self-cure), fluid

Table 1. Physical Properties of Denture-base Resins

Property	Poly(methyl methacrylate)	Vinyl–acrylics
tensile strength, MPa[a]	48–62	51
compressive strength, MPa[a]	75	69–75
elongation, %	1–2	7–10
elastic modulus, MPa[a]	3.8×10^3	2.3×10^3
impact strength, N·m[b]	1050	3150
transverse strength, MPa[a]	41–55	41–55
flexural strength, MPa[a]	83–117	69–110
Knoop hardness	16–22	14–20
polymerization shrinkage, %	6	6
24-h water sorption, %	0.3–0.4	0.07–0.4

[a] To convert MPa to psi, multiply by 145.
[b] N·m = J; To convert J to ft·lbf, multiply by 0.74.

Table 2. Typical Range of Physical Properties of Resin Restoratives

Property	Unfilled PMMA	Microfilled composite	Conventional and filled hybrid composite
inorganic content, wt %		33.2–50.9	70.4–80.2
compressive strength, MPa[a]	55–76	221–330	127–350
tensile strength, MPa[a]	14–28	28–56	28–63
water sorption, mg/cm^2	2.03	0.94–2.20	0.13–0.74
polymerization shrinkage, vol %	5.2–8.0	1.9–5.8	1.2–5.3

[a] To convert MPa to psi, multiply by 145.

Table 5. Estimated U.S. Resin Consumption for Restorative Dental Materials, 1984[a]

Appliance	Units made per yr, 10^6	Wt of resin per unit, g	Total resin used, t
full dentures	4.5	25	112.5
partial dentures	4.1	6	24.6
acrylic teeth	14	0.5	7.0
denture-reliner and repair resins	8.0	4	32.0
filling materials	39	0.15	5.9
others[b]			75.0
Total			*257.0*

[a] Estimates based on 1982 data.
[b] Dental implants, mouth protectors, maxillofacial prostheses, etc.

or pour-type resins, injection-molding resins, and light-cured resins], special-purpose resins, repair resins, denture reliners (eg, resilient liners and tissue conditioners), crown and bridge resins, plastic teeth, mouth protectors, and maxillofacial prosthetic materials.

Restorative Materials

Restorative materials (Tables 2 and 3) include filling materials (eg, resin cements and dental composites), polyelectrolyte-based restoratives (eg, polycarboxylate cements and glass-ionomer cements), sealants, and adhesives.

Dental Impression and Duplication Materials

Impression materials (Table 4) include rigid gels of the reversible hydrogen-bonding type, alginates (irreversible hydrocolloids), and elastomers such as silicones, polysulfides, and polyethers. Duplicating materials are employed to prepare duplicates of the original cast for partial dentures.

Economic Aspects

Estimated U.S. consumption of resins for dental restoratives in 1984 was 257 t (Table 5); an additional 250 t was used as impression materials and in nondisposable and disposable supplies and sundries.

BIBLIOGRAPHY

R. W. Phillips, *Skinner's Science of Dental Materials,* 8th ed., W. B. Saunders, Philadelphia, Pa., 1982.

R. G. Craig, W. J. O'Brien, and J. M. Powers, *Dental Materials: Properties and Manipulation,* 3rd ed., C. V. Mosby Co., St. Louis, Mo., 1980.

Dentists Desk Reference: Materials, Instruments and Equipment, 2nd ed., American Dental Association, Chicago, 1983.

J. F. McCabe in D. C. Smith and D. F. Williams, eds., *Biocompatibility of Dental Materials,* Vol. IV, CRC Press, Inc., Boca Raton, Fla., 1982.

G. M. Brauer
J. M. Antonucci
National Bureau of Standards

Table 3. Typical Range of Physical Properties of Polyelectrolyte-based Dental Cements

Property[a]	Zinc polycarboxylate	Zinc phosphate	Glass ionomer	Silicate
compressive strength, MPa[b]	42–132	69–161	140–229	168–290
diametral tensile strength, MPa[b]	5–16	5–8	7–16	3–15
24-h solubility, wt %	0.04	0.04–3.3	0.1–1.25	0.5–0.7
adhesion to tooth	yes	no	yes	no

[a] Strength 24 h after mixing.
[b] To convert MPa to psi, multiply by 145.

Table 4. Properties of Elastomeric Impression Materials[a]

Property	Polysulfide	Polyether	Silicone	
			Condensation	Addition
handling and setting times	long	short	medium	medium
ease of removal, flexibility	high	low	medium	medium
tear resistance	excellent	fair	good	good
reproduction of detail	very good	very good	very good–excellent	very good–excellent
accuracy	good	good	fair	very good
time of pour	within 1 h	up to several hours	within 1 h	up to several hours

[a] Courtesy of Dental Consultants, Inc.

DEPOLYMERIZATION

Depolymerization is characterized by the scission of the main polymer chain backbone. A polymer molecule can be broken down into fragments by (*1*) random degradation, where bonds in the chain are broken at random; (*2*) depolymerization, where chain scission is the reverse of addition polymerization and monomer units are released successively from a chain end or at a weak link (often called depropagation, unzippering, or unzipping); and (*3*) weak-link degradation, where chain scission occurs at weak links distributed at random along the chain, which are more easily ruptured than normal bonds (Table 1). These types of reactions may occur separately or in conjunction with each other, and are initiated thermally, photochemically, by ultrasonic irradiation, mechanical action, or some other method.

Kinetics

Investigating the kinetics of depolymerization of polymer systems is a means of elucidating the depolymerization mechanism. The two most important mechanisms are the scission of main-chain bonds followed by a reverse of polymerization resulting in high yields of monomer and the evaporation of low molecular weight homologues produced as a result of random degradation of the macromolecules; the mechanisms can occur simultaneously.

In random degradation, the number of chain cleavages is proportional to the number of linkages, ie, to the polymer concentration; if the mass of linear polymer molecules is subject to a random degradation, the variations of the number of bonds N as a function of time t is proportional to the number of bonds N present in the solution at time t:

$$-\frac{dN}{dt} = k_a N \quad (1)$$

where k_a is the rate constant for the random initiation. Integrating equation 1, when at the start of the reaction $t = 0$ and $N = N_0$, gives at time t

$$N = N_0 e^{-k_a t}$$

or

$$\ln \frac{N}{N_0} = -k_a t \quad (2)$$

The total number of monomer units in this system may be expressed as L and

$$L = N_0 + m_0 = N + m_t \quad (3)$$

where m_0 and m_t represent the number of free monomers at the beginning of the reaction and after time $t(m_t > m_0)$. Then

$$\frac{N}{N_0} = \frac{L - m_t}{L - m_0} = \left(1 - \frac{m_t}{L}\right) \Big/ \left(1 - \frac{m_0}{L}\right) \quad (4)$$

The number-average degree of polymerization is expressed as

$$DP = \frac{L}{m} \quad (5)$$

Thus,

$$\frac{N}{N_0} = \frac{1 - 1/DP_t}{1 - 1/DP_0} \quad (6)$$

where DP_0 and DP_t represent the number-average degree of polymerization at the beginning of the reaction and after time t. Substitution of equation 6 into equation 2 yields

$$\ln(1 - 1/DP_t) - \ln(1 - 1/DP_0) = -k_a t \quad (7)$$

Applying the approximation $1 \ll DP$ gives

$$\frac{1}{DP_t} - \frac{1}{DP_0} = k_a t \quad (8)$$

or

$$\frac{1}{M_t} - \frac{1}{M_0} = k_b t$$

where DP_0, DP_t, M_0, and M_t are the degree of polymerization and the molecular weights at the beginning of the reaction and at moment t, respectively, and k_b equals moles of bonds broken per gram per minute. This rate constant k_b can therefore be evaluated for different polymers at different temperatures by measuring the molecular weights during the reaction.

Table 1. Characteristics of Depolymerization

Types of reactions	Initiation points	Structure of typical polymers	Products	Loss of volatile products	Molecular weight decrease
depolymerization[a]	chain ends	1,1-disubstituted vinyl polymers	monomers	rapid	slow
random degradation[b]	random points	condensation polymers, mono- or nonsubstituted vinyl polymers	dispersed mixture of products	slow	rapid
weak-link degradation[b]	weak links	natural polymers	dispersed mixture of products	slow	rapid

[a] Chain reaction (depropagation, unzipping, unzippering, chain depolymerization).
[b] Stepwise reaction.

Thermodynamics and Depolymerization, Polymerization Equilibrium

The concept of a ceiling temperature has been outlined by several research groups. To obtain quantitative information about a depolymerization–polymerization equilibrium, the following steps of reversible addition polymerization are considered:

$$M_n^* \underset{k_p}{\overset{k_d}{\rightleftharpoons}} M_{n-1}^* + M \tag{9}$$

where M_n^* represents the number of active species with a *DP* of n that may add or release a monomer unit, M. The rate constants k_d and k_p represent the depropagation and the propagation reactions, respectively.

At ordinary temperatures, $k_p[M_{n-1}^*][M] \gg k_d[M_n^*]$, and hence the rate of polymerization R_p is

$$R_p = k_p[M_{n-1}^*][M] \tag{10}$$

where $[M_{n-1}^*]$ is the concentration of propagating species, and $[M]$ is the concentration of monomer.

Because $E_d = E_p + \Delta H$, the depolymerization increases more rapidly with increasing temperature. At the ceiling temperature T_c the rate of depropagation becomes equal to that of propagation, regardless of the variation of $[M_n^*]$ with increasing temperature.

$$k_p[M_{n-1}^*][M] = k_d[M_n^*] \tag{11}$$

when the polymer chains are long, $[M_{n-1}^*] = [M_n^*]$. Then

$$k_p[M] = k_d \tag{12}$$

or

$$A_p \exp(-E_p/RT_c)[M] = A_d \exp(-E_d/RT_c)$$

where A_p and A_d are the collision frequency factors for polymerization and depolymerization, respectively. Then

$$T_c = \frac{E_p - E_d}{R\ln(A_p[M]/A_d)} = -\frac{\Delta H}{R\ln(A_p[M]/A_d)} \tag{13}$$

Above the ceiling temperature in the presence of monomer at 101.3 kPa (1 atm) pressure, the polymer active species depolymerize rather than grow.

Many polymers exist even at temperatures considerably above the ceiling temperature because they are in a state metastable equilibrium. A dead polymer that has been removed from the reaction is stable and does not depolymerize unless an active end is produced by cleavage of an end group or at some point along the polymer chain. When such an active site is produced by thermal, chemical, photolytic, or other means, depolymerization follows until the monomer concentration becomes equal to $[M]$ for the particular temperature.

Energetics

For many polymers, depolymerization, characterized by the breaking of the weakest bond, is a function of the bond dissociation energy. The strength of the chemical bond corresponds to the limit of vibrational energy that a molecule may possess without bond rupture. Because vibrational energy is increased by heat, thermal stability is related to the dissociation energy of the various bonds. Thus greater thermal stability requires bonds with greater dissociation energy.

Practical Applications

Depolymerization has a number of practical applications, including polymer modification, preparation of carbonaceous materials, synthesis or recovery of small molecules, and waste treatment and utilization employing polymers.

BIBLIOGRAPHY

L. Reich and S. S. Stivala, *Elements of Polymer Degradation,* McGraw-Hill Inc., New York, 1971.

W. Schnabel, *Polymer Degradation,* Macmillan Publishing Co., Inc., New York, 1982.

H. Sawada, *Thermodynamics of Polymerization,* Marcel Dekker, Inc., New York, 1976.

HIDEO SAWADA
Daicel Chemical Industries, Ltd.

DEPROPAGATION. See DEGRADATION; DEPOLYMERIZATION.

DESIGN. See PRODUCT DESIGN.

DEVOLATILIZATION

Polymerization products frequently require a separation process to remove contaminants, such as unreacted monomers or diluents, eg, solvent, water. Reactor discharges sometimes contain large amounts of volatile components that must be reduced to a content of a few parts per million. Purity requirements are dictated by product applications, eg, food-packaging materials, and ultimate mechanical properties can be adversely affected by residual contaminants.

Because the contaminants in most cases are volatile relative to their polymeric hosts, they are removed from the condensed phase (polymer) by evaporation into a contiguous gas phase. Such separation processes are commonly referred to as devolatilization (DV). The polymer to be devolatilized may be in the form of a melt or particulate solid. Separation is effected by applying a vacuum or by using inert substances, eg, in purging with nitrogen gas or by steam injection.

Equipment

Devolatilization equipment is classified according to the physical state of the polymer, ie, melt or solid, and the transport conditions, ie, high or low contaminant concentration. Equipment includes flash evaporators and vented extruders (Fig. 1) and particulate beds.

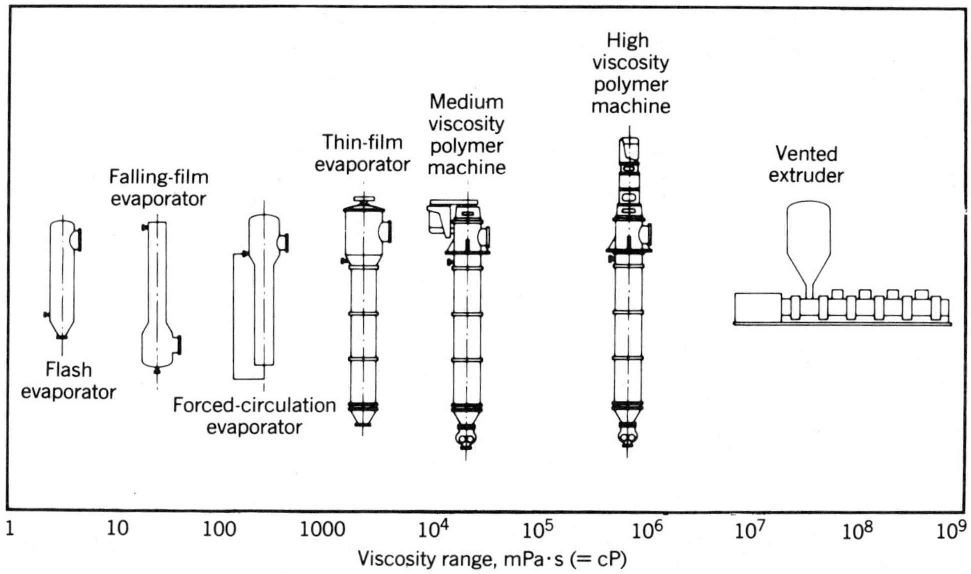

Figure 1. Devolatilization equipment.

BIBLIOGRAPHY

J. A. Biesenberger, ed., *Devolatilization of Polymers,* Hanser, Munich; Macmillan, New York, 1983.

J. A. Biesenberger and D. H. Sebastian, *Principles of Polymerization Engineering,* John Wiley & Sons, Inc., New York, 1983.

Devolatilization of Plastics, VDI-Verlag, Düsseldorf, FRG, 1980.

JOSEPH A. BIESENBERGER
Stevens Institute of Technology

DEXTRAN

Dextrans are a class of polysaccharides (qv) synthesized from sucrose by bacterial enzymes (dextransucrases, glucansucrases, or glucosyltransferases) to give D-glucans with contiguous α-1→6 glucosidic linkages in the main chains and a variable amount of α-1→2, α-1→3, or α-1→4 branch linkages. Dextran has plagued humans as a contaminant in food preparations containing sucrose. It presents a problem in the sugar-refining industry where it plugs process pipes and filters, inhibits crystallization, reduces the yield, and distorts optical rotation measurements used in the standardization and determination of sucrose purity. About 1970, dextran was also recognized to be the principal component of dental plaque and to be involved in the development of dental caries. Dextrans and dextran derivatives, however, have found valuable uses as blood plasma substitutes, molecular sieves, anticoagulants, food additives, explosives, high viscosity gums, and several other applications.

Structure

Dextrans can be divided into three classes based on structural features: class-1 dextrans contain a main chain of consecutive α-1→6-linked glucosyl residues, with branching at positions 2, 3, or 4; class-2 dextrans contain nonconsecutive α-1→3 and α-1→6 linkages and α-1→3 branch linkages; and class-3 dextrans contain consecutive α-1→3 linkages and α-1→6 branch linkages. Class 1 comprises most of the polysaccharides usually thought of as dextrans.

Biosynthesis

In nature dextrans are synthesized from the disaccharide sucrose by enzymes secreted mainly by *Leuconostoc* and *Streptococcus*. Dextran cannot be produced from glucose, mixtures of glucose and fructose, or any other naturally occurring sugar; sucrose is absolutely required. The relatively high energy of the acetal–ketal linkage joining the glucose and fructose moieties of sucrose is used to form glucosyl-enzyme intermediates. The enzyme catalyzes the addition of the glucosyl unit to the reducing end of a second glucosyl unit to form the α-1→6 linkage. The synthesis proceeds by the addition of glucosyl units to the reducing end of the growing dextran chain. The polymerization is terminated by acceptor reactions that release the dextran from the enzyme.

Clinical Dextran

Dextran of relatively low molecular weight can be used as a therapeutic agent in restoring blood volume for mass casualties. Dextran produced by *L. mesenteroides* B-512F is the material of choice for clinical dextran because it has a low degree of antigenicity and a high percentage (95%) of α-1→6 glycosidic linkages.

Derivatives and Uses

Many different types of esters and ethers of dextran provide macromolecules with diverse properties and negative, positive, or neutral charges. Properties depend on the type of substituent, the degree of substitution, and the molecular weight of the dextran.

Cross-linked Dextran. The most widely used dextran derivative is obtained by the reaction of an alkaline solution of dextran with epichlorohydrin to give cross-linked chains. The product is a gel that is used as a molecular sieve and is available commercially as the product Sephadex.

Dextran Sulfate. The sulfate ester of dextran may be prepared by mixing chlorosulfonic acid in pyridine at −10°C, raising the temperature to 60°C, and adding dry, finely powdered dextran. Dextran sulfate has anticoagulant properties similar to, although less effective than, that of heparin, a naturally occurring polysaccharide containing carbohydrate sulfate esters.

Mercaptodextran. Mercaptodextran may be synthesized by thiolating dextran with *N*-acetyl homocysteine thiolactone. High affinity for heavy-metal ions, combined with low toxicity, suggests possible uses in acute, heavy-metal poisoning and in environmental cleanup of heavy-metal contamination.

BIBLIOGRAPHY

P. T. Murphy and R. L. Whistler in R. L. Whistler and J. N. BeMiller, eds., *Industrial Gums, Polysaccharides and Their Derivatives*, 2nd ed., Academic Press Inc., New York, 1973, pp. 526–542.

G. J. Walker in D. J. Manners, ed., *International Review of Biochemistry, Biochemistry of Carbohydrates II*, Vol. 16, University Park Press, Baltimore, 1978, pp. 75–126.

R. M. Alsop, *Prog. Ind. Microbiol.* **18**, 1 (1983).

J. F. Robyt in R. J. Doyle and J. E. Ciardi, eds., *Glucosyltransferases, Glucans, and Dental Caries*, IRL Press, Washington, D.C., 1983, pp. 123–133.

JOHN F. ROBYT
Iowa State University

DEXTRINS.

See POLYSACCHARIDES.

DIACETYLENE POLYMERS

The solid-state reaction of the diacetylene monomer yields a fully conjugated polymer backbone in a single crystal of macroscopic dimensions, several centimeters in some cases. The polymer crystals are highly anisotropic in their electronic properties, essentially one-dimensional, with the π-electron network of the fully conjugated backbone providing intense light absorption in the visible spectral range. Research interest ranges from the fundamental chemistry and structural physics of the reaction process to theoretical and experimental characterization of the unusual electronic properties of the final polymer. Several applications of these materials are of current interest.

The polymerization of diacetylene is lattice-controlled and proceeds by a 1–4 addition reaction. The reaction is initiated by heat, ultraviolet irradiation, pressure, or high energy radiation. The reactivity is controlled by the packing of the diacetylene monomers in the single crystal and can be explained by least-motion arguments.

Electronic Properties of Polydiacetylene Single Crystals

The optical properties of polydiacetylene crystals are dominated by the π–π transition of the conjugated backbone, which results in strong absorption through the visible spectrum. This strong absorption results in metalliclike reflection; the crystals are typically green-gold metallic in appearance.

Polydiacetylenes show substantial photoconductivity, but only in the single-crystal form, which limits application opportunities. Conductivities are in the insulating range but can be increased to the semiconducting range with doping.

The nonlinear optical properties of polydiacetylenes are of considerable interest. The third-order susceptibility (the property governing frequency tripling) of a polydiacetylene crystal is comparable to that of the best nonlinear inorganic semiconductors. Thus, if optical-quality materials can be produced, these polymers could be considered for application in various nonlinear optical devices.

Polydiacetylene Solutions

Most polydiacetylenes are insoluble even in unusual organic solvents and cannot be characterized by normal polymer-characterization techniques. Exceptions to this rule are poly-butoxycarbonylmethylurethanes (BCMU). Solutions of these polymers show dramatic solvent and/or temperature-induced chromic transitions.

Applications

Polydiacetylenes have many interesting applications; time–temperature and radiation indicators are the most promising area at this time. Other potential application areas are nonlinear optics, holographic imaging, and insulator coatings in electronic devices.

The polymers are generally nontoxic. However, monomers should be handled with care as some are toxic and others are skin and eye irritants.

BIBLIOGRAPHY

R. H. Baughman and R. R. Chance, *Ann. N.Y. Acad. Sci.* **313**, 705 (1978).

G. Wegner, *Discuss. Faraday Soc.* **68**, 494 (1980).

R. R. Chance, G. N. Patel, E. A. Turi, and Y. P. Khanna, *J. Am. Chem. Soc.* **100**, 1307, 6644 (1978).

R. R. CHANCE
Allied Corporation

DIALLYL AND RELATED POLYMERS

Allyl or 2-propenyl compounds comprise a very large group of terminally unsaturated organic compounds, a few of which are monomers of outstanding importance; for example, the diallyl carbonate ester CR-39 for optical and glasslike plastics, diallyl phthalates for heat-resistant moldings, and triallyl cyanurate, in low concentrations, as an effective curing and cross-linking agent for a wide range of plastics.

Monoallylic Compounds

Monoallylic compounds do not form homopolymers of high molecular weight under free-radical initiation because of low activation of the double bond and so-called degradative chain transfer (qv) resulting from the high reactivity of hydrogen atoms attached to the third carbon, which tends to terminate chain growth:

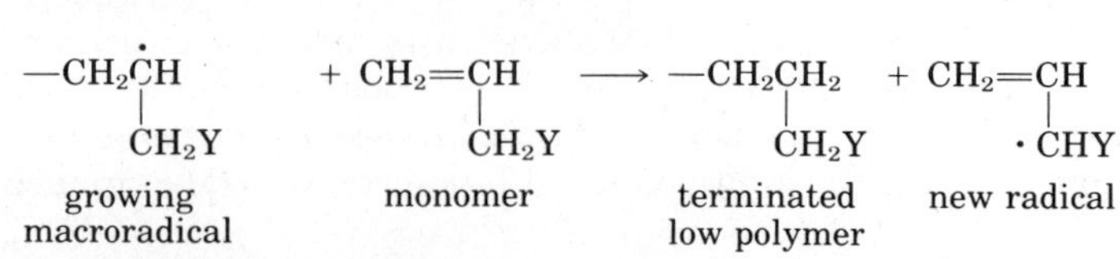

Monovinyl Versus Divinyl Versus Diallyl Polymerization

Typical free-radical polymerizations of monovinyl, divinyl, and diallyl compounds may be compared as follows:

Monovinyl polymerization:

$$R^{\cdot} + n\,M \longrightarrow R\text{—}M\text{—}M\text{—}M\text{—}M\text{—}M\text{—}M\text{—}M\text{----}$$

Divinyl polymerization:

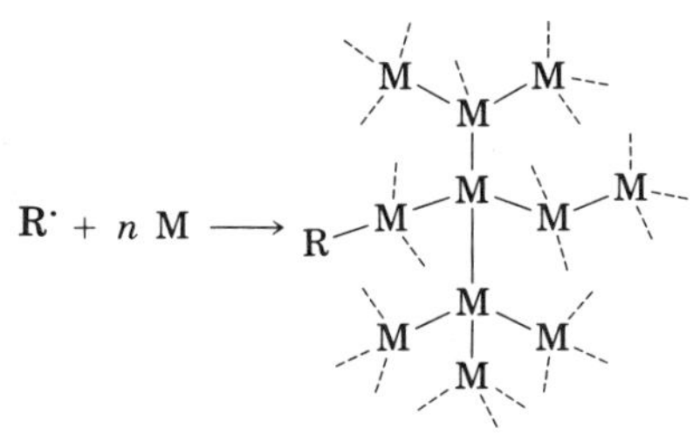

Diallyl polymerization:

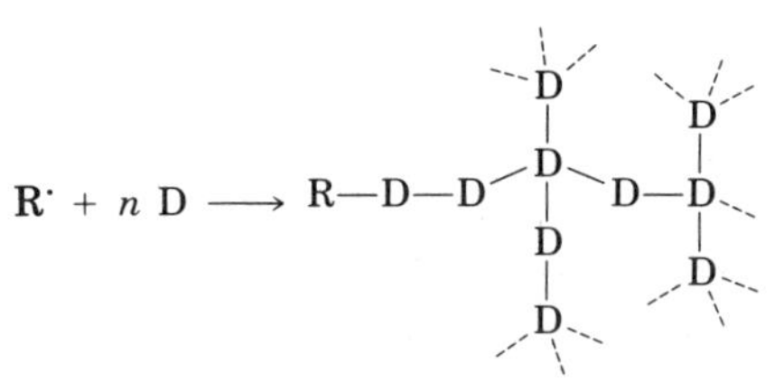

where M and D represent the monomer molecules that become units of polymer structures by converting the double bonds to single bonds.

Diallyl Diglycol Carbonate Castings

DADC Monomers. Reaction of diethylene glycol bis(chloroformate), obtained from the glycol and phosgene, with allyl alcohol in the presence of alkali gives the monomer

$$O(CH_2CH_2OCCl)_2 + 2\,CH_2{=}CHCH_2OH + 2\,NaOH \longrightarrow$$
(C=O in the OCCl group)

$$O(CH_2CH_2OCOCH_2CH{=}CH_2)_2 + 2\,NaCl + 2\,H_2O$$
(C=O in the OCO group)

Reaction of allyl chloroformate and diethylene glycol in the presence of alkali with cooling is another method of preparing the diallyl carbonate ester, CR-39. CR-39 is moderately toxic by ingestion (LD_{50} in rats = 349 mg/kg). Properties of the CR-39 monomer are given in Table 1.

DADC Homopolymerization. Bulk polymerization of CR-39 gives clear, colorless, abrasion-resistant polymer castings that offer advantages over glass and acrylic plastics in optical applications.

Casting CR-39. Sheets, rods, and lens preforms are cast from CR-39 or prepolymer syrup by methods similar to those used for methacrylate ester syrups. Typical properties of homopolymers from CR-39 monomer are given in Table 2.

Polymeric Nuclear-track Detectors. The most important recent application of CR-39 polymer is in solid-state detectors of nuclear particles, including α-particles, fast neutrons, cosmic rays, and ions of elements of atomic number 10 and above. The outstanding sensitivity of the cast polymers in thin sheets and films to diverse types of ionic radiations with a wide range of mass-to-charge ratio has increased applications in nuclear and space sciences, medicine, mining, and ecological research. Tracks from radiation are developed by treatment of the cast polymer with alkali.

Table 1. Typical Properties of Commercial CR-39 Monomer[a]

Property	Value
appearance	clear, colorless liquid
color, APHA	10
odor	none to slight
specific gravity	1.15_4^{20}
refractive index, n_D^{20}	1.452
boiling point at 266 Pa[b], °C	166
melting point (supercooled), °C	−4 to 0
viscosity at 25°C, mm^2/s (= cSt)	15
flash point	
Seta closed cup, °C	173
Cleveland open cup, °C	186
water content, slightly hygroscopic, %	0.1

[a] CR-39 (Columbia Resin 39) is a trade name of PPG Industries, Inc.
[b] To convert Pa to mm Hg, multiply by 0.0075.

Table 2. Physical and Mechanical Properties of CR-39 Homopolymer

Property	Value
density at 20°C, g/cm^3	1.31
tensile strength, MPa[a]	35–41
compressive strength, MPa[a]	155
thermal conductivity, W/(m·K)	0.21
specific heat, kJ/(kg·K)[b]	2.3
linear coefficient of expansion/°C, 25 to 75°C	11.4×10^{-5}
glass transition, °C	85
recommended continuous-use temp, °C	100
recommended 1-h use temp, °C	150

[a] To convert MPa to psi, multiply by 145.
[b] To convert J to cal, divide by 4.184.

Modified Polymers and Copolymers. DADC pure monomer and mixtures with small amounts of comonomers or other additions are commercially available for casting. Monomer formulations are available, including agents for protecting the eyes against uv light. Another grade is designed to absorb infrared radiation, and several modified monomers give copolymers of increased heat resistance and hardness.

Diallyl Phthalates and their Polymerization

The three isomeric diallyl phthalate monomers are colorless liquids of mild odor, low volatility, and relatively slow polymerization in the early stages. At ca 25% conversion the viscous liquid undergoes gelation and polymerization accelerates; however, the last monomer disappears at a slow rate. The monomers are prepared by conventional esterification.

DAP Copolymerization. The diallyl phthalates copolymerize readily with monomers bearing strong electron-attracting groups attached to the ethylenic group. These include maleic anhydride, maleate and fumarate esters, and unsaturated polyesters.

Diallyl Isophthalate. DAIP polymerizes faster than DAP, undergoes less cyclization, and yields cured polymers of better heat resistance, eg, up to ca 200°C. Prepolymer molding materials, such as Dapon M of FMC, are not sticky. Maleic anhydride accelerates polymerization, whereas vinyl isobutyl ether retards it and delays gelations in castings. Copolymers with maleic anhydride are exceptionally hard and tough and may scratch homopolymer surfaces.

Table 3. Properties of DAP Thermoset Moldings

Property	Unfilled	Glass-fiber reinforced	Mineral filled
molding temperature, °C	140–190	144–193	133–193
mold shrinkage, cm/cm		0.0005–0.005	0.002–0.007
tensile strength, MPa[a]	27.6	41–76	34–55
compressive strength, MPa[a]	165	172–241	138–220
specific gravity		1.70–1.96	1.65–1.85
water absorption[b], 24 h at 25°C, %	0.20	0.12–0.35	0.2–0.5

[a] To convert MPa to psi, multiply by 145.
[b] 3-mm thickness.

Applications. The largest use of diallyl phthalate thermoset polymers is in moldings and coatings for electronic devices requiring high reliability under long-term adverse environmental conditions. These devices include electrical connectors and insulators in communication, computer, and aerospace systems. Properties of DAP thermoset moldings are given in Table 3.

Diallyl phthalates are used with glass cloth and roving in tubular ducts, radomes, aircraft, and missile parts of high heat resistance. Diallyl esters find little application in lenses. However, DAP and DAIP can be polymerized by high-energy radiation in lens molds. Coatings of silica and alumina by vaporization give antiglare, scratch-resistant lenses.

Other Diallyl Esters

Although dimethallyl phthalates have been little studied, the ortho isomer has been copolymerized with vinyl acetate at 60°C with bp initiator, and reactivity ratios have been reported. Copolymers of diallyl succinate and unsaturated polyesters cured by x rays provide wear-resistant coatings of MMA dental polymers. Copolymers of diallyl itaconate with *N*-vinylpyrrolidinone and styrene have been proposed as oxygen-permeable contact lenses (qv). Properties of liquid polyfunctional allyl esters are given in Table 4; some of these esters have sharp odors and cause skin irritation.

Table 4. Properties of Some Diallyl (DA) Esters

Diallyl ester[a]	Bp_{Pa}[b]	n_D^{20}	d_4^{20}
oxalate	$107_{1.9}$	1.4460	1.0081
succinate	$94_{0.13}$	1.4507	1.056
sebacate	$164_{0.26}$	1.4550	0.978
triallyl citrate	$130_{0.67\times10^{-3}}$	1.135	

[a] Except where otherwise stated.
[b] To convert Pa to mm Hg, multiply by 0.0075.

Allyl–Vinyl Compounds

Monomers such as allyl methacrylate and diallyl maleate have applications as cross-linking and branching agents selected especially for the different reactivities of their double bonds; some physical properties are given in Table 5. These esters are colorless liquids soluble in most organic liquids but little soluble in water; DAM and DAF have pungent odors and are skin irritants; DAM in oral tests in rats give LD_{50} = 300 mg/kg.

Polyfunctional Allyl Nitrogen Monomers

Triallyl Cyanurate as Cross-linking Agent. Triallyl cyanurate (TAC) or 2,4,6-tris(allyloxy-*s*-triazine) and its isomer triallyl isocyanurate (TAIC) are used as cross-linking agents with comonomers and for after-curing preformed polymers such as olefin copolymers in electrical insulations. Properties of TAC and TAIC are given in Table 6. TAC is moderately toxic by ingestion.

Triallyl Cyanurate Cure of Preformed Polymers. TAC and TAIC are often used in small amounts with vinyl-type and condensation polymers for cured plastics, rubber and adhesive products of high strength, and heat and solvent resistance. In some cases, chemical stability is also improved.

TAIC as a Curing Agent. Triallyl isocyanurate is prepared from an alkali cyanate with allyl chloride. Together with DAP, small amounts of TAIC were used to cure unsaturated polyesters in glass-reinforced thermosets.

Diallyl Ammonium Polymers. *N,N*-Diallyl dimethyl (DADM) ammonium salts are used for the preparation of polyelectrolytes.

Table 6. Properties of TAC and TAIC

Property	TAC	TAIC
melting point, °C	31	24
boiling point, °C	$140_{67\ Pa}$[a]	$126_{40\ Pa}$[a]
density at 30°C, g/cm³	1.1133	1.1720

[a] To convert Pa to mm Hg, multiply by 0.0075.

Table 5. Properties of Allyl Vinyl Monomers

Property	Allyl methacrylate (AMA)	Diallyl maleate (DAM)	Diallyl fumarate (DAF)
structure	CH_3 | CH_2=C—COOA[a]	HC—COOA ‖ HC—COOA[a]	AOOCCH ‖ HC—COOA[a]
boiling point, °C	$55_{4\ kPa}$[b]	$112_{0.53\ kPa}$[b]	$140_{0.40\ kPa}$[b]
density at 25°C, g/cm³	0.930	1.070	1.0516
viscosity, mPa·s (= cP)	13	4.3	3.0

[a] A = allyl.
[b] To convert kPa to mm Hg, multiply by 7.5.

Other Diallyl Compounds

A great number of other polyfunctional allyl compounds have been prepared, especially allyl ethers and allyl ether derivatives of carbohydrates and other polymers. They are made by reaction of alcoholic groups with allyl chloride in the presence of alkali. Polymerizations and copolymerizations generally are slow and incomplete. Most of the products have only limited use in coatings, inks, and specialties.

BIBLIOGRAPHY

C. E. Schildknecht, *Allyl Compounds and Their Polymers (Including Polyolefins)*, Wiley-Interscience, New York, 1973.

C. E. Schildknecht, ed., *Polymerization Processes*, Wiley-Interscience, New York, 1977.

U.S. Pat. 2,273,891 (Feb. 24, 1942), M. A. Pollak, I. E. Muskat, and F. Strain (to PPG Industries, Inc.)

U.S. Pat. 2,318,959 (May 11, 1943), I. E. Muskat and coworkers (to PPG Industries, Inc.).

U.S. Pat. 2,335,962 (Dec. 7, 1943), M. A. Pollak and A. G. Chenicek.

U.S. Pat. 2,320,536 (June 1, 1943), M. A. Pollak, F. Strain, and I. E. Muskat (to PPG Industries, Inc.).

U.S. Pat. 2,332,461 (Oct. 19, 1944), I. E. Muskat and coworkers (to PPG Industries, Inc.).

C. E. Schildknecht
Gettysburg College

DIALLYL MALEATE POLYMERS.

See Diallyl and related polymers.

DIALLYL PHTHALATE POLYMERS.

See Diallyl and related polymers.

DIALYSIS.

See Membranes.

DICHROISM.

See Optical properties; Proteins.

DICUMYL PEROXIDE.

See Peroxy compounds.

DIE DESIGN

In the extrusion (qv) of polymers, the extrusion die that shapes the melt occupies a central position. It must be considered in conjunction with the adjacent takeoff device and the former, or forming jig, which are responsible for maintaining the desired dimensions and at the same time achieving a specified degree of cooling (Fig. 1). In the production of profiles, cable coatings, etc, from vulcanized rubber or cross-linkable ther-

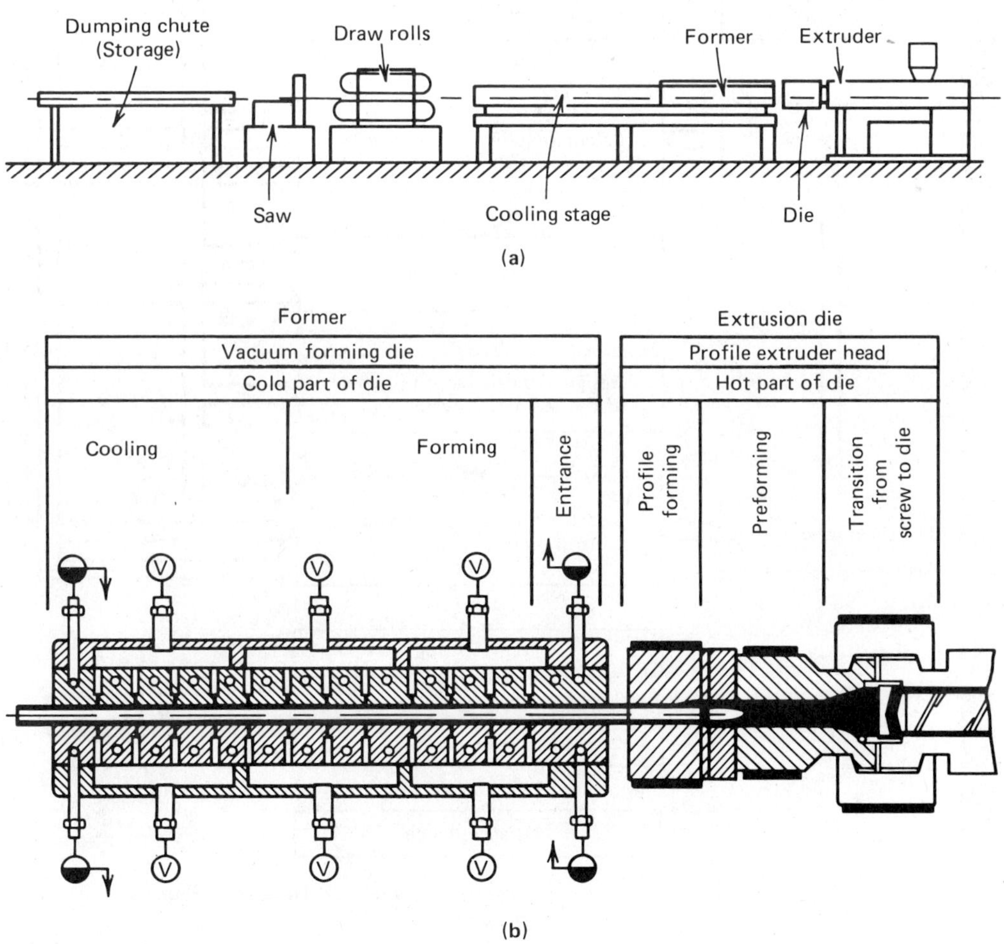

Figure 1. Schematic of an extrusion plant. Courtesy of Carl Hanser Verlag.

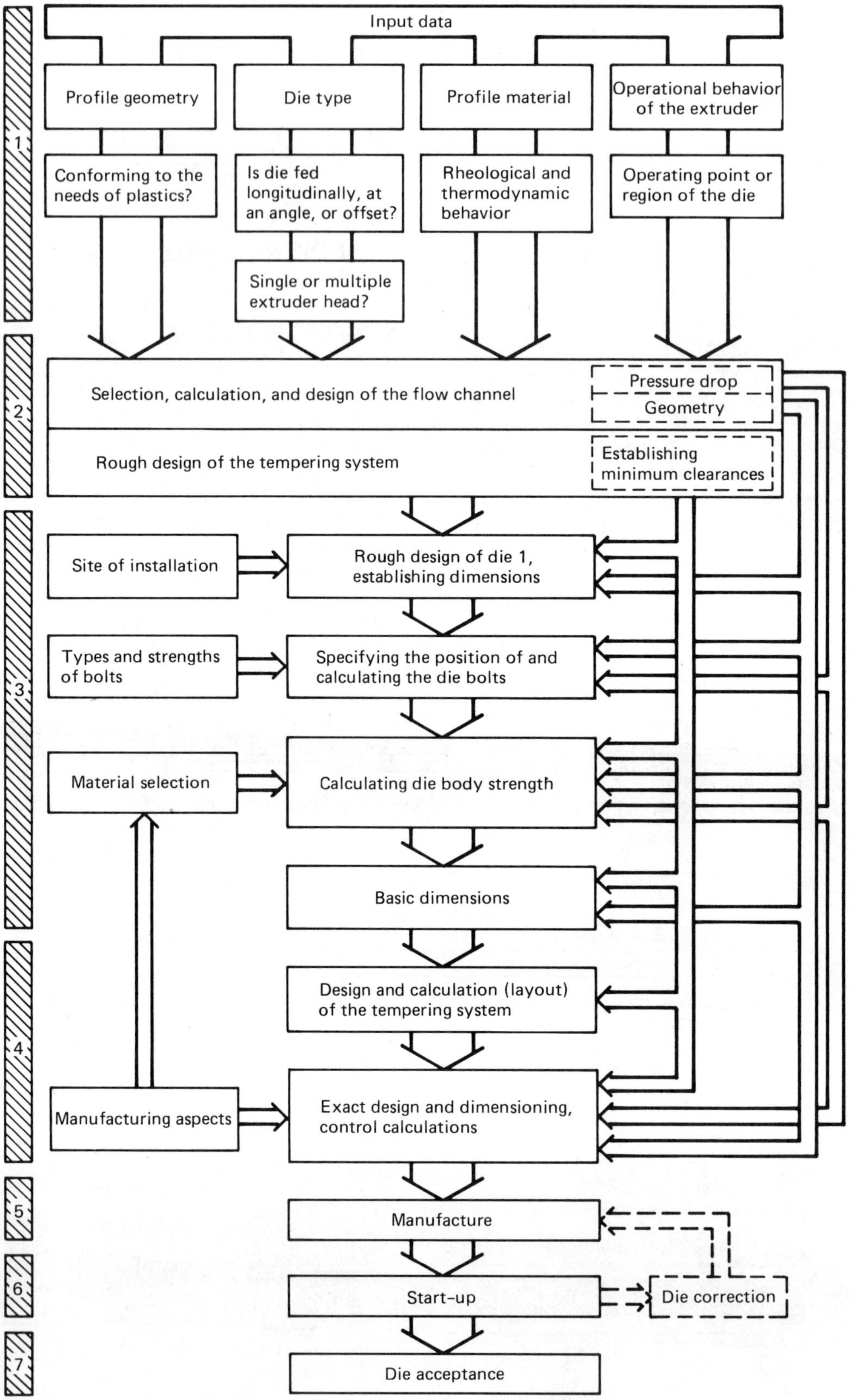

Figure 2. Design of extrusion dies. Courtesy of Carl Hanser Verlag.

moplastics, in-line downstream equipment can also be used as a cross-linking device.

The extruder must deliver a pulsation-free, reproducible homogeneous melt stream, which the dies and the equipment following, eg, the forming jig, form into the semifinished product. The rheological and thermodynamic processes taking place as well as any stretching processes between the die and former have a decisive effect on the quality of the semifinished products. To design the extrusion die and the forming jig or any of the following devices, flow, deformation, and temperature relationships in these parts of the production line must be considered. If any analytical description of the physical processes is selected, the empirical portion of die and forming-stage design can be reduced, because changes in the operating conditions or in the rheological and thermodynamic properties of the polymer can be evaluated directly with regard to shaping, cooling, or cross-linking of the extruded semifinished products. This leads to a more reliable design (Fig. 2).

Rheological considerations include the pressure loss in the die; melt velocity across the die outlet; effect of viscoelastic properties on profile geometry (swelling); and flow velocity and the profile surface (at certain shear rates, a discontinuous change may exist in the viscosity function).

Thermodynamic considerations include maximum temperatures occurring in the melt stream on the basis of heat transmission and dissipation relationships (internal friction of the melt); correlation of dissipation with pressure drop, $\Delta T \sim \frac{\Delta p}{\rho c_p}$; and uniform and controllable temperatures in the die and the forming equipment.

Operational aspects include stiffness of the extrusion die; easy installation and dismantling of die and forming tools; easy cleaning; few but well sealed surface die joints; and detachable, tight connection between extruder and die. As a rule, extruder and extrusion die interact in their operational behavior according to specific characteristics.

Manufacturing costs can be kept low by using materials suitable for machining, polishing, and tempering.

Extrusion Dies

Dies may be classified as single-melt (annular exit cross section, flat-slit exit cross section, profile dies, variable exit cross section) and coextrusion dies.

BIBLIOGRAPHY

W. Michaeli, *Extrusion Dies for Polymers,* Carl Hanser Verlag, Munich, 1984; Ger. ed. 1979.

R. Doyle and L. Gans in J. H. Du Bois and co-workers, ed., *Plastics Mold Engineering Handbook,* Van Nostrand Reinhold Co., Inc., New York, 1978, pp. 467–510.

J. Wortberg, *Die Design for Single and Multilayer Extrusion,* Dissertation, RWTH Aachen, Aachen, FRG, 1978.

U. Masberg, *Use of the Finite Element Method in Extrusion Die Design,* Dissertation, RWTH Aachen, Aachen, FRG, 1981.

B. Gesenhues, *Calculation of Flow in Extrusion Dies With Respect to the Viscoelastic Material Properties,* Dissertation, RWTH Aachen, Aachen, FRG, 1984.

W. MICHAELI
Carl Freudenberg

DIELECTRIC HEATING

Heat transfer to materials that are poor thermal conductors is difficult. Surface overheating or even burning may occur. Dielectric heating can circumvent these problems. This phenomenon occurs when most electrical insulators are placed in a high-frequency electric field. Because the electric field penetrates throughout the material, the heat is generated within, and heat penetration from the surfaces is not necessary.

Dielectric heating follows the laws of electricity. However, the frequency may be 50,000 to 2×10^6 times as high as normal power-line frequencies. Therefore, capacitive reactances are much smaller, and inductive reactances are much larger.

Dielectric-heating generators, or the applicators where heating takes place, are usually equipped with controls to adjust them for the variations in product characteristics, such as dimensions and moisture content, and allow for the normal variations of the electrical characteristics as the temperature rises, the moisture evaporates, or the dimensions change during heating. Although calculations are guides only, they are worth pursuing in many cases.

Theory

Most molecules are polarizable in an electric field (Fig. 1). In any given high-frequency electric field the degree of polarization and the energy required to achieve it depend on the loss factor or dissipation factor of a material. A material that is readily polarized by a weak electric field has a high loss factor and is easy to heat.

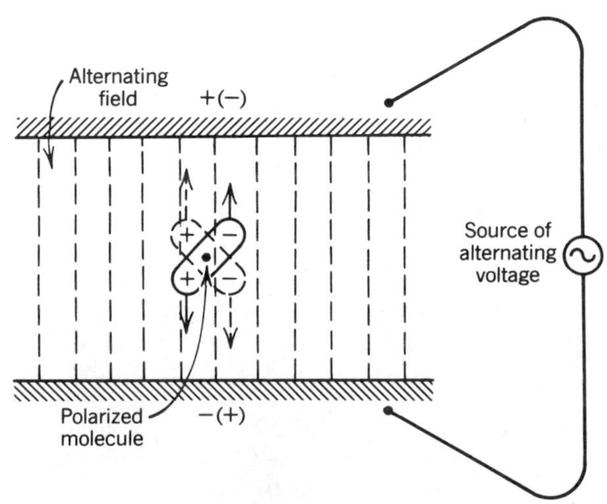

Figure 1. A molecular dipole immersed in a high-frequency electrical field.

Amount of Heat Needed

In determining the amount of heat needed for a process, the following must be taken into account: amounts of heat lost to the surrounding air; heat lost to the electrodes; heat lost to the heating and evaporation of solvents; the basic sensible heat to the product itself; and the amount of heat received from a hot oven or press platen.

Uniformity

The uniformity of heat generation with a dielectric (nonconducting) material depends on several factors:

1. The shape of the electrodes (conductors) forming the field.
2. The shape of the piece being heated, particularly in the direction of the field, i.e., usually perpendicular to parallel-plate electrodes.
3. An air gap between the electrodes and the material.
4. Nearby metallic objects that may distort the field.
5. The edge effect of the electrodes.
6. The uniformity of the internal structure of the dielectric.
7. The generator frequency, especially for large pieces.
8. The dimensions of the piece being heated.
9. The amount and uniformity of external cooling or heating at the surfaces of the insulating material.

The Dielectric-heating Equation

The rate of heat input, and therefore power input, is important. Where the part is sufficiently small that the electric field can be considered to be uniform, the following equation defines the power:

$$P = 2\pi f E^2 C \tan \delta$$

where P = power input in W, f = frequency of the generator in MHz, E = field strength, or voltage between the electrodes in kV, C = capacitance of the material in picofarads (pF), and $\tan \delta$ = the loss factor. The product $\epsilon' \tan \delta$, is often called the dissipation factor or loss index and is a pure number.

Voltage-distribution Controls

Raising the frequency or treating large loads can cause uneven heating, which can, however, be controlled.

Theory. Feeding power to the center improves the voltage distribution because of the shorter distance to the ends of the electrode. The voltage rises toward the ends away from the feed point in all electrodes without special voltage controlling devices.

This normal voltage increase toward the ends of the electrode can be controlled by connecting inductances, or stubs between the electrodes near points of high voltage. Since it is difficult to obtain a high voltage across a low impedance, the lower the inductance, ie, the shorter the stub, the lower the voltage. By proper adjustment of the stub inductance, the voltage at the point of connection of the inductance can be reduced as desired—even below that at the feed point. Adjusting a stub can also alter the voltage at the feed point and may require a generator adjustment to return it to the desired level.

Stubs. Stubs have three functions: to adjust the voltage distribution as discussed; to adjust the impedance of the electrode system to a value that the generator can handle; and to reduce the current in the transmission line and generator output circuitry.

Stubs are often made of strips of copper, aluminum, or brass in widths of 5–20 cm and even wider. For a given inductance requirement, the wider the strip, the longer it must be.

Shielding

Use of ISM frequencies reduces shielding requirements. These frequencies require generators without excessive harmonics in output voltages and currents in order to avoid radiation on non-ISM frequencies. Otherwise, good shielding is required.

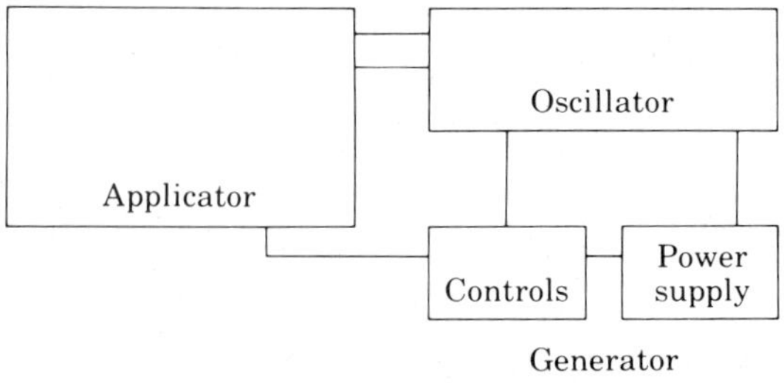

Figure 2. Diagram of a dielectric-heating system.

The manufacturer of generators and applicators usually provides the shielding necessary to meet local requirements.

Equipment

High-frequency heating equipment consists of two sections (Fig. 2). The generator accepts the electrical power from the power source and adapts it to high-frequency output. In the applicator section, the high-frequency fields are established by the electrode system. These two sections are connected by a coaxial transmission line (or sometimes a connecting strap, if within one cabinet).

Applications

Dielectric heating is used in plastics preheating, plastics welding or sealing, woodworking, business forms, and drying many materials.

BIBLIOGRAPHY

F. Clark, *Insulating Materials for Design and Engineering Practice*, Wiley-Interscience, New York, 1962.

A. R. Von Hippel, *Dielectric Materials and Applications*, The Technology Press of M.I.T. and John Wiley & Sons, Inc., New York, 1966.

H. P. Zade, *Heatsealing and High-Frequency Welding of Plastics*, Temple Press Books, Ltd., London, 1959, pp. 301–370.

T. L. Wilson, *Report on Dielectric Heating*, Electric Power Research Institute, Palo Alto, Calif., 1987.

T. LAMONT WILSON
Consultant

DIELECTRIC PROPERTIES. See ELECTRICAL PROPERTIES.

DIELECTRIC RELAXATION

Polarization phenomena in homogeneous dielectric materials may be classified into three categories based on the nature of atomic or molecular origins:

1. Electronic polarization that results from a shift of electron orbits with respect to the center of the atom.
2. Atomic polarization that results from a shift of relative positions, eg, between the carbon and the hydrogen atoms by stretching in response to an electric field.
3. Dipole polarization that results from orientation of permanent dipoles under the influence of the electric field.

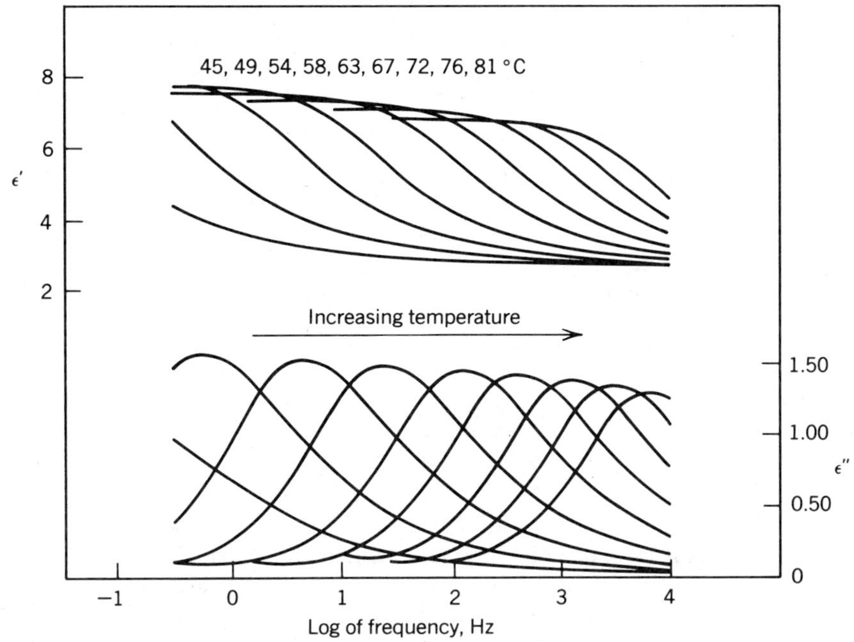

Figure 1. Dispersion for poly(vinyl acetate). The upper part represents in-phase dielectric constant ϵ' and the lower part, the dielectric loss factor ϵ''. Courtesy of G. E. Johnson.

The first and second categories are based on the displacement of positive and negative charges with respect to each other. The third category, on the other hand, is based on the thermal motions of molecular dipoles in some restraining environment; most studies dealing with the dielectric relaxation of polymers are concerned with this type of polarization.

Dielectric materials often contain freely migrating charges such as free ions. The migrations of these charges are directed by the electric field. In a d-c field, the negative ions accumulate ultimately on the positive electrode, and the positive ions on the negative electrode. Polarization by charge migration is called space-charge polarization or ion polarization, depending on the nature of the migrating charges. Polarization by charge migration often occurs simultaneously with polarization by dipole orientation. Distinction between the two similar behaviors is often difficult, and much effort has been spent to separate the two phenomena in experiments as well as in analyses.

The amount of charge building up in response to the imposed electric potential is time-dependent. The speed of polarization depends on the rate of electronic and atomic polarization and permanent dipole orientation. For electronic or atomic polarization, this speed corresponds to the electronic and atomic vibration and is very rapid. These rapid polarizations are termed instantaneous polarizations. In the case of dipole-orientation polarization, however, since reorientation of molecules or molecular segments is involved, the process is interfered with by the friction from the surrounding molecules. The reorientation of dipoles involves motion in viscous media; thus the polarization process in the imposed electrical field is delayed, and an equilibrium is reached only in a relatively long time. When the field is removed, it again takes a long time for polarization to disappear. This phenomenon is called the dielectric relaxation. The time required for the polarization to decay to $1/e$ of the original equilibrium value is called the relaxation time.

Dielectric Relaxation in Polymers

Not only polymers, but many nonpolymers exhibit multiple transitions (loss peaks) and a broad distribution of relaxation times. The dielectric loss factor of poly(vinyl acetate) plotted against logarithmic frequency between 45 and 80°C is shown in Figure 1. The curves represent the relaxation processes in the equilibrium-liquidus state.

Applications

Dielectric data provide valuable information on the nature of molecular motion in polymers. Knowledge of dielectric properties of polymers as well as the related experimental techniques is valuable in all types of research.

BIBLIOGRAPHY

D. W. McCall in R. S. Carter and J. J. Rush, eds., *Molecular Dynamics and Structures of Solids, Special Publication 301,* National Bureau of Standards, Washington, D.C., 1969.

G. Williams, *J. Chem. Soc. Faraday Trans.* **2,** 1045 (1972).

R. F. Field in A. R. von Hippel, ed., *Dielectric Materials and Applications,* MIT Press and John Wiley & Sons, Inc., New York, 1954, Chapt. 3.

R. H. Boyd in F. E. Karasz, ed., *Dielectric Properties of Polymers,* Plenum Publishing Corp., New York, 1971.

R. H. Boyd in R. A. Fava, ed., *Methods in Experimental Physics,* Vol. 16, Academic Press, Inc., New York, 1980, Pt. C.

S. Matsuoka
AT&T Bell Laboratories

DIELS-ALDER POLYMERIZATION.

See Cyclopentadiene polymers; Cyclopolymerization; Ladder and spiro polymers.

DIENE POLYMERS.

See Butadiene derivatives, polymeric; Butadiene polymers; Chloroprene polymers; Cyclopolymerization; Isoprene polymers.

DIE PRESSING. See THERMOFORMING.

DIES. See DIE DESIGN; EXTRUSION; MOLDS.

DIETHYLENE GLYCOL BIS(ALLYL CARBONATE). See DIALLYL AND RELATED POLYMERS.

DIFFERENTIAL THERMAL ANALYSIS. See THERMAL ANALYSIS.

DIFFUSION

The fundamental physical property that describes a diffusion or mass-transfer process is the mutual diffusion coefficient D. This article concerns isothermal diffusion in binary solutions consisting of a monodisperse, amorphous polymer and a low molecular weight penetrant. Results are presented for diffusion in both polymer solutions and in glassy or solid polymers, and it is indicated how results obtained for monodisperse, amorphous polymers can be used in the study of crystalline, branched, or cross-linked polymers and in the investigation of the effects of polydispersity on the diffusion process. However, diffusion in polymer–polymer systems is beyond the scope of this article.

Diffusional Transport

Deborah Number Diagrams. When a penetrant diffuses into a polymer, the polymer molecules rearrange toward a new conformation that is consistent with the concentration of the penetrant. This conformational change toward a new equilibrium state may not take place instantaneously, and the relative rates of the diffusion of penetrant molecules and the relaxation of polymer molecules to new conformations determine the nature of the diffusion process and lead to a wide variety of mass-transfer phenomena. Introduction of a dimensionless group, called the diffusion Deborah number, provides a means of characterizing the various types of diffusion behavior that are observed in polymer–penetrant systems. The introduction of the diffusion Deborah number attempts to place the temperature–concentration diagrams for such systems on a quantitative basis.

Viscous Diffusion. When the polymer–solvent system behaves like a purely viscous fluid, the diffusion process is described by the classical laws of diffusion.

Elastic Diffusion. It is commonly accepted that an elastic diffusion process can be analyzed by using the classical diffusion equation with a diffusion coefficient independent of concentration, even though the system is clearly not a purely viscous fluid mixture.

Viscoelastic Diffusion. When the diffusion Deborah number is on the order of unity, the diffusion process can be classified as viscoelastic, and the mass-transfer process exhibits a number of features that cannot be explained by the classical theory of diffusion. The anomalous or non-Fickian phenomena observed in the study of diffusional transport in polymer–solvent systems have been described in detail. The following examples of anomalous behavior have been observed for differential and integral sorption experiments in the study of penetrant diffusion in polymers: thickness anomalies; anomalous-sorption curves, case-II sorption; and dual-mode sorption.

Diffusion Coefficients

Mutual Diffusion Coefficients. The mutual diffusion coefficient is defined using a constitutive equation for $\boldsymbol{j}_\mathrm{I}^\mathrm{a}$, the mass-diffusion flux of component I relative to an appropriate average velocity. The constitutive equation for the diffusion flux takes the following general form for diffusion in binary mixtures:

$$\boldsymbol{j}_\mathrm{I}^\mathrm{a} = -D\overline{B}\nabla y_1 \tag{1}$$

where $\overline{B}$ is a factor that contains appropriate densities of the system, D is the binary mutual diffusion coefficient, and y_1 is some measure of the concentration of species I. In general, D is a function of concentration, temperature, and pressure.

Self-diffusion Coefficients. The mutual diffusion coefficient is a measure of the rate at which concentration gradients in a mixture are dissipated. On the other hand, self-diffusion coefficients measure the mobility of molecules. A self-diffusion coefficient represents the rate of diffusion of one of the components in a two-component system that has a uniform chemical composition. Inasmuch as the self-diffusion process involves the interchange of labeled and unlabeled molecules, the true mobility of the labeled molecules with respect to a stationary solution is measured since there is no bulk flow.

Measurement of Diffusion Coefficients. Measurement techniques include sorption, permeation, free diffusion, dynamic light scattering, radioactive tracers, and nuclear magnetic resonance.

Equations for Diffusion Coefficients. In the study of the diffusion process for a given polymer-solvent system at a particular temperature and pressure, it is convenient to formulate different theoretical expressions for D, depending on the polymer concentration and molecular weight.

Infinitely Dilute Region. The Kirkwood-Riseman theory provides the best method currently available for predicting the mutual diffusion coefficient at infinite dilution under theta conditions for linear, flexible-chain polymers.

Dilute Region. The dependence of the mutual diffusion coefficient in dilute polymer solutions on the polymer mass density ρ_2 can be expressed in the form

$$D = D_0(1 + k_D\rho_2) \tag{2}$$

where D_0 is the mutual diffusion coefficient at infinite dilution and where the coefficient k_D can be determined from the following equation:

$$k_D = 2A_2M_2 - k_\mathrm{S} - 2\hat{V}_{20} \tag{3}$$

Here, A_2 is the second virial coefficient for a dilute polymer solution, M_2 is the polymer molecular weight, $\hat{V}_{20}$ is the partial specific volume of the polymer in the limit of zero polymer concentration, and k_S is a coefficient in a series expansion for the concentration dependence of the friction coefficient f for a polymer molecule:

$$f = f_0(1 + k_\mathrm{S}\rho_2 + \cdots) \tag{4}$$

where f_0 is the value of f for zero polymer concentration. It is clear from equations 2 and 3 that the nature of the variation of D with concentration depends on the interactions of hydrodynamic, thermodynamic, and volumetric effects in the solution.

Concentrated Region. Transport in the concentrated region is too complicated to analyze rigorously using molecular mechanics because of the numerous degrees of freedom and the complex intramolecular and intermolecular interactions. The most comprehensive molecular theory appears to have been proposed by workers who derived equations for the prediction of mutual diffusion coefficients using the free-volume theory of transport.

Diffusion Characteristics in the Concentrated Region. The strong concentration dependence of D for typical polymer–solvent systems is illustrated in Figure 1, particularly at low solvent concentrations and low temperatures. It is clear that the concentration dependence of D near zero solvent mass fraction becomes less pronounced as the temperature increases. This figure also shows that D assumes a maximum value in the concentration interval.

Diffusion Coefficients for Dissolved Gases. In the limit of very small solvent concentrations, a free-volume equation for the mutual diffusion coefficient can be used to describe the diffusion of a trace of solvent in an amorphous polymer, and thus is applicable to the study of the temperature dependence of D for sparingly soluble materials such as some dissolved gases.

Fickian Diffusion in Glassy Polymers

It is reasonable to expect that the diffusion process is Fickian at sufficiently low penetrant concentrations below the polymer glass transition temperature T_g. It has been argued that free-volume theory should provide an adequate description of transport in glassy polymers, and a theory has been developed for describing the temperature dependence of D below T_g for a system consisting of a trace of solvent and an amorphous polymer. It is an interesting aspect of diffusional transport in glassy polymers that transport below T_g can be significantly enhanced by the extra hole-free volume that is effectively frozen into the polymer in the nonequilibrium or glassy state.

Crystalline, Branched, Cross-linked, and Polydisperse Polymers

Crystalline Polymers. Below the melting point, the analysis of the diffusion process is relatively straightforward, if it is assumed that the polymer crystals are completely impenetrable to the diffusing solvent molecules. In this case, the mutual diffusion coefficient can be computed by multiplying the diffusion coefficient in the amorphous phase by a tortuosity factor.

Branched Polymers. The methods described above for the prediction of the mutual diffusion coefficient and of the solvent self-diffusion coefficient for the concentrated region should also be applicable to branched polymers (qv), at least for low to moderate branching frequencies. However, the previous results for dilute polymer solutions are limited to linear, flexible-chain polymers, since branched and ring polymers and stiff chains introduce complications into the statistical mechanics of polymer chains.

Cross-linked Polymers. The presence of cross-links in a polymer generally leads to decreases in the mutual diffusion and solvent self-diffusion coefficients.

Polydisperse Polymers. Mutual diffusion and solvent self-diffusion coefficients for polymer solutions are essentially independent of molecular weight over most of the concentration

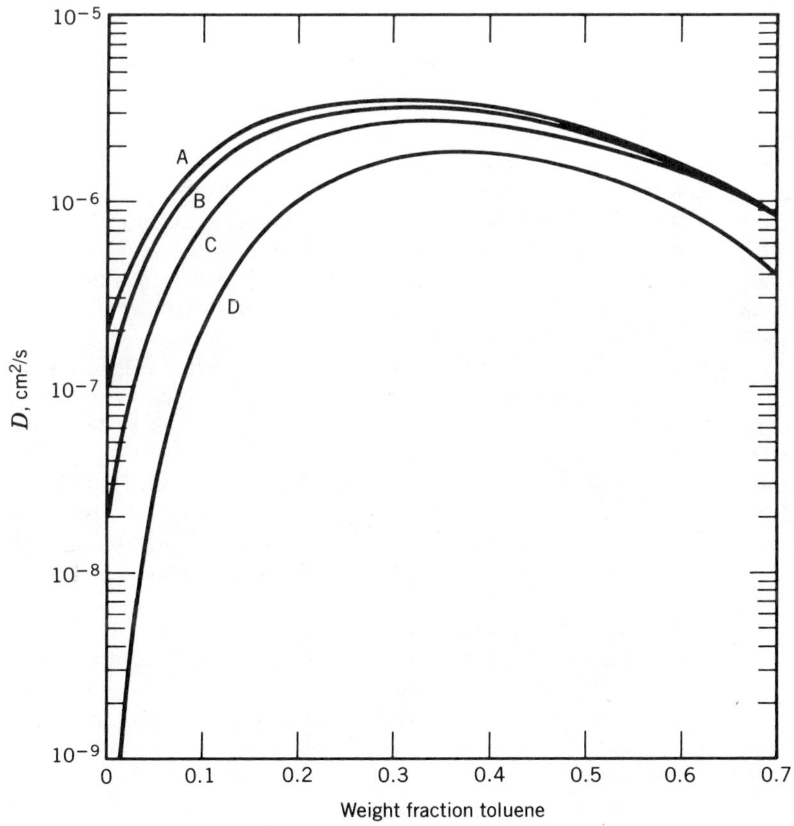

Figure 1. Variation of the mutual diffusion coefficient of the toluene–polystyrene system with temperature and concentration. A, 170°C; B, 160°C; C, 140°C; and D, 110°C.

range, and hence, effects of polymer polydispersity must be considered only for dilute polymer solutions.

BIBLIOGRAPHY

J. S. Vrentas and J. L. Duda, *AIChE J.* **25,** 1 (1979).

H. Fujita in J. Crank and G. S. Park, eds., *Diffusion in Polymers,* Academic Press, Inc., New York, 1968, Chapt. 3.

J. Crank, *The Mathematics of Diffusion,* 2nd ed., Clarendon Press, Oxford, UK, 1975.

R. B. Bird, W. E. Stewart, and E. N. Lightfoot, *Transport Phenomena,* John Wiley & Sons, Inc., New York, 1960.

J. S. Vrentas
J. L. Duda
The Pennsylvania State University

DIFFUSION, POLYMER–POLYMER

When dense macromolecular fluids are processed with changing temperature, pressure, or flow conditions, the establishment of thermal equilibrium may require a large-scale transport of macromolecules by diffusion. Long, flexible polymers relax slowly even in dilute solution; in denser states (concentrated solutions, melts, and blends), their diffusion is slowed down further due to mutual entanglements between the coils. Thus it is a nontrivial question whether in a process there is enough time for gradients in density, concentration, and other properties to fully relax. In fact, some applications of partially incompatible blends make use of the possibility of stopping phase separation in the fluid state by a quench into the amorphous solid state. Mechanical properties can then be controlled via a desired two-phase morphology. Polymer applications, such as optical applications, may require a superior homogeneity in density and composition or both. Clearly, it is important to characterize the diffusion coefficients experimentally, and to predict theoretically their dependence on molecular weight, temperature, and condition. These diffusion coefficients change by many orders of magnitude according to molecular weight. A further slowing of diffusive motions occurs when the temperature approaches the glass-transition temperature. Thus it has been difficult to develop reliable techniques to measure diffusion coefficients over a wide parameter range, and the number of polymers for which precise data are available is limited. Thus theoretical treatments that relate diffusion coefficients phenomenologically to more frequently studied quantities (eg, melt viscosities) are useful. Similarly, the relations between self-diffusion or tracer diffusion coefficients are of interest. These coefficients measure the mean-square displacement of a polymer with time. Cooperative diffusion coefficients, also called interdiffusion or collective diffusion coefficients, measure the concentration deviation of a species from its average concentration with time. However, the theoretical understanding of these problems is rudimentary (see also Macromolecular dynamics).

Definitions and Main Theoretical Predictions

The self-diffusion coefficient D_N of a macromolecule with polymerization index N describes the translational motion of the macromolecule as a whole. It is defined by the mean-square displacement of its center-of-gravity position vector $\mathbf{r}_{CG}(t)$ with time t (Einstein relation) as

$$D_N^* = \lim_{t\to\infty} \langle[\mathbf{r}_{CG}(t + t_0) - \mathbf{r}_{CG}(t_0)]^2\rangle/6t \qquad (1)$$

Here the brackets $\langle\ \rangle$ denote a statistical average, and thermal equilibrium is assumed (there is no dependence on the origin of time t_0). The limit $t \to \infty$ means that the time to be considered in practice must be larger than the largest intrinsic relaxation time τ_N of the polymer, the time scale on which all internal modes have relaxed. *For flexible polymers in solution or melt,* a simple scaling argument relates τ_N, D_N^* and the mean-square gyration radius $\langle R_{\text{gyr}}^2\rangle$ of the coil. Apart from factors of order unity, one has

$$D_N^* \approx \langle R_{\text{gyr}}^2\rangle/\tau_N. \qquad (2)$$

If τ_N can be estimated from the viscoelastic response, equation 2 can be used for an order-of-magnitude estimate of D_N^*.

The main facts about D_N^* in various situations are now summarized. In *dilute solution* coils do not overlap, and their motion is governed by hydrodynamic backflow forces mediated via the solvent (see Solution properties). This yields $D_N^* \sim 1/(\eta_s\langle R_{\text{gyr}}^2\rangle^{1/2})$, where η_s is the solvent viscosity.

In *semidilute solutions* the coils interpenetrate on length scales larger than the so-called screening length (or "blob size") ξ, which is much larger than the size of a monomer σ but much smaller than the size of the coil. For good solvents scaling considerations predict a concentration dependent $\xi \sim \sigma c^{-3/4}$, c being the polymer concentration. If N is not too large, the chains are only weakly entangled with each other, and qualitatively the Rouse model describes the chain dynamics, implying that $D_N^* \sim 1/N$. For large enough N and large enough concentration, however, the entanglements between the chains provide an effective tube of diameter d_T for each chain, in which it moves with a snakelike diffusive motion. The same mechanism is thought to describe polymer motion in polymer melts. It yields $D_N^* \approx \tau_1 d_T^2/N^2$, τ_1 being the characteristic time for reorientation of a single monomer. Note that d_T should have the same concentration dependence as ξ in semidilute solution, but $d_T >> \xi$. This is also true in a melt, where $\xi \approx \sigma$, while d_T, which is also of the order of the typical distance between entanglements, typically is of the order of 3–10 nm.

This *reptation model* implies $\tau_N \sim N^3$. For times $t < \tau_N$, it implies that $\langle \mathbf{r}_{CG}^2(t)\rangle \sim t^x$ with $x = 1/2$: this is an example of "anomalous diffusion". Anomalous diffusion (due to cooperative effects?) is also predicted by some theories to occur in polymer melts near the glass transition.

The above discussion of self-diffusion via reptation (R) is not the whole story, since the constraints forming the tube (entanglements with other chains) are not rigid, because all the surrounding other chains also move: this leads to constraint release (CR), an additional mechanism for diffusion. One has $D_N^* = D_R + D_{CR}$, where D_R is the diffusion constant as estimated above and D_{CR} the diffusion constant due to constraint release. This latter mechanism is particularly important if one considers the selfdiffusion of a high molecular weight polymer in a low molecular weight matrix, but is not fully understood theoretically.

In addition to self-diffusion, in polymer solutions a different diffusion coefficient is needed to describe how inhomogeneities in polymer concentration relax, the so-called cooperative diffusion coefficient D_{coop}. A Fourier component $\delta c_k(t)$ of a concentration fluctuation with wavelength $2\pi/k$ decays with time t as $\delta c_k(t) \sim \exp[-D_{coop}k^2t]$, where D_{coop} is inversely proportional to the solution viscosity and the hydrodynamic screening length ξ_H. It is not yet fully understood under which conditions $\xi_H \sim \xi$.

D_{coop} can be viewed as a special case of an interdiffusion coefficient (polymer–solvent interdiffusion). In binary polymer mixtures (AB), it is also of interest to consider interdiffusion of polymers of species A and species B: One considers the relaxation of a deviation $\delta\Phi(\mathbf{x},t) = \Phi(\mathbf{x},t) - \Phi_0$ of the local volume fraction $\Phi(\mathbf{x},t)$ of species A at point $\mathbf{x}$ from its average value Φ_0 with time t. Again the Fourier component $\delta\Phi_k(t)$ of $\delta\Phi(\mathbf{x},t)$ decays as $\delta\Phi_k(t) \sim \exp[-D_{int}k^2t]$. It is difficult, however, to predict D_{int} for a mixture from the properties of its constituents: both kinetic properties and thermodynamic properties enter. Even in a fully compatible mixture, where the polymers are mixed completely at random, it has been a difficult problem to find approximate relations between D_{int} and the self-diffusion coefficients of the chains contained in the mixture. In the general case, of course, the mixing is not ideal: in the framework of the Flory-Huggins model, the Flory-Huggins parameter χ is nonzero. There is a critical line $\chi_s(\Phi)$, the so called spinodal curve, in the temperature–concentration phase diagram of the mixture. One now predicts $D_{int} \sim \chi_s(\Phi) - \chi$, which implies that for $\chi > \chi_s(\Phi)$ homogeneous mixtures are unstable, since $D_{int} < 0$: the decay of mixtures quenched to such an unstable state by growth of concentration fluctuation (due to $D_{int} < 0$) is called spinodal decomposition and has also been amply observed, since most polymer mixtures are at least partially incompatible. The detailed description of the dynamics of concentration inhomogeneities in polymer blends is an active area of research.

Experimental Techniques

The concentration profile obtained after diffusion of a labeled component across an interface is analyzed in the techniques of infrared scanning, forward recoil spectrometry (fres), Rutherford backscattering (rbs), scanning electron microscopy (sem) and to some extent in small angle neutron and x-ray scattering. The diffusive decay of a pattern (or hologram) produced with photoreactive labels is monitored in the methods of fluorescence redistribution after pattern photobleaching (frapp) and forced Rayleigh scattering (frs). Nuclear spins are used as diffusive probes in the nmr pulsed field gradient technique. The decay of concentration fluctuations is detected in photon correlation spectroscopy (pcs) and pulse induced critical scattering (pics). The range of diffusion coefficients covered by these techniques extends from 10^{-5} to 10^{-17} cm^2/s.

BIBLIOGRAPHY

M. Doi and S. F. Edwards, *The Theory of Polymer Dynamics*, Clarendon Press, Oxford 1986.

H. H. Kausch and M. Tirrell "Polymer Interdiffusion," *Ann. Rev. Mater. Sci.* **19**, 341–77 (1989).

H. Sillescu and D. Ehlich, in J. P. Fouassier and J. F. Rabek, eds., *Applications of Lasers in Polymer Science and Technology*, CRC Press, Boca Raton, in press

K. Binder, *Colloid Polymer Sci.* **265**, 273 (1987).

KURT BINDER
HANS SILLESCU
University of Mainz

DIISOCYANATES. See ISOCYANATE-DERIVED POLYMERS; POLYURETHANES.

DIISOPROPENYLBENZENE POLYMERS

In the past, development of diisopropenylbenzene-based polymers was hindered by lack of a commercial source of the monomers. Introduction of *m*-diisopropenylbenzene to the marketplace by American Cyanamid in 1987 at \$6.05/kg (truckload quantities) should lead to more uses for this monomer and further development of diisopropenylbenzene-based polymers.

Physical Properties

m-Diisopropenylbenzene is a colorless liquid soluble in aromatic hydrocarbons and halogenated hydrocarbons and partially soluble in lower aliphatic alcohols. The para isomer is a white crystalline solid with solubility characteristics similar to those of the meta isomer. Important physical properties are given in Table 1.

Table 1. Physical Properties of Diisopropenylbenzenes

Property	Meta	Para
melting point, °C	−74	64–65
boiling point at 101.3 kPa[a], °C	231	245
density at 20°C, g/mL	0.925	0.965
flash point, °C	93	93

[a] To convert kPa to mm Hg, multiply by 7.5.

Chemical Properties

One key to the chemistry of diisopropenylbenzene is the stability of the tertiary benzylic carbenium ion derived during acid-catalyzed reactions. This stability precludes some carbenium ion reactions, such as the reaction with carbon monoxide and subsequent hydrolysis to give the diacid (Koch-Haff reaction) but allows for regioselective reactions with electron-rich aromatics. Reaction of *p*-diisopropenylbenzene with phenol in the presence of hydrochloric acid gives primarily the *p,p'*-bisphenol (**1**) instead of a random mixture of isomers.

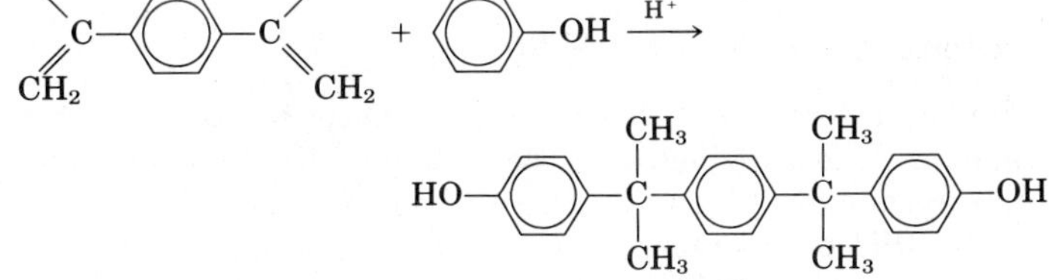

(**1**)

Similar behavior is seen in the reaction of aniline with *p*-diisopropenylbenzene, although a higher temperature is required. *m*-Diisopropenylbenzene also gives *p,p'*-substitution on electron-rich aromatics but the selectivity is not as pronounced as with the para isomer.

Both diisopropenylbenzene isomers readily add inorganic acids, such as hydrochloric acid, isocyanic acid, and isothiocyanic acid. Isocyanic acid and isothiocyanic acid are extremely reactive and are either formed by mixing a strong mineral acid with a slurry of the alkali metal cyanate or thiocyanate in a solution of diisopropenylbenzene, or are preformed and used rapidly.

Peroxy compounds are formed by adding alkylhydroperoxides or hydrogen peroxide to *p*-diisopropenylbenzene. In the former case bis(*t*-butylhydroperoxy)diisopropylbenzene is isolated from the reaction of *p*-diisopropenylbenzene with *t*-butyl hydroperoxide in the presence of a tertiary alkyl halide catalyst. In the latter case hydrogen peroxide reacts with *p*-diisopropenylbenzene to give diisopropylbenzene dihydroperoxide, which undergoes subsequent rearrangement to hydroquinone and acetone in the acidic reaction medium.

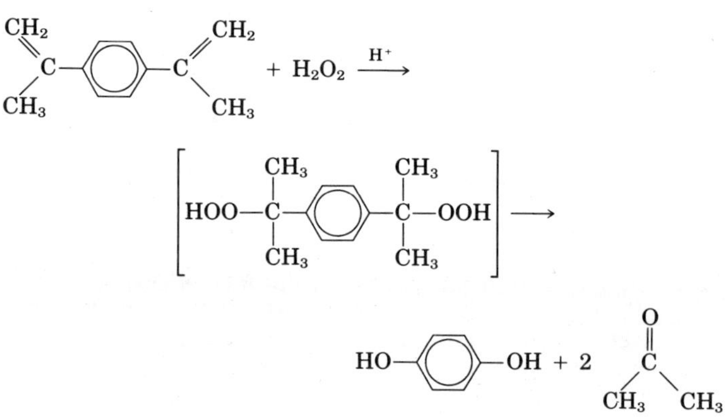

Other reactions of diisopropenylbenzenes closely parallel those of α-methylstyrene. Addition of dichlorocarbene to *m*-diisopropenylbenzene gives a mixture of mono- and bis-*gem*-dichlorocyclopropanes. Partial autooxidation of *p*-diisopropenylbenzene in the presence of transition metal catalysts is reported to give a mixture of *p*-isopropenylacetophenone and *p*-diacetylbenzene. *p*-Isopropenylacetophenone was obtained in 71% selectivity at 46% conversion in the presence of a vanadium oxyacetylacetonate catalyst. Mercaptans add to diisopropenylbenzene to give benzylic sulfides under acidic conditions or phenethyl sulfides under radical conditions.

Monomer Synthesis

The most convenient laboratory synthesis starts with commercial diisopropylbenzene. Benzylic bromination by *N*-bromosuccinimide followed by dehydrohalogenation with pyridine yields diisopropenylbenzene in 70–80% yield. Commercial quantities of monomer are prepared by dehydrogenation of diisopropylbenzene.

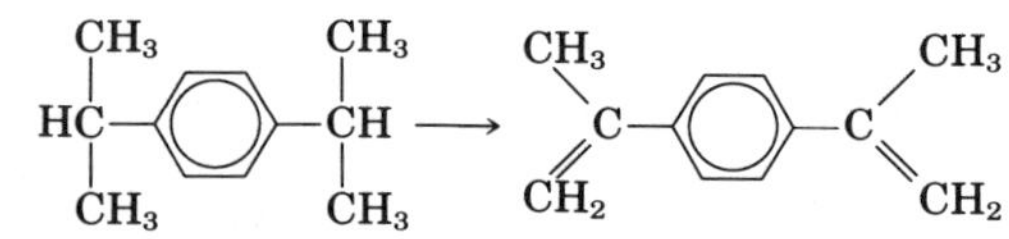

Polymerization

Diisopropenylbenzenes, bis(1-methylethenyl)benzenes, can be polymerized using anionic, cationic, and free-radical initiators. In both anionic and free-radical polymerizations, the pendent isopropenyl groups on the polymer are much less reactive than the isopropenyl groups of the monomer. Thus the pendent isopropenyl groups can act as sites for functionalizing the polymer. Many diisopropenylbenzene-based polymers and copolymers made via cationic polymerization show good oxidative stability for polymers containing aliphatic hydrocarbon moieties. Diisopropenylbenzene can be used to alkylate materials such as phenol, aniline, and isocyanic acid to give new monomers for step-growth polymerization.

Applications

The principal commercial applications of diisopropenylbenzene will most likely utilize derivatives that can be used as components in step-growth polymerization. A variety of functionalized aromatics and inorganic acids can be added to both *meta*- and *para*-diisopropenylbenzene to give difunctional monomers suitable for this type of polymerization. Diisopropenylbenzene derivatives also are used as polymerization initiators.

Health and Safety Factors

American Cyanamid's material safety data sheet states that *m*-diisopropenylbenzene is nonmutagenic in the AMES test. It is not irritating to rabbit eyes in laboratory studies and only causes minimal skin irritation. Flash point data are given in Table 1.

BIBLIOGRAPHY

H. Colvin and J. Muse, *Chemtech* **16**, 500 (1986).

U.S. Pat. 4,528,413 (July 9, 1985), H. Colvin, R. Fiedler, J. Muse, and D. Smith (to The Goodyear Tire & Rubber Co.).

P. Lutz, G. Beinert, and P. Rempp, *Makromol. Chem.* **183,** 2787 (1982).

B. Culbertson and C. Pittman, eds., *New Monomers and Polymers*, Plenum, New York, 1984.

Howard Colvin
Goodyear Chemical Research & Development Division

DILATOMETRY

Dilatometry is concerned with the measurement and study of dimensional changes of matter. Aside from providing data for engineering applications of polymers (including processing), dilatometry is used to study the physical and chemical phenomena associated with dimensional changes, such as melting, crystallization, glass formation, secondary transitions, polymerization, cross-linking reactions, fluid sorption, formation of voids and crazes during mechanical deformation, and others. The important role of pressure in several of these phenomena may require the inclusion of pressure as a variable. It is easiest to measure linear-dimensional changes of solid polymers at atmospheric pressure, using commercially available attachments to thermal analysis systems based on the principle of Figure 1. At the other end of the spectrum of commercially available instruments are pressure dilatometers, capable of measuring volume changes of solid and liquid polymers at pressures to 200 MPa (approx. 28,000 psi).

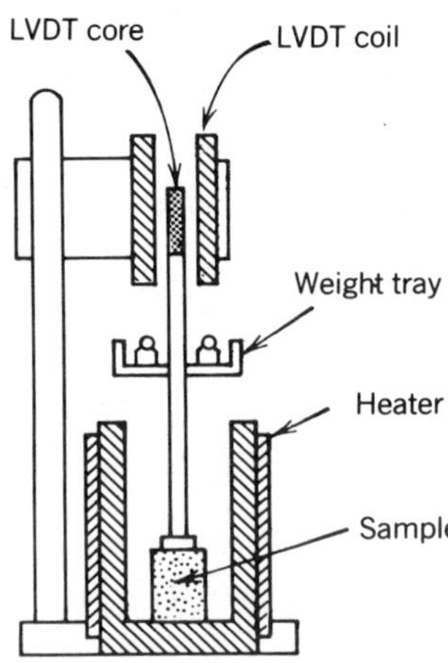

Figure 1. Schematic parts of a dilatometer for measuring linear-dimensional changes of rigid solids. LVDT = linear variable differential transducers.

BIBLIOGRAPHY

P. Zoller, P. Bolli, V. Pahud, and H. Ackermann, *Rev. Sci. Instrum.* **47,** 948 (1976).

J. W. Barlow, *Polym. Eng. Sci.* **18,** 238 (1978).

W. J. Coumans and D. Heikens, *Polymer* **21,** 957 (1980).

PAUL ZOLLER
E.I. du Pont de Nemours & Co., Inc.

DIMENSIONAL STABILITY, FIBERS

With the exception of dimensionally stable inorganic fibers such as glass, aluminum, and carbon fibers (qv), both natural and synthetic polymeric fibers exhibit various types of dimensional instabilities caused by applications of load, temperature, or solvents. Dimensional instabilities are usually detrimental to fiber quality, and numerous studies have been made to reduce the tendency of fibers to undergo reversible and irreversible deformational changes. In some cases, such as shrinkage in hot water or steam, instability is essential for activating latent strains in textured fibers and yarns and is exploited for technological purposes. This article is limited to synthetic fibers.

Morphological Characteristics of Fibers

Drawability is an important characteristic of fiber-forming polymers, ie, they can easily undergo large (several hundred percent) permanent deformation under load at ambient or elevated temperatures. Because of this inherent dimensional instability under load, unoriented polymers are not used in fibrous form.

The required dimensional stability is achieved by a drawing step during which the spun and essentially unoriented fiber is stretched several times its original length to form a dimensionally much more stable, highly uniaxially oriented structure (see also ORIENTATION; ORIENTATION PROCESSES).

Deformation in Tension (Creep)

Creep and recovery after creep are complex phenomena. In most cases, the data obtained at various conditions for a given fiber cannot be presented accurately in a mathematical equation. Within a narrow range of loads and temperatures, the data can sometimes be superimposed to yield master curves that allow prediction of results over a wide range of times, assuming time–temperature and time–load equivalence.

Structural Parameters Affecting Creep and Analysis. For polyethylene, polypropylene, poly(ethylene terephthalate) (PET), and other fibers, creep depends on molecular weight, draw ratio, and treatments affecting morphology, eg, heat setting, radiation, cross-linking, etc.

Critical Stress for Nonrecoverable Creep. With high molecular weight fibers and cross-linked and highly drawn fibers, the recovery is frequently complete up to a critical load above which permanent plastic deformation takes place.

For technological purposes it is important to note that creep can be reduced by increasing draw ratio and molecular weight, cross-linking, and copolymerization. Creep depends also on the manufacturing process of the fiber. The retarding effects are primarily caused by a slowdown of slip in oriented amorphous regions.

Plastic Deformation in Creep Versus Plastic Deformation in Drawing. It was shown that in uniaxially oriented low-density polyethylene, the most likely creep mechanism is interfibrillar slip coupled with chain slippage, a mechanism occurring also during drawing of fibers between intermediate and high draw ratios.

Measurement of Creep and Creep Recovery. Extensional creep and recovery can be measured with the apparatus shown in Figure 1.

Bending Behavior and Recovery

When a polymeric fiber is bent, the inner regions yield under compression. This plastic deformation inhibits the return of the filament after the removal of load into the original configuration. In general, oriented fibers exhibit poor bending recovery, and recovery capabilities decrease with increasing draw ratio. This is a consequence of a large difference between mechanical properties in tension and compression. Modulus and yield stress are much higher in tension than the corresponding values in compression. Bending recovery is important in the

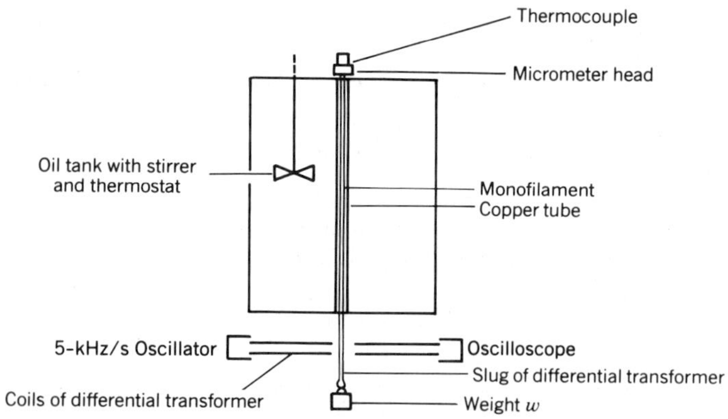

Figure 1. High temperature creep test apparatus. Courtesy of Pergamon Press.

design of crease-resistant fibers, resilience and appearance of carpets, wrinkling of fabrics on washing and drying, etc.

Measurement of Bending Recovery. To measure bending recovery, fiber samples are bent over wires of different diameters. After a certain time, the load is cut off at a distance from the wire, and the fiber is allowed to recover from the bending strain. After a preset recovery time, the bent parts of the sample are cut off, dropped on a black surface or placed between crossed-polarized sheets, and photographed. The recovery angles are measured from the photographs.

Shrinkage

Thermal Shrinkage. At elevated temperatures, thermoplastic fibers contract; near the melting point they assume longitudinal dimensions close to those before drawing. The thermal shrinkage is the result of the contraction of extended-chain tie molecules that assume random coil configurations or crystallize in chain-folded crystals.

Thermal shrinkage is important in the fabrication of composite structures such as tires, V-belts, and hoses, where temperatures can be sufficiently high to cause shrinkage leading to losses in strength and modulus. Exposure of fibers to elevated temperatures also activates latent asymmetric stresses introduced by various crimping processes, ie, a relatively straight fiber or yarn develops crimp upon heating. This characteristic is very important in the manufacture of carpets, stretchable fabrics, hosiery, etc.

Solvent-induced Shrinkage. When polymeric fibers are exposed to solvents, melting points and glass-transition temperature are depressed, resulting in shrinkage.

Solvent-induced shrinkage and solvent interaction with fibers are important in dyeing technologies and cross-linking treatments to improve crease resistance, develop permanent-press finish, or reduce shrinkage on washing (see TEXTILE RESINS).

BIBLIOGRAPHY

A. Peterlin in A. Ciferri and I. M. Ward, eds., *Ultra-High Modulus Polymers,* Applied Science Publishers, Ltd., London, 1979, Chapt. 10.

D. C. Prevorsek, "Recent Advances in Mechanical Properties of Fibers" in *Proceedings International Symposium on Fiber Science and Technology,* Hakone, Japan, August 1985, The Society of Fiber Science and Technology, Japan, 1985.

D. C. Prevorsek and Y. D. Kwon, *J. Macromol. Sci. Phys.* **12,** 447 (1976).

A. Takaku, *J. Appl. Polym. Sci.* **26,** 3565 (1981).

D. C. PREVORSEK
Allied Corporation

DIMENSIONAL STABILITY, PLASTICS. See POLYMER CREEP.

2,3-DIMETHYLBUTADIENE POLYMERS. See BUTADIENE DERIVATIVES, POLYMERIC.

DIMETHYLOLUREA. See AMINO RESINS.

DIOLEFIN POLYMERS. See BUTADIENE DERIVATIVES, POLYMERIC; BUTADIENE POLYMERS; CHLOROPRENE POLYMERS; CYCLOPOLYMERIZATION; ISOPRENE POLYMERS.

DIP COATING. See COATING METHODS.

DISACRYL. See ACROLEIN POLYMERS.

DISPERSING AGENTS. See COLLOIDS; EMULSION POLYMERIZATION.

DISPLACEMENT BLOW MOLDING. See BLOW MOLDING.

DISPOSAL AND REUSE OF PLASTICS

Plastics are common materials in everyday life and, because of their low density, are more visible in wastes than might be indicated by U.S. or worldwide production. Disposal of plastics is discussed without differentiation between thermoplastic and thermosetting materials. Recycling of thermosetting materials is more difficult because of their properties, but they are recycled as fuel and are used sometimes, by grinding, as fillers in new thermosetting materials. In addition, some vulcanized elastomers are recycled (see also RUBBER RECLAIMING).

Sources of Plastic Waste

In the United States, mixed municipal solid waste (MSW) typically contains from 4 to 7 wt % plastics, consisting mainly of polyethylene (PE), polypropylene (PP), polystyrene (PS), and poly(vinyl chloride) (PVC), sometimes in the form of laminates with paper, other plastics, or combined in other composite structures. Small amounts of engineering plastics may be included in the form of kitchen utensils and as various parts and mechanisms.

Construction and demolition wastes contain plastics from electrical conduit and wiring, flooring, plumbing, and insulation. The recycling of all but the metallic components, and items such as bricks, is difficult to achieve; disposal is by landfill, usually in special facilities.

A special source of plastics waste is discarded agricultural mulch (film) (see also AGRICULTURAL APPLICATIONS).

Amounts and Composition of Plastic Waste

Reliable statistics on the amount of plastics in wastes from all sources have not been available for several years. In 1974, ca 4.5×10^6 t of plastic wastes was discarded in the United States, ca 3.6×10^6 t in the European Economic Community, and ca 1.8×10^6 t in Japan. Updated estimates of waste generation are given in Table 1.

Table 1. 1984 Plastics Production and Projected Waste Generation

Type of plastic	Production 10^6 t	Discarded per year, %	Waste 10^6 t	% of total
polyethylene	6.00	65.6	3.94	65.3
polystyrene[a]	2.69	38.1	1.03	17.1
polypropylene	1.86	27.5	0.51	8.5
poly(vinyl chloride)	3.05	17.9	0.55	9.1

[a] Includes PS, ABS, and other copolymers.

Energy Conservation

Energy can be generated by using waste plastics as a source of fuel or chemicals, or conserved by substituting plastics for materials which are more energy intensive in their construction or use. Cumulative energy savings derived from plastics are given in Table 2.

Table 2. Energy Savings Derived from Plastics, 1977–1978

Application	Cumulative energy savings, PJ[a]	Years in the analysis
soft drink containers	12	5
pipe and fittings	1,723	7
blow-molded bottles	155	15
refrigerators and freezers	21	3
household fans	15	2
building insulation[b]	40	3
automobile, light-weight parts	110	11

[a] To convert J to Btu, divide by 1054.

[b] Average annual net energy savings from 30 yr of use of plastic insulation produced in the 3-yr period.

Disposal of Plastics Waste

Sanitary Landfill. The main disposal practice for municipal and other wastes is burial in the ground. A sanitary landfill is a facility operated under strict rules; odors, fires, vermin, insects, leachate, and visual intrusion must be minimized.

Incineration. Plastics alone or mixed with other wastes may be burned for disposal, with or without heat recovery. They may be included with other wastes for the production of a waste-derived fuel. Persistent concerns regarding the incineration of mixed waste, particularly when it contains chlorinated plastics, are the evolution of HCl leading to corrosion, air pollution, and dioxin formation.

Energy Recovery

The fuel value of waste plastics alone or mixed with other wastes is reclaimed by incineration with heat recovery. The waste plastics may be incorporated in some form of waste-derived fuel (such as refused-derived fuel (RDF)) or used by itself as fuel. Waste-derived fuels are used by themselves or as a supplement to coal or wood.

The methods for the preparation of RDF include at least some portion of the plastics fraction of the waste in the product fuel. This raises the heat value and lowers the moisture and ash content. The heat value of RDF, ca 22 MJ/kg (9500 Btu/lb) calculated on a hypothetical moisture and ash-free basis from laboratory measurements, reflects the small contribution of plastics compared to paper and cardboard. Extrusion can be used to densify RDF to *d*-RDF to produce a fuel that can be mixed with lump coal and burned on a stoker grate. Some wastes must be dried before extrusion; addition of plastics may aid the extrusion process and subsequent binding of the *d*-RDF into stable forms.

Energy and Materials Recovery by Chemical Conversion. Because of the large amounts of plastics in Japanese MSW, they have been investigating a two-bed pyrolysis system. The fluidizing medium circulates between the pyrolysis and combustion reactors; the latter generates steam from the medium heat-value gas obtained from RDF by pyrolysis (ca 23 MJ/m^3 or 22,000 Btu/m^3).

Chemical Conversion of Waste PET

The rapidly growing, large-scale use of PET beverage containers, and the generation of waste during the production of PET fibers, film, and textiles, have led to investigation of chemical conversion of these wastes.

Terephthalic acid monomer can be reclaimed from PET wastes by hydrolysis with water at 150–250°C under pressure or by aqueous NaOH. Low molecular weight esters can be formed by transesterification of the PET using catalysts.

Various processes for hydrolyzing PET waste are available, but only alkaline hydrolysis with aqueous ammonia gives a product not needing further purification.

Blends of PET and nylon can be treated with alkali at elevated temperature and pressure. Terephthalic acid is recovered by acidification and filtration; caprolactam (from nylon-6) is recovered as aminocaproic acid. PET fibers are separated from blends with cellulose fibers by carbonizing the cellulose in H_2SO_4.

Terephthalic acid, ethylene glycol, and silver have been recovered from scrap PET-based x-ray film by wet oxidation.

Plastics Recovery and Reuse from Industrial Waste

Plastics can be separated from mixtures of wastes such as cable chopping or x-ray film (PET and cellulose ester bases) by a method similar to froth flotation of minerals. The separation of industrial waste streams is good and the reported properties of the separated plastics indicate that they could be used on a commercial scale. However, this process is not in commercial operation.

Agricultural Mulch. Methods of recycling waste agricultural mulch film were developed in Europe and Japan. The film is shredded and passed through a water wash to remove sand and other foreign matter. The washed film is shredded further and washed again to separate PE from PVC. The plastics are mechanically dewatered and dried and the water effluents are cleaned.

Telephone and Electronic Equipment. Discarded equipment can be reground and remolded for less demanding applications. Plastic waste from electronic assemblies has been burned with waste solvents.

Wire and Cable Insulation. Scrap and discarded wire and cable are processed for recovery of the metallic conductors by comminution (usually in a knife shredder), followed by a vibrating table with air separation, which produces useful metal and a separation of plastics.

Lead-acid Battery Cases. A process for the wet grinding, washing, and air classifying of PP battery cases has been described.

Catering Wastes. Waste containing 35 wt % plastic material plus paper and laminates, glass, and food waste has been treated to recover polystyrene. Paper has been separated from recovered paper–plastics mixtures by acid hydrolysis of the paper portion. Attempts to separate plastics and paper by electrostatic means have been revived in an application to the paper–plastics fraction recovered by the Flakt process.

Recovery of Plastics from Household Waste

In the United States, few attempts have been made to reclaim plastics from MSW mechanically. A froth flotation method was applied to a mixture of hand-sorted articles recovered from the screened heavy fraction of an air classifier. The mixture was chopped into pieces of about 0.5 cm dia and separated into dirt and three fractions: polyolefins (PO); PS, contaminated with unidentified materials; and PVC, contaminated with rubber. Only the PO fraction had physical properties of commercial interest when molded.

Recovery from Plastic Bottles

PET Bottles. PET soft drink containers with base cups of high density PE (HDPE) are recycled to significant extent. The PET bottles reportedly represent 165×10^3 of PET and 60×10^3 of HDPE per year. The market for such bottles is growing worldwide and much attention has been focused on the recycling after voluntary or forced-deposit collection. Reportedly, about 25% of these containers are now returned. The processing train to separate usable PET usually includes shredding, washing (often in hot caustic to remove labels and to pulp paper), sink–float in water to separate the PET from the HDPE, and drying.

The principal markets for recovered PET are fiberfill insulation (for clear resin; colors are rarely accepted), strapping, and chemical conversion to unsaturated polyester molding compounds. The 1981 market and projections to 1988 are given in Table 3.

Table 3. 1980 and Projected Use of Recovered PET

Application	Total PET, 10^3 t/yr	Recovered usage 1980, 10^3 t/yr	Projected recovered usage 1988, 10^3 t/yr
fiberfill	109	4.5	2.3
strapping	5.4	0.5	1.4
unsaturated polyesters	negligible	negligible	<4.5

Reprocessing of Recycled Plastics

Many recycling processes rely on thermal processing (extrusion or injection molding) of recovered plastics. Thus there is the potential of thermal or oxidative degradation of the plastic material. The amount of degradation depends on processing factors and the presence of stabilizers and antioxidants. This degradation may cause discoloration and changes in electrical insulating properties, limiting the applications of the recovered materials.

BIBLIOGRAPHY

Plastic Waste. Resource Recovery and Recycling in Japan, Plastic Waste Management Institute, Tokyo, 1985.

A. Skordilis in M. P. Ferranti and G. L. Ferrero, eds., *Sorting of Household Waste and Thermal Treatment of Waste,* Elsevier Applied Science Publishers, London, 1985.

H. Alter, *Materials Recovery from Municipal Waste. Unit Operations in Practice,* Marcel Dekker, Inc., New York, 1983.

J. Leidner, *Plastics Waste. Recovery of Economic Value,* Marcel Dekker, Inc., New York, 1981.

HARVEY ALTER
U.S. Chamber of Commerce

DISTRIBUTION OF MOLECULAR WEIGHTS.

See FRACTIONATION; MOLECULAR WEIGHT DETERMINATION.

DIVINYL POLYMERS.

See BUTADIENE DERIVATIVES, POLYMERIC; BUTADIENE POLYMERS; CHLOROPRENE POLYMERS; CYCLOPOLYMERIZATION; ISOPRENE POLYMERS.

DONNAN EFFECT.

See COLLOIDS.

DOUBLE REFRACTION.

See OPTICAL PROPERTIES; BIREFRINGENCE.

DRAG REDUCTION

Drag reduction in polymer solutions is the phenomenon whereby extremely dilute solutions of high molecular weight polymers exhibit frictional resistance to flow much lower than the pure solvent.

The principal interest to date has been in attempting to relate the effect to the fluid mechanics of turbulence and turbulent flow. These areas are so complex that progress has been slow, and the understanding of the flow details in polymer solutions is limited.

Polymers of molecular weight above 100,000, of a generally linear structure, and soluble in the liquid of interest, reduce turbulent friction by up to 70–80%. Many natural and synthetic polymers fall into this category.

Turbulent flow is defined here in the engineering sense of

the flow exceeding a critical Reynolds number, which is for pipes:

$$\text{Reynolds number}_{\text{pipe flow}} = \frac{VD}{\nu} > 2{,}300$$

and for an external flow such as over a ship hull:

$$\text{Reynolds number}_{\text{external}} = \frac{VL}{\nu} > 500{,}000$$

where V = average flow velocity, D = pipe diameter, L = axial length, and ν = kinematic viscosity of the polymer solution.

The quantities of polymer required to influence the turbulent friction are surprisingly low; in the laboratory, substantial friction reductions are noticed for polymer concentrations as low as 1 ppm by weight (mg/L) for the most effective substances. More commonly, concentrations in the tens or hundreds ppm give maximum effectiveness.

Synthetic water-soluble polymers such as poly(ethylene oxide) and polyacrylamide are extremely effective drag reducers, as is oil-soluble polyisobutylene. Commercially available highly effective natural polymers include guar (an extract from *Cyamopsis tetragonologus*) and *Xanthamonas* exudates. The requirements for polymer drag reduction include turbulent flow, high molecular weight, good solubility, linearity, and a relatively dilute solution.

Soap and surfactant solutions and suspensions of fibers or other solids also impart drag reduction. The mechanism with these substances is possibly similar to that occurring in polymer solutions. Alkali metal and ammonium soaps have good friction-reducing properties. Some commercially available nonionic surfactants are good drag reducers, as well as suspensions of fibers such as wood pulp, nylon, and asbestos. Dust and sand particles in suspension in both air and water reportedly reduce friction.

The drag-reduction effect seems to depend in part on the stretching of individual molecules by high strain rates in the flow. Elongated particles, ie, the polymer molecules, align in the direction of the principal strain rate and large extensions result. At the same time, a form of strain-rate hardening occurs in which the elongational viscosity becomes very high, and a limit is forced on the strain rate. This limiting strain rate corresponds to a lower strain rate than would occur without the presence of the polymers. The high elongational viscosity appears to play a role in reducing turbulent fluctuations, thus reducing the friction.

Table 1. Commercial Drag-reducing Polymers

Type[a]	Manufacturer
Poly(ethylene oxide)	
Polyox	Union Carbide Corp.
Polyacrylamides	
Separan	Dow Chemical Co.
TRO	Calgon Corp.
Magnafloc	Allied Colloids
Superwater	Berkeley Chemical
Xanthans	
Kelzan	Kelco Co.
Rhodopol	Rhone-Poulenc
Guar	
Jaguar	Stein, Hall & Co., Inc.
Oil-soluble polymers	
CDR	Conoco, Inc.
FLO	ChemLink Petroleum, Inc.

[a] All trade names listed are registered trademarks.

Drag-reducing Polymers

Commercial Sources. Although many polymers exhibit drag-reducing properties, only certain materials are marketed with this purpose in mind (Table 1).

Applications

Polymer drag reduction is used in fire fighting, storm-sewer flow augmentation, and pipelines. The most notable current use is in the trans-Alaska pipeline.

BIBLIOGRAPHY

J. W. Hoyt, *Trans. ASME* **94D,** 258 (1972).

R. H. J. Sellin, J. W. Hoyt, O. Scrivener, and J. Pollert, *J. Hydraul. Res.* **20,** 29, 235 (1982).

R. H. J. Sellin and R. T. Moses, eds., *Proceedings of the 3rd International Conference on Drag Reduction, Bristol, UK,* University of Bristol, Bristol, UK, 1984.

Jack W. Hoyt
San Diego State University

DRAPE FORMING. See Thermoforming.

DRAWING. See Fibers, manufacture.

DRIERS. See Drying.

DRILLING MUD ADDITIVES. See Oil-field applications.

DRUG APPLICATIONS

During the last decade, research in the pharmaceutical industry has been aimed at adjusting concentrations of drugs in the body to the therapeutic need and at increasing the duration of drug action at the target site. Using polymers, better dosage forms have been developed; they are termed controlled-delivery pharmaceutical dosage forms. Polymers utilized in controlled drug-release systems are listed in Table 1.

Components

Drug. The drug is chosen for a number of reasons, including its therapeutic effect, pharmacokinetic behavior, physicochemical characteristics, and the length of its half-life.

Delivery Module. The delivery module regulates drug release from the dosage form according to a predetermined therapeutic program. The module comprises a reservoir to store and protect the drug, a rate controller to establish and main-

Table 1. Polymers Utilized in Controlled Drug-release Systems

Polyesters
poly(glycolic acid)
poly(lactic acid)
glycolic–lactic acid copolymer
polycaprolactone
Polyorthoesters
polydioxyalkyltetrahydrofuran[a]
copolymer of poly(3,9-bis(methylene)-2,4,8,10-tetraoxaspiro(5,5)undecane with 1,6-hexanediol)
Polyacids
olefin–maleic anhydride copolymers
polyacrylates
polymethacrylates
Hydrogels
poly(ethylene glycol monomethacrylate)
gelatin cross-linked with formaldehyde
Cellulose
dextran
Polypeptides
poly(glutamic acid)
glutamic acid–leucine copolymer
Polyaminotriazole
alkyleneaminotriazole copolymers
Albumin beads
albumin cross-linked with glutaraldehyde

[a] Alzamer matrix.

tain the selected pattern of drug administration, an energy source to effect drug release from the dosage form, and a delivery portal. The nature, size, and shape of the delivery module vary according to the design of the dosage form.

Platform. The platform integrates the various elements of the controlled-release dosage form and permits appropriate site placement. It must be reliable and safe, as well as easy to use. The platform's constituents must neither interact with nor affect the drug. Furthermore, they should not be affected by sterilization or packaging processes.

Therapeutic Program. A controlled-release dosage form is characterized by the therapeutic program, which is expressed as the rate of drug release per time unit and the total duration of release.

Sites and Portals

Although the sites of administration for controlled-release drugs are not necessarily those used for conventional drug administration, they are generally the most accessible surfaces or orifices, including skin, eyes, mouth, lungs, gastrointestinal tract, brain, nose, vagina, and rectum. All present barriers between the site of introduction and the blood stream. Injections and infusions create portals by penetrating the skin.

Delivery Methods

Delivery methods include liquid infusion, electrical systems, diffusion, osmosis, and other mass-transport mechanisms, eg, bioerosion.

Liposome Drug-delivery Systems

Liposomes are lipid bilayer structures formed primarily of phospholipids, although other biomolecules may also be involved. Both polar and nonpolar drugs, enzymes, or other substances can be dissolved in the central aqueous compartment of liposomes, from which they diffuse into the systemic circulation or tissue.

Despite their initial promise, results of recent studies suggest that liposomes have only limited use in controlled-release systems. Liposomes are transferred from the systemic circulation only to the liver, spleen, and reticuloendothelial system; they readily concentrate in the latter, where they are degraded by lysosomes. Use of liposomes might be advantageous for the delivery of drugs to these sites.

BIBLIOGRAPHY

J. Urquhart, ed., *Controlled-Release Pharmaceuticals,* American Pharmaceutical Association, Washington, D.C., 1981.

J. R. Robinson, ed., *Sustained and Controlled Release Drug Delivery Systems,* Marcel Dekker, Inc., New York, 1978.

F. E. Yates, H. Benson, R. Buckles, J. Urquhart, and A. Zaffaroni in J. H. U. Brown and J. F. Dickson III, *Advances in Biomedical Engineering,* Academic Press, Inc., New York, 1975.

K. Heilmann, *Therapeutic Systems, Pattern-specific Drug Delivery: Concept and Development,* Georg Thieme Publishers, Stuttgart, 1978.

HARRIET BENSON
ALZA Corporation
BARBARA HARLEY
DNAX Research Institute
EDWARD E. SCHMITT
Cooper Vision, Inc.

DRYING

Drying is a unit operation in which a volatile liquid is separated from a solid or semisolid material by vaporization. In dehydration, vegetable and animal products are dried to less than their natural moisture contents, or chemically bound moisture is removed from chemical compounds. In freeze drying, the wet material is cooled to freeze the liquid; vaporization then proceeds by sublimation of ice. In evaporation, both evaporator feed and product are pumpable fluids; the equipment is different from that employed in drying. The drying of monomers also involves the removal of moisture or other volatile material from fluids. When drying gases, condensable vapors are separated from noncondensable gases by cooling to below the dew points of the condensable fractions, vapor adsorption on solid desiccants, absorption in desiccant liquids, or gas compression and cooling.

Polymer drying is an important industrial operation employed to isolate the salable product from the reaction process. Before melting and reextrusion for molding or spinning, many polymers must be dried again because they are slightly hygroscopic and susceptible to degradation if melted in the presence of moisture. Drying is usually necessary following the three common polymerization processes: solution, suspension, and emulsion polymerization.

In drying, material is transferred from one phase to another. Phase transfer is complicated by the need to transfer heat and mass simultaneously, but in opposite directions. The liquid must be vaporized before it is transferred, and most importantly, the presence of a third, solid phase interferes with heat, liquid, and vapor movement and retards the overall process because of the physical structure of the solid. In most dryers, heat is transferred first to the material surface and then into the interior. Convection, conduction, radiation, and dielectric or microwave heating are the usual external heat-transfer modes.

Many dryers employ more than one of these modes; nonetheless, most industrial dryers are characterized by one that predominates. Mass transfer during drying involves the removal of surface moisture and movement of internal moisture to the surface. Material structure and internal liquid and vapor-flow mechanisms control the drying rate when internal moisture is removed.

The capability of the equipment to handle solids is of greatest importance. The solid must enter one end of the dryer, leave at the other, and not linger too long in passage. If possible, drying should be preceded by crystallization, coagulation, and mechanical separation of solid and liquid in order to minimize the liquid quantity that must be vaporized. Liquid separation without vaporization is always less expensive than drying.

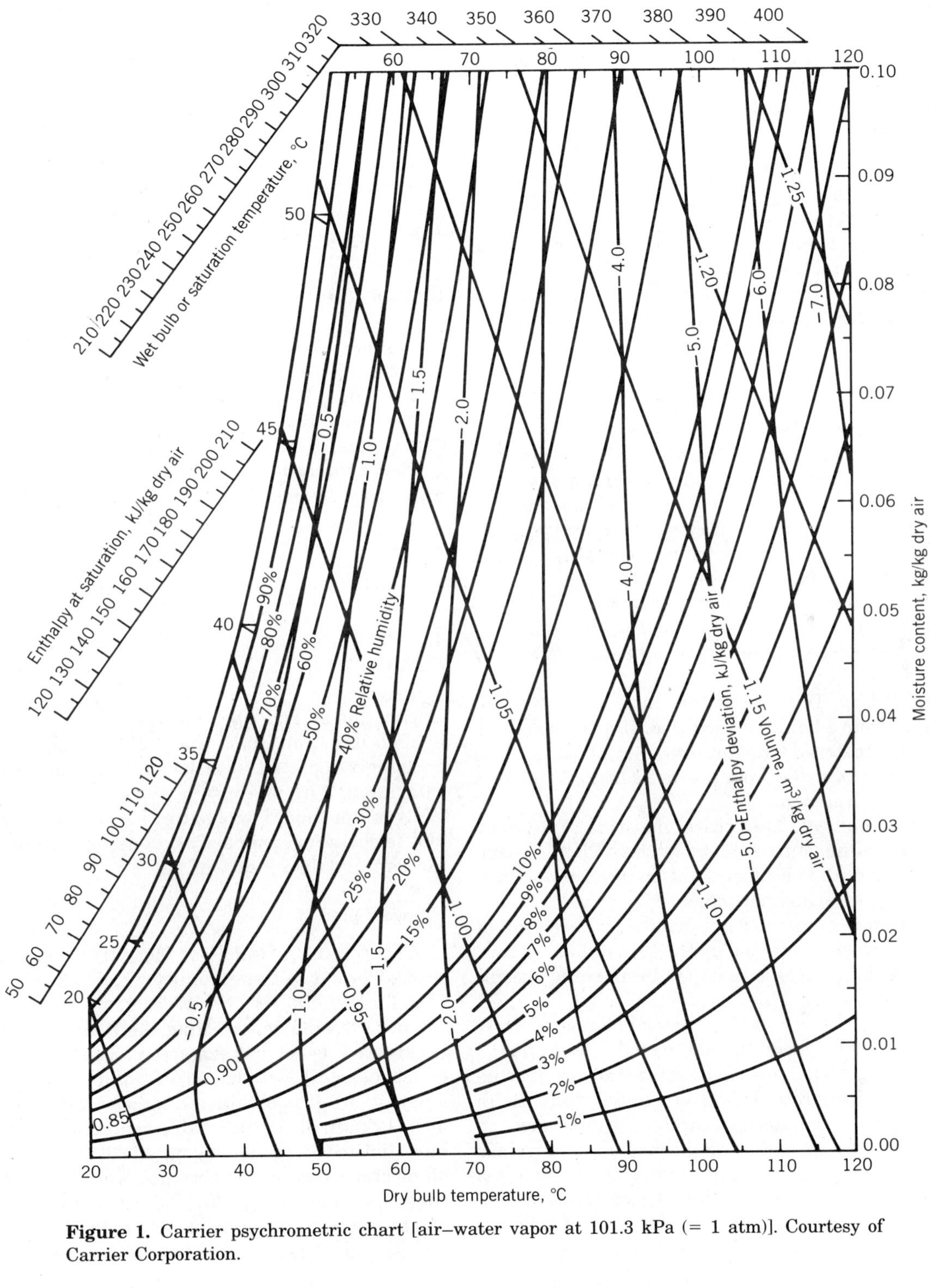

Figure 1. Carrier psychrometric chart [air–water vapor at 101.3 kPa (= 1 atm)]. Courtesy of Carrier Corporation.

Costs depend on many factors, including plant size, materials of construction, and continuous or batch operation. In order of importance, the bases for dryer selection are safety, product moisture and quality, solids-handling capability and versatility, heat- and mass-transfer efficiency, energy availability, labor costs, and capital costs. Continuous dryers are usually more economical than batch dryers. Costs rise rapidly as plant size falls below 500 t per year. Atmospheric-pressure batch dryers are about one-third as costly as batch vacuum dryers, which in turn are about one-third as costly as vacuum freeze dryers. A once-through air dryer is about half as expensive as one employing inert gas recirculation. Energy may be saved by gas recirculation, but the choice is usually a question of safety rather than cost. Microwave and freeze dryers are the most expensive in investment and operating costs when evaluated on a unit-of-evaporation basis.

Psychrometry

Before drying can begin, a wet material must be heated to such a temperature that the vapor pressure of the liquid content exceeds the partial pressure of the corresponding vapor in the surrounding atmosphere. The effect of the atmospheric vapor content of a dryer on drying rate and material temperature is conveniently studied by construction of a psychrometric chart (Fig. 1).

Drying Mechanisms

Although it is sometimes possible to select a suitable drying method simply by evaluating variables such as humidities and temperatures when removing unbound moisture, many polymer drying processes involve removal of bound moisture retained in capillaries among fine particles or moisture actually dissolved in the polymer. A knowledge of internal liquid and vapor mass-transfer mechanisms is essential. These mechanisms are best identified by measuring drying-rate behavior under controlled conditions. A change in material-handling method or any operating variable, such as heating rate, may affect mass transfer.

Constant-rate Drying. During constant-rate drying, vaporization proceeds as from a liquid surface of constant composition and vapor pressure; material structure has no influence. Drying proceeds by diffusion of vapor from the wet surface through a gas film into the environment.

Critical Moisture Content. The critical moisture content, ie, the average material moisture content at the end of the constant-rate drying period, is a function of material properties, the constant-drying rate, and particle size.

Equilibrium Moisture Content. Equilibrium moisture content is a steady-state condition obtained by the gain or loss of moisture when a material is exposed to an environment of specific temperature and humidity. This equilibrium condition, independent of drying method or rate, is a material property. Only hygroscopic materials have equilibrium moisture contents.

Falling-rate Drying. The principal mass-transfer mechanisms controlling falling-rate drying are liquid diffusion in continuous, homogeneous materials; capillarity in porous and fine granular materials; gravity flow in granular materials; vapor diffusion in porous and granular materials; flow caused by shrinkage-induced pressure gradients; and pressure flow of liquid and vapor when a material is heated on one side and vapor escapes from the other.

Capillarity. The outer surface of a porous solid has pore entrances of various sizes. As surface liquid is evaporated during constant-rate drying, a liquid meniscus forms across each pore entrance, and capillary forces are set up by interfacial tension between the liquid and solid. These forces draw liquid from the interior to the outer surface.

Diffusion. Liquid diffuses through materials in which it is soluble. Vapor diffuses in the gas phase when liquid is vaporized in isolated pockets in porous and granular materials. Diffusion-controlled drying is assumed when liquid or vapor flow appears to conform to Fick's second law of diffusion:

$$\frac{\delta c}{\delta \theta} = D_{AB}\frac{\delta^2 c}{\delta z^2}$$

where c = concentration of one component in a two-component phase of A and B; θ = diffusion time; z = distance in the direction of diffusion; and D_{AB} = binary diffusivity of the mixture A–B. This equation is the nonsteady-state diffusion equation in mass-transfer notation. It applies to diffusion in solids, stationary liquids, and stagnant gases.

BIBLIOGRAPHY

R. H. Perry and D. W. Green, *Perry's Chemical Engineers' Handbook,* 6th ed., McGraw-Hill, Inc., New York, 1984, Sect. 20.

R. E. Treybal, *Mass Transfer Operations,* 3rd ed., McGraw-Hill, Inc., New York, 1980.

J. Crank and G. S. Park, *Diffusion in Polymers,* Academic Press, Inc., New York, 1968.

R. B. Keey, *Introduction to Industrial Drying Operations,* Pergamon Press, Inc., Elmsford, N.Y., 1978.

P. Y. McCormick
E.I. du Pont de Nemours & Co., Inc.

DRYING OILS

Drying oils oxidize upon exposure to air from a liquid film to a solid, dry film. Since the 1920s, drying oils have been increasingly replaced by other film-forming materials (see also Coatings). Although use of synthetic drying oils is substantially larger than that of natural drying oils, their consumption also is declining, primarily because oxidation leads to embrittlement and discoloration.

Natural Oils

Occurrence and Isolation. Most drying oils are derived from seeds. The largest volume drying oil, linseed oil, is obtained from flax, *Linum usitatissimum.* Linseed oil is isolated by continuous pressing, ie, expelling, oil from the seed; a further yield is obtained by solvent extraction. The second most important oil, soybean oil, without modification, is a semidrying oil, not a drying oil. It is obtained from soybeans, the seed of *Glycine max* (L) Merrill. Perilla, safflower, sunflower, and walnut oils have limited usage as drying oils. Tung oil, also called wood oil or chinawood oil, is obtained from the seed kernels of the tung tree, *Aleurites fordii.* The only significant animal drying oils are marine fish oils, primarily from menhaden, pilchard, and sardines. Fish oil is isolated by steam treatment of the fish. For use as drying oil, the fraction of saturated acids is reduced by cooling followed by filtering off of the solid, primarily saturated, triglycerides which freeze out.

Composition and Analysis. Naturally occurring drying oils are triesters of glycerol (1,2,3-propanetriol) with mixtures of fatty acids. Their reactivity results from the presence of esters of fatty acids with two or more double bonds separated by single methylene groups or two or more conjugated double bonds.

Typical compositions are given in Table 1.

Table 1. Typical Fatty-acid Composition of Drying Oils, %[a]

Oil	Saturated	Oleic	Linoleic	Linolenic
linseed	9	17	14	60
perilla	7	14	16	63
soybean	16	24	51	9
tung[b]	6	4	8	

[a] Actual compositions can vary greatly.
[b] Also 82% α-eleostearic acid.

Synthetic and Modified Drying Oils

Synthetic and modified drying oils include varnishes, conjugated oils, esters of higher functionality polyols, oils modified with maleic anhydride (maleated oils), and conjugated and nonconjugated drying oils reacting with vinyl monomers.

Economic Aspects

Data on consumption of oils are listed in Table 2.

Table 2. U.S. Consumption of Drying Oils, 10^3 Metric Tons

	Paint[a]		Resins[b,c]		Other[c,d]	
Oil	1972	1982[e]	1972	1982[e]	1972	1982[e]
fish	4.2	2.0	NA	NA	NA	1.8
linseed	73.7	19.3	10.8	9.5	NA	NA
soybean	35.4	17.7	25.0	42.6	NA	22.1
tung	8.2	2.6	1.7	1.5	4.6	2.5

[a] Including varnish, enamel, and similar products, but not drying oils.
[b] Plastics, alkyds, plasticizers.
[c] NA = not available.
[d] Drying oils, glazing and caulking compounds, printing inks.
[e] Oct. 1981–Sept. 1982.

Uses

The largest current industrial use of natural drying and semidrying oils is as raw materials, either directly or as the fatty acids obtained by saponification, in the manufacture of oxidizing alkyds, epoxy esters, urethane oils, and synthetic drying oils. Major uses of natural and synthetic drying oils are coatings and printing inks. Tall-oil fatty acids are also used to manufacture dimer acids.

BIBLIOGRAPHY

A. E. Rheineck and R. O. Austin in R. R. Myers and J. S. Long, eds., *Treatise on Coatings,* Vol. 1, No. 2, Marcel Dekker, Inc., New York, 1968, pp. 181–248.

D. H. Solomon, *The Chemistry of Organic Film Formers,* 2nd ed., Robert E. Drieger Publishing Co., Huntington, N.Y., 1977, pp. 35–74.

D. Swern, ed., *Bailey's Industrial Oil and Fat Products,* Vol. 1, John Wiley & Sons, Inc., New York, 1979, Chapt. 10; Vol. 2, Chapt. 5.

ZENO W. WICKS, JR.
North Dakota State University at Fargo

DYEING

Large numbers of dyes have been synthesized in laboratories around the world, a few hundred of which have found their way to large-scale production and industrial application. However, this process is slowing down as development of new dyes becomes more expensive, and safety, health, and ecological requirements become more stringent.

Dyeing describes the fashioning of a new and often permanent color, especially by impregnating with a dye. It is generally used in connection with fibers and textiles, paper (qv), and leather (qv).

Dyes are usually dissolved or dispersed in a liquid medium before being applied to a substrate onto which they are fixed by chemical or physical means, or both. Due to its suitability, availability, and economy, water is usually the medium used in dye application; however, nonaqueous solvents have been much studied in recent years.

Textile substrates can be classified into cellulosic, protein, and synthetic polymer fibers; the first two categories are hydrophilic and the last is hydrophobic. Cellulosic fibers are vegetable fibers, such as cotton (qv), linen, and jute (see FIBERS, VEGETABLE), and regenerated cellulose fibers such as rayon (qv). Protein fibers comprise wool, silk (qv), and various animal-hair fibers. The synthetic polymeric fibers may be classified according to their ionic properties: nonionic fibers are cellulose acetate and triacetate and conventional polyester fibers; cationic fibers are polyamides; and anionic fibers are modified polyacrylonitriles (see ACRYLIC FIBERS; CELLULOSE ESTERS, ORGANIC; POLYAMIDES, FIBERS; POLYESTERS, FIBERS). Even and economical distribution of a small amount of dye, ie, generally 3 wt % of dry goods, throughout the substrate and fixation of the dye in the substrate are the keys to successful commercial dyeing.

Production of dyeings of acceptable quality requires many auxiliary products and chemicals, such as bleaching agents (see BLEACHING), wetting and penetrating agents, leveling and retarding agents, and lubricating agents. Other agents speed the dyeing process or are used for dispersion, oxidation, reduction, or removal of dyes from poorly dyed textiles. Fastness properties can be improved by chemical treatment.

The U.S. Federal Trade Commission lists 13 classes of dyes. Approximately 20% by volume of all dyes consist of fluorescent whitening agents, 16% of vat dyes, 14% of direct dyes, 13% of disperse dyes, 10% of acid dyes, 6% of basic dyes, and smaller percentages of azoic, fiber-reactive, food, drug, and cosmetic, mordant, solvent, sulfur, and other dyes.

In 1982, approximately 106×10^3 t of dyes valued at about $\$810 \times 10^6$ were consumed in the United States. Approximately 13.5×10^3 t, valued at ca $\$210 \times 10^6$, were imported. It is estimated that perhaps 65% of dyes consumed are used

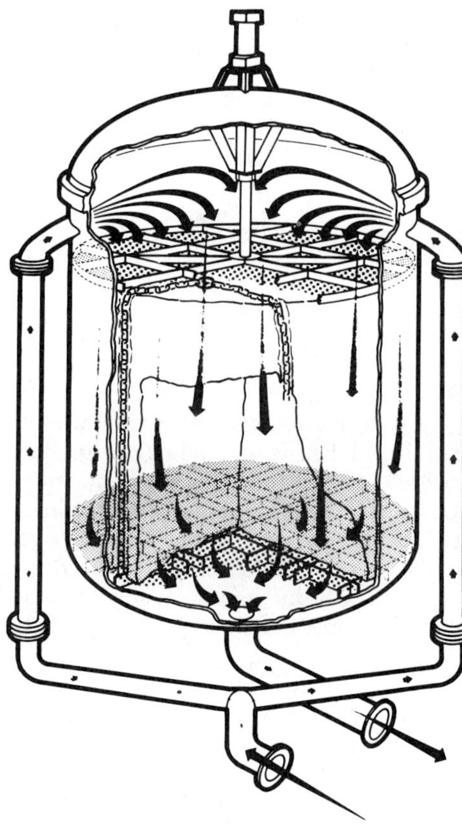

Figure 1. Schematic raw-stock dye machine. Courtesy of Gason County Dyeing Machine Co.

in the textile industry and the remainder for paper, leather, detergents, cosmetics, plastics, food, etc.

Theories

The dyeing process is dynamic. It normally entails changes in the dyeing temperature, dye concentration, bath composition, and fiber structure. Although aspects of the dyeing process can be accurately described with modern theories and related models, a complete description of the dyeing process at the molecular level is not yet available.

Affinity and diffusion are fundamental aspects of the dyeing process. The former describes the force by which the dye is attracted by the fiber, whereas the latter describes the speed with which it travels within the fiber from areas of higher concentration to areas of lower concentration. Alone or together with fiber characteristics, affinity and diffusion determine the speed of dyeing, its temperature dependency, the equilibrium exhaustion of the dye, and some of its fastness properties.

Dyeing processes are either nonionic or ionic. The former refers to the dyeing of synthetic fibers such as polyester, nylon, and triacetate with disperse dyes. It is governed, as a first approximation, by diffusion behavior according to Fick's second law and by a Nernst-type isotherm. The dye is bonded to the fiber by hydrogen bonds, charge-transfer forces, and van der Waals forces. The ionic dyeing process implies the presence of ionic groups in fibers and dyes and an ion-exchange mechanism. Ion exchange is clearly involved in dyeing acrylic fibers with basic dyes, nylon with acid dyes, and wool and silk with acid or direct dyes; the absorption isotherm in these cases is of the Langmuir type or may be a combination of Langmuir and Nernst types.

Machinery

In the application of dyes, three techniques for textiles have been developed: (*1*) the dye liquor is moved as the material is held stationary, eg, raw-stock or package machine dyeing (Fig. 1); (*2*) the textile material is moved without mechanical movement of the liquor, eg, jig dyeing and continuous dyeing involving fabric padding (Fig. 2); (*3*) a combination of (*1*) and (*2*), as exemplified by a Klauder-Weldon skein dye machine (Fig. 3) in which the dye liquor is pumped as the skeins are mechanically turned. Another example is a jet dyeing machine in which both the goods and the liquor are constantly moving.

Figure 2. Jig dye machine. Courtesy of Parrott and Ballentine.

Figure 3. Klauder-Weldon skein dyeing machine. Courtesy of Hussong-Walker-Davis Co.

Equipment Control. Over the years dye and machinery manufacturers have employed any mechanical or electrical equipment that would enable them, day after day, to produce repeatable dyeing of top quality. First, thermometers were installed in dye lines; these soon evolved into thermocouples with remote recording. Other improvements include automatic four-way valves with variable-interval controls, flow controls, pressure recorders, hydraulic and air pressure sets on rollers, pH controls, etc.

Fibers

Acid (Anionic) Dyes. Acid dyes are used for polyamide, wool, silk, modified acrylic, and polypropylene fibers, as well as blends of these fibers with other fibers, such as cotton, rayon, polyester, acrylic, etc.

Basic (Cationic) Dyes. In modern dyeing technology, basic dyes are used for acrylic and modified basic dyeable polyester fibers.

Direct Dyes. Direct dyes, or substantive dyes are used in cellulosic fibers, such as cotton, rayon, and linen, and dyed from a salt bath.

Disperse Dyes. Disperse dyes, introduced in the 1920s, were first used for the dyeing of cellulose acetate fibers. Today, they are used primarily for polyester and triacetate fibers.

Fiber-reactive Dyes. Fiber-reactive dyes are the newest dyestuff group, developed in the last 20 years. Fiber-reactive dyes supplement vat, direct, and naphthol dyes; they are available in a full range of colors of outstanding brilliance and fastness properties. They can be applied to wool, silk, and polyamide, but are used primarily for cellulosic fibers. They are applied in batch processes as well as by semicontinuous and continuous methods.

Sulfur Dyes. The fastness properties of sulfur dyes are between those of diazotized direct dyes and vat dyes. They have better washfastness than diazotized or aftertreated direct dyes, but are generally inferior in this respect to vat dyes.

Azoic Dyes. Azoic dyes, ie, naphthols, are used primarily for cellulosic fibers and for special applications on acetate, polyester, nylon, and silk.

Vat Dyes. Vat dyes are primarily used to produce exceptionally fast coloration on cellulosic fiber, chiefly cotton. However, under proper conditions, they can be used for almost any fiber, eg, silk, wool, acrylic, polyamide, and polyester fibers, with varying degrees of fastness.

Fluorescent Whitening Agents. Fluorescent chemical compounds convert uv radiation to visible radiation at the blue end of the spectrum. Such compounds are known as fluorescent whitening agents (FWAs) or optical brighteners. More than half of all FWAs consumed are used in household detergents: 25% on paper, 10% on textiles, and the rest in plastics and spin-dyeing applications on synthetic fibers.

Other Dyes. Other dyes include dyes developed on the fiber, onium dyes, oxidation dyes, diphenyl black, mordant dyes, natural dyes, and mineral dyes.

Pigments. Pigments are applied continuously from a bath consisting of an oil-in-water emulsion of organic pigments with thermosetting or thermoplastic resin binders. A catalyst may be required, depending on the type of binder used. Light-to-medium depth shades are applied to large quantities of cotton and cotton/polyester blends.

Plastics

Dyes have found wide application in plastics. In 1983, ca 2.5×10^3 t were used for thermoplastic and thermosetting resins. Compared with pigments, their main advantages are high color strength and the ease by which they are incorporated.

Pigments. The terms pigment and dye cannot always be clearly distinguished. Dyes do not always dissolve completely, and can therefore act as pigments. Pigments, on the other hand, can dissolve partially or completely and then act as dyes. Properties such as heat stability and lightfastness of the colorant can be affected by the degree of solution.

Classification. Dyes for plastics are classified in the *Color Index* as published by the Society of Dyers and Colourists according to their class, generic name, and number (CI) under which the chemical structure of the pigment is shown.

Dyes. Plastics dyes include azo dyes, quinoline dyes, perinone dyes, anthraquinone dyes, nigrosine black, and metal complex dyes.

Methods of Incorporation. Dyes are dry-blended with the resin before processing in an injection-molding machine; three groups of equipment are used.

The first group consists of planetary mixers: in one type, one or two paddles rotate while the open container spins around its axis; in the other type, the container is conical and rotates while a screw parallel to the wall lifts the material.

The second group contains tumbling units, which rotate a container, often drums, around its longest axis.

The third group consists of a high speed impeller, rotating at up to 3600 rpm located on the bottom of a cylindrical container. The blades are designed to break up agglomerates and implant them into the resin; powdered resin is usually preferred.

Test Methods for Plastics

Colorants are tested for migration, lightfastness, and heat stability. The Gray scale rating and an extraction test are used also.

Evaluation of Dyes for Textiles

Dyes are evaluated by fastness tests, qualitative and quantitative analysis (paper and thin-layer chromatography, high performance liquid chromatography, spectroscopic methods, and reflectance spectrophotometry).

BIBLIOGRAPHY

C. L. Bird and W. S. Boston, eds., *The Theory of Colouration of Textiles,* The Dyers Company Publications Trust, London, 1975, distributed by The Society of Dyers and Colourists, Bradford, UK.

T. G. Webber, ed., *Coloring of Plastics* (Society of Plastics Engineers Monographs), John Wiley & Sons, Inc., New York, 1979.

H. Saechtling, *International Plastic Handbook,* Hansa Publisher, 1982.

E. R. Trotman, *Dyeing and Chemical Technology of Textile Fibers,* C. Griffin & Co., Ltd., London, 1964.

ROLF G. KUEHNI
Mobay Corporation

H. HENNING BUNGE
Dayglo Corporation

DYES, MACROMOLECULAR

Macromolecular dyes are a group of compounds thought of as intrinsically or structurally colored polymers, ie, possessing inherent properties. Their applications are markedly dependent on their relatively high molar mass. Thus, macromolecular dyes are useful polymers or oligomers as well as useful dyes with high tinctorial strength. They are soluble in the medium in which they are applied. Classifications of macromolecular dyes can be made by chromophoric type (azo-, anthraquinone, etc), physical characteristics (basic dyes, direct dyes, etc), application route (textile dyes, food dyes, paper dyes, hair dyes), method of synthesis (chain-growth, step-growth and ring opening polymerization), attachment to electrophilic or nucleophilic assemblies, and structural characteristics.

Synthetic Strategies

Addition Polymerization Routes. Most addition polymerization leading to macromolecular dyes is radical in character. The monomer dye may be induced to undergo polymerization by chemical initiation, photochemical activation, radiation-induced initiation, and thermal initiation.

Alternating Copolymerization. Here the chromophoric group is involved in the propagation stages of a copolymerization in alternating fashion to give species of type

Pendent Coloring Groups. Color-group attachment to polymer backbones may be represented

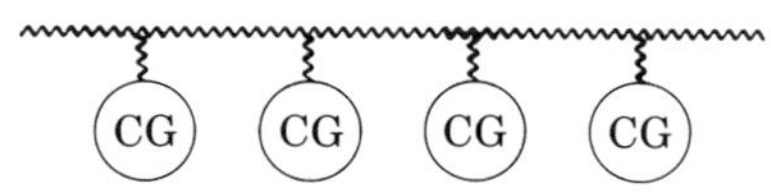

where the coloring group CG is linked by an individual covalent bond or a low molar mass grouping.

Polymer backbones may be electrophilic or nucleophilic in character. Nucleophilic polymers provide the easiest route to macromolecular dyes largely because of the scarcity of double electrophiles which can otherwise induce undesirable cross-linking.

Totally Chromophoric Macromolecular Dyes. In these dyes, the chromophoric centers are linked together as shown by aniline black

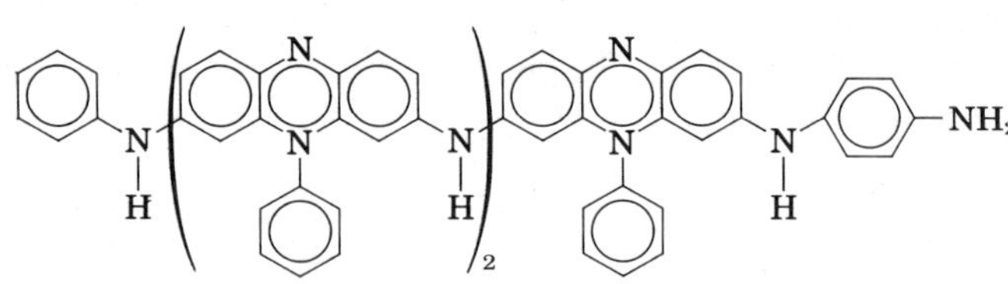

and degraded polyacrylonitrile residues

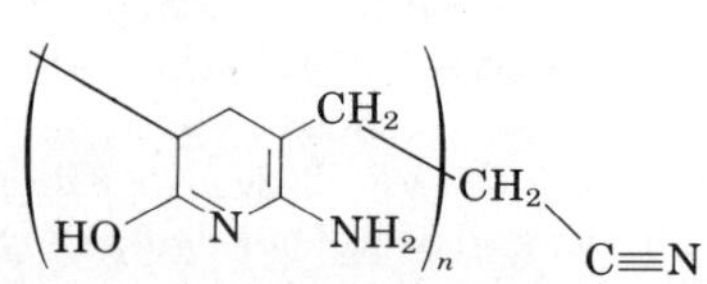

Totally chromophoric macromolecular dyes differ from other types in that main chain scission results in modification of the chromophoric character of the polymeric dye.

Polymerizable Dyes

Addition Routes. Chromophores can frequently be converted into their polymerizable equivalents by incorporation of a vinyl or acrylate function into the molecule. Polymerization produces copolymers or mixtures of homopolymers. There are many examples: anthraquinone–glycidyl methacrylate systems; anthraquinone derivative–vinyl monomer condensation products; reactive vinyl sulfonyl dyes; acrylated azo dyes; vinyl basic dyes; vinylanthraquinone chromophores; and azo dye–glycidyl methacrylate systems.

Step-growth Routes. ***Interfacial Polycondensation.*** Direct dyes with $—NH_2$ or —OH groups and a functionality of two or more can be converted into macromolecular dyes by treatment with dicarboxylic acid chlorides under interfacial conditions. Much of the interest has been in direct azo and bis–azo types.

Novel Polyviologens. These photochromic redox polymers can be designed to have good film forming properties. The chromophore in the viologen system has a radical cation center and is blue/violet.

Miscellaneous Examples. Miscellaneous examples include polymeric vat dyes, the polyvinylamine-eosin system, polymeric phthalocyanines, and coordination macromolecular dyes.

Applications

Development of systems which incorporate macromolecular dyes has been somewhat uncoordinated and based largely on the desired applications. Application areas include textile dyes, paper colorants, food dyes, colorants for cosmetics, fugitive stains for registration purposes, printing, plastics coloration, photoconductive polymers, photographic and reprographic applications, biological and biomedical applications, immunoassay procedures, process efficiency monitoring, fluorescent brighteners, electroconductive polymers, photocell assemblies, colored photoresists, and liquid–crystalline polymeric dye systems.

BIBLIOGRAPHY

D. J. Dawson, *Aldrichimica Acta* **14,** 32 (1981).

P. Brear and J. T. Guthrie, *Polym. Photochem.* **2,** 65 (1982).

I. R. Bellobono, S. Calgari, and E. Selli, *Organic Coatings and Applied Polymer Science Proceedings,* Vol. 46, American Chemical Society, New York, 1981.

R. S. Asquith, H. S. Blair, A. A. Crangle, and E. Riordan, *J. Soc. Dyers Colour.* **93,** 114 (1977).

J. T. GUTHRIE
The University of Leeds

DYNAMIC MECHANICAL PROPERTIES

Dynamic mechanical properties are the mechanical properties of materials as they are deformed under periodic forces. The dynamic modulus, the loss modulus, and a mechanical damping or internal friction express these properties. The dynamic modulus indicates the inherent stiffness of material under dynamic loading conditions. It may be a shear, a tensile, or a flexural modulus, depending on the investigating technique. Dynamic modulus is especially important for engineering structural applications of plastics and composites since the variation of properties is easily determined as a function of temperature and frequency (or time). The polymeric materials under use conditions are often in the dynamic stress and strain field, such as parts of automobiles, aircraft, and aerospace structures. Knowledge of the dynamic moduli and properties of polymeric materials is indispensable for design of these materials.

The mechanical damping or internal friction indicates the amount of energy dissipated as heat during the deformation of the material. The internal friction of material is important as a property index and for environmental and industrial applications. Since noise is radiated by vibration, especially of metallic materials with small internal friction (0.001–0.004), the application of damping materials to the vibrating surface converts the energy into heat, which is dissipated within the damping materials rather than being radiated as airborne noise. Many viscoelastic polymers are good damping materials with high internal friction (0.1–0.3). High damping or internal friction is essential in decreasing the effect of undesirable vibration and in reducing the amplitude of resonance vibrations to safe limits, as well as in playing a key role in all kinds of structures from airplanes to buildings.

The investigation of the dynamic modulus and internal friction over a wide range of temperatures and frequencies has proved to be very useful in studying the structure of high polymers and the variations of properties in relation to performance. These dynamic parameters have been used to determine the glass-transition region, relaxation spectra, degree of crystallinity, molecular orientation, cross-linking, phase separation, structural or morphological changes resulting from processing, and chemical composition of polymer blends, graft polymers, and copolymers. In addition to these structure–property relationships, dynamic mechanical studies have been extended to material composites (qv) and structural systems.

Theory

Dynamic mechanical properties are determined by various instruments measuring the response (deformation) of polymers and their assemblies to periodic forces. Thus, the vibrational parameters—amplitude, frequency, and type of oscillation and wave propagation—become important in this analysis.

In studies of the response of a material to vibrational forces, stress, strain, frequency, and temperature are the key variables. The stress applied to a viscoelastic body results in a linear or nonlinear dynamic response. For example, if the stress varies sinusoidally with time at a given frequency, the strain varies cyclically at the same frequency. If the amplitude of the stress is small enough, the strain is also sinusoidal with time, and the amplitude of the strain is proportional to the amplitude of the stress at a given temperature and frequency, ie, the behavior is linear. However, if the amplitude of the stress is large enough, the strain varies with the same frequency but is nonsinusoidal; it can have high harmonics, showing nonlinear viscoelastic behavior.

The method of sinusoidal excitation and response is very useful for determination of the dynamic mechanical properties of polymeric materials. The applied force and the resulting deformation both vary sinusoidally with time; the rate is usually specified by the frequency f in Hz or $\omega = 2\pi f$ in rad/s. For linear viscoelastic behavior, the strain alternates sinusoidally but is out of phase with the stress. This phase lag results from the time necessary for molecular rearrangements and is associated with relaxation phenomena.

Primary and Secondary Transitions

The dynamic mechanical properties of polymers are usually studied over a wide temperature range (−150–300°C). Typical dynamic mechanical properties of polymers as a function of temperature are illustrated in Figure 1.

In the region where the dynamic-modulus–temperature curve has an inflection point, the internal friction (tan δ) curve goes through a maximum. This dispersion is called the glass-transition region where the dynamic modulus E' changes from approximately 1 GPa (10^{10} dyn/cm^2) in the glassy state to about 1 MPa (10^7 dyn/cm^2) in the soft, rubbery state.

Other relaxation transitions can be found in a glassy state on the lower temperature side of the primary dispersion (the α-peak). These are called secondary dispersions and are usually designated, β, γ, etc, in order of decreasing temperature.

Molecular Motion and Loss Factor

Motions of long chain segments in the polymer structure have a profound effect on the loss factors (tan δ and loss modulus, E'') of the dynamic mechanical properties. The tan δ value at the α-peak (the glass transition) is of greater magnitude than at the dispersion peaks for lower temperatures. It is accompanied by the greatest decrease in dynamic modulus with increasing temperature.

The largest loss peak is generally the α-peak, which is associated with the glass-transition temperature. This transition occurs in the amorphous regions of the polymer with the initiation of the micro-Brownian motion of the molecular chains. The magnitude of the α-peak in the amorphous polymer is much greater than in the semicrystalline polymer, because the chain segments of the amorphous polymer are freer from restraints imposed by crystalline polymers in the glass-transition region.

The glass-transition temperature is affected by a number of chemical and molecular structures. Flexibility of the molecular chain, bulkiness of the side groups attached to the backbone, and molecular polarity affect the value of T_g. The peak height and temperature at which the α-peak occurs are closely related to the glass-transition phenomenon. Therefore, similar interpretations are used for both the α-peak and T_g (see also GLASS TRANSITION).

The magnitude of the β-peak is small in comparison with the α-peak. This dispersion is associated with an amorphous-phase relaxation in most polymers. It is resolved into a double or triple peak and appears to involve local motion in the chain. In the case of amorphous polymers, the β-peak is very broad and may appear as a shoulder of the α-peak. It is associated with motion about the chain backbone of a relatively small number of monomer units or with motion of side groups.

The γ-peak temperature is below the β-peak temperature. In this region, the chain segments are frozen in, and side-group motion is made possible by defects in packing or configuration in the glassy and crystalline states. This γ-peak of polymers can be related to the following general classifications: side-group motions in the amorphous and crystalline phases, end-group rotation, crystalline defects, backbone-chain motions of short segments or groups, and phase separation of impurities or diluents.

Crystallinity and Molecular Orientation

Many polymers, eg, poly(ethylene terephthalate) and nylon, crystallize when cooled from the melt, and chain-folding of the molecules in lamellar crystals takes place. The crystalline aggregates are known as spherulites.

The degree of crystallinity and molecular orientation are affected by heat treatment and stress in the polymer processes.

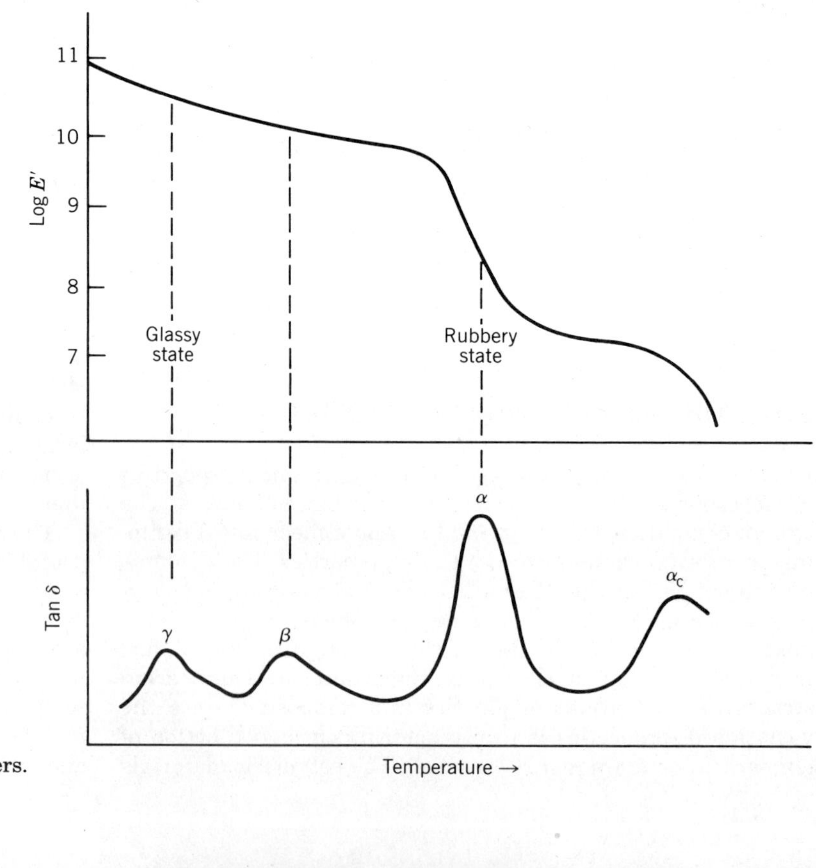

Figure 1. Typical dynamic mechanical properties of polymers.

The dynamic modulus generally increases with increasing degree of crystallinity.

Crystallinity has a great effect on the α-dispersion. Poly(ethylene terephthalate) (PET) is a good example of the α-peak changes due to crystallinity and molecular orientation. Controlling these parameters led to the introduction of a number of new products with a variety of mechanical properties.

Molecular Weight and Cross-linking

Effect of Molecular Weight. Increasing molecular weight has a large effect in the glass-transition range, transforming the behavior from viscous flow to a plateau range of rubberlike behavior. The length of the rubbery plateau region increases with increasing molecular weight.

The loss tangent (tan δ) above T_g is also strongly dependent on molecular weight. The value of the tan δ at the minimum above T_g decreases as the molecular weight increases.

Effect of Cross-linking. The effect of cross-linking on the dynamic properties of the vulcanized rubber is evident above T_g. After the large decrease at T_g, the dynamic modulus of a cross-linked rubber does not decrease further with increasing temperature. The loss tangent is a sensitive indicator of cross-linking. The α-peak (T_g) shifts to higher temperatures with increasing cross-linking, and at temperatures well above T_g, damping decreases with increasing degree of cross-linking. Epoxies and highly cross-linked phenol–formaldehyde resins are examples of polymers with a much higher degree of cross-linking (qv) than normal rubbers.

Polymer Blends and Copolymers

The dynamic mechanical properties of polymer blends (qv) are determined primarily by the mutual solubility of the two polymers. If two polymers are compatible and soluble in one another, the properties of the blend are nearly the same as those of a random copolymer of the same composition (see COMPATIBILITY). However, many polymer mixtures form two phases, due to the insolubility of the components. In this case, the damping-temperature curve shows two peaks; each peak is characteristic of the glass-transition temperature of one of the components. The two steps in the dynamic-modulus–temperature curves are characteristic of an immiscible two-phase system.

The dynamic mechanical properties of copolymers are strongly influenced by the characteristics of the homopolymers and the copolymer composition. In a homogeneous copolymer all molecules have the same chemical composition, and hence should have a single sharp α-transition (T_g). On the other hand, the damping curves of the dynamic properties of heterogeneous copolymers reflect the broadening of the transition region caused by the increase in heterogeneity.

Composite Structures

Composite materials consist of two or more components and phases (see COMPOSITES). The dynamic mechanical properties of composite materials are determined by the properties of the components, the morphology of the system, and the nature of the interface between the phases. Therefore, a great variety of properties can be obtained by varying the structure of the system or the interface properties. An important property of the interface is the degree of adhesive bonding between the phases.

Methods of Determination

An accurate measurement of the response of a polymeric material to vibrational forces requires many instrumentational considerations. A large number of techniques have been developed by using different vibrational principles at various frequencies. The methods are free vibration, resonance-forced vibration, off-resonance forced vibration, and acoustic and wave-pulse propagation techniques. The frequency for each method has a limited range.

In recent years, a wide variety of commercial dynamic testing instruments has become available for operation over a range of testing frequencies. Representative instruments include torsion pendulum, torsional-braid analyzer (TBA), Rheometrics Dynamic Spectrometer, Dynastat, the PL-dynamic mechanical thermal analyzer, B&K complex modulus apparatus, DuPont 981 Dynamic Mechanical Analyzer (DMA), and the Rheovibron and Autovibron. In 1982, the Perkin-Elmer Co. introduced a new DMA module for the TMS-2 thermomechanical analyzer.

BIBLIOGRAPHY

T. Murayama, *Dynamic Mechanical Analysis of Polymeric Materials,* 2nd ed., Elsevier Scientific Publishing Co., Amsterdam, 1982.

J. D. Ferry, *Viscoelastic Properties of Polymers,* 3rd ed., John Wiley & Sons, Inc., New York, 1980.

B. E. Read and G. D. Dean, *The Determination of Dynamic Properties of Polymers and Composites,* John Wiley & Sons, Inc., New York, 1978.

L. E. Nielsen, *Mechanical Properties of Polymers and Composites,* Marcel Dekker, Inc., New York, 1974, 2 vols.

TAKAYUKI MURAYAMA
Monsanto Company

E

EBONITE. See RUBBER, HARD.

ECONOMIC EVALUATION

The abundance of petrochemicals represents a scientific and engineering achievement, as well as an economic success. Each step along the way required a compromise between the economic interests of the producers and the consumers. The price of petrochemicals had to be high enough to induce producers to accept the risks involved in building the necessary plants, but low enough to induce consumers, who have other choices, to buy the products.

Economic evaluations are based on the following basic principles:

1. An economic decision involves a choice between at least two alternatives. One is the proposed investment; the other is the base case, the cash flows that will occur if the proposed investment is rejected. The proposed investment normally provides more benefits, but requires more investment. The base case must be the next best way to invest the money.
2. Economic decisions are based on future cash flows.
3. Only actual cash flows should be counted. Accounting charges that do not reflect an actual flow of cash should be ignored. Depreciation is an example.
4. If possible, a project is divided into steps and each step is analyzed separately.
5. An economic evaluation covers only the economic aspects of a problem.

Evaluation Methods

Discounted cash flow is the best way to perform economic evaluations because it expresses a project's attractiveness as an equivalent interest rate, and permits direct comparison to the cost of money. Discounted cash flow also recognizes the time value of money.

Reliability of Economic Evaluations

Economic evaluations can yield a wide range of results. The mathematics of discounting are rigorous. The problem is that all the cash flows in the analysis are projections and therefore uncertain. The plant might cost more, take longer to build, and cost more to operate than expected. Operating problems or slower sales growth might delay reaching design capacity. Selling price might fall faster than expected. Recessions might cause less than expected capacity utilization during a number of years. Reliable economic evaluations require realistic projections of these cash flows.

BIBLIOGRAPHY

H. Bierman and S. Smidt, *The Capital Budgeting Decision,* 4th ed., Macmillan, Inc., New York, 1975.

R. Brealy and S. Myers, *Principles of Corporate Finance,* McGraw-Hill, Inc., New York, 1981.

J. B. Malloy, *Chem. Ind. London,* 1242 (1971).

D. B. Hertz, "Risk Analysis in Capital Investment," *Harv. Bus. Rev.* **42,** 95 (Jan.–Feb. 1964).

JOHN B. MALLOY
Amoco Corporation

EDUCATION

The bulk of the recent activity centers around the Joint Polymer Education Committee.

Recent Survey

A 1980 survey to which 127 schools responded indicates the current state of polymer science education.

The formal courses were categorized into the following subject areas: (*1*) polymer synthesis, including courses on polymerization and reaction kinetics; (*2*) properties and characterization, including mechanical properties, chemical properties, instrumentation, rheology, and physical chemistry; (*3*) process engineering, involving polymerization techniques, processing, molding, tooling, and product design; (*4*) polymer science courses, including general, basic topics; (*5*) fibers; (*6*) rubber; (*7*) coatings; (*8*) biopolymers; and (*9*) others, which includes topics such as polymers for medicine and optical properties of polymers.

Of 19 bachelor degrees offered by schools, 10 are specifically plastics or polymer degrees, and 9 are in the traditional disciplines with emphasis or specialization in polymers. Finally, there are 12 Associate of Applied Science and certificate degree programs; 10 of these are Associate of Applied Science degrees.

The following is a list of schools offering degrees in polymer science:

Graduate degrees

University of Akron	University of Minnesota
Auburn University	University of Missouri at Kansas City
Carnegie-Mellon University	State University of New York
Case Western Reserve University	North Dakota State University
University of Cincinnati	Pennsylvania State University
University of Connecticut	Polytechnic University
University of Detroit	Princeton University
Drexel University	Rensselaer Polytechnic Institute
University of Florida	University of Southern Mississippi
Georgia Institute of Technology	Stevens Institute of Technology
University of Illinois at Urbana-Chicago	University of Tennessee at Knoxville
Lehigh University	University of Utah
Lowell University	Washington State University
University of Massachusetts	
University of Michigan	

Bachelor degrees	
California Polytechnic State University	North Dakota State University
California State University at Chico	Pennsylvania State University
Case Western Reserve University	Pittsburgh State University
Central Connecticut State College	Purdue University
University of Detroit	University of Southern California
Georgia Institute of Technology	University of Southern Mississippi
University of Lowell	University of Texas at San Antonio
Michigan Technological University	University of Utah
	Western Washington University
	Wright State University

Associate and certificate degrees	
Cincinnati Technical College	State University of New York, Agricultural and Technical College
Ferris State College	Pittsburgh State University
Hennepin Technical Center	Shawnee State Community College
Los Angeles Trade Technical College	
New Jersey Institute of Technology	

Continuing Education

The Education Committee of The Society of Plastics Engineers (SPE) initiated a number of ventures, including frequent surveys of course offerings by U.S. universities. The Plastics Institute of America (PIA) is a nonprofit corporation with headquarters at the Stevens Institute of Technology. Courses offered by PIA are designed to give participants a familiarity and competence within specific areas, eg, polymer additives, extrusion processes, coloring, and packaging. The PIA and the ACS offer a number of audio courses, including topics concerning injection molding, mechanical properties of plastics, and rheology of polymer melts.

International Aspects

Numerous societies worldwide promote polymer education, such as those listed in Table 1.

Table 1. Societies Concerned with Polymer Education

Country	Society
Australia	Polymer Division, Royal Australian Chemical Institute
USSR	Russian Academy of Sciences
U.S.	American Chemical Society; Joint Polymer Education Committee (JPEC); Adhesive and Sealant Council; Plastics Education Foundation; Plastics Institute of America; Stevens Institute of Technology; American Institute of Chemical Engineers
France	Centre National de la Recherche Scientifique
FRG	Fachgruppe Makromolekulare Chemie (FMC)
Italy	Associazione Italiana di Scienca e Technologia delle Macromolecule (AIM)
UK	MACRO Group; Plastics and Rubber Institute
Japan	Society of Polymer Science

BIBLIOGRAPHY

C. Carraher, *Polym. News,* ongoing series.

C. Carraher and R. Deanin, *J. Chem. Educ.* **57,** 436 (1980).

C. Carraher and A. M. Sarquis, eds., *J. Chem. Ed.,* periodic ongoing series.

Charles E. Carraher
Florida Atlantic University

ELASTICITY

Rubberlike

Elasticity is the reversible stress–strain behavior by which a body resists and recovers from deformation produced by a force. This behavior is exhibited by rubberlike materials in a unique and extremely important manner. Unlike metals or glasses, they can undergo very large deformations without rupture (and are thus similar to liquids) and then come back to their original shape (as do solids).

Deformation and Mechanical Testing

When a body is submitted to external stress or pressure, it undergoes a change in shape or volume. Mechanical testing of rubber involves application of a force to a specimen and measurement of the resultant deformation, or application of a deformation and measurement of the required force. The results are expressed in terms of stress and strain and thus are independent of specimen geometry. Stress is the force f per unit original cross-sectional area A, and strain the deformation per unit original dimension. The units of stress are Newtons per square meter (N/m^2) or Pa; strain is dimensionless. The ratio of stress to strain is called the modulus. A material is Hookean when its modulus is independent of strain (typical of a metal spring or wire); elastomers are Hookean only in the range of very small deformations.

The simplest deformations are isotropic compression under hydrostatic pressure, uniaxial extension (or compression), and simple shear. They are used to determine the bulk modulus K, Young's modulus E, and shear modulus G, respectively. Elastomer testing for commercial applications is highly dependent on the method; the conditions, eg, strain rate, temperature; and the shape of the samples. Therefore, mechanical tests have been standardized for uniformity and simplicity by the American Society for Testing and Materials (ASTM) and are codified; for example, ASTM D1456-61 corresponds to strain testing of vulcanized rubber.

Conditions for Rubberlike Elasticity

Conditions include long, highly flexible chains, network structure, and weak interchain interactions.

Differences Between Elastomers and Metals

Elastomers and metals differ greatly with regard to deformation mechanisms. Metals and minerals are formed of atoms arranged in a three-dimensional crystalline lattice, joined by powerful valence forces operating at relatively short range. Deformation of such materials involves changes in the interatomic distances, which requires large forces; hence the elastic modulus of these materials is very high. After a small deformation, slippage between adjacent crystals occurs, at the yield point, and the deformation increases much more rapidly than the stress and becomes irreversible or plastic.

Stretching of elastomers does not involve any significant changes in the interatomic distances, and therefore the forces required are considerably lower. The number of available configurations for a network chain is reduced in the deformation process. After removal of the stress, the specimen recovers its original shape, since this corresponds to the most disordered state.

Extension of Elastomers and Compression of Gases

The extension of elastomers and the compression of gases are associated with a decrease in entropy (disorder). Thus similar behavior is expected in regard to (adiabatic) deformation and heating. The work of deformation of a gas is $dW = -pdV$ where p is the pressure and V the volume. For an elastomer it is $dW = fdL$ where f is the force and L the length; the term $-pdV$ is not taken into account, since there is only a negligible change of rubber volume during stretching. These observations and classical thermodynamic principles can be used to explain Gough's experiments (increase in temperature upon stretching an elastomer).

Statistical Distribution of End-to-end Dimensions of a Polymer Chain

Before treating the statistical properties of a network, the statistics of a single chain must be considered, mainly to establish the relationship between the number of configurations and the deformation. A polymeric chain is constantly changing its configuration by Brownian motion. Statistical methods and idealized models permit calculation of the average properties of such a chain.

The Freely Jointed Chain. This type of idealized chain consisting of n links of length l is represented in Figure 1, where r is the end-to-end distance. When the chain is completely extended, $R = nl$. The chain may assume many configurations, each associated with an end-to-end distance r_i. In statistics, it is equivalent to consider a molecule at different times or an assembly of N molecules at the same time. An average quantity describing the assembly is the mean-square end-to-end distance $\overline{r^2}$ defined by

$$\overline{r^2} = (1/N) \sum_{i=1}^{N} r_i^2 \quad (1)$$

where r_i is the end-to-end distance of the ith chain. The vector r_i is the sum of the link vectors I_j:

$$r_i = \sum_{j=1}^{n} I_j \quad (2)$$

and r_i^2 is the scalar product of r_i with itself:

$$r_i^2 = \sum_{j=1}^{n} I_j^2 + 2 \sum_{k<j} I_k \cdot I_j \quad (3)$$

If the chain is assumed to be freely jointed and volumeless, any two links can assume any orientation with respect to each other. Therefore the second term of equation 3 is zero and

$$r_i^2 = nl^2 \quad (4)$$

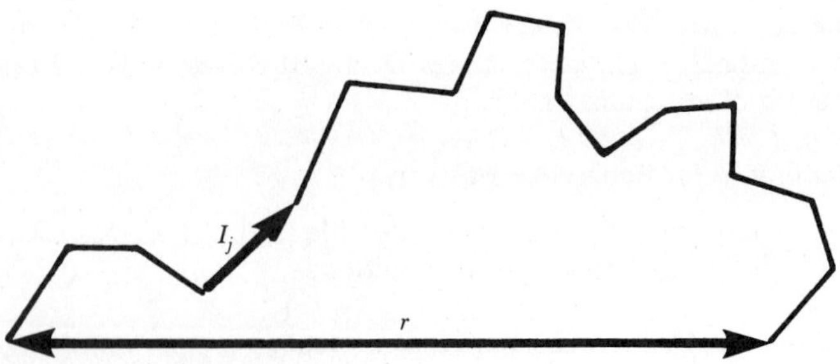

Figure 1. Ideal chain formed with n links of length l. r = End-to-end distance; I_j = link vector.

Chain with Bond-angle Restrictions. Although the chain in the above treatment was assumed to be freely jointed, a real polymer chain has fixed bond angles θ. Therefore, the second term in equation 3 is no longer zero. The scalar product of two vectors $I_i \cdot I_j$ is $l^2 \cos^{|j-i|} \theta$. It can then be shown that

$$\overline{r^2} \simeq nl^2(1 - \cos\theta)/(1 + \cos\theta) \quad (5)$$

In the particular case of a tetrahedrally bonded chain, $\theta = 109.5°$ and $\overline{r^2} = 2nl^2$, twice the value for a freely jointed chain. Thus a real chain is quite different from this ideal representation. Nevertheless, any flexible real chain can be represented by a simple model that is the statistical equivalent of a freely jointed chain. The two conditions are that the real and freely jointed chains have the same mean-square end-to-end distance and the same length at complete extension:

$$\overline{r^2} = \overline{r_e^2} = n_e l_e^2 \quad (6)$$

and

$$R = R_e = n_e l_e \quad (7)$$

Chain with Bond-angle Restrictions and Hindered Rotations. The angle of rotation ϕ of a single bond around the axis formed by the preceding one may be restricted by steric interferences between atoms. When n is large and the average value $\overline{\cos\phi}$ of $\cos\phi$ is not too close to unity, the following relationship can be established:

$$\overline{r^2} = nl^2 \left(\frac{1 - \cos\theta}{1 + \cos\theta}\right)\left(\frac{1 + \overline{\cos\phi}}{1 - \overline{\cos\phi}}\right) \quad (8)$$

When the rotation is not hindered, ie, when $\overline{\cos\phi} = 0$, equation 8 is equivalent to equation 5.

The Equation of State for a Single Polymer Chain. The variation of the Helmholtz free energy is the negative of the work of deformation in isothermal elongation:

$$dA = -dW = fdr \tag{9}$$

with

$$dA = dU - TdS \tag{10}$$

This free energy is then used to calculate the elastic equation of state and the energetic and entropic components of the elastic force.

Classical Thermodynamics of Rubber Elasticity

In experiments concerning the relationships between length, temperature, and force, usually the change of force with temperature at constant length is recorded. It is therefore necessary to extend the thermodynamic treatment of the elasticity. Moreover, the force is not purely entropic, and the energetic contribution carries useful information on the dependence on temperature of the average end-to-end distance of the network chains in the unstrained state. It is therefore important to be able to deduce these quantities from a thermoelastic (force–temperature) experiment.

The energetic and entropic components of the elastic force, f_e and f_s respectively, are

$$f_e = \left(\frac{\partial U}{\partial L}\right)_{T,V} = f - T\left(\frac{\partial f}{\partial T}\right)_{p,\alpha} \tag{11}$$

$$f_s = -T\left(\frac{\partial S}{\partial L}\right)_{T,V} = T\left(\frac{\partial f}{\partial T}\right)_{p,\alpha} \tag{12}$$

Their determination provides much insight into the thermodynamics of the elastic deformation.

Statistical Treatment of Rubberlike Elasticity

A network is an ensemble of macromolecules linked together, each of them rearranging its configurations by Brownian motion. Classical thermodynamics explains the behavior of elastomers with regard to force, temperature, pressure, and volume, but does not give the relationship between the molecular structure of the network and elastic quantities such as the moduli. Therefore, statistical mechanics was introduced in the 1940s, and its theoretical predictions were tested. Because of the complexity of network structures, two models based on affine and phantom networks were studied. The cross-linking points of the affine network are fixed, whereas those of the phantom network can undergo fluctuations independent of their immediate surroundings. The effects of the macroscopic strain are transmitted to the chains through these junctions at which the chains are multiply joined, with the internal state of the network system specified in terms of the positions of the cross-linkages. Deformation transforms the arrangements of these points.

Comparisons with Experimental Results. Stress–strain measurements in uniaxial extension can be compared with the prediction of an affine Gaussian network. The Gaussian relationship in the affine limit is valid, unfortunately only at small deformations. Experimental curves may generally be well represented by adjusting the parameters of a non-Gaussian stress–strain relationship. These disagreements between experiments and the simple predictions of statistical mechanics (qv), however, have led some workers to develop a phenomenological theory of rubberlike elasticity.

Phenomenological Theory. Continuum mechanics is used to describe mathematically the stress–strain relations of elastomers over a wide range in strain. This phenomenological treatment is not based on molecular concepts, but on representations of observed behavior. The main goal is to find an expression for the elastic energy W stored in the system (assumed to be perfectly elastic, isotropic in its undeformed state, and incompressible), analogous to the free energy of the statistical treatment.

Statistical Theory of Real Networks. Affine and phantom networks (qv) are extreme limits. Stress–strain measurements in uniaxial extension have revealed that the behavior of the real networks is between these limits. A theoretical attempt has been made to account for this dependence of reduced stress [f^*] on reciprocal elongation in terms of a gradual transition between the affine and phantom deformations, and a molecular theory has been formulated by Flory and Erman. In this model, the restrictions of junction fluctuations due to neighboring chains are represented by domains of constraints. At small deformations, the stress is enhanced relative to that exhibited by a phantom network. At large strains, or high dilation (swelling), the effects of restrictions on fluctuations vanish and the relationship of stress to strain converges to that for a phantom network. In a later theoretical refinement, the behavior of the network is taken to depend on two parameters. The more important is κ, which measures the severity of entanglement constraints relative to those imposed by the phantom network. Another parameter, ζ, takes into account the nonaffine transformation of the domains of constraints with strain.

Swelling Equilibrium. The isotropic swelling of a cross-linked elastomer by a liquid has two important opposing effects: the increase of mixing entropy of the system because of the presence of the small molecules, and the decrease of configurational entropy of the network chains by dilation. Therefore, an equilibrium degree of swelling is established, which increases as the cross-linking density decreases.

Determination of the Degree of Cross-linking

There are several methods for determining cross-link density in elastomer networks. Stress–strain and swelling equilibrium measurements are the most common. Chemical determination is done by titrating unreacted cross-linking groups. The number of junctions may be counted by nmr spectroscopy or estimated through chain-orientation measurements, eg, fluorescence polarization, infrared dichroism, and birefringence.

Entanglements

Entanglements have been introduced in the later Flory-Erman theory as constraints that restrict junction fluctuations. Another viewpoint considers entanglements to act as physical cross-links, being based in part on the observation that linear polymers of high molecular weight exhibit a storage modulus $G'(\omega)$, which remains relatively constant over a wide range of frequencies. This plateau modulus G_N^0 is independent of chain length for long chains and is insensitive to temperature. Since it varies with the volume fraction of polymer in concentrated solutions, it could be due to pairwise interactions between chains, and a universal law has been proposed for the dependence of G_N^0 on the chemical structure of the polymer. During the cross-linking process, some of such interactions or entanglements could be trapped in the network and act as physical junctions. This conclusion has been tested by irradiation cross-linking of already deformed networks, and subsequent measurement of the dimensional changes.

Cross-linking Processes

Recoverability requires that the chains be linked together by relatively permanent connections. Processes include sulfur vulcanization, peroxide thermolysis, irradiation, and chain-end linking.

Reinforcement by Fillers

In industrial applications, elastomers are typically compounded with carbon blacks, silica and silicates (qv), and clays (see also CARBON). These fillers (qv) reinforce the polymeric matrix and also reduce costs.

Fine particulate fillers increase stiffness and strength of vulcanized rubber. The vulcanized rubber surrounding the filler particles can be considered to be an isotropic elastic continuum with the same elastic properties as the rubber without filler and it can be assumed that the rubber adheres to the filler surface. The following equation gives the dependence of the small-deformation Young's modulus E on the volume concentration c of filler:

$$E = E_0(1 + 2.5c + 14.1c^2) \quad (13)$$

where E_0 is Young's modulus of the matrix. At low filler concentration, the modulus is increased.

Strain-induced Crystallization

Elastomers formed with chains of high structural regularity may crystallize on deformation. Elongation of such elastomeric networks reduces the entropy of its chains, and only a small additional reduction is required for crystallization. This effect, called strain-induced crystallization, is studied by x-ray diffraction, birefringence techniques, and calorimetric measurements. When crystallization occurs in uniaxial stretching, the retraction curve is lower for the stress and higher for the birefringence. It has the very important effect of reinforcing the elastomeric material.

BIBLIOGRAPHY

P. J. Flory, *Principles of Polymer Chemistry,* Cornell University Press, Ithaca, N.Y., 1953, Chapt. XI.

L. R. G. Treloar, *The Physics of Rubber Elasticity,* 3rd ed., Clarendon Press, Oxford, UK, 1975.

J. E. Mark and J. Lal, eds., *Elastomers and Rubber Elasticity,* ACS, Washington, D.C., 1982.

P. J. Flory, *Polym. J.* **17,** 1 (1985); *Brit. Polym. J.* **17,** 96 (1985).

J. P. QUESLEL
J. E. MARK
University of Cincinnati

Hard-elastic Behavior

Hard-elastic behavior is observed in polymers crystallized from a stressed melt with subsequent annealing (qv) and in profusely crazed amorphous polymers. Morphology is similar to fibrils parallel to the elongation direction bridging layers of solid polymer. To date, no commercial applications have been found for these unusual materials, possibly because of creep, which is much higher than in conventional elastomers.

Hard-elastic behavior, which is unexpected in a crystalline polymer, is attributed to a particular morphology, consisting of an aggregate of layers (lamellae) joined by small fibrils. On elongation, this structure does not "neck" like a spherulitic crystalline polymer; instead the fiber diameter remains nearly constant, corresponding to a Poisson's ratio approaching zero.

Mechanical Properties

The mechanical properties of crystalline hard-elastic materials have been reported in greatest detail for polypropylene and acetal copolymer fibers. Other crystalline polymers show similar behavior. Typical mechanical performance of polypropylene is shown in Figure 1. When deformed, the fiber or film exhibits normal Hookean behavior, with a modulus somewhat higher than for a bulk polymer, although substantially lower than for a "drawn" fiber. The fiber yields and increases in stress to a higher level which remains constant to very large extensions.

Amorphous Polymers

After hard-elastic properties were discovered in crystalline polymers, they were observed in two amorphous polymers. The most detailed study has been made of high impact polystyrene (HIPS), a two-phase amorphous polymer. A single-phase amorphous polymer, polycarbonate, shows similar properties. This behavior is produced in glassy amorphous polymers by stretch-

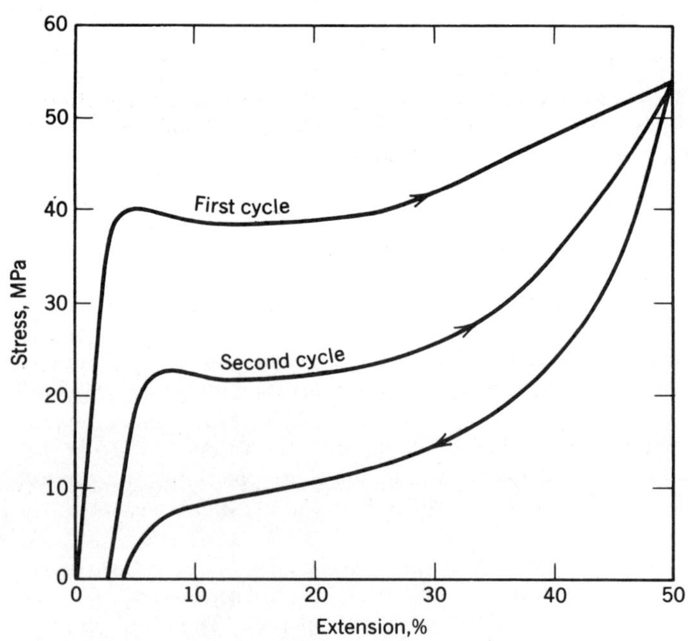

Figure 1. Typical stress–strain recovery behavior for hard-elastic material for first and second strain cycles. To convert MPa to psi, multiply by 145. Courtesy of *Journal of Macromolecular Science, (Part B) Physics.*

ing the material enough to produce profuse crazing. The resulting morphology consists of craze fibrils bridging layers of solid material with surfaces normal to the applied tensile stress. Thus the morphology superficially resembles the lamella-fibril assemblage proposed for crystalline hard-elastic polymers, although the molecular aggregation must be different.

BIBLIOGRAPHY

R. G. Quynn and H. Brody, *J. Macromol. Sci. Phys.* **B4,** 953 (1970); **B5,** 721 (1971).

E. S. Clark in R. W. Lenz and R. S. Stein, eds., *Structure and Properties in Polymer Films,* Plenum Press, New York, 1972, pp. 267–282.

I. K. Park and H. D. Noether, *Colloid Polym. Sci.* **253,** 824 (1975).

A. Moet, I. Palley, and E. Baer, *J. Appl. Phys.* **51,** 5175 (1980).

EDWARD S. CLARK
The University of Tennessee

ELASTOMERIC FIBERS. See FIBERS, ELASTOMERIC.

ELASTOMERS, ACRYLIC. See ACRYLIC ELASTOMERS.

ELASTOMERS, BUTYL RUBBER. See BUTENE POLYMERS.

ELASTOMERS, CHLOROSULFONATED POLYETHYLENE. See ETHYLENE POLYMERS.

ELASTOMERS, ETHYLENE–PROPYLENE RUBBER. See ETHYLENE–PROPYLENE ELASTOMERS.

ELASTOMERS, FLUORINATED. See CHLOROTRIFLUOROETHYLENE POLYMERS; FLUOROCARBON ELASTOMERS.

ELASTOMERS, SYNTHETIC

Elastomers are derived from natural rubber and synthetic polymers with rubberlike properties. They exhibit both long-range deformability on application of stress and complete recovery on removal. These two properties are not shared by lower molecular weight solids, liquids, and gases, and metals, glasses, or ceramics. Rubberlike behavior is exhibited exclusively by polymeric materials consisting predominantly of long-chain molecules, as only such macromolecules have the capacity of forming a sufficiently large number of arrangements in space to be able to alter their end-to-end separations when stress is applied (see ELASTICITY). Elastomers must be above the glass-transition temperature T_g to manifest high local segmental mobility. They must be amorphous in the unstressed state to exhibit high local segmental mobility, the chains must be flexible and intermolecular (interchain) forces must be weak. Elastomers must possess a sufficient number of chemical or physical cross-links to form a continuous network in order to restrain chain sippage (see CROSS-LINKING; NETWORKS). Some triblock copolymers, eg, styrene–butadiene–styrene (SBS), exhibit elastomeric behavior due to two-phase morphology where the polystyrene domains act as thermally reversible physical cross-links as well as fillers. These materials exhibit permanent set when stretched to high elongations and allowed to relax (see ELASTOMERS, THERMOPLASTIC).

Depending on their chemical structure and regularity, the chains of synthetic rubbers may or may not be able to undergo crystallization on stretching. *cis*-1,4-Polyisoprenes prepared with Ziegler-Natta catalysts exhibit the phenomenon of temporary crystallization, but less so than natural rubber (see ISOPRENE POLYMERS; RUBBER, NATURAL). Strain-induced crystallization plays an important role in determining the physicomechanical properties of rubbers, such as tensile strength, tear strength, and fatigue resistance.

A rubber able to crystallize on stretching crystallizes, at least partially, on cooling without stretching. The temperature of optimum crystallization is referred to as its crystalline melting temperature T_m. It is usually below room temperature; otherwise the rubber crystallizes and hardens on storage. This is occasionally observed with natural rubber stored at low temperatures. Copolymers exhibit less tendency to crystallize than homopolymers. Vulcanizates do not undergo strain-induced crystallization and exhibit poor tensile and tear strengths.

Nomenclature and Classification

The nomenclature of synthetic rubbers is based on ASTM D1418(7-9): M denotes rubbers containing a saturated polymethylene chain and appears in many symbols for synthetic rubbers; N represents rubbers containing nitrogen, and O those with oxygen in the polymer backbone; R designates rubbers with an unsaturated carbon chain, natural rubber and synthetic rubbers derived at least partly from conjugated diolefins. Rubbers with silicon in the backbone are designated by Q, and rubbers with sulfur by T. Rubbers with carbon, oxygen, and nitrogen in the macromolecule chain are described by U. Carboxylic groups substituted on the polymer chain are denoted by the letter X and halogen atoms by prefixes B or C for bromine or chlorine. The prefix P denotes the presence of vinylpyridine. The chemical family is represented by prefix letters as follows: A, acrylate rubber; B, butadiene rubber; C, chloroprene rubber; I, isoprene rubber (synthetic); N, acrylonitrile rubber; S, styrene rubber (see also BUTADIENE POLYMERS; ACRYLIC ELASTOMERS; CHLOROPRENE POLYMERS; ETHYLENE–PROPYLENE ELASTOMERS).

Stereospecific Polymerization

As early as 1929, Staudinger and co-workers had recognized that each of the propagation steps in the growth of a polymer chain could give rise to stereoisomerism. By controlling the conditions of polymerization, two types of polymers could be prepared from vinyl isobutyl ether: a tacky, rubberlike amorphous product or a nontacky solid that exhibited a crystalline x-ray diffraction pattern in the unstretched state and a sharp fiber diagram on stretching. Polymer–solvent interactions were different for the crystalline and rubbery polymers.

A new era in the synthetic rubber industry began in the mid-1950s when Goodrich, Goodyear, and Firestone announced the synthesis of high *cis*-1,4-polyisoprene with properties

Table 1. Long-term Rubber Consumption (10^3 t) Forecast by Area and Type[a,b]

	North America		Western Europe		Latin America		Asia and Oceania		Africa and Middle East		Total	
Rubber	1987	1992	1987	1992	1987	1992	1987	1992	1987	1992	1987	1992
styrene–butadiene[c]	839	833	727	745	329	386	674	783	63	72	2,632	2,819
carboxylated styrene–butadiene	447	458	335	370	21	29	125	152	5	6	933	1,015
polybutadiene	422	424	240	257	109	131	270	309	19	21	1,060	1,142
ethylene–propylene–diene	204	216	170	195	11	16	107	126	2	3	494	556
polychloroprene	86	85	81	77	19	23	68	78	5	5	259	268
nitrile	68	66	81	85	13	17	51	66	3	4	216	238
other synthetics[d]	457	479	216	244	39	46	197	223	22	33	931	1,025
Total synthetic	*2523*	*2561*	*1850*	*1973*	*541*	*648*	*1492*	*1737*	*119*	*144*	*6,525*	*7,063*
natural	872	871	945	980	247	296	1568	1804	140	180	3,771	4,131
Total rubber	*3395*	*3432*	*2460*	*2953*	*788*	*944*	*3060*	*3541*	*259*	*324*	*10,296*	*11,914*

[a] Source: International Institute of Synthetic Rubber Producers. Courtesy of *Chemical and Engineering News*, American Chemical Society.
[b] Excluding centrally planned economy countries.
[c] Solid and latex forms of SBR.
[d] Includes polyisoprene and butyl.

matching those of natural rubber. This development had been made possible by utilization of the new catalyst systems discovered by Ziegler in 1953 for producing linear polyethylene under low pressure (see ZIEGLER-NATTA CATALYSTS). At Goodrich–Gulf, isoprene polymerized with a combination of a trialkylaluminum and titanium tetrachloride produced a rubbery material (Ameripol SN), which virtually duplicated the natural rubber molecule in microstructure and physical characteristics (see ISOPRENE POLYMERS).

Economic Aspects

The rubber industry has not been very profitable in the 1980s. However, there seem to be signs for improvement in business of this mature industry, but without a significant increase in demand for its products. Data on production and consumption of rubber are compiled by the International Institute of Synthetic Rubber Producers (IISRP) with 51 synthetic rubber producers from non-Communist countries. These companies produce more than 95% of the world's synthetic rubber.

Synthetic rubber (SR) accounted for 67.9% of total rubber consumption in 1987. The 5-year forecast projects an increase in global consumption from 9.75×10^6 to 10.78×10^6 t. This share of synthetic rubber is expected to remain the same during this period. Styrene-butadiene rubber (SBR) will continue to be the largest-volume SR produced and consumed in non-Communist countries (Table 1), whereas polybutadiene rubber (BR) ranks second.

Following the pattern of many years, tires and tire products constitute 50% of rubber consumption. Overall, the percentage of total synthetic rubber use for tire and tire products in the non-Communist world decreased from 53.5% in 1980 to 50.2% in 1985 and 43.1% in 1987. A similar decrease took place in the percentage of NR consumption. Both SBR and BR remain the workhorses of the tire industry; butyl rubbers (see ISOBUTYLENE POLYMERS) and synthetic polyisoprene are used in inner-tire lining and tire tubes, respectively, and in blends with SBR, BR, and NR.

Product development in elastomers, particularly solution rubbers, is continuing, particularly in the United States, Japan, the FRG, Italy, and the United Kingdom. Increasingly, SBR is being adapted to solution polymerization techniques to provide a combination of low rolling resistance and high wet traction for passenger tire treads, features that have not been attainable by the widely accepted emulsion SBR without sacrificing good wear (abrasion resistance).

BIBLIOGRAPHY

J. Lal and J. E. Mark, eds., *Advances in Elastomers and Rubber Elasticity*, Plenum Press, 1986, pp. 17–36.

Rubbers and Latices Nomenclature, ISO Standard 1629, American National Standards Institute, New York, 1976; D1418-85, ASTM Philadelphia, Pa., 1985.

A. K. Bhowmick and H. L. Stephens, eds., *Handbook of Elastomers: New Developments and Technology*, Marcel Dekker, New York, 1988, Chapt. 16.

Worldwide Rubber Statistics, International Institute of Synthetic Rubber Producers, Inc., Houston, Tex., 1988.

JOGINDER LAL
Lal Associates

ELASTOMERS, THERMOPLASTIC

Thermoplastic elastomers were introduced about 1960. They have many of the physical properties of rubbers, ie, softness, flexibility, and resilience, but in contrast to conventional rubbers, they are processed as thermoplastics. With thermoplastic elastomers, the transition from a processible melt to a solid, rubberlike object is rapid and reversible and takes place upon cooling. Thermoplastic elastomers can be processed on conventional plastics equipment, such as injection molders and extruders; scrap is usually recycled.

Because of increased production and the lower cost of raw material, thermoplastic elastomeric materials are a significant and growing part of the total plastics market. However, because the melt-to-solid transition is reversible, some properties of thermoplastic elastomers, eg, compression set, solvent resistance, and resistance to deformation at high temperatures, are usually not as good as those of the corresponding vulcanizates. Applications of thermoplastic elastomers are therefore

Table 1. Hardnesses, Prices, Specific Gravities, and Productive Capacity for Thermoplastic Elastomers

Elastomer	Shore hardness range	\$/kg[a]	Sp gr[a]	U.S. production capacity, 10^3 t/yr
Styrenic block copolymers				
Pure				
S—B—S	60A–90A	2.00–2.10	0.94	170
S—I—S	30A–40A	2.20–2.30	0.92	
S—EB—S	65A–75A	3.85–4.75	0.91	
Compounded				
S—B—S	35A–75A	1.40–2.40	0.93–1.1	
S—EB—S	30A–50D	2.10–3.85	0.9–1.2	
Other block copolymers				
polyurethane	70A–70D	4.40–5.50	1.05–1.25	35
polyester	35D–75D	5.15–7.15	1.15–1.45	7
polyamide	75A–65D	5.30–10.00	1.0–1.15	—
Copolymer blend				
polypropylene/ethylene–propylene	60A–75D	1.65–3.30	0.90–1.1	35

[a] Excluding fire-retardant grades or highly filled materials for soundproofing.

in areas where these properties are less important, eg, footwear, wire insulation, adhesives, and polymer blending, and not in areas such as automobile tires (qv).

Structure

Thermoplastic elastomers are multiphase compositions in which the phases are intimately dispersed. In many cases, the phases are chemically bonded by block or graft copolymerization. In others, a fine dispersion is apparently sufficient. At least one phase consists of a material that is hard at room temperature but fluid on heating. Another phase consists of a softer material that is rubberlike at RT. A simple structure is an A—B—A block copolymer, where A is a polystyrene segment hard phase and B is an elastomer segment.

Property–Structure Relationships

Molecular Weight. Compared with homopolymers of similar molecular weight, styrenic block copolymers have very high melt viscosities, which increase with increasing molecular weight.

Ratio of Hard to Soft Phases. As expected, the modulus of thermoplastic elastomers increases with the proportion of hard phase.

Elastomer Segment. The choice of elastomer segment affects the properties of block copolymers. In the styrenic block copolymers, those with polyisoprene segments (S—I—S) are softer than corresponding materials with polybutadiene segments (S—B—S). Conversely, those with poly(ethylene–butylene) segments (S—EB—S) are harder but more resistant to degradation. In the polyurethane thermoplastic elastomers, those with polyether elastomer have better hydrolytic stability and low temperature flexibility, whereas those with polyester segments are tougher and have better oil resistance.

Hard Segment. The choice of the hard segment determines the upper service temperature and solvent resistance.

Synthesis

Block copolymers (qv) are synthesized by a variety of methods; most important are sequential polymerization and step growth.

Commercial Production

Commercially, the styrene block copolymers are made by anionic polymerization (qv).

Economic Aspects

The U.S. production capacity for thermoplastic rubbers is estimated at about 250,000 t/yr, and world capacity at 500,000 t/yr. At present, the market is estimated to be divided as follows: styrenic block copolymers, 53%; polypropylene/ethylene–propylene copolymer blends, 29%; thermoplastic polyurethanes, 12%; and thermoplastic polyesters, polyamides and others, 6%. Hardnesses, prices, specific gravities, and production capacities are given in Table 1.

BIBLIOGRAPHY

B. M. Walker and C. P. Rader, eds., *Handbook of Thermoplastic Elastomers,* 2nd ed., Van Nostrand Reinhold, New York, 1988.

N. R. Legge, G. Holden and H. E. Schroeder, eds., *Thermoplastic Elastomers—A Comprehensive Review,* Carl Hanser Verlag, Munich; Oxford University Press, New York, 1987.

GEOFFREY HOLDEN
Shell Development Company

ELECTRICAL BIREFRINGENCE. See ELECTROOPTICAL APPLICATIONS.

ELECTRICAL CONDUCTIVITY. See CONDUCTIVE PLASTICS; ELECTRICALLY CONDUCTIVE POLYMERS; ELECTRICAL PROPERTIES.

ELECTRICAL DISCHARGE MACHINING. See MACHINING.

ELECTRICAL DOUBLE LAYER. See COLLOIDS.

ELECTRICALLY CONDUCTIVE POLYMERS

Polymers with conjugated π-electron backbones display unusual electronic properties such as low energy optical transitions, low ionization potentials, and high electron affinities. The result is a class of polymers that can be oxidized or reduced more easily and more reversibly than conventional polymers. Charge-transfer agents (dopants) effect this oxidation or reduction and in doing so convert an insulating polymer to a conducting polymer with near metallic conductivity in many cases. Conducting polymers represent an important research area with diverse scientific problems of fundamental significance and the potential for commercial applications.

The two key developments in the early stages of conducting polymer research were the discovery of a synthetic route to freestanding, high quality films of polyacetylene (PA) and the discovery that these films could be doped by electron donors or acceptors to conductivity levels (~1000 S/cm) approaching those of some metals. The conductivity of doped polyacetylene is comparable to the highest values achieved in an organic system, ie, those of highly ordered organic charge-transfer crystals. It is the ultimate goal of this research to combine the processibility and other attractive properties of polymers with the electronic properties of metals or semiconductors.

Several different theoretical techniques have been applied to the various conjugated polymers, with varying degrees of success in predicting properties such as ionization potentials, optical spectra, and photoemission spectra. The ionization potential is especially important because it determines whether a particular electron acceptor is capable of ionizing the polymer chain, which is the crucial step in obtaining an acceptor-doped conducting polymer. Much of the theoretical work on doped polymers has concentrated on the charged defects formed on the polymer chain. These defects induce dramatic changes in backbone geometry and electronic structure. Although most of this theoretical work has been concerned with polyacetylene, charged defects in other conducting polymer compositions are now receiving due attention.

Synthesis and Characterization of Conjugated Polymers

Synthesis. There is no singular method for synthesizing polymers that can be transformed into conductors. In designing polymer syntheses, the incorporation of extended π-electron conjugation is of foremost importance. The monomeric starting materials are therefore typically of an aromatic or multiple carbon–carbon bond constitution that must be preserved in the polymer backbone, or achieved in a subsequent step that transforms the initial polymeric product into a conjugated polymer.

Most of the procedures for the syntheses of the neutral conducting polymer precursors are those commonly used in the polymer industry, or are adaptations thereof. The polymerization methods of Ziegler-Natta, Friedel-Crafts, and nucleophilic displacement generate polyacetylene, poly(*p*-phenylene), and poly(*p*-phenylene sulfide), respectively. Other methods include Diels-Alder elimination, Wittig, and electrochemical coupling. Of all of the currently known precursors to conducting polymers, only poly(*p*-phenylene sulfide) is commercially available (Ryton V-1 thermoplastic).

Molecular Weights. The determination of the degree of polymerization (DP) (qv) for most conjugated polymers is, at best, descriptive. The inability to obtain precise data stems largely from the insolubility of these polymers, which precludes the use of conventional methods of determining molecular weight. The insolubility is largely due to the inflexibility of the backbone structures.

Despite the difficulties in determining DP, the correlation of molecular weight with conductivity is important enough to encourage experimentation. A number average molecular weight of 10,500 for polyacetylene is suggested by experiments with ^{14}C and ^{3}H radiolabels.

Electronic Characterization

Doping. The doping of conjugated polymers generates high conductivities, primarily by increasing the carrier concentration. This is accomplished by oxidation or reduction with electron acceptors or donors, respectively. Conductivities achieved span the range from insulators to semiconductors to metals (Fig. 1).

It is characteristic of conducting polymers that the doping process can be reversed, ie, the conducting polymer can be rendered insulating by neutralization back to the uncharged state. This return to neutrality is referred to as compensation.

Electrical Conductivities and Carrier Transport. Conductivities of various conducting polymers are given in Table 1. For a particular polymer, conductivity tends to be relatively insensitive to the identity of the doping agent. The most impor-

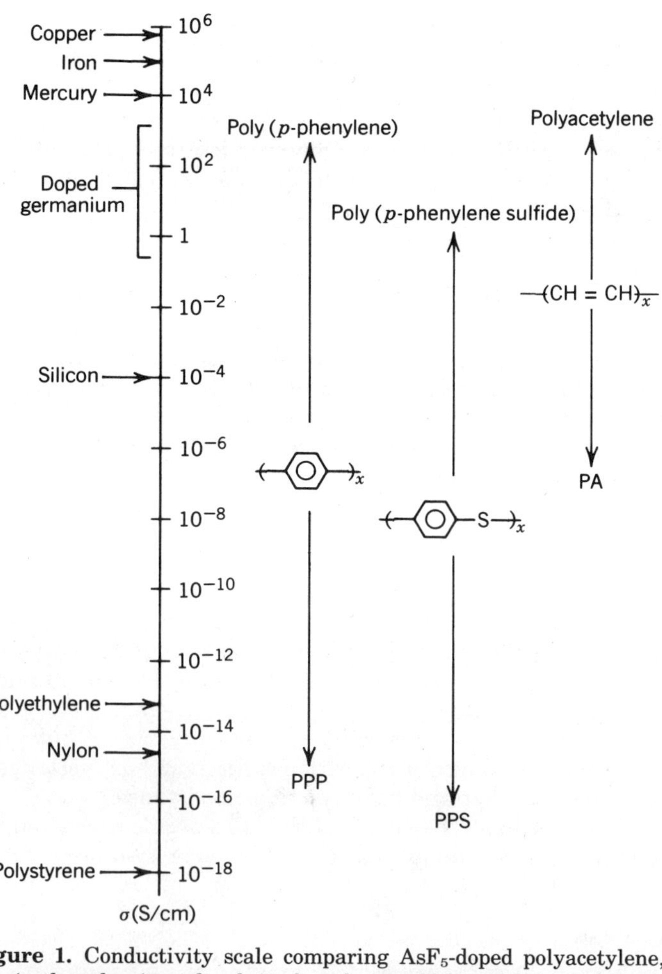

Figure 1. Conductivity scale comparing AsF_5-doped polyacetylene, poly(*p*-phenylene), and poly(*p*-phenylene sulfide) with conventional materials. Conductivity units are Siemens/cm, equivalent to $\Omega^{-1}cm^{-1}$.

Table 1. Electrical Conductivities of Conjugated Polymers with Selected Dopants

		Dopant ion		
Polymer	Abbreviation	Species	Concentration[a]	Conductivity, S/cm
polyacetylene	PA	I_3^-	0.10	550
		AsF_6^-	0.10	1100
		BF_4^- [b]	0.09	100
		K^+ [c]	0.16	50[d]
poly(*p*-phenylene)	PPP	I_3^-		$<10^{-5}$
		AsF_6^-	0.4	500
		BF_4^- [b]	0.2	70
		K^+ [c]	0.6	20
poly(*p*-phenylene sulfide)	PPS	I_3^-		$<10^{-5}$
		AsF_6^-	0.7	5[e]
		BF_4^- [b]		$<10^{-5}$
		K^+		$<10^{-5}$
polypyrrole	PPY	I_3^-		600
		AsF_6^-	0.3	100
		BF_4^- [f]	0.3	100
		K^+		$<10^{-5}$
polythiophene	PTh	I_3^-		0.1
		AsF_6^-	0.24	10
		BF_4^- [f]	0.42	100
		K^+		[g]

[a] Molar concentration of the probable dopant ion on the basis of moles of that ion to moles of the minimum chemical repeat unit of the polymer, ie, CH for PA, C_6H_4 for PPP, C_6H_4S for PPS, C_4H_3N for PPY, and C_4H_2S for PTh.
[b] Doping by exposure to $NO_2^+BF_4^-$.
[c] Doping by exposure to potassium naphthalide solution.
[d] Annealing of donor-doped PA has yielded conductivities up to about 500 S/cm.
[e] Conductivities as high as 100 S/cm have been achieved for PPS/AsF_6^- films deposited from solution.
[f] Electrochemically doped.
[g] Although PTh can be electrochemically reduced, the reduced polymer is apparently unstable.

tant requirement for a dopant is sufficient oxidizing or reducing power to ionize the polymer.

The mechanism of carrier transport in doped conjugated polymers has not been established. It is clear that disorder plays an important role and for that reason variable-range hopping is an attractive concept for further exploration in doped polyacetylene, as well as other conducting polymer systems for which transport experiments have been reported. Metallic islands could also be important, particularly at higher dopant concentrations. Their existence is likely to be controlled by disorder in the starting material and the homogeneity of the doping process. Finally, theoretical explanations for spinless conductivity are qualitative, at best, and require more theoretical work and more detailed experimental studies of the broad spectrum of doped-conjugated polymers that are now available.

Optical Properties. The doping of polyacetylene and other conjugated polymers is accompanied by dramatic changes in optical properties. At low doping, new optical transitions are introduced, most notably at energies lower than the absorption threshold (band gap) of the undoped polymer. Dopant-induced optical transitions grow with dopant concentration at the expense of the absorption band of the undoped polymer.

Conducting Polymer Solutions and Composites

Solutions. To date, three examples of conducting polymer solutions have been reported. The first system reported was AsF_5-doped poly(*p*-phenylene sulfide) in AsF_3 solvent, which demonstrated proof of concept. The second was polycarbazole in I_2 solvent. More recently, doped, substituted polythiophenes have been dissolved in common organic solvents, and conducting films deposited from such solutions. These developments represent a major step forward in the processing of conducting polymers.

Composites. A second route to the processing of conducting polymers has been taken through the development of composites of the conducting plastics. The added components are chosen for their mechanical and chemical properties such as cohesion, adhesion, air stability, and flexibility.

Theory

Theoretical work on conducting polymers has been mainly concerned with radical and ionic sites, referred to as neutral and charged defects, respectively. Most of these studies are carried out on *trans*-polyacetylene because of its unusual ground-state geometry.

Calculations of the electronic structure of conjugated polymers contribute to the understanding of ionization of existing materials and to the prediction of new conducting polymer structures. The key properties of a conjugated polymer in relation to its potential as a precursor to a conducting composition are ionization potential, electron affinity, band gap, and band width.

Applications

The projected commercial applications of conducting polymers are based on the promise of a novel combination of light weight, processibility, and electronic conductivity. Foremost among current commercialization ventures is their application in rechargeable batteries. Conducting polymers have also been proposed as promising new materials in circuitry elements, functioning as passive elements (conducting traces) or active elements (Schottky or *p–n* junction devices). An application with considerable commercial potential is electromagnetic (EMI) shielding.

BIBLIOGRAPHY

T. A. Skotheim, ed., *Handbook of Conducting Polymers,* Vols. 1 and 2, Marcel Dekker, New York, 1986.

H. Shirakawa, E. J. Louis, A. G. MacDiarmid, C. K. Chiang, and A. J. Heeger, *J. Chem. Soc. Chem. Commun.,* 578 (1977).

A. G. MacDiarmid and A. J. Heeger, *Synth. Met.* **1,** 101 (1980).

J. L. Bredas, R. Silbey, D. S. Boudreaux, and R. R. Chance, *J. Am. Chem. Soc.* **105,** 6555 (1983).

JANE E. FROMMER
RONALD R. CHANCE
Allied-Signal Corporation

ELECTRICAL PROPERTIES

Polymers, both uncompounded and compounded, are used in many electrical applications (qv), often as primary electrical insulation. In other cases, such plastics may have primarily a structural function, but also an electrical function (see also ELECTRONICS APPLICATIONS; ELECTROMAGNETIC INTERFERENCE).

The electrical properties of polymers must be understood and taken into account for many different applications. The electrical behavior of insulating materials is influenced by temperature, time, moisture and other contaminants, geometric relationships, mechanical stress and electrodes, and frequency and magnitude of the applied voltage. These factors interact in a complex fashion.

To determine the electrical properties of plastics in a fashion useful for engineering purposes, it is necessary to duplicate conditions of service as closely as possible. However, these conditions are not always known, and can never be completely duplicated in the laboratory. In fact, simple rather than complex and lengthy tests are frequently justified. Electrical properties should be determined as a function of variables such as temperature and time. Time is an important variable in voltage degradation and moisture exposure. Properties should be given in the form of a curve where possible. Some plastics manufacturers supply such data.

Theoretical Aspects

When electrical voltage is applied to a material, current flows. If the voltage is steady (direct), the current is defined by Ohm's law,

$$I = E/R \tag{1}$$

where I = current, A; E = voltage, V; and R = resistance, Ω. Needless to say, such a simple relationship rarely holds in practical situations. Voltage is never absolutely steady and usually varies in cyclic fashion (alternating) with a repetition rate of 0.1–10^{11} Hz (cycles per second) or more. The d-c resistance depends on the dimensions of the material and a number of other factors such as ambient conditions, heat induced by the current flow, and the characteristics of the material.

Materials are classified broadly by resistivity (Table 1). The classes are conductors when hot or wet, or when they contain additives, such as carbon or SiC powder (see also CONDUCTIVE PLASTICS).

Table 1. Classification of Materials Based on Resistivity

Resistivity, Ω·cm[a]	Class
0–10^3	conductor
10^3–10^7	partial conductor or semiconductor[b]
10^5–10^{18} and higher	insulator

[a] Unit resistance between two opposite faces of a 10-mm cube.
[b] Solid-state materials, eg, silicon.

For voltage variations, the concept of capacitance is introduced:

$$C = Q/V$$

where C = capacitance, F; Q = charge, C (A·s); and V = voltage, V. An insulating material (dielectric) can be represented by a combination of resistance and capacitance in several ways (Fig. 1). The more complex and realistic relationship (Fig. **1a**) can be simplified to the equivalent circuits shown in Figures **1b** and **1c**. If voltage is suddenly applied to the series circuit, a current flows (limited by resistance) to charge the capacitance. As voltage builds up across the capacitance, the current decreases. With a good dielectric the series resistance is low. In the parallel representation, on the other hand, the higher the resistance, the better the dielectric and the less current bypassed or leaked.

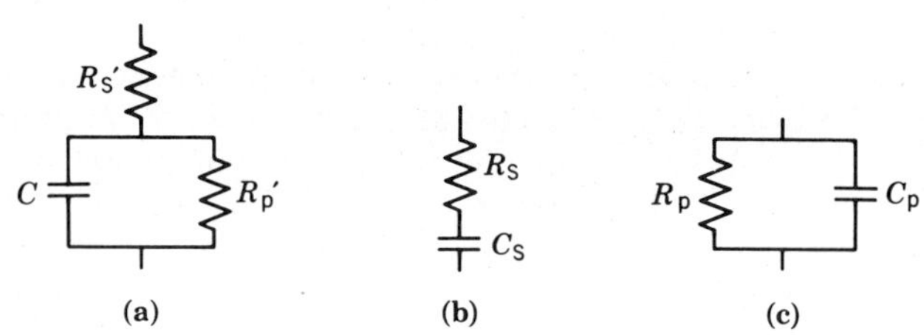

Figure 1. (**a**) Schematic representation of an insulating material (dielectric); (**b**) equivalent series circuit; (**c**) equivalent parallel circuit. C = capacitance, R = resistance; the subscripts p and s refer to parallel and series, respectively.

When an alternating voltage is applied to a perfect dielectric, a current flows that is displaced in time in such a way that it is 90° out of phase with the voltage. Since no insulating material is perfect, the voltage leads the current with time.

In considering the influence of alternating voltage on polymeric insulating materials, it is important to realize that not all of the conductance observed must derive from the migration of charge carriers in the material. An insulating material may contain bound charge or, under the influence of electrical stress, develop additional bound charges of various types that are displaced to a limited extent by the action of the electric field.

Such displacement constitutes an energy-consuming process dependent on temperature and the frequency of the applied voltage. Polarization losses are measured as a-c conductance.

Electric Breakdown

Types of Breakdown. Electric breakdown occurs when the applied voltage can no longer be maintained across the material in a stable fashion without excessive flow of current and physical disruption. Failure of a material, due to the application of a voltage stress, is called voltage breakdown; the voltage gradient at failure is called the dielectric strength, or electric strength, and is usually expressed as kV/mm.

After years of study, the theoretical basis for electric breakdown in solids is still limited. Several types of breakdown in solids are recognized: intrinsic breakdown, breakdown dependent on physical defect, thermal breakdown, discharge-dependent breakdown, failure by electrical treeing, and failure by water (electrochemical) treeing. Conditions and material characteristics determine the type of breakdown that takes place. Actual electrical breakdown may involve two or more types simultaneously or sequentially.

Causes of Breakdown. The electric breakdown of a plastic cannot be defined by a single value. For each application it is necessary to evaluate properties under the conditions of use because service factors may affect the voltage performance of plastics.

Geometry. Thickness, area, shape, and the nature of the conducting contaminants influence the breakdown voltage of plastics.

Thickness and Spacing. The interrelationship of factors that influence voltage breakdown in plastics may be important. The a-c breakdown at room temperature is not linear with thickness and statistical spread of the values is wide; corona and thermal runaway may be involved.

Electric strength depends on thickness and the number of layers of the material. If tests are prolonged, the difference is greater because of corona degradation. On the other hand, a multilayer dielectric may be more resistant to breakdown than a single sheet of equivalent thickness.

Time. Different types of breakdown have different times to failure, as shown in Table 2.

Table 2. Voltage Breakdowns

Approximate time	Description
$0–10^{-8}$ s	space charge or electron-transit time limited
10^{-8}–1 s	electron avalanche, intrinsic breakdown
0.1–10^{3} min	thermal runaway (dielectric heating)
0.1–∞ min	partial discharges (corona) at surfaces or in voids
10^{3}–∞ min	dry and wet treeing
10^{-2}–∞ s	arc (dry) tracking
10^{3}–∞ min	wet and contaminated tracking

Electrical Discharge (Corona). The degradation of plastics by corona may take place over long periods of time up to several years. Corona can occur at the surface of the plastic and in voids or other discontinuities. Geometry, dielectric constant, and gas pressure are among the many factors that influence the start of corona and the resulting degradation.

Electrical Treeing. In treeing, sporadic discharges develop that are intermittent rather than steady like corona. Small channels of degradation develop into a treelike pattern, which may finally bridge the gap between the electrodes (or conductors) and cause failure. The practical importance of treeing was emphasized when failures occurred in buried, polyethylene-insulated cable.

Temperature. Temperature affects the intrinsic, thermal, and discharge-initiated voltage breakdown strength of most plastics.

Thermal Aging. Prolonged thermal aging causes changes that differ for different materials. Loss of weight, eg, plasticizer, embrittlement from oxidation, polymeric cross-linking and scisson, and sometimes acceleration of hydrolytic degradation may all adversely influence performance in service. Thermal aging often improves some "dry" electrical characteristics but crazing and cracking (from aging) may degrade the electrical properties measured after exposure to moisture.

Frequency. Since dielectric heating is directly proportional to the frequency, the tendency for thermal breakdown increases with a rise in frequency.

Moisture. The effects of moisture degradation are more pronounced with resistance, dissipation factor, and capacitance than with breakdown. Functionally, however, the effect of moisture on the electric breakdown may be disastrous. High relative rather than absolute humidity is the causative factor. At higher temperatures the rate of moisture permeation increases. Moisture condensation on plastic surfaces, especially in the presence of ionizable or carbonizing contaminants, may be an important source of degradation. Surface flashover may occur almost instantaneously with moisture condensation, whereas volume breakdown develops after weeks or months.

Tracking from Moisture and Contamination. Electrical failure of a wet and contaminated plastic may progress slowly with the development of surface tracks. The character of these tracks depends on the surface and the contaminant. The tracks are produced by the intense localized heat of multiple tiny arcs (scintillation) that form between more highly conducting areas over less conducting areas on the surface (the wet–dry band process).

Surface Arcs (Arc Resistance). When an arc forms in air between electrodes on a plastic surface, it causes both thermal degradation and low resistivity at the surface. Shifting of the current in the arc from the air to the surface causes further damage.

Variability in Electric Strength. In some cases, normal (Gaussian or more often Weibull) distributions are obtained for breakdown voltage; skewed distributions are common.

Measurement Techniques. ***Electric Breakdown Strength.*** Standardized equipment and procedures for electric breakdown tests are described in ASTM D149. Important aspects are a sine-wave voltage; an appropriate device to measure the test voltage, ie, at the specimen and not only the primary voltage of the testing transformer; a device for increasing the test voltage at a specified, uniform rate; and a provision for rapidly removing the voltage at breakdown.

Corona Tests. A wide variety of methods have been applied to investigate the effect of corona on polymeric materials. Meaningful standardized procedures are difficult to develop, especially for volume corona, ie, in voids and at conducting inclusions. ASTM D2275 is used to measure the corona resistance of plastic.

Treeing Tests. A great deal of effort has led to a plethora of electrical and water treeing tests, each with its advocates. ASTM D3756 describes a test for electrical treeing between a sharply pointed needle and a rounded needle.

The only standardized test for wet-treeing of cable is EIC CS5-79, of the Association of Edison Illuminating Companies.

Tracking Tests. In the United States, four types of tracking tests have been standardized: Dust-and-fog Tracking and Erosion, ASTM D2132; Differential Wet Tracking, ASTM D2302; Inclined-plane Tracking and Erosion, ASTM D2303; and Comparative Tracking Index, ASTM D3638.

Arc Resistance. For dry arc resistance, ASTM D495 uses 12.5 kV (open circuit), which produces an arc between chisel-shaped, tungsten electrodes. The reproducibility of the test results is poor and the correlation of the 10-mA test with service, for which much higher currents are involved, is dubious.

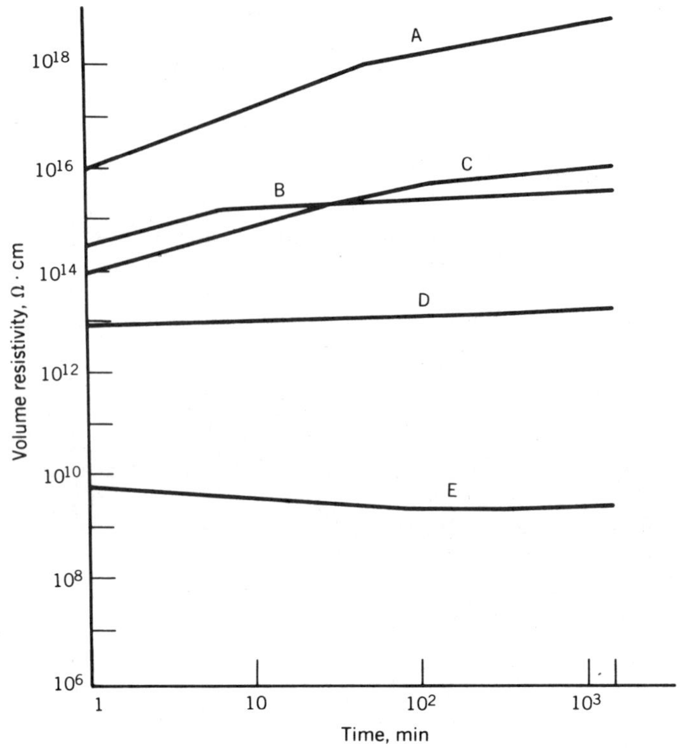

Figure 3. Volume resistivity vs time of electrification. A, cross-linked PE, 23°C; B, cross-linked PE, 100°C; C, dry PVC; D, wet PVC; and E, asbestos fiber-filled epoxy resin.

Resistance, Conductance

In some applications, the plastics resistance or its reciprocal, the conductance, is of functional significance, eg, in electronic circuits for which high resistance or impedance is of operational importance. More often resistance is used to measure indirectly the influence of other factors, such as moisture, which may affect electric strength. Resistance measurements are also useful in studying resin cure, determining the presence of impurities, and investigating the fundamental characteristics of polymeric materials.

Geometry and Homogeneity. Insulation resistance is defined as the d-c resistance of a material or device measured between electrodes or conductors without regard to geometry. Insulation resistance is a combination of surface and volume resistance, as illustrated in Figure 2.

Time. In Figure 3, resistivity is plotted vs time of electrification for materials under several conditions. The change in resistivity with time (polarization) for cross-linked PE is marked, but at 100°C the steady ionic current is large enough and polarization is not apparent. In a similar fashion, absorption is apparent in dry, plasticized, poly (vinyl chloride) (PVC), but masked by ionic conductivity in wet PVC. The ionic conductivity of even a dry, asbestos-filled resin may be so large that the much smaller absorption current is not observed. Thus dielectric polarization provides a useful means of identifying the character and state of polymeric materials, eg, wet, dirty.

Voltage. Resistance measurements are usually made at 100 or 500 V d-c, but for special purposes, eg, to increase sensitivity, the voltage may be increased to 5 kV or even higher.

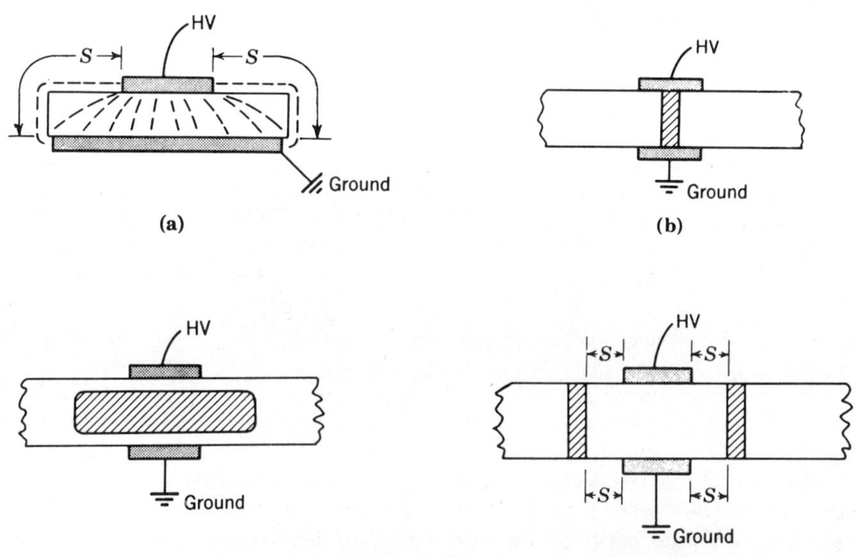

Figure 2. Effect of specimen geometry and homogeneity on surface and volume resistivity: S, path for surface resistance; HV, voltage electrode; flaws are cross-hatched. (**a**) Measures combined surface and volume (plus fringing) resistance; (**b**) primarily volume resistance in low resistance fault between electrodes; (**c**) high volume resistance despite internal low resistance fault between electrodes, and (**d**) surface resistance ($4S$) in series with two low resistance faults outside electrode area.

Temperature. An increase in temperature produces a marked decrease in the volume resistance of plastics.

Thermal Aging. Contrary to widespread opinion, the volume resistance of most plastics continues to increase during thermal aging far beyond the point of mechanical degradation. Consequently, electrical resistance is usually a poor criterion of thermal aging unless combined, for example, with periods of moisture exposure.

Moisture. Measurements of moisture are subject to many difficulties including the precise control of humidity above 90% rh and the long periods of time needed to obtain equilibrium in the test specimens.

The effects of moisture on the electrical resistance of several plastic laminates were studied at General Electric over several years with precise control of temperature and humidity; some of these results are shown in Figure 4.

Measurement Techniques. Techniques for the measurement of resistance (or conductance) are described in ASTM D257.

Alternating-current Characteristics

Although the a-c characteristics of plastics (dielectric constant and dissipation factor) may be more difficult to measure than d-c resistance and breakdown, they provide important theoretical and practical information. Compilations of dielectric constants and dissipation factors have been published for plastics, but are difficult to keep up-to-date. Manufacturers often supply data for specific materials.

Theoretical Aspects. An understanding of the fundamental relationships between dielectric constant and dissipation factor as a function of frequency and temperature is necessary. Electrical properties of polymers depend largely on molecular structure.

The dielectric constant (unit capacitance, permittivity) and dissipation factor, as a function of temperature and frequency, are very different for nonpolar and polar materials. The values of both properties for nonpolar materials, such as PFTE, are nearly constant over a wide frequency range and for temperatures below the softening point. In contrast, for polar materials, such as phenolic resins or plasticized PVC (often referred to as "lossy" materials), the dielectric constant decreases with an increase in frequency, whereas the dissipation factor increases and decreases in a cyclic manner. (Changes with temperature are similar.) It is customary to plot the dielectric constant and the dissipation factor (or its product with the dielectric constant, the loss index), as shown in Figure 5. The absorption peaks (maxima in dissipation factor or loss index) occur at the frequencies at which the dielectric constant changes most rapidly.

Effects of Temperature and Frequency. A dielectric-loss peak within the range of temperature and frequency in use might adversely affect the performance of a plastic at high voltage stress or in some electronic applications requiring a high impedance (or Q). The a-c electrical properties of plastics may also indirectly assess mechanical properties.

Correlation of Physical and Electrical Properties. Since the change in dissipation factor and dielectric constant with temperature and frequency is at least partly determined by molecular structure, it would seem logical that mechanical and electrical properties would be related. A considerable degree of correlation has in fact been found in a number of materials.

Moisture. Like d-c resistance, both dielectric constant and dissipation factor are related to the presence of moisture. Unlike d-c resistance, a-c properties are sensitive to the presence of moisture even in the absence of a continuous low-resistance path. For example, moisture, absorbed in a filler at interfaces between different components or in laminations, can be detected by changes in dissipation factor and capacitance. Moisture may be detected a short distance below the surface. Thus a-c measurements respond more quickly to the presence of moisture and detect moisture penetration before it can influence d-c resistance.

Voltage and Time. From an engineering point of view, dielectric loss in plastics is important because it may be related to electric breakdown. In studying such effects, the dissipation factor $\tan \delta$ is plotted against voltage. The rapid increase in $\tan \delta$ for the wet paper-phenolic resin A is undoubtedly due to dielectric heating and the associated increase in temperature. Dielectric heating probably contributes to the increase in $\tan \delta$ for the drier laminate B, but in this case corona losses are also likely to be involved. The sharp increase in $\tan \delta$ with voltage for the glass–silicone laminate is probably due to the sudden onslaught of corona at the interface between the glass fibers and poorly adhered resin without much, if any, dielectric heating. In low loss materials, such as PE or PTFE, a marked increase in $\tan \delta$ with voltage is due entirely to corona.

Measurement Techniques. The measurement of the dissipation factor and capacitance may be complicated, particularly over a wide range of frequencies. ASTM D150 describes methods applicable to frequencies up to about 10^8 Hz; problems, such as the effect of stray capacitance, arise. At frequencies above 100 MHz, the series resistance and capacitance of the test leads cause measurement problems.

Thermal Aging. The a-c properties usually do not give an indication of the physical embrittlement that leads to the end of life. However, if thermally aged plastics are exposed to moisture, the presence of physical discontinuities or flaws is indicated by an increase in the dissipation factor of the wet specimens.

Engineering Applications of A-c Measurements. In electronic and high voltage applications, the a-c characteristics are often of direct functional importance. Cable insulation should have minimum capacitance (low dielectric constant) with the lowest possible dielectric loss. The single value of a dissipation factor at 1 kHz or 1 MHz, so often supplied by plastics manufacturers, is almost useless for design purposes.

The a-c measurements are particularly useful in detecting dielectric discontinuities, such as voids and delamination, positioned at an angle to the direction of the electric stress; d-c resistance may be more useful in detecting a fault parallel to the electric stress.

Special Electrical Characteristics

Piezoelectric Effects and Electrets. When certain polymers, eg, PTFE and poly(vinylidene fluoride), are placed while hot in a d-c electric field (polarized), and cooled while the d-c field is still applied, stable trapped charges remain. Such permanently polarized materials, called electrets, may possess piezoelectric properties, which are useful in microphones and small loudspeakers.

Static Charge and Triboelectrification. Amber, when rubbed on certain materials such as fur, develops powers of attraction, as has been known for a long time. Several triboelectric series

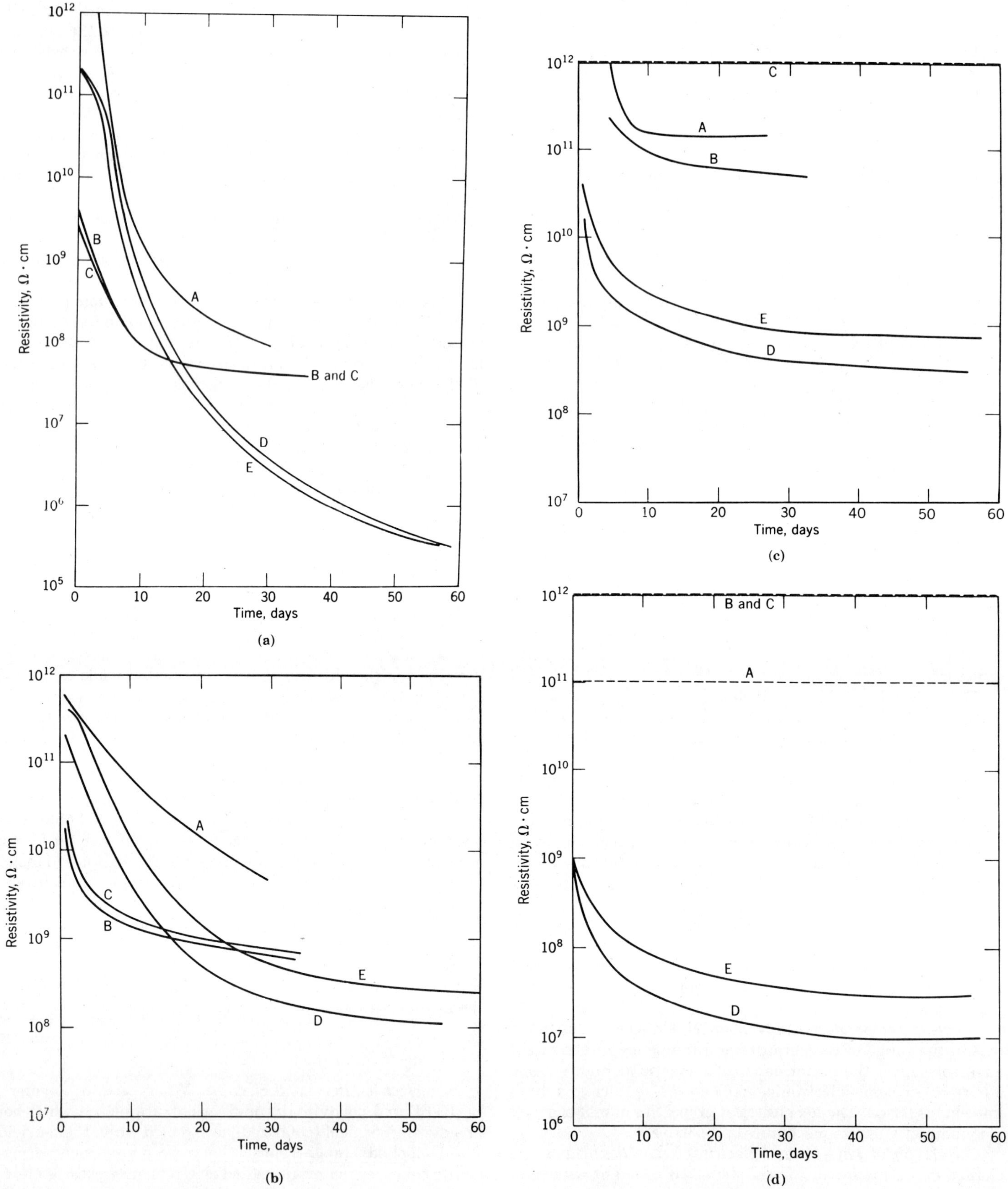

Figure 4. Resistivity of laminate, 6.2 mm thick, at 96% rh and 23°C. A, volume resistivity perpendicular to laminations; B, surface resistivity in machine direction; C, surface resistivity across machine direction; D, volume resistivity along laminations in machine direction; E, volume resistivity along laminations across machine direction. (**a**) Paper–phenolic resin laminate; (**b**) cotton cloth–phenolic resin laminate; (**c**) coarse glass cloth–silicone resin laminate; and (**d**) fine-weave, glass cloth–silicone resin laminate.

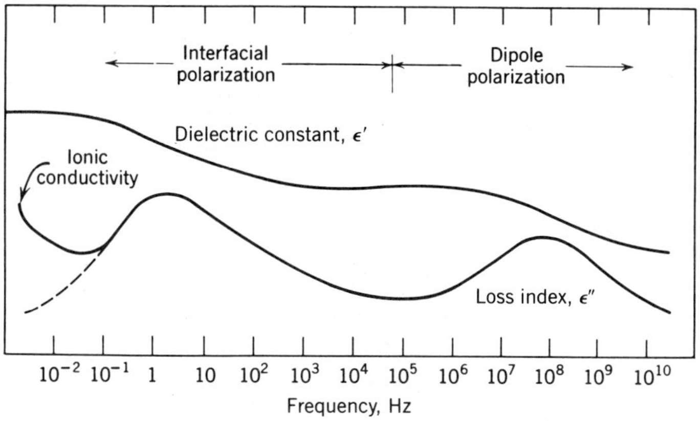

Figure 5. Schematic representation of the dielectric constant and dielectric loss index as a function of frequency.

have been obtained from electrometer measurements of the charge.

Static electricity and surface charge are important in xerography, electrostatic spraying, and precipitation, as well as the adverse effects on the performance of some electronic and computer circuits.

BIBLIOGRAPHY

A. Bradwell, ed., *Electrical Insulation,* Peter Peregrinus Ltd., 1983.

W. Tiller Shugg, *Handbook of Electrical and Electronic Materials,* Van Nostrand Reinhold, New York, 1986.

R. Bartnikas and E. J. McMahon, eds., *Corona Measurement and Interpretation,* ASTM STP 669, American Society for Testing and Materials, Philadelphia, Pa., 1979.

R. M. Eichhorn in R. Bartnikas and R. M. Eichhorn, eds., *Electrical Properties of Solid Insulating Materials: Molecular Structure and Electrical Behavior,* ASTM STP 783, American Society for Testing and Materials, Philadelphia, Pa., 1983.

KENNETH N. MATHES
Consultant

ELECTRICAL SHIELDING.
See ELECTRICAL PROPERTIES; ELECTROMAGNETIC INTERFERENCE.

ELECTRIC INSULATORS.
See ELECTRICAL PROPERTIES.

ELECTROCHEMICAL INITIATION

The passage of current through a solution is always accompanied by chemical changes at the electrodes. Electrons are gained at the cathode, and chemical compounds reduced; electrons are lost at the anode where oxidation occurs. This exchange may produce reactive intermediates, which initiate a chain polymerization. Free-radical or ionic polymerization may thus be initiated by one electron transfer per chain. Alternatively, the initial electron transfer may facilitate coupling reactions that lead to successive chain growth as more electrons are transferred.

The electrochemical method for initiating polymerization is intriguing in its universality and appealing for its simplicity of operation, but can be intimidating because of the possible range of product distributions.

Theory

Faraday's law is the fundamental principle governing electropolymerization. A comparison of product yield charge transferred is a reliable criterion for judging whether a simple reaction path is followed or a complex set of reactions takes place. The occurrence of chain propagation reactions is immediately evident when large numbers of monomer units are polymerized per electron transferred. An examination of molecular weights and of molecular weight distributions provides further insight into reaction mechanism. The variations of molecular weight differ with rate and extent of reaction for ionic, free-radical, and condensation mechanisms, and specific comparison against theoretical models can be made.

Apparatus and Procedure

The design of the apparatus should reflect attempts to maximize efficiency and minimize chemical complications. The uniformity of the electric field is improved by symmetrical geometry based on planar electrodes or concentric cylindrical designs. The spacing between electrodes must be minimized for low losses, particularly in organic solutions. However, the reaction products formed at the counterelectrode may interfere or react with those formed at the working electrode. Utilization of a two-compartment cell with fritted glass or membrane barriers eliminates this interference, but at the cost of increased resistance in the overall cell, with an accompanying loss in power efficiency.

The choice of chemical components is based on the solubility of monomer and polymer in a solution of adequate conductivity. Because of the limited solubility of most monomers and polymers in water, a suitably conducting organic solvent system is needed.

Polymer Coatings

Metals. The coating of metallic substrates by films polymerized directly on their surfaces provides some unusual advantages. The integrity of the films is high. Pinholes tend to be self-healing because the current density is greatest at the exposed portions of the substrate. Cross-linking may be favored by the chemical nature of the monomer, and the resulting films can be insoluble, adherent, and corrosion-resistant.

Graphite Fibers. The modification of the physical properties of graphite fibers by a coating of electropolymerized polymers has attracted considerable attention. To improve bonding between the carbon matrix and the polymeric composite, coating can be generated *in situ* by electropolymerization in aqueous or nonaqueous solvents.

Conducting Polymers

The synthesis of electrically conductive polymers (qv) has emerged as an important application of electropolymerization after it was found that electrochemically oxidized pyrrole gives an electrically conducting polymer. This phenomenon is not confined to pyrrole, but is general to families of heterocyclic aromatic compounds, including thiophene, aniline, furan, azulene, carbazole, and substituted derivatives.

Polymer-modified Electrodes

An important new application of electropolymerization is the formation of polymeric films containing functionally desirable groups. The modification of an electrode surface by adsorbed layers of molecules or ions is the basis for various phenomena in electrochemistry. The effect may be to increase or decrease reaction rates, corrode or passivate electrodes, or influence the morphology of electrodeposition. The locus of electrochemical reactions is the electrode–substrate interface. Modification of the electrode surface can produce a reaction site whose chemical, optical, and electrochemical characteristics are governed by the immobilized molecules. The electrode surface can be designed to favor such diverse functions as photosensitization, inhibition of corrosion, stereoregulation of chirality, and facilitation of fast electron-transfer kinetics.

BIBLIOGRAPHY

A. F. Diaz and J. Bargon in J. A. Skotheim, ed., *Handbook of Conducting Polymers,* Vol. 1, Marcel Dekker, New York, 1986.

B. L. Funt and J. Tanner in N. L. Weinberg, ed., *Techniques of Electroorganic Synthesis Part 2,* Vol. 5, John Wiley & Sons, Inc., New York, 1975.

J. Bargon, S. Mohmand, and R. S. Waltman, *IBM J Res. Dev.* **24,** 330 (1983).

R. W. Murray in A. J. Bard, ed., *Electroanalytical Chemistry,* Vol. 13, Marcel Dekker, Inc., New York, 1984.

B. Lionel Funt
Simon Fraser University

ELECTRODEPOSITION OF COATINGS.

See Coatings, electrodeposition.

ELECTRODES, POLYMER-MODIFIED

A conductor surface is usually modified in order to change the chemical or physical properties of the electrode material, eg, to catalyze a particular electrochemical reaction or to slow the photocorrosion of a semiconductor anode in a photocell. Such problems can be solved by modifying the electrode, ie, by introducing a modifying species at the conductor–solution interface.

To modify electrodes with polymers, two methods are usually employed. In the first method, the electrode is spin- or dip-coated with a solution of a bulk polymer. After the solvent is evaporated, the modified electrode can be used in solvent–electrolyte systems in which the polymer is insoluble.

In the second method, the polymer is formed from a suitable monomer directly on the electrode surface. Advantages are that insoluble, cross-linked polymers can be formed and often the polymer can be covalently linked directly to the electrode surface. Thus polymer solubilities do not restrict the selection of the solvent–electrolyte system.

Charge Transport in Polymers

In any electrochemical application of a polymer-modified electrode, a mechanism for charge to transverse the polymer must exist. The mechanism of charge transport can be electronic conduction, ionic conduction, or a combination of both.

BIBLIOGRAPHY

R. W. Murray, *Acc. Chem. Res.* **13,** 135 (1980).

R. W. Murray in A. J. Bard, ed., *Electroanalytical Chemistry,* Vol. 13, Marcel Dekker, Inc., New York, 1984.

L. R. Faulkner, *Chem. Eng. News* **62,** 28 (1984).

C. Michael Elliott
Colorado State University at Fort Collins

ELECTRODIALYSIS.

See Membranes.

ELECTROKINETIC PHENOMENA.

See Colloids.

ELECTROLESS PLATING.

See Metallizing.

ELECTROMAGNETIC INTERFERENCE

Electromagnetic interference (EMI) in the radio-frequency (r-f) range consists of undesirable r-f signals superimposed on a wanted signal. These signals may originate from a device intended to transmit or from a device not intended to transmit, but with components that are transmitters in the r-f range. The latter group is used in commercial equipment in the United States, and this creates problems with r-f receiving devices. Manufacturers that do not produce r-f-transmitting equipment usually do not test for extraneous emissions. The method for measuring the EMI of a device is described in Appendix C of *FCC Methods of Measurements of Radio Noise Emissions for Computing Devices,* available from the Federal Communications Commission (FCC) Office of Science and Technology. The FCC has traditionally limited its regulation to equipment intended for the radiation of r-f power. However, pressure from agencies whose function depends on clear r-f transmission, as well as the public, has resulted in action by the FCC in the form of regulations limiting the amount of inherent r-f radiation from a device not intended for communication.

Many electronic devices that inherently, but not intentionally, transmit a r-f signal are enclosed in housing made from a polymeric material. Unfilled polymeric materials offer very little attenuation of r-f energy. The restrictions imposed by the FCC led to the production of systems that would enable polymeric materials to attenuate r-f energy by using fillers or by applying a metallic plate to the surface of the plastic part. These systems are incorporated in the device housing or in a component housing within the device. This practice permits compliance with FCC regulations and protection of r-f-sensitive components. The design engineer must have access to data indicating the relative effectiveness of various systems. The system chosen affects the costs.

Theory of R-f Radiation Attenuation

The attenuation of an r-f signal is affected by signal frequency, distance between the source of the r-f signal and the shield (attenuator), and the thickness of the shield.

Metals and carbon are the common materials to attenuate r-f power. Therefore, any system applied to a plastic for the purpose of attenuating r-f energy has a metallic or carbon content, referred to here as a conductor. The conductor must be arranged to provide good electrical continuity within itself and with mating parts. The properties affecting attenuation efficiency are conductivity and permeability. The higher these properties, the better the attenuator. The physical and electrical characteristics of the signal to be attenuated determine how high these values must be in a given situation.

The r-f signal creating EMI is either near field or far field. A near-field condition exists when the distance of signal source to shield is less than the signal wave length λ divided by 2π; a far-field condition exists if this distance is greater than the signal wave length λ divided by 2π.

Application Engineering

The specifications for an attenuating system must have quality-control capability. At present, there are no good quality-control procedures for filled material or filled coatings on a plastic part. Special test equipment, unique to a given application, would be required for a filled plastic or coating. (see also CONDUCTIVE PLASTICS).

BIBLIOGRAPHY

H. W. Ott, *Noise Reduction Techniques in Electronic Systems,* John Wiley & Sons, Inc., New York, 1976.

J. R. Burgoon, Jr., *Insul./Circuits* **16,** 47 (1970).

R. W. Cowdell, "Nomographs Simplify Calculations of Magnetic Shielding Effectiveness," *EDN* (Sept. 1, 1972).

LEROY CHARLES OBERHOLTZER
Rockwell International

ELECTROMAGNETIC SHIELDING.

See ELECTROMAGNETIC INTERFERENCE.

ELECTRON DIFFRACTION

The structure of polymers can be studied by the various techniques of diffraction, including diffraction by x rays, electrons, and neutrons. This article presents the fundamental physics of electron scattering and geometrical diffraction theory. Only single scattering, kinematical scattering, has been considered.

Diffraction Theory

The geometrical nature of diffraction is independent of the type of radiation; thus the theory of diffraction, originally developed for x rays, can be used to describe diffraction by electrons and neutrons. The principal difference between the methods is the physical nature of the scattering of the radiation by atoms, which is conventionally described in terms of the atomic scattering factor f.

Electrons are scattered by the electrostatic potential of the atom, which depends both on the positively charged nucleus and the negatively charged electron cloud surrounding the nucleus. This potential distribution corresponds closely to that of the electron density distribution, but falls off less sharply with increasing distance from the nucleus. Analysis of electron diffraction data thus describes the potential distribution through the crystal; maxima correspond to the locations of atoms.

Wavelength. The experimental observation of electron diffraction was first reported in 1927 and confirmed the dual nature of electrons postulated by de Broglie in 1924. That is, electrons simultaneously possess both particle and wavelike properties.

Atomic Scattering. When a free electron, which may be represented by a plane wave of the form

$$\psi_0 = \exp(2\pi i\mathbf{k}_0\cdot\mathbf{r})$$

interacts with an atom or molecule it may be elastically scattered into a spherical wave of the form

$$\psi_s = 1/r\cdot\exp(2\pi i\mathbf{k}\cdot\mathbf{r})\cdot f_e(u)\cdot\exp(i\delta)$$

where $u = 4\pi/\lambda\cdot\sin\theta$ and 2θ is the angle between the propagation vector of the incident plane wave, $\mathbf{k}_0$, and the direction of an arbitrary vector $\mathbf{r}'$ from the scattering atom to a point on the scattered wave front. $f_e(u)$ is the atomic scattering amplitude and δ is the phase shift introduced by the scattering. This is shown schematically in Figure 1.

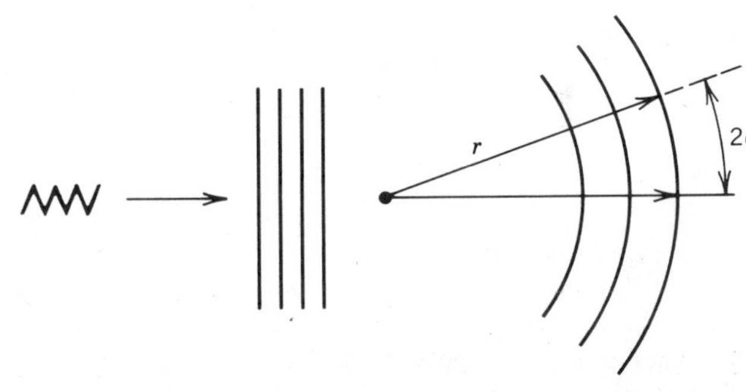

Figure 1. Schematic of electron scattering by an individual atom; ∿∿ is $e^{2\pi i\cdot\mathbf{k}_0\cdot\mathbf{r}}$.

Typical values for the absolute value of the atomic scattering factors are $f_e(\theta) \sim 10^{-8}$ cm for electrons and $f_x(\theta) \sim 10^{-12}$ cm for x rays. Consequently, electrons are scattered much more efficiently than x rays and solid specimens must be thinner to transmit electrons. Since intensities are much greater for electron scattering, it is possible to obtain diffraction patterns using electrons from specimens that give no pattern when using x rays.

Diffraction by a Crystal Lattice

The crystalline state of a material may be defined as a regular three-dimensional arrangement of atoms in space described by a space lattice. The fundamental repeating volume, or unit cell, of the space lattice contains one or more atoms and is defined by three coplanar vectors, $\mathbf{a}_1$, $\mathbf{a}_2$, and $\mathbf{a}_3$. A vector $\mathbf{R}$, defining the location of any atom in the space lattice with respect to some arbitrary origin, often another atom, is given by

$$\mathbf{R} = (x + u)\mathbf{a}_1 + (y + v)\mathbf{a}_2 + (z + w)\mathbf{a}_3$$

where x, y, and z are the fractional coordinates of the atoms within the unit cell and u, v, and w are the integers providing the required translational operations parallel to the unit cell vectors $\mathbf{a}_1$, $\mathbf{a}_2$, and $\mathbf{a}_3$, respectively.

When the incident electron beam is scattered by two atoms of the crystal lattice, the scattered waves will have a phase difference between them equal to $2\pi/\lambda$ times the path difference. For constructive interference between the scattered waves, the path difference must be equal to an integer multiple of 2π. That is

$$2\pi/\lambda\cdot(\text{path difference}) = 2\pi\cdot n$$

which may be written as

$$n\lambda = (\mathbf{k} - \mathbf{k}_0)\cdot\mathbf{R}$$

Considering only scattering within the unit cell the equation may be written

$$n\lambda = (\mathbf{k} - \mathbf{k}_0)\cdot\mathbf{r}$$

where $\mathbf{r} = x\mathbf{a}_1 + y\mathbf{a}_2 + z\mathbf{a}_3$.

The scattering may then be resolved into components along each unit cell vector $\mathbf{a}_1$, where the phase must be an integer value for each. That is

$$h\lambda = (\mathbf{k} - \mathbf{k}_0)\cdot\mathbf{a}_1$$

$$k\lambda = (\mathbf{k} - \mathbf{k}_0)\cdot\mathbf{a}_2$$

$$l\lambda = (\mathbf{k} - \mathbf{k}_0)\cdot\mathbf{a}_3$$

where h, k, and l are integers known as the Miller indices. These equations are known as the Laue equations, which must be satisfied simultaneously for diffraction to proceed.

The three Laue equations can be combined to give

$$(\mathbf{k} - \mathbf{k}_0)/\lambda = h\mathbf{b}_1 + k\mathbf{b}_2 + l\mathbf{b}_3$$

where the $\mathbf{b}_i$ are defined by $\mathbf{a}_i\cdot\mathbf{b}_j = \delta_{i,j}$.

The $\mathbf{b}_i$ define a lattice that is reciprocal to the real space lattice called the reciprocal lattice.

After further manipulation this equation may be written as

$$n\lambda = 2d \sin \theta$$

where d is the distance between the (hkl) lattice planes. This equation is the well-known Bragg law of diffraction. Transmission of electron diffraction patterns from crystalline materials is obtained as spot patterns from single crystals or arrays of only slightly misoriented microcrystals; ring patterns from randomly oriented microcrystals; or some combination indicating a preferred or fiber orientation of a polycrystalline array of microcrystals.

Intensities of Diffraction Patterns

In general terms, the intensity of diffraction maxima is determined by the product of two factors: the structure factor for the unit cell, which is determined by the sum of scattered amplitudes from each atom in the unit cell and their respective phase differences, and the shape factor, which is determined by the physical dimensions of the crystal that is being studied in the microscope.

The location of the Bragg reflections, or diffraction angles, is determined only by the space group of the crystal lattice and not by the arrangement of the individual atoms in the unit cell.

Experimental Considerations

The electron microscopy and diffraction of polymers present unique, but often tedious and difficult problems:

1. Polymers are usually susceptible to radiation damage from the electron beam and may suffer a loss in mass or crystallinity.
2. Many of the standard techniques used for the preparation of metallurgical or biological specimens cannot be used for polymers.
3. Because polymers are composed of low atomic number elements, electron scattering is weak and the required long exposure times increase the risk of radiation damage to the polymeric specimen.

Low Dose Techniques Using the Condenser-objective Lens

In newer transmission electron microscopes (tem), especially those equipped with scanning devices (stem), the objective lens is designed to operate as a much stronger lens with a large pre-specimen component of magnetic flux. Such lenses permit imaging of the final aperture of the illuminating system at the specimen plane. Usually this is the final condenser aperture, or C2 aperture. In this configuration the C2 aperture acts as a field-limiting aperture at the specimen similar to the selected-area diffraction (sad) aperture. However, the C2 aperture is between the specimen and the electron gun and therefore limits the area of the specimen being irradiated. The consequence of this mode of operation, called the condenser-objective mode, is that only the area of the specimen inside the shadow of the C2 aperture is exposed to the electron beam. Thus preliminary setup can be done on one area, and with a small translation of the specimen, a new nonirradiated area for photography can be found.

In another approach, condenser-objective lens is coupled to two sets of deflection coils located both in front of and after specimen. This configuration provides a mechanism for tilting the beam off the optical axis to set up on a non-interesting area of the specimen. A third method, using beam blanking, provides for deflection of the incident beam behind an aperture above the specimen to minimize radiation damage.

The use of a low temperature cryostat for cooling the specimen to near absolute zero temperature has been found to reduce the rate of radiation damage to the specimen.

All these methods are presently being practiced in the study of beam-sensitive materials in the electron microscope.

BIBLIOGRAPHY

P. B. Hirsch, A. Howie, R. B. Nicholson, D. W. Pashley, and M. J. Whelan, *Electron Microscopy of Thin Crystals,* Butterworth Publishers, Ltd., London, 1965.

D. L. Dorset, *J. Electron Microsc. Tech.* **2,** 89 (1985).

D. T. Grubb, *J. Mater. Sci.* **9,** 1715 (1974).

R. H. Geiss, G. B. Street, W. Volksen, and J. Economy, *IBM J. Res. Dev.* **27,** 321 (1983).

ROY H. GEISS
IBM Research

ELECTRONIC HEATING. See DIELECTRIC HEATING.

ELECTRONIC HEAT SEALING. See ADHESION AND BONDING, BONDING.

ELECTRONICS APPLICATIONS

Applications of polymers in the electronics industry are continually increasing. This article deals principally with electronic coatings applications. For discussion of other electronics applications see LITHOGRAPHIC RESISTS; EMBEDDING; ELECTRICALLY CONDUCTIVE POLYMERS; PHOTOCONDUCTIVE POLYMERS; ELECTROMAGNETIC INTERFERENCE, and such polymeric substance articles as POLYIMIDES; EPOXY RESINS; and SILICONES.

Polymeric Encapsulants

The purpose of encapsulation is to protect electronic IC devices from moisture, ionic contaminants such as mobile sodium, potassium, and chloride ions, uv and alpha particle radiation, and other hostile operating environments. In addition, encapsulation enhances the mechanical and physical properties of fragile IC devices and improves manufacturing yields and reliability. The encapsulant must be an ultrapure material with an excellent barrier to moisture and contaminant mobile ions; superior electrical, physical, and mechanical properties; and ease of application and repair in production.

Inorganic polymeric encapsulants consist mainly of silicon dioxide $(SiO_2)_x$ and silicon nitride $(Si_3N_4$ or $SiN)_x$ repeating units (see also PRECERAMIC POLYMERS; SILICATES). These two materials are primarily used as device passivation layers for impurities, mobile ions, moisture, and scratches; as interlayer dielectrics for insulation between conductor layers; and as diffusion and ion implantation masks. Plasma-assisted and thermal chemical vapor deposition (CVD) are commonly used processes for these two inorganic materials.

Organic Encapsulants

Organic polymeric encapsulants are divided into three categories: (*1*) thermosetting polymers, silicones, polyimides, epoxies, silicone-modified polyimides, polyesters (qv), butadiene-styrenes, alkyd resins (qv), allyl esters, and silicone-epoxies; (*2*) thermoplastic polymers, mainly poly(vinyl chloride), polystyrene, polyethylene, fluorocarbon polymers, asphalt, acrylics, tars, Parylene, and recently developed silicone-modified polyimides; and (*3*) the elastomers, silicone rubbers, silicone gels, natural rubbers, and polyurethanes. However, for IC technology applications, only ultrapure silicones, epoxies, and polyimides have been shown to be acceptable IC encapsulants.

BIBLIOGRAPHY

C. P. Wong, *American Chemical Society Symposium Series* **242,** 285 (1984).

C. P. Wong and D. M. Rose, *IEEE Trans. Components Hybrids Manufacturing Technology* **CHMT-6,** 485 (1983).

C. P. Wong, *American Chemical Society Symposium Series* **184,** 171 (1982).

S. M. Sze, ed., *VLSI Technology,* McGraw-Hill, Inc., New York, 1983.

C. P. WONG
AT&T Engineering Research Center

Electronic Compositions

Polymers are used broadly in many electronic compositions, generally formulations composed of the polymer, usually with a solvent, and mixed with particulate ingredients such as metal powders, glass powders, etc. The particulates enhance thermal or electrical conductivity, strength, or some specific requirement such as solder capability. Products include adhesives, thick film compositions, solder pastes, dielectric tapes, and miscellaneous coatings.

The polymers in electronic compositions disperse the particulate ingredients, furnish the proper rheology for the application method, bind the particles together, and furnish adhesion to substrate or device. In some applications the binder is a permanent part of the final cured material, and in others, the binder is burned away.

BIBLIOGRAPHY

P. J. Holmes and R. G. Loasby, *Holmes Handbook of Thick Film Technology,* Electrochemical Publications Ltd., Glasgow, 1976.

J. J. Licari, *Plastic Coatings for Electronics,* McGraw-Hill, Inc., New York, 1976.

ROLF E. FUNER
ICI Americas, Inc.

ELECTRON MICROSCOPY

Electron microscopy (em) is playing an ever-increasing role in the determination of the microstructural state of polymer materials. The resolving power of the scanning electron microscope (sem), a few nanometers under the best circumstances, and that of the transmission electron microscope (tem), which approaches 0.1 nm, are far superior to the light microscope. Since many polymer microstructures lie below the resolution limit of the light microscope, the preference for tem and sem can be easily understood (see also LIGHT MICROSCOPY). Another important development has been the so-called analytical electron microscope (aem), combining tem, sem, x-ray microanalysis, and electron energy loss spectroscopy (eels). The aem is a powerful investigative tool providing the researcher with much detailed information on the composition and type of bonding in addition to morphology (qv). It is likely that such hybrid machines will find increasing utilization in polymer research.

Polymeric materials offer structural variations over 4–5 orders of scale, that is, from the atomic side-by-side packing of chain molecules in the unit cell to the micrometer scale, as in the arrangement of reinforcing fibers in composites. Signals generated by the electron beam and their relation to the specimen thickness are shown in Figure 1. Electron diffraction (qv), eels, and tem require ultrathin specimens, whereas sem and x-ray microanalysis are applicable to any thickness.

TRANSMISSION ELECTRON MICROSCOPE

When studying polymeric specimens, familiarity with diffraction contrast in crystalline materials, mass-thickness contrast in stained amorphous materials, and phase contrast present in all specimens is essential for correct interpretation of tem images and diffraction patterns.

Optics. A cross section of an electron microscope is shown in Figure 2.

Bright-field and Dark-field Imaging. Bright field (bf) is certainly the most commonly used mode of tem imaging; dark field (df) provides a higher contrast image, although of weaker intensity, which is strongly dependent on the sample-electron beam geometry and on the size and location of the objective aperture in the scattering pattern.

Electron Diffraction. In addition to bf and df imaging capabilities, electron diffraction patterns can be easily obtained by simple changes in the lens currents in such a way that the back focal plane of the objective lens is imaged on the phosphor observation screen. For crystalline materials, electron diffraction patterns contain a wealth of information.

Radiation Damage. The main drawback of both tem and sem for the study of polymers is the radiation damage, which results in cross-linking to form disordered amorphous solids or decomposition (chain scission) and mass loss. Therefore, it is vital in any electron microscopy study to take steps to recognize radiation damage and minimize it (see also CROSS-LINKING BY RADIATION; RADIATION-RESISTANT POLYMERS).

Specimen Preparation. For tem, the main requirement is that the specimen be thin enough for the electron beam to pass through readily. An operative contrast mechanism is also required to provide distinction between the microstructural features. Uniform thin polymer specimens are easily solution-cast onto a variety of substrates.

Image Interpretation: Tem Contrast Mechanisms

The incident electrons, after interacting with the specimen, carry information in the amplitude and phase of their wave function. Information from the electrons is usually obtained by employing the squared amplitude of the wave function to produce amplitude contrast (mass–thickness contrast or diffraction contrast). The phase part of the electron wave function may be used to produce changes in intensity (phase contrast). In order to produce a change in intensity from phase differences in the electron wave induced by the object, the phase of a portion of the electron waves must be further shifted by the microscope system. In a light microscope, this shift can be produced by a phase plate that uniformly changes the phase of selected photons (Zernike contrast). In an electron microscope this phase shift is produced by spherical aberration and by defocusing the objective lens. However, defocusing shifts the phases of the electrons in a complicated manner and makes image interpretation more complex.

For polymer specimens, which may consist of complex arrays of small structural units, the image is a two-dimensional projection of the three-dimensional specimen. Thus for phase contrast the projected phase change determines the image, whereas for amplitude contrast imaging of noncrystalline samples, the projected mass-thickness distribution is imaged.

SCANNING ELECTRON MICROSCOPY

Scanning electron microscopes have become commercially available in the last 20 years. Due to the great depth of focus,

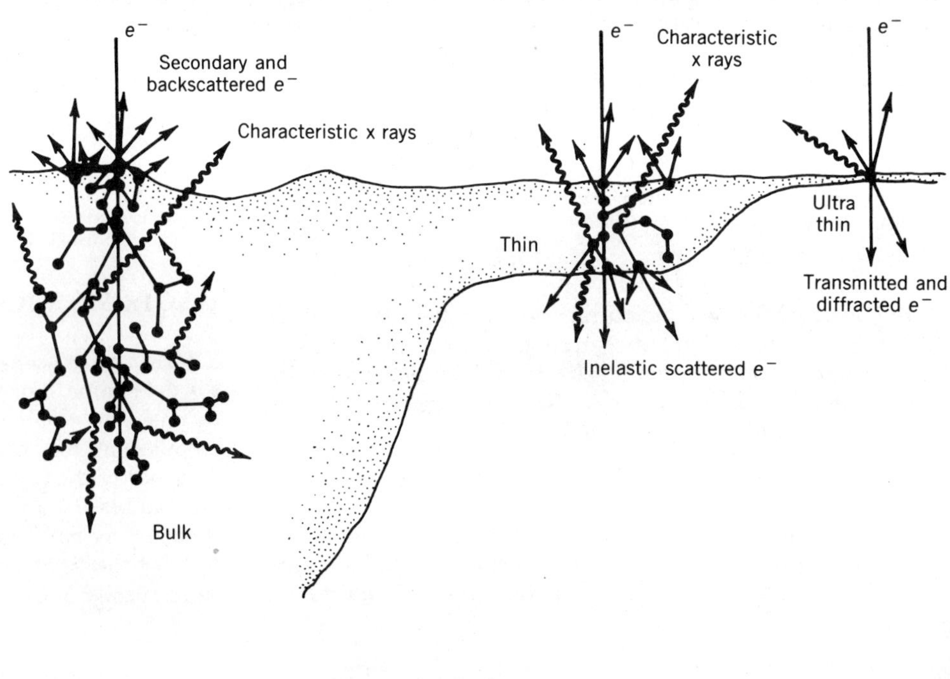

Figure 1. Signal types generated by incident electron beam in specimen areas of varying thickness.

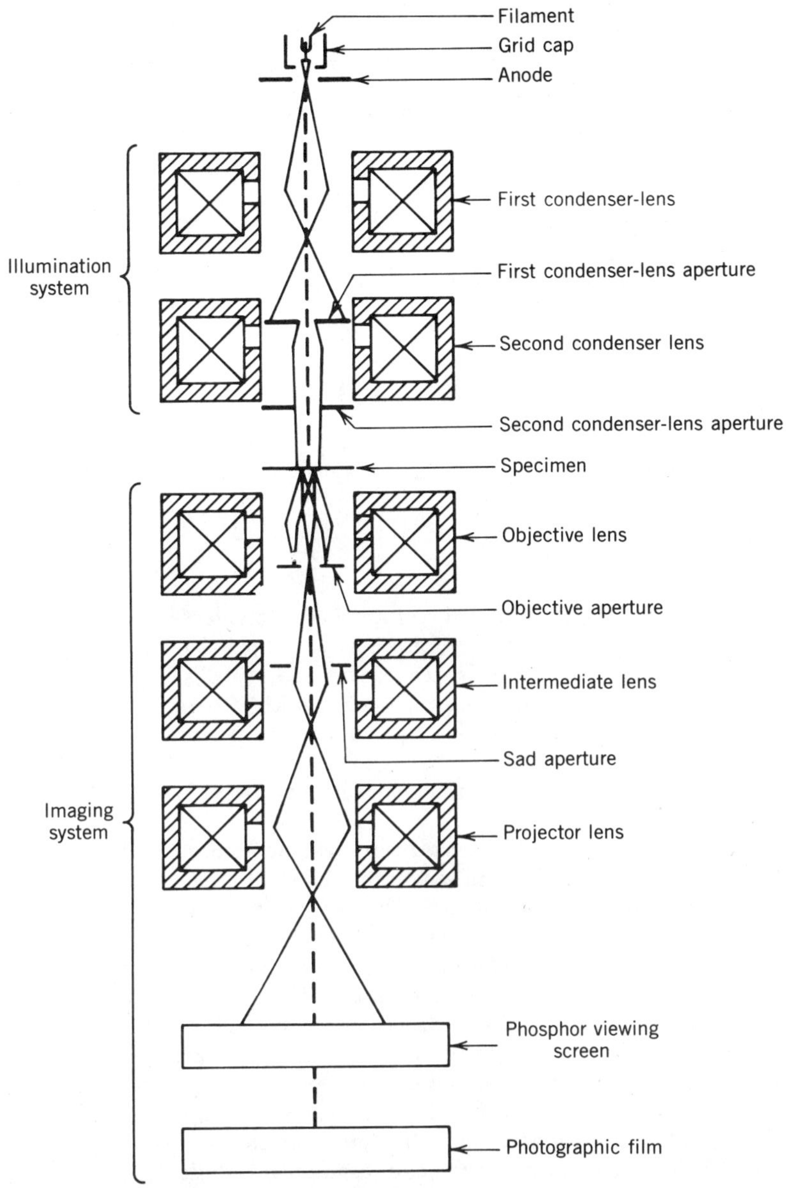

Figure 2. Cross-section of a transmission electron microscope showing the illumination and magnification systems. Sad = selected-area diffraction.

relatively simple image interpretation, and ease of sample preparation, sem is the preferred technique for viewing specimen detail at a resolution well exceeding that of the light microscope. The sem images vividly display the three-dimensional characteristics of the object surface under examination.

Optics. The essential features of a scanning electron microscope are shown in Figure 3.

Secondary-electron Imaging (Sei). In general, sem images are produced by the abundant secondary electrons. Preliminary survey of a sample is normally by secondary electron imaging (sei), even when backscattered-electron imaging (bei) or x-ray energy spectroscopy (xes) is applied.

Backscattered-electron Imaging (Bei). The backscattered electrons, with a much higher energy than secondary electrons, are only slightly deviated by the electric field, which attracts the slowly moving secondary electrons into the secondary-electron detector. Therefore, only backscattered electrons that leave the specimen traveling directly toward the detector contribute to the signal measured. For maximum efficiency, the bei detector should subtend a large solid angle at the specimen, a feature of less importance with the sei detector.

X-ray Microanalysis

For x-ray microanalysis, a detector must be provided that measures either the wavelength or the energy of the x rays generated in the pear-shaped excitation volume of the primary electron beam. The energy dispersive systems (eds) are preferred because they are much faster and can handle all x-ray energies simultaneously. The wavelength dispersive (wds) method is much slower but more precise and provides information down to lower atomic numbers (lower characteristic x-ray energies). Neither technique is particularly attractive for analysis of elements with atomic number below 10, but wds can be used with difficulty down to beryllium. In the wds technique, the x rays entering the detector impinge onto a focusing crystal. The counter is positioned in such a way that it accepts only x rays that are reflected by the focusing crystal in a mir-

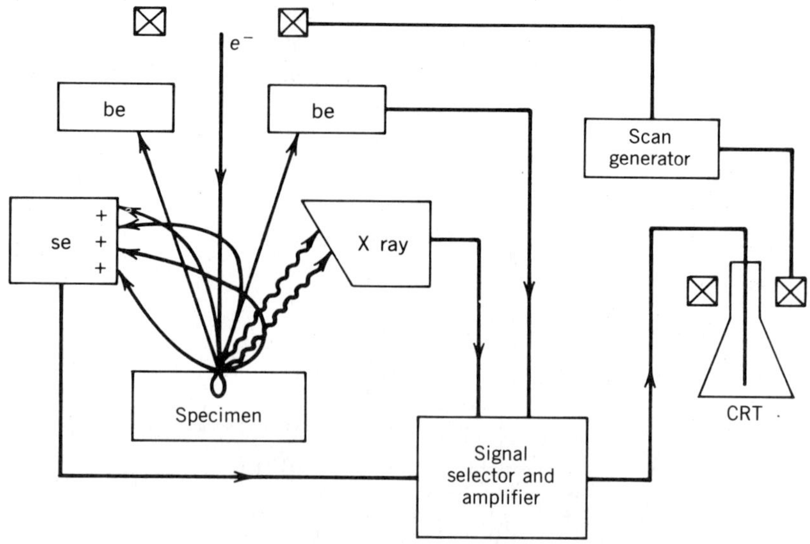

Figure 3. Basic electron optical arrangement in an sem equipped with secondary electron detector (se), backscattered-electron detector (be), and energy dispersive x-ray detector (X ray).

rorlike fashion. A signal is recorded only when the x-ray wavelength satisfies Bragg's law for the focusing crystal atomic plane spacing and the angle of incidence for which the crystal is set. In the eds technique, the x rays enter a solid-state detector and create a number of electron-hole pairs proportional to the energies of the various x rays. This current is swept away by an applied bias voltage to create current pulses, which, after suitable amplification and conversion to voltage pulses, are sent to a pulse-height analyzer where the incoming pulses are sorted by their respective amplitudes (x-ray energies). For both types of measurements, the electron beam can be stationary, set onto a small region for microanalysis, or line or area scanned across a region of chosen dimensions using the scanning coils. If the scanning area mode is chosen, the signal from the x-ray detector may be displayed on the cathode ray tube as in the other imaging modes. In this case, the signal consists of a short pulse of standard strength each time the x-ray count rate for a selected wavelength or energy window exceeds a certain threshold. The resulting image consists of a dot pattern, with the concentration of dots in any particular area dependent on the local abundance of the element.

Specimen Preparation. Perhaps the most attractive feature of sem is the ease of specimen preparation. Since surface topography is normally the prime interest, sample thickness is not critical; only a conductive sample is needed that can be mounted to the goniometer stage.

Surface Preparation. Since sei highlights surface features, any process that serves to emphasize topographical differences enhances contrast. For bei and xes, a flat surface is desired to emphasize compositional rather than topographical contrast.

Radiation Damage. Although electron beam damage is a primary concern in the study of microstructure, the polymeric sample is often examined by sem without regard for possible beam-induced artifacts. The same procedure as used in tem for monitoring irradiation effects should be employed with sem.

Scanning Transmission Electron Microscopy and Analytical Microscopy

The key features of a tem are high resolution imaging and electron-diffraction capabilities. The key feature of a sem is imaging of three-dimensional surfaces. Over the years, x-ray microanalysis, usually accomplished in an electron microprobe instrument, has nearly become a standard sem accessory. An improved hybrid microscope, the analytical scanning transmission microscope (stem), combines all of these features and provides a detailed description of microstructure and composition.

Optics. The basic components of an analytical stem are shown in Figure 4.

Aem of Thin Specimens. Analytical electron microscopy (aem) utilizing stem is a new technique. Applications to polymers are still in the developmental stage, mainly because of possible severe radiation damage. The rewards from aem techniques, such as xes and eels, are nevertheless sufficiently attractive, particularly in biopolymers and cell biology, to justify efforts to deal with radiation-damage limitations.

Xes. X-ray energy spectroscopy on a stem is the same as on a sem, except for the higher accelerating voltage employed and the thinness of the specimen. Because the sample is thin, the x-ray signal generation volume is restricted to a short cylinder defined by the incident beam diameter, permitting neglect of sample absorption and self-fluorescence corrections to the spectra.

Eels. Particularly suited to analysis of light elements, eels is a technique of great promise for polymers; the elements of atomic number 6–9, ie, carbon, nitrogen, oxygen, and fluorine, comprise a very large proportion of the atomic species in synthetic macromolecules. The eels technique depends on the characteristic energy loss suffered by an inelastically scattered electron as it passes through the specimen. There are many inelastic processes, but that of ionization of inner shell electrons provides ready identification of the atomic composition in the probe region.

A crucial limitation for eels is to assure that the inelastically scattered electrons only carry information of the one specific ionization event.

Outlook

Almost every characterization study of polymers now utilizes tem or sem, or both. Stem is still developing a following in the field of polymer research but should find increased usage. Future developments in polymer microscopy will depend on im-

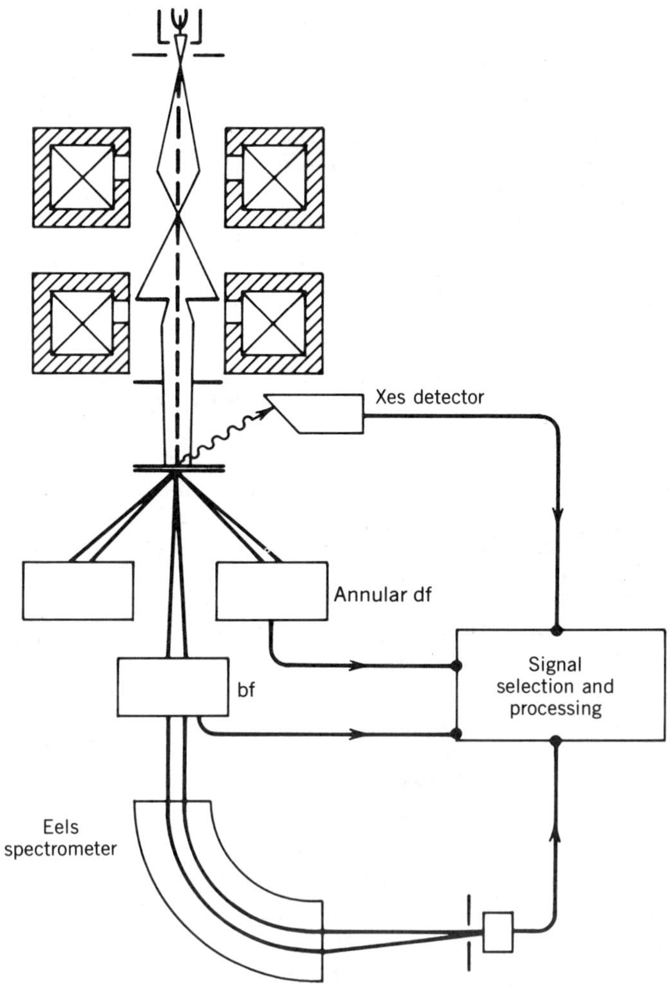

Figure 4. Basic components of an analytical stem.

provements in the instrument and advances in specimen preparation techniques.

BIBLIOGRAPHY

D. L. Misell, *Image Analysis, Enhancement and Interpretation,* Elsevier-North Holland, Amsterdam, 1978.

D. T. Grubb in D. C. Bassett, ed., *Developments in Crystalline Polymers—I,* Applied Science Publishers, London, 1982.

E. L. Thomas in I. H. Hall, ed., *Structure of Crystalline Polymers,* Elsevier Applied Science Publishers, London, 1984.

J. R. White and E. L. Thomas, *Rubber Chem. Tech.* **57,** 457 (1984).

EDWIN L. THOMAS
University of Massachusetts

ELECTRON-SPIN RESONANCE

Electron-spin resonance (esr), also known as electron paramagnetic resonance (epr), is a versatile technique that has found a great many uses in polymer research. In this article, the basic theory of esr is explained and the emphasis is on the use of esr as a tool to prove experimentally the molecular occurrences associated with mechanical damage and failure of polymers.

Fundamental Theory

Electron-spin Transitions. Electron-spin resonance is a form of microwave absorption spectroscopy in which transitions are induced between the Zeeman energy levels arising from the interaction of an assemblage of paramagnetic electrons with an externally applied magnetic field.

The Spectroscopic Splitting Factor. The spectroscopic splitting factor g is defined by the relation $h\nu = g\beta H$. The position center of the spectrum gives the magnetic field H at which resonance occurs and can be used to calculate the g value, which measures the contribution of the spin and orbital motion to its total angular momentum. The ratio of spin magnetic moment $(eh/m_0)\cdot s$ to spin angular momentum S is twice as large as the ratio of orbital magnetic moment $(eh/2lm_0)\cdot l$ to orbital angular moment l, where e is the electronic charge and m_0 its mass. Based on this and classical considerations, a free-electron spin value of $g = 2.0$ is expected. A slight relativistic correction yields $g = 2.002319$ for a completely free spin.

Experimental Considerations. Experimentally, it is convenient to supply incident radiation at a fixed frequency, ν, and then sweep the magnetic field through the range of interest. Energy absorbed by the specimen near resonance may be plotted vs field strength. In general, for sensitivity reasons, a small a-c magnetic field is superimposed on the magnetic field sweep and the resulting signal is phase detected. The result is a "first derivative" spectral line whose intensity is a measure of the number of electrons participating in the resonance absorption and whose shape provides information about the local chemical environment of the electron.

Paramagnetism. In order to exhibit paramagnetic resonance absorption, a specimen must contain unpaired electrons, ie, electrons not interacting with neighboring electrons to such an extent as to be paired with them.

Hyperfine Structure. In addition to its association with the applied external field, the orbit of the electron embraces the nucleus, which itself can possess a magnetic moment and spin. This produces an interaction between the electron and the spin of the nucleus, and the energy levels of the electron are split in accordance with the alignment or opposition of the electron's magnetic moment with respect to that of the nucleus. The magnitude of this splitting interaction depends on the type of atom involved, the phase of the specimen, presence of solvent, and other factors.

In covalently bonded materials, the electron is not associated with a single nucleus; hence, splitting can occur owing to a number of neighboring, shared nuclei. The additional splitting caused by interaction with nuclear spins causes the single esr line to show a more detailed hyperfine structure. A study of this hyperfine splitting provides valuable information about the site of the unpaired electron. It can, for example, be used to identify free radicals produced in polymers by various means.

Secondary-radical Reactions. Care must be exercised in attributing the hyperfine structure observed near room temperature by esr to the radicals originally formed by chain scission. Most polymeric radicals are relatively mobile and through proton migration are able to isomerize to other sites of greater stability. Furthermore, radicals often react with oxygen or other impurities and are converted to other species, eg, peroxy radicals. Consequently, identification of the primary radicals must often be done at cryogenic temperatures, where many radicals are more stable. Upon warming, the primary radical may revert to other radicals, and the spectra may even disappear completely.

Radical-decay Effects. Problems associated with the decay of secondary free radicals may be alleviated by working with more stable radicals.

When chain scission occurs, it may be possible to detect the resulting new end groups by ir spectroscopy.

Polymerization and Degradation Reactions

Most experimental results based on esr are used primarily to corroborate the reaction mechanisms (eg, free-radical polymerization, oxidative degradation) suggested by other analytical tools.

Electron-spin resonance seems to have been more useful in degradation studies than in polymerization, since the degradation mechanism can be increased in intensity until detectably high radical concentrations have been reached.

Mechanical Degradation of Polymers

Mechanical degradation of polymers has been studied extensively by numerous teams. Fracture (qv) is at best a phenomenon which, despite considerable research, is not well understood. The complex morphology of polymers makes it difficult to explain polymer fracture. However, certain models provide helpful insight, and a molecular understanding of fracture should be useful. Esr has been widely used for this purpose.

Grinding of Polymers. Because of the stability of radicals and in order to produce an extensive fracture surface in a small sample, many studies have been conducted on materials degraded by grinding, rasping, milling, or slicing at LN temperatures.

Induced Ordering in Elastomers

Some applications, eg, space or cryogenic seals, require that rubber withstand very low temperatures, where it usually becomes brittle and cracks propagate easily. Applying strain to the rubber before cooling drastically modifies this behavior.

In an engineering sense, these materials toughen. Specimens strained at RT and subsequently fractured at low temperature exhibit strong esr signals, whereas specimens fractured at the same low temperatures without prior straining fail in a brittle manner and show little, if any, detectable radical production during deformation.

It is proposed that prestrain and subsequent cooling produce many small ordered regions throughout the specimen that are comparatively impermeable to cracks. As the load increases, a stress is reached at which microcracks begin to form.

As the deformation increases, more and more microcracks are produced with the associated bond rupture and free-radical production detectable by esr. Eventually, these microcracks start to coalesce and grow to a critical size, after which macroscopic failure ensues. In essence, if somewhat oversimplified, this model suggests that the low temperature ductility of strained rubbers may be explained in terms of the conversion of the rubber into an oriented semicrystalline polymer as a result of temperature and strain.

Filled Elastomers. It appears that esr can be used to investigate mechanical breakdown of filled polymers, at least at low temperatures. This technique can be useful for investigating the effectiveness of fillers and the effect of coupling agents (qv) or other surface treatments on the reinforcing properties of fillers (qv).

Ozone Attack on Elastomers. Ozone cracking of rubber is an economic problem. It is detrimental to serviceability, strength, and appearance and has received considerable experimental attention (see also ANTIOZONANTS).

The esr spectra for rubbers degraded by ozone with time as a parameter are shown in Figure 1.

Oriented Semicrystalline Fibers. The earliest and most extensive esr studies of fracture were on highly oriented polymers, fibers, and films. Although the current understanding of fracture is far from conclusive, there is extensive evidence indicating that bond rupture is a controlling mechanism for fracture of the highly oriented semicrystalline polymers studied.

Attack by Adverse Environments. The effects on residual strength and fracture phenomena of plastics, nylon, and PE exposed to mechanical stress and reactive environments, ie, air, O_3, NO_x, SO_2, and uv radiation, have been studied. The specimens were tested by stressing to failure as received, after stress, and after the specimens were exposed to the environment at various degrees of mechanical stress. Highly oriented and spherulitic forms of semicrystalline PET and polypropylene (PP) are being studied.

Fracture of Aramid Fibers. Fracture studies of semicrystalline thermoplastic fibers led to a study of chain-rupture ki-

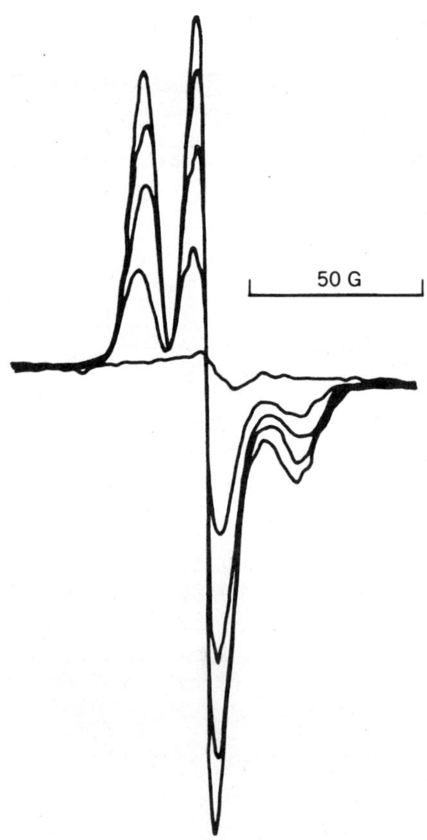

Figure 1. Esr spectra of ozone-degraded NBR (BFGoodrich Hycar 1043). Ozone concentration: 2.8 mg/L. The lowest curve is the residual signal; successively larger ones appear at 5-min intervals after application of 25% strain.

netics in extended chain polymers, such as DuPont's aromatic polyamide fiber, Kevlar. Early efforts to study fracture in Kevlar were hampered by a strong residual esr spectrum always present in the available material. However, more recently, annealing techniques have been described for weakening this signal.

Mechanical Initiation of Energetic Materials

Investigators of the detonation of explosives have speculated on the role of free radicals as mechanical initiators. It has been suggested that free radicals produced by mechanically induced chain scission (similar to those accompanying polymer fracture) act as local hot spots initiating detonation. Esr studies of explosives have provided some, but not conclusive, evidence for this hypothesis.

Conducting Polymers

Most organic polymers are insulators, with some notable exceptions. Conductors have been prepared by the reaction of certain polymers with electron donors or acceptors and have been studied by esr.

Electron Nuclear Double Resonance (endor)

Electron nuclear double resonance (endor) is an elegant technique developed in 1956 that facilitates the determination of hyperfine structure; resolution approaches that of nuclear magnetic resonance (nmr). In conventional esr, measurement of hyperfine coupling constants is hampered by the width of absorption lines and the often small separations of adjacent lines. Endor combines nuclear and electron-spin resonance and provides up to four orders of magnitude of improvement in resolution.

Spin-labeling Techniques

Spin label refers to radicals, often nitroxides, covalently bonded to a polymer molecule. These probes can also be dispersed in a polymer matrix; nitroxides are often used in this approach. Nitroxide labels have been used to study molecular motions in biological systems and to investigate transitions and relaxations in polymers.

BIBLIOGRAPHY

B. Ranby and J. F. Rabek, *ESR Spectroscopy in Polymer Research,* Springer-Verlag, New York, 1977.

J. Sohma and co-workers in P. Kinell, ed., *ESR Applications to Polymer Research,* Almquist-Wiksell, Sodergarn, Sweden, and John Wiley & Sons, Inc., New York, 1973.

M. Igarashi and K. L. DeVries, *Polymer* **24,** 769 (1983).

I. M. Brown, T. C. Sandrecski, and R. J. Morgan, *Polymer* **25,** 737 (1984).

KENNETH DEVRIES
University of Utah

DAVID ROYLANCE
Massachusetts Institute of Technology

ELECTRON-TRANSFER POLYMERS

The term electron-transfer polymer (ETP) denotes polymer systems that transfer or accept electrons from their surroundings.

In this discussion, the term ETP characterizes a polymer bearing only one type of redox system. The term redox implies the chemically reversible electron transfer between the polymer system and reactive ions or molecules. The hydroquinone (QH_2)/quinone (Q) system (eq. 1) shows further that the redox process is connected with a charge compensation leading to electroneutrality. This is realized by donating or accepting ions, in this case, protons.

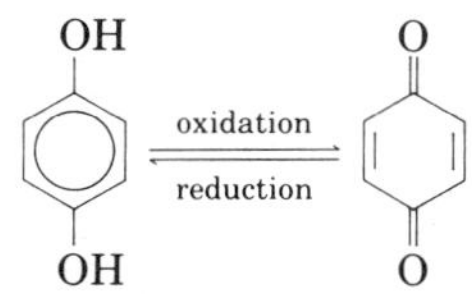

$+ 2\,H^+ + 2\,e^-$ (1)

By this definition, the ETPs represent reversible polymer reagents, in addition to ion exchangers. Furthermore, ETPs are important catalysts.

Preparation

The ETPs are prepared by the methods common in polymer synthesis and polymer functionalization: addition polymerization, condensation polymerization, polymer reactions, redox functions not covalently bound, redox functions and ETPs connected to surfaces via covalent bonds, ETPs used as polymer reagents, and redox ion exchangers.

Properties and Characterization

The most important characteristics of an ETP are redox potential and redox capacity. These parameters allow an estimation of the degree of conversion.

Electrochemical Behavior. Completely soluble polymer redox systems attain a redox equilibrium with an inert electrode. The redox potential is determined by the Nernst equation. This statement is only correct, however, if the redox reactions at the electrode are reversible and free from adsorption. Therefore, both these requirements should be carefully tested by cyclic voltammetry.

Redox Properties. The redox capacity of an ETP defines its reduction or oxidation power measured in milliequivalents (meq) and is referred to as mass (grams of dry product) or volume (milliliters of swollen resin of defined grain size).

Physical Properties. If the redox systems of an ETP are in the same state, the system behaves like a normal macromolecule. For the determination of physical properties such as bulk density, molecular weight, molecular weight distribution, cross-linking, porosity, pore-size distribution, and surface area, standard methods are applied.

Applications

The behavior and the application possibilities of an ETP are determined by the redox potential of its redox systems and its inherently coupled acid–base properties; in addition, they are also strongly influenced by the morphology and molecular structure of the ETP.

The stabilities of the redox functions and the polymer matrix are of particular importance. The acid–base properties of the redox functions are buffered by introducing suitable ionic groups into the matrix. By modifying the synthesis, solubility properties and morphology (grain, sphere, rod, membrane, film, or coating) can be adjusted. In general, ETPs are used as polymer reagents, but recently they have also been applied as redox catalysts and mediators.

BIBLIOGRAPHY

H. G. Cassidy and K. A. Kun, *Oxidation-Reduction Polymers,* Wiley-Interscience, New York, 1965.

A. S. Lindsey in S. Patai, ed., *The Chemistry of the Quinoid Compounds,* John Wiley & Sons, Inc., New York, 1974, Pt. 2, pp. 793–855.

A. V. Kozhevnikov, *Electron-Ion Exchangers,* translated from Russian by R. Koudor, John Wiley & Sons, Inc., New York, 1975.

G. Manecke in E. Seleguy, ed., *Charged and Reactive Polymers,* Vol. 4, D. Reidel Publishing Co., Dodrecht, The Netherlands, 1976, Pt. 2, pp. 173–189.

GEORG MANECKE
Institut für Organische Chemie
Freie Universität Berlin

WINFRIED STORCK
Fritz-Haber Institut der Max-Planck-Gesellschaft

ELECTROOPTICAL APPLICATIONS

Theoretical Development

Electrooptical effects can be generally defined as any change in the optical properties brought about by an external electromagnetic field. The most general expression describing such an effect is given by nonlinear dependence of the polarization on the electric field. At the molecular level, this is given by

$$p_i = p_{0i} + \alpha_{ij}e_j + \beta_{ijk}e_je_k + \gamma_{ijkl}e_je_ke_l + \cdots$$

where p_0 is the permanent polarization of the molecular site, e stands for the local electric field strength, and α, β, and γ (tensors) are the polarizability, first and second hyperpolarizabilities, etc, the various molecular electrical parameters.

Although a large fraction of the research activity in the field of electrooptics is theoretical, a body of experimental research has begun to develop. Attempts have been made to calculate Kerr constants, SHG coefficients, and other relevant nonlinear optical parameters from molecular–structural data. These properties have subsequently been measured for a number of systems.

Experimental Methods and Results

The development of the experimental methods in electrooptics has been closely connected with the technological advancements in design and construction of analytical tools.

Details of time dependence of birefringence as a function of the onset and termination of the electric field could not be studied until high voltage, short-duration electric-pulse generators and fast photodetectors were available. Addition of a photon-counting apparatus with signal averagers and computer control has provided further automation.

Liquid crystalline (LC) polymers (qv), both class I with LC monomer repeat units and class II with mesogenic side groups, represent an important domain of research in polymer electrooptics. The high polarizability and collective orientation properties associated with the liquid crystalline materials result in Kerr coefficients of a much higher order of magnitude than the best known values for ordinary materials. Consequently, various technological applications, eg, display devices, light valves, etc, take advantage of the larger electrooptic effects of liquid crystalline materials.

Possible Applications

For electrooptical or nonlinear optical applications, three properties are required as follows: large electrooptic coefficients, fast response time, and low loss over the wavelength domain of interest. Recent developments in the design and synthesis of novel conjugated and liquid crystalline polymers promise applications in optical communications systems, storage devices, sensors, and display technologies. The superior processing characteristics and molecular design opportunities of polymers may offer substantial advantages over inorganic materials.

A new dimension in recording application has been introduced recently by exploiting a unique characteristic of liquid

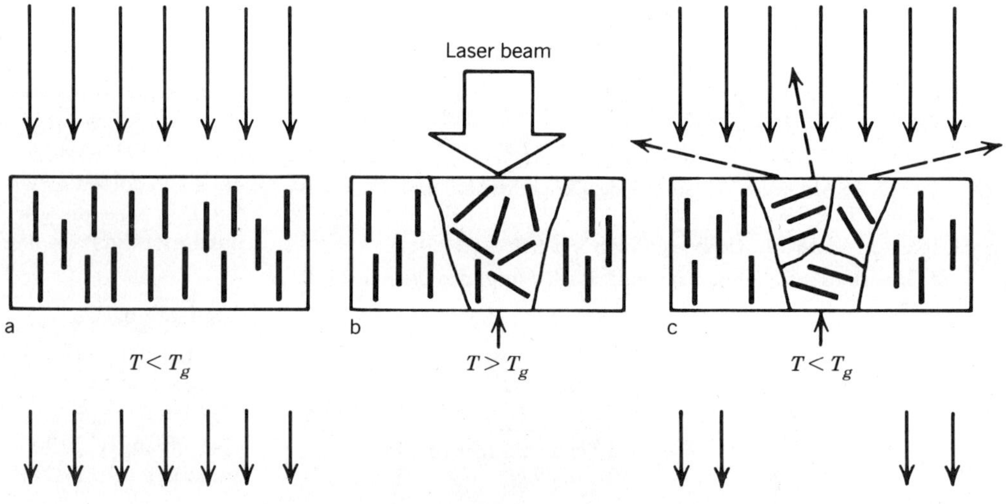

Figure 1. Scheme of thermorecording using a film of an oriented nematic liquid crystalline polymer of class II; for clarity, only the mesogenic units are displayed. T_g is the glass-transition temperature of the backbone polymer. Courtesy of *Polymer Communication.*

crystalline polymers. The field-induced orientation of the side groups of class-II liquid crystalline polymers can be frozen in by cooling the system below the glass-transition temperature T_g of the main polymer backbone. The frozen orientation can be selectively remelted with high-intensity laser beams in such a way that the recording beam is scattered away in the specific regions. In a reverse analogy to a magnetic storage device that utilizes the magnetically oriented domains as the recorded entities, such electrooptical storage devices would employ optically misoriented regions for the identical purpose (Fig. 1).

BIBLIOGRAPHY

C. T. O'Konski, ed., *Molecular Electro-optics,* Marcel Dekker, Inc., New York, 1976.

S. Krause, ed., *Molecular Electro-Optics,* Plenum Press, New York, 1980.

A. Yariv, *Quantum Electronics,* 2nd ed., John Wiley & Sons, Inc., New York, 1967, Chapt. 16.

D. J. Williams, ed., *Nonlinear Optical Properties of Organic Polymeric Materials,* ACS Symposium Series, Washington, D.C., 1983.

MRINAL THAKUR
AT&T Bell Laboratories

SUKANT TRIPATHY
University of Lowell

ELECTROPHORESIS

Electrophoresis refers to the movement of dissolved or suspended charged, i.e., ionic, particles under the influence of an electric d-c field. In the sense commonly used by chemists, the term also applies to techniques used to separate mixtures of ionic components on the basis of differences in their electrophoretic migration. Most applications involve solute macromolecules, although other kinds of charged particles, such as small ions, colloidal and subcellular particles, blood cells, and microorganisms, can also be separated. The rate of migration depends on the charge, size, and shape of the particle, the applied voltage, and the resistance the medium exerts against the migrating particle.

Principles

Electrophoretic Migration. When a particle of effective electrical charge Q is subjected to an electrical-field strength (potential gradient) E, a force of magnitude QE results. This constant force causes acceleration of the particle until the accelerative force is balanced by the frictional resistance f exerted by the medium in which the particle moves. The particle attains a constant velocity of migration (electrophoretic velocity) v, which reflects the difference between the two opposite forces. The equilibrium conditions can be most generally expressed by

$$QE = f = \frac{RT}{D} \cdot v \qquad (1)$$

where D is the diffusion constant of the particle under given conditions, R the Boltzmann (universal gas) constant, and T the absolute temperature. In free solution and for spherical particles, f obeys Stoke's law

$$f = 6\pi r v \eta \qquad (2)$$

where r is the particle radius and η the medium viscosity.

The velocity attained by a charged particle in an electric field of unit strength (1 V/cm) is defined as the mobility m, measured in $\text{cm}^2/(\text{V}\cdot\text{s})$. It is the measured migration distance d of the particle under unit conditions of time t and E:

$$m = \frac{v}{E} = \frac{d}{tE} = \frac{Q}{f} = \frac{ZeD}{RT} \qquad (3)$$

Under given electrophoretic conditions, m is a physical constant that is not equal but only proportional to d. Only under identical conditions of t and E can the values of d be compared directly.

Factors Affecting Migration. The simple definition of m is in practice complicated by a number of interacting factors concerning both the solute and medium that affect the parameters in equations 2 and 3. In the relationship between these factors, E and m are equally valid for both free and supported electrophoresis; however, their manifestation in the individual modes of electrophoresis may differ. In the presence of a supporting medium, a number of additional factors influence m, resolution, and zone sharpness. Limiting factors include counterions, buffer pH and ionic strength, deviation from sphericity of the migrating particles, temperature, Ohmic heat, diffusion, voltage gradients, electroosmotic flow, and interactive effects.

Resolution. In a simplified form, resolution R_s between two zones, or peaks, if a densitometric scan is to be evaluated, is defined as

$$R_s = \frac{d_{\text{ab}}}{0.5(w_{\text{a}} + w_{\text{b}})} \qquad (4)$$

where d_{ab} is the distance between the centers of two zones or peak-to-peak distance and w is the zone width or the peak width measured either at the foot or at the middle of its height.

Techniques

Several criteria can be used to classify electrophoretic techniques, their hybrid forms, and applications. The criteria are difficult to compare and their selection is determined by the purpose of a particular classification. A single comprehensive classification, although possible, would be very unwieldy. The choice of the electrolyte system and its arrangement with respect to the electric field determines the separation principle and thus allows for a simple and systematic classification. Division of electrophoretic techniques by the supporting medium used to stabilize the separation system is popular, as it includes many practical aspects. Techniques include (*1*) classification by method of separation: migration only [zone electrophoresis, moving-boundary electrophoresis, displacement electrophoresis, isoelectric focusing, continuous-flow electrophoresis, and combined methods (discontinuous and two-dimensional electrophoresis)]; migration and interaction (affinity electrophoresis, immunoelectrophoresis, and qualitative immunoelectrophoresis); and (*2*) classification by the method of stabilization: carrier-free electrophoresis and electrophoresis on solid and gel-supporting media.

BIBLIOGRAPHY

Z. Deyl, ed., *Electrophoresis: A Survey of Techniques and Applications,* Vol. 18 of *Journal of Chromatographic Library,* Elsevier Scientific Publishing Co., Amsterdam, 1979, Pt. A.

A. T. Andrews, *Electrophoresis: Theory, Techniques, and Biochemical and Clinical Applications,* Clarendon Press, Oxford, UK, 1981.

C. F. Simpson and M. Whittaker, eds., *Electrophoretic Techniques,* Academic Press, Inc., New York, 1983.

O. Gaal, G. Medgyesi, and L. Vereczkey, *Electrophoresis in the Separation of Biological Macromolecules,* Wiley-Interscience, New York, 1980.

BOREK JANIK
Gelman Sciences, Inc.

ELECTROPLATING. See METALLIZING.

ELECTROSTATIC COATING. See COATING METHODS.

ELEMENTARY ANALYSIS. See CHEMICAL ANALYSIS.

EMBEDDING

Embedment is a useful technique for the protection or decoration of a device or assembly. Components and circuits may be embedded in various types of ceramics and cements, liquids, or granular or powdered solids. In this article, only organic resin-type materials are discussed. Applications range from electrical insulation and protective packaging to mechanical devices. By far the largest use is for electrical and electronic devices. Since the inception of the electrical industry, circuit components have been coated, buried, or otherwise encased in dielectric materials for protection from oxygen, moisture, temperature, electrical flashover, current leakage, salt spray, radiation, solvents, chemicals, microorganisms, mechanical shock, and vibration (see also ELECTRONICS APPLICATIONS).

The terms embedding, casting (qv), potting, molding, impregnation, and encapsulation are often used interchangeably. Generally, embedding implies complete encasement in some uniform external shape; most of the package consists of the embedment material. Encapsulation (Fig. 1), on the other hand, is a coating, and the part is normally dipped in a high viscosity or thixotropic material to obtain a thin, ie, 0.25–12.7-mm-thick coating. The primary protection is provided by a seal, which often imparts mechanical strength.

The casting, mold, and assembly are designed to provide minimum internal stress as the resin shrinks during curing. Potting is similar, except that the shell is not separated from the finished part, as shown in Figure 2. Molding is a technique of embedment in which an object is encased in a plastic that flows into the mold under pressure.

Impregnation is a specialized method of embedding used mainly for coils and transformers in which liquid resin is forced into the interstices of the component.

Materials

Embedding materials include resins (epoxies, silicones, polyurethanes, polyesters, polysulfides, and allylic resins), fillers and extenders (finely ground solids such as silica, quartz, zirconium silicate, calcium carbonate, metals, iron oxide, aluminum oxide, mica, and glass), flexibilizers that reduce the hardness and rigidity of epoxy systems (organic materials that react with the epoxy system, plasticizers and epoxidized vegetable oils), powders used to impregnate and encapsulate coils and windings, foams (epoxy, syntactic, polyurethane, and expandable polystyrene bead foams), decorative and specimen embedments (acrylics or polyester), products with limited utility because of individual weaknesses, eg, low melting temperatures (polystyrenes, polybutadienes, ethylene vinyl acetate (EVA), phenolics, wax, asphalts and tars, and cellulose esters).

Engineering Design

Mechanical Aspects. The stresses that develop between the components and the resin are critical factors in potting and encapsulation. Differentials between the thermal expansion coefficient of the resin and that of the embedded components or housing are a source of design problems. Stresses occur mainly after the cure and are caused by resin shrinkage.

Flexible resin systems produce less internal stress under

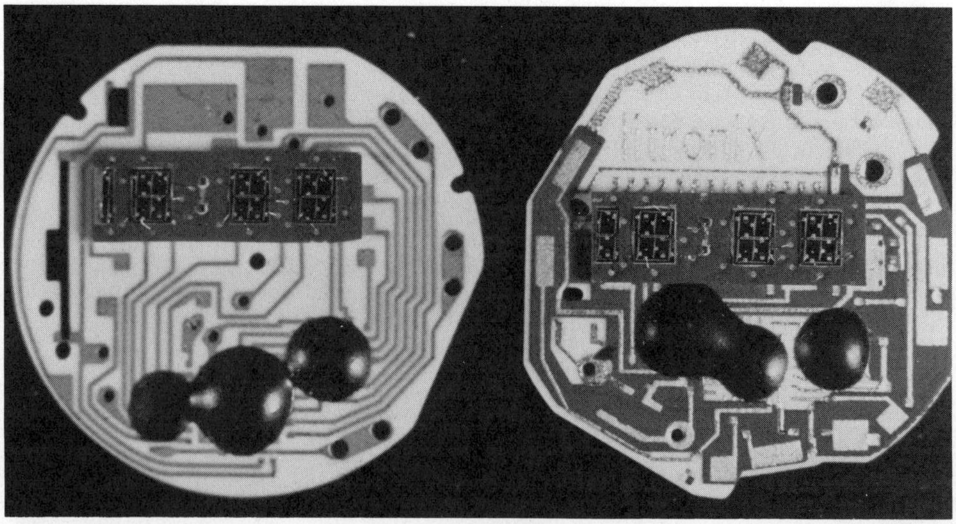

Figure 1. Encapsulated watch circuit. Courtesy of Hysol Div., Dexter Corp.

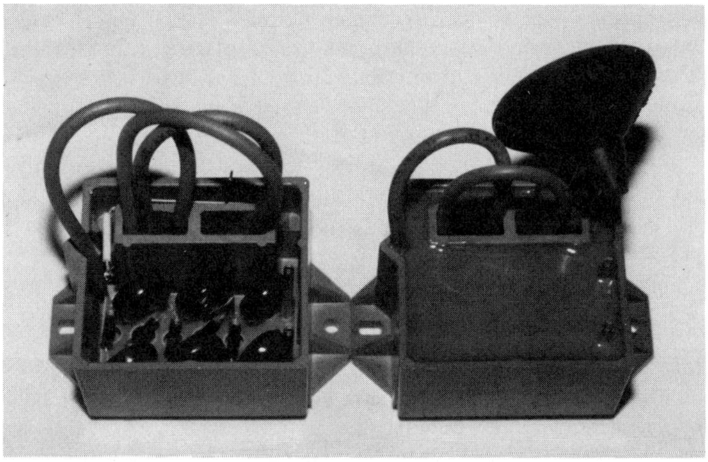

Figure 2. Television voltage multiplier before and after potting. Courtesy of Hysol Div., Dexter Corp.

low temperatures than rigid systems. Generally, the more flexible the system, the weaker the stress, even though the coefficient of linear expansion may be higher than that of a more rigid system.

Viscosity Effects. Viscosity is of crucial importance in embedment resins. When too high, the compound is difficult to mix and pour and does not flow around inserts or components. Encapsulation requires an adherent thixotropic material with an extremely high viscosity, since the part is dipped and cured without a mold or container. For potting and casting, viscosity limits are less important, provided that the mold or container is tight enough to prevent leakage. If impregnation is not required and components are not tightly packed, a viscosity of 1–5 Pa·s (1000–5000 cP) is usually satisfactory.

Exothermic Factors. The embedment of large components entails the risk of runaway exotherms. Excessive stresses and even cracking throughout the casting may result from temperature variations causing localized shrinkage. High temperatures caused by exotherm may volatilize the curing agents and modifiers, causing bubble entrapment. Temperature-sensitive solid-state devices can be damaged and magnet-wire insulation degraded, causing electrical short circuits.

Exothermic reactions can be monitored by a graphic plot of temperature vs time in which the values of gel time, peak temperature, and time-to-peak temperatures are measured.

Electrical Aspects. Embedding resins are good insulators, but it is important to know the values of the other electrical properties as a measure of dielectric isolation of electronic components and electric insulation. Electrical properties (qv) include surface and volume resistivity, arc resistance, dielectric constant and strength, and dissipation and loss factors.

Failure Mechanisms. Failure mechanisms include coefficient mismatch, voids, residual stresses, adhesion, and electromigration.

Processing

Equipment. For precision work, vacuum equipment is ideal for the manufacture of high voltage electronic components. Vacuum impregnation equipment is engineered to dispense precise, metered portions of flowable single- or multicomponent materials into intricate spaces. Dispensing in a vacuum prevents the formation of voids and air entrapment.

Cure Monitoring. The degree and rate of cure are determined by ir analysis and by measuring the changes in electrical properties. The degree of cure of an epoxy resin is determined by monitoring the epoxide linkage, which has an assigned absorption band of 10.93 μm. The progressive disappearance of the epoxide linkage is directly related to the degree of cure (see also CURING).

Molding. Electronic and electrical devices are encapsulated by molding for electrical insulation and protection against physical damage and moisture. Because of their fragility, such devices cannot be rigidly clamped in the mold, although they must be held firmly in place. Resins can be used for compression and transfer molding (qv), injection molding (qv), and reaction injection molding (RIM) (qv).

Molds. ***Materials.*** The embedment mold can be made from wood, metal (machined or cast from low melting alloys), glass, plastics (Teflon), silicone, or almost any material capable of retaining its shape while wetted by the resin and immune to the exotherm temperature of the reaction; metal and plastics are most commonly used.

Design. The design of the mold and the assembly should provide for minimum internal stresses during curing and proper final dimensions after allowing for shrinkage. If removable molds are used, the mold should have a taper or slight draft to facilitate release. A nearly uniform cross section and a smooth surface minimize stresses during cure.

Decapsulation

Embedment materials are selected for their resistance to solvents, chemicals, abrasive handling, and moisture. Repair, quality control, the recovery of metals and devices, and analytical investigation in the case of failure frequently require removal of the embedment material. Methods include mechanical, thermal, plasma, and chemical treatment.

Biomedical Applications

Implantable, embedded electronic devices such as the pacemaker shown in Figure 3 must be biocompatible with the host environment and function during the period of implantation.

Figure 3. Encapsulated pacemaker. Courtesy of Hysol Div., Dexter Corp.

Biocompatibility requires that the material remain biologically inert without any degradation of polymeric material, which would in turn release toxic substances and cause device malfunction.

Present systems employ bisphenol-A type epoxies cured with diethylenetriamine (DETA) or polyurethanes.

BIBLIOGRAPHY

S. M. Lee, *SAMPE J.* **14**(6), Part I, 5 (1978).

S. M. Lee, *SAMPE J.* **15**(1), Part II, 4 (1979).

S. M. Lee, *Epoxy Resins-Science and Technology,* 2nd ed., Marcel Dekker, Inc., New York, Chapt. 9, in press.

W. P. Hudgins III, B. G. Raplee, C. J. Tirman, Jr., and D. A. Brent, *Space Qualified Potting Compounds, AFWAL-TR-82-4005,* Air Force Wright Aeronautics Laboratories, Dayton, Ohio, Feb. 1982.

S. D. Senturia, *Adhes. Soc. Newslett.* **5**(3), 3 (1983).

STUART M. LEE
SRI International

EMBOSSING. See CALENDERING; DECORATING.

EMULSION POLYMERIZATION

Emulsion polymerization is a heterogeneous reaction process in which unsaturated monomers or monomer solutions are dispersed in a continuous phase with the aid of an emulsifier system and polymerized with free-radical initiators. The product, a colloidal dispersion of the polymer or polymer solution, is called a latex. Emulsion polymerization is one of the most important techniques for the commercial production of polymers. The chemistry involves free-radical reactions that are common to many bulk, solution, and suspension polymerization systems, but the physical degree of subdivision of the reaction mixture in the colloidal size range can have a pronounced influence on the course of the polymerization and the characteristics of the product and its performance (see also COLLOIDS).

Monomers polymerized by way of emulsion processes include styrene, butadiene, acrylics, chloroprene, vinyl chloride, vinyl acetate, acrylonitrile, acrylamide, ethylene, and many others.

Reactions

Ingredients for a typical emulsion-polymerization system include the continuous phase, usually water; monomer, monomer mixture, or monomer solution; emulsifiers and stabilizers; initiator system; and sometimes minor ingredients such as chain-transfer agents, buffers, a seed latex, swelling agent, and solvent. The product of the emulsion-polymerization reaction is a colloidal dispersion of polymer particles or polymer–solvent droplets.

The ingredients, the manner in which they are added to the reactor, the reactor design and operating procedures, and post-reaction processing steps all influence product characteristics.

In conventional emulsion polymerization, a hydrophobic monomer, such as styrene, is dispersed in an aqueous continuous phase with an oil-in-water emulsifier. The polymerization is usually carried out with a water-soluble initiator system. The size of the polymer particles in the final latex is generally much smaller than the droplets formed when the monomer is emulsified at the beginning of the reaction.

Temperature. Selection of an initiator system depends mainly on temperature. Higher temperatures permit the use of less reactive systems, such as persulfate, whereas cold polymerizations usually require a redox system.

Pressure. The pressure range is determined by the monomers, the reaction temperature, and the operating procedures. Processes operated above atmospheric pressure generally require complex and costly reactor- and feed-system designs.

Polymerization Rate. The rate of reaction for emulsion homopolymerization can be written

$$R_p = k_p[M]_p[R\cdot]_p = k_p\langle[M]_p\bar{n}\rangle N/N_A$$

where R_p is normally expressed as the rate per volume of continuous phase; $[M]_p$ is the monomer concentration at the reaction site in the polymer particles; $[R\cdot]_p$ is the free-radical concentration per liter of continuous phase; N is the particle-number concentration; N_A is Avogadro's number; $\bar{n}$ is the average number of free radicals per particle, ie, the average among particles of the same size; and $\langle[M]_p\bar{n}\rangle$ is an average of the product $[M]_p\bar{n}$, over the particle-size distribution.

Heat Removal. Removal of the heat of polymerization is often the rate-limiting factor in commercial systems. Circulation of cooling fluids through reactor jackets is the most common method of heat removal from production reactors.

Coagulation and Polymer Formation at the Wall. A good commercial process is based on a "clean" system that is able to run for long periods of time without significant buildup of wall polymer.

Mixing. Agitation promotes heat transfer and mixing (qv) of the ingredients. Rotating impellers generate turbulence within the flowing mixture and cause it to flow. A balance between flow and turbulence is important.

Reactors

A wide variety of both laboratory- and commercial-scale reactors have been designed for emulsion polymerization. They are classified as batch, semibatch, and continuous reactors. Figure 1 is a flow diagram of a process that can be used in either batch or semibatch mode (see REACTOR TECHNOLOGY).

Products

Solids Concentration. High solids concentrations are used for coatings and adhesives and in many formulated products.

Particle-size Distribution. Variations in process design can yield latices with a broad range of size characteristics.

Molecular Weight. Emulsion polymerization can produce high molecular weight polymers at high rates.

Molecular Structure. Cross-linking, branching, and grafting can be significant in emulsion polymerization because the average polymer concentration at the reaction site is higher over most of the reaction cycle. This is advantageous if polymer–radical reactions are desired, as is sometimes the case with grafting. In many processes, however, extensive branching or cross-linking is detrimental to product performance.

Chain-transfer Agents. Chain-transfer agents control molec-

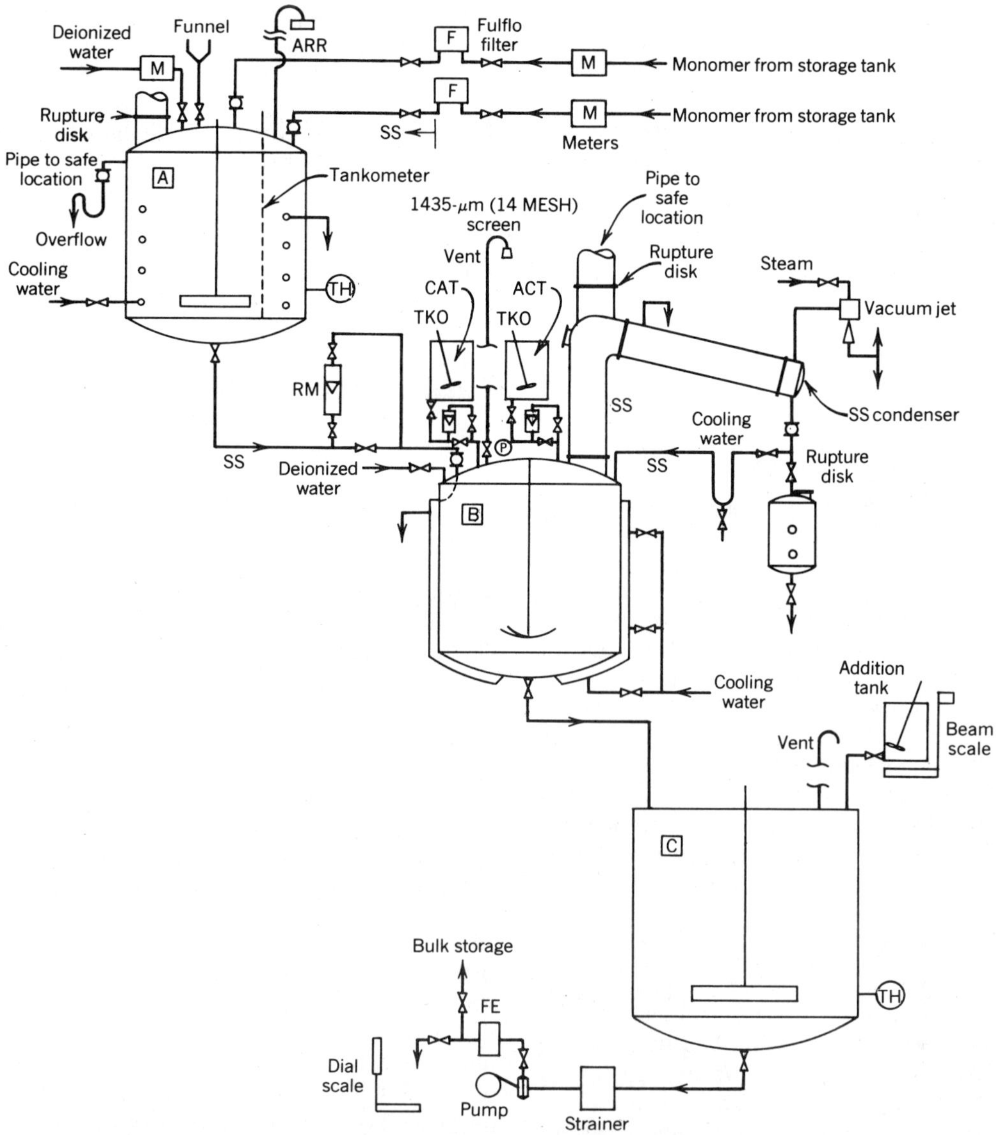

Figure 1. Plant for emulsion polymerization (Rohm & Haas CM-104 (A/cf)). Vessel A is the same size as B; vessel C is 1.25 the size of B. SS = stainless steel; F = filter; M = meter; RM = rotameter; TKO = tankometer; CAT = initiator; ACT = activator; FE = filtration equipment; P = pressure indicator; TH = thermometer; ARR = flame arrester.

ular weight, reactivity, and transport of free radicals in the reaction, and thereby influence the final product properties.

Polymer Composition. In many commercial polymeric products, two or more monomers react to form molecules with various distributions of repeating units. The phenomenon of composition drift during the course of a reaction is important in polymerization, and must be minimized or controlled.

Particle-surface Characteristics. The chemical and physical nature of the surfaces of latex particles can be important in determining stability, application performance, and potential for interfacial chemical reactions. The most common surface groups in conventional latex formulations are adsorbed emulsifiers and initiator fragments attached to the ends of polymer molecules.

Latex Stability. Most stabilization techniques are based on developing suitable surface characteristics, through adsorbed or bound groups, that generate interparticle repulsion forces adequate to prevent particle–particle contact and subsequent coalescence. Stability also depends on particle size, solids concentration, and composition parameters.

Rheology. The flow properties of a latex or a compounded latex product are a complex function of particle-size characteristics, the volume fraction of solids, surface properties, the nature of the continuous phase, and the deformation rate. A typical plot of relative viscosity η vs volume fraction solids is given in Figure 2.

Particle Morphology. Variation in composition of the polymer being formed during polymerization is obviously a significant factor in establishing particle morphology. Other factors include coagulation, monomer polarity, extended particle nucleation, free-radical reactions with polymer molecules, polymer–monomer solubility, and the influence of ingredients such as solvents and blowing agents (qv).

Composition of the Continuous Phase. Most emulsion polym-

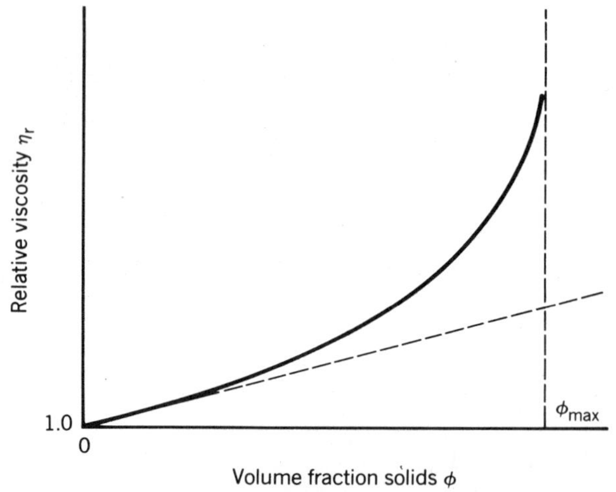

Figure 2. Typical viscosity–concentration curve.

erizations employ initiators that are soluble in the continuous phase. Thus, free radicals generated in this phase may react with dissolved monomer, terminate with other free radicals, or diffuse into loci such as micelles, polymer particles, or monomer drops. The transport of free-radical oligomers, monomers, emulsifiers, chain-transfer agents, and other ingredients within the continuous phase establishes the course of the reaction and the characteristics of the latex particles. The continuous phase also influences rheology, latex stability, and product performance by means of additives such as thickeners and poststabilizers.

Inverse Emulsion Polymerization

In conventional emulsion polymerization, a hydrophobic monomer or monomer mixture is dispersed in a continuous aqueous phase using an oil-in-water emulsifier followed by polymerization with a water-soluble free-radical initiator. In inverse emulsion polymerization, a hydrophilic monomer, usually in aqueous solution, is dispersed in an organic phase using a water-in-oil emulsifier. The polymerization is carried out with oil-soluble or water-soluble initiators to yield a colloidal dispersion of water-swollen polymer particles in oil.

BIBLIOGRAPHY

D. Bassett and A. E. Hamielec, eds., *Emulsion Polymers and Emulsion Polymerization, ACS Symposium Series No. 165,* The American Chemical Society, Washington, D.C., 1981.

I. Piirma, ed., *Emulsion Polymerization,* Academic Press, Inc., New York, 1982.

G. W. Poehlein, R. H. Ottewill, and J. W. Goodwin, *Science and Technology of Polymer Colloids,* Vols. I and II of NATO-ASI Series, Series E: Applied Sciences, No. 67, Martinus Nijhoff Publishers, The Hague, The Netherlands, 1983.

GARY W. POEHLEIN
Georgia Institute of Technology

EMULSIONS. See COLLOIDS; EMULSION POLYMERIZATION; LATICES.

ENAMELS. See COATINGS.

ENCAPSULATION. See EMBEDDING; MICROENCAPSULATION.

END-GROUP ANALYSIS. See CHEMICAL ANALYSIS.

ENERGY APPLICATIONS

Thermal Insulation

Cellular plastics, especially polystyrene, polyurethanes, polyisocyanurates, phenolics, and fire-resistant polymers are widely used as thermal insulation materials and in insulation systems. Currently, their prime use is for building envelope components. Other uses include chemical processing, refrigeration, and cryogenic applications (see also CELLULAR MATERIALS).

Foams based on polyimides, polybenzimidazoles, polypyrrones, polyureas, polyphenylquinoxalines, and phenolic resins produce less smoke, are more fire resistant, and can be used at higher temperatures. These materials are more expensive and are used only for special applications. Poly(vinyl chloride) (PVC) foams are available in small quantities (see also FLAMMABILITY; HEAT-RESISTANT POLYMERS).

Currently, national codes and standards specify building insulation. Building applications account for over 90% of the total volume of cellular plastic materials used for insulation purposes (see also CONSTRUCTION AND BUILDING APPLICATIONS).

Function of Thermal Insulation

The basic mechanisms of heat transmission in thermal insulation are radiation (electromagnetic waves), conduction (atomic or molecular collisions), and convection (fluid motion).

It is the function of thermal insulation to minimize and control the three modes by introducing low emittance–high reflectance barriers to attenuate radiation; a large number of small, low density, low thermal conductivity elements to minimize solid conduction and convection; and by evacuation or inclusion of a high density gas to eliminate or reduce gas conduction and convection.

Although thermal performance is probably the principal influence on the choice of a thermal insulation, the final selection for a particular application depends on the following: suitability for the temperature and environmental conditions; compressive, flexure, shear, and tensile strengths; dimensional stability; shock and vibration resistance; chemical, environmental, and erosion resistance; fire resistance; space limitations; health effects; availability and ease of application; and economics.

Table 1. Typical Properties of Cellular Plastic Materials Used as Thermal Insulation

Property	Polyurethane and polyisocyanurate	Polystyrene	Urea–formaldehyde and urea-based foam	Phenolic resin
density, kg/m³	30–40	12–48	9–18 (dry)	45–60
closed-cell content, %	90	>90	5–70	>90
water-vapor permeability[a], perm in.	2–3	0.5–1[b] 1–4[c]	4–5	<1
water absorption, vol %	2–5	0.02–1[b] 2–4[c]	high, depending on conditions	<2
thermal expansion × 10^{-6} per °C	30–40	30–45		20–40
heat capacity, J/(kg·K)[d]	1500	1200–1300	1500–1800	2000
thermal conductivity, W/(m·K)	0.026[e] 0.020[f]	0.028[b] 0.036[g]	0.035	0.018
fire resistance	combustible	combustible	combustible	combustible
flame spread[a]	30–50[h] 25[i]	5–25	0–25	20–25
smoke development[b]	155–500[h] 55–200[i]	10–40	0–10	5–15
toxicity	toxic gases when burned	CO when burned	may evolve formaldehyde	CO when burned
dimensional stability, vol %	0–12	<1	1–50	<1
upper temperature limit, °C	120	75	210 decomposes	150

[a] Units given in ASTM test.
[b] Extruded.
[c] Molded.
[d] To convert J to cal, divide by 4.184.
[e] Aged unfaced.
[f] Impermeable skins.
[g] Expanded.
[h] Polyurethane.
[i] Polyisocyanurate.

Cellular Plastics as Insulation

A low thermal conductivity polymer fabricated into a low density foam with a multitude of small cells, preferably closed, fulfills many of the above criteria, especially good thermal performance.

The drawbacks of cellular materials include limited temperature of application, poor fire resistance, uncertain dimensional stability, thermal aging and degradation, friability, embrittlement owing to thermal effects and uv light, and possible health hazards.

Properties

The important properties (Table 1) include thermal conductivity, mechanical and physical properties, dimensional stability, fire resistance, and production of toxic gases, especially during combustion. Other criteria are water-vapor permeability, water absorption, and prolonged environmental effects.

Health and Safety Factors

During the course of combustion, cellular plastics can evolve smoke containing toxic gases, eg, cyanides from polyisocyanates.

Urea–formaldehyde might emit free formaldehyde, causing irritation and, possibly, serious illness.

In addition, small amounts of potentially toxic fluorocarbon gases used in manufacture and contained in many cellular plastics can leak slowly into the environment over long periods of time.

BIBLIOGRAPHY

C. J. Benning, *Plastic Foams: The Physics and Chemistry of Product Performance and Process Technology,* Vol. 1, John Wiley & Sons, Inc., New York, 1969.

An Assessment of Thermal Insulation Materials and Systems for Building Applications, DOE Report BNL-50862 UC-95d, U.S. Department of Energy, Washington, D.C., 1978.

R. P. Tye and D. L. McElroy, eds., *ASTM STP 718, Thermal Insulation Performance,* American Society for Testing and Materials, Philadelphia, Pa., 1980, pp. 9–26.

W. C. Turner and J. F. Malloy, *Thermal Insulation Handbook,* R. E. Krieger Publishing Co., Inc., Melbourne, Fla., 1981, pp. 191–275.

The Kline Guide to the Plastics Industry, 2nd ed., Charles H. Kline and Co., Inc., Fairfield, N.J., 1982.

R. P. Tye
Dynatech Scientific Inc.

Solar Applications

The gathering and processing of solar energy is based on large areas that may be in the form of windows, glazings for rooftop collectors, covers for biological reaction ponds, liners for solar ponds, or reflecting surfaces for focusing heliostats or mirrors. In all these applications, polymers rank high on the list of possible materials.

Polymers in Solar Energy Systems

Glazings or cover plates for flat-plate collectors and windows require materials of high transparency. The amount of radia-

Table 1. Thermal and Optical Properties of Cover-plate Materials[a]

			Normal incident transmittance					
Material	Trademark	Index of refraction	Short-wave, $\lambda = 0.4–2.5\ \mu m$	Long-wave, $\lambda = 2.5–4.0\ \mu m$	Thickness, mm	Density, g/cm^3	Specific heat, $J/(g \cdot K)$[b]	Thermal capacity[c], $W/(m^2 \cdot K)$
glass		1.518	0.840	0.020	3.175	2.489	0.75	1.659
fiber-glass-reinforced polyester	Sunlite	1.540	0.870	0.076	0.6350	1.399	1.465	0.361
acrylics	Plexiglas	1.490	0.900	0.020	3.175	1.189	1.465	1.534
polycarbonate	Lexan	1.586	0.840	0.020	3.175	1.199	1.193	1.260
polytetrafluoroethylene	Teflon	1.343	0.960	0.256	0.05080	2.148	1.72	0.036
poly(vinyl fluoride)	Tedlar	1.460	0.920	0.207	0.1016	1.379	1.256	0.049
polyester	Mylar	1.640	0.870	0.178	0.1270	1.394	1.046	0.051
poly(vinylidene fluoride)	Kynar	1.413	0.930	0.230	0.1016	1.770	1.256	0.063
polyethylene		1.500	0.920	0.810	0.1016	0.910	2.302	0.059

[a] These values correspond to the thickness associated with the stated transmittances. They represent thicknesses of commercial films.
[b] To convert J to cal, divide by 4.184.
[c] Thermal capacity = (thickness)(density)(specific heat).

tion reflected from the surfaces and thus lost from transmission is controlled by the index of refraction of the material. The thermal and optical properties of cover-plate or glazing materials are given in Table 1. Fluorinated polymers seem to be particularly well suited for solar applications, where they provide uv stability and weatherability.

BIBLIOGRAPHY

W. F. Carrol and P. Schissel, *Polymers in Solar Technologies: An R&D Strategy, SERI/TR-334-601,* Solar Energy Research Institute, Golden, Colo., July 1980.

B. L. Butler and R. S. Claassen, *Trans. ASME J. Solar Eng.* **102,** 175 (1980).

C. A. Pryde, "Photoaging of Polycarbonates: Effects of Selected Variables on Degradation Pathways." *ACS Symp. Ser. 1* **280** (1985).

Gordon E. Gross
Solar Energy Research Institute

ENERGY CONSERVATION

Energy conservation in the plastics industry can be considered from several viewpoints. Utilization of plastics compared to other materials can reduce the energy required to manufacture or use a particular type of product. A well-known example is the increasing use of plastics in place of metals to reduce the weight of automobiles and thus to reduce fuel consumption. When glass is replaced by plastics in beverage containers, the estimated savings depend on the assumptions made in the calculation, particularly with regard to recycling. It must be decided whether the chemical bond energy of the resin is to be included in the calculation as part of the energy required for production (see also Disposal and reuse of plastics).

Taking into account all polymerization and processing steps, a thermosetting material might be preferred to a thermoplastic to save the energy of injection molding. Reaction injection molding (RIM) requires lower injection pressures and temperatures than injection molding. In addition, the exothermic polymerization provides much of the heat required for curing. Recent advances in technology permit reduction in curing temperatures and times, and this further reduces the energy requirements of the process.

Many products cannot be made by means of reactive processing, and so most plastics products are made from thermoplastic resins. However, the use of new polymerization processes is currently resulting in substantial energy savings. In particular, linear low density polyethylene (LLDPE) produced at low pressures is replacing branched polyethylene, which must be processed at very high pressures. This copolymer reduces both the energy consumption and investment cost of the polymerization process (see Ethylene polymers).

Some resins are designed for low processing energy requirements. New resins have been developed that require lower processing temperatures, thus reducing the energy needed for extrusion or molding.

In a thermoplastic forming plant, energy can be conserved in various ways. Maintenance and repairs are necessary, such as the periodic lubrication of mechanical devices and cleaning of heat-exchange surfaces. The plant layout can be modified while retaining the same basic process, for example, by installing more elaborate control systems and high efficiency extruders. Another alternative is the development of entirely new processes for producing plastics products, for example, the application of dielectric heating to fuse resin particles without bulk melting.

In process modification, the investment costs must be taken into account since the relationship between energy consumption and capital costs is important.

Waste-heat Recovery

Virtually all the energy that flows into a polymer processing operation must be removed as heat. This heat can be recovered and used to provide space heating. In an injection-molding plant, heat is removed mainly from the hydraulic pump and the mold by cooling water of 27–35°C. Most plants use a recirculating water system with a cooling tower, usually located on the roof of the building. A water-to-air fan-coil heat exchanger can transfer this heat to the building for heating.

BIBLIOGRAPHY

R. E. Nunn and K. Ackerman, *Soc. Plast. Eng. Tech. Pap.* **27,** 786 (1981).

N. R. Schott and H. Derby in *ACS Symposium Series, No. 107,* American Chemical Society, Washington, D.C., 1979.

C. I. Chung, E. M. Mount III, and D. E. McClelland in *ACS Symposium Series, No. 107,* American Chemical Society, Washington, D.C., 1979.

J. M. Dealy, *Polym. Eng. Sci.* **22,** 528 (1982).

JOHN M. DEALY
McGill University

ENERGY TRANSFER

Much of the early interest in the photochemistry (qv) of polymers was directed toward the prevention of photodegradation, which has gained in importance with the increasing use of polymeric materials in severe weathering environments, eg, solar collectors. In more recent years, polymer photochemistry has found applications in the electronics industry via photoresists, in which case a photoreactive polymer is desirable (see also LITHOGRAPHIC RESISTS; PHOTOCONDUCTIVE POLYMERS).

In many polymers, for example polystyrene, the absorbing chromophore is present at each repeating unit or, as in certain copolymers, a significant mole fraction of the absorbing unit is attached to the chain. If certain criteria are met, the energy deposited in the initially excited chromophore may be transferred to a nearby chromophore in such a way that the excitation is delocalized. Because there is no driving force for the energy to migrate in any particular direction, it is best thought of as a "random walk" of the excitation on a disordered, nearly one-dimensional lattice. Polymer-bound chromophores acting as energy traps may exist, or most chromophores may self-trap. Thus, the fraction of the polymer coil over which the excitation migrates depends on the average transfer time (τ_t) between chromophores relative to the lifetime of the excited state (τ_E), since the approximate number of transfer steps is τ_E/τ_t, and on the density of trapping groups or the rate of self-trapping. The complex relationship between those characteristics and the average energy migration length is the subject of active theoretical investigation.

Mechanisms

Energy transfer can be represented as a chemical reaction:

$$A^* + B \longrightarrow A + B^* \qquad (1)$$

Species A and B may be chemically and energetically equivalent but they represent different sites in crystal lattices or polymer chains. The process represented in equation 1 is equivalent to a simultaneous downward transition from the $A^* \rightarrow A$ state and an upward transition $B \rightarrow B^*$. The overall probability of this energy transfer step is proportional to the spectral overlap integral:

$$S_n = \int \epsilon_B(\bar{\nu}) I_A^*(\bar{\nu}) \bar{\nu}^{-n} d\bar{\nu} \qquad (2)$$

where $\epsilon_B(\bar{\nu})$ is the molar absorption coefficient of B and $I_A^*(\bar{\nu})$ is the emission spectrum of A* normalized to unit area. The factor $\bar{\nu}^{-n}$ accommodates dielectric effects and depends on the coupling mechanism. The S_n integral is extremely important in the molecular design of polymers for energy transfer because it can govern the choice of chromophores. In the case of energy transfer between equivalent molecules, the S_n integral represents the extent of overlap of the emission and absorption of molecule A. In general, this overlap is most favorable for chromophores that undergo minor geometric change in the excited state.

First-order time-dependent perturbation theory can be used to treat the process of equation 1.

Energy Trapping

Energy trapping refers to the localization of the excitation at some point in space, in this case on the polymer chain. The energy trap may consist of a chemically distinct species to which energy transfer is exothermic:

$$A^* + B \rightleftharpoons A + B^* + \text{heat} \qquad (3)$$

The endothermic reverse reaction has been written explicitly; energy trapping is not necessarily irreversible.

Excited-state Annihilation Processes

Excited-state annihilation processes are well-known. Annihilation occurs if the excited-state molecules diffuse together in solution or if the excitations themselves migrate within a crystal lattice or along a polymer chain. In fact, one of the strongest indications of energy migration is excited-state annihilation.

Special Characteristics of Energy Transfer in Polymers

Every phenomenon discussed up to this point has a counterpart in standard solution or solid-phase molecular photophysics. In addition, polymer photophysics exhibits a number of unique features.

Tacticity and Chain Characteristics. The rate of energy transfer between polymer-bound chromophores is obviously governed by their average separation and orientation.

Copolymers. Copolymers are readily produced in which one chromophore acts as an energy trap or one of the components does not have an absorbing chromophore in the spectral region of interest and, hence, acts as an inert spacer group.

Effect of Molecular Weight. The molecular weight of a polymer affects both the coil density (for a given solvent) and the number of chromophores per coil.

Solvent Effects. Solvents can affect the photophysics of a polymer by excited-state quenching and by affecting coil density.

Films. All energy-transfer phenomena discussed above occur in films, including some of the features peculiar to polymers.

BIBLIOGRAPHY

L. F. Thompson, C. G. Willson, and M. J. Bowden, eds., *ACS Symposium Series, No. 219,* The American Chemical Society, Washington, D.C., 1983.

H. Morawetz and I. Z. Steinberg, eds., *Ann. N.Y. Acad. Sci.* **366** (1981).

J. E. Guillet, *Polymer Photophysics and Photochemistry,* Cambridge University Press, New York, 1985.

D. Phillips, ed., *Polymer Photophysics,* Chapman and Hall, London, 1985.

S. E. WEBBER
University of Texas at Austin

ENGINEERING FIBERS. See FIBERS, ENGINEERING.

ENGINEERING PLASTICS

Engineering plastics are thermoplastics that maintain dimensional stability and most mechanical properties above 100°C and below 0°C. This definition encompasses plastics that can be formed into functional parts that can bear loads and withstand abuse in temperature environments commonly experienced by the traditional engineering materials: wood, metals, glass, and ceramics. Generic resins falling within the scope of this definition include acetals, polyamides (nylons), polyimides, polyetherimides, polyesters, polycarbonates, polyethers, polysulfide polymers, polysulfones, blends or alloys of the foregoing resins, and some examples from other resin types (see also POLYETHERETHERKETONES). Thermoplastic elastomers and commodity resins, such as most styrenics, acrylics, polyolefins, polyurethanes, poly(vinyl chloride)s, and related chlorinated polyolefins, lose mechanical properties below 100°C (see also ELASTOMERS, THERMOPLASTIC). Exceptions include certain acrylonitrile–butadiene–styrene (ABS) polymers that maintain engineering properties in the vicinity of 100°C.

Plastics may be crystalline or amorphous. Acetals, most nylons, some polyesters, poly(arylene sulfide)s, and polyetherketones are crystalline resins. In general, crystalline polymers do not transmit light, whereas amorphous polymers generally do. The mode of processing or treatment can alter apparent morphology. For example, crystalline poly(ethylene terephthalate) can be made into clear film by dimensional orientation. Another property that distinguishes amorphous polymers from crystalline polymers is organic solvent resistance. In applications where solvent resistance is important, such as in gear boxes exposed to oils and greases, a crystalline polymer is the likely choice. On the other hand, for filter bowls, glazing, etc, where clarity is required, an amorphous polymer might be preferred. When treatment of polymers results in morphological change, properties can change as well. Table 1 notes general comparative properties.

Processing

Processing methods for engineering plastics include injection molding, blow molding, extrusion, reaction-injection molding, reactive casting, hot-stamping, and many others. Conceptually, any processing method appropriate to thermoplastic resins can be applied to engineering plastics. In practice, the very properties of engineering plastics such as high heat-distortion temperature (HDT) and stiff melt flow require equipment designed to cope with these more rigorous requirements.

Table 1. Morphology Effects on Properties

Property	Crystalline	Amorphous
organic solvent resistance	high	low
light transmission T	none to low	high
lubricity	high	low
dimensional stability	high[a]	moderate
mold shrinkage	high	low

[a] With a high degree of crystallinity and no water absorption.

Resins

Polyetheretherketone Resin (PEEK). Polyetheretherketone resins are now produced in both the UK and the United States under the trade name Victrex PEEK resin via the displacement reaction of 4,4-difluorodiphenyl ketone by the potassium salt of hydroquinone:

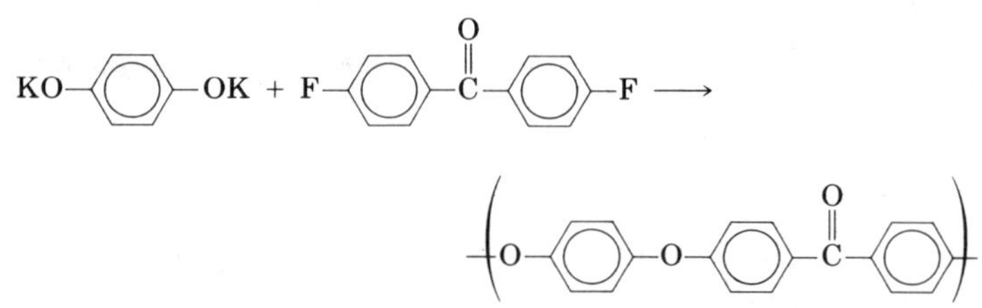

The PEEK resin is gray, crystalline, and has outstanding solvent resistance; its T_g is ca 185°C, and it melts at 288°C.

The unfilled resin has an HDT of 165°C, which can be increased to near its melting point by incorporating glass filler. The resin is thermally stable, and maintains ductility for over one week after being heated to 320°C; it can be kept for years at 200°C. Its hydrolytic stability is excellent. The resin is flame retardant, has low smoke emission, and can be processed at 340–400°C. Recycled material can be safely processed.

The PEEK resin is marketed as neat or filled pellets for injection molding, as powder for coatings, or as preimpregnated fiber sheet and tapes. Applications include parts that can be exposed to high temperature, radiation, or aggressive chemical environments. Aerospace and military uses are prominent.

Polyamideimide Resin (PAI). The first commercial PAI resin, Torlon, is produced by the solution condensation of trimellitic trichloride with methylenedianiline:

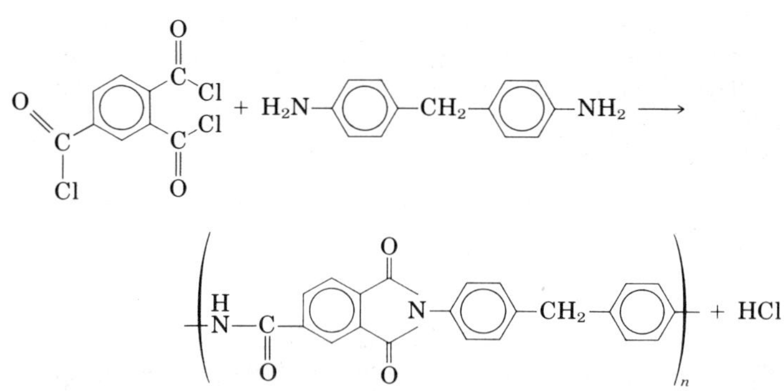

Torlon resin is an amber-to-gray colored amorphous polymer with an HDT of 274°C, flexural modulus of 4600 N/mm^2 (667 × 10^3 psi), and good impact resistance. Although it contains amide linkages, it is dimensionally much more stable under various humidity conditions than standard crystalline nylons. As are aromatic polyamides and imides, Torlon is flame retardant and has low smoke emission. Creep under load is low. Electrical, chemical, hydrolytic, and radiation stability are excellent. Because of these properties, PAI can replace metal in gears, connectors, and other mechanical parts. It is also used in electrical applications where high temperature and chemical resistance are required.

Polysulfone Resins. Commercially important polysulfones (qv) are aromatic, ie, in the generalized formula for the repeating unit

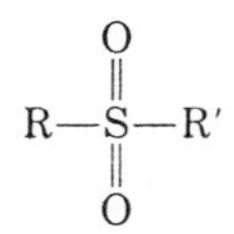

both R and R′ contain aromatic rings and ether linkages. The resins are made by batch processes employing Friedel-Crafts reactions or nucleophilic aromatic substitution.

Polyarylethersulfones are amber-to-yellow colored, transparent, amorphous resins. Because of high HDTs, resistance to thermooxidative conditions, and hydrolytic stability, they are suitable for appliances, electrical and electronic components, aircraft interior parts, and biological and medical devices requiring steam sterilization. Flammability and smoke generation are low. Udel resin has excellent microwave properties and is employed in dishware suitable for microwave ovens. These resins, however, are subject to attack by organic solvents. Melt flow is stiff, limiting the size and complexity of injection molded parts. The most serious problem with these resins is uv light instability. For outdoor use, a uv screen coating is recommended.

Polyimide Resins. Polyimide resins are used when heat and environmental resistance is required. Kapton resin and, more recently, Upilex resin are sold as films, partially polymerized resin coating solutions, and sinterable powders. These resins have extensive applications in the high performance insulation markets (traction motor, wire and cable) and flexible printed circuits. They are not melt reprocessible.

Upjohn's Polyimide 2080 resin, Rhone-Poulenc's Kerinide resin, Mitsubishi's BT resins, and Toshiba's Imidaloy resin are based on bismaleimide technology. Maleic anhydride reacts with a diamine to produce a diimide oligomer. Further polymerization by additional diamine and heat produces the resins:

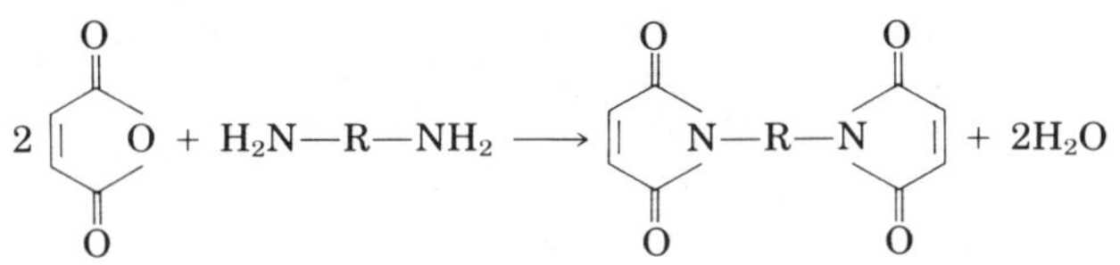

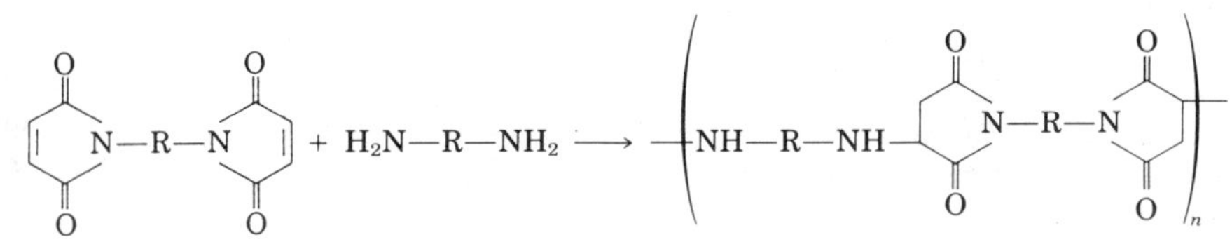

Polyetherimide Resins (PEI). Polyetherimide resins (PEI) are the newest generic engineering plastics. They are produced by an unusual nucleophilic substitution process:

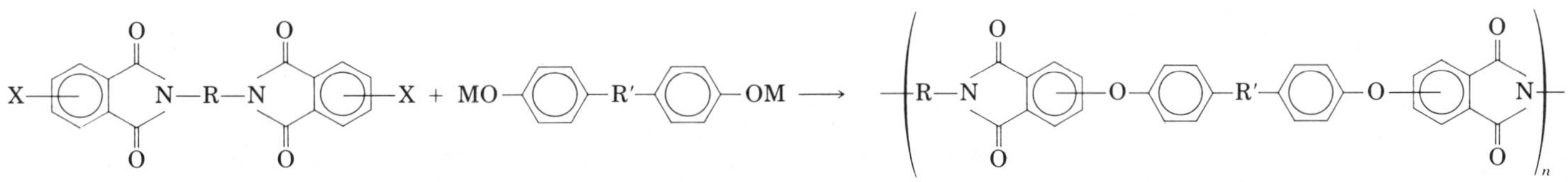

where M = metal, or by a conventional condensation of diamines and dianhydrides:

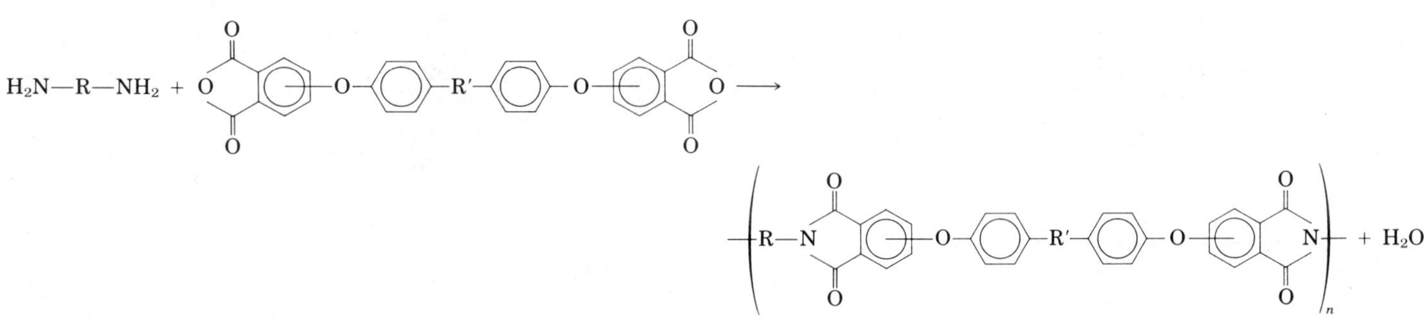

Ultem PEI resins are amber colored and amorphous, with HDT between polyarylate resins and high temperature crystalline polymers, such as PEEK or PAI resins. Ultem resins exhibit high modulus and are stiff yet ductile. Light transmission is low. Despite the high use temperature, they are processible by injection molding, structural foam molding, or extrusion techniques at moderate pressures between 340 and

425°C. They are inherently flame retardant and generate little smoke; dimensional stabilities are excellent. Large flat parts such as circuit boards or hard disks for computers can be injection-molded to maintain critical dimension. Creep properties are excellent; for example, Ultem 1000 maintains its apparent flexural modulus for thousands of hours at a loading of 34 N/mm (4930 psi). Unlike PAI resins, annealing is not required. Chemical resistance is high for an amorphous resin. Ultem resins are unaffected by automotive organic chemicals and aromatic, chlorocarbon, and ester solvents; hydrolytic stability is excellent (see POLYIMIDES).

Ultem PEI resins are used in numerous engineering applications where a combination of heat, chemical, and flame resistance is needed. Typical uses include rigid memory disks, fuse bodies, gears, fans for computer and business equipment, sterilizable surgical appliances, hot combs, brushes, iron skirts, microwave-oven components, aircraft interior components, and high temperature switches, connectors, and circuit boards for industrial and military devices.

Poly(phenylene sulfide) Resins. Poly(phenylene sulfide) resins are manufactured from *p*-dichlorobenzene and sodium sulfide in a dipolar aprotic solvent. The polymer is crystalline with a melting point of 287°C. It is flame retardant and has excellent hydrolytic and chemical resistance. As do most crystalline polymers, the neat material exhibits high mold shrinkage and a tendency for molded parts to warp, which can be corrected by fillers and reinforcement. Although molded parts maintain their mechanical properties during prolonged heating, under some conditions the resin continues to polymerize. This reduces the thermoplastic character of the resin. Recycling should be carefully considered.

Poly(phenylene sulfide) resins are chiefly used for injection molding (see POLY(ARYLENE SULFIDES)S).

Polyester Resins. The general formula for engineering thermoplastic polyester resins is

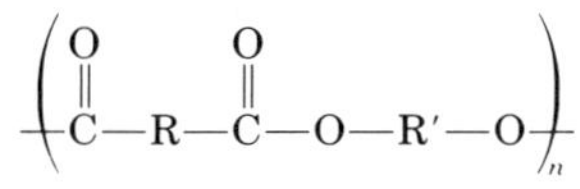

Most polyesters are based on phthalates. They are referred to as aromatic–aliphatic or aromatic according to the copolymerized diol.

Aromatic–Aliphatic Polyester Resins. Unlike most other classes of engineering plastics, which are made by only a few manufacturers, aromatic–aliphatic polyester resins are produced and compounded by several dozen firms.

Low molecular weight polyethylene terephthalate (PET) and poly(butylene terephthalate) (PBT) resins are made by melt processes. For higher molecular weight resins, melt processes followed by solid-state polymerizations are used. Although terephthalic acid can be directly esterified, the most common process involves transesterification of dimethyl terephthalate with ethylene glycol or 1,4-butanediol in the presence of trace amounts of metal ion catalysts.

The engineering applications of PET resins include injection-molded parts, blow-molded bottles, films, and extrusions. The PBT resins are mainly used for injection molding and related applications, although blow moldable grades are now appearing in the marketplace.

The properties given in Table 2 are for cream-color resins with good-to-excellent solvent resistance and good hydrolytic stability, but no clarity (see POLYESTERS, THERMOPLASTIC).

Aromatic Polyester Resins. Commercial aromatic polyester resins or polyarylates are a combination of bisphenol A with isophthalic acid or terephthalic acid. The resins are made commercially by solution polymerization

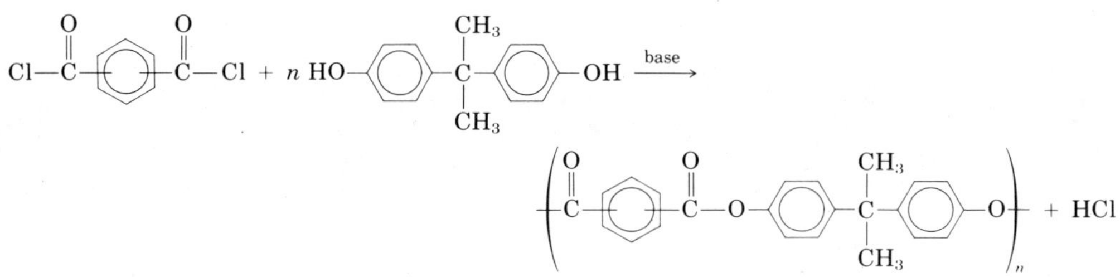

Table 2. Properties of Aromatic–Aliphatic Polyester Resins

Property	ASTM method	Valox 325[a]	Valox DR-51[b]	Rynite 530[c]
HDT at 1.82 N/mm^2 [d], °C	D 648	130	160	224
notched Izod at 20°C, 3.2-mm thickness, J/m[e]	D 256	50	55	90
flex modulus at 20°C, N/mm^2 [d]	D 790	2300	4530	6050
specific gravity	D 792	1.31	1.41	1.56
flammability rating, UL 94			V-O	HB

[a] Neat PBT resin.
[b] PBT resin, 15% glass-reinforced.
[c] PET resin, 30% glass-reinforced.
[d] To convert N/mm^2 to psi, multiply by 145.
[e] To convert J/m to ftlbf/in., divide by 53.38.

or melt transesterification such as

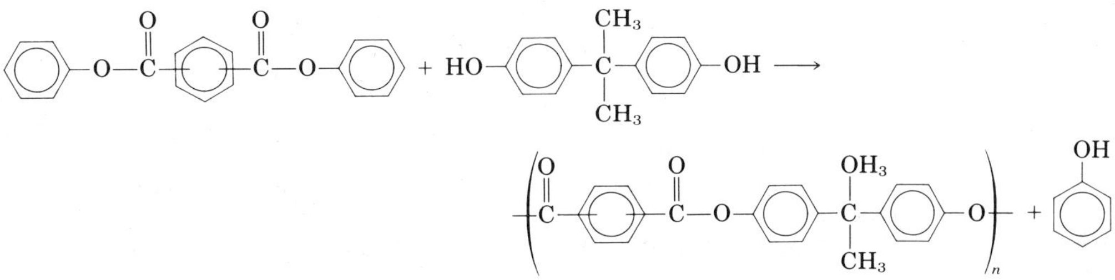

Blends of iso- and terephthalates give amorphous, clear resins, mostly yellow in color. Heat-deflection temperatures depend on the iso-to-terephthalate ratio. These resins are flame retardant; mechanical and electrical properties are similar to those of polycarbonate resins. The notched Izod impacts are lower at 250–300 J/m (4.7–5.6 ftlbf/in.), even in thick sections. Long-term uv radiation stabilities are excellent, but yellowness increases during initial exposure owing to photo-Fries rearrangements.

Polyarylates are sensitive to heat. Although their mechanical properties are not affected very much by heat, their colors darken. Hydrolytic stability and resistance to organic solvents are fair (see POLYESTERS, MAIN-CHAIN AROMATIC).

Polyarylates have been employed for electrical and electronic components, firemen's helmets, and applications requiring higher HDTs than polycarbonate resins. Polyarylates were first used for lighting, especially for small automotive lenses and sodium light outdoor lamps. However, the yellow color and heat-induced darkening have limited sales potential.

Bisphenol-A Polycarbonate Resins. These resins are typically manufactured by interfacial polymerization.

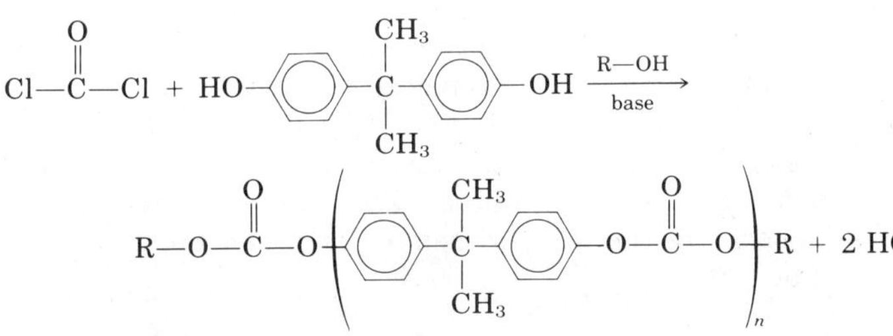

Polycarbonates (PC) (qv) find application in many areas. Sheet products are used for glazing and thermoformed products such as signs, cases, and furniture. PC resin film is suitable for silk-screen printing and is used in illuminated consoles and soft-touch controls. PC resins can be used for packaging as components of multilayer extruded sheet and film and in coextruded blow-molded containers. Injection molding consumes 80% of production. Applications include electrical and electronic parts such as housings, connectors, and fuses, appliance parts, power-tool housings, and parts for the transportation industries ranging from automobile dashboards to headlamps.

Polyester Carbonate Copolymers. Polyester carbonate resins are made by the interfacial process described for standard polycarbonate resins. Polyester carbonates are used primarily in applications requiring 5–26°C higher heat-deflection temperature and better hydrolytic performance than provided by standard polycarbonate resins. Properties are given in Table 3.

Acetal Resins. Acetal resins (qv) are poly(methylene oxide) (polyformaldehyde) homopolymers or formaldehyde copolymerized with aliphatic oxides such as ethylene oxide. The copolymers are produced by the cationic catalyzed reaction of 1,3,5-trioxane and an ethylene oxide.

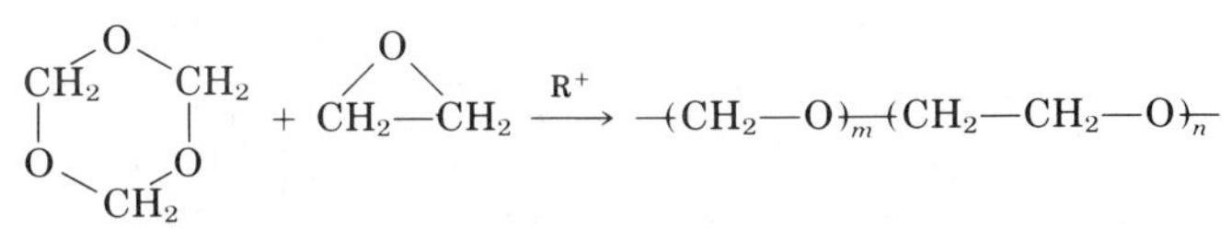

Polyacetals are translucent white and crystalline. They are solvent resistant and self-lubricating. Unlike nylons, they have low moisture absorption and maintain dimensional stability under varying humidity. Melt-flow characteristics are excellent, allowing for short mold cycles. Good fatigue resistance allows for such applications as ski-binding parts. On the other

Table 3. Properties of Polycarbonate (PC) Resins

Property	ASTM method	Lexan 101	Merlon 8320[a]	Lexan 4701[b]
HDT at 1.82 N/mm^2 [c], °C	D 648	132	142	163
notched Izod at 20°C, 3.2-mm thickness, J/m[d]	D 256	600	100	500
flex modulus at 20°C, N/mm^2 [c]	D 790	2300	3380	2300
specific gravity	D 792	1.21	1.27	1.2
flammability rating, UL 94		V-2	V-1	HB
light transmission T, %		85	opaque	85

[a] 20% glass-reinforced.
[b] Polyester carbonate.
[c] To convert N/mm^2 to psi, multiply by 145.
[d] To convert J/m to ftlbf/in., divide by 53.38.

hand, impact resistance is low, which affects toughness. Mold design must allow for shrinkage above 25% for unfilled resins. Specific gravities are high. Effective fire retardants have not been developed, resulting in limited applications.

Acetal resins are processed mainly by injection molding. Consumer articles include cassette cartridges, zippers, and container valves. Automotive parts include mechanical components such as cranks, gears, and bearings. In the industrial and construction areas, valves and connectors are made of acetal resins. Because they are easily chrome-plated, polyacetals can be used for plumbing fixtures.

Nylon Resins. Nylon engineering thermoplastic resins have the polyamide structures shown below.

$$\left(\text{R—NH—}\overset{\text{O}}{\overset{\|}{\text{C}}}\text{—R}'\text{—}\overset{\text{O}}{\overset{\|}{\text{C}}}\text{—NH}\right)_n \qquad \text{—R—}\overset{\text{O}}{\overset{\|}{\text{C}}}\text{—NH—}$$

diacid, diamine-based — lactam-based

Commercial engineering thermoplastic nylons are mainly crystalline resins. Nylon-6,6 is the largest volume resin, followed by nylon-6.

Nylon resins are made by numerous methods ranging from ester amidation to the Schotten-Baumann synthesis. The most commonly used method for making nylon-6,6 and related resins is the heat-induced condensation of monomeric salt complexes. In this process, stoichiometric amounts of diacid and diamine react in water to form salts.

$$n\ \text{H}_2\text{N}(\text{CH}_2)_6\text{NH}_2 + n\ \text{HO—}\overset{\text{O}}{\overset{\|}{\text{C}}}(\text{CH}_2)_4\overset{\text{O}}{\overset{\|}{\text{C}}}\text{—OH} \xrightarrow{\text{H}_2\text{O}}$$

$$n\ [\text{H}_3\overset{+}{\text{N}}(\text{CH}_2)_6\overset{+}{\text{N}}\text{H}_3][^-\text{O—}\overset{\text{O}}{\overset{\|}{\text{C}}}(\text{CH}_2)_4\overset{\text{O}}{\overset{\|}{\text{C}}}\text{—O}^-] \xrightarrow{\text{heat}}$$

$$\text{—}\!\!\left[(\text{CH}_2)_6\text{NH—}\overset{\text{O}}{\overset{\|}{\text{C}}}(\text{CH}_2)_4\overset{\text{O}}{\overset{\|}{\text{C}}}\text{—NH}\right]_n + \text{H}_2\text{O}$$

Physical properties of commercial nylons are shown in Table 4. Crystalline nylons are white and chemically and hydrolytically stable.

Crystalline nylons are processed by injection molding and extrusion. The extrusion products are mostly films. Coated or laminated with moisture-barrier resins, eg, poly(vinylidine chloride) or polyethylene polymers, they are used for meat wrappings and shaped food containers. Injection-molded parts are used for small mechanical, electrical, and construction articles (see POLYAMIDES).

Poly(phenylene ether)–Styrene Resin Alloys. Poly(phenylene ether) resins, composed of phenolic monomers, have very high T_g values. The commercial resins are based primarily on 2,6-dimethylphenol. The resins are produced by oxidative polymerization (qv) in toluene solution over an amine catalyst. Poly(phenylene ether) resins are usually blended with other resins to produce alloys. Styrene is the most common alloying resin. T_g values of the alloys vary inversely with styrenic content. More recently, nylon resin alloys have been marketed.

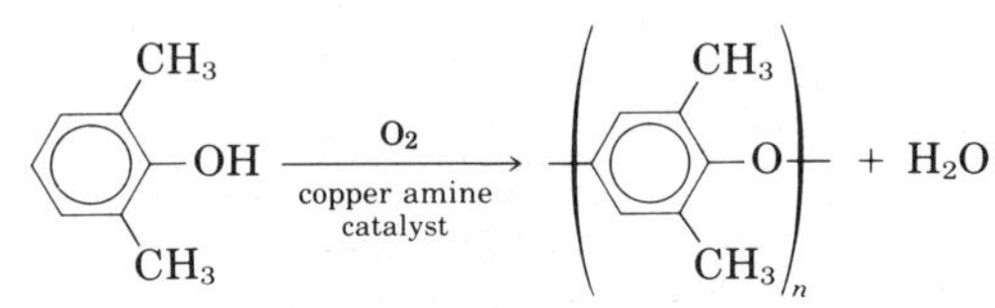

Poly(phenylene ether)–styrene resin alloys are cream colored and are characterized by excellent melt flow. They are easily formed either by an internal heat-activated blowing agent or by in-line injection of a volatile material. Flame retardance is conferred by halogen-free compounds. Electrical properties are outstanding and moisture absorbance is low. With excellent dimensional stabilities, the alloys can be made into very large parts. Although hydrolytic stability is very good, organic solvent resistance is poor, resulting in limited applications such as in large automotive parts. Final finishing must take chemical resistance into consideration. Another problem area is color stability to uv radiation.

Blends. Engineering resin blends, sometimes called engineering resin alloys, are mixtures of engineering resins or mixtures of engineering resins with commodity resins. Poly(phenylene ether)–styrene resin alloys can be considered here, but due to their high production volumes they are classified separately. In the list below, asterisks indicate sales above 900 t/yr.

Engineering Resin with Engineering Resin

Poly(butylene terephthalate)–poly(ethylene terephthalate)*
Polycarbonate–poly(butylene terephthalate)

Table 4. Properties of Nylon Resin

Property	ASTM method	Nylon-6,6 Zytel 101[a]	Nylon-6 Zytel 211[a]	Trogamid T[b]
HDT at 1.82 N/mm^2 [c], °C	D 648	194	129	124
melting point, °C	D 2117	270	228	
notched Izod at 20°C, 3.2-mm thickness, J/m[d]	D 256	53	80	65
flex modulus at 20°C, N/mm^2 [c]	D 790	540–2800	1000	2700
specific gravity	D 792	1.14	1.13	1.12
flammability rating, UL 94		V-2	HB	V-O

[a] As molded; crystalline.
[b] Amorphous yellow resin; light transmission 85–90%.
[c] To convert N/mm^2 to psi, multiply by 145.
[d] To convert J/m to ftlbf/in., divide by 53.38.

Table 5. Properties of Polycarbonate–Poly(butylene terephthalate) Blends

Property	ASTM method	PBT	50–50 blend	PC
HDT at 1.82 N/mm² [a], °C	D 648	130	99	132
notched Izod at 20°C, 3.2-mm thickness, J/m[b]	D 256	50	25	800
flex modulus at 20°C, N/mm² [a]	D 790	2300	2000	2300
specific gravity	D 792	1.31	1.25	1.21
organic solvent resistance		good	fair	poor

[a] To convert N/mm² to psi, multiply by 145.
[b] To convert J/m to ftlbf/in., divide by 53.38.

Polycarbonate–poly(ethylene terephthalate)*
Polycarbonate–polyester carbonate
Polysulfone–poly(ethylene terephthalate)
Polyarylate–nylon
Poly(phenylene oxide)–nylon

Engineering Resin with Other Resins
Polysulfone–ABS
Modified acetal
Modified nylon*
Modified poly(butylene terephthalate)
Polycarbonate–ABS*
Polycarbonate–styrene maleic anhydride
Poly(phenylene oxide)–polystyrene*

Companies produce polymer blends to extend their existing resin capacity and product lines without high capital investment and to achieve property profiles required for specific applications.

The properties of PBT and polycarbonate resins and of a blend of these two resins are given in Table 5.

Outlook

Applications of engineering resins in industry are likely to increase if the conservative estimates of 8–8.5% worldwide annual growth rates are realized. The consumption of 2×10^6 t/yr should double by 1990, mainly because of the promotion of resins tailored for specific uses. Current resins will most likely be modified, as opposed to the marketing of new resins, due to investment considerations.

BIBLIOGRAPHY

M. Grayson, ed., *Kirk-Othmer Encyclopedia of Chemical Technology,* 3rd ed., Wiley-Interscience, New York, 1980.

Handbook of Plastics and Elastomers, McGraw-Hill, Inc., New York, 1975.

Polymer Handbook, 3rd ed., John Wiley & Sons, Inc., New York, 1989.

Modern Plastics Encyclopedia, McGraw-Hill, Inc., New York, latest edition.

DONALD C. CLAGETT
General Electric Company

ENHANCED OIL RECOVERY.

See OIL-FIELD APPLICATIONS.

ENVIRONMENTAL RESISTANCE.

See CHEMICALLY RESISTANT POLYMERS; CRYOGENIC PROPERTIES; WEATHERING.

ENVIRONMENTAL STRESS CRACKING.

See CRAZING; FRACTURE.

ENZYMES

Enzymes are the catalysts of the living cell and because microorganisms are able to synthesize thousands of complex organic molecules, enzymes represent an enormous catalytic potential to the industrial chemist.

Molecular Viewpoint

Enzymes are proteins (qv), which are large organic molecules constituting approximately 50% of the dry weight of cells.

Enzymes, the largest class of proteins, are biocatalysts. Over 2000 different kinds are known, each catalyzing a different type of chemical reaction. Enzymes differ in two ways from other polymers. From a structural point of view, they have very complex and specific three-dimensional conformations. From a functional viewpoint, they are dynamic macromolecules with enormous catalytic power.

Structure. Enzymes, which are polymers of amino acids coupled by peptide linkages, vary in molecular weight from 5×10^3 to 7×10^6. The basic structures of the amino acids, except for proline, are identical and consist of a free carboxyl and a free amino group, both attached to the α carbon. The nature of an amino acid is determined by its side chain, the R group. The R groups can be nonpolar, polar, uncharged, charged, acidic, basic, aromatic, etc, and, as they are not involved in peptide linkages, they are available for other important interactions.

Synthesis in the Cell. Enzymes are synthesized intracellularly by complex organelles of the cell, the ribosomes.

Function. Enzymes are biological catalysts and are extremely well equipped for their task. Compared to catalysts employed by the organic chemist, they differ in their specificity of action: only certain well-defined substrates are acted on and only a single type of reaction takes place without side reactions or by-products.

Another remarkable aspect of enzymes is their enormous catalytic power; they can enhance reaction rates by a factor of from 10^8 to 10^{20}. Furthermore, enzymes can function in dilute aqueous solution under moderate conditions of temperature and pH.

The formation of a stereospecific complex between enzyme and substrate is responsible for the high catalytic activity of enzymes. Every enzyme has an active site composed of a binding site and a catalytic site.

Kinetics

The Michaelis-Menten Equation. The celebrated Michaelis-Menten equation for the kinetics of enzyme action is represented mathematically by

$$r_s = \frac{r_{s,\max} C_s}{K_M + C_s} \tag{1}$$

where r_s in mol/m^3·h is the rate of the reaction catalyzed by the enzyme under consideration; $r_{s,\max}$ in mol/m^3·h is the maximum rate of the forward reaction; C_s is the substrate concentration; and K_M is the Michaelis-Menten saturation constant, which is equal to the substrate concentration at which the rate of reaction reaches half the maximum rate $r_{s,\max}$.

Chemical Reaction Kinetics. The kinetics of enzyme reactions are intimately related to the kinetics of chemical reactions. In a reaction in which compound A is transformed to product B, the transformation does not continue until A is completely exhausted. After a certain time, A and B reach concentrations C_A^* and C_B^*, which become independent of time; ie, a state of thermodynamic equilibrium is reached. The equilibrium constant is

$$K = (C_B^*/C_A^*) \tag{2}$$

The value of K is fixed by the thermodynamic properties of the reactants and depends on the free energy change $\Delta\mu_R^0$ of the reaction:

$$\Delta\mu_R^0 = -RT \ln K \tag{3}$$

Reaction of an Enzyme and a Substrate Forming One Product. The action of an enzyme can be explained by the following kinetic scheme:

$$E + S \underset{k_{-1}}{\overset{k_1}{\rightleftharpoons}} ES \tag{4}$$

$$ES \underset{k_2}{\overset{k_{-2}}{\rightleftharpoons}} E + P \tag{5}$$

where S is the substrate; E is the free enzyme; ES is the complex of the enzyme and the substrate where the substrate is at the active site; P is the product; and k_1, k_{-1}, k_2, and k_{-2} are mass-action-law kinetic constants.

Reversible Inhibition. Reversible inhibition occurs where the activation or inhibition phenomenon is caused by a substance that binds to the enzyme and can dissociate, rendering the enzyme active again. These phenomena can be treated by a classical approach or one based on allosteric controls of enzyme activity.

Activity and Stability. The activity of an enzyme is determined by exogenous parameters that influence the conformation of the enzyme. Optimal conditions for maximal activity are usually not compatible with conditions for maximal stability, and in the case of industrial application the correct balance between activity and stability must be determined to reach satisfactory productivity.

Inactivation. The productivity of industrial bulk enzymes is primarily determined by temperature and to a lesser extent by pH and metal ions. Higher temperatures increase reaction rates and thereby reduce processing time. Most enzymes, however, are readily inactivated by high temperatures.

Effect of pH. An important feature of enzyme application is the dependence of activity and deactivation on pH. In fact, the pH range in which an enzyme can be economically used is one of the prime factors determining its commercial success. At extreme pH, the enzyme may be irreversibly denatured.

Molecular Treatment of Inactivation Phenomena. Enzymes can be inactivated by a covalent or noncovalent modification of the molecule. Noncovalent changes are caused by the incorrect refolding of the denatured form and can be of a polymolecular type (aggregate formation) or a monomolecular type (incorrect structure formation).

Chemical reactions, usually occurring at high temperatures and extreme pH, produce covalent changes such as the oxidation or reduction of S—S linkages; intra- and intermolecular formation of incorrect S—S linkages; hydrolysis of S—S linkages resulting in thiocysteine and dehydroalanine; isopeptide formation such as that between Lys and Asp; the reaction of Lys with reducing sugars; hydrolysis of peptide bonds and of Asn, Gln, and Arg; racemization and β-elimination of amino acids; and proteolytic cleavage, especially of unfolded enzymes.

Units of Enzyme Activity. The official definition of the unit of enzyme activity, as proposed by the Enzyme Commission, is the katal, the amount of enzyme required for the conversion of one mole of substrate to product per second. The specific activity is the number of katals per kilogram of protein.

Applications

Important industrial enzymes and their applications are given in Table 1.

Other Applications. New fields include the manufacture of complex organic compounds such as antibiotics and optically pure amino acids; medicinals; and analytical and research applications, where extremely high selectivity is required.

Economic Aspects

The following subdivision of markets based on the nature of the enzyme application can be distinguished: bulk industrial enzymes, enzymes for organic synthesis, and specialty enzymes.

The bulk industrial enzymes are produced on a large scale and are used in large-scale applications. The enzymes must be either inexpensive or very active. Worldwide sales for 1981 are given in Table 2.

Typically, enzyme costs are about 4% of the total added value, comparable with the catalyst costs in other large-scale operations.

Sources and Production

With some exceptions, eg, calf rennet and papain, industrial enzymes are obtained from microorganisms by highly cost-effective processes.

Production Technology. Production by microorganisms is generally referred to as fermentation. In the surface-culture method, the organism grows on the surface of a solid or semi-solid medium. In the submerged-culture method, the organism is grown in large tanks submersed in a liquid phase in which the culture medium is dissolved or dispersed. The submerged-

Table 1. Survey of Industrial Enzymes

Market and enzyme	Application
Starch	
α-amylase	liquefaction of starch through random hydrolysis of α-1,4-glycosidic linkages at a 90–105°C; dextrins are end product
β-amylase	production of maltose from starch by hydrolysis of α-1,4-glycosidic linkages at the nonreducing ends of starch or dextrin molecules
isoamylase	hydrolysis of α-1,6-glycosidic linkages; in combination with α-amylase for the liquefaction of starch
amyloglucosidase	saccharification of dextrins by hydrolysis of α-1,4-glycosidic linkages at the nonreducing end of the dextrin molecules; end product is glucose; also referred to as glucoamylase
transglucosidase	a transferase that synthesizes α-1,6-glycosidic linkages, produces mainly isomaltose, and is found in some amyloglucosidases as an undesired side activity
acid amylase	an α-amylase with a low pH optimum (pH 4) that in combination with amyloglucosidase improves saccharification yield
pullulanase	hydrolysis of α-1,6-glycosidic linkages in conjunction with amyloglucosidase to obtain higher glucose yield from dextrins
glucose isomerase	isomerization of glucose into fructose; immobilized form used for the production of HFCS
cyclodextrin glucosyltransferase	transfer of glucose from one dextrin molecule to another through hydrolysis and synthesis of α-1,4-glycosidic linkages; used for the production of circular glucose oligomers (α-, β-, or γ-cyclodextrins), which contain 6, 7, or 8 glucose units, respectively
Sucrose	
invertase	hydrolysis of sucrose into glucose and fructose; used for the production of confections, candies, syrups, sweets, and ice cream
fructosyltransferase	transfer of fructose from sucrose to fructose oligomers, after isomerization with glucose isomerase, giving a 1:3 glucose–fructose mixture; can be used to produce fructose-enriched syrups from sucrose
raffinase	decomposition of raffinose to improve yield of sucrose during crystallization process
Inulin	
inulinase	hydrolysis of inulin, a fructose polymer; can be used for the production of glucose–fructose syrups of high fructose content
Dairy	
calf rennet	coagulation of milk protein (casein) for the production of cheese
fungal proteases	replacement of calf rennet in cheese production by acid proteases from *Mucor* species
chymosin	the active component of calf rennet, responsible for the milk clotting; chymosin is produced by microorganisms that have received the genetic information of the original calf enzyme; intended for the replacement of other rennets; product is still under development
β-galactosidase	hydrolysis of lactose into glucose and galactose; improves digestibility and formulating properties of milk and whey products
lipases	hydrolysis of lipids (fat) to enhance cheese flavor; application in conjunction with certain proteases for the production of EMC; also used for the production of modified butter fat and flavor development of cheese
proteases	used in conjunction with lipase for the production of EMC (enzyme modified cheese); also applied in the processing of whey protein to improve whey-derived products
catalase	decomposition of H_2O_2 into H_2O and O_2 in milk, which has been cold-sterilized with hydrogen peroxide
lysozyme	suggested application in cheese production as antibacterial agent; hydrolyzes peptidoglycan, the main constituent of gram-positive bacterial cell walls; principal source, egg-white
sulfydryl oxidase	suggested application for the reduction of the cooked flavor of UHT milk; catalyzes the formation of disulfide bonds in protein
superoxide dismutase	suggested application in preserving the flavor of milk with long shelf life; catalyzes reduction of oxygen-free radicals into H_2O_2; coupled with catalase, prevents lipid oxidation in milk products
Beverages	
pectinase	mixture of pectinolytic enzymes applied in the fruit juice and wine industry to improve the yield, lower viscosity, improve filtration properties, and prevent haze formation; action based on the degradation of methylated poly(galacturonic acid)s
cellulases	hydrolysis of β-1,4-linked glucose polymers; accelerates extraction of juice from fruits; processing aid in the malting stage of beer production
hemicellulase	hydrolysis of mixed pentose polymers; processing aid in juice, wine, and beer production; increases extraction yields, prevents haze formation, and improves filterability
glucanases	degradation of glucans
proteases	plant proteases, especially papain, for haze removal in beer
α-amylase	hydrolysis of starch in fruit juice processing to prevent gel formation
acetolactate decarboxylase	degradation of acetolactate, a precursor of diacetyl produced during beer maturation, imparting an off-flavor; under development
naringinase	removing bitterness from grapefruit juice by hydrolysis of naringin, a glucoside
anthocyanase	suggested for color reduction in white and light-red wines and for the decolorization of fruit juices

Table 1. (*Continued*)

Market and enzyme	Application
glucose oxidase	catalytic for the conversion of glucose into gluconic acid and H_2O_2; application in soft drinks, beer, and wine for the removal of oxygen
hesperidinase	suggested for the clarification of juices by hydrolyses of hesperidine-glycosides
Baking	
α-amylase	limited hydrolysis of starch to produce nonreducing ends for the action of β-amylase
β-amylase	production of maltose used by yeast for CO_2 production; increases the rise of dough
amyloglucosidase	production of glucose to increase CO_2 production by the yeast; rise of dough
proteases	partial hydrolysis of gluten protein; dough conditioner
lipoxygenase	oxidation of polyunsaturated fatty acids leading to the formation of fatty acid hydroperoxides and malonaldehyde; main source, soy flour; bleaching agent
pentosanase	hydrolysis of pentosan (hemicellulose, xylan); antistalling agent
glucose oxidase	production of H_2O_2 through oxygen-dependent oxidation of glucose; bleaching agent
Meat and fish	
papain	tenderization of meat by limited hydrolysis of muscle protein with plant proteases, especially papain, bromelain, and ficin
proteases	recovery of protein from bones and trash fish
Protein extracts	
proteases	modification of protein-rich material, by controlled proteolysis; reverse action used to lengthen peptides (plastein-type reaction)
glucanases	removing carbohydrate material from soy protein
Flavors	
ribonuclease	degradation of ribonucleic acid (RNA) into 5′ nucleoside monophosphates; used for the production of 5′ guanosine monophosphate (GMP), a flavor enhancer
protease	production of yeast extracts and soy sauce
Other foods	
lipases	interesterification of lipids for the upgrading of fat; enzyme-catalyzed conversion of olive oil into cocoa butter
pectinase	acceleration of coffee bean processing
proteases	improving the drying properties of egg products
Detergents	
alkaline proteases	laundry detergents for the removal of protein-containing stains; removal of starchy stains; removal of fatty stains; still under development
α-amylase	
lipase	
Agriculture	
pectinase	suggested to improve the retting of textile crops such as flax and hemp
Alcohol	
α-amylase	starch-degrading enzymes, employed for the generation of yeast-fermentable sugars
cellulase	degradation of cellulosic material in wood and waste products into glucose, which can be fermented to ethanol; trials at pilot-plant level
Pulp and paper	
ligninase	suggested for the degradation of lignin, a main wood component and a waste product of the pulp and paper industry; enzymes from *Phanecoochaete* species are under development
Leather	
proteases	treatment of skins with proteases selected for a high activity on α-keratin
trypsin	use in leather processing; partially degrades protein fibers, softens skin; trypsin is isolated from animal pancreas

Table 2. Industrial Enzyme Products and Sales—1981

Enzyme	Application	Sales, 10^6 $
alkaline protease	detergents	90
α-amylase	starch liquefaction	23
glucoamylase	starch–glucose conversion	26
glucose isomerase	glucose–fructose interconversion	25
rennet		
animal	cheese manufacture	52
microbial	cheese manufacture	28
amylases		18
pectinases	juice clarification	18
trypsin	leather treatment	10
other		20
Total		*310*

Source: Industry sources, F. Eberstadt & Co., Inc. estimates.

culture method is generally preferred as it is more cost effective and allows tighter control of microbial activity.

Formulation. Liquid enzyme preparations, ie, concentrated solutions of purified enzyme, often require a stabilizer and component preventing microbial contamination. Solid preparations may contain various compounding agents. For selected enzymes, special formulation methods are employed, eg, prills or granulates for detergent enzymes.

Chemical Modification. Chemical modification is based on the reactivity of certain chemical agents toward specific amino acids. The effect of modification is unpredictable. Activity might be reduced or lost, and the surface charge might be changed. The characteristics of the modifying agent (toxicity, price, handling) must be taken into account.

Production Organisms. Usually a microbial source of the enzyme is isolated from a soil sample from a suitable environment. Growth in continuous culture is a powerful isolation tool.

The wild type obtained by screening or a directed selection program is often not economical, and a program of mutation and selection is devised to obtain overproducers.

Protein Engineering. Protein engineering is based on the principle that after the primary amino acid sequence of an enzyme is fixed, the secondary, tertiary, and quaternary structure of the enzyme follows. Furthermore, a direct relationship exists between the three-dimensional structure of the enzyme and its activity and stability at a given pH, temperature, and other environmental conditions. If the three-dimensional structure of an enzyme is determined, an alteration by means of recombinant DNA techniques can be suggested that would improve properties.

Health and Safety, and Legal Aspects

Enzymes occurring naturally in the daily diet are nontoxic and inherently safe in industrial applications. Industrial enzymes, however, are not pure and the contaminants may be toxic. Proteins different from body proteins may give rise to allergic reactions. In industrial production these considerations determine the necessary degree of purification and formulation. The manufacture of enzymes for food and feed processing is under governmental control.

The regulation of large-scale production processes employing microorganisms obtained by r-DNA methods is subject to continuing debate.

Outlook

The application of enzymes in industry has grown in the recent past, and no doubt this growth will continue in the near future. New technologies have given impetus to this growth.

BIBLIOGRAPHY

T. Godfrey and J. Reichelt, eds., *Industrial Enzymology,* Nature Press, New York, 1983.

J. A. Roels, *Energetics and Kinetics in Biotechnology,* Elsevier-North Holland, Inc., Amsterdam, The Netherlands, 1983.

L. Stryer, *Biochemistry,* W. H. Freeman & Co. Publishers, San Francisco, Calif., 1981, Chapts. 20, 31.

B. Poldermans
J. A. Roels
Gist-brocades, Inc.

ENZYMES, IMMOBILIZED

Enzymes are natural linear polypeptides that, in their folded states, catalyze biochemical reactions. Enzymes have molecular weights of 10^4–10^5.

The characteristics that make enzymes difficult to use *in vitro* are overcome by immobilizing the enzyme on a solid support. Immobilized enzymes are usually more stable to fluctuations in pH and temperature than their soluble counterparts. When suspended in water, most of these insoluble enzyme derivatives are stable for months, and can, therefore, be used many times as a catalyst. When the products and reactants are soluble, separation is easy since these compounds can simply be rinsed from the insoluble catalyst material. Recycling immobilized enzymes reduces costs, whereas the soluble counterparts have been prohibitively expensive.

Methods of Immobilization

Enzymes can be immobilized on nearly any material. The method of immobilization is dictated by the support and the particular enzyme required for a given application.

Enzyme Purity. The grade or purity of the enzyme material should be selected with careful consideration of the intended use of the immobilized catalyst. Methods of immobilization are outlined in Figure 1.

Polysaccharide Activation. Most enzymes have been immobilized by activation of polysaccharides, where cellulose or a beaded polysaccharide is allowed to react at an alkaline pH with cyanogen bromide. The intermediate reacts immediately with the soluble enzyme.

Cross-linking with Glutaraldehyde. Glutaraldehyde can be utilized to cross-link the enzyme structure intermolecularly.

Characteristics of Immobilized Enzymes

Guidelines for the characteristics of immobilized enzymes have been suggested by The Working Party on Immobilized Biocatalysts within The European Federation of Biotechnology. The recommendations are specific for particulate supports with preparative and industrial applications. Characteristics include support properties, catalyst properties, stability, conformational changes on immobilization, and specificity.

Biochemical Processing

The first step in the development of a commercial process is to select the type of reactor best suited to the process under consideration. Then the enzyme(s) required for the process can be immobilized and tested. The method used for immobilization depends on the enzymes required and the reactor design.

The reactors are either column reactors or tank reactors.

Applications

Immobilized enzymes are used in starch hydrolysis and in the production of high fructose corn syrup. Other examples of im-

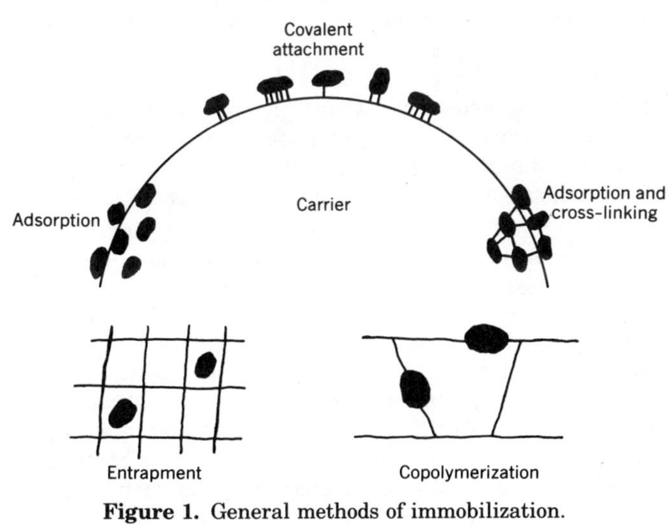

Figure 1. General methods of immobilization.

mobilized enzymes used in commercial applications are lactase, penicillin acylase, and aminoacylase.

Analytical Applications. Immobilized enzymes are also used in such analytical applications as enzyme electrodes and partitioned-enzyme sensors.

BIBLIOGRAPHY

L. Goldstein and G. Manecke in L. B. Wingard, Jr., E. Katchalski-Katzir, and L. Goldstein, eds., *Applied Biochemistry and Bioengineering,* Vol. 1, Academic Press, Inc., New York, 1976.

K. Mosbach, ed., *Immobilized Enzymes,* Methods in Enzymology, Vol. 44, Academic Press, Inc., New York, 1976.

M. H. Keyes in M. Grayson, ed., *Kirk-Othmer Encyclopedia of Chemical Technology,* 3rd ed., Vol. 9, Wiley-Interscience, New York, 1980.

I. Chibata, T. Tosa, and T. Sato in A. Mizrahi and A. L. Van Wezel, eds., *Advances in Biotechnological Processes,* Vol. 1, Alan R. Liss, Inc., New York, 1983.

MELVIN H. KEYES
DAVID ALBERT
Anatrace, Inc.

EPDM. See ETHYLENE–PROPYLENE ELASTOMERS.

EPICHLOROHYDRIN ELASTOMERS. See 1,2-EPOXIDE POLYMERS.

EPISULFIDES. See SULFUR-CONTAINING POLYMERS.

EPITAXIAL PROCESSES

Epitaxial crystallization can be defined, most generally, as the oriented overgrowth of a crystalline phase on the surface of a substrate that is also crystalline.

Epitaxial crystallization has been observed to occur from the vapor phase, melt, and solution. To date, examples of this phenomenon range from the deposition of single atoms through more complex organic molecules to high molecular weight stereoregular polymers.

In general, the action of the substrate is to enhance the nucleation of a second crystalline phase by lowering the free energy of activation by means of the force field near the surface. This force field is a function of the chemical constituents of the substrate and their geometrical arrangement. The substrate acts not only as a heterogeneous nucleation catalyst on which normal crystallization can occur, but at times can induce crystalline arrangements that are rarely observed under ordinary conditions.

In this article, a method is discussed that allows variation of monomer and polymer structure and morphology, applying a technique termed epitaxial polymerization. Monomers are crystallized on crystalline substrates. Lattice matching is a process that allows the variation of monomer structure and morphology, thus yielding different polymer structures and morphologies on polymerization (see also CRYSTALLINE POLYMERS; CRYSTALLIZATION KINETICS; MORPHOLOGY).

Crystallization on Inorganic Substrates

Alkali Halide Substrates. Sodium chloride single crystals were the first and are still the most common interactive templates used as heterogeneous nucleation catalysts for polymer crystallization. In addition to their solubility in water, these crystals represent a series whose faces have easily obtainable long-range two-dimensional periodicity and simple cubic symmetry, and offer the possibility of altering the chemical constitution and lattice parameters in a quantitative fashion.

Parameters. Rules of epitaxy can be defined in a broad sense as the manner in which the morphology and molecular structure of the deposit depend on parameters such as temperature, solution concentration, polymer molecular weight, and solvent and substrate.

Other Inorganic Substrates. Other inorganic substrates include graphite and quartz.

Crystallization on Organic Substrates

Polymer–organic solvent mixtures where both components have similar melting temperatures have been investigated. In addition to possible polymer–solvent eutectics, several examples of polymers and paraffins (dotriacontane and hextriacontane) were found that crystallized epitaxially onto freshly formed solvent crystals. Examples of these systems are mainly linear polyesters solvated by trioxane and aromatic hydrocarbons.

These results encouraged the investigation of the epitaxy of polymers onto organic substrates in a more systematic manner. Of particular interest was the crucial issue of lattice matching, on which epitaxy is thought to rest. Epitaxial crystallization of polyethylene and linear polyesters onto crystals of a homologous series of condensed aromatic hydrocarbons and linear polyphenyls has been studied. Polyethylene and the linear alkyl polyesters, which include a paraffinic segment in their repeat unit, provide a whole class of related yet different paraffinic polymers with varying unit-cell parameters and geometries. In addition, the condensed aromatic hydrocarbons, together with the linear polyphenyls, provide two parallel series of chemically and crystallographically related substrates.

Crystallization on Polymeric Substrates

Polymer–polymer epitaxy can be considered as a specific interfacial interaction between two distinct macromolecular species, as opposed to homoepitaxial behavior as is seen, eg, in hydrodynamically induced crystalline structures (shish kebabs). Because of the lack of macroscopic polymer single-crystal substrates, the interactive template consists of a polymer sample drawn uniaxially to produce a fiberlike structure that, at best, can be expected to possess an average degree of one-dimensional long-range order. Furthermore, as opposed to systems with high cohesive energy densities such as inorganics, ordered polymer substrates are sensitive to temperature.

Polymerization

Epitaxial polymerization is a new phenomenon in the field of macromolecular science. In this process epitaxial crystallization of a monomer on a crystalline substrate is followed by solid-state polymerization.

BIBLIOGRAPHY

J. W. Mathews, ed., *Epitaxial Growth,* Academic Press, Inc., New York, 1975.

S. E. Rickert, J. B. Lando, A. J. Hopfinger, and E. Baer, *Macromolecules* **12,** 1052 (1979).

S. E. Rickert, H. Ishida, J. B. Lando, J. Koenig, and E. Baer, *J. Appl. Phys.* **51,** 5194 (1980).

J. B. Lando, E. Baer, S. E. Rickert, H. Nae, and S. Ching, *ACS Symp. Ser.* **212,** 89 (1983).

G. S. Swei
J. B. Lando
S. E. Rickert
Case Western Reserve University
K. A. Mauritz
University of Southern Mississippi

EPM. See Ethylene–propylene elastomers.

EPOXIDATION. See 1,2-Epoxide polymers; Epoxy resins.

1,2-EPOXIDE POLYMERS

Ethylene Oxide Polymers and Copolymers

The polymeric derivatives of ethylene oxide are divided into two classes that are defined by molecular weight. This division is based on methods of production, commercial nomenclature, and properties and uses. Low molecular weight polymers with an average molecular weight of 200–20,000 are called poly(ethylene glycol)s (PEGs) in this article. They are used in ceramics, cosmetics, lubricants, metalworking, pharmaceuticals, urethanes, rubber chemicals, electronics, and other applications. The upper molecular weight limit is imposed by base-catalyzed side reactions occurring during polymerization.

Polymeric derivatives with molecular weights of 1×10^5–5×10^6 are called ethylene oxide polymers or poly(ethylene oxide) resins. They are used in flocculation, packaging films, acid cleaners, denture adhesives, lubricants, friction reduction, and other applications where water solubility and high viscosity are desired.

Copolymers of propylene oxide and ethylene oxide have achieved commercial importance in such diverse areas as fire-resistant hydraulic fluids, lubricants, quenchers, and polyether polyols for urethanes. These and other applications have grown to a point where the preparation of propylene oxide consumes a significant portion of the total propylene production. Most of propylene oxide production is used for polymer formation.

The properties and uses of the copolymers of ethylene oxide and propylene oxide are related to the properties and uses of their respective homopolymers.

Poly(ethylene glycol)s and methoxy poly(ethylene glycol)s are miscible with water and compatible with a wide variety of formulations. They exhibit low toxicity, good stability, good lubricity, and can be mixed with water or other solvents to give a wide range of viscosities. The higher molecular weight polymers are solids.

The preparation of higher molecular weight ethylene oxide polymers required the development of catalyst systems that function by a different mechanism than that used in the base-catalyzed preparation of poly(ethylene glycol)s. These polymers, identified in Chemical Abstracts as poly(ethylene oxide) resins, are sold under the trademark POLYOX; $—(CH_2—CH_2—O)_n—$ is the repeating unit. The commercial product is a free-flowing white powder in molecular weight grades ranging from about 100×10^3 to about 5×10^6. These water-soluble resins give high viscosity aqueous solutions at low concentrations.

Poly(propylene glycol)s are prepared by the base-catalyzed reaction of water or propylene glycol with propylene oxide. Because of side reactions occurring during the preparation, they are limited to molecular weights below 5000. In contrast to the poly(ethylene glycol)s, these polymers are generally viscous liquids that are insoluble in water.

The properties of copolymers of ethylene oxide and propylene oxide are intermediate between those of the poly(ethylene glycol)s and poly(propylene glycol)s. Water solubility depends on the proportion of ethylene oxide relative to propylene oxide used in manufacture. Higher proportions of ethylene oxide increase the resemblance to a poly(ethylene glycol). Mechanistic conclusions as to the reactions occurring in the preparation of these copolymers are generally drawn by analogy from studies of the relevant homopolymer. The homopolymers are usually the models for the properties of the copolymer. Commercially, poly(propylene glycol)s are often classed with the copolymers because of the similarities of their applications and properties. Inclusion of poly(propylene glycol)s widens the property range of a product line of alkylene oxide-derived polyethers.

The general structure of homopolymers and copolymers of ethylene oxide and propylene oxide is shown below:

$$—\!(CH_2—\underset{}{\overset{R_1}{\overset{|}{C}}}H—O)_x\!—\!(CH_2—\overset{R_2}{\overset{|}{C}}H—O)_y\!—$$

where $R_1 = R_2 = H$	ethylene oxide homopolymer
$R_1 = R_2 = CH_3$	propylene oxide homopolymer
$R_1 = H, R_2 = CH_3$	copolymer

The presence of the ether linkages is responsible for the unique properties of these polymers. The presence of the electron-rich oxygen atoms in the backbone structure of the polymers offers a site for coordination. The water solubility of these polymers is due to hydrogen bonding. This water solubility and the ability of other electron-poor groups to associate with these polymers are critical elements in many of the uses of these polymers.

Monomers

Ethylene oxide and propylene oxide are three-membered ring compounds containing one oxygen atom:

(1) (2)

These cyclic ethers are the first members of a homologous series of alkyl substituted derivatives of (**1**), called oxiranes or epoxides. The strain of the three-membered ring imparts unusual chemical properties and reactivity. Although they are cyclic ethers and have many of the properties of ethers, they are usually considered as a separate class of compounds.

Using the Hantzsch-Widman nomenclature, the three-membered ring containing the oxygen atom is called oxiran.

Ethylene oxide. ***Manufacture.*** Ethylene oxide is produced by silver-catalyzed oxidation of ethylene.

Physical Properties. Ethylene oxide is a colorless gas at room temperature. It condenses at 10.5°C to a colorless, mobile liquid. Physical properties are given in Table 1.

Shipment; Specifications. Ethylene oxide is available in bottles, cylinders, special drums, and insulated tank cars with capacities of 37.85 to 94.63 m^3. Appropriate DOT regulations should be consulted before shipping ethylene oxide.

Commercial ethylene oxide is of high purity, and assay specifications are usually not given.

Ethylene oxide is toxic both in the gaseous and liquid form. The current TWA is 1 ppm. Ethylene oxide should be considered a suspected cancer agent.

Propylene Oxide. Propylene oxide is manufactured by the chlorohydrin process and the oxirane process.

Physical Properties. Propylene oxide is a colorless, flammable liquid. It is miscible with a wide variety of organic solvents. Although soluble in water, it is not completely miscible and can form a two-phase system. The physical properties are given in Table 2.

The current TWA for propylene oxide is 20 ppm. Propylene oxide is a suspected animal carcinogen.

Polymers

Mechanism of Polymerization. The mechanism of this polymerization has been postulated to be a reaction of ethylene oxide with alkaline hydroxide to produce the anion of ethylene glycol. This anion reacts with ethylene oxide to produce the anion of diethylene glycol, which subsequently reacts further in a series of steps to give the following product:

$$KOH + \overset{O}{CH_2—CH_2} \longrightarrow HOCH_2CH_2O^-K^+ \quad (1)$$

$$HOCH_2CH_2O^-K^+ + \overset{O}{CH_2—CH_2} \longrightarrow HOCH_2CH_2OCH_2CH_2O^-K^+ \quad (2)$$

$$HOCH_2CH_2OCH_2CH_2O^-K^+ + n\,\overset{O}{CH_2—CH_2} \longrightarrow H(OCH_2CH_2)_{n+2}O^-K^+ \quad (3)$$

Table 1. Physical Constants of Ethylene Oxide

molecular weight	44.05
freezing point, °C	−112.5
boiling point, °C	10.4
critical pressure, MPa[a]	7.19
critical temperature, °C	195.8
explosive limits in air, vol %	
upper	100.0
lower	3.0
flash point, Tag open cup, °C	<−18.0

[a] To convert MPa to atm, multiply by 9.87.

Table 2. Physical Properties of Propylene Oxide

Property	Value
molecular weight	58.08
freezing point, °C	−112.0
boiling point, °C	34.2
critical pressure, MPa[a]	4.92
critical temperature, °C	209.15
explosive limits in air, vol %	
upper	37.0
lower	2.3
flash point, calculated, °C	<−20.0

[a] To convert MPa to atm, multiply by 9.87.

Preparation. Reactions of ethylene oxide and propylene oxide must be carried out in a safe manner to protect the operator from explosion, fire, or exposure hazards. The ring-opening reactions of these epoxides are exothermic by ca 84 kJ/mol (20 kcal/mol). Storage vessels for the monomers must be carefully protected from contact with catalytic material.

For the commercial-scale preparation of poly(ethylene glycol), poly(propylene glycol), or copolymers of propylene and ethylene oxides, a semibatch stainless-steel reactor equipped with circulating pumps and external heat exchangers is used; both 720 kPa (90 psig) and full vacuum at 160°C can be applied. Agitation is supplied by a turbine-type agitator and a recirculation system. Oxide is fed from measured storage tanks. Water, glycol, or another alcohol containing a catalytic amount of caustic is used to initiate polymer growth. The reaction mixture is heated to 100–200°C and fed the required oxide. The rate of oxide feed is limited by the heat-removal capability of the system or by the pressure limits of the equipment, typically on the order of 520 kPa (60 psig). Unreacted oxide produces a substantial vapor pressure at reaction temperature. Thus heat removal or pressure restrictions limit the feed rate. After the required amount of oxide has been added, the system is held at reaction temperature for a few hours to react with the oxide completely. The reaction is considered complete when, at constant temperature, the pressure remains constant. The product is neutralized.

Commercial Poly(ethylene glycol)s. Poly(ethylene glycol)s (PEGs) are produced by Union Carbide Corporation (Carbowax), Dow (Polyglycol E), Texaco Chemical Company (Jeffox), Olin (PolyG), BASF Wyandotte (Pluracol E), Hodag [poly(ethylene glycol)s], and ICI Americas (ATEG). These compounds are identified as to class and average molecular weight;

eg, Carbowax poly(ethylene glycol) 8000 is a diol product with an average molecular weight of 8000. Methanol as the starter in place of water or ethylene glycol gives methoxy poly(ethylene glycol)s.

Both poly(ethylene glycol)s and methoxy poly(ethylene glycol)s are miscible with water, low in toxicity, and compatible with a wide variety of formulations. They have good stability and lubricity, and can be mixed with water (or other solvents) to a wide range of viscosities.

Specifications of typical poly(ethylene glycol)s are given in Table 3.

Shipping, Storage, and Handling. Low molecular weight liquid poly(ethylene glycol)s should be stored below 50°C, preferably under nitrogen.

The higher molecular weight grades are sold as solids in bags or fiber drums. The material should be stored below its melting or softening temperature in sealed containers.

Applications. Poly(ethylene glycol)s are used in a wide range of applications where water solubility, thermal stability, hydrophilicity, and clean burn-off are important. Commercial poly(ethylene oxide) resins are high molecular weight homopolymers of ethylene oxide sold as free-flowing white powders. They are commercially available in molecular weight grades ranging from 100×10^3 to 5×10^6 produced by Union Carbide Corporation (Polyox), Meisli Chemical Work, Ltd. (AlKox), and Seitetsu Kagasku Co., Ltd. (Peo).

Physical Properties of Solutions. Poly(ethylene oxide)s are miscible in water at room temperature in all proportions. Solutions of the higher molecular weight grades are stringy at concentrations below 1%. Solutions of higher concentrations are elastic gels. At about 20% and higher, the solutions are hard and tough water-plasticized polymers. However, close to the boiling point of water, the water solubility decreases and a cloudy liquid phase forms. This temperature is called the precipitation temperature or cloud point and is dependent on molecular weight and concentration; it is pronounced in ethylene oxide–propylene oxide copolymers. Poly(ethylene oxide) readily dissolves in common organic solvents, increasingly with rising temperature.

Solid State. The melting point increases with molecular weight until about 15,000, where it stabilizes.

The density calculated from structural x-ray data is 1.33 g/cm^3 at 20°C; the experimentally determined value is 1.15–1.26 g/cm^3. The extrapolated melt density data give a value of 1.13 g/cm^3 at 20°C for the amorphous phase.

Association Reaction. The presence of an oxygen atom in every third position of the backbone of ethylene oxide polymers has a significant impact on physical and chemical properties. Water solubility is controlled by hydrogen bonding of the oxygen atoms to water. The electron-rich oxygen also forms association complexes with other compounds. Complexes are formed with phenols, phenolic resins, bromine, iodine, gelatin, liquid sulfonic acids, mercuric salts, tannic acid, poly(acrylic acid), and urea. Because of their tendency to associate, these polymers are used as flocculation agents for colloidal silica, clays, coal fines, and lignins.

Degradation can be prevented by addition of phenothiazine, butyrated hydroxytoluene (BHT), or butyrated hydroxyanisole (BHA) in concentrations of 0.01 to 0.5 wt %.

Shipping, Storage, and Handling. Ethylene oxide polymers are available in 45-kg polyethylene-coated fiber drums. The drums should be stored in a covered area and the containers sealed to exclude moisture. Finely divided poly(ethylene oxide) suspended in air can lead to a dust explosion.

Toxicology. Ethylene oxide polymers exhibit very low oral toxicity.

Applications. Most applications of ethylene oxide polymers are based on water solubility, solution rheology, low toxicity, thermoplasticity, and flocculation ability. They are used in detergents, cosmetic formulations, contact lens solutions, and dental adhesives.

Poly(ethylene oxide) films are used for water-soluble packaging of toxic and difficult-to-handle products.

Commercial Ethylene Oxide–Propylene Oxide Copolymers

Ethylene oxide–propylene oxide copolymers are divided into three broad classes depending on structure and use: polyols, block copolymers, and the so-called "functional fluids."

Polyols for Polyurethane. The reactions of the polyol depend on the hydroxyl group, and the performance of these polyols depends on the functionality and reactivity of the hydroxyl group.

Formation and Reactions. The hydroxyl end group of the polyol reacts with the isocyanate to form the urethane linkage.

Table 3. Specifications of Typical Poly(ethylene glycol)s

Property	Carbowax PEG 400	Carbowax PEG 3350
average molecular weight	390–420	
pH (25°C, 5% aqueous solution)	4.5–7.5	4.5–7.5
clarity		
25% aqueous solution	free from haze or turbidity	[a]
100% material	free from haze or turbidity after standing 1 h at 20°C	
color[b], % max	18	25
sulfated ash, % by weight, max	0.05	0.05
suspended matter	substantially free	melt substantially free from dirt and lint
freezing point, °C		54–57
viscosity at 99°C, mm^2/s (= cSt)		76–110

[a] A 25% solution transmits at least 95% of the light transmitted by the same solution after filtering.
[b] 25% aqueous solution.

The principal influence on reactivity is the steric environment at the end of the molecule. A hydroxyl group derived from ethylene oxide is a primary hydroxyl group; if it is from a propylene oxide, it is mainly a secondary hydroxyl group. Primary alcohols are more reactive in urethane formation than secondary alcohols. The kinetics of the reaction of alcohols with isocyanates is complex, involving a preliminary complexation of the alcohol with the isocyanates, followed by reaction to give the urethane.

Polyethers produced for urethane applications must be treated with special care. Urethane formulations are sensitive to traces of acid or base and must be free of the basic catalyst used in the polyol preparation.

Antioxidants (qv) may be added to suppress air oxidation.

Physical Properties. The physical properties of polyols are dependent on the functionality of the starter, the molecular weight, and the proportions of ethyloxy and propyloxy groups in the polymer.

The commercial polyols are defined by hydroxyl number, viscosity, functionality, and performance, but not often by structure.

Block Copolymers. Block copolymers (qv) are produced by BASF Wyandotte under the trade name Pluronic with molecular weights of 1,100 to over 14,000. A poly(propylene glycol) is prepared first from propylene oxide and a propylene glycol. Ethylene oxide reacts at each end of the molecule to give a block structure that contains 10–80% ethyloxy groups. The poly(propylene glycol) block acts as a hydrophobe, whereas the ethylene oxide ends are hydrophilic. When the length and proportions of these constituents are varied, polymers are produced that range from mobile liquids to solids hard enough to be flaked. Surface-active properties are controlled by the sizes of the hydrophobic and hydrophilic blocks.

Pluronics are low to moderately foaming surfactants.

Functional Fluids

Monohydroxy Polymers of Propylene Oxide. The commercial series of monohydroxy polymers of propylene oxide are linear and based on an alcohol. They are characterized by an alkoxy group at one end and a hydroxy group on the other. Polymers of this type are available in a wide range of molecular weights. The principal manufacturers are Union Carbide Corporation (Ucon LB series) and Olin Corporation (Poly G WI series).

Random Copolymers of Propylene Oxide and Ethylene Oxide. The two commercial classes of random copolymers of ethylene oxide and propylene oxide are linear monols and diols. The monohydroxy polymers have an alkoxy group on one end and a hydroxyl on the other, whereas dihydroxy polymers have a hydroxy group on each end.

Conformation. The Raman spectra show that the structures around O—CH_2, CH_2—CH_2, and CH_2—O bonds are trans, gauche, and trans, respectively.

Solution Properties. The lower molecular weight poly(propylene glycol)s are water soluble but at about 800–900 mol wt they are essentially insoluble in water. The solubility in water is due to the polar hydroxyl end groups of the polymer.

Handling and Storage, Specifications. Poly(propylene glycol)s, random copolymers of propylene oxide and ethylene oxide, and urethane polyols are all liquids and can be stored in carbon-steel tanks. High viscosity fluids are stored in heated tanks. Urethane polymers must be protected from moisture.

Health and Safety. The alcohol- or diol-based polymers of propylene oxide and copolymers of propylene oxide and ethylene oxide are relatively nontoxic. The lower molecular weight species may be slightly toxic when taken orally.

As expected, the block copolymers are also very low in toxicity.

Economic Aspects. Urethane polyol synthesis consumes about 60% of the propylene oxide production. Polyols have had this share of the propylene oxide market for the last decade and are expected to keep it through the 1980s. Polyols, therefore, represent more than 450×10^3 t propylene oxide consumption and more than 45×10^3 t ethylene oxide. Economic growth of polyols through the 1980s is expected to be 6–7%/yr, or even higher. Estimates of the remainder of the poly(alkylene oxide) production are difficult because of inconsistent and inaccurate reporting to the U.S. Tariff Commission and the U.S. International Trade Commission.

Applications. Polyethers are used as quenchers, in metalworking, hydraulic fluids, heat-transfer fluids, lubricants, textile applications, cosmetics, and antifoaming agents.

BIBLIOGRAPHY

F. E. Bailey, Jr. and J. V. Keleske, *Poly(ethylene oxide)*, Academic Press, New York, 1976.

J. Furukawa and T. Saegusa, *Polymerization of Aldehydes and Oxides*, Interscience Publishers, New York, 1963.

UCON Fluids and Lubricants, Booklet SC-818, Union Carbide Corporation, Danbury, Conn., 1987.

A. S. Kastens in N. G. Gaylord, ed., *Polyethers*, Wiley-Interscience, New York, 1963, Pt. 1.

NYE CLINTON
PAUL MATLOCK
Union Carbide Corporation

Propylene Oxide and Higher 1,2-Epoxide Polymers

Polymeric derivatives of propylene oxide (PO) and higher epoxides are very important in polymer chemistry. Low molecular weight homopolymers of PO and copolymers of PO and ethylene oxide (EO) are used in polyurethanes, hydraulic fluids, lubricants, surface active agents, and chemical intermediates. They are known as polyether polyols, polyethers, or polyols. The urethane industry produced 750×10^3 t of polyurethane foam in 1983; its principal components are polyether polyols.

Low molecular weight polyethers are prepared via acid or base catalysis.

High molecular weight polymers are obtained in many cases. The scientific literature contains hundreds of reports of catalysts that polymerize epoxides, PO in particular, to molecular weights of ca 10^6.

Chemistry. The ring-opening polymerization reaction of epoxides can be represented by the following stoichiometric equation:

$$XH_m + n\,\overset{O}{\overbrace{RCH-CH_2}} \longrightarrow X\!\left(CH_2-\underset{R}{\underset{|}{CH}}O\right)_{\!n}\!H$$

where XH_m is an initiator with m reactive hydrogens, such as water, alcohol, amine, phenol, amide, or acid.

Propylene Oxide

Manufacture. Propylene oxide became a leading industrial chemical after World War II when its importance in polyurethanes was recognized. Production, capacity, and price have increased continuously since then except for recession years. Production decreases observed in 1975 and 1980 were a result of the oil embargo of 1973 and 1979, which caused a shortage of raw material and the chlorohydrin process phase-out. At present, only two principal companies supply bulk PO: The Dow Chemical Co. (chlorohydrin process) and ARCO Chemical (peroxidation process).

Optically Active Propylene Oxide. (+)Propylene oxide is dextrorotary and has an *R* configuration. (*R*)-(+)-PO was prepared after the following method:

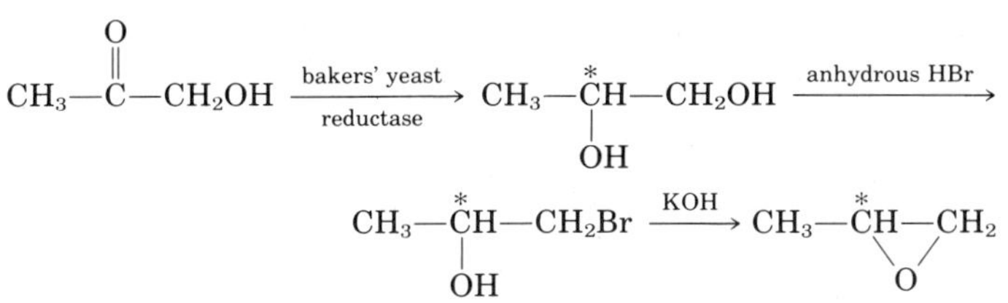

(+)propylene oxide

This method is used to prepare optically pure isopropylethylene oxide and *tert*-butylethylene oxide from the corresponding acetols:

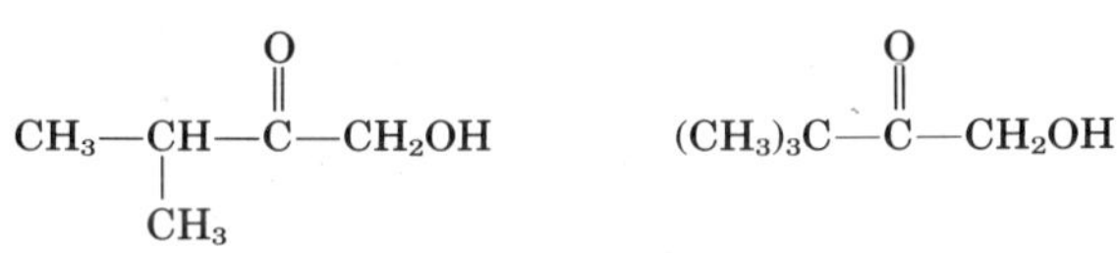

Purification. Commercial PO is refluxed for 2 h over potassium hydroxide and fractionally distilled; the first and last 15% are discarded. Epoxide can also be dried by refluxing over calcium hydride followed by fractional distillation.

The physical properties of propylene oxide, 1-butene oxide (BO), and epichlorohydrin (ECH) are shown in Table 1. Although PO is by far the largest selling epoxide, BO is used in surfactants and ECH in epoxy resins (qv).

Polymerization Mechanisms. Epoxide polymerization can be described in terms of three different mechanisms: anionic (base catalysis), cationic (acid catalysis), and coordinate. The coordinate mechanism combines the features of the first two, since it involves coordination of the monomer oxygen at a Lewis acid catalyst site followed by attack by an alkoxide bound to the site.

Polyols

Manufacture of Polyols. For the manufacture of polyols, continuous tubular or column reactors or semibatch autoclaves are used. In a semibatch operation, initiator and catalyst are charged into the reactor and EO or PO is added continuously until the desired molecular weight is obtained; the reaction is stopped and the product discharged.

Stoichiometry. In the KOH-catalyzed bulk polymerization of PO, conversion is quantitative, permitting calculation of the number-average molecular weight ($\bar{M}_n$). The molecular weight of polyether polyols is usually determined by end-group titration. The result is expressed as mg KOH/g of polyol, known as the hydroxyl number.

$$\text{hydroxyl number} = \frac{\text{functionality} \times 56{,}100}{\text{molecular weight}} = \frac{\text{hydroxyl equivalents} \times 56{,}100}{\text{total charge}}$$

The hydroxyl number is inversely proportional to the molecular weight; 56,100 = mg per equivalent of KOH.

Stabilization. In an oxygen atmosphere, polyols are subject to autoxidative degradation. The products are hydroperoxides,

Table 1. Physical Properties of Propylene Oxide, Butylene Oxide, and Epichlorohydrin

Property	Propylene oxide	Butylene oxide	Epichlorohydrin
molecular weight	58.081	72.108	92.526
boiling point, °C	33.9	63.2	116.11
vapor pressure constants[a]			
A	6.095	6.971	6.596
B	1065.27	1670.3	1587.9
C	226.283		230
freezing point, °C	−104.4	−150	−57.2
coefficient of thermal expansion α[b]	0.00151^{20}	0.00139^{55}	0.00104^{55}
refractive index n_D, at 25°C	1.36322	1.381	1.4358
viscosity η, mPa·s (= cP), at 25°C	0.28	0.40	1.03
heat of vaporization ΔH_v, kJ/mol[c]	28.75^{34}	31.97^{20-63}	37.91^{116}
solubility in water at 20°C, wt %	40.5	5.91	6.58
solubility of water in, at 20°C, wt %	12.8	2.65	1.47
specific gravity at 20°C	0.8304	0.8297	1.1807

[a] The vapor pressure can be calculated from $\log p = A - (B/T)$ or the Antoine equation $\log p = A - [B/(T + C)]$. These equations fail near the critical point; $\log p(\text{kPa}) = \log p(\text{mm Hg}) - 0.875$.

[b] $\alpha = \frac{1}{V}\left(\frac{\partial V}{\partial T}\right)_p = \frac{(d_1/d_2) - 1}{T_2 - T_1}$ The second form of the equation can be used to calculate the density if the temperature difference is small. Superscripts are temperatures, °C.

Table 2. Properties of Isotactic PPO

Property	Value	Remarks
melting point T_m, °C	78	$M_n = 5 \times 10^5$ (isotactic sequence = 55)
	82	multiple melting transitions
	70–75	various
glass transition T_g, °C	-75 ± 2	independent of stereoregularity
x-ray crystal density, g/cm³		
crystalline	1.157	
amorphous	1.002	
heat of fusion $\Delta H_f(T_m)$, kJ/mol[a]	8.4	
entropy $\Delta S_f(T_m)$, J/(mol·K)[a]	23.8	
ord spectra	+	in $CHCl_3$, ether
	−	trifluoroethanol, benzene
specific volume		temperature varied
activity coefficient		in benzene
crystallization kinetics		
viscosity		Stuart–Briegleb model
unperturbed dimensions, nm	8.0	$(\bar{r}^2/M)^{1/2}$

[a] To convert J to cal, divide by 4.184.

which are unstable and break down to aldehydes and carboxylic acids. Stabilizers are usually based on hindered phenols or aromatic amines.

Chemical Properties of Atactic Poly(propylene oxide). The most important chemical property of PPO in polyurethane applications is the primary or secondary hydroxyl end group. Its concentration varies inversely with the molecular weight and is usually given as the hydroxyl number. The secondary hydroxyl group is less reactive than the primary in the reaction with isocyanates. It is the predominant structure in products where the last step is base-catalyzed polymerization of PO.

Physical Properties of Atactic Polyoxypropylene. ***Solubility.*** Polyoxypropylene polyols with a molecular weight below 700 are water soluble.

Density. At low equivalent weight, the specific gravity (density) of polyethers depends on the initiator, at high equivalent weight, on the alkylene oxide. Most commercial products have a specific gravity of 1.005–1.020 at 25°C unless they are of low hydroxyl equivalent weight.

Viscosity. In the molecular weight range of 200–6000, poly(propylene oxide) polyethers are liquids. At low molecular weight, the viscosity is high and is determined by the initiator.

Nuclear Magnetic Resonance Spectra of PPO. The ^1H nmr

Table 3. Properties of Epoxides and Their Homopolymers

Monomer	Boiling point, $°C_{kPa}$[a]	Refractive index, $n_D^{°C}$	Density, g/cm³	Homopolymers T_g, °C	Homopolymers T_m, °C
1-butene oxide	$61_{98.5}$	1.3797^{30}	0.970	−83	76
	63	1.3855^{17}			
trans-2-butene oxide					90,100
cis-2-butene oxide					130,162
butadiene monoxide	66.5			−75	74,79
tert-butylethylene oxide[b]	89–90.5	1.3998^{20}		35	149–150
isobutylene oxide	50.5	1.3667^{30}		−10	173
styrene oxide[c]	$54_{0.5}$	1.5310^{30}	1.118	40	162,176
	$65_{1.1}$	1.5331^{25}			
	$77_{1.6}$				
	$87_{3.1}$				
cyclohexene oxide	129–130	1.4520^{20}			108–126
epichlorohydrin	116	1.4375^{20}	1.28	−25	119–120
1-hexene oxide	$117_{98.6}$	1.4024^{30}	0.925	−68	71
		1.4060^{20}			
1-octene oxide	$61_{2.0}$	1.4163^{30}			
		1.4193^{20}			
1-dodecene oxide	$57_{0.3}$				
allyl glycidyl ether	$49_{1.0}$	1.4310^{30}			
	154	1.4324^{20}			
phenyl glycidyl ether	$91_{0.4}$				203
	245				

[a] At 101.3 kPa unless otherwise noted. To convert kPa to mm Hg, multiply by 7.5.
[b] (*R*)-enantiomer $[\alpha]_D = -18.1°$.
[c] (*R*)-enantiomer $[\alpha]_D = +33.2°$.

spectrum of PPO consists of a broad doublet centered at 1.1 ppm (methyl) and a complex multiplet centered at 3.5 ppm.

Physical Properties of Isotactic PPO. ***Fractionation.*** Polymerization of PPO with alkylaluminum or alkylzinc catalysts gives a mixture of amorphous and crystalline polymers. The isotactic fraction can be separated by taking advantage of the differential solubility in acetone, benzene, methanol, and isooctane.

Properties of isotactic PPO are shown in Table 2.

Viscosity–Molecular Weight Relationship. The solution viscosity is the most often quoted property of a high molecular weight polyether. The limiting viscosity $[\eta]$, also called the intrinsic viscosity, is defined as:

$$[\eta] = \lim_{c \to 0} \frac{\eta - \eta_0}{\eta_0 c}$$

where c = concentration.

Properties of Other Epoxide Polymers. Selected properties of some other epoxides and their homopolymers are shown in Table 3.

Copolymers of Epoxides

A number of block and random epoxide copolymers is known. Reactivity ratios relate the amount of each monomer that is incorporated into the copolymer compared to the amount of each monomer in the feed. The reactivity is defined as

$$\frac{dm_1}{dm_2} = \frac{M_1\,(r_1M_1 + M_2)}{M_2\,(r_2M_2 + M_1)}$$

where m is the observed moles of monomer in the copolymer, M is the number of moles in the feed, and r is the reactivity to be determined, typically at low conversion (see also COPOLYMERIZATION).

Random copolymers of PO and EO, prepared by the simultaneous addition of the epoxides to an alcohol, are offered commercially. The copolymers of PO with monomers containing pendent double bonds, such as allyl glycidyl ether, have been cross-linked to give rubbers with high tensile and tear values, oil resistance, and low temperature flexibility.

Health and Safety

Propylene oxide is highly reactive. It is a primary irritant, a mild protoplasmic poison, and a mild depressant of the central nervous system.

Epichlorohydrin has a high oral toxicity and moderate dermal toxicity. It is also a suspected carcinogen.

The toxicological properties of other simple epoxides are intermediate between PO and epichlorohydrin.

BIBLIOGRAPHY

T. Tsuruta, *J. Polym. Sci. Part D* **10,** 179 (1972).

E. J. Vandenberg, *J. Polym. Sci. Part A-1* **7,** 525 (1969).

Y. Ishii and S. Sakai in K. C. Frisch and S. L. Reegen, eds., *Ring Opening Polymerization,* Marcel Dekker, Inc., New York, 1969, Chapt. 1.

Y.-C. Yen and T.-S. Tsao, *Polyols for Polyurethanes,* Process Economics Report 45A, Stanford Research Institute, Menlo Park, Calif., May 1982.

STEVEN D. GAGNON
BASF Corporation

Polyether Elastomers

Epichlorohydrin Elastomers

Elastomers based on epichlorohydrin (EPI) are important commercial products because their vulcanizates offer an attractive combination of heat resistance, fuel and oil resistance, ozone resistance, and low temperature flexibility. In addition, they provide excellent resistance to vapor permeation by hydrocarbons, fluorocarbons, and air.

Properties of the uncompounded (raw) rubber of these elastomers are listed in Table 1. All have a high specific gravity and exhibit varying degrees of flame resistance, depending on their chlorine content. The EPI homopolymer and EPI–EO copolymer are the most significant commercially and have been designated CO and ECO, respectively, by ASTM.

The solubility parameter for the homopolymer is 9.6–9.8 and for the polymers as a group 11.0 ± 2.0.

Structure. The epichlorohydrin elastomers are linear, amorphous polymers. As an unsymmetrical molecule, EPI can enter the growing chain to produce a head-to-tail, symmetrical monomer sequence or a random head-to-tail, head-to-head product; the commercial homopolymer is 97–99% head-to-tail.

Manufacture. Polymerization is generally carried out in a solution process at 40–130°C utilizing an aromatic solvent such as benzene or toluene. The catalyst systems employed are alkylaluminum–water and alkylaluminum–water–acetylacetone. Chain propagation is believed to follow a cationic, coordination mechanism. The product may be isolated by aqueous coagulation. Alternative polymerization systems in bulk and by means of hydrocarbon slurries have been reported. Commercial manufacture is carried out in a continuous process whereby solvent, catalyst, and monomer are fed to a backmixed reactor maintained at the desired temperature.

Chemical Modification. Epichlorohydrin elastomers readily undergo nucleophilic-substitution reactions at the pendent chloromethyl group, which is the site for many vulcanization systems.

Table 1. Commercial Epichlorohydrin Elastomers[a]

Property	EPI	EPI–EO	EPI–AGE	EPI–EO–AGE
epichlorohydrin, %	100	68	95	63–70
chlorine, %	38.4	26	36.5	24–27
Mooney viscosity[b] at 100°C	45–75	50–110	50–70	45–90

[a] Key: EPI = epichlorohydrin; EO = ethylene oxide; AGE = allyl glydicyl ether.
[b] ASTM D 1646, ML (1 + 4).

Applications for the modified polymers include breaking emulsions, drainage aids in paper manufacture, flame retardants, flocculating agents, selectively permeable membranes, shrink-proofing wool, photosensitive material, and water thickeners.

Epichlorohydrin elastomers have been modified by radical-induced graft polymerization.

Processing and Compounding. Like other vulcanizable elastomers, epichlorohydrin elastomers are compounded with fillers, plasticizers, processing aids, stabilizers, and vulcanizing agents (Table 2). These ingredients may be added on a two-roll mill or in an internal mixer, such as a Banbury mixer.

Curing Systems. The most widely used cross-linking agent for epichlorohydrin elastomers is 2-mercaptoimidazoline, also called ethylene thiourea.

Fillers. Vulcanizates are reinforced by fillers such as carbon black. Various types of alumina and silica are also effective reinforcing fillers. Some of the less expensive, nonreinforcing fillers include calcium carbonate, talc, and some clays (see also FILLERS).

Plasticizers. The most effective plasticizers for epichlorohydrin elastomers are ethers such as TP-90B, di(butoxyethoxyethyl) formal (Thiokol). Incorporation of plasticizers affects tensile strength and hardness and improves processing. Low temperature performance is improved by low molecular weight plasticizers such as dioctyl phthalate (DOP) and TP-90B.

Stabilizers and Antioxidants. During manufacture, antioxidants (qv) of the hindered-phenol type are added for protection during storage. Common heat-stabilizing acid acceptors include red lead oxide, magnesium oxide, and barium carbonate.

Blends. Blends of epichlorohydrin elastomers with other elastomers do not perform well and have not met with commercial success.

Uses. Epichlorohydrin elastomer vulcanizates have a broad operating temperature range and excellent chemical resistance, which makes them ideal for automotive applications.

Propylene Oxide Elastomers

The PO–AGE copolymer has been commercialized and given the ASTM designation GPO and is sold under the trademark PAREL 58 (Hercules, Inc.) with a specific gravity of 1.01 and a brittleness temperature of −90°C. Its swelling characteristics resemble those of polychloroprene (CR), its dynamic properties those of natural rubber (NR). Heat resistance and ozone resistance are superior to those of CR and NR. The uncured rubber is soluble in most aliphatic and aromatic solvents. The commercial product exhibits no crystallinity by dsc or x-ray analysis.

Manufacture. The PO–AGE copolymer is manufactured in the manner described for epichlorohydrin elastomers.

Curing Systems. The only practical curing systems for the propylene oxide elastomers are sulfur-based.

Table 2. Typical Compounding Recipe for Epichlorohydrin Elastomer

Additive	phr
carbon black	70
plasticizer	5
antioxidant	1
processing aid	1
acid acceptor	5
vulcanizing agent	1.5

Fillers. The elastomer is reinforced with carbon-black fillers in a manner similar to other elastomers; however, the interaction is not as strong as with most elastomers, and highly loaded vulcanizates are possible and desirable.

Plasticizers. Plasticizers (qv) are required in PO–AGE elastomers only in highly filled compounds or when softening or viscosity is needed. Aromatic oils such as Sundex 790 (sun oil) and esters such as DOP may be used.

Stabilizers and Antioxidants. The air and ozone resistance of the rubber is improved by the addition of nickel dibutyldithiocarbamate. Heat resistance can be improved by including 2-mercaptobenzimidazole.

Uses. The most important applications of GPO are in automotive engine mounts and under-the-hood suspension bushings where temperatures exceed those where NR is stable.

Safety and Handling

Monomers. The International Agency for Research on Cancer (IARC) has categorized EO as probably carcinogenic to humans. The same group characterized PO as a practical-purpose carcinogenic risk.

All epichlorohydrin monomers can cause severe respiratory, skin, and eye irritation. AGE and EPI also can cause skin sensitization. AGE and EPI are absorbed through the skin and can cause systemic toxic effects. The TLVs for airborne concentrations in the workplace are as follows: AGE, 5 ppm; EPI, 2 ppm; EO, 1 ppm; PO, 20 ppm.

Polymers. A vulcanizate of the EPI–EO copolymer and a synthetic sweat extract of it did not produce skin irritation or sensitization in standard animal tests.

Compound Ingredients. The most commonly used curing agent for epichlorohydrin elastomers, ethylenethiourea (ETU), has been found to be carcinogenic and teratogenic in rats. The common acid acceptors, red lead oxide and barium carbonate, are both toxic by inhalation and ingestion. They are used in industry as dispersions in EPDM.

BIBLIOGRAPHY

U.S. Pats. 3,158,580, 3,158,581, and 3,158,591 (Nov. 24, 1964), E. J. Vandenberg (to Hercules, Inc.).

E. J. Vandenberg, *Rubber Plast. Age* **46,** 1139 (1965).

A. Maeda, *Nippon Gomu Kyokaishi* **53,** 341 (1980).

R. W. BODY
V. L. KYLLINGSTAD
Hercules, Incorporated

1,3-EPOXIDE POLYMERS.

See OXETANE POLYMERS.

EPOXY RESINS

Epoxy resins are characterized by a three-membered ring known as the epoxy, epoxide, oxirane, or ethoxyline group:

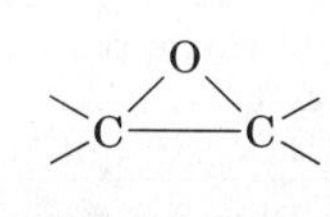

Commercial epoxy resins contain aliphatic, cycloaliphatic, or aromatic backbones. The capability of the epoxy ring to react with a variety of substrates imparts versatility to the resins. Treatment with curing agents gives insoluble and intractable thermoset polymers. In order to facilitate processing and modify cured resin properties, other constituents may be included in the compositions: fillers, solvents, diluents, plasticizers, and accelerators.

Epoxy resins were first offered commercially in 1946 and are now used in a wide variety of industries. Of the 135 metric tons sold in the United States in 1983, 45% were used in protective coatings, and the remainder in structural applications such as laminates and composites, tooling, molding, casting, construction, bonding and adhesives, and others.

High chemical and corrosion resistance, good mechanical and thermal properties, outstanding adhesion to various substrates, low shrinkage upon cure, flexibility, good electrical properties, and ability to be processed under a variety of conditions are characteristics of epoxy resins.

Synthesis

Epoxy resins are prepared by the reaction of compounds containing an active hydrogen group with epichlorohydrin followed by dehydrohalogenation:

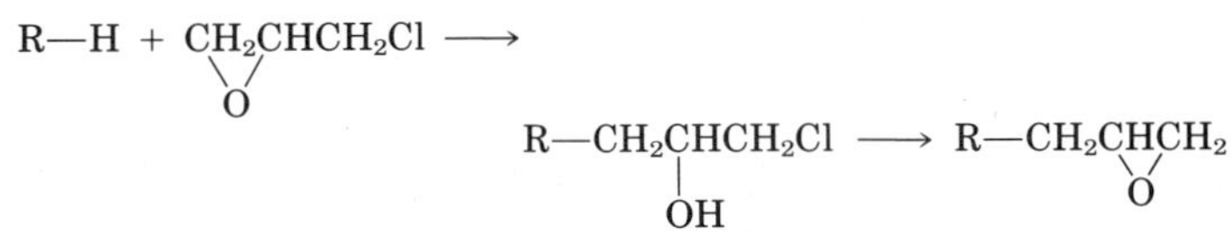

or by direct epoxidation of olefins by peracids.

Liquid Epichlorohydrin–Bisphenol A Resins. The most important intermediate in epoxy resin technology is the liquid reaction product of excess epichlorohydrin and bisphenol A, which can be described as the crude diglycidyl ether of bisphenol A (DGEBPA) (**1**), where the degree of polymerization n is nearly zero ($n \cong 0.2$):

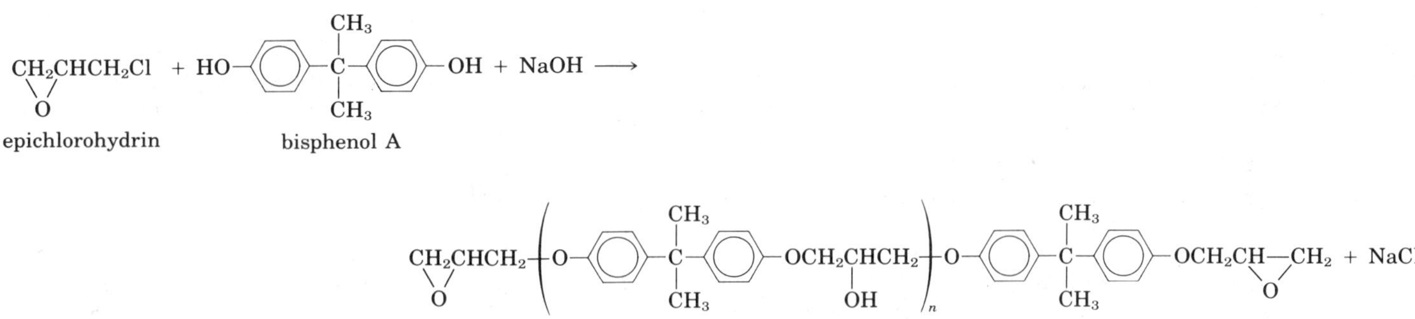

DGEBPA
(1)

Solid Epichlorohydrin–Bisphenol A Resins. High molecular weight solid epoxy resins are characterized by a repeat unit containing a secondary hydroxyl group with degrees of polymerization, ie, n values, ranging from 2 to about 30 in commercial resins; two terminal epoxy groups are theoretically present.

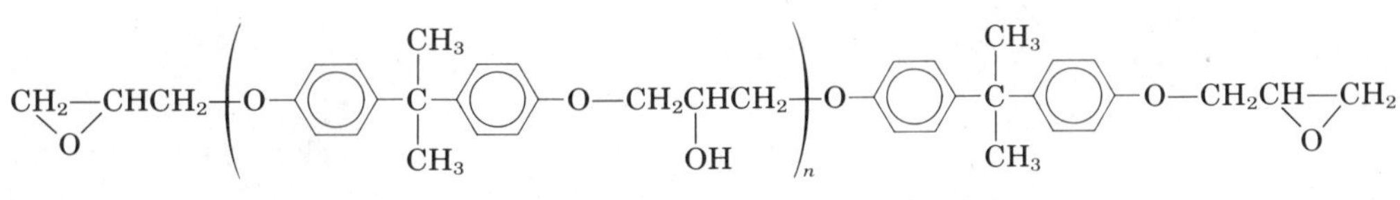

Epoxy resins with varying n values are prepared by two processes. The first is directly from epichlorohydrin, bisphenol A, and a stoichiometric amount of NaOH; this process is sometimes referred to as the taffy process. An alternative method is the chain-extension reaction of liquid epoxy resin with bisphenol A, sometimes referred to as the advancement or fusion process. In the direct process, a calculated excess of epichlorohydrin governs the degree of polymerization. However, preparation of the higher molecular weight species is subject to practical limitations. The effect of epichlorohydrin–bisphenol A (ECH–BPA) ratio for a series of solid resins is shown in Table 1.

Phenoxy Resins. Phenoxy resins are commercially important thermoplastic polymers derived from bisphenols and epichlorohydrin. Their molecular weights are higher, ie, ca 45,000, than those of conventional epoxy resins, ie, 8000 maximum. They lack terminal epoxide functionality and are therefore thermally stable and can be fabricated by conventional thermoforming techniques. The phenoxies have the same repeat unit as advanced epoxy resins and are classified as polyols or polyhydroxy ethers:

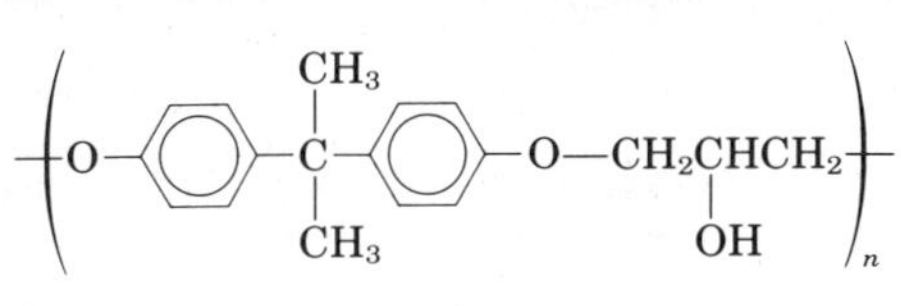

Table 1. Effect of Epichlorohydrin–Bisphenol A Ratio on Resin Properties

Mole ratio ECH:BPA	n value	$\bar{M}_n$	Wpe[a]	Mp, °C[b]
1.57:1.0	2	900	450–525	65–75
1.22:1.0	3.7	1400	870–1025	95–105
1.15:1.0	8.8	2900	1650–2050	125–135
1.11:1.0	12.0	3750	2400–4000	145–155

[a] Weight per epoxide.
[b] Durran's mercury method.

Phenoxy resins are prepared by reaction of high purity bisphenol A with epichlorohydrin in a 1:1 mole ratio or from high purity DGEBPA and bisphenol A in a 1:1 mole ratio. Solution polymerization may be employed to achieve the molecular weight and processability needed. The phenoxides are offered in the form of solutions for coating applications and as granular molding compounds.

Esterification. The esterification of epoxy resins with commercial fatty acids is a well-known process employed for industrial coatings. The carboxylic acids are esterified with the epoxy terminal groups or the pendent hydroxyl sites on the polymer chain:

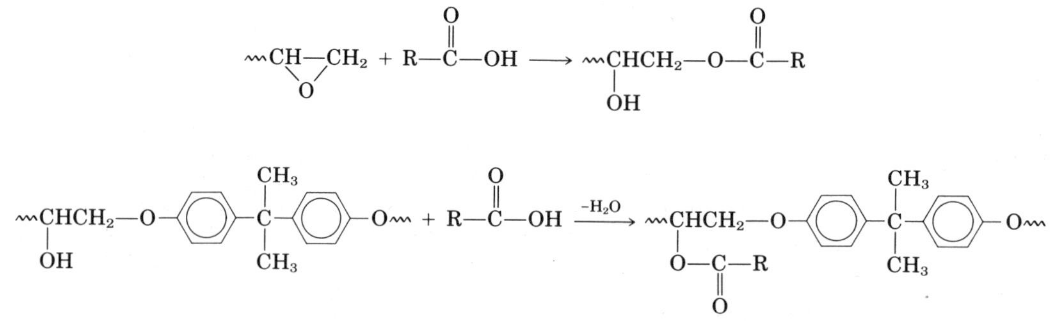

A wide variety of saturated and unsaturated fatty acids is utilized to confer properties useful in air-dried, protective, and decorative coatings.

Multifunctional Epoxy Resins. ***Aromatic Glycidyl Ether Resins.*** Both epoxy phenol novolac resins (EPN) and epoxy cresol novolac resins (ECN) have attained commercial importance. The former are made by glycidylation of the phenol–formaldehyde condensates obtained from acid-catalyzed resinification of phenol and formaldehyde. This produces random ortho- and para-methylene bridges. Properties of epoxy novolacs are given in Table 2.

The *o*-cresol novolac epoxy resins (ECN) and phenol novolac resins are analogous. They are prepared at formaldehyde–*o*-cresol ratios of less than unity; epoxidation is obtained with an excess of epichlorohydrin, as described for the phenol novolac resins (Table 3).

Aromatic Glycidyl Amine Resins. Among the multifunctional epoxy resins containing an aromatic amine backbone, only a few have attained any commercial significance, such as the tetraglycidyl ether of methylene dianiline and the triglycidyl derivative of *p*-aminophenol.

Glycidyl Esters. Glycidyl esters are prepared by the reaction of cycloaliphatic carboxylic acids with epichlorohydrin followed by dehydrohalogenation with caustic:

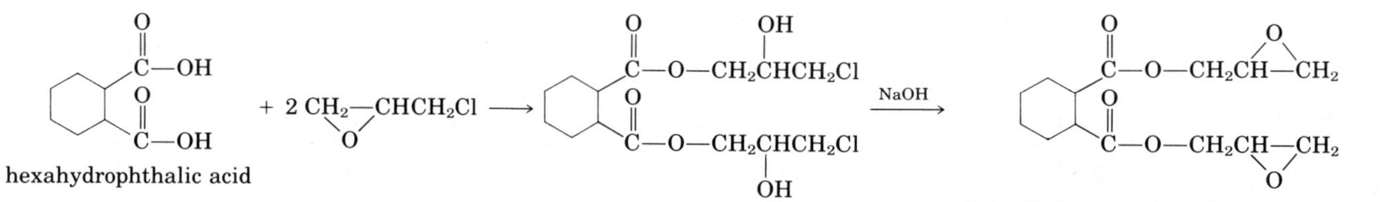

hexahydrophthalic acid

diglycidyl ester of hexahydrophthalic acid

Viscosities of these esters are low, ie, ca 500 mPa (=cP) and reactivities resemble those of standard bisphenol A epichlorohydrin resins. The commercial products are CY 184 (CIBA–GEIGY) and Lekutherm X-100 (Mobay Chemical Co.).

Cycloaliphatic Epoxy Resins and Epoxidized Oils. The preferred industrial route to cycloaliphatic epoxy resins is based on the epoxidation of cycloolefins with peracids, particularly

Table 2. Typical Properties of Epoxy Novolacs

Property	D.E.N. 431[a], EPN 1139[b]	D.E.N. 438[a], EPN 1138[b]	D.E.N. 439[a]	XD 7855[a]
n value	0.2	1.6	1.8	3.5
wpe[c]	175	178	200	185
viscosity[d], mPa·s (= cP)	$1{,}400^{52}$	$35{,}000^{52}$	$3{,}000^{100}$	800^{150}
softening point[e]			53	73
color, Gardner	1	2	2	2

[a] The Dow Chemical Co.
[b] CIBA-GEIGY Corp.
[c] Weight per epoxide.
[d] Superscript indicates temperature of measurement, °C.
[e] Durran's mercury method.

Table 3. Typical Properties of Epoxy Cresol Novolac Resins[a]

Property	ECN 1235	ECN 1273	ECN 1280	ECN 1299
molecular weight	540	1080	1170	1270
wpe[b]	200	225	229	235
softening point, °C	35	73	80	99
epoxide functionality	2.7	4.8	5.1	5.4

[a] CIBA-GEIGY Corp.
[b] Weight per epoxide.

peracetic acid. The costs are low and few side reactions are encountered. Properties are given in Table 4.

Epoxidation of unsaturated fatty acids and their glycerol esters is effected readily by peracids:

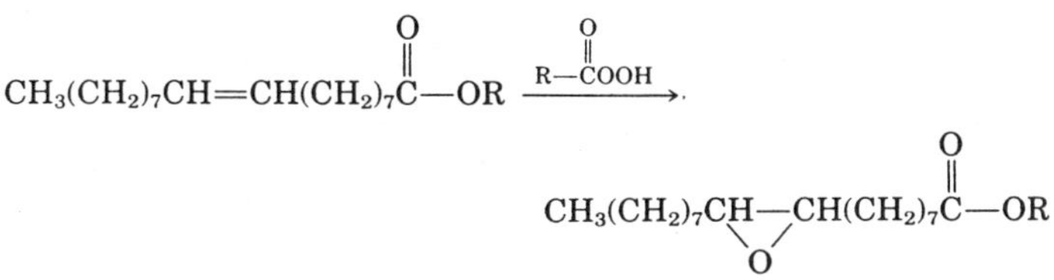

Epoxidized soybean and linseed oils are the most frequently used plasticizers/stabilizers for vinyl resins.

Characterization of Uncured Epoxy Resins

Uncured liquid epoxy resins are mainly characterized by epoxy content, viscosity, color, hydrolyzable chlorine, and volatile content. Less often analyzed are α-glycol content, total chlorine content, ionic chloride, and sodium.

Uncured solid resins are characterized by epoxy content, solution viscosity, melting point, color, and volatile content. Less often quoted are phenolic hydroxyl content, hydrolyzable chlorine, ionic chloride, sodium, and esterification equivalent.

In addition, gpc, high performance liquid chromatography, and other analytical procedures are performed to determine molecular weight, molecular weight distribution, and purity.

Curing Mechanisms

Optimum performance properties are obtained by cross-linking the epoxy resins into a three-dimensional insoluble and infusible network. The resin is treated with a curing agent or hardener; its choice depends on processing methods, curing conditions, and the physical and chemical properties desired.

Curing agents are either catalytic or coreactive. A catalytic curing agent functions as an initiator for epoxy resin homopolymerization, whereas the coreactive curing agent acts as a comonomer in the polymerization process. The curing agent can react with the epoxy and pendent hydroxyl groups on the resin backbone by way of either an anionic or cationic mechanism.

Curing Agents

The choice of a resin and curing agent depends on the application and on handling characteristics, curing temperature and time, use properties, toxicological and environmental limitations, and cost.

In 1981, the U.S. market for epoxy resin curing agents was estimated at 41 × 10^3 t; 126.4 × 10^3 t epoxy resins were sold in the United States (Table 5).

Relationship Between Cured Epoxy Resin Structure and Properties

The curing of a thermoset epoxy resin can be expressed in terms of a time–temperature transformation (TTT) diagram (Fig. 1).

Table 4. Cycloaliphatic Epoxy Resins

Chemical name	Structure	Commercial products	Wpe[a]	Viscosity, mPa·s (= cP) at 25°C[b]
3′,4′-epoxycyclohexylmethyl 3,4-epoxycyclohexanecarboxylate	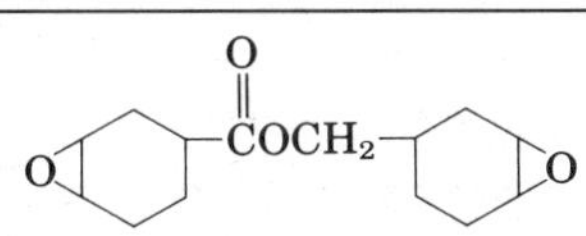	ERL-4221[c] CY-179[d]	131–143	350–450
3,4-epoxycyclohexyloxirane	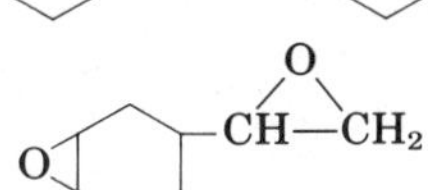	ERL-4206[c]	70–74	15
2-(3′,4′-epoxycyclohexyl)-5,1″-spiro-3″,4″-epoxycyclohexane-1,3-dioxane	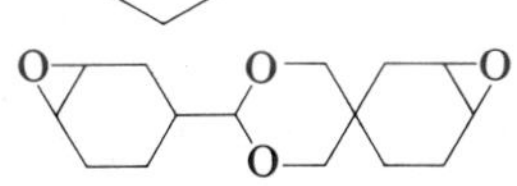	ERL-4234[c]	133–154	7,000–17,000[e]
bis(3,4-epoxycyclohexylmethyl) adipate	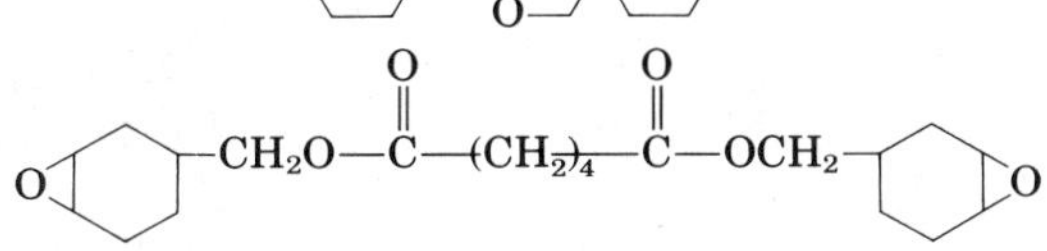	ERL-4299[c]	180–210	550–750

[a] Weight per epoxide.
[b] Unless otherwise stated.
[c] Union Carbide Corp.
[d] CIBA-GEIGY Corp.
[e] At 38°C.

Table 5. U.S. Consumption of Curing Agents for Epoxy Resins, 1981

Curing agents	10^3 t	Percentage of market
polyamides	10.8	27
formaldehyde resins	7.3	18
anhydrides	5.4	13
aliphatic polyamines	5.0	12
aromatic polyamines	3.6	9
polyamide amines	2.7	7
polycarboxylic polyesters	1.8	4
Lewis acids and bases	1.3	3
polysulfides and polymercaptans	1.3	3
phenol novolac resins	0.9	2
cycloaliphatic polyamines	0.9	2
Total	*41*	*100*

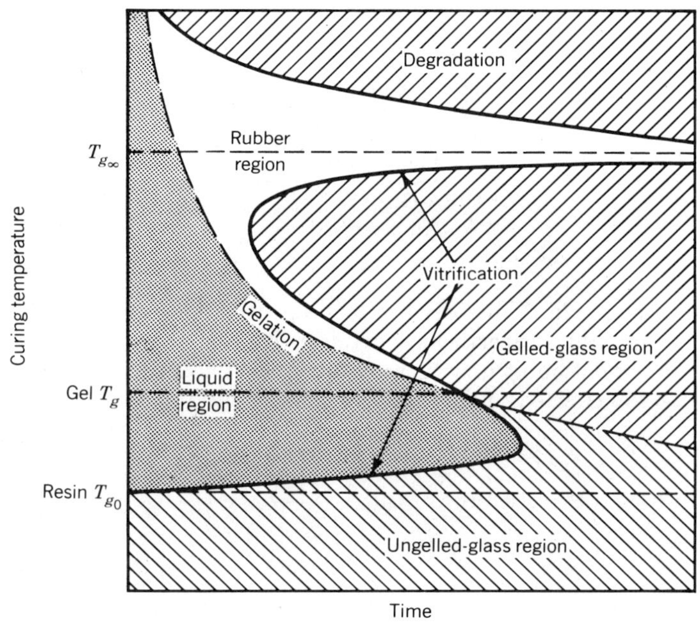

Figure 1. Time–temperature transformation diagram.

The curing cycle is represented by the resin rheology as a function of time and isothermal temperature. The S-shaped gelation curve and the vitrification curve divide the time–temperature plot into four distinct states of the thermosetting-cure process: liquid, gelled rubber, ungelled glass, and gelled glass; T_{g0} is the glass-transition temperature of the unreacted resin mixture, $T_{g\infty}$ the glass-transition temperature of the fully cured resin, and gel T_g the glass-transition temperature of the resin at its gel point. Properties are influenced by factors at the molecular level, such as backbone structures of epoxy resin and curing agent, nature of the covalent bond developed between the epoxy resin and the curing agent during cross-linking, and density and extent of cross-linking, ie, degree of cure.

The extent of cross-linking is a measure of the degree of cure. The most favorable properties are obtained by maximum cross-linking. The curing temperature strongly influences the ultimate cross-link density. Heating increases molecular mobility, resulting in higher cross-link density. The increase in chemical resistance properties after postcure also demonstrates the effects of increased cross-link density.

Modifiers

The behavior and properties of a system can be controlled by diluents (Table 6), fillers (dolomitic marble flour, chalk powder, sand, silica flour, mica flour, slate powder, vermiculite, phenolic microballoons, zircon flour, aluminum powder, chopped glass, asbestos fiber, and aluminum oxide), toughening agents, etc.

Economic Aspects

Epoxy resin sales increased rapidly in the 1960s and continued into the 1970s, as new applications were developed. New developments are expected in the automotive, electronics, and composites industries.

Worldwide capacity for epoxy resins is estimated at over 400×10^3 t/yr; CIBA-GEIGY, Dow, and Shell account for over 70% of this capacity. Trademarks of the principal producers are:

Company	Trade name
Celanese Polymer Specialties	Epi-Rez

Table 6. Commercial Reactive Diluents

Name	Structure
n-butyl glycidyl ether	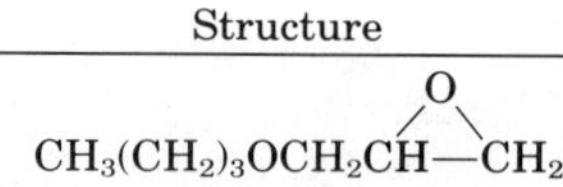
C_{12}–C_{14} aliphatic glycidyl ether	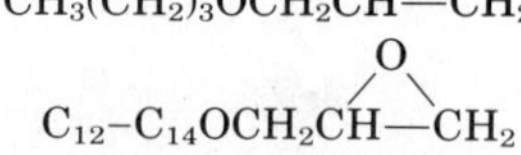
o-cresol glycidyl ether	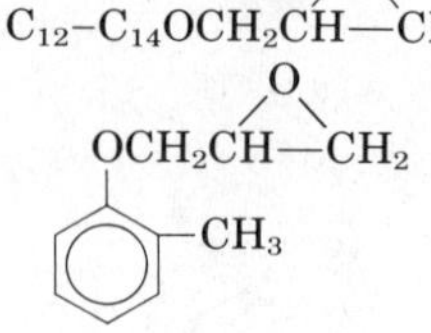
neopentylglycol diglycidyl ether	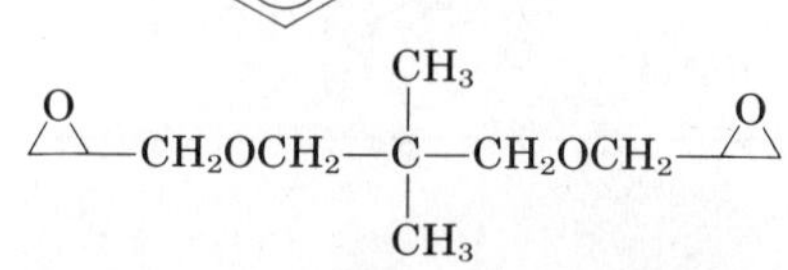
butanediol diglycidyl ether	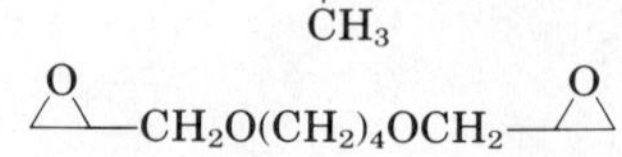

Table 7. U.S. Epoxy Resin Sales, 1979–1983[a], 10^3 t

Market	1979	1980	1981	1982	1983	1984
protective coatings	66.4	62.4	62.4	55.4	59.6	69.6
laminates and composites	27.4	28.5	30.0	26.6	33.5	39.4
tooling, molding, and castings	14.1	11.7	13.1	10.9	12.7	14.1
construction	8.9	8.7	9.3	8.3	8.4	8.2
bonding and adhesives	7.9	8.1	8.0	6.9	7.0	9.1
others	9.9	9.4	9.5	11.4	13.3	13.1
Total, domestic	*134.6*	*128.8*	*132.3*	*119.5*	*134.5*	*153.5*
export	16.2	16.1	18.7	17.6	15.8	18.7
Total	*150.8*	*144.9*	*151.0*	*137.1*	*150.3*	*172.2*

[a] Based on data from SPI, Committee on Resin Statistics, as compiled by Ernst and Ernst.

CIBA-GEIGY Corp.	ARALDITE, ARACAST
The Dow Chemical Co.	D.E.R., D.E.N.
Reichhold Chemical Co.	Epotuf
Shell Chemical Co.	Epon
Union Carbide Corp.	Bakelite

Health and Safety Factors

Handling and safety procedures must be adapted to the individual system and application.

All suppliers provide material safety data sheets that contain the most recent toxicity data. These are the best sources of information.

Applications

Cured epoxy resins exhibit excellent adhesion to a variety of substrates; outstanding chemical and corrosion resistance; excellent electrical insulation; high tensile, flexural, and compressive strengths; thermal stability; a wide range of curing temperatures; and low shrinkage on cure.

The distribution of U.S. consumption of epoxy resins is given in Table 7.

BIBLIOGRAPHY

H. Lee and K. Neville, *Handbook of Epoxy Resins,* McGraw-Hill, Inc., New York, 1967, reprinted 1982.

ACS Symp. Ser. **221** (1983).

The Epoxy Resin Formulators Training Manual, The Society of Plastics Industry, Inc., New York, 1984.

C. A. May and Y. Tanaka, eds., *Epoxy Resins Chemistry and Technology,* Marcel Dekker, Inc., New York, 1973.

LOUIS V. MCADAMS
JOHN A. GANNON
CIBA–GEIGY Corporation

EPR. See ELECTRON-SPIN RESONANCE.

ESCA. See CHARACTERIZATION OF POLYMERS.

ESR. See ELECTRON-SPIN RESONANCE.

ESTERS, CELLULOSE. See CELLULOSE ESTERS, INORGANIC; CELLULOSE ESTERS, ORGANIC.

ESTERS, POLYMERIC. See ALKYD RESINS; POLYESTERS.

ETHERS, CELLULOSE. See CELLULOSE ETHERS.

ETHYL ACRYLATE POLYMERS. See ACRYLIC AND METHACRYLIC ESTER POLYMERS.

ETHYLCELLULOSE. See CELLULOSE ETHERS.

ETHYLENE–ACRYLIC ELASTOMERS. See ACRYLIC ELASTOMERS.

ETHYLENE–CARBON MONOXIDE COPOLYMERS. See OLEFIN–CARBON MONOXIDE COPOLYMERS.

ETHYLENE–ETHYL ACRYLATE COPOLYMERS. See ACRYLIC ELASTOMERS.

ETHYLENEIMINE POLYMERS. See ALKYLENEIMINE POLYMERS.

ETHYLENE–MALEIC ANHYDRIDE COPOLYMERS. See ETHYLENE POLYMERS; MALEIC AND FUMARIC POLYMERS.

ETHYLENE OXIDE POLYMERS. Sce 1,2-EPOXIDE POLYMERS.

ETHYLENE POLYMERS

Introduction

The basic structure of polyethylene (PE) is the chain, —(CH_2—CH_2)$_n$—, which has no substituent groups, ie, branches on the backbone. In commercial PE, n may be from 400 to more than 50,000; alkyl substituents, called short-chain branches, are usually present on the chain backbones.

PE is a partially crystalline solid, somewhat flexible, whose properties are strongly influenced by the relative amounts of crystalline and amorphous phases. The smallest crystalline units, called lamellae, are planar in shape and consist of chains that are perpendicular to the plane and fold back and forth, every 5–15 nm, although sometimes less frequently. Lamellae are interconnected by a few chains that pass from one lamella, through a small amorphous region, to another lamella. The lamellae form much larger spherically shaped units, called spherulites, which are interconnected through amorphous regions. The crystalline phase provides rigidity and a high softening temperature, whereas the amorphous phase, which is relatively soft at service temperatures, provides flexibility and high impact strength. The size of the spherulites greatly affects certain properties.

Unsubstituted PE has a density of 0.960–0.970 g/cm^3 and crystallinity of 70–90%; the melting point of the crystalline phase can be as high as 135°C. Such PE, generally called high density homopolymer, represents only a small proportion of commercial PE. Most polyethylenes contain a varying number of alkyl substituents with 2–8 carbon atoms. These substituent groups cause a reduction in crystallinity (to 40–50%), melting point (to 105–115°C), and density (to 0.910–0.915 g/cm^3). As a result, stiffness is reduced and impact strength is increased. Density is closely related to crystallinity. The density of a pure PE crystal has been calculated as 1.00 g/cm^3, whereas that of an amorphous phase is believed to be about 0.855 g/cm^3 (see also CRYSTALLINITY DETERMINATION).

High density PE (HDPE) includes both homopolymers and copolymers with α-olefins such as 1-butene, 1-hexene, or 1-octene. Copolymerization by transition metal catalysts introduces alkyl substituents into the chains and gives HDPE copolymers of density 0.941–0.958. Linear low density copolymers of density 0.915–0.940 are also obtained by copolymerization, as well as very low density PE of density 0.890–0.915.

In high pressure PE produced by free–radical polymerization, alkyl substituents are formed by rearrangement of the chains during chain growth to give a density of 0.915–0.930 g/cm^3. Polar substituents may also be introduced by copolymerization with polar monomers.

Other important differences in the structure of polyethylenes are average molecular weight and molecular weight distribution. In LDPE, the chains have branches, called long-chain branches, which are as long as the chain backbones before branching occurs.

Controlled variations in these structural parameters result in a broad family of products with wide differences in physical properties, such as softening temperatures, stiffness, hardness, clarity, impact strength, etc. Because of this versatility, PE has become the largest volume commercial polymer in the world. Estimated 1984 production in the United States was 3.8×10^6 t of low and medium density polyethylene and 2.8×10^6 t of high density polyethylene (HDPE). Total production in other countries is somewhat higher.

BIBLIOGRAPHY

J. Agranoff, ed., *Modern Plastics Encyclopedia 1986–1987,* Vol. 63 (10A), McGraw-Hill, Inc., New York, 1986, pp. 52–68, 533, 534.

P. J. Hogan in M. Grayson, ed., *Kirk-Othmer Encyclopedia of Chemical Technology,* 3rd ed., Vol. 16, Wiley-Interscience, New York, 1981, pp. 421–433.

R. N. Short in M. Grayson, ed., *Kirk-Othmer Encyclopedia of Chemical Technology,* 3rd ed., Vol. 16, Wiley-Interscience, New York, 1981, pp. 385–401.

R. Raff and K. W. Doak, eds., *Crystalline Olefin Polymers,* John Wiley & Sons, Inc., New York, 1964, Pt II; 1965, Pt I.

L. I. Chriswell, *Chem. Eng. Prog.,* **79,** 84–92 (1983).

KENNETH W. DOAK
Consultant

Low Density Polyethylene (High Pressure)

Low density polyethylene (LDPE) has a polymethylenelike structure that contains alkyl substituents, or short-chain branches, on the chain backbone. Synthesis by a high pressure, free-radical process involves unique chemistry, whereby branches are formed by molecular rearrangements of the PE backbone during chain growth.

The alkyl groups cannot conveniently fit into the crystal lattice of polyethylene; thus melting point and crystallinity, and therefore density, are reduced. Densities may vary from 0.910 to 0.955 g/cm^3, compared with 0.960–0.970 for linear, ie, unbranched, PE. The homopolymers produced under the reaction conditions available in commercial processes usually have densities of 0.916–0.930 g/cm^3; they are called LDPE. Typical commercial products have a degree of crystallinity of 45–55% and melting points of 105 to 115°C. These products contain 15–25 short-chain branches per 1000 carbon atoms.

In addition to alkyl substituents, long-chain branches as long as the backbone chain are formed by the grafting of polymer chains to inert chains before they can be removed from the reactor. Such chains broaden molecular weight distribution and have a pronounced effect on rheological properties.

Additional substituents can be introduced by copolymerization with polar monomers, such as vinyl acetate or ethyl acrylate, which reduce crystallinity further and improve flexibility and impact strength.

The structural features of LDPE can thus be varied widely by a careful choice of reaction conditions and type and amount of comonomer, resulting in a broad spectrum of physical properties and an extremely wide range of applications.

Homopolymerization

Basic Reactions. Homopolymerization follows the classical equations for free-radical polymerization (qv), although with important deviations. Side reactions leading to the formation of the short branches (alkyl groups), long branches (grafted PE chains), unsaturated groups, particularly vinylidene end groups, and low molecular weight vinylidene compounds are significant in contrast to the polymerization of vinyl monomers such as styrene and methyl methacrylate.

Copolymerization

When two monomers M_1 and M_2, such as ethylene and an α-olefin, are polymerized simultaneously, both may enter the chain to form a copolymer such as:

$$—M_1—M_2—M_2—M_1—M_2—M_1—M_1—M_1—M_2—$$

The numbers of M_1 and M_2 that enter the chain, as well as their arrangement along the chain, are controlled by four propagation reactions. The two types of polymer radicals with different monomer units at the end of the growing chains are designated $M_1\cdot$ and $M_2\cdot$. These radicals may add to M_1 or M_2; however, the relative reactivities of M_1 and M_2 may be different toward the two polymer radicals. The rate of each propagation reaction is determined by the concentrations of polymer radical and monomer and the corresponding propagation constant.

Vinyl acetate has about the same reactivity as ethylene toward both radical types. Copolymers containing 2–28 weight % are widely used commercially. Hydrolysis of vinyl acetate copolymers gives copolymers containing vinyl alcohol. Alkyl acrylates are more reactive than ethylene toward both radical types. Commercial copolymers contain about 20% of ethyl acrylate or methyl acrylate, or 6.5–15% acrylic acid or methacrylic acid. Neutralization of the acid groups forms ionomers.

α-Olefins such as propylene or 1-butene are less reactive than ethylene. Only small amounts can be incorporated by copolymerization because of the adverse effect of these monomers on molecular weight.

Manufacture

The commercial production of PE by the high pressure, free-radical process involves engineering difficulties because of the constraints dictated by the polymerization chemistry. In order to obtain products with the desired molecular characteristics at an economical production rate, the following conditions are required: a pressure of 122–303 MPa (1200–3000 atm); a temperature of 130–350°C; a very short polymerization (residence) time, ie, from 15 s to 2 min, depending on the reactor type; and continuous reactors that allow a high production rate of 10–18 t/h (3 t/h for older reactors). Initiators are oxygen or organic peroxides.

The reactor is either a tubular reactor with an inner diameter of 2.5–6.4 cm and 0.5–1.5 km long, or a stirred autoclave with a capacity as high as ca 0.7 m^3.

A typical commercial process is shown schematically in Figure 1 and includes the following steps: staged compression, introduction of a free-radical initiator, partial conversion of ethylene to PE, separation of ethylene from PE, extrusion of molten polymer, and cooling of ethylene.

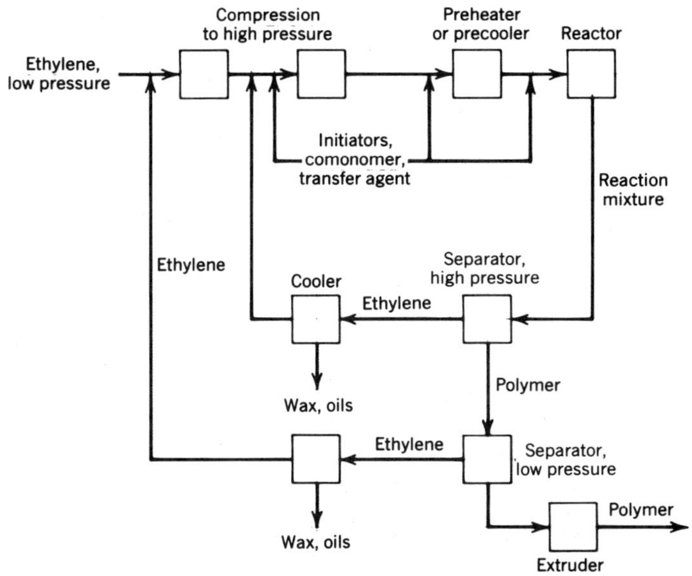

Figure 1. Generalized flow diagram for high–pressure polymerization processes. Tubular process uses preheater; autoclave process may use precooler.

Properties

Physical Properties. LDPE has a unique combination of properties: toughness, high impact strength, low brittleness temperature, flexibility, processibility, film transparency, chemical resistance (especially to polar compounds), low permeability to water, stability, and outstanding electrical properties. The properties shown in Table 1 are given in ranges because of their dependence on density and temperature.

Electrical Properties. The insulating properties of PE compare favorably with those of any known dielectric material. The outstanding electrical properties of LDPE were the basis for its initial commercialization and use as radar insulation during World War II. Today, various types of PE are the principal material used for power and communications cable insulation.

Table 1. Physical Properties of Low and Medium Density High Pressure Polyethylenes

Property	ASTM method	LDPE	MDPE[a]
density, g/cm^3	D 792	0.912–0.925[b]	0.926–0.940[b]
melting temperature, °C		102–112	110–120
refractive index, n_D^{25}	D 542	1.51–1.52	1.52–1.53
tensile yield stress, MPa[c]	D 638	6.2–11.5	10–19
yield elongation, %	D 638	20–40	10–20
tensile modulus, MPa[c]	D 638	102–240	205–370
Shore hardness, durometer	D 676	D40–50	D45–60

[a] Medium density PE.
[b] Arbitrary definition.
[c] To convert MPa to psi, multiply by 145.

Table 2. 1984 U.S. Consumption of LDPE, 10^3 t

Market	All LDPE[a]	LLDPE	EVA[b]
blow molding	21	7	
extrusion coating	286		30
film[c]	2074	764	292
pipe and conduit	36	23	
sheet[c]	11	3	
wire and cable	175	61	15
other extrusion	42		3
injection molding	268	102	18
rotomolding	43	41	
export	432	82	52
adhesives			50
other[d]	345	77	28
Total	*3733*	*1160*	*488*

[a] Includes LLDPE and EVA.
[b] Ethylene–vinyl acetate copolymer.
[c] <300 μm (12 mil).
[d] Includes LDPE and EVA used in blending and compounding.

Rheological Properties. Above its melting point, PE is highly viscous; the viscosity at any temperature depends on average molecular weight, molecular weight distribution (partially related to long-chain branching), and shear rate or shear stress. Molten PE may be classified as a pseudoplastic fluid, ie, its apparent viscosity decreases with increasing shear stress (undergoes shear-thinning), contrary to a classical Newtonian fluid in which the viscosity at constant temperature is relatively independent of shear rate or stress. Shear-thinning increases as long-chain branching increases, and processability is improved. Typically, the melt viscosity at 200°C drops almost two orders of magnitude as the shear rate is increased nearly four orders of magnitude.

Solution Properties. The solution properties of PE are influenced by long-chain branching. As in the molten state at high temperatures and high shear rates, molecules of LDPE are less extended in solution and more compact than those of unbranched PE.

Degradation, Aging, and Stabilization. PE exhibits better heat stability than many other polymers, but undergoes thermal degradation (qv) at high temperatures (eg, above 375°C) in the absence of oxygen, where the bonds in the chain are ruptured; such degradation cannot be prevented. In the presence of oxygen, degradation proceeds slowly at ambient temperature, and becomes more rapid as temperature increases. Certain stabilizers markedly retard oxidative degradation and increase service lifetime.

Products and Uses

The large number of PE uses is the result of the enormous number of variations possible in the molecular structure of homopolymers and polar copolymers. Resins for many types of film and sheet account for over half of the consumption.

The 1984 pattern of U.S. consumption is given in Table 2.

Table 3. U.S. Consumption of PE, 10^3 t

Year	Low density PE				HDPE
	All	LLDPE	EVA	LDPE[a]	
1983	3657	991	463	2203	2594
1984	3733	1160	488	2085	2666

[a] Other high pressure, mostly homopolymer, estimated by difference.

Table 4. U.S. Producers and Capacities of High Pressure PE

Producers	10^3 t
Tubular[a]	
El Paso	185
Exxon	225
Mobil	225
Union Carbide	225
Autoclave	
DuPont	345
Eastman	285
Gulf	410
Autoclave and tubular	
Dow	365
Northern Petrochemicals	460
U.S.I.	370

[a] Some producers listed under tubular have autoclave capacity for specialty copolymers.

Economic Aspects

High pressure PE is a low cost, high volume product. In July 1984, prices were reported as \$0.83–0.94 kg for large volume grades. The U.S. consumption for 1983 and 1984 is shown in Table 3. The consumption outside the United States was probably higher.

The nameplate capacity of the high pressure PE plants in the United States on Jan. 1, 1985 was 3.06×10^6 t. Total capacity outside the United States is believed to be higher. The capacities of the U.S. producers are given in Table 4.

BIBLIOGRAPHY

R. L. Boysen in M. Grayson, ed., *Kirk–Othmer Encyclopedia of Chemical Technology,* 3rd ed., Vol. 16, Wiley-Interscience, New York, 1981, pp. 402–420.

P. Ehrich and G. A. Mortimer, *Adv. Polym. Sci.* **7,** 386–448 (1970).

R. Fava, ed., *Methods of Experimental Physics,* Vol. 16, Academic Press Inc., New York, 1980.

KENNETH W. DOAK
Consultant

Linear Low Density Polyethylene

Linear low density polyethylene (LLDPE) is changing the character of the polyethylene industry. For fifty years, low density polyethylene has been produced at pressures of up to 345 MPa (50,000 psi) and temperatures near 300°C. Now, new technology capable of operating at less than 2 MPa (300 psi) and near 100°C has been developed. This technology has rapidly established itself as a low cost route to low density polyethylene (LDPE). These new products, called LLDPE, have definite advantages over those made by high pressure processes and are revolutionizing the polyethylene industry.

In the past, the polyethylene industry was conveniently classified by product density and process type. Today, polyethylenes are more appropriately described as branched polyethylenes and linear polyethylenes.

Branched polyethylene is made with a free-radical catalyst and contains long-chain branches (LCB). Linear polyethylene is made with a transition metal catalyst and contains no long-chain branching. Both branched and linear polyethylene may contain short-chain branching (SCB).

Molecular Structure

LLDPE, as the name implies, is characterized by linear molecules without long-chain branches. In contrast, LDPE products contain many long-chain branches. Short-chain branches are common to both LDPE and LLDPE and are the primary determinant of resin density for both.

Commercial LLDPE products are available based on propylene, 1-butene, 4-methyl-1-pentene, 1-hexene, 1-octene, and certain mixtures thereof. The selection of a comonomer for LLDPE seems to be based on process compatibility, cost, and product properties.

Chemical Pathways to Polyethylene

Kinetic and mechanistic investigations have shown that distinctly different reaction processes operate at high pressures to produce LDPE, and at low pressures to produce LLDPE. The conventional high pressure process operates by a free-radical mechanism. In polymerization catalyzed by transition metals at low pressures and temperatures, the polymerization occurs by a coordinated anionic mechanism. Propagation occurs by monomer coordination and insertion into a transition metal–carbon bond. These bonds may be generated by alkylation of a transition metal compound ($TiCl_4$) using a metal alkyl (R_3Al). Reduction by an olefin, which occurs with the CrO_3/SiO_2 catalyst, may also provide sites for polymerization. Low valent transition metal compounds, ie, $TiCl_2$, may also function as catalysts.

For linear low density products, ethylene is copolymerized with an alpha-olefin, such as 1-butene, 1-hexene, or 1-octene.

Manufacture of Linear Polyethylene

Processes include low pressure, slurry-phase polyolefin processes (loop-reactor, light-diluent processes; loop-reactor, heavy-diluent processes; stirred-tank, heavy-diluent processes; and liquid-pool processes); solution-phase polyolefin processes (medium pressure process; adiabatic low pressure process; cooled low pressure process); modified high pressure polyolefin processes; and low pressure, gas-phase polyolefin processes.

Product Capabilities. Polyethylene is characterized in terms of melt index and density. Melt index increases as weight-average molecular weight decreases. The melt indexes of commercial linear polyethylenes range from about 0.01 g/10 min to several hundred grams per 10 minutes. This corresponds to a weight-average molecular weight between 300,000 and 30,000.

The markets and applications for linear polyethylenes have developed over the years as a result of product properties and fabrication know-how (Fig. 1).

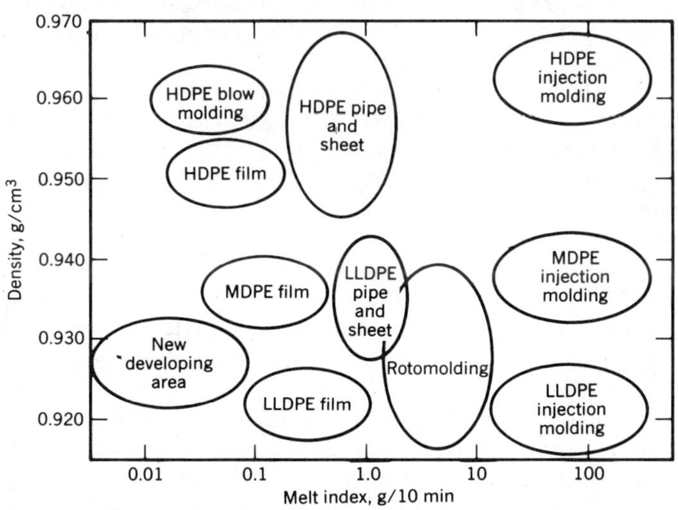

Figure 1. Linear polyethylene markets: melt index vs density.

Table 1. 1988 Nameplate Capacities of U.S. Polyethylene Producers, 10^3 t

Producer	Total	LDPE	HDPE	LLDPE/HDPE[a]
Allied	545		545	
Amoco	182		182	
Chevron	680	420	260	
Dow	985	350	195	440
DuPont	364	364		
Eastman	300	300		
El Paso	205	205		
Exxon	546	264		282
Hoechst	130		130	
Mobil	452	232		220
Occidental	518		518	
Phillips	660		660	
Soltex	522		552	
Union Carbide	995	250		745
Quantum	1490	811	564	115
Westlake	100	100		
Total	*8674*	*3296*	*3606*	*1802*

[a] Company makes HDPE but does not disclose the amount of LLDPE–HDPE swing capacity dedicated to HDPE.

Economic Aspects

Worldwide polyethylene demand in 1988 was 28×10^6 t. It is estimated that this industry will continue to grow at an average annual rate of 4–5%, reaching 38×10^6 t by 1995. In 1988, LLDPE accounted for about 33% of total U.S. LDPE sales.

Various sources estimate that by 1995, LLDPE will account for as much as 50% of the U.S. LDPE market. The estimated nameplate capabilities of U.S. polyethylene producers as of January 1, 1988 are given in Table 1.

Prices vary widely, depending on, but not limited to, ethylene cost, energy cost, and supply and demand (Table 2).

Table 2. U.S. Polyethylene Price Ranges, 1988

Product	Market price, $/kg
LDPE	0.92–1.12
LLDPE	
butene based	0.92–1.15
hexane based	1.06–1.28
HDPE	0.95–1.23

Table 3. Blown-film[a] Properties of LLDPE Copolymers

Property	Ethylene–butene	Ethylene–hexene	Ethylene–octene
melt index, g/10 min	1.0	1.0	1.0
density, g/cm^3	0.918	0.918	0.919
dart impact, g	150	165	180
Elmendorf tear, N/mm[b]			
MD[c]	59	125	125
TD[d]	133	220	275
tensile strength, MPa[e]			
MD[c]	38	38	43
TD[d]	31	32	34
elongation, %			
MD[c]	620	570	550
TD[d]	760	790	660

[a] 37-μm film at 2:1 BUR (blow-up ratio).
[b] To convert N/mm to lb/in., divide by 0.175.
[c] Machine direction.
[d] Transverse direction.
[e] To convert MPa to psi, multiply by 145.

Properties and Applications

Polyethylene is semicrystalline, ie, it contains both crystalline and amorphous regions. Its density correlates with crystallinity. Linear polyethylene, with a density of 0.915–0.970 g/cm^3, exhibits a crystallinity of about 25–75 vol %.

The LDPE/LLDPE markets typically are divided into five applications areas: film (the largest) (Table 3), injection molding, wire and cable, rotomolding, and others.

Health and Safety Factors

Manufacturers of LLDPE usually provide information regarding health and safety in Product Information Sheets or Material Safety Data Sheets. Issues discussed include storage and handling; health and safety; combustibility; disposal; precautionary labeling; and FDA status.

Polyethylene resins are among the most inert commercial polymers and in normal handling constitute no hazard from skin contact or ingestion.

Polyethylene resins can be disposed of by landfill or incineration. With properly controlled industrial, commercial, or municipal incineration, particulate or gaseous discharge into the air can be maintained within allowable levels.

BIBLIOGRAPHY

Polyolefins Through the 80s—A Time of Change, SRI International, Menlo Park, Calif., 1983.

Pap. Film Foil Converter **58,** 49 (1984).

Plast. World **41,** 92 (1983).

Mod. Plast. **63,** 29 (1986).

David E. James
Union Carbide Corporation

High Density Polyethylene

By historical definition, high density polyethylene (HDPE) is the product of ethylene polymerization with density 0.940 g/cm^3 and higher; it includes both ethylene homopolymers and copolymers with higher alpha olefins.

Molecular Structure

Chains. Linear polyethylene (PE) is a predominantly regular polymer with the chemical composition of polymethylene $(CH_2)_n$. Its name reflects the principal method of production: ethylene polymerization by various mechanisms. HDPE molecules always contain a small number of chain branches that are deliberately introduced by copolymerization of ethylene with alpha olefins, or that are present in the chains as a result of side reactions intrinsic to a particular polymerization mechanism. The most regular linear structure is obtained with chromium oxide-based catalysts.

Branches. For copolymerization with alpha olefins, the structure of branches is determined by the type of olefin:

$$-\!\!\left(CH_2CH_2CH_2\underset{\displaystyle CH_3}{\underset{|}{C}}HCH_2CH_2\right)\!\!-$$

ethylene–propylene copolymers

$$-\!\!\left(CH_2CH_2CH_2\underset{\displaystyle C_2H_5}{\underset{|}{C}}HCH_2CH_2\right)\!\!-$$

ethylene–1-butene copolymers

$$-\!\!\left(CH_2CH_2CH_2\underset{\displaystyle C_4H_9}{\underset{|}{C}}HCH_2CH_2\right)\!\!-$$

ethylene–1-hexene copolymers

The number of branches depends on polymerization conditions, including olefin concentration; it is determined by ^{13}C-nmr or ir methods.

Chain Ends. Usually one of the chain ends in a given HDPE molecule is a methyl group: $—CH_2CH_2CH_2CH_3$; the other chain end is either a similar saturated group if hydrogen is used as a chain-transfer agent in the polymerization, or it contains a double bond.

Crystalline Structure

The only stable local chain conformation of HDPE at low temperature is all-trans (tt), corresponding to the flat zigzag chain configuration.

The principal crystalline form of linear PE is orthorhombic (space symmetry group P_{nam}), the same as that of linear paraffins, which can be regarded as HDPE homologues.

Morphology. The principal morphological units of PE crystallizing under typical conditions from the melt are spherulites, very small anisotropic spheres (ca 1–5 μm) visible under high magnification with polarizers.

Phase Transitions and Relaxation Phenomena. Relaxation studies of semicrystalline HDPE using dynamic mechanical or electrical methods reveal three temperature ranges in which significant relaxation phenomena occur: the α peak in the 30–80°C range, the β peak in the 0 to −60°C range, and the γ peak in the −100 to −140°C range.

Determination of the glass-transition temperature T_g for linear PE, as for all highly crystalline polymers, is controversial.

Crystallization Kinetics. HDPE crystallizes rapidly. The kinetics of crystallization in the initial stages is described by the Avrami-Kholomogorov equation:

$$X_t/X_\infty = 1 - \exp(-(kt)^n)$$

where X_t and X_∞ are polymer crystallinites at time t and its limit crystallinity, respectively, and k and n are parameters that depend on temperature, molecular weight, and branching.

Orientation. Articles made of HDPE, such as films, filaments, pipes, and injection-molded articles, exhibit some degree of molecular and crystal orientation (qv). This orientation develops either spontaneously during melt flow and crystallization or is introduced deliberately by stretching. The degree of orientation can be measured by x ray, polarization spectroscopy, acoustical methods, or birefringence.

Physical Properties

Optical Properties. Because of the high crystallinity of HDPE and the difference in index of refraction between crystalline and amorphous phases, thin HDPE films are translucent, with significantly lower transparency than LDPE. Additionally, the complex morphological structure of HDPE, such as boundaries between spherulites, further decreases transparency. The ultraviolet transmission limit of HDPE is ca 230 nm. When stretched, HDPE film becomes highly birefringent. The specific birefringences for crystalline and amorphous phases of HDPE are 0.059 and 0.28, respectively. These parameters are useful for measuring orientation.

Physical, thermal, electrical, and mechanical properties are listed in Table 1.

Chemical Properties and Reactivity

Linear PE is a mixture of predominantly saturated linear hydrocarbons. In general, it exhibits low chemical reactivity.

At room temperature, HDPE is not soluble in any known solvent, although several solvents (ie, xylenes) have a swelling effect. However, certain binary solution mixtures including CS_2 dissolve HDPE at as low as 30–40°C.

Degradation. HDPE is relatively stable to heat. Chemical processes at high temperature in an inert medium or in vacuum result in breakage and cross-linking of polymer chains. This reaction is noticeable above 290–300°C and is similar to thermocracking of linear hydrocarbons.

At elevated temperatures, oxygen attacks the macromolecules. Thermooxidation is an important pretreatment step in commercial processes involving surface dyeing and printing. The poor adhesion of dyes to HDPE can be improved by chemical modification of the surface, such as thermooxidation by treatment with an open flame or in a strong electric field.

Photooxidative degradation of HDPE, ie, degradation in the presence of oxygen and light, resembles thermooxidative degradation but proceeds at low temperatures.

Photooxidative degradation results in rapid aging, development of surface cracks, brittleness, and change in color, as well as deterioration of mechanical and dielectrical properties. Photooxidative degradation is prevented by light stabilizers that afford protection by absorbing uv radiation.

Catalysts

At present, two classes of catalysts are used for the commercial manufacture of HDPE: those based on chromium oxide, eg, Phillips catalysts, and those based on Ziegler-Natta systems.

Table 1. Physical, Thermal, Electrical, and Mechanical Properties of HDPE

Property	Highly Linear	Low degree of branching[a]
Physical		
density, g/cm³	0.962–0.968	0.950–0.960
refractive index, n_D^{25}	1.54	1.53
Thermal		
melting point, °C	128–135	125–132
brittleness temp, °C	−140 to −70	−140 to −70
heat resistance temp, °C	ca 122	ca 120
thermal conductivity, W/(m·K)	0.46–0.52	0.42–0.44
heat of combustion, kJ/g[b]	46.0	46.0
Electrical		
dielectric constant at 1 MHz	2.3–2.4	2.2–2.4
dielectric strength, kV/mm	45–55	45–55
Mechanical		
yield point, MPa[c]	28–40	25–35
tensile modulus, MPa[c]	900–1200	800–900
tensile strength, MPa[c]	25–45	20–40
shear strength, MPa[c]	20–38	20–36
elongation, % at yield point	5–8	10–12
hardness, Brinell, MPa[c]	60–70	50–60

[a] 2–3 CH_3 per 1000 carbons.
[b] To convert J to cal, divide by 4.184.
[c] To convert MPa to psi, multiply by 145.

Other catalysts based on other transition metals are known, including scandium, titanium, vanadium, chromium, cobalt, nickel, zirconium, niobium, molybdenum, tungsten, palladium, rhodium, and ruthenium, and even lanthanides and actinides, although these have not achieved commercial significance.

Polymerization Processes

The technologies available for HDPE manufacture are solution, slurry or suspension, and gas-phase polymerization. Each has its advantages and disadvantages.

Because solution viscosity increases rapidly with molecular weight, solution polymerization, also called the solution form process, is employed primarily for low molecular weight PE. Slurry (suspension) polymerization is the oldest and most widely used method because of process-engineering maturity and flexibility. This technology is the only one capable of producing the full range of HDPE grades, from low molecular weight waxes to ultrahigh molecular weight HDPE. Gas-phase polymerization, the newest technology, is increasingly utilized because of the low investment requirements and operating costs.

Rheology and Processing

Typically, HDPE is molten twice during its lifetime: during pelletization and during fabrication. At both times, antioxidants or other means are utilized to prevent oxidation.

Rheology of PE Melt. The HDPE melt of molecular weight >10,000 is a non-Newtonian liquid. Its effective viscosity is reduced sharply, up to several thousand times, when the melt flow speed is increased. Such behavior is typical at temperatures of 140–200°C, whereas at ca 300°C, HDPE is a Newtonian liquid. The deviation of the melt flow from Newtonian behavior is usually presented graphically as an upward deviation of the flow curve, plotting log of shear stress vs log of shear rate. Non-Newtonian behavior of the HDPE melt at very low shear rates allows correlation of melt viscosity with molecular weight. The reason for this viscosity anomaly is disentanglement of polymer coils at high shear rate and their partial orientation in the flow direction.

Processing. Low melting point and high chemical stability facilitate the processing of HDPE by conventional techniques, ie, injection molding, blow molding, and extrusion (see also BLOW MOLDING; INJECTION MOLDING).

Economic Aspects

Anticipated and projected capacities of high density polyethylene are shown in Table 2. U.S. producers are shown in Table 3.

Table 2. World Capacity of HDPE, 10^6 t/yr

Region	1988	1990
Europe	3.99	4.28
United States	3.93	4.85
Far East and Australia	2.41	2.73
South America	0.47	0.47
Canada	0.43	0.46
Middle East and Africa	0.36	0.36
Total	*11.59*	*13.15*

Table 3. U.S. Producers of HDPE, 1990

Company	Capacity, 10^3 t
Quantum Chemical	938
Phillips	809
Occidental	704
Soltex	636
Allied	545
Exxon	324
Dow	257
Mobil	230
American Hoechst	164
Union Carbide	114
Chevron	110
Himont (Montedison)	21

Health and Safety Factors

Due to its inertness and absence of polyethylene toxicity, PE is employed on a large scale in food and drug packaging applications.

Recently, HDPE has been used in prosthetic devices, eg, hip or knee joint replacements. Such applications underscore polyethylene's safety.

Polyethylene does present a health hazard during combustion, pyrolysis, or thermocutting when smoke, fume, or toxic decomposition products are formed.

Uses

Blow-molding products are the single largest use of HDPE resins, with packaging applications accounting for the greatest volume. The blow-molding market represents ca 41% of total HDPE consumption.

Injection-molding products are the second largest application, with approximately 30% of the HDPE market. Specific uses include housewares, toys, food containers, pails, crates, and cases.

Films are a rapidly growing application for HDPE resins, replacing paper and glass. Films account for about 7% of the total HDPE market.

Pipe (qv) and conduit represent another important application, amounting to about 10% of total HDPE products. This use is anticipated to grow 6% per annum.

Low molecular weight HDPE with molecular weight of several thousand (waxes) is widely used for paper coatings, spray coatings, emulsions, printing inks, crayons, and wax polishes.

BIBLIOGRAPHY

H. V. Boenig, *Polyolefins: Structure and Properties,* Elsevier Publishing Co., Amsterdam, 1966.

R. S. Stain in B. Ke, ed., *Newer Methods of Polymer Characterization,* Wiley-Interscience, New York, 1964.

J. Boor, Jr., *Ziegler-Natta Catalysts and Polymerizations,* Academic Press, Inc., New York, 1979.

J. M. McKelvey, *Polymer Processing,* John Wiley & Sons, Inc., New York, 1962.

DAVID L. BEACH
Chevron Chemical Co.

YURY V. KISSIN
Mobil Chemical Co.

Ultrahigh Molecular Weight Polyethylene

Ultrahigh molecular weight polyethylene (UHMWPE) is a high density polyethylene with a molecular weight over 3×10^6. Chemically, high density polyethylene (HDPE) and UHMWPE are identical; both are straight-chain linear polymers. The molecular weight of conventional HDPE is rarely above 50,000 whereas ASTM definitions require a molecular weight of UHMWPE above 3×10^6; it can be as high as 6×10^6 (ASTM D 4020). Crystallinity of UHMWPE is typically about 45%.

The long molecular chain of UHMWPE confers the following unique combination of properties: abrasion resistance greater than that of any other thermoplastic; the highest impact toughness of any plastic, even at cryogenic temperatures; good corrosion resistance and excellent environmental stress-crack resistance; resistance to cyclical fatigue; noise- and shock-abatement properties; low coefficient of surface friction; nonstick surface; and resistance to radiation. In addition, UHMWPE has been approved by the FDA, USDA, and National Bureau of Standards sanctions for pure water and food handling.

The polymerization process for UHMWPE employs a Ziegler catalyst similar to that for conventional HDPE. The process can be batch or continuous. Most UHMWPE is produced by a slurry process, but both solution and bulk processes are applicable.

Formulation

Unlike other thermoplastics, UHMWPE can be thermally processed without additives or stabilizers because the long polar chains are resistant to scission, and even when broken can have a mol wt $>10^6$. However, fillers and modifiers are used. Graphite fibers, powdered metals, and glass fibers and beads improve stiffness, decrease deformation under load, and increase heat-deflection temperatures. Because of superior properties, UHMWPE can compete with rigid engineering plastics (qv). Except for graphite and glass fibers, most additives lower abrasion resistance and impact strength.

Processing

Because of its extremely high molecular weight, UHMW resin cannot be readily processed by thermoplastic melt-processing techniques such as injection molding (qv), blow molding (qv), thermoforming (qv), or screw extrusion (qv). Nevertheless, UHMW resin can be processed by compression molding into sheet, block, and precision parts; ram extrusion into board, rods, pipe, and profiles; and forging into complex parts such as gears, wheels, sprockets, and cams.

Applications

The principal applications of UHMWPE are in chemical processing, the food and beverage industries, foundries, the lumber industry, bulk materials handling, medical implants, mining, mineral processing, paper, recreational equipment, textiles, and transportation. In Italy and the United States, UHMWPE is used for highly efficient battery separators. A recent application is in very strong, light fibers made by gel spinning.

UHMWPE is produced in the United States by HIMONT U.S.A., Inc. (1900 UHMW) and American Hoechst (Hostalen Gur), in Japan by Ashai (Sunfine-U), and in the FRG. Worldwide capacity is approximately 27,000–34,000 t; worldwide consumption is about 22,650 t.

BIBLIOGRAPHY

Abrasion-Resistant 1900 UHMW Polymer, Properties and Fabrication, Bulletin No. 500-657, Hercules, Inc., Wilmington, Del., Mar. 1982.

UHMW Polymer Modification, Bulletin No. P-129, Hercules, Inc., Wilmington, Del., Mar. 1985.

J. J. Coughlan
D. P. Hug
HIMONT U.S.A., Incorporated

Modified Polyethylenes

CHLORINATED POLYETHYLENE

Of the many chemical reactions used to modify polyethylene, chlorination and chlorosulfonation are the most important commercial methods. The properties of chlorinated polyethylene (CPE) depend on the properties of the starting material and on the amount and distribution of the chlorine introduced.

Chemistry of Preparation

In the chlorination of polyethylene, a chlorine atom is substituted for a hydrogen atom on the polymer chain, accompanied by the formation of HCl:

$$-CH_2CH_2CH_2- + Cl_2 \longrightarrow -CH_2\underset{\displaystyle Cl}{\underset{|}{C}}HCH_2- + HCl$$

This occurs by a free-radical mechanism, usually catalyzed by ultraviolet light or initiators:

$$Cl_2 \xrightarrow{h\nu} 2\,Cl\cdot$$

Chlorination Processes

Chlorination processes include solution (solvent) processes and processes without solvent (eg, aqueous suspension and anhydrous processes).

Properties

The hardness (qv) of CPE varies widely. Commercial products are soft, with a chlorine content of 23–48%. The tensile properties of flexible CPE polymers are often compared to those of flexible PVC, but the stress–strain characteristics are different (see also Vinyl chloride polymers).

Effect of Chlorine Distribution. A CPE polymer exhibiting a very small heat of fusion can be produced with the addition of less than 25% Cl in hindered chlorination. Block chlorination may give a product of measurable heat of fusion with a chlorine content above 57%.

Effect of Parent Polyethylene. Higher molecular weight polyethylene yields a CPE with higher viscosity and tensile strength.

Compounding

Thermoplastic Compositions. Components include stabilizers, antioxidants, lubricants, plasticizers, and fillers and pigments.

Blends with Other Polymers. Little true dissolution takes place between CPE and other polymers. The miscibility with polycaprolactones, ethylene–acrylic rubbers, and a few other polymers appears to be strongly influenced by chlorine content and distribution. More typically, a separate phase forms when CPE is combined with other polymers. Nevertheless, such compositions often have useful properties.

Elastomers. A variety of mechanisms can be used to cross-link CPE elastomers. Peroxide or free-radical vulcanization (qv) is used in many commercial applications. The peroxide cure creates numerous carbon–carbon cross-links, resulting in superior compression set and heat, ozone, and chemical resistance.

Unlike the peroxide cure system, the dimercaptothidiazole system is compatible with the inexpensive aromatic oils and is not sensitive to antioxidants, offering significant economic advantages compared to peroxide cure. There is, however, some loss in compression set and heat resistance.

Elastomeric CPE cross-linked by electron-beam and gamma radiation also is in commercial use.

Processing and Fabrication

Thermoplastics. Usually, CPE is in the form of fine, soft particles, which can absorb atmospheric moisture, organic solvent vapors, and liquid. The tendency to stick becomes worse as liquids are absorbed; an inert anticaking additive is often present. The soft CPE powder also fuses easily. Amorphous polymers can usually be banded on a roll mill below 80°C. For good dispersion, temperatures as high as 200°C and above may be needed to blend CPE with a resinous material such as PVC or ABS. When CPE is the base polymer in a CPE–PVC blend, best dispersion is obtained when a low molecular weight PVC is used.

Processes for CPE include dry blends, Banbury mixing, extrusion, and molding.

Elastomers. The first important step in the fabrication and processing of CPE elastomers is good dispersion in the polymer matrix. Internal mixers of the Banbury type are commonly used; the solids are added first, followed by liquids and CPE. Cooling, load factor of 70%, and drop temperature of 95–102°C are the conditions that produce satisfactory CPE elastomers. The product is cooled on a two-roll mill and cut into strips or pelletized for calendering, extrusion, or molding.

Health and Safety Factors

Chlorinated polyethylene polymers have a very low toxicity; their primary hazard appears to be HCl evolution at high temperatures. CPE reportedly causes only minor eye and skin irritation.

Applications

The 1985 worldwide production capacity for CPE was estimated to be above 50,000 t per year.

Thermoplastics. The modification of rigid PVC is one of the largest single uses of CPE. CPE-based polymers are used in highly filled compounds, industrial sheeting, wire and cable, and solution applications.

Elastomers. Hydraulic, automotive, and industrial hoses made from CPE elastomers exhibit excellent resistance to heat, ozone, oil, and chemicals.

BIBLIOGRAPHY

W. L. Young and E. D. Serdensky, *paper presented at the Regional Technical Conference, Miami, Fla., 1966,* Society of Plastics Engineers, Greenwich, Conn., 1966, Val. Sect. 8.

Dow CPE Resins: Basics, Product Bulletin, Form No. 305-895-281, The Dow Chemical Co., Plaquemine, La., 1984.

R. R. Blanchard and C. N. Burnell, *SPE J.* **24**(1), 74 (1968).

R. R. Blanchard, *J. Elastomers Plast.* **6**(1), 3 (1974).

W. L. Young
R. R. Blanchard
The Dow Chemical Company

CHLOROSULFONATED POLYETHYLENE

Chlorosulfonated polyethylene (CSM) is a family of synthetic elastomers prepared by the simultaneous chlorination and sulfonation of polyethylene (PE) in homogeneous solution. This reaction can be represented by

$$-(CH_2)_n- + (b + c)Cl_2 + c\,SO_2 \xrightarrow{\text{initiator}} -(CH_2)_a(\underset{\displaystyle Cl}{\underset{|}{CH}})_b-(\underset{\displaystyle SO_2Cl}{\underset{|}{CH}})_c- + HCl$$

where a free-radical initiator or light is used. Because the reaction occurs in homogeneous solution, the three types of

Table 1. Commercial CSM Polymers

Hypalon[a]	Chlorine, %	Sulfur, %	Specific gravity	Mooney viscosity, ML_4[b]	Comment
45	24	1.0	1.08	37	
623	24	1.0	1.08	21	easy processing grade
20	29	1.2	1.14	28	branched
LD-999	35	1.0	1.18	30	
40S	35	1.0	1.18	45	
40	35	1.0	1.18	56	general-purpose elastomer
4085	36	1.0	1.20	97	
30	43	1.1	1.27	30	branched, low solution viscosity
48S	43	1.0	1.26	62	
48	43	1.0	1.26	78	high fluid resistance

[a] DuPont trademark.
[b] Large rotor, 4 min (ASTM D 1646).

groups in the product are distributed nearly randomly along the polymer chain. In a typical elastomeric product containing 36% Cl and 1% S, they are present in the ratio $a:b:c = 108:32:1$.

The introduction of chlorine and sulfur dioxide onto the PE molecule breaks up the crystalline segments, changing a thermoplastic material into an amorphous polymer that behaves as an elastomer and can be thermoset by way of the sulfonyl chloride groups. Incorporation of chlorine also increases oil resistance. Commercial chlorosulfonated polyethylene (Hypalon) offers the advantages of superior resistance to oil, ozone, and heat compared with standard synthetic rubbers. The optimum chlorine content for an amorphous elastomer is the minimum amount required to destroy crystallinity completely. Lower chlorine concentrations give a product with a higher glass-transition temperature, which therefore is less rubbery but offers increased resistance to oils and nonpolar solvents. Chlorinated PE prepared in solution is very similar to CSM except that the lack of a cure site makes it more difficult to cure and allows less compounding latitude.

Commercial Synthetic Elastomers

Commercial chlorosulfonated polyethylenes are shown in Table 1.

BIBLIOGRAPHY

U.S. Pat. 2,212,786 (Aug. 27, 1940), D. M. McQueen (to E.I. du Pont de Nemours & Co., Inc.).

S. C. Ostrowski, *Hypalon R Bulletin HP–210.1(R2),* E.I. du Pont de Nemours & Co., Inc., Wilmington, Del., Apr. 1984.

I. C. Dupuis, *Hypalon R Bulletin HP–110.1,* E.I. du Pont de Nemours & Co., Inc., Wilmington, Del., Sept. 1984.

GERALD D. ANDREWS
ROBERT L. DAWSON
E.I. du Pont de Nemours & Co., Inc.

ETHYLENE–PROPYLENE COPOLYMERS. See

ETHYLENE–PROPYLENE ELASTOMERS.

ETHYLENE–PROPYLENE–DIENE ELASTOMERS. See

ETHYLENE–PROPYLENE ELASTOMERS.

ETHYLENE–PROPYLENE ELASTOMERS

Ethylene–propylene copolymers and ethylene–propylene terpolymers are specialty elastomers. An elastomer is a polymeric material that rapidly recovers its shape after removal of a strain of at least 50% and whose entropically derived equilibrium modulus increases with temperature. It is usually a synthetic polymer; the term rubber usually applies to the natural product. These specialty elastomers possess some particular qualities suitable for certain applications. EPM and EPDM, hereafter together called EPDM, have two particular characteristics that make them unique: chemical resistance, especially to ozone, and the ability to tolerate high concentrations of filler and oil with retention of good physical properties. Thus rubber compounding is very cost effective. Compounds for long-lasting automobile radiator hose are formulated with only ca 10% elastomer, yet can be extended several hundred percent and possess significant tensile strength (>10 MPa). Roof sheeting can be prepared with a service life of over 30 yr.

Because of such useful properties, EPDM production capacity and consumption have steadily increased since its introduction in the early 1960s, following the discoveries of Ziegler and Natta, despite the fact that global synthetic rubber production has decreased since the oil crises of the early 1970s. Currently, EPDM is third or fourth of the synthetic elastomers in total tonnage produced, substantially behind only the general-purpose tire elastomers, styrene–butadiene copolymers, and polybutadiene. In 1984, world capacity was ca 500×10^3 t.

At present, 11 manufacturers in the noncommunist trade area produce over 150 EPDM grades. Manufacturers and their trade names are shown in Table 1. Automotive applications are the single largest use for these elastomers, although industrial construction, plastics alloy, and electrical and other uses account for about half of the market.

The basic building blocks are ethylene and propylene, which may or may not be combined with a third, or even fourth, monomer to provide olefinic sites along the backbone.

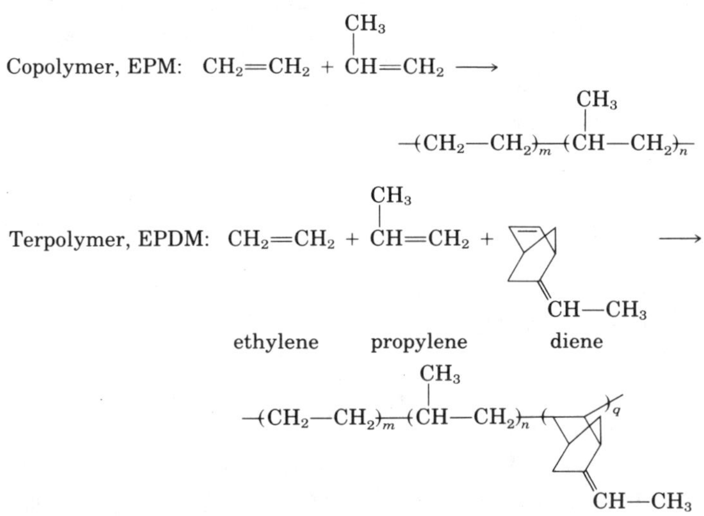

Olefinic sites improve cross-linking response with peroxides and permit direct sulfur cure.

Properties of Monomers

Ethylene and propylene, the two principal components of EPDM, are normally used as polymerization grades obtained from petrochemical refineries, and therefore most elastomer plants are located near refineries. In certain cases, a significant percentage, eg, 10%, of propane may be present in the propylene. Both are gases at ambient temperature and pressure; propylene is easily liquefied under pressure. Monomer properties are given in Table 2.

Properties of the Polymer EPDM

Except for a small mole fraction of olefin, EPDM is a saturated hydrocarbon with a random-coil configuration. The mechanical properties depend on the glass-transition temperature, which serves as the reference point for Williams-Landel-Ferry behavior. Properties also depend on crystallinity. The degree of crystallinity and the T_g depend on average composition and compositional and sequence distribution. Molecular weight and

Table 1. Ethylene–Propylene Elastomer Manufacturers

Company	Trade name	Plant location[a]
Bunawerke Huls	Buna	Marl, FRG
Copolymer	Epsyn	Addis, La.
DuPont	Nordel	Beaumont, Texas
Dutch State Mines	Keltan	Beek, Holland
Exxon	Vistalon	Baton Rouge, La. Notre Dame de Gravenchon, France
Japan Synthetic Rubber	JSR	Yokkaichi, Japan
Mitsui	Mitsui	Ichiham, Japan
Montedison	Dutral	Ferrara, Italy[b]
Polysar	Epcar	Orange, Texas[b]
Sumitomo	Esprene	Anegasaki, Japan
Uniroyal	Royalene	Geismar, La.
USSR	SKEP or SKEPT	USSR

[a] Solution process unless otherwise indicated.
[b] Slurry process.

its distribution and long-chain branching influence rheological and processing characteristics at elevated temperatures. Some properties are given in Table 3.

Composition and Sequence Distribution. The ratio of ethylene to propylene in the polymer is important for a number of reasons. Glass-transition temperature and degree of crystallinity, as well as the oxidative stability and propensity for chain scission during free-radical cross-linking, vary with ethylene content. Oxidative stability and resistance to scission increase as the ethylene content increases and the number of methine hydrogens decreases.

The composition may vary from molecule to molecule, or along a given molecule. In inter- and intramolecularly homogeneous polymers, the arrangement of monomer units along the chain can vary at a given average composition. For EPDM prepared in continuous backmixed flow reactors, at constant monomer concentrations in the reacting medium, the compositional heterogeneity and sequence distribution are determined by the catalyst, solvent, and temperature.

Glass Transition. The glass-transition temperature of EPDM is a function of the ethylene and diene content. In most commercial grades, it is between −55 and −60°C when measured at a slow heating rate and low frequency. However, at high diene or low ethylene, content, it can be 10–20°C higher.

Crystallinity. Although most commercial EPDM is amorphous at 20°C, several grades may attain crystallinity up to 30%. All grades that contain above 40 mol % ethylene develop crystallinity below ambient temperature if annealed for several hours.

Solution Properties. The solution properties for EPDM are important because it is manufactured in solution, polymer characterization relies heavily on solution properties, and certain applications involve solutions. The solubility parameter of EPDM is ca 16 $J^{1/2}/cm^{3/2}$ (8 $cal^{1/2}/cm^{3/2}$). At atmospheric pressure and temperature, it is soluble in aliphatic hydrocarbons of C_5 and above, low polarity halogenated and unhalogenated aromatic and naphthenic hydrocarbons, and tetrahydrofuran. It is not soluble in aldehydes, ketones, carboxylic acids, esters, or dioxane. Solubility is a function of molecular weight and ethylene content, the latter as related to crystallinity. All but the lowest ethylene samples exhibit association phenomena in dilute solution to 80°C. Filtration and characterization operations should be carried out above that temperature unless association effects are desired.

Molecular Weight and Distribution. After a specific reactor and residence time are chosen, the characteristics of the polymer are primarily determined by the catalyst. Catalyst selection contributes to the wide variety of specialty EPDM products. Details of MWD are best determined by gpc, which should be performed under conditions appropriate for PE.

Branching. Long-chain branching adds further variety to these polymers. Even in copolymers, end-group copolymerization can lead to trifunctional branch points. Branching is more important in ter- and higher polymerizations where diene is present.

The characterization of branching is linked to the MWD of the polymer.

Gel Formation. Certain commercial polymers contain gel

Table 2. Properties of Monomers

Property	Ethylene	Propylene	Ethylidene–norbornene	Dicyclopentadiene	1,4-Hexadiene
molecular weight	28.0	42.1	120.2	132.2	82.1
mp, °C	−169	−185	−80	32[a]	
bp at 101.3 kPa[b], °C	−104	−49	147.6	170	65
specific gravity			0.895^{20}	$0.97^{15.5}$	0.7^{20}
viscosity, liquid, mPa·s (= cP)[c]	0.07^{-18}		1.1^{20}	1.5^{38}	
heat of vaporization, kJ/kg[c,d]	320^{-29}	410^{-29}	300^{bp}		

[a] Pure endo.
[b] To convert kPa to atm, multiply by 0.01.
[c] Superscripts indicate temperature, °C.
[d] To convert J to cal, divide by 4.184.

Table 3. Physical Constants of Ethylene–Propylene Elastomers[a]

Constant	Value
density, g/cm^3	0.854
expansivity, per °C	
$(1/V)(dV/dT)_P$	8.8×10^{-4}[b]
	7.5×10^{-4}
$\beta = (1/L)(dL/dT)_P$	2.5×10^{-4}
specific heat C_p, J/(g·K)[c]	2.22
thermal conductivity, kW/(m·K)	1757
refractive index n_D at 23°C	1.4740
dielectric constant at 10^3 Hz	2.5
dissipation factor at 10^3 Hz	0.2
compressibility $\beta = (1/V)(dV/dP)_T$, per 101 kPa[d]	5.4×10^{-5}
diffusion constant $D_{(0)}$, cm^2/s, at 23°C	
benzene	6.1×10^{-8}
n-hexane	4.1×10^{-8}
CH_2Cl_2	1.3×10^{-7}
water-permeability coefficient at 37.5°C, $\frac{\text{mol}}{\text{m·s·TPa}}$[e]	0.15

[a] 0.4–0.6 mol fraction ethylene; amorphous samples at ca 23°C unless otherwise stated.
[b] Estimated from PE.
[c] To convert J to cal, divide by 4.184.
[d] To convert kPa to atm, divide by 101.
[e] To convert $\frac{\text{mol}}{\text{m·s·TPa}}$ to $\frac{\text{cc·cm}}{\text{cm}^2\text{·s·atm}}$, divide by 4.4×10^4.

which arises in the polymerization medium or in recovery operations. Gels are often highly cross-linked particles that do not disperse well in compounding operations. They can be found by extruding the elastomer through a screen pack and subsequent microscopic examination. Gel analysis is difficult and gel formation should be avoided.

Rheological Properties. EPDM has high plateau modulus, about 1.5 MPa, and correspondingly low (3800) critical entanglement molecular weight in its viscosity–molecular weight relationship. The high modulus and low propensity for crystallization under stress give EPDM low tack. A wide variety of MWDs, long-chain branching, and in some cases crystallinity lead to correspondingly varied rheological responses. The melts are very non-Newtonian, and at constant shear stress the activation energy of flow is about 9 kcal/mol for linear EPDM.

Vulcanizate Properties. Due to its high plateau modulus EPDM tolerates high loadings of filler and oil. It generally behaves as a non-crystallizing elastomer with properties more like SBR than NR. A comparison with other specialty elastomers is given in Table 4.

Properties such as aging, compression set, and cure reversion largely depend on the cross-linking agent. For the best heat resistance, peroxides or radiation curing are used, which produce carbon–carbon bonds. For good aging properties and network stability, the stronger monosulfide links are preferred. For high strength applications, polysulfide links are preferred.

If the ethylene content of the polymer is high enough to produce crystallinity at rest or upon stretching, the mechanical properties are enhanced. Crystallinity provides hardness, modulus, and strength.

Table 5. Catalysts for EPDM Polymerization

Group I–III metal	Transition metal
$(C_2H_5)_2AlCl$	VCl_4
$(i\text{-}C_4H_9)_2AlCl$	$(CH_3COCH_2COCH_3)_3V$
$(C_2H_5)_3Al$	$VOCl_3$
$(C_2H_5)_2Zn$	$TiCl_4$[a]
$(C_2H_5)_2Mg$	$TiCl_3$[a]

[a] On various supports, eg, $MgCl_2$.

Polymerization and Processing

Polymerization Catalysts. Only Ziegler-Natta catalysts are used for EPDM polymerization.

Components include a Group I–III organometallic compound and a Group IV–VII transition-metal compound. Combinations that show activity are countless; some of the more commonly used components are shown in Table 5.

Manufacturing Processes. Commercial production is carried out in plants of capacities of ca 50,000 t/yr. Polymerization proceeds by solution or slurry processes. Both are operated con-

Table 4. Properties of Vulcanized Elastomers

Property	EPDM	Butyl	Chloroprene
density, gum, g/cm^3	0.86	0.92	1.23
hardness, Shore A	35–90	30–95	40–90
tensile strength, reinforced, MPa[a]	20	22	28
tear strength, kN/m[b]	15–45	25–35	35–45
upper temperature limit, °C	150–175	120–175	100–150
brittleness point, °C	−60	−60	−40
dielectric constant	3.0	3.5	6.7
resistance[c] to			
weather	E	G	G
ozone	E	G	G
acids and alkalies	E	G	E
oils and solvents	F	P	G
steam	E	G	G
gas permeation	P	E	G
ASTM D 2000	BA–DA,BC	BA,CA	BC,BE

[a] To convert MPa to psi, multiply by 145.
[b] To convert kN/m to ppi, divide by 0.175.
[c] Key: E = excellent; G = good; F = fair; P = poor.

tinuously in liquid-phase back-mixed reactors. Most manufacturers use a solution process in hexane; only Polysar and Montedison use propylene slurry.

Health and Safety Factors

Ethylene–propylene elastomers are not toxic and are flammable only after heating to thermal-decomposition temperatures, eg, ≥300°C. Certain grades have received FDA approval for contact with the skin and in pharmaceutical applications. The most likely contaminants are residual monomer, polymerization diluent, catalyst residues, slurry, antifoam or deashing aids, scale from vessels, trash, and antioxidant; none of these are highly toxic.

Threshold limit values (TLV) for airborne concentrations in the work environment, as OSHA requires, are defined by ACGIH.

BIBLIOGRAPHY

L. Corbelli in *Developments in Rubber Technology and Synthetic Rubbers,* Vol. 2, Applied Science Publishers, New York, 1981, Chapt. 4.

F. P. Baldwin and G. Ver Strate, *Rubber Chem. Technol.* **45,** 709 (1972).

S. Cesca, *Macromol. Rev.* **10,** 1 (1975).

GARY VER STRATE
Exxon Chemical Company

ETHYLENE–SULFIDE DIOXIDE COPOLYMERS.

See OLEFIN–SULFUR DIOXIDE COPOLYMERS.

ETHYLENE–SULFIDE POLYMERS.

See POLYSULFIDE POLYMERS.

ETHYLENESULFONIC ACID POLYMERS

Ethylenesulfonic (or ethenesulfonic) acid is one of the few readily available monomers that can be polymerized to a strongly acid polyelectrolyte (see POLYELECTROLYTES). The sodium salt is manufactured in the United States by Air Products & Chemicals, Inc., and by Alcolac, Inc., and probably in the FRG by Farbwerke Hoechst AG. The monomers are classified as low volume specialty products.

Sodium ethylenesulfonate is sold as a weakly alkaline, 25% aqueous solution, either unstabilized or containing a small amount of hydroquinone monomethyl ether. It contains 1–3% of inorganic salts and has a specific gravity (20°C/4°C) of about 1.2.

Polymerization

Ethylenesulfonic acid and its derivatives have been polymerized and copolymerized to linear polymers by various free-radical catalysts.

$$n\ CH_2{=}CHSO_3Na \xrightarrow{\text{initiator}} -\!\!\left(CH_2\underset{\displaystyle SO_3Na}{\underset{|}{CH}}\right)_{\!n}\!\!-$$

The solvent depends on the particular derivative; the free acid and its salts are usually polymerized in aqueous solution, whereas other derivatives are polymerized in bulk, solution, or emulsion.

Properties of Polymers

The alkali metal and ammonium salts of poly(ethylenesulfonic acid) are water-soluble, glassy, brittle solids; the alkaline earth and heavy metal salts are insoluble in water. The alkali metal salts do not lose water of hydration completely even after prolonged heating above 100°C in vacuum, and they absorb moisture on exposure to the atmosphere. Wet polymer dried in contact with glass adheres to it tenaciously.

Uses

Although there are numerous references to the use of ethylenesulfonate homopolymers and copolymers, few appear to have been commercialized. Most important is the application as a dyeing site in acrylic fibers (qv); it shares this market with styrenesulfonate and allyl- or methallylsulfonate.

BIBLIOGRAPHY

D. S. Breslow and R. R. Hough, *J. Am. Chem. Soc.* **79,** 5000 (1957).

H. Distler, *Angew. Chem.* **77,** 291 (1965); *Angew. Chem. Int. Ed. Engl.* **4,** 300 (1965).

W. Kern and R. C. Schulz in E. Muller, ed., *Houben–Weyl, Methoden der Organischen Chemie,* 4th ed., Vol. 14, Georg Thieme Verlag, Stuttgart, FRG, 1961, Pt. 1, pp. 1097–1106.

D. S. Breslow and A. Kutner, *J. Polym. Sci.* **27,** 295 (1958); **38,** 274 (1959).

DAVID S. BRESLOW
D. S. Breslow Associates

ETHYLENEUREA.

See AMINO RESINS.

ETHYLENE–VINYL ACETATE COPOLYMERS.

See ETHYLENE POLYMERS.

ETHYL FUMARATE.

See MALEIC AND FUMARIC POLYMERS.

2-ETHYLHEXYL ACRYLATE.

See ACRYLIC AND METHACRYLIC ESTER POLYMERS.

ETHYLHYDROXYETHYLCELLULOSE.

See CELLULOSE ETHERS.

ETHYL MALEATE.

See MALEIC AND FUMARIC POLYMERS.

ETHYL METHACRYLATE.

See ACRYLIC AND METHACRYLIC ESTER POLYMERS.

ETHYNYLATION.

See ACETYLENIC POLYMERS.

ETHYNYL-TERMINATED OLIGOMERS.

See ACETYLENE-TERMINATED PREPOLYMERS.

EXPANDED MATERIALS.

See BLOWING AGENTS; CELLULAR MATERIALS.

EXTENDERS.

See FILLERS.

EXTRUSION

In typical plastics extrusion processing, a viscous melt under pressure is forced through a shaping die in a continuous stream. The feedstock may enter the extrusion device in the molten state, but more commonly it consists of solid particles that must be subjected in the extruder to melting, mixing, and pressurization. The solid feed may be in the form of pellets, powder, beads, flake, or reground material. It may also be a combination of any of these with or without liquid additives, all of which may be premixed or fed separately at one or more feed ports. Before the existence of the thermoplastics industry, extrusion devices were applied to rubber, clay, and ground meat. Today extrusion has been extended to other food processing, machines designed to injection-mold plastics, and blow-mold plastic bottles. Extruders of some type are required to process almost all thermoplastics. Many plastics are first extruded in polymer manufacturing plants and in compounding operations before reaching the final extrusion device used to make fabricated products.

Most plastics extruders incorporate a single screw rotating in a horizontal cylindrical barrel with an entry port mounted over one end and a shaping die mounted at the discharge end. The largest capacity single-screw extruder known, with a 600-mm-dia screw, is designed to extrude 29 t/h; the smallest models, for torque rheometers, have a capacity of 5 kg/h. Very small extruders are used in laboratories and commercially for special applications such as extruding a thin narrow red stripe for decoration on polypropylene soda straws. Most U.S. single-screw extruders are sized in diameter by inches (in the range of 5.08–25.4 cm dia).

In general, twin-screw extruders and other multiple-rotor devices are more expensive than conventional single-screw extruders. Their use is confined primarily to tasks that cannot be performed easily. For both single-screw and twin-screw extruders the screw design profoundly affects the performance capabilities of the extruder. A typical simple screw design for a solid-fed single screw extruder must convey the plastic entering the screw into a heated compacted environment, where the shear force developed by the rotating screw melts the plastic and mixes it to a reasonably uniform temperature while pressurizing the melt and pumping it uniformly through the die. Because of highly sophisticated advances in screw design and control features, the single-screw extruder is becoming increasingly versatile. Nevertheless, new developments in combining plastics and incorporating various additives are placing more demands than ever on twin-screw extruders, thus assuring their expanded use. Both single-screw and twin-screw extruders can be designed with one or more vents to remove volatiles and with additional feed ports downstream of the first feed port. Sometimes extruders are arranged in tandem or as coextruders to produce different layers of molten polymers through the same die. At the extruder exits, melt filtration may be provided, usually by screen packs. Occasionally a static mixer is placed in an adapter tube leading to the die, and sometimes a gear pump may precede the die to assure a more consistent output rate. The possibilities for die design are almost unlimited and cover many applications, such as compounding (coloring and blending of melts, additives, and fillers), cross-head extrusion (wire, garden hose), sheet, film, coatings, pipe, rod, profiles, and coextruded products.

Single-screw Processing

Power and Energy Requirements. The theoretical energy requirement for heating and melting the polymer and pumping it through the die is given by the following equation:

$$P_T = \dot{m} \int_{T_1}^{T_2} C_p \, dT + \dot{m} \, v \, \Delta P$$

where P_T = total net energy input to polymer, $\dot{m}$ = mass flow rate of polymer; T_1,T_2 = polymer feed and exit temperature, respectively; C_p = specific heat capacity of polymer; v = specific volume of the polymer melt; and ΔP = screw discharge pressure.

Commercial Single-screw Extruders

Single-screw extruders are designed for versatility. The principal functional element is the helical screw rotating in a close-fitting cylindrical barrel. The d-c drive motor is coupled to a double-reduction gear box to reduce the rpm while increasing torque sufficiently to operate the screw under normal processing conditions. The screw requires a heavy-duty thrust bearing to absorb the backward thrust as the screw develops the pressure required to force the polymer melt through the die. Although the extruder shown in Figure 1 is equipped with vacuum venting and a water cooling system in combination with heating elements for barrel temperature control, many extruders are sold without vents and with air cooling.

Screw Design

Screw design (qv) is at the heart of the extrusion process, which suffers if the screw design is not well suited to the application. Before 1967, most screws were of conventional single-stage and two-stage design, and theoretical attention was directed to optimizing the length and depth of conventional feed, transition, and metering sections in relation to processing requirements. Some screws were built with different pitches and multiple flights, and occasionally mixing pins, ring barriers, undercut spiral barriers, and Dulmage sections (many parallel interrupted mixing flights), plus a few other novelties were employed. In a notable study on mixing screw design, a highly sensitive oscillograph recorder traced temperature and pressure. It was demonstrated that incorporation of a Union Carbide mixing section in a conventional screw extruding polyethylene at good output rates could reduce temperature and pressure fluctuations entering the die to about ±0.25%. Other mixing sections were developed, and since that time, high performance screws have been used more and more.

Dies and Auxiliary Equipment

Die Design. The extrudate entering the die is typically a polymer melt that can be characterized rheologically. Conse-

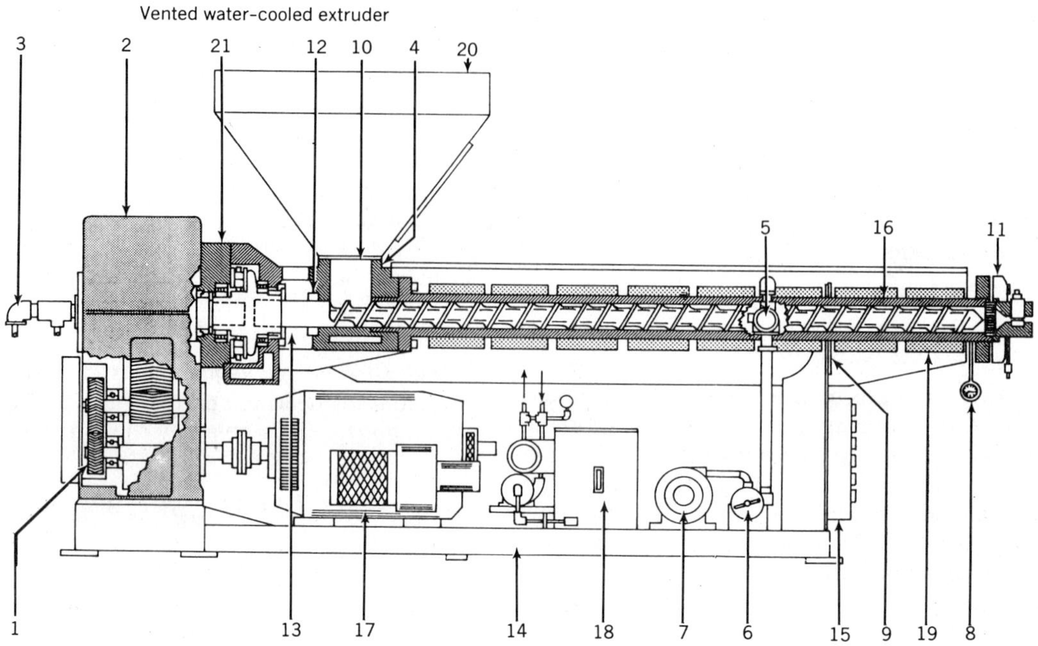

Figure 1. Vented, water-cooled extruder. 1, Change gears; 2, gear reducer; 3, rotary union; 4, feed throat; 5, side vent; 6, trap; 7, vacuum pump; 8, pressure measurement; 9, front barrel support; 10, large rectangular feed opening; 11, hinged swing gate; 12, rear seal; 13, large air spacer; 14, extruder base; 15, electrical box; 16, heavy wall cylinder; 17, motor; 18, barrel water cooling; 19, barrel heating and cooling; 20, hopper; and 21, thrust bearing assembly. Courtesy of HPM Corporation.

quently classic theoretical equations for flow through various sections of the die and adapter aid die design (qv) and indicate die performance at different throughput rates and with different polymer melts. Usually, a good blown PE film die can be used for a wide range of thicknesses. Likewise, a good high impact polystyrene sheet die can be used for a wide range of thicknesses. Nevertheless, fundamental calculations are invaluable in achieving the best flow uniformity at low enough extruder head pressure to avoid undesirably high melt temperature. For coextruded products, control of flow through the die can be especially important to preserve the uniformity of layer thickness.

The Poiseuille equation for Newtonian flow gives the following relationship between flow and pressure through the die.

$$P = K\mu Q$$

where Q = volumetric flow rate, μ = melt viscosity, and K = die resistance.

Downstream Auxiliary Equipment. For quality products, such as synthetic fibers and transparent sheet and film, polymer melt filtration is very important. Simple melt-filtration procedures employ coarse mesh screen packs mounted in front of a breaker plate at the end of the extruder barrel to separate metal scraps, dirt, or other solids. Contaminants are more likely to be present if reprocessed scrap resin is incorporated into the feedstock.

Fine screen packs are sometimes used that remove solid particles to below 50 μm, but supports of a coarser backup screen are required to prevent breakage. Several screen layers in combination are called a screen pack; these are employed in a fixed position ahead of a breaker plate at the end of the barrel or in a separate screen-changer apparatus.

A static mixer inserted into an elongated die adapter may improve extrudate uniformity, promote the mixing of additives, and equalize the melt temperature.

Coextrusion

Coextrusion of multiple layers of different materials through the same die lips is employed for plastic products such as sheet, film, coatings, profiles, pipe, and foam-core products. In a multimanifold die, the polymer melt streams are confined to individual flow channels until they are joined near the die-lip exit, shortly before or directly at the primary die land. In the feedblock method, the melt streams are combined and arranged to flow together in layers moving in laminar flow as they enter the die body. The adapter combining the melt streams ahead of the die is known as the feedblock. The layers must spread laterally while thinning down as they move through the die because the profile of the die lips is much more extended than that of the feedblock, which has an internal diameter or width of only a few centimeters. The rheological properties of the layers must be reasonably well matched, and feedblock and die must be carefully designed to preserve layer thicknesses as the melt moves downstream.

Multimanifold dies are used commercially for blown-film applications, eg, for five-layer barrier film structures. For flat sheet, the feedblock method is preferred even for seven-layer impermeable barrier structures. Some flat sheet, flat film, and extrusion coatings are made with multiple manifold dies, but usually only two- and three-layer structures. Both methods are applied to pipe extrusion.

The most promising new application of coextrusion with excellent growth potential is in barrier structures for food and beverage packaging.

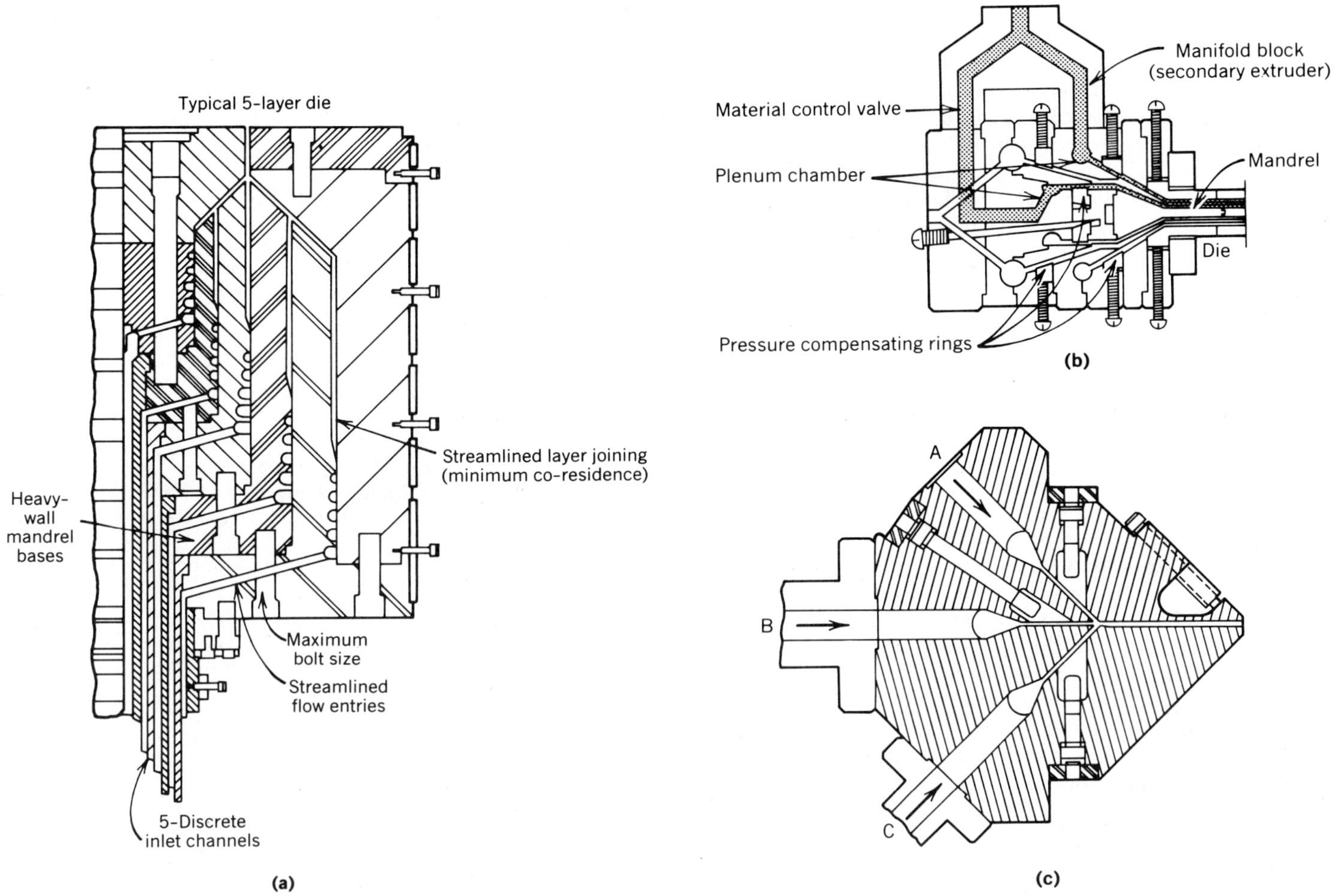

Figure 2. Multiple manifold dies. (**a**) For film; (**b**) for pipe. Courtesy of *Modern Plastics*. (**c**) For sheet. Courtesy of *Plastics Technology*.

Currently, coextrusion is also used for two- and three-layer structures for encapsulating scrap, coloration of outer layers, uv-resistant layers on siding, foam cores, and layers to promote heat sealability and grease resistance.

Examples of multimanifold dies for blown film, flat sheet or film, and pipe are shown in Figure 2.

An example of a feedblock for sheet production is shown in Figure 3.

Commercial Extruder Operation

Evaluation of Performance. Frequently an increase in output is required without loss in quality, or quality must be improved

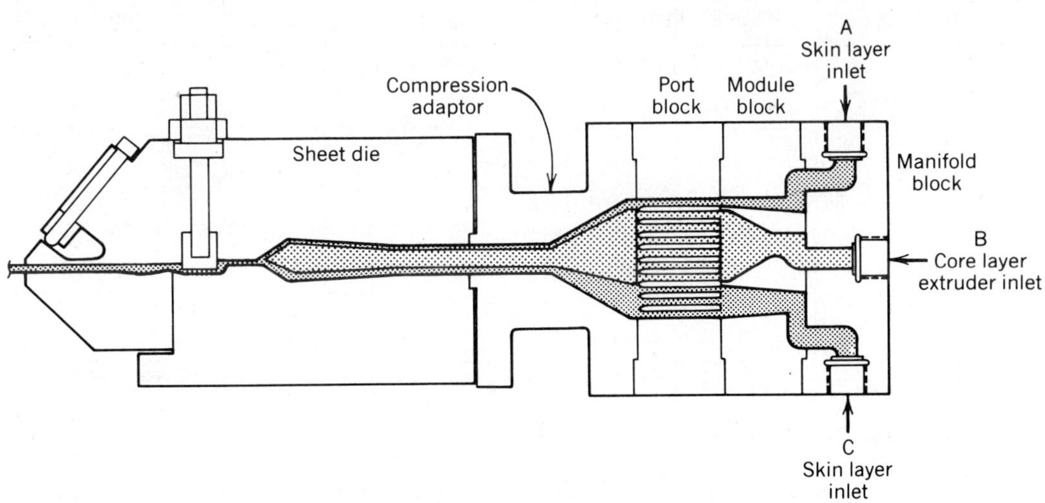

Figure 3. Modular sheet coextrusion tooling assembly. Courtesy of HPM Corporation.

without loss of output. In such cases, the extrusion process may deteriorate, and this must be corrected. Sometimes, a new product or formulation must be adapted to an available extruder. Typical extrusion problems are given in Table 1.

Operating Practices and Safety. In designing modern extruders, safety features are incorporated to protect the operator from electrical, mechanical, and thermal hazards. Signs are attached at strategic locations to warn against potential dangers. Where extruders are properly maintained and operated by trained safety-conscious personnel, extrusion plants have an excellent safety record. Specific safety instructions should be tailored to each production situation by qualified engineers.

Commercial Applications. Extruded products include blown and cast film, sheet, pipe and tubing, rod and profiles, coated wire and cable, coated paper products, and monofilaments and fiber.

Twin-screw Extruders

Although twin-screw extruders are used less than single-screw extruders, they are widely employed for difficult compounding applications, devolatilization, and other special operations, and for extruding finished products from viscous polymers with limited heat stability. Of the four basic types, only three are available commercially, including corotating and counterrotating intermeshing twin screws; nonintermeshing twin screws are offered only with counterrotation. A distinction should also be made between fully intermeshing and partially intermeshing systems and between open- and closed-chamber types, as shown in Figure 4.

Table 1. Typical Extrusion Problems

Nonuniform feedstock
variations within lot, between lots, and between sources
variation in percentage and nature of regrind components
variations in moisture content, additives, premixing, predrying, etc
Solids conveying
bridging in hopper causing erratic feed
density variations or low density feed
poor filling of feed flights for powder feed at higher screw speeds
poor frictional characteristic in feed section
poor design of feeding system or screw feed section
Melting and metering
poor screw design for the application
poor barrel temperature profile
screw rpm exceeding extruder melting capacity
worn screw or barrel
fouled cooling system, burned out heaters
inadequate extruder size, drive hp, or rpm range
inadequate or poorly maintained machine controls
inadequate or excessive pressure drop through screen pack and die
vent bleed
Melt quality
melt pressure and temperature fluctuations
output surge product dimensional fluctuations
incomplete melting, excessive melt temperature
poor distributive or dispersive mixing
contaminants, degradation, plugged screen, etc
poor die design, adjustment, or temperature settings

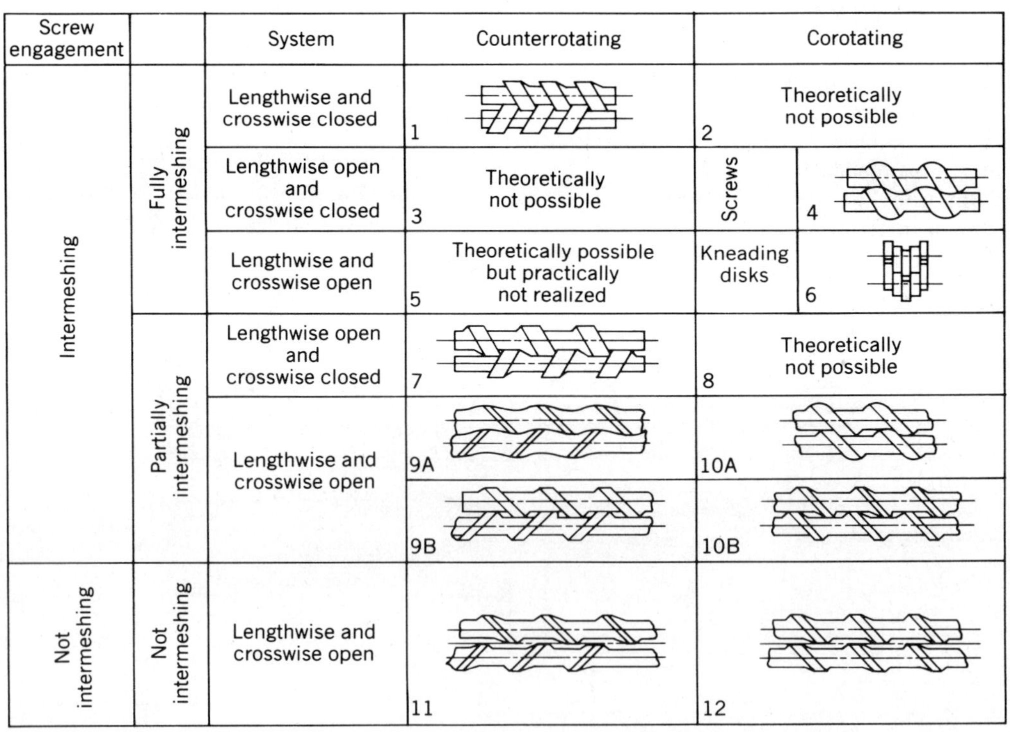

Figure 4. Twin screw design concepts. Courtesy of Delia Associates.

Other Commercial Extrusion Devices

Other single-rotor and multirotor devices are also used in the extrusion processing of polymers, particularly for compounding, although single-screw tandem-extruders have been developed that give far more mixing than conventional screws.

BIBLIOGRAPHY

Z. Tadmor and I. Klein, *Principles of Plasticating Extrusion,* Van Nostrand Reinhold, New York, 1970.

E. C. Bernhardt, *Processing of Thermoplastic Materials,* Reinhold Publishing Corp., New York, 1959.

K. Eise, H. Werner, H. Herrmann, and U. Burkhardt, *Advances in Polymer Technology* **1** (Apr. 1981).

G. Schenkel, *Plastics Extrusion and Technology Theory,* American Elsevier Publishing Co., New York.

GEORGE A. KRUDER
Consultant

F

FABRICS.

See TEXTILE RESINS.

FABRICS, COATED

Desirable properties of a fabric can be supplemented by coating it with a polymer. The fabric provides tensile, tear, and elongation control. The coating reduces porosity, protects the fabric from the chemical and physical environment, and, in some cases, provides decorative effects. In 1983, the U.S. sales of coated, filled, bonded, and laminated fabrics totaled $1.4 billion.

Textile Materials

Coated fabrics are designed for specific applications. Prime considerations are the chemical environment (water, acid or base, solvents), the physical environment (temperature, stress, flexing), and cost.

Fiber. Fibers include cotton, nylon, polyester, polypropylene (PP), rayon, glass, and high strength fibers.

Yarn. Yarn is produced as mono- or multifilament (synthetic) or staple (natural or chopped filament).

Fabric Construction. The properties of the substrate are determined by the fiber properties and fabric construction. In 1981, 70% of the fabric coated or filled in the United States was woven, 20% was nonwoven, and 10% knitted.

Finishing. In woven and knitted fabrics, the yarns almost always need to be sized or lubricated. Before coating, the sizes and lubricants must be removed.

Coating Materials

The polymer portion of the coating material provides many of the required properties. Considerations in selecting the polymer include chemical environment, mechanical requirements, costs, and processing characteristics. Coating materials include poly(vinyl chloride) (PVC) (used most for the coating of fabrics), polyurethane, and elastomers.

Processing

Processing is governed by the properties of the substrate and the coating. The viscosity of the latter must permit flow around the yarn or fiber surface. In calendering or extrusion, the compound is fluidized by pressure and heat. In casting, viscosity is reduced by solution or dispersion (Fig. 1). After coating, the fabric should be flat and smooth.

Impregnation. Impregnation is used in coated fabrics to penetrate the interstices of the textile. Binders, adhesion primers, and flame retardants are incorporated in the fabric by this process.

Decorative Coated Fabrics. Wall covering, upholstery, and apparel are examples of decorative coated fabrics. Inks are applied with one or more gravure printers to correct color or add patterns. Relief patterns are obtained by applying heat and pressure with embossing rolls. The outermost layer has a slip finish, a hard, thin layer which confers a final gloss adjustment, imparts the right feel to the product, and protects the ink coat from abrasion.

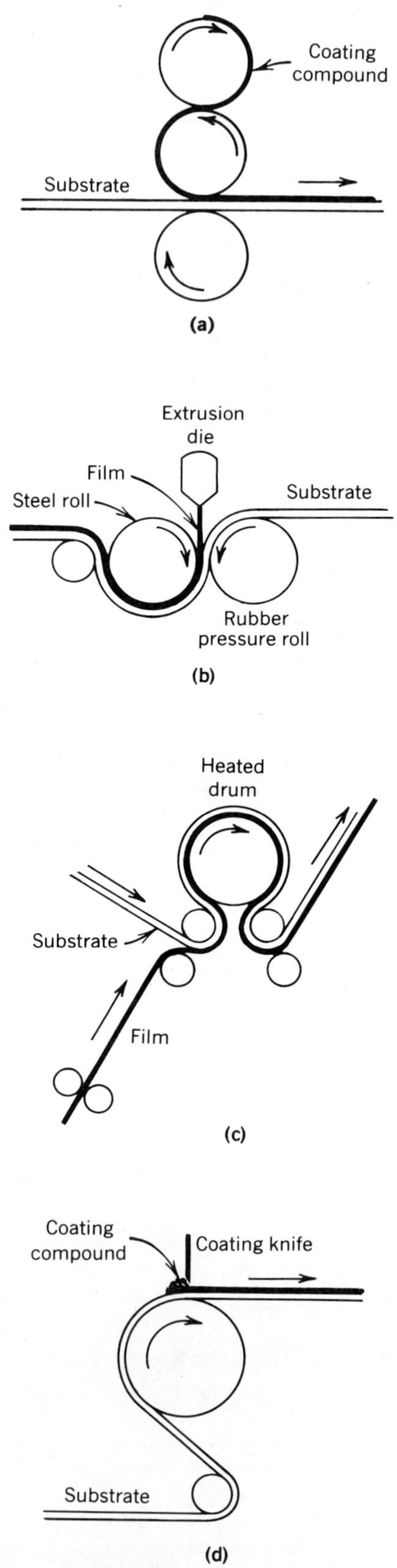

Figure 1. (**a**) Simple calender coating; (**b**) extrusion coating; (**c**) lamination coating; (**d**) casting: knife-over-roll coating.

Industrial Coated Fabrics. Although appearance is not as important in industrial applications as it is for consumer products, the same aftertreatment techniques are used.

Applications

About 75% of coated fabrics are used in applications that would be considered decorative.

BIBLIOGRAPHY

Summary of Trade and Tariff Information; Coated, Filled, Bonded, and Laminated Fabrics, USITC Publication 841, U.S. International Commission, Washington, D.C., Oct. 1982; June 1985 Supplement.

1983 Annual Book of ASTM Standards, The American Society for Testing and Materials, Philadelphia, Pa., 1983.

D. C. Miles and J. H. Briston, *Polymer Technology,* Chemical Publishing Co., Inc., New York, 1979.

L. I. Nass, ed., *Encyclopedia of PVC,* Vols. 1–3, Marcel Dekker, Inc., New York, 1977.

FRED N. TEUMAC
Reeves Brothers, Inc.

FABRICS, FINISHING.

See TEXTILE RESINS.

FABRICS, NONWOVEN.

See NONWOVEN FABRICS.

FAILURE MECHANISMS.

See CRAZING; FATIGUE; FRACTURE; ABRASION AND WEAR.

FATIGUE

Lifetime Curves

Numerous studies have shown that the cyclic lifetime of a test bar or component varies inversely with the magnitude of the applied cyclic stress, strain, or deflection. Various test procedures have been developed to obtain mechanical property information useful in fatigue design analysis. Long ago, Wohler developed a procedure in which the specimen, notched or unnotched, is subjected to a cyclic stress (S) and the number of loading cycles (N) to cause failure is noted. Examples of S–N fatigue curves are shown in Figure 1. The cyclic lifetime of a given polymer decreases with increasing applied stress, consistent with greater damage accumulation per cycle at higher stress.

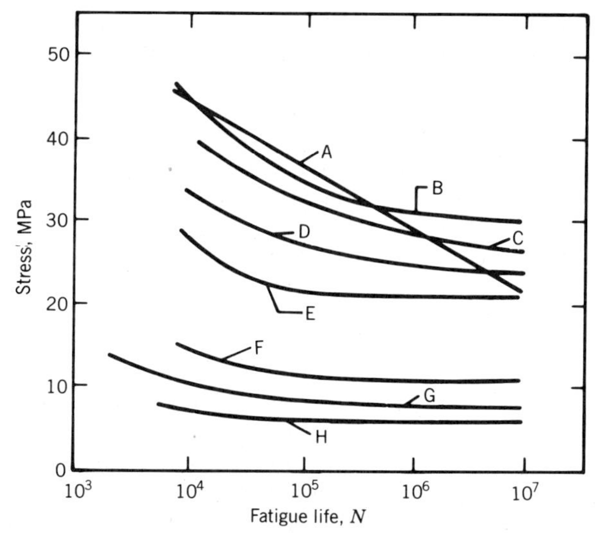

Figure 1. Stress vs cyclic life plots for typical plastics. Cyclic lifetime N_1 corresponds to stress S_1. A, polyurea; B, diallyl phthalate resin; C, alkyd resin; D, cellulose acetate; E, acrylic resin; F, polypropylene; G, polyethylene; H, polytetrafluoroethylene. To convert MPa to psi, multiply by 145.

Hysteretic Heating. The fatigue ranking of the plastics in Figure 1 may not measure the relative intrinsic fatigue resistance of these materials. These test results can be applied in design only when all design factors, including magnitude and mode of the stress, part size and shape, temperature, heat transfer conditions, cyclic frequency, and environmental conditions, are comparable to the test conditions.

Specimen "failure" takes place by either of two modes. Crack nucleation and subsequent growth across the gauge section of a specimen may produce failure. Alternatively, some materials may become overheated as a result of the accumulation of hysteretric energy generated during the loading cycles. Such energy dissipation produces a significant temperature rise within the specimen, which in turn causes the elastic modulus of the material to decrease.

Cyclic Strain Fatigue. Some experiments have been conducted to evaluate the fatigue response of engineering plastics under constant strain amplitude test conditions involving fully reversed tension–compression loading conditions. During the course of such testing, the height of the hysteresis loops may either increase or decrease, reflecting conditions of cyclic strain hardening and softening, respectively. Both changes in material properties have been identified for a large number of metallic alloy systems. In crystalline, amorphous, and composite engineering polymers, however, only cyclic strain softening has been reported.

Crack Initiation

So far cyclic fatigue lifetimes have been discussed as the number of cycles both to initiate and propagate a defect to some critical length. Hence,

$$N_T = N_i + N_p$$

where N_T = total fatigue life, N_i = number of cycles to initiate a crack, and N_p = number of cycles to propagate a crack to a critical length necessary for fracture.

Crack Propagation

Engineering plastic components may contain material defects, eg, porosity and inclusions, and may also suffer damage as a result of fabrication and service. These defects may not cause immediate fracture under the intended design stress, but these cracks may grow slowly to critical dimensions under the influence of cyclic loading. For this reason, the kinetics of fatigue crack propagation (FCP) must be adequately identified. Fatigue crack propagation data for amorphous, semicrystalline, and multiphase polymeric solids are available in the literature and may be obtained from various test specimens. Beginning with a mechanically sharpened notch, cyclic loads are applied and the resulting change in crack length is recorded as a function of the number of load cycles. Such crack-length monitoring may involve either direct visual examination or indirect compliance methods. The growth rate increases with increasing crack length and stress (Fig. 2) as described by a relationship of the form

$$da/dN \propto f(\sigma,a)$$

where a is crack length. It has been postulated that the stress intensity factor, itself a function of stress and crack length, is the overall controlling factor in the FCP process (see FRACTURE).

Cyclic Frequency. Although hysteretic heating explains a decrease in fatigue resistance with increasing cyclic frequency in unnotched polymer test samples, the fatigue resistance of notched polymers is exceedingly complex. FCP rates can either decrease, remain unchanged, or increase with increasing test frequency (Fig. 3). A correlation has been found between the FCP frequency sensitivity factor (FSF) in polymers and the frequency of movement of the main chain segments responsible for generating the transition peak at room temperature.

Test Temperature. Based on the viscoelastic response of engineering plastics, the FCP response of these materials is expected to depend on test temperature.

Effects of Chemistry in Homogeneous Polymers

Molecular Weight and Distribution. Molecular weight and distribution profoundly affect the thermal and mechanical properties of polymers. Thus the tensile strength of many polymers is very low or negligible when the average M is low (10^4–10^5), but increases rapidly above M_0, a characteristic value of molecular weight, and then levels off. When $M > M_0$, the relationship between strength and molecular weight may often be expressed by an empirical equation of the form

$$\sigma_B = A - B/\overline{M}_x$$

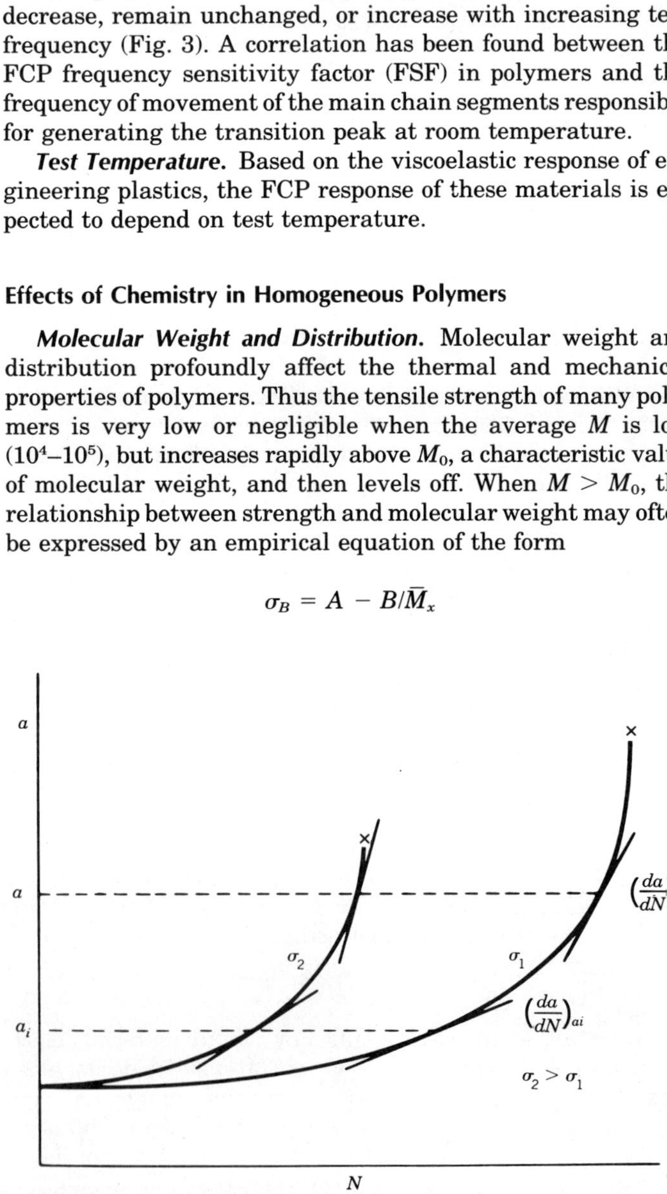

Figure 2. Crack length (a) increases with increasing number of loading cycles (N). Crack growth rate increases with increasing crack length and stress level. × signifies failure.

where σ_B is the tensile strength, A and B are constants, and $\overline{M}_x$ is an average molecular weight.

However, in spite of the association of high fatigue resistance with high values of M, the sensitivity to molecular weight is much greater in fatigue than in monotonic loading. Thus generalities about fatigue resistance of generic polymers must be viewed cautiously, for rankings depend strongly on molecular weight as well as on structure.

Cross-linking. Few controlled studies on the effects of cross-linking on fatigue exist. When irradiation was used to increase $\overline{M}_w$ in polystyrene, keeping $\overline{M}_n$ constant, the fatigue life was little affected by the total dose applied, both up to the gel point and beyond. The FCP resistance of cross-linked polystyrene is lower than that of polystyrene or rubber-modified polystyrene. A typical epoxy resin is much less resistant to FCP than thermoplastic polymers. This observation is generally consistent with the ability of cross-linking to inhibit segmental mobility, crazing, and consequent energy dissipation.

Effect of Diluents. Diluents, eg, plasticizers, residual solvents or monomers, or absorbed environmental agents, greatly affect the viscoelastic response of a polymer and related properties such as damping, modulus, and yield strength. The resultant changes in properties can therefore affect fatigue behavior whether the diluents are dispersed uniformly in the bulk or only at the surface.

The effects of lower stiffness differ depending on whether the testing is conducted under constant stress or constant deflection conditions. In the first case, fatigue behavior may deteriorate due to the increase in strain experienced by a more compliant, ie, less stiff, specimen. In the latter case, the more compliant specimens, which experience lower stresses, may exhibit superior performance.

Surface Modification. Previous examples involve essentially homogeneous polymer–diluent systems. Interesting effects with practical consequences also occur when unnotched samples are exposed to various environmental media. Because fatigue failure in such cases usually involves initiation at the surface, such effects are to be expected.

Deleterious effects of atmospheric exposure on mechanical properties are well known. Surface oxidation in rubbers results in embrittlement, surface cracking, enhanced fatigue crack propagation, and reduced fatigue life. Even in a ductile plastic, the formation of brittle surface layers and cracks invariably facilitates the onset of crack propagation under static or fatigue loading.

Although liquid environments often induce environmental stress crazing or crazing in glassy plastics, not all liquids are deleterious to fatigue performance, and some are beneficial.

Thermal History. This is an important parameter in determining viscoelastic and mechanical behavior. For example, in amorphous polymers, quenching from above T_g increases the free volume and hence facilitates relaxation, while annealing below T_g decreases free volume and enthalpy, increases the yield stress, and decreases fracture toughness (see also AGING). With crystalline polymers, the morphology (qv) depends very much on the balance between nucleation and growth corresponding to the rate and pattern of cooling from the melt (see CRYSTALLIZATION KINETICS).

Crystallinity and Morphology. As a class, semicrystalline polymers tend to exhibit superior fatigue resistance, relative to that of amorphous polymers, both in terms of S–N behavior and fatigue crack propagation rates, as long as tests are con-

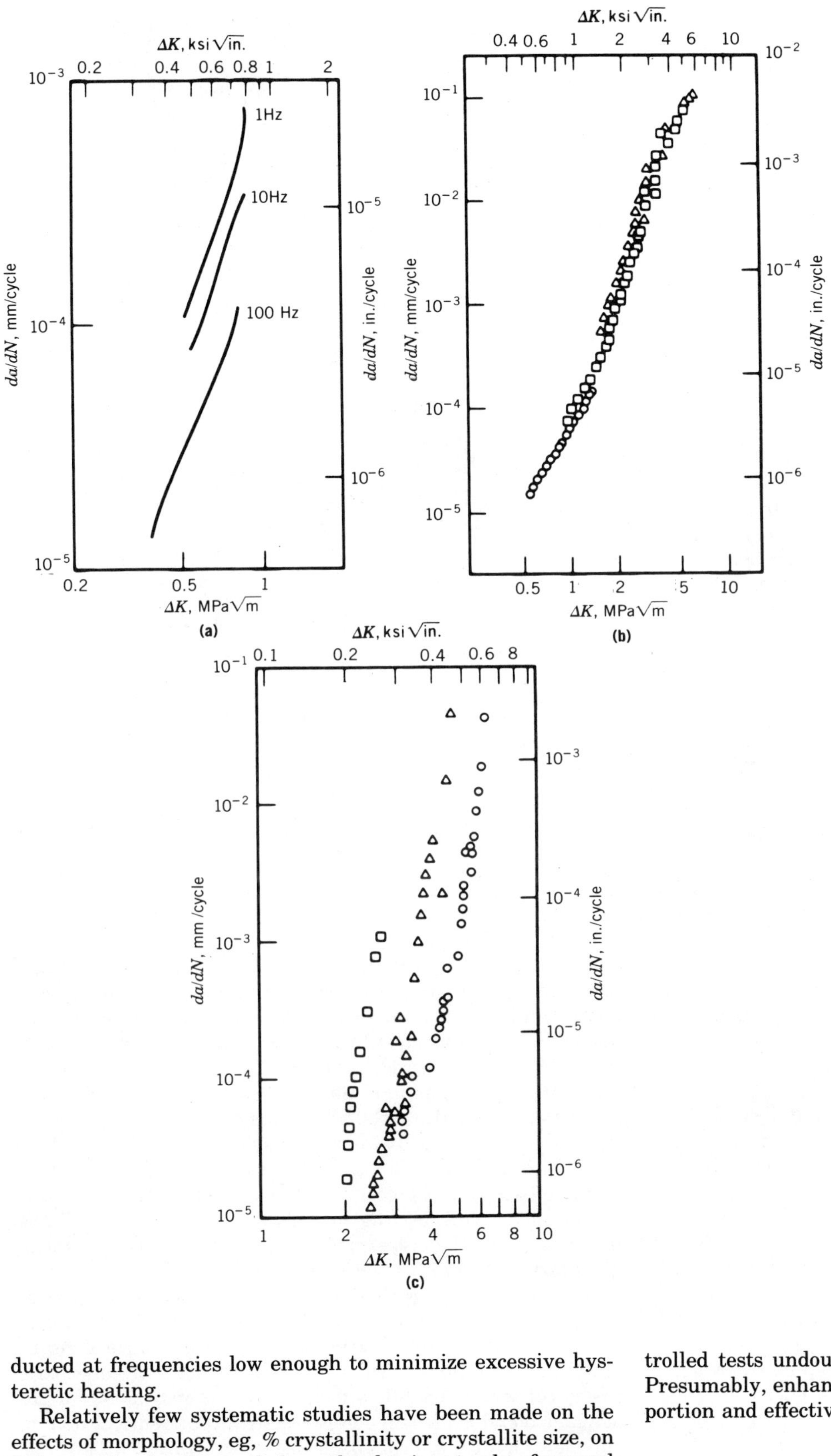

Figure 3. Effect of cyclic frequency on FCP in (**a**) PVC; (**b**) polycarbonate: ○, 100 Hz; □, 10 Hz; △, 1 Hz; (**c**) Zytel ST 801: ○, 1 Hz; △, 10 Hz; □, 30 Hz.

ducted at frequencies low enough to minimize excessive hysteretic heating.

Relatively few systematic studies have been made on the effects of morphology, eg, % crystallinity or crystallite size, on fatigue behavior. Nevertheless, the dominant role of several antioxidants in increasing fatigue life has been attributed to their ability to nucleate smaller and more uniform spherulites. In addition to specific effects of the fine-structured morphology, the associated reduction in hysteretic heating in stress-controlled tests undoubtedly also helps to increase fatigue life. Presumably, enhanced nucleation rates also increase the proportion and effectiveness of tie molecules.

Design Concepts

The fatigue life of an engineering component depends markedly on the size of the existing defect, the magnitude of the stresses, and the characteristics of the component material.

Crack Propagation Calculations. Fatigue-life calculations are required to design a new part with a safe operating service life under cyclic loading conditions. Such computation can be performed by the following equation (where $\Delta K = Y \Delta \sigma\sqrt{a}$) between the starting and final flaw sizes as the limits of the integration.

$$da/dN = A\ \Delta K^m$$

When the geometrical correction factor Y does not change within the limit of integration for the case of a circular flaw where $K = 2/\pi\ \sigma\sqrt{\pi a}$, the cyclic life is given by

$$N_f = \frac{2}{(m-2)AY^m\Delta\sigma^m}\left(\frac{1}{a_0(m-2)/2} - \frac{1}{a_f(m-2)/2}\right) \quad \text{for } m \neq 2$$

Fracture Surface Morphology

Macrofractography. Macroscopic examination of numerous service failures generated by repeated loading reveals distinct fracture surface markings. Typically, the fracture surface is flat, indicating the absence of appreciable gross plastic deformation during service life. The absence of observable plastic flow belies the accumulation of irreversible damage through fatigue crack formation. These flat fracture surfaces generally contain parallel sets of lines, referred to as "clamshell markings," "arrest lines," and/or "beach markings," attributed to different periods of crack extension, during the intermittent operation of a pump in a liquid transport system. Different fatigue processes in engineering plastics also give rise to various colors and textures on the fatigue fracture surfaces. Stress whitening has been observed on the fatigue fracture surfaces of a number of engineering plastics, particularly semicrystalline polymers such as polymides, polyoxymethylene, chlorinated polyether, and poly(vinylidiene fluoride). Such whitening is believed to result from refractive index changes occurring as a result of molecular-chain reorientation and spherulite breakdown in the slowly advancing fatigue crack-tip plastic zone.

Microfractography. The fatigue-fracture-surface micromorphology of engineering plastics is complex. At relatively high ΔK levels, many polymers reveal the presence of fatigue striations that correspond to the successive location of the advancing crack front after each loading cycle (Fig. 4**a**); the width of each striation, therefore, represents the amount of crack-front incremental advance resulting from the prior load excursion. The microscopic measurement of the fatigue-crack-propagation rate, ie, striation width, is in excellent agreement with macroscopically determined crack growth rates (Fig. 5).

Another set of parallel bands has been identified on fatigue fracture surfaces (Fig. 4**b**). These bands reflect discrete crack advance increments that develop following several hundred to several thousand loading cycles of total crack arrest. Calculations show that the width of these discontinuous growth bands corresponds to the dimension of the crack tip plastic zone as formulated by the Dugdale model.

Polymer Blends

Effects of rubbery phases on the fatigue behavior of a rigid polymer can be very complex. As with unmodified polymers, high values of $\Delta\sigma$ or frequency may generate enough hysteretic heating to induce failure by softening (Fig. 6); lower values may lead to failure by the initiation and propagation of a crack in a nominally brittle manner. Even with the latter, thermal effects at and beyond the crack tip can greatly modify FCP behavior. While softening due to cyclic damage, with or without hysteretic heating, is typical of most polymers, a rubbery phase can cause several specific effects, significant decreases in yield stress and modulus, and a greater susceptibility to creep. Moreover, the reduction in ultimate elongation associated with cyclic softening may be greater in rubber-modified polymers. Thermal effects, which may be beneficial or deleterious, are likely to be more intense, at least with cycling between fixed stress limits, for the loss compliance (D'') may be increased by the presence of rubber. Also, rubber modification may permit cycling to higher values of load and ΔK, thus increasing the propensity for hysteretic heating as cracks grow further before fracture.

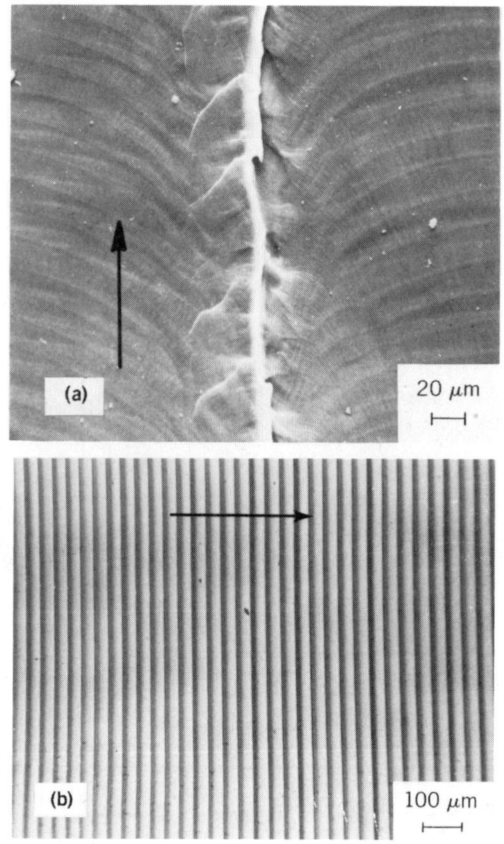

Figure 4. (**a**) Fatigue striations in polysulfone corresponding to the incremental advance of a crack as a result of a single loading cycle. Arrow indicates crack growth direction. (**b**) Discontinuous growth bands in poly(vinyl chloride). Arrow indicates crack growth direction.

Thus it is not surprising to find exceptional diversity of fatigue behavior in rubber-modified systems. Although results can often be explained in terms of a plausible balance of mechanistic factors, unequivocal quantitative conclusions about the role of rubber are usually difficult because the behavior of the polymer blend is seldom compared with the identical matrix polymer, and loading conditions used have varied widely.

Rigid Particle Systems

Fillers (silica, calcium carbonate, and mica) are commonly added to polymeric matrices to increase modulus and dimensional

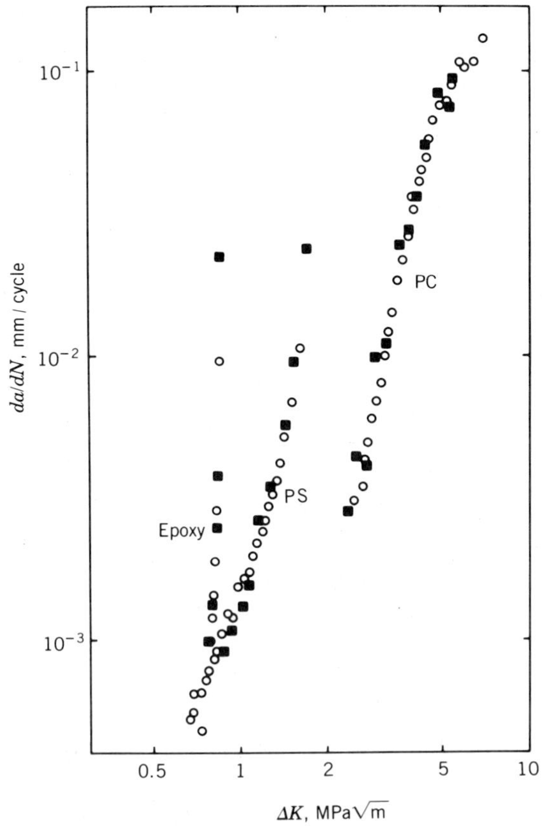

Figure 5. Macroscopic (○) vs microscopic (■) fatigue crack growth in epoxy, polystyrene, and polycarbonate.

stability, and sometimes to increase wear resistance and decrease the coefficient of friction. Preformed rigid polymer particles are also incorporated into casting resins for orthopedic or dental restorations. The effects of filler on ultimate fracture behavior is less predictable and is dependent on the filler content, interfacial adhesion, and temperature and ductility of the matrix. Thus, depending on the system, the value of $\mathcal{G}_c$ (strain

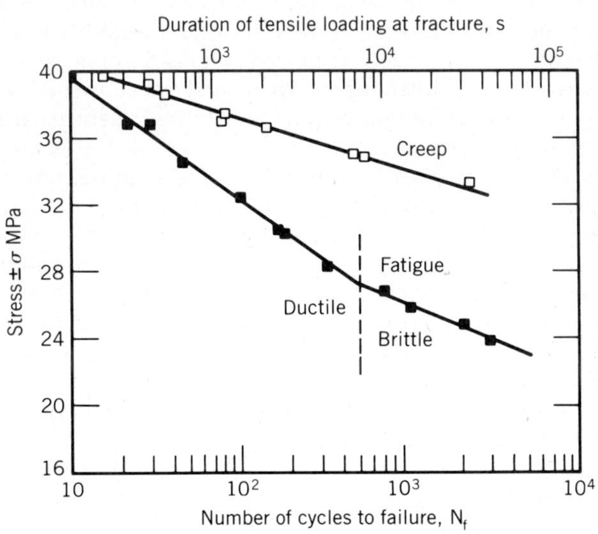

Figure 6. Comparison of S–N curve under square-wave loading with creep-rupture data for typical ABS. Note that the creep-rupture time scale is equivalent to that of the fatigue cycle scale in terms of time under tensile load. To convert MPa to psi, multiply by 145.

energy release rate) may increase, decrease, or exhibit a maximum as filler content is increased. This complex behavior reflects competition between several factors, cracking of the subsurface, increase in surface roughness and modulus, ability of particles to pin the crack, and ductility from variations in matrix constraint at the interface. In the case of K_c, values may increase or decrease with filler content. Although much remains to be learned about the balance of factors mentioned, typical particulate fillers often are deleterious to strength and toughness.

Fibrous Composites

As the shape of a high modulus phase dispersed in polymer is elongated from a sphere to a fiber, the ability of the dispersed phase to bear load is progressively increased. At a given concentration of uniformly dispersed reinforcement, several matrix-related properties are usually improved: stiffness, strength, resistance to creep and fatigue, and heat-distortion temperature; fracture energy is also often increased, especially with long fibers. Maximum effects are achieved with long or continuous fibers, but significant improvements are conferred by short fibers as well. In practice, fibers such as glass, boron, carbon, and high modulus organic materials, aromatic polyamides (aramids), are typical, often by themselves but also in combination with each other, or with particulate fillers (as in sheet-molding compounds). A wide variety of geometrical arrangements are used, including randomly and uniaxially oriented fibers, cross-plied laminates, and woven structures. Compositions range from those in which the matrix is the dominant component to those in which the matrix serves to bond and protect an aggregation of fibers (see also COMPOSITES).

Other Composite Systems

Fatigue of adhesive joints is of considerable practical importance. As with other composite systems, a fracture mechanics approach can be useful; the rate of crack growth may be described by the energy-based equivalent of the Paris law (ie, replacing ΔK with $\Delta\mathcal{G}$). The locus of failure (center of bond vs interface) depends on the material, the frequency, and the environment, although the Paris law variant has been reported to hold in all cases.

The fatigue behavior of portland cement concretes has been studied to a limited extent. Significant improvements in fatigue performance, noted with concretes impregnated with PMMA, have been attributed to improved toughness and interfacial bonding between the concrete constituents (see also CEMENT ADDITIVES).

Recently, the incorporation of low concentrations (0.2–2 vol %) of metallic glass ribbons has been shown to reduce fatigue crack growth rates in an epoxy significantly. The reduction is greatest when the width of the ribbon is perpendicular to the crack growth direction; however, when the width is parallel, absolute values of da/dN for the composite are lower by an order of magnitude.

BIBLIOGRAPHY

R. W. Hertzberg and J. A. Manson, *Fatigue of Engineering Plastics*, Academic Press, Inc., Orlando, Fla., 1980.

A. J. Kinloch and R. J. Young, *Fracture Behavior of Polymers*, Applied Science Publishers, New York, 1983.

R. W. Hertzberg, *Deformation and Fracture Mechanics of Engineering Materials,* 2nd ed., John Wiley & Sons, Inc., New York, 1983.

Fatigue in Polymers, Plastics and Rubber Institute, London, UK, 1983.

RICHARD W. HERTZBERG
JOHN A. MANSON
Lehigh University

FELTS. See NONWOVEN FABRICS.

FEP. See TETRAFLUOROETHYLENE POLYMERS.

FERROCENE POLYMERS. See ORGANOMETALLIC POLYMERS.

FIBERBOARD. See COMPOSITION BOARD.

FIBER OPTICS. See FIBERS, OPTICAL.

FIBERS

A fiber may be defined as a flexible, macroscopically homogeneous body having a high ratio of length to width and a small cross section. A large segment of the world's agricultural activity is concerned with the growth and harvesting of natural fibers. The production of synthetic fibers is an important activity of the worldwide chemical industry. The textile and paper industries are the primary converters of fibers into products with properties directly related to the unique characteristics of fibers. The paper industry uses a natural cellulosic fiber derived from wood almost exclusively, and the textile industry uses both natural and synthetic fibers in the manufacture of a variety of products.

Manufacturing Methods. Synthetic fiber manufacture is based on three common methods of fiber formation or spinning. The three principal methods are melt spinning, dry spinning, and wet spinning, although there are many variations and combinations of these basic processes (see also FIBERS, MANUFACTURE).

Fiber Consumption Trends

Increasing worldwide production of cotton and synthetic fibers between 1920 and 1980 is shown in Figure 1.

The rapid increase in worldwide fiber production reflects not only the increasing world population but also increasing per capita consumption of textile fibers. World population has increased from 2.5 billion people in 1950 to about 5.0 billion in 1987, an average annual population increase of about 2%. The average per capita fiber consumption has increased from about 5.6 kg in 1965 to ca 7.0 kg in 1980.

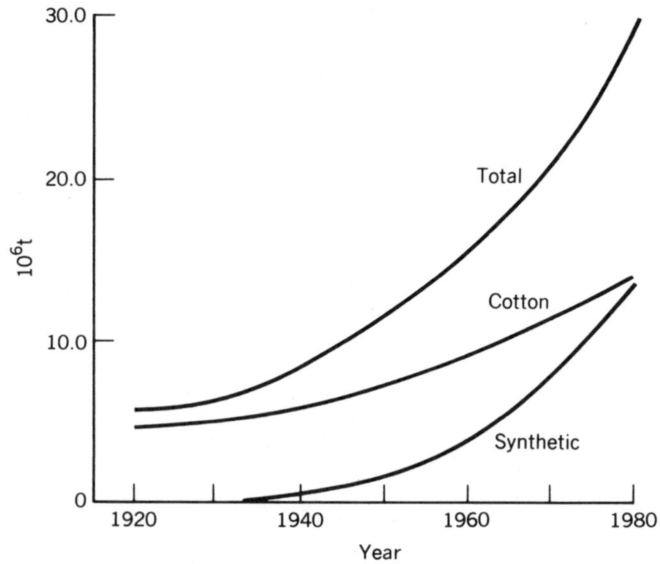

Figure 1. Worldwide fiber production trends.

Types of Fibrous Materials

Fibers are used in the manufacture of a wide range of products, generically referred to as fibrous materials. The properties of these materials depend on the properties of the fibers and on the geometric arrangement of the component fibers in the structure.

Several types of fibrous materials can be distinguished on the basis of fiber organization. One extreme is an isotropic assembly in which the fibers are arranged in a completely random fashion with no preferred orientation in any of the three principal spatial axes. Alternatively, and more commonly, fibrous materials have their fibers arranged in well-defined spatial patterns.

Textile Yarns. Textile yarns and several preliminary linear structures, such as roving and sliver, are typical of structures with a high degree of fiber orientation with respect to the principal axis of the material.

Textile yarns are produced from staple (finite length) fibers by a combination of processing steps referred to collectively as yarn spinning. After preliminary fiber alignment, the fibers are locked together by twisting the structure to form the spun yarn, which is continuous in length and remarkably uniform.

Textile yarns produced from continuous filament synthetic fibers are called multifilament yarns and are characterized by nearly complete filament alignment and parallelization with respect to the yarn axis.

Textile Fabrics. Yarns are used principally in the formation of textile fabrics, either by weaving or knitting.

Nonwoven Fabrics. The term "nonwoven" simply suggests a textile material that has been produced by means other than weaving, but these materials really represent a unique class of fibrous structure. In nonwoven fabrics (qv), the fibers are processed directly into a planar sheetlike fabric structure, bypassing the intermediate one-dimensional yarn state, and then are either bonded chemically or interlocked mechanically (or both) to achieve a cohesive fabric.

Fiber Properties

The properties of textile fibers may be conveniently divided into three categories: geometric, physical, and chemical, as shown in Table 1.

Table 1. Classification of Fiber Properties

Geometric	Physical	Chemical
Length	*Density*	*Response to*
average value	linear	acids
distribution	bulk	alkalies
Cross section	*Thermal*	oxidation
average value	melting point	reduction
distribution	transitions	heat
shape	conductivity	*Sorption*
Crimp	*Optical*	moisture
frequency	birefringence	dyes
form	refractive index	*Swelling*
	luster and color	anisotropy
	Electrical	
	resistivity	
	dielectric constant	
	Surface	
	roughness	
	friction	
	Mechanical	
	tension	
	compression	
	torsion	
	bending	
	shear	

Structure of Fiber-forming Polymers

Fiber structure is described at three levels of molecular organization, each relating to certain aspects of fiber behavior and properties. Of first importance is the organochemical structure defining the chemical composition and molecular structure of the repeating unit in the polymer, and also the nature of the polymeric link. This primary level of molecular structure is directly related to chemical properties, dyeability, moisture sorption, swelling characteristics, and indirectly related to all physical properties. The macromolecular level of structure describes the chain length, length distribution, and stiffness, as well as molecular size and shape. The supramolecular organization is the arrangement of the polymer chains in three-dimensional space. The physical properties of fibers are strongly influenced by the organization of polymeric chains into crystalline and noncrystalline domains, and the disposition of these domains with respect to each other. In the case of natural fibers, a further level of structural organization related to the natural growth and development of the fiber must be considered. This morphology is quite complex, with fibrils, microfibrils, and similar structural subunits frequently surrounded by a matrix material in a composite configuration.

All polymeric fibers that are useful in textile applications are semicrystalline, irreversibly oriented polymers, ie, the polymeric chains are partially ordered into regions or domains with near-perfect registry where the laws of x-ray diffraction are obeyed. In other regions of the fiber, the molecular chains or segments of chains are not perfectly ordered and may approach random coil configurations. This is a simple statement of the two-phase crystalline–amorphous model of the structure of semicrystalline polymers that has been used to describe fiber structure and to interpret fiber properties.

Vegetable Fibers

Cotton is the best known and most extensively used vegetable fiber in the textile industry, but there are hundreds of other fibers derived from plants and used as textile fibers. They are classified according to the part of the plant from which the fiber is obtained (see also FIBERS, VEGETABLE).

Seed hairs—cotton, kapok, coir.
Stem or bast fibers—flax, hemp, ramie, jute.
Leaf fibers—sisal, abaca (Manila hemp).

The chemical composition of some vegetable fibers is shown in Table 2. The principal component in these vegetable or plant fibers is cellulose; hence, these fibers are frequently called natural cellulosic fibers. Cellulose is the most abundant natural polymer. Its chemical structure is poly(β-1→4-D-anhydroglucopyranose).

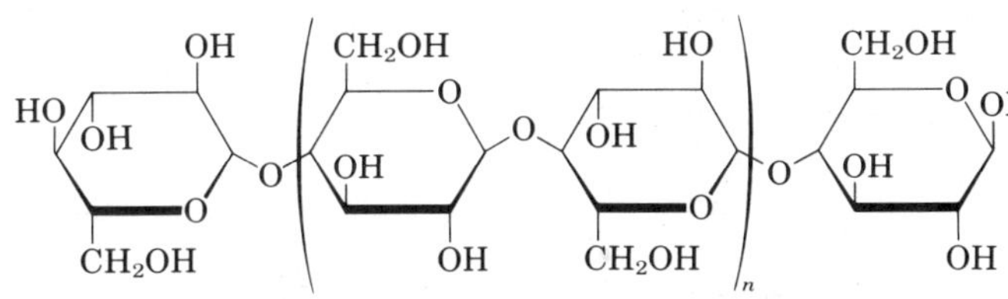

degree of polymerization = $2n + 2$
cellulose

Because of its prevalence in natural vegetable fibers, the supramolecular structure of cellulose is critically important to understanding its properties, particularly for cotton, which in its purified state is essentially entirely cellulose. The inherent chain stiffness of cellulose and its hydrogen bonding capacity allows extensive interaction between neighboring cellulose chains. In the native state these interactions lead to well developed crystalline regions, and the natural cellulosic fibers are therefore among the most highly crystalline fibers. Sche-

Table 2. Chemical Composition of Common Vegetable Fibers

Fiber	Cellulose	Hemicellulose	Pectin	Lignin	Extractives
cotton	94.0	2.0	2.0		2.0
kapok	43.2	32.4	6.6	15.1	
flax	71.2	18.5	2.0	2.2	6.0
hemp	74.3	17.9	0.9	3.7	3.1
ramie	76.2	14.5	2.1	0.7	6.4
jute	71.5	13.3	0.2	13.1	1.8
abaca	70.0	21.8	0.5	5.7	1.8
sisal	73.2	13.3	0.9	11.0	1.6

matic representations of the micellar, continuous, fringed micellar, and fringed fibrillar structures are shown in Figure 2.

Cotton. ***Fiber Growth and Structure.*** Cotton (qv) is indigenous to both the Eastern and Western hemispheres and is grown commercially in nearly every nation of the world. Cotton refers to the seed hair on the plant, as well as the plant itself, which belongs to the genus Gossypium of the tribe hibisceae.

The growth and development of cotton fibers is complex and follows several distinct and identifiable stages. Cotton fibers emerge as the hair covering on the surface of the seeds in a cotton boll several days after the plant flowers. Each seed produces 10,000 or more fibers, and there are several seeds in each boll. Each fiber is a single biological cell. A fiber (lint) consists of a thin tubular membrane called the primary wall, which encloses the nucleus and the living protoplasm. There is a thin cuticle layer covering the primary cell wall that consists of waxes, pectins, and some proteinaceous material.

The convolutions formed when the cotton fiber dries and collapses after boll opening are unique and characteristic of cotton fibers. The helical structure of the fiber's secondary wall causes the collapse to form convolutions, which alternate in their direction because of the helical reversals. These convolutions allow easy extension of cotton fibers under very low stress.

Fiber Properties and Processing. After the boll has opened and dried in the field, the cotton is harvested by hand or machine picking. The process of removing the fiber or cotton lint from the seed surface and separating it from the seed is known as ginning. After ginning the cotton lint contains considerable quantities of dirt, trash, and leaf matter that is at least partly removed by mechanical cleaning before baling and shipping. The seed yields cottonseed oil by crushing and pressing operations after ginning.

Cotton is the purest form of native cellulose and contains only small quantities of hemicelluloses, pectin, and surface oils and waxes. The noncellulosic components are quite easily removed by scouring in dilute alkaline solutions at elevated temperatures. In normal textile processing this purification is never performed until the fibers have been spun into yarn. The natural oils and waxes act as efficient spinning lubricants and protect the fiber from chemical or physical damage. Often, as part of desizing, the noncellulosic components are not removed until after weaving. The noncellulosic components must be removed before dyeing and other chemical-finishing treatments to permit the penetration of the reagents into the interior of the fiber.

Fineness, defined in terms of cross-sectional area, linear density, or maturity, also affects cotton's processibility and utility.

Cotton fibers are classified as stiff, strong, and modestly extensible (Table 3).

Table 3. Typical Properties of Cotton Fibers

Property	Value
tenacity at break, N/tex[a]	
65% rh	0.26–0.44
wet	0.26–0.53
extension at break, %	
65% rh	4–10
wet	5–13
elastic modulus, N/tex[a]	
65% rh	3.5–7.9
wet	2.6–5.3
moisture regain at 65% rh, %	7.0–8.0
specific gravity	1.54
approx. volumetric swelling in water, %	40

[a] To convert N/tex to g-f/den, multiply by 11.3.

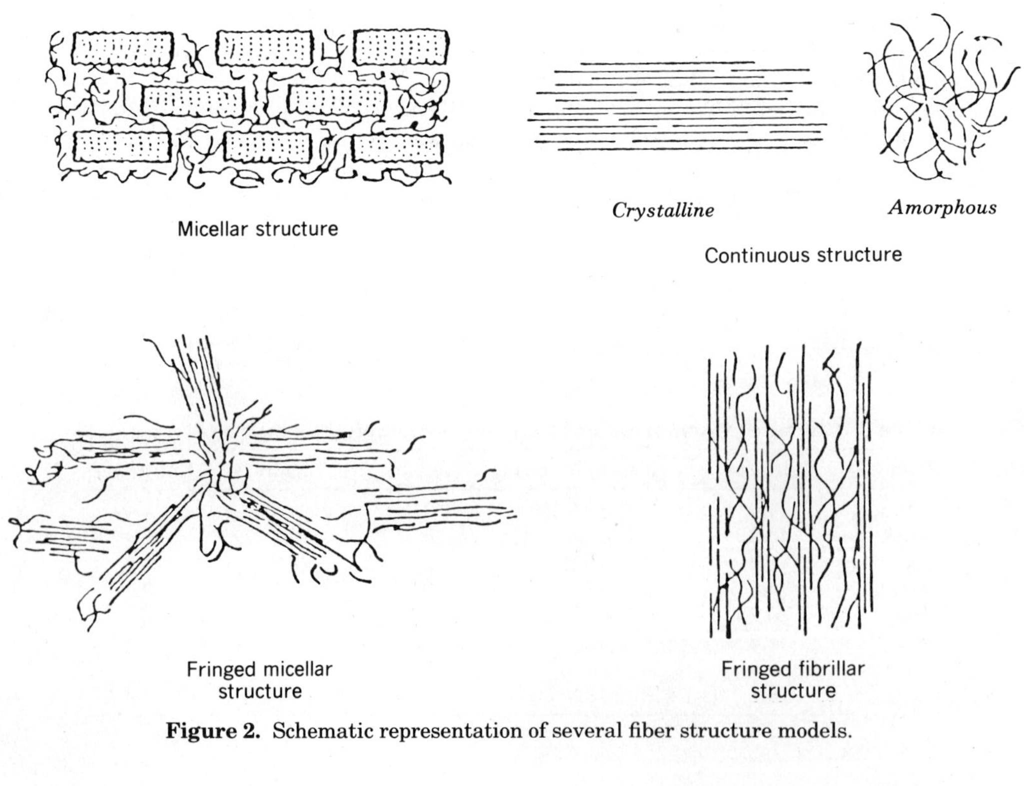

Figure 2. Schematic representation of several fiber structure models.

Cotton may be dyed with a variety of dyes available in a full spectrum of colors with good fastness properties.

Mercerization. Cotton is frequently mercerized, ie, exposed to aqueous solutions of NaOH. Mercerization is not only a severe intracrystalline swelling treatment, but also a recrystallization process that involves a polymorphic transformation of the native cellulose I crystal structure to the regenerated cellulose II crystalline modification. The transformation may involve a progressive shift of cellulose chain sheets within crystallites from a quarter-staggered relation to complete correspondence. The primary purpose of mercerizing cotton fabrics is the improvement of luster resulting from the smoother fiber surface. Increased fabric strength and sorptivity also result.

Bast Fibers. Bast fibers (eg, flax, hemp, jute, ramie) are derived from the stem of a wide number of plants. The fibers used in commercial processing are aggregates of the individual or ultimate fibers, and the dimensions of these aggregates depend to a large extent on the cleaning and preparation methods used. Even the dimensions of the ultimate fibers, or ultimate cells as they are most commonly called, vary a great deal (Table 4).

Unlike a cotton fiber, which is a single biological cell, even the ultimate cells (fibers) of the bast fibers are not single cells, but rather they are composed of many biological cells having a chemical composition (cellulose) and morphology similar to that of cotton. These biological cells are held together by intercellular material composed of hemicellulose, pectin, and lignin (Table 2). Thus they are similar to wood fibers with a morphology that may be depicted as in Figure 3. The secondary walls are fibrillar in nature, just as in cotton, and the fibrils are helically disposed although with a smaller helical angle and no structural reversals. The sense of the helix is characteristic for some of the fibers in this group as indicated in Table 4. Also, these fibers are not convoluted so that the spiral angle is not distorted with respect to the fiber axis as in cotton. As a result, these ultimate fibers or cells are generally stronger, stiffer, and less extensible than cotton. However, in commercial processing it is not the ultimate fibers or cells that are used, but rather larger and longer aggregates of ultimate fibers. The mechanical properties of the aggregate bast fibers may be significantly different from those of the ultimate fibers. X-ray diffraction shows that the cellulose in these fibers has a higher degree of order or crystallinity compared with cotton.

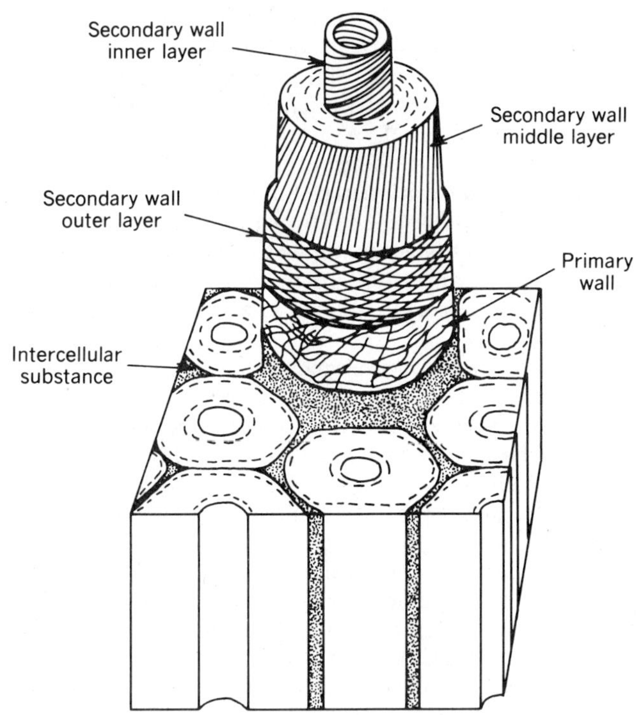

Figure 3. Schematic representation of the fine structure of a wood fiber.

Animal Fibers

The hair covering of animals provides several fibers of great interest to the textile industry. The word wool is used to describe the hair covering of animals; however, from a textile point of view it should be reserved for the fleece of sheep and lambs. In addition to wool the following animal fibers are used in textile applications, primarily as specialty fibers: mohair, cashmere, common goat hair, camel hair, llama hair, alpaca hair, and vicuna wool. Silk is an important animal fiber that differs in structure and properties from the other fibers in this group. Human hair, though never used in normal textile applications, is of interest as a structurally and functionally related material. Animal fibers, with the exception of silk, belong to a class of proteins known as keratin (qv). Keratins are natural cellular systems of fibrous proteins that serve as a protective outer barrier for the higher vertebrates. They are the main components of the outer layer of skin and of appendages, eg, hooves, scales, hair, and feathers. Keratin fibers have a complex morphology consisting of at least two major structural components, ie, the cuticle and the cortex, and certain coarse fibers possess a third structural component, a central medulla. All three components grow in a concentric fashion out of the germinal layer of the hair follicle, and their relative contribution to the total mass depends on the animal species and on the location within the fleece. Another important component of the total keratin fiber morphology is the cell membrane complex. This complex contains a cementing substance called the δ-layer, which holds together the individual cells of both the cuticle and the cortex.

Mechanical Properties. The mechanical properties of keratin fibers reflect the complex molecular and supramolecular struc-

Table 4. Morphological Structure of Ultimate Bast Fibers

Property	Cotton	Flax	Jute	Ramie	Hemp
mean length, mm	25	10–50	2.5	25–250	20
mean diameter, mm	0.012–0.025	0.023	0.015	0.050	0.022
spirality	*S/Z*	*S*			*Z*
spiral angle	23°	←——	—— 5°–10° ——	——	——→

tures of these fibers. When a keratin fiber is subjected to tensile deformation, the stress–strain curve may be interpreted as the response to the deformation of the microfibril and its surrounding matrix. The stress–strain curve reveals three distinctly different regions (Fig. 4).

Supercontraction. Supercontraction of keratin fibers is equivalent to denaturation in globular proteins. A keratin fiber supercontracts by 20–40% of its original length in hydrogen-bond-breaking systems, eg, aqueous phenol solutions, alkali halide solutions, and thiocyanate solutions at elevated temperatures. This supercontraction is a two-stage process. The first stage is completely reversible and presumably occurs by the disruption of hydrogen bonds in easily accessible regions of the fiber. The second stage is not reversible and involves deep-seated structural rearrangements, including the interchange mechanism with free sulfhydryl. This is indicated by the fact that partially reduced fibers supercontract much more quickly than do corresponding untreated fibers; the increased rate is due to the higher SH-group concentration and not the decrease in disulfide cross-link density.

Wool is a hygroscopic fiber and absorbs moisture from the atmosphere. Its regain at standard conditions of humidity and temperature (65% rh and 21°C) is about 15%, depending on type and previous history. Some physical properties of wool fibers are summarized in Table 5. Compared to other fibers, wool can be classified as relatively weak and extensible. Its unique property is high elastic recovery, particularly at elongations up to about 25%.

Silk. Commercial silk is obtained from the cultivated or domesticated silkworm, ie, the caterpillar of a small moth belonging to the species *Bombyx mori*. There are also several wild caterpillars harvested for their silk, the principal one being Tussah silk. Well over one-half of the world's silk supply comes from Japan; other countries with significant silk productions are China, Brazil, the Soviet Union, India, and Korea.

The protein of the silk filament is fibroin. Its structure is much simpler than that of keratin, although some 17 amino acids have been identified (see SILK).

Individual silk filaments are triangular in cross section, smooth-surfaced, and partially transparent. Their average diameter is approximately 10 μm. Under standard conditions the tenacity at break ranges from 0.26 to 0.44 N/tex. The wet strength is ca 80–90% of that under standard conditions. The filaments have an elongation to break of 20–25% dry and approximately 30% when wet. The elastic modulus of silk filaments is the highest of all animal fibers, ranging from 7.1 to 10.6 N/tex under standard conditions. Silk has a specific gravity of 1.25 and an equilibrium moisture regain of 11.0% at 65% rh and 21°C. As a protein fiber, silk can be readily dyed with a variety of dyestuffs and is seriously degraded and eventually dissolved by acids and alkalies. It is completely resistant to the common organic solvents.

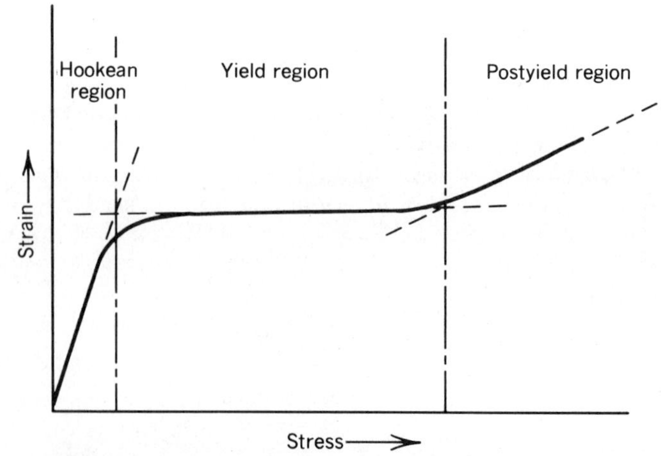

Figure 4. Stress–strain curve of a single keratin fiber.

Table 5. Typical Properties of Wool Fibers

Property	Value
tenacity at break, N/tex[a]	
65% rh	0.09–0.15
wet	0.07–0.14
extension at break, %	
65% rh	25–35
wet	25–50
elastic modulus, N/tex[a]	
65% rh	2.2–3.1
wet	1.8–2.6
moisture regain at 65% rh, %	15–17
specific gravity	1.30
approx. volumetric swelling in water, %	35

[a] To convert N/tex to g-f/den, multiply by 11.3.

Regenerated Cellulose Fibers

Regenerated cellulosic fibers, known generically as rayon (qv), were the first synthetic fibers to receive wide recognition and general acceptance. Rayon fibers are the most extensively used synthetic textile fibers manufactured from a naturally occurring polymer. The key steps in manufacture are purification of the cellulose raw material, cellulose solubilization, extrusion of the cellulose solution, cellulose regeneration by precipitation or chemical reconversion, and mechanical drawing or stretching to develop fiber structure. The most common method of rayon manufacture is the viscose process that accounts for more than 90% of the rayon produced. Other methods are the cuprammonium process and the saponified cellulose-acetate process. Methods of rayon manufacture involving direct dissolution of cellulose in a suitable solvent, bypassing chemical conversion of cellulose to a derivative, are being sought.

The source of cellulose for rayon manufacture is either relatively low molecular weight wood pulp or higher molecular weight cotton linters. The choice depends on the desired properties, and on the availability and price of the two materials.

Properties of Regenerated Cellulose Fibers. Because of the many variations possible in the viscose process and other manufacturing processes, fibers with widely different properties may be obtained. Rayon fibers have a less well developed crystalline structure than natural cellulosic fibers. The regenerated cellulose in rayon is in the form of the cellulose II crystalline polymorph, and unless special efforts are made rayon fibers do not have a fibrillar morphology.

Regenerated cellulose fibers are hydrophilic, with an equilibrium moisture regain of 10–14% in a standard atmosphere of 65% rh and 21°C.

Table 6. Typical Properties of Regenerated Cellulose Fibers

Property	Viscose process	Cuprammonium process	Regenerated acetate	Polynosic fibers
tenacity at break, N/tex[a]				
65% rh, 21°C	0.12–0.33	0.14–0.19	0.58–0.66	0.25–0.33
wet	0.07–0.21	0.09–0.12	0.45–0.54	0.21–0.29
extension at break, %				
65% rh, 21°C	10–30	10–17	6.0	7–10
wet	15–40	20–25	7.0	9–12
elastic modulus, N/tex[a]				
65% rh, 21°C	3.7–5.8	4.1	12.4–14.9	3.7–5.8
wet	1.2–3.3	3.3	8.26–10.7	2.9–4.1
moisture regain at 65% rh, %	12–16	11.0	10–12	10–12
specific gravity	1.50–1.54	1.54	1.52	1.54
approx. volumetric swelling in water, %	45–82	100–134	18–26	40

[a] To convert N/tex to g-f/den, multiply by 11.3.

The mechanical properties vary over a wide range and depend primarily on the techniques of cellulose regeneration and the degree of stretching during manufacture, as indicated in Table 6.

Regenerated Protein Fibers

The success of regenerated cellulose fibers suggests that regeneration of naturally occurring vegetable or animal protein into fiber form is technologically feasible and economically attractive. Unfortunately, some fundamental differences between cellulose molecules and protein chains prevent the regenerated protein fibers from achieving the success of their cellulosic counterparts. When a naturally occurring protein is solubilized, thereby disrupting the protein tertiary structure, the protein chain lacks some of the basic requirements for fiber formation.

Despite these inherent shortcomings, several attempts have been made to produce commercial fibers from natural proteins. These fibers, generically known as azlon fibers, have found brief acceptance, but none has ever been a significant commercial textile material. The proteins from milk, corn, soya, beans, and peanuts have been isolated and regenerated into fibers by wet-spinning processes.

The regenerated protein fibers are relatively weak and extensible (Table 7).

Cellulose-derivative Fibers

These fibers, based on acetyl derivatives of cellulose and called acetate fibers, are of two types: regular "secondary" acetate with a degree of substitution somewhat less than 2.4, and cellulose triacetate, in which essentially all three of the hydroxyl groups on each anhydroglucose unit are acetylated (see CELLULOSE ESTERS, ORGANIC, FIBERS). Acetate fibers are produced by the dry-spinning process.

There are many similarities in the properties of the acetate and triacetate fibers, but each has its own unique characteristics (Table 8).

Table 7. Properties of Regenerated Protein Fibers

Property	Value
tenacity at break, N/tex[a]	
65% rh	0.07–0.11
wet	0.03–0.06
extension at break, %	
65% rh	30–40
wet	40–70
elastic modulus, N/tex[a]	
65% rh	2.2
moisture regain at 65% rh, %	10–15
specific gravity	1.2–1.3

[a] To convert N/tex to g-f/den, multiply by 11.3.

Polyamide Fibers

These synthetic polymers have repeating units connected by amide linkages that are formed by the condensation of a dibasic acid with a diamine or by self-condensation of a lactam (see POLYAMIDES, FIBERS). Polyamide fibers based on aliphatic monomers are generically called nylons; those with monomers that are at least 85% aromatic are referred to as aramids.

Nylons. ***Manufacture.*** Nylon filaments are manufactured by the melt-spinning system.

Structure. Both nylon-6 and nylon-6,6 are semicrystalline polymers with several possible crystalline polymorphs.

Properties. The mechanical properties of nylon-6 fibers depend on processing, drawing, and the nature of heat setting.

The data shown in Table 9 are typical, although high tenacity nylon filaments for industrial uses have been made with breaking tenacities of ca 0.9 N/tex (10 g-f/den) under standard

Table 8. Typical Properties of Cellulose Acetate Fibers

Property	Secondary acetate	Triacetate
tenacity at break, N/tex[a]		
65% rh, 21°C	0.09–0.13	0.11–0.19
wet	0.06–0.10	0.07–0.12
extension at break, %		
65% rh, 21°C	25–40	20–28
wet	30–45	30–40
elastic modulus, N/tex[a]		
65% rh, 21°C	2.2–3.5	3.1–3.5
moisture regain at 65% rh, %	6.4	4.5
specific gravity	1.30–1.35	1.32
approx. volumetric swelling in water, %	10–30	

[a] To convert N/tex to g-f/den, multiply by 11.3.

Table 9. Typical Properties of Nylon Fibers

Property	Continuous filament	Staple
tenacity at break, N/tex[a]		
65% rh, 21°C	0.40–0.71	0.35–0.44
wet	0.35–0.62	0.31–0.40
extension at break, %		
65% rh, 21°C	15–30	30–45
wet	20–40	30–50
elastic modulus, N/tex[a]		
65% rh, 21°C	3.5	3.5
moisture regain at 65% rh, %	4.0–4.5	
specific gravity	1.14	
approx. volumetric swelling in water, %	2–10	

[a] To convert N/tex to g-f/den, multiply by 11.3.

conditions. Nylon fibers are used in many apparel, household, and industrial applications, especially where high strength is required along with high resilience; carpets constitute one of the principal markets for nylon yarns. Since fatigue resistance of nylon fibers is high, these fibers are also excellent as tire reinforcements, particularly for heavy-duty truck tires.

Aramids. Many aromatic polyamide polymers have been synthesized, but only a few have found general use as fiber-forming materials. Aramid fibers are characterized either by flame resistance and thermal stability or by high strength and stiffness. Combinations of these high performance properties are also possible. In general, flame resistance and thermal stability are characteristic of the *m*-isomers, whereas high strength and stiffness are associated with the *p*-isomers. Physical properties of the principal commercial aramid fibers are summarized in Table 10 (see POLYAMIDES, AROMATIC).

Polyester Fibers

Synthetic fibers, formed from a synthetic polymer composed of at least 85% by weight of an ester of a dihydric alcohol and terephthalic acid, are known as polyesters. The most common polyester fibers have ethylene glycol as the dihydric alcohol. In one polyester fiber type, 1,4-dimethylolcyclohexane is used.

As in the case of polyamide fibers, high speed spinning is beginning to replace the traditional two-step spinning and drawing process.

The properties of polyester fibers are summarized in Table 11 (see POLYESTERS, FIBERS).

Dyeing of polyester fibers is difficult because of the lack of hydrophilic sites and the inherent stiffness of polyester molecules. New techniques of dyeing had to be developed for full-shade range with adequate fastness. These methods are high temperature dyeing, carrier dyeing, and the Thermosol method of dye fixation.

Table 10. Typical Physical Properties of Aramid Fibers

Property	Kevlar	Nomex
specific gravity	1.44	1.38
tenacity at break, N/tex[a]		
65% rh, 21°C	1.8–2.0	0.39–0.49
wet	1.8–2.0	0.26–0.35
extensibility, %		
65% rh, 21°C	2.5–4.0	20–30
wet	2.5–4.0	20–30
elastic modulus, N/tex[a]		
65% rh, 21°C	44–71	6.2–7.9
wet	44–71	4.4–6.2
equilibrium–moisture regain, 65% rh, %	4.5	8.0
melting point, °C	decomp. 460	380

[a] To convert N/tex to g-f/den, multiply by 11.3.

Table 11. Typical Properties of Polyester Fibers

Property	Continuous filament	Staple
tenacity at break, N/tex[a]		
65% rh, 21°C	0.35–0.53	0.31–0.44
wet	0.35–0.53	0.31–0.44
extension at break, %		
65% rh, 21°C	15–30	25–45
wet	15–30	25–45
elastic modulus, 65% rh, 21°C, N/tex[a]	7.9	7.9
moisture regain at 65% rh, %	0.4	0.4
specific gravity	1.38	1.38
approx. volumetric swelling in water, %	none	none

[a] To convert N/tex to g-f/den, multiply by 11.3.

Acrylic and Modacrylic Fibers

Acrylic fibers are long-chain polymers composed of at least 85% by weight acrylonitrile units. A modacrylic fiber has less than 85%, but at least 35% by weight, of acrylonitrile units. Polyacrylonitrile is formed by the addition polymerization of acrylonitrile. In most commercial acrylic fibers, various quantities of other vinyl monomers are copolymerized with acrylonitrile. They confer specific chemical and physical properties, and include such monomers as vinyl acetate, vinyl chloride, styrene, vinylpyridine, acrylic esters, and acrylamide. In commercial modacrylic fibers the comonomer is either vinyl chloride or vinylidene chloride. These fibers are manufactured by either dry- or wet-spinning processes. In dry spinning dimethylformamide is a common solvent.

Typical properties of acrylic fibers (qv) are shown in Table 12.

Table 12. Typical Properties of Acrylic Fibers

Property	Continuous filament	Staple
tenacity at break, N/tex[a]		
65% rh, 21°C	0.40–0.44	0.22–0.26
wet	0.35–0.40	0.18–0.26
extension at break, %		
65% rh, 21°C	15–20	25–35
wet	20–30	35–45
elastic modulus, 65% rh, 21°C, N/tex[a]	5.3–6.2	2.2–3.5
moisture regain at 65% rh, %	1.6–2.0	1.8–2.5
specific gravity	1.17	1.17
approx. volumetric swelling in water, %	slight	2–5

[a] To convert N/tex to g-f/den, multiply by 11.3.

The dyeing of acrylic fibers is done mostly with cationic or basic dyes at temperatures above the boiling point of water.

As a general class, acrylic fibers do not have well developed crystalline structures, although there are strong dipolar interactions of the polymer chains through the nitrile groups.

Acrylic fibers have good resistance to chemical and microbial attack.

Several commercial modacrylic fibers are produced primarily because of their flame resistance resulting from their high halogen content.

Olefin Fibers

Olefin fibers (qv) include polypropylene, polyethylene, poly(1-butene), poly(1-pentene), poly(3-methyl-1-butene), poly(4-methyl-1-pentene), poly(4-methyl-1-hexene), and polystyrene (Table 13).

Spandex Fibers

An elastic fiber may be defined as one that recovers completely from long-range deformations immediately upon removal of the deforming force. The fiber-forming substance in spandex fibers is a long-chain, synthetic polymer composed at least 85% by weight of a segmented polyurethane.

Spandex fibers can be formed as continuous filaments by traditional dry- and wet-spinning processes.

Spandex fibers have high extensibilities and low elastic moduli, and very high elastic recoveries from large deformations (Table 14) (see FIBERS, ELASTOMERIC).

Carbon Fibers

Because these fibers combine low density with exceptional mechanical properties, they have found increasing use as reinforcing elements in fiber-reinforced composites. Carbon fibers, along with certain aramid and glass fibers, provide strength and durability to engineering thermosetting and thermoplastic resins to make composites one of the most exciting materials of the future. The fibers consist of small crystallites of "turbostratic" graphite, which is structurally similar to crystalline graphite.

Carbon fibers typically are fabricated from continuous precursor fibers by a three-stage procedure. Initially, the precursor fibers are stabilized at low temperatures to prevent fusion or melting in later stages. Noncarbon elements are then eliminated during a carbonization heat-treatment step; and finally, a high temperature graphitization stage enhances the mechanical properties of the carbon fibers.

Fiber Properties. The physical properties of carbon fibers depend on internal structure, which in turn depends on the precursor and on processing conditions. A typical high performance carbon fiber may have a diameter of 8 μm, a specific gravity of 1.95, an elastic modulus of 390 GN/m^2, and a tensile strength of 2.2 GN/m^2.

Table 13. Typical Properties of Polyolefin Fibers

Property	Polypropylene	Polyethylene
tenacity at break, N/tex[a]		
65% rh, 21°C	0.44–0.79	0.18–0.35
wet	0.44–0.79	0.18–0.35
extension at break, %		
65% rh, 21°C	15–30	20–40
wet	15–30	20–40
elastic modulus, N/tex[a]		
65% rh, 21°C	2.6–4.0	
wet	2.6–4.0	
moisture regain		
at 65% rh, %	0	0
specific gravity	0.90	0.92–0.96
volumetric swelling in water	none	none
melting point, °C	165	115–135

[a] To convert N/tex to g-f/den, multiply by 11.3.

Table 14. Typical Properties of Elastic Fibers at 65% rh, 21°C

Property	Natural rubber	Spandex
tenacity at break, mN/tex[a]	26	79
extension at break, %	500	600
elastic modulus, mN/tex[a]	1.76	4.4
recovery from 100% extension, %	97	95
moisture regain at 65% rh, %		1.3
specific gravity		1.21

[a] To convert mN/tex to g-f/den, multiply by 0.0113.

Vinyl Fibers

Two principal fiber types fall under the general category of vinyl fibers. Those fibers that contain at least 85% vinyl chloride by weight are generically referred to as vinyon fibers; those that are composed of at least 50% vinyl alcohol by weight are referred to as vinal fibers. Other fibers in this category are based on vinylidene chloride or contain fluorine.

Vinyon Fibers. Poly(vinyl chloride) fibers, generally containing about 10% vinyl acetate units, are produced by dry spinning using mixed solvents such as acetone–benzene.

Fluorine-containing Fibers. The only important fiber in this category is that made from polytetrafluoroethylene, the addition polymer of tetrafluoroethylene. Since the polymer is insoluble in all known solvents, the fiber cannot be wet- or dry-spun, and its extremely high melting point makes melt spinning impossible without degradation.

Other Fibers

Other organic synthetic fibers include Novoloid fibers, sulfur or poly(phenylene sulfide) fibers, PBI fibers, and polymer blend fibers.

Inorganic Fibers

Asbestos. Asbestos is a generic name for a number of minerals that occur naturally in fibrous form. Over 75% of commercial asbestos is obtained from Canada.

Products manufactured from asbestos fibers have been used primarily in industrial applications when heat resistance is required in combination with mechanical stability. Because asbestos has been identified as a carcinogen, its use is now strictly regulated and severely restricted.

Glass. Glass fiber is the only inorganic synthetic fiber that is used extensively in the textile industry, particularly for industrial products and household items such as drapery material. Continuous glass filaments are manufactured by the melt-spinning process. The main ingredients of a typical glass are silica (sand), limestone, aluminum hydroxide, soda ash, and borax.

Metallic Fibers. Metallic yarns are used primarily for decorative purposes. Metallic fibers are also manufactured in continuous form by repeated attenuation and drawing through diamond and tungsten dies. It is possible to weave traditional textile fabrics from steel filaments.

Other Inorganic Fibers. A number of fibers have been developed with inorganic lattice structures. Inorganic fibers (qv) are being made from alumina, silicon carbide, boron nitride, and boron carbide.

BIBLIOGRAPHY

B. C. Goswami, J. G. Martindale, and F. L. Scardino, *Textile Yarns: Technology, Structure and Applications,* John Wiley & Sons, Inc., New York, 1977.

J. W. S. Hearle in *Fundamentals of Structure and Mechanics,* Vol. 1, Halsted Press, John Wiley & Sons, Inc., New York, 1982.

L. Rebenfeld, *J. Appl. Polym. Sci. Part C* **(9)**, 91 (1965).

M. Lewin and E. M. Pearce, eds., *Fiber Chemistry,* Marcel Dekker, Inc., New York, 1985.

LUDWIG REBENFELD
Textile Research Institute

FIBERS, ELASTOMERIC

Elastomeric fibers consist of polymers with a main glass-transition temperature well below room temperature. This criterion excludes some fibers with elastic properties, eg, crimped hard fibers, but allows inclusion of biconstituent fibers where one of the constituents is an elastomer.

The principal materials for elastomeric fibers are polyisoprenes (natural rubber) and segmented polyurethanes (spandex).

Synthetic rubbers and nonurethane segmented elastomers are used in limited amounts for elastomeric fibers, where they offer special advantages.

The principal spandex fiber producers are given in Table 1.

Chemical Composition and Structure

Rubber. High quality natural rubber or *cis*-1,4-polyisoprene is converted into a cross-linked elastomer for elastomeric fibers by compounding with vulcanizing agents, stabilizers, antioxidants, and pigments, followed by curing at elevated temperature. Degradation is reduced by using latex for extrusion. In the final product, the molecular weight between cross-links is on the order of 3,000–10,000.

Cross-linking takes place during sulfur vulcanization. Other agents include accelerators and activators. These are chosen to minimize discoloration and staining in washing operations.

Spandex. Elastomers used for spandex are block copolymers made by a one- or two-step reaction of diisocyanates with hydroxyl-terminated, low molecular weight polymers (macroglycols) and diamines or glycols (chain extenders). Their general formula can be written as

$$\left[\left(\text{O—MG—O—CO—NH—DI—NH—CO}\right)_{m}\left(\text{NH—CE—NH—CO—NH—DI—NH—CO}\right)_{n}\right]_{z}$$

or

$$\left[\left(\text{O—MG—O—CO—NH—DI—NH—CO}\right)_{m}\left(\text{O—CE—O—CO—NH—DI—CO}\right)_{n}\right]_{z}$$

where MG, CE, and DI represent the residues of macroglycol, chain extender, and diisocyanate, respectively. In practice, m and n are kept at about 2, by the operation of the polymerization, and z is controlled with a monofunctional chain terminator to give molecular weights on the order of 50,000–100,000.

The hard segments in segmented polyurethanes interact with one another and behave like cross-links. This was a fundamental discovery, one that led to the development and commercialization of spandex fibers. It allowed elastomeric fibers to be made by economical solution or melt-spinning operations

Table 1. Spandex Producers[a]

Spinning method	Producer	Country	Trade name(s)
dry	DuPont and subsidiaries	United States	Lycra, Opelon, Likra
dry	Farbenfabriken Bayer	FRG	Dorlastan
dry	Asahi Chemical Industry	Japan	Roica
dry	Toyobo	Japan	Espa
dry	Tae Kwang	Korea	Acelan
wet	Fujibo	Japan	Fujibo Spandex
wet	F.I.E.S.	Italy	Lynel
wet	Gomelast	Venezuela	Spandaven
reaction	Globe Manufacturing	United States	Glospan, Cleerspan
melt	Nisshinbo	Japan	Mobilon
melt	Kanebo	Japan	Loobell
melt/biconstituent	Kanebo	Japan	Eslon
melt/biconstituent	Toray Industries	Japan	Shevelina

[a] Listed by spinning method in descending order of capacity.

without the cross-linking step required in conventional rubber processing. Properties were surprisingly good, with a modulus about twice that of natural rubber.

Commercial spandex fibers contain various additives to modify luster, protect the fiber from degradation and discoloration, or provide dye sites.

Other Elastomeric Fibers. Besides natural rubber, synthetic rubbers such as neoprene and nitrile rubber are used in cut rubber thread or strips where solvent resistance is required, as in dry cleaning. Besides spandex, other segmented elastomers, such as those with a polyether soft segment and a poly(tetramethylene terephthalate) hard segment, have been developed; these are melt-spun into threads.

Biconstituent Elastomeric Fibers. These bicomponent fibers are usually melt-spun. Their hard-fiber component may be a polyamide or polyester, whereas the elastomeric component is a segmented polymer with polyurethane or polyester hard segments.

Fiber Manufacture

Cut Rubber. To produce cut rubber thread, rubber is milled and compounded with vulcanizing agents, antioxidants, and pigments in the usual manner and calendered into sheets for vulcanization. Multiple sheets can be vulcanized together in layers. The vulcanized sheets are cut into threads with an approximately square cross section or into strips with a rectangular cross section. Properties can be controlled by the vulcanization recipe affecting cross-link density and by fillers.

Extruded Rubber. A latex mixture is forced through a capillary tube into an acid bath, where it is coagulated; the threadlike coagulum is pulled from the bath, and washed, dried, and cured continuously.

Spandex. *Polymerization.* Spandex polymer is usually made in three steps: synthesis of the macroglycol, reaction with excess diisocyanate, and chain extension.

Additives. In solution spinning, additives, including pigments and stabilizers, are mixed with solvent and added as a slurry to the polymer solution. In solvent milling (wet grinding), a small amount of urethane polymer may be added to increase viscosity and slurry stability. In melt or reaction spinning, additives are mixed with the macroglycol, and their possible reaction with diisocyanates should be taken into account.

Dry Spinning. Most spandex is produced by dry spinning, a solution process that allows economical preparation of yarn with superior properties due to urea hard segments.

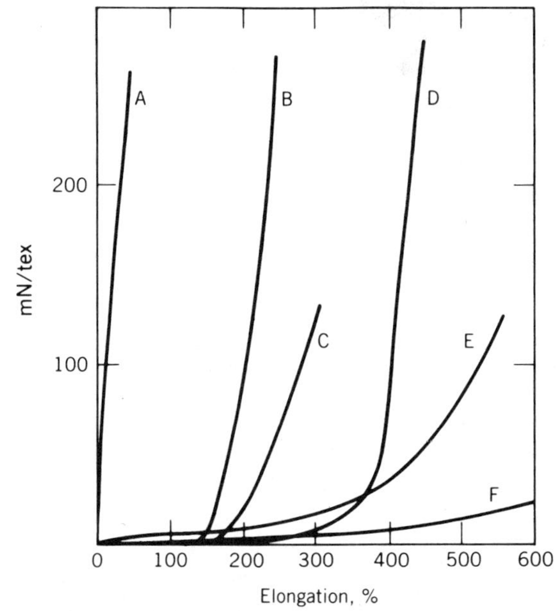

Figure 1. Stress–strain curves: A, textile nylon; B, bicomponent nylon; C, biconstituent elastomeric fiber; D, textured nylon; E, spandex; F, extruded rubber. To convert N/tex to g/den, multiply by 11.33.

Wet and Reaction Spinning. In these processes, a bath is used to coagulate or chain-extend the polymer to form spandex.

Melt Spinning. Only a few spandex producers use a melt-spinning process.

Physical Properties

Yarn Size. The size of spandex yarns has customarily been expressed in denier, ie, grams per 9000 m. The SI unit, however, is the tex or grams per 1000 meter.

Stress–strain Properties. Simple stress–strain curves of elastomeric fibers, as shown in Figure 1, give information on the force needed to stretch the fiber for the first time and the tenacity and elongation at break. These properties are important for textile processing, as they determine knitting tensions, resistance to breaks, and available stretch.

To measure retractive force, called "power," cyclic tests are used where the fiber is stretched a number of times and the retractive force measured on the return of the last cycle.

Properties of several elastomeric fibers are given in Table 2.

Table 2. Physical Properties of Elastomeric Fibers Cycled to 300% Elongation

Fiber	Tex[a]	Tenacity, mN/tex[b]	Elongation at break, %	5th cycle unload power, mN/tex[b] At 100% elongation	At 200% elongation	Set after 5 cycles, %
Spandex						
dry-spun	4.4	90	456	1.59	2.58	23
wet-spun	6.7	53	500	0.81	1.61	26
reaction-spun	7.8	42	517	1.24	1.93	24
melt-spun	4.2	67	381	1.26	2.05	23
Rubber						
extruded	88.0	24	590	0.70	0.90	12

[a] To convert tex to den, multiply by 9.

[b] To convert mN/tex to g/den, multiply by 0.0113.

Biconstituent elastomeric and other bicomponent fibers generally lack elongation for comparison with elastomeric fibers in these tests.

Set. After cycling to 300% elongation, the elastomeric fibers do not immediately return to their original length as the retractive force drops to zero at 10–20% elongation. This set is not permanent, because a simple relaxed heat treatment, eg, in boiling water, restores the fiber practically to its original length.

Permanent Set. A different type of set, which is more permanent, can be achieved by heat-setting spandex, which is not cross-linked, in fabric form.

Fabric Testing. The elastic properties of one-way stretch fabrics are usually determined by sewing a strip into a loop and measuring elastic power on cycling and stretchability, ie, elongation under very high stress.

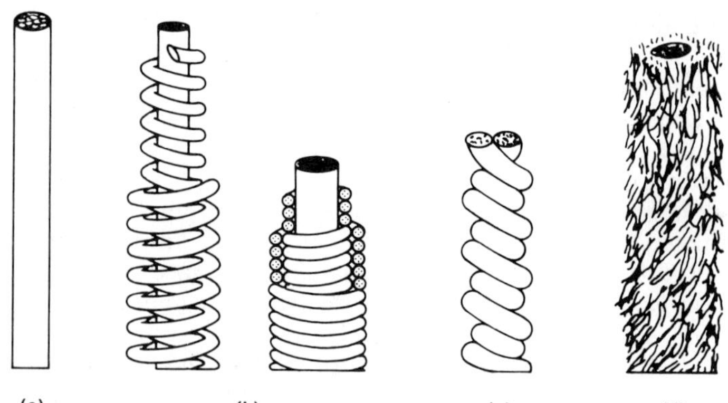

Figure 2. Forms of elastomeric yarn: (**a**) bare; (**b**) double-covered (shown during covering and relaxed, respectively); (**c**) single-covered or core-plied; (**d**) core-spun.

Chemical Properties

Color. With the help of phenolic and amine-type stabilizers acting as antioxidants, the whiteness retention of the best spandex fibers approaches that of hard fibers. The color of rubber, on the other hand, is poorer than that of spandex because of the effects of vulcanization.

Oxidative Degradation. The oxidative attack by heat, light, atmospheric contaminants, and chlorine degrades the polymer by a free-radical chain reaction. Even in the presence of antioxidants, rubber is subject to attack by ozone, which has little effect on spandex. Spandex with polyether soft segment is more sensitive to oxidative degradation than spandex with polyester soft segment, but with an effective stabilizer, spandex has a significantly longer stretch life than rubber.

Hydrolytic and Mildew Resistance. Hydrolytic stability is important in acidic dyeing and for mildew resistance in swimwear. Rubber and polyether-based spandex have excellent resistance under these conditions, whereas polyester-based spandex is more sensitive.

Solvent Resistance. Solvent stability is important in scouring and dry cleaning. Rubber swells in hydrocarbon solvents, which have little effect on spandex. Perchloroethylene swells spandex considerably, but after evaporation the physical properties are not affected.

Dyeing. Although rubber cannot be dyed, spandex has an affinity for dispersed and acid dyes, including the chromed and metallized dyes.

Processing

In commercial applications, elastomeric fibers are always used in combination with hard fibers. Either the elastomeric fiber is used bare in a woven or knit structure, or it is first combined with hard fiber by covering, core-spinning, or core-plying (Fig. 2).

After incorporation in the fabric, heat-settable spandex may undergo further changes during heat-setting, boarding, or molding. The purpose of these treatments is to set fabric dimensions, stretch, and power; stabilize the fabric against wrinkling in subsequent finishing operations; or shape finished garments such as socks and hosiery.

Uses

Elastomeric yarns are used mainly in clothing, such as hosiery, swimwear, foundations, underwear and narrow fabrics, where they provide stretch and comfort. Other uses include medical supplies, parachutes, golf balls, automotive seats, and upholstered furniture.

BIBLIOGRAPHY

G. H. R. Weiss, *Nat. Rubber Technol.* **10,** 80 (1979).

M. Couper in M. Lewin and J. Preston, eds., *High Technology Fibers,* Marcel Dekker, Inc., New York, 1985, pp. 51–85.

K. Ichikawa in K. Ashida and K. C. Frisch, eds., *International Progress in Urethanes,* Vol. 3, Technomic Publishing Co., Lancaster, Pa., 1981, pp. 241–259.

K. H. Wolf, *Text. Prax.* **1981,** 839 (1981).

A. J. Ultee
E.I. du Pont de Nemours & Company, Inc.

FIBERS, ENGINEERING

Earth Sciences

Agriculture. Product development of fibrous structures for the agricultural market involves mostly sheet structures. Key properties are light, air, and moisture permeability; the fiber plays a limited role in determining these characteristics. Fiber properties are more important when cost effectiveness is considered in terms of weight, strength, and dimensional stability. Also, fibers are critical in determining uv stability, moisture pickup, and thermal-use limits of the sheet. Fibers in current use and suppliers of sheet structures are listed in Table 1 by agricultural applications and fiber type.

Soil Engineering. In soil engineering, fibrous structures are used for separation, drainage, reinforcement, and filtration. When so used, the structures are called geotextiles. Tables 2 and 3 list applications and properties of fibers used in geotextiles. Geotextile selection for a project is based on improved design performance such as more efficient installation, reduced maintenance, increased life, or a substantial cost savings compared to alternatives.

Water Purification, Separation. Reverse osmosis, the production of pure water from saline water utilizing pressure as the driving mechanism, is one of the major markets for hollow-

Table 1. Fibers in Agricultural Applications

Use	Fiber	Manufacturer
row covers	polyester	Reemay, Inc., Unitika, Freudenberg
	polypropylene	Kimberly Clark, Sodoca, Freudenberg
	nylon	Chicopee
	polyamide	Beghin-sey
tobacco seed	polyester	Reemay, Inc.
bed covers	polypropylene	Kimberly Clark
	nylon	Chicopee
	cotton	Chicopee
turf covers	polyester	Reemay, Inc., Hoechst
weed barrier	polyester	Hoechst
	polypropylene	Reemay, Inc., Phillips, Dewitt

fiber membranes. Other areas of application of hollow fibers are ultrafiltration, the separation of dissolved species based on size differences, and the separation of gases. The use of membranes for these separations is the most energy-efficient method known. The greatest advantage of fibrous membranes over other membrane forms is their extremely high surface area per device unit volume. A wide variety of polymers with differing physical and chemical properties can be spun as hollow fibers.

Types of hollow fibers used are shown in Table 4.

Biomedical Applications

Fibers have been used for many years in medical applications such as dressing, sanitary pads, tapes, and operating-room apparel. Application of engineering fibers include sutures, surgical implants, prosthetic devices, and medical equipment. Engineering fibers are also used for separation and purification such as kidney dialysis, and they all play important roles in emerging biotechnology applications such as substrates for large-scale cell culture. These fibers aid in maintaining and improving the quality of life.

Fiber Properties. Fibers possess certain advantages applicable to biomedical uses. The surface-to-volume ratio is very large, which is important in surface dominated applications such as separations. Engineering fibers can also be processed into knitted, woven, or hollow structures that provide the varying degrees of porosity and strength needed in biomedical applications such as synthetic blood vessels, artificial ligaments, and controlled drug release.

Many engineering fibers are used within the body because of their good mechanical properties, long-term stability, and nontoxic character. These fibers are also biocompatible; that is, they do not produce any significant changes in the surrounding tissue over long periods of time. Biocompatibility of the fibers depends on the type of polymer from which they are made, on processing additives, and on the surface properties of the fiber. Woven or knitted fibers may approximate the stress–strain behavior of natural tissue more closely than other structures and thus offer greater compatibility. This is of particular importance in applications such as the replacement of natural blood vessels.

Fiber Types. Both natural and synthetic engineering fibers are used in the biomedical field (Table 5).

Table 2. Uses for Geotextile Fabrics

Application	% of all uses	Function of fabric
stabilization	28	separation and reinforcement
primary road	26	reinforcement and water-proofing
secondary road	13	separation and reinforcement
drainage	13	drainage and separation
erosion control	10	drainage, filtration, and separation
silt fence/other	7	filtration
railroad tracks	4	reinforcement and separation

Table 3. Fibers Used in Geotextiles

Fiber	Strength, N/tex[a]
aramid	2.2
polyamide	0.53–0.88
polyester	0.62–0.88
polypropylene	0.31–0.62
polyethylene	0.31–0.62

[a] To convert N/tex to g-f/denier, multiply by 11.3.

Advanced Composite Materials

In order to meet the requirements of advanced military systems, a number of low density fibers have been produced that exhibit both high specific stiffness and high specific strength.

High Performance Fibers. Carbon and silicon are the principal constituent elements of commercially available high performance fibers. These fibers can be grouped into three broad categories; carbon, organic resins, and inorganic compounds. Within each group, several classes of high performance fiber materials have been developed that satisfy the basic criteria of low density, high strength, and high stiffness in varying degrees.

Carbon Fibers. Carbon fibers (qv) are currently the predominant high strength, high modulus fiber used in the manufacture of advanced composite material. These fibers are made by the pyrolytic degradation of a fibrous organic precursor. Representative properties are given in Table 6.

Most commercially available carbon fibers are obtained by the pyrolysis of polyacrylonitrile (PAN).

Low strain-to-failure characteristics have been the principal limitation of standard PAN-derived carbon fibers, but high strain carbon fibers are now available that exhibit tensile elongations on the order of 1.7–1.8%.

Organic Fibers. The measured tensile strength and modulus of most natural and synthetic organic fibers are significantly lower than the theoretical values of the maximum tensile strength (>20 GPa) and modulus (>100 GPa) based on the force constants of bonds in the polymer chain and crystal extensions observed by x rays on stressed fibers. A number of polymeric fibers are now commercially available whose strength and modulus more closely approach these theoretical values and are high enough for these materials to be considered as reinforcements for advanced composites. The materials include primarily aromatic polyamides, but also ultrahigh molecular weight polyolefins and certain classes of polyesters. Compared to textile fibers, these high strength, high modulus materials all exhibit a high degree of molecular orientation and chain alignment (see also HIGH MODULUS POLYMERS; POLYAMIDES, ARO-

Table 4. Types of Fibers Used in Purification

Polymer	Hydrolytic stability	Oxidative stability	Biological stability	Acceptable pH range
aramid	excellent	poor	excellent	4.0–11.0
cellulose acetate	poor	fair	poor	4.5–7.5

Table 5. Engineering Fibers Used in Biomedical Applications

Type	Examples
natural fibers	collagen, silk, cellulose, chitin
synthetic organic fibers	polyester, polyamide, polypropylene, polyurethane, polytetrafluoroethylene, polyethylene, poly(vinyl alcohol), polyacrylonitrile, poly(glycolic acid), poly(lactic acid), regenerated cellulose, polydimethylsiloxane, aramid
inorganic fibers	carbon, graphite, alumina, glass

MATIC; POLYESTERS, MAINCHAIN AROMATIC). In the early 1970s, DuPont introduced Kevlar aramid fibers, which are based on para-substituted aromatic polyamides, most notably poly(*p*-phenyleneterephthalamide)(PPD-T).

Several types of Kevlar fibers offer different combinations of properties to satisfy the requirements of a variety of uses. Selected properties of Kevlar 49 and Kevlar 29, two products used in plastics reinforcements, are presented in Table 7.

Inorganic Fibers. Even though the densities of inorganic fibers are significantly higher than those of carbon and polymer fibers, these materials are of interest in advanced composites because of their resistance to high temperatures and their compatibility with metal and ceramic matrices. Inorganic fibers (qv) of commercial interest are based on silica (Table 8), alumina, and silicon carbide (Table 9) (see also PRECERAMIC POLYMERS).

Matrix Materials. The matrix of an advanced composite material is the continuous phase that binds the fibers together so they can act in concert. The main characteristics required of a matrix material are chemical compatibility with the reinforcing fibers, compatibility with the manufacturing methods used to fabricate the desired advanced composite component, and environmental stability under conditions of use of the advanced composite component.

The matrix material has the primary influence on mechanical properties of interlaminar shear strength, and compression and flexural strength, especially at high temperatures, and also dictates the processibility of the composite and its environmental resistance.

Resin-matrix materials dominate the technology and the commercial use of advanced composite materials.

Carbon-, ceramic-, or metal-matrix composites are being developed principally for use in high temperature environments hostile to resin-matrix materials or in applications that require characteristics such as electrical conductivity or the toughness of metals.

Epoxy resins are by far the most important polymeric-matrix material because of the good balance of properties obtained with relative ease of handling and processing. Thermoplastic resins have not been used as matrix materials for continuous-fiber reinforced composites, in spite of their greater ease of processing, because of their lower creep resistance at elevated temperatures. Furthermore, thermoplastics as a class tend to be more susceptible to solvent attack than thermosets. However, thermoplastic materials that exhibit good high temperature properties and reasonable solvent resistance have become available, eg, aromatic polyesters, poly(phenylene ether), poly(phenylene sulfide), polysulfone, aromatic polyetheretherketone (PEEK), and polyetherimides, polyesterimides, or po-

Table 6. Representative Properties of Standard PAN-based Carbon Fibers

Designation properties	Type I high strength	Type II high modulus	Type III ultrahigh modulus
carbon content, wt %	92–94	>99	>99.9
specific gravity	1.7–1.8	1.8–1.9	1.9–2.1
filament diameter, μm	7–8	7–8	8–9
tensile modulus, GPa	220–250	340–380	520–550
tensile strength, GPa	2.5–3.5	2.2–2.4	1.8–1.9
tensile elongation, %	1.2–1.4	0.6–0.7	0.3–0.4
toughness, MPa	20	7.5	3
electrical resistivity, $\mu\Omega$/m	15–18	9–10	6–7
longitudinal coefficient of thermal expansion, 10 m/(m·K)	−0.5	−0.7	−0.9 (est)
range of available tow counts, thousand fibers/bundle	1–320	1–12	<1
price range, 1985 $/kg	40–300	90–1000	>300

Table 7. Representative Properties of Kevlar *p*-Aramid Fibers

Properties	Kevlar 29, standard	Kevlar 49, standard	Kevlar 49, improved
specific gravity	1.44	1.44	1.44
filament diameter			
tex[a]	0.17	0.17–0.25	0.13
μm	13	13–16	11
tensile modulus, GPa	62[b]	131	124
tensile strength, GPa			
unimpregnated yarn[c]	2.8	2.7	
impregnated strand[d]	3.8	3.8	4.1
tensile elongation, %	4.0	2.8	3.3
toughness, MPa	87	55	68
electrical resistivity, μΩ/m			
longitudinal coefficient of thermal expansion, 10^{-6} m/(m·K)		−2	
range of available tow counts, thousand filaments/strand	0.13–10	0.13–2	4

[a] To convert tex to denier, multiply by 9.
[b] Initial modulus.
[c] As per ASTM Test D 885.
[d] As per ASTM Test D 2343.

lyamideimides. These materials all contain an aromatic backbone, but differ in the way the aryl groups are linked.

Processing and Fabrication Technology. Most fibers used in reinforced advanced composite material have a diameter of 5–20 μm.

The manufacture of advanced composite components entails surrounding the fibers with the matrix material in a nonrigid state, arranging the reinforcing fibers in selected patterns and orientations, solidifying the matrix material under constrained conditions that prevent fibers from moving during solidification, and integrating the advanced composite into the finished component or system.

Properties of Composite Structures. A unique feature of composite materials is the directional dependence or anisotropy of the various properties. This feature is intuitively clear for a unidirectional lamina: properties along the direction of the fiber reflect the higher performance characteristics of the fiber whereas properties perpendicular to the fiber tend toward the behavior of the matrix material. This directional dependence introduces the opportunity to tailor the material to specific applications. The alignment of stiff and strong fibers along anticipated load paths can yield highly efficient structures.

The major applications of advanced composite materials are: aerospace equipment (satellite structures, rocket component, military aircraft, rotorcraft, business aircraft, secondary structures of commercial transports), marine equipment (spars for racing boats, hulls for small craft, propeller shafts), ground transportation equipment (racing-vehicle components, limited-production automobile body and chassis parts, springs and drive shafts of heavy-duty truck engine components), industrial and scientific equipment (process piping, oilfield structures, robotic components, high speed machinery, safety equipment including ballistic protection, communications antennae,

Table 8. Representative Properties of Some Inorganic Oxide Fibers

Properties	E-glass	S-glass S2-glass	Quartz	Ceramic	Alumina
specific gravity	2.60	2.49	2.2	2.7	3.25
filament diameter, μm	9(*)	9(*)	10	8–12	17
range of available tow counts, thousands filaments/strand	0.2–2	0.2	0.3	0.39	1
price range, 1985 $/kg	1	6–30	100–400	180	760
Properties at ambient					
tensile modulus, GPa	72	87	69	152	200
tensile strength, GPa	3.45	4.6	0.9	1.5	1.5
tensile elongation, %	4.8	5.4	1.3	1	0.8
toughness, MPa	83	124	5.9	7.7	5.6
dielectric constant 10e10 Hz	6.1	5.2	4.5	4.7	
longitudinal coefficient of thermal expansion, 10^{-6} m/(m·K)	5	5.6			
High temperature properties					
melt temperature, °C	1260		1650	1800	
max use temperature, °C	600	760	900	1200	
90% tensile strength retention temperature, °C	220	275		1100	1250

Table 9. Representative Properties of SiC Fibers

Properties	Chemical vapor deposition[a]	Pyrolysis[b]
specific gravity	3	2.55
filament diameter, μm	140	10–20
range of available tow counts, thousand filaments/strand	not applicable	0.5
price range, 1985 $/kg	1100	500
Properties at ambient		
tensile modulus, GPa	430	190
tensile strength, GPa	2.4	2
tensile elongation, %	0.56	1.1
toughness, MPa	13	11
electrical resistivity, μΩ/m		0.1
longitudinal coefficient of thermal expansion, 10^{-6} m/(m·K)		3.1
High temperature properties		
melt temperature, °C	2700[c]	2700[c]
max use temperature, °C	1150	
90% tensile strength retention temperature, °C	800	300

[a] Fiber supplied by Avco.
[b] Fiber supplied by Nippon Carbon Co./Dow-Corning.
[c] Sublime.

electronic-circuit boards, medical devices), and consumer products (sporting goods, musical instruments).

Industrial Applications

Industrial applications of engineering fibers include rubber goods, eg, tires and V-belts, ropes and cables, friction products and gaskets, and ballistics applications.

Electro-optical Uses

Engineering fibers are making contributions to the transmission of information, the miniaturization of computer components, and the efficient handling of electrical power.

Optics. The largest use of optical fibers is the transmission of information by light, as a direct substitute for transmission

Table 10. Reinforcing Fibers for Printed-circuit Boards

Properties	E-glass fiber	*p*-Aramid fiber	Quartz fiber
Bulk			
density, g/cm^3	2.54	1.44	2.2
Electrical			
dielectric constant, 1 MHz	6.32	3.6	3.7
dissipation factor, 1 MHz	0.005		0.0001
volume resistivity, Ω-cm	1×10^{15}	1×10^{15}	1×10^{15}
moisture regain, %			
50% rh	neg	3.2	neg
75% rh	neg	4.8	neg
90% rh	neg	5.7	neg
Mechanical			
tensile stiffness, GPa[a], at 20°C	72	110	69
tensile strength, GPa[a]	3.5	3.6	0.9
elongation, %	4.6	2.9	1.3
Thermal			
coefficient of thermal expansion	5.0	−5.0	0.5
softening point, °C	841		1670
decomposition temperature, °C		500	
thermal conductivity, W/(m·K)[b]	10.4	NA	
Chemical Resistance			
acids	excellent	fair	excellent
bases	excellent	fair	excellent
organics	excellent	excellent	excellent
Other			
drillable	excellent		
clean hole	excellent	clean when plasma etched	excellent
drill bit wear	acceptable	acceptable	poor
punchable	unacceptable	unacceptable	unacceptable
PWB relative cost[c], $/15 m	1	20	20

[a] To convert GPa to psi, multiply by 145,000.
[b] To convert W/(m·K) to (Btu·in)/(h·ft^2·°F), multiply by 6.94.
[c] Cellulose paper relative cost, $0.5/15 m.

Table 11. Typical Properties[a] of Fibrous Electrical Insulation Materials[b]

Material	Specific gravity	Thickness, mm	Elongation, %	Tensile strength, MPa[c]
Inorganic				
asbestos paper (with no binder)	0.72	0.088		0.90
glass cloth (with epoxy resin)	1.3–1.45	0.12–0.25	2.3	MD: 110–250
			2.4	CD: 66–131
Organic				
cellulosic, standard kraft paper	0.68–1.35	0.25	3.5	MD: 90–145
			7.5	CD: 19–55
aramid paper	1.01	0.25	21	
(calender, no binder)			18	
polyester paper (with epoxy resin)	1.5	0.38	24	MD: 46
			5–24	CD: 30
laminates (mat/film/mat)	0.97	0.25	42	
			46	

[a] Not for establishing specifications.
[b] Source: published product data and DuPont test results.
[c] To convert MPa to psi, multiply by 145.

of electronic signals over metal wires, but optical fibers have many other uses based on their unique ability to transmit light. For example, they are used in decorative objects and signs.

Electronics. Fibers are making substantial contributions to the miniaturization of electronic computer components. A printed-circuit or wiring board (PCB or PWB) functions primarily as an electrically insulating substrate on which various electrical or electronic components (resistors, capacitors, integrated circuits, etc.) are mounted, connected by relatively thin copper lines. PCBs eliminate bulky wiring, increase reliability of circuitry and interconnections, and provide miniaturization advantages. PCBs, mostly composites, are found in computers, telecommunications equipment, and electronics for military, automotive, industrial, and consumer uses. Table 10 compares properties of the three reinforcing fibers that are generally recognized as meeting the requirements for use in rigid printed-circuit boards.

Electrical Insulation. Inorganic fibers used for electrical insulation are asbestos and glass. Predominant synthetic fibers are polyesters and aromatic polyamides (aramids).

In most cases, fibers are used in fabricated form such as woven or nonwoven fabric or paper with or without resin impregnation. There are endless variations and combinations of fibrous, film, and resin systems. Table 11 compares the important properties of selected fibrous electrical insulation materials in their most widely used form.

BIBLIOGRAPHY

Geotextile Engineering Manual, Federal Highway Administration/National Highway Institute, 1984.

D. Annis and co-workers in E. Chiellini and P. Glusit, eds., *Polymers in Medicine,* Plenum Publishing Corp., New York, 1983.

D. Tanner, A. K. Dhingra, and J. J. Pigliacampi, *J. Met.* **38**(3), 21 (1986).

Analysis of the Market for PWB's and Related Materials for the Year 1982, Institute for Interconnecting and Packaging of Electronic Circuits, 1982 Program of the Technology/Marketing Research Council, 1982.

ASHOK K. DHINGRA
HERBERT G. LAUTERBACH
E.I. du Pont de Nemours & Co., Inc.

FIBERS, INORGANIC.

See INORGANIC FIBERS.

FIBERS, MANUFACTURE

Polyester, nylon, acrylic, polypropylene, and viscose fibers are today the principal commodity products (see also POLYAMIDES, FIBERS; POLYESTERS, FIBERS; OLEFIN FIBERS). Various higher priced specialty fibers are manufactured on a smaller scale. These include the elastane (spandex) fibers, which possess elastomeric properties, and the aramid fibers, which possess high strength and modulus or high chemical and flame resistance (see also FIBERS, ENGINEERING; FIBERS, ELASTOMERIC).

Man-made fibers, which include those made from natural organic polymers as well as synthetic organic polymers, are mostly produced by a process of extrusion through fine holes that is known as spinning, a word derived from and still applied to the formation of yarns from fibers of any kind. Since the natural polymers and their derivatives undergo decomposition at or below their melting points, they can be converted into a form suitable for spinning only by dissolution. Consequently, the two processes for extruding polymer solutions, known as wet and dry spinning, were developed well before the third process, melt spinning.

Usually the product obtained from the spinning process is not directly suitable for commercial use. If the polymer has undergone little or no orientation during extrusion, the resulting fiber is relatively weak or undergoes irreversible extension. In such cases the filaments are stretched or drawn immediately after extrusion or as a separate operation to increase tensile strength. They may then be subjected to further operations, according to the final usage. Thermal setting and thermal relaxation processes provide dimensional stability;

twisting and interlacing provide interfilament cohesion; texturing provides a voluminous yarn; crimping and cutting provide staple products similar in length and processing behavior to natural fibers.

Continuous-filament yarn, which consists of a small number of roughly parallel, continuous, individual filaments of unlimited length held together by a slight twist or by intermingling, is usually packaged as a tube or cone. Staple fiber is made up of a very large number of discontinuous, randomly oriented, individual fibers normally shipped in a box or bale. It is usually subjected to a series of processes, culminating in textile spinning to yarn. The precursor of staple fiber is tow, which consists of a large number of roughly parallel, continuous filaments. It is converted by cutting or breaking into staple fiber or directly into a top or sliver, intermediate stages between staple fiber and yarn. In the latter case the filaments remain parallel.

Substantial amounts of fiber are also sold as monofilament, which is a single filament yarn of substantially greater diameter than those present in continuous filament and staple yarn, and as tapes, fibrillated tapes, and slit-film products. In addition, certain types of so-called nonwoven fabric are directly formed from continuous filaments without isolation of a yarn. Products of this type include melded and spun-bonded fabrics.

Individual polymers and the fibers derived from them are dealt with in other articles in this Encyclopedia (see also FIBERS; POLYAMIDES, FIBERS; POLYESTERS, FIBERS). Polymers of natural origin that have been converted into fibers by manufacturing processes include cellulose (derived mainly from wood pulp), which is the source of cupro, viscose, modal, acetate, and triacetate fibers; alginic acid from seaweed, used to spin alginate fibers; and arachin, zein, and casein, from peanuts, maize, and milk, respectively, and soya-bean protein, which have each been the basis of protein-fiber manufacture. Synthetic polymers for fiber production are made by melt, solution, and suspension techniques; addition and condensation polymerization; chain-growth and step-growth processes; free-radical, anionic, and cationic mechanisms; and homogeneous and heterogeneous methods. Continuous polymerization processes are then used to produce polymers for commodity fibers. Many synthetic fibers are based on copolymers that contain units introduced to modify physical or chemical properties such as dyeability, solubility, or flame retardancy.

Spinning Processes

The most commonly used spinning processes are melt spinning, dry spinning, and wet spinning.

Melt Spinning. Melt spinning was developed in the late 1930s for nylon-6 and nylon-6,6 and is now used for several fibers, including polyamides, polyesters, and polyolefins. In this process molten polymer is forced by an extruder or gear pump through the fine holes in a spinneret and solidified by passage through cool air to a driven roll that draws the fibers away from the face of the spinneret (Fig. 1).

Dry Spinning. In dry spinning, a polymer solution is extruded from a spinneret through a zone in which the solvent is rapidly evaporated, leaving filaments that are wound up at speeds up to about 1000 m/min (Fig. 2).

Wet Spinning. In wet spinning, a polymer solution is extruded from a spinneret into a nonsolvent that coagulates (ie, precipitates) the polymer (Fig. 3). In some cases, such as the viscose process, both the solution and the precipitation stages involve chemical reaction. Conventional wet spinning is by far the slowest in linear velocity of the principal processes, but since it permits very short distances between the holes in the spinneret face, a single spinneret may carry a very large number of holes. Hence, the productivity from a single spinning position may be very high.

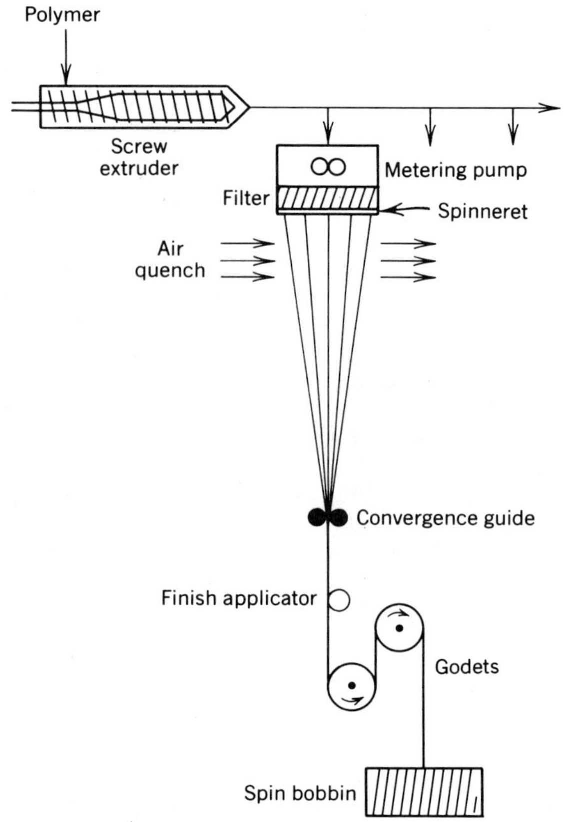

Figure 1. Typical melt-spinning arrangement.

Other types of fiber-forming process include: reaction spinning; dispersion, emulsion, and suspension spinning; fusion-melt spinning; phase-separation spinning; and gel spinning.

Numerous techniques for producing fibers without using a spinneret have been devised, including centrifugal spinning, electrostatic spinning, and tack spinning.

Fiber-orientation Processes

Processes for manufacturing fibers usually convert an isotropic polymer into a highly anisotropic form. The macromolecules are constrained, resulting in main chains more nearly parallel to the axis of the fiber. The more closely the chains parallel the axis, the more highly oriented the fiber is said to be.

Yarns and tows are usually drawn to orient them by stretching them between two rolls or two sets of rolls. The individual filaments thus extend irreversibly to several times their original length. The ratio of the stretched length to the unstretched length, which is controlled by the ratio of the surface speeds of the rolls, is known as the draw ratio. In many cases heat is applied to the filaments just before or during stretching in order to facilitate the process.

In recent years, modified spinning processes that introduce orientation into the yarns at the extrusion stage have been developed for some fibers. In the case of melt-spun fibers this is achieved by using very high wind-up speeds. Some of these

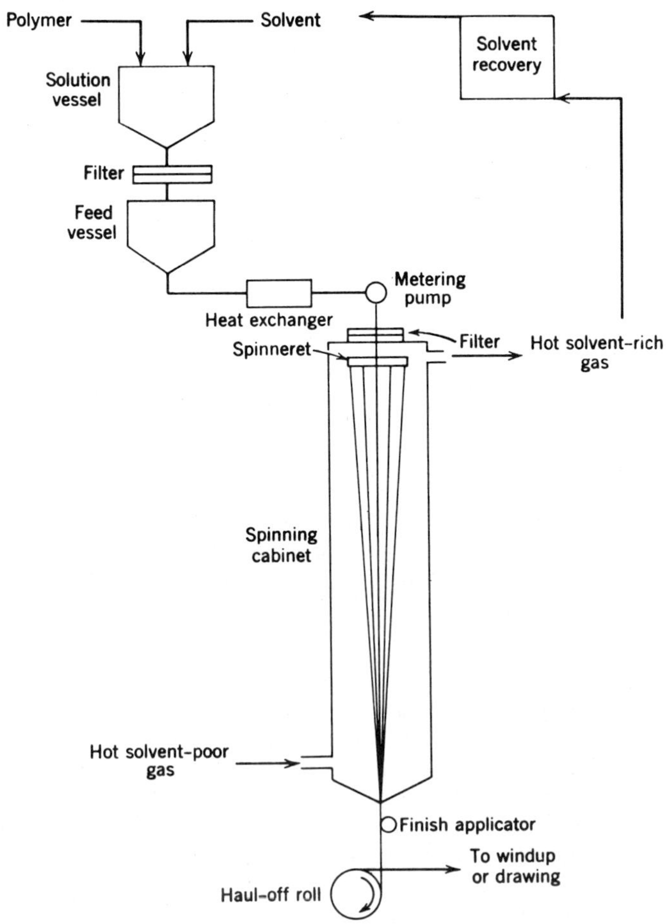

Figure 2. Typical dry-spinning arrangement.

processes provide a partially oriented yarn, which requires only a small draw ratio to complete the orientation, whereas others provide a fully oriented yarn directly without further drawing.

Twisting and Intermingling

Continuous-filament yarns that lack cohesion between filaments serve poorly in textile manufacturing operations because of their tendency to form a ribbon over surfaces such as guides and because individual filaments may become sufficiently separated from the rest to snag and break. Traditionally, cohesion is provided by introducing twist into the filament bundle. Conventional up-twisting and down-twisting processes, as practiced generally in the textile industry, are frequently applied to filament-yarn production. In draw-twisting, the drawing stage is integrated sequentially with down-twisting. Introduction of true twist by methods such as this is slow and expensive and for many products has been replaced by the much faster process of draw-winding. The most important method of introducing cohesion into draw-wound yarns is intermingling, which employs air jets to produce turbulent zones in which individual filaments in a yarn are slightly interlaced. Intermingling is sometimes used as an ancillary to other processes such as draw texturing, where it creates yarn cohesion and also generates specialty products such as heather yarns. Here two or more yarns are combined in the intermingling jet (comingling). When filament yarns of higher twist or folded yarns are required, the two-for-one twisting technique is employed.

Textured Yarns

The extruded and drawn synthetic yarns consist of straight and parallel filaments. Texturing introduces permanent distortions, crimp, loops, coils, or crinkles without destroying the essential continuity of the filaments. This improves the texture of the yarns and extends their range of applications.

Some texturing processes also introduce stretch properties. The most important texturing processes for continuous-filament yarns arc false-twist texturing, air-jet (entanglement) texturing, stuffer-box texturing and jet texturing (using hot air or steam in the jet). The first two are usually applied to relatively light yarns for apparel use, and the last two to heavier yarns for carpets and upholstery. False-twist texturing is almost invariably combined with drawing, sometimes sequentially, but more often in the process known as simultaneous draw texturing in which drawing and texturing processes are carried out within the same zone.

Tow, Top, and Staple Fiber

Tow for staple fiber manufacture passes after drawing through a number of ancillary processes, which may include washing,

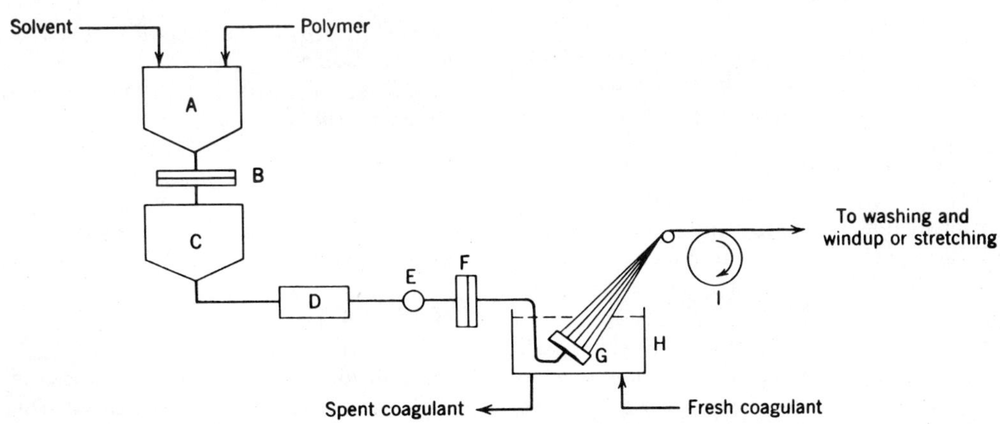

Figure 3. Typical wet-spinning arrangement: A, solution vessel; B and F, filters; C, feed vessel; D, heat exchanger; E, metering pump; G, spinneret; H, spin bath containing coagulant; I, haul-off roll.

finish application, heat-setting, dyeing, drying, crimping, cutting or breaking, and baling. Crimping is usually carried out by the shifter-box method. Tow is often cut to staple fiber by forcing it against the sharp edges of blades arranged around the circumference of a rotating cutting reel. The cut fiber is baled and passed to the manufacturer for staple-yarn production. It may then be processed alone or in blends with natural fibers. In direct crush-cut or stretch-break conversion systems, tow is directly reduced into staple in the form of a sliver or top, without losing the intrinsic parallel arrangement of the filaments of the original tow.

Finishes

Handling and processing of fibers and yarns are improved by the application of finishes, typically consisting of a mixture of components, including a lubricant, such as hydrocarbon oil or a long-chain aliphatic ester, an emulsifier, an antistat, a bactericide, and an antioxidant. Finish is generally applied during the spinning process before windup in order to reduce friction against guides, improve interfilament cohesion, and eliminate filament separation arising from electrostatic charge development. More finish may be applied later to modify yarn behavior during processes, such as stuffer-box crimping, or to improve the handling in textile processing.

Manufacture of tapes, slit-film and split-film products for carpet backings, geotextiles, ropes, twines, and many other uses is widespread, with a large number of relatively small plants. The production of split film is based on the fact that uniaxially drawn film, especially if very highly drawn, splits much more readily parallel to the direction of orientation than transversely. Polypropylene is easy to fibrillate in this way and is the most important polymer in this field, although high density polyethylene is also widely used.

The film is extruded by conventional hot die (with water quench or chill roll) or tubular film technique, slit into tapelike strips, drawn to produce orientation, fibrillated, and wound up.

Coloration and Delustering

The yarn, fabric, or other product is usually dyed after fiber manufacture. This procedure readily accommodates the variety of color and fashion changes and minimizes inventories and distribution costs earlier in the process. Nevertheless, fiber is frequently colored during manufacture when dye sites are lacking (eg, in polypropylene), when the dyeing rate is low (eg, aramid fibers), or when high fastness to washing, rubbing, solvents, or light and large quantities of colors are required. Mass coloration of solution-spun fibers, called dope dyeing, requires dye-stuffs that can be dispersed or dissolved in the spinning solution before extrusion. Mass coloration of melt-spun fibers requires thermal stability of the dyestuff up to about 300°C. The color most frequently introduced by mass coloration is black, usually by applying special grades of carbon black. The incorporation of delusterant is usually not classified as coloration. Nearly all synthetic fibers intended for textile use are delustered by addition of a white pigment of high refractive index.

The term multiconstituent (biconstituent, etc) covers all multiphase fibers (Fig. 4) with definable phase boundaries made from more than one constituent. The two main classes are those where the constituents are combined before extrusion by feeding each to a defined point in the cross section in such a way that each remains continuous along the length of the filament and those where the constituents are blended before extrusion in such a way that one or more is dispersed in a random way.

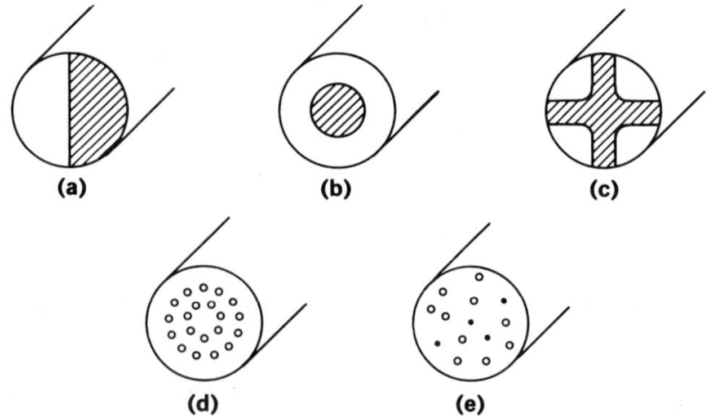

Figure 4. Multiconstituent fibers: (**a**) side–side; (**b**) sheath-core; (**c**) multiple-segment; (**d**) islands-in-the-sea; (**e**) matrix-fibril.

Inorganic Fibers

The most important inorganic fiber is glass, which is produced by melt-spinning processes in both continuous-filament and staple forms. Refractory fibers are produced from alumina–silica by centrifugal spinning and by air- and steam-blast techniques. Refractory fibers of still higher melting point are produced from aluminum and zirconium oxychlorides, which are incorporated into an organic carrier fiber such as viscose or poly(vinyl alcohol). Boron, boron nitride, silicon carbide, and silicon nitride fibers are also produced commercially.

Production of man-made organic fibers in the world exceeds 20 million t per year, of which synthetic fibers represent over 80% and polyester fiber alone about 40%. Production of staple and tow exceeds that of continuous-filament yarns and monofilaments, and approaches 60% of the total synthetic fiber production.

BIBLIOGRAPHY

M. Grayson, ed., *Encyclopedia of Textiles, Fibers and Nonwoven Fabrics,* Wiley-Interscience, New York, 1984.

M. Lewin and E. M. Pearce, eds., *Handbook of Fiber Science and Technology, Vol. IV: Fiber Chemistry,* Marcel Dekker, Inc., New York, 1985.

F. Happey, ed., *Applied Fiber Science,* Vols. 1–3, Academic Press, London, 1977–1979.

B. von Falkai, ed., *Synthesefasern,* Verlag Chemie GmbH, Weinheim, Germany, 1981.

J. E. McIntyre
M. J. Denton
University of Leeds

FIBERS, OPTICAL

A transmission source for fiber optics, as it is known today, had its inception in 1960 with the discovery of the laser. At that time, glass fibers had a transmission loss of ca 1000 dB/km. In 1966, scientists at Standard Telecommunications Labora-

Table 1. Composition and Properties of Cable Components

Fiber coating	Filling compound	Jacket compound
Requirements		
primary		
low modulus down to −40°C, thermally and hydrolytically stable, adhesion to glass, no H_2 evolution	low modulus down to −40°C, hydrophobic, no flow at elevated temperature[a], no H_2 evolution	same as conventional copper cable
secondary		
tough, abrasion resistant, thermally and hydrolytically stable, no H_2 evolution		
Composition		
primary		
silicone[b], rubber hot melt, or acrylate[c]	oil-extended rubber, greases, polybutylene	polyethylene, plasticized PVC, thermoplastic polyurethane, etc
secondary		
nylon or acrylate[c]		

[a] Dependent on design application.
[b] Thermally or uv curable.
[c] Uv curable.

tories in the UK proposed that a clad fiber with a transmission loss under 20 dB/km was possible if scattering and absorption losses from impurities were reduced. A clad fiber consists of a doped silica core surrounded by a silica cladding. In 1970, Corning Glass Works produced the first such fiber with less than 20 dB/km transmission loss.

The current state of the art has been made possible by advances in glass technology, transmission and detection devices, and methods of packaging glass fibers. In the last area, polymers are commonly applied. The packaging of glass fibers is necessary to preserve fiber strength and prevent so-called microbending-induced transmission loss.

The most common solution to the microbending problem is the application of two polymer coatings, a soft inner (primary) coating for microbending protection and an outer (secondary) coating of a stiffer material providing toughness and abrasion resistance.

The types of polymeric materials used for the cable components and important property requirements are given in Table 1. Fiber-coating materials have received more attention than jacketing compounds. For the latter, commercially available materials used in copper cable are generally employed.

The current market trend is toward lower transmission loss. Some of the improvement in loss performance associated with microbending loss is obtained through innovative engineering of the polymeric materials used in the cable.

Coating Technology

Glass optical fibers require protective polymer coatings in order to preserve the inherent strength of the glass and provide the fiber with low microbending susceptibility. Since contact of the newly drawn glass fiber with any solid surface results in the immediate formation of microflaws that reduce fiber strength, polymer coatings must be applied as thin liquids that solidify by cooling or curing. The viscosity of the liquid coating is usually less than 10 Pa·s (100 P).

Waterproofing

In light-guide cable, water accelerates the static fatigue failure of the fiber as compared with a low or ambient humidity state and, if it freezes, induces microbending loss. Optical-fiber cable must be protected against water by pressurized air or by a filler.

Fiber Packaging

Sets of fibers are packaged within the cable in stranded, loose-tube, or V-grooved units or ribbons.

Fiber Joining

Fibers are joined by permanent splices and by connectors. The connectors are designed to be repeatedly connected and disconnected. Both methods require alignment of the two fiber cores.

Economic Aspects and Future Applications

Fiber optics has now penetrated significantly into the long-distance telecommunications network and will continue to do so. Growth for new communication cable must eventually come in other areas such as the subscriber loop.

BIBLIOGRAPHY

L. L. Blyler, F. V. DiMarcello, and J. C. Williams, *Technical Digest of OFC (Optical Fiber Communications) '85,* Optical Society of America, Washington, D.C., 1985.

P. Halley, *Fibre-Optic Systems,* John Wiley & Sons, Inc., New York, 1987.

Alvin C. Levy
Carl R. Taylor
AT&T Network Systems

FIBERS, VEGETABLE

Fibers for commercial and domestic use are broadly classified as natural or synthetic. The natural fibers are vegetable, animal, or mineral in origin. Vegetable fibers are derived from plants. Their principal chemical component is cellulose, and

Table 1. Vegetable Fibers of Commercial Interest

Commercial name	Other name, source	Botanical name of plant	Growing area
Bast fibers	*soft fibers*		
China jute	Abutilon	*Abutilon theophrasti*	China
flax		*Linum usitatissimum*	north and south temperate zones
hemp		*Cannabis sativa*	all temperate zones
jute		*Corchorus capsularis; C. olitorius*	India
kenaf		*Hibiscus cannabinus*	India, Iran, USSR, South America
ramie		*Boehmeira nivea*	China, Japan, United States
roselle		*Hibiscus sabdarifa*	Brazil, Indonesia (Java)
sunn		*Crotalaria juncea*	India
urena	cadillo	*Urena lobata*	Zaire, Brazil
Leaf fibers	*hard fibers*		
abaca		*Musa textilis*	Borneo, Philippines, Sumatra
cantala	Manila maguey	*Agave cantala*	Philippines, Indonesia
caroa		*Neoglaziovia variegata*	Brazil
henequen		*Agave fourcroydes*	Australia, Cuba, Mexico
istle		*Agave* (various species)	Mexico
mauritius		*Furcraea gigantea*	Brazil, Mauritius, Venezuela, tropics
phormium		*Phormium tenax*	Argentina, Chile, New Zealand
sansevieria	bowstring hemp	*Sansevieria* (entire genus)	Africa, Asia, South America
sisal		*Agave sisalana*	Haiti, Java, Mexico, South Africa
Seed-hair fibers			
cotton		*Gossypium* sp	United States, Asia, Africa
kapok		*Ceiba pentranda*	tropics
milkweed floss		*Chorisia* sp	North America
Other fibers			
broom root	roots	*Muhlenbergia macroura*	Mexico
broom corn	flower head	*Sorghum vulgare technicum*	United States
coir	coconut husk fiber	*Cocos nucifera*	tropics
crin vegetal	palm leaf segments	*Chamaerops humilis*	North Africa
palmyra palm	palm leaf stem	*Brossus flabellifera*	India
piassava	palm leaf base fibers	*Attalea funifera*	Brazil
raffia	palm leaf segments	*Raphia raffia*	East Africa

therefore they are also called cellulosic fibers. Vegetable fibers are classified according to their source in plants into bast fibers from the bark section of certain plants, called soft fibers for textile use; leaf fibers, called hard fibers, mainly used in cordage; seed-hair fibers, the source of cotton (qv), the most important vegetable fiber; and palm fibers, also called brush fibers, from various sections of the plant. Commercially important fibers are given in Table 1.

Properties

Chemical Composition. The chemical composition of the principal commercial vegetable fibers is given in Table 2.

Physical Properties. Bast and leaf fibers are stronger (higher tensile strength and modulus of elasticity) but lower in elongation (extensibility) than cotton. Vegetable fibers are stiffer but less tough than synthetic fibers. Kapok and coir are relatively low in strength; the former is known for its buoyancy. Textile-related properties of some bast fibers and polyester are given in Table 3.

Economic Aspects

The principal bast and leaf fibers are produced in yields of 1–5%, with some exceptions, on a green plant basis. U.S. imports of jute, istle, kapok, sisal, and other fibers have dropped 70–90%, whereas abaca imports remained stable except for a 40% drop from 1980 to 1983. These market trends reflect the 1970s recession and development of synthetic fibers.

Table 2. Chemical Compositions of Vegetable Fibers, wt %

Fiber	Cellulose	Hemicellulose	Pectins	Lignin	Extractives
Bast fibers					
flax	71.2	18.6	2.0	2.2	6.0
jute	71.5	13.4	0.2	8.1	1.8
ramie	76.2	14.6	2.1	0.7	6.4
Leaf fibers					
abaca	70.1	21.8	0.6	5.7	1.8
sisal	73.1	13.3	0.9	11.0	1.6
Seed-hair fibers					
cotton	92.9	2.6	2.6		1.9

Table 3. Textile-associated Properties of Cotton, Flax, and Polyester

Property	Cotton	Flax	Polyester
density, g/cm^3	1.54–1.5	1.50	1.22–1.35
moisture regain, %	8–11	12	0.4–0.8
fiber tenacity, mN/tex[a]	260–440	230–240	180–790
elongation, dry, %	3–10	2.7–3.3	18–67
recovery from 2% elongation, %	75	65	85–97
length–width ratio	1400	1200	

[a] To convert N/tex to g/den, multiply by 11.33.

Uses

Vegetable fibers have application in a broad range of fibrous products, including textiles and woven goods, cordage and twines, stuffing and upholstery materials, brushes, and paper.

BIBLIOGRAPHY

AATCC Technical Manual, Vol. 62, American Association of Textile Chemists and Colorists, Research Triangle Park, N.C., 1987.

B. P. Corbman, *Textiles, Fiber to Fabric,* 6th ed., McGraw-Hill, Inc., New York, 1983.

M. L. Joseph, *Textile Science Introductory,* 5th ed., Holt, Rinehart, and Winston, New York, 1986.

JOHN N. MCGOVERN
University of Wisconsin-Madison

FILAMENT WINDING

Filament winding is a process for the fabrication of reinforced composites in which a series of continuous resin-impregnated filaments are applied to a mandrel according to a predetermined geometrical relationship. The continuous reinforcement, locked within the resin matrix, carries more load than a discontinuous reinforcement. In addition, controlled fiber alignment allows more reinforcement in a given volume and consequently higher directional strength (Fig. 1). In filament winding, directional strength and stiffness can be varied, permitting efficient material utilization. No more strength than is actually required must be provided in any one direction, an advantage lacking in isotropic materials. This provides greater directional strength at similar weight or similar strength at less weight. Filament-wound composites have the same characteristics as other reinforced composites: corrosion resistance, high dielectric strength, low thermal expansion, and good impact resistance. When a number of these characteristics are required, a filament-wound composite may be the only alternative, or it may be the most cost-effective one (see also COMPOSITES, FABRICATION).

The simplest form of filament winding is shown in Figure 2. A series of reinforcements is drawn through a resin bath mounted on a traversing carriage. While the mandrel rotates about its central axis, the carriage traverses a number of circuits from one end of the mandrel to the other end, until the desired wall thickness is built up. The composite is cured on the mandrel, which is subsequently extracted from the wound product, or the mandrel may remain as an integral element of the part.

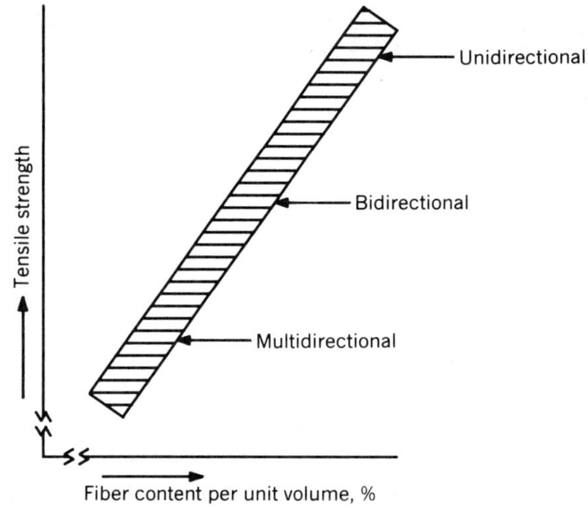

Figure 1. Tensile strength vs fiber content and orientation. Unidirectional alignment allows more reinforcement in a given volume and higher directional strength. In multidirectional alignment strengths are equal in all directions.

In principle, almost any continuous reinforcement is suitable for filament winding. The most commonly used at the present time is fiber glass, primarily E glass and S glass. Thermosetting resins, such as polyesters and epoxies, are the most commonly used resin matrices.

The development of computer-controlled filament-winding equipment combined with continuing advances in fibers and resins has expanded the filament-winding process into an increasing range of applications. A manufacturing unit consisting of a computer-controlled filament winder, a curing station, an extraction station, and secondary finishing operations can produce a range of reinforced composites with minimum lead times and minimum inventory.

Applications

Whereas the early applications were for corrosion-resistant storage tanks and lightweight, high strength rocket motor cases, recent applications have covered a much broader product range

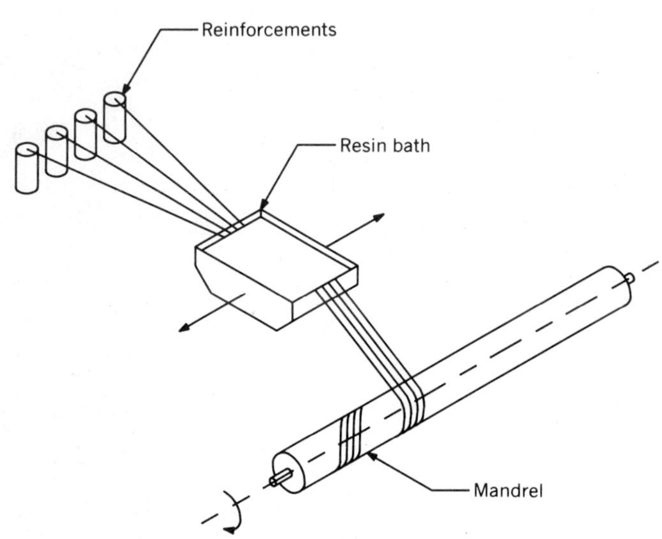

Figure 2. Simple filament winding.

in the following areas: automotive, aviation and aerospace, marine, pipes, tanks, and pressure vessels.

BIBLIOGRAPHY

A. M. Shibley in G. Lubin, ed., *Handbook of Composites,* Van Nostrand Reinhold Company, New York, 1982, pp. 449–478.

Publication 5-CR-6516—Filament Winding with Type 30 Roving, Owens Corning Fiberglas Corp., Toledo, Ohio, May 1974.

Technical Service Bulletin TS-600—Filament Winding, PPG Industries, Pittsburgh, Pa., 1984.

M. M. Schwartz, *Composite Materials Handbook,* McGraw-Hill Book Company, New York, 1984, pp. 4.52–4.62.

RAYMOND C. HAYES
McClean-Anderson, Inc.

FILLERS

Fillers, or extenders as they are called in the coatings industry, are finely divided solids added to polymer systems to improve properties or reduce cost. Fillers can be minerals, metallic powders, organic by-products, or synthetic inorganic compounds. They cover a broad range of particle sizes and shapes and may have undergone surface treatments. Although most fillers are solids, a few silicate types contain air voids to reduce effective filler density.

Fillers include calcium carbonate (the most widely used filler for plastics, caulks, sealants, and coatings) (Table 1), kaolin, talc, alumina trihydrate, and organic fillers (eg, ground by-products and waste products based on wood, shells, corn, and rice).

Morphology

There are almost as many types of filler geometry and morphology as there are types of fillers. Some particles are amorphous and approximately round because of wet or dry grinding, which removes sharp corners and edges. Some fillers, such as metallic powders, mica, or talc are thin plates. A few, such as wollastonite and California talcs, tend to be needle-shaped. Others, such as fly-ash floats and spray-dried products, are almost spherical; common shapes are shown in Figure 1.

Loading and Packing

The ability of filler particles to be packed into a polymer matrix is of fundamental importance to the paint, plastics, rubber, and related industries. The loading and packing efficiency of a given filler is closely related to morphology, particle-size distribution, and oil absorption.

Table 1. Properties and Applications of Calcium Carbonates

Physical form	Maximum free-moisture, %	Mean particle size, μm	Applications
fine wet ground	0.20	3.0	caulks and sealants, paints
coarse wet ground	0.20	12.5	latex- and solvent-based paints
fine dry ground	0.20	6.5	rubber and adhesives, vinyls
coarse dry ground	0.20	18.0	texture paints, putties
surface-modified ultrafine ground[a]	0.30	0.7	trade sales and industrial solvent-based coatings
coarse ground limestone	0.15	74.0	traffic paint, flooring

[a] pH = 8.5–9.5.

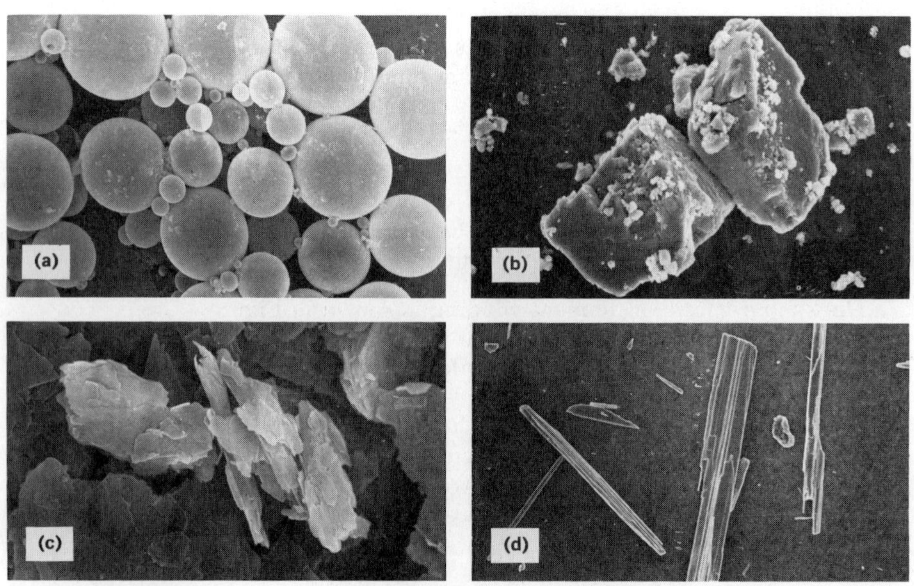

Figure 1. Common filler particles: (**a**) sphere (magnification 502×); (**b**) irregular (magnification 1,510×); (**c**) flake (magnification 10,100×); (**d**) fiber (magnification 304×). The specific fillers shown are (**a**) hollow glass bubbles; (**b**) calcium carbonate; (**c**) talc; and (**d**) wollastonite.

Applications

Fillers are used in plastics; caulks, sealants, putties, and adhesives; coatings and inks; and paper.

Economic Aspects

Calcium carbonate is the dominant filler in terms of tonnage in the plastics, paint, and caulks–sealants–putties industries; kaolin leads the way in the rubber and adhesives markets.

Growth-rate projections for fillers and extenders are in the 6–10% range over the next decade. The total filler consumption is thus expected to approach 7×10^6 t by 1990.

BIBLIOGRAPHY

F. J. Washabaugh in Joan Agranoff, ed., *Modern Plastics Encyclopedia 1984–1985,* Vol. 61 (10A), McGraw-Hill, Inc., New York, 1984.

H. S. Katz and J. V. Milewski, *Handbook of Fillers and Reinforcements for Plastics,* 1st ed., Van Nostrand Reinhold Co., New York, 1978, pp. 79–182.

W. S. Stoy and T. G. Brown, *Mod. Paint Coatings* **72,** 162 (1982).

E. M. Dannenberg, *Rubber Chem. Technol.* **55,** 860 (1982).

W. S. Stoy
F. J. Washabaugh
Engelhard Corporation

FILMS

Films are planar forms of plastic, thick enough to be self-supporting but thin enough to be flexed, folded, or creased without cracking. Film thickness depends on applications and manufacturing; in most applications it is below 125 μm.

The use of plastic films has shown phenomenal growth during the past three decades. Numerous technical breakthroughs in polymerization and processing techniques resulted in com-

Table 1. U.S. Film Producers and Converters

Company	Principal products	Comment
DuPont	cellophane	
	polyester	
	aramid	
	polyimide	exclusive product
Eastman Kodak	polyester	essentially all captive
	cellulose triacetate	
Celanese	polyester	
Grace	EVA[a], composites	
	irradiated LDPE[b] shrink films	
Hercules	OPP[c]	
ICI	polyester	
	OPP[c]	
Mobil	OPP[c]	
	LDPE[b]	
Olin	cellophane	
3M	polyester	all captive
	OPP[c]	
Allied Chemical	nylon	
	fluoropolymer	
American Can	LDPE[b], composites	leading suppliers of packaging material
	HDPE[d], coextruded	
Bemis	OPP[c]	
Crown Zellerbach	cast PP	leading converter
	LDPE[b], coextruded and composites	
Dixico	composites	leading converter
Dow Chemical	PVDC[e]	
	polystyrene	
Ethyl	LDPE[b]	
	cast PP	
Rexham	composites	converter
Printpak	HDPE[d], coextruded composites	leading converter
St. Regis Paper	LDPE[b], composites	converter, leading producer of industrial and consumer bags
Union Carbide	LDPE[b], coextruded	

[a] Ethylene–vinyl acetate copolymer.
[b] Low density polyethylene.
[c] Oriented polypropylene.
[d] High density polyethylene.
[e] Poly(vinylidene chloride).

Table 2. Properties of Packaging Films

Films	Gauge, μm	Yield[a], m^2/N^b	Specific gravity	Refractive index	Haze, %	Ultimate tensile strength, MPa^c	Elongation, %
cellophane	20–48	2.82	1.40–1.55	1.54	1–2	49–111	25–50
cellulose acetate	12–500	3.04	1.28–1.31	1.48	2	78–113	15–70
nylon-6	12–100	3.55	1.129		1–4	69–111	250–400
polycarbonate	12–100	3.34	1.20		1–2	60	75–105
HDPE	10–75	4.20	0.94–0.965		5–85	20–52	25–300
LDPE	20–100	4.34	0.91–0.93		7–30	13–19	220–600
ionomer	25–100	4.27	0.94			20–34	300–400
polyester, PET	12–180	2.90	1.23–1.395	1.65	2–4	167–275	30–120
polyimide	25–125	2.82	1.42	1.78		167	70
polypropylene	12–75	4.43	0.9–0.91	1.5	1–2	127–226	50–250
polystyrene	19–100	3.76	1.05	1.6	1–2	59–78	10–20
nonrigid PVC	19–100	3.11	1.20–1.45		2–6	9.8	60–200
Saran	12–50	2.43	1.64–1.71		2–5	49–11	30–60

[a] Per 25 μm.
[b] To convert m^2/N to m^2/kgf, multiply by 9.807.
[c] To convert MPa to kgf/cm^2, multiply by 10.2.
[d] To two significant figures, GPa = 10^4 kgf/cm^2.
[e] To convert N/m to gf/cm, multiply by 1.02.

positions and properties designed for flexible and rigid packaging, industrial applications, and other uses. In addition, manufacturing costs were greatly reduced.

In 1984, U.S. consumption of plastic films was about 2.95×10^6 t valued at >$\$6.3 \times 10^9$. Commodity products, low density polyethylene (LDPE), and poly(vinyl chloride) (PVC) accounted for about 2.15×10^6 t; the rest were specialty films.

Specialty films include the higher priced, high performance films of coated, coextruded, and laminated versions of oriented polyester (OPET), oriented polypropylene (OPP), cellophane, oriented nylon (ON), oriented polystyrene (OPS), and vinylidene chloride copolymers (PVDC).

The U.S. film industry comprises film producers and film converters (Table 1) of varying technical capabilities.

Specialty films and their coated, coextruded, and laminated compositions are well suited for meeting the packaging requirements of food items in terms of processing, protection, barrier properties, consumer appeal, and government regulations. Other packaging applications include bags and wrappers for textiles, records, and tapes (see also PACKAGING MATERIALS, FLEXIBLE).

Properties

Properties of the principal film materials are given in Table 2.

BIBLIOGRAPHY

J. E. Peters, ed., *Packaging Encyclopedia,* Vol. 30, No. 4, Cahners Publishing Co., Boston, Mass., 1985.

C. J. Benning, *Plastic Films for Packaging,* Technomic Publishing Co., Inc., Lancaster, Pa., 1983.

C. R. Oswin, *Plastic Films and Packaging,* John Wiley & Sons, Inc., New York, 1985.

R. A. Priemom, *Annual Book of ASTM Standards,* American Society for Testing and Materials, Philadelphia, Pa., 1983.

H. C. Park
A. M. Nahmias
Mobil Chemical Company

FILMS, MANUFACTURE

Virtually every thermoplastic material can be used in film form. Most thermoplastic films are prepared by conventional extrusion techniques based on calendering or solvent casting or blown-film or tenter-film systems. In the selection of a film for a particular application, the properties of the thermoplastic materials must be considered in view of the application. Thermal properties (Table 1), molecular characteristics, and crystallinity of the polymer affect processing and film properties. Additives (Table 2) influence extrusion and orientation processes and improve film properties.

Manufacturing processes include extrusion (coextrusion, casting, uniaxial orientation, and biaxial orientation), solvent-solution casting (coagulation and evaporation), rolling, and calendering.

Finishing operations include coating and embossing.

BIBLIOGRAPHY

J. Briston, *Plastics Films,* 2nd ed., Longmans, Inc., New York, 1983.

C. J. Benning, *Plastic Films for Packaging,* Technomic Publishing Co., Inc., Lancaster, Pa., 1983.

J. Stěpěk with V. Ducháček, D. Curda, J. Horáček, and M. Šípek, *Polymers as Materials for Packaging,* Ellis Horwood Series in Poly-

Modulus, GPa[d]	Stiffness[a], N/m[e]	Tear strength, N/mm[f]	Water-vapor transmission rate[g], $\frac{\mu mol}{m^2 \cdot s}$ [h]	Gas permeability[i], $\frac{mol}{m \cdot s \cdot PPa}$ [j] O_2	CO_2	Dielectric strength, V/25 µm	Volume resistivity, Ω/cm³
1.4–3.1	6.9–25	0.8–3.9	2.9–10	0.086	0.13	2000–2500	10^{11}
2.4–3.1	9.8–18	1.6–3.9	64–209	0.23–0.3	1.7–2.0	5000	10^{16}
0.6–0.8		7.8–20	30–69	0.065	0.25	700–800	
2.5		7.8–9.5	50–60	0.6–0.8	3.0–3.4	1000	10^{16}
0.7–1.0	3.9–7.8	5.9–118	2.9–4	0.36	1.2	500	10^{16}
0.1–0.25	0.98–1.96	39–157	7.7–12	1.1	4.8	4700	10^{16}
0.15–0.4	1.96–3.9	5.9–9.8	16–23	0.77–0.91	0.77–0.9		10^{13}
3.4–5.5	15.7	7.8	16–19	0.006–0.012	0.033–0.04	7000	10^{18}
3.1		3.1	54	0.05	0.091	7200	10^{18}
2.5–4.9	4.9–10.8	2.4–3.9	3.5–45	0.21–0.3	0.9–1.1	3000	10^{16}
2.7–3.7	19.6	1.96	45–100	0.62	1.7	2500	10^{16}
1.0–2.0	2.9–18	24–39	9.6–30	0.3	1.9	250–1000	10^{11}–10^{14}
1.1–1.2	3.9–5.9	3.9–39	1–2.9	0.0016–0.0022	0.0077–0.012	3000–5000	10^{12}–10^{15}

[f] To convert N/mm to gf/25 mm, multiply by 2.549.
[g] At 37°C, 90% rh, per 25 µm.
[h] To convert $\frac{\mu mol}{m^2 \cdot s}$ to $\frac{g}{m^2 \cdot d}$, divide by 0.643.
[i] At 25°C, 0% rh, per 25 µm.
[j] To convert $\frac{mol}{m \cdot s \cdot PPa}$ to $\frac{cc}{m^2 \cdot d \cdot atm}$, multiply by 7695.

Table 1. Thermal Properties of Film Polymers[a]

Class	Polymer	Common symbol	Glass-transition temp T_g, °C	Melting temp T_m, °C	T_{dec}[b], °C
polyolefin	high density polyethylene	HDPE	−(120–140)	130–137	335–450
	low density polyethylene	LDPE	−(95–130)	106–115	335–450
	polypropylene	PP	−10	170	328–410
	ethylene–propylene copolymer	EP	−30	128–145	328–410
	poly(1-butene)	PB	−24	126	288–425
	poly(4-methylpentene)	TPX	30	235	335
	polytetrafluoroethylene	PTFE	27	330	508–538
vinyl polymers	polystyrene	PS	85–105	240	285–440
	poly(vinyl acetate)	PVAC	75–105	240	213–325
	poly(vinyl chloride)	PVC	75–105	212–310	200–300
	poly(vinylidene chloride)	PVDC	−18	210	225–275
	poly(vinyl fluoride)	PVF	40	200–235	372–480
	polyacrylonitrile	PAN	97	320	250–280
	poly(methyl methacrylate)	PMMA	65	160	170–300
polyethers	polyoxymethylene	POM	−(85–110)	181	100
	poly(phenylene sulfone)		220	300	310
	cellulose triacetate	LA	105	306	250–310
polyesters	poly(ethylene terephthalate)	PET	81	265	283–306
	polycarbonate	PC	147	225	420–620
polyamides	polycaprolactam[c]		40–87	220	310–380
	polyhexamethyleneadipamide[d]		50	265	310–380
	polyundecanoamide[e]		46	194	310–380
	polyimide		310–365		310–380

[a] Temperature quoted represents values for specific grades of commercial materials and may not be representative of all materials of this classification.
[b] Decomposition temperature.
[c] Nylon-6.
[d] Nylon-6,6.
[e] Nylon-11.

Table 2. Film Additives

Additive function	Supplier	Trade name	Chemical name or class
antioxidants	Goodrich	Goodrite 3114	phenolic
	Koppers	BHT	butylated hydroxytoluene
	Koppers	DBPC	di-*tert*-butyl-*p*-cresol
	Mobil		trialkylhydrogen phosphites
	Mobil		dialkylhydrogen phosphites
	CIBA-GEIGY	Irganox	high molecular weight hindered phenols
		Poly TDP-2000	hydroquinone derivative
heat stabilizers	Cardinal	CC-7711	organotin
	American Cyanamid	Cyastab	lead-based carbonates, sulfates, phthalates
			phosphites
	Ferro		barium–cadmium
			cadmium–zinc
	Dart and Kraft		antimony mercaptides
	Uniroyal Chemical		*N,N′*-diphenyl-*p*-phenylene diamine
uv stabilizers	American Cyanamid	Cyasorb	benzophenone
			benzotriazole
	Borg-Warner	Spinuvex	hindered amine (HALS)[a]
	Borg-Warner	Weston 619	solid phosphite
	CIBA-GEIGY	Tinuvin	hindered amine (HALS)[a]
slip agents	Witco	Kenamide	fatty acid amides
	Dow Corning	Dow 200 fluid	polydimethylsiloxane
	Allied	A-C Polyethylene	PE wax
		Admer	polyolefin
fillers, antiblock agents		Atomite	calcium carbonate
		DiCaperl Hp-510	glass bubble
	North Georgia Minerals & Chemicals		aluminum trihydrate
			celestite
			barytes
			talc
	James River	Solka-Floc	cellulose
	Manville	Celite	diatomaceous earth
	Manville	Micro-Cel	hydrous calcium silicate
		Suzorite	mica
		Wollastonite	calcium metasilicate
		Glomax	calcined kaolin clay
		Kaopaque	delaminated aluminum silicate

[a] HALS = Hindered amine light stabilizer.

mer Science and Technology, John Wiley & Sons, Inc., New York, 1987.

H. C. PARK
E. M. MOUNT III
Mobil Chemical Company

FILMS, MULTILAYER

Multilayer coextrusion of thermoplastic film and sheet began only twenty years ago and has developed into an important plastic fabrication process, providing significant growth opportunities for the plastics industry. Coextruded multilayer plastics are challenging such traditional materials as metals, glass, paper, and textiles.

The attraction of coextrusion is both economic and technical. It is a single-step process starting with two or more plastic materials that are simultaneously extruded and shaped in a single die to form a multilayer sheet or film. Thus coextrusion avoids the costs and complexities of conventional multistep lamination and coating processes, where individual plies must be made separately, primed, coated, and laminated. Coextrusion readily allows manufacture of products with layers thinner than can be made and handled as individual plies. Consequently, only the necessary thickness of a high performance polymer is used to meet particular specifications of the product.

Layers may be used to place colors, bury recycle, screen ultraviolet (uv) radiation, minimize die-face buildup, and control film-surface properties. Additives, such as antiblock, antislip, and antistatic agents, can be placed in specific layer positions. High melt strength layers can carry low melt strength materials during fabrication.

The largest market for coextruded films and sheets is in packaging applications, eg, two- or three-layer films for trash bags or five-to-nine-layer structures for flexible and semirigid packages. Growing applications for coextrusion are in automotive, construction, appliance, sanitary ware, and recreational vehicle markets.

Methods of Coextrusion

Coextruded films are produced by a tubular-blown film process (Fig. 1), and a flat-die, chill-roll casting process (Fig. 2).

Coextrusion dies include tubular-blown film dies (circular dies), flat dies (slit dies), feedblock/single manifold dies, and combined feedblock/multimanifold dies.

Rheological Considerations

Polymer rheology is important for designing coextrusion dies and feedblocks. Polymer-flow characteristics also influence the selection of polymeric materials for coextruded products. Layered melt streams flowing through a coextrusion die can become unstable, leading to layer nonuniformities and even intermixing of layers under certain conditions. The causes of these instabilities are related to non-Newtonian flow properties of polymers and viscoelastic interactions (see also VISCOELASTICITY).

Several types of flow instabilities have been observed in coextrusion including interface distortion from viscosity mismatch and elastic instability.

Polymers for Coextruded Films

Polymers are chosen for individual layers to achieve specific combinations of properties, eg, mechanical strength and stiffness, gas and water-vapor barrier, oil and grease barrier, heat seal-hot tack, adhesion, optics, formability, machinability, and economics. Individual layers may be pure polymers or blends, sometimes with regrind or recycled scrap. The polymer layers are often formulated with color; antislip, antiblock, and antistatic processing aids; and stabilizer additives to enhance layer characteristics. Adhesive polymers are used as tie layers to bond dissimilar polymers that do not normally adhere to each other.

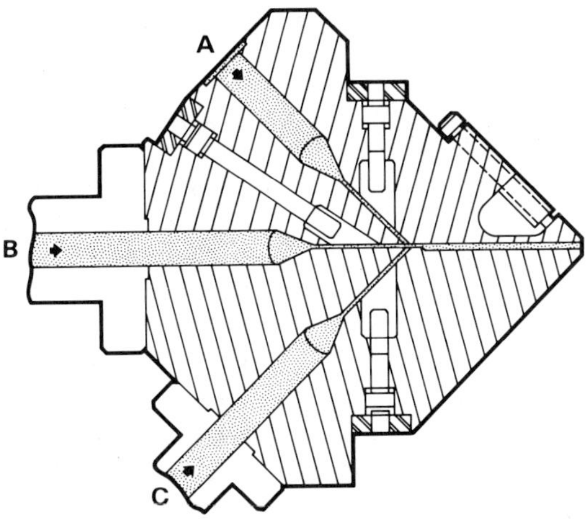

Figure 2. Cross-sectional view of three-layer internal-combining multimanifold flat film or sheet die.

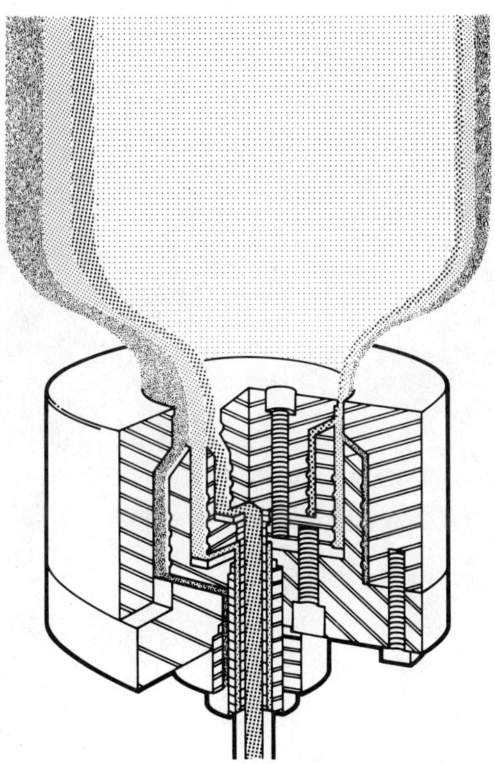

Figure 1. A three-layer blown film die.

Common polymers used in coextrusion applications are listed below with their abbreviations.

Name	Abbreviation
low density polyethylene	LDPE
medium density polyethylene	MDPE
high density polyethylene	HDPE
high molecular weight, high density polyethylene	HMW-HDPE
linear low density polyethylene	LLDPE
linear medium density polyethylene	LMDPE
ethylene–vinyl acetate copolymer	EVA
ethylene–acrylic acid copolymer	EAA
ethylene–methyl acrylate copolymer	EMA
ethylene–methacrylate acid salts	ionomer
hydrolyzed ethylene–vinyl acetate	EVOH
polyamide	nylon
poly(vinyl chloride)	PVC
poly(vinylidene chloride) copolymers	PVDC
polypropylene	PP
polybutylene	PB
poly(ethylene terephthalate)	PET
polycarbonate	PC
polystyrene	PS
impact polystyrene	IPS
acrylonitrile–methyl acrylate copolymer	AN/MA
acrylonitrile copolymers	AN
acrylonitrile–butadiene–styrene polymers	ABS
styrene–butadiene block copolymer	SB

BIBLIOGRAPHY

D. Djordjevic, *TAPPI/PLC Conference Proceedings,* 1984, pp. 67–75.

A. Toporek, *TAPPI Coextrusion Seminar,* 1983, pp. 23–28.

W. J. Schrenk and T. Alfrey in D. R. Paul and S. Newman, eds., *Polymer Blends,* Academic Press, New York, 1978, Chapt. 15.

C. D. Han, Y. J. Kim, and H. B. Chin, *Polym. Eng. Rev.* **4,** 3 (July 1984).

WALTER J. SCHRENK
EARL W. VEAZEY
The Dow Chemical Company

FINE ARTS

The exploration of polymers has been an important aspect of artistic creation. Today polymeric materials are available in various physical forms: liquid monomers that can be polymerized at room temperature in open molds; powders and beads that can be melted, extruded, and expanded; sheets that can be sawed, glued, and heat-formed; and blocks that can be carved.

Polymers are also used as adhesives, coatings, impregnating materials, binders, and varnishes. By combining polymers designed for industrial applications, the most extraordinary shapes can be created. All kinds of objects and even landscapes can be embedded in polyethylene (PE), polypropylene (PP), or woven nylon (Fig. 1). Poly(vinyl chloride) (PVC) dissolved in an organic solvent can be sprayed over a wire armature to obtain three-dimensional objects, and a vacuum film-forming technique can produce works of art.

Painting Media

The 20th century has seen natural paints increasingly replaced by synthetic polymers. Some paints are specially made to allow the artist to paint in a traditional way with modern materials.

Polymers as Artists' Media. The most commonly used polymer for artist paints is a copolymer of methyl methacrylate and ethyl acrylate in aqueous dispersion (Primal AC 33 and AC 634, Rohm & Haas; Plextol B 500, Rohm). Paints based on this polymer are stable in acid and alkaline media and resist dimensional changes of the support. Small amounts of organic colorants, sometimes used to obtain certain tints, may affect light stability.

Table 1. Formulation for White Acrylic Paint

Ingredient	Amount, g
titanium dioxide	14
barium sulfate	12
kaolin	2
calcium carbonate	23
diatomaceous earth	2
carboxymethylcellulose	1
water	20
poly(ammonium acrylate)	0.03
acrylic dispersion	22
5% sodium hydroxide solution	1
antifoaming agent	1
diethylene glycol	1.77
disinfectant	0.10

The typical formula for a white acrylic paint, given in Table 1, can be colored with pigments ground in water.

Although acrylics are usually preferred, some paints are based on pure or plasticized poly(vinyl acetate) (PVAC) (Mowilith 30 or 40, Hoechst; Bakelite; AYAF, Union Carbide) and copolymers of vinyl acetate with dibutyl maleate (Mowilith DM 2 and DM 4, Hoechst). Such paints are less weather resistant and less stable on alkaline supports, such as fresh plaster walls. The formula of a PVAC paint is given in Table 2.

Collages. Collages are made by gluing various materials

Figure 1. *Valley Curtain, Rifle, Colorado,* 1970–1972 by Christo. A large curtain (18,580 m^2) of woven nylon fabric suspended at a width of 400 m and a height ranging from 55 m at the center to 111 m at each end. Photograph by H. Shunk. Copyright Christo 1972.

Table 2. Poly(vinyl acetate) (PVAC) Paint

Ingredient	Amount, g
water[a]	29
fungicide	1.5
sodium hexametaphosphate[a]	0.1
antifoaming agent[a]	1.5
20% hydroxyethylcellulose solution[a]	20
pigment[a]	400
fillers[a]	
TiO_2	225
$CaCO_3$	90
$BaSO_4$	70
kaolin	20
ethylene glycol	17
butyl glycol acetate[b]	7.5
poly(vinyl acetate) dispersion[b]	365
sodium benzoate[b]	4

[a] These ingredients are combined in a homogeneous suspension.
[b] These ingredients are added to the suspension. The pH is adjusted to 7.5–8 with 20% ammonium hydroxide.

such as paper, fabric, sand, or ground or crushed plastics on a canvas. The adhesive is a low molecular weight PVAC, known as all-purpose glue, or a PVAC dispersion, known as white wood glue. Large objects are glued with epoxy resins.

Sculptures. Traditionally, three-dimensional articles are made by carving wood or stone or by casting a metal in a mold. Plastics such as polyesters, epoxies, acrylics, and foams offer more possibilities: they can be carved and molded, laminated, coated, heat-formed, and extruded.

Polyester. Plaster molds are made from living models and copied in polyester reinforced with glass mat. Tinting is done by mixing pigments in the casting mixture or by painting, after hardening, with acrylic paint.

Epoxy Resin. Epoxy resins (qv) resemble polyester. Higher cost is compensated for by improved stability and water resistance and less shrinkage.

Acrylics. Acrylics, available since 1936, are the materials favored by most sculptors. The monomer, methyl methacrylate, can be polymerized *in situ,* or the polymer, in powder or bead form, can be cast at elevated temperature; however, the ready-formed sheet is the preferred form.

Foams. Foams, e.g. polyurethane foams and polystyrene, can be shaped into sculptures.

Art Restoration and Repair

Today, polymers play an important role in conservation and experimental treatments. In restorations, synthetic polymers are employed as adhesives, protective coatings, and binders, and for consolidation. Long-term stability is required.

BIBLIOGRAPHY

N. Roukes, *Sculptures in Plastics,* Watson-Guptill Publications, New York, 1978.

Dokumenta 7, D & V Paul Diericks, Kassel, 1981.

W. Rotzel, *Object Kunst,* Verlag M. DuMont, Kohn, 1972.

N. S. Brommelle, E. M. Peye, P. Smith, and G. Thomson, "Adhesives and Consolidants," *Preprints of the IIC-Paris Congress, Sept. 2–8, 1984,* International Institute for Conservation of Historic and Artistic Works, London, 1984.

E. De Witte
Koninklijk Instituut voor het Kunstpatrimonium

FINISHING

Finishing is generally defined as "the act or process of completing" (*Webster's Third New International Dictionary,* G. & C. Merriam Co., Springfield, Mass., 1961). To the polymer engineer this may mean the process by which raw polymer in the reactor becomes the final polymer product (see Recovery and isolation techniques).

More often finishing means decorating (qv), polishing, buffing, flash removal, etc, during plastics processing (qv). It also refers to the application of finishes, ie, of decorative or protective surface coatings (see Coating methods; Coatings). Finishes include organic materials, such as alkyd resins (qv) or plastisols, as well as inorganic coatings (see, eg, Metallizing). In the textile field, finishing refers not only to decorating [dyeing (qv), printing] but also to chemical treatments that impart such properties as crease resistance or water repellancy. See Textile resins.

FIRE-RESISTANT MATERIALS. See Flammability.

FIRE RETARDANCY. See Flammability.

FLAME RETARDANTS. See Flammability.

FLAME SPRAYING. See Coating methods.

FLAMMABILITY

The combustible material in most fatal fires is polymeric in nature. Whether natural or synthetic, these solid fuels account for 94% of lives lost, 80% of reported injuries, and 94% of property lost in residential fires in the United States. As a result, numerous tests have been devised for flammability and many codes and regulations cite them. Not coincidentally, significant effort has been devoted to improve the fire resistance of polymers. Such progress relies on understanding the nature of fire, fuel properties and how to measure them, and the mechanisms of polymer degradation.

Fire

Incidence. Incidence and severity of fires vary according to location, community size, and economics. Although residential fires comprise only 25% of the total, they account for 80% of the deaths, 65% of the injuries, and 46% of the property losses. Within residences, beds and upholstered furniture are the main sources of ignition. Cigarettes and heating equipment, especially space heaters, are the most frequent sources of ignition in fatal fires. Two-thirds of the fire deaths occur at a distance from the fire source; four-fifths are caused by inhalation of toxic smoke and not by burns.

Combustion Processes. ***Pyrolysis and Gasification.*** When subjected to a sufficient heat flux for a sufficiently long time, all organic polymeric materials undergo thermal degradation. In the absence of oxidation, these processes are referred to as pyrolysis, which is usually endothermic. In the presence of oxygen, the overall process is referred to as oxidative pyrolysis or thermal oxidative degradation. This can be endothermic or exothermic. The susceptibility of a polymer to thermal degradation is determined by its chemical constitution and affected by the presence of fillers, pigments, stabilizers, plasticizers, and flame retardants. The relative importance of thermal oxidative degradation under flaming combustion conditions is much less well defined. This is especially so for the potentially important surface oxidation reactions and their effects on the rates of mass consumption and flame spread.

For the downward burning of polyolefins, surface oxidation provides sufficient energy for fuel gasification. However, no more than one-fifth of the needed energy is thus generated. Little oxygen is bound to the surface of a variety of polymers during downward burning. Oxygen does not seem to affect the surface pyrolysis of polystyrenes, but during transient poly(methyl methacrylate) (PMMA) pyrolysis, it affects the degree of polymerization in the condensed phase.

Ignition. As the volatilized polymer fragments leave the surface, they mix with the ambient air. That gas mixture is heated by the internal energy of the fuel gases and their absorption of infrared energy from the pyrolyzing energy source. The latter is especially important if that source is a remote flame. As the temperature rises, the gases interact with oxygen, at first slowly, but at a rate that increases exponentially with temperature. At some point, the exothermic gas-phase reaction is sufficiently fast, generating enough energy to supply more fuel, that combustion is sustained. This is the autoignition temperature, which depends on the chemical nature of the gases and the mixture with oxygen.

Ignition and combustion are dominated by free-radical reactions. Thus addition or removal of radicals can affect the likelihood of a fire.

Another important parameter, especially for solid substrates, is the critical radiant flux. As the energy density incident on the polymer surface increases, gasification and thus ignition are accelerated. Critical radiant fluxes for some common materials are given in Table 1.

Table 1. Minimum Radiant Flux for Ignition of Common Materials

Material	Flux, kW/m^2
polyethylene	19
poly(methyl methacrylate)	18
polyoxymethylene	17
polypropylene	20
polystyrene	29
polytetrafluoroethylene	43
polyurethane foam	16–34
poly(vinyl chloride)	21

Flaming. Most of the destruction associated with fires occurs during the flaming phase. The visible plumes above the burning fuel result from rapid, often incomplete combustion at temperatures near 1200°C. In a building fire, the energy released further buckles nonflammable components, such as wallboard and structural steel, opening new avenues for fire growth.

Smoldering. Porous materials may undergo a different mode of combustion. Unlike flaming, smoldering is a surface phenomenon. It occurs in open-cell foam cushions; cellulosic wall insulation; and quiescent, lit cigarettes. It is a slow process, often taking hours to run its course.

The hazards of smoldering are twofold. First, a person in close proximity to the smoldering material is likely to inhale substantial amounts of noxious carbon monoxide from a combustion producing too little heat or smoke to cause alarm. Meanwhile, the energy balance in smoldering is delicate. A slight closing of the pores can snuff the fire. A small breeze over it can create the second hazard, a transition to a flaming fire.

Extinguishment. Left to itself, a fire burns until energy feedback from the flames is insufficient to sustain a flammable mixture of vapors. The fire is said to self-extinguish. This may result from total consumption of material or of available oxygen. It may also be a consequence of chemical changes within the substrate, eg, the evolution of flame inhibitors from an additive to the polymer.

Fires are most frequently extinguished with water, which has a considerable sensible heat and heat of vaporization. It absorbs much energy and reduces the fuel temperature; to assure suppression, large quantities are often required. In experiments, however, room fires have been extinguished with less than 2 L of water. Optimizing such results in actual fires is a current research and technological priority.

Hazards. Hazards include flames, heat, loss of visibility, and smoke toxicity. (See Table 2 for sample fire toxicological data.)

Limiting Hazard. Fire professionals have adopted the concept of the limiting hazard. For a given location in the burning building, one of the above hazards becomes life-threatening first. If that hazard is frequently the limiting hazard, it must be eliminated first. Information on which to base this assessment is currently being gathered from mathematical models and fire data.

Flammability Parameters. ***Rate of Heat (Energy) Release.*** The rate of heat or energy release measures the contribution of a substrate to a fire.

Mass-loss Rate. The mass-loss rate does not directly affect fire growth. It relates, however, the results from one test to another or from one test to a fire-growth model.

Ease of Ignition. Resistance to ignition reduces hazard. Ease of ignition can be measured by the time of onset of flaming or smoldering from a fixed-energy source. It can also be considered as the energy needed to start combustion in a fixed time period.

Ignition Temperatures. At some temperature, a polymer decomposes rapidly enough to generate a flammable mixture. If no external ignition source is available, this is referred to as the autoignition temperature. If a flame or spark is present, piloted ignition occurs; the result is the analogue of the flash point for liquids. Representative ignition temperatures are given in Table 3.

Table 2. LC_{50} Values for Some Common Polymers[a]

Material	Flaming, mg/L		Nonflaming, mg/L	
	Mean	95% confidence limits	Mean	95% confidence limits
ABS	19.3	16.7–22.3	30.9	21.2–45.0
modacrylic fiber	4.4	3.9–5.0	5.3	4.0–7.1
poly(phenyl sulfone)	19.8	14.8–26.5	9.5	9.1–10.1
polystyrene	38.9	37.9–39.9	>40.0	
polytetrafluoroethylene	0.045	0.039–0.054	0.045	0.017–0.120
poly(vinyl chloride)	17.3	14.8–20.2	20.0	
with zinc ferrocyanide	15		11.3	
Douglas fir	36.0	21.1–50.8	22.8	13.4–32.2

[a] NBS test method; 30 min plus 14 d.

Flame-spread Rate. This measures the spread of flames, the most conspicuous feature of fire growth, and thus the fire hazard.

Soot Yield. The particulates and aerosols generated from polymer combustion affect visibility and smoke-detector performance and cause discomfort. Emission of these condensed species is measured as the fraction of combusted fuel appearing as airborne condensate and as the light attentuation caused by the particles.

Materials

Thermal Degradation of Polymers. When subjected long enough to a sufficient heat flux, organic polymeric materials undergo thermal degradation. In a large-scale fire, or a downward-propagating candlelike flame, little oxygen reaches the polymer surface, and the generation of volatile fuel gases is controlled by pyrolytic, thermal degradation. Consequently, susceptibility of the polymer to thermal degradation is determined by chemical constitution and affected by the presence of fillers, pigments, stabilizers, plasticizers, and flame retardants.

Different polymers generate different volatile products in varying concentrations over different temperature ranges. The chemical steps leading to the generation of volatiles may differ in electronic nature, eg, radical vs ionic; however, the process models are independent of these variables. The three schemes common to most polymer degradations are described below:

1. Random chain cleavage followed by chain unzipping is characterized by high monomer yields in the degradation products and a slow decrease in the molecular weight of the polymer.
2. Random chain cleavage followed by further random chain cleavage is characterized by low monomer yields in the degradation products and a rapid drop in molecular weight.
3. An intrachain chemical reaction is followed by a cross-linking reaction and formation of a carbonaceous residue, or random chain cleavage. This process generates a relatively high yield of volatiles from the intrachain reaction, produces very little monomer, and produces no or a very slight loss in molecular weight during the initial stages of degradation.

In some cases, several of these schemes occur simultaneously, depending on sample size, heating rate, pyrolysis temperature, environment, and additives (see DEGRADATION; DEPOLYMERIZATION).

Thermally Stable Polymers. Although polymer degradation schemes can be greatly altered by the presence of comonomers, side-chain substituents, and other chemical constitutional factors, the ultimate thermal stability is determined by the relative strengths of the main-chain bonds. Many additives and comonomers employed as flame retardants are thermally labile; as a result the thermal stability of the polymer system is reduced. In order to reduce the observed effects of flame-retardant additives on the thermal stability of polymeric materials, more thermally stable and hence inherently fire-resistant polymers are of great interest to fire-safety experts.

There are two general strategies which have been used to produce more thermally stable polymers. The first is illustrated by the highly cross-linked three-dimensional network structure of an amine cured epoxy. Rupture of bonds does not generate combustible fuel gases, and carbonization is promoted. Ladder polymers, in which main chains are bonded together at each repeat unit by a cross-link, serve the same purpose. These principles are illustrated in Figure 1 for an aramid (single chain thermally stable polymer), a pyrrone (a ladder polymer), and an amine-cured aromatic epoxy (three-dimensional network polymer) (see also LADDER AND SPIRO POLYMERS).

Another strategy is to increase the strength of the main-chain bonds. Resonance stabilization can be achieved by utilizing aromatic and heterocyclic ring structures with resonance stabilization energies as high as 16.7 kJ/mol (70 kcal/mol) (Fig. 1).

Table 3. Average Surface Temperatures at Ignition

Material	Temperature, °C	
	No pilot	Pilot
cardboard	636	322
newspaper	292	283
pine	488	408
black rubber	653	421
canvas	[a]	335
polytetrafluoroethylene	660	
polyurethane foam	456–579	
poly(vinyl chloride)	507	
polystyrene	573	
polyethylene	488	
polypropylene	498	
poly(methyl methacrylate)	478	

[a] No ignition.

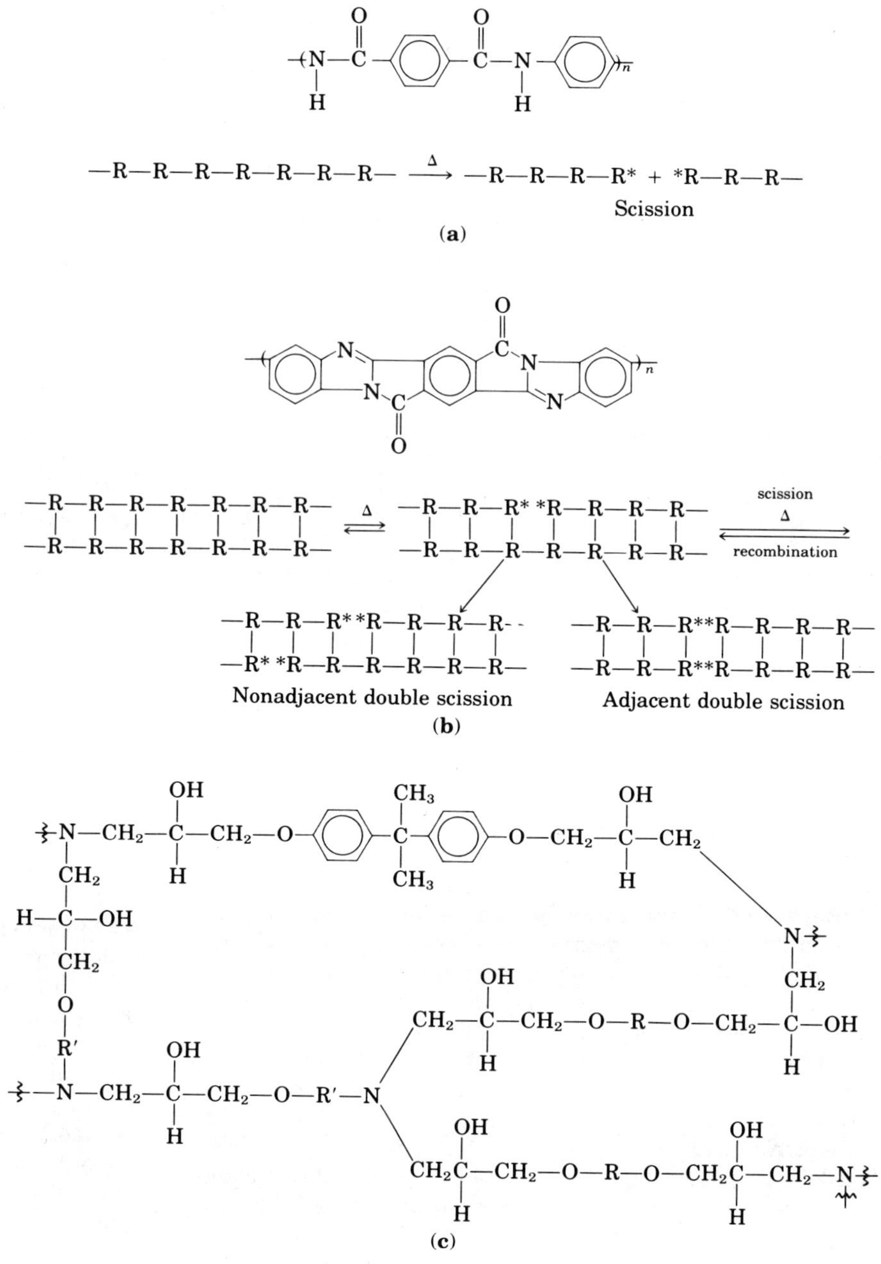

Figure 1. Thermally stable polymers: (**a**) single chain: aromatic amide (polybenzamide); (**b**) ladder chain: pyrrone; (**c**) three-dimensional network: amine-cured epoxy.

The structures of several classes of thermally stable polymers are given in Table 4.

Flame Retardant Strategies. A flame retardant interferes with the chemistry and physics of combustion in such a way that the flammability of a combustible material is changed either in a specified use or under a specified set of conditions. The flame retardant must meet one or more of the following criteria.

1. The flame retardant or its decomposition products is intended to volatilize simultaneously with the gases generated by the thermal degradation of the substrate and inhibit the vapor-phase combustion of the fuel gases.
2. The flame retardant is intended to alter the pathway of the substrate's thermal degradation by providing a low energy process that promotes solid-state reactions leading to the carbonization of the material over those generating combustible gases.
3. The flame retardant is intended to form a protective coating that insulates the substrate from the thermal energy that would otherwise result in degradation and the generation of fuel gases.

Solid-state Chemistry. In flaming combustion, the generation of combustible gases is a critical component in the continued propagation of the flame. Therefore, the most effective strategy for flammability reduction would be to interfere with the mechanism of degradation in such a way as to prevent formation of volatile fuel gases. Flame retardants that act in this manner are referred to as condensed-phase active.

Thermal Barriers. Generation of volatile gases can also be prevented by protecting the surface from the heat of the flame

Table 4. Thermally Stable Polymer Structures

Polymer class	Structural type	Structure	Decomposition temp, °C[a]
poly(p-phenylene)	aromatic	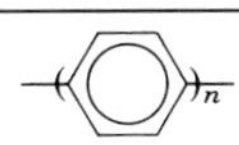	660
polybenzimidazole	step ladder	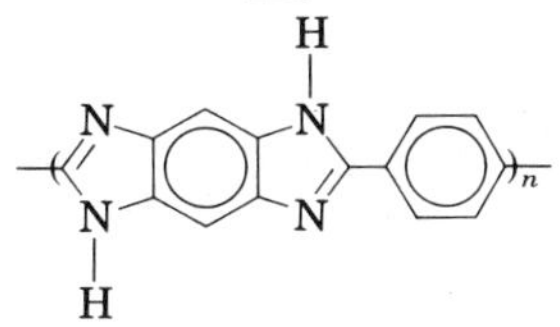	650
polypyrrone	ladder	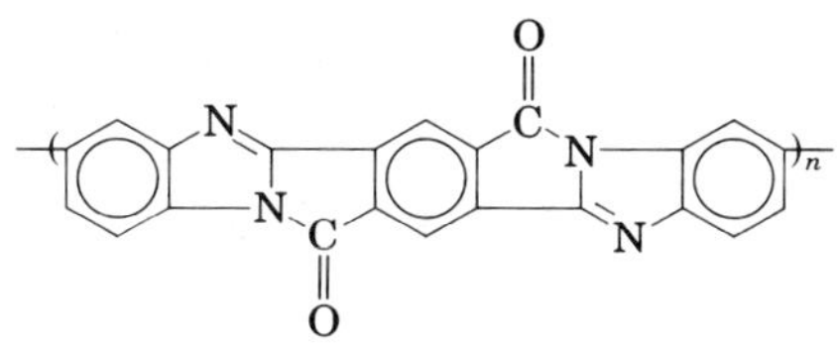	660
polybenzamide	aromatic	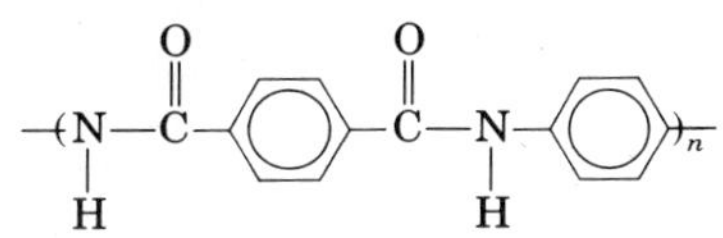	500
polyester	aromatic	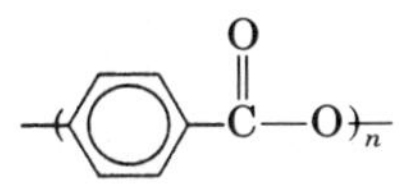	480

[a] Extrapolated onset of weight-loss temperature by tga in an inert atmosphere.

by a thermal barrier such as an intumescent coating. Their mechanism of action is largely independent of the formulation.

Thermal Sinks. Inert fillers and additives can act as thermal sinks. The inert fillers replace potential fuel, ie, the solid polymer, with inorganic materials such as talc, calcium carbonate, crushed marble, or clays.

Vapor-phase Chemistry. Where coatings are not suitable and no facile route toward altering the solid-state chemistry is available, flames are inhibited by interference with the chemistry of the combustion processes. With respect to linear, thermoplastic, polymeric materials, this strategy is most commonly employed. Additives altering the flame chemistry are referred to as vapor-phase-active retardants.

Smoldering. Boric acid and phosphorous compounds are effective, but their mode of action has not been elucidated. Sulfur has been used to inhibit smoldering.

Synergism. When two or more active flame retardants are employed in a single formulation, their effects on flame retardance may be described as additive, synergistic, or antagonistic. With additive effects, the efficiency of the total system is equal to the sum of the efficiencies of the individual components. Synergistic effects occur when the efficiency of the system is significantly higher than the additive efficiencies of the individual components under the same conditions.

Effectiveness and Efficiency. Flame retardants may be evaluated on the basis of effectiveness, the performance of a formulation under a specified set of test conditions, or on the basis of efficiency, the effect of a formulation on thermal or combustion parameters. Ease of ignition and extinguishment can also be measured. Performance criteria reflect degrees of flammability suitable for a specific use.

Flame Retardant Additives. In 1984, over 206 t of flame retardants were used in the United States (Table 5). Despite the large number of different flame-retardant additives, ca 90% of the commercial retardants currently used belong to one of six chemical classes: alumina trihydrate (ATH); organochlorine compounds; organobromine compounds; organophosphorus (including halogenated phosphorus compounds); antimony oxides; and boron compounds.

Test Methods for Fire Performance

Various test methods for fire performance are cited in government regulations. The test result may be pass/fail or a quantitative value. A test reflects a response of a material to a specific fire environment. Thus different tests may not give the same ranking of materials.

Table 5. U.S. Consumption of Flame Retardants, 1984

Type	10^3 t
Additives	
alumina trihydrates	92
antimony oxides	15
boron compounds	5
bromine compounds	15
chlorine compounds	15
phosphorus compounds	17
halogenated phosphorus compounds	11
others	8
Reactives	
epoxy intermediates	11
polycarbonate intermediates	3
polyester intermediates	5
flexible urethane intermediates	1
rigid urethane intermediates	4
others	4

BIBLIOGRAPHY

D. Drysdale, *An Introduction to Fire Dynamics,* John Wiley & Sons, Inc., New York, 1985.

J. W. Lyons, *The Chemistry and Uses of Fire Retardants,* John Wiley & Sons, Inc., New York, 1970.

National Materials Advisory Board, *Fire Safety Aspects of Polymeric Materials, Publ. # NMAB 318-1,* National Academy of Sciences, Washington, D.C., 1977.

National Fire Codes, The National Fire Protection Association, Quincy, Mass., published annually.

RICHARD G. GANN
RICHARD A. DIPERT
U.S. National Bureau of Standards
MICHAEL J. DREWS
Clemson University

FLASH DEVOLATILIZATION

Like other mass transfer processes, devolatilization requires the generation of surface area, in this case between the liquid polymer phase and the gas phase. Various forms of devolatilization are distinguished by the mechanism of surface generation. A sophisticated technique known as flash devolatilization generates surface through homogeneous nucleation, boiling, and subsequent two-phase flow.

The superior scale-up characteristics of flash devolatilization suggest its use as a primary means for removing solvents and unreacted monomers from polymer melts. Practical results show that this is indeed possible.

Equilibrium Calculations

Suppose that the back-pressure valve in Figure 1 is adjusted so that no boiling occurs within the shell-and-tube heat exchanger. Also assume that the heat-transfer area downstream from the valve is small. Then the flash is approximately adiabatic. T_{in} and P_{out} should be known because they are independently adjustable using the heat exchanger and vacuum system. Also known is the inlet solvent volume fraction ϕ_{in}. The two dependent variables, T_{out} and ϕ_{out}, must be determined. A heat balance gives

$$T_{out} = T_{in} - \frac{v}{C_p}\lambda \tag{1}$$

where v is the weight fraction of the entering stream that is vaporized, λ is the heat of vaporization, and C_p is the specific heat of the polymer solution. For simplicity λ and C_p are assumed to be constant, although equation 1 can be replaced by an ordinary differential equation should λ and C_p be strong functions of temperature or composition.

The material balance gives

$$\phi_{in} = v + (1 - v)\phi_{out} \tag{2}$$

where, for simplicity, weight and volume fractions are assumed equivalent. Finally, an equilibrium relationship is needed. The Flory-Huggins form is normally used for polymer solutions:

$$\frac{P_{out}}{P_{sat}} = \phi_{out} \exp\left[\left(1 - \frac{1}{m}\right)(1 - \phi_{out}) + \chi(1 - \phi_{out})^2\right] \tag{3}$$

where m is the degree of polymerization and χ is the Flory-Huggins interaction coefficient (typically 0.45 for a good solvent). The vapor pressure of the pure solvent P_{sat} is a function of T_{out} and is given by a correlation such as the Antoine or Lee-Kesler equation. The system described by equations 1–3 contains three unknowns (T_{out}, ϕ_{out}, and v) and is solved simultaneously.

High Solids Flashes

A typical application for flash devolatilization is the removal of 10–15% styrene–ethylbenzene from polystyrene. This can be achieved in an adiabatic, equilibrium flash:

	Pressure	Temperature	Solvent Fraction
Inlet	>1.8 MPa	241°C	15%
Outlet	1.3 kPa	220°C	500 ppm

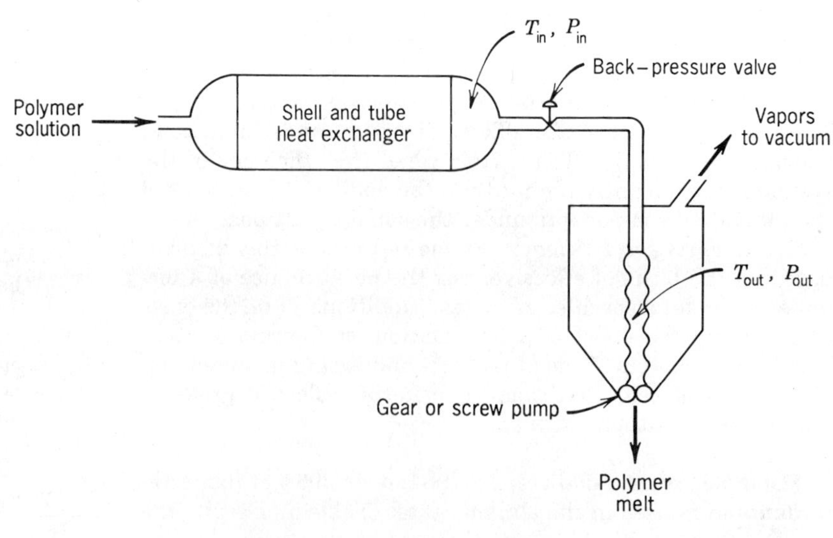

Figure 1. Flash devolatilization apparatus.

The secret to achieving a near-equilibrium flash is the generation of surface area through foaming. Vapor bubbles nucleate, grow, and coalesce into an open-celled foam provided T_{out} exceeds the glass-transition temperature of the polymer.

Induced Foaming

In induced foaming, a relatively short, melt-fed, single-screw extruder is used to heat and pump the incoming, partially devolatilized polymer. The extruder is also used to mix the foaming agent, typically water, into the polymer. Upon subsequent exposure to vacuum, the foaming agent expands and creates an open-celled foam.

Deep Adiabatic Flashes

A rather different mechanism applies to a very deep flash, say from 5–10% polymer to 50% or more. The vapor volume generated in such a flash completely disrupts the foam structure so that the partially devolatilized polymer is dispersed in the vapor as fine droplets. Equilibrium is closely approached in these flashes. Deep adiabatic flashes have found utility in a process known as compositional quenching.

BIBLIOGRAPHY

J. A. Biesenberger and D. H. Sebastian, *Principles of Polymerization Engineering,* John Wiley & Sons, Inc., New York, 1983.

R. Mallikarjun and E. B. Nauman, *Polym. Plast. Tech. Eng.* **28,** 137 (1989).

E. B. Nauman and co-workers, *Chem. Eng. Commun.* **66,** 29 (1988).

E. B. Nauman
Rensselaer Polytechnic Institute

FLAX. See Fibers, vegetable.

FLEX CRACKING. See Antioxidants; antiozonants.

FLOCCULATION

The process of flocculation is the gathering together or aggregation of small masses, usually in a liquid medium, into larger masses called flocs. Although this definition is generally applicable, various other definitions or refinements, especially when used in conjunction with the term coagulation, may be found in the literature.

Small suspended particles or colloids impede the separation of solids and liquids, resulting, for example, in excessively long sedimentation or filtration times. Increasing the particle size by flocculation alleviates these difficulties. The objectives of flocculation are to allow rapid separation or drainage of the liquid phase from a suspended solid phase, minimize the solids remaining in the liquid phase, and maximize the solids content in the solid phase (dewatering).

In order for flocculation to occur, chemicals called flocculants or coagulants are introduced to the solid suspension. Most of these chemicals are water-soluble polymers.

The principal practical applications of flocculation are in raw and waste-water clarification, sludge dewatering, mineral processing, and papermaking.

Flocculants

Inorganic Flocculants. The inorganic flocculants, more frequently referred to as coagulants, are based on the hydrolyzable salts of aluminum and iron, mainly aluminum sulfate (filter alum or alum, $Al_2(SO_4)_3$) and ferric chloride, $FeCl_3$.

Polymeric Flocculants. Polymeric flocculants are water-soluble, organic macromolecules capable of destabilizing and flocculating suspended matter. Water solubility is conferred by ionic and nonionic polar functional groups. Water-soluble polymers (qv) may be synthesized or derived from natural sources; usage depends on the application. For water treatment, synthetic polymers are used almost exclusively. In the pulp and paper industry, on the other hand, the consumption of natural polymers is nearly twice that of synthetic polymers.

Polymeric flocculants are distinguished by charge type, charge density, and molecular weight. Charge type provides the basis for classification into anionic, cationic, and nonionic polymers. Anionic and cationic polymers are known as polyelectrolytes (qv), but this term is often applied to nonionic polymers as well. The charge density, defined as the percentage of monomer units bearing a charge, may range from 100 mol % for homopolymers prepared from charged monomers to none for nonionic polymers. Molecular weights are at least 1×10^4 g/mol and as high as 1×10^7 (Table 1).

Synthetic Polymers. *Based on Polyacrylamide.* More than half of all synthetic polymeric flocculants and all those of high and ultrahigh molecular weight are copolymers in which the principal monomer is acrylamide. 43.8% of acrylamide flocculants are anionic, 48.8% are cationic, and 7.4% are nonionic acrylamide homopolymers (see Acrylamide polymers).

Quaternary Ammonium Types. Approximately 18% of the polymeric flocculants are homopolymers of the quaternary ammonium monomers, which are also used in the preparation of the cationic acrylamide copolymers. These homopolymers are low to medium molecular weight polyelectrolytes with 100% charge density.

Flocculants Based on Polyamines. Polyamine flocculants are cationic polymers of low to medium molecular weight, frequently referred to as coagulants. They are prepared from amines and epichlorohydrin (EPI) or ethylene dichloride (see Polyamines and Polyquaternary ammonium salts).

Natural Polymer Flocculants. Most natural polymeric flocculants, ie, 97%, consist of starch and its derivatives. Starch is a nonionic polymer of glucose and is obtained principally from corn.

Physical Forms. Low to medium molecular weight polymers are most frequently sold as concentrated aqueous solutions

Table 1. Molecular Weight Definitions

Description	Range (g/mol)
low	10^4–10^5
medium	10^5–10^6
high	1×10^6–5×10^6
ultrahigh	above 5×10^6

containing 10–50 wt % polymer. The high molecular weight polymers are supplied in solid, gel, and emulsion form. The emulsion polymers are also referred to as latex or liquid polymers, with concentrations of 25–30% polymer.

Mechanisms

Flocculation mechanisms include charge neutralization, bridging, electrostatic patch mechanism, and sweep-floc mechanism.

Kinetics

The kinetics of polymer flocculation depend on several physical processes. On addition of a flocculant, the polymer must be distributed throughout the system and must become adsorbed onto the colloidal particles, and the particles with adsorbed polymer must collide for aggregation to occur. Collision resulting from simple Brownian motion is called perikinetic flocculation, while orthokinetic flocculation results when aggregation is promoted by externally induced velocity gradients. By definition, before flocculation can occur, the repulsive forces caused by double-layer interactions must be reduced, allowing particle collisions. As these repulsive forces are overcome to the extent that aggregation occurs with every collision, rapid flocculation results. If sufficient surface charge exists, particles are repelled on approach unless they have sufficient kinetic energy to overcome the potential energy barrier. This is called slow flocculation.

Physical Effects

The rate, extent, and mechanism of flocculation are affected by particles, flocculants, and solution.

Particle properties of concern are concentration and size. Relevant solution properties include ionic strength and pH effects.

Health and Safety Factors

Since many polymer products currently on the market are used in potable applications or in the fabrication of paper products used in the food industry, they must be approved by the Department of the Environment, UK; the U.S. Environmental Protection Agency; and the U.S. Food and Drug Administration.

There is no evidence that synthetic polymers are toxic. Some polymer solutions, however, can cause minor skin and eye irritation.

BIBLIOGRAPHY

C. A. Finch, ed., *Chemistry and Technology of Water Soluble Polymers,* Plenum Press, New York, 1983.

J. Bratby, *Coagulation and Flocculation,* Uplands Press, Inc., Croydon, UK, 1980, Chapt. 1. Contains the most complete account of the title subject.

J. Lyklema in K. J. Ives, ed., *The Scientific Basis of Flocculation,* Sijtjoff & Noordhoff International Publishers, Alpen aan den Rijh, The Netherlands, 1978, pp. 3–36.

W. L. K. Schwoyer, ed., *Polyelectrolytes for Water and Wastewater Treatment,* CRC Press, Inc., Boca Raton, Fla., 1981.

GERARD R. ROSE
MICHAEL R. ST. JOHN
Nalco Chemical Company

FLOORING MATERIALS

Flooring materials containing polymers include resilient sheet goods and tile, prefinished and acrylic impregnated hardwood, and seamless, monolithic, and troweled surface coatings. The principal area of use is resilient flooring, where total U.S. resin consumption is over 300,000 t per year. The largest portion of this is poly(vinyl chloride) (PVC) homopolymer or copolymers containing vinyl acetate. Significant amounts of acrylics, polyurethanes, and butadiene–styrene resins are also used.

Resilient flooring is produced by at least 50 manufacturers in 14 countries throughout the world. The 1983 worldwide production figures for vinyl-sheet flooring, according to U.S. International Trade Commission estimates, are shown in Table 1.

Resilient Floor Coverings

Resilient floor coverings include vinyl sheet, vinyl tile, linoleum, asphalt tile, cork tile, and rubber tile. Vinyl sheet and vinyl tile, based on PVC and poly(vinyl chloride-*co*-vinyl acetate), represent the greatest part of the market. Linoleum and asphalt tile are mainly of historical interest. Cork and rubber-based products represent only a small fraction of the flooring market.

Vinyl-sheet Flooring. Vinyl-sheet flooring is produced in widths up to 4.5 m in two types. The first is a solid PVC composite sheet that may be covered with a clear protective surface, ie, wear layer. A plastisol is cast onto a removable backing or, in the case of narrow-width products, a calendering (qv) process is used. The second and more widely used type has a backing such as a polymer-impregnated mineral felt or a glass–synthetic fiber mat. The backing material is coated with a plastisol, which usually contains a blowing agent to produce a foam structure. The foam provides a cushioning effect and enhances the embossing-design possibilities. The composite is heated sufficiently to gel the plastisol and enable printing of the surface with a design. The surface is then covered with a clear PVC layer. For superior gloss retention and abrasion resistance, a polyurethane topcoat is often applied. The structure of this so-called no-wax flooring is illustrated in Figure 1.

Other Materials. Linoleum and asphalt are mainly of historical interest as flooring materials. Rubber tile is used primarily in commercial and industrial applications.

Prefinished Hardwood Flooring

Acrylic–Wood Flooring. The impregnation of wood with methyl methacrylate monomer, followed by polymerization, gives a flooring that is harder and more durable than wood. The aesthetics of the wood grain is enhanced by the polymer treatment.

Table 1. Vinyl-sheet Flooring Production

Area	10^3 m^2
United States	130,000
Europe	84,000
Japan	50,000
Australia	8,000
Korea	5,000

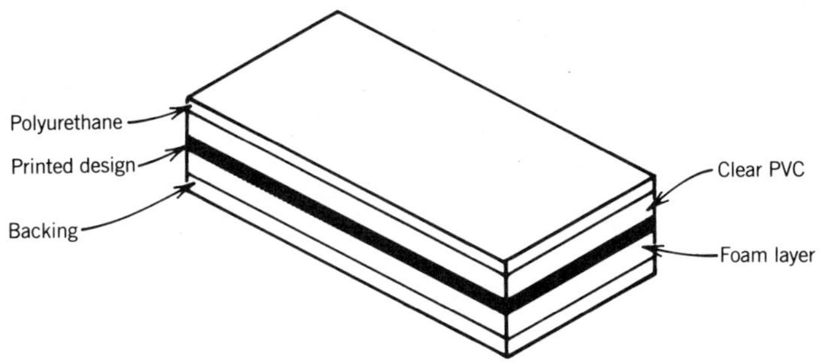

Figure 1. Structure of no-wax vinyl flooring.

In addition, resistance to abrasion, discoloration, and water spotting is increased (see WOOD, POLYMER-IMPREGNATED).

Seamless Flooring. Seamless flooring, also called monolithic flooring, makes up 1% of the noncarpet flooring market. Polymeric systems used in seamless flooring include epoxy resins, polyurethanes, and polyesters.

BIBLIOGRAPHY

U.S. Pat. 3,293,094 (Dec. 20, 1966), R. F. Nairn, J. C. Harkins, Jr., F. E. Ehrenfeld, Jr., and H. Tarlow (to Congoleum Corp.).

Can. Pat. 1,079,133 (June 10, 1980), R. Conger and V. Pogozelski (to Congoleum Corp.).

E. A. Peterson, C. J. Schmidle, and R. Barclay, *Radiat. Curing* **5**(1), 11 (Feb. 1978).

U.S. Pat. 4,041,200 (Aug. 9, 1977), A. G. Boranian and W. Timm (to GAF Corp.).

CLAUDE J. SCHMIDLE
Congoleum Corporation
ROBERT P. CONGER
ARMEN BORANIAN
Consultants

FLOOR POLISHES

Floor polish provides a thin, temporary coating that confers an attractive, clean, well-cared-for appearance. A polished floor can withstand the abusive wear of foot traffic, and the useful life of the flooring beneath the polish is extended. Scientific studies have repeatedly demonstrated the protective effect of floor polishes.

The floor-polish market is divided into household floor polishes and industrial and institutional floor polishes. Reliable estimates place the annual sales volume (in 1985 U.S. dollars) of household polishes at $\$110 \times 10^6$ for the United States and $\$450 \times 10^6$ worldwide. Best estimates place the annual sales volume of industrial and institutional floor polishes at $\$390 \times 10^6$ for the United States and $\$1.7 \times 10^9$ worldwide.

Ingredients

In order to provide both attractive appearance and protection, polishes must meet stringent performance requirements. Each performance property imposes constraints on the raw materials used to prepare the polish.

Ingredients include polymer latex, alkali-soluble resin, wax, and additives (eg, leveling aids, plasticizers, coalescing solvent, and metal ions).

Household Floor Polishes

Household floor polishes are formulated to produce good appearance, but they must also be easily removable. Typical formulations are given in Table 1. Household polishes fall into two general categories: dry-bright polishes and self-stripping polishes.

Industrial and Institutional Floor Polishes

Industrial polishes are generally formulated to minimize the time required for routine maintenance (Table 1). This includes maximum resistance to water and detergents as well as the properties that lead to good durability, eg, resistance to abrasive wear, scuffs, scratches, and heel marks. Industrial polishes are applied with a mop and removed by machine.

Industrial and institutional polishes fall into the following categories: dry-bright nonbuffable polishes, semi-buffable polishes, buffable polishes, sealers, spray-buffing polishes, and polishes for high speed burnishing.

BIBLIOGRAPHY

Performance of No-Wax Flooring in the Institutional and Industrial Environment, Booz, Allen & Hamilton, Inc. for Chemical Specialties Manufacturers Association (CSMA), Washington, D.C., Nov. 1979.

R. E. Zdanowski, *Soap Chem. Spec.* **36**(11), 78, 109 (Nov. 1960).

J. Brady, *Soap Cosmet. Chem. Spec.* **52**(12), 61, 72 (Dec. 1976).

L. P. Johnson and W. E. Draper in G. Illmann, *Wax Technology,* Chemical Specialties Manufacturers Association, Washington, D.C., 1975, pp. 22–26.

GORDON K. GREGORY
S. C. Johnson & Son, Inc.

Table 1. Typical Floor-polish Formulations[a], pph

Ingredient	1	2	3	4	5	6	7
polymer latex, 15% NV[b]	70	80	10	55	29	0	95
alkali-soluble resin, 15% NV[b]	20	20	80	22	8	25	5
wax emulsion, 15% NV[b]	10	0	10	25	63	74	0
coalescing solvent	1.0	4	1.0	1.5	0.9	0	1
plasticizer	0.4	0.8	0.5	0.4	0.2	0	0.4
surfactant	0.3	0.3	0.8	0.3	0.2	2.5	0

[a] Key: 1 = conventional dry-bright; 2 = clear, transparent liquid; 3 = self-stripping; 4 = semibuffable; 5 = buffable; 6 = wash wax; 7 = sealer.
[b] Nonvolatile.

FLOW.
See RHEOLOGY.

FLOW BIREFRINGENCE.
See BIREFRINGENCE; OPTICAL PROPERTIES.

FLUIDIZED-BED COATING.
See COATING METHODS.

FLUORINATED ETHYLENE–PROPYLENE COPOLYMERS.
See TETRAFLUOROETHYLENE POLYMERS.

FLUORINATED PLASTICS.
See TETRAFLUOROETHYLENE POLYMERS; VINYLIDENE FLUORIDE POLYMERS.

FLUORINE-CONTAINING POLYMERS

The substitution of fluorine for hydrogen in organic polymer structures has produced remarkable materials, including plastics, elastomers, and fluids, that are used in environments which degrade most other known polymers.

The various classes of materials are covered in several Encyclopedia articles: CHLOROTRIFLUOROETHYLENE POLYMERS; FLUOROCARBON ELASTOMERS; TETRAFLUOROETHYLENE POLYMERS; VINYL FLUORIDE POLYMERS; and VINYLIDENE FLUORIDE POLYMERS.

FLUOROCARBON ELASTOMERS

Fluorocarbon elastomers are synthetic, noncrystalline polymers that exhibit elastomeric properties when vulcanized. They are designed for demanding service applications in hostile environments characterized by broad temperature ranges and contact with chemicals, oils, or fuels.

Table 2. Properties of Fluoroelastomers

Property	Value
Physical properties	
tensile strength, MPa[a]	8.96–18.62
100% modulus, MPa[a]	2.07–15.17
elongation at break, %	100–500
hardness range	
Shore A	50–95
compression set, %	
70 h at 200°C	10–30
specific gravity of gumstock	1.80–1.86
brittle point[b], °C	−18 to −50
thermal degradation temperature, °C	400–550
General characteristics	
gas permeability	very low
flammability	self-extinguishing or nonburning[c]
radiation resistance	good to fair
abrasion resistance	good and satisfactory for most uses
weatherability and ozone resistance	outstanding[d]

[a] To convert MPa to psi, multiply by 145.
[b] Highly dependent on material grade.
[c] When properly formulated.
[d] Unaffected after 200-h exposure to 150-ppm ozone.

The principal commercial fluorocarbon elastomers are given in Table 1. Of the approximately 5000 t consumed worldwide per year, 40% is used in the United States, 30% in Europe, and 15% in Japan; 1985 prices were $33–242/kg.

Properties

The characteristics of vulcanizates prepared from commercially available fluorocarbon elastomer gumstocks are given in Table 2.

Table 1. Commercial Fluorocarbon Elastomers

Copolymer	Trademark	Supplier
poly(vinylidene fluoride-*co*-hexafluoropropylene)	Dai-el	Daikin
	Fluorel	3M
	Tecnoflon	Montedison
	Viton	DuPont
poly(vinylidene fluoride-*co*-hexafluoropropylene) plus cure-site monomer	Fluorel[a]	3M
poly(vinylidene fluoride-*co*-hexafluoropropylene-*co*-tetrafluoroethylene)	Dai-el	Daikin
	Fluorel	3M
	Tecnoflon	Montedison
	Viton	DuPont
poly(vinylidene fluoride-*co*-hexafluoropropylene-*co*-tetrafluoroethylene) plus cure-site monomer	Dai-el[a]	Daikin
	Fluorel[a]	3M
	Viton[a]	DuPont
poly[vinylidene fluoride-*co*-tetrafluoroethylene-*co*-perfluoro(methyl vinyl ether)] plus cure-site monomer	Viton[a]	DuPont
poly[tetrafluoroethylene-*co*-perfluoro(methyl vinyl ether)] plus cure-site monomer	Kalrez	DuPont
poly(tetrafluoroethylene-*co*-propylene)	Aflas[a]	Asahi Glass
poly(vinylidene fluoride-*co*-chlorotrifluoroethylene)	Kel-F[a]	3M

[a] Peroxide-curable.

Table 3. Typical Fluorocarbon Elastomer Compound

Component	phr
elastomer	100
magnesium oxide or calcium hydroxide	6–20
filler, reinforcing or nonreinforcing	0–60
accelerators or curatives	0–6
process aids	0–2

Manufacture

The elastomers listed in Table 1 are typically prepared by high pressure, free-radical, aqueous emulsion polymerization. The initiators are organic or inorganic peroxy compounds, such as ammonium persulfate, and the emulsifying agent is usually a fluorinated acid soap.

Processing

Compounding. Compared with the large number of ingredients required in a conventional rubber recipe, fluorocarbon elastomer compounding seems simple (Table 3) (see RUBBER CHEMICALS).

Mixing. Fluorocarbon elastomers may be compounded by any standard rubber technique.

Preforming. Formulations are easily preformed at cool barrel temperatures with a screw- or piston-type extruder (Barwell). Gums must be in the proper viscosity range and scorching must be avoided.

Molding. Both compression and injection molding are generally used. Flow requirements and cure rheology are different for both, and care should be taken in selecting the correct grade.

Extrusion. Extrusion techniques are used for tubing, hose, O-ring cord, preforms, and shaped gaskets.

Postcuring. Postcuring for 16–24 h at 200–260°C develops maximum physical properties, tensile strength, and compression-set resistance.

Health and Safety

Under normal handling conditions, fluorocarbon elastomers are nontoxic and nonirritating. Safe handling procedures are provided by manufacturers on request.

Uses

About 45% of U.S. fluorocarbon elastomer consumption is in ground transportation. Other industries include petroleum and petrochemicals, pollution control, and industrial hydraulic and pneumatic applications.

BIBLIOGRAPHY

"Fluoroelastomers" in D. A. Stivers, ed., *The Vanderbilt Rubber Handbook,* R. T. Vanderbilt Co., New York, 1978.

M. Morton, ed., *Rubber Technology,* Van Nostrand Reinhold Co., New York, 1973, pp. 407–439.

D. Apotheker, J. B. Finlay, P. J. Krusic, and A. L. Logothetis, *Rubber Chem. Technol.* **55,** 1004 (1982).

ACS Symp. Ser. **260** (1984).

MICHAEL M. LYNN
ALLAN T. WORM
3M Company

FLUOROCARBON POLYMERS.

See FLUOROCARBON ELASTOMERS; TETRAFLUOROETHYLENE POLYMERS.

FOAM CONTROL.

See ANTIFOAMING AGENTS.

FOAMED MATERIALS.

See CELLULAR MATERIALS.

FOAMING AGENTS.

See BLOWING AGENTS.

FOOD APPLICATIONS

Foods are highly complex chemical matrices containing a variety of polymers. Proteins (qv) and starch, two of the three principal nutritive food components, are polymers. Many foods, whether natural or fabricated, contain significant amounts of nonnutritive polymers providing important fiber (roughage). In addition, polymers determine to a large extent the form and texture of foods. These characteristics, together with flavor and color, determine the attractiveness of food.

Polymers are not absorbed by the human body except in very small amounts (see BIOLOGICAL ACTIVITY). The size of polymer molecules prevents their diffusion across the membranes of the gastrointestinal mucosa. However, the gastrointestinal tract contains digestive enzymes which break down certain types of polymers to monomeric units that can be absorbed.

Food technologists routinely use nonnutritive and nutritive polymers in the fabrication of food products. Most nonnutritive polymers added directly to food are hydrophilic colloids (hydrocolloids, also referred to as gums), usually polysaccharides. Nonnutritive food polymers are derived from many natural and synthetic sources. The natural polymers and their derivatives are obtained from plant exudates (arabic, ghatti, karaya, and tragacanth gums), seeds (guar and locust-bean gums), seaweed extracts (agar, alginate, carrageenan, and furcellaran), cellulose, and wood (arabinogalactan) and by microbial fermentation (xanthan gum and bakers' yeast glycan). Synthetic polymers include polydextrose and anoxomer.

Nonnutritive hydrocolloids perform functions such as gelation, thickening, texture modification, inhibition of crystallization, stabilization of emulsions and suspensions, surface-tension control, and encapsulation of flavor oils.

In food processing, acrylamide-based resins are used as flocculants in the production of sugars. Dimethylamine–epichlorohydrin copolymer is also used in sugar processing. Vinylpyrrolidinone polymers are employed as clarifying agents for vinegar, beer, and wine. Dimethylpolysiloxane, poly(ethylene glycol), and propylene glycol alginate are used as defoaming agents. Ion-exchange polymers (qv) are used in the preparation of foods and the treatment of the potable water added during processing. Enzymes (qv), free or immobilized, play an important role; eg, insoluble glucose isomerase preparations are used in the production of high fructose corn syrup.

Another area is food protection. Coumarone–indene resin and oxidized polyethylene are applied directly to certain fruits and vegetables to prevent moisture loss, thereby extending storage life. Food packaging is of course the most important use of polymers for food protection (see also PACKAGING MA-

TERIALS). A special application of polymers is in chewing gum, which is legally considered food.

The application of a particular polymer in food is determined by its ability to accomplish a certain technical effect, legal status and safety, and market considerations such as price and consumer acceptance. Consumption of nonnutritive polymers in food applications is most likely in excess of 3×10^4 t/yr. Under the U.S. Food, Drug, and Cosmetic Act, polymers added to food are considered food additives or generally recognized as safe (GRAS) substances.

Naturally Occurring Polysaccharides

Naturally occurring polysaccharides (qv) include xanthan (qv) gum arabic, and carrageenan (see also GUMS, INDUSTRIAL).

Synthetic Polymers and Modified Polysaccharides

Synthetic polymers and modified polysaccharides include polymers derived from cellulose, polydextrose, anoxomer, and polyvinylpyrrolidinone (see *N*-VINYLAMIDE POLYMERS).

Food Protection

Most fruits and vegetables lose moisture through transpiration after they are harvested. Consequently, for many years coating formulations containing hydrophobic materials have been applied to harvested fruits and vegetables for protection.

BIBLIOGRAPHY

M. Glicksman, ed., *Food Hydrocolloids,* Vols. I and II, CRC Press, Inc., Boca Raton, Fla., 1982 and 1983.

R. Davidson, ed., *Handbook of Water-soluble Gums and Resins,* McGraw-Hill, Inc., New York, 1980.

R. Whistler, ed., *Industrial Gums (Polysaccharides and Their Derivatives),* 2nd ed., Academic Press, Inc., New York, 1973.

M. Glicksman, *Gum Technology in the Food Industry,* Academic Press, Inc., New York, 1969.

TERRY C. TROXELL
PATRICIA S. SCHWARTZ
U.S. Food and Drug Administration

FORENSIC ASPECTS

Forensic science is the collective term for a hybrid applied science embracing any of the natural, engineering, or medical sciences that may be called upon for assistance in the administration of criminal or civil justice. This article is concerned principally with the utilization of polymers as items of trace evidence. The small size of samples usually available makes their examination nonroutine. A significant part of the forensic examination of polymeric substances is concerned with factors other than those deriving from the polymer itself because criminalistics is distinguished from all other sciences by its attempts to individualize.

It has only recently become possible to identify polymers by instrumental methods applied to microsamples. Even with this advantage, and in spite of the many possible modifications in polymer composition, it is the trace components that serve to individualize these substances and distinguish them from their close relatives.

Polymers are usually of interest for one of the following reasons.

Trace Evidence. Polymers may be crushed to powder, melted, left behind or carried away, burned, or broken.

Failures. Polymeric materials, like others, fail. Such failures are most often studied with regard to civil cases involving liability litigation, although they occasionally have bearing on a criminal case.

Burned Polymers. Most cases are civil, eg, the flammability of various fabrics or plastics and their suitability for a particular use.

Evidence Collection

Sample poverty is the usual rule in legal investigations. The sample available for study is frequently quite small and, in most cases, the requirement that the entire sample not be consumed in analysis places a further burden on the analyst and the techniques or instruments utilized. As a result, microscopy is the obvious first choice for most forensic examinations. The ability to locate, collect, manipulate, identify, and compare microscopic particles of trace evidence characterizes the successful forensic microscopist.

Analysis

Microscopical methods are most commonly employed because they allow the examination of minute specimens and are usually nondestructive. Other equipment less frequently used because larger samples are required include Instron testers, instruments for thermal analysis (dta, dsc, tga), and apparatus for determining molecular weights. Micromodifications of some of these instruments and techniques have been described and may have occasional significance.

Fibers

In the forensic study of fibers, four questions are commonly asked:

1. What is the identity of the fiber?
2. Could the questioned fiber have the same source as the known?
3. How was the fiber broken?
4. Who manufactured the fiber or fabric?

The first question can almost always be answered, no matter how little fiber is available. The other three are often more difficult.

A frequent problem is identification and comparison of fibers transferred from an item of clothing, carpet, or vehicle to a suspect, victim, or crime scene. In these examinations, the analyst receives evidence fibers and fibers from a suspect source that must be compared to decide whether or not the questioned and known fibers could have a common origin. As part of this comparison, the fibers must be classified according to their generic type. Tests designed to distinguish the fibers from each other are used. If at the conclusion of the tests the fibers cannot be distinguished, they are reported as nondistinguishable and it can be said that they may have a common origin. However, at best, fibers provide class-type evidence; they are not as individual as fingerprints or blood. The study of fiber fracture becomes important whenever fibers fail in use.

The possibility of determining the manufacturer of a single fiber is impossible in most cases but has been done successfully on occasion. The cooperation of fiber manufacturers and a comprehensive collection of known fibers are essential for this type of determination (see IDENTIFICATION, FIBERS).

Paint

Paint is usually submitted to the forensic scientist to be compared with paint from a known source. In the United States, most of these examinations involve automotive finishes in relation to automotive accidents and hit-and-run cases. House paints may be examined in connection with burglaries, and paint from tools in cases involving industrial sabotage. Tests are made to measure different properties in an attempt to distinguish two samples from each other.

Plastics and Rubber

Particles of plastics or rubber are often seen in debris recovered from crime scenes, victims' and suspects' clothing, etc. Occasionally, these have forensic significance and must be identified and compared. Rubber particles mounted on a microscope slide are recognized by the fact that they can be squashed when pressure is applied to the coverlip with a needle and recover fully when the pressure is released. Plastics are more easily recognized by an experienced eye by ruling out other types of particles.

Particles of plastic are identified by microscopy, solubility, melting point, microchemical tests, and micro-ftir spectrophotometry. Rubber is more difficult to identify by type using microscopical methods owing to the usual presence of pigments and fillers. The most useful technique found to date for these particles is pyrolysis, followed by ftir spectrophotometry of the pyrolyzate (see also IDENTIFICATION, ELASTOMERS).

BIBLIOGRAPHY

R. Saferstein, ed., *Forensic Science Handbook II,* Prentice-Hall International, Inc., Englewood Cliffs, N.J., 1986.

D. A. Hemsley, *The Light Microscopy of Synthetic Polymers,* Oxford University Press and The Royal Microscopical Society, London, 1984.

P. Echlin, ed., *Analysis of Organic and Biological Surfaces,* John Wiley & Sons, Inc., New York, 1984.

SKIP PALENIK
Walter C. McCrone Associates, Inc.

FORGING. See COLD FORMING.

FORMALDEHYDE POLYMERS. See ACETAL RESINS; AMINO RESINS; PHENOLIC RESINS.

FORMING. See CASTING; EXTRUSION; MACHINING; THERMOFORMING.

FOSSIL RESINS. See RESINS, NATURAL.

FOUNDRY RESINS

The foundry industry uses several procedures for casting metal parts: permanent mold casting, centrifugal casting, and sand casting. Permanent molds are made of metal, graphite, or a ceramic. In sand casting, molds and cores are used. Cores are required for hollow castings and must be removed after the metal has solidified. Cores are used in both sand and permanent molds.

Binders were developed to strengthen the cores, which are the most fragile part of a mold assembly.

This article is limited to the recently developed synthetic organic resin-based binders, which, because of their versatility and performance, have resulted in significant labor and energy savings and higher foundry productivity. In practice, these binders are mixed with sand, the mixes are compressed into the desired shape of the mold or core, and the binders are hardened, ie, cured, by chemical or thermal reactions to fixate the shapes. Typically, ca 0.7–4.0 parts (usually 1–2 parts) of binder are added to 100 parts of sand.

Core and Mold-making Processes

Five types of fabrication processes are used by the foundry industry to produce molds and cores from sand and organic binders. In the no-bake and the cold-box processes, the binder is cured at room temperature, while in the shell-molding, hot-box, and oven-bake processes, heat cures are applied. Selection of the process and type of binder depends upon the size and number of cores or molds required, production rates, and equipment.

Binder Resins

The foundry industry employs phenolic resins, furan resins, and unsaturated resins. In addition, various blends, such as furan–phenolic resin combination, are commonly used. Certain types of the phenolic and unsaturated resins used with polymeric isocyanates are applied as two-component (polyurethane-forming) binder systems.

Phenolic Resins. Phenolic resins (qv) are the most widely used resin binders in the foundry industry. They are produced by polycondensation of phenols with formaldehyde. The three types (Table 1) differ in catalyst and mole ratio of reactants used in their preparation. Furthermore, they have different molecular structures and reactivities and require different curing agents.

Furan Resins. Furan resin-based binders are widely used in no-bake, hot-box, and cold-box processes. Furfuryl alcohol is blended with preformed resins, including furfuryl alcohol homopolymers, or polymerized by acid catalysis with formaldehyde, ketones, phenol, or urea (see FURAN POLYMERS).

The great variety of furan binders available provides widely differing performance characteristics for use in different applications. Water content may be as high as 15% and nitrogen content as high as 10% in inexpensive resins made with urea. The choice of a specific binder depends on the type of metal to be cast.

Unsaturated Resins. Binders based on unsaturated resins are used in the production of foundry molds and cores. These in-

Table 1. Phenolic Resins for Foundry Applications

	Novolacs	Resoles	Polybenzylic ethers
ratio of phenol to formaldehyde	>1	<1	<1
pH conditions	acidic	basic	neutral, metal salts, anhydrous
thermal behavior	thermoplastic	thermosetting	thermosetting
foundry process	shell process	hot-box process no-bake process	cold-box process no-bake process

clude the core oils based on drying oil, alkyd–isocyanate binders, and vinyl-unsaturated urethanes. Unlike the phenolic or furan-based binders, these systems are cured primarily by free-radical mechanisms.

BIBLIOGRAPHY

P. R. Carey and G. P. Sturtz, *Foundry Manage. Technol.* **114**(2), 22 (1986).

A. Knop and L. A. Pilato, *Phenolic Resins,* Springer-Verlag, New York, 1985, pp. 24–58, 256–267.

W. J. McKillip and E. Sherman in M. Grayson, ed., *Kirk-Othmer Encyclopedia of Chemical Technology,* 3rd ed., Vol. 11, John Wiley & Sons, Inc., New York, 1980, pp. 499–527.

H. J. Langer, W. R. Dunnavant, and H. Kögler, *Giesserei* **69,** 150 (1982).

HEIMO J. LANGER
WILLIAM R. DUNNAVANT
Ashland Chemical Company

FRACTALS

Fractal geometry is concerned with the quantitative description of complex structures and the ways in which these structures transform under a change of length scales. The most simple fractals are self-similar or scale invariant structures that are invariant to an isotropic change of length scales. They "look the same" under different magnifications. Familiar examples include coastlines, clouds, and biological structures such as the vascular system and nerve cells. The concepts of fractal geometry were developed during the past two to three decades, primarily by B. B. Mandelbrot. During the past few years these ideas have become widely disseminated and applied to a very broad range of materials and phenomena. Applications to the physical sciences have been described in a number of books, reviews, and conference proceedings.

Considerable interest has developed in fractal structures that grow under nonequilibrium conditions, much of it beginning with the diffusion-limited aggregation (DLA) model in which particles are added, one at a time, to a growing cluster or aggregate of particles via random walk trajectories originating far from the growing cluster. This model stimulated interest in a variety of nonequilibrium growth and aggregation models that lead to the formation of fractal structures. In contrast, the most important models for polymer structures are (appropriately) equilibrium models. Thus the theoretical apparatus of equilibrium statistical mechanics (qv) can be brought to bear to develop an understanding of the asymptotic (high molecular weight) scaling properties of polymers. In general, the models for polymer structure are much better understood than nonequilibrium models such as DLA or diffusion-limited cluster–cluster aggregation.

This article concerns primarily the structure of polymers at the molecular level. However, structures on a large length scale formed by processes such as fracture, phase separation, aggregation, deposition, degradation, etc, often have a fractal geometry. In most cases, nonequilibrium models must be used to describe the evolution of the morphologies that result from these processes.

Fractal Geometry

Some of the basic concepts of fractal geometry needed for a discussion of polymers are introduced here. Additional information can be found in the references given.

Self-similar Fractals. Figure 1 shows four stages in the construction of a self-similar fractal. In the first stage, 5 discs are joined to form a small cross (Fig. **1a**). Five of these crosses are then joined to form a larger cross containing 25 discs (Fig. **1b**). At each stage in this construction process the mass (number of discs) is increased by a factor of 5 and the overall size (spatial extent) of the structure is increased by a factor of 3. The structure can also be extended to smaller length scales by replacing each disc of radius d_0 by a cross consisting of five discs of radius $d_0/3$, etc. If these construction processes are extended to infinitesimally small and infinitely large length scales, a self-similar fractal is generated that can be mapped exactly onto itself after uniform dilation or contraction by a factor of 3 (or 3^n). If the structure is dilated by a factor of 3, then 5 replicas of the structure are needed to cover the dilated structure and the fractal dimensionality D is given by

$$D = \log 5/\log 3 \simeq 1.4649 \tag{1}$$

This is in accord with intuitive ideas concerning dimensionality in ordinary (Euclidean) objects.

For physical realizations of simple self-similar fractals the fractal dimensionality can also be thought of in terms of the scaling relationship between mass M and length ℓ

$$M \sim \ell^{D_m} \tag{2}$$

Here ℓ is a characteristic length that describes the overall spatial extent of the structure. The radius of gyration R_g or the maximum radius could equally well be used for the length ℓ. The exponent D_m in equation 2 is often called the mass fractal dimensionality. Many of the approaches that are used to measure D for structures generated by experiments or computer simulations are based on equation 2. Another approach is to cover the fractal with a series of "grids" whose elements have sides of length ϵ (volume $= \epsilon^d$ where d is the dimensionality

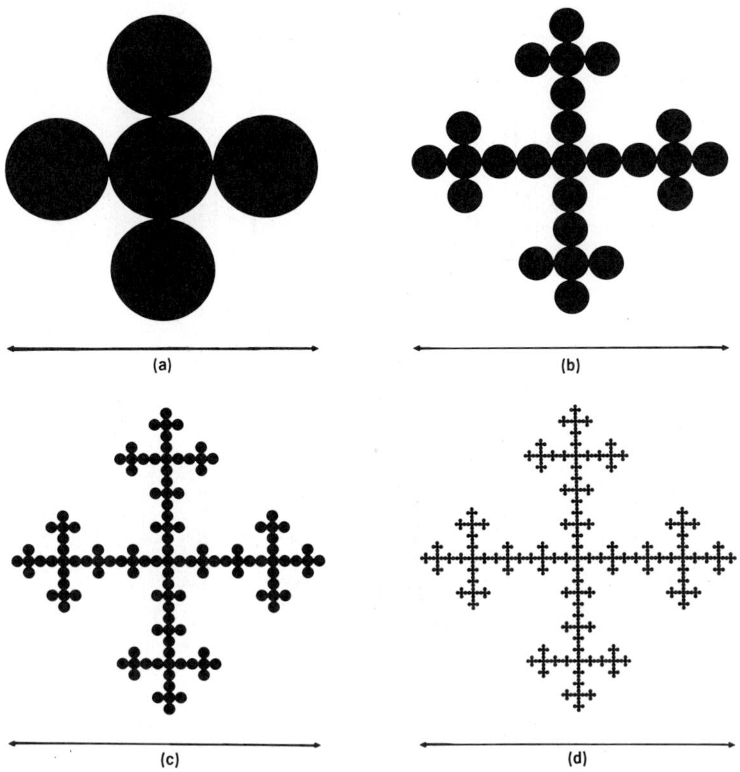

Figure 1. Four stages in the construction of a self-similar fractal. At each stage the mass (number of discs) increases by a factor of 5 while the diameter increases by a factor of 3. The fractal dimensionality is given by $D = \log(5)/\log(3) = 1.4649$. (**a**) 3 Diameters; (**b**) 9 diameters; (**c**) 27 diameters; (**d**) 81 diameters.

of the embedding space in which the fractal exists). The number of grid elements that are occupied depends on ϵ according to the power law

$$N(\epsilon) \sim \epsilon^{-D_\epsilon} \tag{3}$$

For self-similar fractals $D_\epsilon = D_m = D$ where D is the all-purpose fractal dimensionality.

Self-affine Fractals. Many of the important models for the structure of polymers are self-similar fractals that are invariant to an isotropic change in length scales. However, a broader range of systems exhibits different scaling behavior in different "directions." Fractals with these properties are said to be self-affine.

Measurement of Fractal Dimensionalities. Many approaches to the measurement of fractal dimensionalities are based on equation 2. For example, the dependence of log R_g on log s can be measured in a growth or aggregation process. Here s is the size (mass) of the structure. Because of corrections to the asymptotic scaling behavior, the slope of this curve provides an estimate of the effective fractal dimensionality $\{D_{\text{eff}} = [d(\log R_g)/d(\log s)]^{-1}\}$ that is, in general, different from the asymptotic ($s \to \infty$) fractal dimensionality. If sufficient data are evaluated to reduce the statistical uncertainties, it is useful to explore the dependence of D_{eff} (the effective fractal dimensionality) on s to obtain an estimate of the magnitude of the corrections to scaling. In favorable cases, it may be possible to fit the dependence of R_g on s by a two term power law

$$R_g \sim s^{\beta}(1 + As^{-\gamma}) \tag{4}$$

Uncertainties due to corrections to scaling can be reduced by extending the simulation or experiments to a larger range of length scales, or by reducing statistical uncertainties by using a large number of samples and carrying out a systematic series analysis based on equation 4 or a more general form for the corrections to scaling.

Applications to Polymers

The concepts of fractal geometry can be used to describe the structure of polymer molecules and to develop an understanding of their properties. Very often the application of these ideas leads to a more simple and more intuitive understanding than alternative approaches.

BIBLIOGRAPHY

B. B. Mandelbrot, *The Fractal Geometry of Nature,* W. H. Freeman and Co., New York, 1982.

J. Feder, *Fractals,* Plenum, New York, 1988.

T. Vicsek, *Fractal Growth Phenomena,* World Scientific, Singapore, 1988.

D. Avnir, ed., *The Fractal Approach to Heterogeneous Chemistry: Surfaces, Colloids, Polymers,* J. Wiley & Sons, Ltd., Chichester, UK, 1989.

PAUL MEAKIN
E.I. du Pont de Nemours & Co., Inc.

FRACTIONATION

Polymers are usually mixtures of homologues differing in molecular weight. In some homopolymers, differences in branching may be present. In copolymers there are differences in chemical composition. Many graft and block copolymers may contain varying amounts of homopolymers. These heterogeneities affect the polymer properties. Fractionation is a tool used to determine the distribution of molecular weight and the heterogeneous structure in a polymer. It is also used in the preparation of homogeneous polymers that may not be directly obtainable by polymerization (see also MOLECULAR WEIGHT DETERMINATION).

The necessity of fractionating polymers resulted in the development of many useful fractionation methods. In batch fractionation and column-elution fractionation, actual fractions are collected and analyzed individually; hence these methods are slower. Turbidimetric methods, ultracentrifugation, and modern chromatographic methods do not require the collection of fractions and are faster. The faster chromatographic methods are usually preferred, although the slower methods survive and are occasionally more convenient.

A copolymer can be fractionated according to molecular weight or chemical composition. Determination of the two different distributions by two separate fractionation experiments often gives only a limited knowledge of the true heterogeneity of the sample. For a complete understanding of two-dimensional distribution, the fractions from one experiment are collected and each fraction is fractionated further by a different experiment. Such a procedure is called cross fractionation. When branching is present, cross fractionation is required to determine the two-dimensional distribution of molecular weight and degree of branching.

Batch Fractionation

Batch fractionation is based on the partition of polymer molecules in two liquid phases. The polymer is dissolved first to give a dilute solution. The solution is separated into two phases by the addition of a nonsolvent or by lowering the temperature. The partition of the polymer into the two phases depends on molecular weight or chemical composition.

Two batch fractionation processes are used: partial precipitation or successive precipitation fractionation, and coacervate extraction or successive solution fractionation. Refractionation schemes can be applied to both processes to improve the monodispersity of the fractions.

The coalescence of phases in batch fractionation may take a long time. The fractionation vessel must be immersed in a constant-temperature bath or surrounded by refluxing vapors. An apparatus suitable for the fractionation of polyethylene at elevated temperatures is shown in Figure 1.

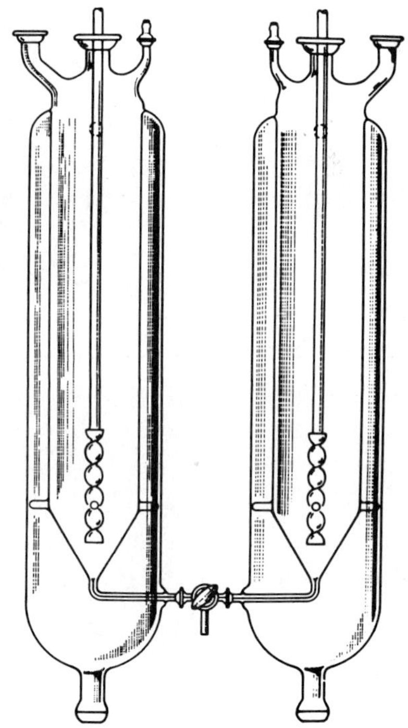

Figure 1. Vessels for high temperature batch fractionation.

Batch fractionations are slow and tedious. Characterization of the fractions is also time-consuming. The construction of the molecular weight distribution from the fractionation data is not very precise. However, when a polymer contains functional groups that tend to be adsorbed by the packings of gel-permeation chromatographic columns or when a polymer is suspected of containing gels that might plug a costly chromatographic column, batch fractionation is very useful.

Column-elution Fractionation

Column-elution fractionation was conceived by Destreux and Spiegel as a speedy method for the fractionation of low density polyethylene. The polyethylene, deposited on a support of Celite, is eluted with toluene at successively higher temperatures. Other workers used solvent and nonsolvent mixtures for elution at the same temperature. The waiting time for the phases to coalesce and settle into layers is eliminated, and the fractions may be collected within a few hours.

A typical apparatus for column-elution fractionation based on the design of Francis and co-workers is used to fractionate polypropylene.

Column-elution fractionation is faster than batch fractionation but the time consuming characterization is still required. It is not nearly as fast or precise as gel-permeation chromatography (gpc). Column-elution fractionation is used for the determination of molecular weight distribution only when gpc is not successful, for example, for gels containing high density polyethylene.

Turbidimetric Methods

Turbidimetric titration is a form of partial precipitation fractionation. A nonsolvent is added continuously to a dilute solution of the polymer. The turbidity created by the precipitation or phase separation is measured by a photometer.

The molecular weight distribution of the polymer is derived from the turbidity generated as a function of solvent composition. Turbidimetric titration is much faster than batch and column-elution fractionations since individual fractions are not recovered and characterized. However, the method is not quantitatively reliable because of the uncertainty between the turbidity and the amount of polymer precipitated and other complications.

Ultracentrifugation

In the sedimentation-velocity method, the rate of sedimentation is measured. In equilibrium centrifugation and density-gradient centrifugation, the sedimentation process is allowed to come to equilibrium with the back diffusion process. A commercial instrument suitable for all three ultracentrifugation methods is the Beckman Spinco model E analytical ultracentrifuge.

Ultracentrifugation (qv) today is mainly a tool for the characterization of biopolymers in aqueous solutions. Its use for hydrophobic flexible polymers has been largely superseded by gel-permeation chromatography. Nevertheless, sedimentation velocity is a powerful method for the determination of branching and molecular weight distribution of high molecular weight polymers if a suitable solvent is available.

Gel-permeation Chromatography (Size-exclusion Chromatography)

In gel-permeation chromatography (gpc), polymers of different molecular weight are separated in a column packed with porous cross-linked gels (see also CHROMATOGRAPHY). When a polymer solution is eluted, the species of higher molecular weight, which permeate the porous gels to a lesser degree than the species of lower molecular weight, are eluted out first. In gpc, the gels are packed tightly in columns, and sample solutions are injected into the eluting solvent stream at the column inlet. The eluting solvent, being pumped at a near constant rate, generates a highly reproducible relation between molecular weight and retention volume. Because of the reproducibility and the speed of operation, gpc is the principal fractionation method for analytical purposes. Preparative gpc instruments are also available commercially. With aqueous polymer solutions, the technique is called aqueous gpc. The designation size-exclusion chromatography was adopted later to reflect the nature of the method.

Thin-layer Chromatography

Thin-layer chromatography (tlc), introduced as a form of adsorption chromatography, is a common analytical tool for organic chemists and biochemists. The stationary phase is a thin layer of absorbent, about 0.25 mm thick. The sample is deposited as a small spot on the absorbent. The eluent solvent, called developer, is the mobile phase. The developer is allowed to migrate past the sample spot, and the ratio of the rate of migration of a sample to the rate of migration of the developer, defined as the R_f value, is determined. If the species in the polymer sample have different R_f values, fractionation is achieved. Polymeric samples are separated by three thin-layer processes: adsorption–desorption, molecular sieving or exclusion, and phase separation or precipitation. Partition and ion exchange, the two other thin-layer processes, are effective only in the separation of low molecular weight compounds.

Thin-layer chromatographic plates are commercially available. The most useful stationary phases for polymer fractionation are silica gels activated (heated) to varying degrees of dehydration.

Adsorption Chromatography

The development of quantitative data in tlc is cumbersome and tedious. For higher precision, better reproducibility, and the possibility of automated analysis, chromatographic methods utilizing packed columns are more desirable. Gel-permeation chromatography is the column version of tlc molecular sieving. Adsorption chromatography is the column version of tlc adsorption–desorption and the phase-separation processes.

Theory on adsorption of polymers shows that adsorption (qv) gives good molecular weight separation only for oligomer and low molecular weight polymer samples. Therefore, adsorption chromatography is used mainly for the determination of chemical-composition distribution of copolymers. Experimentally, adsorption chromatography for polymers does not differ greatly from conventional liquid chromatography.

Field-flow Fractionation

In field-flow fractionation (fff), separation takes place in a thin open channel of rectangular cross section. The flowing fluid assumes a laminar-flow profile with the highest velocity at the center of the channel and the slowest near the surface of the wall. An external field is applied perpendicular to the flow direction to push the solutes into narrow layers against the wall on the opposite side. The movement of a solute species in the transverse direction depends on its degree of interaction with the applied field and its tendency to diffuse away from the region of higher concentration. The solutes, which distribute in positions nearest the fastest moving center layer, are eluted first. A detector at the outlet records the concentration of the solute as a function of time (a factoram). Despite some promising results, fff is not used very widely.

Foam Fractionation

Foam fractionation has been used for the characterization of poly(vinyl alcohol). Foam bubbles are collected from an aqueous solution of the polymer and analyzed. The separation mechanism is based on the adsorption of solute at the bubble surfaces.

Fractionation of Copolymers

Batch and column-elution fractionation can be used to fractionate copolymers according to molecular weight or chemical composition, depending on the choices of solvents. Since the solvent interaction parameters are related to Hildebrand's solubility parameters, the latter are useful guides for selecting solvents. Cloud-point titration is used often to screen solvents experimentally.

In ultracentrifugation only density-gradient centrifugation is useful for analyzing copolymer chemical-composition distribution. In gel-permeation chromatography a second detector is required to determine the concentration of copolymers. Uv and infrared spectrophotometers are common second detectors used with the standard differential refractometer detector. A stop-flow technique is used to allow the determination of the complete infrared spectrum at any elution volume.

With the help of a multidetector in gpc, the variation of chemical composition can be determined as a function of elution volume but not the chemical-composition distribution itself. For the determination of chemical-composition distributions, tlc and adsorption chromatography are preferred.

To gain a detailed knowledge of the two-dimensional distribution of molecular weight and chemical composition, cross fractionation is required, where fractions obtained from one fractionation are used as the starting samples of a second fractionation. If the first fractionation is separating mainly according to molecular weight, the second fractionation should separate mainly according to chemical composition. Cross fractionation can be tedious when batch fractionation is used for both fractionations.

Chromatographic cross fractionation is less tedious, as shown by studies using tlc and adsorption columns of dried silica or two-coupled gpc. In the latter case, the solvent used in the second gpc promotes adsorption and phase separation; therefore the fractionation in the second gpc is more according to chemical composition. This technique is termed orthogonal chromatography. Gpc can also be combined with adsorption chromatography. Conditions for a successful fractionation of copolymers are often critical.

Fractionation of Branched Polymers

Two types of branches are found in polymers: short-chain branches with lengths from one to several carbons long and long-chain branches with lengths of the same order of magnitude as those of the main backbone chains.

Short branches impede crystallization and cause a decrease in melting point. Based on this effect, fractionation of short-chain branching in low density polyethylene is achieved by the technique of temperature-rising elution fractionation. From a column-elution apparatus packed with Chromasorb, and xylene as the elution solvent, starting at 50°C and rising at a rate of 4°C/h to 90°C, the fractions obtained differed mainly in the degree of short-chain branching. Temperature-rising elution fractionation has been combined with gpc for automatic cross fractionation of low density polyethylene. A two-dimensional distribution of short-chain branching and molecular weight has been determined.

Long-chain branching is characterized by dimensional differences in dilute solutions. A parameter g, defined as the ratio of mean square radii of the branched molecule and a linear molecule of the same molecular weight, is used to correlate the degree of long-chain branches. This parameter can be determined from intrinsic viscosities for narrow distribution samples. Several studies on broad distribution branched polymers were made, first by batch fractionation to obtain narrow molecular weight distribution fractions and then by performing the tedious double experiments to characterize the polymer in each of the fractions. A faster method is the use of gpc for the fractionation and on-line dual detectors to perform the double characterization.

Another method of determining the distribution of branching in polymers is by combining gpc and sedimentation-velocity measurements. In gpc, branched molecules, smaller in size, are eluted later as if they were linear polymers of lower molecular weights. In sedimentation measurements, the denser and smaller branched molecules settle faster and behave like linear polymers of higher molecular weights. This method has a higher sensitivity for the detection of long-chain branching than the dual-detector gpc methods.

BIBLIOGRAPHY

L. H. Tung, ed., *Fractionation of Synthetic Polymers,* Marcel Dekker, Inc., New York, 1977.

M. J. R. Cantow, ed., *Polymer Fractionation,* Academic Press, Inc., New York, 1967.

B. Ke, ed., *Newer Methods of Polymer Characterization,* Wiley-Interscience, New York, 1964.

W. W. Yau, J. J. Kirkland, and D. D. Bly, *Modern Size-Exclusion Liquid Chromatography,* John Wiley & Sons, Inc., New York, 1979.

L. H. TUNG
The Dow Chemical Company

FRACTURE

Definition

Fracture is defined as a stress-biased material disintegration through the formation of new surfaces within a body. The brittle fracture of a plastic knife and the rupture of an inflated elastomeric balloon are typical examples. Fracture is synonymous with rupture and breakage but not with failure. The latter term is more general and also encompasses nonmechanical breakdown through heat (thermal failure) or environmental degradation (chemical attack, irradiation). For fracture to occur, it is generally necessary that a specimen be subjected to mechanical loads, that it deform (viscoelastically), and that some (molecular) damage develops which lead to material disintegration.

Fracture Phenomena

Fracture phenomena include brittle fracture at low temperatures, high loading rates, or high strain energy; dry crazing, stress-whitening, and environmental stress cracking; creep crazing (fissurelike brittle rupture); and fatigue. Yielding, necking, creeping, and flowing also contribute to polymer fracture. Localized fracture events also influence abrasion, erosion, and wear of a material. (See also CRAZING; POLYMER CREEP; ABRASION AND WEAR.)

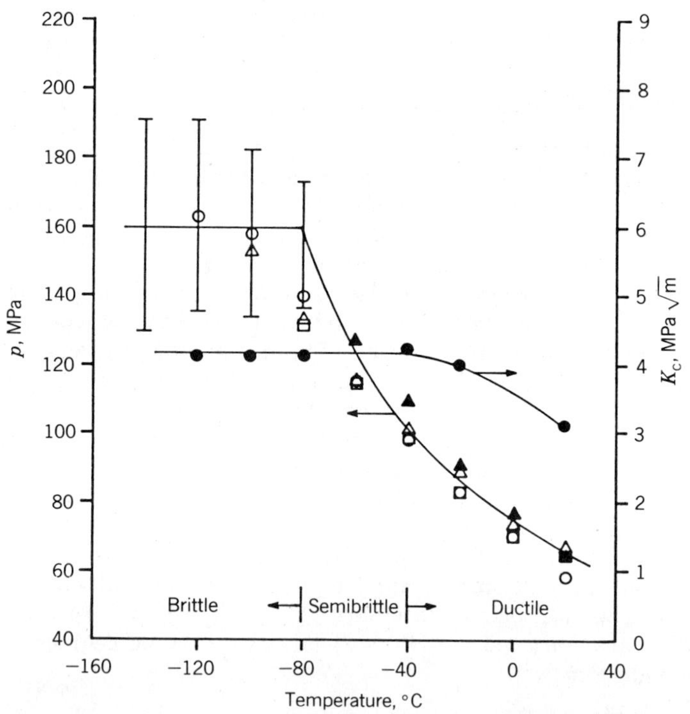

Figure 1. Fracture stress in tension (T) or bending (three-point bending) of unplasticized PVC. The vertical bars indicate the maximum scatter band. Courtesy of S. Hashemi, London.

		specimen width, mm	specimen depth, mm
tension	●	6	10
	△	3	12.5
	□	6	15
3-point bending	▲	12.5	15
	○	6	50

Fracture through cracking typically shows two to four phases. In the first, defect initiation or activation, there is no visible crack growth. In the second phase, a deformation zone or a crack starts growing "thermally"; depending on crack length, stress, and material characteristics, the initially stable crack growth can turn into unstable crack propagation, which is generally equivalent to instantaneous specimen failure. Fracture research and engineering design, therefore, are directed towards a better understanding and control of the nature of material defects, the behavior of a crack or other defect in a loaded body, and the influence of external variables (state of stress and strain; temperature; environment) on crack initiation and propagation.

Molecular and Microscopic Theories

Kinetics. As opposed to continuum-mechanical theories, molecular theories of fracture recognize the presence of discrete particles or elements forming the material. These theories relate breakage, displacement, and reformation of these elements with deformation, defect development, and fracture of the structured material. Some of these theories share the assumption that macroscopic failure is a rate process, that the basic fracture events are controlled by thermally activated breakages of secondary or primary bonds, and that the accumulation of these events leads to crack formation or breakdown of the loaded sample. These events are sometimes defined in a general manner as damage and are not experimentally related to particular morphological changes or to the existence of microcracks. In contrast, morphological theories are based on the physical evidence of molecular damage obtained from spectroscopic or scattering methods.

In a structured material several different mechanisms can simultaneously contribute to failure; the prevailing mechanism may change with stress, temperature, and/or state of defect development. Excellent examples are given in Figures 1 and 2. The kinetics of the failure mechanisms "creep in shear," "chain disentanglement," and "thermal degradation" are clearly different, leading to well pronounced transitions in the fracture behavior.

Chain Scission and Disentanglement. Molecular failure through chain scission occurs in a loaded polymer if molecular segments are strained to rupture rather than to disengage by slippage. Chain scission or rupture indicates the attainment of extremely large molecular stresses. It can be achieved if chains are heavily entangled (high molecular weight) or strongly interacting through chemical cross-links or physical bonds (hydrogen bonds, crystal lamellae in semicrystalline fibers, perfect alignment of molecular chains in extended chain crystals) or if the rate of loading (considerably) surpasses the rate of disentanglement of a given molecular segment. The transition from disentanglement of chains (at low rates of loading and/or at high temperatures) to chain scission occurs, for instance, in the breakdown of craze fibrils, in long-time loading of semicrystalline thermoplastics, or at the tip of a growing crack. This phenomenon has been studied in great detail by microstructural analysis (sem, tem, esr, ir, mol wt analysis) and in correlation with deformation and fracture behavior.

Structural Implications. The important influence of chain length on polymer strength and toughness has already been

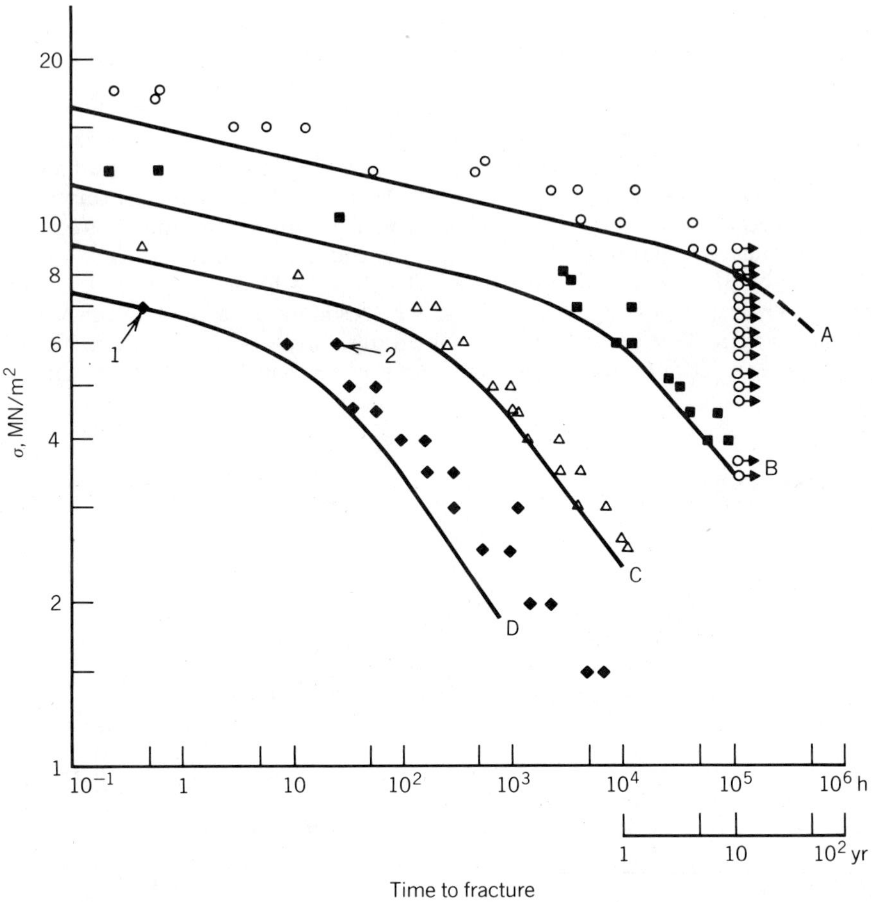

Figure 2. Times to failure of high density polyethylene (HDPE) water pipes under internal pressure p at different stresses and temperatures: A, 20°C; B, 40°C; C, 60°C; D, 80°C. 1 = ductile failure; 2 = creep crazing. Circumferential stress $\sigma = d_m/2s$, where d_m = average diameter and s = wall thickness. To convert MN/m² to psi, multiply by 145.

indicated. Other influences on fracture strength include molecular mobility; chain branching, tacticity, and degree of chain orientation and crystallinity; morphology (spherulitic or fibrillar or highly chain extended structure); and cross-link density. The important influence of second components (incompatible phases, reinforcing agents) should also be mentioned (see COMPOSITES; POLYMER BLENDS).

Defects. Through the stress concentration that they cause, material defects are at the origin of crazes and cracks. Defects strongly influence the absolute value of the strength of a specimen and its time-to-failure; the irregularity in size, shape, position, and nature of the defects generally is responsible for the observed scatter of the ultimate properties (see Figures 1 and 2). Quantification of this effect is one of the major aims of (linear elastic) fracture mechanics.

Intrinsic Crazing and Stress-whitening. At appropriate stress levels ordinary or extrinsic crazes are initiated. In amorphous single-phase polymers these crazes are preferentially initiated at the sample surface through foreign particles, fingerprints, or surface grooves (see CRAZING; for multiphase polymers see POLYMER BLENDS). Such crazes can grow into cracks. Conversely, notches or cracks in stressed glassy polymers are often preceded by crazes, whose lengths and rates of growth are a (calculable) function of the stress concentration. Stress-whitening caused by intrinsic crazing at higher sample extension ratios has been reported for a few cases of amorphous single-phase polymers. Strong evidence indicates that the initiation of intrinsic crazes within the total sample volume is governed by a temperature- and stress-activated instability of the entanglement network frozen-in during the glass transition.

Environment. Liquid and even some gaseous environments can lead to surface plastification and microcracking, especially of stressed polymers. The specific environmental problems of the different classes of polymers are summarized in Table 1.

Fracture Mechanics

Aims and Restrictions of Linear Elastic Fracture Mechanics as Applied to Polymers. Fracture mechanics describes the behavior of cracks or other defects when a body is loaded. A crack constitutes a discontinuity within a body. Fracture mechanics describes the conditions under which it grows or propagates, creating new surface area. At the boundary of the discontinuity, very high local stresses and strains are usually observed, and in the limit of a sharp crack in an elastic body, these may theoretically be infinite or singular. There are two approaches to describing fracture. In the first, the balance of energy is discussed, some of which is released from the elastic deformation by the creation of the new crack area and some of which is absorbed by the area created. In the second, the local stress field around the crack tip is considered and characterized by some parameter at fracture. The energy approach is taken here, and characterizing parameters are used when convenient.

In general, fracture criteria are material properties, whereas stability criteria are functions of the body concerned. Fracture mechanics is a macro theory, being concerned with the overall behavior of a body. The particular material behavior is considered only in connection with fracture criteria or deformation processes involved. Much of the theory is thus not material specific and applies equally to all materials. It is based on linear elastic fracture mechanics (LEFM), which assumes linear elastic deformation and very local energy absorption at the crack tip. The theory has been extended to plastically deforming materials, which undergo rate-independent, irreversible deformation. Neither of these assumptions is strictly true for polymers, since all are somewhat viscoelastic and exhibit time dependence, giving behavior intermediate between these limiting states. However, much of the behavior observed in the fracture of polymers can be described in terms of LEFM, and plasticity modifications are of value in dealing with ductile polymer fracture. It is possible to modify the analysis to encompass time effects; this is necessary for some phenomena, eg, slow crack growth.

Toughness. The toughness of a material may be defined in terms of its resistance or R curve, as shown in Figure 3, where R is the energy per unit area necessary to produce the new fracture surface. For a crack of length a_0 and in a plate of uniform thickness B, the R curve may be expressed in terms of the crack growth Δa. Up to some value of R, there is no growth and the line is vertical. However, at point A, there is crack-growth initiation at $R = R_c$. Subsequent growth may be described, as in line A, with R increasing as the crack extends. This usually involves the increase of the local plastic flow in a zone of size r_y around the crack tip. If the extent of this zone, which represents the energy absorption, remains the same, R remains at R_c, a constant as shown in line B.

Energy-release Rate. The driving force for fracture comes from the energy G that is released by the growth of the crack. This growth violates continuity within the body and results in the final energy state being lower than the initial; the differ-

Table 1. Problematic Environments of Polymers

Environment	Polymer class
organic liquids and vapors	amorphous polymers
hydrocarbons, tensioactive agents (detergents)	polyolefins
alcohols, hydrocarbons	poly(methyl methacrylate)
humidity, acid- or base-catalyzed	polymers with heteroatoms in backbone, ie, polysaccharides, polyethers, polyesters, polyamides, polyurethanes
Lewis acids, zinc, copper	polyamides
ozone	unsaturated rubbers
sunlight	polymers with photosensitive groups, ie, polyacetals, polysulfone, polyphenylene
biodegradation	short-chain polymers, nitrogen-containing polymers, polyesters

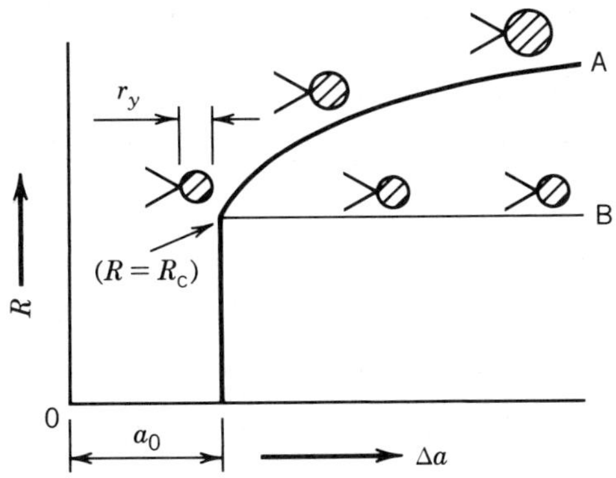

Figure 3. The resistance or R curve. See text for significance of curves A and B.

ence is released and available to overcome the crack resistance R. At crack initiation $G = G_c = R_c$.

Stress-intensity Factor. The stresses around the crack tip, as shown in Figure 4, may be expressed in the form of a series:

$$\sigma = \frac{Kf(\theta)}{\sqrt{2\pi r}} + A(\theta) + B(\theta) \cdot r^{1/2} + \cdots$$

where f, A, B, etc, are functions of the angle θ and r is the distance from the crack tip. In a state of plane stress (thin plates) K is related to G by $K^2 = EG$. For an infinitely sharp crack in a thin plate under uniaxial stress σ_0 one obtains

$$K = \sigma_0\sqrt{\pi a_0}$$

K therefore designates the combined efforts of mechanical stress and crack length. Typically a cracked polymer specimen will show no crack growth if K is smaller than a threshold value K_{Ic}; slow, thermal crack growth at $K > K_{Ic}$; rapid, instable crack propagation if K is larger than a material-dependent, characteristic value K_{cc}.

Plastic Zone and Size Effects. The elastic analysis predicts infinite stresses at the crack tip, but local yielding prevents this from happening. Nevertheless, LEFM can be applied as long as the local yielding remains restricted (inner zone in Figure 5). The outer, larger zone characterizes the elastic field; it corresponds to some fraction of the sample dimensions. Generally LEFM can be applied if all sample dimensions are larger than 2.5 $(K/\sigma_y)^2$ where σ_y is the yield stress.

Plasticity Effects. Most polymers show some degree of plasticity, which can have an important influence on fracture behavior. This is clearly seen at higher temperatures and low loading rates where pronounced plasticity may develop. To apply LEFM, it is stipulated that plastic-zone size r_y be very small, but it is often difficult to make measurements under

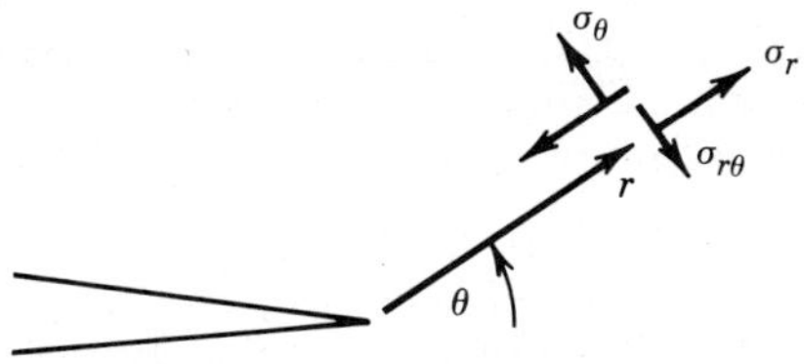

Figure 4. Local stresses around a crack tip.

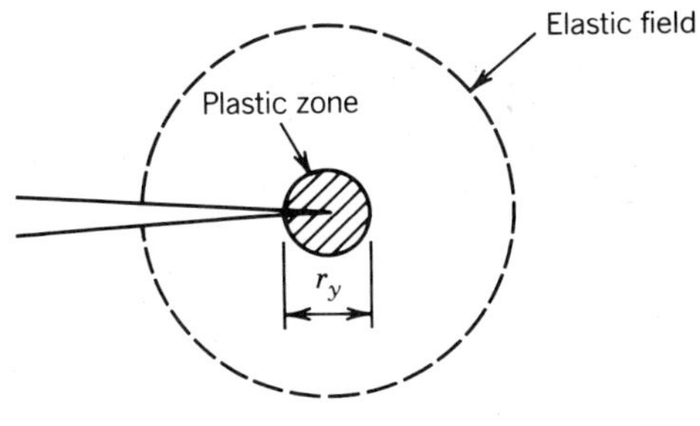

Figure 5. Elastic field and plastic zone; $\sigma = \frac{K}{\sqrt{2\pi r}} \cdot f(\theta)$.

conditions where this is true because of size limitations. Even when it is not true, it is possible to find a fracture resistance R, but it is not G_c in this case but a "corrected toughness" termed J_c. In effect, the crack length is increased by

$$r_y - \frac{1}{2\pi}\frac{EJ}{\sigma_y^2}$$

to compensate for the modification of the elastic stress field by the plasticity:

$$J = \frac{\pi\sigma^2}{E}(a + r_y) = \frac{\pi\sigma^2}{E}\left(a + \frac{1}{2\pi} \cdot \frac{EJ}{\sigma_y^2}\right)$$

$$\frac{J}{G} = \left[1 - \frac{1}{2}\left(\frac{\sigma}{\sigma_y}\right)^2\right]^{-1} \quad \text{where } G = \pi\sigma^2\, a/E$$

High Rate Effects. The behavior of polymers at high testing rates is of considerable importance since their viscoelastic nature can result in substantial changes. For fracture, the changes in yield stress are of particular importance since increases result in a tendency to brittle fracture. Such concerns are reflected in the large number of impact tests employed in polymer testing; it is useful to employ fracture mechanics in such evaluations (see IMPACT RESISTANCE).

Characterization and Test Methods. Fracture as a phenomenon is of great importance and its occurrence must be controlled. Fracture mechanics has been utilized mainly as an analytical method to characterize materials and as a design method to ensure component integrity via the relation

$$K_{Ic}^2 = EG_c = Y^2 \cdot \sigma^2 a$$

Toughness and related properties are characterized by K_{Ic}, G_c, or J_c. In the basic test of fracture mechanics, a cracked specimen is loaded slowly and the onset of crack growth (at load P_0) is observed. All fracture-mechanics tests use prenotched specimens, since the analysis is concerned with the behavior of notches or cracks. The nature of the initial notch is of great importance since the measured toughness can be sensitive to its sharpness. Ideally, a natural crack should be grown in a prenotching test, often at low temperatures, but these tests are frequently not easy to perform. The notch must be carefully formed with a very sharp cutter to avoid crazing, crack branching, or blunting; fatiguing is time-consuming (see FATIGUE). The critical stress intensity factor in a three point bending test is obtained as

$$K_{Ic} = P_0 sf(a/w)bw^{1/2}$$

where s = span between supports, b and w = specimen thickness and width, and $f\,(a/w)$ is a (tabulated) geometrical correction function.

Design Methods. Applications of fracture mechanics in design are based on the assumption that all bodies contain imperfections (of size a_I). The operating conditions are defined as those under which such flaws do not make the component unsafe or unserviceable. Uniaxial stress criteria are of the form

$$\sigma_0 < K_{Ic}\sqrt[2]{\pi a_I}$$

BIBLIOGRAPHY

H. H. Kausch, *Polymer Fracture,* 2nd ed., Springer-Verlag, Berlin, 1987.

J. G. Williams, *Fracture Mechanics of Polymers,* Ellis Horwood, Ltd., Chichester, UK, 1984.

A. J. Kinloch and R. J. Young, *Fracture Behavior of Polymers,* Appl. Sci. Publ., London, 1983.

H. H. Kausch, ed., *Advances in Polymer Science,* Vols. 52 and 53: *Crazing in Polymers,* Springer-Verlag, Berlin, 1983.

H. H. Kausch
École Polytechnique Fédérale de Lausanne

J. G. Williams
Imperial College of Science, Technology and Medicine

FREE-RADICAL POLYMERIZATION.

See Radical polymerization.

FREEZING-POINT DEPRESSION.

See Molecular weight determination.

FRICTION.

See Abrasion and wear; Drag reduction; Interfacial properties.

FRIEDEL-CRAFTS CATALYSTS.

See Cationic polymerization; Ziegler-Natta catalysts.

FUMARIC POLYMERS.

See Maleic and fumaric polymers.

FUNGICIDES AND FUNGISTATS.

See Biocides.

FURAN POLYMERS

Macromolecules prepared from furan compounds form one of the rare families of synthetic polymers not derived from petroleum. Their origin is the biomass. Renewable vegetable resources can be exploited through simple chemical transformations to manufacture two furan derivatives: furfural (2-furan carboxyaldehyde) and 5-hydroxymethylfurfural. These processes can also yield 5-methylfurfural and 5-chloromethylfurfural.

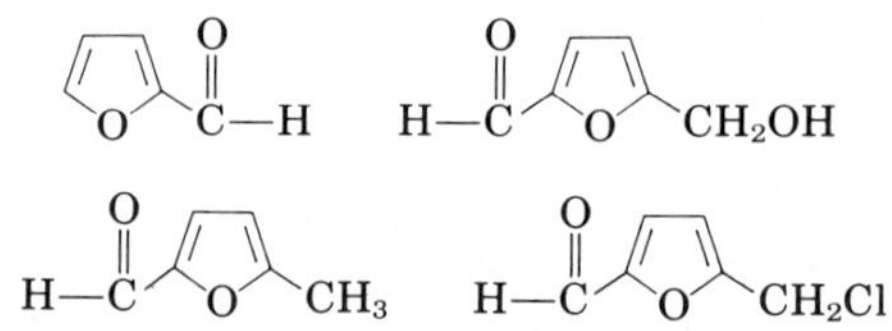

Recent interest in wood and other biomass sources of chemicals has renewed interest in furan compounds.

There are many sources for furfural production because most agricultural wastes and woods contain sufficient amounts of pentosans to justify industrial exploitation by one of the well-established and relatively simple technologies.

Furan polymers are varied and include macromolecules in which the furanic character has been destroyed; the furan ring is maintained as part of the polymer backbone; the furan ring is pendent to the chain; linear and/or branched structures prevail; or cross-linking is important.

Thermosetting Resins

Many formulations containing furan derivatives undergo cross-linking on heating, usually in the presence of a catalyst. Furfural and furfuryl alcohol are by far the most frequently used compounds in these resins, which vary from simple mixtures, involving essentially only the furan compound, to very elaborate ones. The complexity of the latter does not allow a proposal for the mechanism of polycondensation or for the structure of intermediates and products.

Furfural Compositions. ***Furfural Alone.*** Furfural resinifies to give a black cross-linked product when heated at 100–250°C in an inert atmosphere.

Furfural–Phenols. Novolac-type resins prepared solely from furfural and phenol or with formaldehyde or a phenol–formaldehyde prepolymer added to the mixture have been used for decades as molding compounds.

Furfural–Acetone.

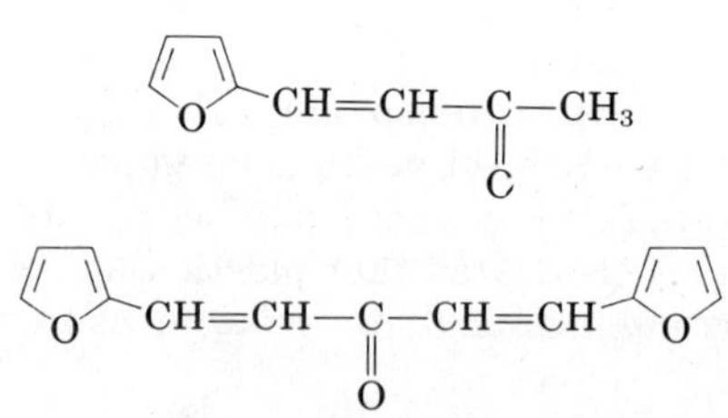

The anionic polymerization of mono- or difurfurylidene acetone gives a thermoplastic material, while treatment of these compounds with acidic catalysts leads to cross-linked black products.

Furfuryl Alcohol Compositions. Resins derived from furfuryl alcohol are by far the most important polymers of the furan series in terms of industrial output and applications. Progress has been achieved in understanding the resinification of furfuryl alcohol through specific studies of both the soluble intermediates and the solid cross-linked final materials. From this recent evidence and previous work, a chemically consistent resinification mechanism involves first the condensation of methylol groups with C-5 hydrogen atoms to give predominantly linear chains:

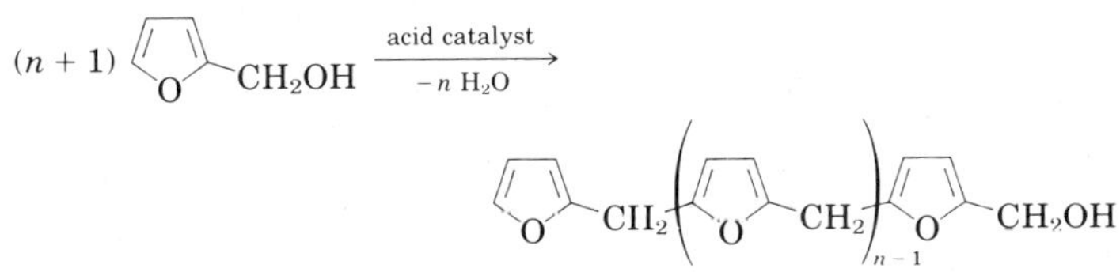

followed by branching at $—CH_2—$ (cross-linking) and polyconjugation along the chains (brown-black color).

Uses. Applications of furfuryl alcohol resins span a wide range of technologies. Materials with different properties are prepared from furfuryl alcohol alone or with coreactants such as furfural, formaldehyde, phenols, urea, and various prepolymers. Furfuryl alcohol resins are used in metal-casting cores and molds, low flammability and low smoke composites and foams, carbonaceous products, polymer concretes, wood adhesives and binders, and sand consolidation and formation plugging (see FOUNDRY RESINS).

Other Thermosetting Compositions. Other thermosetting compositions include 2,5-bis(hydroxymethy)furan, and hydroxymethylfurfural.

Thermoplastic Polymers from Furfural

The decarbonylation of furfural and homologues to give furan and alkylfurans has been widely investigated. Furan and methylfurans polymerize through the ring unsaturations in the presence of acids.

On the other hand, furfural and homologues, 2-alkenylfurans, furfurylidene acetone and homologues, 2-furyl vinyl ketone, furfuryl methacrylate and homologues, vinyl 2-furoate and homologues, furfuryl vinyl ether and homologues, and furyloxirane and homologues polymerize and copolymerize through the corresponding function external to the ring with radical, anionic, cationic, and stereospecific initiators giving mostly linear macromolecules with furan rings pendent to the chains.

The dual dienic–aromatic character of the furan ring is revealed by the side reactions encountered in radical and cationic polymerizations, respectively.

Thermoplastic Polymers from Hydroxymethylfurfural

The macromolecular materials described in this section arise from polycondensation reactions involving at least one bifunctional furan monomer derived from 5-hydroxymethylfurfural; the furan ring is therefore always part of the chain backbone. Such polymers include polyesters, polyamides and polyhydrazides, polyurethanes, and polymeric Schiff bases.

Chemical Modification of Furan Polymers

Modification of polymers bearing furyl groups by specific reactions has provided a series of new materials, some of which possess remarkable properties. Polymers cross-linked by a bis(dienophile) can be reverted to linear structures by heating.

BIBLIOGRAPHY

W. J. McKillip and E. Sherman in M. Grayson, ed., *Kirk-Othmer Encyclopedia of Chemical Technology,* 3rd ed., Vol. 11, John Wiley & Sons, Inc., New York, 1981, pp. 499–527.

C. W. Bird and G. H. W. Cheeseman, eds., *Comprehensive Heterocyclic Chemistry,* Vol. 4, Pergamon Press, Oxford, UK, 1984, Chapts. 1, 2, 3, 10, 11, and 12.

A. Gandini, *Adv. Polym. Sci.* **25,** 47 (1977).

ALESSANDRO GANDINI
Institut National Polytechnique de Grenoble (EFP)

FURFURAL. See AMINO RESINS; FURAN POLYMERS.

FURFURAL POLYMERS. See FURAN POLYMERS.

FURFURYL ALCOHOL. See FURAN POLYMERS.

FURNITURE

Plastics in Furniture Construction

Applications are given in Table 1. ABS is used primarily for modern-style furniture to provide high gloss and bright colors. Melamines and phenolics are used mostly for stain resistance of tables and dresser tops, where printed surfaces match the color and texture of surrounding wood surfaces. Phenolic resin is also used as a binder for wood chips used in the manufacture of plywood (see also AMINO RESINS; LAMINATES).

Simulated Wood Parts

Both PS and polyurethane can be fabricated into rigid parts that are easy to decorate to simulate the appearance of carved wood.

Decoration of Simulated Wood Parts

Before finishing, the resin part is cleaned to remove contaminants, which are mostly mold release agents. Silicone release agents cannot be removed and should not be used. To obtain a good color match, wood and plastic parts must be finished at the same time.

Table 1. Typical Applications of Plastics in Furniture

Resin	Methods of fabrication	Applications
acrylics	forming; injection molding; machining	see-through furniture; lamps
ABS	rotational and expansion casting; injection molding	chair and sofa frames; cast chair shells; decorations; modern furniture
alkyds	spray or brush coating	combined with other varnish or enamel resins
cellulosics	spray or brush coating	lacquer coatings
epoxies	casting; lamination; compression molding	imitation marble tabletops; decorative parts and trim; bonding adhesive; coating; molds
polyamides	injection molding	drawer guides, bearings, wheels; fiber reinforcement
phenolics	compression molding; lamination	bonding agent for plywood; molded decorative panels
polyesters	casting; compression molding; reinforced-plastics molding; coating; machining	tabletops and bases; imitation slate and marble; decorative fronts, trim, and panels; molded chairs; modern furniture
polyethylene	rotational molding; blow molding; injection molding	molded chairs; drawer guides; institutional seating; door tracks
polypropylene	injection molding; extrusion; blow molding	molded institutional chairs; shelving; decorative trim
polystyrene	injection molding; structural foam molding; extrusion	wood-grain-patterned chair parts; panels; trim; mirror frames; drum tables; pedestals; other decorative and structural parts
polyurethane		
rigid	casting; expansion molding; RIM	chair frames and shells; sofa frames; decorative trim and components; decorative and structural parts
flexible	foam molding by slabstock extrusion or injection	upholstery cushioning; modern furniture
coating	spray or spread coating	coating of upholstery fabrics
poly(vinyl acetate)	spray or brush coating	coating; adhesive
PVC	lamination; rotational casting; extrusion; injection molding; calendering; spread coating	veneers and wood-grain-patterned film for lamination; other decorative and structural parts
amino	compression molding; laminating	urea adhesives for plywood, particle board; melamine laminates for decorative tabletops

BIBLIOGRAPHY

B. J. Breberg, *Molded Plastics Components in Furniture Conference, SPI, National Association of Furniture Manufacturers, Southern Furniture Manufacturers' Association, Washington, D.C., Sept. 14, 1967,* The Society of the Plastics Industry, Inc., New York, 1967.

W. C. Kuhlke and N. E. Frailey, *Finishing of Polystyrene Molded Furniture Components, Technical Bulletin SC:68-46,* Shell Chemical Co., Houston, Texas, Oct. 1968.

W. H. Parks, *SPI Conference, Washington, D.C., Sept. 4, 1967,* The Society of the Plastics Industry, Inc., New York, 1967.

William C. Kuhlke
Ray Giornelli
Shell Chemical Company

G

GAMMA GLOBULIN. See PROTEINS.

GAS CHROMATOGRAPHY. See CHROMATOGRAPHY.

GAS PERMEABILITY. See BARRIER POLYMERS; DIFFUSION; MEMBRANES.

GAS-PHASE POLYMERIZATION

Since the late 1960s, designers have envisioned the ideal process for polymerization of olefins: the introduction of gaseous monomer and catalyst into a continuous reactor, followed by removal of dry, free-flowing polymer powder ready to be pelletized and sold. Chemists developed highly specialized catalysts that make gas-phase polymerization possible.

Engineering

Common to the new technology processes are polymerization kinetics, heat transfer, mass transfer, and dynamics of powder mixing and transport. While previous processes required a reactor design that accommodated the first three, in a gas-phase reactor, powder dynamics must also be considered. With better than 98% reliability, a single catalyst particle produces a single powder particle. The catalyst is activated by contact with a cocatalyst. The monomer migrates and is adsorbed on the catalytic surface in the correct orientation (especially for propylene). The addition reaction is highly exothermic, 3.83 MJ/kg (917 kcal/kg) for ethylene, and this heat must be transferred from the heterogeneous solid catalyst to a gas stream, a boiling liquid monomer, or other cooler solid particles which rapidly remove the heat from the reactor. Failure to transfer this heat in a rapid, controlled fashion produces a fused, agglomerated mass accompanied by uncontrollably rising temperatures. Control of the operation cannot be recovered if these lumps reach the size of a fist. The reactor must then be shut down with emergency procedures and cleared of lumps. Gas–powder reactors require well-designed thermal sensors and controls.

The management of heat transfer has been resolved in several ways. In a vertical stirred bed of particles, the gas is recirculated through an external heat exchanger and cooled (ethylene process) or partially condensed (propylene process). In a horizontal stirred bed, the flow of powder from one end to the other approaches plug flow, with cooling in each zone by sprays of condensed propylene.

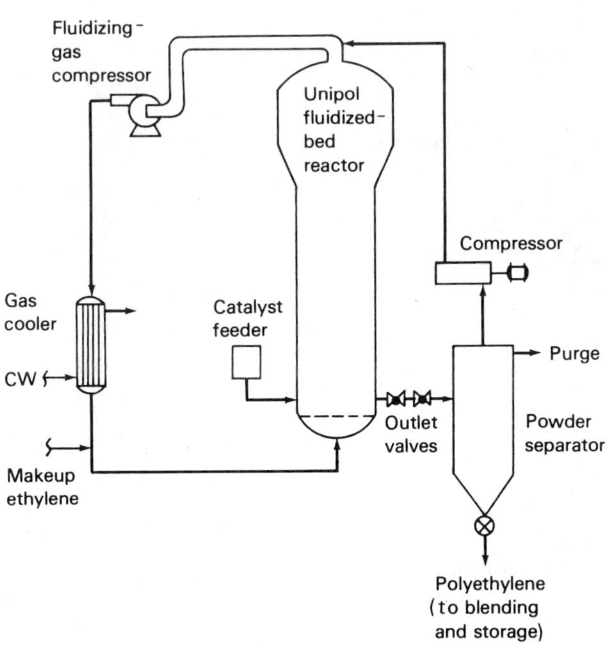

Figure 1. Unipol PE process.

Processes

The Unipol process for HDPE or LLDPE uses a fluidized-bed reactor (Fig. 1). Use of the process is growing rapidly because of its simplicity and favorable economics. Its success depends heavily on the development of a highly active and selective catalyst. The first such catalyst consisted of chromium compounds and optional modifiers on dehydrated silica; other catalysts are also employed.

Polypropylene. ***Stirred-bed Reactors.*** The activity of a transition-metal catalyst is significantly lower for making isotactic PP than for HDPE. Addition of modifiers to improve the stereospecificity to over 97% isotactic also tends to lower catalyst activity for PP, resulting in a productivity of 1.5–2.0 kg/g titanium. Therefore, early gas-phase reactors for PP were usually designed to provide 4–6 h contact time, considerably more than for HDPE. An early commercial gas-phase PP plant, built by BASF, used vertical stirred beds for reactors.

In a horizontal stirred-bed reactor, catalyst and modifiers are fed near one end of the bed, while the product is removed from the other end. This configuration greatly reduces axial dispersion and short-circuiting of powder and produces a higher growth ratio of polymer per catalyst particle in a given residence time.

Fluidized-bed Reactor. The fluidized-bed process for polyethylene has been adapted by Union Carbide to produce isotactic PP. This process, called Unipol PP, started-up commercially in 1985 and has since been licensed to a number of other companies.

Ethylene–Propylene (E–P) Copolymers. Gas-phase polymerization is a boon to the manufacture of E–P copolymer because there is no bulk liquid present to extract the soluble fraction. However, E–P copolymer powders tend to soften more easily with increasing temperature and are more sticky than either homopolymer.

BIBLIOGRAPHY

J. F. Ross and W. A. Bowles, *Oil Gas J.* **82,** 82 (Nov. 5, 1984).
Oil Gas J. **82,** 92 (Jan. 9, 1984).

N. F. Brockmeier and J. B. Rogan, *Ind. Eng. Chem. Prod. Res. Dev.* **24,** 278 (1985).

K. Y. Choi and W. H. Ray, *J. Appl. Polym. Sci.* **30,** 1065 (1985).

Norman F. Brockmeier
Amoco Chemicals Company

GELATIN

Gelatins are high molecular weight polypeptides derived from collagen, the primary protein component of animal connective tissues, such as bone, skin, and tendon.

Conversion of Collagen to Gelatin

Collagen Molecular Structure. The collagen molecule consists of three helical peptide chains (alpha chains) held in close, parallel association; they form a rigid rod with dimensions of 1.4×300 nm.

Transformation of Soluble Collagen to Gelatin. Collagen is transformed into gelatin by denaturation and physical and chemical degradation. Denaturation, particularly thermal denaturation, is central to any conversion process.

Commercial Processes

Although gelatins can be manufactured reproducibly with only slight degradation, they are a class of materials with a wide range of properties, such as color, clarity, impurities, molecular weight average and distribution, isoelectric pH, and gel rigidity. The term gelatin is often applied indiscriminately to the whole class but should be understood to have only a generic meaning.

Commercial gelatins are not derived from soluble collagen but from the more abundant bone and skin collagen fibrils,

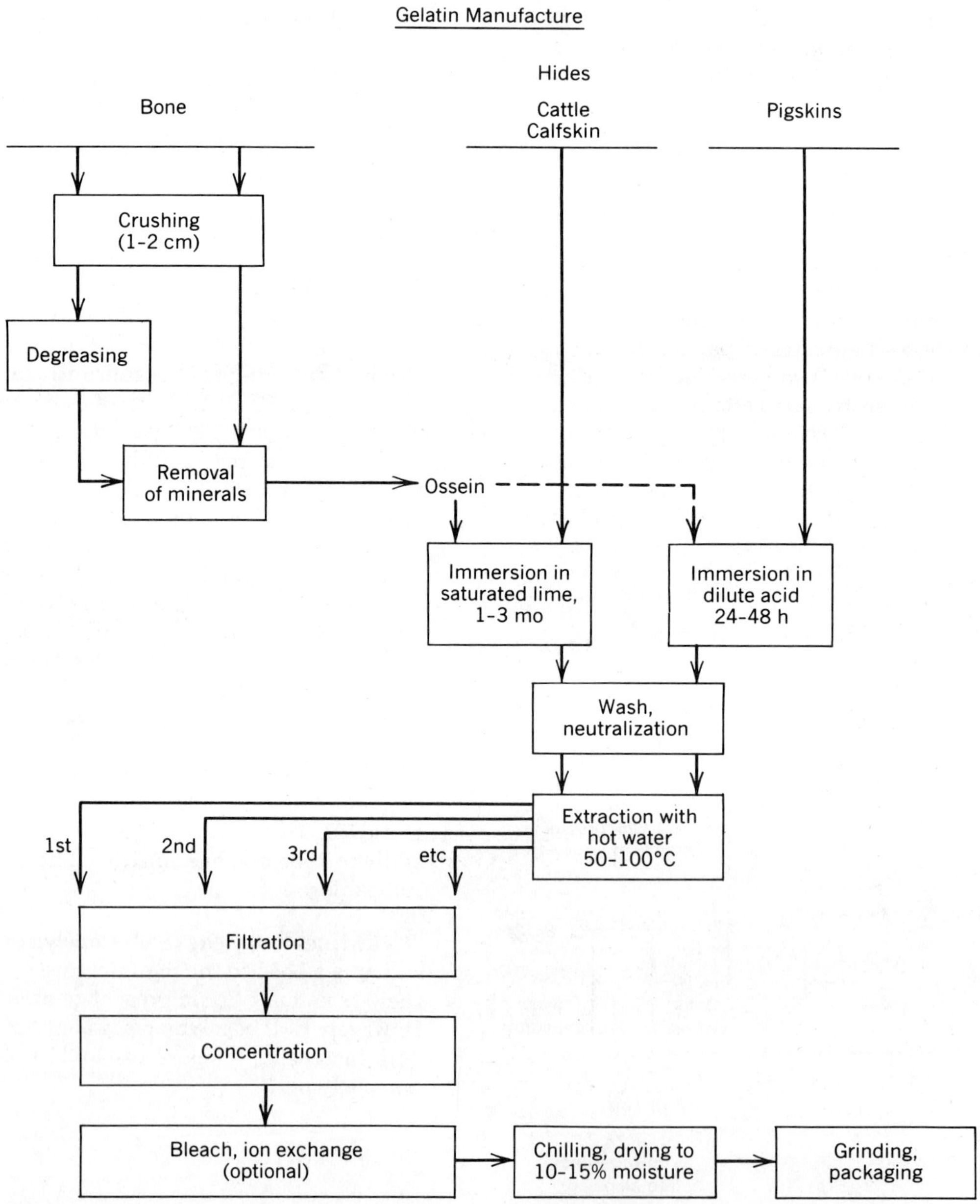

Figure 1. Gelatin manufacture.

which are stabilized by Schiff base and aldol formation. Manufacturing processes employ lyotropic and hydrolytic treatments with aqueous solutions of acid or base, which swell the collagen and ultimately solubilize it. The different types of gelatin result from combinations of raw materials, extent and type of raw-stock treatment (pretreatment), and conditions used for extraction. Typical schemes for gelatin production are shown in Figure 1. The challenge in gelatin manufacturing is to produce materials that satisfy many chemical and physical criteria and to employ economical processes that avoid excessive degradation.

Raw Materials. Most vertebrate animal collagen occurs in skin (hide) and bone, the principal raw stocks for gelatin production. Cattle and pigs are the principal commercial sources, but other mammalian species and fish are also used. Raw stock is selected on the basis of cost, availability, processing, the desired gelatin properties, and intended uses. Certain process–raw-stock combinations are more successful than others in terms of economy and product properties.

Pretreatment. Pretreatment procedures swell and soften the hides or ossein in preparation for denaturation and extraction. Virtually all gelatin is produced by lime or acid pretreatment, but alternative techniques employ proteolytic enzymes, hydrogen peroxide, and a sequence of acid and alkali treatments.

After pretreatment, the swollen collagen is denatured and extracted by hot water.

Extraction. The washed, nearly neutral pretreated stock is transferred to stainless steel extraction tanks provided with temperature controls. The collagen is extracted with hot water, causing denaturation (disruption of the helical conformation), additional hydrolysis, and solubilization of the gelatin. Usually, the gelatin is extracted in stages of increasing time and temperature. At the end of a stage, termed a cook, the extraction liquor may contain 3–10% gelatin and is drained off. Hot water is added to the remaining stock, and extraction continues at a higher temperature. The hot water may circulate through the stock during extraction.

Gelatins from successive extractions have different physical and chemical properties. The first extract is obtained at the lowest temperature and invariably has the highest Bloom value (gel strength), which is the common measure of gelatin quality. Extractions at near-neutral pH cause Bloom values to decrease steadily with extraction number, whereas extraction at lower pH produces more constant values. Color and turbidity increase in later extracts, particularly with ossein gelatins, whereas viscosity often passes through a maximum at a variable point. Typical changes in gel strength and viscosity values are shown in Figure 2.

Postextraction Processing. The dilute extraction liquors are processed rapidly above 40°C to avoid gelling and minimize thermal or biological degradation. High standards of cleanliness must be maintained, and the equipment should be sterilized frequently. The details of most of these processes vary according to the manufacturer and also depend on the intended gelatin use. Processes may include clarification, concentration, deionization, bleaching, and drying.

Chemical Modification. The amino and hydroxy side groups provide the sites for acylation. These reactions reduce the pI

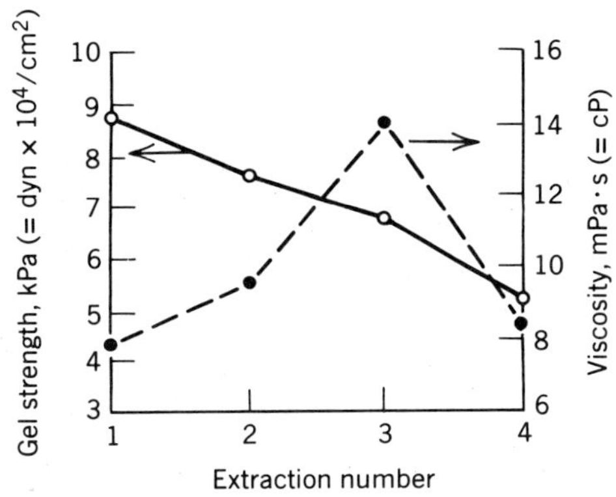

Figure 2. Gel strength and viscosity values vs extraction number for the gelatins obtained in a typical extraction of limed ossein. Extraction pH = 6.4; extraction temperature increased with extraction number.

and may cause coacervation when the pH is near the isoionic point; the phase separation is reversible by adjusting the pH.

Physical and Chemical Testing and Standards

Certain standard properties (see Table 1) such as Bloom value, viscosity, moisture, and ash are specified by the manufacturer, but additional testing may be required, particularly with photographic applications of gelatins.

Amino Acid Composition. Gelatins are composed of essentially the same amino acids as their collagen source and thus have a high glycine content (33 mol %) and also two unusual amino acids: hydroxyproline (10 mol %) and hydroxylysine (0.5 mol %). Cysteine and tryptophan are absent (except perhaps in very small amounts as impurities). All amide side-groups of collagen are hydrolyzed to the carboxylic acids when gelatin is prepared by the alkali-pretreatment process.

Color and Clarity. The yellow color and turbidity of commercial gelatins have been attributed to mucoprotein impurities; both color and turbidity are highest in the later extracts.

Table 1. Typical Property Values of Gelatin

Property	Value
density, dry (g/cm^3)	1.35
refractive index, dry, at 546.1 nm	1.54
weight-average mol wt, $\overline{M}_w$	1×10^5 to $>5 \times 10^5$
number-average mol wt, $\overline{M}_n$	5×10^4 to 1.5×10^5
viscosity at 6.67 wt %, 60°C, mPa·s (= cP)	5–10
gel rigidity, g Bloom	75–330
specific optical rotation at 589 nm[a], degrees	−134 to −140
refractive increment at 300 nm, *dn/dc*	0.187
glass transition, dry, °C	217
moisture, wt % at 50% rh	9–15
ash, wt %	0–3
isoionic pH	
type B	4.8–5.2
type A	7–9
nitrogen, wt %	
type B	18.15
type A	18.30
hexose sugars, wt %	0.45–0.65

[a] These specific optical rotations are typical values found by the author at 45°C. Published values sometimes are 5–10% lower, possibly a result of moisture and ash.

Properties of Gelatin Solutions, Gels, and Solids

Solutions and Gels. ***Molecular Size.*** Dissolved gelatin molecules are in a conformationally disordered sol state above 40°C, which should be regarded as a minimum temperature for molecular characterization and certain aspects of gelatin use. Lower temperatures may be used in the presence of denaturing solutes in high concentrations such as urea, guanidinium ion, and certain neutral salts. As in other polymers, the intrinsic viscosity $[\eta]$ of gelatin is a function of molecular size and shape.

Hydrolysis. Reduction of molecular weight by hydrolytic cleavage of covalent cross-linkages or peptide bonds inevitably occurs in solutions; hydrolysis rate is lowest near neutrality (pH 5 to 7).

Polyampholyte Properties. The polyampholyte character of gelatin is demonstrated in Table 2, which gives mole percentages and pK_a for the weakly acidic and basic amino acid residues. These enable a gel to function as a Donnan membrane and to act as a cation or an anion exchanger, depending on pH.

Colloid Protective Ability. The ability of gelatin to stabilize colloids is exemplified by its Zygmondy number (measuring stability of gold sols), which is the highest of any biopolymer. This results from adsorption at solid–aqueous interfaces in a loop-and-train conformation.

Gel Formation. Below 35–40°C, gelatin molecules slowly change conformation and aggregate, but gel formation depends on concentration and temperature (see also GELS). Aggregation of the molecules occurs during a refolding process that forms quasi-crystalline structures similar to the native collagen triple-helix. Aggregation does not cause gelation when solutions below about 0.5 wt % are chilled, but with higher concentrations, network junctions form, leading to a gel state. The gel modulus tends to increase slowly over very long times, and usually a true equilibrium value is not attained.

Organic Solvents. Gelatin is soluble in a number of organic solvents, including acetic acid, trifluoroethanol, formamide, ethylene glycol, glycerol, and dimethyl sulfoxide.

Solid State. Anhydrous gelatin has a density of 1.34 g/cm^3; however, if it is heat-dried below 1% moisture, it tends to cross-link covalently and become insoluble. Aqueous gelatin dries to a semicrystalline solid, which is clear, tough, and brittle, swellable in cold water, and completely soluble in hot water in the absence of cross-linking.

Table 2. Abundance and pK_a Values of the Ionizable Side Groups of Gelatin

Amino acid	Gelatin, mmol/g	pK_a
Anionic		
aspartic acid	0.50 (0.32)[a]	4–4.5
glutamic acid	0.78 (0.52)[a]	4.5
tyrosine	0.011–0.044	10
α-COOH	0.011	3.6
Cationic		
lysine	0.30	10–10.4
hydroxylysine	0.054	9.5
arginine	0.53	>12
histidine	0.044	6.5–7
α-NH_2	0.011	7.8

[a] Values in parentheses are for type-A gelatin.

Solid-state properties are dominated by the thermal history and the presence of diluents, notably water. Bloom value and setting and drying conditions determine the amount, thermal stability, and orientation of crystallites in the solid.

Uses

The principal uses of gelatin are categorized as edible (55%), pharmaceutical (25%), photographic (15%), and other technical uses (5%).

Economic Aspects

The industry appears to experience a steady annual growth of a few percent. Recent U.S. prices for edible gelatin are $3–4/kg. Pharmaceutical gelatin prices are somewhat higher, and photographic grades, currently near $8/kg, are the highest.

The U.S. gelatin production is estimated at 30,000 t/y; worldwide production is 125×10^3 t/yr. Most gelatin is manufactured in Europe.

BIBLIOGRAPHY

L. Stryer, *Biochemistry,* 2nd ed., W. H. Freeman and Co., San Francisco, Calif., 1982, pp. 185–204.

A. G. Ward and A. Courts, eds., *The Science and Technology of Gelatin,* Academic Press, Inc., New York, 1977.

A. Veis, *The Macromolecular Chemistry of Gelatin,* Academic Press, Inc., New York, 1964.

B. Idson and E. Braswell in E. M. Marek, S. G. Stewart, and C. O. Chichester, eds., *Advances in Food Research,* Academic Press, Inc., New York, 1957, pp. 235–338.

PHILIP I. ROSE
Eastman Kodak Company

GEL EFFECT

The gel effect, also known as the Trommsdorff or the Trommsdorff–Norrish effect, is the autogenous increase in rate of polymerization and molecular weight observed in media of high viscosity. It appears to be related to decreased diffusion of polymer chains and, therefore, to reduced rates of termination with increasing conversion. See BULK POLYMERIZATION; RADICAL POLYMERIZATION. See also articles on specific cases, such as ACRYLONITRILE POLYMERS; ACRYLIC AND METHACRYLIC ESTER POLYMERS.

GEL FILTRATION.

See CHROMATOGRAPHY; FRACTIONATION.

GEL-PERMEATION CHROMATOGRAPHY.

See CHROMATOGRAPHY; FRACTIONATION.

GEL POINT

A polymer at its gel point (GP), the critical gel, is in a transition state between liquid and solid. Its molecular weight distribution is infinitely broad ($M_w/M_n \to \infty$) and molecules range from the smallest unreacted oligomer to the infinite cluster. The molecular motions are correlated over large distances but the critical gel has no intrinsic size scale. The polymer reaches the GP at a critical extent of reaction, $p \rightarrow p_c$. The liquid polymer before the GP, $p < p_c$, is called a sol because it is soluble in good solvents. The solid polymer beyond the GP, $p_c < p$ is called a gel, which is not soluble, even in a good solvent. However, low molecular weight molecules (sol fraction) are still extractable. The polymer at the gel point is in a critical state and may, therefore, be called the critical gel to distinguish it from the various materials that are commonly called gel.

Polymers around the gel point are just beginning to be explored. They are of interest for a broad spectrum of applications such as gel processing, reactive processing (gel as intermediate state), and the development of new polymeric materials (adhesives, absorbants, porous catalysts, vibration dampers, membranes, etc). They are also important outside the polymer field, for example, in food technology or in blood clotting.

The intermediate state of the critical gel is expressed in its high adhesion strength (tackiness), which is combined with extreme ductility and fragility when subjected to large strain. The adhesion behavior must be accounted for when designing experiments with gels. It also suggests future applications of gels as adhesives.

The principal differences between chemical and physical gels lie in the lifetime and the functionality of the network junctions. Chemical bonds are considered to be permanent while the physical junctions have finite lifetimes. Defining GP of a physical gel is difficult because its molecular weight is finite even if it forms infinite clusters and it is soluble even beyond GP. Therefore, it is common to use mechanical properties to define GP, and a variety of GP definitions are in use. The functionality of the chemical junction is clearly defined by the chemistry of cross-linking (low functionalities of 3 or 4, in most systems), but the functionality of physical junctions strongly depends on sample preparation. It can be very high, possibly 50 or higher.

A typical evolution of equilibrium mechanical properties during gelation is schematically shown in Figure 1. The initial ($p \rightarrow p_c$) liquid system has a steady shear viscosity that grows with reaction time as the average molecular weight increases. In the approach to GP, the viscosity diverges (ie, no steady state can be reached in flow at constant shear rate $\dot{\gamma}$) but the equilibrium modulus is still zero because the stress in a deformed critical gel can still relax completely. In the immediate vicinity of GP

$$|p - p_c| \ll 1 \tag{1}$$

the steady-state measurements fail, because the relaxation time diverges to infinity and a steady state cannot be reached any more. Beyond p_c, the equilibrium modulus starts to grow until it reaches its full value when the reaction is brought to completion ($p \rightarrow 1$). All viscoelastic liquid states of this polymer are in between the Newtonian liquid ($p = 0$) and the critical gel ($p = p_c$). Equivalently, all viscoelastic solids are in between the critical gel and the Hookean solid ($p = 1$). The rheology at each of the three limiting states is universal and of extreme simplicity.

Instead of the extent of reaction, the stoichiometric ratio is often chosen as independent variable of a chemical gelation

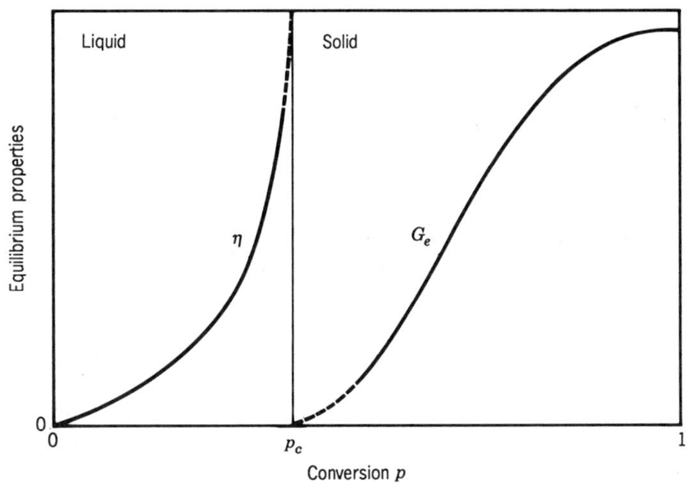

Figure 1. Evolution of steady-state mechanical properties of a cross-linking polymer as a function of cross-linking time (schematic). Representative properties are the steady shear viscosity for the liquid state (sol) and the equilibrium modulus for the solid state (gel).

experiment. Critical stoichiometry results in a gel after the reaction is complete.

Rheological Properties of the Critical Gel

At the onset of long-range connectivity, $p = p_c$, materials exhibit a very simple, critical rheological behavior on an equal footing with the Newtonian liquid and the Hookean solid. Linear viscoelasticity has been found to reduce to a simple power law form for the linear relaxation modulus

$$G(t) = St^{-n}; \qquad p = p_c \tag{2}$$

or a relaxation time spectrum

$$H(\lambda)\, d \ln \lambda = (S/\Gamma(n))\lambda^{-n}\, d \ln \lambda \tag{3}$$

The range of the power law is not yet explored for many polymers, but the lower time limit seems to be determined by the chain between cross links or by the glass length. The upper time limit depends on sample size.

Introduced into a general constitutive equation for linear viscoelasticity, this power law results in the constitutive equation for critical gels, the Winter-Chambon gel equation for the stress τ as a function of deformation rate history $\dot{\gamma}(t')$

$$\tau(t) = S \int_{-\infty}^{t} (t - t')^{-n} \dot{\gamma}(t')\, dt'; \qquad p = p_c \tag{4}$$

which seems to be a universal property at the gel point. The gel equation predicts all known rheological properties of critical gels, such as infinite viscosity and zero equilibrium modulus. For dynamic mechanical experiments, it gives power laws for the storage and loss moduli with a phase angle $\delta = \pi n/2$.

Prediction of Gel Point

Mean field theories are able to predict the critical conversion p_c, even if they might be inaccurate in their description of the structural details of the network at GP. The threshold p_c depends on the functionality of the cross-links and on the stoichiometric ratio. New methods of measuring the gel point can be derived from the gel equation.

BIBLIOGRAPHY

D. Stauffer, A. Coniglio, and M. Adam, *Adv. Polym. Sci.* **44,** 74 (1982); H. E. Stanley, *Introduction to Phase Transitions and Critical Phenomena,* 2nd ed., Oxford University Press, New York, 1985.

H. H. Winter and F. Chambon, *J. Rheol.* **30,** 367 (1986).

F. Chambon and H. H. Winter, *J. Rheol.* **31,** 683 (1987).

E. E. Holly, S. K. Venkataraman, F. Chambon, and H. H. Winter, *J. Non-Newtonian Fluid Mech.* **27,** 17–26 (1988).

H. H. WINTER
University of Massachusetts

GELS

A gel is a cross-linked polymer network swollen in a liquid medium. Its properties depend strongly on the interaction of these two components. The liquid prevents the polymer network from collapsing into a compact mass, and the network, in turn, retains the liquid (see also NETWORKS).

Gels are important intermediates in the manufacture of polymers such as rubber, plastics, glues, films, and membranes. They are used as an absorbent in disposable diapers, for water purification, and in chromatography and electrophoresis, where molecules are separated according to the speed with which they pass through or are expelled by the pores. Gels are widely used as implants in plastic surgery and for artificial and soft contact lenses.

Gels are characterized by their equilibrium, dynamic, and kinetic properties. These properties depend on the gel state represented by osmotic pressure, temperature, solvent composition, and degree of swelling. In particular, the distance of the gel state from the phase boundary crucially influences all properties.

The phases that can be taken by a gel are sol, gaslike gel, and liquidlike gel. At the critical point of the gel–gel phase transition, the two gel phases become indistinguishable. Gel–sol transitions have been studied, but gel–gel phase transition is not fully understood. Most experimental and theoretical characterizations of gels have been carried out in the gel–gel phase regime, where the two phases cannot be distinguished.

It is important to understand the processes by which the gel is formed or dissolved. The process of gel–sol transition has significant effects on the structure and properties of the formed gel.

Sol–Gel Transition

The polymer network of a gel can be formed in various ways. In a typical condensation polymerization, a network is formed by polymerizing bifunctional units and polyfunctional units, as shown in Figure 1. Gels are also classified by the strength of the cross-linkages.

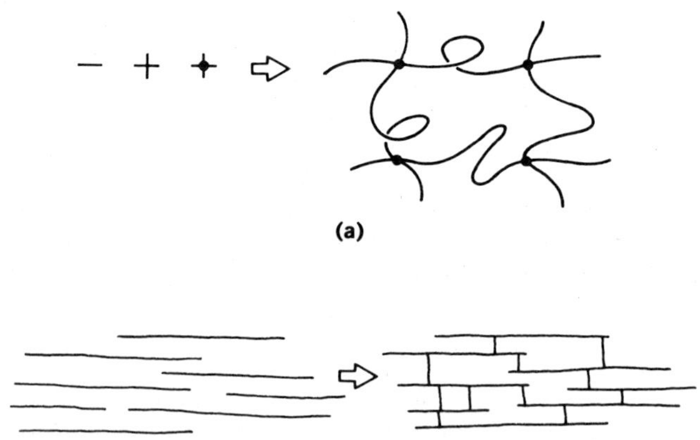

(a)

(b)

Figure 1. Network formation. (**a**) In typical condensation polymerization, bifunctional units are connected to form long chain molecules; polyfunctional units serve as cross-links. (**b**) The polymer network can also be formed by cross-linking polymer chains, which are already formed from bifunctional units.

Equilibrium Properties

Phase Properties. In order to characterize a physical system, its phases must be known. The system at each phase should be characterized in terms of its mechanical, optical, thermal, and other properties. The phase diagram and the transitions among the phases give information concerning the interaction of molecules.

Volume Phase Transition. The volume change at the transition increases with the degree of hydrolysis. Volume phase transition also occurs when the temperature is varied. At higher temperatures, the gel is swollen; at lower temperatures, it is collapsed.

Equation of State of Ionic Gels. The equation of state of gels relates osmotic pressure, polymer-network concentration, temperature, and solvent composition. Most phase properties can be theoretically deduced from it.

Effect of pH and Salt Ions on Swelling. Reversible swelling and contraction of ionic gels respond to changes of pH and ion compositions. The cycle of swelling–contraction phenomena was used to develop a mechanochemical transducer. Although large volume changes were observed in the gel on varying pH or ionic composition, no discontinuous volume transition was observed.

Collapse of Gels Under Electric Field. Phase transition can be induced by applying an electric field across the gel. When a gel is placed in a 50% acetone–water mixture and held between two platinum electrodes and an electric field of ca 500 mV/cm is applied, the gel undergoes a discontinuous volume change. At the critical position, the portion of the gel from that point to the positive electrode collapses and the remainder is swollen. The collapsed part swells again when the electric field is turned off. The physical mechanism involved in this phenomenon is not yet understood.

Universal Feature of Volume Phase Transition. According to the theory presented above, any gel may undergo the discontinuous volume transition if the polymer chain is sufficiently ionized or stiff. This universal feature has been confirmed with several different artificial and natural gels. The theory predicts that the volume transition should not depend on the sign of the charge of counterions. It has been demonstrated that positively ionized gels undergo volume transitions similar to those of negatively charged gels.

Conformation of Polymers Constituting a Gel. In order to evaluate the validity of the equation of state independently from the swelling equilibrium, the conformation of polymer chains incorporated within a polymer network must be known. Conformation is measured by observing neutron scattering in a small number of deuterium-labeled chains incorporated within a polymer network.

Inhomogeneities in Polymer Network. Density fluctuations of cross-linkings may affect mechanical properties and the spatial resolution in electrophoresis and chromatography. The inhomogeneities of a gel may be explained by phase transition, where the gel has two phases with different polymer concentrations that create inhomogeneities.

Mechanical Properties

The mechanical properties of a gel are typically represented by the stress–strain relationship. Both swelling and mechanical properties have equilibrium, dynamic, and kinetic aspects. Each can be determined by the frequency-dependent stress–strain relationship.

Dynamic Properties

Collective Diffusion of Polymer Networks. The polymer network of a gel is constantly in random thermal motion, giving rise to the space and time fluctuations of the polymer density. Such density fluctuations reveal the fundamental properties of the gel and can be observed by light-scattering or neutron-scattering techniques.

Light and Neutron Scattering from Collective Modes. The collective diffusion modes give rise to light or neutron scattering. The space–time correlation function of the scattered-light electric field is proportional to the spatial Fourier transform of the time-correlation function of the density fluctuations.

Critical Behavior. The temperature dependence of the light scattering reveals the critical behavior of the gels. As the temperature approaches the critical point, the intensity of scattered light increases to infinity and the relaxation time slows down to zero.

Nmr Measurements of Local Motions. In order to understand the molecular state and motions of polymer segments and solvent molecules in gels, the local conformational fluctuations of polymer networks in a gel have been studied by nmr.

Kinetics of Gel Swelling and Shrinking

In spite of its importance both in science and technology, the kinetics of the volume change of gels is not fully understood. It was demonstrated experimentally and theoretically, however, that the swelling process of a gel is determined by the diffusion of the polymer network.

Novel Systems

Extended applications of gel phenomena are anticipated. Gels made of photosensitive, conductive, or magnetic polymers will provide exciting subjects to explore. Gels consisting of rodlike polymers show interesting coupling phenomena between electrical and mechanical properties (see also ELECTRICALLY CONDUCTIVE POLYMERS; LIQUID CRYSTALLINE POLYMERS; PHOTOCONDUCTIVE POLYMERS).

BIBLIOGRAPHY

P. J. Flory, *Principles of Polymer Chemistry,* Cornell University Press, Ithaca, New York, 1953.

P. G. deGennes, *Scaling Concepts of Polymer Physics,* Cornell University Press, Ithaca, New York, 1979.

T. Tanaka, *Sci. Am.* **244,** 100 (1981); *Phys. Rev. Lett.* **40,** 820 (1978).

T. TANAKA
Massachusetts Institute of Technology

GENETIC ENGINEERING.

See POLYNUCLEOTIDES; PROTEINS, BIOSYNTHESIS.

GLASS FIBER.

See INORGANIC FIBERS.

GLASS TRANSITION

When a polymeric material is cooled from the liquid or rubber state, it becomes much stiffer as it goes through a certain temperature range. This stiffening is the result of one of two possible events: crystallization or glass transition. For crystallization to occur, the polymer molecules must be sufficiently regular along their length to allow formation of crystalline lattices and cooling rate must be slow enough for the crystallization process to take place before the molecular motions become too sluggish. When the polymer fails to crystallize for either reason, the amorphous, liquidlike structure of the polymer is retained, but the molecular motion becomes frozen and the material turns into a glass. Such a glass transition occurs over a finite temperature interval, but is still realized abruptly enough to merit the term "transition." The glass transition can be recognized by the change in many of the properties of the material, the most important from a practical point of view being the increase in the modulus of the material by several orders of magnitude.

Glass formation can be achieved with many low molecular weight materials and with certain metallic alloys by special preparation techniques such as rapid quenching. With polymers, the opportunities for irregularity along the chain are numerous and the crystallization rate is inherently slow; as a result the formation of the glassy state is a more common occurrence. Thus the glass state is a state of polymeric materials equal in importance to the semicrystalline state and the rubbery state (see AMORPHOUS POLYMERS). Also, the melting temperature T_m and the glass-transition temperature T_g are the two most important parameters of a given polymer that characterize its properties over a wide temperature range.

Nature of the Glassy State

The volume–temperature relationship of a typical polymer is depicted in Figure 1. Upon crystallization at temperatures moderately below T_m, the specific volume of the polymer decreases significantly in comparison to that of the amorphous polymer. If crystallization is prevented, either because of irregularities in the polymer chain or because of a rapid cooling rate, the material undergoes a glass transition manifested not by a change in specific volume but by a change in the thermal expansion coefficient, α. If the enthalpy of the polymer is measured by calorimetry and plotted against temperature, the resulting enthalpy–temperature relationship would be very similar to the plot in Figure 1. In any first-order transition, such as melting or boiling, the volume and entropy, which are both

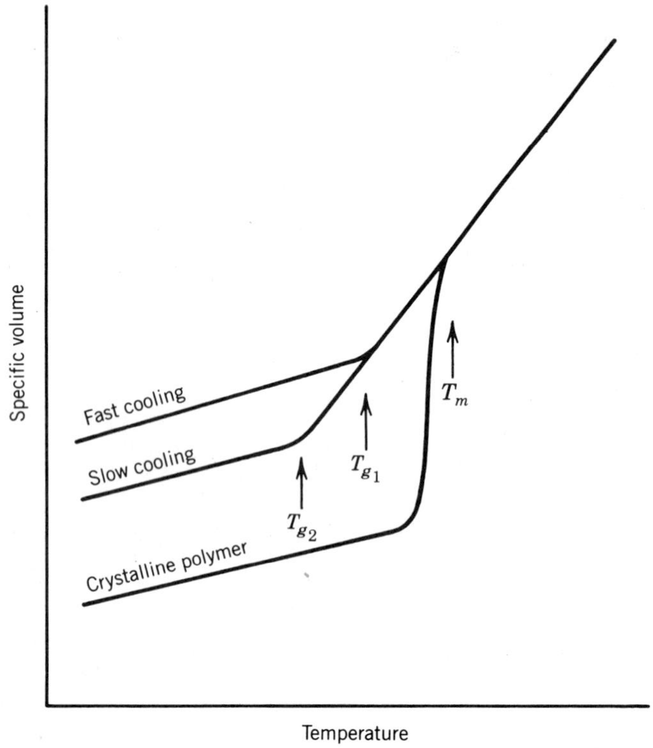

Figure 1. Schematic illustration of the volume–temperature relationship of a typical polymer. When the polymer is prevented from crystallization, it is brought to temperatures below T_m in an amorphous state and is then turned into a glass at the glass-transition temperature T_{g1} or T_{g2}, which depends on the cooling rate.

first derivatives of the free energy, undergo a discontinuous change. In the second-order transition, the second derivatives of the free energy, such as α and heat capacity, undergo a discontinuous change. The glass transition gives the appearance of the second-order transition in this sense. The glass-transition phenomenon as observed ordinarily is, however, a nonequilibrium phenomenon.

Glass-transition Temperature

The values of T_g reported for the same polymer often differ greatly, sometimes as much as 10–20°C, because the glass transition occurs over a temperature range rather than at a single, sharply defined temperature, and because the observed T_g varies somewhat, depending on the method of measurement used and on the thermal history of the sample.

Effect of Chemical Structure and Chain Flexibility. It can be stated that polymers with flexible backbone chains have low T_g and those with stiff chains have high T_g.

Correlation Between T_m and T_g. There exists a fairly good correlation between the T_m and T_g of a large number of polymers. A useful rule of thumb is that the ratio T_g/T_m is ½ for symmetrical polymers, ie, those containing a main-chain atom having two identical substituents, and ⅔ for unsymmetrical polymers.

Effect of Molecular Weight. For high molecular-weight polymers, T_g is essentially independent of molecular weight, but as the polymer chain length becomes shorter the T_g is likely to decrease appreciably. This lowering of T_g is essentially an end effect; the two ends of the polymer chain are able to move about more freely than a segment in the chain interior.

Effect of Pressure. The application of pressure compresses the volume of a liquid and consequently its free volume is reduced. This retards molecular mobility, and the glass transition occurs at a higher temperature.

Copolymers and Polymer Blends. For random copolymers and compatible polymer blends, only a single T_g, usually lying intermediate between the T_g of the corresponding pure homopolymers, is observed. For block copolymers with mutually incompatible blocks, the microdomains formed by the different blocks exhibit different T_g, and for incompatible polymer blends separate T_g values are also observed.

Measurement of T_g

T_g can be measured by dilatometry (qv), refractive index, differential scanning calorimetry, dynamic mechanical measurement, and dielectric measurement.

BIBLIOGRAPHY

R. O. Davies and G. O. Jones, *Adv. Phys.* **2,** 370 (1953); *Proc. R. Soc. London Ser. A* **217,** 26 (1953).

M. L. Williams, R. F. Landel, and J. D. Ferry, *J. Am. Chem. Soc.* **77,** 3701 (1955).

L. C. E. Struik, *Physical Aging in Amorphous Polymers and Other Materials,* Elsevier North-Holland, Inc., New York, 1978.

R. Simha and R. F. Boyer, *J. Chem. Phys.* **37,** 1003 (1962).

Ryong-Joon Roe
University of Cincinnati

GLASSY STATE.

See Amorphous polymers; Atomistic modeling, polymeric glasses; Glass transition; Viscoelasticity.

GLOBULINS.

See Proteins.

GLOW DISCHARGE.

See Coating methods.

GLUE.

See Adhesive compositions.

GLUEING.

See Adhesion and bonding.

GLYCERYL PHTHALATE.

See Alkyd resins.

GLYCIDYL METHACRYLATE.

See Acrylic and methacrylic ester polymers.

GLYCOGEN

Glycogen is a branched polymer consisting of linear segments of α-D-glucopyranosyl units joined by α-D-(1→4) glycosidic linkages with α-D-(1→6) glycosidic linkages joining the segments at the branch points.

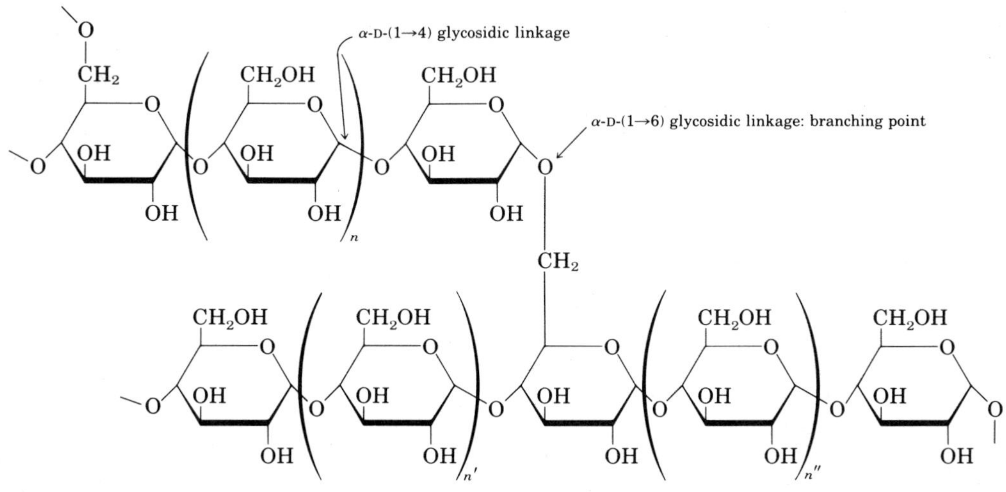

glycogen

Occurrence

Glycogen is the principal carbohydrate food-reserve material in animals. It is found in both vertebrates and invertebrates.

Biosynthesis

Glycogenesis begins on a protein molecule that acts as the primer. The polysaccharide is attached to the protein via a glycosidic linkage to the phenolic hydroxyl group of a tyrosine unit.

Isolation and Purification

Several methods have been used for the isolation of glycogen. Extraction with hot, concentrated alkali is still used, although this method is known to cause extensive degradation. Milder extraction techniques with cold trichloroacetic acid solutions, methyl sulfoxide, or water–phenol mixtures give higher molecular weight glycogens that are therefore considered to be more like native glycogen.

Structure and Properties

Glycogen is an amorphous polymer. Considering its very high molecular weight, the polymer is highly soluble and exhibits a fairly ideal hydrodynamic behavior. All polysaccharides are polydisperse; glycogen is especially so because it exists in a dynamic state. It serves to store D-glucose units until they are needed as an energy source and serves as a buffering molecule to control the concentration of blood glucose. As a result, besides organismic and tissue differences in the molecular weights of glycogen molecules, the dynamic state produces a range of molecular weights, even within a given tissue. The range depends on the metabolic state of the tissue. Glycogen occurs in tissues as single spherical particles termed β-particles and as aggregations of β-particles called α-particles.

Uses and Functions

Glycogen functions as a reserve source of energy that is stored in certain cells. The amount of stored glycogen increases when exogenous supplies of energy are abundant; it is used as an energy source when these supplies are insufficient to meet the current needs of the cell or organism.

BIBLIOGRAPHY

L. D. Kennedy, B. R. Kirkman, J. Lomako, I. R. Rodriguez, and W. J. Whelan in M. C. Berman, W. Gevers, and L. H. Opie, eds., *Membranes and Muscle,* IRL Press, Oxford, 1985, pp. 65–84.

B. R. Kirkman and W. J. Whelan, *FEBS Lett.* **194,** 6–11 (1986).

R. Geddes in G. O. Aspinall, ed., *The Polysaccharides,* Vol. 3, Academic Press, Orlando, Fl., 1985, pp. 283–336.

D. W. James, Jr., J. Preiss, and A. D. Elbein in G. O. Aspinall, ed., *The Polysaccharides,* Vol. 3, Academic Press, Orlando, Fl., 1985, pp. 173–196.

JAMES N. BEMILLER
Purdue University

GRAFT COPOLYMERS

A graft copolymer is a polymer comprising molecules with one or more species of block connected to the main chain as side chains, and having constitutional or configurational features that differ from those in the main chain, exclusive of branch points. In a graft copolymer, the distinguishing feature of the side chains is constitutional; ie, the side chains comprise units derived from at least one species of monomer different from those that supply the units of the main chain. The simplest case of a graft copolymer can be represented by the structure

```
—AAAAAAAXAAAAAAA—
        |
        B
        B
        B
        B
        B
        B
        B
        |
```

The A monomer units are referred to as the main chain or backbone, the sequence of B units is the side chain or graft, and X is the unit in the backbone to which the graft is attached. The following configurations are possible: backbone and side chains may both be homopolymeric, the backbone may be homopolymeric and the side chains copolymeric or vice versa, or both backbone and side chains may be copolymeric but of different chemical compositions. Branching in one or more stages and cross-linking may also occur. Cross-linked structures usually cannot be fully characterized because they are insoluble and frequently infusible.

Graft polymerization is a common method for modifying polymer properties. Because the main chain and the branch chain are usually thermodynamically incompatible, most graft copolymers can be classified as multiphase polymers in the solid state, analogous to polymer blends, block copolymers, and interpenetrating polymer networks. Because immiscible phases are joined by covalent bonds, analogous to block polymers, a limited range of composition-sensitive phase and morphological behavior is expected. Microphase-separated graft polymers can exhibit many of the unique thermal and mechanical properties observed in block copolymers, including thermoplastic elasticity (see ELASTOMERS, THERMOPLASTIC). Since the morphology of heterophase polymers can be affected by the casting solvent and the nature of its interaction with the polymer blocks, the physical properties are also expected to depend on the casting solvent. These factors are responsible for the technological importance of polymeric materials such as high-impact polystyrene (HIPS) (see STYRENE POLYMERS), ABS resins (acrylonitrile–butadiene–styrene graft copolymers), and cellulose graft copolymers (qv).

Free-radical polymerization methods are the oldest and most widely used procedures for the synthesis of graft polymers, because they are relatively simple. However, they usually give heterogeneous materials that are difficult to characterize. More recent methods have been developed to produce graft polymers with relatively well-defined structures.

Special Properties of the Backbone and Branches

Historically, graft copolymers have been prepared by polymerization of a monomer in the presence of a preformed backbone. The monomer can be polymerized by any of the traditional modes of polymerization. Backbones for free-radical graft copolymerization require the presence of an atom or group that can be abstracted or displaced by another radical, by radiation of sufficient intensity, or by mechanical degradation. Backbones for ionic or condensation polymerization require a reactive site or functional group capable of participating in specific chemical reactions.

Graft copolymers can be prepared by copolymerizing preformed branches with the monomer constituting the major portion of the backbone. The branch must have an end group capable of copolymerizing with the monomer by the mode of polymerization being used.

Methods of Preparation

Free-radical grafting procedures involve radical chain transfer, and/or addition reactions to the polymer backbone in the presence of polymerizing monomer. These processes are used to synthesize commercial multiphase polymeric materials, such as high-impact polystyrene (HIPS), acrylonitrile–butadiene–styrene (ABS), and methyl methacrylate–butadiene–styrene (MBS).

Selective Generation of Backbone Initiation Sites. Unwanted homopolymer formation from the grafting monomer can be minimized using "grafting-from" methods, which selectively generate initiation sites for graft chain growth on the backbone polymer chain.

Chemical modification or copolymerization introduces reactive monomer units into the backbone polymers and redox reactions generate radical initiation sites singly on the backbone polymer.

Ionic grafting methods offer distinct advantages for the selective generation of backbone initiation sites since numerous initiation methods can be used to generate single grafting sites selectively. Selective cationic grafting is limited by the types of cationically polymerizable monomers and by chain transfer.

Coupling of Preformed Polymers with Polymer Backbones. Coupling reactions provide an excellent method for the synthesis of graft copolymers with well-defined structures. In this "grafting-onto" method, a suitable reactive functional group on the backbone polymer chain is coupled with a preformed polymer chain containing an antagonistic, reactive functional group on its end. Ionic polymerization methods are most suitable for preparation of the grafting branch, since polymers with controlled structures and reactive chain-end functionality can be prepared.

Macromonomer Procedures. A macromonomer is an oligomer with an end group that can copolymerize with monomers to form comb-type graft copolymers with pendent, preformed polymer chains. Macromonomers can be prepared by ionic and free-radical methods. The resulting macromonomer can be copolymerized with a backbone-forming comonomer using free-radical, cationic, anionic, or insertion polymerization methods.

Characterization

It must be assumed that the products of intended graft copolymerizations are mixtures of graft copolymer, unreacted backbone, and homopolymer of the graft. Sometimes careful analysis reveals only mixtures of homopolymers. The first step in the characterization of a graft copolymer is the determination of whether any graft copolymerization has occurred and if so, how much.

Purity. Most purification methods are based on solubility differences between the graft copolymer and impurities.

Surface Grafting. Characterization of surface-grafted copolymers uses methods such as uv, attenuated total reflectance (atr) ir, x-ray photoelectron spectroscopy (esca), and contact-angle measurement.

Molecular Structure. Complete characterization of a graft copolymer requires knowledge of the molecular weight and molecular weight distribution of both backbone and branches, of the number and location of the branches on each macromolecule, and of the chemical composition of both backbone and branches.

Properties and Uses

Physical properties of specific graft copolymers of commercial importance, eg, HIPS, ABS, and cellulose graft copolymers, appear in articles about those substances (see STYRENE POLYMERS; ACRYLONITRILE–BUTADIENE–STYRENE POLYMERS). Graft copolymers are used for reinforcement of natural rubber, modification of rubber-based adhesives, in fibers, and in films.

Economic Aspects

Branched copolymers are specialty polymers and as such are relatively expensive. Nevertheless, branching can lead to desirable properties, such as processing ease. Branched copolymers combine vastly different properties in a single molecule and thereby provide a means of blending and introducing special properties such as dyeability, crease resistance, or water repellency that are not inherent in the parent backbone.

It is likely that the utility of graft copolymers will continue to expand as more is learned about their synthesis and unique properties.

BIBLIOGRAPHY

H. A. Battaerd and G. W. Tregear, *Graft Copolymers,* Wiley-Interscience, New York, 1967.

R. J. Ceresa, ed., *Block and Graft Copolymerization,* Vols. 1 and 2, Wiley-Interscience, New York, 1973 and 1976.

L. H. Sperling, ed., *Recent Advances in Polymer Blends, Grafts, and Blocks,* Plenum Press, New York, 1974.

P. Dreyfuss
Michigan Molecular Institute
R. P. Quirk
University of Akron

GRAPHIC ARTS. See Decorating; Fine arts.

GRAPHITE. See Carbon.

GRAPHITE FIBERS. See Carbon fibers.

GRAVURE PRINTING. See Decorating; Printing ink vehicles.

GROUP-TRANSFER POLYMERIZATION

Group-transfer polymerization (GTP) is a new method of polymer formation. It involves the repeated addition of monomer to a growing polymer chain end that carries a reactive silyl ketene acetal group. During the addition, the silyl group transfers to incoming monomer, regenerating a new ketene acetal function ready for reaction with more monomer, hence the name group-transfer polymerization. The process is illustrated below for methyl methacrylate monomer.

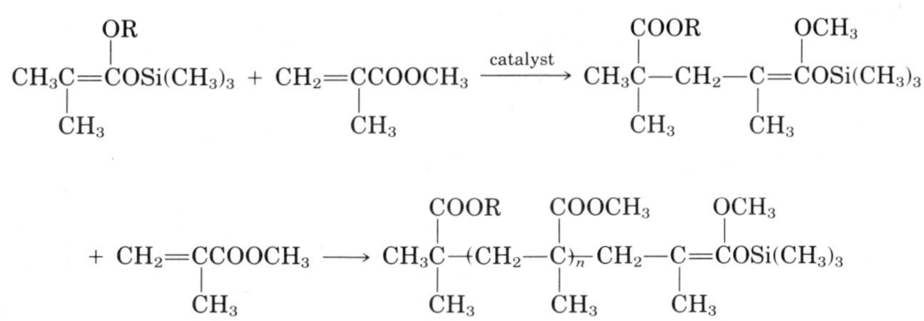

Group-transfer polymerization provides "living" polymer rapidly at room temperature and offers new dimensions in the construction and design of polymer chains. A living polymer grows if new monomer is added. This technique is used to control molecular weight, end groups, pendent functionality, and chain configuration. Polymerization control is essential for construction of polymers that are used as catalysts, and to separate mixtures, conduct electricity, and deliver drugs. Polymers produced by GTP include methyl methacrylate, ethyl acrylate, butyl acrylate, and 2-methacryloxyethyl acrylate.

GTP is a safe operation. A runaway polymerization can be quickly quenched with protonic solvent. Since polymerization goes to completion, no unwanted toxic monomer remains. The silicon group on the living end after hydroxylation is removed as inactive siloxane. The use of highly toxic cyanide and azide as catalysts for GTP may present a problem. Small amounts of bifluoride are likely to be scavenged during work-up.

Conditions and Typical Procedure

Because of the strict exclusion of active hydrogen compounds, the polymerization must be conducted under anhydrous conditions. Oxygen does not interfere. Typical solvents include toluene, tetrahydrofuran, 1,2-dimethoxyethane, acetonitrile, and *N,N*-dimethylformamide for nucleophilic catalysis, and toluene, dichloromethane, and 1,2-dichloroethane for Lewis acid catalysis.

Temperatures may range from −100°C to ca 150°C, but 0–50°C is preferred. For acrylates, temperatures at or below 0°C give best results.

In a typical procedure, methyl methacrylate (360 mmol) is added slowly to a solution of 1-methoxy-1-(trimethylsiloxy)-2-methyl-1-propene (8.5 mmol) and tris(dimethylamino)sulfonium bifluoride ($TASHF_2$) (0.01 mmol) in dry tetrahydrofuran (20 mL) at room temperature with exclusion of moisture. After completion of the exothermic reaction (about 1 h), methanol (3 mL) is added, and the solution is evaporated. Yield of poly(methyl methacrylate) (PMMA) is quantitative; $\overline{M}_n$ 4300 (theory, 4343), $\overline{M}_w$ 5300, dispersity $D = \overline{M}_w/\overline{M}_n = 1.24$, as determined by gel-permeation chromatography (gpc).

Initiators. Since in any living polymerization the growing chain ends are the initiators for the remainder of the polymerization, the GTP initiator should be similar in structure to the living ends. The living ends of methacrylates are 1-alkoxy-1-(trimethylsiloxy)-2-methyl-1-alkenes, and the best initiators have this structure. They are made by silylation of ester enolates with alkyl silicon halides.

Catalysts. The initiator is activated by nucleophilic catalysts such as soluble fluorides, bifluorides, azides, and cyanides.

Polymer Architecture

Molecular Weight. Although molecular weight can be readily controlled in the 1,000–20,000 range, higher molecular weights are difficult to obtain. With highly pure monomers, solvent, catalysts, and initiators, polymer in the 100,000–200,000 molecular weight range can be prepared.

Tacticity. The tacticity of PMMA made by GTP is independent of the type of solvent used but dependent on the reaction temperature and nature of the catalyst.

Random Copolymers. Random copolymers are made by adding a mixture of monomers of the same family, all methacrylate or all acrylate, to the initiator and catalyst.

Block Copolymers. Block copolymers within the same family form easily. A new monomer is added when the first one is depleted. An AB block copolymer with methacrylate and acrylate segments is prepared by polymerizing the less reactive methacrylate monomer first and then adding the acrylate monomer.

Applications

Because of the stringent reaction conditions and requirement for pure, dry monomers and solvents, GTP is likely to be more expensive than traditional free-radical processes.

Possible areas of application are high-performance automobile finishes, ie, finishes that can be applied at higher concentration, reducing pollution; finishes that cure at relatively low temperature, thus saving energy; and finishes that last longer, up to ten years. Other areas where cost is less important are fabrication of silicon chips and coatings for optical fibers.

BIBLIOGRAPHY

O. W. Webster, W. R. Hertler, D. Y. Sogah, W. B. Farnham, and T. V. RajanBabu, *J. Am. Chem. Soc.* **105,** 5706 (1983).

W. R. Hertler, D. Y. Sogah, O. W. Webster, and B. M. Trost, *Macromolecules* **17,** 1417 (1984).

D. Y. Sogah and W. B. Farnham, *Seventh International Symposium on Organosilicon Chemistry, Kyoto, Japan, Sept. 9, 1984,* Ellis Horwood Limited, Chichester, UK, 1985.

T. H. Maugh II, *Science* **222,** 39 (1983).

D. Y. Sogah, W. R. Hertler, O. W. Webster, and G. M. Cohen, *J. Am. Chem. Soc.* **20,** 1473 (1987).

O. W. WEBSTER
E. I. du Pont de Nemours & Co., Inc.

GROUT, CHEMICAL. See MINING APPLICATIONS.

GUANAMINE–ALDEHYDE POLYMERS. See AMINO RESINS.

GUMS, INDUSTRIAL

Gums are polymeric substances that, in an appropriate solvent or swelling agent, form highly viscous dispersions or gels at low dry-substance content. This group of polymers is commonly divided into the following classes: water-soluble gums, rubbers and elastomers, resins, oleoresins and balsams, and high molecular weight hydrocarbons.

The term "industrial gums" usually refers to polysaccharides and their derivatives. These polysaccharides of commerce are water-soluble polymers that produce viscous aqueous dispersions and are classified both by source (Table 1) and by structure (Table 2) (see also POLYSACCHARIDES; WATER-SOLUBLE POLYMERS).

The usefulness of industrial gums is based on their capacity to thicken and/or gel aqueous solutions and otherwise to control water. Because all gums modify (control) the flow of aqueous solutions, dispersions, and suspensions, the choice of gum for a particular application often depends on its secondary characteristics. These applications include utilization as adhesives, binders, bodying agents, bulking agents, crystallization inhibitors, clarifying agents, cloud agents, emulsifying agents, emulsion stabilizers, encapsulating agents, film formers, flocculating agents, foam stabilizers, gelling materials, mold-release agents, protective colloids, suspending agents, suspension stabilizers, swelling agents, syneresis inhibitors, texturing agents, whipping agents, in water absorption and binding, and in coatings. Most gums are not true emulsification agents, but contribute to emulsion and suspension stability by increasing the at-rest viscosity of the aqueous phase. Some also exhibit protective colloid action. These properties are manifestations of both the chemical structure and the shape of the gum molecules in a solvent or swelling agent.

General Properties. Gums are tasteless, odorless, colorless, and nontoxic. All gums are subject to microbiological attack and can be depolymerized by acid- and enzyme-catalyzed hydrolysis of the glycosidic (acetal) linkages joining the monomeric (saccharide) units.

Rheological Properties. In general, the rheology of gum solutions is non-Newtonian and pseudoplastic; they exhibit shear

Table 1. Classification of Commercial Polysaccharides by Source

Class	Examples
seed gums	corn starch, guar gum, locust bean gum
tuber and root gums	potato starch, tapioca starch
seaweed extracts	algin, carrageenan, agar
plant extracts	pectin
exudate gums	gum arabic
fermentation (microbial) gums	xanthan (qv), dextran (qv)
derived gums	carboxymethylcellulose, hydroxyalkylmethylcellulose, methylcellulose, starch acetate, starch phosphate, hydroxyethylstarch, hydroxypropylstarch, oxidized starches, dextrinized starches

Table 2. Classification of Commercial Polysaccharides by Structure

	Examples
Classification by shape	
linear	algin, starch amylose and its derivatives, carrageenan, pectin, cellulose derivatives
branched	
regular short branches	guar gum, xanthan, locust bean gum
branch-on-branch structures	starch amylopectin and its derivatives, gum arabic
Classification by monomeric units	
homoglycan (1 monosaccharide)	starch amylose and amylopectin, dextran, pectin
diheteroglycan (2 monosaccharides)	guar gum, locust bean gum
triheteroglycan (3 monosaccharides)	xanthan
tetraheteroglycan (4 monosaccharides)	gum arabic
Classification by charge	
neutral	guar gum, methylcellulose, starch acetate, hydroxyethylstarch, locust bean gum
anionic	algin, pectin, xanthan, carboxymethylcellulose, gum arabic

thinning (see SOLUTION RHEOLOGY; RHEOLOGICAL MEASUREMENTS). Variables that affect the rheology of gum solutions include the following: polymer composition, polymer size, gum concentration, shear rate, temperature, pH, and the concentration of organic solvents, salts, sequestrants, and other cosolutes.

Starches and Starch Derivatives

The production and use of starches and starch derivatives exceed that of all industrial gums combined. Starch use permeates the entire economy for two reasons; corn starch, in particular, is abundantly available and quite inexpensive, and starch occurs in granules. Other gums often control rheology better, coat better, or perform better; but starch has a much lower cost, and its granular nature makes handling relatively easy.

Starch granules are found in chloroplasts, storage organs, seed, roots, tubers, bulbs, stems, and fruit of green plants. Commercial starch is obtained from cereal grains [corn, sorghum (milo), rice, and wheat], tubers and roots [potato, cassava (manioc, tapioca starch), and arrowroot], and stems [sago palm].

Relative properties of native starches are given in Table 3.

Starches are modified and derivatized to improve specific properties. In general, derivatization can increase solution and gel clarity, freeze-thaw stability, and peak viscosity, improve water binding, and reduce the gelatinization temperature and the tendency to retrograde. Combinations of substitutions are used to obtain desired properties for specific applications. The useful properties of starches, which can be modified and controlled by various treatments, are their ability to thicken aqueous systems; produce a suspension of insoluble granules without thickening until the slurry is heated; create semisolid gels on cooling of pastes; form strong, adhesive films and coatings; act as a binder; and disperse and suspend fats, oils, and particulate matter. In addition, starch pastes and gels provide important textural qualities to prepared foods.

Other principal industrial gums can be classified as neutral or acidic (anionic).

Neutral Gums. Commercial guar and locust bean gums are the ground endosperm of the seeds of the respective legumes. The purified polysaccharides belong to the class galactomannans. Addition of hydroxyethyl, hydroxypropyl, methyl, and hydroxypropylmethyl ether groups to cellulose makes it into a water-soluble gum.

Anionic Gums. *Gums with Carboxylate Groups.* Algins, which are extracted from closely related species of brown algae, are salts or esters of alginic acid. Carboxymethylcellulose (CMC) is the sodium salt of the carboxymethyl ether of cellulose.

Pectins are mixtures of water-soluble polysaccharides of varying methyl ester content and degree of neutralization. Xanthan is the extracellular polysaccharide produced in a fermentation medium by *Xanthomonas campestris*. Gum arabic exudes from cracks or incisions in the bark of *Acacia* trees.

Gums with Sulfate Half-ester Groups. Kappa-, lambda-, and iota-carrageenans and agar are sulfated polysaccharides extracted from species of red algae.

Table 3. Properties of Native Starches

Starch	Amylose, %	Gelatinization temperature, °C	Relative viscosity	Paste rheology (body)	Clarity of pastes	Tendency to gel	Relative shear stability
corn	27	62–70	low	short	turbid	strong	stable
waxy maize	<2	63–72	high	long	slow turbid	weak	unstable
high amylose corn	50–70	>100, for 70%	low	short	opaque	very strong	stable
tapioca	17–23	52–64	high	long	translucent	medium	unstable
potato	17–23	58–62	very high	very long	translucent	weak	unstable

BIBLIOGRAPHY

R. L. Whistler and J. N. BeMiller, eds., *Industrial Gums*, 2nd ed., Academic Press, Inc., Orlando, Fla., 1973; 3rd ed., 1988.

R. L. Whistler, J. N. BeMiller, and E. F. Paschall, eds., *Starch: Chemistry and Technology*, 2nd ed., Academic Press, Inc., Orlando, Fla., 1984.

JAMES N. BEMILLER
Purdue University

GUMS, NATURAL. See GUMS, INDUSTRIAL; POLYSACCHARIDES.

GUNCOTTON. See CELLULOSE ESTERS, INORGANIC.

GUTTA-PERCHA. See RUBBER, NATURAL.

H

HAIR. See KERATIN.

HALOACRYLIC ESTER POLYMERS. See ACRYLIC AND METHACRYLIC ESTER POLYMERS.

HAND LAY-UP. See COMPOSITES, FABRICATION; REINFORCED PLASTICS.

HARDBOARD. See COMPOSITION BOARD.

HARD ELASTIC POLYMERS. See ELASTICITY, HARD-ELASTIC BEHAVIOR.

HARDENERS. See CURING; DRYING OILS.

HARDNESS

Hardness, or the resistance to local deformation, defines the ease with which a material can be indented, drilled, sawed, or abraded (see also ABRASION AND WEAR). It involves a complex combination of properties (elastic modulus, yield strength, and strain-hardening capacity).

Hardness is measured by static penetration of the material with a standard indenter exerting a known force, by dynamic rebound of a standard indenter of known mass dropped from a standard height, or by scratching with a standard pointed tool exerting a known force.

Hardness can be important for assessing product quality and uniformity and is used in identification, classification, and quality control. Hardness tests provide a rapid evaluation of variations in mechanical properties affected by changes in chemical or processing conditions, heat treatment, microstructure, and aging. Morphological and textural changes in crystalline polymers can be detected by hardness tests.

Table 1. Hardness Values Derived from the Ideal Shear Strength of a van der Waals Solid and Experimental Vickers Hardness Data

Polymer	Calculated H, MPa[a]	Experimental H, MPa[a]
polyethylene	150–170	50–125
nylon-6	113	90
polystyrene[b]	94	129–170
polyoxyethylene	159	170
polyoxymethylene	93	70
selenium	37	
poly(*trans*-1,4-isoprene)	87	

[a] 9.807 MPa = 9.807 N/mm^2 = kgf/mm^2.
[b] Isotactic.

Correlation of Hardness with Other Mechanical Properties

The hardness value for a pyramid indenter producing plastic flow is approximately three times the yield stress, $H \simeq 3Y$. This is the basic relation between indentation hardness and bulk properties. It is, however, only applicable to an ideally plastic solid showing no elastic strains.

Commercial Polymers

Hardness should be intimately related to the critical stresses required for plastic deformation of the polymer crystal. Thus applied compressive and shear stress must overcome the cohesive force between the molecules. Calculated and experimental values of crystal hardness for various polymers are listed in Table 1. The discrepancies are due to the complex microstructure of the polymer solids. Hardness is also a reliable index of water diffusion in certain polymers.

Microstructure and Hardness

Macroscopic unoriented polymers show a lamellar morphology, with crystalline lamellae intercalated by amorphous regions. Semicrystalline plastics such as polyethylene can be regarded as composite materials containing distinct compliant (disordered) and hard crystalline elements. It is further known that the crystal thickness decrease is related to the expansion of the unit cell. In case of low-density polyethylene, the decrease of crystal hardness has thus been correlated with the increasing expansion of the lattice unit cell, caused by the inclusion of a fraction of chain defects within the crystals.

Surface Hardening of Polymers by Chemical Attack

The controlled exposure of polymers to etching atmospheres greatly strengthens the mechanical properties of the surface, yielding hardness approaching that of some metals, eg, lead–antimony.

Microindentation Anisotropy of Oriented Polymers

The correlation between molecular orientation and hardness can be applied to the characterization of drawn materials. The influence of orientation on hardness has been demonstrated for poly(ethylene terephthalate) films by using a steel ball (r = 1.5 mm; P = 0.15 N or 0.015 kgf). The difference between the main axes of the elliptical impression increases with draw ratio. The influence of draw ratio on the anisotropy of indentation after load removal has been investigated in polyethylene, polyoxymethylene, polystyrene, polypropylene, poly(vinyl chloride), and polyamides, among others.

Test Methods

Tests for hardness include Brinell, Vickers, Knoop, Rockwell, scleroscopy, and the scratch-hardness test.

BIBLIOGRAPHY

J. Bowmann and M. Bevis, *Colloid Polym. Sci.* **255,** 954 (1977).

F. J. Baltá-Calleja, *Adv. Polym. Sci.* **66,** 117 (1985).

P. Eyerer and G. Lang, *Kunststoffe* **52,** 222 (1972).

Annual Book of ASTM Standards Part 10, American Society for Testing and Materials, Philadelphia, Pa., 1978.

F. J. Baltá-Calleja
J. Martinez-Salazar
D. R. Rueda
Institute for Structure of Matter
CSIC, Madrid, Spain

HARD RUBBER. See Rubber, hard.

HDPE. See Ethylene polymers.

HEAD-TO-HEAD POLYMERS

In addition polymerization, where polymers are obtained from α-olefins or vinyl or acrylic monomers, the monomer units are exclusively or predominantly arranged in a head-to-tail (H-T) fashion with the side groups placed on alternating carbon atoms.

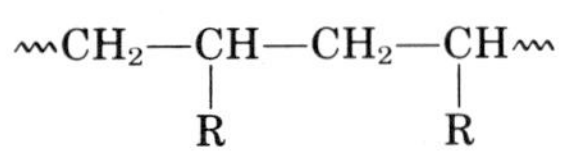

H-T linkage 1,3

H-H linkages are present in addition polymers prepared by radical polymerization by recombination of the growing polymer radicals.

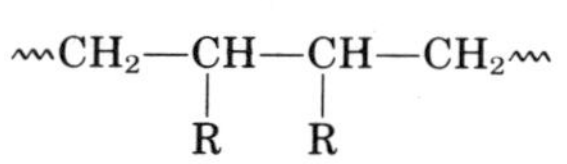

H-H linkage 1,2

Because H-H polymers cannot be made directly from α-olefins, or vinyl or acrylic monomers, indirect synthetic methods have been developed. In this article, only those polymers that have a polyethylene backbone chain are discussed.

H-H Poly(α-olefins)

Diene Polymerization and Hydrogenation of Diene Polymers. The 1,4-polymerization of 2,3-disubstituted-1,3-butadienes followed by hydrogenation of the double bond gives H-H polyolefins with hydrogen atoms on the substituted carbon atoms.

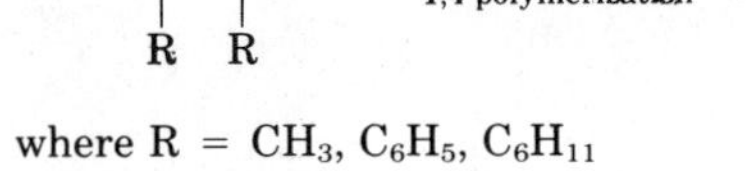

where R = CH_3, C_6H_5, C_6H_{11}

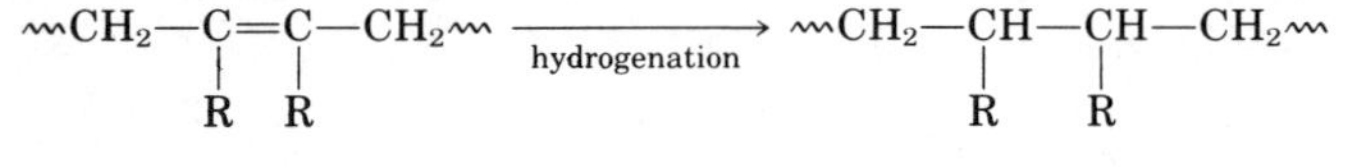

H-H Polypropylene (PP). H-H PP is synthesized from 1,4-poly(2,3-dimethyl-1,3-butadiene) (PDMB) by catalytic hydrogenation of the double bond. The key to this reaction is the polymerization of 2,3-dimethyl-1,3-butadiene (DMB) with 100% 1,4-linkages in the polymer. Some time ago H-H PP was prepared by alternating copolymerization of ethylene with a large excess of 2-butene.

H-H Polystyrene (PS). H-H PS is prepared by bulk polymerization of 2,3-diphenyl-1,3-butadiene (DPB) with azobisisobutyronitrile (AIBN) as radical initiator. Hydrogenation of 1,4-poly(2,3-diphenyl-1,3-butadiene) (PDPB) is best carried out by chemical reduction with potassium and ethanol as reducing agent for the stilbene-type double bond of the polymer chain.

H-H Polyvinylcyclohexane (PVCH). H-H PVCH is prepared from H-H PS or PDPB by catalytic hydrogenation with a heterogeneous noble metal catalyst such as palladium-charcoal. Higher temperatures are required and 30% catalyst based on the polymer is needed.

1,4-Dibromobutane Polymerization. The H-H polymers of gem-disubstituted monomers cannot be made by catalytic hydrogenation. They are prepared by polycondensation of 2,2,3,3-tetrasubstituted-1,4-dibromobutanes.

H-H Polyisobutylene (PIB). H-H PIB is prepared from an intermediate polymer that has either an H-H or a T-T linkage. Prepolymers with an H-H linkage are obtained by cationic polymerization of 2,5-dimethyl-2,4-hexadiene, commonly called 1,1,4,4-tetramethyl-1,3-butadiene (TMB). For the formation of a prepolymer with a T-T linkage of fairly high molecular weight, 1,4-dibromo-2,2,3,3-tetramethylbutane is polymerized.

H-H Poly(vinyl-1-naphthalene) (P1VN) and Poly(vinyl-2-naphthalene) (P2VN). The two polyvinylnaphthalenes are synthesized from 1,4-dibromo-2,3-di(naphthyl)butane or 1,4-dibromo-2,3-di(2-naphthyl)butane by polycondensation using the Yamamoto method.

Poly(vinyl halides)

H-H Poly(vinyl chloride) (PVC). Early work on the chlorination of *cis*- and *trans*-1,4-poly(1,3-butadienes) in trichloroacetonitrile at 0°C with a ferric chloride catalyst gave cross-linked polymers. More recent work on the chlorination of *cis*-1,4-poly(1,3-butadiene) of high molecular weight with less than 0.4% 1,2-linkages gave H-H PVC of high molecular weight.

Poly(vinyl bromide) (PVB). PVB has been prepared from *cis*-1,4-poly(1,3-butadiene) by bromination in carbon tetrachloride or chloroform solution at elevated temperature in a 0.25% polymer solution; the brominated products from either geometric isomer have very similar properties.

Polyacrylates

H-H Poly(methyl acrylate) (PMA). H-H PMA is prepared from an alternating copolymer of maleic anhydride and ethylene that is esterified with methanol and a sulfuric acid catalyst, followed by diazomethane.

Other H-H Polyacrylates Substituted at the α-Carbon. H-H poly(methyl crotonate) is prepared similarly to H-H PMA. The H-H poly(methyl cinnamate) is prepared from the alternating copolymer of stilbene and maleic anhydride.

Attempts to Prepare H-H Poly(methyl methacrylate) (PMMA). Attempts were made to prepare pure H-H poly(methyl methacrylate), but with limited success. They include cyclopolymerization (qv) of methacrylic anhydride or dimethacrylimides followed by hydrolysis and esterification. Another attempt was the alternating copolymerization of 2,3-dimethylmaleic anhydride with ethylene.

Poly(allyl alcohol) and Derivatives. H-H poly(allyl alcohol) (H-H PAA) is prepared from H-H PMA by lithium aluminum hydride reduction in tetrahydrofuran. The resulting H-H PAA is acetylated to H-H poly(allyl acetate) (H-H PAAc) or benzoylated to H-H poly(allyl benzoate) (H-H PABz).

Properties

Spectral Properties. The infrared spectra of H-H polymers are similar to those of H-T polymers. The nmr spectra, both ^{1}H and ^{13}C nmr, are different and allow the identification of H-H polymers or H-H linkages in polymers.

Thermal Behavior. *Glass-transition Temperature.* Glass-transition temperatures, as determined by differential scanning calorimetry (dsc), are given in Table 1.

Thermal Degradation. The maximum degradation-rate temperature (MDT) of H-H and H-T polymers is given in Table 2.

Solution Behavior. The solution behavior of some of the H-H polymers has been investigated. It has been found that the unperturbed dimension of H-H PP in solution is smaller than that of atactic H-T PP.

The theta temperature of H-H PST in cyclohexane is lower than that of H-T PST. The unperturbed dimension of H-H PS is larger than that of isotactic and atactic H-T PS, which indicates that the phenyl substituent in the 1,2-position (H-H placement) decreases the flexibility of the polymer chain.

Blending Behavior. The blending behavior of H-H polymers has been studied. Blends of erythro-predominant and threo-predominant H-H PP indicate that H-H polymers of slightly different stereochemistry are completely miscible as determined by dsc. H-H and H-T PPs of comparable molecular weights are also completely miscible over the entire range of compositions.

Table 1. Glass-transition Temperatures of H-H and H-T Polymers[a]

	T_g, °C	
Polymer	H-H	H-T
polypropylene	−39	−17
polyisobutylene	87	−61
polystyrene	97	98
polyvinylcyclohexane	88	138
poly(vinyl-1-naphthalene)	170	162
poly(vinyl-2-naphthalene)	146	
poly(vinyl chloride)	91	83
poly(vinyl bromide)	76	85
poly(methyl acrylate)	40	12
poly(methyl crotonate)	107	80
poly(methyl cinnamate)	210	190
poly(methyl methacrylate)	160–170 est	100
poly(allyl alcohol)[b]	105	95
poly(allyl acetate)[b]	50	[c]
poly(allyl benzoate)[b]	95–100	95

[a] Heating rate, 20°C/min, except where otherwise stated.
[b] Heating rate, 10°C/min, softening temperature.
[c] Rubbery at room temperature.

Table 2. Thermal Degradation Behavior of H-H and H-T Polymers

	Maximum degradation rate temperature, °C	
Polymer	H-H	H-T
polypropylene	468	479
polyisobutylene	315	385
polystyrene	397	398
polyvinylcyclohexane	433	441
poly(methyl acrylate)	418	419
poly(methyl crotonate)	383	410
poly(methyl cinnamate)	376	410
poly(vinyl alcohol)	435	449
poly(vinyl acetate)	402	411
poly(vinyl benzoate)	399	412

In contrast, H-H and H-T PIB are not miscible, as demonstrated with blends of molecular weights of approximately 10,000.

BIBLIOGRAPHY

O. Vogl, *J. Macromol. Sci. Chem.* **A21,** 1725 (1984).

O. Vogl, *J. Macromol. Sci. Chem.* **A22,** 541 (1985).

O. Vogl and M. Kryszewski, *Dresden Polymer Discussions 2* **8,** 51 (1986).

O. Vogl, M. Malanga, and W. Berger, *ACS Symposium Series* **4,** 35 (1985).

OTTO VOGL
Polytechnic University
STEVEN GROSSMAN
University of Lowell

HEAT AGING. See ANTIOXIDANTS; ANTIOZONANTS.

HEATING, DIELECTRIC. See DIELECTRIC HEATING.

HEAT-RESISTANT POLYMERS

Heat-resistant polymers are used in many applications, such as insulators for microelectronic components, sealants for fuel tanks in high-speed aircraft, coatings on cookware, binders in brake systems, and structural components in space vehicles.

Heat-resistant resins are defined as materials that retain mechanical properties for thousands of hours at 230°C, hundreds of hours at 300°C, minutes at 540°C, or seconds up to 760°C. These temperatures may be encountered during the manufacture or use of the product.

Requirements for Heat Resistance

Heat resistance is the capacity of a material to retain useful properties for a stated period of time at elevated temperatures (230°C) under defined conditions, such as pressure or vacuum, mechanical load, radiation, and chemical or electrical influences at temperatures ranging from cryogenic to above 500°C (see also CRYOGENIC PROPERTIES). Both reversible and irreversible change can occur. Reversible changes occur primarily as a function of T_g, which for high temperature structural polymers must be above 230°C.

The chemical factors that influence heat resistance include primary bond strength, secondary or van der Waals bonding forces, hydrogen bonding, resonance stabilization, mechanism of bond cleavage, molecular symmetry (structure regularity), rigid intrachain structure, cross-linking, and branching. The physical factors include molecular weight and molecular weight distribution, close packing (crystallinity), molecular (dipolar) interactions, and purity. The primary bond strength is the single most important factor contributing to the heat resistance.

Determination of Heat Resistance

Heat resistance is assessed by thermogravimetric analysis (tga), isothermogravimetric analysis (itga), differential scanning calorimetry (dsc), dynamic mechanical analysis (rheovibron or torsional-braid analyzer), thermal mechanical analysis (expansion, penetration), and dielectrometry. Heating rate, sample form, atmosphere, frequency, load, and thermal history affect the test results.

Polymers with High Heat Resistance

The characteristics that contribute to the thermal stability also make it more difficult to process, namely, poor solubility and high melt viscosity. The evolution to melt-processible forms of heat-resistant polymers is illustrated by polyimides and to a lesser extent by the polyquinoxalines, polyquinolines, and polybenzimidazoles. Processibility requirements for polymers potentially useful as fibers or ribbons having all-para structures and commonly referred to as extended chain and rigid rodlike polymers are different. Several of these polymers are processed from solution (generally in strong acids such as methane sulfonic acid) to form fibers and ribbons; others can be molded under high temperature and pressures to form complex shapes.

Polyimides. Polyimides (qv) derived from long-chain aliphatic diamines, which remain in a melt under the polymerization conditions, are usually tractable and of high molecular weight. An improved route (eq. 1) has been reported where an aromatic dianhydride (pyromellitic dianhydride) and an aromatic diamine (4,4′-oxydianiline) in a polar solvent (N,N-dimethylacetamide, DMA) form a high molecular weight poly(amide–acid), also called polyamic acid. The poly(amide–acid) can be cyclodehydrated by chemical (acetic anhydride and pyridine) or thermal means to the polyimide.

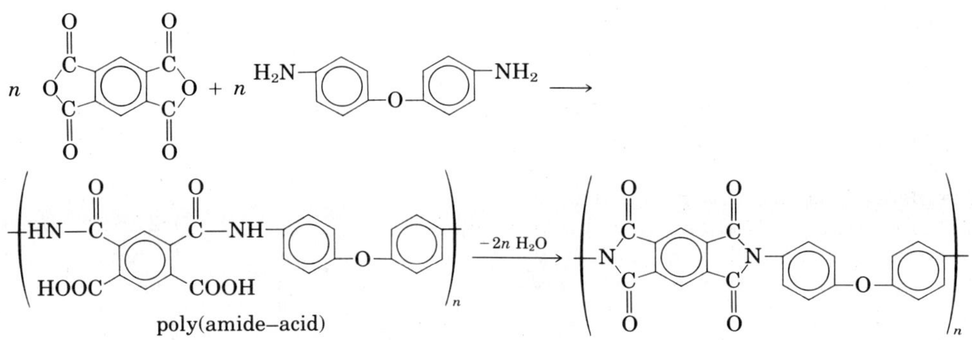

(1)

The properties of polyimides are controlled by chemical structure, molecular weight, and process conditions (thermal history). When assessing structure–property relationships, the method and conditions of determining T_g and the thermal history should be considered.

Process conditions affect properties as shown in Table 1; tensile strength and tensile modulus increase dramatically on hot drawing.

Several materials derived from a poly(amide–acid), such as Kapton film or Vespel moldings, are commercially available and are thought to have the chemical structure shown in equation 1. Kapton has excellent mechanical properties (Table 2).

The processibility of PI was improved by eliminating the hydrolytically unstable, volatile-producing poly(amide–acid). In the early 1970s, imide oligomers terminated with norbor-

Table 1. Thin-film Properties of LARC Polyimide at 25°C

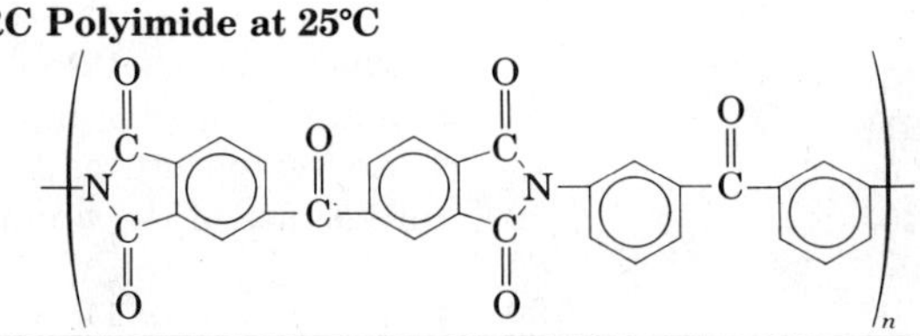

Process conditions	Tensile strength[a], MPa[b]	Tensile modulus[a], GPa[c]	Elongation[a], %
dried 1 h at 100, 200, and 300°C in air	111.0	3.25	6.6
same as above but drawn 50% at 260°C before testing	169.6	4.65	6.7

[a] ASTM D 882-647.
[b] To convert MPa to psi, multiply by 145.
[c] To convert GPa to psi, multiply by 145,000.

Table 2. Properties of Kapton-type H Film

	Test temperature, °C				
Property[a]	−195	23	200	300	450
ultimate tensile strength, MPa[b]	241	172	117	96.5	41.4
yield point at 3%, MPa[b]		69	41		
tensile modulus, GPa[c]	3.5	3.0	1.86		
ultimate elongation, %	2	70	90	110	110
density[d], g/cm^3		1.42			

[a] ASTM D 822-64T except where otherwise stated.
[b] To convert MPa to psi, multiply by 145.
[c] To convert GPa to psi, multiply by 145,000.
[d] ASTM D 1505-63T.

nene and acetylenic groups were developed for structural applications.

Several imide copolymers, eg, esters, phenylquinoxalines, and amides, have been reported, but the Torlon polyamide-imides (Amoco Chemical Corp.) have received the greatest attention and are currently used in many applications. Polyimides in general are used where heat resistance is required. Various forms for use as adhesives, composite matrices, coatings, films, and moldings are commercially available.

Polyquinoxalines. The polyquinoxalines (qv) (PQ) are prepared by the reaction of aromatic bis(*o*-diamines), such as 3,3′,4,4′-tetraaminobiphenyl (TAB), and aromatic bisglyoxal hydrates, such as 4,4′-oxydiphenylenediglyoxal dihydrate, in *m*-cresol at 80–100°C.

Polyphenylquinoxalines (PPQ), first reported in 1967, are easier to make than PQ and offer superior solubility, processability, and thermooxidative stability. High molecular weight PPQs are readily synthesized as shown in equation 2 from aromatic bis(*o*-diamines), such as TAB, and bis(phenyl-*α*-diketones), such as 4,4′-oxydibenzil (ODB), in *m*-cresol at ambient temperature.

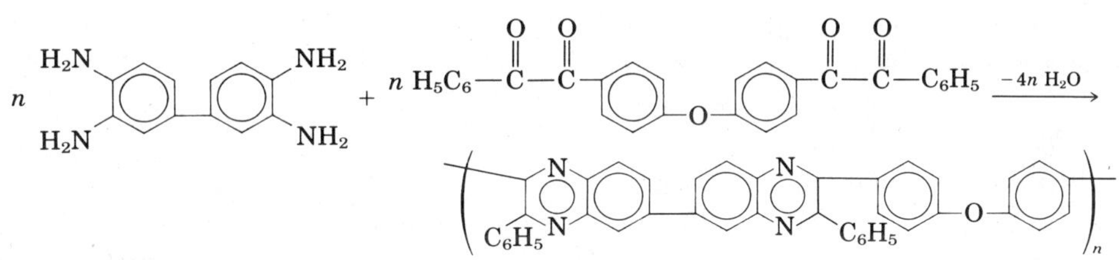

(2)

The PPQs have excellent high temperature adhesive, film, and composite properties.

Polyquinolines. Polyquinolines (qv) and polyphenylquinolines are synthesized by routes similar to those for polyanthrazolines and polyisoanthrazolines. In the synthesis of polyquinolines or polyphenylquinolines, a bis(*o*-aminoketone) [eg, 4,4′-bis(2-aminobenzoyl) diphenyl ether] reacts at 130°C with aromatic diacetyl (eg, 4,4′-diacetyldiphenyl ether) and di(phenylacetyl) [eg, 4,4′-di(phenylacetyl)diphenyl ether] monomers, respectively, in a 3:1 mixture of *m*-cresol and polyphosphoric acid (PPA).

Polybenzimidazoles. Polybenzimidazoles (qv) (PBI) were initially synthesized in 1961 by melt polycondensation of TAB with diphenyl esters of aromatic dicarboxylic acids as shown in equation 3.

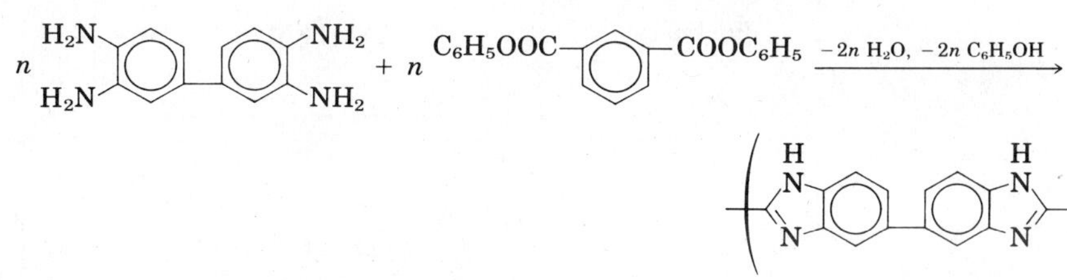

(3)

Other aromatic bis(*o*-diamines) have also been employed.

Polybenzimidazoles are processed into low-to-medium density foams that have excellent strength retention after 100 h at 316°C. The foams, like other forms of PBI, are resistant to thermal shock from cryogenic to very high temperatures.

Rigid Rodlike Polymers. High temperature, rigid rodlike polymers, as exemplified by polybenzothiazoles (PBT), polybenzoxazoles (PBO), benzimidazobenzophenanthroline-type ladder polymer (BBL), and poly(4-hydroxybenzoic acid) (poly-4-benzoate), are being developed primarily as high temperature reinforcements, eg, fibers and ribbons. Poly-4-benzoate is also considered as molding material. The PBTs and PBOs are lyotropic liquid crystalline polymers with high thermooxidative stability due to their morphology and high molecular weight. PBTs with intrinsic viscosities of 30.3 dL/g were prepared in PPA at 190–200°C by the polycondensation of 1,4-dimercapto-2,5-diaminobenzene dihydrochloride with terephthalic acid.

Similar to PBTs, PBOs are prepared in PPA by the polycondensation of 4,6-diaminoresorcinol dihydrochloride with terephthalic acid or terephthaloyl chloride. PBO fiber properties include a modulus of 84.6 GPa (711 g/den), tenacity of 570 MPa (4.8 g/den), and elongation of 0.7% after a 425°C heat treatment (see POLYBENZOTHIAZOLES AND POLYBENZOXAZOLES).

Outlook

Virtually every thermally stable heterocyclic polymer that can be prepared through the formation of the ring has been reported. It is unlikely any significant improvement in the thermal stability of organic polymers will be obtained. Polymers with outstanding heat resistance include the following:

- Kapton, with excellent heat resistance and wide industrial use.
- PI adhesive with no strength loss at 232°C after 37,000 h at 232°C in air.
- PMR-15 graphite filament reinforced composites with good retention of mechanical properties at 316°C after hundreds of hours at 316°C in air.
- PQ adhesive with high strength at 371°C after 50 h at 371°C in air.
- PPQ film with high thermooxidative stability and weight loss of ~1% at 316°C in air after 1,100 h.
- PBI composites with useful mechanical properties for short time at temperatures >538°C.
- PIQ composites with excellent retention of mechanical properties at 371°C.

BIBLIOGRAPHY

J. P. Critchley, G. J. Knight, and W. W. Wright, *Heat-resistant Polymers,* Plenum Press, New York, 1984.

M. I. Bessonov, M. M. Koton, V. V. Kudryavtsev, and L. A. Laius, *Polyimides—Thermally Stable Polymers,* Consultants Bureau, Division of Plenum Publishing Corp., New York, 1987.

P. E. Cassidy, *Thermally Stable Polymers,* Marcel Dekker, Inc., New York, 1980.

W. D. Weber and M. R. Gupta, *Recent Advances in Polyimide Science and Technology,* Society of Plastics Engineers, Inc., Poughkeepsie, New York, 1987.

P. M. HERGENROTHER
NASA Langley Research Center

HEAT SEALING. See ADHESION AND BONDING.

HEAT TRANSFER

Fundamental Principles

Transfer of heat within solid, liquid, or gaseous media or from one medium to another can occur by conduction, convection, or radiation. It can be assumed that in molten plastic bodies no further flow processes take place when the cooling process starts, ie, shaping is completed. These bodies, such as injection moldings or extrudates, generally have much smaller thickness than length and breadth and may thus be regarded as being of a flat plate shape or as made up of flat plates. Because the temperature gradients over the thickness are much greater than in the other directions, a unidimensional problem is created.

Equation of Transient Heat Conduction. If the progress of a semifinished product is observed from a viewpoint next to the plant, a steady state normally prevails at each point. To calculate the temperature field in the extrudate, the system of coordinates is convected by a volume element of the extrudate. Viewed from this angle, the process is a transient one. The cooling of a molding part, on the other hand, is always to be regarded as a transient problem.

Temperature Equalization. If a body with uniform internal temperature distribution is brought into an environment of different temperature, temperature equalization commences. Only conduction heat transfer occurs inside the body (no heat sources).

Calculation of Transient Temperature Fields. The parameters required for the calculation of transient temperature fields are given in Figure 1. The material is assumed to be homogeneous and isotropic.

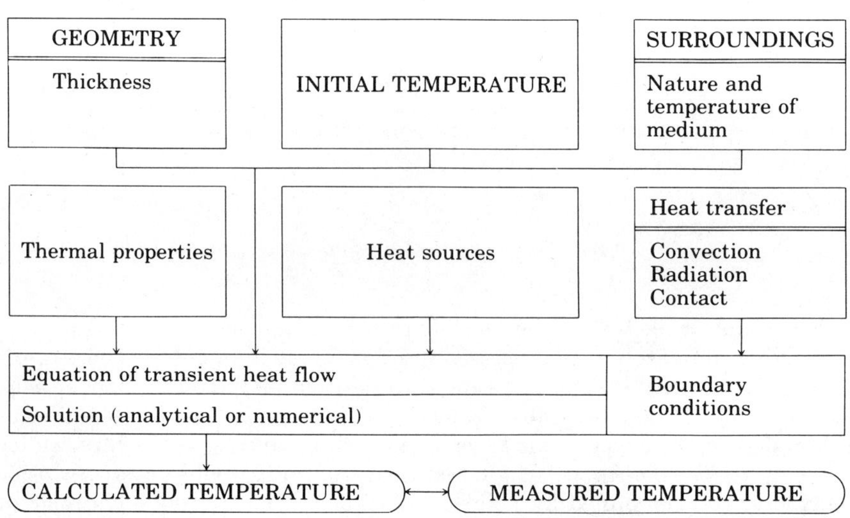

Figure 1. Parameters required for the calculation of transient temperature fields.

The thickness strongly affects the result and hence the thermal shrinkage should also be considered.

Thermal Properties. The most important influence on the thermal properties of a material is temperature.

In some semicrystalline polymers, the degree of crystallization, and hence the thermal properties at a given temperature, are highly dependent on the cooling rate.

Heat-transfer Coefficients. *Convection.* The heat-transfer coefficient α can be presented by the dimensionless Nusselt number:

$$\mathrm{Nu} = \frac{\alpha \cdot l}{\lambda}$$

The characteristic length l of the body is arbitrary in some cases and must always be indicated. In contrast to the Biot number, here λ is the thermal conductivity of the flowing medium.

Radiation. The radiative heat-transfer coefficient is calculated using the Stefan-Boltzmann law. The material value can be set at 0.95 for plastics that are not transparent to radiation. The reflection factor is 0.04–0.06.

Contact. The contact heat-transfer coefficient α_{cont} can only be established by the comparison of measurement and calculation.

Heating and Cooling of Semifinished Products and Molded Parts

Flat Plate. Assuming an infinite flat plate for describing heat transfer, only heat flow through the plate is observed. This method can also be applied to extruded film and sheet since the temperature gradients in the other directions can always be neglected. This then gives the simple case of unidimensional heat transfer in a rectangular coordinate system for which analytical solutions and diagrams are available. This type of solution can also be used for cylinders and spheres.

Other Shapes. Thick-walled hollow cylinders, such as pipes with a ratio of wall thickness to internal radius of approximately 1:1, can be described numerically by cylinder coordinates. Pellets may be regarded as cylinders or spheres, depending on their shape.

Extruded profiles not corresponding to the simple geometries mentioned above require a two-dimensional calculation of the temperature fields using the finite-difference or finite-element method.

Multilayer moldings, composite materials, multilayer semifinished products, and cables with single or multilayer insulation generally constitute cases of unidimensional heat transfer, for which, however, numerical calculation methods are required.

Cooling of Extruded Semifinished Products. After being shaped in the die, extrudates are usually cooled in different sections of the downstream equipment, although they may also be heated to be stretched or cross-linked. Downstream equipment should be divided into sections on the basis of the heat applied or removed (Fig. 2).

Heat Transfer from Polymerizations

Polymerization reactions are almost exclusively exothermic, and usually the considerable amounts of heat must be removed. Heat can be removed from a reactor through external jacketing, immersed coils, overhead condensers, or external loops. The coolant is usually water. Frequently, more than one means of cooling is used.

BIBLIOGRAPHY

H. H. Winter, *Adv. Heat Transfer* **13,** 205 (1977).

S. Middleman, *Fundamentals of Polymer Processing,* McGraw-Hill, Inc., New York, 1977.

J. L. Throne, *Plastics Process Engineering,* Marcel Dekker, Inc., New York, 1979.

W. Dietz, *Colloid Polym. Sci.* **255,** 755 (1977).

EDMUND HABERSTROH
Institut für Kunststoffverarbeitung
Technische Hochschule Aachen

HEAT-TRANSFER LABELING.

See DECORATING.

HELIX–COIL TRANSITIONS

When all bond lengths and bond angles are fixed, a linear polymer forms a helix if successive backbone dihedral angles exhibit a repeating pattern. The simplest pattern is one in

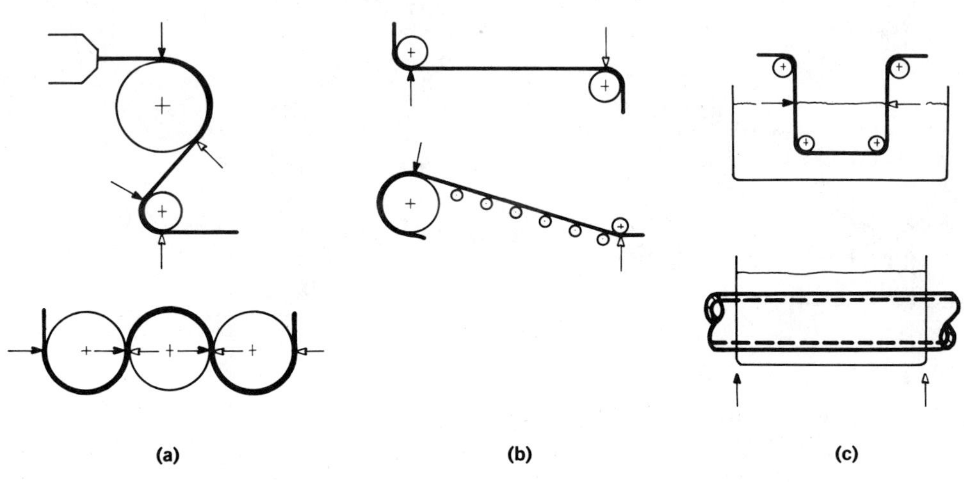

Figure 2. Cooling-section elements of extrusion lines: (**a**) roller cooling; (**b**) air cooling; (**c**) water cooling. ↓ and ↓ indicate start and finish of section, respectively.

which the conformation at every bond is the same. More complicated repeating patterns arise when the monomer units contribute different types of bonds to the backbone.

Short-range interactions are sufficient for stabilization of a helical conformation in some poly(amino acid)s. Interactions of somewhat longer range play the dominant role in stabilization of the better known α-helix (Fig. 1).

The α-helix–coil transition and the related β-structure–coil transition serve as useful models for the denaturation of globular proteins. Studies of these transitions in synthetic poly(amino acid)s provide information about the interactions that stabilize native protein conformations.

Experimental Observations

Methods. Experiments can provide three different types of information: specification of the helix as single or multichain, measurement of the fraction of the monomer units in helix and coil states, and characterization of the overall dimensions of the polymer. The single or multichain nature of the helix is revealed by the behavior of the molecular weight in the helix–coil transition region. The fraction of the monomer units in α-helices is usually deduced from measurements of optical activity. The helix content can also be deduced from the titration curve if there are ionizable groups on the side chains. The overall shape of the chain can be investigated with light-scattering and hydrodynamic methods.

Conclusions. The results of numerous experimental studies of helix–coil transitions in homopolymers of amino acids can be summarized by several conclusions. The amino acid residue must satisfy certain covalent requirements if there are to be conditions under which the α-helix is stable. Any feature that interferes with the formation of the hydrogen bond from the C=O of residue i to the N—H of residue $i + 4$ prevents helix formation. The substitution of alkyl groups for the amide hydrogen atom prevents formation of an α-helix in polysarcosine and poly(L-proline). The presence of competing hydrogen-bonding groups near the backbone also discourages formation of the α-helix. For this reason, a methyl side chain supports helix formation, but hydroxymethyl does not. A more subtle covalent requirement is that alternative helices that might compete with the α-helix must be suppressed. Although polyglycine does not experience adverse interactions in an α-helix, it has never been observed in this conformation. It prefers another helix, known as polyglycine Form II. The polyglycine Form II structure is sterically inaccessible when side chains larger than a hydrogen atom are present.

Copolymers in Protein Studies. Many proteins contain α-helix segments. Therefore, there is special interest in the helix–coil transitions in water of polymers formed from the 20 amino acid residues usually found in proteins. However, most of the homopolymers are not sufficiently soluble in water to permit investigation by the usual techniques; thus, interest has centered on copolymers.

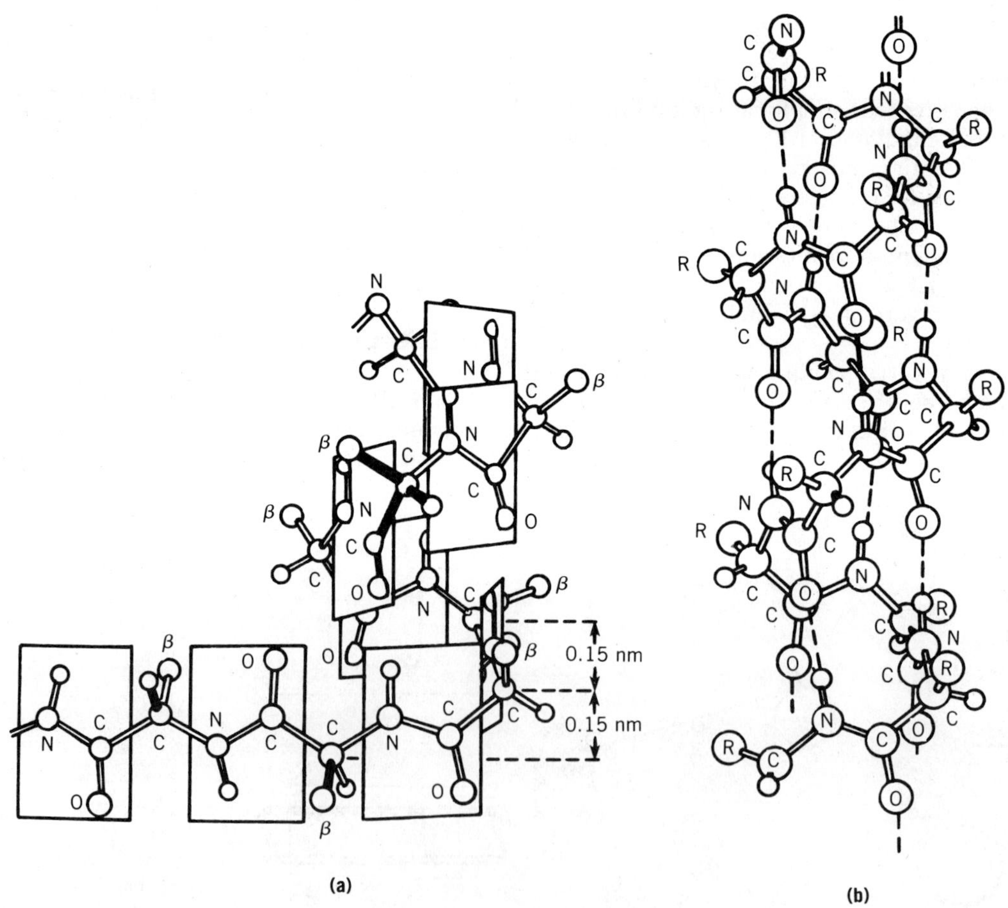

Figure 1. Formation of (**b**) right-handed α-helix from (**a**) a linear poly(amino acid). Courtesy of Springer-Verlag.

Theory

Homopolymer. Various theories have been developed to account for the experimental observations and the dependence of the helix content on molecular weight and temperature in a given solvent, and to provide information about the interactions that stabilize the helix. These theories, taking cognizance of the special geometric features that characterize the α-helix (Fig. 1), fall into two categories: one deals with the interactions in the helix and coil, and the other is concerned with enumeration of the allowable states of the system. The latter type of theory provides phenomenological parameters that reflect the interactions in the helix and coil.

Copolymers. Because of the insolubility of most homopolymers in water, it is not possible to obtain the (helix) nucleation parameter σ and the equilibrium constant $s(T)$ for the naturally occurring amino acids from experiments on homopolymers. Random copolymers are the easiest type of copolymers to synthesize, by mixing the *N*-carboxyanhydrides of the monomers. Thus water-soluble random copolymers are prepared from a water-soluble host and a guest monomer, which is, in turn, each of the 20 naturally occurring residues. Useful hosts are poly(hydroxypropyl-L-glutamine) (PHPG) and poly(hydroxybutyl-L-glutamine) (PHBG). The theory for the helix–coil transition in a random copolymer can be formulated on a similar basis to that for the homopolymer.

Application of Copolymer Theory. The copolymer theory was tested initially on random copolymers of hydroxypropylglutamine and hydroxybutylglutamine. Since experiments on the homopolymers, PHPG and PHBG, yielded values of σ and $s(T)$ for both types of residues, it was possible to use these homopolymer parameters in the foregoing theory for a random copolymer to compute the transition curves for random copolymers of these two residues. The computed curves agreed reasonably well with experimental transition data for the random copolymers.

BIBLIOGRAPHY

D. Poland and H. A. Scheraga, *Qt. Rev. Biophys.* **10,** 239 (1977).

B. H. Zimm and J. K. Bragg, *J. Chem. Phys.* **31**, 526 (1959).

A. Kidera, M. Mochizuki, R. Hasegawa, T. Hayashi, H. Sato, A. Nakajima, R. A. Fredrickson, S. P. Powers, S. Lee, and H. A. Scheraga, *Macromolecules* **16,** 162 (1983).

M. Sueki, S. Lee, S. P. Powers, J. B. Denton, Y. Konishi, and H. A. Scheraga, *Macromolecules* **17,** 148 (1984).

HAROLD A. SCHERAGA
Cornell University

WAYNE L. MATTICE
Louisiana State University

HEMICELLULOSE.
See CELLULOSE; POLYSACCHARIDES.

HEMP.
See FIBERS, VEGETABLE.

HETEROGENEOUS POLYMERIZATION.
See COORDINATE POLYMERIZATION; EMULSION POLYMERIZATION; RADICAL POLYMERIZATION; STEP-REACTION POLYMERIZATION.

HETEROPOLYMERIZATION.
See COPOLYMERIZATION.

HEVEA.
See RUBBER, NATURAL.

HEXAFLUOROPROPYLENE.
See TETRAFLUOROETHYLENE POLYMERS.

HEXAMETHYLENE ADIPAMIDE POLYMERS.
See POLYAMIDES.

HEXAMETHYLENE DIISOCYANATE.
See POLYURETHANES.

HEXAMETHYLENE SEBACAMIDE POLYMERS.
See POLYAMIDES.

HEXENE POLYMERS.
See OLEFIN POLYMERS.

HIGH-DENSITY POLYETHYLENE.
See ETHYLENE POLYMERS.

HIGH-ENERGY RADIATION.
See CROSS-LINKING WITH RADIATION.

HIGH-FREQUENCY HEATING.
See DIELECTRIC HEATING.

HIGH-IMPACT PLASTICS.
See IMPACT RESISTANCE.

HIGH MODULUS POLYMERS

High modulus polymers are potential materials of choice in applications requiring high mechanical properties and light weight. Figure 1, a display of specific modulus and specific strength of high performance materials, demonstrates that high modulus polymers show a distinct advantage over metals and ceramics on a property–weight basis. However, the penetration of polymers into these applications has been relatively modest for several reasons:

1. Only the aramid fibers, Kevlar (DuPont) and Arenka (Akso), have a significant history of commercial availability.
2. The thermotropic copolyesters, Vectra (Celanese) and Xydar (Dart), and high modulus polyethylene fibers, Spectra 900 (Allied), were not introduced commercially until 1985.
3. Because they are new materials, design criteria and a performance expectation experience base are lacking.

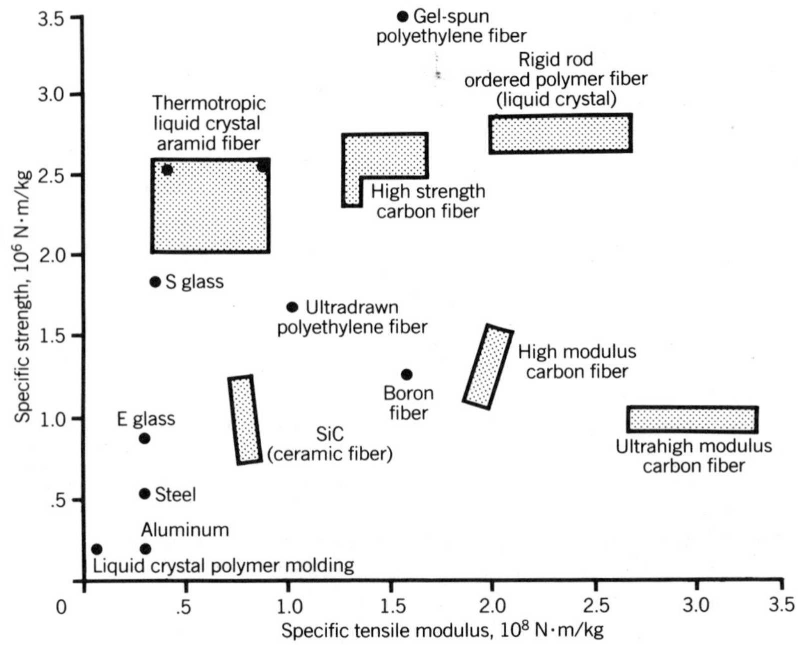

Figure 1. Comparison of specific strength vs specific modulus for a variety of high-performance materials. Specific properties are normalized by material density (Pa or N/m^2 divided by kg/m^3).

4. With the exception of the thermotropic copolyesters, high modulus polymers are available only as fibers and hence must be used in composite form to produce complex three-dimensional parts.

These polymers are highly anisotropic, and the high modulus is achieved only in the direction of molecular chain orientation. Properties normal to the molecular axis tend to be an order of magnitude or more lower than the longitudinal values, resulting in low compressive and/or shear properties. Compensation for the low shear and compressive properties is accomplished through composite design concepts.

In this article, modulus is taken to mean the tensile modulus as defined by the slope of the initial linear portion of the load extension response (stress–strain curve) of a specimen deformed at room temperature.

There is no single, all-encompassing consistent definition of high modulus polymer. In this article the practical but imprecise "higher than typical" criteria will be employed. Polymeric products with moduli of >70 GPa (10 × 10^6 psi) can be produced through a variety of technologies. In general, the tensile strength of polymers tends to follow the same molecular and structural principles as the modulus. Strength development is complicated, however, by the effects of molecular weight and the statistical probabilities of finding flaws or stress concentrators leading to failure of the specimen (see also MECHANICAL PROPERTIES).

Production

Two fundamentally different approaches to the production of high modulus polymers have emerged. One is the transformation of conventional, flexible-chain random-coil polymers into almost perfectly oriented, extended-chain structures in the solid state. The most successful example of this approach is gel spinning of high molecular weight polyethylene. Other examples include solid-state extrusion, die drawing, and superdrawing. These methods are extensions of fiber processing.

The second approach is the molecular design of stiff, planar molecules that exhibit liquid-crystal behavior in solution (lyotropic) or in the melt (thermotropic) and form highly oriented, extended-chain structures in the solid state. Three important examples are poly(*p*-phenyleneterephthalamide), the lyotropic aramid commercialized first by DuPont as Kevlar; ordered polymers, a group of ultrastiff lyotropic molecules that exhibit the highest modulus values yet observed in organic polymers; and thermotropic, aromatic copolyesters (see POLYESTERS, MAINCHAIN AROMATIC).

High Modulus Structures from Conventional Polymers

The production of high modulus articles from conventional polymers involves fiber-spinning technology in which the control of molecular orientation allows control of mechanical properties of the final fiber. Molecular orientation may be imparted to a fiber during the extrusion (spinning) process or after spinning in a solid-state orientation (drawing) process, and will be permanent if conditions include a faster orientation time than polymer relaxation time (see also ORIENTATION PROCESSES).

Liquid Crystalline Polymers

The liquid crystalline nature of stiff polymer molecules in solution was predicted by Flory in 1947, and the concept was verified by the synthesis of aromatic polyamides (aramids).

In addition to the main-chain nematics, which are inherently high modulus in the solid state, mesogenic polymers exhibiting smectic and cholesteric phases have been synthesized.

Four categories can be recognized:

Rodlike

aromatic polyamides, esters, azomethines, benzobisoxazoles

Helical

polypeptides, nucleotides, cellulosics

Side-chain mesogenic (comb polymers)

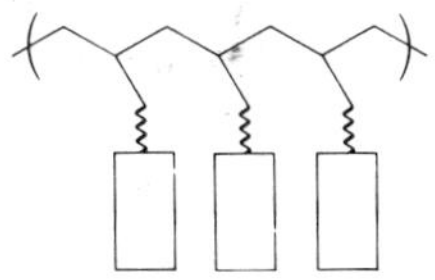

Block copolymers with alternating rigid and flexible units

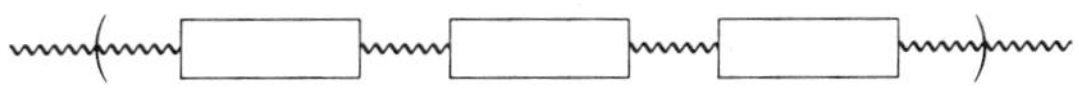

Polymers exhibiting nematic behavior are highly aromatic, utilize linear linkages of restricted rotation, eg, esters, amides, or azomethines, and exhibit a high aspect ratio and planar conformations in the melt or solution, rather than the random coils typically associated with polymer molecules.

The advantages of nematic liquid-crystalline polymers include both processing ease and solid-state properties. Tractability is a problem in liquid-crystal polymers; the molecular structures are so perfect that crystals that form in the solid state decompose before melting. The two predominant approaches to achieving tractability in liquid-crystal polymers are solution processing (lyotropic systems) and copolymerization to lower the melting point to a usable range (thermotropic polymers).

Thermotropic Polymers. Thermotropic copolyesters are perhaps the most versatile class of high modulus, mesogenic polymers amenable to fabrication.

In 1985, both the Dart division of Kraft, Inc. and the Celanese Corporation announced decisions to proceed toward commercialization of their thermotropic copolyester compositions, primarily as plastics designed for injection-molding and shaped-extrusion applications. The Dart material is similar to the Excel polymer developed by Carborundum in the 1970s. The Celanese polymer is a proprietary composition developed internally. Both companies offer a broad range of compounded resins, designed for a variety of specific uses. The properties of these resins are summarized and contrasted in Table 1.

Lyotropic Polymers

Stiff, essentially linear polymers, which decompose before melting, must necessarily be processed into useful products from solution. Considerations of mass transfer during solution processing require that the extrudates possess at least one very thin dimension; hence, these processing methods are applicable to fiber, or perhaps film, formation only. A typical polymer of this type is the aramid, poly(*p*-phenyleneterephthalamide) (PPT), commercialized by DuPont as Kevlar (see POLYAMIDES, AROMATIC). In solution, PPT is lyotropic; with the proper choice of solvent (commonly 100% sulfuric acid concentration) and molecular weight, these polymers form nematic liquid crystals.

Polymer Blends and Molecular Composites

An interesting approach to high modulus polymers is the blending of stiff and flexible polymers to create *in situ* composites. Blends range from ca 80/20–20/80 stiff/flexible. Properties follow the rule of mixtures; phase separation and poor adhesion between phases are clearly evident. Systems investigated include both lyotropic and thermotropic polymers as the stiff phase. Most descriptions of these materials appear in the patent literature.

Molecular composites use truly rodlike polymer, eg, poly(*p*-phenylene benzobisthiazole) to reinforce a flexible polymer such as a nylon in the same way that carbon or glass fiber is used to reinforce engineering plastics.

Applications

The most important market sectors include high-performance composites (for aerospace, sporting goods, etc), ballistic protection devices, tire reinforcement, and ropes and cables. High modulus polymers are also used in electronics, telecommunications, hostile environment applications, and other specialty markets where cost is not critical.

BIBLIOGRAPHY

A. E. Zachariades and R. S. Porter, eds., *The Strength and Stiffness of Polymers,* Marcel Dekker, Inc., New York, 1983.

Table 1. Injection-molded Properties of Thermotropic Copolyesters

Property	Vectra[a]	Zydar[b]
tensile strength, MPa[c]	140–240	120
tensile modulus, GPa[d]	9.6–38	9.6
break elongation, %	1.2–6.9	
flexural strength, MPa[c]	150–30	130
flexural modulus, GPa[d]	9–34	
notched Izod, J/m[e]	53–534	240–300
dielectric constant	2.6–3.3	2.9–3.7
dissipation factor	0.03–0.004	0.04–0.006
arc resistance, s	63–180	138
dielectric strength, 1.6 mm, kV/mm	31–43	31
melting point, °C	275–330	421
HDT, 1.8 MPa[c], °C	180–240	337–355

[a] Celanese.
[b] Dart.
[c] To convert MPa to psi, multiply by 145.
[d] To convert GPa to psi, multiply by 145,000.
[e] To convert J/m to ft·lb/in., divide by 53.4.

G. W. Calundann and M. Jaffe, *Proc. Robert A. Welch Found. Conf. Chem. Res.* **26,** 247 (1982).

A. Cifferri and I. M. Ward, eds., *Ultra-High Modulus Polymers,* Applied Science Publishers, Barking, UK, 1979.

MICHAEL JAFFE
Hoechst-Celanese

HIGH-SPEED TESTING.

See TEST METHODS.

HIGH-TEMPERATURE STABILITY.

See HEAT-RESISTANT POLYMERS.

HISTORY OF POLYMER SCIENCE

Origin of the Term Polymer

In 1826, Faraday reported the fractionation of a liquid that separated from illuminating gas under pressure. The most volatile component was a gas with the elemental composition of ethylene but with twice the density. This was puzzling because it was believed that elemental composition defined a compound.

The substance, now known as butene, was at first called by Berzelius an isomer of ethylene; later, he suggested the term polymer for any compound with a molecular weight that was a multiple of the molecular weight of another compound with the same elemental composition. The first addition polymerizations were reported for chloral in 1832 and for styrene in 1839, but the nature of these reactions was not understood.

A lecture, delivered by Berthelot to the Chemical Society of Paris in 1863, contains the first general discussion of polymerization in the literature. Berthelot isolated the dimer, trimer, and tetramer of pentene and, from combustion data, estimated the heat of dimerization and tetramerization. He observed that "all compounds able to add hydrogen, hydrogen chloride, water, must be able to add molecules identical to themselves" and since the unsaturation is preserved in such an addition, he believed there was no limit to the chain length that might be produced by repetition of the process. He discovered that BF_3 reacts with a monomer, which induces neighboring molecules to polymerize with more heat evolution, so that "once the process is started it spreads like a fire."

Early Studies of Natural Macromolecules

Long before the purposeful preparation of synthetic polymers, chemists faced the problem of the nature of cellulose, starch, rubber, and proteins. Rubber proved especially important for the development of early ideas, being the only natural macromolecule that apparently could be reconstituted after a pyrolysis to small molecules. Faraday reported in 1826 that rubber is a hydrocarbon with a C:H ratio of 6.812. Williams emphasized that "the composition of caoutchouc coincides with that of isoprene" and that heat disrupts "a polymeric body into substances having a simple relation to the parent hydrocarbon." Harries believed that rubber consisted of dimethylcyclooctadiene residues associated by "secondary valence bonds" between the carbon–carbon double bonds. This interpretation was challenged by Pickles, who noted that bromination, which eliminates the double bonds, does not appear to degrade rubber. He proposed for the first time that molecules of Hevea rubber are made by 1,4-addition of isoprene residues (see also ISOPRENE POLYMERS).

In 1811, Gay-Lussac and Thenard pointed out that the elemental composition of certain natural products corresponds to an equal number of carbon and water molecules and suggested the term "carbohydrates." The hydrolysis of starch, the most intensively studied carbohydrate, to "a sugary syrup" was described in 1815, a year after the discovery of the starch–iodine color reaction.

In 1839, extraction of wood with dilute nitric acid yielded a carbohydrate residue that was called cellulose. The early literature refers to "celluloses" that were considered to represent isomers or higher aggregation states of starch.

The hydrolysis of proteins to amino acids was introduced in 1873; progress, however, became significant only when Emil Fischer developed methods for amino acid separation and condensation to polypeptide chains containing as many as 18 amino acid residues, which he thought to be similar to protein molecules. In the early 20th century, electrophoresis (qv) led to the characterization of proteins by their isoelectric points. At the same time, the denaturation of protein was first studied.

The Beginnings of the Polymer Industry

In 1907, Hofmann presented a plan to attempt rubber synthesis to the management of Farbenfabriken in Elberfeld and obtained financial backing for ten years. He first tried to synthesize dimethylcyclooctadiene, which, according to Harries, was the building block of rubber. When this effort failed, he tried, again without success, to polymerize isoprene as described by Bouchardat. At this point, he concluded that it was necessary to synthesize very pure isoprene and devised a six-step synthesis starting with *p*-cresol. In 1912, Farbenfabriken obtained two patents for the polymerization of isoprene in bulk and in emulsion to a product described optimistically as "indistinguishable from natural rubber." During the same period, the polymerization of dienes by alkali metals was discovered almost simultaneously in England and Germany.

In 1862, it was found that silk is soluble in ammoniacal copper solution and solidifies after extrusion through spinnerets. Nitrocellulose fibers were patented in 1883 and applied to textiles by de Chardonnet. The first plastic was nitrocellulose plasticized by camphor in 1870.

The acid-catalyzed condensation of phenol with formaldehyde was first reported in 1872. The industrial production of Bakelite dates from 1910 (see PHENOLIC RESINS).

X-ray Crystallography

In 1917, optical investigation of denitrated nitrocellulose films revealed "cylindrical crystal aggregates." X-ray diffraction showed cellulose, silk, and asbestos to be crystalline, and sharp x-ray diffraction patterns were discovered from bundles of ramie and flax fibers.

The first interpretation of a diffraction diagram in terms of chain molecules was proposed for cellulose in 1926, but encountered skepticism since it contradicted chemical evidence that cellulose is comprised of cellobiose residues. However, Meyer and Mark showed that the crystallographic and chemical evidence are in agreement; they also proposed the structure of silk fibroin.

Other synthetic polymers studied crystallographically before World War II include polychloroprene, polyethylene, and polyisobutene, which was the first polymer for which a helical structure was suggested. A crystallographic study of gelatin gels led to the concept of fringed micelles, according to which polymer chain molecules pass through crystalline and amorphous regions.

Staudinger's Work on Macromolecules

Around 1920, Staudinger, who had previously gained recognition for his work on ketenes and aliphatic diazo compounds, decided to concentrate on "high molecular compounds." His first publication in this field asserted, without experimental proof, that Hevea rubber, polystyrene, and polyoxymethylene consist of long-chain molecules rather than colloidal aggregates of small molecules, as was widely believed. The use of synthetic polymers as models for biological macromolecules was one of Staudinger's most fruitful innovations.

In a monograph published in 1932, Staudinger explained that the inability of polymers such as polystyrene to crystallize was due to their lack of stereoregularity. Their amorphous state accounted for their solubility, which others had claimed to be incompatible with very high molecular weight.

The Ultracentrifuge

In 1925, Svedberg and Fahreus constructed a centrifuge generating 5,000 times the force of terrestrial gravitation and showed that the equilibrium solute distribution in a hemoglobin solution corresponded to particles of a unique size with a molecular weight of 68,000. This was the first unambiguous demonstration that proteins are not colloidal aggregates, but contain molecules of very high molecular weight.

Polycondensation

The work of Carothers led to a profound change in the understanding of polycondensations. Starting in 1928 in the DuPont laboratories, he used polycondensations to obtain the longest possible chain molecules. He realized that formation of cyclic monomers or dimers may compete with the production of chain molecules and determined the dependence of the stability of cyclic compounds on ring size by the yield of cyclic esters and anhydrides in the thermal decomposition of polyesters and polyanhydrides.

When Flory joined Carothers' laboratory, he derived an expression for the distribution of chain lengths in polycondensation, establishing the independence of reactivity of a functional group from the size of the molecules.

Free-radical Polymerization

In his first paper, "On Polymerization," Staudinger wrote that "numerous molecules assemble until an equilibrium between the large molecules . . . is obtained." Initially, it was suggested that chains of vinyl monomers grow by hydrogen migration.

$$\mathrm{CH_3\overset{R}{\overset{|}{C}}H(CH_2\overset{R}{\overset{|}{C}}H)_{\mathit{n}}CH{=}CHR + CH_2{=}CHR \longrightarrow CH_3\overset{R}{\overset{|}{C}}H(CH_2\overset{R}{\overset{|}{C}}H)_{\mathit{n}+1}CH{=}CHR}$$

Although Staudinger and Kohlschutter assumed "free valences" at the ends of propagating chains, papers published in the early 1930s mostly referred vaguely to "activation of monomers." The existence of free radicals generated from peroxides was still controversial as late as 1937.

For a long time it was widely believed that the length of polymer chains was limited by a decrease of reactivity with an increase in chain length. Flory introduced the concept of chain transfer; others showed that the increase in polymerization rate in some vinyl polymerizations at higher monomer conversion is due to reduced chain termination in very viscous media.

In the first published copolymerization study, it was reported that maleic anhydride and stilbene copolymerize, even though neither substance polymerizes alone. Later, Marvel proved copolymer structures by chemical degradation. Staudinger concluded that there is no relation between the ease with which a monomer homopolymerizes and copolymerizes.

When a theory of the dependence of a binary copolymer on the composition of the monomer mixture was first developed, an error was made in assuming that the relative number of chains terminating with the two monomer residues equals the relative concentration of these residues in the copolymer. Later, it was shown that the copolymer composition can be predicted from reactivity ratios of the propagating radicals with the monomers.

A theory of chain-transfer agents was formulated in 1945. Formation of a graft polymer was first accomplished by polymerization of *p*-chlorostyrene in the presence of poly(methyl methacrylate).

Polymerizations in which water-insoluble monomers were dispersed in aqueous media led to patents claiming polymers of finely divided insoluble solids as "protective colloids," preventing the coalescence of the polymerizing droplets. The high rates in emulsion polymerization were explained in the work on synthetic rubber during World War II.

The first detailed study of the polymerization of a crystalline vinyl monomer was reported in 1960. The depolymerization (qv) of vinyl polymers was shown to be a chain reaction. In the presence of free radicals, polymerization and depolymerization lead to the same equilibrium state.

Ionic and Coordinate Polymerization

Berthelot described polymerization catalyzed by acids more than a century ago, and at about the same time the base-catalyzed polymerization of styrene was discovered.

Anionic polymerization became important with the discovery of the alkali metal-catalyzed conversion of dienes to rubbers. In 1934, the polymerization of butadiene initiated by butyllithium led Ziegler and co-workers to the conclusion that both 1,2 and 1,4 additions take place.

The nature of cationic polymerization was first suggested in 1934 and the role of water as a co-catalyst was demonstrated in 1946.

Because the addition of ethylene to alkyllithium compounds produced insoluble LiH when olefins were split off, Ziegler began to study the addition of ethylene to $LiAlH_4$. Following a series of experiments yielding unexpected results, Ziegler found that ethylene could be polymerized at atmospheric pressure and ambient temperature in a heterogeneous catalytic system to give unbranched chains. In describing the sequence of events that led to his epochal discovery, Ziegler stressed in his Nobel Prize address that he never intended to perform a search for a new polyethylene synthesis.

In 1954, Natta attempted the polymerization of propylene by the Ziegler process, hoping that a rubbery product would be obtained. When the resulting polymer was fractionated, a highly crystalline portion was isolated. A short note claimed the synthesis of crystalline polypropylene, polybutene, and polystyrene and specified that the repeat period of polypropylene contains three monomer residues. These polymers were assigned a stereoregular structure for which the designation "isotactic" was proposed.

As early as 1955, Natta's group reported chain conformations in isotactic vinyl polymer crystals and proved the versatility of coordination polymerization by the preparation of the first syndiotactic polymer, of butadienes with a prevalent 1,4-cis- or 1,4-trans-monomer addition, and of an amorphous high molecular weight rubbery polypropylene.

Ring-opening Polymerization

Trialkyloxonium salts were discovered in the 1930s, and later used as catalysts for the polymerization of tetrahydrofuran. This was surprising because five-membered rings were not expected to polymerize.

In 1906, Leuchs discovered that small amounts of water lead to evolution of CO_2 from glycine *N*-carboxyanhydride with formation of a water-insoluble product later characterized as a glycine polypeptide.

Proteins and Nucleic Acids

The discovery that crystals of globular proteins yield highly resolved x-ray diffraction patterns proved that they must contain molecules having not only the same amino acid sequence but also a closely defined chain conformation. For some time, it was believed that the amino acid sequence exhibited periodicities, but this was disproved when Sanger determined the amino acid sequence of insulin, a protein containing an 18-member and a 30-member chain with one intrachain and two interchain disulfide bonds.

In 1951, Pauling and co-workers proposed a helical structure with intramolecular hydrogen bonds for the α-form of fibrous proteins and predicted that this α-helix would be an important structural feature of globular proteins. The suggestion that the folding of the polypeptide chain in native globular protein minimizes the free energy and is determined by the amino acid sequence was substantiated by the recovery of the native form after protein denaturation.

Determination of the structure of hemoglobin by x-ray crystallography was made possible by the discovery that heavy metal atoms could be attached to specific sites of the protein molecule without a change in the crystal structure, and by the development of powerful computers. Low-resolution structures of myoglobin and hemoglobin were published in 1958 and 1960.

In 1944 it was observed that exposure of bacteria to a cell-free solution of deoxyribonucleic acid (DNA) from another bacterial strain produces a change in their hereditary character. Chargaff found that the DNAs from different organisms have different nucleotide compositions, and that in any given DNA the nucleotides are unevenly distributed along the chain.

After Chargaff's finding that the adenine–thymine and the guanine–cytosine ratios are always close to unity, Watson and Crick proposed the DNA double helix and conjectured that in DNA replication each strand of the double helix acts as a template governing the nucleotide sequence in a newly synthesized complementary strand.

Spectroscopy

The utility of uv spectroscopy in the study of styrene–butadiene copolymers was demonstrated in 1946. Infrared absorption spectroscopy was successfully applied to the estimation of the degree of branching and content of carbonyl groups in low-density polyethylene, and to the quantitative determination of the relative frequency of 1,2- and 1,4-*cis*- and 1,4-*trans*-butadiene incorporation into polymer and copolymers. Ir spectroscopy provided an easy method for distinguishing the α and β forms of polypeptides and showed that intramolecular hydrogen bonding is retained in solution. Nuclear magnetic resonance (nmr) spectroscopy and electron spin resonance (esr) spectroscopy found valuable applications in polymer studies.

Polymer Solutions

Staudinger's theory of intrinsic viscosity was based on the assumption that chain molecules behave like rigid rods. Assuming free rotation around the bonds in chain molecules, Kuhn derived the probability distribution of their end-to-end displacement and concluded that intrinsic viscosities $[\eta]$ can be understood if the solvent is trapped within the coiled macromolecule and the chain is expanded beyond the predicted dimension due to an excluded volume effect. Later, a study of the hydrodynamic behavior of macroscopic randomly coiled model chains showed that the dependence of $[\eta]$ on chain length reflects the extent to which a solvent can pass through the coil. It was assumed that this is also the case with chain molecules; the excluded volume effect was not considered. Flory's analysis of the effect of polymer–solvent interaction on the dimensions of flexible chain molecules proved that under most conditions of interest, chain molecules behave like impermeable coils whose dimensions can be deduced from $[\eta]$ values.

Polymers in Bulk

The glass transition (qv) in polymers was first studied on a polyisobutene sample with M_n 4900, which led to the conclusion that T_g corresponds to a viscosity of ca 10 TPa (10^{13} P), as in fluids composed of small molecules. However, later studies showed that T_g approaches an asymptotic limit with increasing chain length, being governed by the free volume rather than the melt viscosity.

The prediction that the conformation of molecular chains in a polymer crystal corresponds to the energy minimum of the isolated chain was confirmed for most, but not all, cases. Polymer single crystals and the folding of chain molecules in these crystals were discovered almost simultaneously in three laboratories. This folding explained the previously observed long periodicities unrelated to unit-cell dimensions and the structure of polymer spherulites.

Early treatments of the mechanical behavior of polymers were summarized in monographs published in 1948 and 1952. The activation energy for viscous flow was found to be independent of the length of high polymer chains, suggesting that motion of chain segments is involved. At constant temperature, plots of the logarithm of melt viscosity against the logarithm of chain length exhibited an abrupt change of slope, which was

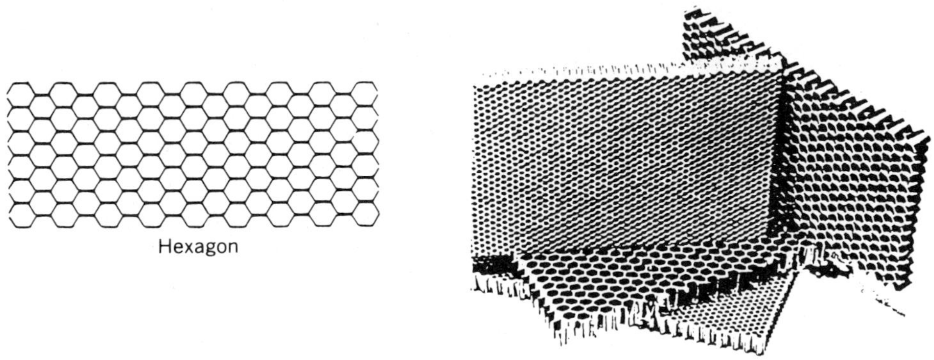

Figure 1. Honeycombs.

interpreted as marking the point at which chain entanglements control the flow rate.

The viscoelasticity of a polymer was first studied in connection with the creep recovery of polystyrene fibers. The behavior of plasticized poly(vinyl chloride) demonstrated the applicability of the Boltzmann superposition principle. The dielectric dispersion of this material and the viscoelasticity of rubber showed that polymeric materials must be characterized by a broad distribution of relaxation times.

BIBLIOGRAPHY

H. Morawetz, *Polymers: The Origin and Growth of a Science,* John Wiley & Sons, Inc., New York, 1985.

P. J. Flory, *Principles of Polymer Chemistry,* Cornell University Press, Ithaca, N.Y., 1953, Chapt. 1.

H. Morawetz, *Angew. Chem. Internat. Ed.* **25,** 93 (1967).

HERBERT MORAWETZ
Polytechnic University

HOBBING. See MOLDS.

HOLLOW-FIBER MEMBRANES. See FIBERS, ENGINEERING; MEMBRANES.

HOLOCELLULOSE. See CELLULOSE; WOOD.

HOMOELEMENT POLYMERS. See INORGANIC POLYMERS.

HOMOGENEOUS POLYMERIZATION. See ANIONIC POLYMERIZATION; BULK POLYMERIZATION; CATIONIC POLYMERIZATION; COORDINATE POLYMERIZATION; RADICAL POLYMERIZATION; SOLUTION POLYMERIZATION; STEP-REACTION POLYMERIZATION.

HOMOPOLYMERIZATION

The IUPAC defines a homopolymer as a polymer derived from one species of monomer ["Basic Definitions of Terms Relating to Polymers," *Pure Appl. Chem.* **40,** 479–491 (1974)]. Homopolymerization is then polymerization in which a homopolymer is formed.

HONEYCOMBS

Honeycomb consists of very thin sheets attached to form cells (Fig. 1). The basic cell shape is the hexagon; however, there are many other configurations, such as overexpanded, reinforced, flex-core, and square. Honeycomb can be manufactured from just about any thin-sheet material. Some of the more common metallic cores are aluminum and stainless steel, while fiber glass, Nomex, and Kraft paper are very popular nonmetallic core materials.

Manufacture

There are two major methods in manufacturing honeycomb: the expansion process and the corrugation process, as shown in Figures 2 and 3.

The expansion procedure for metallic cores consists of printing adhesive lines on foil sheets, cutting and stacking them, curing the adhesive in a press, sawing slices, and finally ex-

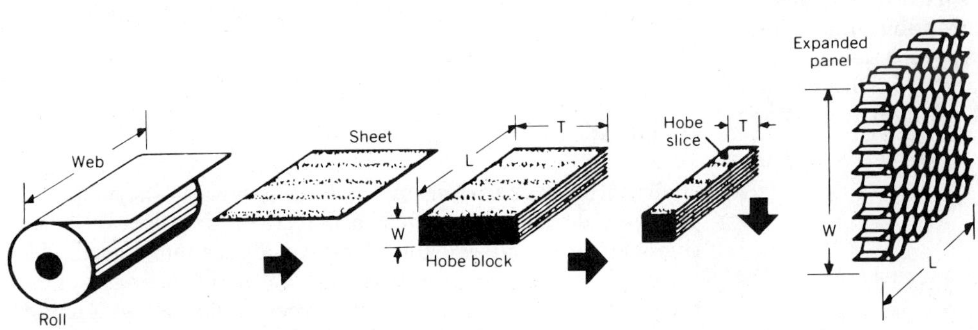

Figure 2. Expansion process.

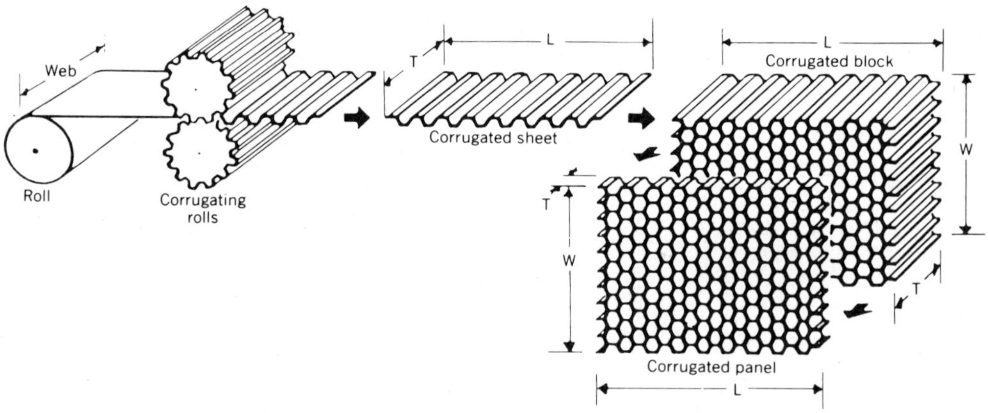

Figure 3. Corrugation process.

Table 1. Cell Size and Densities of Honeycombs

Core	Cell size, mm	Density, kg/m³	Process or property
aluminum	1.59–25.4	16–192	expanded process
aluminum	3.18–6.35	192–881	corrugation process
stainless steel	3.18–12.7	32–320	high temperature resistance
fiber glass	3.18–9.53	16–192	nonmetallic
Nomex	3.1–9.53	16–144	impact resistant
Kraft paper	6.35–38.1	16–144	low cost

panding the Hobe slices. This procedure is slightly different for nonmetallic honeycomb. Here the core does not retain its shape after expansion; consequently, it must be held expanded in a rack and the slight amount of resin in the web material heat-set in an oven. Then the block is dipped in a liquid resin and oven-cured. This dipping–curing cycle is repeated until the honeycomb block is at the desired density.

In the corrugation method, the sheets are first corrugated to the desired shape and adhesive is applied to the nodes. The sheets are then stacked, cured in an oven, and cut into slices. Instead of adhesive being used to connect the nodes, some cores, such as stainless steel have brazed or spot-welded nodes. Cell sizes and density ranges of common core types are given in Table 1.

Applications

For structural applications, honeycomb is used in conjunction with thin facing sheets attached by adhesive bonding, brazing, or diffusion bonding. The resulting structure is extremely lightweight with the highest stiffness-to-weight and strength-to-weight ratios possible. In designs of this type, it is assumed that the facings take the bending and axial loads and that the core takes the shear load, provides shear rigidity, and stabilizes the skins.

These honeycomb sandwich panels are used mainly in secondary structural aircraft elements where minimum weight, aerodynamic smoothness, and high fatigue resistance are required. They are also used in flooring panels, partitions, military shelters, and other applications where light weight and stiffness are desired.

BIBLIOGRAPHY

Mil HDBK-23A, Structural Sandwich Composites, 1968.

Mil Std-401B, Sandwich Construction and Core Materials, 1967.

TB 120, Mechanical Properties of Hexcel Honeycomb, Hexcel Corp., Dublin, Calif., 1980.

TSB 124, The Basics on Bonded Sandwich Structures, Hexcel Corp., Dublin, Calif., 1982.

T. N. Bitzer
Hexcel Corporation

HORTICULTURAL APPLICATIONS.

See Agricultural applications.

HOT SHAPING.

See Thermoforming.

HOT STAMPING.

See Decorating.

HYBRID POLYMERS.

See Polymer blends.

HYBRID SHEETS.

See Composites.

HYDRAZIDES, POLYMERIC.

See Polyhydrazides and polyoxadiazoles.

HYDROCARBON RESINS

Hydrocarbon resins are low molecular weight thermoplastic polymers derived from cracked petroleum distillates, turpentine fractions, coal tar, and a variety of pure monomers. They are used in adhesives, printing inks, hot-melt coatings, rubber, protective coatings, flooring, and other applications to modify other materials; they are only rarely used alone. The average molecular weight is usually below 2000, and physical forms range from viscous liquids to hard, brittle solids. Hard resins are sold commercially in flaked form, solid in drums, or in solution according to customer demand. Some resins are also available in the form of anionic, cationic, or nonionic aqueous emulsions. Colors range from water-white resins produced from pure vinylaromatic monomers, to pale yellow and amber, to dark brown inexpensive petroleum by-product resins.

Terpene resins are similar to petroleum resins and have been used since the mid-1930s, particularly in adhesives. They are polymers of α-pinene, β-pinene, and d-limonene of citrus origin, or dipentene mixtures from sulfate turpentine. Terpene–phenol resins are also produced.

Resins from pure monomers are a recent development, and are made from styrene, α-methylstyrene, vinyltoluene, isobutylene, and compounds with related structures. They are very light in color, frequently water-white. Dicyclopentadiene is an important monomer obtained from cracked petroleum streams.

Properties

Certain properties such as the softening point, color, color stability, or solubility ("solubility contours") determine the performance characteristics of the resin. Physical properties are related to more fundamental properties, eg, molecular weight distribution and aliphatic vs aromatic character. Manufacturers frequently strive to tailor-make a resin for a specific need by addition of selected comonomers and adjustments in polymerization conditions.

Coumarone–Indene or Coal Tar Resins

Sulfuric acid was first used to catalyze the formation of coumarone–indene resins, but higher melting, lighter-colored resins can be obtained with aluminum chloride or boron trifluoride and its complexes as initiator.

Coumarone–indene resins with softening points 100°C and higher have been used in coatings with film-forming materials such as drying oils. They have also been employed as process aids and pigment-dispersing agents in the compounding of natural and synthetic rubber. BF_3-polymerized terpolymers of α-methylstyrene, indene, and vinyltoluene with softening points of 92–108°C have been recommended for alkyd-based coatings because of improved compatibility compared with resins based only on α-methylstyrene and vinyltoluene (see also ALKYD RESINS). Polyindene resins reportedly improve the flexural modulus and processibility of high molecular weight poly(vinyl chloride), polystyrene, and other thermoplastic resins.

Table 1. Properties of a Typical C_5 Aliphatic Olefin–Diolefin Resin

color	
Gardner scale, 50% in toluene	8
USDA Rosin scale	WG
softening point, ring and ball, °C	100
refractive index, n_D^{20}	1.53
specific gravity at 15.5°C	0.98
specific heat, 0–100°C, kJ/kg	11.06
bromine number	45
saponification number	<1
acid number	<1

Table 2. Properties of a Typical High Melting Aromatic Petroleum Resin

Property	Value
color, Gardner (50% in toluene)	11
softening point, ring and ball, °C	140
specific gravity, 25°C	1.08
acid number	<1
bromine number	21
melt viscosity, Pa·s[a]	
at 245°C	0.1
at 205°C	1.0
at 180°C	10

[a] To convert Pa·s to poise, multiply by 10.

Petroleum Resins

Petroleum resins are low molecular weight, thermoplastic hydrocarbon resins derived from cracked petroleum fractions. They are distinguished from higher molecular weight polymers, eg, polyethylene, polypropylene, or polystyrene, which are made from almost pure starting materials. They have essentially replaced coumarone–indene resins in all applications. Large volume uses for petroleum resins are in the rubber industry, and in printing inks, adhesives, and coatings.

C_5 Aliphatic Petroleum Resins. Some properties of a typical C_5 aliphatic olefin–diolefin resin are given in Table 1.

C_9 Aromatic Petroleum Resins. Some properties of typical high melting C_9 aromatic petroleum resins are given in Table 2.

Dicyclopentadiene Resins. Some properties of a typical cyclic diolefin resin are given in Table 3.

Terpene Resins

Terpene resins are obtained by polymerizing various terpene hydrocarbons, most of which occur naturally.

Polymerization. The most important terpene resins are those made from α-pinene, β-pinene, and limonene or dipentene by cationic polymerization. Less important terpenes include β-phellandrene, myrcene, 3-carene, camphene, terpinolene, and various terpinene isomers.

Copolymers. Terpenes have been copolymerized with other terpenes and with a variety of nonterpene monomers. A process for the preparation of resins with low softening points (0–40°C) by $AlCl_3$ polymerization of a blend of β-pinene, dipentene, and terpene oligomers (oily dimers and trimers) has been disclosed.

Resins from Pure Monomers

Thermoplastic resins from high purity monomers, eg, styrene, substituted styrenes, isobutylene, and the like, have colors that are Gardner 1 or lighter, and are often water-white. Styrene

Table 3. Properties of a Typical Cyclic Diolefin Resin

Property	Value
color, Gardner (50% in toluene)	11
softening point, ring and ball, °C	100
specific gravity, 15.5°C	1.11
acid number	<1
bromine number	55
iodine number	180
refractive index, 25°C, n_D^{20}	1.58

and alkylsubstituted styrenes such as vinyltoluene and α-methylstyrene respond well to Friedel-Crafts polymerization initiators. For example, a process for continuous production of α-methylstyrene homopolymers using boron trifluoride, ethylaluminum dichloride, or diethylaluminum chloride as initiator have been described.

Uses

Hydrocarbon resins are used in a wide variety of materials including adhesives, printing inks, hot-melt coatings, rubber articles, caulks and sealants, floor coverings, textiles, paints and varnishes, and plastics and chewing gum. They are rarely the sole component in a particular application, but rather are compounded with elastomers, plastics, waxes, alkyds, and oils to provide special properties for a particular use.

For convenience of handling, hydrocarbon resins are available in various physical forms including solid, flake, crushed, powder, bead, molten, water emulsion or dispersion, and solution in solvents or oils.

BIBLIOGRAPHY

J. F. Holohan, Jr., J. Y. Penn, and W. A. Vredenburgh in M. Grayson, ed., *Kirk-Othmer Encyclopedia of Chemical Technology,* 3rd edition, Vol. 12, John Wiley & Sons, Inc., New York, 1980, pp. 852–869.

WALTER VREDENBURGH
KENDALL F. FOLEY
ARTHUR N. SCARLATTI
Hercules, Inc.

HYDROCHLORIDE RUBBER. See RUBBER DERIVATIVES.

HYDROGELS

Hydrogels, or water-containing gels, are polymers characterized by hydrophilicity and insolubility in water. In water, they swell to an equilibrium volume, but preserve their shape. The hydrophilicity is due to the presence of water-solubilizing groups, such as —OH, —COOH, —CONH—, —SO_3H, etc. The insolubility and stability of shape are due to the presence of a three-dimensional network. The swollen state results from a balance between the dispersing forces acting on hydrated chains and cohesive forces that do not prevent the penetration of water into the network. Cohesive forces are most often due to covalent cross-linking. Others are electrostatic, hydrophobic, or dipole–dipole in character. The degree and nature of cross-linking and the tacticity and crystallinity of the polymer are responsible for its characteristics in the swollen state. The ability to imbibe water and ions without the loss of shape or mechanical strength is valuable in many natural hydrogels, such as those found in muscles, tendons, and cartilage.

Natural hydrogels are used in pulp and paper production, artificial silk, cellulosic membranes, and biomedical applications. Synthetic hydrogels are used in prosthetic materials, soft lenses, and membranes for controlled drug release because of their compatibility with living tissue (see also CONTACT LENSES; DRUG APPLICATIONS). The gels are fixed to supports by coating, grafting, or chemical modification.

Classes

Hydrogels Based on Methacrylic and Acrylic Esters. The principal monomer is

where R = H or CH_3. To form a gel, the monomer must be copolymerized with a cross-linking agent, most often ethylene dimethacrylate (EDMA), formerly called ethylene glycol dimethacrylate (EGDMA).

This class of hydrogels includes not only hydrogels based on hydroxyalkyl methacrylates or acrylates but also copolymers of hydroxyalkyl methacrylates with enhanced hydrophilicity, copolymers with ionogenic comonomers, copolymers with low hydrophilicity and superior mechanical properties, copolymers with hydrophilic and hydrophobic constituents, copolymers incorporating presynthesized macromers or oligomers, and hydroxyalkyl methacrylates and copolymers with ordered structures, as well as hydrogels not based on hydroxyalkyl esters of methacrylic acid, eg, polymers based on 2-(*N*,*N*-dimethylamino)ethyl methacrylate (DMAEMA).

Acrylamide (Methacrylamide) Hydrogels. Acrylamide-based hydrogels differ significantly from those based on 2-hydroxyethyl methacrylate.

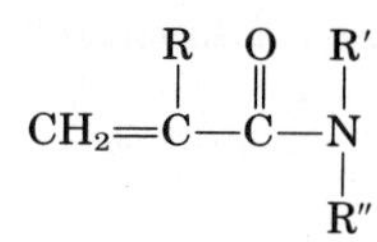

They are not limited in respect to water content, but they lack hydrolytic stability. This disadvantage is not present if R′ and R″ are alkyl or hydroxyalkyl. Acrylamide hydrogels are usually cross-linked with *N*,*N*′-methylenebisacrylamide (MBAA). This class of hydrogels includes hydrogels with unsubstituted amide groups, acrylamide hydrogels with ionogenic comonomers, and hydrogels with substituted amide groups.

Hydrogels Based on *N*-Vinyl-2-pyrrolidinone. *N*-vinyl-2-pyrrolidinone (NVP), a lactam, is soluble in water and many other solvents.

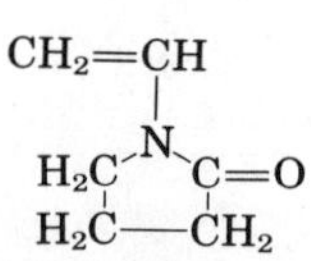

Its strong interaction with water creates advantages and disadvantages with respect to hydrogel formation. Homogeneous cross-linked polyNVP is difficult to prepare and low in mechanical strength. Cross-linking agents include *N*,*N*′-methylenebisacrylamide, methylenebis(4-phenyl isocyanate), ethylene dimethacrylate, divinylbenzene, and allyl methacrylate.

This class of hydrogels includes *N*-vinyl-2-pyrrolidinone copolymers containing less hydrophilic comonomers, *N*-vinyl-2-pyrrolidinone copolymers containing ionizable groups, *N*-vinyl-2-pyrrolidinone copolymers containing macromers or oligomers, and *N*-vinyl-2-pyrrolidinone copolymers containing hydrophobic comonomers.

Polyelectrolyte Complexes and Charged Hydrogels. ***Complexes with Ionizable Groups of Opposite Signs on Polymer Chains.*** Polyelectrolyte complexes (qv) represent only one type of hydrogel cross-linked by ionic bonds. These bonds exist in polyelectrolyte complexes between pairs of charge sites on anionic and cationic polyelectrolyte chains. Polyelectrolyte complexes are easily prepared with an excess of one ionic species or by changing the ratio of anionic and cationic components. These hydrogels swell in water, but are prevented from dissolving by the ionic bonds.

Polyelectrolytes or Ionomers Cross-linked with Polyvalent Ions. Some complexes contain only one type of polyelectrolyte chain, which is ionically bound by small polyvalent ions. This type of polyelectrolyte complex usually contains poly(acrylic acid) or poly(methacrylic acid), or their copolymers with acrylamide or methacrylamide.

Charged Composite Hydrogels. Charged gels are also prepared as polymer composites containing a nonionic matrix, a charged component carrying ionizable groups, and a cross-linking agent. Old formulations used collodion as the matrix and 3:1 alcohol–ether as solvent. Composite gel membranes made of polymeric matrix materials, such as polyacrylonitrile, poly(vinylidene fluoride), poly(vinylidene chloride), polyoxyphenylene, and polyesters, etc, with poly(styrenesulfonic acid) or poly(1-vinyl-3-methylimidazolium iodide), are a more recent development.

Covalently Cross-linked Polyelectrolyte Networks. Pure poly(acrylic acid) and poly(methacrylic acid) networks cross-linked with ethylene dimethacrylate, as well as partly hydrolyzed polyacrylonitrile or poly(methyl acrylate-*co*-vinyl acetate-*co*-ethylene-*co*-divinylbenzene) etc, were studied.

Charged Hydrogels Containing Presynthesized Sequences. Copolymers of hydrophilic monomers with presynthesized oligomers or macromers are increasing in importance, as are charged gels. Acrylic acid is copolymerized in alkaline media with α,ω-(2,3-epoxypropyl) polyoxyethylene to yield hydrogels with extraordinary water absorption.

Cross-linked Poly(vinyl alcohol) Gels. Poly(vinyl alcohol) (PVA) is a water-soluble film-forming polymer with corresponding applications. A hydrogel can be formed by cross-linking it covalently; it contains no domains capable of hydrophobic interactions or dissociation. Nevertheless, because of its polar character, dissolved PVA can be stabilized by treatment with Cr^{3+}.

There are many possibilities for covalently cross-linking PVA. Some of the simplest methods use formaldehyde or glutaraldehyde (see VINYL ALCOHOL POLYMERS).

Hydrogel Surface Coatings. Hydrogel surface coatings are obtained by etching a polymer surface, by dipping an object into a solution of hydrogel, which becomes a fixed film, or by grafting hydrophilic substances onto hydrophobic objects chemically or by irradiation.

Heterochain Hydrogels. In the hydrogel systems described above, vinyl groups give rise to the main polymer chain. Heteroatoms occur only in side chains and cross-links. Hydrogels containing heterochain segments are among the most promising with respect to mechanical properties, eg, sulfonated polysulfone, polyether–urethane hydrogels.

Polymerization and Processing

Hydrogels are usually formed by copolymerization and cross-linking.

Hydrogel soft contact lenses must be given an extremely precise shape. Dry gels (xerogels) can be machined on a lathe, or gels can be polymerized by spin-casting with simultaneous cross-linking.

Properties

The most important property of hydrogels is the ability to imbibe water. The resulting osmotic swelling is opposed by the elastic contractility of the stretched hydrogen network.

Health and Safety Factors

When monomeric hydroxyalkyl acrylates are handled, contact with eyes and skin should be avoided. If it occurs, washing with water usually prevents ill effects. Acrylates are absorbed through the skin in toxic amounts. Polymerization and cross-linking suppress toxicity. Monomeric acrylamide and *N,N'*-methylenebisacrylamide are eye and skin irritants. They are neurotoxic and should be handled in a ventilated area with rubber gloves and a dust respirator. Methacrylamides and substituted methacrylamides are less toxic or nontoxic; methacrylamide and acrylamide polymers are nontoxic. Acrylonitrile is a neurotoxin, and inhalation, body contact, and ingestion must be avoided.

BIBLIOGRAPHY

J. D. Andrade, ed., *Hydrogels for Medical and Related Applications, ACS Symposium Series 31,* American Chemical Society, Washington, D.C., 1976.

B. D. Ratner in D. F. Williams, ed., *Biocompatibility of Clinical Implant Materials,* Vol. II, CRC Press, Inc., Boca Raton, Fla., 1981, pp. 145–175.

S. W. Kim, R. Petersen, and J. Feijen in E. Ariens, ed., *Drug Design,* Vol. X, Academic Press, Inc., New York, 1980, pp. 193–250.

M. F. Refojo, *Surv. Ophthalmology* **26,** 257 (1982).

V. KŮDELA
Institute of Macromolecular Chemistry
Czechoslovak Academy of Sciences

HYDROGENATION

Hydrogenation of polymers is an important method for generating new products, as well as for aiding in polymer characterization and molecular understanding. Hydrogenated materials are typically more stable and their spectra are easier to interpret than those of unhydrogenated analogues. The most commonly hydrogenated (reduced) polymers are those with unsaturated or halogenated backbones.

Hydrogenation of Unsaturated Polymers

Products. ***Homopolymers.*** The structures and properties of hydrogenated polymers depend on the precursor polymers and the extent of hydrogenation. Polymer crystallinities tend to increase with increasing hydrogenation, but glass-transition temperatures increase or decrease, depending on the starting material. Thermal-oxidative stability increases with the extent of hydrogenation and the loss of double bonds.

Hydrogenation of elastomeric 1,4-polybutadiene (1,4-PBD) gives a crystalline polyethylene plastic.

$$-\!(CH_2-CH=CH-CH_2)\!\!\frac{}{2}\!_n \xrightarrow{H_2} -\!(CH_2CH_2CH_2CH_2)\!\!\frac{}{2}\!_n$$

The properties of hydrogenated polybutadiene and low-density polyethylene are compared in Table 1.

Copolymers. Hydrogenation of copolymers, especially block copolymers, is a means of extending the performance range of copolymers. The hydrogenated copolymers are either hydrogenated diene–diene copolymers or hydrogenated diene–aromatic copolymers.

Functional Polymers. The hydrogenation of functional unsaturated polymers presents special problems because many functional groups coordinate with and poison organometallic hydrogenation catalysts or are reduced. In extreme cases, poisoning prevents hydrogenation; in other cases, catalyst efficiencies and polymer conversions are adversely affected.

Processes. Saturated polymers are obtained from unsaturated polymers by catalytic (homogeneous, heterogeneous) and noncatalytic (diimide, hydroboration) methods. Such methods differ in yield, selectivity, side reactions, and catalyst poisoning and removal.

Equipment. For low-pressure hydrogenations, Parr hydrogenators are preferred (Parr Instrument Company, Moline, Illinois). For high-pressure operations, heated thick-walled steel vessels are used. Fixed-bed reactors are designed for hydrogenation of large volumes.

Reduction of Halopolymers

Chloropolymers are hydrogenated by reductive dehalogenation with tri-*n*-butyltin hydride. Dichlorocarbene derivatives of polydienes have been reduced to cyclopropane adducts by this method.

Table 1. Properties of Hydrogenated Polybutadiene and Low-density Polyethylene[a]

Property	Hydrogenated polybutadiene	Low-density polyethylene
tensile strength, MPa[b]	16.2	13.1
elongation, %	750	600
stiffness modulus, MPa[b]	103.4	137.9
brittle point, °C	<−73	<−73
impact strength, Izod	[c]	[c]
refractive index	1.50	1.51

[a] Courtesy of the American Chemical Society.
[b] To convert MPa to psi, multiply by 145.
[c] Does not break.

Characterization and Structure Elucidation

Hydrogenation facilitates the nmr analysis of the sequence distribution of polydiene polymers.

Reductive dehalogenation with (n-C_4H_9)SnH and (n-C_4H_9)SnD has made the elucidation of the microstructure of PVC possible. These methods have also facilitated the study of chain transfer in the polymerization of vinyl chloride monomer.

Hydrogenation is also employed to prepare model polymers, which are used for the study of molecular theories and rheology.

Health and Safety Factors

The primary hazard in the use of liquid or gaseous hydrogen is its extreme flammability in oxygen or air. Solvents for hydrogenation should be pure and inert to the catalysts, precursor polymers, and products, and must not form peroxides.

Economic Aspects

Hydrogenation is commercially competitive for specialized products, especially those not available by direct polymerization. Examples are certain block and functional polymers, Kraton G (Shell), Zeon (Nippon), and hydrogenated NBR (Bayer).

BIBLIOGRAPHY

P. N. Rylander, *Catalytic Hydrogenation in Organic Synthesis,* Academic Press, New York, 1979.

M. Friefelder, *Catalytic Hydrogenation in Organic Synthesis—Procedure and Commentary,* John Wiley & Sons, Inc., New York, 1978.

B. R. Janes, *Homogeneous Hydrogenation in Organic Chemistry,* Reidel Publishing Company, Dorderecht-Holland, 1976.

D. N. Schulz, S. R. Turner, and M. A. Golub, *Rubber Chem. Technol.* **55,** 809 (1982).

DONALD N. SCHULZ
Exxon Research & Engineering Company

HYDROGEN-TRANSFER POLYMERIZATION. See ISOMERIZATION POLYMERIZATION.

HYDROPEROXIDES. See PEROXY COMPOUNDS.

HYDROXYACID POLYMERS. See POLYESTERS.

HYDROXYALKYL ACRYLATE. See ACRYLIC AND METHACRYLIC ESTER POLYMERS.

HYDROXYBENZOIC ACID POLYMERS. See HEAT-RESISTANT POLYMERS.

HYDROXYETHYLCELLULOSE. See CELLULOSE ETHERS.

HYDROXYPROPYLCELLULOSE. See CELLULOSE ETHERS.

I

IDENTIFICATION

Elastomers

Elastomers are comprised of natural rubber and synthetic polymers with rubberlike properties. They may be formulated to produce rubbers with certain characteristic properties, including high resilience; high tensile strength and elongation; resistance to tear, flexing, freezing, and abrasion; and low permanent set. Elastomer composition can range from relatively uncomplicated homopolymer elastomers such as natural rubber to very complex multimonomer copolymers, EPDM elastomers (see also RUBBER, NATURAL; BUTADIENE POLYMERS; ELASTOMERS, SYNTHETIC). The analytical procedures used for the identification of these polymers range from a relatively simple color or flame test to complex, sophisticated procedures using nmr, ir, and pyrolysis–gas chromatography–mass spectrometry (py/gc/ms). No single analytical technique can solve all of the identification problems that might be encountered, especially in rubber blends (see also CHARACTERIZATION OF POLYMERS; CHEMICAL ANALYSIS).

Table 3. Principal Infrared Bands for Selected Elastomers

Elastomer	Principal band, cm^{-1}
polybutadiene	909, 962
ethylene–propylene	1370, 909, 885
butyl rubber	1370, 1390
butadiene–acrylonitrile	2220, 1610, 1590
isoprene–natural rubber	1370, 885
styrene–butadiene	990, 909, 775, 699

Physical and Chemical Tests

Physical and chemical tests include visual examination, specific gravity (Table 1), flame tests (Table 2), elemental analysis, and color tests.

Instrumental Techniques

Instrumental techniques include infrared spectroscopy (Table 3), pyrolysis–gas chromatography–mass spectrometry, and nuclear magnetic resonance.

BIBLIOGRAPHY

J. Haslam, H. A. Willis, and D. C. M. Squirrel, *Identification and Analysis of Plastics,* Heyden & Son Inc., Philadelphia, Pa., 1981.

Table 1. Appearance and Specific Gravity of Elastomers

Elastomer	Appearance	Specific gravity
polybutadiene	water-white, slightly yellow, opaque	0.91–0.93
poly(styrene-*co*-butadiene)	water-white, slightly yellow, opaque	0.94
natural rubber	dark brown, opaque	0.91–0.95
polyisoprene	water-white to slightly yellow, opaque	0.91
polyacrylonitrile	yellowish to dark brown, opaque	0.95
polyisobutylene	water-white, transparent	0.85
poly(ethylene-*co*-propylene-*co*-diene)	white, translucent	0.85
polychloroprene	white, opaque	0.95
polydimethylsiloxane	white, opaque	0.98
polyurethanes	yellow to dark brown, opaque	1.20

Table 2. Burning Characteristics of Elastomers

Elastomer	Flame characteristics	Odor
polybutadiene	clear flame, sometimes yellow	hydrocarbon
poly(styrene-*co*-butadiene)	smoky sooty, yellow	strong styrene
natural rubber	smoky, yellow	hydrocarbon
polyacrylonitrile	smoky, yellow	strong, trace of bitter almonds
butyl rubber	smoky, yellow	hydrocarbon
EPDM	clear	waxy, similar to burning candle
chloroprene	yellow, extinguishes on removal from flame	strong odor of hydrochloric acid
silicone	slight smoke leaves white residues of SiO_2	
polyurethanes	some burn, some melt, and some burn after melting	strong smell of hydrocarbons and amines

1985 Annual Book of ASTM Standards, Vol. 09.01, American Society for Testing and Materials, Philadelphia, Pa., 1985, Sect. 9.

ASTM D 2702-81, "Determination of Infrared Absorption Characteristics," *1985 Annual Book of ASTM Standards,* Vol. 09.01, American Society for Testing and Materials, Philadelphia, Pa., 1985.

ASTM D 3677-83, "Rubber Identification by Infrared Spectroscopy," *1985 Annual Book of ASTM Standards,* Vol. 09.01, American Society for Testing and Materials, Philadelphia, Pa., 1985.

D. O. Hummel, *Infrared Analysis of Polymers, Resins and Additives. An Atlas,* 2nd ed., Wiley-Interscience, New York, 1978, Parts I and II.

GINO J. FRISONE
General Tire Company

Fibers

The ever-increasing number of types of natural and synthetic fibers poses problems of identification in the textile industry (see also FIBERS). Textile manufacturers must be able to identify fibers to avoid putting the wrong fiber in a blend, to duplicate a construction when specifications are missing, to confirm or deny implications regarding their product, to recognize their fibers vs a competitor's, etc. (see also FORENSIC ASPECTS).

Because there are many ways to identify fibers, there is no routine method available to the textile analyst. Fibers may be subjected to burning tests, solubility tests, selective dye tests, microscopical examination, spectroscopy, etc. The methods might be classified as follows:

Microscopical
longitudinal appearance
cross-sectional appearance
refractive index
birefringence

Chemical
burning
solubility
chemical reaction
  dyeing
  other
differential thermal analysis (dta)

Physical
specific gravity
melting point
moisture regain
infrared spectrum
ultraviolet spectrum
spectroscopy
mechanical properties
  breaking strength
  modulus of elasticity
  other

BIBLIOGRAPHY

Textile-Fibers, Zippers, 1984 Annual Book of ASTM Standards, Vol. 07.02, The American Society for Testing and Materials, Philadelphia, Pa., 1985.

T. G. Rochow and E. G. Rochow, *An Introduction to Microscopy by Means of Light, Electrons, X-Rays, or Ultrasound,* Plenum Press, New York, 1979.

AATCC Technical Manual, The American Association of Textile Chemists and Colorists, Research Triangle Park, N.C., 1985.

Identification of Textile Materials, 7th ed., The Textile Institute, Manchester, UK, 1975.

PAUL A. TUCKER
North Carolina State University

Plastics

The general analytical procedure begins with preliminary tests of solubility (Table 1), density (Table 2), and thermal behavior (Table 3). These tests may be conclusive, or they may suggest appropriate specific tests. Identification of heteroatoms is a useful indicator of plastic type. Specific tests confirm preliminary conclusions.

Testing for Heteroatoms

Heteroatoms are those elements present in plastic in addition to carbon and hydrogen, eg, nitrogen, sulfur, chlorine, fluorine, silicon, and sometimes phosphorus. The Lassaigne method is usually used for the qualitative determination of nitrogen, sulfur, and chlorine (see also CHEMICAL ANALYSIS).

Table 1. Plastics Dissolved by Selected Solvents

Water	Tetrahydrofuran (THF)	Boiling xylene	Formic acid
polyacrylamide	all uncross-linked polymers[a]	polyolefins	polyamides
poly(vinyl alcohol)		styrene polymers	poly(vinyl alcohol) derivatives
poly(vinyl methyl ether)		vinyl chloride polymers	urea– and melamine–formaldehyde condensates (uncured)
poly(ethylene oxide)		polyacrylates	
polyvinylpyrrolidinone		polytrifluorochloroethylene	
poly(maleic anhydride)			

[a] Except polyolefins, polyfluorohydrocarbons, polyacrylamide, polyformaldehyde, polyamides, polyurethanes, urea and melamine resins.

Table 2. Approximate Densities of Plastics

Density, g/cm³	Material	Density, g/cm³	Material
0.80	silicone rubber[a]	1.18–1.24	cellulose propionate
0.83	polymethylpentene	1.19–1.35	plasticized PVC[b]
0.85–0.92	polypropylene	1.20–1.22	polycarbonate[c]
0.89–0.93	high-pressure (low-density) polyethylene (LDPE)	1.20–1.26	cross-linked polyurethanes
0.91–0.92	poly(1-butene)	1.26–1.28	phenol–formaldehyde resins[d]
0.91–0.93	polyisobutylene	1.21–1.31	poly(vinyl alcohol)
0.92–1.0	natural rubber	1.25–1.35	cellulose acetate
0.94–0.98	low-pressure (high-density) polyethylene (HDPE)	1.30–1.41	phenol–formaldehyde resins[e]
1.01–1.04	nylon-12	1.3–1.4	poly(vinyl fluoride)
1.03–1.05	nylon-11	1.34–1.40	celluloid[f]
1.04–1.06	acrylonitrile–butadiene–styrene polymers (ABS)	1.38–1.41	poly(ethylene terephthalate)
1.04–1.08	polystyrene	1.38–1.41	rigid PVC
1.05–1.07	poly(phenylene oxide)	1.41–1.43	polyoxymethylene[g]
1.06–1.10	styrene–acrylonitrile copolymers	1.47–1.52	urea– and melamine–formaldehyde resins[h]
1.07–1.09	nylon-6,10		
1.12–1.15	nylon-6	1.47–1.55	chlorinated PVC
1.13–1.16	nylon-6,6	1.5–2.0	phenoplasts and aminoplasts[i]
1.1–1.4	epoxy resins, unsaturated polyester resins	1.7–1.8	poly(vinylidene fluoride)
1.14–1.17	polyacrylonitrile	1.8–2.3	polyester and epoxy resins[j]
1.15–1.25	cellulose acetobutyrate	1.86–1.88	poly(vinylidene chloride)
1.16–1.20	poly(methyl methacrylate)	2.1–2.2	polytrifluorochloroethylene
1.17–1.20	poly(vinyl acetate)	2.1–2.3	polytetrafluoroethylene

[a] Silica filled, up to 1.25 g/cm³.
[b] Approximately 40% plasticizer.
[c] Based on bisphenol A.
[d] Unfilled.
[e] Filled with organic materials, eg, paper, fabrics.
[f] Cellulose nitrate and camphor.
[g] Polyformaldehyde.
[h] With organic fillers.
[i] With inorganic fillers.
[j] Filled with glass fibers.

Table 3. Softening and Melting Ranges of Thermoplastics

Thermoplastic	Softening or melting range, °C
poly(vinyl acetate)	35–85
polystyrene	70–115
poly(vinyl chloride)	75–90 (softens)
polyethylene, density 0.92 g/cm³	~110
density 0.94 g/cm³	~120
density 0.96 g/cm³	~130
poly(1-butene)	125–135
poly(vinylidene chloride)	115–140 (softens)
poly(methyl methacrylate)	120–160
cellulose acetate	125–175
polyacrylonitrile	130–150 (softens)
polyoxymethylene	165–185
polypropylene	160–170
nylon-12	170–180
nylon-11	180–190
polytrifluorochloroethylene	200–220
nylon-6,10	210–220
nylon-6	215–225
polycarbonate	220–230
poly(4-methyl-1-pentene)	240
nylon-6,6	250–260
poly(ethylene terephthalate)	250–260

Fluorine. The solution is acidified with dilute hydrochloric or acetic acid and 1 N calcium chloride solution is added. A gel-like precipitate of calcium fluoride indicates the presence of fluorine.

Phosphorus. Upon adding a solution of ammonium molybdate to a solution acidified with nitric acid, a precipitate is obtained after heating for approximately 1 min.

Silicon. A mixture of 30–50 mg of plastic sample, 100 mg of dry sodium carbonate, and 10 mg of sodium peroxide in a small platinum or nickel crucible is melted slowly over a flame. After cooling, a few drops of water are added, and the solution is brought quickly to a boil and neutralized with dilute nitric acid. A drop of ammonium molybdate solution is added and the mixture heated nearly to boiling. After cooling, 1 drop of benzidine solution is added, and then 1 drop of saturated aqueous sodium acetate solution. A blue color indicates silicon.

General Identification Reactions

Qualitative tests that allow ready identification of plastic material categories have been developed. These include the *p*-dimethylaminobenzaldehyde test, the Gibbs indophenol test, and the formaldehyde test.

Specific Plastic Tests

Polyolefins. The simplest method of identification of polyolefins is by infrared absorption spectroscopy.

Polystyrene. Polystyrene and most styrene-containing copolymers can be identified by placing a specimen in a test tube, adding 4 drops of fuming nitric acid, and evaporating the acid without polymer decomposition.

Poly(methyl methacrylate). Identification of poly(methyl methacrylate) is based on its relatively easy depolymerization to the monomers, which is accomplished conveniently by heating a 0.5 g sample in a test tube with ca 0.5 g dry sand.

Polyacrylonitrile. Identification of acrylonitrile polymers is based on their evolution of HCN upon heating.

Poly(vinyl acetate). Polymers containing vinyl acetate can be recognized by the fact that they produce acetic acid on thermal decomposition.

Poly(vinyl alcohol). Highly saponified poly(vinyl alcohol)s are insoluble in organic solvents, but soluble in water and formamide. Others are identified by reaction with iodine.

Polymers Containing Chlorine. Poly(vinyl chloride), poly(vinylidene chloride), chlorinated rubber, chlorinated polyolefins, and polychloroprene can be identified by using a color reaction with pyridine.

Polyoxymethylene. Polyoxymethylenes produce formaldehyde on heating.

Polycarbonates. For polycarbonates containing bisphenol A, the color reaction with *p*-dimethylaminobenzaldehyde or the Gibbs indophenol test are used.

Polyamides. In some cases melting point determination permits distinction among different polyamides. Polyamides can also be recognized through color reactions with *p*-dimethylaminobenzaldehyde.

Polyurethanes. On pyrolysis, polyurethanes re-form the isocyanates used in their synthesis to some extent.

Phenolics. Phenolic resins may be identified in the Gibbs indophenol test and the bound formaldehyde may be detected with chromotropic acid.

Amino Resins. Amino resins contain nitrogen and bound formaldehyde that can be identified using chromotropic acid.

Epoxy Resins. Epoxy resins give a positive reaction for phenol according to the Gibbs indophenol test from the presence of bisphenol A.

Polyesters. The categories of polyesters should be distinguished as unsaturated polyesters and saturated aliphatic and aromatic polyesters. The acid components of most unsaturated polyesters are maleic, fumaric, phthalic, succinic, or adipic acid.

Cellulose Derivatives. A suspected sample of cellulose is dissolved or suspended in acetone, 2–3 drops of a 2% solution of α-naphthol in ethanol are added and a layer of concentrated sulfuric acid is carefully introduced underneath. At the phase boundary, a red to red-brown ring forms. In the presence of cellulose nitrate, a green ring forms. Sugars and lignin produce interference. Differentiation between cellulose acetate and cellulose acetobutyrate is usually done by smelling the vapors produced by dry heating of the sample.

Silicones. These polymers are identified by testing for the element silicon (see SILICONES; POLYSILANES AND POLYCARBOSILANES).

Rubberlike Plastics. Butyl rubber, polyisobutylene with some isoprene repeat units, can be identified with mercury(II) oxide. Polybutadiene and polyisoprene contain double bonds, which can be detected by Wijs solution, made by dissolving iodine monochloride in glacial acetic acid. The Burchfield color reaction also differentiates rubbers.

BIBLIOGRAPHY

A. Krause, A. Lange, and M. Ezrin, *Plastics Analysis Guide,* Hanser Publishers, Munich, Vienna, New York, 1983; distributed in the U.S. and Canada by Macmillan Publishing Co.

D. Braun, *Simple Methods for Identification of Plastics,* Carl Hanser Verlag, Munchen, Wien, 1982; H. Saechtling, *Plastics Identification Table.*

J. Haslam, H. A. Willis, and D. C. M. Squirrel, *Identification and Analysis of Plastics,* 2nd ed., Butterworth Publishers, Ltd., London, 1972; paperback reprint edition, Heyden & Son, Ltd., London and Philadelphia, 1980.

D. O. Hummel and F. Scholl, *Atlas of Polymer and Plastics Analysis,* 2nd ed., 3 Vols., Carl Hanser Verlag/Verlag Chemie, Munich/Weinheim, 1978, 1981, 1982, 1984.

DIETRICH BRAUN
Deutsches Kunststoff-Institut

IMIDES, POLYMERIC. See POLYIMIDES.

IMINES. See ALKYLENEIMINE POLYMERS.

IMMUNOCHEMISTRY. See ANTIBODIES AND ANTIGENS, POLYMER-BOUND.

IMPACT RESISTANCE

Impact resistance is a measure of the ability of a material or structure therefrom to withstand the application of a sudden load without failure. The impact resistance of a structure is therefore a complex function of geometry, mode of loading, load application rate, environment, material properties, and, quite important, the definition of failure.

The issue of stress geometry or mode is all important: a material highly resistant to failure in one mode of loading can fail catastrophically in another. Therefore, the impact resistance of a structure is relevant only with respect to a particular mode of loading. A soundly designed structure can fail unexpectedly if the geometry of a structural component is altered by the gradual change of shape, material wear, or crack growth, or if the mode of loading is changed, as when a force is applied in an unforeseen manner. Thus it is often necessary to test the material or structure under conditions more severe than those of actual use. It is also prudent to follow certain well-established design guidelines.

Impact often implies a high rate, but the type of failure usually associated with impact can also occur under apparently mild conditions, even at low rates of load. When the rate of strain energy accumulation is higher than can be dissipated by the material in the vicinity of a crack or a flaw, unstable fracture can occur, often entailing crack speeds of hundreds of meters per second. Higher speeds are more likely to cause unstable fracture, other conditions being equal, for then even small flaws, less severe geometries, or milder temperatures can become critical. This is partly due to the viscoelastic nature of polymers. Under truly high speed conditions, when inertial effects are involved, the deformation and fracture behavior is complicated by wave propagation and constraint effects.

In contrast to impact resistance, the fracture toughness of a solid polymer is much better defined for a given set of environmental conditions (see also FRACTURE). Toughness is improved by the incorporation of a soft, elastomeric phase into the rigid polymer matrix. The art and science of material modification for the purpose of toughening is extremely complex. Enhancement of toughness improves impact resistance.

The performance of a material or a structure depends on local geometrical variations and processing conditions, such as cooling rate, shear stress, and melt-flow paths, which result in orientation and residual stress. A structure produced by a given process may possess significant morphological differences from test specimens. Ultimately, it may be unavoidable to test the impact resistance of manufactured components.

Since the impact toughness of a material, as it is commonly called, depends on the technique of measurement, the testing instruments and the technique must be considered first.

Testing Machines and Techniques

Standard Test Methods. Many impact tests are sufficiently standardized to have ASTM designations (see Table 1).

Special Test Machines. Hydraulically and pneumatically driven machines have found increasing use in research and quality control laboratories; ASTM standards have not yet been established. These machines require higher operator skills and are more expensive, but offer better control and produce more information.

Product Testing. The specimens are usually specially prepared for falling weight tests, swinging-pendulum tests, tensile impact tests, and others. However, frequently the fabricated articles perform differently. The conventional forms of impact testing are more suitably applied as control or material tests. This limitation has become more serious in recent years with the introduction of new polymers. The fabricating process can influence impact strength. A typical example is the effect of residual flow orientation in thin-walled injection moldings. The flow properties of the material during the filling of the mold may be more important than the basic impact properties of the unoriented material in governing the impact behavior of the thin-wall moldings. Specimen geometry is also important because at shock loading, the rate of stress buildup is governed by the article shape. Therefore, the impact strength of the finished product should be tested in addition to the usual control and quality tests.

Flexed-plate Impact. In flexed-plate impact, a ball, dart, or striker is driven by gravity or pressure into a thin plate supported by or clamped over an annular area. The nose of the dart is usually hemispherical, though sometimes square. A flexed-plate impact test may correspond closely to conditions of actual use. However, geometric factors, such as specimen thickness and dart-to-annular opening ratio, cannot be readily standardized; furthermore, the geometry itself can cause a change in the deformation mode.

Flexed-beam Impact. Flexed-beam impact tests are sometimes carried out on unnotched specimens, although notched specimens are preferred. The Izod and Charpy testing geometries are different; in the German Charpy test a square notch is used rather than a V-shaped. Regardless of the geometry, the notch creates a severe stress concentration. Many materials that are tough in the tensile and flexed-plate impact tests fail because of brittleness in notched flexed-beam tests.

Fracture Toughness from Flexed-beam Impact Tests. The so-called impact toughness values obtained from Izod and Charpy tests are not material properties because they depend on specimen thickness, notch depth, notch radius, and other factors. Even the mode of failure is affected by a changing specimen geometry. These values are therefore useful only for the comparison of different materials and are useless for design calculations. Yet it is desirable to have fracture toughness values that are material properties taken under impact conditions. The fracture toughness can be determined from the impact energy if the geometry of the specimen is known.

Impact-resistant Materials

Impact resistance is a relative term. It depends on structural parameters as much as it does on intrinsic material toughness.

Table 1. Standard Test for Impact Resistance

Test	Designation	Description
brittleness temperature	ASTM D 746	The temperature is determined at which plastics and elastomers exhibit brittle failure under impact.
falling weight	ASTM D 3029	Impact resistance indicated by energy to break or crack rigid plastics by means of a falling weight (tup[a]); constant height and variable weight are recommended.
falling weight	ASTM D 1709	Similar to D 3029 but for measuring impact resistance of polyethylene film by free-falling dart.
falling weight	ASTM D 2444	For impact resistance of thermoplastic pipe and fittings by falling weight (tup)[a].
fracture toughness	ASTM E 399	For plane-strain fracture toughness.
high rate stress/strain (tension)	ASTM D 2289	Area under stress–strain curve corresponds to measure of impact resistance at testing speeds up to 254 m/min.
Izod impact	ASTM D 256[b]	Energy to break a notched (cantilever beam) specimen upon impact by a pendulum. Notch tends to promote brittle failure. Unnotched impact strength is obtained by reversing the notched specimen in the vise. Notch sensitivity can be determined by using Method D.
tensile impact	ASTM D 1822	Recommended for plastic materials too flexible, too thin, or too rigid to be tested by ASTM D 256. Measures energy to break by "shock in tension" imparted by a swinging pendulum.

[a] Designation of the weight.
[b] Includes Charpy impact testing.

Table 2. Typical Notched Izod Impact and Drop-weight Impact Toughness

Thermoplastic	Notched Izod, J/m[a]		Falling dart[b], N·m[c]	
	22°C	−40°C[d]	22°C	−40°C[d]
acetal	53–53			
nylon, amorphous	587–1121	213–480[e]		
nylon-6[f]	53–81			
impact modified nylon-6[f]	961	294[e]	65	57[e]
nylon-6,6[f]	27–64			
toughened nylon-6,6[f]	907–1174			
33% glass[f]	219			
nylon/ABS	998	96[e]	73	35[e]
nylon/PPO	213	133[e]	51	40
polyarylate	294	213		
polycarbonate	641–854	133–598		
polycarbonate/ABS	454–641	267–614[e]	38–61	30–46[e]
polycarbonate/PBT[f,g]	694–854	160–641	56	49[e]
polycarbonate/PET	907	694[e]	57	50[e]
polyester, aromatic (LCP)[h]	53–534			
thermoplastic polyester, PBT[g]	43–53			
impact modified PBT[g]	801–854	747–801[e]	43–54	43[e]
impact modified PBT/PET	801–854		54	
30% glass-reinforced PET, thermoplastic polyester	101	96		
toughened 30% glass PET	139–235	123–160		
polyetheretherketone	59–81			
polyetherimide	32–53		34–36	
polyethersulfone	53–81			
poly(phenylene oxide) modified	267–534	133–187	15–34	3–11
poly(phenylene sulfide)	27–75			
poly(phthalate carbonate)	534			
polysulfone	69	64		
polyurethane, engineering thermoplastic	53–213			

[a] To convert J/m to ft·lbf/in., divide by 53.38.
[b] Gardner or falling dart.
[c] To convert N·m (J) to in.-lb, divide by 0.113.
[d] Unless otherwise stated.
[e] At −29°C.
[f] Dry as molded.
[g] Poly(butylene terephthalate).
[h] LCP indicates liquid crystalline polymer.

Impact resistance is sometimes gained at the cost of reduced modulus and strength, and perhaps fatigue resistance. The impact toughness of commercial plastics is shown in Table 2.

BIBLIOGRAPHY

A. J. Kinloch and R. J. Young, *Fracture Behavior of Polymers,* Applied Science Publishers, London, 1983. This is a particularly comprehensive treatise and contains much up-to-date (early 1980s) information.

J. G. Williams, *Stress Analysis of Polymers,* 2nd ed., Halsted Press, Chichester, 1980. Presents fundamental mechanics using relatively simple mathematics. The last chapter is particularly relevant.

C. B. Bucknall, *Toughened Plastics,* Applied Science Publishers, London, 1977. Contains mid-1970s state-of-the-art information on the science and technology of impact-modified plastics.

S. Turner, *Mechanical Testing of Plastics,* 2nd ed., Longman Inc., New York, 1983.

ALBERT F. YEE
University of Michigan

INDENE–COUMARONE RESINS. See HYDROCARBON RESINS.

INDUSTRIAL GUMS. See GUMS, INDUSTRIAL.

INDUSTRIAL MICROBIOLOGY. See ENZYMES.

INDUSTRIAL POLYMERIZATION PROCEDURES. See POLYMERIZATION PROCEDURES, INDUSTRIAL.

INFINITE NETWORKS. See CROSS-LINKING; GELS; INTERPENETRATING POLYMER NETWORKS; NETWORKS.

INFRARED-ABSORPTION SPECTROSCOPY

Infrared spectroscopy is an old and familiar friend to the organic and polymer chemist. It involves the vibrational energy levels of molecules but often provides only part of the complete vibrational spectrum. Other so-called vibrational modes appear in the Raman spectrum although, in molecules having a degree of symmetry, certain modes can also be optically inactive. The selection rules governing this behavior are well known and the appearance of modes in the ir and Raman can be predicted from the symmetry properties of the normal modes relative to those of the molecule through the methods of group theory.

By the early 1970s, the use of vibrational spectroscopy for obtaining novel insight into polymer structure appeared to have peaked. Indeed, this seemed true of ir spectroscopy as a general tool for the characterization of organic compounds. Other techniques, particularly nmr, provided far more detailed information. However, the discovery of Fourier transform infrared (ftir) spectrometers changed this situation.

This technique has a number of advantages over conventional dispersive spectroscopy, not the least of which is higher optical throughput. It was also initially claimed to be much faster. The key to the renaissance of ir spectroscopy was the subsequent development in the speed and power of small, affordable computers. This has allowed the application of various advanced programs that are capable of providing new information.

Origin of the Vibrational Spectrum

Theory. A change in the total energy of a molecule occurring on interaction with electromagnetic radiation may be observed as the absorption spectrum of the material. To describe this interaction and formulate a useful mathematical model, certain assumptions are usually made. The total energy of a molecule consists of contributions from the rotational, vibrational, electronic, and electromagnetic spin energies. This total energy can be approximated to a sum of the individual components and any interaction treated as a perturbation. The separation of the electronic and nuclear motions, known as the Born–Oppenheimer approximation, depends on the large difference in mass between the electrons and nuclei. Since the former are much lighter, they have relatively greater velocities and their motion can be treated by assuming fixed positions of the nuclei. Conversely, small nuclear oscillations (compared with interatomic distance) occur in an essentially averaged electron distribution. The change in energy of these nuclear vibrations on interaction with radiation of suitable frequency is the origin of the vibrational spectrum. This energy change is, of course, quantized. In addition, an infrared-absorption band or Raman scattering band nearly always corresponds to discrete vibrational transitions in the ground electronic state. Absorption of visible or uv light is usually required to produce changes in electronic energy.

Since molecular vibrations cannot be observed directly, a model of the system is required in order to describe these normal modes. The nuclei are considered to be point masses and the forces acting between them springs that obey Hooke's law. The motion of each atom is assumed to be simple harmonic.

The normal modes of a molecule can be calculated by the methods of classical mechanics. Although an approximation, it can be shown that in the limit of harmonic motion the classical solution corresponds to that obtained by quantum mechanics. The equations are set up by assuming that displacements are opposed by "springs" linking the atoms. Each spring has a characteristic force constant, and the collection of such interatomic forces for a particular molecule comprises a force field. Usually, the force constants are defined in terms of internal coordinates, those forces resisting the stretching, bending, and torsions around bonds. This description is intuitively satisfying from a chemical viewpoint and also allows calculation of the frequencies and the form of the normal modes of complex systems using the Wilson *GF* matrix method.

Conditions for Infrared Absorption. A basic and inherent condition for ir absorption is that the frequency of the absorbed radiation must correspond to the frequency of a normal mode of vibration and, hence, a transition between vibrational energy levels. This condition is not by itself sufficient, and additional so-called selection rules are required to determine activity. An understanding of the selection rules can only be attained through the methods or theories capable of successfully describing the interaction of radiation with matter.

Ir absorption is simply described by classical electromagnetic theory; an oscillating dipole is an emitter or absorber of radiation. Consequently, the periodic variation of the dipole moment of a vibrating molecule results in the absorption or emission of radiation of the same frequency as that of the oscillation of the dipole moment. The requirement of a change in dipole moment with molecular vibration is fundamental. Certain normal modes do not result in such a change.

The Vibrations of Chain Molecules. At first glance, the enormous number of atoms present in a polymer chain would seem to make the calculation of its vibrational frequencies impossible. However, it can be shown that for a theoretically infinite polymer chain in a regular, well-defined conformation, the problem can be reduced by symmetry to the determination of the normal modes of the translational repeat unit. In effect, an ordered polymer chain can be considered to be a one-dimensional crystal or linear lattice. This is so, physically, because the intermolecular forces between chains packed in a three-dimensional lattice are usually an order of magnitude less than the intramolecular interactions between repeat units in the same chain. Consequently, their effect on the vibrational spectrum can be neglected or treated as a perturbation.

Instrumentation

A large number of dispersive instruments are still used in chemical laboratories. Although inferior in optical throughput, flexibility, and data manipulation, they have the advantage of being inexpensive. Today, however, a computerized dispersive spectrometer is similar in price to a good desktop Fourier transform instrument.

A limiting feature of infrared spectroscopy and other spectroscopic techniques is the incident energy at the detector per unit time. Dispersive instruments, as the name implies, use either grating or prisms to separate the radiation into its component wavelengths. Resolution is obtained by a set of slits. The narrow openings of the entrance and exit slits, required for high resolution, severely limit the amount of energy reaching the detector. Consequently, if good signal-to-noise ratios are to be achieved, slow scan rates and hence, high measurement times per frequency interval are required. Interferometers have two advantages; high energy throughput and multiplexing. Various types of interferometer are now in use, but the general principles of the technique are clearly illustrated from the simplest and still most widely used design, shown in Figure 1. This instrument consists of two mirrors, one fixed and one movable, at right angles to one another. A beam splitter bisects the angles between these two mirrors. There are no slits to limit the incoming energy and the energy throughput is usually 80–200 times greater than that of a dispersive instrument, depending on resolution.

The principal disadvantage of ftir is that usually an un-

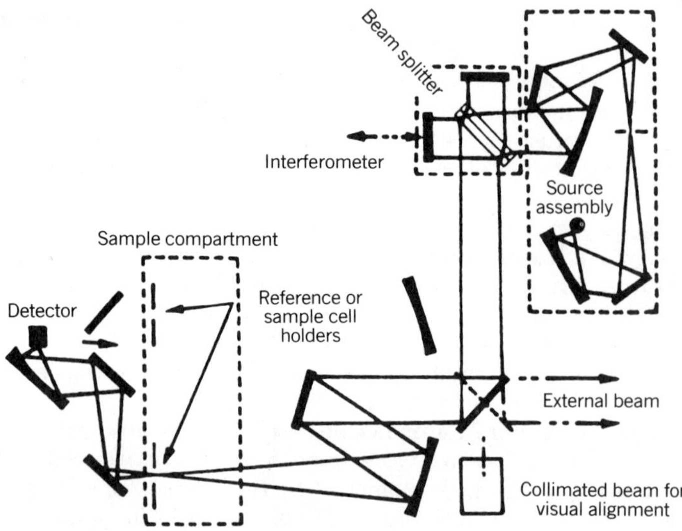

Figure 1. Schematic representation of an optical bench of a Fourier transform infrared (ftir) spectrometer. Courtesy of Digilab, Inc.

recognizable interferogram rather than a spectrum is produced. However, a number of advances allowed the introduction of turnkey commercial instruments. Typically, these consist of a number of basic components: an interferometer, minicomputer (often including a processor for specific calculations in addition to the central processing unit), a magnetic disk storage device, and various input–output interfaces and peripheral devices (terminals, plotters, etc). Software is included, but there are various programs developed by specific laboratories.

Quantitative Analysis

The quantitative analysis of polymers in transmission studies is based on the Beer-Lambert law:

$$A = abc$$

where A is the intensity (in absorbance, not transmittance) of the band of interest; integrated areas or peak heights can be used, but the units change accordingly; a is the absorptivity or extinction coefficient, b is the sample thickness, and c is the concentration of the component of interest. For multicomponent systems, the absorption of any particular band can be written

$$A = \sum_{i=1}^{n} a_i c_i b$$

assuming there are no interactions between the i components.

The general theory for diffuse reflectance at scattering layers within powdered samples is similar in form to the Beer-Lambert law and (for an infinitely thick layer) can be written as

$$f(R_\infty) = (1 - R_\infty)^2/2R_\infty = k/s$$

where R_∞ is the absolute reflectance of the layer, s a scattering coefficient, and k the molar absorption coefficient.

Data-processing Techniques

The principal advantages of the advances in instrumentation that have occurred in the last ten years are a direct result of various new data-handling techniques. The most widely used program (and conceptually the simplest) has proved to be the spectral subtraction routine. This allows spectra to be multiplied by a variable-scale parameter before subtraction. The magnitude of the scale parameter is usually altered on a trial-and-error basis until the desired difference spectrum is obtained. One application is the separation of crystalline and amorphous spectral components. For polymers with readily identifiable amorphous bands, this procedure is straightforward; the scaling parameters are altered until the amorphous bands in the difference spectrum are subtracted to the baseline. In many instances, however, bands of the components to be separated overlap and the absorptions of weakly absorbing species are completely hidden by the bands of a more strongly absorbing component. The subjective judgment of a composite spectrum is a potential source of large errors.

Characterization of Polymeric Materials

Infrared spectroscopy can be used to characterize polymeric materials at different levels of complexity. Most commonly, ir spectroscopy is a rapid and easy method for qualitative identification of principal components. Today, identification of polymeric materials by modern computer-assisted instruments has become a (relatively) trivial task. Libraries of standard spectra stored in the computer memory and used in conjunction with search routines have greatly simplified the process.

Ir spectroscopy may also be employed to characterize the structure of polymeric materials, the overall chemical composition of the polymer chain, and the distribution of the individual units. Thus it is feasible to obtain information concerning the nature and concentration of structural and conformational units present in the polymer chain. This, in turn, leads to consideration of the ir spectral features observed when a polymer orients or crystallizes. Symmetry plays an important role in the observed spectra of ordered systems, making the spectrum sensitive to the conformation of polymer chains. Ir spectroscopy may also be used to study the changes occurring on chemical modification, degradation, and oxidation of polymers. At the present time, polymeric materials of industrial significance are rarely composed of a single polymer. Considerable advances have been made in the characterization of multicomponent polymer systems by ir spectroscopy. Finally, ir spectroscopy is employed in conjunction with computer calculations. Such studies advance the understanding of the origin of the vibrational spectra of polymers.

BIBLIOGRAPHY

E. B. Wilson, Jr., J. C. Decius, and P. C. Cross, *Molecular Vibrations*, McGraw-Hill, Inc., New York, 1955.

P. C. Painter, M. M. Coleman, and J. L. Koenig, *The Theory of Vibrational Spectroscopy and Its Application to Polymeric Materials*, John Wiley & Sons, Inc., New York, 1982.

J. L. Koenig, *Adv. Polym. Sci.* **54,** 87 (1983).

T. Hirschfeld in J. R. Ferraro and L. J. Basile, eds., *Fourier Transform*

Infrared Spectroscopy: Applications to Chemical Systems, Vol. 4, Academic Press, Inc., New York, 1978.

M. M. COLEMAN
P. C. PAINTER
The Pennsylvania State University

INHIBITION. See RADICAL POLYMERIZATION.

INIFERS. See BLOCK COPOLYMERS; TELECHELIC POLYMERS.

INITIATION

Various methods are employed to initiate polymerization, and reactions are often classified thereby. Depending on the method used to effect initiation, polymerizations may be said to be catalyzed, thermal, enzymatic, electrochemical (see ELECTROCHEMICAL INITIATION), mechanochemical (see MECHANOCHEMICAL REACTIONS), or radiation-induced. The plasma state is also used for initiation (see PLASMA POLYMERIZATION). The word catalyst is often loosely applied to initiation. A distinction should be made between true catalysts and initiators; the latter are incorporated into the polymer molecule or otherwise consumed during the reaction. If a photon source is employed, whether in the presence or absence of chemical sensitizers or initiators, the term photopolymerization (qv) is used. See also AZO COMPOUNDS; CHAIN-REACTION POLYMERIZATION; KINETICS OF POLYMERIZATION; PEROXY COMPOUNDS; RADICAL POLYMERIZATION; STEP-REACTION POLYMERIZATION.

INJECTION MOLDING

Injection molding is an important processing technique for converting thermoplastic and thermosetting materials into final products. In 1989, approximately 5.4×10^6 t (20%) of the 25.2×10^6 t of thermoplastics sold in the United States were injection-molded.

In the United States, ca 80,000 injection molding (IM) machines are used in about 8,000 plants. They tend to cluster in groups of multiples of twelve, which is all a foreman and crew

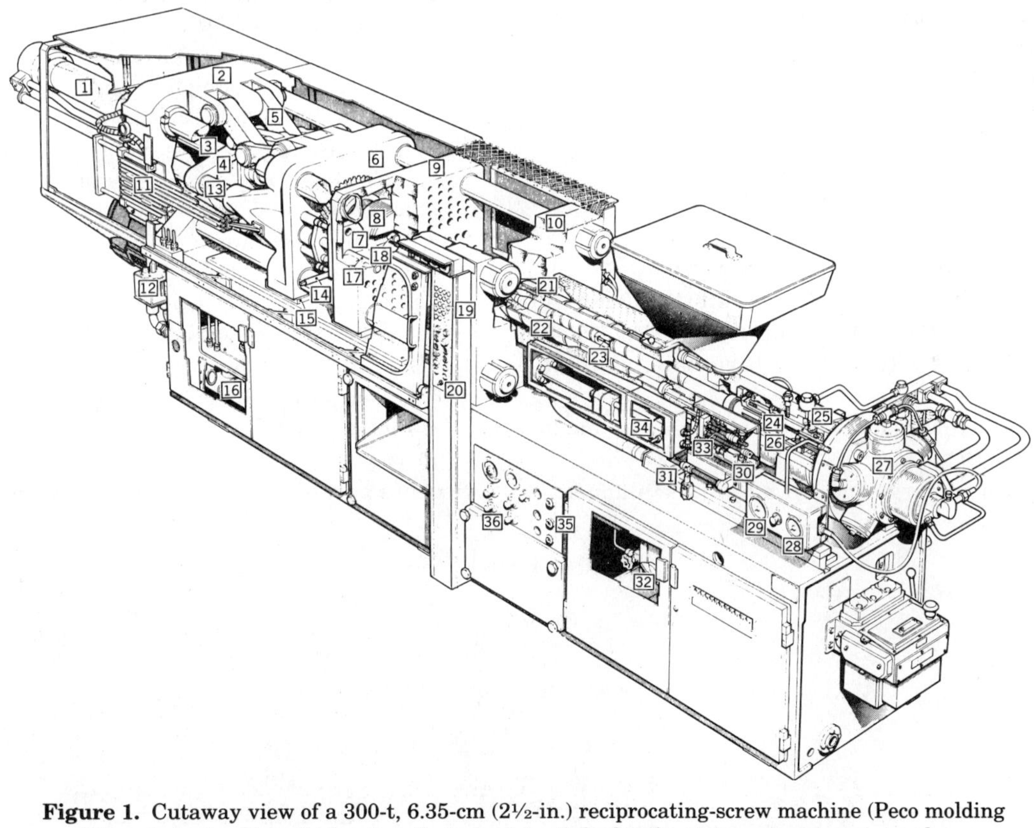

Figure 1. Cutaway view of a 300-t, 6.35-cm (2½-in.) reciprocating-screw machine (Peco molding machine). 1, hydraulic cylinder; 2, tail stock plate; 3, hydraulic piston extension; 4, toggle crosshead; 5, toggle link; 6, moving back plate; 7, ejector plate; 8, mold height-adjustment screw; 9, moving platen; 10, stationary platen; 11, limit switches and stops; 12, lubrication pump; 13, toggle crosshead guide bar; 14, mold height-adjustment mechanism; 15, moving-plate support pad; 16, hydraulic tank; 17, ejector bar; 18, hydraulic ejector; 19, solenoid indicator lights; 20, manual-control panel; 21, injection heating cylinder; 22, screw; 23, air tube and bore; 24, screw coupling; 25, bearing; 26, hydraulic-motor drive shaft; 27, hydraulic motor; 28, screw-speed indicator; 29, injection pressure gauge; 30, shot-volume control mechanism; 31, retraction stroke-limit switch; 32, screw-speed control; 33, injection secondary-pressure control; 34, hydraulic injection cylinder; 35, water valves; and 36, hydraulic controls.

can handle efficiently. About 95% of the machines are reciprocating (Fig. 1) or two-stage screw (Fig. 2); the remainder are plunger machines. In 1988, 60% of the IM machines were imported.

Injection molding produces parts in large volume at high production rates. Labor costs per unit are low and the process can be automated. The parts require little or no finishing and many different surface finishes and colors are available. The same article can be molded in different materials on the same equipment. Close tolerances can be maintained. Parts can be molded in a combination of plastic and fillers, eg, glass, asbestos, talc, and carbon. Metallic and nonmetallic pieces can be inserted.

The process permits the manufacture of very small parts, which are almost impossible to fabricate in quantity by other methods. Scrap losses are minimal as runners, gates, and rejects can be reground and reused. Aside from die casting, it is the only commercial process where the scrap can be reused immediately by regrinding and remolding at the machine. Energy requirements are low, which will become increasingly important as petroleum sources diminish.

Screw Plasticizing. In a screw, the melting of the plastic is caused by the shearing action of the screw, which converts the mechanical energy of the screw drive into heat energy. The heat is applied directly to the material. This and the mixing action of the screw contribute to plasticization, which offers the following advantages:

High shearing rates lower the viscosity.

The flow is nonlaminar.

Good mixing homogenizes the melt.

The residence time in the cylinder is approximately three shots compared to 8–10 shots of a plunger machine.

Since most of the heat is supplied directly into the material, the cycle can be delayed for a longer period before purging.

The method can be used with heat-sensitive materials, such as PVC.

The action of the screw reduces material holdup and subsequent degradation.

The screw is easier to purge and clean than the plunger.

Reciprocating-screw System. In a reciprocating screw (in-line screw), the material is fed from the hopper, plasticized in the screw, and forced past a one-way valve at the injection end of the screw. The material accumulates in front of the screw, forcing back the screw, its drive and motor, and the piston(s) of the hydraulic injection cylinder(s). The return oil from these cylinders passes through the back-pressure control valve into the tank. This valve controls the mixing in the cylinder. The higher the pressure setting, the more complete the mixing.

Screw–Plunger System. In the screw pot or two-stage screw shown in Figure 2, a fixed screw is used for plasticizing. A reciprocating screw would permit continual operation of the screw throughout the whole cycle.

The reciprocating screw and screw pot are both preplasticizing systems. They differ in the location of the pot, which is in front of the reciprocating screw and is a separate cylinder in the two-stage machine. Most machines are of the reciprocating type, but screw-pot equipment offers significant advantages:

Because the screw does not act as the injection ram, lighter bearings can be used.

The extruder barrel need only be strong enough to maintain the pressure of the material during plasticization, which is rarely over 34.5 MPa (5000 psi).

The connection between the two stages can be a ball check valve, which is troublefree and easy to maintain. It presents minimum flow resistance.

The small clearances between the plunger and barrel of the second stage help in degassing the material and give more accurate shot control.

The connection between the two stages results in better mixing of the melt.

All the material passes the full flights of the screw, receiving the same amount of heat.

In a two-stage machine, the screw pumps only against the injection ram, which is floating in oil in the hydraulic cylinder. Therefore, the shot size control is considerably more accurate in the two-stage machine.

Because part of the energy input is used to push the heavy

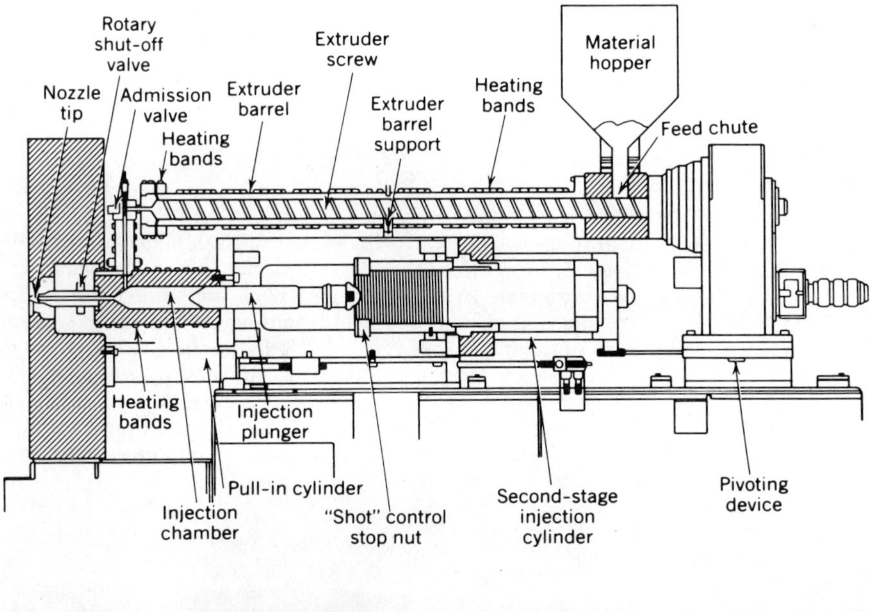

Figure 2. Two-stage screw-plunger machine. Courtesy of HPM Division of Koehring Company.

carriage back, an in-line machine gives slightly less output per unit input.

Control of injection speed and pressure is better in screw-pot equipment, and a larger area can be molded. Extremely high injection pressures are available.

The size of a pot in front of a reciprocating screw is limited by the length of the feed section. In a two-stage machine, there is no limitation. The two-stage machine, however, requires two cylinders and two sets of heat controls. It is more difficult to clean and to set up and materials sensitive to high heat cannot be processed. Furthermore, it takes up more space than a reciprocating screw, and it is more expensive. Cylinders for molding thermosets and rubber are designed only for reciprocating screws.

Intrusion Molding. For the molding of heavy sections or when the shooting capacity of the machine is not adequate, intrusion molding is used where the screw turns continuously, filling the cavity directly. When the cavity is filled, a cushion is extruded in front of the plunger, which then comes forward to supply the needed injection pressure.

Thermoset Molding. Thermoset materials can be molded on reciprocating screw machines. The screw is designed differently and the barrel is usually heated with hot water. The material cannot be permitted to cool in the cylinder as it will solidify. If it does, the screw must be removed and the thermosetting material chipped out.

Injection Molds

A high quality mold is absolutely essential for a profitable molding operation. An excellent mold is one that produces the designed part time after time without interruptions or deviation in cycle, with minimum downtime for repair.

The hot plasticized material is injected into the mold where it is maintained under pressure. When the plastic material has sufficiently solidified, the machine opens, the mold halves separate, and the plastic pieces are ejected.

The quality of the part and its cost of manufacture are strongly influenced by mold design, construction, and excellence of workmanship (see MOLDS).

Computer-assisted Moldmaking. Computer-assisted design (CAD) and computer assisted manufacture (CAM) are being rapidly adopted by the U.S. plastics industry. The former is extremely helpful in the design of a part. Views, cross sections, projections, changes in size and color, and mechanical and thermal analyses are easily made by the computer (see also COMPUTER-ASSISTED ENGINEERING).

Mold Base. The steel parts that enclose the cavities and cores are called the mold base, mold frame, mold set, die base, die set or shoe.

Mold Types. The injection mold is identified descriptively by a combination of some of the following terms:

Parting line
regular
irregular
two-plate mold
three-plate mold

Material
steel
stainless steel
prehardened steel
hardened steel
beryllium copper
aluminum
brass
epoxy

Surface
chrome-plated
electroless nickel
etched
sand blasted
EDM

Number of cavities

Methods of manufacture
machined
hobbed
gravity cast
pressure cast
electroplated
EDM (spark erosion)

Runner system
hot
insulated
cold (for thermosets)
stack

Gating
edge
restricted (pin-pointed)
submarine
sprue
ring
diaphragm
tab
flash
fan
multiple

Ejection
stripper ring
stripper plate
unscrewing
cam
removable insert
hydraulic core pull
pneumatic core pull
knockout pins

Materials for Cavities and Cores. The principal material of construction for molds is steel, followed by beryllium copper alloy. Brass, aluminum, and steel-filled epoxy are also used.

Surface Finish. The surface finish of a mold affects appearance, ejectability, and cost. It is specified by comparing it with six different finishes. This is an SPE/SPI standard that is available from SPI.

Equipment for Fabricating Cavities and Cores. Toolroom equipment is used for machining mold bases, cores, cavities, pins, blocks, and other parts. Fabrication can be assisted by electronically punched tape and CAM.

Hobbing. Hobbing is a method of cold forming metal. The term is used in plastics for the high pressure cold displacement of one material by another.

Pressure Castings. Beryllium–copper cavities and cores are fabricated by pressure casting or, more accurately, hot hobbing. The hob is larger than the finished plastic part. After hobbing, the alloy shrinks as it cools just as plastic shrinks after molding. The hob is made of a good-quality, hot, working die steel that does not deform under the temperatures and pressures of casting. For one or two cavities, a different alloy of beryllium can be used for the hob. Beryllium is also gravity cast.

Gravity Casting. Gravity casting is readily adaptable for injection molds. Any metal can be cast, especially with the Shaw process.

Electroforming. A master, sometimes called a mandrel, is an exact reverse of the cavity. On it is plated ca 38 mm (0.150 in.) of a nickel cobalt compound at the rate of ca 0.1 mm/8 h (0.004 in./8 h). On top of this, copper is electroplated, which is harder than mold steel. The rustproof finished cavity has good dimensional stability and high thermal conductivity and is very precise within 0.0025 mm (0.0001 in.). Electroforming is primarily used when other methods are difficult and expensive such as in gears, pen barrels, and reflectors.

Duplicating. This process is mechanical reproduction by means of cutting tools that are guided by a master, proportional in size to the desired finished parts.

Spark Erosion. Steel is easily removed by an electrical-discharge machine (EDM). An electrode, made of carbon or copper or other conducting material, is made in the reverse shape of the part to be produced. The steel and electrode are immersed in a circulating solution that serves to flush away the eroded material and cool the workpiece. When a-c power is rectified and charged into a capacitor system, the discharge between the electrode and the cavity creates a spark that erodes the steel.

Parting Line. When a mold closes, core and cavity meet, producing an air space into which the plastic is injected. This junction appears on the molded piece as a line, called the parting line.

The selection of the parting line is influenced by the shape of the piece, method of fabrication, tapers, method of ejection, type of mold, aesthetic considerations, postmolding operations, venting, wall thickness, orientation, number of cavities, and location and type of gating.

Venting. Since injection of hot plastic into the mold displaces air, vents are required. Vents are usually ground on the parting line. Their size depends on the nature of the material and the size of the cavity. Inadequate venting may burn the material or cause flash.

Ejection. The molded part is ejected by knockout (KO) pins, sleeves, stripper rings, stripper sleeves, stripper plates, air, or a combination of these methods. The quality of the molded piece is influenced by the method of ejection. Undercuts are pieces of mold material that obstruct ejection.

Gates. The gate is the connection between the runner and the molded part. It must permit enough material to enter and fill the cavity, plus material lost through shrinkage.

Temperature Control. For consistent molding, strict mold temperature control is required. Refrigerated water and hot water to 93°C are needed at each machine. Electrical-heating units are employed to control the temperature of the mold. They include bands, cartridges, and strip heaters controlled by pyrometers.

Automation. Automation is expensive. It requires high quality machinery, controls, and molds, as well as trained personnel and managerial skill. It can be a highly satisfactory and economical operation.

High Quality Parts. A well-designed and built mold should provide high quality parts. Quality-control specifications are used as a guide.

Molecular Aspects of Injection Molding and Physical Properties of Molded Parts

The polymeric materials used in the molding process may be completely amorphous, or partly amorphous and partly crystalline. Varying amounts of thermal and frictional or shear energy are required to provide the plastic flow conditions necessary for good moldings. The viscosity of the melt must be sufficiently low to allow the injection pressure to force the plastic into the cold mold.

As the temperature increases, the energy absorbed by the polymer chains causes vibrational, rotational, and translational or segmental motion of the polymer molecules. This essentially Brownian motion results in a random molecular arrangement.

A unidirectional force is applied to a polymer by the injection plunger. It moves away from the force, filling the cavity. If the force is applied gradually so that the Brownian movement will overcome the orienting forces caused by the flow, the mass of the polymer moves with a rate proportional to the applied stress. This is termed Newtonian flow (see also RHEOLOGICAL MEASUREMENTS).

As the molecules move more rapidly under the influence of higher pressures in the injection process, the chains untangle, using up varying amounts of the energy from the driving force. This does two things. First, it slows down the acceleration compared to an untangled material that would not require any energy for untangling.

Second, the chains move so rapidly that the orientations cannot be removed by the Brownian movement. As the molecules separate they move farther apart. The forces holding them together include van der Waal's forces, which decrease as the sixth power of the distance. Therefore, the increased shear rate (speed) is no longer proportional to the shear stress (force). This is non-Newtonian flow, which is characteristic of injection-molded polymer behavior.

Since good molding requires constant viscosity, the speed of the injection plunger must be accurately controlled. This is best done with computer-controlled machines with feedback systems.

As the flow rate increases, it reaches its final stage with maximum molecular orientation and separation and no further untangling. Then an increase in shear stress proportionally increases the shear rate and the material again acts as a Newtonian fluid. Injection molding is not done in this shear rate range.

Health and Safety Factors

Molding machines are hazardous; some have caused fatalities and severe injuries. The safety gate should be large enough and high enough to prevent body contact with the platen area when the platens are closing. On smaller machines a cover is essential. Both front and rear gates should be mechanically, electrically, and hydraulically interlocked to prevent the operation of the machine if open.

BIBLIOGRAPHY

I. I. Rubin, *Injection Molding, Theory and Practice,* Wiley-Interscience, New York, 1972.

R. C. W. Pye, *Injection Mold Design,* George Goodwin, Ltd., 2nd ed., London, UK, 1978.

H. Gastrow, *Injection Molds: 102 Proven Designs,* Carl Hanser Verlag, Munich, FRG (Macmillan Publishing Co., Inc., New York), 1983.

L. Sors, Bardocz, and Radnoti, *Plastic Molds and Dies,* Van Nostrand Reinhold Co., 1981, pp. 133–204.

G. Menges and P. Mohren, *Anleitung fur den Bau von Spritzgiewerkzeugen* (in German), Carl Hanser Verlag, Munich, FRG, 1974.

I. I. Rubin, *Plast. Eng.,* Part 1, 16 (Feb. 1980); Part 2, 48 (Mar. 1980).

IRVIN I. RUBIN
Robinson Plastic Corporation

INK. See PRINTING-INK VEHICLES.

INORGANIC FIBERS

Inorganic fibers exhibit high temperature resistance, high modulus, hardness, brittleness, and chemical resistance. Starting with the successful development of fiber glass, a large family of inorganic fibers has been developed in the past two decades. They are classified into textile and insulating fibers. Textile fibers are made in discontinuous form (staple) and continuous form (filaments), which are converted into yarn, twines, cords, felt, tapes, and fabric structures. Insulating fibers are usually in the form of mats or boards used for protection against heat, fire, cold, and noise.

Size and Shapes

In continuous form, fibers exist in single strand monofilament or multiple strand multifilament. In discontinuous form staple fibers range in length from 12.5 to 38 mm (short staple); a length of greater than 76 mm is considered a long staple. A short staple fiber is suitable for cotton textile machines, whereas the long staples require wool textile machines.

Inorganic fibers have diameters ranging from less than one micrometer, as in the case of asbestos, to over 100 μm (Fig. 1). The cross-sectional shape of fibers affects the flexural rigidity and the surface area. The bending stiffness of fibers also relates to cross-sectional shape; for example, a triangular cross section produces a higher bending stiffness than a circular fiber of the same cross-sectional area.

Tensile modulus is a material property associated with each type of fiber. With moduli >40 GPa (5.8×10^6 psi) inorganic fibers are characterized by high modulus, which contributes to their high bending stiffness.

Structure and Properties of Inorganic Fibers

Inorganic fibers can be classified according to the source of origin or by their internal structure. Asbestos is an example of a naturally occurring inorganic fiber, and fiber glass an example of a synthetic inorganic fiber. Inorganic fibers can be classified as amorphous, single crystal, polycrystalline, and multiphase.

The various structural features of inorganic fibers reflect the complex interaction of fiber chemistry and the fiber formation technique. Amorphous inorganic fibers such as fiber

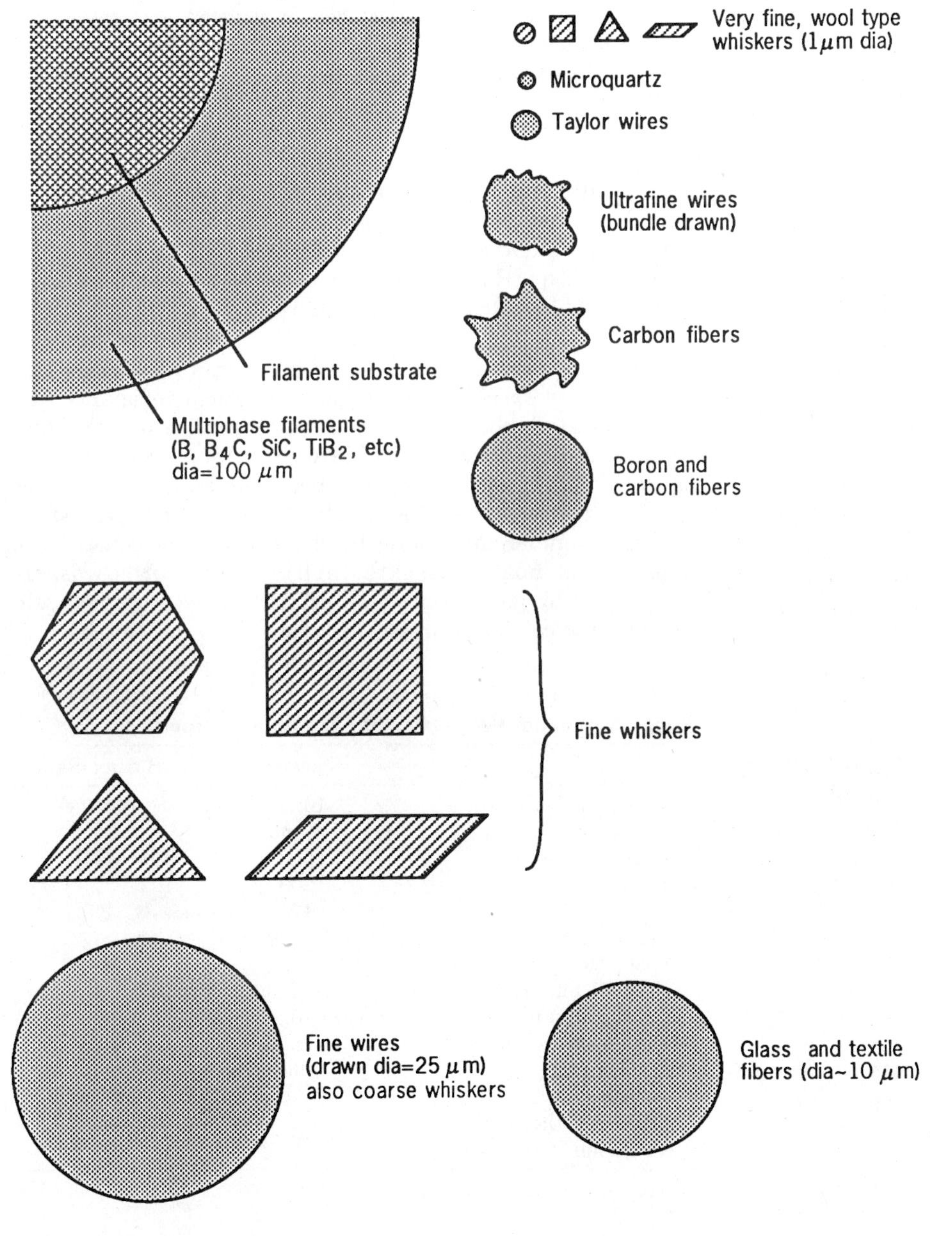

Figure 1. Cross-sectional dimensions for various fiber reinforcements. ▩, Continuous fibers; ▨, discontinuous fibers.

glass and fused silica fibers are manufactured by melt spinning. With a 3-D atomic bonding structure glass fibers have no long range molecular order or orientation. Although they possess relatively high tensile strength, their modulus is among the lowest of inorganic fibers, depending on the thermal history of the fiber. On the other hand, single crystal fibers such as whiskers are chemically pure and highly ordered. This results in strength approaching the theoretical limit, making them the strongest of all fibers.

The polycrystalline fibers include ceramic fibers such as alumina and mullite; metallic fibers such as stainless steel; and graphite fibers. Ceramic and metallic fibers are three-dimensional isotropic structures; graphite fibers are two-dimensional stratified structures. The bonding in the ceramic and graphite fibers is covalent, twice as strong as the metallic bonds in the metallic fibers. With a high molecular orientation, paracrystallinity, and strong covalent bonds, the graphite fibers have an extremely high modulus. Because of the sheetlike structure in graphite fibers, held together by van der Waals forces that are one to two orders of magnitude lower than the covalent bonds in the fiber axis, graphite fibers have weak transverse strength. Ceramic and metallic fibers are isotropic, with ceramic fibers having higher strength and melting temperature. The limiting factor for the strength potential of polycrystalline fibers is their dependence on imperfections and grain size. Prolonged exposure of these fibers to elevated temperatures promotes grain growth, weakening the fiber.

Monofilament boron and silicon carbide are examples of multiphase fibers, which are usually produced by chemical vapor deposition on a fine filament substrate. With the ability to tailor core and sheath structures, multiphase filaments have a good balance of strength and modulus and are thermally more stable than polycrystalline fibers. Diameters are much larger than those of other inorganic fibers. With the unique morphological features, inorganic materials in fibrous forms exhibit high specific strength, modulus, and thermal stability (Fig. 2).

The high specific properties of the inorganic fibers makes them excellent candidate reinforcements for composites. In the design and modeling of composite structures, knowledge of the fundamental elastic constants of the fibers is essential. A system of microtesters has been developed for the characterization of the elastic constants from single fibers.

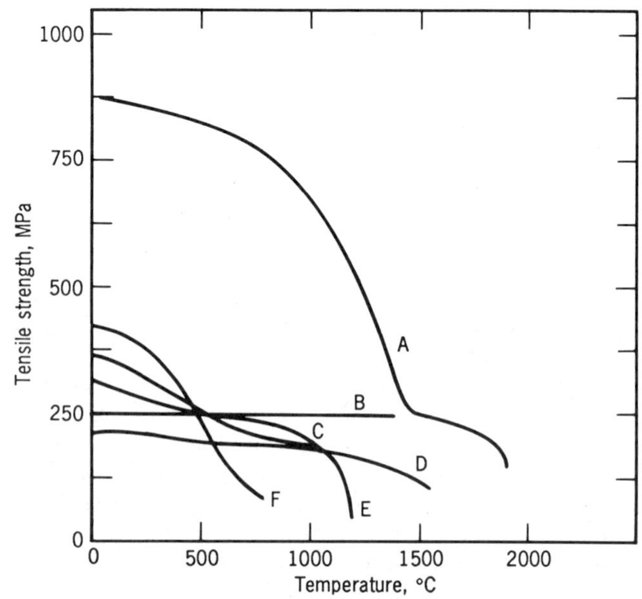

Figure 2. Strength of inorganic fibers at elevated temperature (continuous filament except for Al_2O_3 whiskers). A, Al_2O_3; B, carbon; C, tungsten; D, SiC; E, boron; F, S-glass. To convert MPa to psi, multiply by 145.

Glass Fibers

Glass is an amorphous material obtained by supercooling the molten glass, so that no ordered regions are formed. Chemically, glass is a combination of metallic oxide with silica. Morphologically, glass is a three-dimensional network, an aggregate of tetrahedron networks of silicon and oxygen with metallic ions or inorganic elements randomly distributed among the networks. Typical physical and mechanical properties of E, C, S, and S-2 fiber glass are given in Table 1.

Glass fibers are extensively used as plastic reinforcement because they are inexpensive to produce and possess high strength, low specific gravity, good chemical resistance, and good insulating characteristics. Printed circuits on glass-fiber reinforced high pressure laminates, containing either woven or wet-process glass fibers, have become standard in the electrical industries. Glass-fiber products are also used in residential and commercial roofing markets and other construction applications. Boat hulls, skis, vaulting poles, surfboards, and other sporting goods have been primarily made of glass-reinforced plastics. Glass-fiber fabrics are used in filter bags for

Table 1. Average Properties of Glass Single Filaments

Property	E glass	C glass	S glass	S-2 glass
specific gravity	2.54	2.49	2.48	2.49
tensile strength, MPa[a]				
at 22.2°C	3448	3310	4586	4586
at 371.1°C	2621		4448	
at 537.8°C	1724		2414	
tensile modulus, GPa[b]	72	69	86	87
elongation at 22°C, %	4.8	4.8	5.7	5.4

[a] To convert MPa to psi, multiply by 145.
[b] To convert GPa to psi, multiply by 145,000.

Table 2. Typical Properties of Nicalon SiC Fibers

	Nicalon	SCS monofilament
filament diameter, μm	10–15	140–142
cross section	round	
filaments, yarn	500	
count, tex	200	
density, g/cm³	2.55	3.0
strength, GPa[a]		
at 20°C, as produced	2	3.4–4.5
at 1400°C in argon	<1	
at 1400°C in oxygen	<0.5	
tensile modulus, GPa[a]	26	428
creep strain at 1300°C, 0.6 GPa[a], 20 h %	4.5	
coefficient of thermal expansion × 10^{-6} per °C	3.1	4.8

[a] To convert GPa to psi, multiply by 145,000.

Table 3. Typical Properties of Silicon Nitride Fiber

Property	Value
tensile strength, MPa[a]	2500
tensile modulus, GPa[b]	300
filament diameter, μm	10
density, g/cm^3	2.5
maximum temperature for continuous use, °C	1200
crystal structure	amorphous

[a] To convert MPa to psi, multiply by 145.
[b] To convert GPa to psi, multiply by 145,000.

air-pollution control. The rail and road transport industry as well as the aerospace industry consume large quantities of fiber glass-reinforced plastics.

Quartz and Silica Fibers

Silica fibers are smooth-surfaced, glasslike fibers, with a near-round cross section. They are spun from silicon dioxide, which may be pure or contain a small amount of other materials. Silica fibers can be produced indirectly from glass filaments from which all constituents other than silica have been removed or through spinning a viscous filament that contains a high amount of silica. The organic materials are burned away, leaving a porous silica filament.

Quartz fibers are made from natural quartz crystals by softening quartz rods in an oxy-hydrogen flame and drawing the rods into filaments. Because high purity quartz crystals are rare, the cost of quartz fiber is considerably higher than that of fiber glass and high silica fibers.

Quartz filaments have assumed a new role as high temperature resistant fibers, and are produced in considerable quantities for high temperature and corrosion-resistance applications. They are widely used as filtration and insulation materials at temperatures above those of mineral silicate fibers. Quartz and silica fiber reinforced composites are used in jet aircraft, rocket nozzles, nose cones, and reentry heat shields for spacecraft. Silica (G) fibers are principally used in chemical engineering, high temperature electrical insulation, and high temperature thermal insulation.

Ceramic Fibers

Ceramic fibers are polycrystalline refractory materials composed of metal oxides, metal carbides, metal nitrides, and their mixtures. Starting materials include aluminum silicate, alumina, zirconia, boron nitride, silicon nitride, thoria, aluminum nitride, silicon nitride, potassium titanate, and more recent polymeric materials such as polycarbosilane. Compared to glass fiber, ceramic fibers have higher thermal and chemical resistance, and higher oxidation resistance than carbon and graphite fibers. The densities of ceramic fibers are higher than those of glass, carbon, and graphite fibers (Tables 2–4).

Boron Fibers

Boron fibers are produced by thermal decomposition of a boron hydride or by reduction of a boron halide. The structure of boron fibers depends on the deposition temperature; below 1300°C an amorphous structure is obtained, whereas at higher temperatures, crystalline boron is formed. Between 1400 and 1500°C the boron is deposited in the β-tetragonal form, and above 1500°C in the β-rhombohedral form.

The surface of a boron fiber looks like a corncob structure, consisting of modules separated by boundaries. The cross section reveals boron, tungsten, and a zone containing W_2B, WB, W_2B_5, and WB_4 where boron has diffused into the tungsten substrate.

Boron is particularly suitable for aerospace application where low density, high strength, and high modulus are important.

Table 5. Properties of Boron Fibers

Property	Value
fiber diameter, μm	124–142
tensile strength, GPa[a], (average)	3.5
coefficient of variation, %	
average	15
maximum	20
modulus, GPa[a]	400
density, g/cm^3	2.58
melting point, °C	2100

[a] To convert GPa to psi, multiply by 145,000.

Table 4. Typical Properties of Alumina Fibers

Fiber	Producer	Diameter, μm	Density, g/cm^3	Tensile strength, GPa[a]	Tensile modulus, GPa[a]	Maximum use temperature, °C
FP	DuPont	15.2–25.4	3.71	1.3	345	2045
Sumica	Sumitomo Chemical Co.	1.7	3.2	1.8	210	
Saffil RF grade	Imperial Chemical Industries, Ltd.	3	3.29	2.0	295	1600
Nextel 312	3M	9.9–11.9	2.71	1.3–1.7	150	1205
Nextel 440	3M	9.9–11.9	3.10	1.3–2.0	205–240	
Fiberfrax	Standard Oil Engineered Materials Co.	2–12	2.60	1.0–1.7	105	1795
Kaowool[b]	Babcock & Wilcox Co.	11	2.7	1.7	150	1425
PRD-166	DuPont	21	4.2	2.1	380	

[a] To convert GPa to psi, multiply by 145,000.
[b] Continuous fiber.

Table 6. Properties of Metal and Ceramic Whiskers

Material	Melting point, °C	Density, g/cm^3	Tensile strength, MPa[a]	Modulus, GPa[b]
iron	1520	7.89	13,117	290
copper	1066	8.89	2,938	97
silver	944	10.39	1,759	117
silicon	1392	2.33	3,821	159
Al_2O_3	2038	3.97	2,759	428
BeO	2571	2.86	13,103	345
B_4C	2427	2.53	13,793	483
SiC	2688	3.19	20,690	483
Si_3N_4	1899	3.19	13,793	310
graphite	3649	1.67	19,655	703

[a] To convert MPa to psi, multiply by 145.
[b] To convert GPa to psi, multiply by 145,000.

Table 7. Properties of Commercial High Modulus Whiskers

	Supplier	Diameter, μm	Density, g/cm^3	Tensile strength, MPa[a]	Elastic modulus, GPa[b]	Cost, $/kg
Al_2O_3	Thermokinetic Fibers, Inc.	1–10	3.97	4,138–24,138	552–1,034	16,500
Al_2O_3	Horizons, Inc.	0.5–2.0	3.97	6,897–20,690	414–483	18,700
SiC	Thermokinetic Fibers, Inc.	1–10	3.22	13,793–41,379	552–1,034	25,300
SiC	Carborundum Co.	0.5–0.3	3.22	20,690	483	440
Si_3N_4	Thermokinetic Fibers, Inc.	1–10	3.19	4,828–10,345	276	
Al_2O_3	Tyco Laboratories, Inc.	38–500	3.97	2,552–4,310	462	

[a] To convert MPa to psi, multiply by 145.
[b] To convert GPa to psi, multiply by 145,000.

Boron filament is also fabricated in a prepreg tape form with epoxy resin. The tape can be used for composite structures. Boron fiber composites are used in various aircrafts and in the space shuttle. It is also used as reinforcement in sporting goods. Properties are given in Table 5.

Whiskers

Whiskers are metallic and nonmetallic filamentary single crystals of ultrahigh strength (7 GPa or 10^6 psi) and micrometer-size widths or diameters. Their extremely high strength is attributed to a near perfect crystal structure, chemically pure nature, and fine diameters that minimize defects. Whiskers are the only high modulus fibers that retain elongation in the 1–4% range. In addition, whiskers exhibit a much higher resistance to fracture (toughness) than all other types of reinforcing fiber. More than 100 materials, including metal, oxides, carbides, nitrides, graphite, and organic compounds have been prepared as whiskers. Properties are given in Tables 6 and 7.

Fabrication. In general, the growth of whiskers results from recrystallization (primary) or a decrease in supersaturation.

Applications. The unique ultrahigh strength and stiffness of whiskers are most valuable for structural applications, especially aerospace designs, where the strength-to-weight ratio is of primary importance. Specific applications include high temperature turbine engine components, such as blades or vanes, and aircraft and helicopter components. In dental applications whisker-reinforced composites are used for porcelain, cements, resins, and alloys. Whiskers alone are used in insulation, paper, felts, filters, and dipoles for light modulations.

Metal Fibers

Metal fibers can be grouped into whiskers, metal wools (of which steel wool is a common example), and filament length fibers. Staple length fibers are produced by cutting the fibers produced by the last two categories. Commercial metal wool is produced by shaving metal stock (in wire, tube, or sheet form) with sharp serrated blades to the desired fineness.

Metals in wire or filament form exhibit high, consistent strength. Beryllium, steel, and tungsten are among the most important ones. Beryllium has extremely low density and high modulus. Tungsten wires have high density and were originally developed for lamps. Because of its refractory properties, tungsten wire is used as a reinforcement in some nickel- and cobalt-based superalloys. Steel wire is a common reinforcement material for concrete and tires. Mechanical properties of commercial steel fibers are given in Table 8.

Applications. Steel fiber is used in carpets to reduce static electricity. Metal fabrics produced with metal yarns by tra-

Table 8. Mechanical Properties of Steel Fibers

		Tensile strength, MPa[a]	Modulus, GPa[b]	Elongation, %
Brunsmet	304 ss	1380–2070	186–207	0.9–1.5
Riotec-Gr		900		
drawn	203 ss	760		
steel		2760	200	
Naslon		1880–3140	240	1–2

[a] To convert MPa to psi, multiply by 145.
[b] To convert GPa to psi, multiply by 145,000.

Table 9. Physical Properties of the Various Asbestos Fibers[a]

Property	Source[b]	Chrysotile	Anthophyllite	Amosite	Crocidolite	Tremolite	Actinolite
tensile strength, MPa[c]	1	5683	28	1979	6041	6.9–55	6.9
flexibility		high	poor	good	good	poor	poor
specific gravity, g/cm^3	1	2.55	2.85–3.1	3.43	3.37	2.9–3.2	3.0–3.2
hardness, mohs	2	2.5–4.0	5.5–6.0	5.5–6.0	4	5.5	6
specific heat, kJ[d]/g·K	2	1.11	0.88	0.81	0.84	0.89	0.91
fusion point, °C		1521	1468	1398	1193	1315	1393

[a] Reproduced with permission of the Asbestos Textile Institute Handbook of Asbestos Textiles.
[b] 1, Cape Asbestos Fibres Ltd.; 2, Badoleet (1960).
[c] To convert MPa to psi, multiply by 145.
[d] To convert kJ to kcal, divide by 4.184.

Table 10. Properties of Fiber Substitutes for Asbestos

Fiber	Limiting oxygen index, %	Service temperature, °C	Filament diameter, μm	Specific gravity, g/cm^3	Resistance to acids	Resistance to alkali	Cost $/kg
asbestos	100	649	<1	2.5–3.4	good	excellent	2.2–4.4
glass fiber[a]	100	482	3.5–9.0	2.54	good	good	2.2–6.6
ceramic fiber	100	1204	12+	2.70	excellent	excellent	110
HM aramid	29	204	12+	1.44	fair	fair	17.6
aramid	29	260	12+	1.4	poor	good	17.6
PBI	38	399	12+	1.4	excellent	good	83.6
acrylic	18	177	12+	1.4	good	fair	2.2–4.4
OPF[b]	60	260	10+	1.4	good	poor	13.2–17.6

[a] Utilizing alkali-resistant (AR) glass.
[b] Oxidized polyacrylonitrile fiber.

ditional textile methods (weaving, knitting, and braiding) are used as mold release and for high temperature gaskets. Metal wools and felts are used as heat-, sound-, and vibration-insulating materials. New applications of metal fiber are in the composite area. Metals in fiber or yarn forms are used as reinforcement in resin, ceramic, glass, and metal matrices.

Asbestos Fibers

Asbestos is a mineral fiber that occurs in nature and is derived from serpentine and amphiboles. To obtain the fibers, the rock is mined or removed, and sent for crushing and extraction at a refinery.

The commercial value of asbestos depends largely on its incombustibility and its unique fibrous structure. The fibers can be separated into finer fibers and process well even in wet processes. Other valuable properties include resistance to heat and chemical attack; properties are given in Table 9. Asbestos can be fabricated in different textile forms, from fiber to various types of yarn, braid, and fabric.

Impregnated with metal, ceramics, and resin, asbestos fibers increase the strength and durability of the composites. Asbestos–aluminum materials are widely used as heat insulators in industrial installations and as protective clothing.

Long-term exposure to asbestos can be hazardous. These fibers are extremely fine, and inhalation of dust increases the risk of lung cancer, mesothelioma, and asbestosis, a common disease to asbestos workers.

Because of the extremely hazardous nature of asbestos fibers, suppliers and users have been searching for replacements, but no single fiber exhibits all the characteristics of asbestos. Commercial substitutive fibers are shown in Table 10. The most common replacement is fiber glass, which, like asbestos, has a specific gravity >2.5 g/cm^3 and is not flammable.

BIBLIOGRAPHY

P. Bracke, H. Schurmans, and J. Verhoest, *Inorganic Fibers and Composite Materials,* PO Applied Technology Series, Vol. 3, Pergamon International Information Corp., 1984.

J. G. Mohr and W. P. Rowe, *Fiber Glass,* Van Nostrand Reinhold Co., Inc., New York, 1987.

Engineered Materials Handbook, Vol. 1, *Composites,* ASM International, 1987.

Weeton, Peters, and Thomas, eds., *Engineer's Guide to Composite Materials,* American Society for Metals, Ohio, 1987.

Frank K. Ko
Ping Fang
Drexel University

INORGANIC POLYMERS

This article is confined to descriptions of homoelement, inorganic polymeric compositions. Coverage of other inorganic polymers may be found under the titles Coordination polymers, Organometallic polymers, Phosphorus-containing polymers, Polyphosphazenes, Polysilanes and Polycarbosilanes, Polysulfide polymers, Preceramic polymers, Silicates, Silicones, and Sulfur-containing polymers.

If the term polymer refers to a structure capable of infinite extension, elements known to form homoelement polymers are restricted to the main-group elements of the carbon, nitrogen, and oxygen families (groups 14, 15, and 16) and mercury. Most of the elements of these main-group families, in at least one allotropic form, exist in high molecular weight forms but are not considered polymers.

Related to the structures of elemental forms are Zintl-ion phases, which may be broadly defined as ions of about 4–20 atoms of the heavier group-14, -15, or -16 elements, usually forming cage and cluster structures. The anionic forms are often formed by reduction of the element by an active metal, such as sodium or potassium with cation complexation by a polyether or cryptand macrocycle. The cationic forms are usually obtained in superacid media. Exceptions to the usually discrete and highly condensed structures found for Zintl phases are examples relevant to this discussion: the room-temperature oxidation of sulfur by AsF_5 in a 2:1 mixture of SO_2 and SO_2ClF produces the S_{19}^{2+} ion consisting of two seven-membered rings joined by a five-atom chain. Since the process for converting cyclo-$S_8 \rightarrow S_{19}^{2+}$ is complex, other products, some possibly polymeric, may be available by this or similar routes.

Potentially interesting anionic forms are typified by $SrSb_2$, where antimony is found in infinitely extended, zig-zag anionic chains. Related structures are known in several other alkali and alkaline-earth metal arsenides and antimonides (stibinides). Reactions of these anionic forms with carbocation synthons are possible sources of organometallic polymers (qv).

An important influence of polymer formation and stability is the possibility for ring–chain equilibria, which is a function of the central element and its substitution. The existence of small rings rather than long chains is one of the principal constraints of inorganic polymer chemistry (see MACROCYCLIC POLYMERS; OLIGOMERS; POLYPHOSPHAZENES).

The periodic relationships are important in the preferences of elements for ring or chain structures. If the first-row elements are excluded, no members of groups 11, 12, or 13 form neutral homoatomic ring structures. The group-14 elements show well-developed catenate chemistries and little preference for rings or chains. In groups 15 and 16, however, a clear trend exists: the tendency to form chain structures in preference to rings increases dramatically as a family is descended. The trend is clearly seen in the group-16 elements thermodynamically stable at room temperature, ie, S_8 (ring), Se_∞ (polymer), and Te_∞ (polymer) and in the monoorgano group-15 catenates: $(RP)_x$, few chain structures; $(RAs)_x$, examples of both rings and chains; and $(RSb)_x$, only two confirmed examples of ring structures, but numerous examples of poorly characterized high molecular weight phases.

Group 14

The homoelement chemistry of silicon, germanium, tin, and lead derives from the addition of the R_2E unit (where E = element). The steady decline in homoelement bond energies as the carbon family is descended leads to the conclusion that the only commercially interesting homoelement polymers that can be obtained from group-14 elements other than carbon are silicon polymers.

Group 15

Perhaps the greatest diversity in noncarbon, homoelement polymers is observed in nitrogen, phosphorus, arsenic, antimony, and bismuth.

If one valence of a trivalent group-15 atom is blocked, eg, by an organic group, the remaining valence positions provide a functionality identical to elemental chalcogens.

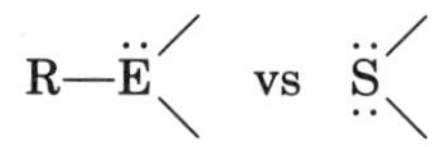
(E = P, As, or Sb)

The analogy has a strong chemical basis as well, ie, many reactions of CH_3AsH_2 strongly resemble those of H_2S; similarly $C_6H_5PCl_2$ resembles SCl_2.

At ambient temperatures and in the absence of low valent transition metal compounds, most organic groups attached to phosphorus or arsenic are inert. The exchange of organic groups is, however, a constant constraint in the syntheses of antimony or bismuth compounds. The homonuclear single-bond energies for group-15 elements are at a maximum for phosphorus for reasons more likely to be related to the inherent weakness of the N—N bond than to the strength for P—P bonds. The similar bond strengths across the second period suggest an efficiency of bond formation determined largely by atom size and the similarity of orbitals involved in bond formation. The usual trend of decreasing bond energies as a family is descended is found.

The potential-energy surfaces characterizing the chemistry of these elements become smoother with increasing atomic weight. Also, as a rough rule, the velocity of a substitution reaction increases approximately several thousandfold from one central element to the next heavier element. Consequently, although most compounds containing P—P bonds are discrete and characterizable, their As and Sb analogues frequently exist only as average assemblages of parts in continual dynamic reorganization.

Nitrogen Polymers. Although chains of up to eight nitrogen atoms have been claimed, and monosubstituted, unsaturated cyclopentamers are known and well characterized, saturated catenates built of R—N< units have not been prepared and would undoubtedly be even less stable than their unsaturated analogues.

Phosphorus Polymers. A number of cyclic, oligomeric organopolyphosphines are known, but no confirmed P—P bonded, high molecular weight compounds exist, and no systematic studies of the potential for such polymeric compositions have been undertaken. Solid products of empirical formula RP accompanying the preparation of polyphosphines have been reported, suggesting the presence of polymeric, possibly open-chain materials. It has been proposed that two of the high-melting crystalline forms of polyphenylcyclopolyphosphine may be polymeric. Both forms, however, decompose to the cyclopentamer at their melting points.

Arsenic Polymers. Thousands of organoarsenic materials, probably with extended As—As bonded structures, have been reported. The confirmed polymeric compositions are limited to a single system $(CH_3As—AsCH_3)_x$, which might better be considered as a weakly associated, extended array ladder structure. The structure is composed of normal 0.25-nm As—As single bonds forming the ladder "rungs," and long 0.29-nm As—As half-bonds forming the "uprights."

Antimony Polymers. Unlike phosphorus and arsenic compounds, primary organoantimony compounds do not readily form stable ring structures. However, an orange-red crystalline compound has been reported, analyzing as $(C_6H_5Sb)_6 \cdot C_6H_6$. The presence of a cyclohexastibine structure has been crystallographically confirmed for the 1,4-dioxane adduct. RSb units have been condensed to characterizable high molecular weight polymers by the reaction of primary alkylstibines, $RSbH_2$, with halogen sources such as I_2, ICl, CI_4, $(CH_3)_2AsI$, organosilicon chlorides, and HCl.

Bismuth Polymers. Very few compounds containing Bi—Bi bonds have been reported. Methylbismuthine is thermally unstable above −78°C; its decomposition yields BiH_3 and $Bi(CH_3)_3$, not $(CH_3Bi)_x$. A few dibismuthines have been prepared, several of which are strongly thermochromic. Reduction of phenylbismuth halides with $LiBH_4$ or $LiAlH_4$ produces a black, ether-insoluble polymeric substance described as $(C_6H_5Bi)_x$.

Mercury Polymers. Oxidation of Hg by AsF_5 or SbF_5 in SO_2 gives $Hg_{2.82}AsF_6$ or $Hg_{2.90}SbF_6$ as a crystalline lattice of the octahedral AsF_6^- (or SbF_6^-) anions in which nonintersecting channels are occupied by nearly linear mercury chains. Two newer mercury chain structures have been reported, $Hg_{2.88}NbF_6$ and $Hg_{2.88}TaF_6$. Both have the same golden metallic appearance as the AsF_6^- and SbF_6^- compounds and all are isomorphous (tetragonal, $I4_1/amd$).

BIBLIOGRAPHY

A. L. Rheingold, ed., *Homoatomic Rings, Chains and Macromolecules of the Main-Group Elements,* Elsevier, Amsterdam, 1977.

D. G. Adolphson, J. D. Corbett, and D. J. Merryman, *J. Am. Chem. Soc.* **99,** 7163 (1977); H. Schäfer, B. Eisenmann, and W. Müller, *Angew. Chem. Int. Ed. Eng.* **9,** 694 (1973); R. G. Teller, L. J. Krause, and R. C. Haushalter, *Inorg. Chem.* **22,** 1809 (1983).

R. J. Gillispie, *Chem. Soc. Rev.,* 315 (1979); R. C. Burns, R. J. Gillispie, and J. F. Sawyer, *Inorg. Chem.* **19,** 1423 (1980).

F. G. A. Stone and W. A. G. Graham, *Inorganic Polymers,* Academic Press, New York, 1972.

A. Rheingold
University of Delaware

INSERTION POLYMERIZATION

High molecular weight linear polymers can be obtained from unsaturated monomers by successive addition to free radicals or ions located at the end of the growing chain or from cyclic monomers by ring-opening polymerization. In the case of unsaturated hydrocarbons, it became evident, particularly from the work of Ziegler and his group and of Natta and co-workers, that polymerization occurs also in the presence of compounds containing highly reactive metal–carbon bonds with a substantially covalent character, eg., aluminum–carbon or transition-metal–carbon bonds:

$$MtR + CH_2{=}CH_2 \longrightarrow Mt{-}CH_2{-}CH_2R$$

$$\text{where } Mt = Al, Ti, V, Co, \text{etc}$$

These polymerizations involve a stepwise insertion of the monomer between the metal atom and the last carbon atom of the growing chain, as proven in many cases by the nature of the end groups of the macromolecules. In some cases, under mild conditions, no termination occurs; each metal–carbon bond produces a single macromolecule and the process can be considered a living polymerization. If chain termination occurs, it usually involves the formation of a new metal–hydrogen or metal–carbon bond which can start a new polymer chain by monomer insertion. Therefore, when termination occurs, the polymerization can be considered a catalytic reaction; each catalytic center produces a large number of macromolecules.

The inorganic moiety of the active organometallic group exerts a large influence on the polymerization process, including its regio- and stereoselectivity. The process has been called polyinsertion or insertion polymerization. The same type of mechanism has been proposed for the polymerization of polar monomers with transition metals.

Another type of insertion polymerization is the synthesis of polymers from cycloolefins via metathesis, which involves a transition-metal–carbon double bond as the catalytic center and a metallacyclobutane derivative as intermediate. This type of polyinsertion is discussed in the articles Metathesis polymerization and Polyalkenamers.

Much less investigated is the insertion of monomers with carbenelike structure into essentially covalent metal–carbon bonds; examples of this type of insertion polymerization are the synthesis of polyiminomethylenes with nickel catalysts, the synthesis of polyalkylidenes, and the copolymerization of carbon monoxide and ethylene with palladium or nickel catalysts.

For conjugated diolefins, chemoselectivity, that is 1,4- vs 1,2-polymerization or, for 2-substituted diolefins, vs 3,4-polymerization, adds complications. These polymerizations can be carried out in the presence of free radicals, free ions, ion couples, or substantially covalent organometallic bonds. Regioselectivity in polymerization of diolefinic hydrocarbons has been little investigated; stereoselectivity appears to be very high in the polymerization catalyzed by organometallic compounds of transition metals and, in the case of isoprene, by lithium alkyls.

Chemoselectivity in the transition-metal-catalyzed polymerization of diene seems in many cases to be controlled mainly by the inorganic moiety of the organometallic bond. Stereoselectivity is mainly controlled by the stereochemistry of the allyl complex formed between the last monomeric unit of the growing chain and the transition-metal atom of the catalytic centers.

The polymerization of other unsaturated monomers, allenes and acetylenes catalyzed by main-group or transition-metal compounds containing organometallic bonds, is believed to take place by inserting the monomer into a substantially covalent organometallic bond (see also Acetylenic polymers).

The exceptional regio- and stereoselectivity observed in insertion polymerization of α-olefins has spurred the investigation of the influence of the counterion in ionic polymerization of monoolefins containing functional groups. Remarkable results have been obtained particularly in anionic polymerization. In some cases, not only the configuration of stereogenic sites in the polymer chain, but also the conformation of the main chain, can be controlled under conditions in which unsolvated ion couples are active sites in polymerization. Diolefinic monomers containing ester groups have also been investigated and chemo-, regio-, and stereoselectivity have been achieved under conditions in which unsolvated ion couples should be present.

General Features of Insertion Reactions

In polymer synthesis, examples of insertion reactions into substantially covalent metal–carbon bonds include insertion of carbenes and molecules with carbenelike structure (:C≡O:, R—N≡C:)

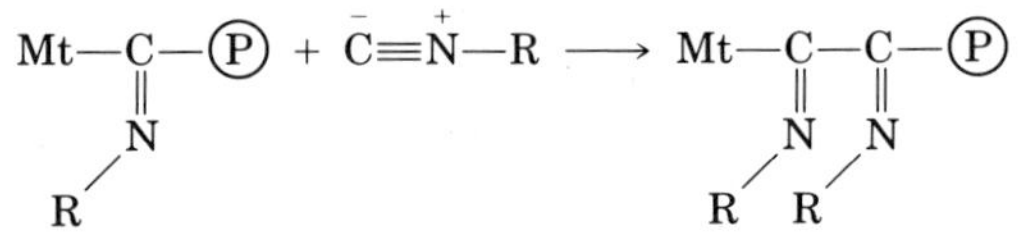

where Mt = a metal complex

of olefinic double bonds,

$$\mathrm{Mt—CH_2—CH_2—(P) + CH_2{=}CH_2 \longrightarrow Mt—CH_2—CH_2—CH_2—CH_2—(P)}$$

of conjugated or cumulated double bonds,

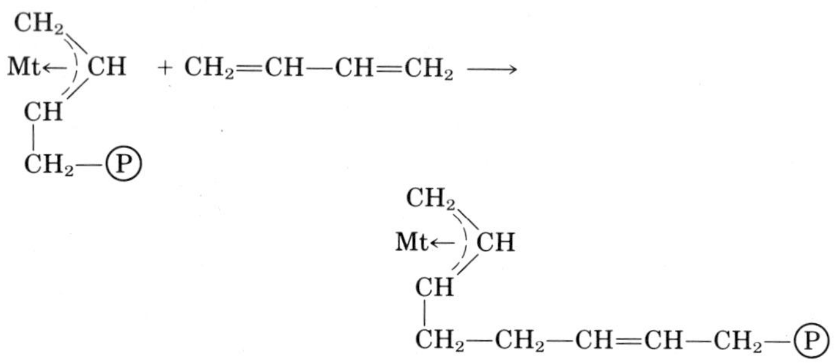

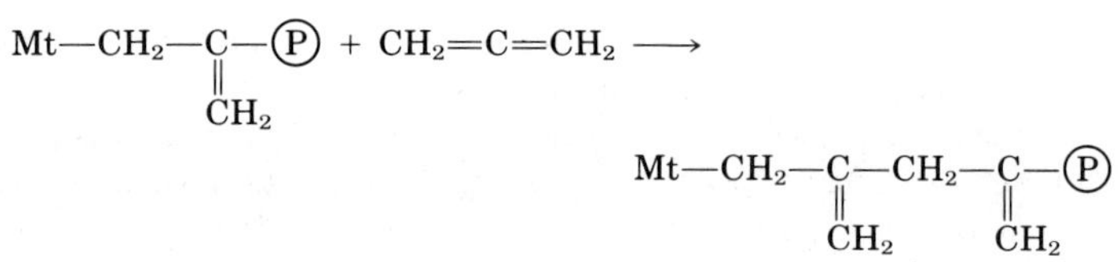

and of triple bonds

$$\mathrm{Mt—CH{=}CH—(P) + CH{\equiv}CH \longrightarrow Mt—CH{=}CH—CH{=}CH—(P)}$$

These types of insertion reactions are well known in organometallic chemistry and the mechanisms have been discussed in detail.

A general feature of these insertion reactions into metal–carbon bonds is the presence of a vacant coordination site at the metal which allows the approach of the monomer to the metal atom and, in some cases, the formation of an intermediate π-complex.

Contrary to the situation in radical polymerization, the formation of the active species is usually not kinetically relevant. Some of the carbon–metal bonds active in polymerizations are relatively stable, and in some cases compounds containing such bond types have been isolated. For the α-olefins the rate of insertion of the first monomer molecule is of the same order of magnitude as the rate of the subsequent insertion steps. However, for diolefin polyinsertion starting with metal alkyls, the first step, consisting of the formation of an η^3-allyl group, can occur at a rate very different from that of the chain growth reaction.

In many cases the metal–carbon bonds are not stable and must be formed immediately before the start of the polymerization or in the presence of the monomer, by alkylation of a transition-metal compound with a stable metal alkyl. When the complex containing the active metal–carbon bond is stable, the kinetics in a homogeneous phase are first order with respect to both the monomer and the organometallic compound. When heterogeneous catalysts are used, different reaction orders are sometimes observed. Propagation rate constants in a homogeneous phase can be very different depending on the organometallic compound used. In general, for ethylene they are lower than those found in radical processes. An exception is the system $Cp_2Ti(CH_3)_2/[Al(CH_3)O]_n$, (where Cp = cyclopentadienyl), which appears to be much more active than all the other organometallic catalysts.

When heterogeneous catalysts are used, propagation rate constants are difficult to determine because the concentration of the active centers can only be roughly estimated and in many cases it decreases with time. In principle, in contrast to radical polymerization, the polyinsertion can have a living character; under certain conditions the average molecular weight increases with time and a Poisson distribution of the molecular weights has been found in the synthesized polymers. However, with transition metals under the usual polymerization conditions, numerous chain-termination reactions occur depending on the type of catalytic system and on the type of monomer used. With heterogeneous systems the molecular weight distribution is very broad. This is probably due to the existence of catalytic sites with different structure; however, other explanations (eg, diffusion-controlled polymerization rate) have been proposed and may affect molecular weight distribution under suitable conditions.

The actual mechanism of the monomer insertion has not been investigated in depth and can be different depending on the monomer and catalytic system used. For monomers with carbenelike structure, a coordination of the monomer to the metal atom bearing the growing chain, followed by migration of the growing chain to the coordinated group, as proved for the insertion of CO into manganese–carbon bonds, is very likely. For the olefins and acetylenes a nucleophilic attack of the growing chain on a monomer occupying a coordination site around the metal has been postulated. Transition states with some charge separation are more widely accepted than completely concerted four-center transition states. Transition states for the 1,4- and 1,2- or 3,4-insertions of conjugated diolefins, as well as for allenes, have been discussed very little.

Monomers with Carbenelike Structures

The research on the polymerization of monomers having a carbenelike structure has concentrated on the polymerization of methylenes and alkylidenes, isonitriles, and carbon monoxide. In all cases, the results indicate that polymerization or copolymerization occur according to an insertion mechanism (see also POLYISOCYANIDES; OLEFIN–CARBON MONOXIDE COPOLYMERS).

Insertion Polymerization of Alkenes

After the 1953 discovery of the "Metallorganische Mischkatalysatoren" by Ziegler and co-workers, olefin polymerization underwent a very rapid development. The insertion nature of the polymerization carried out with catalyst prepared from titanium derivatives and aluminum alkyls was clearly proved by Natta and co-workers investigating the terminal groups of the macromolecular chains, in the case of propylene and ethylene polymerization. A typical catalyst combination was triethylaluminum and either titanium tetrachloride or titanium trichloride. Subsequently, an enormous variety of mixtures of this general type were prepared and used in polymerization, and these are commonly referred to as catalysts of the Ziegler-Natta type. The interest in Ziegler-Natta catalysts resides in their remarkable stereospecificity.

The hundreds of catalytic systems used for olefin polyinsertion are not easy to classify because, in most cases, the exact nature of the catalytic centers is obscure and only the catalyst precursors are known. Table 1 is an attempt to group the catalytic systems according to the nature of the catalyst precursors and to the solubility of the catalytic systems in the reaction medium.

Olefins Used in Insertion Polymerization. Insertion polymerization of olefinic hydrocarbons is highly substrate-selective; in fact, terminal linear α-olefins are polymerized with few exceptions, but 1,1-disubstituted α-olefins are neither homo- nor copolymerized with other monoolefins. Internal olefins do not homopolymerize but give alternating copolymers with ethylene. Among cycloolefins, only cyclobutenes and norbornene have been successfully homopolymerized by double-bond opening and yield highly crystalline polycyclic polymers. Very poor results have been obtained with cyclopentene and 3-methylcylohexene. Nonconjugated diolefins can be cyclopolymerized by using $TiCl_4/Al(C_2H_5)_3$ catalytic systems. 1,6-Heptadiene yields a linear polymer consisting mostly of *cis*-methylene-1-cyclopentyl units.

Copolymerization. Insertion polymerization can yield random, block, and alternating copolymers. Although the well-known Alfrey-Goldfinger approach has been applied to copolymerizations carried out both with heterogeneous and homogeneous catalysts, this approach is justified only in the second case where the catalytically active centers are believed to have substantially similar structure. In the case of heterogeneous catalysts, different catalytic centers display different substrate selectivity. Therefore, different reactivity ratios are expected for the different catalytic sites and no meaningful extrapolation concerning the relative monomer reactivity can be attempted on the basis of the average r_1 and r_2 values measured.

Industrial Significance. Out of the large number of olefin polymers obtained by insertion polymerization, several are produced commercially on enormous scale [high density polyethylene (HDPE), linear low density polyethylene (LLDPE), isotactic polypropylene (PP), ethylene–propylene copolymers (EPR)] and others in lower quantities [isotactic poly(1-butene), isotactic poly(4-methyl-1-pentene)] (Table 2) (see ETHYLENE POLYMERS; PROPYLENE POLYMERS; ETHYLENE–PROPYLENE ELASTOMERS; POLYBUTENES; 4-METHYL-1-PENTENE POLYMERS).

Table 2. World Polyolefins Capacity (1983)

Polymer	10^3 t
LDPE	13,700
HDPE	7,670
HDPE/LLDPE	4,400
PP	8,900
EPR	530
poly(1-butene)	23
poly(4-methyl-1-pentene)	6

Table 1. Types of Catalysts for the Insertion Polymerization of Olefins

Nature of the catalyst precursor	Solubility in the reaction medium	Examples
organometallic compounds or hydrides of nontransition metals	soluble	LiR; BeR_2; AlR_3
organometallic compounds or hydrides of transition metals	soluble	Cp'_2LuCH_3; Cp'_2NdH; $Cp'_2Sc\eta^3\text{-}C_3H_5$[a,b]
	insoluble, supported on inorganic supports	Nph_4Zr on Al_2O_3[c]
catalysts obtained by reaction of a transition-metal derivative with a main-group metal-alkyl derivative (Ziegler-Natta catalysts)	soluble	$Cp_2TiCl_2/(—Al(CH_3)—O—)_n$[d] $VOCl_3/AlCl(C_2H_5)_2$
	insoluble	$TiCl_3/AlR_2Cl$; VCl_3/AlR_2Cl
	insoluble, supported on inorganic supports	$TiCl_4/MgCl_2/AlR_3$
catalysts obtained by reaction of a transition-metal derivative with a transition-metal alkyl	insoluble	$TiCl_3/TiCl_3CH_3$; $Cp_2^dTiR_2/TiCl_4$
catalytic systems obtained from components with no metal–carbon or metal–hydrogen bonds	insoluble	$TiCl_2$; $TiCl_3/NR_3$
	insoluble, supported on inorganic supports	$CrO_3/SiO_2/Al_2O_3$ CoO/MoO_3 reduced with H_2

[a] Cp′ = pentamethylcyclopentadienyl.

[b] η^3-C_3H_5 designates the allyl groups in which all three carbon atoms are bound to the metal.

[c] Nph = neophyl, ie, $(CH_3)_2C(C_6H_5)—CH_2—$ (with C_6H_5 attached to the C).

[d] Cp = cyclopentadienyl.

Polymerization of Conjugated Diolefins

Isomerism Phenomena. Isomerism in polymers of conjugated diolefins is more complex than in vinyl polymers. Butadiene can be polymerized to yield 1,4- or 1,2-monomeric units. The 1,4-units, $—CH_2—CH{=}CH—CH_2—$, contain an internal double bond that can be cis or trans. Therefore, two types of stereoregular 1,4-polybutadienes are possible, namely *cis*-1,4- and *trans*-1,4-polybutadiene. The 1,2-units contain a tertiary carbon atom which can assume two opposite configurations. Therefore, 1,2-polybutadiene shows the same type of isomerism as vinyl polymers, ie, three types of polymers can exist, isotactic, syndiotactic, and atactic.

In polymers from dienes longer than butadiene, ie, $CH_2{=}CH—CH{=}CHR$, R adds to the complexity. Every unit derived from this monomer, whatever its structure, contains two sites of isomerism. The 1,4-units contain an internal double bond, which can be cis or trans, and an asymmetric carbon atom. Therefore, a cis-1,4- or a trans-1,4-unit of this type can exist in two enantiomeric forms. Consequently, whereas *cis*-1,4- or *trans*-1,4-polybutadiene can have only one stereoisomeric form, for the polymers derived from $CH_2{=}CH—CH{=}CHR$, three different cis-1,4- or trans-1,4-stereoisomers are possible, ie, isotactic, syndiotactic, or atactic. Monomers of this type can also give a 1,2- and a 3,4-unit. A 1,2-unit contains a tertiary carbon atom and a double bond, which can be cis or trans. Consequently, monomers $CH_2{=}CH—CH{=}CHR$ can give four types of stereoregular polymers with a 1,2-structure: two isotactic polymers, which differ in the configuration of the double bonds in the side group (all cis in one polymer, all trans in the other), and, analogously, two syndiotactic polymers. A 3,4-unit contains two tertiary carbon atoms. Hence, three types of stereoregular polymers with a 3,4-structure are possible: disyndiotactic, threodiisotactic, and erythrodiisotactic. In summary, a monomer $CH_2{=}CH—CH{=}CHR$ can give, in principle, eleven types of stereoregular polymers. Obviously, 1,4-disubstituted butadienes show even greater complexity (see also STEREOREGULAR LINEAR POLYMERS).

Diolefin Polymers. Transition-metal catalysts are effective for the polymerization of various types of 1,3-dienes:

$$CH_2{=}CH—CH{=}CH_2,\ CH_2{=}CR—CH{=}CH_2,\ CH_2{=}CH—CH{=}CHR$$
$$CH_2{=}CR—CR{=}CH_2,\ CH_2{=}CR—CH{=}CHR,\ CH_2{=}CH—CH{=}CR_2$$
$$CH_2{=}CH—CR{=}CHR \quad \text{and} \quad CHR{=}CH—CH{=}CHR$$

where R = alkyl or aryl. Table 3 reports the types of stereoregular polymers obtained with the most common catalyst systems from butadiene, isoprene, and pentadiene.

Transition-metal Catalysts. Catalysts based on transition metals now available for diolefin polymerization can be divided into three classes as in Table 4.

Polymerization Stages. ***The First Stage.*** The initial stage of the polymerization of butadiene with $(\eta^3\text{-crotylNiI})_2$ has been elucidated by proton nmr. The results show that the insertion of the first butadiene molecule occurs between the metal and the allyl group with reformation of an η^3-butenyl group.

Growth Reaction. The initial stages of the polymerization of butadiene and of monomers of the type $CH_2{=}CR—CH{=}CH_2$

Table 3. Polymer Structures Obtained from Butadiene, Isoprene, and Pentadiene with Common Transition-metal Catalysts

Catalyst system	Butadiene	Isoprene	Pentadiene
AlR_3–$TiCl_4$[a]	65–70% cis-1,4 25–30% trans-1,4 5% 1,2	97% cis-1,4	ca 65–70% cis-1,4
AlR_3–TiI_4[b]	95% cis-1,4		
AlR_3–$TiCl_3$	trans-1,4(cryst) + amorphous products	trans-1,4(cryst) + amorphous products	trans-1,4-isotactic + amorphous products
AlR_3–VCl_3	trans-1,4	trans-1,4	trans-1,4-isotactic
AlR_3–VCl_4[c]	trans-1,4	trans-1,4	
AlR_3–$Ti(OR)_4$[d]	1,2-syndiotactic (low crystallinity)	predominantly 3,4 + minor amounts of cis-1,4-units	cis-1,4-isotactic (ca 60% of the crude products)
$Al(C_2H_5)_3$–$V(acac)_3$[e]	1,2-syndiotactic		
$Al(C_2H_5)_3$–$Cr(acac)_3$[e,f]	1,2-syndiotactic		
$Al(C_2H_5)_3$–$Cr(CNC_6H_5)_6$[f]	1,2-isotactic + amorphous products		
AlR_2Cl–$Co(acac)_2$[e]–H_2O	98% cis-1,4	74% cis-1,4 + 26% 3,4	cis-1,4-syndiotactic + amorphous products
AlR_3–$CoCl_2{\cdot}2Py$[g]	1,2-syndiotactic		
$(R_2Al)_2O$–$CoCl_2{\cdot}2Py$[g]	1,2-syndiotactic		
$Al(C_2H_5)_3$–$Co(acac)_3$–CS_2	1,2-syndiotactic		
$Al(C_2H_5)_2Cl$–$Ni(acac)_2$	85–90% cis-1,4		
AlR_3–$Ni(OCOR)_2$–$BF_3{\cdot}O(C_2H_5)_2$	97% cis-1,4		cis-1,4-syndiotactic
$Al(C_2H_5)_2Cl$–$Nd(OCOR)_3$–AlR_3	98–99% cis-1,4	95% cis-1,4	cis-1,4-isotactic

[a] Al/Ti ratio = 1.
[b] Or $TiCl_2I_2$.
[c] Or $VOCl_3$.
[d] Al/Ti ratio = 7.
[e] Acetylacetonate.
[f] Al/Cr ratio = 6.
[g] Pyridine.

Table 4. Transition-metal Catalysts for the Polymerization of 1,3-Dialkenes

Catalyst precursor	Solubility in the reaction medium	Examples[a,b]
Ziegler-Natta	insoluble	AlR_3–$TiCl_4$; AlR_3–$TiCl_3$
		AlR_3–VCl_4; AlR_3–VCl_3
	soluble	AlR_3–$Ti(OR)_4$; AlR_3–$V(acac)_3$; AlR_3–$Cr(acac)_3$; AlR_2Cl–$Co(acac)_3$
allyl derivatives of transition metals	soluble	$(\eta^3\text{-}C_3H_5)NiX$, where X = Cl, Br, I, $OCOCF_3$
		$(\eta^3\text{-}C_3H_5)_3Cr$
	insoluble	$(\eta^3\text{-}C_3H_5)NiCl + TiCl_4$
precursor with no preformed	soluble	$Rh(NO_3)_3$
		$AlCl_3$–$CoCl_2$
metal–carbon bonds	insoluble	$Ni(PCl_3)_4$–$TiCl_4$

[a] Acac designates acetylacetonate.
[b] η^3-C_3H_5 designates the allyl groups in which all three carbon atoms are bound to the metal.

(R = CH_3, C_2H_5, i-C_3H_7, t-C_4H_9) with $(\eta^3\text{-crotylN}\pi)_2$ have been studied by nmr. The growth reaction occurs by monomer insertion into the η^3-allyl bond between the transition metal and the last polymerized unit. A new η^3-allyl bond is reformed after each insertion.

Termination. The several processes that can bring about the termination of a growing chain have been studied mostly for the polymerization of butadiene and isoprene to cis-1,4-polymers because of the practical interest of these polymers. Depending on the catalyst used and the operative conditions, termination reactions include transfer between the growing chain and the monomer and/or the aluminum alkyl, transfer with a suitable compound introduced in the reaction medium, and monomolecular or bimolecular terminations.

Kinetic Aspects. In most kinetic studies of butadiene and isoprene polymerization, the rate is first order with respect to both monomer and catalyst concentration; different dependencies have also been observed. Reported results are sometimes difficult to compare, because the experimental conditions are not identical.

Chemoselectivity and Stereoselectivity of the Insertion Reaction. The chemoselectivity of the polymerization, ie, the formation of 1,4- vs 1,2-monomeric units, is determined by the fact that a η^3-butenyl group has two reactive points, C-1 and C-3. Reaction of the incoming monomer at C-1 gives rise to the formation of a 1,4-monomeric unit, whereas reaction at C-3 gives rise to a 1,2-unit:

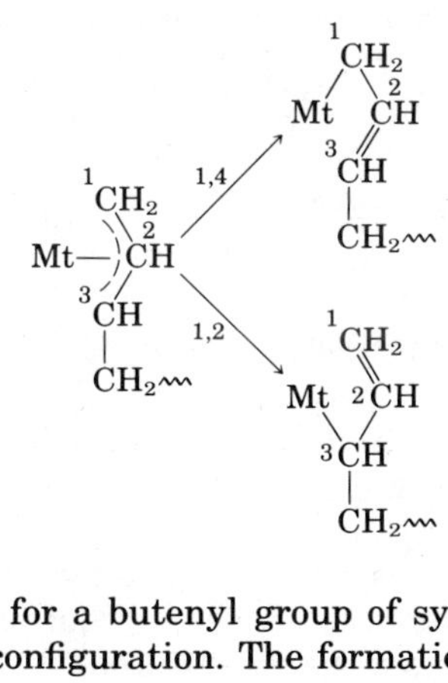

This holds both for a butenyl group of syn configuration (as shown) or anti configuration. The formation of a cis-1,4- or a trans-1,4-unit is related to the anti or syn structure of the butenyl group

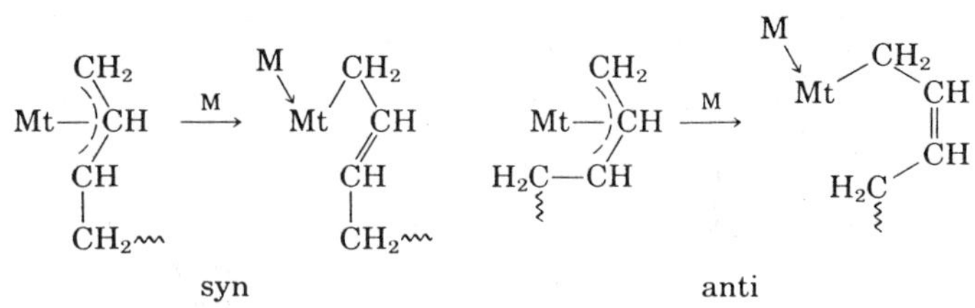

where M = monomer

Various tacticities obtain for 1,2 and 1,4 structures according to previous descriptions.

Copolymerization of Dienes. Several copolymerizations of dienes with transition-metal catalysts have been reported. For most of them the monomer reactivity ratios have been determined and, in most cases, the microstructures of the units of each comonomer are the same as in the homopolymers. The butadiene–isoprene copolymerization, in particular, has been studied.

Industrial Significance of Conjugated Diolefin Polymerization. *cis*-Polybutadiene is currently produced on an industrial scale and occupies a well-defined position in the complex market of elastomeric products. *cis*-Polyisoprene is produced in limited amounts, because it is not price-competitive with natural rubber (see ISOPRENE POLYMERS).

Propylene–butadiene alternating copolymers and butadiene–pentadiene copolymers have been produced in pilot plants, but no large-scale production has followed. *trans*-Polyisoprene has been manufactured in the UK, since the natural polymer, balata, is expensive and in limited supply. Transition-metal-catalyzed polymerization of 1,3-dienes has provided a variety of new products, but from a practical standpoint, it has not matched the importance of olefin polymerization.

Polymerization of Cumulated Diolefins

Allene has been polymerized to high molecular weight linear polymers by various catalysts derived from aluminum alkyls and compounds of Ti, V, Cr, Mn, Fe, Co, and Ni. Crystalline and amorphous polymers have been produced, in various ratios, with each of the catalysts used. The crystalline polymers

Table 5. Polymerization of Monomers Containing Heteroatoms with Ziegler-Natta Catalysts

Monomer	Catalyst	Polymer structure
$CH_2{=}CH{-}CH_2{-}Si(CH_3)_3$	$TiCl_4$–$Al(C_2H_5)_3$	crystalline isotactic
$CH_2{=}CH{-}SiH_2(C_2H_5)$	VCl_3–$Al(C_2H_5)_3$	crystalline isotactic
$CH_2{=}CH{-}(CH_2)_n{-}N(i\text{-}C_3H_7)_2$, where $n = 2,3$	δ-$TiCl_3{\cdot}0{,}3AlCl_3(i\text{-}C_4H_9)_2AlCl$	crystalline
N-alkenylcarbazole	γ-$TiCl_3$–$(C_2H_5)_2AlCl$	crystalline isotactic
$CH_2{=}CH{-}(CH_2)_3{-}X{-}Si(CH_3)_3$, where X = O or i-C_3H_7N	δ-$TiCl_3{\cdot}0{,}3AlCl_3(i\text{-}C_4H_9)_2AlCl$	crystalline[a]
$CH_2{=}CH{-}(CH_2)_9\,I$	$TiCl_3$–$(C_2H_5)_2AlCl$	amorphous
p-methoxystyrene	$TiCl_4$–$Al(i\text{-}C_4H_9)_3$	amorphous
halogenated styrenes	$TiCl_4$–$Al(i\text{-}C_4H_9)_3$	[b]

[a] X = O; after hydrolysis a crystalline polyalcohol has been obtained.
[b] *o*-Fluorostyrene: amorphous; *p*-fluorostyrene: crystalline.

consist predominantly of units of the form $-CH_2-\underset{\underset{CH_2}{\|}}{C}-$ whereas the amorphous polymers also contain methine units with pendent vinyl groups and units $-CH{=}CH{-}CH_2-$, exclusively of cis configuration.

Acetylene Polymerization

Acetylene polymerization initiated by Ziegler-Natta catalysts is believed to occur through insertion of the monomer into a Ti—C bond. Acetylene was first polymerized by Natta and coworkers by the $Ti(OC_4H_9)_4$–$Al(C_2H_5)_3$ catalyst to a crystalline polymer having a predominantly trans enchainment of the conjugated double-bond sequences. The polymer has increasingly higher cis content as the polymerization temperature is decreased and an 88% *cis*-polyacetylene is produced at 195 K by the same catalyst.

Monomer insertion into a $Ti{-}CH_3$ bond has been proven by nmr for the Cp_2TiCl_2—$(CH_3)_2AlCl$-catalyzed polymerization of phenylacetylene. The microstructure of polyacetylene suggests a cis insertion mechanism leading to a cis–transoid structure at low temperature. The increasing trans content with increasing polymerization temperature has been explained by a cis–trans, thermally activated isomerization, which probably takes place before crystallization of the chain segment.

A large number of monosubstituted acetylenes have been polymerized including alkyl-, alkenyl-, aryl- and aryloxyacetylenes. Also, optically active alkynes have been polymerized yielding optically active polyacetylenes (see ACETYLENIC POLYMERS).

Monomers Containing Heteroatoms

Insertion polymerization of monomers containing heteroatoms has been less investigated than polyinsertion of unsaturated hydrocarbons. Three groups of monomers can be considered: monomers in which the heteroatoms are not electronically interacting with the double bond to be polymerized, monomers in which groups containing the heteroatoms are directly bound or conjugated with the olefinic double bonds, and monomers in which unsaturated bonds between carbon atoms and heteroatoms are polymerized.

Because of the very high reactivity of the organometallic complexes that polymerize olefins, heteroatoms usually hinder the insertion reaction into carbon–metal bonds. However, some monomers containing heteroatoms have been homopolymerized and/or copolymerized with olefins, using typical Ziegler-Natta catalysts, by minimizing the interaction between the heteroatom and the catalyst (Table 5). This has been accomplished by various methods: the heteroatom in the monomer and the metal–carbon bond in the catalyst precursors are sterically shielded; the electron-donating character of the heteroatom is lowered by attaching particular groups to it; the heteroatom has a low reactivity; the heteroatom is complexed with aluminum alkyls or with the metal alkyl precursor of the catalyst; the heteroatom is separated from the double bond by more than one methylene group.

BIBLIOGRAPHY

K. Ziegler, *Angew. Chem.* **64**, 323 (1952); G. Natta, P. Pino, G. Mazzanti, U. Giannini, E. Mantica, and M. Peraldo, *Chim. Ind. Milan* **39,** 19 (1957); G. Natta, P. Pino, E. Mantica, F. Danusso, G. Mazzanti, and M. Peraldo, *Chim. Ind. Milan* **38,** 124 (1956).

H. G. Elias, *Makromoleküle,* Hütig and Wepf Verlag, Basel, FRG, 1971.

T. Keii and K. Soga, eds., *Catalytic Polymerization of Olefins,* Elsevier Science Publishing Co., New York, 1986.

R. P. Quirk, H. H. Hsieh, G. B. Klingensmith, and P. S. T. Tait, eds., *Transition Metal Catalyzed Polymerizations, Part A,* Harwood Academic Publishers, London, New York, 1983.

INSULATION.

See ACOUSTIC PROPERTIES; CELLULAR MATERIALS; ELECTRICAL PROPERTIES; ENERGY APPLICATIONS.

INSULATION BOARD.

See COMPOSITION BOARD.

INSULATORS.

See ACOUSTIC PROPERTIES; ELECTRICAL PROPERTIES; ELECTRICALLY CONDUCTING POLYMERS.

INSULIN.

See PROTEINS.

INTERFACIAL POLYMERIZATION

Between 1944 and 1960, techniques were gradually developed for the preparation of condensation polymers at low temperature (Table 1). The interfacial process employs water and an organic liquid, whereas the solution method uses a single liquid medium. These processes provide the required speed, need only simple equipment, and are widely used.

Table 1. Interfacial Condensation Polymerizations

Polymer	Date
urethanes	1944–45
urethanes	1954–55
amides, urethanes	1952
amides	1953
carbonates[a]	1954
amides	1955
urethanes	1956
aromatic esters[a]	1957
aromatic sulfonates	1960
amides and others	1958
aromatic amides	1959
nylon-rope experiment	1959

[a] Early use of phase-transfer agents.

In interfacial polycondensation, the two fast-reacting intermediates are dissolved in a pair of immiscible liquids, one of which is preferably water. The water phase contains the diamine or diol and usually an inorganic base to activate any diol and neutralize the by-product acid. The other phase consists of the diacid chloride and an organic liquid such as dichloromethane, toluene, or hexane. Polymer formation takes place at or near the liquid–liquid interface when the two solutions are brought in contact or stirred together. By this process and its variations, thousands of polymers have been synthesized. Some of its features are shown in Table 2.

Polycondensation without Stirring

If the two solutions for an interfacial polycondensation are brought together without stirring and if the polymer is not soluble in the organic liquid, certain polymers form as thin films at the interface. Under the right conditions, this film is coherent and fairly strong. It has a high molecular weight and can be pulled from the area of the interface as a continuous collapsed tube. The experiment is called the nylon rope trick.

Mechanism of Polymer Formation. Many water-insensitive polymers seem to form at the interface or in the organic phase near the interface. The aqueous phase acts as a reservoir for one type of intermediate and the acid acceptor and extracts any by-product acid from the polymerization zone.

Applications. The film or rope of collapsed film can be shredded with a blender or other device to yield fibrous particles (fibrids), which form tough paper or that can be used to bind other chopped fibers into paper. Interfacial polycondensation has been used to form cell-like capsules containing enzymes and other materials for use in artificial kidneys, catalysis, and ion exchange. Membranes have been prepared for use in reverse osmosis.

Table 2. Interfacial Polycondensation Processes

Conditions	
time	5 min to several hours
temperature	0 to 100°C
pressure	101.3 kPa (1 atm)
equipment	simple; often open but may be closed for gaseous reactants
stirring	low-to-high speed
Polymer characteristics	
structure	wide range possible
molecular weight	low to high
copolymers	some limitations
Product recovery	
polymer isolation	may need nonsolvent precipitant
	filtration, washing, and drying
polymer yield	low to high
by-products and solvents	recovery or disposal required

Polycondensation with Stirring

Polymerization with stirring offers more versatility. It may be applied to polymers that precipitate rapidly or to those that form metastable or stable solutions; conditions may be adjusted to the reaction. For rapid reactions with polymer precipitation, a laboratory blender at high speed is effective for small-scale preparations. Slower reactions with the polymer in solution may be stirred at moderate speeds. Although fast polymerizations are completed in a few minutes or as soon as the two phases have been thoroughly mixed, the process is also applied to reactions that require up to several hours at elevated temperatures.

Polymerization with Water-immiscible Solvents. The mechanism of polymer formation in a stirred system is the same as that for an unstirred system. The polymer may form a precipitate or remain in solution in the organic phase, as determined by the polymer solubility. Many reactions used in interfacial polycondensation have kinetic rates estimated to be 10^2–10^6 L/(mol·s).

Polymerization in Water-miscible Solvents. Completely or partially water-miscible solvents can be used for polyamides, polyesters, polyurethanes, polysulfonamides, and in particular for systems with a diacid chloride component of relatively high hydrolytic stability. Hydrolysis is reduced by use of weakly alkaline acid acceptors, eg, sodium carbonate.

Vapor–Liquid Interfacial Polycondensation. The gas– or vapor–liquid process is useful when one of the intermediates is a vapor at ambient or slightly higher temperatures. It is mostly employed in the preparation of the polycarbonates with phosgene, but it is also useful for the preparation of polyureas from phosgene and a diamine and for polyoxamides from oxalyl chloride. The reactant vapor is introduced into the system in an inert gaseous diluent. The liquid system can be the aqueous phase alone or may also contain an immiscible organic solvent.

Copolymerization. The formation of completely random copolymers in interfacial processes is difficult because conditions interfere with the uniform introduction of two or more similar intermediates into the growing polymer chains.

Water-insoluble Reactants. Occasionally a reactant, which would normally be placed in the aqueous phase, has little or no water solubility. In this case, for example, a diamine such as bis(4-aminocyclohexyl)methane is dissolved in a portion of the organic solvent added to the aqueous phase.

Polycondensation in Two-phase Nonaqueous Systems. Nylon-6,10 film can be prepared in xylene–ethylene glycol. Other polyamides are prepared in stirred systems with glycol, glycerol, and a cyclic sulfone instead of water. Several nonaqueous, two-phase media have been evaluated for preparation of organometallic polymers.

Polycondensation Without Organic Solvents. When the diacid chloride or the analogous reactant is a vapor or a liquid, the polymer can be prepared at an aqueous interface. Thus polycarbonates are prepared from phosgene vapor and aqueous alkaline solutions of bisphenols.

Phase-transfer Agents. Phase-transfer agents (PTA), also called catalysts or accelerators, are used in the synthesis of small molecules. Their addition to a two-phase system comprised of a water-immiscible solvent and water promote a chemical reaction by enhancing the transfer of the aqueous intermediate across the interfacial boundary into the presence of the complementary intermediate. Addition of phase-transfer agents increases the rate of polymer formation and results in higher molecular weight and yield. Polyethylene glycols; tertiary aliphatic amines; quaternary ammonium, phosphonium, arsonium, and sulfonium compounds; crown ethers; and similar substances act as phase transfer agents.

Applications. Interfacial polycondensation with stirring is widely applied in research on polymer synthesis. Polycarbonates are manufactured commercially by this process. Interfacial polycondensation in water-miscible solvents is particularly useful in the preparation of many aromatic polyamides, which are frequently obtained with low molecular weight when water-immiscible solvents are used.

Health and Safety Factors

Interfacial polycondensations are no more hazardous than most simple laboratory procedures. The usual care should be observed in the storage, handling, and disposal of solvents and reagents.

BIBLIOGRAPHY

P. W. Morgan, *Condensation Polymers by Interfacial and Solution Methods,* Wiley-Interscience, New York, 1965.

F. Millich and C. E. Carraher, Jr., eds., *Interfacial Synthesis,* Vols. 1 and 2, Marcel Dekker, New York, 1977.

L. J. Mathias and C. E. Carraher, eds., *Crown Ethers and Phase Transfer Catalysis in Polymer Science,* Plenum Press, New York, 1984.

J. K. Stille and T. W. Campbell, eds., *Condensation Monomers,* John Wiley & Sons, Inc., New York, 1972.

PAUL W. MORGAN
Consultant

INTERFACIAL PROPERTIES

The two most common properties associated with interphases are surface tension and interfacial tension. The surface tension is a controlling factor in processes involving wetting or coating operations (see also ADHESION AND BONDING). Polymer–polymer interfacial tension and structure greatly affect the mechanical properties of adhesive joints, multiphase block copolymers and polymer blends and composites or laminates. In addition, interfacial tension is a primary factor in controlling rheological properties, stability, and domain size in polymer–polymer dispersions.

The technological importance of surface-interfacial tension and morphology has led to development of numerous thermodynamic theories of the surface and interfacial properties of polymeric materials and experimental methods for their characterization. In this article, some of the salient features of polymeric interphase behavior and the associated theories are reviewed, in particular, the general behavior of vapor–liquid and liquid–liquid polymeric interphases. For these systems, an extensive body of knowledge exists pertaining to interfacial properties such as surface and interfacial tension and their dependence on polymer structure and environment. In addition, equilibrium theories that predict the interfacial properties, surface tension, interfacial tension, and interphase thickness have been developed. These theories have also been applied to the interfacial properties of block copolymers. Block copolymers are amenable to controlled synthesis and serve as convenient model systems through which the effects of interphase structure on properties have been ascertained and can be effectively illustrated.

Thermodynamics of Interphases

In discussing the classical thermodynamics of surfaces or interfaces, it is convenient to employ the concept of an interphase as a boundary region S between coexisting bulk phases, α and β, as pictured in Figure 1. The bulk phases are isotropic in their properties, whereas the properties in the interphase change gradually between those of the two bulk phases. In all planes parallel to the plane of the interphase, parallel to AA′ or BB′, the forces arise only from hydrostatic pressure P and are equivalent. In contrast, the force balance for planes that cross the interphase, perpendicular to AA′, includes an additional term due to the interfacial tension γ. This force is associated with the anisotropic nature of intermolecular forces that result from the concentration gradient within the interphase.

The influence of the interfacial tension on the thermodynamics can be illustrated by considering the work W performed on the interphase when an additional interphase is formed. If the interphase volume increases by a factor dV^s, ie, a thickness increase of $d\tau$ and an area increase of dA^s, the force balance leads to

$$\begin{aligned} W &= -PA^s d\tau - (P\tau - \gamma)dA^s \\ \text{or } W &= -PdV^s + \gamma dA^s \end{aligned} \tag{1}$$

where τ is the thickness of the interphase, A^s its area, and V^s its volume. This result differs from the familiar relation for bulk phases in the appearance of the interfacial tension force. Incorporation of this expression into the first and second laws of thermodynamics for multiconstituent systems yields the classic relation for the internal energy change for an interphase

$$dU = TdS - PdV^s + \gamma dA^s + \sum_i \mu_i \, dN_i \tag{2}$$

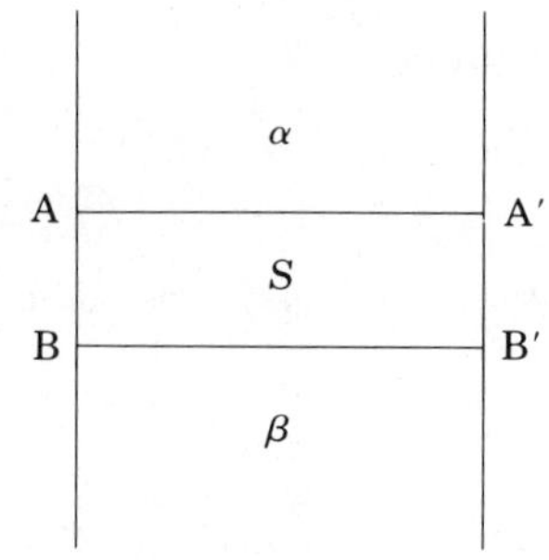

Figure 1. Definition of an interphase. Courtesy of Ellis Horwood, Ltd.

where T is the thermodynamic temperature and S the entropy; N_i and μ_i are the number of moles and the chemical potential, respectively, for the ith constituent.

Vapor–Liquid Interfaces

Homopolymers. The single most accessible experimental parameter for the characterization of interphase thermodynamics is interfacial tension. In the case of vapor–liquid interfaces, it is usually referred to as the surface tension. A number of empirical relations correlate these data. The most general of these is the Macleod relation

$$\gamma = \gamma^0 \rho^n \tag{3}$$

which was developed originally for small molecule liquids. The effects of molecular weight and temperature are contained solely within the mass density term ρ. The constant γ^0 is related only to the chemical structure and can be written in terms of the parachor $\mathcal{P}$:

$$\gamma = \left(\frac{\mathcal{P}}{m}\right)^n (\rho_\ell - \rho_g) \tag{4}$$

where the subscripts ℓ and g refer to liquid and gas, respectively. The power-law exponent n is 4 for small molecule liquids and m is the molecular weight. When applied to polymers, ρ_g is assumed to be negligible; m refers to the formula weight of the monomer and n is between 3 and 4.

The temperature dependence can be derived by differentiation of equation 3 to give

$$-d\gamma/dT = n\alpha_T\gamma \tag{5}$$

where α_T is the thermal-expansion coefficient. The surface tension of polymers is usually a linear function of temperature.

The Macleod equation also provides a framework for understanding the effect of molecular weight on homopolymer surface tension. Free volume changes due to end groups have been used to account for the molecular weight dependence of bulk polymer properties. In the case of mass density, the familiar Fox-Flory relation can be incorporated into equation 3 to give

$$\gamma^{1/4} = \gamma_\infty^{1/4} - \frac{k_s}{M_n} \tag{6}$$

where the subscript ∞ refers to the infinite molecular weight value and M_n is the number average molecular weight. The constant k_s is a function of the molar volumes and parachors for the monomer repeat unit and end groups. The applicability of this relation is confirmed in Figure 2, which also shows, however, that the relation

$$\gamma = \gamma_\infty - \frac{k_e}{M_n^{2/3}} \tag{7}$$

applies equally well to the results. The constant k_e is

$$k_e \simeq (\gamma_\infty - \gamma_e)(2m)^{2/3} \tag{8}$$

with γ_e reflecting the surface tension of the end groups.

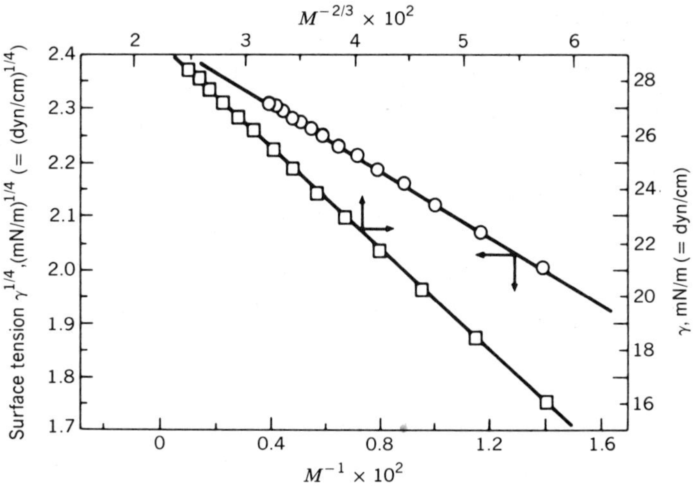

Figure 2. Variation of the surface tension of n-alkanes with molecular weight: □ refers to upper and right scale; ○ refers to lower and left scale. Courtesy of Marcel Dekker, Inc.

By combining the generalized density-gradient theory of inhomogeneous fluids and the compressible-lattice-fluid theory of polymers, a microscopic theory was developed that describes both bulk and surface properties. The reduced surface tension is shown to be

$$\tilde{\gamma} = \frac{\gamma}{\gamma^*} = 2\int_{\tilde{\rho}_g}^{\tilde{\rho}_\ell} (\tilde{\kappa}\Delta\tilde{a})^{1/2} d\tilde{\rho} \tag{9}$$

where γ^* is a characteristic surface tension.

Multiconstituent and Multiphase Polymer Systems. In multicomponent systems such as copolymers, polymer blends, and solutions, the low-energy component tends to segregate preferentially or adsorb at the surface.

Liquid–Liquid Interphases

Immiscible Binary Homopolymer Blends. The term interfacial tension is used here to differentiate liquid–liquid interfacial properties from those of the gas–liquid interface. The general behavior of homopolymer–homopolymer melt interfacial tension is similar in many respects to that discussed for homopolymer melt surface tension. The interfacial tension decreases linearly with temperature, as shown in Figure 3. The temperature coefficients $d\gamma/dT$ are smaller than the corresponding values for surface tension, and typically fall below -0.03 mN/m (= dyn/cm) per °C. The molecular weight dependence usually follows the empirical $M_n^{-2/3}$ dependence represented by equation 7.

Multicomponent Systems—The Effect of Additives. The effective interfacial tensions of binary homopolymer blends may be altered appreciably by adding different types of materials that behave as surface-active agents.

Solid Interfaces

The description and characterization of interfaces containing solid materials present considerable difficulties from both an experimental and theoretical viewpoint. Whereas newly formed liquid surfaces and interfaces eventually assume equilibrium conformations, the same is not true for solid surfaces. The state

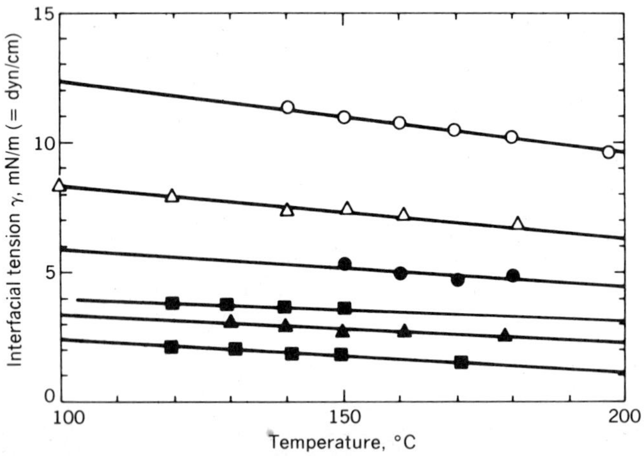

Figure 3. Interfacial tension vs temperature for immiscible binary homopolymer mixtures: ○, poly(vinyl acetate)–linear polyethylene; △, poly(vinyl acetate)–polyisobutylene; ●, poly(*n*-butyl methacrylate)–linear polyethylene; □, polychloroprene–branched polyethylene; ▲, poly(*n*-butyl methacrylate)–poly(vinyl acetate); ■, poly(methyl methacrylate)–poly(*n*-butyl methacrylate). Courtesy of Marcel Dekker, Inc.

of solid surfaces is therefore thermodynamically metastable and cannot be described by the classical thermodynamics discussed previously. Furthermore, the surface energy of solids is found to be generally anisotropic, dependent on the orientation of crystal-growth faces, spatial variations in surface structure and composition, and surface roughness. In addition, the characterization of solid surfaces is severely hindered by the lack of direct methods for measuring surface and interfacial tension. The behavior of polymers at solid surfaces is important in a number of problems, including coatings, adhesion, wetting, adsorption, and filled composites.

Block Copolymer Structure. A number of thermodynamic models for block copolymer morphology have been proposed. In more refined methods, the concept of a finite interphase region of intermediate composition was applied and the interfacial-composition profile calculated. The latter theories employ the diffusion-equation approach, as described earlier, and required adoption of a microdomain geometry, spherical, cylindrical, or lamellar. The primary focus is the calculation of the microdomain repeat period and interphase thickness, parameters that may be measured experimentally. Some theories yield an interphase thickness identical to that derived for polymer blends.

An extremely novel approach to the study of polymer interfaces has been the examination of so-called tapered block copolymers. These are tri-block copolymers wherein the central block is a random or tapered copolymer comprised of the two end-block components. The effective interphase thickness can be controlled by varying the relative proportion of this central block. The microdomain repeat distance D is unaffected by the variable interphase size and follows the $D \sim M_n^{2/3}$ dependence observed for similar styrene–isoprene copolymers. The interphase thickness becomes larger with increasing random copolymer content. Calculations of interphase thicknesses from electron micrographs and small-angle x-ray scattering experiments agree well and document a linear correspondence between interphase thickness and midblock content.

The Effect of Interphase Structure on Block Copolymer Properties. There is a strong body of evidence that the structure and nature of the interphase region affect the thermal, mechanical, and rheological properties of block copolymers. The most systematic study was a series of experiments performed on block copolymers with a variable interphase content. The materials examined are styrene–isoprene block copolymers containing a midblock of styrene–isoprene random copolymer. They are designated SI-*XX*, where *XX* is the weight percent of the interblock. The effect of the variable interphase on the dynamic mechanical properties is demonstrated in Figure 4.

Experimental Methods

Surface and Interfacial Tension. A great number of methods have been developed for measuring polymer surface and interfacial tension. The most appropriate methods for polymeric liquids are the Wilhelmy plate method and the drop-profile methods. The surface tension of solid polymers cannot be measured directly, and several indirect methods have been proposed. These methods invoke various restrictive assumptions and usually require extrapolation from the liquid state.

The intercept of the plot at $\cos\theta = 1$ gives γ_c. In practice, γ_c varies with testing fluids, and Zisman plots are complicated by scatter and curvature. The other methods described give more consistent and reliable results and are preferred.

Interfacial Structure. *Scattering Methods.* Scattering methods are particularly useful in studying block copolymer interfaces.

Other Methods. Microscopy has been used in several investigations of interphase structure. Nuclear magnetic resonance techniques rely on the discrimination of mobility differences to provide an indirect characterization of interphases. Infrared methods applicable to surface characterization can be used to

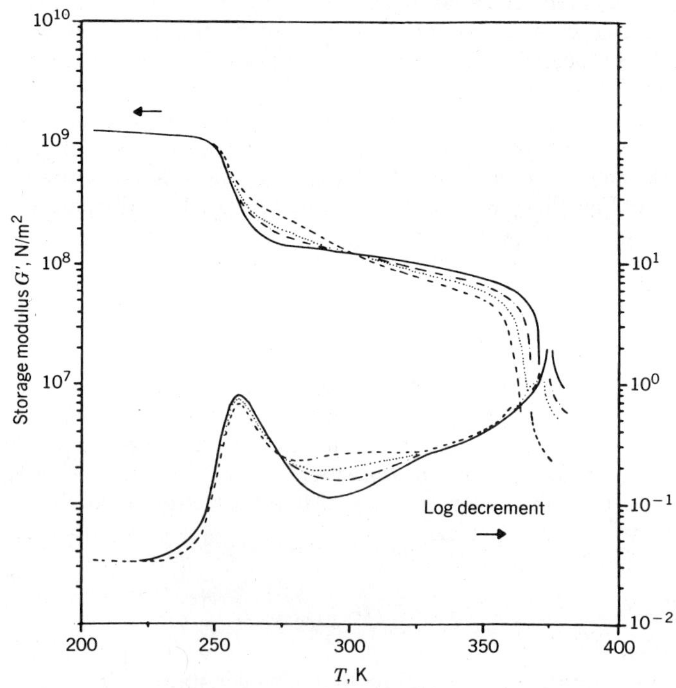

Figure 4. Dynamic mechanical properties of poly(isoprene-*b*-(isoprene-*co*-styrene)-*b*-styrene) copolymers: —, SI-0; -·-·-, SI-10; ····, SI-20; ---, SI-33, where numbers represent weight percent of the interblock. Courtesy of Hüthig & Wepf Verlag.

determine surface chemical structure, composition depth profiles (by angle-resolved measurements), and surface orientation of polymer chains.

BIBLIOGRAPHY

M. J. Jaycock and G. D. Parfitt, *Chemistry of Interfaces,* Ellis Horwood, Ltd., Chichester, UK, 1981.

K. Solc, ed., *Polymer Compatibility and Incompatibility: Principles and Practice,* Vol. 2, MMI Symposium Series, Michigan Molecular Institute, Harwood Cooper Station, N.Y., 1982.

S. Wu, *Polymer Interface and Adhesion,* Marcel Dekker, Inc., New York, 1982.

K. L. Mittal, ed., *Physiochemical Aspects of Polymer Surfaces,* Vol. 1, Plenum Publishing Corp., New York, 1983.

JEFFREY T. KOBERSTEIN
The University of Connecticut

INTERNALLY COLORED COPOLYMERS. See COLORANTS; DYES, MACROMOLECULAR.

INTERNAL PLASTICIZER. See PLASTICIZERS.

INTERPENETRATING POLYMER NETWORKS

An interpenetrating polymer network (IPN) is an intimate combination of two polymers both in network form, at least one of which is synthesized or cross-linked in the immediate presence of the other. Unlike chemical blends, there are no induced covalent bonds between the two polymers; monomer A reacts only with other molecules of monomer A, as does B. A schematic representation of an ideal IPN is shown in Figure 1. In sequential synthesis, polymer A is swollen in the presence of monomer B and a cross-linking agent, and B is polymerized. In simultaneous synthesis, monomers A and B are cross-linked and polymerized by way of noninterfering modes. Thus, in addition to mechanical blending and copolymerization, IPNs represent a third mechanism by which different polymers can be combined.

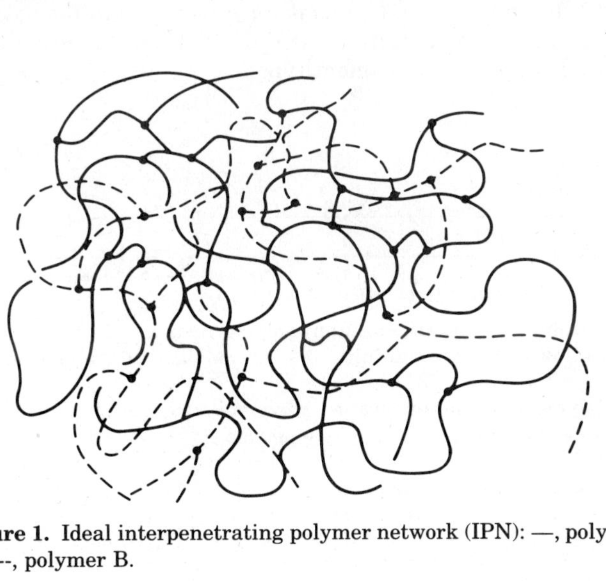

Figure 1. Ideal interpenetrating polymer network (IPN): —, polymer A; ----, polymer B.

The chemical and physical combination of two or more structurally dissimiliar polymers provides a convenient route for the modification of properties to meet specific needs. It facilitates processing and may impart flexibility, tensile and impact strength, chemical resistance, weatherability, flammability resistance, or other properties. The physical properties of the combined polymers depend on the properties of the polymers and the way they are combined.

An IPN is distinguished from simple polymer blends, blocks, or grafts in two ways: it swells but does not dissolve in solvents, and creep and flow are suppressed.

In sequential interpenetrating polymerization, monomer I is combined with cross-linking agent and initiator to form network I. Network I is then swollen in monomer II containing cross-linking agent and initiator to form network II (Fig. 2**a**). Simultaneous interpenetrating networks (SINs) (Fig. 2**b**) are synthesized by combining monomers or linear prepolymers of the two monomers together with their respective cross-linking agents and catalysts, in bulk (melt), solution, or dispersion. The individual monomers are polymerized by chain or stepwise polymerization. Reaction between the polymers is usually prevented due to the different polymerization modes. If one of the two polymers is in network form (cross-linked) and the other a linear polymer, not cross-linked, a semi-IPN results (Fig. 2**c**). If the polymerizations are sequential in time, four semi-IPNs may be distinguished. If polymer I is cross-linked and polymer

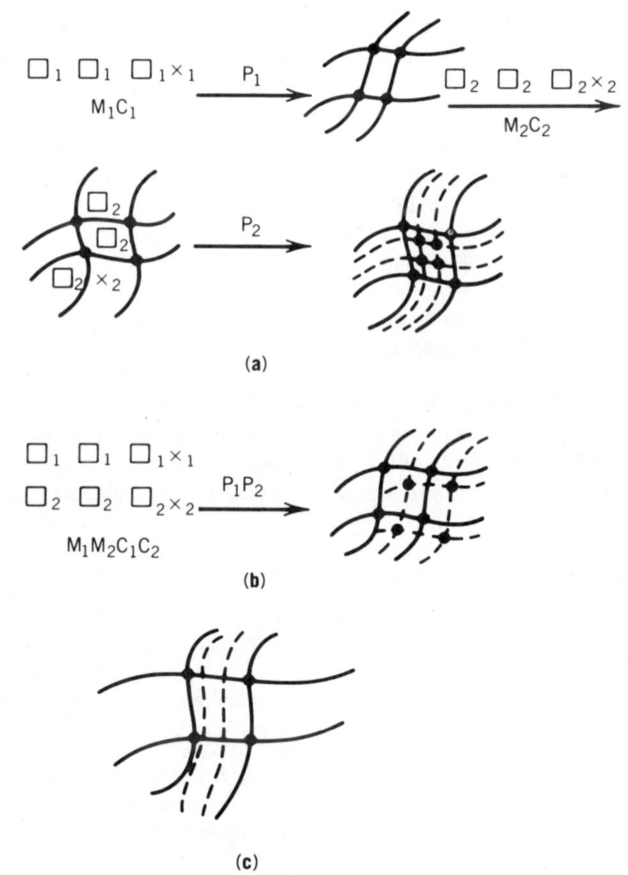

Figure 2. Synthesis of IPNs: (**a**) sequential IPN; (**b**) simultaneous interpenetrating network (SIN); (**c**) semi-IPN. Network I, solid lines; network II, dotted lines; cross-link sites, filled circles. M = monomer; C = catalyst; × = cross-linking agent; P = polymerization.

II is linear, the product is called a semi-I IPN. If polymer I is linear and polymer II is cross-linked, a semi-II IPN results. For simultaneous polymerization, only two semi-IPNs can be distinguished. The different categories of IPNs are shown in Table 1.

Interpenetrating polymerization is a mode of blending two or more polymers to produce a mixture in which phase separation is not as extensive as it would be otherwise. It is the only way of combining cross-linked polymers. Normal blending or mixing results in a multiphase morphology due to the well-known thermodynamic incompatibility of polymers (see COMPATIBILITY; MICROPHASE STRUCTURE). This incompatibility arises from the relatively small gain in entropy on mixing the polymers due to contiguity restrictions imposed by their large chain length. However, if mixing takes place on a lower molecular weight level and polymerization is achieved simultaneously with cross-linking, phase separation may be kinetically controlled, since the entanglements have been made permanent by cross-linking. In other words, phase separation cannot occur without breaking covalent bonds.

Synthesis

Sequential IPNs. The term sequential refers to the time order of polymerization. Polymer I and its multifunctional cross-linking agent, divinylbenzene, are polymerized first. Polymer I is usually an elastomer because of its ease and rapidity of swelling. Above the T_g, extreme care must be taken to prevent solvent-type stress cracking. Monomer II is then swollen into the polymer I network and cross-linked via its initiator and corresponding cross-linking agents. The cross-link density in polymer I limits the extent of swelling. Sequential IPNs include poly(ethyl acrylate) with polystyrene, poly(methyl methacrylate), poly(styrene-*co*-methyl methacrylate), poly(*n*-butyl acrylate) with polystyrene, styrene–butadiene with polystyrene, castor oil with polystyrene, filled sequential IPNs, anionically polymerized sequential IPNs, polybutadiene with polystyrene, latex interpenetrating elastomeric networks, latex interpenetrating polymer networks, poly(vinyl chloride) with poly(butadiene-*co*-acrylonitrile), and latex semi-IPNs (commercial ABS plastic).

Simultaneous IPNs. Simultaneous IPN (SIN) synthesis refers to the mixing of monomers (or prepolymers), linear polymers, cross-linking agents, and catalysts, followed, usually immediately, by simultaneous polymerization by noninterfering modes. The advantage of SINs is ease of processing; for example, polyurethane–epoxy SINs are used in reaction injection molding (RIM) applications. Simultaneous IPNs include several different preparations of polyurethane with poly(methyl methacrylate), polyurethane with acrylics, polyurethane with epoxy or unsaturated polyester, polyurethane with polystyrene, polyurethane with poly(methyl methacrylate-*co*-methacrylic acid), polyurethane with polyacrylate or epoxy, castor oil with polystyrene, vernonia oil with polystyrene, epoxidized triglyceride oils with polystyrene, polyurethane with polyester, polyurethane with polysiloxane, butyl rubber with poly(methyl acrylate), poly(butyl acrylate), or poly(nonyl acrylate), epoxy with acrylics, polyurethane with polyester or epoxy, grafted SINs, and topotactic polymerization.

Thermoplastic IPNs. Although the IPNs previously described contain covalent cross-links, physical cross-linking modes may also be used. In ABA triblock copolymer, the glassy end blocks usually form a discrete phase in the rubbery phase, providing physical cross-links. In the formation of ionomers, at about 5% ionic groups such as sodium methacrylate, segregation of the ionic groups into concentrated domains occurs. Crystalline segments of amorphous polymers serve as physical cross-links. The whitish haze in polyethylene is an example of a crystalline portion of an amorphous polymer. All these thermoplastic IPNs behave as thermoplastics at elevated temperature and can be melt-processed (see also ELASTOMERS, THERMOPLASTIC). Thermoplastic IPNs include poly(styrene-*b*-ethylene-*co*-butylene-*b*-styrene), polyester with polycarbonate, functional silicone fluids with thermoplastic resin, thermoplastics with bicyanates, and polystyrene–isoprene triblock copolymer with polystyrene.

Morphology and Glass Transition

The IPNs synthesized to date show varying degrees of phase separation. Phases vary in size, shape, sharpness at the interfaces, and degree of continuity. The morphology of IPNs, controlled by molecular orientation, is related to physical properties such as modulus and glass transition and provides insight into the latter. Transmission electron microscopy (tem), an important technique used to visualize the component phases of IPNs, involves staining of double bonds or other groups such as urethanes by osmium tetroxide. The morphology of IPNs is controlled by chemical compatibility (eg, solubility parameters), interfacial tension, cross-link densities, kinetics of polymerization, and component types.

Table 1. Classification of IPNs

Category	Definition
full IPN	any material containing two or more polymers in which there are no induced cross-links between the individual polymers
sequential IPN (SIPN)	polymer A is swollen in monomer B, its cross-linking agent, and initiator, polymerizing B *in situ*
simultaneous IPN (SIN)	monomers A and B, and their respective cross-linking agents and initiators, are polymerized simultaneously by way of noninterfering modes
thermoplastic IPN	a two-polymer IPN in which the individual polymers are thermoplastics; polymers may contain physical cross-links, eg, ionomers that join two or more chains together or may be phase-separated single polymer systems
semi-I IPN	sequential IPN in which polymer I is cross-linked and polymer II linear
semi-II IPN	sequential IPN in which polymer I is linear and polymer II cross-linked
pseudo-IPN	simultaneous IPN in which one polymer is in network form, ie, cross-linked, and the other linear

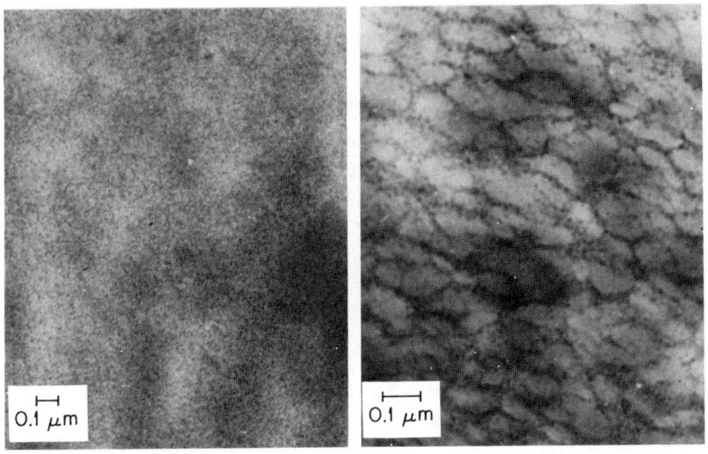

(a) (b)

Figure 3. Electron micrographs of IPNs: (**a**) 75/25 poly(ethyl acrylate)/poly(methyl methacrylate); (**b**) 50/50 poly(ethyl acrylate)/polystyrene. A small amount of butadiene was copolymerized with the poly(ethyl acrylate) to aid in osmium tetroxide staining. Courtesy of *Macromolecules.*

All amorphous polymers have a glass–rubber (T_g) transition temperature depending on the chemical nature of the polymer (see GLASS TRANSITION). In incompatible or phase-separated IPNs and other similar two- (or more) polymer systems, each phase retains its T_g. The transitions may be broadened or shifted by more or less mixing in IPNs. In the limit of total compatibility (complete mixing), only one transition is observed, usually spanning the glass transitions of the component polymers.

The glass transitions of compatible IPNs can be calculated from the following equations:

$$\frac{1}{T_g} = \frac{W_1}{T_{g1}} + \frac{W_2}{T_{g2}}$$

$$T_g = W_1T_{g1} + W_2T_{g2}$$

where $W_1(W_2)$ and $T_{g1}(T_{g2})$ are the weight fractions and glass transitions of polymers I and II, respectively. The glass transitions of compatible IPNs shift inward from the individual component. In some cases, they are broadened so much that the peaks overlap. The glass transition temperature increases with increasing cross-link density.

Sequential IPNs. The differences in the compatibility of sequential IPNs are illustrated in Figure 3. The isomeric poly(ethyl acrylate)–PMMA IPNs (Fig. 3**a**) produced dispersed domains smaller than 10 mm (fine structure) because of their similar solubility parameters. This gives a ΔG_M slightly larger than zero (very compatible), which prevents phase separation until high conversion of PMMA.

A much less compatible poly(ethyl acrylate)–PS IPN (Fig. 3**b**) produces a cellular structure with 100-nm domains. Phase separation occurs early in the polymerization, forming cellular regions rich in PS and styrene monomer and later leading to a fine structure of PS domains in the poly(ethyl acrylate) cell wall.

Simultaneous IPNs. Morphology has been studied as a function of mutual solubility for polyurethane-based SINs. The morphology of a polyurethane–PMMA SIN is shown in Figure 4**a**; the components have very similar solubility parameters. The polyurethane–polystyrene SINs in Figure 4**b**, with different solubility parameters, exhibit more phase separation. Unlike sequential IPNs, both networks form at the same time but not necessarily at the same rate. However, these figures do not imply that SINs of simultaneous gelation always have complete molecular mixing or dual-phase continuity.

Physical Properties

The effect of filler on the behavior of the dielectric loss factor tan δ was studied over a temperature range of -140 to 20°C. The tan δ_{max}, corresponding to the dipole segmental losses of the cross-linked polyurethane, shifts from -10 to 0°C as the silica content is increased to 50% or when the divinylbenzene concentration in the PS is increased. The drop in the height of the tan δ indicates reduced chain mobility in the polyurethane network. The corresponding average reduction in density, in conjunction with reduced chain mobility, indicates that the PS exerts a hydrostatic pressure on the polyurethane, probably in connection with the filler. Therefore, the PS polymerization locus may be concentrated on the polymer I–filler interface.

As the tan δ peak is reduced, less energy is transferred from an a-c power source to the polymer in the -10 to 0°C range. The reduction of dielectric heating and concomitant reduction

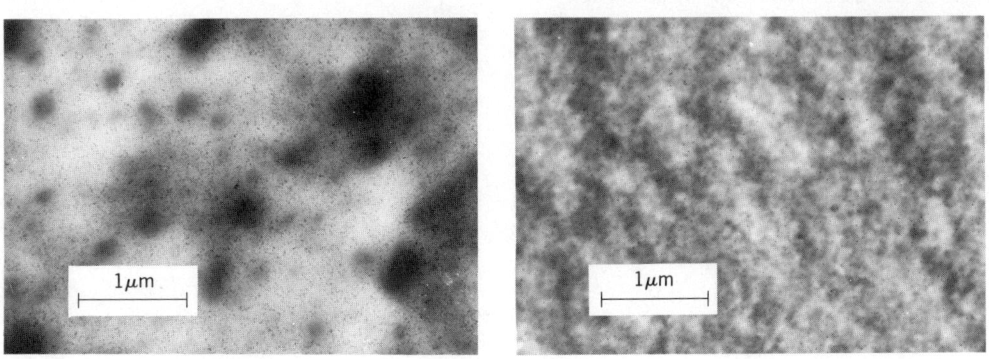

(a) (b)

Figure 4. Morphology of simultaneous interpenetrating networks (SINs): (**a**) electron micrograph of SIN of 75/25 polyurethane/poly(methyl methacrylate) (polyurethane stained with osmium tetroxide); (**b**) 75/25 polyurethane/PS. (**a**) Courtesy of *Macromolecules;* (**b**) courtesy of *Polymer Engineering and Science.*

in power loss by using filler could prove useful in many applications.

The density of polyurethane–PS IPNs prepared under high pressure shows an abrupt change between atmospheric pressure and 250 MPa (2500 atm), where it remains constant at 1.088 g/cm^3. The calculated average density based on volume change of mixing occurs due to interpenetration. These same trends hold for polyurethane–PMMA SINs, but with higher overall densities.

The ultimate tensile properties of poly(urethane–urea)–polyacrylate latex IPNs include a minimum of the ultimate tensile strength reached at 20% polyacrylate, probably because polyacrylate acts initially as a plasticizer, reducing the hydrogen bonding within the poly(urethane–urea) network. At 75% polyacrylate, the material behaves like a rubber toughened plastic, with ultimate tensile strengths higher than either homopolymer network.

To be useful as reinforced elastomers, graft polymers or IPNs should have well-dispersed elastomeric phase domains in the range of 50–500 nm and a T_g of the elastomer below −40°C.

Charged groups affect the tensile properties of polyurethane–acrylic–epoxy ternary IPNs. IPNs exhibit better mechanical properties than blends of the constituent polymers. The high degree of interpenetration and good compatibility in the former case results in an experimental tensile strength much higher than that calculated.

Mechanical property variations in epoxy–poly(*n*-butyl acrylate) simultaneous IPNs (SINs) were studied with regard to epoxy prereaction time and the concentration of acrylic initiator and cross-linking agent, DEGDM. All are tougher than the parent epoxy, and the initial slope, Young's modulus, is less because of the rubber phase, poly(*n*-butyl acrylate), in the material.

Shore A hardness, modulus, tensile strength, and thermal-degradation properties in polyurethane–PMMA SINs were studied. The combination of improved tensile properties of SINs over the semi-SINs and linear blends, along with smaller domain sizes, demonstrates interpenetration.

Thermal gravimetric analysis (tga) shows that IPNs, regardless of composition, have better thermal properties than the corresponding homopolymers. This phenomenon can be explained in terms other than interpenetration, since all IPNs perform almost equally. The more easily degradable polymer, PMMA, absorbs free radicals and stabilizes the less easily degradable polymer, polyurethane. The close juxtaposition of the IPNs allows the PMMA and its degradation products to act as free-radical scavengers for the polyurethane.

Lap shear and peel strength, important properties in coatings and adhesive applications, are enhanced by interpenetration, eg, polyurethane–epoxy SINs and semi-SINs.

Tensile strength and elongation increase significantly in 80/20 polyurethane–acrylic IPNs. The pseudo- or semi-IPNs and linear blends do not show this enhancement. The enhancement in physical properties is most likely due to the higher degree of mixing and greater adhesion between the dispersed and continuous phases in the IPNs.

A study of the reaction kinetics of polyurethane–polyester SINs in bulk polymerization showed that interaction is high, even though there is no chemical bonding between the two polymer networks.

Applications

Many types of IPNs, SINs, semi-IPNs, etc, are available in commercial products, although they are rarely identified as such. The term interpenetrating polymer network was coined before the full significance of phase separation was recognized. Since most IPNs are phase-separated, the term interpenetrating phases may more accurately describe some systems. However, dual-phase continuity is not required in the definition of an IPN. Many IPNs are physically cross-linked networks or thermoplastic IPNs. They are used in sheet-molding compounds (SMC), dental fillings, leather IPNs, noise and vibration damping, rimplast thermoplastic IPNs, dicyanate pseudo-IPNs (semi-IPNs), and Kraton thermoplastic IPN.

BIBLIOGRAPHY

H. L. Frisch, K. C. Frisch, and D. Klempner, *Polym. Eng. Sci.* **14**, 646 (1974).

L. H. Sperling, *Polym. Eng. Sci.* **24**, 1 (1984).

K. C. Frisch, D. Klempner, and H. L. Frisch, *Polym. Eng. Sci.* **22**, 1143 (1982).

S. C. Kim, D. Klempner, K. C. Frisch, and H. L. Frisch, *J. Appl. Polym. Sci.* **21**, 1289 (1977).

D. KLEMPNER
University of Detroit

L. BERKOWSKI
BASF Corporation

INTERPOLYMERIZATION. See COPOLYMERIZATION.

ION-EXCHANGE POLYMERS

Ions are charged species that carry a positive or a negative charge; in this state, they are stable and are attracted to the oppositely charged species. Because the activation energy for exchange of ions is small in liquid media, 21 kJ/mol (5 kcal/mol), ions are very mobile when solvated and move with relative ease from one oppositely charged ion to another. The size of the ion, its propensity to solvate and carry large numbers of solvent molecules, and its environment greatly influence its mobility.

Synthesis

Most ion-exchange resins are sold in bead form and are made by suspension polymerization (qv), also known as dispersion or pearl polymerization. All organic polymers used as ion exchangers, including condensation polymers, can be made by suspension polymerization. However, most, if not all, of the condensation resins based on the reaction of phenol with formaldehyde are sold as irregularly shaped particles. These resins are manufactured by unagitated bulk polymerization followed by fracturing and grinding of the fragments into smaller particles. The particles are sieved to provide the desired range of particle sizes. Condensation resins based on the reaction of a difunctional halocarbon with a polyamine, the condensation of polyethyleneimine or diethylenetriamine with epichlorohydrin, are made in bead form by a reverse-phase suspension polymerization.

Suspension polymerization of monomers to give cross-linked resins in a finely divided state is a straightforward technique. Upon cross-linking, polymers quickly become intractable unless the bulk polymer is reduced to smaller sized particles. Size reduction improves manageability, as well as the kinetics of subsequent operations, and is usually the first step in the synthesis and manufacture of ion-exchange resins.

Building polymers in finely divided form is readily accomplished by suspending the mixture of monomers containing the initiator in a continuous fluidizing phase in which the monomers are not soluble or only partially miscible. For nonpolar, hydrophobic monomers, styrene and divinylbenzene, water is preferred. For water-soluble polar monomers, nonpolar, hydrophobic liquids with low volatility are preferred, such as chlorobenzene, dichlorobenzene, *m*- and *p*-xylenes, C_{10}–C_{12} paraffinic hydrocarbons, methylcyclohexane, white mineral oil, kerosene, and ultrasene.

Particle size, particle-size distribution, and the stability of the monomer droplets depend on the interfacial energy between the two liquid phases; viscosity of each phase; density difference between the two phases; reactor geometry; agitator geometry; baffling in the reactor; rpm of agitator; droplet stability imparted by the protective colloid; and intrinsic nature and properties of the dispersing agents.

The most commonly used cross-linking monomer is divinylbenzene. The functionalization of the cross-linked polymers is accounted for by the following reactions: sulfonation, phosphonation, phosphination, chloromethylation, amino-methylation, aminolysis, and hydrolysis.

The active site on anion-exchange resins is an amino or a quaternary ammonium group. Anion-exchange resins are prepared by the following methods:

1. Chloromethylation of aromatic copolymers followed by aminolysis of the chloromethyl group.
2. Aminomethylation of aromatic copolymers.
3. Aminolysis of nitrile and ester groups of aliphatic polymers.
4. Condensation polymerization with polyfunctional amines.

Classification

Synthetic ion-exchange resins are classified by the physical and chemical characteristics of both the polymeric matrix and the functional group. Cation exchangers possess a fixed negative charge on the resin matrix and exchange cations such as Ca^{2+}, Na^{+}, H^{+}, etc. Similarly, anion exchangers possess a positive charge on the functional group or are positively charged during the ion-exchange reaction where anions, such as HCO_3^-, SO_4^{2-}, OH^-, and others are exchanged. Structural characteristics are shown in Figure 1.

Equipment

Vessels designed for cocurrent operation are the simplest and least expensive. Both loading and regeneration is downward through the bed.

Vessels designed for countercurrent operation are more complex but give better overall performance and more efficient regeneration (less regenerant) and less leakage. The loading flow is usually down through the resin bed, whereas regeneration is in the opposite direction. A "stratabed" vessel is very similar to a countercurrent system.

A mixed bed consists of anion and cation resins, which are intimately mixed during the loading cycle. Following loading, the resins are separated by backwashing into two distinct layers with the anion resin on top. Regeneration of the two resins must contact the cation and anion resins with acid and base without cross-contamination of resins or regenerants. Frequently a collector is placed at the interface of the cation and anion resins.

Contactors. ***Batch.*** In batch reactors, the entire electrolyte solution contacts a mass of ion-exchange resin. A single equilibrium stage is followed by phase separation, eg, filtration.

Column. In fixed-bed columns, fresh electrolyte solution continuously contacts the ion-exchange resin in a vertical vessel. Column contactors offer multiple-stage equilibrium reactions during a single pass. Deeper beds and slower-service flow rates increase the number of stages and drive the exchange equilibrium toward completion. Column contactors are highly efficient and are used for virtually all commercial ion-exchange operations.

Semicontinuous and Continuous. Commercial semicontinuous and continuous contactors employ countercurrent moving-bed systems. Engineering simplicity is sacrificed for processing efficiency both in terms of resin utilization and chemical requirements. Among the semicontinuous packed-bed types, the Higgins contactor and the Asahi contactor are the most widely used by industry.

Properties

Ion-exchange resins consist of a polymer matrix that carries functional moieties. The functional groups serve as active, fixed sites where chemical exchange occurs with charged mobile ions. Mobile counterions with charges opposite those on the matrix provide electroneutrality. The fixed ionic groups determine the ion-exchange properties. The nature of the solvent (solution phase) strongly influences performance because of hydrophilic or hydrophobic interactions and diffusion. The polymeric matrix and the chemical cross-links provide both rigidity and control of elasticity and affect the chemical properties.

Ion-exchange Capacity. Exchange capacity is the single most important characteristic of ion-exchange resins. Maximum exchange capacity implies a high degree of functional ionic sites. Among the current commercial resins, those based on weak bases and weak acids exhibit higher exchange capacities than those based on strong bases or strong acids.

Moisture-retention Capacity. Equilibrium moisture retention defines the volume occupied by water when the resin is fully swollen. This parameter reflects distances between polymer chains within the resin. The size of molecules that can penetrate the resin matrix is determined by the moisture-retention capacity.

Density. Density is an important property of ion-exchange resins since most applications require resin particles denser than the fluid. Polymer matrix, functionality, and ionic form (counterion) all affect product density.

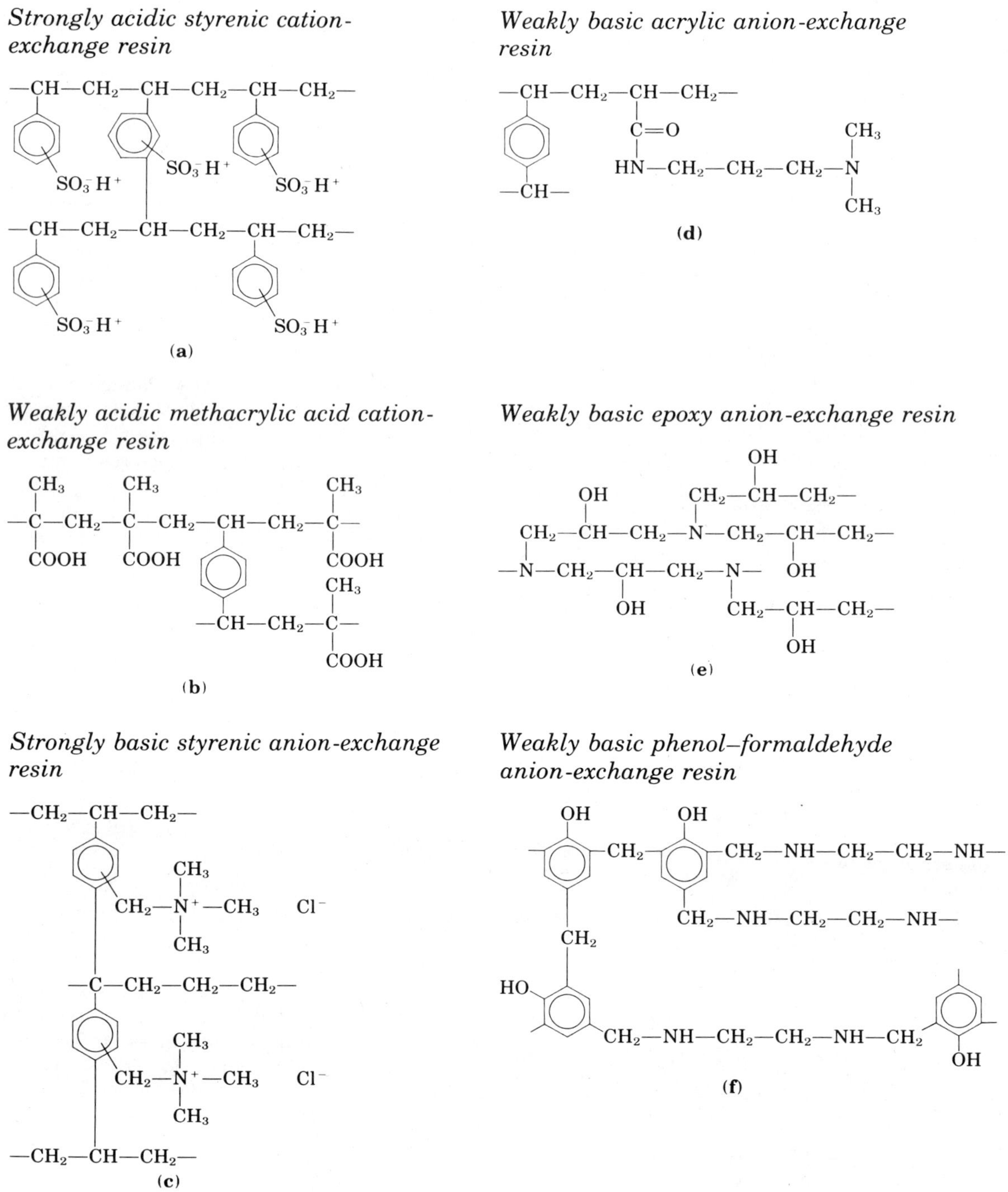

Figure 1. Structures of ion-exchange resins.

Hydraulic Behavior. Ion-exchange operations are controlled by the hydraulic pressure drop and the bed expansion.

Particle Size. Commercial ion-exchange resins exhibit a spectrum of particle sizes in the range of 0.3 to 1.2 mm (16–50 mesh). Exchange reaction rate (kinetics), pressure drop, and hydraulic expansion depend directly on resin particle size.

Porosity. The gel-phase microporosity associated with water-retention capacity relates to interchain distances or pore sizes. Microporous resins exhibit single-phase homogeneous gel structures wherein counterions diffuse uniformly. Pore sizes, as measured by this diffusivity, vary with cross-linking and the ionic strength of the solution.

Swelling. Swelling equilibria in ion-exchange resins balance solvation forces against matrix elasticity. Swelling is favored by polar solvents, low cross-linking, high exchange capacity, large, strongly solvated counterions, low valence counterions, and low ionic strength solutions.

Stability. The stability of ion-exchange resins determines their length of service and ultimately their utility. Osmotic degradation, dissolution, oxidation, thermal degradation, and radiation offer pathways to instability.

Fouling. Certain contaminants in feedwaters foul ion exchangers by irreversible sorption or precipitation within the resin beads. Contaminants include hydrous metal oxides (iron,

Table 1. U.S. Manufacturers of Ion-exchange Resins

Company	Trade name	Manufacturing location(s)
Dow Chemical	Dowex	Midland, Michigan
Sybron Chemical	Ionac	Birmingham, New Jersey
Rohm & Haas	Amberlite	Philadelphia, Pennsylvania
	Duolite	Redwood City, California

manganese, copper, aluminum, magnesium), calcium sulfate or carbonate, oils, greases, and suspended matter. Iron compounds are the most frequent contaminants of cation exchangers.

Chemical Equilibria

The nature of the chemically active exchange groups determines the overall chemical behavior of ion-exchange resins. Acid–base titration curves provide the best evidence for chemical equilibria. The equilibrium constants reflect the acid–base strength of the exchange group. Physical structure (matrix, type, cross-linking, porosity) influences exchange equilibria to a lesser degree.

Applications

Ion-exchange resins perform one or more of four functions on electrolyte solutions: transformation, removal, concentration, and fractionation. Ion-exchange resins find application in a wide variety of industrial, domestic, governmental, and laboratory operations. Water treatment in fixed-bed columns including industrial and domestic softening, as well as deionization, constitutes 75% of U.S. ion-exchange resin consumption.

Health and Safety

Ion-exchange resins pose little or no health hazard to users. An exception to this rule involves systems where toxic ions are exchanged and possibly concentrated by the resin.

Physical properties associated with ion-exchange resins can result in hazards if proper handling practices are not followed. Ion-exchange resins in their regenerated forms represent solid polymeric forms of strong acids and strong bases. Moreover, moderate concentration (1–2 *N*) mineral acids and alkalies are used for regeneration of these resins. Users need to take the normal precautions against exposure and reactivity.

Economic Aspects

Ion-exchange resins are used in all industrialized countries. Major production facilities exist in the United States (Table 1), Japan, the USSR, France, Italy, the FRG, and the United Kingdom. Worldwide annual production of synthetic ion-exchange resins probably exceeds 1.4×10^5 m^3 annually.

BIBLIOGRAPHY

K. Dorfner, *Ion Exchangers,* Ann Arbor Science Publishers, Inc., Ann Arbor, Mich., 1972.

R. Kunin, *Ion Exchange Resins,* R. E. Krieger Publishing Co., Huntington, N.Y., 1972.

J. A. Marinsky and Y. Marcus, *Ion Exchange and Solvent Extraction,* Vols. 1–9, Marcel Dekker, Inc., New York, 1966–1985.

D. Naden and N. Streat, *Ion Exchange Technology,* Ellis Horwood Series of the Society of The Chemical Industry, John Wiley & Sons, Inc., New York, 1984.

ROBERT L. ALBRIGHT
PETER A. YARNELL
Rohm and Haas Company

IONIC POLYMERIZATION.

See ANIONIC POLYMERIZATION; CATIONIC POLYMERIZATION; COORDINATE POLYMERIZATION; INSERTION POLYMERIZATION.

IONIC POLYMERS

Ionic polymers represent a broad classification of polymers with inorganic salt groups attached to the polymer chain. The combination of low ionic content and low polarity backbone results in a class of commercial polymers that continues to be of scientific interest. These ionic polymers are commonly referred to as ionomers. The terms ionomers and ionic polymers are employed interchangeably here to describe these materials.

Unlike homogeneous polymer systems, pendent ionic groups interact to form ion-rich aggregates contained in the nonpolar polymer matrix. The resulting ionic interactions strongly influence polymer properties and applications, which have made this a fertile area for research and development. Typical ionomer structures can be depicted as follows:

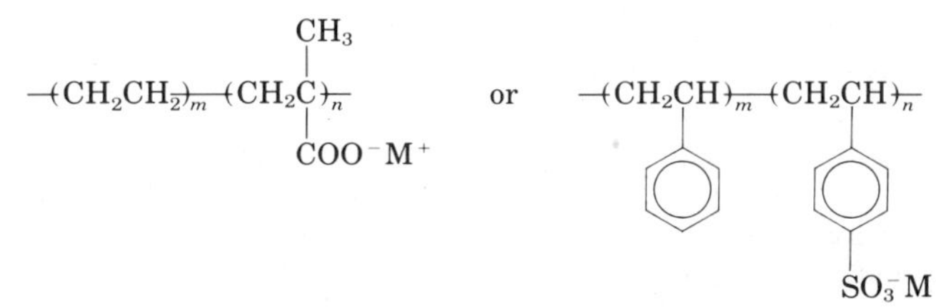

The ratio of *m*/*n* is usually on the order of 10 to 100, reflecting a low overall content of ionic groups. It is generally accepted that the ionic groups interact or associate to form ion-rich regions in the polymer matrix.

Ionic content, degree of neutralization, and type of cation dominate the properties of the resultant polymer system. With this range of variables, the spectrum of polymer properties within the ionomer family is extremely broad.

Ionic polymers are usually prepared by copolymerization of a functionalized monomer with an olefinic monomer or by direct functionalization of a preformed polymer. Typically, carboxyl-containing polymers are obtained by copolymerization of acrylic acid or methacrylic acid with ethylene, styrene, or similar comonomers by free-radical copolymerization. The resulting copolymer is available as the free acid, which can be neutralized to the desired degree with metal hydroxides, acetates, and similar salts.

A second route to ionic polymers involves modification of a preformed polymer. This route is most useful when only a low degree of functionalization is desired. It is especially suitable for sulfonate-containing polymers.

Examples of commercial and experimental ionic polymers are given in Table 1.

Table 1. Commercial and Experimental Ionomers

Polymer system	Trade name	Manufacturer	Applications
Commercial			
ethylene–methacrylic acid copolymer	Surlyn	DuPont	modified thermoplastic
butadiene–acrylic acid copolymer	Hycar	Goodrich	high green-strength elastomer
perfluorosulfonate ionomers	Nafion	DuPont	multiple membranes
perfluorocarboxylate ionomers	Flemion	Ashai Glass	chloro-alkali membrane
telechelic polybutadiene	Hycar	Goodrich	specialty uses
sulfonated ethylene–propylene–diene terpolymer	Ionic Elastomer[a]	Uniroyal	thermoplastic elastomer
Experimental			
styrene–acrylic acid copolymer			model ionomer system
sulfonated polystyrene			model ionomer system
sulfonated polypentenamer			model ionomer system
telechelic polyisobutylene sulfonated ionomers			model ionomer system
alkyl methacrylate–sulfonate copolymers			high green-strength elastomer
styrene-based polyampholytes			
acid–amine ionomers			

[a] Developmental stage.

Structure

A large amount of structural data on ionic polymers has accumulated over the past twenty years. An important question deals with the state of aggregation of the ionic species in the nonpolar polymer matrix. A number of models have been proposed, and this is an area of active research.

Analytical methods applied to the structure elucidation include small angle x-ray scattering, small angle neutron scattering, transmission electron microscopy, infrared spectroscopy, Raman spectroscopy, Mössbauer spectroscopy, and direct measurements of mechanical and dielectric properties.

Above a critical ion concentration, two types of ion aggregates have been proposed: multiplets and clusters. The former are considered to be small isolated polar groups dispersed throughout the polymer matrix. The larger aggregates have a lower ionic interaction but are significantly larger and are defined as clusters.

Models. The state of ionic aggregates in both bulk polymers and hydrocarbon solutions is the subject of intensive research. No single model can adequately describe the wide range of ionomer structures that are now emerging. Most systems are intermediate between the homogenous aggregate and the phase-separated cluster, depending on backbone polarity and ionic functionality.

Properties

The poly(ethylene-*co*-methacrylic acid)s introduced by DuPont in 1964 were the first ionomers whose physical properties were recognized as a consequence of phase-separated ion-rich domains. Although the DuPont products were based on methacrylic acid, poly(ethylene-*co*-acrylic acid) behaves similarly over a range of neutralization levels (Table 2).

A growing body of evidence suggests that the characteristic ionomer properties can be attributed to ionic aggregation. Furthermore, the presence of relatively small amounts of metal carboxylate or sulfonate groups pendent to a polymer chain clearly has profound effects on polymer properties. Specific properties, such as glass transition, rubbery modulus above the glass transition, dynamic mechanical behavior, melt rheol-

Table 2. Properties of Poly(ethylene-*co*-acrylic acid) Salts

	Percent neutralized	Melt index, 10 g/min		Secant modulus, 1% extension, MPa[b]	Ultimate tensile strength, MPa[b]	Elongation at break, %
		at 298 kPa[a]	at ca 3 MPa[b]			
control material, 14.8 wt % acrylic acid	0	67	2.57	48.26	14.8	470
sodium salt	12.0	12.2	256	230	21.7	420
	30.0	3.9	92	540	27.6	330
	47.5	1.0	30	293	31.7	310
	66.0	0.3	7.6	273	33.0	280
potassium salt	8.0	16.3	360	202	21.0	470
	25.0	4.5	110	362	25.5	410
	51.0	2.7	49	341	30.6	370
	63.0	0.57	15	308	34.4	390
lithium salt	12.0	18.7	442	181	21.7	410
	28.5	5.2	116	337	26.5	350
	52.5	1.4	38	334	28.2	260
	67.5	0.2	5.4	254	31.7	250

[a] To convert kPa to psi, multiply by 0.145.
[b] To convert MPa to psi, multiply by 145.

ogy, relaxation behavior, dielectric properties, and solution behavior are modified by even small amounts of ionic groups.

Cluster Formation and Physical Properties. The rheological behavior of ionomers offers convincing evidence that cluster formation or microphase separation has a significant impact on properties. With poly(styrene-*co*-methacrylic acid) ionomers, for example, stress–relaxation data at low ionic content (6 mol %) give satisfactory time–temperature curves. In effect, these results indicate a low degree of cross-linking, which may be attributed to multiplet formation.

Similar stress–relaxation data for samples of higher ion concentration (7.7 mol %) provide a different picture of ion aggregation. Attempts to apply conventional time–temperature superposition by shifting the modulus–time curves were unsuccessful. In other words, a single master curve could not be generated that would describe the viscoelastic response over the entire region of time and temperature. The critical ion concentration for a breakdown in time–temperature superposition for the ionomer based on styrene–methacrylic acid is ca 6 mol %. The current interpretation is that superposition does not apply, because at the higher ionic contents a second relaxation mechanism occurs due to a change in the structure.

A significant difference in water absorption is observed for polymers of different ion contents. The impact of clustering or some form of microphase separation on ionomer properties seems to be significant, and an enhanced degree of water absorption appears conclusive.

Dynamic Mechanical Properties. The dynamic mechanical properties of ionomer systems provide definitive evidence that the salt forms of these systems are dramatically different from the acid form or the parent polymers. Specifically, the neutralized forms of these materials generally display a rubbery plateau in the modulus–temperature curves that is not present in the parent materials.

In other systems, such as polystyrene-based ionomers, dynamic mechanical behavior shows evidence of microphase separation. A single loss-tangent peak is observed for the acid or parent polymer; two such peaks of high intensity appear on neutralization of the copolymer acid. This phenomenon has been interpreted as the result of two glass-transition temperatures: one is normally observed for the polymer matrix, the T_g of polystyrene of the ionomers based on polystyrene; the other T_g is related to the phase-separated regions containing a high percentage of ionic material.

Plasticization

The plasticization of conventional polymers by fugitive and nonfugitive additives is well known. In the case of ionomers, plasticization can have an additional feature, since either the polymer backbone or the inorganic salt groups respond to these additives in a selective manner. This is reflected in polymer properties in several ways. For example, the addition of 5 wt % glycerol as a plasticizer in sulfonated polystyrene (S-PS) reduces the melt viscosity by a factor of 1000 at elevated temperatures.

Solution Behavior

Ionic polymers that are not water-soluble because of low ionic functionality behave quite differently from conventional polyelectrolytes when dissolved in low polarity solvents. In effect, the same electrostatic forces present in bulk ionomers can persist in nonpolar solvents and induce ion-pair associations.

At high polymer concentrations, the reduced viscosity for the ionomers is substantially higher than that for polystyrene, but at low concentrations the situation is reversed. Interassociations increase the viscosity values at high polymer concentrations; however, intraassociations decrease viscosity at low concentrations, as compared to nonionic polymers. Although this behavior broadly typifies strongly associating polymer systems, it is particularly appropriate for ionic polymers dissolved in low polarity solvents.

Elastomers

Elastomers include butadiene–methacrylic acid ionomers and sulfonated ethylene–propylene–diene terpolymers.

Thermoplastics

Thermoplastics include crystalline ethylene–methacrylic acid ionomers, amorphous polystyrene ionomers, and sulfonated derivative polystyrene.

Health and Safety Factors

Information relating to toxicity of specific ionomers can be obtained from the manufacturers.

Uses

Most ionomer applications exploit several key characteristics (eg, enhanced elastomeric green strength, enhanced melt viscosity, toughness, and outstanding abrasion resistance), which can be attributed to ionic aggregation or cluster formation or the interaction of polar groups with ionic aggregates. Ionic polymers are used in packaging films, sporting goods, impact modifiers, adhesives, footwear, garden hoses, and membrane applications, among other applications.

BIBLIOGRAPHY

W. J. MacKnight and T. R. Earnest, *J. Macromol. Sci. Polym. Technol. Rev.* **16**, 41 (1981).

A. Eisenberg and M. King, *Ion-Containing Polymers,* Academic Press, Inc., New York, 1977.

W. J. MacKnight and R. D. Lundberg, *Rubber Chem. Technol.* **57,** 652 (1984).

M. Pineri and A. Eisenberg, *Structure and Properties of Ionomers,* D. Reidel Publishing Co., Dordrecht, The Netherlands, 1987.

ROBERT D. LUNDBERG
Exxon Chemical Company

ION IMPLANTATION

Ion implantation, a process whereby a solid substrate is bombarded with a beam of energetic ions that come to a stop at a certain depth and become "implanted" in the substrate, is a well known technique in the electronics industry for introducing precisely controlled amounts of chemical species or impurities into semiconductors to modify their electronic and physical properties and for fabricating integrated circuits and other microelectronics devices. Ion implantation is of growing interest in polymer science and engineering because of its demonstrated capability to modify the molecular structure, morphological structure, and physical properties of polymers. Modifications include the dramatic increase of electrical conductivity of normally insulating polymer films to semiconducting and near-metallic levels as a result of ion implantations. There is significant reduction of the degree of crystallinity of ion-implanted semicrystalline polymers. Changes in the index of refraction, optical transmission and reflection, and other optical properties of polymer films follow ion implantation. Magnetic properties may be introduced in polymer films by implantation of ferromagnetic species such as iron. Adhesion properties change. There is a substantial reduction in film thickness and significant increase in density as a result of volume and density changes accompanying ion implantation of polymer films. Changes in solubility as a result of irradiation-induced chain scission or chain cross-linking occur. Structural changes, such as increase in the degree of conjugation and the carbon content, ionization, and amounts of free radicals present in a sample, result from ion implantation.

The growing commercial applications of polymer thin films in microelectronics also require knowledge of the characteristics of ion-implanted polymers. For example, the use of photoresists and polyimides as high resolution implant masks for device fabrication requires knowledge of the penetration depth or range of various ions in the polymers. The differential solubility that arises from ion-beam irradiation of polymer films and consequent chain scission or chain cross-linking is being explored as the basis of ion-beam lithography for fabrication of microelectronic devices.

Techniques

Equipment and Procedure. Ion implantation is a conceptually simple method. A beam of ions accelerated through a voltage of about 10^3–10^7 V is focused and steered onto the surface of a target material. The energetic ions penetrate the substrate and come to rest at a mean depth R_P, which depends primarily on the atomic number of the ions Z_1 and the energy of the incident ion beam. The theory of ion implantation is primarily concerned with the problems of the interaction of energetic ion beams with target materials, and hence the mechanisms of ion energy loss and the prediction of the implanted ion density or depth profile. The basic components of equipment needed to implement this technique are an ion source, ion-accelerating electrodes, analyzing magnet for mass separation, electrostatic beam focusing and scanning optics, and a vacuum chamber to hold the target.

Spatial Distribution of Ions in Implanted Solids. Of paramount theoretical and practical importance in ion implantation is the density (n, ions/cm^3)–depth (x) profile, $n(x)$, where x is the penetration depth normal to the surface. The depth distribution determines the physical properties of the implanted solid, especially electronic and electrical properties.

Ion-implanted Polymers

Numerous polymers, ranging from new materials to commercially available films, have been ion implanted and studied. A summary of most of the currently known ion-implanted polymer results and some of the implantation conditions are shown in Table 1.

Table 1. Ion Implantation of Polymers

Polymer	Ion(s) implanted	Ion-beam energy, keV	Comments
polyacetylene $(CH)_x$	F, Br, Cl, I, Pd, Na	6–90	*p*-type and *n*-type conductors; *p*–*n* junctions
poly(sulfur nitride) $(SN)_x$	Br, Pd	25–46	conducting material
poly(*p*-phenylene sulfide) (PPS)	F, Cl, Br, Kr, As, N, Li, I	100–200; 320–50,000 (I, F)	conducting materials; *p*-type conductor (Br)
poly(2,6-dimethylphenylene oxide) (PPO)	N, Br, Kr, As	200	conductivity
polyacrylonitrile (PAN)	Br, N, As, Kr	130–200	*n*-type conductor (Br)
poly(4-phenyl-2,6-quinoline)	Ar	200	*n*-type conductor; esr signal
benzimidazobenzophenanthroline ladder polymer (BBL)	Ar, B, Kr	200	conductivity; patterned conducting lines
polydiacetylenes	Na, As	150–200	conducting material; spin concentration
poly(ethylene terephthalate)[a] (PET)	O, H, F, B	25–250; 6000 (O)	crystallinity changes; conductivity; morphology
poly(methyl methacrylate) (PMMA)	H, He, Li, Be, B, C, Ar	5–300	ion implant mask; depth profiles
polyimides[b]	H, Ar, Ne, Cu, Zn, N_2	150–300; 3000 (N_2)	ion implant mask; optical reflectivity; morphology; conductivity
polyethylene	Fe, Cr, Cu, Ag	25	magnetic properties; structural changes
poly(vinylidene fluoride) (PVDF)	Fe, H, O, C, Ni	25–6000	magnetic properties; depth profile
photoresists (AZ-1370; AZ-1350; HNR-120; AZ-4000)	H, B, Ne, Ar, Cl	50–200; 15,000 (Cl)	implant mask; depth profile; conducting material
poly(*p*-xylylene)	Ar	40	optical properties; refractive index
polystyrene	Ar, Kr, Xe	25–250	density changes; structural changes
polyphthalocyanines (PcAlF)	I	100–200	optical properties; depth profile (rbs)
poly(vinyl chloride) (PVC)	He, Ar	2000	conductivity
nylon	Fe, Cr, Cu, Ag	25	structural changes; magnetic properties
poly(*N*-vinylcarbazole) (PVK)	I	30–80	conductivity; depth profile (sims)
poly(*p*-phenylene) (PPP)	H_2	300	conducting materials; depth profile (sims)

[a] Mylar film.

[b] Kapton film designated XU-218, PI-2250, PIQ-13.

BIBLIOGRAPHY

H. Ryssel and I. Ruge, *Ion Implantation*, John Wiley & Sons, Inc., New York, 1986. An excellent introductory book with emphasis on the experimental technique and procedure of ion implantation with many schematic and photographic illustrations of equipment.

G. K. Hubler, O. W. Holland, C. R. Clayton, and C. W. White, eds., *Ion Implantation and Ion Beam Processing of Materials, Materials Research Society Symp. Proc.*, Vol. 27, North Holland, New York, 1984.

S. A. Jenekhe and S. J. Tibbetts, *J. Polym. Sci., Part B* **26**, 201 (1988).

C. M. Balik, M. A. Said, and J. D. Carlson, *J. Polym. Sci. Part B* **25**, 817 (1987).

SAMSON A. JENEKHE
University of Rochester

IONOMERS.

See IONIC POLYMERS.

ISOBUTYLENE POLYMERS

Isobutylene polymers of commercial importance include homopolymers and isobutylene copolymers with small amounts of diolefin such as isoprene. Chlorinated and brominated derivatives also enjoy significant sales. The monomer was of interest from the early days of synthetic polymer research when Friedel-Crafts catalysts were used to prepare elastic materials.

Monomers

Manufacture. The principal monomers used in the preparation of polyisobutylene and butyl rubbers are 2-methylpropene (isobutylene) and 2-methyl-1,3-butadiene (isoprene). Physical properties of pure isobutylene and isoprene are given in Table 1.

Monomer Purity. Production of the high molecular weight, elastomeric grades of polyisobutylene and butyl rubber requires isobutylene of 99.5 wt % purity. Polymerization units commonly include fractionation facilities to dry and further purify the isobutylene. Impurities include propylene, *n*-butenes, pentenes, isobutane, water, and oxygenated organic compounds. Isoprene of 98 wt % purity is desired for butyl rubber manufacture.

Polymerization

Chemistry of Isobutylene Polymerization. The mechanism of cationic isobutylene polymerization with Lewis acids is highly complex. Friedel-Crafts-type Lewis acids at low temperature give an extremely high polymerization rate. In the first step, the initiation reaction, a carbenium ion–counteranion pair is formed between a Lewis acid coinitiator, an initiator, and the monomer (see also CATIONIC POLYMERIZATION). Typical Lewis acid coinitiators include $AlCl_3$, $RAlCl_2$, BF_3, etc, whereas initiators are Brønsted acids such as HCl, RCOOH, H_2O, or alkyl halides, eg, $(CH_3)_3CCl$.

Initiation is followed by propagation reactions where monomer units add to the carbenium ion forming the chain. These reactions are fast and highly exothermic. The chemistry of the propagation is affected by the temperature, polarity of the solvent, and counteranions. The propagation proceeds until chain transfer or termination occurs.

In chain transfer (qv), the carbenium ion chain end reacts with another species (chain-transfer agent) to terminate the chain and yield a new carbenium ion; this step controls the molecular weight. The isobutylene itself is an effective chain-transfer agent. Other chain-transfer reactions involve counteranions or species with an unshared electron pair, RX, solvents, and olefins. Since the activation energy of chain transfer is much larger than that of the propagation, the molecular weight of polyisobutylene strongly depends on the polymerization temperature.

Termination is the irreversible destruction of the propagating carbenium ion and discontinuance of the kinetic chain. Termination reactions include the collapse of the propagating carbenium ion–counteranion pair, hydride extraction from olefin and formation of dormant or stable allylic carbenium ion, destruction of the carbenium ion by nucleophiles, eg, alcohols, amines, or alkoxides.

Laboratory Preparation. Polyisobutylene and isobutylene copolymers are prepared by cationic polymerization with Lewis acids in halogenated solvents (CH_3Cl, CH_2Cl_2) at low temperature (−40°C or lower). A typical laboratory batch polymerization is conducted under dry nitrogen in a stirred temperature-controlled bath.

Table 1. Physical Properties of Isobutylene and Isoprene

Property	Isobutylene	Isoprene
boiling point at 101.3 kPa[a], °C	−6.90	34.07
melting point at 101.3 kPa[a], °C	−140.35	−145.95
heat capacity at 25°C, J/(mol·K)[b]		
gas	88.1	102.7
liquid	128.0	151.1
heat of formation at 25°C, kJ/mol[b]		
gas	−17.1	75.8
liquid	−37.7	49.4
density, kg/m³ at 20°C	594.2	681.0

[a] To convert kPa to mm Hg, multiply by 7.5.
[b] To convert J to cal, divide by 4.184.

Manufacturing Processes

Polyisobutylene. Commercial polyisobutylene polymers range from low molecular weight semiliquids (M_v of 50,000–90,000) to high molecular weight elastomers (M_v = 500,000). The low-to-medium molecular weight polymers are produced by polymerization of high purity isobutylene in isobutane or hexane diluent at −10 to −40°C, using aluminum chloride or boron trifluoride as catalyst (coinitiator). Producers include Exxon Chemical Company (U.S.), BASF (FRG), and Nisseki (Japan).

Polybutenes. Polybutenes are produced in solution by copolymerizing isobutylene and a portion of the *n*-butenes contained in refinery streams over aluminum chloride or boron trifluoride catalyst. Commercial polybutenes are liquids with number-average molecular weights from 230 to 2700. Different manufacturers employ similar processing sequences, although specific equipment features and operating conditions differ.

Butyl Rubber. The bulk of the world production of butyl rubber is made by a slurry process with aluminum chloride at −98 to −90°C in methyl chloride diluent.

Halogenated Butyl Rubber. Both brominated and chlorinated butyl rubber are produced in commercial quantities. The halogenation is carried out in hydrocarbon solution using elemental chlorine or bromine in a 1:1 molar ratio with enchained isoprene in the butyl feedstock. Both reactions are fast; chlorination is faster. More than one halogen atom per olefin unit can be introduced. However, the reaction rates for excess halogen are lower and some chain fragmentation occurs.

Properties of Polymers

Structure. *Polyisobutylene.* Polyisobutylene is an amorphous polymer with a glass-transition temperature of ca −70°C. The polymers have no asymmetric carbon atoms and crystallize on extension if they are high in molecular weight. X-ray diffraction studies indicate a helical conformation that repeats every eight monomer units. The isobutylene units appear to be almost exclusively in a head-to-tail configuration.

Butyl Rubber. In butyl rubber isoprene is enchained by 1,4-addition in the trans configuration.

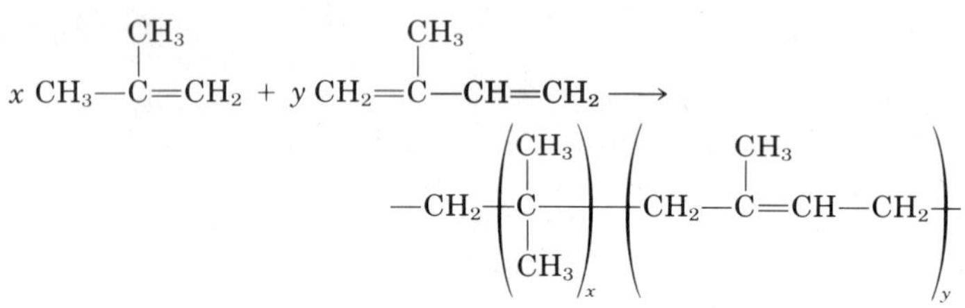

Halogenated Butyl Rubbers. The structures of both chlorinated and brominated butyl rubber are not what would be expected based on halogenation of trisubstituted alkenes. The methyl groups adjacent to the unsaturation in butyl rubber prevent halogen addition across the double bond. The predominant structure shown below has been determined by ^{13}C nmr.

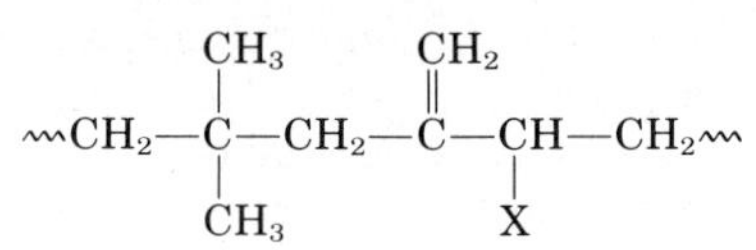

Backbone structure appears to be largely unaffected by halogenation.

Physical Properties. The physical properties of butyl rubber are given in Table 2. The physical properties of polyisobutylene and chlorinated and brominated butyl rubber are quite similar.

Chemical Properties. Polyisobutylene and butyl rubber have the chemical resistance expected of saturated hydrocarbons. Oxidative degradation is slow and the materials may be further protected by antioxidants, eg, hindered phenols. Degradation of polyisobutylene by radiation can be reduced and cross-linking of halobutyl rubber can be accomplished by the addition of divinylbenzene or mercaptans that react with the free radicals produced.

In butyl rubber, the hydrogen atoms α to the double bond permit vulcanization into a cross-linked network with sulfur and accelerators.

Solubility. Amorphous polyisobutylene is readily soluble in nonpolar liquids. The solution properties of polyisobutylene, butyl rubber, and halogenated butyl rubber are very similar. Cyclohexane is an excellent solvent, benzene a moderate solvent, and dioxane a nonsolvent for polyisobutylene polymers.

Viscosity–Molecular Weight Relationships. The relationship between intrinsic viscosity $[\eta]$ and molecular weight for isobutylene polymers in diisobutylene is

$$[\eta]_{20°C} = 3.6 \times 10^{-4} M^{0.64} \qquad 10^4 \leqq M$$

$$[\eta]_{20°C} = 1.78 \times 10^{-2} M^{0.465} \qquad M \leqq 10^4$$

Data on other solvents have been tabulated.

Processing of Polymers

Compounding. Polyisobutylene-based elastomers, like other rubbers, are frequently compounded and vulcanized to optimize processing and properties. The polymer is selected for its vulcanized physical and rheological properties and its cross-linking sites. Grades and compounding ingredients are recommended by polymer manufacturers. Compounding ingredients include fillers, processing aids, stabilizers, and cure systems.

Stabilizers. Antioxidants prevent autooxidation during finishing steps, storage, and compounding. Halogenated butyl rubber must be stabilized with calcium stearate against dehydrohalogenation, which is caused by elevated temperature, as well as acid.

Antiozonants improve the resistance of butyl and halobutyl rubber vulcanizates to ozone cracking. The low segmental mo-

Table 2. Physical Properties of Butyl Rubber and Polyisobutylene

Property	Value	Composition[a]
density, g/cm^3	0.917	B
	1.130	CBV
coefficient of volume expansion	560×10^{-6}	BV
$(1/V)(V/T)$, K^{-1}	460×10^{-6}	CBV
glass-transition temperature, °C	−75 to −67	B
heat capacity, C_p, kJ/(kg·K)[b]	1.95	B
	1.85	BV
thermal conductivity, W/(m·K)	0.130	BV
	0.230	CBV
refractive index, n_D	1.5081	B
	1.5077	P

[a] B = butyl rubber.
BV = vulcanized butyl rubber.
CBV = vulcanized butyl rubber with 50 phr carbon black.
P = polyisobutylene.
[b] To convert J to cal, divide by 4.184.

Table 3. Annual Polybutene Production Capacity

Producer	10^3 metric tons
Amoco Chemical Company	
Texas City, Texas	110
Whiting, Indiana	75
Antwerp, Belgium	25
Chevron Chemical Company	
Richmond, California	25
Cosden Petroleum Company	
Big Spring, Texas	8
Exxon Chemical Company	
Baytown, Texas	75[a]
Bayway, New Jersey	44[a]
Koln, FRG	35
Lubrizol	
Deer Park, Texas	70
Rouen, France	30
Petrofina	
Montreal, Canada	12
Polybutenos Argentinos	
Ensenada, Argentina	9
BASF	
Ludwigshafen, FRG	20[a]
British Petroleum Company	
Grangemouth, UK	50
Naphtachemie, France	45
Nipon Sekiyu Kagaku, Ltd.	
Kawasaki, Japan	14

[a] Includes capacity for polyisobutylene at about 10,000 t/yr.

bility in butyl rubber apparently limits the crack growth rate; however, antiozonants, such as *N,N'*-dioctyl-*p*-phenylenediamine, are effective even at high concentrations of plasticizer.

Vulcanization

Vulcanization introduces a chemical cross-link approximately every 250 carbon atoms along the polymer chain, producing a molecular network. The olefin sites in butyl rubber and the allylic halogen sites in halogenated butyl provide the necessary reactivity. Butyl rubber is cross-linked by sulfur and organic accelerators, phenolic resins, bisazidoformates, and quinone derivatives. Halogenated butyls are cross-linked by the same materials, as well as by zinc oxide, bismaleimides, diamines, peroxides, and dithiols.

Health and Safety Factors

Polyisobutylene and isobutylene–isoprene copolymers are considered to have no chronic hazard associated with exposure under normal industrial use.

Table 4. 1985 U.S. Sales Prices for Isobutylene Polymers

	\$/kg[a]
butyl rubber	2.00
chlorobutyl rubber	2.09
bromobutyl rubber	2.20
polyisobutylene	
high mol-wt elastomer	2.24
low mol-wt semiliquid	2.53
polybutene	0.80

[a] FOB plant.

Economic Aspects

Manufacturers of polybutene and polyisobutylene are given in Table 3; plant locations and estimates of production capacity are included The listed 1985 U.S. sales prices for isobutylene polymers are given in Table 4.

BIBLIOGRAPHY

E. G. M. Tornqvist in J. P. Kennedy and E. G. M. Tornqvist, eds., *Polymer Chemistry of Synthetic Elastomers,* Part 1, John Wiley & Sons, Inc., New York, 1968.

J. P. Kennedy and M. Marechal, *Carbocationic Polymerization,* John Wiley & Sons, Inc., New York, 1982.

F. P. Baldwin and R. H. Schatz in M. Grayson, ed., *Kirk-Othmer Encyclopedia of Chemical Technology,* 3rd ed., Vol. 8, John Wiley & Sons, Inc., New York, 1979.

EDWARD N. KRESGE
RALPH H. SCHATZ
H-C WANG
Exxon Chemical Company

ISOCYANATE-DERIVED POLYMERS

Polymeric isocyanates are undistilled products made by the phosgenation of oligomeric amines obtained in the acid-catalyzed reaction of aniline and formaldehyde. In contrast, polyisocyanates are prepared by the homopolymerization of monoisocyanates.

Condensation polymers obtained from diisocyanates and suitable substrates with the loss of carbon dioxide are of commercial importance.

Monomers

Preparation. Mono- and diisocyanates are manufactured on a commercial scale by the phosgenation of amines.

$$\mathrm{RNH_2 + COCl_2 \xrightarrow{-HCl} RNHCOCl \xrightarrow{-HCl} RNCO}$$

Health and Safety Aspects. The highly reactive isocyanates must be handled with caution. The toxicities of toluene diisocyanate (TDI) and 4,4'-methylenebis(phenyl isocyanate) (MDI) have been extensively investigated. Some sensitization and asthmalike reactions can occur on exposure to TDI or MDI. Exposure limits have been set by OSHA at a 20-ppb limit for an 8-h TWA. Experiments with isocyanates should be conducted in a well-ventilated fume hood.

Homopolymerization

Monoisocyanates. Anionic homopolymerization of monoisocyanates, like that of olefins, takes place by addition across the C=N double bond to form nylon-1 polymers, as shown by their carbonyl absorption at 1710 cm^{-1}. The mechanism of the homopolymerization is shown below (B = base).

initiation

$$\mathrm{RN{=}C{=}O \xrightarrow{B} R\overset{-}{N}{-}\overset{\overset{\displaystyle O}{\|}}{C}{-}B}$$

propagation

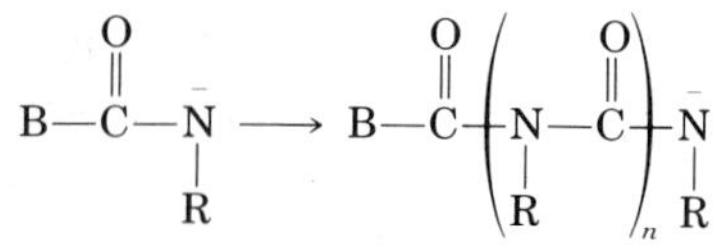

termination

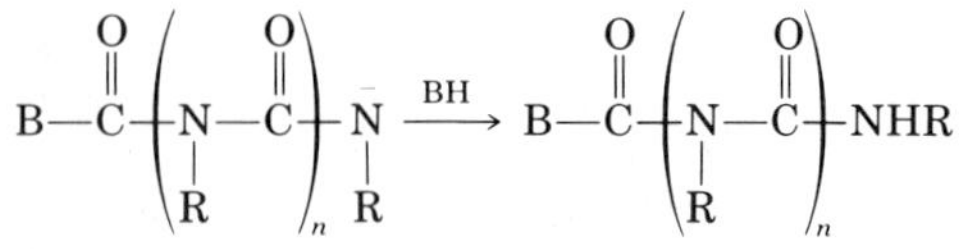

Diisocyanates. Homopolymerization of diisocyanates at low temperatures in dimethylformamide (DMF), initiated by sodium cyanide in DMF, produces high molecular weight crosslinked polyisocyanates.

Copolymerization

With Olefins. Copolymers of *n*-butyl isocyanate with olefins are prepared by anionic copolymerization in a mixture of toluene and tetrahydrofuran at −80°C, using sodium biphenyl as the initiator.

With Ketenes. The copolymerization of isocyanate with ketenes is best conducted in DMF with sodium cyanide as the initiator. This method gives copolymers of dimethylketene and isocyanates and diphenylketene and phenyl isocyanate.

With Aldehydes. Phenyl isocyanate is copolymerized with chloral in DMF at −68°C with potassium cyanate as the initiator.

With Ethylene Oxide. The copolymerization of phenyl isocyanate and ethylene oxide at −78°C uses triethylaluminum as the initiator.

Alkyl isocyanates do not copolymerize with ethylene oxide.

Cyclopolymerization

Homopolymers derived from unsaturated monoisocyanates, such as vinyl isocyanate, undergo further cyclization on heating in DMA, treatment with azoisobutyronitrile, or exposure to uv light to produce ladder polymers with enhanced thermal stability.

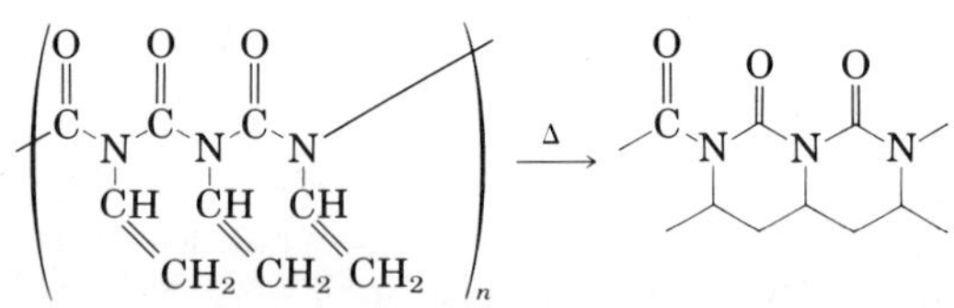

Cyclopolymerization also occurs during treatment of 1,2-diisocyanatoethane, 1,2-diisocyanatopropane, or 1,3-diisocyanatopropane with sodium cyanide in DMF.

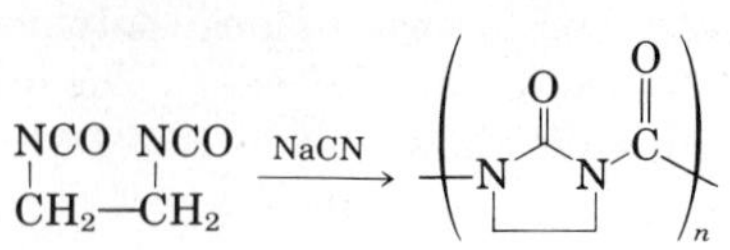

Similar cyclopolymers are obtained from 1,2,3-triisocyanatopropane and 1,2- and 1,3-diisocyanatocyclohexane.

Condensation Polymers

Polycarbodiimides. Heating of diisocyanates and polymeric isocyanates in the presence of phospholene–oxide catalysts produces polycarbodiimides.

Polyamides. Polyamides are prepared from diisocyanates and dicarboxylic acids in inert solvents with carbon dioxide generation.

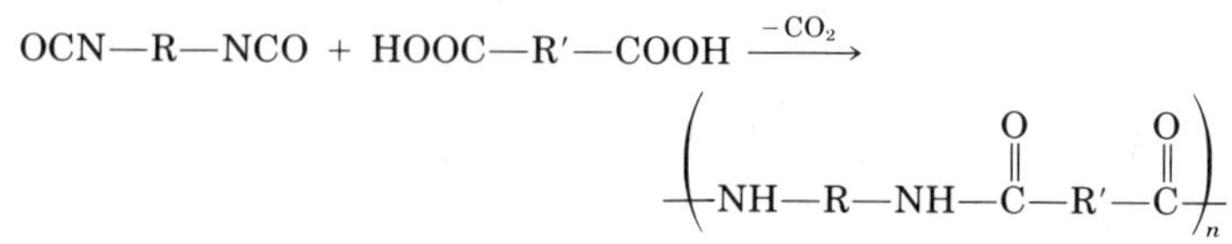

Polyamideimides. The reaction of diisocyanates with anhydrides of tricarboxylic acids, such as trimellitic anhydride, produces polyamideimides.

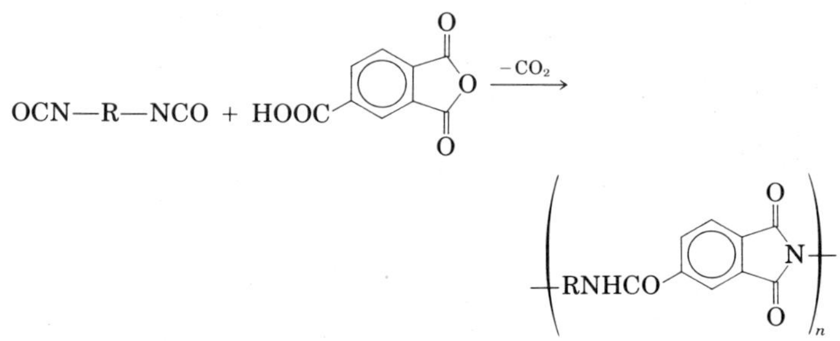

Polyimides. Linear polyimides are prepared from diisocyanates and dianhydrides in polar solvents.

Addition–Condensation Polymers

Poly(parabanic acid)s. The reaction of MDI with hydrogen cyanide gives linear cyanoformamidines that cyclize to produce polyiminoimidazolidinones, which are hydrolyzed to poly(parabanic acid)s.

Polyhydantoins. The reaction of diisocyanates with polyfunctional glycinates produces polyhydantoins.

Polyoxazolinediones. Heterocyclic polymers can also be obtained from α-hydroxy acids, or α-hydroxy acid esters and diisocyanates.

Polyhydrouracils. In the production of urea polymers from diisocyanates, amines are more reactive than hydroxy compounds. Polyhydrouracils are a good example of the products of this reaction. Thus aromatic diisocyanates and *N,N'*-biscarbomethoxyethyl-1,6-hexanediamine in a polar solvent produce linear ureas that cyclize on heating in polyphosphoric acid.

Polybenzoxazinediones. The reaction of di-*o*-hydroxyarene dicarboxylates with diisocyanates in dimethyl sulfoxide (DMSO), in the presence of a tertiary amine catalyst, produces thermally stable polybenzoxazinediones.

Polyquinazolinediones. The reaction of diisocyanates with bisanthranilic acids gives high molecular weight poly(urea acids) that cyclize to polyquinazolinediones.

BIBLIOGRAPHY

H. Ulrich, *J. Polym. Sci. Macromol. Rev.* **11,** 93 (1976).

A. A. R. Sayigh, H. Ulrich, and W. J. Farrissey, Jr. in J. K. Stille and

T. W. Campbell, eds., *Condensation Monomers,* Wiley-Interscience, New York, 1972, pp. 369–476.
I. Carney, *Toxicology of Isocyanates,* International Isocyanate Institute, Inc., Parsippany, N.J., 1980.
V. E. Shashoua, W. Sweeny, and R. F. Tietz, *J. Am. Chem. Soc.* **82,** 866 (1959).

HENRI ULRICH
The Dow Chemical Co.

ISOMERIZATION POLYMERIZATION

Isomerization polymerizations are mainly polyaddition reactions in which the propagating species rearranges to energetically preferred structures before propagation in such a manner that the repeat unit of the polymer does not possess the structure of the original monomer.

$$nA \longrightarrow -(B)_n-$$

where A = monomer and B = repeat structure. These polymerizations involve the rearrangement of the monomeric unit or the propagating species. The former case implies that monomer isomerization precedes the propagation step, whereas the latter is due to isomerization occurring during propagation. Recently, isomerization polycondensations were described.

Energetic factors are usually considered responsible for the occurrence of isomerization processes, but kinetic, steric, and other effects may play a role, including the type and number of active species, the characteristics of the reaction medium and the chain propagating groups, and the presence of reactive groups on the polymer chain. Frequently, rearrangements are partial and not selective.

A polyfunctional molecule may react in a variety of ways, including ring opening followed by rearrangement or group transfer, carbene formation, and irregular tail-to-head (T-H) addition, etc. The resulting structures may be very complex, eg, cyclopolymerization of dienes and trienes. The isomerization process is thoroughly understood only in a few cases. Polymerizations are known to be controlled by monomer isomerization when the monomers are internal olefins. The determination of isomerized units has been facilitated by the use of ^{13}C-nmr spectroscopy for elucidating the microstructure of complicated macromolecules resulting from partial rearrangements.

Isomerization polymerization is best understood in the case of cationic processes; less is known about anionic, free-radical, and condensation polymerization. Isomerization polymerization of olefins catalyzed by transition metals may have industrial interest.

Monomer-isomerization Polymerization

Isomerization preceding polymerization is conceptually simple, since the two steps are independent, but an exact description of monomer-isomerization polymerization (MIP) is not easily given. Isomerization gives rise to a new monomer, whereas chain growth proceeds by conventional processes, which explains the final macromolecular structure. The identification of the isomerization route, which involves positional and geometric but not skeletal rearrangements, and the determination of the concentration of the isomerized units may be difficult if isomerization and polymerization occur simultaneously. The new molecules are usually in equilibrium with the original ones and their rate of formation determines the polymerization rate. The majority of MIPs were observed in homopolymerizations and copolymerizations induced by coordination, ionic, and free-radical (a few cases) catalysts. The principal monomers involved have been olefins, diolefins, *N,N*-dialkylalkenylamides, alkenylamides, acrylic acid, hydroxyalkyl acrylates, diethyl methylene malonate, and dialkyl maleates.

Isomerization During Polymerization

When isomerization takes place after the attack of the monomer, it involves the active species formed and the rearrangement is then essential to propagation. These are cases of true isomerization polymerization (IP) that occur mainly by bond rearrangements (electron migration) or by atom and group migration, possibly in the form of ions.

Bond rearrangements can occur inter- or intramolecularly at one or two reactive sites in a linear or cyclic structure, eg, α,ω-dienes, α,ω-dialdehydes, or olefins containing a strained ring. Alternatively, the first step can consist of the ring opening of a suitable monomer followed by rearrangement of the initial structure and formation of new functional groups.

Olefins. Complicated structures are usually formed during the cationic polymerization of linear olefins because of the lack of preferred rearrangements that do occur when the monomers contain tertiary or quaternary carbon atoms. Bases for synthetic lubricants are obtainable by cationic IP of internal olefins. Also, cyclic olefins undergo IP.

Bifunctional Monomers. Certain bifunctional monomers that form six-membered rings undergo intra–intermolecular polymerization, ie, cyclopolymerization. Examples are α,ω-dienes and α,ω-dialdehydes. Rearrangements also occur during the transannular polymerization of norbornadiene derivatives and alkylideneorbornenes. Similarly, norbornene, alkenylnorbornenes, and dicyclopentadiene undergo the Wagner-Meerwein isomerization of the norbornyl skeleton (exo–exo, syn–syn units) owing to the tendency of the norbornyl cation to rearrange. These rearrangements occur just before the propagation step, after the attack by the electrophilic species. However, norbornene derivatives are also polymerized by coordination catalysts via double-bond or ring opening.

Cyclic Monomers. *Homopolymerization.* The mechanism of cationic oligomerization of cyclic monomers (different from olefins) has been elucidated for ethylene oxide and isobutylene oxide. For isobutylene oxide, dioxolane-type dimer and trioxane-type trimer are mainly formed at room temperature, where hydride transfer occurs easily. At low temperature, polymer formation is accompanied by formation of dioxane-type dimer (backbiting termination). Other investigated monomers include oxiranes, alkoxy phospholanes, chloromethylthiirane, chlorothietane, cyclic imino ethers (oxazolines, oxazines, and their alkyl or aryl derivatives), and cyclic ethylene ketene acetals. Ionic and zwitterionic mechanisms are mainly involved in these homopolymerizations and copolymerizations.

Isomerization in Polycondensations

Structural irregularities arising during polycondensation as a consequence of monomer or active species isomerization are much less favored than in polyaddition reactions and therefore are seldom described. However, modern tools for elucidating polymer microstructures have stimulated investigations into structural isomerism in polycondensates.

Due to the complexity of polycondensation systems, the orientation of the units was mainly studied. In the simplest polycondensation involving a single unsymmetrical monomer XabX (X is a leaving group), four units can be formed: -a/a-, -a/b-, -b/a-, and -b/b-, although the possibility of distinguishing between the second and third units is negligible.

The degree of order can be measured by a parameter s that describes the probability for two adjacent unsymmetrical units to point in the same direction. Another parameter K_m describes the degree of microheterogeneity and is equal to the degree of randomness B.

The value of s, which is defined by the relationship $s = |a/b|/(|a/a| + |a/b| + |b/b|)$, is 1 when the polymer units have the same orientation (head-to-tail linkage), whereas it is 0 for a perfectly alternating structure. For a random chain, where no structural order exists, $s = 0.5$. These theoretical treatments were tested with bipolymer systems such as $XabX^- = YccY$. Polyesters with few -accb- units ($s \simeq 0.2$) or -acca-/-bccb- units ($s \simeq 0.6$) were obtained from terephthalic acid and unsymmetrical diols. Polyamides containing sequences $(\text{-accabccb-})_{n/2}$ with $n > 100$, were prepared from ethylenediamine and an unsymmetrical complex diacid. Also, polyureas having $(\frac{1}{2}n)$ $(\text{-accabccb-})_n$ or $(1/n)$ $(\text{-accb-})_n$ or random structure were recently obtained.

BIBLIOGRAPHY

M. A. Schmuck, P. Pino, and U. W. Suter, *Macromolecules* **18,** 823 (1985).

J. P. Kennedy, *Cationic Polymerization of Olefins: A Critical Inventory,* John Wiley & Sons, Inc., New York, 1975.

T. Saegusa, *Angew. Chem.* **16,** 826 (1977); T. Saegusa and S. Kobayashi, *Pure Appl. Chem.* **50,** 281 (1978); *J. Polym. Sci. Polym. Symp.* **62,** 79 (1978); S. Kobayashi and T. Saegusa in K. J. Ivin and T. Saegusa, eds., *Ring Opening Polymerization,* Vol. 2, Elsevier Applied Science Publishers, Essex, UK, 1984, Chapt. 11.

S. Cesca, A. Priola, and G. Santi, *Polym. Lett.* **8,** 573 (1970); G. Ferraris, C. Corino, A. Priola, and S. Cesca, *Macromolecules* **10,** 188 (1977); **13,** 1104 (1980); **12,** 404 (1979); A. Priola, C. Corno, and S. Cesca, *Macromolecules* **13,** 1110 (1980); **14,** 475 (1981); C. Corno, A. Priola, and S. Cesca, *Macromolecules* **13,** 1099, 1314 (1980).

SEBASTIANO CESCA
ENICHEM SpA

ISOPIESTIC METHODS.

See MOLECULAR WEIGHT DETERMINATION.

ISOPRENE POLYMERS

Physical Properties of the Monomer

In 1860, isoprene was isolated from products of the destructive distillation of rubber. The correct structural formula was proposed later.

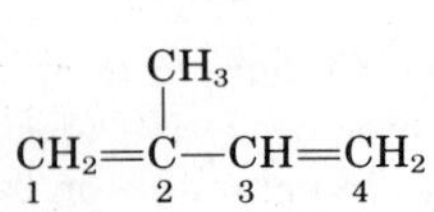

Table 1. Physical Properties of Isoprene

Property	Value
molecular weight	68.11
boiling point at 101.3 kPa[a], °C	34.067
freezing point, °C	−145.95
density, g/cm^3	
at 20°C	0.68095
at 30°C	0.67074
refractive index n_D^{20}, 25°C	1.41852
flash point, °C	−48
heat of formation at 25°C, kJ/mol[b]	
liquid	−49.40
gas	−75.78
free energy of formation at 25°C, kJ/mol[b]	−146.0
coefficient of expansion[c]	0.0016
surface tension, mN/m (= dyn/cm)	
at 20°C	18.22

[a] To convert kPa to atm, divide by 101.3.
[b] To convert J to cal, divide by 4.184.
[c] Between −20.6 and 21.1°C.

Isoprene or 2-methyl-1,3-butadiene is a colorless, volatile liquid that is soluble in most common hydrocarbons and virtually insoluble in water. Typical properties are given in Table 1.

Chemical Properties of the Monomer

Isoprene is a reactive diene that undergoes cycloaddition, electrophilic, nucleophilic, and coordinated catalytic addition, and thermal dimerization.

Manufacture of the Monomer

Isoprene is obtained mainly by extraction from hydrocarbon streams formed by the cracking of naphtha or gas oil. Until the mid-1970s, isoprene was produced synthetically. However, with the availability of C_5 streams, extraction processes have proved to be more economical. Synthetic routes are currently not in commercial use in the United States.

Properties of Polymers

Polymer Structure. Depending on the catalyst and conditions, isoprene may undergo 1,2-, 3,4-, or 1,4-addition polymerization leading to several isomeric structures:

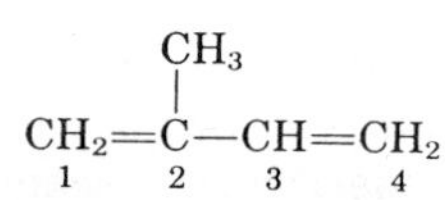

isoprene monomer

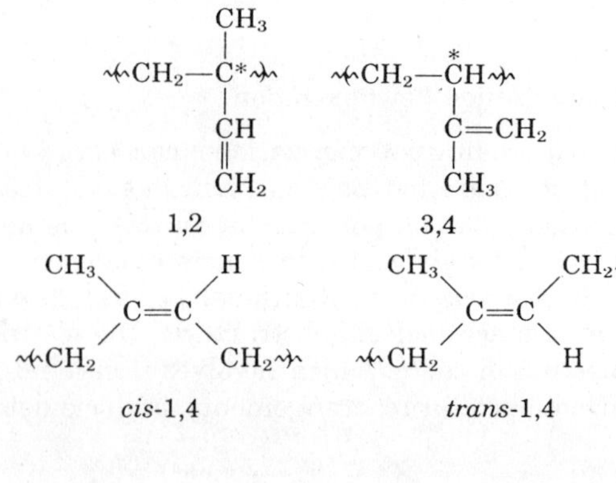

1,2 3,4

cis-1,4 *trans*-1,4

In the 1,2 and 3,4 additions, an asymmetric carbon atom is formed (marked by an asterisk) that has an *R* or *S* (D or L) configuration. No optical activity is observed since equal numbers of *R* and *S* configurations are produced. Syntheses involving asymmetric induction result in optical activity. The arrangement of the *R* and *S* configurations along the polymer chains leads to diastereomeric isomers called isotactic, syndiotactic, and atactic (see also MICROSTRUCTURE).

—*R*—*R*—*R*—*R*—*R*—*R*— (isotactic) —*R*—*S*—*R*—*S*—*R*—*S* (syndiotactic) —*R*—*R*—*S*—*S*—*S*—*R*—*R*—*S*— (atactic)

In isotactic polymers all monomer units have the same configuration in which the asymmetric carbon atoms have side chains containing the same regular configuration shown below.

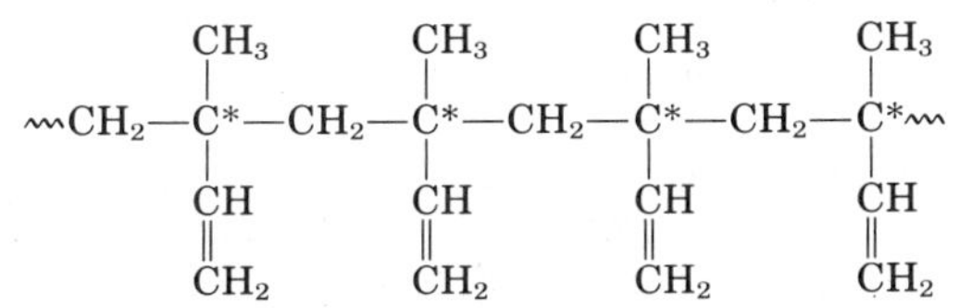

isotactic 1,2-polyisoprene

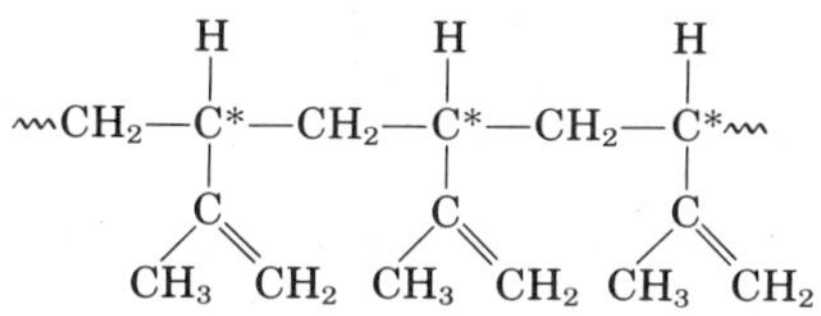

isotactic 3,4-polyisoprene

In syndiotactic polymer, the monomer units have alternating *R* and *S* configurations in which the asymmetric carbon atoms have side chains containing regular alternations.

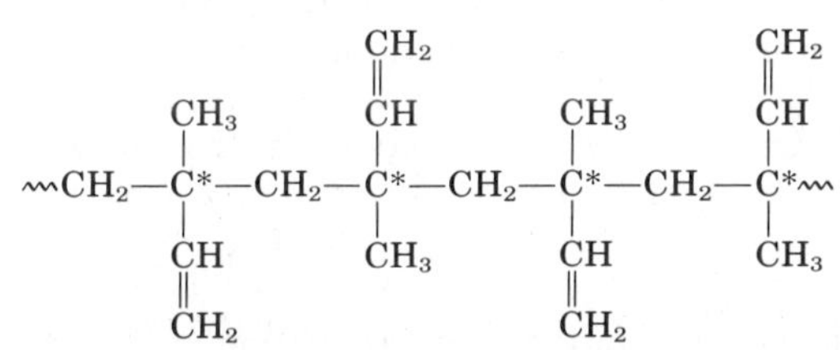

syndiotactic 1,2-polyisoprene

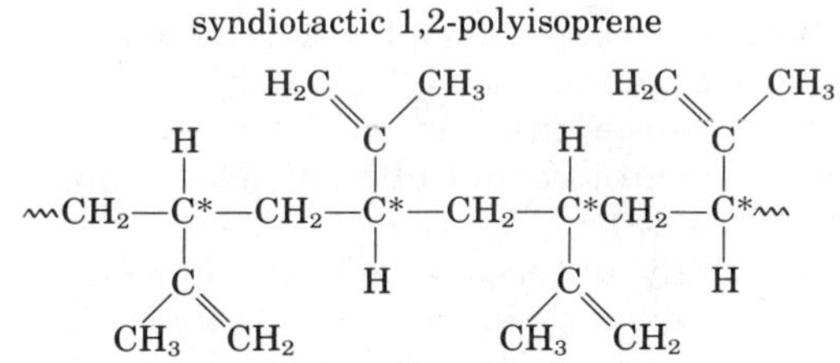

syndiotactic 3,4-polyisoprene

For the atactic polymer, a random succession of *R* and *S* configurations occurs along the polymer chain.

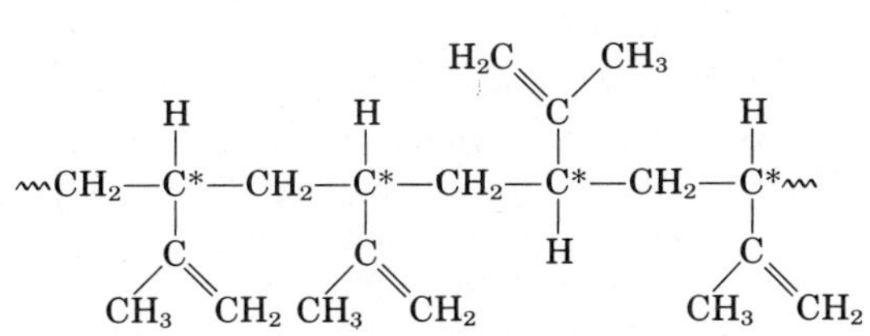

atactic 3,4-polyisoprene

The different arrangements and configurations examined thus far result in eight possible isomeric polyisoprenes. However, the only three known isomeric polyisoprenes of high stereoregularity are *cis*-1,4-polyisoprene, *trans*-1,4-polyisoprene, and 3,4-polyisoprene.

For crystallinity a very high degree of structural stereoregularity is required.

Polyisoprene with more than 90% cis-1,4-microstructure is rubbery, whereas material with >90% trans-1,4 structure is crystalline.

Molecular Weight. Polyisoprene molecular weights are determined by the usual methods, including osmotic pressure for number-average molecular weight M_n, light scattering and ultracentrifugation for weight-average molecular weight M_w, and intrinsic viscosity for viscosity-average molecular weight M_v.

Molecular Weight Distribution. The ratio $\overline{M}_w/\overline{M}_n$ is a useful parameter for defining the polydispersity of molecular weight distributions. Today molecular weight distribution is determined by gel-permeation chromatography (gpc). Automated instruments are commercially available.

Gel and Branching. The amounts of gel and branching in *cis*-1,4-polyisoprenes depend on the catalyst system used in their preparation. Microgels with particles of colloidal size are determined for polyisoprenes by a combination of centrifugation and ultrafiltration.

Glass-transition Temperature. The flexibility of amorphous polymers is greatly reduced as they are cooled below a characteristic transition temperature known as the glass-transition temperature (T_g). Below the T_g, there is no segmental motion. Elastomers, eg, polyisoprene, are used above the T_g where glassy amorphous polymers become rubberlike because of the onset of segmental motion.

The T_g of polyisoprene rubbers is determined by differential scanning calorimetry (dsc) and differential thermal analysis (dta).

Rheology. Rheology is the study of deformation and flow behavior of materials. The rheological behavior of rubbers is important in understanding relationships to polymer structure and rubber-processing behavior.

Like most polymers, polyisoprene exhibits non-Newtonian flow behavior at the shear rates normally encountered in rubber-processing equipment. Non-Newtonian behavior is characterized by a dependence of viscosity on shear rate or shear stress.

According to the ASTM definition, the relative ease of handling of a raw or compounded rubber in rubber machinery is called processibility. Processibility depends to a large extent on viscoelastic properties.

Chemical Modification of Polymers

Chemical modification of polymers is a convenient way of altering physical and mechanical properties. Highly unsaturated polymers such as polyisoprene are easily modified because the olefinic double bonds are readily available. Their high reactivity with a variety of chemical substances offers advantages in chemical modification and vulcanization. However, reaction with oxygen and ozone causes degradation and loss of physical and mechanical properties.

Chemical modifications include hydrohalogenation, halogenation, oxidation, ozonolysis, epoxidation, hydrogenation, carbene additions, and cyclization.

Polymerization

Catalysts. Several catalysts are important for the commercial polymerization of isoprene. For *trans*-1,4-polyisoprene, a coordination catalyst consisting of a vanadium salt and an alkylaluminum is utilized. For *cis*-1,4-polyisoprene, three catalyst systems are employed: alkyllithiums, a coordination catalyst consisting of titanium and trialkylaluminum, and a coordination catalyst consisting of titanium tetrachloride and an alane (AlH_3). Goodyear Tire and Rubber Co. began commercial production of *cis*-1,4-polyisoprene in 1962 at Beaumont, Tex. with a catalyst derived from trialkylaluminum and titanium tetrachloride. This plant is currently the sole U.S. producer of *cis*-1,4-polyisoprene rubber.

Anionic Polymerization. Polymers of uniform chain length and predictable molecular weight are only produced by the anionic-polymerization mechanism. The anionic mechanism is characterized by living, growing chains and control of the stereoisomeric placement of the incoming monomer units. The phenomenal growth of anionic polymerization was spurred by the discovery that lithium metal and its organic compounds are capable of polymerizing isoprene to a high (>90%) cis-1,4 microstructure as determined by ir analysis. This polyisoprene has almost the same chain structure as natural Hevea rubber. Furthermore, it was discovered that nonterminating chain addition is possible with vinyl polymers.

Sodium naphthalene solutions initiate the homogeneous anionic polymerization of styrene and produce living polymers with chains that show no tendency to terminate growth as long as monomer is present.

Initiation. Compared to propagation, the reaction between some organolithiums and isoprene in hydrocarbon solvents is slow, partly because of the strong association of these organolithiums in hydrocarbon solvents.

Some semiquantitative conclusions have been reached with regard to the reaction rates of various alkyllithium species in the initiation of isoprene in hydrocarbon solvents. Rates of reaction are in the following order: menthyllithium > *sec*-butyllithium > isopropyllithium > *tert*-butyllithium > *n*-butyllithium > ethyllithium. The order of reactivity in hydrocarbon solvents is toluene > benzene > *n*-hexane > cyclohexane.

Propagation. Kinetic studies on propagation show that the propagation reaction follows a first-order dependence on monomer concentration in hydrocarbon and ether solvents. For the alkyllithium polymerization of isoprene in ethers or amines, the propagation reaction exhibits first-order dependence on the concentration of growing chains (initiator charge). The propagation kinetics of isoprene are rather complex. Studies on dienes in aliphatic and aromatic solvents show kinetic orders of between one fourth and one sixth.

Molecular Weight and Molecular Weight Distribution. The absence of spontaneous termination in many homogeneous anionic polymerizations allows the synthesis of polyisoprenes with very narrow molecular weight distributions when the initiation and propagation rates are of the same order of magnitude (see MONODISPERSE POLYMERS). Under these conditions, the molecular weight distribution approaches the Poisson distribution

$$P_j = e^{-x}x^{j-1}/(j-1)!$$

where P_j = the number fraction of j-mers, and x = the number of monomers reacted per initiator molecule.

To obtain predictable molecular weight and narrow molecular weight distributions from anionic polymerization, the following conditions are necessary. Terminating impurities such as moisture must be excluded. The initiation rate must be comparable to the propagation rate. The polymerization media must be homogeneous during both the initiation and propagation steps.

Amine Modification. The anionic polymerization of isoprene can be carried out in the presence of N,N,N',N'-tetramethylethylenediamine (TMEDA), increasing the polymerization rate and the 3,4-microstructural content of the resulting polymer; a plateau is reached for the ratio of TMEDA/living species = 4 (ca 70% 3,4-addition). In anionic polymerization of isoprene in cyclohexane by oligoisoprenyllithium complexed with TMEDA or pentamethyldiethylenetriamine (PMDT) the propagation rate can increase or decrease depending on the concentration range. This result is similar to that found for styrene.

Free-radical Polymerization. Isoprene does not polymerize readily under free-radical conditions in bulk or solution presumably due to the high mutual termination of growing radicals. However, emulsion polymerization at 50°C with a potassium persulfate initiator gives a 75% conversion to polyisoprene in 15 h with an $[\eta]$ of 1.15 dL/g.

Cationic Polymerization. Polymerization of isoprene by cationic catalysts is extremely complex and can cause cyclization. The copolymerization of isobutylene with a small amount of isoprene provides the unsaturation for sulfur vulcanization. This process is used commercially for butyl rubber (see ISOBUTYLENE POLYMERS).

Isoprene readily undergoes cationic polymerization with conventional Lewis acids in chlorinated solvents at low temperature. At low conversion, low molecular weight products are obtained. At high conversion, the products are cross-linked, insoluble resins. The soluble products have mainly trans-1,4-microstructure and exhibit less than the theoretical unsaturation.

Polymer Structure

Microstructure. The stereoregularity of polyisoprene in bulk and in different hydrocarbon and polar solvents initiated by a constant, low concentration of *n*-butyllithium initiator at 25°C has been studied. The concentration of alkyllithium initiator was kept constant to study the solvent effect on polyisoprene microstructure. The amounts of 3,4- and 1,4-microstructure are determined by ^{1}H nmr spectra, and the cis-1,4- and trans-1,4-microstructures are determined from ^{13}C nmr spectra. The effect of initiator concentration and solvent on the microstructure of polyisoprenes in this particular study is comparable to that obtained in another study.

Model Polymers. The polyisoprenes synthesized by homogeneous anionic polymerization possess very narrow molecular weight distributions. In addition, they are free of branching and may be prepared with the desired microstructure. After hydrogenation these polymers are ideal models for molecular theories, eg, the rotational isomeric state (RIS).

Block Copolymers. Shell Chemical Co. commercialized thermoplastic elastomers in 1965 under the trade name Kraton (see ELASTOMERS, THERMOPLASTIC). These elastomers are composed of linear-block copolymers of styrene and butadiene or isoprene. The rigid polystyrene end segments are joined by an elastomeric polyisoprene center block. The polystyrene segments associate with each other to form large aggregates called domains. The S—I—S block copolymers are triblock polymers of styrene–isoprene–styrene. Since the two rigid polystyrene blocks form hard plastic domains that anchor the central rubbery polyisoprene block, the raw polymer behaves like a cured elastomer at room temperature. When heated above the T_g of polystyrene, over 100°C, the styrene segments lose rigidity and the block copolymer behaves like an uncured elastomer. Therefore, a product, eg, a shoe sole, may be injection-molded at high temperature without a curing step. A typical synthetic route to these block copolymers is given below where only the 1,4-microstructure is shown for polyisoprene.

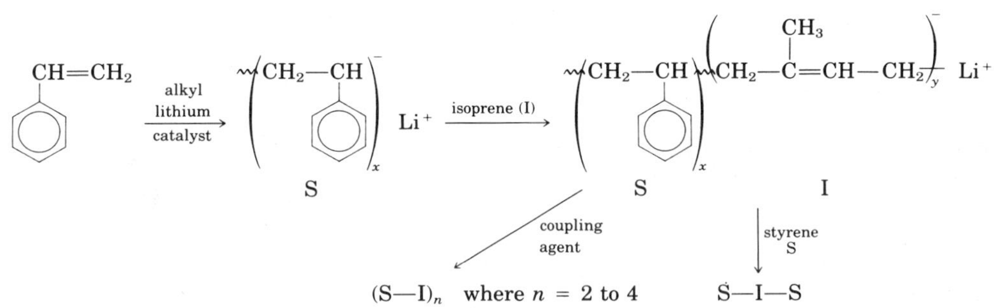

Although more difficult to prepare than random copolymers, block copolymers are not hampered by the comonomer reactivity-ratio restrictions of the random copolymers.

Amorphous Block Copolymers. The early block copolymers were of the styrene–isoprene–styrene, S—I—S, or ABA type. They were synthesized by anionic polymerization by a three-stage sequential addition process using a monofunctional initiator, a difunctional initiator process, a coupling process, or a tapered-block process.

Crystalline Block Copolymers. Block copolymers of the A—B—A type with an elastic center block and crystalline end blocks are prepared as follows. Triblock polymers of ethylene sulfide–isoprene–ethylene sulfide and thiacyclobutane–isoprene–thiacyclobutane are synthesized with a difunctional initiator, 1,4-dilithio-1,1,4,4-tetraphenylbutane. The two ring opening systems operate by different mechanisms: ethylene sulfide through a thiolate species, and thiacyclobutane through a carbanionic chain end, respectively.

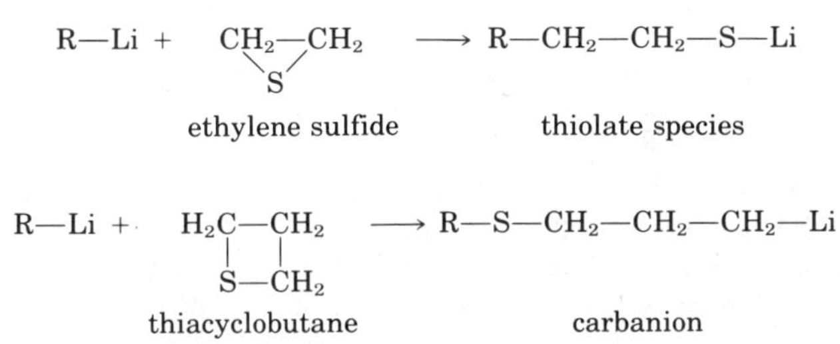

Table 2. Microstructure of Polyisoprenes

Initiator, mol/L	Solvent	Temperature, °C	Mol %[a] cis-1,4	trans-1,4	1,2	3,4
n-C_4H_9Li						
5×10^{-6}	none	25	98	0	0	2
5.3×10^{-6}	n-hexane	25	90	6	0	4
5×10^{-6}	benzene	25	72	20	0	8
4.8×10^{-6}	diethyl ether	25	2	38	10	50
4.9×10^{-6}	THF	25	0	11	28	61
sec-C_4H_9Li						
8×10^{-6}	none	20	96	0	0	4
3×10^{-3}	none	20	77	18	0	5
10^{-5}	n-hexane	20	86	11	0	3
10^{-2}	n-hexane	20	70	25	0	5
4×10^{-5}	benzene	20	70	24	0	6
9×10^{-3}	benzene	20	69	25	0	6

[a] Determined by nmr.

These block copolymers have crystalline end segments that reinforce the polymers in a manner similar to the amorphous or glassy-styrene domains. Ethylene sulfide–isoprene–ethylene sulfide triblock copolymers have a spherulitic texture.

Hydrogenated Block Copolymers. Hydrogenation of the initially prepared block copolymer gives useful products. Butadiene–isoprene–butadiene and butadiene–isoprene–styrene block copolymers are prepared by sequential anionic polymerization with *sec*-butyllithium initiator in cyclohexane. The butadiene moieties are then selectively hydrogenated with cobalt 2-ethylhexanoate–triethylaluminum catalyst.

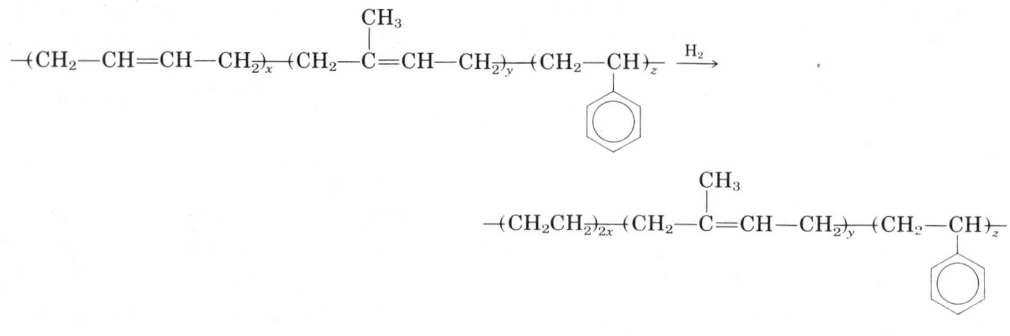

The polyethylene–polyisoprene–polyethylene block copolymers exhibit maximum ultimate tensile strength of ca 15.2 Mpa (2200 psi) between 66 and 75 wt % polyethylene at 600 to 730% elongation; they are tough, clear, rubbery materials. Block copolymers of ethylene–isoprene–styrene have soft rubbery characteristics with low ultimate tensile strengths of ca 8.2 MPa (1200 psi). Complete hydrogenation of an S—I—S (styrene–isoprene–styrene) block copolymer with 17 wt % styrene using 10% palladium-on-carbon catalyst gives a vinylcyclohexane–ethylene–propylene vinylcyclohexane block copolymer (VCH–EP–VCH). Hydrogenation of the polystyrene segments increases the glass-transition temperature of the hard segment by ca 40–50°C.

Processing of Polymers

Synthetic polyisoprene is processed on conventional rubber machinery. The compounding technology is similar to that for natural rubber. Polyisoprene is mixed with compounding ingredients, eg, sulfur, accelerators, antioxidants, fillers (carbon black, clays, or silicates), activators, and plasticizers. The ingredients are incorporated into the raw polyisoprene rubber with a Banbury (internal) mixer or a mill. After mixing, the rubber compounds are calendered, molded, extruded, or fabricated into the desired shape and vulcanized at ca 150°C. Synthetic polyisoprene usually processes much like premasticated natural rubber.

Isoprene Copolymers

Isoprene–Isobutylene Copolymers. Butyl rubber is an elastomer with very low permeability to gases and higher stability against ozone or oxygen attack than natural rubber or styrene–butadiene rubber (SBR). It is prepared by cationic copolymerization of isobutylene with small amounts (1.5–4.5%) of isoprene using an aluminum trichloride catalyst in methyl chloride solution at −100°C. The isoprene units provide the unsaturation necessary for vulcanization. The copolymer is insoluble in the reaction mixture and emerges from the reactor as a fine slurry. Commercial butyl rubbers are random copolymers containing ca 98.5 mol % isobutylene and ca 1.5 mol % isoprene. The isoprene microstructure is mainly trans-1,4 as confirmed by nmr and by the fact that cationic polymerization of isoprene results predominantly in *trans*-1,4-polyisoprene. The copolymer repeat unit can be represented.

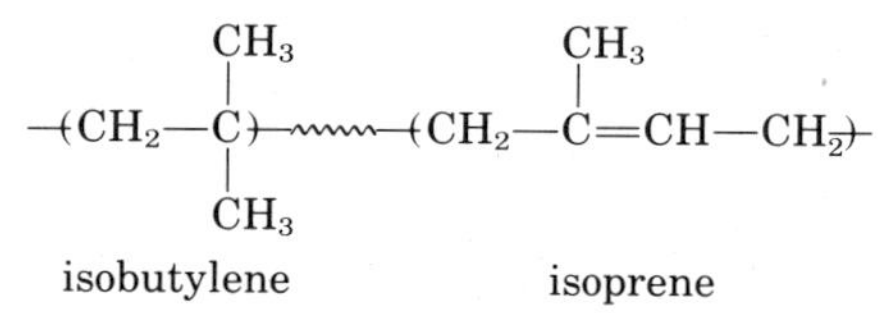

Isoprene–Styrene Copolymers. Anionic copolymerization of isoprene and styrene is readily initiated with alkyllithiums. However, monomer activity is reversed in hydrocarbon solution in which isoprene is more active than styrene; in homopolymerization styrene is the more active monomer. This gives essentially block copolymers, whereas a change in copolymerization characteristics in the presence of small amounts of polar solvent can result in essentially random copolymers. Further evidence of the inversion phenomenon with alkyllithiums in hydrocarbons in the copolymerization of styrene and isoprene is that the propagating polymer chains are rich in isoprene until late in the reaction when suddenly styrene incorporation increases (inversion).

Isoprene–Butadiene Copolymers. *Alkyllithium Catalysts.* 1,3-Butadiene was copolymerized with ^{14}C-labeled isoprene in hexane solution with butyllithium at 50°C to determine the copolymer composition. The reactivity ratios for butadiene and isoprene are 3.38 and 0.47, respectively, indicating that butadiene is the more active monomer. In alkyllithium-initiated homopolymerization, isoprene is more reactive than butadiene, whereas in the alkyllithium copolymerization of a butadiene and isoprene mixture, butadiene polymerizes first. This is the same inversion phenomenon in hydrocarbon solvents that is observed for isoprene and styrene.

Coordination Catalysts. The copolymerization of 1,3-butadiene and isoprene was first studied with a homogeneous diethylaluminum chloride–cobalt acetylacetonate catalyst $(C_2H_5)_2AlCl\text{-}Co(acac)_2$ in toluene at −20 to 50°C. The isoprene composition ranges from 3% to 90% in the mixture when all monomer is initially charged. Triisobutylaluminum–titanium tetrachloride $(i\text{-}C_4H_9)_3Al\text{–}TiCl_4$ and diisobutylaluminum chloride–cobalt chloride–ethanol $(i\text{-}C_4H_9)_2AlCl\text{–}CoCl_2\text{–}C_2H_5OH$

were also investigated. Actual copolymers are formed and the reactivity ratios are 1 for both monomers with the $(i\text{-}C_4H_9)_3Al$–$TiCl_4$ catalyst. For the cobalt catalyst, the reactivity ratio of butadiene is 2 and of isoprene is 0.9. The microstructure of the butadiene units in the copolymer is the same as in its homopolymer (see BUTADIENE POLYMERS).

Ethylene–Isoprene Alternating Copolymer. A new ethylene–isoprene stereospecific rubber is prepared by alternating copolymerization with an $(RO)_2VOCl$–triisobutylaluminum catalyst. The catalytic activity increased significantly with increasing β-substitution of the alkyl group R in the vanadium portion of the catalyst. With a $[(CH_3)_3C—CH_2—O]_2$ VOCl–triisobutylaluminum catalyst, a 51.2/48.8 (molar ratio) ethylene–isoprene copolymer is obtained with ~98% of 1,4-microstructure, 85/15 trans-1,4/cis-1,4, and a T_g of −85°C. The crystallization properties have also been examined.

trans-1,4-Polyisoprene

trans-1,4-Polyisoprene is crystalline and very hard at room temperature and exhibits good tensile strength in the unvulcanized state. Synthetic *trans*-1,4-polyisoprene melts at ca 60°C and has a T_g of ca −60°C. It can be processed in conventional rubber-fabrication machinery above its melting temperature. Mooney viscosity is ca 30 (ML-4 at 100°C); processing characteristics and Mooney viscosity are similar to those of natural balata. *trans*-1,4-Polyisoprene can be extruded, calendered, injection-molded, and compression-molded. It can be compounded with fillers and used in blends with other polymers.

Economic Aspects

In 1986, synthetic *cis*-1,4-polyisoprene (Natsyn) was manufactured on a commercial scale in the United States only by Goodyear Tire and Rubber Co. at Beaumont, Tex. This plant has an annual capacity of ca 60,000 t. In 1982, approximately 65% of the synthetic *cis*-1,4-polyisoprene consumed in the United States was used in tires and tire products. *trans*-1,4-Polyisoprene is not produced commercially in the United States, but is produced in Japan.

Economic data for synthetic polyisoprene in the United States are given in Table 3; production figures are proprietary. Figures for noncommunist countries' production of polyisoprene are shown in Table 4. Figures are not available from the USSR or other Communist-bloc countries, but production is believed to be roughly three times that of the free world. The prices for synthetic *cis*-1,4-polyisoprene (Natsyn) (~97% cis-1,4-microstructure) are given in Table 5. In 1980 the price of imported *trans*-1,4-polyisoprene in the United States was ca $13.23–14.33/kg.

Table 3. U.S. Synthetic Polyisoprene Economic Data, 10^3 t

		Consumption			
Year	Production	trans-1,4[a]	cis-1,4	Imports	Exports
1970	122		81.3		
1972	133.8		113.5		
1974	85.2	1.02	80.2	0.52	6.6
1976	74.4		62.1	0.5	6.7
1978		1.25[b]	75.8	10.1	4.5
1980			62.6	11.1	7.2
1982	40[b]		42[c]	5.3	5.0
1983				4.3	4.5

[a] Includes balata and gutta percha.
[b] Estimated.

Table 4. Synthetic Polyisoprene Production in Noncommunist Countries

Year	10^3 t
1974	220
1976	210
1978	223
1980	221
1982	128

Specifications, Standards, and Test Methods

Product specifications for solid synthetic *cis*-1,4-polyisoprene include analyses for volatile matter, extractables, ash, and Mooney viscosity at 100°C. Manufacturers may also specify the Monsanto Rheometer data for vulcanizates prepared by a standard recipe.

Health and Safety Factors

The flash point of isoprene is −54°C by the Tag-open-cup (TOC) method. Isoprene is classified as a flammable liquid by the U.S. Department of Transportation. In the absence of inhibitors, isoprene forms dangerous peroxides in air. Isoprene is a moderate irritant to the skin, eyes, and mucous membranes.

Isoprene is found in the vapor phase of tobacco smoke, as a precursor of polycyclic aromatics, demonstrated by thermal condensations at 450–700°C. Isoprene occurs in many biological systems, mammalian tissue, and plant bacteria. The isoprene unit is an important building block for lipids, steroids, terpenoids, and natural products, eg, latex, which is the raw material of natural rubber.

Uses

cis-1,4-Polyisoprene is used in tires and tire products, molded and mechanical goods, belts, gaskets, foam rubber, rubber sheeting, rubber bands, bottle nipples, footwear and sporting goods, gloves, sealants, and caulking. Because of the close resemblance in structure and properties, synthetic *cis*-1,4-polyisoprene and natural rubber have similar applications.

trans-1,4-Polyisoprene, a crystalline thermoplastic, resists abrasion, cutting, and scuffing, and is used mainly in golf-ball covers, orthopedic devices, and splints, and occasionally in transmission belts, cable coverings, and adhesives. *trans*-Polyisoprene offers an unusual combination of properties for medical uses such as prosthetics, splints, braces, casts, and attachments for artificial limbs.

Table 5. Prices of Synthetic *cis*-1,4-Polyisoprene[a,b]

	$/kg
Natsyn 2200	1.70
Natsyn 2210, low viscosity	1.70
Natsyn 2205, low gel	1.81

[a] As of November 1985; Data from Goodyear Tire and Rubber Co.
[b] Carload or truckload, FOB at Beaumont, Tex., 34 kg.

Some isoprene is used in the production of butyl rubber by Exxon Chemical Company at Baytown, Tex., and Baton Rouge, La., and by Polysar, Inc. Block copolymers such as styrene–isoprene–styrene are used in adhesives and footwear.

BIBLIOGRAPHY

E. Ceausescu, *Stereospecific Polymerization of Isoprene,* Pergamon Press Inc., New York, 1983.

M. Morton, *Anionic Polymerization: Principles and Practice,* Academic Press, New York, 1983.

W. M. Saltman, ed., *The Stereo Rubbers,* Wiley-Interscience, New York, 1977.

E. Schoenberg, H. A. Marsh, S. J. Walters, and W. M. Saltman, *Rubber Chem. Technol.* **52,** 533 (1979).

MICHAEL L. SENYEK
Goodyear Tire and Rubber Co.

ISOTACTIC POLYMERS.

See MICROSTRUCTURE; STEREOREGULAR POLYMERS.

J

JET MOLDING. See Injection molding.

JOINTS, ADHESIVE. See Adhesion and bonding.

JUTE. See Fibers, vegetable.

K

KAPOK. See FIBERS, VEGETABLE.

KERATIN

Fibrous proteins in vertebrates have essentially mechanical functions. Keratins are a special form of fibrous proteins providing protective outer animal coverings, eg, fur, wool, hair, and feathers, and external appendages, such as claws, horns, nails, and quills. Keratins are durable, insoluble, chemically unreactive, and suited to exposure to severe environmental conditions. They are pliable and recover with little loss of properties after repeated deformations. A high content of the sulfur-containing dimeric amino acid cystine distinguishes keratins from other fibrous proteins. Cystine stabilizes the keratin structure.

The outer covering of vertebrates consists of two distinct strata, the dermis and the epidermis. The dermis is a network composed mainly of proteins, collagen, elastin, and reticulin fibrils that merges gradually into the subcutaneous tissue. The epidermis consists of a number of layers. In the germinal layer, which is closest to the dermis, potential keratin cells are produced. These cells migrate through zones where protein synthesis and hardening (keratinization) occur to the outermost layer of dead tissue, producing skin keratin and the keratins of such appendages as the scales of reptiles, the feathers of birds, and the hooves, claws, nails, horns, hair, and wool of mammals. Keratin also is the main constituent of certain types of skin tumors.

Hair and wool are produced in fiber follicles. The cortical cells form the bulk of a hair or wool fiber. The cells start growing at the bottom of the follicle, which is bulbous and contains the germinal matrix where cell division occurs. The cells complete their elongation immediately above the bulb. Keratinization, the process of stabilizing the elongated cortical cells, occurs from the top of the bulb and is completed before the fiber protrudes through the surface of the skin. Structure stabilization of the fiber involves the formation of disulfide bonds by oxidation of thiol groups (S—H) present in the materials immediately above the bulb.

The term α-keratin (and β-keratin) refers to the typical wide-angle x-ray diffraction pattern obtained from these fibers. Fibrous keratins, eg, hair, fur, and wool, all have an external cuticle layer covering the cortex, with a central axial medulla often present in the coarse fibers (Fig. 1). The cuticle consists of layers of scales, each $\sim 0.5\ \mu m$ thick in human hair and nearly 1-μm thick in wool. The scales in all types of hair form a ratchetlike structure, resulting in a directional friction effect that plays a vital role in the entanglement of fibers.

Chemistry

Chemical Composition. The bulk of material forming α-keratins consists of polypeptide chains

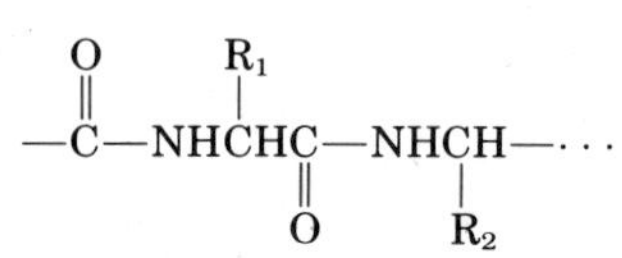

where R_1, R_2 are the side chains of the amino acids; 20 different compositions occur in keratin in varying proportions.

As in all proteins, a large variety of hydrogen bonds and van der Waals interactions exist between and within the chains. Important for the properties of α-keratin fibers are the coulombic interactions, also referred to as salt links, that exist between the charged side chains of lysine, arginine, and histidine and the acidic side chains of glutamic and aspartic acids.

The amino acids are determined after acid hydrolysis of the polypeptide chain. In native keratins, the degradation of some amino acids is catalyzed by metal ions.

Keratin Derivatives. Many different protein fractions have been extracted from keratin. In biosynthesis, the fibers are formed in the keratinization step where thiols are oxidized to disulfides, forming a covalent link between polypeptide chains.

When extracting proteins from keratin, the protein fractions must be solubilized by breaking the disulfide bonds. The unstable groups that are formed, such as thiol, must be blocked to obtain stable derivatives.

The reduction of α-keratin fibers produces three distinct fractions: low sulfur *S*-carboxymethyl kerateine A (SCMKA), high sulfur *S*-carboxymethylkerateine B (SCMKB), and high Gly-Tyr proteins rich in glycine and tyrosine. The SCMKA proteins have an α-helix content of ca 50% and a molecular weight of from 45×10^3 to 60×10^3.

Chemical Modification of Keratin Fibers. The reactive groups, mainly the side chains, may be used to modify keratin fibers. For example, attachment of iodine to the tyrosine residue increases the intensity of certain x-ray reflections. The mechanical properties of fibers are changed by rupturing disulfide

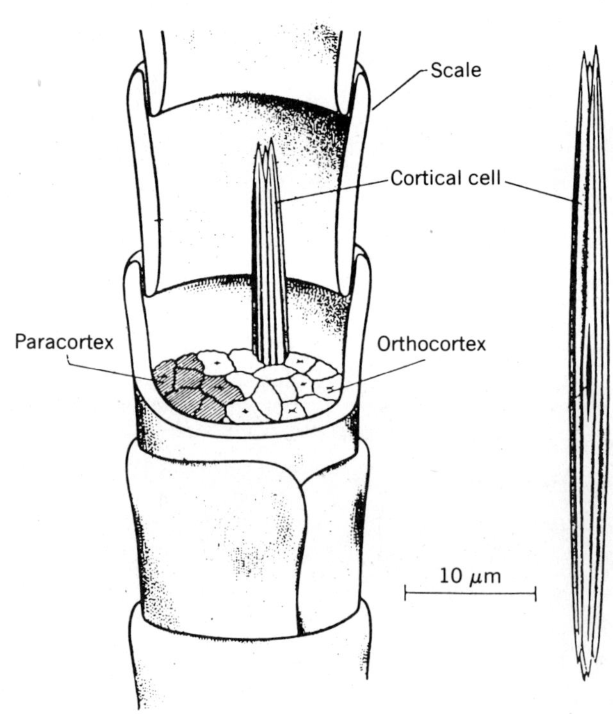

Figure 1. The histological components of a Merino wool fiber as revealed by the light microscope.

bonds and by modification of coulombic interactions of acidic and basic side chains. Introducing cross-linking reagents, eg, formaldehyde and nitrophenyl esters, also affects mechanical properties. Water content change produced by chemical change also modifies the mechanical properties.

Keratin–Water System

Water is a highly polar, extremely mobile and penetrating molecule, dimensionally equivalent to little more than an atom of oxygen. In the keratin structure, it interacts with negatively and positively charged groups, such as carboxy and amine groups of the polypeptide chains. A water molecule can form up to four hydrogen bonds with other water molecules or hydrogen-bonding groups, eg, amide, carboxyl, and hydroxyl. Water produces a continuous three-dimensional network of interactions within the keratin structure, providing the keratin–water system with electrical conductivity.

The structural location of the water molecules is of prime importance in understanding the physical and mechanical properties of α-keratin fibers. Water affects the mechanical properties of fibers, acting as a plasticizer and reducing fiber stiffness by a factor of about three from dry to wet. The swelling of hair fibers due to the water absorption is confined primarily to the radial direction, a result of the basic two-phase structure of the keratin cortex.

The high volume swelling of the matrix indicates that the water is not primarily absorbed into a matrix of a highly cross-linked network formed by polypeptide chains and disulfide bonds, but rather that the material of the matrix is in the form of high sulfur globular protein material and external water. The nature of the wet matrix has been further elucidated by a number of workers who have studied the dependence of water sorption and mechanical properties of α-keratin fibers on the protein content of the matrix. An increase of matrix high sulfur protein reduces the water sorption of the matrix, by displacement of water by the protein, and the stiffness of the matrix increases with decreasing water content.

Mechanical Properties of α-Keratin Fibers

Longitudinal Mechanical Properties. The longitudinal mechanical properties of α-keratin fibers vary markedly with temperature, humidity, and time. The stress–strain behavior can be considered in terms of three distinct regions of strain (Fig. 2). If a fiber is progressively extended under constant temperature and humidity conditions, once it is straightened, which may involve considerable strain, the stress–strain curve (up to a few percent strain) exhibits a nearly linear region usually referred to as the Hookean region (Fig. 2). Further extension results in the yield region of low stress increase with strain, followed with further extension by a stiffer postyield region.

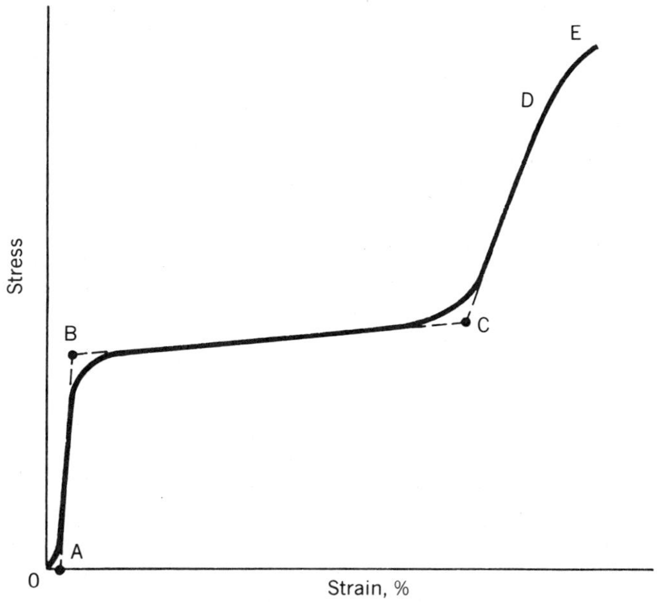

Figure 2. A typical stress–strain curve of an α-keratin fiber showing the Hookean region AB, yield region BC, and postyield region CDE.

Keratin Fiber on Extension

Mechanically, only two distinct phases of keratin can be identified. The matrix M, which mechanically is sensitive to the presence of water, includes the material between the microfibrils and the material whose mechanical properties are indistinguishable from that of the matrix, as defined by the transmission electron microscope. The "mechanical" microfibril phase C, which is unaffected by water, corresponds to the organized α-helical material, which is about 50% of the microfibril visible and defined by the electron microscope. Side chains from some α-helical components of the microfibrils interact with the matrix and the distinction of these into one or the other phase is difficult, and some water penetration of the microfibrils into the nonhelical structure may occur. The mechanical events occurring in each of the two phases during fiber extension are shown in Table 1. These events are compatible with the evidence available and lead to an understanding of the mechanical properties of α-keratin fibers.

Swelling, Contraction, and Set. The longitudinal stability of α-keratin fibers is primarily a result of long, intact, organized microfibrils oriented in the fiber direction. The α-helical ropes provide the main crystalline components of the microfibrils and maintain the longitudinal stability, which is minimally affected by the presence of water. The effect of other swelling agents on the stability of α-keratin fibers can be expressed best in terms of the two phases consisting of microfibrils and matrix.

Table 1. Molecular Events Within the Microfibrils and Matrix of a Keratin Fiber in Water Reflected by Mechanical Properties

Region of extension	Phase C	Phase M[a]
Hookean	α-helices strained	hydrogen-bonded water network plus globular proteins in gel state
yield	$\alpha \rightarrow \beta$	during extension the extending structure in the sol state
postyield	$\alpha \rightarrow \beta$[b]	extending structure in the sol state[b]

[a] See text for definition.

[b] Associated covalent network under strain with possible irrecoverable bond breakdown.

If the matrix and the microfibrils are swollen and the α-helices are randomized, eg, fibers in concentrated aqueous solutions of lithium bromide at molarities ≥6.6, the mechanical stiffness is drastically reduced, corresponding to that of a purely elastomeric solid. The wide-angle x-ray diffraction α-pattern and the optical birefringence have both disappeared in agreement with the stucture now consisting of randomized polypeptide chains. The length of the fiber is greatly reduced, and the fiber is said to be supercontracted.

If an α-keratin fiber is held extended in the yield or postyield regions in water or steam at 100°C for sufficiently long periods, the released fiber, instead of contracting, retains some of the extended length, depending on the environment (dry or wet) and the temperature. The fiber is said to be set at the extended length. The process of setting is of commercial importance for wool fibers and human hair.

Imparting set involves not only relatively high extensions as above but also application of chemical and physical techniques at low extensions and a variety of distortions. Both breakdown and reformation of the disulfide bonds occur in many of these techniques. The basic nature of set as a physical process is common for all fibers.

BIBLIOGRAPHY

R. D. B. Fraser, T. P. MacRae, and G. E. Rogers, *Keratins, Their Composition, Structure and Biosynthesis,* Charles C. Thomas, Springfield, Ill., 1972.

W. G. Crewther, L. M. Dowling, P. M. Steinert, and D. A. D. Parry, *Int. J. Biol. Macromol.* **5,** 267 (1983).

R. D. B. Fraser, L. N. Jones, T. P. MacRae, E. Suzuki, and P. A. Tulloch in R. Srinivasan, ed., paper presented at *International Symposium on Biomolecular Structure, Conformation, Function, and Evolution, Madras, India,* 1978, pp. 133–152.

M. Feughelman, *J. Soc. Cosmet. Chem.* **33,** 385 (1982).

MAX FEUGHELMAN
University of New South Wales

KERR EFFECT.

See ELECTROOPTICAL APPLICATIONS.

KETONE POLYMERS

Ketones are generally regarded as stable materials. Unlike aldehydes, thioaldehydes, and thioketones, they do not normally give cyclic trimers or linear polymers. Polyketones have been made from ketenes. The only polyketone of commercial importance is PEEK (see POLYETHERETHERKETONES).

KINETIC CHAIN LENGTH.

See CHAIN-REACTION POLYMERIZATION; DEGREE OF POLYMERIZATION; KINETICS OF POLYMERIZATION.

KINETICS OF POLYMERIZATION

The kinetics of a reaction constitutes one of the most important outward manifestations of the mechanism. This is particularly true for polymerization reactions, in contrast to those involving other simple substances, because the average molecular weight and the molecular weight distribution of the polymer produced are kinetic observations, just as much as are the rates and orders of reactions. That is, both are consequences of the basic features of the mechanism.

All polymerization processes consist of a number of basic reaction steps that contribute to the overall kinetics. This includes a propagation step; indeed, this reaction essentially is polymerization, the linking of a large number of units into a long sequence. Many polymerizations have a distinct initiation step due to the action of a catalyst or initiator, but in others initiation is distinguishable in character from subsequent propagations solely by the size of the contributing molecules. Most, but by no means all, polymerization processes also have a termination step that permanently ends the growth of molecules. From a kinetic point of view, true kinetic termination must be distinguished from transfer or exchange processes (see CHAIN TRANSFER). Exchange processes influence the size of individual molecules without necessarily having any effect on the rate of reaction; by termination, a reaction that not only brings molecular growth to an end but also tends to reduce the concentration of some species essential for the reaction's continuation is implied. Hence, termination is a factor in the determination of overall rate. Although termination and transfer processes are different with respect to rate of reaction, they are essentially equivalent in considerations of the polymer's degree of polymerization.

Further discussion of kinetics in general terms requires that a distinction be made between two broad classes of polymerization reactions, each of which is described in more detail elsewhere in the Encyclopedia. These are step-reaction polymerization and chain-reaction polymerization, whose nomenclature corresponds closely but not exactly to the older classes "condensation polymerization" and "addition polymerization," respectively.

As the names suggest, the chain-reaction class depends on the repeated reaction of an active intermediate (radical, ion, or polar complex), whereas the step-reaction class involves a course consisting of the repetition of a simple reaction, often a condensation, between molecules of gradually increasing size. A number of broad differences in kinetic characteristics between the two reaction types are given below:

Step reaction	Chain reaction
high molecular weight material only produced at high conversion	high molecular weight material produced in earliest stages
propagation steps involve species of all molecular sizes	each propagation step involves a monomer unit
monomer concentration diminishes rapidly in early stages	monomer remains as the most plentiful single molecular size until essentially the end
after early stages, propagation is almost entirely reaction between polymer molecules	polymer molecules do not react with one another once formed

BIBLIOGRAPHY

C. H. Bamford, W. G. Barb, A. D. Jenkins, and P. F. Onyon, *The Kinetics of Vinyl Polymerization by Radical Mechanisms,* Butterworth & Co. Publishers, Ltd., London, UK, 1958, Chapt. 1.

A. D. Jenkins in A. D. Jenkins and A. Ledwith, eds., *Reactivity, Mechanism and Structure in Polymer Chemistry,* John Wiley & Sons, Inc., New York, 1974, Chapt. 4.

P. E. M. Allen and C. R. Patrick, *Kinetics and Mechanisms of Polymerization Reactions,* Ellis Horwood, Ltd., Chichester, UK, 1974, Chapt. 7.

AUBREY D. JENKINS
The University of Sussex

KNIFE COATING. See COATING METHODS.

L

LABORATORY POLYMERIZATION PROCEDURES. See POLYMERIZATION PROCEDURES, LABORATORY.

LAC. See RESINS, NATURAL.

LACQUERS. See COATINGS.

LACTAM POLYMERS. See POLYAMIDES; *N*-VINYL AMIDE POLYMERS.

LACTONES. See POLYESTERS.

LADDER AND SPIRO POLYMERS

As science and technology progress, there is an ever-increasing demand for materials that are resistant to all types of degradation: thermal, oxidative, photochemical, mechanical, radiative, and biochemical. Ladder and spiro polymers, special cases of double-stranded polymers, are therefore of particular interest because, theoretically, they should possess increased resistance to any type of degradation as compared to conventional polymers (see also CYCLOPOLYMERIZATION; DEGRADATION; HEAT-RESISTANT POLYMERS).

Many properties of polymers, such as tensile strength and toughness, are directly related to the molecular weight of the material. Any reaction producing a cleavage that results in a decrease in molecular weight will adversely affect these properties. In a linear polymer that is composed only of single bonds, such as polyethylene, polystyrene, and polyamides, any break in the main polymer chain will reduce the number-average molecular weight to half the original value, as indicated below.

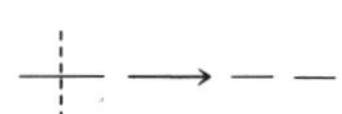

Ladder polymers (so named because a schematic diagram of their molecular structure resembles a ladder) and spiro polymers behave somewhat differently. These polymers are also sometimes called "double-stranded," "ordered linear network," and "nonrandom cross-linked" polymers. In addition to their resistance to degradation, many regular ladder and spiro polymers, because of their stiffness and restriction to rotation, exhibit liquid crystalline behavior. Highly conjugated ladder polymers, because of their planarity, have very interesting electrical properties.

ladder polymer

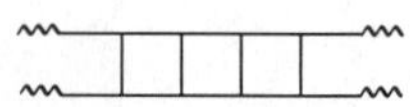

spiro polymer

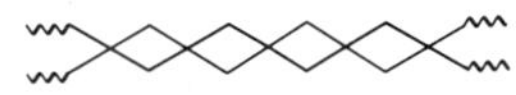

Both ladder and spiro polymers can undergo a cleavage of any single bond in the chain without a decrease in molecular weight and resulting deterioration in properties; two cleavages must occur in the same ring before a decrease in molecular weight occurs. If the breaking of single bonds occurs at random along the chain, many cleavages might take place before two occurred in the same ring. Therefore, it can be expected that a ladder or spiro polymer will be more resistant to mechanical, chemical, and thermal degradation than its single-chain counterpart.

Even partial ladder polymers, where every twelfth or sixth bond is single, are somewhat more stable than the corresponding single-chain polymer. However, much of the resistance of the polymer to molecular-weight decrease has been lost. Thus, having a complete, or perfect, ladder structure is critical to achieving maximum stability. In practice, only a few synthetic approaches to ladder polymers ensure the formation of perfect ladder structures.

Experimental Verification of Theory

The validity of the theoretical conclusions has been verified in part by studies of the basic hydrolysis of the ladder polymer from 2-vinyl-1,3-butadiene and the cyclic dimer of tetramethylene fumarate (**1**). A plot of the molecular weight of the resulting product as a function of the extent of hydrolysis of this polyester ladder polymer (**1**) is given in Figure 1 together with data from a related linear polyester, poly(tetramethylene adipate). The ladder polymer is indeed more resistant to a decrease in molecular weight as hydrolysis proceeds.

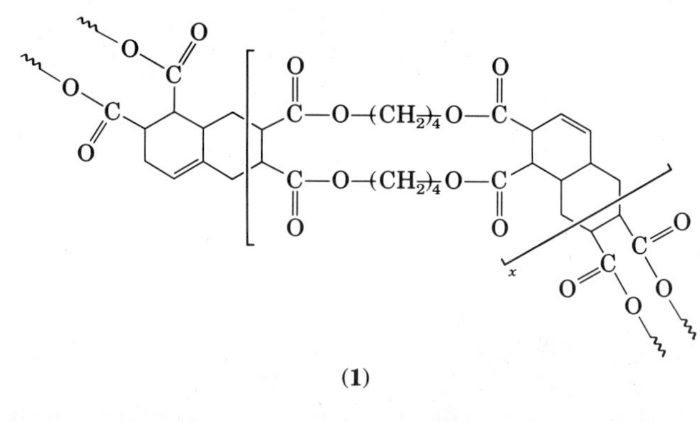

(**1**)

Problems in Making a Complete or "Perfect" Ladder Polymer

From the above discussion, the importance of having a perfect ladder polymer is quite obvious. However, in practice, it is often difficult or impossible to prepare a perfect or nearly perfect ladder structure. The reasons are manifold. First, many ladder polymers are insoluble in common solvents, at least partly because of the inherent lack of rotation in the double-stranded chain, and precipitate from solution before reactions are complete. In the solid state, many partial or step-ladder structures have high glass-transition temperatures, and as ring closure

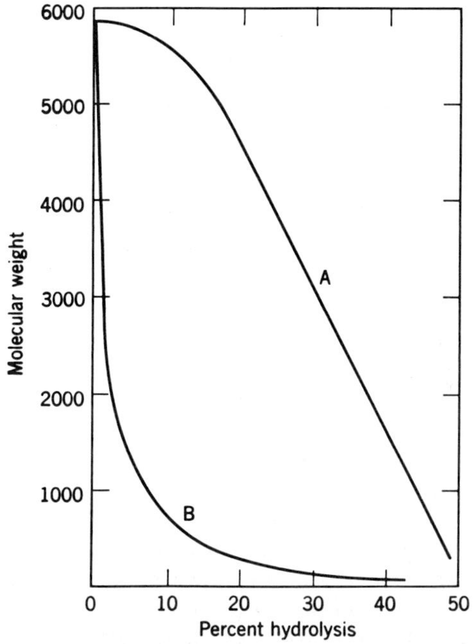

Figure 1. Molecular weight as a function of extent of hydrolysis. A, ladder polymer; B, poly(tetramethylene adipate).

progresses, the glass transition increases further. As the temperature is raised to keep the molten polymer above the glass transition, side reactions often occur.

In a sense a ladder polymer can be considered a nonrandomly cross-linked polymer. If the reactions involved in the ring closure are at all reversible, prolonged heating to complete the reactions may cause some random cross-linking rearrangement or redistribution to occur to give an insoluble cross-linked material.

Finally, there are several problems associated with a process in which one of the sides of the ladder is formed first and then the second side is formed in a subsequent process. If the second step involves a chain process, it is essentially impossible to find an external reagent that would begin at one end of the chain and zip up all the side groups to form a complete ladder structure.

Even if the ring closure is designed to involve adjacent pairs of reactive groups, there may be a statistical problem. Thus in the example in Figure 2, if there is free rotation of the group holding the reactive groups A and B, and A can only react with B, not all of the rings can be closed.

Thus it appears that the most desirable type of reaction for the formation of a complete or perfect ladder is one in which both sides of the ladder are either formed simultaneously or no reaction occurs. If the reaction is at all reversible, it is desirable that both sides of the ladder structure are broken at the same time. Such reactions include the Diels-Alder reaction and 1,3-dipolar addition reactions. Similarly, for the formation of complete or perfect spiro polymers, ketal formation appears suitable because the intermediate hemiketal is usually unstable.

Ladder Polymers

Early Developments. The early work in the field of double-chain polymers has been reviewed by many authors in many languages. The effect of introducing rings into a polymer backbone on its chemical and physical properties has been recognized for some time. In 1926 the self-condensation of cyclopentadiene to a low molecular weight ladder polymer (**2**) was reported (see also CYCLOPENTADIENE POLYMERS).

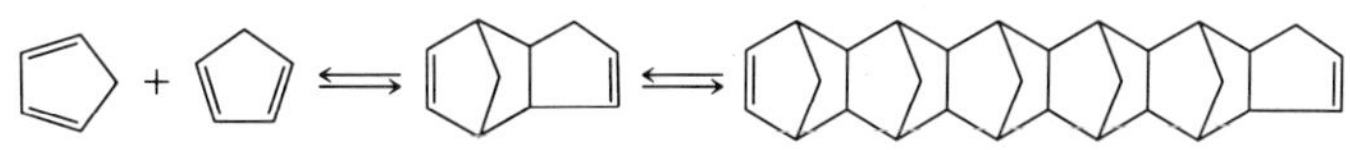

(**2**)

However, because of the low reactivity of double bonds in the adducts and the strain in the fused rings, the highest molecular weight material isolated was the hexamer. Also in 1926, researchers reinvestigated the cyclization of natural rubber in which the rubbery material is converted to a plastic with an acid catalyst. Because the resulting structure is complex, a variety of structures have been proposed. Earlier workers proposed a structure containing monocyclic derivatives, but more recent workers have proposed the formation of a fused polycyclic system or a partial or incomplete ladder structure.

Polyacrylonitrile Cyclization. Undoubtedly the most studied nitrile cyclization is that involving polyacrylonitrile (PAN). When PAN is heated to 250–275°C in air, it undergoes simultaneous cyclization and oxidation to "Black Orlon." When this material is heated in the flame of a blowtorch, it merely glows red but maintains its shape. Treatment at 900°C for 3 h is necessary to destroy this material.
of Black Orlon has been given.

The formation of carbon fibers (qv) is now a big business and the principal starting material for these fibers is either PAN or copolymers of acrylonitrile with 0.5–5.0% (mainly 0.5–2.0%) of methyl acrylate, methyl methacrylate, itaconic acid, or methacrylic acid. These fibers are stabilized by heating in air to convert the linear fibers into a highly condensed cyclized ladder polymer. The next step involves a rapid thermal pyrolysis under tension in an inert atmosphere, usually high purity nitrogen, at 1200–2000°C. In the 350–1300°C temperature range, a thermally stable pyridinoid structure is formed, which subsequently collapses into a stacked-ring carbon structure to form carbon fibers. The overall yield from the PAN is about 50%. Heat treatment under tension above 2500°C for a short time in argon or nitrogen can be used to produce ultrahigh modulus products, 520 GPa (7.5 × 10^7 psi), and higher modulus.

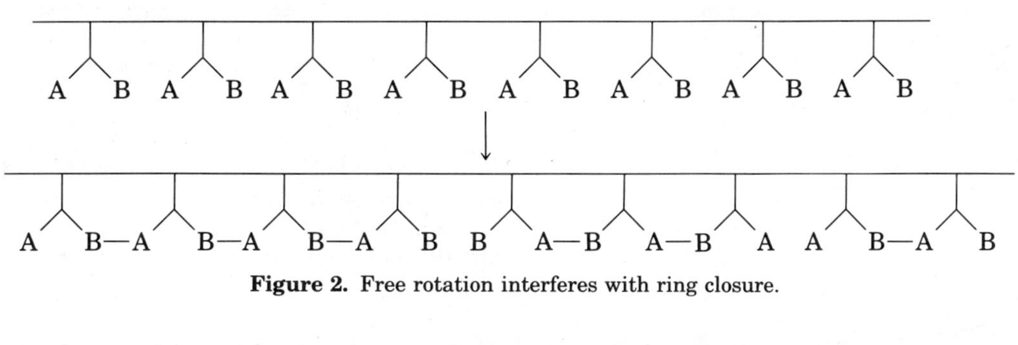

Figure 2. Free rotation interferes with ring closure.

Carbocyclic Ladder Polymers. Probably the first nearly perfect ladder polymer (**3**) was prepared from 2-vinylbutadiene and *p*-benzoquinone in 1954.

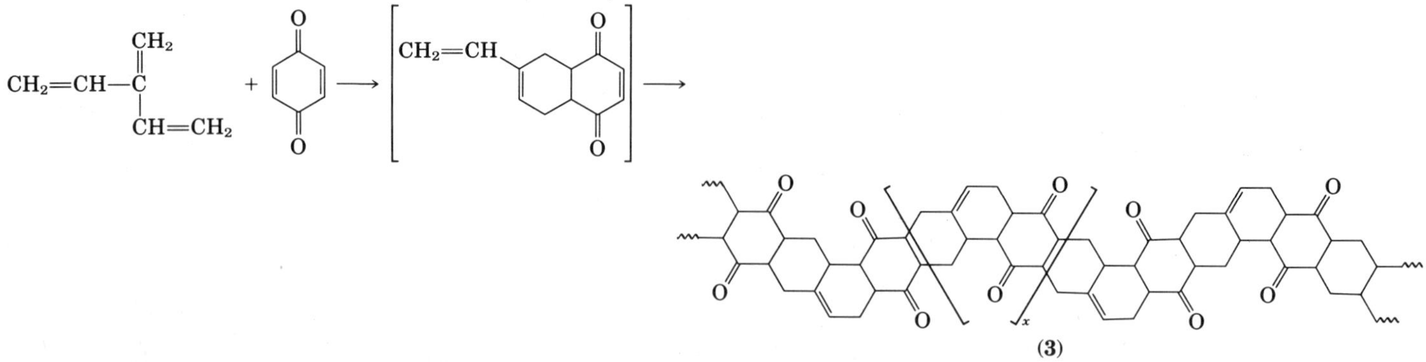

(**3**)

The Diels-Alder reaction is ideally suited for the preparation of ladder polymers because both "rungs" of the ladder are closed simultaneously and are not governed by statistical considerations. This Diels-Alder polymer (**3**) was reported at the time to be a crystalline brown powder that was infusible and insoluble in all common solvents. It was later shown, however, to be soluble in 2-hexafluoropropanol and to have a molecular weight of 7000. It decomposes on heating to form a carbon residue with a 37% loss in weight at 900°C. A similar polymer (**4**) can be prepared from 1,2,4,5-tetramethylenecylcohexane and *p*-benzoquinone.

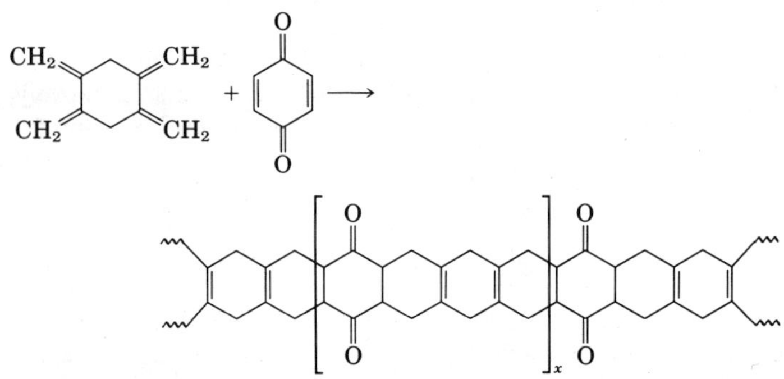

(**4**)

Because there at first appeared to be no way to fabricate or characterize these materials, they attracted very little interest, but solvent availability allows fabrication. Both polymers (**3**) and (**4**) can be deoxygenated to the corresponding hydrocarbon polymer by the following process:

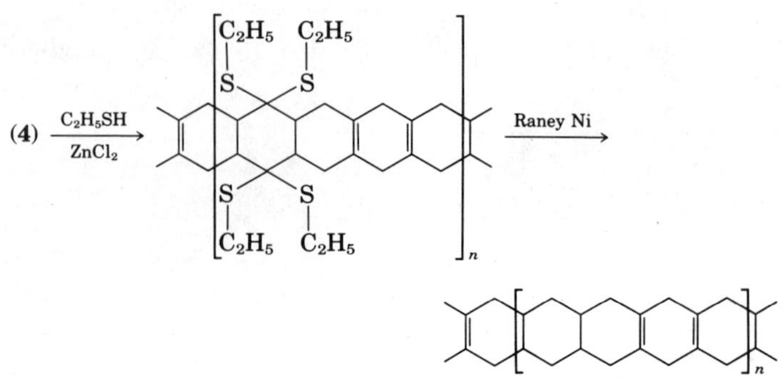

(**5**)

In addition to cyclized natural rubber, which has been reviewed, a large variety of diene polymers have been studied as potential sources of ladder polymers.

Ladders from Acetylenic Monomers. Acetylene-containing polymers have been cyclized to products that have been described as ladder polymers without much structural determination. For example, vinylacetylene and hexen-1-yne-3 have been polymerized with acid catalysts to a presumed ladder polymer (**6**) containing conjugated double bonds; (**6**) is stable in air at ≥400°C.

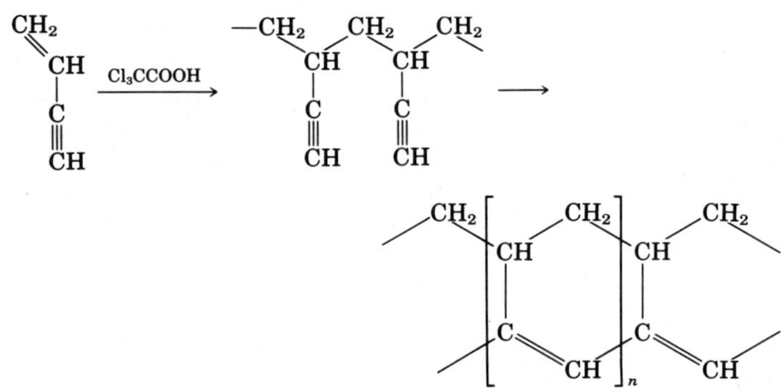

(**6**)

Highly Conjugated Carbocyclic Ladder Polymers. A number of workers have synthesized highly conjugated ladder polymers that are of interest as good electrical conductors. For example, heating of the dianhydride (**7**) gives the polyperinaphthalene (**8**).

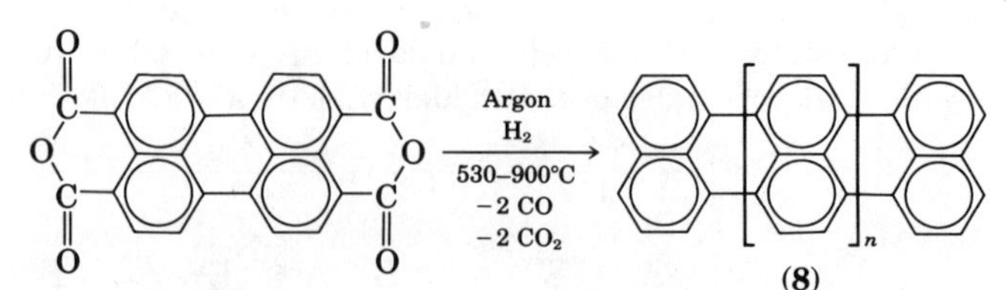

(**7**) (**8**)

If the pyrolysis is carried out below 650°C, the polymer (**8**) is a semiconductor if undoped (see ELECTRICALLY CONDUCTIVE POLYMERS). At 650–800°C the polymer changes to acquire a metallic behavior. Its x-ray and Raman spectra show that the high temperature material is more regular. If the polymer (**8**) is doped with iodine, $FeCl_3$, or $NbCl_2$, its conductivity at room temperature greatly increases.

Polyacenes decrease in stability as the length of the acene sequence increases. Several workers have made calculations in several other conjugated systems. Thus predictions based on theoretical calculations of the relative stabilities of various ladder polymers related to biphenylene have been made.

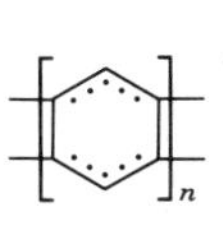

(**9**)

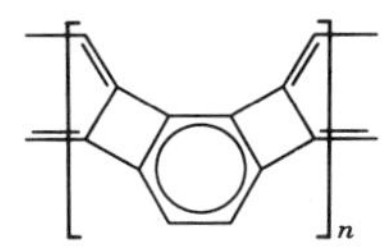

(**10**)

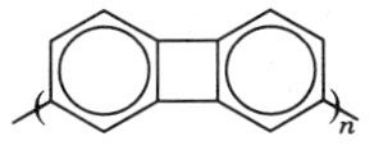

(**11**)

Of the three polymers (**9**), (**10**), and (**11**), (**10**) had the highest ionization potential, the highest band gap, and fairly narrow band width. A structure related to (**9**) was proposed for the uncharacterized, aromatic yellow-red polymer obtained from 1,2,4,5-tetrabromobenzene by a Wurtz-Fittig-type reaction by use of a sodium–potassium alloy.

Ladder Polyquinones. The condensation of an aromatic dianhydride with an activated aromatic ring in a Friedel-Crafts condensation gives a black, insoluble, infusible material for which structure (**12**) has been proposed, although with little evidence.

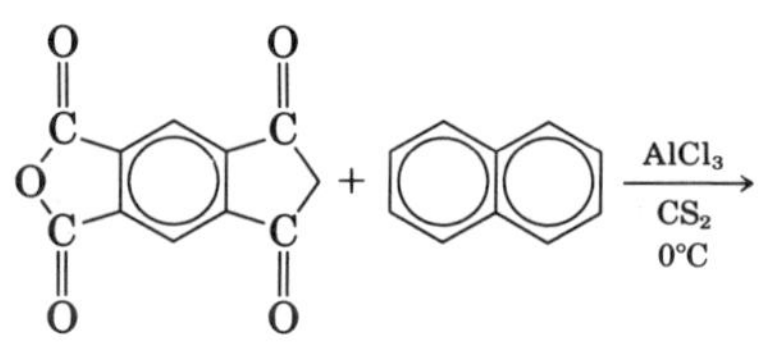

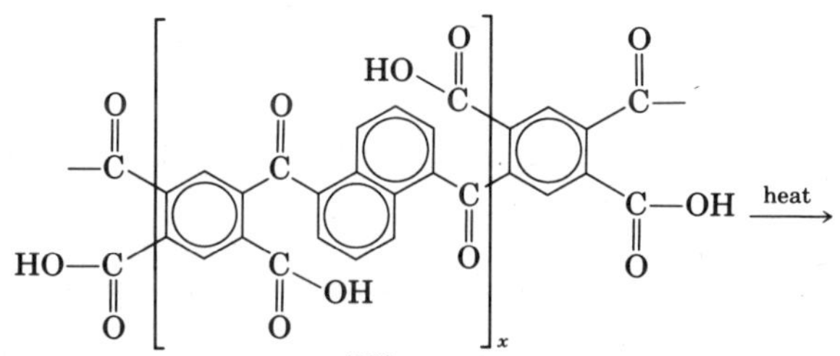

(**13**)

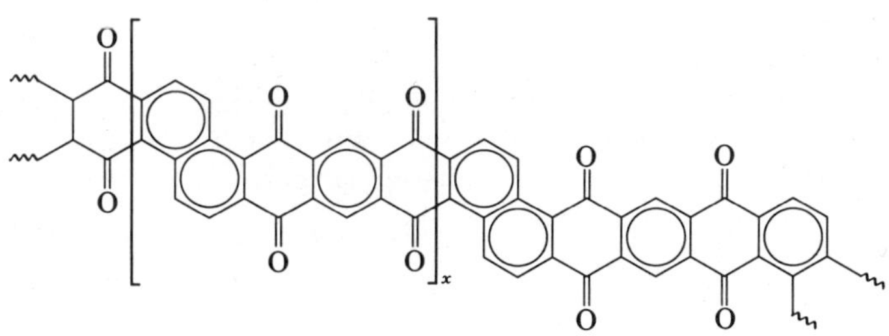

(**12**)

More recently workers have established the structure of the ketonic acid (**13**) by infrared spectroscopy and have studied model compounds to give some evidence for the structure of (**12**). The rate of ring closure of (**13**) to (**12**) decreased after 50–60% conversion because of increased rigidity of the polymer as the quinones are formed. These materials are photoconductive with a high concentration of paramagnetic centers. Several workers have made similar polymers from pyromellitic anhydride and fluorene to obtain soluble ladder polyquinones.

Silicon-containing Ladder Polymers. The first soluble and perfect or complete ladder (**14**) was reported in 1960. Equilibrated phenylsilsesquioxane polymers (**14**) have a highly ordered structure consisting of cis-anti-cis fused cyclotetrasiloxane rings.

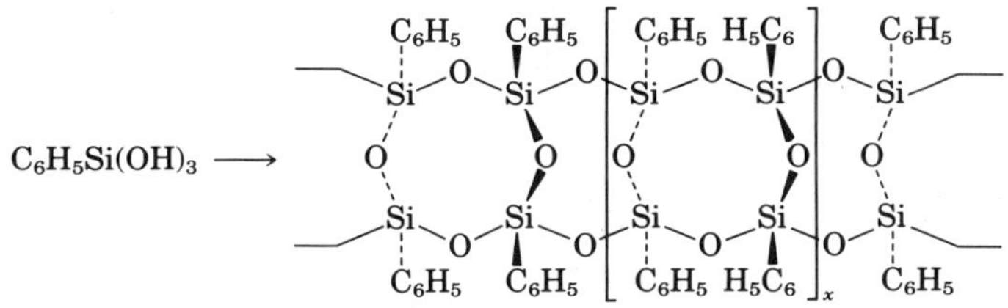

(**14**)

However, nonequilibrated trifunctional silicon-containing condensates were found to have, instead, moderately orderly bead-chain structures composed of polycyclic blocks linked through single bonds (**15**), where some of the beads may have the structure (**16a**) or (**16b**).

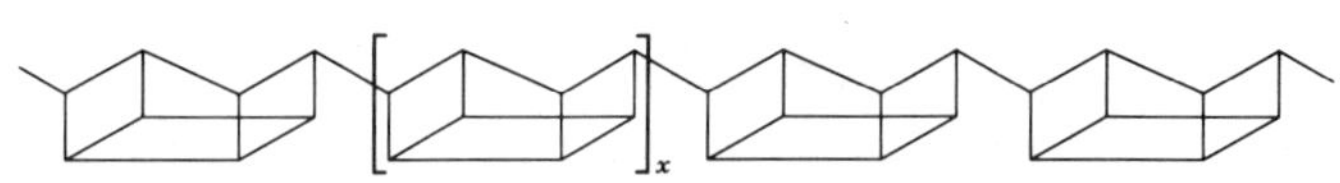

(**15**)

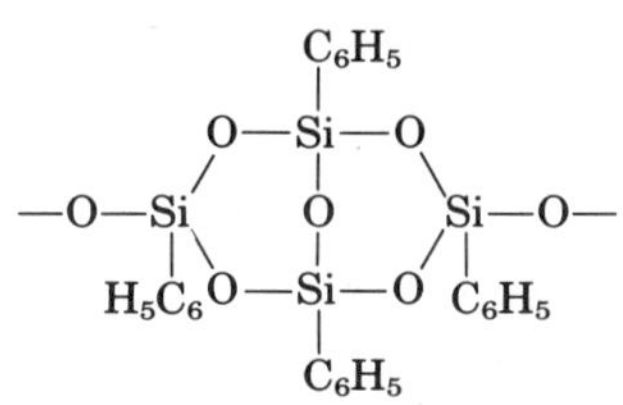

(**16a**)

or

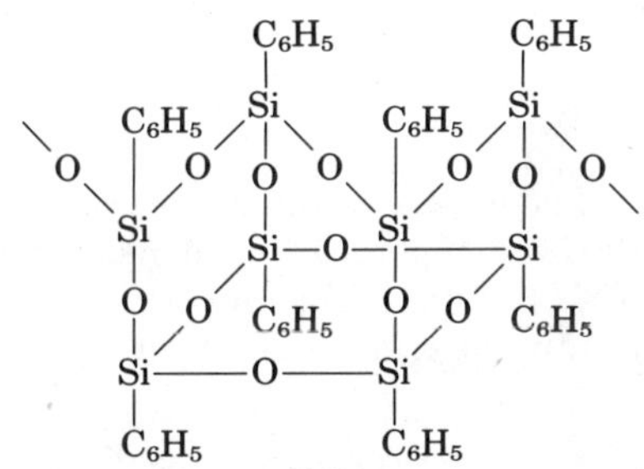

(**16b**)

The fact that the two forms (**14**) and (**15**) are interconvertible and the reactions are reversible points out a difficulty in preparing perfect or complete ladder polymers by a process that involves equilibration. In order to duplicate the synthesis of a particular polymer, the conditions must be followed exactly. A typical commercial use of these siloxanes is in coatings that are cured at high temperatures. Under these conditions it is extremely unlikely that the ladder structure survives. Regardless of the structure of the final coating material, the silsesquioxanes have excellent oxidative and thermal resistance as well as good electrical and mechanical properties.

The poly(phenylsilsesquioxane) (**14**) although infusible, is soluble in a number of solvents, such as toluene, chloroform, and acetone. It thus can be cast into films and spun into fibers. This material has excellent high temperature stability but because it is somewhat brittle and hydrolytically sensitive, it has not been put into commercial production.

Numerous workers have studied the solution behavior of the ladder polyphenylsilsesquioxane (PPSQ) (**14**). The solution properties of the phenylsilsesquioxane ladder polymer indicate that the polymer is not a rigid rod in solution. Solutions of the silane ladder (**14**) have solution properties qualitatively very similar to the properties of flexible chain polymers with one exception; the characteristic time $\tau = \eta_0 R_0$ is longer and more dependent on molecular weight. Data on $[\eta]$, second virial coefficient A_2, and the mean-square radius of gyration as functions of molecular weight, temperature, and solvent lead to the conclusion that the conformation of PPSQ can be modeled with a wormlike chain model with a persistence length ρ of 7.4 nm.

On the other hand, several workers have used viscosity data to indicate that polyphenylsilsesquioxane (**14**) is imperfect and highly branched. Measurements of intrinsic viscosities, molecular weights, radii of gyration, second virial coefficients, and optical anisotropies indicated that not all the rings in (**14**) were closed and the polymer was branched.

There is also some disagreement concerning the conformation of the ladder part of the PPSQ (**14**). It was originally thought that the bulk of the rigid phenylsilsesquioxane groups coupled with the requirement of maintaining tetrahedral bond angles and the preferred bond angle of 155° for the silicon–oxygen–silicon bonds ruled out all regular structures except the cis-syndiotactic ladder form. However, the local conformations of $(C_6H_5SiO_{1.5})_n$ and $(CH_3SiO_{1.5})_n$ were determined by classic semiempirical conformational analysis. Because the calculated values exceeded the experimental ones, it was suggested that the experimentally observed chain flexibility was related to defects in the structural regularity.

Because the polyphenylsilsesquioxane (**14**) is semicrystalline, the structure of the crystalline portion has been determined by x-ray diffraction. X-ray structures of a series of model compounds, such as hydroxy-terminated octaphenyloctasilsesquioxane, have been determined. The models indicate that the crystallinity decreases with increasing molecular weight.

A variety of copolymers have been prepared by cohydrolysis of two alkyl or aryltrichlorosilanes. A series of chlorinated PPSQ ladder polymers in which the chlorine content varied from 0 to 39.9% were made and their properties determined. Also, a variety of ladder silsesquioxane polymers containing reactive side groups have been described.

Even though heating usually rearranges the ladder structure of a polysilsesquioxane, a large number of patents and publications claim the use of such materials for baked-on coatings for corrosion prevention or for insulation in microelectronics. The commercially available Glass Resins GR 100, GR 650, and GR 950 (ladder-type silicone oligomers) are common starting materials. The resulting coatings are resistant to cracking and oxidation.

In addition to the polysilsesquioxanes many other silicon-containing ladder polymers have been reported. Some difunctional silicon has been incorporated into the polysilsesquioxane structure by the process described in the equation below, where R = CH_3, C_6H_5, CH=CH_2:

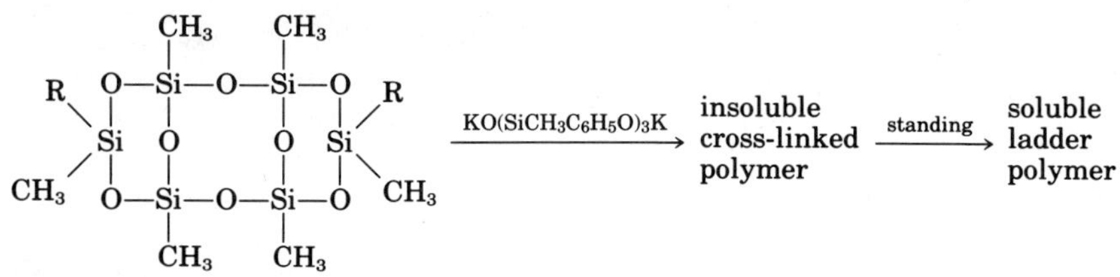

Ladderlike cyclolinear branched polymers were reported from the low temperature condensation of $[C_6H_5—Si(OH)_2O]_3Al$ and $[C_6H_5Si(OH)_2O]_2Zn$. Several interesting ladder polymers containing nitrogen and silicon also have been reported.

Heteroarylene Nitrogen-containing Ladder Polymers. The highly conjugated heteroarylenes are of great interest not only for their high thermal and oxidative stability but also for their electrical properties.

Polypyrrones. Poly(benzoylenebenzimidazole) or poly(imidazopyrrolone) or "pyrrones" represent one of the most studied classes of heteroarylene ladder polymers.

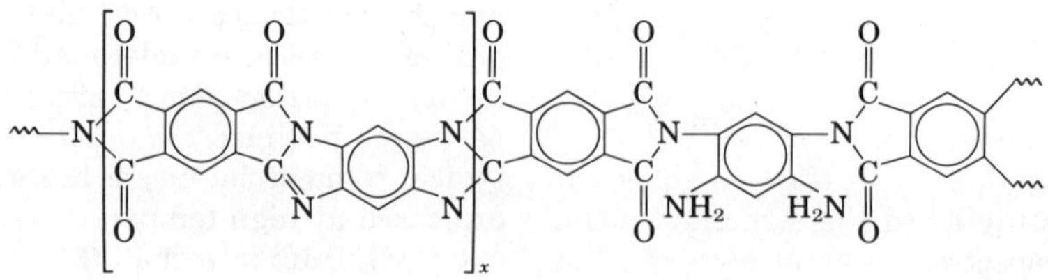

(17)

Although the pyrrone (**17**) undergoes only negligible weight loss up to 600°C, there is indication that not all the rings are closed. In this case statistical factors are not a barrier but the low solubility of the partially converted ladder polymer is a problem. The very properties that are desirable in many final products make the synthesis and fabrication of ladder polymers extremely difficult. If the formation of the ladder polymer is carried out in solution, the product tends to precipitate at a relatively low molecular weight or with an incomplete ladder structure. If the reaction is carried out in bulk, the reaction mixture becomes immobile before a high molecular weight complete ladder structure is obtained. For many of the same reasons that make their synthesis difficult, these infusible materials with low solubility are also difficult to fabricate. One technique that has been only partially successful involves the fabrication of an intermediate nonladder or partial ladder polymer to the desired shape, followed by reaction to complete the ladder structure. This technique has been successful for partial or step-ladder polymers.

Because the two fused five-membered rings in (**17**) were somewhat strained, a pyrrone-containing fused six- and five-membered-rings (**18**) was prepared by heating 1,4,5,8-naphthalene tetracarboxylic acid and 1,2,4,5-tetraaminobenzene in PPA for 1 h at 75°C and then 20 h at 180–190°C.

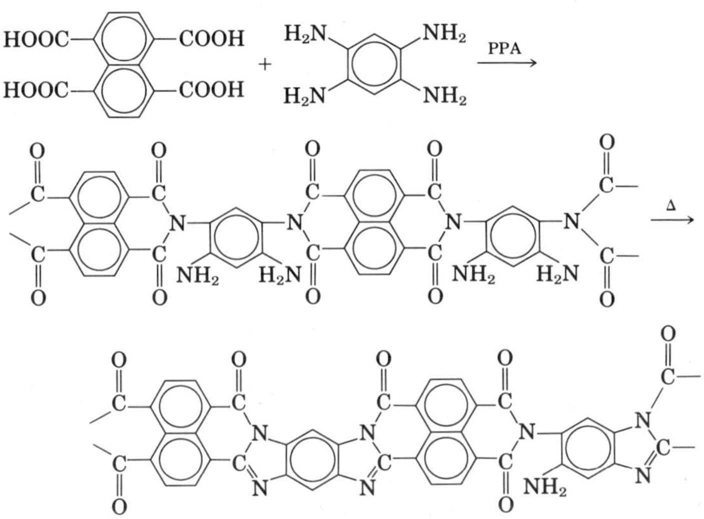

(**18**)

A number of workers have studied the ring-closing reaction of the BBL (**18**) polymer. The thermal stability of fibers made from BBL with only 65% cyclic structure (determined by ir) in the presence of H_3PO_4 was studied. The thermal stability was highest in fiber containing 0.15–0.40% phosphorus, which apparently catalyzes the ring closure.

A principal breakthrough in processing and fabricating articles from BBL (**18**) occurred when it was discovered that BBL forms microscopic sheets on precipitation from strong acids and these solid particles could be made to coalesce into strong films. The coalescence of discrete particles of solid matter takes place without going through a melt phase. As a result this process obviates processing BBL (**18**) in strong acid and thus makes possible the coating of metals. Thus the BBL ladder polymer (**18**) can be formed into tough, durable films by collecting suspensions of the polymer obtained from acid reprecipitations on a fritted glass funnel. After drying, the polymer can be removed as a film, with the thickness dependent on the amount of material used. Such a film has a tensile strength of 66 MPa (9,600 psi) as compared to tensile strength of 110 MPa (16,000 psi) obtained from the same polymer when cast from methanesulfonic acid solution.

Polyquinoxalines. Two groups of workers prepared ladder polyquinoxaline essentially simultaneously. Model quinoxalines were studied and several syntheses of ladder polymers, eg, (**19**) and (**20**), were reported.

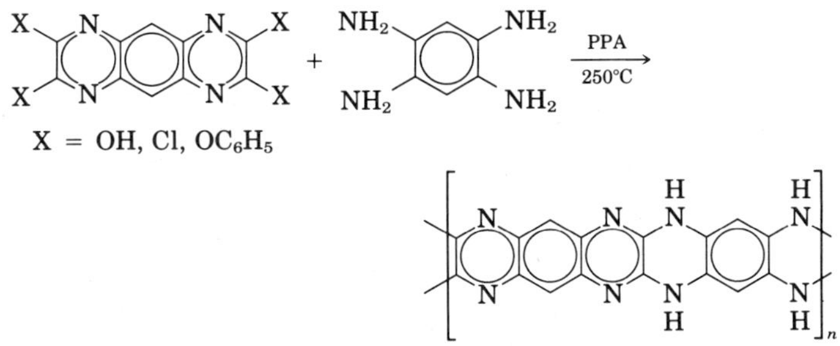

(**19**)

In a tga determination (**19**) [η_{inh} = 2.5 dL/g (CH_3SO_3H)] had an initial weight loss at 350°C. A similar synthesis of polyfluoflavin (**20**) was developed.

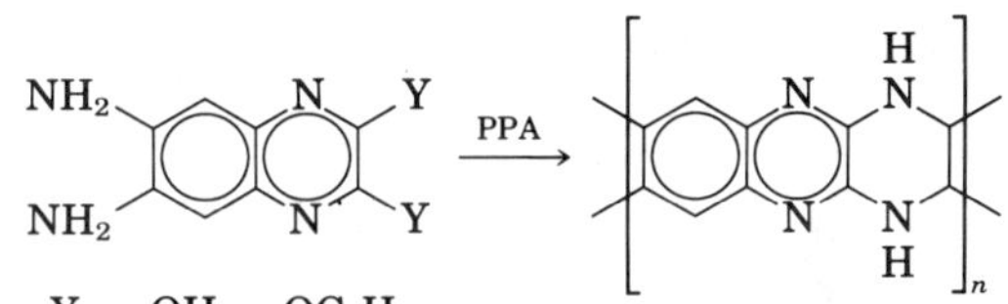

(**20**)

Miscellaneous Nitrogen-containing Aromatic Ladder Polymers. Closely related to the polyquinoxalines are the ladder polymers such as (**21**), which is a green powder, [η] = 0.86 dL/g (in H_2SO_4).

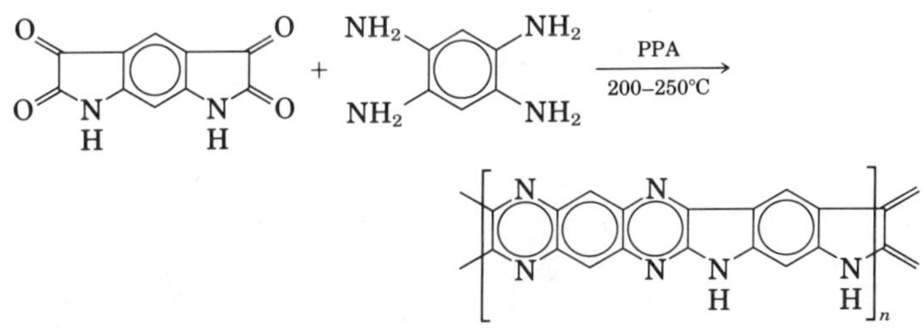

(**21**)

A ladder polymer (**23**) may be made from the condensation of the diaminodicyanopyrazine (**22**) according to the following:

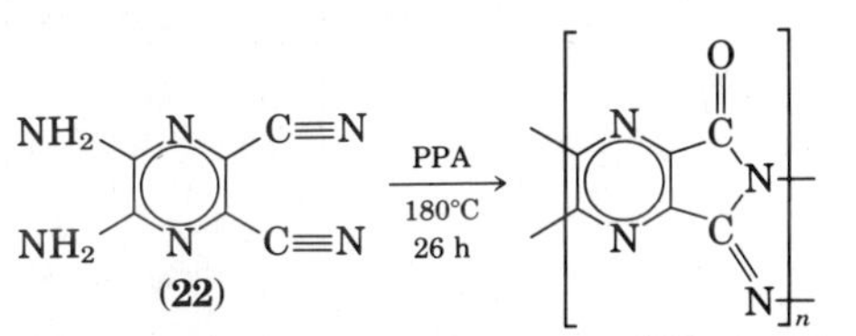

(**23**)

The actual structure varies depending on the conditions.

A poly(iminoquinazolinepyrrolone) (**24**) decomposes in air at 400–450°C and has a tensile strength of 147–167 MPa (1500–1700 kg/cm^2).

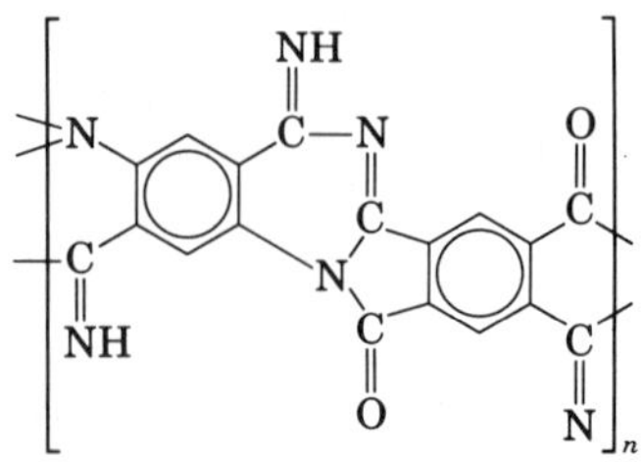

(24)

Poly[4-(3-pyridyl)-8-methyl-2,3,6,7-quinolino)] (25) has been synthesized by the following reactions.

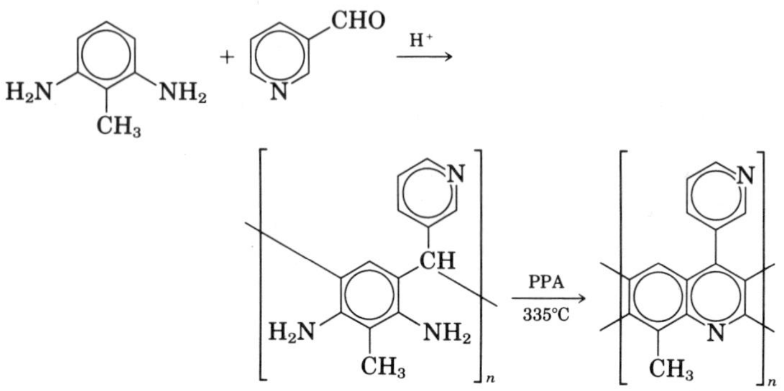

(25)

The soluble ladder polymer (25) undoped had a conductivity of 2.1×10^{-5} S/cm, but when doped with I_2, the conductivity increased three orders of magnitude to 0.017 S/cm.

In a reinvestigation of some older work an interesting conjugated aromatic ladder polymer (26) and its coordination complex (27) were prepared.

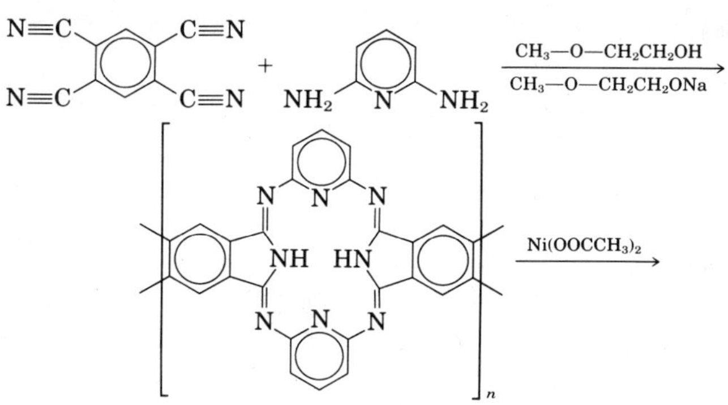

(26)

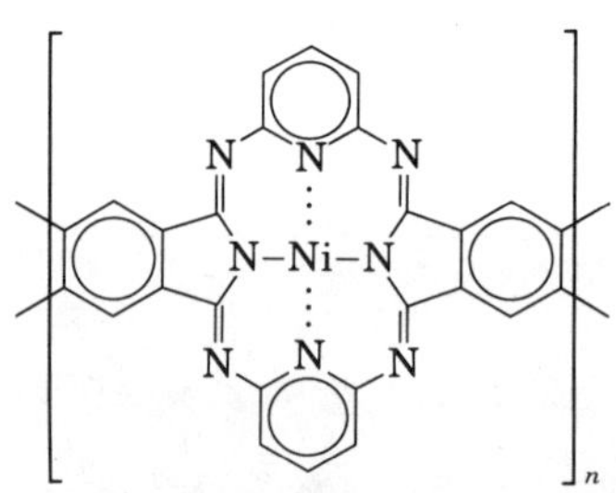

(27)

Because only about 50% of the theoretical amount of ammonia was liberated, polymer (26) must contain some unclosed rings.

Tetracyanobenzene was condensed with 2,6-diaminophthaloyldinitrile to obtain a very complex aromatic ladder polymer containing isoindole, benzimidazole, and quinazoline rings in the backbone. A series of aromatic ladder polymers containing carbazole–quinone groupings have been reported by many workers.

Heteroarylene Oxygen-containing Ladder Polymers. ***Polyphenoxazines.*** In 1966, a two-step synthesis for the polyphenoxazines (28) was developed as indicated in the following, where X is Cl or OH.

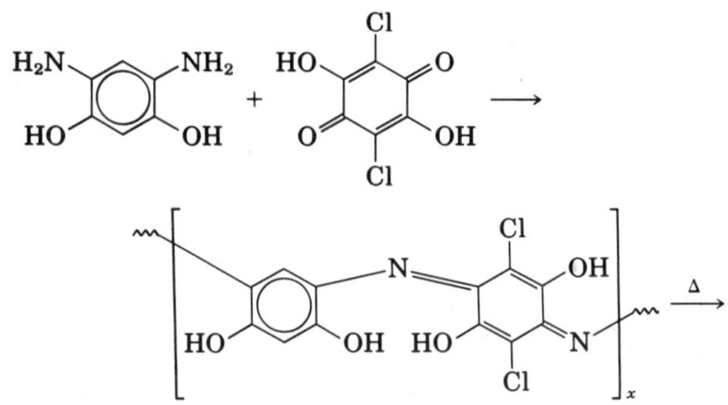

(29)

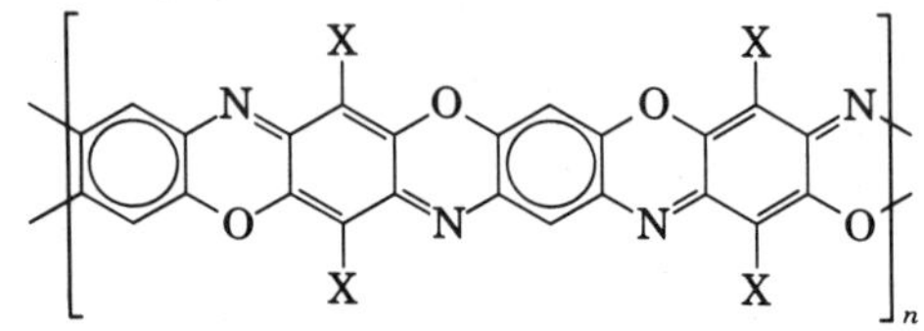

(28)

The unsubstituted polyphenoxazine (30) was prepared in a similar manner from either the 2,5-dihydroxybenzoquinone (31) or its diacetate.

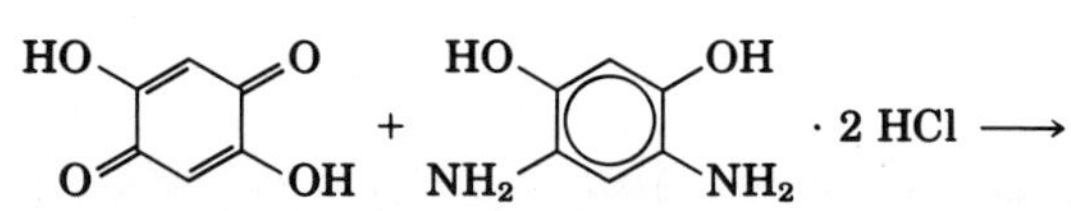

(31)

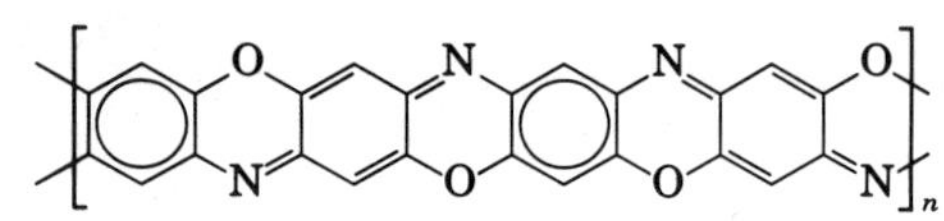

(30)

The two-step procedure permitted the production of a high molecular weight soluble intermediate (29). However, the poor thermal stability of these materials may have been due to incomplete ladder formation.

A synthesis for a hybrid polyoxazine and polyquinoxaline (32) from either the tetrahydroxy derivative (33) or its tetrachloro derivative was developed.

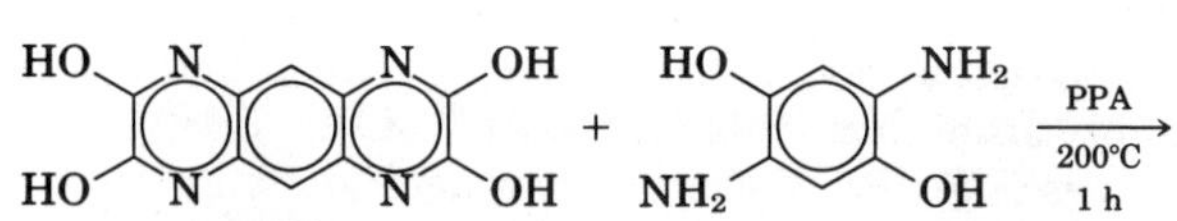

(33)

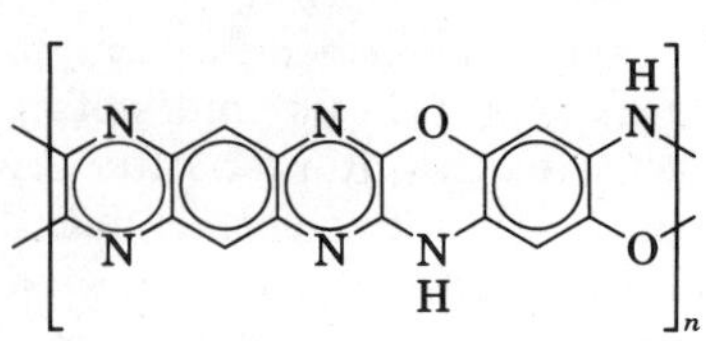

(32)

The ladder polymer (32) is only 60% soluble in concentrated H_2SO_4 with the soluble fraction having an η_{inh} of 0.85 dL/g.

Polydioxines. The synthesis of an aromatic ladder polymer

(**34**) containing the dioxine ring was reported by the following reaction:

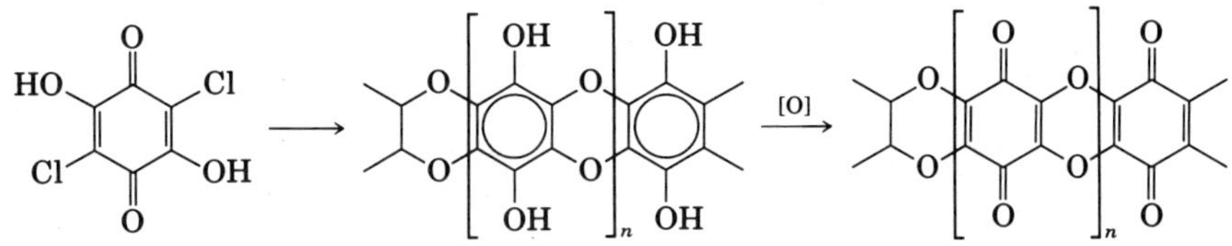

(**34**) (**35**)

Oxidation of (**34**) produced the corresponding polyquinone (**35**) containing the dioxine ring system.

Several hybrid dioxine and quinoxaline aromatic ladder polymers such as (**36**) and (**37**) were prepared:

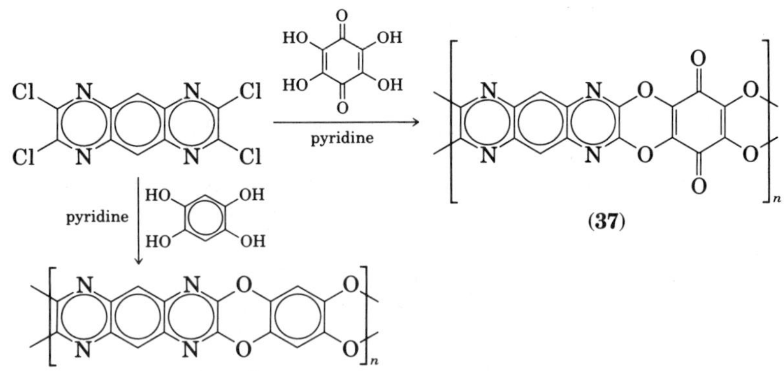

(**36**)

The ladder polymer (**36**) was black with an η_{inh} of 0.03 dL/g in H_2SO_4; (**37**) was brown with η_{inh} = 0.11 dL/g.

A xanthene-containing polymer containing short fused ladder sequences was prepared by the condensation of hydroquinone with pyromellitic dianhydride containing a small amount of phthalic anhydride. The xanthene-containing material shows semiconducting properties. The reaction of dibenzofuran with pyromellitic dianhydride is claimed to produce an aromatic ladder polymer containing the furan ring.

Heteroarylene Sulfur-containing Ladder Polymers. ***Polythiazines.*** In 1967 a hybrid polymer (**38**) containing both thiazine and diazine rings was prepared by the following reaction.

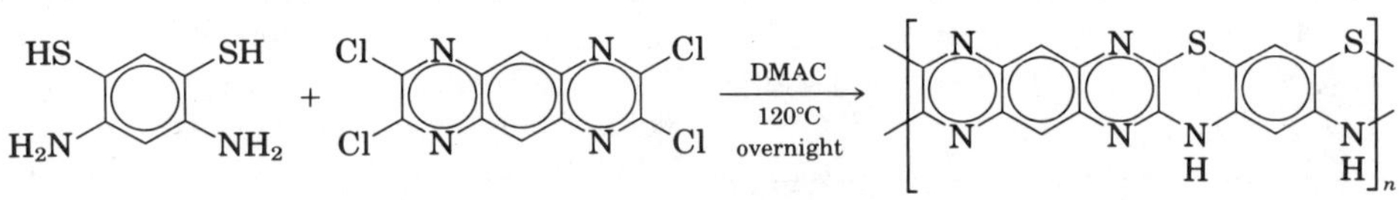

(**38**)

The ladder polymer (**38**) was a black powder with a η_{inh} of 1.19 dL/g (H_2SO_4). The related tetrahydroxy compound could also be used in the synthesis. Later, a polythiazine ladder (**39**) containing recurring anthraquinone units was prepared by this process:

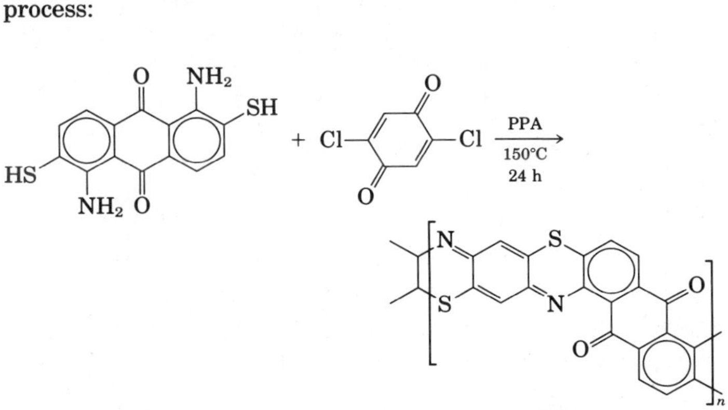

(**39**)

Combining the reagents from the previous two syntheses led to preparation of the polyphenothiazine (**40**).

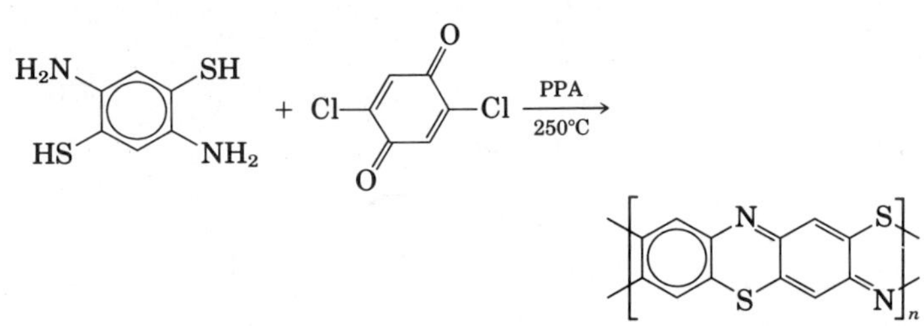

(**40**)

The black insoluble amorphous polymer (**40**) when prepared in polyphosphoric acid had a $\sigma = 4 \times 10^{-6}$ ohm^{-1} cm^{-1}. However, when (**40**) was prepared in dimethylacetamide, it had a σ of only 5×10^{-10} ohm^{-1} cm^{-1}, presumably because of a poor ladder structure and low molecular weight. When the ladder (**40**) was doped with $ClSO_3H$, the σ increased to 2.1×10^{-2} $ohm^{-1}\,cm^{-1}$. (**40**) has a large third-order optical susceptibility.

Polydithianes. In 1948, a series of fused-ring polymers were prepared from sulfur and aromatic hydrocarbons. Although the sulfur content varied from 29.6 to 45.7% and the molecular weights were low, the structure was proposed to be (**41**)

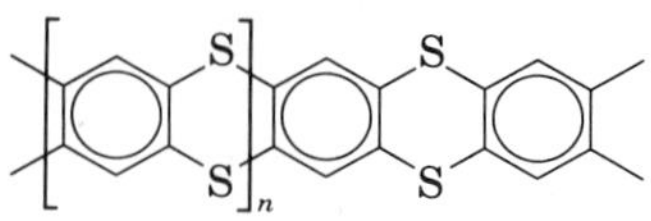

(**41**)

Over the years a number of other workers have also treated aromatic hydrocarbons and sulfur or aromatic sulfides or disulfides with aluminum chloride to obtain polydithianes.

Polybenzothiophenes. A conducting polybenzothiophene (**42**) was produced from poly(phenyl sulfide) by heavy doping with SO_3.

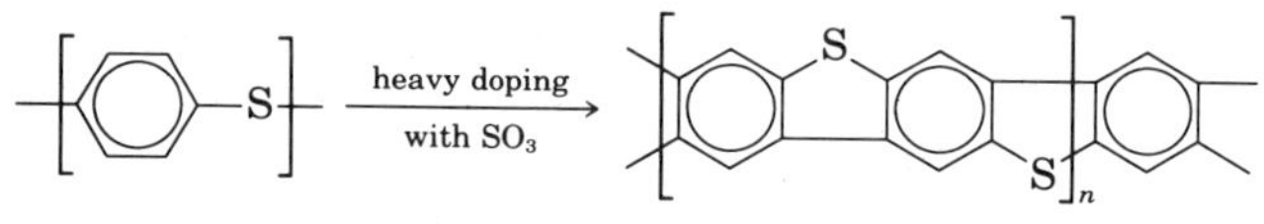

(**42**)

Additional cross-links via sulfuryl groups restricted conjugation and, therefore, the conductivity.

The acylation of dibenzothiophene with pyromellitic dianhydride should produce an aromatic ladder polymer containing benzothiophene units.

Heating carbon diselenide under pressure gives a number of intermediates. If the heating is carried out long enough, the final result is an aromatic ladder polymer containing selenium bridges. Selenium analogues of the polydithianequinones and polybenzothiophene have also been reported.

Nonaromatic Heterocyclic Ladder Polymers. ***Nonaromatic Oxygen-containing Ladders.*** Derivatives of polyacrolein (**43**) can be considered as incomplete ladder polymers containing oxygen (see also ACROLEIN POLYMERS).

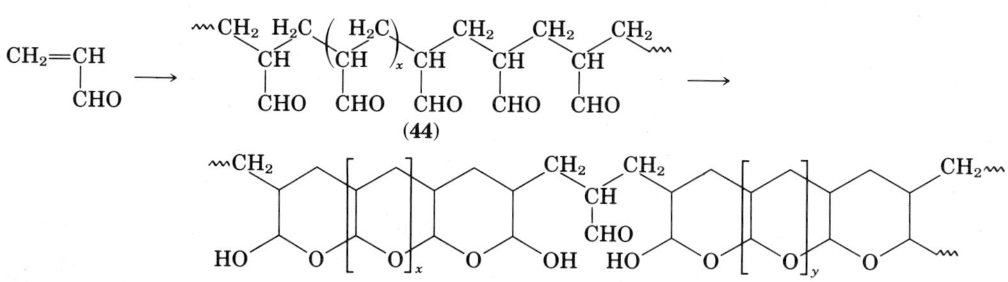

(**44**) (**43**)

When acrolein is radically initiated, polymerization occurs through the vinyl double bond to give a poly(vinyl aldehyde) (**44**), with degrees of polymerization in the range of 100–20,000. The aldehyde groups of (**44**) are partly converted to acetals to give (**43**), containing sequences of two to four fused tetrahydropyran rings.

In an attempt to produce a ladder polymer by zipping up the side group of a preformed polymer, allyl methacrylate was polymerized anionically to give the linear polymer (**45**). When (**45**) was treated with benzoyl peroxide at 80°C in very dilute solution, a partial ladder polymer (**46**) was obtained in which >80% of the double bonds were consumed without gelation.

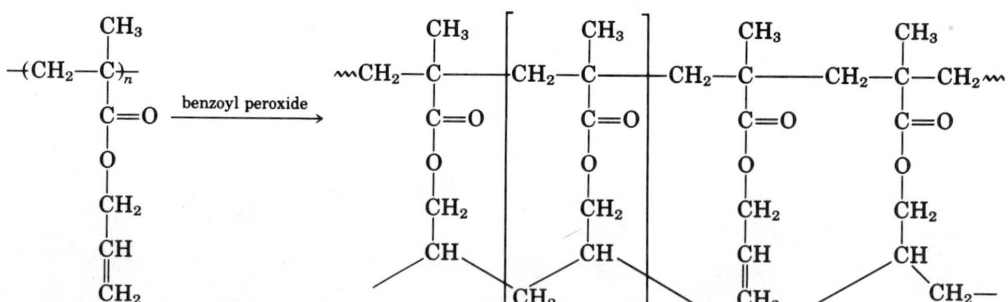

(**45**) (**46**)

A similar result was reported for vinyl methacrylate.

Nonaromatic Nitrogen-containing Ladders. In order to facilitate the preparation of a nearly perfect ladder polymer by eliminating the statistical factor in ring closure, a polymer in which the groups were paired in such a way that only one type of ring closure was theoretically possible was synthesized.

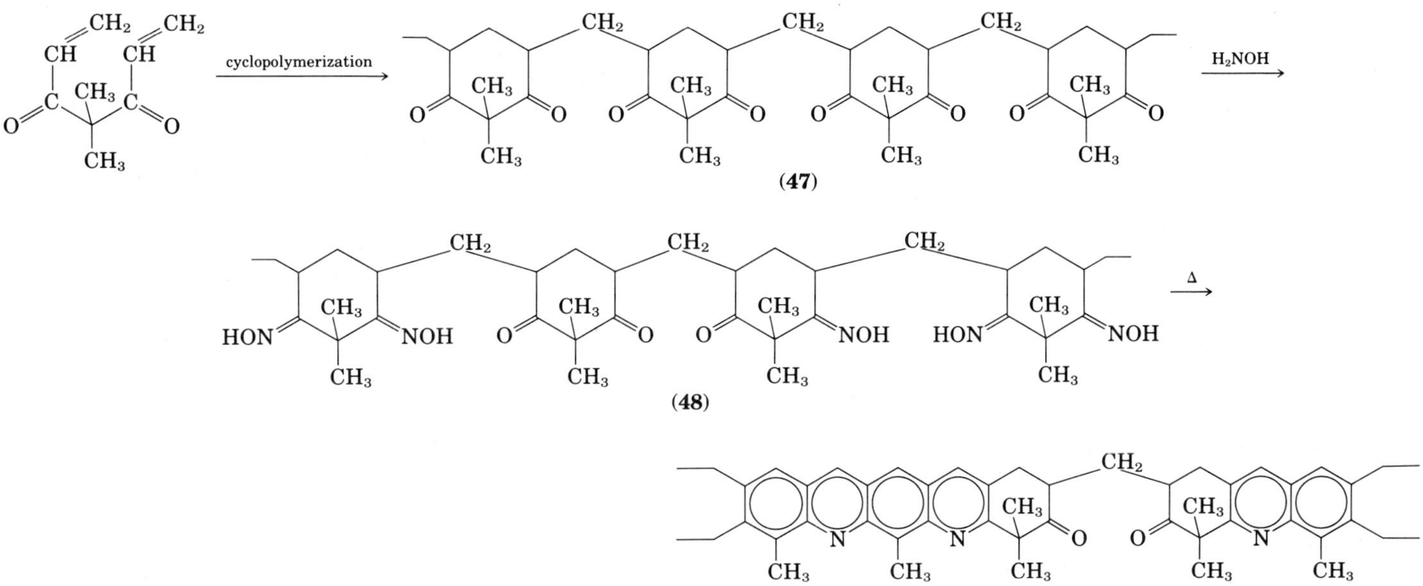

(47) (48) (49)

Unfortunately, the oximation reaction proceeded to only about 90% so that the final ladder polymer contained only about 10 units in the fused-ring system. A similar result was obtained with the unsubstituted diacrylylmethane.

The preparation of ladder polymer (**50**) from vinyl isocyanate (**51**) has been investigated; these polymers were prepared by two methods:

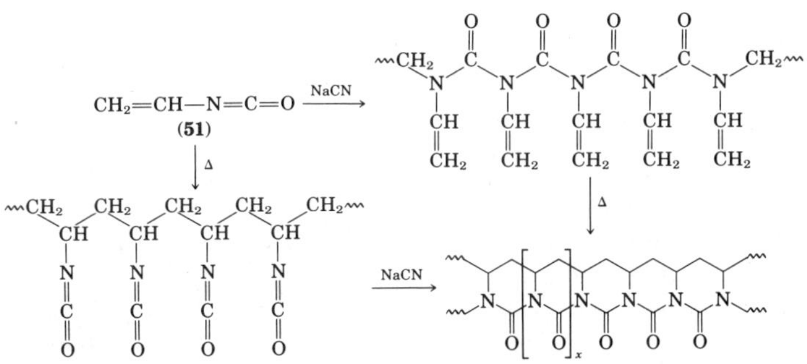

(**51**) (**50**)

In both cases either a cross-linked network was obtained, if the final reaction was forced, or a soluble partial ladder polymer was obtained in which about 15% of the side groups had not reacted (see also ISOCYANATE POLYMERS). Both of these attempts illustrate the statistical barrier to preparing a complete or perfect ladder polymer.

Theoretically, the rungs of the ladder do not have to be covalent bonds. Polymerization of 4-vinylpyridine in the presence of poly(styrenesulfonic acid), poly(vinylsulfonic acid), or poly(methacrylic acid) yields insoluble polymeric salts consisting of stoichiometric quantities of polyvinylpyridine and the polyacid. Thus the complex (**52**) obtained when solutions of poly(methacrylic acid) and polyvinylpyridine were mixed has an infrared spectrum almost identical to that obtained by the reaction of 4-vinylpyridine with poly(methacrylic acid). Even though a ladder structure can be written for the product (**52**), the ladder structure would be imperfect because there is little driving force to hold two chains always paired and parallel. Thus there would be a tendency to rearrange and form cross-links.

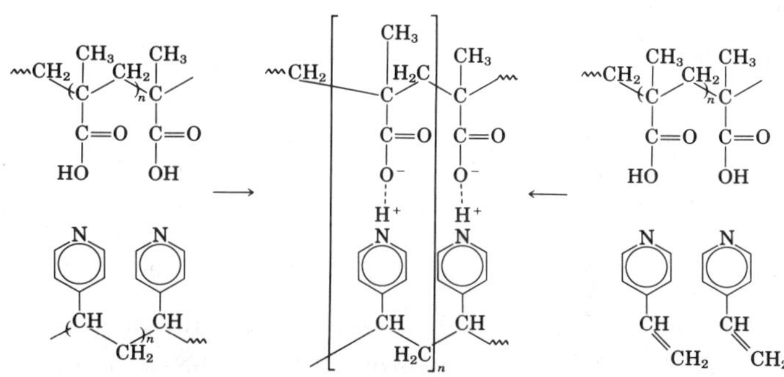

(**52**)

Hydrogen bonding is often strong enough to form the rungs of a ladder polymer. In fact the ultimate ladder structure is the double helix first proposed by Watson and Crick for polynucleotides (qv). There are many natural polymers for which a ladder polymer structure has been proposed.

Nonaromatic Organometallic Ladder Polymers. Methylarsine, which is produced from methyldihydrogenarsine and methyldiiodoarsine, has been shown to be a ladder polymer (**53**).

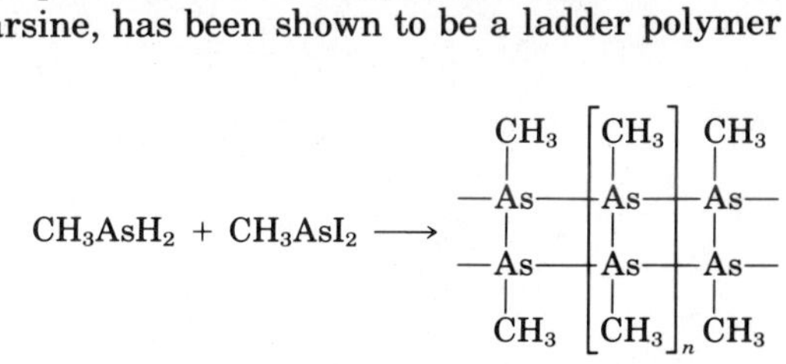

(**53**)

The sides of that ladder have long 0.29 nm arsenic–arsenic half-bonds whereas the rungs of the ladder have normal 0.24 nm arsenic–arsenic single bonds. It has a semiconducting property that is intrinsic (thermal activation of about 1.5 eV) at 273–500 K and extrinsic below 273 K. In the near ir it is a responsive photoconductor.

Treatment of diacetylferrocene (**53**) with calcium carbide yields a product that was assigned the conjugated ladder polymer structure (**54**):

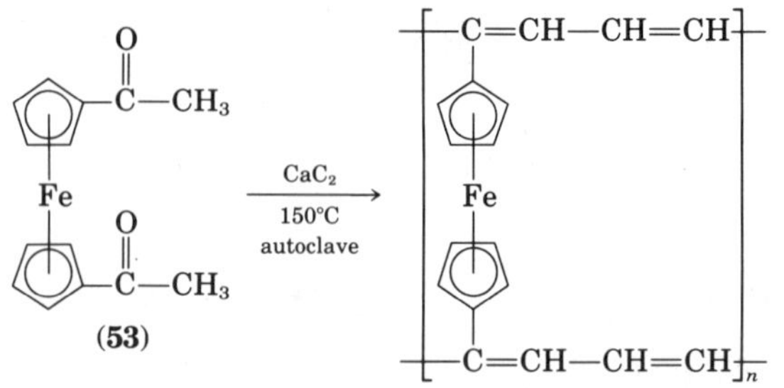

(**53**) (**54**)

When divinylferrocene was polymerized, a soluble polymer was obtained that was considered to be possibly a ladder polymer. Polymerization of the ferrocene derivative containing the two silicon dienes (**55**) gave a product that was assigned the ladder structure (**56**).

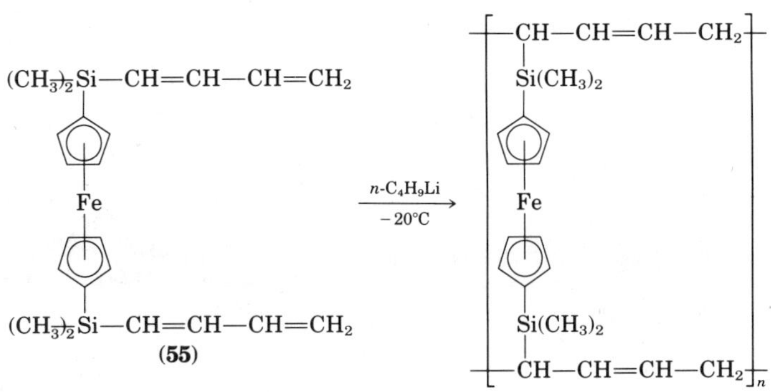

(**55**) (**56**)

Solid-state polymerization of a mercury bis-diacetylene derivative either by ultraviolet radiation or thermally gives the proposed ladder polymer structure (**57**):

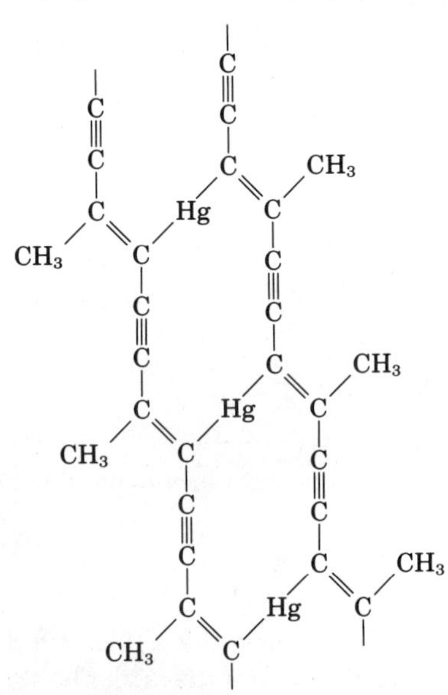

(**57**)

Spiro Polymers

A discussion of the inorganic spiro polymers is beyond the scope of this article and the coverage is restricted to organic spiro polymers and those spiro polymers that involve the coordination of polyfunctional organic moieties to metal ions.

In order to prepare a complete or perfect spiro polymer it is desirable to utilize a reaction in which both sides of the spiro ring are formed simultaneously or nearly simultaneously or no ring is formed. Ketal formation meets this requirement because it either produces a stable spiro ketal ring or no new bond at all. The condensation proceeds stepwise, but because the hemiketal intermediate (**58**) is unstable, no condensation would be expected to take place until the stable ketal (**59**) is formed.

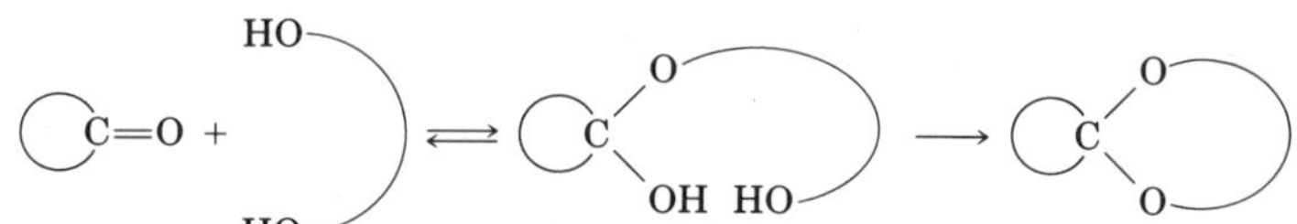

(**58**) (**59**)

By use of this reaction, the first completely organic spiro polymer reported was the polyspiroketal (**60**).

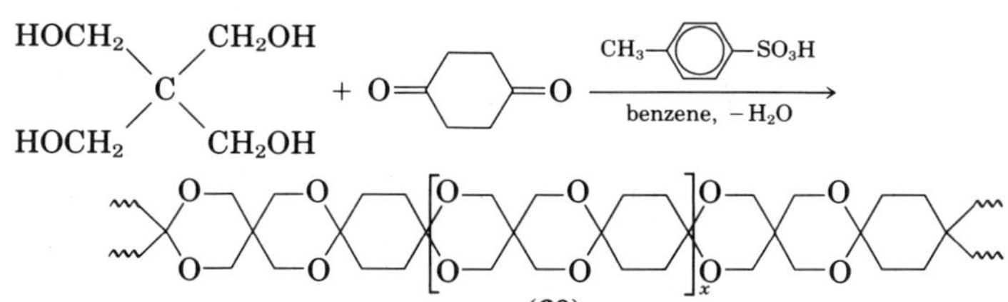

(**60**)

This polymer is a white crystalline solid, soluble in hexafluoro-2-propanol but insoluble in other organic solvents. X-ray analysis indicates that (**60**) is highly crystalline. The polyketal is infusible, decomposing at 440°C, about 150°C higher than related linear polyethers. The molecular weight can be controlled by the addition of cyclohexanone as a chain stopper to produce a low molecular weight soluble polymer. The use of a flexible 18-membered cyclic diketone in place of the cyclohexanedione produces a more flexible and more soluble analogue.

When the condensation polymerization was carried out with the tetrol (**61**), which contains a hindered ketone, the adjacent bulky groups prevented involvement of this keto group in the reaction to give the spiro polymer (**62**), which is soluble in many common solvents.

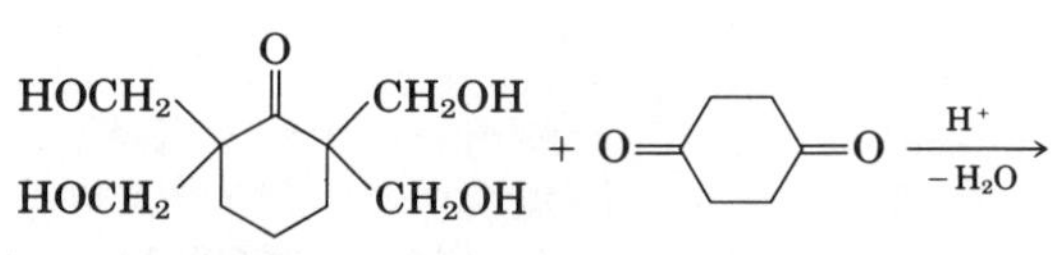

(**61**)

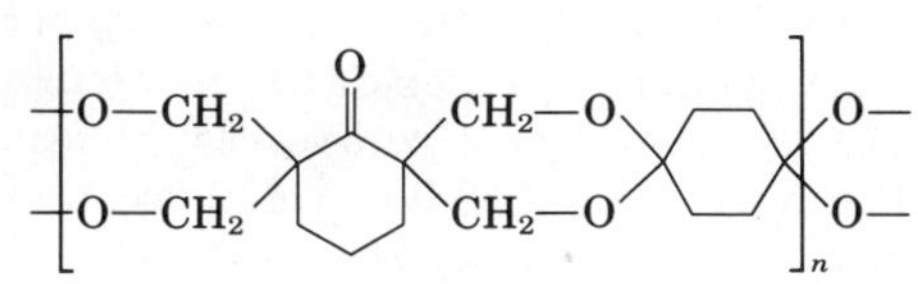

(**62**)

The examples discussed above all contain six-membered rings for the ketal formation. However, five- and seven-membered rings work just as well.

Similar polymers were prepared by replacing the cyclohexanedione with 2,7-diketotetrahydronaphthalene. Even eight-membered rings will work in this polymerization if the conditions are right. Thiols also work equally well in a similar condensation.

An interesting spiro polyester (**63**) can be produced by heating the dihydroxydiacid (**64**).

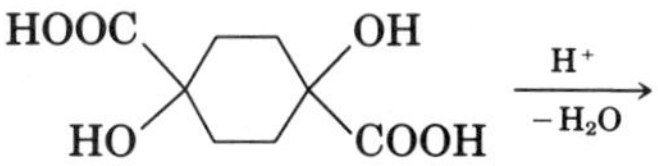

(**64**)

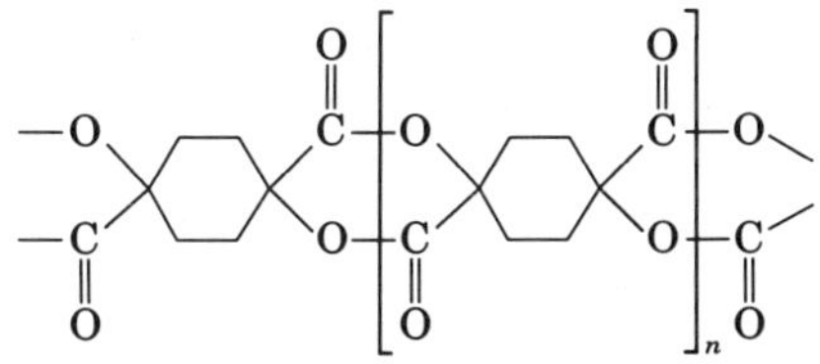

(**63**)

The spiro glycolide (**63**) is soluble in hexafluoro-2-propanol but decomposes at 400°C without melting. Thus all the spiro polymers discussed above have restricted rotation and do not melt but decompose first. The synthesis of a spiro polyamide, which is analogous to that producing the spiro polyester (**63**), has been reported. A German patent reports a series of spiro polyesters formed by the esterification of several tetraacids with tetraols.

A hybrid spiro–ladder polymer (**65**) containing both a ladder polyester segment and a spiro polyketal segment was prepared by a Diels-Alder reaction.

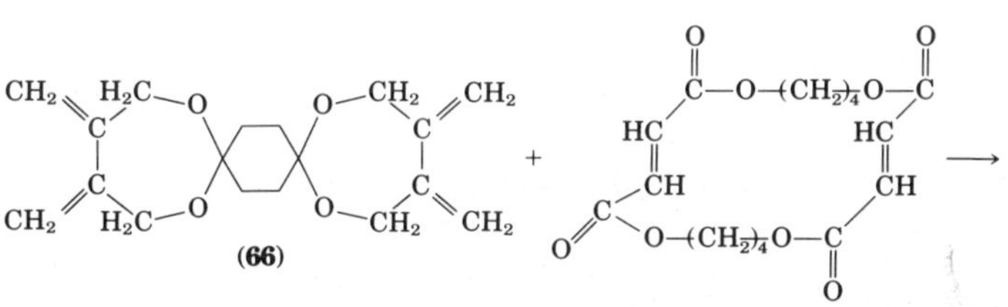

(**66**)

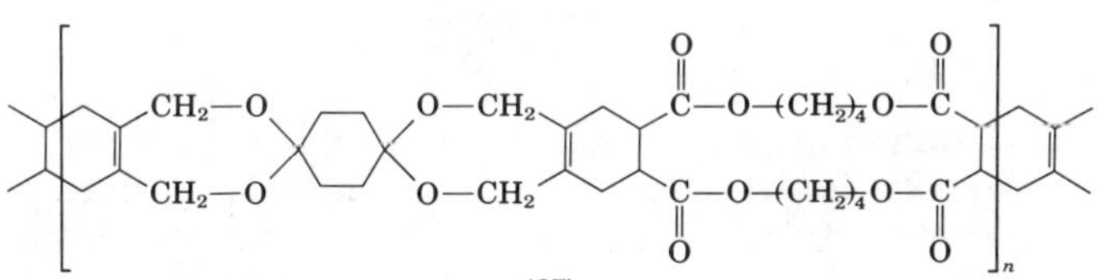

(**65**)

Coordination Spiro Polymers. Because many metallic ions have coordination numbers of four or six, they are ideally suited for the production of spiro polymers. However, often the added polarity renders the spiro polymer intractable and the resulting low solubility causes the polymer to precipitate at low molecular weights. In this section only a few representative examples are given, many of which are exceptions to the general rule just discussed (see also COORDINATION POLYMERS; ORGANOMETALLIC POLYMERS).

Coordination Involving Oxygen Atoms. Some of the more interesting spiro polymers are the soluble and fusible poly(metal phosphinates) to which structure (**67**) has been assigned. This random "hybrid" polymer has a molecular weight of over 10,000 and melts at 150°C to an amorphous material that remains flexible at −78°C.

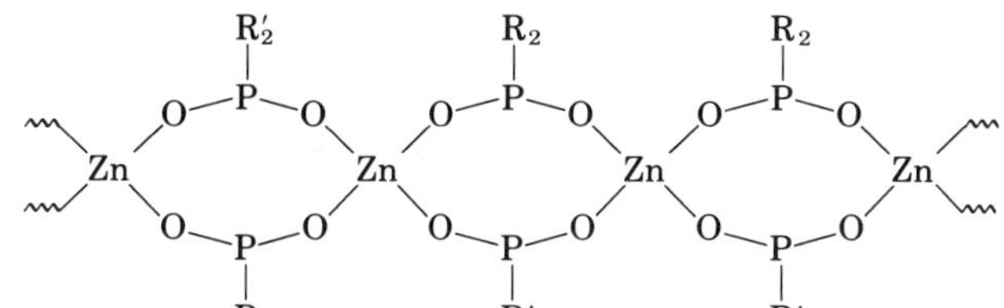

(**67**)

Coordination Involving Nitrogen Atoms. Complexes of the bisphenol oxime (**68**) with Cu^{2+}, Ni^{2+}, and Pd^{2+} produce spiro polymers (**69**).

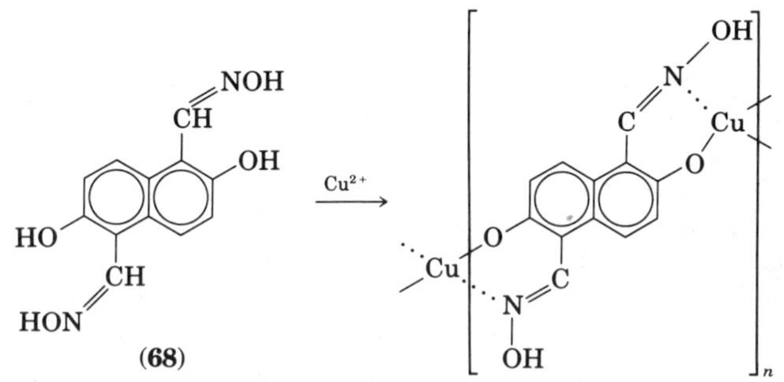

(**68**) (**69**)

Coordination Involving Sulfur Atoms. Several workers have reported complexes with thioamides to produce spiro polymers (**70**).

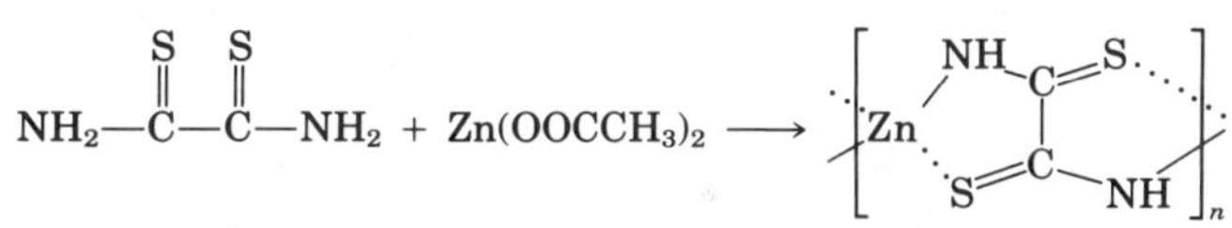

(**70**)

1,2,4,5-Tetramercaptobenzene reacts with nickel salts to produce the spiro polymer (**71**), which can be easily oxidized to (**72**).

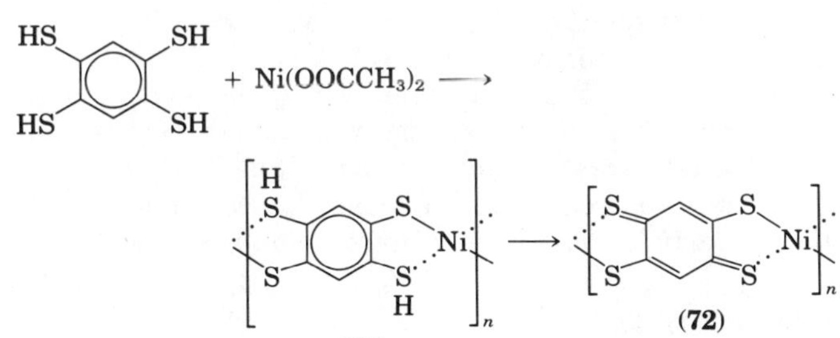

(**71**) (**72**)

The conductivities for the spiro polymer from tetrathiooxalate and a nickel(II) salt range from 5 to 20 S/cm while the copper and palladium complexes are up to 1 S/cm.

More Complex Polymers

Although this discussion has been centered on double-chain polymers, more complex structures can be conceived, eg, polymers with a triple or a quadruple chain, parquet polymers, or a cylindrical molecule. One obvious extension of nonrandom cross-linked polymers is a sheet-like polymeric array. Graphite, for example, has been assigned a structure consisting of layers of highly condensed aromatic rings (**73**).

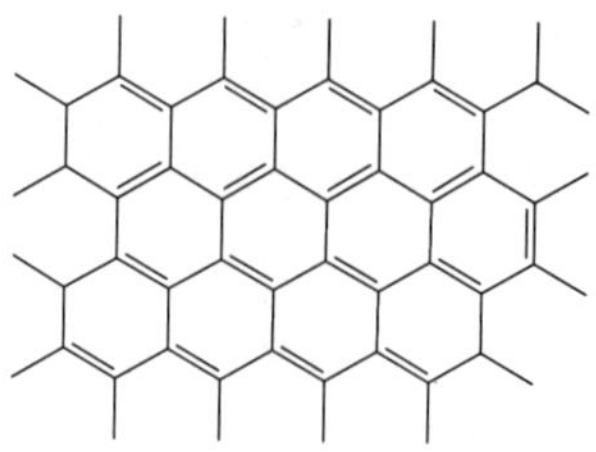

(**73**)

Physical Properties

From the very nature of ladder and spiro polymers several *a priori* predictions can be made concerning some of their physical properties. Because there are no single bonds present that can act as swivels, these polymers are extremely stiff. The only flexibility of the polymer chains is the flipping and bending of the rings. If the rings are large and flexible, the polymer will have some mobility; if the rings are small or rigid and aromatic, the chain will be relatively inflexible. Most of the complete ladder and spiro polymers reported to date have been essentially infusible; they all decompose before they melt. In general, the presence of rings in a polymer chain raises the melting point; the effect is at its maximum in a ladder or spiro polymer.

It is obvious that the synthesis and fabrication of ladder and spiro polymers demand some new techniques and technology. However, enough is known at this time to predict that these double-chain polymers will be important in the future for the preparation of chemically, mechanically, and thermally stable materials.

Proposed Nomenclature

A committee of the American Chemical Society has proposed a system of polymer nomenclature based on structure rather than on the identity of the monomer, ie, poly(bivalent radical) instead of poly(monomer) (see NOMENCLATURE OF POLYMERS).

Ladder and spiro polymers are named in an analogous manner, but the relation of the four free valences is denoted by pairs of locants separated by a colon; pairs of locants imply a linear double-stranded polymer. If both bivalent and tetravalent radicals are present in the structure, the mer is depicted with the tetravalent radical component at the left end of the unit. This proposed system of nomenclature is under study by the Macromolecular Nomenclature Commission of IUPAC.

BIBLIOGRAPHY

C. Arnold, Jr., *Macromol. Rev.* **14,** 265 (1979).

V. V. Korshak and A. L. Rusanov, *J. Macromol. Sci. Rev. Macromol. Chem.* **C21,** 275 (1981).

A. V. Pocius, *Rubber Chem. Technol.* **58,** 622 (1985).

G. B. Butler and J. E. Kresta, eds., *Cyclopolymerization and Polymers with Chain-Ring Structures,* American Chemical Society, Washington, D.C., 1982.

WILLIAM J. BAILEY
University of Maryland

LAMINATES

Laminates consist of layers, or laminae, bonded together by suitable binders. Laminae are usually materials that are readily available in continuous-sheet form, such as paper or woven fabrics. The binders are synthetic resins, predominantly phenolic resins, that are solvent-coated or impregnated into the base laminae. After drying, several laminae are stacked and the entire mass is consolidated under heat and pressure to form a rigid sheet or panel used for its mechanical, electrical, chemical, or aesthetic qualities.

In industry, laminates are used for their electrical properties, impact strength, wearing quality, or chemical resistance. They are used in electronic applications, electrical switches, gears, etc. Although extremely important functionally, they are seldom conspicuous.

Laminates consist of a reinforcing fibrous web extending through a matrix of synthetic resinous material. The choice of web and resin determines the properties of the laminate and, to a great extent, the process used for manufacture (see also REINFORCED PLASTICS; COMPOSITES).

Laminating Resins

Phenolic resins play a significant role in both decorative and industrial laminates. To meet a wide range of product requirements, such as strength, flexibility, durability, hardness, etc, many other types have been used either as portions of the product or as homogeneous laminates. These include substituted phenolics, ureas, melamines, polyesters, epoxides, silicones, and acrylics.

Laminating Bases

A laminate base is a sheetlike material that may be treated with resin and consolidated by heat and pressure. Typical bases are papers, woven fabrics, mats, felts, and similar materials. The base is usually in the form of a continuous roll, although wood veneers are a notable exception. Laminating bases are classified by the form (paper, woven fabric, etc) or the constituent, which may vary from absorbent organic cellulose to nonabsorbent inorganic glass.

Manufacture

The laminating process consists of treating, collating, and pressing. In the first step, the base is brought into contact with the resin solution by impregnation or coating; the solvent is removed by drying. The treated material is cut into sheets that are placed on each other in a parallel arrangement; this is called collating. Heat and pressure are applied to consolidate the sheets into a permanently bonded and infusible unitary mass; this step is called pressing, or molding. Finishing operations complete the procedure.

Properties and Standards

The National Electrical Manufacturers Association (NEMA) publishes complete information on the properties of and test methods for both industrial and decorative laminates. Property values of the various types are given in Table 1.

Copper-clad Laminates. Grades X, XXP, XXXPC, FR-2, FR-3, FR-4, FR-5, FR-6, G-10, G-11, CEM-1, and CEM-3 laminates

Table 1. Performance Properties of Decorative Laminates[a]

Types and grade	Nominal thickness, mm	Minimum resistance: Wear, cycles	Minimum resistance: Impact, mm	Maximum dimensional change[b]: % MD	Maximum dimensional change[b]: % CD	Minimum resistance: Boiling water[c]	Minimum resistance: High temperature[c]	Minimum resistance: Radiant heat, s
General-purpose								
GP 50	1.270	400	1270	0.5	0.9	NE	SL	125
GP 38	0.965	400	889	0.6	1.0	NE	SL	100
GP 28[d]	0.711	200	508	0.7	1.2	NE	SL	80
GP 20[d]	0.508	200	381	0.8	1.3	NE	SL	60
Postforming								
PF 42	1.067	400	762	1.1	1.4	SL	SL	100
PF 30[d]	0.762	300	508	1.1	1.4	SL	SL	80
Cabinet liner								
CL 20	0.508	50	254	1.2	2.0	M	M	
Backer								
BK 20	0.508							
BK 50	1.270							
Specific-purpose								
SP 125	3.175	400	1905	0.3	0.7	NE	SL	200
SP 62	1.575	400	1397	0.5	0.9	NE	SL	150
High wear								
HW 120	3.048	3000	1905	0.3	0.7	NE	SL	200
HW 80	2.032	3000	1016	0.4	0.8	NE	SL	175
HW 62	1.575	3000	889	0.5	0.9	NE	SL	150
Fire-rated								
FR 62	1.575	400	1397	0.5	0.9	NE	SL	125
FR 50	1.270	400	1143	0.5	0.9	NE	SL	75
FR 32	0.813	300	508	0.7	1.2	NE	SL	50

[a] Test procedures according to NEMA standards.
[b] MD = machine direction; CD = cross direction.
[c] Rating system: NE = no effect; SL = slight effect; M = moderate effect; X = severe effect (NEMA Standard 12-31-1980).
[d] Normally specified for vertical use.

Table 2. Typical Properties of Industrial Laminates Meeting NEMA Standard

Properties[a]	X	XXXPC	C	L	AA	G-3	N-1	FR-4
modulus of elasticity, in tension, GPa[b]								
lengthwise	13.1	6.9	6.9	8.3		13.8	2.8	17.2
crosswise	9.7	5.3	6.2	6.9		11.7	2.8	13.8
tensile strength, MPa[c]								
lengthwise	138	85	69	90	83	159	59	276
crosswise	110	66	55	82	69	138	55	241
compressive strength, MPa[c]	248	173	255	241	262	345		414
Rockwell hardness, M scale	M-110	M-95	M-103	M-105	M-111	M-100	M-105	M-115
electrical strength[d], V/mm								
short time, 10^3	27.6	25.6	7.9	7.9	8.9	27.6	23.6	27.6
step by step, 10^3	19.6	17.7	5.9	4.7	5.3	19.7	17.7	17.7
insulation resistance[e], MΩ		50,000					50,000	200,000
specific gravity	1,360	1,300	1,360	1,350	1,555	1,650	1,150	1,850
coefficient of lengthwise thermal expansion per °C	2.0×10^{-5}	2.0×10^{-5}	2.0×10^{-5}	2.0×10^{-5}	1.5×10^{-5}	1.8×10	3.0×10^{-5}	1.0×10^{-5}
thermal rating, continuous, °C	130	125	125	125	140	170		140

[a] As determined by standard NEMA tests; all tests performed at Condition A, ie, as received, unless otherwise noted.
[b] To convert GPa to psi, multiply by 145,000.
[c] To convert MPa to psi, multiply by 145.
[d] Perpendicular to laminations 1 g thick.
[e] 96 h at 35°C and 90% rh.

are used with copper foil bonded to one or both surfaces and are intended primarily for printed wiring boards but not for multilayer applications.

Epoxy–Glass Laminates. Epoxy-resin-impregnated glass cloth in the semicured state is cured for bonding the individual circuit layers of multilayer printed wiring boards.

Polyester–Glass Laminates. Grades GPO-1, GPO-2, and GPO-3 polyester glass-mat sheet laminates are intended for both mechanical and electrical applications. Properties of industrial laminates are given in Table 2.

Applications

The number of applications of industrial laminates is almost endless. They are used mainly for electrical insulation, mechanical applications, and as decorative laminates.

BIBLIOGRAPHY

R. W. Martin, *The Chemistry of Phenolic Resins,* John Wiley & Sons, Inc., New York, 1956, Chapts. 4 and 5.

G. L. Booth, *Coating Equipment and Processes,* Lockwood Publishing, New York, 1970.

S. A. Giddings
Formica Corporation

LATICES

A latex is conveniently defined as a stable colloidal dispersion of a polymeric substance in an aqueous medium. The polymer particles are usually approximately spherical and of typical colloidal dimensions; particle diameters range from ca 30 nm to 500 nm. The dispersion is relatively concentrated; the volume fraction of contained polymer usually ranges from 0.40 to 0.70. The dispersion medium is usually a dilute aqueous solution containing substances such as electrolytes, surface-active compounds, hydrophilic polymers, and, in the case of synthetic latices, initiator residues. In recent years, there has also been some industrial interest in stable colloidal dispersions of polymers in nonaqueous media. Although outside the above definition of a latex, they are nevertheless sometimes referred to as nonaqueous latices; polymer dispersions that conform to the above definition are then known as aqueous latices. The preferred plural of latex is latices, but the alternative latexes is widely used in North America. Polymer latices are usually white mobile liquids whose fluidities, by comparison with those of typical polymer solutions, belie the high concentrations of polymers that they usually contain.

The number of types of commercial latices is legion. Other types have been available in the past, or have been prepared in the course of developing commercial products or as scientific curiosities. Most latices are anionic, ie, the particles carry negative electric charges. Important examples of rubber latices include natural rubber latex in various forms, styrene–butadiene rubber latices, acrylonitrile–butadiene rubber latices, and polychloroprene latices. Important examples of resin latices include vinyl acetate polymer and copolymer latices, and latices containing polymers derived from various acrylic monomers.

Characteristics of Latices

Characteristics relating to the colloidal aspects of latices include volume fraction of contained polymer, particle shape and size, particle-size distribution, the nature of the contained polymer, particle morphology, molecular weight and molecular weight distribution, gel content and cross-link concentration of the contained polymer, and physical characteristics of the contained polymer, such as glass-transition temperature, crystallinity, and mechanical properties, notably those relating to viscoelastic behavior.

The characteristics pertaining to the dispersion medium include the nature of its principal component, and the natures and concentrations of dissolved substances. The characteristics pertaining to the interfacial region between the particles and the dispersion medium include the following: the nature and concentration of functional groups that may be chemically bound at the surface of the particles; the extent to which such groups are ionized; the nature and concentration of adsorbed species that may be present in the interfacial region; and the extent to which molecules of dispersion medium are firmly bound to the surface of the particles because of the presence of the adsorbed species or because of the nature of the polymer particle itself.

Colloid Aspects of Latices

Stability. Latices are a subgroup of colloid systems known as lyophobic sols. The distinguishing feature of latices is that the dispersed phase is polymeric in nature. The factors determining the colloid stability of latices are those determining the colloid stability of any lyophobic sol. The colloid stability is determined by the balance between attractive and repulsive forces affecting two particles as they approach each other.

Destabilization. Although polymer latices must be of sufficient colloidal stability to persist almost unchanged at normal temperature for periods of months, if not years, for many applications they should also be capable of controlled destabilization at certain stages of processing. In some applications, destabilization may be achieved by evaporation of the dispersion medium. The latex particles are thereby constrained in close proximity with each other. Coalescence of the particles tends to occur if they are sufficiently soft at the temperature at which evaporation occurs. In other applications, a chemical influence (eg, chemical agencies such as acidic substances, water-soluble salts of heavy metals, water-miscible organic liquids, polymer-miscible organic liquids, cationic surfactants, heat-sensitizing coagulants, and delayed-action coagulants) may be employed to facilitate destabilization by reducing the height of the interparticle barrier to flocculation. Physical agencies other than mere removal of dispersion medium by evaporation are also able to reduce colloid stability, and some use is also made of these agencies in latex technology.

Natural Rubber Latex

Natural rubber latex originates as the sap of the *Hevea brasiliensis* tree. Rubber plantations are found in many tropical countries, including Malaysia, Thailand, Indonesia, Vietnam, India, Sri Lanka, and parts of Africa.

The *Hevea brasiliensis* is a large and rather sparse tree. The latex is found in cells known as latex vessels that form an interconnecting mantle of tubular cells that surround the tree in the inner cortex, the region between the cambium layer and the bark. The latex is obtained from the tree, usually early in the morning, by a process called tapping. The tree is wounded, and the latex flows out spontaneously; the flow continues for a few hours.

As the latex effluxes from the tree, it is collected in a cup attached to the tree and is then brought to a local collecting station where it is bulked with other latex. It is then transported to a processing factory where, if intended for use as latex rather than for the manufacture of dry rubber, it is concentrated.

Constitution of Fresh Natural Rubber Latex. The chemical composition of fresh natural rubber latex is complex. The principal components are water and rubber hydrocarbon. Rubber hydrocarbon, which typically comprises about 33 wt % of the latex, is a highly linear and stereospecific *cis*-1,4-polyisoprene with an average molecular weight of ca 5×10^5. It is present in the form of colloidally stable particles, approximately spherical, with diameters of 50–1300 nm. In addition, fresh natural rubber latex contains small amounts of proteins, lipids (notably phospholipids), carbohydrates, free amino acids, other organic acids, metals in various chemical combinations, and inorganic cations.

Preservation. Short-term preservation ensures that the latex remains in a liquid condition for a few days. Long-term preservation includes measures that will keep the latex intact and free of decay during transportation to and storage in the country of final use. The preservative protects the latex against spontaneous coagulation and putrefaction by suppressing the activity and growth of microorganisms, enhancing colloid stability, and deactivating trace metals by sequestration in solution or precipitation as insoluble salts. The most widely used preservative for natural rubber latex is ammonia.

Concentration. The crude latex is concentrated to a rubber content of 60% or more to reduce transportation costs and enhance suitability for various latex processes. Product uniformity is also improved, partly because of the partial removal of nonrubber constituents and partly because of latex bulking. Concentration can be achieved by evaporation, creaming, centrifugation, and electrodecantation. Of these processes, centrifugation is the most important.

Ammonia-preserved Concentrate. The composition of ammonia-preserved natural rubber latex concentrate differs in several important respects from that of fresh natural rubber latex. Some of these differences are a consequence of the alkalinity, pH ca 10, imparted by the ammonia; others result from the concentration process.

Biosynthesis of Natural Rubber Hydrocarbon. Much interest has been shown in the biochemical pathway by which *cis*-1,4-polyisoprene is produced in the *Hevea brasiliensis* tree. It is not produced by the addition polymerization of isoprene, but by the condensation polymerization of isopentenyl pyrophosphate.

Synthetic Latices

Emulsion Polymerization. Synthetic latices are produced from monomers by emulsion polymerization. A typical aqueous emulsion polymerization is carried out by gently agitating water, monomer (usually sparingly soluble in water), water-soluble initiator, and surfactant. The mixture is maintained at such a temperature that the initiator decomposes to generate a sufficient concentration of free radicals for the initiation of polymerization. The polymerization reaction occurs primarily within a large number, typically ca $10^{15}/cm^3$, of separate loci. These loci, which eventually become the latex particles, are so numerous that at any one time they contain at most only a small number of propagating chains, and often only one. Therefore, the free-radical polymerization is said to be compartmentalized within isolated reaction loci. The conversion time curve for a typical emulsion polymerization shows three stages, or intervals. In Interval I, the reaction loci are nucleated; in Interval II, compartmentalized free-radical polymerization occurs in the loci while monomer droplets are still present as a separate phase. In Interval III, monomer droplets are no longer present as a separate phase, and the unreacted monomer is now entirely imbibed within the growing latex particles.

Agglomeration and Concentration. In some cases, a synthetic latex must be concentrated before use. This is especially necessary for general-purpose styrene–butadiene rubber (SBR) latices that are intended to compete with concentrated natural rubber latex in the manufacture of latex foam rubber. Although some interest has been shown in concentration by creaming, the usual method is evaporation. If it is not possible to concentrate the crude product of the emulsion polymerization sufficiently without forming an intractable viscous paste, as in the case of SBR latex, the latex is partially agglomerated in a controlled manner to increase the average particle size and broaden the particle-size distribution. The concentrated product is of lower viscosity than would have been the case had the latex not been partially agglomerated.

Processes for the controlled agglomeration of synthetic latices can be classified as follows: solvent addition; partial soap neutralization; electrolyte addition; freezing; mechanical agitation; and addition of hydrophilic colloids. Freezing and mechanical agitation are the most widely used of these processes.

Synthetic Rubber Latices, Nonfunctionalized. Styrene–butadiene rubber latices are produced in large quantities as competitors for ammonia-preserved natural rubber latex concentrates in certain applications, notably in the production of foam rubber. Other synthetic rubber latices in this class are those of acrylonitrile–butadiene rubbers, polychloroprene rubbers, and rubbery acrylic polymers. They are used in applications requiring a rubber with reduced tendency to swell in hydrocarbon oils.

Synthetic Plastic Latices, Nonfunctionalized. The wide range of synthetic latices of this type is based mainly on copolymers of vinyl acetate and other vinyl esters, and on nonelastomeric acrylic polymers. Such latices find application as surface coatings, binders, adhesives, impregnants, etc.

Functionalized Synthetic Latices. Any type of synthetic latex can be modified to contain a small amount, ie, up to 5 mol %, of units to which are attached reactive functional groups such as carboxylic acid, hydroxyl, sulfonic acid, amine, amide, or *N*-methylol amide groups. Such latices are referred to as functionalized synthetic latices. The best known are the carboxylated synthetic latices; their polymers contain small amounts of carboxylic acid groups.

Carboxylated synthetic latices of the styrene–butadiene copolymer type are commonly used in applications such as paper coating, textile impregnation, and carpet backing. These latices exhibit superior mechanical and chemical stability and enhanced tolerance to mineral fillers.

Specialty Synthetic Latices. Most synthetic latices for industrial applications are anionic. Cationic synthetic latices, the particles of which carry positive electric charges, are used for specialized applications. Such latices can be prepared using cationic surfactants in conventional emulsion polymerizations, and also using copolymerizable cationic monomers.

Artificial Latices

Several general-purpose synthetic rubbers, notably stereoregular polybutadienes, stereoregular polyisoprenes, ethylene–propylene elastomers, and butyl (isobutylene–isoprene) rubber, are not manufactured by emulsion polymerization (see also ISOBUTYLENE POLYMERS; ISOPRENE POLYMERS). This is also true of certain specialty synthetic rubbers, such as chlorosulfonated polyethylenes, polyurethane rubbers, epichlorohydrin rubbers, and silicone rubbers (see also ETHYLENE POLYMERS, MODIFIED POLYETHYLENES; POLYURETHANES; 1,2-EPOXIDE POLYMERS, POLYETHER ELASTOMERS; SILICONES). Latices of rubbers such as these are produced by dispersing the bulk rubber in an aqueous medium. However, such latices are expensive and usually contain an unacceptably high ratio of colloid stabilizer to polymer.

Chemically-modified Latices

Chemically-modified latices are latices of polymers that have been subjected to chemical modification while still in latex form. Although most polymers are susceptible to chemical modification in latex form, in practice such interest has been confined to natural rubber latex.

Prevulcanized Natural Rubber Latex. The polymer contained in natural rubber latex and certain other rubber latices can be cross-linked without prior coagulation. The product is, in effect, a latex of vulcanized rubber. Natural rubber latex is usually prevulcanized by heating with dispersions of sulfur and an accelerator, such as zinc diethyldithiocarbamate, to 50–80°C.

Compounding and Applications

Latices frequently must be compounded before use. Some compounding ingredients, notably stabilizers and thickeners, are used with almost all types of latices and in diverse applications. The stabilizers are either surfactants or water-soluble hydrocolloids. Water-soluble hydrocolloids of high molecular weight are also used as thickeners. The latices of glassy polymers are frequently compounded with additives such as plasticizers, pigments, and fillers. Plasticizers are invariably liquids that are miscible with the polymer and very sparingly soluble in water. They may be permanent or fugitive; the latter are volatile and eventually lost from the solid phase that is deposited from the latex. Latices of rubbery polymers are often compounded with vulcanizing ingredients that cause the solid phase deposited from the latex to become cross-linked when suitable conditions, usually elevated temperatures, are established. The most common vulcanizing agent for unsaturated hydrocarbon rubbery polymers is sulfur; very fast accelerators, notably metal dialkyldithiocarbamates and metal alkylxanthates, are also required. The latices of rubbery polymers are often compounded with antioxidants, usually amines or hindered phenols. Plasticizers, pigments, and fillers are also compounded into rubber latices. Latices of glassy polymers are used as reinforcing fillers to enhance tensile strength, tear strength, and puncture strength of vulcanized films. Rubber latices can also be compounded with water-miscible condensation resins to increase the modulus and tensile strength of vulcanized films.

Latex foam rubber is produced by foaming a compounded rubber latex by mechanical or chemical means, and then converting the latex phase to a rubbery continuum. Thin-walled rubber articles, such as gloves, condoms, and balloons, are produced by latex dipping. Other uses for rubber latices include application as binders for nonwoven carpets and bonded-fiber fabrics; paper treatment, of which coating is the most important; water-based adhesives; in the manufacture of rubber thread by the extrusion of latex through an orifice into a coagulant bath; in the binding of hair to form a resilient cushioning material; as can sealants; as bitumen modifiers; as concrete modifiers; in combination with monomethyl sodium siliconate as a treatment for buildings suffering from dampness; and in the dipping of tire cords used in the manufacture of pneumatic tires to promote adhesion between the cords and the rubber matrix in which the cords are embedded. Another major use for the latices of glassy polymers is as surface coatings.

BIBLIOGRAPHY

D. C. Blackley, *High Polymer Latices: Their Science and Technology,* 2 Vols., Maclaren & Sons, Ltd. and Applied Science Publishers, Ltd., London, UK, 1966.

K. O. Calvert, ed., *Polymer Latices and Their Applications,* Applied Science Publishers, Ltd., London, UK, 1982.

H. Warson, *The Applications of Synthetic Resin Emulsions,* Ernest Benn, Ltd., London, UK, 1972.

R. Buscall, T. Corner, and J. F. Stageman, *Polymer Colloids,* Applied Science Publishers, Ltd., London, UK, 1985.

D. C. BLACKLEY
London School of Polymer
Technology

LAYUP. See REINFORCED PLASTICS.

LDPE. See ETHYLENE POLYMERS.

LEATHER

Prehistoric humans learned to make leather by preserving hides and skins with juices extracted from plants. From that time on, leather has provided footwear, garments, shelter, and other articles affording protection, comfort, durability, and aesthetic properties available from no other natural product. Despite many attempts, no synthetic material has been produced to date to match all of the properties of leather. The demand for leather exceeds the supply and the principal function of substitutes is to fill those needs for which the supply of leather is inadequate.

Large variations can be found in the physical properties of leather made from different species of animal hides or skins. Similarly, variations in processing of a type of hide can also produce very different properties for specific end uses. Even when leather is manufactured for a single use, eg, shoe uppers, a variation in properties is desirable to allow for differences between casual, military, work, or dress shoes. Values for various physical properties of cattlehide shoe upper leather are shown in Table 1. Heat resistance and shrinkage depend on moisture content; leather without any moisture decomposes at 160–165°C.

The most important property of leather is its unmatched ability to provide comfort in footwear, partly because of its unique ability to conform to the shape of the foot. The other unique property of leather is its ability to dissipate moisture produced by perspiration. The mechanism of the transmission of water vapor through leather is a function of the material rather than a linear diffusion process.

Raw Materials

Most hides and skins used as raw material for leather manufacture are by-products of the meat industry. The supply of raw material bears no relationship to the demand for leather and leather goods. Although the supply of hides and skins worldwide has been increasing, their availability on the world market has decreased. Many countries do not permit the free export of hides and skins, preferring to encourage the export of higher value leather and leather goods. In the United States the demand for red meat has declined over the past decade and with it the supply of cattlehides. Decreasing raw material supplies and increasing growth of the consumer goods industries, demanding more leather for footwear, gloves, luggage, handbags, small personal leather goods, upholstery, and garments, have combined to make cattlehides a valuable commodity.

Chemical Composition. Fresh cattlehides, like most biological materials, contain 65–70% water, 30–35% dry substance, and <1% ash. The dry substance is largely made up of the fibrous proteins collagen (qv), keratin (qv), elastin, and reticulin.

Microstructure. The skin forms a tough, protective, thermal blanket on the living animal and provides the raw material for leather. In general, mammalian skins are composed of a dense, interwoven, fibrous mat in the center (dermis, including grain and corium); a thin, cellular outer layer (epidermis) with protruding hairs; and a loose, fatty, inner (subcutaneous) layer attached to underlying muscles. The looseness of the inner layer facilitates mechanical removal (flaying) of the skin in the slaughterhouse.

Table 1. Physical Properties of Shoe Upper Leather

Property	Range	Representative value
tensile strength, MPa[a]	15.26–37.48	27.6
elongation at break, %	29.5–73.0	40
stitch tear strength, N/cm[b]	1280–2275	1751
tongue tear strength, N/cm[b]	226–961	525
thickness, mm	1.5–2.4	1.8
bursting strength, kN/cm[b]	1.10–24.5	17.5
grain cracking, N/cm[b]	525–1489	1051
wet shrinkage temperature[c], °C	96–120	100
apparent density, g/cm³	0.6–0.9	0.75
real density, g/cm³	1.4–1.6	1.5
flexibility (Flexometer)		
bending length, cm	6–9	7
flexural rigidity, mg/cm	10,000–50,000	20,000
bending modulus, MPa[a]	19.7–68.9	34.5
compression modulus, MPa[a]		0.345
cold resistance[d], K		92[e]
temper of chrome side leather (butt)		
flexibility factor, g		400
recovery, %		57
resilience, %		21

[a] To convert MPa to psi, multiply by 145.
[b] To convert N/cm to lbf/ft, divide by 14.6.
[c] Chrome-tanned leather values.
[d] Ordinarily limited by finish-crack or lubricant hardening properties.
[e] Workable without cracking in the absence of finish or fat liquor.

Manufacture and Processing

Conversion of a raw hide into leather takes place in three distinct phases. Historically, the first phase takes place in the beam house, where the hair is removed and the hide prepared for tanning. The beam house is so named because this part of the process was carried out by hand with the skin draped over a wooden beam. Tanning, the second phase, was originally done outdoors in pits containing vegetable tannin extracts. This area was called the tan yard. Today leather is tanned in large drums or mixers, containing hundreds of hides, with chromium as the primary tanning agent. Only a small percentage is tanned with vegetable tannins. Finishing, the last phase, was developed much later to protect the grain surface of the hide and enhance its appearance. Although methods have changed greatly, areas called the beam house and tan yard still exist in modern tanneries. Additional technical information about the leather manufacturing process can be found in the general references that describe modern tanning practices in detail.

Hides can be processed into leather immediately on removal from the animal, although this is not generally done. Once a hide or skin is removed from an animal, deterioration begins immediately. Efforts to preserve the hide and prevent this deterioration from affecting the leather must be started within a few hours after slaughter. The hide is usually cured in brine or salt, which provides a satisfactory preservation from the time it is taken from the animal until the tanner is ready to begin processing it into leather. Salt curing has several disadvantages that are related to effluent pollution and the corrosive nature of concentrated salt solutions. As a result, a wide variety of alternative preservation methods have been investigated; however, none has been adopted commercially on a large scale.

Salt-cured hides, as received by the tanner, contain as much as 14% of their weight as salt. A fresh hide contains ca 64% water; after brine curing it contains ca 45% water and 41% protein, only two-thirds of which is leather-making collagen. The tanner first removes the salt and rehydrates the fibers by soaking the hide. A detergent usually is added to speed hydration. The soaking procedure also removes water-soluble proteins and washes the hide free of manure and dirt. The soaking stage is followed immediately by the hair removal process.

Chrome Tanning. The primary function of a tanning agent is to stabilize the collagen fibers so that they are no longer biodegradable. At the turn of the century almost all leather was tanned with vegetable extracts. In contrast, 95% of all the leather manufactured in the United States today is chrome-tanned.

Vegetable Tanning. Sole leather, by far the main use for vegetable-tanned leather, has been replaced by synthetics in all but the most expensive shoes. Vegetable tanning produces a fullness and resilience characteristic of only this type of tannage. It has excellent molding characteristics allowing, for example, adaptation to the shape of the foot. Vegetable-tanned leathers have good strength and dimensional stability and are still used in power-transmission belts. Because of their hydrophilicity, vegetable-tanned shoe linings absorb perspiration efficiently.

Other Tanning Agents. Chrome and vegetable tannages have the greatest commercial importance. Other tannages provide a pretreatment or posttreatment of chrome or vegetable tanned leather and are employed to confer specific properties to the final leather product. They include minerals other than chrome, synthetic tannages, oil tannages, and aldehyde tannages.

Posttreatment. It is easier to control the final properties of leather at the stage of processing where color, retans, and fatliquors (CRF) are added, than at any other. Leather can be prepared with physical properties ranging from soft and supple for high fashion clothing to firm and stiff for use in the manufacture of men's work shoes and boots. Through the use of different retanning materials, dyes, and fat liquors, a wide variety of leathers are available to meet the demands of the market.

Drying. After all of the wet operations have been completed, excess water is removed and the reactions of hide with the tanning agents, fat liquors, and dyes is completed. The water is removed first from the surface and then from within the leather fibers by controlled diffusion.

Finishing. Uniformity is of prime importance in any leather product. Individual skins vary in thickness, angle of fiber weave, and natural defects, not only from animal to animal, but from area to area in one piece of skin. The leather manufacturer minimizes these differences by splitting and shaving the leathers to uniform thicknesses, tanning and filling the voids in looser-structured flanks, and using mordants to obtain level dyeing. However, it is the finishing process, with its application of natural or synthetic polymers and of colorants, within and on the surface of the leather, that produces the uniformity, appearance characteristics, and resistance to scuffing and abrasion required for a commercial product.

By-products. The weight of finished grain leather constitutes only 20–30% of the weight of the raw hide, depending on the thickness of the leather and the hide selection. The remaining 70% consists of either by-products that can be sold or wastes requiring disposal. Over the years, as markets have changed, some wastes have become by-products and various by-products have become wastes. The largest by-product by weight is the "split," which may account for 30–50% of the incoming hide. If the hide is split after liming, the split can be used to make collagen casings and other potential food uses, cosmetics, and medical materials such as burn dressings and sutures. If the hide is split in the blue, its use is limited to split leather manufacture. Although this market fluctuates, the split is almost always a salable by-product. The only other saleable by-products are the fleshings which can be rendered to produce oils and grease.

BIBLIOGRAPHY

F. O'Flaherty, W. T. Roddy, and R. M. Lollar, eds., *Chemistry and Technology of Leather,* Vols. I–IV, Reinhold Publishing Corp., New York, 1956–1965.

J. J. Tancous, *Skin, Hide and Leather Defects,* Leather Industries of America Laboratory, Cincinnati, Ohio, 1969.

T. C. Thorstensen, *Practical Leather Technology,* Robert E. Krieger Publishing Co., Inc., Huntington, New York, 1975.

J. H. Sharphouse, *Leather Technicians Handbook,* Leather Producers' Association, London, 1971.

DAVID G. BAILEY
Eastern Regional Research Center
U.S. Department of Agriculture

LEATHERLIKE MATERIALS

After World War II, with the development of poly(vinyl chloride), nylon, polyester, polyurethane, and other polymers, leather imitations began to be employed successfully in shoes, bags, clothing, and home furnishings. Because leather is a by-product of the meat industry, the market for the latter significantly affects the price and supply of the former. With the significant increase in world population and its tendency toward urbanization, the demand for shoes and clothing, the two principal applications of leather, continues to increase, requiring increased production of leather and its substitutes.

Poromerics simulate leather to such an extent that they are nearly identical in appearance, physical properties, and processibility. This rather recent material enjoyed quick success in all leather-use areas.

Materials

Coated Fabrics. Leather substitutes are designed to imitate the appearance of leather with its surface grain. This is accomplished by coating substances that are capable of forming a uniform film. This requirement was first met by plasticized poly(vinyl chloride) (PVC), which became available following World War II. When plasticized PVC is coated onto a substrate, a leatherlike material, termed vinyl-coated fabric, is produced.

Vinyl-coated fabrics exhibit high density, extremely low water-vapor permeability, cold touch, poor flex endurance, and plasticizer migration. However, they have good scratch resistance and colorability and are inexpensive.

Urethane-coated fabrics, developed in the 1960s, were an improvement. This material is prepared by applying a coat of polyurethane to woven or knit fabric. With a glass-transition temperature below 0°C, polyurethane is flexible at room temperature without a plasticizer. Furthermore, its molecular structure allows water-vapor permeability. In addition, the solvents normally used for polyurethane (PU) permit coagulation by a nonsolvent with formation of a porous structure, thereby increasing flexibility and water-vapor permeability. Ordinary urethane-coated fabrics are manufactured by drying a cast polyurethane solution to form a film that is laminated onto the substrate. Significant improvement in appearance, feel, and grain break is achieved by using a brushed fabric as the substrate that is laminated with a cast polyurethane film (Fig. 1). Alternatively, an organic-solvent solution of polyurethane is applied to a brushed woven fabric that is immersed in a nonsolvent bath for coagulation.

Poromerics. Poromerics, also referred to as synthetic leather, were developed in the 1960s and early 1970s as an improve-

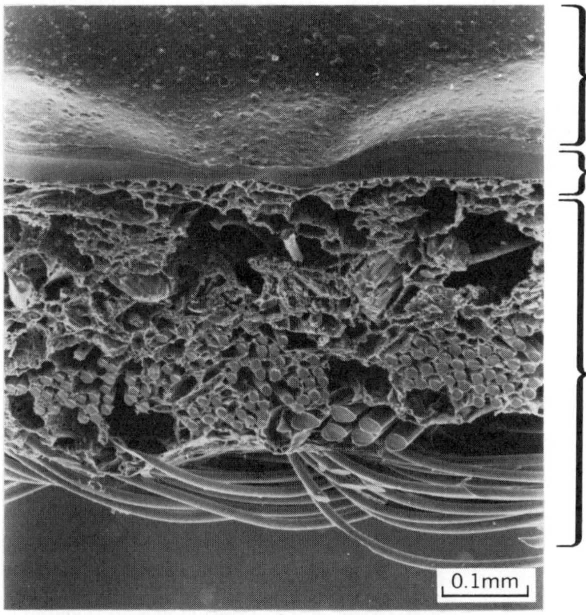

Figure 1. Cross-sectional view of urethane-coated fabric.

ment over coated fabrics. Although coated fabrics provide a good imitation of leather, their application is limited by the properties of the knit or woven substrates. Poromerics utilize a nonwoven fabric that is impregnated with a polymer, thereby creating a substrate resembling leather, eg, Clarino (Kuraray Co., Ltd., Fig. 2).

Manufacture

Coated Fabrics. Solid and foam vinyl-coated fabrics are made by calendering (qv), extrusion (qv), or coating.

Poromerics. In poromerics, nonwoven fabrics are used as substrate. Fiber as fine as 0.03 tex (0.3 den), and sometimes less, is frequently used to imitate the fibrillated collagen fibers of leather, thereby attaining the soft feel and good appearance essential for clothing applications. The nonwoven fabric is impregnated with a polymer to increase density and improve softness, mechanical properties, and processibility. The coating layer in poromerics that corresponds to the grain layer of leather is prepared with polyurethane. Poromerics are manufactured by combining the technologies of fiber and nonwoven fabrics manufacture with those of polyurethane preparation and coagulation.

Finishing

Finishing is an important process in the manufacture of leatherlike materials. Surfaces are either grained or sueded.

Fabrics coated with PVC can be colored by adding pigments to the PVC compound or paste. After the coating layer is bonded to the substrate, it is softened by ir irradiation and grained with cold embossing rolls. Coating with a mixture of poly(methyl methacrylate) and PVC adjusts luster and prevents blocking. Surface patterns are provided by gravure rolls.

Figure 2. Cross-sectional view of Clarino.

Table 1. Physical Properties of Leather and Leatherlike Materials

Product substrate	Weight, g/m^2	Density apparent, g/cm^3	Tensile strength, MPa[a] MD[c]	Tensile strength, MPa[a] TD[c]	Tear strength, kN/m[b] MD[c]	Tear strength, kN/m[b] TD[c]	Water-vapor permeability, $mg/(cm^2 \cdot h)$	Flex endurance, 10^3
Clarino[d], nonwoven	525	0.35	4.9	7.2	36	33	2.9	>1000
Clarino F[d], nonwoven	618	0.43	11.6	14.3	56	46	5.0	>1000
Sofrina[d], nonwoven	225	0.47	12.5	13.7	102	82	5.4	>1000
Suedemark[d], nonwoven	220	0.40	12.7	12.7	80	68	15.0	>1000
vinyl								
solid knit	641	0.99	12.1	6.6	29	47	0.4	60
foamed knit	1045	0.76	4.6	3.3	24	21	0.3	150
polyurethane								
cast woven	372	0.42	12.1	6.9	44	86	1.3	400
coagulated woven	713	0.38	5.2	3.1	29	49	1.2	350
leather								
calf	768	0.65	19.0	14.0	53	49	10.1	>1000
side	994	0.69	26.4	11.5	74	88	9.2	>1000

[a] To convert MPa to kgf/cm^2, multiply by 10.2; to psi, multiply by 145.
[b] To convert kN/m to ppi, multiply by 5.71.
[c] MD = machine direction; TD = transverse direction.
[d] Registered trademark of Kuraray Co., Ltd.

Urethane-coated fabrics, on the other hand, whose grain appearance is obtained from release paper and the casting of color coating, are finished by applying polyurethane finishing agents; gravure rolls adjust luster and touch. Through use of the pattern embossed on the release paper, urethane-coated fabric can be produced in various surface patterns and shades by changing the thickness of the top coating. Patterns in additional colors require gravure rolls or spray guns. Some high grade urethane-coated fabrics can be dyed to deep shades.

Poromerics are given a grain finish by embossing the coating layer with heated rollers engraved with the desired pattern. The surface is then color-printed with gravure rolls, spraying, or dyeing. Embossing and coloring affect appearance and properties. Excessive embossing and coloring may damage the porous structure of the coating layer by creating a thick surface layer that impairs water-vapor permeability, flex endurance, and grain break. Gloss or patent-leather effects are obtained by finishing the surface to a mirror finish.

Sueded leatherlike materials are made in two ways: by coating the material's surface with a porous layer, and by napping the substrate's surface before dyeing, producing a fibrous material. The former method gives a suedelike appearance by sanding the surface with a macroporous sponge or by embossing or tearing the porous structure at the intermediate stage. Sueded poromerics, produced by napping urethane-impregnated nonwoven fabrics made of fine polyester or nylon fibers, are dyed and used for apparel, shoes, upholstery, and home furnishings.

Properties

The properties of leather substitutes depend on the polymer used for substrate and coating layer. Feel, handling, and grain break are affected by the construction. The physical properties of leather substitutes are given in Table 1.

Economic Aspects and Applications

Leatherlike materials are made of petrochemicals, except for cotton, which is often used as the substrate material. Petroleum supplies and prices are subject to great variations. Prices began to fall in 1985, after years of being high.

The production of coated fabrics in the Far East shows an increase, but production in other areas of the world shows no growth since 1978. High quality poromerics are produced mainly in Japan due to highly advanced techniques, resulting in a steady increase in annual production.

Applications vary according to quality and price. Vinyl-coated fabrics, for example, are used extensively for automotive seats, upholstery, sandals, casual shoes, clothes, and bags. Urethane-coated fabrics are used mainly for shoes, handbags, and apparel. In the production of coated fabrics, 70–73% is consumed for industrial materials and upholstery, 15% for apparel, and 12–15% for footwear. In the case of poromerics, 40–50% of production is used for various types of shoes, 35–47% for clothes, and 20–25% for bags, balls, upholstery, cases, and belts.

BIBLIOGRAPHY

K. Nagoshi in K. Ashida and K. C. Frisch, eds., *International Progress in Urethanes,* Vol. 3, Technomic Publishing Co., Inc., Lancaster, Pa., 1981, pp. 193–217.

O. Fukushima, *J. Coated Fabr.* **5,** 3 (1975).

K. Nagoshi
Kuraray Company, Ltd.

LEUCHS' ANHYDRIDE. See Polypeptides.

LEVANS. See Dextran; Polysaccharides.

LIGHT ABSORBERS. See Stabilization.

LIGHT MICROSCOPY

For illumination of an object, light microscopes utilize a spectrum of electromagnetic radiations that include the visible spectrum from violet (400–455 nm) to red (650–700 nm), and the invisible uv (ca 200–400 nm) and ir spectra. The invisible parts are of minor importance except in photomicrography and in the utilization of the property of certain objects to fluoresce in the uv to emit visible light. Other kinds of microscopes, such as those employing electrons, x rays, or ultrasound, also use optical parts. Therefore it is ambiguous to refer to light microscopes as optical microscopes (see ELECTRON MICROSCOPY).

In the polymer industry, light microscopy is used for quality control and problem solving, and to investigate compositions, properties, structures, and morphology. In patent application or litigation, microscopical data are important. In research, microscopical data often reveal variations in crystal structure. Microscopes also provide a useful tool in education.

Microscopical Attributes

Resolution. The fineness of detail revealed by an optical device is termed resolution. More specifically, it can be defined as the separation of two points of the object on two different receptors, separated by at least one receptor, on the retina of the eye.

The numerical aperture defines the theoretical (limiting) resolving power for a given wavelength of the illuminating beam.

Contrast. The ability to differentiate the structural components of an object by different intensity levels between black and white or between colors is called contrast perception. Thus contrast is as important as resolution. Unless there is contrast between the spaces and points, they are not visible, even though resolved. Separation of transparent portions by opaque portions or different colors enhances contrast. Contrast can also be enhanced by staining. Herzberg's stain (iodine in KI, $ZnCl_2$, and H_2O) differentiates among cellulosic fibers and tissues. Iodine-based reagents are also useful for starch identification and in estimating the degree of dextrinization. Contrast between structural portions in polymeric fibers, moldings, crystals, and foils can also be enhanced by means of polarized light.

Optical Staining. Colored disks, called Rheinberg disks or filters, inserted in the substage condenser, provide optical staining. Many patterns of optical staining are possible.

Correction of Objective Lenses for Refractive Aberrations. These include spherical aberration, coma, astigmatism, field curvature, and distortion.

Spherical aberration occurs even with a single truly spherical surface illuminated with parallel, monochromatic light. Rays closer to the lens center necessarily travel farther through the refractive solid and are focused at a greater distance from the lens. This is corrected by combining a single or double convex lens with a concave-aberration lens of proper shape (Fig. 1) and refractive index to add spherical aberration of equal but opposite direction.

Coma is the aberration of rays from any off-axis point of an object. Coma is so called because the image of a sharp point is shaped like a blurred comet. This is due to the greater magnification by rays through the periphery of a lens rather than through the center. The correction is similar to that for spherical aberration.

Astigmatism occurs in a lens that is not a product of rotation. Since microscopical lenses are ground and polished by rotation, astigmatism is not a problem with light microscopes. However, if an observer has astigmatism of the eye, the corrective eyeglasses should be worn also while using a microscope.

Distortion usually does not occur with monobjective light microscopes, but stereobinocular microscopes might image a square as a barrel or a pin cushion.

Depth of Focus. The thickness of the image space in acceptable focus is defined as the depth of focus. It is inversely proportional to the numerical aperture. A great depth of focus (low numerical aperture, NA) is required for micromanipulation. On the other hand, short depth of focus (high NA) is an advantage in optical sectioning when studying the spatial arrangements of structures.

Depth of Field. The thickness of the object space in focus is defined as the depth of field. For resinography, composite materials are prepared in sections ca 30 nm thick. Such specimens require depth of field in the microscopical objective.

Illumination. Kinds and extents of illumination are very important in microscopy. As indicated in Figure 2, the substage condenser offers light transmitted through the object. The three types of illumination by reflected light are also shown in Figure 2.

Reflection is used for an opaque matrix. Orthogonal (grazing) illumination (Fig. 2) is provided by a condenser with its axis at right angles to the axis of the objective.

Radiation. The microscopical resolution d, ie, the distance between two points discerned as separate, depends on the wavelength λ and the numerical aperture NA of the objective, where $d = 0.6/\text{NA}$. Both achromatic and apochromatic objectives can be corrected for green light. Therefore, a green filter is used in black and white photomicrography to provide the sharpest image.

Anisotropy in the Object. An optically anisotropic object has visible properties that vary with changing direction through the specimen. Optical anisotropy is detected microscopically by polarizing the light before it reaches the condenser.

Magnification. In microscopy, some magnification is necessary to place two adjacent but separate points in the object onto two eye receptors, which are separated by at least one other receptor. Magnification a thousand times the numerical aperture of the objective (1000 NA) is considered both useful and convenient. Any additional magnification reveals no more information and affects focus.

Field of View. The visible portion of the specimen, called the field of view, originates in the objective. It is first reduced by the diameter of the microscope tube or by its diaphragm. The field of view is bounded by the diaphragm in the focal plane of the eyepiece (Fig. 1).

Antiglare Devices. Glare, which consists of harsh, bright light, does not contribute to image formation and interferes with contrast. Application of a low refracting, transparent coating on lens-to-air surfaces counteracts glare.

Working Distance. This is the depth of space between the surface of the specimen and the front surface of the objective's lens system. A stereomicroscope provides sufficient space under its objective for needle, tweezers, micromanipulator, or hot or cold stage.

Temperature Control. A cold stage permits observation during freezing of an emulsion or suspension of a polymer. A hot stage is indispensable in studying crystallization, melting points, and glass transitions at varying rates of heating and cooling.

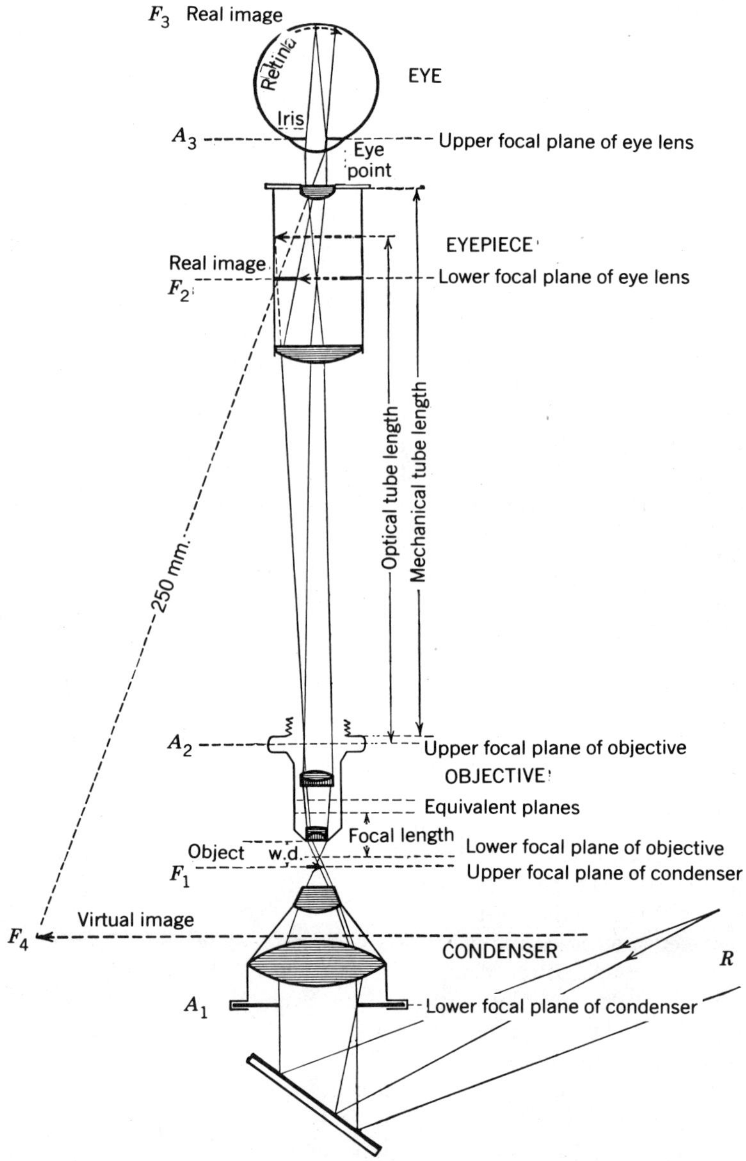

Figure 1. Optical systems of the light microscope.

Specimen

Structure. Microscopical manifestation of the internal structure is commonly correlated with optical properties that are usually determined by means of polarized light.

Refractive Index. The refractive index can be determined by light microscopy. In anisotropic crystals and fibers the refractive index varies with the direction of polarized light through the structure.

Morphology. The smaller the size and the more complex the shape of an object, the greater the required resolving power, contrast, aberration correction, and other attributes contributing to visibility.

Behavior. Changes in properties of a specimen with time, temperature, radiation, illumination, or other conditions can be very revealing.

Preparation. The best preliminary examination of a specimen (as received) is by stereoscopic microscope; reflected light, as well as transmitted illumination, should be employed. If the sample is heterogeneous, the apparently relevant part should be separated and impurities removed. A minimum of preparation is usually advisable. The kind and extent of preparation for light microscopical examination depends on the nature of the polymeric sample: particulate, fibrous, fabricated, textured, molded, laminated, composited, or otherwise produced.

Photomicrography

A photomicrograph is an image developed on a sensitized surface from radiation emanating from a microscope. The radiation may be visible or invisible light, including ir and uv.

Photomicrographs are intended to describe objects and illustrate observations for records, reports, publications, specifications, patents, and other communications. They are relied on as factual, and compete with drawings, diagrams, patterns, and spectra. Motion pictures or time-lapse studies are often indicated.

Light Microscopes

Light microscopes include simple microscopes (magnifiers), handheld compound microscopes, stereomicroscopes, metallographs

Figure 2. Principal types of illumination. Directional beams lie on or within a principal cone. Symmetrical types are cones within the principal cones.

and other inverted microscopes, and near-field scanning optical microscopes.

BIBLIOGRAPHY

T. G. Rochow and E. G. Rochow, *Introduction to Microscopy by Means of Light, Electrons, X-rays, or Ultrasound,* Plenum Publishing Corp., New York, 1978.

R. P. Loveland, *Photomicrography,* Vols. 1 and 2, John Wiley & Sons, Inc., New York, 1970; reprinted by R. E. Krieger Publishing Co., Inc., Melbourne, Fla., 1981.

W. C. McCrone, L. B. McCrone, and J. G. Delly, *Polarized Light Microscopy,* 4th ed., Ann Arbor Science, Ann Arbor, Mich., 1984.

T. G. Rochow
North Carolina State University

LIGHT SCATTERING

Study of the scattering (qv) of electromagnetic radiation from polymers and their solutions embraces x-ray, neutron, and light scattering; some features of the scattering from these diverse sources can be treated with a common formalism. The molecular properties elucidated include intramolecular properties, eg, molecular weight ($\bar{M}_w$), root-mean-square radius of gyration (R_G), translational diffusion constant (D_T^0), and others obtained by extrapolation of measured quantities to infinite dilution. Data on solutions at low concentrations yield information on intermolecular properties such as the second virial coefficient A_2, and the mutual-diffusion constant D_M.

A useful measure of the concentration c (weight per volume) is the product, $[\eta]c$, where the intrinsic viscosity $[\eta]$ is defined as the limiting value of $\partial \ln \eta/\partial c$ at infinite dilution, with η the solution viscosity. The product $[\eta]c$ provides a measure of the intermolecular overlap of the polymer solute inasmuch as $[\eta]M/N_A$ may be considered a hydrodynamic volume per solute molecule, with cN_A/M the number of chains per unit volume. Thus low concentrations are those for which $[\eta]c \ll 1$, and intermolecular interactions are small or suppressed in the limit as $[\eta]c$ goes to zero. For moderately concentrated solutions, $[\eta]c \approx 1$–10, and the effects of intermolecular interactions are important.

Theory

Much of the discussion of light scattering is concerned with interference effects between light rays scattered from different parts of the sample; these may arise from different parts of a single solute molecule or from parts of different molecules. The interference between rays scattered from elements i and j depends on the vector distance $\mathbf{r}_{ij} = \mathbf{r}_i - \mathbf{r}_j$, where $\mathbf{r}_i$ and $\mathbf{r}_j$ are the vectors from an arbitrary reference to elements i and j, respectively. For elastic or quasielastic scattering, a reciprocal dimension defined as

$$q = (4\pi/\lambda)\sin\theta/2 \qquad (1)$$

with θ the scattering angle and λ the wavelength of light in the medium, appears in a natural way when considering interference effects.

Since digital photon-counting methods find increasing use in light scattering and are the preferred method for quasielastic light scattering studies, the presentation here is cast in terms of measurement of the number of photon counts $n(q,t;\Delta\tau)$ measured at time t over a time interval $\Delta\tau$. This information may be presented in terms of the probability density $p_1(n)$ for the count rate, or the bivariant probability density $p_2(n,n')$, where n and n' signify $n(q,t;\Delta\tau)$ and $n(q,t';\Delta\tau)$, respectively, with $t' = t + \tau$. The average intensity $I(q)$ of the elastic scattering is independent of t, and proportional to $\Delta\tau\langle n(q,\Delta\tau)\rangle$, where

$$\langle n(q,\Delta\tau)\rangle = \int n\, p_1(n)dn \qquad (2)$$

It is useful to express $\langle n(q,\Delta\tau)\rangle$ and $I(q)$ in terms of a quantity $\mathbf{R}(q)$ called the Rayleigh ratio, defined as

$$\mathbf{R}(q) = R^2 I(q)/VI_{\text{inc}} = R^2\langle n(q,\Delta\tau)\rangle/Vn_{\text{inc}} \qquad (3)$$

where R is the distance between the scattering centers and the photomultiplier, V, the scattering volume, I_{inc}, the intensity of the incident beam, and n_{inc}, the count rate obtained with the incident beam at the $\Delta\tau$ used to determine $\langle n(q,\Delta\tau)\rangle$. In most cases, the excess Rayleigh ratio is equal to the difference of the Rayleigh ratios for solution and solvent; $\mathbf{R}(q)$, etc, refers to the excess quantity unless otherwise specified.

The photon-count distribution may also be used to obtain the photon-count correlation function,

$$G^{(2)}(q,\tau;\Delta\tau) = \int nn'p_2(n,n')dndn' \qquad (4)$$

$$g^{(2)}(q,\tau;\Delta\tau) = \frac{G^{(2)}(q,\tau;\Delta\tau)}{G^{(2)}(q,\infty;\Delta\tau)} \qquad (5)$$

Spatial averaging of the scattered rays from the scattering volume gives

$$g^{(2)}(q,\tau;\Delta\tau,A) = 1 + f(A)[g^{(2)}(q,\tau;\Delta\tau,0) - 1] \qquad (6)$$

The factor $f(A)$ may be considered to be an instrument parameter dependent on the specific optical arrangements of the detector optics; typically, $f(A)$ is in the range of 0.1–0.9. In most cases,

$$g^{(2)}(q,\tau;\Delta\tau,0) - 1 = [p(q,\Delta\tau)g^{(1)}(q,\tau)]^2 \qquad (7)$$

Scattering at Infinite Dilution. ***Static Scattering at Zero Scattering Angle.*** The ratio $[\mathbf{R}(0)/c]^0$ of the Rayleigh ratio divided by the concentration, extrapolated to zero angle and infinite dilution, is discussed below for several cases. For isotropic scatters,

$$[\mathbf{R}(0)/c]^0 = K'(2\pi N_A)^2 \sum_{\nu}^{C} w_\nu(n_\nu\alpha_\nu)^2 M_\nu^{-2} \frac{RT}{(\partial\Pi/\partial c)_\nu^0} \qquad (8)$$

Here Π is the osmotic pressure, λ_0 the wavelength of the incident light in vacuum, $w_\nu = c_\nu/c$ the weight fraction of component ν with c the total solute concentration, and $n_\nu\alpha_\nu$, is given by

$$n_\nu\alpha_\nu = \left(\frac{\tilde{n}_s}{2\pi N_A}\right)\psi_\nu M_\nu \qquad (9)$$

$$\psi_\nu = \left(\frac{\partial\tilde{n}}{\partial c_\nu}\right)_w^0 \qquad (10)$$

Here, $\tilde{n}$ is the refractive index, and the subscript s denotes the solvent. Equation 8, obtained by Debye, reveals the connection between the scattering and the osmotically dampened concentration fluctuations. Since

$$\lim_{c=0} \Pi_\nu = RT\frac{c_\nu}{M_\nu} \qquad (11)$$

equation 8 reduces to

$$[\mathbf{R}(0)/c]^0 = K'\tilde{n}_s^2\left(\frac{\partial\tilde{n}}{\partial c}\right)_w^2 M_{\text{LS}} \qquad (12)$$

where

$$K' = 4\pi^2/N_A\lambda_0^4 \qquad (13)$$

$$\left(\frac{\partial\tilde{n}}{\partial c}\right)_w = \sum_{\nu}^{C} w_\nu\psi_\nu \qquad (14)$$

$$M_{\text{LS}} = \left(\frac{\partial\tilde{n}}{\partial c}\right)_w^{-2} \sum_{\nu}^{C} \psi_\nu^2 w_\nu M_\nu \qquad (15)$$

with $(\partial\tilde{n}/\partial c)_w$ the refractive index increment determined at constant temperature, pressure, and composition w of solute.

If the solute comprises a homologous series with ψ_ν independent of ν and equal to $(\partial\tilde{n}/\partial c)_w$,

$$M_{\text{LS}} = \sum_{\nu}^{C} w_\nu M_\nu = \bar{M}_w \qquad (16)$$

If the system contains multiple solvent components,

$$M_{\text{LS}} = \frac{(\partial\tilde{n}/\partial c)_\mu^2}{(\partial\tilde{n}/\partial c)_w^2}\bar{M}_w \qquad (17)$$

where $(\partial\tilde{n}/\partial c)_\mu$ is the refractive index increment determined at constant temperature, pressure, and at osmotic equilibrium of the solvent components in the solution and the solute-free solvent mixture. Since $(\partial\tilde{n}/\partial c)_\mu$ is conveniently measured by differential refractometry on solutions dialyzed against the solvent mixture, $\bar{M}_w$ may be obtained for solutions in mixed solvents with little more effort than is required with a single-component solvent.

For a copolymer comprising scattering elements with refractive index increment ψ_A or ψ_B and molecular weight m_A and m_B, respectively, with $W_{A,\nu}$ and $W_{B,\nu} = 1 - W_{A,\nu}$ the weight fractions of the two monomers in component ν, and

$$\Delta w_\nu = w_{A,\nu} - w_A = w_B - w_{B,\nu} \tag{18}$$

then

$$M_{LS} = \bar{M}_w(1 + 2pY + qY^2) \tag{19}$$

Y is given by equation 20:

$$Y = \left[\frac{\psi_A - \psi_B}{(\partial n/\partial c)_w}\right] \tag{20}$$

and

$$p = \sum_\nu^C w_\nu M_\nu \Delta w_\nu / \bar{M}_w \tag{21}$$

$$q = \sum_\nu^C w_\nu M_\nu (\Delta w_\nu)^2 / \bar{M}_w \tag{22}$$

For a strictly alternating copolymer or a copolymer with random placements of the monomers, $\Delta w_\nu = 0$ and $M_{LS} = \bar{M}_w$.

Static Scattering at Arbitrary Angle. For isotropic scatterers, $[\mathbf{R}(q)/c]^0$ is expressed in terms of the product $M_{LS}P_X^0(q)$, where $P_X^0(q)$, defined below, accounts for interference effects among the rays scattered from a solute. For a system with C-solute molecules, each characterized by a number n_ν of scattering elements, the expression for $[\mathbf{R}(q)/c]^0$ for an optically isotropic molecule, with $(\partial n/\partial c)_w \neq 0$, is

$$[\mathbf{R}(q)/c]^0 = K'\tilde{n}_s^2 \left(\frac{\partial \tilde{n}}{\partial c}\right)_w^2 M_{LS} P_X^0(q) \tag{23}$$

where M_{LS} is given by equation 15 (or for copolymers by equations 19, 21, and 22), and

$$P_X^0(q) = \frac{\sum_\nu^C w_\nu M_\nu^{-1} \left(\sum_j^{n_\nu}\sum_k \psi_j\psi_k m_j m_k \left\langle \frac{\sin q|r_{jk}|}{q|r_{jk}|}\right\rangle\right)}{\sum_\nu^C w_\nu M_\nu^{-1}\left(\sum_j^{n_\nu} \psi_j m_j\right)^2} \tag{24}$$

For a homopolymer with m_i and ψ_i identical for all scattering elements, $M_{LS} = \bar{M}_w$ (eq. 16), and equation 24 becomes

$$P_X^0(q) = \frac{1}{\bar{M}_w}\sum_\nu^C w_\nu M_\nu P_\nu^0(q) \tag{25}$$

$$P_\nu^0(q) = \frac{1}{n_\nu^2}\sum_j^{n_\nu}\sum_k \left\langle \frac{\sin q|r_{jk}|}{q|r_{jk}|}\right\rangle \tag{26}$$

$$P_X^0(q) = 1 - \tfrac{1}{3}R_{G,X}^2 q^2 + \cdots \tag{27}$$

$$R_{G,X}^2 = \frac{\sum_\nu^C w_\nu M_\nu R_{G,\nu}^2}{\bar{M}_w} \tag{28}$$

$$R_{G,\nu}^2 = \frac{1}{2n_\nu^2}\sum_j^{n_\nu}\sum_k \langle r_{jk}^2\rangle \tag{29}$$

where $R_{G,\nu}$ is the rms radius of gyration of component ν. Examples of R_G for models of typical interest are given in Table 1.

Table 1. Radius of Gyration and Hydrodynamic Radius for Some Model Structures

	Characteristic lengths	R_G^2	R_H/R_G
persistent coil	L = contour length	$\frac{L\rho}{3}S(\rho/L)$	$3^{1/2}\left(1+\frac{L}{4\rho}\right)^{1/2}\left(\frac{27L}{16\rho}+\ln^2\frac{3L}{2d}\right)^{+1/2 a}$
	ρ = persistence length		
	d = diameter		
coil ($\rho/L \ll 1$)	$b = 2\rho$ = step length	$\frac{L\rho}{3} = \frac{nb^2}{6}$	$\frac{2}{3}\left(\frac{9\pi^{1/2}X}{6+16X}\right)^b$
	$L = nb$		
rod ($\rho/L \gg 1$)	L = rod length		
	d = rod diameter	$\frac{L^2}{12}$	$\sqrt{3}/\ln(3L/2d)$
disk	R = radius	$\frac{1}{2}R^2$	$\sqrt{2}/\pi$
sphere	R = radius	$\frac{3}{5}R^2$	$\sqrt{5/3}$
spheroid	$2R_1$ = length of unique axis		$\left[\frac{5(p^2-1)}{2+p^2}\right]^{1/2}\frac{1}{\ln\{p[1+(1+p^{-2})^{1/2}]\}}$;
	$2R_2$ = length of transverse axis	$\left(\frac{2+p^2}{5}\right)R_1^2$	$\left(\frac{1-p^2}{2+p^2}\right)^{1/2}\frac{1}{\arctan(p^{-2}-1)^{1/2}}$; $p<1$
	$p = R_2/R_1$		

[a] An approximate relation.
[b] $\pi^{3/2}X = (L/R_G)(\zeta/6\pi\eta_s d)$ with ζ the friction factor per step length.

The ensemble average $\langle(\sin q|r_{jk}|)q|r_{jk}|\rangle$ in equation 26 may be completed for several chain models. Results for models with n_ν identical scattering elements that find frequent use give

coil: $$P^0_\nu(q) = \frac{2}{u^2_\nu}\,[u_\nu - 1 + \exp(-u_\nu)] \quad (30)$$

rod: $$P^0_\nu(q) = \frac{2}{x^2_\nu}\,[x_\nu Si(x_\nu) - 1 + \cos x_\nu] \quad (31)$$

disk: $$P^0_\nu(q) = \frac{2}{z^2_\nu}\,[1 - J_1(2z_\nu)/z_\nu] \quad (32)$$

sphere: $$P^0_\nu(q) = \frac{9}{y^6_\nu}\,(\sin y_\nu - y_\nu \cos y_\nu)^2 \quad (33)$$

for random-flight coils, thin rods, thin disks, and spheres, respectively. In these expressions, $u_\nu = R^2_{G,\nu}q^2, x^2_\nu = 12u_\nu$, $z^2_\nu = 2u_\nu$, and $y^2_\nu = 5u_\nu/3$. The Bessel function of 1st order $J_1(z)$ and the sine integral

$$\mathrm{Si}(z) = \int_0^z \frac{\sin t}{t}\,dt \quad (34)$$

may be found in standard tabulations; $P^0_\nu(q)$ is shown as a function of $qR_{G,\nu}$ for several models in Figure 1.

Expansion of equation 30 as a power series for $[P^0_\nu(q)]^{-1}$ or $[P^0_\nu(q)]^{-1/2}$ in q^2 gives relations that are more nearly linear for small q^2 than is equation 27:

$$[P^0_\nu(q)]^{-1} = 1 + 1/3u_\nu + 1/36u^2_\nu + 0(u^3_\nu) \quad (35)$$

$$[P^0_\nu(q)]^{-1/2} = 1 + 1/6u_\nu + 0(u^3_\nu) \quad (36)$$

Thus, experimental data for $[\mathbf{R}(q)/c]^0$ vs q^2 are frequently analyzed by using plots of $[c/\mathbf{R}(q)]^0$ or $\{[c/\mathbf{R}(q)]^0\}^{1/2}$ vs q^2.

Dynamic Scattering at Arbitrary Angle. If solutions are sufficiently dilute that intermolecular correlations may be ignored, the function $g^{(1)}_\nu(q,\tau)$ for the νth solute is given by

$$g^{(1)}_\nu(q,\tau) = \exp(-\gamma_\nu\tau) \quad (37)$$

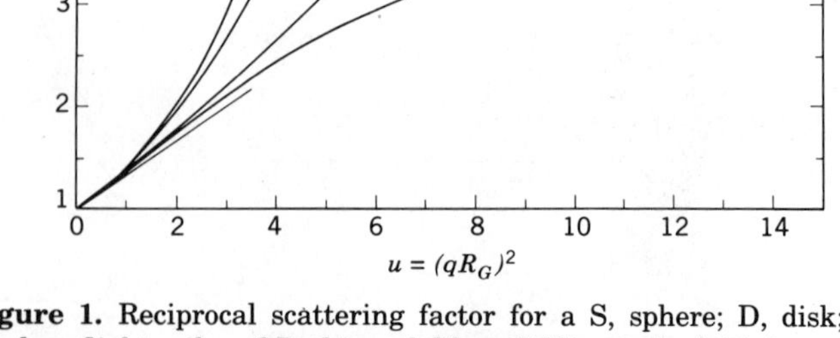

Figure 1. Reciprocal scattering factor for a S, sphere; D, disk; C, random-flight coil; and R, thin rod. The solid line is the initial tangent for all curves; –·–· indicates the asymptotic behavior for the coil for large u.

where D^0_T is the diffusion constant for center of mass translation and $\gamma_\nu = q^2 D^0_{T,\nu}$ plays the role of an inverse coherence time. It is frequently useful to express $D^0_{T,\nu}$ in terms of a hydrodynamic radius $R_{H,\nu}$

$$R_{H,\nu} = kT/6\pi\eta_s D^0_{T,\nu} \quad (38)$$

where η_s = solvent viscosity. The relation between $R_{H,\nu}$ and $R_{G,\nu}$ for several models is given in Table 1.

The translational diffusion constant (at infinite dilution) may be expressed in terms of a molecular-friction coefficient Ξ^0_ν and the osmotic pressure by the relation

$$D^0_{T,\nu} = \frac{M_\nu}{N_A \Xi^0_\nu}\left(\frac{\partial\Pi}{\partial c}\right)^0_\nu = \frac{kT}{\Xi^0_\nu} \quad (39)$$

For $\Delta\tau$ in the range usually preferred to study $g^{(2)}(q,\tau;\Delta\tau,A)$, $p(q,\Delta\tau)$ is essentially unity, ie, $\gamma\Delta\tau < 1$,

$$g^{(1)}(q,\tau) = \sum_\nu^C r_\nu \exp(-\gamma_\nu k\Delta\tau) \quad (40)$$

where r_ν is the fraction of the scattered light associated with $g^{(1)}_\nu(q,\tau)$.

The method of cumulants is based on the expansion

$$\ln\{[g^{(2)}(q,\tau;0) - 1]^{1/2}\} = K_1\tau + \frac{1}{2!}K_2\tau^2 + \cdots \quad (41)$$

$$K_i = (-1)^i \left\{\frac{\partial^i \ln[g^{(2)}(q,\tau;0) - 1]^{1/2}}{\partial\tau^i}\right\}_{\tau=0} \quad (42)$$

Thus with equation 40

$$K_1 = \sum_\nu^C r_\nu\gamma_\nu \quad (43)$$

$$K_2 = \sum_\nu^C r_\nu\gamma^2_\nu - K^2_1 \quad (44)$$

For example, $K_1 = q^2 D^0_T$ and $K_2 = 0$ if the solute is monodisperse.

If the molecules are large enough that $qR_{G,\nu}$ exceeds about unity, and if they have asymmetric instantaneous conformations, correlations among the scattering elements comprising one molecule result in a more complicated representation for $g^{(1)}_\nu(q,\tau)$

$$g^{(1)}_\nu(q,\tau) = \exp(-\Omega_\nu\tau)D(\Omega_\nu\tau) \quad (45)$$

where Ω_ν and $D(\Omega_\nu\tau)$ have the properties

$$\lim_{q=0}\Omega_\nu = \gamma_\nu = q^2 D^0_{T,\nu} = \frac{kT}{6\pi\eta_s}\frac{u_\nu}{(R_{H,\nu}R^2_{G,\nu})} \quad (46)$$

$$\lim_{q=0} D(\Omega_\nu\tau) = \lim_{\tau=0} D(\Omega_\nu\tau) = 1 \quad (47)$$

with $u_\nu = (qR_{G,\nu})^2$. Thus with equation 45, generally the second cumulant $K_2 \neq 0$, even for a monodisperse solute.

For several models for small u_ν, $g^{(1)}_\nu(q,\tau)$ may be represented by equation 45 with $\Omega_\nu = \gamma_\nu$ and

$$D(\Omega_\nu\tau) = 1 - u^2_\nu s^{(2)}_\nu(u_\nu)[1 - \exp(-k^{(2)}_\nu\gamma_\nu u^{-1}_\nu\tau)] + \cdots \quad (48)$$

For rodlike chains $s_\nu^{(2)}(0) = 1/45$ and $\tau_\nu^{(2)} = (6D^0_{R,\nu})^{-1}$, with $D^0_{R,\nu}$ the rotational diffusion constant. For flexible coil chains, $s_\nu^{(2)}(0) = 1/45$, and $\tau_\nu^{(2)}$ is the time constant for the longest relaxation mode.

Scattering at Arbitrary Concentration. ***Static Scattering at Zero Scattering Angle.*** *Behavior for Small* $[\eta]c$. For a monodisperse solute,

$$\frac{Kc}{\mathbf{R}(0)} = \frac{1}{M}[1 + 2A_2Mc + 3q_3(A_2Mc)^2 + \cdots] \qquad (49)$$

$$\left[\frac{Kc}{\mathbf{R}(0)}\right]^{1/2} = \left(\frac{1}{M}\right)^{1/2}[1 + A_2Mc + \tfrac{1}{2}(3q_3 - 1)(A_2Mc)^2 + \cdots] \qquad (50)$$

where $q_3 = A_3/A_2^2M$. If $q_3 \approx 1/3$, $[Kc/\mathbf{R}(0)]^{1/2}$ is more nearly a linear function of c than $Kc/\mathbf{R}(0)$.

It is convenient to express A_2 in terms of the excluded volume per molecule U defined as

$$A_2M = \frac{N_A}{M}\cdot\frac{U}{2} \qquad (51)$$

For particles with a hard-core potential, U, which represents the volume excluded by one particle to all other particles, may be calculated exactly based on geometric considerations. Such calculations have been completed for spheroids, disks, and rods.

More realistic intermolecular potentials exhibit both repulsive and attractive terms. Thus U, or A_2, may be zero or negative, in contrast to the case for the hard-core potential for which $U > 0$. With flexible chain polymers, it is useful to express A_2M in the form

$$A_2M = \frac{N_AR_G^3}{M}\Psi\left(\frac{U}{R_G^3}\right) \qquad (52)$$

where $\Psi(x)$ is a function with the properties

$$\lim_{x=0} \Psi(x) = 0 \qquad (53)$$

$$\lim_{x=\infty} \Psi(x) = \Psi_\infty \qquad (54)$$

with $\Psi_\infty \approx 6$–7 for linear flexible chain polymers.

For heterodispersed systems in which $\Psi_\nu = \Psi$ for all ν,

$$\mathbf{R}(0) = Kc\bar{M}_w(1 + 2A_2^R\bar{M}_wc + \cdots)^{-1} \qquad (55)$$

$$A_2^R = \frac{1}{\bar{M}_w^2}\sum_\nu^C\sum_{\mu \neq \nu}^C w_\nu w_\mu M_\nu M_\mu B^0_{\nu\mu} \qquad (56)$$

Behavior for $[\eta]c > 1$. If $c[\eta] > 1$, the power-series expansions in the former case are not applicable.

For flexible chain polymers in systems for which U/R_G^3 is large, A_2Mc forms a natural dimensionless variable for $M\Pi/RTc$, or $\mathbf{R}(0)/KcM$. For A_2Mc in the range 1–10, it is expected that Π should depend on c, but not on M. Thus retaining the form based on the dimensionless variable A_2Mc,

$$\frac{\Pi M}{RTc} = P(A_2Mc) \qquad (57)$$

the condition $\partial\Pi/\partial M = 0$ requires that $(\partial \ln P/\partial \ln M) - 1 = 0$. Making use of the relation $\partial \ln A_2/\partial \ln M = \alpha - 1$ with $a = \partial \ln [\eta]/\partial \ln M$, this condition gives the result

$$1 + p = \frac{\partial \ln P(A_2Mc)}{\partial \ln A_2Mc} = \frac{1}{a} \qquad (58)$$

For large A_2Mc, for which a tends to $4/5$, p goes to $1/4$. The power-series expansion valid for small A_2Mc and the power-law representation may be combined in the semiempirical relation

$$\frac{\Pi M}{RTc} = 1 + Z\left(1 + \frac{q_3}{p}Z\right)^p \qquad (59)$$

where $Z = A_2Mc$. Numerically similar relations have been derived using renormalization group methods. This relation provides a semiempirical representation for Π over the range from infinite dilution to large $[\eta]c$, but should not be used if $c/\hat{\rho}$ is larger than ca 0.5 (with $\hat{\rho}$ the solute density), as then $\Pi M/RTc$ may not scale with A_2Mc.

With equation 60,

$$\mathbf{R}(0) = K\frac{RTc}{(\partial\Pi/\partial c)} \qquad (60)$$

this expression for Π gives

$$\mathbf{R}(0) = KcM\left\{\left[1 + 2Z\left(1 + \frac{q_3}{p}2Z\right)^P\right]H(Z)\right\}^{-1} \qquad (61)$$

where $H(Z) \approx 1$.

Dynamic Scattering at Arbitrary Angle. *Behavior for* $[\eta]c \ll 1$ *and* $qR_G \ll 1$. In the limit as q goes to zero, for the behavior at infinite dilution, equation 62

$$S(q,\tau) = \sum_i\sum_j^\eta \langle \exp i\mathbf{q}\cdot[\mathbf{r}_i(0) - \mathbf{r}_j(\tau)]\rangle \qquad (62)$$

involves only the center of mass translation, with

$$g^{(1)}(g,\tau) = \exp - \gamma\tau \qquad (63)$$

$$\gamma = q^2D_M(c) \qquad (64)$$

However, the mutual diffusion constant depends on c (eg, compare eq. 39):

$$D_M(c) = \frac{M}{N_A\Xi(c)}\cdot\frac{\partial\Pi}{\partial c}(1 - \phi) \qquad (65)$$

with ϕ the volume fraction of solute and $\Xi(c)$ the friction factor at concentration c. Of course, $D_M(c)$ reduces to D_T^0 at infinite dilution. For $[\eta]c \ll 1$, $\Gamma(c)$ may be represented by

$$\Gamma(c) = 2\Gamma_2 + 3\Gamma_3c + \cdots \qquad (66)$$

and $\Xi(c)$ may be expanded in a power series in $[\eta]c$:

$$\Xi(c) = \Xi^0[1 + k_\Xi[\eta]c + \cdots] \qquad (67)$$

where $\Xi^0 = kT/D_T^0$. Combination of these relations gives

$$D_M(c) = D_T^0[1 + k_D[\eta]c + \cdots](1 - \phi) \quad (68)$$

$$k_D = \frac{2A_2M}{[\eta]} - k_\Xi \quad (69)$$

In general, k_Ξ has the form

$$k_\Xi = c_1 + c_2A_2M/[\eta] \quad (70)$$

where c_1 and c_2 depend on the nature of the solute. For flexible chain polymers, one treatment gives $c_1 \approx 0.2$ and $c_2 \approx 1.2$.

BIBLIOGRAPHY

B. J. Berne and R. Pecora, *Dynamic Light Scattering,* Wiley-Interscience, New York, 1976.

B. Chu, *Laser Light Scattering,* Academic Press, Inc., New York, 1974.

M. Kerker, ed., *Electromagnetic Scattering,* ICES I, Pergamon Press, Inc., Elmsford, N.Y., 1963.

D. McIntyre and F. Gornick, eds., *Light Scattering From Dilute Solutions,* Gordon and Breach Science Publishers, Inc., New York, 1964.

GUY C. BERRY
Carnegie-Mellon University

LIGHT STABILITY OF POLYMERS. See DEGRADATION; STABILIZATION; WEATHERING.

LIGNIN

Lignin is an amorphous, aromatic biopolymer second in natural abundance only to cellulose. The biosphere is estimated to contain 3×10^{11} t of lignin with an annual biosynthesis rate of ca 2×10^{10} t. Based on pulp- and paper-production statistics, about 50×10^6 t/yr of lignin is produced from woody plants at pulp mills worldwide.

Lignin is defined as a plant constituent that responds in a certain manner to an applied stain or as a substance that yields certain degradation products upon chemical treatment under defined conditions. Since lignin has characteristic uv-light-absorbing properties, it can also be detected by uv-light microscopy. The second criterion for positive lignin identification is based on a degradative treatment of the lignified plant tissue.

Plants whose tissue show positive response to any one of the above tests have two specialized functions: the conduction of solutes over extended distances and mechanical support. Lignification is thus distinctly associated with the development of a water-conduction (vascular) system and solar-energy collection devices (leaves, needles, and a crown) above ground level.

Lignin is not necessarily uniform in its chemical structure or morphological distribution. Although the chemical structure of lignin may significantly vary with its location within the cell wall, lignin is generally divided into two classes: guaiacyl and guaiacyl–syringyl lignin. This designation refers to the presence of different types of aromatic ring structures in lignin:

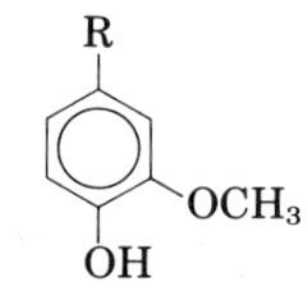

guaiacyl
(3-methoxy-4-hydroxyphenyl)
substituent;
if R = CHO, vanillin

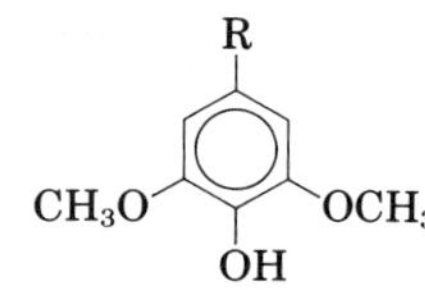

syringyl
(3,5-dimethoxy-4-hydroxyphenyl)
substituent;
if R = CHO, syringaldehyde

where R is normally an oxygenated C_3-substituent in lignin giving the basic repeat unit a C_9-configuration.

Biosynthesis

Early studies on lignin biosynthesis benefited from the development of methods for preparing a ligninlike polymer in vitro, the so-called dehydrogenation polymer (DHP). In this effort, lignin precursors were pictured as being synthesized enzymatically in the cambium of the growing plant. Phenolic precursors, partly because of their toxicity to the organism and partly because of poor solubility in water, needed to be converted enzymatically to a glycoside. In softwoods, the principal glycosidic precursor, coniferin, was detected in the first eight to ten cell layers inside the cambium. Since coniferin concentration decreases at this point and the lignin content increases, it was concluded that lignin was the product of (random) diffusion of monomeric building blocks over the distance of several cell layers and that lignin itself had a random poly(oxyphenylpropane) structure. A more-or-less uniform mixture of guaiacyl, syringyl, and *p*-hydroxyphenyl units was expected for hardwood lignins.

This concept of lignin biosynthesis needed revision in light of new experimental evidence, which contradicted the theory of intercellular transport of lignin precursors. A revised concept of liquification holds that each individual cell controls the biosynthesis of lignin precursors and the rate of their delivery to the respective lignifying cell wall.

Lignin precursors, phenylpropane units, are generated through the shikimate and cinnamic acid pathways. The reactions along these pathways are precisely controlled by a number of different enzymes. The activity of enzymes needed in lignin formation is negligible until the carbohydrate framework has been formed.

Polymerization of Lignin Precursors. Coniferyl and sinapyl alcohols, the reduction products of ferulic and sinapic acids, are the monomeric precursors of lignin. A ligninlike polymer forms by the enzymic dehydrogenation of coniferyl alcohol. The reaction proceeds by free-radical coupling and the process is initiated by peroxidase and H_2O_2. These alcohols, or mixtures of both, may be dehydrogenated by a peroxidase–H_2O_2 system to form a ligninlike polymer by a free-radical coupling mechanism.

Lignification Control. The potential for controlling the lignification process is being investigated. The benefits of control are obvious because of the enormous energy devoted to separating lignin from carbohydrates during pulping. With the goal of improving the efficiency of pulping, ie, delignification, the biosynthesis of lignin might be modified. First, the total lignin content of the tree might be reduced. This operation would be limited by the physical and mechanical requirements of the tree. The second approach would be to modify the types of linkages present in the lignin macromolecule, since certain intermonomer linkages are more easily cleaved, under pulping conditions, than others.

Structure

The process of dehydrogenative polymerization of monomeric lignin precursors has been studied *in vitro* using (crude) isolated enzyme fractions from mushrooms. Numerous dimers, trimers, and oligomers have been isolated from this mixture, and the bonds that they represent were identified in isolated polymeric lignin as well. Hence, strong evidence has been presented in favor of a random polymerization process of phenoxy radicals and of the fact that this variety of isomeric phenoxy radicals, so-called mesomers, produces a wide variety of intermonomer linkages. This explains the complexity of the chemical structure of lignin in terms of functionality and interunit linking.

Although the difference between coniferyl alcohol and sinapyl alcohol is minor, a methoxy group at the ring 5-position, the structural difference between guaiacyl and guaiacyl–syringyl lignin is more dramatic. Syringyl structures, with an extra methoxyl group, cannot form linkages at the 5-position, which account for about 20–25% of the linkages found in softwood lignin. In comparison, the 5-position is involved in only 10% of the linkages in hardwood lignin. The consequence of the OCH_3 substitution in position 5 of the syringyl lignin is that hardwood lignin has a greater abundance of alkyl–aryl ethers, which are hydrolyzable by pulping lignins, and fewer phenolic hydroxyl groups. As a result, hardwood lignin is more easily degraded (by alkaline and acidic pulping lignin) to smaller molecules with higher phenolic OH contents, which are more soluble in alkali.

In accordance with the physiological need of the plant for lignin as a reinforcing polymeric matrix, the distribution of lignin across cell wall and middle lamella is nonuniform. Furthermore, different anatomical regions are lignified at different stages of the life cycle of the living cell. It is then not surprising that differences have been found in the chemical structure of lignin in different anatomical and morphological regions. This is particularly apparent for the distribution of guaiacyl and syringyl residues in the lignin of hardwood species. Table 1 illustrates the case for white birch.

Nonuniformities are indicated on biochemical, physical, and chemical grounds, but a striking uniformity has been revealed in experiments exploring the morphology of lignin in the cell wall. Lamellae, or nodes, ca 7–10 nm in size are dispersed throughout the cell wall in a regular pattern. This regularity suggests that whatever the differences in chemical composition, they are not significant enough to disturb the preferred orientation of lignin in the S_2 layer.

The macromolecular structure of lignin in wood is the subject of controversy. In one view, lignin exists as a (linear or branched) molecule whose solubility is constrained by a cage of cell-wall material from which it can escape only after the opening of micropores, allowing lignin to diffuse into solution. The second perspective interprets lignin as a gel (according to the Flory-Stockmayer theory) with a dissolution behavior entirely consistent with network behavior and independent of association phenomena with carbohydrates. It sees delignification as a rupture of tetrafunctional branch sites in wood. Both views ignore the possibility that the amorphous component of wood may be representative of a lignin–carbohydrate (block?) copolymer or a semi-interpenetrating polymer network (SIPN).

Table 1. Distribution of Guaiacyl and Syringyl Residues in Lignin in White Birch[a]

Morphological region[b]	Guaiacyl:syringyl Bromination and tem–edxa[c]	Guaiacyl:syringyl Uv spectral analysis
fiber S_2	12:88	syringyl
vessel S_2	88:12	guaiacyl
ray parenchyma S	49:51	syringyl
$ML_{cc(F/F)}$	91:9	50:50
$ML_{cc(F/V)}$	80:20	guaiacyl
$ML_{cc(R/R)}$	100:0	50:50
$ML_{cc(R/R)}$	88:12	50:50

[a] Courtesy of Academic Press, Inc.
[b] S = secondary wall region; ML = middle lamella; subscript cc = cell corners; F/F = fiber/fiber; F/V = fiber/vessel; R/R = ray/ray.
[c] tem = transmission electron microscopy; edxa = energy dispersive x-ray attachment.

Isolation and Purification

By virtue of its role in woody tissue, lignin is intimately associated with polysaccharides. Although many methods are available for the isolation and purification of lignin (Table 2), none satisfies the requirements of completeness, 100% yield, and structural authenticity. A compromise must be reached between yield and chemical modification.

Lignin preparations are often associated with both inorganic and organic impurities. Inorganic impurities are often peculiar to the particular isolation procedure and may reflect the composition of the reaction vessels used during the separation from wood. Organic impurities are mostly of carbohydrate or extractive nature. Purification employs partitioning lignin between different solvents or membrane filtration. Various methods are available for determining contamination; methoxyl content may be used as an expression of purity.

All lignins, regardless of method of isolation and purification, are polydisperse mixtures of molecules that differ in size and often in chemical structure. This heterogeneity has been established for industrial lignin sulfonates and kraft lignin.

Processing

Lignin exists as a polymer in wood. Chemical processing of wood usually aims at separating a liquid (soluble) lignin fraction from a solid, cellulose-rich fiber component. Delignification technologies have been widely used in industrial pulp and paper production. The removal of lignin from wood in the form of a water- or alkali-soluble derivative constitutes an important industrial technology. Its chemistry has been intensively researched during the past several decades.

Isolated lignins may be derived in different forms from several industrial sources, primarily pulping (chemical delignification) processes which employ acidic, neutral, or alkaline conditions.

Table 2. Isolation of Lignin[a]

Method	Reagents	Names	Approximate yield, % lignin	Selected property	Chemical characteristics
Lignin as residue					
acid hydrolysis of polysaccharides	H_2SO_4 HCl, HF, CF_3COOH	acid-hydrolysis lignin	>80	mostly insoluble	condensed, especially at elevated temperatures
oxidation of polysaccharides	$Na_3H_2IO_6$	periodate lignin	>80		some demethylation
hydrolysis/dissolution of polysaccharides	$NaOH–H_2SO_4/Cu(NH_3)_4(OH)_2$	cuoxam lignin	>80		some hydrolysis
Lignin by dissolution					
mechanical disintegration, also in combination with celluloytic enzymes	aqueous dioxane	MWL[b], CEL[c]	25–50	soluble polyether with high molecular weight	unchanged; some carbohydrate association
mild acid hydrolysis of lignin–carbohydrate bonds	alcohol, dioxane, phenol, or acetic acid with acid catalyst	organosolv lignin	50–<100	low molecular weight oligophenol, thermoplastic	hydrolytically degraded
steam hydrolysis of lignin–carbohydrate bonds	steam- and *in situ*-generated acetic acid	steam-explosion or autohydrolysis lignin		very similar to organosolv lignins	
alkaline hydrolysis of weak ether bonds	NaOH and Na_2S or catalyst	kraft or sulfate lignin	80	soluble in alkali, insoluble in most organic solvents	degradation and modification
Lignin derivative					
sulfonation	$SO_2–H_2O$ and base	sulfite lignin	80	high molecular weight, water-soluble	some hydrolysis, sulfonate groups
thioglycolation	$HSCH_2COOH–HCl$	thioglycolic acid lignin	≤100	alkali-soluble, low molecular weight	hydrolytic degradation, thioglycolic acid groups

[a] Courtesy of Walter de Gruyter.
[b] Milled wood lignin (MWL).
[c] Cellulolytic enzyme lignin (CEL).

Isolated lignins from industrial sources may differ greatly in their chemical composition and properties, which influences downstream polymer processing. This processing may be directed toward depolymerization to low molecular weight chemicals or liquid fuels, to the application of isolated lignins as fillers in various polymer systems, or to the use of chemically modified lignins as polymeric components in copolymers.

Chemically Modified Copolymers. All lignin used today for nonfuel applications is used in the form of chemically modified copolymers. This modification is sometimes a result of the particular isolation procedure (ie, from wood by sulfite pulping) or may be in the form of a chemically modified isolated polymer. Although an almost endless array of chemical modifications of lignin has been explored, only a few have achieved industrial application, mostly as a constituent of water-soluble polymeric systems. Polymeric lignin can be used in water-soluble polymeric systems as a surface-active and viscosity-control agent and in structural network polymers to enhance the amorphous modulus.

Economic Aspects

Most chemical wood-processing technologies deal with the chemical fractionation of wood into its structural components, cellulose (fibers), hemicellulose (or polyoses), and lignin. Since cellulose is by far the most important wood-derived chemical, the production of lignin is closely tied to pulp production. With pulp production at approximately 120 × 10^6 t/yr worldwide, the production of lignin can be safely estimated above 50 × 10^6 t/yr. Production figures for the United States are ca half of world production. Annual U.S. sales of lignin in the mid-1980s amounted to 700 × 10^3 t at a market value of \$200 × 10^6.

Although lignin is generated in concentrated, although not pure, form by pulp-producing industries, most remains unisolated. Only 2% of the lignin contained in the spent pulping liquors of pulp mills worldwide is separated for by-product purposes; the remainder is discarded or incinerated. The almost universal practice of spent-liquor combustion in kraft (alkaline) pulping employs lignin as the fuel needed for the recovery of inorganic pulping chemicals from dilute aqueous solution by evaporation.

The demand for lignin sulfonates has increased from ca 250 × 10^3 t in 1970 to ca 500 × 10^3 t in 1980. In 1984, lignin sulfonates ranked first among industrial surfactants. Because industrial lignin chemistry has developed out of the need to dispose of water-soluble lignin sulfonates from sulfite pulping, ca 95% of all lignin products today are lignin sulfonates and only ca 5% (or 35,000 t/yr) are derived from kraft pulping. However, the price differential between kraft lignin and lignin sulfonates is ca 3:1. The 23 application areas for lignin products by six major producers are all polymeric in nature. The potential size of U.S. lignin markets that might be developed by the year 2000 from biomass-conversion technologies has been estimated at ca 5×10^6 t/yr. Low molecular weight chemicals, obtained by degradative depolymerization, also hold economic promise.

Classification and Shipping

Isolated lignins are normally classified, according to their genetic and procedural origin, into lignin sulfonates (hardwood or softwood), kraft (sulfate or thio) lignin, milled wood lignin (MWL), etc. Other classifications may be based on method of purification and post-treatment.

For shipping purposes, lignin products are designated as lignin liquor or lignin pitch. Lignin products are shipped as free-flowing powders in bags or as solutions and suspensions of 40–55% solids content.

Uses

Lignin uses fall into five categories: dispersants, emulsion stabilizers, sequestrants, binders, and resin coreactants; fillers are a potential application (particularly in rubber tires). Uses involve water- or alkali-soluble products and all (except resin coreactants) are based on sulfonated derivatives, mainly lignin sulfonates.

BIBLIOGRAPHY

K. Freudenberg and A. C. Neish, *Constitution and Biosynthesis of Lignin,* Springer-Verlag, New York, 1968.

K. V. Sarkanen and C. H. Ludwig, eds., *Lignins: Occurrence, Formation, Structure and Reactions,* Wiley-Interscience, New York, 1971.

T. Higuchi, *Wood Res.* **66,** 1 (1980).

S. Saka and D. A. I. Goring, *Biosynthesis and Biodegradation of Wood Components,* Academic Press, Inc., New York, 1985, Chapt. 3, pp. 51–62.

WOLFGANG G. GLASSER
STEPHEN S. KELLEY
Virginia Polytechnic Institute and State University

LIGNOCELLULOSE.
See LIGNIN; WOOD.

LIGNOSULFONATES.
See LIGNIN.

LINEAR POLYETHYLENE.
See ETHYLENE POLYMERS.

LINEAR POLYMERS.
See STEREOREGULAR LINEAR POLYMERS.

LINOLEUM.
See FLOORING MATERIALS.

LINSEED OIL.
See DRYING OILS.

LINTERS.
See COTTON.

LIPOSOMES.
See VESICLES.

LIQUID CRYSTALLINE POLYMERS

Low molecular weight liquid crystalline (LC) compounds have been known for about 100 years. However, LC polymers have attained prominence only in the last 20 years, owing largely to the discovery of the aramids [eg, poly(*p*-phenyleneterephthalamide) (PPD-T), the polyamide from *p*-phenylenediamine and terephthalic acid] (**1**) and thermotropic polyesters [eg, copolymers of *p*-hydroxybenzoic acid (HBA), *p,p'*-biphenol, and terephthalic acid] (**2**), and the commercial development of strong fibers and unique plastics from these materials.

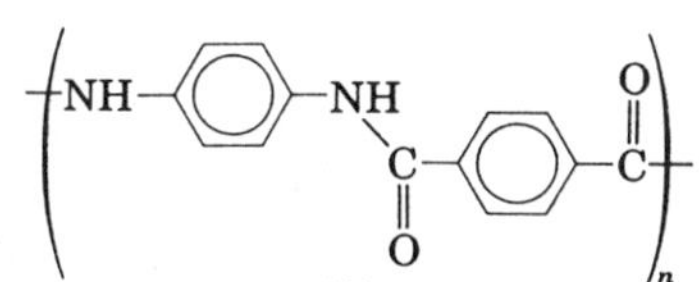

(**1**)

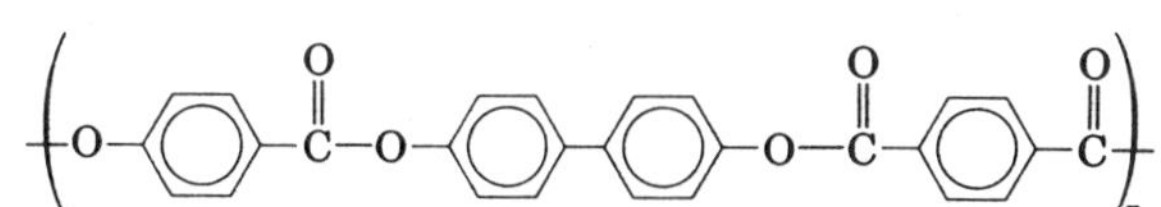

(**2**)

Such polymers may exhibit LC order either in solution (lyotropic liquid crystals) or in the melt (thermotropic liquid crystals). Specific structural features of LC polymers usually involve a succession of para-oriented ring structures to give a stiff chain with a high axial ratio (ratio of length of molecule to its width, aspect ratio) x (main chain LC polymers), or a rodlike ring containing side chains on, eg, vinyl polymer chains (side chain LC polymers).

The common structural feature of low molecular weight LC compounds is asymmetry of molecular shape, manifested either as rods characterized by a uniaxial order with an axial ratio usually greater than three or by thin platelets with biaxial order. Some illustrative structures follow:

Nematic:

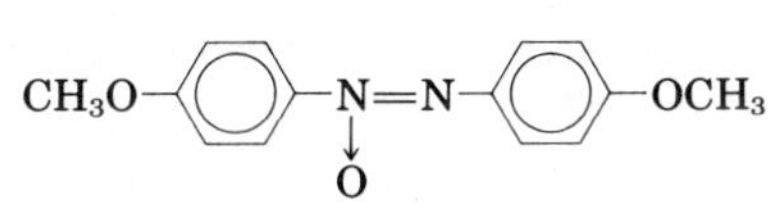

p-azoxyanisole

Smectic:

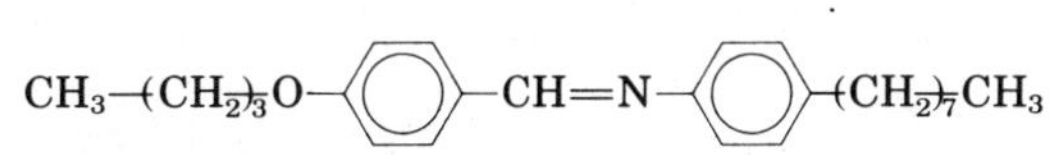

N-(*p*-butoxybenzylidene)-4-octylaniline

Cholesteric:

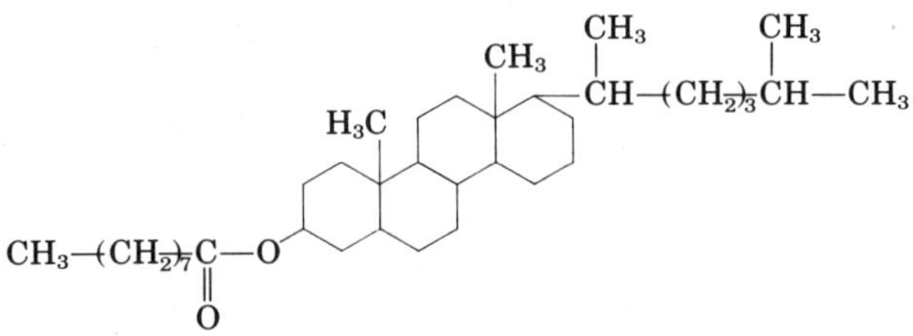

cholesteryl nonanoate

Discotic:

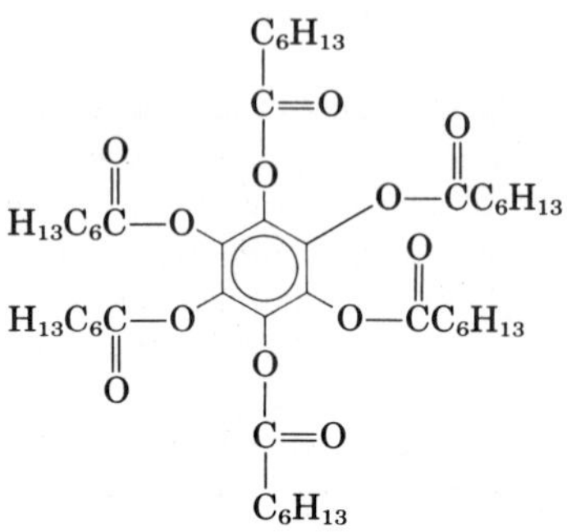

benzene hexa-*n*-heptanoate

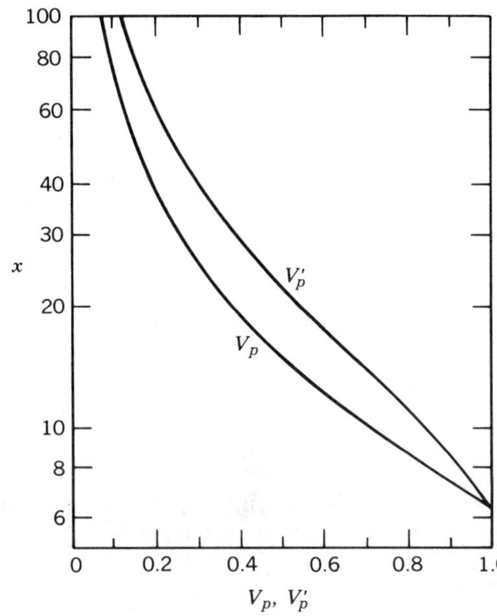

Figure 2. Compositions of coexisting isotropic and anisotropic phases expressed in volume fractions V_p and V_p', respectively, as functions of the axial ratio x of hard rods in athermal solutions. Courtesy of Springer-Verlag.

The types of ordered structures that LC materials form may be represented as shown in Figure 1. These ordered LC structures give birefringent solutions and melts and have been studied by optical methods with polarized light.

Theory

The formation of liquid crystals is solely a consequence of molecular asymmetry. It is not due to intermolecular attractions, but is dominated by intermolecular repulsions, the fact that units in two molecules cannot occupy the same space. There is, therefore, in the case of polymers, a limit to the number of rodlike chains that can be accommodated in random arrangement in solution, or to the axial ratio of chains that can exist randomly in the melt. When this limiting concentration is exceeded in solution, a liquid crystalline phase forms. When a critical axial ratio is exceeded in the melt, a liquid crystalline melt forms (Fig. 2).

In the separated LC phase, the molecules align parallel to each other in domains to accommodate the higher concentration or molecular weight. As overall concentration increases further, an additional ordered phase is formed at the expense of the random or isotropic phase until the volume fraction V_p' is reached. In melts, molecular weight differences alone account for LC phase formation due to a higher axial ratio of molecules in the anisotropic phase.

The Flory lattice theory is widely applicable for quantitative treatment of phase-separation phenomena in solutions and melts of rodlike polymers. Athermal solutions of hard rods of axial ratio x reach the threshold volume fraction V_p^*, resulting in the appearance of a stable anisotropic phase in accordance with

$$V_p^* \approx (8/x)[1 - (2/x)] \approx 8/x$$

at high values of x.

Since phase separation depends on both polymer volume concentration V_p and axial ratio x in solution, and axial ratio alone in the melt, rodlike polymers with a distribution of molecular weights (usual case) are partitioned between phases

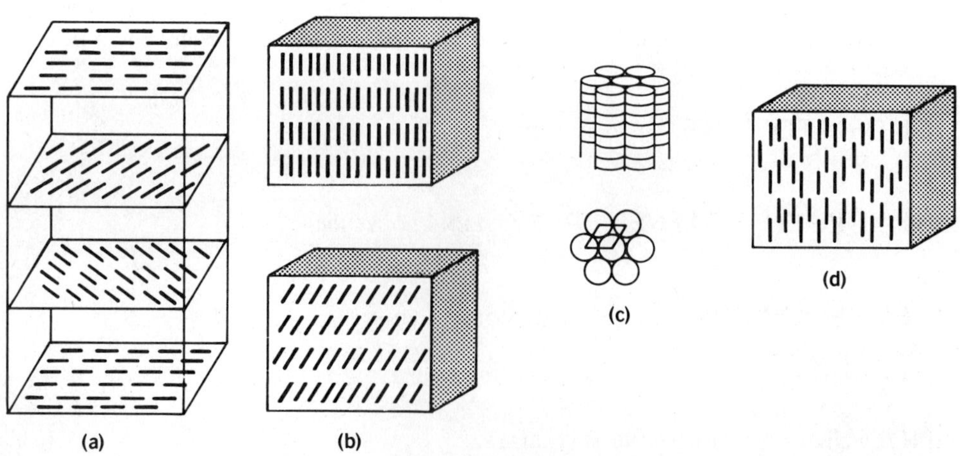

Figure 1. Mesophasic structure: (**a**) cholesteric; (**b**) smectic A (top) and smectic C (bottom); (**c**) discotic; (**d**) nematic. Courtesy of The American Chemical Society and Springer-Verlag.

with the higher molecular weight molecules preferentially populating the liquid crystalline phase.

Rodlike and Flexible Chain Mixtures. Theory predicts that anisotropic phases tolerate only a negligible concentration of flexible random-coil molecules, whereas isotropic phases tolerate a fair concentration of rodlike molecules. This obvious consequence of the lattice treatment method was confirmed experimentally. In rejecting the random-coil polymer, the LC phase shows a principal feature of crystalline materials in general, rejection of a foreign substance, thus justifying the term liquid crystal.

Semiflexible Chains. Most rodlike polymers are not truly rods. Even aramids, which experimentally verify Flory's theory for rods, depart slightly from linearity and, as molecular weight approaches infinity, can be considered semiflexible. Thus complete rodlike character is not required for LC phase formation. However, as flexibility increases, higher solution concentrations are required to achieve the critical V_p^* for phase separation. Therefore, for cellulose derivatives (Table 1), V_p (critical) ranges from 0.25 to 0.39, depending on the solvent and molecular weight M. These values are higher than for the stiffer aramids of comparable or lower M ($V_p < 0.1$).

Intermolecular Interactions. Most rigid polymers discussed here, eg, aramids, require strong polymer solvent interaction to achieve the concentrated solutions necessary for exceeding V_p^*.

Orientation-dependent Interactions and Semiflexible Chains. Polymer–polymer interactions are important in determining the stability of LC phases if these forces are orientation-dependent. Groups in polymer chains with polarizabilities higher along their bond axes than perpendicular thereto can contribute to the stability of the LC state. The interactions are calculated in terms of segmental contacts and are secondary and additive to the effect of molecular asymmetry as measured by chain extension or chain stiffness in promoting LC phase formation. The net effect of the interaction is to lower the critical axial ratio needed for LC phase separation. These interactions can be the dominant forces in low molecular weight compounds and play an important role in semiflexible long-chain polymers, particularly in the melt.

Monte Carlo lattice treatment demonstrates that semiflexibility alone does not lead to LC phase separation. It is necessary to invoke local orientation-dependent intermolecular interactions to obtain a nematic transition. In semiflexible chains, eg, poly(γ-benzyl-L-glutamate) (PBLG), that exhibit either rigid or coil-like configurations, formation of an anisotropic phase is accompanied by a pronounced induced rigidity in which the ordered phase contains chains substantially more rigid than those populating the coexisting isotropic phase.

Persistence Length. According to the Flory lattice theory, an axial ratio of only 6.42 (Fig. 2) is required for a polymer to be stiff enough to form an LC melt. In solution, a higher axial ratio is required. Molecular weight M is usually a lesser factor. For the achievement of good mechanical properties, the M of a rodlike polymer is always high enough to reach the axial ratio required for formation of an LC phase. Another measure of the rodlike character, or straightness, of a chain is the persistence length, the vector sum of the projections of the $(n + i)$th segments where $i = 0$ to ∞ on the nth segment (chosen at random) of a very long chain. The more rodlike a chain, the more the $(n + i)$th bond remembers the direction of the nth bond. The persistence length can be calculated from chain-conformation data or derived from measured polymer solution properties.

Lyotropic Systems

The important criteria to be satisfied for formation of a LC polymer phase in solution are that the polymer has a high

Table 1. Cellulose Derivatives Forming Lyotropic Solutions

Derivative	DS[a]	Solvents
Esters		
triacetate		chlorophenol–dichloromethane, *m*-cresol, dichloroacetic acid, difluoroacetic acid, formic acid–1,2-dichloroethane, trifluoroacetic acid, trifluoroacetic acid–dichloromethane
		acetic acid, acetone, aniline, *m*-cresol, dimethylacetamide, methyl acetate, nitromethane, 3-picoline, 4-picoline, pyridine, tetrachloroethane
		dichloroacetic acid, dichloromethane, trifluoroacetic acid, trifluoroacetic acid–dichloromethane
acetate	2.41	acetic acid, chloroform–formic acid, *p*-chlorophenol, dichloroacetic acid, diethylacetamide, dimethylacetamide, dimethyl sulfoxide, dioxane, ethylene sulfite (cyclic), phosphoric acid, trifluoroacetic acid
acetate–butyrate		acetic acid, dichloromethane–methanol, dimethylacetamide, formic acid, methyl ethyl ketone, phenol
nitrate		γ-butyrolactone, cyclopentanone, diethyl oxalate, dimethylacetamide, dimethyl carbonate, dimethyl cyanamide, dimethylformamide, dimethyl maleate, dimethyl sulfoxide, *N*-methyl-2-pyrrolidinone, 3-pentanone, pyridine
		acetone, 2-hexanone, methyl acetate, methyl ethyl ketone, 4-methyl-2-pentanone, methyl propyl ketone, 3-pentanone, *n*-pentyl acetate, *n*-propyl acetate
sulfate[b]	1.76	water
Ethers		
ethyl	2.2–2.4	acetic acid, carbon tetrachloride, chlorophenol–dichloromethane, *m*-cresol, dichloromethane–methanol, trifluoroacetic acid, trifluoroethanol
		chloroform, 1,1,2-trichloroethane
	2.5	acetic acid, benzene–methanol, chlorophenol–dichloromethane, *m*-cresol, dichloromethane, formic acid, phenol, toluene–methanol, trifluoroacetic acid
		acetic acid, dichloroacetic acid
hydroxymethyl		dimethyl sulfoxide

[a] DS is number of substituents per glucose unit, ie, degree of substitution (qv).
[b] Sodium salt.

Table 2. Polyamides That Form Liquid Crystalline Solutions

Diamine	Diacid	Solvent[a], %	Concentration[b], wt %
1,4-phenylene	fumaric	[c]	27[d]
chloro-1,4-phenylene	fumaric	96	31
4,4′-benzanilide	*trans,trans*-muconic	96	20–30
1,4-phenylene	mesaconic	100	25[e]
1,4-phenylene	*trans*-1,4-cyclohexylene	100.6	14.3
chloro-1,4-phenylene	*trans*-1,4-cyclohexylene	100.6	14.4
1,4-phenylene	1,4-dimethyl-*trans*-1,4-cyclohexylene	100.6	10
1,4-phenylene	2,5-pyridine	96	10
chloro-1,4-phenylene	2,5-pyridine	100	17
3,3′-dimethyl-4,4′-biphenylene	2,5-pyridine	100	14.6
1,4-phenylene	4,4′-stilbene	100	12
chloro-1,4-phenylene	4,4′-stilbene	99.6	20
1,4-phenylene	4,4′-azobenzene	97	12
4,4′-azobenzene	4,4′-azobenzene	99.9	18
1,4-phenylene	4,4′-azoxybenzene	101.5	12
4,4′-azobenzene	4,4′-azoxybenzene	101.2	12
1,4-cyclohexylene	4,4′-azobenzene	100	10
4,4′-azobenzene	terephthalic	96	12
3,8-phenanthridinone	terephthalic	99	>7
4,4′-biphenylene	terephthalic	101.0	12.9
4,4′-biphenylene	4,4′-bibenzoic	96	insoluble
1,4-phenylene	4,4′-bibenzoic	100.3	10.9
1,4-phenylene	4,4′-terephenylene	96	insoluble
1,4-phenylene	2,6-naphthalic	98.7	10
1,5-naphthylene	terephthalic	101.3	10
3,3′-dimethyl-4,4′-biphenylene	terephthalic	101	12
3,3′-dimethoxy-4,4′-biphenylene	terephthalic	101	12
3,3′-dimethoxy-4,4′-biphenylene	4,4′-bibenzoic	101	12

[a] Solvent is H_2SO_4 unless otherwise noted.
[b] Concentration is not the critical concentration; temperature is 25–30°C unless otherwise noted.
[c] FSO_3H.
[d] Above 50°C.
[e] Above 80°C.

persistence length (structure-dependent) and sufficient solubility to exceed the threshold volume concentration for LC phase separation (solvent, temperature-dependent). Polymers that form lyotropic solutions include polypeptides, eg, PBLG, aromatic polyamides (Table 2), polyoxamides and polyhydrazides (Table 3), poly(amide-hydrazide)s, polyesters, polyazomethines, polymers with heterocyclic units (Table 4), polyisocyanides, polyisocyanates, polyorganophosphazenes, metal polyynes, and cellulose and cellulose derivatives (Table 1).

Thermotropic Systems

Thermotropic systems require a melting point below that at which polymer degradation and decomposition occur. Many LC polymers and copolymers fulfill this condition because the stiffness requirement (Fig. 2) is much less severe in the melt. Aromatic polyesters are prime candidates because of their high thermal stability and high axial ratio or persistence length. The melting point may be depressed below the decomposition temperature by copolymerization, ring substitution, and use of flexible spacer groups between stiff mesogenic units while maintaining overall stiffness sufficient for LC formation. Examples are shown in Tables 5 and 6 to indicate the diversity of structure possible in main chain thermotropic polymers.

Polymers with mesogenic side chains usually are thermotropic. Low molecular weight mesogens are attached to backbone carbon chains or siloxane chains. Although the LC properties of the polymers parallel those of the low molecular weight analogues, the differences are due mainly to (*1*) high melt viscosity, which impedes structural rearrangements effected by external fields, and (*2*) the interaction of the polymer backbone chain, which tends to assume a random-coil configura-

Table 3. Liquid Crystalline Solutions of Poly(terephthalic hydrazide) in Aqueous Organic Bases

	Approximate solution range[a], wt %	
Base	Base	Polymer
tetramethylammonium hydroxide	3–30	6–27
tetraethylammonium hydroxide	7.5–25	7.4–26
diethylamine	10–15	7.5–9
piperidine	10	7.5–9

[a] Determined at 27°C on polymer with inherent viscosities of 3.2–4.5 dL/g.

Table 4. Lyotropic Solutions of Polyheterocyclic Compounds

Compound	Structure	Lyotropic solution
poly(1,4-phenylene-2,6-benzobisimidazole)	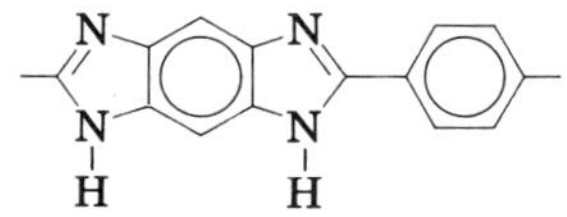	methanesulfonic acid
poly(1,4-phenylene-2,6-benzobisoxazole) (PBO)	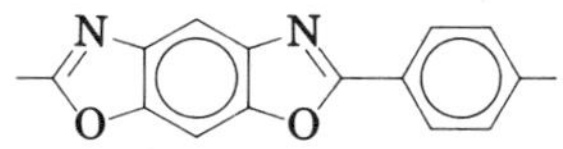	methanesulfonic acid chlorosulfonic acid 100% sulfuric acid
poly(1,4-phenylene-2,6-benzobisthiazole) (PBT)	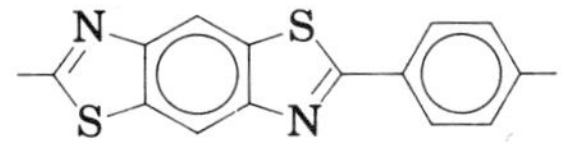	5–10% in polyphosphoric acid methanesulfonic acid
poly[2,6-(1,4-phenylene)-4-phenylquinoline]	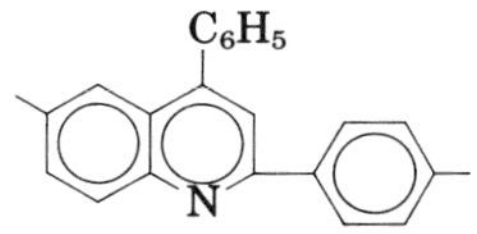	1.0–1.5% in *m*-cresol–di-*m*-cresyl phosphate
poly[1,1′-(4,4′-biphenylene)-6,6′-bis(4-phenylquinoline)]	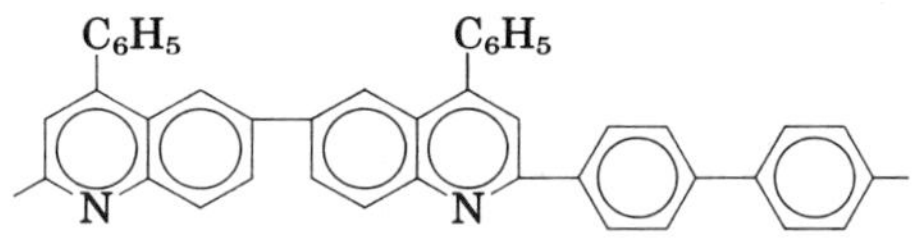	>9% in *m*-cresol–di-*m*-cresyl phosphate
poly[2,5(6)-benzimidazole] (AB–PBI)	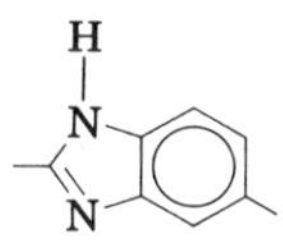	methanesulfonic acid
poly(1,4-phenylene-1,3,4-oxadiazole)	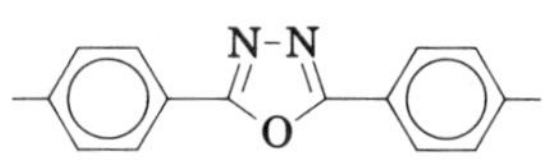	none

tion, and the mesogenic groups, which try to organize in LC phases. This interaction is reduced by inserting spacer groups between the backbone chain and the mesogenic groups. Both nematic and smectic melts may be obtained by variation of the polymer structure (Table 7).

Synthesis

Step-growth polymers are prepared by melt or low temperature polycondensation. Choice of reactants and their purity as well as polymerization conditions are quite critical for attainment of optimum high molecular weight. Single crystals with a unique morphology are made by solid-state polymerization of monomers to obtain the simplest AB aramid, poly(1,4-benzamide), and aromatic polyester, poly(1,4-benzoate).

Polymers with mesogenic side chains are prepared by polymerizing a monomer containing an activated vinyl double bond attached by way of a spacer group to a mesogenic group. Methacrylates and acrylates are convenient starting materials. Another method is by addition of a vinyl mesogenic compound to a silyl group on a preformed polymer.

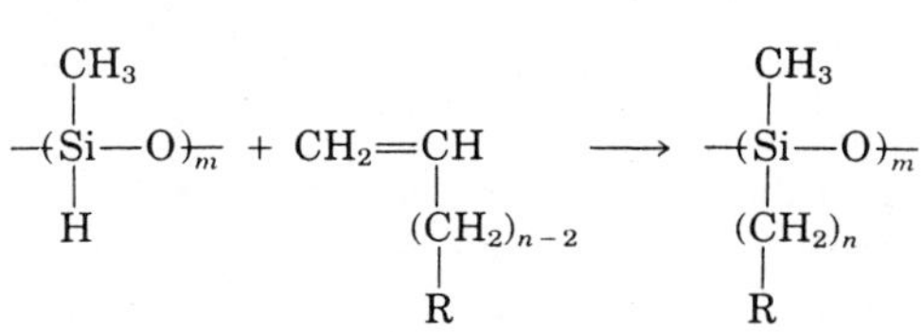

where R is a mesogenic group and $-(CH_2)_n-$ a spacer group.

Properties and Processing

Polymer Structure and Morphology. Thermotropic main-chain polymer melts generally are nematic. Phases may be identified with the aid of a polarizing microscope, but some polymeric mesophases do not exhibit recognizable textures. Mutual miscibility with known mesogens is at times useful for phase identification, as is x-ray diffraction.

Aramid fibers spun from lyotropic solutions and extruded moldings from thermotropic melts have a similar morphology. The structure of the solution or melt is frozen into the solid state. It consists of a hierarchy of structures from microfibrils (0.05 μm in diameter) to fibrils (0.5 μm) to macrofibrils (0.5 mm) in a highly oriented skin. The organization is much higher in the fiber.

Rheology and Processing. The rheology of LC systems is complex. The liquid crystalline melt shows a yield stress at low stress values and an initial rapid decrease in viscosity with increasing shear rate unlike poly(ethylene terephthalate) (PET), which is essentially Newtonian at low shear rates. Under normal processing conditions, therefore, the bulk viscosity of LC polyesters can be two orders of magnitude less than that of PET. The viscosity is a sensitive function of shear history as well. Because of the long relaxation time of the LC systems, a more fluid melt is formed by preshearing or initially heating to a higher temperature. In fiber spinning, a die swell near zero, which can be negative at lower temperatures, is observed. Fiber and molding properties depend markedly on processing conditions.

Table 5. Polymers Forming Thermotropic Melts

Structural units	Mole ratios	T_m, °C
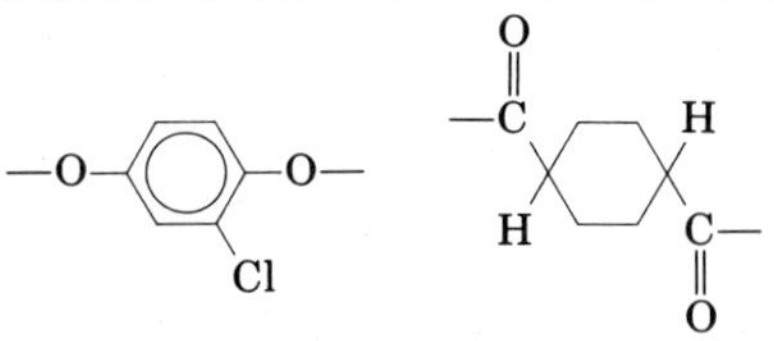 (trans)	1:1	305
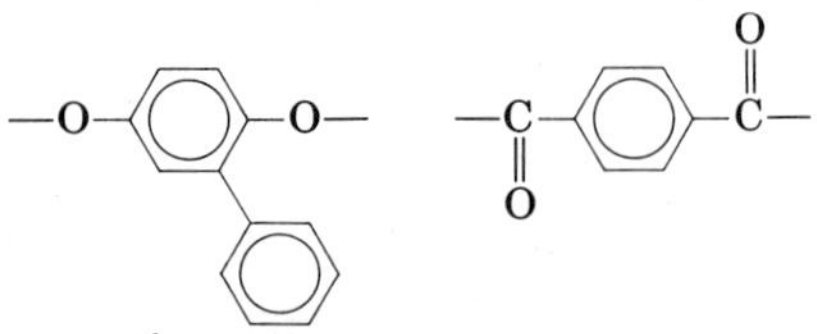	1:1	340
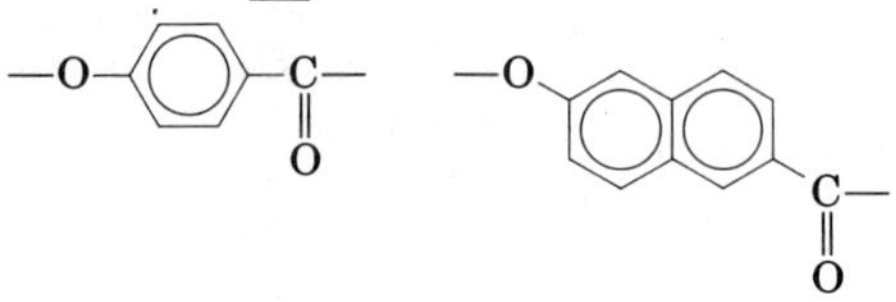	3:1	302
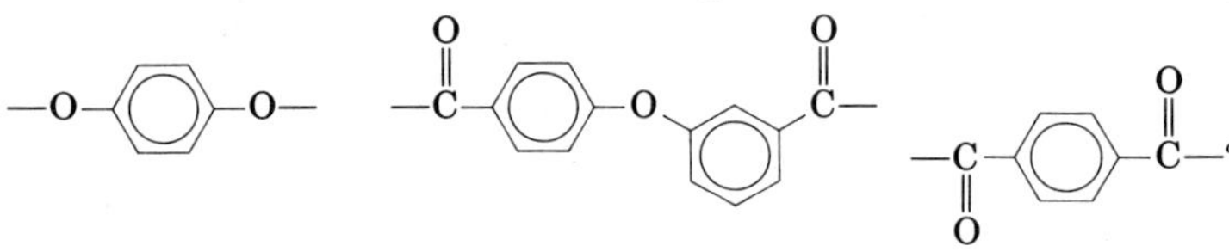 [a]	5:3.5:1.5	
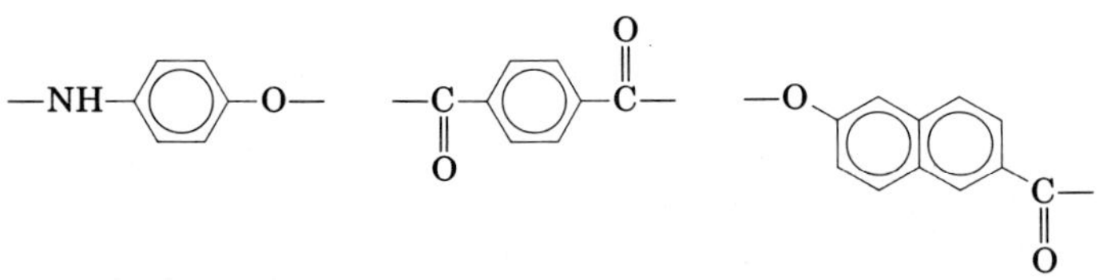	2:2:6	280
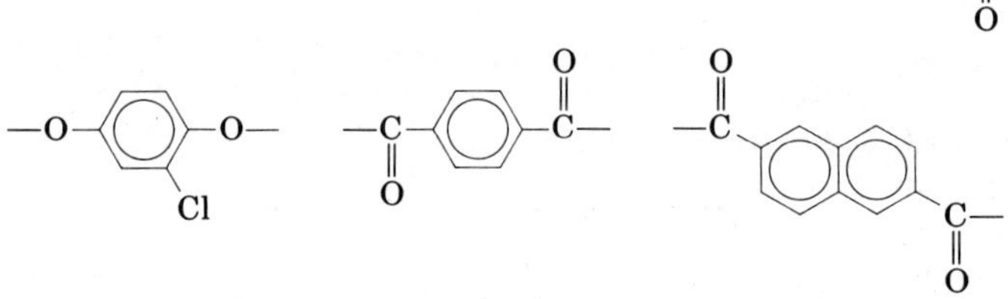	1:0.7:0.3	302
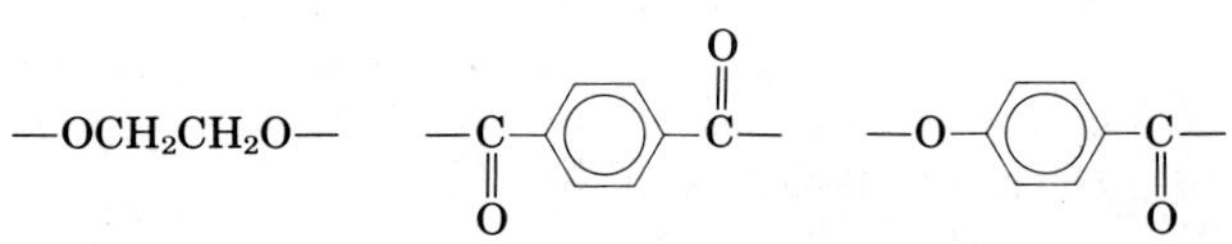	2:2:6	260
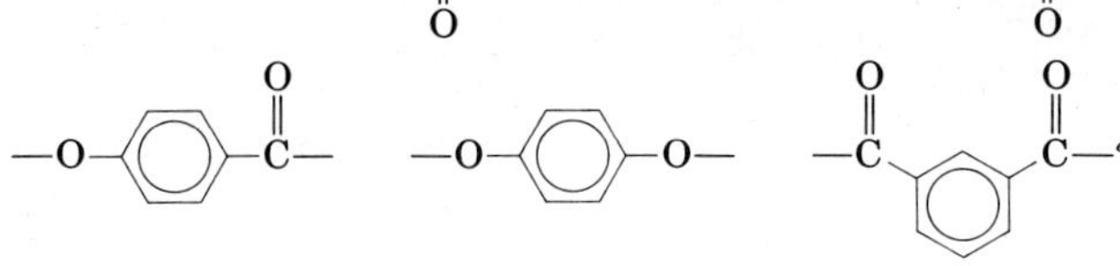 [a]	43:28.5:28.5	304
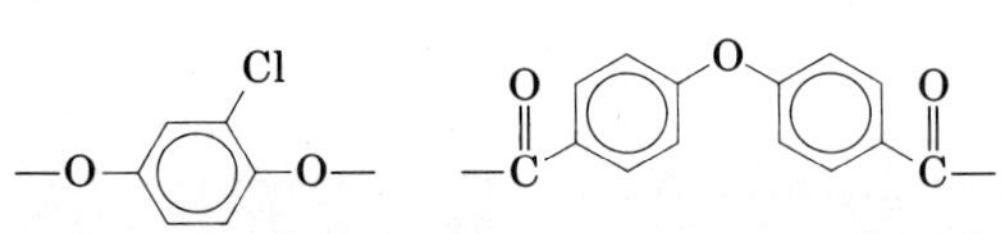	1:1	339
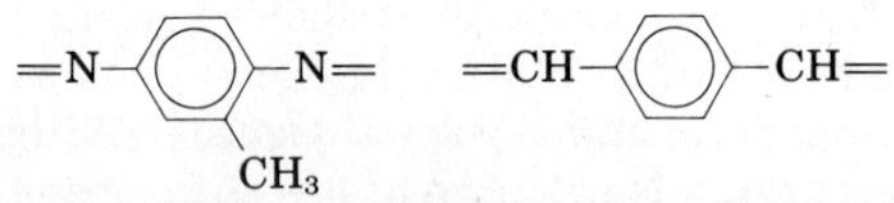	1:1	255

[a] If an extended conformation can be stabilized in the solid polymer, modulus is maintained despite a flexible or kinked unit.

Structure–Property Relationships. As few as two para-linked aromatic rings (axial ratio about three) suffice for a mesogenic unit in thermotropic copolymers. The mesogenic group may consist of a combination of aromatic and cycloaliphatic rings. The rings may be substituted by alkyl, aryl, or halogen groups, provided the substituents are not too large.

All types of LC phases have been observed in LC polymers, including cholesteric and discotic melt phases. At longer spacer lengths in both main-chain and side-chain LC polymers, the LC phases shift from nematic to smectic (Table 7). At even longer spacer lengths, liquid crystallinity may not appear.

Table 6. LC Polymers With Mesogenic and Spacer Groups

Mesogenic unit	Spacer unit	n
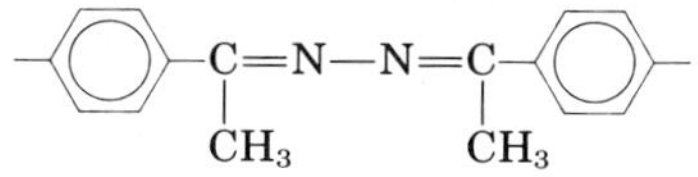	$-\mathrm{OC(=O)O}-(\mathrm{CH_2})_n-\mathrm{OC(=O)O}-$	6,8,10, 12
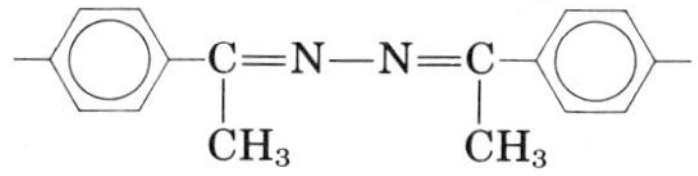	$-\mathrm{OC(=O)}-(\mathrm{CH_2})_n-\mathrm{C(=O)O}-$	8,10,12
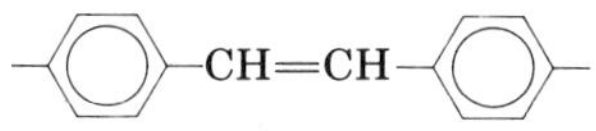	$-\mathrm{C(=O)}-\mathrm{O}-(\mathrm{CH_2})_n-\mathrm{O}-\mathrm{C(=O)}-$	5,6, 10, 4–10
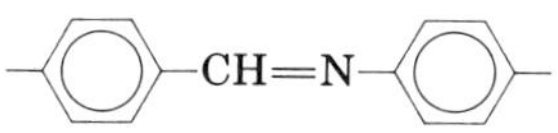	$-\mathrm{OC(=O)}-(\mathrm{CH_2})_n-\mathrm{C(=O)O}-$	10
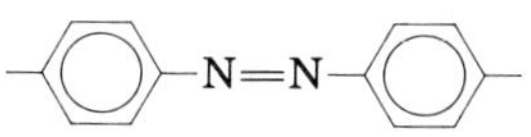	$-\mathrm{O}-(\mathrm{CH_2CH_2O})_n-$	2–4
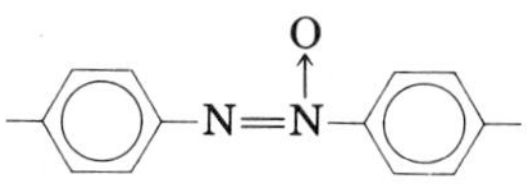	$-\mathrm{C(=O)}-\mathrm{O}-(\mathrm{CH_2CH_2O})_n-\mathrm{C(=O)}-$	2–4
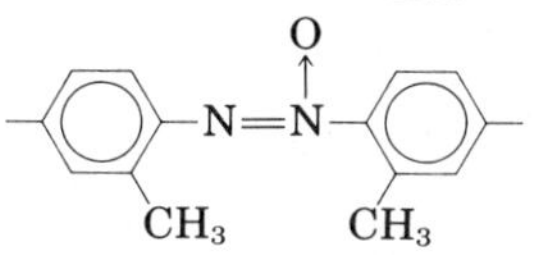	$-\mathrm{OC(=O)}-(\mathrm{CH_2})_n-\mathrm{C(=O)O}-$	10
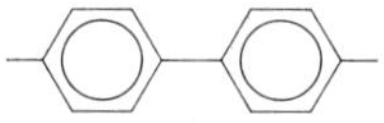	$-\mathrm{OC(=O)}-(\mathrm{CH_2})_n-\mathrm{C(=O)O}-$	5,6,8, 10,12
	$-\mathrm{C(=O)}-\mathrm{O}-(\mathrm{CH_2})_n-\mathrm{O}-\mathrm{C(=O)}-$	2,4–6, 10
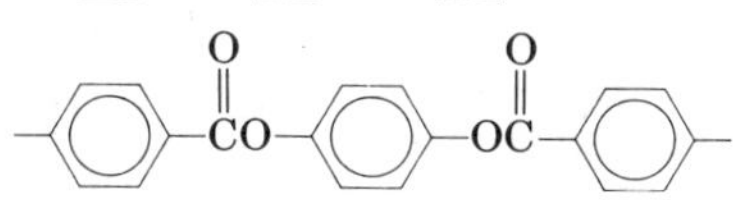	$-\mathrm{O}-(\mathrm{CH_2CH_2O})_n-$	4
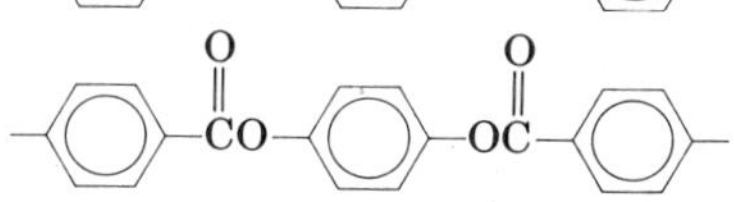	$-\mathrm{O}-(\mathrm{CH_2})_3-(\mathrm{Si(CH_3)_2O})_n-\mathrm{Si(CH_3)_2}-(\mathrm{CH_2})_3-\mathrm{O}-$	2,3,5
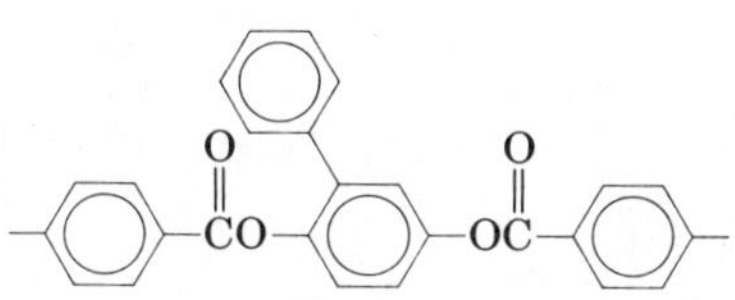	$-\mathrm{O}-(\mathrm{CH_2})_n-\mathrm{O}-$	10
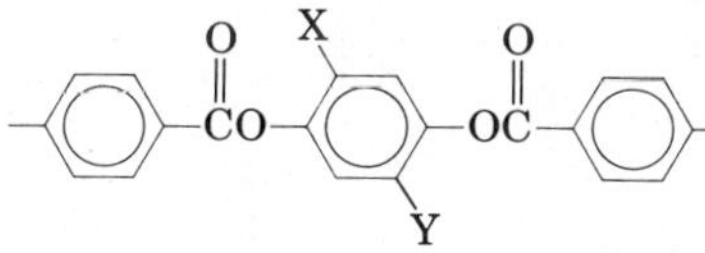 [a]	$-\mathrm{O}-(\mathrm{CH_2})_n-\mathrm{O}-$	10
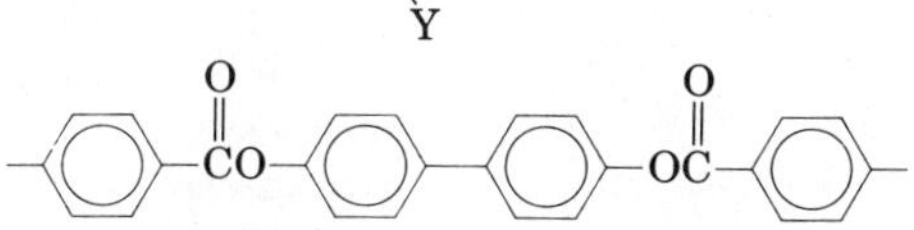	$-\mathrm{O}-(\mathrm{CH_2})_n-\mathrm{O}-$	9–11
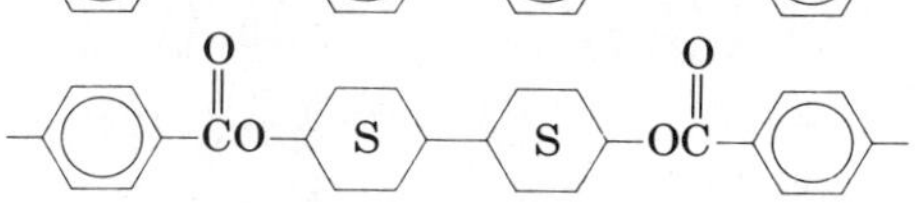	$-\mathrm{O}-(\mathrm{CH_2})_3-(\mathrm{Si(CH_3)_2O})_n-\mathrm{Si(CH_3)_2}-(\mathrm{CH_2})_3-\mathrm{O}-$	2,3,5

[a] X = H, Cl; Y = H, Cl, Br, CH_3.

Table 7. Smectic Polymers Obtained by Lengthening the Segments

Polymer	$\Delta \tilde{H}_{LC-i}$, J/g[a]	Phase transitions[b] at K
Segment A (end group)		
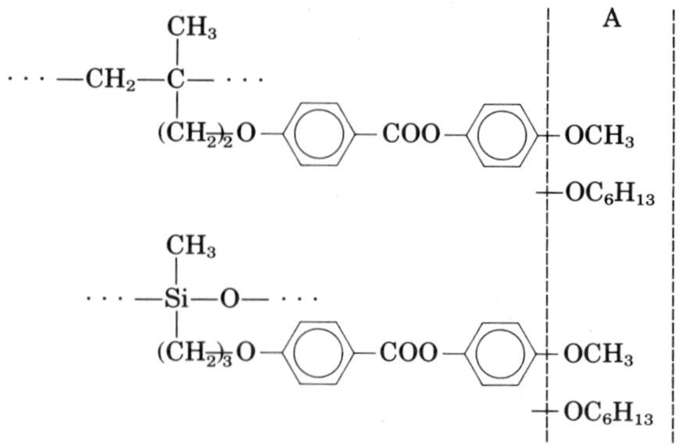		
—OCH$_3$ (C backbone, $(CH_2)_2$)	2.3	g 369 n 394 i
—OC$_6$H$_{13}$ (C backbone, $(CH_2)_2$)	11.5	g 410 s 451 i
—OCH$_3$ (Si–O backbone, $(CH_2)_3$)	2.2	g 288 n 334 i
—OC$_6$H$_{13}$ (Si–O backbone, $(CH_2)_3$)	11.6	g 288 s 385 i
Segment B (spacer group)		
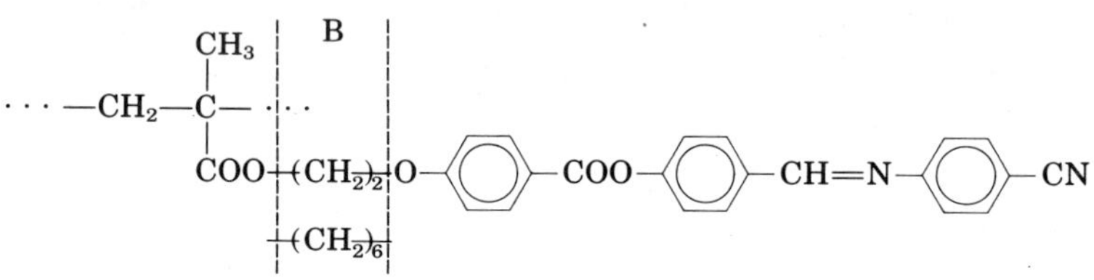		
$(CH_2)_2$		g 361 n 580 i
$(CH_2)_6$		g 324 s 607 i

[a] To convert J to cal, divide by 4.184.
[b] g = glassy; n = nematic; i = isotropic; s = smectic.

Table 8. Properties of Xydar and Vectra Resins

Property	Xydar[a]	Vectra[b]
tensile strength, MPa[c]	126	138–241
tensile modulus, GPa[d]	8.3	9.6–38
elongation, %	5	1.2–5.5
flexural strength, MPa[c]	131	151–296
flexural modulus, GPa[d]	13.1	8.9–34.4
Izod impact strength, J/m[e]		
notched	208	53–534
unnotched	454	
density, g/cm^3	1.35	
T_m or Vicat softening point, °C	358	275–330
heat-deflection temperature at 1.8 MPa[c], °C	337	180–240
coefficient of thermal expansion, ppm, °C		
flow		0–27
transverse		27–54
mold shrinkage		0.006
gamma radiation, 5 MGy		no effect
microwave radiation		transparent
flammability (UL 94)	V-0	V-0
limiting oxygen index, %	42	35–50
dielectric strength, V/μm, 1.6 mm	31	31–43
arc resistance, s	138	63–146
processing, stock temperature, °C	360–398	285–325
mold temperature, °C		50–150
production capacity, t/yr	10,000	230

[a] SRT-500; Dart and Kraft.
[b] Range of properties for several compositions; Hoechst-Celanese.
[c] To convert MPa to psi, multiply by 145.
[d] To convert GPa to psi, multiply by 145,000.
[e] To convert J/m to ft·lbf/in., divide by 53.38.

Applications

Fibers. LC polymers are used to make aramid fibers (Kevlar), poly(*p*-phenylenebenzobisthiazole) (PBT) fibers, aromatic polyester fibers, and fibers from polymers with alternating mesogens and flexible spacer units. These fibers often exhibit exceptionally high strength and tensile modulus. They have a variety of uses including rubber reinforcement, rigid composites for boat hulls and aerospace applications, protective clothing, ballistic vests, and for asbestos replacement in brake linings.

Films. Since aramids and other lyotropic polymers are spun from high boiling solvents, eg, dialkylamides containing salts, the process does not lend itself easily to film formation.

Thermotropic melt aromatic polyesters give films with superior tensile properties, especially in the machine direction. Values of tenacity and modulus are higher than for PET films. More balanced properties are afforded by two-way stretching, and by cross-lamination of films or sheets. Polymer films are useful in the information industry and as a coating for optical fibers. Toughness and low linear coefficient of thermal expansion are additional important properties.

Molecular composite films have been made from LC PBT and a more flexible type benzimidazole polymer. Properties are superior to those of LC thermotropic polyesters.

Molded Plastics. Because of their liquid crystallinity, Xydar and Vectra LC resins represent a new dimension in plastics (Table 8). They have superior tensile properties comparable to glass-filled PET, good electrical properties, and ease of processing. Melt viscosity is low and not sensitive to temperature at processing shear rates. Vectra uses standard injection-molding equipment; Xydar requires some modification because of a higher melting point. The latter has excellent high temperature performance. Transparency to microwaves and high heat-deflection temperature made Xydar highly suitable for microwave ovenware. Good electrical properties and precision moldability due to low mold shrinkage make LC polymers highly suitable for electrical and electronics applications. Chemical resistance is important in molded tower packing for chemical distillation columns where a single Vectra molding replaces several expensive metal parts.

Deficiencies include weak weld lines, poor compressive strength, and poor transverse properties caused by anisotropy. Fillers such as talc (50%), glass fiber, and metal powder improve these properties.

Information-storage Devices. Highly oriented LC polymer films with mesogenic side chains can be used to store information. For example, in the LC melt, held between two conductive glass plates, the side chains are oriented homeotropically by an electric field to produce a transparent film. The electric field is then turned off and a focused laser beam used to imprint the film by heating selected areas above the clearing temperature. These heated isotropic areas scatter light when the cooled film is viewed. The "writing" is stored indefinitely at temperatures below the polymer glass-transition temperature but may be erased by heating or applying an alternating electric field (see also OPTICAL DATA STORAGE).

BIBLIOGRAPHY

M. Gordon, ed., *Liquid Crystal Polymers I,* Vol. 59, and *Liquid Crystal Polymers II/III,* Vols. 60/61 of *Advances in Polymer Science Series,* Springer-Verlag, New York, 1984.

A. Ciferri, W. R. Krigbaum, and R. B. Meyer, eds., *Polymer Liquid Crystals,* Academic Press, Inc., New York, 1982.

H. Finkelmann, W. Helfrich, and G. Heppke, eds., *Liquid Crystals of One- and Two-dimensional Order,* Springer-Verlag, Berlin, 1980.

S. L. KWOLEK
Consultant
P. W. MORGAN
Consultant
J. R. SCHAEFGEN
Consultant

LITERATURE OF POLYMERS

Since the publication of the first edition of this Encyclopedia 20 years ago there has been a tremendous expansion of polymer literature, both in terms of increased volume of publications and new types of materials and services, as well as new methods of handling the increased mass of information. Books have continued to be published at the pace of the 1960s, with a large number of new titles each year, but the subjects covered have shown a noticeable trend toward narrower and more specialized fields, targeting specific-interest groups. This change is also reflected in new journals covering specific aspects of polymer science and engineering, instead of the entire polymer field as does the *Journal of Polymer Science.* Patents as part of published literature have gained prominence during this period. By means of computer and advanced data-processing methods, patent information has become a most effective tool in opening up a wealth of technical information. The patent literature is considered to be equal to books and journals as regards the value of information provided (see PATENT INFORMATION).

Another form of literature source that has gained widespread acceptance during this period is the database. By definition, a database is a collection of computer-readable records that can be searched rapidly to generate information requested. Most of the technical literature has been converted into the machine-readable form, on magnetic tapes or disks, and can be searched electronically on a computer. There are 2805 databases listed in the new edition of *Computer-readable Databases* (American Library Association, Chicago, Ill., 1985) available for lease, license, or purchase. Information can be retrieved from a database by subscribing or purchasing the magnetic tapes or disks from the database producer and processing them on an in-house computer, or by subscribing to the service of one of the numerous information vendors (614 listed in the above book) and accessing their database collection on a terminal or microcomputer via long-distance telephone; this is called online searching. This type of information retrieval, however, requires considerable skill and training and should be performed by professionals.

Conferences and Symposia

Conferences and symposia constitute an important source of polymer literature. Every year hundreds of papers are presented at meetings, providing up-to-date coverage of all fields in polymer science and engineering. These papers either report the latest findings or offer state-of-the-art surveys. Most meetings are organized by professional societies, eg, the American Chemical Society and IUPAC (International Union of Pure and Applied Chemistry), or trade associations, and most provide abstracts or preprints before the meetings. The well-known polymer preprints are: *ACS Polymer Preprints, ACS Polymeric Materials Science and Engineering Proceedings, IUPAC Symposium on Macromolecular Chemistry, IUPAC Microsymposia on Macromolecules,* and *Deutsche Kunststofftagung, Preprints.*

Full or modified papers of the conferences are sometimes published after the meeting as proceedings in one of several forms: a separate issue in a polymer journal or symposium series; the proceedings of the conference; or a book. The important polymer proceedings publications are: *ACS Symposium Series; ACS Advances in Chemistry Series; Journal of Applied Polymer Science, Applied Polymer Symposia; Journal of Polymer Science, Polymer Symposia Edition; Die Makromolekulare Chemie, Makromolekulare Symposia; Midland Macromolecular Monographs; Pure and Applied Chemistry; Society of Plastics Engineers;* and *Annual Technical Conference Papers.*

Many society-sponsored meetings are not published via an official channel. A number are edited and published as individual books by commercial publishing houses, such as Plenum Press, Academic Press, Inc., and others, or by the society itself. A considerable number of meeting papers are scattered among many journals around the world, and many never appear in book form. The best sources for locating published proceedings are: *Directory of Published Proceedings Series SEMT,* Interdock Corp., White Plains, N.Y.; *Index of Conference Publications Received by BLLD,* British Library Lending Division; *Index to Scientific and Technical Proceedings,* Institute for Scientific Information, Philadelphia, Pa.; and *Proceedings in Print,* Proceedings in Print, Inc., Manalapan, N.J.

Of equal importance is information on forthcoming meetings. In addition to the polymer and chemical journals, such as *Polymer News, Chemical and Engineering News (C&E News), Polymer Preprints, Plastics Engineering,* and *RAPRA Abstracts,* which carry calenders of meetings, a few general lists provide good coverage: *World List of Future International Meetings in Science, Technology, Agriculture and Medicine,* Library of Congress; *World Meetings: U.S. and Canada,* World Meetings Information Center, Chestnut Hill, Mass.; *World Meetings: Outside U.S. and Canada,* World Meetings Information Center, Chestnut Hill, Mass.; and *Scientific Meetings,* Special Libraries Association, New York.

Theses and Dissertations

Other sources of primary information are theses and dissertations, which are reports of research work undertaken in universities to fulfill the requirements of graduate degrees, namely, masters and doctorates. By tradition, this type of research deals with a subject or an aspect of a subject not previously investigated, and the information has never before been published. Some polymer theses and dissertations are published in journals and then by *Chemical Abstracts;* a significant portion, however, never appears in the open literature. However, such work can be located in several published directories. Although there is no specific listing for the polymer field, the general listings can be searched by subject. The U.S. directories are *Master Theses in the Pure and Applied Sciences Accepted by Colleges and Universities of the United States and Canada* (Plenum Press, New York, 1957–); *Master Abstracts* (University Microfilm International (UMI), 1962–); *American Doctoral Dissertations* (UMI 1933/34–); *Dissertation Abstracts International* (UMI, 1938–); and *Comprehensive Dissertation Index (CDI)* (Xerox UMI, 1973).

Theses of foreign origin can be located in the following national lists: *Index to Theses Accepted for Higher Degrees in the Universities of Great Britain and Ireland* (ASLIB, London, 1953–); *Catalogue des Theses de Doctorate Soutenues Devant les Universités Françaises* (1984/85–); *Jahresverzeichnes der Deutschen Hochschulschriften* (VEB Verlag für Buch und Biblio Thekswesen, Leipzig); *Katalog Kandidatskikh i Doktorskikh Dissertatsii Postupivshikh v Biblioteku imeni V. I. Lenina i Gosudarstuennuyu Tsenytral'nuyu Nauchnuyu Medisinskuyu Biblioteku* (Moscow, 1958–); *Canadian Theses* (National Library of Canada, Ottawa, 1961–); *European Dissertation Abstracts* (Foundation for European University and Research Documentation, Bern, Switzerland); and *Dissertations Abstracts International* (Section C, European Abstracts, UMI). For research work currently in progress on campuses, the best source is the *ACS Directory of Graduate Research.*

Patents

The primary source of most patents is the weekly gazette of countries that announce either the granting of legal protection to inventions, as in the United States, or the filing of unexamined patent applications, as in the quick-issue countries, eg, FRG, France, Japan, etc. These weekly disclosures are the earliest records of new technology. As rights of invention apply only to the country issuing the patent, patent applications are usually sent to many countries for legal protection. This has led to the worldwide multilingual coverage of patent literature. The tremendous task of procuring and tracking the huge volume of multilingual records has given rise to commercial patent services that provide abstracts and indexes in English for the weekly patent records reported in the world. In the polymer area, the principal patent services are Derwent Publications *World Patent Index Gazette (WPI), Central Patent Index (CPI), World Patent Abstracts (WPA), Alerting Bulletins, Basic Abstracts Journal,* and *CPI Section A: PLASDOC; Chemical Abstracts (CA);* IFI/Plenum Data Co. (*IFI Comprehensive Database to United States Chemical Patents, Weekly Patent Profile, Patent Intelligence and Technology Report,* and *IFI Assignee Index*); and International Patent Documentation Center (INPADOC) (*Patent-Family Service* and *INPADOC Patent Gazette (IPG).*

Abstracts and Indexes

Abstracts and indexes are the best substitute for original documents that are not readily available in many cases. They cover a multitude of sources and subjects and are quickly delivered to the readers a few weeks to a few months after the first publication date. Although some services do not supply abstracts, they usually provide bibliographic data to the original materials, and many can be searched retrospectively through their cumulative indexes. In the polymer field, the most important abstracting service is *Chemical Abstracts (CA)*, published by the Chemical Abstracts Service, a division of the American Chemical Society. The *CA* database is available for online searching through the commercial systems DIALOG (Lockheed), SDC (System Development Corp.), BRS (Bibliographic Retrieval Service), and STN International (formerly CAS Online). Most industrial and academic libraries now access these databases, although the CA printed form is still the most heavily used on campuses.

On par with *CA* in terms of size and coverage, but not of usage, are the *Refretivnyi Zhurnal Khimicheskoe (RZK)* (in Russian) and the *Bulletin Signaletique (BS)* (in French). *RZK* is published by the All-Union Institute of Scientific and Technical Information, USSR Academy of Sciences, Moscow. *Bulletin Signaletique Polymère* (discontinued in 1983), like the *CA* Macromolecular Section, is one of the 50 sections of the *BS* published by the Centre National de la Recherche Scientifique, Paris. On a smaller scale is the *German Literatur-Schneldienst Kunststoffe, Kautschuk, Fasern,* published monthly by Deutsche Kunststoff-Institut, Darmstadt, FRG, and *RAPRA Abstracts* published by the Rubber and Plastics Research Association, London.

Reviews

The high volume of published information in polymer science and technology has created a strong demand for a compressed and digested form of literature by which scientists can keep abreast of their field of interest. Reviews serve an important function in supplying critical evaluation of literature in each area. They are published in hardbound editions, review series, review journals, and as articles scattered in different journals. The principal hardbound polymer review series are *Polymer Reviews* (Wiley, New York, 1958–); *Macromolecular Chemistry,* specialist periodical report (Royal Society of Chemistry, London, 1980–); *Advances in Polymer Science* (Springer-Verlag, Berlin, 1958–); Developments Series (Applied Science, London); and *Polymer Yearbook* (Harwood Academic, New York).

The following journals are devoted to review papers only: *Journal of Polymer Science, Macromolecular Reviews* (Wiley, New York); *Journal of Macromolecular Science [Part C]: Reviews in Macromolecular Chemistry* (Dekker, New York); *Polymer-Plastics Technology and Engineering* (Dekker, New York); *Polymer Engineering Review* (Freund, London); *Progress in Polymer Science* (Pergamon, Elmsford, N.Y.); *Progress in Colloid and Polymer Science* (Steinkopff, Darmstadt, FRG); *Progress in Organic Coatings* (Elsevier Sequoia, Lausanne, Switzerland); *Reports on Progress in Polymer Physics in Japan* (Fakujutsu Bunken Fukyukai, Tokyo); *Rubber Review,* annual issue on *Rubber Chemistry and Technology* (ACS Rubber Division, Akron, Ohio).

In addition, review articles on polymers are published in numerous polymer and general chemical journals. The best source is the *CA Review Index,* which contains between 25,000 and 30,000 reviews a year, and can be searched online via the DIALOG system.

Journals

Polymer journals have a relatively short history; very few titles go back as far as the 1940s, even though polymer literature had been published long before that in scientific and engineering publications. Today, the information explosion has produced hundreds of periodicals devoted to various aspects of polymer science and technology, but a significant number of polymer papers are still scattered among journals in the fields of chemistry, physics, and materials science and engineering. Most basic polymer journals were started in the 1960s, a period during which many new journals were published, such as *Macromolecules* (1968), *Journal of Macromolecular Science* (1967), *Polymer* (1961, UK), *Polymer Journal* (1970, Japan), *European Polymer Journal* (1965), and when existing journals, such as *Die Makromolekulare Chemie* (1943) and *Journal of Polymer Science* (1946), underwent substantial structural changes to provide increased specialization by division and subdivision. The *Journal of Applied Polymer Science* was spun off from the *Journal of Polymer Science* in 1959 and *Angewandte Makromolekulare Chemie* from *Die Makromolekulare Chemie* in 1967. In 1963, the *Journal of Polymer Science* was divided into three parts: *General, Polymer Letters,* and *Polymer Symposia.* In 1966, *Part A, General* was further subdivided into the *Polymer Chemistry* and *Polymer Physics* editions. *Applied Polymer Symposia* was added to the *Journal of Applied Polymer Science* in 1965, and *British Polymer Journal* was spun off from the *Journal of Applied Chemistry* in 1969. These numerous changes reflect a time of intense activity in the polymer industry as a whole.

Books

General Reference Works. The most familiar desk-top reference book is the *Polymer Handbook* (J. Brandrup and E. H. Immergut, eds., 3rd ed., Wiley, New York, 1989). It is an invaluable single-volume compilation of constants, properties, and tables of synthetic polymers and derivatives. *Modern Plastics Encyclopedia* (McGraw-Hill, New York) is an annual reference work that provides reviews on materials and processes in addition to properties and technical data. The *Encyclopedia of Plastics, Polymers and Resins* (M. and I. Ash, Chemical Publishing Co., New York, 1982–1984, 3 vols.) is a dictionary of polymer and plastics trademark products, containing information on composition, general, mechanical, and thermal properties, curing conditions, and toxicity and handling. Among encyclopedias, *Kunststoff-Handbuch* (R. Vieweg and L. Goerden, eds, Carl Hanser, Munich, 1963–1974, 12 vols., in German) is a comprehensive work covering preparation, properties, and applications of a wide range of polymers. The annual *Encyclopédie Française des Matières Plastiques* (Les Publicateurs Techniques Association, Paris, in French) is comparable to the *Modern Plastics Encyclopedia.* The *Enciklopedia Polimerov* (W. A. Kargin, ed., Sovietskaya Enciklopedia Publ., Moscow, 1972, 3 vols., in Russian) is less known in this country because of the language barrier. And, of course, there is the 19-volume *Encyclopedia of Polymer Science and Engineering.*

BIBLIOGRAPHY

J. T. Lee in S. M. Lee, ed., *Encyclopedia of Composites,* VCH, Inc., New York, 1989 or 1990.

J. E. Lotz in M. Grayson, ed., *Kirk-Othmer Encyclopedia of Chemical Technology,* 3rd ed., Vol. 16, Wiley-Interscience, New York, 1981, pp. 889–945.

R. E. Maizell, *How to Find Chemical Information,* 2nd ed., Wiley-Interscience, New York, 1987.

Y. Wolman, *Chemical Information,* John Wiley & Sons, Inc., New York, 1983.

JULIA T. LEE
Michigan Molecular Institute

LITHOGRAPHIC RESISTS

Most commercial printing is accomplished with lithographic materials based on simple inking–noninking systems. A polymer image is formed by masked exposure to light, followed by the dissolution of the unwanted polymer. The polymer adsorbs ink and the substrate rejects it. The ink image is then transferred to paper. For many applications, however, inking selectivity is not sufficient for image transfer. A relief image must be created on the plate by etching the uncovered areas of the substrate. The developed polymer image protects underlying areas during etching. These imaging polymers are called resists because of their resistance to etching.

Historically, lithography has been associated with the printing industry. However, the manufacture of printed circuit boards, microelectronic chips, and other photofabricated products also involves the use of photosensitive polymers to form images. Each industrial application requires a specific set of polymers, sensitizers, and developers, but the fundamental concepts of image formation and transfer remain the same. The process depicted in Figure 1 may be used to create a printing plate, a copper circuit pattern connecting various electrical components, or, on a microscopic scale, may be used to create the gates and interconnects of a semiconductor microcircuit.

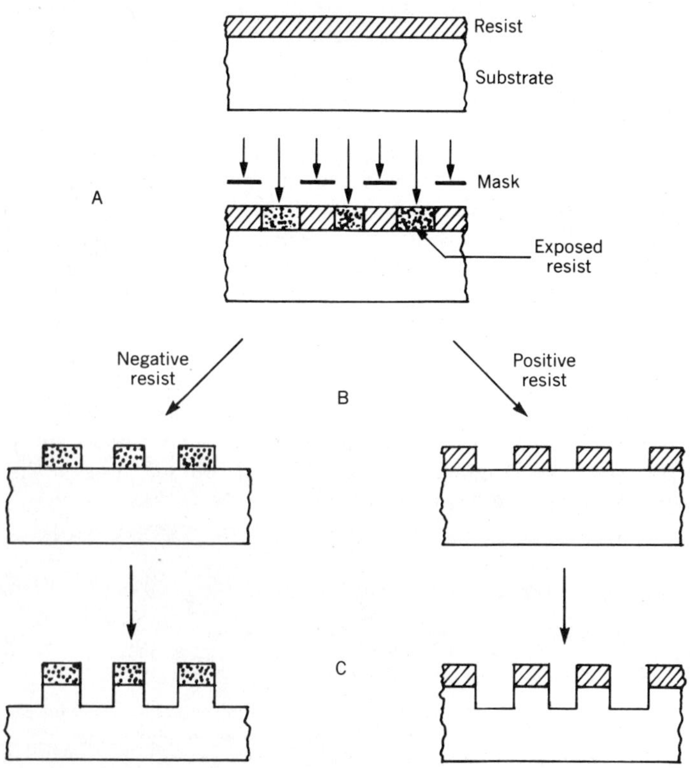

Figure 1. Negative and positive imaging: A, exposure; B, development; C, etching.

Image formation involving polymers usually requires changing the solubility of the exposed polymer. In negative systems, where the image is the reverse of the mask or template, the exposed area becomes less soluble or insoluble in the developer. Solubility is usually reduced by photocross-linking reactions between polymer chains or by the polymerization of the monomers in the coating (see also PHOTOPOLYMERIZATION). With a positive photoresist the image is the same as on the mask. Following exposure, the exposed area is more soluble and is removed during development. Increased solubility can be the result of polymer degradation, or the photochemical formation of a more soluble chemical group in the resist. The common methods for producing polymer relief images are given below.

Negative images: exposed areas become insoluble
photocross-linking, including dimerization
photoinitiated chain-growth polymerization
photocondensation polymerization
photofunctional change

Positive images: exposed areas become soluble
photofunctional change
polymer degradation

A lithographic resist is a carefully formulated mixture of components, each of which has specific functional requirements. The total system includes coating solvents, polymers, energy-sensitive components, and developers, as well as an expensive array of equipment to expose and process the images in a controlled environment. Depending on the application, resists may transfer their developed image via ink as in printing, or may be used as a mask for the acid etch of copper in circuit boards, or, in microelectronics, as a mask during ion implantation or plasma etching. Each of these image transfer techniques requires a unique resist material optimized for the requirements of its application. Historically, most commercial lithographic applications have employed negative resists. The current trend is toward the use of positive resists to achieve improved resolution and to reduce dependence on organic solvents and developers.

Two photochemistries have dominated microresist applications. Bisazides (and now monoazides) are the basis for almost all negative near-uv resists. On exposure, nitrenes are formed that react with the binder polymer (typically cyclized polyisoprene) to cross-link the exposed regions. Development with an organic solvent leaves the cross-linked polymer as the image of the mask. Positive systems take advantage of the photoreaction of naphthoquinone diazides to form indene carboxylic acids. In the unexposed areas, the naphthoquinone diazide acts to inhibit dissolution of the novolac binder in an aqueous base developer. The exposed areas, containing the acid, are more soluble, and their removal results in a positive image.

While these two photochemical systems remain the mainstay of the industry, they are slowly giving way to other resist systems that are capable of either finer resolution or better resistance during device fabrication. Two major areas of research in the past few years have been the synthesis of polymers tailored to high energy exposure sources, or the creation of resists and other coatings used in a multilayer approach.

Various high energy resists have been optimized for use with electron beams, ion beams, x rays, or excimer lasers, and have demonstrated better resolution (smaller geometries) than can be achieved with conventional near-uv polymers because they are not limited by the diffraction of light at submicrometer dimensions. Multilayer schemes are needed because a single resist layer is incapable of simultaneously achieving all the process requirements of surface smoothing, control of critical dimensions, and resistance to chemical or plasma etching at resolutions of 1 μm or less. A multilayer system can address each of these concerns separately with appropriate materials. Both approaches (high energy and multilayer) have demonstrated resolutions of well below 0.5 μm.

Electron-beam resists are frequently used for making fine resolution master masks for microlithography, and are commercially available. However, the long times required to "write" the device circuitry with a beam of electrons has precluded its widespread use in manufacturing. The most common *e*-beam materials are poly(butene sulfone) (positive resist) and COP (a copolymer containing glycidyl methacrylate used as a negative resist). Systems using methyl methacrylate and its analogues are widely studied, but are generally too slow or too poor in chemical resistance for industrial applications. Many of these *e*-beam resists are also used for x-ray and ion-beam lithography. For x-ray resists, certain elements are frequently added to absorb the characteristic x rays from a particular source. Ion-beam and excimer lasers are still in the initial stages of development for microlithographic applications. These tools can produce images by the normal route, but are also capable of making ablative images, that is, the exposed areas are literally exploded from the surface of the device. Wet chemical development is not required to remove unwanted polymer. Ablative exposure mechanisms require less specific polymers to work, and the excimer laser, in particular, may have a large impact on the future direction of microlithography.

The second general approach to achieving smaller geometries is via multilayer systems. Multilayer schemes generally begin with a thick layer to planarize the surface of a device that may already have several layers of features on it. The photosensitive resist coated on top of this smoothing layer can be very thin and uniform, and thereby capable of high resolution. The key to multilayers is finding a development approach that will transfer the image exactly through both the imaging and the nonimaging layers to the surface of the device. This often requires a combination of exposures in the near and deep uv to harden the resist layers, or the use of a silicon containing layer that is then used as a mask in an oxygen plasma etch. Other possible layers in a multilayer scheme include an antireflection coating (ARC) and a contrast enhancing layer (CEL). Although they operate in very different ways, both ARCs and CELs contain dyes that are intended to improve the contrast of the exposed image, and thereby enhance the resolution of the device features.

Combinations of improved exposure tools and advanced resist chemistry will ensure the use of optical lithography well into the next decade. The use of photopolymers is expanding in the areas of microlithography and printed circuit boards. Growth in these areas is driven by the need for high resolution, high performance materials. Only moderate expansion in the use of photopolymers for photomachining is anticipated. The printing industry will continue to expand its use of high production flexible polymer printing plates.

BIBLIOGRAPHY

L. F. Thompson, C. G. Willson, and M. J. Bowden, *Introduction to Microlithography,* ACS Symposium Series, 219, American Chemical Society, Washington, D.C., 1983.

W. S. DeForest, *Photoresist Materials and Processes,* McGraw-Hill, New York, 1975.

W. Schnabel and H. Sotobayashi, *Prog. Polym. Sci.* **9,** 297 (1983).

E. A. Chandross, E. Reichmanis, C. W. Wilkins, Jr., and R. L. Hartless, *Solid State Technol.* **24,** 81 (1981).

G. N. Taylor, T. M. Wolf, and L. E. Stillwagon, *Solid State Technol.* **27,** 145 (1984).

R. W. Blevins
R. C. Daly
S. R. Turner
Eastman Kodak Co.

LIVING POLYMER SYSTEMS

The term living polymers is often used to describe systems in which active centers remain after complete polymerization, so that a new batch of monomer subsequently added will add to the existing chains and increase their degree of polymerization (qv). These systems are well-documented in anionic polymerization (qv). Indeed, in the article of that name living polymerization is treated as a normal feature of Anionic polymerization. In living polymerization the mechanism is comprised of chain initiation and chain propagation with no termination reactions [or chain transfer (qv)]. Each initiator molecule produces one growing chain, so that the degree of polymerization at the end of polymerization will simply be M_0/I_0 if initiation occurs by addition to the monomer, or $2M_0/I_0$ for radical anion initiation, where M_0 and I_0 are the initial monomer and initiator concentrations. If the rate of chain initiation is rapid compared to that of chain propagation, then all chains start at the same time and, apart from a small statistical factor, add monomer equally. Highly monodisperse polymers (qv) with a Poisson distribution of molecular weights form.

More recently, living systems other than anionic have been developed. This article presents various living polymerizations, classified largely according to the nature of the propagating species. The first section begins with a brief discussion of the requirements for a truly living system.

Living Cationic Ring-opening Polymerization

According to the original Szwarc definition several criteria must be met simultaneously in order for a system to be called living. In the living system, active species are formed relatively fast (compared to their further propagation) and most of these species stay active until the monomer is consumed. This description, translated into the language of measurable quantities, requires that

1. $k_i \geq k_p$, ie, the rate constant of initiation is at least as large as the rate constant of propagation;
2. $[C^*]$ = constant, ie, the concentration of active species is constant; the rate constants of termination and transfer are close to zero (k_t, $k_{tr} \sim 0$).

These requirements, if simultaneously met in the studied system, have the following consequences:

1. $\ln [M]_0/[M]_t = k_p[C^*]t$, ie, the polymerization follows first-order kinetics with respect to monomer concentration;
2. $\overline{DP}_n = ([M]_0 - [M]_t)/[C^*]$, ie, the number-average degree of polymerization is proportional to the monomer conversion;
3. The molecular weight distribution corresponds to the Poisson distribution.

If these points are strictly observed, there is hardly any system in the cationic field (both vinyl and ring opening) qualifying as living polymer. The closest candidates are cyclic amines described in the next section. In systems involving these monomers initiation is fast, propagation is irreversible, transfer and termination are virtually absent.

Systems with Irreversible Polymerization, Fast Initiation, and No Transfer or Termination. Chain transfer to polymer, a general phenomenon in cationic ring-opening polymerization, may be eliminated when the polymer chain is substituted with bulky groups, providing steric hindrance to the reaction of the heteroatom in the chain with active species (both intermolecularly and intramolecularly). This idea, coupled with the high nucleophilicity of the nitrogen atom, resulting in high initiation rate and relatively high stability of the active species (ammonium cations), makes some cyclic amines obvious candidates for living cationic ring-opening polymerization.

N-t-Butylaziridine, 1,3,3-trimethylazetidine, and conidine provide living systems. They offer the advantage of the high ring strain of three- and four-membered rings, resulting in irreversible polymerization. Thus there is no MWD broadening due to the reversibility of propagation. Because of high nucleophilicity of the heteroatom, initiation with a number of initiators is fast as compared to propagation; the system may still be simplified by reaction of monomer with an equimolar quantity of initiator, $(C_2H_5)_3O^+BF_4^-$, isolation of the corresponding ammonium salt, and use of it as initiator. In such a case $k_i \approx k_p$. This approach has been used for the polymerization of 1,3,3-trimethylazetidine and conidine. Because of the stability of active species, the quaternary ammonium ions, the systems are not sensitive to impurities. Polymerizations may be carried out even in methanol as a solvent and are not affected by water. The quaternary ammonium ions do not recombine with counterions (anions), thereby avoiding termination. Furthermore, the presence of bulky substituents effectively inhibits chain transfer to polymer.

Reactions in Cationic Ring-opening Polymerization Disturbing the Living Character of the System

Slow Initiation. The necessary prerequisite for a living system is that the inequality $k_i \geq k_p$ holds. Otherwise, the polymer chains do not start growing simultaneously, resulting in a broadening of the molecular weight distribution.

Reversible Propagation. Ring-opening polymerization of monomers with more than four atoms in the ring proceeds as a highly reversible process. Propagation is represented as follows:

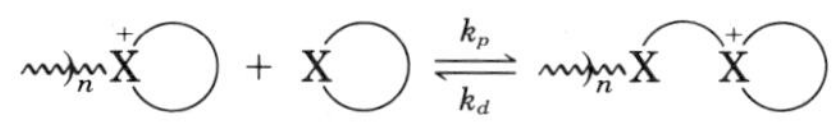

The most comprehensively studied system in which the reversibility of polymerization distorts the living character is the polymerization of the five-membered cyclic ether tetrahydrofuran (THF) (see TETRAHYDROFURAN POLYMERS).

Irreversible Chain Transfer to Polymer. When there is no sufficient steric hindrance for the reactions involving heteroatoms in the chain, formation of polymeric onium ions may lead to irreversible termination, as shown for a series of *N*-substituted aziridines, where a correlation between the size of the substituent and k_p/k_{tr} ratio has been found, namely the larger the substituent the smaller the proportion of transfer.

Chain Transfer to Polymer Followed by Chain Scission (Fast Reinitiation). Chain transfer to polymer, leading to formation of the unreactive polymeric onium ions, is equivalent to termination of both kinetic and material chains. In a number of other systems polymeric onium ions may, however, participate in further propagation reactions with rate constants comparable to those of the desired propagation. This leads to broadening of MWD. Cationic polymerization of cyclic acetals belongs to such a category (see ACETAL RESINS).

Chain Transfer to Monomer. In contrast to the polymerization of vinyl monomers, chain transfer to monomer in cationic ring-opening polymerization occurs in only a limited number of systems. Some monomers have exocyclic groups, linked to the heteroatom bearing the charge through a bond similar to the bonds to be broken in the ring. This is the case with cyclic amines and cyclic phosphates or phosphites. It has been shown for cyclic six-membered phosphate that transfer involves release of the exocyclic group from the growing species.

Recombination with Counterions. If the counterions are unstable in the given system, they may break down and provide an anion giving a covalent bond with growing macrocation. This is a termination step.

Strongly nucleophilic monomers, as a rule, give weakly electrophilic growing species and vice versa. Thus the following general rule can be formulated: in the polymerization of strongly nucleophilic monomers (eg, cyclic amines) even relatively nucleophilic counterions (Br^-) can be used without causing termination. In the polymerization of weakly nucleophilic monomers (eg, cyclic acetals) the same counterions would lead to termination; to avoid termination, stable complex anions such as SbF_6^- or AsF_6^- must be used as counterions.

Some counterions may react with cationic-active species reversibly, giving covalent species. Anions such as ClO_4^- or $CF_3SO_3^-$ or halide anions belong to this category.

General Characteristics of Cationic Ring-opening Polymerization

Living systems allow the study of the propagation step as the only chemical process, not disturbed by initiation, termination, and transfer reactions.

Thermodynamics of the Polymerization. In real systems, the monomer–polymer, solvent–monomer, and solvent–polymer interactions may contribute significantly to the overall change of the thermodynamic parameters. This effect is especially pronounced for the polymerization of heterocyclic monomers because of their basic and solvating properties.

Structure of Active Species. Active species in the cationic ring-opening polymerization are mostly onium: ammonium, oxonium, sulfonium, and phosphonium ions. These ions have been observed directly by nmr and their concentration determined in a number of systems.

Propagation-rate Constants. The characteristic feature of cationic ring-opening polymerization is the equal reactivity of ions and ion pairs at least in propagation. This is explained by strong solvation of both ions and ion pairs with monomer, overshadowing the differences in electrostatic effects between these species. Nevertheless, ions do not propagate merely in the clusters created by monomer and polymer segments, usually the strongest solvating agent present, and the presence of solvent molecules in the immediate vicinity of the active species has been clearly demonstrated from the kinetic solvent effects. Large dependences of the activation parameters on the solvent solvation power have been observed. The higher the solvent solvation power, the higher ΔH_p.

Propagation Mechanism. Most of the phenomena discussed above are characteristic for the systems in which propagation proceeds as a reaction between charged (or polar covalent) species, located at the end of the growing macromolecule, with a monomer molecule:

$$\sim\!\sim M^* + M \longrightarrow \sim\!\sim M\text{—}M^*$$

For most onium ions, the actual propagation step proceeds as a borderline S_N2 reaction. Thus, in the transition state, bond formation is less advanced than bond breaking, eg, for THF,

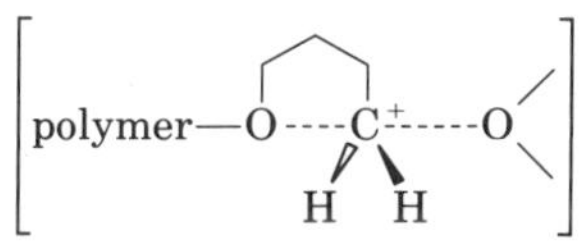

Another mechanism may operate with propagation proceeding as a reaction of the cationically activated monomer with a nucleophilic group located at the end of the growing chain. This mechanism the present authors term the activated monomer (AM) mechanism. The AM mechanism has mostly been studied in the polymerization of cyclic ethers.

Thus, polymerization of oxiranes catalyzed by protonic acid in the presence of alcohols proceeds, according to this mechanism:

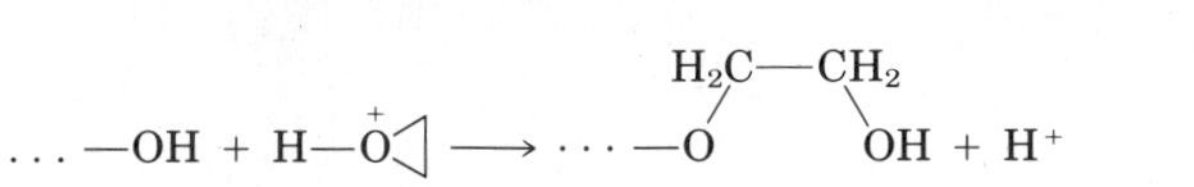

where H^+ denotes a proton exchanging fast between all bases present in the system.

In this process the growing chain end is a hydroxyl group. Thus termination and chain transfer to polymer are mostly avoided as there is no charge on the macromolecule itself. As long as the AM mechanism predominantly operates in the system, polymerization proceeds according to living polymerization with no transfer or termination.

Synthetic Aspects

Systems with long living propagating active species retain activity long after the polymerization is completed, and offer a possibility of preparing functionalized polymers by reaction of active species with an added terminating agent or block copolymers by addition and polymerization of the other monomer.

Functionalized Polymers. Polymers of THF, containing hydroxyl groups at both ends, are important precursors for the synthesis of polyurethanes and polyetheresters or polyetheramides, where they constitute soft segments. Poly-THF diols are made by protic acid-initiated polymerization. The head hydroxyl group is formed in the initiation reaction and the tail hydroxyl group by the terminating reaction (after reaching equilibrium conversion) with water, which reacts with living active species (ionic or covalent). The products are strictly bifunctional, which is proof that all macromolecules are, before termination, fitted with active (ionic and covalent) chain ends. Polymerization of THF can be terminated with other nucleophiles, leading to quantitative formation of the following end groups:

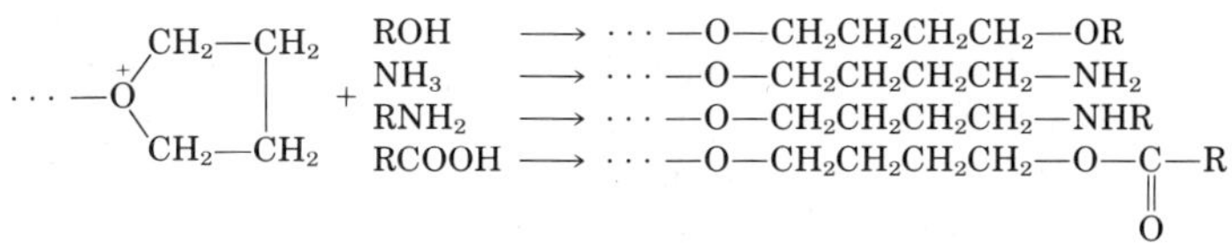

Quantitative termination of THF polymerization with phenolate (end capping) or tertiary phosphine (ion trapping) has been used to determine the concentration of active species.

After development of difunctional and multifunctional initiators, polymers containing two or more well-defined end groups (star polymers) have been prepared.

Block Copolymers. An important feature of living anionic polymerization is the possibility of preparing well-defined block copolymers by sequential polymerization. This approach was also used in cationic ring-opening polymerization, but none of the systems studied gave quantitatively well-defined block copolymers free of the corresponding homopolymers. Nevertheless block copolymers have been prepared by coupling of mono- and dicationically ended polymers, eg, poly-THF, with mono- or dianionically ended ones. In some of these systems, modification of active species prior to coupling is required.

BIBLIOGRAPHY

M. Szwarc, *Living Polymers and Electron Transfer Processes,* Wiley-Interscience, New York, 1968.

K. J. Ivin and T. Saegusa, eds., *Ring Opening Polymerization,* Vols. 1 and 2, Elsevier Applied Science Publishers, Barking, UK, 1984.

S. Penczek, P. Kubisa, and K. Matyjaszewski, *Cationic Ring-Opening Polymerization, Adv. Polym. Sci.* **37,** 1 (1980) (Vol. 1); **68/69,** 1 (1985) (Vol. 2).

B. A. Rosenberg, W. I. Irzhak, and N. S. Enikolopian, *Interchain Reactions in Polymers,* Chimia, Moscow, 1975.

STANISLAW PENCZEK
PRZEMYSLAW KUBISA
Polish Academy of Sciences

Living Cationic Polymerization, Alkenyl

Living cationic polymerization encompasses two principal fields: (*1*) living polymerization of vinyl and related alkenyl compounds (vinyl ethers, styrenes, isobutylene, etc) via a carbocationic propagating species; and (*2*) ring-opening living polymerization of heterocycles (cyclic ethers and amines, oxazolines, etc) via an onium intermediate (see RING-OPENING POLYMERIZATION). The latter category, first found in 1965 for tetrahydrofuran, has been well documented and falls outside the scope of this article (see also TETRAHYDROFURAN POLYMERS).

In contrast, the living carbocationic polymerization of vinyl monomers has long been regarded as extremely difficult, although not impossible, in view of the inherent instability of the growing carbocations that tend to undergo chain transfer, termination, isomerization, and other undesirable side reactions under acidic conditions. It was in fact not until the early 1980s that well-defined living cationic processes became available for vinyl monomers, although there had been heralding reports on the formation of long-lived polymers with acid initiators.

Vinyl Ethers

Alkyl Vinyl Ethers. In general, primary and secondary alkyl vinyl ethers ($CH_2{=}CH{-}OR$; R = alkyl) can be polymerized by the HI/I_2 and related initiating systems into well-defined living polymers with controlled molecular weights and near monodisperse (Poisson) molecular weight distributions (MWD) ($\bar{M}_w/\bar{M}_n \leq 1.1$). The reported examples include methyl, ethyl, isopropyl, *n*-butyl, isobutyl, cyclohexyl, *n*-hexadecyl, benzyl, and 2-chloroethyl vinyl ethers, among which the isobutyl derivative (IBVE) has been studied in most detail. No living processes have been found for aryl vinyl ethers, which are often subject to side reactions under acidic conditions.

The most important factor for realizing living vinyl ether polymerizations is the judicious selection of initiators or initiating systems. The following four systems have proved effective: hydrogen iodide/iodine; hydrogen iodide/zinc halide; ethylaluminum dichloride/ester; and ethylaluminum dichloride/ether. For the latter two, water or cationogens may be employed as coinitiators.

Propenyl Ethers. Living cationic polymerizations of these β-methyl derivatives ($CH_3{-}CH{=}CH{-}OR$; R = alkyl; *cis* and *trans*) of vinyl ethers are feasible by use of the HI/I_2 initiating system in toluene at −40°C; examples for ethyl and isobutyl propenyl ethers have been reported.

Trialkylsilyl Vinyl Ethers. On treatment with an aldehyde or its precursor as initiator, *t*-butyldimethylsilyl vinyl ether and related silyl vinyl ethers ($CH_2{=}CH{-}OSiR_3$; R = alkyl) undergo living polymerization at room temperature or −60°C in the presence of a Lewis acid catalyst.

N-Vinylcarbazole

In contrast to vinyl ethers, *N*-vinylcarbazole can be polymerized quantitatively by hydrogen iodide alone in toluene at −40°C to afford living polymers. Another method to prepare living poly(*N*-vinylcarbazole) uses iodine as initiator at −78°C in methylene chloride containing a trace amount of tetra-*n*-butylammonium iodide. The ammonium iodide or a common ion salt is required to suppress the ionic dissociation of the growing end in the relatively polar solvent.

Styrene Derivatives

p-Methoxystyrene. The HI/ZnI_2 system initiates living polymerization of this reactive styrene derivative in toluene at −15 or even +25°C.

p-Methylstyrene. *p*-Methylstyrene is less reactive than the *p*-methoxy derivative, so that it cannot be polymerized by the hydrogen iodide-based initiating systems. Instead, its long-lived polymers can be prepared with acetyl perchlorate as initiator, which generates the perchlorate counteranion that is less nucleophilic than the iodide anion.

Isobutylene

Living cationic polymerization of isobutylene has been achieved with use of the *t*-alkyl acetate/BCl_3 and tertiary alkyl ether/BCl_3 initiating systems in various solvents (methyl chloride, methylene chloride, etc) at −30 to −80°C. The resulting polymers have fairly narrow MWDs ($\bar{M}_w/\bar{M}_n$ = 1.1–3.0) and their $\bar{M}_n$s are controlled by the feed ratio of the monomer to the ester or ether initiator.

Synthetic Applications

As in the case of anionic living polymerization, living cationic polymerization provides a wide variety of methods to prepare polymers with a controlled structure, molecular weight, and functionality. Among these applications, four areas are of particular importance: near monodisperse polymers; pendent-functionalized polymers; end-functionalized polymers; and block polymers.

End-functionalized Polymers. The living cationic polymerization of vinyl ethers initiated by the HI/I_2 system offers two routes for the synthesis of end-functionalized polymers, as shown in equations 1 and 2:

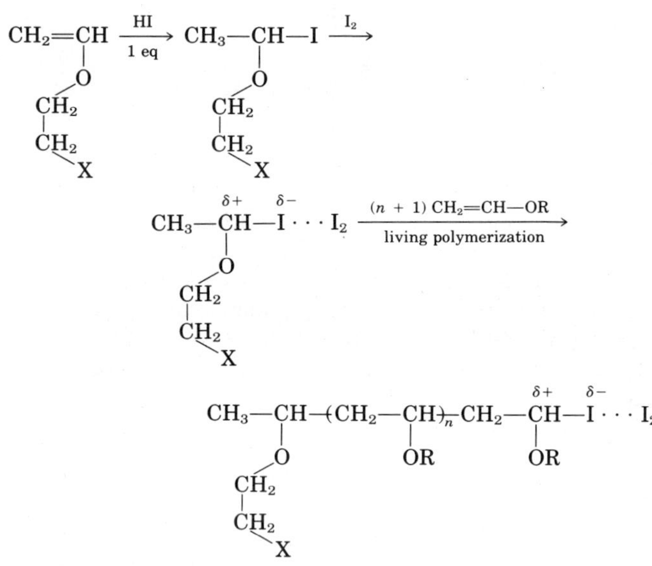
(1)

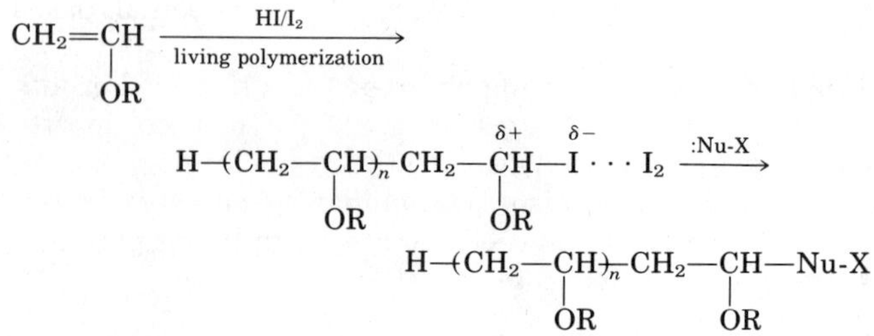
(2)

Block Polymers. Sequential living polymerization of two or more monomers gives block copolymers (qv), and this is equally applicable to living cationic polymerization. The following AB- (**1–4**) and ABA-type (**5**) block polymers have been prepared by the living polymerization of vinyl ethers (and *p*-methoxystyrene in some cases) initiated by HI/I_2 and related systems. Most of these polymers have narrow MWDs ($\bar{M}_w/\bar{M}_n = 1.1$–$1.3$) and moderate molecular weights ($\bar{M}_n = 4 \times 10^3 - 3 \times 10^4$).

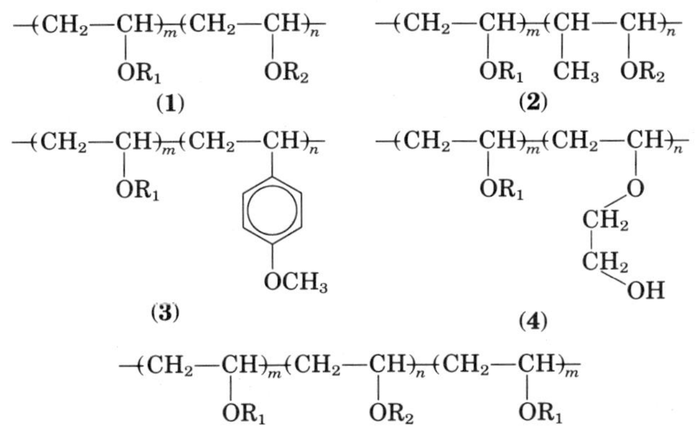

(**1**) (**2**) (**3**) (**4**) (**5**)

($R_1, R_2 = C_1$-C_{16} alkyl)

BIBLIOGRAPHY

J. P. Kennedy and E. Marechal, *Carbocationic Polymerization,* John Wiley & Sons, New York, 1982.

T. Higashimura and M. Sawamoto, *Makromol. Chem. Suppl.* **12,** 153 (1985); M. Sawamoto and T. Higashimura, *Chem. Macromol. Symp.* **3,** 83 (1986); T. Higashimura, S. Aoshima, and M. Sawamoto, *Chem. Macromol. Symp.* **3,** 99 (1986); Ibid., **13/14,** 457 (1988). M. Sawamoto, S. Aoshima, and T. Higashimura, *Makromol. Chem. Macromol. Symp.* **13/14,** 513 (1988).

T. Higashimura and M. Sawamoto, *Adv. Polym. Sci.* **64,** 49 (1984).

T. Higashimura and M. Sawamoto in *Comprehensive Polymer Science,* Vol. 3, Part I, Pergamon, 1989, Chapt. 42.

M. SAWAMOTO
T. HIGASHIMURA
Kyoto University

Epoxides and Lactones, New Developments

Ring-opening polymerizations of epoxides and lactones take place by a variety of anionic, coordinate anionic, and cationic initiators or catalysts to give polyethers and polyesters, respectively.

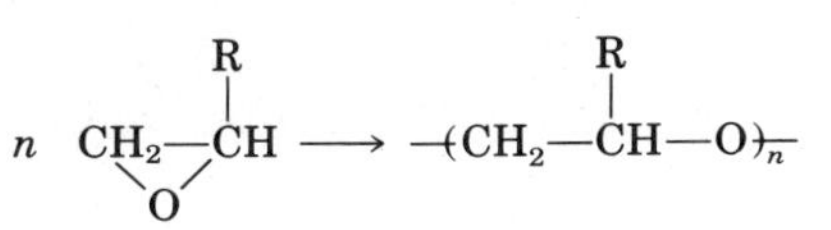

(1)

(**1**)

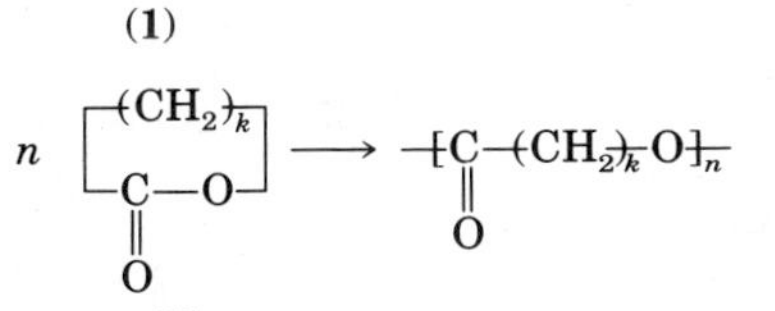

(2)

(**2**)

The polymerization of epoxyethane [(**1**), R = H] initiated by alkali metal alkoxide is one of the first successful examples of the living polymerization of heterocyclic monomers, in which the number-average degree of polymerization ($\overline{DP}$) of the polymer is equal to the mole ratio of the monomer to the initiator, and the ratio of the weight- and number-average molecular weights ($\bar{M}_w/\bar{M}_n$) is close to unity. Further addition of epoxyethane to the polymerization system at 100% conversion results in clear increase of the molecular weight of polymer, retaining the narrow molecular weight distribution. However, this system is not applicable to the living polymerization of substituted epoxides such as 1,2-epoxypropane [(**1**), R = CH_3], because proton abstraction from the monomer by the growing alkoxy species occurs, and the formation of a low molecular weight polymer with terminal unsaturation results (eq. 3).

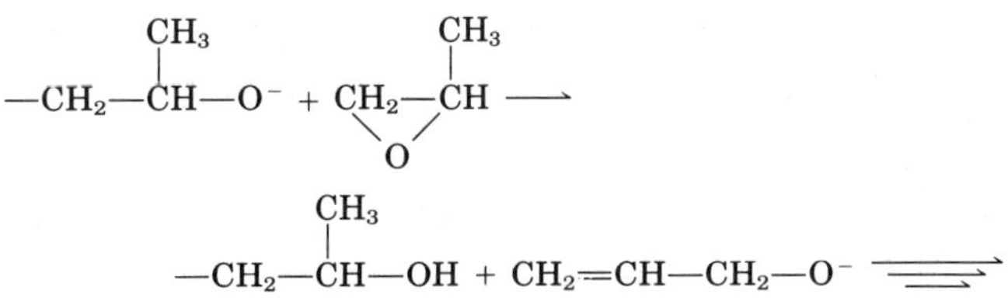

(3)

Polymerizations of lactones by nucleophilic initiators (X—M) may be classified into two types according to the modes of ring fission: alkyl–oxygen and acyl–oxygen cleavage to give a metal carboxylate and a metal alkoxide, respectively, as the growing species. The polymerizations of four-membered lactones often suffer from this multiplicity of growing species.

Possible side reactions in the anionic polymerizations of lactones are proton abstraction from the active carbon–hydrogen group adjacent to the carbonyl moiety of the monomer and intramolecular and intermolecular transesterifications. The former leads to the formation of polymers with terminal unsaturation, and the latter two lead to the formation of cyclic oligomers and chain reshuffling of the polymers, respectively. Broadening of the molecular weight distribution of polymers is caused by these unfavorable reactions, which are related to the nucleophilic nature of the initiators.

Aluminum Porphyrins as Novel Initiators

The novel initiator system of aluminum porphyrins [(TPP)AlX, (**3**)] brings about living polymerizations and copolymerizations of a wide variety of monomers, and a variety of novel block copolymers with tailored block lengths can be easily synthesized thereby.

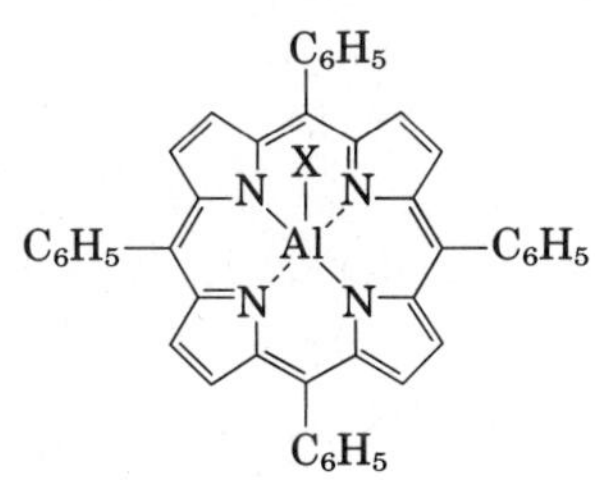

(**3**)

Epoxides. The polymerizations of epoxides with aluminum porphyrins [(TPP)AlX, X = Cl, OOCR, OAr, OR] as initiators give polyethers of very narrow molecular weight distribution.

The growing species of the polymerization of epoxide initiated by aluminum porphyrin is a (porphinato)aluminum alkoxide (**4**), which is formed by the insertion of the monomer into the central aluminum atom–axial group bond of the initiator (eq. 4).

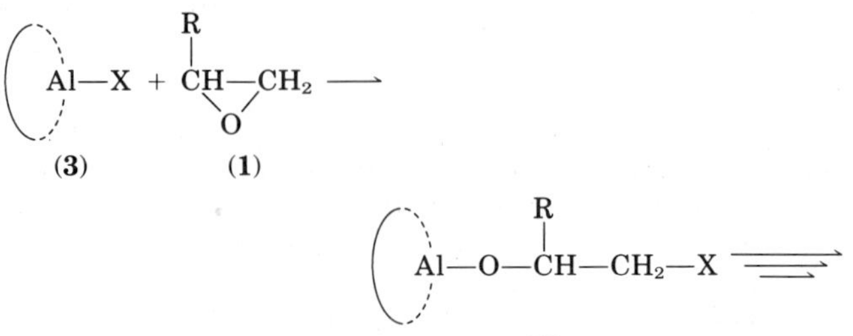

(3) (1) (4) (4)

The copolymerizations between different epoxides by (TPP)AlCl also give copolymers of narrow molecular weight distribution. The copolymerizations of epoxides with carbon dioxide and cyclic anhydride, respectively, by aluminum porphyrin take place to give copolymers with narrow molecular weight distribution, although rich in the units from epoxides. On the other hand, alternating copolymers, polycarbonates and polyesters, are formed with narrow molecular weight distribution when (TPP)AlX as initiator is combined with quaternary organic salts such as ammonium and phosphonium salts.

Lactones. Aluminum porphyrins such as (TPP)AlCl and (TPP)AlOOCR are also effective for the living polymerizations of four-membered lactones such as β-propiolactone [(**2**), $k = 2$] and β-butyrolactone to give the corresponding polyesters of uniform, controlled molecular weights. On the other hand, the polymerizations of six- and seven-membered lactones such as δ-valerolactone and ϵ-caprolactone [(**2**), $k = 3, 4$] hardly take place by either (TPP)AlCl or (TPP)AlOOCR as initiator, but proceed with living character by (TPP)AlOR via ring fission of the monomer at the acyl–oxygen bond, producing a (porphinato)aluminum alkoxide (**5**) as the growing species (eq. 5).

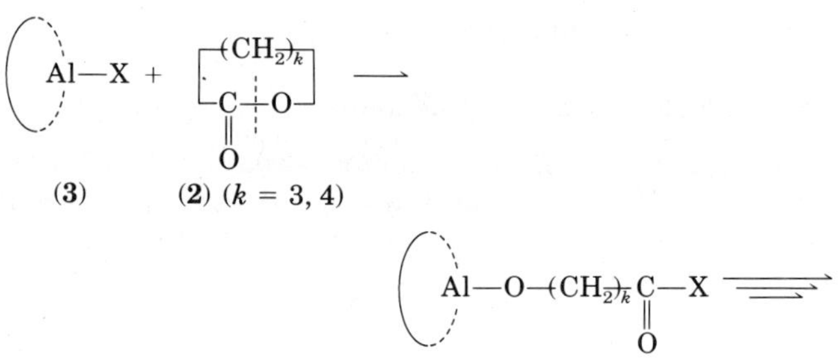

(3) (2) ($k = 3, 4$) (5) (5)

Accordingly, the living polymer of (**4**) successfully initiates the polymerizations of these monomers to give the corresponding polyether–polyester block copolymers of very narrow molecular weight distribution. Similarly, the living polymerization of lactide takes place by using (TPP)AlOR or (**4**) as initiator, in which the monomer is also opened at the acyl–oxygen bond to give a (porphinato)aluminum alkoxide such as the species (TPP) Al—O—CH(CH_3)—CO— as the growing chain.

Sequential Addition, Ring-opening Living Polymerizations. An alkyl–aluminum porphyrin such as (TPP)$AlCH_3$ [(**3**), X = CH_3] brings about the living polymerization of alkyl methacrylates (**6**) at 30°C under irradiation of visible light to give polymers of controlled molecular weight with narrow distribution. The polymerization proceeds by the conjugate addition of the monomer to the aluminum–methyl bond of the initiator to give a (porphinato)aluminum enolate (**7**) as the growing species (eq. 6).

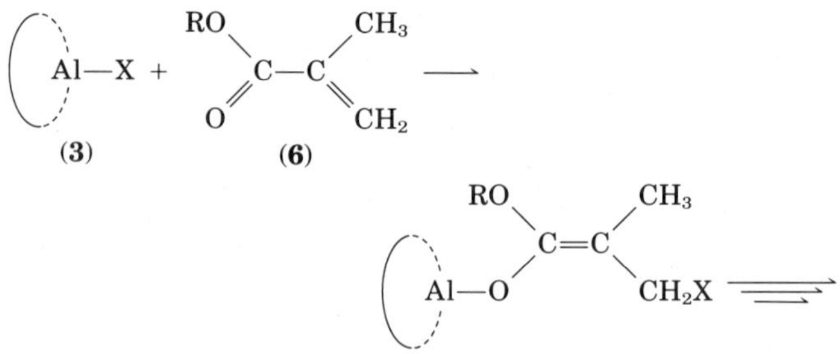

(3) (6) (7) (6)

The ring-opening polymerizations of epoxides and lactones successively take place from the growing species (**7**) to give novel polyvinyl–polyether and polyester block copolymers of controlled block lengths with narrow distribution, respectively. This unprecedented achievement is due to the wide applicability of the system of aluminum porphyrins as excellent initiators for living polymerizations.

Immortal Polymerization. In the case of the polymerizations of epoxides and lactones initiated by aluminum porphyrins, polymers of uniform molecular weights are formed even in the presence of protic chain-transfer agents that immediately react with the growing species. Furthermore, the number of polymer molecules is not limited by that of the initiator molecules but can be increased by the amount of protic compounds added. The polymerization of epoxides by aluminum porphyrins in the presence of protic compounds is called "immortal" polymerization, in the sense that the termination reaction is absent. The immortal nature of polymerization is due to the unusually high nucleophilicity of (TPP)AlX (X = Cl, RCOO) compared to the other usual metal salts, taking into account the example of polymerization of epoxides by sodium alkoxide, which is terminated by hydrogen chloride as a result of the formation of NaCl, an inert salt. A manganese porphyrin such as (TPP)MnOOCR is also an excellent initiator for the immortal polymerization of epoxides.

BIBLIOGRAPHY

S. Inoue and T. Aida in K. J. Ivin and T. Saegusa, eds., *Ring-Opening Polymerization,* Vol. 1, Elsevier Applied Science Publishers, Ltd., Barking, UK, 1984, pp. 185–298.

S. Inoue and T. Aida in J. E. McGrath, ed., *ACS Symp. Ser.,* Vol. 286, American Chemical Society, Washington, D.C., U.S.A., 1985, pp. 137–146.

S. Inoue and T. Aida in G. C. Eastmond, A. Ledwith, S. Russo, and P. Sigwalt, eds., *Comprehensive Polymer Sci.,* Vol. 3, Pergamon Press, Oxford, U.S.A., 1989, pp. 553–569.

SHOHEI INOUE
TAKUZO AIDA
University of Tokyo

Olefin Metathesis

The evolution of the ring-opening metathesis polymerization (ROMP) of strained cyclic olefins, from a curious side reaction observed in Ziegler-Natta polymerizations to well defined living systems, is a remarkable story. This work, which has recently culminated in the development of these living catalyst systems, epitomizes the goals of mechanistic organometallic chemistry. Progress in this area began with early work that led from the proposal of carbenes as the active propagating species to their identification, demonstration of activity, and eventual synthesis and use as active, well-controlled metathesis catalysts.

Several groups have isolated well-characterized transition metal metallacycles and alkylidenes that possess, when utilized as ROMP catalysts, the requisite characteristics of living systems. These well-characterized systems are discussed in this article. The mechanism of olefin metathesis involves the interconversion of a carbene and a metallacycle, either of which may be the stable chain-carrying species.

Metallacycles

Titanacyclobutanes. β-Substituted titanacyclobutanes derived from the reaction of the Tebbe reagent with olefins undergo only degenerate metathesis of terminal olefins. The unproductive nature of this reaction is the result of the steric influence of the cyclopentadienyl rings that direct the olefin's substituent to the β-position of the forming metallacycle.

To promote productive metathesis it is necessary to design systems in which, upon retro-cleavage of the metallacycle, only a destabilized α-carbene is formed. This may be accomplished by the incorporation of strained rings into the initial metallacycle. This added ring strain provides a thermodynamic bias that drives the reaction into the productive cleavage manifold.

The metallacycle (**2**)–norbornene polymerization system has several unique features that put it in the class of living systems: the presence of only one propagating species in solution; fast initiation relative to propagation; and the suppression of back-biting reactions that constitute chain-transfer and depolymerization steps. These desirable features are a function of both the catalyst and the monomer substrate. Relaxation of any one of these features would result in a broadened molecular weight distribution, and in some cases, lead to the loss of the living nature of the system.

Kinetics. The rate of polymerization of norbornene using (**2**) shows the reaction to be zero-order in monomer. This is consistent with a mechanism that has a rate-limiting unimolecular cleavage of the propagating metallacycle followed by rapid trapping of the subsequent carbene with norbornene (Fig. 1).

The utility of the living titanacyclobutane-mediated polymerizations has been demonstrated not only for the synthesis of monodisperse polymers of controlled molecular weights, but for the synthesis of well-defined di- and tri-block copolymers and in polymer end-capping processes as well.

Tantallacyclobutanes. Metallacycle complexes analogous to the titanium cases cited above can be prepared from the reaction of $Ta(CH\text{-}t\text{-}C_4H_9)(OR)_3(THF)$ complexes (where OR = 2,6-diisopropylphenoxide or 2,6-dimethylphenoxide) and norbornene at −30°C (eq. 1).

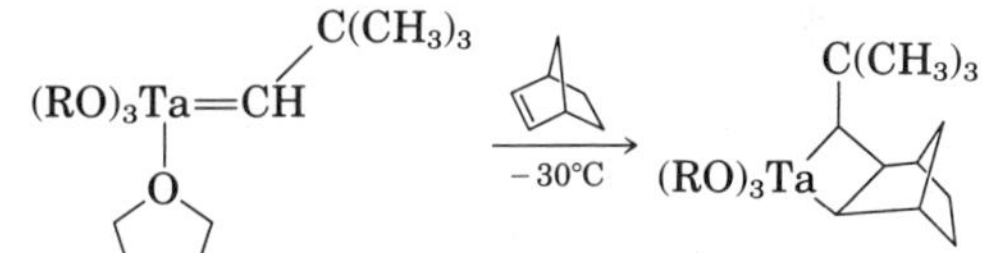

(**4**) (1)

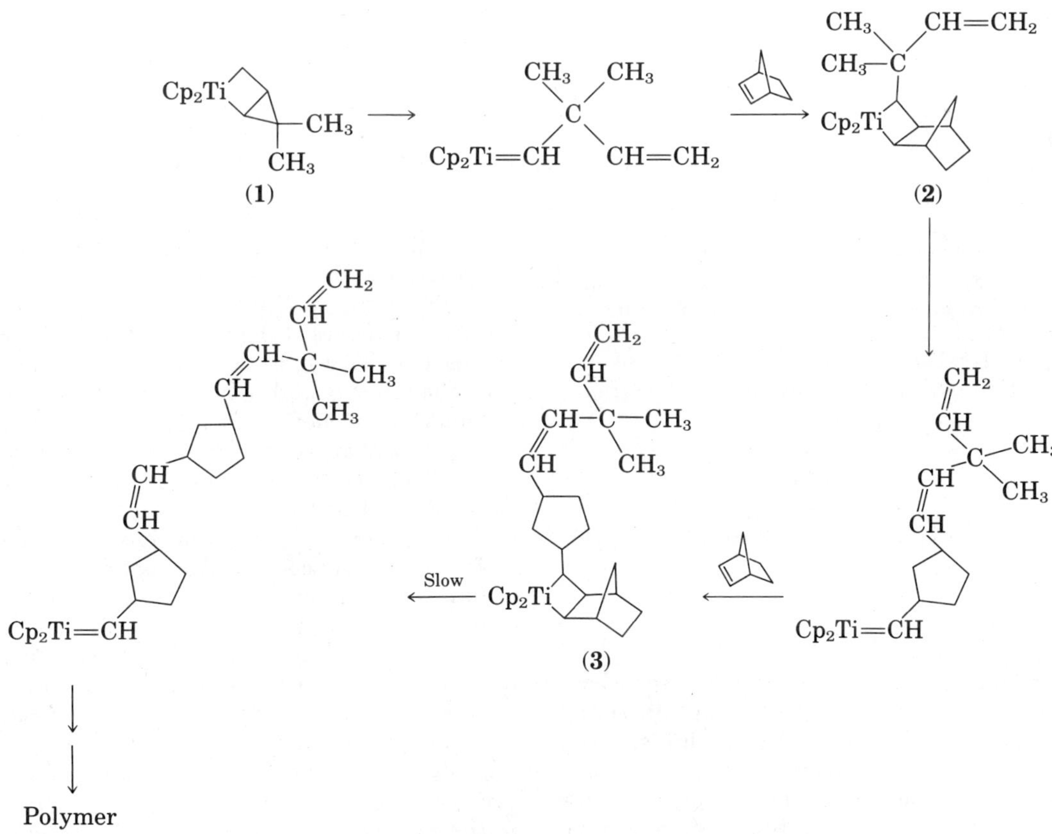

Figure 1. Mechanism of norbornene polymerization using (**2**).

Alkylidenes

As a complement to the stable metallacyclobutane catalysts, a series of stable alkylidene catalysts have been prepared and shown to be active living polymerization catalysts. The complex $W(CH\text{-}t\text{-}C_4H_9)(NAr)(O\text{-}t\text{-}C_4H_9)_2$ (where Ar = 2,6-diisopropylphenyl) reacts with 50–200 equivalents of norbornene in toluene at 25°C, followed by addition of benzaldehyde as an end-capping agent, to provide polymers in which the major component has a molecular weight proportional to the number of equivalents of norbornene consumed, and dispersities of approximately 1.05. A variable amount (10–20%) of high molecular weight polymer is also produced. Again, it is thought that the high molecular weight material forms via an extremely active, as yet unidentified, tungsten–carbene species present during the polymerization.

Additional synthetic utility can be introduced by changing the metal in the above alkylidene complexes from tungsten to molybdenum. Molybdenum, being less oxophilic than tungsten, allows for the polymerization of monomers containing mildly reactive functionality such as esters without appreciable catalyst deactivation.

A variety of isolated pentacoordinate tungsten–carbene complexes are known to be active metathesis catalysts. At least one of these systems has been proposed to be living based primarily on ^1H nmr identification of the propagating alkylidene.

An important extension of the tungsten system $W(CH\text{-}t\text{-}C_4H_9)(OCH_2\text{-}t\text{-}C_4H_9)_2Br_2/GaBr_3$ in which the conversion of the initiating metal–carbene complex into the initial metallacyclobutane complex by the addition of a first equivalent of norbornene is directly observed at low temperature by ^1H nmr. Subsequent orthogonal cleavage of the metallacyclobutane and its reformation by the addition of further equivalents of norbornene is also observed. The direct observation of these primary steps may lead to a better understanding of both the kinetic and thermodynamic factors that balance during these propagating steps.

BIBLIOGRAPHY

K. J. Ivin, *Olefin Metathesis,* Academic Press, London, 1983.

P. J. Flory, *J. Am. Chem. Soc.* **62,** 1561 (1940); M. Szwarc, *Carbanions, Living Polymers, and Electron Transfer Processes,* Wiley-Interscience, New York, 1968; M. Szwarc in *Advances in Polymer Science,* Vol. 49, Springer-Verlag, Berlin, 1983.

L. R. Gilliom and R. H. Grubbs, *J. Am. Chem. Soc.* **108,** 733 (1986).

R. H. Grubbs and W. Tumas, *Science,* **243,** 907 (1989).

ROBERT GRUBBS
BRUCE M. NOVAK
California Institute of Technology

Radical Polymerization

Polymerization systems that involve propagating free radicals are likely to involve very short-lived intermediates. Free-radical termination reactions are very fast, being diffusion-controlled even for the smallest radicals. The pronounced inclination of free radicals to terminate offers a severe challenge to chemists who would hope to develop living free-radical polymerization systems. However, approaches have been developed that afford the polymer chemist with many of the synthetic advantages that living polymerization systems provide. These include the synthesis of polymers with narrow molecular weight distributions, block copolymers, and polymers with star or radial block architectures.

Two approaches have been followed in efforts to develop living radical polymerization systems. The first uses physical methods to prevent radicals from contacting each other, thereby preventing their termination. The second, sometimes called pseudo-living polymerization, involves repeated reinitiation of polymer chains by thermolysis or photolysis of weak linkages that are built into the chains.

Limiting the Mutual Access of Radicals

Since radicals generally terminate by reaction with other radicals, any process that limits radical recombination reduces the rate of termination. In the extreme, such processes can result in living polymerization behavior.

The rates of radical-termination reactions are very sensitive to viscosity. In the case of methyl methacrylate polymerization, the termination rate constant varies over six orders of magnitude during the course of the reaction and is inversely proportional to the viscosity of the polymerization medium. If radicals are generated in sufficiently viscous media, they can remain active for long periods.

Precipitation Polymerization. It was recognized as early as 1941 that it was possible for polymer particles formed during polymerization reactions to contain trapped radicals, which could remain so for months. This is an extreme example of the viscosity effect just discussed.

Template Polymerization. Poly(acrylic acid) and poly(methacrylic acid) form complexes with a large number of polymers, including poly(*N*-vinylpyrrolidinone), poly(ethylene glycol), polyvinylpyridine, and polyethyleneimine. When these complexes are formed from growing polymer radicals and a co-complexing polymer, the mobility of the radicals is decreased substantially, termination rate constants are reduced, and living polymer behavior may be possible (see TEMPLATE POLYMERIZATION).

Polymers with Rigid Backbones. Although the termination rate constant for methyl acrylate polymerization is on the order of 10^7 L/mol·s, the corresponding value for diisopropyl fumarate is 0.84 L/mol·s. This very low value for diisopropyl fumarate, a consequence of the rigidity of poly(diisopropyl fumarate) chains, enables it to be polymerized to high molecular weight, even though it is highly hindered.

Emulsion Polymerization. Electron-spin resonance measurements show that polymer radicals can exist in polymer latices (qv) for long periods. According to the simplest model for emulsion polymerization (qv), one half of the monomer–polymer particles present at any given time in an emulsion polymerization system can contain propagating polymer radicals. Such radicals are terminated by the entry of other, generally small, radicals into the monomer–polymer particles. When generation of new radicals stops, those that remain in the monomer–polymer particles can live for long periods, can continue to grow as monomer diffuses into them, or can add to different monomers to form block copolymers. The key to obtaining living polymerization behavior is to control the radical initiation step. This has been attempted in two ways: by photochemical initiation, and by removing components that are required for initiation from the system.

Polymerization in Inclusion Compounds. A somewhat specialized but effective method for hindering radical-termination reactions and obtaining living polymerization behavior is to conduct polymerization in clathrate channels (see POLYMERIZATION IN CLATHRATES).

Employment of Radical Complexes. There is evidence that oxygen radicals such as the *t*-butoxy and *t*-butylperoxy radicals can be stabilized by forming complexes with Co^{3+} ions and that the reactivity of chlorine atoms is reduced by complexation with benzene. Therefore, there has been hope that a way may exist of converting propagating polymer radicals into complexes that would be reluctant or unable to participate in termination reactions.

Repeated Reinitiation of Polymer Chains

Instead of attempting to eliminate termination reactions as a means of achieving living radical polymerization behavior, the termination step may be accepted as inevitable and compromise solutions that provide for polymer chains to be reinitiated repeatedly during the course of a polymerization process may be attempted. This more easily achieved objective results in polymerizations with the characteristics of those that occur without termination, but without the same precision of structure control.

There are two general methods for achieving repeated reinitiation: use of multifunctional initiators, and pseudo-living radical polymerization.

Plasma-initiated Induced Polymerization. Living radical polymerization behavior has been claimed for polymerizations that occur slowly at room temperature in the dark after monomers (mostly MMA) have been exposed for brief periods (30–60 s) to microwave discharge. Conversion and molecular weight increase steadily with time for periods up to 45 days. Molecular weights as high as 10^7 have been obtained. The conversions are generally low (~5%), except for polymers that precipitate during polymerization. The process is used to prepare a variety of block and statistical copolymers.

BIBLIOGRAPHY

T. Otsu, *Makromol. Chem. Macromol. Symp.* **10/11,** 235 (1987).

C. I. Simonescu, E. Comanita, M. Pastravanu, and S. Duminitriu, *Prog. Polym. Sci.* **12,** 1 (1986).

S. R. Turner and R. W. Blevins, *Polym. Prepr. Am. Chem. Soc. Div. Polym. Chem.* **29**(2), 6 (1988).

C. I. Simonescu and B. C. Simonescu, *Polym. Prepr. Am. Chem. Soc. Div. Polym. Chem.* **29**(2), 75 (1988).

H. J. HARWOOD
University of Akron

Ziegler-Natta Polymerization

Recently, several examples of living insertion (coordination) polymerization by transition metal catalysts have been reported. Monomers such as α-olefins, cyclic olefins, and substituted acetylenes have been used. This article reviews recent progress in living insertion polymerization (qv) and the synthesis of tailor-made polymers.

Synthesis of Living Polyolefins

Syndiotactic polymerization of propylene using homogeneous catalyst systems based on VCl_4 and AlR_2Cl in the presence or absence of a weak Lewis base such as anisole have been investigated. For polymerization with anisole at −78°C, in 1967 it was found that both the polymerization rate and average molecular weight of the resulting polymer increased linearly as a function of time (Fig. 1). This may have been the first discovery of living polymerization of olefins promoted by Ziegler-Natta catalysts.

The system has been extended. Experiments carried out in toluene solvent with catalyst systems composed of $V(acac)_3$ and AlR_2X have led to the following conclusions: (*1*) living polymerization of propylene takes place at low temperatures, below −65°C; (*2*) dialkylaluminum monohalides such as $Al(i\text{-}C_4H_9)_2Cl$, $Al(n\text{-}C_3H_7)_2Cl$, $Al(C_2H_5)_2Cl$, and $Al(C_2H_5)_2Br$ act as effective cocatalysts for living polymerization; (*3*) the formation of active centers is instantaneous; (*4*) the distribution of the molecular weights of polypropylene is very narrow ($M_w/M_n < 1.2$).

The heterogeneous system composed of $TiCl_3$ (essentially free of aluminum) and $[(CH_3C)_5]_2Ti(CH_3)_2$ also catalyzes the living polymerization of propylene. Contrary to the vanadium systems, the pentamethylcyclopentadienyl catalyst gives highly isotactic polypropylene.

Terminally Functionalized Polyolefiins. The reaction of living polyolefins with additives is not only important for the synthesis of terminally functionalized polymers but also for understanding the reactivity and structure of living polymer ends.

The vanadium–carbon bond of living polypropylene reacts instantaneously with I_2 to give an iodine-bonded polypropylene in approximately 100% yield. The ^1H and ^{13}C nmr analyses of iodine-bonded polypropylene (**2**) have revealed the structure of the living polypropylene ends as (**1**):

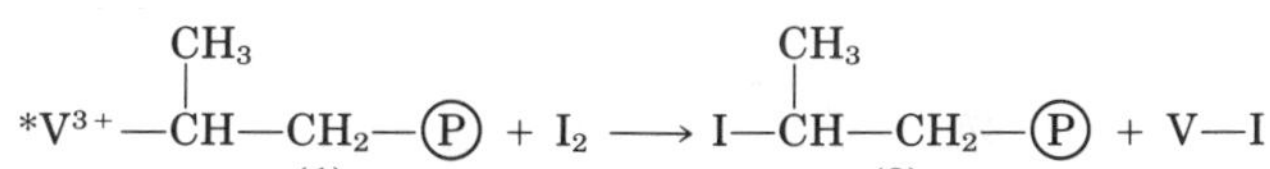

(**1**) (**2**)

where V^{3+} and (P) represent an active vanadium center and polypropylene chain, respectively. It was previously demonstrated in syndiotactic polymerization that propylene insertion occurs into a V—C bond of the secondary (or 2–1) type. The

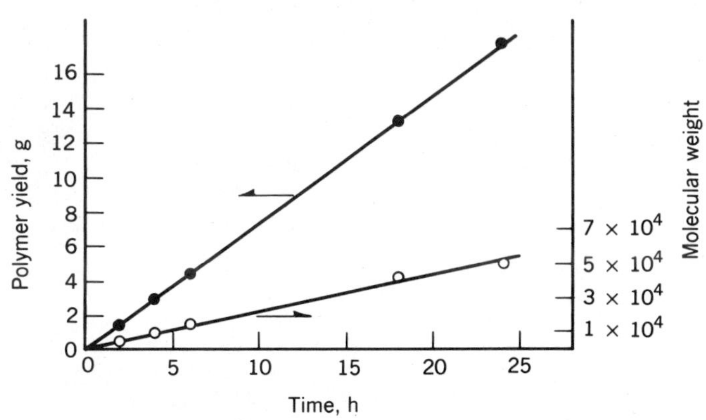

Figure 1. Change in polymer yield and molecular weight with polymerization time. Polymerization conditions: T = −78°C, C_3H_6 = 40 g, *n*-heptane = 50 mL, $VCl_4 = 0.5 \times 10^{-3}$ mol, anisole = 5×10^{-3} mol, $Al(C_2H_5)_2Cl = 15 \times 10^{-3}$ mol; ○ = mol wt, ● = polymer yield.

same conclusion has been reached from spectroscopic studies using ir and nmr. The end structure (**2**) of iodine-functional polypropylene gives the same conclusion.

Block Copolymers. Thermoplastic elastomers having the general formula $(AB)_nA$, where each A is a crystalline isotactic polypropylene block, each B is a random ethylene–propylene copolymer block, and n is an integer from one to about 12, have been prepared by using the $TiCl_3$–$[(CH_3C)_5]_2Ti(CH_3)_2$ catalyst system. Several types of well-defined block copolymers have been prepared with the vanadium systems by using the following methods: sequential polymerization of propylene and ethylene; transformation of living polypropylene ends to radical or cationic ends that initiate the polymerization of polar monomers; and a coupling reaction between iodine-functional polypropylene and living polystyrene anion.

Living Polymerization of Acetylenes

The living polymerization of substituted acetylenes such as 1-chloro-1-octyne, *o*-(trimethyl)phenylethyne, etc by $MoCl_5$- and $MoOCl_4$-based catalysts has been reported.

Living Polymerization of Cyclic Olefins

Ring-opening methathesis polymerization of cyclic olefins has also been extensively studied, and several ring-opened polymers such as polynorbornene are currently produced on an industrial scale (see LIVING POLYMER SYSTEMS, OLEFIN METATHESIS).

BIBLIOGRAPHY

A. Zambelli, G. Natta, I. Pasquon, and R. Signorini, *J. Polymer Sci. Part C* **16,** 2455 (1967).

Y. Doi and T. Keii, *Adv. Polym. Sci.* **73/74,** 201 (1986).

U. S. Pat. 4,408,019 (1983), H. W. Blunt (to Hercules Inc.).

L. R. Gilliom and R. H. Grubbs, *J. Am. Chem. Soc.* **108,** 733 (1986).

TOMINAGA KEII
YOSHIHARU DOI
KAZUO SOGA
Tokyo Institute of Technology

LLDPE.
See ETHYLENE POLYMERS.

LOW DENSITY POLYETHYLENE.
See ETHYLENE POLYMERS.

LOW PRESSURE MOLDING.
See REINFORCED PLASTICS.

LOW TEMPERATURE PROPERTIES.
See CRYOGENIC PROPERTIES.

LUBRICANTS.
See RELEASE AGENTS.

M

MACHINING

Plastic parts are usually produced by molding processes that make good use of the rheological characteristics of plastics, but additional machining operations are frequently required. Plastic parts used in precision machinery, electronics, and optics are produced by machining instead of molding processes because of the need for high accuracy, smooth surface, and lower cost. Supplementary machining procedures may be required in finishing operations, as shown in Table 1. Machining methods used on plastic parts are shown in Table 2.

Plastics are machined by techniques developed for wood or metals. However, phenomena are observed in plastics machining that differ from those of metal or wood machining and that are based on characteristics peculiar to plastics. Plastics are available in many types and grades and some are difficult to machine. Machined surfaces of plastics are often rough and cracked or covered with undesirable tool marks. Excessive frictional heat that develops may cause burning and a further finishing operation may be required. Cutting edges of tools wear out more readily.

Cutting with a Tool

Cutting with a tool is a controlled fracture. When a cutting tool advances into the workpiece along a given axis or cutting direction, a cutting force R' acts on the tool at the rake surface; R' is the sum of the normal force F_n and the tangential or friction force F_f. The action of R' on the work material is R; it deforms and fractures the work material.

The type of chips obtained with a given plastic varies with the conditions of fracture, such as depth of cut, cutting speed, temperature, and tool geometry. The cutting process is controlled by these factors.

Chip Formation

Continuous Chip. Chip deformation is essentially elastic, and chip thickness is nearly equal to the depth of cut. Continuous-flow formation is produced by high elastic deformation rather than plastic fracture; it does not occur in metal cutting. With this type of chip, fluctuations in cutting force and surface roughness are low and the accuracy of the machined part is fairly good. Undercut or overcut tends to occur because of high elastic deformation. Material that produces this type of chip has good machinability. However, it is only produced by certain cutting conditions. Shear intervals are relatively large, producing discontinuous chips. Surface roughness is excessive and the accuracy of the machined part inadequate. Conditions producing this type of chip should be avoided.

"Sticky" cutting action in polystyrene produces a discontinuous complex-type chip (Fig. 1). Dimensional accuracy of the machined part is poor and excessive surface roughness is due to the large stresses during cutting. Cutting conditions producing this type of chip are to be avoided.

Observations of the type of chip during cutting provide important information on how the cutting is proceeding in turning, shaping, diamond-cutting, drilling, milling, or sawing. It is important to select tool designs, cutting speeds, and depths of cut that will produce the smooth, continuous chips associated with clean, precisely machined surfaces.

Table 1. Supplementary Machining Operations Used to Finish Plastic Products

Processing method	Purpose of machining operation	Types of machining used
compression, transfer, injection, and blow molding	degating, deflashing, polishing	cutting off, buffing, tumbling, filing, sanding
extrusion	cut lengths of extrudate	cutting off
laminating	cut sheets to size, deflashing edges	cutting off
vacuum forming	polish cut edges, trim parts to size	cutting off, sanding, filing

Table 2. Machining Methods Used for Plastic Parts

Parts	Methods
bearing, roller	turning, diamond-cutting, milling, drilling, shaping
button	turning, drilling
cam	turning, copy turning
dial and scale	engraving, sandblasting
electronic parts	diamond-cutting
gear	turning, milling, gear shaving, broaching
liner and brake lining	cutting off, shaping, planing, milling
optical parts	diamond-cutting
pipe and rod	cutting off, turning, threading
plates for ceilings and panels	cutting off, drilling, tapping
tape for PTFE	peeling

Figure 1. Discontinuous complex-type chip formation in polystyrene. Cutting speed, 0.2 m/min; depth of cut, 0.25 mm; rake angle, −30°.

Critical Rake Angle

Deformation in the material surrounding the tool must be controlled or compensated for, to obtain maximum dimensional accuracy and minimum distortion after machining. Type and magnitude of strain vary with the rake angle of the tool. With a positive rake angle, the material tends to be pulled up during cutting, whereas with a zero or negative angle, it is pushed down. Thus the rake angle affects the accuracy of the cut.

Elastic recovery during machining interferes with the cutting action because the expansion of compressed material causes the friction between the recovered cut surface and the relief surface of a tool to increase. In addition to generating heat, this friction causes tool wear. Elastic recovery makes the holes smaller in diameter than those of the drill used and makes the cut strips of material pinch the saw used to cut them.

This value has been defined as the critical rake angle. In most cases, it is also the optimum rake angle of a single-point tool for cutting plastics to obtain the highest accuracy because the direction of the resultant cutting force coincides exactly with the direction of cutting. Under these conditions, the material is deformed neither upward nor downward during cutting. The critical rake angle depends on material, cutting speed, and depth of cut. It is also influenced by the chip friction on the rake surface and must be determined for each plastic separately. The rake angle of a cutting tool must be at the critical value or higher.

The grinding technique for producing and maintaining a sharp edge on the cutting tool is also important, since the magnitude and direction of the resulting cutting force components vary with the shape of tool edge and the progress of wear. When cutting rigid poly(vinyl chloride), the force components F_c and F_t both increase with increasing roundness of the cutting tool. A tool wears least if it has the proper rake angle and a sharp edge.

Ultraprecision Diamond Turning

In fly-cutting operations, a diamond tool is fixed in a tool holder 37 mm off the center of an air-bearing spindle. The tool holder can be dynamically balanced, and a natural diamond is used as the cutting tool. The maximum work dimension that can be cut with this apparatus is 60 mm. Depth of cut ranges from 1 to 30 μm, feeding speed from 7.5 to 45 mm/min, and spindle rotation speed from 500 to 3000 rpm.

The distinctive feature of the ultraprecision air spindle is the spherical rotating shaft, which gives better spherical accuracy than a cylindrical shaft. A sphere is mounted on one shaft end, and a journal bearing on the other. The journal bearing is installed inside a semisphere that is used for highly accurate alignment during assembly. The spindle radial loads are supported by both the spherical and journal bearings, the axial loads only by the spherical bearing. Spindle shaft elongation caused by heat generation extends to the journal-bearing side and does not influence journal-bearing performance.

Several types of natural diamond tools with nose radii of 0.5–5.0 mm and rake angles of −10 to 10° are used for cutting plastics.

Polycarbonate. The surface of polycarbonate cut with a diamond tool is strongly affected by the conditions, especially the feeding rates. Feeding marks disappear with decreasing feeding rates.

Allyl Diglycol Carbonate. Surface roughness for allyl diglycol carbonate decreases with decreasing feeding rate. However, these values are always higher than those of the geometrically calculated surface roughness R_{th}.

Poly(methyl methacrylate). Surface roughness decreases with decreasing feeding rate. Optical grade surface roughness of ca 5 nm can be obtained at a cutting speed of 730 m/min, depth of cut of 10 μm, and feeding rate of 3 μm per revolution. PMMA is the plastic most suitable for optical parts. To obtain a high quality smooth surface, the cutting tool must be carefully polished, especially to sharpen the cutting edge.

Multiple-point Tool

Selections of cutting conditions for multiple-point tool operations, drilling, milling, or circular sawing, are more difficult.

Depth of Cut. The tooth depth of cut depends on the feed rate and the cutting speed, which significantly affect the surface quality and dimensional accuracy. In saws, these variables determine the number of teeth or cutting edges per tool rather than the rake angles on the cutting edges or teeth. The geometry of each individual saw tooth is governed by the phenomena described previously.

Drilling

The behavior in drilling is similar to that in cutting with a circular saw. Drilling is done by revolving and simultaneously feeding the drill into the material. The depth of cut per tooth d is calculated from equation 1, using terms defined in Figure 2.

$$d = \frac{s}{n} \sin \frac{\theta}{2} = \frac{1}{n} \frac{v}{N} \sin \frac{\theta}{2} \qquad (1)$$

Thus the tooth depth of cut is directly proportional to the feeding speed v and inversely proportional to the rotational speed N of the drill.

Multilayer Printed Circuit Boards. For the drilling of printed circuit boards, a drill, punch, laser beam, electron beam, or liquid jet is required. A drill gives the best surface quality of holes and workability.

A printed circuit board is a composite of epoxy resin and glass cloth over copper. A cemented carbide drill may produce

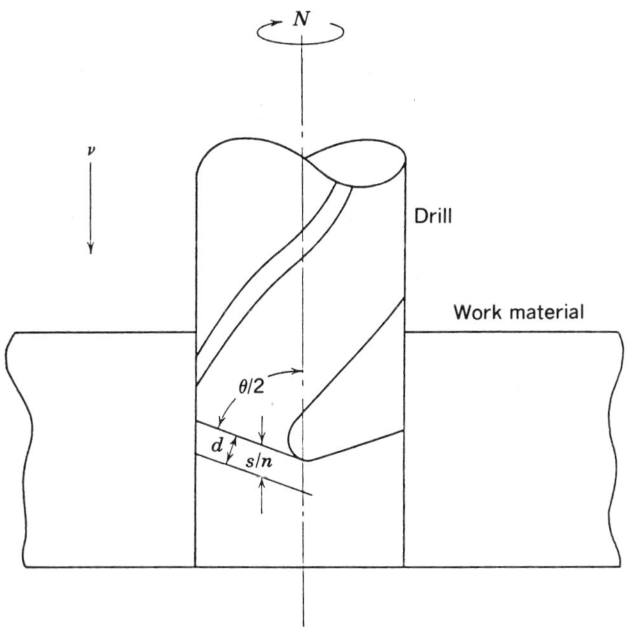

Figure 2. Twist drill; s, feeding distance per one revolution of drill; v, feeding speed; N, rotational drill speed; n, number of teeth; θ, point angle of drill.

a rough inner surface, an overcut of copper layer called an "epoxy smear," and swelling or burning of the copper around the hole's edges.

Glass–epoxy laminates are easily swelled, cracked, or damaged. Drilling requirements differ from those for other plastics because of their nonhomogeneous structure. A multilayer printed circuit board is drilled perpendicular to the direction of lamination. Drilling at right angles to the laminate causes swelling or cracking. Swelling indicates partial delamination adjacent to the drilled hole and separation of the layers.

Cutting Off

Most processes for producing plastic sheet yield products with rough edges that must be trimmed or cut to the desired sizes. The speed and efficiency of the cutting-off operation affect production rate and product cost. Therefore, cutting tools and techniques are of considerable interest to the plastics industry. Efficiency is governed by speed of operation, surface qualities, degradation, accuracy, and roughness, and the cost by kerf loss, power consumption, and labor.

Machinability

Machinability is the ease with which a material can be cut. Criteria for machinability include tool life, cut-surface finish, and the power required to make the cut. Good machinability requires little power and produces smooth cut surfaces with little tool wear. A machinability factor η_1 can be expressed by

$$\eta_1 = \frac{V_m}{V_w \times W \times R} \tag{2}$$

where V_m is the cut volume per unit time, V_w the volume of material worn from the tool per unit time, W the net power consumption during cutting, and R the cut-surface roughness, usually in μm; η_1 increases as tool wear, power consumption, and surface roughness decrease.

When a continuous chip is formed during machining, surface roughness is low. Machinability improves as the formation of continuous chips increases; R is then no longer a factor, and the machinability factor η_2 is given by

$$\eta_2 = \frac{V_m}{V_w \times W} \tag{3}$$

If tool wear is also negligible, the machinability factor becomes

$$\eta_3 = \frac{V_m}{W} \tag{4}$$

Thus machinability improves as a specific cutting force decreases, since W is a reciprocal of the specific cutting force. The maximum depth of cut that produces continuous chips is a basis for measuring machinability. The greater the depth, the higher the efficiency for cutting smooth and accurate parts when a continuous chip is formed.

Maximum Depth of Cut. The maximum depth of cut that can produce a continuous chip is shown in Figure 3 for various plastics.

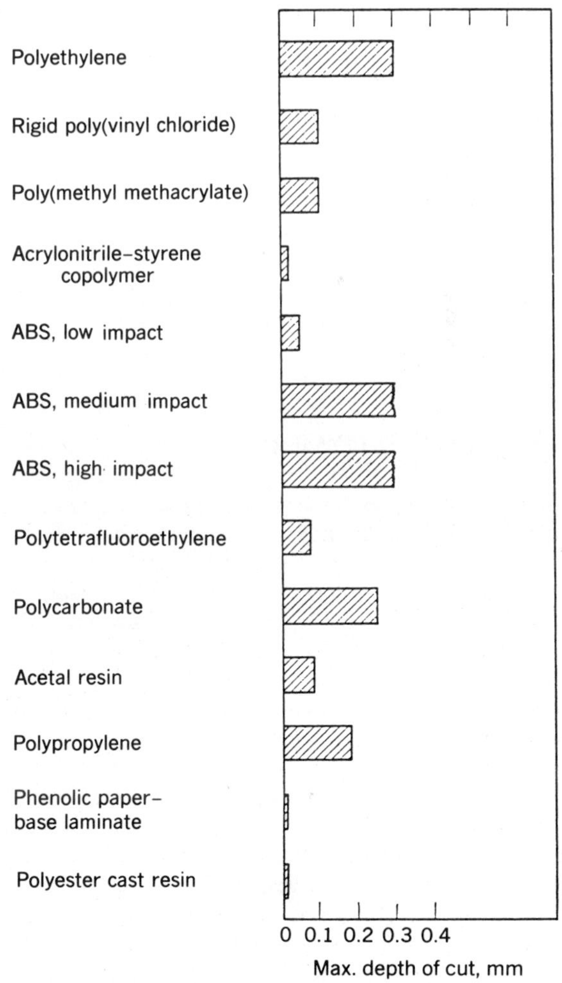

Figure 3. Maximum depth of cut that can produce continuous chips using a tool with a rake angle of 0° at a cutting speed of 400 m/min.

Table 3. Machinability Grade of Plastics

Machinability grade	Thermosetting plastics	Thermoplastics
excellent		polyethylene ABS, medium and high impact polytetrafluoroethylene polyamide, nylon-6 and -6,10 polycarbonate
good to excellent		rigid poly(vinyl chloride) acrylic resin ABS, low impact acetal resin polypropylene polyamide, nylon-6,6
good		acrylonitrile–styrene copolymer
fair	phenolic paper-base and cloth-base laminate toluene paper-base laminate	polystyrene
poor	melamine paper-base laminate phenolic glass-base laminate polyester cast resin epoxy cast resin	

Grades. The machinability grades for several plastics are given in Table 3.

BIBLIOGRAPHY

A. Kobayashi, *Mod. Plast.* **40,** 11 (July–Aug. 1963); **41,** 1 (Sept.–Dec. 1963); **41,** 5 (Jan. 1964).

A. Kobayashi and K. Saito in B. Rosen, ed., *Fracture Processes in Polymeric Solids,* Wiley-Interscience, New York, 1964, pp. 489–516.

A. Kobayashi, *Machining of Plastics,* McGraw-Hill Inc., New York, 1967; R. E. Krieger Publishing Co., Inc., New York, 1980.

A. Kobayashi, *SME Technical Paper MR80-214,* Society of Manufacturing Engineers, Dearborn, Mich., 1980.

Akira Kobayashi
Ibaraki Polytechnic College

MACROCYCLIC POLYMERS

Macromolecules are commonly represented as long, flexible, randomly coiled chains that may be branched. Some cyclic structures may be present, although in small amounts. Ring–chain equilibria occur, however, in a number of polycondensation systems, implying the presence of functional links in the polymer chain. Some biological macromolecules also exhibit cyclic structures.

Cyclic polymer samples covering a broad range of molecular weights, and exhibiting low polydispersity have been prepared by ring–chain equilibration followed by fractionation, or by end-to-end cyclization followed by removal of the products formed by chain extension. The influence of cyclization on morphology and dimensions of macromolecules is studied by comparing their solution behavior with that of their linear counterparts.

Because of lower hydrodynamic volume and higher segment densities, macrocyclic polymers exhibit lower limiting viscosity numbers, lower frictional coefficients, and higher size-exclusion chromatography (sec) elution volumes than their linear homologues.

The lack of end groups influences melt viscosity and thermomechanical behavior in the bulk. Because the molecular weight dependence of the conventional glass-transition temperature is assumed to originate from end group effects, the T_g of macrocyclic compounds is expected to be independent of molecular weight.

Cyclization Methods

The synthesis of cyclic macromolecules is based either on the existence of a ring–chain equilibrium in a polymer system, or on end-to-end cyclization, starting from a polymer chain with reactive chemical functions at both ends. The latter reaction should be carried out at high dilution to favor intramolecular coupling. Careful fractionation is required to remove the "polycondensate" formed simultaneously.

Solution Behavior of Macrocyclic Polymers

Cyclic polymers have, of course, no chain ends. The mean square radius of gyration, defined as the average of the mean square distance between each segment and the center of mass, is used to characterize the molecular dimensions.

Limiting-viscosity Number. The limiting-viscosity numbers $[\eta]_L$ and $[\eta]_C$ have been measured on linear and cyclic polymers, respectively, in θ-solvents, as well as in good solvents.

The values obtained on polydimethylsiloxane (PDMS) samples agree with the expected values, although the range of molecular weights covered was not very broad. The ratio $[\eta]_C/[\eta]_L$ is on the order of 0.66 in poor solvents and somewhat lower in good solvents (Fig. 1); the exponents of the viscosity laws for linear and cyclic PDMS are almost the same.

The unperturbed dimensions K of a cyclic macromolecule can be determined from viscosity measurements in a good solvent over a range of molecular weights, as in the case of linear macromolecules.

Translational Diffusion Constants and Sedimentation Constants. For translational diffusion measurements on linear and cyclic PDMS samples, both the classical boundary spreading and quasielastic light scattering techniques are used. The ratio of the frictional coefficients $f = kT/D$ for cyclic and linear

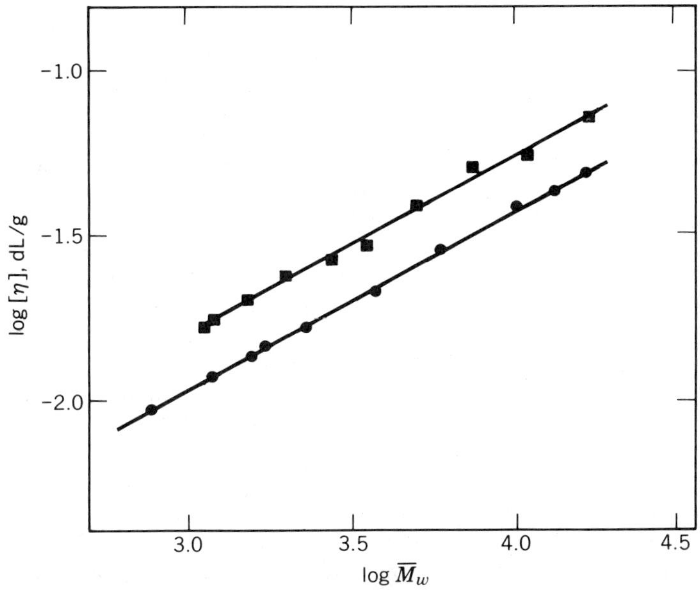

Figure 1. Logarithmic plots of the limiting viscosity number vs molecular weight for linear (■) and cyclic (●) polydimethylsiloxane fractions measured in butanone at 20°C.

polymers is 0.83, confirming theoretical expectations. This indicates that the hydrodynamic radius of a cyclic polymer in a good solvent is about 15% smaller than that of its linear homologue.

Coil Expansion Near the θ-point. The θ-temperature of cyclic polystyrene (PS) is ca 6°C lower in cyclohexane, indicating that some volume effect is present in cyclic polymers at a temperature at which none is evident in linear homologues. Topological interactions, or enhanced segment density, might be the reason for this phenomenon.

Size-exclusion Chromatography (sec). The hydrodynamic volume of the solute molecules determines chromatographic retention in sec columns. It can thus be anticipated that the elution volume of a cyclic polymer should be higher than that of the linear polymer of the same molecular weight; this was confirmed experimentally on cyclic PDMS and on cyclic PS.

BIBLIOGRAPHY

J. A. Semlyen, *Cyclic Polymers,* Elsevier Science Publishing Co., Inc., New York, 1986; *Adv. Polym. Sci.* **21,** 43 (1976).

G. Hild, C. Strazielle, and P. Rempp, *Eur. Polym. J.* **19,** 721 (1983).

J. Roovers and P. M. Toporowski, *Macromolecules* **16,** 843 (1983).

J. Roovers, *J. Polym. Sci. Polym. Phys. Ed.* **23,** 1117 (1985).

G. Hadziioannou and co-workers, *Macromolecules* **20,** 493 (1987).

C. J. C. Edwards, R. F. T. Stepto, and J. A. Semlyen, *Polymer* **23,** 869 (1982).

PAUL REMPP
CLAUDE STRAZIELLE
PIERRE LUTZ
Institut Charles Sadron

MACROMERS

Macromers are polymers of molecular weight ranging from several hundred to tens of thousands, with a functional group F at the chain end that can further polymerize:

∿∿∿F

where F = —CH=CH$_2$, —CHCH$_2$ (with O bridging, epoxide), —CH(COOH)—COOH, —CH(OH)—OH, —CH(NH$_2$)—NH$_2$

The term macromer was originally a trademark registered by Milkovich of PPG Industries (an abbreviation of macromolecular monomer, macromonomer), but is now a common noun. The functional group F may be any polymerizable group. Vinyl groups are most typical, but polymers with dicarboxylic acid, diol, or diamine end groups are used in step-growth polymerizations. Macromers give easy access to well-defined graft copolymers.

Synthesis of Graft Copolymers

Graft copolymers can be synthesized by several methods. The reaction between two polymers occurs typically between a living polymer anion F$_2$ and a reactive functional group F$_1$ in preformed polymer, as shown in Figure 1.

A monomer can be grafted on a preformed polymer by thermally initiating the polymerization of the monomer at the functional groups (I) on the preformed polymer; an active species may be produced by irradiation (Fig. 2). In the syntheses of graft copolymers by the copolymerization of a macromer with a comonomer, the macromer forms the branches and the comonomer the backbone chain in a graft copolymer (Fig. 3). This is the method of choice.

Polymerization Reactivity

Macromers are introduced as branches into graft copolymers by copolymerization. Vinyl-polymerizable macromers are usually difficult to homopolymerize by radical initiation because of the low concentration of double bonds. Chain-transfer reactions cannot be ignored even in solvents with a chain-transfer constant value in the range of 10^{-5}, which usually can be used for polymerization. Anionic polymerization gives higher molecular weights. In copolymerization with low molecular weight comonomers, the polymerizable functions in macromers

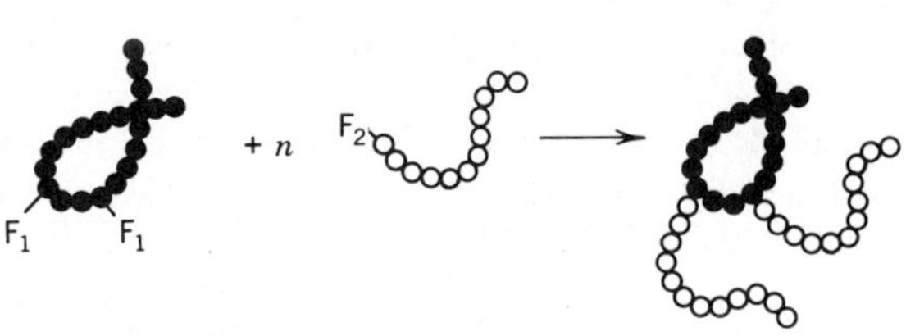

Figure 1. Synthesis of a graft copolymer from two polymers: ●, prepolymer; ○, living polymer anion. F$_2$ represents an anionic site.

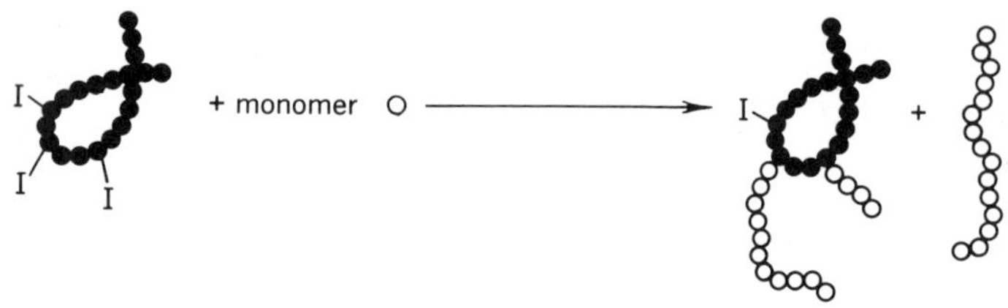

Figure 2. Grafting on a preformed polymer by thermal initiation or irradiation: ●, prepolymer; ○, monomer.

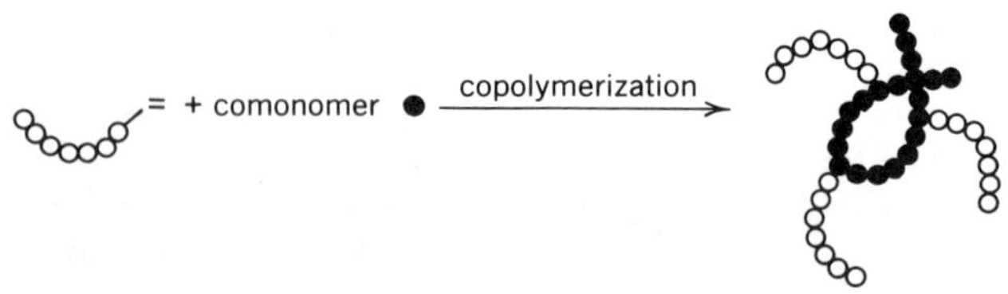

Figure 3. Graft copolymerization with a macromer: ○, macromer; ●, comonomer.

behave normally, although in some cases problems concerning the homogeneity of the system are seen. The reactivity r_M of macromer in copolymerization can be treated by the Mayo-Lewis equation:

$$\frac{d[A]}{d[M]} = \frac{[A]}{[M]} \cdot \frac{r_A[A] + [M]}{[A] + r_M[M]}$$

where $[M]$ and $[A]$ indicate the concentrations of a macromer and a comonomer, respectively.

Characterization

The principal problems in characterizing macromers are purification and quantitative determination of functionality in high molecular weight macromers. Low molecular weights facilitate the use of ordinary techniques. Terminal-group functionality can be determined by uv or ft nmr methods. Molecular weight data can be obtained by osmometry and gel-permeation chromatography. In this way, a variety of well-defined macromers becomes accessible.

Industrial Applications

Graft copolymers are surface-active and can be used as compatibilizers for polymer blends, surface-modifying agents, thermoplastics, medical materials, and other applications.

BIBLIOGRAPHY

Y. Kawakami, *Kagaku* **37,** 588 (1982).

P. Rempp, P. Lutz, P. Masson, and E. Franta, *Makromol. Chem. Suppl.* **8,** 3 (1984).

R. Asami and M. Takaki, *Makromol. Chem. Suppl.* **12,** 163 (1985).

K. Ito, H. Tsuchida, A. Hayashi, T. Kitano, E. Yamada, and T. Matsumoto, *Polym. J.* **17,** 827 (1985).

YUHSUKE KAWAKAMI
Nagoya University

MACROMOLECULAR DYNAMICS

The dynamics of macromolecules are those time-dependent configurational changes that occur as a result of thermal equilibration processes of the macromolecules with their surroundings. The kinetics of these microscopic rearrangements determine the macroscopic flow and viscoelastic properties of polymer solutions and melts, as well as the transport properties of the polymeric chain themselves and many related phenomena.

Because of their large size there is a hierarchy of dimensions within macromolecules, each of which exhibits motions on a different time scale, as shown in Figure 1 for a simple —CH_2— backbone chain. On the scale of Figure **1a**, it is the rotation about backbone and side-group bonds that is important. The time scales are determined by energy barriers to bond rotation and are typically on the order of 10^{-11} s. In Figure **1b** and **c**, the spatial scales are characteristic of larger segments and of the molecule as a whole, respectively; these change their configurations over longer times. Motions at all scales are interdependent within a single molecule, being connected not only along the chain contour but also via hydrodynamic interactions, and the resulting detailed dynamic behavior can be very complicated. The fact that many of the properties of interest in polymer solutions or melts depend on relaxations at long times on the scale of Figure **1c** facilitates the theoretical understanding of macromolecular dynamics. For configurational rearrangement on these scales, the local dynamic behavior on the scale of Figure **1a**, which is the main problem for the case of simple (monomeric) liquids, is averaged out, and can be accounted for by some mean frictional coefficient per unit length of the chain; the value of this parameter must be determined empirically for a given system, but once this has been done, the dynamic properties of that system, at a given temperature, will depend only on the overall size of the polymer chain and on its concentration.

Polymer Dynamics and the Entanglement Effect

The basic features manifested by entangled polymer solutions or melts (or uncross-linked rubbers) are illustrated by a simple experiment. If such an entangled system is at equilibrium and is subjected to a fixed, steady stress σ at some instant, eg, in a Couette viscometer, its subsequent deformation may be monitored by measuring the strain $e(t)$ as a function of time t. In Figure 2, the variation with time of the quantity $J(t) = e(t)/\sigma$, known as the creep compliance, is schematically shown.

At very short times, typically $t < 10^{-10}$ s, all polymers exhibit a glassy region in common with simple, monomeric liquids when only small-scale motion is possible. At longer times, the behavior depends on the degree of polymerization (DP) or N of the polymer being stressed: for short chains, when $N < N_e$ (the entanglement DP), the polymer sample goes from a glasslike state, through a so-called transition region, directly to liquidlike flow. N_e differs somewhat from polymer to polymer, but is typically on the order of 200–300 backbone units in vinyl polymers. For $N > N_e$ a plateaulike region in the compliance appears, up to some terminal time τ_t, and beyond this the polymer flows like a liquid, with the strain increasing linearly (and irrecoverably) with time. In models of entangled chains, the terminal time τ_t is identified with the time for the polymer molecules to disentangle from each other, ie, it signals

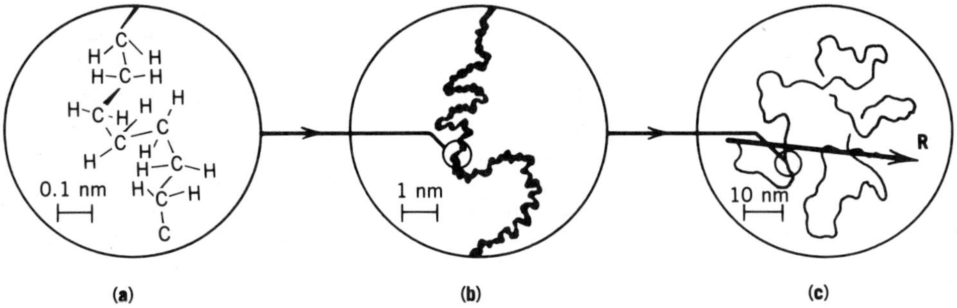

Figure 1. The spatial scales in a randomly coiled polyethylene molecule. (**a**) Monomeric scale; (**b**) segmental scale; (**c**) macromolecular scale.

the breakdown of the transient network or rubberlike structure. A large body of experiments shows that the conventional steady-flow viscosity $\eta \propto \tau_t$. The constant of proportionality is the inverse of the plateau compliance J_0 and may be thought of as the elastic modulus of the transient network existing over the rubbery plateau time. In the case of a large number of entangled, linear polymer melts, it has been found that both the viscosity η and the terminal time τ_t vary rapidly with the degree of polymerization of the polymer

$$\eta,\ \tau_t \propto N^{3.4}$$

Thus the effect of entanglements on both the relaxation (as manifested by τ_t) and the flow properties (through η) of polymer melts is very marked.

Quantitatively,

$$\tau_t = \tau_{R1} N^{3.4}$$

where τ_{R1} is comparable to the characteristic rotational relaxation time for a single monomer ($N = 1$) in a liquid of monomers. This time is typically 10^{-11} s; since high polymers can reach degrees of polymerization $N \simeq 10^4$, the overall molecular relaxation process can take on the order of tens of seconds for long entangled chains, a distinctly macroscopic time, which is also manifested in the extremely high viscosities of entangled melts.

Theory

Linear Chains. The detailed solution to the problem of the dynamic behavior of many entangled chains is extremely complex and all theoretical treatments have involved some form of approximation to account for the entanglement phenomenon. The most successful of these accounts for the fact that entangled chainlike molecules in a network cannot cross through fixed obstacles, eg, permanent cross-links, by considering them confined inside a virtual tube or pipe defined by the locus of contacts of the polymer with the noncrossable obstacles, ie, within the tube the polymer can move freely, but it may not cross through the walls of the tube. In this model the motion takes place by random curvilinear or reptilelike motion (called reptation) that does not violate the entanglement criterion (Fig. 3).

The original model was applied to the dynamics of flexible chains in fixed networks such as gels, and it has also been extended to the case of entangled melts. In this case, the tube about any molecule is defined by the locus of its entanglements with neighboring chains. Furthermore, the disengagement time can be identified with the terminal time. It is the time required for a polymer chain to disentangle completely from its surroundings and so lose all correlation with its original configuration. The reptation model provides the basis for the currently accepted molecular theories of the dynamics and viscoelastic properties of entangled polymers, and its qualitative and quantitative predictions have had considerable suc-

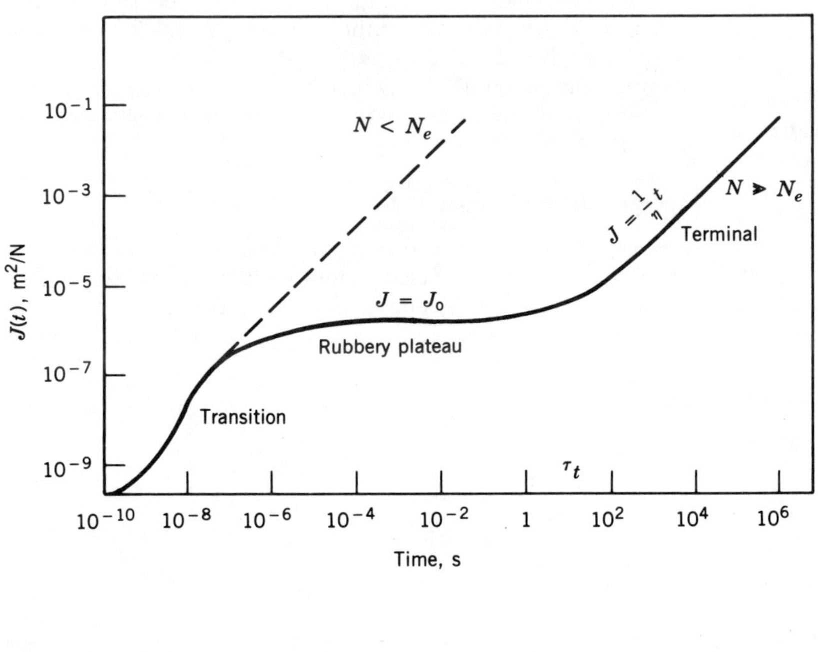

Figure 2. Creep compliance $J(t)$ of flexible polymer melts. The broken line is the behavior at low N ($<N_e$) values; the solid line is for highly entangled polymers ($N \gg N_e$). To convert m²/N to cm²/dyn, divide by 10.

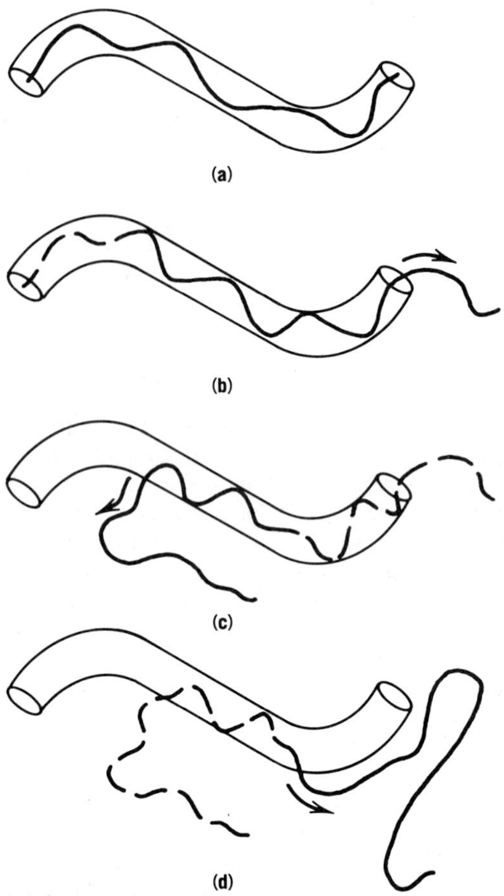

Figure 3. In the reptation model, a chain laterally constrained in a fixed tube (or fixed obstacle matrix) moves by random back-and-forth curvilinear motion with its ends free to choose new directions. (**a**) Initial configuration, $t = 0$; (**b**), (**c**) subsequent motion with progressively less of the chain contour confined to the initial constraining tube; (**d**) $t = \tau_d$ (average time for disengagement). The chain has completely escaped its initial tube, and its new conformation is uncorrelated with its conformation in (**a**).

cess. It also serves as the starting point for improved theoretical calculations of these properties.

Nonlinear Chains. Although the reptation-in-tubes model provides an attractive and largely successful solution for the translation and relaxation of entangled linear polymers, it must be modified for the case of nonlinear chains such as branched or cyclic polymers. For star-branched polymers, a reptative motion within enclosing tubes involves doubling up of two or more arms. This process is strongly resisted by the molecule as it requires considerable entropic (or elastic) energy to double up substantial portions of its arms in this way. It is energetically cheaper for an entangled star-branched molecule to renew its configuration by retracting an arm so that it retraces its path along the confining tube (its entangling neighbors) to the position of the center monomer; it may then extend out again to adopt any new configuration uncorrelated with its original one. The time required for such an arm relaxation process may again be identified with the terminal relaxation time. The dynamics of processes such as the arm retraction shown in Figure 4 is considerably more sluggish than a straightforward reptation process (for the same chain length) because the additional free energy required to attain the doubled-up configuration of Figure **4b** before relaxation (as in Fig. **4c**) can take place, implies an exponentially low probability for such configurations.

Commercial polymers are frequently branched at random, in a treelike fashion, and cyclic (ringlike) polymer chains have also been synthesized. The dynamic behavior of these may also be modeled in terms of the tube concept. The essential idea in all cases of an entangled nonlinear molecule is that in order for the polymer as a whole to relax it must pass through configurational states which are quasi-linear.

Tube Renewal. The discussion of theoretical models of entangled chains in terms of reptation within tubes has so far assumed that the entangling environment about any chain, and hence the tube about it, are fixed in time. This is not necessarily true in the case of an uncross-linked polymer melt. In these cases, entangling neighbors of the chain of interest, which define the tube about it, may themselves reptate away; whenever one of these diffuses away a constraint about the enclosed molecule momentarily relaxes, and the tube at that point may distort slightly. Through a series of such small steps the tube about the molecule may itself renew its configuration, and this provides an additional mechanism by which the enclosed molecule relaxes, over and above the curvilinear reptation within the tube.

Experimental Methods. The molecular properties of interest (which control the bulk behavior) are the spectrum of relaxation times of entangled molecules and the closely related diffusion coefficient. These two are intimately connected, both being controlled by friction of the macromolecules with their environment and the need to overcome, in moving, the constraining effect of neighboring chains with which they are entangled.

The various experimental approaches have focused not only on measuring the absolute values of these parameters, but also on the way in which they vary with polymer size, polymer concentration in entangled solutions, and molecular structure, eg, linear or nonlinear (see also DIFFUSION; TRANSPORT PROPERTIES).

Experimental Indications

Overall, the experimental evidence from studies of both the longest relaxations (and closely related steady flow viscosities), and especially the diffusion of linear and nonlinear polymers in entangled solutions and melts, supports the idea that these

Figure 4. A mechanism for arm relaxation and center-of-mass diffusion by retraction of one of the arms, (**a**) and (**b**), and its subsequent unfolding, as shown in (**c**). Such a process requires a doubling up of one arm only.

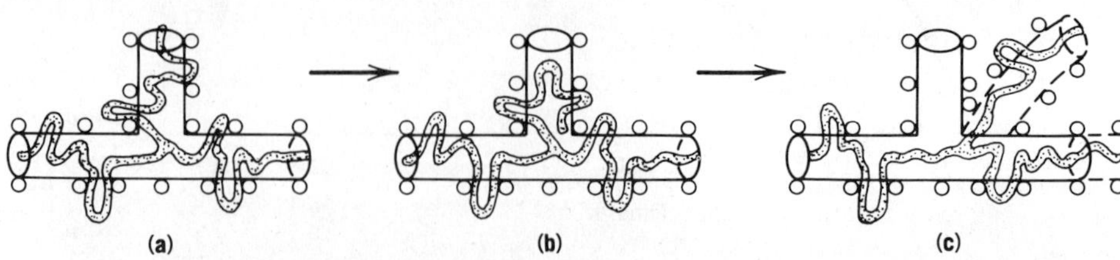

processes proceed via curvilinear motion (reptation) of the entangled molecules. Although some serious discrepancies are still unexplained (notably the experimental 3.4 power-law variation with N of η and τ, compared with the theoretical third-power prediction), there is, by and large, both qualitative and quantitative agreement between the data and the predictions of detailed theories based on the reptation concept. These, therefore, currently provide the most satisfactory theoretical framework for the understanding of the dynamics of entangled polymers at the molecular level. They open the prospect for controlled design of polymers (with respect to their size and topology) to yield optimal material properties for applications controlled by the dynamics of the entangled chains.

Applications

The dynamics of macromolecules influence all properties of polymeric materials where motions of the molecules are involved. These include viscoelasticity and stickiness, bonding between polymers, polymer crystallization, crazing (qv) in polymer glasses, diffusion-controlled reactions, and diffusion through fixed networks.

BIBLIOGRAPHY

J. D. Ferry, *Viscoelastic Properties of Polymers,* 3rd ed., John Wiley & Sons, Inc., New York, 1980.

M. Doi and S. F. Edwards, *Polymer Dynamics,* Oxford University Press, London, 1986.

P. G. de Gennes, *Scaling Concepts in Polymer Physics,* Cornell University Press, Ithaca, N.Y., 1979.

J. Klein, *Macromolecules* **19,** 105 (1986).

JACOB KLEIN
The Weizmann Institute

MAGNETIC POLYMERS

Ferromagnetism is a well known phenomenon in metallic and inorganic materials. Not until recently have organic materials been discovered to be ferromagnetic. Very little work has been done on organic ferromagnets, including polymers. This article reviews the current status of studies on organic ferromagnets with particular emphasis on polymeric materials.

Magnetic Properties

Diamagnetism is a phenomenon based on a change of the electron path; hence, all materials have associated diamagnetic components. Paramagnetism, ferromagnetism, and antiferromagnetism are all phenomena based on spin axis orientation.

Iron-containing Organic Ferromagnets

Ferromagnetism has been observed for pure metals, alloys, inorganic salts, and others, but it has not been observed in purely or mostly organic substances until recently. However, as early as 1957, the synthesis of a series of heterocyclic molecules including poly(2,6-pyridinediylmethylidenenitrilohexamethylene nitromethylidene) (PPH) was reported. This polymer is easily synthesized by the condensation reaction of 2,6-pyridinecarboxaldehyde with 1,6-hexanediamine. When PPH reacts with $FeSO_4 \cdot 7H_2O$ in an aqueous solution, the white powder instantaneously turns a bluish-purple color. This substance is metamagnetic. In the ferromagnetic phase, the material has a net magnetic moment of 1.95×10^{-10} Wb·m ($155{,}000 \times 10^{-6}$ cgs). This iron-containing polymer (**1**) is insoluble in common organic solvents and does not melt below 300°C. Compositional analysis suggests that each iron is surrounded by six nitrogen atoms.

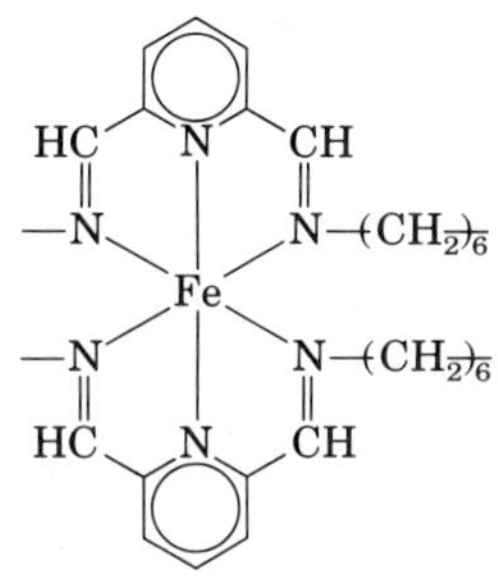

(**1**)

Some iron-containing molecular complexes have also been studied. One-dimensional decamethylferrocinium, 7,7,8,8-tetracyano-*p*-quinodimethanide, $\{Fe[\eta^5\text{—}(CCH_3)_5]_2\}^+$ $(TCNQ)^-$, is reported to be metamagnetic and exists as a ferromagnet on application of an external field >119 kA/m (1500 Oe). On the other hand, a similar molecule, decamethylferrocinium tetracyanoethylenide, $Fe[\eta^5\text{—}(CCH_3)_5]_2{}^+$ $(TCNE)^-$, is ferromagnetic.

Purely Organic Ferromagnets

Whereas iron-containing organic molecules have been demonstrated to be ferromagnetic, the synthesis of noniron-containing organic ferromagnets has been difficult.

Attempts to synthesize organic ferromagnets have apparently only begun in recent years, eg, preparation of a ferromagnetic aromatic tetracarbene as shown in Figure 1. The esr spectra obtained in the range 1.8–150 K and the temperature dependence of the total signal intensity indicate a nonet structure in the ground state.

The magnetic properties of the polycarbenes, including the temperature dependence of the paramagnetic susceptibility, have been further studied in detail.

From a plot of χ^{-1} vs T, the effective magnetic moment of

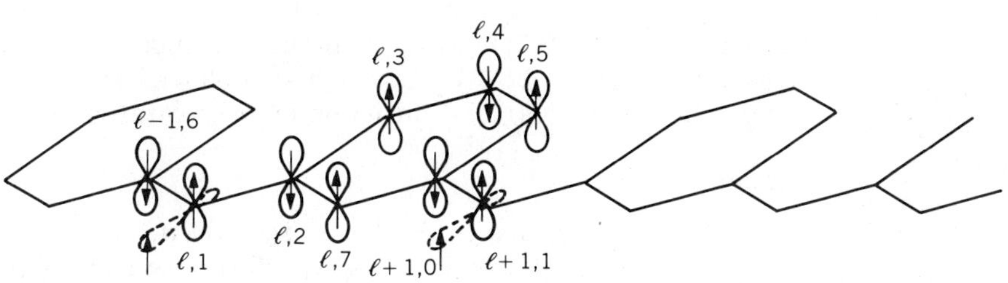

Figure 1. Ferromagnetic polycarbene. Orbitals shown by solid lines are π-orbitals; dotted line orbitals are nonbonding.

$\mu_{eff} = 9.08\,\mu_B$ with a local spin of $S = 8.1$ was calculated. The magnetization curve obtained at 2.1 K shows complete saturation in a magnetic field greater than 2.5 T. When the size of the fine metal particles is below 3.0 nm in diameter, a single domain structure is expected. If the anisotropic energy is large, these fine particles exhibit magnetic hysteresis, behaving like tiny ferromagnets. Particles with small anisotropy, on the other hand, can be magnetized at any direction without exhibiting any coercive force. In the latter case, the electron spins in the particle are aligned by the ferromagnetic exchange interaction; however, these particles behave independently of each other, like a paramagnet. This is called superparamagnetism. The polycarbenes studied can be regarded as organic molecular superparamagnets.

It has been reported that *s*-triaminobenzene polymerized by iodine in solution exhibits ferromagnetism. A few percent of the total material synthesized showed especially strong magnetic properties.

Another ferromagnetic polymer was reported when a stable biradical, 1,4-bis-(2,2,6,6-tetramethyl-4-oxy-4-piperidyl-1-oxyl)-butadiyne (BIPO) was polymerized.

Theory of Organic Ferromagnets

The origin of ferromagnetism is poorly understood. Two extreme theories proposed by Heisenberg and Block exist, which are synonymous with the valence bond method and the molecular orbital method for the study of molecular electronic structure, respectively. The Heisenberg theory is based on electron exchange between localized electronic systems, and Block's approach is based on collective electron theory. Until recently, ferromagnetic substances have only been known for metals or metallic oxides containing *d*-electrons. Now, however, the possibility of organic ferromagnets has been suggested.

BIBLIOGRAPHY

R. Friend, *Nature London* **326,** 335 (1987).

T. Sugano, M. Nomura, K. Awaga, P. L. Kieng, T. Ohta, and M. Kinoshita, *Bull. Chem. Soc. Jpn.* **59,** 2615 (1986).

Y. Teki, T. Takui, K. Itoh, H. Iwamura, and K. Kobayashi, *J. Am. Chem. Soc.* **105,** 3722 (1983).

Yu. V. Korshak, T. V. Medvedeva, A. A. Ovchinikov, and V. N. Spector, *Nature London* **326,** 370 (1987).

HATSUO ISHIDA
Case Western Reserve University

MALEIC AND FUMARIC POLYMERS

The polyfunctional monomers maleic anhydride (2,5-furandione or *cis*-butenedioic anhydride), maleic acid ((*Z*)-2-butenedioic acid), and fumaric acid ((*E*)-2-butenedioic acid) are chemicals of worldwide commercial importance. The presence of both a reactive double bond and anhydride or acid moieties confers unique properties to maleic anhydride (MAN,**1**), maleic acid (MA,**2**), and fumaric acid (FA,**3**), providing opportunities for applications in chain- and step-growth polymerization systems. The largest use of these monomers is in the manufacture of unsaturated polyesters, where both the unsaturation and the carboxylic acid groups are utilized.

Maleic anhydride and maleic acid are also important raw materials for a host of other products, including alkyd resins (qv), surface coatings, adhesives, lubricant additives, plasticizers, copolymers, and agricultural chemicals. Fumaric acid occurs in many plants and is named after the genus *fumaria.* It is used in food additives, quick-setting inks, paper-sizing agents, and lacquers.

Maleic anhydride, first produced by dehydration of maleic acid, is second only to phthalic anhydride in commercial importance. Production has grown from 2×10^3 t in 1940 to ca 163×10^3 t in 1983.

Maleic acid can be isomerized to its trans isomer, fumaric acid, by a variety of techniques; for commercial production, catalytic isomerization starting with purified maleic anhydride is preferred.

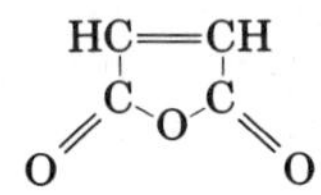

(1)

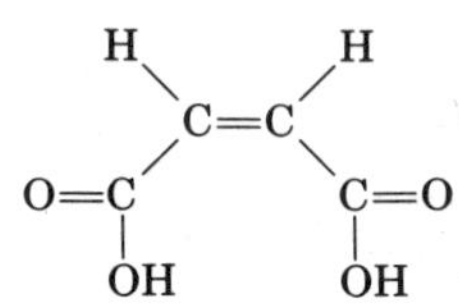

(2)

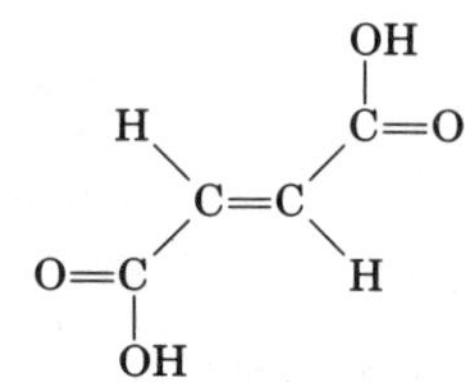

(3)

Monomers

Physical Properties. Physical properties of MAN (mol wt 98.06), MA (mol wt 116.07), and FA are given in Table 1.

Chemical Properties. The conjugated C=C and C=O bonds may participate individually in chemical reactions or both may react. Since MAN is susceptible to both ring-opening and double-bond reactions, a clear understanding of the chemistry involved is required for polymerization studies. Union Carbide Corp. uses MAN in a proprietary process to produce butanediol, γ-butyrolactone, and tetrahydrofuran. Alkyl maleates and fumarates have been reviewed.

Manufacture. The predominant commercial route to MAN is the vapor-phase oxidation of hydrocarbons, eg, benzene, *n*-butane, or a C_4-hydrocarbon mixture, over a solid catalyst.

Health, Safety, and Toxicology. Maleic anhydride has an oral LD_{50} for rats of 481 mg/kg; the subcutaneous LD_{50} is 610 mg/kg. By ingestion in single doses MAN is considered slightly toxic.

Maleic anhydride is an intense local irritant, especially where moist tissue promotes hydrolysis. It is regarded as a sensitizer of the skin and respiratory tract with a recommended workplace TLV of 0.25 ppm (1 mg/m^3). Left in contact with the skin, MAN can cause severe pain.

Polymerization

The presence of both an electron-acceptor double bond and an anhydride moiety confers unique properties to MAN, allowing it to be used in both addition- and condensation-type polymerizations. In 1961, it was shown that MAN could be homopolymerized with uv radiation, in the presence of radical initiators, anionically with various bases, electrochemically, and under shock waves. Polymerization methods include initiation by radiation, free-radical initiation (the preferred route for commercial production), and initiation by ionic catalysts.

Polymer Systems

MAN Polymer. *Properties.* Maleic anhydride polymer prepared by radical initiators is a creamy white, low molecular

Table 1. Physical Properties

Property	MAN	MA	FA
mp, °C	52.8	138–139[a]	287
bp, °C	202	~138, dec	290[a]
specific gravity			
at 20°C, solid	1.480	1.590	1.635
at 70°C, molten	1.3		
molar volume		81	79
heat of formation, kJ/mol[b]	−470.41	−790.57	−811.03
free energy of formation, kJ/mol[b]		−625.09	−655.63
heat of combustion, MJ/mol[b]	−1.390	−1.359	−1.335
heat of hydrogenation, kJ/mol[b]		−153.2	−130.2
heat capacity, kJ/(kg·K)			
solid	1.21		
molten	1.67		
heat of vaporization, kJ/mol[b]	54.8		
heat of fusion, kJ/mol[b]	13.65		
heat of solution, kJ/mol[b]		18.6	
heat of hydrolysis, kJ/mol[b,c]	−34.9		
heat of neutralization, kJ/mol[b]	126.9		
flash point, °C			
open cup	110		
closed cup	102		
flammable limits, vol %			
lower	1.4–3.4		
upper	7.1		
autoignition t, °C	477		
vapor density (air = 1)	3.38		
space group, nm	$P2_12_12_1$	$P2_1/C$	
a	0.7180	0.747	
b	1.123	1.015	
c	0.539	0.765	
dipole moment at 25°C, 10^{-30} C·m in dioxane	13.2	10.6	8.17
viscosity, mPa·s (= cP)[d]			
at 60°C	1.61		
at 90°C	1.07		
at 150°C	0.60		
pH, water solutions			
1×10^{-2} *M*			2.42
5×10^{-3} *M*			2.62
1×10^{-4} *M*			3.10
acid dissociation constants		$pK_{a1} = 1.83$ $pK_{a2} = 6.07$	$pK_{a1} = 3.03$ $pK_{a2} = 4.44$
polarographic half-wave potential, V	−0.72[d,e]		

[a] Sublimes at 165°C, 0.23 kPa (1.7 mm Hg).
[b] To convert J to cal, divide by 4.184.
[c] Hydrolyzes slowly.
[d] Absolute viscosity at 70°C = 1.5 mPa·s (= cP).
[e] Various reduction potentials reported with first (E_{p1} = 1.10) and second (E_{p2} = −1.30) cyclic voltammetric reduction potentials probably most usable.

weight material that is soluble in water, dilute alkali, ketones, acetonitrile, lower alcohols, esters, and nitroalkanes.

The infrared spectrum of poly(MAN), analogous to the monomer, exhibits strong carbonyl peaks at 1859 and 1880 cm^{-1}; the polymer carbonyl peaks are broader. The nmr spectrum (DMSO-*d*6) exhibits a broad resonance at 5.9 τ for the CH residue, and a resonance at 3.7 τ attributable to the polymer with carboxylic acid groups; the polymer decomposes at its melting point (320°C).

Poly(maleic acid) is readily obtained by hydrolysis of poly(MAN). It is a polyelectrolyte with a high potential charge on every carbon atom on the polymer backbone. The polymer normally behaves as a monobasic polyacid on potentiometric titration. Intramolecular hydrogen bonding between adjacent groups has been offered as an explanation for this behavior.

Uses. The patent literature suggests a large number of potential applications for MAN and MA polymers, including water-treatment chemicals (detergent additive, boiler-scale inhibitors, heavy metal chelating and removal agents). With the increasing concern over phosphate detergents, a substantial market could develop for MAN polymer or copolymers.

MA and FA Polymerization. Nippon Oil and Fats Co. claims to have solved the problems of fumaric acid polymerization. Their polymer is claimed to have superior heat stability and to be useful for optical applications and separation membranes. In contrast, MA can be easily polymerized in water in the presence of $K_2S_2O_8$ and poly(*N*-vinylpyrrolidinone). The polymer–MA complex, which promotes polymerization, decomposes at ca 100–130°C. Low molecular weight MA polymer is manufactured by polymerization of monoalkali or monoammonium maleates in aqueous solution with peroxide initiators. The method provides efficient control of molecular weights and high yield of polymer.

MAN Random Addition Copolymerizations. Maleic anhydride is easily copolymerized with vinyl comonomers. In fact, many 1,2-disubstituted monomers that normally fail in radical polymerization yield copolymers with MAN having superior polarity and rigidity, and enhanced T_g and functionality. The MAN functionality on the polymer backbone promotes or improves hydrophilicity, adhesion, dyeability, heat distortion, and compatibility with other polymers and provides a cross-linking moiety. Copolymers containing MAN are useful in coatings, adhesives, plastics, and fibers.

Vinyl-substituted Aromatic Monomers. Styrene is the most important monomer copolymerized with MAN. Polymerization of the styrene–MAN pair is highly exothermic unless controlled with diluents, cooling, and careful initiator selection. The reaction is most often run in solution with peroxide initiators at 50–150°C with highest rates under heterogeneous conditions.

Other vinyl-substituted aromatic monomers, eg, α-methylstyrene, *p*-methylstyrene, vinyltoluene, *p*-*t*-butylstyrene, vinylnaphthalene, vinylanthracene, *p*-arylsulfoxyaminostyrene, and *p*-styrenesulfonic acid, have also been copolymerized with MAN via radical initiation. Only poly(styrene-*co*-MAN) has achieved commercial importance.

Vinyl Halides. Vinyl chloride (VC) readily copolymerizes with MAN and MA esters with free-radical initiators. Dialkyl maleates, such as the diethyl ester, also copolymerize readily with VC, entering the polymer chain at a faster rate so that the MA ester content of the polymer is higher than that in the feed. Copolymers with ca 20% dimethyl maleate have very good processing properties. In the late 1980s, production of VC–maleate copolymers ceased in the United States.

Vinyl Esters. Vinyl acetate (VA) readily undergoes radical copolymerization with MAN. Other vinyl esters, eg, vinyl abietate, vinyl stearate, vinyl perfluorobutyrate, and vinyl 3-(2,5-di-*tert*-butyl-4-hydroxyphenyl) propionate, easily copolymerize with MAN, maleates, and fumarates. Copolymers of ethoxyalkyl maleates are used as surfactants for vinyl acetate latex. Photoinitiated copolymerization of VA–MAN gives good yields of random copolymer.

Acrylic Acids. Radical-initiated copolymerization of acrylic acid (AA) and methacrylic acid (MAA) with MAN have received little attention. When MAN concentration exceeds the acrylic or methacrylic acid concentration in the feed, only 1:1 alternating copolymers are observed.

Acrylic and Methacrylic Esters. Acrylic ester–MAN mixtures polymerize in solution with peroxides or azobisisobutyronitrile initiators; the highest rates of copolymerization are observed for heterogeneous systems.

Maleic anhydride copolymerizes easily with acrylates or methacrylates in benzene or dioxane, with or without photoinitiation. Free-radical, bulk polymerization of alkyl methacrylates may be achieved with good control at 80–140°C, by first charging the reactor with 5–15 wt % dialkyl maleates or fumarates. Typical methyl methacrylate–MAN copolymerization with radical initiators produces random copolymers.

Acrylamides. Maleic anhydride and acrylamide have been copolymerized in the solid phase with gamma rays (^{60}Co source) and in polar solvents with free-radical initiators; molecular weights of the copolymers have been investigated.

Olefins. Olefin–MAN free-radical copolymerizations usually give 1:1 alternating copolymers; under certain conditions random, nonequimolar copolymers are produced.

Acrylonitrile and Acrolein. Free-radical and uv-initiated MAN–acrylonitrile bulk and solution copolymerizations have been studied; random, nonequimolar polymers are obtained.

Acrolein–MAN and α-ethylacrolein–MAN mixtures are readily copolymerized with free-radical initiators. High molecular weight copolymers with 10–45 wt % MAN are insoluble in water, inherent viscosities are in the range of 0.5–5.0 dL/g, and molecular weights are between 1×10^5 and 3×10^6.

Allyl Monomers. Polyfunctional allyl compounds, eg, allyl ethers, allyl malonates, allyl carbonates, triallyl cyanurate, and triallyl isocyanurate, give nonequimolar copolymers with MAN. Most of these materials are cross-linked plastics; they have not been well characterized.

Vinyl Ketones. Solution copolymerization of methyl and phenyl vinyl ketone with MAN gives good-to-fair yields of nonequimolar copolymers. With peroxide initiators in aromatic solvents at 50–135°C, yields of 14–44% were obtained.

Other Monomers. Maleic anhydride copolymerizes readily with vinylsulfonic acid, methallylsulfonic acid, and 2-propene sulfonate in the presence of free-radical initiators giving water-soluble, nonequimolar copolymers.

Random Copolymer Properties. Of all the random MAN copolymers prepared with <50% MAN, only the poly(styrene-*co*-MAN) materials have achieved commercial importance. Characteristics and solubility parameters (δ) have been determined for several styrene–MAN (SMA) copolymer compositions (Table 2).

Random Terpolymerizations. Under azeotropic conditions the monomer feed mixture and copolymer composition remain constant during the course of the copolymerization reaction. For most binary systems that do not contain MAN, the problem of azeotropy is easily solved. Where MAN is involved, complications arise from the difficulty of MAN homopolymerization and the tendency of MAN to undergo 1:1 alternating copolymerization with a large number of electron-donor monomers. The problem is even more complicated when MAN is copolymerized with two or more monomers. Where azeotropic copolymers may be formed, an azeotropic copolymer index has been formulated to predict homogeneous terpolymerizations.

Terpolymers Containing Styrene or Olefins. The styrene–MAN monomer pair copolymerizes with numerous other vinyl monomers, usually in solution with typical free-radical initiators.

Terpolymers Containing Vinyl Acetate or Acrylates. Vinyl acetate–MAN and several other vinyl ester–MAN pairs copolymerize with radical initiators to give nonequimolar copolymers. Most of these copolymerizations have been with vinyl chloride.

Alkyl acrylate–MAN and alkyl methacrylate–MAN pairs have been copolymerized with each other and with a large

Table 2. Physical Properties of Styrene-*co*-MAN Polymers

	MAN, mol %				
Property[a]	0	5	18	25	33
$\bar{M}_n \times 10^3$ by osmometry	87.5	82	75.4	79.6	75.2
$\bar{M}_n \times 10^3$ by gpc	90.4	82	84.0	79.6	65.1
$\bar{M}_w \times 10^3$ by gpc	195.1	175.2	184.3	177.6	146.4
$\bar{M}_w/\bar{M}_n$ by gpc	2.16	2.14	2.20	2.26	2.25
n, 10% in 2-butanone	4.06	3.92	4.07	3.96	4.19
solubility parameter δ	9.18	9.314	9.729	9.874	9.907
T_g, °C		106			155
dta maximum endotherm[b], °C	95	111	131	133	153

[a] Combinations of tetrahydrofuran (THF)/H_2O, THF/α-pinene, and acetone/*tert*-butyl alcohol used for fractionations.
[b] At 48% MA, the dta maximum occurs at 160°C.

number of other monomers; methyl acrylate and methyl methacrylate pairs with MAN are used.

Alternating Addition Copolymerizations. Even though MAN does not easily homopolymerize, it has been copolymerized with a large number of electron-donor monomers to obtain 1:1 alternating copolymers (see also COPOLYMERS, ALTERNATING). Derivatives, such as alkyl- or halogen-substituted MAN, also function as electron-acceptor monomers, undergoing alternating copolymerization with select electron-donor monomers. Maleimide and isoimide derivatives of MAN, as well as fumarates and maleates, are also known to undergo alternating copolymerization.

Allyl Monomer–MAN Copolymers. Homopolymerizations of allyl compounds and MAN give only low molecular weight materials because of degradative chain-transfer processes. However, high molecular weight and good yields of MAN–allyl acetate copolymer are obtained by conventional radical polymerization; the cross propagation step competes fairly effectively with allyl termination. A large number of allyl carboxylates and other allyl-type monomers have been copolymerized with MAN by a radical process.

Vinyl Ether–MAN Copolymers. Alkyl and aryl vinyl ethers, as well as unsaturated cyclic ethers, undergo radical copolymerization with MAN under mild conditions, giving 1:1 alternating copolymers.

Vinyl Ester–MAN Copolymers. Vinyl acetate–MAN copolymerizations and copolymers have been investigated. At one time Monsanto produced and marketed these copolymers as part of their Lytron resins series.

Olefin–MAN Copolymers. The copolymerizations of monoolefins, conjugated dienes, cyclic olefins, cyclic dienes, bicyclic olefins, and bicyclic dienes with MAN are initiated by free radicals, uv light, and ionizing radiation, giving 1:1 structures. The lower α-olefins copolymerize easily; procedures are described for equimolar and nonequimolar compositions. Cyclic and long-chain olefins copolymerize slowly. Monsanto marketed ethylene–MAN (EMA) copolymers and Gulf Corp. marketed a polyanhydride from a C_{18} α-olefin.

Vinyl Aromatic Monomer–MAN Copolymers. Since the very early patents on the copolymerization of electron-rich styrene with electron-poor MAN, considerable work has been published on this monomer pair. However, copolymerization between styrene and dialkyl maleates is poor. The free-radical copolymerization system of styrene and MAN is one of the best examples of an alternating copolymerization system. With a large excess of styrene, some polystyrene is also formed. The copolymerization reaction is strongly exothermic and is controlled by the initiator, solvents, and temperature. Polystyrene peroxide formation accelerates the reaction.

Acrylic Monomer–MAN Copolymers. Good yields of acrylic acid–MAN copolymers are obtained with excess MAN; otherwise acrylic acid homopolymer is produced along with 1:1 copolymer. The corresponding fumaric acid–acrylic acid copolymer is also known.

Vinyl Heterocyclic Monomer–MAN Copolymers. Vinyl heterocyclic compounds such as *N*-ethyl-2-vinylcarbazole, *N*-vinylpyrrolidinone, *N*-vinylcarbazole, *N*-vinylphthalimide, *N*-vinylsuccinimide, *N*-vinylcaprolactam, 3-vinyloxyethyl-5,5-dimethylhydantoin, and vinylenecarbonate, readily undergo free-radical initiated copolymerization with MAN, giving 1:1 alternating copolymers. The free-radical initiated copolymerization of *N*-vinylcarbazole with diethyl fumarate possibly produces a 1:1 copolymer.

MAN Copolymer Properties. Copolymerization with maleic anhydride improves the physiocochemical properties of polymers by providing increased polarity, rigidity, glass-transition temperature, and functionality. MAN, MA, or FA based functionality promotes hydrophilicity, adhesion, dyeability, compatibility, and higher heat distortion and provides a reactive group for cross-linking or other reactions.

Styrene Copolymers. Styrene–MAN (SMA) copolymers are divided into two types. Type I materials are brittle, low molecular weight materials with a high MAN content soluble in weakly alkaline solutions. Typical of these are Monsanto's Scriptset and ARCO's SMA series. Type II materials are mainly styrene–MAN terpolymers with low (5–15%) MAN content containing a third comonomer. These terpolymers are of high molecular weight and are useful in engineering applications. Typical are Dylark (ARCO) and Cadon (Monsanto) polymers.

Olefin Copolymers. Ethylene–MAN copolymers were first marketed by Monsanto under the trademark EMA. In Europe the copolymer is called Malethamer or VINAC. EMA materials are available in a range of molecular weights as the anhydride, free acid, or sodium derivative. Other olefin-*co*-MAN or MA resins include those of 1-octadecene (PA-18 by Gulf Oil Chemicals), isobutylene (ISOBA M by Kuraray Co., Ltd.), and butadiene (Maldene by Borg-Warner Corp.). At one time Gulf offered PA-6, PA-10, and PA-14 α-olefin-MAN copolymers, indicating the number of carbon atoms in the olefin chains.

Physical properties of EMA resins are shown in Table 3.

Table 3. Properties of EMA Resins[a]

	Anhydride	Acid	Na^+ salt
density, kg/m^3	320	594	400
pH of 1% aq soln	2.35[b]	2.35	6
softening point, °C	170	155	
decomposition temp, °C	247	250	
mp, °C	235	217	

[a] White powder.
[b] Dissolves with hydrolysis to free acid in ca 2–3 h at 80°C or 10–15 min at 95°C.

These copolymers are readily soluble in acetone, pyridine, dimethylformamide, and molten MAN, and insoluble in most other common organic solvents. The anhydrides readily react with amines or alcohols, forming ester or amide polymers soluble in polar organic solvents. Salts of EMA materials are soluble only in water.

Vinyl Ether Copolymers. Vinyl alkyl ethers copolymerize readily with MAN, with free-radical initiators, yielding predominantly 1:1 copolymer and traces of homopolymer. GAF Corp. markets the methyl vinyl ether-*co*-MAN polymer as the Gantrez AN resin series in several grades (AN-119, 139, 149, and 169), and as the free-acid (hydrolyzed) (Gantrez S), half-ester (ES-225, 335, and 425), and half-amide copolymers. Gantrez-type resins (Table 4) are compatible with a wide variety of other polymeric materials. In Europe, ICI markets the copolymer as Viscofres. Gantrez AN resins are readily soluble in polar solvents and insoluble in aliphatic or aromatic hydrocarbons.

Vinyl Ester Copolymers. The water-soluble vinyl acetate–MAN copolymers (VA–MAN) were marketed by Monsanto under the trademark Lytron 897, 898, and 899. Monsanto Benex resin, Amoco Drilling Aid 420 and 421, and Baroid (National Lead) Xtend resins are vinyl acetate–MA copolymers for oil-field applications; in Europe these resins are named Povimalya.

Applications

MAN copolymers of styrene, olefins, vinyl acetate, and alkyl vinyl ethers fill the need for polymeric materials with low viscosity, high polarity, water solubility, and high reactive functionality.

Styrene Copolymers. The principal uses for low molecular weight, alkali-soluble, SMA-type polymers containing a high MAN-to-styrene ratio (1:1) are in paper and textile sizing, floor polishes, and rug shampoos.

In 1983 fabrication of one-piece instrument panels and headliners for automotive applications accounted for ca 70% of all modified SMA resins, mostly Dylark resins, which are a blend of SMA and high impact polystryrene (HIPS). Dylark has also captured a good share of the resin market for business machines, appliance housings, water tumblers, cutlery, coffee makers, and other small appliances. The main market for Cadon is in business machines and appliances. The current volume of SMA-type polymers is less than that of ABS; the annual growth rate is about 10%.

Olefin Copolymers. Olefin–MAN polymeric materials have a low annual sales volume. EMA resins and their derivatives are used as dispersants, thickeners, suspending agents, adhesives, modifiers and additives, detergent builders, and paper and textile sizing agents. ISOBAM polymers are used as suspending agents, adhesive modifiers, paint dispersants, and paper and textile sizing agents. Long-chain α-olefin-*co*-MAN materials such as PA-18 are most useful in oil additives, wax extenders for paper, paper and textile sizing, and cosmetics dispersants.

Maldene is a chemical intermediate for products useful in textile, adhesives, detergent, and cleaning applications. Butadiene and isoprene–MAN copolymers are also useful in coatings, surfactants, sealants, adhesives, tackifiers, paper and textile sizing, and molding and laminating applications.

Vinyl Ether Copolymers. The methyl vinyl ether–MAN copolymers marketed by GAF Corp. as Gantrez AN resins (Table 4) are used in hair sprays, shampoos, cosmetics, and other protective colloids.

The divinyl ether–MAN copolymer (Pyran) has been evaluated experimentally in the treatment of cancer. It is said to function as a adjuvant to surgery, radiation, or chemotherapy, and to promote interferon production.

Vinyl Ester and Other Copolymers. Vinyl acetate–MAN copolymers, also called VAMA resins, have been explored as extenders for bentonite mud and as flocculants for drilling clays. However, no market has developed for these materials. The acrylic acid-*co*-MAN and methacrylic acid-*co*-MAN polymers hold promise for market development as dispersants and controlled-release agents.

MAN Graft Copolymers

Methods for grafting MAN or MAN copolymers to the backbone of both saturated and unsaturated polymers has been extensively studied to improve physicochemical properties by providing polarity to promote hydrophilicity, adhesion, and dyeability. The pendent anhydride moiety provides functionality for cross-linking and other chemical modifications, promotes compatibility with other polymers and fillers, and improves heat resistance. In rubber chemistry, grafting of MAN has been investigated for producing materials with improved green strength, vulcanization, and fatigue properties, and resistance to oxygen. The techniques explored for grafting MAN to polymer backbones include Diels-Alder, free-radical, and "ene" reactions.

Table 4. Gantrez AN Resins[a]

Property	Value
η_{sp}[b]	
AN-119	0.1–0.5
AN-139	1.0–1.4
AN-149	2.6–3.5
softening point, °C	200–225
T_g, °C	133[c]
sp gr	1.37
bulk density, kg/m^3	320
pH of 5% aq soln (free acid)	~2

[a] Methyl vinyl ether-*co*-MAN resins; white fluffy powder.
[b] Solutions 1.0 g/100 mL methyl ethyl ketone at 25°C.
[c] The T_g values for the ethyl vinyl ether-*co*-MAN and butyl vinyl ether-*co*-MAN polymer, respectively, are reported as 68 and 56°C.

BIBLIOGRAPHY

B. C. Trivedi and B. M. Culbertson, *Maleic Anhydride,* Plenum Press, New York, 1982.

W. D. Robinson in M. Grayson, ed., *Kirk-Othmer Encyclopedia of Chemical Technology,* Vol. 14, John Wiley & Sons, Inc., New York, 1981, pp. 770–793.

K. Lohbeck and H. Haferkorn, *Ullmanns Encyclopedia of Technical Chemistry,* Vol. 16, Verlag Chemie, New York, 1978, pp. 407–414.

B. M. Culbertson, *Catalysis Today* **1**(5), 609 (Oct. 1987).

BILL M. CULBERTSON
Ashland Chemical Company

MALEIC ACID ANHYDRIDE.

See MALEIC AND FUMARIC POLYMERS; POLYESTERS, UNSATURATED.

MAN-MADE FIBERS.

See FIBERS.

MARINE APPLICATIONS

Long-term performance, rather than initial cost, governs the selection of a material for marine use. Thus it is not surprising that polymers are used in a wide variety of marine applications. They remain stable for a long time in the severe marine environment, reduce weight and volume, retain their appearance, require little maintenance or lubrication, and resist water, corrosion, rot, and fungus growth.

Many applications of polymeric materials are unique to seagoing vessels; others find their way into shore applications.

Structural and Mechanical Applications

Reinforced plastics are used for numerous topside applications to achieve reduction of weight, freedom from corrosion, retention of color and gloss, and decreased maintenance. Masts, antennae, fairings, sails, deck houses, and other components for pleasure craft are frequently constructed from polymeric materials. Plastics reinforced with filament or cloth are used in hull construction, especially for pleasure boats.

Buoys are manufactured from natural or synthetic rubber or from plastics. Buoys, fenders, floats, wharves, and pilings may be coated with a polyurethane elastomer to increase wear resistance and cushion impacts. Syntactic foams, created from hollow glass or plastic microballoons bonded with epoxy resin, are used to provide buoyancy in deep-submergence craft. Polyurethane and polystyrene foams are used in buoys, floats, and pontoons. Elastomers find many applications in underwater acoustic systems, including vibration isolators and dampers, sound reflectors, and acoustic absorbers (see ACOUSTIC PROPERTIES). Polyesters are used for pump bodies and impellers. Gears, springs, packing glands, panels, moldings, seals, and gaskets are frequently made from polymeric materials. Plastics are also used in piping and ventilating systems.

Preservation and Appearance

Unless coated, steel and aluminum degrade rapidly in the marine environment. Preservation is essential to maintain function and appearance.

Coatings. The most widely used marine coatings are two-component epoxy–polyamide systems. Although a variety of epoxy resins can be employed, the epoxy is usually a liquid or solid resin derived from epichlorohydrin and bisphenol A. The polyamide resin is manufactured from dimers of unsaturated fatty acids and ethylene diamine or its higher analogues.

Sealants. Polymers of bis(ethyleneoxy)methane react with sodium disulfide to produce polymers containing disulfide linkages and terminal mercaptan groups. These are known as polysulfide polymers and are used in sealants to inhibit corrosion.

Comfort and Safety

Plastics are used for cups, plates, trays, furniture, and tabletops. Stair treads, cove moldings, kick plates, and floorings are made of wear-resistant polymeric materials. Plastics are used for shower compartments, wash basins, and racks and fixtures because they resist moisture and biological attack and are easy to maintain. Polyurethane foams are used for flame-resistant matresses and cushions. A nonflammable fluorosilicone foam is produced in place as a fire barrier around pipes and cables that penetrate decks or bulkheads.

Plastic-coated fabrics in garments improve comfort and safety, and offer protection against subzero temperatures and moisture. Life vests contain closed cell polyurethane foams for buoyancy; ear protectors are made from an impact-resistant polymer lined with plastic foam.

Electrical and Electronics Applications

Various polymers are used in electrical construction and insulation. Printed circuit boards manufactured from epoxy or phenolic resins are used in almost all types of electronic equipment. Acetal resins find widespread use in marine applications because they resist high humidity and immersion in water. Antennae and radomes frequently are constructed of fiber glass-reinforced epoxy resins.

BIBLIOGRAPHY

Proceedings of the First International Symposium on Polymers in a Marine Environment, *Transactions of the Institute of Marine Engineers,* Vol. 97, Conference 2, 1985.

Proceedings of the Second International Symposium on Polymers in a Marine Environment, *Transactions of the Institute of Marine Engineers,* Vol. 100, Conference 2, 1988.

W. B. Moniz, J. R. Griffith, R. F. Brady, and C. M. Thompson, *Memorandum Report 5517,* Naval Research Laboratory, Washington, D.C., March 6, 1985.

ROBERT F. BRADY, JR.
U.S. Naval Research Laboratory

MARKET EVALUATION

The Market for Plastics

The thermoplastic resins market can be divided into three product groups according to the price–volume relationship (Table 1).

The performance distinctions among various plastic resins are not always clearly delineated. In Figure 1 the axis of the price vs performance curve represents the value options occupied at specific locations by each resin. In the overlapping areas, more than one material offers the same combination of price and performance. These are regions of interpolymer competition; one material can be substituted for another. Inter-

Table 1. Consumption and Prices of Plastics, 1984[a,b]

Resin	Price, \$/kg[c]	Volume, 10^3 t	Number of producers[d]
Commodity			
polyethylene	0.75	6530	17
polypropylene	0.84	2189	12
polystyrene	0.97	1743	15
poly(vinyl chloride)	0.68	3090	13
ABS	1.78	499	3
thermoplastic polyester	1.26	390	8
Engineering			
nylon	3.39	177	11
acetal resin	3.61	46.8	2
poly(phenylene oxide)		89.1	1
polycarbonate	3.56	123	2
Specialty			
poly(phenylene sulfide)	4.33[e]	10	1
polysulfone	8.40[e]	13	1
polyethersulfone	14.30[e]	na	1
poly(amide–imide)	42.46[e]	na	5

[a] Source: The Society of the Plastics Industry Committee on Resins Statistics as compiled by Ernst & Whinney, except as noted.
[b] In 1985 and 1986 the relative volumes for commodity and engineering plastics remained approximately the same; values fluctuated about ±7%. Thermoplastic polyester is the only significant exception; volume rose to 654×10^3 t.
[c] Prices of commodity resins follow oil prices and have fallen since 1984.
[d] Since 1984, consolidation of producers of commodity resins; increased numbers of producers of engineering and specialty resins.
[e] Manufacturer's list price; bulk, general-purpose grade.

polymer competition is increasing because new polymers have been developed, performance boundaries have been extended, advances in materials technology continue, and price competition between materials exists. Also, new processes, the expiration of patents, and the diffusion of technology affect competition.

Principles of Market Evaluation

Reasons. A market evaluation is often performed to understand or assess the commercial potential for a new or improved material. Commercialization implies wide sales of the material for profit. The following questions can be answered with the help of market evaluation:

What is the market potential (volume) for this material?
What is the price potential for this material given the volume targets?
Will revenues generated cover all associated costs and yield an adequate profit?
What are the prospects of competition?
What is a reasonable estimate of achievable market share?

There are many more questions to be answered. It is the task of the market manager to devise and implement a strategy that assures sales and profitability.

Precision. Historical or current market evaluations can be highly precise, depending on the effort and money expended in their development. And long-term forecasts are, of course, less dependable than short-term forecasts.

Need. The focus of a market evaluation is the purchase decision. Usually one or two key individuals have the ultimate responsibility to authorize a purchase. These individuals are the focus of the seller's efforts.

Determination of Demand. The plastics markets belong to a segment of the economy known as the chemical economy. This economy is both vast and invisible. Very few of its products are sold directly to the government or households. Nonetheless, over 80% of chemical products end up in consumer preparations purchased by households.

Stages. The first step in conducting a market evaluation for plastics is identifying the source and extent of the final product demand. The final demand is composed of numerous individual purchase decisions and resides in one or more of the four sectors of the economy: household spending, government spending, private and business spending, and exports. The final demand for products incorporating a plastic produces a derivative demand for the resin or monomer within the chemical economy.

After the sources of final demand for plastics have been identified, it is necessary to sketch a comprehensive flow chart of incorporating the plastic into end products. Then the magnitude of final demand for each end product is assessed and the derived demand for the plastic calculated.

Measurement of Derived Demand. When the final demand for an end product is known, a derived demand for its components can be calculated from the input–output relationship at each production stage.

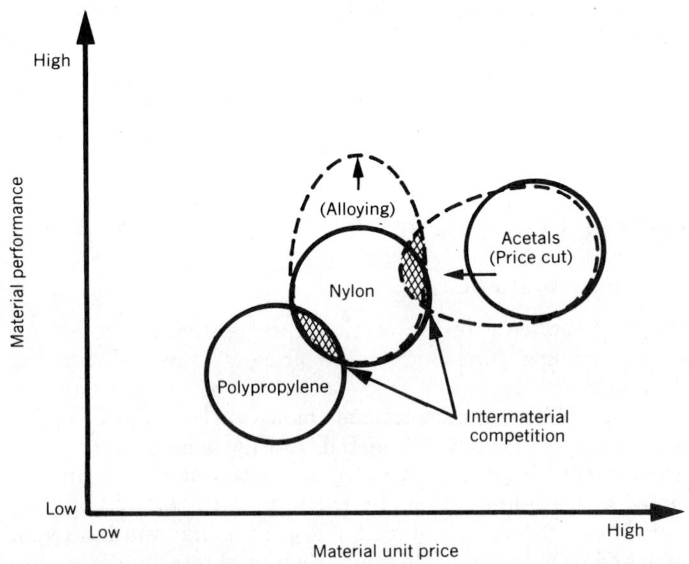

Figure 1. Price–performance for plastics.

Characterization of Demand

Product. The freedom, flexibility, or limitations a buyer has in purchasing or using a product depend on its character. Some products are commodities that buyers perceive as equal, regardless of the specific suppliers. Gasoline, cement, and liner-grade polyethylene are such commodities. Buyers can easily change suppliers; this decision is usually based on price. Specialty products are more closely associated with the specific firm or company manufacturing them.

Perishable goods cannot be stored for long periods of time; their demand tends to be more consistent, since accumulation of inventories rarely results. Deferrable purchases represent nonessential items, discretionary purchases, or repairable goods that can be reconditioned in lieu of replacement. And demand for certain items varies in a predictable, time-dependent fashion. Seasonal goods are items whose consumption is usually weather-dependent.

Microeconomics. Microeconomic factors influencing the final demand for a product include its price, the so-called price elasticity of demand, the price of associated goods, and the income of purchasers. The price elasticity of demand defines the percentage change in demand quantity that accompanies a 1% change in price:

$$\text{price elasticity of demand} = P/E = \frac{\dfrac{\text{demand change}}{\text{demand}}}{\dfrac{\text{price change}}{\text{price}}}$$

Products with few substitutes, eg gasoline, tend to be price-inelastic ($P/E < 1$) within reasonable limits of price changes. Price-inelastic goods are also called price-insensitive. Price increases reduce demand only marginally, whereas large price decreases do not immediately stimulate demand. Price-elastic products ($P/E > 1$) usually have many substitutes or are discretionary purchases. The availability of substitutes allows changing from one product to another as price rises.

Macroeconomics. Macroeconomic considerations affect the consumption or demand for certain goods, that are typically larger purchases, eg, automobiles, houses, etc. These goods are highly dependent on the chemical economy for a number of critical raw materials, plastics, resins, coatings, rubber, foams, fibers, and paints. In turn, the chemical economy is highly dependent on demand for these finished goods to sustain a derived demand for these raw materials.

Buyer Concentration and Power. Buyer concentration describes the percentage of total demand accounted for by a single purchaser. In the fragmented household market, a single household accounts for very little of the total demand from this sector. Other markets, however, are much more concentrated in terms of buyers. An extreme example is the market for military goods. The U.S. Department of Defense can determine the fortunes of an entire industry based on its decision to proceed with or cancel a major defense contract.

Intermaterial Competition. Intermaterial competition is the dynamic, ongoing process of substitution within every market.

Substitution

Substitution is the process of one product or service supplanting another in performing a particular function for a buyer. The degree of substitution, the number of purchase decisions that can be shifted from an existing product to a new substitute, is one of two factors that ultimately determine market demand for polymers. The other factor is the number of new market applications that can be justified. Disposable packaging and recreational goods represent markets in which plastic materials have created new applications.

Identification of Substitutes. A substitute is a product or service that performs the same function as the typically used product or service.

Other Methods. Other methods of substitution must also be considered, eg, elimination, usage-rate reduction, recycling, reusing and reconditioning, backward integration, and upstream substitution.

Economics of Substitution. Successful substitution occurs when the value–price relationship of a substitute relative to the existing material is economically beneficial to the buyer. The benefit must also exceed the cost of changing to a new material.

Sources of Market Information

Performing a market evaluation requires a wealth of information, both qualitative and quantitative. Although some of this information may be gained through a search of published sources, the most timely and usually the most valuable information can only be developed from direct contact and personal interviews.

Directories of Information Sources. Guides published by Harvard Business School include *Business Reference Sources; Business Intelligence and Strategic Planning;* and *Business Forecasting for the 1980s and Beyond.* A chemical industries information source is the *Management Information Guide #29.*

BIBLIOGRAPHY

O. Eckstein, ed., Foundation of Modern Economics Series, Prentice-Hall, Inc., Englewood Cliffs, N.J. A series of 13 concise books surveying the principal branches of economics, including *Prices and Markets, National Income Analysis, Labor Economics,* and *International Economics.*

J. Wei, F. Russell, and M. Swartzlander, *The Structure of the Chemical Processing Industries: Function and Economics (Chemical Engineering Series).* Provides an excellent discussion of the chemical economy, including sections on economics, accounting, industry structure, and highlights on specific chemical products.

M. Porter, *Competitive Advantage,* The Free Press, New York, 1985.

Chem. Week, published weekly.

KARL D. LOOS
JAMES A. FINNEGAN
Arthur D. Little, Inc.

MASS POLYMERIZATION. See BULK POLYMERIZATION.

MASS SPECTROMETRY

The primary and very significant impact of mass spectrometry (ms) on polymer analysis has arisen from its use in the characterization of localized polymer structure by examination of thermal degradation or pyrolysis (py) products. While the mass range of most common analyzers is typically on the order of just 1000 daltons, analyzer mass range has not been the main obstacle to the application of mass spectrometry to intact polymers. Instead, the principal hindrance arises from the requirement for isolated, gas-phase ions. The alleviation of this problem and improved degradative analytical methods account to a large degree for the rapid growth of polymer ms.

Procedure

Mass spectrometry utilizes the effects of electromagnetic fields to control the trajectory of isolated ions and thereby determine their mass-to-charge ratio (m/z). The experiment can be considered in four stages, each of which introduces constraints on potential applications. In the first stage, sampling, the analyte must generally be vaporized or otherwise isolated from its matrix on a molecular scale. Thermally labile, nonvolatile samples will often be seriously degraded when heated for vaporization. In many cases, thermal degradation products are usefully related to small-scale structural features of large molecules, eg, monomer or functional group identities may be retained; this is the basis for pyrolysis ms techniques.

The second ms step, ionization, is most often accomplished by electron impact (ei). Interpretation of ei fragmentation patterns supplies the main source of structural information in conventional ms. However, the information content is usually small for spectra of thermally labile, nonvolatile materials. The newer, so-called soft ionization techniques (many of which combine sampling and ionization or reverse their order) have solved this problem in many instances. The success of soft ionization, in fact, has spurred development of collision-based methods for recovering structural information from intact molecular ions; this is achieved by promoting their fragmentation after a preliminary ms separation step (ms–ms experiments).

After ionization and sampling, ions are accelerated to kinetic energies compatible with the mass analyzer in use (tens to thousands of eV). The charge dependence of this acceleration coupled with subsequent mass-discriminating interactions with applied magnetic or electric fields account for the *m/z* (rather than simple mass) response of virtually all mass spectrometers in the third (separation) stage of the ms experiment.

The final step, apart from data interpretation, is detection. Many of the difficulties in quantitating ms data arise from the mass, resolution, and energy dependence of the response of detectors in common use. For the most demanding quantitative applications, one or more internal standards that include isotopically labeled analyte molecules may be employed to correct for variations in sensitivity. Such standards, however, are seldom available for polymer samples.

Sampling. In many ms applications, the distinction between sampling and ionization is blurred. Experiments in which a separate sampling step can be discerned include those using gas inlets, direct-insertion probes, rapid heating, pyrolysis, gas chromatography, and liquid chromatography. Polymer applications of gas inlets and direct insertion probes (typically operated up to 300°C) are generally limited to analysis of additives and residual volatiles (unreacted monomer and solvent).

Ionization. Several ionization techniques developed in the 1980s could also be classified as sampling techniques, because they rely on low energy solution chemical processes (electrolyte dissociation or ion association with polar analytes) for sample ionization, generating so-called preformed ions that are subsequently coaxed into the spectrometer ion source region by various means. Although these newer techniques have extended applications to so-called middle molecules (mol wt 1,000–10,000), electron impact and chemical ionization remain the mainstay of ms.

Electron Impact (ei). Seventy-eV electrons from a heated filament are regularly employed for ms ionization. Fragmentation is generally extensive, although diagnostic, revealing structural information. For many labile molecules, the intact molecular radical cation $M^{+\cdot}$ is not detected. Fragmentation can be reduced at the cost of poorer reproducibility by lowering the electron energy. Chemical derivatization is frequently utilized to increase the vapor pressure of the sample.

Chemical Ionization (ci). The source region in ci is enclosed with only a small leak to the spectrometer, allowing maintenance of relatively high pressures of a reagent gas, typically 133 Pa (1 torr). Electron energies of up to several hundred volts are employed. These energies are dissipated primarily in reactive collisions with the reagent gas, which is present in large excess; sample ionization occurs predominantly through subsequent collisions with the ionized reagent. Because internal-energy transfer in these collisions is relatively inefficient, fragmentation in ci is normally appreciably less than that observed in ei, despite the higher initial electron energies involved. Nonvolatile solids can be accommodated in desorption ci (dci) by placing the sample on the inert probe tip of a direct-insertion probe positioned within the ci plasma.

Photoionization (pi). Single- or multiphoton photoionization is used with a variety of sample-introduction methods, including conventional molecular beams and, more recently, thermospray (ts). Tunable radiation, usually lasers, can afford remarkably selective photoionization for both molecular and atomic analysis.

Field Ionization (fi). As molecules pass through regions of extremely high electric field strength (10^7–10^8 V/cm), electrons can be removed by a very gentle quantum-mechanical tunneling mechanism, resulting in the generation of molecular ions with little fragmentation (but low abundance due to the small cross section for tunneling). High fields are created by applying 2–20 kV between a cathode and a sharp needle, wire, or blade anode positioned about 1 mm away. The field strength is enhanced by growing dendrites (whiskers) on the anode, producing an activated emitter. The sample must be vaporizable, limiting polymer applications to characterization of residual volatiles, volatile additives, and pyrolysis products.

Field Desorption (fd). Volatility requirements are reduced in fd by depositing a sample solution or suspension directly on an activated emitter analogous to that used in fi, then allowing the solvent to evaporate. The emitter must be heated during analysis, presumably to promote molecule migration to the high field sites at the dendrite tips. Heating induces fractional distillation, requiring repeated scanning and integration of several spectra for accurate characterization of sample mixtures. For most organic substances, a best anode temperature for maximum molecular ion currents is reached between 20°C and 300°C. At higher temperatures, fragmentation becomes more extensive; pyrolysis fdms has been reported. The ability to sample high molecular weight material without excessive fragmentation has made fd the most extensively used tool in nondegradative polymer ms.

Secondary-ion Mass Spectrometry (sims). In sims, a solid sample is bombarded with ions accelerated to ca 2–20 keV. The sample is usually deposited as a thin layer on a metal (often silver) substrate. With low primary ion current densities (on the order of 10^{-9} A/cm^2), each bombarding particle encounters a fresh, undamaged surface, resulting in molecular sputtering known as static, organic, or molecular sims. Some fragmentation is evident, even under the lowest dosage conditions. As current densities increase, sample damage becomes more severe, and fragments become much more abundant than molecular ions. In the high current dynamic sims mode, erosion of the sample surface permits depth profiling, although these applications are generally limited to investigations of elemental (atomic) distributions. Sims experiments can also provide spatially resolved ($<1\ \mu$m) chemical information about the sample surface. A serious complication in applying sims or imaging sims to polymers and other insulators can arise from the buildup of charge on the sample surface. This can repel incoming primary ions and distort the kinetic energy of the secondary ions of interest, complicating the mass analysis. Thin conducting overlayers and low-energy electron flood guns have been used to combat these difficulties.

Fast Atom Bombardment (fab). By neutralizing the primary ions used in sims, a beam of energetic (fast) atoms can be generated for a sims-type experiment with less charging than in conventional sims. Charging can be further reduced if the sample is introduced in a conductive liquid matrix. Furthermore, beam damage inflicted by primary-particle impact can be healed by the flow of the solution, restoring fresh sample to the surface. The liquid matrix also prompts the preformation of ions by low energy solution chemical reactions. The preformed ions can be sputtered from the surface by overcoming solvation energies that, although appreciable for most ionic species, are generally much lower than the lattice energies of ionic solids. In fact, solvent evaporation from sputtered ions furnishes a means of reducing their internal energy, thereby decreasing fragmentation.

Because the liquid sample is bombarded within the evacuated spectrometer ion source, the matrix cannot be volatile. Glycerol has good solvent properties and low volatility, but other matrices are superior in certain applications.

Plasma Desorption (pd). The energy of the primary bombarding particle in pdms reaches the MeV range by using ions from a particle accelerator or products from the spontaneous nuclear fission of ^{252}Cf. The sample is usually deposited as a thin film on a thin foil or polymer substrate. In common with fab, the addition of alkali salts improves performance for polar compounds, presumably by promoting the formation of cationized quasi-molecular ions. Plasma desorption ms has been particularly successful with high mass (10,000–20,000 daltons) samples.

Laser Desorption (ld). Most ld applications employ pulsed, focused lasers to promote desorption of ions and neutral particles from a thin film sample. Attainable power densities depend on the wavelength of irradiation, the most common being 10.6 μm (CO_2 lasers) and 265 nm (frequency-quadrupled Nd-YAG lasers). At shorter wavelengths, extreme focusing is possible (to μm dimensions), allowing acquisition of spatially resolved spectra (laser microprobe mass analyzer; lamma).

Electrohydrodynamic Ionization (eh). Electrohydrodynamic ionization uses the electric field at the tip of a biased (several kV) capillary emitter to promote the desorption of ions directly from a liquid meniscus. As a result, eh spectra contain almost exclusively preformed ions representative of the composition and solution chemistry of the sample. Sampling is accomplished with minimal increase in internal energy, and there is virtually no fragmentation.

Separation. Virtually every type of mass analyzer has been applied to various aspects of polymer characterization. The choice of analyzer depends on needed resolution, mass range, scan speed, sensitivity, and compatibility with ion energies associated with different sampling and ionization techniques.

Time-of-flight (tof) measurements offer fast response (<msec) for pulsed ionization systems (eg, ld and pd), with high mass range but generally low resolution. Ion energies are typically in the keV range, and resolution improves when the spread in ion energies is reduced. Quadrupole instruments relying on superimposed d-c and r-f electric fields also offer fast scan speeds, with moderate resolution over a relatively limited mass range (typically up to 500 or 1000 daltons). Ion kinetic energies must be relatively low (usually 1–15 eV), but wide relative distributions within this constraint can be tolerated. Magnetic sector-based instruments rely on ion interactions with essentially d-c magnetic and electric fields. keV kinetic energies are required, and high resolution (up to 10^5) can be obtained if an energy filter (electrostatic analyzer) is employed to reduce the spread of energies. A mass range up to about 5,000–10,000 is common. Laminated magnets offer scan rates up to at least 0.1 s per decade of mass. Nonlinear scan laws necessitate careful calibration. In ion–cyclotron resonance (icr) or fourier transform (ft) ms, d-c magnetic and r-f electric fields are superimposed to induce coherent "cyclotron" motion of low energy (<10 eV) ions in a small (cm) high vacuum (<100 mPa or 10^{-8} torr) cell. Commercial ftms instruments can measure the mass-dependent cyclotron frequency of ions of several different masses simultaneously, providing a mass spectrum very rapidly (ms). Over a narrow mass range, extremely high (10^6) resolution is attainable, especially at low mass. Alternatively, a wide mass range can be examined, with some loss in both resolution and sensitivity (especially at higher mass).

Mass Spectrometry–Mass Spectrometry (ms–ms). Application of ms to direct characterization of mixtures is difficult owing to the complexity of spectra obtained through conventional ionization, eg, ei and ci. These constraints can be relieved by employing a mass spectrometer to isolate an ion attributable to one component of a mixture and a second mass spectrometer to obtain a fragment spectrum from that ion. A collision cell between the spectrometers promotes fragmentation of the molecular ions through collisional activation, using neutral gas targets or a collision surface. A short chromatographic column may be retained in the inlet for cleanup of volatile samples. Sequential steps of mass analysis interspersed with collisions can be accomplished with a single ft or ion trap (a quadrupole variant) spectrometer.

Detection. Probably the simplest and oldest ms detector is photographic film. In instruments that focus several ions simultaneously on the exit plane, film offers the advantage of multichannel detection, with integration for signal averaging. Unfortunately, the sensitivity and dynamic range of most emulsions are limited. Furthermore, data manipulation is more cumbersome than with electronic detectors (electron multipliers), which are more common. Multichannel (array) electronic detectors are becoming commercially available. The limited mass range for simultaneous detection by arrays will probably expand with growing applications.

Data Treatment. Mass-spectral data are usually presented graphically or numerically as a collection of m/z–intensity data pairs. These may be collected and processed digitally with a computer-based data system, or manually from analogue recordings of mass spectra. In chromatography–ms, time may provide a third dimension; for example, total ion current may be plotted vs time to generate a (reconstructed) chromatogram that can provide estimates of chromatographic retention times. Alternatively, single-ion chromatograms can be plotted for the signal at one or more m/z of interest; thus ions characteristic of a particular molecule or organic functional group can be monitored. This can provide an extremely sensitive indication of the presence of specific species. A full mass spectrum of each chromatographic peak can often be obtained.

Three aspects of the mass spectra of heavy ions require special consideration during data interpretation. First, mass defects can become substantial for large organic molecules. The difference between nominal mass and exact mass (eg, 1 vs 1.007825 for ^1H) is cumulative, reaching 6 daltons for the monoisotopic mass of a polystyrene 100-mer. While the necessary correction is easily calculated, theoretical ion intensity even at the corrected mass will be small (relative to other ions in an isotope cluster) for ions of mass above 10,000; incorporation of minor isotopes (especially ^{13}C) becomes highly probable for heavy organics, so that the most abundant ion in an isotope cluster will have a mass significantly different from (usually heavier than) the monoisotopic mass. Corrections for isotopic abundance are also relatively straightforward, but using the average mass (eg, C = 12.01115) can often provide a useful molecular weight estimate based on mass spectral data. Unfortunately, the interpretation is further complicated by hydrogen scrambling—proton, hydrogen, and hydride transfer among sample molecules that can be extensive during ionization of large organic molecules. In practical experiments at very high mass, distinctions between peaks due to varying isotopes and those due to hydrogen scrambling may blur, given limited resolution. Even when peaks differing by one unit in nominal mass can be resolved, each of these may have important contributions from a number of isobaric (same nominal mass) ions. A resolution of several million would be required to distinguish among heavy ions containing ^{13}C, ^2H, or two ^1H's. Fortunately, errors in molecular weight estimates due to these limitations are small in relative terms (<1%) at high mass.

Applications

Despite the progress in soft ionization techniques outlined above, most ms applications in polymer characterization involve degradative analysis. This reflects both the power of py–ms for providing qualitative structural information and the persistent constraints on ms, the utility of which may now be said to extend to intact middle molecules but not yet beyond.

Synthetic Polymer Surfaces. Desorption ionization (di) experiments (sims, fab, ld, and lamma) have been employed successfully as probes of the molecular structure and chemistry of synthetic polymer surfaces, with depth profiling in some instances. These applications include studies of structure, additives, and spatial chemical analysis.

Bulk Synthetic Polymers. *Degradative Methods.* The degradative analysis of synthetic polymers comprises the most extensive body of polymer ms applications. Mechanistic interpretation of pyrograms (py gas chromatograms) and py mass spectra serves two distinct analytical purposes. Such studies have been valuable probes of the combustion chemistry of materials of known structure, furnishing insight, for example, into means of fire retardancy and the pyrolytic toxicity of commercial polymeric fabrics. For applications to new materials, the background of studies with polymers of known structure aids in deriving microstructural (eg, tacticity and branching) and compositional (eg, monomers, additives, and copolymer ratios) information. In both applications, attractive features of py–ms and py–gc–ms include their high potential sensitivity (milligram or even microgram samples are suitable), their compatibility with intractable materials that cannot be dissolved for conventional nmr or ir analysis, and (particularly for py–ms) the relatively short data-acquisition times involved.

Molecular Weight Distributions. For direct mass-spectral characterization of molecular weight distributions (MWDs), various soft ionization techniques are used to generate a characteristic molecular (or more often quasi-molecular) ion from each component in a mixture of *n*-mers. Polymer–ion interactions may also be probed.

Natural Polymers. *Degradative Methods.* The use of py–ms for the characterization of natural polymers and natural products in general has been extensive. Diverse applications include structure–property correlation; fossil fuel exploration and exploitation; bacterial fingerprinting; food science; and fuel-conservation technology.

Sequencing. In sequencing studies, the extended mass range accessible with di has substantially reduced the microchemical manipulations necessary before ms analysis, facilitating direct sequencing and verification and correction of protein sequences determined from standard DNA sequencing schemes.

BIBLIOGRAPHY

H.-R. Schulten and R. P. Lattimer, *Mass Spectrom. Rev.* **3**, 231 (1984).

J. T. Watson, *Introduction to Mass Spectrometry,* Raven Press, New York, 1985.

K. J. Voorhees, ed., *Analytical Pyrolysis: Techniques and Applications,* Butterworth & Co. (Publishers) Ltd., Kent, UK, 1984.

K. L. Busch, G. L. Glish, and S. A. McLuckey, *Mass Spectrometry/Mass Spectrometry,* VCH Publishers, Inc., New York, 1989.

KELSEY D. COOK
University of Tennessee

MATERIALS, SELECTION

Plastics are selected and designed by the same easy and practical methods as steel, aluminum, wood, glass, ceramic, and other materials. Many different products have been designed from plastics for over a century; they usually provide a cost advantage over other materials. Since 1980, more plastics have been used worldwide on a volume basis than any other material except wood and concrete. Before the end of this century, more plastics will be used on a weight basis, excluding wood and concrete; the United States uses about one third of total world production.

Plastics offer the opportunity to optimize design by focusing on material composition and orientation, as well as structural geometry. The interrelationship between materials selection (including reinforcement orientation), performance, shape, cost, and processing is more important for plastics than for other materials, where the designer is limited to specific forms or profiles that are bent or welded.

Materials selection is based on design process requirements. Data concerning requirements are available from many sources, including manufacturers, or are developed by the materials selector. The behavior of plastics strongly depends on the time of load application, history of loading, and temperature, temperature cycles, and environmental conditions.

The range of mechanical properties of plastics is illustrated in Figure 1. Certain plastics provide higher strength and stiffness (modulus of elasticity) than any other material, eg, steel, aluminum, etc. Since plastics properties vary widely, a logical selection process must be followed to meet performance requirements that relate to cost factors. Plastics may be made hard, elastic, rubbery, tough, crystal-clear, opaque, strong, stiff, weather resistant, or electrically conductive, and can exhibit almost any property.

Although over 10,000 different plastics are available, only a few hundred are used in large quantities. Reinforced or composite plastics incorporate fibers or filaments of glass, carbon–graphite, aramids, boron, and single crystals and whiskers, as well as flakes. Since no one plastic (or any other material) can meet all property requirements for all products, the designer or materials selector can combine different plastics; if desired, they can be kept as separate materials and then used in coextrusion or coinjection.

Types of Plastics

Plastics are also called polymers, resins, reinforced plastics, and composites, each having its own connotation. All plastics flow at some stage; they are soft and pliable and can be shaped, usually by the application of heat, pressure, or both, into desired forms. Some can be cast without pressure, and some can be resoftened and rehardened by heating and cooling; others cannot be softened after they are hardened.

Thermoplastics. Thermoplastic materials soften repeatedly when heated and harden when cooled. At high enough temperature, they melt, and at low enough temperatures some become brittle. They consist of monomeric units forming chains. These chains consist of rigid segments and flexible joints that are bent and coiled rather than straight. They may be continuous chains or branched. The chains are directly attached to each other by weak short-range secondary valence or van der Waal's forces. At any temperature above absolute zero, segments of the chains are in continual motion, coiling, uncoiling, and twisting.

Thermosets. Thermoset plastics pass through one soft plastic stage and then harden irreversibly. Like thermoplastics, they develop a time–temperature profile during polymerization; in the soft plastic stage, they can be shaped. During polymerization, monomer molecules are cross-linked to form large molecules. Softening by heating destroys these linkages.

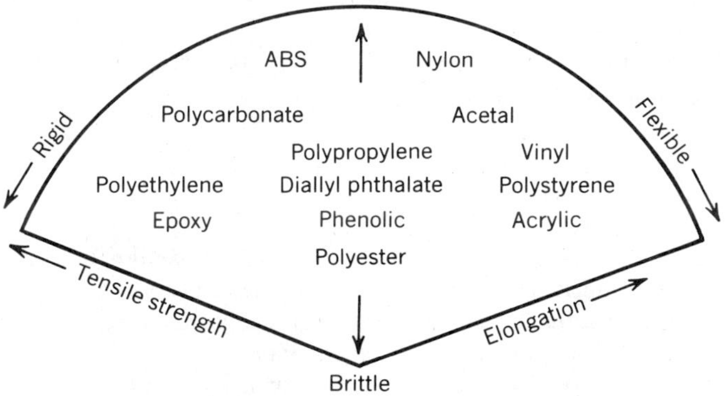

Figure 1. Range in mechanical plastics properties; with formulation changes (additives, fillers, reinforcements, alloying, etc), positions change.

Behavior of Plastics

Plastics are classified according to behavior, which varies widely and depends on time, temperature, and environment. The degree and duration of stress and strain affect structural response and behavior.

Both nonstructural and structural characteristics influence costs. Some plastics are compatible with other structural materials.

Viscoelasticity. An important property is viscoelastic behavior under different loading and environmental conditions. In viscous liquids and elastic solids, deformation is proportional to the applied force. Because of the wide range of viscoelastic behavior, some plastics are extremely flexible, whereas others provide the strongest materials available; some operate at extremely low temperatures and some at extremely high temperatures.

Creep and Stress Relaxation. Due to the viscoelasticity of plastics, these properties are highly dependent on temperature and time. Loading with respect to time is important.

Effect of Temperature

The stress response of plastics is much more temperature-dependent than that of metals or ceramics; four principal regions of viscoelastic behavior can be identified (Fig. 2).

Processing Considerations

Conventional plastics are available in sheets, I-beams, and other forms, like most materials, but offer the advantage of processibility, which allows the design of complex shapes not possible with other materials. The choice of fabricating process depends on the nature and requirements of the materials, the properties desired in the finished product, the cost of the process, speed, and the volume to be produced. Some processes can be used with many kinds of plastics; others are specialized. The fabrication process markedly influences the structural properties of a plastic or composite. About 36% of all plastics are processed by extruders, 32% by injection molding, 10% by calenders, 5% by coating, 3% by compression, 2% as powder, and 10% by other methods. Injection-molding machines are used four times as frequently as extruders.

Material Guidelines

The formability of plastics into almost any conceivable shape permits infinitely varied design stages, including a wide spec-

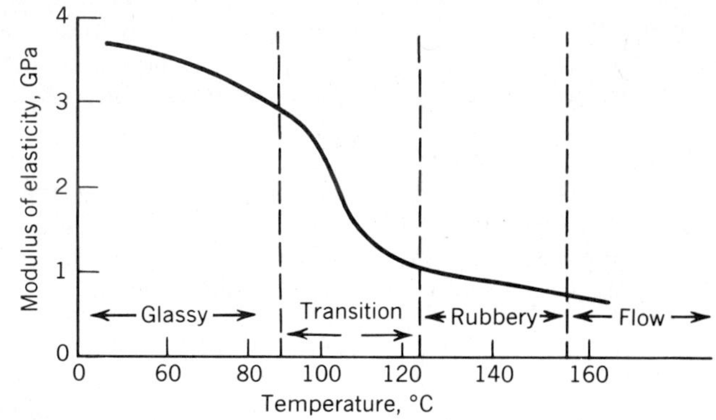

Figure 2. Plastics modulus of elasticity vs temperature. To convert GPa to psi, multiply by 145,000.

trum of strength properties, especially stiffness and bending resistance. In all materials, elementary strength of material theory (related to the moment of inertia I) demonstrates that some shapes resist deformation resulting from external loads or stresses better than others. This is due to the fact that deformation in beam and sheet sections depends on the product of the modulus (E) and the second moment of inertia, commonly expressed as EI. Physical performance can be changed by varying the moment of inertia, the modulus, or both.

To meet stiffness requirements by increasing thickness is expensive because large quantities of material must be used. Furthermore, the long heating and cooling times affect the process economics by increasing cycle times.

Ribs. Stiffness is obtained by ribbing, double curvature, or shaping. Although these methods increase costs, long production runs offset these investments by reducing molding cycles and raw-material consumption. Integral skin foam cores further increase stiffness.

Shape and Stiffness. In most cases, a basic beam structure can be used in the design of plastics parts. Plastic hollow-channel and I- and T-shapes designed with large radii and other flow considerations rather than sharp corners are more efficient on a weight basis because they use less material and provide a high second moment of inertia. The moments of inertia of such simple sections and hence stresses and deflections can be calculated using simple theories.

Part Thickness. With few exceptions, wall thickness is fundamental to design. Except in purely aesthetic parts, a suitable thickness specification is based on an analysis of the performance requirements and the strength, stiffness, and impact properties of the material.

Thickness Tolerance. To ensure that the minimum amount of material is used, it is important to select a material that can be repeatedly processed to meet the required tolerances. Tolerances that can be met range from 5% for 0.5 mm thick, 1% for 7.5 mm, 0.5% for 25 mm, to 0.25% for 12.5 cm, and so forth. Economical production requires that tolerances not be specified any tighter than necessary. However, after production, it is most economical to mold as tightly as possible. Many plastics change dimensions after molding because of temperature, humidity, or load changes. In certain plastics, heat treatment can significantly reduce or even eliminate these changes.

Material and Design Parameters

In the practical engineering approach, most plastic products are required to withstand a mechanical load without static or dynamic loads, as in appliance housings, containers, and similar applications. Property requirements could include mechanical strength and stiffness, chemical resistance, heat resistance, and electrical and thermal insulation. Short-term, conventional static tests generally suffice.

When a part is exposed to load for a long period, the design engineer is provided with long-term property data, eg, creep, fatigue, tensile strength, and others. These data provide the basis for common stress analysis. In the advanced engineering approach, data obtained over the years on plastic materials are used.

Testing and Quality Control

Quantitative test criteria for the design and material selection of plastic parts are available. Data-sheet properties are to be used only for initial screening of materials; the elastic modulus is usually less affected by variables than is strength. These data can be meaningful if they meet performance requirements and process conditions. Structural properties must be obtained on materials made by the process used in the manufacture of the components. Data generally include static and dynamic loadings.

The most widely used characterization of a plastic for initial screening and final material selection is the stress–strain relationship. These data show the presence or absence of a yield point, the slope of the initial curve, which is the elastic modulus in tension and relates to stiffness, and resistance to deformation. The mechanical properties observed depend on the testing rate. Plastics are usually tested in tension, flexure, compression, torsion, and at impact speeds.

Predicting Performance

Structural failure can be avoided by predicting material performance. Advanced computer methods have been developed for calculating stresses in complex structures. These methods have replaced former simplified models of materials. Early comprehensive analysis permits determination of the effects of temperature, loading rate, environment, and material defects on structural reliability. This information is supported by stress–strain-behavior data collected in actual materials evaluations.

BIBLIOGRAPHY

D. V. Rosato and D. V. Rosato, *Injection Molding Handbook,* Van Nostrand Reinhold Co., New York, 1985.

G. Lubin, *Handbook of Composites,* Van Nostrand Reinhold Co., New York, 1982, pp. 6–10.

M. A. Dorgham and D. V. Rosato, *Designing with Plastics and Advanced Plastic Composites,* Interscience Enterprises, Geneva, 1986.

D. V. Rosato and D. V. Rosato, *Blow Molding Handbook,* Carl Hanser Verlag, Munich, 1988.

DONALD V. ROSATO
University of Lowell
Borg-Warner Chemicals

MATRIX POLYMERIZATION.

See TEMPLATE POLYMERIZATION.

MECHANICAL DEGRADATION.

See DEGRADATION; ABRASION AND WEAR.

MECHANICAL PROPERTIES

The properties considered here include rubberlike behavior and viscoelasticity (qv), and range from small strain static and dynamic studies to the yield and fracture behavior of both amorphous and semicrystalline polymers.

In rubberlike elasticity (qv) the strains are large and can be considered to occur instantaneously as a consequence of chain uncoiling. Unhindered bond rotation can occur only above the glass-transition temperature T_g, the value of which in many cases can be related to a free-volume model. The main theoretical approaches are: finite strain elasticity, which involves an extension of the small strain stress–strain relations to finite deformation usually by the development of a strain-energy function; or a statistical theory modeled on a network of molecular chains.

Most polymers are, in contrast, viscoelastic, and show a delayed component of strain even at small deformations. Linear viscoelasticity, ie, total deformation proportional to stress, and stress relaxation proportional to strain, can be modeled by combinations of Hookean springs and Newtonian dashpots. Such models throw no direct light on molecular structure, but can be used to derive retardation and relaxation spectra from simple experiments performed over a range of temperatures, applying the principle of time–temperature superposition or equivalence. A particularly useful concept is that of the Boltzmann superposition principle: creep is a function of the entire loading history, with each loading step making an independent contribution.

Dynamic investigations into the temperature dependence of viscoelasticity indicate damping maxima that relate to molecular relaxation processes. The effect of crystallinity on the curtailment of molecular mobility in the glass transition is shown in Figure 1. The simplest treatment of temperature dependence is in terms of transition state (or barrier) theories, but for amorphous polymers, the experimental superposition of compliance–frequency plots at different temperatures through a shift along the logarithmic frequency axis is best accounted for in terms of the change in free volume with temperature [the Williams-Landel-Ferry (WLF) equation].

Normal mode theories, originally derived for dilute solutions, have been used to predict the relaxation spectrum for amorphous polymers. A more recent approach, associated particularly with the names of de Gennes, Doi, and Edwards, describes the dynamics of entangled polymers by assuming that each molecular chain moves within a well-defined tube; the motions are called reptation because of their snakelike nature (see MACROMOLECULAR DYNAMICS).

Relaxation transitions have customarily been labeled α,β,γ, etc, with decreasing temperature, irrespective of their molecular origin. Comparative studies on amorphous polymers that differ only slightly have been used to interpret these relaxations in terms of molecular motions and particular chemical groups attached to the molecular chains (Fig. 2). In a few crys-

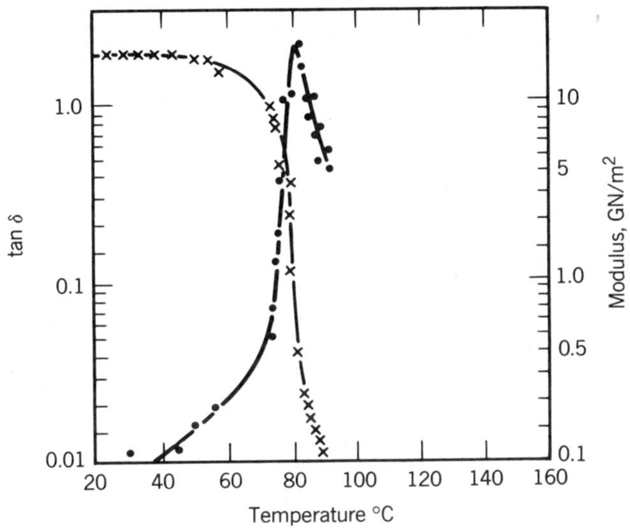

(a)

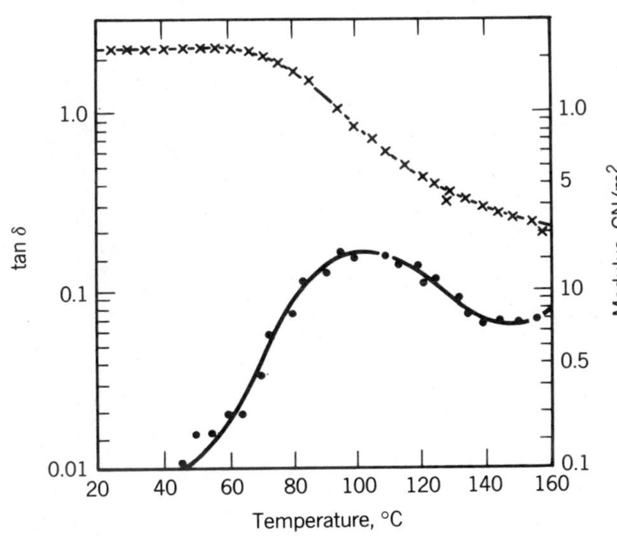

(b)

Figure 1. Tensile modulus and loss factor tan δ for (**a**) unoriented amorphous poly(ethylene terephthalate) and (**b**) unoriented crystalline poly(ethylene terephthalate) as a function of temperature at ~1.2 Hz. To convert GN/m² to psi, multiply by 145,000.

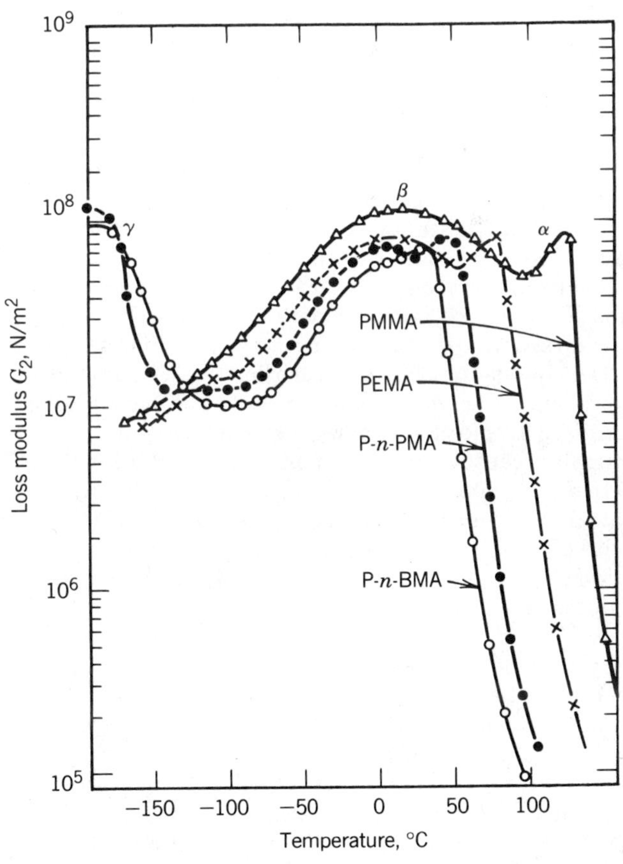

Figure 2. Temperature dependence of loss modulus G_2 for poly(methyl methacrylate) (PMMA), poly(ethyl methacrylate) (PEMA), poly(*n*-propyl methacrylate) (P-*n*-PMA), and poly(*n*-butyl methacrylate) (P-*n*-BMA). To convert N/m² to psi, multiply by 1.45×10^{-4}.

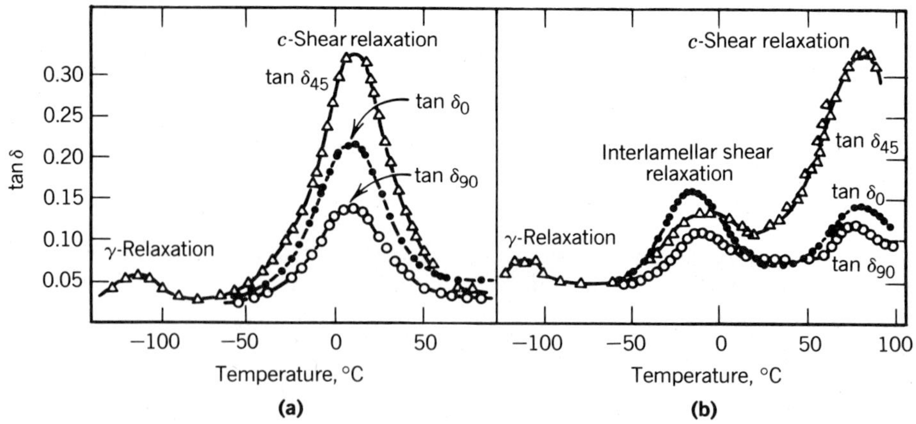

Figure 3. Temperature dependence of tan δ in three directions in (**a**) cold-drawn and (**b**) cold-drawn and annealed sheets of low density polyethylene (LDPE) at ~500 Hz.

talline polymers, particularly oriented polyethylene, some of the relaxations have been assigned to molecular processes (Fig. 3) but, in general, interpretation is still at the speculative stage, and is much influenced by the structural model selected.

The Boltzmann superposition principle is appropriate only at small strains, as viscoelasticity is generally nonlinear. Because there is no comprehensive mathematical representation that provides physical insights, two contrasting approaches have generally been adopted, those of the engineer and the rheologist. In the former, performance is predicted from a limited number of initial experiments using empirical relationships between stress, strain, and time; the latter extends the formal descriptions of linear viscoelasticity using single and multiple integral representations. Multiaxial deformation, which is of considerable technological importance, adds further complexity.

In many commercial applications of polymers, anisotropic behavior is observed, and the low strain elastic or viscoelastic properties must be described in terms of a generalized Hooke's law: five compliance (or stiffness) constants or functions are required for materials with fiber orientation and nine when the materials are biaxially oriented as is usual in polymer films. For instance, stiffness measured at various angles to the draw direction may change in completely different ways (Fig. 4). The change in behavior with orientation in a range of both amorphous and crystalline polymers has been simulated by an essentially single-phase aggregate model due to Ward, in which stretching causes the alignment of preexisting units. Other polymers, such as high density (or linear) polyethylene, that show a clear lamellar structure, are best modeled as a composite of linked crystalline and amorphous regions, using an approach pioneered by Takayanagi. Polyethylene of ultra-high modulus has been modeled either as an array of needlelike crystalline fibrils in a more compliant matrix, or as a stacking of crystal blocks linked by crystalline bridges passing through interspersed amorphous layers.

At high enough strains, dependent on both temperature and strain rate, ductile polymers yield, exhibiting a characteristic necking and drawing process; indeed, textile fibers are useful only when they have been subjected to stabilized drawing. When oriented polymers are extended in a direction different from that of the initial draw, the deformation may be concentrated in a narrow deformation band (Fig. 5).

Yielding has been interpreted as a shear-activated rate process, modified by the hydrostatic component of stress. Such theories imply that stress changes the rate of deformation until it equals the imposed strain rate. A contrasting theory by Argon assumes yield to be nucleation controlled. An important point is that cold drawing is no longer considered to arise through strain softening caused by an appreciable temperature rise at the neck. Furthermore, even in semicrystalline polymers, where plastic deformation causes drastic reorganization at the morphological level, the drawing behavior is considered to involve deformation of a molecular network to its limiting extensibility.

Griffith fracture theory in terms of linear elastic fracture mechanics has been applied to provide a framework for understanding the brittle fracture of polymers in terms of cracks spreading from existing flaws. High measured values of frac-

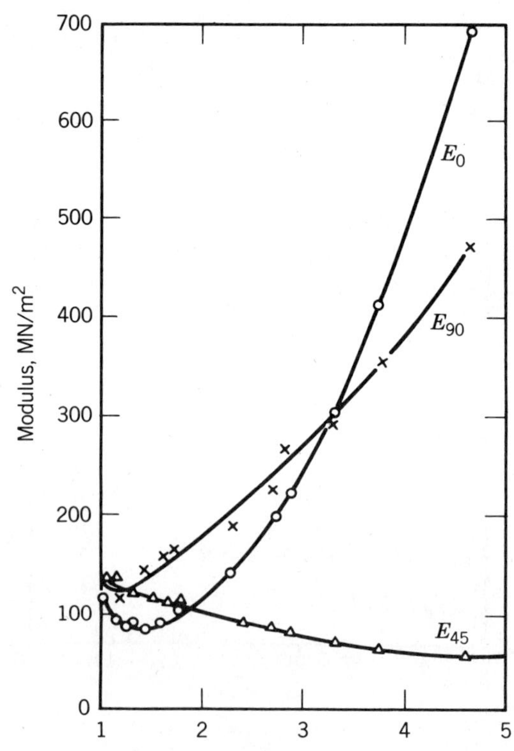

Figure 4. Moduli at various angles to draw direction in cold-drawn sheets of low density polyethylene (LDPE) at room temperature. To convert MN/m^2 to psi, multiply by 145.

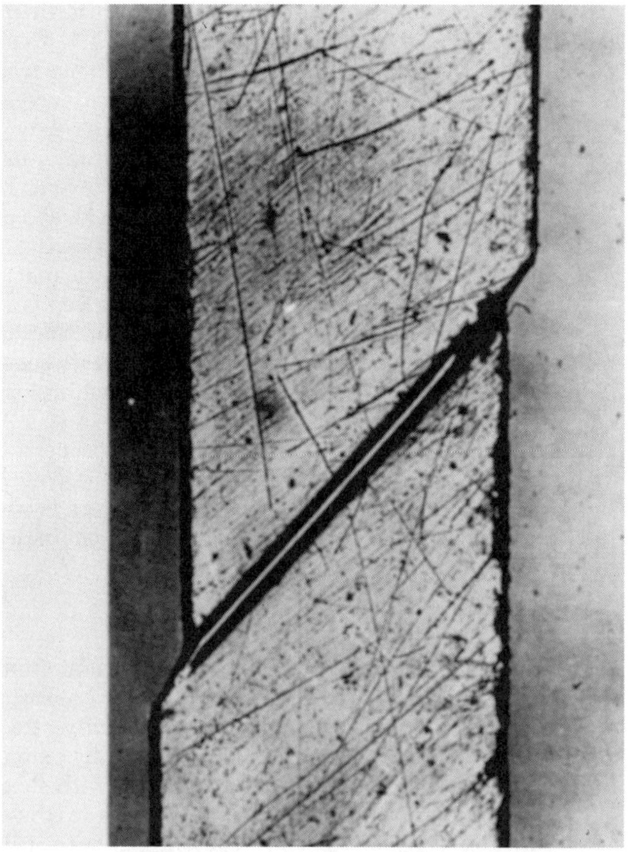

Figure 5. Deformation band in an oriented poly(ethylene terephthalate) (PET) sheet.

ture surface energy indicate that work is expended in the alignment of polymer chains ahead of the crack, giving rise to a wedge of porous material known as a craze. The brittle–ductile transition can be considered in terms of competition between the differently activated processes of crazing (qv) and yielding. There are only tentative suggestions of the way in which fracture behavior is related to molecular processes, but a generalized theory of fracture (qv) mechanics may provide a starting point.

Over the past 30 years polymer mechanics has made startling progress from the earlier applications of empirical relations, but much remains to be done, particularly for anisotropic semicrystalline polymers, before mechanical properties can be related unambiguously to processes at the underlying molecular level.

BIBLIOGRAPHY

J. D. Ferry, *Viscoelastic Properties of Polymers,* 3rd ed., John Wiley & Sons, Inc., New York, 1980.

N. G. McCrum, C. P. Buckley, and C. B. Bucknell, *Principles of Polymer Engineering,* Oxford University Press, Oxford and New York, 1988.

S. Turner, *Mechanical Testing of Plastics,* 2nd ed., Godwin, Harlow and Longman, New York, 1983.

I. M. Ward, *Mechanical Properties of Solid Polymers,* 2nd ed., John Wiley & Sons, Chichester and New York, 1983.

D. W. Hadley
University of Reading
I. M. Ward
University of Leeds

MECHANOCHEMICAL REACTIONS

Stress-induced reactions of polymers, a field commonly called mechanochemistry, are the object of intensive research. These reactions are observed in a great variety of disciplines and industries. Stress-induced reactions must be considered when handling high molecular weight polymers. With increased molecular weight, the conversion of the mechanical energy applied to a polymer is shifted increasingly from viscous to elastic, from thermal dissipation toward molecular energy storage. This energy storage leads first to changes in supramolecular structure, then to molecular bond rotations, chain orientation and elongation, bond bending and extension, and ultimately to the rupture of chemical bonds. Macroradicals are generated and molecular topology is changed, ie, molecular weight, its distribution, and branching and cross-linking. Molecular topologies more resistant to energy storage are created and restrict the reactions caused by the stress conditions. In terms of molecular weight, this limit decreases for more severe mechanical deformations and for poorer solvents (Table 1). The limit increases for tests in good solvents and for polymers with long-chain branching. For extensive polymer degradation, the limiting (final) molecular weight and distribution are independent of the initial values of molecular weight, M_{in}.

Bond rupture can also occur in the mechanics of solid (rigid) polymers. Solids may be considered as semicrystalline or amorphous polymers when they are below their glass temperature or highly cross-linked. The modes of stress reactions in the solid state may be classified into four categories: comminution (grinding, milling, crushing); machining (drilling, cutting, slicing); stretching; and other means (fatigue, tear, and wear tests).

Polymer Reactions

Most of the available data are on the stress reactions of amorphous polymer chains. These results have been obtained on both dilute and concentrated solutions; the latter include melts. These two categories are defined in terms of reactions of individual polymer molecules vs reactions among highly entangled polymer chains.

Table 1. Influences on Mechanochemical Reactions[a]

Increases in	Temperature	M_{in}	Initial viscosity	Shear intensity	Radical scavenger
reaction rate	↓	↑	↑	↑	→
extent of reaction	↓	↑	↑	↑	↓
M_{lim}	↑	→	→	↓	↑

[a] ↑, increases; ↓, decreases; →, no effect.

Dilute-solution theories for stress reaction are based on the rotation and elongation of chains by hydrodynamic forces. With very high molecular weight polymers, self-entanglement may contribute to stress reactions, even in dilute solution.

For entangled polymers, a related view leads to a different result. Imposition of shear on entangled polymers places tension on the temporary network constructed of interchain entanglements. The maximum chain tension, and thus the potential rupture site, is projected to occur midway between network points leading to a rupture process that is random along chains.

The two reaction regimes described above and a third less-observed degradation route differ in changes in molecular weight distribution resulting from stress reaction. In the third case, short chain fragments produced by stress are observed. These fragments have molecular weights in the range of chain-entanglement lengths.

The modes of bond rupture depend on both compositional and stress conditions. Most are radical processes. In decreasing order of importance, polymer-chain rupture reactions are random, exhibit midchain preference or short-chain segments. An additional major degradation process, unzippering, the reverse of vinyl polymerization, seems not to be significant under most conditions of stress reactions.

Deformation

The storage of strain within main-chain bonds leads to stress reaction. The direct measure of the increased energy storage is difficult, although potentially possible by spectroscopy. The extent of reaction is more commonly measured by less direct determinations of stress reaction, such as by changes in molecular weight.

Rheological measurements can be used to determine the tendency of stress reaction. Shear rather than tension or compression is usually considered.

Correlation of Stress. The applied and calculable shear stress is more closely correlated with the extent of stress reaction than the shear rate $\dot{\gamma}$. The stress $\tau = \dot{\gamma}\eta_a$, by Newton's law, provides the apparent viscosity η_a at reaction conditions. Other shear-rate variables may be considered for correlation with the extent of stress reaction. These include the rate of energy input $\dot{\gamma}^2\eta_a$ and various reduced forms, modified from theories of linear viscoelasticity.

Molecular Weight Effects. Most mechanochemical processes reduce molecular weight. Only in a few cases, eg, shear of rubbers and polyolefins under oxygen-starved conditions, can cross-linking and branching reactions predominate, leading to a molecular weight increase.

Property Modification. Mechanochemical processes can modify the chemical and physical structure of polymers, resulting in changes in the melt rheology and mechanical properties of solids. New functional groups may be formed along or at the chain ends.

Effects of Temperature. The temperature of mechanochemical reactions affects the physical state of the polymer, its properties (elasticity and viscosity), and the bond-rupture mechanism.

Reaction Media. Gaseous media can be divided into three classes: those incapable of reacting with macroradicals; those capable of reacting with the radicals generated; and those capable of reacting with radicals and controlling the secondary reactions. Nitrogen, vacuum, and air have been used for the cold mastication of natural rubber. In the presence of oxygen, chain scission reduces the molecular weight. In an inert atmosphere (nitrogen or vacuum), branching results in cross-linking and gelation; a small amount of oxygen, 0.1%, prevents gel formation. This accounts for discrepancies among test results. The relative importance of competitive and successive reactions depends on the chemical structure of the rubber and the reactivity of macroradicals toward active groups in the chain.

Radical Acceptors. The incorporation of radical acceptors influences polymer reactions induced during processing. The radical character of the oxygen molecule leads to reactions comparable with those of scavengers. Radical acceptors are usually nonpolymeric, exhibiting greater diffusion than polymer chains. Their presence prevents reactions between macroradicals and polymer chains leading to branching, cross-linking, and modification of properties.

The use of radical acceptors allows control of properties by avoiding undesirable secondary reactions during polymer processing and use. Incorporation of so-called peptizers is widely used in the rubber industry, resulting in linear chain-degradation products.

Polymer States

Solid State. Modes of stress reaction include comminution, machining, uniaxial stress, fatigue, and wear and abrasion.

Rubber State. Highly elastic polymers are readily attacked by shear where chain displacement is restricted by extensive molecular entanglements. Internal stresses rise with shear to a point where they may exceed covalent bond strengths even without chemical cross-links. This reaction has been utilized by passing natural rubber between spiked rollers for softening and transformation into a coherent mass. The method, called mastication, allows for easy dispersion of fillers and additives (accelerators, vulcanizers, and antioxidants). This technique is less important for synthetic rubbers because molecular weight can be regulated during polymerization.

Molten State. High temperature processing has been applied in the extrusion of polyolefins and in the modification of homopolymers by producing block and graft copolymers during processing. At processing conditions, degradation is almost exclusively thermal and possibly oxidative; others conclude that degradation is mainly mechanical. Although degradation is a thermal process, a reduction in temperature is needed for reactions provided by the mechanical energy stored by shear within the polymer chains. This corresponds to a shear-induced change in the potential energy function for thermal bond rupture. The effect is beyond any temperature rise resulting from mechanical working.

Solutions. Polymer solutions of high concentration range are more sensitive than dilute solutions to mechanochemistry. Entangled systems usually appear rubbery, being both highly viscous and elastic; the solvent is usually absorbed by the polymer. Homogeneous solutions may require a long time to prepare; stirring can lead to mechanochemical reactions.

When polymer solutions are agitated, extruded, or sheared, polymer and solvent interact, resulting in deformation and extension. When the force is removed, the polymer molecules return to an equilibrium and relaxed state. Strong forces cause chain rupture, and macroradicals (reaction sites) are created.

Copolymer Synthesis

Recently, demands have increased for polymers with complex compositions. Properties can be obtained by combining polymers in true solutions or heterophase systems. Block and graft polymerization permit a combination of different properties, eg, heat resistance and elasticity, or hydrophobic and hydrophilic properties. Stress can generate these systems.

Methods of mechanical syntheses include comminution, vibromilling, mastication, extrusion, stirring, ultrasonic irradiation, freezing and thawing, high voltage discharge, and swelling. These methods have been employed to synthesize block and graft copolymers using, as initiator, the radical formed by the mechanical scission of macromolecules.

BIBLIOGRAPHY

R. S. Porter and A. Casale, *Polym. Eng. Sci.* **25,** 129 (1985).

A. Casale and R. S. Porter, *Polymer Stress Reactions,* Vols. 1 and 2, Academic Press Inc., New York, 1978 and 1979.

H. H. Kausch, *Polymer Fracture,* Springer-Verlag, New York, 1978.

N. K. Baramboin in W. F. Watson, ed., *Mechanochemistry of Polymers* (translated from Russian by R. J. Moseley), Rubber and Plastic Research Association of Great Britain, Maclaren, Ltd., London, 1964.

ROGER S. PORTER
University of Massachusetts

ANTONIO CASALE
Snia Tecnopolimeri SpA

MEDICAL APPLICATIONS

The first medical applications of polymers made use of commercially available plastics, adapted as necessary. Although the science and technology of polymers for biomaterials application is still at a relatively early stage of development, progress has been dramatic. Medical polymers must adhere to very rigid standards, and must be nontoxic, noncarcinogenic, biocompatible, and in no way injurious to the biological environment (see also DRUG APPLICATIONS).

Artificial organs and prostheses constitute some of the most valuable remedies in present medical armamentaria. They have contributed immeasurably to the welfare and health of the human race. Although the production of medical polymers is small compared with other industrial plastics, the number of individuals who benefit from these uses is very large.

The 1983 U.S. applications are given in Table 1. Disposable dialyzers that use a variety of polymeric membranes are an example of a biomaterial that has become a clinically significant commercial product.

Clinical Applications

The medical applications of polymers are given in Table 2.

Table 1. Medical Applications and Implants in the United States, 1983

Applications	Number per year
heart valves	30,000
pacemakers	150,000
vascular grafts	150,000
oxygenators	200,000
dialysis procedures	10,000,000
dental implants	20,000
breast prostheses	100,000
retinal surgery	50,000
insulin infusion pumps	5,000

Table 2. Medical Uses of Polymers

Polymer	Abbreviation/trade names	Applications
poly(vinyl chloride)	PVC	disposable medical articles; extracorporeal devices; hemodialysis or hemoperfusion blood tubing line; cardiac catheters; blood bags and iv infusion bags; endotracheal tubes; surgical tape; sheet oxygenators; artificial heart; blood pump; artificial limb materials; packaging materials
polyethylene	PE	catheters; film substrates; tubing, connectors, and bottles; plastic surgery implants; oxygen tents; packaging material
ultrahigh molecular weight polyethylene	UHMWPE/RCH 1000	acetabulum in total hip prostheses; artificial knee prostheses
polypropylene	PP	disposable syringe; membrane for membrane oxygenator; connectors, housings, bottles; finger-joint prostheses; indwelling iv cannulae; unabsorbable sutures; packaging materials
polystyrene	PS	monodisperse latex for preparing diagnosis kits; calibration for microscopes; packaging materials; disposable laboratory articles, eg, test tubes

Table 2. (*Continued*)

Polymer	Abbreviation/trade names	Applications
carboxylated PS		monodisperse latex for preparing diagnostic kits
styrene–acrylonitrile copolymer	SAN	housing for hollow-fiber hemodialyzer; connectors, components
silicone oil		defoaming agent for blood oxygenator; plastic surgery implants; mammary augmentation
silicone rubber	Silastic	hydrocephalus shunts; catheters of all types; roller pump tubing (heart–lung machine); membrane for membrane oxygenator; artificial skin for burn dressing; plastic surgery implants; artificial heart, heart-assist pump materials; drug-release system; atrioventricular shunts; ear prostheses; facial prostheses; artificial heart valve prosthesis (Star Edwards); tendon and finger-joint repair; tracheal prostheses; bladder prostheses; bladder patch intestine patch; dura-mater prostheses; retinal-detachment treatment; impressing material; heart pacemaker leads
polycarbonate	PC/Lexan, Nuclepore	housings, connectors; membrane for oxygenator and hemodialyzer; potential packaging material; plasmapheresis membrane
polyester	Dacron, Mylar	vascular graft prostheses; Dacron mesh, fixation devices for tissue; hernia repair; patches for heart, bladder, arteries; suture
polytetrafluoroethylene	PTFE/Teflon, Gore-Tex, Impra	vascular graft prostheses; heart patch; stapes prosthesis (in tympanoplasty); retinal-detachment treatment
fluorinated ethylene—propylene copolymer	FEP	dual lumen subclavian cannula
polyolefin rubber	Hexsyn[a]	artificial heart pump material
polyurethane	PU/Avcothane[b], Biomer[c]	artificial heart pump material and heart assist; balloon for IABP; heart-valve prostheses; vascular graft prostheses; tubing; coating for blood compatibility
poly(vinyl alcohol)	PVA	drug-release system
poly(acrylonitrile)	PAN	membrane for dialysis
polyacrylamide	PA	drug-release system; diagnostics, eg, in electrophoresis
polyacrolein	PAL	microbeads for diagnosis use; cell culture
polysulfone		membrane for dialysis or ultrafiltration; oxygenator
regenerated cellulose	Cuprophane	membrane for dialysis
cellulose acetate		membrane for hollow-fiber dialyzers
poly(methyl methacrylate)	PMMA	bone cement; artificial teeth, implanted teeth, denture material, dental fillings; bone prostheses, cranial bone replacement; intraocular lens; membrane for dialysis or ultrafiltration

Table 2. (*Continued*)

Polymer	Abbreviation/trade names	Applications
methyl methacrylate–styrene copolymer	MMA–S	bone cement
poly(hydroxyethyl methacrylate)[d]	polyHEMA/Hydron[e], Soflens[e]	soft contact lens, 39% H_2O; denture soft liners; burn dressing; active carbon encapsulating material
poly(hydroxyethyl methacrylate-*co*-vinylpyrrolidinone-*co*-ethylene dimethacrylate)	HydroCurve[e]	soft contact lenses 45% H_2O
poly(hydroxyethyl methacrylate-*co*-vinylpyrrolidinone-*co*-ethylenebis(oxyethylene) dimethacrylate)	Hydralens[e]	47% H_2O
poly(hydroxyethyl methacrylate-*co*-methacrylic acid-*co*-ethylene dimethacrylate)	Tresoft[e]	48% H_2O
poly(vinylpyrrolidinone-*g*-hydroxyethyl methacrylate)	Softcon[e]	55% H_2O
poly(hydroxyethyl methacrylate-*co*-3% methacrylic acid)		65% H_2O
poly(hydroxyethyl methacrylate-*co*-methacrylic acid-*co*-vinylpyrrolidinone	Perma Lens[e]	68% H_2O
poly(vinylpyrrolidinone-*co*-methyl methacrylate-*co*-allyl methacrylate-*co*-ethylene dimethacrylate)	Sauflon-70, Sauflon-85[e]	70% H_2O 85% H_2O
poly(*N*-vinyl-2-pyrrolidinone) interpolymer	Hydromer	coatings for catheters; extremely low coefficient of friction when wet
poly(alkyl-α-cyanoacrylate)		tissue adhesives

[a] Hexsyn rubber (Goodyear Corp.) is a terpolymer of 95% poly(1-hexene) and 5% of a mixture of 4-methyl-1,4,-hexadiene and 5-methyl-1,4-hexadiene in a 3:2 ratio.
[b] Avcothane-51 (Arco Corp.) is a block copolymer of 90% polyether urethane and 10% polydimethylsiloxane; it is available as a ca 15 wt % solution of a 2:1 mixture of THF and 1,4-dioxane.
[c] Biomer (Ethicon, Inc.) is a segmented polyether polyurethane; it is available as a 30 wt % solution in dimethylacetamide.
[d] With a small amount of ethylene dimethacrylate.
[e] Manufacturers: Hydron, National Patent Development Corp.; Soflens, Bausch & Lomb, Inc.; HydroCurve, Soft Lenses, Inc.; Hydralens, Ophthalmos, Inc.; Tresoft, Urocon International; Softcon, American Optical Corp.; Perma Lens, Global Vision, Ltd. (UK); Sauflon, Sauflon International.

Requirements for Biomedical Polymers

A biomedical polymer should possess appropriate mechanical and physical properties; tissue, cell, and blood compatibility (minimum histotoxicity, noncarcinogenicity); appropriate aging property in the body (stability or degradability); sterilizability; and shelf stability. It must be readily available, easily processed, and of reasonable cost. In special cases, eg, bone cements and tissue adhesives, exotherm during polymerization should be minimal to avoid tissue trauma.

Mechanical and physical properties of polymers continue to be the subject of intensive research, as they relate to end uses for prostheses, slow release of drugs, diagnostic related tests, and synthetic biopolymers.

BIBLIOGRAPHY

T. Akutsu, N. Yamamoto, M. A. Serrato, J. Denning, and M. A. Drummond in C. G. Gebelein and F. K. Koblitz, eds., *Biomedical and Dental Applications of Polymers,* Vol. 14 of *Polymer Science and Technology,* Plenum Press, New York, 1981.

J. W. Hodge, Jr. and C. W. R. Wade in M. Szycher and W. J. Robinson, eds., *Synthetic Biomedical Polymers: Concept and Applications,* Technomic Publishing Co., Inc., Westport, Conn., 1980.

Medical Polymers: Chemical Problems, 17th Microsymposium on Macromolecules, Institute of Macromolecular Chemistry, Prague, Aug. 15–18, 1977, as reported in *J. Polym. Sci. Polym. Symp.* **66** (1979).

E. P. Goldberg and A. Nakajima, *Biomedical Polymers: Polymeric Materials and Pharmaceuticals for Biomedical Use,* Academic Press, Inc., New York, 1980.

B. David Halpern
Polysciences, Inc.

Y.-C. Tong
Tianjin Bureau of Chemical Industry

MELAMINE–ALDEHYDE RESINS. See AMINO RESINS.

MELAMINE–THIOUREA RESINS. See AMINO RESINS.

MELAMINE–UREA RESINS. See AMINO RESINS.

MELT EXTRUSION. See EXTRUSION.

MELT FLOW. See RHEOLOGICAL MEASUREMENTS.

MELT FRACTURE. See EXTRUSION; RHEOLOGICAL MEASUREMENTS.

MELT INDEX. See RHEOLOGICAL MEASUREMENTS.

MELT VISCOSITY. See RHEOLOGICAL MEASUREMENTS.

MEMBRANES

A membrane is a thin barrier that permits selective mass transport. Thus the term permselectivity was coined to distinguish between a thin, nonpermeable film or layer, and a permselective membrane. The membrane can be made of an inorganic or organic, synthetic or biological product. A frame of reference for the synthetic membrane is the biological membrane, which provides the semipermeable boundaries of living cells.

As permselective barriers, synthetic membranes have been employed in a variety of applications and can be classified according to their utilization, configuration, composition, allowable mode of transport, and permselectivity (Tables 1 and 2, Fig. 1).

Configuration and Morphology

Membranes are used in four principal configurations: sheets, tubes, hollow fibers, and capsules (or microcapsules). However, there is no conceptual limitation for the configuration of a membrane, and membranes can be shaped in almost any geometrical form desired.

The membrane morphology dictates the mode of permeation and separation, as shown in Figure 2. The basic morphologies are isotropic (dense or porous) and anisotropic with a tight surface extending from a highly porous wall structure. The tight surface can be a dense, selective skin, permitting only diffusive transport or a porous skin, allowing viscous flow of the permeate, as in conventional ultrafiltration. Membrane separation is achieved by manipulation of these basic morphologies.

The semipermeability of porous membranes is based on the spatial cross section of the permeating species, that is, small molecules exhibit a higher permeation rate than large molecules, and linear chain molecules permeate better than those with a globular shape. The semipermeability of the dense membrane or a membrane that exhibits anisotropic morphology with a dense skin is obtained primarily through a solution-diffusion mechanism.

Composite membranes consist of ultrathin semipermeable layers supported by highly porous substrates with minimum resistance to mass-transport. In this class of membranes the properties of two or more different materials are combined to yield a desired product.

Table 1. Synthetic Membrane Applications and Processes

Special applications as permselective barriers	Membranes as permselective barriers	Membrane reactor
ion-specific electrodes controlled release tissue-culture growth biosensors	dialysis microfiltration ultrafiltration reverse osmosis osmotic pumping pervaporation gaseous separation electrodialysis Donnan dialysis membrane distillation steam filtration thermoosmosis	immobilization of catalyst and enzyme within the membrane

Principles of Permeation and Separation

Permeation. In most membrane processes, mass transport through a nonporous polymer membrane is caused by a chemical potential gradient or, in more general terms, by the difference in the free energy of the permeate between membrane upstream (feed-side) and downstream (product-side) interfaces. Correspondingly, the mass transport through a membrane can be described by the following equation:

$$J = KF'/l \tag{1}$$

where J is the permeate flux, F' the resultant driving force, l the membrane resistance to transport expressed often as the membrane-effective thickness, and K a constant.

The most common mode of transport through a dense neutral membrane can be described by a simplified solution-diffusion mechanism where the chemical potential gradient materializes as a concentration gradient. In such cases, the flux-per-unit area of the membrane can be expressed by Fick's first law

$$J = -Ddc/dx \tag{2}$$

where D is the diffusion coefficient and dc/dx the concentration gradient. Accordingly, the flux can be described by

$$J = D(C_1^m - C_2^m)l \tag{3}$$

where C_1^m and C_2^m represent the concentration of the permeate in the membrane upstream and downstream interfaces. For vapors at very low activities and for gases for which a linear relation between solute concentration and pressure exists, the solubilities can be described by Henry's law constants (for a solute in the liquid phase, the solubility can be described by the partition coefficient)

$$C = Sp \quad \text{or} \quad C_m = SC_s \tag{4}$$

Table 2. Commercial Polymer Membranes

Material	Abbrev	Processes[a]
cellulose acetate	CA	MF, UF, RO, D, G
cellulose triacetate	CTA	MF, UF, RO, G
CA–triacetate blend		RO, D, G
cellulose esters, mixed		MF, D
cellulose nitrate		MF
cellulose, regenerated		MF, UF, D
gelatin		MF
polyamide, aromatic		MF, UF, RO, D
polyimide		UF, RO
polybenzimidazole	PBI	RO
polybenzimidiazolone	PBIL	RO
polyacrylonitrile	PAN	UF, D
PAN–poly(vinyl chloride) copolymer	PAN-PVC	MF, UF
PAN–methallyl sulfonate copolymer		D
polysulfone		MF, UF, D, G
poly(dimethylphenylene oxide)	PPO	UF, G
polycarbonate[b]		MF
polyester[b]		MF
polytetrafluoroethylene	PTFE	MF
poly(vinylidene fluoride)	PVF_2	UF, MF
polypropylene	PP	MF
polyelectrolyte complexes		UF
poly(methyl methacrylate)[c]	PMMA	UF, D
polydimethylsiloxane	PDMS	G

[a] MF = microfiltration, UF = ultrafiltration, RO = reverse osmosis, D = dialysis, and G = gas separation.
[b] Track-etched.
[c] Stereotopic complexes.

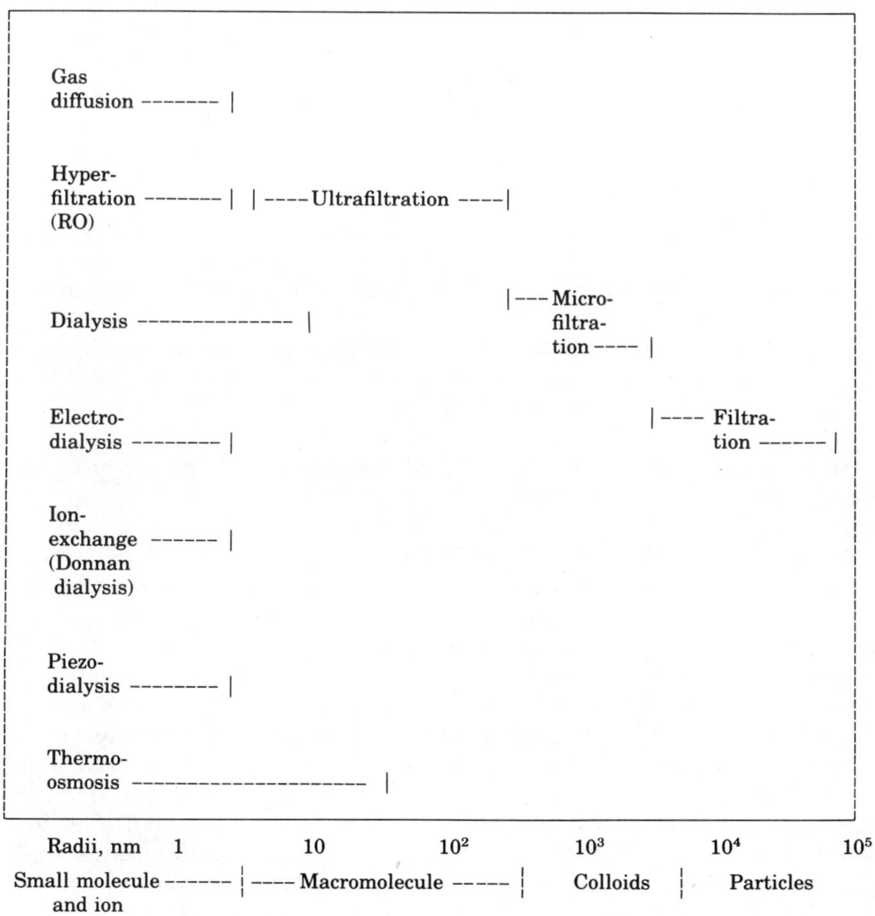

Figure 1. Size range of permeating species in membrane processes.

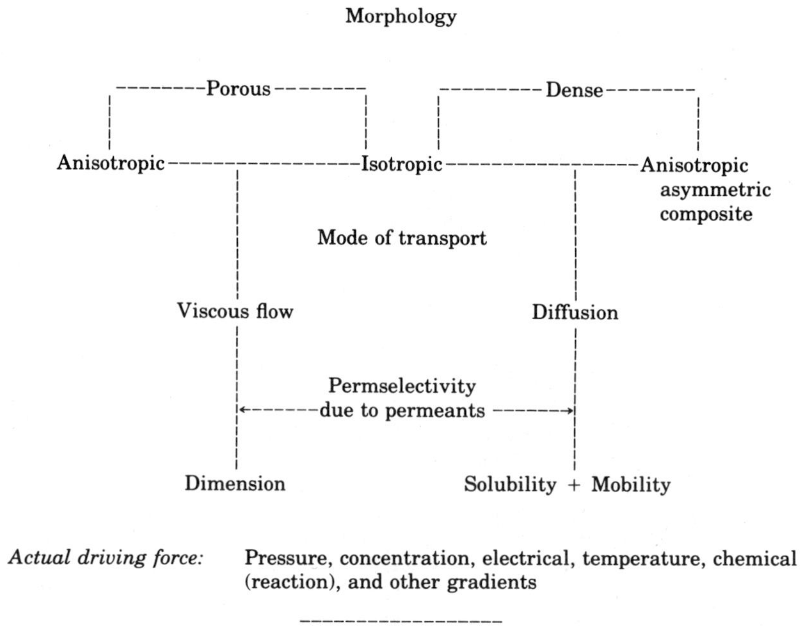

Figure 2. Configuration, morphology, transport, permselectivity, and driving forces employed in membrane science and technology.

where C and C_m represent the solubility of the permeant in the membrane, p the partial pressure of the permeant in the gas phase, C_s the concentration of the permeant in the solution, and S the solubility constant. Combining equations 3 and 4 yields convenient flux equations

$$J = DS(p_1 - p_2)/l \tag{5}$$

$$J = DS(C_1 - C_2)/l \tag{6}$$

where p and C represent the partial pressure and the concentration of the permeating species in the feed and product, respectively. These flux equations are also used for membrane systems where the interpretation and measurement of the solubility and diffusion constants are more complex and difficult, as in membranes made of glassy polymers. Therefore, it is convenient to describe the membrane-transport characteristic in terms of a permeability coefficient

$$P = DS \tag{7}$$

The permeability coefficient P is commonly used for comparing different membranes for a given separation. It is easily determined from steady-state permeation experiments. Nevertheless, any improvement sought in the inherent performance of a given membrane requires the determination of the sometimes complex relationship between D and S, obtained by performing a set of well-documented sorption and diffusion experiments (see DIFFUSION).

For many applications, membrane permeability is too low and the separation process becomes feasible only when the membrane resistance to transport (eq. 1) reaches a certain minimum. This can be attained for a given P value by reducing the thickness of the permselective barrier, which for many materials is below the level allowable for mechanical self-support. The solution to this problem was found in the structures and morphologies of anisotropic membranes that support permselective barriers of $<0.1\ \mu m$ (Figs. 2–4).

The general flux equation for a pressure-driven process can be written as

$$J = K(p_1 - p_2)/l \tag{8}$$

where K is the permeability coefficient (equal to P for gas or vapor separation) and is often designated as the hydraulic permeability Lp in liquid permeation processes.

Separation. Membrane performance is commonly assessed in terms of permeation rates and separation factors. For a feed stream composed of components A and B, the separation factor $\alpha_{A/B}$ is defined as

$$\alpha_{A/B} = (X_A/X_B)^{P}/(X_A/X_B)^{F} \tag{9}$$

where X_A^F, X_B^F, X_A^P, X_B^P denote the concentration of components A and B in the feed solution F and the product P; X is expressed in the relevant unit to the system, ie, mol %, wt %, or volume fraction.

The separation of solvent from solute can be described by

$$\text{separation (\%)} = (1 - C_P/C_F)\ 100 \tag{10}$$

where C_P and C_F are the concentration of the solute in the product and in the feed, respectively.

Materials

Membranes are prepared from almost every available material. In large-scale production, commercial thermoplastics and cellulosics are primarily used (Table 2).

The selection process of materials for the construction of membranes requires a thorough understanding of the structure–property relationship, which affects segregative mass transport. Hitherto most permselective membranes have been prepared through an empirical process of trial and error. In this regard, membrane science has been outpaced by membrane technology. Consequently, polymer membranes (as those listed in Table 2) are effectively being employed in spite of the

Figure 3. Cross section of an anisotropic membrane.

fact that, at the molecular level, only limited fundamental knowledge is available regarding the formation and functioning of a membrane. Difficulties concerning the material selection process are frequently due to the elusive nature of the amorphous phase in polymers. Mass transport in neutral, dense polymer membranes takes place only through the amorphous phase, where solubility and mobility of a permeating species largely depend on the local arrangement of the polymer chains above and below the glass-transition temperature. In the same vein, semicrystalline polymers are often considered heterogeneous material because the nature of molecular diffusion varies through the pure amorphous phase and within the crys-

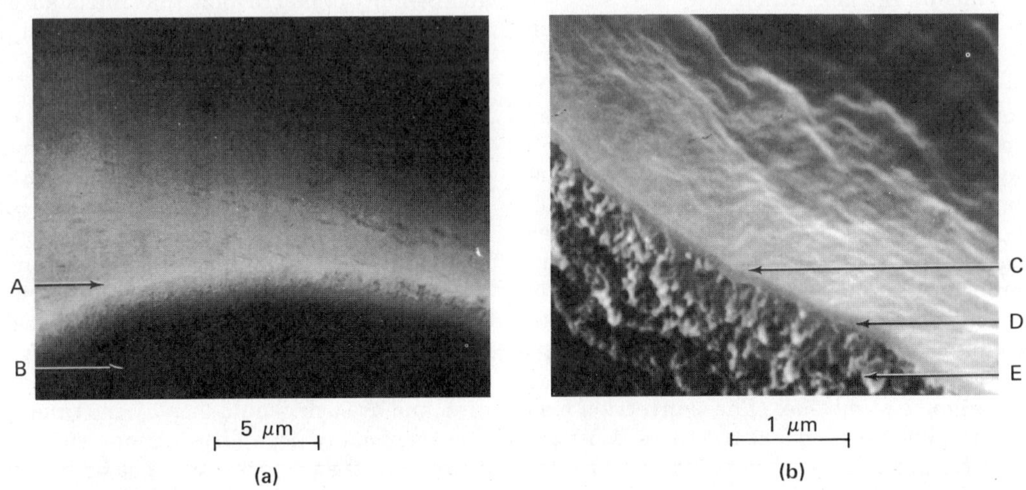

Figure 4. Composite hollow-fiber membranes. (**a**) Polysulfone hollow fiber coated with furan resin. A and B denote furan resin surface and porous support, respectively. (**b**) Cross section of composite hollow fiber (PEI/TDI coated on polysulfone matrix). C, D, and E denote tightly cross-linked surface, "gutter" gel layer, and porous support, respectively. Both fibers were developed for reverse-osmosis application.

talline–amorphous interfaces. The interaction of the diffusing species with the functional groups that are pendent to, or part of, the polymer chains largely depends on the distribution of these groups in the amorphous phase. Consequently, materials that otherwise are viewed as homogeneous would be considered highly heterogeneous when judged on permeate diffusion pathway (see also TRANSPORT PROPERTIES).

Processing

Most solid polymer membranes are produced by conventional extrusion processes. The fabrication of a membrane is a multistage procedure. Membranes initially formed by extrusion or other methods are submitted to a variety of posttreatments to establish permselectivity and durability. All starting materials must meet a set of stringent criteria, and the components employed in casting or spinning formulations must be consistent in every batch preparation. Commercial polymers, even when obtained from the same manufacturer, often display unacceptable inconsistencies resulting from batch-type production.

Phase-inversion Process. Most of the membranes used in microfiltration, dialysis, reverse-osmosis, and gas-separation processes are prepared by the wet extrusion process. Homogeneous solutions of the polymers in solvent mixtures are prepared and phase inversion is achieved by solvent evaporation, exposure to excess of nonsolvent, and thermal gelation. These methods can be applied to a large number of polymers as well as inorganic materials. The phase-separation phenomenon can be characterized by thermodynamic and kinetic parameters. The rate at which phase separation takes place determines whether the structure of a precipitating membrane is the result of nucleation or spinodal decomposition. It also determines the degree of crystallinity and the stresses left in the membrane matrix as a result of a fast transformation of the nascent membrane from a highly plasticized state to an amorphous glass. The glass-transition temperature progressively shifts toward that of a pure polymer throughout the coagulation process. A methodology that combines all the elements involved in the phase-inversion process into scientific formalism has yet to be developed. Phase diagrams like those shown in Figure 5 serve as practical tools in membrane fabrication.

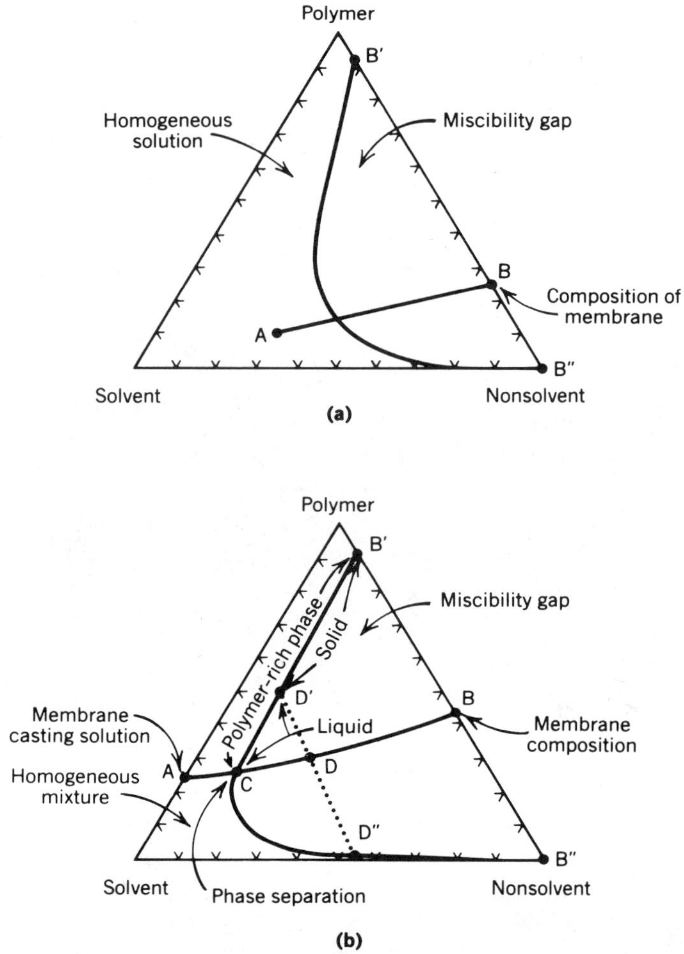

Figure 5. Phase diagrams showing the formation of a microporous membrane by (**a**) evaporation of a solvent from a three-component polymer mixture exhibiting a miscibility gap at certain conditions of temperature and composition; (**b**) addition of a nonsolvent to a homogeneous polymer solution.

Anisotropic Membranes. Early attempts in the fabrication of anisotropic membranes via phase inversion, using wet-casting or spinning, were all empirical trial and error, and yielded the following set of working rules. The solubility parameters of the casting composition should be near the solubility boundary facing the coagulating medium. The loss of volatile components, when evaporation precedes coagulation in a nonsolvent, should move the composition out of the solubility area rather than into it. The solid content at the solution boundary must be high in order to cause a rapid transition from sol to gel, and all components of the system should be miscible with the coagulating medium (nonsolvent). These rules, however, are not binding.

Substructure of Anisotropic Membranes. Phase inversion may lead to a variety of morphologies and substructures. Most anisotropic membranes consist of dense skin resting on nodular (micellar) layers 0.1–2 μm thick (Fig. 6). From this structure evolves the porous matrix from which the remainder of the membrane is composed. The porous structure can be weblike, an open-cell foam, or coral structure. Depending on the dope formulation and casting conditions, the membrane may exhibit an assembly of macrovoids that intrude from the surface deep into the porous structure. Bottom skin frequently forms on the side of the membrane, which does not interface with the coagulant.

The nodular structure was first shown to exist beneath the skin of anisotropic cellulose acetate membranes; ion-etching the surface revealed an assembly of closed-packed spheres, 18.8 nm in dia. Depending on the polymer composition, nodules in the range of 12–1000 mm have been reported. The fusion of the nodules in the membrane's upper zone is largely responsible for permselective properties of the membrane.

Bottom Skin. The origin of the bottom skin is different from that of the surface skin. The bottom surface of the membrane is usually an integral extension of the highly porous substructure, and as such should display the same porous morphology. However, many dope formulations yield a highly structured, dense skin that may or may not be porous. Bottom-skin formation is largely due to the interaction between the casting surface and the components of the dope mixture. Polymers with different degrees of adherence to the casting surface (eg, glass, polyethylene, and Mylar) yield skin layers with a wide range of characteristics.

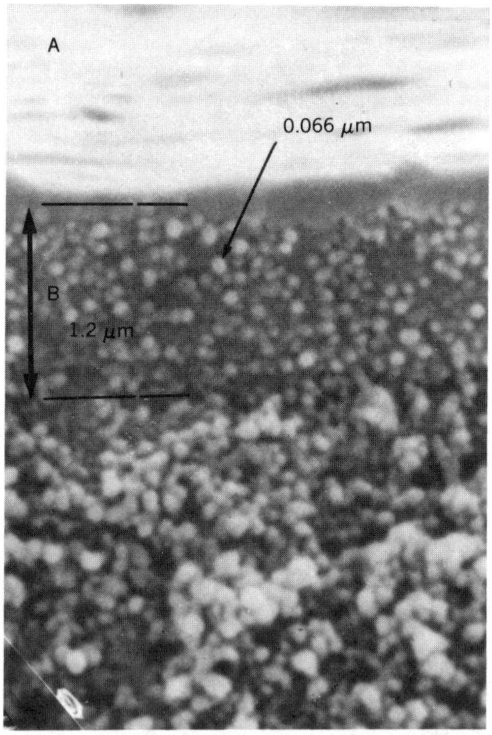

Figure 6. Cross section of the surface of a polysulfone hollow fiber showing nodular zone of 1.2 μm and well-defined nodules (dia = 0.066 μm): A, polydimethylsiloxane coating; B, nodular zone.

Macrovoids. Macrovoids seem to be the result of coagulant penetration through the nascent membrane interface during the first stages of coagulation, following the submersion of the cast layer (or the spun extrudate) into the coagulating bath.

Isotropic Porous Membranes. Isotropic porous membranes include stretched semicrystalline film, nucleation-track membranes, and sintered-particle membranes.

Polymer Blends and Alloys

Dense and anisotropic (asymmetric) alloy membranes with dense skin can be produced from a homogeneous blend of different polymers. Polymer blending is a simpler, more feasible technology than methods that compound different polymer segments via copolymerization and the formation of interpenetrating matrices. Homogeneous polymer blends (qv) consist of two (or more) polymers miscible at the molecular level and combining the properties of the components to yield a distinct new material.

Dynamic Membranes

Concentration–polarization phenomena take place with high flux membranes or if a strong affinity between the solute and the membrane material exists. These phenomena, which usually impair mass transport through synthetic membranes, are used to produce dynamic membranes.

Dynamic membranes are prepared by *in situ* deposition of a colloidal slurry or a low concentration of polymers on the surface of filtration or ultrafiltration membranes. Organic or inorganic membranes as well as combinations can be used.

Facilitated and Coupled Transport Membranes, Liquid Membranes

The terms facilitated and coupled transport refer to transport of a permeant through a membrane via a carrier mechanism (Fig. 7**b**). The membrane phase is impermeable to the permeant and contains a free, mobile carrier that does not dissolve (or evaporate) into the feed and product streams. In facilitated transport, the carrier, which can interact with the permeant at the membrane–liquid interface, extracts the permeant into the membrane phase. The permeant coupled with the carrier diffuses through the membrane and is eventually released at the downstream interface.

In coupled-transport mechanism (Fig. 7**c**), a chemical reaction takes place at the interfaces of both membranes and mass flux occurs in both directions.

Biomimetic Membranes

Biological membranes are composed primarily of lipids and proteins. The lipids vary in composition and generally consist of a polar head group and nonpolar aliphatic tails (Fig. 8). This structure confers surface activity. In aqueous environments, the lipids align in tail-to-tail continuous bilayers, with the polar head groups in contact with the water phase. In models such as Figure 8, the globular-shaped proteins extend throughout the lipid bilayer. Water, gases, protons, and hydrophobic ionic molecules are transported through both proteins and bilayer. Specific permeations occur throughout the variety of proteins embedded in the bilayer.

The principal transport mechanisms identified in the biological membrane are passive and facilitated. The first is driven by a random diffusion of the permeate. The second includes carrier, channel (via the alignment of cylindrical protein), and gate-mediated transports.

Membrane-mimetic chemistry is concerned with the development of processes inspired by the biological membrane. Numerous surfactants that can form micelles, monolayers, organized multilayers, bilayer lipid membranes, and vesicles (qv) have been studied. From those, families of polymerizable surfactants have been evolved to provide the building blocks necessary for synthetic chemists to mimic the biological membrane.

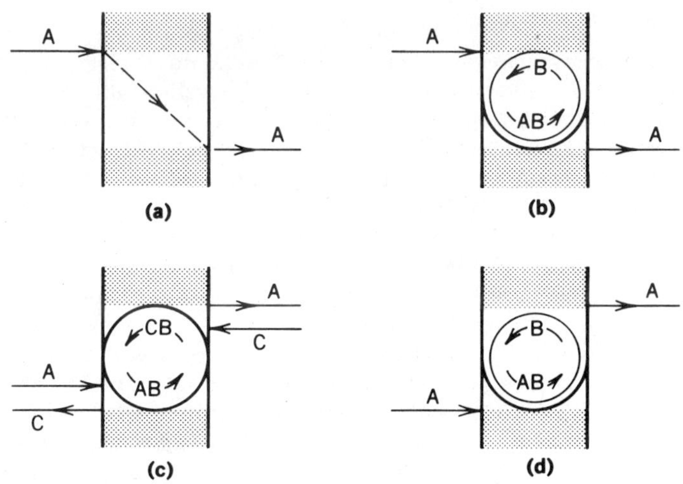

Figure 7. Membrane-transport mechanisms: (**a**) diffusive transport; (**b**) facilitated transport (carrier mechanism); (**c**) coupled transport; and (**d**) facilitated active transport.

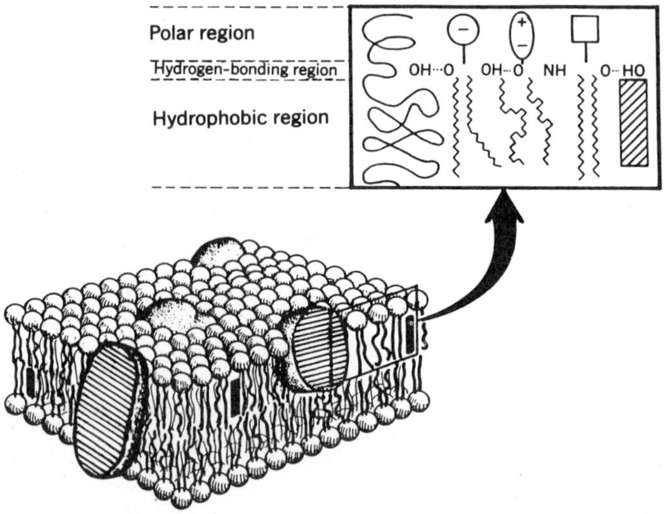

Figure 8. Biological membrane showing proteins embedded in the lipid bilayer.

These types of membranes are being developed for potential application in medicine and industry. Of these, most notable is the potential restoration of damaged cellular membranes in the living system and the production of artificial cells.

Ion-exchange Membranes

Ion-exchange membranes are composed of polymers with cation- or anion-exchange groups. The former usually contain carboxylic or sulfonic acid groups; the latter contain quaternary groups, primarily quaternary ammonium salts. These membranes are used in electrodialysis, Donnan dialysis, reverse osmosis, pervaporation, and other applications.

The ion-exchange membrane is characterized by high electric conductivity and higher permeability for the counterions than for the co-ions; the latter comprise the ionic species with the same charge sign as the exchanger framework. The co-ions are prevented from entering the membrane as long as their chemical potential in the bulk solution is lower than that within the membrane.

Liquid Separation

Organic Liquids. The separation of liquid mixtures by membrane techniques is closely related to the well-established extractive distillation and liquid–liquid extraction processes. In contrast to extraction in membrane separation, extraction and stripping occur concurrently, providing the driving force for steady-state mass transport. A thin, amorphous, dense polymer film is brought into contact with the liquid mixture (feed), and the permeant dissolves and permeates through the membrane by a diffusion mechanism along a concentration gradient. The permeate is stripped from the membrane at the downstream side. The process continues by maintaining a low chemical potential for the permeates at the product side. The methods employed have a large impact on the separation efficiency, eg, separation factor, permeation rate, and energy consumption.

Filtration-related Membranes. Microfiltration (MF), ultrafiltration (UF), and hyperfiltration (ie, reverse osmosis) are well-established membrane technologies for the treatment of aqueous solutions and water purification. Microporous membranes prepared from cellulose esters have been in use since the beginning of the century. Progress in this field was slow until the development of the cellulose acetate anisotropic (asymmetric) reverse-osmosis membrane. The processes mentioned above are all pressure-driven.

Hemodialysis and Hemofiltration

The removal of low molecular weight metabolic waste, such as urea, creatinine, and other toxic substances from the blood of uremia patients is perhaps the largest single application of separation membranes. The process, designated hemodialysis, is driven by a concentration gradient of blood solutes (up to 6000 mol wt) being dialyzed against a physiological solution. Because hemodialysis does not rectify all the problems associated with the removal of metabolites from blood, a complementary process, hemofiltration, was developed. In this process, solutes up to 20,000 mol wt are removed via an ultrafiltration membrane. Hemodialysis and hemofiltration are conceptually different processes. The first is based on diffusive transport and the second on a convective transmembrane flux generated by mild hydraulic-pressure differences across the membrane.

Other Membranes for Therapeutic Application. Microporous membranes are used in plasmapheresis and as oxygenators in the heart–lung machine. Low density polyethylene and propylene in the form of flat-sheet and hollow-fiber membranes were developed for these applications. Other materials commonly used in plasmapheresis are cellulose acetate, polycarbonate, and polysulfone.

Gas Separation

Membranes are employed in the separation of H_2 from supercritical gases such as N_2, CO, and CH_4; the separation of CO_2 and water vapor from natural gas; and the enrichment of air by nitrogen or oxygen. The removal of H_2S and other contaminants from natural gas streams has been marginally successful.

The fundamentals of gas separation are based on the permeability equation 7. At low pressures, membrane selectivity is closely related to the ideal separation factor, which is partitioned between the preferential sorption, expressed by S_A/S_B and the ratio of the mobilities D_A/D_B. Hence the ideal separation is expressed as

$$\alpha_{AB} = P_A/P_B = (S_A/S_B)\cdot(D_A/D_B) \tag{11}$$

Membrane Systems

In the laboratory, research and development of membrane units is conducted with flow areas $<1\ m^2$ with an overall flux range of 0–50 kg/($m^2\cdot h$). Therefore, a very large membrane area is required for the scale-up of processes for the production of 100 t per day or for seawater desalination of $12 \times 10^3\ m^3/d$. Such units may require 10^3–$10^7\ m^2$ of membrane area, and unit configurations with a high membrane area-to-volume ratio are needed. This is referred to as the packing density of the membrane system. Theoretically, the highest packing density can be achieved with hollow-fiber configurations. However, the type of membrane used depends on the separation, fluid-dynamic considerations, durability, maintenance costs, and other factors.

Currently four types of membrane modules are used: plate-and-frame, spiral-wound, tube-in-shell, and hollow-fiber. Performance characteristics are shown in Table 3.

Table 3. Performance Characteristics of Membrane Module Configurations

Characteristics	Plate-and-frame	Spiral-wound	Tube-in-shell	Hollow-fiber
packing density, m^2/m^3	200–400	300–900	150–300	10,000–75,000
permeate flux, m^3/m^2 per day	0.3–1.0	0.3–1.0	0.3–1.0	0.004–0.1
flux density, m^3/m^3 per day	60–400	90–900	45–300	36–2,400
feed channel dia, mm	5	1.3	13	0.1–1
method of replacement	sheets	module assembly	tubes	entire module
replacement labor	high	medium	high	medium
pressure drop				
product side	medium	medium	low	high[a]
feed side	medium	medium	high	low[a]
concentration polarization	high	medium	high	low
suspended solids buildup	low–medium	medium–high	low	high

[a] Permeation into the fiber bore.

BIBLIOGRAPHY

R. B. Kesting, *Synthetic Polymeric Membranes,* 2nd ed., Wiley-Interscience, New York, 1985.

I. Cabasso in M. Grayson, ed., *Kirk-Othmer Encyclopedia of Chemical Technology,* 3rd ed., Vol. 12, John Wiley & Sons, Inc., New York, 1980.

S. L. Matson, J. Lopez, and J. A. Quinn, *Chem. Eng. Sci.* **38,** 503 (1983).

V. T. Stannett, W. J. Koros, D. R. Paul, H. K. Lonsdale, and R. W. Baker, *Adv. Polym. Sci.* **32,** 69 (1979).

ISRAEL CABASSO
State University of New York-ESF
at Syracuse

MERCERIZATION. See COTTON.

MESOMERIC POLYMERS. See LIQUID CRYSTALLINE POLYMERS.

METAL-CONTAINING POLYMERS. See CHELATE-FORMING POLYMERS; COORDINATION POLYMERS; INORGANIC POLYMERS; ORGANOMETALLIC POLYMERS.

METAL FIBERS. See INORGANIC FIBERS.

METALLIC DRIERS. See DRYING OILS.

METALLIZING

Plastics can be coated with metals for decorative or functional purposes, or both. The most important commercial process is electroless plating, which includes plating on plastics for automotive and decorative uses and the related technologies for printed circuits. Flame or arc spraying is mainly used for functional purposes such as radio-frequency-interference (rfi) shielding. Sputtering and vacuum metallizing are widely used, often in continuous coating of plastic films where they are the predominant technologies.

Electroless Plating

Theory of Electroless Deposition. The formation of a continuous metallic film on a nonconductive plastic surface without the use of an electric current is called electroless plating. It can be defined as the controlled reduction of a solution of a metal ion by a dissolved reducing agent at a catalytic interface to give a coherent film. Most electroless-plating phenomena have analogues to classical electroplating. The simplest electroplating baths consist of a solution of a soluble metal salt. Electrons are supplied to the conductive metal surface, where electron transfer to and reduction of the dissolved metal ions occur. These simple baths are rarely satisfactory. Additives are required to control conductivity, pH, crystal structure, throwing power, and other conditions.

Electroless Copper. Electroless copper was introduced in the mid-1950s. Today, a great variety is available commercially. Formaldehyde is usually the reducing agent, copper sulfate (occasionally nitrate or chloride) the metal salt, and sodium hydroxide is used to control pH. The complexers may be tartrate, ethylenediaminetetraacetic acid (EDTA), tetrakis(2-hydroxypropyl)ethylenediamine, nitrilotriacetic acid (NTA), or other strong chelate. Numerous proprietary stabilizers, eg, sulfur compounds, nitrogen heterocycles, and cyanides, are used. These formulated baths differ in deposition rate, ease of treatment, stability, bath life, copper color and ductility, operating temperature, and component concentration. Most have been developed for specific processes; all deposit nearly pure copper metal.

Electroless Nickel. Sodium hypophosphite is the reducing agent, nickel sulfate or chloride the metal salt, and ammonia the buffer. Mild complexes, eg, citrate or glycolate, and trace quantities of proprietary stabilizers, eg, heavy metals or sulfur compounds, are used; temperatures are 20–50°C. A large variety of electroless nickel baths for plastics coating are available from commercial suppliers. Most deposit a low phosphorus (1–5%)–nickel alloy.

Electroless Silver. Silver was the first metal used for metallizing nonconductors, but is rarely used today because of high cost, low stability, and inability to plate a thick deposit autocatalytically. Formulations given in handbooks and patents typically consist of ammoniacal silver nitrate solutions and formaldehyde or a reducing sugar.

Electroless silver has only minor uses today. Most applications are for silvering glass or metallizing record masters. A unique application is the use of silver-plated nylon cloth as a bacteriostatic wound dressing. A tiny current applied to the cloth causes slow silver dissolution, which acts as a bactericide.

Plating on Plastics. Plating on plastics consists of the following steps: etching, etch neutralization, catalyst application, catalyst activation, and plating. Most commercial applications, except rfi shielding, use the initial deposit as a base for subsequent electrolytic plating. The initial copper or nickel layer is coated with successive layers of electrolytic copper, nickel, and chromium. The exact types and thicknesses of metal used are determined by the end use of the part, eg, automotive exterior, decorative, plumbing, and others.

Chemistry. The plating of plastics depends on a precise series of chemical steps to clean, roughen, and catalyze the surface before plating. These steps are critical for formation of an adherent, continuous electroless coating and for optimum durability after electrolytic plating.

Equipment. An electroless plating line consists of a series of lead-lined (for the etch) or plastic-lined tanks equipped with filters and heaters, separated by rinse tanks.

Applications and Economic Aspects. Automatic processing lines allow low cost production of plated parts even though equipment and solution control cost can be high; rejects are typically less than 5% of production. These can be stripped and replated, or reground and remolded. Total chemical costs are typically $0.65–1.30/m^2.

The largest application for electroless plating is for automotive parts such as grilles and headlamp bezels. This market segment has dropped in size because cars are smaller and painted parts have replaced plated finishes.

Health and Safety. Electroless-plating operations must comply with current EPA and local waste-treatment guidelines. The most troublesome chemicals are electroless nickel and copper and the chromic acid etchant.

Printed Circuit Boards. Most of present printed-circuit-board (PCB) production uses the subtractive process. In the simplest version a thin copper foil is laminated to a nonconductor, holes are fabricated, and the unwanted copper etched off. These single sided boards do not require plating. Now known as print-and-etch, this version is used for the most simple printed circuit boards.

The need to install more components on one board led to the development of double-sided boards with copper laminated to both sides of a nonconducting support. The first commercial boards using electroless plating for connection between the sides were made by Motorola in 1953. An unclad plastic board was coated with silver, a reverse resist applied, and copper electroplated to the required circuit thickness. The resist and excess silver were removed. This process is used today, with copper replacing the silver, as the semiadditive process.

Chemistry. Many of the chemicals and process steps used for PCB processing are similar to those used in the plating on plastics (POP) industry, although electroless copper is used exclusively. Processing variations depend on the type of board, its composition, and the technology used. The surface must be clean for uniform coverage, surface roughening, and copper deposition. Copper-clad boards are given an acid or alkaline cleaning step. The copper layers are then microetched with acidic persulfate or hydrogen peroxide–sulfuric acid solutions to promote adhesion of the electroless copper. Multilayer boards receive a precleaning step to eliminate resin smear on the internal copper connections. The precleaning step may use plasma etching, concentrated sulfuric acid, alkaline potassium permanganate, or concentrated chromic acid.

Further processing steps are needed before the printed circuit board is finished. These include electroplating, imaging, resist removal, copper etching, and other steps.

Equipment. Electroless-plating equipment for printed circuit production is very similar to that used for plating of plastics. The exact baths and sequences differ, but the same basic principles apply.

Application Methods. Printed circuit production is outlined in Table 1. Additional steps used for multilayer production before processing with any of the other techniques include application of black or red oxide to foil; application of photoresist or screened etch resist; imaging and development of photoresist; copper etching; removal of resist; assembling of layers, checking registration, and lamination; drilling; and removal of drill smear on copper inner layers, followed by subtractive, semiadditive, or additive processing sequence. Rinses and additional cleaning, neutralizing, light etching, and developing steps are required. Other steps may include a liquid- or dry-film resist, plastic etchants and micro- and final-metal etchants, panel or pattern plating, and use of flash or heavy electroless copper.

Economic Aspects. Printed-circuit-board production is a multibillion-dollar business in the United States alone. An estimated 2000 shops are involved in production. They range from small companies specializing in one aspect of production to huge organizations containing both research and production facilities and capable of producing any type of board. U.S. production of printed circuit boards is forecast to reach $\$13.7 \times 10^9$ by 1993, an increase from $\$3.5 \times 10^9$ in 1983; electroless plating is only a small part of this cost.

Health and Safety. The same EPA and health rules as for plating of plastics apply here. Additional waste treatment is due to rinses from tin–lead plating baths, solder cleaners, metal etchants, and resist strippers. Other hazards may be caused by molten-solder reflow steps and component soldering.

Metal-spraying

There are several ways to provide a metal film on a plastic part through the use of atomized metal at atmospheric pressure. The temperature of the part being coated is controlled by the application rate. This process is widely used for the

Table 1. Printed-circuit-board Processing

Subtractive	Semiadditive	Additive
drilling	drilling	drilling
cleaning	epoxy etching	epoxy etching
copper microetching	catalyzing	catalyzing
catalyzing	accelerating	accelerating
accelerating	electroless copper deposition, 2.5 μm	drying
electroless copper deposition	drying	image resist
dry and image resist, optional	image resist	electroless copper deposition, 35 μm
electrolytic copper deposition, 25 μm	electrolytic copper deposition, 35 μm	strip resist
electrolytic deposition of Sn or Sn–Pb	electrolytic deposition of Sn or Sn–Pb	immersion Sn or Sn–Pb
strip resist	strip resist	
copper-foil etching	electroless copper etching	

protection of metal against corrosion, especially for *in situ* protection of structural members. Spray heads using a flame or an electrical arc to melt metal wires or powders directly are much more convenient and are now the only types used on a large scale in the United States. The principal metal used for spraying of plastics at the present time is zinc; aluminum and copper are also used (Fig. 1).

Equipment. The equipment is relatively simple: a spray head for feeding the metal wire; electrical contacts for maintaining an arc or gas orifices for the flame; compressed air; and fixtures for holding the parts to be coated.

Application. A great advantage of zinc arc spray is that it can be applied to almost any plastic. It is also suitable for prototypes and small lots of materials. It is less suited for very small parts and parts with blind holes or complex interior surfaces, or where warpage is a problem. Zinc flame spray is normally applied to a thickness of 0.05–0.10 mm.

Economic Aspects. Zinc arc spray is an inexpensive process in terms of equipment and raw materials; only 55–110 g Zn/m^2 is required for a standard 0.05–0.10-mm thickness. Manpower is more expensive.

Figure 1. Zinc arc-spray head. Courtesy of TAFA, Inc.

The largest market for flame-spray coatings on plastic at present is for rfi shielding, a relatively new concern in the United States. Environmental pollution is minimal and only vaporized zinc particles need to be removed from the air.

Health and Safety. Zinc arc- or flame-spray equipment is hardly more hazardous than a welding torch, and only safety goggles and gloves are required. Safety aspects emphasize reduction of noise and vapor inhalation.

BIBLIOGRAPHY

G. A. Krulik, *J. Chem. Ed.* **55,** 361 (1978).

G. G. Gawrilov, *Chemical (Electroless) Nickel-Plating,* Portcullis Press, Redhill, UK, 1979.

M. E. Pole-Baker, *Print. Circuits Fab.* **7,** 26 (Dec. 1984); **8,** 103 (Jan. 1985); **53** (Feb. 1985).

C. F. Coombs, Jr., *Printed Circuits Handbook,* 2nd ed., McGraw-Hill Inc., New York, 1979.

GERALD A. KRULIK
Morton Thiokol, Inc.

Sputtering

Physical sputtering is a process whereby material is dislodged and ejected, primarily in atomic form, from the surface of a solid or liquid by the momentum exchange arising from energetic particle bombardment. Sputter deposition is a vacuum coating process wherein substrates intercept the flux sputtered from a source of coating material called a target. The bombarding species are generally ions of a heavy inert gas, usually argon. In most industrial coating processes, the target is the cathode electrode of a low pressure glow discharge, as shown in Figure 1. In sputter cleaning, sputtering is used to remove contamination from a material surface.

The most striking characteristic of the sputtering process is its universality. Since the coating material is passed into the vapor phase by an atomic-scale momentum exchange, almost any substrate can be coated with almost any material. Direct-current glow discharges are used to sputter metals. Discharges driven at radio frequencies (rf) are used for semiconductors and insulators. Films containing almost every element in the periodic table have been prepared by sputtering. Sput-

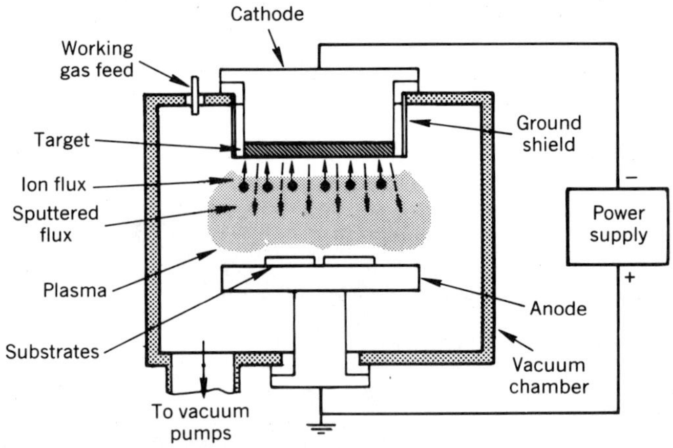

Figure 1. Planar diode glow-discharge sputtering device. Courtesy of the University of Illinois.

tered coatings generally have the same composition as the target. Sputtering can be adapted to hybrid processes capable of producing an almost unlimited range of unique materials.

In chemical sputtering, the bombarding particles produce at the target surface an unstable compound that passes into the gas phase. This technique is important in plasma-etching applications in the electronics industry.

Basic Process

The sputtering-momentum exchange, which arises from the energetic particle bombardment, occurs by a sequence of collisions involving the atoms within the target and culminates with the ejection of surface atoms. The sputtered atoms are ejected with relatively high energies, which increase with the target atomic mass; average values are between 10 and 40 eV. For polycrystalline target materials used in coating deposition, the atoms are ejected in a near-random direction, eg, a cosine distribution.

Sputtering Yield. Sputtering yield, which may be defined as the average number of atoms ejected per incident ion, depends on the target material and on the identity, energy, and angle of incidence of the bombarding-particle species.

Multicomponent Targets. One of the most important features of sputtering is that, after an equilibration period, the overall flux of material leaving a target of complex composition is identical in composition to the target, provided that the target is cooled sufficiently to suppress internal diffusion and that it does not decompose.

Sputtered Species. In most situations, the sputtered species are primarily atoms, but atomic clusters and positive and negative ions also occur; molecular fractions can account for a significant portion of the species emitted.

Ion Reflection. A significant fraction of the bombarding particles is reflected at the target surface if its atomic mass is less than that of the target atoms. In most cases, the reflected ion is neutralized.

Recoil Implantation. If not reflected, the ion is entrapped in the target. Many surface atoms are driven into the target by forward-sputtering, recoil implantation.

Glow Discharge. Ions are used in sputtering because they are easily accelerated in an electric field. They are created within a glow discharge and may then be accelerated, as ion-beam sources, or the glow discharge itself may be used as a sputtering source.

Bombarding Ion Energies. When ion beam sources are used, the ion energies are typically between a few hundred eV and several keV. The energies of the ions incident on the cathode electrode of a glow discharge depend on the collision processes that occur within the plasma sheath. In the case of planar-diode devices, the sheath voltage drop may be several thousand volts. However, the operating pressures in these devices are higher, and charge-exchange collisions within the sheath cause the bombarding flux actually to consist of neutral atoms and ions with average energies of a few hundred volts.

R-f Sputtering. Direct-current methods cannot be used to sputter nonconducting targets because of charge accumulation on the target surface. This difficulty is overcome by operating at r-f frequencies. R-f sputtering is versatile, and a single apparatus can be used to sputter metals, semiconductors, and insulators; frequencies above 1 MHz are required. Most apparatuses operate at the frequency of 13.56 MHz, allocated by the FCC for industrial–scientific purposes.

Equipment

The sputter-deposition system shown in Figure 2 consists of a vacuum chamber with a high vacuum of 130 μPa–130 nPa (10^{-6}–10^{-9} mm Hg) or ultrahigh vacuum (<130 nPa or 10^{-9} mm Hg) pumping system. The high vacuum pumps are of the oil-diffusion, turbomolecular, titanium-sublimation, or cryopump types. Thermocouple and ionization gauges are used to measure the vacuum in the chamber. Gases are introduced through needle valves or automatic flow controllers. Capacitive manometer-type absolute-pressure gauges measure gas pressure. If toxic gases are used, scrubbers are placed in the pump exhaust line. The sputtering source is driven by an appropriate power supply. Ports or a vacuum interlock are provided for introducing the substrates. Provision is made for controlling the substrate temperature during deposition. Other control instrumentation may include a quartz-crystal deposition-rate monitor, a mass spectrometer for observing gas composition, and an optical spectrometer for observing plasma radiation.

Ion Plating. Ion plating is a generic term applied to deposition processes in which the substrate surface and the depositing film are subjected to bombardment by a flux of energetic particles, usually ions, that cause changes in the properties of the interfacial region and the coating.

Operation

Target Fabrication. Targets can be fabricated from commercial plate and tube stock, vacuum cast high purity metals, and hot-pressed powders by a variety of methods.

Predeposition Pumping. Sputtering chambers usually operate at pressures of 3×10^{-3}–3×10^{-5} Pa (2×10^{-5}–2×10^{-7} mm Hg), depending on the application. A mechanical pump evacuates the chamber to ca 10 Pa (0.1 mm Hg) as soon as the substrates are loaded. The gate valve is then opened, and pumping is continued with the high vacuum pumps.

Deposition. Coating properties depend on the deposition procedures. Since every detail is important, automatic control systems are being used more and more; the details of procedures are usually proprietary.

Reactive Sputtering. In reactive sputtering, at least one of the coating species enters through the gas phase. This has the

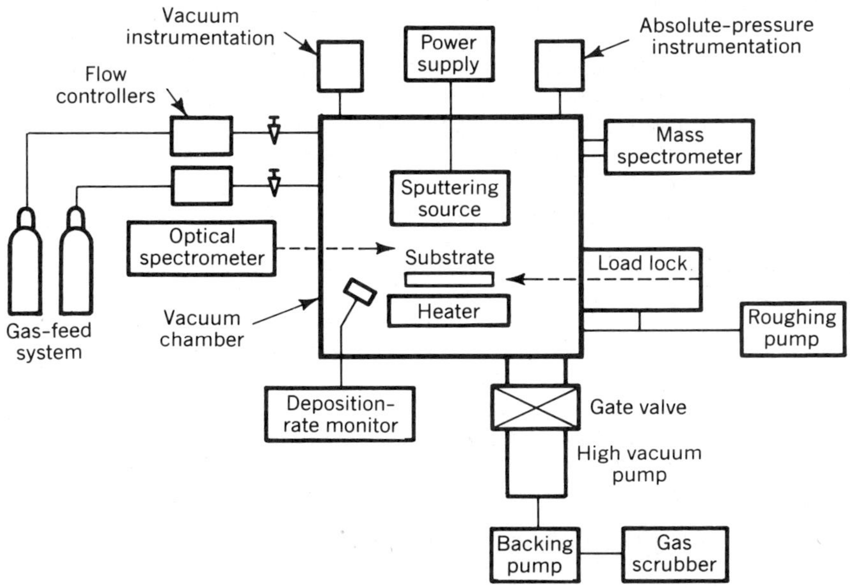

Figure 2. Overall sputter-deposition systems. Courtesy of the University of Illinois.

advantages that complex compounds can be deposited using easily made metal targets, poorly conducting compounds can be deposited by using d-c power, and coatings with graded compositions are easily formed.

Coating Properties

Coating Structure. A diagram summarizing the structure of metal coatings that have been deposited using magnetron sources is shown in Figure 3 as a function of the argon working gas pressure and the substrate temperature T relative to the melting point T_m of the coating material. The zones result from the interaction of the basic processes that occur during deposition, coating flux transport, and surface and bulk diffusion. The indicated trends also apply to other types of sputtering sources and to other materials.

Internal Stress. When a coating is applied to a substrate, stresses usually develop within the coating and at the interface, consisting primarily of superimposed thermal and intrinsic stresses.

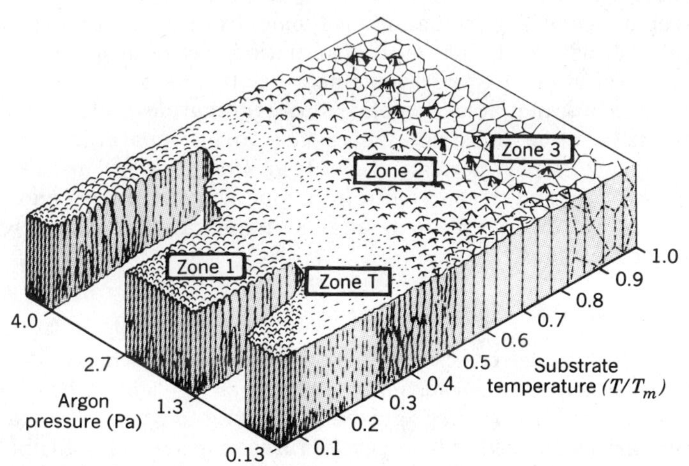

Figure 3. Zone diagram showing influence of argon gas pressure and substrate temperature T relative to coating material melting point T_m on the structure of metal coatings deposited by magnetron sputtering; 1.3 Pa = 10^{-2} mm Hg.

Adhesion. The mechanisms of adhesion between metal coatings and organic polymers are typical of those that are generally observed, classed as mechanical or interlocking, weak boundary layer, chemical, and electrostatic.

Applications and Economic Aspects

Sputtering, like other vacuum coating processes, requires expensive equipment, which can be justified for large-scale production or where coating properties are crucial and cost is of secondary importance. On the other hand, sputtering is reliable and lends itself to automatic control, permitting economics of scale that make even decorative coatings feasible.

Most sputtering applications have been in the electronics industry, where they are used for metals, insulators, and semiconductors; sputtering has become the principal method for metallizing silicon wafers.

BIBLIOGRAPHY

J. A. Thornton and D. W. Hoffman in D. Satas, ed., *Plastics Finishing and Decoration,* Van Nostrand Reinhold Co., Inc., New York, 1985, pp. 345–371.

B. Chapman, *Glow Discharge Processes in Sputtering and Plasma Etching,* John Wiley & Sons, Inc., New York, 1980.

R. Behrisch, ed., *Sputtering by Particle Bombardment I* and *Sputtering by Particle Bombardment II,* Vols. 47 and 52 of Topics in Applied Physics, Springer-Verlag, New York, 1981 and 1983. Excellent reviews on physics of sputtering by leading authorities in the field.

R. V. Stuart, *Vacuum Technology, Thin Films and Sputtering: An Introduction,* Academic Press, Inc., New York, 1983. One of the few texts written at the technician level.

JOHN A. THORNTON
University of Illinois

Vacuum Metallizing

Vacuum metallization is the thermal evaporation and subsequent condensation of a metal onto a substrate in vacuum. The metal is primarily aluminum and the substrate plastic most often styrene, ABS, and acrylic resins.

Decorative applications account for ca 80% of metallizing. However, since about 1980, the functional sector of the industry has been growing at a faster rate.

Decorative Applications

Coating Equipment and Processing. Base- and topcoats can be applied by a simple tool such as a hand sprayer, an elaborate automated spray robot, or a flow coater. Although dipping and spraying are common coating techniques, flow coating is used at most installations.

Coating materials in decorative metallizing are usually called lacquers. Lacquers are usually dried in a convection oven. Typical lacquers are based on nitrocellulose or other cellulosics, as well as acrylic and vinyl resins.

Metallizing Equipment. For industrial vacuum metallizing, horizontal cylinders are operated in batch mode. Vacuum chambers range in size from small laboratory models (610 mm dia × 610 mm length) to very large production equipment (2134 mm dia × 2794 mm length). The chamber is evacuated down to the pressure of aluminum vaporization (6.7×10^{-2} Pa or 0.5 μm Hg) by a series of vacuum pumps. A rotary piston pump is used first, followed by an intermediate blower pump, and finally by a diffusion pump.

Process

The following steps are used in a typical process in which the objective is to obtain a first surface silverlike finish:

1. The parts are loaded on the work reel.
2. The basecoat is applied.
3. The basecoat is dried in a forced-air oven for 45 min at 65.5°C.
4. The coated reels are loaded on an evaporation carriage.
5. The aluminum is evaporated at 6.7×10^{-2} Pa (0.05 μm Hg).
6. The metallized reels are unloaded.
7. The topcoat is applied.
8. The topcoat is dried in a forced-air oven for 45 min at 54.5°C.
9. The parts are unloaded from the work reel.

Discoloration occurs if the product of the vaporization time and current is too high, the load is not large enough, or part placement is not optimal. Firing too long or at an excessively high current produces burnings.

Functional Applications

The oldest industrial functional use of vacuum metallizing is probably the evaporation of aluminum for optical reflection, eg, the reflector of the standard automotive seal-beam headlight. The newer functional vacuum coating processes have been utilized mostly in the electronics industry. Aluminum evaporation is also used for compact audio and video disks.

Nonmetal Vacuum Deposition. Both silicon monoxide (SiO) and magnesium fluoride (MgF_2) are processed in standard industrial batch metallizing systems as previously described.

BIBLIOGRAPHY

A. Roth, *Vacuum Technology,* North-Holland Co., New York, 1976. A detailed treatise on vacuum pumps.

L. I. Maissel and R. Glang, *Handbook of Thin Film Technology,* McGraw-Hill Inc., New York, 1970. A thorough treatment of low pressure gas kinetics.

U.S. Pat. 4,544,571 (Oct. 1, 1985), W. J. Miller (to Stokes Division-Pennwalt Corp.).

C. Kirkland, *Plast. Technol.* **31,** 73 (Apr. 1985).

WALTER J. MILLER
Pennwalt Corporation

METALLOCENE POLYMERS.

See ORGANOMETALLIC POLYMERS.

METAL-VAPOR SYNTHESIS

Metal-vapor synthesis (MVS) is the use of atoms of the transition metals, lanthanoids, and actinoids as reagents in the preparation of organic, organometallic, coordination, and metal-cluster compounds and catalyst materials. The technique also employs coordinatively unsaturated molecular high temperature gas-phase species, eg, metal oxides, halides, and nitrides, as reagents. Vapor formed at high temperatures is combined under vacuum with a coreactant at a cold surface where products may be formed and isolated. Reduced steric restrictions and the high potential energy for reaction implicit in the free state of the atom often lead to the very low activation-energy requirements for these atom–molecule reactions. The introduction of the carbon-vapor reactor initiated the use of high temperature species for synthesis. This concept was extended to transition-metal atoms. The extensive MVS literature recognizes the kinetic and thermodynamic advantages of this technique in providing more direct or higher-yield routes to existing compounds and making available new compounds and materials. Applications of MVS in polymer science are just emerging. The methodology has been used primarily on a research scale to explore synthesis of organometallic monomers; monomer polymerization induced by high temperature vapors; preparation of organometallic polymers by direct reaction of a polymer with metal atoms; concatenation of organic molecules by metal atoms to give organometallic oligomers; use of polymers as host matrices for metal-atom aggregation or mediation of metal-atom–substrate reactions; and preparation and screening of polymerization catalysts. Metal-vapor methods can be expected to contribute in unique ways to the chemical or physical incorporation of single or aggregate metal centers into composites.

Experimental Methods

MVS techniques can be distinguished on the basis of whether the reaction vessel is stationary (Fig. 1) or rotatable. The procedures are further distinguished according to whether all vapors are condensed at low temperatures to give a solid, fusible film (static MVS) or whether metal atoms are deposited directly into dynamically mixed neat liquids, solutions, or suspensions of solids (liquid-phase MVS). Given the same starting materials, the products obtained may reflect the choice of technique

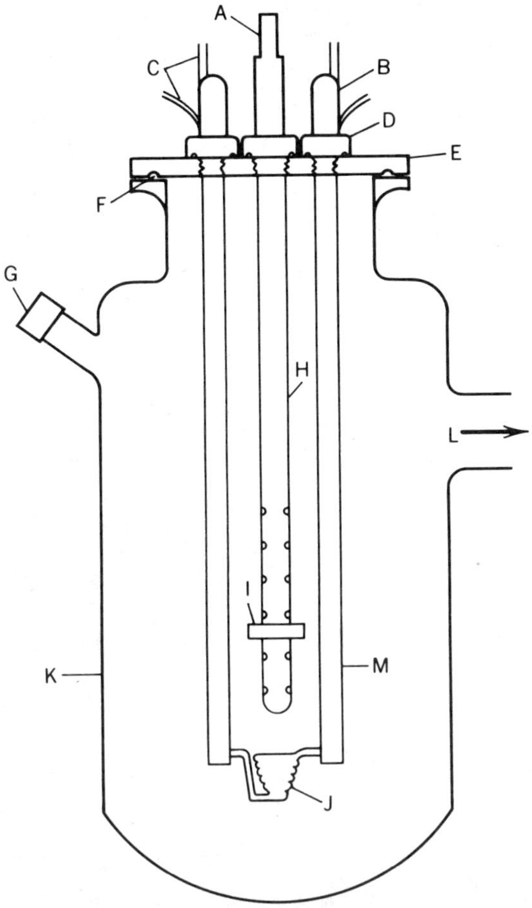

Figure 1. A simple static glass reactor for resistance vaporization of metals. A, cocondensate inlet tube; B, electrical contact for resistive electrode; C, cooling-water connections; D, O-ring seal; E, Plexiglas flange; F, gasket; G, inlet for product extraction; H, perforated cocondensate inlet tube for gas dispersion; I, quartz-crystal microbalance for monitoring metal-atom deposition rate; J, crucible; K, Pyrex reaction vessel; L, to vacuum station; M, water-cooled electrode.

or procedure. Except in the preparation of metal aggregates, the facile self-association reactions of the high temperature species are suppressed to favor atom–coreactant interactions; hence, high pumping speeds are required with sensitive strategies for mixing reagents and efficient furnace and refrigeration designs.

BIBLIOGRAPHY

J. R. Blackboro and D. Young, *Metal Vapor Synthesis in Organometallic Chemistry,* Springer-Verlag, Berlin, 1979.

K. J. Klabunde, *Chemistry of Free Atoms and Particles,* Academic Press, Inc., New York, 1980.

Mark P. Andrews
AT&T Bell Laboratories

METATHESIS POLYMERIZATION

Metathesis polymerization describes the ring-opening polymerization of cycloalkenes initiated by olefin metathesis catalysts. These reactions were discovered around 1960 for monomers such as cyclopentene and norbornene (bicyclo[2.2.1]hept-2-ene) using catalyst systems such as MoO_3/Al_2O_3, $TiCl_4/LiAlR_4$, and $WCl_6/(C_2H_5)_3Al$ at room temperature.

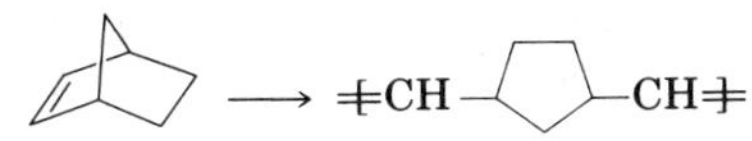

It is now known that these reactions are propagated by metal–carbene complexes via intermediate metallacyclobutane complexes.

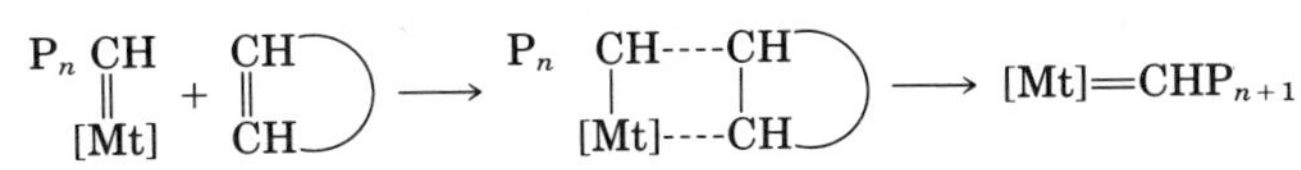

Thus metathesis polymerization is not only a distinctive type of addition polymerization; it is also an important part of transition-metal–carbene chemistry, and related to transition-metal–alkyl chemistry and Ziegler-Natta polymerization.

Four types of equilibria may be encountered in ring-opening polymerization of cycloalkenes: monomer–polymer equilibria, ring–chain equilibria, cis–trans equilibria, and equilibria between chains of different lengths. In ring–chain equilibria, rings of different sizes are in equilibrium with linear polymers of different lengths.

The products of ring-opening cycloalkene polymerization are known as polyalkenamers (qv), some of which are of industrial importance. Cyclooctene and norbornene polymers have applications as specialty elastomers, and the polymerization of *endo*-dicyclopentadiene can be used for reaction injection molding (qv) (RIM).

Catalyst Systems

Hundreds of compounds of the transition elements, ranging from Group IV (Ti) to Group VIII (Ir), have been employed either alone or in conjunction with a cocatalyst to initiate metathesis polymerization. The most common catalyst systems are based on compounds of molybdenum or tungsten. A third component, such as water, an alcohol, oxygen, or any oxygen-containing compound, often causes a marked increase in activity.

The range of effective catalysts for a given monomer is governed to some extent by the exothermicity of the reaction. Thus the release of strain energy allows the metathesis polymerization of norbornene and cyclobutene to be readily initiated by a wide range of catalyst systems including those based on Ti, V, Ru, Os, and Ir compounds. Less strained cycloalkenes, such as cyclopentene, require more active catalyst systems.

Catalyst systems are of three types. The first are metal–carbene complexes, such as $(C_6H_5)_2C{=}W(CO)_5$. The second type contains an alkyl or allyl group in one of the components, eg, $C_2H_5AlCl_2$ or $(CH_3)_4Sn$, from which a carbene ligand is generated through reaction with the transition-metal compound. In the third type the components lack a preformed carbene or alkyl group, and a carbene ligand must be formed by interaction of the monomer itself with the transition-metal center. With certain catalyst systems it is possible to detect both the propagating metal–carbene complex and its metallacyclobutane precursor by ^{1}H and ^{13}C nmr.

Unsubstituted Monoenes. High-cis cycloalkene polymers usually have a blocky cis–trans double-bond distribution, which indicates that two kinetically distinct species propagate the reaction.

A rather remarkable phenomenon is observed when the cyclic oligomers of cyclododecene are submitted to mass spectral analysis. If the sample is held at 150°C the main component reaching the mass spectrometer is the tetramer (C_{48}). As the temperature is raised to 315°C, progressively heavier molecules (C_{60} to C_{144}) reach the ionization chamber and the maximum intensity moves toward higher mass numbers. However, at 330° and 370° there is a marked jump in the relative intensity of the hexamer and heptamer peaks. Similar observations have been made on fully hydrogenated cyclic oligomers. It is especially significant that no monomeric species are detected at the higher temperatures. These results may be interpreted in terms of the presence of catenanes in the oligomeric products. These may be formed by two successive backbiting reactions, as in Figure 1.

Substituted Monenes. A ring substituent is expected to have an adverse effect on the free energy of polymerization, ie, to make it less negative. It is not surprising, therefore, that 1-methylcyclopentene and 3-isopropylcyclopentene do not polymerize, although 3-methylcyclopentene does so. Substituents in C_4, C_8, and larger rings are usually tolerated in any position. It is surprising to find that 1-methyl-(*Z*)-cyclooctene will not polymerize, but 1-methyl-(*E*)-cyclooctene will do so.

Cyclic Polyenes. Cyclopentadiene gives a high yield of powdery amorphous polymer with $WCl_6/AlBr_3$ as catalyst; it also gives a soluble polymer ($\bar{M}_n$ = 2400) with WCl_6 in toluene solution, and copolymerizes with cyclopentene and norbornene. None of these products has been closely characterized. Most higher cyclic polyenes undergo metathesis polymerization.

Bicyclic Alkenes. All the bicycloalkenes listed in Table 1 can be polymerized by metathesis catalysts.

Copolymers of Cycloalkenes. There are several ways of using the olefin metathesis reaction to generate copolymers. According to the method used, the copolymers may have either statistical or block structures. Methods include direct copolymerization, successive addition of monomers, cross-metathesis of unsaturated polymers, and partial conversion of homopolymers.

Cross-metathesis of Acrylic Olefins with Monomers and Polymers. Acyclic olefins act as chain-transfer agents in the metathesis polymerization of cycloalkenes. In small quantities they effect a moderate reduction in molecular weight of the product. If used in amounts comparable with the cycloalkenes, the result is a telomer (see also TELOMERIZATION). When the olefin bears functional groups, such as esters at both ends, the final product may be a telechelic polymer (qv).

The same end result is achieved by submitting an unsaturated polymer to a metathesis catalyst in the presence of an acyclic olefin. The ultimate degradation product, obtained by using an excess of symmetrical acyclic olefin, is an adduct of the monomer repeat unit and the olefin.

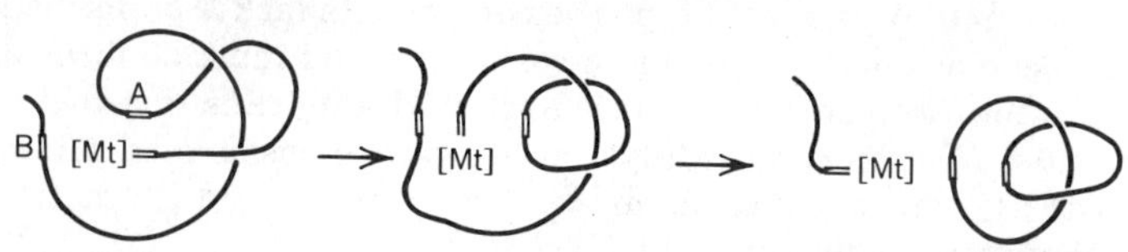

Figure 1. Mechanism for formation of catenanes by successive intramolecular metathesis reactions at double bonds A and B.

Table 1. Ring-opening Polymerization of Some Bicycloalkenes

Unsaturated ring size	Ring system	Compound
4	[4.2.0]	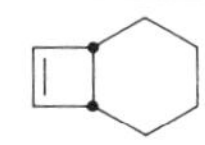
4	[3.2.0]	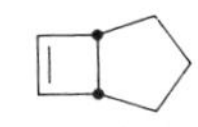
4,5	[3.2.0][a]	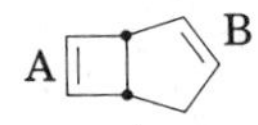
5,6	[2.2.1]	
5,5	[2.2.1]	
5,6	[4.3.0][b]	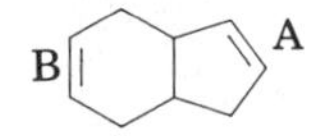
6,6	[2.2.2]	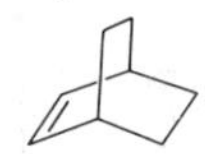
7	[5.1.0]	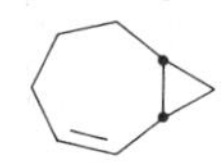
8	[6.1.0]	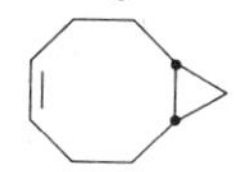
12	[10.1.0]	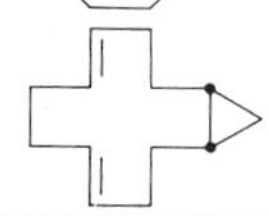

[a] Polymerizes by ring opening at A.
[b] Polymerizes reversibly; polymerization probably occurs by ring opening only at A.

Acetylenes. The metal–carbene complexes, for example, $C_6H_5(CH_3O)C{=}W(CO)_5$ and $(C_6H_5)_2C{=}W(CO)_5$, cause phenylacetylene and alkylacetylenes to polymerize to high molecular weight polymers, just as they do with cycloalkenes. Other olefin metathesis catalysts, notably $WCl_6/(C_6H_5)_4Sn$ and $MoCl_5/(C_6H_5)_4Sn$, also bring about polymerization of acetylenes. It is assumed that propagation occurs through metal–carbene complexes, the initiation being represented as follows:

$$\begin{matrix}[Mt]{=}CHR^1 \\ + \\ R^3C{\equiv}CR^2\end{matrix} \longrightarrow \left[\begin{matrix}[Mt]{-}CHR^1 \\ | \quad\ \ | \\ R^3C{=}CR^2\end{matrix}\right] \longrightarrow \begin{matrix}[Mt] \quad CHR^1 \\ \| \quad\ \ \| \\ R^3C{-}CR^2\end{matrix}$$

This mechanism requires that the original triple bond becomes a single bond in the polymer chain in contrast to the Ziegler-Natta mechanism of acetylene polymerization, whereby the triple bond becomes a double bond.

BIBLIOGRAPHY

K. J. Ivin, *Olefin Metathesis,* Academic Press, Inc., London, 1983.

V. Drăgutan, A. T. Balaban, and M. Dimonie, *Olefin Metathesis and Ring-Opening Polymerization of Cycloolefins,* Editura Academiei/Wiley, Bucharest, Rumania, 1985.

R. R. Schrock, J. Feldman, L. F. Cannizzo, and R. H. Grubbs, *Macromolecules* **20,** 1169 (1987).

R. H. Grubbs in G. Wilkinson, ed., *Comprehensive Organometallic Chemistry,* Vol. 8, Pergamon Press, Inc., Oxford, UK, 1982.

KENNETH J. IVIN
The Queen's University of Belfast

METHACRYLAMIDE POLYMERS. See ACRYLAMIDE POLYMERS.

METHACRYLATE POLYMERS. See ACRYLIC AND METHACRYLIC ESTER POLYMERS.

METHACRYLIC ACID POLYMERS. See ACRYLIC AND METHACRYLIC ACID POLYMERS.

METHACRYLIC ESTER POLYMERS. See ACRYLIC AND METHACRYLIC ESTER POLYMERS.

METHACRYLONITRILE POLYMERS

High methacrylonitrile (2-cyanopropylene), $\mathrm{CH_2{=}\overset{\overset{CH_3}{|}}{C}{-}CN}$, polymers are readily made using free-radical or ionic catalysis. Copolymers are prepared by free-radical methods over wide ranges of composition, resulting in materials with extremely divergent properties. In general, the inclusion of methacrylonitrile in a copolymer enhances both its solvent resistance and barrier and thermal properties. Polymers containing methacrylonitrile have been employed industrially in coatings cast from either solvent or latex as well as in solvent-resistant elastomers. These same polymers are utilized in thermoplastic materials, especially in packaging because of their excellent barrier properties. Methacrylonitrile is also a significant component of many polymeric photoresist compositions in the electronics industry.

Monomer

Physical Properties. Pure methacrylonitrile is a colorless liquid with a pungent, sweet odor; its properties are summarized in Table 1. Methacrylonitrile is soluble in most organic sol-

Table 1. Physical Properties of Methacrylonitrile

Property	Value
molecular weight	67.09
boiling point, °C	
at 101.3 kPa[a]	90.3
at 100.6 kPa[a]	89.2
boiling point of water azeotrope	
16 wt % of H_2O, °C	77
melting point, °C	−35.8
specific gravity[b]	
20°C	0.8001 (vacuum), 0.7998
25°C	0.805
refractive index	
n_D^{20}	1.4007, 1.4002, 1.40131, 1.4003
n_D^{23}	1.3961
n_D^{25}	1.3977, 1.398
n_α^{20}, n_β^{20}	1.39801, 1.40804
latent heat of vaporization, kJ[c]/mol at bp	31.8
vapor pressure, kPa[a]	
at 0°C	2.133
at 10°C	3.796
at 30°C	10.47
at 50°C	24.88
at 70°C	52.54
at 90°C	100.5
at 100°C	134.9
$\log p\ (\mathrm{kPa})^d = 6.105 - \dfrac{1274.959}{t(^\circ\mathrm{C}) + 220.734}$	
dp/dt = 0.3248°C per kPa[a] at the boiling point	
coefficient of expansion	
20–25°C, per °C	0.00133
viscosity, at 20°C, mPa·s (= cP)	0.392
surface tension, at 20°C, mN/m (= dyn/cm)	24.4

[a] To convert kPa to mm Hg, multiply by 7.5.
[b] d_4^T.
[c] To convert kJ to kcal, divide by 4.184.
[d] To convert log p (kPa) to log p (mm Hg), add 0.875.

vents. The solubilities of methacrylonitrile in water and of water in methacrylonitrile are given in Table 2.

Chemical Properties. Methacrylonitrile possesses three important functional groups: the nitrile (CN) group, the activated double bond (C═C), and the allylic system (C═C—CH_3).

Reactions of the Nitrile Group. Electrophiles (electron-seeking reagents) attack the lone pair of electrons on the nitrile nitrogen, whereas nucleophiles (electron-rich reagents) attack the nitrile carbon.

Reactions of the Double Bond. Methacrylonitrile can add electrophiles, nucleophiles, and free radicals, and undergoes cycloaddition reactions.

Reactions of the Allylic System. The allylic hydrogens of methacrylonitrile become labile at elevated temperatures. Besides thermal additions, the methyl group of methacrylonitrile can be substituted by halogens, and condensed with sulfur dioxide and ammonia to give 4-cyanoisothiazole and with sulfur to produce 4-cyanotrithione. Furthermore, the catalytic isomerization of methacrylonitrile at 200–500°C yields *cis*- and *trans*-crotononitrile.

Manufacture of Monomer. Processes based on acetone cyanohydrin, isobutyronitrile, chlorinated isobutyronitrile, methallylamine, and isobutylene have been utilized in the pilot- or commercial-scale production of methacrylonitrile at different times since 1940. The acetone cyanohydrin and isobutylene routes are presently the only processes used commercially.

Monomer Uses. Methacrylonitrile is an intermediate for the production of methacrylamide, methacrylic acid, and its esters. Commercial utilization of this route to the monomers is dependent on large-volume methacrylonitrile production. Cyanoisopropylation is the largest potential use for methacrylonitrile outside the polymer area.

Polymerization

Methacrylontrile is stable to thermal initiation in the absence of air or oxygen. In the presence of free-radical or ionic catalysts and proper polymerization conditions, methacrylonitrile readily forms addition polymers via the vinyl group. The free-radical polymerization of methacrylonitrile generally proceeds more slowly than that of acrylonitrile and yields an amorphous material. Although the homopolymer exhibits some of the solvent resistance characteristic of polyacrylonitrile, this thermoplastic is soluble in ketones and other highly polar solvents. Among the initiators used are benzoyl peroxide, potassium peroxydisulfate, α,α'-azobisisobutyronitrile (AIBN), and various redox systems, such as potassium peroxydisulfate/sodium metabisulfite. Free-radical polymerizations of methacrylonitrile have been carried out in bulk, emulsion, and solution systems.

Unusual properties can be produced in polymethacrylonitrile through the use of ionic and coordination catalysts. Natta and co-workers reported that diethylberyllium and diethylmagnesium are highly efficient catalysts for preparing stereospecific polymethacrylonitrile. The melting points of these materials are much higher than that of amorphous polymethacrylonitrile, which softens in the range of 100–130°C; they are insoluble in solvents which attack noncrystalline polymethacrylonitrile. Anionic catalysts, such as n-butyllithium, are also used to prepare these crystalline materials.

Table 2. Methacrylonitrile–Water Solubilities

Temp, °C	Water in methacrylonitrile, wt %	Methacrylonitrile in water, wt %
0	1.06	2.89
10	1.32	2.67
20	1.62	2.57
30	1.98	2.54
40	2.38	2.59
50	2.83	2.69
60		2.83

Irradiation Polymerization. Radiation-initiated polymerization of methacrylonitrile proceeds by a free-radical mechanism at 10 to −40°C and by an ionic mechanism at very low temperatures of −40 to −196°C with a radiation source of cobalt-60 at a dose of 1–3.5 kGy/h (100–350 krad/h). All of the polymers produced are soluble in acetone. The amount of syndiotactic sequences increases in the solid-state polymerizations as the temperature decreases.

Copolymerization

Many monomers are capable of forming copolymers with methacrylonitrile under free-radical catalysis. Bulk, solution, and emulsion systems have been used. The most significant commercial application of methacrylonitrile in the near future will probably be in the area of copolymers. Most important are those with acrylonitrile, styrene, methyl methacrylate, and methyl acrylate.

Copolymers of methacrylonitrile and butadiene range from rubbery materials to thermoplastics as the methacrylonitrile concentration in the copolymer increases. Butadiene–methacrylonitrile elastomers are comparable to commercial butadiene–acrylonitrile rubber; at selected methacrylonitrile compositions these copolymers may possess advantages over butadiene–acrylonitrile materials. Alternating copolymers prepared in the presence of Lewis acids reportedly have better properties than random copolymers.

Graft Copolymerizations. The most common commercial grafting reactions are of a thermoplastic material onto an elastomeric polymer to produce impact-resistant thermoplastics of the ABS type (see GRAFT COPOLYMERS; ACRYLONITRILE–BUTADIENE–STYRENE POLYMERS). These products generally have good weather resistance and excellent solvent and barrier properties.

Polymethacrylonitrile has also been grafted onto a variety of other polymers by various methods. Graft copolymers of cellulose and polymethacrylonitrile, catalyzed by cerium salts, are resistant to heat and microorganisms (see CELLULOSE, GRAFT COPOLYMERS). Anionic graft polymerization of methacrylonitrile on alkali metal alkoxide derivatives of starch and dextrin yields products with more densely packed grafted side chains possessing lower molecular weights compared to those obtained by free-radical graft polymerization. High grafting efficiency is achieved in the anionic graft polymerization of methacrylonitrile onto nylon-6. Electron-beam radiation has been used in grafting reactions on wool.

Polymer Properties

Solubility and Solvent Resistance. Although polymethacrylonitrile is generally more soluble than polyacrylonitrile, it does possess considerable solvent resistance. Polymethacrylonitrile is unaffected by most common solvents, acids, and water; it is attacked by polar solvents and decomposed by concentrated alkalies and hot dilute alkali.

Molecular Weight Relationships. The dependence of the dilute solution viscosity of polymethacrylonitrile on molecular weight was first published for a study in acetone at 20°C, where $[\eta]$ is in dL/g.

$$\overline{\mathrm{DP}}_n = 0.92\,[\eta]^{1.81}$$

Another study in dimethylformamide at 29.2°C gave the Mark-Houwink form:

$$[\eta] = 3.06 \times 10^{-3}\,[M_n]^{0.503}$$

Both of these relationships were based on osmotic-pressure measurements (number average molecular weights) for primary calibration.

Transitions in Polymethacrylonitrile. The glass-transition temperature of amorphous polymethacrylonitrile prepared by free-radical methods has been observed to be 120°C. Values of 115–123°C have been recorded by differential thermal analysis, depending on analytical procedures and sample preparation.

The stereoregular polymethacrylonitriles prepared with organometallic catalysts have melting points in the range 180–260°C, which are highly dependent on the conditions of polymerization and the purification of the polymer. The more highly crystalline materials have melting points above 200°C.

Degradation and Color Formation. Polymethacrylonitrile readily undergoes depolymerization. Its ceiling temperature of 145°C is 10°C lower than that of poly(methyl methacrylate). Rapid heating above this temperature converts the polymer to monomer. Consequently, it is employed as a positive resist with high resolution capability.

Polymethacrylonitrile tends to discolor when heated, especially in the presence of air or oxygen, and in solution in the presence of strong bases. The color development proceeds from light yellow through orange to dark red and finally approaches black.

Crystal Structure and Tacticity. Conversion to poly(methyl methacrylate) and subsequent analysis by nmr indicated almost completely random configurations, independent of the polymerization conditions studied. A later report indicated that polymethacrylonitrile prepared using free-radical initiators has a higher syndiotacticity than isotacticity.

It was suggested that crystalline polymethacrylonitrile exists in two modifications. X-ray diffraction analysis of oriented polymethacrylonitrile indicates modification I indexed as a pseudohexagonal lattice [$a = 0.903$ nm and $c = 0.687$ nm (fiber identity period)]. The modification II material is indexed with a monoclinic lattice [$a = 1.35$ nm, $b = 0.771$ nm, $c = 0.762$ nm (fiber identity period), and $\beta = 97°40'$].

Bulk Properties. The properties of a compression-molded polymethacrylonitrile resin are given in Table 3.

Processing

Injection molding, thermoforming, injection blow molding, extrusion, and extrusion blow orientation (film) can be used to prepare objects of polymethacrylonitrile and copolymers high in methacrylonitrile content. Care must be taken to avoid overheating to prevent thermal discoloration reactions and depolymerization to monomer.

Applications

The commercial uses of polymethacrylonitrile depend on the properties of the nitrile group and the advantages of the methyl group in the α-position. The nitrile group is among the most polar organic functional groups found in polymers. This polarity provides matrix immobility, which leads to solvent resistance, barrier to passage of gases, adhesion to polar surfaces, and cohesion for film strength.

Table 3. Properties of Polymethacrylonitrile

Test	ASTM method	Value
Mechanical properties		
flexural strength, MPa[a]	D790	83–97
flexural modulus, GPa[b]	D790	3.86–4.48
tensile strength, MPa[a]	D638	55–69
elongation, %	D638	2–3
Rockwell hardness, M scale	D785	95
impact resistance, notched Izod, J/m[c]	D256	21
Thermal properties		
heat-deflection temp, °C, 1.8 MPa[a]	D648	97–100
Electrical properties		
dielectric constant	D150	
60 Hz		4.14
dissipation factor	D150	
60 Hz		0.046
volume resistivity, Ω·cm	D149	1.14×10^{16}
Barrier properties		
water-vapor transmission, μmol/(m²·s)[d]	E96, Procedure E	20
oxygen permeability, mol/(m·s·EPa)[e]	D1434-63 Method M	0.7

[a] To convert MPa to psi, multiply by 145.
[b] To convert GPa to psi, multiply by 145,000.
[c] To convert J/m to ftlb/in., divide by 53.38.
[d] To convert μmol/(m²·s) to g·mil/(100 in.²·d), divide by 10.
[e] To convert mol/(m·s·EPa) to cc·mil/(100 in.²·d·atm), divide by 2.

Methacrylonitrile compositions are used in metal coatings for outdoor applications, where color stability and weatherability are important. The gas-barrier properties of methacrylonitrile polymers have led to their use for packaging resins. Polymethacrylonitrile also has been made into practical films and fibers. Methacrylonitrile is a component in many polymer-resist formulations; it is often utilized along with various acrylate and methacrylate comonomers. Methacrylonitrile–methacrylic acid copolymers are positive electron-beam resists with good sensitivity, submicrometer resolution, high thermal stability, and dry-etch resistance.

Nitrile rubber latices with excellent initial color and color stability, high tensile strength, solvent resistance, and good compatibility with other latices have been commercialized. Dip goods and adhesives for nonwoven fabrics are among the applications developed.

Dispersions of selected methacrylonitrile copolymers immobilize enzymes useful for food manufacture and pharmaceutical production. Copolymers are also used as aqueous printing components.

Health and Safety Factors

Methacrylonitrile, like acrylonitrile and certain other aliphatic nitriles, is toxic by ingestion, by vapor inhalation, and by liquid and vapor absorption through intact skin and mucous membranes.

BIBLIOGRAPHY

J. Brandrup and E. H. Immergut, eds., *Polymer Handbook*, 3rd ed., John Wiley & Sons, Inc., New York, 1989.

U.S. Pat. 3,565,876 (1971), L. E. Ball and H. R. Musser (to The Standard Oil Co. of Ohio); U.S. Pat. 4,473,506 (1984), J. D. Burrington, J. F. Brazdil, and R. K. Grasselli (to The Standard Oil Co. of Ohio).

U.S. Pat. 4,417,034 (1983), O. W. Webster (to E.I. du Pont de Nemours & Co., Inc.); U.S. Pat. 4,414,372 (1983), W. B. Farnham and D. Y. Sogah (to E.I. du Pont de Nemours & Co., Inc.).

Ger. Pat. 1,817,156 (1970), G. Schroeder, W. Gaenzler, and W. Bitscg (to Rohm & Haas GmbH); Jpn. Pat. 80 56,112 (1980) (to Asahi Chemical Industry Co., Ltd.).

LAWRENCE E. BALL
BENEDICT S. CURATOLO
The Standard Oil Company

METHYL ACRYLATE POLYMERS.

See ACRYLIC AND METHACRYLIC ESTER POLYMERS.

2-METHYL-1,3-BUTADIENE POLYMERS.

See ISOPRENE POLYMERS.

METHYLBUTENE POLYMERS.

See BUTENE POLYMERS.

METHYLCELLULOSE.

See CELLULOSE ETHERS.

METHYL 2-CHLOROACRYLATE POLYMERS.

See ACRYLIC AND METHACRYLIC ESTER POLYMERS.

N,N'-METHYLENEBISACRYLAMIDE POLYMERS.

See ACRYLAMIDE POLYMERS.

METHYL METHACRYLATE–ACRYLONITRILE–BUTADIENE–STYRENE COPOLYMERS.

See ACRYLONITRILE–BUTADIENE–STYRENE COPOLYMERS.

METHYL METHACRYLATE–BUTADIENE–STYRENE TERPOLYMERS.

See ACRYLONITRILE–BUTADIENE–STYRENE COPOLYMERS.

METHYL METHACRYLATE POLYMERS.

See ACRYLIC AND METHACRYLIC ESTER POLYMERS.

N-METHYLOLACRYLAMIDE.

See ACRYLAMIDE POLYMERS.

METHYLOLUREA.

See AMINO RESINS.

4-METHYL-1-PENTENE POLYMERS

4-Methyl-1-pentene can be polymerized by two mechanisms: with Ziegler-Natta catalysts or with cationic catalysts. The first method produces highly regular, partially crystalline products that have various applications. Cationic polymerization of 4-methyl-1-pentene yields amorphous polymers of irregular structure.

Commercial production of crystalline poly(4-methyl-1-pentene) (PMP) was started in the UK in 1968 by ICI. PMP is a general-purpose plastic with high thermal and chemical stability, good mechanical and electrical characteristics, and high light transparency. Currently, it is produced commercially by Mitsui Petrochemical Industries in Japan by license from ICI, at an annual production of ca 6000 metric tons. The 1986 price was $4.84–5.72/kg.

Monomer

4-Methyl-1-pentene, $CH_2{=}CH{-}CH_2{-}CH(CH_3)_2$, is a colorless, light, flammable liquid. It is an irritant and, in high concentrations, a narcotic. Because of a low flash point, this chemical compound represents a significant fire hazard when exposed to heat, flame, or oxidizing agents. The chemical properties are typical for an α-olefin: it participates in addition reactions, oxidations, complex formation with transition-metal derivatives, polymerizations, and copolymerizations with other olefins.

Polymer

Molecular and Crystalline Structure. Poly(4-methyl-1-pentene) made with heterogeneous Ziegler-Natta catalysts has a regular head-to-tail structure with a repeating monomer unit:

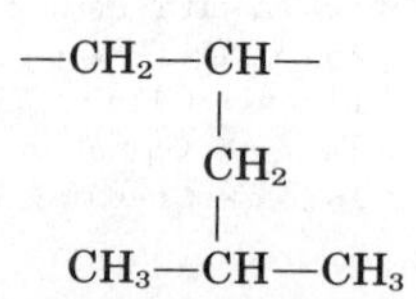

Commercial PMP is a random copolymer of 4-methyl-1-pentene and 3–5 wt % of an α-olefin with a linear alkyl chain C_6–C_{16}.

Physical Properties. Physical, thermal, mechanical, and electrical properties of commercial crystalline PMP appear in Table 1.

Health and Safety Factors

Poly(4-methyl-1-pentene) is an inert, nontoxic material that usually represents no health hazard.

Uses

The applications of PMP are based on high optical transparency, excellent dielectric characteristics, high thermal stability, and chemical resistance. The manufacture of medical equipment consumes ca 40% of production and includes hypodermic syringes, needle hubs, blood collection and transfusion equipment, pacemaker parts, urine collection tubes, blood analysis cells, breast pumps, and respiration equipment. High PMP transparency in the visible and near uv ranges, combined with thermal stability, chemical resistance, and rigidity, makes it acceptable as medical and chemical laboratory equipment.

Table 1. Properties of Crystalline PMP

Property	ASTM method	Value
Physical		
density at 20°C, g/cm³	D792	0.830–0.835
water absorption, at saturation, %	D570	0.01
refractive index, n_D^{20}		1.463
Thermal		
melting point, °C		235–240
heat of fusion, J/g[a]		61.65
Vicat softening point, °C	D1525	173–180
glass transition, °C		
homopolymer		ca 50
copolymer		ca 30
deflection temperature under flexural load, °C		
0.46 MPa[b]	D686	80–90
1.82 MPa[b]	D686	48–50
coefficient of linear expansion per °C	D696	1.17×10^{-4}
thermal conductivity, W/(m·K)	C177	0.167
specific heat capacity, J/(g·K)[a]	C351	2.18
Electrical		
dielectric constant at 25°C, 10^2–10^6 Hz	D150	2.12
volume resistivity, Ω·cm	D257	$> 10^{16}$
dielectric strength, kV/mm	D149	63–65
dissipation factor at 10 MHz	D150	1.5×10^{-4}
Mechanical		
tensile yield strength, MPa[b]	D638	23–28
tensile break strength, MPa[b]	D638	17–20
elongation at break, %		10–25
bending strength, MPa[b]	D790	25–35
tensile modulus, MPa[b]	D638	1500–2000
flexural modulus, MPa[b]	D790	1300–1800
compression modulus, MPa[b]	D695	800–1200
notch–impact strength, kJ/m[c]	D256	100–200
Rockwell hardness		L80–93

[a] To convert J to cal, divide by 4.184.
[b] To convert MPa to psi, multiply by 145.
[c] To convert kJ/m to ft·lb/in., divide by 53.39.

Thermal and chemical resistance provides the basis for PMP applications in chemical and metallurgical industries. Additional applications include films, transparent sheets, tubes, rods, filaments, and fibers. Highly transparent PMP fibers have excellent potential for use in fiber optic applications. PMP is also utilized for wire and cable coating and for film and paper coatings with good release properties.

BIBLIOGRAPHY

J. Boor, Jr., *Ziegler-Natta Catalysts and Polymerizations,* Academic Press, Inc., New York, 1979.

Y. V. Kissin, *Isospecific Polymerization of Olefins with Heterogeneous Ziegler-Natta Catalysts,* Springer-Verlag, New York, 1985.

H. Ohi in R. J. Martin, ed., *Modern Plastics Encyclopedia,* Vol. 63 (10A), McGraw-Hill Inc., New York, 1986–1987.

YURY V. KISSIN
Mobil Chemical Company

METHYL RUBBER. See BUTADIENE DERIVATIVES, POLYMERIC; BUTADIENE POLYMERS.

METHYL VINYL KETONE POLYMERS. See VINYL KETONE POLYMERS.

MICELLES. See COLLOIDS; EMULSION POLYMERIZATION; VESICLES.

MICROBIAL POLYSACCHARIDES. See DEXTRAN; POLYSACCHARIDES.

MICROEMULSION POLYMERIZATION

During the past 50 years emulsion polymerization (qv) has become an important process for the production of industrial polymers in the form of polymer colloids (qv) or latices (qv), which are the bases of adhesives, paints, polishes, and other coatings.

Structure of Microemulsions

In contrast to emulsions, which are opaque, unstable, and macrodisperse systems of 1–10-μm droplet size, microemulsions are thermodynamically stable and transparent oil-water systems stabilized by an interfacial layer of surface-active agents. These structures can exist either alone or in equilibrium with water and/or oil phases in excess. It is now well established that in water- and oil-rich regions, microemulsions consist of uniform and spherical droplets dispersed in a continuous medium. The particle size is ca 10^{-2} μm, which explains the optical transparency of these systems. The droplets can be swollen with oil and dispersed in water (direct or oil–water microemulsions), or swollen with water and dispersed in oil (inverse or water–oil microemulsions).

Polymerization in Oil–Water Microemulsions

Most of the investigations reported up to now refer to polymerization of styrene in microemulsions containing an alcohol as a cosurfactant. If the amount of styrene is low, ie, 1–2%, there is no evidence of phase separation in the course of polymerization. Once polymerized, these systems appear quite stable to dilution. When the styrene content exceeds a few percent, phase separation or polymer precipitation is usually encountered at various stages of the reaction.

Polymerization in Water–Oil Microemulsions

A strong incentive to study these systems comes from the numerous practical applications of high molecular weight water-soluble polymers (qv), eg, polyacrylamide and derivatives (see ACRYLAMIDE POLYMERS). These are widely used for flocculation (qv) of colloidal dispersions, sewage treatment, coatings, adhesives, and as thickeners in enhanced oil recovery.

The polymerization of a water-soluble monomer in a microemulsion medium rather than in a conventional emulsion can overcome problems of inverse emulsion polymerization, eg, instability of the latices resulting in rapid sedimentation of the polymer and a broad size distribution of final particles. The choice of the system to be polymerized, especially the nature of the surfactant, is of critical importance since it controls the ultimate characteristics of the latices. Both anionic and nonionic surfactants have been investigated.

Polymerization in the Continuous Phase in Water–Oil Microemulsions

Some attempts have been made to use the continuous phase of microemulsions to prepare solid polymers capable of encapsulating the dispersed phase, eg, water. This would allow inclusion of materials in the dispersed phase that would otherwise be insoluble in the polymer. Phase separation occurs, however, in polymerizing the continuous phase of microemulsions. This instability must be attributed to the complex effect of entropic increase of free energy as a result of conformational limitations and/or the limited solubilities of polymer–water and polymer–cosurfactant.

BIBLIOGRAPHY

A. M. Bellocq and co-workers, *Adv. Colloid Interface Sci.* **20,** 167 (1984).

A. Jayakrishnan and D. O. Shah, *J. Polym. Sci. Polym. Lett. Ed.* **22,** 31 (1984).

J. O. Stoffer and T. Bone, *J. Dispersion Sci. Technol.* **1,** 37 (1980).

F. Candau, Z. Zekhnini, and J. P. Durand, *J. Colloid Interface Sci.* **114**(2), 398 (1986).

FRANÇOISE CANDAU
Institut Charles Sadron

MICROENCAPSULATION

Microencapsulation is the envelopment of small solid particles, liquid droplets, or gas bubbles with a coating. Many terms are used to describe the contents of a microcapsule, eg, active agent, actives, core material, fill, internal phase (IP), nucleus, and payload. The coating material used to form the capsule is called a wall, shell, coating, carrier, or membrane. It may be an organic polymer, hydrocolloid, sugar, wax, fat, metal, or inorganic oxide (Table 1). The unique feature of a microcapsule is its size, ie, between 1 and 1000 μm.

Microcapsules can have a wide range of geometries and structures (Fig. 1). A given active agent can be encapsulated by different materials. A capsule is characterized by particle size and size distribution, particle geometry, actives content and distribution, release mechanism, storage stability, and other parameters.

Manufacturing costs are based on the coating material, solvent, equipment, and labor. Coating-material prices vary greatly, and as a rule, the cheapest acceptable material is used. Coatings that can be applied without solvent or with water are preferred. Environmental and safety regulations greatly increase the cost of processes that use volatile organic solvents. Processes requiring only simple equipment and minimal labor are preferred.

Table 1. Formation of Commercial Microcapsules

Coating material	Chemical class	Encapsulation process	Applications	Regulatory status
acacia (gum arabic)	polysaccharide	spray drying	food flavorings	edible
gelatin	protein	spray drying	vitamins	edible
ethylcellulose	cellulose ether	polymer–polymer phase separation or Wurster air suspension	pharmaceuticals	edible
polyurea	cross-linked polymer	interfacial or *in situ* polymerization	pesticides	nonedible
aminoplasts[a]	cross-linked polymer	*in situ* polymerization	carbonless paper; perfume	nonedible
maltodextrins	low molecular weight carbohydrate	spray drying or pressure extrusion	food flavorings	edible
sucrose	carbohydrate	spray drying or pressure extrusion	food flavorings	edible
hydrogenated vegetable oils	glycerides	Wurster air suspension	food additives	edible

[a] Amino resins.

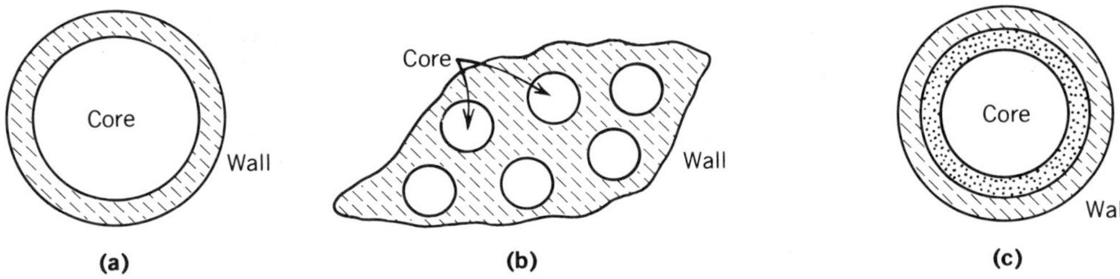

Figure 1. Microcapsule structures: (**a**) mononuclear capsule; (**b**) microcapsule aggregate; (**c**) double-wall structure.

A classification of encapsulation processes is given below:

Category I	Category II
Liquid suspending media	**Gas suspending media**
Water-in-oil Oil-in-water	**Solvent Solvent-free**
complex coacervation	spray drying and congealing
polymer–polymer incompatibility	fluidized-bed and Wurster process
interfacial and *in situ* polymerization	coextrusion
desolvation	vacuum coating
solvent evaporation from emulsions	gelatin
gelation	electrostatic deposition
pressure extrusion	

In Category I, a liquid is used as the suspending medium throughout. Emulsification or dispersion of two or more immiscible phases is a key step in these processes. In Category II, a gas is the suspending medium in which the capsules are formed, and atomization of a liquid phase is the key step.

Applications

Microcapsule-based products are used in the graphic arts, pharmaceutical, food and pesticide industries, and for specialty products. The following organizations manufacture capsules on a contract basis or accept research contracts and development work: Battelle Memorial Institute, CDS (Controlled Delivery Systems), IGI Industries, Inc., Ronald T. Dodge, Eurand America, Southern Research Institute, The Coating Place, Inc., and Southwest Research Institute.

Graphic Arts. Carbonless copy paper is by far the largest single application of microcapsules. The three-part business form illustrates the principle on which carbonless paper is based (Fig. 2).

Pharmaceuticals. Microencapsulation provides the pharmaceutical industry with a range of new dosage forms. Orally administered capsules conceal unpleasant taste, prolong release, provide enteric release, improve stability of incompatible drug mixtures and resistance to oxidation, reduce volatility, and convert a liquid drug to a powder form (see also DRUG APPLICATIONS).

Food Additives. A number of food additives are available in encapsulated form (Table 2).

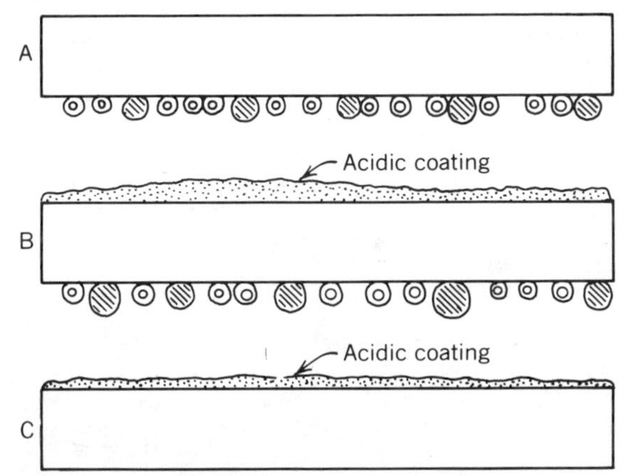

Figure 2. Cross-section of a three-part business form prepared from carbonless paper. A–C represent sheets 1–3 respectively, ◎, Microcapsules; ⊘, starch.

Table 2. Encapsulated Food Ingredients

Encapsulated product	Active agent	Producer
Borden Caps	flavors	Borden, Inc.
Durarome	flavors	MCP Foods
Durkote	powders	Durkee Industrial Div., SCM Corp.
Durstay	citrus flavors	Durkee Industrial Div., SCM Corp.
Spice-Sol-E	spice extracts	Durkee Industrial Div., SCM Corp.
Flav-O-Lok	flavors	PFW Div., Hercules, Inc.
In-Cap		
Nu-Spice	spice and vegetable extracts	Griffith Laboratories, Inc.
Permastabil	citrus flavors	Sunkist Growers, Inc.
Cap-Shure	citric and ascorbic acid	Balchem Corp.

Table 3. Encapsulated Pesticides and Herbicides

Trade name	Active ingredient	Manufacturer
Penncap-M	methyl parathion	Pennwalt Corp.
Knox-Out 2FM	diazinon	Pennwalt Corp.
Capsolane	EPTC	Stauffer Chemical Co.
Surpass 45	vernolate	Stauffer Chemical Co.
Fonofos Seed Treatment	fonofos	Stauffer Chemical Co.
Altosid SR-10	methoprene	Zoecon Corp.
Sectrol	pyrethins	3M Co.
Dursban ME	dursban	The Dow Chemical Co.
Lasso Micro-Tech	alachlor	Monsanto Co.

Pesticides. Microcapsules in pesticide formulations (Table 3) prolong activity and reduce mammalian toxicity, volatilization losses, phytotoxicity, environmental degradation, and soil movement.

Specialties. Scented strips are used to promote perfumes. A strip of scent-loaded capsules is coated on paper sheets which are used as promotional inserts. Live organisms are entrapped in gel beads to protect transplanted cells, eg, islets of Langerhans, from the immune system of recipients.

Other microencapsule applications include cosmetics, fish nutrients, animal-feed additives, veterinary products, plant-growth stimulants, fertilizers, adhesives, and biomedical applications; the last application area appears most interesting.

BIBLIOGRAPHY

P. B. Deasy, *Microencapsulation and Related Drug Processes,* Marcel Dekker, Inc., New York, 1984.

A. Kondo, *Microcapsule Processing and Technology,* Marcel Dekker, Inc., New York, 1979.

C. Thies, *CRC Crit. Rev. Biomed. Eng.* **8,** 335 (1983).

T. Kondo, ed. *Microencapsulation: New Techniques and Applications,* Techno, Tokyo, 1979.

CURT THIES
Washington University

MICROMECHANICAL MEASUREMENTS

In order to determine the strength and endurance of a material under stress, it is necessary to characterize its mechanical behavior. Moduli, strain, strength, and toughness are measured macroscopically. These parameters are useful for materials selection and design, but they are not easily transferable because of a lack of understanding of the mechanisms of deformation and fracture. This is particularly true of polymers because of their viscoelastic behavior. The fracture behavior

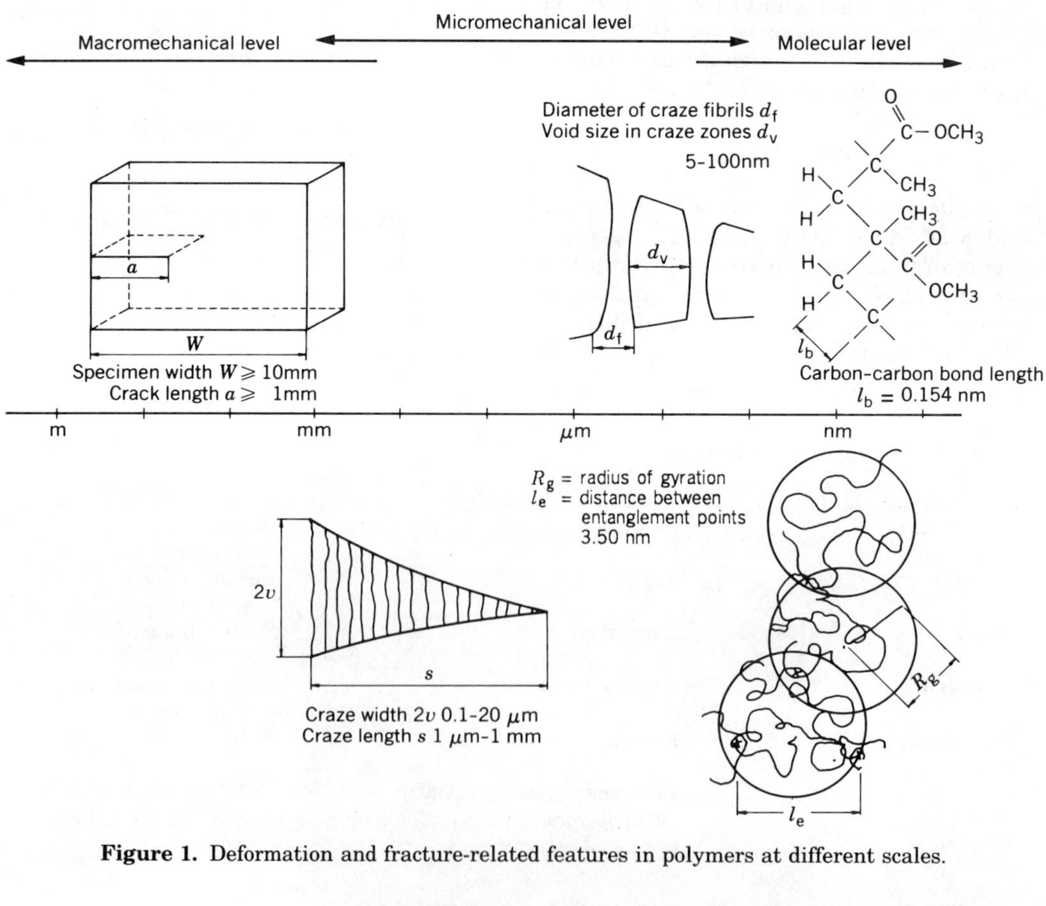

Figure 1. Deformation and fracture-related features in polymers at different scales.

of materials, especially macroscopically brittle materials, is governed by the microscopic mechanisms operating in a heterogeneous zone at the crack tip or stress-raising flaw. In order to supplement macromechanical investigations and advance knowledge of the fracture process, micromechanical measurements of the deformation zone are required to determine local stresses and strains. The required measurements depend to a large extent on the complexity of the model used in the subsequent analysis.

In thermoplastics, craze zones are the most important microscopic features around a crack tip governing strength behavior (see also CRAZING). Fracture (qv) is preceded by the formation of a craze zone, which is a wedge-shaped region spanned by oriented microfibrils. The investigation of craze properties on the micromechanical level covers orders of sizes between ca 10 nm to several hundred μm and is hence situated between the macromechanical and the molecular levels, as can be seen in Figure 1. Measurements in this microregion are adequate for analyses using craze-micromechanics models, which include the elastic properties of the continuum.

Experimental methods of craze-zone measurement include interference optics, diffraction techniques, scanning electron microscopy (sem), and transmission electron microscopy (tem).

Measured dimensions have been analyzed, correlating them with strength and deformation parameters in order to further the understanding of material behavior on the microscopic scale. For example, the fracture energy can be related to the maximum length of craze fibrils ($2v$) and the engineering stress along the craze zone. Analytical methods applied include the Dugdale model, and mathematical and numerical methods.

BIBLIOGRAPHY

W. Döll in H. H. Kausch, ed., "Crazing in Polymers," *Adv. Polym. Sci.* **52–53,** 105 (1983).

E. J. Kramer, *ibid.,* p. 1.

H. R. Brown, *Materials Science Reports* **2,** 315 (1987).

L. Bevan, W. Döll, and L. Könczöl, *J. Polym. Sci. Polym. Phys. Ed.* **24,** 2433 (1986).

W. DÖLL
L. KÖNCZÖL
Fraunhofer-Institut für Werkstoffmechanik
L. BEVAN
North East London Polytechnic

MICROPHASE STRUCTURE

Polymer systems containing two or more components are known as multicomponent polymer materials, polymer alloys, or polymer blends. The following generic terms include all combinations of two or more polymers: graft copolymers, block copolymers, interpenetrating polymer networks (IPNs), and AB-cross-linked polymers, as well as various polymer blends.

Multicomponent polymer materials are an important class of engineering materials applied as impact-resistant plastics, tires, resilient rugs, knife handles, rubber shoe soles, and ion-exchange resins. They exhibit outstanding properties because controlled phase separation frequently leads to synergistic behavior. This article discusses morphologies achieved on phase separation, especially aspects of phase inversion and cocontinuous phases.

Thermodynamics of Phase Separation

The terms compatible and incompatible have been used to indicate a single phase or multiphase mixture of two or more polymers, respectively (see COMPATIBILITY). However, these terms are also used by polymer technologists to indicate satisfactory or unsatisfactory behavior on blending, such as good adhesion between the constituents, improved mechanical behavior, or simply ease of mixing. More recently, the terms miscibility and immiscibility have been suggested to strengthen the thermodynamic implications (see MISCIBILITY).

Characteristics of Phase-separated Systems

There are two simple tests to distinguish a polymeric material as homogeneous or phase-separated. First, if the material is milk white or hazy, then two or more phases exist. The whitish appearance arises because of the difference in refractive indexes of the two phases, which causes intense light scattering.

The second method of characterization employs the glass–rubber transition phenomenon. Each polymer has a glass-transition temperature (T_g), and a phase-separated polymer material typically exhibits two T_g values. If mixed polymers happen to have identical glass-transition temperatures, this test of homogeneity fails.

A white product or observation of two T_g values, or both, easily identifies a multiphase material, but it is harder to be sure that a material contains only one phase. Other helpful characterization information is monomer composition, ir and nmr spectra, etc. The most powerful experiments involve electron microscopy, the principal tool in elucidating polymer blend morphology.

Sample Preparation for Microscopy

Cutting and staining techniques are important in three general methods of observing samples: light microscopy, transmission electron microscopy (tem), and scanning electron microscopy (sem).

The development of the osmium tetroxide staining technique greatly extended observational possibilities. OsO_4 stains and hardens polymers with double bonds:

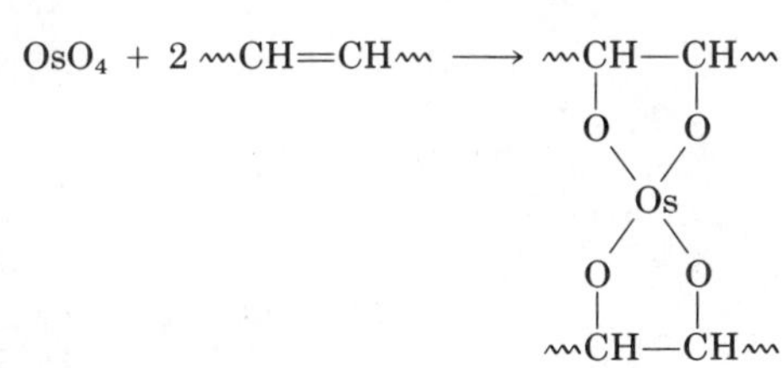

Another method involves staining carboxyl groups, and is especially effective for esters such as poly(n-butyl acrylate). The polymer is treated with hydrazine or hydroxylamine, then with osmium tetroxide in the absence of water. Each technique can be used to observe multicomponent morphologies.

Two-phased Morphologies

Important aspects of phase separation include the size and shape of the phases formed, their internal structure, the extent of molecular mixing between the two components, the adhesion between the two phases, and the average distance between phase domains. The method of preparation of polymer blends affects these parameters.

Polymer Blend Preparation. In a simple laboratory method for preparing polymer blends, solutions of two polymers in separate containers are mixed together and the solvent is evaporated. This method leads to an uncontrolled morphology and consequently is rarely used industrially. A more controlled simple method of polymer blending in the melt state involves equipment such as rollers or extruders. A more sophisticated method for preparing high impact polystyrene involves the solution graft copolymer process.

PS–PB Multiphase Polymers. Several known morphologies of polystyrene–polybutadiene materials are illustrated in Figure 1.

ABS Polymers. Utilization of core–shell latex technology brings about a related morphology in acrylonitrile–butadiene–styrene (ABS) polymers; the final product is a modified styrenic material. First, a seed latex of butadiene and acrylonitrile is polymerized to form the core, polymer I. Then, a mixture of styrene and acrylonitrile monomers is added, and the polymerization continued to form polymer II, a styrene–acrylonitrile (SAN) copolymer. Some of the styrene and acrylonitrile solution dissolves in polymer I, and the remainder forms a shell around the exterior. On polymerization, the morphology shown in Figure 2 is obtained.

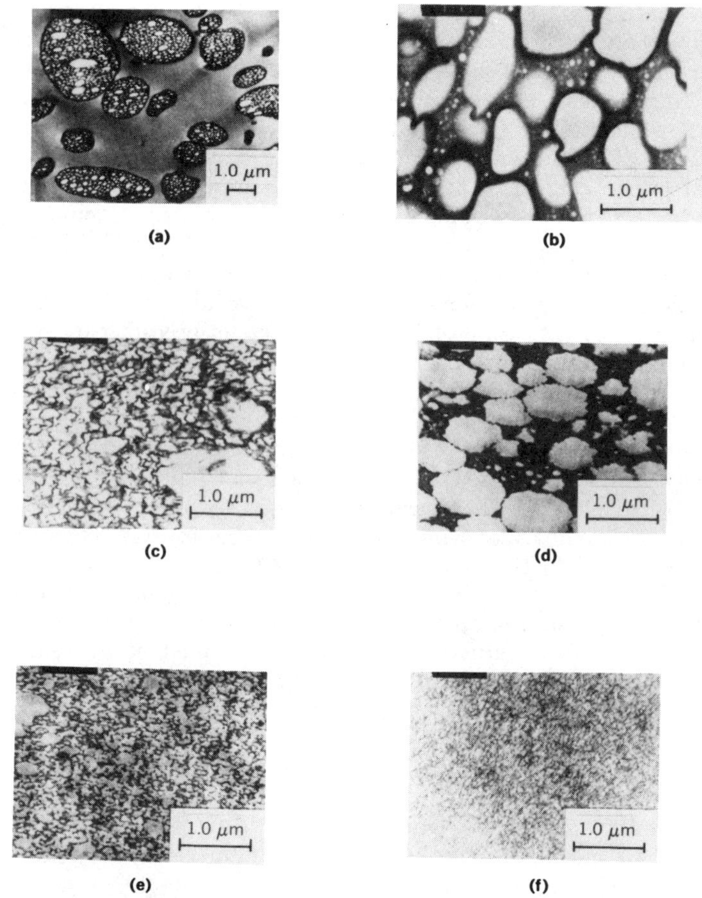

Figure 1. Six morphologies of graft copolymers and IPNs of polybutadiene and polystyrene. (**a**) High impact polystyrene (phase inverted); (**b**) graft, a consequence of not stirring (no phase inversion); (**c**) *cross*-polybutadiene-*inter*-polystyrene semi-I IPN; (**d**) polybutadiene–*inter*-*cross*-polystyrene semi-II IPN; (**e**), (**f**) full IPNs, *cross*-polybutadiene–*inter*-*cross*-polystyrene; (**e**) 0.1% cross-linking; (**f**) 0.2% cross-linking.

Block Copolymers

Most block copolymers phase separate. Because the blocks are attached to one another, the phases are constrained to be no larger than the molecular dimensions. If they were much larger or smaller, either significant rarifications or densifications would have to be present, with considerable increases in free energy of formation. The structure of such domains is illustrated in Figure 3. The individual blocks assume nearly random coil conformations within each domain, and the junction between the two blocks is at the interface.

Three typical morphological structures are known among block copolymers: spheres, cylinders, and alternating lamellae. Which one will appear is governed by the ratio of the molecular

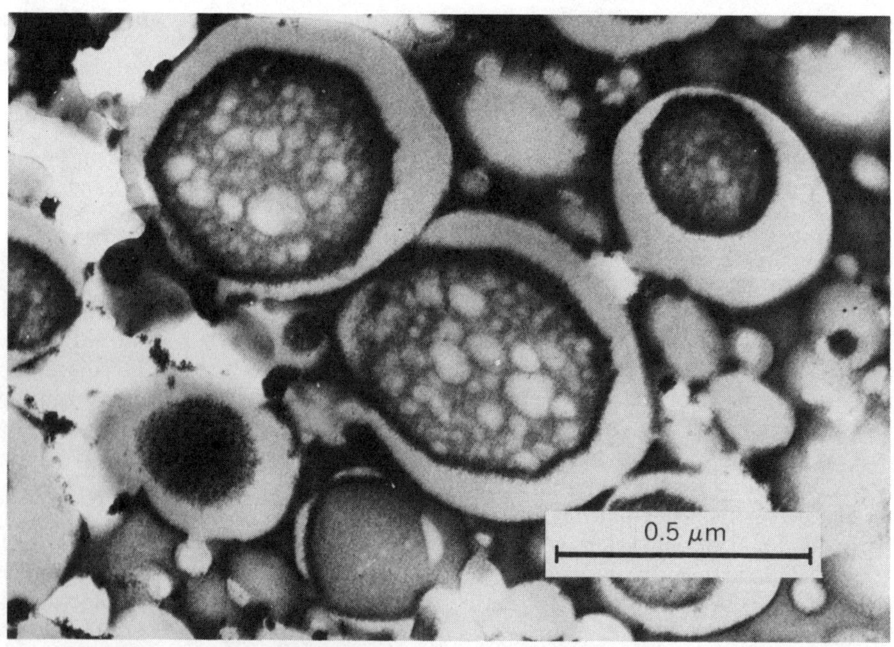

Figure 2. Transmission electron micrograph of an ABS latex, showing the core–shell morphology.

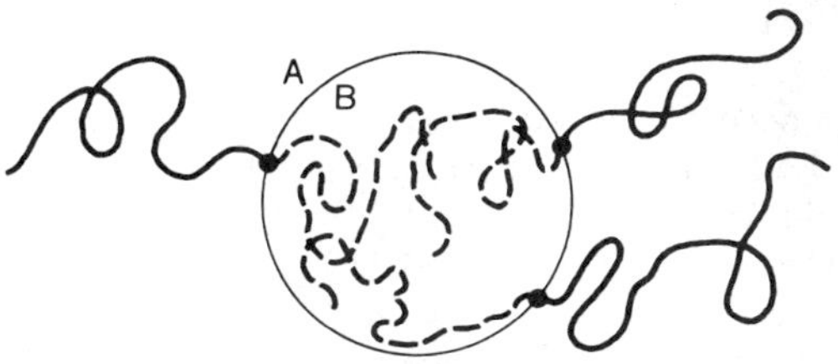

Figure 3. The Meier model of a block copolymer domain.

weights of the blocks, which determines the free energy of the system.

Onion Ring Morphologies

Block copolymers and AB-cross-linked polymers made by limited conterminous grafting can form unusual morphologies resembling onion rings in the presence of the homopolymer of one of the components. Isolated concentric rings of copolymer are formed, surrounded by homopolymer. The materials are prepared by a free-radical process in such a way that no B-homopolymer is formed and that the cross-linking index is sufficiently low that no gel is formed. All of the chemical species are derived from an ABA block copolymer type of structure.

The onion ring structure develops when macroscopic phase separation initially produces droplets of the minor phase of block copolymer. Copolymer molecules in the vicinity of the droplet surface preferentially have their A-block in the A-rich phase, while the B-blocks form a B-rich phase on the interior. This layering persists to the middle of the droplet, resulting in the onion ring structure. The onion ring morphology is really a topological variation of the alternating lamellar structure with curved, closed structure formation.

Phase Inversion

During the synthesis of high impact polystyrene, phase inversion occurs. At the beginning the rubber-rich phase is continuous, but at the end the plastic-rich phase is continuous. An analogy can be made to an oil-in-water suspension, which under certain conditions can be made into a water-in-oil suspension. In both of these cases, properties differ greatly, depending on which phase is continuous.

The most important factors determining phase continuity include the melt viscosity η of each phase, and its phase volume ϕ. Qualitatively, the phase with the larger volume fraction will be continuous, all other factors being equal. Similarly, the phase with the lower viscosity tends to be continuous, because it flows around the higher viscosity phase. It is often observed that regions of phase inversion occur near where the relation η_1/η_2 times ϕ_2/ϕ_1 is unity.

Cocontinuous Phases

A phase is a definite part of a system, homogeneous throughout, and physically separated from other phases by distinct boundaries. The nearly synonymous terms interpenetrating phases, interlocking phases, cocontinuous phases, and dual-phase continuity are used to describe two phases that are each continuous in space, such that each contacts all portions of the sample space.

Topologically, there are two independent ways of forming equal cocontinuous phases: with alternating lamellae or with interpenetrating phases. Alternating lamellae possess cocontinuity only in two dimensions; ideally both phases are discontinuous in the third dimension. In the case of interpenetrating phases, each phase maintains continuity in all three dimensions. An example of cocontinuous phases in three di-

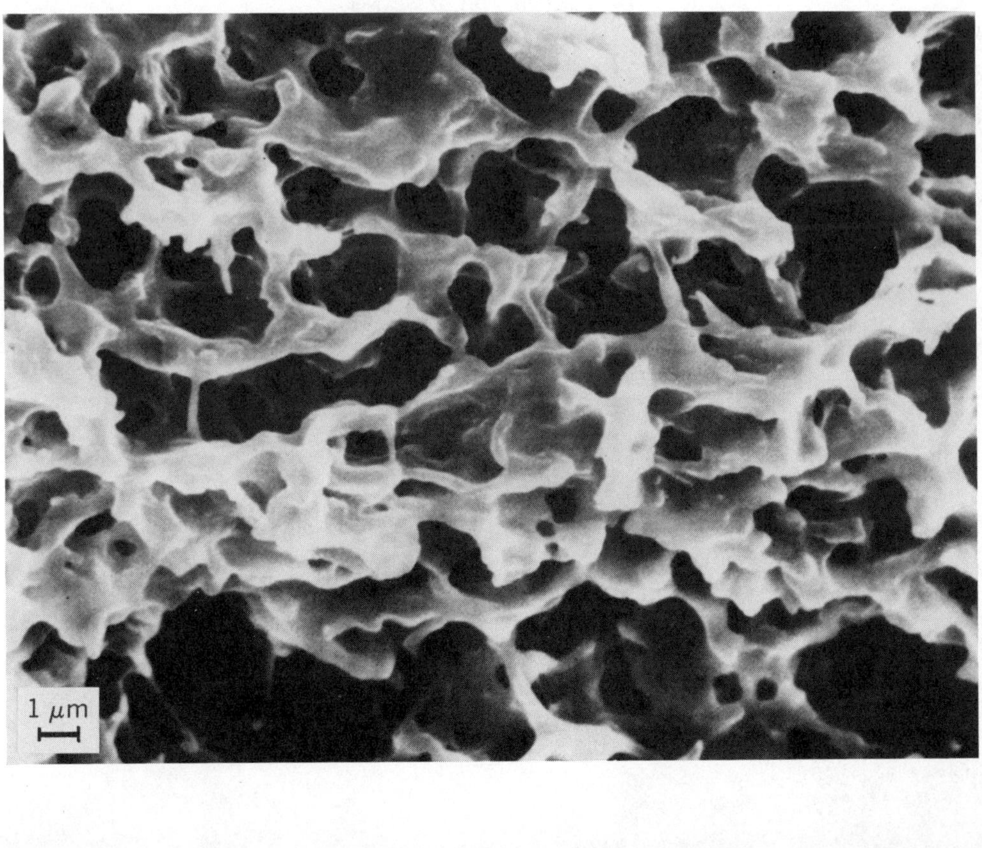

Figure 4. Scanning electron micrograph of a 70/30 blend of EPM and polypropylene. The EPM phase was extracted with heptane, leaving the polypropylene.

mensions is an air filter and the air that flows through it. A Maxwell demon can traverse the entire system in either phase.

Polypropylene Blends. Important systems utilizing the phase inversion technique are polypropylene blended with ethylene–propylene elastomers (EPM or EPDM). The morphology of a cocontinuous polypropylene–EPM blend is shown in Figure 4.

Block Copolymers in Thermoplastic IPNs. In the case of block copolymers forming one of the components of a thermoplastic IPN, the block copolymer may show no measurable true shear flow and this feature can be used to form cocontinuous phases by a method called templating. The block copolymer, in this gel-like state, fragments along shear planes during mixing with the other, immiscible polymer, forming a skeletal continuous structure. The second component polymer takes up what is called the "antitropic" continuous phase structure.

Interpenetrating Polymer Networks. Interpenetrating elastomer networks (IENs) are made by mixing two cross-linkable latices, followed by curing reactions. If one of the polymers flows more easily than the other, the polymer with the higher viscosity remains more or less spherical within the other. However, when the higher viscosity latex particles can be made to touch, a different type of dual-phase continuity arises.

BIBLIOGRAPHY

L. H. Sperling in D. R. Paul and L. H. Sperling, eds., *Multicomponent Polymer Materials, Adv. Chem. Ser. 211,* American Chemical Society, Washington, D.C., 1986.

J. A. Manson and L. H. Sperling, *Polymer Blends and Composites,* Plenum Press, New York, 1976.

L. H. Sperling, *Interpenetrating Polymer Networks and Related Materials,* Plenum Press, New York, 1981.

D. R. Paul and S. Newman, eds., *Polymer Blends,* Vols. I and II, Academic Press, Inc., New York, 1978.

L. H. SPERLING
Lehigh University

MICROSCOPY. See ELECTRON MICROSCOPY; LIGHT MICROSCOPY.

MICROSPHERES

Microspheres are spherical particles, typically less than 200 μm, which provide unique properties to plastics when used as filler. They differ from nonspherical fillers (qv) (flakes, platelets, granules, fibers, whiskers, and irregular-shaped powders) in several important ways. To begin with, the ratio of surface area to volume is the smallest for any type of particle. At equivalent volume loadings, spheres have the smallest surface area in contact with the resin, and, therefore, the viscous drag is minimal. Resin viscosity increases slowly with microsphere loading, thus making higher loadings possible with good flow properties.

Microspheres are isotropic, ie, their properties are distributed uniformly over their spherical shape. As a result, their properties do not depend on the direction in which they are measured. In polymer applications, this means that internal stresses during cure can be distributed evenly, thereby reducing the tendency toward cracking. During actual use, external stress is evenly distributed; hence, the orientation of the stress in relation to the orientation of the fillers has no effect on

Table 1. Composition and Physical Properties of Solid Microspheres

	A-glass	Ceramic
Composition, wt%		
SiO_2	72	50–58
Al_2O_3		25–30
Na_2O	14	
CaO	10	6–10
MgO	3	
Na_2O, K_2O		1–4
Al_2O_3,K_2O,Li_2O	1	
other oxides		1–5
Properties		
true particle density, g/cm^3	2.5	2.4
bulk density, g/cm^3	1.3	1.3
particle size, μm		
range	10–1000	5–60
average	30–50	10–30

Table 2. Composition and Physical Properties of Hollow Microspheres[a]

Properties	Eccospheres MC-37, sodium borosilicate glass	Eccospheres SI, high silica glass, 98% SiO_2	Q-Cel 300, sodium borosilicate	Ucar BJO-0930, phenolic
true particle density, g/cm^3	0.36	0.25	0.21	0.23
bulk density, g/cm^3	0.23	0.15	0.12	0.10
particle size, μm				
range	5–150	20–180	10–180	5–150
average	40	80	65	45
hydrostatic strength, survivors at 10.3 MPa[b], vol %	96	55	16	
thermal conductivity, W/(m·K)	0.06	0.05	0.04	
melting temperature, °C	480	980	450	150

[a] Information was obtained from company technical literature: Eccospheres: Emerson & Cuming, Inc. (Apr. 1980); Q-Cel: Philadelphia Quartz (1983); Ucar: Union Carbide (Sept. 1982).

[b] To convert MPa to psi, multiply by 145.

failure. Because of the filler's spherical shape, the surface of the polymer is smoother than fibrous or irregular-shaped fillers.

Microspheres are either solid (Table 1) or hollow (Table 2). The solid ones most commonly used as reinforcement in plastics are made of glass or ceramic. Carbon, graphite, zinc, and plastics, such as acrylics, poly(vinyl chloride), and polystyrene, are employed in special applications. Hollow microspheres are made of glass, ceramic, carbon, or resins, eg, phenolics, poly(vinylidene chloride), or polystyrene. Glass microspheres are by far the most popular hollow fillers. Mixed with a polymerizable resin, hollow microspheres form a special lightweight material known as syntactic foam.

BIBLIOGRAPHY

H. S. Katz and J. V. Milewski, *Handbook of Fillers and Reinforcements for Plastics,* Van Nostrand Reinhold Co., New York, 1978, pp. 301–310.

U.S. Pat. 3,138,444 (June 23, 1964), C. E. Searight, H. W. Smalley, and H. L. Robinson (to Ferro Corp.).

U.S. Pat. 2,978,339 (Apr. 4, 1961), F. Veatch, H. E. Alford, and R. D. Croft (to Standard Oil Co.).

JOHN F. PLUMMER
Emerson & Cuming, Inc.

MICROSTRUCTURE

Molecular Structure

The term microstructure is used principally to describe specific repeat-unit structures, sequencing of repeat units, average compositions, and composition distributions. The average degree of polymerization, although not usually included when defining microstructure, is also necessary to describe polymer molecular structure. Therefore, molecular weight, molecular weight distribution, and the polymer-chain skeletal structure, ie, topology, must be included with detailed repeat-unit microstructure to define polymer molecular structure completely. Polymer microstructure is a subdivision of the broader topic of polymer structure, which also includes, for example, solid- and liquid-state structures, solution characteristics, and chain conformation.

In addition to establishing the structure and concentration of each type of repeat unit in a polymer chain and the degree of polymerization, analyses of repeat-unit structure must yield information about isomeric arrangements, that is, stereochemical configuration and modes of monomer-unit addition. Sequencing of repeat units must also be established before structure–property relationships can be well understood.

Polymer microstructure, therefore, should be defined from the following viewpoints:

structure of repeat unit(s),
repeat-unit chirality,
skeletal topographical structure,
end groups and degree of polymerization, and
sequencing of repeat units.

Because of some fundamental differences in polymer structures produced by their mode of preparation, polymer molecules are often classified by the routes through which they are obtained, as chain or step-growth polymers (see CHAIN REACTION POLYMERIZATION; STEP-REACTION POLYMERIZATION).

Structure Determination. For structural determinations of polymers from step-growth processes, spectroscopic techniques or a characterization of products formed after hydrolyses of functional groups are used. Spectroscopic tools, particularly infrared and nuclear magnetic resonance, are among the foremost methods employed in the characterization of polymers prepared by chain-growth polymerization.

Mode of Addition. Although homopolymers are synthesized from only one kind of monomer unit, the repeat-unit structures may not be uniform, particularly for chain-growth polymerizations. Differences in modes of addition and chain-transfer reactions during polymerization may lead to complex chain structures. Such homopolymers may resemble copolymers or terpolymers depending on the number of isomeric repeat structures possible.

Chirality. Complex microstructures occur for homopolymers prepared from olefins that produce repeat units with side-chain groups. Differences in the chirality of the repeat units may occur when the olefin monomers link head-to-tail to form the polymer chain.

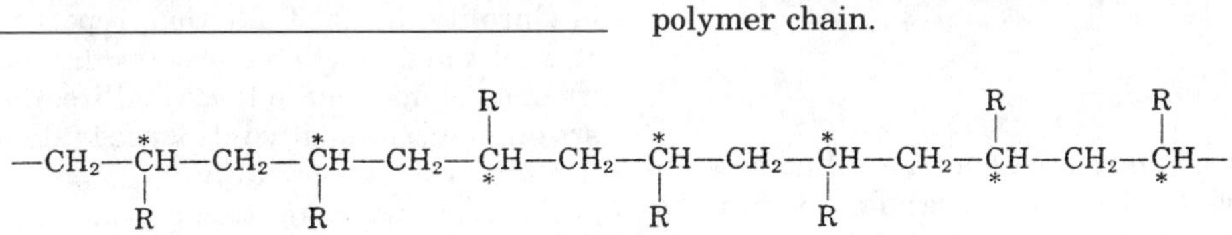

The * designates chiral carbons and the R groups are placed up or down to denote different stereochemical relationships. Chirality significantly affects physical properties.

Copolymers. Two kinds of monomer units can be combined to form several characteristic arrangements of repeat units:

$$x\,A + y\,B \longrightarrow A{-}A{-}A{-}A{-}A{-}A{-}A{-}A{-}A{-}B{-}B{-}B{-}B{-}B{-}B{-}B{-}$$

block copolymer

$$\longrightarrow A{-}A{-}A{-}A{-}B{-}A{-}A{-}A{-}B{-}B{-}A{-}A{-}A{-}A{-}B{-}B{-}B{-}$$

random copolymer

$$\longrightarrow A{-}B{-}A{-}B{-}A{-}B{-}A{-}B{-}A{-}B{-}A{-}B{-}A{-}B{-}A{-}B{-}A{-}$$

alternating copolymer

Few polymers form perfectly random, block, or alternating copolymers, but the predominant structure is so described. Complexity of microstructure is increased if differences arise from chirality or the modes of monomer addition. The microstructure of terpolymers can be exceedingly complex because of the various sequence arrangements analogous to copolymers and differences arising from chirality and modes of addition.

Topography. The overall topographical structure of a polymer chain can be either linear, branched, or cross-linked. Polymers with branched skeletal structures may be described as either stars or combs. A "Y" structure, which is the simplest form of a star polymer, is a prevalent form of skeletal branching. Comb polymers typically have branches of approximately the same length (see BRANCHED POLYMERS; GRAFT COPOLYMERS).

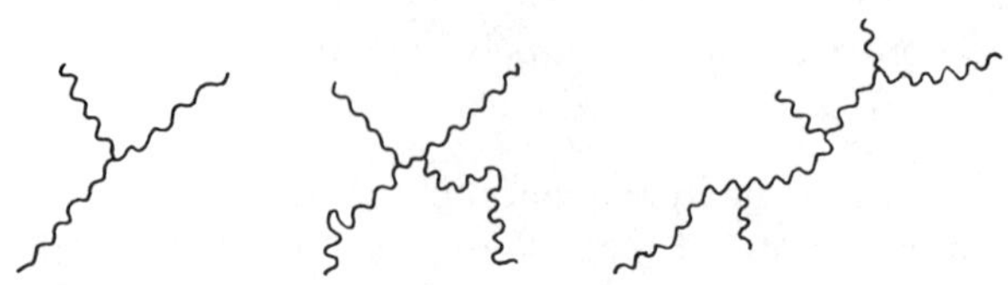

Y star comb

Cross-linked polymers have a network structure and are difficult to characterize because of insolubility, which often precludes solution characterization.

End Groups. Many chain-growth polymers have end groups arising from the initiator used in polymerization. For example, the diol used to initiate ring-opening reactions may lead to an end group that differs from the repeat-unit structure. Chain transfer also determines the nature of end groups. The importance of end groups as part of polymer microstructure is illustrated in telechelic polymers (qv) having functional groups. Linking reactions via the functional groups allow a variety of polymers from segmented blocks to cross-linked networks to be synthesized.

Average Molecular Structure. Polymer molecules are necessarily represented by average structures because of the various structural possibilities produced by repeat-unit arrangements, modes of addition, chirality, and differences in chain lengths. Diverse molecular systems may even require more than one type of average to describe the polymer microstructure adequately.

Molecular Weight

Polymer chains in any given sample may not all possess the same lengths; hence, a determination of molecular weight yields an average molecular weight for a particular polymer system. Average molecular weight determinations are based on the total numbers of molecules or the relative sizes of the polymer molecules. Thus there are three common averages, the number average, the weight average, and the z average. Gel-permeation chromatography or size-exclusion chromatography (sec) is a useful analytical technique because each of the molecular weight averages can be calculated from a single-column elution curve.

Polymer-chain Conformation in Amorphous and Semicrystalline Materials

Conformation, solid-state properties, and ultimate physical properties of polymers are intimately related to polymer microstructure. Polymers are useful materials because of the properties they contribute in a particular application. Polymers are extruded, drawn, blown, pressed, and molded to make articles. Each of these processes can be designed to enhance valuable physical properties, but the range of properties attainable in any particular application is related to specific and desirable polymer microstructures.

Polymer-chain conformation is related to both crystallinity and viscoelastic properties. The term "conformation" indicates polymer three-dimensional macromolecular structure, which is a manifestation of local conformations arising from torsions around main-chain single bonds.

A given polymer-chain segment may exist in either a crystalline or an amorphous state, depending on previous thermal history. Some order must develop within chain conformations for crystallization to occur. Some polymer-chain segments fold neatly into a crystalline lattice during crystallization, but other parts of the same chain may remain in an amorphous state leading to a solid-state structure containing both crystalline and amorphous regions, layered within a larger, often spherulitic, superstructure. Consequently, crystallizable polymers are often referred to as semicrystalline.

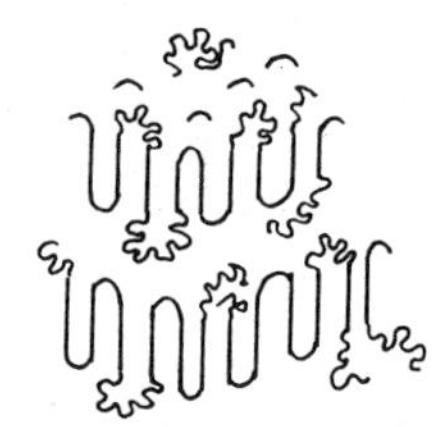

semicrystalline polymer

Amorphous polymers have largely disordered conformations, short co-repeat-unit sequence lengths, and irregular stereochemical sequences. By comparison, crystalline polymers have regularly ordered sequences, that is, either long uninterrupted repeat-unit sequence lengths, or regular co-repeat-unit or chiral sequence patterns.

From thermal history, which can be utilized to control crystallinity, the range of crystallinity attainable for a particular polymer or copolymer is related to the polymer microstructure. Any chain disruption in the form of a new unit or any change in chirality, mode of addition, repeat-unit inversion, branch, or unplanned functional group will reduce the innate capability of a polymer chain to crystallize. Consequently, new units are often intentionally introduced into linear polymers to control the available crystallinity.

Another factor affecting polymer crystallinity is molecular weight. High molecular weight chains are intertwined and less mobile and do not crystallize as rapidly or as fully as polymers of lower molecular weight under quiescent conditions.

Overall polymer dimensions also depend on whether the macromolecule is linear or long-chain branched. Significant changes in both physical properties and processing characteristics result from the presence of long-chain branches.

An understanding of polymer-melt characteristics and the relationship to structure is essential for optimizing processing conditions. Molecular weight, the molecular weight distribution, and topology are key factors determining the melt rheology of macromolecules; polymer composition is not nearly as important in establishing polymer-melt processing behavior as is the relative size of the molecules and degree of long-chain branching.

Measuring Polymer-chain Microstructure

The first step in determining polymer microstructure is to identify the structure of the repeat units. The repeat-unit identity must include not only the generic type but also the isomeric structure, which encompasses both the manner of addition of the original monomer unit and its chiral form as they are linked in the polymer chain. Spectroscopic techniques, such as infrared and nuclear magnetic resonance, have been valuable for establishing repeat-unit identities and isomeric forms.

Establishing the overall chain topology is difficult because the number of individual branch sites is extremely low compared to the total number of repeat units. Indirect methods based on properties associated with conformation, such as intrinsic viscosity and light-scattering techniques combined with size-exclusion chromatography, have been the most successful because long-chain branches have dramatic effects on conformation.

Sequence Determinations in Copolymers

Copolymers are of considerable interest commercially because performance can be controlled by the nature and amount of comonomer inclusions. For example, copolymers from the same comonomers may be either rubbers or semicrystalline plastics depending on the amount and the manner in which the comonomers are incorporated. Knowledge of copolymer-sequence structure is therefore important in establishing structure–property relationships.

Sequence structures can only be identified through a determination of various connecting-unit concentrations, through relationships with specific physical properties arising from certain chain architectures, or through knowledge of reactivity ratios or transition probabilities (see COPOLYMERIZATION).

The best approach to defining copolymer structure is to examine initially only the possible arrangements of the two units, defined simply as "A" and "B." Any copolymer consisting of repeat units that do not possess differences in chirality or modes of addition of the comonomers will be defined completely.

Chiral Sequences in Homopolymers

Monosubstituted vinyl homopolymers are chiral because a substituent appears on every other carbon atom of the polymer-chain backbone; in the strictest sense, these methine carbons are asymmetric carbon atoms. It has been recommended that asymmetric carbons be called "chirotopic" and nonasymmetric carbons "achirotopic," and that any isomeric molecular-bonding arrangements leading only to different senses of chirality are stereogenic, regardless of whether individual carbon atoms composing the structures are chirotopic or achirotopic. Accordingly, each methine carbon atom of a monosubstituted vinyl polymer is chirotopic because it is bonded to four different structural groups, but within long chains, the substituents represented by each end of the chain become equivalent and the methine carbon atoms are essentially achirotopic. Regardless of whether the methine carbons in a vinyl polymer are considered chirotopic or achirotopic, they are still stereogenic. An interior sequence of a typical vinyl chain illustrates this concept of vinyl polymer chirality:

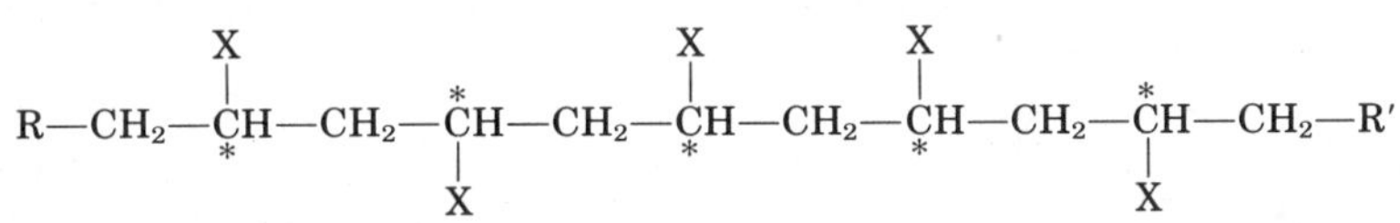

where X can be $—CH_3$, $—C_2H_5$, —Cl, —Br, —F, $—C_6H_5$, etc, and a * designates chirotopic carbon atoms. If R and R′ are identical, the only methine carbon that is not chirotopic in the strictest sense of the word is the center methine carbon.

The structural considerations discussed in this article relate primarily to intrachain compositional relationships. Average structures, even in the case of sequence distributions, have been used to describe complex and diverse chain structures. Compositional distributions describing interchain structures are equally important. Eventually, polymer microstructures will be routinely defined in terms of chain punctuation and sequence distributions as a function of both molecular weight and composition.

BIBLIOGRAPHY

F. A. Bovey, *Chain Structure and Conformation of Macromolecules,* Academic Press, New York, 1982, Chapt. 3.

J. C. Randall, *Polymer Sequence Determination: Carbon-13 NMR Method,* Academic Press, New York, 1977.

F. A. Bovey, *Polymer Conformation and Configuration,* Academic Press, New York, 1969.

J. C. Randall, "A Review of High Resolution Liquid ^{13}C NMR of Ethylene Based Polymers," *J. Macromol. Sci. Rev. Macromol. Chem.,* C **29**(2–3) 201–317 (1989).

J. C. RANDALL
Exxon Chemical Company

MICROTACTICITY.

See STEREOREGULAR POLYMERS.

MILDEW PROOFING.

See BIOCIDES.

MINING APPLICATIONS

Mineral Processing

A wide variety of hydrophilic polymers, both synthetic and natural, is used in the mineral processing industries principally as flocculants and flotation depressants, with the former having more importance. Other applications include use as dispersants, pelletizing binders, and ion exchangers. In 1982 flocculants accounted for more than half the total usage of polymers in mining applications. The trend toward beneficiation of leaner deposits, which require ever-finer grinding, coupled with environmental and energy factors, has increased the usage of polymeric flocculants (see also FLOCCULATION).

Table 1. U.S. Suppliers of Synthetic Polymers for Mining Applications

Company	Trade name	Type
Allied Colloids	Percol	polyacrylamides, polyacrylates
American Cyanamid	Superfloc	polyacrylamides, polyamines, polyammonium quaternary salts
Betz	Betz	polyacrylamides, polyammonium quaternary salts
Calgon	Calgon	polyacrylamides, polyamines, polyammonium quaternary salts
Calloway Chemical	Jayfloc	polyacrylamides
Dow Chemical	Separan	polyacrylamides
Henkel	Nopcofloc, Floxan	polyacrylamides, polyamines
Hi-Tek Polymers	Polyhall	polyacrylamides, polyamines, polyammonium quaternary salts
Nalco	Nalco	polyacrylamides, polyamines, polyammonium quaternary salts
POLYPURE	Clarifloc	polyacrylamides, polyamines, polyammonium quaternary salts
Union Carbide	Polyox, Ucarfloc	poly(ethylene oxide)

Organic Flocculants and Related Reagents

Organic flocculating agents are water-soluble polymers capable of aggregating solid particles suspended in an aqueous medium. They facilitate solid–liquid separation in filtration, centrifugation, and gravity settling. Aggregation of mineral suspensions by polymers is achieved either by partially neutralizing the surface charge on the particle, or by bridging the polymer between the particles. Other modes of action can also be involved. The mechanisms of flocculation are discussed in greater detail in the literature. Special aggregation techniques are selective flocculation and pellet flocculation; both find limited application at the present time.

Naturally occurring organic polymers, eg, starch and animal glue, were the only settling aids used in mineral processing until 1951, when synthetic water-soluble polymers were introduced to the industry, ushering in a new era of solid–liquid separation by polymeric chemical reagents. Synthetic polymers are classified as anionic, nonionic, and cationic according to the electrical charge of the molecule.

Of the natural products, starch is still the most widely used polymer in mineral processing, followed by guar gum and locust-bean gum. All three polysaccharides are essentially nonionic in nature (see also GUMS, INDUSTRIAL; POLYSACCHARIDES).

Synthetic polymeric flocculants and related mining reagents have simple structures compared to those of natural products. Most synthetic polymers used in mining are based on acrylic monomers. They may be neutral, like polyacrylamide; positively charged, such as copolymers of acrylamide and (meth)acryloxyethyltrimethylammonium salts; or negatively charged, eg, acrylic acid–acrylamide copolymers and hydrolyzed polyacrylamides. The last group (mol wt $1–20 \times 10^6$), comprises the most important group of synthetic polymeric flocculants in mineral processing.

The principal U.S. manufacturers of synthetic water-soluble flocculants are given in Table 1.

BIBLIOGRAPHY

P. Somasundaran and B. M. Moudgil eds., *Reagents in Mineral Technology,* Surfactant Science Series, Vol. 27, Marcel Dekker, Inc. New York, 1988.

F. Halverson and H. P. Panzer, "Flocculating Agents" in M. Grayson, ed., *Kirk-Othmer Encyclopedia of Chemical Technology,* Vol. 10, John Wiley & Sons, New York, 1980, pp. 489–523.

R. M. Goodman, "Dispersants" in M. Grayson, ed., *Kirk-Othmer Encyclopedia of Chemical Technology,* Vol. 7, John Wiley & Sons, Inc., 1979, pp. 833–848.

HOWARD I. HEITNER
TERENCE FOSTER
HANS P. PANZER
American Cyanamid Company

Chemical Grouts

Chemical grouts (Table 1) are injected into soil, rock, or concrete in order to alter the physical properties of the grouted mass. A chemical grout is defined as a true solution (free of suspended solid particles) that is used for grouting purposes. Grouting is used to restrict or reroute the flow of water through a formation or to improve its strength. Grouting is often applied in the construction or rehabilitation of dams, buildings, tunnels, shafts, and other structures, as well as in mines.

Inorganic Grouts

In the United States silicate-based grouts have been estimated to have 75–90% of the chemical grout market and almost all of the market for structural applications; a small but growing percentage is held by urethanes, not including rock-bolting usage. Acrylamides and acrylates account for the remainder of the market, primarily in seepage control.

Organic Grouts

Organic reactants can be used as catalysts for sodium silicate; esters are preferred in Europe and glyoxal in Japan. Acrylamide grouts, AM-9, introduced in the 1950s, contained acrylamide and methylenebisacrylamide, a cross-linking agent. Marketed by the American Cyanamid Co., production was discontinued in 1978 because of possible health problems associated with the monomers. Currently, other acrylamide-based grouts are commercially available: Rocagil from France, and Sumisoil and Nitto SS from Japan.

Acrylate grouts, which first appeared in the early 1980s, are marketed by Geochem (AC-400) and Celtite as concentrated solutions and contain the same catalyst system as the acrylamides. Acrylate and acrylamide grouts have similar properties.

Table 1. Chemical Grouts

Material	Corrosivity or toxicity	Viscosity	Strength
Silicates			
Joosten process	low	high	high
Siroc	medium	medium	medium–high
silicate–bicarbonate	low	medium	low
Lignosulfites			
terra firma	high	medium	low
blox-all	high	medium	low
Phenoplasts			
terranier	medium	medium	low
geoseal	medium	medium	low
Aminoplasts			
Herculox	medium	medium	high
Cyanaloc	medium	medium	high
Acrylamides			
AV-100	high	low	low
Rocagel BT	high	low	low
Nitto-SS	high	low	low
Polyacrylamide			
Injectite 80	low	high	low
Acrylate			
AC-400	low	low	low
Polyurethane			
CR-250	high	high	high

BIBLIOGRAPHY

Bibliography on Chemical Grouting, Vol. 92, No. SMI, Soil Mechanics and Foundations Division of American Society of Civil Engineers, New York, 1966, pp. 39–66.

G. W. Clough and co-workers, *Development of Design Procedures for Stabilized Soil Support Systems for Soft Ground Tunneling: Vol. 2, Preliminary Results, U.S. D.O.T., TST 77-74,* Office of University Research of U.S. Dept. of Transportation, Washington, D.C., 1978.

Evaluation of Available Materials, U.S. D.O.T., FHWA-RD-77-50 and 54, Federal Highway Administration, Washington, D.C., 1977.

R. H. Karol, *Chemical Grouting,* Marcel Dekker, Inc., New York, 1983.

REUBEN H. KAROL
Rutgers University

MISCIBILITY

In the search for new polymeric materials, the mixing together of two or more polymers has attracted much interest as a means of arriving at new property combinations without having to synthesize novel structures. A major feature encountered when two polymers are mixed is that in the majority of combinations the components tend to phase-separate to form heterogeneous mixtures that do not exhibit enhanced properties. Only in a limited number of cases do amorphous polymers blend to form one-phase mixtures. The term miscible will be used here to describe a mixture of two or more polymers that form a single-phase (solid or liquid) solution. Whereas this appears to be a simple and clear-cut definition, which should lend itself to a precise thermodynamic description, where polymers are concerned, the existence of a single phase may be a relative rather than an absolute concept and can depend on the method used in its identification.

Criteria for Miscibility

Several methods can be used to determine miscibility in polymer blends (qv) and it is useful to recognize the strengths and limitations of each of the ones that follow.

Optical Clarity. When a film is prepared from a miscible blend, it is usually found to be optically clear. This technique is not always reliable and transparent films can be prepared from immiscible systems if very thin films are examined or if the two polymers have similar or equal refractive indexes (see OPTICAL PROPERTIES).

Glass-transition Temperature. Perhaps the most commonly used and widely accepted criterion for miscibility is the existence of a single glass-transition temperature, T_g, for a miscible blend.

Nuclear Magnetic Resonance. For a more refined measurement of miscibility and assurance of mixing at the segmental level, nmr or excimer fluorescence may be the answer. This technique may be too sensitive for most practical purposes (see NUCLEAR MAGNETIC RESONANCE).

Excimer Fluorescence. An excimer is an excited molecular complex formed between two identical aromatic rings, one of which is in a singlet excited state. By attaching an active unit to one polymer chain in low concentration, miscibility with another polymer can be assessed in the following way. If I_M is the emission intensity from an isolated ring and I_D that from the excimer, then a measure of the ratio (I_D/I_M) indicates the extent of miscibility. Thus $(I_D/I_M) \rightarrow 0$ shows that the two polymers are well dispersed and intermixed, but when the system is immiscible the ratio becomes larger, indicating that the polymer containing the active ring is close to similar structures, thereby allowing the excimer to form.

Basic Thermodynamics

The miscibility of a mixture is determined by the Gibbs free energy of mixing ΔG_M, which is related to the entropic ΔS_M

and enthalpic ΔH_M components through the simple relation

$$\Delta G_M = \Delta H_M - T\,\Delta S_M \qquad (1)$$

If the system is to be miscible then ΔG_M should be negative. However, this criterion is not totally adequate as can be illustrated by examining the behavior of ΔG_M as a function of mixture composition. First, equation 1 should be recast in a more amenable form following the treatment developed by Flory and Huggins for polymer–solvent mixtures, which is based on a rigid lattice model (see SOLUTION PROPERTIES). Although this theory is inadequate in many respects, it has proved remarkably versatile when used to describe phase equilibria in polymer solutions. In its simplest form it can be applied also to polymer–polymer mixtures by introducing the concept of a reference segment volume V_R that can be approximated by the smallest polymer repeat unit. For practical purposes this is often taken to be ~100 cm³/mol.

For such mixtures the entropy term in equation 1 is expanded into a combinatorial entropy of mixing term using the volume fraction ϕ_i as the concentration term:

$$\Delta S_M = -R\left(\frac{V}{V_R}\right)\left\{\frac{\phi_1}{m_1}\ln\phi_1 + \frac{\phi_2}{m_2}\ln\phi_2\right\} \qquad (2)$$

where m_i is the number of segments per chain molecule (ie, the degree of polymerization) of the ith component relative to the reference volume.

The enthalpy of mixing can be expressed as a dimensionless interaction parameter per segment of polymer, χ, based on the van Laar form:

$$\Delta H_M = RT\left(\frac{V}{V_R}\right)\chi_{ij}\phi_i\phi_j \qquad (3)$$

This yields the final equation for the free energy of mixing

$$\Delta G_M = \frac{RTV}{V_R}\left\{\frac{\phi_1}{m_1}\ln\phi_1 + \frac{\phi_2}{m_2}\ln\phi_2 + \chi_{12}\phi_1\phi_2\right\} \qquad (4)$$

where V is the total molar volume of the two components.

The free energy change ΔG_M can now be plotted as a function of ϕ_i for different χ_{ij} (or m_i) values and, for a pair of polymers of equal molecular weight, a series of curves is obtained that can be represented schematically as shown in Figure 1. For curve 1, ΔG_M is always negative and the system is miscible over all compositions, whereas curve 4 represents an immiscible system in which ΔG_M is always positive. In the situation represented by curve 3, again ΔG_M is always negative but as the composition changes there is a reversal of curvature in the middle of the composition range. This means that when the mixture is at point A it can lower its free energy by splitting into two phases of compositions B and B′ where there are free energy minima. This leads to a more rigorous thermodynamic definition of a miscible system, which at constant temperature and pressure is

$$\left[\frac{\partial^2\,\Delta G_M}{\partial\phi_2^2}\right]_{T,P} > 0 \qquad (5)$$

Thus a mixture is stable in one phase, only as long as this condition prevails.

Phase Equilibria

It is instructive to examine first the phase equilibria detected in mixtures of low molecular weight compounds and in solutions of polymers dissolved in small molecule liquids. These are represented schematically in Figure 2 where the hatched areas represent two-phase regions (see also PHASE TRANSITIONS).

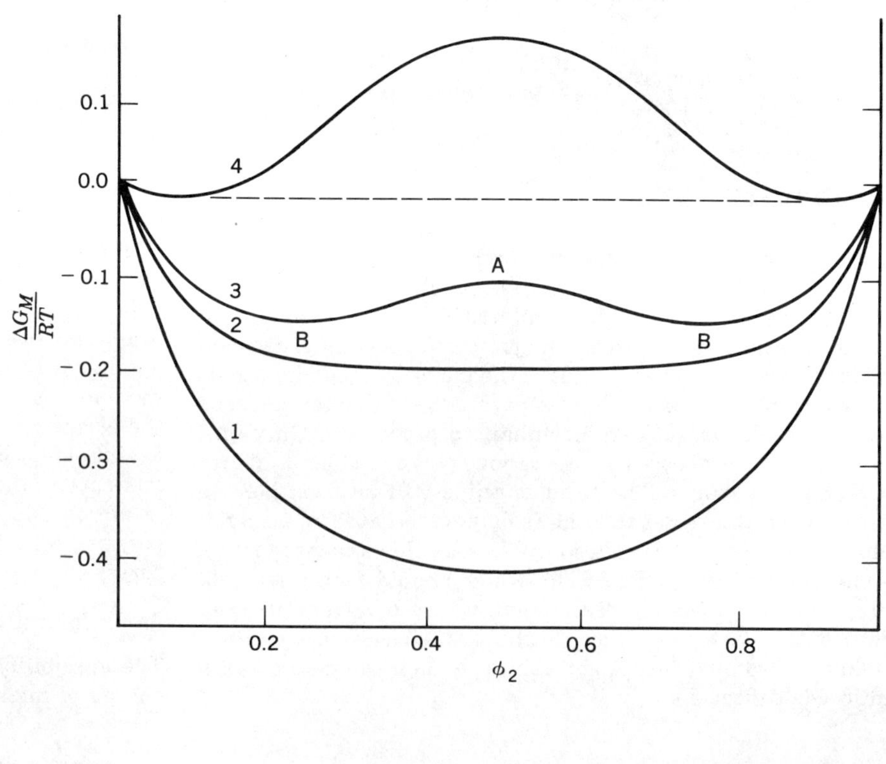

Figure 1. Free energy change ΔG plotted as a function of mixture composition for two molecular species of equal molecular weight. The curves are derived from equation 4 for different values of the interaction parameter and represent various degrees of miscibility as described in the text.

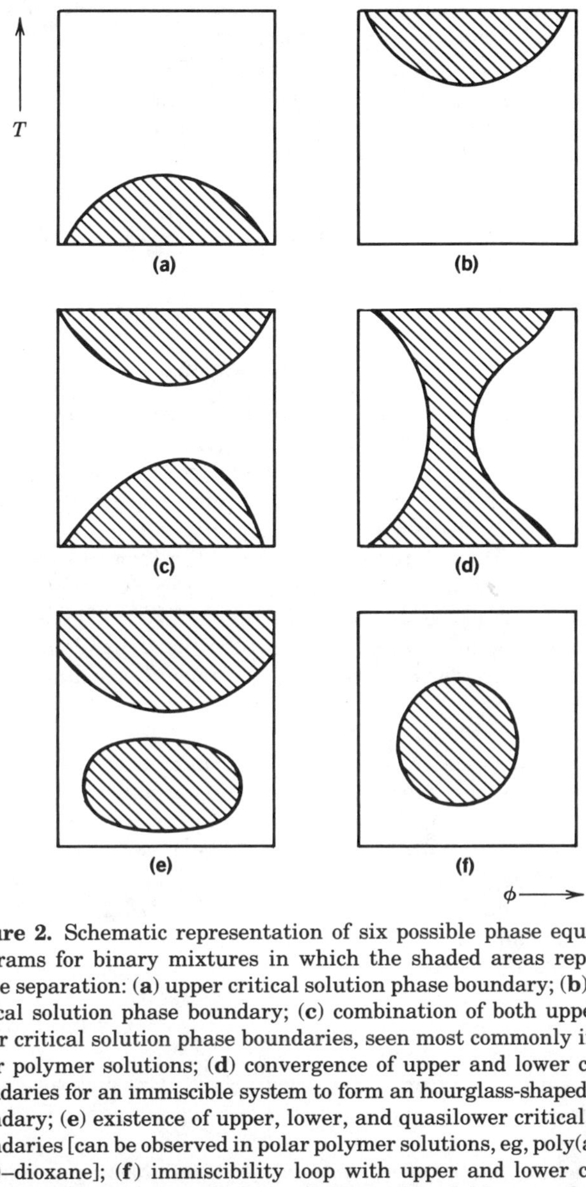

Figure 2. Schematic representation of six possible phase equilibria diagrams for binary mixtures in which the shaded areas represent phase separation: (**a**) upper critical solution phase boundary; (**b**) lower critical solution phase boundary; (**c**) combination of both upper and lower critical solution phase boundaries, seen most commonly in nonpolar polymer solutions; (**d**) convergence of upper and lower critical boundaries for an immiscible system to form an hourglass-shaped phase boundary; (**e**) existence of upper, lower, and quasilower critical phase boundaries [can be observed in polar polymer solutions, eg, poly(acrylic acid)–dioxane]; (**f**) immiscibility loop with upper and lower critical phase boundaries inverted (cf part **c**). Examples of this type of system are nicotine–water and poly(vinyl alcohol)–water.

On cooling a single-phase polymer solution, often a temperature is reached at which phase separation is observed; Figure 2**a** represents a system in which there is an upper critical solution temperature (UCST). The UCST represents the maximum temperature (and composition) at which the system begins to separate into two phases. This will depend on the solution composition; if the phase separation temperature T_c is measured for several concentrations a cloud point curve can be constructed that defines the limits of miscibility in the system. This cloud point curve, or binodal curve, is actually the locus of the points of the common tangents to the free energy composition curves measured at different temperatures that are extrapolated to the temperature–composition plane of the phase diagram. The curve then defines the limits of metastability in the mixture.

In miscible polymer–polymer blends, a lower critical phase boundary can often be detected when the mixture is heated and the polymer components phase separate. Thus the phase behavior depicted in Figure 2**c** is typical of many polymer blends.

Quasi-binary Systems with $\Delta H_M \geq 0$

The majority of polymer pairs exhibit an endothermic heat of mixing when blended together, which does not favor the formation of a single amorphous phase. Values of ΔH_M, which indicate whether pairs may be miscible, can be estimated for nonpolar components by using the Hildebrand approach to regular solutions, which introduces the concept of the solubility parameter δ. This is related to the cohesive energy density defined as the energy of vaporization ΔE per unit volume V of material and is expressed as

$$\delta = \left(\frac{\Delta E}{V}\right)^{1/2} \tag{6}$$

in units of $(J/m^3)^{1/2}$ $[(cal/cm^3)^{1/2}]$.

By assigning to each polymer a value of δ_i, the heat of mixing is derived from the equality

$$\Delta H_M = V(\delta_1 - \delta_2)^2\phi_1\phi_2 \tag{7}$$

where ϕ_i is the component volume fraction and for a polymer–polymer mixture

$$\chi_{12} = \frac{V_R}{RT}(\delta_1 - \delta_2)^2 \tag{8}$$

Indirect experimental methods must be used to calculate δ for a polymer; alternatively, there are several group contribution methods that can be applied. Some contribution to miscibility can come from the combinatorial entropy of mixing (as reflected in the component molecular weight) but mixtures involving high molecular weights can only be obtained if ΔH_M is very small or negative.

Systems with $\Delta H_M < 0$

Polystyrene–poly(α-methylstyrene) mixtures highlight the fact that even when the components of the mixture are chemically similar ($\delta_1 \approx \delta_2$), miscible systems either cannot be obtained or are limited to relatively low molecular weights when $\Delta H_M > 0$. Rather than seeking chemically similar pairs, more success is likely if chemically dissimilar combinations are examined, particularly if there is the possibility of specific interactions occurring between the components that will result in an exothermic heat of mixing. A number of these interactions have been identified and found to enhance polymer miscibility, eg, hydrogen bonding, ionic and dipolar interactions, π-electron interactions, and charge-transfer complexes.

Homopolymer–Copolymer Blends

It has recently been reported that miscible blends can be obtained in mixtures that do not appear to exhibit obvious specific interactions. These usually involve blends where one or more of the components is a statistical copolymer and miscibility is restricted to mixtures falling within well-defined limits of copolymer composition (the miscibility window).

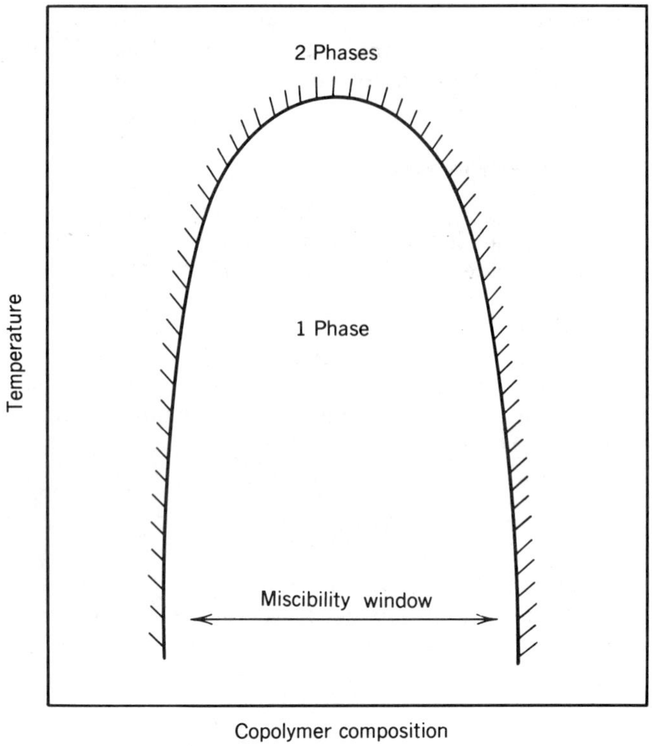

Figure 3. A "miscibility window" in a polymer blend.

Among the best documented examples are miscible blends of poly(*o*-chlorostyrene-*co*-*p*-chlorostyrene) with both poly(2,6-dimethylphenylene oxide) (PPO) and polystyrene, and poly(methyl methacrylate) (PMMA) with poly(styrene-*co*-acrylonitrile) (SAN).

The "miscibility window" can be represented on a temperature–composition plot (Fig. 3), but it is also possible to calculate miscibility regions in the composition plane if χ_{ij} parameters are known for each of the segment–segment interactions in the system. This can most easily be achieved by adopting a modified Flory-Huggins approach to the problem to calculate the overall interaction parameter for a blend (χ_{blend}), which can be expressed as a summation of each segmental χ_{ij} and the appropriate composition term for the copolymer w_{ij} where $0 \leq w_{ij} \leq 1$, and

$$\chi_{\text{blend}} = \sum_{ij} w_{ij}\chi_{ij} \quad (9)$$

In general, for a binary mixture of two statistical copolymers (A_xB_{1-x}) and (C_yD_{1-y}), where x and y define the copolymer compositions, the interaction parameter for the blend is given by

$$\chi_{\text{blend}} = xy\chi_{AC} + y(1-x)\chi_{BC} + x(1-y)\chi_{AD} + (1-x)(1-y)\chi_{BD} - x(1-x)\chi_{AB} - y(1-y)\chi_{CD} \quad (10)$$

BIBLIOGRAPHY

D. R. Paul and S. Newman, eds., *Polymer Blends,* Academic Press, New York, 1978.

O. Olabisi, L. M. Robeson, and M. T. Shaw, *Polymer-Polymer Miscibility,* Academic Press, New York, 1969.

D. Patterson, *Rubber Chem. Technol.* **40,** 1 (1967).

J. M. G. Cowie in L. A. Kleintjens and P. J. Lemstra, eds., *Integration of Fundamental Polymer Science and Technology,* Vol. 1, p. 138, Elsevier Appl. Science Publishers, 1986.

J. M. G. Cowie
Heriot-Watt University

MIXING

The Principles of Laminar Fluid–Fluid Mixing

In polymer processing, mixing significantly affects material properties, processibility, and cost. Polymers are blended with other polymers to produce selective barriers to air, moisture, or certain light frequencies. Various reinforcing materials are incorporated to increase moduli or impact toughness. Some processes require chain extenders or stabilizers to control a reaction. Additives are included to improve flame retardance or corrosive resistance. Thermal homogenization is needed for flow control of materials with temperature-dependent viscosities and to prevent hot spots where thermal degradation can occur.

In polymer processing mixing is often performed in a continuous mode such as extrusion. The materials are fed through a hopper into the heated barrel of an extruder within which one or more turning screws push the materials through. They are compressed, melted, mixed, and driven through a die.

The principles of melting and pumping in polymer processing are reasonably well understood, permitting the rational design of efficient equipment such as extruders. However, no quantitative theoretical treatment can accurately predict the efficiency of a particular processor. The traditional solution to mixing problems has been to increase the energy input to the process (eg, by lengthening an extruder). With the increasing importance of mixing to polymer processing, greater understanding of mixing is needed to design more efficient processing equipment.

Characterization of Mixing and Mixtures. Mixing is the intermingling by mechanical action of two or more components. Mixing involves two distinct physical phenomena, dispersion and blending. When one or more of the components has a cohesive nature, such as agglomerated particles or droplets held together by interfacial tension, mechanical stress must be applied to reduce the scale of the particles or droplets. This process is called intensive mixing or dispersion, and is governed by the history of fluid mechanical stresses in a mixture. In the absence of cohesive barriers to mixing, as in the thermal homogenization of a polymer melt or in the mixing of thermodynamically miscible fluids, mixing is governed by the history of deformation imparted to the fluid, ie, the strain. This process is called extensive mixing, distributive mixing, or blending.

Mixing

Direct measurements of mixing include two statistical measures of mixing quality, the scale and intensity of segregation that serve as a point of reference in many studies.

Mixing in polymer melt processing is primarily the reduction of scales of segregation between fluids. The scale of a polymer mixture is typically described by an average striation thickness or the amount of the interfacial area. Many indexes have been proposed in recent years to quantify the mixed state; most of these are related to interfacial area or striation thickness distributions.

Decoupling Flow and Mixing. Most mixing problems of practical interest involve systems in which flow and mixing are coupled, but analysis of the simpler decoupled idealizations give valuable insight. The following analysis of mixing kinematics is based on the assumption that flow and mixing are decoupled. Important mechanisms of mixing in laminar flow are present in systems in which the component distribution does not affect fluid mechanics. Much has been learned about the importance of mixing of elongational flows and of interface orientation through the study of decoupled systems.

The Kinematics of Interfacial Growth. The first fundamental study of mixing in laminar flow was based on the interfacial area growth in simple shear. The deformation of an arbitrarily oriented interfacial element within a fluid undergoing simple shear was studied. The growth of the interfacial area was found to be a function of the magnitude of the shear and the initial orientation of the element.

This relationship has been applied frequently to laminar mixing for a large unidirectional shear. For large shear, the expression reduces to

$$\frac{A_f}{A_i} = S \cos \alpha \tag{1}$$

where A_i and A_f are the initial and final interfacial areas, S is the magnitude of the shear strain, and α is the angle defining the initial orientation of the element relative to the shearing plane.

Interfacial growth in shear is important because for many purposes simple shear has been a good approximation to actual melt processing flows; it illustrates the roles of strain and orientation in mixing.

Mixing as a Function of Strain

Mixing depends on the general strain tensor that can be expressed as principal elongation ratio and principal directions. Elongation ratios can be easily visualized as the principal axes of the ellipsoid resulting from the deformation of a unit sphere.

BIBLIOGRAPHY

H. Aref, *J. Fluid Mech.* **143** (1984).

F. Mokhtarian and L. Erwin, *Polym. Eng. Sci.* **23**(2), 49 (1983).

J. Y. Oldshue, *Fluid Mixing Technology,* McGraw-Hill Book Co., New York, 1983.

LEWIS ERWIN
Kraft Company

MODACRYLIC FIBERS.

See ACRYLIC FIBERS.

MODELING, POLYMER.

See POLYMER MODELING.

MODEL NETWORKS.

See ELASTICITY; NETWORKS.

MODEL POLYMERS

Model polymers, also referred to as tailor-made macromolecules, were originally defined as polymer species in which the linear structure of the molecules and their average size are defined unambiguously, and in which the fluctuations in molecular weight (and in composition) are very low. A broader definition of model polymers has progressively emerged, as macromolecules exhibiting more complex architectures attracted increasing interest, because of their specific properties and potential applications. Examples of such polymeric species include: ω-functional polymers, branched macromolecules, block copolymers, graft copolymers, and polymeric networks with elastically effective chains of known average length (Fig. 1). In all these cases, emphasis is placed on the accurate control of the molecular structure of the species formed, rather than on low polydispersity in molecular weight or low compositional heterogeneity. See also BLOCK COPOLYMERS; BRANCHED POLYMERS; DENDRITIC POLYMERS; GRAFT COPOLYMERS; MACROCYCLIC POLYMERS; MACROMERS.

The need for linear model polymers originates from the development of specific characterization techniques, especially of those requiring calibration, such as viscometry, or size-exclusion chromatography (sec or gpc). The importance of model polymers as calibration species was enhanced further when the relations between molecular structure and macroscopic properties had become one of the major issues of polymer science and engineering.

The polymerization methods commonly used in industry or in the laboratory to synthesize linear polymers or copolymers generally do not involve control of molecular structure. The average molecular weights cannot be chosen beforehand, molecular weight distributions are usually rather broad, and unwanted branching sometimes occurs. Fluctuations in composition must be taken into account if two monomers are involved. More elaborate structures, such as controlled branching, chemical linking of homopolymeric blocks, or functionalization at chain ends, are generally inaccessible by the classic polymerization procedures.

For these reasons, specific methods have been designed for the preparation of model polymers, allowing proper control of the parameters governing chain growth. These methods, sometimes called macromolecular engineering, have been developed to synthesize structurally controlled polymers, especially those not accessible otherwise.

Synthesis

The chief requirement for polymerizations to be applicable to model polymer synthesis is the survival of the active site at the chain end, throughout the process. This condition implies the absence of spontaneous termination and transfer, so that the reaction scheme merely comprises initiation and propagation:

Initiation $\quad R^* + M \longrightarrow R\text{—}M^*$

Propagation $\quad R\text{—}M_{i-1} - M^* + M \longrightarrow R\text{—}M_i - M^*$

where M is the monomer, * the active site, and R the radical originating from the initiator.

If all sites are formed at the same time and remain active until total conversion has been attained, the number-average

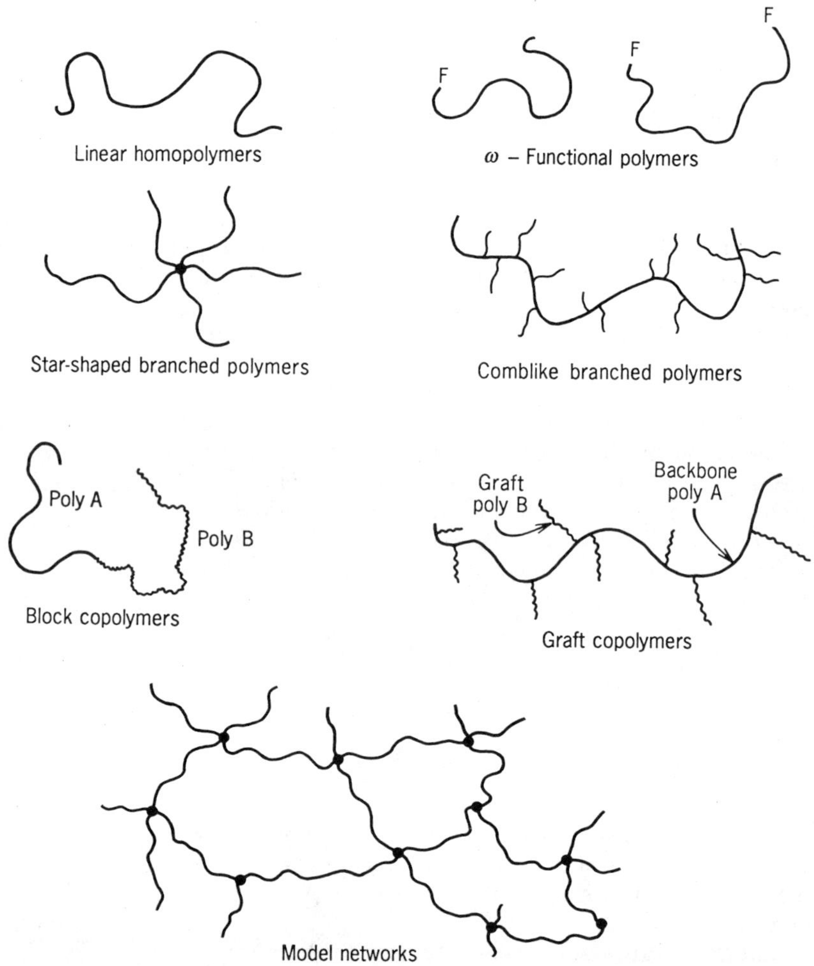

Figure 1. Model polymers.

degree of polymerization of the sample is given by the mole ratio of monomer to initiator. Under such conditions, the molecular-weight distribution is very narrow (Poisson-like). The theoretical polydispersity index tends toward unity as the molecular weight increases. In the absence of transfer and termination reactions, the polymer molecules formed carry active sites at the chain end, which can react to functionalize chain ends by induced deactivation of the sites; in coupling with an appropriate bifunctional reagent, added in stoichiometric amount; as an initiator for the polymerization of an additional amount of monomer, whereby the already existing chains are extended; and as a possible initiator for the polymerization of another monomer, leading to the formation of a block copolymer. Such polymerizations are referred to as living because of the survival of the active sites. They have been studied in great detail (see LIVING POLYMER SYSTEMS).

BIBLIOGRAPHY

P. Rempp, E. Franta, and J. E. Herz, *Adv. Polym. Sci.,* **86,** 145 (1988).

P. Rempp, E. Franta, and J. Herz in J. E. McGrath, ed., *ACS Symposium Series* **166,** 59 (1981).

J. P. Kennedy and E. Marechal, *Carbocationic Polymerization,* John Wiley & Sons, Inc., New York, 1982.

D. Y. Sogah, W. R. Hertler, O. W. Webster, and G. M. Cohen, *Macromolecules* **20,** 1473 (1987).

PAUL REMPP
JEAN E. HERZ
Institut Charles Sadron

MODIFIERS. See CHAIN TRANSFER; EMULSION POLYMERIZATION.

MOLDING. See MOLDS.

MOLDING, BAG. See REINFORCED PLASTICS.

MOLD-RELEASE AGENTS. See RELEASE AGENTS.

MOLDS

Molds are the core of the complex system required to produce plastic molded products. The quality and accuracy of the prod-

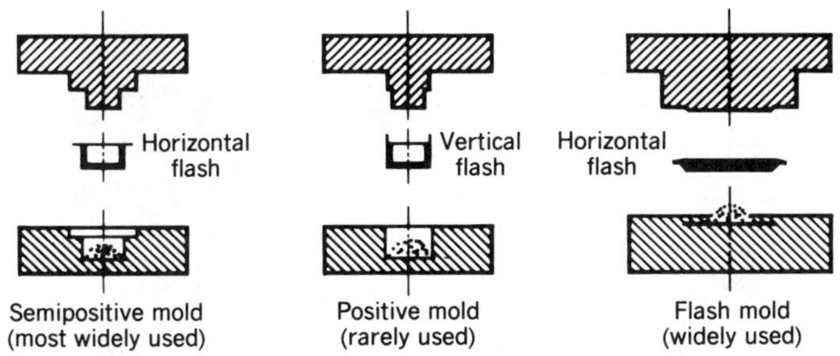

Figure 1. The three basic types of compression molds.

uct as well as the productivity of the system depend primarily on the mold construction. Although there have been few changes in the principle of molding plastics since its inception, the design of molds has matured, and has become a recognized branch of engineering.

Loading the Cavity Space

Compression Molding. The plastic is loaded into the open mold that must part in a horizontal plane requiring a vertical machine. The raw plastic occupies a greater volume than the cured material; as the mold is closed, the plastic is compressed and fills the cavity under pressure. A slight excess of material is necessary to ensure complete filling. This excess is squeezed out at the parting line in the form of so-called flash that is discarded.

The three basic positive, semipositive, and flash molds are shown in Figure 1. Most molds used are flash molds (for records, dinnerware, and other flat articles) or semipositive molds (for technical products).

Injection Molding. The plastic is injected into the cavity space of the closed mold. The plastic is made soft (plastic) and flowable in the injection unit, today almost exclusively an extruder, and forced into the cavity by the action of the extruder itself (screw injection or one-stage injection) or by using an injection cylinder (or pot) filled from the extruder and used to inject the plastic, so-called two-stage injection. Thermosets are heated to just below the curing temperature before injection into a heated mold where the plastic is cured. Thermoplastics are heated to increase flowability for easy and fast injection into a (relatively) cold mold where the plastic solidifies. The machines are mostly horizontal, with a vertical parting plane; this permits easier access to the mold and dropping of the finished products. Today, most molding operations are based on injection molding (Fig. 2).

Injection–Compression Molding. This is a combination of the two previous methods (Fig. 3). Injection–compression molding is used occasionally for large, flat products that require excellent surface definition, such as record grooves.

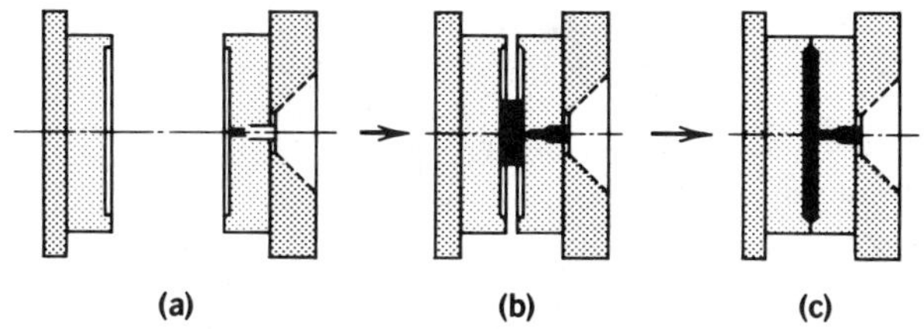

Figure 3. Injection compression molding. (**a**) Mold open; (**b**) mold almost closed, plastic injected; (**c**) mold closed, plastic compressed.

Materials

Thermosets. Thermosets undergo a chemical change when cured in the mold, similar to cement, which becomes concrete after pouring; the cured material cannot be reused as molding material.

Thermoplastics. Thermoplastics do not change significantly during the transformation from powder or pellets to finished products. The material is heated in the injection unit to make it flowable; after injection it is cooled in the cavities to make it rigid again before ejection.

Mold Types

Prototype Molds. For engineering evaluation, such as performance of the product in its intended use, a prototype mold must be designed with the essential features planned for the production mold, such as runners, gating and gate location, venting, and cooling. With little additional cost for ejection, such a mold, even though designed for prototyping, should perform in a manner similar to the production mold, but with one or two cavities only.

Production Molds. Production molds differ generally only by the method of product and runner ejection. A semiautomatic (SA) mold requires an operator to remove the product from the mold and to ensure that the mold is clean before initiating the next cycle. Today, however, almost all molds are designed so that the assistance of an operator is not required to remove the product. Better ejection designs and mold finishes eliminate the need for mold spray, and automatic lubrication of the mold can be connected with the lubrication system of the machine.

Characteristics and Mold Elements

Ducting the Plastic into the Cavities. In compression molding no ducting is necessary; the plastic is placed into the open cavities, before the mold closes.

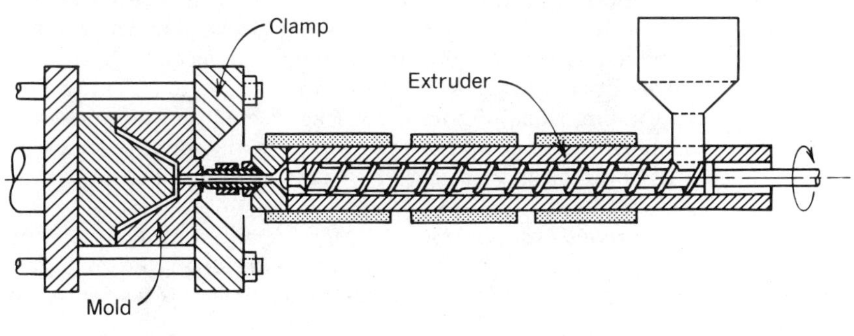

Figure 2. Injection molding.

Transfer Molding. For thermosets, transfer molding is employed. For single-cavity molds, a transfer cylinder or pot is located above the cavity; the runner is a short drop or sprue from the pot into the cavity. The plastic is placed into the pot, and after the mold has closed, the piston descends and pushes the plastic into the cavity. The cull, a thin, molded disk at the end of the piston, must be keyed to the piston so that it can be manually removed after injection.

Injection Molding. There are several methods of ducting the plastic from the machine nozzle to the cavities; they evolved from the previously described transfer methods. However, because the plastic in the runners can generally be reused, the waste problem is never serious; the system of runners that are cooled and ejected with the products (cold runners) allows a simple mold construction, and many molds in operation today are of this design; sprue gating, two-plate, and three-plate systems are employed.

Contamination of Material. Any small, foreign particle, such as paper scraps, tobacco, metal chips, and others, that may have entered the plastic, usually associated with poor housekeeping, may stop at the gate and prevent the cavity from filling. However, the loss of production is limited to this one shot. The impurity is molded into the cold runner and ejected with it. The next cycle finds the gate open. Thus, both two-plate and three-plate molds are not affected by contamination.

Runnerless Molding. This is a misnomer. The runners in runnerless molding remain fluid as the mold opens to permit product ejection. The runners freeze or cure only after the machine is shut down. Runnerless molding saves the cost of handling and regrinding scrap, thereby avoiding a common source of contamination of the plastic, and reduces noise and dust; cycle time is shortened by eliminating the often thick sprues from machine nozzle to runner; the need is eliminated to reuse an (often high) percentage of the plastic that may affect the quality of the product; injection capacity of the machine is increased because there is no need to plastify and inject that portion of the injected plastic (shot) that is not used to make products, and that is recycled or scrapped; and the energy is saved to first plasticize and then cool this extra amount of plastic.

For runnerless thermoplastic molding, two different systems are available: insulated runners that permit runnerless molding once the mold is running on cycle, but that present some difficulties in start-up, and true hotrunners, with heated runners so designed that the mold can be started up easily and safely, provided certain rules are observed.

Hotrunner Systems. In the earlier, European system, a hotrunner plate (manifold) is heated with cartridge heaters, sandwiched between a cavity plate and a backing plate (or machine platen). It depends on thick insulating plates to insulate the hot manifold from the rest of the mold and the machine. This system is well suited for small molds but not successful for larger ones.

In the American system (Fig. 4), the heated system is separated from the cool surroundings by insulating them with air space, having minimal surface contact between cold and hot masses.

Contamination. Contamination is probably the main reason preventing many molders from using hotrunners. Because the gates are small, any foreign matter that cannot pass through them blocks the passage of subsequent injections, with the loss of output from the blocked cavity. Unless the hotrunner system is so designed that the gate and nozzle are easily accessible within the machine, the cleaning of a plugged gate may interrupt production. Even in systems where the cleaning of the gate within the machine is easy, the work stoppage may last half an hour or more, with resulting loss of production.

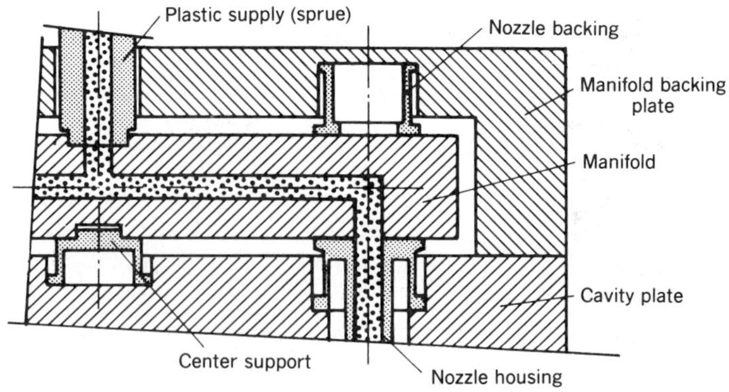

Figure 4. American hotrunner system.

Color Changes. Cold runner molds do not present a problem because the mold is cleaned after every cycle. With insulated runner molds, the runner can be removed after the mold has cooled down, so that the mold is clean for a new color run.

Color can also be changed without stopping the machine: the hot core of the insulated runner is replaced after about 10 shots with the new color material, while the surrounding, colder material of the old color remains without affecting the new product.

Runnerless Molding of Thermosets (Warm Runners). This technology is based on applying the hotrunner method for thermoplastics to the injection molding of thermosets. The extruder heats the plastic to below its curing temperature. It is held as close as possible to this temperature in the duct from machine nozzle to gates. While the plastic within the hot mold cavities is cured (hardened), the warm runners must remain viscous without curing to permit the next injection. The principal mold-design problems are balancing of temperatures, insulating hot cavities from warm runners, and different expansion of mold plates because of the different temperatures.

Balancing of Runners. Balancing refers to a layout of runners to ensure that the injection pressure in a multicavity mold is the same at all cavities. Balancing is important to reduce residence time of plastic in hotrunners and to warm runners to prevent degradation of the plastic.

Product Ejection. With fully automatic molds, 100% reliability of product ejection from the mold is required. No operator is present to pull the product off the core or to ensure that the mold is clean before initiating the next cycle. The basic methods have not changed for many years, such as ejector pins and sleeves, stripper plates (or rings), and air ejection. Some molds use combinations of these, such as ejector pins and sleeves, strippers and pins, and strippers or pins and air. Ejection may be in one motion or staggered (stepwise), eg, pins first, then stripper, etc.

Controlled Product Removal. More recently, random ejection has been used less and less. Molds are designed to incorporate ejection methods that ensure the product never loses orientation.

Venting. The air inside the mold must be released as the plastic enters the cavity space. Improvements in mold designs and the increased use of strippers or air ejection, combined with rapid injection, made the location and shape of vents more important. Today, vent design is an important part of mold engineering, and vents are machined at the time the mold is built. The designer analyzes the probable flow path and provides as many vents as possible for fast, uniform filling of the cavity.

Mold Cooling and Heating. A processing temperature is needed high enough for easy filling, but cold enough for rapid cooling. Mold cooling occurs by normal heat transfer. Cooling efficiency depends on factors such as the temperature difference ΔT_1 between coolant entering and leaving, the coolant volume passing through the channels, its speed, the Reynolds number indicating laminar (poor) or turbulent (good) flow, and the chemistry of the coolant. By feeding all available data into the computer and varying the parameters under control of the designer, such as diameters, location, and lengths of bores, number of cooling circuits, distances from molding surfaces, and mold materials, the engineer can selcct the most advantageous design.

Alignment of Mold Halves. To ensure that the mold halves met without lateral shifting, leader pins and bushings were used almost universally. They are still used extensively, but where high accuracy of alignment is required, leader pins are assisted or replaced by taper locks.

Specialty Molds

Specialty molds include stack molds, insert molds, two-color molds, injection-blow molds, and co-injection molds.

BIBLIOGRAPHY

G. Menges and P. Mohren, *Anleitung für den Bau von Spritzgiess-werkzeugen,* C. Hanser Verlag, München, 1974. Also available in English.

I. I. Rubin, *Injection Molding Theory and Practice,* John Wiley & Sons, Inc., New York, 1972.

K. Stoeckhert, *Moldmaking Handbook for the Plastics Engineer,* C. Hanser Publishers, MacMillan Publishing Co., New York, 1983.

R. H. Bebb, *Plastics Mould Design, Vol. 1 (Compression and Transfer Molding),* The Plastics Institute, Iliffe Books Ltd., London, UK, 1962.

HERBERT REES
Consultant

MOLECULAR DYNAMICS.

See MACROMOLECULAR DYNAMICS.

MOLECULAR GRAPHICS

Traditionally, chemists have used the well-known Kendrew and Corey, Pauling, and Koltun (CPK) wire and stick models to visualize the structure of molecules. It was extremely time consuming to build such models and with the advent of computer-aided molecular graphics (CAMG) systems, research utilizing these models has increased. The field of molecular graphics as applied to polymer science has been developed since the late 1970s. Development in CAMG has mainly been in the areas of hardware, software, and techniques in using these.

Hardware

The CAMG systems developed involve the user as an integral part; a typical system is shown in Figure 1.

Output. Output devices perform tasks such as providing feedback of the effects of input actions and effectively rendering the information in the molecular model (visualization).

Computational Hardware. Computational hardware or the compute server is made up of two processors, a general purpose processor and a graphics processor. The general purpose processor, eg, a VAX (Digital Equipment Corp.) computer, communicates with the graphics processor, eg, an E&S PS390, through a slow asynchronous line (~128 Kbytes/s) or a fast, eg, ethernet (~1 Mbyte/s) communications link.

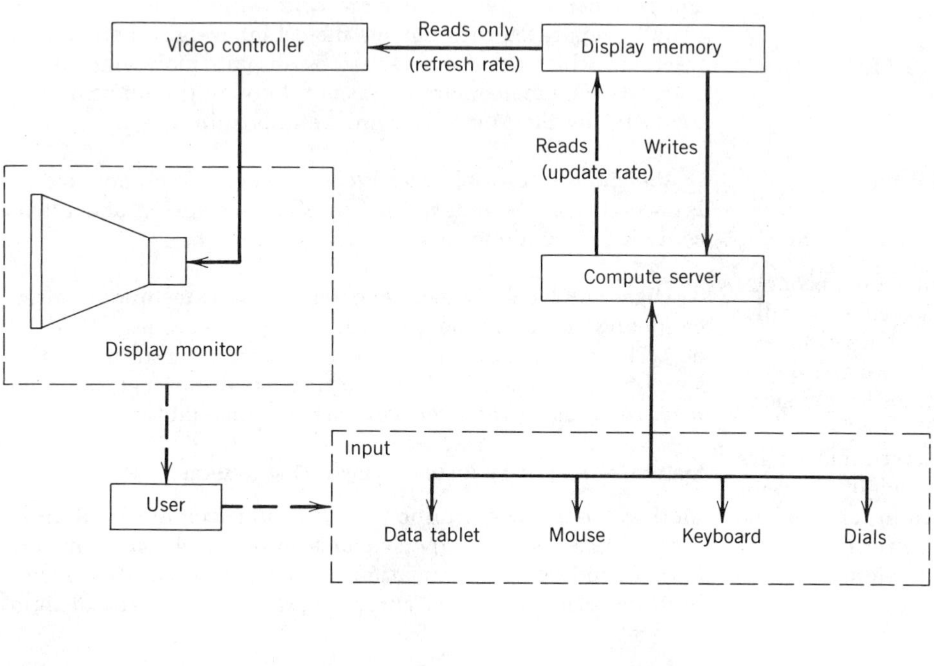

Figure 1. Generic CAMG system.

Software

Software packages are available for dealing with polymers, but mainly biopolymers; these include SYBYL (Tripos Associates), INSIGHT (Biosym Technologies), QUANTA (Polygen Corp.), CHEMX (Chemical Design Ltd.), and others. The software produces representations of ribbon structures, space-filling structures, molecular surfaces, and molecular orbital diagrams.

Techniques

Molecular graphics is used extensively as a tool in biotechnology.

Biopolymers. The various advances of graphics in biopolymers (proteins and peptides, nucleic acids, and sugars) led to a complete description of 3-d structural analysis of macromolecules.

Catalysts and Inorganic Polymers. Modeling has been used for zeolites, which are crystalline aluminosilicates whose 3-d framework consists of corner-sharing SiO and AlO tetrahedrons.

Synthetic Polymers. The prediction of properties of synthetic polymers computationally has been developed since the late 1960s. As the cost of industrial research has risen and as polymers have moved increasingly into more specialized high performance areas, interest in molecular graphics applications to synthetic polymers has increased dramatically.

BIBLIOGRAPHY

W. M. Newman and R. F. Sproul, *Principles of Interactive Computer Graphics,* 2nd ed., McGraw-Hill, Inc., New York, 1979.

R. Fletterick and M. Zoller, *Computer Graphics and Molecular Modeling,* Cold Spring Harbor Laboratory, Cold Spring Harbor, New York, 1986.

C. R. A. Catlow, *J. Computer-Aided Molecular Design* **2,** 255 (1988).

R. C. Cook, *Computer Applications in Polymer Science II, ACS Symp. Ser.,* Vol. 404, American Chemical Society, Washington, D.C., 1989.

ROY J. VAZ
Tripos Associates
NEELA SRINIVASAN
University of North Carolina

MOLECULAR WEIGHT AVERAGE.

See MOLECULAR WEIGHT DETERMINATION.

MOLECULAR WEIGHT DETERMINATION

Molecular Weight Averages and Distributions

Most polymeric materials are composed of mixtures of molecules of various sizes. This distribution of molecular weights is caused by the statistical nature of the polymerization process. A complete description of the molecular weight distribution of a polymer is necessary to understand its physical, rheological, and mechanical properties.

Certain techniques for molecular weight determination are only capable of yielding one of the molecular weight averages of the distribution. These averages are defined in terms of the molecular weight M_i and the number of moles n_i or the weight w_i of the component molecules. The molecular weight averages are defined by equations 1–5.

Number-average molecular weight:

$$\bar{M}_n = \frac{\Sigma n_i M_i}{\Sigma n_i} = \frac{\Sigma w_i}{\Sigma w_i / M_i} \quad (1)$$

Weight-average molecular weight:

$$\bar{M}_w = \frac{\Sigma n_i M_i^2}{\Sigma n_i M_i} = \frac{\Sigma w_i M_i}{\Sigma w_i} \quad (2)$$

z-Average molecular weight:

$$\bar{M}_z = \frac{\Sigma n_i M_i^3}{\Sigma n_i M_i^2} = \frac{\Sigma w_i M_i^2}{\Sigma w_i M_i} \quad (3)$$

(z + 1)-Average molecular weight:

$$\bar{M}_{z+1} = \frac{\Sigma n_i M_i^4}{\Sigma n_i M_i^3} = \frac{\Sigma w_i M_i^3}{\Sigma w_i M_i^2} \quad (4)$$

Viscosity-average molecular weight:

$$\bar{M}_v = \left[\frac{\Sigma n_i M_i^{1+\alpha}}{\Sigma n_i M_i}\right]^{1/\alpha} = \left[\frac{\Sigma w_i M_i^{\alpha}}{\Sigma w_i}\right]^{1/\alpha} \quad (5)$$

Molecular Weight Distribution Functions

Various mathematical functions have been employed to describe the distribution of molecular weights. Some of the more common functions are shown in Table 1 in terms of the mole fraction X.

Determination of Molecular Weight

Number-average Molecular Weight. Number-average molecular weight can be determined by end group analysis and measurement of colligative properties (ebulliometry, cryoscopy, membrane osmometry).

Viscosity-average Molecular Weight. The viscosity of dilute polymer solutions may be related to the molecular weight of the polymer by the appropriate calibration. The polymer is usually separated into narrow molecular weight distribution fractions, which are characterized by absolute molecular weight methods. The molecular weight is related to the intrinsic viscosity $[\eta]$ by the Mark-Houwink relationship.

Weight-average Molecular Weight. Weight-average molecular weight can be determined by light scattering and ultracentrifuge techniques.

Higher Molecular Weight Averages. Classic techniques often cannot be used for higher molecular weight averages ($\bar{M}_z$, $\bar{M}_{z+1}$, etc). These are usually measured by determining the molecular weight distribution by fractionation or chromatography and back-calculating the averages from the distribution.

Determination of Molecular Weight Distribution

Methods used to determine the molecular weight distribution include fractionation (qv); various forms of chromatography (qv), including gel-permeation or size-exclusion, thin layer, high performance liquid chromatography, supercritical fluid

Table 1. Molecular Weight Distribution Functions

Name	Function	Comments
Gaussian	$X(M) = \frac{1}{\sigma_n(2\pi)^{1/2}} \exp\left[\frac{(M - M_m)^2}{2\sigma_n^2}\right]$	M_m = median value equal to $\overline{M}_n$
log-normal	$X(M) = \frac{1}{\sigma_n(2\pi)^{1/2}} \exp\left[\frac{(\ln M - \ln M_m)^2}{2\sigma_n^2}\right]$	M_m = geometric mean
Poisson[a]	$X(M) = \frac{\nu^{M-1}}{\Gamma(M)} \exp(-\nu)$	$\nu = \overline{M}_n - 1$
Flory-Schulz[b]	$X(M) = \frac{\beta^{k+1} M^{k-1} \overline{M}_n}{\Gamma(k+1)} \exp(-\beta M)$	k = degree of coupling $\beta = \frac{k}{\overline{M}_n}$

[a] $\Gamma(M)$ = γ-function.
[b] $\Gamma(k + 1)$ = γ-function.

chromatography, phase-distribution chromatography, field-flow fractionation; electrophoresis (qv); and atomization of polymer solutions.

BIBLIOGRAPHY

L. H. Peebles, *Molecular Weight Distributions in Polymers,* Wiley-Interscience, New York, 1971.

N. C. Billingham, *Molar Mass Measurements in Polymer Science,* Halsted Press, a division of John Wiley & Sons, Inc., New York, 1977, Chapt. 9.

L. H. Tung, ed., *Fractionation of Synthetic Polymers,* Marcel Dekker, Inc., New York, 1967.

A. R. Cooper, ed., *Determination of Molecular Weight,* Chemical Analysis Series, Vol. 103, John Wiley & Sons, Inc., New York, 1989.

ANTHONY R. COOPER
Lockheed Missiles & Space Company, Inc.

MOLECULAR WEIGHT DISTRIBUTION.

See FRACTIONATION.

MONODISPERSE POLYMERS

The preparation of homopolymers and block copolymers with near monodisperse (uniform) molecular-weight distributions is of both academic and commercial value. The absence, in certain cases, of a spontaneous termination step during propagation allows for the preparation of linear and star-shaped polymers. This facilitates studies in block polymer morphology, chain dynamics, diffusion (both tracer and self-diffusion), unperturbed chain dimensions, and rheology. These investigations have benefited from the narrow distribution of chain lengths present in the samples. The materials evaluated have been prepared by anionic polymerization of dienes and by hydrogenation of these polydienes.

In any consideration of molecular weight distributions resulting from termination-free systems, the potential influence of the relative rates of initiation and propagation must be recognized. Quantitative relations have been developed between the molecular weight distribution and the ratio of propagation to initiation rates. Even in the extreme case where $k_p/k_i = 10^6$, the resultant x_w/x_n ratio does not exceed ~1.4. Termination-free polymerizations, in which the active center has two forms in dynamic equilibrium with each exhibiting its own propagation rate have been examined. This approach has been used to develop a method for dissecting k_p into the contributions from the various ionic structures of the propagating center in polar media from an analysis of the resultant molecular weight distributions. Any desired polydispersity in molecular weights is obtained by controlling the addition of initiator during a nonterminating polymerization.

BIBLIOGRAPHY

P. J. Flory, *J. Am. Chem. Soc.* **62,** 1561 (1940).

B. D. Coleman and T. G. Fox, *J. Am. Chem. Soc.* **85,** 1241 (1963).

G. V. Schulz, *Ber. der Bunsenges.* **78,** 1064 (1974).

T. A. Orofino and F. Wenger, *J. Chem. Phys.* **35,** 532 (1961).

LEWIS J. FETTERS
Exxon Research and Engineering Company

MONOMER

A monomer is defined as a compound consisting of molecules, each of which can provide one or more constitutional units of a polymer (or oligomer). For example, the monomer molecule styrene (—$CH{=}CH_2$) provides the constituent unit —CH—CH_2—. The monomer molecule $CH_2{=}CH{-}CH{=}CH_2$

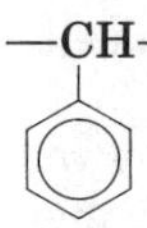

can provide the constitutional units

—$CH_2CH{=}CHCH_2$— or —$CHCH_2$— with pendant $CH{=}CH_2$

The monomer molecule CH_2N_2 can provide the constitutional unit —CH_2—.

A monomer molecule of δ-valerolactone 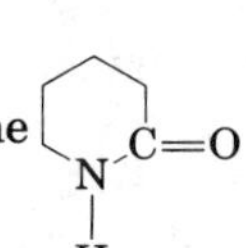 can provide the constitutional unit

$$-\mathrm{NH(CH_2)_4}\overset{\mathrm{O}}{\overset{\|}{\mathrm{C}}}-$$

By a condensation process, the monomer molecules hexamethylene diamine $H_2N(CH_2)_6NH_2$ and adipoyl chloride $ClCO(CH_2)_4COCl$ can lead to the regular polymer molecule $-[NH(CH_2)_6NHCO(CH_2)_4CO]_n-$ which contains the constitutional units

$$-\mathrm{NH(CH_2)_6NH}- \quad \text{and} \quad -\overset{\mathrm{O}}{\overset{\|}{\mathrm{C}}}\mathrm{(CH_2)_4}\overset{\mathrm{O}}{\overset{\|}{\mathrm{C}}}-$$

A monomeric unit or "mer" is the largest constitutional unit contributed by a single monomer molecule in a polymerization process.

BIBLIOGRAPHY

"Basic Definitions of Terms Relating to Polymers," IUPAC, *Pure Appl. Chem.* **40,** 475 (1974).

MONOMER REACTIVITY RATIOS. See COPOLYMERIZATION.

MORPHOLOGY

The morphology of polymers is concerned with the shape, arrangement, and function of crystals alone or embedded in the solid. Polymers are either truly homogeneous amorphous or heterogeneous semicrystalline solids. Even single crystals obtained from dilute solution have a lower density than the ideal crystal lattice, thus inviting the notion of a finite amorphous component. Completely crystalline polymers are obtainable only from crystalline oligomers by polymerization in the solid state.

The emphasis on technical applications reduces the importance of the morphology of single crystals. Of technical importance are molded materials and highly oriented polymers that, because of the low density ρ of polymers, yield in one direction much higher specific elastic modulus E/ρ and strength σ/ρ than metals.

Under special conditions, the completely amorphous material may partially crystallize and be transformed into the semicrystalline polymer. The coexistence of crystalline and amorphous components is observed in all synthetic polymers that are not obtained by polymerization in the solid state of crystalline monomers or cyclic oligomers. The amorphous state is often called the liquid state, although it may differ quite appreciably from a normal low molecular weight liquid. The dominance of the long, randomly coiled linear macromolecules with the enormous anisotropy of their force field and the presence of crystals strongly affects the behavior of the chains in the amorphous component. Virtually all amorphous chain sections have one end (cilia) or both ends (folds, two different crystals connecting tie molecules) fixed in the crystal lattice. Hence the properties of the polymer chains in the transition region between the crystals and the amorphous layers are substantially different from those in truly amorphous material without any crystals or strongly sorbing solids.

In the absence of crystals, the material is fully amorphous like a true rubber or glass. These two states are separated by the glass-transition temperature T_g. Chain mobility is pronounced in the rubber state but almost absent in the glassy state. This influences the time dependence of different effects, although the conformations of the macromolecules in both states are rather similar.

Crystal Models. In the fringed micelle model, it is assumed that bundles of macromolecules form crystals as long as the parallelism of the chains in the bundle is not too disturbed by separation and inclusion of part of the chain in other crystals independent of the first one. Because there is no uniformity in the crystal length, this model cannot explain the existence of a repeat distance L in small-angle x-ray scattering (saxs). If each densely packed chain in the laterally extended crystal enters and remains in the amorphous layer after a single passage, the density of the quasi-amorphous phase with random-chain conformation ought to be at least twice the density of the crystals with parallel chains. These two deviations from the experimental facts generally exclude the micelle model. The folded-chain crystal model assumes that the crystalizable part of the macromolecule is in crystals of finite thickness. The regular and irregular folds, the free ends, and the tie molecules (TM) connecting different crystals form the amorphous component. In the extreme switchboard model of the crystal, more than half the chains in the lamella reenter at random the same crystal and the rest pass to the next crystal, whereas in the adjacent reentry model all molecules adjacently enter the same crystal. Very likely the truth lies somewhere between the two extremes. In fibrillar material, taut TM (TTM) connect the blocks while a substantial part of the rest of the chains folds back in the same crystalline block. The former model is too dense in the amorphous layers while the latter is free of this deficency.

Composite Structure of the Semicrystalline Solid. In semicrystalline solids, the polymer chains in the amorphous part are intimately connected with the crystals and cannot act independently. The crystalline and the amorphous components depend on each other in a way that is different from the situation in the nonpolymeric composites. The chain mobility is limited by the crystals or other solids where the chains are anchored. This causes a volume change during uniaxial extension or compression in total disagreement with the infinite rubber model in which the volume remains constant with deformation. The inflow of amorphous molecules from the sides tries to achieve constancy. In a polymeric semicrystalline solid the amorphous chains cannot move as far in the long and thin amorphous layers between the crystals. Consequently, the deformation of the semicrystalline solid polymer must include in the elastic modulus of the amorphous component both moduli of rubber, the large bulk modulus K that considers the volume changes and the small shear modulus G without any volume change.

Crystallization. Chains crystallize if they are sufficiently regular and not too hampered in the liquid state by entanglements. Hence a normally polydisperse medium molecular weight (<100,000) highly regular PE without side chains crystallizes very easily.

The static mechanical properties, elastic modulus E and

strength σ, are usually measured in a tensile field. Both depend on the rate of force application and the measurement time, but also on the time the sample has relaxed after the last treatment. The microhardness measurements with an indenter determine the dimensions of the depression a certain load has made on the surface of the specimen. The method is rapid and measures the plastic resistance and its anisotropy on the surface of the sample. The mechanical modulus of wave propagation may be used to detect the mechanical anisotropy of the sample at the frequency of the wave.

Microscopy. Experimental methods include light microscopy, acoustic microscopy, electron microscopy (em), scanning em (sem), transmission em (tem), scanning tem (stem), and so-called decoration with heavy metal atoms to reveal rough surfaces.

Scattering. Scattering techniques include wide-angle x-ray scattering (waxs), small-angle x-ray scattering (saxs), small-angle neutron scattering (sans), small-angle light scattering (sals), and the laser speckle interferometer.

Spectroscopy. Spectroscopic methods include infrared (ir) and Raman spectroscopy. Differential scanning calorimetry (dsc) mainly tells about melting and crystallization in the sample.

Crystals from Dilute Solution

Nucleation of Single Crystals. First, a nucleus is formed that grows and yields the lamellae. In the presence of a large number of primary nuclei, small crystals tend to develop. Fewer nuclei give large, uniform crystals. The origin of nuclei in crystallization from dilute solution is uncertain. Because crystallization may occur at a much higher temperature than is theoretically considered possible by homogeneous nuclei, it must be concluded that invisible heterogeneous nuclei, most probably fragments of the original crystals or small dust particles, exist in solution above the so-called dissolution temperature.

Single-crystal Growth. Subsequent to primary nucleation, the crystal grows into a two-dimensional lamella by epitaxial deposition of chains. The crystal grows from the edge of the nucleus with the same structure of unit cell. The same applies to the heterogeneous nucleus, even if the unit cells of the nucleus and those of the deposited chains differ. The long-chain linear molecules adopt a folded configuration in the crystals.

Folding profoundly influences the developing geometry of the lamellar crystals. Well-defined, lateral growth-shapes reflect the symmetry of the unit cell. As indicated in Figure 1, the lateral growth faces in the lozengelike crystal of PE are {110} faces, whereas in the truncated type, the longer faces are {110} and the shorter faces are {100} (Fig. 2). In the case of the hexagonal unit cell of polyoxymethylene (POM), the simplest lamellar crystal is in the shape of a hexagon.

Irregular Lateral-growth Habit. In some cases the lamella does not exhibit simple geometrical shapes. The periphery of the single crystal is rounded or serrated on such a fine scale that it is difficult to ascertain the occurrence of crystallographic microfaceting. The polymers from whose solution single crystals are grown have a irregular shape and may have similar energy levels of different lateral plans and folds that depend on the concentration and temperature of the solution and on the molecular weight of the polymer. Modes of crystallization or folding are easily changed. With irregular chains, such effects are the consequence of low tacticity, eg, in polyacryloniltrile (PAN), polychlorotrifluoroethylene (PCTFE), and poly(vinyl chloride); side chains in branched PE; random side groups in chlorinated, brominated, or sulfochlorinated PE; and random copolymers of ethylene with propylene or 1-butene.

Single-crystal Mats. These are obtained by precipitation of single crystals from dilute solution (Fig. 3). Individual lamellae are formed with few connections from highly dilute solutions. In such a case, the mat is extremely brittle with little lateral cohesion.

Lamellae. The thickness $L = D + l$ of single lamella tends to be uniform at any supercooling $\Delta T = T_m - T$, of which it is a sensitive function if the surface energy σ_e is the same. Here T_m is the melting point or the solution point of an infinite crystal in a solvent at 101.3 kPa (1 atm). This is indicated by both em and saxs. Exactly speaking the crystal thickness D with the high E_c, measured by lam, ought to be equal because l depends uniquely on the highest temperature to which the crystal was exposed. Any temperature changes during crystal growth are reflected in the thickness D or L. Abrupt temperature changes produce small steps.

Attachment of Heavy Atoms to Folds. Initially chlorine and bromine atoms or ions do not enter the crystal structure. They react with amorphous chains up to the interface with the crystals, ie, with the loops and cilia on the surface of the crystals. Thus they make the lamellar surfaces impermeable. Because

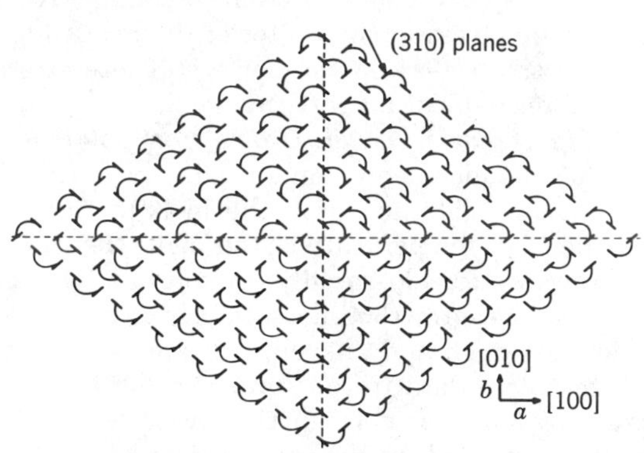

Figure 1. Regular folds with adjacent reentry showing the "sectorization" of a single crystal of PE. Courtesy of *Journal of Applied Physics.*

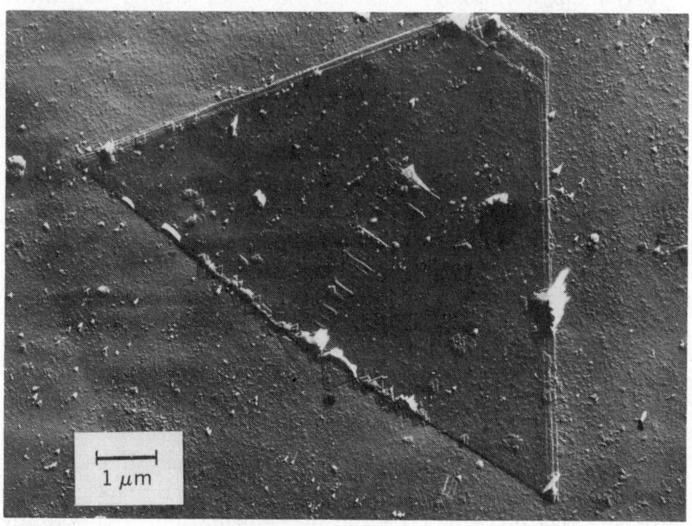

Figure 2. Trapezoidal PE single crystal obtained from truncated pyramidal shape of lamella as formed after the pyramid broke along a diagonal. Courtesy of *Journal of Applied Physics.*

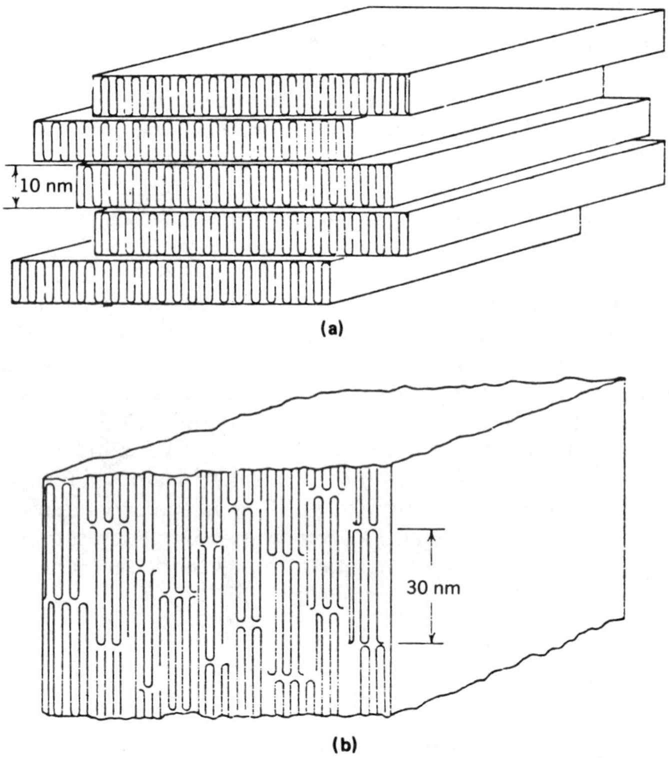

Figure 3. Oriented single-crystal mats of PE (**a**) as grown and (**b**) after annealing. Courtesy of *Journal of Applied Physics*.

they are heavier than the replaced H atoms, they shift the lam frequency to smaller values. Additional details are seen with ir, such as the formation of double bonds and the attachment of the second chlorine or bromine atom on the same monomer. At a later stage, the attacking atoms enter the crystals primarily from the side faces that are not chlorinated or brominated. The crystalline core is thus divided into the penetrated part and the part at a distance from the side faces. The attached atoms are too large to be included in the crystal lattice of an eventually recrystallized solid.

Pyramidal Morphology. A number of observations lead to the conclusion that PE crystals in suspension are not in a planar but a pyramidal form. This is equally applicable to other polymers crystallizing in the lamellar habit. In most cases, analogous features are observed. The most obvious and frequent morphological feature appearing on dry crystals is a central pleat and sometimes, as in the case of a truncated crystal broken in half across the *b* axis, a trapezoid-shaped structure (Fig. 2). Corrugations that can be related to the special type of pyramidal crystal often appear and do not disappear for up to several months after removal of the solvent.

Twinning. Growth twins, in which two identical crystals grow with a specific mirror relationship to each other, are frequently found in polymer crystallization. In complexity, they are intermediate between dendrites and lozenges. They may play an important role in the development of grosser morphological structures, but this point has not yet been investigated. Planar twinning has mainly been recorded in PE in association with spiral growth of lamellae by deformation of the single crystals or with growth-determined mirror planes.

Multilayers. Multilayered crystals very likely result from spiral growth of single lamellae. Crystals with such sharp fold surfaces would not be expected to exhibit any rotation; they should fit into a good lattice register. This has been demonstrated by dislocation networks of low molecular weight fractions of PE on such double crystals.

Annealing. Using the Moire technique, em experiments on bilayered single crystals on a carbon substrate show that the dislocation density increases by a factor of about 10^3 after annealing at 95°C; this may be due to chain sliding. For dry, single-crystal mats, saxs at RT does not indicate any change in L by annealing at 90°C. The T_A is too low.

Dissolution. Dissolution by the same solvent occurs if the temperature is above the original formation temperature of the crystals. Dissolution is actually melting with the added effects of polymer–solvent interaction. The penetration by the solvent proceeds from the lateral surfaces of the crystal and some internal crystal defects. Holes are formed inside the crystal because the areas around the defects are dissolved. The undissolved parts of the original crystal serve as epitaxial nuclei for the recrystallization with larger thickness of dissolved macromolecules. Gradually the interior of the crystal disappears completely, leaving a picture-frame morphology. The increase in thickness during such an annealing experiment is a consequence of dissolution and epitaxial recrystallization.

Effect of Substrate. Nucleation of polymer crystals on solid surfaces affects orientation. In epitaxy, the electrical field of the unit cell of the substrate rules the mode of deposition.

Crystals from Concentrated Solution and Melt

The most common mode of crystallization from melt and concentrated solution is the formation of spherulites. With very few primary nuclei, eg, at slight supercooling with complete exclusion of heterogeneous nuclei, the spherulites may grow to large dimensions before they impinge on each other. Extensive supercooling, quenching, or dense heterogeneous nucleation produces many primary nuclei. The spherulites are small and poorly developed. Only randomly oriented stacks of parallel lamellae can be observed. High molecular weight slows the crystallization because the melt viscosity increases with molecular weight to the 3.4th power. A monodisperse high molecular weight sample yields only randomly oriented stacks of lamellae of different thickness.

Crystallization increases density and reduces the volume. In molds, the crystallization starts at the surfaces where the melt is in contact with the cooler parts. Crystallization may be normal although columnar crystallization is preferred adjacent to the cold surface. The interior crystallizes subsequently with an appreciable reduction in volume. As a result there is a tensile stress in the center of the molded specimen and a compressive stress on the surface. These stress fields influence the mechanical properties.

Relationship of Single Crystals Grown from Solution to Melt-grown Crystals. Folded-chain lamellae in the form of single crystals cannot be obtained from the melt in their original shape. At the high temperature of the melt, the thickness of lamellae grows extremely rapidly. In solution, the slow nucleation affects the growth time so much that the resulting lamellae are too thick and uneven. Only recently was it possible to extrapolate data to the start of solidification, $t = 0$, and to prove by lam that ν' is equal in crystals obtained from dilute solution and melt at the same supercooling.

Extended-chain Crystals. Under high pressure the polymers, especially PE, may crystallize from the melt in the form of so-called extended-chain crystals whose average lamella thickness of 150 nm is much larger than the usual L of 20 nm and more comparable to the extended macromolecule length. Such crystals are not usually made of fully extended chains, but contain folds that are less numerous the thicker the crystals and the lower the molecular weight.

Cross-linked Polymers. Because the cross-links tend to be excluded from the crystal lattice, a sufficient length of segments between consecutive cross-links is needed for crystallization. As a consequence of the numerous excluded cross-links, the lamellae are thinner than in a normal LPE, full of defects, and with less contrast between the crystals and the quasi-amorphous layers. The spherulites are smaller, about 20–30 μm, as in the case of highly branched material. A grosser structure, found earlier, is an artifact caused by crystallization of shorter sections of the cross-linked material obtained by oxidizing the solid.

Lamella Structure

Simple spatial considerations require that each macromolecule occupy a larger volume in the amorphous, rather than in the crystalline region. If the chain in the amorphous region has no preference for one direction, if its orientation is completely randomized, the number of amorphous chains per cross section is halved. This applies equally to macromolecules crystallizing from dilute solution or from the melt. Thus with the empirical knowledge that chain folding occurs, two models for the individual crystals from the melt were proposed. In one the folds are sharp and the macromolecule reenters the lattice adjacent to the emerging chain, and in the other the folds are irregular and any reentrance is at random (switchboard model). No experimental data are in complete agreement with either theory. The switchboard model, without some nearby reentry in the same crystal, yields too high a density of the amorphous layers where the chains are perpendicular to the surface; the adjacent reentry model without the tie molecules and cilia does not yield a thick enough amorphous layer.

Oriented Polymers

Most semicrystalline polymers, particularly those produced commercially, are partially oriented; that is, their chains have an overall alignment that may impart to the bulk polymer certain advantageous properties, eg, increased mechanical strength or dielectric polarizability. Molecular orientation, whether arising from crystallization under stress or deformation of a solidified polymer, or in naturally occurring oriented crystalline polymers such as cellulose or keratin, is always associated with an orientational morphology.

The chain orientation in a semicrystalline polymer is against the equilibrium, which in large domains requires complete randomness. The sample is stable only at sufficiently low temperatures that extend chain- and crystal-relaxation times longer than the time of the experiment, which may extend over a few years. The oriented material is not absolutely stable. Increasing the temperature increases the mobility of the macromolecules and crystals; the chains move perceptibly toward their equilibrium position.

Uniaxial Deformation of Solidified Polymers. Highly oriented material is obtained by drawing or extrusion below the melting point. The first method requires a sample with sufficient longitudinal cohesion by which it resists the uniaxial elongation force in the stretching device.

Extrusion is substantially improved by the split-billet technique. The cylindrical billet is cut in half and the sample is inserted. This permits the drawing of incoherent samples into thin sheets, for instance, dried gels and not yet annealed single-crystal mats.

Deformation of Bulk Polymer. Plastic deformation by drawing or extrusion is the best method to obtain highly oriented samples from unoriented solid material. According to the microfibrillar model of fibrous structure and its formation, the wide, thin lamellae are destroyed and the sheared-off or melted and recrystallized blocks are incorporated in highly aligned, long and narrow microfibrils.

Single-crystal Drawing. To elucidate the structural aspects of oriented polymers, the deformation process of single crystals was investigated. These crystals are the strongest constituent of semicrystalline polymers. The initial stages of the deformation of PE single crystals were homogeneous up to ca 15% elongation, accomplished by twinning or phase changes. If the crystalline chains firmly adhere to the strained, deforming substrate, whole lamellar segments tilt and slip toward the draw direction. Free crystals are transformed into microfibrils without any detectable chain tilt and slip. The orientation of the chains in the crystal lattice and in the amorphous layers between the crystalline blocks is established during the formation of microfibrils and does not improve substantially during later drawing.

Drawing of the Fibrous Structure. The drawing of the fiber based on longitudinal displacement of fibrils instead of the displacement of microfibrils requires less motion of the microfibrils. The shear forces on each strongly displaced fibril result in a weaker shear motion of microfibrils. The fibrils are more skewed, but the bulk draw ratio is not increased. The nearly linear increase in the interfibrillar TTM fraction per amorphous area yields a similar increase in the axial elastic modulus. The forces on each microfibril are shear forces that do not affect the structure of the axially strong microfibril, particularly the longitudinal alternation of amorphous layers and crystalline blocks. However, these forces displace the folds and straight sections of the chains in the crystalline blocks to such an extent that the axial projection of the differences in the electron density of the microfibrils continuously decreases. This effect levels off and finally completely cancels the meridional maximum of the saxs; no long period may be deduced any more from it. If annealing effects are excluded, the long period L of the microfibril does not change during the drawing of the fibrous structure.

BIBLIOGRAPHY

I. H. Hall, ed., *Structure of Crystalline Polymers,* Elsevier Applied Science Publishers, Ltd., Barking, UK, 1984.

D. C. Bassett, *Developments in Crystalline Polymers,* Elsevier Applied Science Publishers, Ltd., Barking, UK, 1982.

D. C. Bassett, *Principles of Polymer Morphology,* Cambridge University Press, Cambridge, UK, 1981.

A. Keller in F. Ciardelli and P. Giusti, eds., *Structural Order in Polymers,* Pergamon Press, Oxford, UK, 1981.

ANTON PETERLIN
Consultant

N

NAPHTHENATES. See DRYING OILS.

NATURAL FIBERS. See COTTON; FIBERS, VEGETABLE; SILK.

NATURAL GUMS. See GUMS, INDUSTRIAL; POLYSACCHARIDES.

NATURAL POLYMERS. See BIOPOLYMERS.

NATURAL RESINS. See RESINS, NATURAL.

NATURAL RUBBER. See RUBBER, NATURAL.

NEOPRENE. See CHLOROPRENE POLYMERS.

NETWORKS

Polymeric networks, systems of interconnected macromolecular chains, exhibit the property of high extensibility coupled with the capacity for full recovery that is implied by the term "rubber elasticity." Although the constitutional units comprising various networks may differ widely, their mechanical properties have much in common; this is especially evident in their stress–strain relationships.

The junctions that confer recoverability on the network ordinarily are provided by covalent bonds. In some instances, however, physical combinations of chains, mediated, for example, by crystallites, may serve the same function as the permanent chemical interlinkages. Thus rubber elasticity assumes importance beyond the narrow limits by which it is commonly designated. It is operative in the swelling of polymeric networks, the deformation of semicrystalline polymers and the viscoelastic behavior of linear polymers under flow in the liquid or amorphous state, and it is essential to the functions of elastic proteins and muscle. The theory of rubber elasticity is centrally important to much of polymer science.

The density of junctions, or cross-links, in elastomeric networks is generally low, typically 0.05–0.2 mol/kg. Yet, the degree of interlinking places them far beyond the gel point. It follows that imperfections in the form of chains attached to the network at only one end are usually of minor consequence. To characterize the constitution of the network in a way that comprehends network imperfections, the cycle rank of the network defined below turns out to be the appropriate measure of connectivity.

A feature of networks, second in importance only to the degree of connectivity, is the copious interpenetration of the chains comprising them. Chains emanating from a given junction of the network are embedded in a maze of other chains and junctions. Most of the latter are only remotely topologically related to the junction considered. This fact, often overlooked, has an important bearing on the properties of elastomeric networks. It precludes their treatment in terms of a lattice, albeit a disordered one, consisting of junctions related to their spatial neighbors by connecting chains (see also INTERPENETRATING POLYMER NETWORKS).

Molecular Theory

The basic premise of the molecular theory of elasticity in networks asserts that the stress in a typical strained network originates within the molecular chains of the structure; contributions from interactions between the chains are negligible. This premise finds direct support in elasticity measurements on polymeric networks. The temperature coefficient of the stress at fixed strain and its constancy with dilution are especially significant in this connection.

The Isolated Chain

The chains in elastomeric networks typically consist of 100–1000 skeletal bonds. For free chains of this length, the function describing the distribution of end-to-end chain vectors **r** is Gaussian in good approximation.

Network Structure and Topology

A polymer network may be characterized by the number μ_J of its junctions, their functionality ϕ (or average functionality $\bar{\phi}$), and by the number ν_{ends} of chain ends. The number of chains in the network, including those with only one end attached is

$$\nu = \tfrac{1}{2}(\mu_J\phi + \nu_{\text{ends}}) \qquad (1)$$

The effective number ν_e of chains is less than ν as a result of the imperfections caused by free chain ends. For a perfect network where $\nu_{\text{ends}} = 0$,

$$\nu_e = \nu = \mu_J\phi/2 \qquad (2)$$

Affine Networks

The high degree of interpenetration in elastomeric networks and the fact that each junction is located in an environment dominated by chains and junctions whose structural relation to the junction considered is remote lend credence to the assumption universally adopted in the earliest theories of rubber elasticity that the positions of the junctions are approximately affine in the macroscopic strain.

Phantom Networks

The theory of James and Guth addresses networks of Gaussian chains for which the only action is to deliver contractile forces (proportional to their displacement lengths r) at the junctions to which they are attached. The chains have no other material properties; they may pass through one another freely and they are not subject to the volume exclusion requirements of real molecular systems. Being free of constraints by neighboring chains, the junctions of the phantom network thus described undergo displacements that are affected only by their connections to the network and not at all by their immediate surroundings.

Fluctuations in Real Networks

Inasmuch as the basic premise that the stored elastic free energy resides within the chains is fully validated, observed departures from the form of the stress–strain relationship prescribed implicate the relationships assumed, or deduced, between the macroscopic strain and the distribution of chain vectors as the source of the discrepancy. The assumption that the latter are affine in the strain makes no allowance whatsoever for excursions of the junctions from their mean positions. Fluctuations of this nature are implicit in the molecular mobility essential to the high compliance exhibited by elastomers. On the other hand, the large, unimpeded fluctuations deduced for phantom networks may be curtailed severely by the profusion of chains in which each junction is embedded. The restrictions on fluctuation considered here are operative at equilibrium. They are unrelated to the kinetic inhibitions that impede time-dependent relaxation processes.

Considerations such as these led to the suggestion that real networks may behave in a manner between the two extremes. It was suggested further that a shift in proximity to these respective extremes might be expected with strain, as phantom behavior is more closely approached with elongation or dilation. Inasmuch as the factor of proportionality to the strain function is smaller according to phantom theory than for affine theory, the observed departures from these theories might thus be explained.

Network Junctions Subject to Strain-dependent Constraints

The model adopted for the purpose of giving quantitative expression to the ideas described above is shown in Figure 1. Point A represents the mean position of the chosen junction in the hypothetical phantom state of the network. The radius of the large dashed circle centered at A represents the root-mean-square fluctuation $\langle(\Delta R)^2\rangle_{ph}^{1/2}$ around this position in the phantom state. The domain of constraints due to entanglements with surrounding real chains and to their steric requirements are represented by the smaller dashed circle centered at B and separated from A by $\bar{s}$. It is located as if the constraints were suddenly imposed at an instant during which a random excursion of the junction around its mean position A carried it to point B. (The manner in which the network was actually formed is irrelevant.) After the constraints have been established, the mean position of the junction is at point C removed from A by $\Delta\bar{R}$. In other words, under the combined influences of its connections with the network, the phantom network forces, and the constraints, the mean position of the junction in the unstrained, real network is at C. The instantaneous position of the junction happens to be at point D, which is outside the domain of constraints by neighbors, but inside the domain representing fluctuations of the phantom network. Both of the domain boundaries are diffuse rather than rigid; hence, the junction may wander beyond either of them, although the probability of its doing so diminishes with the distances from their centers. For simplicity and without significant sacrifice of accuracy, the action of the domain of constraint is taken to be a Gaussian function of the distance Δs of the junction from B, just as the action of the (phantom) network is a Gaussian function of ΔR.

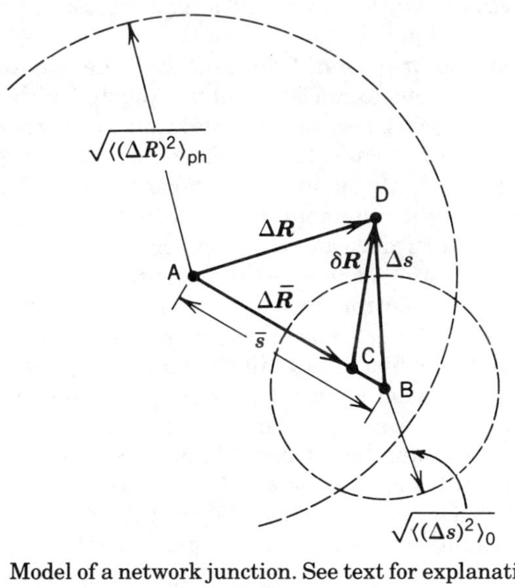

Figure 1. Model of a network junction. See text for explanation. Courtesy of the American Chemical Society.

BIBLIOGRAPHY

P. J. Flory and B. Erman, *Macromolecules* **15,** 800 (1982).

P. J. Flory, *Br. Polym. J.* **17,** 96 (1985).

P. J. Flory, *Polym. J.* **17,** 1 (1985).

J. E. Mark and B. Erman, *Rubberlike Elasticity. A Molecular Primer,* Wiley-Interscience, New York, 1988.

PAUL J. FLORY
Stanford University

NEUTRON SCATTERING

Neutron scattering has found numerous applications in the fields of biology, polymer science, physical chemistry, materials science, metallurgy, colloids, and solid-state physics. In polymer science the majority of these applications involve small angle neutron scattering (sans), which gives information on the time averaged structure and conformation of the molecules. In view of the large numbers of nonspecialists who are increasingly using neutron scattering, there is the need to provide rapid access to the method without unnecessary detail and mathematical rigor. This article is intended to meet this need for the sans technique. Although there has been less work on other subcategories of neutron scattering (i.e., inelastic scattering, spin-echo, etc.), such experiments provide valuable new information on the *motion* of polymer molecules.

Theory

Energy and Momentum Transfer. In the context of this article, scattering means the deflection of a neutron beam from its original direction by interaction with the nuclei of polymer or solvent molecules in a sample. In a scattering experiment a proportion of the incident neutrons is scattered and the remaining fraction is transmitted through the sample. The intensity of the scattered neutrons is measured as a function of the scattering angle and energy.

The kinetic energy of a typical neutron, wavelength $\lambda = 0.53\,\text{nm}$, is $4.7 \times 10^{-22}\,\text{J}$ or 3 meV. Such energies are much lower than electromagnetic radiation and are of the same order as the vibrational and diffusional energies of molecular systems. Exchanges of energy between the incident particle neutron and molecule give rise to inelastic scattering, which depends on the dynamics of the system studied. Although the angular dependence of the scattering of both x rays and neutrons is easily measured, the energies of molecular vibrations (~3 meV) are much lower than incident x-ray photons (~10 keV) and thus energy transfers are difficult to detect for x-ray scattering. In contrast, the energy transfers resulting from neutron scattering are easily resolved and permit the elucidation of dynamic processes. Neutrons are thus a unique probe for studying the condensed state, as they simultaneously have both the appropriate wavelength and energy to investigate the structure and dynamics of molecular systems in general, including polymers.

For an incident neutron of wavelength λ_0 and velocity v_0 that is scattered through an angle 2θ in an inelastic process, this results in a final wavelength λ and velocity v, and the energy gained by the target (and lost by the neutron) is given by

$$\Delta E = \frac{m}{2}(v^2 - v_0^2) = \frac{\hbar^2}{2m}(k^2 - k_0^2) = \hbar\omega \tag{1}$$

where $\mathbf{k}_0$ and $\mathbf{k}$ are the initial and final wave vectors ($k = 2\pi/\lambda$). The momentum transfer is

$$\hbar\mathbf{Q} = \hbar(\mathbf{k} - \mathbf{k}_0) \tag{2}$$

$$\hbar|\mathbf{Q}| = \hbar(k^2 + k_0^2 - 2kk_0 \cos 2\theta)^{1/2} \tag{3}$$

If energy is transferred in the scattering process ($\Delta E \neq 0$), the process is termed inelastic. If no energy change takes place ($\Delta E = \hbar\omega = 0$, $\lambda = \lambda_0$), the scattering is termed elastic and

$$|\mathbf{Q}| = \frac{4\pi}{\lambda}\sin\theta \tag{4}$$

If ΔE is small compared to the incident neutron energy ($|\Delta E| << E_0$) the scattering is termed quasielastic. Most neutron scattering measurements on polymers have involved neutrons scattered at small values of the momentum transfer ($Q \to 0$). This type of measurement is usually referred to as small-angle (rather than small Q) neutron scattering, though the terms are equivalent for long wavelengths ($\lambda_0 > 0.4\,\text{nm}$). It is easily seen that for long wavelengths, $Q \to 0$ implies $k \to k_0$ and the scattering is predominantly elastic, as any neutron scattered with a large energy transfer ΔE could not satisfy both energy and momentum conservation at small Q.

Scattering Length and Cross Section. Scattering theory is usually developed by considering a single atom that is fixed at the origin and hence cannot accept energy from the neutron. The interaction between the neutron and nucleus is known to be short-ranged (~10^{-6} nm) compared to the wavelength of the neutron (~0.5 nm). Because of this, the scattering can contain only zero angular momentum components. This has the important consequence that the scattering is isotropic for slow neutrons and there is not an angle-dependent form factor as in the case of x rays. Thus if an incident plane wave of neutrons is described by a wave function of unit density

$$\psi_0 = e^{ik_0z} \tag{5}$$

the scattered wave from a fixed nucleus is spherically symmetrical and of the form

$$\psi_1 = -\frac{b}{r}e^{ikr} \tag{6}$$

The quantity b has the dimensions of length and is called the scattering length. Although the value of b is in principle dependent on the incident neutron energy E_0, the variation of energies normally encountered in neutron scattering studies is negligible and b may be regarded as real and a (known) constant for a given nucleus (isotope). The scattered neutrons may be envisaged as originating from a sphere of radius r centered on the nucleus, and, using the scattering length, a scattering cross section σ for the nucleus may be defined by

$$\sigma = \frac{\text{number scattered neutrons/s}}{\text{incident neutron flux}} \tag{7}$$

and hence the single-atom cross section is given by

$$\sigma = 4\pi b^2 \tag{8}$$

It can be seen from equation 8 that σ has the dimensions of area.

To a first approximation the cross section may be regarded as the effective area that the target nucleus presents to the incident beam of neutrons for the elastic scattering process. The cross section defined in equation 8 is usually called the bound-atom cross section, as the nucleus is considered fixed at the origin. However, where the atom is free to recoil, eg, in the gaseous state, the cross section applicable to this state is called the free-atom cross section. The bound-atom cross section is generally relevant to polymer studies, which are virtually always conducted on samples of macroscopic dimensions in the solid or liquid state.

Coherent and Incoherent Cross Sections. The magnitude of b varies from nucleus to nucleus and is typically on the order of 10^{-12} cm. This gives rise to the usual unit for a cross section, called a barn ($10^{-24}\,\text{cm}^2$). Unlike the x-ray scattering factor f, which increases with the atomic number of the atom, there is no general trend throughout the periodic table in the values of b, which vary from isotope to isotope and from nucleus to nucleus of the same isotope if it has nonzero spin.

Because the neutron has spin ½, it can interact with a nucleus of spin I to form one of two compound nuclei with spins ($I \pm 1/2$), each of which has a different scattering length, b^+ or b^-, which is associated with the spin-up or spin-down states. For a given spin state J, the number of orientations is $(2J + 1)$ and thus the number of possible orientations for the compound spin states of $(I + 1/2)$ and $(I - 1/2)$ are $2(I + 1)$ and $2I$, respectively. The total number of spin states is $2(2I + 1)$, and as the probabilities of each state are equal, the statistical weights are $\dfrac{I+1}{2I+1}$ and $\dfrac{I}{2I+1}$, respectively.

The average (coherent) scattering length is

$$\langle b\rangle = \frac{I+1}{2I+1}\,b^{+} + \frac{I}{2I+1}\,b^{-} \tag{9}$$

where the brackets ⟨ ⟩ represent a thermal average over the spin-state population. A coherent cross section can be defined for each isotope by

$$\sigma_{\text{coh}} = 4\pi\langle b\rangle^2 \tag{10}$$

whereas the total-scattering cross section is given by

$$\sigma_{\text{tot}} = 4\pi\langle b^2\rangle \tag{11}$$

The difference between the two is the incoherent cross section σ_{inc}, which is given by

$$\sigma_{\text{tot}} - \sigma_{\text{coh}} = \sigma_{\text{inc}} = 4\pi\{\langle b^2\rangle - \langle b\rangle^2\} \tag{12}$$

If the isotope has no spin, then $\langle b^2\rangle = \langle b\rangle^2$ as $\langle b\rangle = b$ and there is no incoherent scattering. Only the coherent-scattering cross section contains information on interference effects arising from spatial correlations of the nuclei in the system, ie, the structure of the sample. The incoherent cross section contains no information on interference effects and forms an isotropic (flat) background, which must be subtracted in sans structural investigations. It does, however, contain information on the motion of single atoms (particularly hydrogen), which may be investigated via energy analysis of the scattered beam. Although most of the atoms encountered in neutron scattering from polymers are mainly coherent scatterers (eg, carbon, oxygen), there is one important exception. In the case of hydrogen (^{1}H), the spin-up and spin-down scattering lengths have opposite sign ($b^+ = 1.080 \times 10^{-12}$ cm; $b^- = -4.737 \times 10^{-12}$ cm), and as $I = 1/2$,

$$\sigma_{\text{coh}} = 1.76 \times 10^{-24}\ \text{cm}^2 \tag{13}$$

$$\sigma_{\text{tot}} = 81.5 \times 10^{-24}\ \text{cm}^2 \tag{14}$$

$$\sigma_{\text{inc}} = 79.7 \times 10^{-24}\ \text{cm}^2 \tag{15}$$

For photons, there is no strict analogue of incoherent scattering of neutrons due to nonzero spin in the scattering nucleus. Compton scattering that occurs for x rays is similar in that it contains no information on interference effects, ie, the structure of the sample, and forms a background that must be subtracted off.

Table 1 gives the cross sections and scattering lengths for atoms commonly encountered in synthetic and natural polymers. These cross sections refer to bound protons and neglect inelastic effects arising from interchange of energy with the neutron. For coherent scattering that is a collective effect arising from the interference of scattered waves over a large correlation volume, this approximation is reasonable, especially at low Q where recoil effects are small. However, for incoherent scattering, which depends on the uncorrelated motion of individual atoms, inelastic effects become increasingly important for long wavelength neutrons with the result that the H-incoherent cross section, and hence the sample transmission is a function of both the incident neutron energy and sample temperature. It may be seen from Table 1 that there is a large difference in the coherent scattering length between deuterium and hydrogen and that the latter value is actually negative. This arises from a change of phase of the scattered wave and results in a marked difference in scattering power (contrast) between polymer molecules synthesized with deuterium atoms or hydrogen atoms along the chain.

The basic experiment consists of an incident neutron beam, energy E_0, which is scattered by an assembly of sample nuclei into solid angle $d\Omega$ with energy change dE recorded by a neutron detector. The double-differential-scattering cross section for unit volume of sample, $d^2\Sigma/d\Omega dE$, is defined as the number of neutrons scattered per second into a solid angle $d\Omega$ with energy change dE, divided by the incident neutron flux (neutrons per second per unit area). In this article, σ denotes the cross section of a single nucleus and Σ denotes an assembly of nuclei, except where it denotes the standard summation sign. For such an assembly the double-differential-scattering cross section is given by standard scattering theory as

$$\frac{d^2\Sigma}{d\Omega dE} = \frac{k}{2\pi k_0}\int_{-\infty}^{+\infty} dt\, \exp(-i\omega t)\left\langle \sum_{ij} b_i{*}b_j F_{ij}(\mathbf{Q},t)\right\rangle$$
$$= \frac{k}{k_0}\,S(Q,\omega) \tag{16}$$

where

$$F_{ij}(\mathbf{Q},t) = \exp\,[-i\mathbf{Q}\cdot\mathbf{R}_i(0)]\,\exp\,[i\mathbf{Q}\cdot\mathbf{R}_j(t)] \tag{17}$$

Table 1. Bound-atom Scattering Lengths and Cross Sections for Elements in Synthetic and Natural Polymers

Atom	Nucleus	Scattering length b_{coh}, 10^{-12} cm	Coherent cross section σ_{coh}, 10^{-24} cm^2 ($\sigma_{\text{coh}} = 4\pi b^2$)	Incoherent cross section σ_{inc}, 10^{-24} cm^2
hydrogen	^{1}H	−0.374	1.76	79.7
deuterium	^{2}H (D)	0.667	5.59	2.01
carbon	^{12}C	0.665	5.56	0
nitrogen	^{14}N	0.930	11.10	0
oxygen	^{16}O	0.580	4.23	0
silicon	^{28}Si	0.415	2.16	0
chlorine	Cl[a]	0.958	11.53	5.9
fluorine	^{19}F	0.566	4.03	~0

[a] Values are for the naturally occurring element and are an average over the mixture of isotopes.

In equation 16 the symbol $*$ denotes a complex conjugate and $\langle\ \rangle$ denotes a thermal average over all configurations of scatterers at position vectors $\mathbf{R}(t)$ at time t. $S(Q,\omega)$ is called the scattering law or scattering function. Equation 16 may be separated into coherent and incoherent components of the cross-section

$$\frac{d^2\Sigma^{\text{coh}}}{d\Omega dE} = \frac{k}{2\pi k_0}\int_{-\infty}^{+\infty} dt \exp(-i\omega t)\left\langle \sum_{ij} \langle b_i\rangle * \langle b_j\rangle F_{ij}(\mathbf{Q},t)\right\rangle \quad (18)$$

and

$$\frac{d^2\Sigma^{\text{inc}}}{d\Omega dE} = \frac{k}{2\pi k_0}\int_{-\infty}^{+\infty} dt \exp(-i\omega t)\left\langle \sum_{i} |b_i - \langle b\rangle_i|^2 F_{ii}(\mathbf{Q},t)\right\rangle \quad (19)$$

From equations 18 and 19 it may be seen that coherent scattering contains information on the correlations between different nuclei and hence gives information on relative spatial arrangement of atoms in the system (eg, structure) and its time dependence. The incoherent cross section, on the other hand, contains information on correlations from the same nucleus and hence gives information on the time dependence of the motion of an individual atom (eg, vibration, diffusion, etc).

Because the scattering law separates into coherent and incoherent components, which may be further subdivided into elastic and inelastic and quasielastic processes, the type of information that may be obtained from neutron scattering from polymers subdivides into the general categories shown in Table 2. The majority of neutron scattering experiments undertaken on polymers fall into the category of sans which is an example of predominantly coherent elastic scattering and the first applications of this technique to polymer science were made in the early 1970s in Europe where the first small-angle scattering cameras were developed. The construction of such instruments in the United States began in the late 1970s and by the end of the 1980s, over 30 laboratories and institutes throughout the world had constructed or initiated sans instrumentation.

Small-angle Neutron Scattering

High Concentration Labeling. In experiments designed to determine the chain conformation in bulk polymer by sans, no energy discrimination is employed and the detector integrates over all energies. The scattered intensity $I(Q)$ is measured as a function of angle, and for isotropic (nonoriented) samples Q is a scalar quantity. The differential cross section is obtained by integrating equation 18 over ω, and noting that this operates on the phase factor to give a delta function in time. Furthermore, for the typical elements contained in polymers, the scattering lengths may be treated as real, dropping the complex conjugate. Where the element present consists predominantly of a single isotope, $\langle b\rangle$ may be replaced by b; eg, for naturally occurring carbon $\langle b\rangle = b = 0.665 \times 10^{-12}$ cm. With these simplifications equation 18 becomes

$$\frac{d\Sigma^{\text{coh}}}{d\Omega}(Q) = \left\langle \sum_{ij} b_i b_j \exp\left[i\mathbf{Q}\cdot(R_j - R_i)\right]\right\rangle \quad (20)$$

Nuclear cross sections have the dimensions of area, and as sample cross sections are normalized to unit volume, $d\Sigma(Q)/d\Omega$ has the dimensions of inverse length and is typically given in units of cm^{-1}.

For a bulk polymer sample with N molecules per unit volume of pure unlabeled component, a coherent scattering length of a monomer unit can be defined

$$a_{\text{H}} = \sum_k b_k \quad (21)$$

where the summation runs over all the atoms in an unlabeled monomer unit and a similar equation may be written for the coherent scattering length of a labeled monomer unit a_{D}. If the two polymers are blended together so that X_{H} equals the mole fraction of unlabeled components and X_{D} is the mole fraction of the labeled polymer component the cross section is given by

$$\frac{d\Sigma}{d\Omega}(Q) = X_{\text{D}}X_{\text{H}}(a_{\text{H}} - a_{\text{D}})^2 NZ^2 P(Q) \quad (22)$$

where $P(Q)$ is the intrachain signal that originates from monomer pairs belonging to the same polymer chain and is called the form factor of the molecule ($P(Q) = 1$). Equation 22 is based on the assumptions that deuteration of the hydrogenous molecule has a negligible effect on the monomer–monomer interactions, and that both chains have the same polymerization index Z and the same number of molecules per unit volume (N). This shows that the form of the scattering curve in this case is governed by the single-chain form factor $P(Q)$. The mole fraction of each component modulates the scattered intensity with the maximum coherent scattering of the blend occurring at a 50:50 mixture of the two components. Thus $P(Q)$ may be obtained from the measured coherent intensity at labeling levels up to 50%. Although equation 22 is essentially the same formula derived by von Laue for random binary alloys, the result was not appreciated in the earliest sans studies of bulk

Table 2. Information Obtained from Neutron-scattering Experiments on Polymers

Type of scattering	Energy change: Elastic, $\Delta E = 0$	Quasielastic, $\lvert\Delta E\rvert << E_0$	Inelastic, $\Delta E \neq 0$
coherent	sans chain configuration in bulk and solution, polymer compatibility, orientation mechanisms wide-angle diffraction crystal structure	molecular dynamics in the bulk and in solution via Dopper broadening of the elastic peak	elastic constants of crystalline polymers via phonon-dispersion curves
incoherent		effective diffusion for segmental motion in the bulk, activation energy	side-group vibrational frequencies and rotation barriers, intermolecular potentials

polymers and concentrated solutions. These studies relied on analogies with light and x-ray scattering, where the limit of zero concentration was required to eliminate interchain interference. The scattered signal was assumed to be proportional to c, the concentration (g/cm^3) of labeled (deuterated) polymer, and this assumption has been shown to be a reasonable approximation in this limit:

$$\frac{d\Sigma}{d\Omega}(Q) = c(a_H - a_D)^2 \bar{M}_{wD} \frac{N_A}{m_D^2} P(Q) \qquad (23)$$

where N_A is Avogadro's number (6.023×10^{23}), m_D is the molecular weight of a repeat unit of the deuterated polymer, and $\bar{M}_{wD}$ is the (weight-averaged) molecular weight of a deuterated chain.

The quantity $(a_H - a_D)^2$ is related to the difference in scattering power between labeled and unlabeled chains and is called the contrast factor. In general, radiation incident on a medium whose scattering power is independent of position is scattered only into the forward direction ($2\theta = 0$). For every volume element S that scatters radiation through an angle $2\theta > 0$, there is another volume element S' that scatters exactly (180°) out of phase, and therefore all scattering cancels unless the scattering power is different at S and S', ie, fluctuates from point to point in the sample. By analogy with x-ray scattering, which is caused by fluctuations in electron density, neutron scattering arises from differences in scattering length density (SLD), which is defined as the sum of coherent scattering lengths over all atoms lying in a given volume δV, divided by δV. For partially labeled polymer blends, the SLD is given by the coherent scattering length (eq. 21) divided by the monomer volume. The coherent cross section of a system of uniform scattering length density is zero, although fluctuations may be introduced by means of isotopic substitution, thus giving rise to a finite cross section that is proportional to $(a_H - a_D)^2$. In order to produce contrast observable by x-ray scattering that can be used to give direct information on $P(Q)$ it is necessary to change the electron density of the labeled chain. Such experiments have been performed by labeling polystyrene chains statistically with iodine atoms and by this means concentrated solutions and bulk polymers were investigated by x-ray scattering. This type of labeling method relies on changing the chain chemistry and in general produces a greater perturbation on the chain trajectory than deuterium labeling methods. The method seems to give reasonable results after extrapolation to infinite dilution of the iodine-labeled molecules, but it seems unlikely that labeling levels of up to 50% could be used with this approach.

The contrast variation methods that have found wide application in biology can sometimes be used to remove a component of the scattering by matching its scattering power with that of the medium in which it is dispersed, and thus removing the fluctuation giving rise to the scattering. This principle may be illustrated by an example suggested by Prof. D. Engelman of Yale University. Two tubes (A and B) each contain two Pyrex beads embedded in glass wool, which has a lower refractive index than the Pyrex. Tube A is filled with a solvent that has the same refractive index as the glass wool. When light shines on the tubes, only the Pyrex beads are visible in tube A because the electron density and hence the scattering power of the glass wool has been matched with that of the solvent, thus eliminating this component of the scattering and making the wool transparent to light. In tube B, both the beads and glass wool scatter light, but only the glass wool can be seen because it dominates the scattering.

The parameter used to describe the overall size of a polymer chain is the radius of gyration R_g, which may be derived from equation 22 or 23 by expanding $P(Q)$ in a power series for low $Q (Q < R_g^{-1})$ and plotting $d\Sigma^{-1}(Q)/d\Omega$ vs Q^2. Alternatively, these parameters may be obtained by plotting $\ln d\Sigma(Q)/d\Omega$ vs Q^2 at low Q. These types of plot are conventionally referred to as Zimm and Guinier plots, respectively, and the former is generally used for investigating polymer configurations, as it has been found to be linear over a wider Q range.

The first measurements in the bulk and concentrated solution were generally performed in the limit of low relative labeling and extrapolated to zero concentration. In this range

$$\frac{d\Sigma^{-1}}{d\Omega}(Q) = \frac{m_D^2}{C(a_H - a_D)^2 \bar{M}_{wD} N_A}\left[1 + \frac{Q^2 R_g^2}{3} + \cdots\right] \qquad (24)$$

Thus R_g and $\bar{M}_{wD}$ may be derived from the slope and $Q = 0$ intercept of such a plot. The realization that the same information could be obtained with greater accuracy at much higher levels of labeling was made and verified independently by several groups. The variation of R_g with the concentration of labeled molecules in amorphous polycarbonate is shown in Figure 1; the measured values are independent of the level of labeling. Similarly, it is shown in Figure 2 that the extrapolated $Q = 0$ cross section $d\Sigma(0)/d\Omega$ is proportional to the product $X_H X_D$ for amorphous polystyrene, as expected from equation 22.

Amorphous Polymers. The first convincing demonstration of the power of the sans technique was made in the field of bulk amorphous polymers. It is well known that there have been several theoretical approaches to the molecular conformation in these systems, based on the unperturbed Gaussian (random) coil due to Flory and coworkers, and the meander or bundle models, where a large fraction of the polymer molecules are envisioned in quasiparallel arrangement. In addition, a collapsed-coil model was advanced for some systems. Before the development of the sans technique there was no way of measuring the molecular conformation in bulk polymers and this led to intense debate on this issue in the literature. The first

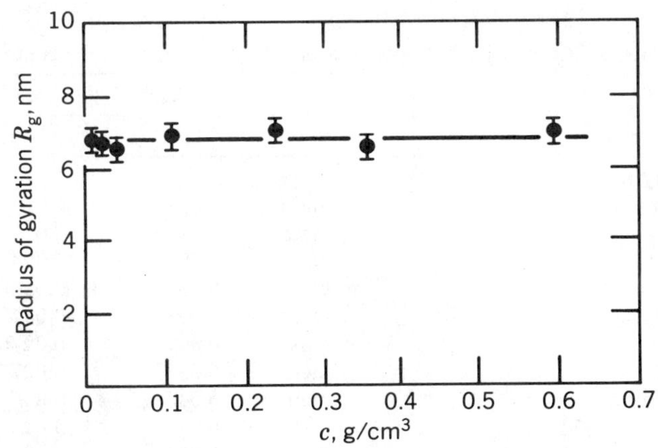

Figure 1. Radius of gyration R_g for molecules in amorphous polycarbonate vs the concentration of labeled (deuterated) polymer. $\bar{M}_w$ (sans) = 20,100; $\bar{M}_w$ (viscometry) = 20,300; $\bar{M}_w/\bar{M}_n$ = 1.79. Courtesy of *Polymer Bulletin*.

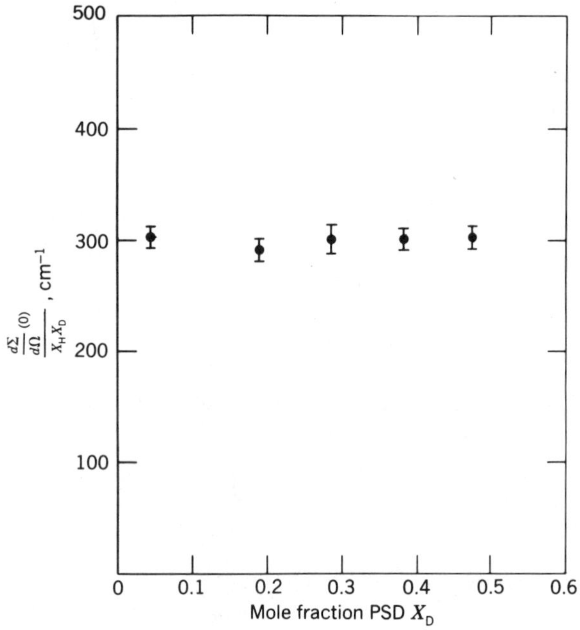

Figure 2. Variation of $1/X_H X_D\, d\Sigma/d\Omega(0)$ with X_D for blends of deuterated (PSD) and normal (hydrogenous) polystyrene. Courtesy of *Polymer.*

suggestion to use the contrast between deuterated and normal (hydrogenous) molecules to provide a direct determination of $P(Q)$ was made in the late 1960s and the method was first demonstrated in principle in the early 1970s.

According to the random-coil model, R_g^w should be proportional to $\bar{M}_w^{1/2}$, where $\bar{M}_w$ is the (weight-averaged) molecular weight, with the same constant of proportionality in the bulk as in an ideal θ-solvent. It may be seen from Table 3 that in general this prediction holds remarkably well for amorphous polymers and that there is close agreement between data on the same polymer studied independently by different groups.

Although these results have given impressive support to this model, they are not in themselves conclusive, because it was subsequently shown for crystalline polymers that R_g is similar for molecules in the molten (amorphous) and solid (crystalline) states. Thus the finding that molecules exhibit the unperturbed dimensions in the molten or glassy amorphous state does not in itself rule out significant parallelism for an appreciable fraction of the molecules. To test how far the local molecular configuration, as opposed to the overall R_g, is described by the various models, measurements have been extended to higher values of Q. To first order the scattered intensity at a given Q is sensitive to fluctuations in the scattering length density on a distance scale $D \sim 2\pi/Q$. Thus as Q increases, the scattering is increasingly determined by the local chain configuration. This configuration may be calculated for the random-coil model using rotational isomeric statistics, and hence the scattered intensity may be estimated numerically and compared with the experiment. This is accomplished by measuring the scattering in the intermediate-angle range ($1 < Q < 6\ \text{nm}^{-1}$), which is sensitive to the local configuration of the chain over distances ~1–5 nm. Figure 3 shows the intermediate-angle neutron scattering (ians) data for molten polyethylene at $T = 150°\text{C}$ compared to the rotational isomeric state (RIS) calculation of Yoon and Flory and the Debye model for a coil with a Gaussian distribution of chain elements. The data are plotted as $Q^2\, d\Sigma(Q)/d\Omega$ vs Q because this representation enhances the scattering at higher Q and facilitates comparison with different models. It may be seen from Figure 3 that $d\Sigma(Q)/d\Omega$ varies as Q^{-2} in this region, leading to a plateau in the plot, which is closely fitted by both the RIS calculation and Gaussian coil function. Similarly, the ians data for atactic polystyrene are consistent with the Debye model both in the melt and glassy states. Similar comparisons in the intermediate Q range have also been made with polyisobutylene, poly(methyl methacrylate), and polycarbonate, and in each case reasonable agreement was achieved with the rotational isomeric theory. To the author's knowledge there are no major discrepancies between theory and the scattering experiments that might be indicative of quasiparallel packing of polymer chains.

Table 3. Molecular Dimensions in Bulk Amorphous Polymers

		$[(R_g^w)^2/\bar{M}_w]^{1/2}$, nm/g$^{1/2}$	
Polymer	State	Bulk	Solvent
poly(methyl methacrylate)			
atactic	glass	0.027	0.025
	glass	0.025	0.025
syndiotactic	glass	0.029	0.024
isotactic	glass	0.030	0.028
polystyrene, atactic	glass	0.0280	0.0275
	glass	0.0275	0.0275
	melt	0.0280	0.0275
polyethylene	melt	0.045	0.045
	melt	0.046	0.045
poly(vinyl chloride)	glass	0.040	0.037
polyisobutylene	glass	0.031	0.030
poly(ethylene terephthalate)	glass	0.039	0.039–0.042
polybutadiene	melt	0.035	0.034–0.042

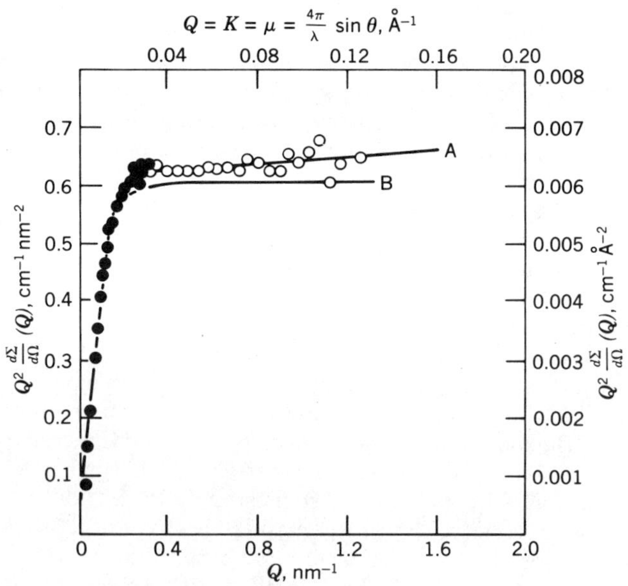

Figure 3. Kratky plot for molten polyethylene at 150°C. A, Yoon and Flory rotational isomeric calculation (1976); B, Debye, Gaussian coil. $\frac{d\Sigma}{d\Omega}(Q) = \frac{2}{u^2}\frac{d\Sigma}{d\Omega}(0)[u - 1 + e^{-u}]$, $u = Q^2R_g^2$. Courtesy of Butterworth & Co. (Publishers) Ltd.

The Crystalline State. The arrangement of molecular chains within the lamellae of semicrystalline polymers has long been disputed, and sans has provided new information on this subject. Semicrystalline polymers exhibit a lamellar morphology for material crystallized both from the melt and from dilute solution. The thickness of lamellae is typically 10–50 nm with amorphous polymer interspersed between the crystalline regions. The molecular chains are at an angle (0–45°) to the lamellar normals and have lengths much greater than the lamellar thickness, thus traversing one or more lamellae several times. Based on considerations of density conservation at the crystal–amorphous boundary, Flory demonstrated that a considerable fraction (~0.5) of chains must return to the same crystal. What is in question is whether the molecule returns with predominantly adjacent or with random reentry to the crystallite of origin.

In general, sans experiments reveal that the radius of gyration R_g of molecules remains unchanged on crystallization from the melt and hence has a $\bar{M}_w^{1/2}$ dependence in both the molten and crystalline states. This indicates that the molecules crystallize with a similar distribution of mass elements to that possessed in the melt and hence is distributed over several lamellae in the crystalline state. For solution-crystallized material, however, the radius of gyration is relatively independent of molecular weight, and is usually markedly reduced from the dimensions in dilute solutions, indicating a much more compact configuration. These measurements of R_g were made at low Q and contain no information on the mutual arrangement of stems, ie, straight sections of a chain traversing a crystalline lamella. This is best examined by experiments in the intermediate-angle range, which is sensitive to the correlation of stems over distances of 1–5 nm. This type of measurement has been made for several systems and compared with a variety of model calculations that simulate the chain trajectory.

Figure 4 shows ians data for quench-crystallized polyethylene, where the scattering function $F_n(Q)$ is defined by

$$F_n(Q) = (n + 1)Q^2P(Q) = (n + 1)Q^2 \frac{\frac{d\Sigma}{d\Omega}(Q)}{\frac{d\Sigma}{d\Omega}(0)} \qquad (25)$$

where n is the number of bonds in the chain. There is reasonable consistency between the data from several groups that are compared with model calculations based on Monte Carlo statistics as a function of the probability (p_{ar}) that the stem folds adjacently along the ⟨110⟩ plane. It may be seen that the model leads to a Q^{-2} dependence for $d\Sigma(Q)/d\Omega$ and hence a plateau in the plot. The plateau levels differ by a factor ~2 for the extremes of random ($p_{ar} = 0$) and adjacent ($p_{ar} = 1$) reentry, and it was concluded that the ians data were inconsistent with regular folding ($p_{ar} < 0.3$).

In a careful review of the scattering from both quenched and slowly cooled materials, Sadler concluded that the degree of adjacency within the rows is ca 30–50% for melt-grown crystals. Although the precise definition of p_{ar} varies between different models, it seems that probabilities of this order are inconsistent with regular folding of an appreciable number (>4) of stems in one crystallographic plane. Longer sequences of adjacent stems would lead to an observable modulation of the wide-angle neutron scattering pattern ($Q > 6\ \text{nm}^{-1}$), and such

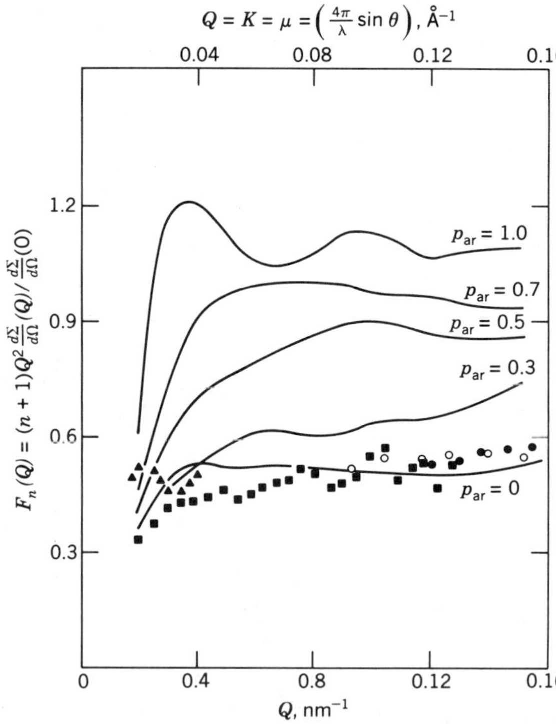

Figure 4. Calculation of $F_n(Q)$ as a function of the probability of adjacent folding (p_{ar}) (number of bonds in chain n = 2000). Various symbols for different studies. Courtesy of IUPAC.

modulations are not observed for melt-crystallized polyethylene. However, a large proportion of folds are relatively close and the folds that are not adjacent are "near" and rarely involve stem separations greater than three nearest neighbors. Although some significant disagreement still persists and further work is in progress, it seems generally agreed that the neutron data rule out the possibility that a typical molecule is regularly folded in one crystallographic plane over many stems without interruption. This model had gained widespread support over the previous decades for both melt and solution crystallized material. Similarly the extremes of random configurations have been ruled out with a large fraction of molecules usually folding in near reentry within a few nearest neighbors.

There are many other important areas where the sans technique has been applied in polymer science (e.g., semidilute and concentrated solutions, polymer blends, block copolymers, the response of individual molecules to deformation and swelling, diffusion, relaxation, the structure of latices, etc.). All of these studies have been based on the assumption that the interaction parameter between labeled and unlabeled molecules of the same species χ_{HD} is zero. This assumption has also been implicit in previous sans studies, although calculations have indicated that χ_{HD} may be finite and on the order of 10^{-4}–10^{-3}. This has led to the suggestion that for sufficiently high molecular weights, demixing could occur in mixtures of deuterated and hydrogenous molecules of the same species, and that this could lead to the measurement of χ_{HD}. Such experiments have been performed by Bates and coworkers who examined binary mixtures of deuterium-labeled and unlabeled 1,4-polybutadienes. By regarding the labeled and unlabeled molecules as different species with volume fractions ϕ_H and ϕ_D, respectively, the inter-

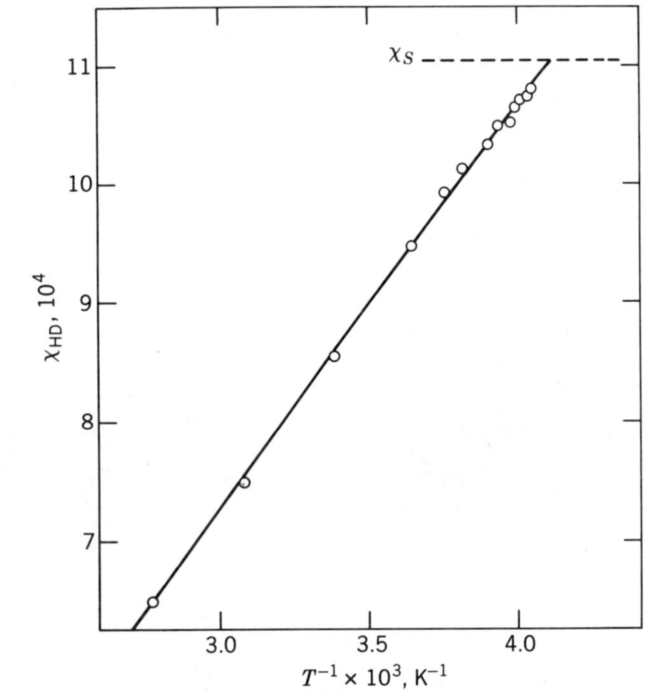

Figure 5. Plot of χ_{HD} vs T^{-1} for blend of deuterated (ϕ_D = 0.31) and hydrogenous (ϕ_H = 0.69) polybutadienes; $\chi = \frac{0.326}{T} - 2.3 \times 10^{-4}$. Courtesy of the American Physical Society.

action parameter χ_{HD} between them may be estimated by fitting the random phase approximation (RPA) of deGennes to the measured scattering data.

Figure 5 shows a plot of χ_{HD} as a function of inverse temperature (K^{-1}), which reaches the value on the spinodal at T_S = 242.1 ± 0.7 K. The critical exponents exhibit the mean-field values and at room temperature (T = 296 K) the value of $\chi_{HD} = 8.8 \times 10^{-4}$. Similar measurement in binary mixtures of deuterated and protonated polystyrenes, polyvinylethylenes, polyethylethylenes, and polydimethylsiloxanes corroborate the prediction of a universal isotope effect, which has been shown to be a consequence of zero-point motion in conjunction with the anharmonicity of the interatomic potential. χ values on the order of 10^{-4}–10^{-3} are usually much smaller than the values measured between different species. However, the correction for this effect should be considered when these methods are applied to the measurement of smaller χ parameters. Similarly, the effect of a finite χ_{HD} does not invalidate the vast majority of investigations using the sans techniques, as segregation effects are important only in high molecular weight systems and have generally not been observed.

The examples discussed herein are indicative of the information on the time averaged structure of polymers that can be gained on the distance scale of 1–100 nm by small-angle scattering techniques. Similarly, inelastic and quasi-elastic neutron scattering have given a great deal of new information on the motion of molecules or segments in a wide variety of polymer systems.

BIBLIOGRAPHY

G. E. Bacon, *Neutron Diffraction,* Clarendon Press, Oxford, UK, 1971; W. Marshall and S. W. Lovesey, *Theory of Thermal Neutron Scattering,* Clarendon Press, Oxford, UK, 1971.

G. D. Wignall, "The Physics of Photons and Neutrons with Applications of Deuterium Labeling Methods to Polymers," p. 27 in *Scattering, Deformation and Fracture in Polymers,* Vol. 79, ed. by G. D. Wignall, B. Crist, T. P. Russell, and E. L. Thomas, Materials Research Society, Pittsburgh, 1987.

J. S. Higgins and A. Maconnachie in K. Skold and D. L. Price, eds., *Neutron Scattering,* Volume 23, Academic Press Inc., New York, 1987, p. 287.

F. S. Bates, G. D. Wignall, and W. C. Koehler, *Phys. Rev.* Lett. **55,** 2425 (1985).

D. Sadler in I. Hall ed., *The Structure of Crystalline Polymers,* Elsevier Applied Science Publishers, Ltd., Barking, UK, 1983, p. 125.

G. D. Wignall
Oak Ridge National Laboratory

NEWTONIAN FLOW. See Viscoelasticity.

NITRILE RUBBER. See Butadiene polymers.

NITROCELLULOSE. See Cellulose esters, inorganic.

NITROETHYLENE POLYMERS. See Anionic polymerization.

NITROSO POLYMERS

The term "nitroso polymers" has been used for polymers with the structure —N—O—(C)$_m$— as the repeating unit (nitroso rubbers). Discovery of a novel polymeric system having the azo-N,N'-dioxide —(N(→O)=N(→O))— structure in the repeating unit, corresponding to the true dimeric form of nitroso compounds, required broadening the definition. Also, polymerization of nitrosobenzene and 1,4-dinitrosobenzene has been achieved and polymers with pendent nitroso groups have been prepared.

Nitroso Rubbers

Trifluoronitrosomethane (CF_3NO) and tetrafluoroethylene (C_2F_4), in the absence of light and air, react at 20°C to give a mixture of perfluoro-2-methyl-1,2-oxazetidine (**1**) and a polymeric oil having the form of structure (**2**):

(**1**) (**2**)

The oxazetidine is the predominant product at ambient temperatures and above. However, below 0°C, a 1:1 alternating elastomeric copolymer having structure (**2**) is obtained in high yield. This nitroso rubber is nonflammable, is completely amorphous, has commendable low temperature flexibility (T_g = −51°C), and has excellent resistance to protic solvents, oxidizing agents, and ozone. Despite this combination of properties, nitroso rubbers have found only limited and highly specialized applications because of the unavailability of an efficient vulcanization system, poor resistance to heat and electromagnetic radiation, and degradation to extremely toxic products.

The only nitroso polymers of any commercial importance so far are the copolymers of trifluoronitrosomethane and tetrafluoroethylene and terpolymers that include small quantities of functional nitroso compounds as a third monomer, eg, 4-nitrosohexafluorobutyric acid, to provide cross-linking sites.

Polymerization. CF_3NO does not homopolymerize, but reacts spontaneously on mixing with an equimolar quantity of C_2F_4 to give a 1:1 alternating copolymer at temperatures below 0°C. Preparative methods for bulk and suspension polymerization have been reported.

Mechanism. Copolymerization of CF_3NO and C_2F_4 probably proceeds by a free-radical mechanism. However, a precise initiating step has not been elucidated.

Structure. A copolymer of CF_3NO and C_2F_4 prepared at −78°C was shown by mass spectrometry to have a structure similar to that of (**2**). However, high resolution ^{19}F nmr studies on a polymer prepared at −25°C by a suspension method indicated the presence of the following type of minor irregular repeat units:

$$-\!\!\left[ON(CF_3)CF_2CF_2N(CF_3)O\right]\!\!- \qquad -\!\!\left(O-CF_2CF_2-O\right)\!\!-$$

The reaction variables, particularly temperature, influence the amount of irregular orientation.

Properties. The properties of the raw copolymer of CF_3NO and C_2F_4 are shown in Table 1.

Poly(azoalkylene-*N,N′*-dioxide)

Bifunctional nitrosoalkanes with suitable structural features form stable poly(azoalkylene-*N,N′*-dioxide) (poly(nitrosolkane)) polymers, as shown below

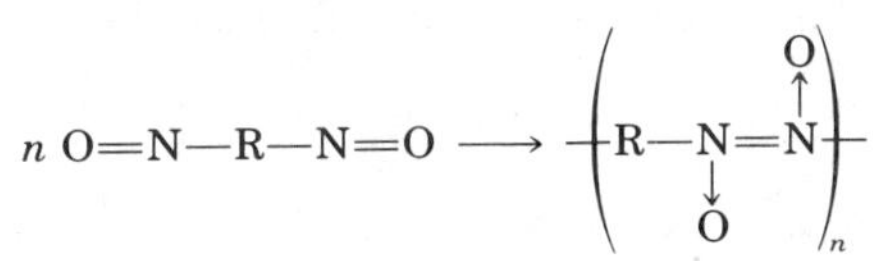

The polymerization apparently proceeds by dimerization of nitroso groups.

Aromatic Nitroso Polymers

Various condensation polymers have been synthesized by the reaction of monomers containing nitroso groups, such as dimeric 4-nitroso-3,5-dichlorobenzoyl chloride with diols or diamines. These polymers have $-\overset{\overset{O}{\uparrow}}{N}=\underset{\underset{O}{\downarrow}}{N}-$ linkages in the chain backbone. The nitrosation of numerous additional polymers for the purpose of cross-linking by dimerization of nitroso groups has also been described. Such cross-links are reversible because they are thermally labile.

Table 1. Properties of Unvulcanized Copolymer of CF_3NO and C_2F_4

Property	Value	
Mark-Houwink constants[a] in	*K*	*a*
trichlorotrifluoroethane (35°C)	2.80×10^{-4}	0.51
perfluorotributylamine (25°C)	8.77×10^{-5}	0.66
solubility and solution properties	soluble in fluorocarbons	
specific gravity	1.93	
n_D^{25}	1.3170	
T_g, °C	−51	
flammability	nonflammable; will not flame	
crystal structure[b]	amorphous	

[a] $[\eta] = KM^a$.
[b] By x-ray diffraction.

Health and Safety Information

LC_{50} for C_2F_4 is 40,000 ppm/4 h. This compound has been reported to detonate unpredictably, particularly if uninhibited. Nitroso rubbers are not known to be toxic, but their degradation products, ie, CF_2O, CF_3NCF_2, are extremely toxic. All nitroso compounds should be handled with care.

BIBLIOGRAPHY

C. E. Griffen and R. N. Haszeldine, *Proc. Chem. Soc.* **1959,** 369; *J. Chem. Soc.* **1960,** 1398.

G. H. Crawford, *J. Polym. Sci.* **45,** 259 (1960).

G. A. Morneau, P. I. Roth, and A. R. Shultz, *J. Polym. Sci.* **55,** 609 (1961).

D. K. Dandge and L. G. Donaruma in B. M. Culbertson and C. U. Pittman, Jr., eds., *New Monomers and Polymers,* Vol. 25 of *Polymer Science and Technology Reviews,* Plenum Press, New York, 1984.

D. K. DANDGE
ST & E, Inc.

L. G. DONARUMA
Macrochem Corporation

NOMENCLATURE

Source-based Nomenclature

Traditionally, polymers have been named by attaching the prefix poly to the name of the real or assumed monomer (the "source") from which it is derived. Thus polystyrene is the polymer made from styrene and will often be found in an index under "styrene, polymer of." When the name of the monomer consists of two or more words, parentheses should be used, as in poly(vinyl acetate), poly(methyl methacrylate), poly(sodium styrenesulfonate), etc. Source-based nomenclature can be easily adapted to the naming of copolymers, eg, poly(styrene-*co*-acrylonitrile).

A block copolymer containing 75 mass % of polybutadiene and 25 mass % of polystyrene is polybutadiene-*block*-polystyrene (0.75:0.25 *w*) or *block*-copoly(butadiene/styrene) (75:25 mass %).

Structure-based Nomenclature

For organic polymers that are regular, ie, have only one species of constitutional unit in a single sequential arrangement, and consist only of single strands, the IUPAC has promulgated a structure-based system of naming polymers.

The steps involved in naming the constitutional repeating unit are (*1*) identification of the unit, taking into account the kinds of atoms in the main chain and the location of substituents, (*2*) orientation of the unit, and (*3*) naming of the unit.

Table 1. Examples of Systematic Structure-based Names for Polymers[a]

Structure	Structure-based name	Common (source-based) name
$-(CH(CH_3)CH_2)_n-$	poly(propylene)	polypropylene
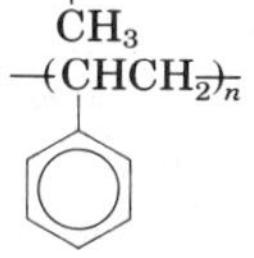	poly(1-phenylethylene)	polystyrene
$-(CH(OOCCH_3)CH_2)_n-$	poly(1-acetoxyethylene)	poly(vinyl acetate)
$-(C(CH_3)(COOCH_3)-CH_2)_n-$	poly[1-(methoxycarbonyl)-1-methylethylene]	poly(methyl methacrylate)
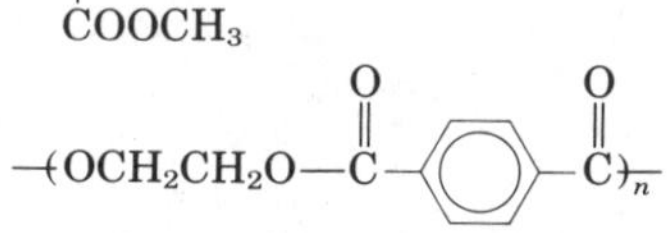	poly(oxyethyleneoxyterephthaloyl)	poly(ethylene terephthalate)

[a] Courtesy of *Pure and Applied Chemistry*.

Examples of names for some common polymers are given in Table 1.

BIBLIOGRAPHY

IUPAC, *Compendium of Macromolecular Nomenclature* (*The Purple Book*), Blackwell Scientific Publications, Oxford, 1990.

NORBERT M. BIKALES
National Science Foundation
Secretary (1978–1987), IUPAC Commission on Macromolecular Nomenclature

NONDESTRUCTIVE TESTING

Nondestructive testing (NDT) involves determining the usefulness of a specimen for a given purpose by any means that do not decrease that usefulness. Familiar NDT methods include visual testing, radiography, ultrasonic testing, penetrant testing, acoustic or tap testing, acoustic emission testing, leak testing, thermal testing, magnetic, electrical, and eddy-current testing, and composition testing by x-ray fluorescence or neutron activation analysis. These and many other less familiar methods are used to evaluate material properties, composition, or structure. Serviceability is related to the detection of defects (discontinuities judged to harm the usefulness of the specimen) such as cracks, lack of bonding between components, incorrect composition, inclusions, voids, and improper thickness or other structural property.

Nondestructive testing is used for quality control, process control, and reliability assurance for raw materials, fabrication processes, in-service monitoring, and maintenance programs. Such testing finds use in predicting maintenance and serviceable life as well as life extension for a variety of materials, coatings, components, assemblies, and operating systems. Proper use can help prevent premature failure, reduce manufacturing and service costs, minimize liability risks and costs, and increase reliability and safety for the users of the simplest materials to complex systems such as aircraft.

Large corporations, regulatory agencies, and consumers such as the Department of Defense have recognized the importance of specifying and using NDT in the fabrication and use of the complex systems they produce, acquire, or regulate. This is particularly true for expensive facilities and systems or circumstances in which failures during service are potentially disastrous. Nondestructive testing programs have been adopted to provide public, corporate, and governmental protection from the costs and penalties of failures. These programs include industrial-independent NDT programs, code-required programs, and special governmental agency programs. Governmental agencies using special NDT programs for polymeric components and systems include the Department of Defense, NASA, and the Federal Aviation Administration.

BIBLIOGRAPHY

Materials Evaluation, American Society for Nondestructive Testing, Columbus, Ohio 43228 (published monthly).

F. A. Iddings, *Survey of Nondestructive Evaluation,* Electric Power Research Institute (EPRI), EPRI NDE Center, Charlotte, N.C. 1986.

Howard E. Boyer, ed., *Metals Handbook,* 8th Edition, Volume 11, "Nondestructive Inspection and Quality Control," American Society for Metals, Metals Park, Ohio 44073.

R. C. McMaster, ed., *Nondestructive Testing Handbook,* The Ronald Press Co., New York, 1959 and 1963; American Society for Nondestructive Testing, Columbus, Ohio 43228, 1977 and 1979.

FRANK IDDINGS
Southwest Research Institute

NONAQUEOUS DISPERSIONS. See COATINGS; EMULSION POLYMERIZATION; LATICES.

NONCOMBUSTIBLE FABRICS. See FLAMMABILITY.

NON-NEWTONIAN FLOW.
See VISCOELASTICITY.

NONWOVEN FABRICS

Survey

Nonwoven fabrics are porous, textilelike materials, usually in flat sheet form, composed primarily or entirely of fibers assembled in webs. These fabrics, also called bonded fabrics, formed fabrics, or engineered fabrics, are manufactured by processes other than spinning, weaving, or knitting. The thickness of the sheets may vary from 25 μm to several centimeters and the weight from 10 g/m^2 to 1 kg/m^2. A sheet may resemble paper or a woven or knitted fabric in appearance or may have a unique texture or pattern. It may be as compact and crisp as paper or supple and drapable as a conventional textile; it may be resilient or limp. Its tensile properties may be barely self-sustaining or so high that it is impossible to tear, abrade, or damage the sheet by hand. The fiber components, one or several types, may be natural or synthetic, from 1–3 mm long to endless. The tensile properties may depend on frictional forces or a film-forming polymeric additive functioning as an adhesive binder. All or some of the fibers may be welded by heat or solvent. A scrim, gauze, netting yarn, or other conventional sheet material may be added to one or both faces, or embedded within as reinforcement. The nonwoven fabric may be incorporated as a component in a composite structure.

According to a recent definition, textiles and nonwoven fabrics are classified into three groups: fiber-web structures, netlike structures, and multiplex structures. In fiber-web structures, fibers or filaments are the fundamental units arranged into a web and bonded in such a manner that the bond-to-bond distances are greater than 50–100 times the fiber diameter. This definition clearly excludes paper, wherein bond-to-bond distances are on the order of the fiber diameter. However, this definition also excludes some nonwoven fabrics heavily impregnated with binder, where bond-to-bond distances are less than 50–100 times the fiber diameter.

In contrast to conventional textiles, the basic structure of all nonwovens is a web of fibers; thus the basic element is the single fiber. The individual fibers are arranged more or less randomly (Fig. 1). Tensile, stress–strain, and tactile properties are imparted to the web by an adhesive or other physicochemical bonding, by fiber-to-fiber friction created by entanglement, and reinforcement by yarns, scrims, nettings, films, foams, and additional unbonded webs.

The technology for making nonwoven fabrics is based on the following primary elements: fibers of various lengths and diameters; a web geometrically arranged according to the method of forming and processing; the bonding within the web by adhesive, or mechanical–frictional forces created by fiber contact or entanglement; and reinforcement and composites, eg, yarns, scrims, films, nettings, and unbonded webs. Variations of the several elements in combination allows for the enormous range of nonwoven fiber types.

Fibers

In the present context, as applied to regenerated cellulose and synthetic fibers, staple fibers are of relatively uniform length, ca 1.3–10.2 cm, with a regular crimp. Regenerated and other extruded fibers are endless as formed. They are cut during the manufacturing process to a specified length to meet a processing or market need. Extruded fibers are also produced as continuous filaments without crimp.

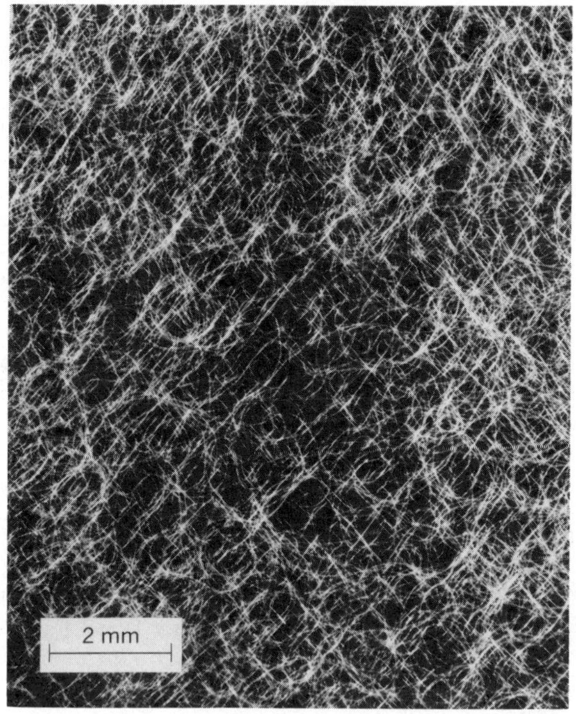

Figure 1. Polyester air-laid web with random fiber distribution. Courtesy of Chicopee.

The fibers, as defined by chemical composition and as a result of their mechanical properties, determine the ultimate properties of the fabric. Web structure and bonding maximize inherent fiber characteristics.

All natural fibers that are commercially available can be formed into webs. In practice, wood pulp, which is far shorter than staple length, is the only natural fiber used in large amounts because of its high water absorbency, bulk, and low cost. Regenerated fibers include viscose rayon and cellulose acetate (see FIBERS IN CELLULOSE ESTERS, ORGANIC). Among the synthetic fibers, poly(ethylene terephthalate) (PET), nylon-6 and -6,6, polypropylene (PP), and poly(vinyl alcohol) fibers are employed in nonwoven fabric manufacture.

Manufacturers in the UK (ICI), Japan (Chisso), and the United States (BASF) have developed bicomponent fibers especially for nonwoven materials. These fibers contain a low melting and a high melting component. The former is usually the outer layer and the latter is the core; occasionally, the two components are side-by-side.

Length. Natural fibers, such as wood pulp and cotton, are characterized by specific length. Synthetic and regenerated fibers are made by extrusion and can be prepared in any length; they are usually furnished in precision-cut, uniform length. The length is usually chosen for convenience in the manufacturing process.

Diameter. Regenerated and synthetic fibers are produced in a wide range of diameters for both conventional textiles and nonwoven materials. The fiber diameter has a great effect on the properties of the finished fabric and must be chosen carefully. Fiber diameter influences the fabric hand, bonding properties, cover or opacity, strength and resilience, web thickness, filtering properties, and processing equipment requirements.

Crimp. Crimp refers to a three-dimensional wave along the fiber, identified as crimps per centimeter. In the finished nonwoven material, crimp imparts a third dimension and improves resilience, appearance, and feel. Fibers intended to be made into webs by wet or papermaking processes are not crimped, because this would promote fiber entanglement in the water-dispersion step. Absence of crimp contributes to the flat, dense properties typical of wet-formed nonwovens.

Cross Section. Synthetic fibers, eg, polyester, nylon, and polypropylene, are extruded in a variety of shapes. Some noncircular shapes are better for bonding efficiency and fabric softness, but round cross-sectional fibers are commonly used in nonwoven fabric.

Spin-finish. Spin-finish refers to the lubricant that is added in small amounts (ca 0.2 wt %) to the surface of extruded fibers to enhance processibility.

Draw Ratio. Synthetic fibers for nonwoven use are drawn during extrusion to the same extent as for general textile use except in special cases involving PET and PP to make them more suitable as bonding fibers.

Web Formation

All nonwoven fabrics are based on a fibrous web. The characteristics of the fiber web profoundly affect the physical properties of the final product. These characteristics depend largely on fiber geometry, which is determined by the mode of web formation. Fiber geometry includes the predominant fiber direction whether oriented or random, fiber shape (straight, hooked, or curled), the extent of interfiber engagement or entanglement, crimp, and z-direction compaction (web-density control). Web characteristics are also influenced by fiber diameter, length, web weight, and chemical and mechanical properties of the polymer.

The choice of method for forming the web is determined by fiber length. Initially, the methods for forming webs from staple-length fibers were based on the textile-carding process, whereas web formation from short fibers was based on papermaking technologies. Although those technologies are still in use, new methods have been developed. For example, webs are formed from long, virtually endless filaments directly from bulk polymer; both web and fibers are produced simultaneously. The latter processes are known as spun-bond and melt-blown.

Bonding

The bonding of fibers gives the strength to the web and influences other properties (see also ADHESION AND BONDING). Both adhesive and mechanical means are used. Mechanical bonding employs the engagement of fibers by frictional forces.

Bonding can also be achieved by chemical reaction, formation of covalent bonds between binder and fibers. Some large-scale latex bonding of rayon is enhanced by excess cross-linking agent in the latex, eg, *N*-methylolacrylamide. In a laboratory experiment, a binder was applied to a web of cellulose fibers and formed covalent bonds, using fiber-reactive dye technology.

Adhesive Bonding. Adhesive bonding techniques include latex bonding (by impregnation, spraying, foam application or printing in a pattern), thermal bonding, and solution bonding. Thermal bonding, using low melting fibers or granules, has become important in recent years. It is practiced by heating components in a through-air oven or an embossing calender.

Mechanical Bonding. Mechanical bonding techniques include needle punching and hydroentangling, commonly known as the spin-lace process. Hydroentangling is the latest and possibly the most important of all nonwoven technologies.

Reinforcements and Composites

Reinforcements and composites are expected to play an important role in nonwoven fabrics. Reinforcements were first used in yarn scrims (Kimberly-Clark) and plastic netting (Chicopee) within an adhesively bonded base web. The reinforcements increased the tensile force to break, but did not have the same stress–strain properties as the nonwoven web. Composites are also made by hydroentangling components, as in surgical fabrics.

Stitch-bonded Fabrics. Laboratories supported by the government of the GDR and Czechoslovakia have developed a group of nonwoven fabrics made by extensive overstitching or knitting of yarns into fiber webs.

Manufacturing Process

The following technologies are used for the manufacture of nonwoven fabric.

Dry-formed, Carded, and Bonded. The webs are formed from staple fibers by carding or air-laying, bonded overall or in a pattern with latex or other waterborne adhesives.

Thermal-bonded. Dry-formed webs of staple fibers, or continuous filament spun-bond webs, are bonded with fusible fibers. Thermal bonding is also carried out on webs composed entirely of fusible fibers.

Air-laid. Wood-pulp fibers, with or without added staple fibers, are bonded with latex or similar adhesives.

Wet-formed. Short fibers, wood-pulp, and short regenerated or synthetic fibers are formed into a web by processes derived from papermaking technology, followed by bonding with latex or thermal binders.

Spun-bonded. Webs composed of long filaments with normal textile diameter are formed directly from bulk polymer and are usually thermally bonded.

Melt-blown (microfiber). Webs of long, extremely fine-diameter fibers are formed directly from bulk polymer and are usually bonded by hot embossing processes.

Hydroentangled. Dry-formed webs, usually from staple textile fibers, are mechanically entangled by multiple fine, high pressure water jets, in most cases without adhesive binder.

Needle-punched. Fibers are mechanically entangled by multiple reciprocating banks of barbed needles.

Laminated. Different layers are combined in composite or reinforced fabric by an adhesive, thermal fusion, or entanglement.

Stitch-bonded. Staple-fiber webs are reinforced or mechanically entrapped by yarns stitched or knitted through the webs.

Economic Aspects and Uses

Nonwoven fabrics can be classified into disposables and durable. The former are used once and are exemplified by facings for disposable diapers, sanitary napkins, surgical supplies, and

Table 1. Economic Data of the U.S. Nonwoven Industry, $\$10^6$ [a]

Category	1974	1984	Growth to 1990[b]
disposable[c]			
diaper cover stock	75	215	360
sanitary napkins, underpads, adult diapers	35	65	
roll goods, surgical gowns, packs	33	160	285
wiping cloths, roll towels	32	145	285
durable[c]			
interlinings	65	115	
coated fabrics	35	150	
geotextile and roofing		100	245
total disposables	*230*	*870*	*1610*
total durable	*355*	*690*	*1310*
products incorporating nonwovens			
disposable baby diapers	560	2700	
sanitary napkins		975	
fabricated medical products	175	845	1630
industrial–institutional disposables	90	600	1300
process			
staple-fiber dry-formed webs	300	300	29[d]
spun-bonded	140	415	32[d]
wet-laid	70	180	11[d]
spun-laced		100	11[d]
melt-blown		75	8[d]
air-laid pulp		55	5[d]
raw material consumption			
staple fibers		440	570
polyester		150	
polypropylene		120	
rayon		120	
polymer for spun-bond–melt-blown		315	575

[a] Unless otherwise stated.
[b] Estimated.
[c] Not inclusive.
[d] Percentage of total.

other medical applications; absorbency or repellency is the main characteristic.

Durable nonwoven fabrics include garment interlinings, fabrics used in automobiles, substrates for vinyl-coated upholstery and luggage, home furnishings such as mattress covers and draperies (stitch-through technology), geotextiles (civil engineering fabrics), floppy disk liners, indoor and outdoor carpets (needle-punched), wall coverings, garment insulation and fiber fill, architectural roofing fabrics, and suedelike composite structures.

The United States is the largest producer and consumer of nonwoven fabrics in the world, accounting for approximately 60% of world production; sales figures are given in Table 1.

BIBLIOGRAPHY

B. M. Lichstein, ed., *The Nonwovens Handbook,* INDA, New York, 1988.

J. Lunenschloss and W. Albrecht, *Nonwoven Bonded Fabrics,* Halsted Press, a division of John Wiley & Sons, Inc., New York, 1985.

INDA Technical Symposia Papers, Association of the Nonwoven Fabrics Industry, New York, 1973 to present.

Annual Symposia Papers, Clemson Nonwoven Fabrics Forum, Clemson University, Clemson, S.C., 1969 to present.

ARTHUR DRELICH
Chicopee

Spun-bonded

In 1984 the production of spun-bonded fabrics in the United States reached a record 135×10^3 t or approximately 25% of the total nonwoven fabrics produced. Spun-bonded fabrics are distinguished from other nonwoven fabrics by a one-step manufacturing procedure, which provides a complete chemical-to-fabric route in some instances and polymer-to-fabric in others. In either instance the process integrates the spinning, laydown, consolidation, and bonding of continuous filaments to form a fabric.

The area of largest growth has been disposable diaper coverstock, which accounts for approximately 50% of the U.S. coverstock market. Forecasts for the future growth of spun-bonded fabrics continue to be favorable as consumption in both durable and disposable areas continues to grow (see Table 1).

Fabric Characteristics

Structure. Spun-bonded fabrics are filament sheets made by an integrated process of spinning, attenuation, deposition, bonding, and winding into rolls. The fabrics are up to 5.2 m wide and usually not less than 3.0 m in order to facilitate productivity. Fiber sizes range from 0.1 to 50 dtex (0.09–45.0 den), although a range of 1.5–20 dtex (1.36–18.00 den) is most common. A combination of thickness, fiber denier, and number of fibers per unit area determines the fabric basis weight or weight

Table 1. U.S. Growth of Nonwoven Products, 1983–1987, 10^6 Dollars (Point-of-use Value)[a]

Application	1983	1987
consumer disposables		
baby diapers	2300	4000
adult diapers	25	300
wipes and roll towels	250	500
fabric-in-drier softeners	225	300
surgical packs	200	350
surgical gowns	60	150
surgical masks, caps, shoe covers	70	125
hospital bed underpads	80	115
medical dressings	70	150
industrial disposables		
filtration media	300	600
wipe cloths	100	250
other	50	75
durables		
interlining-interfacing (apparel)	125	200
coated and laminated products	100	100
home furnishings	100	180
carpet backing (primary, secondary)	125	175
geotextiles, civil engineering	NA[b]	NA[b]
roofing	100	200
other	100	150
total	*4380*	*7920*

[a] Data from Maran Marketing Co., Inc. (New York).
[b] NA = not available.

per unit area, which ranges from 10 to 800 g/m^2, typically 17–180 g/m^2.

The method of bonding greatly affects the thickness of the sheet and other characteristics. Fiber webs bonded by calendering are thinner than needle-punched webs because calendering compresses the structure through pressure, whereas needle punching moves fibers from the xy plane of the fabric into the z-direction (thickness).

Composition. The method of fabric manufacture determines the sheet characteristics, whereas the polymer determines the intrinsic properties. Properties such as fiber density, temperature resistance, chemical and light stability, ease of coloration, surface energies, and others are a function of the base polymer.

Most spun-bonded fabrics are based on isotactic polypropylene or polyester. Small quantities are made from nylon-6,6, and increasing amounts from high density polyethylene (HDPE). The properties of fibers made from different polymers are given in Table 2. Linear low density polyethylene (LLDPE) is increasingly used as base polymer because it gives a softer fabric.

Properties. Spun-bonded fabrics are characterized by tensile, tear, and burst strength; elongation to break; weight; thickness; porosity; and stability to heat and chemicals. These properties reflect fabric composition and structure. Typical physical properties are given in Table 3.

Spinning and Web Formation

Spin bonding combines fiber spinning with web formation. In the traditional methods the fiber is first spun and collected, and then converted into a fabric by a separate process. In spin bonding the bonding device is placed in line with spinning (Fig. 1). In some arrangements the web is bonded in a separate step, which first appears to be less efficient. However, this arrangement is more flexible if more than one type of bonding is required for the same web.

The spinning process is similar to the production of continuous-filament yarns and uses similar extruder conditions for a given polymer. Fibers are formed as the molten polymer exits the spinnerets and is quenched by cool air. The objective of the process is to produce a wide web (eg, 3 m), and, therefore, many spinnerets are placed side by side to generate sufficient fibers across the width. The grouping of spinnerets is often called a block or bank. In commercial production two or more blocks are used in tandem in order to increase the coverage of fibers.

Bonding

Many methods can be used to bond the fibers in the spun web. Although most procedures were developed for nonwoven staple fibers, three were adapted for continuous filaments: mechanical needling, thermal bonding, and chemical bonding. The last two bond large regions (area bonding) or small regions (point bonding) of the web by fusion or adhesion of fibers.

Melt-blown Fabrics

The fibers of melt-blown fabrics are composed of discontinuous filaments and are smaller than those of spun-bonded fabrics. Although melt-blown fabrics are not generally referred to as spun-bonded because of discontinuous filaments, the integration of spinning, attenuation, laydown, and bonding during the production of melt-blown webs describes a process traditionally defined as spun bonding. Fibers produced by melt blowing are very fine with a typical diameter of 3 μm, which is smaller by an order of magnitude than the smallest traditional spun-bonded fiber. The webs are weak and easily distorted, because the fibers are extremely fine and largely unoriented. Most commercial products are made of polyester or high melt-flow polypropylene, but other thermoplastic polymers have been used.

Table 2. Properties of Fibers

Polymer	Breaking tenacity, N/tex[a]	Elongation, %	Specific gravity	Moisture regain[b], % at 21°C, 65% rh	Melting range, °C
polyester	0.17–0.84	12–150	1.38	0.4	248–260
nylon-6,6	0.26–0.88	15–70	1.14	4.0	248–260
polypropylene	0.22–0.48	20–100	0.91	~0.0	162–171

[a] To convert N/tex to gf/den, multiply by 11.3.
[b] Equilibrium water content.

Table 3. Typical Physical Properties of Spun-bonded Products

Product	Basis weight, g/m^2	Thickness, mm	Tensile strength[a], N[b]	Tear strength, N[b]	Mullen burst, kPa[c]	Bonding method
Accord	69		144 MD 175 XD	36 MD 41 XD	324	point thermal
Bidim	150		495	279	1550	needle-punch
Cerex	34	0.14	182 MD[d] 116 XD	40 MD[e] 32 XD	240	chemically induced area
Colback	100	0.6	300/5 cm[f]	120[g]		area thermal (sheath core)
Corovin	75		130[f]	15[h]		point thermal
Lutradur	84	0.44	275[d] 297 XD	86[e] MD 90 XD	600[i]	copolymer area thermal
Polyfelt	137		585[d]	225[j]	1450[k]	needle-punch
Reemay	68	0.29	225[d] MD 180 XD	45[j] MD 50 XD	331[k]	copolymer area thermal
Terram	137	0.7	850[d]	250[l]	1100[m]	area thermal (sheath core)
Trevira	155		630[n] MD 495 XD	270[j] MD 248 XD	1520[k]	needle-punch
Typar	137	0.38	650[o] MD 740 XD	345[p] MD 355 XD	1210[n]	undrawn segments–area thermal
Tyvek	54	0.15	4.6 MD 5.1 XD	4.5 MD 4.5 XD		area and point thermal

[a] MD = machine direction; XD = transverse direction.
[b] To convert from N to pound-force, divide by 4.448.
[c] To convert kPa to psi, multiply by 0.145.
[d] ASTM D1117-1682.
[e] ASTM D2263.
[f] DIN 53857.
[g] DIN 53363.
[h] DIN 53356.
[i] ASTM D774.
[j] ASTM D1117.
[k] ASTM D3786.
[l] BS4303-1968.
[m] BS4768-1972.
[n] ASTM D751.
[o] ASTM D1682.
[p] ASTM D4533.

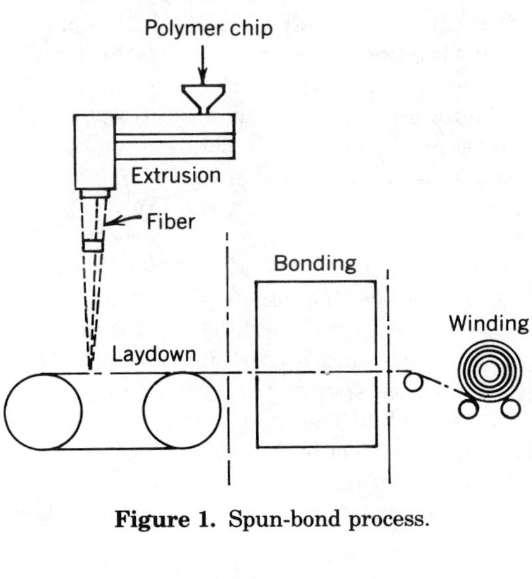

Figure 1. Spun-bond process.

Flashspun Fabrics

The production of spun-bonded webs by flash spinning is a radical departure from the conventional melt-spinning methods. In flash spinninig, a 10–15% solution of, for example, high density polyethylene in trichlorofluoromethane or methylene chloride is heated to ca 200°C and pressurized to 4500 kPa (650 psi) or more. The pressurized vessel is connected to a spinneret containing a single hole. When the pressurized solution is permitted to expand rapidly through the single hole, the low boiling solvent is instantaneously flashed off, leaving a three-dimensional film-fibril network referred to as a plexifilament. The film thickness is 4 μm or less; the three-dimensionality results from the cross-linking interconnection of the subdenier fibers. Thus a multitude of individual but interconnected fibers are created from a single-hole spinneret.

Flash spinning is the most complex and difficult method of manufacturing spun-bonded fabrics because of the need to spin heated pressurized solutions under precise conditions. The result, however, is a unique structure that shows great promise.

Although the process is almost 30 years old, it is used only by one company.

BIBLIOGRAPHY

U.S. Pat. 3,338,992 (Aug. 29, 1967), G. A. Kinney (to E. I. du Pont de Nemours & Co., Inc.).

U.S. Pat. 3,692,618 (Sept. 19, 1972), O. Dorschner, F. Carduck, and C. Storkebaum (to Metallgesellschaft AG).

H. Kulter in J. Lunenschloss and W. Albrecht, eds., *Non-Woven Bonded Fabrics*, Ellis Horwood, Ltd., Chichester, UK, 1985.

RONALD L. SMORADA
E. I. du Pont de Nemours & Co., Inc.
Reemay, Inc.

NORBORNADIENE.

See ETHYLENE–PROPYLENE ELASTOMERS.

NOVOLOID FIBERS.

See PHENOLIC RESINS.

NOVOLACS.

See PHENOLIC RESINS.

NUCLEAR MAGNETIC RESONANCE

Nuclear magnetic resonance (nmr) spectroscopy is a method of great interest and importance for the study of polymers. Nmr measurements of polymers in solution can be used to determine the identity of unknown polymers, to establish stereochemical configuration and geometrical isomerization, to quantify chain branching, to assess conformational preference and flexibility, and to determine copolymer composition. In the solid state, nmr experiments can assess motional dynamics, chain conformations, phase separation, the effects of plasticizers, and the degree of orientation.

The Nmr Phenomenon

Magnetic Energy Levels and Transitions. When the spin quantum number I of a nucleus is nonzero, the nucleus possesses a magnetic moment. This condition is met if the mass number and atomic number are not both even. Examples of nmr-active nuclei include ^{13}C and ^{1}H, both of which have a spin I of $1/2$. When placed in a magnetic field of strength B_0, nuclei with nonzero I occupy quantized magnetic energy levels, called Zeeman levels, the number of which is equal to $2I + 1$. The relative populations of the Zeeman levels are normally given by a Boltzmann distribution. All nuclei with spin greater than $1/2$ have nonspherical distributions of positive charge and therefore also possess electric quadrupole moments.

The separation of Zeeman levels is given by

$$\Delta E = h\nu_0 = \mu B_0/I \tag{1}$$

For spin-$1/2$ nuclei, there are two energy levels for which

$$\Delta E = h\nu_0 = 2\mu B_0 \tag{2}$$

For spin-1 nuclei, there are three energy levels separated by μB_0; μ is the magnetic moment of the nucleus.

Transitions between energy levels can be made to occur by means of a resonant radio-frequency (r-f) field B_1 of frequency ν_0. The resonance condition for any nucleus may be expressed as

$$\nu_0 = \frac{\omega_0}{2\pi} = \frac{\gamma B_0}{2\pi} \tag{3}$$

where ω_0 is the frequency in rad/s and γ is the magnetogyric ratio, equal to $2\pi\mu/Ih$.

An alternative way of picturing the resonance phenomenon arises from the fact that when placed in a magnetic field, a nucleus undergoes Larmor precession about the field direction at a rate given by ν_0 or ω_0 in the above equations. Transitions between energy levels occur when the frequency of the r-f fields equals the Larmor precession frequency.

The Chemical Shift. Resonance occurs at slightly different frequencies for each chemically different nucleus, depending on its chemical binding and position in a molecule. This variation is caused by the cloud of electrons about each nucleus, which shields the nucleus against the magnetic field, thus requiring a slightly lower value of ν_0 to achieve resonance than for a bare proton. For example, protons attached to or near electronegative groups such as OH, OR, OCOR, COOR, and halogens experience a lower density of shielding electrons and resonate at higher ν_0. Protons farther removed from such groups, as in hydrocarbon chains, resonate at lower ν_0. Similar structural dependence is observed for ^{13}C nuclei. These variations are termed chemical shifts and are commonly expressed in relation to the resonance of tetramethylsilane (TMS) as the zero of reference. The total range of proton chemical shifts in organic compounds is on the order of 10 ppm, eg, ca 1 kHz in a magnetic field of 2.34 T (23,400 G). For ^{13}C nuclei and all other magnetic nuclei, the range is much greater, at least 200 ppm; this is the principal reason for the great interest in the study of polymers by ^{13}C nmr.

For any nucleus, the separations of chemically shifted resonances, expressed in Hz, are proportional to B_0; when expressed in ppm, they are independent of B_0. An important advantage of high field magnets is the greater resolution of peaks and finer discrimination of structural features of polymer chains that they make possible. The sensitivity of the chemical shift to the nearby electronic environment and the high spectral resolution attainable in solution make nmr spectroscopy the method of choice for establishing polymer microstructure.

The electronic screening of nuclei is actually anisotropic and must be expressed as a tensor. In solution, motional averaging produces the isotropic chemical shift value, which is the trace of the tensor. In solids, however, the chemical shift depends on the orientation of the molecule with respect to the magnetic field direction. Single crystals and oriented fibers display sharp lines whose frequenices (chemical shifts) vary as the molecule is rotated in the magnetic field. The principal values of the shift tensor, σ_{11}, σ_{22}, and σ_{33} specify its magnitude, whereas the direction cosines specify the orientation of the principal values with respect to the principal axis system (molecular framework) of the molecule. In glassy polymers or unoriented powders, the contributions from all possible orientations of the molecule must be considered. The resultant chemical shift anisotropy powder pattern illustrates two types of chemical shift anisotropy, axially symmetric and axially asymmetric. In solid polymers the partial collapse of the chemical shift anisotropy pattern is a valuable indicator of molecular motion.

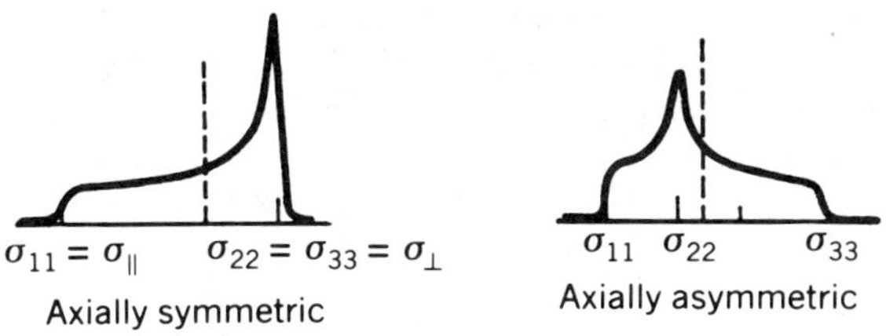

Direct Dipole–Dipole Coupling. In most substances, protons contribute to local fields and are sufficiently numerous to have a marked effect not only on proton spectra but on those of other nuclei. For example, the local field at a ^{13}C nucleus is given by

$$B_{loc} = \pm \frac{h}{4\pi} \lambda_H \frac{(3 \cos^2 \theta_{C-H} - 1)}{r^3_{C-H}} \tag{4}$$

The plus-or-minus sign indicates that the local field may add to or subtract from B_0, depending on whether a neighboring proton dipole is aligned with or against the direction of B_0, an almost equal probability. The angle θ_{C-H} is the angle that the C–H bond vector makes with the magnetic field direction, B_0, and r is the internuclear C–H distance. As is the case with the chemical shift, contributions to the dipole–dipole interaction from all orientations of the molecules must be considered in a solid polymer. This produces a broadening of spectra of solids. The dipole–dipole coupling of protons in solution however, is usually averaged to zero by rapid molecular tumbling and segmental motion. Because the dipolar coupling is orientation dependent, the study and interpretation of dipolar-broadened spectra, particularly of polymer protons, yield valuable insights into molecular motion in solid materials.

Indirect Nuclear Coupling. Magnetic nuclei may transmit information to each other concerning their spin states not only directly through space but also through the intervening covalent bonds. This is indirect or scalar nuclear coupling, also known as J coupling. Rapid tumbling of the molecule does not average this interaction to zero. If a nucleus has n sufficiently close, equivalently coupled spin-½ neighbors, its resonance will be split into $n + 1$ lines corresponding to the spin states of the neighboring group of spins. Intensities are given by simple statistical considerations and are therefore proportional to the coefficients of the binomial expansion. Thus one neighboring spin splits the observed resonance to a doublet, two produce a 1:2:1 triplet, three a 1:3:3:1 quartet, and so on. The size of the coupling is denoted by a coupling constant J and is expressed in Hz; to first order, it is independent of B_0. Because the size of the coupling for protons on adjacent carbons depends on the dihedral angle between them, J can be used to establish conformations of polymer chains.

Spin–Lattice Relaxation. A spin system may be readily disturbed from its equilibrium state. Before the sample is placed in the magnet, the spin states are equally populated. In the magnetic field, the spins establish a new equilibrium population, given by the Boltzmann distribution. The time to establish this equilibrium is termed the spin–lattice relaxation time, or T_1. The populations may be equalized by an appropriate pulse of resonant r-f energy, and it is even possible by this means to invert the populations. Once equalized or inverted, the rate of return to equilibrium can be monitored. Figure 1 shows vector diagrams illustrating relaxation processes. Because spin–lattice relaxation affects the magnetization along the field direction, it is also sometimes called longitudinal relaxation. The term lattice denotes all the molecules in the sample and is used for both fluids and solids. The lattice serves as a heat sink in the equilibration, but there must be a link by which thermal energy may be exchanged. This link is provided by molecular motion. Molecular motion can be characterized by a correlation time τ_c, which defines the time a molecule can be considered to be in a particular motional state before it moves to the next state. Relaxation in polymers generally occurs by dipole–dipole interaction. For isotropic rotation and dipolar relaxation of a ^{13}C nucleus by ^1H, the T_1 is related to the correlation time by:

$$\frac{1}{T_1} = \frac{N \gamma_H^2 \gamma_C^2 h^2}{r_{C-H}} \cdot \tau_c \tag{5}$$

where N is the number of directly bonded protons.

Spin–Spin Relaxation. Nuclear spins are not merely small static magnetic dipoles; even in a rigid solid, they are precessing about the direction of B_0. Such a precessing moment

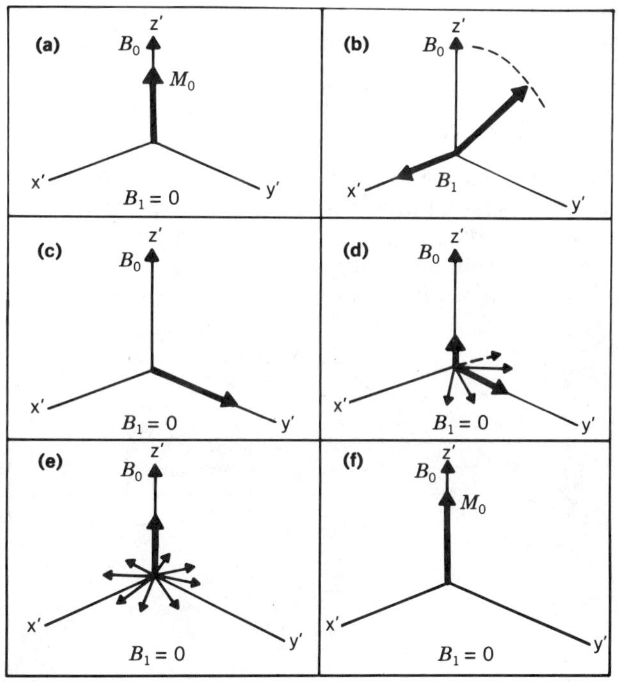

Figure 1. Vector diagrams illustrating the behavior of magnetization under the influence of a radio-frequency field, B_1. This diagram is simplified by assuming a coordinate system that rotates with B_1 (and at resonance with the net magnetization, M_0). (**a**) The net magnetization M_0 is aligned along the magnetic-field direction. (**b**) and (**c**) An r-f field B_1 is applied perpendicular to B_0. The duration of B_1 is sufficient to tip the net magnetization by 90°. (**d**) and (**e**) The spins begin to relax in the $x'y'$ plane by spin–spin (T_2) processes and in the z' direction by spin–lattice (T_1) processes. (**f**) The equilibrium magnetization is reestablished along B_0.

may be resolved into static and rotating components. The rotating component is the right type of magnetic field to induce transitions in a neighboring nucleus precessing at the same frequency. If such a spin flip occurs, the nuclei exchange magnetic energy states with no overall change in the energy of the system but with a shortening of the spin lifetime of each. This introduces an uncertainty broadening, in conformity with the Heisenberg principle, of the same form and magnitude as the broadening by the static components. Both effects are included in the quantity T_2, which is the characteristic time for spin–spin relaxation. This process is also also termed transverse relaxation, and is taken as the inverse of the line width:

$$T_2 = \frac{1}{\pi\delta\nu} \qquad (6)$$

Nuclear Overhauser Enhancement. The nuclear Overhauser enhancement, or NOE, arises from dipole–dipole relaxation, and is observed when a continuous resonant r-f field is applied to one spin of a dipolar-coupled system. If, for example, the protons attached to a particular carbon are irradiated, the intensity of the carbon resonance will increase by a factor of η (η is the NOE enhancement factor, or NOEF) to produce a signal of intensity $1 + \eta$ ($1 + \eta$ is defined as the NOE). The maximum possible η is given by $\frac{\gamma_S}{2\gamma_I}$, where γ_S and γ_I are the magnetogyric ratios of the S and I spins. When protons are decoupled from ^{13}C, the maximum $^1H—^{13}C$ NOE is 1 + 1.98, or almost 3. For homonuclear $^1H—^1H$ dipolar interactions, the maximum η is 50%. Because the NOE arises from dipole–dipole interactions, it can be used as a convenient measure of distance.

Experimental Methods

Continuous-wave and Pulsed Fourier Transform Nmr. Figure 2 shows a block diagram of a modern nmr spectrometer. The heart of the instrument is its magnet and r-f circuit. In one type of magnet and r-f coil combination the magnetic field is generated by a solenoid bathed in liquid helium so that its electrical resistance is zero, ie, it is superconductive. The r-f coil is of the Helmholz design. Electromagnets and permanent magnets are also commonly used for nmr spectroscopy. For electromagnet and permanent magnet systems, r-f coils of solenoid design may also be employed.

The magnetic field B_0 removes the degeneracy of the spin energy levels, and the electromagnetic r-f field B_1 excites transitions between these energy levels. The resonance condition occurs when the frequency of the r-f field B_1, in MHz, is exactly equal to the Larmor frequency of the nucleus under observation. The actual method for achieving resonance is more complicated than it first appears. Because most molecules have more than one type of magnetically nonequivalent group, eg, a methyl and a methine, a single sample has multiple Larmor frequencies, perhaps fractions of a part per million (ppm) apart, corresponding to each of the chemical shifts. Any excitation method must cover all of the Larmor frequencies in the sample. The two principal methods used to achieve the necessary resonant condition, continuous-wave (cw) or Fourier transform (ft), are conceptually different. In the cw method, the resonant condition is met for each nucleus by sweeping either the B_1 radio frequency or the B_0 field. Each magnetically equivalent nucleus in the sample is successively brought into resonance by the sweeping process. When a particular nucleus comes into resonance, it induces a voltage in the pickup coil. This signal, microvolts or less in magnitude, is amplified and detected directly in the frequency domain and recorded in a plot of voltage (or intensity) vs frequency.

In contrast, the ft method involves detection in the time domain and subsequent Fourier transformation into the frequency domain. A short burst or pulse of r-f energy, nominally monochromatic at ν_0, is used to excite all of the Larmor frequencies in a sample simultaneously. The chief advantage of pulsed Fourier transform techniques is in time savings. The data can be collected all at once, rather than from a slow sweep of the field.

Decoupling in Solution. The smaller of the two sources of nuclear coupling is the scalar, or J coupling, that arises from indirect spin–spin coupling. If the larger dipolar coupling has been averaged to zero by rapid molecular tumbling in solution,

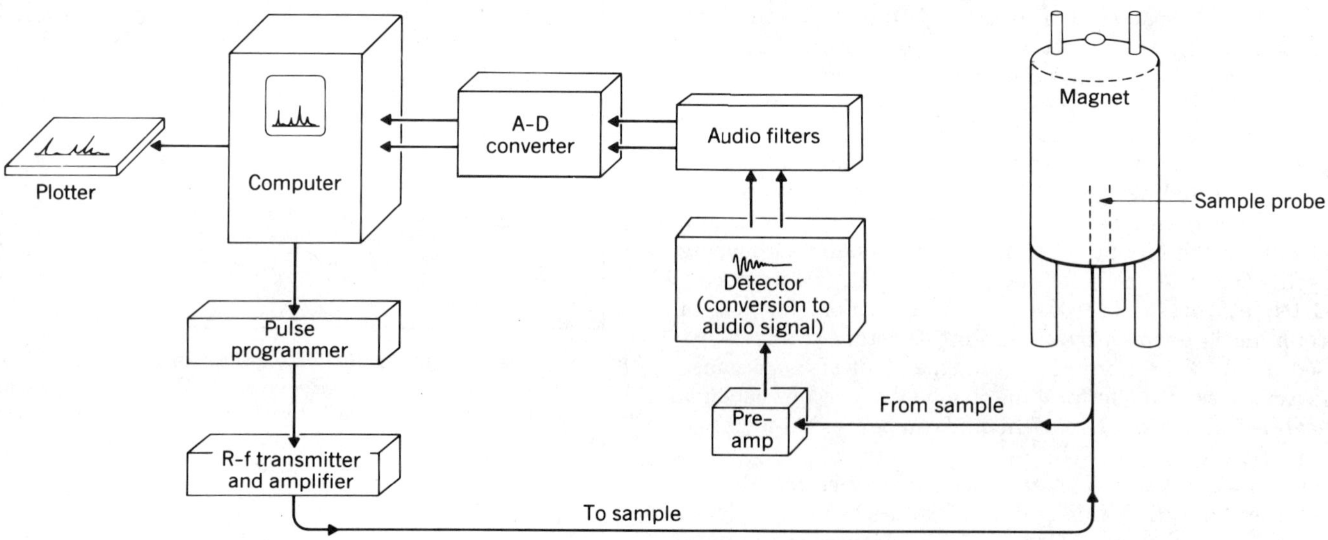

Figure 2. Block diagram of an nmr spectrometer, showing the main components necessary to observe a signal. (The lock and decoupler are not shown.)

substantial further simplification of ^{13}C spectra is obtained on removing the J coupling by applying an additional r-f field B_2 resonating at the frequency of the proton spins. This auxiliary field causes the proton spins to oscillate rapidly compared to the weak (ca 200 Hz) spin–spin interactions, thereby decoupling their influence.

Special Techniques for Obtaining Nmr Spectra of Solid Polymers. Molecular motion in most solid polymers is generally so slow that the dipolar couplings, normally averaged to zero by segmental motion of polymers in solution, are quite large (ca 50 kHz). Dipolar decoupling, or high power proton decoupling, is used to remove the effects of dipolar coupling in solids. It is analogous to the removal of the scalar J couplings; however, the strength in frequency of the applied auxiliary field must be greater than the dipole–dipole interaction. Once the dipole–dipole interaction in solids has been removed by dipolar decoupling, the largest remaining source of line broadening is the chemical shift anisotropy. Although the width of the chemical shift powder pattern usually obscures detailed structural information, it contains considerable angular-dependent information that is particularly useful when determining the exact nature of anisotropic molecular motion.

Broadening caused by the chemical shift anisotropy can be removed by a technique known as magic angle spinning in which the sample is mechanically rotated about an axis forming the "magic angle" (54.7°) with respect to the external magnetic field B_0. Rapid mechanical rotation causes the direction cosines, which specify the orientation of the chemical shift tensor in the molecular frame, to become time dependent in the rotor period. The expression for the chemical shift σ becomes:

$$\sigma = \tfrac{1}{2}\sin^2\beta(\sigma_{11} + \sigma_{22} + \sigma_{33}) + \tfrac{1}{2}(3\cos^2\beta - 1) \times (\text{functions of direction cosines}) + (\text{time-dependent terms}) \quad (7)$$

where β is the angle between the magnetic field and the rotation axis, and σ_{11}, σ_{22}, and σ_{33} are the principal values of the chemical shift tensor. When β is 54.7°, the first term on the right is one-third times the sum of the diagonal elements of the chemical-shift tensor, which is precisely the isotropic chemical shift. The $(3\cos^2\beta - 1)$ term is zero at the magic angle. (The time-dependent terms in the above equation produce spinning sidebands.) Thus magic angle spinning in conjunction with dipolar decoupling permits attainment of high resolution spectra of solid polymers.

Although high resolution spectra of solids can be obtained, there is often limited molecular motion, leading to inefficient spin–lattice relaxation. Cross-polarization is a technique that takes advantage of the fact that generally all of the protons in a solid have the same T_1 value because of a process called spin diffusion, in which all the protons communicate efficiently with each other. The proton T_1 is usually short compared with the carbon T_1. (In this discussion, cross-polarization from protons to carbons is emphasized, but these concepts apply to polarization between any abundant and rare spin pair.) Cross-polarization works by effectively forcing an overlap of the proton and carbon energies in the rotating frame. Thus energy transfer between protons and ^{13}C nuclei with differing Larmor frequencies can be made to occur when

$$\gamma_C B_{1C} = \gamma_H B_{1H} \quad (8)$$

This equation, called the Hartmann-Hahn condition, results in a match of the rotating-frame energies for ^1H and ^{13}C. Because γ_H is four times γ_C this match occurs when the strength of the applied carbon field B_{1C} is four times the strength of the applied proton field B_{1H}. When the proton and carbon rotating-frame energy levels match, polarization is transferred from the abundant spin (protons) to the rare spin (carbons). The advantages of cross-polarization are twofold. Because polarization is transferred from the protons to the carbons, it is the shorter T_1 of the protons that dictates the repetition rate for signal averaging. In addition, the carbons obtain a significant enhancement in signal intensity from the cross-polarization experiment.

Observation of Nuclear Relaxation. Nmr relaxation measurements on polymers are particularly useful for understanding the complex molecular motions of polymer chains, both in solution and in the solid state. Essentially all relaxation in polymers is produced by molecular motion, which gives rise to fluctuating local dipolar fields. There are four primary types of nuclear relaxation: T_1, T_2, NOE, and $T_{1\rho}$. T_1, T_2, and NOE have been defined above. The $T_{1\rho}$ describes the loss of spin magnetization in the rotating frame, and is sensitive to slower motions. The motional time scales probed by these relaxation measurements are summarized in Table 1 where they are compared with other nmr methods for determining molecular motion.

Deuterium Quadrupole Nmr. The deuterium nucleus is particularly useful for exploring the details of molecular motion in solid polymers. It has a spin of 1 and therefore exhibits three equally spaced energy levels when placed in a magnetic field. Deuterium is a quadrupolar nucleus, and interaction of the quadrupole moment with the molecular electric field gradient at the nucleus produces a substantial perturbation of these energy levels. The perturbation is actually so large in solids that the dipolar, chemical shift, and J coupling interactions

Table 1. Nmr Experiments Useful for Determining Molecular Motion

Experiment	Most informative time scale, s	Comments
T_1	10^{-6}–10^{-10}	measured by inversion–recovery or progressive saturation; useful for solutions and solids
NOE	10^{-8}–10^{-10}	most useful for solutions; gives information on fraction of relaxation that is dipolar
$T_{1\rho}$ (C)	10^{-2}–10^{-4}	for solids
narrowing of chemical-shift powder pattern	10^{-1}–10^{-3}	measured by off-axis spinning
deuterium nmr line shapes	10^{-4}–10^{-7}	requires specific labeling of sample; for solids
T_2	10^{-5}–10^{-8}	most useful for solutions

can be ignored. The magnitude of the quadrupolar splitting is given by

$$\Delta\nu_q = \frac{3e^2qQ}{4h}(3\cos^2\theta - 1) \tag{9}$$

where the charge on the electron is e, eqQ is the quadrupole coupling constant, θ is the angle that a particular carbon–deuterium bond vector makes with the external magnetic field, and h is Planck's constant.

In a powder, where all orientations of carbon–deuterium bond vectors are present, the actual spectrum is composed of two overlapping, axially symmetric powder patterns, one from each of the transitions. This pattern is often called a Pake doublet. These theoretical patterns are shown in Figure 3**a**. The left-most figure shows two axially symmetric patterns, one of which is crosshatched. The $\perp$ and $\parallel$ frequencies occur when the C–D bonds are perpendicular and parallel to the field, respectively. The right part of Figure 3**a** shows the smoothed line shape. For a static C–D bond, $\Delta\nu_q$ or d is ca 130 kHz. Figure 3**b** shows the expected pattern for a rotating methyl group and Figure 3**c** the pattern for a two-site hop through the tetrahedral angle. Figures 3**d** and **e** illustrate the fact that solid-state deuterium nmr spectroscopy can readily distinguish aromatic rings that are flipping by 180° about their 1,4-phenylene axes from those undergoing unrestricted rotation. Thus the solid state deuterium powder pattern is extremely useful for determining the details of motions in the solid state.

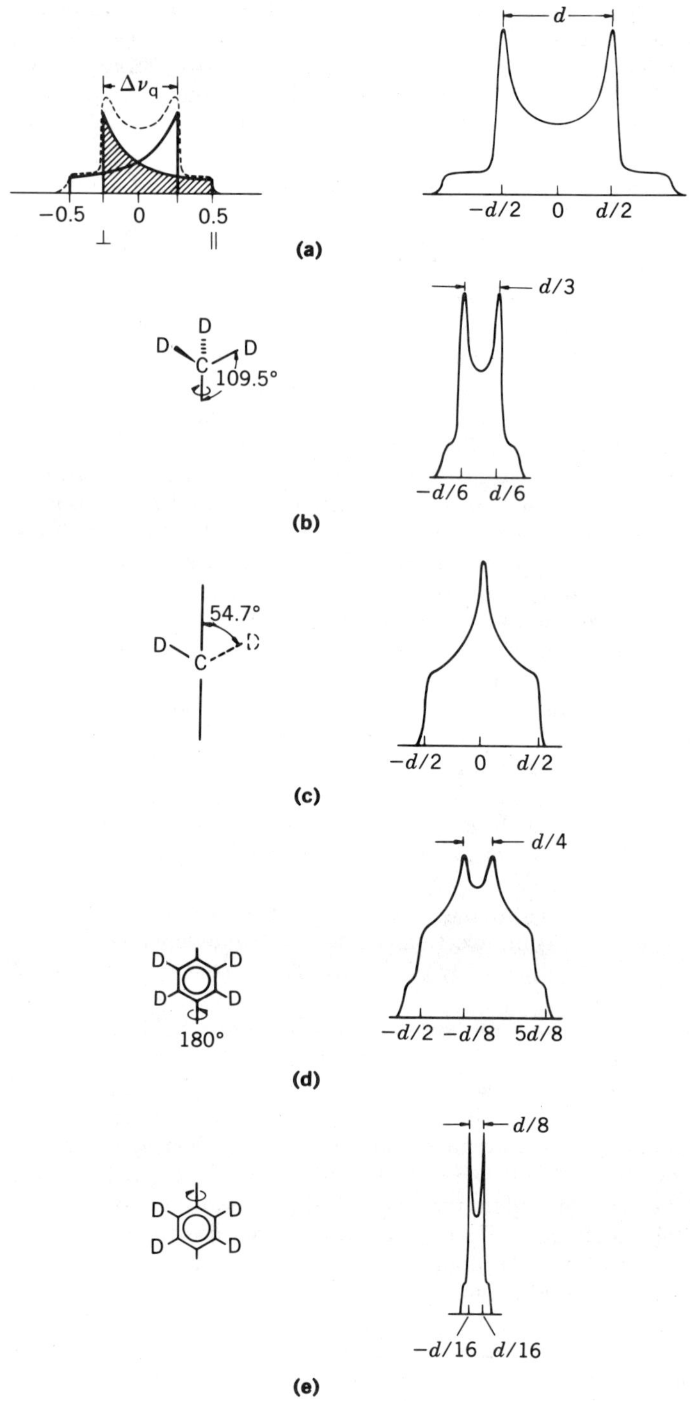

Figure 3. Theoretical solid-state deuterium nmr powder patterns. The static pattern is shown in (**a**) $\Delta\nu_q$, or d, is ca 130 kHz. (**b**) Rotating methyl; (**c**) two-site hop; (**d**) 180° phenyl-ring flip; (**e**) free diffusion of phenyl ring.

Applications to Polymers

Microstructure. Nmr is particularly informative concerning chain microstructure (qv), revealing details seen by no other technique. Figure 4 shows the 500 MHz proton spectra of poly(methyl methacrylate) (PMMA) prepared with a free-radical initiator (Fig. 4**a**) and with fluorenyllithium in toluene, an anionic initiator (Fig. 4**b**). The profound effect of the nature of the initiator is evident in the marked differences between these spectra. Proton nmr can provide absolute stereochemical information concerning polymer chains without recourse to x-ray or other methods.

Geometrical Isomerism. Geometrical isomerism in diene polymers may be observed and measured by nmr. In addition to the cis and trans structures found in natural isoprene polymers formed by 1,4-addition, chain propagation of diene monomers may also proceed by incorporation of the monomer through one double bond (1,2-addition) to produce isotactic or syndiotactic forms. Each of these structures gives distinctive peaks; the intensities of the peaks can be analyzed to determine the amount of each isomer in the chain.

The olefinic carbons are only moderately sensitive to geometrical isomerism, but the methylene carbons are highly sensitive because they are substantially more shielded (ca 8 ppm) in the cis structures. Although chains of mixed structure exhibit complex spectra because of sequence effects, these spectra can be analyzed and the distribution of cis, trans, and 1,2 units can be determined.

Branching. Branches in vinyl polymers may be introduced deliberately by employing diene or divinyl comonomers, or by processes that are under less specific control and involve chain transfer. Branches are of particular importance in polyethylene because their presence reduces the melting point and extent of crystallinity. Infrared spectroscopy indicates that high pressure polyethylene has unusually large numbers of methyl groups, normally expected only at chain ends. When combined with molecular weight measurements, these results indicate that there are many more ends than molecules or, in other words, that the chains contain short branches. Carbon-13 nmr can supply details of the types and distribution of these branches because of the sensitivity of ^{13}C chemical shifts to such structural variables (Fig. 5).

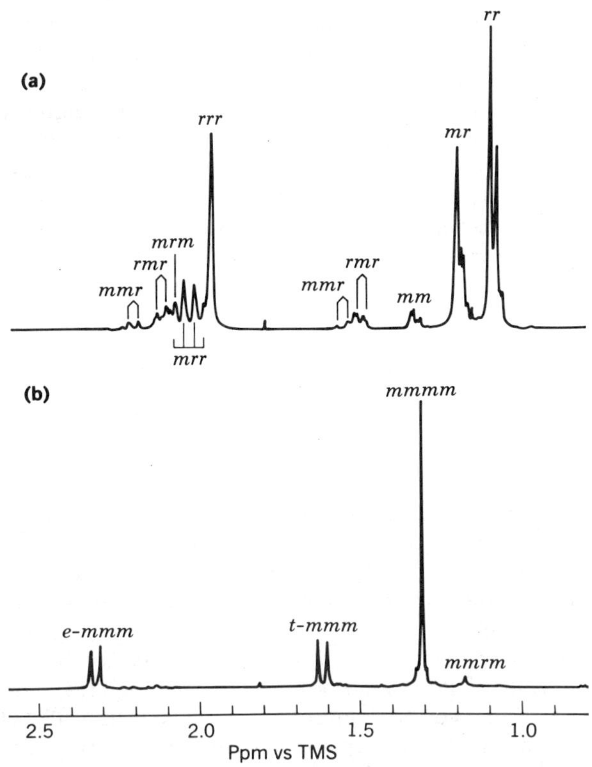

Figure 4. 500-MHz proton nmr spectra of (**a**) syndiotactic and (**b**) isotactic PMMA (the —OCH_3 peaks are not shown): *e* and *t* denote erythro and threo, respectively.

Copolymer Structure. Copolymers are broadly divided into three types: random, block, and graft. Block copolymers (qv) and graft copolymers (qv) contain relatively long sequences of one monomer bonded to similar sequences of another. Their overall composition is usually known from their method of synthesis and they do not present microstructural problems essentially different from those of homopolymers. The random type, in which two or more types of comonomer units are present in each chain, presents more complex problems.

The composition of copolymers of vinyl and diene monomers are not in general the same as that of the monomer mixtures from which they are formed and cannot be deduced from the homopolymerization rates of the monomers involved. Traditionally, the determination of reactivity ratios, which provide important information concerning the behavior of monomers and growing chains, has required the determination of the overall comonomer composition of copolymers prepared from a series of feed ratios. Elemental analysis is most commonly used. A number of computational and graphic methods are employed to do this. The theoretical treatment that predicts overall composition also predicts the frequency of occurrence of comonomer sequences (see COPOLYMERIZATION).

Comonomer sequences can be readily measured by nmr, by which it is also possible to observe copolymer stereochemistry and the presence of anomalous units. Deviations from the simplest model, in which only the terminal residue of a growing chain determines its reactivity, can be detected; the effect, if any, of the penultimate unit, may be clearly observed. By nmr sequence measurements, reactivity ratios can be determined from only a single copolymer provided the feed ratio is known. It may still be desirable to observe a range of compositions to assist in resonance assignments, but it is not essential in principle. Figure 6 shows the proton nmr spectra of poly(vinylidene chloride) and polyisobutylene and a copolymer containing 70 mol % of vinylidene chloride. The reactivity ratios may be calculated from the dyad resonances. Analysis of the spectrum shown in Figure 6**c** shows that the vinylidene chloride radicals prefer to add vinylidene chloride and that chains terminating in isobutylene units have only a very small tendency to add another isobutylene. More searching tests of the propagation mechanism can be obtained from consideration of tetrad intensities.

Chain Motion in Macromolecules. Measurements of nuclear relaxation furnish a powerful means for observing motions of macromolecules in two regimes: polymer solutions, in which chain motion is relatively fast, and polymers in the solid state, where motional freedom may vary from fast to essentially nil. Synthetic polymers are useful mainly for their solid-state properties, so observation of the solid state may be of greater prac-

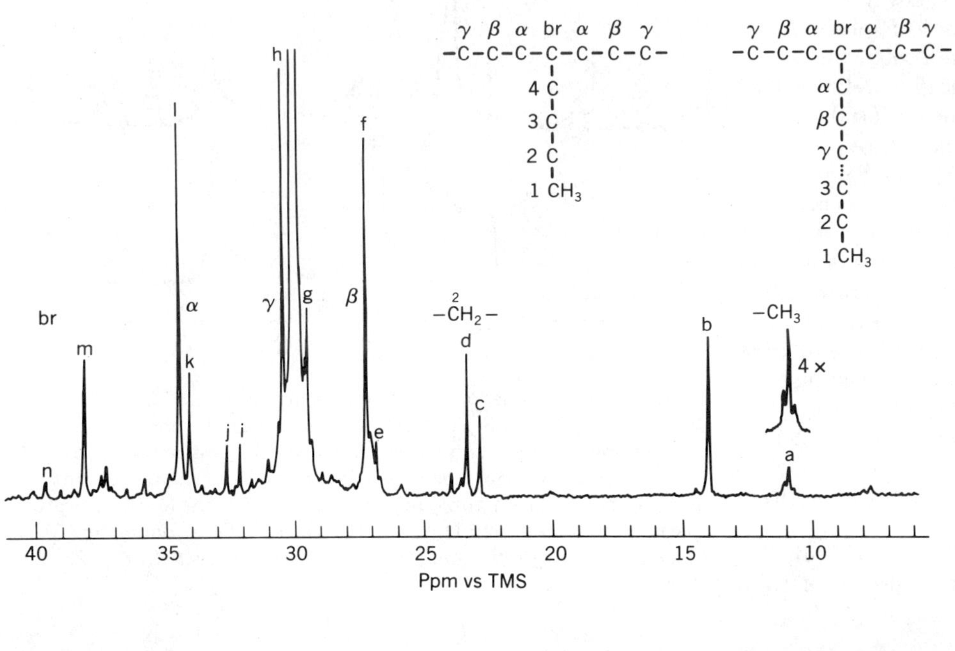

Figure 5. The 50.3-MHz ^{13}C nmr spectrum of high pressure PE observed in 5% (wt/vol) solution in 1,2,4-trichlorobenzene at 110°C: a, CH_3 of an ethyl branch; b, CH_3 of a butyl, amyl, or longer branch; c, penultimate carbon of an amyl or longer branch; d, penultimate carbon of a butyl branch (γ to the br carbon); e, fourth carbon of an amyl branch (β to br carbon) and methylene of an ethyl branch (α to br carbon); f, main-chain β-C from an ethyl, butyl, or amyl branch and β-C of a long branch; g, third carbon of a butyl branch (β to br carbon); h, main-chain γ-C from an ethyl, butyl, or amyl branch and γ-C of a long branch; i, third carbon of a long-chain branch; j, third carbon of an amyl branch (γ to br carbon); k, fourth carbon of a butyl branch (α to br carbon) and main-chain α-C from an ethyl branch; l, main-chain α-C from a butyl or amyl branch and α-C of an amyl branch; m, branched carbon for butyl, amyl, or longer; n, branched carbon for ethyl branch; br = branched.

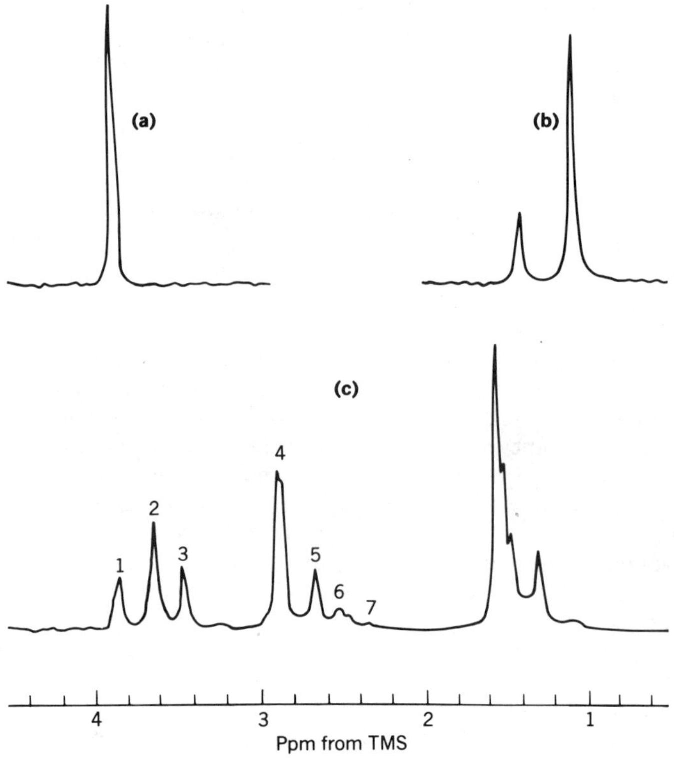

Figure 6. The 60-MHz proton spectra of (**a**) poly(vinylidene chloride); (**b**) polyisobutylene; (**c**) a vinylidene chloride (m_1):isobutylene (m_2) copolymer containing 70 mol % m_1. Peaks are identified with monomer tetrad sequences: 1, $m_1m_1m_1m_1$; 2, $m_1m_1m_1m_2$; 3, $m_2m_1m_1m_2$; 4, $m_1m_1m_2m_1$; 5, $m_2m_1m_2m_1$; 6, $m_1m_1m_2m_2$; 7, $m_2m_1m_2m_2$. Courtesy of *Kolloid-Zeitschrift.*

tical importance. However, solution measurements are often more revealing of intramolecular influences, divorced from the effects of neighboring chains.

Most synthetic polymers in solution give fairly well resolved proton and ^{13}C spectra and thus must enjoy considerable segmental freedom, because rigid molecules of comparable molecular weight would give very much broader resonances. Line widths are independent of molecular weight down to the oligomer level and are very nearly independent of solution concentration and viscosity up to 25–30%. Even solutions so viscous that they barely pour from the nmr tube give well-resolved spectra, another indication of the dominance of local motion. Nevertheless, polymer spectra usually appear more poorly resolved than those of small-molecule counterparts. Some of this line broadening, in polymers of irregular structure, represents an envelope of chemical shifts. The true solution proton line width for many polymers at elevated temperatures is usually 3–5 Hz, whereas ^{13}C resonances are somewhat narrower, typically 1–2 Hz. The motion of flexible polymers in solution may be elucidated by spin–spin (T_2) or spin–lattice (T_1) relaxation measurements. Usually such motion is regarded as being composed of segmental motion and overall tumbling. Various factors influence this motion; these include the effect of side chains, the effect of heteroatoms, double bonds, and stereochemistry. Relaxation times for a wide variety of polymers in solution have been compiled.

Chain Orientation in the Solid State by ^{13}C Nmr. With dipolar decoupling and magic angle spinning, nmr spectra of solid polymers can be obtained in which many of the individual resonances are observed. Although the peaks are generally too broad for microstructural characterization, they are generally sufficiently well resolved to measure copolymer composition and conformational and packing effects. Solid state ^{13}C spectra are also useful for determining orientation in fibers. For example, from x-ray studies it is known that the main chain in poly(ethylene terephthalate) (PET) fibers is oriented in such a way that the aromatic rings are not colinear with the fiber draw direction:

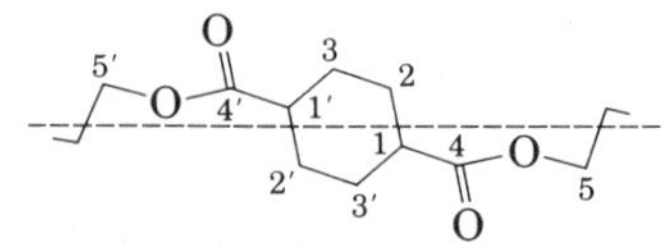

Fundamental questions arise concerning the degree of orientation of the polymer chains and the possible changes in orientation that may occur when the fibers undergo shrinkage due to excessive heat. Solid-state ^{13}C nmr is useful for examining such materials because of the orientation dependence of the chemical shift.

If the chain molecules in the fibers were perfectly oriented the material should produce a single line for each magnetically distinct carbon even without magic angle spinning. The chemical shifts of these lines depend on the angle that the fiber axis makes with the magnetic field. In the ^{13}C nmr spectra of an oriented PET sample (Fig. 7), the peaks are assigned, from right to left, to C5, C3, C2, C1, and C4. These spectra are obtained with the fiber draw axis parallel to the magnetic field. Spectra such as these can be interpreted to indicate overall orientational distributions in the crystalline regions. Because the ortho carbons (C2 and C3) reside in slightly different environments, their peaks are distinct at low temperature. These two peaks undergo a type of coalescence phenomenon at elevated temperatures, where it is observed that the rings undergo rapid 180° flips, thereby time-averaging the frequencies of the two peaks.

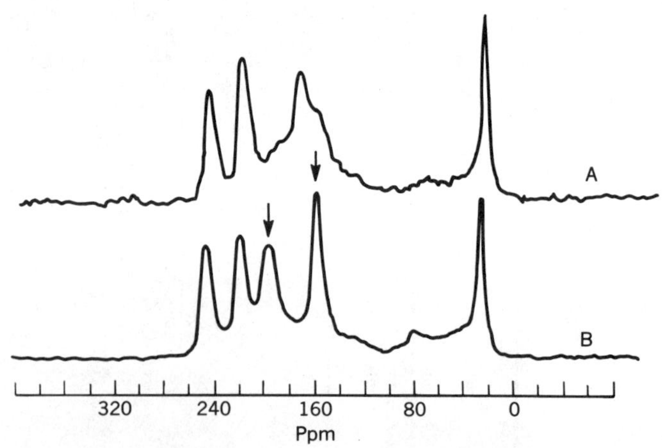

Figure 7. Solid-state ^{13}C nmr spectra of oriented PET fibers. The draw direction of the fibers is parallel to the magnetic field. Spectra (obtained without magic angle spinning) are shown for two temperatures: A, 137°C; B, −105°C. The resonances for the ortho aromatic carbons (C2 and C3) (marked by arrows) appear separately in the low temperature spectrum (adapted from D. L. van der Hart, G. G. A. Böhm, and V. D. Machel, *Polym. Prepr. Am. Chem. Soc. Div. Polym. Chem.* **22,** 261 (1981).

Chain Motion in the Solid State by ¹³C Nmr. Hytrel is a thermoplastic elastomer. It is a segmental copolymer composed of poly(butylene terephthalate) "hard" segments and poly(tetramethylene ether) "soft" segments. The soft segments impart elastomeric character to the polymer, whereas the hard segments form thermally reversible noncovalent networks. Molecular motion and phase mixing in a series of Hytrel copolyesters can be studied by solid state ^{13}C nmr. Local motion can be measured at each chemically different site in the copolymer. Complete analysis is possible because each chemically different site produces a separate peak, whose nmr parameters, including line width and relaxation times, can be measured.

For example, it is observed that the soft-segment carbons (ie, those of the polyether groups) in a wide range of Hytrel samples give nmr spectra without requiring the solid-state nmr techniques of dipolar decoupling, cross-polarization, or magic angle spinning. This suggests that the hard- and soft-segment carbons are phase separated and that the soft-segment carbons are mobile. The relaxation times (T_1 and NOE) of the soft-segment polyether carbons are independent of the mole fraction of hard segments, showing that the rates of the average local segmental motions of the soft phase are not affected by the hard segments. The relaxation times of the soft-segment carbons may be analyzed using a broad distribution of correlation times because a single correlation time model does not adequately fit the data. The line widths for the soft-segment carbons are a linear function of the average hard-block length. This finding is consistent with the presence of a large interfacial area in which the angular range of the excursions of the soft-segment carbons is affected by the presence of the hard blocks. Taken together, these results suggest that most of the hard segment is unaffected by the presence of soft-segment material.

Chain Motion in the Solid State by ²H Nmr. Deuterium nmr line shapes (Fig. 8) of selectively deuterated polymers are useful for observing molecular-level details of dynamical processes. The advantages of deuterium nmr are readily apparent in the study of complex copolymer materials, where one phase can be selectively labeled, thereby simplifying the spectra. For example, solid-state deuterium nmr measurements have been used to address the nature of phase separation, the amount of interfacial material, and the role of hard-segment material in the interface of the specifically labeled polyurethane based on diphenylmethane-4,4′-diisocyanate, chain-extended with butanediol:

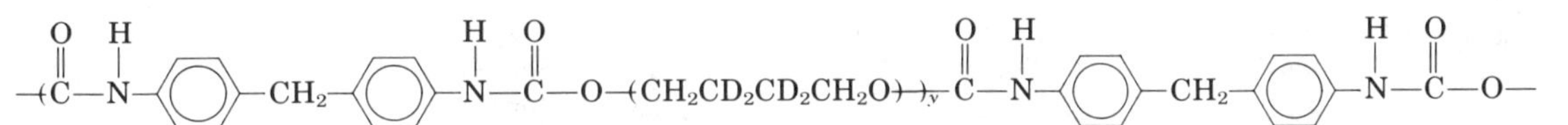

hard segment

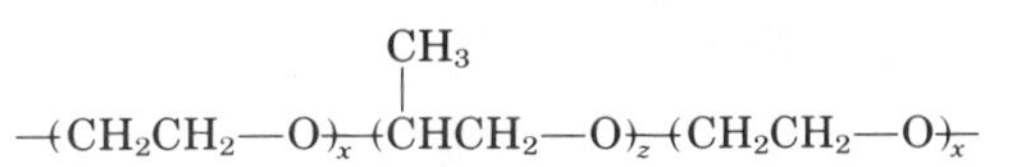

soft segment

In these samples, only the central methylene carbons of the butanediol group are labeled. Solid-state deuterium nmr spectra are shown in Figure 8. Because only the hard segment is labeled, the two-component nature of these spectra must arise from hard segments that reside in two different dynamical environments. Computer subtraction of appropriate amounts of the spectra in the center column from those in the left column produce the difference spectra shown in the right column. The sharp component, whose width remains constant, comprises 14, 20, and 50% of the signal intensity for the 70, 60, and 50 wt % hard samples, respectively.

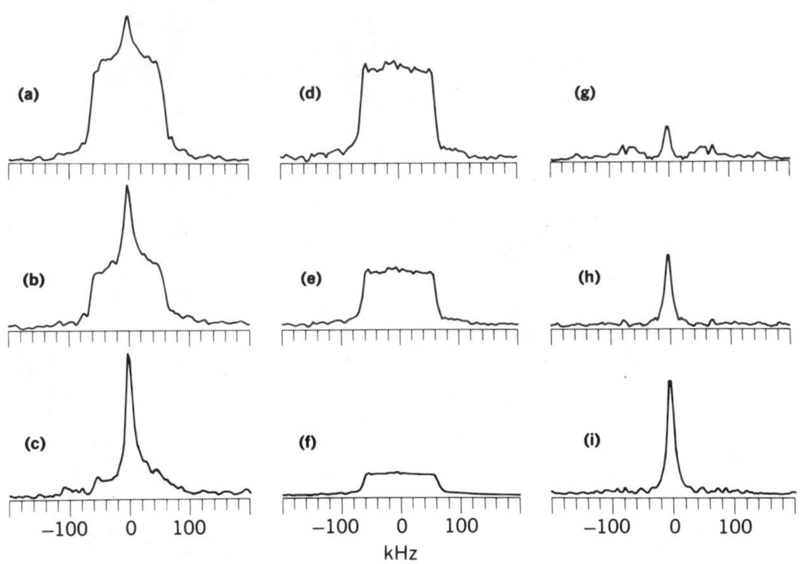

Figure 8. Solid-state deuterium nmr spectra of hard-segment labeled polyurethanes (see text for structural formula) obtained at 55.26 MHz and 22°C: (**a**) 70 wt % hard; (**b**) 60 wt % hard; (**c**) 50 wt % hard; (**d**)–(**f**) 100% hard at different vertical gains; (**g**)–(**i**) subtraction of spectrum in middle column from that in left column.

The unambiguous result from the subtraction process suggests that the central cores of the hard-segment-rich microdomains are composed of material that behaves in a dynamical sense in a manner similar to that of the all-hard-segment material. On the basis of additional small-angle x-ray scattering (saxs) results, the sharp component is attributed to hard segments residing in interfacial regions. The narrow line widths of the solid deuterium nmr spectra for these interfacial hard segments indicate that they undergo nearly isotropic motion and suggest that there are few long-lived interurethane hydrogen-bonding interactions in this region of the copolymer.

This example shows that solid-state deuterium nmr can provide highly selective information, in this case about the interface in a segmented copolymer. Appropriate analysis of the deuterium nmr lineshapes and their temperature dependence can also yield details of both the rate and angular range of molecular motion.

BIBLIOGRAPHY

F. A. Bovey, *High Resolution NMR of Macromolecules,* Academic Press, Inc., New York, 1972.

F. A. Bovey and L. W. Jelinski, *Chain Structure and Conformation of Macromolecules,* Academic Press, New York, 1982.

R. A. Komoroski, ed., *High Resolution NMR of Synthetic Polymers in Bulk,* VCH Publishers, Deerfield Beach, Fla., 1986.

A. E. Tonelli and F. C. Schilling, *Acc. Chem. Res.* **14,** 233 (1981).

V. J. McBrierty, *Magn. Reson. Rev.* **8,** 165 (1983); L. W. Jelinski, *Ann. Rev. Mater. Sci.* **15,** 358 (1985); H. W. Spiess, *Adv. Polym. Sci.* **66,** 23 (1985).

F. A. Bovey
L. W. Jelinski
AT&T Bell Laboratories

NUCLEAR RADIATION.

See RADIATION-RESISTANT POLYMERS.

NUCLEIC ACIDS.

See POLYNUCLEOTIDES.

NUCLEOSIDES.

See POLYNUCLEOTIDES.

NUCLEOTIDES.

See POLYNUCLEOTIDES.

NUMBER-AVERAGE MOLECULAR WEIGHT.

See MOLECULAR WEIGHT DETERMINATION.

NYLON.

See POLYAMIDES; POLYAMIDES, FIBERS; POLYAMIDES, PLASTICS.

O

OFFSET PRINTING.
See DECORATING.

OIL-FIELD APPLICATIONS

Polymers are used extensively in the various phases of drilling, operating, and maintaining oil and gas wells, including drilling, cementing, work-over and stimulation operations, and even the abandonment of wells. Application of polymers in enhanced oil recovery (EOR) processes prolongs the economic lifetime of oil fields. Over 600 polymer products are marketed for these applications.

Both water-soluble and oil-soluble polymers are employed in oil-field technology. The principal applications of water-sol-

Table 1. Properties of Water-soluble Polymers[a]

Polymer	Cost[b]	Viscosity[c], mPa·s (= cP)	Shear stability[d]	Salt tolerance[e]	Acid stability[f]	Enzyme stability[f]	Residue in broken gel[g]	Applications
guar gum	1	34	3	C	NS	NS	R	drilling fluids, spacers, friction reduction in stimulation, fracturing and lost circulation (cross-linked guar gum gels), and fluid-loss additives
hydroxypropyl guar	1	36	3	C	NS	NS	R	drilling fluids, spacers, completion and work-over fluids, friction reducer in fracturing, fracturing and lost circulation (cross-linked HPG gels), and fluid-loss additive
CMC	2	55	3	IC	NS	NS	RF	drilling fluids
HEC	2	37	3	C	NS	NS	RF	fluid-loss additive for cementing spacers, completion and work-over fluids, fracturing fluids, friction reduction in stimulation, and enhanced recovery
CMHEC	2	32	3	C	NS	NS	RF	fluid-loss additive and retarder for cementing spacers, gelling weak acids, temporary diverting agents in fracturing (cross-linked gels)
xanthan gum	4	34	1	C	MS	MS	RF	drilling, completion, fracturing, and enhanced oil recovery fluids
polyacrylamide, partially hydrolyzed	3	34	2	IC	S	S	RF	fluid-loss additive in cementing, drilling fluids, friction reduction in fracturing, enhanced oil recovery, and scale inhibitor
copolymer of acrylamide	3	25	2	MC	S	S	RF	fluid-loss additive in cementing, drilling fluids, friction reduction in fracturing, enhanced oil recovery, water–oil ratio reduction, improvement of injection profile (cross-linked)

[a] Courtesy of the Society of Petroleum Engineers.
[b] The order of cost per kg of polymer, increasing from 1 to 4. The price varies with the supplier and the quantity.
[c] Solution of 4 g/L at 300 rpm.
[d] The order of shear stability, increasing from 1 to 3. Xanthan gum solution viscosity is not reduced permanently by shear.
[e] C = compatible, IC = incompatible, and MC = moderately compatible.
[f] NS = not stable, MS = moderately stable, and S = stable.
[g] R = residue present and RF = residue-free.

Table 2. Properties of Sand-consolidation Resins

Resin	Well closing, h	Formation temperature, °C	Compressive strength of consolidated sand, MPa[a]	Permeability retention, %
phenol–formaldehyde	10–48	30–93	20.6	50–70
epoxy	2–24	10–120	34.4–48.2	67
furan	1–6	4–>204	20.6	90

[a] To convert MPa to psi, multiply by 145.

uble polymers arise from their ability to alter the rheological properties of aqueous fluids used in drilling, cementing, well completion, sand control, fracturing, acidizing, and enhanced oil recovery. Cross-linked polymer systems, including water-soluble, oil-soluble, and water-dispersible systems, reduce rock permeability in a controlled manner. These systems find applications in cementing, sand-control processes, and sealing of high permeability zones. Both *in situ* polymerization and *in situ* cross-linking may be employed in sealing high permeability zones. Both organic and inorganic polymers may be used in a variety of lesser well known applications, eg, in formation-damage control and as dispersing agents. The organic polymers may be synthesized from petrochemicals or derived from plants or microorganisms. They cover a wide range of molecular weights, chemical structures, and costs.

Oil-field Technology

Products and techniques used in oil-field technology include drilling fluids, cement slurries and spacers, completion and work-over fluids, hydraulic fracturing, acidizing, treatment for water intrusion into production wells, and enhanced oil recovery.

Water-soluble Polymers

In oil-field applications, the most important property of a water-soluble polymer is its solution viscosity, defined as the ratio of shear stress S_s to shear rate S_r. Shear stress is the force per unit cross-sectional area applied to the polymer solution. Shear rate is the displacement of the polymer solution divided by the height of the solution affected by the shear stress. At 20°C, water has a viscosity of 1.0 mPa·s (= cP). Properties and applications of these polymers are given in Table 1.

Sodium Silicate Gels. Sodium silicate gels are used to seal channels, vugs, and high permeability rock zones in production wells. In injection wells, *in situ* gelation of sodium silicate improves the injection profile and stops water leakage during drilling operations by plugging highly permeable channels.

Lignosulfonates. Lignosulfonates are employed as drilling-fluid dispersants and to prolong cement-slurry set time. *In situ* gelation to increase the volumetric sweep efficiency of waterfloods and EOR processes is potentially the largest volume oil-field application of lignosulfonates. They appear to offer the best potential of the cross-linked polymer systems for in-depth plugging of highly permeable channels or zones.

Cationic Organic Polymers. Applications of polymers containing quaternary ammonium groups in the repeat unit are based mainly on their adsorption characteristics and their properties as organic salts. These cationic organic polymers (COPs) adsorb rapidly and tenaciously to mineral surfaces, particularly clays in an ion-exchange process. The multiplicity of cationic sites on a single polymer chain causes the permanency of adsorption observed for many COPs (see POLYAMINES AND POLYQUATERNARY AMMONIUM SALTS).

Oil-soluble Polymers

Furan, Epoxy, and Phenol Resins. The cementation of poorly consolidated rock formations by cross-linking certain organic polymers adsorbed on rock surfaces can be particularly effective in minimizing sand production from poorly consolidated formations. This effectiveness accounts for their continued use, despite the relatively high cost of this sand-control method. The most commonly used plastics for this application are furan and furfuryl alcohol resins, epoxy resins, and phenol–formaldehyde resins. The properties of the most common sand-consolidation resins are given in Table 2.

BIBLIOGRAPHY

J. Chatterji and J. K. Borchardt, *J. Pet. Technol.* **33,** 2042 (1981).

G. V. Chilingarian and P. Vorabutr, *Drilling and Drilling Fluids,* Elsevier Science Publishing Co., Inc., New York, 1981.

T. O. Allan and A. P. Roberts, *Production Operations,* Vol. 1, 2nd ed., Oil & Gas Consultants, Inc., Tulsa, Okla., 1982.

T. O. Allen and A. P. Roberts, *Production Operations,* Vol. 2, 2nd ed., Oil & Gas Consultants, Inc., Tulsa, Okla., 1982.

JOHN K. BORCHARDT
Shell Development Company

OILS. See DRYING OILS; SILICONES.

OLEATES. See DRYING OILS.

OLEFIN–CARBON MONOXIDE COPOLYMERS

Carbon monoxide can be copolymerized with various olefins by free-radical processes; the greatest interest has been in copolymers and terpolymers with ethylene. The copolymers of ethylene and carbon monoxide fall into two categories depending on composition. Those containing small amounts of carbon monoxide have properties similar to low density polyethylene, except for a greatly increased sensitivity to photodegradation. Compositions close to 1:1 copolymer have much

Table 1. Melting Point vs % Carbon Monoxide

% carbon monoxide	Melting point, °C
27.5	111
35.1	122
38.7	126
40.6	147
43.9	181
46.3	203
49.0	235
49.8	242

higher melting points and improved mechanical properties. Terpolymers of ethylene, vinyl acetate, and carbon monoxide are used as permanent plasticizers for poly(vinyl chloride). Other vinyl and acrylic compounds can also be used in terpolymers with ethylene and carbon monoxide.

Copolymerization

The free-radical copolymerization may be carried out in bulk or in inert organic solvents, eg, benzene, cyclohexane, or saturated aliphatic hydrocarbons, as water is not a suitable medium for polymerization.

Structure of the Copolymers

The ratio of ethylene and carbon monoxide is determined by elemental analysis. As in polyethylene, the chains in crystals of ethylene–carbon monoxide copolymers have a planar zigzag conformation.

Properties of the Copolymers

The dependence of the melting point on the percent carbon monoxide is shown in Table 1. Up to about 40% carbon monoxide the melting point is similar to or slightly lower than that of a polyethylene made by a similar free-radical process. Between 40 and 50% carbon monoxide the melting point rises sharply. One of the most striking consequences of incorporating keto groups in polyethylene is its increased sensitivity to photodegradation by ultraviolet light.

Uses

Packaging materials discarded across the countryside may produce a legislative mandate for biodegradability or photodegradability. Copolymers of ethylene with a few percent carbon monoxide can meet this demand.

BIBLIOGRAPHY

U.S. Pat. 2,495,286 (Jan. 24, 1950), M. M. Brubaker (to E. I. du Pont de Nemours & Co., Inc.).

H. W. Starkweather, *J. Polym. Sci. Polym. Phys. Ed.* **15,** 247 (1977).

G. H. Hartley and J. E. Guillet, *Macromolecules* **1,** 165 (1968).

C. F. Hammer in D. R. Paul and S. Newman, eds., *Polymer Blends,* Vol. 2, Academic Press, Inc., Orlando, Fla., 1978.

HOWARD STARKWEATHER
E. I. du Pont de Nemours & Co., Inc.

OLEFIN FIBERS

Polyolefin fibers, also called olefin fibers, are defined as "manufactured fibers in which the fiber-forming substance is any long-chain synthetic polymer of at least 85% by weight of ethylene, propylene, or other olefin units." A number of crystalline polyolefins are capable of forming fibers (Table 1), but only polypropylene (PP) and, to a much lesser extent, polyethylene (PE) are of practical importance. Crystalline olefin polymers are hydrophobic and resistant to most solvents. These properties impart good environmental qualities and resistance to staining, but impede dyeing. Crystalline olefin polymers, having melting points lower than PP, have limited use in textile applications. With the exception of poly(4-methyl-1-pentene), the high cost of the higher melting polyolefins limits their applications. Furthermore, the low T_g of polyolefins may result in compressive creep. In spite of these drawbacks, polyolefin fibers are one of the fastest growing segments of the synthetic fiber market. The advantages of PP fibers are low cost and light weight. Improvements in stabilization, coloration, and texturing of this versatile fiber led to substantial volume growth in the 1970s, which continues at an even higher rate owing to recent advances in fiber-forming and textile technology.

Manufacture

For many years, the primary technology employed for the manufacture of PP filaments was melt spinning, similar to the methods employed for polyester and polyamide fibers. A number of texturizing operations were developed to impart characteristics of interest to carpet and upholstery manufacture. More recently, systems have been developed that are tailor-made for PP, eg, the short-spin system for low volume staple production. The newer spin systems take account of the fact that polyolefins, because of their high molecular weight and

Table 1. Properties of Crystalline Polyolefins and Other Fiber-forming Polymers

Polymer[a]	T_g, °C	T_m, °C	Softening temperature, °C	Thermal-degradation temperature, °C
high density polyethylene (HDPE)	−120	130	125	
i-polypropylene	−20	170	165	290
i-poly(1-butene)	−25	128		
i-poly(3-methyl-1-butene)		315		
i-poly(4-methyl-1-pentene)	18	250	244	
poly(ethylene terephthalate) (PET)	70	265	235	400
nylon-6,6	50	264	248	360

[a] i = isotactic.

broad distribution, have much higher viscoelasticity than traditional melt-spun polymers such as polyester and nylon.

The basic process of melt spinning is illustrated in Figure 1. The polymer resin and ingredients, primarily stabilizers, pigments, and rheological modifiers, are fed into a screw extruder, melted, and extruded through a fine-diameter spinneret. A metering pump and a mixing device are usually installed in front of the spinneret to ensure the uniform mixing and delivery rate necessary to produce uniform drawdown at high speeds. The fiber is pulled through a quench chamber by a take-up device and routed to downstream finishing operations.

Draw. Polyolefin fibers are usually drawn to increase orientation and modify the physical properties of the fiber further.

Texturing. The final step in olefin-fiber production is texturing; the method depends primarily on the application. For carpet and upholstery, the fiber is usually bulked, a procedure in which heated fiber is deposited on a moving screen through a hot-air or steam jet.

Slit-film Fiber. A substantial volume of olefin fiber is produced by slit-film or film-to-fiber technology. The fibers are used primarily for carpet backing or rope and cordage applications.

An olefin film is usually produced and, as in melt spinning, the morphology and composition of the film determine the processing characteristics. Fibers may be produced by cutting or slitting a film, or by mechanical or chemomechanical fibrillation. In the first case, the fiber is oriented by stretching before or after the cutting process. In the latter case, film anisotropy is created by stretching before fibrillation.

Bicomponent Fibers. In these fibers, two incompatible polyolefins, often PP and PE, are spun together to give a fiber with a side-by-side arrangement of the two materials. The lower melting polymer can melt and form adhesive bonds to other fibers; the higher melting component causes the fiber to retain some of its textile characteristics.

"Spurted" and Melt-blown Fibers. Unusual olefin fibers are produced by new spinning techniques. A variety of directly formed nonwovens with excellent filtration characteristics are made by melt-blown processes originally developed by Exxon Corp. producing very fine, submicrometer filaments. Pulplike olefin fibers are produced by a high pressure spurting process developed by Hercules, Inc. and Solvay, Inc.

High Strength Fibers. A new, high modulus commercial PE fiber with properties approaching those of aramid and graphite fibers is prepared by gel spinning. Higher tensile strengths are currently available only from gel spinning or dilute fibrillar crystal growth.

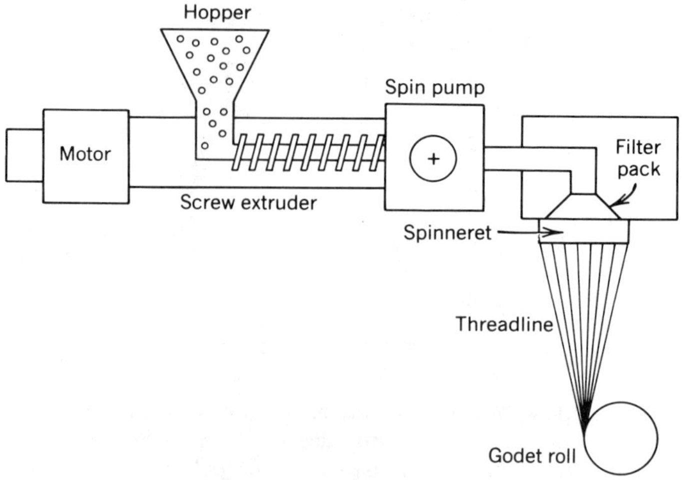

Figure 1. Melt-spinning process.

Hard-elastic Fibers. Hard-elastic fibers are prepared by annealing a moderately oriented spun yarn at high temperature under tension. They are prepared from a variety of olefin polymers, acetal copolymers, and polypivalolactone.

Properties

Physical. Physical properties of commercial olefin fibers and those of other synthetic fibers are given in Table 2. Of particular interest are the lower densities, allowing much lighter weight fiber at a specified size or coverage, and the low moisture absorption.

Tensile. As with all polymers, tensile properties are a function of molecular weight, morphology, and testing conditions. Increasing strain rate and lower temperature result in higher breaking stresses at lower elongation, consistent with the general viscoelastic behavior of polymeric materials. Under the same testing conditions, higher molecular weight at constant draw ratios results in higher tensile strength.

Creep, Stress Relaxation, Elastic Recovery. Olefin fibers exhibit creep and undergo stress relaxation. Higher molecular weight and high orientation reduce creep. At a fixed molecular

Table 2. Physical Properties of Commercial Fibers[a]

Polymer	Standard tenacity, GPa[b]	Breaking elongation, %	Modulus, GPa[b]	Density, kg/m^3	Moisture regain at 21°C and 65% rh, %
polyester	0.37–0.73	13–40	2.1–3.7	1380	0.4
carbon	3.1	1	227	1730	
nylon	0.23–0.60	25–65	0.5–2.4	1130	4–5
rayon	0.25–0.42	8–30	0.8–5.3	1500	11–13
acetate	0.14–0.16	25–45	0.41–0.64	1320	6
acrylic	0.22–0.27	35–55	0.51–1.02	1160	1.5
olefin	0.16–0.44	20–200	0.24–3.22	910	0.01
glass	4.6	5.3–5.7	89	2490	
aramid	2.8	2.5–4.0	113	1440	4.5–7
fluorocarbon	0.18–0.74	5–140	0.18–1.48	2100	
polybenzimidazole	0.33–0.38	25–30	1.14–1.52	1430	15

[a] Courtesy of *Textile World*.
[b] To convert GPa to psi, multiply by 145,000.

weight, the stress–relaxation modulus is higher for a highly crystalline sample prepared by slow cooling than for a smectic sample prepared by rapid quench. Annealing the smectic sample slightly raises the relaxation modulus, but not to the degree present in the fiber prepared by slower cooling.

Chemical. The hydrocarbon nature of olefin fiber, lacking any polarity, imparts high hydrophobicity and consequently resistance to soiling or staining by polar materials, a property important in carpet and upholstery applications.

Thermal and Oxidative Stability. The thermal and oxidative stability of polyolefins are poor.

Uv Degradation. Polyolefins are subject to light-induced degradation; PE is more resistant than PP. In any fiber application, stabilization against light is necessary.

Flammability. Most polyolefins can be made fire retardant with a stabilizer, usually a bromine-containing organic compound, and a synergist, such as antimony oxide. However, the required loadings are usually too high for fibers to be spun.

Dyeing. Because of their hydrophobic nature, olefin fibers are difficult to dye. Acid-dyeable olefin fibers have been developed by blending copolymers such as vinyltoluene-*co*-α-methylstyrene. A broad variety of polymeric dye sites have been blended with PP; nitrogen-containing copolymers are the most favored. Dyeable blends are expensive and create problems in spinning fibers with fine filaments for apparel application where dyeing is important. Hence, olefin fibers are usually colored by pigment blending during manufacture.

Economic Aspects

Olefin fibers are used for home furnishings and industrial applications. The former include carpets and upholstery, and the latter rope, geotextiles (see FIBERS, ENGINEERING), and disposable and nondisposable nonwovens. Traditionally, PP fiber has had limited use in apparel because of the lower melting temperature. However, this market is increasing as manufacturers learn to take advantage of the wicking properties (moisture transport), as, for example, in lightweight sportswear.

BIBLIOGRAPHY

S. Al-Malaika and G. Scott in N. S. Allen, ed., *Degradation and Stabilization of Polyolefins,* Elsevier Applied Science Publishers, Ltd., Barking, UK, 1983, Chapt. 6.

M. Ahmed, *Polypropylene Fibers: Science and Technology,* Elsevier Science Publishing Co., Inc., New York, 1982, pp. 344–346.

L. M. LANDOLL
Hercules, Inc.

OLEFIN POLYMERS

This article is devoted to polymeric materials derived from higher olefins with vinyl double bonds, ie, olefins with carbon atom numbers five and higher, containing linear or branched alkyl groups as well as vinyl aromatic compounds. Polymers of higher α-olefins are produced with heterogeneous Ziegler-Natta catalysts. Some higher olefins can also be polymerized with cationic catalysts to materials with isomerized structures.

Monomers

The monomers of greatest interest are those available as byproducts from commercial processes such as refining of petroleum products, products of thermal cracking of hydrocarbons, and ethylene oligomers. Olefins with branched alkyl groups are also produced by catalytic dehydration of the corresponding alcohols. Careful purification, in particular, removal of water and oxygenated compounds, is essential when employing Ziegler-Natta catalysts.

Polymers

Physical Properties. Physical properties of isotactic polymers of higher α-olefins are given in Table 1. All isotactic polymers have the helix conformation in the crystalline state.

Chemical Properties. The chemical properties of polymers of higher olefins resemble those of polypropylene and poly(1-butene).

Polymerization

Ziegler-Natta Catalysts. Polymers of higher α-olefins produced by polymerization with heterogeneous Ziegler-Natta catalysts have highly regular head-to-tail structures. Reactivities of higher α-olefins largely depend on the substituents with strong steric effects of alkyl groups at the olefin double bond being evident. The measure of olefin reactivity, the reactivity ratio r_2, decreases with increasing carbon number, becoming essentially constant after 1-dodecene. If the alkyl group in a higher olefin contains a second double bond, a special copolymerization reaction can take place giving polymers containing cyclized monomer units.

Cationic Catalysts. Higher α-olefins are also polymerized with cationic catalysts. Olefins containing linear alkyl groups are oligomerized to oils of irregular structure. However, α-olefins with alkyl groups branched in the α-position to the double bond can be polymerized to high molecular weight polymers with regular structures due to a uniform isomerization polymerization (qv).

Processing

Processing of isotactic polymers of higher α-olefins is similar to the processing of other isotactic polymers, eg, polypropylene, poly(1-butene), and poly(4-methyl-1-pentene).

Economic Aspects

The polymers of higher olefins discussed in this article are noncommercial and detailed economics associated with production are not available.

Health and Safety Factors

Polymers of higher α-olefins are expected to be nontoxic based on structural and compositional similarities to polypropylene, poly(1-butene), and poly(4-methyl-1-pentene). The main potential health hazards are associated with residual monomer, antioxidants, and catalyst residues.

Uses

The lower molecular weight liquid oligomers of higher α-olefins have found applications as synthetic base oils in the formu-

Table 1. Properties of Isotactic Polyolefins

Polymer	Melting point, °C	Crystal type	Density, g/cm³ Crystalline	Density, g/cm³ Average	T_g, °C
poly(1-butene)[a]	126–142	rhombohedral	0.95	0.915	−20 to −25
poly(1-butene)[b]	122–130	tetragonal	0.96	0.90	
poly(1-pentene)[a]	80	monoclinic	0.92	0.87	−40 to −55
poly(1-pentene)[b]	75–80	pseudoorthorhombic	0.90		
poly(3-methyl-1-butene)[b]	350	monoclinic	0.93	0.90	
poly(1-hexene)[c]		monoclinic	0.73	0.84	−50
poly(3-methyl-1-pentene)[c]	200				20–50
poly(4-methyl-1-pentene)[c]	235–240	tetragonal	0.813	0.838	20–50
poly(1-heptene)[a]	18				−30
poly(4-methyl-1-hexene)[c]	188–200	tetragonal	0.845	0.86	
poly(5-methyl-1-hexene)[a]	110–130	monoclinic	0.84	0.85	
poly(1-octene)	~20				−65 to −45
poly(5-methyl-1-heptene)[a]		tetragonal			
polyvinylcyclohexane[b]	385	tetragonal	0.95	0.93	80
polystyrene[a]	213–240	rhombohedral	1.12	1.08	80–100
poly(*o*-methylstyrene)[b]	>360	tetragonal			
poly(*m*-methylstyrene)[a]	215	tetragonal	1.01		
poly (α-vinylnaphthalene)[b]	~360	tetragonal	1.125		

[a] Helix 3_1.
[b] Helix 4_1.
[c] Helix 7_2.

lation of various lubricants such as transformer oils, transmission and crankcase fluids, and the like. The true polymers of higher α-olefins, however, have found only limited application.

BIBLIOGRAPHY

H. Tadokoro, *Structure of Crystalline Polymers,* John Wiley & Sons, Inc., New York, 1979.

Y. V. Kissin, *Isospecific Polymerization of Olefins with Heterogeneous Ziegler-Natta Catalysts,* Springer-Verlag, New York, 1985.

J. Boor, Jr., *Ziegler-Natta Catalysts and Polymerizations,* Academic Press, Inc., Orlando, Fla., 1979.

J. P. Kennedy, *Cationic Polymerization of Olefins: A Critical Review,* John Wiley & Sons, Inc., New York, 1975.

DAVID L. BEACH
Chevron Chemical Company
YURY V. KISSIN
Mobil Chemical Company

OLEFIN–SULFUR DIOXIDE COPOLYMERS

Olefin–sulfur dioxide copolymers [poly(olefin sulfone)s and poly(alkylene sulfone)s] are obtained by the free-radical reaction of olefins with sulfur dioxide. In the principal class, the hydrocarbon and sulfonyl residues alternate, as in poly(propylene sulfone) (**1**); its International Union of Pure and Applied Chemistry (IUPAC) systematic name is poly(sulfonyl-1-methylethylene). Such copolymers are also formed from ethylene, olefins with two or more substituents, cyclic olefins, and olefins with heteroatoms in the side chain or ring. In a second class, a nonalkyl substituent is linked directly to the olefinic carbon, and the sulfonyl groups comprise less than half the residues, eg, poly(vinyl chloride sulfone) (**2**) with $m > 1$ (typically 2).

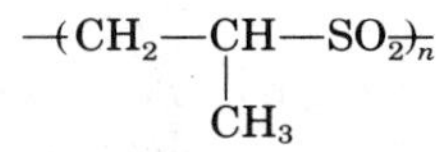
(**1**)

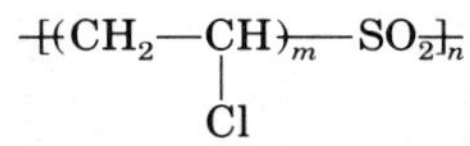
(**2**)

Polysulfones also form from conjugated and nonconjugated dienes and from alkynes, although not from acetylene. More than one olefinic structure may be incorporated into a single macromolecule. A second method of preparation is by polysulfide oxidation. The term "polysulfone" usually refers to engineering plastics, such as poly(aryl ether sulfone)s (**3**) (see POLYSULFONES).

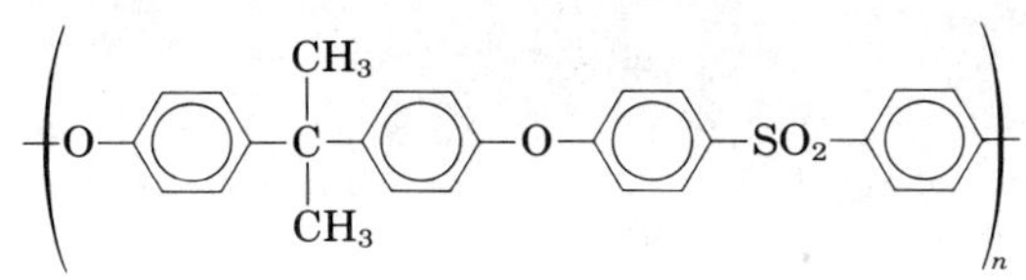
(**3**)

Monomers

Sulfur Dioxide. The odor threshold of SO_2 (0.3–2.5 ppm) is close to the recommended TLV of 2 ppm and the maximum workplace concentration of 5 ppm. Short exposure to concentrations over 400 ppm may be lethal. The most susceptible tissues are the mucous membranes of the eye and respiratory tract where sulfurous acid may form. Prolonged low concentration exposure may result in bronchitis and pulmonary edema. Physical properties for SO_2 are given in Table 1.

Thermodynamics of Poly(olefin sulfone) Formation

At the ceiling temperature T_c, the free energy of a mixture of SO_2 and an olefin is the same as the free energy of the 1:1

Table 1. Physical Constants of Sulfur Dioxide[a]

Property	Value
melting point, °C	−75.52
boiling point, °C	−10.08
enthalpy of fusion, kJ/mol[b]	8.241
enthalpy of vaporization, kJ/mol[b]	24.94
vapor pressure, kN/m^2[c]	
at −44°C	16.1
at 20°C	327

[a] The following equations have been proposed for the liquid state:

$$\rho = 1430 - 2.486\,t - 2.63 \times 10^{-3}\,t^2 - 5.591 \times 10^{-5}\,t^3 + 8.1 \times 10^{-8}\,t^4 \text{ kg/m}^3$$
$$\epsilon = 9.512 \exp(-6.676 \times 10^3\,T)$$
$$\log P\ (\text{kN/m}^2) = 15.2003 - 1867.5/T - 0.015865\,T + 0.000015574\,T^2$$

where t = °C and T = K.

[b] To convert J to cal, divide by 4.184.

[c] To convert kN/m^2 to mm Hg, multiply by 7.5.

copolymer. The polymer forms on initiation only below the T_c; above the T_c the polymer is unstable. For a particular feed of monomers and solvent, T_c is independent of initiator and may be determined by extrapolating dilatometric or gravimetric measurements made over a range of temperatures below T_c of rates of polymerization or of molecular weight. The relationship between standard enthalpy and entropy changes of polymerization, monomer activities, and T_c is

$$T_c = \Delta H_p/[\Delta S^0 + R \ln (a_{SO_2}a_M)] \qquad (1)$$

This relation usually holds, regardless of the polymer solubility in the feed, but is inadequate for the polysulfones of isobutene and 3-methyl-1-butene. Plots of $\ln(a_{SO_2}a_M)$ vs $1/T$ have two branches. Perhaps for conformational reasons the solid polymer obtained from a polar, SO_2-rich feed differs from that obtained from a less polar, olefin-rich feed.

Mechanism

Charge-transfer complexes are sometimes considered to play an important role in the formation of olefin–sulfur dioxide copolymers, particularly at low temperatures. The formation of 1:1 copolymers might also be the consequence of the radical of one monomer reacting with the other monomer (in Lewis-Mayo terms, $r_1 = r_2 = 0$) or of complicated radical dissociation processes taking place. The spontaneous initiations with cyclopentene, cyclohexene, or bicyclo[2.2.1]-2-heptene are associated with traces of peroxide impurities functioning as hydroperoxide initiators.

Microstructure

The ir spectra of polysulfones display the strong bands of the asymmetric and symmetric SO_2 stretches near 1320 and 1140 cm^{-1}, as well as the other low frequency bands associated with the sulfone group identified from studies on small alkyl sulfones. Solution ^{13}C nmr spectroscopy shows poly(1-olefin sulfone)s to be atactic.

Preparation

Polysulfones are normally formed from an olefin and sulfur dioxide in the liquid phase, many free-radical sources being suitable for initiation. A solvent such as chloroform promotes miscibility of the two monomers. Following uv or high energy irradiation, copolymerization has also been carried out in the solid state with isobutene, butadiene, vinyl acetate, and styrene at low temperatures and bicyclo[2.2.1]-2-heptene at ambient temperatures. Emulsion and suspension methods have been described.

Properties

Solution Properties and Chain Dynamics. Poly(methylene sulfone) and poly(propylene sulfone), where the sulfone group forms a large proportion of the repeat unit, dissolve in polar solvents, eg, concentrated sulfuric acid or dimethyl sulfoxide. Other polysulfones, such as those formed from 1-olefins and cyclic olefins, dissolve in common solvents, such as acetone, chloroform, and benzene. The polysulfones of the 1-olefins with long alkyl groups dissolve in hydrocarbon fuel oils and, if made with a small proportion of ionizing heteroatomic groups, cannot be extracted into water. Chain dynamics and conformation are controlled by strong electrostatic interactions between adjacent sulfone dipoles in a unique manner.

Physical Properties. Crystallinity has been readily detected by x-ray diffraction studies in the poly(ethylene sulfone)s of variable composition, in poly(butadiene sulfone), and in polysulfones with three, four, five, or six main-chain carbon atoms in the repeat unit. Density and gas-permeability values suggest that the C_{18} 1-olefin copolymers may have ordered side chains.

The densities of polysulfones are given in Table 2. The density increases with increasing sulfone to hydrocarbon ratio. The glass-transition temperatures are shown in Table 3 and melt temperatures in Table 4. Liquid crystalline polysulfones have mesogenic groups in the side chain. Poly(propylene sulfone) can be molded between 180 and 200°C, and poly(1-butene sulfone) between 125 and 180°C.

Table 2. Densities of Polysulfones

Olefin	ρ, kg/m^3
propylene	1457
1-hexene	1220[a]
1-octadecene	990
1-butene	1245
1-hexadecene	990[a]
isobutene	1406

[a] The density varies smoothly with chain length between 1-hexene and 1-hexadecene.

Table 3. Glass-transition Temperatures of Polysulfones

Olefin	T_g, °C
1-butene	81–95
bicyclo[2.2.1]-2-hepene	117
1-hexadecene	77
cyclopentene	82
styrene, 2:1	180–200
1-butene–cyclopentene[a]	57

[a] 50% of each.

Table 4. Melting Points of Polysulfones[a]

Olefin	T_m, °C
ethylene[b]	135
1-butene	150–160
1-hexene	76
cyclohexene	200–205
bicyclo[2.2.1]-2-heptene	240–290
norbornene acetate	240–245
propylene	280
isobutene	230
1-octadecene	45
cis,cis-1,5-cyclooctadiene	260
norbornene nitrile	275–280

[a] Melting is often accompanied by decomposition. For noncrystalline polymers, T_m is the temperature at which the polymer melts and leaves a trail when moved across a hot metal bar.
[b] 31% SO_2.

Stability and Degradation. The polysulfones of ethylene, propylene, butadiene, and other dienes as well as the poly(methylene sulfone)s dissolve in strong acids and are resistant to concentrated sulfuric acid. The thermal instability of poly(alkyl sulfone)s has been attributed to the weakness of the carbon–sulfur bond (240 kJ/mol or 57.3 kcal/mol compared to 345 kJ/mol or 82.4 kcal/mol for the carbon–carbon bond) and to a decomposition mechanism, depending on the presence of hydrogen atoms β to the sulfonyl group.

Consistent with their high G scission values for γ-irradiation is the sensitivity of olefin–sulfur dioxide copolymers to high energy electrons. Doses on the order of $1–4 \times 10^{-2}$ C/m^2 reduce the molecular weight by main-chain scission, permitting solution development of patterns in thin films. This treatment has been commercialized for poly(1-butene sulfone). Because of a low ceiling temperature, poly(2-methyl-1-pentene sulfone) films disintegrate faster than other polysulfones in a process termed "vapor development."

Uses

Thermal instability near the molding temperature has prevented olefin–sulfur dioxide copolymers from being used as bulk thermoplastics, but certain specialty uses have been developed. The most significant use at present exploits sensitivity to high energy radiation. Poly(1-butene sulfone) (PBS) was developed at Bell Laboratories, now known as AT&T Bell Laboratories, as a resist in the electron-beam fabrication of chromium photomasks using electron-beam exposure. It is available as a filtered solution ready for spin coating at a cost of approximately $350/L (Mead Chemical Co., Rolla, Mo. and Chisso Corp., Japan). Current interest lies in improving the resistance of polysulfones to dry etching and in developing polymers with liquid crystalline side chains.

BIBLIOGRAPHY

F. S. Dainton and K. J. Ivin, *Q. Rev. London* **12,** 61 (1958).

K. J. Ivin and J. B. Rose, *Adv. Macromol. Chem.* **1,** 335 (1968).

M. J. Bowden and J. H. O'Donnell in N. Grassie, ed., *Developments in Polymer Degradation,* Vol. 6, Elsevier Applied Science Publishers, Ltd., Barking, UK, 1985.

ALLAN H. FAWCETT
The Queen's University of Belfast

OLIGOMERS

The International Union of Pure and Applied Chemistry (IUPAC) defines an oligomer as a substance composed of molecules containing a few of one or more species of atoms or groups of atoms (constitutional units) repetitively linked to each other. This does not specify an absolute degree of polymerization or molecular weight that distinguishes an oligomer from a polymer, but the IUPAC definition further states that the physical properties of an oligomer vary with the addition or removal of one or a few of the constitutional units from its molecules. This structure–property definition is perhaps the most meaningful definition of an oligomer.

Oligomers are used to study polymerization mechanisms and the dynamics and equilibria of cyclization and are of enormous importance for the development of the host–guest complex chemistry or bioorganic chemistry. The primary reasons for studying oligomers include the use of oligomers as models for understanding polymerization mechanisms, as intermediates in preparative polymer chemistry, and as models for studying the structure–property relationships of polymers.

Models for Polymerization Mechanisms

Anionic Polymerization. ***Vinylic Monomers.*** The oligomerization of vinyl monomers has been studied to understand the stereochemistry of polymerization and termination reactions. In particular, comparisons have been made of the stereochemistry of 2-vinylpyridine and 4-vinylpyridine polymerizations by consecutive monomer additions to a tetrahydrofuran (THF) solution of an alkali metal salt of 2-ethylpyridine or 4-ethylpyridine, followed by termination with methyl iodide.

These studies demonstrate that the geometry of the carbanion site can be an important factor in the methylene stereochemistry of anionic oligomerization and polymerization of vinyl monomers. Acrylic monomers and vinyl phenyl sulfoxide have also been studied.

Heterocyclic Monomers. Most anionic ring-opening polymerizations are nucleophilic substitution reactions, eg,

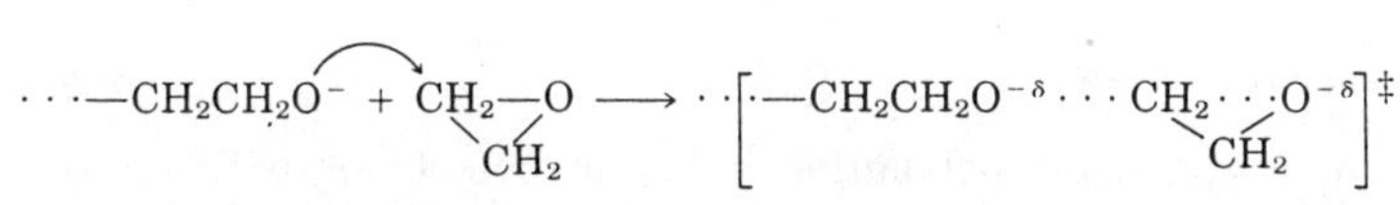

For some heterocyclic monomers the unique chemical structure of the growing species follows unequivocally from the monomer structure. The bond to the heteroatom is the most susceptible to chemical change. However, isomeric structures must be taken into account in the case of ambident monomers.

A method has been developed that allows the determination of the chemical structure of active centers of a growing oligomer. This method is based on oligomeric anion-capping with

$ClP(O)(OC_6H_5)_2$, followed by quantitative analysis of the resulting phosphate end groups by $^{31}P(^1H)$ nmr spectroscopy.

$$RX^-, Mt^+ + ClP(O)(OC_6H_5)_2 \longrightarrow RXP(O)(OC_6H_5)_2 + MtCl$$

where X = heteroatom and Mt = metal

Cationic Polymerization. ***Heterocyclic Monomers.*** Cationic ring-opening polymerizations are also nucleophilic substitution reactions, eg,

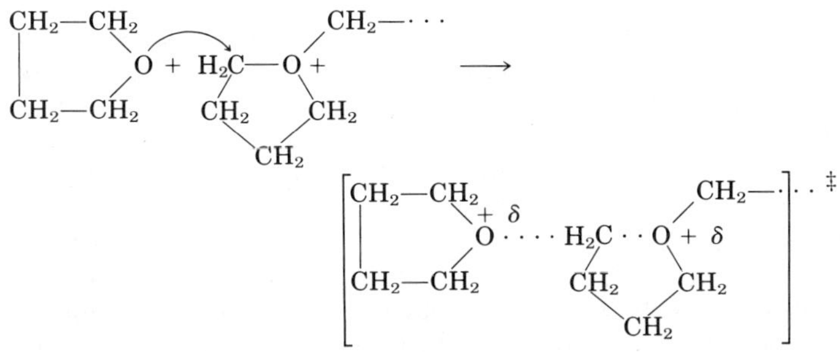

Two experiments are frequently employed to follow the cationic polymerization of heterocyclic monomers. The first is based on the polymerization of a low molar ratio of monomer and initiator directly in an nmr sample tube. Trapping experiments such as those described for the anionic polymerization of heterocyclic monomers have been designed to determine the structure of the growing cations by using R_3P as a capping reagent.

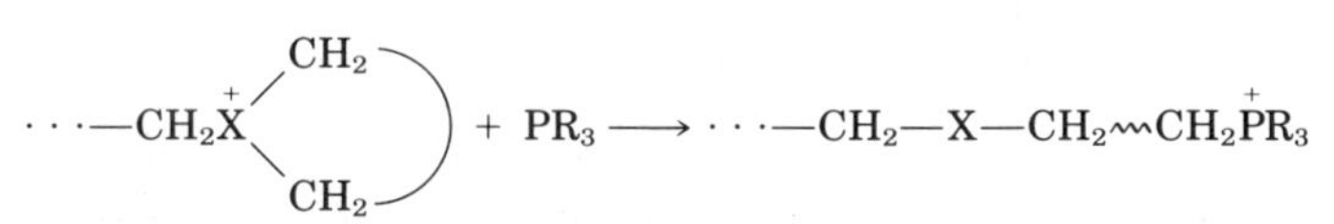

Radical Polymerization. The most widely accepted pathway for the spontaneous generation of radicals in the thermal polymerization of styrene was proposed in 1961 and is outlined here:

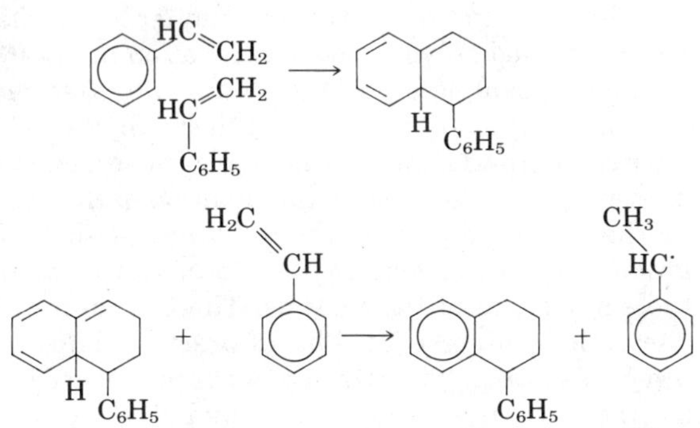

(2) (3)

Although ample evidence for the transient formation of the Diels-Alder dimer (**1**) exists, its reaction with styrene and the identity of the radicals that initiate polymerization have remained speculative and controversial. The definitive experiment confirming this mechanism was based on the retardation of the propagation step in the thermal polymerization of styrene with $FeCl_3$ in DMF. The resulting oligomers containing between 6 and 14 monomer units were characterized by ^1H nmr spectroscopy and it was shown that radicals (**2**) and (**3**) are indeed responsible for the initiation of the polymerization.

Intermediates in Preparative Polymer Chemistry

Poly(ethylene oxide) Derivatives. The synthetic methods used to make functional derivatives of poly(ethylene oxide) (PEO) and their applications have been extensively reviewed (see also 1,2-EPOXIDE POLYMERS). The cyclic PEO oligomers, known as crown ethers, were first prepared in 1937. However, their ability to complex cationic substrates, which makes them useful catalysts and phase-transfer reagents, was not recognized until 1967. These materials have been reviewed. Most crown ethers are synthesized in rather high yields by template-assisted Williamson etherifications, which organize the intermediate(s) in the cyclization step. However, the cyclic oligomers of PEO can also be made by oligomerization reactions of ethylene oxide, although not in high yields.

Polyethylene Derivatives. ***Oligomers with Two Functional Chain Ends.*** Telechelic polyethylene (PE) oligomers have been obtained by the selective oxidative degradation of PE single crystals at the amorphous chain folds using either fuming nitric acid or ozone (see also TELECHELIC POLYMERS). Dicarboxylic-acid-terminated paraffins are obtained with lengths corresponding to the crystal thickness (approximately 100 carbon atoms).

Supported Reagents and Catalysts. Ethylene oligomers with molecular weights higher than 1000 are insoluble in organic solvents at 25°C, but form solutions in a variety of solvents at 90–100°C. Based on this solubility behavior a novel concept for polymer-supported reagents and catalysts based on functionalized oligomers of ethylene has been developed. At room temperature they become insoluble and can be separated from the reaction products by centrifugation or filtration. An additional property of ethylene oligomers is their ability to become entrapped quantitatively in PE powders by coprecipitation with high density polyethylene (HDPE). Therefore, these functionalized ethylene oligomers behave either as catalysts supported on soluble polymers at high temperature or as catalysts supported on insoluble polymers at low temperature.

Models for Structure–Property Relationships

Crystallization and Chain Folding. ***Polyethylene Oligomers.*** One of the most extensive uses of oligomers and short polymer chains has been in studies of chain folding, which is the standard mode of polymer crystallization. That is, oligomers help to answer questions such as at what chain length does chain folding set in? What are the lengths of the straight chain segments or stems? Where are the chain ends located and what is their effect on chain folding? And what is the nature of the fold? Cyclic oligomers have been used to study the last question; linear oligomers and short polymer chains have been used to consider all four problems.

Polydimethylsiloxane Oligomers. In contrast to the cyclic alkanes, comparison of the solid-state ^{13}C and ^{29}Si nmr spectra of cyclooctamethyltetrasiloxane with its solution ^{13}C nmr spectra demonstrates that the conformational mobility of such polar oligomers is significantly determined by the molecular packing in the crystal, and the solid–solid transition (265 K) is only observed in solid-state experiments and not in solution.

The thermal behavior (dsc) of both cyclic (5–329 structural units) and linear polydimethylsiloxanes (2–343 structural units) have been studied. In addition to melting and the solid–solid transition, a T_g and a cold crystallization exotherm were observed in cyclic and linear oligomers with greater than 42 and 33 structural units, respectively. Furthermore, the temperatures of these transitions were much lower than those of octamethyltetrasiloxane.

Poly(ethylene oxide) Oligomers. The effect of polydispersity on crystallinity was studied by comparing the monodisperse nonamer and pentadecamer of ethylene oxide with "polydisperse" (1.05–1.10) PEO (number-average degrees of polymerization from 9 to 68). All samples were lamellar with a 7/2 helical structure. The monodisperse glycols were highly crystalline with their chain ends incorporated in the crystalline array, but the chain ends in the polydisperse samples were increasingly incorporated in the noncrystalline surface area as chain length increased.

Poly(tetramethylene oxide) Oligomers. As with most other oligomers the oligomers of poly(tetramethylene oxide) or polytetrahydrofuran have been used to calculate the melting temperature of the corresponding polymer of infinite chain length (419.5 K) (see TETRAHYDROFURAN POLYMERS).

Vitrification. The physical properties of polymers and oligomers vary with the degree of polymerization. Because polymeric end groups have a greater influence on properties as molecular weight decreases, PS samples have been used to test the assumption that the T_g of anionically and thermally initiated polymers coincide at high molecular weight, but differ significantly at low molecular weights where the concentration of chain ends is much higher. However, the same dependence of T_g on molecular weight for both types of polymer at all molecular weights was found. Because this is true even for the oligomeric samples and lower molecular weight polymers, vitrification must not be governed by free volume (see also CURING; GLASS TRANSITION).

Segmented Thermoplastic Elastomers. There is interest in tailor-making model oligomers of the "hard" and "soft" segments of thermoplastic elastomers because it is desirable to control the combined thermoplastic processing properties and elastomeric mechanical properties of the materials, especially as they are influenced by chain-length distributions (see ELASTOMERS, THERMOPLASTIC).

Thermal Stability of Poly(methacrylate). Poly(methyl methacrylate) (PMMA) prepared by an anionic mechanism is more thermally stable than that made by a free-radical process. The thermal degradation of PMMA by a free-radical mechanism occurs in two distinct stages. Approximately 50% of the sample degrades between 170 and 250°C; the remainder requires temperatures well in excess of 300°C.

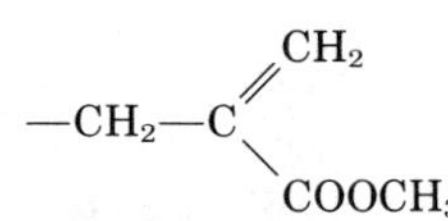

(4)

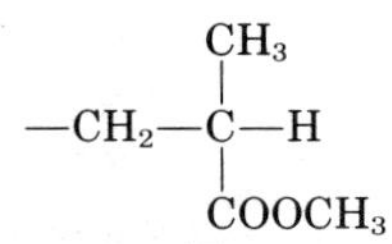

(5)

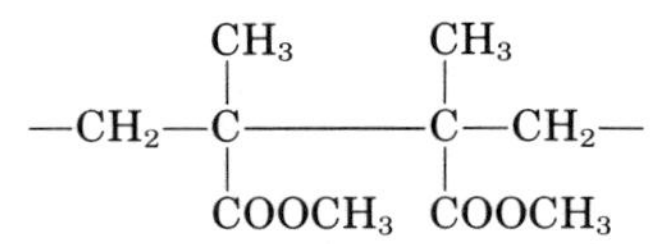

(6)

In order to interpret the thermal stability of PMMA, the oligomers (**7**), (**8**), and (**10**) have been synthesized as models for the polymers containing end groups or linkages formed by the different termination mechanisms:

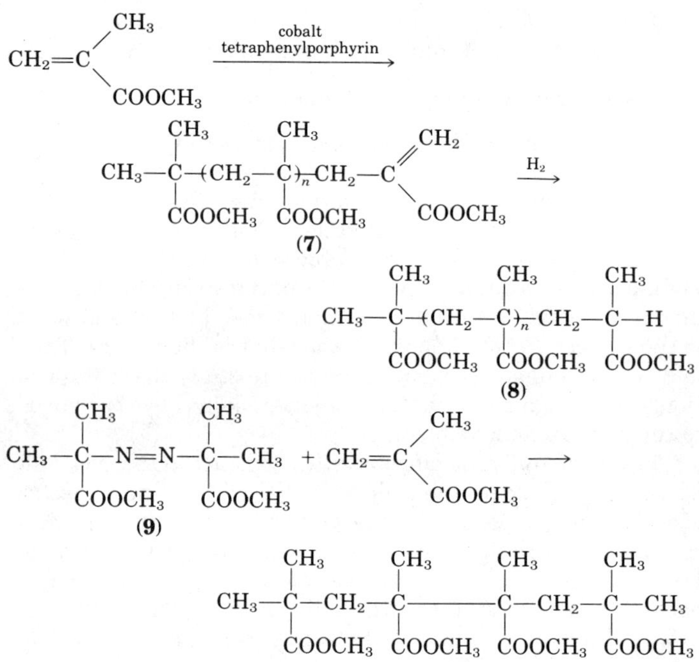

The decomposition temperatures for the individual oligomers are 250°C for (**7**), corresponding to structure (**4**); 300°C for (**8**), which is like (**5**); and 190°C for (**10**), corresponding to (**6**). These decomposition temperatures can clearly explain the thermal behavior of radically and anionically formed PMMA. The anionically obtained polymer has a structure similar to (**5**), whereas the radically obtained polymer is a combination of (**4**)–(**6**).

Within a given sample of PMMA obtained by free-radical polymerization, longer chains are more thermally labile than their shorter counterparts. Most of the higher molecular weight molecules are expected to arise by combination of the growing chains and hence would contain the weak head-to-head linkage. Degradation of these chains, therefore, will be observed during the early stage of thermolysis. These results are not only of theoretical interest but also of practical importance, because even a radical polymerization can be performed in such a way that the amount of structures (**4**) and (**6**) are controlled and therefore the thermal stability of the resulting polymer designed.

Conducting Oligomers. The answer to the question of whether an interchain or intrachain process is the dominant one responsible for electrical conductivity in conducting polymers comes from the study of the electrical conductivity of polymers as a function of molecular weight (see also ELECTRICALLY CONDUCTING POLYMERS).

Thermotropic Liquid Crystalline Oligomers. Thermotropic liquid crystalline polymers containing calamitic or rodlike mesogenic units can be classified as side chain or main chain. Side-chain liquid crystalline polymers have the mesogenic units attached as side groups from the polymer backbone. The mobility of the mesogenic side groups is decoupled from the mobility of the polymer backbone through a flexible spacer. Main-chain liquid crystalline polymers have the mesogenic groups in the main chain and they are coupled either through flexible or rigid spacers. Side-chain liquid crystalline oligomers were synthesized in order to obtain a more detailed understanding of the change in phase behavior in going from the monomer to the polymer.

Polymer Blends. Theoretical models have predicted that any miscible polymer blend should exhibit a lower critical solution temperature (LCST) and that the LCST should decrease as the molecular weight of one of the two polymers increases. The theoretical prediction concerning the dependence of the LCST on molecular weight has been experimentally demonstrated by measuring the LCST of blends of oligomers with different molecular weights with a polymer having a constant molecular weight (see also POLYMER BLENDS).

BIBLIOGRAPHY

P. Rempp and E. Franta, *Adv. Polym. Sci.* **58,** 1 (1984).

F. W. Harris and H. J. Spinelli, *ACS Symp. Ser.*, 282 (1985).

J. Macromol. Sci. Chem. **21,** 8–9 (1984).

E. J. Goethals, *Adv. Polym. Sci.* **23,** 103 (1977).

VIRGIL PERCEC
COLEEN PUGH
Case Western Reserve University

ONE-ELECTRON-INITIATED POLYMERIZATION. See RADICAL POLYMERIZATION.

OPEN-CELL FOAM. See CELLULAR MATERIALS.

OPTICAL BLEACHES. See PAPER.

OPTICAL BRIGHTENING AGENTS. See PAPER.

OPTICAL DATA STORAGE

Optical storage is the emerging technology for storage, processing, and retrieval of vast quantities of data. In various formats, it is suitable for applications where mass replication of predetermined data is required, long-term (over 10 years) archiving, recording of legally nonalterable records, and where infinite erasability and rerecording are required. Because optical storage provides very high data density (up to 1 gigabyte per side on a 130-mm disk), removability from the drive, random access, and low cost per bit, it will be used to supplement and replace magnetic and micrographic storage devices.

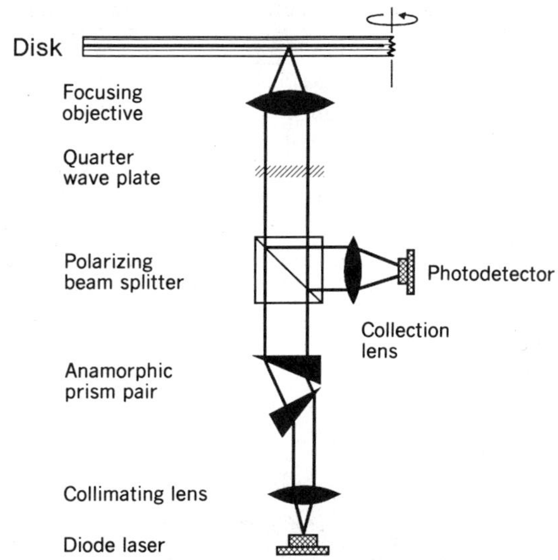

Figure 1. Optical recording system.

This remarkable advance in data storage is possible by using a modulated laser beam as the recording head. A continuous laser beam functions as the playback probe. Information storage occurs in a recording layer on the surface of a support or substrate. During playback or reading, differences in the optical properties of the disk, which correspond to the encoded information, modulate the reflected read beam to produce a signal that corresponds to the input data.

In replicated, read-only memory (ROM) systems such as audio disks, information, in the form of surface pits, is manufactured in place. Modulation of the read beam is achieved by optical interference resulting from the depth of the pits. The surface of the disk is uniformly highly reflective. In the case of a user-recordable disk, the recording process changes the optical properties of the recording layer. These changes may be physical or chemical, but in all practical systems they must be completed in tens of nanoseconds and must be immediately detectable, ie, they must not require a subsequent developing step. This type of recording, usually called direct read-after-write (DRAW), is necessary if data recorded are to be immediately verified.

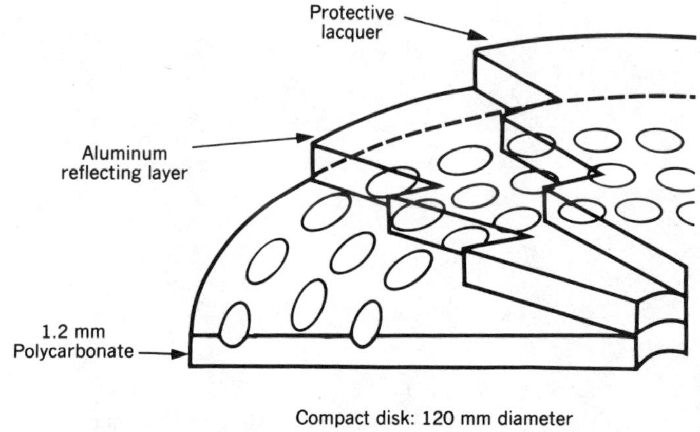

Figure 2. Read-only disk.

Table 1. Comparison of Substrate Materials[a]

Property	Glass	Polycarbonate	PMMA	Epoxy	Olefin
melt viscosity	NA	0	0	NA	X
birefringence	XX	X	XX	XX	XX
optical clarity	XX	X	XX	XX	XX
solvent resistance	XX	0	X	XX	X
thermal properties	XX	X	0	XX	U
moisture regain	XX	X	0	X	XX
cost	0	XX	XX	0	U

[a] XX = excellent, X = good, 0 = poor, U = unknown, and NA = not applicable.

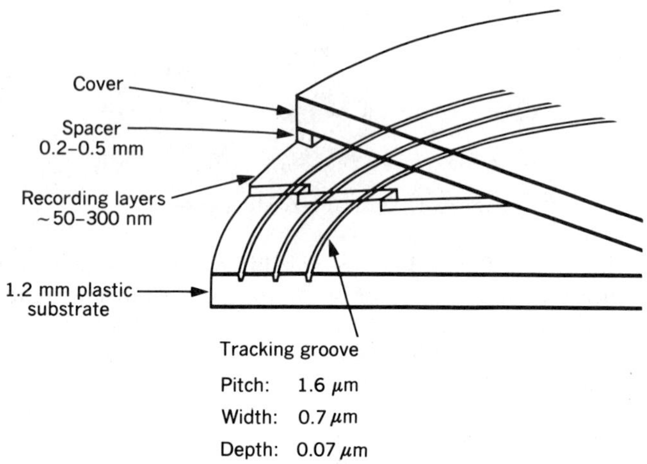

Figure 3. Recordable air sandwich disk.

Table 2. Optical Characteristics of Recording Layer[a]

Baseline reflectivity, R_b	Value
blank area, %	10–90
uniformity, % (relative)	±12
changes, %	
for processes with increased reflectivity	>130
for processes with decreased reflectivity	<60

[a] Adapted from International Standards Organization Draft Standard for 130-mm disks, supplemented with data representing standard industry practice.

Almost all optical recording phenomena depend on the recording layer absorbing the recording radiation, followed by conversion of the absorbed energy to heat, which allows a change in the recording layer to occur. A specialized case, magneto-optical recording, is a major focus of many development efforts. Irradiation of appropriate amorphous metallic layers results in local heating above the threshold temperature required for reversal of magnetic orientation by an external field. Accompanying the reversal of magnetic field is a change in the angle of reflected plane-polarized light (the Kerr effect). Thus reading of data is done by purely optical means, whereas recording is by a combination of optical (thermal) and magnetic means. This phenomenon is infinitely reversible.

Several key features of optical recording that dictate the properties of polymers used in this technology are shown in Figure 1. To make the optical design more compact, detection is by reflection not by transmission, and to allow convenient detection of the reflected beam polarized light is used to permit separation of the incoming and returning beams. An additional feature is the use of preformatted disks, ie, disks with tracking, sector, track address information, etc, manufactured into the surface of the disk in the form of grooves and pits. The optical disk is a component of the optical path, because the data layer is enclosed by protective layers. In practice, the substrate itself is generally the layer chosen to be part of the optical path.

Protection of the data or recording layer is necessary because the spot size from the focused beam is small, less than 1 μm. In read-only disks, a simple construction gives this protection, as shown in Figure 2. The substrate, having replicated datum pits, is coated with a reflective aluminum layer, which is subsequently protected with a uv-curable lacquer that is in direct contact with the aluminized information surface. In the case of recordable layers in which no surface deformation oc-

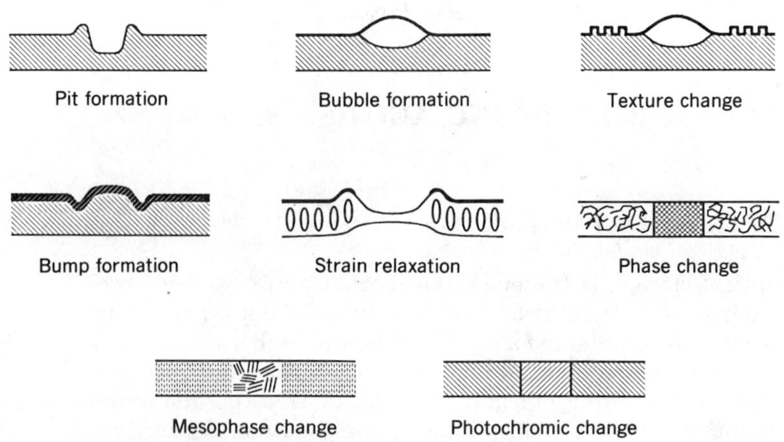

Figure 4. Optical recording mechanisms.

curs, a similar type of protection can be used. However, in the case of surface deformation recording, an in-contact layer usually results in such high energy requirements for recording data that recording at acceptable frequencies is not possible with commercial diode lasers. Special silicone elastomers have been claimed as suitable in contact overcoats for pit-forming media, but in commercial media noncontact protection is used in the form of an air sandwich, as shown in Figure 3.

In optical data storage technology polymers are used as photoresists for mastering (see LITHOGRAPHIC RESISTS); transparent resins for substrates (Table 1); uv-cured lacquers for protective layers; adhesives, photosets, and thermosets for the assembly of the disk structure; and specialty applications in active recording layers (Table 2).

Several material systems with different marking mechanisms have been explored for optical storage applications (Fig. 4).

BIBLIOGRAPHY

J. E. Kuder, *J. Imaging Sci.* **32,** 51 (1988).

H. Ryu, K. Akagane and H. Kori, *Chem. Econ. Eng. Rev.* **18**(12), 26 (1986).

G. Kaempf, *Polym. J.* **19,** 257 (1987).

H. Urrows and E. Urrows, *Opt. Inform. Sys.* **8,** 202 (1988).

R. SIDNEY JONES
JAMES E. KUDER
Hoechst Celanese Corporation

OPTICALLY ACTIVE POLYMERS

In general, the term "polymer" indicates synthetic macromolecular systems; accordingly, this article deals with optically active synthetic polymers only. Only those molecules with dissymmetric or chiral structures, those without inversion and reflection-symmetry elements, possess optical activity. Both low molecular weight compounds and macromolecules must obey the same symmetry rules, and the preparation of optically active polymers presents problems similar to those encountered in traditional organic chemistry.

Measurement of Chiroptical Properties of Polymers

When a linearly polarized light beam passes through an optically active medium, an elliptically polarized wave is generated with the main axis rotated by an angle α (optical rotation) with respect to the original polarization plane. The arc tangent of the minor to the main axis of the elliptical vibration gives ψ (the ellipticity angle). A medium can show a nonvanishing optical rotation and ellipticity angle for chemical, physical, or both reasons.

The instruments used for measuring α and ψ, polarimeters and dichrographs, respectively, are substantially the same as for low molecular weight compounds, but treatment of experimental results and the type of information obtained are clearly different. In most cases chemists are interested in obtaining indications at the molecular level, and this is better achieved by measuring chiroptical properties in dilute solutions where possible artifacts arising from orientation, aggregation, and supermolecular order, such as those occurring in liquid crystals and solid crystalline materials, are not observed. Most measurements on polymers have been performed in solution, although a few examples of measurement in the solid state are also known.

Structure and Optical Activity

In order to show measurable optical activity, a polymer must consist of chiral macromolecules with one-handedness appreciably predominating.

Optical activity is used as an analytical tool to relate molecular structure to chiroptical properties. In many cases, these properties are very different for macromolecules and low molecular weight compounds resembling the structure of the monomeric residue. Interpretation of these differences allows development of a picture of the conformation in solution, which is difficult to obtain by other techniques.

Synthesis

The synthesis of optically active polymers is based mostly on the introduction in the macromolecules of stereogenic centers, such as asymmetric carbon atoms, with a single absolute configuration. A few cases are known in which it has been possible to isolate conformations of a single-handedness from macromolecules having achiral primary structure. In macromolecular chemistry, the preparation of an optically active product does not necessarily coincide with the polymerization process, which must then be followed by resolution.

Synthetic methods require an optically active agent as the monomer, as the catalyst, or as the separation support, the chiral discrimination being then of chemical origin. In some cases natural optically active molecules can be used as starting materials or asymmetric systems, such as enzymes or bacteria that are capable of chiral discrimination.

Polymerization of Optically Active Monomers. *1-Olefins.* Terminal alkenes exhibit a typical relationship between chiroptical properties and both primary and secondary structure. Closely following the discovery of Ziegler-Natta catalysts for the highly stereospecific polymerization of propylene, several optically active 1-olefins were polymerized in the presence of these catalysts.

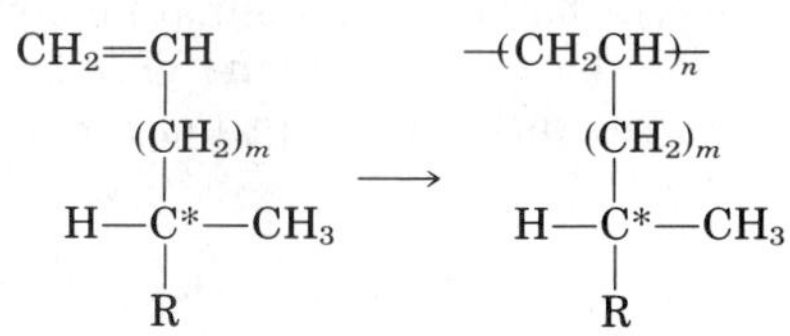

(1) (2)

where $m = 0, 1, 2, 3$ and $R = -C_2H_5$ or $-(CH_2)_3CH(CH_3)_2$

As a result of systematic studies, the following observations were made regarding the relationships between structure and optical rotation in solution:

1. The optical rotation absolute value increases asymptotically with increasing isotacticity.

2. The molar optical rotation per monomeric residue $[\phi]_\gamma^\tau$ depends on the monomer structure and in particular on the distance of the asymmetric carbon atom from the double bond. For $m = 0$ or 1, the values of $[\phi]_\gamma^\tau$ are much larger in absolute value than for monomeric structural analogues. This difference is reduced with $m = 2$ and becomes practically negligible for $m = 3$.
3. Polymers of structure (**2**) regardless of the values of m or the nature of R, show plain ord curves between 650 and 200 nm; extrapolation of these data by the one-term Drude equation allows evaluation of the wavelength of the lowest energy, optically active electronic transition. This is located nearly in the same position as for chiral paraffins. Recent studies in the vacuum region, 200–140 nm, have shown a dichroic band centered at about 160 nm, which can account for the rotatory properties of poly(1-olefin)s up to the visible region, in perfect agreement with ord studies.
4. At $m = 0$ or 1, $[\phi]_\gamma^\tau$ increases in absolute value with increasing enantiomeric purity of the polymerized monomer, reaching the asymptotic value at ca 70%. However, a linear dependence is observed when $m = 2$ or 3.

Other Unsaturated Monomers. The unsaturation in optically active monomer is usually due to a carbon–carbon double bond, in vinyl and acrylic or methacrylic derivatives, whereas there are few examples of dienes, acetylene derivatives, and monomers containing multiple bonds between a carbon atom and a heteroatom. All mechanisms of chain-growth polymerization give optically active polymers, depending on the type of monomer.

Copolymers of Optically Active and Achiral Unsaturated Monomers. In some copolymers made up of an optically active monomer and an achiral monomer, the nonchiral monomer units assume a predominant chirality due to insertion in the copolymer macromolecules. This dissymmetric effect can involve isolated units as well as sequences of the originally nonoptically active moieties. The presence of units derived from the optically active monomer is absolutely necessary to stabilize dissymmetric conformations in sequences of achiral monomer units.

Cyclic Monomers. Through ring-opening polymerization, chiral cyclic monomers give rise to macromolecules with main-chain tertiary carbon atoms, asymmetric owing to the different nature of the four atoms or groups of atoms attached to them (**3**). In vinyl or acrylic polymers, each main-chain tertiary carbon atom is flanked by two methylene groups and its asymmetry depends on the different length of the two chain fragments attached to it (**4**).

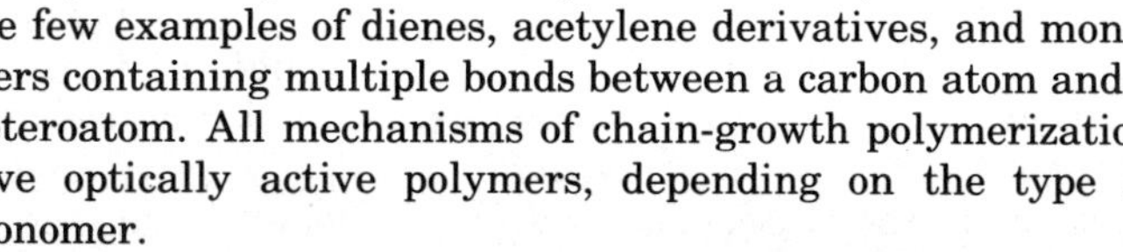

(**3a**) (**3b**)

(**3**)

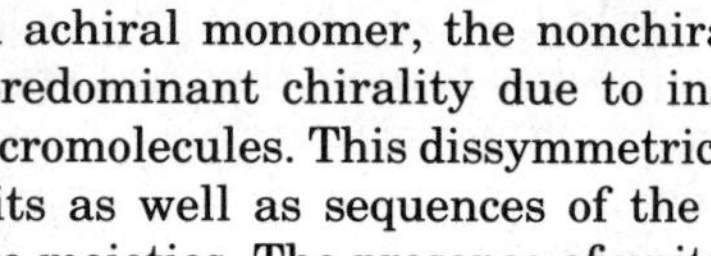

(**4**)

In the isotactic polymers (**3**), stereoregularity and optical activity are directly related as chains and have no mirror-symmetry elements. Clearly, an equimolar mixture of (**3a**) and (**3b**) is optically inactive because of intermolecular compensation and can be considered a racemate.

Monomers for Step-growth Polymerization. Optically active polymers can be prepared by step-growth polymerization, starting with one enantiomer of a compound containing the two functionalities needed for chain formation. If compounds with different functionality, such as a diacid and a diamine, are polymerized, the chirality can be in either or in both (**5**).

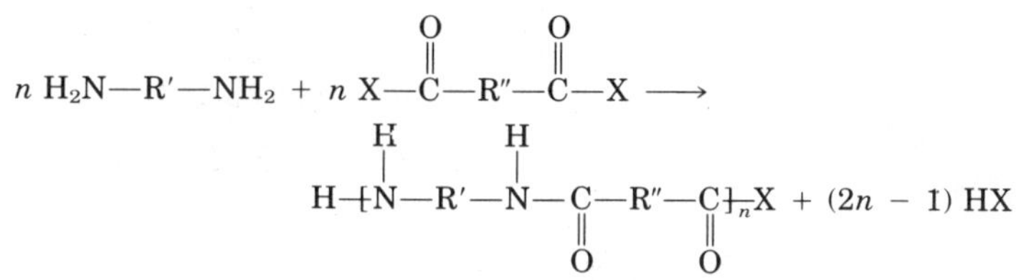

(**5**)

where R′ or R″ contains an asymmetric carbon atom of a single absolute configuration.

Because of the increasing interest in condensation polymers in applications as engineering polymers and liquid crystalline polymers, synthetic approaches have been developed, increasing the number of structures available. Consequently, the number of optically active polymers prepared in this way has also been substantially increased.

Polymerization of Racemic Monomers. The preparation of racemic monomers is generally much less costly than that of pure enantiomers and they afford economical and convenient starting materials. However, optically active polymers can be obtained from racemic monomers only when the two enantiomers are polymerized at different rates and the conversion is not complete (stereoelective or asymmetric-selective polymerization). If the catalyst or initiator polymerizes the two enantiomers at the same rate, optically active polymers can finally be obtained by the separation of macromolecules differing in the content of monomeric units derived from the left- and right-handed monomer.

Stereoelective and stereoselective processes have been demonstrated in the polymerization of both unsaturated and cyclic racemic monomers. The former process uses optically active initiators and requires the control of conversion whereas the latter requires a separation method according to chirality.

Asymmetric Polymerization of Prochiral Monomers. Prochiral monomers, in spite of having a symmetry plane, give rise to chiral monomeric residues. Asymmetric polymerization is a process that, starting with a prochiral monomer, gives macromolecules consisting predominantly of units with a single chirality; in this case, the polymer shows appreciable optical rotation. An optically active agent is needed to control the stereochemistry of the polymerization process.

Polymers with Rigid Chiral Conformation. Few polymers show optical activity of purely conformational origin because, in the absence of chiral groups, right- and left-handed structures tend to reach an equimolar equilibrium. Sequences of achiral monomer units can be placed in conformations of a single handedness, but only when included in a copolymer macromolecule with optically active comonomer residues. On the other hand, homopolymers with one chiral conformation may maintain it because of backbone rigidity strongly restricting rotation around bonds. Clearly, an optically active agent is necessary either as a catalyst to build up chains of only one chiral conformation or as a separation medium to resolve the racemic mixture consisting of macromolecules with left- and right-handed conformations.

Introduction of Optically Active Groups in Polymers. Optically active polymers can also be prepared by the attachment through chemical reaction of optically active groups to macromolecules. The macromolecule usually acts as a matrix and only in a few cases has the polymer become intrinsically optically active.

Applications

Initially, optically active synthetic polymers were investigated with the aim of obtaining information on polymerization stereochemistry and macromolecular conformation. The former topic is related to the understanding of stereospecific polymerization, whereas the latter has contributed to the analysis of ord and cd spectra of complex naturally occurring macromolecules. In addition, optically active polymers have found application in several specialized fields as chiral reagents and catalysts for asymmetric synthesis; packing materials of chromatographic columns for enantiomer separation; for the separation of racemates on helical poly(triphenylmethyl methacrylate); and chiral materials for the preparation of cholesteric liquid crystal polymers.

BIBLIOGRAPHY

P. Pino, F. Ciardelli, and M. Zandomeneghi, *Ann. Rev. Phys. Chem.* **21,** 561 (1970).

E. Sélégny, ed., *Optically Active Polymers,* D. Reidel Publishing Co., Dordrecht, The Netherlands, 1979.

F. Ciardelli, M. Aglietto, C. Carlini, E. Chiellini, and R. Solaro, *Pure Appl. Chem.* **54,** 521 (1982).

P. Pino, A. Oschwald, F. Ciardelli, C. Carlini, and E. Chiellini in J. C. W. Chien, ed., *Coordination Polymerization,* Academic Press, Inc., Orlando, Fla., 1975.

FRANCESCO CIARDELLI
Università di Pisa

OPTICAL MICROSCOPY.

See LIGHT MICROSCOPY.

OPTICAL PROPERTIES

Plastics offer several advantages in optical applications. Manufacturing costs are low because precision injection molding produces a product that needs no finishing operations and no grinding or polishing. The ability to design mountings into moldings or to produce multiple-lens arrays reduces the assembly costs of optical products. Most plastics have a higher impact strength than silicate glass, and when they fracture, the fragments are less likely to damage the eyes. The density of plastics is less than half that of glass, reducing the weight of spectacles. Some plastics are sufficiently flexible, especially when thin, to allow formation of soft contact lenses, flexible mirrors, and other products. Some plastics have a greater transmissibility than silicate glass in ir or uv parts of the spectrum.

On the other hand, certain disadvantages prevent plastics from being used in applications such as car windshields. Abrasion resistance is low. Various coatings can improve scratch resistance, but the yield stress is low. Heat tolerance is low. Even the glass-transition temperatures above 250°C are below the range of the glass transition of silicate glass (>500°C). The temperature coefficients of refractive index are high. Optical plastics are mostly homopolymers; it is not possible to adjust the refractive index or dispersion by adding metal oxides as to silicate glass. The variation in refractive index from point to point is greater than for optical glass by an order of magnitude.

The most important optical plastic is poly(methyl methacrylate) (PMMA) because of its superior weathering and scratch-resistance properties (Table 1). It is produced in two forms: a sheet that has been polymerized between glass plates with an extremely high weight-average molecular weight ($\bar{M}_w > 2 \times 10^6$) and a lower molecular weight form ($\simeq 1.5 \times 10^5$) that can be injection molded (see ACRYLIC AND METHACRYLIC ESTER POLYMERS; CASTING).

Injection molding is widely used to make plastic optical components from thermoplastics. With modern microprocessor control of injection pressures, speeds, and temperatures, highly reproducible parts are obtained; linear dimensions can be held to $\pm 8\ \mu m$ for small parts. However, the molded part is always smaller than the steel mold.

Table 1. Properties of the Optical Plastics

Polymer	Abbreviation	n_D	Abbe value	Density, kg/m^3	Thermal expansion, 10^{-6}/°C	T_{max}, °C[a]
poly(methyl methacrylate)	PMMA	1.491	57.2	1190	65	90
polystyrene	PS	1.590	30.9	1100	70	80
polycarbonate	PC	1.586	29.8	1200	70	120
allyl diglycol carbonate	CR-39[b]	1.499	57.8	1320	120	100
poly(styrene-*co*-acrylonitrile)[c]	SAN	1.571	35.3	1070	70	95
poly(styrene-*co*-methyl methacrylate)[c]	NAS	1.562	34.7	1140	65	95
poly(4-methyl-1-pentene)[c]	PMP	1.466	56.4	830	117	180

[a] Maximum use temperature.
[b] Columbia Resin CR-39 (Pittsburgh Plate Glass).
[c] Only small amounts are used.

The optical properties of plastic film used for packaging, photographic products, and other applications also depend on the method of production. Film of glassy polymers with little molecular orientation are produced by extruding the melt onto steel rolls (see FILMS, MANUFACTURE). This does not take advantage of the potential strength of semicrystalline film and is only used when dimensional change on overheating must be avoided. Most film is biaxially stretched while being cooled from the melt, eg, blown polyethylene film where the velocity and diameter of the tubular bubble are increased several-fold, or while being reheated from the glassy state, eg, poly(ethylene terephthalate) film or polypropylene film. These semicrystalline polymers have greatly increased tensile strengths in the plane of the films as a result of the biaxial orientation. Optical clarity can be improved by keeping the size of the crystalline microstructure as small as possible, by crystallization at a low temperature.

Properties

Refractive Index. The refractive index n of an isotropic material is defined in terms of the speed of light c in that material, compared with the speed of light $c = 3.00 \times 10^8$ m/s in a vacuum:

$$c = c_0/n$$

The refractive index is a decreasing function of the wavelength of the light. This is a consequence of the frequency of visible light lying between that required for electronic transitions, causing absorption in the uv region, and that for atomic vibrations, which causes absorption in the ir region. The refractive index is measured at a number of standard wavelengths, such as the F, D, and C lines in atomic emission spectra (Table 2).

Effect of Chemical Structure. Under the influence of an electric field, the electrons in the chemical bonds of a polymer chain slightly shift their average position. Both the effective electric field E and the induced dipole P are vector quantities linearly related by the bond polarizability α, which is a second-rank tensor.

$$P = \alpha E$$

Nonlinear optical phenomena, the Raman effect, require E^2 or E^3 terms on the right-hand side of this equation, but they are outside the scope of this article.

In general, the polarizability of the chemical bonds is anisotropic; it is greater along the bond than it is at a 90° angle to it. On the other hand, the random orientation distribution of the chemical bonds in a glassy polymer indicates an isotropic refractive index. The mean polarizability of a chemical bond is defined in terms of the diagonal terms of α.

$$\overline{\alpha} = \tfrac{1}{3}(\alpha_{11} + \alpha_{22} + \alpha_{33})$$

The Lorentz-Lorenz formula relates the mean refractive index n to the mean polarizability of the ith kind of chemical bond and to the number N_i of such bonds per unit volume:

$$\frac{n^2 - 1}{n^2 + 1} = \frac{\Sigma N_i \overline{\alpha}_i}{3\epsilon_0}$$

where ϵ_0 is the permittivity of free space $= 8.85 \times 10^{-12}$ F/m. In order to achieve a high refractive index, the polymer must contain chemical bonds that have a high mean polarizability and yet do not occupy too high a volume. In general, the refractive index increases in the following bond sequence:

C—F, C═O, C—H, 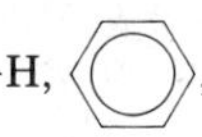, C—Cl

but differences in density may also affect the refractive-index ranking.

Transmittance. The intensity of light is reduced during passage through a plastic in two ways: first, there is a reflection at the air–plastic surfaces caused by change in refractive index and second, there are absorption processes that occur in the plastic; light scattering also plays a role. For light normally incident on an interface between materials 1 and 2 with refractive indexes n_1 and n_2, the reflectivity (defined as the ratio of reflected to incident light intensity) is

$$R = \frac{(n_1 - n_2)^2}{(n_1 + n_2)^2}$$

Internal Absorption. Internal absorption processes cause the transmitted intensity I to fall exponentially as the path length L increases through the plastic.

$$I = (1 - R)^2 I_0 \exp(-AL)$$

where I_0 is the incident intensity and A is a constant. The transmittance T is defined as the ratio I/I_0.

Plastics can be colored with inorganic pigments or organic

Table 2. Refractive Indexes and Related Properties

	PMMA	PS	PC	NAS
line and wavelength, nm				
F, 486	1.497(8)	1.604(1)	1.593(4)	1.574(4)
D, 589	1.491(7)	1.590(3)	1.586(0)	1.504
C, 651	1.489(2)	1.584(9)	1.576	1.558(3)
Abbe value	57.2	30.8	34.0	35
dispersion D^a, $\times 10^{-2}$	1.75	3.25	2.97	2.85
linear-expansion coefficient, $\times 10^{-5}$/°C	6.5	6.3	6.8	6.5
dn/dT, $\times 10^{-5}$/°C	−8.5	−12.0	−14.3	−14.0

[a] $D = \dfrac{n_F - n_C}{n_D - 1}$.

dyes. To maintain good optical transmission, soluble dyes must be used.

Transmission becomes critical when plastics are considered for use in fiber-optics systems (see FIBERS, OPTICAL). There have been great advances in the preparation of high purity, silica-based glasses for fiber-optics communications over long distances.

Polarizing filters are another type of absorbing material. Sheet polarizers typically transmit 50% of the intensity of light polarized in the optimum direction but only 0.0002% of the intensity of light polarized at 90° to this.

Photoelasticity. When elastic stresses in an amorphous solid cause the refractive index to become anisotropic in a reversible way, this is referred to as the photoelastic effect. If, however, the stresses exceed the yield stress and the refraction index changes are permanent, this is called the photoplastic effect. Two-dimensional models cut from a sheet of a photoelastic material have been widely used for stress analysis. This technique provides a visualization of high stress regions, and prototype designs can be modified before being put into production. The equipment used to load and observe the models is referred to as a polariscope. Applications of photoelasticity, particularly to crack growth and three-dimensional problems, have been described. Recent advances in computer-based techniques, such as finite-element analysis, have made the photoelastic technique less attractive. The more complex techniques, such as the stress freezing of three-dimensional photoelasticity experiments, are relatively easy to perform and photographs of the stress patterns provide powerful visual aids. It is vital to understand the phenomena causing photoelastic fringes when examining transparent plastic moldings with polarized light, in order to avoid confusion with effects caused by permanent molecular orientation, or by residual stresses.

Two types of fringe can be observed when a sheet of photoelastic model material is viewed in a polariscope. Isochromatic fringes appear colored if the model is illuminated with white light. These are contours of maximum shear stress τ_m, which is related to the principal stresses σ_1 and σ_2 by

$$2\tau_m = \sigma_1 - \sigma_2$$

An ideal material will have no isochromatic fringes when unloaded and will exhibit on the order of 10 fringes before the elastic limit is reached. Cast epoxy thermoset sheet 5 mm thick is a suitable material, and the contour levels occur at intervals of 1 or 2 MN/m^2. Isoclinic fringes always appear black. This is the locus of points where the axes of principal stress have the same inclination to a reference direction.

The pattern of isochromatic and isoclinic fringes is easiest to interpret either if the point of interest occurs at the surface of the model or if standard features such as straight cracks or parallel-side beams occur. In general, the individual principal stresses at an interior point can only be found either by a further experimental measurement such as the thickness strain, which is proportional to $\sigma_1 + \sigma_2$, or by measuring the isochromatic fringe order and isoclinic angle at a closely spaced square grid of points using an automated polariscope. Numerical integration of the data is then necessary.

The term "birefringence" is used for the magnitude of the difference between the largest refractive index, n and the smallest refractive index n_2 in the material, the values being measured with plane polarized light. In addition to the birefringence caused by elastically loading a model, other causes are nonequilibrium effects in rapidly cooled plastic products.

Separation of Residual-stress Birefringence from Orientation Birefringence. When transparent glassy plastics are injection molded and the product examined in a polariscope, there may be some doubt about the origin of the isochromatic fringes. Because no external forces are applied to the article, the conventional photoelastic effect is not present. However, the stresses in the product do not have to be zero because the average tensile stress on some cross section is zero. The most common type of residual-stress distribution is shown in Figure 1; the outer layers, which have solidified first, are under compressive stress, balanced by the tensile stresses in the interior. Consequently, the integral

$$\int_0^t \sigma_{xx} dy = 0$$

for the stresses along the line AA′. These residual stresses cause birefringence because the plastic is photoelastic. The other cause of birefringence that may be present is molecular orientation, due to rubberlike deformation of the molecules in the flowing melt. This molecular orientation does not cause stresses in the glassy state because the molecules are no longer free to change their shapes. If the isochromatic fringes in the product change their positions when saw cuts are made, then part of the birefringence is a result of residual stress.

Photoplasticity. In certain stress-analysis problems, limited yielding occurs in a body that otherwise remains elastic. The most common examples are the events at the tip of a crack and those that are due to concentrated surface loads as in hardness testing. As most thermoset model materials fail in a brittle manner in tension, the only suitable model materials are thermoplastics such as polycarbonate. However, many difficulties have hindered the widespread use of photoplastic modeling. The shape of the stress–strain curve after yield must be the same in the model material as in the metal that it simulates. Polycarbonate exhibits an upper-yield point in a tensile test and then cold draws at a 20% lower stress until a strain of ~100% is reached; this is quite unlike the behavior of any metal. Nevertheless, an isotropic polycarbonate sheet of ca 0.2-mm thickness can be used to illustrate a model of yielding at a crack tip.

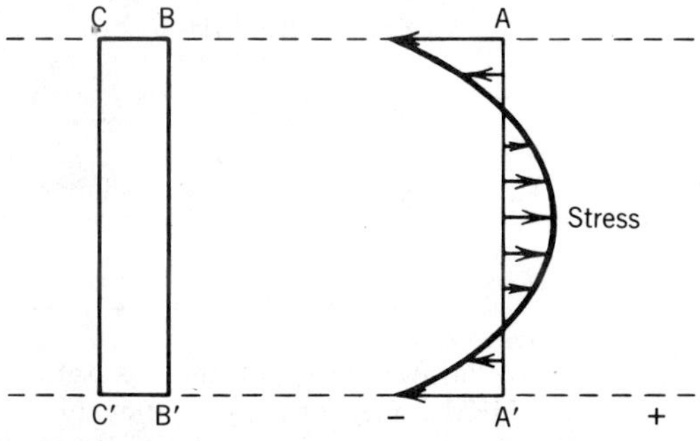

Figure 1. Distribution of residual stresses parallel to the surface of a sheet of plastic that has been cooled rapidly during extrusion or injection molding. If the sheet is cut into a slice as shown by the lines BB′ and CC′, the residual stresses in the x-direction are reduced by at least 95%.

Light Scattering. When a collimated beam of light passes through or is reflected from a plastic sheet or film, scattering causes that light to deviate in direction without any reduction in the total light intensity (see also LIGHT SCATTERING). The light intensity is reduced by pigments or dyes. If light is scattered through an angle exceeding 90°, the forward light intensity is reduced. Scattering as a result of surface imperfections is dealt with separately from scattering due to internal microstructural inhomogeneities, because the mechanisms for scattering are different.

Several common terms are given specific meanings when used in connection with light scattering. "Gloss" refers to the reflected light intensity as a percentage of that from an ideal surface. The specular gloss is measured within a specified angular range of the ideal reflected angle *r*; a standard, polished, black glass of refractive index 1.567 has a specular gloss of 100%. "Haze" refers to the integrated, transmitted intensity for all light that deviates by an angle $\beta > 2.5°$, as a percentage of the incident-light intensity. "Clarity" is defined in terms of the image degradation of standard printed scales viewed through the plastic sheet. "Transparency" refers to the transmitted intensity for all light that deviates by an angle $\theta < 2.5°$, as a percentage of the incident-light intensity. These definitions are more precise when the test equipment is specified in the appropriate national standards.

Applications

Applications for which optical properties of plastics are important include glazing and barrier films (Table 3), horticultural glazing, headlight lenses, aircraft windows, electronic and information technology, and plastic lenses (Tables 4 and 5).

Table 3. Properties of Plastic Sheet for Outdoor Applications

Property	PMMA	PC	10-mm twin-wall PC	PVC	PVC + wire mesh
light transmission of 3-mm sheet, %	92	88	70–80	85	72
notched Izod impact strength, J/m[a]	15–20	650–950		20–500	
service temperature range, °C	−30 to 80	−50 to 120		−30 to 55	−30 to 55
weathering	good	average	average	average	average
flammability, surface spread of flame[b]	Class 3	Class 1	Class 1	Class 1	Class 1
cost relative to PVC per m^2	1.5	2.5	1.3	1.0	2.4

[a] To convert J/m to ft·bf/in., divide by 53.38.
[b] BS 476:Pt 7.

Table 4. Properties of Ophthalmic Lens Materials

Material	Refractive index n_D	Density, kg/m^3	Weight of lens, g	Impact strength, J[a]
Crown glass, annealed	1.525	2.53	25.7	0.04
PMMA	1.495	1.19	12.3	
CR-39	1.498	1.32	13.6	0.4
PC, coated	1.596	1.20	11.7	22

[a] To convert J to ft·lbf, divide by 1.36.

Table 5. Contact Lens Materials

Materials	Abbreviations	Refractive index n_D	Water content, %
hydrophilic			
poly(hydroxylethyl methacrylate)	HEMA	1.43–4.44	≤40
HEMA–poly(vinyl pyrrolidinone)	HEMA–PVP	1.43	≤72
PMMA–ethylene glycol dimethacrylic acid	PMMA–EGDMA	1.43	≤80
hydrophobic			
PMMA		1.49	0
cellulose acetate butyrate	CAB	1.44	2
polydimethylsiloxane	PDMS	1.44	0

BIBLIOGRAPHY

The Handbook of Plastics Optics, 2nd ed., U.S. Precision Lens, Inc., Cincinnati, Ohio, 1983.

S. A. Paipetis and G. S. Holister, eds., *Photoelasticity in Engineering Practice,* Elsevier Applied Science Publishers, Ltd., Barking, UK, 1985.

N. J. Mills, *Plastics,* E. Arnold, London, 1986.

G. H. Meeten, ed., *Optical Properties of Polymers,* Elsevier Applied Science Publishers, Ltd., Barking, UK, 1986.

N. J. Mills
University of Birmingham, UK

OPTICAL ROTATORY DISPERSION. See OPTICALLY ACTIVE POLYMERS.

ORGANOALUMINUM COMPOUNDS. See ZIEGLER-NATTA CATALYSTS.

ORGANOLITHIUM COMPOUNDS. See ANIONIC POLYMERIZATION.

ORGANOMETALLIC CATALYSTS. See ANIONIC POLYMERIZATION; COORDINATE POLYMERIZATION; ZIEGLER-NATTA CATALYSTS.

ORGANOMETALLIC POLYMERS

Macromolecules generally contain fewer than 10 different elements (carbon, hydrogen, nitrogen, oxygen, sulfur, phosphorus, halides). There are available well over 40 additional elements, many metals, which can be built into polymers. Metals often can exist in several different oxidation states. Geometries and reactivities depend on such factors as the nature and extent of substitution. Potentially then, a large number of polymer families exist in theory, limited only by intention and imagination. Organometallic polymers are those containing metals either in the backbone chains or pendent to them. The metals may be connected to the polymer by σ or π bonds to carbon, coordination bonds to elements containing free electron pairs, or, less commonly, by σ or π bonds to other elements. Organometallic polymers can be prepared by addition, condensation, coordination, ring-opening, and other common modes of polymerization.

Classification. Organometallic polymers can be classified into condensation polymers, addition polymers, coordination polymers, and others based on their mode of preparation.

Applications

Organometallic polymers have many current and potential uses. Poly tris[5,5′-bis(3-acrylyl-1-propoxycarbonyl)-2,2′-bipyridine]ruthenium is one of a series of polymerized electrochromic compounds that have been reported to go through a range of seven different colors, from orange to cherry red. This series of polymeric complexes may be useful in generating multicolor displays in electronic display panels. Tributyltin methacrylates have found widespread use in antifouling paints, which slowly release or ablate tributyltin hydroxide.

Commercial colorants, dyes or pigments, often exhibit poor migration fastness, bleeding, blooming, or plateout. Polymer-bound organometallic colorants might reduce these difficulties.

Polymeric metal phosphinates having single-, double-, or triple-bridged structures (**1**)–(**3**) (Fig. 1) have been made using aluminum, beryllium, cobalt, chromium, nickel, titanium, and tin. They form films with thermal stabilities to 450°C and the chromium(III) polyphosphinates have been used as thickening agents for silicone greases to improve their high pressure physical properties.

Organometallic polymers are being investigated as medicinal agents. Polymer matrices containing europium chelates have also been used as the active species in the electronic energy-transfer processes of lasers.

One of the most vigorously studied areas of application is the use of polymer supports for homogeneous catalysts, so that fixed-catalyst beds can be used in flow processes, while the chemical selectivity inherent in the homogeneous organometallic catalyst is maintained.

There are applications in the area of functionalized electrodes for electrocatalysis, photoelectrocatalysis, photovoltaic cells, and specialized electrode and sensor development. Several organometallic polymers have been used as preheat shields for targets in inertial-confinement nuclear fusion.

Condensation Polymers

Condensation reactions exhibit several characteristics such as expulsion of a smaller molecule (often H_2O or HX) leading to a repeat unit containing fewer atoms than the sum of the atoms in reactant molecules. The reaction site can be at a metal atom,

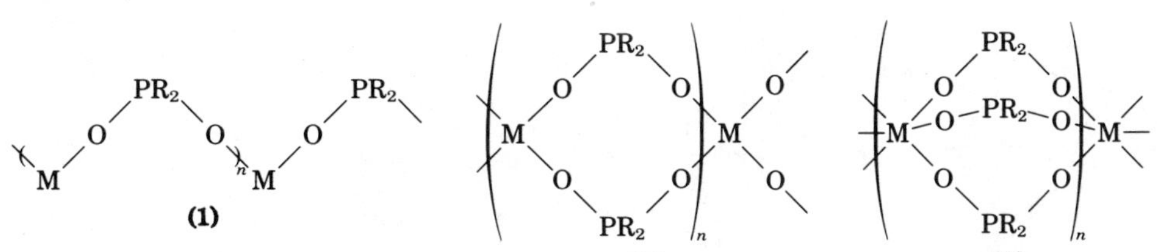

Figure 1. Polymeric metal phosphinates.

adjacent to the metal, or further removed from the metal atom as in (**6**).

$$R_2MX_2 + H_2NRNH_2 \longrightarrow -\!\!(\underset{R}{\overset{R}{M}}-\overset{H}{N}-R-\overset{H}{N})\!\!- + 2\,HX$$

(**4**) (**5**)

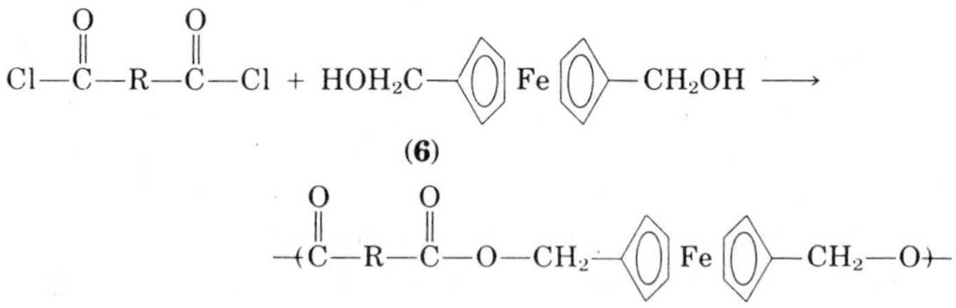

(**6**)

(**7**)

There is often no clear distinction between coordination and condensation reactions. Here, reactions directly involving the metal already chemically bonded to an organic moiety, such as (**4**), will be considered condensation reactions, whereas reactions involving simple metal ions and metal ion oxides undergoing chelations and electrostatic attractions will be covered in "Coordination Polymers," below.

Condensations at the Metal Atom. Many organometallic halides possess a high degree of covalent character and can behave as organic acid derivatives in many reactions:

$$RCOCl + H_2O \longrightarrow RCOOH + HCl$$

$$-\overset{|}{\underset{|}{M}}Cl + H_2O \longrightarrow -\overset{|}{\underset{|}{M}}-OH + HCl$$

Thus many organometallic polycondensations are extensions of organic polyesterification reactions.

$$Cl\overset{O}{\overset{\|}{C}}R\overset{O}{\overset{\|}{C}}Cl + HOR'OH \longrightarrow -\!\!(\overset{O}{\overset{\|}{C}}R\overset{O}{\overset{\|}{C}}OR'O)\!\!- + 2\,HCl$$

$$R_2MCl_2 + HOR'OH \longrightarrow -\!\!(\underset{R}{\overset{R}{M}}-OR'O)\!\!- + 2\,HCl$$

Polymers can be formed by reaction of a variety of Lewis bases with organometallic Lewis acids. An organometallic compound may also be condensed with suitable natural polymers, eg, dextran, cellulose, and wool, and synthetic polymers, eg, poly(vinyl alcohol) and polyethyleneimine, that contain appropriate reactive sites.

$$-\!\!(CH_2-\underset{OH}{CH})\!\!- + R_3SnX \longrightarrow -\!\!(CH_2-\underset{\underset{SnR_3}{O}}{CH})\!\!- + HX$$

Reaction Media. Because many of the reactants and products are thermally sensitive, low temperature solution and interfacial condensation systems have typically been employed. Most solution systems are inferior to the interfacial condensation systems.

Water-based solution systems are attractive for water-stable organometallic reactants and reactants that form usable aqueous species. For reactants that are not water soluble or that hydrolyze forming inert products, nonaqueous solution systems have been developed.

Biological Activity. Condensation organometallic polymers with biological activity have been made. The activity derives from the total polymer chain or through controlled release of monomeric fragments.

Phthalocyanine Polymers. Shish kabob macromolecules with the stacked metal phthalocyanine structure (**8**) (M = silicon, germanium) have been made (Fig. 2).

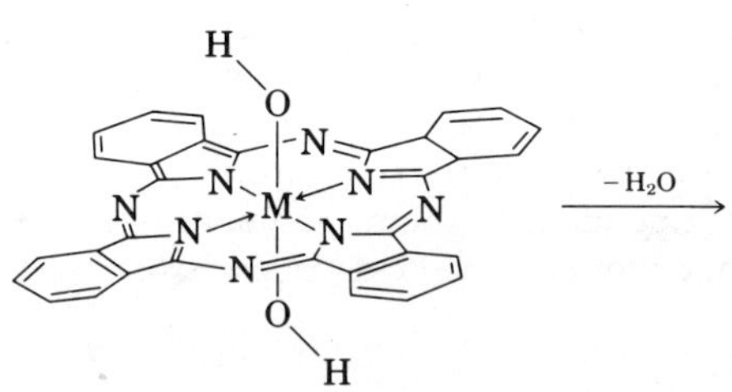

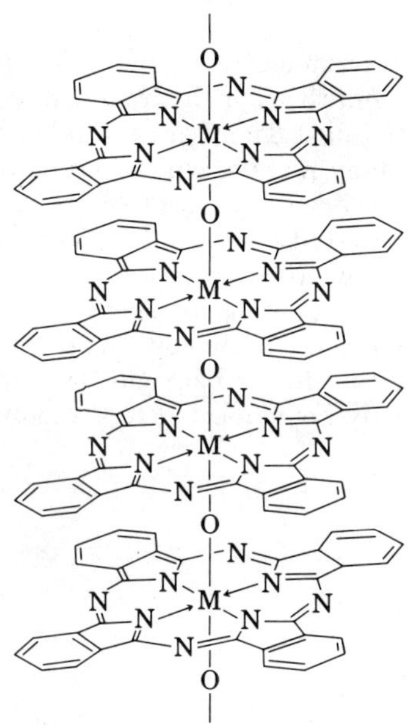

where M = Si, Ge, Sn

(**8**)

Figure 2. Formation of metallophthalocyanines.

Condensations away from the Metal Atom. Organometallic monomers, in which the metal is not the site of condensation, have been polymerized in wide variety. In many of these cases the metal is in the backbone, however. For example, melt-phase polytransesterifications of 1,1′-bis(carbethoxy)cobalticinium hexafluorophosphate (**9**) and diols such as 1,10-decandediol or bis(hydroxymethyl)benzene give fairly high molecular weight ionic polyesters (**10**). Polyamides (**11**) have been prepared by reaction of aromatic diamines with 1,1′-dicarboxycobalticenium chloride in molten antimony trichloride.

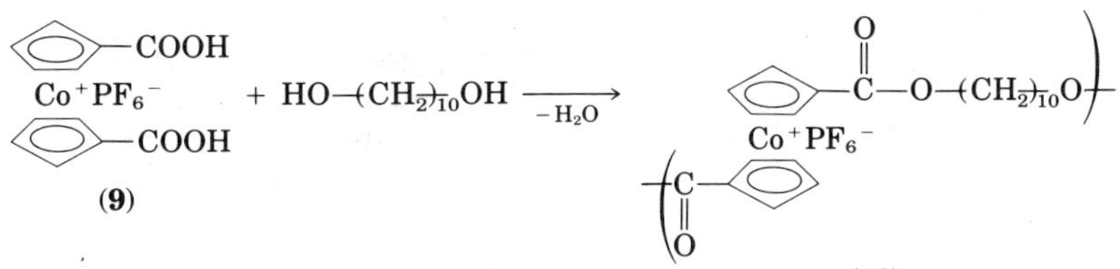

(**9**) (**10**)

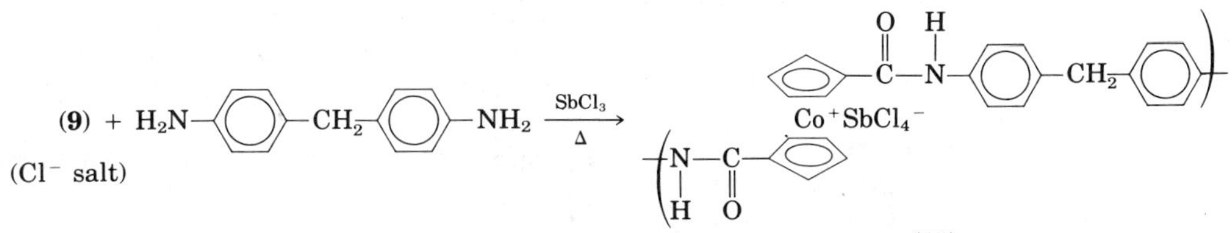

(**11**)

Ferrocene diepoxide derivatives have been made in good yields and incorporated into epoxy resins.

Many ferrocene-containing polyesters, polyamides, and polyurethanes have been reported in which classic polycondensation methods have been employed. Interfacial polycondensations of 1,1′-bis(β-aminoethyl)ferrocene with diacid chlorides and diisocyanates give new polyamides and polyureas.

Poly(ferrocenylphosphine oxides) and sulfides result from the reaction of phenyldichlorophosphonic acid or phenylphosphonothioic dichloride with ferrocene in the presence of $ZnCl_2$. However, cleavage of the rings from iron also occurs and cycloalkyl-bridged ferrocene units are formed within the polymer. A boron-containing poly(benzborimidazoline) (**12**) was synthesized from ferrocene-1,1′-diboronic acid and 3,3′-diaminobenzidine as part of an effort to produce polymers of superior thermal stability.

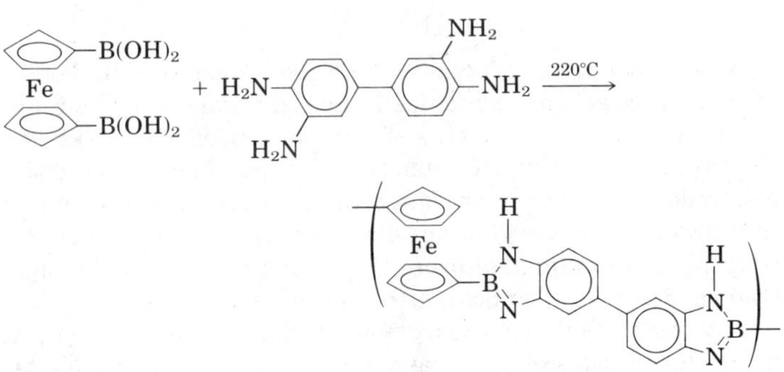

(**12**)

In other attempts to increase thermal stability while achieving processibility, silicone polymers containing ferrocene backbone groups were studied.

Coordination Polymers

Coordination polymers have long been known. Tanning leather and generating select-colored pigments depend on the coordination of metal ions. Many coordination polymers have unknown and/or irregular structures. Coordination polymers can be prepared by a number of routes; the three most common are:

1. Metal coordination complexes polymerize through attached functional groups; polymer formation is a condensation or addition reaction, eg,

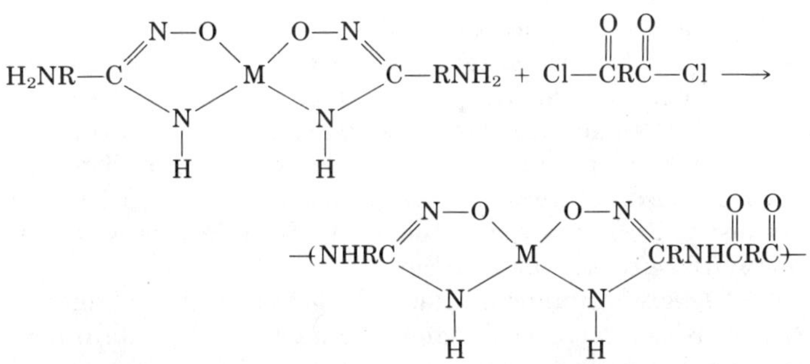

2. Preformed polymers containing chelating groups coordinate metal ions, eg,

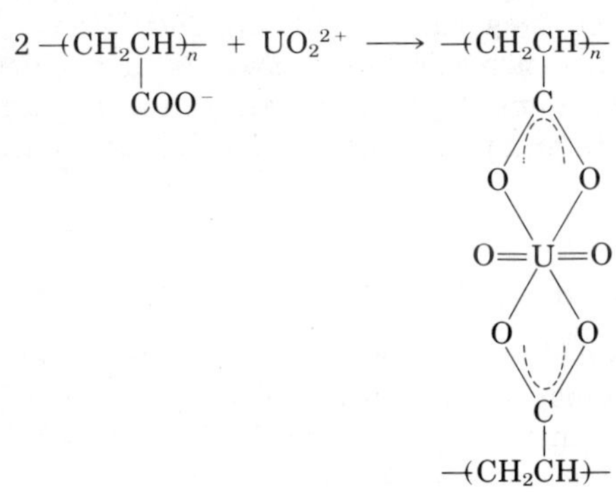

3. Metal ions react with ligands capable of coordinating to

two metal atoms; metal-to-donor coordination builds chains, eg,

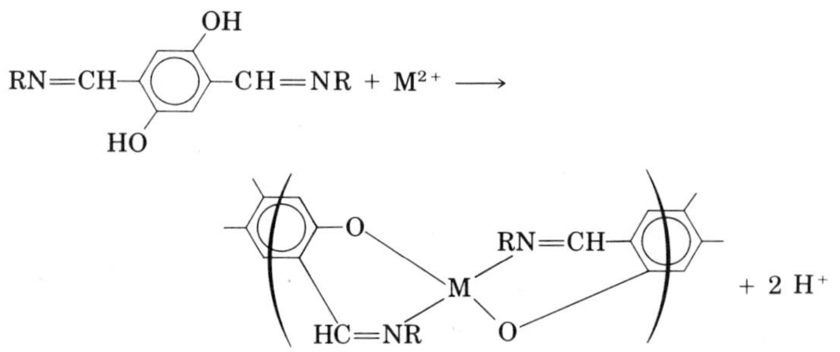

Ionomers are an important class of commercial polymers formed through route 2.

Semiconductivity. Mixed-valence, transition-metal complexes of the dianion of oxalic acid act as semiconductors. The iron(II) oxalates are linear-chain complexes. Oxidation of $Fe(C_2O_4)(H_2O)_2$ with Br_2 gives $Fe(C_2O_4)(H_2O)_{2.4}Br_{0.6}$, a semiconducting polymer with conductivity at RT of 10^{-4} S/cm. The reaction of $Fe(C_2H_2O_4)(H_2O)_2$ with I_2 results in the formation of a complex with a conductivity six orders of magnitude greater than that of the unoxidized polymer.

Phthalocyanine Polymers. Shish kabob-type phthalocyanine polymers with alternating metal and pyrazine units, or with alternating metal and oxygen atoms in the backbone, have been previously mentioned. In general, metal complexes of macrocyclic ligands, such as phthalocyanine, crystallize in stacks to afford linear chains that have the ability to conduct electricity along the stack axis.

Other Stacked Complexes. Stacked metal–chain complexes without macrocyclic ligands are also known. Tetracyanoplatinate-stacked complexes resemble the metal phthalocyanine polymeric structure, except that coordinating macrocycles do not chelate each metal atom. Instead, stacks of square-planar $Pt(CN)_4$ units are formed with anions between the stacks. The cyano groups surrounding Pt are staggered with respect to those coordinating the adjacent platinum atoms. These well-studied materials are one-dimensional conductors with conductivities greater than 1 S/cm.

Poly(yne)s. An unusual class of organometallic polymers is the transition metal–poly(yne) polymers, which contain transition metals bound to carbon within the main chain and which might vaguely be considered coordination polymers. The platinum and palladium copolymers of this class have rodlike structures. This linear structure results from the all-trans configurations at the platinum and palladium atoms. All of these polymers form nematic liquid crystals in trichloroethylene, and some of them align their main chain in a direction perpendicular to an applied magnetic field; others align parallel to the field.

Addition Organometallic Polymers

A large number of vinyl organometallic monomers have been prepared, homopolymerized, and also copolymerized with other classic vinyl monomers. The synthesis and polymerization of many organometallic vinyl monomers has been reviewed. The variety of vinyl organometallic structures includes vinyl groups attached to cyclopentadienyl rings, dienes, phenyl rings, and ether or ester functions; these functions are bound to the metal. Such monomers are organometallic analogues of styrene, vinyl ethers, acrylates, and methacrylates. The vinyl groups may be directly attached to metals:

$$(CH_3)_3SiCH{=}CH_2 \qquad (CH_3)_3GeCH{=}CH_2 \qquad (CH_3)_3SnCH{=}CH_2$$

Preparation of Monomers. The methods of monomer preparation include the Friedel-Crafts method, the Wittig method, use of alkenylcyclopentadienide salts, ligand displacement, and the use of acrylic and methacrylic esters.

Polymerization of Vinyl Organometallic Monomers. A transition metal with its various readily available oxidation states and large steric bulk may be expected to exert unusual electronic and steric effects during polymerization. The homopolymerization of vinylferrocene has been initiated by radical, cationic, coordination, and Ziegler-Natta initiators.

Derivatization of Preformed Polymers. ***Catalytic Applications.*** Many preformed vinyl polymers and resins have been derivatized with organometallic functions to make polymer-bound catalysts. This topic has been extensively reviewed (see also CATALYSTS, POLYMER-SUPPORTED).

Selectivity of Catalysts. Steric effects of polymer matrices and the barriers they impose to diffusion of reagents to the attached catalytic site have been employed to modify the regioselectivity of homogeneous hydrogenations catalyzed by $RhCl[P(C_6H_5)_3]_3$.

Multistep Catalytic Reactions. Two or more anchored catalyst centers have been used to carry out sequential multistep catalytic reactions. The products from the first catalytic reaction become reagents for the second reaction. An example is the sequential catalytic cyclooligomerization–selective hydrogenation reaction and the cyclooligomerization–hydroformylation sequences illustrated in Figure 3 where butadiene is the starting material. The polymers used were styrene–divinylbenzene resins functionalized with $-P(C_6H_5)_2$ groups that ligated the metal complexes.

Miscellaneous Vinyl Polymers. The use of modified vinyl polymers with covalently or ionically bound organometallic functions is of interest in the area of polymer-coated electrodes (see ELECTRODES, POLYMER-MODIFIED). Such polymers protect the electrodes from oxygen or from a variety of corrosive reactions that occur when carrying out electrocatalysis or electrochemistry in surrounding solutions or when promoting hole–electron separation at semiconductor surfaces.

A variety of other vinylorganometallic polymers have been made. Ferrocenylacetylene was polymerized using Ziegler-Natta, radical, and cationic initiators to give polyferrocenylacetylene.

Of particular interest are polyacrylates containing the tetraphenylporphinatosilver(I), tetraphenylporphinatocobalt(II), and chloro(tetraphenylporphinato)iron(III) groups as side chains. Unlike the stacked shish kabob structure, it seems the pendent units could be isolated. However, uv–visible and esr spectroscopy and magnetic susceptibility measurements show an antiferromagnetic interaction between metal ions in the silver, cobalt, and iron cases. It was thought that these polymers might exhibit ferromagnetic properties. This possi-

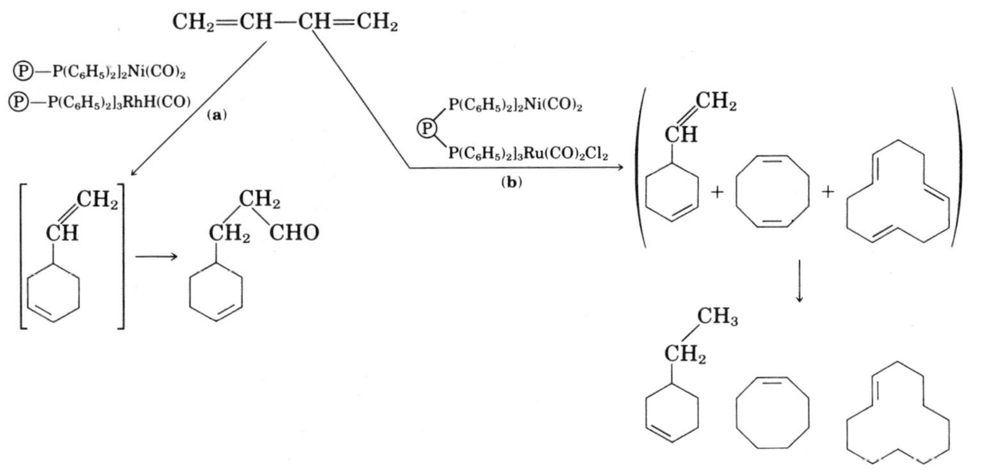

Figure 3. Multistep catalytic reaction. (**a**) The reaction conditions are 115°C under CO for 24 h, then H_2/CO at 1.7 MPa (250 psi) and 70°C; (**b**) the reaction conditions are 100°C for 24 h, then H_2 at 1 MPa (150 psi) and 160°C.

bility exists because electron-exchange interactions between metal ions may give ferromagnetic substances. Thus the antiferromagnetic interaction, depicted in (**13**), may reduce the chances of observing desirable magnetic properties (see also MAGNETIC POLYMERS).

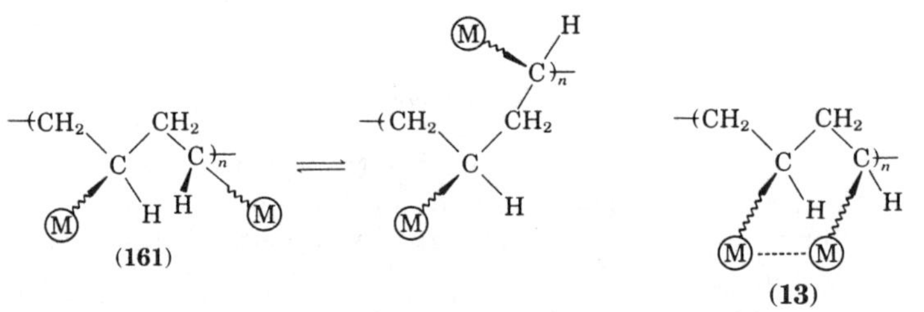

where ⁓Ⓜ represents the carboxy metal group (—COOⓂ) and Ⓜ is a 5-(4-substituted phenyl)-10,15,20-triphenylporphinatometal group

Polysilanes

Polysilanes are a new class of organometallic polymers that have an all-silicon backbone. Polysilanes serve as excellent reactive-ion-etching barriers for bilevel-resist applications because a protective layer of SiO_2 is formed during exposure to oxygen plasma (see LITHOGRAPHIC RESISTS). The λ_{max} depends on the degree of polymerization leading to a nonlinear bleaching photochemistry. This unique "bleaching latency" is useful in contrast-enhanced lithography. At 3M Co., chemists have found that polysilanes have extraordinary activity as polymerization catalysts for making high molecular weight vinyl polymers, and polysilanes are being used as precursors for β-silicon carbide fibers at the Air Force Materials Laboratory.

BIBLIOGRAPHY

J. E. Sheats, C. E. Carraher, Jr., and C. U. Pittman, Jr., eds., *Metal-containing Polymer Systems,* Plenum Publishing Corp., New York, 1985, pp. 1–523.

C. U. Pittman, Jr., in G. Wilkinson, F. G. A. Stone, and E. W. Abel, eds., *Comprehensive Organometallic Chemistry,* Vol. 8, Pergamon Press, Oxford, UK, 1982, Chapt. 55, pp. 553–608.

F. R. Hartley, *Supported Metal Complexes,* D. Reidel Publishers, Dordrecht, The Netherlands, 1985, pp. 1–318.

C. U. Pittman, Jr., in E. I. Becker and M. Tsutsui, eds., *Organometallic Reactions,* Vol. 6, Marcel Dekker, Inc., New York, 1977, pp. 1–62.

CHARLES U. PITTMAN, JR.
Mississippi State University

CHARLES E. CARRAHER, JR.
Florida Atlantic University

JOHN R. REYNOLDS
University of Texas-Arlington

ORGANOPHOSPHORUS POLYMERS.

See PHOSPHORUS-CONTAINING POLYMERS.

ORGANOSOL.

See VINYL CHLORIDE POLYMERS.

ORIENTATION

The orientation of macromolecules in fabricated and naturally occuring polymers plays an important role in determining their performance, ranging from mechanical to optical characteristics. No laboratory for characterizing polymeric materials can be considered complete if it does not contain tools for characterizing orientation.

Representation of Orientation

Uniaxial Orientation. The orientation of chains in polymers is expressed in terms of the angle ϕ between the symmetry axis and the segments of the polymer chains (Fig. 1). The distribution function $N(\phi)$ is defined to represent the distribution of segments possessing an angle ϕ with the axis. This function can be normalized:

$$2\pi \int_0^{\pi} N(\phi) \sin \phi \, d\phi = 1$$

In a semicrystalline polymer, it is possible to distinguish between f_{H_c}, the orientation factor of the crystalline phase, and $f_{H_{am}}$, the orientation factor of the amorphous regions.

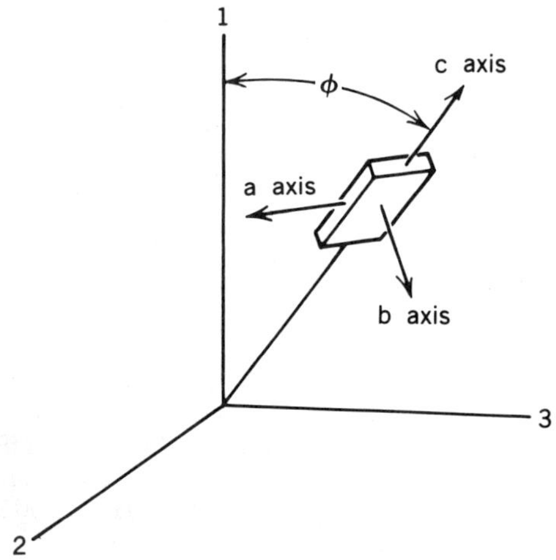

Figure 1. Uniaxial orientation.

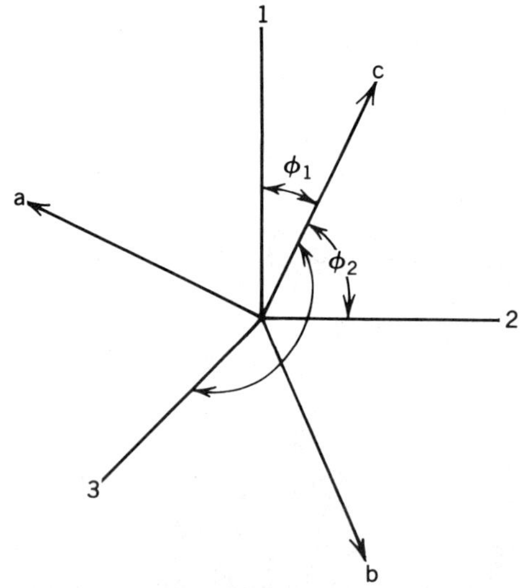

Figure 2. Biaxial orientation representation in terms of angles ϕ_1 and ϕ_2.

A second approach to the specification of orientation in polymers is based on the anisotropy of the polarizability tensor α_{ij}:

$$f_H = \frac{\alpha_{11} - \alpha_{22}}{\Delta\alpha_0} = \frac{\alpha_{11} - \alpha_{33}}{\Delta\alpha_0}$$

where $\Delta\alpha_0$ is the value of the difference in polarizability along and perpendicular to polymer chains, and α_{11}, α_{22}, and α_{33} are the values of the polarizability tensor along, 1, and perpendicular, 2 and 3, to the symmetry axis. The relation of f_H to $\cos^2 \phi$ may be derived by coordinate transformations.

Biaxial Orientation. Biaxial orientation in polymers can be represented in several ways, eg, by Euler's angles. This involves the angle ϕ with respect to the first symmetry axis and a second latitude. A different set of angles was proposed by Wilchinsky in 1963 and later expanded (Fig. 2). These involve the above-mentioned angle ϕ, denoted here by ϕ_1, and a second angle ϕ_2 defined from an orthogonal axis.

Methods of Measurement

Birefringence. Optically, solids are categorized in three fundamental symmetry systems: cubic (isotropic), uniaxial, and biaxial. These systems possess one, two, or three principal refractive indexes, respectively. In other words, in cubic or isotropic systems, the optical indicatrix is a sphere ($\eta_1 = \eta_2 = \eta_3$); in a uniaxial system, the indicatrix is an ellipsoid of revolution ($\eta_1 \neq \eta_2 = \eta_3$); and in a biaxial case (ie, orthorhombic), it is a general ellipsoid ($\eta_1 \neq \eta_2 \neq \eta_3$). The polymeric materials produced by processes such as compression molding, melt spinning, biaxial stretching, film blowing, and rolling possess one of these three general optical symmetries.

Birefringence (qv) is a classic method of measurement of orientation. The refractive index represents the slowing of the progress of an electromagnetic wave through a material because of interaction of the wave with polarizable molecules.

Dichroism. Dichroism is a measure of anisotropy of absorption of electromagnetic waves in a medium. The absorption of the electromagnetic waves occurs when the frequency of electromagnetic waves is equal to the natural vibration frequency of electrons. The transition-moment vector defines this natural vibration. In the infrared wavelength range, these vibrations are associated with the molecular motions. The absorbance at a given frequency is defined as

$$A = \log_{10} [I_0/I_\alpha(\mathbf{M}\cdot\mathbf{E}\cdot)^2]$$

where I_0 and I_α are the incident and transmitted intensity; **M** is the transition-moment vector associated with a certain vibration mode, and **E** is the electrical-field vector of the incoming beam.

The two classes of dichroic investigations reported in the literature involve visible dichroism using dyes, and dichroism of chain segments in the infrared range. The latter is widely used. Visible dichroism is hampered by the unknown rules governing the orientation of absorbed dye molecules with respect to polymer chains.

In certain polymers, notably polyethylene, infrared absorbing groups coincide with the crystallographic axes. It is thus possible to establish the orientation of the individual crystallographic axes in these materials.

Dichroism is a much more powerful technique than birefringence. The latter measures mean values over all the components and phases present, whereas dichroism is specific to particular absorption bands. It can thus be applied to different groups on the same molecule, to different segments in a copolymer, or to different components of a polymer blend.

X-ray Diffraction. The wide-angle x-ray diffraction technique is sensitive to the diffraction caused by crystalline regions (Fig. 3). This technique is useful to obtain the orientation of a given direction of unit cell (generally *a*, *b*, and *c* axes) (see X-RAY DIFFRACTION).

Other Techniques. Almost all studies of orientation in polymer systems use birefringence, dichroism, or wide-angle x-ray diffraction, although other methods have also been developed, eg, fluorescence, Raman scattering, and broadline nmr (see NUCLEAR MAGNETIC RESONANCE; RAMAN SPECTROSCOPY).

Influence of Orientation on Properties

Orientation affects mechanical, optical, and electrical properties. Uniaxial orientation has a strong influence on mechan-

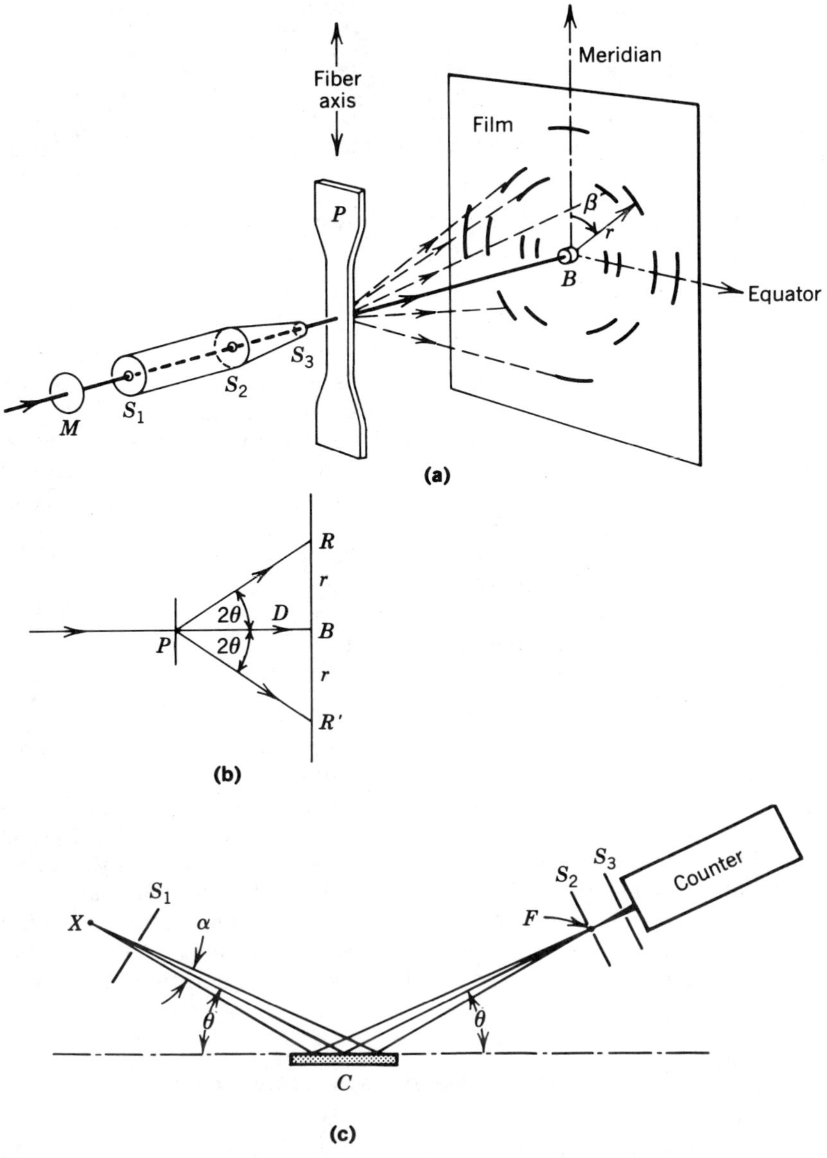

Figure 3. Wide-angle x-ray diffraction technique. (**a**), (**b**) Flat film technique; (**c**) diffractometer technique.

ical properties. At low orientation, modulus and tensile strength correlate with either crystalline or amorphous orientation. After the crystallites are fully oriented, Young's modulus and tensile strength often increase further with additional stretching. During uniaxial stretching, birefringence continues to increase.

Uniaxially oriented samples are frequently brittle in the transverse direction. Introducing biaxial orientation produces materials with increased elongation to break in all directions in the plane of the film. Similar transverse toughening characteristics have been observed in other polymers.

ORIENTATION PROCESSES

Development of Process Technology

Products with "designed-in" orientation are increasingly important in the polymer industry. They include filaments with uniaxial orientation and films and bottles with biaxial orientation. A variety of molding operations have been developed in which multiaxial orientation is built in during the process. Uniaxial orientation results in high tensile modulus and high tensile strength in all directions, whereas equal biaxial orientation results in equal properties in all directions in the film plane. Biaxial orientation in bottles results in balancing the mechanical properties between the bottle-axis direction and the hoop direction.

Efforts toward orientation processing has led to new molding technology. A process for extruding, reheating, and inflating tubes produced orientation. In an injection-molding process the core of an annular mold rotates to produce a helically oriented molded part (Fig. 4). The DuPont work on biaxially oriented poly(ethylene terephthalate) involved the molding of an isotropic glassy parison that was heated and inflated with air in a mold to form a biaxially oriented bottle. This gave rise to a new technology called stretch blow molding (qv). The Corpoplast process is shown in Figure 5 (see also EXTRUSION; INJECTION MOLDING).

New polymers have been biaxially oriented by older technologies. Tentering-frame and double-bubble technologies have been used to make a biaxially oriented polypropylene film.

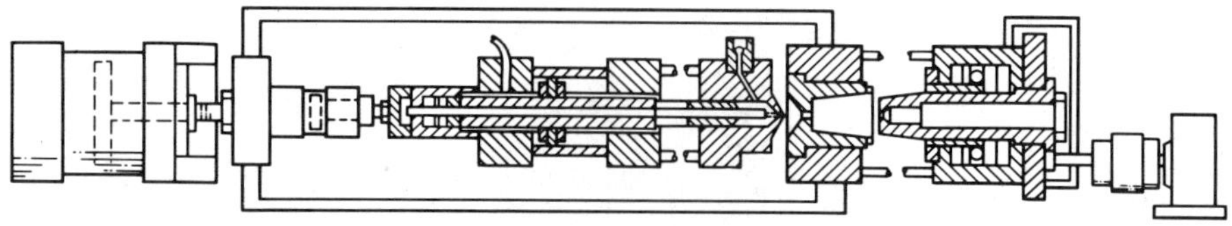

Figure 4. Rotational technology for biaxial orientation, injection molding.

Orientation and Process Conditions

In the processing of thermoplastics the degree of orientation appears to be related to the stress field at the time of solidification and the accompanying thermal history. This behavior is due to the fact that flowing polymer melts consisting of flexible polymer chains obey the rheooptical law between birefringence (qv) and stress, ie, birefringence $(n_i - n_j)$ is proportional to the principal stress differences $(\sigma_i - \sigma_j)$:

$$n_i - n_j = C(\sigma_i - \sigma_j)$$

where C is the stress-optical constant.

Fibers

Fibers are usually processed in such a manner as to produce uniaxial orientation along the filament axis with the chains oriented in the axial direction. This results in high Young's modulus and tensile strength, although often poor lateral properties. These are usually less important, as the primary tensions in structural members are in the axial direction.

Melt and Solution Spinning. Fiber-processing operations can be divided into a shaping step, where a polymer melt or solution is transformed into a continuous filament, and an orienting step, in which filaments are drawn in a state of uniaxial tension (see FIBERS, MANUFACTURE).

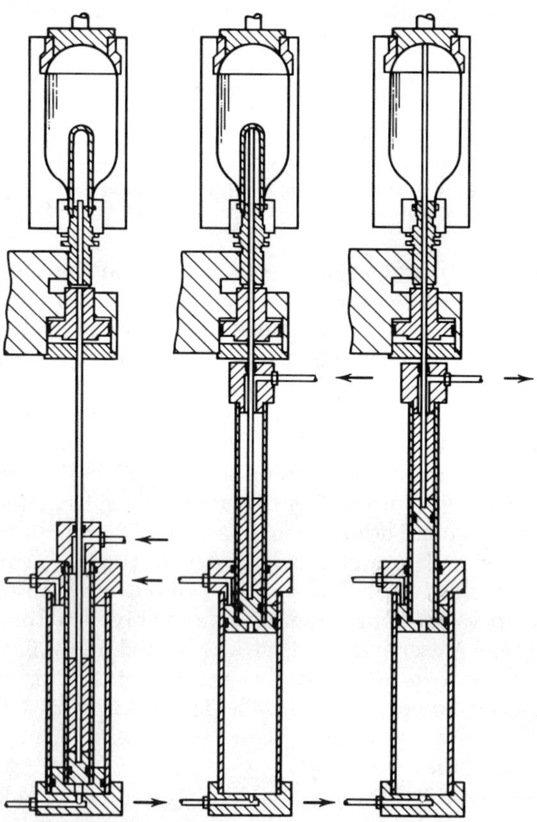

Figure 5. Corpoplast process.

Films

Films are produced in such a manner as to yield systems with approximately equal properties in the film plane. High uniaxial orientation impairs transverse properties. Such uniaxially oriented films frequently fibrillate and are used as intermediates to produce continuous filaments. The degree of orientation and biaxiality in films varies with the process technology used (see FILMS, MANUFACTURE).

As with filaments, film production has two stages. First, the polymer melt is shaped by being forced through a slit or annular die of narrow dimension. In the second step, the desired biaxial orientation is produced.

Molding

Traditionally, molding processes have been used to shape objects, but rarely to induce orientation. For a high degree of orientation, a two-step process is required. First, a parison is injection molded, which may be heated above its softening temperature and reshaped. In stretch blow molding, the injection-molding tubes are deformed into the desired shape, usually a bottle (Fig. 5), resulting in a higher degree of biaxial orientation. Another procedure employs molds with rotating members; it is used to introduce hoop orientation into cups.

Other Orientation Processes

Other processes include solid-state extrusion and gel processing.

BIBLIOGRAPHY

D. W. Hadley and I. M. Ward in I. M. Ward, ed., *Structure and Properties of Oriented Polymers,* Wiley-Interscience, New York, 1975.

U.S. Pat. 3,311,679 (1967), E. J. Moore; B. E. Read and R. S. Stein, *Macromolecules* **1,** 116 (1968).

G. V. Vinogradov, A. I. Isayev, D. A. Mustafaev, and Y. Y. Podolsky, *J. Appl. Polym. Sci.* **22,** 665 (1978).

K. Oda, J. L. White, and E. S. Clark, *Polym. Eng. Sci.* **18,** 53 (1978).

JAMES L. WHITE
MUKERREM CAKMAK
University of Akron

OSMOMETRY

The principles of a membrane osmometer are illustrated in Figure 1. At constant temperature T, if a solvent α and a solution β consisting of the solvent and a polymeric (macromolecular) solute are separated by a membrane SM, permeable to solvent only (the semipermeable membrane), then solvent α flows across SM to the solution phase β until the hydrostatic

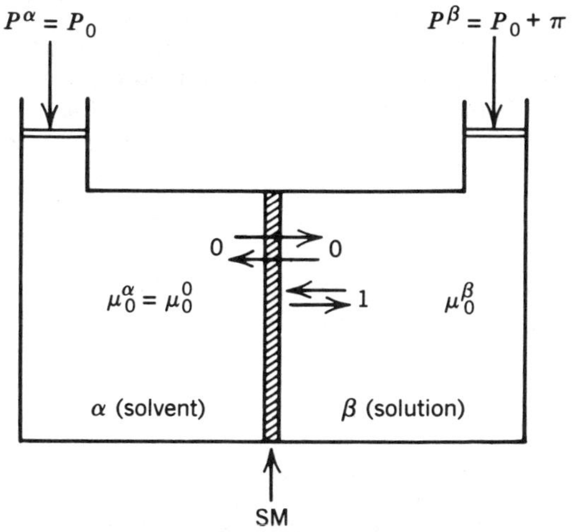

Figure 1. Principle of a membrane osmometer. Component 0 is the solvent; component 1 is a macromolecular solute retained by the semipermeable membrane (SM), which separates the solution β from the solvent α. At constant temperature T, when the pressure difference between the two sides is π, no solvent flows from side α to β, and osmotic equilibrium is attained. At osmotic equilibrium $\mu_0^0 = \mu^{\beta}$.

pressure in side β becomes sufficiently large to prevent the flow of solvent from side α (see also MEMBRANES). The flow of solvent from α to β can be prevented by increasing the pressure on side β from its initial value P_0 to $P_0 + \pi$, giving a pressure difference of π. This excess pressure π is called the osmotic pressure. The same effect can be achieved by lowering the pressure on the solvent side α from P_0 to $P_0 - \pi$. Thus as long as the pressure difference between the two sides is π, no solvent flows from α (where the solvent has a higher chemical potential) to β (where the chemical potential of the solvent is lower). If the pressure on side β is increased beyond $P_0 + \pi$, solvent flows from β to α; this is sometimes referred to as reverse osmosis and is being considered in the purification of brackish water.

Osmotic pressure is a colligative property that depends on the number of solute particles. For nonassociating solutes it provides the molecular weight M if the sample is homogeneous, or the number-average molecular weight $\bar{M}_n$ if the sample is heterogeneous. In addition, these measurements provide information about the nature of the solution, whether it is ideal or nonideal, and whether self-association occurs. It is one of the older methods for studying polymeric solutions. In recent years, the development of high speed membrane osmometers, in which measurements can be performed in minutes instead of hours or days, has revolutionized the technique. Osmotic-pressure measurements are generally restricted to macromolecules with a molecular weight >8000 g/mol and $<1 \times 10^6$ g/mol. For lower molecular weight solutes, vapor-pressure osmometry, cryoscopy, ebulliometry, or sedimentation equilibrium should be used (see also MOLECULAR WEIGHT DETERMINATION).

Osometry provides a means of obtaining the activity a_0 or the osmotic coefficient g of the solvent. For monodisperse, nonassociating solutes it gives the molecular weight M and the nonideal term B. For heterogeneous or polydisperse solutes, osmometry provides the number-average molecular weight $\bar{M}_n$ of the polymeric solute as well as the osmotic-pressure second virial coefficient B_0, a measure of the nonideality of the system. In fact, osmometry and other colligative methods (cryosopy, ebulliometry, and vapor-pressure osmometry) are the only thermodynamic methods that yield an unambiguous average molecular weight $\bar{M}_n$, the number-average molecular weight. Membrane osmometry can be applied to solutions that contain homogeneous or heterogeneous, nonassociating solutes; solutions in which the solute undergoes self-association; and solutions containing two different solutes in which a mixed association occurs.

Sensitivity and Significance of Measurements

Osmometry is an absolute method that is fast, reliable, and simple to use. For best results, solvents and solutions must be degassed before use. Membrane osmometry is more sensitive than vapor-pressure lowering, vapor-pressure osmometry, freezing point depression (cryoscopy), or boiling-point elevation (ebulliometry).

The values of M or $\bar{M}_n$ obtained by osmometry agree well with those obtained by other methods, as seen in Table 1 for proteins and dextran.

Table 1. Molecular Weight Determined by Membrane Osmometry and Other Methods

Substance	Molecular weight	Method
ovalbumin	$(4.295 \pm 0.04) \times 10^4$	mo[a]
ovalbumin	$(4.30 \pm 0.17) \times 10^4$	se[b]
lysozyme	$(15.1 \pm 0.45) \times 10^3$	mo[a]
lysozyme	$(14.4 \pm 0.10) \times 10^3$	se[b]
dextran T-70[c]	4.30×10^{4}[d]	mo[a]
dextran T-70	4.10×10^{4}[d]	ega[e]
canine high density lipoprotein	2.13×10^5	mo[a]
canine high density lipoprotein	2.14×10^5	se[b]

[a] Membrane osmometry.
[b] Sedimentation equilibrium.
[c] Pharmacia Dextran T-70 Lot 7981.
[d] Number-average molecular weight.
[e] End-group analysis.

Modern Instrumentation

In the United States, two high speed membrane osmometers are available: the Wescan (formerly Melabs) (Wescan Instruments, Santa Clara, Calif.) and the Knauer (UIC, Inc., Joliet, Ill.). Both utilize the strain gauge.

Membranes

For aqueous solutions, cellulose acetate membranes are usually employed, but any dialysis or ultrafiltration membrane can be used. The latter may be more retentive and take longer to reach equilibrium. The membranes should be conditioned in solvent or buffer and degassed before use while still in the solvent. For organic solvents, gel cellulose or cellophane membranes are preferred.

BIBLIOGRAPHY

M. P. Tombs and A. R. Peacocke, *The Osmotic Pressure of Macromolecules*, Oxford University Press, London, 1974.

E. T. Adams, Jr., P. J. Wan, and E. F. Crawford, *Methods Enzymol.* **48**, 69 (1987).

E. T. Adams, Jr., *Biochemistry* **4,** 1655 (1965).
E. T. Adams, Jr., *J. Lipid Res.* **27,** 1233 (1986).
B. W. McCarty and E. T. Adams, Jr., *Biophys. Chem.* **28,** 149 (1987).

E. T. ADAMS, JR.
Texas A&M University

OVALBUMIN. See PROTEINS.

N-OXACRYLAMIDES. See ACRYLAMIDE POLYMERS.

OXACYCLOBUTANE POLYMERS. See OXETANE POLYMERS.

OXAZOLINES. See CYCLIC IMINO ETHERS, POLYMERIZATION.

OXETANE POLYMERS

Oxetane is a name commonly used for the unsubstituted cyclic ether with a four-membered ring and the structure:

$$\begin{array}{cc} CH_2\text{—} & O \\ | & | \\ CH_2\text{—} & CH_2 \end{array}$$

(1)

This monomer is also known by other names including trimethylene oxide, oxacyclobutane, and 1,3-epoxypropane. The parent ether polymerizes readily as a result of ring strain associated with the small ring size. A variety of substituted oxetanes have been synthesized and many of them also polymerize readily (Table 1).

Monomers

Properties. More than a hundred different oxetanes have been synthesized; a few selected examples are given in Table 1.

Polymerization

The ring-opening polymerization of oxetane is primarily accomplished by cationic polymerization methods. Insertion polymerization, anionic polymerization, and with certain special oxetanes, a free-radical ring-opening polymerization have also been used. In certain cases, aluminum-based catalysts impose a degree of microstructural control on the polymerization process that is not obtainable with a pure cationic system.

Properties of Polymers

A large number of polyoxetanes have been prepared and partially characterized. The properties of the polymers vary greatly with the symmetry, bulk, and polarity of the substituents on the chain (Table 1). Many copolymers of oxetanes with cyclic ethers have been studied. The properties of block copolymers of oxetanes with tetrahydrofuran resemble those of thermo-

Table 1. Properties of Selected Oxetanes

Monomer	Structure	Abbreviation	Boiling point, °C[a]	Melting point, °C	Polymer T_m, °C
oxetane		OX	47.8	−99	35
2-methyloxetane	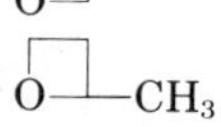	2MOX	60		amorphous
3-methyloxetane	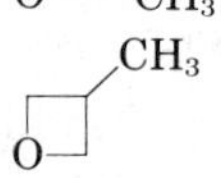	3MOX	67		amorphous
3-azidooxetane	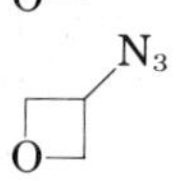	AZOX	120–130[b]		low molecular weight oil
3-nitrooxetane	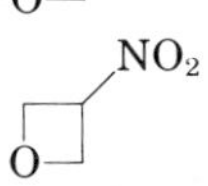	NIOX	77[c]		
3,3-dimethyloxetane	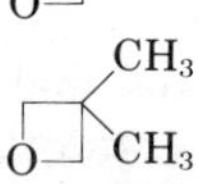	DMOX	80		47
3,3-dinitrooxetane	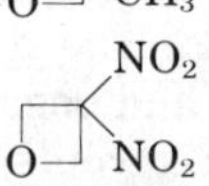	DNOX		70, 71	200–202
3,3-bis(chloromethyl)oxetane	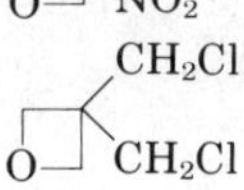	BCMO	91[d]	19	170
3,3-bis(azidomethyl)oxetane	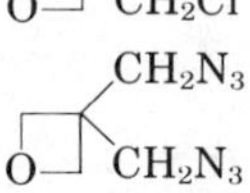	BAMO	liquid density 160–170		75–78

[a] At 101.3 kPa unless otherwise noted. To convert kPa to mm Hg, multiply by 7.5.
[b] 0.93–1.3 kPa.
[c] 0.067–0.13 kPa.
[d] 2.3 kPa.

plastic elastomers (see ELASTOMERS, THERMOPLASTIC). Extensive physical property measurements have been reported only for poly[3,3-bis(chloromethyl)oxetane] (Table 2), the one polyoxetane that was available commercially for a time.

Table 2. Typical Properties of Poly[3,3-bis(chloromethyl)oxetane]

Property	Value
melting temperature, °C	α form, 186
	β form, 180
glass-transition temperature, °C	7–32[a]
density, 25°C, g/cm^3	
amorphous	1.386
crystalline	1.47
compression modulus, 6.9 MPa[b]	130
tensile strength, MPa[b]	41.4
elongation at break, %	35–150
heat-distortion temperature, °C	
1.82 MPa[b]	85–93
0.48 MPa[b]	149
Izod impact (notched), 228°C, J/m[c]	27
Rockwell hardness	R100
flexural strength, MPa[b]	34.5
thermal expansion, 10^{-5}/°C	8
volume resistivity, Ω·cm	1×10^{16}
dielectric strength, V/μm	16
dielectric constant, 10^6 Hz	3
power factor, 10^6 Hz	0.011
flammability	self-extinguishing
water absorption, % (24 h)	0.1
mold shrinkage, cm/cm	0.005

[a] Depends on method of determination.
[b] To convert MPa to psi, multiply by 145.
[c] To convert J/m to ftlbf/in., divide by 53.38.

Economic Aspects

At the present time, polyoxetanes are not commercially available. Consequently, few oxetane monomers are available and those only in laboratory-size quantities; the monomers are expensive. The price varies with the substituent, but all are at least several dollars per gram.

Health and Safety Factors

Very little has been reported about the toxicity and hazards of handling the oxetanes and their polymers. Both properties are influenced markedly by the substituents. Oxetane itself is a low boiling flammable organic liquid and should be handled with care like any other low boiling and flammable liquid. Details of its toxicity are unknown, although one report states that it may be narcotic in high concentrations. BCMO can evolve toxic fumes on decomposition and can be dangerous in disasters. The azido- and nitro-containing monomers (Table 1) are explosive and should never be distilled. Special precautions are required for their handling.

BIBLIOGRAPHY

S. Penczek, P. Kubisa, and K. Matyjaszewski, *Adv. Polym. Sci.* **68/69,** 1 (1985).

P. Dreyfuss and M. P. Dreyfuss in C. H. Bamford and C. F. H. Tipper, eds., *Comprehensive Chemical Kinetics,* Vol. 15 of *Non-Radical Polymerization,* Elsevier, Amsterdam, The Netherlands 1976, Chapt. 4, pp. 259–330.

E. J. Vandenberg and A. E. Robinson in E. J. Vandenberg, ed., *Polyethers, ACS Symp. Ser.* **6,** American Chemical Society, Washington, D.C., 1975, Chapt. 7, pp. 101–119.

T. Saegusa and S. Kobayashi, *Prog. Polym. Sci. Jpn.* **6,** 107 (1973).

M. P. DREYFUSS
P. DREYFUSS
Michigan Molecular Institute

OXIDATION.

See ANTIOXIDANTS; ANTIOZONANTS; DEGRADATION.

OXIDATION OF POLYMERS.

See DEGRADATION.

OXIDATIVE POLYMERIZATION

The polymerization of monomers by oxidative means has been known for many years. Probably the most familiar example of a commercial polymer prepared by this technique is poly(phenylene oxide) (PPO), originally synthesized at General Electric in 1959.

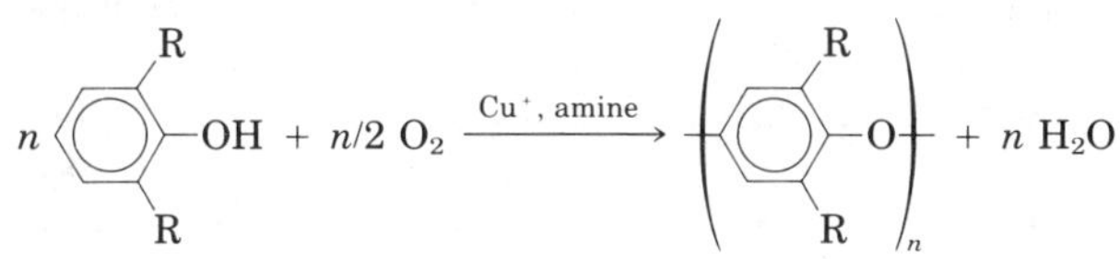

Oxidative polymerization has also been carried out successfully with aromatic amines, thiols, dialkynes, homocyclic and heterocyclic aromatic nuclei, and other miscellaneous monomers.

Dialkynes

In 1869, it was discovered that the cuprous derivative of phenylacetylene undergoes a smooth coupling in air to afford 1,4-diphenyl-1,3-butadiyne. Subsequently, isolation of the cuprous salt was found to be unnecessary.

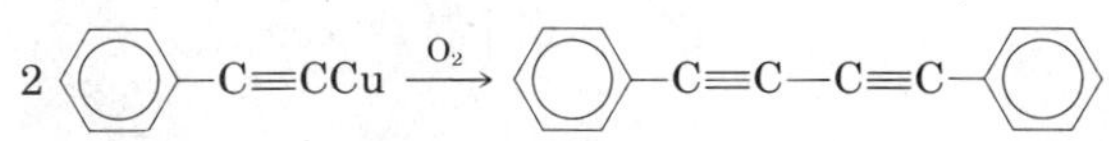

A wide variety of terminal diynes have been transformed successfully into polymers under oxidative conditions. Copolymerization has also been carried out. Low molecular weight individual oligomers in the series $R(C{\equiv}C)_nR'$ (where n = 4–10, 12; R, R′ = H, C_6H_5, t-C_4H_9, mesityl) have been prepared via coupling of triethylsilyl-protected terminal diynes, followed by alkaline desilylation.

Mechanism of Coupling. The mechanism of oxidative coupling of alkynes remains unclear. The active catalyst appears to be an amine complex of a copper salt. Although best results have been obtained with cuprous salts, under coupling conditions, cuprous ion should be rapidly oxidized to cupric ion. A three-step mechanism has been proposed for the reaction.

However, this mechanism does not account for the observed catalysis by cuprous ion in basic media.

$$RC{\equiv}CH \xrightarrow{\text{base}} RC{\equiv}C^-$$

$$RC{\equiv}C^- + Cu^{2+} \longrightarrow RC{\equiv}C\cdot + Cu^+$$

$$2\ RC{\equiv}C\cdot \longrightarrow RC{\equiv}C{-}C{\equiv}CR$$

This effect may be attributed to formation of an alkyne–copper complex, which facilitates ionization of the terminal alkyne.

$$RC{\equiv}CH + Cu^+ \longrightarrow \underset{\vdots \atop Cu^+}{RC{\equiv}CH}$$

Uses. Diyne polymers have not attained the commercial status of poly(phenylene oxide)s. A facile cross-linking reaction brought about thermally or photochemically provides the main interest.

Aromatic Nuclei

Benzene. Two decades ago, the facile conversion of benzene to an insoluble brown powder by treatment with aluminum chloride–cupric chloride at mild temperatures was reported.

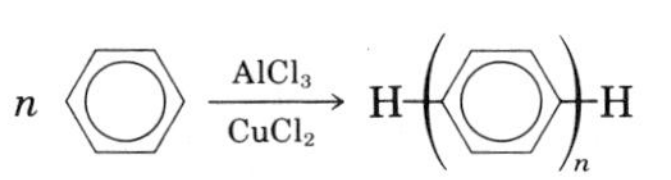

The structure of the polymerization product was assigned that of poly(*p*-phenylene) (PPP) on the basis of elemental analysis, ir spectrum, x-ray diffraction pattern, pyrolysis products, oxidative degradation, insolubility, color, and thermal stability. Polyphenylenes have been prepared by several other routes, including coupling of aryl halides via the Ullmann and Fittig reactions, free-radical arylations of aromatic substrates, and aromatization of cycloaliphatic precursors. Significant alternative routes have been reported and polyphenylenes have been reviewed.

Other Homocyclic Aromatic Hydrocarbons. Several other homocyclic aromatic molecules have been converted to polymeric materials via oxidative polymerization. Toluene yields a polymer with a poly(*o*-phenylene) backbone and possibly some para structure on treatment with $AlCl_3$—$CuCl_2$.

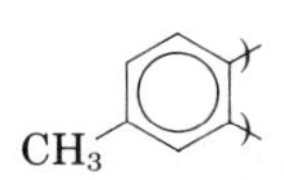

Biphenyl and terphenyl mainly yield sexiphenyl on treatment with $AlCl_3$—$CuCl_2$ in *o*-dichlorobenzene at temperatures <80°C. At higher temperatures (80–155°C), soluble, processible polymers with *m*- and *p*-phenylene linkages are produced.

Treatment of naphthalene with aluminum chloride–cupric chloride or ferric chloride–water yields a product mixture containing binaphthyls, ternaphthyls, and higher molecular weight material. Fluorene, phenanthrene, and pyrene principally convert to dehydrodimers and trimers when treated at room temperature with aluminum chloride–cupric chloride. A soluble polymer with M_n of 4800 can be obtained from polymerization of *trans*-stilbene with $AlCl_3$—$CuCl_2$ in carbon disulfide solvent.

Aryl Halides. Chlorobenzene and fluorobenzene produce oligomers with backbones composed of *o*-phenylene units.

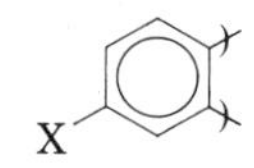

where X = Cl, F

Phenols, Thiols, Ethers, and Nitriles. Phenol is not polymerized by aluminum chloride–cupric chloride, presumably because of deactivation of the $AlCl_3$ by the phenolic hydroxyl. However, $FeCl_3$ successfully couples phenol, catechol, and hydroquinone. Dehydropolycondensation of thiophenol with $MoCl_5$ or WCl_6 results in the formation of a polyphenylene with retention of the SH groups. High molecular weight linear polymers have been obtained through oxidation of di-1-naphthoxylalkanes with $FeCl_3$ in nitrobenzene. 1,4-Dimethoxybenzene is converted to low molecular weight poly(2,5-dimethoxy-1,4-phenylene) by oxidative polymerization with $AlCl_3$–$CuCl_2$ in nitrobenzene.

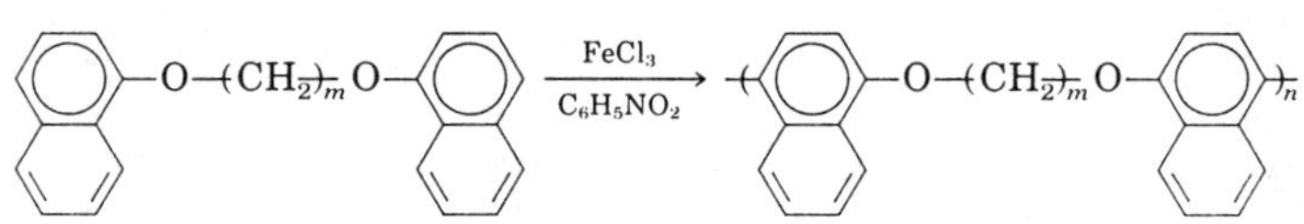

where m = 2–7

Bifunctional *p*-tolylcyanoacetic esters are oxidatively coupled with O_2/[Cu(OH)TMEDA]$_2$Cl$_2$ in $CHCl_3$ to yield high molecular weight polymers. Monofunctional phenylcyanoacetic esters, under similar conditions, give oligomeric species with benzylic carbon to *p*-aromatic carbon coupling.

Amines. Diphenylamine has been converted to a polyamine resin by ferric chloride, aluminum chloride–cupric chloride, or at 200–260°C, by molecular oxygen. Aniline is oxidized to polyaniline (aniline black) by electrochemical or chemical means. The polymer consists of alternating quinonediimine and phenylene moieties.

Heterocyclic Aromatic Molecules. Relatively few reports describe the direct coupling of heteroaromatic substrates by chemical means. Dehydrocoupling of thiophene by $MoCl_5$ or WCl_6 gave products in which the original ring structure was preserved. Chlorinated monomers and dithienyl were also studied. Coupling of benzothiophene is accomplished with molybdenum pentachloride or tungsten chloride. Recently, in connection with an increasing interest in conducting organic polymers, many studies of electrochemical coupling of heteroaromatic nuclei have been reported. Details of electrochemical polymerization may be found in a recent review.

Uses. There is a wide range of potential commercial applications for PPP. Currently, the most prominent of them is the use of PPP as a conductive polymer (following doping) (see ELECTRICALLY CONDUCTING POLYMERS).

BIBLIOGRAPHY

G. Wegner, *NATO Conf. Ser. [Ser.] 6,* **1** (*Mol. Met.*), 209 (1978).

P. Kovacic and M. B. Jones, *Chem. Rev.* **87,** 357 (1987).

A. F. Diaz and J. Bargon in T. Skotheim, ed., *Handbook on Conjugated*

Electrically Conducting Polymers, Marcel Dekker, Inc., New York, 1986, Chapt. 3.

Martin B. Jones
University of North Dakota
Peter Kovacic
University of Wisconsin-Milwaukee

OXIRANES. See 1,2-Epoxide polymers.

OZONE REACTIONS. See Antioxidants; Antiozonants.

P

PACKAGING MATERIALS

Flexible Materials

Packaging protects contained products against air, water, relative humidity, odor, microorganisms, insects, light, compression, and impact. Simultaneously, packaging aids in retaining desirable product constituents such as color, flavor, form, and structure.

Plastic packaging materials may be classified into rigid and flexible, with considerable overlap. The dividing line between rigid and flexible material is determined by caliper or gauge, with thicknesses <0.25 mm generally regarded as being in the flexible category. Flexible packaging is composed of both single-layer and multilayer structures. The latter may be further subdivided into laminated, coated, and coextruded, or combinations of these.

Table 1. Properties of Packaging Materials[a]

	Specific gravity	Tensile strength, MPa[b]	Heat-seal range, °C	Water-vapor transmission rate[c], μmol/($m^2\cdot s$)[d]	Gas transmission[e], nmol/($m^2\cdot s$)[f]	Use temperature, °C	
						Maximum	Minimum
cellophane							
nitrocellulose, coated	1.44	124 (MD) 62 (TD)	93–176	5	16	depends on rh	
Saran, both sides coated	1.44	124 (MD) 62 (TD)	93–176	4.5	0.4	depends on rh	
fluorocarbon	2.2	34–69	176–204	0.25–0.5	56–120	148	−45
ionomer	0.94–0.96	21–69	87–204	13–21	1,800–3,870	71	−73
nylon							
uncoated[g]	1.13–1.14	48–124	176–260	240–260	21	176–232	−59
Saran-coated, one side	1.13–1.14	48–124	176–260	2	4	93	−40
polyester, PET							
uncoated	1.35–1.39	172–228		13	40	204	−62
Saran-coated, one side	1.4	179–214	135–204	6[i]	3.2	coating softens at 82	−51
metallized[h]	1.35–1.39	172–228		0.3–1.4	0.32	204	−62
polyethylene							
LDPE[g]	0.910–0.925	6.9–24	121–176	12	2,000–6,700	65	−51
MDPE[g]	0.926–0.940	14–35	126–154	5–10	1,300–2,700	82–104	−51
HDPE[g]	0.941–0.965	21–52	135–154	3–6.5	260–2,000	121	−51
LLDPE[g]	0.915–0.935	24–55	121–176	12	2,000–6,700	76–82	−51
LDPE/12% EVA copolymer	0.94	21–35	93–148	39	4,100–5,200	60	−51
polypropylene							
nonoriented	0.88–0.90	21–62	162–204	5–6.5	670–3,300	121	not recommended at low temp
oriented	0.905	172–207	requires coating or additive	3–4	880	121	−51
oriented and metallized[h]	0.905	131 (MD) 220 (TD)	requires coating	1	24–80	121	−51
poly(vinyl chloride)	1.21–1.37	14–110	121–176	>40	40–12,000	93[j]	[j]
poly(vinylidene chloride)[g]	1.64–1.71	55–138	121–148	0.5–3	0.64–14	softens ca 82	above −18

[a] Transparent unless otherwise stated; good resistance to grease and oils.
[b] To convert MPa to psi, multiply by 145.
[c] 2.5-μm gauge film, 38°C, 90% rh.
[d] To convert to g/(100 in.$^2\cdot$d), divide by 10.
[e] 25-μm gauge film, 23°C, 0% rh, 101.3 kPa.
[f] To convert to cc/(100 in.$^2\cdot$d), divide by 8.
[g] Transparent to translucent.
[h] Opaque.
[i] 12.5-μm gauge.
[j] Depends on plasticizer.

The U.S. Flexible Packaging Industry

In 1986 packaging in the United States was a \$60+ billion industry, growing at a real rate of 3% annually or slightly faster than the national economy. On a value basis, about 8% is represented by flexible packaging, growing at a rate of ~6%. More than half of flexible packaging is used for food, ca 10% for health and beauty aids, 5% for drugs and pharmaceuticals, and the remainder for a range of products such as electronics, toys, hardware, tobacco, stationery, and paper goods.

More than 1000 firms participate in the flexible packaging industry, including the small number that produce polymer resins and other raw materials such as paper and aluminum. At the next level are the manufacturers that process the plastic resin into film by blowing or slot casting, a unit operation often assumed by the converter.

The processes of slitting, printing, coating, laminating, coextruding, and fabricating into preformed pouches and bags are usually performed by converters. Almost all converters perform printing, but only about 200 laminate, coat, or coextrude.

The output of the converter, usually in roll form but sometimes in bag or pouch form, is marketed to packagers. Product manufacturers employ the materials to wrap or otherwise contain their products. Some flexible packaging, such as that for fresh meat and produce, is sold directly to retailers.

Film Materials

The properties of film packaging materials are given in Table 1.

Cellulosics. Cellophane is not a true plastic because it is derived from cellulose. In 1987 cellophane was being manufactured in the United States only by one company. Less than 40,000 t, valued at ca \$3.50/kg, were manufactured in or imported into the United States. A fraction of cellophane is reinforced with oriented polypropylene (see also CELLOPHANE).

Plastics. Plastic flexible packaging materials are thermoplastic, reversibly fluid at high temperatures, and solid at RT (Table 1). In its various forms, over 1.4×10^6 t of polyethylene film are produced annually in the United States for packaging alone. In addition, large quantities of most plastic films are used for nonpackaging applications. The major plastic films used in packaging in 1987 included nylon used for thermoforming web; polyester, a dimensionally stable, tough material; poly(vinyl chloride) (PVC), used for high permeability applications such as produce wraps; and oriented polypropylene (OPP), which, except for polyethylene, is the most important packaging film in volume and growth. OPP is an excellent moisture barrier with good clarity and fair stiffness, and is used in snack and bakery goods packaging and as a laminant. Plastic materials may be modified by copolymerization, additives in the blend, and alloying as well as surface treatment and coating.

Coextrusions

The use of film coextrusions in packaging increased rapidly in the late 1970s and 1980s; well over 113.5×10^3 t were used in the United States in 1985. Applications are expected to increase in the future.

In coextrusion two or more thermoplastic resin melts are extruded simultaneously from the same die. Although variants on this basic mechanism are in commercial use, the single multilayer die from multiple extruders is conceptually sound. Coextrusion permits an intimate combination of materials in precisely the quantities required to function. Incompatible plastic materials are bonded with thermoplastic adhesives.

Coextruded films may be made by extrusion-blowing or slot-casting of two or three layers (AB or ABA). Recently, seven- and nine-layer blown-film equipment has been developed.

BIBLIOGRAPHY

M. Bakker, ed., *The Wiley Encyclopedia of Packaging Technology,* John Wiley & Sons, Inc., New York, 1986.

C. J. Benning, *Plastic Films for Packaging,* Technomic Publishing Co., Inc., Lancaster, Pa., 1983.

F. Paine and H. Paine, *Principles of Food Packaging,* Leonard Hill, Ltd., London, 1983.

S. Sacharow and A. L. Brody, *Packaging: An Introduction,* Harcourt Brace Jovanovich, Orlando, Fla., 1987.

AARON L. BRODY
Schotland Business Research, Inc.

Rigid Containers

About 4% of the total U.S. plastic market is used to make bottles and jars. Jars have openings approximately the same as those of the body, whereas the neck diameter of the bottle is significantly smaller than that of the body. More than 17 billion plastic bottles and jars were produced annually between 1983 and 1986; distribution is given in Table 1. The volume has been increasing rapidly as more than 8 billion plastic soft-drink bottles were used in 1987. Additionally, plastic bottles were introduced for pourable salad dressings and jars for condiments and spices.

Although rigid and semirigid plastic bottles and jars were made before 1950, production increased sharply after the mid-1960s with the replacement of glass bottles by plastic for household bleach and pasteurized milk in 3.785-L (1-gal) sizes; the development of plastic bottles for liquid detergents followed. Early plastic bottles and jars were mostly fabricated by extrusion blow molding of high density polyethylene (HDPE).

Multilayer blown bottle technology, developed in Japan before 1974, allowed the incorporation of high oxygen-barrier ethylene–vinyl alcohol copolymers (EVOH) for containers for oxygen-sensitive foods. This technology has been expanded to

Table 1. Distribution of Plastic Bottles, 1986

Application	Percent
food and beverages	49
milk	16
food	6
soft drinks	21
other beverages	6
household chemicals	17
industrial chemicals	2
automotive and marine applications	3
toiletries and cosmetics	15
medicinal and health	14
other	<1

Table 2. Properties of Plastics for Rigid and Semirigid Packaging Applications

	Nylons	Polyesters	Low density polyethylene	High density polyethylene	Polypropylene	Polystyrene	Poly(vinyl chloride)
physical							
density, g/cm^3	1–1.4	1.37	0.91–0.93	0.95–0.96	0.9	1.04–1.07	1.3–1.5
transparency	no	yes	no	no	no	yes	yes and no
melt temperature, °C	210–220	254–257	106–115	130–137	168	100–105	75–105
water absorption, g, in 24 h	1.3–1.9	0.08–0.09	<0.01	<0.01	0.01	0.01	0.04
mechanical							
tensile strength, MPa[a]	62.0–82.7	68.9	6.2–17.2	20–37.2	33–37.9	41.3–51.7	34.4–55.1
tensile modulus, MPa[a]	1170–2760	2760–4140	138–186		1100–1520	3100	2410
Izod impact, J[b]	1.6–5.4	1.0		0.5–19	0.5–2.9	0.4	0.6–27
Rockwell hardness	120	68–78			80–102	60–75	
thermal							
maximum use temperature, °C	82–149	79	82–100	79–121	107–149	66–77	66–79
heat-deflection temperature at 180 MPa[a], °C	66–49	116	38–49	43–54	52–60	82–104	54–79
barrier							
gas	excellent	good	poor	poor	poor	poor	variable
water-vapor	poor	good	excellent	excellent	excellent	poor	variable

[a] To convert MPa to psi, multiply by 145.
[b] To convert J to ft·lbf, divide by 1.355.

a variety of narrow- and wide-mouth food containers, with the prospect of displacing glass packaging.

As much as 20% of extrusion blow-molded bottles are self-manufactured by dairies and liquid-detergent packagers. Perhaps more than 1000 merchant bottle blowers operate in the United States, employing extrusion and injection blow molding.

The U.S. plastic container industry aims to replace the almost 40 billion glass bottles and jars used annually in the domestic market by plastic. Plastic bottles, jars, and thermoformed containers are classified as semirigid and injection-molded articles are usually regarded as rigid.

Fabrication Processes

Fabrication processes for rigid plastics include compression molding, thermoforming, sheet extrusion, steam-chest expansion, three-dimensional packaging (cartons, spin welding, blow molding, extrusion blow molding, stretch blow molding, and injection–stretch blow molding), heat-set molding, the blow mold–fill–seal system, barrier enhancement, and monolayer modifications.

Rigid and Semirigid Plastic Materials

Plastic materials for the fabrication of packaging materials by thermoforming and by injection, compression, and blow molding are usually thermoplastic. A small quantity of thermosets is employed for bottle closures; thermoset PET is employed for compression-molded food trays suitable for microwave and conventional ovens. Properties of typical plastic packaging materials are given in Table 2.

BIBLIOGRAPHY

Modern Plastics Encyclopedia, McGraw-Hill Inc., New York, 1985.

Opportunities for Barrier Plastics in Packaging, Schotland Business Research, Inc., Princeton, N.J., 1986.

J. Frados, *Plastics Engineering Handbook,* Van Nostrand Reinhold Co., Inc., New York, 1976.

S. Sacharow and A. L. Brody, *Packaging: An Introduction,* Harcourt Brace Jovanovich, Orlando, Fla., 1987.

Aaron L. Brody
Schotland Business Research, Inc.

Medical Supplies

Changes brought about by single-use, disposable medical devices, prepackaged and sterilized by the manufacturer, include the development of new packaging materials and forms. The primary requirements for packaging of disposables are protection of the products, sterility maintenance, and ease of use. In order to satisfy these criteria, new materials such as Tyvek by DuPont and special heat-sealable coatings have been utilized. Thermoformable plastics such as polystyrene (PS), poly(vinyl chloride) (PVC), acrylics, polyesters, polypropylene (PP), and polyethyenes (PE) are used extensively in tray or blister-type packaging. Film extrusions or laminations, alone or in combination with papers or foil, meet the needs of flexible packages.

Sterilization methods, process conditions, shelf-life requirements, product resistance, and protection from environmental conditions are some of the factors to be considered in the selection or development of medical packaging materials. It is expected that the uses for polymeric materials in the packaging of medical products will continue to grow with the continued development of and concern for modern health-care practices.

Material Characteristics

Three categories of materials are commonly used to satisfy the basic requirements of the primary package forms: pouches, bags, and trays. Flexible materials, ie, papers, Tyvek, films, composite material laminations, may be used in pouches or bags, or as lidding materials on trays. Semirigid materials, such as various plastic sheet products, typically over 0.127 mm (5 mils) thick, are used in preformed trays or may be run on form–fill–seal equipment. Sealants or adhesives are generally applied to the surface of a flexible material and enable the sealing of different packaging structures together, usually with heat and pressure.

Flexible Material Properties. Important properties to consider include barrier properties, strength, flexibility, peelability, safety, sealability, printability, processibility, stability, and aesthetics.

Tyvek. The spun-bonded material Tyvek has several attributes that make it very close to an ideal sterile packaging material. High strength makes it a clean, nonparticulating component of many peelable packages. Tyvek is resistant to delamination, and even if it delaminates, its continuous-fiber structure prevents the release of particulate material. The high porosity of this HDPE material makes it ideal for ETO sterilization. It has temperature limitations due to its thermoplastic nature, but may also be used in steam sterilization with suitable adhesives and well-controlled sterilization conditions (120°C).

Films, Laminations, and Coextrusions. Through laminations, coextrusions, or coatings, film constructions may be tailored to perform a wide variety of specialized functions in medical packaging (Table 1).

Semirigid Materials. *Design.* The versatility of thermoplastic makes possible an almost unlimited variety of three-dimensional packages. Designs can include features that follow the contours of the device or hold several components in one unit. The development of the thermoformed plastic tray with peelable Tyvek lid was a key factor in the widespread use of custom procedure kits in which several components of a particular surgical procedure are contained in one tray. Other built-in features of thermoformed trays include compartments, snap-in lugs, embossed surfaces, hinges, stacking lugs, or various types of recesses.

Material Selection. Functional, economic, and aesthetic factors determine the selection of plastics for a particular application.

Sealants and Adhesives. Sealants must provide adhesion, integrity, and ease of opening. Adhesive enables two or more components to bond. Adhesives are necessary with materials that have little or no affinity for each other. Integrity implies that the package components will hold together throughout stages of packaging, sterilization, shipping, handling, and storage. The medical device industry values the functional benefits of cleanly peelable, easy-opening packaging. This feature is primarily attributable to improved adhesives in conjunction with high strength substrates.

Table 1. Laminated and Coextruded Film Constructions and Characteristics

Film layers	Properties and applications
nylon–PE	thermoformable packaging
polyester–PE–EVA	peelable, heat-seals to plain Tyvek
polyester–Surlyn[a]	wide seal range, high hot tacks, strong seals
polyester–nylon–PE	high flexibility and puncture resistance
polyester–PP	steam sterilization packaging
polyester–nylon–PP	high temperature resistance and flexibility
polyester–PVDC[b]–PE	oxygen or moisture barrier
metallized polyester–PE	light resistance and improved barrier
polyester–foil–PE	high barrier and puncture resistance
PET–foil–PET–Surlyn	high barrier and chemical resistance
PC–PE–EVA	high clarity, strength, flexibility

[a] PE copolymer ionomer resin.
[b] Poly(vinylidene chloride).

Processing and Functional Properties

Typically, a thermoplastic such as PE goes through a temperature-flow cycle during which softening occurs incrementally with increasing temperature until the coating becomes molten. When the heat source is removed, the thermoplastic returns gradually to its original state, but not necessarily at the same rate as shown by its softening curve. This behavior can be predicted to some extent on the basis of certain known physical properties of the material, such as melt index and molecular weight. Peelable heat-seal coatings are usually mixtures of polymers and various modifiers designed to provide a balance of functional characteristics, eg, blocking, hot tack, bond strength, stress fatigue, and cling.

Figure 1 illustrates the entire range of the heat-seal characterization cycle. Such an analysis can be a useful tool in the evaluation and selection of heat-seal packaging materials.

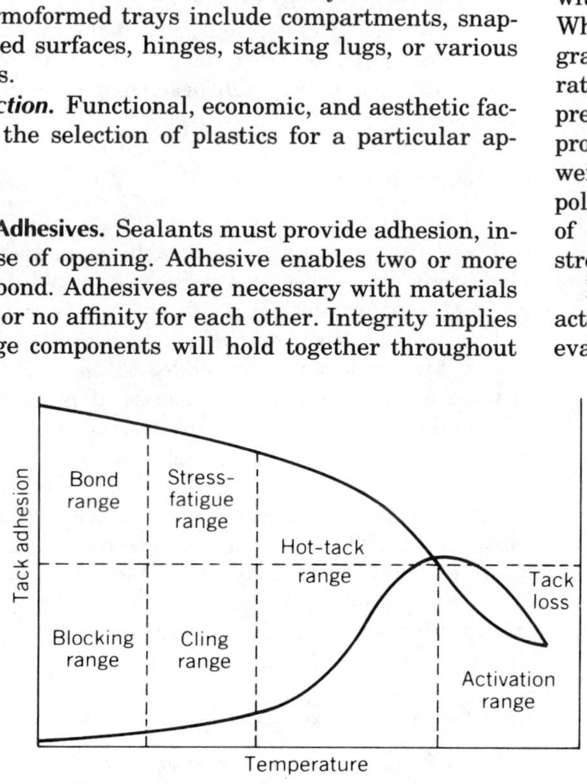

Figure 1. Heat-seal characterization cycle.

BIBLIOGRAPHY

C. D. Marotta, *Med. Dev. Diag. Mag.* **3** (May 1981).

A. Friedman and C. D. Marotta, *Med. Dev. Diag. Mag.* **6** (Apr. and June 1984).

C. D. Marotta, *Med. Dev. Diag. Mag.* **6** (Dec. 1984).

C. D. Marotta, *Med. Dev. Diag. Mag.* **2** (Sept. 1980).

CARL D. MAROTTA
Innovation Technology, Inc.

PAD TRANSFER PRINTING. See DECORATING.

PAPER

Paper consists of sheet materials that are comprised of bonded, small discrete fibers. The fibers are usually cellulosic in nature and are held together by chemical bonds which, most probably, are hydrogen bonds (see CELLULOSE). The fibers are formed into a sheet on a fine moving screen from a dilute water suspension. The word paper is derived from papyrus, a sheet made in ancient times by pressing together very thin strips of an Egyptian reed, *Cyperus papyrus*.

Paper is made in a wide variety of types and grades to serve many functions. Writing and printing papers constitute ca 30% of the total production. The balance, except for tissue and toweling, is used primarily for packaging (see also PACKAGING MATERIALS). Paperboard differs from paper in that it is thicker, heavier, and less flexible.

Fibrous Raw Materials

The main components used in the manufacture of paper products are listed in Table 1. More than 95% of the base material is a fibrous pulp with over 90% originating from wood.

Physical Properties

Most physical properties of paper depend on direction. For example, strength is greater if measured in the machine direction (MD), the direction of manufacture, than in the cross-machine direction (CD). For paper made on a Fourdrinier paper machine, the ratio of the two values varies from about 1.3 to 3.5. An even greater anisotropy is observed if either of the in-plane values is compared to the out-of-plane strength.

Paper may be considered an orthotropic material, one possessing three mutually perpendicular symmetry planes. The three principal directions are the machine, cross-machine, and thickness directions. The three-dimensional, orthotropic nature of paper is most easily observed in the elastic properties. It is possible to easily measure seven of the nine elastic moduli necessary to describe paper as an orthotropic material, using sound-wave propagation methods. Table 2 gives some of the elastic moduli for several commercial papers and laboratory sheets (handsheets) measured this way.

Paper is a viscoelastic material exhibiting creep and stress relaxation phenomena. The mechanical and most other properties, therefore, are highly sensitive to the temperature and the moisture content of the paper.

Papers are specified in terms of their basis weight, the mass in grams per square meter. The thickness or caliper is the thickness of a single sheet measured under specified conditions. The apparent density of paper is defined as basis weight divided by caliper. It is, therefore, the density of the air and fiber (and water) composite. The tensile strength is the force per unit width (area) parallel to the plane of the sheet that is required to produce failure in a specimen of specified width (and thickness) and length under specified conditions of loading. Stretch is the strain at failure resulting from the application of a tensile load applied under specified conditions. Tensile-energy absorption (TEA) or toughness is the work done on the specimen in straining it to failure. Ring crush is a measure of the edge-compressive strength, typically measured on linerboard and corrugating medium. The bursting strength is the hydrostatic pressure required to rupture a specimen when it is tested in a specified instrument under specified conditions. Tear strength or the internal-tear resistance is the average force required to tear a single sheet of paper under standardized conditions in which a small cut is put in the specimen prior to tearing. Stiffness is related to bending resistance. It is most commonly measured by determining the force required to produce a given deflection or by measuring the deflection produced by a given load when the paper specimen is supported rigidly at one end and the deflecting force is applied at the free end. Folding endurance refers to the number of folds a paper can withstand before failure. Typical strength properties for some commercial and laboratory papers are given in Table 3. The common optical properties of paper are brightness, color, opacity, transparency, and gloss.

Table 1. Main Components of Paper and Paperboard, wt %[a]

	World	North America
wood pulp		
mechanical and semichemical	20.8	21.5
unbleached kraft chemical	18.5	25.2
white chemical[b]	26.6	28.6
waste fiber	25.3	20.3
nonwood fibers	4.2	0.9
fillers and pigments	4.6	3.5

[a] 1980 data.
[b] Includes unbleached sulfite.

Chemical Properties

The chemical composition of paper is determined by the types of fibers used (pulping and bleaching methods) and by any nonfibrous substances incorporated in or applied to the paper during the papermaking or subsequent converting operations. In most papers, the chemical composition largely reflects nonfibrous materials that were added to the paper to achieve the desired physical, optical, or electrical properties. Examples of chemicals and resultant properties are dyes and optical brighteners to enhance appearance, resins to impart wet strength, rosin or starch size to reduce penetration of aqueous liquids, pigment coatings to provide a smooth surface for printing, mineral fillers to increase opacity, polymers applied by saturation or extrusion to impart mechanical or barrier properties, and cationic polyelectrolytes and resistive polymers used in the interior and on the surface, respectively, of papers for dielectric recording (see also PAPER ADDITIVES AND RESINS).

Manufacture and Processing

Stock Preparation. Stock preparation includes the operations undertaken to prepare the pulp and additive mixture (furnish)

Table 2. Elastic Properties[a]

Material	Apparent density ρ, kg/m³	Three-dimensional bulk stiffness, GPa[b]								
		C_{11}	C_{22}	C_{33}	C_{12}	C_{13}	C_{23}	C_{44}	C_{55}	C_{66}
carton stock	780	8.01	3.84	0.042	1.36	0.092	0.91	0.099	0.137	2.04
linerboard	691	8.12	3.32	0.032	1.19	0.113	0.082	0.104	0.129	1.80
corrugating medium	538									1.24
bleached-kraft softwood handsheets[c]	721	10.9	6.40	0.172				0.290	0.343	3.09
	673	16.6	2.78	0.073				0.151	0.260	2.28

[a] All tests were carried out at 50% rh; 23°C. The subscripts refer to the direction of measurement, the first normal to surface, the second in the direction of force.
[b] To convert GPa to psi, multiply by 145,000.
[c] At different densities.

Table 3. Strength Properties[a]

Sample	Apparent density, kg/m³	Breaking length, km			Stretch, %		Compressive strength[b], MPa[c]		Tear resistance, MN[d]	
		MD	CD	ZD	MD	CD	MD	CD	MD	CD
linerboard	721	7.22	3.71				8.28	4.83		
kraft envelope	695	8.42	4.32		1.2	2.6			863	912
rag bond	687	4.9	3.47		1.8	4.7			637	637
newsprint	596	3.65	1.84		1.1	1.4			118	226
bleached-softwood kraft handsheet[e]	721	9.11	5.28	0.066						
	673	14.3	3.15	0.056	3.2	3.7	23.1	15.1		

[a] MD = machine direction; CD = cross-machine direction; ZD = thickness direction.
[b] Short-span test.
[c] To convert MPa to psi, multiply by 145.
[d] To convert MN to pound-force, multiply by 225,000.
[e] At different densities.

for the papermaking process. Papermaking pulps are most conveniently handled as aqueous slurries. They can be transported, measured, subjected to desired mechanical treatments, and mixed with nonfibrous additives before delivery to the paper machine. In the case of adjacent pulping and papermaking operations, pulps are usually delivered to the paper mill in slush form directly from the pulp mill. Purchased pulps and recycled paper are typically received as dry sheets or laps, or in bales, and must be slushed before use. The objective of slushing is to separate the fibers and disperse them in water with a minimum of mechanical work in order not to alter the fiber properties.

Paper-machine Operations. Continuous-sheet forming and drying came into use ca 1800; the cylinder machine and the Fourdrinier were employed. The former is especially suited to multi-ply sheets. A series of cylinders covered with wire mesh are mounted in vats containing the fiber slurry. As the cylinder revolves, water drains inward through the screen and the paper web is formed on the outside. The wet web is removed at the top of the cylinder onto previously formed paper layers carried by a felt. The sheet with a felt then passes through press rolls for water removal and then over steam-heated cylindrical dryer drums for drying. The Fourdrinier (Fig. 1) basically consists of an endless screen belt, termed the wire, supported by various devices that enhance drainage. The fiber slurry is introduced at one end from a headbox and the slurry loses water as it progresses down the wire, thereby forming the sheet. It then passes to presses and dryers as in the cylinder machine. At the end of the dryer section, many machines are equipped with

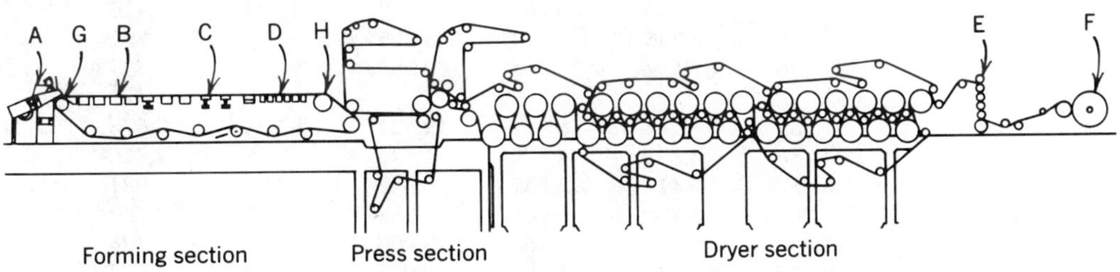

Figure 1. Fourdrinier paper machine with A, headbox; B, Fourdrinier wet end with foil boxes; C, wet and D, dry suction boxes, pickup, and closed transfer of web through the press section and dryer section; E, calender; and F, reel; G and H are the breast roll and couch roll, respectively. Courtesy of Beloit Corp.

Table 4. 1984 World Production of Pulp and Paper and Board, 10^3 t

Country	Pulp	Paper and paperboard
United States	55,530	69,477
Canada	23,679	16,638
Japan	16,415	24,624
People's Rep. China	8,350	12,645
U.S.S.R.	10,065	10,750
Fed. Rep. Germany	2,358	10,576
Finland	9,001	8,653
World total	160,551	226,357

a calender, a stack of loaded rolls through which the paper is threaded to produce the desired surface finish and caliper (sheet thickness), before winding on the reel.

Subsequent to stock preparation and dilution, the paper furnish is usually fed to the paper machine through screens or centrifugal cleaners to remove contaminants, eg, dirt and fiber bundles. The stock then enters a flowspreader, such as a tapered header, which provides a uniform stream the width of the paper machine. In all machines, except vat-type designs, the slurry then enters a headbox, whose function is to deliver to the wire a uniformly and properly dispersed flow of stock. Uniformity is a key consideration in the various paper-machine operations. Hence, the headbox must deliver a continuous, uniform flow rate of constant dilution and fiber makeup at all points across the machine, up to 9.14 m (30 ft) wide.

Filling and Loading. Materials, mineral pigments for filling and loading, may be added to the pulp slurry to make the papermaking furnish (see PAPER ADDITIVES AND RESINS). Pigments are used in varying amounts, depending on the paper grade, and may comprise 2–40 wt % of the final sheet. Fillers can improve brightness, opacity, softness, smoothness, and ink receptivity. They almost invariably reduce the degree of sizing and the strength of the sheet. Other additives include sizing and coloring.

Sheet Processing. ***Forming.*** The Fourdrinier (Fig. 1) is still the most common paper machine used today. The paper between the breast roll and couch roll is dewatered by drainage through the single wire under the influence of table rolls driven by the wire, foils, and suction boxes. Newer machine designs may pinch the slurry between two wires or may have multiple headboxes to produce multilayer papers.

The classical forming section for producing boxboard and other multi-ply grades is the vat or cylinder machine. Because of the limited drainage capacities of the vats, and the method of carrying the sheet, vat machines are limited to low speeds. Modern versions of the vat machine employ cast and drilled shells, suction boxes, and pressurized headboxes. These additions increase the drainage capacity and facilitate water handling, and the formers can be placed on top of the carrying felt for higher speed operation.

Pressing. The sheet leaving the forming section of either type of machine has typically been dewatered to near 20% solids. By passing the web through mechanical rotary presses, the solids content can be increased to 35–45%, while compacting the sheet and generally improving properties.

Drying. At 35–50% solids, additional water removal by conventional mechanical means is not feasible and evaporative drying must be employed. This is a costly process and often the bottleneck of papermaking. The dryer section most commonly consists of a series of steam-heated cylinders.

Converting. Almost all paper is converted by further treatment after manufacture. Operations may include coating, calendering, supercalendering embossing, impregnating, saturating, laminating, printing, and the forming of special shapes and sizes, eg, bags and boxes.

Economic Aspects

The paper industry in the United States employs about 660,000 people in 391 pulp mills, 884 paper and board mills, and about 4,500 converting plants. In 1984 the United States produced nearly 36% of the total world production of pulp and almost 33% of the total world production of paper and paperboard (Table 4).

The U.S. paper and board capacity for 1982 and 1983 and forecasts through 1987 are given in Table 5. The estimated total paper and board capacity increase from 1983 to 1987 is about 1.7% per year.

Table 5. United States Paper and Board Capacity, 10^3 t

	1984	1985	1986	1987	1988[a]	1989[a]	1990[a]	1991[a]
Paper								
newsprint	5,851	5,788	5,804	6,006	6,115	6,301	6,877	7,117
printing and writing	19,890	20,463	20,980	22,077	23,064	23,971	24,838	25,887
packaging and other	6,088	5,985	5,758	5,492	5,394	5,410	5,445	5,531
tissue	5,276	5,345	5,472	5,505	5,626	5,757	5,946	6,096
Total	*37,106*	*37,581*	*38,014*	*39,080*	*40,199*	*41,439*	*43,106*	*44,631*
Paperboard								
unbleached kraft packaging	17,834	18,144	18,622	19,421	20,174	20,675	21,439	22,597
bleached packaging	4,185	4,370	4,576	4,735	4,679	4,715	4,815	4,985
semichemical	5,252	5,411	5,557	5,682	5,796	5,987	6,160	6,369
semichemical	8,722	8,795	9,008	9,267	9,468	9,752	9,907	9,983
Total	*35,992*	*36,720*	*37,763*	*39,106*	*40,116*	*41,128*	*42,321*	*43,933*
Construction	2,676	2,161	1,990	2,001	1,933	1,993	2,002	2,013
Total paper and board	*75,943*	*76,622*	*77,921*	*80,341*	*82,435*	*84,685*	*87,553*	*90,701*

[a] Forecast.

Environmental Issues and Mill Efficiency

Federal laws affecting the pulp and paper industry include the Clean Air Act of 1970, Clean Water Act of 1974, Resource and Recovery Act of 1976, Toxic Substances Control Act of 1977, Occupational Safety and Health Act of 1970, Federal Hazardous Substances Act, and Federal Insecticide, Fungicide, and Rodenticide Act of 1972. Recent requirements have resulted in increased lead time for expansion of basic production facilities and have affected the nature of industrial growth. Removal of the few percent of pollutants from paper-plant sources could be prohibitively expensive.

Initially, the concern was for general pollutants, eg, biochemical oxygen demand (BOD) and total suspended solids (TSS) in water and particulates, total reduced sulfur (TRS), and sulfur dioxide (SO_2) in air. Concern for particularly toxic chemicals, eg, chlorinated organic compounds, has increased.

Water quality is of continuing high priority to the pulp and paper industry and to the public. With the growing concern over hazardous materials, the disposal of sludges as landfill is being questioned and alternative solutions must be found. Wide acceptance of incineration as a disposal method for problematic sludges will give concentrated ash, requiring special disposal methods.

BIBLIOGRAPHY

D. C. Smith, *History of Papermaking in the United States 1691–1969,* Lockwood Trade Journal Co., New York, 1970.

R. E. Mark, ed., *Handbook of Physical and Mechanical Testing of Paper and Paperboard,* Vol. 1, Marcel Dekker, Inc., New York, 1983.

J. P. Casey, *Pulp and Paper Chemistry and Chemical Technology,* 3rd ed., Vol. 1, Wiley-Interscience, New York, 1980.

Statistics of Paper, Paperboard, and Wood Pulp (data through 1984), American Paper Institute, New York, 1985.

GARY A. BAUM
EARL W. MALCOLM
JOHN D. SINKEY
JACK D. HULTMAN
DWIGHT B. EASTY
HARDEV S. DUGAL
The Institute of Paper Chemistry

PAPER ADDITIVES AND RESINS

This article discusses the chemicals, polymers, and resins used in the papermaking and paper-coating processes. In 1985 sales of all paper chemicals totaled $3.8 billion. These sales are predicted to grow at a rate of 2.7%/yr to $4.2 billion in 1990. The fastest growth rate is predicted for surface-treatment (3%/yr) and wet-end additives (3.2%/yr), in near agreement with another study, which predicts a growth rate of 3.5–4% for specialty paper chemicals. Pulping chemicals are expected to have the slowest growth rate (1.5%/yr).

Additives are described in terms of their point of use in the papermaking process, because usually they are specifically designed to be used only in one location on the paper machine. Three main categories are wet-end additives, surface-treatment additives, and paper-coating additives. A distinction is made between wet-end additives designed to improve the papermaking process (processing aids) and those designed to modify the properties of the paper (functional additives).

WET-END ADDITIVES

These chemicals are added to the pulp slurry prior to sheet formation and drying.

Processing Aids

Processing aids include: retention aids (papermaker's alum or hydrated aluminum sulfate, low molecular weight ($M = 10^3$–10^5) resins containing amine groups, cationic starch, polyacrylamide-based bridging polymers, and poly(ethylene oxide) (PEO)); drainage aids (polyethyleneimine and its derivatives); formation aids and wet-web additives (water-soluble polymers such as polyacrylamide, poly(ethylene oxide), and natural gums such as guar and locust bean gum); defoamers (paste, solid, or brick consisting primarily of high solids (~60–70%) dispersions of fatty acids, fatty alcohols, fatty esters, and often, mineral oil, and liquid defoamers based on an oil such as mineral oil or kerosene); pitch dispersants (surfactants, talc, or synthetic polyolefin pulps); creping aids (adhesives such as animal glues, aminopolyamide resins and aminopolyamide–epichlorohydrin resins); biocides and slimicides; and release agents (silicone oils and paraffin oils).

Functional Additives

Dry-strength Resins. Considering that it is made up of a collection of short (0.1–0.4 cm) rigid fibers, a sheet of paper has considerable strength. A bleached kraft paper sample having a basis weight of 65 g/m² has a tensile strength of approximately 35 N/cm (20 lb per inch width). The exact mechanism by which paper develops dry strength is not well known. Some of the strength is due to the strength of the fibers themselves, but most of the dry strength is from bonding between the fibers. Several types of bonding have been proposed, including covalent bonds, mechanical entanglement of fibers, ionic bonds, hydrogen bonds, and van der Waals forces. The fact that waterleaf paper loses almost all of its strength when soaked in water suggests that the likely possibilities are ionic and hydrogen bonding. Interfiber bonding is probably due to hydrogen bonding between cellulose hydroxyls on adjacent pulp fibers. Dry-strength resins include natural gums, starch, carboxymethylcellulose, and acrylamide polymers, which act primarily to enhance fiber bonding.

Wet-strength Resins. The earliest method for developing wet-strength was to fuse the cellulose fibers to one another, either by heating the paper to very high temperature or by partially solubilizing the paper with sulfuric acid. The latter is the method by which parchment paper is produced. The first chemical method for improving wet strength involved the use of formaldehyde, which can form methylene bridges between hydroxyl groups on the surface of cellulose fibers. Wet-strength resins include urea–formaldehyde resins, melamine–formaldehyde resins, aminopolyamide–epichlorohydrin resins, polyamine–epichlorohydrin resins, glyoxal resins, polymeric amines, and chitosan.

With the exception of the polymeric amines, all effective wet-strength resins undergo cross-linking reactions and some react with functionality on the surface of cellulose pulp fibers. Two mechanisms for wet-strength development have been proposed. The protection theory suggests that the resin forms a cross-linked network around existing fiber–fiber bonds and prevents swelling and disruption of these bonds. The new bond theory proposes that the wet-strength resin creates new fiber–fiber bonds that are covalent and thus are not broken on exposure to water.

Paper containing wet-strength resin is more difficult to repulp than nonwet-strength paper. Therefore, the repulpability properties of wet-strength resins must be considered when choosing a resin.

Sizing Agents. Paper is a porous structure consisting of hydrogen-bonded cellulosic fibers. Cellulose is a hydrophilic polymer with a strong affinity for water. At 50% rh, cellulose absorbs approximately 5 wt % water. The porous-hydrophilic nature is useful for applications requiring rapid absorption of water (eg, tissue and paper towels). However, there are a number of applications where wetting and absorption are undesirable properties for paper. Examples are printing and writing grades designed to be printed by aqueous inks, and liquid containers such as milk cartons and paper cups. In these grades the wettability of the paper is decreased through the use of chemicals called sizes or sizing agents. A paper size is defined as an additive that controls the penetration of liquids, especially water, into paper. Sizing agents are all amphipathic in nature and include rosin sizes, cellulose-reactive sizes (alkyl ketene dimer and alkenyl succinic anhydride), wax emulsions, and fluorochemical sizes. The efficiencies of various sizing agents are given in Table 1. The sizing mechanism is complex and general for each type of size. In each case, the additive must be retained in the sheet, distributed to cover the fiber surface, oriented so that the hydrophobic (water-repelling) portion is away from the fiber, and anchored in place.

Fillers. Opacity is an important property for paper, particularly for grades that are printed on both sides of the sheet and in which show-through must be minimized. Newsprint has high opacity due to the use of high levels of groundwood, a high surface area, and highly light-scattering pulp. For lower surface-area pulps it is necessary to use a particulate filler to increase light scattering and opacity. Mineral pigments are the most commonly used fillers, including kaolin clay, calcium carbonate, titanium dioxide, silica, hydrated alumina, and talc. In recent years attempts have been made to use synthetic polymers as fillers for paper. Fillers based on polystyrene latices and on urea–formaldehyde particles have been evaluated.

Table 1. Efficiency of Sizing Agents

Agent	Typical usage, kg/t of paper
rosin-based	
ordinary rosin	15–20
fortified rosin	5–7.5
high free-rosin emulsion	2–4
synthetic	
alkyl ketene dimer emulsion	1–2
cationic alkyl ketene dimer emulsion	1–2
theoretical minimum	
rosin-based	0.5–1.5

SURFACE TREATMENTS

In some cases it is not possible to add a paper chemical to the dilute pulp slurry and achieve the desired paper property. Hence, a number of functional additives mentioned are also applied to the surface of preformed paper, usually after it has been dried.

Several techniques for the surface application of chemicals to paper are available. The most common involves the use of a size press, a device consisting of two rolls forming a nip, which are flooded with a solution or dispersion of the chemical additive. Paper, usually dried below equilibrium-moisture content, is passed through the nip between the rolls to apply the additive. The paper is then dried in a second dryer section. Surface treatments include strength additives (poly(vinyl alcohol), starch, carboxymethylcellulose, and natural gums), sizing agents (wax emulsions or blends of these emulsions with rosin size, styrene–maleic anhydride copolymers, polyurethane resins, fluorochemical sizes, and chromium complexes of fatty acids), and various processing chemicals such as deformers. By far the most common surface additive is starch or modified starch.

PAPER COATING

Chemical additives may be placed onto the surface of paper in a special applicating device (ie, the coater). The paper coating is usually applied at relatively high solids compared to the surface treatments previously described. Both aqueous and nonaqueous coatings are used with both on-paper-machine and off-paper-machine application. Generally, the coating is not intended to strike deeply into the paper web.

The coating applied to paper is called a coating color. Generally, a coating color contains 40–65 wt % solids, the major component of which is pigment. Binders are used at relatively low levels, 5–30 parts by weight per 100 parts pigment. Other paper-coating additives include rheology modifiers, defoamers, dispersants, and lubricants.

BIBLIOGRAPHY

J. P. Casey, *Pulp and Paper Chemistry and Technology,* 3rd ed., Vols. II and III, Wiley-Interscience, New York, 1980–1981.

R. W. Davison in W. F. Reynolds, ed., *Dry Strength Additives,* TAPPI Press, Atlanta, Ga., 1980.

J. W. Swanson, ed., *Internal Sizing of Paper and Paperboard, TAPPI Monogr.* **33,** 2 (1971).

G. G. Spence, D. H. Dumas, J. C. Gast, S. M. Ehrhardt, and D. B. Evans in R. W. Johnson and E. Fritz, eds., *Fatty Acids,* Marcel Dekker, Inc., New York, 1989.

GAVIN G. SPENCE
Hercules Incorporated

PAPER CHROMATOGRAPHY. See CHROMATOGRAPHY.

PAPER COATINGS. See PAPER ADDITIVES AND RESINS.

PARISON. See BLOW MOLDING.

PARQUET POLYMERS. See COORDINATION POLYMERS.

PARTICLE BOARD. See COMPOSITION BOARD.

PARTING AGENTS. See RELEASE AGENTS.

PARTITION CHROMATOGRAPHY. See CHROMATOGRAPHY.

PARTS REMOVAL. See ROBOTICS.

PATENT INFORMATION

To a first approximation patents are the literature of technology. A patent is a legal document conferring on its owner the right to exclude others from practicing an invention; making, using, or selling a particular product; or carrying out a particular process. In return for this limited monopoly the inventor must fully disclose the details of the invention, and this information becomes part of the collective knowledge of humankind to serve as a spur to further invention and innovation.

The potential value of a patent monopoly more often than not causes inventors and their employers to file for patent protection, although there are some technological advances whose substance is not disclosed but is rather maintained as a trade secret. Consequently, a very substantial proportion of the published information on advances in all technologies, including the technology of polymers, is found in patent documents.

The number of patents that deal with polymers is large as reflected by statistics from the *Chemical Patents Index* (CPI) produced by Derwent Publications, Ltd. of London. The CPI is the most comprehensive abstracting and indexing system for patents with a broad definition of what constitutes a chemical patent and what constitutes a polymer-related patent. Since 1981 its polymer section, PLASDOC, has included nearly one-third (31.7%) of all chemical patents in the CPI, an annual total that reached about 45,000 new basic patents in 1985. These patents cover all aspects of polymers, from monomer preparation and purification, through polymer synthesis, compounding, and processing, to uses (Fig. 1), and this scope of information must be dealt with by information services that cover polymer patents. The range is enormous, encompassing not only chemistry and chemical engineering but also mechanical engineering, electrical engineering, physics, and other disciplines.

Three information services are especially important in the area of polymer patents: Derwent Publications, Ltd; Chemical Abstracts Service (CAS); and IFI/Plenum Data Co. The first two play a principal part in both current awareness and retrospective information retrieval, whereas IFI/Plenum is mainly involved in retrospective retrieval. There are other information services that include polymer patent information but do not provide any special information capability regarding patents. Thus the INPADOC data base produced by the International Patent Documentation Center in Vienna includes bibliographic information on all patents from virtually all countries, but has limited subject-based retrieval capability. The APIPAT data base produced by the American Petroleum Institute has a specialized polymer retrieval system, but is limited to polymers of direct interest to the petroleum industry.

- Monomer preparation, purification
- Polymerization catalyst systems; initiators; modifiers
- Polymerization processes
- Polymer structure, composition
- Compounding agents, techniques
- Polymer modification reactions, processes
- Fabrication techniques, equipment
- Polymer shapes, forms
- Molecular and physical properties
- Uses

Figure 1. Scope of polymer patent information.

BIBLIOGRAPHY

J. B. Gambrell, C. M. Cox, and P. E. Kreiger in M. Grayson, ed., *Kirk-Othmer Encyclopedia of Chemical Technology,* 3rd ed., Vol. 16, John Wiley & Sons, Inc., New York, 1981, pp. 851–889.

J. W. Lotz in M. Grayson, ed., *Kirk-Othmer Encyclopedia of Chemical Technology,* 3rd ed., Vol. 16, John Wiley & Sons, Inc., New York, 1981, pp. 889–945.

S. M. Kaback in J. M. Bernard, ed., *Computer Handling of Generic Chemical Structures,* Gower, Aldershot, Hampshire, UK, 1984, pp. 38–65.

S. M. Kaback, *J. Chem. Inf. Comput. Sci.* **25,** 371 (1985).

STUART M. KABACK
Exxon Research and Engineering Company

PBT. See POLYESTERS, THERMOPLASTIC.

PEARL POLYMERIZATION. See SUSPENSION POLYMERIZATION.

PECTIC SUBSTANCES. See POLYSACCHARIDES.

PEEK. See POLYETHERETHERKETONE.

PELLETIZING

In the production of thermoplastics, the primary products frequently leave the manufacturing facility in the form of pellets. Incorporation of antioxidants, stabilizers, colors, fillers, reinforcing agents, and other additives requires that the polymer be in the melt stage to ensure uniform distribution. After incorporation the melt is pelletized into uniform sizes and shapes to facilitate handling.

Polymerization products are frequently in powder form, with the raw material already compounded. Uniform pelletized form offers the following processing advantages: simpler feeding system with fewer feeders, dust-free handling, and easier cleaning between changes of feedstock; fewer unsatisfactory products, because of uniformity of size and homogeneity of feed; and greater extrusion capacity and lower shipping costs as a result of higher feed-bulk density.

Offsetting these advantages may be a higher heat exposure of the product (an additional melting step), as well as the additional manufacturing costs associated with pelletizing. Molten reactor products must be pelletized.

The process involves pressure development for forced flow through a die. The extrudate is cooled to solidification and then cut into pellets, or the molten extrudate is cut as it emerges from the die and the pellets are subsequently cooled. In the latter case, both cutting and cooling may be done in air or water, or cutting may be done in air, followed by quenching in water.

Equipment

Pelletizing equipment can be classified as follows:

Quenching, solidifying, and cutting	Cutting, quenching, and solidifying die-face pelletizers
dicers	dry-face pelletizers
strand pelletizers	water-ring pelletizers
	underwater pelletizers
	centrifugal pelletizers
	rotary-knife pelletizers

Each type of pelletizer has its function in plastics compounding, depending on the throughput rate required, the kind of compounding equipment used upstream, the material being processed, and the pellet shape required.

BIBLIOGRAPHY

H. Herrmann, ed., *Granulieren von Thermoplastischen Kunststoffen*, VDI-Verlag GmbH, Dusseldorf, FRG, 1974.

D. B. Todd
APV Chemical Machinery Inc.

PENTADIENE POLYMERS

Polymers containing pentadiene are used as resins for adhesives and in the USSR for liquid rubbers. The monomer is usually a mixture of *cis*-1,3-pentadiene and *trans*-1,3-pentadiene, also called piperylene. It may also be regarded as 1-methyl-1,3-butadiene to stress its relationship to other butadiene derivatives.

Products based on cationic copolymerization of pentadiene with olefinic components of the same refinery streams appeared in the 1950s. U.S. commercial production of these resins for adhesives and tackifiers in 1986 used about 65×10^6 kg of pentadiene to produce 110×10^6 kg of resins.

In the USSR, synthetic low molecular weight polypentadienes (SKPs) were available in the early 1960s. Soviet production of pentadiene polymers may exceed that of the rest of the world.

Monomers

Physical properties of the monomers are given in Table 1, along with those of the most significant homologues.

Isomers and Isomerization. Unlike the other commercial diolefins, 1,3-pentadiene has two separable geometric isomers in addition to the more readily interconverted cisoid and transoid conformations that exist for all conjugated dienes. Therefore, it is more precise to refer to the pentadienes than to pentadiene. Commercial pentadiene or piperylene is a mixture in which *trans*-1,3-pentadiene predominates, but the proportions vary. Typical cracking conditions produce an equilibrium ratio of about 65:35 trans:cis.

Isomerization of 1,3-pentadienes is important to consider because it provides methods of purification and separation, reduces wastage when one isomer is inert, and sometimes occurs under the conditions of polymerization.

Sources. A polymer-grade pentadiene isomer mixture is obtained in steam-cracking operations while producing ethylene and propylene from petroleum feed stocks. With the increasing demand for by-products such as butadiene and aromatics, crackers are now usually managed to produce saleable pentadienes as well. In addition, a C-5 fraction may be taken from the waste that would be used for fuel after the removal of butadiene.

Recovery and Purification. For cationic pentadiene–olefin copolymerization, the 40% stream (piperylene concentrate) can be employed without further enrichment. Other polymerization systems require a higher grade, which is obtained by another fractionation increasing the pentadiene concentration to as high as 95%. However, the boiling points of cyclopentene (44.1°C), cyclopentadiene (41°C), and 1-pentyne (39.7°C) are so close to those of the pentadienes that they are not removed by fractional distillation. In most polymerization systems based on transition or alkali metals, trace amounts of these particular C-5 contaminants impede the polymerization.

Gel-permeation chromatography and liquid chromatogra-

Table 1. Properties of Diene Monomers[a]

Compound	Boiling point at 101.3 kPa[b], °C	Density at 20°C, g/mL	Refractive index
cis-1,3-pentadiene	44.086	0.6916	1.43634
trans-1,3-pentadiene	42.032	0.6771	1.43008
1,3-butadiene	−4.41	0.6211	1.4292^{25}
isoprene	34.067	0.68095	1.42194

[a] From BUTADIENE DERIVATIVES, POLYMERIC.
[b] To convert kPa to mm Hg, multiply by 7.5.

Table 2. Behavior of Pentadiene Geometric Isomers with Polymerization Initiators

	Polymerization		Polymer structure[a]; %		Relative polymerization rate	
Catalyst	Cis	Trans	Cis	Trans	Cis	Trans
cationic	+	+	mixed	mixed	high	low
anionic, C_4H_9Li	+	+	*trans*-1,4; 61	*trans*-1,4; 54	low	moderate
free radical	+	+	*trans*-1,4; 70	*trans*-1,4; 70	moderate	low
Alfin	+	+	*trans*-1,4; 85	*trans*-1,4; 85	moderate	low
transition metals						
V	+	+	*iso-trans*-1,4; 98	*iso-trans*-1,4; 98	high	high
Ti	−[b]	+		*iso-cis*-1,4; 87		low
Co, H_2O	−	+		*syndio-cis*-1,4; 90		moderate
Co	−	+		*syndio-trans*-1,2; 90		low
Ce or Nd	−	+		*iso-cis*-1,4; 80		moderate
Fe	+	+	*iso-cis*-1,4; 99	*syndio-trans*-1,2; 90	moderate	high
Cr	+	+	(*iso-*)*cis*-1,4; 98	*cis*-1.4-; 75	moderate	moderate

[a] *Iso-trans* is isotactic trans stereoregular polymer; *syndio-trans* is syndiotactic trans stereoregular polymer.
[b] The cis isomer isomerized slowly to trans isomer.

phy can be used to separate *cis-* from *trans*-pentadiene and recover each isomer 99% pure. Industrially, azeotropic or extractive distillation is preferred.

Polymerization

Table 2 summarizes the polymerization behaviors of *cis-* vs *trans*-pentadiene under different initiator conditions. The very high isotactic *cis*-1,4-polypentadienes prepared with either iron or chromium have properties potentially useful in tire rubbers. The chromium initiator is the only one known that polymerizes both *cis-* and *trans*-pentadiene monomers to quite high *cis*-rubbers.

Copolymers with Other Conjugated Dienes. The trans copolymers of butadiene–30 ± 5% pentadiene, made with a soluble vanadium initiator, are the most likely Ziegler-Natta copolymers to be of potential industrial utility. Copolymers of butadiene–pentadiene (*cis-* and *trans*-1,4) have been made with rare earth and uranium initiators; with these initiators and soluble vanadium only *trans*-pentadiene participates.

Industrial Applications

Cationic pentadiene–olefin copolymers are used primarily in adhesives; they are a significant component (30–50% by weight) in typical formulations for pressure-sensitive adhesives (annual U.S. consumption ca 160 × 10^6 kg). Pentadiene resins improve the thermal stability of the adhesive and improve the adhesion to various substrates ranging from metal to fiber.

Pentadiene and its polymers can serve as vehicles for carrying reactive polar materials such as maleic anhydride into polymers. Epoxy resins can be made from liquid polypentadienes. Epoxidation of polypentadiene in the presence of alcohols improves the pot life of polymeric formulations. Epoxy–oligopentadiene improves wear properties in tire-tread rubbers.

Hydrogenation, phenolation, sulfurization, and chlorination of polypentadiene have been patented. Styrene and acrylonitrile have been grafted to polypentadiene to obtain a high impact polymer. Polypentadiene grafted to polyester fiber improves fiber–rubber adhesion. Terpolymers of pentadiene–acrylonitrile–methacrylic acid have been made for leather-finishing compositions. Various blends have been made with chlorinated PVC resin to stabilize the PVC, with low molecular weight acrylate for protective coatings, with nitrile rubber to improve mechanical properties, with ethylene–vinyl acetate (EVA) for resin-to-steel adhesives, and with concrete.

BIBLIOGRAPHY

G. Natta and L. Porri in J. P. Kennedy and E. G. Tornqvist, eds., *Polymer Chemistry of Synthetic Elastomers (High Polymers 23),* John Wiley & Sons, Inc., New York, 1968, pp. 597–678; L. E. Forman, pp. 535–553.

B. J. Davis, *paper presented at the Technical Association of the Pulp and Paper Industry (TAPPI) Hot Melt Short Course,* June 1979, p. 112.

M. Bruzzone, A. Carbonaro, and L. Gargani, *Rubber Chem. Technology* **51,** 907 (1978).

T. L. Hanlon, K. C. Kauffman, R. W. Kavchok, and R. G. Bauer, *Appl. Polym. Symp.,* (26), 61 (1975).

J. Neil Henderson
Consultant

M. C. Throckmorton
Consultant

PEROXY COMPOUNDS

Peroxy compounds containing the O—O linkage encompass a variety of organic and inorganic derivatives. Some of these provide free-radical and epoxidation sources in the polymer industry. The ultimate source of the O—O linkage is oxygen, obtained by direct oxidation or reactions of compounds with peroxidic materials such as ozone, hydrogen peroxide, or other organic and inorganic peroxides. Air oxidation of many synthetic and natural noninhibited organic compounds produces organic peroxide intermediates manifested by rancidity in fats, loss of activity in certain vitamins, gum formation in lubricating oils, prepolymerization of some vinyl monomers, biological processes, and buildup of dangerous peroxides in certain vinyl monomers and solvents, eg, ethers. Buildup of peroxides can be the cause of explosions during distillations or long-term storage of such solvents and monomers.

In the polymer field, peroxides are widely used as radical sources to initiate vinyl monomer (co)polymerizations, telomerizations, for resin and elastomer cure and cross-linking, and to modify polymers by grafting or controlled degradation (shear thinning). Other free-radical-initiated processes include autoxidation, halogenation, additions to aromatic compounds, and anti-Markownikoff additions to terminal olefins.

The largest volume of uses of peroxides in the polymer industry is in epoxidation and the production of phenol and acetone from the acid-catalyzed decomposition of cumene hydroperoxide. These are nonfree-radical applications. Peroxide manufacturers now offer over 65 different peroxides in more than 200 formulations designed for the specific needs of the polymer and plastics industries.

Properties

Although some commercial peroxide initiators are available in high assay (90–99%) liquid or solid form, many are formulated into solutions, pastes, and dispersions or extended with solid fillers. Formulation is necessary in most cases for safety and ease of handling.

The classes of commercial organic peroxide initiators are shown in Table 1; nonpromoted temperature-activity ranges are expressed in 10-h half-life ($t_{1/2}$) temperatures, the temperature at which the peroxides are 50% decomposed in 10 hours. The activity of peroxide free-radical initiators varies from below room temperature to above 100°C, depending on structure.

Some organic peroxide types, diacyl peroxides and peroxyesters, exhibit a wide variation in temperature activity depending on structure, whereas the activity of peroxydicarbonates is independent of structure. Monoperoxycarbonates, diperoxyketals, and dialkyl peroxides show a moderate variation in activity with differing structures. Only one sulfonyl peroxide is produced commercially, namely, acetyl cyclohexanesulfonyl peroxide ($t_{1/2}$ = 10 h at 40 °C in trichloroethylene (TCE), 0.2 *M*) as a 30% solution in phthalate ester solvent. It is used exclusively as an initiator for vinyl chloride polymerization. Hydroperoxides, ketone peroxides, and peroxydisulfates are primarily used in combination with promoters in redox systems; consequently, half-life temperatures are not too significant for most applications.

Peroxycarboxylic acids are mainly used as epoxidizing and bleaching agents and not as free-radical initiators. Commercial products include peracetic acid as a 40% solution in acetic acid and solid *m*-chloroperbenzoic acid. Peroxyformic acid, generated *in situ*, is used extensively for epoxidation.

The activity of commercial organic peroxide initiators is related to three factors: the relative stability of the radicals formed, ie, the more stable the radical, the less stable the

Table 1. Commercial Peroxides

Peroxide type	Structure	10-h $t_{1/2}$, °C[a]
diacyl peroxides	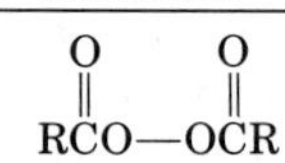	20–75
acetyl alkylsulfonyl peroxides	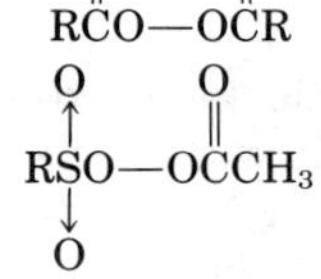	32–42
dialkyl peroxydicarbonates	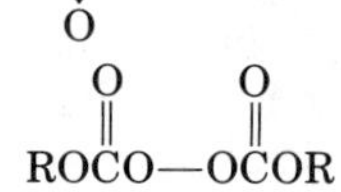	49–51
tert-alkyl peroxyesters	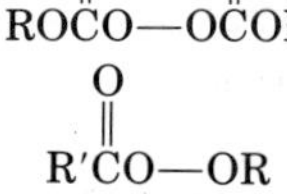	49–107
OO-tert-alkyl *O*-alkyl monoperoxycarbonates	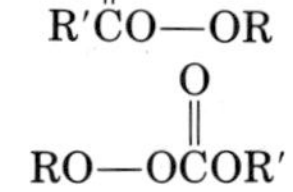	90–100
di(*tert*-alkylperoxy)ketals	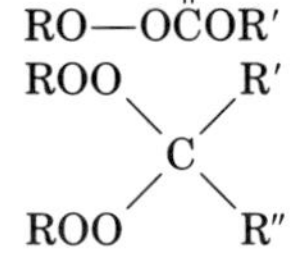	92–115
di(*tert*-alkyl) peroxides	ROOR′	117–133
tert-alkyl hydroperoxides	ROOH	133–172
ketone peroxides	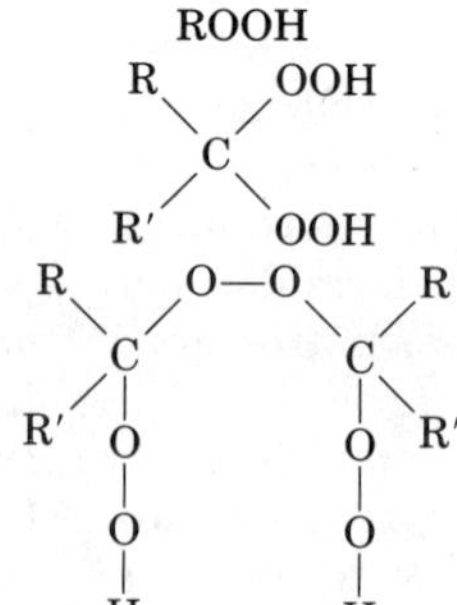 and other structures	

[a] Temperature at which $t_{1/2}$ = 10 h.

peroxide; steric factors, ie, highly strained peroxides are less stable because decomposition relieves steric strain; and electronic effects where electron-donating functions destabilize and electron-withdrawing functions stabilize peroxides. These factors are manifested in the E and A factors in the Arrhenius first-order equation:

$$k = A \times \exp(-E/RT)$$

The initial radicals are generated in a so-called cage of solvent or monomer molecules, as depicted by the brackets in the following equations:

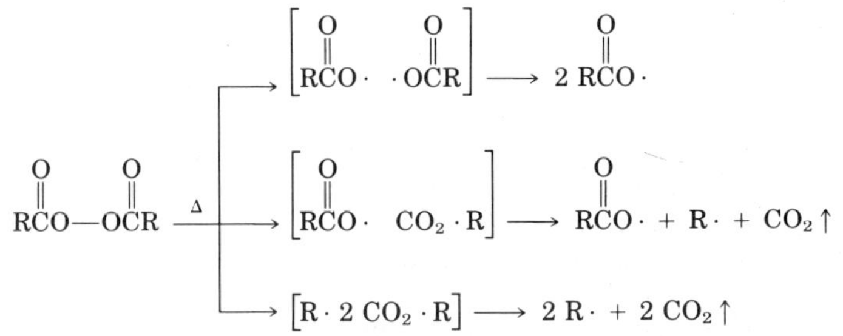

Radical–cage interactions occur and play an important role in peroxide initiator efficiency. Only radicals that diffuse out of or react with the cage molecules are effective in initiating radical reactions such as polymerizations. Single-bond initiators are very efficient because radical–cage recombination regenerates the peroxide. With multiple-bond initiators, radical–cage interactions result in the formation of stable products by radical coupling or disproportionation:

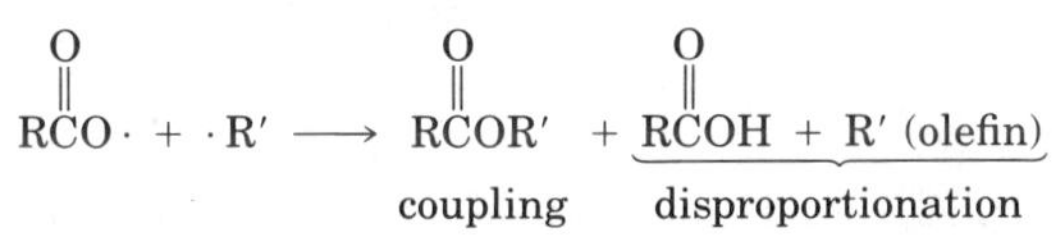

The ratio of radical–cage interation to diffusion is a measure of the efficiency of the peroxide and varies with peroxide structure and decomposition environment; more cage interactions occur in highly viscous systems.

Preparation

Diacyl peroxides are produced by the reaction of an acid chloride or anhydride with sodium peroxide or hydrogen peroxide in the presence of a base. *tert*-Alkyl peroxyesters are manufactured from acid chlorides or anhydrides and the sodium or potassium salts of *tert*-alkyl hydroperoxides. *OO-tert*-Alkyl *O*-alkyl monoperoxycarbonates are produced similarly, employing alkyl chloroformates. Dialkyl diperoxyketals (*gem*-diperoxides) are prepared from a ketone and a *tert*-alkyl hydroperoxide in the presence of an acid. Di-*tert*-alkyl peroxides are formed from a *tert*-alkyl hydroperoxide and a sulfated *tert*-alkyl alcohol or suitably branched olefin. Dialkyl peroxydicarbonates are manufactured by treating alkyl chloroformates with sodium peroxide. *tert*-Alkyl hydroperoxides are prepared by the reaction of a sulfated *tert*-alkyl alcohol or olefin with hydrogen peroxide or by autoxidation of a branched hydrocarbon. Ketone peroxides are produced by treating a ketone with hydrogen peroxide under acidic catalysis.

Economic Aspects

The 1985 U.S. consumption of organic peroxide initiators was 18,000 t. The principal U.S. producers of organic peroxide initiators include Lucidol Division, Pennwalt Corp. (all types); Akzo Chemie America-Noury Chemicals (most types); U.S. Peroxygen Division, Witco Chemical Corp. (most types); Catalysts Resources, Inc., Phillips Petroleum Co. (several types); Hercules, Inc. (dialkyl peroxides, hydroperoxides); PPG Industries, Inc. (peroxycarbonates, sulfonyl peroxide); Norac Co. (BPO, hydroperoxides, ketone peroxides); Reichhold Chemicals, Inc. (BPO, ketone peroxides, hydroperoxides); Freeman Resins, Catalyst Division (ketone peroxides); ARCO Chemical Co. and Aristech Chemical Corp. (hydroperoxydes).

Inorganic peroxides and peroxyacids are produced by FMC Corp. (hydrogen peroxide, peracetic acid, peroxydisulfates), E. I. du Pont de Nemours & Co., Inc., Ashland Chemical Co., and Interox America (hydrogen peroxide, peroxydisulfates).

Health and Safety Factors

Although many commercial organic peroxides are characterized by low acute toxicity, most are irritants to the skin and eyes; the lower molecular weight peroxides containing the hydroperoxide function (—OOH) are especially irritating. The U.S. Department of Health and Human Services lists nine commercial organic peroxides on its suspected carcinogenic list and several are undergoing testing in the National Toxicology Program.

Polymeric Peroxides

Polymeric peroxides can be classified into polyperoxides where the peroxy group is a repeating unit in the main polymer or oligomer chain, and peroxy-containing polymers or oligomers where the peroxy group is chemically bound to a polymer or oligomer as a nonrepeating unit, such as an end group, a connecting group, or a random side chain.

Polyperoxides form in certain solvents and monomers under poor storage and handling conditions. In the presence of oxygen, vinyl monomers (eg, styrene, α-methylstyrene, butadiene, acrylate and methacrylate esters, vinyl chloride, acrylonitrile, methacrylonitrile, vinyl acetate) can form polyperoxides, $-(OOCH_2CHX)_n-$, by copolymerization with oxygen.

Pure polyperoxides are viscous liquids or amorphous solids, usually containing as many as 10 repeating units. They undergo potentially dangerous explosive decomposition if heated pure or in high concentration.

Polyperoxides are also prepared by polycondensation techniques utilizing difunctional peroxides such as hydrogen peroxide, 2,5-dihydroperoxy-2,5-dimethylhexane, and peroxy-containing diacid chlorides with other suitable difunctional comonomers. Peroxy-containing polymers have been prepared by a variety of methods such as autoxidation, peroxy-acylation, and grafting of polymers and introduction of peroxy-functions during polymerizations. Many such polymeric peroxides have been used to prepare block and graft copolymers.

BIBLIOGRAPHY

C. S. Sheppard and O. L. Mageli in M. Grayson, ed., *Kirk-Othmer Encyclopedia of Chemical Technology,* 3rd ed., Vol. 17, Wiley-Interscience, New York, 1982, pp. 27–90.

S. Patai, ed., *The Chemistry of Peroxides,* John Wiley & Sons, Inc., New York, 1984.

D. Swern, ed., *Organic Peroxides,* Vols. I, II, and III, Wiley-Interscience, New York, 1970, 1971, and 1972.

CHESTER S. SHEPPARD
Pennwalt Corporation

PESTICIDES. See BIOCIDES.

PETROLEUM ADDITIVES

Most petroleum-derived fuels and lubricants are formulated with additives. The largest use of polymeric additives is in the formulation of lubricant additives. The additives used in fuels are in most cases the same as or derived from lubricant additives. Lubricant additives are used to improve viscosity characteristics and performance. The total world market for lubricant additives is $\$2.5 \times 10^9$–$2.7 \times 10^9$ annually. Growth in sales was less than 1% per year in North America from 1975 to 1985 and about 2% per year in other parts of the world.

Viscosity Improvers

The most prominent additive component of a multigraded mineral oil is a high molecular weight, polymeric viscosity improver. The principal function of the viscosity improver is to minimize viscosity variations over a wide temperature range.

Composition. Viscosity improvers are oil-soluble polymers with molecular weights ranging from 2000 to 10^6, depending on application (Table 1).

Preparation. Polyisobutylene is prepared by polymerizing isobutylene at low temperatures (ca − 70°C) using a Friedel-Crafts-type catalyst in hydrocarbon solvent. The alkyl methacrylate copolymers are prepared by the free-radical-initiated polymerization of mixed alkyl methacrylates. Ethylene–propylene copolymers are prepared by copolymerization of ethylene and propylene in a hydrocarbon solvent with a vanadium-based Ziegler-Natta catalyst. Hydrogenated diene–styrene copolymers are prepared by anionic copolymerization of styrene with isoprene or 1,3-butadiene in a hydrocarbon solvent with butyllithium as initiator. Maleic anhydride–styrene copolymers are prepared by the free-radical copolymerization of styrene and maleic anhydride.

Mode of Action. The performance of viscosity improvers depends on the behavior of the polymer in oil solution. As the polymer dissolves in oil, the viscosity of the fluid increases, depending primarily on the volume occupied by the polymer molecules and associated solvent molecules.

Uses. The principal use of viscosity improvers is in the formulation of multigrade crankcase oils.

Table 1. Commercial Viscosity Improvers

Compound
ethylene-propylene copolymers
hydrogenated diene-styrene copolymers
poly(alkyl methacrylate) copolymers
maleic anhydride-styrene copolymers
polyisobutylene
poly(vinyl acetate-alkyl fumarate) copolymers

Dispersants

Dispersants are added to lubricating oils to suspend contaminants and oil-insoluble material and to prevent deposit formation. Excessive deposits inhibit mechanical functioning or lubricant flow. In fuels, dispersants at low concentrations prevent deposits from forming; at high concentrations they remove formed deposits. They also are used to stabilize emulsions. The largest application is in crankcase engine lubricants where three types are used. Dispersants include succinimides, succinic esters, and Mannich dispersants.

Pour-point Depressants

All mineral oils contain paraffin wax and eventually solidify on chilling. This point of nonfluidity, the pour point, is defined as the lowest temperature at which the oil pours or flows when chilled under prescribed conditions (ASTM D97-66). The pour point is lowered by addition of a pour-point depressant to the formulation. Pour-point depressants are used primarily in crankcase lubricants for winter service. A common characteristic of commercial pour-point depressants (Table 2) is the presence of side chains containing 10–22 carbon atoms; the length of the side chain depends on the application.

Table 2. Polymeric Pour-point Depressants

Compounds
alkyl methacrylate copolymers
α-olefin copolymers
vinyl ester copolymers
alkylated polystyrene
maleic anhydride-styrene copolymers
alkylated naphthalenes

Mode of Action. Pour-point depressants (ASTM D2500-66) function by enhancing the nucleation and directing or retarding the growth of wax crystals with decreasing temperature.

Antifoaming Agents

Most lubricating oils foam to some extent as a result of agitation and aeration in operating engines, transmissions, and other lubricant applications. In most lubricant applications, the antifoaming agent (qv) is polydimethylsiloxane.

BIBLIOGRAPHY

T. R. Erdman, *SAE Soc. Automot. Eng. Tech. Pap. 810015,* 1981.

R. C. Tupa and C. J. Dorer, *SAE Soc. Automat. Eng. Tech. Pap. 841211 (SP-603),* 1984.

R. J. Kopko, "Wax Crystal Modifiers for Solvent Dewaxing Processes, FL 83-82," *NPRA Fuels and Lubricants Symposium Meeting,* Houston, Tex., 1983, National Petroleum Refiners Association, Washington, D.C.

E. R. Booser, *Handbook of Lubrication,* Vol. I, CRC Press, Inc., Boca Raton, Fla., 1983, pp. 3–44.

STANLEY K. BACZEK
WILLIAM B. CHAMBERLIN
The Lubrizol Corporation

PHASE TRANSFORMATIONS

Phase changes, such as melting of a crystal or precipitation of a polymer solution, have attracted a great deal of theoretical and experimental attention. In thermodynamic equilibrium a system may be composed of one or several physically distinct macroscopic homogeneous parts, called phases, which are separated from one another by well-defined interfaces. These phases are determined by several thermodynamic parameters, such as temperature, pressure, and electric and magnetic fields. By continuously varying the parameters it is possible to induce the transformation of the system from one phase to another.

From a thermodynamic point of view two types of phase transitions may be distinguished. The transition in which the two phases coexist in equilibrium at the transformation point are called first-order phase transitions. Melting of crystals, isotropic-nematic phase transitions, or microphase-separation transitions in block copolymers are good illustrations. First-order phase transitions are accompanied by a sharp discontinuity in entropy and a latent heat of transformation. More generally, one or more first derivatives of the relevant thermodynamic potential, such as the magnetization in a ferromagnet, the density in a liquid–gas system, or the concentration in a binary mixture, change discontinuously at the transition. In polymer systems this simple thermodynamic picture may often be complicated by very slow kinetics of the phase-separation process, structural imperfections, or polydispersity effects.

Much more universal is another kind of phase transformation, called continuous or second-order transition, for which the first derivatives of thermodynamic potentials (entropy, density, magnetization, electric polarization) remain continuous, but the higher-order derivatives (specific heat, compressibility, magnetic or electric polarization) diverge or change discontinuously at the transition point. Typical examples of continuous transitions are the paramagnetic-ferromagnetic transition at the Curie point, the paramagnetic-antiferromagnetic transition at the Neel point, or the superfluidity transition at the γ-line. The point at which such continuous transition occurs is called a critical point.

Other examples of continuous-phase transitions in polymer systems include volume-change transition in ionic gels (see also GELS) or recently demonstrated ferroelectric-paraelectric transition in copolymers of vinylidene fluoride and trifluoroethylene. Among the continuous-phase transitions occuring in polymeric material, the phase-separation transition in solutions and mixtures has attracted the most theoretical and experimental attention (see also MICROPHASE STRUCTURE).

Polymer–Solvent Phase Diagrams

Polymer solutions exhibit a great variety of phase-separation transitions. The most familiar case is that in which the system is miscible in all proportions at high temperatures and separates into two coexisting homogeneous phases (polymer-rich and solvent-rich) as the temperature T is lowered through the transition point. If the solution is prepared at a critical concentration $c = c_k$, then, as T decreases, a continuous transition from a single-phase region into the two-phase domain occurs at the upper critical solution temperature (UCST) $T = T_c$. Such situations may be found for most θ-solvents; a classical example is the solution of polystyrene in cyclohexane.

Many alkene polymers, when dissolved in aliphatic or aromatic solvents, exhibit a lower critical solution temperature (LCST), they separate when the temperature is increased. Some of these nonpolar solutions, eg, the polyisobutylene–benzene system, show both UCST and LCST. The latter occurs at elevated temperatures between the normal boiling point and the liquid–vapor critical point of the pure solvent. The higher the molecular weight, the lower the LCST. The existence of LCST for low molecular weight nonpolar mixtures is highly unusual. In polymer solutions at high temperatures, the pure solvent is compressible compared to the polymer; the system may exhibit a negative volume of mixing. In such situations the total system entropy may increase when phase separation occurs. In polymer solutions phases separate because the combinatorial entropy of mixing is very small.

Several statistical models, such as equation-of-state theories or lattice fluid theory, have been developed to describe phase diagrams of nonpolar polymer solutions. These theories take into account the finite compressibility of the solutions and explain the existence of both (lower critical-solution temperature) LCST and UCST. They predict the values of critical temperatures and the heat and volume of mixing.

Critical Properties of Binary Solutions

The essential features of the phase separation in polymer solutions have been described by mean-field theories such as the Flory-Huggins lattice model. These models predict the molecular weight dependence of the critical concentration and temperature as well as a detailed behavior of thermodynamic quantities near the critical point. However, the mean-field theories ignore or take into account only in a very approximate manner spatial-concentration fluctuations. These fluctuations may play a crucial role in the vicinity of the critical point. Consequently, beyond the classical mean-field approximation a more adequate description based on modern renormalization group and scaling ideas is required. This theoretical effort is promoted by careful experimental studies of critical phenomena in θ-solvents.

Critical Phenomena in Binary Polymer Mixtures

Segregation effects in polymer blends (qv) are extremely important. Most polymer pairs are completely immiscible in the experimentally accessible temperature range. At the origin of this incompatibility of polymer chains is a strong repulsion between unlike molecules, even when the repulsion between unlike monomers is relatively weak. The consolute points can be observed under three conditions:

1. In low molecular weight polymers the entropy of mixing can be sufficiently strong to induce mixing at high temperatures. At lower temperatures the van der Waals attractions between similar monomers dominate and the phases separate. The upper critical solution temperature may be thus observed. Examples of such systems include mixtures of polystyrene and polybutadiene chains of molecular weight of a few thousand.
2. When specific interactions between monomers are present, the polymers may have a favorable heat of mixing at room temperature. Separation can often be induced on heating, and the LCST may be observed. Phase separation on heating may be attributed to either unfavor-

able volume-of-mixing changes caused by different thermal expansions of component polymers or to highly orientational interactions (weak hydrogen bonds) weakening at high temperatures. A typical example of a blend showing the LCST behavior is polystyrene–poly(vinyl methyl ether).

3. When the heat of mixing is very low, a UCST may be observed even for high molecular weight polymer mixtures. This situation can occur when two statistical copolymer chains of very similar composition are mixed. A UCST is also observed for mixtures of normal (protonated) and perdeuterated 1,4-polybutadienes of high molecular weight.

The phase transition in polymer blends is expected to exhibit unique critical properties very different from those observed in normal binary mixtures or even in polymer solutions. In fact, it may be argued that critical behavior in polymer mixtures should be correctly described by classical mean-field exponents rather than Ising exponents, except very near the critical point.

Phase transition in polymer solutions and mixtures demonstrates clearly that the critical behavior of polymer systems may be well described employing the concepts developed for phase transitions in low molecular weight systems. However, polymer systems show some unique features.

BIBLIOGRAPHY

S. K. Ma, *Modern Theory of Critical Phenomena,* Benjamin/Cummings, New York, 1976.

A. Z. Patashinskii and V. L. Pokrovskii, *Fluctuation Theory of Phase Transitions,* Pergamon Press, Oxford, UK, 1979.

C. Comb and M. S. Green, eds., *Phase Transitions and Critical Phenomena,* Vols. 1–8, Academic Press, Inc., New York, 1972–1984.

L. Leibler
Université Pierre et Marie Curie

PHENOL–FORMALDEHYDE RESINS.

See Phenolic resins.

PHENOLIC RESINS

Phenolic resins encompass a wide variety of polymeric substances; each polymer is composed of a multitude of structures, and a variety of raw materials and catalysts are used. Phenolic resins are employed in a wide range of applications from commodity construction materials to high technology applications in electronics and aerospace. Generally, but certainly not exclusively, thermosetting in nature, phenolic resins provide numerous challenges in the areas of synthesis, characterization, production, product development, and quality control.

The early phenolic resins consisted of self-curing, resole-type products made with excess formaldehyde and novolacs, which are thermoplastic in nature and require a hardener. The early products produced by General Bakelite were used in molded parts, insulating varnishes, laminated sheets, and industrial coatings. These areas remain important applications, but have been joined by numerous others, just as the number of producers has increased to more than 25 in the United States and more than 60 worldwide.

Table 1. Properties of Phenol

Property	Value
mol wt	94.1
mp, °C	40.9
bp, °C	181.8
flash point, °C	79.0
autoignition temperature, °C	605.0
explosive limits	2–10 vol %
vapor pressure at 20°C, Pa[a]	20

[a] To convert Pa to mm Hg, multiply by 7.5.

Monomers

Phenol. Phenol (C_6H_5OH) is the monomer or raw material used in the largest quantity to make phenolic resins. As a solid with a low melting point (Table 1), phenol is usually stored and handled in liquid form at 50–60°C.

The most widely used process for the production of phenol is the cumene process developed and licensed in the United States by Allied Chemical Corp. (now Allied-Signal). Other commercial processes for making phenol include the Raschig process, which uses chlorobenzene as the starting material, and the toluene process via a benzoic acid intermediate. In the United States, 35–40% of the phenol produced is used for phenolic resins.

Substituted Phenols. Although substituted phenols are used for specialty resins, phenol itself is used in the largest volume. The substituted phenols are typically alkylated phenols made from phenol and the corresponding α-olefin with acid catalysis. The incorporation of alkyl phenols into the resin reduces reactivity, hardness, cross-link density, and color formation, and increases solubility in nonpolar solvents, flexibility, and compatibility with natural oils.

Formaldehyde. Formaldehyde, principally in aqueous form, is used almost exclusively in the production of phenolic resins, regardless of the type of phenol.

Other Aldehydes. The higher aldehydes react with phenol in much the same manner as formaldehyde, although at much lower rates.

Hexamethylenetetramine (hexa). Hexa, a complex molecule with an adamantane-type structure, is prepared from formaldehyde and ammonia and can be considered a latent source of formaldehyde.

Other Reactants. Other reactants are used in smaller amounts to provide phenolic resins with specific properties, especially coatings applications. Aniline can be incorporated into both resoles and novolacs. Other materials include rosin (abietic acid), dicyclopentadiene, unsaturated oils, such as tung oil and linseed oil, and polyvalent cations for cross-linking.

Polymerization

Phenolic resins are prepared with strong acid (Table 2) or alkaline catalysts (Table 3). Occasionally, weak or Lewis acids, such as zinc acetate, are used for specialty resins.

Manufacture

The final state of a phenolic resin varies dramatically from thermoplastic to thermoset and from solid to liquid, including

Table 2. Novolac Resin Properties

Property	Catalyst	
	Acid	Zn acetate[a]
formaldehyde–phenol molar ratio	0.75	0.60
nmr analysis		
2,2′, %	6	45
2,4′, %	73	45
4,4′, %	21	10
gpc analysis		
phenol, %	4	7
M_n	900	550
M_w	7300	1800
water, %	1.1	1.9
T_g, °C	65	48
gel time, s	75	25

[a] High ortho.

Table 3. Properties of Resole Resins

Property	Catalyst	
	NaOH	Hexa
concentration, pph	3	10
formaldehyde–phenol ratio	2.0	1.5
water solubility, %	100	swells
gpc analysis		
phenol, %	6	8
M_n	280	900
M_w	500	3000
T_g, °C	35	47
gel time, s	65	110

solutions and dispersions. The critical equipment parameters, assuming material compatibility, are reactor size, cooling capacity, and agitation. With a bulk process, resole resins, in neat or concentrated form, must be produced in small batches (ca 2–9.5 m^3) in order to maintain control of the reaction and obtain a uniform product. On the other hand, if the product contains a large amount of water, such as liquid plywood-adhesive resins, very large reactors (19 m^3) can be used. Melt-stable products, like novolacs, can be prepared in large batches (19–38 m^3), provided the exotherms can be controlled.

Batch processes for most phenolic resins employ the equipment similar to that shown in Figure 1. Liquid reactants are metered into the stirred reaction vessel through weigh tanks, whereas solid reactants such as bisphenol A and $Ba(OH)_2$ present handling problems. Facilities are provided to carry out the reaction under a vacuum or an inert gas.

Materials of Construction. Compatibility of the materials of construction and the process chemicals is extremely important. The reactors are usually made of stainless steel alloys. Copper is avoided because of the possible presence of amines.

Novolac Resins. In a conventional novolac process molten phenol is placed into the reactor, followed by the precise amount of acid catalyst. The formaldehyde solution is added near 90°C at a formaldehyde-to-phenol molar ratio of 0.75–0.85. For safety reasons slow continuous or stepwise addition of formaldehyde is preferred over adding the entire charge at once. Reaction enthalpy has been reported to be above 80 kJ/mol (19 kcal/mol). The heat of reaction is removed by refluxing the water combined with the formaldehyde or by using a small amount of a volatile solvent such as toluene. Toluene and xylene are used for azeotropic distillation. Following decantation, the toluene or xylene is returned to the reactor.

High Ortho Novolacs. The process for high ortho novolacs is similar to that used for those catalyzed by strong acid; however, the catalyst is frequently zinc acetate.

Resoles. Like the novolac processes, a typical resole process consists of reaction, dehydration, and finishing.

Phenolic Dispersions. These systems are resin-in-water and

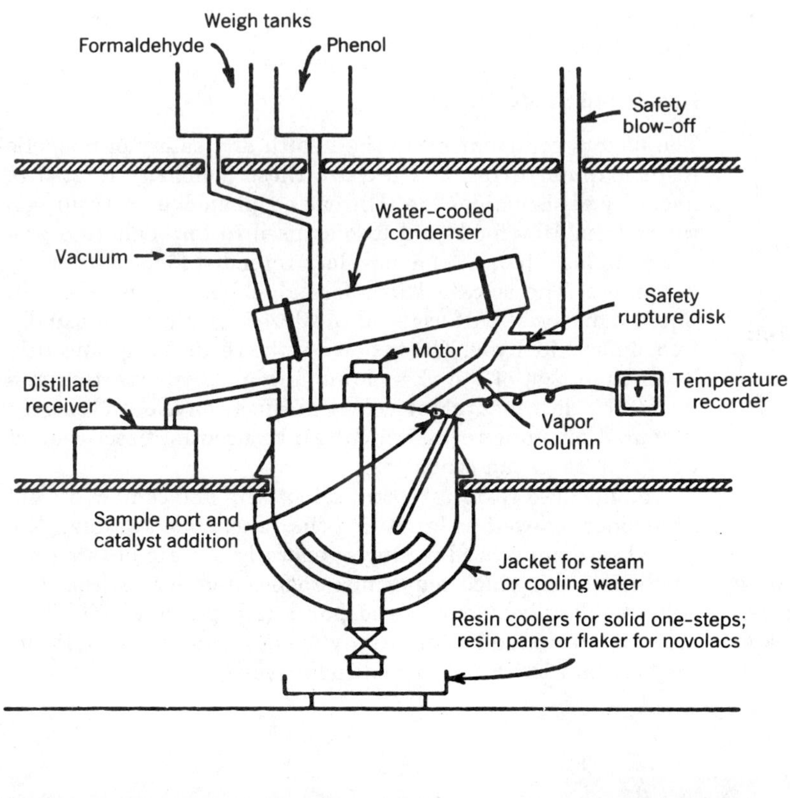

Figure 1. Typical phenolic resin production unit.

resin-in-nonsolvent mixtures, in which the resin exists as discrete particles. Particle size ranges from 0.1 to 2 μm for stable dispersions and up to 100 μm for dispersions requiring constant agitation.

In the postdispersion process the solid phenolic resin is added to a mixture of water, cosolvent, and dispersant at very high shear mixing, possibly with heating. The cosolvent, frequently an alcohol or glycol ether, heat softens the resin, permitting small particles to form. On cooling, the resin particles stabilized by dispersant and perhaps thickener, harden and resist settling and agglomeration. Both resole and novolac resins have been made by this process.

The *in situ* process is simpler with less material handling. The process is analogous to the free-radical catalyzed emulsion polymerization of unsaturated monomers, but the particles formed are rarely submicrometer in size. This process has been used only for resole resins.

Spray-dried Resins. Spray drying produces resins in particulate form. Spray drying a resole solution containing a blowing agent produces phenolic microballoons. Spray drying also produces uncross-linked resins. The principal application for this type of product is believed to be wood binding, especially for waferboard applications.

Cure

A typical resin has an initial molecular weight of 150 to perhaps 1500. For systems of unsubstituted phenols, the final cross-link density is 150–300 atomic mass units (amu) per cross-link. In other words, 25–75% of the ring-joining reactions occur during the cure phase.

Resoles. The advancement and cure of resole resins follow reaction steps similar to those described previously for resin preparation; the pH is 9 or higher and reaction temperature does not exceed 180°C. In particular, methylol groups condense with other methylols to give dibenzyl ethers and react at the ortho and para position on the phenol to give diphenylmethylenes. In addition, dibenzyl ethers eliminate formaldehyde to give diphenylmethanes.

Novolacs. Novolac resins are typically cured with 5–15% hexamethylenetetramine as the cross-linking agent. The reaction mechanism and reactive intermediates have been studied extensively. It is not known how the hexamethylenetetramine is broken down or how stable the benzylamine intermediates are. Nevertheless, the early studies showed that as much as 75% of nitrogen is chemically bound. More recent studies of resin cure have made use of tga, dta, gc, and ir. They confirm that the cure is at least two-step in nature.

Decomposition. Above 250°C cured phenolic resins decompose; for example, the dibenzyl ethers disproportionate.

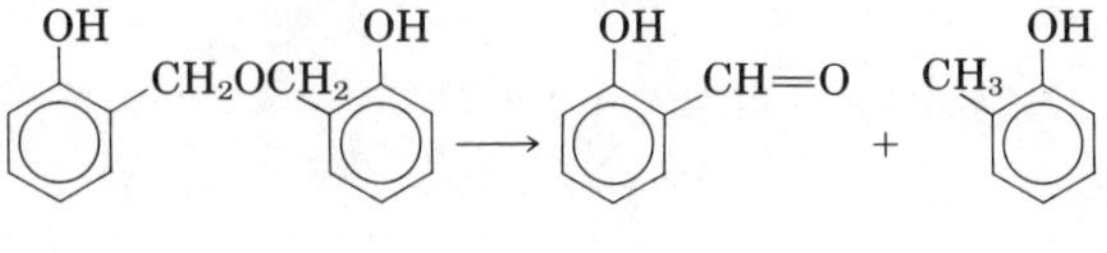

The aldehyde group is rapidly oxidized to the corresponding carboxylic acid. In an analogous reaction in hexa-cured novolacs, tribenzylamines are decomposed into cresols and azomethines, which cause yellowing.

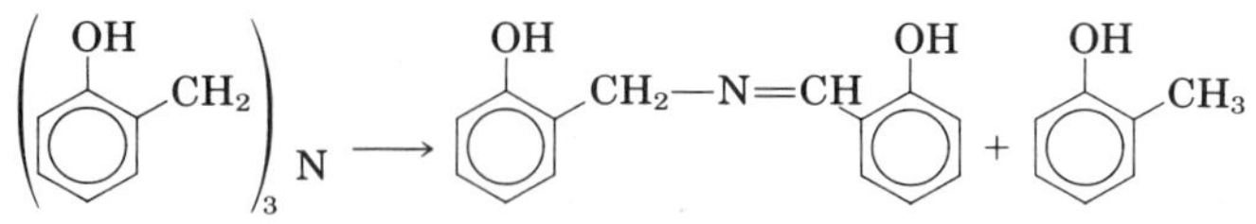

Substantial decomposition of phenolic resins begins above 300°C. In the presence of oxygen the methylene bridging group is converted to a hydroperoxide, which yields alcohols and a ketone on decomposition.

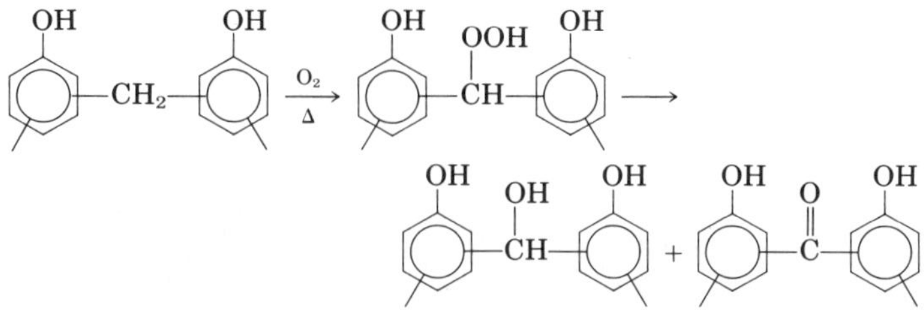

The ketone is especially susceptible to random chain scission. Decomposition continues up to ca 600°C; the by-products are mostly water, CO,CO_2, and phenols.

Analysis and Characterization

The principal techniques for determining the microstructure of phenolic resins include mass spectroscopy, proton and ^{13}C nmr spectroscopy, as well as gc, lc, and gpc. The softening and curing processes of phenolic resins are effectively studied with thermal and mechanical technical techniques, such as tga, dsc, and dynamic mechanical analysis (dma). Ir and electron spectroscopy are also employed.

Control Tests. Numerous chemical and physical tests are used in the manufacture of phenolic resins to assure correct properties of the finished resins. Additional quality-control tests are made on the final product and to monitor the resins storage. Properties tested include the refractive index, viscosity, the solid content, miscibility, specific gravity, the melting point, the gel-time, resin flow, particle size, flash point, and autoignition temperature.

Health and Safety Factors

The factors contributing to the health and safety of phenolic resin manufacturing and use are those primarily related to phenol and formaldehyde. Unreacted phenol in a resin can range from 10% for liquid resoles used in impregnation processes to less than 1% for novolacs intended for use as epoxy hardeners. The so-called free formaldehyde can be 2–4% in liquid adhesives. Sulfuric acid-catalyzed novolacs are usually free of formaldehyde. The toxicity of the resins is significantly lower than that of the phenol and formaldehyde starting materials. No detrimental toxicological effects have been reported for cured phenolic resin, which can be used in direct contact with food as in can coatings.

The uncured resins are skin sensitizers and contact should be avoided, as well as breathing the vapor, mist, or dust. Novolac-based pulverized products generally contain hexamethylenetetramine, which may cause rashes and dermatitis. Phenolic molding compounds and pulverized phenolic adhesives must be controlled as potentially explosive dusts. In addition, they contain irritating or toxic additives.

Table 4. U.S. Phenolic Resin Consumption by Use, 10^3 t

Use	1980	1981	1982	1983	1984	1985	1986
molding compounds	112	113	87	105	108	95	89
bonding and adhesive resins							
laminating	44	61	58	75	68	65	76
coated and bonded abrasives	15	17	11	13	10	7	8
friction materials	16	13	11	15	16	12	12
insulation materials	123	164	158	163	140	191	191
foundry and shell molding	36	37	25	32	37	35	23
plywood	202	504	454	595	600	570	578
fibrous and granulated wood	40	91	66	81	84	99	107
protective coatings	11	12	8	9	10	10	10
other applications	79	57	44	53	59	107	114
Total	*678*	*1069*	*927*	*1141*	*1130*	*1191*	*1210*

Wastewater. Phenol is a toxic pollutant to the waterways with an acute toxicity to fish of about 5 mg/L. Chlorination of water gives chlorophenols, which impart objectionable odor and taste at 0.01 mg/L. Biochemical degradation is most frequently used to treat waste water containing phenol.

Economic Aspects

In 1986 worldwide consumption of phenolic resins exceeded 2.7×10^6 t; slightly less than half of the total volume was produced in the United States. The largest volume application is in plywood adhesives. As shown in Table 4, this area accounts for ca 47% of U.S. consumption.

As a mature industry U.S. production and application of phenolic resins has paralleled the growth in the GNP. The compounded annual growth from 1975 to 1985 was approximately 9%. A list of the larger producers of phenolic resins in the United States follows.

U.S. phenolic resin manufacturers

Acme Resins Co.	Pioneer Plastics Corp.
American Hoechst Corp.	Plaslok Corp.
Ametek, Haveg Div.	Plastics Engineering Co.
Ashland Chemical Co.	Polymer Applications, Inc.
Bendix Corp.	Polyrez Co., Inc.
Borden, Inc.	Raymark Industries, Inc.
Chembond Corp.	Reichhold Chemicals, Inc.
Formica Corp.	Rogers Corp.
Georgia-Pacific Corp.	Union Carbide Corp.
Inmont Corp.	Uniroyal Chemical Div., Uniroyal, Inc.
Ironsides Resins, Inc.	Valite Div., Valentine Sugars, Inc.
Koppers Co.	Van Leer Containers, Inc.
Owens-Corning Fiberglas Corp.	
Perkins Industries, Inc.	

Prices vary. In 1986 the average commodity phenolic resins was priced at \$0.66/kg. Liquid-laminating grades were ca \$0.44/kg and thermal-insulation liquid grades were 0.66/kg. Since the raw materials of phenolic resins are derived from crude oil and natural gas, the prices of phenolic resins depend on the price of these resources.

Applications

Applications for phenolic resins include coatings, dispersions, adhesives, carbonless copy paper, molding compounds, abrasives, friction materials, foundry resins, laminates, air and oil filters, battery separators, wood bonding, fiber bonding, composites, reaction injection molding, foam, hollow spheres, and fibers.

BIBLIOGRAPHY

R. W. Martin, *The Chemistry of Phenolic Resins,* John Wiley & Sons, Inc., New York, 1956.

N. J. L. Megson, *Phenolic Resin Chemistry,* Butterworth & Co. (Publishers) Ltd., Kent, UK, 1958.

J. F. Walker, "Formaldehyde," *ACS Monogr.* **159,** Van Nostrand Reinhold Co., Inc., New York, 1964.

A. Knop and L. Pilato, *Phenolic Resins,* Springer-Verlag, Berlin, 1985.

PETER W. KOPF
Arthur D. Little, Inc.

PHENOPLASTS.

See PHENOLIC RESINS.

PHENOXY RESINS.

See EPOXY RESINS.

PHENYL ETHER–FORMALDEHYDE RESINS.

See PHENOLIC RESINS.

PHENYLENE OXIDE POLYMERS.

See POLY(PHENYLENE ETHER)S.

PHENYL SULFIDE POLYMERS.

See POLY(ARYLENE SULFIDE)S.

PHOSPHATES.

See PHOSPHORUS-CONTAINING POLYMERS.

PHOSPHONITRILIC CHLORIDE POLYMERS.

See PHOSPHORUS-CONTAINING POLYMERS; POLYPHOSPHAZENES.

PHOSPHORUS-CONTAINING POLYMERS

The only valid generalization about phosphorus polymers is that they tend to be flame retardant. The flame-retardant effect depends heavily on the phosphorus content, but structural effects are also significant.

Other interesting features of phosphorus polymers are adhesion to metals, metal ion-binding characteristics, and increased polarity. Plasticizing properties are contributed by the P—O—C linkage. Phosphorus polymers with acid groups are useful for ion exchange, adhesion, and scale inhibition. Some form water-soluble coatings used as primers in metal protection and for photolithographic plates. Dental applications have been reported and medical uses (control of bone growth) are under study. Phosphoproteins are of interest as food ingredients. Academic interest has been stimulated by the many types of bonding characteristics of the phosphorus atom and by the relationship of certain phosphorus polymers to natural products, such as the nucleic acids (see POLYNUCLEOTIDES).

Polymers based principally on phosphorus-containing components are relatively expensive and have therefore not attained much commercial significance. On the other hand, the modification of conventional polymers with small amounts of phosphorus comonomers or oligomers to impart flame retardancy or improve other properties has become commercially significant.

Polymeric Forms of Elemental Phosphorus

Phosphorus occurs in various allotropic and polymeric forms. Heating white phosphorus (P_4) at 270–400 °C, preferably with a catalyst, produces red phosphorus, a stable, nontoxic, high melting, insoluble polymeric solid. Red phosphorus (often stabilized by additives) has been used for many decades as the igniting agent for the striking surface of safety matches.

The stability and low toxicity of red phosphorus, in contrast to the white allotrope, which is highly flammable and toxic, make it a useful flame-retardant additive. It is used commercially, particularly in Europe, in molded nylon electrical parts. Other applications for red phosphorus are in epoxy and phenolic circuit boards. To prevent decomposition to the toxic and highly flammable white form, red phosphorus is stabilized with additives and/or encapsulated with a thermoset resin.

Inorganic Phosphorus Polymers

Inorganic polyphosphates are covalently linked polymers. Long chains of repeating phosphate units are formed by eliminating the elements of water from adjacent orthophosphate units. Polyphosphoric acid itself has found utility mainly as a supported catalyst in the petroleum industry for alkylation, olefin hydration, polymerization, and isomerization, and for syntheses of fine chemicals and dyes.

Heating alkali metal or alkaline earth metal dihydrogen phosphates produces polymeric salts (cyclic metaphosphates and linear polyphosphates) and cross-linked polyphosphates (ultraphosphates), depending on temperature and the presence of other ingredients. This complex group of polymers includes materials with glasslike as well as thermoplastic and thermoset characteristics; some are useful as binders for metals and ceramics. Calcium sodium metaphosphate fiber was shown to be useful as an asbestos substitute.

Heating of ammonium phosphates under an atmosphere of ammonia or in the presence of urea produces ammonium polyphosphate. The insoluble form of ammonium polyphosphate is used as a flame-retardant additive for plastics and as the latent acid component in intumescent paints and mastics.

Condensed phosphoramides with linear, cyclic, or cross-linked structures are produced by the reaction of $POCl_3$ with ammonia; on further heating $(PON)_x$ is formed. An amorphous substance usually written as P_3N_5, but probably a polymer as shown by various spectroscopic methods, is an excellent electrical insulator and is used as a dielectric layer in solid-state electronic circuitry; it is usually made by vapor-phase deposition from ammonia and phosphine.

Organic Phosphorus Polymers

Phosphines. Polymeric phosphines exhibit strong metal-binding properties. Nonpolymeric phosphines, in particular triphenylphosphine, are employed as ligands for cobalt and rhodium in hydroformylation catalysts used in plasticizer manufacture; analogous polymeric phosphines have been made and shown to be catalytic, but have not been commercialized.

Phosphine Oxides. The outstanding property of polymeric phosphine oxides is their stability. Polymeric phosphine oxides are made by a variety of condensation- and addition-polymerization methods. An important commercial phosphine oxide polymer is a cross-linked finish formed on cotton by treatment with tetrakis(hydroxymethyl)phosphonium salts followed by treatment with ammonia and hydrogen peroxide.

Phosphinites and Phosphonites. In common with phosphines these classes of phosphorus structures have unshared electron pairs on the phosphorus. They tend to be unstable to oxidation and are good metal binders. Consequently, they have been prepared as catalyst ligands and ion-extraction reagents. An oligomeric phosphonite prepared from phenylphosphonous dichloride has been claimed as an antioxidant synergist.

Phosphites. Nonpolymeric phosphites are employed as the peroxide-decomposing components of the antioxidant systems of polyolefins and diene rubbers. Polymeric analogues have lower volatility; an example is shown below.

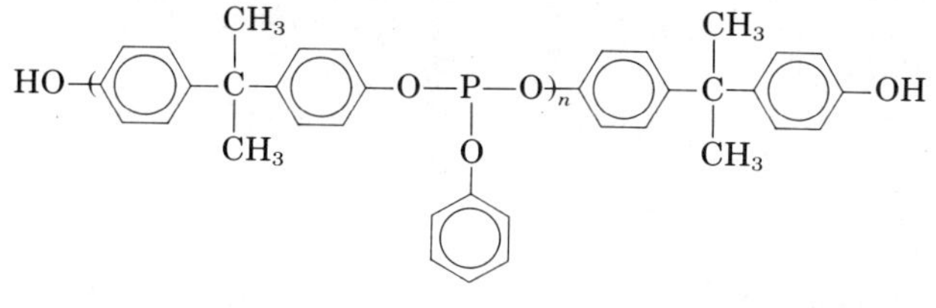

Polyphosphites have been prepared by transesterification of trialkyl phosphite with diols such as dipropylene glycol.

Phosphinates. Phosphinate structures confer flame retardancy to polyester fibers. Polyphosphinates are made by the reaction of an alkylene dihalide with a diphosphonite or by the self-condensation of a haloalkylphosphonite:

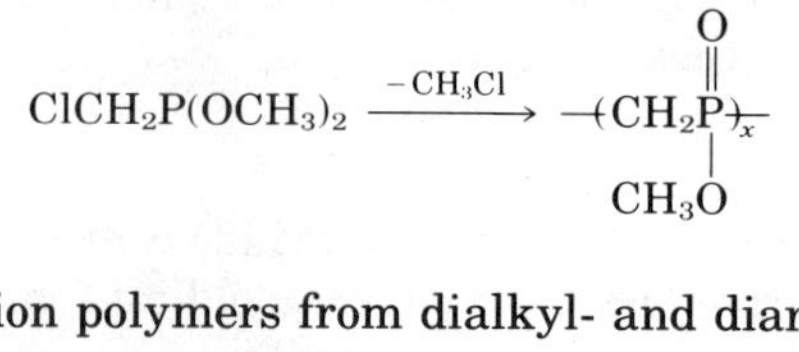

Coordination polymers from dialkyl- and diarylphosphinic

acid and various metals have been investigated because of high temperature stability.

Phosphonates. The principal synthetic routes include addition and condensation methods. Although a large number of vinyl and diene phosphonate monomers have been described in the literature, only bis(2-chloroethyl) vinylphosphonate (Akzo's Fyrol Bis-Beta monomer) is available in commercial quantities. In the 1970s, in response to children's sleepwear regulations promulgated under the Federal Flammable Fabrics Act, a number of water-soluble phosphonate monomers were designed, principally for cotton flame-retardant finishing, such as the following example:

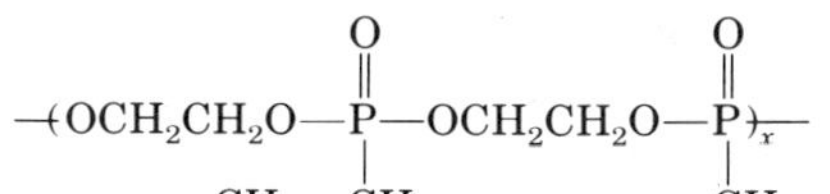

Polyphosphonates have been made from phosphonyl dichlorides or by transesterification from diphenyl phosphonates and a wide variety of diols. Those made from bisphenols have high thermal stability. The following was used commercially as a flame-retardant additive for polyester fiber:

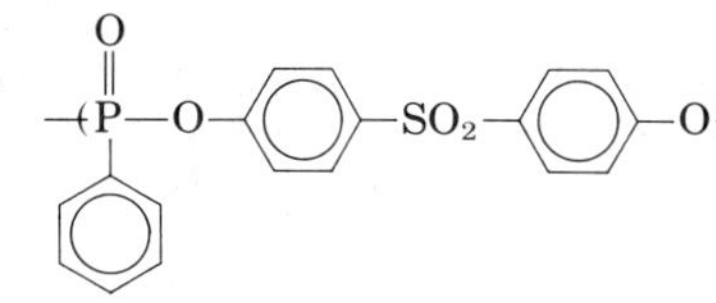

Phosphates. Polymeric phosphates are used for the preparation of flame-retardant polymers, ion-exchange resins, and models for natural biopolymers. Oligomeric phosphates have been used commercially as flame retardants in urethane foams and thermoset resins, such as the following:

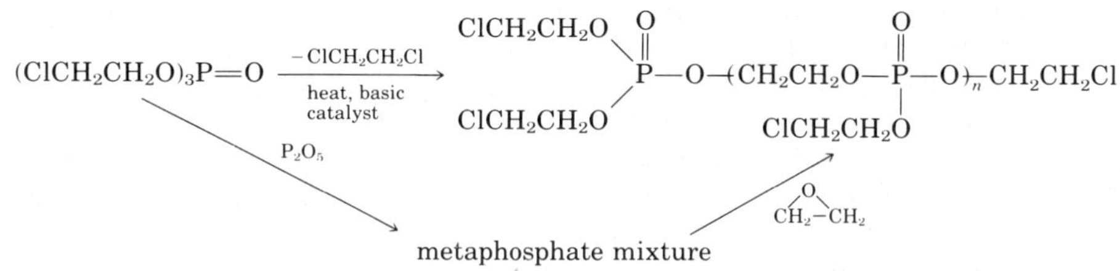

Phosphorus–Nitrogen Polymers. The most important of such polymers are the polyphosphazenes (qv). Research on polymeric phosphonamides, phosphoramides, and phosphorimides has led to no commercial products as yet. Many of these polymers are hydrolytically unstable.

Phosphorus-modified Plastics, Resins, and Textiles

Usually only small quantities of phosphorus are introduced into a polymer to achieve desired properties such as flame retardancy.

Polyolefins. Polyethylene can be phosphorylated with phosphorus trichloride and oxygen in a free-radical chain reaction. Hydrolysis gives a polymeric phosphonic acid, which has good adhesion to metals. It also inhibits dental caries and undesirable bone growth. Similarly, ethylene–propylene copolymer is chlorophosphonylated and converted to the phosphonic acid sodium salt, which shows the reversible cross-linking characteristics of an ionomer with improved tensile strength and tensile modulus (see IONIC POLYMERS).

Vinyl and Acrylic Polymers. Vinyl and acrylic polymers containing phosphorus (usually for flame retardancy or improved adhesion) are prepared from a phosphorylated monomer or by treating the polymer with a phosphorus compound.

Thermoplastic Polyesters. Cocondensation with a phosphorus acid, diol, or diester introduces a small amount of phosphorus into polyesters for flame retardancy, color stability, electrical properties, or antistatic properties. Hoechst and Toyo Spinning (Toyobo) have increased the flame retardancy of poly(ethylene terephthalate) fibers by incorporating small amounts of phosphinates. The Hoechst technology places the phosphinate group in the main chain.

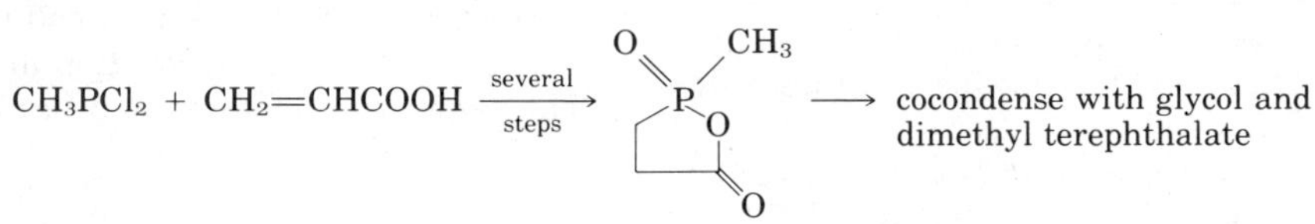

Polyester Resins (Thermoset). Many patents describe polyesters containing an alkylene oxide reaction product with a phosphonic or phosphoric acid to promote flame retardance; some patents describe transesterifying a neutral or acid phosphate directly into the reaction mixture; however, none of these resins appear to be commercial.

Polyamides. Aromatic polyamides made with small amounts of a phosphonate-modified diamine have been investigated at NASA.

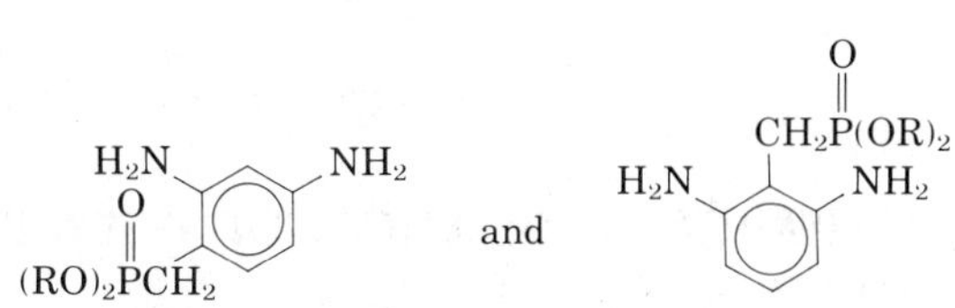

Epoxy Resins. In the combustion of epoxy resins, phosphorus inhibits the evolution of combustible gases. Phosphorus can be introduced into epoxy resins by reaction of the epoxy group with a phosphorus acid group, generating a β-hydroxyalkyl phosphate or phosphonate linkage.

Polyurethanes and Isocyanurate-modified Polyurethanes. Phosphorus flame retardants, both additive and reactive, are used commercially in polyurethanes. Rigid urethane foams employed in thermal insulation have made use of a phosphonate diol or a phosphate diol, which are commercially available.

Phenolic and Amino Resins. Numerous patents describe incorporation of phosphorus to increase flame retardancy. Paper impregnated with phenolic or amino resins, such as that used in automotive air filters, is treated with oligomeric phosphates or phosphonates.

Cellulosic Resins. The direct phosphorylation of cellulose has been studied extensively, particularly to prepare flame-retardant cotton fabric or paper and ion-exchange materials. Heating cellulose with phosphoric acid results in little phosphorylation and much degradation. However, in the presence of urea or dicyandiamide, the degree of phosphorylation is sufficient to impart self-extinguishing properties while retaining strength. Such processes have been employed for semidurable, flame-retardant finishing of draperies. Durable flame-retardant finishes have made use of various phosphonates, which are curable to cross-linked polymers. Highly durable finishes are based on tetrakis(hydroxymethyl)phosphonium salts reacted on the fabric with ammonia to produce a cross-linked structure mainly having phosphine oxide linkages. Wood has been commercially treated for flame retardancy by aminoplast–phosphoric acid compositions, which form cross-linked amino resin structures that bind to the cellulose.

Phosphorylated Proteins. Naturally occurring phosphorus-containing proteins include such important examples as casein. Phosphorylated proteins have been synthesized by modification of nonphosphorus proteins to improve gel-forming and water-binding properties.

BIBLIOGRAPHY

E. D. Weil, "Flame Retardants (Phosphorus Compounds)" in M. Grayson, ed., *Kirk-Othmer Encyclopedia of Chemical Technology,* 3rd ed., Vol. 10, Wiley-Interscience, New York, 1980; "Phosphorus-based Flame Retardants" in M. Lewin, S. M. Atlas, and E. M. Pearce, eds., *Flame-retardant Polymeric Materials,* Vol. 2, Plenum Publishing Corp., New York, 1978, Chapt. 4, pp. 103–128.

M. Sander and E. Steininger, *J. Macromol. Sci. Rev. Macromol. Chem. Part C* **1,** 1 (1967); **2,** 1 (1968).

S. V. Vinogradova and M. A. Anoreva, *Usp. Obl. Sint. Elementorg. Polim.* **97** (1980). A helpful guide to the Russian literature with 291 Refs. to 1976.

S. L. Vail, D. J. Daigle, and A. W. Frank, *Text. Res. J.* **52**(11), 671, 678, 738 (1982); D. J. Daigle and A. W. Frank, *Text. Res. J.* **52**(12), 751 (1982).

EDWARD D. WEIL
Polytechnic University, Brooklyn

PHOTOCHEMICAL OXIDATION OF POLYMERS.

See DEGRADATION.

PHOTOCHEMICAL POLYMERIZATION.

See PHOTOPOLYMERIZATION.

PHOTOCHEMISTRY

The absorption of light by polymers produces noticeable physical and chemical changes. Many commercial plastics are discolored and lose mechanical strength when exposed to uv light, particularly in the presence of oxygen. Photodegradation and photooxidation severely limit the usefulness of untreated plastics exposed in sunlight. On the other hand, the photoinstability offers advantages in waste disposal. Plastics have been developed that decompose rapidly on exposure to uv light; these materials are used as packaging materials and film for agricultural mulches. In other applications, photodegradation increases costs, and photostabilization is required to enhance lifetimes for such outdoor uses as all-weather siding and automotive applications, as well as in highly specialized functions such as in solar energy panels and heat exchangers.

The effects of light on polymers play an important role in a number of specialized technical applications. Polymers insolubilized via photocross-linking are used as durable coatings and negative-working lithographic resists in printing plates and integrated circuits. Solubilization by light-induced depolymerization or chain scission provides the basis for positive-working lithographic resists. Solventless inks based on uv-curable resins fulfill the government regulations limiting the use of volatile solvents in the workplace. Photoconductivity is the principle behind photocopying. Photovoltaic effects in semiconducting polymers are being investigated with the goal of developing solar cells for the generation of electricity from sunlight. All of these applications require in-depth knowledge of the photophysics and photochemistry of polymers.

The interaction of light with biopolymers is also based on photochemistry. The chemistry of vision and photosynthesis, and the light-induced DNA damage that can cause skin cancer, all involve light absorption by large molecules (see also CHARGE-TRANSFER COMPLEXES; DEGRADATION; ELECTRONICS APPLICATIONS; ENERGY TRANSFER; LITHOGRAPHIC RESISTS; PHOTOCONDUCTIVE POLYMERS; PHOTOPOLYMERIZATION; RADIATION-RESISTANT POLYMERS; STABILIZATION; VINYLCARBAZOLE POLYMERS; VINYL KETONE POLYMERS; WEATHERING).

Basic Concepts

The basic processes of photochemistry are the same for polymers and small molecules. The Grotthus-Draper law states that no photochemical reactions can occur unless a photon of light is absorbed. This means, eg, that many commercial plastics transparent in the near uv can undergo photodegradation only as a result of the absorption of light by impurities.

The absorption of light is described by the Beer-Lambert law:

$$A = -\log T/100 = \epsilon bc \quad (1)$$

where A is the sample absorbance, T % transmission, ϵ the molar extinction coefficient, b the sample thickness, and c the molar concentration of chromophores, ie, light-absorbing functional groups. The Beer-Lambert law can be used to determine the penetration depth of light into a polymer sample. If b is defined as the path length at which 90% of the light of a particular wavelength has been absorbed ($A = 1$), rearranging equation 1 gives

$$b \text{ (in } \mu\text{m)} = 10^4/\epsilon c \quad (2)$$

Equation 2 applies to polymers in solution and in the solid state as long as the chromophores are distributed uniformly. In the case of solid polymers the exact calculation of c requires knowledge of the bulk density. In polymers with high chromophore concentrations or strongly absorbing chromophores, b is small and photochemistry is a surface phenomenon.

Table 1 gives some polymer absorption maxima and extinction coefficients. The σ-bond framework of hydrocarbon polymers does not absorb at wavelengths long enough to intercept solar uv. Other chromophores such as C=O, ROOH, and aromatic groups are necessary to extend the absorption of such polymers into the solar uv range, but the search for photoresist materials with high sensitivity to light below 300 nm has prompted renewed interest in the photochemistry of simple polymers.

Properties of Excited States. An energy-level (Jablonski) diagram for a ketone, a common chromophore in polymers reveals several intramolecular processes competing with photochemical reactions, including reemission of a photon as fluorescence or phosphorescence, radiationless decay to the ground state, and crossing from one excited state to another. The term photophysics is used to describe photoprocesses that do not lead to the formation of new chemical products.

The ground states of nearly all organic compounds have all electron spins paired. Absorption of a photon promotes an electron from this singlet state S_0 to a higher energy singlet state $S_1, S_2 \ldots S_n$, numbered in order of increasing energy above the ground state. A change in the spin state of an electronically excited molecule is called intersystem crossing; it gives rise to triplet species $T_1, T_2 \ldots T_n$, with two unpaired spins. A triplet state is always lower in energy than the corresponding singlet state.

Singlet states may emit light and return to the ground state. This fluorescence is a fast process, occurring on a 1–500-ns time scale. The analogous emission process from a triplet state is called phosphorescence. Because direct emission from T_1 to S_0 involves a change of electron spin, phosphorescence is a forbidden process and occurs much more slowly than fluorescence.

Quantum Yields and Lifetimes. The efficiency of deactivation of an excited molecule by any process A is described by a quantum yield ϕ_A, defined as

$$\phi_A = \frac{\text{moles of excited states depleted by process } A}{\text{moles of photons absorbed}}$$

For a primary process $0 < \phi_A < 1$. Photochemical reactions frequently proceed by a complex sequence of mechanistic steps. In this case the quantum yield of formation of product B is defined as

$$\phi_B = \frac{\text{moles of product formed}}{\text{moles of photons absorbed by starting material}}$$

The quantum yield ϕ_B differs from ϕ_A in having values greater than unity in certain chain reactions, such as photoinitiated free-radical polymerizations and polymer photooxidation.

An excited state has a lifetime τ given by

$$\tau = \left(k_e + \sum_i k_i + \sum_j k_j[Q_j]\right)^{-1} \quad (3)$$

In equation 3 k_i and k_j are the rate constants for intra- and intermolecular deactivation processes, respectively, k_e is the rate constant for emission of light, and $[Q_j]$ is the concentration of species j capable of quenching the excited state. The quantum yield ϕ_A and the rate constant k_A of a process A are related as follows to the lifetime of the excited state:

$$\phi_A = k_A\tau \text{ for an intramolecular process} \quad (4)$$

$$\phi_A = k_A\tau[Q] \text{ for an intermolecular process involving another molecule } Q \quad (5)$$

Quenching of Excited States. Quenching plays an important role in photochemistry. Polymers can be stabilized against photodegradation if quenching competes with reaction of an excited state, such as a ketone n–π^* triplet. On the other hand, if energy transfer to another chromophore introduces a new photoreaction, sensitized photodegradation can occur.

Excited states can be quenched by several mechanisms, including electronic-energy transfer, excimer or exciplex formation, electron transfer, proton transfer, or enhanced intersystem crossing by interaction with a molecule such as methyl iodide, which contains a heavy atom. Whatever the quenching mechanism, the result is a reduction in the quantum yield of any competing photochemical process.

Excimers and Exciplexes. Many photochemical reactions proceed by way of exciplexes, complexes formed between an excited molecule D* and a quencher Q as shown in equation 6.

$$\begin{array}{ccc} D^* + Q & \underset{k_{-1}}{\overset{k_1}{\rightleftharpoons}} & [DQ]^* \longrightarrow \text{products} \\ \downarrow k_d & & \downarrow k_c \\ D_0 & & D_0 + Q \end{array} \quad (6)$$

The quencher can also be a ground-state D species, in which case the symmetric complex [DD]* is called an excimer.

Electron Transfer and Proton Transfer. The excited states of molecules often have dramatically different acid-base and redox properties from their ground states. The energy of the absorbed proton can drive processes that would be highly endothermic from the ground state. Photochemical electron transfer:

$$D + A \xrightarrow{h\nu} D^{\dot{+}} + A^{\dot{-}} \quad (7)$$

which does not occur in the ground state, may occur at the

Table 1. Absorption λ of Maxima and Extinction Coefficients ϵ of Some Polymers

Polymer	λ_{max}, nm	ϵ, $M^{-1}cm^{-1}$[a]
polyethylene	140	
poly(methyl methylacrylate)	214	170
poly(methyl vinyl ketone)	290	18
polystyrene	269	175
	262	222
poly(2-isopropenylnaphthalene)	323.5	242
	280	4550
poly(N-vinylcarbazole)	345	4500
	293	12900

[a] In moles of repeating units.

diffusion-controlled rate in polar solvents following excitation of the donor or acceptor. Proton transfer or acid dissociation can be greatly enhanced by light absorption:

$$B\text{–H} \xrightarrow{h\nu} BH^* \longrightarrow B{:}^- + H^+ \qquad (8)$$

$$B{:} + H^+ \xrightarrow{h\nu} B{:}^* + H^+ \longrightarrow BH^+ \qquad (9)$$

Photoreactions

Polymer photochemistry has evolved into a broad discipline with important consequences in numerous technical areas, from photolithography to the outdoor weathering of plastics. Much of small-molecule photochemistry has its parallel in polymer systems. The differences between polymer and small-molecule photochemistry arise mainly from the greater kinetic and mechanistic complexity of polymer systems. Future developments should include the photochemistry of larger and more complex chromophores, as the variety of polymers in use continues to increase, and an extension to the realm between conventional photochemistry and radiation chemistry, reflecting the use of polymers in space and in enhanced-resolution microlithography. Photoreactions include bond-breaking reactions, those involving ketones, dimerizations and cycloadditions, rearrangements and isomerizations, and reactions with singlet oxygen.

Photodegradation

The photooxidative degradation of polymers involves both photochemical and thermal reactions. The primary function of photochemistry is to generate free radicals in the initiation step. Such initiating reactions include direct and sensitized photolysis of hydroperoxides and peroxides, H-atom abstraction by the triplet states of aldehydes and ketones, and other σ-bond homolyses occurring with light of wavelengths below 290 nm. These photochemical intermediates yield additional hydroperoxide via free-radical chain reactions, as shown for polypropylene:

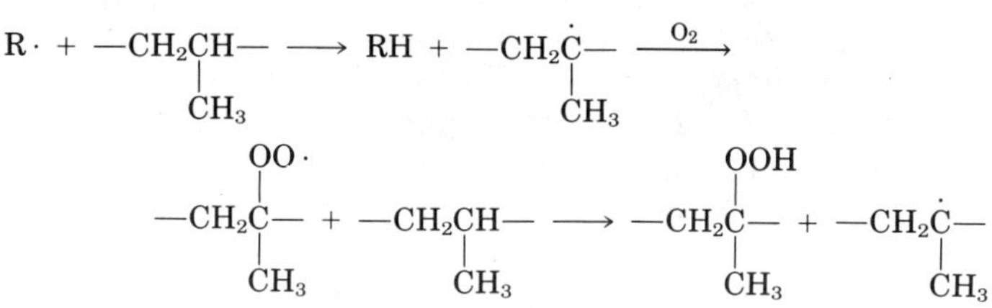

Following their initial formation during autooxidation, hydroperoxides and their decomposition products continue to be photolyzed in proportions dictated by their relative extinction coefficients and by the efficiency of sensitization. The sensitizer is frequently TiO_2 or ZnO, two widely used pigments in plastics.

Polymers and Small Molecules

Despite many similarities, small-molecule and polymer photochemistry have important differences, especially in kinetics.

Kinetics of Chain Scission and Cross-linking. For a chain-scission reaction occurring at random along a polymer chain, the evolution of the molecular weight is described by

$$M_n(0)/M_n(t) - 1 = S = \phi_{cs} I_0 t \qquad (10)$$

where S is the average number of chain breaks per initial chain, ϕ_{cs} the quantum efficiency of chain scission, I_0 the amount of light absorbed per unit time, and t the irradiation time; $M_n(0)$ and $M_n(t)$ are the number-average molecular weights at times 0 and t, respectively. Examination of equation 10 shows that only a few chain breaks are required to convert a polymer from a useful material with a degree of polymerization (DP) of 1,000, eg, to a useless material with DP < 200. Thus polymer photochemistry differs from small-molecule photochemistry in the fact that critical changes in material properties can occur at less than 1% reaction of the photosensitive groups.

Molecular Mobility. For a photoreaction to occur in a solid polymer, the molecule must be sufficiently mobile or the photoreaction must proceed without motion of the surrounding polymer. Molecular motion must occur on the time scale of the excited-state lifetime. Limited molecular mobility persists in amorphous polymers below the glass-transition temperature; only large-scale movement of main-chain segments is frozen out at T_g. As a result many photoreactions, requiring only rotation around individual bonds or diffusion of molecules small relative to the mean free-volume element, occur well below T_g.

Effects of Inhomogeneous Chromophore Distributions. A further complication in polymer photochemistry is the absence of a uniform distribution of chromophores. Figure 1 compares the distributions of chromophores present in a fluid solution of a small molecule, a polymer solution, and bulk polymer after photoreaction has proceeded around randomly distributed initiating sites. In the polymer systems (Fig. **1b** and **c**), the average separation between a chromophore and its nearest neighbor is considerably smaller than in the isotropic solution.

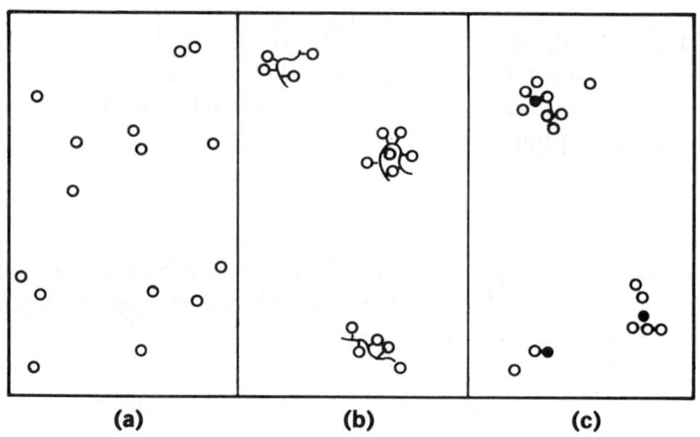

Figure 1. Local concentration effects. Chromophore distribution for (**a**) a dilute solution of a small molecule; (**b**) a dilute solution of a polymer; and (**c**) a polymer film following photooxidation at randomly distributed light-absorbing sites.

Effect of Kinetic Inhomogeneity. In polymers, the inhomogeneous distribution of chromophores and restrictions on molecular mobility result in differences in molecular environment persisting throughout the excited-state lifetime. As a result the excited states are not described by a single lifetime τ but rather by a superposition of lifetimes.

BIBLIOGRAPHY

J. Guillet, *Polymer Photophysics and Photochemistry,* Cambridge University Press, Cambridge, UK, 1985.

J. F. McKellar and N. S. Allen, *Photochemistry of Man-made Polymers,* Applied Science Publishers, London, 1979.

J. F. Rabek, *Experimental Methods in Photochemistry and Photophysics,* John Wiley & Sons, Inc., New York, 1982, Pts. 1 and 2.

D. Phillips, ed., *Polymer Photophysics,* Chapman and Hall, London, 1985.

DAVID A. HOLDEN
University of Waterloo

PHOTOCONDUCTIVE POLYMERS

Photoconductivity is defined as an increase of conductivity caused by radiation. Photoconductive polymers, according to this definition, are polymers that are insulating or poorly conductive in the darkness and more conductive when illuminated.

Electronically photoconductive polymers are those polymers in which the carriers of electricity are produced (generated) in the bulk of the material by the action of light, and which, at the same time, support migration of the carriers. The carriers of electricity can be photogenerated extrinsically, in an adjacent photoconductor, and injected into the polymer. In this case the polymer acts merely as a charge-transporting polymer.

Photoconductive or charge-transporting polymers can be *p*-type or *n*-type, depending on the main (majority) carrier. Only in special cases can polymers be bipolar or capable of transporting both holes and electrons. Most known photoconductive or charge-transporting polymers are *p*-type; only holes can move.

These definitions lead to the concept of an ideal photoconductive polymer, which contains no free charge carriers, is sandwiched between blocking electrodes preventing injection of holes and electrons in the darkness, has high quantum efficiency of photogeneration of carriers, allows migration of photogenerated charges with maximum speed, is bipolar, does not lose the mobile carriers by recombination or trapping, and no carriers stay immobile in the bulk of the material. The real photoconductive polymers are, however, far from ideal.

The only photoconductive polymers studied in detail are those containing the carbazole moiety, such as polyvinylcarbazole (PVK). The lack of detailed studies on photoconductive polymers other than PVK is noteworthy in view of the commercial importance of organic photoconductors, which are essential in the multibillion-dollar xerographic-copier, laser-printer, and duplicator industries.

The practical polymeric photoconductors are based on two types of systems, structurally and conceptually different, but similar in the mechanism of charge-carrier generation and transport. In one class the activity is built into the polymer structure itself; the pendent groups, the in-chain groups, or the chain itself absorbs light and supports charge migration. The second class consists of inactive binder polymers containing high concentrations of active chromophores or charge-transport molecules. These so-called molecularly doped polymers are widely used in xerographic photoreceptors. The molecularly doped polymers have one advantage over truly photoconductive polymers: they combine the useful mechanical properties of the host-binder polymer with the efficiency of carrier generation or transport via the small molecules.

The process of photoconduction can be separated into the following steps:

1. Absorption of radiation.
2. Generation of charge carriers.
3. Separation of carriers.
4. Injection of carriers.
5. Diffusion and migration.
6. Immobilization in traps.
7. Release from traps.
8. Recombination of carriers.
9. Space charge, space-charge-limited current (SCLC).

Materials

Only polymers capable of both producing the charge carriers upon exposure to light and transporting them through the bulk are photoconductive. Polymers that do not absorb the incident light but accept charges generated in an adjacent material are merely transport materials or electronic semiconductors. All known and studied transporting polymers are photoconductive in uv light, and, conversely, all truly photoconductive polymers can act as transporting media for charges generated extrinsically. Thus the term photoconductive must be related to the particular wavelength of incident light.

Photoconductive polymers can be *p*-type (hole-transporting), *n*-type (electron-transporting), or bipolar (capable of transporting both holes and electrons). Typically, bipolarity can be accomplished by adding electron-transporting small molecules such as trinitrofluorenone (TNF) to a donorlike, hole-transporting polymer such as PVK. All practical photoconductive charge-transporting polymers are *p*-type. Practical electron-transporting polymers are notably absent in the literature.

Polymers with Polyconjugated Multiple Bonds. Many photoconductive polymers and oligomers with conjugated C═C, C═N, and C≡C bonds have been reported in the patent literature, but in most cases no fundamental parameters are available.

Polymers with Large Polynuclear Aromatic Groups. Polymers with pendent or in-chain, large, aromatic groups, such as anthracene, pyrene, chrysene, and acridine, display some degree of photoconductivity; they usually absorb light only in the uv range. No application has been found for polymers of this group.

Polymers with Aromatic Amine Groups. Most known photoconductive and charge transporting polymers contain an aromatic amine group:

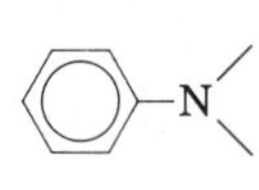

Most practical systems and all organic, charge-transporting materials used in electrophotographic photoreceptors, eg, PVK, belong in this category.

Inorganic Polymers. Photoconductivity is also displayed by polymers with elements other than carbon in the main chain. For example, TNF-doped polyphenylmethylsilylene, a silicon-backbone polymer, is photoconductive in visible light and the polymer itself is photoconductive in the uv range. Holes are the main (majority) carriers with mobilities of $\sim 10^{-4}$ cm^2/(V·s).

Photoconductivity has also been detected in polymers prepared by glow-discharge polymerization of tetramethylsilane, tetramethyltin, and tetramethylgermanium.

Preparation

Photoconductivity is found in many polymer types and classes. The active groups can be pendent or within the chain. Consequently, there is no single or preferred method of synthesis. The known photoconductive polymers are prepared by free-radical, cationic, and anionic polymerizations, ring-opening polymerizations, polycondensations, attachments of active groups to polymers, Wurtz-like polycondensation reactions, and other methods.

Applications

The commercial utilization of photoconductive polymers has been very limited. The only area of utilization is electrophotography. Most organic photoreceptors are based on the illustration in Figure 1, where the photoreceptor is charged first by corona in the dark, then exposed to the image reflected from the original. Light reflected from the paper penetrates the polymeric transport layer and generates charges in the adjacent photoconductor. The charges migrate through the transport layer to neutralize the opposite charges at the surface. The dark areas of the print remain charged. Following generation of the latent image on the surface, the image is developed on the photoreceptor and transferred to a sheet of paper. The photoconductor is cleaned of the ink and any remaining electrostatic pattern is removed by an erase lamp. This completes the print cycle and the photoconductor is ready to be recharged.

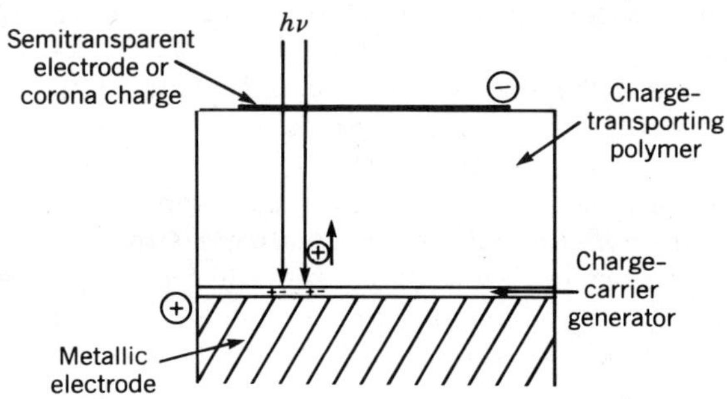

Figure 1. A layered photoconductor in which charge photogeneration is separated from charge transport.

BIBLIOGRAPHY

J. M. Pearson and M. Stolka, *Poly(N-vinylcarboazole)*, Gordon and Breach, New York, 1981.

J. Mort and G. Pfister, eds., *Electronic Properties of Polymers,* John Wiley & Sons, Inc., New York, 1982.

M. Stolka, J. F. Yanus, and D. M. Pai, *J. Phys. Chem.* **88,** 4707 (1984).

J. Mort and D. M. Pai, eds., *Photoconductivity and Related Phenomena,* Elsevier Science Publishing Co., Inc., New York, 1976.

MILAN STOLKA
Xerox Corporation

PHOTOCROSS-LINKING. See PHOTOPOLYMERIZATION.

PHOTOCURING. See CURING; PHOTOPOLYMERIZATION.

PHOTODEGRADATION. See DEGRADATION.

PHOTOELASTICITY. See BIREFRINGENCE; OPTICAL PROPERTIES.

PHOTOGRAPHIC APPLICATIONS

Since the rudimentary beginning of photography more than 100 years ago, the impressive growth of the technology, which accounts for the state of the art today, is due to advances primarily in solid-state silver halide techniques and organic chemistry.

Flexible Film Supports

The properties of photographic film supports are given in Table 1. In addition, a finished photographic support often includes an antistatic layer and an adhesion-promoting sublayer.

Gelatin

Ever since the invention of the gelatin-AgBr emulsion by Maddox in 1871, gelatin has been used as the hydrophilic binder (or vehicle) in virtually every silver halide photographic product. Gelatin (qv) is flexible, colorless, transparent, chill-setable, and abrasion-resistant. Furthermore, it is important that the resultant film swells during processing to allow access of the imaging chemistry to the processing chemicals.

Water-soluble Synthetic Vehicles

The outstanding physical properties of gelatin derive from its tertiary structure and inter- and intrachain hydrogen bonding. Efforts to replace gelatin with a synthetic vehicle have failed to produce a fully satisfactory material. Nevertheless, synthetic vehicles play important roles as partial gelatin replacements in specialized applications.

Peptization of silver halide is a key role of gelatin, where the applicability of synthetic vehicles has received attention. Peptizers influence the crystalline morphology of silver halide grains, grain size and distribution, and accessibility of the grain surface to various additives and processing chemicals. Effective synthetic peptizers for silver halide contain sulfide or quaternary ammonium functional groups and solubilizing comonomers.

Water-soluble synthetic vehicles are often used as carriers of photographically useful chemicals. These can be polymer-bound color couplers, hardeners, development inhibitors, oxidized developer scavengers, or other reactive species.

Water-insoluble Vehicles

Latex or emulsion polymers are employed in modern photographic products, where they improve physical properties in some cases and act as carriers for hydrophobic additives in

Table 1. Applications and Properties of Sensitized Photographic Film Supports

	Cellulose triacetate	PET	Polycarbonate	Polystyrene
applications	graphic arts aerial motion picture amateur roll x ray microfilms	graphic arts aerial industrial x ray professional sheet	graphic arts aerial	graphic arts
water absorbed at 21°C and 50% rh, %	1.5	0.5	0.3	<0.1
yield strength, MPa[a]	81	95	69	72
tensile strength at break, MPa[a]	107	173	81	72
tensile elongation at break, %	35	110	110	3
Young's modulus, MPa[a]	3,800	4,500	2,400	3,100
toughness, MPa[a]	31	152	76	2
tear propagation strength, g/μm[b]	0.39	1.57	0.79	0.16
thermal coefficient of expansion	0.003	0.001	0.004	0.0035
humidity coefficient of expansion	0.005–0.010	0.002–0.004	0.002–0.004	0.001–0.003
processing size changes, %	<−0.1	+0.03−0.03	+0.03−0.03	+0.03−0.03
uv transmittance λ at 10% transmittance, nm	270	315	300	290
refractive index (in plane of sheet)	1.48	1.65	1.58	1.59

[a] To convert MPa to psi, multiply by 145.
[b] To convert g/μm to g/mil, multiply by 25.4

others. The latex or any residual components of its preparation must be photographically inert. If there is an effect, it must be desirable and reproducible. Latices are always a dispersed phase, and gelatin is the continuous phase.

Latices have been employed in several applications in which physical property modification was advantageous or essential. Properties that can be affected include dimensional stability, flexibility, drying rate (coating rate), cracking, abrasion resistance, and differential swelling.

Instant Photography

In the early 1970s the role of polymers in photography took on special significance with the development of litter-free, integral, instant-imaging systems such as Polaroid SX-70 instant color film or Kodak instant color film PR10. These systems required, under ambient conditions, control of processes formerly included in photofinishing. Polymers were utilized as thickeners, mordants, supports, and for opacification, antireflection, pH control, interimage control, and temperature compensation.

Polymer Reagents

Differential layer hardening is made possible by employing a layer-specific water-soluble polymeric hardener that does not diffuse. Polymeric reagents are also used as color couplers, which must remain in the desired layer. Redox polymers, such as polyvinylhydroquinone, have been utilized to remove stains (see ELECTRON-TRANSFER POLYMERS). Other polymeric reagents include uv absorbers, peptizing agents, antifoggants, chemical sensitizers, silver scavengers, and reactive mordants.

Static-discharge Control

Cellulose acetate and PET are more stable, stronger, and less flammable than the former film base cellulose nitrate. Static fires are virtually nonexistent, but the forces that cause static buildup have increased. Polyanions, polycations, and ionene systems are utilized in a binder which provides flexibility, improves adhesion, imparts stability, and acts as a moisture sump.

On cinefilm, another tactic has been followed. Carbon black is coated on the backside of the film by an organic solvent with an alkali-soluble binder. The carbon provides excellent antistatic properties.

Permanent thin films containing organic or inorganic semiconductors are highly conductive independent of humidity. The semiconductor-containing layer is optimized for conductivity and an overcoat layer is provided for other physical properties.

BIBLIOGRAPHY

P. Z. Adelstein, ed., SPSE: *Handbook of Photographic Science and Engineering,* John Wiley & Sons, Inc., New York, 1973.

T. H. James, ed., *The Theory of the Photographic Process,* 4th ed., Macmillan, New York, 1977.

G. A. Campbell and J. L. R. Williams, *Polym. News* **8,** 294 (1983).

V. K. Walworth, "Color Photography, Instant" in M. Grayson, ed., *Encyclopedia of Chemical Technology,* Vol. 6, 3rd ed., John Wiley & Sons, Inc., New York, 1979, pp. 646–682.

EUGENE ARMOUR
G. A. CAMPBELL
D. A. UPSON
Eastman Kodak Company

PHOTOPOLYMERIZATION

Photopolymerization, the utilization of electromagnetic radiation (or light) as the energy source for polymerization of functional monomers, oligomers, and polymers, is the basis of important commercial processes with broad applicability, including photoimaging and uv curing of coatings and inks. These processes require that light is absorbed by the system and utilized to effect the formation of new chemical bonds. Photopolymerization may be achieved by the utilization of photoinitiators, photocross-linking agents, and photocross-linkable polymers. Photocross-linking agents and photocross-linkable polymers are used primarily in photoimaging applications; photoinitiators are more broadly used in both photoimaging and curing processes.

Photoinitiators

Photoinitiators absorb light in the uv–visible spectral range (250–450 nm) and convert this light energy into chemical energy in the form of reactive intermediates, such as free radicals and reactive cations, which subsequently initiate polymerization of functional monomers and oligomers. Monofunctional monomers form linear polymers, whereas multifunctional monomers give three-dimensional cross-linked networks. In photoimaging and curing applications, cross-linking imparts solvent resistance. Cross-linked polymer networks are more durable, which is important for protective coatings.

Light absorption by the photoinitiator requires that an emission line from the light source overlaps with an absorption band of the photoinitiator. In this event, the photoinitiator (PI) is raised to an electronically excited state PI* by promotion of an electron to a higher energy orbital, as shown below:

$$\text{PI} + \text{light} \longrightarrow \text{PI}^*$$

The lifetime of PI* is short, generally less than 10^{-6} s. During this time, PI* may be partitioned between several processes including (*1*) decay back to PI (with emission of light and/or heat), (*2*) excited-state quenching by O_2, monomer, or other quenching agent, and (*3*) a chemical reaction yielding the initiator species, such as I· or I^+:

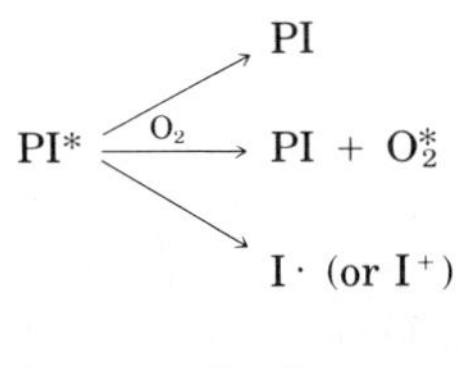

The initiator species are further partitioned between reactions that either lead or do not lead to initiation. When the initiator species are free radicals (I·), reaction with oxygen produces peroxy radicals (IO_2·), which do not initiate polymerization efficiently, thereby reducing the fraction of initiating radicals that react with monomer (M):

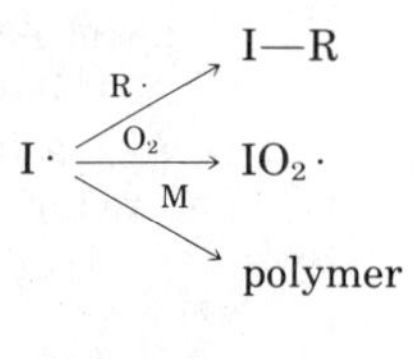

The rate of initiation (R_i) of photopolymerization is proportional to the fraction of PI* that yields initiator species times the fraction of initiator species that initiate polymerization. Thus, the reaction of O_2 with PI* and I· reduces R_i, resulting in air-inhibition, which occurs primarily at the air interface where the concentration of O_2 is highest.

The problem of air-inhibition, which results in retardation of surface-cure, may be eliminated by conducting photopolymerization in vacuo or in an inert atmosphere such as N_2. Under these conditions, lower intensity light sources and lower PI concentrations may also be utilized.

Lower PI concentration tends to give more uniform polymerization throughout a film as well as lower cost, since the PI is expensive. Nonuniform polymerization may lead to stresses in films that result in poor mechanical properties and poor adhesion. On the other hand, providing an inert N_2 atmosphere tends to complicate the curing process, which must be taken into account.

Photoinitiators for Radical Polymerization. Photoinitiators for radical polymerization generally function by intramolecular bond cleavage, notably α-cleavage of an aromatic ketone group in the photoinitiator, or intermolecular H abstraction from an H donor by the photoinitiator, notably benzophenone, benzil, and quinone derivatives.

It has been suggested that photoinitiators undergoing intramolecular bond cleavage be classified as type PI_1, since initiator radicals are produced by a unimolecular process. Photoinitiators that undergo intermolecular H abstraction may correspondingly be classified as type PI_2, since a bimolecular reaction is involved.

Photoinitiators for Cationic Polymerization. Thermally stable photoinitiators for cationic polymerization are triarylsulfonium and diaryliodonium salts with complex metal halide anions.

$$Ar_3S^+\,X^- \qquad Ar_2I^+\,X^-$$

$$\text{where } X^- = SbF_6^-, AsF_6^-, PF_6^-$$

Irradiation of these salts results in cleavage of Ar—S (or Ar—I) bonds to yield reactive radical cations.

$$Ar_3S^+\,X^- \xrightarrow{h\nu} Ar_2S^{+\cdot}\,X^- + Ar\cdot$$

$$Ar_2I^+\,X^- \xrightarrow{h\nu} ArI^{+\cdot}\,X^- + Ar\cdot$$

Direct evidence for this photocleavage has been obtained from laser-flash photolysis studies. The radical cations were found to be highly reactive with nucleophiles, including cyclohexene oxide. Thus the radical cations may directly initiate polymerization of epoxides, which is supported by polymer end-group analysis.

A newer class of photoinitiators for cationic polymerization are mixed arene cyclopentadienyl metal salts of complex metal halide anions, such as (η^6-benzene) (η^5-cyclopentadienyl) Fe(II) hexafluorophosphate. A mechanism for photoinitiated polymerization of epoxides is shown in the following:

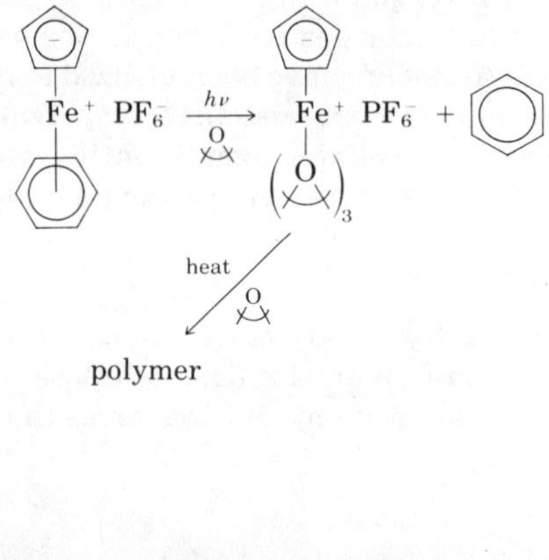

Photosensitization. The spectral response of aryl sulfonium and iodonium salts may be extended into the 300–400-nm region (as well as into the visible) by utilization of photosensitizers (PS). Possible mechanisms are given below.

Energy transfer:

$$\mathrm{PS} \xrightarrow{h\nu} \mathrm{PS}^*$$

$$\mathrm{PS}^* + \mathrm{Ar_3S^+}\,(\mathrm{Ar_2I^+}) \longrightarrow \mathrm{PS} + (\mathrm{Ar_3S^+})^* \text{ or } (\mathrm{Ar_2I^+})^*$$

$$(\mathrm{Ar_3S^+})^* \text{ or } (\mathrm{Ar_2I^+})^* \longrightarrow \mathrm{Ar_2S^{+\cdot}}\,(\mathrm{ArI^{+\cdot}}) + \mathrm{Ar}\cdot$$

$$\mathrm{Ar_2S^{+\cdot}}\,(\mathrm{ArI^{+\cdot}}) + \underset{\diagup\diagdown}{\mathrm{O}} \longrightarrow \text{polymer}$$

Electron transfer:

$$\mathrm{PS} \xrightarrow{h\nu} \mathrm{PS}^*$$

$$\mathrm{PS}^* + \mathrm{Ar_3S^+}\,(\mathrm{Ar_2I^+}) \longrightarrow \mathrm{PS^{+\cdot}} + \mathrm{Ar_3S^\cdot}\,(\mathrm{Ar_2I^\cdot})$$

$$\mathrm{PS^{+\cdot}} + \underset{\diagup\diagdown}{\mathrm{O}} \longrightarrow \text{polymer}$$

$$\mathrm{Ar_3S^\cdot}\,(\mathrm{Ar_2I^\cdot}) \longrightarrow \mathrm{Ar_2S}\,(\mathrm{ArI}) + \mathrm{Ar}\cdot$$

$$\mathrm{Ar_3S^\cdot}\,(\mathrm{Ar_2I^\cdot}) + \mathrm{PS^{+\cdot}} \longrightarrow \mathrm{PS} + \mathrm{Ar_3S^+}\,(\mathrm{Ar_2I^+})$$

Photoinitiator Design. Photobleaching should be considered in the design of photoinitiators. It appears desirable that the principal chromophoric group of the PI responsible for the long wavelength absorption is reduced during photogeneration of the initiator. This results in bleaching, which increases the efficiency of initiator photogeneration, since photoinitiator fragments do not compete with light absorption by the PI. In addition, weatherability (or exterior durability) of coated products is improved because of lower absorptivity following photopolymerization.

Photocross-linking Agents

An important class of photocross-linking agents are bisarylazides, such as

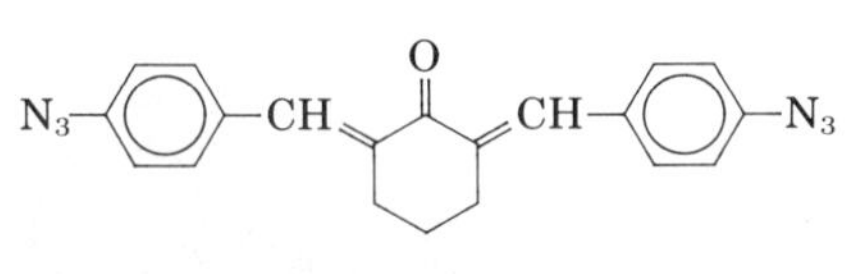

The synthesis and utilization of bisazido compounds in photoimaging applications have been reviewed.

Photocross-linkable Polymers

Polymers that can be "photocross-linked" possess chromophoric groups that undergo light-induced chemical bonding with each other. The conceptually simplest systems in this group, and apparently the most important commercially, are polymers that undergo photodimerization.

The chromophoric groups may be in the polymer backbone or pendent, as exemplified by a backbone chalcone group, and poly(vinyl cinnamate), respectively.

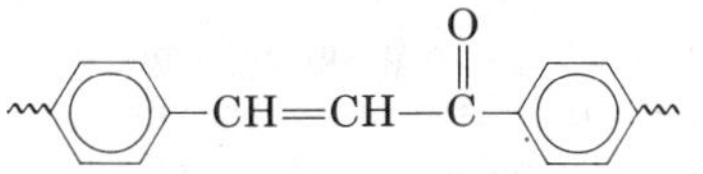

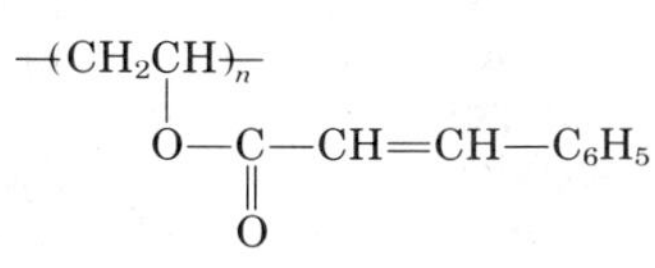

Alternatively, cinnamate groups may be present in the backbone, just as chalcone groups may be pendent.

Uv Curing and Related Processes

Uv curing of coatings and printing inks is based on the photoinitiation of radical and cationic polymerization. Photoinitiated radical polymerization of acrylate resins (monomers and oligomers) is widely employed commercially. Photogeneration of initiator radicals by α-cleavage (PI_1) or H abstraction (PI_2) is followed by reaction with the acrylate to produce a new radical that reacts with a second acrylate, etc.

$$\mathrm{PI} \xrightarrow{h\nu} \mathrm{I}\cdot$$

$$\mathrm{I}\cdot + \mathrm{CH_2{=}CH(COOR)} \longrightarrow \mathrm{I{-}CH_2{-}CH(COOR)}\cdot \xrightarrow{\mathrm{CH_2{=}CHCOOR}}$$

$$\mathrm{I{-}CH_2{-}CH(COOR){-}CH_2{-}CH(COOR)}\cdot \xrightarrow{\mathrm{CH_2{=}CHCOOR}} \text{linear polymer}$$

The utilization of monoacrylates yields a linear polymer. However, in uv curing a cross-linked polymer is produced from mixtures of mono- and multi-acrylates. The properties of the coatings, such as hardness, flexibility, abrasion resistance, and solvent resistance, are determined by the composition of the acrylate resins.

Limitations of uv-curing technology include the utilization of oddly shaped substrates and the curing of opaque coatings. Weatherability or durability requires special consideration, since conventional light stabilizers tend to interfere with the uv-curing process. Latent light stabilizers that are activated by light absorption may alleviate this problem.

For curing opaque acrylate films, electron beams or curtains may be used, a process known as EB curing; uv and EB curing are complementary technologies. In the latter the depth of electron penetration is not limited by opacity, but is principally dependent on the density of the composition and may be controlled by the accelerating voltage of the EB processor.

Photoimaging and Related Techniques

Photoimaging is based on exposure of a photopolymerizable system to light through a pattern of transparent and opaque regions, such as a photographic negative. Polymerization and cross-linking occur below the transparent regions, resulting in a latent image of insolubilized polymer in the exposed areas and soluble (uncross-linked) polymer in the unexposed areas. Development of the latent image may be accomplished by employing a solvent to dissolve the polymer in the unexposed areas selectively. The remaining cross-linked polymer serves as a barrier (or resist) against a subsequent process, such as etching or implantation. The photopolymerizable composition is called a photoresist. The process, known as photolithography, is used for integrated circuits (IC), printed circuits (PC), and printing plates. Related technologies are electron-beam, x-ray, and ion-beam lithography, which utilize the corresponding resists.

BIBLIOGRAPHY

S. P. Pappas, ed., *UV Curing Science and Technology,* Technology Marketing Corp., Norwalk, Conn., 1978; Vol. II, 1985.

C. G. Roffey, *Photopolymerization of Surface Coatings,* Wiley-Interscience, New York, 1982.

J. V. Crivello, *Adv. Polym. Sci.* **62,** 1 (1984).

H. J. Hageman, *Progr. Org. Coat.* **13,** 123 (1985).

S. PETER PAPPAS
North Dakota State University

PHOTORESISTS. See LITHOGRAPHIC RESISTS; PHOTOPOLYMERIZATION.

PHTHALIC ACIDS AND DERIVATIVES. See ALKYD RESINS.

PHTHALOCYANINE POLYMERS

The phthalocyanine macrocycle, although complex in size and structure, is synthetically amenable to many variations and imparts novel, unique, and useful properties to molecular substances.

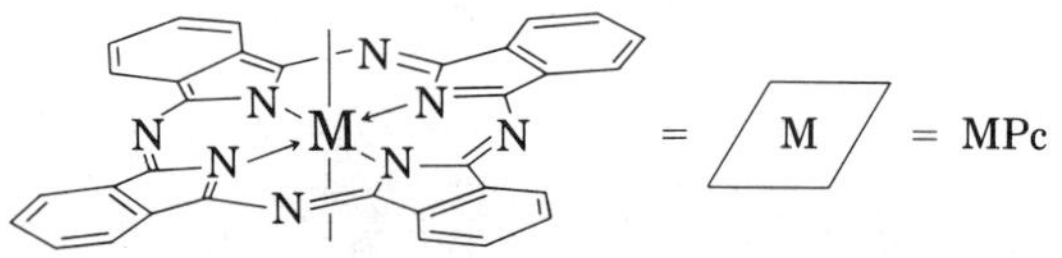

(1)

where M = 2 H, transition and nontransition metals

Monomeric compounds of this type are currently produced in multimillion kilogram quantities for colorant, catalytic, semiconductor, and a variety of other commercial applications. Incorporation of the phthalocyanine structure as a main-chain component in polymeric materials and investigation of resulting properties are objectives of extensive research.

Phthalocyanine polymers may be divided into two structural groups: those in which the polymeric linkage is through the axial bonding sites of a completed metal (**2**) and those in which the polymeric linkage is through the substituent positions at the periphery of the phthalocyanine ring (**3**).

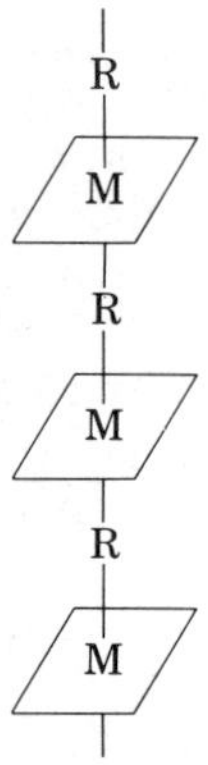

(2)

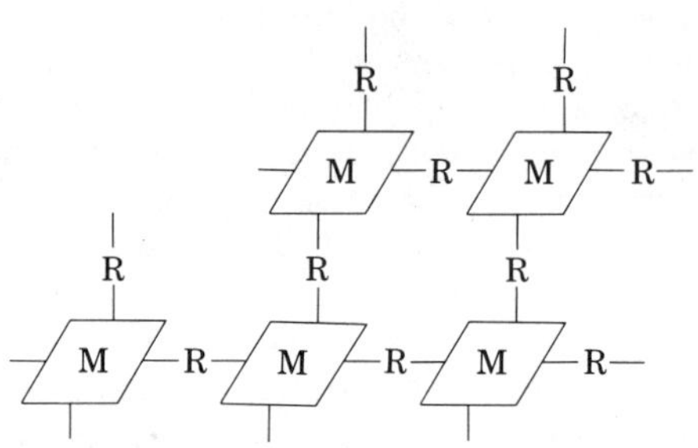

(3)

Axially linked phthalocyanine polymers are linear, well-characterized structures and are currently being studied for their electrical properties. Peripherally linked phthalocyanine polymers are much less well characterized. They are structurally more complicated and are being investigated for thermal, catalytic, and electrical properties. Reviews of phthalocyanine polymers summarize the literature until 1983 (see also COORDINATION POLYMERS; ORGANOMETALLIC POLYMERS).

BIBLIOGRAPHY

F. H. Moser and A. L. Thomas, *The Phthalocyanines,* Vols. I and II, CRC Press, Inc., Boca Raton, Fla., 1983.

D. Wohrle, *Adv. Polym. Sci.* **50,** 45 (1983).

A. A. Berlin and A. I. Sherle, *Russ. Chem. Rev.* **48,** 1125 (1979).

ARTHUR W. SNOW
JAMES R. GRIFFITH
Naval Research Laboratory

PHYSIOLOGICAL PROPERTIES. See BIOLOGICAL ACTIVITY; BIOMATERIALS; DRUG APPLICATIONS; FOOD APPLICATIONS; MEDICAL APPLICATIONS.

PIGMENT BINDERS. See COATINGS; PAPER ADDITIVES AND RESINS.

PIGMENTED COATINGS. See COATINGS.

PIGMENTS. See COLORANTS.

PINENES. See HYDROCARBON RESINS.

PIPE

Plastics were first used for pipe in the mid-1930s. They were more widely used by the mid-1950s as a result of advances in materials, formulation, and screw-extrusion technology. Since then the industry has grown rapidly and, in the United States, its dollar value is now ranked second to steel pipe. Plastics have surpassed clay, copper, asbestos-cement, aluminum, iron, and concrete. The current annual rate of U.S. consumption of plastics for pipe is ca 1.7×10^6 t, ca 80% of which is poly(vinyl chloride) (PVC). This material combines strength, stiffness, and toughness with economy, making it the most cost-effective choice for pressure and nonpressure applications.

Polyethylene (PE) follows PVC with ~16% of the market. Features include flexibility, toughness, and heat fusibility. The remainder of the plastics pipe consumption, or ca 4%, is principally accounted for by acrylonitrile-butadiene-styrene (ABS), chlorinated poly(vinyl chloride) (CPVC), and polybutylene (PB).

Although reinforced thermosetting resin pipe represents a small portion of the market, it is the product of choice for many special applications. Corrosion resistance, toughness, and ease of handling and installation contribute to the growing acceptance of plastic piping.

Commercial Products and Joining Methods

Plastic piping is available in an impressive array of materials, constructions, diameters, wall thicknesses, and lengths.

A variety of methods are available for joining plastic pipe. The characteristics of the material govern the choices. For example, the solvent-resistant polyolefins (PE, PB, and PP) cannot be solvent-cemented, vinyls (PVC and CPVC) and ABS are difficult to heat-fuse, and thermosets cannot be heat-fused or solvent-cemented. The choice for a given material is further limited by performance, installation, and maintenance requirements, as well as by the availability of fittings and tools. For applications requiring maximum joint strength and chemical resistance, the following bonding methods yield strong, monolithic types of joints: heat fusion for PE, PB, PP, and PVDF; solvent cementing for ABS, PVC, and CPVC; and adhesive bonding for reinforced thermosetting resin. Joint design is typically of the socket- (also called bell-) and -spigot type. Squarely cut butt ends of pipes or fittings can be heat-fused. Butt fusion is efficient and practical for joints of pipes with diameters > 10 cm.

Manufacture

Thermoplastic Piping. Thermoplastic pipe and fittings are usually made from a base polymer containing small amounts of additives, such as stabilizers, antioxidants, lubricants, colorants, modifiers, extrusion aids, and fillers.

Thermoplastic pipe is manufactured by extrusion from precompounded pellets, or a dry blend of powdered resin and required ingredients. Extrusion by the last method is more economical than by the first method. Practically all PVC pipe is now extruded from dry blends. Special extrusion equipment that homogenizes the material is required.

Reinforced Thermosetting Resin Pipe. In reinforced thermosetting resin pipe (RTRP), a reinforcing material is embedded in or surrounded by cured thermosetting resin. Fiber glass in the form of continuous filament, chopped fibers, or mats is the most frequently used reinforcing material. Granular or platelet fillers, thixotropic agents, pigments, or dyes may be included. Fiber glass-reinforced pipe (FRP) is by far the most widely used. Reinforced plastic mortar pipe (RPMP) consists of layers of thermosetting resin-sand aggregate mixtures between layers of resin-fiber glass reinforcements. In another construction, the sand is replaced by glass microspheres. In general, RTRP is produced by filament winding (qv) or centrifugal casting.

Properties and Tests

An important characteristic of plastic pipe is resistance to galvanic chemical attacks, such as those causing corrosion of metals by the action of water, soils, mild acids, bases, and other electrolytes. Other important features are ease of handling, durability, and economy. For many applications, plastic piping is the most cost-effective choice.

Typical physical, mechanical, and thermal properties of thermoplastic piping materials are given in Table 1; those of glass-fiber-reinforced, thermosetting resin pipe wall constructions are given in Table 2. The actual values of these limits and other properties may vary according to the material composition or wall construction.

Because the behavior of plastics is influenced by temperature, environment, and the duration of loading, the values of strength and stiffness obtained by short-term testing (Tables 1 and 2) are not suitable for long-term service design. The limits of durability under the anticipated-use conditions should be established.

Inertness to Drinking Water and Other Fluids. The base resins from which plastic pipe is made have been shown to be inert to drinking water. However, the pipe composition may contain additives that, if leached into the water, could affect toxicological safety, taste, or odor. Therefore, all products used in drinking-water installations must be tested and approved by special laboratories. The usual criteria for this evaluation are the minimum extraction, and taste and odor requirements given in National Sanitation Foundation Standard 14, *Plastic Piping System Component and Related Materials.*

Materials

Materials used in plastic pipe include poly(vinyl chloride), chlorinated poly(vinyl chloride) (CPVC), polyethylene (PE), cross-linked polyethylene (PEX), polybutylene (PB), polypro-

Table 1. Properties of Principal Thermoplastic Piping Materials[a]

Property at 23°C	ASTM test method	ABS	PVC	CPVC	PE	PB	PP	PVDF
specific gravity	D792	1.08	1.36	1.54	0.95	0.92	0.92	1.76
tensile strength, MPa[b]	D638	48	48	55	22	29	34	48
tensile modulus, GPa[c]	D638	2.1	2.5	2.9	0.9	0.4	1.4	1.5
impact strength, Izod, J/m[d] of notch	D256	210	270	80	>500	>500	100	200
coefficient of linear expansion, per °C, $\times 10^{-5}$	D696	10.8	9.0	6.3	16	13	8.0	13
thermal conductivity, W/(m·K)	C177	0.19	0.19	0.14	0.46	0.22	0.17	0.22
specific heat, kJ/(kg·K)[e]		1.4	1.0	0.8	2.3	1.9	1.9	1.2
approx operating limit[f], °C								
under pressure		70	50	90	60	90	65	140
not under pressure		80	54	100	80	100	90	150

[a] The actual properties are determined by the resin and additives used.
[b] To convert MPa to psi, multiply by 145.
[c] To convert GPa to psi, multiply by 145,000.
[d] To convert J/m to ftlbf/in., divide by 53.38.
[e] To convert J to cal, divide by 4.184.
[f] The actual maximum operating-temperature limit depends on the composition and use conditions.

Table 2. Properties of Glass-fiber-reinforced Thermosetting Resin Pipe[a]

Property at 23°C	ASTM test method	Cast polyester	Filament-wound Epoxy	Filament-wound Vinyl ester	Plastics mortar[b]
specific gravity	D792	1.6	1.9	1.9	1.7–2.2
tensile strength, MPa[c]	D2105				
hoop direction		240	140–350	140–350	100–400
axial direction		170	14–70	14–70	20–140
tensile modulus, GPa[d]	D2105				
hoop direction		10	10–28	10–28	10–25
axial direction		10	6–14	6–14	6–12
coefficient of linear expansion per °C, $\times 10^{-5}$	D696				
hoop direction		2.3	0.9–1.3	0.9–1.4	1.8
axial direction		2.3	1.8–3.2	1.8–3.2	2.7
thermal conductivity, W/(m·K)	C177	0.12	0.2–0.6	0.2–0.6	0.2
approx temp limit, °C[e]					
under pressure		95	115	105	60
not under pressure		115	150	120	95

[a] The actual properties are determined by the resin, additives, reinforcement, and wall construction.
[b] Polyester.
[c] To convert MPa to psi, multiply by 145.
[d] To convert GPa to psi, multiply by 145,000.
[e] The maximum operating-temperature limit depends on the specific pipe construction and the use conditions, including nature of environment.

pylene (PP), acrylonitrile–butadiene–styrene (ABS), poly(vinylidene fluoride) (PVDF), and polyacetal.

Reinforced Thermosetting Resin Pipe

Materials used in RTRP include epoxies, polyesters, vinyl esters, and furans.

BIBLIOGRAPHY

Standards for Plastics Piping, PPI TR-5, Plastics Pipe Institute, New York, 1988.

Plastic Pipe Assessment Study, American Gas Association, Arlington, Va., 1981–1983.

J. S. Murphy, *Trends in the Use of Plastics Pipe,* 1980–1984, Techline Industrial Data Services, Ltd., Surrey, UK.

Stan Mruk
Plastics Pipe Institute

PLANT DESIGN

Polymer synthesis plants resemble chemical synthesis plants. Polymer processing plants, which convert solid or liquid polymers made in polymer synthesis plants into useful objects, resemble plants in which mechanical goods are fabricated. This article assumes that the reader is familiar with chemical plant design or mechanical goods plant design and concentrates on those features of polymer synthesis and polymer processing plant design that are peculiar to polymer plants.

Unique Problems. The special problems of polymer plant design derive mostly from a few of the unusual physical properties of polymers. The designer should be aware of these unusual properties (eg, shear thinning and thermal degradation) and their implications for design. In particular, the designer should become familiar with the power law model of polymer viscosity and its use.

Quantitative Design. Except for the chemistry of polymer synthesis and stabilization, the polymer manufacturing industry has drawn very little on polymer science in a quantitative way for design and operation of its plants because the calculations necessary to apply polymer science in plant design were judged too complex and time-consuming to be practical. The advent of sophisticated, readily available computer programs to do process-type calculations is changing this situation. It may be confidently expected that within the life of this edition of the Encyclopedia practice in designing polymer plants will become much more quantitative with favorable consequences to costs, polymer product quality, and plant reliability.

Site Development

The geographic location selected for a plant is most important to its economic success. Access to markets, access to raw materials, transportation costs, local taxes, labor costs, labor quality, and waste disposal costs all play a large part in determining whether a new plant will be successful. Once a plant site is selected, the first step in designing a production unit is thinking through the site development plan. To make a site development plan, one should consider *inter alia* production unit block size; flows; utilities needs; shop, office, and laboratory requirements; neutralizing objectional features of production units; roads, pipe racks, sewers, fire protection; drainage; safety problems; outside hazards; zoning, building codes; waste treatment; and initial cost minimization.

Polymer Synthesis Plants

Polymer synthesis plants are usually built outdoors for reasons of economy and safety. Only operations in which polymers might become contaminated, such as filtration or pelletizing, are placed indoors. The population density in a synthesis plant is usually low compared with polymer processing plants (see also POLYMERIZATION PROCEDURES, INDUSRTRIAL).

Design decisions need to take into account the equipment and processing steps specific to the plants' operation, ie, type of polymerization reactor, reaction vessels, agitation system, reaction runaway, and reactor venting, reactor discharge, devolatilization, melt filtration, filtration and washing, drying polymer solids, pelletizing, polymer pellet handling and storage, and waste disposal.

Instrumentation and Control of Polymer Synthesis Plants. The instrumentation of polymer synthesis plants is much like the instrumentation used in organic chemicals synthesis plants. The common variables measured and controlled are temperature, pressure, flow, level, and power. The instruments used to measure temperature, pressure, and flow of polymer melt streams are necessarily different because of the high viscosity of polymer melts and solutions.

Safety. Polymer synthesis plants hold many potential hazards for personnel and property. Polymer synthesis plant designers should look carefully into the problem presented by their specific process and devise designs appropriate to the process in order to arrive at safe and efficient plants.

Polymer Processing Plants

The operations in polymer processing plants—compounding (qv), extrusion (qv), injection molding (qv), film blowing, fiber spinning, embossing, thermoforming (qv), decorating (qv), regrinding, etc—resemble operations in plants that make mechanical goods (see also BLOW MOLDING; CALENDERING; COMPRESSION AND TRANSFER MOLDING; FIBERS, MANUFACTURE; FILMS, MANUFACTURE).

Because the polymer processing industry is highly competitive, plants must be highly productive, labor efficient, and economical of raw materials (produce little scrap). They should meet quality requirements consistently while being economical of capital in order to succeed. They must be safe and meet standards of good practice and all applicable codes and regulations. The population density in polymer processing plants is usually high. Particular attention should be paid to providing a safe, comfortable, hazard-free working environment.

Ideally, raw materials should arrive at one side of a processing plant from suppliers or from a warehouse and flow directly through the processing machines to a warehousing or packing and shipping area on the opposite side of the plant. Perhaps the most important aspect of processing plant design is development of a superior plant layout.

A common sequence of operations is raw materials handling and distribution, compounding, pellet drying, forming (extrusion, injection molding, etc), finishing (sprue and flash removal, etc), inspection, warehousing, packing, and shipping. A regrind operation to reuse plastic materials, sprues, reject, flashing, etc, is usually required.

BIBLIOGRAPHY

J. A. Biesenberger and D. H. Sebastian, *Principles of Polymerization Engineering,* John Wiley & Sons, Inc., New York, 1983.

Z. Tadmor and C. G. Gogos, *Principles of Polymer Processing,* John Wiley & Sons, Inc., New York, 1979.

J. M. McKelvey, *Polymer Processing,* John Wiley & Sons, Inc., New York, 1961.

C. Rauwendaal, *Polymer Extrusion,* Carl Hanser Verlag, Munich, 1986.

DON E. CARTER
Monsanto Company

PLASMA POLYMERIZATION

Polymerization of molecules, inorganic or organic, saturated or unsaturated, occurs in various plasma environments, such as plasma jets, electron beams, ion beams, corona discharges, laser-induced plasmas, and nonequilibrium glow discharges at low and high pressures and low and high temperatures.

In this review the polymerization from vapor-phase molecules in a nonequilibrium plasma at low temperatures and pressures is discussed. Ion–molecule and radical–molecule mechanisms are chiefly responsible for the dissociation of neutral species and subsequent polymerization. Introduction of inert gases as carriers enhances these reactions. The presence of very small amounts of water or halogens may increase the polymerization rate, in some cases to more than 50 times, as does additional electromagnetic radiation such as laser beams.

Distinction is made between plasma polymerization and plasma deposition. However, the nomenclature in the literature differs sometimes in the interpretation. The term plasma polymerization is generally applied to a process occurring in surface film formation, whereas the term deposition is more often applied to deposition of powder particles formed in the gas phase. However, some researchers refer to plasma polymerization as the polymerization that occurs at high rates in the gas phase resulting in powder, and to deposition as any sorption occurring on the surface, with each sorption step being a deposition step. Deposition occurs when an impinging particle fails to bounce back from a colliding surface. The sticking or deposition coefficient can be defined as the number of particles deposited divided by the total number of impinging particles.

The plasma usually employed in laboratories and commercial reactors consists of a low pressure glow discharge of positively and negatively charged species, cations, anions, electrons, free radicals, excited and neutral species, and electromagnetic radiation. The number density of positively charged ions equals the number of negatively charged particles, resulting in an electrically neutral plasma. An important aspect of such low temperature, low pressure plasma is the nonequilibrium state of this glow discharge, ie, while the ions and neutral molecules are at room or elevated temperature, the electrons exist at very high temperatures; at the pressures typically used, the electron temperature may be as high as 20,000–30,000 K. Therefore, the nonequilibrium plasma has generally the following range of properties: electron temperature $\approx$ 1–10 eV, electron density $\approx 10^9$–$10^{12}/cm^3$ and pressure $\approx$ 13.3–133 Pa (0.1–1 mm Hg).

Comparison of the energies in a glow discharge with the typical bond energies in molecules shows that the energies of species in such a plasma are sufficient to break covalent bonds in molecules exposed to this environment. Plasmas are rich sources of electromagnetic radiation, particularly in the uv and vacuum uv, and are frequently used as sources in those regions. In fact, an entirely new plasma phenomenon occurred with the production of a plasma in a gas by the action of light alone.

This technique is effectively employed in modern plasma reactions.

Many films can be deposited in a plasma at low temperatures, whereas conventional vapor deposition would require prohibitively high temperatures. Recent advances in technology and apparatus design resulted in higher deposition rates in clearly defined zones and yielded films of higher quality.

Kinetics of Polymerization

The primary processes in the generation of ions involve electron impact and photoionization. Similarly, free radicals may be generated by electron impact, thermally and by photolysis. Ionization processes also take place via collisions of metastable species and molecules or atoms. These processes are referred to as Penning ionization and are treated in terms of metastable–normal and metastable–metastable collisions resulting in ionization.

Because of ion–molecule reactions with the discharge gas of the plasma or with trace impurities, primary ions readily transform into secondary ions through a series of steps before termination. They may then have low enough heat of formation for reaction with neutral components, are further broken up to form free-radical ions, leave through the reactor, or are lost by recombinations. The following basic reactions control deposition of films in a glow discharge: electron-impact reactions, inelastic collisions between heavy particles, and heterogeneous reactions.

Polymerization of Organic and Inorganic Monomers

In the deposition of aliphatic and aromatic polymer films in a plasma, all monomers, saturated or unsaturated, can be made to polymerize in a glow discharge, including those that are not polymerized by any of the conventional techniques. In general, insoluble material is deposited or formed on a surface as a film. Under certain conditions of glow discharge, however, soluble materials of lower molecular weight may be obtained or, under other conditions, the polymer is obtained in powder form. Among unusual conditions, polymerization of methane in a glow discharge at liquid air temperature gives polymethylene.

Polymerization is usually initiated and partially propagated in the vapor phase. As propagation proceeds the growing polymer cannot remain in the vapor phase and diffuses onto a surface. However, primary decomposition products in the gas phase, such as ions and free radicals, may simultaneously interact with surfaces inside the reactor that have been activated by the impact of the glow discharge to form ions or free radicals from the surface molecules of substrates, followed by propagation steps on that surface.

It is characteristic of plasma that virtually any organic or organometallic vaporizable species can be polymerized. In a hydrocarbon plasma a number of ions have been identified, including CH_3^+, $C_2H_2^+$, $C_2H_3^+$, $C_2H_4^+$, and $C_2H_5^+$; the prevailing species is $C_2H_3^+$. In r-f plasma deposition of hydrocarbons the rates decrease in the order $C_2H_6 > C_3H_8 > C_4H_{10}$ for these monomers; the deposition rates also decrease with increasing flow rate. The conventional initiation, propagation, and termination steps may be preceded by an excitation step, permitting the excited species to engage in so-called Penning dissociation or Penning ionization and deexcitation to issue electromagnetic radiation. Plasma films are usually highly cross-linked and resistant to higher temperatures, abrasion, and chemicals; adhesion to substrates is high.

BIBLIOGRAPHY

H. Yasuda, *Plasma Polymerization,* Academic Press, Inc., Orlando, Fla., 1986.

H. V. Boenig, *Plasma Science and Technology,* Cornell University Press, Ithaca, N.Y., 1982.

H. V. Boenig, *Advances in Low Temperature Plasma Chemistry and Technology Applications,* Technomic Publishing Co., Inc., Lancaster, Pa., 1984.

A. T. Bell in J. R. Hollahan and A. T. Bell, eds., *Techniques and Applications of Plasma Chemistry,* John Wiley & Sons, Inc., New York, 1979.

Y. Catherine, *Proceedings of the Symposium on Plasma Processes,* Electrochemical Society, Princeton, N.J., 1985.

HERMAN V. BOENIG
Research Institute of Plasma Chemistry and Technology

PLASTIC

A plastic is a material that contains as an essential ingredient one or more organic polymeric substances of large molecular weight, is solid in its finished state, and, at some stage in its manufacture or processing into finished articles, can be shaped by flow. Rubber, textiles, adhesives, and paint, which may in some cases meet this definition, are not considered plastics (see also POLYMER).

BIBLIOGRAPHY

"Standard Definitions of Terms Relating to Plastics," ASTM D883-86b, American Society for Testing and Materials, Philadelphia, 1986.

PLASTICIZERS

A plasticizer is incorporated in a material to increase its workability, flexibility, or distensibility (ASTM D883). Addition of plasticizer may lower the melt viscosity, the second-order transition temperature, or the elastic modulus of the plastic. For a plasticizer to be effective with any polymeric material, the two must be intimately mixed. With organic resins, this is done most commonly by heating until the resin dissolves in the plasticizer or the plasticizer dissolves in the resin. The resultant mixture is then formed and cooled. Alternatively, resin and plasticizer may be mixed by dissolution in a common solvent without heat, followed by removal of the solvent by evaporation. Plasticizer effects are apparent when a high molecular weight poly(vinyl chloride) homopolymer without plasticizer or stabilizer is fluxed on a hot two-roll mill. At 160°C, the resin particles are unchanged and fall like sand between the rolls. At slightly higher temperatures, the resin softens and forms a sheet. It also will decompose from excessive heat. When the sheet is removed and cooled, it is brown to black, horny to brittle, corrosive from liberated hydrogen chloride, and worthless. If, on the other hand, a small amount of a plasticizer such as bis(2-ethylhexyl) phthalate (dioctyl phthalate, DOP) is added when the resin is poured onto the roll mill at 160°C, the resin softens and fuses; the mass becomes homogeneous and forms a sheet around the hotter or the faster moving of the two rolls. If the sheet is removed quickly and cooled, it is clear, colorless to faintly yellow, noncorrosive, and semirigid to very flexible, depending on the amount of plasticizer used.

Plasticizers usually are moderately high molecular weight organic liquids or, occasionally, low melting solids. Most commonly they are esters of dicarboxylic acids or phosphoric acid. Other types include hydrocarbons, halogenated hydrocarbons, ethers, polyesters, polyglycols, and sulfonamides. The principal use of plasticizer is in thermoplastic resins; PVC because of its unique molecular structure, which accepts a wide range of plasticizers but limits their effects by an unusual array of crystal systems, reflects the largest use area for organic plasticizers. Many other thermoplastic resins are suited to beneficial modification by plasticizers. Thermosetting resins used in the field of surface coatings are modified by plasticizers to affect flow and leveling during application and to improve impact and check resistance of the dried film. Some, such as phenolics, are plasticized to improve forming during an intermediate processing stage and impact resistance of the final product.

Nomenclature

The rules of nomenclature do not lend themselves conveniently to many plasticizers. In plasticizer terminology iso- denotes a highly branched alkyl radical, usually from the oxo process; usually a mixture of isomers is meant.

The name *N*-ethyl (*o,p*)-toluenesulfonamide refers to a mixture of ortho and para isomers. A name such as (*n*-octyl, *n*-decyl) phthalate denotes the statistical mixture obtained when phthalic anhydride is completely esterified with approximately equimolar quantities of *n*-octyl and *n*-decyl alcohols.

Physical Properties and Test Methods

For quality control in either plasticizer manufacture or purchase, a number of properties are routinely measured. Most common, and frequently the only ones measured in purchasing, are specific gravity and refractive index (Table 1).

Table 1 includes three other types of more complex physical properties: solubility parameters δ, a thermodynamic property; dielectric constant ϵ; and the hydrogen-bonding index γ. These are related to plasticizer compatibility and performance and will be discussed later.

Mechanism of Plasticizer Action

Four general theories have been proposed to account for the effects that plasticizers have on certain resins. According to the *lubricity theory,* the rigidity or the resistance of a resin to deformation results from intermolecular friction. The plasticizer acts as a lubricant to facilitate movement of the resin macromolecules over each other and provides internal lubricity.

The *gel theory,* which was devised for amorphous polymers, contends that their resistance to deformation results from an internal, three-dimensional honeycomb structure or gel. This gel is formed by loose attachments, which occur at intervals along the polymer chains. The dimensions of the cells in a stiff or brittle resin are small because the points of attachment are close. The action of a plasticizer on a resin with many points of attachment along the polymer chains is breakage of the attachment and masking of the centers of force, which hold the polymer chains together, by selectively solvating the polymer at these points. This produces much the same result as if fewer points for attachment were provided on the macromolecules in the first place.

It is assumed that solvents or plasticizers of different classes are attracted to the resin macromolecule by forces of different magnitudes and that none of them are bound permanently when they are attached. Instead, there is a continuous exchange whereby one plasticizer molecule becomes attached to a given active group or force center only to be dislodged and to be replaced by another. This results in a dynamic equilibrium between solvation and desolvation, by which a certain fraction of the force centers of the polymer chains are masked by solvent or plasticizer under a given set of conditions such as plasticizer concentration, temperature, and pressure. This solvation–desolvation with accompanying aggregation–disaggregation of resin molecules is basically the *mechanistic theory.*

The *free volume theory* is related to the molecular theory, which assumes that there is nothing but free space between molecules. Because free volume increases with increased molecular motion, a disproportionate amount of free volume is associated with the chain end groups in a polymer system. Increasing the concentration of chain ends by decreasing the molecular weight, increases the free volume. Addition of flexible side chains produces a similar result. All of these effects can be used for internal plasticization, and the free volume is geographically fixed with regard to the polymer. However, addition of a smaller molecule adds a larger free volume at any location for the amount of material added; this is external plasticization. The size and shape of the molecule that is added and the nature of its atoms and groups of atoms (ie, nonpolar, polar, hydrogen bonding or not, and dense or light) determines how it functions as a plasticizer. In all cases, the free volume increases with added heat; this may result in "plastication."

Processing Aids versus End-use Plasticizers. Plasticizers are frequently used to aid in processing. Easier processing can be achieved by better internal lubricity, reduced sticking of resin to metal equipment, lower hot-melt viscosity, or improved resin stability. At times a plasticizer can aid processing in all of these ways simultaneously. An idealized response of a resin to plasticizers is as processing aid or as end-use plasticizer. The processing aid is intended to lower the processing temperature without changing the softening temperature or upper limit of usefulness of the final article. The end-use plasticizer must lower the softening point T_g, which is the lower limit of usefulness, and not affect the flow temperature. This produces a broadening of the elastomeric range desired from the plasticizer.

Plasticization. A stepwise hypothesis has been suggested. Initially there is a rapid, irreversible plasticizer uptake as plasticizer molecules enter porous areas of the resin and adhere there. Then absorption occurs in two steps. The first, with a low to moderate activation energy, implies a simple diffusion process. The second absorption step, a more complex diffusion process, involves higher activation energies and it appears that polymer molecules are disentangling and separating. At this point, most or all of the plasticizer is incorporated in the resin and slight volume changes have taken place, but plasticization is not complete. Plasticizer is present as clusters of molecules between clusters or bundles of polymer segments or molecules. As energy is applied to the system, a marked change occurs:

Table 1. Plasticizer Properties and Compatibility Predictors[a]

Plasticizers	Specific gravity at 25°C	n_D^{25}	Boiling range, °C/kPa[b]	Pour point[c] (or melting point), °C	Solubility parameter[d], $(J/cm^3)^{1/2}$ δ	Dielectric constant at 10 kHz ϵ	H-bond index γ
adipic acid derivatives							
heptyl, nonyl adipate	0.920	1.444	224/1.23	−13 (p)	17.38	4.15	
bis(2-ethylhexyl) adipate (DOA)	0.924	1.446	214/0.66	−65 (p)	17.30	4.13	6.3
bis(2-butoxyethyl) adipate	0.997[e]	1.442	208–215/0.53	−34 (c)	17.98	5.46	
azelaic acid derivative							
bis(2-ethylhexyl) azelate (DOZ)	0.916	1.446	237/0.66	−63 (p)	17.26	4.04	
benzoic acid derivatives							
dipropylene glycol dibenzoate	1.129	1.528	225–235/0.66	−40 (f)	19.66	7.44	7.4
citric acid derivatives							
tri-*n*-butyl citrate	1.042	1.443	170/0.13	−62 (p)	18.49	7.62	
tri-*n*-butyl acetylcitrate	1.048	1.441	173/0.13	−59 (p)	18.45	4.42	
epoxy derivatives							
epoxidized soy oil (ESO)	0.993	1.471	>150 (dec)/0.66	−1 (p)	ca 18.20	5.47	10.2
2-ethylhexyl epoxy tallate	0.920	1.458	>215 (dec)/0.66	−22 (p)	ca 17.58		9.9
glycol derivatives							
triethylene glycol di(2-ethylbutyrate)	0.995[e]	1.441	196/0.66	−65 (p)	17.65	6.59	
hydrocarbons and derivatives							
hydrogenated terphenyls HB-40	1.004	1.567	340–396/101	−26 (p)	18.45	2.53	2.7
chlorinated paraffin (52 wt % chlorine) Cereclor S-52	1.230		>260 (dec)/101		18.71	8.59	2.5
oleic acid derivatives							
butyl oleate	0.865	1.451	190–230/0.87	−27 (p)	17.16	4.0	
phosphoric acid derivatives							
tributyl phosphate	0.975	1.426	289/101	<−100	17.50	7.96	11.7
tri(2-ethylhexyl)phosphate (TOF)	0.926[e]	1.441	220/0.66	−90 (p)	16.83	4.90	13
tributoxyethyl phosphate	1.020	1.434	215–228/0.53	−94 (p)	17.53	8.7	
triphenyl phosphate (TPP)	1.268 (60°C)	1.550 (60°C)	370/101	49 (m)[f]	20.65	9.6	
tricresyl phosphate (TCP)	1.165	1.556	265/0.66	<−35 (c)	20.16	7.20	18.8
t-butylphenyl diphenyl phosphate	1.180	1.555		−25 (p)	20.82	7.33	10.0
2-ethylhexyl diphenyl phosphate	1.091	1.508	230/0.66	−54 (p)	19.02	7.33	10.8
phthalic acid derivatives							
dimethyl phthalate (DMP)	1.190	1.513	284/101	0.0 (c)	21.37	8.49	
dibutyl phthalate (DBP)	1.046	1.493	340/101	−40 (p)	19.24	6.42	5.7–6.3
di(heptyl, nonyl, undecyl) phthalate	0.969	1.482	252/1.33	<−50 (p)	18.24	4.76	5.6
bis(2-ethylhexyl) phthalate (DOP)	0.982	1.485	231/0.66	−47 (p)	18.06	5.19	5.6
di(*n*-hexyl, *n*-octyl, *n*-decyl) phthalate DNODP-610	0.978[e]	1.483	177–259/0.53	−48 (p)		4.70	
diisononyl phthalate (DINP)	0.973[e]	1.486				4.67	
diisodecyl phthalate (DIDP)	0.966	1.483	255/0.66	−46 (p)	17.50	4.43	
diundecyl phthalate (DUP)	0.954	1.481	262/1.33	+2 (c)	17.94	4.43	5.8
ditridecyl phthalate (DTDP)	0.952	1.484	285/0.47	−37 (p)	17.28	4.05	
butyl benzyl phthalate (BBP)	1.119	1.538	370/101	−45 (p)	20.20	6.47	5.7
7-(2,6,6,8-tetramethyl-4-oxa-3-oxo-nonyl) benzyl phthalate Santicizer 278	1.097	1.519	243/1.33	−6 (p)	19.12	5.86	5.7
dicyclohexyl phthalate (DCHP)	solid		212–218/0.66	63 (m)[f]	19.61	6.42[g]	
polyesters							
adipic acid polyester (mol wt 6000) Paraplex G40	1.15	1.471		<−15 (f)			
adipic acid polyester (mol wt 2000) Drapex 334F	1.083	1.465		4 (p)	19.02	6.02	6.5
azelaic acid polyester (mol wt 850) Plastolein 9720	1.03	1.460		−8 (p)			
ricinoleic acid derivatives							
methyl ricinoleate	0.925	1.462	170/0.13	−30 (p)	17.42	5.03	strong
castor oil (90 wt % glyceryl ricinoleate)	0.963	1.475		−12 (p)	18.16		18.7
sebacic acid derivatives							
di-*n*-butyl sebacate (DBS)	0.933	1.439	349/101	−11 (f)	17.75		
bis(2-ethylhexyl) sebacate (DOS)	0.911	1.451	256/0.66	−48 (m)	17.28	3.88	

Table 1. (Continued)

Plasticizers	Specific gravity at 25°C	n_D^{25}	Boiling range, °C/kPa[b]	Pour point[c] (or melting point), °C	Solubility parameter[d], $(J/cm^3)^{1/2}$ δ	Dielectric constant at 10 kHz ϵ	H-bond index γ
stearic acid derivatives							
n-butyl stearate	0.855	1.433	220–225/3.33	27 (m)	16.79	3.11	
sulfonic acid derivatives							
N-ethyl (*o,p*)-toluenesulfonamide	1.190	1.540	340/101	ca 0.0	29.43	36.5	(N) 8.5 (O) 23.4
alkyl sulfonic acid ester of phenol and cresol, Mesamoll	1.05[e]	1.499		<−10	19.73	10.6	
terpenes and derivatives							
camphor	0.990	1.5462	200 (sub)	174[f]	18.32	11.35	
hydrogenated methyl ester of rosin, Hercolyn D	1.02	1.52	362/101		17.57		
trimellitic acid derivatives							
tris(2-ethylhexyl) trimellitate (TOTM)	0.987	1.4823	260/0.13	−46 (p)	18.40	4.7	

[a] Data for commercial products.
[b] To convert kPa to mm Hg, multiply by 7.5.
[c] Pour point (p), melting point (m), double melting point (m2), freezing point (f), crystallization point (c).
[d] To convert $(J/cm^3)^{1/2}$ to $(cal/cm^3)^{1/2}$, multiply by 0.489.
[e] Measurements at 20°C.
[f] Crystalline solid; its performance as the sole plasticizer varies with its degree of crystallinity in the resin.
[g] Super-cooled liquid.

the volume remains the same but the dielectric constant rises significantly, indicating that the polymer molecules are no longer held together rigidly but are free to move as if the polymer were rubbery rather than glassy. Thus the plasticizer molecules have penetrated the clusters or bundles of polymer molecules and plasticization essentially is complete. After heating this mixture, a toughening structure forms upon cooling and may result from van der Waals forces, hydrogen bonding, or crystallization of polymer segments.

Antiplasticization. Sometimes, when plasticizers are added to polymers, they increase modulus and tensile strength and decrease elongation. Such antiplasticization results most frequently when small amounts of plasticizer are used. Although conventional plasticizers at low concentrations are effective processing aids for rigid PVC, the product is unacceptably brittle.

The reasons for this behavior have been debated and various theories have been compiled in some detail. Increased order in these embrittled systems has been attributed to resin crystallization, which results from increased freedom of motion when a small amount of plasticizer is present. Only after a large amount of plasticizer is added do the crystallites redissolve sufficiently for plasticization to occur.

Compatibility

To those concerned with development and production of plastics materials, eg, resins, solvents, and plasticizers, compatibility (qv) is the ability of two or more substances to mix with each other to form a homogeneous composition of useful plastic properties. To some it means exactly the opposite, so a rubber gasket that may be used with a hydraulic fluid without absorbing it and swelling is "compatible" with that fluid. This is "functional compatibility." We will use the first definition unless so stated.

Compatibility of a plasticizer with any resin is a function of its chemical makeup. The number, kind, and arrangement of atoms govern the forces involved. All influence the energy of vaporization, solubility parameter, dipole moment, hydrogen bonding, viscosity, etc, but not in the same proportions. Measurement of any one of these properties gives one viewpoint of the plasticizer molecule, which may then be related to its compatibility with any given resin. Two or more analyses from wisely chosen properties can, theoretically, be used to pinpoint the compatibility behavior of a plasticizer with any given resin. Because the dielectric constant ϵ is a measure of these intermolecular forces and their effect on the molecule's ability to adapt itself to the electrical forces of its environment, the solubility parameter and the dielectric constant may be combined as indicators of plasticizer compatibility.

Interaction Parameter. The theory of Flory and Huggins is based on a statistical-mechanical treatment of a lattice model of polymer solutions. According to theory, the free energy of mixing of a polymer solution is given by

$$\Delta G = RT(n_1 \ln v_1 + n_2 \ln v_2 + \chi n_1 v_2)$$

where ΔG is the free energy of mixing; R is the gas constant; T is the absolute temperature; n_1 and n_2 are the number of moles of solvent and solute, respectively; v_1 and v_2 are the volume fractions of solvent and solute, respectively; and χ is the interaction parameter.

If ΔG is negative, the polymer and solvent forms a solution or, presumably, the polymer and plasticizer are compatible. The interaction parameter is a semiempirical constant, which adjusts for nonideality of the heat and entropy of mixing. In reality, it is not truly a constant for it may vary significantly with, eg, concentration. Nevertheless, the concept has proved sufficiently useful that many equations involving χ have been developed so that various types of measurement can be correlated with polymer–solvent or polymer–plasticizer compatibility. These include measurement of osmotic pressures of dilute solutions, equilibrium swelling, vapor pressures, viscosities, stress–strain behavior of swollen polymers, and depression of melting or freezing points.

Primary and Secondary Plasticizers. It is not at all uncommon for plasticizers to have considerable but incomplete compatibility with a given resin. Plasticizers that are highly compatible with a given resin are primary plasticizers; those that are not are secondary plasticizers. The distinction is hard to quantify, but if a plasticizer can be used alone at high enough concentration to achieve the desired properties without exuding, it is a primary plasticizer.

Temperature Dependence. The effect of temperature on the mechanism of plasticization has been explained according to the mechanistic theory. With polymer–polymer forces tending toward aggregation of polymer macromolecules and resultant syneresis or exudation of plasticizer, and with polymer–plasticizer forces tending toward disaggregation and solvation, the system is in dynamic equilibrium at a given temperature. As the temperature changes, the magnitudes of the effects of these forces change but not to the same degree. With more energy, ie, higher temperature, it is easier to separate polymer–polymer bonds, but it also is easier to separate polymer–plasticizer bonds, which favors aggregation. At lower temperatures, both separations are more difficult.

Depending on the relative magnitude of these changes, a given liquid may be a better solvent for a polymer when hot rather than when cold and another liquid may be a better solvent when cold rather than when hot.

External Versus Internal Plasticization

The solvation–desolvation equilibrium is characteristic of all plasticized resins. Because plasticizer molecules are not attached to the resin by primary bonds, the plasticizer may be lost by evaporation, migration, or extraction.

To circumvent these shortcomings, copolymers were developed in an effort to make products with properties otherwise like a plasticized resin. Thus small to moderate amounts of vinyl acetate were polymerized with vinyl chloride to make poly(vinyl chloride-*co*-vinyl acetate), eg, 95:5 or 85:15 vinyl chloride to vinyl acetate. Such copolymers may be softer than the related polymers, such as poly(vinyl chloride). The term "internal plasticization" was coined to describe these processes, and "external plasticization" then had to be coined for plasticization when necessary for clarification.

Many internally plasticized products have been developed, but none have the properties of the plasticized resin. They are better in extraction and migration resistance and in volatile loss, because the flexibilizing side chains, etc, are an integral part of the polymer. But because they lack the mobility of an added plasticizer they have poorer tear strengths and usually do not have as broad a useful temperature range as a truly plasticized polymer. They also lack the ease of formulation to achieve a wide range of properties that may be achieved by plasticization.

General-purpose Plasticizers

Raw material sources and brief descriptions of syntheses have been given elsewhere. Phthalate esters, particularly dialkyl phthalates, have dominated plasticizer technology since the 1930s. In one good comparison of their performance, they are called the "work horse plasticizers". For many years, bis(2-ethylhexyl) phthalate (DOP) was the accepted industry standard for a general-purpose plasticizer for PVC and is a benchmark for comparison of other plasticizers.

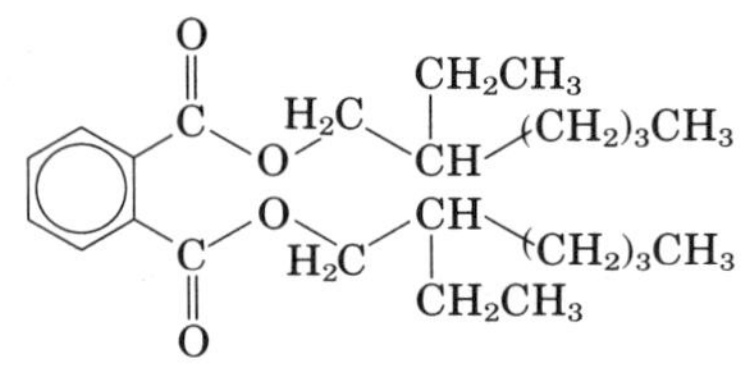

bis(2-ethylhexyl) phthalate (DOP)

Its all-around performance, eg, compatibility with the resin, efficiency in flexibilizing, low volatility, and resistance to extraction are so good that it alone has accounted for one-fourth of the total plasticizer production. More highly branched alcohol phthalates were developed concurrently with the oxo process.

Various diisoalkyl phthalates have accounted for another quarter of the market. And the linear and near-linear phthalates account for approximately another one-fourth of the total market. Collectively, these dialkyl phthalates are so widely usable they are considered general-purpose plasticizers. Because plasticization of PVC consumes ca 80% of plasticizer production, most plasticizers are compared to DOP for price-performance relationships.

Specialty Plasticizers

Specialty plasticizers include benzyl phthalates, glycol benzoates, trimellitates, hydrocarbons and petroleum products, flame-retardant plasticizers (chlorinated paraffins and phosphate esters), low temperature plasticizers (adipates, azelates, sebacates, and tri(2-ethylhexyl) phosphate), polymeric plasticizers, and epoxy plasticizers.

Poly(vinyl chloride)

Modification of PVC by plasticizers may be considered the standard for plasticization. PVC is unique in its acceptance of large amounts of plasticizers with gradual change in physical properties from a rigid solid to a soft gel or viscous liquid (see also VINYL CHLORIDE POLYMERS). Most other plasticized resins undergo this transformation rapidly with either a slight increase of plasticizer content or a slight increase in temperature. Poly(vinyl chloride) has a helical structure with repeating units of $C_{28}H_{42}Cl_{14}$. Application of Zhurkov's law, ie, that there must be one molecule of plasticizer to block each polar group on the polymer chain, to the helical structure of PVC reflects the need for 45 phr of DOP for complete plasticization. The helical structure and its ability to maintain some kind of molecular order, as shown by disappearance and reappearance of the crystallites in PVC solutions, must in part account for such unique response. In general, plasticizers for PVC and poly(vinyl chloride-*co*-vinyl acetate) resins are esters of aliphatic and aromatic dicarboxylic and tricarboxylic acids and organic phosphates. Secondary plasticizers include high molecular weight alkylaromatic hydrocarbons, chlorinated aliphatic hydrocarbons, and occasionally other types.

Methods of Plasticization. There are five principal ways in which plasticizers are incorporated into resins such as PVC: hot compounding, dry blending, plastisols, solutions, and latices. The first methods of incorporating the plasticizer into the resin require energy input by heating and, frequently, vigorous mixing. In the solvent method, the free energy of the resin–solvent system and the small molecular size of the solvent permit intimate mixing of plasticizer and resin without heating. In the method for latices, a variety of forces combine to promote plasticization.

Fusion. The effect of plasticizer on the ease of fusion of resin–plasticizer blends is economically as well as technically important in determining the best processing conditions. In a dynamic fusion test in a torque rheometer with a heating bath temperature of 67°C, fusion of a DOP–PVC blend required 19.4 min, whereas that of a comparable blend with butyl benzyl phthalate required only 3.8 min. When the rheometer is heated during processing, the relative fusion temperatures of plasticizers may be determined. Less heat and mechanical energy is necessary for fusion of PVC with DOP than with diisodecyl adipate. Butyl benzyl phthalate requires even less than DOP.

A list of plasticizer families in order of decreasing ease of fusion is phosphates, phthalates, isophthalates, terephthalates, adipates, azelates, and sebacates. Ease of fusion without shear tends to increase with decreasing molecular size and increasing polarity up to a dielectric constant of ca 8.

Vinyl Foam. Plasticized vinyl foam can be produced by mechanical frothing, incorporation of air or other gas in the plastisol prior to fusion, or by chemical blowing, decomposition of an additive by heat during processing to form gas bubbles.

Modification of Physical Properties. Plasticizer properties and performance in poly(vinyl chloride) are listed in Tables 1 and 2. The main variables are the kind of plasticizer and its concentration. Because of antiplasticization at low plasticizer concentration, for most resin–plasticizer combinations at room temperature, a plasticizer threshold concentration must be passed before the normal plasticizer effects on physical properties are observed. The plasticization threshold is at about 10 wt % for DOP at room temperature but is lower for linear aliphatic plasticizers and higher for more aromatic and more polar plasticizers.

Efficiency. Plasticizer efficiency is a measure of its ability to bring about a desired effect. The more plasticizer required to produce the effect, the less efficient it is. Efficiency factors are devised to compare one plasticizer to another; such factors must be obtained by measurements of each desired physical property. More viscous plasticizers are less efficient than less viscous ones. A high proportion of compatibility-inducing groups in the molecule lowers efficiency. However, with linear polyester plasticizers, their high molecular weight leads to high viscosity but because of their predominantly aliphatic linear structures, they react almost segment by segment and are more efficient than otherwise would be predicted.

Mixtures of Plasticizers. In the modification of physical properties, a blend of two or three plasticizers usually provides effects that are essentially the weighted average of the effects produced by the individual plasticizers. This presumes the plasticizers are mutually compatible with each other and are sufficiently compatible with the polymer. Occasionally plasticizers are sufficiently different that they are not miscible in all proportions.

Effect on Resin Chemical Properties. *Heat Stability.* Incorporation of plasticizers into PVC tends to inhibit thermal degradation during processing. Increased lubricity and decreased hot melt viscosity are important, but chemical inhibition, particularly with epoxy plasticizers, is a major factor.

Oxidative Stability. During processing, conditions are favorable for initiating autooxidation. Plasticizers with structures particularly susceptible to autooxidation, such as tertiary hydrogens, may contribute to early color development and other undesired effects. Antioxidants are commonly used to control these problems.

Latent Degradation. Sometimes the unwanted effects do not appear during or immediately after processing but are evident after further standing, particularly after storage in a hot warehouse. Because the hydroperoxides produced by oxidation of plasticizers during processing are colorless and odorless, no harm is observed until they begin to decompose and oxidize further. The aldehydes, ketones, free acids, and resinous products may have an objectionable odor. The product may be covered with brownish splotches or may be sticky from exuding plasticizer. Such delayed effects are those of latent degradation, because the damage occurred during processing. Again, antioxidants control the problem.

Light Stability. In general, plasticizers improve light stability of PVC when they are used in plasticizing amounts. But effects can range from very beneficial (linear phthalates) to quite harmful (triaryl phosphates alone), depending on compatibility, concentration, and total formulation; uv absorbers are beneficial.

Ionizing Radiation. Plasticized PVC is subjected to β-rays and γ-rays industrially for three major purposes: cross-linking of the polymer, sterilization, and use of materials in environments of high radiant energy. By use of DOP, linear phthalates and adipates with proper stabilization, problems of degradation of PVC with liberation of HCl, stiffening and darkening of the composition are overcome.

Flammability. Because of its high chlorine content, rigid PVC burns with difficulty. Plasticizers add fuel value, but flammability of plasticized PVC is heavily dependent on type and concentration of plasticizer present. Phthalates, adipates, and polyesters contribute to burning; triaryl and alkylaryl phosphates inhibit burning of plasticized PVC and can be blended with other types of plasticizer to achieve an acceptable balance of flame resistance and physical properties.

Corrosion Resistance. Corrosion of PVC is governed largely by the plasticizer. When films of plasticized PVC were exposed to 12 environments ranging in corrosiveness from distilled water through solutions of various acids, bases, and salts, phthalate-plasticized resins were least attacked. Phosphates, adipates, and polymerics were less corrosion resistant. Secondary plasticizers are sometimes beneficial.

Biological Attack. Most commercial synthetic polymers are inert to biological systems. On occasion, however, they may be attacked by microbiological agents such as bacteria and fungi, and macrobiological agents such as termites, insects, and rodents. Microbiological attack is to a large extent chemical, whereas macrobiological attack is primarily physical.

Plasticizers impart to the resin system their own biological properties. The extent of their influence is governed by their compatibility with the resin, ease of extraction, migration, and volatility in addition to their innate biological properties.

Fungal and Bacterial Attack. Some plasticizers are utilized

Table 2. Plasticizer Performance in Poly(vinyl chloride)[a,b]

Plasticizer	Volatility[c] plasticizer loss, wt %	T_f[d], °C	Water extraction[e], wt % loss	Kerosene extraction[f], wt % loss	Hardness[g], Shore A
adipic acid derivatives					
heptyl nonyl adipate	8.5	−71	0.15	>70	66
bis(2-ethylhexyl) adipate	13.9	−66	0.10	>70	65
diisodecyl adipate	3.4	−63	0.07	>70	74
azelaic acid derivatives					
bis(2-ethylhexyl) azelate	5.7	−67.2	0.06	73	70
benzoic acid derivatives					
dipropylene glycol dibenzoate	9.4	−12	0.45	2.7	71
citric acid derivatives					
tri-*n*-butyl citrate	16.5	−38.8	0.43	8	66
tri-*n*-butyl acetyl citrate	17.8	−36	0.09		73
epoxy derivatives					
epoxidized soybean oil	0.9	−28.2	0.0	1.08	75
2-ethylhexyl epoxy tallate	1.5	−58	0.13	81	64
hydrocarbons					
hydrogenated terphenyls[h] 1:1 with DOP	27.6	−30.4	0.35	57	73
chlorinated paraffins, Cereclor S-52	4.5	−12.0	0.61	2.7	89
phosphoric acid derivatives					
tris(2-ethylhexyl) phosphate	11.0	−70.0	0.21	74	75
tricresyl phosphate	1.0	−10	0.03	2.4	79
t-butylphenyl diphenyl phosphate	3.1	−9.4	0.07	1.3	80
2-ethylhexyl diphenyl phosphate	7.4	−39	0.06	7.3	65
phthalic acid derivatives					
dibutyl phthalate	44.0	−41.0	0.25	8.3	62
diisohexyl phthalate	14.8	−39.8	0.01	6.9	67
di(heptyl nonyl undecyl) phthalate	1.7	−48	0.02	>70	71
bis(2-ethylhexyl) phthalate	4.5	−39	0.01	44.3	68
di(*n*-hexyl, *n*-octyl, *n*-decyl) phthalate, DNODP-610	2.1	−50.1	0.01	63.5	71
diisononyl phthalate	1.6	−37	0.07	77	73
diisodecyl phthalate	1.3	−37	0.02	74	76
diundecyl phthalate	0.9	−52	0.05	>70	77
ditridecyl phthalate	0.8	−37.4	0.07	88	83
butyl benzyl phthalate	7.7	−24	0.09	3.5	62
high molecular weight benzyl phthalate, Santicizer 278	1.1	−0.3	0.02	0.6	85
dicyclohexyl phthalate[i]	3.3	+16.7	0.03	0.3	99
polyesters					
adipic acid polyester (mol wt 6000), Paraplex-G40	1.6 (6 d)	−9.6	5.1[j]		85
adipic acid polyester (mol wt 2000), Drapex 334F	2.8 (6 d)	−21.8	11.6[j]	1	74
azelaic acid polyester (mol wt 850) Plastolein 9720	5.1 (6 d)	−30.4	21.6[j]	3.8	72
sebacic acid derivatives					
bis(2-ethylhexyl) sebacate	4.2	−69.1	0.02	70	73
sulfonic acid derivatives					
alkyl sulfonic acid ester of phenol and cresol, Mesamoll	5.3	−31.4	0.03	4.8	72
trimellitic acid derivatives					
tris(2-ethylhexyl) trimellitate	0.5	−32.7	0.0	75	72

[a] Data for commercial products rather than pure compounds.
[b] Performance data are for 40 wt % (67 phr) in a medium molecular weight homopolymer PVC; 1 mm thick PVC sheet.
[c] After 24 h or 6 d as noted at 87°C over activated carbon; ASTM D1203-67.
[d] Low temperature flex. Clash and Berg test; ASTM D1043-72.
[e] After 24 h at 50°C; ASTM D1239-55. Percent loss of total specimen.
[f] After 24 h at 23°C; ASTM D1239-55. Percent loss based on plasticizer weight.
[g] Shore A penetrometer; ASTM D676-58T.
[h] Secondary plasticizer; used in a 50:50 blend with DOP and evaluated at 67 phr total plasticizer.
[i] Crystalline solid; its performance as the sole plasticizer in PVC varies with its degree of crystallinity in the resin.
[j] Test on polymerics was run at 50°C for 4 days in 1% Ivory Soap solution instead of pure water; ASTM D1239-55.

readily by microorganisms as sources of carbon. Others are sufficiently resistant to be useful in warm, moist, or wet environments. Microorganisms are quite selective in the chemical structures they metabolize, eg, linear rather than branched alkyl groups. Proper choice of resin and additives with use of a biocide if needed can control the problem.

Insects, Termites, or Rodents. Plasticization increases the probability of attack by insects, termites, and rodents. Protection may be achieved by incorporating a pesticide in the composition.

Environmental Effects. *Exudation Under Pressure.* A plasticizer that is compatible with PVC under ordinary conditions may exude in use under slight pressure. Proper formulation for the given use avoids the problem.

Effect of Humidity on Compatibility. Some plasticizers, polyesters more than monomeric plasticizers, that are compatible under normal conditions, exude in humid and, especially, hot and humid environments. Exudation is not usually a result of plasticizer hydrolysis but can be accounted for by formation of an incompatible water–plasticizer blend.

Volatility. Early plasticizers for PVC such as dibutyl phthalate were so volatile at ordinary use temperatures that shrinkage and stiffening of finished vinyl articles was common. With increase of molecular weight to higher phthalates, trimellitates, and polyesters, most volatility problems were eliminated. However, volatility is still a concern with each new high temperature application.

Extraction. Because the plasticizer is not held permanently to the resin, it may be extracted partially by oils, waxes, water, soapy water, and other agents. Extraction by any environmental agent can usually be controlled by a suitable plasticizer.

Migration to Other Materials. If plasticized PVC is in contact with other polymeric materials, the resistance at the interfaces is not too high, and the plasticizer is compatible with the second polymer, the plasticizer may migrate from the PVC to the other polymer. The rate of diffusion depends in part on the shape of the plasticizer molecule. Migration may soften or, in severe cases, destroy a surface coating. Such problems are usually controlled by proper choice of plasticizer.

Migration from Other Materials. Migration of trace amounts of other plasticizers from surface coatings or adhesives usually does not noticeably change plasticized PVC. Problems arise when the material migrating into the PVC composition has an obviously undesirable property, eg, staining. Almost invariably the problem can be solved by understanding solubility properties and judicious formulation.

Other Resins and Elastomers

Thermoplastics. Many thermoplastic resins other than PVC are modified for specific uses either with a plasticizer or by restructuring the polymer to make it more flexible, soft, or resilient and then frequently adding a little plasticizer.

Because of structural differences among resins, a plasticizer for one resin may not be compatible with another or, if compatible, may provide little or no plasticizing action. Presence or absence of functional groups in the resin, polarity of the groups present, steric hindrance, and molecular weight are factors in the acceptance or rejection of a plasticizer by a resin. These factors influence the kind and degree of crystallinity and other less well-defined structures that help control plasticizer acceptance and limit the extent of plasticizer activity by toughening action. Plasticization takes place in amorphous regions or at crystal imperfections. Resins that are completely amorphous, eg, ordinary polystyrene, accept large amounts of plasticizers; however, if they have no toughening structure, they rapidly form gums and viscous liquids with increasing plasticizer content. If they are to be used as surface coatings, adhesives, caulks, etc, they usually are improved by at least small amounts of plasticizers. Hot-melt adhesives benefit from solid plasticizers that melt just below the temperature of application and revert to solids or highly inefficient supercooled liquid plasticizers after the bond is formed.

Thermosets. Thermosetting resins may at times benefit from plasticizers for improved pot life and penetration of laminating paper. As there is no sharp distinction between thermoplastic and thermosetting resins, there is no discontinuity in plasticizer technology between them.

Acrylics. Plasticizers are incorporated in acrylics primarily to produce tough, flexible coatings. This necessitates production of a large number of base polymers, usually copolymers of ethyl and butyl acrylates and methacrylates with methyl methacrylate and occasionally with methacrylic acid, etc. BBP is used widely for acrylics because of its excellent compatibility, good but not excessive softening, good adhesion, and good weather resistance. Higher molecular weight benzyl phthalates are characterized by lower volatility and good reflow for bake-sand-bake requirements. Heptyl nonyl adipate provides good low temperature properties and adjustment to swelling and shrinking on wood. Because of its high molecular weight, sucrose acetate isobutyrate inhibits migration and damage of a substrate. DBS is very efficient for use at very low concentrations; is characterized by low viscosity and, therefore, facilitates spraying; and does not excessively soften most plastic substrates. ESO provides a hard top coat and resists plasticizer migration to the PVC substrate and from the substrate into the top coat.

Aminoplasts. Melamine resins, and to some extent urea–formaldehyde resins, may be improved by some small quantities of plasticizer. Phosphate improves gloss, flexibility, and flow and adds flame resistance. Organic phosphates plus ammonium phosphates are used in intumescent coatings (see AMINO RESINS).

Cellulosics. Cellulose nitrate, even with its high degree of crystallinity and high melting point, accepts large amounts of plasticizers because its melting point is easily depressed. Organic esters and ethers of cellulose are compatible with large amounts of plasticizer. An excellent review of plasticizers for cellulosics has been given. Cellulosics are used in plastics, coatings, adhesives, film and sheeting, and fibers. Plasticizers are used to reduce processing temperatures, improve impact resistance, increase flexibility and resistance to cracking, and improve light and heat stability (see CELLULOSE ESTERS, INORGANIC).

Epoxy Resins. Modifiers, diluents, or plasticizers for epoxy resins like those for phenolics, lower the viscosity of the uncross-linked resin for greater ease in application of surface coatings and adhesives and production of prepregnated resin-coated cloth. They accelerate casting and potting. They also may add moderate flexibility to the final product by either internal or external plasticization. Nonreactive plasticizers, eg, phosphates and phthalates at 25 phr, can reduce viscosity from 12.8 Pa·s (128 P) to 1.5–3.0 Pa·s (15–30 P). The final resin is softer but is characterized by poorer impact resistance.

The nonreactive plasticizers lower the extraction resistance and chemical resistance of epoxy resins. They promote resin acceptance of larger amounts of filler. Dibutyl phthalate at 6 wt % improves mechanical properties and weatherability but with a sacrifice in heat-distortion temperature.

Reactive plasticizers are low molecular weight epoxy compounds usually with one epoxy group per molecule. As such, they are difunctional rather than polyfunctional, as the resin epoxides are. They lengthen the polymer segments between cross-links and produce a slight softening and flexibilizing effect with improved impact strength. Among those used are phenyl glycidyl ether, butyl glycidyl ether, allyl glycidyl ether, styrene oxide, and epichlorohydrin. Some of these tend to be skin irritants.

Fluoroplastics and Elastomers. Polytetrafluoroethylene (PTFE) is the most highly fluorinated of the fluoroplastics; it does not melt during processing, which is by sintering, but changes into a gel above 330°C. During the sintering process, volatile solvents or lubricants and even water may be used to aid in molding or surface coating. The lower fluorinated and chlorofluorinated resins accept more conventional plasticizers, usually as aids to processing.

Phenolics. Condensation of phenol and formaldehyde results in short-chain segments with a very high degree of cross-linking. There is little opportunity for the polymer to deform under load, particularly under impact, without separating resin domains or breaking primary valence bonds. Addition of either reactive or nonreactive plasticizers initially or at an intermediate stage can facilitate processing and can alter final physical properties beneficially.

Polyacrylonitrile. Acrylonitrile polymers (qv) are used mainly as textile fibers and, in this form, require little or no external plasticization (see ACRYLIC FIBERS). Polyacrylonitrile's tendency to cyclize and stiffen on heating makes plasticization difficult. It can be successfully plasticized with high polarity plasticizers such as *N,N*-bis(2-cyanoethyl)-*o,p*-toluenesulfonamide to give flexible, leathery sheets.

Polyamides. Nylons are among the best known highly crystalline polymers, although some recent ones appear amorphous. When used as fibers, poly(hexamethylene-*co*-adipic acid) and polycaprolactam, nylon-6,6 and nylon-6, respectively, and related polyamides frequently benefit from 2–5 phr of a very polar, hydrogen-bonding plasticizer as a flow aid, which retards degradation of the resin and speeds up processing. *N*-Ethyl (*o,p*)-toluenesulfonamide is most commonly used; at times the unsubstituted *o,p*-toluenesulfonamide is beneficial. Flexible filament, sheeting, tubing, and especially polyamide coatings and adhesives may require more plasticization. Nylon also is toughened when plasticized by traces of water, anhydrous ammonia, and other polar solvents.

Polycarbonates. Polycarbonates (PC), synthesized from bisphenol A diglycidyl ether, are characterized by a high hot-melt viscosity, which makes fabrication difficult. Incorporation of 0.2–5 wt % of a suitable plasticizer, eg, an aliphatic carbonate, aids fabrication without seriously modifying final properties of the resin. Small amounts of plasticizers facilitate dispersion of pigments or fillers into the resin and improve wetout of pigments by the resin.

Polyesters, Thermoplastic. Poly(ethylene terephthalate) (PET) and related polyesters, the so-called linear polyesters, are crystalline polymers and are difficult to plasticize. They are used for high strength film and textile fibers and, as such, they include little or no external plasticizers. Poly(ethylene terephthalate) is plasticized by water in a manner similar to that of nylon, but to a lesser degree. Plasticization is employed briefly to achieve greatly increased speed and intensity of dyeing, by addition of certain low molecular weight plasticizers or dye carriers that are emulsified in the dye bath. Those that are insoluble in water but soluble in the resin work best. High volatility is desirable, because plasticization is wanted only long enough for dye diffusion; then the plasticizer is removed by evaporation. Many plasticizers have been used, including butyl benzoate, *o*-phenylphenol, *p*-phenylphenol, and biphenyl.

Polyesters, Thermosetting. Butyl benzyl phthalate at about 10 phr is used to improve machinability of polyester castings and avoid chipping. Plasticizers, including low molecular weight polyesters and liquid nitrile rubber at 9 to 17 phr, improve flexibility and toughness of laminates and avoid edge chipping during deflashing of fiber-reinforced parts. Phosphate plasticizers add flame retardance.

Polyolefins. Interest in plasticizers for polyethylene persists, but their use results in general reduction of physical properties. Compatibility of typical vinyl plasticizers with low density polyethylene usually is about 0–2 wt %. Hydrocarbons, partially hydrogenated terphenyls and certain mineral oils, are compatible up to about 20 wt % (25 phr). Chlorinated paraffins are used in conjunction with antimony oxide to impart flame retardance. Small amounts, ca 0.25 wt %, of 2-ethylhexyl diphenyl phosphate may improve oxidation resistance and inhibit odor formation in polyethylene (see also ETHYLENE POLYMERS).

Polyurethanes. Flexible polyurethanes, or polyurethane elastomers, accept plasticizers with almost predictable changes in physical properties. The elastomer, already resilient in its own right, is made more resilient by addition of plasticizer.

The moisture content of the plasticizer should be controlled carefully to avoid a slight drift in hardness upon aging. Plasticizers such as tri(2-ethylhexyl) phosphate and alkylaryl phosphates give some flame retardance with good to excellent low temperature properties. *N*-Ethyl *o,p*-toluenesulfonamide increases the cure rate. To prevent overheating from polymerization of large castings or of large foamed products a plasticizer or other liquid is at times added as a heat sink.

Rigid Polyurethanes. Any large increase in flexibility or lowering of T_g would be undesirable; however, small additions, ie, 2–5 phr, of BBP can produce a foam of equal density with higher compressive strength, finer cell structure, and tougher cell wall. The resulting foam is stronger, insulates better, and is resistant to environmental degradation.

Poly(vinyl alcohol). Poly(vinyl alcohol) (PVA) is plasticized with glycerol, sorbitol, and other polyols and small amounts of alkyl diaryl phosphates. The sulfonamides also will plasticize it. On excessive heating it will cross-link, and the plasticizer will exude (see VINYL ALCOHOL POLYMERS).

Poly(vinyl butyral). Poly(vinyl butyral), PVBu, is compatible with many plasticizers. Its complex structure, as if it were a terpolymer of vinyl butyral, vinyl acetate, and vinyl alcohol, permits attraction to a wide variety of plasticizers; triethylene glycol bis(2-ethylbutyrate) was for decades the plasticizer of choice for auto safety glass interliner (see VINYL ACETAL POLYMERS).

Poly(vinyl formal). Poly(vinyl formal) (PVFo), used in adhesives, does not receive plasticizers as readily as poly(vinyl butyral). Its usual high PVA content demands strong hydrogen bonding plasticizers and solvents.

Poly(vinylidene chloride). Poly(vinylidene chloride) and its copolymers are blended with moderate amounts of plasticizers that contribute high but controlled blocking action in temporary food wraps. Ethyl phthalyl ethyl glycolate, tributyl acetylcitrate, and 2-ethylhexyl diphenyl phosphate are useful.

Poly(N-vinylpyrrolidinone). Polyvinylpyrrolidinone (PVP), a water-soluble resin, is plasticized with phosphates, polyols, dibutyl tartrate, or lower phthalates for use in hair sprays, adhesives, and water-soluble or removable lacquers. Sulfonamides, including low molecular weight sulfonamide–formaldehyde resins, improve adhesion to glass. In addition, the resin helps give a tack-free PVP coating or ink on glass or ceramic wafers.

Rubber and Related Elastomers. Natural rubber, because of its high initial molecular weight, is not easily fabricated into commercially useful articles (see RUBBER, NATURAL). It is common practice to masticate raw rubber on a roll mill or, more commonly, in a Banbury mixer to lower the molecular weight by breaking the polymer chains. The product is more plastic and less elastic, or has less nerve, than raw rubber. Low molecular weight fragments, which are produced from mixing, plasticize the remaining high polymer so that it processes easily and may be readily fabricated. Usually it is more economical to use a petroleum oil for at least part of the plasticizer rather than degrade more expensive rubber to produce the plasticizer *in situ*.

Plasticizers. Addition of plasticizers during mixing can reduce hot-melt viscosity and retard chain cleavage. Because most elastomers are comprised of long hydrocarbon segments and are largely amorphous, they accept heavy petroleum oils, coal tars, and other predominantly hydrocarbon products, which are less expensive than the esters that commonly are used with PVC. Ester plasticizers, eg, linear dialkyl phthalates, adipates, and sebacates, impart unusually good low temperature properties in SBR; thus they are used, eg, in tires designed for arctic regions. For most other uses they are too expensive.

Silicones. Silicone fluids and silicone rubber are used for their high temperature resistance and their relative inertness to many chemicals and physical environments, eg, on gaskets that must resist excessive swelling. Their solubility parameters of about 7.3 (Small's method) or 7.0–9.5 in nonhydrogen-bonded solvents to 9.5–11.4 in strongly hydrogen-bonded solvents indicate some hydrocarbons and some polar, hydrogen-bonding solvents might attack them.

A prominent preservative for renewing the appearance of plasticized PVC and other plastics and rubbers combines polydimethylsiloxane with glycols and glycerol in an emulsion. It thus makes use of incompatibilities to achieve "functional compatibility." The polysiloxane is not attacked by the polyols, and the PVC and its plasticizers are not attacked and swollen or extracted by either, so it can clean and protect (see SILICONES).

Polymerizable Plasticizers. There has long been interest in polymerizable plasticizers that would permit use of plastisol techniques to make rigid products. Polyglycol dimethacrylates were among the first to appear promising, followed by 1,3-butylene glycol dimethacrylate, and trimethylolpropane bis(methacrylate), (2-ethyl-2-hydroxymethyl-1,3-propanediol dimethacrylate), and other cross-linking esters. All of these can be used with paste PVC resins to yield satisfactory plastisols. Formulation and processing of PVC with polymerizable plasticizers requires unusual expertise to develop the best possible properties. Lowered viscosity and increased hardness, modulus, and tensile strength can be achieved by replacing increasing amounts of primary plasticizer, DOP, or polyester. Tear and abrasion resistance first increase and then decrease with increasing methacrylate content. A peroxide, eg, di-*t*-butyl peroxide, is required at about 1 phr to polymerize the plasticizer as the resin fuses. The resin requires stabilization, and the stabilizers fight the polymerization initiator. Improper processing can result in severe odor and color problems as well as poor hardness or low impact strength. Polymerizable plasticizers are especially useful in coatings where their use permits production of 100% solids coating materials (no volatile solvents) for application to steel, cloth, wood, paper, or as a top coat for vinyl tile to improve stain and abrasion resistance.

Specialized Systems

Many special systems require the use of two or more polymers, two or more plasticizers, and a host of other additives. Frequently their formulations have grown up and become increasingly complex through the years. Some special systems are new and simple. Specialized systems include liquid color, adhesives (see also ADHESIVE COMPOSITIONS; PRESSURE SENSITIVE ADHESIVES), surface coatings, and polymer blends and block copolymers.

Economic Aspects

Approximately 400 plasticizers are available commercially. The rate of introduction into the market of new plasticizers has decreased significantly in the last two decades and is characteristic of maturing plasticizer developments. Of the 400 available plasticizers, perhaps 100 are of significant commercial value. Table 3 reflects recent production of the more common plasticizer types manufactured in the United States. The decline is a combination of changing styles in, eg, floor coverings and automobile interiors, and worldwide economic trends, and to a large extent the development of competitive products for large market areas carved out and developed by plasticized PVC.

Health and Safety Factors, Toxicology

In plasticized PVC, the resin, its stabilizer, and the plasticizer must be essentially nontoxic when used, eg, in film or bottles

Table 3. U.S. Plasticizer Production, Thousands of Metric Tons

	1965	1975	1978	1986
phthalates	308	421	572	515
phosphates	30	44	53	43
adipates	22	22	31	60
polymerics	18	17	24	33
trimellitates	1	9	15	24
epoxy				50
others[a]	108	113	252	39
Total	487	626	947	764

[a] Includes epoxy plasticizers, citrates, stearates, benzoates, and glycol derivatives except that epoxies are listed separately for 1986.

for food packaging or in tubing for maple syrup, milk, or carbonated water. It also must be nonsensitizing to skin, when used for disposable hospital aprons and sheets. Most plasticizers used in PVC are characterized by low skin sensitization, eye irritation, and other untoward effects. (Certain chlorinated and nitrated plasticizers and *o*-cresyl phosphates have long been removed from the market because of their irritating or toxic effects.)

DOP and heptyl, nonyl, undecyl phthalates are degraded quickly in the river die-away test and the semicontinuous activated sludge test, although not as quickly as the control, which is a biodegradable sodium alkylbenzenesulfonate. Broad geographical surveillance has enabled monitoring of rivers and bays across the United States for presence or absence of three benzyl phthalates and di(heptyl, nonyl, undecyl) phthalate. The Environment Protection Agency (EPA) regulations regarding potential hazardous waste in the plastics industry have been reviewed briefly. New tests and new regulations are being devised and required for improved safety in waste sites, including the potentially broad TCLP, or toxicity characteristic leaching procedure.

Material Safety Data Sheets, available from plasticizer manufacturers, should be consulted for up-to-date information on safe handling and toxicity of plasticizers.

BIBLIOGRAPHY

N. W. Touchette and J. K. Sears in R. A. Meyers, ed., *Encyclopedia of Physical Science and Technology,* Vol. 10, Academic Press, Orlando, Fla., 1987, pp. 753–769.

J. K. Sears and J. R. Darby in L. I. Nass and C. R. Heiberger, eds., *Encyclopedia of PVC,* 2nd ed., Vol. 1, Marcel Dekker, Inc., New York, Chapt. 9, 1986.

J. K. Sears and J. R. Darby, *The Technology of Plasticizers,* Wiley-Interscience, New York, 1982.

P. F. Bruins, ed., *Plasticizer Technology,* Reinhold, New York, 1965.

J. K. SEARS
N. W. TOUCHETTE
Consultants

PLASTICS PROCESSING

A plastic is classified as a thermoplastic or thermosetting resin, depending on the effect of heat. Thermoplastic resins, when heated, soften and flow as liquids; when cooled, they solidify. These changes on heating and cooling can be repeated. Almost without exception, thermoplastic products are made by melting thermoplastic compounds, shaping the melt, and cooling it while maintaining the shape. Thermosetting resins liquefy when heated and solidify with continued heating; the polymer cross-links during heating. These materials are formulated for controlled chemical change during molding, and cross-linking is permanent. Thus once molded, a thermoset part cannot be reheated and molded again. However, thermoplastic articles can be reworked.

Plastic raw materials are characterized by properties important for processing and those important for the intended product. Many grades of raw materials differ only in processing properties.

Thermoplastic Resins

Almost 90% of the resins produced are thermoplastics. Although a number of chemically different kinds of thermoplastics are on the market, they can be divided into two broad classes: amorphous and crystalline.

Thermoplastics are usually processed as received from the manufacturer. Additives are not required; some may have undesirable effects. Colored resins are available, but frequently it is more convenient and economical to buy uncolored resins and blend them with color concentrates. This avoids handling dusty pigments and ensures uniform color distribution. Colorants (qv) must, however, be compatible with the resin.

Extrusion. Extrusion (qv) is employed for the processing of approximately 50% of all resins. In the extrusion line shown in Figure 1, the resin is melted in the extruder and pumped through a die. The die forms the melt into a shape closely related to that of the final product. The melt may be drawn to a thinner cross section in an air gap. It is then cooled and its shape is maintained. The solid plastic enters a puller characterized by a shape complementary to that of the plastic. The puller draws melt from the die and through the cooling system. It is followed by the cutter, which slices the product to length or winds it into a roll or coil. Extrusion is a continuous process, and products of the same cross section are produced in infinite length, including film, sheeting, pipe, tubing, profiles, coated wire, and coated paper.

Molding. ***Injection Molding.*** In injection molding (qv), a molten thermoplastic is injected under high pressure into a steel mold. After the plastic solidifies, the mold is opened and a part in the shape of the mold cavity is removed. The machine consists of an injection unit and a clamp unit (Fig. 2). The injection unit melts the plastic and injects it into the mold. The clamp unit opens, closes, and holds the mold closed against the pressure of the melt. The size of an injection-molding machine is described by the capacity of the clamp unit, ranging from 50 to 4000 t. A machine with a given clamp unit can be supplied with a variety of injection units. Most injection units are extruders specially built for injection-molding machines. The size is described by the shot capacity, which is the maximum volume of melt that can be injected in a cycle.

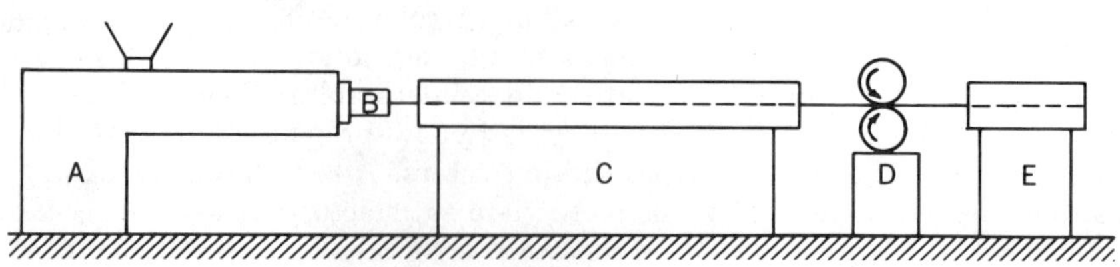

Figure 1. Extrusion line: A, extruder; B, die; C, sizer cooler (water trough); D, pull rolls; E, cutter (coiler). Courtesy of Society of Plastics Engineers.

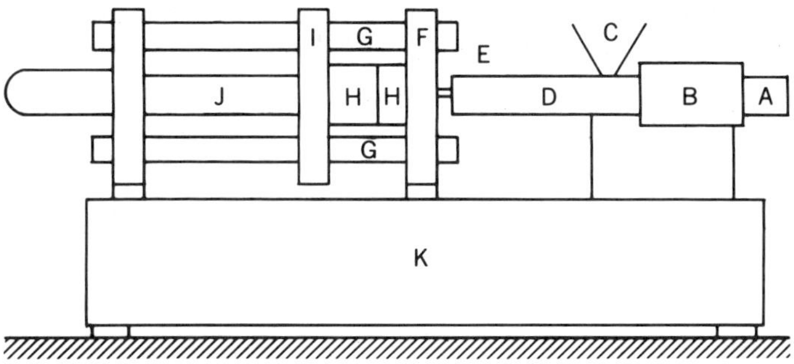

Figure 2. Injection-molding machine: A, hydraulic motor for turning the screw; B, hydraulic cylinder and piston allowing the screw to reciprocate about three diameters; C, hopper; D, injection cylinder (a single-screw extruder); E, nozzle; F, fixed platen; G, tie rods; H, mold; I, movable platen; J, hydraulic cylinder and piston used to move the movable platen and to supply the force needed to keep the mold closed; K, machine base.

Structural Foam. Structural-foam molding is a modified injection-molding process for large articles with 20–40% reduction in density. It is most frequently used for polyethylene and high impact polystyrene. Some modifications to the resin, machine, and mold are required, and a blowing agent (qv) is added to the resin. A chemical blowing agent, which releases gas when heated, is commonly used (see STRUCTURAL FOAMS).

Blow Molding. Blow molding (qv) is the most common process for making hollow articles such as bottles. A single production line for one product is often used on a large scale. In Figure 3**a** a molten tube of resin is extruded from a die above an open mold. In Figure 3**b** the mold is closed around the tube or parison, and the bottom of the tube is pinched together by the mold. Air under pressure is fed through the die into the tube, which expands to fill the mold, and the part is cooled as it is held under internal air pressure. Figure 3**c** shows the open mold with the part falling free.

Rotational Molding. Hollow articles and large, complex shapes are made by rotational molding (qv), usually from polyethylene. The resin is in the form of a fine powder. A measured quantity is placed inside an aluminum mold and the mold is heated in an oven and rotated at low speed. The resin melts and coats the inside of the mold. The mold is then cooled by water spray and the part solidifies, duplicating the inside of the mold.

A rotomolding machine has three long arms extending from a central driving mechanism; each arm rotates several molds in two planes. The arms are moved from one process station to the next, from unloading and loading to heating and cooling. The molds are usually made of cast aluminum, but sheet metal is also used. Rotation of the mold surface during heating or

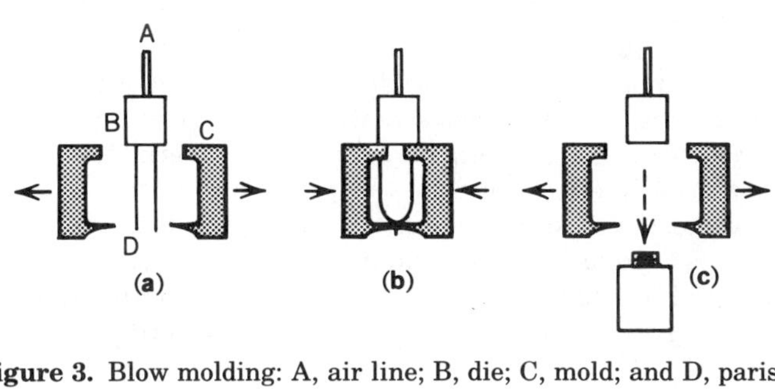

Figure 3. Blow molding: A, air line; B, die; C, mold; and D, parison.

cooling results in uniform wall thickness. Common rotomolded products include drums, tanks, boxes, furniture, hobby horses, and dolls. The process is particularly suitable for large tanks and boxes.

Thermoforming. Thermoforming (qv) is a process for converting a plastic sheet into an article, eg, a sign or a tray for packaging meat. When a vacuum is used the process is called vacuum forming. Sheet is clamped in a frame and is exposed to radiant electric heaters. It softens to a formable condition and is moved over a mold. The molten resin is sucked against the mold by vacuum and cooled. Excess plastic is trimmed and recycled. In some plants, sheet extrusion and vacuum forming are integrated into a continuous process.

Cast Acrylic Sheeting. For more than 50 years cast acrylic sheeting has been made by polymerization of methyl methacrylate in a cell assembled from two glass plates and a flexible gasket. Heating for a few hours until polymerization is complete gives sheets with excellent optical properties. Sheeting (qv) is also made by continuous casting; a monomer–polymer–catalyst mixture is fed onto a stainless steel belt on which the polymerization is completed. These products are used for glazing and thermoforming (see CASTING).

Polystyrene Bead Processing. Molding expandable polystyrene gives foamed products such as insulation board, shapes for packaging, and coffee cups. Such processes are also called bead molding or steam molding. For each type of product there is a preferred bead size and blowing agent concentration. The beads are preexpanded by heating with steam or hot air. Then they are stored for a number of hours in order to let the air diffuse into the cells of the foam. Live steam expands the beads further and fuses the foamed particles. After storage and conditioning for several days, the blocks are cut into boards. Shapes are made in a special press and mold, into which the preexpanded beads are transferred by guns; the filled mold is heated by steam and cooled by a water spray.

Foamed-sheet Extrusion. Foamed polystyrene sheet is used for thermoforming egg cartons, meat trays, and fast-food containers; two extruders are used.

Compression Molding. Compression molding of thermosetting resins is the oldest process of the plastics industry. The equipment consists of a vertical hydraulic press with platens for mold attachment. Mold cavities are hardened and highly polished, similar to those of an injection mold. The mold is hot for curing the compound. A compressed cake of granular resin is placed in the mold, and the mold is closed. The resin melts and pressure forces the liquefied material to fill the cavity. Continued heating cures the resin within a minute or two, and the part is removed from the mold. Compression molding is employed for phenol–formaldehyde (phenolic molding compounds), urea–formaldehyde, melamine–formaldehyde resins (amino molding compounds) (see AMINO RESINS; PHENOLIC RESINS), and polyester resins (see COMPRESSION AND TRANSFER MOLDING).

Calendering. Calenders are used for making rubber sheeting and similar products from semirigid and flexible PVC and ABS with thicknesses of 0.05–0.75 mm. A calender has four heavy, large steel rolls, which are usually assembled in the inverted

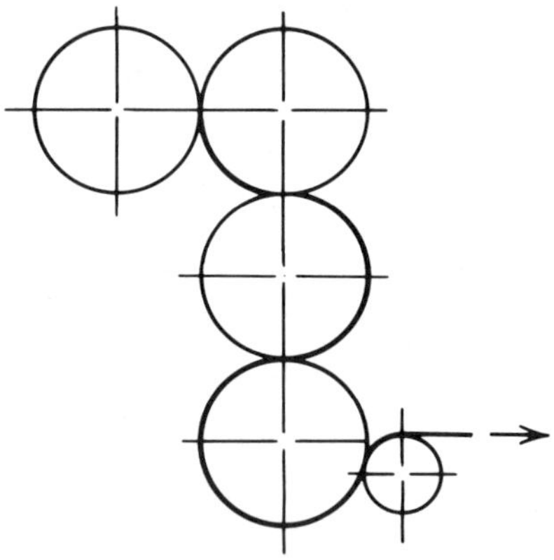

Figure 4. A four-roll, inverted L calender.

L configuration shown in Figure 4. The molten resin is worked and metered at the three nips between the rolls, and the product is characterized by uniform thickness (see CALENDERING).

BIBLIOGRAPHY

R. M. Ogorkiewicz, ed., *Thermoplastics: Effects of Processing,* CRC Press, Cleveland, Ohio, 1969.

R. J. Crawford, *Plastics Engineering,* Pergamon Press, Elmsford, N.Y., 1981.

J. M. McKelvey, *Polymer Processing,* John Wiley & Sons, Inc., New York, 1972.

J. L. Thorne, *Plastics Process Engineering,* Marcel Dekker, Inc., New York, 1979.

PAUL N. RICHARDSON
E.I. du Pont de Nemours & Co., Inc.

PLASTICS REINFORCED. See REINFORCED PLASTICS.

PLASTICS TOOLING. See TOOLING WITH PLASTICS.

PLASTISOLS AND PLASTIGELS. See VINYL CHLORIDE POLYMERS.

PLATING. See METALLIZING.

PLEINOMERS. See OLIGOMERS.

POLARIMETRY. See OPTICALLY ACTIVE POLYMERS.

POLYACETALDEHYDE. See ALDEHYDE POLYMERS.

POLYACETALS

Polymers that arise by bond opening and polymerization of the carbonyl group of an aldehyde are often called polyacetals. The most important commercial materials so formed are homopolymers or copolymers of formaldehyde.

$$n\ H_2C{=}O \longrightarrow -\!\!(H_2C{-}O)_n\!\!-$$

$$n\ CH_2{=}O + x\ \overset{O}{\triangle} \longrightarrow -\!\!(OCH_2CH_2O)_x\!\!-\!\!(CH_2{-}O)_n\!\!-$$

where $n >>> x$

These polymers are usually termed acetal resins (qv). Formaldehyde polymers or acetal resins were commercialized 25 years ago. The general designation aldehyde polymers (qv) in this Encyclopedia refers to polymers of higher aliphatic aldehydes and haloaldehydes.

By analogy to organic reactions and terminology, it would be preferable to restrict the term polyacetal to reaction products of polyols and aldehydes.

$$HO\!-\!(CH_2)_x\!-\!OH + RCHO \longrightarrow HO\!-\![(CH_2)_x\!-\!O\underset{\displaystyle R}{\underset{|}{C}}HO]_n\!-\!(CH_2)_x\!-\!OH$$

Such materials were investigated in the early 1960s, but have never been commercialized. Consequently, the current usage of polyacetal as a synonym for acetal resin or aldehyde polymer is likely to persist.

POLYALKENAMERS

Metathesis polymers, designated polyalkenamers, are obtained by the ring-opening polymerization of cycloolefins catalyzed by transition metals. The alternative designation, polyalkenylene, is being used more and more and is in accordance with IUPAC-recommended nomenclature; but the term polyalkenamer reflects the initial terminology of the early pioneers Natta and Dall'Asta in 1964. Ring-opening polymerization leads to a linear repeat unit and retention of the number of double bonds present in the original monomer. Although two seemingly equivalent representations of the polymerization reaction have been used, the first of the following equations more accurately reflects the chemical process:

$$\underset{\smile R \smile}{C{=}C} \longrightarrow \begin{cases} {=}\!\!\!\!+\!C{-}R{-}C\!+\!\!\!\!{=} \\ +\!C{=}C{-}R\!+ \end{cases}$$

Ring-opening proceeds by the facile cleavage and reformation of the double bond, a remarkable process involving the strongest bond in the molecule. The general reaction has been designated the olefin metathesis reaction, and has been exhaustively reviewed.

The rapid progress made during the last 20 years in understanding the metathesis reaction has made it possible to exercise considerable control over the polymerization process and direct it effectively for the preparation of well defined polymers. As is often the case, successful commercialization becomes dependent on economic factors such as cost and availability of the monomers. With the gradual maturing of this technology, three cycloolefin polymers have recently been introduced into the marketplace, whereas several others have been explored extensively.

Monomers

Despite the great diversity of cycloolefins that have been polymerized, most monomers studied in industrial laboratories have been derived at least in part from butadiene and cyclopentadiene. Refinery by-product streams have been the most economical source of these two low cost materials, which are produced during the manufacture of ethylene. Cyclopentadiene (CPD) is hazardous to store as the monomer, but is easily recovered as dicyclopentadiene by thermal dimerization.

Butadiene is the source of a variety of polymerizable cycloolefins. The dimerization of butadiene to 1,5-cyclooctadiene (COD) was first described in 1953.

Norbornene and several of its polymerized derivatives are made directly by the Diels-Alder condensation of CPD with ethylene or its derivatives.

Polymerization

Metathesis polymerization (qv) is a unique hybrid of addition and condensation polymerization. The chain-growth process is clearly a step-growth addition reaction, wherein monomer units are inserted between the catalysts and the attached polymer chain. This process is typically initiated by a carbene species generated from catalyst components. The vinyl carbons are shown below with at least one H attached, since 1,1-disubstituted vinyl groups are inert toward metathesis.

Initiation:

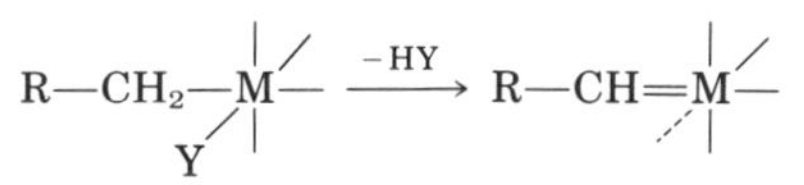

Propagation:

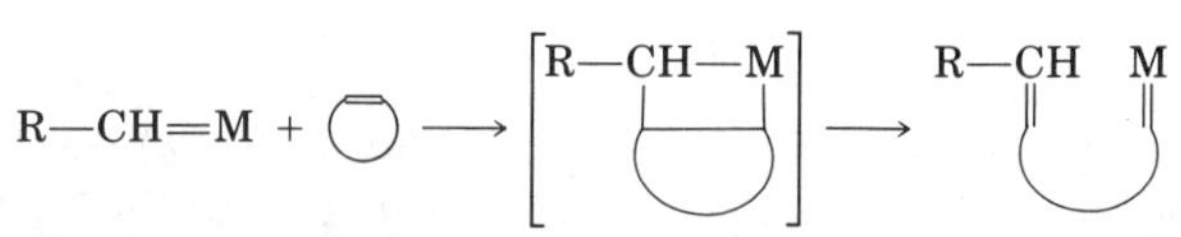

The unusual characteristic of hydrocarbon metathesis polymerizations is the retention of the original reactive moiety, the double bond, along the polymer chain. Thus these enchained monomer units continue to undergo successive metathesis reactions throughout the course of polymerizations and a series of reactions occurs similar to polycondensation processes. These include formation of macrocyclic oligomers from high molecular weight polymer, formation of monomer from polymer when thermodynamically permissable, interchain redistribution reactions that lead to most probable molecular weight distributions as well as randomization of monomer sequence distributions, and chain scission with monofunctional agents (acyclic monoolefins).

Properties

Structure Characterization. Polynorbornene was analyzed by ozonolysis to establish the structure of the repeat unit and verify ring-opening polymerization. Similarly, ozonolysis was used to characterize the basic repeat unit in polypentenamer and polyoctenamer. Other techniques include analysis of the macrocyclic oligomer fractions, which are a normal coproduct of methathesis polymerizations, infrared spectroscopy, and ^{1}H and ^{13}C nmr.

Solution Properties. The simple linear polyalkenamers are soluble in low polarity aliphatic, cycloaliphatic, aromatic, and halogenated hydrocarbons. Crystalline stereoregular polymers may require elevated temperatures to achieve solubilization. Polymers from polar monomers such as 5-norbornenecarbonitrile require highly polar solvents, and chlorinated solvents are often employed for polymerization since they provide polarity without affecting catalyst activity.

Solid-state Properties. The simple linear polyalkenamers have the potential to be elastomeric by virtue of their low glass-transition temperatures T_gs, whereas polymers from some substituted cycloolefins and nearly all of the polycyclic monomers have T_gs above room temperature and thus are thermoplastics. However, the thermoplastic behavior of some stereoregular polymers may be the result of crystallization. Since many polyalkenamers derived from simple cycloolefins exhibit crystallinity, a property that plays an important role in most applications, this phenomenon has been widely studied for the homologous series of linear polymers.

Commercial Products

Norsorex (from norbornene) (Table 1), the first polyalkenamer to be manufactured, is often used in elastomer applications requiring softness and low resiliency combined with strength and low permanent set. Vestenamer (from cyclooctene) (Table 2) is a low melting, crystalline polymer that improves pro-

Table 1. Properties of Norsorex Polynorbornene

Property	Value
microstructure, % trans	80
appearance	white amorphous powder
mol wt	>2,000,000
T_g[a], °C	35
refractive index	1.534
density, g/cm^3	
bulk	0.977
(apparent)	0.30
solubility parameter, MPa$^{1/2}$[b]	19.8
heat capacity, J/(g·K)[c]	2.09
thermal conductivity, W/(m·K)	0.285
heat of formation, J/g[c]	218

[a] By dsc.
[b] To convert MPa$^{1/2}$ to (cal/cm^3)$^{1/2}$, divide by 2.046.
[c] To convert J to cal, divide by 4.184.

Table 2. Properties of Vestenamer Polyoctenamer

Property	Vestenamer 8012	Vestenamer 6213
mol wt	50,000–60,000	ca. 60,000
microstructure, % trans	80	63
crystallinity, %	33	low
T_m[a], °C	55	[b]
T_g[a], °C	−75	
relative viscosity number	120	130
Mooney viscosity at 100°C	60	62

[a] By dsc.
[b] T_m estimated near 30°C.

cessing of rubber compounds at elevated temperatures and provides good dimensional stability of uncured parts. Meton (from dicyclopentadiene) (Table 3) utilizes RIM (reaction injection molding) technology to convert monomer directly to finished, molded plastic parts which have good strength and impact resistance. Several other polymer systems have been explored for commerciallization.

Table 3. Properties of Meton DCPD Resin

Property	Orientation	Unfilled	20% milled glass[a]
flexural modulus, MPa[b]	parallel	1790–2070	2620–2900
	perpendicular		2480
flexural strength, MPa[b]	parallel	62	62–76
	perpendicular		62
tensile modulus, MPa[b]		1620	
tensile strength, MPa[b]	parallel	34	31
tensile elongation, %	parallel	80	25
T_g, °C		90[c]	
coefficient of linear expansion	parallel	37	17
mold shrinkage, cm/cm	parallel	0.035	0.008–0.014
	perpendicular		0.012–0.021

[a] 1.5-mm fibers, Owens Corning OCF 737.
[b] To convert MPa to psi, multiply by 145.
[c] Under some conditions, T_g = 119°C.

Polyalkenamers Under Development

A few polyalkenamers are the subject of extensive development programs, but have not yet been commercialized. These polymers exhibit unusual properties, but uncertainties regarding monomer availability, marketing, and other economic factors are delaying production. They include *trans*-polypentenamer (TPA), cyclorene rubber, and telene.

BIBLIOGRAPHY

V. Dragutan, A. T. Balaban, and M. Dimonie, *Olefin Metathesis and Ring-Opening Polymerization of Cyclo-Olefins,* Editura Academiei, Bucharest, and John Wiley & Sons, Ltd., Chichester, UK, 1985.

K. J. Ivin, *Olefin Metathesis,* Academic Press, Ltd., London, 1983.

N. Calderon, *J. Macromol. Sci. Rev. Macromol. Chem* **C7**(1), 105 (1972).

C. Tosi, F. Ciampelli, and G. Dall'Asta, *J. Polym. Sci. Polym. Phys. Ed.* **11,** 529 (1973).

EILERT A. OFSTEAD
Goodyear Tire & Rubber Company

POLYAMIDES

Polyamides are polymers that contain recurring amide groups as integral parts of the main polymer chains. Linear polyamides are formed as the products of condensation of bifunctional monomers. With amino acid monomers, such as ϵ-aminocaproic acid, the polyamides are called AB types, A representing amine groups and B carboxyl groups. If the polyamides are fomed by condensation of diamines and dibasic acids, they are called AABB types. Polyamides are frequently referred to as nylons, a generic term. Although they are generally considered to be condensation polymers, polyamides may also be formed by addition polymerization. This method of preparation is especially important for some AB polyamides where the monomers are cyclic lactams, such as ϵ-caprolactam or pyrrolidinone.

Typical formulas of linear polyamides are represented by structure (**1**) or, in the case of self-condensation of an amino acid, by structure (**2**), where R and R′ represent chains between functional groups in the reactants and *n* represents the degree of polymerization or the number of repeating units in the polymer chain.

$$H_2NRNH(\overset{O}{\overset{\|}{C}}R'\overset{O}{\overset{\|}{C}}NHRNH)_n\overset{O}{\overset{\|}{C}}R'COOH \qquad H_2NR\overset{O}{\overset{\|}{C}}(NHR\overset{O}{\overset{\|}{C}})_nNHRCOOH$$

(**1**) (**2**)

Reactions of numerous combinations of diacids, diamines, and amino acids have been reported, and copolymers containing various proportions of these compounds have been prepared. The chains between functional units in the reactants may comprise linear or branched aliphatic hydrocarbons, or alicyclic or aromatic rings. They may contain heteroatoms such as oxygen, sulfur, and nitrogen. When the amide groups of the polyamide are connected to aromatic rings at both ends, the polymers are referred to as aramids, a generic term established in the early 1970s (see POLYAMIDES, AROMATIC). Polyfunctional reactants, such as triamines, tetraamines, and tribasic acids, lead to the formation of branched, cross-linked, or network polymers. Thus it is apparent that it is possible to synthesize a tremendous number and variety of polyamides.

Nylons are used in the production of synthetic fibers and engineering resins. Over the years, improvements have been made in the performance and usefulness of these products by various engineering innovations and by polymer modifications such as the introduction of reinforcing fillers and toughening agents. Entirely new levels of strength, modulus, and thermal stability have been achieved with aramids and heterocyclic polyamides. In the 1970s the production of nylon fibers slowed down, reaching ca 1.1 x 10^6 t in the United States in 1984 and ca 3.3 x 10^6 t worldwide. The market share for nylon fibers vs the total of all noncellulosic fiber production has declined from 50% in 1966 to a constant 30–32%, primarily because of the more rapid growth of polyester fibers in the period 1966–1975. The main markets for nylon fibers include carpets, industrial yarns, and tire reinforcements, and various apparel applications, such as hosiery. Another important use is in engineering resins.

Synthetic Methods

Although polyamides have been prepared by many routes, only melt polycondensation, ring-opening polymerization, and low temperature solution polymerization (usually for aramids) have gained commercial importance (see also INTERFACIAL POLYMERIZATION). Methods include direct amidation, reaction of acid chlorides with amines, amidation via phosphorylation, ring-opening polymerization, addition of amines to activated double bonds, polymerization of isocyanates, reaction of formaldehyde with dinitriles, reaction of dicarboxylic acids and diisocyanates, reaction of carbon suboxide and diamines, and reaction between diazlactones (2,2′-bisoxolones) and diamines.

Principles of Polyamidation

A polyamide is formed by the melt condensation of an amine with a carboxylic acid to form an amide group and water. The functional groups at the end of a polymer chain are referred to as end groups.

$$—COOH + —NH_2 \rightleftharpoons —CONH— + H_2O$$

Molecular Weight. In the preparation of polyamides molecular weight and its control are of considerable concern.

Molecular Weight Distribution. In linear condensation polymers with a given $\overline{DP}_n$, all chains do not have the same length. There is a rather broad distribution of chain lengths, called the most probable distribution, in which the ratio of weight average degree of polymerization (DP_w) is about 2 but may be somewhat less for AB polyamides (eg, nylon-6) containing bifunctional stabilizers.

Polyamidation Equilibrium. The equilibrium constant for the polyamidation reaction is given by

$$\frac{[CONH][H_2O]}{[COOH][NH_2]} = K_1 = B \exp(-\Delta H_a/RT)$$

where B is the temperature-independent part of the equilibrium constant and ΔH_a the enthalpy change of the reaction. The formation of polymer is exothermic (ie, ΔH_a is negative).

Effects of End-group Imbalance. For equilibrated polymer with a fixed end-group product [COOH][NH_2], the precise values of the individual end-group concentrations are determined by the degree of imbalance between the ends [COOH − NH_2].

Cyclic Oligomers. In addition to the equilibrium considerations discussed, another equilibrium exists between polymer and cyclic oligomers. For AABB polymers of commercial interest, the concentrations of these oligomers (cyclic monomer dimer, etc) are small (1–2%) because the smallest possible ring is relatively large (eg, 14-membered in nylon-6,6; 18 in nylon-6,10). On the other hand, with AB polymers the concentration of cyclic oligomers can reach proportions of great practical significance.

Kinetics. The published research on the kinetics of amidation appears to be contradictory. However, a consistent pattern emerges from these results. Conversions up to ca 90% indicate that the reaction obeys second-order kinetics and is not accelerated by catalysts but at higher conversions, the reaction is third order, catalyzed by carbonyl groups.

Solid-phase Polymerization. The polyamide condensation reaction can occur quite rapidly at temperatures below the melting point of the polymer. In a study of the polyamidation equilibrium in the solid phase, a shift to higher molecular weights at a given steam pressure was observed in contrast to a molten polymer of the same amide content polymerized at the same temperature. These results were explained by assuming that end groups are essentially excluded from the crystalline regions of the polymer and that the equilibrium in the amorphous region is the same as for a completely amorphous or molten polymer at the same temperature.

Amide Interchange. When two amides are melt-blended, interchange reactions lead to the formation of block copolymers, followed after several hours by random copolymers.

Low Temperature Methods. The preparation of polyamides at low temperature by the reaction of diamines and diacid chlorides is an important route to high melting polyamides, including aramids. Both interfacial and solution polymerization are employed although the latter is of greater commercial significance. At low temperature, acid chlorides react rapidly with free acids and high molecular weights are easily attained for polymers that are unstable at their melting temperatures or that have extremely low reactivity in the absence of acid chloride groups. In order to obtain high molecular weight polymer by low temperature polymerization, the reactivity of the acid chloride must be maintained by excluding water, selecting a reaction medium that keeps the polymer swollen or in solution, employing pure intermediates, cooling the exothermic reaction polymerization mixture, and maintaining a rapid mixing rate in interfacial polymerization.

Lactam Polymerization. In the preparation of nylon-6, the monomer most commonly used is ε-caprolactam rather than ε-aminocaproic acid. Pure, dry caprolactam does not polymerize when heated for as long as 200 h. Amines and carboxylic acids in combination catalyze polymerization; neither is effective alone. Examples of such catalysts are ε-aminocaproic acid, the salts of diamines and dicarboxylic acids (eg, nylon-6,6 salt) or water that hydrolyzes the monomer to produce these end groups. Kinetic studies by various workers have shed considerable light on the mechanism of caprolactam polymerization.

When pure caprolactam containing only water is heated at elevated temperatures (eg, 265 °C), initially, little ring opening occurs. However, both amine and carboxyl-group concentrations increase gradually as a result of monomer hydrolysis. This is accompanied by a progressive increase in the ring-opening rate. A carboxyl-catalyzed addition of an amine end to the cyclic monomer causes the polymer chain to grow. This reaction also applies to cyclic oligomers. The amine and carboxyl groups increase with time to a maximum and then decrease as monomer content is reduced to equilibrium.

Anionic Polymerization of Caprolactam. Cyclic amides, such as ε-caprolactam, are polymerized with the aid of alkaline catalysts. With alkali metals or metal hydrides, the polymerization is characterized by an induction period that decreases with increasing temperature and is ca 5 min at 265 °C. It is followed by an extremely rapid polymerization (accompanied by a corresponding increase in temperature), which is complete in a few minutes at 265 °C. At this point the monomer concentration reaches normal equilibrium and polymer viscosity is at a maximum. Subsequently, the viscosity falls off rapidly, probably as a result of a redistribution of molecular weights from a broad to a most probable distribution. This may result from interchange of the reactive end groups with chain-amide groups or from continuing polymerization and depolymerization. In general, the polymer molecular weight usually decreases with increasing initiator concentration, provided that the water content of the monomer is very low.

Low Temperature Bulk Polymerization of Caprolactam. The equilibrium monomer and oligomer contents in the base-catalyzed and hydrolytic processes are the same. However, when polymers are prepared at temperatures $<T_m$, the monomer content is much lower than would be expected from extrapolation of the monomer content vs temperature relationship, at temperatures $>T_m$, to these lower temperatures. This also applies to anionic polymerization, indicating that the cyclic monomer is excluded from the crystalline regions of the polymer, similar to the end groups in solid-phase polyamidation.

The extraordinarily low monomer content of the resulting polymer is valuable from the technological point of view, and solid moldings have been made by this procedure. The absence of significant by-product formation in anionic polymerization is crucial for this type of process.

Structural Effects. The size of the lactam ring affects polymerizability. Eight- and nine-membered ring compounds polymerize; about half of the seven-membered ring compounds polymerize; of the six-membered rings, only 2,5-diketopiperazine polymerizes; and 2-pyrrolidinone is the only five-membered ring compound that polymerizes, but low temperatures are required to provide acceptable equilibrium monomer content.

Nylon-1. Although N-substitution usually prevents lactam polymerization, the effect is overcome with a sufficiently high ring strain. This has been demonstrated by the ready polymerization of *N*-alkyl isocyanates (two-membered lactams) to *N*-alkyl nylon-1 with molecular weights (M_w) of 400,000. The anionic polymerization was initiated by strong bases below −20 °C.

Structure-Property Relationships

Many data dealing with physical properties of condensation polymers and covering a wide variety of structures have been accumulated since the early work of Carothers and co-workers. Attempts have been made to systematize the data and correlate polymer structure with properties such as melting temperature T_m, glass-transition temperature T_g, solubility, water absorption, and modulus.

Melting Temperatures. The T_m (in K) of a polymer depends on the heat of fusion ΔH_u and entropy of fusion ΔS_u of the repeat unit of the polymer chain in the crystal.

$$T_m = \frac{\Delta H_u}{\Delta S_u}$$

Polyamides melt at increasingly higher temperatures as the concentration of amide linkages and ΔH_u increase.

Effects of Lateral Substituents. Since the high T_m and crystallinity of a polyamide such as nylon-6,6 are a result of strong polar forces between amide linkages, it is not surprising that the introduction of hydrocarbon side chains on the carbon skeleton of the polymer chain introduces lateral disorder by forcing the chains apart and interfering with the intermolecular forces between the amide groups. When nylon-6,6 is substituted on the α-carbon atom of the dibasic acid with hydrocarbon groups of varying size, the melting point decreases as the size of the substitutent increases.

Influence of Heteroatoms. Substitution of an oxygen atom for a methylene group in the polyamide main chain increases chain flexibility and lowers the $T_m \sim 40$ °C for each oxygen atom introduced. Crystallinity and modulus are markedly lowered, whereas moisture absorption and solubility are increased. Polymers with several ether linkages (eg, nylon-3O2O3,6) are even soluble in water.

Introduction of sulfide linkages increases flexibility and reduces modulus. On the other hand, sulfone linkages raise the melting point, modulus, and crystallinity as well as water absorption, eg, from 2–3% to 7–10%.

Effects of Cyclic Structures. Replacement of an aliphatic section of a polyamide by a cyclic segment reduces chain flexibility, but raises the T_m because of a decrease in the entropy of melting. Introduction of aromatic rings also tends to increase modulus and reduce solubility.

Effects of Ring Orientation in Aliphatic-Aromatic Polyamides. Symmetry has a strong effect on the melting point of ring-containing polyamides. For example, replacement of a para-substituted ring by a meta-substituted ring in aliphatic-aromatic polyamides reduces the T_m.

Effects of Cycloaliphatic Rings on Melting Points. The effect of *trans*-1,4-cyclohexane rings on the melting point is similar to that of aromatic rings. Polyamides from *trans*-1,4-cyclohexanebis(methylamine) melt somewhat higher than the corresponding polyamides from *p*-xylene-α,α'-diamine.

Piperazine Polyamides. N,N'-Disubstituted diamines yield very low melting, rubbery polymers. On the other hand, piperazine polyamides, which also have no NH groups, are relatively high melting because the rigid ring contributes to a low entropy of fusion.

Effects of Copolymerization on Melting Points. Irregularity of monomer-group spacing depresses the T_m of a polyamide. This occurs in random copolymerization, where the T_ms of copolymers of nylon-6,10 and nylon-6,6 are plotted against copolymer composition. The lowest T_m (190 °C) is reached in the vicinity of the 70/30 composition. Lower T_ms reflect the reduction in average segment length.

Glass-transition Temperature. The glass-transition temperature T_g of a polymer is determined by the properties of the amorphous regions and by the degree of crystallinity. For polyamides, an increase in rh and water absorption tends to reduce T_g.

Crystallinity. Symmetrical, hydrogen-bonded linear polyamides are generally highly crystalline and owe their excellent mechanical properties, in part, to this property. Yield stress, tensile strength, modulus, hardness, and abrasion resistance increase with increasing crystallinity, whereas water absorption and impact strength decrease.

Degradation

Thermal Degradation. The rate and type of thermal degradation of polyamides depend on temperature and polymer structure and whether the heating is being carried out in the presence or absence of oxygen. The volatile products of degradation of nylon-6,6 and -6 at 305 °C in the absence of oxygen are carbon dioxide, water, ammonia, hexamethyleneimine, *n*-heptylamine, and methylamine. About half the nitrogen is eliminated as ammonia. For many other polyamides, degradation is thought to be related to homolytic cleavage of the C—N bond with formation of a double bond and an amide or a nitrile group.

Thermal Oxidation. Heating polyamides in the presence of oxygen results in chain scission and cross-linking. Nylon-6,6 becomes brittle at 250 °C after 2 h, but at 70 °C only after 2 yr. The degradation is due to a free-radical chain reaction involving the carbon α to the amide nitrogen; it is generally diffusion-controlled.

Photodegradation. Exposure to sunlight or other sources of uv radiation causes a deterioration in the properties of polyamides, especially polyamide fibers. The changes include a reduction in molecular weight and thus in tensile or flex strength and elongation to break. The concentration of oxygen governs the extent of degradation.

Ionizing Radiation. Irradiation of polyamides with high energy electrons (eg, 2 MeV) or γ-rays produces cross-linking and degradation. The latter is greatly accelerated in air. When nylon-6,6 is irradiated in air, rapid strength loss occurs; in addition, at high radiation doses, crystallinity decreases. Diffusion of oxygen into irradiated polymers has a significant influence, since the solubility of oxygen in the polymer is not sufficient to account for much degradation.

Chemical Reactions

Hydrolysis. Most nylon fibers and plastics are very slowly affected by water even at the boiling point. They are generally stable in aqueous media over the pH range of 5–13, but can degrade in more acidic environments.

Substitution of the Amide Hydrogen. Heating of polyamides in an autoclave with ethylene oxide gives hydroxyethylated polyamides with polyether chains attached to the nitrogen. These are characterized by high elasticity and vapor permeability. Ethylene carbonate and polyamides form block copolymers; the amine and carboxyl end groups as well as the amide NH groups react.

Ionizing Radiation. Ionizing radiation of polyamides produces free radicals with a characteristic color (yellow for typical aliphatic polyamides). It is generally believed that the free radicals are on the carbon atom α to the amide nitrogen. These radicals can disappear by reaction with each other or with oxygen in diffusion-controlled reactions.

Radiation Grafting. Low molecular weight saturated compounds, such as amines or organic chlorides, can be grafted to polyamides by high energy radiation but with relatively low

Table 1. Properties of Nylon-6,6 and Nylon-6,12 Resins[a]

Property	Nylon-6,6 (Zytel 101)	Nylon-6,12 (Zytel 158L)
tensile strength, MPa[b]	77.2	60.7
yield stress, MPa[b]	58.6	51.0
elongation at break, %	≥300	≥300
elongation at yield, %	40	25
shear strength (dry), MPa[b]	66.2	55.8
flexural modulus, MPa[b]	1207	1241
deflection temp (dry), °C		
455 kPa[c]	235	180
1820 kPa[c]	90	90
brittleness temp, °C	−65	−109
Izod impact strength, J/m[d]	112	75
melting point[e], °C	255	212
coefficient of linear thermal expansion (dry)	8.1×10^{-5}	9×10^{-5}
specific heat, J/(g·K)[f]	1.67	1.67
thermal conductivity, W/(m·K)	0.25	0.22
dielectric constant, 10^3 Hz	7.0	5.3
dissipation factor, 10^3 Hz	0.2	0.15
flammability, oxygen index	31	28
sp gr	1.14	1.06
water absorption at saturation, %	8.5	3.0

[a] Equilibrated at 23°C and 50% rh.
[b] To convert MPa to psi, multiply by 145.
[c] To convert kPa to mm Hg, multiply by 7.5.
[d] To convert J/m to ftlbf/in., divide by 53.38.
[e] Fisher-Johns apparatus.
[f] To convert J to cal, divide by 4.184.

Table 2. Physical Properties of Nylon-6[a]

Property	Nylon-6
tensile strength, MPa[b]	70–83
yield stress, MPa[b]	44
elongation at break, %	300
shear strength (dry), MPa[b]	59
flexural modulus, MPa[b]	970
Izod impact strength, J/m[c]	270
heat distortion temp, °C	
455 kPa[d]	145–205
1820 kPa[d]	65–80
melting point, °C	223
sp gr	1.12–1.15
refractive index, unoriented	1.53
water absorption at saturation, %	9–11
dielectric constant, 10^3 Hz	8.3
dissipation factor, 10^3 Hz	0.20
specific heat (dry), J/(g·K)[e]	1.6

[a] Conditioned at 23°C and 50% rh.
[b] To convert MPa to psi, multiply by 145.
[c] To convert J/m to ftlbf/in., divide by 53.38.
[d] To convert kPa to mm Hg, multiply by 7.5.
[e] To convert J to cal, divide by 4.184.

Table 3. Melt Temperatures of Homologous AB Polyamides

Nylon	T_m, °C
3	>320
3,3-dimethyl-3	250
4,4-dimethyl-3	296
4	260
6	223
7	233
8	200
9	209
10	188
11	190
12	179

Table 4. Properties of Nylon-11 and -12[a]

Property	Nylon-11	Nylon-12
tensile strength, MPa[b]	54	52
yield stress, MPa[b]	34	41
elongation at break, %	330	250
elongation at yield, %	22	20
modulus, MPa[b]	1730	2200
deflection temp, °C		
460 kPa[c]	149	140
1850 kPa[c]	54	51
sp gr	1.04	1.01
water absorption (long-term), %	1.1	1.55
dielectric constant, 10^6 Hz	4.0	3.0
dissipation factor, 10^6 Hz	0.03	0.03
specific heat, J/(g·K)[d]	1.26	1.26
thermal conductivity, W/(m·K)	0.19	0.19
coefficient of linear thermal expansion, °C^{-1}	9.9×10^{-5}	9.9×10^{-5}

[a] Conditioned at 23°C and 50% rh.
[b] To convert MPa to psi, multiply by 145.
[c] To convert kPa to mm Hg, multiply by 7.5.
[d] To convert J to cal, divide by 4.184.

Table 5. Nylon-4,6 Resin Properties

Property	Stanyl (4,6) Dry	Stanyl (4,6) 50% rh	Zytel 101 (6,6) Dry	Zytel 101 (6,6) 50% rh
tensile strength, MPa[a]	80	65	83	77
yield stress, MPa[a]	95	55	83	59
elongation at break, %	50	300	60	≥300
flexural modulus, MPa[a]	3100	1000	2800	1200
sp gr	1.18		1.14	1.14
melting point, °C	290		270	
mold shrinkage, %	1.5–2.0		1.5	
water absorption, equilibrium				
50% rh	3		1.2	
100% rh	12		8.5	

[a] To convert MPa to psi, multiply by 145.

efficiencies. Since the concentration of the modifier in the polymer is usually <10%, the radiation produces most of the free radicals on the polymeric substrate, on the methylene group adjacent to the nitrogen. These start the grafting process by abstracting a hydrogen or chlorine atom from the modifier, which attaches itself to a substrate-free radical. The maximum graft that can be obtained is equal to half the radical concentration in the amorphous regions of the polymer. Thus high radiation doses of 0.2–0.8 MGy (20–80 Mrad) are required to obtain a significant modification. Much higher grafting efficiencies are obtained with polymerizable vinyl monomers such as acrylic acid.

Commercial and Other Polyamides

Typical aliphatic polyamides include nylon-6,6 (Table 1), nylon-6 (Table 2), nylon-6,12 (Table 1), nylon-4, other AB polyamides (Table 3), nylon-3 polypropiolactams, nylon-7, nylon-8, nylon-11 (Table 4), nylon-12 (Table 4), nylon-4,6 (Table 5), and nylon-4,2 (see POLYAMIDES, FIBERS for a description of nylon nomenclature).

Ring-containing Polyamides. Ring-containing polyamides include aliphatic–aromatic polyamides (Tables 6 and 7), PACM-12 polyamide, and aromatic polyamides (aramids).

Polyamide Blends and Block Copolymers. Amide interchange in blends of compatible polyamides (eg, nylon-6,6 and -6) is slow, and a true random copolymer forms in more than 8 h at 285 °C. Melt blends of nylon-6,6 with aliphatic–aromatic polyamides with higher T_g, such as 6,I (I = isophthalic acid) or MPD-6 (MPD = *m*-phenylenediamine), give block copolymers with degrees of interchange ≤3% (based on the ring-containing polyamide). These have a T_g higher than that of nylon-6,6.

Polyamides from Long-Chain Fatty Acids. Another class of polyamide is based on condensation products of diamines or triamines, such as ethylenediamine or diethylenetriamine, with high molecular weight dibasic acids or esters. The latter are obtained from thermal polymerization of a diene acid or ester such as linoleic acid or lineolates. These polyamides vary from tacky to high melting (175 °C) resins and are used as heat-sealing adhesives and moisture-barrier coatings for paper, wood, cellophane, and other substances.

BIBLIOGRAPHY

D. B. Jacobs and J. Zimmerman in C. E. Schildknecht and I. Skeist, eds., *Polymerization Processes,* John Wiley & Sons, Inc., New York, 1977, Chapt. 12.

P. W. Morgan, *Condensation Polymers by Interfacial and Solution Methods,* Polym. Rev. Ser. X, John Wiley & Sons, Inc., New York, 1965.

Table 6. Properties of Nylon-6,T and Nylon-6,6

Fiber property	Nylon-6,T	Nylon-6,6
T_m, °C	370	265
density, g/cm^3	1.21	1.14
T_g, °C	180	60
water-content equilibrium at 70% rh, %	4.5	4.5
tensile strength, MPa[a]	425	490
elongation to break, %	35	57
modulus, MPa[a]	4250	4000
strength retained after 5 h in air, %		
at 150°C	100	85
at 185°C	100	40
at 220°C	60	0

[a] To convert MPa to psi, multiply by 145.

Table 7. Properties of Nylon-6,I Resin[a]

Property	Nylon-6,I
tensile strength, MPa[b]	58–65
yield stress, MPa[b]	85–105
elongation at break, %	20–80
tensile modulus, MPa[b]	2800
flexural modulus, MPa[b]	3200
flexural strength, MPa[b]	95
T_m, °C	180–210
thermal coefficient of expansion, K^{-1}	6.6×10^{-5}
refractive index	1.59
dielectric constant, 100 Hz	3.8
dielectric loss, 100 Hz	0.02
T_g, °C	120–130
density, g/cm^3	1.185

[a] As-molded.
[b] To convert MPa to psi, multiply by 145.

H. K. Reimschuessel, *J. Polym. Sci. Macromol. Rev.* **12,** 65 (1977).

M. I. Kohan, ed., *Nylon Plastics,* John Wiley & Sons, Inc., New York, 1973.

JOSEPH ZIMMERMAN
Consultant

POLYAMIDES, AROMATIC

Aromatic polyamides are formed by reactions that lead to the formation of amide linkages between aromatic rings. In practice, aromatic diamines and aromatic diacid chlorides are employed in an amide solvent. From these polymers, fibers of exceptional heat and flame resistance and high tensile strength and modulus are produced. Because fibers of aromatic polyamides exhibit unusual physical properties, the generic term aramid was adopted in 1974 by the U.S. Federal Trade Commission for designating fibers of the aromatic polyamide type and was defined as follows: aramid, a manufactured fiber in which the fiber-forming substance is a long-chain synthetic polyamide in which at least 85% of the amide ⟨CO—NH⟩ linkages are attached directly to two aromatic rings. The aramids in commercial use have 100% of the amide linkages attached directly to two aromatic rings.

There is no recognized systematic nomenclature for aromatic polyamides as exists for aliphatic polyamides. However, acronyms or initial letters of the various monomers are frequently combined with the initials of the diamine moieties preceding those of the diacid moieties. The letter T and I are commonly accepted for terephthalamide and isophthalamide, respectively. Using this shorthand, Nomex is based on MPD-I, poly(*m*-phenylene isophthalamide) and Kevlar is based on PPD-T, poly(*p*-phenylene terephthalamide).

Properties

Aramid fibers do not melt in the conventional sense because decomposition usually occurs simultaneously. An endothermic peak in the differential thermal analysis thermogram for aramid fibers can be obtained with values above 400 °C and as high as 550 °C. Glass-transition temperatures T_g range from ca 250 to >400 °C.

Aramid fibers are characterized by medium-to-ultrahigh tensile strength, medium-to-low elongation, and moderately high-to-ultrahigh modulus. Most of these fibers have been reported to be either highly crystalline or crystallizable with densities for crystalline fibers ranging from ca 1.35 to 1.45 g/cm^3; hot-drawn fibers have a density of ca 1.38 g/cm^3 to 1.45 g/cm^3. Fibers from PPD-T, eg, Kevlar, have a density of ca 1.45 g/cm^3.

Heat- and flame-resistant aramid fibers contain a high proportion of meta-oriented phenylene rings, whereas ultrahigh strength-high modulus fibers contain mainly para-oriented phenylene rings. The meta-oriented polymers are considered to be chain-folding polymers. On the other hand, small-angle x-ray diffraction studies of the rodlike para-oriented polymers show no evidence of chain folding.

In order to predict long-term performance at elevated temperatures, tensile properties are evaluated after heat-aging at intermediate temperatures. Heat-aging performance in air, ie, thermooxidative stability, is a severe test because of the high surface-to-volume ratio of the fibers. The effect of heat-aging at 300 °C in air shows Nomex to have a useful life of only about 40 h at 500°C, but at 250 °C in air, about 1400 h. At 177 °C this fiber retains 80% of its strength for several thousand hours in air.

Electrical Properties. Aromatic polyamides have high volume resistivities and high dielectric strengths that are retained at elevated temperatures. Accordingly, they have potential as high temperature dielectrics, particularly in motors and transformers.

Resistance to Chemicals and Radiation. The resistance of aramid fibers to chemicals is excellent. These fibers are much more resistant to acid than nylon-6,6 fibers but not as acid-resistant as polyester fibers, except at elevated temperatures, particularly in the range of pH 4–8. Resistance to strong base is comparable to that of nylon-6,6 fibers; hydrolytic stability is superior to that of polyester and comparable to that of nylon-6,6.

Dyeability. Aramid fibers are exceedingly difficult to dye by conventional techniques because of their very high T_g.

Ultrahigh Strength-High Modulus Aramid Fibers. Although heat-resistant rodlike aramids have very high melting points, >500 °C with decomposition, they are not particularly useful as heat-resistant fibers. Fibers from the rodlike polymers have rather low elongations, and they lose elongation rapidly at elevated temperature, resulting in brittleness. For short exposure, strength retention at elevated temperatures is excellent, eg, Kevlar exhibits tenacity of >0.88 N/tex (10 gf/den) at 300 °C. The chemical resistance of Kevlar is similar to that of Nomex.

Methods of Preparation

The usual methods for the preparation of aliphatic polyamides are unsuited to the preparation of high molecular weight aromatic polyamides. However, two synthetic methods are available for the preparation of medium-to-high molecular weight polymer: low temperature polycondensation (<100 °C) and direct polycondensation in solution using phosphites, especially in the presence of metal salts. The principal low temperature methods are interfacial polycondensation and solution polycondensation.

A–B Polyamides. Low temperature polycondensation of reactive A–B monomers produces high molecular weight A–B polyamides. Polymers derived from *m*- or *p*-aminobenzoic acid have been prepared from the *m*- or *p*-aminobenzoyl chloride hydrochloride by the addition of a base. The *m*- and *p*-aminobenzoyl chlorides are synthesized from aminobenzoic acids and thionyl chloride to yield the thionylaminobenzoyl chlorides, which react with HCl under anhydrous conditions.

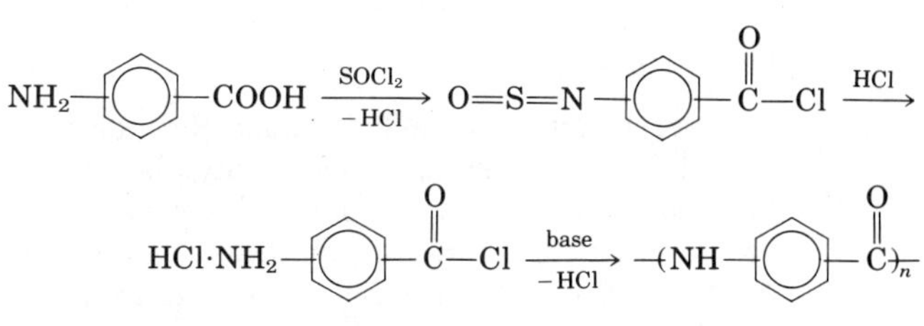

(1)

AA–BB Polyamides. These polyamides are readily prepared from aromatic diamines and diacid chlorides.

$$NH_2{-}Ar{-}NH_2 + Cl{-}\overset{O}{\overset{\|}{C}}{-}Ar'{-}\overset{O}{\overset{\|}{C}}{-}Cl \longrightarrow {-}\!\!(NH{-}Ar{-}NH{-}\overset{O}{\overset{\|}{C}}{-}Ar'{-}\overset{O}{\overset{\|}{C}})\!\!{-} \qquad (2)$$

Ar and Ar′ = 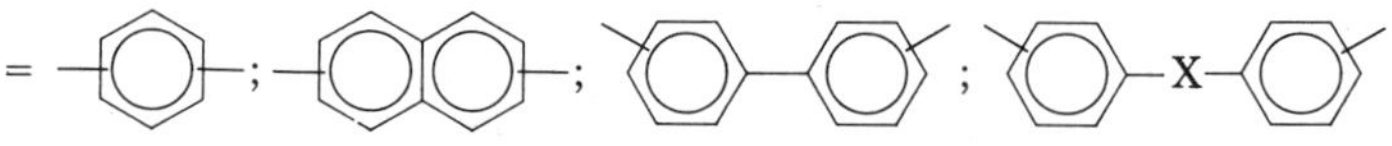

$X = O, S, CO, SO_2, CH_2, {-}\!(CH_2)_n\!{-}, C(CH_3)_2$

Polyamides with phenylene rings bridged by short alkylene chains are included because their properties resemble those of aromatic polymers. Polymers containing pendent methyl or chloro groups on the phenylene units are also included.

Low temperature solution polycondensation is preferred to interfacial polycondensation, as shown by the numerous reports of high molecular weight, fiber-forming polymers of widely varying structure. However, aromatic polyamides usually do not remain in solution when prepared by solution polymerization unless salts, such as lithium or calcium chloride, are present, and a basic solvent, eg, DMAC or NMP, is used. The products of low temperature solution polycondensation may be used in a convenient, economical, and direct process for the spinning of fibers.

Production

Aromatic polyamide fibers are spun from solution. Dry- and wet-spinning techniques appear to be equally satisfactory for polymers soluble in organic solvents, although properties may vary with the spinning method. The wet method usually requires a higher molecular weight, 60,000, for the best tensile properties. Polyamides prepared by solution polycondensation can be conveniently spun from the resulting solution, eliminating the need for isolating and redissolving the polymer. Spinning the polymer solution as prepared directly is necessary for polymers that do not redissolve (or that redissolve only with difficulty) in organic solvents after isolation in the dry state and that are not stable in strong acid solvents. Polymers prepared by interfacial polymerization must be dissolved before spinning.

Economic Aspects

Although the annual production of aramid fibers is not high, the total dollar volume is substantial and rising. There is not strong competition, as with commodity fibers, and profitability is good and should continue to be so. Because of rather high monomer and processing costs coupled with low volume production, aramid fibers are relatively expensive and probably will remain so compared with other large volume synthetic fibers.

Health and Safety Factors

Inhalation studies at the DuPont Haskell Laboratory for Toxicology and Industrial Medicine showed hexamethylphosphoric triamide (HPT), used in the preparation of rodlike polyamides, eg, PPD-T, to be carcinogenic in rats. Accordingly, any work with HPT should be carried out in an efficient hood and contact with the skin must be avoided. In commercial production of PPD-T, amide solvents (eg, NMP) containing $CaCl_2$ are used.

For Nomex, and especially for Kevlar, as with any fine, fibrous material, inhalation of the "fly" generated in certain textile operations should be avoided. Because of its unusual high strength and cutting resistance, caution should be exercised in splicing, handling, or cutting Kevlar.

Applications

Common applications exploit the following properties; heat-resistance, flame resistance, dimensional stability, ultrahigh strength, high modulus, and permselective properties.

BIBLIOGRAPHY

W. B. Black and J. Preston in H. F. Mark, S. M. Atlas, and F. Cernia, eds., *Man-Made Fibers: Science and Technology,* Vol. 2, Wiley-Interscience, New York, 1968, pp. 297–364.

W. B. Black and J. Preston, eds., *High-Modulus Wholly Aromatic Fibers,* Marcel Dekker, Inc., New York, 1973.

J. Preston, "Synthesis and Properties of Rod-Like Condensation Polymers" in A. Blumstein, ed., *Liquid Crystalline Order in Polymers,* Academic Press, Inc., New York, 1978, pp. 141–166.

M. Jaffe and R. S. Jones in M. Levin and J. Preston, eds., *High Technology Fibers,* Pt. A (series: M. Lewin, ed., *Handbook of Fiber Science and Technology,* Vol. III), Marcel Dekker, Inc., New York, 1985, pp. 349–392.

J. PRESTON
Research Triangle Institute

POLYAMIDES, FIBERS

In polyamide fibers monomer units are joined by amide groups prepared from diamines and dicarboxylic acids, $(CONH{-}R{-}NHCO{-}R')_n$ or from lactams, ${-}\!(RCONH)_n\!{-}$. When R or R′ are aliphatic, alicyclic, or aromatic mixtures in such a manner that <85% of the amide linkages are attached directly to two aromatic rings, the polyamides are known as nylon. When at least 85% of the amide linkages are attached directly to aromatic rings, the fibers are generically called aramids. This article is a survey of the nylon fibers; for aramids, see POLYAMIDES, AROMATIC.

The nomenclature commonly designates the number of carbon atoms in the repeating monomer unit. Thus the polyamide from caprolactam is known as nylon-6 and that from hexamethylenediamine and adipic acid as nylon-6,6. In the case of nylons from diamines plus dicarboxylic acids, the first number designates the diamine used. Thus the polymer from hexamethylenediamine and sebacic acid is called nylon-6,10.

Nylon was the first synthetic fiber of importance. Although a polyamide had been prepared in 1899, it was the fundamental research of Carothers at DuPont more than 30 years later that led to nylon fibers.

The world production of nylon fibers has grown from about 1,500 t in 1940, to 55,000 t in 1950, to 3×10^6 t in 1980. Production has remained more or less constant since that time. The production in the United States was ca 1.09×10^6 t in 1984, essentially all from nylon-6,6 plus nylon-6. Estimates indicate ~65% of the U.S. capacity is nylon-6,6, but ~60% of world capacity is nylon-6 (1985 Monsanto Co. estimates).

Of the large-volume synthetic fibers, nylon usually is the most expensive when compared to fibers of the same filament strength and in the same form, eg, as staple or as textured-filament yarn. This is to be expected because the costs of the polymers increase in the series polypropylene < acrylic < poly(ethylene terephthalate) < nylon-6 or nylon-6,6. In 1986 nylon filament yarn of 1.7–2.2 tex (15–20 den) sold for ~\$5.28–6.60/kg, nylon tire yarn of 140 tex (1260 den) was priced at ~\$3.52–3.96/kg, and nylon carpet staple of 1.6–2.0 tex (15–18 den per filament) was priced at ~\$2.20–3.08/kg. Nylon has a lower specific gravity than acrylics or polyester and this, plus special properties, make it fully competitive or even preferred in many applications.

Preparation of Polyamides

Raw Materials. The raw materials used in the production of commercial polyamides are ε-caprolactam, hexamethylenediamine, and adipic acid (see also POLYAMIDES).

Polymerization. Nylon polymers are produced by the melt polycondensation of a diamine and a dicarboxylic acid, or the ring-opening polymerization of caprolactam. Many other methods have been used in the laboratory. Manufacture of nylon-6 and -6,6, and many others have been studied on a smaller scale; wet spinning has been reviewed (see also FIBERS, MANUFACTURE).

In melt-spinning, molten polymer is delivered from an extruder or a metal-grid melter (an early commercial process), or directly from a polymerizer through a filter to a meter pump. This pump accurately meters polymer to the pack, a combination of a small filter and spinneret. Upon leaving the spinneret, the molten filaments pass into a vertical chimney where they are air-cooled and solidify. At the same time they are stretched to the desired diameter. Finish is applied and the yarn is wound on a bobbin.

Next, the fibers are heated and stretched (drawn) to 30–600% of their original length, depending on spinning speed. This results in oriented crystallization and increased tensile strength. At this stage the fiber or yarn (a bundle of fibers) may be suitable for some applications requiring straight (flat) filaments. Where special properties such as very low shrinkage may be required, as in tire yarn, a heat-relaxation step may be added.

For apparel or carpeting applications, a textured yarn is preferred. Texturing introduces crimp, whereby the straight filaments are given a twisted, coiled, or randomly kinked structure. A yarn made up of such filaments is softer and more open in structure and provides fabrics more pleasing to the touch. Combination of two or more of the above steps into consecutive processes, such as spin-draw or spin-draw-texture, reduces manufacturing costs.

The processes described thus far produce continuous-filament yarn. Another type of product is staple, which is obtained by cutting continuous-filament yarn into 3–20-cm lengths. Staple can be processed like natural fibers, such as wool and cotton, with which it can be blended.

Modified Cross Sections. Conventional spinneret orifices are circular and produce a fiber that is round in cross section. Improved machining methods, advances in fiber-production technology, and the study of the effect of profiled fibers on fabric characteristics has increased the importance of noncircular (modified cross section) filaments.

The cross section of the fiber is an important variable in designing the fiber properties for a specific use. For example, if maximum strength is required, as in tire cord, a round cross section with the fewest possible surface flaws is preferred. For antisoil carpet fibers, the cross section should be free from crevices that could trap dirt particles, so round, or squares with rounded edges, or triangles with rounded edges may be preferred. Since flat surfaces enhance light reflection, a triangular cross section may be desired for high luster, as in some upholstery fabrics. An elongated cross section, such as an ellipse, enhances bulk and covers more than round fibers. Some commercial carpet yarns are produced from hollow fibers in which several tiny holes run parallel to the fiber axis. The internal air-polymer interfaces thus created help to hide dirt.

Finishes. The main functions of a fiber finish are to provide yarn cohesion and lubricity. Lubricity is important because many types of surfaces are involved, both in manufacture and processing. Yarn cohesion is essential for the yarn to move as a compact structure without filaments protruding from the bundle. Static electricity, a common cause of poor yarn cohesion, can be alleviated by an antistatic finish. These finishes are usually an emulsion or a water-soluble mixture of lubricants and antistatic agent; wetting agents promote spreading. Mineral oils and esters or ethylene oxide adducts of long-chain alcohols are frequently used as lubricants; other additives include nonionic polyether wetting agents, antistatic agents, and emulsifiers.

Additives. Additives are included in the nylon system before or during polymerization in order to influence the properties and nature of the product. They include delustering agents, spin-dyeing agents, antistatic additives, and flame-retarding additives.

Properties of Nylon-6 and -6,6

The uses of fiber depend on its properties. The dynamic growth of nylon in the textile industry is due mainly to a combination of unique and inherent properties that are not available in any other fiber. Fibers are judged commercially by a combination of properties and costs. Aesthetics, style, and perceived value are difficult to measure quantitatively. In apparel material these properties may include fabric softness, drape, sound caused by movement, light reflection, and comfort. In carpets they may include brightness of color and firmness or plushness of the carpet. Strength, abrasion resistance, recovery from deformation, creep, and resistance to environmental changes are easier to measure by conventional methods.

A nylon fiber is partially crystalline; thus it is composed of

Table 1. Physical Properties of Nylon-6 and -6,6 Continuous-filament Yarns[a]

Property	Normal tenacity	High tenacity
tenacity, N/tex[b]	0.4–0.6	0.75–0.84
wet	0.36–0.51	0.64–0.7
loop	0.40–0.49	0.57–0.69
knot	0.35–0.42	0.52–0.72
tensile strength, MPa[c]	580–635	855–952
elongation at break, %		
conditioned	23–43	12–17
wet	28–41	14–21
tensile modulus, conditioned, N/tex[b]	2.2–3.1	2.9–4.1
average toughness, N/tex[b]	0.07–0.14	0.06–0.14
moisture regain at 21°C, %		
65% rh	4.5	3.3–4.5
95% rh	7.1–7.8	7.8

[a] Conditioned at 65% rh and 21°C.
[b] To convert N/tex to gf/den, multiply by 11.33.
[c] To convert MPa to psi, multiply by 145.

many crystallites, each surrounded by an amorphous region. Physical properties are controlled to a very large extent by the percent crystallinity, the average crystallite size, the crystalline orientation relative to the fiber axis, and the amorphous orientation.

When a nylon fiber is extended, the crystallites, acting as physical cross-links, oppose the extension. Hydrogen bonding and van der Waals attractions in the amorphous regions also oppose the extension but more weakly. These forces acting together provide a short-range elasticity that permits nylon to recover nearly completely from elongations of ca 5–8%, if the extension is maintained for short times. Elastic recovery in nylon is outstanding.

The melting point determines thermal resistance, safe ironing temperature, and heat-setting temperature. The glass-transition temperature greatly influences initial modulus, tensile-strain recovery, work of recovery, drape and hand, wash-and-wear characteristics, comfort, resilience, and ease and cost of dyeing. Moisture absorption affects the drying rate, comfort, and hand or feel of the fabric. Density determines meter per kilogram of fiber and affects material cost per meter, and ease and comfort in wearing (weight).

Tensile Properties. The tensile properties of nylon-6 and -6,6 yarns are shown in Table 1.

The conditions of extension and recovery (load, rate of extension and recovery, time of extension, etc) affect the quantitative comparisons of nylon with other fibers. In comparison with other fibers, nylon yarns have an outstanding degree of elasticity, recovering well from high loads and extensions. The outstanding elastic recovery of the nylons is responsible for the cling or fit in women's hosiery and for their extensive market acceptance.

Temperature and Moisture Effects. The thermal behavior of nylon fiber and the interrelated effects of moisture have a pronounced influence on the physical properties of the yarn and its products. They are of basic importance in fiber manufacture and conversion into fabric, and in use.

The role of water is largely that of a very powerful plasticizer, affecting the mobility of the amorphous regions. It is to be expected that water content of the fiber is very important in the rate of crystallization during spinning, the tensile strength and modulus of the fiber in use, and the dyeing of fabrics. Thermal properties of nylon-6 and -6,6 are given in Table 2.

Stabilization. Polyamide fibers are easily stabilized against light and other degrading conditions. Stabilizers include manganese salts and additives such as hypophosphorous acids, phosphites, and phosphates. In the presence of light, anatase TiO_2, a delusterant, results in autocatalytic formation of a peroxide, which degrades polyamides in the absence of stabilizers. Copper salts are the preferred stabilizers against degradation by heat and light.

Electrical Properties. Electrical conductance of dry nylon is very low, as might be expected from its low moisture regain. The insulating properties of nylon are manifested in the readiness with which it accumulates static electrical charges. These charges are not readily dissipated, especially at very low rh. The problem in carpets is eliminated by the incorporation of a very small percentage of conductive filaments.

Staple Properties. The use of nylon staple, rather than continuous filament, stems from a need to satisfy the demand for

Table 2. Thermal Properties of Nylon-6 and -6,6, °C

	Temp, °C	
Property	Nylon-6	Nylon-6,6
mp	215;220	250
zero strength[a]	232	240
max setting temperature	190	225
softening point	170	235
starts to become plastic	160	220
crit temperature		
in air	163	158[b]
	93[c]	130[c]
in steam	137	140
max ironing temperature	150	180
optimum setting (steam)	128	130
max wash temperature		
set by saturated steam	60	71
set dry	30	60

[a] Heat-stabilized tire yarn, 1.6-kg load.
[b] Heat-stabilized tire yarn.
[c] Unstabilized yarn.

a "soft and warm fabric hand." Even more important is the aim to combine some of the advantageous properties of nylon with those of other staple fibers, natural as well as regenerated cellulosics. The outstanding contributions of nylon staple to carpet, apparel, and upholstery fabrics are abrasion resistance and low moisture absorption. The former increases the wear life of the textile structure, the latter contributes to faster drying.

Dyeability

Nylon has a marked affinity for virtually every dye class, such as disperse, acid, premetallized, chrome, and direct dyes, and certain naphthol combinations (see DYEING). Disperse dyes are frequently used because of their easy applications, and acid dyes for their bright colors. To the dyer differences between nylon-6 and -6,6 are less significant than fiber and fabric uniformity.

Effect of Fiber Characteristics. Chemical structure, crystallinity, molecular orientation, and fabric preparation determine the rate and extent of nylon dyeability. Nylon-6 and -6,6 differ in the arrangement of the carbonyl and NH groups between the hydrocarbon chains of the polymer. When the molecules of the noncrystalline portions of the extruded fiber are aligned (oriented) by drawing, the nylon-6 molecules have a more random, open structure than nylon-6,6. The rate of diffusion of dyes into this open structure is higher. Thus nylon-6 is more easily dyed than nylon-6,6; copolymer modifications of the latter increase dyeability.

Effect of Dyebath. Many methods have been proposed to ensure uniform dyeing. Uniformity is most easily achieved when an equilibrium is established between dye diffusion in and out of the fiber and the dye remaining in the bath. Under this condition, nonuniform fabrics are dyed to a level shade; disperse dyes are unsurpassed for this purpose.

Modified Nylon-6 and -6,6 Fibers

Bicomponent and Biconstituent Fibers. Bicomponent fibers consist of two polymers of the same generic family, eg, nylon-6 and -6,10. Biconstituent fibers are made from components that are not of the same generic family, eg, nylon-6,6 and poly(ethylene terephthalate). The components may be in two continuous phases arranged in a side-by-side or sheath-and-core configuration, or one may be dispersed in the other (island-in-the-sea configuration). The compositions of early commercial bicomponent fibers are shown in Table 3.

Block Copolymers. Nylon block copolymers are prepared by combination in the melt of two or more homopolyamides, reaction of a diamine or diacid monomer containing an amide linkage with another diamine or diacid, reaction of a complex molecule such as bisoxazolone with a diamine, to produce a wide range of multiple amide sequences along the chain, and reaction of a diisocyanate and a dicarboxylic acid.

Random Copolymers. Copolymers are used for the modification of nylon-6,6 fibers to alter their dyeing behavior.

Graft Polymers. Graft polymers in the polyamide field have been studied extensively but have not been commercialized despite indications of improvements in physical properties.

Other Nylons

Qiana. The only nylon of significant commercial use other than nylon-6 and -6,6 has been sold under the trademark Qiana. The polymer for this fiber was produced by the condensation polymerization of bis(4-aminocyclohexyl)methane and dodecanedioic acid. The diamine occurs in several isomeric forms; a 20% cis-trans, 80% trans-trans diamine is believed to have been used in the polymer. Typical yarn properties are shown in Table 4. These yarns have been withdrawn from the market in recent years.

Table 4. Properties of Qiana Nylon Yarns

Property	Value
mp, °C	280
glass temp, °C	
dry	~190
wet	~85
density, g/cm^3	1.03
tenacity, N/tex[a]	0.26–0.29
elongation, %	2.1–2.6
shrinkage in boiling water, %	4–10
moisture regain (at 65% rh), %	2.5

[a] To convert N/tex to gf/den, multiply by 11.33.

Experimental Nylons. In an effort to improve the wrinkle resistance of nylon fabrics and the flatspotting performance of bias and bias-belted tires, a number of additional polyamides with higher T_g than nylon-6 and -6,6 have been tested. Typical properties are shown in Table 5. Another group of nylons has been tested extensively in recent years, possibly for general use in apparel and carpets (Table 6).

Applications

The principal commercial uses of nylon are in carpets, apparel, tire reinforcement, and industrial applications. The consumption in the United States in 1984 was ca 710,000 t for carpets, 140,000 t for apparel, 85,000 t for tire cord, and 105,000 t for home furnishings and other industrial uses.

Table 3. Commercial Bicomponent Fibers

Type	Component A: Copolymer	Component A: Composition	Component B, homopolymer
sheath-and-core	6,6–6,10	50:50	6,6
side-by-side	6,6–6,I[a]	80:20	6,6
side-by-side	6,6–6	80:20	6,6
sheath-and-core	6–6,T[b]	40:60	6

[a] 6,I = hexamethyleneisophthalamide.
[b] 6,T = hexamethyleneterephthalamide.

Table 5. Dimensionally Stable Experimental Nylons

Diamine[a]	Diacid	Mp, °C[b]	T_g[b], °C	Tenacity, N/tex[c]	Elongation, %	Evaluation
PACM	dodecanedioic	298	~220	0.78	13	no flatspotting in tires
CBMA	dodecanedioic	280	175	0.8	13–16	no flatspotting in tires
CBMA	sebacic	291		0.6	14	no flatspotting in tires
CBEA	dodecanedioic	254	135	0.35–0.44	20–30	
MDA	sebacic	270	160	0.48	11	no flatspotting in tires
PXD	sebacic	290	158	0.44–0.53	12–15	tire flatspotting not satisfactory
HMD	terephthalic	370	180	0.22–0.44	30–45	mp is too high for melt spinning
HMD	terephthalic–isophthalic, 65/35	310	165	0.3–0.66	8–40	no flatspotting in tires
TMD	adipic	~319		~0.8		may need copolymerization for processibility

[a]PACM = bis(*p*-aminocyclohexyl)methane; CBMA = *p*-cyclohexanebis(methylamine); CBEA = *p*-cyclohexanebis(ethylamine); HMD = hexamethylenediamine; MDA = 4,4′-methylenedianiline(bis(*p*-aminophenyl)methane); PXD = *p*-xylylenediamine; TMD = tetramethylenediamine.
[b] Dry; T_g measured by Vibron analysis.
[c] To convert N/tex to gf/den, multiply by 11.33.

Table 6. Other Experimental Nylons

Polyamide	Mp, °C	T_g, °C	Tenacity, N/tex[a]	Moisture regain, %	Evaluation
poly(*m*-xylylene adipamide)	243	90	0.26–0.44	4.0	hot-wet properties and light stability inferior to nylon-6 and -6,6
nylon-4	262	72	0.22–0.40	7.3	excessive property loss when wet
nylon-7	235	60	0.26–0.40	2.8	
nylon-11	192	57	0.26–0.44	1.3	
nylon-12	175	40	0.22–0.40	2.0	mp is too low

[a] To convert N/tex to gf/den, multiply by 11.33.

BIBLIOGRAPHY

W. E. Morton and J. W. S. Hearle, *Physical Properties of Textile Fibers*, Butterworth & Co. (Publishers) Ltd., Kent, UK, 1962, Chapts. 15 and 16.

V. V. Korshak and T. M. Frunze, *Synthetic Heterochain Polyamides*, Israel Program for Scientific Translations, Jerusalem, 1964.

H. F. Mark, S. M. Atlas, and E. Cernia, eds., *Man-made Fibers: Science and Technology*, Wiley-Interscience, New York, Vol. 1, 1967, pp. 241–374.

E. M. Hicks, Jr. and co-workers, *Text. Prog.* **3**, 1 (1971).

James H. Saunders
Consultant

POLYAMIDES, PLASTICS

When Carothers introduced polyamides almost 50 years ago, he could have hardly envisioned the magnitude of the product line he created. From basic fiber origins in 1938, the polyamide family has grown into one of immense proportions and versatility. Nylon is a generic term for any long-chain synthetic, polymeric amide in which recurring amide groups are integral to the main polymer chain; the terms nylon and polyamide are used interchangeably here. There is a wide choice of starting materials from which polyamides can be synthesized. The two primary methods for manufacture are condensation of a diamine and a dibasic acid or their equivalents, and polymerization of monomeric substances.

The nylons are identified by a simple numerical system; one number indicates that the product was prepared from a single monomeric substance and represents the number of carbon atoms in the linear chain of the recurring polymer unit. When two reactants are used in the polymer manufacture, they are represented by two numbers separated by a comma. The first refers to the number of carbon atoms in the diamine and the second to the number of carbon atoms in the dibasic acid.

The difference in the numbers of carbon atoms between the amide groups results in widely varying physical and mechanical properties. As can be readily seen, the theoretical potential for polyamide types is as wide as numerical statistics allow. In practice, comparatively few nylons are available commercially. The more common types of polyamide plastics are shown in Table 1. Within each type of nylon, a multitude of grades have been developed for specific market and process purposes.

Manufacturing and Processing

Polyamides are most commonly processed in the molten state. Nylons are hygroscopic and hydrolytically unstable and must be dry when processed. The materials are packaged in moistureproof containers at moisture contents suitable for processing. Processing methods include injection molding, rotomolding, reaction injection molding (RIM), powder coating, and machining.

Table 1. Properties of Nylons, Dry as Molded

Property	Nylon-6	Nylon-6,6	Nylon-11	Nylon-12	Nylon-6,9	Nylon-6,12	ASTM test method
specific gravity	1.13	1.14	1.04	1.02	1.09	1.07	D792
water absorption, wt %							D570
24 h	1.6	1.2	0.3	0.25	0.5	0.5	
equilibrium at 50% rh	2.7	2.5	0.8	0.70	1.8	1.4	
saturation	9.5	8.5	1.9	1.5	4.5	3.0	
melting point, °C	215	255	194	179	205	217	
tensile yield strength, kPa[a]	8.1×10^4	8.3×10^4	5.5×10^4	5.5×10^4	5.5×10^4	5.5×10^4	D638
elongation at break, %	50–150	40–80	200	200	125	150	D638
flexural modulus, kPa[a]	2.8×10^6	2.8×10^6	1.2×10^6	1.1×10^6	2.0×10^6	2.0×10^6	D790
Izod impact strength, J/m[b] of notch	55–65	53–64	40–68	95	58	58	D256
Rockwell hardness, R scale	119	121	108	107	111	114	D785
deflection temp under load, °C							D648
at 455 kPa[c]	185	235	150	150	150	165	
at 1850 kPa[c]	65	75	55	55	55	82	
dielectric strength, kV/mm							D149
short-time	17	24	16.7	18	24	16	
step-by-step	15	11		16	20		
dielectric constant							D150
at 60 Hz	3.8	4.0	3.7	4.2	3.7	4.0	
at 10^3 Hz	3.7	3.9	3.7	3.8	3.6	4.0	
at 10^6 Hz	3.4	3.5	3.1	3.1	3.3	3.5	
starting material(s)	caprolactam	hexamethylene-diamine and adipic acid	11-aminoun-decanoic acid	lauro-lactam	hexamethylene-diamine and azelaic acid	hexamethylene-diamine and dodecanedioic acid	

[a] To convert kPa to psi, multiply by 0.145.
[b] To convert J/m to ftlbf/in., divide by 53.38.
[c] To convert kPa to atm, divide by 101.3.

Additives and Modifications

Additives are incorporated for a variety of reasons: lubricants improve mold release and extrusion characteristics; heat stabilizers increase thermal range; nucleation alters the rate and type of crystallization; and dyes and pigments produce coloration. Varying the acid-to-amine ratio of the end groups affects hydrolytic stability and dye absorption as well as adhesion characteristics. The above additives are used at very low levels and have minor effects on properties. Other additives include plasticizers, modifiers, reinforcement, flame retardants, and regenerated nylon.

Environmental Effects

Heat. Polyamides undergo molecular weight and property losses when exposed to elevated temperatures; the degradation depends on time and temperature.

Ultraviolet Radiation. Similar to most plastic materials, the surface appearance and mechanical properties of polyamides are adversely affected by extended exposure to uv light; toughness and ductility deteriorate rapidly.

Moisture. Nylons absorb moisture, accompanied by changes in mechanical properties and dimensions (Table 2).

Chemical Resistance. Polyamides, in general, exhibit excel-

Table 2. Effect of Moisture on Mechanical Properties

	Nylon-6		Nylon-6,6		Nylon-11		Nylon-12	
Property	Dry	50% rh	Dry	50% rh	Dry	50% rh	Dry	50% rh
tensile strength, MPa[b]	81	45	83	58	55	46	55	48
flexural modulus, MPa[b]	2800	970	2800	1200	1200	640	1100	600
notched Izod impact strength, J/m[c]	55	215	53	113	40		95	

[a] To convert MPa to psi, multiply by 145.
[b] To convert J/m to ft·lbf/in., divide by 53.38.

lent chemical resistance; nylon-6 and -6,6 show no significant differences and are resistant to organic solvents, refrigerants, and hydrocarbons. They are soluble in phenol and formic acid and are chemically attacked by strong acids and oxidizing agents (see CHEMICALLY RESISTANT POLYMERS).

Economic Aspects

Polyamide plastics are the largest volume engineering plastics, with 1984 worldwide sales of 460,000 t.

The 1983 worldwide consumption of polyamide plastics was 415,000 t. Consumption varies widely among the United States, Western Europe, and Japan. In Europe and Japan nylon-6 is the dominant type, utilizing plants and technologies evolved after World War II. In the United States, nylon-6,6 was first developed and still dominates the market.

Table 3. U.S. Polyamide Consumption, 1985

Market	Metric tons
Extruded products	
filaments	7,000
film	14,000
sheet, rod, tube	10,000
wire and cable	9,000
Molded products	
appliances	5,000
consumer products	16,000
electrical and electronic	19,000
industrial	16,000
transportation	52,000
Other	14,000
Total	*162,000*

Applications

Injection-molded applications comprise more than 70% of the polyamide plastics consumed in the United States. In 1985, over 40% was used in the transportation industry (Table 3).

BIBLIOGRAPHY

M. I. Kohan, *Nylon Plastics,* John Wiley & Sons, Inc., New York, 1973.

Capron Nylon Homopolymers for Molding and Extrusion, CL-100-884, Allied Corp., Morristown, N.J., and other Allied technical bulletins.

Zytel Nylon Resins, General Guide to Products and Properties, E-42267, E. I. du Pont de Nemours & Co., Inc., Wilmington, Del.

R. J. WELGOS
Allied Chemical Corporation

POLYAMIDES FROM FATTY ACIDS

Polyamides based on fatty acids are condensation products of fatty dibasic acids (C-13, C-19, C-21, and C-36) and di- and polyfunctional amines. The commercial products are largely based on dimer acid (C-36) and less on C-21 dibasic acid. Other fatty polyamides are based on brassylic acid (C-13) and carboxystearic acid (C-19).

Polyamides from Dimer Acids

The term dimer acids is applied to the dicarboxylic acids formed by the reaction of two or more C-18 unsaturated fatty acids such as oleic and linoleic acids. Commercial dimer acid products are mixtures of C-36 dibasic acids containing some trimer (C-54), higher oligomers, and also small amounts of monomer acids (C-18 saturated, unsaturated, and structurally modified fatty acids).

Tall oil fatty acids (TOFA) are usually the feedstock for the production of dimer acids. However, oleic acid and soya fatty acids are used when a different predominant dimer structure is required.

Preparation of Polyamides. Polyamides are reactive liquids or thermoplastic solids and may be categorized by molecular weight, reactivity, and solubility. Most solid polyamides are neutral polymers from the 1:1 stoichiometric condensation reaction of linear diamines, dimerized fatty acids, and monobasic or dibasic acid modifiers, eg,

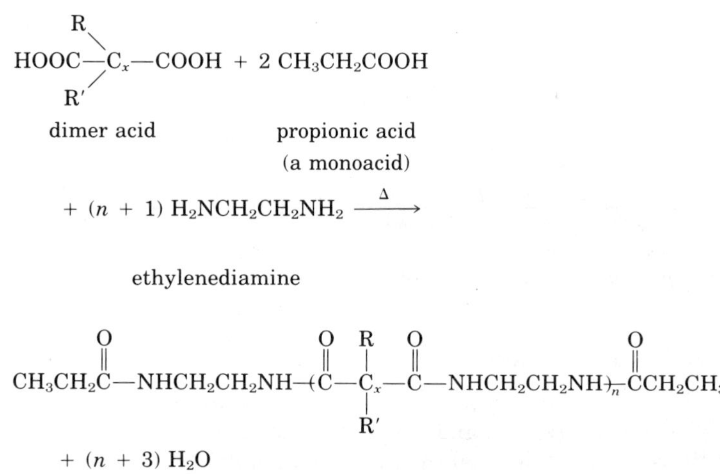

where R, R′ C_x designates a variously substituted chain of carbons of length x. Liquid polyamides are oligomers formulated to have amine reactivity for reaction with epoxy resins.

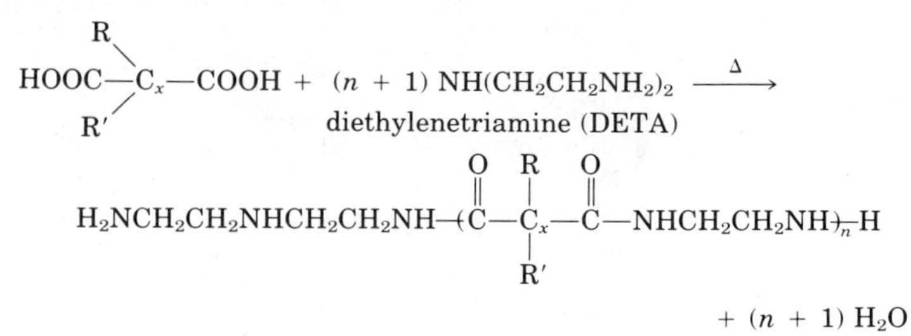

Process. Typical batchwise processing varies with formulations and procedures to meet special requirements.

Table 1. Manufacturers of Dimer-based Polyamide Resins

Company	Applications[a]	Trade name
United States		
AZS Corp.	C,I	Azamide
Bostik	A	Thermogrip
Celanese	C	Epi-cure
Emery	A,I	Emerez
Fuller	A	
Henkel	A,C,I	Versamid Macromelt
Hysol	A	Hysol
Lawter	I	Polymid
Pacific Anchor	A,C	Ancamide
Reichhold	A,C	Epotuf
Sun Chemical	I	SunKem
Sylvachem	C,I	Cropolamid
Union Camp	A,C,I	Uni-Rez
Terrell	A	Terlan
R.T. Vanderbilt	I	Van-Amid
Foreign		
Chem-Plast	A	
Cray Valley	I	
Croda	I	
Dainippon Ink	C,I	Luckamide
Emser	A	Griltex
Fuji	C,I	Tohmide
Hatrick Chem.	C,I	
Henkel	A,C,I	
Huels	A	
Indusguima	C,I	
Kao Soap	I	Leomide
Mitsui Toatsu	C	Epoky-H
Pacific Anchor	A,C	Wolfamide
Rilsan	A	
Sanwa	C	Sanmide
Sanyo	I	Polymide
Schering	A,C,I	Eurelon
Toto		Goodmide
Unichema	A,C,I	
Victor Wolf	I	Wolfamide

[a] A, hot-melt adhesives; C, curing agents; I, inks.

Environmental and Safety Considerations. The reactor room, which may house weigh tanks, filters, condensate receivers, and drum-filling facilities, must be designed considering the potential for leaks, spills, and vapor emissions of flammable and toxic materials. All motors, switches, lighting, and instruments should be explosion proof. All equipment must be electrically grounded, and a fire protection system is required; multiple escape doors must be accessible. Operating personnel must be provided with head, eye, and body protective gear, fire-fighting equipment, respirators, and safety showers and eyewashes.

Applications

Polyamides meeting the special needs of inks, coatings, and adhesives manufacturers are available from several domestic and foreign suppliers (Table 1). Typical properties of commercial polyamide resins are given in Table 2.

BIBLIOGRAPHY

L. V. Berman and M. L. Loeb in E. C. Leonard, ed., *The Dimer Acids*, Humko Sheffield Chemical, Memphis, Tenn., 1975.

R. W. Johnson in E. H. Pryde, ed., *Fatty Acids*, American Oil Chemists' Society, Champaign, Ill., 1979.

Polyamide Resins and Curing Agents, Union Camp Corp., Savannah, Ga., Apr. 1983.

R. W. Johnson
J. M. Valdespino
R. L. Gordon
G. E. Miller
R. W. Kight
Union Camp Corporation

POLYAMINES AND POLYQUATERNARY AMMONIUM SALTS

Most synthetic cationic polyelectrolytes are polyamines and polyquaternary ammonium salts (hereafter called polyquats),

Table 2. Typical Properties of Commercial Polyamide Resins

Manufacturer	Viscosity[a], mPa·s (=cP)	Softening point[b], °C	Acid number	Amine number	Applications
Union Camp					
Uni-Rez 2100	950^{150}			90	coatings and adhesives
Uni-Rez 2800	450^{25}			450	coatings and adhesives
Uni-Rez 2221	290^{145}	105	3	2	high solids inks
Uni-Rez 2931	$3,200^{160}$	110	2	4	inks and adhesives
Uni-Rez 2641	$8,500^{190}$	138			vinyl-bonding adhesives
Uni-Rez 2654	$10,000^{190}$	100	1	4	flexible adhesive
Henkel					
V-750	800^{160}	112	1	4	inks
MM-6239	$32,000^{190}$	129	1	7	inks and adhesives
MM-6902		208			colorless adhesives
Emery					
E-1533	$2,900^{160}$	100	2	4	inks

[a] Brookfield; superscripts indicate temperature in °C.
[b] Ring-and-ball method, ASTM E28-67.

although non-nitrogen-based cationic species are known. Polyamines and polyquats can be prepared by free-radical chain polymerization, epoxide-addition reactions, condensation polymerization, and reactions on polymer backbones (see also ACRYLAMIDE POLYMERS; FLOCCULATION; WATER-SOLUBLE POLYMERS).

The cationic character of polyamines depends on the pH; more strictly speaking, the extent of protonation is determined by the pK_b of the amine and the pH of the medium.

Chain-reaction Synthesis

Synthesis of Vinyl Monomers. ***Acrylic and Methacrylic Esters.*** Of the many structural variations possible the most widely used monomers are the methacrylic esters of dimethylaminoethanol. The amine monomers are usually prepared by transesterification of the methyl esters.

Acrylic and Methacrylic Amides. The polymers produced from cationic amide monomers are more stable to hydrolysis than the esters and are therefore of commercial interest. The most widely used aminoamide and quaternary ammonium amide monomers are prepared by the addition of *N*,*N*-dimethyl-1,3-propanediamine to methacrylic acid or its ester, followed by decomposition of the resultant aminopropionamide.

Allylamines. The most important commercial allylamine, diallyldimethylammonium chloride (DADMAC, DMDAAC, or DADM) is prepared by the reaction of allyl chloride with dimethylamine:

$$2\ CH_2{=}CHCH_2Cl + (CH_3)_2NH \xrightarrow{\text{base}} (CH_2{=}CHCH_2)_2\overset{+}{N}(CH_3)_2Cl^-$$

Vinylbenzyl Derivatives. The only commercial monomer of this type, vinylbenzyltrimethylammonium chloride, is prepared from vinylbenzyl chloride and trimethylamine.

Polymerization of Vinyl Monomers

The water solubility of the monomers, comonomers, and polymers determines, to a large extent, the polymerization conditions for these systems and water is thus the solvent of choice for solution polymerizations (see ACRYLAMIDE POLYMERS). Water is also the disperse phase in water-in-oil (ie, inverse) emulsion polymer systems. Inverse emulsion polymers are obtained by polymerization of the monomers dissolved in the aqueous micelles dispersed in the hydrocarbon continuous phase or by dispersion of polymers in the continuous phase, with or without water. In precipitation polymerization the monomer or mixture of monomers is dissolved in water and a cosolvent, in which the monomer is soluble but the polymer is not. Water is also used as a solvent for monomer and polymer in the preparation of cationic polymers by suspension polymerization, where the final form of the polymer is a spherical bead.

Chain-reaction polymerizations use free-radical initiators such as persulfate salts to provide free radicals thermally or by redox initiation. Other water-soluble initiators include peroxides and hydroperoxides and azo compounds. Cationic polymers are prepared by chain reaction at high concentrations and to a high conversion for economic reasons. In addition, a low monomer content is desirable, since most free monomers are more toxic than the polymers.

Homopolymerization. The most commonly employed monomer is DADMAC. It polymerizes by free-radical initiation via an alternating intra–intermolecular propagation mechanism, ie, cyclopolymerization (qv).

Copolymerization. Acrylamide is the comonomer in almost all commercial copolymers of amine and quaternary ammonium monomers.

Vinylamine Polymers. The monomeric vinylamine has apparently not been isolated. Attempts to prepare it by dehydrobromination of 2-bromoethylamine give ethyleneimine, and attempts at hydrolysis of vinyl isocyanate give acetaldehyde.

$$CH_2{=}CHNH_2 \rightleftharpoons CH_3CH{=}NH \xrightleftharpoons{H_2O} CH_3CH{=}O + NH_3$$

Vinylamine polymers, however, are well known and have been prepared by a number of methods. The earliest preparation was from *N*-vinylphthalimide:

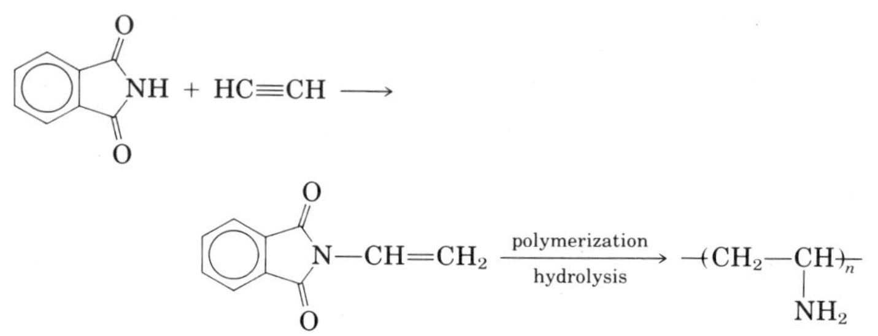

Other methods have employed a Hofmann reaction on polyacrylamide, using basic hypochlorite:

$$\text{-(}CH_2\underset{\substack{|\\ C{=}O\\ |\\ NH_2}}{CH}\text{)}_n\text{-} \xrightarrow[NaOH]{NaOCl} \text{-(}CH_2\underset{\substack{|\\ NH_2}}{CH}\text{)}_n\text{-}$$

The Schmidt reaction with hydrazoic acid has been used to convert polymeric carboxylic acids to amines.

These syntheses do not appear to be of commercial significance. A route from *N*-vinylacetamide has been of interest as a source of polyvinylamine and as a starting point for polymeric dyes (see DYES, MACROMOLECULAR).

A practical laboratory synthesis of poly(vinylamine hydrochloride) starts with conversion of acrylic acid to the acrylyl azide via the acid chloride. The azide is decomposed to vinyl isocyanate, which reacts *in situ* with *t*-butyl alcohol to give *N*-vinyl-*t*-butylcarbamate, followed by polymerization and hydrolysis. Polyvinylamine itself is difficult to isolate because of its tendency to react with CO_2. The acid salts, however, appear to be stable indefinitely.

Polymeric Starting Material

An early commercial polyamine was prepared from poly(acrylic acid) and ethyleneimine. Polyacrylonitrile and ethylenediamine give polyvinylimidazoline:

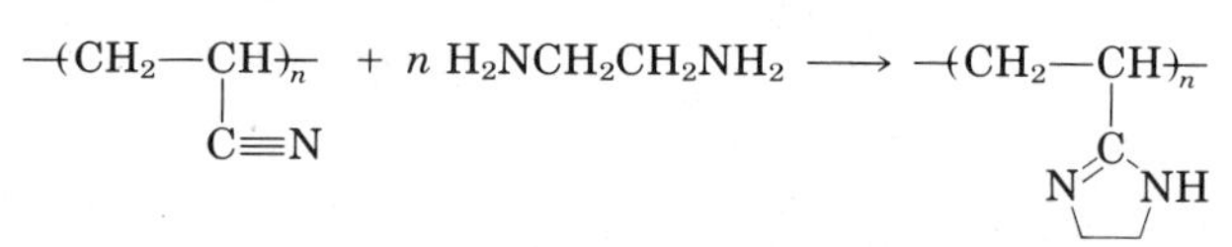

With propylenediamine a polymeric tetrahydropyrimidine is

obtained. A third, widely used polyamine is prepared by the Mannich reaction of formaldehyde and dimethylamine with polyacrylamide:

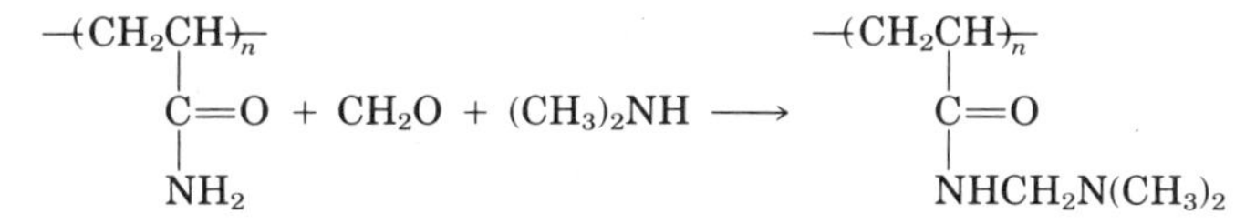

Step-reaction Synthesis

Chain-reaction synthesis gives linear polyamines and polyquats with high molecular weights, whereas linear step-reaction polymers from difunctional amines and epihalohydrins or dihalides rarely have molecular weights above 10,000. Higher molecular weights are achieved at the expense of linearity. High molecular weights in the millions, as are commonly found with cationic polyacrylamides, are rarely obtained commercially by step-reaction synthesis. The extensive branching necessary to reach these high molecular weights usually results in cross-linking, giving insoluble polymers, which are generally not useful.

With regard to the functionality of the amine, each nitrogen is potentially a quaternary ammonium function, ammonia can be tetrafunctional, dimethylamine difunctional, ethylenediamine hexafunctional, and so on. The functionality of the amines is one of the factors that influences molecular weight and degree of branching. With increasing functionality of the amine, higher branching and molecular weight are achievable. Conversely, to reach a particular molecular weight requires less of the amine as the functionality increases.

Polyamines have been prepared by step reaction between polyfunctional amines and epihalohydrins or dihalides.

Uses

Polyamines and polyquats find application by virtue of their cationicity in uses involving interactions with anionically charged colloidal particles (including biological species) in aqueous media in nature. In many instances the applications can be classified by structural type. For example, the high charge density, low molecular weight polyamines and polyquats prepared by step-growth polymerization are useful in the flocculation of particulate matter from turbid natural waters. High molecular weight cationic acrylamide copolymers are employed as pigment retention aids in the manufacture of fine paper. In certain applications, such as sludge dewatering, both high charge density and high molecular weight are required. Other applications include use as filtration aids, emulsion breakers, biocides, and antistatic agents, and in ore processing, enhanced oil recovery, and cosmetics (see MINING APPLICATIONS; OIL-FIELD APPLICATIONS; COSMETIC APPLICATIONS).

BIBLIOGRAPHY

E. Goethals, ed., *Polymeric Amines and Ammonium Salts (IUPAC)*, Pergamon Press, Elmsford, N.Y., 1980.

N. VORCHHEIMER
Betz Laboratories

POLYAMINOTRIAZOLES

Polyaminotriazoles are synthesized by the polycondensation of dihydrazides of dicarboxylic acids or of the diesters of dicarboxylic acids with hydrazine. The reactions of bis(imino ether) or dinitrile with hydrazine offered another route. In this article, the term polyaminotriazole is used to designate the poly(4-amino-1,2,4-triazole)s.

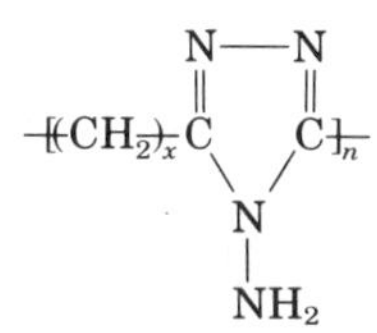

Polyaminotriazole is characterized by high crystallinity, a strong basicity due to the exocyclic amino group, strong affinity for acid dyes, and good fiber-forming properties. Polyaminotriazole can be incorporated in copolymers or blended with polyolefins, polyesters, and polyoxymethylene (POM) to improve the dyeability and thermal stability of the matrix polymers.

Fundamental data on polyaminotriazoles were reported in 1953–1954. Synthesis, chemical and physical properties, and uses have been reviewed since. A few patents were issued in the early 1970s. Polyaminotriazole fiber and yarn have not been produced on a commercial scale so far.

Properties

The chemical properties of polyaminotriazoles are dominated by the aminotriazole structure. The melting temperature of

Table 1. Physical Properties of Fibers[a]

Property	Poly(ethylene terephthalate)	Polyoctamethylene-aminotriazole	Nylon-6,6
specific gravity	1.38	1.12	1.14
dry tenacity, N/tex[b]	0.44	0.37	0.44
breaking elongation, %	20	20	23
tenacity (wet–dry), %	100	81	85
moisture at 65% rh, %	0.4	3.9	4.2
modulus of elasticity, N/tex[b]	7.9	5.6	2.4
shrinkage in boiling water, %	7	8	7
melting point, °C	260	256	256
glass-transition temperature, °C	67	88	57

[a] Medium tenacity.
[b] To convert N/tex to gf/den, multiply by 11.33.

polyaminotriazoles with alkyl groups in the backbone shows the normal zigzag depression with increasing numbers of methylene groups.

Polyoctamethyleneaminotriazole is soluble in 25% methanolic calcium chloride, concentrated aqueous sodium salicylate, and various organic solvents. It is insoluble in water, although the aminotriazole ring increases water solubility. Polyaminotriazole is stable to hydrolysis.

Fiber Properties. The physical properties of polyoctamethyleneaminotriazole fibers are compared in Table 1 with those of nylon-6,6 and poly(ethylene terephthalate) (PET). The dynamic behavior of polyoctamethyleneaminotriazole yarn resembles that of polyester more than that of polyamide, especially with respect to initial stress-strain and extension recovery.

BIBLIOGRAPHY

R. W. Moncrieff, *Text. Manuf.* **79,** 559 (Nov. 1953).

R. W. Moncrieff, *Text. Manuf.* **79,** 635 (Dec. 1953).

R. W. Moncrieff, *Text. Manuf.* **80,** 33 (Jan. 1954).

J. W. Fisher, *J. Appl. Chem. London* **4,** 212 (1954).

TADATOMI NISHIKUBO
Kanagawa University

MASAKI HASEGAWA
University of Tokyo

POLY(ω-AMINOUNDECANOIC ACID).

See POLYAMIDES.

POLYAMPHOLYTES

Macromolecules capable of possessing both positively and negatively charged moieties are commonly known as polyampholytes. The presence of ionic units along the polymer chain has a distinct effect on the solution and solid-state properties of this class of polymeric material. Coulombic attractions between oppositely charged sites afford formation of inter- and intramolecular ionic interactions that are stronger than van der Waals forces, yet weaker than covalent bonds. In the solid state an increase in ionic character results in a corresponding increase in tensile strength and glass-transition temperature, T_g, whereas in solution several factors affect the hydrodynamic volume of the ampholytic polymer chain. The type and concentration of the added salt, pH, ionic nature, and charge-density all influence the dilute solution properties of ampholytic polymers. In the absence of a simple electrolyte in aqueous solution, polyampholytes exhibit behavior ranging from highly water-swollen, insoluble hydrogels to water-soluble substances. The degree to which these effects are observed can be modified by incorporating a neutral monomer into the macromolecular chain, resulting in ampholytic ionomers.

Monomers

Acidic and basic vinyl monomers can be mixed in varying proportions as strong or weak acids with strong or weak bases. The principal materials studied include poly(aminocarboxylic acid)s and copolymers of vinylpyridines or dialkylaminoethyl (meth)acrylates with (meth)acrylic acid. The ionic nature of this class of ampholytic monomer pair depends on the pH of the solution where, at the isoelectric point, ampholytic character can occur.

Polymers

The earliest investigations into the formation of polyampholytes focused on the random vinyl copolymerization of acidic and basic monomers, as in the case of vinylpyridines with (meth)acrylic acid. Soon afterward, other methods were developed based on the chemical modification of preformed polymers in order to improve the structural regularity of the acidic and basic functional groups located along the polymer chain. More recently, vinyl ampholytic monomers containing no non-polymerizable ions were prepared. The location of the cationic and anionic units may be on different vinyl moieties, as with ion-pair comonomers, or on the same pendent group, as in the traditional sulfobetaine inner slats. In either case, each charge is a counterion to an oppositely charged unit and each vinyl moiety is individually polymerizable.

Properties. *Amphoteric Polymers.* It was suggested in 1947 that the behavior of synthetic ampholytic polymers could shed light on the electrochemical solution behavior of their biological analogues, the proteins, since they possess certain similarities. Best known are the existence of an isoelectric point with reversal of electrophoretic mobility, an increased intrinsic viscosity at lower and higher pH values, and an electroviscous effect.

Polymeric Inner Salts. The solution properties of another class of ampholytic polymer, polysulfobetaine inner salts, have been the subject of several studies. These substances exhibit hydrogel characteristics and are insoluble in water, with the unusual exception of three compounds. Their insolubility in water can be accounted for by inter- and intramolecular ionic interactions of the cationic and anionic charges on each pendent group. This is in close analogy with scrambled salt polyelectrolyte complexes consisting of ionic cross-links between two polyions of opposite charges. The hydrogel behavior is a typical case of swelling in a cross-linked material until the osmotic forces, which tend to draw the solvent into the polymer, are balanced by the elastic forces of the network. These osmotic forces are not sufficiently strong to rupture the ionic cross-links; as a result, the polymer becomes swollen with solvent.

Polymeric Ion-pair Comonomers. A class of ampholytic materials similar to those discussed above have been obtained by homo- and copolymerization of ion-pair comonomers. For the copolymers of ion-pair comonomers with a neutral monomer in which ampholytic ionomers were prepared, a linear relationship was found between the T_g and the mol % of ionic character. This was also observed in copolymers of MPTMA-AMPS with styrene and *n*-butyl acrylate, and with a variety of methacrylates.

The rigidity of the ionomer is enhanced by the presence of ionic interactions. This is reflected in an increased tensile strength with increasing ionic content, and a decrease in the average impact strength of the ionomer.

In solution the polymers derived from ion-pair comonomers show diverse properties. Homopolymers display viscosity increases in aqueous salt solutions. Similarly, the viscosity of ampholytic ionomers increases in polar organic solvents containing salts. The viscosity of hydrophobic, ampholytic ionomers in nonpolar–polar solvent mixtures increases with temperature.

Applications

The zwitterion nature of ampholytic polymers provides features that are useful in environmental and industrial applications, eg, as variable-charge, ion-exchange membranes (as a function of pH) (see ION-EXCHANGE RESINS; MEMBRANES).

Ampholytic polymers are used in sewage treatment, as soil conditioners, and in the recovery of minerals from aqueous suspensions. In the displacement of viscous oils from underground reservoirs, addition of water-soluble polymers to the flood water increases the viscosity and the sweep efficiency.

Ampholytic polymers are employed in shampoos, hair conditioners, as paper fortifiers, and as pigment-retention aids (see also PAPER ADDITIVES AND RESINS; COSMETIC APPLICATIONS). They are also used as standards for isoelectric focusing of proteins and as substitutes for proteins and gelatin substrates of photographic film (see also GELATIN; PHOTOGRAPHIC APPLICATIONS).

BIBLIOGRAPHY

E. Goethals, ed., *Polymeric Amines and Ammonium Salts* (IUPAC), Pergamon Press, Elmsford, N.Y., 1980.

J. C. Salamone, T. Nagabhushanam, A. C. Watterson, T. Stone, Q. Anwaruddin, S. B. Clough, and S. J. Shu, *IUPAC Macromolecular Symposium No. 28,* IUPAC, Oxford, UK, 1982; J. C Salamone, *Macromolecular Preprints '89,* Oxford, UK, 1989.

JOSEPH C. SALAMONE
WILLIAM C. RICE
University of Lowell

POLYANHYDRIDES

Research in this class of polymers was revitalized by researchers at MIT in the early 1980s in developing biodegradable polymeric biomaterials, especially for applications in controlled drug delivery.

Polycondensation

In the Melt. The great majority of polyanhydrides have been prepared by the melt-polycondensation technique. As shown in equation 1, starting with a dicarboxylic monomer, a prepolymer consisting of a mixed anhydride of the diacid with acetic acid is formed. The polymer is obtained by subsequently heating the prepolymer under vacuum to eliminate the acetic anhydride by-product.

$$\text{HOOC—R—COOH} \xrightarrow{(CH_3CO)_2O} H_3C—\overset{O}{\overset{\|}{C}}—O—\overset{O}{\overset{\|}{C}}—R—\overset{O}{\overset{\|}{C}}—O—\overset{O}{\overset{\|}{C}}—CH_3 + CH_3COOH \xrightarrow[\text{vacuum}]{\Delta}$$
$$—\!\!\left(R—\overset{O}{\overset{\|}{C}}—O—\overset{O}{\overset{\|}{C}}\right)\!\!— + (CH_3CO)_2O \qquad (1)$$

HOOC—R—COOH ⟶ α-anhydride (linear, mol wt 5,000)

molecular distillation ↘

β-anhydride (cyclic dimer distillate) + ω-anhydride (20,000 residue) (2)

↓ standing

γ-anhydride (cyclic, 5,000)

Representative polyanhydrides obtainable from the melt-condensation technique are shown in Table 1. A variety of copolymers have also been obtained from these structures. Purification of the prepolymer, such as recrystallization, is a crucial step in increasing the molecular weight of the polymer. Table 2 shows the high molecular weights obtainable under optimum conditions.

In Solution. Since melt-polycondensation requires high temperature, a milder reaction condition is desirable for heat-sensitive monomers. Several solution polymerization schemes have therefore been proposed. Polyanhydride formation can be effected at room temperature by a dehydrochlorination between a diacid chloride and a dicarboxylic acid (eq. 3).

$$\text{HOOC—R—COOH} + Cl—\overset{O}{\overset{\|}{C}}—R'—\overset{O}{\overset{\|}{C}}—Cl \xrightarrow{\text{base}}$$
$$—\!\!\left(R—\overset{O}{\overset{\|}{C}}—O—\overset{O}{\overset{\|}{C}}—R'—\overset{O}{\overset{\|}{C}}—O—\overset{O}{\overset{\|}{C}}\right)\!\!— + \text{base·HCl} \qquad (3)$$

Interfacial Polycondensation. Given the success of the interfacial polymerization (qv) technique in preparing polyamides, polyesters, and polycarbonates, it is natural to attempt the polyanhydride polymerization similarly. The results, using sebacic acid as the model monomer, confirmed the feasibility of this approach. However, the molecular weights obtained have been less than satisfying due to hydrolytic decomposition of the diacid chloride.

Polymer Properties

Thermal. Almost all polyanhydrides show some degree of crystallinity as manifested by their crystalline melting points. X-ray diffraction analysis of the aromatic polyanhydrides show discrete Debye-Scherrer rings and sharp powder diffraction peaks, indicating the high degree of crystallinity. Typical spherulitic structures are seen in the poly[bis(*p*-carboxyphenoxy)alkane anhydrides].

The melting points of a large number of polyanhydrides have been determined. As expected, aromatic polyanhydrides melt at higher temperatures than do the aliphatics.

According to the copolymer melting relation introduced by

Table 1. Representative Polyanhydrides, $-\!(\mathrm{R{-}C({=}O){-}O{-}C({=}O)})\!-$, Obtained by Melt-polycondensation

R	
$-(CH_2)_m-$	m = 4–12
$-HC{=}CH-$	
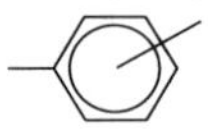	ortho, meta, and para
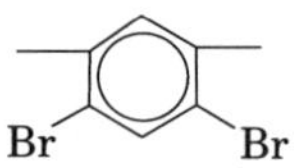	
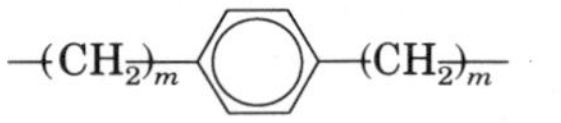	m = 1, 2
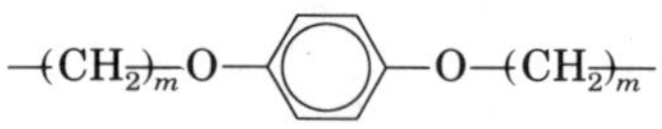	m = 1, 2
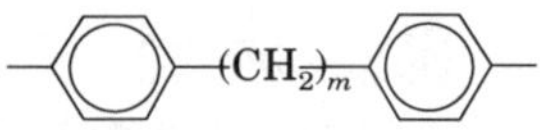	m = 1–6
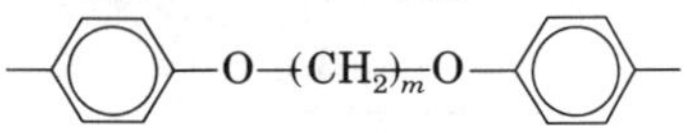	m = 1–10
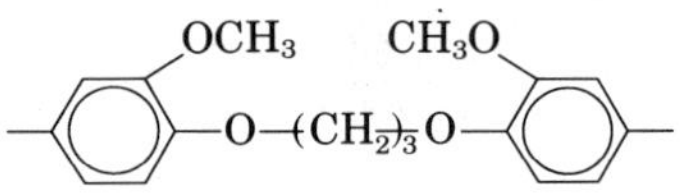	
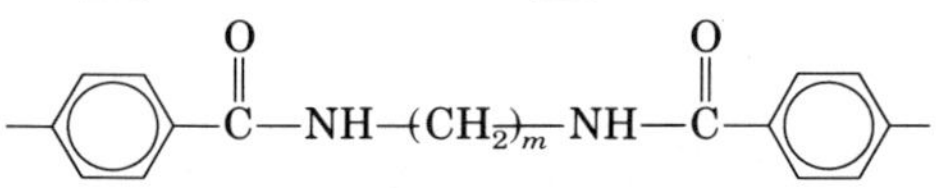	m = 1–3
$-(CH_2)_2S-(CH_2)_2S-(CH_2)_2$	
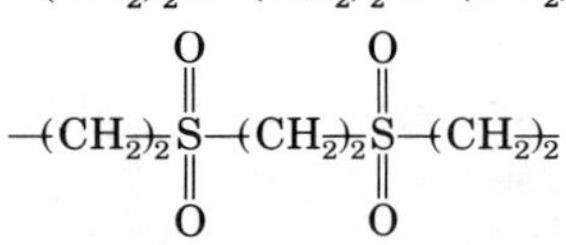	
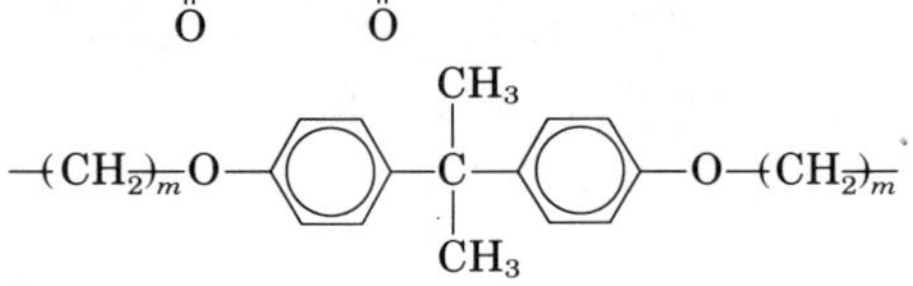	m = 1, 2
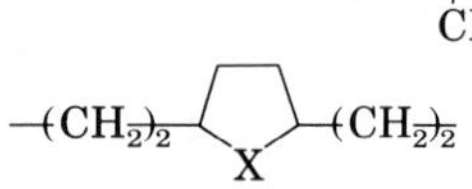	X = O, N—CH_3
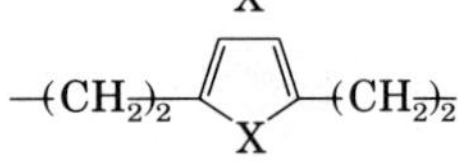	X = O, S, N—CH_3
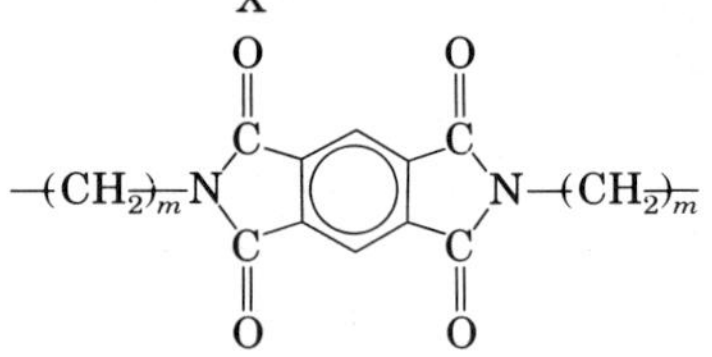	m = 1, 2, 3, 5
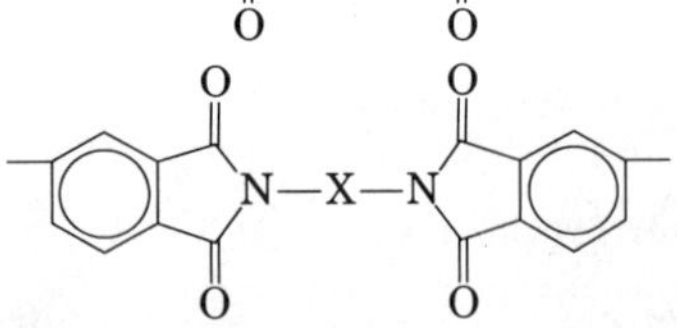	X = $-(CH_2)_m-$, m = 2,6; 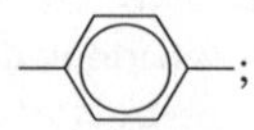; 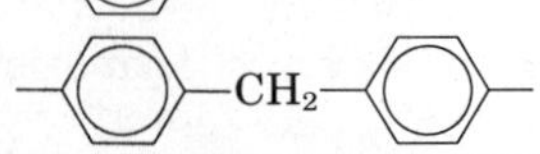

Flory, the melting temperature of a random copolymer can be expressed by equation 4.

$$(1/T_m) - (1/T_m^0) = -(R/H_u)\ln X_A \qquad (4)$$

where T_m is the melting temperature for complete disappearance of crystallinity, T_m^0 is the melting temperature for the pure polymer composed exclusively of A, H_u is the heat of fusion per repeating unit, and X is the mole fraction. This equation holds for the mixed polyanhydrides.

It has been suggested that the principal factor controlling melting of polyanhydrides is the molar cohesion energy per

Table 2. Polyanhydrides Obtained by Melt-Polycondensation[a] of Carefully Purified Prepolymers

Polymer[b]		Molecular weight[c] M_w	M_n	Polydispersity	Viscosity[d] $[\eta]$, dL/g	Mp, °C
PSA		137,300	21,400	6.41	0.92	82
P(CPP–SA)	20:80	116,800	18,200	6.42	0.92	72–74
P(CPP–SA)	35:65	87,900	17,500	5.02	0.86	126
P(CPP–SA)	50:50	37,200	12,900	2.88	0.62	138–142
PCPH		28,900	10,800	2.68	0.53	142
P(CPP–DD)	20:80	125,900	26,850	6.03	1.16	74–76
PPDP		42,300	4,340	9.72	0.37	100
P(IPh–SA)	20:80	118,000	23,600	6.02	0.95	72
P(IPh–SA)	50:50	68,100	15,500	4.39	0.66	145–150
PDD		122,500	25,200	6.38	1.12	94

[a] Polymerization at 180°C for 90 min.
[b] SA, sebacic acid; CPP, (*p*-carboxyphenoxy)propane; CPH, (*p*-carboxyphenoxy)hexane; DD, dodecanedioic acid; PDP, 1,4-phenylene dipropionic acid; IPh, isophthalic acid.
[c] Gpc; calibrated with polystyrene standards.
[d] Chloroform, 23°C.

chain unit. Aliphatic polyanhydrides melt at lower temperatures than polymethylene, in which the anhydride group is replaced by the methylene group. There is also a general decrease of melting point as the proportion of anhydride groups in the chain increases in spite of the increase of intermolecular forces by the polar anhydride groups. These two observations lead to the proposal that there is a bond in the anhydride group around which rotation is considerably easier than in a CH_2 chain.

Mechanical. Little data exist on the mechanical properties of polyanhydrides. Poly[1,2-bis(*p*-carboxyphenoxy)ethane anhydride] fibers obtained in a melt-spinning process showed a tensile strength of 392 MPa (40 kg/mm^2) with an elongation of 17.2% and a Young's modulus of 4.95 GPa (505 kg/mm^2). More recently the effects of copolymer composition and molecular weight on the tensile strength of bis(*p*-carboxyphenoxy)propane and sebacic acid (CPP-SA) have been examined. Increasing either the CPP content in the copolymer or the molecular weight increases the tensile strength.

Solubility. The majority of the polyanhydrides studied dissolve in common organic solvents such as dichloromethane and chloroform. This includes all aliphatic and heterocyclic structures. For the highly aromatic structures such as phthaloyl, bis(*p*-carboxyphenoxy)alkanes, and the imide and amide-containing backbones, more polar solvents such as dimethylformamide or *m*-cresol are needed.

Chemical. It is generally perceived that polyanhydride is hydrolyzed due to its reactivity toward nucleophiles. Aliphatic polyanhydrides are known to decompose on standing. Aromatic polyanhydrides are surprisingly much more resistant to hydrolysis. Because of interest in using polyanhydrides as biodegradable biomaterials, recent hydrolytic studies have been concentrated on *in vitro* conditions, ie, 0.1 *M* pH 7.4 phosphate buffer at 37°C. The degradation rates of polyanhydrides span a wide range, from approximately 0.2 μg/cm^2 per h for poly[bis(carboxyphenoxy)hexane anhydride] to 500 μg/cm^2 per h for poly(fumaric anhydride). In general the decomposition rate decreases with the aromaticity and hydrophobicity of the backbone.

Although hydrolysis of monomeric anhydrides is both acid- and base-catalyzed, it is generally observed that the hydrolytic degradation rate of polyanhydrides increases with pH; hydrolysis of polyanhydrides at neutrality is faster than under acidic conditions.

In the solid state and at room temperature, the polyanhydride is unlikely to react with other compounds. However, under reactive conditions, the polyanhydrides might interact with nucleophiles other than the hydroxide ion.

Biological. Preliminary data on poly[bis(*p*-carboxyphenoxy)propane anhydride], poly(terephthalic anhydride), and their copolymers with sebacic acid show that they have good biocompatibility.

Applications

Since the anhydride linkage is moisture-sensitive, there were no initial commercial applications for polyanhydrides. However, as biodegradable materials for medical applications, the hydrolytic instability of the anhydride bond is an intrinsic advantage and these polyanhydrides have been developed into biodegradable drug carriers. Aside from polymeric controlled drug delivery, the polyanhydrides may also be attractive materials for temporary scaffolds and barriers, such as sutures (qv), bioabsorbable prostheses, and vascular grafts. They may also be useful as biodegradable plastics for bottles or bags. A unique feature of these polyanhydrides is their tendency to undergo surface erosion. Heterogeneous hydrolysis theoretically would better preserve the mechanical strength and physical integrity of the matrix during biodegradation, which is highly desirable in terms of predictable performance (see CONTROLLED RELEASE TECHNOLOGY).

Besides being used as a matrix for controlled drug delivery, polyanhydrides have also been formulated into microspheres (qv) and microcapsules.

BIBLIOGRAPHY

N. Yoda, *J. Polym. Sci. Part A* **1**, 1323 (1963).
R. J. Cottler and M. Matzner, *Chem. Week.* **63**, 113 (1969).

K. W. Leong, Y. Kost, E. Mathiowitz, and R. Langer, *Biomaterials* **7**, 364 (1986).

A. Domb and R. Langer, *Macromolecules* **22**, 2117 (1989).

KAM LEONG
Johns Hopkins University
A. DOMB
E. RON
R. LANGER
Massachusetts Institute of Technology

POLYARYLATES.

See POLYESTERS, MAINCHAIN AROMATIC.

POLY(ARYLENE ETHER)S.

See POLY(PHENYLENE ETHER).

POLY(ARYLENE SULFIDE)S

Although poly(*p*-phenylene sulfide), frequently referred to as poly(phenylene sulfide) or PPS, is classified as a new engineering plastic, it is by no means a new material. In fact, together with the phenolics, it was one of the first synthetic resins. Structurally, the polymer backbone of PPS is composed of a series of alternating aromatic rings and sulfur atoms. The aromatic rings are disubstituted in the para positions (*p*-phenylene) and the sulfur atoms are present as divalent moieties (sulfide), giving rise to the name poly(*p*-phenylene sulfide).

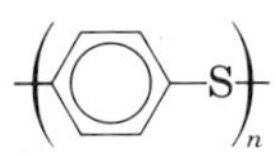

PPS

Properties

PPS, isolated as an off-white powder from the Phillips process, is a linear material of modest molecular weight. Although exact molecular weight determinations by conventional techniques are complicated because of solubility characteristics, dilute-solution light-scattering and gpc studies (in chloronaphthalene at 220°C) indicate a molecular weight near 18,000 (ie, $n \sim 230$). The melt viscosity of this material is low [3–5 Pa·s (30–50 P)] measured at low shear rates. According to differential thermal analysis (dta), amorphous PPS possesses a broad melting point at 285 °C and a moderate glass-transition temperature of 85 °C.

Curing Changes. The most important characteristic of PPS is its ability to change upon heating in the presence of oxygen. This change generally represents alterations in properties and characteristics as a result of increasing molecular weight and has been associated with several descriptive terms such as curing, cross-linking, and chain extension. Although much is known about this change, the phenomena are complex in nature and the complete chemistry is not fully understood. Investigations employing model systems indicate that varying degrees of oxidation, cross-linking, and chain extension are involved. In this article this change is referred to as curing.

The progress of the curing reaction can be conveniently followed by monitoring melt-viscosity changes. As curing occurs, molecular weight and melt viscosity increase in a gradual manner.

General Characteristics. In addition to its thermosetting–thermoplastic characteristics, PPS exhibits good thermal stability, insolubility, resistance to chemical environments, and inherent flame resistance.

Unpigmented moldings are characterized by dark color, hardness, high rigidity, and metallic ring. Unfilled moldings of low molecular weight, cross-linked polymer possess good tensile strength, moderate flexural properties, low elongation, and low impact strength. Because of its affinity for fillers and reinforcing fibers, the properties of PPS can easily be enhanced. Glass reinforcement produces injection- and compression-molding compounds with high tensile strength, high flexural modulus, good flexural strength, high heat-deflection temperature (HDT), low elongation, and moderate impact strength. Pigments and mineral fillers produce a myriad of injection-molding products with properties intermediate between unfilled and 40–50% glass-filled compounds (Table 1). PPS and its compounds exhibit good overall electrical (insulative) properties (Table 2).

Because of its aromatic structure, PPS is a strong absorber of uv energy and subject to some loss in properties during weathering. A small amount of carbon black provides a system with excellent uv resistance. Radiation effects (neutron and γ) are negligible.

Crystallinity. PPS, as prepared by the Phillips process, is highly crystalline. X-ray diffraction analysis allowed the assignment of detailed crystalline structure parameters. Other studies have correlated changes in polymer properties with crystallinity contents and crystallization rate.

Manufacture and Processing

Polymer Production. PPS has been produced by the Phillips Petroleum Co. since 1973 in a commercial facility in Borger, Tex., which initially had an annual rated capacity of 2.7×10^3 t. In 1985, plant capacity was increased to 5.4×10^3 t/yr. Ryton PPS is produced from *p*-dichlorobenzene and sodium sulfide in a polar solvent at elevated temperatures.

Injection Molding. Molding grades of PPS show an excellent affinity for pigments and reinforcing fillers, such as minerals and glass fibers. Because of the substantial enhancement of mechanical properties and the resulting cost-effectiveness, PPS injection-molding resins almost always contain substantial amounts of these materials. The filled compounds mold with an ease similar to that of easily processable thermoplastics such as polypropylene. The only significant difference in processing PPS is the slightly higher melt temperature required. PPS compounds can be injection-molded between 302 and 357 °C.

Compression Molding. Materials for compression molding may be produced by curing the polymer blended with a filler or reinforcing agent.

Free Sintering. Heavily cured PPS compounds can be fabricated by free-sintering techniques.

Economic Aspects

PPS is available commercially from Phillips Petroleum Co. in a variety of forms. Powder grades designed for slurry coating formulations, powder coatings, and feedstocks for custom com-

Table 1. Properties of PPS Compounds[a]

Property	Unfilled	40 wt % glass-filled	Glass–mineral-filled[b]	Test method[c]
density, g/cm^3	1.35	1.6	1.8–2.0	D1505
tensile strength, MPa[d]	65.5	121.0	74	D638
elongation, %	1.6	1.25	0.54	D638
flexural modulus, MPa[d]	3,800	12,000	15,000	D790
flexural strength, MPa[d]	96	160	100	D790
Izod impact, J/m[e]				D256
notched	16	58	31	
unnotched	100	175	101	
compressive strength, MPa[d]	110	145	110	D695
heat-deflection temp (at 1.8 MPa[d]), °C	135	243 (>260[f])	244 (>260[f])	D648
specific heat, J/(g·K)[g]	1.09	1.05		
Rockwell hardness	R-120	R-123	R-121	D785
thermal conductivity, W/(m·K)[h,i]	0.288	0.288		
coefficient of linear thermal exp, 10^{-5}/°C	4.9	4	2.8	
flammability	V-0	V-0/5V	V-0/5V	UL 94
oxygen index	44	46.5	53	D2863

[a] Injection-molded, 135°C mold.
[b] Properties vary slightly, depending on filler–glass ratio.
[c] ASTM unless noted otherwise.
[d] To convert MPa to psi, multiply by 145.
[e] To convert J/m to ftlbf/in., divide by 53.38.
[f] After heating for 4 h at 260°C.
[g] To convert J to cal, divide by 4.184.
[h] −30°C to 30°C determined by thermal mechanical analysis.
[i] To convert W/(m·K) to (Btu·in.)/(h·ft²·°F), multiply by 6.94.

Table 2. Electrical Properties of PPS Compounds

Property	40 wt % glass-filled	Glass–mineral-filled	Test method[a]
dielectric strength, kV/mm[b]	17.7	13.4–15.7	D149
dielectric constant, 25°C			D150
1 kHz	3.9	4.6–6.6	
1 MHz	3.8	4.3–6.1	
dissipation factor, 25°C			D150
1 kHz	0.0014	0.01–0.08	
1 MHz	0.0014	0.01–0.02	
volume resistivity, Ω·cm	4.5×10^{16}	$1–3 \times 10^{15}$	D257
arc resistance, s	34	116–182	D495
arc track rate, mm/s	8.4	0–2	UL 746A
comparative track rate, V	180	160–235	UL 746A
UL temperature index, °C[c]	200–220	200–240	UL 746B

[a] ASTM unless noted otherwise.
[b] Transformer oil, 500 V/s rate of increase, 1.6–3.2-mm thickness.
[c] Indexes vary, depending on requirement and compound composition.

pounding and molding applications are available in a broad spectrum of molecular weights and melt viscosities. The 1987 prices of these powders were $9.35–10.40/kg in large quantities. Likewise, a myriad of injection-molding compounds are available, containing glass reinforcement, mineral fillers, pigments, lubricating additives, and mold-release agents. The 1987 prices for these compounds ranged from a low of $3.20/kg for glass–mineral-filled to $7.30/kg for glass-reinforced compound.

Health and Safety

Ryton PPS compounds have been approved for applications where safety is of the utmost importance; certain grades also comply with government regulations for contact with food.

Toxicology. In acute toxicity tests the maximum amounts of ground coating fed produced no effects in albino rats or mongrel dogs. It was also demonstrated that PPS dust is inert and presents no hazard to health from long-term inhalation exposure.

Pyrolysis. Special precautions should be observed when molding or curing Ryton above 371 °C, producing sulfur dioxide and carbonyl sulfide, which are considered irritants to the mucous membranes. As with other plastic materials, adequate ventilation of the molding area is recommended when injection molding PPS compounds.

Copolymers and Modified Polymers

The process developed at Phillips Petroleum Co. can be applied to the preparation of an entire family of arylene sulfide poly-

Table 3. *m*- and *p*-PPS Copolymers

% para	T_g, °C	T_m, °C	Solubility in tetrahydrofuran, wt %
100	83	283	<0.1
75	68	205	<0.1
50[a]	49		>20
25[a]	27		42
0[a]	15		0.5

[a] These polymers were amorphous and softened at 90–100°C.

mers by substituting other polyhalogenated aromatics for all or part of the *p*-dichlorobenzene.

A series of copolymers is based on mixtures of *m*- and *p*-dichlorobenzene and sodium sulfide (Table 3). As the meta content increases, the melting point and glass-transition temperature of the polymers decrease; above 50% meta content the polymers no longer crystallize. Polymers with higher melting points and glass-transition temperatures than PPS can also be produced.

A modification of the basic Phillips process permits the production of a high molecular weight polymer without a curing step. This resin is suited for fiber and film extrusion as well as injection-molding applications; a carboxylate moderates the polymerization.

Applications

Because of its unique combination of properties and the fact that it can be made in a wide range of melt viscosities, PPS finds applications ranging from large, thick-walled mechanical parts to small, delicate electrical components. Most end products are fabricated by injection molding. Electrical, electronic, and mechanical applications constitute the largest marketing areas for PPS compounds. The appliance and automotive industries also use large quantities.

Small amounts of polymers are used in sintering and compression molding for prototype part production and testing; they are also used where good lubricity is required. Compounds designed for these applications employ a low melt-flow resin combined with glass or carbon reinforcement fibers, fillers, and lubricants such as PTFE or molybdenum sulfide. Typical applications include bearings, roller elements, and seals, as well as rods, tubes, electrical components, valves, pistons, rings, and so on.

Coatings. Significant amounts of PPS resin are used in specialty coatings where it provides thermal stability, chemical resistance, and release characteristics (with incorporation of PTFE).

Composites. Reinforcement of PPS with short glass or carbon fibers provides a cost-effective method for enhancing mechanical properties. The combination of PPS with long- (2.5 cm to continuous) fiber reinforcement produces thermoplastic composites exhibiting a unique combination of strength and toughness. Composite materials containing glass (mat, fabric, and unidirectional), carbon (mat, fabric, and unidirectional), and Kevlar (fabric and unidirectional) fibers are now available (Table 4). These materials will compete for the automotive, appliance, power equipment, aircraft, military hardware, and sporting equipment industries.

Table 4. Thermoplastic PPS Composites

Properties	Stampable sheet	Prepreg and laminates
fiber reinforcement	glass, carbon	glass, carbon, Kevlar
fiber form	random mat	fabric, unidirectional
fiber loading, wt %	20–40	40–70
strength, MPa[a]	90–206	206–1725
modulus, GPa[b]	6–14	14–138
impact, J/m[c], unnotched	530–1325	635–1590

[a] To convert MPa to psi, multiply by 145.
[b] To convert GPa to psi, multiply by 1.45×10^6.
[c] To convert J/m to ftlbf/in., divide by 53.38.

BIBLIOGRAPHY

H. W. Hill, Jr. in R. B. Seymour and G. S. Kirshenbaum, eds., *High Performance Polymers: Their Origin and Development,* Elsevier Science Publishing Co., New York, 1986, pp. 135–148.

H. W Hill, Jr. and J. N. Short, *Chemtech* **2,** 481 (1972).

H. W. Hill, Jr., *Ind. Eng. Chem. Prod. Res. Develop.* **18,** 252 (Dec. 1979).

J. E. O'Connor, A. Y. Lou, and D. G. Brady, *ASC Conference on Composite Materials,* Oct. 1986.

H. WAYNE HILL, JR.
Hill Associates

D. G. BRADY
Phillips Petroleum Company

POLY(ARYLENE SULFONE)S.

See POLYSULFONES.

POLYAZINES.

See AZO POLYMERS.

POLYAZLACTONES

Azlactones are the cyclic anhydrides of *N*-acrylamino acids. Generally, the term azlactone refers to those compounds derived from α-amino acids, although some higher homologues are known.

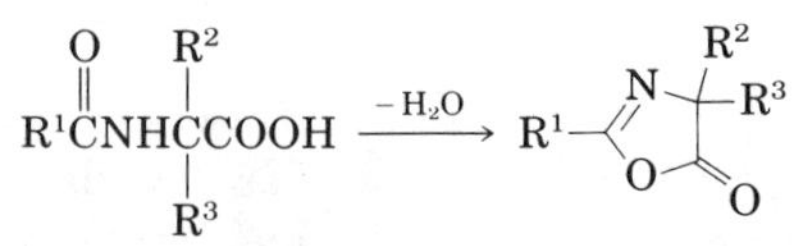

An important reaction of azlactones is the ring-opening addition that occurs with various nucleophiles.

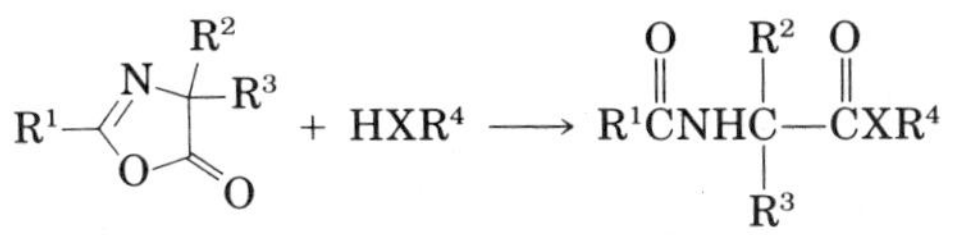

where X = O, S, NR^5

Monomers

A wide variety of azlactone-functional monomers have been synthesized and characterized. Bisazlactones (**1**) are useful in step-growth polymerization and cross-linking reactions. Alkenyl azlactones (**2**) are readily polymerized by free-radical techniques to produce azlactone-functional addition polymers and copolymers, and also are utilized for the preparation of novel acrylamide-functional monomers and new bis- and multiazlactone-functional compounds.

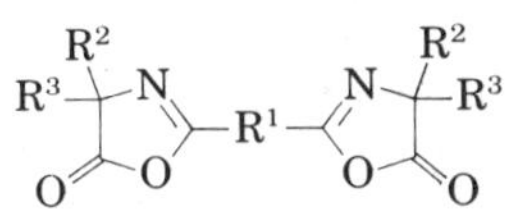

(1)

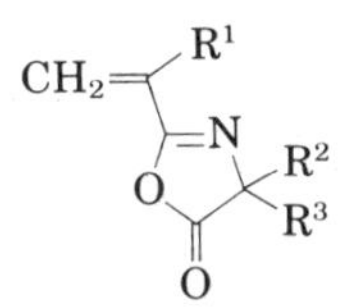

(2)

Preparation. In nearly all azlactone syntheses, the first step is the N-acylation of an appropriate α-amino acid. Generally, the amino acid is dissolved in aqueous alkali and treated with the acid chloride (optionally dissolved in a solvent) under interfacial-reaction conditions (Schotten–Baumann reaction); acidification yields the corresponding N-acylated amino acid) (see INTERFACIAL POLYMERIZATION). Methods for the cyclodehydration of these intermediates to the azlactone monomers differ principally in the nature of the dehydrating agent utilized.

Physical Properties. Most alkenyl azlactones are liquids at RT. Being relatively nonpolar compounds, they are generally soluble in organic solvents, including hydrocarbons, but are insoluble in water. An important consequence of this is that they are miscible with most addition-polymerizable monomers. In contrast, bisazlactone monomers are all crystalline, relatively high melting. Solubilities in many instances are limited to the more polar organic solvents; this is particularly true of the aromatic-linked bisazlactones.

Chemical Properties. The azlactone ring is readily attacked by a variety of nucleophiles to provide ring-opened addition products.

Polymerization

Step-growth Polymerization. Formation of step-growth polymers from bisazlactones was first reported in 1943. Soluble, fiber-forming polyamides and polyesteramides were briefly described, as was the use of bisazlactones as cross-linkers for cellulose acetate, poly(vinyl acetal)s, and casein. The polymers were not well characterized, but it was noted that reaction with amines is particularly rapid whereas alcohols may require an acidic or basic catalyst. Subsequently, more detailed studies on the preparation and characterization of a variety of polyamides have appeared.

Chain-growth Polymerization. The free-radical polymerizability of an alkenyl azlactone was first reported in 1959. Subsequently, homo- and copolymerization studies with a variety of monomers have been reported. Disubstitution in the 4 position is apparently necessary in order to obtain high molecular weight polymers. Rearrangement to pseudoazlactones and chain-transfer reactions compete when hydrogen atoms occupy the 4 position.

Ring-opening Polymerization. The ring-opening homo- and copolymerization of azlactones unsubstituted in the 2 position have been reported. These polymerizations presumably proceed via zwitterionic intermediates and produce polymers containing *N*-formyl-α-peptide units (see also ZWITTERIONIC POLYMERIZATION).

Properties and Applications

In addition to those utilized in copolymerization studies, a variety of other comonomers have been incorporated into azlactone-containing copolymers. In many instances, the alkenyl azlactone is incorporated in minor amounts (<20%) and therefore contributes little to the physical properties of the resultant copolymer. In general, physical properties can be manipulated by proper choice of comonomers and polymerization conditions, as is commonly done in free-radical polymerization. The main utility of azlactone-containing copolymers arises from the presence of the electrophilic azlactone ring as a site for polymer modification.

Health and Safety Factors

Although no studies have been published concerning the toxicological properties of azlactone monomers or polymers, some initial safety evaluations have been conducted at 3M for vinyldimethylazlactone. This monomer is a moderate to extreme eye irritant and a mild skin irritant. Both effects are alleviated by washing with water after contact. Negative results have been obtained in the Ames test and the cell transformation assay.

BIBLIOGRAPHY

Y. S. Rao and R. Filler in I. J. Turchi, ed., *Chemistry of Heterocyclic Compounds,* Vol. 45, John Wiley & Sons, Inc., New York, 1986, pp. 361–729.

C. S. Cleaver and B. C. Pratt, *J. Am. Chem. Soc.* **77,** 1544 (1955).

K. Hubner, F. Kollinsky, G. Markert, and H. Pennewiss, *Angew. Makromol. Chem.* **11,** 109 (1970).

JERALD K. RASMUSSEN
STEVEN HEILMANN
LARRY KREPSKI
3M Company

POLY(AZOARYLENE)S.

See AZO POLYMERS.

POLYAZO COMPOUNDS.

See AZO POLYMERS.

POLYAZOMETHINES.

See AZO POLYMERS.

POLYBENZIMIDAZOLES

Polybenzimidazoles are a class of linear polymers whose repeat unit contains a benzimidazole moiety. The acronym PBI is commonly used for the poly[2,2′-(*m*-phenylene)-5,5′-bibenzimidazole], which is commercially available primarily from Hoechst Celanese Corp. (Charlotte, N.C.).

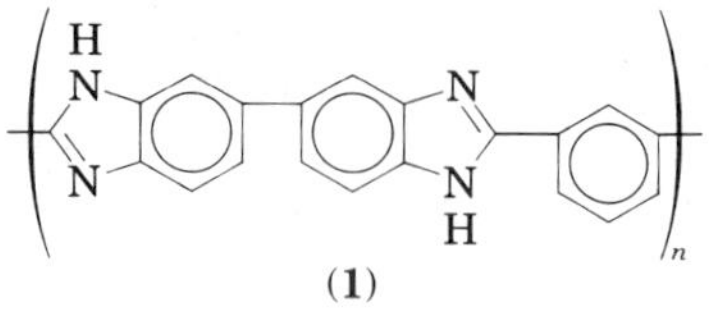

(1)

As a polymer of high temperature stability, nonflammability, and high chemical resistance, PBI can be made into a fiber with excellent textile and tactile performance, exhibiting properties not found in other synthetic fibers. It is being used as a replacement for asbestos, in fire-blocking applications, thermal-protective clothing, and numerous other high temperature- and chemical-resistant applications. The polymer also shows promise in other configurations such as film, foam, paper, adhesive, ion-exchange resins, and microporous adsorbent beads.

Polymer Synthesis

Polybenzimidazoles are prepared by condensation of tetraamino compounds with diacids. In all over 18 tetraamines and 50 diacids have been used.

Polymerization. The polymerization of PBI can be carried out in poly(phosphoric acid), in a hot molten nonsolvent diluent such as sulfolane or diphenyl sulfone, or by a solid-state polymerization. The poly(phosphoric acid) procedure provides the advantage of a moderate temperature (180 °C), homogeneous-solution polymerization. It is an excellent route for preparing laboratory samples of linear, high molecular weight polymer. The disadvantages are the necessity of working at low solid contents (3–5%) and a multiple-step isolation procedure (precipitation, neutralization, and washing), which make this route unsuitable for large-scale production. Likewise, molten diphenyl sulfone, as an inert, heat-transfer medium, is suitable for producing high molecular weight polymers of a wide variety of compositions, but its removal by filtration and solvent extraction eliminates this route for the same reasons.

The most practical process for engineering scale-up is obviously solid-state polymerization. This route offers the advantage of handling solids that can be polymerized in vacuum or at atmospheric pressure to give a product that can be dissolved directly to an extrudable dope. Both glassware and 316-stainless steel reactors can be used in a two-stage polymerization at atmospheric pressure.

Characterization

Polymer characterization includes the determination of inherent viscosity, solution filterability, and molecular weight distribution of the polymer. The inherent viscosity is determined routinely only on the second-stage product at 25 °C and a polymer concentration of 0.4 g in 100 mL of 97.0 ± 0.1% sulfuric acid. Earlier values (0.19–0.25 dL/g) bear no relationship to the final values.

Solubility in DMA. An assessment of the solubility of PBI in dimethylacetamide (DMA) shows that the polymer can be separated into bulk fractions by varying the solution temperatures.

Fiber Formation

A dry-spinning process converts PBI into fiber. A polymer solution or spinning dope is extruded through small holes and the solvent is evaporated away from the fiber on its route to the winder. The spun fibers are then subjected to a series of operations, such as washing, drawing, and acid treatment, before final take-up (Figure 1). The highest dope quality is required for dry spinning.

Properties

Thermal Properties. Torsional-braid analysis confirmed three distinct transition regions: a relaxation process at ~−70 °C, a transition without modulus change at ~310 °C, and a T_g at ~430 °C. The T_g can be increased to 500 °C by annealing in nitrogen at 500 °C for 200 min. The increase has been associated with completion of ring closure, increase in molecular weight, and cross-linking. The low temperature transition is indicative of toughness at low temperatures, which is confirmed by resistance to pulverization at −70 °C. At temperatures as low as −196 °C, the polymer does not embrittle and a PBI adhesive exhibits tensile shear strengths of 39.3 MPa (5700 psi) at −252 °C.

The polymer is almost completely amorphous; crystallinity can be induced only under specific conditions. In almost all cases, it behaves as an amorphous polymer of very high T_g.

Both stabilized and unstabilized PBI behave similarly in thermal decomposition. In nitrogen or air, stabilized PBI loses SO_2 between 420 and 500 °C, followed by degradation of the unstabilized PBI. By optimizing the heat-stabilization process, the onset of SO_2 loss can be delayed until 480 °C and the difference between the two variants becomes even less.

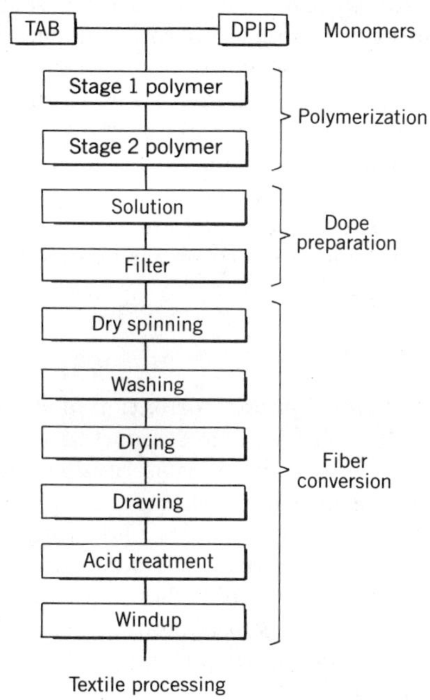

Figure 1. Fiber formation.

Fiber. The FTC has given the name PBI as the generic name for PBI fiber. Properties of unstabilized and stabilized PBI fibers are given in Table 1. The stabilized variant is sold by Hoechst Celanese as a staple fiber. An increase in fiber density occurs during the stabilization treatment as a result of the incorporation of 5–6% sulfur. It accounts for the changes in tensile strength and modulus, which are expressed as N/tex (gf/den) based on the weight of a length of 1000 (9000) m (see FIBERS).

Flammability. Because of the low rate of gas evolution in air and the relative inertness of the off-gases produced, PBI is intrinsically nonflammable in air.

Properties at Elevated Temperatures. The useful life of PBI fibers at elevated temperatures or on exposure to actinic-degradation conditions depends on the amount of oxygen in the environment. In inert atmosphere or vacuum, no significant aging effects have been observed, even at 350 °C after 300 h. Nomex I and Kevlar had lost 57 and 86%, respectively, of their initial strengths at 300 °C after 300 h. The degradation of PBI fibers between 200 and 600 °C in air follows a first-order rate law for both strength and weight retention.

Chemical Stability. Acid-stabilized PBI fiber exhibits outstanding stability, including hydrolytic stability in severe environments. Fiber strength is unaffected by exposure to organic liquids at 86 °C for 168 h, although some are solvents for the precursor polymer.

Wear Comfort. PBI fibers exhibit a high moisture regain, which is enhanced by the acid-stabilization process. This high moisture regain, together with the soft hand of PBI, contributes to wear comfort. In subjective wearer trials, PBI is perceived at least as comfortable as cotton and rayon.

Applications

In 1983 a 450-t/yr facility was put into operation by Hoechst Celanese. Consumption is mainly based on textile staple fiber, providing a combination of nonflammability, high moisture regain, comfort, and textile processibility.

Staple Fiber. Staple fiber is used for protective apparel, flight suits, and aircraft furnishings.

Specialty Applications. Forms other than fibers can be processed directly from a PBI solution. These specialty forms, in various stages of commercial application, utilize the nonflammability, thermal performance, and chemical properties of PBI. Applications include electrical–electronic, fabricated structures, and product separations (hollow-fiber and flat membranes).

Table 1. Physical Properties of PBI Fibers

Properties	Stabilized	Unstabilized
filament fineness, tex[a]	0.16	0.16
tenacity, N/tex[b]	2.3	2.7
breaking elongation, %	30	30
initial modulus, N/tex[b]	39.6	79.2
density, g/cm^3	1.43	1.39
moisture regain at 20°C, 65% rh, %	15	13
shrinkage in air at 204°C, %	<1	>2
flame-test shrinkage, %	6	50
shrinkage at 400–500°C[c]	4	10

[a] To convert tex to den, multiply by 9.
[b] To convert N/tex to gf/den, multiply by 11.33.
[c] Thermal mechanical analysis.

BIBLIOGRAPHY

K. Hoffman, *Imidazole and Its Derivatives,* Interscience Publishers, Inc., New York, 1953.

P. E. Cassidy, *Thermally Stable Polymers,* Marcel Dekker, Inc., New York, 1980.

Celanese PBI Bulletin, Celanese Fibers Marketing Co., Charlotte, N.C., 1980.

ALAN BUCKLEY
D. E. STUETZ
Hoechst Celanese Research Division
G. A. SERAD
Hoechst Celanese Specialty Operations

POLYBENZOTHIAZOLES AND POLYBENZOXAZOLES

Major advances in the theory, synthesis, and processing of rigid polymers in the past 15 years have raised polybenzothiazoles (PBTs) and polybenzoxazoles (PBOs) from the status of obscure laboratory curiosities to a new class of environmentally resistant, high strength, high modulus materials, receiving much attention from both academic and industrial communities.

These PBTs and PBOs are members of the benzazole family of polymers, which have main-chain, repeat-unit structures containing one or more aromatic, heterocyclic moieties shown in Figure 1, where Z is sulfur or oxygen. The backbone bonds that link the heterocyclic units in each repeat unit are positioned regularly, thereby defining a specific catenation angle, the angle between the two exocyclic bonds of each rigid unit. In wholly aromatic systems these exocyclic bonds are formally single bonds, but may have appreciable double-bond character if adjacent rings are coplanar. Resonance structures of this type contribute to increased backbone rigidity, which is characteristic of PBTs and PBOs. The high degree of molecular

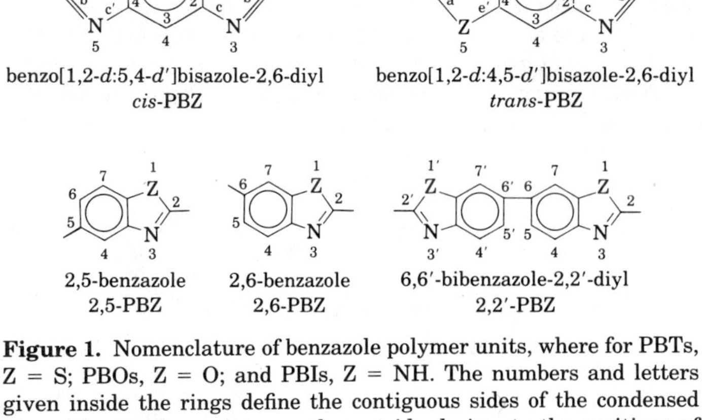

Figure 1. Nomenclature of benzazole polymer units, where for PBTs, Z = S; PBOs, Z = O; and PBIs, Z = NH. The numbers and letters given inside the rings define the contiguous sides of the condensed rings; the numbers given on the outside designate the positions of catenation.

rigidity imparted to the main chain by the various units dominates nearly every aspect of the synthesis, chemistry, processibility, and properties of PBTs and PBOs.

The rigid-rod, molecular-composite concept is distinct from self-reinforced composites. Rigid-rod, molecular-composite technology was first described in connection with the preparation of polymer alloys in which rigid-rod polybenzazoles act as reinforcement in a coil-like polymer matrix. The analogy was drawn to chopped-fiber composites with the potential of improved properties, depending on the higher axial ratio of the reinforcement and virtual elimination of stresses related to the fiber-matrix interface. The Class 1 polymers described here are ideal candidates for this technology because they maintain rigid-rod conformation at the low concentrations required to achieve compatibility of rod and coil in solution.

Classification

Polybenzothiazoles and polybenzoxazoles can be classified into four groups based on differences in the degree of chain extension in solution. These differences can be related directly to the minimum or critical concentration required to form stable, liquid crystalline (nematic) phases. Because the degree of chain extension in solution is related directly to the inherent rigidity of the polymer structure, the different classes exhibit distinct differences in solid-state physical properties, particularly tensile modulus. The chain extension of the rigid-rod polymers of Class 1 is described by the axis ratio x used by Flory in his theory for the phase behavior of athermal rodlike particles. Polymers with ideal rigid-rod structures can be characterized simply by an x, which represents a ratio of the average contour length L to the width d of the backbone. Most polymers, however, are better described as having chain conformations between that of an ideal rod and a random coil. Such polymers that form nematic phases comprise Class 2 homopolymers and Class 3 copolymers and are termed extended-chain or wormlike polymers. They are characterized by a persistence length p, which is a dilute-solution property defined as the average length of backbone rectilinearity; x can then be given by p/d. When the persistence length is insufficient to cause liquid crystalline behavior at any achievable solution concentration, the polymers are called nonmesogenic and are described as Class 4 polymers.

Chemistry of Benzoxazole and Benzothiazole Formation

The benzoxazole or benzothiazole ring structure is formed by the condensation of a carboxylic acid (or its derivatives) with an *o*-aminophenol or an *o*-aminothiophenol, respectively; the two functional groups are referred to as A and B, respectively. An AA monomer is a dicarboxylic acid, and BB and AB designations have corresponding meanings.

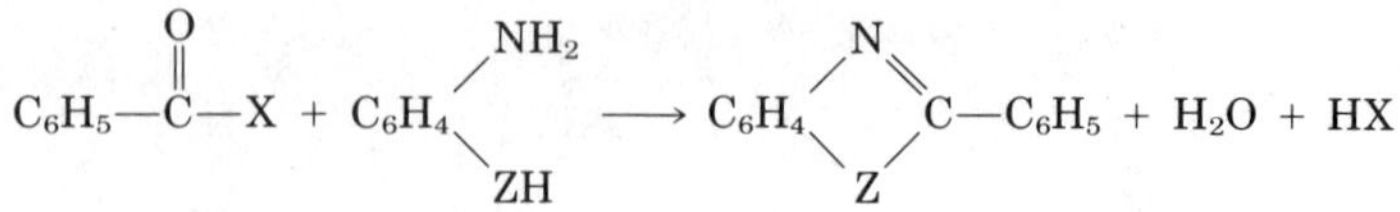

where X = OH, Cl, NH_2, OC_6H_5, etc

Z = S, O

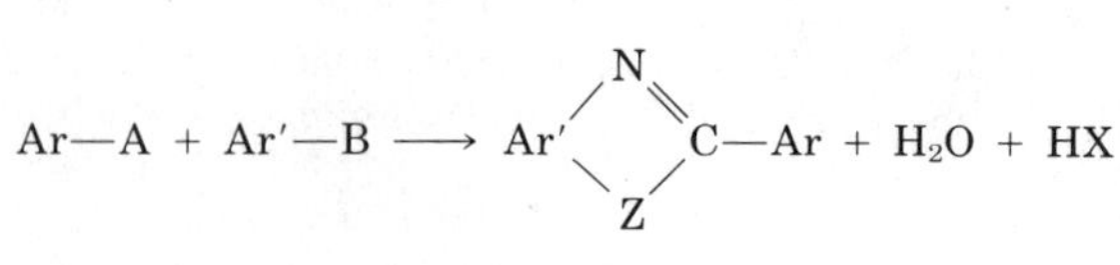

The reaction between A and B occurs in the melt or in a solvent. The choice, especially for polycondensations, depends ultimately on maintaining functional-group reactivity to a high degree of conversion. Both the physical state of the high molecular weight polymer and the rate of the desired condensation relative to the rate of side reactions are prime considerations.

Monomers. The PBTs and PBOs are prepared by condensing AA monomers (eg, terepthalic acid [TA]) with BB monomers (eg, 2,5-diamine-1,4-benzenedithiol-dihydrochloride [DABDT] or 4,6-diamino-1, 3-benzenediol dihydrochloride [DABDO]) or by condensing AB monomers (eg, 3-amino-4-hydroxybenzoic acid hydrochloride [AHBA] or 3-mercapto-4-aminobenzoic acid hydrochloride [MABA]) alone.

Distinct positional isomers, in which the relative positions of the heteroatoms in each repeat unit are the same, are obtained for polymers of benzoxazole, benzothiazole, benzobisoxazole, and benzobisthiazole. This structural regularity distinguishes these polymers from polymers of benzimidazole and benzobisimidazole, which are members of the same class of benzazole polymers, because the symmetry of the ortho diamino functional group results in random placement of the NH moiety (see POLYBENZIMIDAZOLES).

Solution Polymerization

The products of a solution polymerization are obtained either in isolated or in solution form. The choice of solvent is based on polymer solubility to maintain the reactivity of the functional groups until the desired molecular weight is achieved. When a processible solution is the goal, solution viscosity, polymer concentration, and polymer morphology must be considered.

Both PBTs and PBOs can be prepared directly as processible liquid crystalline solutions by solution methods. The most extensively used solvent, PPA, acts as solvent, catalyst, and dehydrating agent. The key to successful polycondensation is maintaining the effectiveness of the PPA while its composition changes on hydrolysis by the condensation by-product. This is accomplished by a technique, which is called the P_2O_5 adjustment method, of adding P_2O_5 to the polymerization medium based on both the concentration of the by-product of condensation and the initial composition of the PPA.

Composition of PPA. Poly(phosphoric acid)s are a continuous series of mixtures of condensed phosphoric acid oligomers of the general formula

$H_{n+2}P_nO_{3n+1}$ or 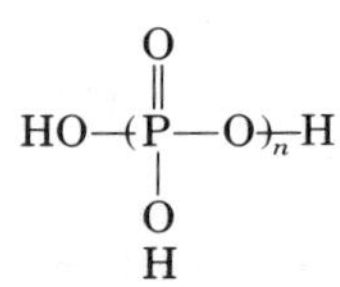

where n can be above 14; the average value of n depends on the ratio of water to phosphorus pentoxide (P_2O_5). The various compositions of PPA are described by either the H_3PO_4 or the P_2O_5 content.

Syntheses

Rigid-rod Homopolymers. ***Poly(p-phenylene-2,6-benzo[1,2-d:4,5-d']bisthiazole).*** *trans*-PBT was first prepared in 1977 by the polycondensation of 2,5-diamino-1,4-benzeneditiol dihydrochloride (DABDT) and terephthalic acid in PPA. With PPA as the polymerization solvent, the achievable molecular weight depends on the polymer concentration, C, developing during polymerization, and the P_2O_5 content, P_3, at the end. The highest molecular weight is obtained if P_3 is >83% and the polymerization is conducted in the nematic phase. For the latter the minimum polymer concentration required for rigid-rod polymers is 5 wt %; polymerization >10 wt % offers additional advantages.

Poly(p-phenylene-2,3-benzol[1,2-d:5,4-d']bisoxazole). The initial attempts to prepare *cis*-PBO from DABDO and TA or TPC in the isotropic phase by conventional PPA methods gave polymers with $[\eta] < 4$ dL/g. Although these low molecular weight samples could be redissolved in methanesulfonic acid to give nematic phases, they could not be processed into high strength fibers. High molecular weight *cis*-PBO was not synthesized until it was discovered that polymerization at high concentration had significant advantages.

Substituted Rigid-rod Polymers. Class 1 polymers have been prepared with various aromatic groups substituted for the backbone hydrogens. This research is aimed at obtaining structure–property relationships resulting in improved solubility or in fibers with increased compressive strength.

Mesogenic Homopolymers. ***Poly(2,6-benzothiazole) (2,6-PBT).*** This polymer was first prepared in the mid-1960s with $[\eta]$ below 2 dL/g. Later, the synthesis of 2,6-PBT in PPA was tried with the hope of providing a high molecular weight matrix resin that would be chemically compatible with PBT fiber. Attempts conducted in the semidilute region (5–10 wt % 2,6-PBT) gave isotropic polymerization media, and the highest $[\eta]$ achieved was ca 4 dL/g. The P_2O_5 adjustment method allows nematic-phase polymerizations in the concentration region of 13–20 wt % and $[\eta]$ of > 17 dL/g can be achieved.

Poly(2,5-benzoxazole) (2,5-PBO). Polymerization of 3-amino-4-hydroxybenzoic acid in PPA to give 2,5-PBO requires concentrations higher than 14.5 wt % to achieve the nematic phase. The slightly higher critical concentration for 2,5-PBO relative to 2,6-PBT can be ascribed to the smaller catenation angle in each repeat unit, which leads to a smaller persistence length in dilute solution.

Mesogenic Copolymers. ***Articulated Rod Polymers.*** Class 3 polymers have been prepared containing various percentages of main-chain articulation or swivel units derived from flexible diacids. In early efforts small amounts of aromatic diacid monomers with ether linkages were substituted for terephthalic acid. The articulated polymers formed cast films under the same conditions that resulted in extensive cracking when the nonarticulated analogue was used.

Ordered Block Copolymers. Block copolymers have been synthesized with an ABA block structure, where the B block is *trans*-PBT and the A block is poly(2,5-benzimidazole). Carboxy-terminated *trans*-PBT polymers with $[\eta]$ of 11–24 dL/g were prepared and allowed to react with 3,4-diaminobenzoic acid (DABA) in dilute PPA. The composition of the *trans*-PBT in the block copolymer (25–50%) was independent of the average length of the rigid segment by varying the amount of DABA added.

Random Block Copolymers. The higher concentrations obtainable with the P_2O_5 adjustment methods allow the preparation of nematic solutions of random block polymers. In these preparations two separate stoichiometric, high concentration polymerization mixtures were blended at intermediate and different degrees of conversion.

Nonmesogenic Polymers. Although early Class 4 polymers demonstrated high thermal stability, the $[\eta]$s were low compared with current materials. The processing alternatives from isotropic solutions for these polymers do not result in properties that distinguish them from other amorphous high temperature polymers. If the expected further development of the technology of PBTs and PBOs makes the appropriate monomers more available, this class of polymers may be of interest.

Properties

Wholly aromatic PBTs and PBOs are among the most thermally and thermooxidatively stable organic polymers known. Molecular structure, molecular weight, and morphology affect stabilities.

Critical oxygen concentration tests show *trans*-PBT and *cis*-PBO to be noncombustible. These polymers char, but do not support combustion. After the fibers have been dried at 250–300 °C, the moisture regain is <1 wt %. The PBTs and PBOs are highly resistant to hydrolysis, acid chemical attack, bases, solvents, electron bombardment, and laser radiation. Resistance to molecular oxygen is particularly important for space applications. The uv stability is outstanding. Most recently, the third-order nonlinear optical susceptibility of *trans*-PBT has been determined to be an order of magnitude higher than carbon disulfide and the response time is subpicosecond, indicating an electronic $\chi^{(3)}$ process.

Solution Properties. The PBTs and PBOs are soluble only in strong protic acids such as PPA, methanesulfonic acid (MSA), chlorosulfonic acid (CSA), trifluoroacetic acid (TFA), and 100% sulfuric acid. $[\eta]$ or η_{inh} measurements are commonly reported for solutions in MSA at 30 °C in dL/g.

Fabrication Processes and Properties of Fabricated Polymers

The synthesis procedures described above provide PBTs and PBOs in strong acid solution, their most processible state. To obtain a useful form of the polymer, the acid solvent must be removed by producing an intermediate shape with a high surface area and coagulating the polymer rapidly by the action of a nonsolvent. The polymer proceeds rapidly from a protonated dispersed state to an unprotonated condensed state. Such a process is called wet extrusion, and all initial fabrication methods of PBTs and PBOs are variations of this process. Variations include dry-jet wet spinning of fibers and films.

PBT and PBO Fibers. The first attempt to dry-jet wet-spin a PBT or PBO into fiber used a nematic MSA solution of *cis*-PBO with $\bar{M}_n < 10{,}000$. Mechanical properties of *trans*-PBT fibers are given in Table 1.

PBT and PBO Films. Films of *trans*-PBT have been prepared from the PPA-*trans*-PBT reaction mixture by extrusion, using a flat die with a variable air gap length, leading onto a carrier roll and into a water congulation bath.

Fiber and Film Structure. The structures, morphologies, and properties of spun and heat-treated *trans*-PBT fibers have been elucidated by electron diffraction and wide-angle x-ray diffraction (waxd). In the spun fibers the phenylene ring is ~20° out-of-plane with the benzobisthiazole moiety, which is in agreement with the structure of the model compound and the lowest energy conformation calculations. The smallest element of supramolecular order in spun *trans*-PBT fiber is seen as bundles composed of ~20 chains measuring ~2 nm laterally. This structure resembles a frozen nematic or two-dimensional crystallite without registry of repeat units. The absence of (hkl) off-axis reflections is consistent with this degree of axial disorder.

Table 1. Properties of *trans*-PBT Filaments

Property	As-spun spin stretch			Heat-treated	
	Low[a]	Medium	High[b]	Routine	Ultimate
strength[c]					
N/tex[d]	1.6	1.5	1.6	2.2	2.6
MPa[e]	2345	2276	2345	3490	4186
elongation to break, %	7.1	4.8	2.4	1.3	1.4
modulus[f]					
N/tex[d]	11.9	43.2	106	194	212
GPa[g]	17.9	65.5	159	303	331
density, g/cm^3		1.47–1.53		1.54–1.60	

[a] High dtex (den) filament.
[b] Low dtex (den) filament.
[c] Averages of five 2.5-cm breaks.
[d] To convert N/tex to gf/den, multiply by 11.33.
[e] To convert MPa to psi, multiply by 145.
[f] Averages of five 25-cm breaks.
[g] To convert GPa to psi, multiply by 145,000.

PBT and PBO Fiber-reinforced Composites. Information on the fabrication and properties of conventional composites reinforced with PBT or PBO fibers has been limited. Epoxy composites have been reported where the tensile strength and modulus of the fiber are preserved. DuPont workers reported properties of unidirectional epoxy composites with a 60% loading of various fibers (Table 2).

Rigid-rod Molecular Composites: Fabrication, Structure, and Properties

The processing technology for so-called rigid-rod molecular composites is based on the theory that the ultimate reinforcing elements in a composite material are individual molecules, and the important figure of merit is the molecule axis ratio. A blend or alloy of a Class 1 PBT or PBO entangled in a coil-like polymer in theory supplies such a composite system with high figures of merit with regard to reinforcement, if the rods and coils could be obtained in the molecularly dispersed form. Rodlike and coil-like polymers are thermodynamically incompatible above a critical solution concentration C_{cr} described by Flory. To achieve a homogeneous ternary composition comprised of rodlike molecules, coil-like molecules, and a solvent, the concentration of both polymers must be below C_{cr}. These solutions, therefore, provide the precursor to the solid-state, molecularly dispersed blend, which is the goal in this technology. To achieve such a blend, the precursor solution must be frozen by the nonequilibrium process of wet extrusion. Class 1 polymers should have high figures of merit for reinforcement because their persistence lengths are independent of concentration, unlike Class 2 and 3 polymers, which have relatively short persistence lengths at the required low concentration ($C < C_{cr}$).

Dry-jet wet-spinning techniques have been applied to isotropic solutions of Class 1 polymers and various host polymers. In this case sufficient work can be placed on the solution to cause alignment of the rigid-rod component in the flow direction. Studies of the characteristics of the precursor solution and of the morphology of rigid-rod, molecular-composite film have shown that a molecular dispersion of the rigid-rod molecules (*trans*-PBT) in the host matrix (poly(2,5(6)benzimidazole), AB PBI) can be achieved by processing at solution concentrations below C_{cr}. A common orientation of the reinforcement can be accomplished by wet extrusion and suitable treatment.

The importance of fabrication below C_{cr} is quite evident from the data in Table 3. The samples processed from solutions with concentrations higher than C_{cr} separate into two distinct phases characterized by 0.1–4-μm ellipsoidal particles that mainly contain crystallites of *trans*-PBT dispersed in a ductile matrix that is mostly AB PBI. The transfer of the rigid-rod polymer

Table 2. Properties of Unidirectional Epoxy Composite[a] with 60% Fiber Loading

Properties	Fiber: Heat-treated *trans*-PBT	Kevlar-49	Graphite
Tensile			
strength, MPa[b]	1900	1380	2100
modulus, GPa[c]	190	75.8	150
strain, %	1.0	1.8	1.5
Poisson's ratio	0.4	0.34	0.27
Compressive			
strength, MPa[b]	200	275	1400
modulus, GPa[c]	190	75.8	125
strain, %	0.12	0.4	1.1

[a] Hercules 3501-6.
[b] To convert MPa to psi, multiply by 145.
[c] To convert GPa to psi, multiply by 145,000.

Table 3. Mechanical Properties for *trans*-PBT, AB PBI, and 30% *trans*-PBT–70% AB PBI Fiber and Film

Sample	Modulus, GPa[a]	Tensile strength, MPa[b]	Elongation to break, %
Neat fibers			
trans-PBT	300	3000	1.1
AB PBI	36	1100	5.2
30% trans-PBT–70% AB PBI			
$C > C_{cr}$ fiber[c]	11	310	13
$C > C_{cr}$ film[c]	1.1	35	5.6
$C < C_{cr}$ fiber[d]	120	1300	1.4
$C < C_{cr}$ film[d]	88	920	2.4

[a] To convert GPa to psi, multiply by 145,000.
[b] To convert MPa to psi, multiply by 145.
[c] Oven-dried.
[d] Heat-treated at elevated temperature.

properties to the material is expected to be poor for this morphology. The samples processed from homogeneous solutions were found to contain crystallites of both *trans*-PBT and AB PBI and lateral dimensions of the ordered regions no larger than 3 nm.

The effects of the molecular weights of both the rigid-rod polymer and the matrix polymer on the processing of rigid-rod molecular composites have been reported. The molecular weight of the matrix polymer must be above a certain value based on entanglement theory, but deficiency in the matrix molecular weight can be compensated for by increasing the rigid-rod molecular weight. If the molecular weights required for sufficient entanglement are employed, the application of shear rates as high as 14,000 s^{-1} does not cause phase separation.

Applications and Outlook

Nonstructural applications may take the form of woven fabrics, fibrids, membranes, and cables. High modulus PBT and PBO yarns are excellent candidates for ballistic-protection fabrics and panels because of the high energy absorption and rapid dissipation of the impact by the fibrillar morphology.

PBT and PBO fibers and films should find applications in space structures, where high specific modulus, resistance to molecular oxygen, and a low coefficient of thermal expansion are important. Such applications might include solar arrays, tethers, antennas, pressure vessels, rocket-motor casings, space-structure frames and joints, impact panels, and others.

Scale-up development of the production of PBT and PBO materials, which is currently in progress, gives a twofold promise for the future application of these materials. The availability of fibers and films will accelerate the evaluation for specific applications. The full-scale process, from raw materials to monomer to polymer to end-product, will come into specific focus. With the markets identified and the cost of production projected with greater confidence, the decision to proceed with full-scale commercial production can be made.

BIBLIOGRAPHY

S. R. Allen, A. G. Filippov, R. J. Farris, E. L. Thomas, C.-P. Wong, G. C. Berry, and E. C. Chenevey, *Macromolecules* **4,** 1135 (1981).

U.S. Pat. 4,533, 693 (Aug. 6, 1985), J. F. Wolfe, P. D. Sybert, and J. R. Sybert (to SRI International).

R. C. Evers, F. E. Arnold, and T. E. Helminiak, *Macromolecules* **4,** 925 (1981).

D. R. Wiff, S. Timms, T. E. Helminiak, and W.-F. Hwang, *Polym. Eng. Sci.* **27**(6), 424 (1987).

JAMES F. WOLFE
SRI International

POLYBENZOXAZINONES

Polymers of the benzoxazinone class provide outstanding thermal and hydrolytic stability.

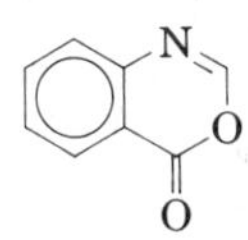

(1)

They are prepared by melt polymerization, cyclopolycondensation, or solution polymerization in poly(phosphoric acid) (PPA). Polybenzoxazinones are sparingly soluble in conventional organic solvents and difficult to process. Synthetic efforts are oriented toward modification of the basic system in order to improve solubility and processibility without the loss of thermal and mechanical stability. A number of new monomers have been synthesized and developed.

Properties

The properties of some polybenzoxazinones are shown in Table 1.

BIBLIOGRAPHY

P. E. Cassidy, *Thermally Stable Polymers: Synthesis and Properties,* Marcel Dekker, Inc., New York, 1980.

G. Neamtu and M. Brama, *Angew. Macromol. Chem.* **103,** 19 (1982).

J. Gopal and M. Srinivasan, *J. Polym. Sci., Polym. Chem. Ed.* **24,** 1577 (1986).

Table 1. Physical Properties and Thermal Behavior of Polybenzoxazinone

Ar in poly(amic acid)	Yield, %	η_{inh}^{25}, dL/g	Density, g/cm³	10% weight loss, °C	Temp[a], °C
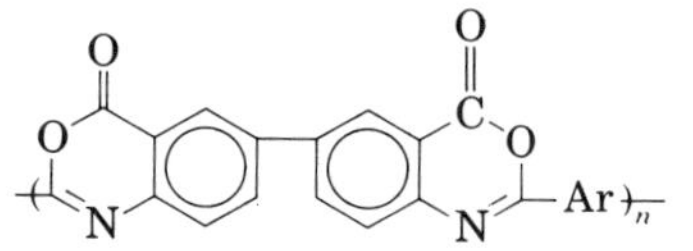					
Ar = 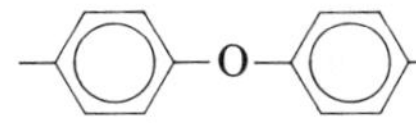	67	0.16	0.95	388	421
Ar = 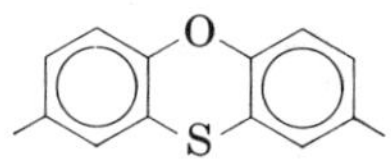	60	0.15	1.12	472	566
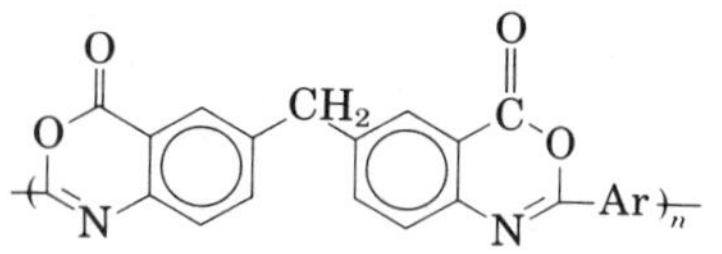					
Ar = 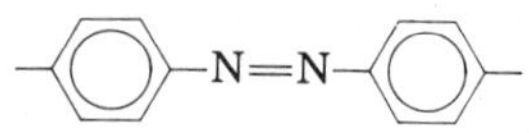	52	0.35	1.54	430	420
Ar = 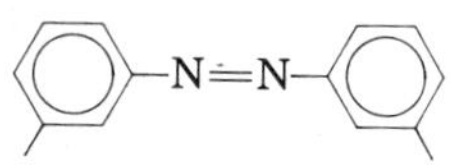	55	0.28	1.52	410	429
Ar = $-CH_2C_6H_4-N{=}N-C_6H_4CH_2-$	60	0.15	1.57	495	447
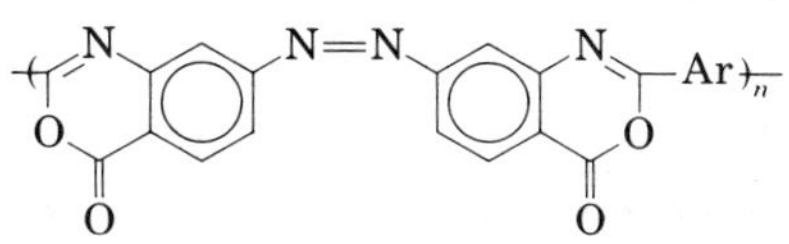					
p-phenyl	65	0.112	1.401	325	420
m-phenyl	68	0.124	1.508	311	405
4,4′-biphenyl	58	0.115	1.397	311	430
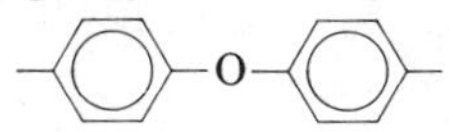	62	0.119	1.377	327	440
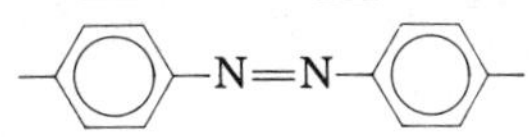	58	0.105	1.517	375	409

[a] Temperature of maximum exothermal peak observed in dta, where rate = 10°C/min.

S. Prema and M. Srinivasan, *J. Polym. Sci., Polym. Chem. Ed.* **24**, 2563 (1986).

M. Srinivasan
Indian Institute of Technology

POLYBITHIAZOLES

Polybithiazoles (or polydithiazoles) are polymers obtained from polycondensation reactions that generate a bithiazole group (**1**) in the main chain (Table 1).

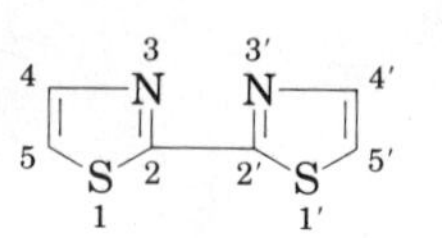

(**1**)

The poly(4,4′-disubstituted-2,2′-bithiazole)s (**2**) are prepared from the polymerization of dithiooxamide (**3**) (ethanedithioamide, dithionoxamide, or rubeanic acid) with difunctional halomethyl ketones (**4**).

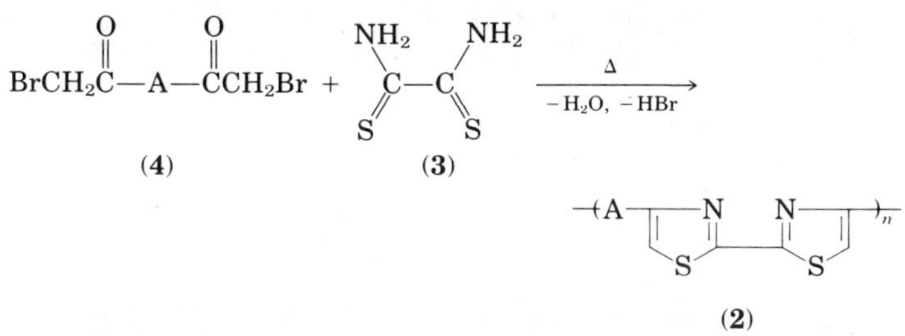

Because difunctional halomethyl ketones can be easily synthesized, a variety of bithiazole-containing polymers can be prepared. The polymerizations proceed rapidly and in moderate-to-high yields. In addition, polyamides and aromatic azo polymers with the bithiazole unit in the polymer backbone have been prepared. These bithiazole-containing polymers have moderate thermal and poor uv stability.

BIBLIOGRAPHY

P. Karrer, P. Leiser, and W. Graf, *Helv. Chim. Acta* **27**, 624 (1944).

G. Bischoff, O. Weber, and H. Erlenmeyer, *Helv. Chim. Acta* **27**, 947 (1944).

Table 1. Properties of Polybithiazoles

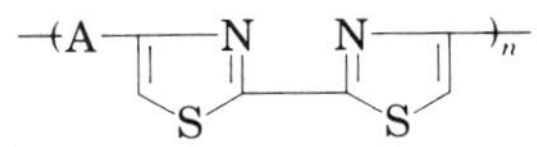

A	Polymer	Mp[a], °C	Yield[b], %	η_{inh}[c], dL/g	M_n[d]
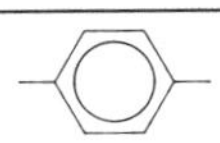	**(5)**	[e]	88	0.23	12,000
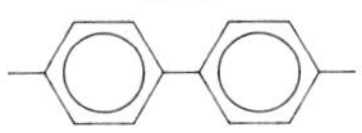	**(6)**	[e]	79	0.24	84,000
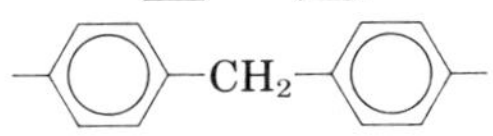	**(7)**	[e]	78	0.47	11,000
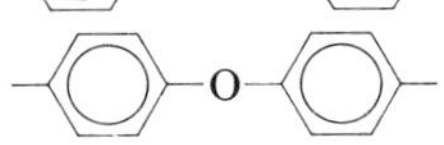	**(8)**	[e]	79	0.38	80,000[g]
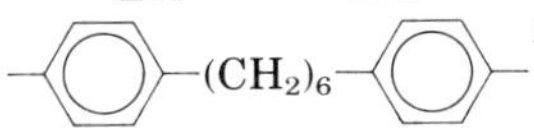 [f]	**(9)**				
fraction A		440–448	56	0.40	80,000[g]
fraction B		238–252	15	0.1	12,000
fraction C		218–235	18	0.1	11,000

[a] Capillary melting point.
[b] Yield of purified polymer.
[c] At 25°C and a concentration of 0.25 g/100 mL in concentrated sulfuric acid.
[d] Number-average molecular weight based on bromine content.
[e] No melting was observed below 520°C.
[f] Fractionated from DMF.
[g] Based on 0.1% Br; Beilstein test indicates less than 0.1% Br.

D. T. Longone and H. H. Un, *J. Polymer Sci. Part A* **3**, 3117 (1965).
W. C. Sheehan, T. B. Cole, and L. G. Picklesimer, *J. Polym. Sci. Part A* **3**, 1443 (1965).

PAUL D. SYBERT
General Electric Company

POLYBLENDS. See POLYMER BLENDS.

POLYBUTADIENE. See BUTADIENE POLYMERS.

POLY(BUTADIENE-*co*-ACRYLONITRILE). See BUTADIENE POLYMERS.

POLY(BUTADIENE-*co*-STYRENE). See BUTADIENE POLYMERS.

POLYBUTENE. See BUTENE POLYMERS.

POLY(BUTYLENE TEREPHTHLATE). See POLYESTERS, THERMOPLASTIC.

POLYCAPROAMIDE. See POLYAMIDES.

POLYCAPROLACTAM. See POLYAMIDES.

POLYCARBODIIMIDES. See ISOCYANATE-DERIVED POLYMERS.

POLYCARBONATES

ALIPHATIC POLYCARBONATES

Aliphatic polycarbonates are much less important than aromatic polycarbonates and are therefore discussed only briefly here.

Preparation. Starting materials for aliphatic polycarbonates are diols and carbonates. Aliphatic polycarbonates may also be prepared from 1,3-dioxan-2-ones obtained by thermal depolymerization of the corresponding polycarbonates. Current methods for the preparation of aliphatic polycarbonates include transesterification of diols with lower dialkyl carbonates, dioxolanones, or diphenyl carbonate in the presence of catalysts such as alkali metal, tin, and titanium compounds.

Properties. The molecular weights of linear aliphatic polycarbonates are process-dependent and are between 500 and 5000, which is adequate for most applications. Polycarbonates with molecular weights up to ~30,000 are obtained by transesterification, whereas those with a molecular weight >50,000 are obtained by polymerization of carbonates possessing six-membered rings.

As would be expected, the starting monomers influence the properties of the polymers. The melting range extends well over 300 °C, beginning with the polycarbonate of diethylene glycol (liquid), through that of hexane-1,6-diol (mp 60 °C) and that of 2,2-dimethylpropane-1,3-diol (mp 109 °C), to that of *trans*-tetramethylcyclobutanediol (mp > 360 °C). These melting points are lower than the melting points of aromatic polycarbonates. Thermal stability is also lower because the aliphatic carbonate group may undergo ester pyrolysis, which can be used to synthesize ω-alkenols.

Uses. In contrast to aromatic polycarbonates, aliphatic polycarbonates are rarely used as thermoplastics. In the form of block cocondensates, they modify the properties of other polycondensates. Aliphatic polycarbonates containing hydroxy groups are widely used as the diol component for the synthesis of polyurethanes and polyurethane-urea elastomers.

The cross-linked polymers of bis(allylcarbonates)s prepared by free-radical polymerization, particularly those based on diethylene glycol, deserve special attention. Because of their high optical transparency, scratch resistance, water resistance, and good mechanical properties, they have become an alternative to glass in the optical industry.

AROMATIC POLYCARBONATES

Aromatic polycarbonates are usually prepared from bisphenols and phosgene by interfacial polymerization.

Bisphenols

Properties. The bisphenols melt between 100 and 200 °C and are virtually insoluble in water. Solubility in organic solvents is determined by the substituents. The boiling points of the bisphenols are very high and depend on the size of the molecule and the polarity. For this reason and because many decompose, bisphenols are seldom distilled.

Chemical Properties. The chemistry of the bisphenols is determined by the phenolic OH groups, the aromatic nuclei, and the bridge segment. Hence, they primarily undergo the same

Table 1. Properties of Injection-molding-grade Bisphenol A Polycarbonate (Makrolon 2800)

Property	Value	ASTM test method
General		
density, g/cm^3	1.20	D792
water absorption	0.12	D570
at 24-h immersion, %		
equilibrium, %	0.34	D570
Optical		
light transmittance at 550 nm, %	87	D1003
haze, %	1.3	D1003
Thermal		
heat deflection at 1.82 MPa[a], °C	131	D648
specific heat, kJ/(kg·K)[b]	1.172	C351
linear thermal expansion per °C	70×10^{-6}	D696
thermal conductivity, W/(m·K)	0.193	C177
brittle temp, °C	−101	D746
flame retardancy		
1.6-mm thickness, UL standard	94 V-2	UL 94
oxygen index, %	25	D2863
Electrical, 3.2-mm thickness		
dielectric strength, mV/m	>16	D149
dielectric constant		D150
at 60 Hz	2.96	
at 1000 Hz	2.92	
dissipation factor		D150
at 60 Hz	0.0006	
at 1000 Hz	0.0090	
vol resistivity, Ω·cm	$>10 \times 10^{16}$	D257
Mechanical		
tensile strength, MPa[a]		D638
yield	62	
ultimate	70	
elongation at yield, %	8	D638
elongation at break, %	120	
tensile modulus, MPa[a]	2300	D638
flexural strength at 5% strain, MPa[a]	86	D790
flexural modulus, MPa[a]	2200	D790
compressive strength at yield, MPa[a]	72	D695
notched Izod impact, J/m[c]		D256
3.2-mm thickness	850	
6.4-mm thickness	120	
tensile impact, 1.6-mm thickness, kJ/m^2[d]	735	D1822

[a] To convert MPa to psi, multiply by 145.
[b] To convert J to cal, divide by 4.184.
[c] To convert J/m to ftlbf/in., divide by 53.38.
[d] To convert kJ/m^2 to $ftlbf/in.^2$, divide by 2.10.

Table 2. Properties of Easy-flow Polycarbonates

Property	Test method	Easy-flow grades[a]		Standard PC	
melt-flow index at 300°C/1.2 kg, g/10 min	DIN 53,735[b] ISO 1133	32–35	17–20	17–20	10–12
spiral flow at 300°C, cm		54	38	38	33
notched impact strength, Charpy, kJ/m²[c]	DIN 53,453[b] ISO R 179	20	40	20	40
elongation at break after 400 h thermal aging at 100°C, %	DIN 53,455[b] ISO R 527	10	50	10	30

[a] Based on polycarbonate with higher branched alkylphenyl end groups.
[b] Deutsche Industrie Norm, FRG.
[c] To convert kJ/m² to ftlbf/in.², divide by 2.10.

reactions as similarly substituted monophenols, esterification and etherification at the OH group, and substitution and hydrogenation of the aromatic nucleus. Because of their bifunctionality, they are useful in the synthesis of high molecular weight linear polyesters and polyethers.

Synthesis. Bisphenols are prepared by the condensation of phenols with ketones, aldehydes, or hydroxymethylphenols, or by the reaction of diols or dihalides with phenols.

Toxicology. Bisphenols are much less toxic than phenol itself. Some bisphenols are used as antioxidants in cosmetics and foodstuffs.

Preparation of Aromatic Polycarbonates

Synthesis of Aromatic Polycarbonates. Aromatic polycarbonates are prepared by the reaction of bisphenols with carbonic acid derivatives. 2,2-Bis(4-hydroxyphenyl)propane (bisphenol A) is usually the bisphenol and phosgene or diphenyl carbonate, the carbonic acid derivative. Aromatic polycarbonates may be prepared by transesterification, solution polymerization, and, most often by interfacial polymerization. Phosgene-free synthesis is still in the experimental stage.

The arrangement of the condensed monomers in the polymer may be linear, branched, or macrocyclic and can be controlled by the reaction conditions. Chain termination is effected with monofunctional monomers, eg, phenol, which are able to regulate the molecular weight and influence the properties of the polymer. Branching can be achieved by using polyfunctional compounds.

Bisphenol A Polycarbonates

Properties. The physical properties of standard bisphenol A (BPA) polycarbonate are given in Table 1. Polycarbonates are insoluble in water, alcohols, organic acids, and aliphatic and cycloaliphatic hydrocarbons. Preferred solvents are chloroform, 1,2-dichloroethane, tetrachloroethane, and methylene chloride. Swelling agents include benzene, chlorobenzene (which, however, dissolves polycarbonates at high temperature), ethyl acetate, acetone, acetonitrile, and carbon tetrachloride. Aliphatic and aromatic hydrocarbons promote crazing of stressed molded samples. Care must be exercised in the choice of solvents for painting and coating operations.

Molecular Weight. Depending on the polymerization process, homopolycarbonates or copolycarbonates of bisphenol A have mean molecular weights between 20,000 and 200,000.

Branched Bisphenol A Polycarbonates. The incorporation of small amounts of trifunctional or polyfunctional monomers gives branched polycarbonates; large amounts give cross-linked polycarbonates. The boundary between branched and cross-linked polycarbonates depends primarily on solubility.

Modification. Modification includes photostabilization, thermal stabilization, stabilization to hydrolysis, improved flow behavior (Table 2), increased structural viscosity, improved release properties, improved flame retardance, and glass-fiber reinforcement (Fig. 1).

Foamed Polycarbonates. Polycarbonates can be foamed with physical or chemical blowing agents during the processing procedure, producing a cellular core and a compact outer skin. Polycarbonate foam exhibits high heat-distortion resistance, impact strength, and rigidity coupled with good flame retardance and high electrical resistivity as well as freedom from stress and a lack of sink marks (Table 3).

Polycarbonate foam is mainly employed for large, heavy moldings, such as housings for office equipment and lamps, roofs of jeeps, and others (see STRUCTURAL FOAMS).

Uses. Polycarbonates are used in electrical applications, household and consumer articles, automotive applications, photographic and optical equipment, sporting goods, construction, medical applications, packaging, laser-optical data-storage systems (Fig. 2), and films.

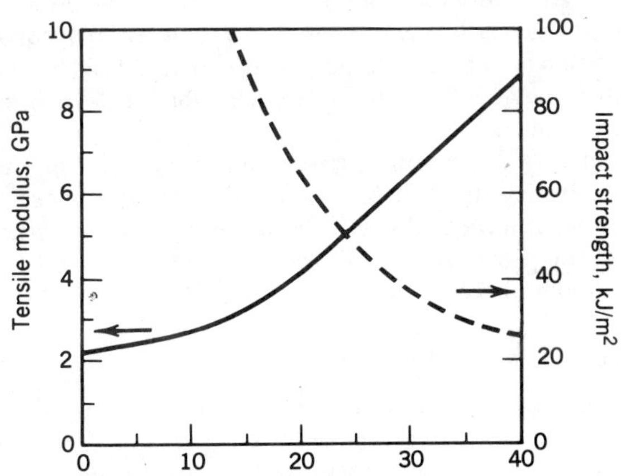

Figure 1. Tensile modulus (——) and impact strength (---) vs fiber concentration for glass-fiber-reinforced BPA polycarbonate. To convert GPa to psi, multiply by 145,000. To convert kJ/m² to ftlbf/in.², divide by 2.10.

Table 3. Properties of Foamed Polycarbonate Makrolon SF 800

Property[a]	Test method[b]	Value
tensile strength, MPa[c]	DIN 53,455	>35
elongation at break, %		5
tensile modulus of elasticity, MPa[c]	DIN 53,457	2,000
flexural strength at 5% peripheral fiber elongation, MPa[c]		
at 23°C	DIN 53,452	70
at 60°C	DIN 53,452	60
flexural modulus of elasticity, MPa[c]		
at 23°C	DIN 53,457	2,300
at 60°C	DIN 53,457	2,000
impact strength, kJ/m^2 [d]		
at +23°C	DIN 53,453	40
at −20°C	DIN 53,453	40
Izod notched impact strength, method E, J/m[e]		
at +20°C	ASTM D256	350
at −20°C	ASTM D256	350
heat-distortion resistance, °C		
at 1.81 MPa[c]	DIN 53,461	127
at 0.45 MPa[c]	DIN 53,461	137
Vicat softening temp VST/B	DIN 53,460	140
UL flammability test, 6.4 mm	UL 94	VO

[a] The mechanical properties were measured on 6.4-mm thick moldings.
[b] DIN = Deutsche Industrie Norm, FRG.
[c] To convert MPa to psi, multiply by 145.
[d] To convert kJ/m^2 to ftlbf/in.^{2}, divide by 2.10.
[e] To convert J/m to ftlbf/in., divide by 53.38.

Processing. Bisphenol A polycarbonate must be dried before processing to <0.2% moisture. In modern fresh-air, high-speed dryers, a drying of 2–5 h is sufficient. It may be processed by injection molding and extrusion.

Economic Aspects. The above-average growth rates recorded worldwide for polycarbonate show no signs of slowing down. Following Bayer, General Electric, Anic, ATO, and Japanese producers (Mitsubishi, Tdemitsu, Teijin), Dow entered the market at the beginning of 1985. In 1984, world polycarbonate consumption reached 320,000 t, and consumption for 1992 is estimated above 500,000 t. Distribution for Western Europe is given in Table 4.

Table 4. Distribution of Polycarbonate Consumption in Western Europe, %

Use	1982	1983
electrical industry	59	55
disks	18	20
automotive industry	6	6
packaging and medicine	5	5
office equipment	2	3
helmets	2	2
other uses	8	9

The factors determining these markets are a combination of transparency, heat-distortion resistance, toughness, flame retardancy, and physiological acceptability; broad spectrum of use; and the possibility of tailoring products with copolymers, blends, and additives.

Modified Aromatic Polycarbonates

Polycarbonates from o,o,o′,o′-Tetramethyl-substituted Bisphenols. *Synthesis.* High molecular weight polycarbonates are synthesized from *o,o,o′,o′*-tetramethyl-substituted bisphenols by a modified interfacial polymerization or by melt transesterification. Tertiary amine catalysts are used in high concentrations.

Properties. The properties of polycarbonates obtained from *o,o,o′,o′*-tetramethyl-substituted bisphenols differ substantially from those of the polycarbonates obtained from nonsubstituted bisphenols. Table 5 compares the properties of TMBPA polycarbonate with those of the polycarbonate obtained from bisphenol A and those of polystyrene.

Blends. The TMBPA polycarbonates can be mixed with bisphenol A polycarbonates to give a single phase. The transparent blends possess properties intermediate between those of the individual polycarbonates, depending on the ratio of components.

Other Bisphenols. The reaction of the three isomeric dihydroxybenzenes with phosgene in pyridine gives the simplest aromatic polycarbonates. Hydroquinone and resorcinol give low molecular weight polycondensates. The resorcinol polycarbonate is glassy amorphous and melts at 200 °C; the hydroquinone polymer is crystalline, insoluble, and infusible.

Industrially useful polycarbonates are formed from monomers of the bisphenol type, where phenolic radicals in the para position are bridged via C, O, S, or alkylene. Particularly important are the alkylidene and cycloalkylidene bisphenols, which are readily obtained by condensation of phenols with aldehydes or ketones. Some bisphenols are used as homopolymeric additives in blends with bisphenol A polycarbonate (eg, oligomers based on tetrabromobisphenol A as a flame retardant) or they are subjected, as comonomers, to a copolymerization with bis-

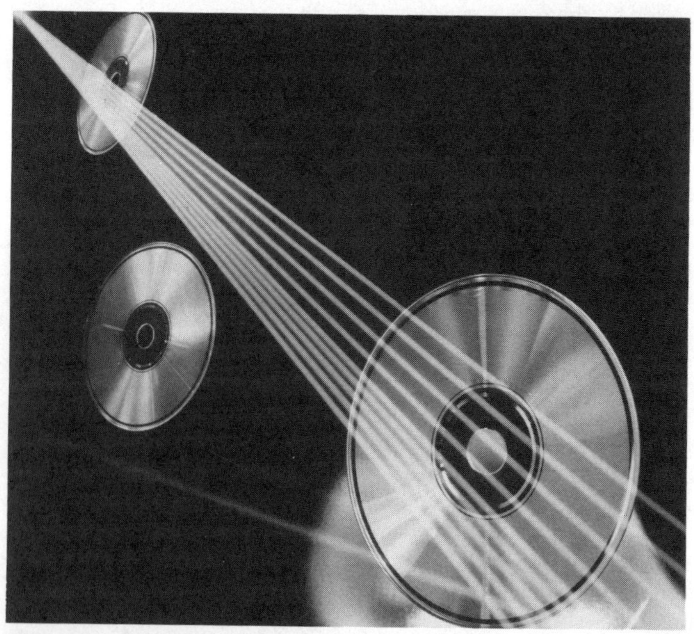

Figure 2. Compact disks from easy-flow polycarbonate (Makrolon CD 2000).

Table 5. Properties of TMBPA Polycarbonate, BPA Polycarbonate, and Polystyrene

Property	Test method, DIN[a]	TMBPA PC, η_r = 1.30[b]	BPA PC, η_r = 130[b]	PS, η_r = 1.579[b]
glass-transition temp, °C		203	150	106
density, g/cm^3	53,479	1.08	1.20	1.05
refractive index		1.546	1.586	1.593
tensile modulus, MPa[c]	53,457	2,400	2,200	3,300
tensile strength, MPa[c]	53,455	80	65	
impact strength, kJ/m^2[d]	53,453	45	no break	20
elongation at break, %	53,455	50	>100	3

[a] Deutsche Industrie Norm, FRG.
[b] η_r determined in a solution of 0.5 g polymer in 100 mL CH_2Cl_2 at 25°C.
[c] To convert MPa to psi, multiply by 145.
[d] To convert kJ/m^2 to ftlbf/in.^{2}, divide by 2.10.

phenol A incorporated randomly as well as in the form of block copolymers.

Copolycarbonates. Copolycarbonates obtained from bisphenol A and tetrabromobisphenol A (TBBPA) and tetrachlorobisphenol (TCBPA) exhibit flame retardancy.

Polycarbonates and copolycarbonates based on 2,2-bis-(4-hydroxyphenyl)-1,1-dichloroethylene (bisphenol C) have an impact strength similar to that of bisphenol A polycarbonate, high transparency, and good flame retardancy. They produce a small amount of fumes and give decomposition products of low toxicity.

Block Copolycarbonates. By incorporating chemically different segments into the polycarbonate chain, the properties of aromatic polycarbonates can be modified. Block copolycarbonates are prepared in the homogeneous phase in an inert solvent or by interfacial polymerization.

The molecular structure of the segmented polycarbonates is determined by the structure of the polycarbonate. Accordingly, three-block structures A—B—A or multiblock structures $(A—B)_n$ are preferred. They are prepared from monofunctional or bifunctional oligomers. By bonding prepolymers that react alternately, eg, reaction of equimolar amounts of hydroxyl-terminated oligomers and chloroformate-terminated oligomers, it is possible to obtain block copolycarbonates with an exactly alternating sequence.

On the other hand, the phosgenation of different hydroxyl-terminated species (oligomers, bisphenols) leads to irregular, random coupling. A completely random structure is only obtained when the hydroxyl-terminated components have the same reactivity toward phosgene. When the components differ in reactivity, fairly long sequences of the same structure are formed. Phosgenation of bisphenols of different acidities gives a block structure.

Block copolycarbonates include aromatic–aromatic polycarbonate-*b*-polycarbonates polyester-*b*-polycarbonates, polysiloxane-*b*-polycarbonates, and poly(methyl methacrylate)-*b*-polycarbonates.

Polyester Carbonates. Aromatic polyester carbonates give products of great heat-distortion resistance and improved low temperature toughness and aging resistance. In addition to a carbonic acid derivative, comonomers used for the preparation are derivatives of one or more aromatic dicarboxylic acids or hydroxycarboxylic acids, and in some cases aliphatic or aromatic diols, in addition to BPA.

The production of polyester carbonates is based on aromatic polycarbonates and aromatic polyesters. Methods include

Table 6. Properties of Polyester Carbonates

Property	Test method[a]	PC	Polyester carbonate 30 wt % ester	50 wt % ester	80 wt % ester	Aromatic polyester
tensile strength, MPa[b]	ISO R 527, DIN 53,455	75	70	66	60	62
elongation at break, %		120	100	80	50	56
flexural strength, MPa[b]	ISO 178, DIN 53,452	95	77	71	66	62
tensile modulus, MPa[b]	DIN 53,457	2,300	2,400	2,300	2,200	2,100
notched impact strength, kJ/m^2[c]	ISO R 179,					
at 23 °C	DIN 53,753	45	35	32	28	22
Vicat softening temp, °C	ISO 306, DIN 53,460	150	159	170	182	192

[a] DIN = Deutsche Industrie Norm, FRG; VDE = Verein Deutscher Elektroingenieure, FRG.
[b] To convert MPa to psi, multiply by 145.
[c] To convert kJ/m^2 to ftlbf/in.^{2}, divide by 2.10.

transesterification of bisphenols with aromatic esters and diphenyl carbonate or polycarbonates with aromatic polyesters in the melt, the reaction of bisphenols with aromatic acid chlorides and phosgene in the homogeneous phase, or by interfacial polymerization.

Properties. Properties are given in Table 6. Blending with polymers with a low melt viscosity improves the processibility of the polyester carbonates. Heat-distortion resistance and low temperature toughness of other thermoplastics are improved by blending with polyester carbonates; graft polymers are also used.

Producers. The following companies offer polyester carbonates:

Company	Product name
General Electric Co.	Lexan-PPC
Mitsubishi Chemical Industries, Ltd., Allied Chemical Corp.	Copec
Dow Chemical Co.	XP 73
Farbenfabriken Bayer AG	Apec

Polyphosphonatocarbonates. Flame-retardant polycarbonates are prepared by the incorporation of phosphonate structures. In modified interfacial methods, bisphenolates are treated with phosphonyl dichlorides and phosgene in the presence of onium compounds, eg, triethylbenzylammonium chloride.

Properties. Cocondensation of phosphonate structures with the polycarbonate chain gives a halogen-free, flame-resistant thermoplastic material (Table 7).

Table 7. Properties of a Polyphosphonatocarbonate[a]

Property	Value
mol wt	16,200
phosphorus, %	3.1
mechanical properties	
impact strength, kJ/m^2 [b]	no break
tensile modulus, MPa[c]	2,500
flexural modulus, MPa[c]	2,400
glass-transition temperature, °C	148
Vicat-B softening temperature, °C	142
flame retardancy, UL 94, 1.5-mm thickness	VO

[a] Phosphonate:bisphenol A:4,4′-dioxydiphenyl = 1:3:1.
[b] To convert kJ/m^2 to ftlbf/in.^{2}, divide by 2.10.
[c] To convert MPa to psi, multiply by 145.

Bisphenol A Polycarbonate Blends

Current world consumption of bisphenol A polycarbonate blends is estimated above 20,000 t. The blends are mostly BPA PC-polyethylene systems, BPA PC-ABS blends, and modified BPA PC-poly(alkylene terephthalate) blends.

Polyester Blends. *Properties.* Heating BPA PC-polyester blends may lead to transesterification. Melts of BPA PC-poly(butylene terephthalate) (PBT) blends give a four-component mixture. The transesterification takes place via a direct ester–ester exchange reaction; acidolysis or alcoholysis play a minor role. The ester exchange is catalyzed by titanium compounds present in the PBT.

Blends of bisphenol A polycarbonate and poly(alkylene terephthalate), mainly those containing poly(butylene terephthalate) and poly(ethylene terephthalate), are less sensitive to stress cracking but more brittle than bisphenol A polycarbonate. Modification with elastomers compensates for the decrease in the toughness of the blends. Such materials are characterized by low temperature toughness, heat-distortion resistance, stability to weathering, resistance to gasolines, dimensional stability, reliable processibility, capability of being coated, and weldability.

Blends of copolyester and BPA PC are transparent and possess desirable physical properties; however, toughness is not optimum. These blends possess better resistance to chemicals, especially resistance to lubricating greases, oils, and plasticizers, than most amorphous materials. Grades that conform to the FDA and USP specifications for medical applications show little tendency to yellow during sterilization with γ-radiation.

Flame retardant formulations are also available (UL 94 VO at 3.2 mm).

Processing and Uses. Drying prior to processing at 100–110 °C is required. During processing melt temperatures of 270–290 °C should not be exceeded, and the residence times should be as short as possible. The melts of the blends possess a lower viscosity than those of standard polycarbonate injection-molding grades and also exhibit a more pronounced structural viscosity. This permits processing by extrusion.

Bisphenol A polycarbonate-poly(alkylene terephthalate) blends have a wide range of uses in automoive, electrical–electronics, mechanical, and sport applications.

ABS Blends. *Properties.* Blends based on bisphenol A polycarbonate and ABS have become increasingly important over the last 15 years. By varying the polycarbonate and ABS grades and proportions, it has been possible to close the technological gap between ABS and bisphenol A polycarbonate. The mechanical and thermal properties of three typical bisphenol A polycarbonate-ABS blends are given in Table 8.

Processing. The BPA PC-ABS blends are processed by conventional methods.

Uses. Bisphenol A polycarbonate-ABS blends are widely used for automobile components, such as flaps for glove compartments, ventilation ports, cladding for steering columns, and dashboards, spoilers, wheel caps, and protective side trims. Because of good electrical insulating properties, blends are also used for production of plugs, sockets, and housings for lamps and safety cutouts. Flame-retardant grades are used in office equipment.

Polyolefin Blends. Among the polycarbonate–polyolefin blends, only the polycarbonate–polyethylene blends have acquired importance. These blends are opaque, are tougher than bisphenol A polycarbonate, stress crack less, and have higher resistance to certain chemicals.

Bisphenol A polycarbonate–polyethylene blends are processed at the temperature usually employed for bisphenol A polycarbonate. Both injection molding and extrusion are used. Typical uses include protective headgear, sporting goods, bobbins for the textile industry, and automobile bumpers.

Styrene-Maleic Anhydride Copolymer Blends. Bisphenol A polycarbonate–styrene copolymer blends are heat-resistant thermoplastics with the toughness of polycarbonate and the processing behavior of styrene copolymers.

The products are suitable for the production of housings, computer components, and interior and exterior automobile components. Transparent formulations can be used for pot lids and projector lenses. Other blends include blends with sty-

Table 8. Physical Properties of PC–ABS Blends (Bayblend)

Typical physical property	Test method	Type I	Type II	Type III
General				
density, g/cm^3	DIN 53,479 ISO R 1183	1.12	1.15	1.16
water-absorption, long-term, %		0.2	0.2	0.2
Mechanical				
tensile stress at yield, MPa[a]	DIN 53,455 ISO R 527	45	50	55
tensile strain at yield, %	DIN 53,455 ISO R 527	4	4.5	5
tensile modulus, MPa[a]	DIN 53,457	1,950	2,100	2,200
flexural strength, MPa[a]	DIN 53,452 ISO 178	75	80	85
impact strength, kJ/m^2[b] at +23°C	DIN 53,453 ISO R 179	not broken	not broken	not broken
notched impact strength, kJ/m^2[b]	DIN 53,453			
Thermal				
Vicat softening temp, °C	DIN 53,460	120	130	137
	ISO 906	110	120	129
flammability rating[c] at 1.6 mm	UL 94	HB	HB	HB

[a] To convert MPa to psi, multiply by 145.
[b] To convert kJ/m^2 to ftlbf/in.^{2}, divide by 2.10.
[c] HB = horizontal burning.

rene–ethylene–butylene–styrene block copolymers (SEBS), and elastomer-modified bisphenol A polycarbonates.

BIBLIOGRAPHY

H. Schnell, *Polym. Rev.* **9** (1964).

H. Vernaleken in F. Millich and C. E. Carraher, Jr., eds. *Interfacial Synthesis,* Vol. 11, Marcel Dekker, Inc., New York, 1977, pp. 65–124.

D. Fox, "Polycarbonates" in M. Grayson, ed., *Kirk-Othmer Encyclopedia of Chemical Technology,* 3rd ed., Vol. 18, Wiley-Interscience, New York, 1982, pp. 479–494.

G. Peilstöcker in R. Vieweg and L. Goerden, eds., *Kunststoff-Handbuch,* Vol. 8, Carl Hanser Verlag, Munich, FRG, 1973.

D. Freitag
U. Grigo
P. R. Müller
W. Nouvertné
Farbenfabriken Bayer AG

POLYCONDENSATION

Polycondensation or condensation polymerization (qv) is defined by IUPAC as polymerization by a repeated condensation process, ie, with elimination of simple molecules. This is in keeping with the preferred usage in organic chemistry wherein a condensation reaction must involve concurrent formation of small molecules.

Because of mechanistic similarities, the term polycondensation has been incorrectly applied to such reactions as polyurethane formation from isocyanates, which is not accompanied by formation of small molecules. Polyurethane formation and similar polymerization reactions in which no elimination of small molecules occurs should more properly be called polyadditions, as they already are in languages other than English (see Step-reaction polymerization).

Polyaddition should not be confused with addition polymerization (qv), a term that is often applied quite differently.

BIBLIOGRAPHY

Pure Appl. Chem. **40,** 477 (1974).

V. Gold, K. Loening, A. McNaught, and P. Sehmi, *Compendium of Chemical Terminology/IUPAC Recommendations,* Blackwell Scientific Publications, Inc., Palo Alto, Calif., 1987.

POLY(DIMETHYLPHENYLENE OXIDE).

See Poly(phenylene ether).

POLYDISPERSITY.

See Molecular weight determination.

POLYDITHIAZOLES.

See Polybithiazoles.

POLYELECTROLYTE COMPLEXES

Polyelectrolyte, or polysalt, complexes are formed when macromolecules of opposite charge are allowed to interact. The interaction usually involves a polymeric acid or its salt with a polymeric base or its salt. Depending on a variety of factors, it may cause the system to separate into a dilute phase and a concentrated complex coacervate phase, or it may result in a more-or-less compact precipitate or gel. The complexes can also remain in solution. Electrostatic interactions constitute the main attractive forces, but hydrogen bonding, ion dipole forces, and hydrophobic interactions frequently play a significant role in determining the ultimate structures.

Complex Coacervates

The early studies on polyelectrolyte complexes dealt mostly with natural polymers of comparatively low charge density. It was particularly noteworthy that liquid coacervates were formed upon mixing natural polymers carrying opposite charges. A number of interesting phenomena on the gum arabic–gelatin system were described, including phase separation.

In many studies of complex coacervates, weakly acidic or basic biopolymers are used. The number of charges on these polyampholytes depends on the pH, whereas the extent of complex coacervation depends on both pH and the mixing ratio of the two polymers. The degree of coacervation is at a maximum when the polyions are present in electrically equivalent quantities.

Increased charge density on the two macromolecules enhances the intensity of coacervation θ and increases the salt tolerance. The parameters θ and ρ, the degree of coacervation or fraction of polymer in the concentrated coacervate phase, as well as the volume and concentration of the coacervate, are complex functions of the salt concentration.

Molecular weight heterogeneity of the reacting polyions is another variable in coacervate formation. In mixtures of symmetrical deionized gelatins of pI 5 and pI 9 where microsalt was removed, the self-suppression of coacervate formation is significant for unfractionated gelatin samples. The phase-separation process selectively accumulates the higher molecular weight components in the coacervate phase, but salt and self-suppression have the effect of increasing the higher molecular weight components in the equilibrium phase.

Synthetic Polymers

The more stable polyelectrolyte complexes are usually formed by the interaction of polyacids and polybases with pK values below 2.0. In contrast to the highly swollen, often liquid complex coacervates, they form more-or-less compact precipitates or gels by the stoichiometric interaction of two polyelectrolytes, which in their ionized form have a comparatively high charge density.

In most systems, insoluble polyelectrolyte complexes, when made under appropriate conditions, exhibit 1:1 stoichiometry in charged groups, independent of the charge density on the macromolecules and the structure of their backbones. This is especially true for linear polymers with easily accessible ionic groups. Examples are the systems polyvinylbenzyltrimethylammonium chloride (PVBTACl) with poly(styrene sulfonate) (NaPSS) or polyacrylate; ionene-type polycations with NaPSS or polyphosphates; quaternized poly(4-vinylpyridine) with NaPSS; and polydimethyldiallylammonium with poly(vinyl sulfonate) or sodium carboxymethylcellulose (NaCMC) at pH 7.

In some systems, nonstoichiometric precipitates are formed as a result of pH changes during complex formation or when very low or very high polymer concentrations are used. Poor accessibility of ionic groups and branching can also cause deviations from 1:1 stoichiometry.

Observation of small spherical particles with a narrow size distribution of 20–40 nm suggested that formation of a polyelectrolyte complex does not result from random cross-linking, which would result in large clusters or gel-like structures. The initial ion-paring process was described as a rotation of extended polymer segments of opposite charge in which the proper positions for the formation of stable ionic cross-links are established. The process is followed by local realignments to adjust for short-range errors during the ion pairing.

The formation of polyelectrolyte complexes is usually, but not always, independent of the initial mixing ratio of the two macromolecules and of the order of mixing. Unlike the mixing of a polyacid and a polybase, where neutralization results in an exothermic reaction, the binding between a polycation and a polyanion is nearly athermal. The driving force of the complex formation is largely entropic owing to the release of microions.

The actual structure of the polyelectrolyte complex has been described as one in between that of a "scrambled-egg" model and a ladderlike structure. The actual structure depends on the structure of the two polymers (eg, the distance between the polymer-bound charges), presence of microsalt, nature of the solvent, and temperature.

Polyelectrolyte complexes formed from macromolecules of high charge density are compact precipitates that are usually insoluble even at high salt concentrations. However, they can often be solubilized in a ternary solvent system of water, a polar organic solvent, and a strongly ionized simple electrolyte.

A variety of methods are employed in the study of polyelectrolyte complex formation, including nephelometry, conductance, potentiometry, electrophoresis, pH measurements, and calorimetry. Structure and properties of polyelectrolyte complexes can be studied by elemental analysis, solubility and swelling measurements, spectroscopic techniques, differential scanning calorimetry (dsc), electron microscopy, and other methods.

The relationship between complex stability and chain length was studied in complexes of poly(methacrylic acid) and oligocations of the ionene type $-\!\!\![CH_2CH_2N^+(CH_3)_2]_n$, where $1 < n < 6$. Stability constants K for the reaction

$$\text{PMAA} + \text{cation} \rightleftharpoons \text{complex} + \text{H}^+$$

increase exponentially with n, according to the relationship

$$K = \text{A} \exp (\text{B}n)$$

where A and B are constants. The K values increase from 110 M^{-1} for $n + 1$ to 13,000 M^{-1} for $n + 6$. Complexes of weak polyacids such as poly(acrylic acid), poly(methacrylic acid), and poly(itaconic acid) with ionenes of the (poly($N,N,N'N'$-tetramethyl-N-p-xylene-N'-alkylene diammonium) type, ie, $-\!\!\![N^+(CH_3)_2\text{-}(CH_2)_n\ N^+(CH_3)_2CH_2C_6H_4CH_2]_n$ have compositions of polycation to poly(carboxylic acid) that equal the degree of ionization of the polyacid.

Poly(dimethydiallylammonium chloride) (PDMDAAC) has frequently been employed as the polycation component in a number of polyelectrolyte complexes. It forms stoichiometric 1:1 complexes with, among others, poly(vinylsulfonate), sodium carboxymethylcelluiose (between pH 5 and 10), and with cellulose sulfate.

Neutral polymers with ion-chelating ligands can be converted into polyelectrolytes when dissolved in salt-containing solutions. They can then form complexes with polyelectrolytes of opposite charge. A typical example is the neutral poly(crown ether)s such as polyvinylbenzo-18-crown-6, a water-soluble polymer. It forms insoluble complexes with carboxymethylcellulose, polyacrylates, polysulfonates, and polynucleotides but only in the presence of a sufficient amount of crown ether-complexable cations such as K^+, Cs^+, Tl^+, and Ba^{2+}. Maximum precipitation occurs when sufficient cations are bound to the crown ether polymer to reach electroneutrality with the added polyanion.

In nonaqueous media complications arise as a result of site binding of counterions to the polymer-bound charges. A high population of ion pairs on a chain can result in strong coiling due to intramolecular interactions, whereas intermolecular interactions cause low solubility of polyelectrolytes in low polarity solvents. Weakly charged copolymers of acrylonitrile and 2-methylvinylpyridine (quaternized with methyl iodide) have been complexed with the copolymer of acrylonitrile and the sodium salt of methallyl sulfonate in dimethylformamide as solvent.

Natural Polymers

Formation of polyelectrolyte complexes is fundamentally no different for natural and synthetic polymers. The former are of course important in living systems where interactions between charged polymers play an important role in the buildup of membranes and other biological structures. Studies include the effects of polyanions on the helix-coil transitions in poly(L-lysine hydrogen bromide), and the formation of complexes between polysaccharides such as glycol–chitosan and heparin or other polyanions.

Crystalline Complexes

Rapid formation and precipitation of polyelectrolyte complexes yield amorphous products. Especially in more concentrated polyelectrolyte solutions, random ion pairing between charged groups of different polymer chains results in the formation of a scrambled-egg-type polymer structure. In more dilute solutions, the initially formed complexes frequently rearrange with time to more stable structures with a higher degree of order. Apparently, the presence of salt, age of the solution, and molecular weight of the polyanion, coupled with time-dependent processes, determine the ultimate morphology of the polyelectrolyte complex.

Template polymerization of ionic monomers on a polyelectrolyte provides a more effective path to crystalline polyelectrolyte complexes. Salts of vinyl sulfonate, acrylate, methacrylate, and styrene sulfonate were polymerized on a matrix of the ionene (**1**) (DP $\approx$ 10):

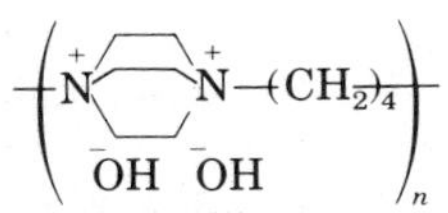

(**1**)

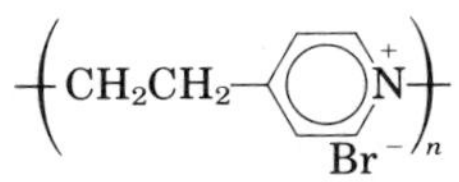

(**2**)

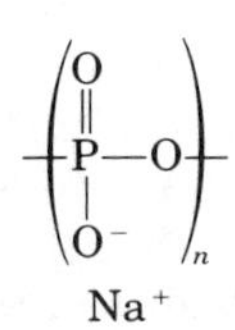

(**3**)

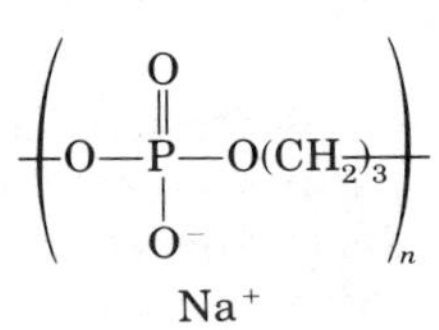

(**4**)

The polyelectrolyte ionene complex with poly(vinyl sulfonate) (PVS) is highly crystalline, but less crystallinity is found with poly(acrylic acid), and none with poly(styrene sulfonate). The size of the monomer counterion bound to the ionene polycation is apparently critical. Crystalline complexes of the ionene poly(1,4-pyridiniumdiylethylene) (**2**) and polyphosphate (**3**) were obtained when the latter polyanion was used as template for the polymerization of 4-vinylpyridine (VP).

Soluble Complexes

Under certain conditions, stable, soluble polyelectrolyte complexes can be formed. Addition of sodium poly(styrene sulfonate) to the ionene poly(*N*,*N*,*N*′,*N*′-tetramethyl-*N*-*p*-xylylenepropylenediammonium dichloride) produces first a precipitate of the 1:1 complex, which is gradually solubilized by excess NaPSS until the solution is completely homogeneous at a ratio of NaPSS to ionene = 3.

A 1:1 complex also precipitates upon mixing NaPSS and poly(*N*-ethyl-4-vinylpyridinium bromide) (PVPQ), but neither an excess of the polycation nor of the polyanion can redissolve the complex. It has been concluded that soluble complexes are only stable with integral-type polyions, such as ionenes, where charges located on the backbone of the polymer are less accessible for pairing with the opposite polyion than in the pendent-type polyions, such as PVPQ.

Properties

Because of their wide use as membranes for dialysis, ultrafiltration, and other solute separation processes, the moisture sorption and water content of polyelectrolyte complexes are of special interest. Permeabilities and permselectivities of solutes through hydrogels made from these complexes are controlled by these properties. Water sorption depends on the chemical structure of the polymer components, charge density, ratio of cationic to anionic charge, pH, presence of microions, and the procedures employed in the complex preparation. The water molecules in the polymer matrix may be free or strongly bound.

The high conductivity and dielectric loss of a number of polyelectrolyte complexes are useful properties in certain applications. Water content and temperature can profoundly affect dielectric loss. Mechanical properties, such as tensile strength, modulus, and elongation at break, are sensitive to charge density, water content, and other factors.

Applications

The best-known applications are membranes for dialysis and ultrafiltration, consisting of PVBTACl and PSS developed by the Amicon Corp. Materials with reported antithrombogenic properties are complexes of PVBTACI and PSS, especially those with a slight excess of sulfonate charges. Complex coacervates of gelatin and gum arabic are used in microencapsulation applications such as carbonless copy paper. Other applications include soil stabilization, the coating of paper, the treatment of cellulose pulp, and flocculation processes.

BIBLIOGRAPHY

A. S. Michaels, *Ind. Eng. Chem.* **57,** 32 (1965).

A. Veis, ed., *Biological Polyelectrolytes,* Marcel Dekker, Inc., New York, 1970.

V. A. Kabanov and A. B. Zezin, *Pure Appl. Chem.* **56**, 343 (1984).
B. Philipp, W. Dawydoff, and K. J. Linow, *Z. Chem.* **22**, 1 (1982).

JOHANNES SMID
DARYLE FISH
State University of New York at Syracuse

POLYELECTROLYTES

The term polyelectrolyte denotes a class of macromolecular compounds, which, when dissolved in a suitable polar solvent (generally water), spontaneously acquire or can be made to acquire a large number of elementary charges distributed along the macromolecular chain. In the former case and when the charge that appears spontaneously has its maximum value, these macromolecules are termed strong polyelectrolytes by analogy with low molar-mass (LMM) electrolytes and, in other cases, weak polyelectrolytes. This distinction is, however, subject to experimental conditions. The number of elementary charges borne by a fully charged polyelectrolyte should be of the same order as the number of monomeric units (degree of polymerization). In most cases, the charges are all of the same sign, although the electrolyte groups on the chain need not be identical. In principle, polyelectrolytes bearing groups with opposite charges also exist or can be prepared.

When in solution and charged, each polyelectrolyte is accompanied by an equivalent amount of LMM ions of opposite sign, the counterions. If, as is often the case, LMM electrolytes are added to a solution of polyelectrolytes, several kinds of counterions can be present as well as several kinds of coions, ions of the same sign as the charge on the macromolecular chain.

It may be expected, and experiments confirm, that polyelectrolytes exhibit properties related to their macromolecular as well as to their electrolytic nature. There is, however, a certain interference between both types of properties leading to a specific behavior of polyelectrolyte solutions. This specific behavior can be understood to a large extent by considering only this interference, without taking into account the exact composition and structure of the macromolecules. This justifies a separate treatment and study of polyelectrolytes and polyelectrolyte solutions. The starting point is the experimental fact that polyelectrolytes in the uncharged state behave like other macromolecules, but that perturbations appear upon charging. Already at a low charge, the specific properties just mentioned appear. Some of these properties and their dependence on the charge of the macromolecules and other factors are discussed below. Addition of a large excess of LMM electrolyte (usually a salt) leads to a reduction of the specific charge effects and, to a certain extent, a recovery of the properties of neutral macromolecules. However, the presence of the third component, the salt, the other two being solvent and macromolecules, must be taken into account.

Classification

Polyelectrolytes may be classified in different ways. As for neutral macromolecules, it is possible to distinguish between natural macromolecules (biopolymers), synthetic macromolecules, and chemically modified biopolymers. Examples are DNA, poly(acrylic acid), and carboxymethylcellulose, respectively. In an analogous way, polyelectrolytes may be grouped into linear, branched, and cross-linked chains (the latter including eventually polyelectrolyte networks) or into homo- and copolymers. On the other hand, classification may be based on electrochemical considerations, distinguishing between polyacids, polybases, and polysalts.

AQUEOUS POLYELECTROLYTE SOLUTIONS

As for neutral macromolecules, the solutions of polyelectrolytes are characterized by the increase of their viscosity with respect to that of the pure solvent η_s. It is very easy to change the relative viscosity increment $\eta_i \equiv (\eta - \eta_s)/\eta_s$ of a polyelectrolyte solution over a wide range of values by changing one of several parameters, such as temperature T, composition of solvent, particularly the LMM electrolyte concentration, and, in the case of a weak polyelectrolyte, its degree of charge β, the fraction of the total number of electrolyte groups that are charged. Only in the case of an excess of LMM salt is it possible to determine the intrinsic viscosity that qualitatively shows the same dependence on the parameters mentioned.

The intrinsic viscosity of a polymer can be related to the average dimensions of the individual macromolecules. Therefore, the changes in $[\eta]$ observed with the degree of charge (or degree of dissociation) of a weak polyelectrolyte at constant c_s or with the concentration of LMM electrolyte at constant degree of charge reflect changes in the average dimensions. These changes can be related to the charge repulsions of the like charges fixed on a given macromolecular chain and their (partial) screening by the small ions in the solution monitored by the concentration of the added salt. The strong dependence of the average dimensions of polyelectrolytes on their linear charge density and on the ionic strength of the solution can also be demonstrated with the help of other properties, eg, by the direct examination of the mean square radius of gyration obtained by light-scattering measurements.

The influence on electrochemical properties of the fixation of the electrolyte groups on a macromolecular chain can be shown by the titration curves of weak polyacids such as PAA, which differ from those of LMM weak acids by a continuous decrease of the apparent dissociation constant K and its strong dependence on the LMM salt concentration. This can be explained by the increasing difficulty of charging an electrochemical group on a chain with increasing charge.

Different empirical expressions for the titration curves of weak polyelectrolytes have been proposed. The so-called extended Henderson–Hasselbalch equation has been widely used:

$$\mathrm{pH} = \mathrm{p}K_a + n \log (\alpha/1 - \alpha)$$

where the two parameters K_a and $n > 1$ are found to be independent of the degree of dissociation α over a more or less wide range of α values around $\alpha = 0.5$, but depend rather strongly on c_s (for a weak LMM acid $n \to 1$ and $K_a \to K_o$). Other empirical expressions are available in different ranges of values.

For some polyelectrolytes, the change of pK with degree of dissociation shows a maximum and a minimum (Fig. 1). Such a nonmonotonous dependence on the charge of the chain is generally attributable to more-or-less discontinuous changes

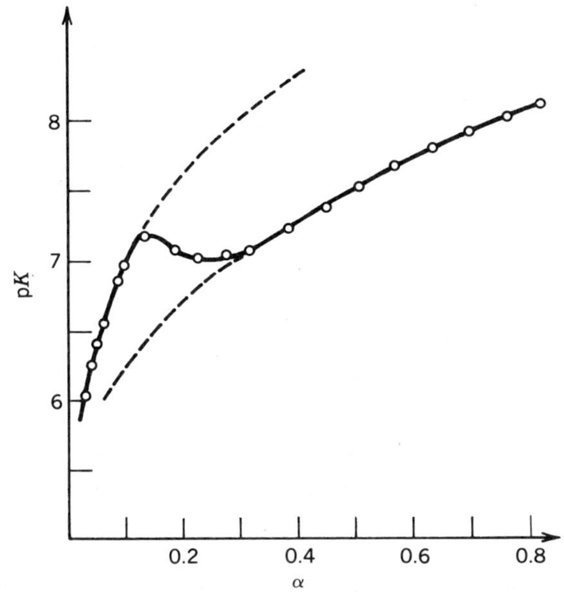

Figure 1. The change of pK with degree of dissociation for poly(methacrylic acid) (PMA) in water at 5 °C showing a conformational transition in the range $0.15 < \alpha < 0.30$. Courtesy of the American Chemical Society.

in the average conformation of the macromolecule in the course of the charging. The helix-coil transition occurring during the titration of poly(glutamic acid) (PGA) is the classical example of such a noncontinuous titration curve; the transition can also be observed by the dependence of other physical quantities on the degree of dissociation.

PMA, which also exhibits such a peculiar titration curve in contrast to PAA, is a good example, showing that, even for charged macromolecules under certain conditions, effects other than electrostatic ones must be taken into account in order to understand the details of the solution properties. In such cases, the exact chemical structure of the macromolecules is more important than when Coulomb interactions are primarily responsible for the polyelectrolyte behavior.

In solutions containing charged macromolecules and small ions, the number and the nature of the thermodynamic components must be chosen with some care, as is also the case for solutions of electrolytes. The thermodynamic components do not necessarily correspond to the species effectively present in solution if some of the latter are charged. Thermodynamics requires that components be defined as electrically neutral entities. In a polyelectrolyte solution in which the macromolecule is (partially) charged, the thermodynamic component is the entity formed by the polyion M^{z-} and its counterions, eg, Na^+, MNa_z; the other is the solvent. Its chemical potential μMNa_z may be formally related to theoretically defined (but not experimentally accessible) single-ion chemical potentials of polyion μ_M and counterion μ_{Na}.

$$\mu_{MNaz} = \mu_M + Z\mu_{Na}$$

In solutions containing both charged macromolecules and LMM electrolyte, the choice of components is not unique. If polyelectrolyte and LMM electrolyte have a common ion, the simplest choice is to consider the polyelectrolyte entity (polyion and counterion) and the LMM electrolyte as thermodynamic components.

Because of their specific properties, polyelectrolytes find a wide range of applications. Both natural polyelectrolytes, mainly polysaccharides, and synthetic ones are used. The former have been utilized for a long time as thickeners or gelling agents in foods, pharmaceuticals, and cosmetics, whereas the latter have been applied for the same purpose in latex paints, eg, where they also act as dispersants. The influence of polyelectrolytes on the rheological properties of solutions and suspensions (or slurries) has also been widely exploited. They can be used in oil-well drilling muds and tertiary oil-recovery processes. Polyelectrolytes are widely employed as water conditioners and flocculating agents for the treatment of wastes and slimes. Their extensive applications in cross-linked form, as ion-exchange resins or electrochemically active membranes, is becoming increasingly important. However, knowledge of polyelectrolyte behavior and polyelectrolyte properties in many of these applications has remained mostly empirical. In many applications, the choice of the polyelectrolyte is also determined by the availability and chemical composition of the macromolecule, which is responsible for its rigidity and biodegradability.

Characterization

The usual methods for the determination of average molar masses, molar-mass distribution, or other characteristic molecular weight quantities can be applied to polyelectrolytes. If they can be dissolved in a nonionizing solvent, the characterization is straightforward, as the polyelectrolyte is not transformed into a charged macromolecule. Such an approach is generally limited to weak polyacids or polybases.

Characterization in aqueous solutions is also possible, but must proceed with some care, particularly for solutions containing (partially) charged macromolecules. For weak acids and bases, the charge on the polyelectrolytes may be suppressed by adding to the solution an excess of strong LMM base or acid, respectively, with eventually LMM inert salt present. Otherwise the characterization has always been to proceed in the presence of an excess of LMM salt.

The number-average molar mass can be determined by osmometry or related thermodynamic methods, if in general the molar mass does not exceed an upper limit, depending on the sensitivity of the equipment used or, for osmometry, if the nonuniform (heterodisperse) sample does not contain low molar fractions permeating the membrane.

The mass-average (or weight-average) molar mass can be determined with the help of Rayleigh (average-intensity) light-scattering measurements. Here, even more care than for osmometry is mandatory. For transforming light-scattering results into the derivative of osmotic pressure $(\partial\Delta\Pi/\partial c)$, the refractive-index increment must be measured at constant chemical potential of water and LMM inert salt, $(\partial\tilde{n}/\partial c)_{\mu 1,\mu 3} = (\partial\tilde{n}/\partial e)\mu$.

Another experimental technique is based on the sedimentation equilibrium of a polyelectrolyte-LMM salt solution in the ultracentrifuge but the analysis of experimental data is difficult for polydisperse samples. A much more frequently used technique is based on viscosity measurements, but it needs calibration. Probably the simplest way to determine the molar-mass distribution of polyelectrolyte samples would be gel-permeation chromatography (gpc), also called size-exclusion chromatography (sec), which for aqueous solutions may present, however, technical problems.

In the characterization of polyelectrolytes, the number of chargeable groups per macromolecule is important (or the maximum value of the average charge per monomeric unit). This refers particularly to copolymers of irregular composition or to polyelectrolytes that have been prepared by chemical modification of nonpolyelectrolytes. Although many analytical methods are available, the simplest ones for polyacids or polybases are potentiometric or conductometric titrations.

It is usually more difficult to determine the number of charged groups per macromolecular chain on a salt of a strong polyacid or strong polybase, which cannot be titrated directly. In that case, the polyelectrolyte must be dissolved in pure (deionized) water and, eventually, the amount of counterion determined directly.

Equilibrium Properties

Equilibrium properties as well as dynamic properties of flexible polyelectrolytes in solution may depend on the presence of excess inert LMM electrolyte. This may be due to the influence of the added electrolyte on the charge interactions involving the polyelectrolyte chain but also by the presence of a third component in the solution or, in slightly different terms, by the influence of the "solvent composition" on the behavior of the polyelectrolytes. The properties of polyelectrolyte solutions in the presence of an excess of LMM salt are in many respects related to those of solutions containing neutral macromolecules.

The Donnan distribution of LMM inert salt between the "inner" and the "outer" solution in equilibrium across a membrane impermeable to the macromolecule is a typical property of polyelectrolyte solutions containing charged macromolecules. It directly follows from the equality of the chemical potential of the LMM salt in the two solutions at equilibrium. If the (small) influence of the osmotic pressure on the value of the standard chemical potential of the salt is neglected, the equilibrium condition takes the following form, in the case where LMM has one ion in common with the polyelectrolyte.

$$c_s(c_s + Zc_m) = c_2'^2 G^2$$

$$G^2 = (y'_\pm)_3^2/(y_\pm)_3^2$$

where $(y_\pm)_3$ and $(y'_\pm)_3$ are the mean-activity coefficients (on the molarity scale) of the LMM salt in the inner and outer solution, respectively.

The Donnan distribution results clearly demonstrate that the mean-activity coefficient of the LMM inert electrolyte in an aqueous polyelectrolyte solution is considerably lowered with respect to its value in the polyelectrolyte-free solution. This lowering of $(y_\pm)_3$ has been directly investigated by potentiometric determinations, usually involving electrochemical cells with liquid junctions. The values of the mean-activity coefficients thus determined have been found to be quite reliable, justifying the assumption that the polyelectrolytes do not significantly influence the diffusion potentials involved in the measured electromotive force (EMF). The value of the ratio $(y_\pm)_3/(y_\pm)_3^0$ decreases rapidly with increasing y (the ratio of the concentration of polyelectrolyte to concentration of LMM salt), from unity at $c_m = 0$ to values of 0.4 or less (Fig. 2).

Compared with the systems discussed previously, systems

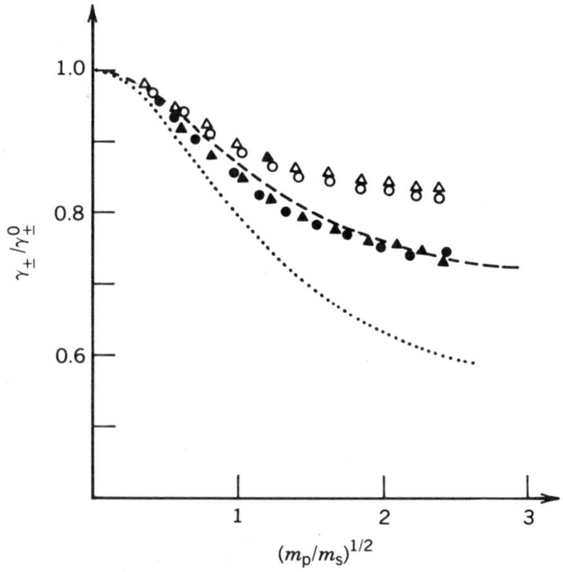

Figure 2. Measured mean-activity coefficient $\gamma_\pm$ on the molality scale) of NaCl in aqueous Na pectate solutions vs the mean-activity coefficient $\gamma_\pm^0$ of NaCl in water vs the ratio $m_p : m_s$ for two pectate sample corresponding to $\lambda = 0.82$ (open symbols) and $\lambda = 1.63$ (filled symbols). Circles refer to solutions with $m_p = 0.01$ mol/kg and triangles to $m_p = 0.025$ mol/kg. The dashed curves represent the theoretical ratios according to Manning's theory. Courtesy of the American Chemical Society.

containing no LMM electrolytes are characterized by properties deviating in an even more striking way from those of solutions containing uncharged macromolecules or only LMM electrolytes. However, the behavior with respect to solutions with LMM salt may also be qualitatively different. No Donnan distribution is, of course, observed in a dialysis experiment involving a macromolecular impermeable membrane. The counterions of the polyelectrolyte solution cannot be distributed on both sides of the membrane as a direct consequence of the electroneutrality condition, which must be obeyed in both solutions at equilibrium. Osmotic pressure and light scattering no longer supply direct information about the molar mass of a polyelectrolyte if present in a charged state in the solution.

Polyelectrolytes without LMM electrolytes exhibit stronger deviations from ideal behavior than their analogues with LMM inert salt. This can be understood by the reduced electrostatic screening of the charges fixed on the macromolecular chains in the former systems.

Theoretical Treatment of Equilibrium Properties

The theoretical understanding of polyelectrolyte behavior is much further advanced for equilibrium than for nonequilibrium properties; the latter involves considerations on the dynamics of the (flexible) charged macromolecular chains, which at present are still difficult to handle. Some of the theoretical developments concerning the equilibrium properties, particularly in recent years, have not only provided clues for the understanding of the nonequilibrium properties, but have also served to stimulate further experimental investigation.

The statistical physics of charged macromolecular chains and their counterions in a polar solvent, with or without LMM inert electrolyte, are very complicated. The asymmetry between the small, simple ions and the polyion on the one hand, and the rather numerous conformational possibilities of the chains, strongly dependent on the electrostatic and perhaps also nonelectrostatic interactions between all particles in solution, on the other, seem to make a rigorous statistical treatment of polyelectrolyte systems nearly impossible. Therefore, simpler treatments, in which models involving many approximations are used, must be considered. Many of these theoretical treatments have been inspired by the Debye–Hückel theory of ordinary electrolyte solutions or make use of the result obtained by this treatment. They are, in general, based on the explicit neglect of other than electrostatic interactions, the treatment of the solvent as a continuum (and therefore also the omission of specific solvation effects), the use of a fixed (average) conformation of the macromolecule, and a limitation of interactions to those between one single macromolecule and the small ions in the solution (no intermacromolecular interactions are taken into account).

This last condition means that only electroneutral systems containing one macromolecule are being considered. Various boundary conditions have, however, been used, such as a chain in an infinitely large system (infinitely dilute solution) or a chain within an electroneutral cell. The volume of the latter is determined by the average volume per macromolecule, assumed to be much larger than the volume occupied by the macromolecule itself (dilute solution of very small macromolecular volume fraction).

In all treatments, the rather large electrostatic potential arising around the macromolecule from the charges fixed thereon determines the distribution of the small ions in the solution, in particular the distribution of the counterions. This potential may be evaluated more or less explicitly, following the Debye–Hückel treatment, or may be postulated to be of a certain simple form. In the first theoretical attempts, the polyion was represented as a permeable sphere containing, distributed over its total volume, a continuous but not necessarily uniform charge density due to the macromolecular charges. However, this model was found to be unsatisfactory, principally because of the assumption of the continuous fixed-charge-distribution. This gives rise to a practically constant electrostatic potential and consequently to a nearly uniform mobile ion concentration inside the sphere. There are, however, good reasons to believe that this is not the case inside a macromolecular coil, where the electrostatic potential varies considerably from one volume element to another, depending on its distance to the nearest charged chain segments as compared with the average screening distance (roughly on the order of the Debye length κ^{-1}, which is inversely proportional to the square root of the ionic strength of the solution).

A more appropriate model, which takes into account the specific chain character of a polyelectrolyte neglected in the former, is the rodlike model. The polyion is represented as a cylinder, of radius a and length L, with $L \gg a$. Although in the absence of LMM electrolyte and at very low concentrations the average conformation of the charged macromolecule tends to become rodlike, this is not necessarily the case under other conditions. Nevertheless, whenever the screening length of the electrostatic forces is sufficiently large with respect to the average distance A between successive charges fixed on the chain, the macromolecule may be considered to be locally stiff enough, because of the repulsion of the like fixed charges, to justify the cylindrical model on that scale. This model is a sufficient approximation insofar as properties are concerned, which, in dilute solutions, have been found to be independent or only very slightly dependent on the length of the chain or its molar mass; they may thus be thought to depend on the local charge density and the local distribution of the small ions around the polyion. The simplest theoretical treatment of polyelectrolyte solutions without added LMM electrolytes based on a cylindrical model is the association or two-phases model.

A sounder but mathematically more difficult theoretical treatment is based on the solution of the Poisson–Boltzmann (PB) equation for the electrostatic potential around the polyion. This is the application to polyelectrolyte solution of the theoretical treatment of LMM electrolytes by Debye and Hückel.

The cylindrical model is a good approximation for discussing polyelectrolyte properties assumed to be independent or only weakly dependent on the average dimensions, or properties under conditions where the average conformation is close to the fully stretched chain. It cannot be used to understand the dependence of the average dimension of a charged macromolecule on its linear charge density or on the ionic strength of the solution.

Recently, several new methods have been proposed. Their models take into account the influence of charge interactions on the local stiffness of the macromolecular chain, including eventually torsional effects as well as excluded-volume effects. They are, strictly taken, valid only in solutions containing an excess of LMM inert electrolyte. The simplest model is that of the so-called wormlike chain, a continuous-chain model in which all structural features are smoothed out.

In all these theoretical considerations, no intermacromolecular interactions have been taken into account. The effects of charge density and ionic strength are, strictly taken, only applicable to solutions sufficiently dilute in polyelectrolytes, where concentration effects should be small. Treatment of higher order concentration effects remains much more difficult. Such effects have been investigated experimentally only recently. Some semiquantitative insight may be gained by using scaling concepts, which have been successfully applied to systems containing uncharged macromolecules.

Although better insight has been gained into the field of polyelectrolyte theory, entirely convincing quantitative explanations for many aspects of polyelectrolyte behavior are still lacking. This is particularly true for solutions without excess LMM salt (especially when $0 < c_s \leqq Zc_m$), which are difficult to treat theoretically. New methods in which the interaction with solvent molecules are more explicitly introduced than has yet been possible are badly needed to increase understanding of solvent effects.

Nonequilibrium Properties

Nonequilibrium properties of polyelectrolyte solutions have been investigated using a large number of methods, of which viscosimetric and conductometric measurements are the oldest applied. In contrast to equilibrium, where a single theoretical treatment, independent of the property studied, may be expected to account for the observed effects, nonequilibrium behavior strongly depends on the response of the system to the external perturbation causing the departure from equilibrium. A theoretical interpretation for all nonequilibrium properties is not yet available, especially since the dynamic behavior of the more-or-less flexible polyelectrolyte chains is still far from being clearly understood. A more pragmatic treatment, tailored to the nonequilibrium conditions explored, is generally used for the different properties studied.

The formal framework of nonequilibrium thermodynamics remains applicable to polyelectrolyte solutions if proper consideration is given to all the forces present in such systems. This kind of formal treatment is extremely helpful in determining the physical significance of experimentally obtained quantities and in defining relations between them. Many equations that have been derived have been formulated in the so-called friction-coefficient formalism.

Viscosity. The relative viscosity increment η_2 of a polyelectrolyte solution depends on the solvent, temperature, macromolecular mass, concentration, linear charge density, and ionic strength of the solution. Most experimental investigations have been performed on aqueous solutions, but several studies on solutions with other polar solvents have been published. At constant temperature, solvent, and LMM salt concentration, the viscosity increment of a given polyelectrolyte in water usually increases with increasing linear charge density of the chain, as is observed upon neutralizing weak polyacids. This increase tends to level off as neutralization progresses. It can be observed at all LMM salt concentrations, including $c_s = 0$, but the change is small at high c_s. The curve η_i vs α' may reflect the occurrence of a conformational transition, since the change in relative viscosity increment is directly related to the change in the average dimensions of the polymer.

If LMM inert salt is in excess, η_i/c usually increases linearly with c at constant temperature and c_s, and can be extrapolated to $c = 0$ to yield the intrinsic viscosity $[\eta]$. The latter decreases with increasing c_s and depends strongly on the molar mass of the polyelectrolyte.

Viscosity results for polyelectrolyte solutions without LMM inert electrolyte or for which $c_s \ll Zc_m$ are difficult to understand. At constant temperature and constant linear charge density, the reduced viscosity number η_i/c exhibits a peculiar concentration dependence that impedes the determinations of the intrinsic viscosity $[\eta]$ for such solutions. Sometimes even more complicated curves have been observed with, at very low concentration, an additional minimum and subsequent increase, but such results should be considered very critically in view of possible interference by other phenomena due to adsorption and impurities. The decrease of η_i/c with decreasing c below the first maximum, as observed with some but not all polyelectrolyte solutions investigated, could be due to such secondary effects, including the additional screening of the macromolecular charges by the self-dissociation of water. However, at these low concentrations, highly accurate measurements and control of disturbing side effects remain very difficult.

Isothermal Diffusion. All kinds of different diffusion measurements have been performed on polyelectrolyte solutions. In the presence of macroscopic concentration gradients, diffusion of electroneutral components takes place. Diffusion is observed on the macroscopic scale where electroneutrality is observed locally. Nevertheless, with the help of special experimental techniques, the motion of the separate charged species may eventually be followed.

Various diffusion coefficients of polyelectrolyte solutions with or without an excess LMM inert electrolyte (with a common counterion) have been determined experimentally. They include experiments with labeled counterions, concentration-gradient measurements, and quasielastic light scattering.

They may yield different kinds of diffusion coefficients. The cooperative diffusion coefficients as, eg, determined by quasielastic light scattering, reveal various concentration regimes as well as the influence of increasing LMM salt concentration. At present it is still difficult to derive a complete theory of polyion and small ion diffusion in polyelectrolyte solutions because of the complexity of the problem in view of all the molecular forces involved. Experiments on the counterion diffusion coefficient generally lead to the conclusion that a fraction of the counterions is diffusing together with the polyions.

Electric Transport. A small, constant electric field applied to a polyelectrolyte solution generates an electric current subject to the transport of the polyions and the small ions through the solution. As for ordinary electrolyte solutions, the conductance or specific and equivalent conductivity are the directly accessible properties that reflect this electric transport; they are easily measured. Related properties are the transference numbers and electrophoresis velocity. Complications arise if the applied field is not time-independent but varies periodically, giving rise to conductivity dispersion and a contribution to dielectric dispersion. Experimental investigations of the electric conductance of polyelectrolyte solutions in the presence of a time-independent field (or of a low frequency field assumed to have the same effect) have been rare.

The formal, thermodynamic treatment of conductivity and related phenomena in polyelectrolyte solutions is possible, although several versions, sometimes apparently in conflict, are available because different ways can be followed and different formalisms used. All this applies to polyelectrolyte solutions with and without LMM electrolytes.

In the absence of LMM salt, the conductivity of polyelectrolyte solutions again reveals association of counterions. This association also strongly influences the transference number of poly- and counterions. The concentration dependence of these equivalent conductances at constant charge density is not monotonous but is not yet well understood.

The situation is not much better for polyelectrolyte solutions containing an excess of LMM salt for which several experimental quantities may be studied. The interpretation of the experimental results of polyelectrolyte–LMM salt solutions is seriously hampered by the lack of any theory that takes into account the coil-like average conformation of the polyions in most of the systems investigated and the non-negligible interactions between the charged macromolecules.

Dielectric Properties. Dielectric properties as revealed in a small, periodically varying, electric field are not only interesting but also striking, particularly for aqueous polyelectrolyte solutions in the absence of LMM salt. The electric-permittivity increment with respect to the pure solvent $\Delta\epsilon$ can easily exceed values as large as several hundred or even thousand units, depending on the molar mass, concentration, and charge density of the macromolecule. Moreover, the increment may decrease rapidly with increasing frequency of the applied field and disappear in the region near 100 MHz. This is in contrast to observations of solutions of uncharged macromolecules or LMM electrolytes. Measurements are, however, difficult, particularly at low frequencies where they require special equipment, generally not commercially available, and where the influence of artifacts is often hard to avoid. Most of these techniques have a limited sensitivity for detecting $\Delta\epsilon$ values, which imposes a lower limit to the polyelectrolyte concentrations. They cannot easily deal with systems presenting a conductivity exceeding $10^{-1}\ \Omega^{-1}/\text{m}$, thus imposing a severe upper constraint on the polyelectrolyte concentration.

Ion Binding

Many of the observed properties of polyelectrolyte solutions are consistent with the assumption that a fraction of the stoichiometric charge of the macroion is reduced by the presence of counterions in the close neighborhood of the macromolecular chain. These counterions are termed associated counterions or sometimes bound counterions. They are closely related to the condensed counterions, although there are differences that follow from the way the concepts are introduced. No direct evidence for association is generally available, but some experimental results seem to justify certain aspects of association, eg, the drag exerted on the counterions by the macroions moving in a gradient of electric or chemical potentials. Nevertheless, association as introduced in the two-phase model or in the corresponding condensation method are operationally defined concepts and not concepts resulting from a theoretical analysis of the systems under investigation. As a consequence, to derive the fraction of associated counterions from experimental results, it is necessary to introduce additional assumptions that may sometimes be questionable. Thus conductivity or related properties of polyelectrolyte solutions can be interpreted assuming association. To calculate from such measurements numerical values for the effective charge of the macroion or the effective fraction of associated counterions, it must be postulated that the latter have no mobility relative to the macromolecule. This assumption has, however, been questioned in view of other properties, such as the anisotropy of the conductance in a field of flow or an alternating electric field, and the high electric polarizability of the macroions.

This may explain why the fraction of bound counterions, derived from different experimental sources but referring to the same systems, usually does not coincide, though often their order of magnitude does. The discrepancies generally exceed what is to be expected, considering the experimental accuracy, and are to be attributed to flaws in the theoretical background underlying the calculation of these values. Nevertheless the concept remains qualitatively very useful and should not be abandoned as long as no better and satisfactory theoretical treatment is available.

BIBLIOGRAPHY

H. Eisenberg, *Biological Macromolecules and Polyelectrolytes in Solution,* Clarendon Press, Oxford, UK, 1976.

T. Odijk in K. H. Bennemann, ed., *Ionic Liquids, Molten Salts, Polyelectrolytes,* Springer-Verlag, Berlin, 1982, p. 184.

M. Mandel and T. Odijk, *Ann. Rev. Phys. Chem.* **35,** 75 (1984).

H. Morawetz, *Macromolecules in Solution,* 2nd ed., John Wiley & Sons, Inc., New York, 1974.

A. Rembaum and E. Selegny, eds., *Polyelectrolytes and their Applications,* D. Reidel Publishing Co., Dordrecht, the Netherlands, 1975.

MICHEL MANDEL
University of Leiden

POLYESTERS

Polyesters are heterochain macromolecular substances characterized by the presence of carboxylate ester groups in the repeating units of their main chains. This article deals primarily with linear homo- and copolyesters derived, whether actually or notionally, from the self-polycondensation of hydroxycarboxylic acids or the polycondensation of dicarboxylic acids with dihydroxy compounds. The first approach yields AB-type products having a unidirectional orientation of carboxylate ester groups in the molecular chain; the second leads to AA/BB-type polymers with the ester groups recurring alternately in their oxycarbonyl and carbonyloxy forms. In homopolyesters the carboxylate ester groups are spaced regularly along the chains, but in copolyesters their separations may vary with the nature of the intervening skeletal groups.

Through variations in molecular architecture and modulation by copolymerization, linear polyesters are obtained with a wide range of properties and uses. Predominant in terms of volume and product value are those based on poly(ethylene terephthalate) (PET) [poly(oxy-1,2-ethanediyloxycarbonyl-1,4-phenylenecarbonyl)] (**1**), where $x = 2$, long established as the basis of fibers, films, and molding plastics, and currently in rapid expansion as the source of widely used containers for liquids. Poly(butylene terephthalate) (PBT) [poly(tetramethylene terephthalate)], poly(oxy-1,4-butanediyloxycarbonyl-1,4-phenylenecarbonyl) (**1**), where $x = 4$, also finds substantial use as a molding plastic and molecular component of polyetherester thermoplastic block copolymer elastomers; the alicyclic ring-containing polyester (**2**) and its copolymers are also manufactured for fiber and plastics uses. In general, these materials have molecular weights in excess of $10–15 \times 10^3$.

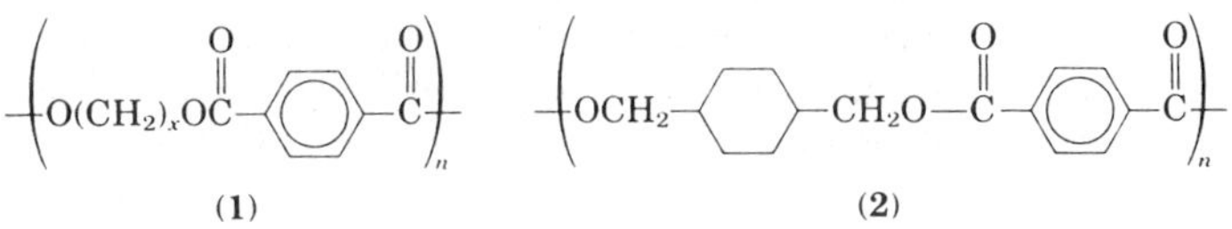

(**1**) (**2**)

Aliphatic polyesters of technical importance are typified by the poly(alkylene adipate)s (**3**), where $x = 2,4$, or 6, $y = 4$, and their copolymers, and by poly(ϵ-caprolactone) {poly[oxy(1-oxo-1,6-hexanediyl)]} (**4**), where $x = 5$. Those with molecular weights up to 5,000 and with hydroxyl groups at both ends of the molecules are employed as macrodiols for the synthesis of elastomeric polyurethanes. Others find use as nonmigrating plasticizers for PVC, and the polyglycolide {poly[1-oxy(1-oxoethanediyl)]}, (**4**), where $x = 1$, forms the basis of bioabsorbable surgical sutures (qv).

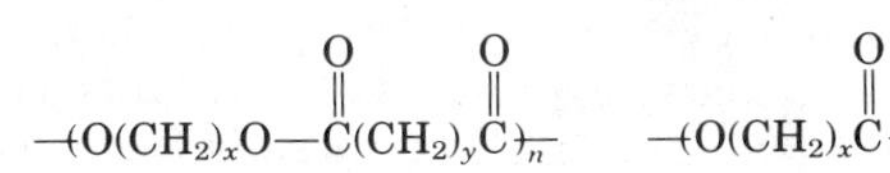

(**3**) (**4**)

Recent developments have occurred with high temperature-resistant polyarylates, the term used to denote polyesters of dihydric phenols, and with products of unusually high mechanical properties; representatives of both groups are now emerging in the market (see LIQUID CRYSTALLINE POLYMERS; POLYESTERS, MAINCHAIN AROMATIC). The span of properties and the commercial value provided by polyesters is thus considerable (Table 1).

Monomers

The starting reactants for the manufacture of the wide range of commercially available polyesters are products mostly of downstream petrochemicals operations given here only in outline. Fuller descriptions are available in the literature.

Table 1. U.S. Production and Approximate Prices of Polyester Products[a]

Class of product	Production per year, 10^3 t 1965	1975	1979	1983	1985	1983 unit price, \$/kg
alkyd resins	269	306	343	327	na	1.36
unsaturated polyesters	148	365	537	536	555	1.43
polycarbonate	7	39	109	110	133	3.65
polyester fibers	205	828	1367	1018	na	1.58–1.80
polyester plastics and film	23	35	268	478[b]	651	1.19 (PET) 1.94 (PBT)
polyester plasticizers	18	17	26	26	na	2.05
polyester-based thermoplastic elastomers[b]			6	10	12	2.42–3.08

[a] Most data from U.S. International Trade Commission Statistics; *Text. Organon.* is another good source.
[b] Of which PET comprised 407×10^3 t.

The important diols ethylene glycol (1,2-ethanediol) and propylene glycol (1,2-propanediol) are derived from ethene and propene, respectively, by way of the corresponding cyclic oxides. ε-Caprolactone (2-oxepanone), the precursor monomer for (**4**), where $x = 5$, is derived from cyclohexane through oxidation of its ketone, cyclohexanone, with peroxyacids, and bisphenol A [4,4′-(1-methyl-ethylidene)bisphenol] is formed by the acid-catalyzed condensation of acetone with phenol. The benzene-based manufacturing routes to nylon-6,6 via cyclohexane or phenol are the source of adipic acid (hexanedioic acid) and the by-product glutaric acid (pentanedioic acid), which are used for some polyester plasticizers. Benzene is also the starting point for the manufacture of maleic anhydride (*cis*-2-butenedicarboxylic anhydride), whereas sebacic acid is derived from vegetable oils. In addition to its major outlet for the preparation of unsaturated polyester resins, maleic anhydride is also the basis of one manufacturing route to 1,4-butanediol, other routes being from 1,3-butadiene and acetylene (see also MALEIC AND FUMARIC ACID POLYMERS). 1,6-Hexanediol is obtained by the hydrogenation of adipoyl compounds and of ε-caprolactone.

Terephthalic Acid. The preeminence of terephthalate polyester products (**1**) and (**2**) became possible only with the evolution of technically and economically acceptable manufacturing processes for terephthalic acid (1,4-benzenedicarboxylic acid), which was not a commercially significant substance at the time of PET's discovery. The acid is made by the oxidation of *p*-xylene (1,4-dimethyl benzene), a product of the catalytic reforming of naphtha.

Purity. The emphasis on purity of the reactants for polyester production has two aspects. First, there is the need to avoid monofunctional impurities that limit chain growth in polymerization. Second, product quality in the final polymer is affected by even small proportions of impurities, whether organic or inorganic, which can cause process difficulties during polymerization and drastically affect the color, thermal and photochemical stability, and long-term durability of the polymers, especially in the form of fibers. Reactant purities are therefore generally at least 99.0% and often 99.5% or better.

Polymer Properties

The properties of polyesters are determined by the proportion of carboxylate ester groups in their structure and by the geometry, polarity, and segmental mobility of their repeating units. Since their intermolecular interactions are not especially strong, the properties of polyesters are more sensitive to variations in structure than are those of strongly interacting polymers such as polyamides or polyurethanes, and hence span a wider range than for these more polar materials.

Linear Acyclic Polyesters. Polyesters of types (**3**) and (**4**) with values of x and $y \geq 2$ are colorless crystalline solids with crystal melting temperatures (T_m) mostly within the rather narrow range of 40–90°C, glass-transition temperatures (T_g) −70 to −30°C, and densities 0.9–1.3 g/cm^3. At low molecular weights (< 5,000), they are brittle waxes, increasing in mechanical strength at higher chain lengths. Generally, they are readily soluble in liquid ketones, amides, and phenols; moderately soluble in chlorohydrocarbons, 1,4-dioxane, aromatic hydrocarbons, and tetrahydrofuran; poorly soluble in the lower alcohols; and insoluble in water. They are substantially resistant to oxidation by air or ozone in normal conditions, but are degraded rapidly by ammonia, hydrazine, warm alkali solutions, and primary or secondary amines that cleave the ester linkages liberating hydroxyl groups and forming salt or amide derivatives of the carboxyl functions. The hydrophobic character of polyesters provides some physical barrier to chemical attack by water or dilute mineral acids at ordinary temperatures, but hydrolysis occurs at higher temperatures or when in contact with steam; the polymers are also prone to solvolysis by stronger organic acids such as formic acid. Some protection against hydrolysis is afforded by end-group blocking or by the addition of carbodiimides. The thermal stability of linear aliphatic polyesters is generally excellent up to ca 200°C.

Substituted Acyclic Polyesters. The properties of aliphatic polyesters are markedly affected by changes in chemical structure that influence the geometrical regularity, polar character, or segmental mobility of the molecular chains.

Alkyl Group Branches. When one or more hydrogen atoms in the repeating units of (**3**) and (**4**) are replaced by alkyl groups, the intermolecular separation of adjacent chain backbones is increased and there is often a reduction in the inherent symmetry of the system leading to a lowering of T_m if crystallinity is retained, and frequently to the loss of crystalline character.

Fluorine-containing Polyesters. The polyesters represented by the structure $-\!\!\left[OCH_2(CF_2)_xCH_2O-\overset{O}{\overset{\|}{C}}(CH_2)_y\overset{O}{\overset{\|}{C}}\right]_n\!\!-$ with $x = 3$ or 4 and $y = 3$–8 are crystalline solids with melting temperatures between 30 and 60°C and brittle temperatures, which may equate to T_g, between −50 and −65°C. These properties are similar to those of the hydrogen analogues, but the substances are claimed to have enhanced high temperature stability.

Heteroatomic Groups in the Chain. When divalent heteroatomic groups such as —O—, —S—, or —N(CH$_3$) are inserted into the alkylene segments of aliphatic polyesters, a discontinuity is caused in the conformational pattern, and the energy barrier to rotation around the bond connecting the heteroatom to adjacent CH$_2$ groups may be lowered compared with that for CH$_2$—CH$_2$. The usual effect is to lower T_m markedly, and often the tendency to crystallize, as compared with all-methylene analogues. In contrast, the insertion of a strongly dipolar group into the chain can substantially increase the intermolecular forces and hence the T_m.

Unsaturation in the Chain. When —CH$\overset{a}{=}$CH$\overset{b}{—}$ replaces the saturated sequence —CH$_2\overset{a'}{—}$CH$_2\overset{b'}{—}$ in the repeating unit, the central bond a is irrotatable as compared with a' (thus reducing the number of configurations available to the chain), but the adjacent bond b has a lower energy barrier to rotation than bond b' in the saturated system. These opposing features are further complicated by the different geometries that may be assumed by the double bond. In general, *cis*-vinylene groups interrupt the uniformity of structure and packing of linear chains, leading to lowered melting temperatures or to noncrystalline character.

Unsaturation lateral to the main chain has less effect on the motions of the latter and is better compared with normal alkyl substitution. Thus the polyesters (**5**) are crystalline, with T_m of 60–90°C (lower than the saturated analogue) when R = H, and 190–200°C when R = CH$_3$.

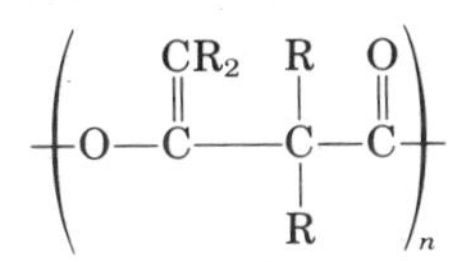

(**5**)

Ring-containing Polyesters. The discovery of PET and the attribution of its properties to the effects of the phenylene ring spurred the study of numerous other ring-containing polyesters. Their characteristics depend on the structure, symmetry, and conformational features of the cyclic units, and on the relative proportions of cyclic and acyclic components in the repeating units. As a broad guide, when the cyclic components, which may be aromatic, alicyclic, or heterocyclic, predominate, then the polyesters, if crystalline, have melting temperatures considerably higher than those typical of acyclic polyesters. The glass-transition temperatures are also higher, whether the materials are crystalline or amorphous. The effect is due to the restriction of chain motion in the ring as compared with an acyclic unit of similar length.

Sulfur-containing Analogues of Polyesters. Polyesters containing the carbothioate linking group S—C(=O) in place of O—C(=O) have been derived from dithiols and bis(thiophenol)s in combination with aliphatic and aromatic dicarboxylic acids. Their melting temperatures are usually 20–100°C higher than those of corresponding oxycarboxylates. This order is reversed in the lower AB-type carbothioate polyesters —[SCH$_2$C(=O)]$_n$—, —[SCH(CH$_3$)C(=O)]$_n$—, and —[SCH$_2$C(CH$_3$)$_2$C(=O)]$_n$—, although poly(δ-caprothiolactone) —[S(CH$_2$)$_5$C(=O)]$_n$— (T_m = 106°C) is higher melting than its oxyanalogue. They appear to have relatively poor thermal stability, but, because of their potential for chelating to metal surfaces, some members have been proposed for use as corrosion-inhibiting coatings for steel.

Copolyesters. For binary random copolyesters composed of structural units A and B, which respectively form crystalline homopolymers $[A]_n$ and $[B]_n$, the normal finding is that the T_m of $[A]_n$ is reduced as B units are incorporated into the chain and correspondingly when A units are incorporated into $[B]_n$. The T_m vs composition diagram for the whole range of compositions is usually V- or U-shaped, with its minimum, termed the eutectic, usually displaced from the equimolar composition of the components toward the side of the lower melting homopolymer. The glass-transition temperatures, however, fall monotonically between those of $[A]_n$ and $[B]_n$. An uncommon pattern is found in stereocopolymers of (**2**) containing *cis*- and *trans*-1,4-cyclohexanedimethylene groups, where both the melting and the glass-transition temperatures increase progressively from the values for the cis homopolymer to those for the trans.

Depending on the structures involved, crystallizability may be preserved across the whole range of composition or there may be a central zone where the copolyesters are amorphous. However, in general, the crystallinity is reduced by random copolymerization, leading to materials of greater plasticity and a wider range of solubilities than in the homopolymers.

Copolymerization thus provides a means of tailoring the characteristics of polyesters to desired values, although it should be remembered that modification of some particular property may also alter others, not always desirably.

Molecular Features of Polyester Properties

The major uses of linear polyesters are for the materials in the solid state and considerable investigation has been devoted to elucidating the molecular basis of their properties and mechanical behavior in this condition. For crystallizing polyesters, the present understanding is based largely on crystal structures determined by x-ray diffraction studies of oriented fibers and films or by electron diffraction with single crystals, supplemented by spectroscopic investigations, thermal analysis, and conformational energy calculations.

If the melting temperature T_m is defined as that temperature where the ordered (crystalline) and disordered (amorphous liquid) phases of the material coexist in equilibrium, then the relationship between T_m, the enthalpy of fusion ΔH_f, and the entropy increase on fusion ΔS_f is given by $T_m = \Delta H_f/\Delta S_f$, such that changes of T_m with structure reflect differences for different polyesters in the balance between the thermodynamic quantities. The presence of ring structures in the chains decreases ΔS_f, and that of strongly polar groups increases ΔH_f, both tending to raise T_m. Conversely, in-chain groups of low polarity, eg, ether links, increase ΔS_f and hence lower the melting temperature relative to their methylene analogues.

Commercially Important Terephthalate Polyesters

The merits of polyester fibers are their relatively high melting and glass-transition temperatures, insensitivity to moisture and common solvents, and the wide range of mechanical properties attainable by variations of molecular weight, orientation, and crystallinity, giving characteristics suitable at one extreme for uses in apparel (eg, in blends with wool or cellulosic fibers), curtains, upholstery, and fiberfill, and at the other, for industrial applications such as sewing threads, tire cords, and filter fabrics (see POLYESTERS, FIBERS). The primary mechanical properties of PET fibers are within the ranges shown in Table 2.

Table 2. Mechanical Properties of PET Fibers

Property	Textile-use fibers	High tenacity types
tenacity, dN/tex[a]	3.7–5.0	7.5–8.2
elongation at break, %	24–45	8–11
tensile modulus, dN/tex[a]	35–60	83–110

[a] To convert dN/tex to g/den, multiply by 1.133.

Although PET is the principal polyester fiber material, fibers of poly(1,4-cyclohexanedimethylene terephthalate) are produced by Eastman Chemical Products, Inc. as the Kodel 200 series and Kodaire tubular cross-section fiberfill, and PBT fibers are being introduced by Celanese Fibers in the United States and by the Teijin, Toray, and Unitika concerns in Japan.

Polyester films are manufactured from PET by melt extrusion and processing into many different grades and clarities, with gauges ranging from 6 to 350 μm and with various extents of uniaxial or biaxial orientation (see POLYESTERS, FILMS). Major brand names include Mylar (DuPont), Melinex (ICI), Hostaphan (Kalle-Hoechst), and Celanar (Celanese Plastics).

The range of uses is large, encompassing photographic and drafting-film applications; bases for magnetic, video, and computer tapes; electrical insulation and membrane switches; surfacing for GRP sheet; packaging (especially in metallized and printable forms, which may themselves be laminated with poly(vinylidene chloride) or polyolefins); "bag-in-box" wine packs; and decorative products. The outstanding barrier properties of PET find particular application in the higher molecular weight grades ($\overline{M}_n$ up to 26,000) sold increasingly since the mid-1970s under such names as Melinar PET (ICI), Polyclear (Hoechst), and Kodapak (Eastman Chemical) for injection stretch blow molding to form high performance bottles for carbonated soft drinks, wines, beers, and spirits, as well as for a miscellany of industrial liquid products (see BARRIER POLYMERS; RIGID CONTAINERS in PACKAGING MATERIALS).

Consumption of engineering plastics grades of PET and PBT in the United States has increased from ca 24,000 t in 1978 to 47,000 t in 1985. The materials, which have high strength and rigidity coupled with good surface hardness and gloss, find application in precision moldings for electrical and electronic devices, domestic and office appliances, and automotive parts. A comparison between the properties of crystalline PET and PBT moldings, unfilled and reinforced, is given in Table 3.

Structure. The chemical structures of PET and PBT are not described completely by the linear formula (**1**), where $x = 2$ and 4. When prepared by melt polymerization, each polymer contains as natural constituents ~1.5% of a family of cyclic alkylene terephthalate oligomers (**6**), with values of n up to 7–9, although composed predominantly of the cyclic trimer in the case of PET and of about equal amounts of cyclic dimer and trimer in PBT.

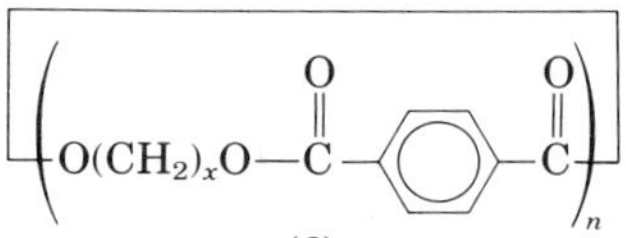

(**6**)

Transitions. The glass-transition temperatures of both polymers depend on the degree of crystallinity and on the technique of measurement. The values reported for PET are in the range of 70–120°C, and for PBT from 22 to 80°C.

Conformation. Characterization of the conformations in the amorphous and crystalline states and their interconversions have been made by ir and other forms of vibrational spectroscopy.

Thermal Degradation. The response of the polyesters to thermal, chemical, and photochemical influences is important for many aspects of their preparation, fabrication, and use. In crystalline form and at ambient or moderately elevated temperatures, PET and PBT are generally resistant toward dilute aqueous mineral acids, nonbasic salts, and many common organic compounds. However, they are prone to attack by oxidizing reagents and particularly by aqueous alkalies. Because of low moisture absorbency, the polyesters are essentially unaffected by water at ordinary temperatures, but at higher temperatures water or humid atmospheres hydrolyze the ester groups to OH and COOH, the latter being autocatalytic for further hydrolysis.

At high temperatures, eg, during the relatively long time periods involved in melt polymerization, PET and PBT undergo thermal degradation, inherently by the β-scission reaction of equation 1, but with differences of consequential detail in the two cases.

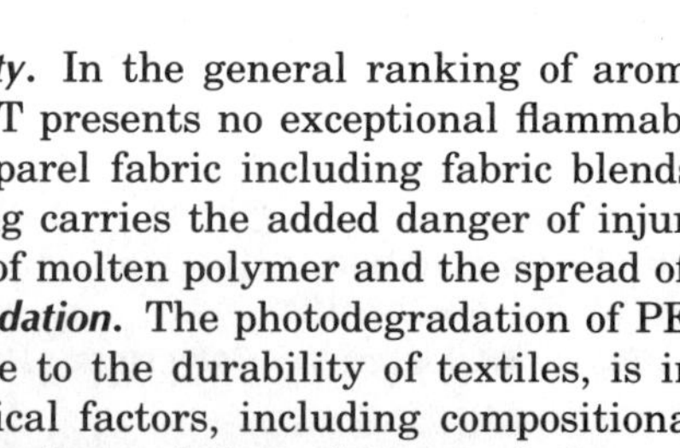 (1)

(**7**)

Flammability. In the general ranking of aromatic organic polymers, PET presents no exceptional flammability hazard, but as an apparel fabric including fabric blends with other fibers, burning carries the added danger of injury caused by the dripping of molten polymer and the spread of fire.

Photodegradation. The photodegradation of PET, of particular relevance to the durability of textiles, is influenced by several practical factors, including compositional deviations resulting from thermal degradation and the presence of diglycol or other copolymerized linkages, the types of catalysts and delustrants used in manufacture, and the spectral range and conditions of environmental exposure.

Synthesis

Many types of reactions can be used to synthesize linear polyesters, but the most important, conceptually and practically, are those where the molecular chain is built up by the formation of an ester linkage at each step. Approaches of more limited utility include those wherein monomers containing

Table 3. Properties of Commercial[a] Molding Plastics Grades of PET and PBT[b]

	Arnite A (PET)		Arnite T (PBT)	
Property	A04 900, unfilled	AV2 360S, 33% glass fiber	T06 200, unfilled	TV4 260S, 30% glass fiber
sp gr, d_4^{23}	1.375	1.73	1.30	1.69
Thermal property				
T_m, °C	255	255	223	223
specific heat at 40°C, kJ/(kg·K)[c]	1.2	0.9	1.35	1.15
thermal conductivity, W/(K·m)	0.29	0.33	0.21	0.24
coefficient of linear thermal expansion, K^{-1}				
direction of flow	7×10^{-5}	2×10^{-5}	7×10^{-5}	$4–5 \times 10^{-5}$
cross direction	7×10^{-5}	$6–7 \times 10^{-5}$	7×10^{-5}	$7–8 \times 10^{-5}$
deflection temperature under load (1.8 MN/m^2)[d], °C	80	230	65	205
Vicat softening point (10 N)[e], °C	235	235	210	215
Mechanical property				
tensile strength, MN/m^2 [d]	81	165	52	135
yield strength, MN/m^2 [d]	81		52	
elongation at yield, %	4		4	
elongation at break, %	70	2	200	2
tensile modulus, MN/m^2 [d]	2,800	11,500	2,600	10,200
flexural modulus, MN/m^2 [d]	3,000	11,500	2,500	9,900
Charpy notched impact strength, kJ/m^2 [f]	3	7.5	3.5	8
Rockwell hardness, scale M	85	102	80	87
Electrical property				
breakdown voltage[g], kV	130	95	>90	76
dielectric strength[g], MV/m	60	45	>45	47
volume resistivity, Ω·m	2×10^{14}	2×10^{14}	5×10^{14}	7×10^{14}
surface resistivity, Ω	6×10^{14}	8×10^{14}	5×10^{13}	3×10^{13}
dielectric constant ϵ_r				
at 10^3 Hz	3.3	3.9	3.0	3.7
at 10^6 Hz	3.3	3.8	2.8	3.6
dissipation factor tan δ				
at 10^3 Hz	0.007	0.005	0.002	0.002
at 10^6 Hz	0.021	0.015	0.017	0.011
flammability, UL 94	94 HB	94 V-O[h]	94 HB	94 V-O[i]
Mold shrinkage, %				
flow direction	1.5–1.6	0.10–0.15[j]	2.0	0.2–0.3
cross direction	1.7		2.0	1.5–1.6[j]

[a] AKZO products.
[b] All at ordinary temperatures unless otherwise indicated.
[c] To convert kJ to kcal, divide by 4.184.
[d] To convert MN/m^2 to psi, multiply by 145.
[e] To convert N to kgf, divide by 9.807.
[f] To convert kJ/m^2 to ftlbf/in.^{2}, divide by 2.10.
[g] 2–2.5-mm sheet.
[h] 1.5-mm thickness.
[i] ≥1.5-mm thickness.
[j] 2 mm-thickness.

preformed ester groups are polymerized by reactions other than esterification.

Syntheses based on ester-group formation are of two main types: step-polyesterification reactions of bifunctional reactants and ring-opening polymerizations of cyclic esters.

Manufacturing Processes for Terephthalate Polyesters

Ester-interchange Route. In the long-established batch procedure, a mixture of dimethyl terephthalate with 2.1–2.2 molar proportions of the diol is heated with a catalyst at temperatures rising from 150 to ca 210°C. Methanol is evolved as a vapor and is removed through a condenser system, the slight excess of diol facilitating its departure at this stage. The immediate product, essentially a bis(2-hydroxyalkyl) terephthalate, is passed to a polycondensation reactor operating at a higher temperature and under reduced pressure. The reactants must be kept molten so that reaction can proceed and the polymer can be eventually discharged from the vessel; the second-stage reaction is therefore carried out at ca 270–280°C (240°C in the case of PBT) with pressures of ca 66–133 Pa (0.5–1 mm Hg). The course of polymerization is followed by torque meters connected with the stirrers, which indicate increases in the viscosity of the melt. The reaction is interrupted when the viscosity reaches the desired value and the product is discharged under nitrogen pressure into a cooling system where it is solidified and then reduced to granules.

Direct Esterification Process. Direct esterification processes for PET manufacture developed from the finding that, in the presence of preformed ethylene terephthalate oligomers, terephthalic acid readily reacts with ethylene glycol at high temperatures and under autogenous pressures. Aside from the economic advantages of direct esterification and continuous operation, the process is claimed to give polyester of low carboxyl end group and low diglycol-linkage content, and to be free of the undesirable side effects incurred by the use of ester-exchange catalysts in the dimethyl terephthalate route.

Characterization

Because of the multiplicity of types and grades of polyester products, there are no uniformly applicable specifications or testing standards. Individual materials are mostly appraised on the basis of criteria related to the end uses.

Special Aspects

Polyester Block Copolymers. The distinctive features of block copolymers are due to the separation of their constituent segments into more or less distinct microphases that, to a considerable extent, contribute their individual characteristics to the behavior of the whole material, thus leading to permanent combinations of properties not displayed by the respective homopolymers alone or as blends. Such a combination occurs in random, multiblock, polyester-based thermoplastic elastomers and elastoplastics composed of hard blocks or segments of high T_m and/or T_g, which provide the resistance to viscous flow, alternating with soft blocks of low T_g (and, if crystallizable, low T_m), usually with molecular weights of 1,000–2,000.

Polyester Graft Copolymers. The modification of PET fibers with grafted chains from monomers such as acrylic acid and its esters, vinyl acetate, and vinyl acetate and vinylpyridines has been claimed variously to improve the dyeability, hydrophilicity, and soiling resistance. The grafting is achieved by treating the fibers with the monomers under the influence of free-radical initiators, γ-rays, or glow discharges. Specific effects can also be introduced, as in the grafting of poly(vinyl phosphonate) chains to confer flame retardancy and the after-treatment of grafted polyvinylpyridine with quaternizing agents to facilitate the dispersal of static electrical charges. In general, the extents of these modifications are limited to the superficial layers of the treated materials.

Polyester Blends. Although polymer blends may display morphological heterogeneity when one or more of the components is crystalline, their constituent species are usually deemed to be miscible at the molecular level, at least in certain ranges of composition. Blending mainly affects the viscoelastic properties of the materials and, broadly considered, affords a means of modifying and tailoring the mechanical and thermal characteristics.

Liquid Crystalline Polyesters. Although birefringence had been observed during the 1960s in melts and precrystallized samples of substituted poly(*p*-phenylene terephthalate)s, its significance emerged only a decade later with the finding that certain homo- and copolyesters formed mesomorphic (liquid crystalline) melts, the anisotropy of which could be retained when the substances crystallized on cooling. The key observations originated with α,α′-dimethylbenzalazine polyesters (**8**), where x = 6, 8, or 10 on the one hand and with ethylene terephthlate–4-oxybenzoyl copolyesters (**9**) on the other. These have served both directly and as archetypes for a prolific volume of subsequent work.

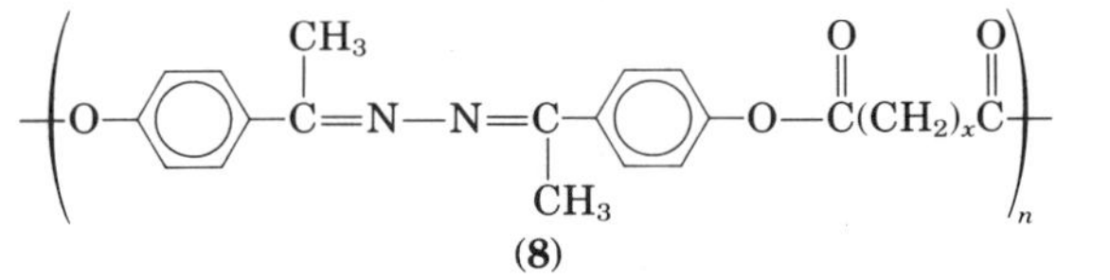

(**8**)

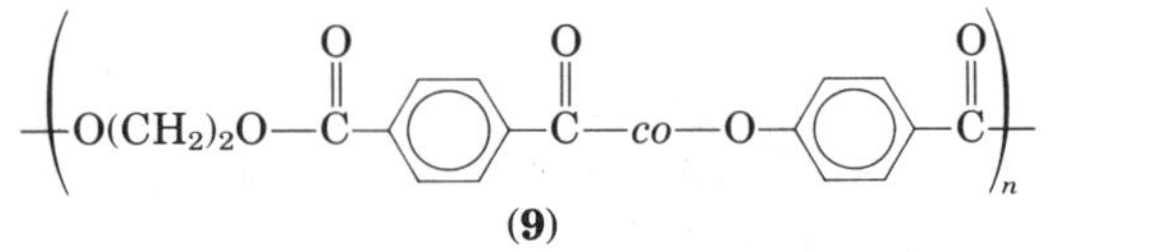

(**9**)

Bioabsorbable Polyesters. Recognition of the potential of polyesters as the basis of sutures of uniform properties, capable of *in vivo* resorption at a rate matching that of wound healing and free of the adverse features of the prior materials dates from the 1960s. Probably because of the relationship to naturally occurring substances, initial studies were of filaments made from polyesters of lactic acid, (**10**),

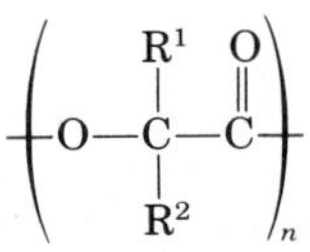

(**10**)

where R^1 = CH_3 and R^2 = H, but their absorption rate was unacceptably low. The desired characteristics were later found in sutures of polyglycolide, (**10**) where $R^1 = R^2$ = H, and its copolymers with ca 10% of lactyl units. As Dexon (American Cyanamid) and Polyglactin 910 and Vicryl XLG (Ethicon), these materials have found widespread acceptance for surgical use. The worldwide market had reached ca $30 × 10^6 by the late 1970s.

Photocross-linkable Polyesters. Polyesters such as (**11**) and (**12**) containing suitably activated sequences of conjugated double bonds undergo insolubilizing cross-linking reactions on exposure to actinic radiation. By tailoring the structural detail of the chromophores, sensitivity can be provided to chosen wavelengths in the uv or the visible spectral regions. Applications of this principle are probable in photoresists and laser imaging, and, in the case of (**13**), for the formation of network cryptate resins with selective binding power for cations.

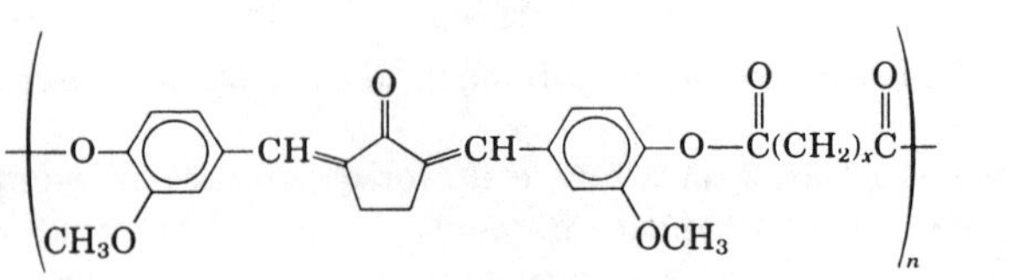

(**11**)

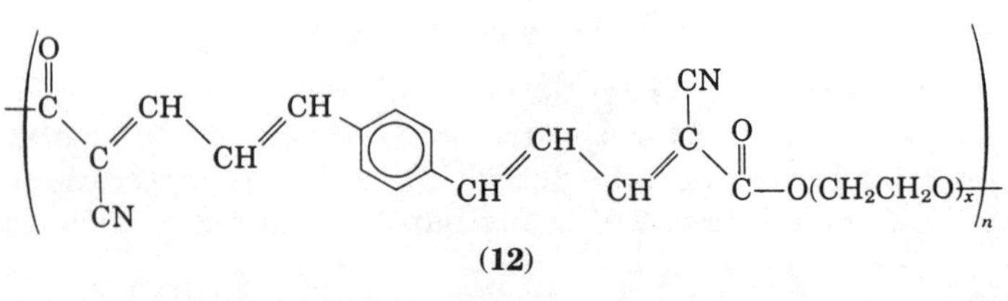

(**12**)

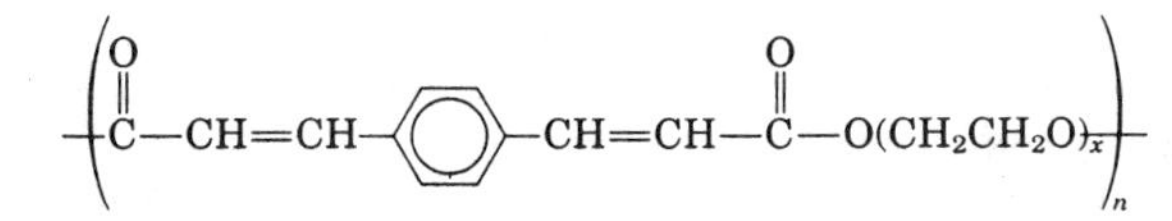

(13)

Polyesteramides. Modification of the properties of polyesters by incorporation of hydrogen-bonded amide groups has been extensively investigated. The materials are either random copolyesteramides in which the proportion of ester to amide groups can be varied through the whole range of compositions, or alternating or patterned polyesteramides with regularly recurring successions of the characteristic groups in the main chain.

BIBLIOGRAPHY

H. Mark and G. S. Whitby, eds., *Collected Papers of Wallace Hume Carothers on High Polymeric Substances,* Interscience Publishers, Inc., New York, 1940.

I. Goodman in R. N. Haward, ed., *Developments in Polymerization-2,* Applied Science Publishers, London, 1979, Chapt. 4.

I. Goodman, ed., *Developments in Block Copolymers-1,* Elsevier Applied Science Publishers, Ltd., Barking, UK, 1982.

Adv. Polym. Sci. **59,** Springer-Verlag, New York, 1984, Parts I–III.

ISAAC GOODMAN
University of Bradford, UK

POLYESTERS, ELASTOMERIC

Thermoplastic elastomers possess many of the characteristics of cured elastomers and yet can be processed as easily as glassy or semicrystalline plastics. Since conventional processing of elastomers is energy- and labor-intensive, elastomers that can be readily processed by techniques such as injection molding represent an important technological advance.

Elastomers are highly viscous fluids converted to useful articles by cross-links, which tie the array of macromolecules into an infinite network. These articles exhibit useful properties such as maintenance of shape at increasing temperature and elastic deformation on elongation or compression. In natural rubber and traditional synthetic elastomers, these cross-links are chemical bonds. In thermoplastic elastomers, these bonds are replaced by thermally labile tie points held together by physical forces. These junctions may be glassy, crystalline, or even hydrogen-bonded molecules or ionic associations. According to elastomer theory, to behave as a thermoplastic elastomer the molecules must contain two types of units, blocks, or chain segments: one is an amorphous or fluid segment above its glass-transition temperature T_g in the normally useful range of the elastomer. These units are referred to as soft or rubbery segments or blocks, since they impart elastomeric character to the copolymer. The other element, the hard block, is the portion capable of intermolecular association with other hard blocks. This block must be solid in the desired use range to convey dimensional stability to the array of molecules. At high temperatures, the physical bonds dissociate, allowing the polymer to soften and flow under shear as does a thermoplastic material. To ensure formation of a three-dimensional network, each macromolecule must, on the average, contain at least two such hard blocks. These hard and soft blocks may be arranged in various ways, randomized or ordered (see also BLOCK COPOLYMERS; ELASTOMERS, THERMOPLASTIC; GRAFT COPOLYMERS). The manner in which they are ordered affects the physical and mechanical properties of the material.

Development of Copolyetheresters

High molecular weight polyesters derived from terephthalic acid (T) and ethylene glycol are of great technical significance as fibers and films. By 1950, up to 30 wt % of poly(ethylene oxide glycol)s had been incorporated in these fibers to improve dyeability, and some elastic properties had been observed along with a significant drop in T_g. Elastic properties were also observed in other segmented copolyesters.

When poly(butylene terephthalate) (4GT) was employed as hard segment in a copolyester, the product crystallized or hardened so quickly from the melt that it appeared to meet the goal of an easily processible thermoplastic elastomer. This same property has led to the widespread use of 4GT, also called PBT, as an engineering plastic.

These new segmented copolyesters exhibited an extraordinary combination of elasticity, toughness, low temperature flexibility, and strength at 150°C. They were quickly commercialized under the trademark Hytrel polyester elastomer (DuPont) in 1972 and with related products are now of great commercial importance as engineering-type thermoplastic elastomers.

Polymer Synthesis

Polyetherester block copolymers are usually prepared by a two-stage melt transesterification involving an aromatic dicarboxylic acid or preferably its dimethyl ester, a low molecular weight poly(alkylene glycol ether) and a short chain diol. The starting materials are commercially available: terephthalic acid and its isomers, and the poly(alkylene oxide glycol)s from ethylene oxide (PO2), propylene oxide (PO3), tetramethylene oxide (PO4), and their copolymers. The most important short-chain diols are ethylene glycol (2G) and tetramethylene glycol (4G); butene diol and hexane diol are also used. The resulting copolyester structures are

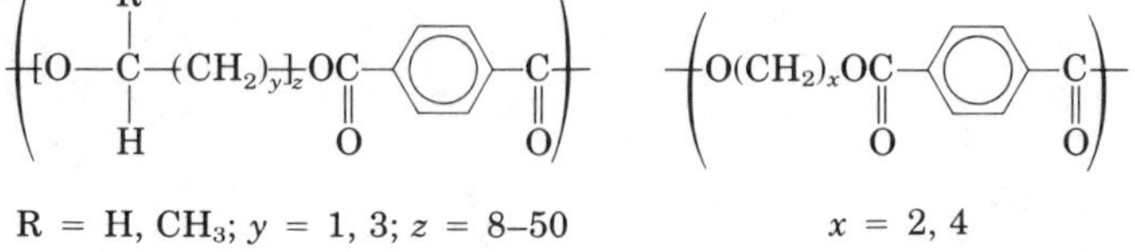

R = H, CH_3; y = 1, 3; z = 8–50 — soft

x = 2, 4 — hard

Molecular Weight and Block-length Distributions

The polymer, prepared as described above, may be considered a condensation polymer formed by randomly joining, head-to-tail, poly(oxytetramethylene terephthalate) (PO4T) soft segments and poly(tetramethylene terephthalate) (4GT) hard segments:

$$-(4GT)_x\text{–PO4T–}(4GT)_y\text{–PO4T–}(4GT)_z\text{–PO4T}-$$

hard soft hard soft hard soft

Gel-permeation chromatography (gpc) shows that the polymer conforms to the geometric molecular weight distribution expected from such an equilibrium condensation reaction. Figure 1 shows the gpc curve for a polymer composed of 58 wt % tetramethylene terephthalate units and 42 wt % poly(oxytetramethylene glycol terephthalate) units. This material has an inherent viscosity of 0.15 L/g, corresponding to an M_n of ca 25,000.

Manufacture

A commercial manufacturing process can be carried out batchwise or in continuous-flow units. The same equipment is used as in the manufacture of polyethylene and poly(butylene terephthalate)s for fibers, plastics, and film. The reaction system consists of three heated feed vessels for the starting materials, an agitated ester-interchange reactor fitted for a short column and a condenser for removal of methanol, and an evacuated, agitated polycondensation reactor for removal of excess tetramethylene glycol.

Properties

Depending on the concentration of the hard 4GT segments, the products range from hazy to translucent solids, rubbery to elastoplastic in character. They exhibit unusual mechanical properties and microphase structure (Table 1).

Morphology. The polymers are of two-phase structure with crystalline and amorphous domains. The T_g is between −70 and 20°C, characteristic of the amorphous phase; the melt transition T_m of the crystalline tetramethylene terephthalate (4GT) phase is in the region up to ca 220°C. Both T_g and T_m are functions of the hard-segment content.

Although it is clear that the copolyesters exhibit two transitions, T_g and T_m, the morphological details have been the subject of speculation, particularly because there are many possibilities. These morphologies arise from changes in the ratios and molecular weights of hard and soft segments and

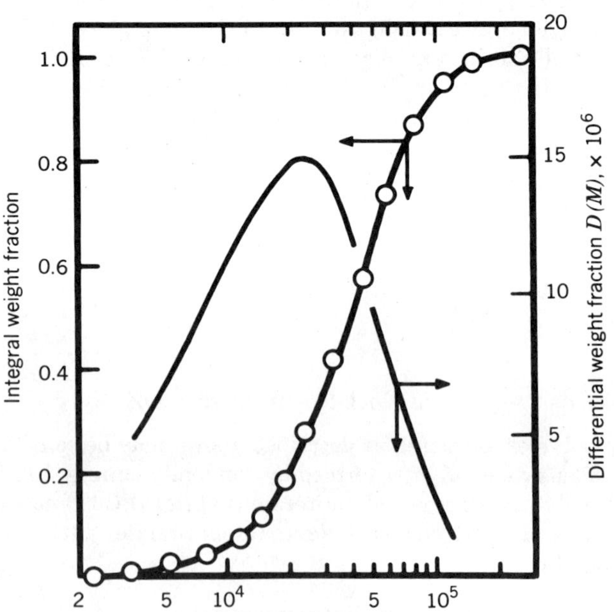

Figure 1. Molecular weight distribution of polymer containing 58 wt % 4GT; inherent viscosity = 0.15 L/g.

from variations in sample preparation and mechanical and thermal history. A lamellar-type morphology exists in which crystalline lamellae form a virtually continuous network throughout the amorphous phase. Under some conditions, other structures such as dendrites or spherulites are developed, but under other conditions they are not observed.

Effect of Molecular Weight. As might be expected, physical properties such as modulus and tensile and tear strengths strongly depend on molecular weight. These materials have the same percentage of crystalline material and therefore similar moduli at low elongations. As the elongation increases to the point where the crystalline network is disrupted, the moduli begin to differ, increasing with increasing molecular weight. The material is responding as an elastomer, and tensile strength also increases with molecular weight. The same dependence on molecular weight is found for tear strength.

Effect of Hard-segment Concentration. The higher the tetramethylene terephthalate content of the polymer, the higher the modulus and tear strength. Polymers containing more than 30% hard segments are tough and retain some strength at elevated temperatures (retention of strength increases with hard segment concentration).

Mechanical Properties. Copolyesters can be designed with a wide range of tensile and mechanical properties. This is particularly useful in engineering-type applications, where functional performance depends on the reliability of properties. To illustrate this point, four commercial Hytrel polymers, based on poly(tetramethylene oxide) (mol wt 1000) and tetramethylene terephthalate, are shown in Table 2.

Other characteristics of the copolyesters include utility over a wide temperature range (−40 to 180°C), excellent fatigue resistance, high impact strength, superior wear (abrasion) and fluid resistance, and good electrical properties at low temperatures.

Oxidative and Environmental Stability. The stability of copolyetheresters to light, heat, and oxidation is inferior to that of polyethylene and poly(tetramethylene terephthalate). Because of the presence of the poly(alkylene oxide) chains, they are subject to photo- and thermal free-radical-promoted oxidation. They are stabilized by hindered polyphenols and secondary amines at ca 1% (see also ANTIOXIDANTS).

Copolyesters with Different Hard and Soft Segments

Hard Segments. The hard or ester segments influence ease of synthesis, molecular weight, crystallization rate, elastic recovery, and tensile and tear strengths. The 4G-based polymer with 50% hard segment is outstanding in ease of synthesis, rapid crystallization from the melt, high melting point, and tensile and tear strengths. The 2G-based polymer, although higher melting, crystallizes slowly from the melt.

The crystallization rates of these polymers are extremely important, particularly in high speed processes such as injection molding. Tetramethylene terephthalate-based thermoplastic elastomers crystallize rapidly, whereas polymers based on hard segments such as ethylene terephthalate crystallize slowly.

Incorporation of 1,4-cyclohexanedimethanol (CHDM) in an elastomeric copolyester fiber results in a high melting point and good crystallization characteristics. The higher melting point of CHDM terephthalate enables the production of copolymers with acceptable melting points in the presence of

Table 1. Mechanical Properties[a]

Property	ASTM method	Short-chain ester in copolyester, %[b]			
		50	60	75	80
hardness, Shore D	D2240	40	55	63	72
specific gravity	D792	1.17	1.20	1.22	1.25
melting point, °C	D3418				
peak of endotherm		145	201	201	217
melt complete		170	220	214	232
tensile strength, MPa[c]	D638	30	44	44	52
ultimate elongation, %	D638	560	560	440	420
flexural modulus, MPa[c]	D790	48	207	345	517
resilience, Bashore, %		62	53	43	na[d]
compression set, 22 h at 70°C, constant, at 9.3 MPa[c], %	D395A	27	4	2	2
tear strength, kN/m[e]	D624				
die B		110	164	185	238
shear strength, MPa[c]	D732	24	37	48	52

[a] Products based on poly(tetramethylene oxide), mol wt = 1000, and poly(tetramethylene terephthalate).
[b] Approximate.
[c] To convert MPa to psi, multiply by 145.
[d] na = not available.
[e] To convert kN/m to lbf/in., divide by 0.175.

greater amounts of soft segments. These properties have led to the use of CHDM in the copolyester elastomer Ecdel (Eastman Kodak Co.).

Changes in the structure of the acid in the hard segment have more pronounced effects. All of the variations show that the properties desired in a high strength elastomer are influenced by other factors in addition to the high melting point of the hard segment.

Soft Segments. The hard segment influences primarily mechanical and processing properties, whereas the soft segment affects physical and chemical properties. Resistance to heat, light, oxygen, and water, as well as low temperature properties and tear strength, are strongly dependent on the chemical structure, chain length, and concentration of the soft segment; melting point, hardness, and modulus are less dependent.

Swelling in oil and water is essentially the same for the hydrophobic PO4- and PO3-based materials, whereas the hydrophilic nature of the PO2-based polymer results in a large increase in water swell and large decrease in oil swell. Other properties may be affected by a change in composition.

Polyester Soft Segments. Copolyesters with ester soft segments are used less than those with polyethers, but have a significant effect on some important properties.

Copolyesters are products with randomly alternating hard and soft segments, which can involve either single or segmental esters. The lower molecular weight soft segment from dimer acid (ca 565 vs 1000–2000 for a polyether glycol) leads to block structure and phase separation, which are less pronounced

Table 2. Hytrel Copolyesters

Short-chain ester concentration, %	Hardness, Shore D	T_m, °C
40	40	170
60	55	210
75	63	213
80	72	218

because the molar concentration of soft segment is increased. Thus to achieve the same hardness (55D) with dimer acid, a higher 4GT concentration is needed, eg, 64% vs 58% in the copolyetherester.

In contrast to the above simple ester-based products, segmented copolyesters with polyester soft segments are not easily prepared by polycondensation. Complete randomization of the esters results in short hard segments and a loss of mechanical properties. To avoid randomization, the preformed ester blocks are coupled by kinetically controlled ester interchange between hard and soft polyester blocks or by chemical coupling of hard and soft polyester segments.

Properties of copolyetheresters and copolyesters of comparable hardness are given in Table 3.

Processing of Copolyesters

Copolyesters are true thermoplastic polymers, with sharp melting points, viscosities more sensitive to temperature change than many common plastics, and rapid crystallization except in the softer grades. The apparent melt viscosity decreases only to a limited extent with shear rate, especially at low shear rates (10^{-1} to 10^2) where the melt behaves like a Newtonian fluid with excellent flow properties. For this reason, copolyesters are useful in low shear applications, such as calendering, melt casting, rotational molding, and impregnation of porous substrates. At high shear rates, characteristic of injection molding and extrusion, the effect of shear rate on viscosity is more pronounced. Increase in temperature lowers the viscosity: this effect increases with hard-segment content and crystallinity.

Copolyesters are processed by most conventional techniques, including extrusion, injection molding, blow molding, rotomolding, thermoforming, melt casting, and machining, as well as powder-spray coating. Good melt stabilities and rapid crystallization rates provide excellent processing properties and versatility. The sharp melting points increase with increasing hardness and crystallinity. In view of their polar character,

Table 3. Properties[a] of Copolyetheresters and Copolyesters

Property	A[b]	B[c]	C[d]	D[e]
melting point T_m, °C	202	195	216	211
glass-transition temperature T_g, °C	−53	−34		−33
melt index, g/10 min	$5.8^{220°C}$	$5.4^{220°C}$	$6.8^{240°C}$	$7.3^{240°C}$
hardness, Shore D	55	56	62	63
tensile strength, MPa[f]	44.1	41.4	40.8	47.6
elongation at break, %	650	620	300–600	510
M_5[g] at 25°C, MPa[f]	6.45	6.48	12	12.2
brittle point, °C	−70	−70	−64	−70
torsional modulus, MPa[f] at 25°C			143	116

[a] Measured by standard ASTM methods.
[b] Copolyetherester, 58% 4GT.
[c] Copolyester from dimer acid, 64% 4GT.
[d] Copolyester from caprolactone, Pelprene S 3000, 75% 4GT.
[e] Copolyetherester, 75% 4GT.
[f] To convert MPa to psi, multiply by 145.
[g] The M notations are moduli at elongations of 5, 10, 100, and 300%, as shown.

copolyesters are slightly hygroscopic. They are stable to hydrolysis under ordinary conditions, but moisture content should be below 0.1%.

Health and Safety Factors

Polyester elastomers are closely related to fibrous and plastic polyesters, which are not hazardous under ordinary conditions and which decompose slowly under processing conditions.

In handling these products, especially dusts and fines, normal precautions to prevent static electricity and spills should be used. Compounding with acid materials should be avoided since they promote decomposition.

During processing, workers should be protected from burns, "blow-back," fire, and inhalation of vapors.

Applications

The broad spectrum of properties offered by these products has led to their acceptance in a wide variety of applications, replacing rubbers and plastics. Although the cost on a weight basis is higher, performance is superior and processing less expensive. Excellent low temperature flexibility, abrasion resistance, and flex-fatigue characteristics make them well suited for such uses as low pressure tires for off-the-road vehicles, such as lawn mowers, golf carts, garden tractors, and snowmobiles. Oil and chemical resistance, together with flexibility and abrasion resistance, make them a logical choice for cable jacketing and hose applications, both as tube and cover. They have replaced rubbers in "boots" for front-wheel-drive cars and steering "boots," where high flexibility is required. Various types of powdered Hytrel are used as protective coatings on metal parts of all sizes and complexities; these may be applied by fluidized-bed coating, electrostatic deposition, and flame and plasma spray. These polymers are also widely used with or without tensile-reinforcing cords as power-transmission belts and roll covers, and in a wide range of molded goods, such as diaphragms, gaskets, seals, O-rings, plugs, and pads. In these uses, they have far surpassed many conventional plastics and rubbers in properties or value-in-use.

Economic Aspects and Products

Many varieties of copolyester elastomers are produced in the United States, Europe, and Japan. The total capacity is difficult to estimate because the manufacturers have not released specific figures. Production capacity of copolyester manufacturers is probably close to 30×10^3–45×10^3 metric tons, with DuPont the principal producer.

BIBLIOGRAPHY

G. K. Hoeschele and W. K. Witsiepe, *Angew. Makromol. Chem.* **29/30,** 267 (1973).

B. M. Walker, *Handbook of Thermoplastic Elastomers,* Van Nostrand Reinhold Co., Inc., New York, 1979, pp. 103–215.

I. Goodman, ed., *Developments in Block Copolymers,* Applied Science Publishers, Ltd., London, 1982.

N. R. Legge, G. Holden, and H. Schroeder, *Thermoplastic Elastomers, Research and Development,* Carl Hanser Verlag, Munich, 1987; see especially Chapt. 8.

HERMAN SCHROEDER
Schroeder Scientific Services, Inc.
RICHARD J. CELLA
E.I. du Pont de Nemours & Co., Inc.

POLYESTERS, FIBERS

A polyester fiber is "a manufactured fiber in which the fiber-forming substance is any long chain synthetic polymer composed of at least 85 wt % of an ester of a dihydric alcohol (HOROH) and terephthalic acid (p-$HOOCC_6H_4COOH$)." The most widely used polyester fiber is made from the linear polymer poly(ethylene terephthalate) (PET) (**1**), and therefore polyester has become synonymous with PET.

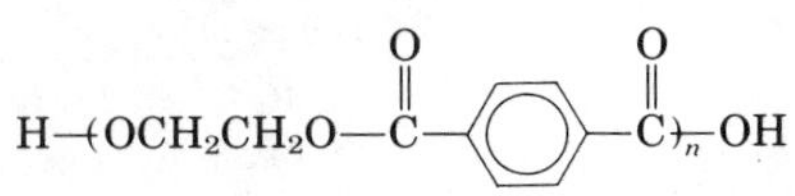

(**1**)

Polyester fibers first became commercially available in the United States in 1953. Worldwide and U.S. production expanded enormously in the 1960s and 1970s; worldwide production reached 6×10^6 metric tons in 1984. Although the rate of increase has fallen slightly, the manufacture of PET fiber continues to be the fastest-growing industry worldwide. This trend is projected to continue as polyester fiber manufacture is selected as an industry by developing countries for future growth. This enormous growth is attributed to the great versatility of polyester and its importance to the world's economy. The United States remains the largest producer and consumer.

A thermoplastic polyester can be selectively engineered at all the basic processing steps: polymerization, fiber formation, and fabrication, which allows the manufacture of a broad range of fiber products for the textile-apparel, home-furnishing, industrial, tire, and carpet markets. Staple blends with cotton for permanent-press garments comprise the largest single use. In the early 1970s, textile-filament yarn for apparel use grew very rapidly and, in addition, polyester became the dominant tire reinforcement in the United States. Other areas such as fiberfill and nonwovens continue to emerge and grow. High strength, high modulus, low shrinkage, heat-set ability, lightfastness, and chemical resistance account for the great versatility of PET (see also FIBERS).

Properties

Chemical Composition. Although many derivatives have been investigated, the linear homopolymer PET, which is a condensation polymer of terephthalic acid (TA) (benzene-1,4-dicarboxylic acid) or its dimethyl ester (DMT), and ethylene glycol (EG), remains the dominant composition for commercial polyester fibers.

Number-average molecular weights of ca 15,000 for PET are required for useful textile-fiber properties. Lower molecular weights give staple fibers of low tendency to pill (unsightly small "fuzz balls" on the fiber surface), whereas higher molecular weights give high strength fibers for industrial applications. Most apparel products contain a delustrant in varying amounts (usually TiO_2 in quantities up to 2%) to provide whiteness and reduce shine. Specialty chemicals, such as optical brightners and colored pigments, are frequently added.

Some commercial PET fibers contain small amounts of comonomers to confer specific properties. Adipic acid, isophthalic acid, and diethylene glycol enhance dyeability with disperse dyes by increasing the amorphous content of the fiber to allow easier access for the dye. Sites for adhesion of ionic dyes, such as cationic dyes, are created by adding the sodium salt of 5-sulfoisophthalic acid. Flame retardancy can be imparted by adding bromine or phosphorous compounds, which have combustion-quenching effects. Hygroscopic properties may be obtained by adding high water-absorptive materials such as poly(ethylnene glycol) (PEG) and organic sulfonates. Antistatic properties are improved by the addition of charge dissipating materials such as PEG, carbon, and metal. Pill resistance can be achieved by cross-linking agents, which reduce tensile properties. However, the addition of comonomers tends to reduce fiber strength and thermal stability, which is undesirable in most cases.

Structural Composition. Polyester fibers may be considered to be composed of crystalline, oriented semicrystalline, and noncrystalline (amorphous) regions. The structural unit (unit cell) for PET has been determined to be triclinic by x-ray diffraction techniques; it contains one repeating unit. Percentage crystallinity and molecular orientation relate to tensile strength and shrinkage.

Relation of Structure, Property, and Processing. Properties of polyester fibers are strongly affected by fiber structure. An oriented semicrystalline structure is required to impart the desired physical properties. Fiber structure, which strongly influences the applicability of polyester fibers, depends heavily on the processing parameters in the fiber-formation steps. The structure is determined by the spinning speed (threadline-stress), hot-drawing (stretching), stress-relaxation, and heat-setting (stabilization) processes employed.

Mechanical Properties. Polyester fibers have a number of desirable physical properties, which depend on the method of manufacture and the molecular weight of the polymer. As the degree of stretch is increased (yielding higher crystallinity and molecular orientation), so are properties such as tensile strength and the initial Young modulus. At the same time ultimate extensibility, elongation, is usually reduced. An increase in molecular weight further increases tensile strength, modulus, and elongation. Typical physical and mechanical properties of PET fibers are given in Table 1, and stress–strain curves in Figure 1.

Chemical Properties. Polyesters have good resistance to weak mineral acids, even at boiling temperatures, and to most strong acids at room temperature, but are dissolved with partial decomposition by concentrated sulfuric acid. Hydrolysis is highly dependent on temperature.

Strong alkalies, such as caustic soda, etch the fiber surface and reduce the polyester's strength. Ammonia and other organic bases, such as methylamine, initially penetrate the structure through the noncrystalline regions, causing degradation of the ester linkages and a loss in physical properties.

Polyesters display excellent resistance to oxidizing agents, such as conventional textile bleaches, and are resistant to cleaning solvents and surfactants. PET is insoluble in most solvents except for some polyhalogenated acetic acids and phenols. Polyester fibers have a low moisture regain of ca 0.4%.

Optical Properties. PET has the clear, shiny, optical properties characteristic of many thermoplastics, providing a bright, sparkle effect that is desirable for some end uses, such as silk-like apparel. Where this effect is not desired, particulate materials are added to obtain the desired optical properties.

Thermal Properties. The thermal properties of PET fibers depend on the method of manufacture. The dta and tma for fibers spun at different speeds show peaks corresponding to the glass-transition, crystallization, and melting regions, depending on the amorphous and crystalline content. The curves for 6000 m/min and above are characteristic of drawn-fiber curves. The glass-transition range is usually around 75°C; crystallization and melting ranges are around 130 and 260°C, respectively.

Thermal degradation of PET proceeds by a molecular mechanism with random chain scission at the ester links, although a radical mechanism has also been proposed.

Dyeing. Because of its rigid structure and lack of reactive dyesites, PET absorbs very little dye in conventional dye systems. This is particularly true for the highly crystalline (highly drawn), high tenacity, high modulus fibers. Polyester fibers are

Table 1. Physical Properties of Polyester Fibers

	Filament yarn		Staple and tow	
Property	Regular tenacity[a]	High tenacity[b]	Regular tenacity[c]	High tenacity[d]
breaking tenacity[e], N/tex[f]	0.35–0.5	0.62–0.85	0.35–0.47	0.48–0.61
breaking elongation, %	24–50	10–20	35–65	17–40
elastic recovery at 5% elongation, %	88–93	90	75–85	75–85
initial modulus, N/tex[f]	6.6–8.8	10.2–10.6	2.2–3.5	4.0–4.9
specific gravity	1.38	1.39	1.38	1.38
moisture regain[g], %	0.4	0.4	0.4	0.4
melting temperature, °C	258–263	258–263	258–263	258–263

[a] Textile-filament yarns for woven and knit fabrics.
[b] Tire cord and high strength, high modulus industrial yarns.
[c] Regular staple for 100% polyester fabrics, carpet yarn, fiberfill, and blends with cellulosic blends or wool.
[d] High strength, high modulus staple for industrial applications, sewing thread, and cellulosic blends.
[e] Standard measurements are conducted in air at 65% rh and 22°C.
[f] To convert N/tex to g/den, multiply by 11.33.
[g] The equilibrium moisture content of the fibers at 21°C and 65% rh.

therefore dyed almost exclusively with disperse dyes, which act primarily through controlled dispersion with only weak hydrogen bonding of the dye molecule to the fiber.

Other Properties. Polyester fibers display good lightfastness; long-term degradation appears to be initiated by uv radiation.

Although PET is flammable, the fabric usually melts and drops away instead of spreading the flame. However, in blends containing cotton, which supports combustion, PET fibers burn.

Polyester has good oxidative and thermal resistance. The resistance to mildew, aging, and abrasion generally is excellent.

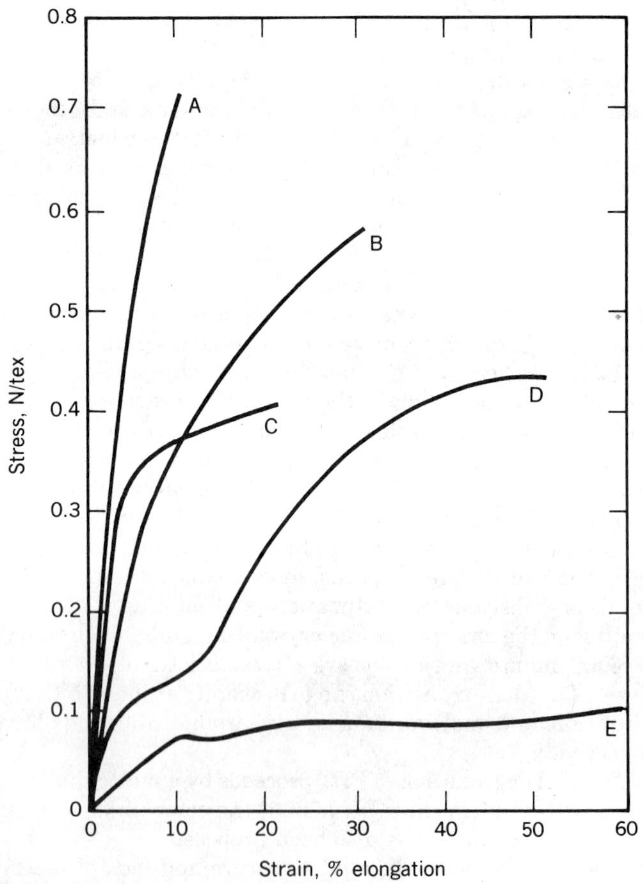

Figure 1. Typical stress–strain curves for polyester fibers: A, high tenacity filament; B, high tenacity staple; C, regular tenacity filament; D, regular tenacity staple; E, POY filament. To convert N/tex to g/den, multiply by 11.33.

Polymer Manufacture

Terephthalic acid (TA) or its dimethyl ester (DMT) and ethylene glycol form a diester monomer that is polymerized to PET. The polymer is melted and extruded (spun) through a spinneret, forming filaments that are solidified by cooling in a current of air. The spun fiber is drawn by heating and stretching the filaments to several times their original length, forming an oriented semicrystalline structure and imparting the desired physical properties. The fiber structure has a strong influence on the applicability of the fibers and depends heavily on the process parameters (Fig. 2).

Polyester polymer is produced commercially in two steps: monomer formation by ester interchange of DMT with glycol or esterification of TA with glycol, followed by polycondensation by removing excess glycol to promote chain extension. This is accomplished commercially on a simple two-vessel batch process or on larger-scale multi-vessel continuous-polymerization systems (Fig. 3).

Spinning Processes. Polyester is thermoplastic and may be converted to fibers by melt spinning from chip (Fig. 4) or directly from the continuous polymerizer. Properties and parameters important in the process are shown in Table 2. Melt-spinning allows great versatility in control of fiber diameter, structure, and cross-sectional shape.

Rheology. In extrusion through the spinneret, shearing flow in the capillary hole is the dominant type of fluid deformation in PET processing. On leaving the spinneret capillary, the filaments exhibit some degree of extrudate swell associated with the relaxation of molecular orientation that develops in the viscoelastic melt as a result of shearing. In the region just below the spinneret, a transition occurs from the shearing flow in the spinneret to the extensional flow field of the spinning threadline.

Determining the ability of fibers to form threads from polymer solutions and melts is a fundamental problem. Fibers are

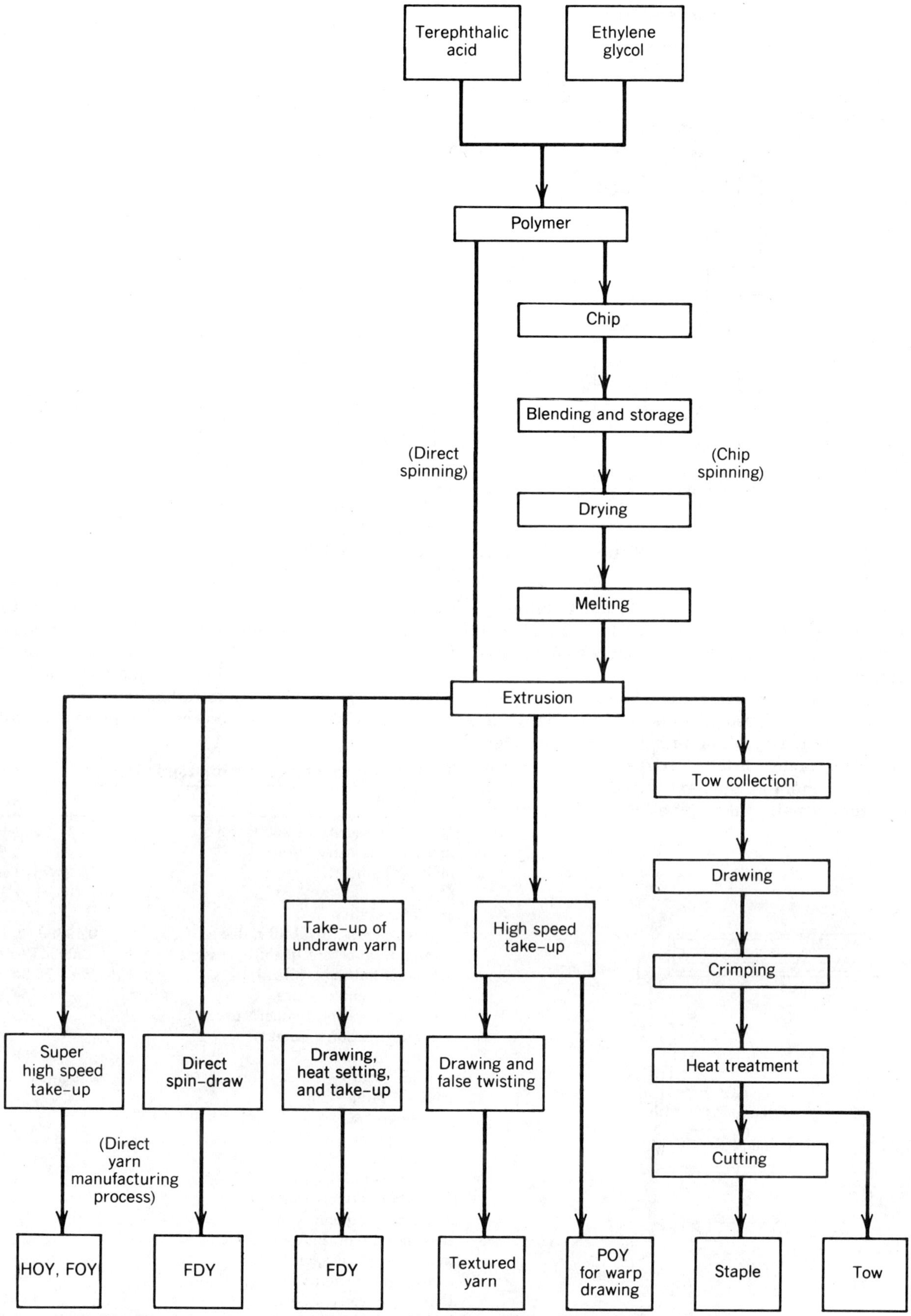

Figure 2. Polyester fiber manufacturing processes. HOY = highly oriented spun yarn; FOY = fully oriented spun yarn; POY = partially oriented spun yarn; FDY = fully drawn yarn.

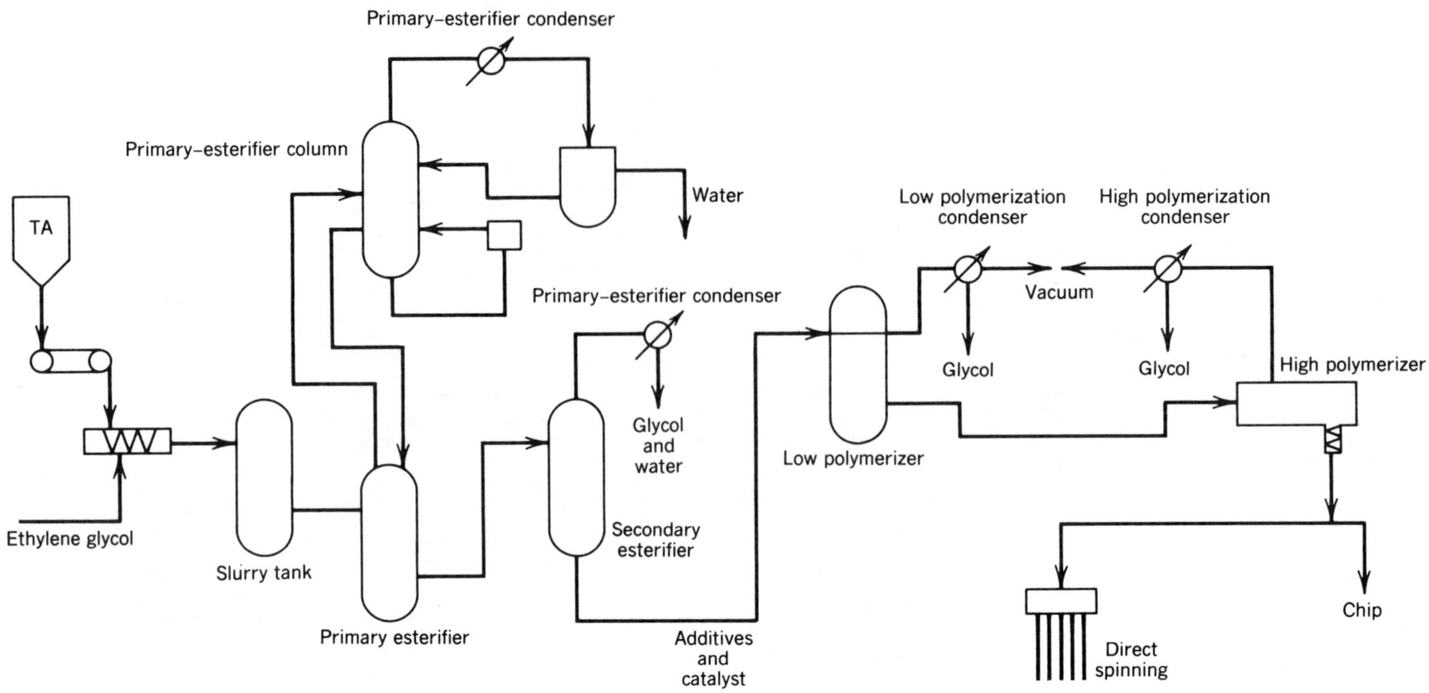

Figure 3. Continuous polymerization of PET.

difficult to form from any polymer melt or solution. Capillarity (surface-tension-related disturbances), ductile failure, and fracture are important. The conditions under which steady-state melt spinning can exist are limited.

Threadline Dynamics and Structure. The molecular orientation that develops in spun polyester fiber is a function of threadline stress and is affected by all spinning-process variables. The first group of variables includes extrusion temperature, dimensions and number of the spinneret orifices, mass outflow, length of the spinning path, take-up velocity, cooling conditions, and the composition, structure, and physical behavior of the material. The second group includes extrusion velocity, filament diameter, and the take-up:extrusion velocity ratio. The third group includes the resulting variables that are determined by the primary variables and the kinematics and dynamics of the spinning process. These include take-up ten-

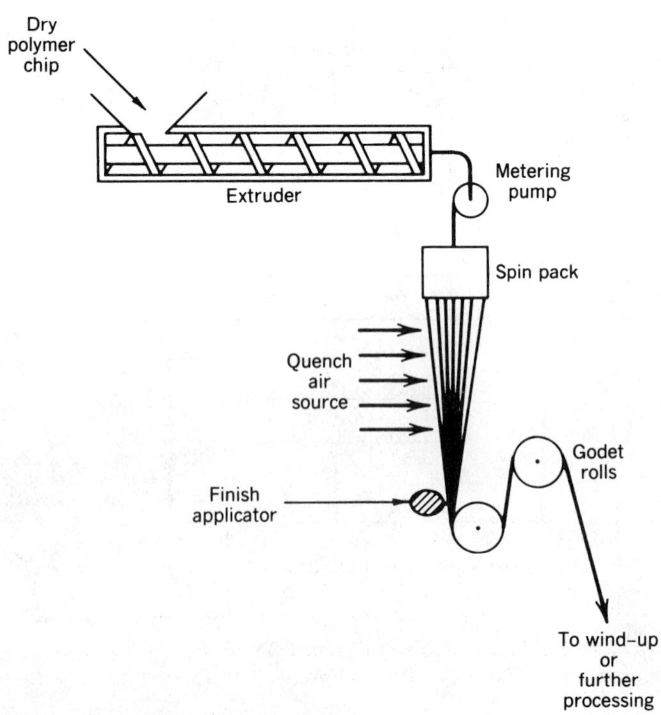

Figure 4. Melt-spinning process for PET.

Table 2. Parameters for Spinning PET

Parameter	Value
absolute viscosity at 280°C, Pa·s[a]	265 (±2.5)
intrinsic viscosity, dL/g	0.629
melting point, °C	254 (±0.5)
$\bar{M}_w/\bar{M}_n$	2.1
crystallizability	
kinetic crystallizability, deg/s	0.46399 (±6.6 × 10^{-3})
maximum crystallizability rate, s^{-1}	0.00947 (±4 × 10^{-5})
time of half-crystallization, s	73.2 (±0.2)
melting temperature, °C	254
glass-transition temperature, °C	66
crystallization rate at °C[b]	
maximum	205
half-width	49, 168, 217
melt spinning	
spinneret	
temperature, °C[c]	286
dia, mm[d]	0.5
orifice height, mm	4
number of orifices[e]	24
spinning path, cm	180
temperature of the cooling medium, °C	20
mass output rate, cm^3/min	0.249–2.184
average extrusion velocity, m/min	1.269–11.131
take-up velocity, m/min	400–3500

[a] 1 Pa·s = 10 P.
[b] Peaks of thermomechanical analysis curves.
[c] 285°C at take-up velocity of 3500 m/min.
[d] 0.4 mm at take-up velocity of 3500 m/min.
[e] 30 at take-up velocity of 3500 m/min.

sion and stress, ultimate filament temperature distribution, and the texture and physical properties of the spun fibers.

The categories of spun polyester yarn (LOY, MOY, POY, HOY, FOY) are differentiated by the degree of orientation and hence by the amount that they can be stretched. The first four categories have enough elongation left to produce fully drawn yarn (FDY).

Drawing Theory. Drawing is plastic deformation of a polymer related to the stretching of the molecular network and the influence of thermally activated processes. A stable network seems to exist that can explain the molecular orientation achieved in hot drawing and that allows for a limit to the extensibility in cold drawing. The change in deformational behavior that occurs at the glass-transition temperature cannot be explained.

Drawing Practice. Different drawing processes are used on a commercial basis. These include draw–twisting, spin–draw, draw–texturing, draw–spinning, and draw–warping processes for filament yarns, and a staple fiber drawing process. A drawing aid is usually provided in the form of a heated pin, plate, or roll, or a hot-water bath or spray. Its purpose is to raise the fiber above its second-order glass-transition temperature and localize the draw point or draw zone. For polyester, the draw zone can extend over a distance of a few millimeters (for filament) to a few centimeters (for staple). In some cases, the filaments are heated before the stress gradient is applied; in others, the filaments are heated during the application of the drawing stress. The change in length that occurs during drawing is usually controlled by stretching the fibers between rolls or sets of rolls. The drawing may be done in one or more stages, depending on the final properties required in the drawn fibers.

Staple-yarn Manufacture. A typical bale of staple fiber contains about 272 kg of 3.8-cm long, 1.6-dtex (1.5-den) filaments, ca 40×10^9 fibers. The cotton spinning system is most often used to convert these fibers into yarn. In the conventional ring-spinning process, the fibers are straightened, aligned, drafted (decreasing the fiber-bundle diameter), and twisted together to form the yarn. After opening the bales, the fibers are carded, cleaned (when cotton is used), and blended. Carding produces a continuous web of fibers that are removed from the card and condensed into a continuous sliver. The card sliver is processed on a draw frame, where several slivers are blended together and further fiber alignment takes place. If polyester and cotton are carded separately, they can be blended together at the draw frame. The next stage is the production of roving where twist is inserted and further fiber alignment and attenuation of the sliver diameter occur. Finally, the ring-spinning operation inserts more twist, and the yarn diameter is attenuated to provide the final yarn size. The final processing step is usually a rewinding of the spun yarn onto larger packages. Carding, drawing, roving, ring-spinning, and rewinding are separate operations.

Lubrication. Lubricants, finishes, or oils may be applied at many stages of polyester fiber manufacture. Finish is usually applied early in the spinning process after the fibers have solidified and have been cooled below 100°C. In the staple process, additional lubricants (finishes) may be applied. In some downstream yarn-manufacturing processes, oils or waxes are applied.

The main function of the lubricants is to ensure uniform and uninterrupted processing without breakage of filaments or threadlines.

Since polyester is an insulator, some charge-transfer and static generation takes place on contact. Finishes containing hydrophilic, polar, or ionic materials dissipate static. Finishes also provide cohesion in processing zero-twist HOY, FDY, or POY yarns to prevent stray filaments from leaving the yarn bundle.

Fabric Manufacture

Many methods exist to convert polyester filament or staple yarns into fabrics that are in turn dyed, finished, cut, and sewn into articles. Fabrics are formed by weaving, weft knitting, and warp knitting, as well as braiding, lace making, and felt making. Nonconventional methods such as tufting and nonwoven techniques are also used.

Woven fabrics are used for shirts, trousers, and sheets; weft-knit fabrics for men's underwear, T-shirts, and skirts; warp-knit fabrics in ladies' lingerie, nightwear, and dresses; and tufted fabrics in carpets and sometimes as wall coverings. Nonwovens are used for disposable articles (see NONWOVEN FABRICS).

The yarns, as manufactured, often need further processing (Table 3) by weaving and warp-knitting methods, where the yarns must be presented simultaneously in the form of a warp sheet. Yarns for tufting are often twisted, heat-set, and rewound before being placed in the tufting machine.

Coloration. Before dyeing and stabilization against shrinkage and creasing, polyester–cotton blends are prepared by heat setting, singeing, desizing, scouring, bleaching, and mercerizing.

Polyester polymer in the form of drawn fiber contains amorphous, oriented, and crystalline areas. The crystalline areas cannot be dyed. The oriented areas, formed during fiber processing, are difficult to dye from an aqueous bath at atmospheric conditions. Most of the dyeing, therefore, takes place in the amorphous fiber regions.

The rate of dyeing polyester depends on the temperature, time, and thermal history of the fiber. Because of the hydrophobic nature of polyester and its lack of dye sites, dyeing from an aqueous bath requires temperatures above 100°C or the use of a dyeing accelerant or carrier.

Finishing. Fabrics of 100% polyester often are finished with softeners to improve appearance and facilitate high speed sewing. Because of its hydrophobic nature, polyester absorbs oily stains and acts as dirt scavenger in laundering. This can be effectively prevented by applying soil-release finishes based on fluorocarbons, acrylic emulsions, or polyoxyethylene. Fabrics of polyester–cotton blends are finished with urea–formaldehyde resins to confer permanent press.

Products

Polyester fibers serve almost all segments of the textile-apparel, home-furnishings, industrial, tire, and carpet markets (Table 4). Within each line, a broad range of products can be engineered to meet special requirements.

Economic Aspects

Since the beginning of polyester fiber production in the 1940s, when TA was a rare and expensive chemical, world production of these fibers has risen to ca 6×10^6 t in 1985, and is expected to continue to rise significantly. The U.S. share, currently at ca 1.5×10^6 t/yr, is declining. Polyester already accounts for ca 50% of all synthetic fibers. Factors pushing polyester staple and filament growths to high levels are favorable interfiber economics due to increased volumes, expanding applications and new products, and substitution for cotton and other fibers.

Table 3. Yarn Preparation for Fabric Formation

Method	Flat yarn	Textured yarn	Spun yarn
Weaving			
filling			
rewinding and waxing	no	no	yes[a]
quilling[b]	yes	yes	yes
warp			
rewinding and waxing	no	no	yes
warping	yes	yes	yes
slashing	yes	yes[c]	yes
tyeing in	yes	yes	yes
Knitting			
weft			
rewinding and waxing	no	no	no[a]
steaming	no	no	yes
warp			
rewinding and waxing	no	no	yes[d]
beaming (warping)	yes	yes	yes
threading-in	yes	yes	yes
Tufting			
rewinding and waxing	not used	no	yes
twisting		some	some
heat setting		some	some

[a] Mainly ring-spun and some open-end-spun.
[b] Shuttle looms only.
[c] Except for compacted or twisted yarns.
[d] Only a small amount of spun yarn is warp-knitted.

Worldwide, these factors have resulted in the emergence of many new fiber producers, particularly in developing countries.

Most of the growth in polyester has occurred in Asia. In the last 10 years, Asia has rapidly expanded its capacity in both staple and filament and collectively has overtaken U.S. production.

Price and supply of raw material feedstocks will be important factors in the future position of polyester fibers. PET requires only a small portion, <1%, of the petroleum stream, but competes with gasoline octane boosters for the supply of *p*-xylene. Nonetheless, feedstock prices tend to increase at the same rate as oil prices, which has been volatile since the 1974 oil crisis.

Health and Safety Factors

Insofar as health and environmental effects are concerned, PET is classified among the least hazardous materials. It is chemically stable and virtually nonbiodegradable.

Table 4. Polyester Processes and Products

Product	Spinning speed (abbreviation)	Drawing and crimping	Applications
staple and tow	low (LOY)	large draw frames and crimpers	100% and blends, apparel, underwear, carpet, upholstery, industrial, nonwovens, sleeping bags, pillows, mattresses, drapes and curtains, sheets, filters, towels, sewing thread
industrial filament	low (LOY)	draw–twist or integrated with spinning	high tex[a,b]: tires, conveyor belts, seat belts, hose, ropes, cords, coated fabrics, dryer belts, V belts low tex[b,c]: sewing thread, sailcloth, coated fabrics
textile filament, flat and textured	medium (POY)	draw–twist or integrated with spinning for fully drawn yarn, draw–texture	apparel, outerwear, sportswear, automotive, upholstery, drapes and curtains, sheets, sewing thread, underwear, loungewear and sleepwear, linings, narrow wovens
	high (HOY)	integrated with spinning, draw–texture	
	very high (FOY)	direct use, texture	

[a] >50.
[b] To convert tex to den, divide by 0.111.
[c] <50.

BIBLIOGRAPHY

G. W. Davis and E. S. Hill in M. Grayson, ed., *Kirk-Othmer Encyclopedia of Chemical Technology,* 3rd ed., Vol. 18, Wiley-Interscience, New York, 1982.

O. Pajgrt and B. Reichstadter, *Textile Science and Technology, Processing of Polyester Fibers,* Vol. 2, Elsevier North-Holland, Amsterdam, The Netherlands, 1979.

A. Ziabicki and H. Kawai, eds., *High-Speed Fiber Spinning,* Wiley-Interscience, New York, 1984.

H. Ludewig, *Polyester Fibers, Chemistry and Technology,* Wiley-Interscience, New York, 1979.

GERALD W. DAVIS
JAMES R. TALBOT
Hoechst Celanese Corporation

POLYESTERS, FILMS

In the early 1940s, a semicrystalline, high melting, processible polyester, poly(ethylene terephthalate) (PET), was discovered in the UK. It offered great promise as a melt-spun oriented fiber. Soon after this discovery, Imperial Chemical Industries, Inc. began the development of films based on this polymer; this development continued at the ICI licensees in the United States, the FRG, France, and Japan. Of the nearly 500,000 t of PET film manufactured worldwide in 1984, approximately 60% was sold to the general merchant market and 40% was used by captive producers (Table 1), primarily for photographic, reprographic, and magnetic tape products. The total market value of PET films used worldwide in 1984 was about $\$1.75 \times 10^9$ (Table 2).

Others types of aliphatic polyesters also form useful films. However, none possess both the excellent properties and low raw material cost of PET.

Great care in design and operation of PET film plants is required to produce the high degree of purity required for videotape, floppy disks, capacitors, and photographic products. Modern plants operate under clean conditions similar to those used for the manufacture of precise electronic or medical components. Automated process monitoring and control are employed to ensure consistent product quality.

Table 2. Worldwide Capacity and Markets for PET Film in 1984

	Production capacity, %	Market, %
United States	42	50
Western Europe	26	23
Japan	25	19
other countries	7	8

Structure

PET is a linear condensation polymer with the repeating unit shown in (**1**).

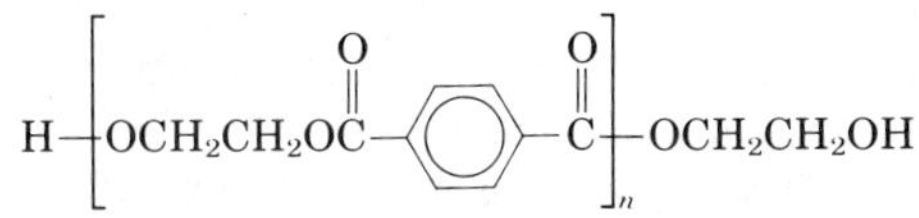

(1)

It is made from approximately equimolar amounts of ethylene glycol and terephthalic acid or dimethyl terephthalate, depending on the polymerization process. The same polymer is used in fibers and stretchblown plastic bottles.

PET is a semicrystalline polymer. Its molecular structure is described by the fringed-micelle model, where the polymer exists as an imperfect two-phase system of interconnected crystalline and amorphous domains.

The usual sequential biaxial process used for the manufacture of most commercial PET films consists of four steps, each of which influences the final molecular structure of the film:

1. Melt extrusion and quenching.
2. Drawing in the longitudinal or machine direction (MD).
3. Drawing in the transverse direction (TD).
4. Heat setting.

Table 1. Merchant and Captive Producers of PET Films

Company	Trademark	Country
merchant producers		
DuPont	Mylar	United States, Luxembourg
ICI	Melinex	United Kingdom, the Netherlands, United States, Brazil
Hoechst	Hostaphan	United States, FRG
Toray	Lumirror	Japan
Teijin	Tetoron	Japan
Diafoil	Diafoil	Japan
Rhône Poulenc	Terphane	France, Brazil
Toyobo	Espet	Japan
captive producers		
Agfa Gevaert		Belgium
Eastman Kodak		United States
3M		United States, Italy
Fuji		Japan
DuPont[a]		United States

[a] DuPont produces PET film in two separate divisions: one supplies products to the merchant markets, and the other manufactures film for its own photographic and graphic arts businesses.

Table 3. Properties of PET Film-grade Polymer

Property	Typical value	Test method
intrinsic viscosity, mL/g	6–8	in dichloroacetic acid solution at 25°C
carboxyl end groups, mol/10^6 g	20–35	titration
oligomer content, wt %	1.3–1.8	boiling xylene, 24-h reflux
density[a], g/cm^3	1.333	density gradient column
M_w/M_n	1.5–1.8	gel-permeation chromatography
refractive index[a], n_D^{25}	1.576	Abbé refractometer
glass-transition temperature, °C	70	differential scanning calorimetry
melt viscosity at 280°C, Pa·s[b]	200–400	

[a] Amorphous sheet.
[b] 1 Pa·s = 10 P.

Other methods employ simultaneous drawing, blown film or bubble processes, and multiple draw (MD–TD–MD or MD–TD–MD–TD) tensilized film processes.

The mechanical properties of biaxially oriented PET are strongly influenced by the structure of the amorphous regions. Several different methods have been described for characterizing these regions, including measurement of birefringence, x-ray scattering, and nmr spectroscopy.

Properties

The properties of the PET typically used for the manufacture of polyester film are shown in Table 3.

The thermal properties of PET film are given in Table 4.

The mechanical properties of a balanced PET film are shown in Table 5. Its excellent tensile properties allow for its use in a wide variety of film-processing equipment.

Table 4. Thermal Properties of Hostaphan PET Film

Property	Typical value
melting point, °C[a]	256
service temperature, °C	−70 to 150
coefficient of thermal expansion at 20–50°C, K^{-1}	20×10^{-6}
specific heat at 20°C, J/(kg·K)[b]	1300
coefficient of thermal conductivity up to 80°C, W/(m·K)	0.13
shrinkage[c], %	0–2

[a] ASTM D2117
[b] To convert J to cal, divide by 4.184.
[c] 30 min at 150°C.

Polyester film has excellent electrical insulating properties, as shown in Table 6, and is certified as a Class B (130°C) insulating material. Because of its chemical inertness, shown by its resistance to solvents, it can be used in processes or environments involving solvents and coatings.

Manufacture

The manufacture of polyester film involves three steps: polymer manufacture, film manufacture, and the auxiliary processing or post-treatment. Each of these steps may be modified to tailor the final film properties to meet different application requirements. For example, catalysts and additives determine film transparency and surface roughness; the stretching ratio, heat-setting temperature, and other process conditions control mechanical properties and dimensional stability; and surface treatments affect the adhesive and antistatic properties.

Polymer. The basic manufacturing process of PET is nearly identical for fibers, filaments, bottles, and films. The main route is via transesterification of dimethyl terephthalate with ethylene glycol, followed by polycondensation.

Film. PET films are formed by extrusion casting the molten polymer, followed by biaxial stretching and heat setting. This process is almost exclusively carried out with linear slot dies combined with a system of cooling and heating, as well as guiding rolls and tenter frames.

Auxiliary Treatments. The myriad of uses to which polyester films are put have resulted in the need for modification to provide special properties, such as slip, adhesion, sealability, abrasion resistance, antistatic behavior, release, and weatherability. Auxiliary treatments include in-line coating, off-line coating, etching, electrical discharge, and other high energy treatments.

Table 5. Mechanical Properties of 23-μm Hostaphan PET Balanced Film

Property	Typical value	ASTM test method
ultimate tensile strength, MD and TD, N/mm^2 [a]	225	D882-67:100%/min
ultimate elongation, MD and TD, %	120	D882-67:100%/min
stress at 3% elongation, MD and TD, N/mm^2 [a]	86	D882-67:100%/min
stress at 5% elongation, MD and TD, N/mm^2 [a]	100	D882-67:100%/min
tensile modulus (tangent), MD and TD N/mm^2 [a]	4900	D882-67:10%/min
tear strength, initiation (Graves), MD and TD, g/μm	24	D1004-67
Elmendorf tear strength, propagation, MD and TD, g/μm	0.6	D1922-67
coefficient of friction[b]	0.45	D1894[c]

[a] To convert N/mm^2 to psi, multiply by 145.
[b] Kinetic.
[c] Film-to-film.

Table 6. Electrical Properties of 23-μm Hostaphan PET Film

Property	Typical value	Test method	Test conditions
dielectric strength, V	7500	ASTM D149-75	23°C, 50% rh, 60 Hz
dielectric constant	3.3	ASTM D150-74	25°C, 0.1 kHz
dissipation factor	0.003	ASTM D150-74	25°C, 0.1 kHz
volume resistivity, Ω·cm	10^{18}	ASTM D257-76	25°C
surface resistivity, Ω/square	10^{12}	ASTM 257-76	23°C, 80% rh

Applications

PET film is classified as an engineering film because of its excellent thermal, mechanical, optical, and electrical properties. In both price and performance, polyester films are between the lower price commodity films, such as polyethylene and polypropylene, and the much higher price specialty films, such as polyimides and polyfluorocarbons. This combination of properties and reasonable cost has resulted in a rapid expansion of the volume and application areas for polyester film. Usage has grown from a few thousand tons per year during the mid-1950s, when it was introduced to the U.S. market by DuPont, to a worldwide use of about 500,000 t/yr in 1984.

In all, there are more than 50 specific application areas for PET film (Table 7).

Table 7. Applications for PET Films

photographic	hot stamping
x-ray base	metallic foils
35-mm slides	wood grain
photographic films	decorative
instant photography	electrical
phototool	capacitor dielectric
photoresist	wire and cable insulation
magnetic media	printed and laminated circuits
computer tape	motor and generator insulation
audio tape	membrane touch switches
video tape	liquid crystal display laminations
instrumentation tape	others
floppy disks	labels and decals
magnetic cards	solar control
reprographic	copier belts
original microfilm	stationery supplies
duplicate microfilm	typewriter ribbons
microfiche	computer printer ribbons
engineering reproductions	pressure-sensitive tapes
drafting films	duct insulation
packaging	laminated products
food wrap	drum heads
boil-in-bag	release film
metallized barrier	fiber glass panel carrier film
pharmaceutical	
medical devices	
retort pouch	

Other Polyester Films

Most polyester films are made of PET, but other polyesters or copolyesters are also used, including 1,4-cyclohexanedimethylene terephthalate, poly(ethylene isophthalate), PET blended with 2–35% poly(butylene terephthalate) resin, and aromatic copolyesters containing 2,6-naphthalene dicarboxylic acid.

BIBLIOGRAPHY

U.S. Pat. 2,823,421 (Feb. 18, 1958), A. C. Scarlett (to E.I. du Pont de Nemours & Co., Inc.).

W. Seifried, *Kunstst.* **75**, 10 (1985).

General Properties, Hostaphan Polyester Film, American Hoechst Corp., 1981.

EBERHARD WERNER
SIEGFRIED JANOCHA
Hoechst AG

MICHAEL J. HOPPER
KENNETH J. MACKENZIE
Hoechst Celanese Corp.

POLYESTERS, MAINCHAIN AROMATIC

The concept of molecular order in a fluid phase was initially envisioned at the end of the last century, and was based on the observations that cholesterol and its derivatives and other compounds, such as *p*-azoxyphenetole and ammonium oleate, had two melting points. That is, a liquid phase existed between the melting point of the crystalline phase and the isotropic, nonturbid liquid. The turbid phase was, and still is, interchangeably referred to as a liquid crystalline or mesomorphic phase, and because it was recognized that this thermodynamically stable phase was temperature dependent, it is also referred to as thermotropic. Some materials exhibit mesomorphic behavior as a function of their concentration in solution; these are called lyotropic liquid crystals.

Over the last 100 years liquid crystal research has had two principal thrusts: identifying and cataloging a myriad of phases with various degrees of molecular order ranging from the least-ordered nematic to highly ordered "plastic crystals" and elucidating the molecular basis of this phenomenon and establishing molecular structure–property relationships essential to practical applications.

Thermotropic Polyesters

The first observation of a liquid crystalline melt in a polymeric material was made by two groups independently. The Eastman group was attempting to increase the use temperature of poly(ethylene terephthalate) by incorporating *p*-hydroxybenzoic acid in various ratios. The most striking result is exhibited by a copolymer containing 60 mol % of the *p*-hydroxy compound. An injection-molded (260°C) sample of this polymer has a fourfold increase in tensile strength and a nearly sixfold increase in flexural modulus compared to the parent polymer and represents the first example of a high strength unreinforced engineering plastic.

This work stimulated an intense effort to find examples of completely rodlike, wholly aromatic, thermotropic polyesters with even better fiber properties but lower thermal transitions. In order to achieve the latter, researchers focused on molecular structural modifications that would diminish interchain association by disrupting interchain correlations. The use of several comonomers in various ratios has limited effect in this regard, but a fairly successful approach to attaining this goal is exemplified by work at Celanese and Eastman Kodak. The key structural element of this work is the 2,6-dicarboxy-, 2,6-dihydroxy-, or 2,6-hydroxy- carboxy-substituted naphthalene moiety.

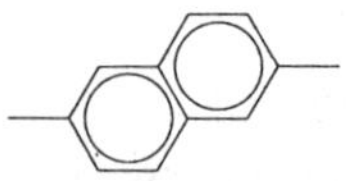

2,6-naphthalene

Because the positions of substitution are colinear, this comonomer does not disrupt the parallelism of the backbone but rather creates a "linear offset." When combined with other comonomers, such as terephthalic acid and hydroquinones, the liquid crystalline transition temperatures for some examples are as low as 240–260°C. Injection molding can be accomplished at 360°C, which generates fibers with engineering properties in the same range as *p*-hydroxybenzoic acid modified poly(ethylene terephthalate).

A more significant effect was realized by placing bulky substituents on the phenyl ring of a hydroquinone comonomer. Dramatic effects were also realized by substituting chlorines, bromines, phenyls, or cyclohexanes on both the acid and the hydroquinone. In general, the thermotropic transition temperatures dropped to about 200–250°C.

Other attempts to disrupt chain packing and lower thermotropic transition temperatures include using long-chain alkyls (C_6 to C_{12}), branched alkyls, and cyclic alkyls as side-chain substituents and substituted biphenyls.

Almost all of the molecular manipulations that generate thermally accessible transitions in the polyesters and polyesteramides are completely ineffective in the polyamide class.

Solubility and Solution Behavior

In general, the solubility of the aromatic polyamides is limited to strong protonic acid and salt-containing amide solvents. The discovery of the cosolvating ability of LiCl was first described in detail in the DuPont patents on Kevlar, and a review of this phenomenon has been presented.

The wholly aromatic polyesters are highly crystalline and are even more difficult to solubilize than the polyamides. Polyesteramides are fairly soluble in salt-containing amide solvents, but the most soluble members of this class contain the substituted biphenyls.

Rigidity is one of the primary prerequisites for liquid crystallinity in polymers. Equilibrium rigidity in solution is one of the ways by which this molecular property is measured, and it serves as a basis of comparison between polymer classes. A relative measure of chain stiffness between polymers can be most easily obtained by plotting the intrinsic viscosity vs the molecular weight; these two quantities can be related to each other by the well-known Mark-Houwink equation.

A much more precise basis of comparison of rigidity for different classes of polymers is the persistence length, which can be obtained by several techniques, the most accurate being light scattering (qv).

In spite of the slightly flexible nature of the polyesters in solution, in the molten state these materials behave as expected of rigid polymers. The notion that the condensed phase induces rigidity and enhances alignment in a shear field has much support.

Synthesis of Polymers

In the laboratory there are a wide variety of choices for generating high molecular weight polymers of all three types (polyesters, polyamides, polyesteramides). The most important considerations are monomers and solvent purity. Because acid chlorides or some equally reactive water-sensitive monomers are frequently used, all reactants, including the reaction vessel, must be scrupulously dry and care must be taken to perform all weighings as accurately as possible and all manipulations as carefully as possible while maintaining anhydrous conditions.

On an industrial scale, polyesters are frequently synthesized from a diacid or diester, instead of a diacid chloride, and diacetates of aromatic diols. In some examples, high molecular weights are obtained via transesterification in a melt condensation. The reaction is driven to completion by removing the low boiling alcohol, ester, or acid.

Engineering Properties

The first DuPont polyamide (Kevlar) patents appeared in the early 1970s. As evidenced by the large number of subsequent patents and papers, this work stimulated worldwide interest in both academic and industrial laboratories. The goal of this intense research activity is focused on understanding molecular parameters and optimizing processing conditions in order to capitalize on the unique opportunities presented by all classes of rigid, rodlike polymers.

The realization of high strength, high modulus fibers, films, and molded parts is intimately connected to the ability to control and ultimately maximize molecular and crystallite orientation. Most mainchain liquid crystalline polymers (qv) exhibit nematic (one-dimensional) order in the molten state and crystalline (three-dimensional) order in the solid state. A useful relationship for specifying the degree of order is the Hermans orientation function:

$$S(\text{or } f) = (\tfrac{1}{2})\,(3\cos^2\theta - 1)$$

The letters S and f are referred to as the order parameter and the orientation factor, respectively. These values range from 0 for completely unoriented molecular species to 1.0 for perfect unidirectional orientation and −0.5 for perfect alignment perpendicular to the orientation direction. Measurement of the degree of order can be accomplished by a number of spectroscopic techniques, including infrared, uv–visible, nmr, x-ray, neutron scattering, and birefringence.

From the standpoint of rheology, three regions of order can be identified by viscosity–shear measurements. At low shear, the domains rotate, slip past each other, and tumble, resulting in a very small viscosity decrease and low order parameter. At higher shear rates, smaller domains merge to form larger domains, but the correlation between them remains low; the viscosity tends to level off in this shear regime as does S. At even higher shear rates the large domains coalesce, the order increases significantly, and, most importantly, the viscosity drops by several decades; this non-Newtonian behavior is known as thixotropy. The low viscosity superdomain is stable in the shear field, particularly at high concentrations in lyotropic solutions or in the thermotropic phase.

An extensive investigation and development effort was undertaken by DuPont for the commercialization of polyaramides. The greatest degree of orientation of liquid crystalline solutions of these materials obtains by means of a process known as dry-jet wet-spinning. The resulting stiff fibers exhibit typical tensile strengths around 3.0–3.5 GPa (5×10^5 psi) and modulus values of approximately 62–124 GPa ($9.0–18.0 \times 10^6$ psi).

Because thermotropic liquid crystals do not require solvent, they offer a great advantage over lyotropics. These materials can be melt spun into fibers, and extruded or injection molded into large objects having high strength. The non-Newtonian flow characteristics uniquely allow these materials to be used in molds with thin walls and complex shapes. However, care must be taken in establishing spinning and molding characteristics and mold design because examination of fibers via birefringence, sem, and waxs have shown significantly higher orientation near the outer surface than in the core.

BIBLIOGRAPHY

L. L. Chapoy, ed., *Recent Advances in Liquid Crystalline Polymers,* Elsevier Applied Science Publishers, Ltd., Barking, UK, 1985.

A. E. Zachariades and R. S. Porter, eds., *The Strength and Stiffness of Polymers,* Marcel Dekker, Inc., New York, 1983.

A. Blumstein, ed., *Liquid Crystalline Order in Polymers,* Academic Press, Inc., New York, 1978.

A. Ciferri and I. M. Ward, eds., *Ultra-High Modulus Polymers,* Elsevier Applied Science Publishers, Ltd., Barking, UK, 1979.

Review articles in *Molecular Crystal Liquid Crystal,* 169 (1989).

RUSSELL A. GAUDIANA
Polaroid Corporation

POLYESTERS, THERMOPLASTIC

Polyesters belong to the broad generic class of organic high molecular weight condensation polymers and are characterized by ester groups $-\!\!\left(\overset{\overset{\displaystyle O}{\|}}{C}-O\right)\!\!-$ distributed either regularly or randomly along the main molecular chain. Whereas many types of polyesters are known, only a few have been commercialized as thermoplastic polyesters (TPE). These few possess the proper combination of economics, properties, and processing characteristics.

The commercial growth of TPEs has been very rapid among the thermoplastic resins because of their broad range of properties, including high heat-distortion temperature, high rigidity and hardness, good mechanical strength and toughness, excellent mold flow, good hydrolytic stability, and favorable economics. The reinforced grades are additionally characterized by greater stiffness, good mechanical strength at elevated temperature, and low tendency to cold flow, all without sacrificing a large degree of impact strength. TPEs have been developed for numerous novel applications, such as engineering materials replacing steel, aluminum, and other metals.

There has been a dramatic increase in the global growth of TPEs from under 500 t/yr in 1970 to over 500,000 t in 1985, and this growth was forecasted to continue at an annualized rate of 14% for five years. A significant portion of this growth is fueled by the dramatic development of poly(ethylene terephthalate) (PET) beverage bottles produced via reheat blowmolding technology.

Linear polyesters can be classified as aliphatic, partly aromatic, or aromatic polyesters. The first class incorporates polyesters derived from aliphatic dicarboxylic acids and aliphatic glycols. These materials are important technologically, but have not achieved importance as TPEs because of their lower melting point and poorer hydrolytic stability.

Partly aromatic polyesters are obtained from aliphatic glycols and aromatic dicarboxylic acids or esters, especially poly(alkylene terephthalate)s and, more specifically, PET and poly(butylene terephthalate) (PBT).

An amorphous, transparent, injection-moldable TPE based on poly(1,4-cyclohexylenemethylene terephthalate-*co*-isophthalate) (PCTI) was introduced soon after PBT. The partly aromatic category also includes the elastomeric TPE derived from a combination of polyoxytetramethylene glycol, 1,4-butanediol, and dimethyl terephthalate.

Aromatic polyesters have all of their ester linkages attached to aromatic rings. These materials are also referred to as polyarylates. Some may be crystalline, but the more important ones are clear, amorphous thermoplastics that combine clarity, high heat-deflection temperatures, high impact strength, and good electrical properties with inherent uv stability and flame resistance.

Chemical Compositions

Although some aliphatic polyesters are used as plasticizers, and hydroxy-terminated aliphatic polyester polyols as prepolymers in urethane technology, the main reason for the development of TPEs is the availability of partly aromatic polyesters, which are injection-moldable and extrudable with ease of fabrication, resulting in tough, durable parts. These fabricated parts must require minimal secondary finishing, provide a good profile of mechanical and electrical properties, and withstand hostile environments. Modifications have been extensively realized by blending, incorporation of small amounts of stabilizers (<1 wt %), mold-release agents, flame retardants, impact modifiers, and large amounts of fillers (eg, glass, up to 55 wt %), minerals, pigments, and other polymers. Technologically, the two most important types of polyesters are based on PET and PBT, which between them have over 95% of the TPE market for injection-moldable parts.

Present product emphasis is for TPEs with high heat resistance, as well as greater strength and modulus. These properties can be increased by additives and mineral reinforcement, but quite often result in lower impact strength. Improvements in impact strength may be obtained through blending with a variety of polymers including aromatic polycarbonates, polyurethanes, thermoplastic elastomeric copolyetheresters, and a variety of low T_g elastomeric methacrylate polymers.

A widely used practice for TPE modification is the incorporation of surface-modified glass fibers to realize enhanced properties, such as greater strength and modulus, and higher heat-deflection temperature, which translate into a significant increase in Underwriters' Laboratory (UL) temperature index.

A third type of commercialized TPE is based on the condensation of 1,4-cyclohexanedimethanol with mixtures of dimethyl iso- and terephthalates to produce the polymer structure, poly(1,4-cyclohexylenedimethylene terephthalates-*co*-isophthalate) (PCTI).

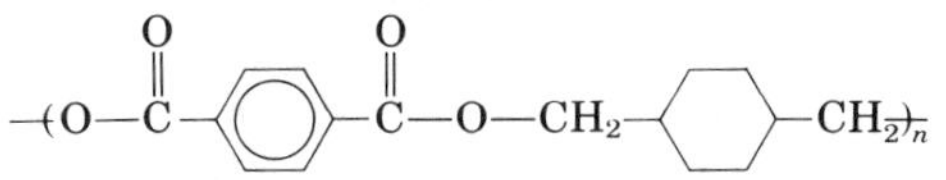

These technologically important TPE's based on 1,4-cyclohexanedimethanol are amorphous, but with poorer solvent resistance than semicrystalline TPEs. As a result, their principal applications are not for injection-molding resins, but because of their clarity, toughness, and good barrier properties, they have found broader usage in bottles and films.

One important approach that has been used to modify the properties of TPEs involves copolymerization with other glycols and/or other acid esters (eg, with PCTI). Thus the incorporation of small amounts of diethylene glycol and polyglycol has produced polymers with increased rates of crystallization.

Another TPE modification that has been commercialized is the development of polyester carbonate random copolymers. These materials are based on bisphenol-A iso- and terephthalate copolymerized with bisphenol-A carbonate. The combination of polycarbonate toughness, clarity, and processibility with polyester chemical and heat resistance has been achieved in these copolymers. Applications requiring polycarbonatelike properties with increased heat-distortion temperature have included safety items such as fire helmets, automotive fuses, and autoclavable medical and biological devices.

Manufacture

Raw Materials. Ethylene glycol (EG) with terephthalic acid (TPA) or dimethyl terephthalate (DMT) is the principal diol used in the condensation to produce PET. 1,4-Butanediol is the basic diol used for PBT.

Terephthalic acid has been prepared commercially from *p*-xylene by nitric acid oxidation, cobalt-catalyzed air oxidation, and the intermediate formation of *p*-toluic acid. The preferred method is air oxidation at ca 200°C and 14.2 MPa (14 atm) in the presence of a cobalt bromide catalyst. TPA is then esterified with methanol to give DMT, which is easily purified by recrystallization or distillation.

Polymerization. All three classes of TPEs (PET, PBT, and PCTI) are prepared commercially in two steps: transesterification or direct esterification to form a diol diester and low molecular weight oligomers, and a melt polycondensation step, which itself may be carried out in several stages to achieve very high molecular weights.

The crystalline TPEs (PET and PBT) possess excellent physical, mechanical, electrical, and thermal properties. Typical properties of unreinforced and glass-filled PET and PBT are summarized in Table 1.

Chemical Resistance. Since PBT and PET are crystalline, they are stable in a wide range of chemicals. The polymers are resistant to water, weak acids and bases, ketones, alcohols, glycols, ethers, aliphatic hydrocarbons at room temperature. These resins are also resistant to gasoline, motor oil, and transmission and brake fluids at temperatures up to 60°C. Use of strong bases at any temperature or in aqueous medium above 50°C is not recommended.

Few solvents will dissolve the crystalline TPEs at room or elevated temperature. Solvents such as hexafluoro-2-propanol, hexafluoroacetone sesquihydrate, and trifluoroacetic acid are hazardous or toxic. Mixed solvents are commonly used for solution viscosity, including 60:40 wt/wt phenol–*sym*-tetrachloroethane at elevated temperatures.

Weathering. In general, TPE resins have limited weather resistance, which can be made acceptable by adding stabilizers. Pigmented products, primarily black, are recommended for improved outdoor stability. Stabilization of TPE materials against weathering damage can be achieved by adding uv-screening agents, absorbers, and quenchers.

Processing

An important attribute of TPEs is their ability to be processed easily to yield a wide variety of molded parts, extruded shapes, sheeting, film, monofilament, or fibers. Processing methods include injection molding, extrusion, and blow molding.

The fabrication of TPEs presents few problems, especially for crystalline PET and PBT. These resins have little tendency to stick to tools because of their high melting points. They can be easily sawed, drilled, turned, milled, filed, threaded, tapped, and reamed. Processes include machining, welding, and coating and stamping.

Recycling. The rising cost of TPE polymers and their prevalent consumer uses place greater emphasis on scrap and recycling of used resins. The scrap parts from TPEs (eg, sprues and runners) are reground and blended with virgin resin up to 25 wt % for injection molding without significant loss of properties. Scrap can also be used to make parts where optical clarity is not important. The recycling of PET bottles is important due to the huge volume of scrap. Presently in the United States, nine states have bottle-deposit laws, and by 1991, there might be a national bottle-return law. There is a continuing need for efficient, cost-effective recovery and recycling systems that can turn scrap bottles into useful products. Recycled PET cannot be used to make bottles. Most of it goes into low-end strapping. Three of the principal resin suppliers, DuPont, Goodyear Tire and Rubber Co., and Eastman Kodak Co., have their own recycling systems, and several machinery producers offer both technology and recycling equipment.

Additives

The compatibility of TPEs with many fillers, polymer additives, and modifying agents is a major reason for the market expansion of injection-moldable polyesters. Many mineral additives are commonly used as fillers, including glass fiber, glass beads, calcium carbonate, talc, silica, wollastonite, clay, mica, calcium sulfate fiber, carbon powder and fiber, and aluminum trihydrate. Other additives include aramid fiber, metal flakes and fibers, organic and inorganic flame retardants, impact modifiers, antistatic agents, uv stabilizers, internal lubricants, color additives, pigments, and blowing agents to produce TPE foams.

Many new developments involving polyester blends and alloys have occurred in the last six years. The most significant commercial alloys from TPEs are PBT with polycarbonate, PET with polycarbonate, PBT with PET, PBT with elastomers, PET with poly(methyl methacrylate), PET with elastomers,

Table 1. Typical Properties of Crystalline Thermoplastic Polyesters

		Glass fiber concentration, wt %				
		PET		PBT		
Properties	Test method[a]	0	45	0	30	50
Physical						
specific gravity	D792	1.37	1.69	1.31	1.53	1.70
water absorption at 23°C, %	D570					
after 24 h		0.08	0.04	0.08	0.06	0.05
at equilibrium		0.60	0.45	0.34	0.26	
Rockwell hardness	D785	M106	M100	R117 M68	R118 M90	M90
Taber abrasion (CS-17) 1000 g, mg/1000 cycles	D1044	3	7	9	19	
coefficient of friction	D1894					
against self			0.17	0.17	0.15	
against metals			0.20	0.13	0.19	
Mechanical						
tensile strength, MPa[b]	D638	53	193	52	117	152
tensile elongation, %		300	2	300	4	3
flexural strength, MPa[b]	D790	114	283	90	193	262
compressive strength, MPa[b]	D695	128	179	90	124	131
shear strength, MPa[b]	D732	59	86	53	61	
flexural modulus, MPa[b]	D790	2,830	13,800	2,340	7,580	
Izod impact strength, J/m[c]	D256					
notched						
at 23°C		43	128	53	96	117
at −40°C		32	123	37	85	
unnotched						
at 23°C		no break	370	no break	800	
at −40°C		no break	370	no break	747	
continuous stress deformation under 13.8 MPa[b] load at 50°C, 24 h, %	D620		0.60[d]	2.6	1.60	
fatigue endurance limit, 10^7 cycles, MPa[b]	D671	24	58	20		
Thermal and flammability						
heat-deflection temp, °C	D648					
at 0.46 MPa[b]		115	250	155	216	221
at 1.82 MPa[b]		85	226	55	210	210
coefficient of linear thermal expansion, $\times 10^{-5}$/°C	D696	7.2	2.30	12.8	2.5	
thermal conductivity, W/(m·K)	C177		0.31	0.16	0.21	
UL temperature index	UL08HR2					
electrical and mechanical with impact, °C		140	140	120	140	
mechanical without impact, °C		140	140	140	140	
oxygen index, %	D2863	21	20	20.6	18.2	18.5
flammability rating, cm/min	D635	SB[e]		3.3	2.5	2.5
Electrical						
vol resistivity, Ω·cm $\times 10^{16}$	D257	3.5	0.1	4.0	3.2	
dielectric strength (short term), kV/mm	D149	23.6	29.6	23.2	24.8	
dielectric constant	D150					
100 Hz			4.0	3.30	3.80	
10^6 Hz		3.4	3.9	3.10	3.70	
dissipation factor	D150					
100 Hz			0.005	0.002	0.002	0.002
10^6 Hz		0.02	0.012	0.02	0.02	0.02
arc resistance, s	D495		126	184	146	

[a] ASTM, unless otherwise noted.
[b] To convert MPa to psi, multiply by 145.
[c] To convert J/m to ftlbf/in., divide by 53.38.
[d] Measurement under 27.6 MPa load.
[e] Slow burning.

Table 2. Principal Manufacturers of PBT

Company	Trade name
United States	
Celanese Corp.[a]	Celanex
GAF Corp.[a]	Gafite
General Electric Co.	Valox
Mobay Chemical Co.	Pocan
Europe	
Akzo Plastics BV	Arnite
BASF AG	Ultradur
Bayer AG	Pocan
Chemische Werke Huls, AG	Vestodur
CIBA-GEIGY Corp.	Crastin
Dynamit-Nobel AG	PTMT
Montedison	Pibiter
Japan	
Dainippon Ink & Chemicals, Inc.	Planac
Mitsubishi Chemical Industries, Ltd.	Novadur
Mitsubishi Rayon Co., Ltd.	Tufpet
Mitsubishi Petrochemical Industries, Ltd.	FR-PMT
Teijin, Ltd.	PBT
Toray Industries, Inc.	PBT

[a] Celanese Corp. acquired GAF's PBT business in 1986.

and PET with polysulfone. Blend technology is becoming increasingly more important in the development of new polymer plastic formulations. A key factor is the ability of polymer-blend technology to satisfy end-use requirements in the most economical manner (see POLYMER BLENDS).

Economic Aspects

Today PBT is the most important TPE for injection molding, accounting for over 80% of total TPE volume, excluding bottles, fibers, and films. The principal manufacturers of PBT in the United States, Europe, and Japan are listed in Table 2.

Health and Safety Factors

The importance of plastics in food packaging has been growing significantly and it has been estimated that over 20% of food packaging materials are now plastic. For the last 20 years, films and coatings of PET have been approved by the FDA for use with food. In 1973 this approval was extended for containers and bottles used for food.

PBT resins can be used for nonalcoholic food-contact applications under specified conditions.

Uses

Engineering TPEs that can be injection-molded are used in many industries, including electrical and electronics, automotive, housewares, lighting, power tools, material handling, sporting goods, and plumbing.

The pattern of U.S. consumption of TPE engineering-grade resins in various major markets for 1981–1985 is summarized in Table 3.

BIBLIOGRAPHY

D. B. G. Jaquiss, W. F. H. Borman, and R. W. Campbell in M. Grayson, ed., *Kirk-Othmer Encyclopedia of Chemical Technology,* 3rd ed., Vol. 18, John Wiley & Sons, Inc., New York, 1982.

N. J. Brozenick in *Modern Plastics Encyclopedia,* McGraw-Hill, Inc., New York, 1986–1987.

W. K. Witsiepe, *Adv. Chem. Ser.* **129,** (1973).

J. R. Wolfe, Jr. in Dale J. Meier, ed., *Block Copolymers: Science and Technology,* MMI Press, Harwood Academic Publishers, New York, 1983, pp. 145–166.

JALANDAR Y. JADHAV
SIMON W. KANTOR
University of Massachusetts

POLYESTERS, UNSATURATED

Unsaturated polyester resins are step-growth polymers formed by the interaction of stoichiometric mixtures of unsaturated and saturated dibasic acids or anhydrides with dihydric alcohols or oxides. The unsaturated acid component is fundamental to the reactivity of the low molecular weight polymers formed and is derived primarily from 1,2-olefinic dibasic acids such as maleic acid or anhydride.

The commercial resins have established a significant position within the plastics industry. They are typically solutions of unsaturated polyester polymers dissolved in unsaturated coreactant liquid monomers such as styrene to enhance reactivity and processibility. Free-radical catalysts initiate cross-linking reactions involving the unsaturated polyester polymer and the unsaturated coreactant monomer, rapidly transforming the low viscosity resin into a rigid thermoset plastic state, comprising a three-dimensional polymer network.

The degree of unsaturation in both the polyester polymer and liquid monomer determines the complexity and physical characteristics of the cross-linked network, although properties such as hardness, flexibility, heat resistance, fire retardance, and others can be modified by substitution with glycols and saturated dibasic acids into the polyester polymer backbone.

Although other unsaturated monomers can be more effective in enhancing specific properties of the cross-linked plastic,

Table 3. U.S. Consumption of Thermoplastic Polyesters in Typical Markets from 1981 to 1985

	Consumption, 1000 t				
Market	1981	1982	1983	1984	1985
appliances	2	3	4	5	5.5
electrical–electronics	9	9	10	18.2	17.3
plumbing–hardware	3	4	5	2.7	2.7
transportation	7	6	7	13.6	14.5
other	2	3	4	7.3	7.3
Total	*23*	*25*	*30*	*46.8*	*47.3*

styrene, in accordance with general cost–performance criteria, is the principal coreactant monomer in commercial formulations.

In the late 1980s, the consumption of unsaturated polyester resins in North America exceeded 500 × 10^3 t annually; consumption in Asia is similar. Polyester products are used in many markets of commercial significance, ranging from automotive to construction and electrical applications. Structural applications consume ca 80% of production; the remainder is used without fibrous reinforcements, but usually extended with inexpensive fillers primarily for producing cast items.

Raw Materials

Petrochemicals supply most functional derivatives used for the manufacture of polyester resins. Lower glycols are derived almost exclusively by the oxidation of ethylene or propylene. Aromatic dibasic acids are produced by the oxidation of xylenes and maleic anhydride by the oxidation of benzene. Styrene monomer is manufactured on a large scale from benzene and ethylene. Initially, unsaturated polyesters were simple compositions of propylene glycol, phthalic anhydride, and maleic anhydride. Dissolution of this viscous precursor in styrene monomer produced resins of low viscosity that were adaptable to simple fabricating processes under ambient conditions. This classical formulation has remained essentially unchanged and is typical of most industrial processes. This family of unsaturated polyester resins is known as (ortho)phthalic resins or general-purpose resins in view of their chemical derivation and wide adaptability and performance characteristics. Even though commercially successful in a wide variety of applications, the phthalic resins exhibit certain limitations with regard to heat resistance, fire retardancy, resistance to chemical attack, and processibility.

By a process of substitution, new families of polyester resins have evolved with improved and specialized properties, characteristic of certain organic constituents. Aromatic derivatives such as isophthalic and terephthalic acid or diols derived from bisphenol A provide a higher degree of hardness, rigidity, and enhanced thermal characteristics. Aliphatic constituents such as adipic acid, 1,4-butanediol, and diethylene glycol yield soft, pliable products. Property modification is influenced by the number of methylene or oxyethylene units separating the reactive functionality. Other characteristics are derived from reactive halogenated compounds, which impart flame resistance.

In view of the wide choice of functional components, unsaturated polyester resins can be tailor-made with specific and exclusive properties. Within economic constraints, most unsaturated polyester resins in commercial applications include different types of glycols, dibasic acids, and monomers (Table 1).

Manufacturing Processes

Polyester resins derived from glycols and dibasic acids are manufactured in heated steel reactor vessels with moderate agitation and overhead condenser systems to collect the aqueous by-products. To prevent discoloration caused by oxygen, the reactor is continuously purged with fresh nitrogen or carbon dioxide. Methods include the fusion-melt process, the isophthalic resin process, the oxide process, the epoxy dimethacrylate process, the dicyclopentadiene process, and terephthalate transesterification.

Cross-linking Mechanism

Catalysts and Initiator Systems. The resin is transformed from a stabilized liquid into a rigid plastic state; free-radical catalysts, eg, ketone hydroperoxides or benzoyl peroxide, initiate a controlled cross-linking reaction involving the fumarate polymer and styrene monomer.

Gelation and Cure. Initially, the free radicals produced by the redox systems are neutralized by quinone stabilizers present in the resin system. After an induction period, the free radicals initiate cross-linking polymerization involving the fumarate groups in the polyester chains and the unsaturated styrene monomer. This so-called gelation time is defined as the period between the addition of the initiator system to the

Table 1. Ingredients of Polyester Resin Formulations in Descending Order of Commercial Use

Glycol	Dibasic acid or anhydride	Unsaturated acid or anhydride	Unsaturated monomer
propylene glycol	phthalic anhydride	maleic anhydride	styrene
ethylene glycol	isophthalic acid	fumaric acid	vinyltoluene
diethylene glycol	adipic acid	methacrylic acid	methyl methacrylate
neopentyl glycol	chlorendic anhydride[a]	acrylic acid	diallyl phthalate
dipropylene glycol	tetrabromophthalic anhydride	itaconic acid	α-methylstyrene
dibromoneopentyl glycol	tetrahydrophthalic anhydride		triallyl cyanurate
bisphenol dipropoxy ether	terephthalic acid		divinylbenzene
propylene oxide	tetrachlorophthalic anhydride		
2,2,4-trimethylpentane-1,3-diol			
tetrabromobisphenol dipropoxy ether			
1,4-butanediol	glutaric acid		
dicyclopentadiene hydroxyl adducts	cyclopentadiene–maleic anhydride[b]		

[a] 1,4,5,6,7,7-Hexachlorobicyclo(2,2,1)-5-heptene-2,3-dicarboxylic anhydride.
[b] Diels-Alder adduct.

liquid resin and the appearance of a soft intractable gel, where 1–2% of the polymer has been cross-linked. Cross-linking proceeds rapidly and establishes the structure of the three-dimensional network in which polymers and monomer are immobile. The corresponding decrease in termination rate constant and net increase in free-radical concentration leads to an exponential jump in styrene-initiated copolymer growth. The gel effect gradually transforms the mass to a rubbery transition phase. As cross-linking and copolymer formation accelerate, heat is generated and a rigid plastic forms. The activation energy of the gelation and cross-linking is estimated to be 50 kJ/mol (12 kcal/mol), and the overall reaction rate constant $2.7 \times 10^9 \exp(-17{,}000/RT)\ \text{min}^{-1}$.

The rate of the gel-structure development depends on the free radicals available after neutralization of the inhibitor system. The complexity of the gel increases as the degree of cross-linking progresses, and enters a rubbery transition state of surprising friability. The cure or development of hardness can be followed with a Shore D penetrometer.

Suppression of Heat Generation. Heat generated during cross-linking is influenced by the fumarate and the styrene content. In small batches, heat is generated in a controlled manner, but in larger batches the heat generated can cause discoloration or cracking. Modifiers can be added to lower the cross-linking rate. α-Methylstyrene was used initially, but has been replaced by copper naphthenate.

Shrinkage. The formation of the cross-linked network is accompanied by some volume contraction (Table 2). The fumarate content has less influence over shrinkage than the monomer, but polymers with high fumarate concentration crack as the more rigid structure attempts to shrink. Fillers and inorganic extenders reduce the shrinkage of large polyester castings, eliminating internal voids and cracking.

Table 2. Volume Shrinkage During Cure

Monomer	Shrinkage, %
styrene	17
vinyltoluene	12.6
diallyl phthalate	11.8
methyl methacrylate	21
vinyl acetate	27
t-butylstyrene	7

Air Inhibition. The free-radical reactions involving styrene and unsaturated polyester polymers are suppressed by oxygen, which lowers the cross-linking activity in the exposed surface of the gelling mass. Phthalic resins are more susceptible to air inhibition than isophthalic resins, and parts fabricated in open molds have a characteristic tacky surface. Although tackiness can be overcome by postcuring, the surface remains undercured and sensitive to increased attack from water and solvents. In general-purpose applications, small amounts of paraffin wax control air inhibition by forming a monomolecular surface layer that excludes oxygen. The wax layer must be removed before the article is painted or bonded. Wax is also used in wood finishes; it is usually removed, followed by buffing and polishing. Wax-free coating systems are based on diethylene glycol compositions incorporating tetrahydrophthalic anhydride, which promotes air drying. In contrast, resins containing acrylate monomers or reactive acrylate groups, eg, epoxy dimethacrylates, are sensitive to air inhibition and do not respond to conventional cobalt–hydroperoxide initiator systems. Surface cures can be promoted by tertiary aromatic amines, benzoyl peroxide catalysts, or redox systems containing high cobalt concentrations.

Performance Characteristics

Unmodified, cured, unsaturated polyester resins have low strength and are prone to stress cracking and brittle fracture. In structural applications, reinforcing fibers or fillers are required. For coatings of furniture and electrical components, reinforcements are not needed, but application in this market has been limited. Performance tests are usually made on clear unmodified resins (0.3-cm castings) and on the reinforced plastic composite, usually as laminates or moldings. Laboratory studies of formulations are made on clear castings of the unmodified resins.

Mechanical Properties. Tests for mechanical properties on samples machined from flat cast sheets often give misleading results because of sensitivity to brittle fracture. Stresses created during cure can lead to anomalous results, and extensive annealing treatment is required to eliminate such problems.

Glass fiber, Kevlar, and carbon fiber reinforcements improve both the tensile- and flexural-strength characteristics of the reinforced plastic composite laminate (Table 3). Increasing the reinforcement to the maximum packing volume eliminates

Table 3. Mechanical Properties of Polyester Resin Castings and Glass-mat-reinforced Laminates

	Phthalic laminating resin		Isophthalic corrosion-resistant resin		Vinyl ester resin	
Property	Casting	Laminate	Casting	Laminate	Casting	Laminate
glass fiber mat, %	0	30	0	30	0	30
flexural strength, MPa[a]	82	172	90	193	103	190
flexural modulus, MPa[a]	4137	5517	4827	6206	4482	6896
tensile strength, MPa[a]	41	89	62	103	82	124
tensile modulus, MPa[a]	4482	4827	4482	8275	5172	8275
tensile elongation, %	1.8	1.4	2.1	1.6	4.5	1.8
heat-distortion temperature, °C	70	150	100	>200	110	>200

[a] To convert MPa to psi, multiply by 145.

differences in both tensile and flexural properties between widely varying polyester compositions.

Thermal Properties. The highly cross-linked nature of the three-dimensional polymer network enhances the resistance of the plastic to softening and deformation at high temperatures. However, because of the high content of aromatic vinyl groups, the cross-linked plastic is susceptible to thermooxidative decomposition, which reduces the long-term-use temperature.

Thermooxidative Properties. Polyesters are used in high temperature electrical or corrosion applications, frequently above their HDT. Although fillers and glass fiber reinforcements improve the structural behavior of the composite at high temperature, the resins at some point begin to dissociate chemically. The decomposition is accelerated by autooxidative reactions, giving by-products such as benzaldehyde, benzene, and acetaldehyde. The cross-linked resins, irrespective of the composition of the polyester polymer or the vinyl monomer, undergo spontaneous decomposition near 300°C, even in the absence of oxygen. This is a characteristic of all vinyl copolymers, which are degraded into monomeric units at high temperatures, unlike nonvinyl cross-linked thermosets, such as the phenolic and epoxy resins, which tend to carbonize.

Dielectric Properties. The low dipolar nature of the unmodified cross-linked plastic contributes to the dielectric strength and surface resistivity characteristic of polyester resins. At high voltage or high current differential, however, the plastic fails due to arcing or tracking, caused by the charring of the polymer surface into a conductive carbonaceous residue. Hydrated fillers such as alumina trihydrate suppress char formation and increase resistance to failure.

Flammability. High styrene content increases flammability with an emission of black sooty smoke. Flammability can be reduced by substituting the dibasic acid by chlorendic anhydride or tetrabromophthalic anhydride or the glycol component by dibromoneopentyl glycol. Polyester derived from tetrahydrophthalic anhydride can be directly brominated in solution.

Corrosion Resistance. The three-dimensional cross-link network gives rise to the thermoset characteristics of polyester plastic to resist attack by corrosive chemicals, including weak alkalies, strong acids, and nonpolar solvents. The ability to resist chemical attack is related to the steric factors that protect the polar ester groups within the cross-linked network. Phthalic resins perform very poorly in alkaline media and disintegrate faster than isophthalic resins; polyesters based on derivatives such as epoxy dimethacrylate or dipropoxylated bisphenol fumarate offer the maximum resistance to saponification.

In strongly alkaline environments, the glass fiber is susceptible to attack, and filaments within the composite act as alkali conduits. Unmodified castings of compositions that allow only surface chemical attack undergo complete disintegration when reinforced. Polyester resins are not recommended for service in strong alkali because glass fiber reinforcement promotes ester disintegration by saponification.

Weathering. Cross-linked polyester resins survive outdoor exposure for more than 30 years, although some discoloration and loss of strength occurs.

Fabrication Processes

The ability of unsaturated polyester resins to undergo controlled transformation from liquid resin to a hard infusible thermoset plastic at ambient temperatures has facilitated the development of fabricating techniques. Glass fiber-reinforcing agents improve the strength properties, enabling composites to compete in applications traditionally using materials such as wood, concrete, and welded steel components. FRP remains unchallenged in large fabricated articles, largely because of low tooling and fabrication costs, and the fact that it can be formed without heat and pressure. Although this remains primarily true for large low volume products, the molding process can be accelerated at higher temperatures, and higher pressures facilitate the flow of resins within the glass fiber reinforcements. Mechanized processes have evolved rapidly for large volume production rates required by certain industrial and commercial markets.

Other less advanced techniques for cast items employ fillers instead of fibrous reinforcements at ambient temperatures. These techniques are used for synthetic marble products, mine-bolt adhesives, and automotive body patching. Fabrication processes include manual methods, spray-up technique, gel coats, molds, resin-transfer molding, compression molding, molding compounds, pultrusion, filament winding, and casting and coating.

Health and Safety Factors

The health and safety aspects of polyester resins relate principally to the FRP process, plant emissions, and the airborne chemicals and particulates that can be inhaled. Styrene, a major resin component, can readily volatilize from open molds and become a hazard in the workplace (see STYRENE POLYMERS). Efficient ventilation in the workplace must be provided by blowers and spray booths.

BIBLIOGRAPHY

H. V. Boenig, *Unsaturated Polyesters: Structure and Properties,* Elsevier Publishing Company, Amsterdam, Neth., 1964.

JEFFREY SELLEY
Consultant

POLYESTERS, WATER-EXTENDED

The low cost and versatility of polyester resins has been further enhanced by special polyesters that allow the use of water as an extender (filler). The use of water as an extender offers advantages over conventional fillers such as low density, high temperature control (fast cure), easy mixing, and lower raw-material cost. These resins form an emulsion with water, providing improved physical properties. Water-extendable polyester emulsions can be thought of as microcellular, water-filled foams.

As a filler, water reduces the cost of the final product. It also acts as a heat sink, facilitating the casting and curing of thick articles without high temperature and cracking problems. A water-in-oil emulsion is prepared where the resin is the continuous phase, and small-diameter (average 2 mm) water droplets are the discontinuous phase. Before curing, the emulsion is a low viscosity liquid that flows easily into the mold, filling it completely. The emulsion gels and cures quickly.

A variety of mold materials have been used, such as latex

rubber, poly(vinyl chloride) hot melts, urethane, and silicone rubber.

Emulsion Preparation

Water-extended polyester emulsions are prepared batchwise with a high speed mixer or continuously with specially designed metering equipment. For top-quality castings, uniform, small water droplets are required. Microscopic examination shows no agglomeration of water during the storage or cure of the emulsion.

Glass-reinforced Applications

A number of new applications have been developed based on the unique handling properties of water-extended polyesters and glass fibers deposited with the same "chopper gun" used with polyester laminating resins. The water moderates the heat of reaction, allowing rapid cure times without overheating. This characteristic, combined with rapid fiber wetting, allows thick laminates to be applied in one operation. Rapid dehydration due to special resins and the glass fibers results in a lightweight part. Physical properties are inferior to those of conventional fiber glass laminates. Energy dissipation is excellent and the composites are satisfactory for many applications.

Reinforcement of Acrylics. The low temperature, fast-cure, single-operation characteristics of glass-reinforced, water-extended polyester are useful in reinforcing thermoformed acrylic. These characteristics, along with good adhesion to acrylic and low light transmission, result in significant labor reduction and cost savings.

Properties

The properties of cured composites vary with water content. At 50% water, the compressive strength approaches that of concrete, whereas the tensile strength is much higher. Increasing the water content to 70% reduces the physical strength to one-fourth of that at 50% water. Additional increases in water reduce strength even further, giving low density, void-filling materials. Properties are given in Tables 1 and 2.

Table 1. Typical Properties of Water-extended Polyesters

Property	50% H_2O	60% H_2O
tensile strength, MPa[a]	17.2	11.7
tensile modulus, MPa[a]	813	558
elongation, %	6.0	3.9
heat-distribution temp, °C	57	56
flexural strength, MPa[a]	33.1	20.7
flexural modulus, MPa[a]	833.5	572

[a] To convert MPa to psi, multiply by 145.

Shrinkage. Castings shrink 1–2% as they cure in the mold. Cure shrinkage, common to all unsaturated polyesters, is almost complete at the time the part cools to room temperature and is completed within 24 h. However, unlike conventional polyesters, shrinkage is extended over a longer time as the water gradually leaves the part.

Freezing. Articles subjected to freeze–thaw cycles may exhibit cracking.

Table 2. Properties of Glass-reinforced, Water-extended Polyester[a]

Property	Value
tensile strength, MPa[b]	84.7
tensile modulus, MPa[b]	393
tensile elongation, %	2.7
flexural strength, MPa[b]	66.8
flexural modulus, MPa[b]	310

[a] 1 part resin, 1 part water, 1 part aluminum trihydrate (filler), reinforced by 22% three-plies Owens Corning fiber (OCF) 721 mat.

[b] To convert MPa to psi, multiply by 145.

Fillers and Pigmentation. Addition of solid fillers does not give acceptable results. Calcium carbonate can be used to increase the weight. Colored castings can be made by mixing a pigment dispersion with the resin.

BIBLIOGRAPHY

U.S. Pat. reissue 27,444 (July 18, 1972), G. Will (to Ashland Oil, Inc.).

B. E. Cooper, *Mold Materials for Water-extended Polyesters,* Ashland Oil, Inc., Ashland, Ky.

Aropol WEP Resins Bulletin No. 1437, Ashland Oil, Inc., Ashland, Ky., 1981.

Ashland Mixer Bulletin No. 1464, Ashland Oil, Inc., Ashland, Ky., 1981.

ROBERT E. CARPENTER
Ashland Chemical Company

POLYESTERS, WATER-SOLUBLE

Since the 1970s, water-soluble polyesters have found increased use in diversified forms and have become mature in the coatings industry. They are also used in textile treatments, adhesives, paper coatings, reinforcements, and inks. Many synonyms are used in the literature, such as waterborne, water-dispersible, water-reducible, water-dilutable, and others. When modified with unsaturated fatty acids, they are frequently known as water-soluble alkyds. The increase in consumption is primarily due to governmental regulation of solvent emissions to reduce air pollution. Ease of application and cleanup, low odor, and nonflammability are additional advantages of coatings based on these polymers.

Water-soluble polyesters are prepared by introducing hydrophilic groups into the polymer chain or in side chains. Block polymers of polyesters and a hydrophilic polymer such as poly(ethylene oxide) can be used. These polyesters are supplied in water or are diluted with water on site. They are frequently combined with neutralizing or salt-forming agents and water-miscible cosolvents. Water-soluble polyesters are usually converted to a water-insoluble form after application and water evaporation. Low molecular weight polyesters are converted to high molecular weights and the hydrophilic groups are converted to hydrophobic form by condensation with water-dispersed coreactants, such as urea or melamine–formaldehyde resins; water-dispersed phenolic, epoxy, blocked isocyanate, or polyaziridine functional resins; and, sometimes, inorganic agents. Water-soluble alkyd coating films can be cross-linked with metallic driers on exposure to air.

The monomers used in the preparation of water-soluble polyesters are usually the same as those used in polyesters and alkyds, except for minor amounts of special intermediates, eg, trimellitic anhydride, to confer the water solubility. The properties and performance of these polymers are similar to those of the parent polyesters (see also POLYESTERS; POLYESTERS, UNSATURATED; COATINGS; ALKYD RESINS).

The largest consumers of water-soluble polyesters are the surface coatings and related industries. Water-soluble polyesters and alkyds also find use in paints, stains, coil and wire coatings, and appliance and electrodeposition topcoats (see also COATING METHODS; COATINGS, ELECTRODEPOSITION).

Type I

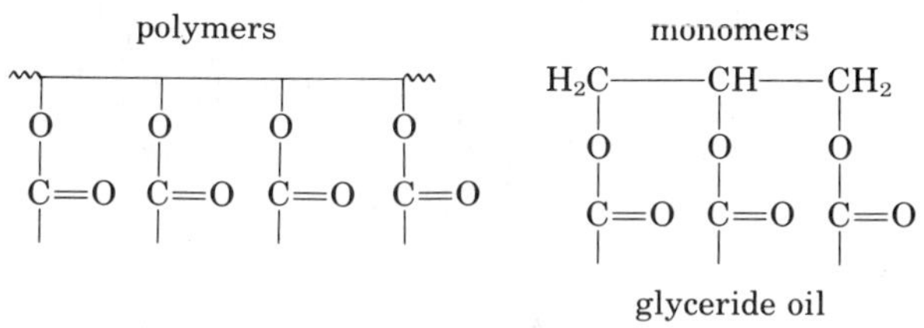

Type II

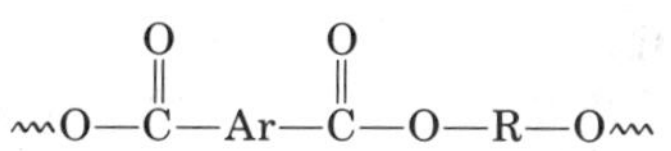

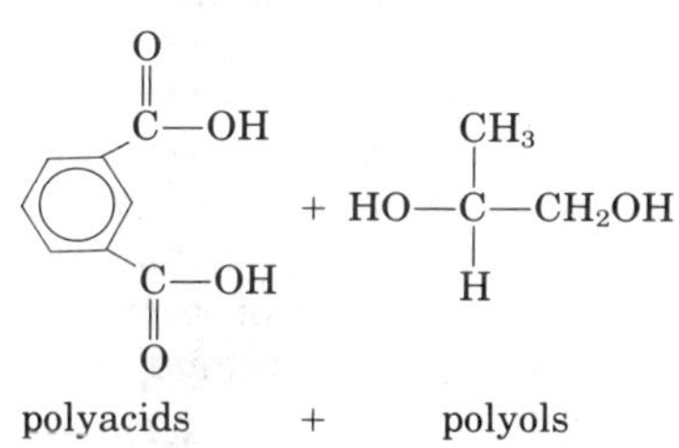

Water-dispersed Oils and Alkyds

Numerous water-soluble paint systems are based on resinous or drying-oil compositions emulsified with natural colloids.

These esters are prepared by shifting the esterification equilibrium at high temperature by removing the water produced. The ester groups are, of course, sensitive to hydrolysis. The polyesters are dispersed and stored in an aqueous solution. Even today, with the market dominated by acrylic and vinyl acetate emulsions, patents are still being issued on improvements to these oil-based paint systems such as for rheological control.

Water-soluble Polyester Compositions

These polyesters can be divided into two types according to whether the ester linkages are in the side chains (Type I) or in the main chain (Type II). In Type I the source of the ester linkage is frequently a fatty acid, imparting an alkyd nature to the polymer.

Type I polyesters include maleic anhydride-treated polyesters, rosin esters treated with maleic anhydride, and epoxy or styrene-allyl alcohol copolymer esters. Type II polyesters include alkyds modified with poly(ethylene oxide), polyesters or alkyds modified with trimellitic anhydride, oil-free thermosetting polyesters, dimethylolpropionic acid polyesters, 5-sulfoisophthalic acid polyesters, and polyesterimides and polyesteramides.

Performance

The replacement of organic solvents by water has several disadvantages, such as the essentially fixed evaporation rate of water at a given temperature and dependence on rh. The high surface tension of water requires additives to facilitate substrate wetting and prevent "crawling" of films. A "cratered" film appears when airborne particulate contaminants impinge on the film, disturbing evaporation and causing localized flow problems.

Neutralizing agents, such as fast evaporating amines, create odor problems. The use of amines also enhances color problems in conjunction with auto-oxidative drying of fatty acids.

Residual carboxyl content increases sensitivity to uv light, alkali, water, and corrosion. Thermosetting resins can be used to remove these carboxyls and cross-link the film.

BIBLIOGRAPHY

E. T. Turpin, *J. Paint Technol.* **47**(602), 40 (1975).

D. H. Solomon, *The Chemistry of Organic Film Formers,* John Wiley & Sons, Inc., New York, 1967, Chapt. II. See also 2nd ed.

H. V. Schenk, H. Spoor, and M. Marx, *Prog. Org. Coat.* **7**(1), 1 (1979).

P. Oldring and G. Hayward, *Resins for Surface Coatings,* Vol. 3, SITA Technology, Inc., London, 1987, Chapt. XI.

ROY C. WILLIAMS
NL Chemicals, Inc.

POLYETHERETHERKETONES

Polyetheretherketone (PEEK) is produced only by ICI Americas, Inc. in production facilities in the United States and the UK under the trade name Victrex. Since its introduction into the marketplace, both sales and production of the polymer have approximately doubled each year. The product is available as powder or crystalline extruded pellets in a range of grades of varying molecular weight and viscosity.

The conventional processibility of PEEK has resulted in a wide range of modified products incorporating fibrous fillers such as glass or carbon, or particulate fillers, eg, milled glass, micas, etc. These compounds are available from both ICI and a number of custom-compounding operators throughout the world.

Typical industries using the polymer are the electrical (cable insulation), automotive (bearings, piston skirts), aerospace (structural components), oil (valves, pump, seals), and chemical industries (compressor valve plates, pH meters). PEEK and new polyaromatics based on PEEK, such as polyetherketone (PEK), are engineering plastics increasingly in demand by high performance polymer users.

PEEK polymer is covered broadly by Union Carbide Corp. patents dating from 1963, all of which expired in 1984 with the exception of those of the United States, where the patents do not expire until 1995.

Polymer Composition

Polyetheretherketone has been described as a tough, crystalline thermoplastic made up of the repeat unit

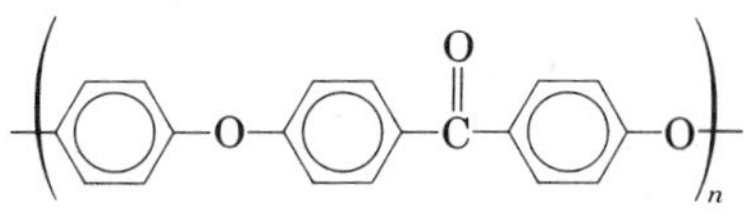

Polymer Properties

The properties of the natural injection-molding resin are described in this section. Reinforcement of PEEK with fibers or particulate fillers affects properties significantly. The natural grade of PEEK is ca 35% crystalline, has a density of 1.32 g/mL, $T_m = 334°C$, $T_g = 143°C$, and shows a mold shrinkage of ca 1%.

Mechanical Properties. Table 1 shows the mechanical properties of natural PEEK.

Thermal Properties. The high oxidative and thermal stability of PEEK is a direct consequence of the molecular structure, which allows substantial electron overlap along the molecule (Table 2).

Electrical Properties. PEEK's electrical properties are substantially unchanged at temperatures up to 200°C. This balance of properties makes PEEK suitable for all electrical applications with the possible exception of those where electrical loss is a highly critical consideration.

Chemical Resistance. There are no common room-temperature solvents for PEEK.

Table 1. Mechanical Properties of PEEK[a]

Property	Value
flexural modulus, GPa[b]	3.7
at 100°C	3.6
at 300°C	0.3
tensile strength, MPa[c]	92.0
at 100°C	50.0
at 300°C	10.0
elongation at break, %	50
Izod impact strength (unnotched)	no break
shear strength, MPa[c]	95
Rockwell hardness, R scale	126

[a] At 23°C, unless otherwise noted.
[b] To convert GPa to psi, multiply by 145,000.
[c] To convert MPa to psi, multiply by 145.

Table 2. Thermal and Flammability Properties of PEEK[a]

Property	Value
Thermal	
heat-distortion temp, °C	156°C
coefficient of thermal expansion	
at 23°C	47×10^{-6}
specific heat, J/(g·K)[b]	1.34
thermal conductivity, W/(m·K)[c]	2.5×10^{-1}
Flammability	
limiting oxygen index, %	35
NBS smoke chamber, nonflaming	1.5
toxicity index[d], 10-g sample	1
acid gas release	undetectable
Burning behavior	
heat of combustion, kJ/g[b]	29
flash-ignition temp, °C	575

[a] Natural material.
[b] To convert J to cal, divide by 4.184.
[c] ASTM C177; to convert W/(m·K) to J/(cm·s·°C), divide by 418.
[d] UK MD Test NES 7.13.

Health and Safety

PEEK is an inert polymer under normal storage conditions and there are no known dust hazards in the handling of fully compounded granules as supplied by ICI.

BIBLIOGRAPHY

Victrex PEEK: A Guide to Grades for Injection Molding UK2/0785, ICI Americas, Inc., Wilmington, Del., 1986.

Victrex PEEK: Health and Safety Information VKT 2/0585, ICI Americas, Inc., Wilmington, Del., 1986.

RENNIE MAY
ICI Americas, Inc.

POLYETHERS

The term "polyethers" generally refers only to those polymers in which the ether linkage (—C—O—C—) is in the backbone. Polymers with pendent ether groups, as in vinyl ether polymers (qv), are therefore excluded. Because the numerous classes of polyethers differ greatly in methods of preparation and properties, they are discussed separately. See ACETAL RESINS; ALDEHYDE POLYMERS; 1,2-EPOXIDE POLYMERS; EPOXY RESINS; FURAN POLYMERS; KETONE POLYMERS; OXETANE POLYMERS; POLYETHERETHERKETONES; POLY(PHENYLENE ETHER); POLYACETALS; TETRAHYDROFURAN POLYMERS. Many important commercial copolymers incorporate polyether linkages (see also POLYESTERS, ELASTOMERIC; POLYIMIDES; POLYSULFONES; POLYURETHANES).

POLYETHYLENE. See ETHYLENE POLYMERS.

POLY(ETHYLENE OXIDE). See 1,2-EPOXIDE POLYMERS.

POLY(ETHYLENE TEREPHTHALATE). See POLYESTERS.

POLYETHYLENEIMINE. See ALKYLENIMINE POLYMERS.

POLYFORMALDEHYDE. See ACETAL RESINS.

POLYHYDRAZIDES AND POLYOXADIAZOLES

A polyhydrazide is a condensation product containing the hydrazide group —CONHNH— in the main chain, obtained by the reaction between hydrazine and a dicarboxylic acid derivative. Significant developments have taken place in recent years, including the synthesis of several new more thermally stable systems; studies of liquid crystalline behavior of these polymers; and evaluation of their fiber- and film-forming properties. However, full commercial exploitation of polyhydrazides and polyoxadiazoles has not been realized, even though their use as fibers, films, and membranes has been known for quite some time and their applications evaluated in certain cases. This may be due to the need for costly amide solvents for processing.

Polyhydrazides

Preparation. In addition to hydrazine, alkanedihydrazines, and phenylenedihydrazines, a range of dihydrazide monomers can be used to obtain different polyhydrazides. These dihydrazides can be prepared by reaction between one equivalent of an alkyl ester and three to four equivalents of hydrazine hydrate in a suitable solvent, eg, methanol, toluene, or *m*-xylene, depending on the diester used.

Table 1. Fiber, Yarn, and Film Properties of Polyhydrazides

Polymer structure	η_{inh}[a], dL/g	Tensile strength, N/tex[b]	Elongation, %	Modulus, N/tex[b]
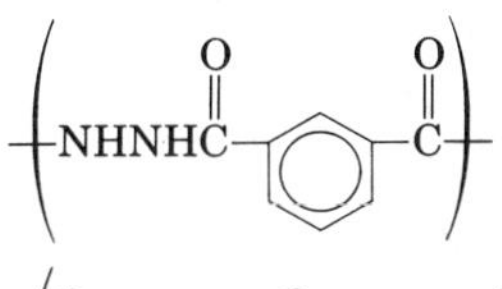	1.00	0.46	14	10.1
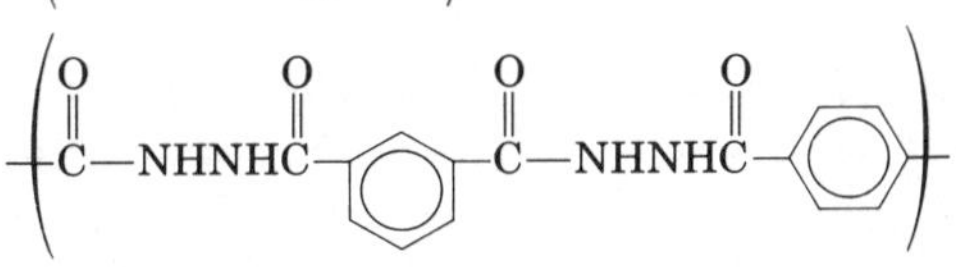	1.03	0.53	8	13.3
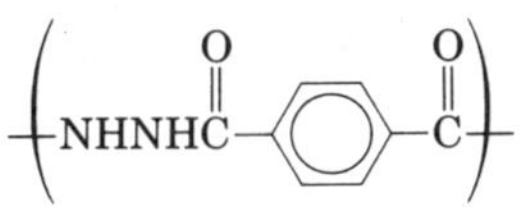	8.28[c]	0.49	15.5	13.9
	8.3[c]	0.54	2.5	28.8
	5.1[d,e,f]	1.12	7.4	32.0
	1.67[f]	0.53	1.3	43.2
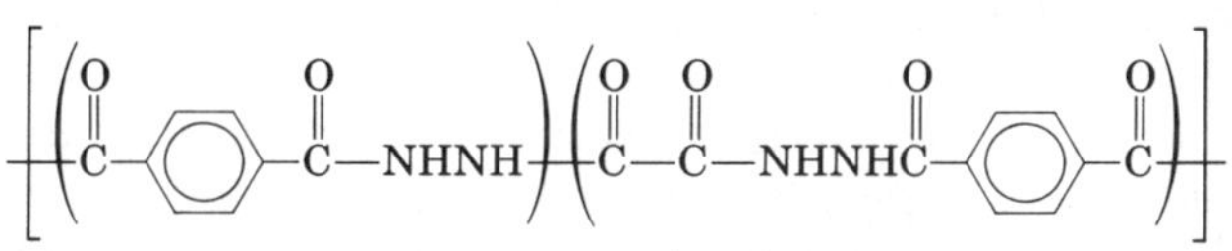		1.38	5.1	42.6
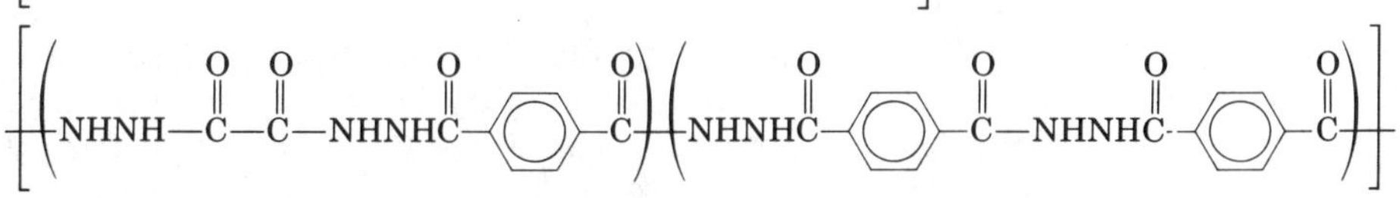	8.4[c]	1.25	4	
	5.8–7.3[c]	1.91	6	39.1
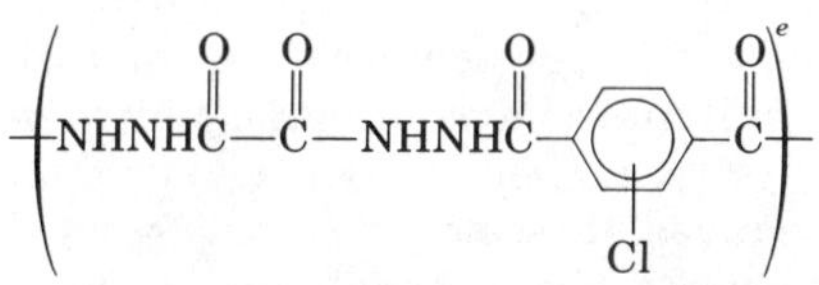	2.40	0.97	5.3	30.5
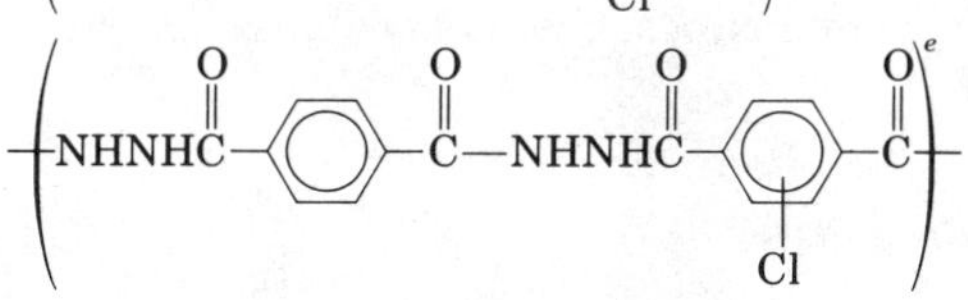	1.44	0.28	2.8	20.0
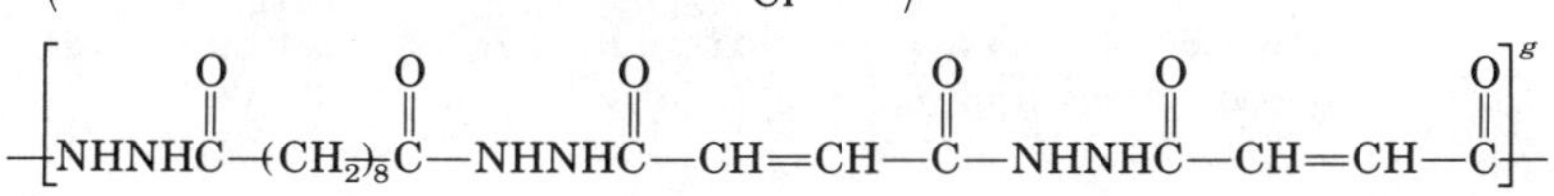	0.58[h]	0.028	5	2.1
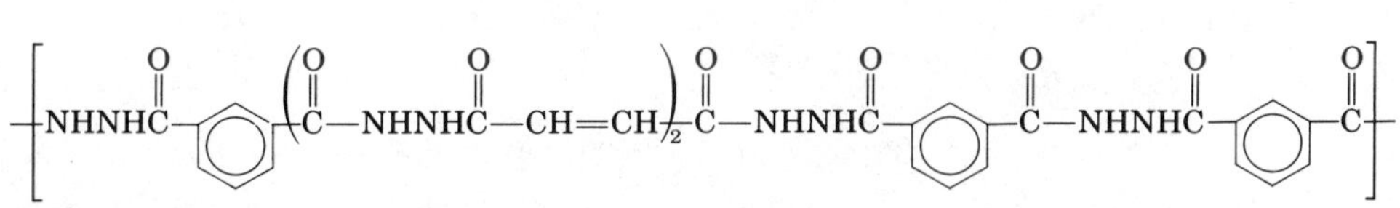	0.36[h]	0.037	1–7	2.1

[a] In DMSO, unless otherwise specified.
[b] To convert N/tex to gf/den, multiply by 11.33.
[c] In dimethylacetamide.
[d] In diethylamine.
[e] Liquid crystalline.
[f] Heat-treated.
[g] Film.
[h] Reduced viscosity $\eta/_{red}$.

A preferred technique for high molecular weight polyhydrazides containing the —R_1CONHNHCOR_2— group is low temperature solution polycondensation under anhydrous conditions. Polyhydrazides containing the —R_1CONHNHR_2— group in the main chain can be prepared by condensing diacid chlorides with alkanedihydrazines or aryldihydrazines.

High temperature solution polycondensation techniques can also be used to obtain polyhydrazides. The phosphorylation technique, developed for the synthesis of polyamides directly from dicarboxylic acids and diamines, has been extended to the synthesis of polyhydrazides as well.

Properties. Table 1 shows some important polyhydrazides and their properties.

Applications. Polyhydrazide fibers have good physical characteristics and can be used as fibers, films, and also in certain catalytic processes.

Polyamide–Hydrazides. ***Preparation.*** Polyamide–hydrazides are usually prepared from dicarboxylic acid chlorides and aminohydrazides, such as *p*-aminobenzhydrazide, by the low temperature solution polycondensation method. They may also be synthesized by condensation of dihydrazides containing an amide linkage with dicarboxylic acid chlorides or dicarboxylic acid chlorides containing an amide linkage with dihydrazides.

Properties. Most of the polymers in this class are soluble in dimethyl sulfoxide, which can be used for spinning, although DMSO is not a suitable solvent for polymerization. These polymers do not have pure melting points. These transitions are obscured by the cyclodehydration reaction that results in polyamide–oxadiazoles.

Applications. The X-500 class of fibers derived from *p*-aminobenzhydrazides have excellent mechanical properties. Depending on their tensile properties, they have been classified as PABH-T(E), PABH-T(T), and PABH-T(G). PABH-T(E) fibers can absorb high energy over an irreversible higher elongation prior to break, a property of importance in certain impact-absorbing devices, such as for single-use aircraft arresters and safety harnesses.

Polyimide–Hydrazides. ***Preparation.*** Polyimide–hydrazides can be prepared by low temperature solution polymerization from diacid chlorides containing preformed imide rings and dihydrazides, or from dihydrazides containing preformed imide rings and diacid chlorides. On heating with polyphosphoric acid, these polymers give polyimide–oxadiazoles.

Properties. These polymers are soluble in dimethylacetamide and *N*-methylpyrrolidinone; some are soluble in DMSO. The specific viscosities of polyimide–hydrazides reported are in the 0.62–1.34 range. They are brittle, difficult to form into film or fibers, and have poor solubility in organic solvents. Their heat resistivity is not ideal in that decomposition occurs above 300°C. They are reported to possess good thermal-insulating properties.

Polyoxadiazoles

Polymers containing oxadiazole units are known to have excellent thermal and hydrolytic stability, in addition to their capacity to form fibers, films, and membranes. Both poly(1,3,4-oxadiazole)s and poly(1,2,4-oxadiazole)s are known and a class of *N*-methylated oxadiazoles has been developed.

Poly(1,3,4-oxadiazole)s. ***Preparation.*** Poly(1,3,4-oxadiazole)s can be prepared by the cyclodehydration of polyhydrazides containing the —CONHNHCO— group.

Properties. Poly(1,3,4-oxadiazole)s have densities of 1.2–1.4 g/cm^3, indicating a lower level of three-dimensional order than polyhydrazides. Poly(1,3,4-oxadiazole)s are thermally more stable than polyhydrazides and have good stability toward hydrolytic attack.

Studies of *p*-phenylene oxadiazole–*N*-methylhydrazide copolymer fibers indicate that these high strength, high modulus fibers are attractive for reinforcement.

Poly(1,3,4-oxadiazole)s containing phenoxaphosphine show self-extinguishing behavior.

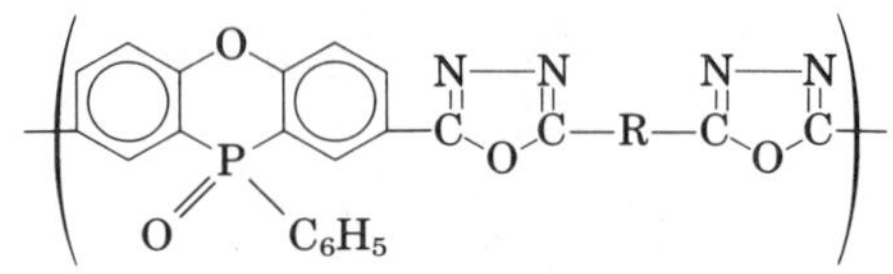

Applications. Poly(1,3,4-oxadiazole) fibers have a combination of properties, ie, good strength and stiffness, good fatigue resistance, and low density. These fibers were found to be competitive in performance when compared to other reinforcing agents such as glass, steel, and commercial poly(phenylene terephthalamide) fibers.

Poly(1,2,4-oxadiazole)s. Poly(1,2,4-oxadiazole)s can be prepared by the thermal cyclodehydration of linear poly(*o*-acylamidoxime)s at 150–300°C.

The properties of poly(1,2,4-oxadiazole)s have not been investigated in detail when compared to poly(1,3,4-oxadiazole)s. The all-aromatic poly(1,2,4-oxadiazole)s are soluble only in strong acids, such as trifluoroacetic and sulfuric acid, and show extremely good hydrolytic stability in both acidic and basic media and good resistance to uv-catalyzed oxidative attack. Although these polymers have high softening temperatures and are reported to be stable at elevated temperatures, no quantitative data regarding their high temperature stability are available. Poly(1,2,4-oxadiazole)s have lower melting temperatures than their precursors, namely, poly(*o*-acylamidoxime)s. When suitably compounded, these polymers may be used as adhesives, caulking compounds, channel sealants, and fuel-tank liners.

Health and Safety Considerations

There have not been any reports of health risks associated with polyhydrazides and polyoxadiazoles.

BIBLIOGRAPHY

A. H. Frazer and F. T. Wallenberger, *J. Polym. Sci. Part A* **2,** 1147 (1964).

J. Preston, W. B. Black, and W. L. Hofferbert, Jr., *J. Macromol. Sci. Chem.* **7,** 45 (1973).

J. Preston, H. S. Morgan, and W. B. Black, *J. Macromol. Sci. Chem.* **7,** 325 (1973).

J. Preston and W. B. Black, *1976 Witco Award Symposium Honoring P. W. Morgan,* ACS Meeting, New York, 1976.

H. C. Bach, F. Dobinson, K. R. Lea, and J. H. Saunders, *J. Appl. Polym. Sci.* **23,** 2125 (1979).

M. J. NANJAN
Hindustan Photo Films
Manufacturing Company, Ltd.

POLYIMIDAZOLE CATALYSTS

The energy crisis and environmental pollution are serious problems that were first encountered in the 1970s and that still affect many aspects of chemistry and the chemical industry. Since then, the development of energy-saving catalysts to promote chemical reactions efficiently under mild conditions and the search for nonpolluting catalysts that are specific or selective for particular reactions and substrates have been pursued. The mechanism of biopolymer catalysis by enzymes has attracted much attention from polymer scientists for the design of enzyme mimetic catalysts. Many enzymes possess histidine residues in the active site. The basic and nucleophilic imidazole ring, a component of histidine, participates in hydrolytic reactions of esters and amides and in the coordination of coenzyme and prosthetic groups.

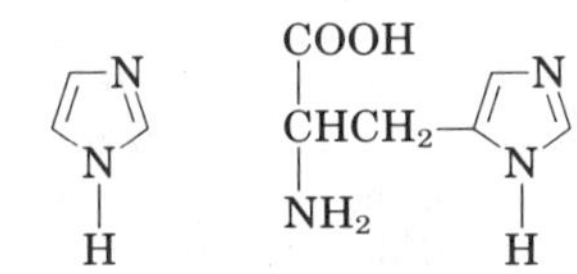

imidazole histidine

However, the catalytic activity of the imidazolyl group involved in enzyme reactions is strongly enhanced by the intramolecular interaction with other functional groups in the environment provided by the polypeptide. Therefore, it seems desirable to develop highly efficient and specific catalysts by designing polymers that carry imidazole groups.

Functional groups, other than the imidazole groups, are also found in the active site of enzymes. However, concentrated research efforts, mainly with imidazole derivatives, stem from the fact that imidazole undergoes a tautomeric isomerization according to pH. Versatile catalytic properties are exhibited by the combination of these species.

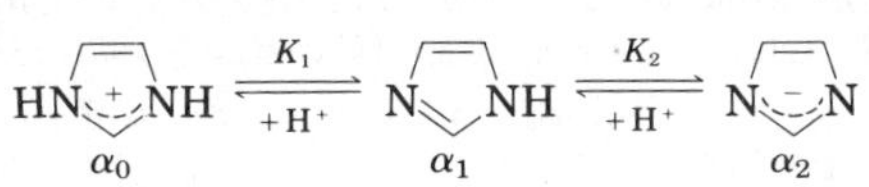

Flexible Catalysts

Polymer catalysts containing the imidazolyl group can be prepared by polymerization or copolymerization of a vinyl compound carrying the imidazolyl group or by the introduction of imidazolyl groups to a preformed polymer by suitable chemical reactions.

One system investigated was the solvolytic reaction of *p*-nitrophenyl acetate with poly[4-vinylimidazole] and the corresponding small catalyst imidazole. Here, studies indicate that the neutral and anionic imidazolyl groups cooperate in the catalytic reaction. The mechanism of the cooperative catalysis proposed is shown in Figure 1. The rate constants in the same reaction catalyzed by poly(*N*-vinylimidazole) and poly(*N*-vinyl-2-methylimidazole), which are polymers of N-substituted imidazole that do not produce anionic imidazolyl functions even in higher pH regions, are not dependent on pH. This observation verifies the participation of anionic imidazolyl groups in the efficient catalysis by poly[4-vinylimidazole]. The small molecule catalyst imidazole does not induce acceleration under the conditions in which anionic imidazoles are produced. The cooperation of neutral and anionic imidazole groups is only likely intramolecularly along the flexible chain of the vinyl polymer.

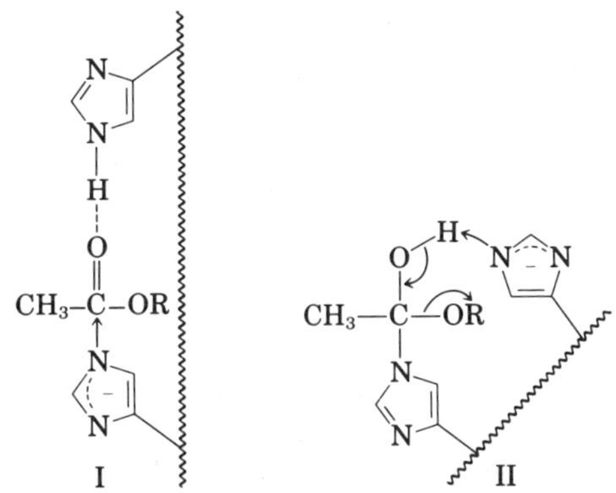

Figure 1. Proposed mechanism for the intramolecular cooperation between neutral and anionic imidazolyl functions.

Rigid Catalysts

The simplest way of preparing polyimidazole catalysts should be the synthesis of histidine-containing polypeptides. Polypeptide chains are more rigid than chains of vinyl polymers and assume a regular structure according to conditions. Since the catalytic activity of enzymes essentially depends on the higher order structure of polypeptide chains, histidine-containing polypeptides and cyclic peptides are promising enzyme models. In fact, they were the first polyimidazole catalysts investigated in this field of polymer chemistry.

Molecular Assemblies Containing Polyimidazoles

The enhancement of reaction rate and enantiomer selectivity in mixed micelles consisting of detergents, detergents containing imidazolyl groups, and amphipathic substrates have been reported by many researchers.

Future Prospects

Purely synthetic approaches to enzyme mimetic polyimidazole catalysts seem difficult. Alternatively, modification of biological polymers will be pursued along four different lines:

1. genetic manipulation of histidine-containing enzymes;
2. chemical modification of histidine-containing enzymes;
3. noncovalent interaction of histidine-containing enzymes;
4. chemical mutation of biologically active substances.

BIBLIOGRAPHY

C. G. Overberger and J. C. Salamone, *Acc. Chem. Res.* **2,** 217 (1969).
Y. Imanishi, *Bioactive Polymer Systems,* Plenum Publishing Corp., New York, 1985, pp. 435–511.
Y. Imanishi, *Adv. Polym. Sci.* **20,** 1 (1976).

J. H. Fendler and E. J. Fendler, *Catalysis in Micellar and Macromolecular Systems,* Academic Press, Inc., New York, 1975, pp. 104–193.

YUKIO IMANISHI
Kyoto University

POLYIMIDAZOLES

Imidazole (glyoxaline) is a five-membered heterocyclic compound, a 1,3-diazole, which may be synthesized, with difficulty, by a number of routes. Preparation from an α-dicarbonyl compound, formamide or ammonia, and an aldehyde is probably the most common method. The amphoteric base occurs widely in nature in the amino acid histidine and its derivative histamine. Polymers containing this functionality have received considerable attention, mostly concerned with the use of imidazole polymers as enzyme analogues, acyltransfer reagents, and catalysts for the hydrolysis of activated esters. Imidazole polymers have, however, shown utility in a wide range of applications.

Heme complexes of imidazole polymers have received considerable attention in studies of reversible oxygen binding, and transition-metal complexes of polyimidazoles, particularly those with Cu(II) and Co(III) have been examined extensively, ostensibly because of the importance of the metal–imidazole linkage in metalloproteins. The dye-binding characteristics of imidazole polymers have led to applications as photographic dye mordants and to the enhancement of dyeability of fibers. Photoconductivity in donor–acceptor complexes of polyvinylimidazole with chloranil, tetracyanoquinodimethane (TCNQ), benzoquinone, and others have been examined, as has conductivity in polymeric imidazolium–TCNQ complexes. Imidazole and imidazolium polymers have also seen application as ion-exchange membranes and resins, and hydrogels.

Less studied applications include additives for lubricating oils, anticorrosive coatings for copper, photosensitized polymer complexes for lithography, and encapsulants for temperature-sensitive and pH-sensitive materials. Polyvinylimidazole complexes with $CdCl_2$ have even been examined as antitumor agents.

The difficulty of the synthesis of vinylimidazole monomers has no doubt limited the commercial exploitation of imidazole-containing polymers. Imidazole polymers have nevertheless become a unique class of functional polymers employed in a wide range of applications.

Monomers

Vinyl substituents can be bonded to the imidazole ring at the 1, 2, or 4 positions.

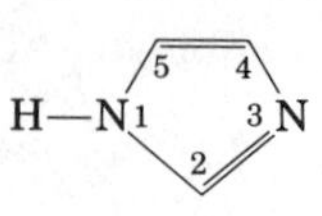

The three isomeric vinyl monomers are known. The 4 and 5 positions are equivalent in imidazole compounds in which nitrogen is not substituted.

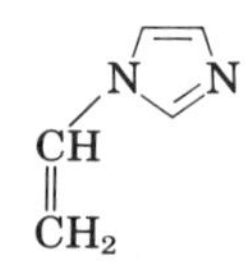
1-vinylimidazole

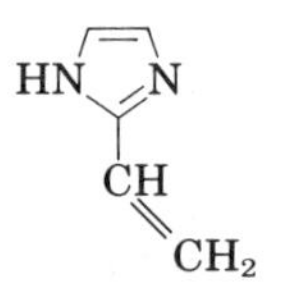
2-vinylimidazole

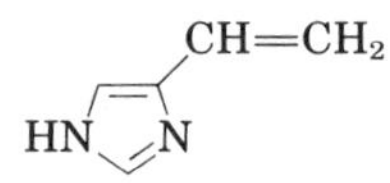
4(5)-vinylimidazole

Homopolymerization of Vinylimidazoles

Studies of the homopolymerization of the three isomeric vinylimidazoles are limited. 1-Vinylimidazole has received the most attention and is typically polymerized by free-radical initiators. The free-radical polymerization of 1-vinylimidazole follows classic solution polymerization kinetics. At high monomer concentrations, however, the homopolymerization appears to be retarded by degradative transfer to monomer. As a result of protonation–deprotonation equilibria, the rate of polymerization of 1-vinylimidazole in water is strongly pH dependent. 2-Vinylimidazoles and 4(5)-vinylimidazoles also undergo facile free-radical homopolymerization. Studies of their homopolymerization characteristics however are scarce.

Copolymerization of Vinylimidazoles

Numerous reports in the literature refer to the copolymerization of vinylimidazoles. 1-Vinylimidazoles have been copolymerized with vinyl acetate, vinyl butyl ether, maleic anhydride, monomethylmaleimide, maleic acid, and others.

Other Imidazole Polymers

There are a number of polymers containing imidazole functionality that are not derived from vinyl monomers. These include condensation polymers, such as poly(L-histidine), which is conveniently prepared by the polymerization of 1-benzyl-*N*-carboxy-L-histidine.

BIBLIOGRAPHY

M. R. Grimmett, "Advances in Imidazole Chemistry" in A. R. Katritzky and A. J. Boulton, eds., *Advances in Heterocyclic Chemistry,* Vols. 12 and 27, Academic Press, Inc., New York, 1970 and 1980.

C. G. Overberger, T. W. Smith, and K. W. Dixon, *J. Polym. Sci. Polym. Symp.* **50,** 1 (1975).

T. M. Handel, H. L. Cohen, and J. S. Tan, *Macromolecules* **18**(6), 1200 (1985).

THOMAS SMITH
Xerox Corporation

POLYIMIDES

Polyimides are condensation polymers derived from bifunctional carboxylic acid anhydrides and primary diamines. They contain the imide structure —CO—NR—CO— as a linear or heterocyclic unit along the main chain of the polymer backbone:

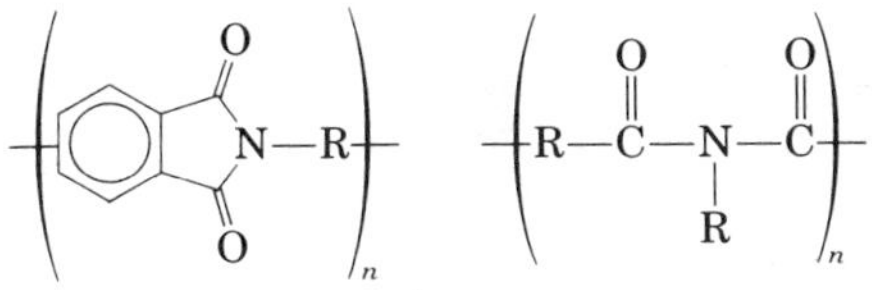

Aromatic, heterocyclic polyimides exhibit outstanding mechanical properties and excellent thermal and oxidative stability and are of major commercial and industrial importance. These materials are widely used in place of metals and glass in high performance applications throughout the electrical, electronics, automotive, aerospace, and packaging industries. Polyimides are available in many forms, including pellets, film, extruded sheet, raw stock, and in solution.

Polyimides are considered specialty plastics because of their outstanding high performance engineering properties. As such, they are priced well above commodity polymers such as polyethylene and polystyrene. Many monomers used to prepare polyimides are also specialty chemicals. Therefore, the price of commercial polyimides covers a wide range, \$8.80–110.00/kg. Polyimides that require special processing or that are provided in a finished form such as Kapton film or machined parts of Vespel are more expensive than powdered or pelletized resin.

Polymerization

Polyimides are produced by condensation polymerization, which is highly dependent on chemical and physical variables. The former include monomer purity, reactivity, and stoichiometry; chain stopper; solvent purity; catalysts; and side reactions. Physical variables include the solubility of monomers, intermediates, and catalysts; the melting point of monomers and intermediates; reaction temperature; and time.

The polymerization method, catalysts, solvents, reaction time, and temperature are usually chosen according to the monomers being used and the physical properties of the desired polymer.

Aliphatic Diamines. Polyimides are prepared from aliphatic diamines and aromatic tetracarboxylic acids in a multistep sequence in the melt. Preparation from aromatic tetracarboxylic dianhydrides is a much more versatile method. Mixing the diamine and dianhydride (20–40%) in DMAC or NMP at ambient temperature produces a solution of poly(amic acid) oligomers. Such solutions are commonly used to cast films or coatings.

Aromatic Diamines. The polymerization reaction is carried out in much the same fashion as for aliphatic diamines and dianhydrides. However, aromatic diamines are less reactive than aliphatic diamines, and longer reaction times are needed. The reaction rate depends mostly on the solvent.

Bisphenols and Dinitrobisimides. The chemistry used in the condensation of bisphenols with dinitrobisimides involves a nucleophilic aromatic substitution reaction in which the polymer chain is generated by the formation of successive aromatic ether bonds. The polymers derived by this route are known as polyetherimides.

Table 1. Thermal Stability[a] of Polypyromellitimides

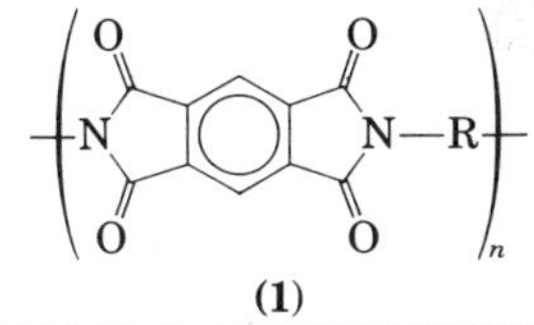

(1)

	Decomposition onset, °C	
R	In nitrogen	In air
1,3-benzene	530	318
4,4′-methylenediphenyl	485	240
hexamethylene	370	285

[a] Determined by dta.

Properties

Thermal Properties. Polyimides derived from aromatic diamines and aromatic dianhydrides exhibit outstanding thermal stability. For example, the thermal decomposition temperatures for typical aromatic and aliphatic polypyromellitimides (**1**) are given in Table 1.

Solvent Resistance. Polyimides exhibit good hydrolytic stability. Aromatic polypyromellitimides are insoluble in most organic solvents. Strong acids, such as concentrated sulfuric or nitric acids, dissolve these materials. Aromatic polyetherimides, on the other hand, display a range of solubilities in a variety of solvents, including chlorinated hydrocarbons such as methylene chloride and chlorobenzene; phenols such as *m*-cresol and *o*-chlorophenol; and dipolar aprotic solvents such as DMF and DMAC. Solvent resistance is measured as the time required to produce surface cracking under a given stress loading. Solvent resistance of polyimides can be improved by cross-linking.

Although polyimides burn, the self-extinguishing behavior of most aromatic polyimides makes them ideal candidates for applications in the transportation and construction industries. Polyimides derived from highly purified monomers have been prepared and range in color from bright yellow to nearly colorless, depending on the structure. Polymers prepared from technical-grade monomers, especially aromatic diamines, may be highly colored. Polyimides may also be translucent or opaque. Mechanical and electrical properties are given in Tables 2–4; properties may vary widely.

Processing

Aromatic polypyromellitimides are not processible in the molten state and are usually processed from poly(amic acid) pre-

Table 2. Mechanical Properties of Polyimides

Property	Ultem 1000	Vespel	Polyimide 2080
tensile strength, MPa[a]	105	87	111
tensile elongation at break, %	60	6.5	10
flexural strength, MPa[a]	145	76	200
flexural modulus, MPa[a]	3300	2902	3306

[a] To convert MPa to psi, multiply by 145.

Table 3. Electrical Properties of Polyimides

Property	Ultem 1000	Ultem 5000	Ultem 6000
dielectric strength[a], kV/mm	28	28	30
dielectric constant at 1 kHz	3.15	2.92	3.0
dissipation factor at 1 kHz	0.001	0.001	0.001
volume resistivity[a], Ω·m	6.7×10^{15}	$>1.1 \times 10^{15}$	1.0×10^{15}

[a] 1.5-mm thickness.

Table 4. Mechanical Properties of Filled Polyetherimides

		Glass reinforcement, %		
Property	Ultem 1000	10	20	30
tensile strength, MPa[a]	105	114	139	169
flexural strength, MPa[a]	145	193	207	228
flexural modulus, MPa[a]	3300	4950	6675	8745
tensile elongation at break, %	6	6	3	3

[a] To convert MPa to psi, multiply by 145.

Table 5. Polyimides for Aerospace Applications Curable by Addition Mechanism

Monomers	Unsaturated end cap	Examples
methylenedianiline, benzophenone dianhydride	nadic imide	PMR-15 LARC-13
methylenedianiline, maleic anhydride	maleimide	Kerimid 601
methylenedianiline, benzophenone dianhydride	phenylacetylene	ATPI Thermid

cursors in dipolar aprotic solvents to yield thin films. Aromatic polyetherimides are easily processed in the melt by conventional injection-, extrusion-, and compression-molding equipment. Processing amorphous thermoplastic materials such as polyimides in the melt requires a decomposition temperature higher than the melt-processing temperature, which in turn must be at least 50–100°C above the T_g of the polymer.

In most nonmelt processes, the polyimide is handled as a powder or poly(amic acid) solution. A combination of heat and pressure or vacuum is used to form the polymer into the desired shape in the fully imidized form. The choice of processing method is determined by the properties of the particular polyimide and the desired product. Many polyimide resins offer numerous processing options to the user.

Modified Polyimides

A large number of structurally modified polyimides have been developed in recent years for specific applications, particularly composites for the aerospace industry, where the need for materials with high temperature capabilities for prolonged periods and resistance to aggressive solvents is crucial; other necessary properties include strength and low flammability. To meet these demanding requirements, several polyimides curable by an addition mechanism have been developed and commercialized (Table 5). These systems are commonly referred to as PMR (polymerization of monomer reactants) or bismaleimide systems. They contain low molecular weight oligomers end-capped with unsaturated functional groups such as olefins or acetylenes (see also ACETYLENE-TERMINATED PREPOLYMERS).

Health and Safety

Prudent handling of all monomers and solvents used in the synthesis of polyimides should be practiced. Dianhydride monomers are skin and eye irritants. Several aromatic diamines used in polyimide synthesis are suspected carcinogens and should be handled with caution; eye and skin contact should be avoided.

BIBLIOGRAPHY

A. K. St. Clair, T. L. St. Clair, W. S. Slemp, and K. S. Ezzell, *Proceedings of the 2nd International Conference on Polyimides,* American Chemical Society, Washington, D.C., 1985.

K. L. Mittal, *Polyimides,* Vols. 1 and 2, Plenum Press, New York, 1982.

J. W. VERBICKY, JR.
General Electric Company

POLY(ISOBUTYLENE-*co*-ISOPRENE).

See ISOBUTYLENE POLYMERS; ISOPRENE POLYMERS.

POLYISOCYANIDES

The first successful syntheses and characterizations of high molecular weight polyisocyanides were shown to occur by cationically catalyzed, repeated α,α-addition under controlled conditions:

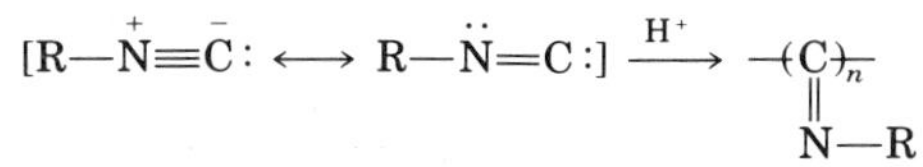

(1) (2)

The Schiff base structure (**2**) assigned to this class of polymers was substantiated by ir and x-ray powder diagrams.

Many polyisocyanides with a large variety of side chains are now available as a result of studies that primarily evolved two catalytic procedures capable of addition propagation with good kinetic chain lengths. The characterization of high molecular weight polyisocyanides was achieved only after the successful synthesis of the first few uncommonly soluble high polymers. Much of this characterization was carried out on poly(α-phenylethyl isocyanide) in solution and in the solid state. Polyisocyanides have interesting properties with regard to solubility, viscosity, cryoscopy, optical rotation and helicity, catalysis, and liquid crystals. The exploitation of combinations of some of the remarkable polymer properties in applications of industrial and commercial interest is imminent.

Polyisocyanides, constituted as they are by unsaturation and large pendent groups at each carbon atom of the chain backbone, promise unusual properties of exceptional kind or degree. Polyisocyanides ideally serve fundamental theoretical study of rigid-rod macromolecules and of helices. Their configuration is uniquely different from that of condensation and vinyl polymers and was deduced from structural considerations, viscosity, and the elevated values of molar rotation, $M_D^{27°} = 500°\ cm^2/g$. Polyisocyanides provide helical rigid rods of highest integrity and solution stability, such that polyisocyanides are mixtures of resolvable left-handed and right-handed helices; chromatographic resolution of racemic poly(t-butyl isocyanide) has been achieved. The Mark-Houwink relationship has been further refined by the concept of isohydrodynamic isoviscosity as an outgrowth of polyisocyanide viscometry.

Monomers

Preparation. Convenient laboratory preparations of monomers involve dehydration of formamides by means of several chemical reagents (Q), such as $ClCOOCCl_3$/t-amine, $POCl_3$, or the Vilsmeier reagent $(CH_3)_2NCHO/SOCl_2$.

$$R{-}NH_2 + HCOOC_2H_5 \longrightarrow R{-}NH{-}CHO \xrightarrow{Q} RN{\equiv}C: + Q{\cdot}H_2O$$

Isocyanides are also available by way of other routes, such as by removal of sulfur or oxygen from isothiocyanates and isocyanates, respectively, and by modification of isocyanides by substitution reactions.

Reactions. Isocyanides undergo α,α-additions leading to compounds of the structural type (**3**), or subsequently to that of type (**4**) by rearrangement if (**3**) is relatively unstable; the first step relates to polymerization, wherein

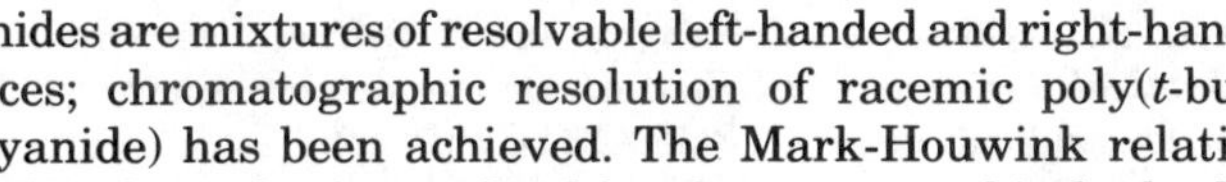

(3) (4)

X and Y–Z become end groups of a catanated chain of carbons.

Isocyanides can be thermally rearranged to nitriles usually at temperatures above 180°C. In one report of a free-radical reaction of an isocyanide, cyclohexyl isocyanide is said to react with benzoyl peroxide to give α,α-addition products.

Oxidation with mercuric oxide converts isocyanides to isocyanates. Heating isocyanides with elemental sulfur forms isothiocyanates.

The most common reaction of isocyanides with transition-metal halides is the formation of coordination complexes. Isocyanides also stabilize cations and higher oxidation states of transition metals. Isocyanide compounds of the following metals have been described: boron, aluminum, copper, silver, gold, zinc, cadmium, mercury, chromium, molybdenum, tungsten, manganese, rhenium, iron, ruthenium, cobalt, rhodium, iridium, nickel, palladium, platinum, and the lanthanides.

Controlled Polyisocyanide Synthesis. Poly(α-phenylethyl isocyanide) provided the first linear high polymer and the first polymer to be characterized by membrane osmometry and light scattering. Tan or yellow preparations of the polymer of M_n = 25,000–270,000 have been produced by a catalyst system that includes a substrate that provides a high surface area, ie, finely ground glass; an acidic substance, sodium bisulfate or concentrated sulfuric acid, coated over the substrate and an inert solvent in which the polymer is insoluble. A free-radical source, eg, air, oxygen, or benzoyl peroxide, although not essential, has been shown to increase the polymerization rate and slightly reduce the molecular weight of poly(α-phenylethyl isocyanide). This catalyst system is usually employed at temperatures of 25–80°C and has successfully initiated the homopolymerization of *sec*-butyl- and n-butyl, n-hexyl-, o-tolyl-, and many other aliphatic and aromatic isocyanides, as well as optically active isocyanides.

A variation of the basic procedure is to produce a fine dispersion of sulfuric acid droplets in n-heptane (in a high speed blender) during inoculation by addition of monomer.

Isocyanides, which have coordinating side groups as does carbylhistamine, are polymerized by nickel chloride in the presence of one equivalent of zinc chloride to block the coordinating sites of the imidazole nucleus competitively.

Thus polyisocyanides are conveniently available by means of protonic and metal ion procedures of initiation, which have in common cationic initiation, use of liquid media in which the polymer product precipitates, propagation proceeding with autogenic stereoregulation, similar requirements of temperature and time of reaction, and similar yields and molecular weights. For highest molecular weights and fewer restrictions, the acid-coated, ground-glass procedure is recommended.

Oligomerization. Several ways of producing oligoisocyanides have been investigated. α-Diimines can be made by condensation of amines with carbonyl compounds, but this method is not easily extended to the preparation of vicinal triimines due to the formation of other products. By analogy to the alkazone reaction between 1-methyl-1-phenylhydrazine and carbohydrates, a reaction was developed that yielded a polyvicinal pentimine from dextrose and α-phenylethylamine. Polymerizations of isocyanides with high chain transfer yield chromatographically resolvable mixtures of vicinal oligoisocyanides. Benzene solvent, saturated with gaseous hydrogen chloride, primarily yields a distribution peaking with tetramer from t-butyl isocyanide. The cationic polymerization of isocyanides takes place with some primary, many secondary, and even with tertiary monomers.

Copolymerization. Some patents have described the reactions of isocyanides and diisocyanides with carboxylic acids. Filaments, films, coatings, and molding powders may be produced from hexamethylene α,ω-diisocyanides and α,ω-dicarboxylic acids. Lacquers, adhesives, and textile auxiliaries have also been prepared by modifying polymers containing pendent carboxylic acid groups with monoisocyanides. The products are polyamides and do not contain the imine repeating unit of structure.

Equimolar mixtures of acetaldehyde and cyclohexyl isocyanide were allowed to react in the presence of $BF_3{\cdot}O(C_2H_5)_2$ at -78°C for two days. Cyclic oligomers and material of presumed imidate structure were obtained.

Mixed isocyanides and diazomethane with isocyanides have been cationically copolymerized using $BF_3{\cdot}O(C_2H_5)_2$ at -78°C in ether solution. In an investigation of the $BF_3{\cdot}O(C_2H_5)_2$-catalyzed copolymerization of binary mixtures of cyclohexyl (M_1) with several arylisocyanides at -40°C, it was found that cyclohexyl isocyanide polymerized more easily than the aryl isocyanides.

Some optically active copolymers have been synthesized using achiral aryl isocyanides in asymmetric induction studies.

Properties

Low solubilities often correlate with high melting points, since both properties are affected by limited entropy gain on dispersal, which is very much a factor with the restricted, tightly wound helical conformation of polyisocyanides. Thermogravimetric data reveal decomposition rather than melting for many polyisocyanides.

Literature reports of anomalous freezing point depression describe systems in which the solute is ineffective in reducing the solvent freezing point. In contrast poly(α-phenylethyl isocyanide) produces an enormous depression on the freezing point of benzene, increasing with dilution until at zero concentration a limiting value of the apparent M_n of 66 is reached for a sample of $M_{osm} = 34{,}800$.

Polyisocyanides of α-phenylethyl, *o*-tolyl, *n*-butyl, *n*-octyl, and benzyl isocyanides have been characterized through x-ray diffraction patterns. A proton nmr of poly(α-phenylethyl isocyanide) has been published, and ^{13}C nmr spectra have been described for the neutral and acidified forms of *n*-butyl, *n*-octyl, and benzyl polyisocyanides.

Applications

There has been no commercial exploitation of polyisocyanides, although a patent was issued on poly(cyclohexyl isocyanide) from which compression-molded disks were prepared at 225°C and a materials patent has been issued on some poly(vinyl isocyanide)s. Ion-transporting vesicles structured from polyisocyanides are currently being studied as biomembrane models.

BIBLIOGRAPHY

F. Millich and R. G. Sinclair, *Polym. Prepr. Am. Chem. Soc. Div. Polym. Chem.* **6,** 736 (1965); *16th Midwest Chemistry Conference,* Nov. 19–20, 1964, Kansas City, Mo., p. 30; *J. Polym. Sci. Part C* **22,** 33 (1968).

F. Millich, E. W. Hellmuth, and S. Y. Huang, *J. Polym. Sci. Polym. Chem. Ed.* **13,** 2143 (1975).

F. Millich and G. K. Baker, *Macromolecules* **2,** 122 (1969).

A. J. M. van Beijnen, R. J. M. Nolte, J. W. Zwikker, and W. Drenth, *J. Mol. Catal.* **4,** 427 (1978).

FRANK MILLICH
University of Missouri-Kansas City

POLYISOPRENE.

See ISOPRENE POLYMERS; RUBBER, NATURAL.

POLYKETANILS.

See AZO POLYMERS.

POLYKETAZINES.

See AZO POLYMERS.

POLYKETONES.

See KETONE POLYMERS.

POLYMER

A polymer is a substance composed of molecules characterized by the multiple repetition of one or more species of atoms or groups of atoms (constitutional units) linked to each other in amounts sufficient to provide a set of properties that do not vary markedly with the addition or removal of one or a few of the constitutional units.

BIBLIOGRAPHY

Pure Appl. Chem. **40,** 477 (1974).

V. Gold, K. Loening, A. McNaught, and P. Sehmi, *Compendium of Chemical Terminology/IUPAC Recommendations,* Blackwell Scientific Publications, Inc., Palo Alto, Calif., 1987.

POLYMER BLENDS

The scientific and commercial progress in the area of polymer blends during the past two decades has been tremendous and was driven by the realizations that new molecules are not always required to meet needs for new materials and that blending can usually be implemented more rapidly and economically than the development of new chemistry. However, successful blend technology requires a sound scientific basis.

The question of whether two polymers are miscible is of paramount importance. A decade ago, very few examples of miscible blends were known. Since that time a very large number of polymer pairs have been found to be miscible and only selected examples and broad categories of structures are discussed here.

Equilibrium Phase Behavior

At equilibrium, a mixture of two amorphous polymers may exist as a single phase of intimately mixed segments of the two macromolecular components or separate into two distinct phases consisting primarily of the individual components. Which of these occurs is dictated by the same principles of solution thermodynamics governing the phase behavior of mixtures of low molecular weight liquids, with some quantitative differences arising from the higher molar masses in the case of polymers. The polymer literature often refers to the molecularly mixed blends as compatible, but there is growing preference to use the more scientifically precise term, miscible, to avoid confusion with other uses of the former term (see COMPATIBILITY). It is important to note that amorphous polymers form glasses upon sufficient cooling and that a homogeneous, or miscible, polymer blend exhibits a single, composition-dependent glass-transition temperature T_g, whereas an immiscible blend has separate glass transitions associated with each phase. If one of the polymers in a blend can crystallize, a separate crystalline phase of that component can form even when the two polymers are miscible (see MISCIBILITY).

Experimental Techniques

A variety of experimental techniques are used to prepare and characterize polymer blends. Some of the more important ones focus on blend preparation, transition behavior, sorption, analogue calorimetry, spectroscopy, and scattering.

Effect of Molecular Structure on Polymer–Polymer Interactions

The literature contains extensive reports on investigations of the equilibrium phase behavior for an enormous number of polymer–polymer pairs. The number of blends known to be miscible has grown so rapidly over the past decade that it is more instructive to attempt to understand these observations in terms of the molecular structures of the components rather than to catalog them. A simple binary interaction model is the framework for this discussion, since the heat of mixing, the primary governing factor, is determined by a combination of interactions between units in each of the polymers being mixed. This approach offers the potential for making *a priori* predictions about the phase behavior of a mixture of two polymers based on observations for other blends or from calorimetric data for low molecular weight analogues. This model envisions each polymer as being comprised of structural units. The interaction between unlike units i and j within the blend or in the polymer itself is characterized by the product of an interaction energy density B_{ij} and the volume fractions (or mole or surface fractions) of these units, regardless of to what they may be covalently bonded. Thus, the heat of mixing a volume V_A of polymer A and a volume V_B of polymer B is given by

$$\Delta H_{mix} = (V_A + V_B) \sum_{i>j}^{\text{blend}} B_{ij}\phi_i\phi_j - V_A \sum_{k>l}^{A} B_{kl}\phi'_k\phi'_l - V_B \sum_{m>n}^{B} B_{mn}\phi''_m\phi''_n \quad (1)$$

where each summation is over unlike pairs without double counting, ϕ_i gives the concentration of units i in the blend, ϕ'_k the concentration of units k in A, and ϕ''_m the concentration of units m in B. There are many ways to subdivide a polymer into units, but for most instances here these units are taken as monomer residues or repeat units.

Homopolymer A–Homopolymer B. For this case, if 1 and 2 denote the monomer residues of the two homopolymers A and B, respectively, equation 1 reduces to

$$B = B_{AB} = B_{12}$$

In the limit of high molecular weights, A and B are miscible only if mixing is exothermic, ie, $B < 0$, which requires obviously that B_{12} must be negative.

Copolymer A–Homopolymer B. Many examples are known where a random copolymer A, comprised of monomers 1 and 2, is miscible with a homopolymer B, comprised of monomer 3, even though neither homopolymer 1 nor 2 is miscible with homopolymer 3. Figure 1 shows a detailed illustration for blends of poly(phenylene oxide) (PPO) with copolymers of *o*-chlorostyrene and *p*-chlorostyrene. The PPO blends with these copolymers exhibit LCST behavior, and Figure 1 shows temperature regions where blends containing 60% PPO have one or two phases, depending on copolymer composition. Neither homopolymer forms miscible blends with PPO, but certain copolymers are single-phase to rather high temperatures. Other systems show similar behavior.

The binary interaction model offers a simple explanation for the increased likelihood of random copolymers forming miscible blends with other polymers. The overall interaction parameter for such blends is

$$B = B_{13}\phi'_1 + B_{23}\phi'_2 - B_{12}\phi'_1\phi'_2 \quad (2)$$

where the ϕ'_i describe the copolymer composition, and $\phi'_1 + \phi'_2 = 1$.

Copolymer A–Copolymer B. When copolymer A contains monomers 1 and 2 and copolymer B contains monomers 3 and

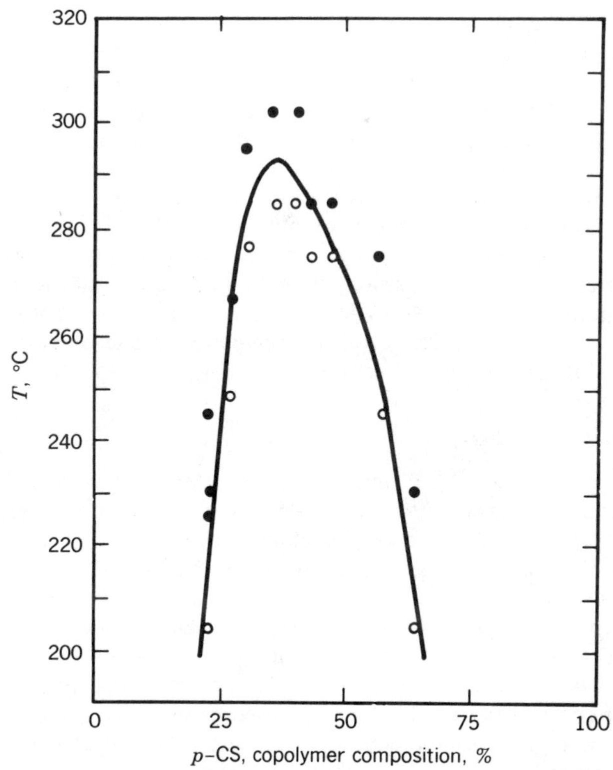

Figure 1. Phase behavior of blends containing 60% poly(phenylene oxide) (PPO) with copolymers of *o*-chlorostyrene (*o*-CS) and *p*-chlorostyrene (*p*-CS) as a function of copolymer composition: ○, one phase; ●, two phases. Courtesy of Butterworth Publishers, Ltd.

4, equation 1 gives as the most general result

$$B = B_{13}\phi_1'\phi_3'' + B_{14}\phi_1'\phi_4'' + B_{23}\phi_2'\phi_3'' + B_{24}\phi_2'\phi_4'' - B_{12}\phi_1'\phi_2' - B_{34}\phi_3''\phi_4'' \quad (3)$$

Clearly, there is a double opportunity for miscibility driven by intramolecular repulsion stemming from the last two terms even when all B_{ij}s are positive.

Spatial Structural Effects. In the binary interaction model used here, it is assumed that mixtures may be modeled by interactions between groups without regard to details about the spatial placement of these groups on the molecules. Clearly, this approximation is not adequate for all situations. In some cases stereoregularity has an effect on the phase behavior of blends, which, of course, is not accounted for in this model. Aliphatic polyesters with branched units in their repeat structures differ very little from those without branches in blends with PVC. Similar branching causes phase separation upon heating for blends with the polyhydroxy ether of bisphenol A and complete immiscibility in blends with bisphenol A polycarbonate. Causes for these branching effects may include shielding of certain groups from interactions, inductive effects, and increased intermolecular spacings.

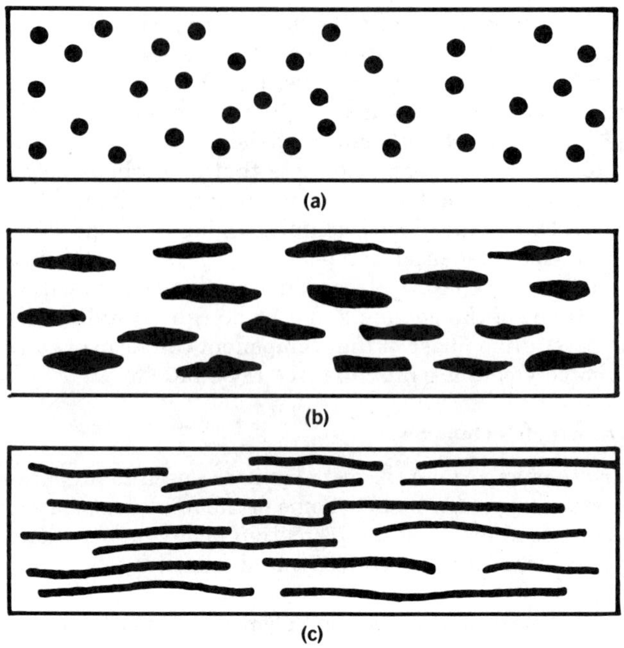

Figure 2. Different types of dispersion of a polymer (dark regions) in the matrix of an immiscible polymer. The spherical droplets (**a**) are progressively extended into platelets (**b**) (biaxial) or fibrils (**c**) (uniaxial) and by deformation.

Blend Morphology

Although many miscible polymer pairs have been identified, most combinations of polymers are immisicible. In many instances, phase-separated blends are preferred for achieving useful results. For example, such polymer–polymer composites yield materials whose stiffness can be adjusted, in principle, to any value between those of the component polymers. However, tailoring blends to achieve this or other characteristics requires, among other things, control over the spatial arrangement or morphology of the phases and some degree of stability once formed. These arrangements may consist of one phase dispersed as simple spheres in a matrix of the other polymer, as shown in Figure 2**a**. On the other hand, the dispersed phase may take the form of fibrils or platelets with varying aspect ratios, as shown in Figure 2**c** and 2**b**, respectively. Another distinct morphology consists of both phases simultaneously having a continuous character or an interpenetrating network of phases (Fig. 3). The typical dimensions of the phases are important in all these morphologies.

Chemical Reactions

Polymers undergo the usual chemical reactions seen in low molecular weight materials. Frequently, this chemical reactivity results in a loss of molecular weight and reduction of properties. Undesirable reactions include those that can cleave the polymer backbone, such as the free-radical chain reactions encountered when thermally processing materials containing abstractable hydrogen atoms, elimination reactions associated with thermally unstable materials such as PVC, and hydrolytic reactions encountered when processing moisture-containing polyesters. Numerous techniques are available for controlling the rates of these undesirable reactions.

On the other hand, beneficial reactions may occur. Ester-exchange reactions can convert immiscible blends of polyesters to single-phase materials. For example, blends of PA with poly(ethylene terephthalate) (PET) or PC are both initially immiscible, but become single-phase mixtures during melt processing, as determined by optical transparency and a single T_g. Polyolefins and other polymers can be chemically modified by grafting unsaturated monomers containing the desired functionality to the polymer melt. Property improvements can result from either improved physical interactions or additional chemical reactions between the compo..ents. For example, grafting of unsaturated acids and esters to PE or PP or ethylene–propylene copolymers produces materials suitable for impact modification of nylons. These materials, as well as maleic anhydride-grafted olefins and other ester grafts, probably undergo amide–ester exchange reactions when mixed with nylon to provide the adhesion necessary for efficient utilization of the olefin modifier.

Properties

In the simplest cases, the properties of polymer blends reflect a composition-weighted average of the properties possessed by the components. In general, however, the property dependence

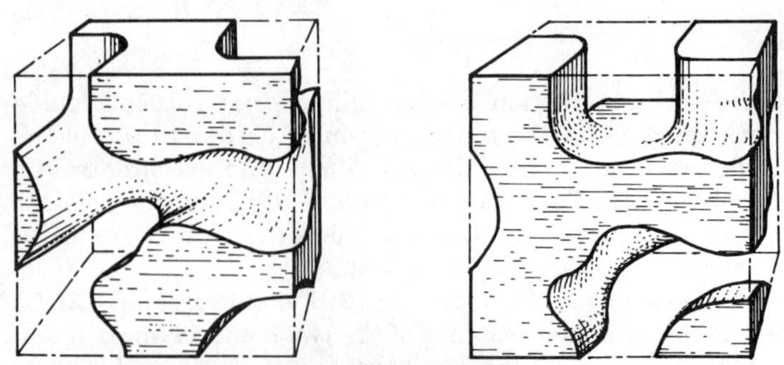

Figure 3. Conceptual illustration of an interpenetrating network of phases by showing the two interlocking materials separated from one another. Courtesy of Carl Hanser Verlag.

on composition varies in a complex way with the particular property, the nature of the components (glassy, rubbery, or semicrystalline), the thermodynamic state of the blend (miscible or immiscible), and its mechanical state (whether its molecules and phases are oriented by the shaping of the material for testing). The amorphous portions of the blend components for a miscible pair form a single phase. The property dependence on composition for such systems is different from that for immiscible mixtures, where each component in the blend is present as a separate phase. The mechanical properties of immiscible mixtures, particularly those related to ductility, are related not only to those of the separate phases but also to the degree of adhesion between the phases. Given all these complexities and possibilities, no simple "rules of mixtures" are adequate to describe the property behavior of blend products.

Miscible Blends. Miscible binary blends of amorphous polymers have only one phase, morphologically the simplest case, and their properties are easily related to those of the blend components. Commercially important examples of this subclass include PPO–PS, PC–CPTI, PVC–nitrile rubber, and PVC–MeSAN (α-methylstyrene–acrylonitrile copolymer) blends. The glass-transition temperature T_g is the primary thermal transition for these blends, and it varies monotonically with composition, following equations such as the Fox equation and

Table 1. Mechanical Properties of Polystyrene and Rubber-toughened Polystyrene

Property	PS	HIPS
tensile modulus, GPa[a]	3.5	1.6
tensile strength at yield, MPa[b]	[c]	17.5
tensile elongation at yield, %	[c]	2
tensile strength at break, MPa[b]	54	21
tensile elongation at break, %	2.1	40
notched Izod impact strength, J/m[d]	21	96
heat-deflection temperature, °C	100	96
light transmission	clear	opaque

[a] To convert GPa to psi, multiply by 145,000.
[b] To convert MPa to psi, multiply by 145.
[c] Does not yield.
[d] To convert J/m to ftlbf/in., divide by 53.38.

Table 2. Commercial Blends of Miscible Components

System	Advantages	Applications
PVC–nitrile rubber	permanent plasticization of PVC, improved processibility	wire and cable insulation, food-contact service
PVC and related copolymers and terpolymers	permanent plasticization, improved processibility	pond liners, automobile interiors, shoes
PVC–MeSAN	increased HDT, improved processibility	rigid siding
PVC–chlorinated PE[a]	permanent plasticizer	wire and cable insulation, food-contact service, pond liners, automobile interiors, shoes
PPO–HIPS	better processibility and toughness than PPO, better HDT than PS	appliance components, business-machine housings
ABS–MeSAN	better HDT than ABS	appliance components
ABS–SMA	increased HDT	automotive applications
PMMA–PVF_2	better chemical and uv resistance than PMMA, better clarity than PVF_2	outdoor film
PEMA–PVF_2	good weatherability, clarity, and chemical resistance	decorative stripes on automobiles
PBT–PET	lower cost, better gloss, and flexibility than PET	electrical and electronic applications, brake and fuel lines
PP–poly(1-butene)	better crystallization rate of polybutene	film
PC–CPTI	better processibility, and better hydrolytic and γ-radiation resistance than PC	sterilizable applications
PBT–phenoxy	improved filler acceptance, cross-linking	electrical and electronic applications

[a] More than 40% chlorine.

others, which were originally written to describe the composition dependence of T_g for copolymers.

Immiscible Blends. An immisicible mixture of polymers shows multiple amorphous phases as determined, eg, by the presence of multiple glass-transition temperatures.

For example, in amorphous immisicible binary mixtures, the stiffness, strength, and HDT properties of the principal component largely determine the properties of the blend. This contrasts with the more nearly linear composition dependence shown for miscible amorphous binary systems and suggests some advantage in forming immiscible blends.

The permeation behavior in immiscible blends is greatly influenced by phase morphology. Blends with lamellar morphologies can exhibit outstanding barrier properties. Such a blend is used to replace metal and glass in certain packaging applications.

Toughened Polymers

Improvement of toughness is a frequent reason for blending. Usually, a small amount of an elastomer is incorporated as a discrete particulate second phase. The properties of such plastics are balanced; the significant improvement of the impact strength is offset by only a small reduction of modulus and tensile strength. This mechanical behavior, not predictable from additivity of polymer properties, is due to the heterophase nature of the blends. The high modulus, brittle polymer remains as a matrix; its tensile properties are only moderately affected by the rubbery dispersed phase, except for the improvement of toughness by the enhanced crazing or shear-yielding mechanisms caused by the rubber phase. Toughened polymers include high impact polystyrene (HIPS) (Table 1), acrylonitrile–butadiene–styrene, poly(vinyl chloride), poly(2,6-dimethyl-1,4-phenylene oxide) blends (PPO), polypropylene, nylon, and epoxy resins.

Commercial Blends and Applications

The 1983 production of commercial alloys and blends (not including rubber-toughened products such as HIPS and ABS) has been estimated to be ca 540×10^6 t or 3% of the total polymer production. It is growing at a rate higher than the

Table 3. Commercial Blends of Immiscible Components

System	Property advantages	Applications
PVC–ABS	better processibility and toughness than PVC, better fire retardancy than ABS	mass-transit interiors, appliance housings
PVC–acrylic	impact-modified, similar to PVC–ABS	mass-transit interiors, appliance housings
PVC–chlorinated PE[a]	better impact than PVC	pipe and siding
PC–ABS	better toughness and HDT than ABS, better processibility and lower cost than PC	appliance and business-machine housings, automotive components
PSF–ABS	similar to PC–ABS, composition can be electroplated, lower cost than polysulfone (PSF)	plumbing fixtures, food-service trays, appliances
PC–PE	better flow and energy absorption than PC	automotive applications
PC–PET	better chemical resistance and processibility and lower cost than PC	tubing, auto bumpers, business-machine housings
PC–PBT	better solvent resistance and processibility than PC	tubing, auto bumpers, business-machinc housings
PET–PMMA	lower cost than PMMA, lower warp and shrink than PET	electrical and electronic applications
PC–SMA	impact-modified, better toughness and ductility than SMA, better retention of properties upon aging at high temperature and lower cost than PC	automotive applications, appliances, cookware
PP–EPDM	better impact and toughness than PP	wire and cable insulation, auto bumpers, hose and gaskets
PE–ethylene copolymers	better chemical resistance, impact, and toughness than PE	film
nylon–ethylene copolymers	better toughness and impact	transport containers, sports equipment

[a] Less than 40% chlorine.

10% growth rate of the entire polymer industry. Exact figures for blends are hard to obtain because manufacturers do not disclose production figures for specific products, but the growth rate of products based on polymer blends has been estimated to be between 11 and 23%/yr. Of all patents in the area, 20–25% pertain to blends.

Tables 2 and 3 give the key properties and applications for miscible and immiscible commercial blend systems. Most of these blends contain only two main components, although many are impact-modified. Research on blends containing more than two main components promises more sophisticated commercial blends in the future. Commercial blends are nearly evenly divided between miscible and immiscible binaries, although the production tonnage is higher for the immiscible systems.

Poor elongation and impact-strength properties are related to poor stress transfer between the phases of the immiscible blend. Since low molecular attractive forces between the blend components are responsible for the immiscible-phase behavior, low attractive or adhesive forces between phases are to be expected and it is believed that this poor interfacial adhesion causes premature failure under stress as a result of the usual crack-opening mechanisms.

Blend ductility can be improved by using the rheological forces associated with flow processing to change the morphology of the dispersed phase from spheres to lamellar or ribbon-like structures with high surface areas and aspect ratios that are oriented in the load-bearing direction. Polymer–polymer adhesion plays a significant role in determining the ductility-related properties of immiscible blends. Better adhesion between ductile and brittle components improves ductility.

The properties of immiscible mixtures may be improved if the interfacial zone can be increased or strengthened by incorporating an interfacially active block or graft copolymer compatibilizer (Fig. 4). In its simplest form, the compatibilizer

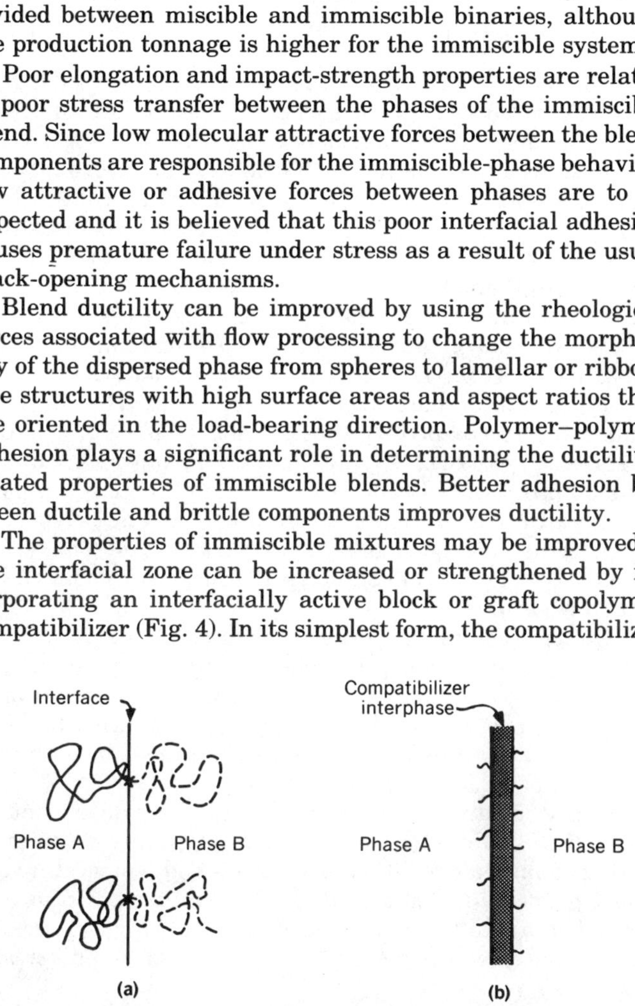

Figure 4. (**a**) Ideal configuration of a block copolymer at the interface between polymer phases A and B. (**b**) Formation of an interphase between phases A and B promoted by a compatibilizer. Courtesy of Carl Hanser Verlag.

has block or graft segments that are chemically identical to those in the respective phases, although nonidentical segments, which are miscible or partially miscible in the respective phases, should work equally well to improve interfacial adhesion and blend properties, permit a finer dispersion during mixing, and provide a measure of stability against gross segregation.

BIBLIOGRAPHY

O. Olabisi, L. M. Robeson, and M. T. Shaw, *Polymer–Polymer Miscibility,* Academic Press, Inc., New York, 1979.

D. R. Paul and J. W. Barlow, *J. Macromol. Sci. Rev. Macromol. Chem.* **18,** 109 (1980).

D. R. Paul and S. Newman, eds., *Polymer Blends,* Vols. I and II, Academic Press, Inc., New York, 1978.

D. J. Walsh, J. S. Higgins, and A. Maconnachie, eds., *Polymer Blends and Mixtures, NATO ASI Series, Series I, Applied Sciences, No. 89,* Martinus Nijhoff Publishers, Dordrecht, The Netherlands, 1985.

D.R. PAUL
J. W. BARLOW
H. KESKKULA
University of Texas

POLYMER CONCRETE

Concrete is the most widely used construction material. Its low cost, convenience, and compressive strength are the principal reasons for its universal acceptance. However, it has shortcomings, most of which are attributable to the Portland cement binder. Shortcomings include freeze–thaw deterioration, destruction by corrosive chemicals, and poor tensile strength (see also CEMENT ADDITIVES).

Polymer concrete is a composite material consisting of selectively graded aggregates with an organic resin binder. The hardening or curing action is created by chemical initiators and promoters, or thermal catalytic and radiation-induced polymerization. The resulting material has qualities far superior to conventional concrete. It is impervious to liquids, resists most corrosive liquids, has excellent dielectric properties and several times the compressive strength of Portland cement concrete, and can be made to resemble granite and marble. These qualities make polymer concrete a viable substitute for many materials that are now used in the architectural, construction, electrical, and industrial areas.

Polymer concretes have been utilized in Europe and Japan for more than 20 years and are now being accepted in the United States. Current applications include utility vaults, precast drains, industrial block, synthetic marble floor tiles, median barriers, piling jackets, bridge-parapet panels, and architectural cladding.

Properties

Material properties for polymer–aggregate systems are compared to high strength Portland cement concrete in Table 1.

Fabrication

Polymer–concrete composites are fabricated in the same way as Portland cement concrete and can be precast or cast-in-place.

Economic Aspects

The cost of polymer–concrete composite systems is a function of resin type and content, aggregate, filler, and amount and type of additives, eg, reinforcing fibers, silane coupling agents, and wetting agents.

The 1986 cost of redimix Portland cement concrete was ca $41.8/m^3$, with a typical density of 88 kg/m^3. The 1986 cost of polymer concrete at approximately the same density was $190–304/m^3$.

Health and Safety

The monomers used as binders in polymer–concrete composites are usually volatile, combustible, and toxic liquids. Attention

Table 1. Material Properties for Several Polymer–Aggregate Systems Compared to Those of Conventional Concrete

Material properties	Polymer–aggregate system I	III	V	High strength conventional concrete
density, kg/m^3	2050	2170	2350	2430
compressive strength, N/mm^{2a}	145.6	126.3	74.2	34.5
flexural strength, N/mm^{2a}	32.6	35.2	14.9	4.7
splitting tensile strength, N/mm^{2a}	13.3	15.2	10.8	3.4
modulus of elasticity, kN/mm^2	10.1	22.1	19.0	23.5
polymer, %	20	14.5	8.3	
max size of aggregate, mm	0.38	2.4	12.7	12.7

[a] To convert N/mm^2 (MPa) to psi, multiply by 145.

to basic safety requirements ensures safe use of polymer–concrete materials.

BIBLIOGRAPHY

J. T. Dikeou, M. Steinberg, L. E. Kukacka, and co-workers, *Concrete-polymer Materials, 3rd Topical Report, USBR REC-ERC-71-6 and BNL 50275 (T-602),* U.S. Bureau of Reclamation (DOI), Washington, D.C., and Brookhaven National Laboratory, Upton, N.Y., Jan. 1971.

B. W. Staynes, "Epoxide Resin Concrete—A Structural Material," in *Proceedings of the 1st International Congress on Polymers in Concrete,* Concrete Construction Publications, Inc., Addison, Ill., 1976.

Y. Ohama, "Development of Concrete-polymer Materials in Japan," in *Proceedings of the 2nd International Congress on Polymers in Concrete,* University of Texas at Austin, Concrete Construction Publications, Inc., Addison, Ill., 1978.

L. Czarnecki, *The Status of Polymer Concrete,* Concrete Int.: Design & Construction, Detroit, Mich., July 1985.

L. C. Muszynski
Lone Star Industries, Inc.

POLYMER CREEP

The term "creep" describes the time-dependent deformation or strain of a material resulting from a sustained load, generally described as an applied stress and stress state. Most solids creep at high temperatures and/or stresses, but polymers are particularly prone to creep as a result of the relatively weak intermolecular forces acting between segmental units. Creep in polymers is the result of complex molecular processes that involve cooperative segmental motions leading to chain backbone conformational changes. In this regard, creep is similar to other rheological processes (see RHEOLOGICAL MEASUREMENT; VISCOELASTICITY).

Engineering applications of polymeric solids require that some degree of structural rigidity be maintained throughout the product's design lifetime. The ability of a material to withstand loads without excessive creep is a major factor in materials selection and design.

Many variables control creep in polymers, but some generalizations can be made about the creep response of the various classes of polymeric materials. Thermoplastics tend to be susceptible to creep, though the degree of creep depends on the individual material and is a function of the applied stress level, stress state, temperature, degree of crystallinity if semicrystalline, prior history of vitrification if glassy, molecular weight, plasticizer concentration, and filler content. In contrast, highly cross-linked thermosets such as epoxy resins are relatively creep-resistant, though they may creep when subjected to adverse conditions, including high temperature and moisture. Lightly cross-linked elastomers exhibit primary creep, but undergo secondary creep only when chain scission or cross-link breakdown and reformation allow chain slippage to occur. Labile cross-link effects depend on the specific chemical nature of the cross-links, the environment, and the stress level.

Polymers that might be unsuitable for structural applications because of excessive long-time creep can often be modified by the addition of glass fibers or other fillers. The resultant reinforced material or FRP (fiber-reinforced plastic) may display greatly reduced creep rates, improved temperature limits, and enhanced stiffness and strength. Many commercial polymers are available in reinforced grades, including polypropylene, polyethylene, nylon, and polycarbonates. These materials are processable as conventional thermoplastics (see REINFORCED THERMOPLASTICS).

Polymer creep generally varies with stress state; uniaxial tension, uniaxial compression, shear, hydrostatic pressure, or some complex combination of stresses may be imposed on a material, depending on the application. In addition to stress state, the stress–time history is important in determining the amount of creep at any particular time.

Creep is a kinetic process and is strongly dependent on temperature. Laboratory creep data are typically obtained at constant temperature. When both stress and temperature are time-dependent, the ability to predict creep rates from isothermal data is difficult at best and commonly involves theory, the validity of which has not been shown to be universally applicable.

Creep mechanisms in polymeric solids may be characterized as either homogeneous or inhomogeneous. Inhomogeneous creep can be defined as that which causes a modification of the material properties in subsequent tests, eg, reapplication of stress to obtain another creep curve.

Homogeneous polymer creep is prevalent in melts, solutions, amorphous rubbers, and glasses, at stress levels for which the theory of linear viscoelasticity applies. It is doubtful that semicrystalline polymers ever creep by a homogeneous mechanism; these materials appear to behave nonlinearly even at low stresses.

Theory

Definition of Creep Compliance Function. The simplest creep experiment involves step loading a material to a fixed stress, then measuring the strain as a function of time. The creep compliance D is defined as

$$D(\sigma,t) = \epsilon(\sigma,t)/\sigma \qquad (1)$$

where $\epsilon(\sigma,t)$ is the strain resulting from the stress applied at time zero and held constant thereafter. The state of stress under which the creep test is conducted is indicated by the symbol used for creep compliance in equation 1: D for uniaxial tension or compression, J for shear, B for hydrostatic pressure. These functions are, in general, different and not interchangeable. The remainder of the theory section considers the so-called one-dimensional theory; hence, state of stress is irrelevant.

Linear Theory and Creep Recovery. Most applications involve complex stress histories; unloading (or stress removal) followed by reloading is frequently encountered. A simple loading and unloading sequence is depicted in Figure 1**a**. The creep strain developed from 0 to t_1 is a function only of the load applied at $t = 0$, so for linear viscoelastic behavior the strain developed is

$$\epsilon(t) = \sigma D(t) \qquad \text{for} \qquad 0 < t < t_1 \qquad (2)$$

According to the Boltzmann theory of linear viscoelasticity, the strain after time t_1 is given by the superposition of a creep curve due to stress (which no longer acts) and another creep curve, beginning at time t_1 (a shift in time) and due to a negative stress $-\sigma$. The net stress is zero, but the strain is given by

$$\epsilon(t) = \sigma[D(t) - D(t - t_1)] \qquad \text{for} \qquad t > t_1 \qquad (3)$$

and is referred to as the creep recovery function. Both the creep and creep recovery curves are shown in Figure 1**b**.

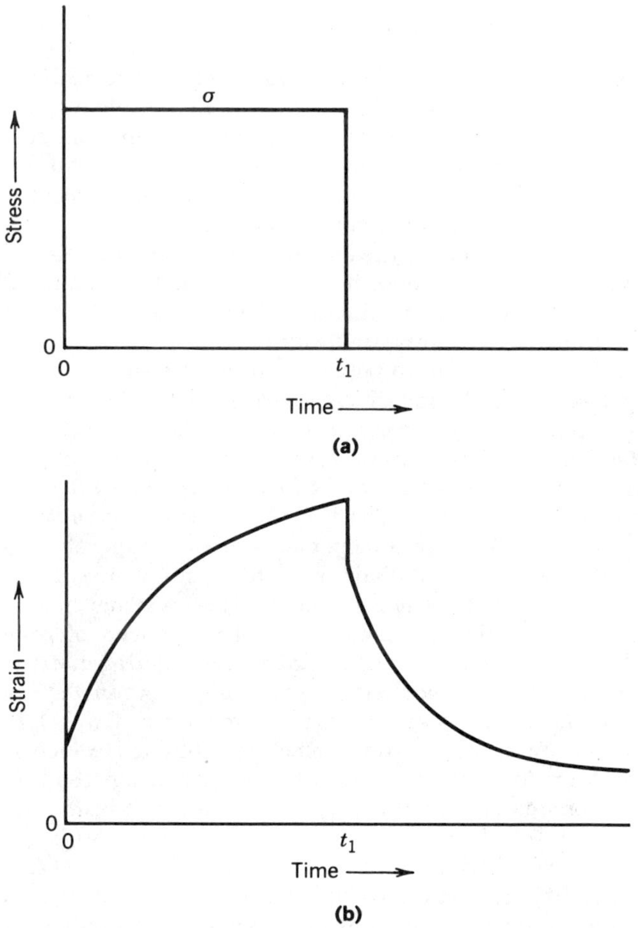

Figure 1. (**a**) Stress–time history used to define creep recovery behavior; (**b**) strain response to the stress history given in (**a**), showing the creep and recovery behavior.

Time–Temperature Superposition. Temperature effects on polymeric solid creep are often critical to the application. Creep rates can become excessive with relatively small increases in temperature, yet the onset of accelerated creep may occur at a time beyond the range of the data base. It is therefore important to understand the general features of TTS, the validity of which and applicability to amorphous polymers in the glass-transition temperature region are well documented.

From a phenomenological viewpoint, the TTS concept states that the viscoelastic response (eg, creep compliance) of a material at some temperature T and time t is equivalent to the response at another temperature T_0 and a different time ξ. The ratio of times ξ/t is a function only of the temperature ratio T/T_0 (independent of time). This implies that a change in temperature serves to change all material time scales uniformly, eg, an increase in temperature contracts (shortens) all times equally. Stated mathematically,

$$D(T,t) = D(T_0, \xi) \qquad (4)$$

where

$$\xi/t = A_T(T) \qquad (5)$$

and where $A_T(T)$ is the time–temperature shift factor and $A_T(T_0) = 1$.

In terms of the retardation-time spectra, TTS implies that the spectrum $H(\tau)$ has been shifted on the log τ axis without change of shape. Similarly, the creep curve at temperature T is the same as the creep curve at temperature T_0, except that it has been shifted on the log time axis by an amount log A_T. The inverse of this procedure forms a basis for accelerated creep testing. A creep curve is constructed at some specific temperature T_0 from segments of creep curves obtained at various temperatures but over a limited time range.

Nonlinear Theories. Nonlinear viscoelastic behavior manifests itself as a failure to conform to one or both of the required conditions previously noted: (*1*) cause and isochronous effect do not scale in a linear manner, and (*2*) superposition of effects from individual causes do not equal the effect from the compound set of causes. The consequences of nonlinear creep are significant in that each stress history studied results in a unique response (strain) function, not readily related to other stress histories. The failure of Boltzmann superposition coupled with the possible inapplicability of TTS effectively negate any theoretical and quantitative basis for accelerated testing. This has led to considerable experimental and theoretical work on nonlinear creep behavior.

BIBLIOGRAPHY

J. D. Ferry, *Viscoelastic Properties of Polymers*, 3rd ed., John Wiley & Sons, Inc., New York, 1980.

J. M. Schultz, ed., *Properties of Solid Polymeric Materials, Treatise on Materials Science and Technology,* Academic Press, Inc., New York, 1977, Pt. B.

J. Gittus, *Creep, Viscoelasticity and Creep Fracture in Solids,* John Wiley & Sons, Inc., New York, 1975.

W. N. Findley, J. S. Lai, and I. Onaran, *Creep and Relaxation of Nonlinear Viscoelastic Materials,* North-Holland, New York, 1976.

S. S. Sternstein
C. S. Van Buskirk
Rensselaer Polytechnic Institute

POLYMERIZATION IN CLATHRATES

Clathrates are crystalline addition compounds in which two components, called host and guest, are linked by weak forces essentially steric in origin, related to the shape and dimension and not to the chemical nature of the molecules. This distinguishes clathrates or inclusion compounds from other adducts, such as hydrates, solvates, or coordination compounds, where hydrogen bonding, ion–dipole, or other specific interactions play an important role in the stabilization of the crystal structure.

The polymerization of unsaturated monomers included as guests in channel-like clathrates represents an important example of polymerization in organized systems. This kind of polymerization is often referred to as inclusion, channel, canal, or tunnel polymerization, and has a wide field of application with regard to the number and type of host compounds and monomers. It can afford polymers with a high degree of constitutional, stereochemical, and supermolecular order.

Inclusion polymerization differs from the solid-state polymerization of crystalline monomers on two main points, ie, the temperature at which the process can be carried out is far higher than the melting point of the pure monomer, and the wider choice of suitable monomers that are no longer confined to easily crystallizing molecules, but can be extended to common monomers that are usually liquid or gaseous at polymerization temperatures.

Extensive research was conducted to determine the scope and limitations of this unusual polymerization method and on the properties of the polymers obtained. Interest has been concentrated on the host–monomer relationship, the radical nature and conformation of the active chain end, the stereochemical control and asymmetric induction, the macroconformation of the native polymer, and the study of the inclusion copolymerization. The main elements controlling polymerization in clathrates are known: by choosing a suitable host–guest pair and polymerization conditions, polymers are obtained endowed with high chain regularity and crystallinity and, when possible, with optical activity; moreover, an extended-chain morphology can be induced. These features are useful in the synthesis of polymer samples of known structure to be used as reference compounds for spectroscopic assignments or for the study of structure–property relationships.

According to the proposed nomenclature, the clathrates most commonly used in inclusion polymerization are defined as binary (or ternary) tubulate clathrates. The different topology of the cavities existing in cage, channel, or layer clathrates is mainly determined by the shape of the host; trigonal or hexagonal molecules often give rise to cage or channel clathrates. The rigidity of the cavity walls, and consequently the selectivity of the inclusion phenomenon, is related to the nature of the interaction of the host molecules with each other.

From the point of view of the phase rule, clathrates are similar to the other addition compounds, such as hydrates, solvates, and racemic compounds. Several classes can be considered, depending on the stability of the empty crystal structure, in the absence of guest, and on the miscibility of clathrate and pure host in the crystal phase. When the empty structure is unstable and does not form mixed crystals with the solid host, as in the cases of urea, thiourea, perhydrotriphenylene (PHTP), deoxycholic acid (DCA), apocholic acid (ACA), and tris(*o*-phenylenedioxy)-cyclotriphosphazene (TPP), clathrates with volatile guests have a definite decomposition pressure, which obeys the Clapeyron law.

The occurrence of polymerization in the channels of the host–monomer clathrate was demonstrated by x-ray diffraction and phase-diagram studies. Experiments carried out in an irradiated matrix show that the polymerization rate suddenly drops to zero when certain limits of pressure or temperature are exceeded. From the examination of the corresponding phase diagrams, the conclusion can be drawn that the range of the pressure and temperature in which polymerization takes place coincides with that of the existence of clathrates with monomers. In other words, polymerization occurs only when the monomeric adduct actually forms, and not in the presence of a simple mixture of host and monomer.

Polymerization is usually initiated by γ-radiation and proceeds through a living radical mechanism. The most interesting examples of polymers obtained are given in Table 1.

Topochemical Control of the Polymer Structure

Control of Structure. The occurrence of polymerization in the clathrate cavities has a marked effect on the polymer structure. Every channel should be considered as an isolated reaction environment in which only linear polymers can form. When using difunctional monomers, such as divinylbenzene, soluble noncross-linked polymers are obtained.

Butadiene and substituted dienes give rise to different polymers depending on the host: in urea, thiourea, PHTP, and TPP, pure 1,4-polymers are produced; in DCA and ACA both 1,2 and 1,4 monomeric units are formed.

As a rule, head-to-tail polymers are produced, but in some cases head-to-head and tail-to-tail are observed. This phenomenon is particularly important in the case of isoprene.

Tacticity. Highly isotactic 1,4-trans polymers are obtained from *trans*-penta-1,3-diene and *trans*-2-methylpenta-1,3-diene by using PHTP as host. The ^{13}C nmr spectrum of polypentadiene obtained from trans monomers at room temperature presents no evidence of *r* diads; a higher content of *r* diads is revealed in the spectra of polymers synthesized from cis monomer and/or at higher temperatures. The tacticity of poly(2-methylpentadiene) was determined after hydrogenation to hemiisotactic polypropylene; a value higher than 96% was measured. Polymers with a variable amount of vinyl, 1,4-cis and trans units, and *m*/*r* stereodefects are obtained by inclusion in DCA or ACA according to the polymerization conditions (host, monomer, temperature, presence of an extraneous guest, etc).

In general, vinyl monomers give rise to polymers with rather low tacticity; isotactic, syndiotactic, or atactic polymers are described, but only few quantitative data are presented.

Optical Activity. The availability of chiral hosts in a pure enantiomeric form (PHTP, DCA, ACA) made it possible to perform asymmetric polymerization in clathrates. Suitable

Table 1. Polymerization in Clathrates

Host	Monomer	Polymer structure
thiourea	2,3-dimethylbutadiene	1,4-trans, crystalline
	other substituted butadienes	1,4
	vinylidene chloride, several monomer pairs[a]	statistical copolymers
urea	butadiene	1,4-trans, crystalline
	vinyl chloride	partly syndiotactic
	acrylonitrile	partly isotactic
(±) PHTP	ethylene	linear polymer
	propylene	partly syndiotactic
	butadiene	1,4-trans, crystalline
	trans-2-methylpentadiene	1,4-trans, highly isotactic, crystalline
	other substituted butadienes	1,4-trans
	several monomer pairs[b]	block copolymers
	several monomer pairs[a]	statistical copolymers
(−) PHTP	*trans*-pentadiene	1,4-trans, highly isotactic, crystalline, optically active
	isoprene + *trans*-pentadiene[b]	optically active block copolymer
(+) DCA	butadiene	mainly 1,4-trans
	cis-pentadiene	mainly 1,4-trans, atactic, optically active
	other substituted butadienes	mainly 1,4-trans
(+) ACA	*trans*-pentadiene	mainly 1,4-trans, optically active
	trans-2-methylpentadiene	mainly 1,4-trans, optically active
	other substituted butadienes	1,4-trans
TPP	vinyl chloride	mainly syndiotactic
	butadiene	1,4-trans, crystalline
	divinylbenzene	linear polymer, not cross-linked
	acrylonitrile	mainly isotactic
	methyl methacrylate	mainly syndiotactic
	several monomer pairs[a]	statistical copolymers

[a] The two monomers are included together at the same time.
[b] The second monomer replaces the former after partial polymerization.

monomers are pentadiene and 2-methylpentadiene, two planar achiral compounds that can be transformed into chiral 1,4-polymers. *trans*-1,3-Pentadiene included in (−)PHTP gives an optically active polymer with very high stereoregularity but with low optical purity (7%); on the other hand, polymerization of *cis*-1,3-pentadiene included in DCA gives a polymer with higher rotatory power but with lower stereoregularity. Highly optically active poly(2-methylpentadiene) was obtained in DCA and ACA.

Asymmetric polymerization in clathrates was the first example of asymmetric synthesis occurring inside a crystalline phase; moreover, it represents the only case of asymmetric radical polymerization known. Asymmetric induction occurs by a through-space and not by a through-bond mechanism. It is the chiral shape of the channel and not the presence of a chiral initiator that is responsible for the preferential formation of polymeric chains of a given configuration.

Macroconformation of Native Polymer. Some crystalline polymers obtained by inclusion polymerization show a particular morphology strictly related to the topochemical nature of the process. In a perfect crystal, cavities are not connected to each other and the stack of monomers existing inside a given channel transforms in a nonfolded polymer chain, parallel to but independent of the chains existing in adjacent channels. Thus, inclusion polymerization generates under mild operating conditions polymers with an extended-chain macroconformation, a result generally obtained by using extremely severe conditions, eg, crystallization under very high pressure.

Copolymerization. The living nature of channel polymerization makes possible the synthesis of block copolymers by subsequent inclusion and polymerization of two monomers. Simultaneous inclusion of monomers leads to statistical copolymers. This phenomenon has a much wider occurrence than first thought. The monomer distribution can be determined by ^{13}C nmr: monomers similar in shape give rise to a Bernoullian distribution; in other cases, a tendency toward the formation of blocks of like units is revealed.

BIBLIOGRAPHY

J. F. Brown and D. M. White, *J. Am. Chem. Soc.* **82,** 5671 (1960).

K. Takemoto and M. Miyata, *J. Macromol. Sci., Rev. Macromol. Chem.,* **18,** 83 (1980).

M. Farina in J. L. Atwood, J. E. D. Davis, and D. D. MacNicol, eds., *Inclusion Compounds,* Vol. 3, Academic Press, Inc., London, 1984, p. 297.

M. Farina, G. Di Silvestro, and P. Sozzani in D. J. Sandman, ed., *Crystallographic Ordered Polymers, ACS Symposium Series* **337,** American Chemical Society, Washington, D.C., 1987, p. 79.

MARIO FARINA
GIUSEPPE DI SILVESTRO
Università di Milano
Italy

POLYMERIZATION PROCEDURES, INDUSTRIAL

Polymerizations have certain characteristics that present challenging problems to scale up to industrial quantities. The most significant characteristics to consider are the exothermic nature of polymerizations, potential for large increases in the viscosity of the polymerization mass, low diffusivities, and low thermal conductivities of the polymerization mass.

Technical factors that must be studied in the design of industrial polymerization facilities include the size and geometry of the reactor, temperature control, accuracy of raw material charge, heat removal, agitation, exclusion of air or other contaminants from the polymerization, control of the degree of polymerization, effect on the distribution of molecular weight, removal of the polymer from the reactor, separation of the polymer from the reaction mass, recovery and, if feasible, recycling of residual monomers and solvents, and cleaning of the equipment.

Design of the industrial facilities is always under economic and environmental constraints. Economic factors include the choice of optimum equipment size and geometry to minimize operating costs, selection of materials of construction resistant to corrosion and to formation of an adherent polymer deposit, design of equipment and piping systems for safe use and for ease of dismantling, and the choice of polymerization conditions that best avoid excessive costs and excessive number of operations in the process. Design of the industrial facilities must include provisions for containing a possible runaway polymerization and must meet all OSHA and EPA regulations. These agencies provide information in publications on best available technology (BAT), such as the EPA's New Source Performance Standards on Polymer Plants.

General Reaction Characteristics

Most commercial polymers have an average molecular weight of ca 10^4 to 10^7 and consist of a mixture of macromolecules with a distribution of molecular weights. Melts and solutions of such large molecules are very viscous. Polymer molecules are produced by bond-forming reactions (polymerization) of small molecules with the potential of proceeding indefinitely, and almost all polymerizations are exothermic. The heat of reaction must be removed to maintain isothermal conditions.

Classification of Polymerizations. Polymerizations can be classified according to the bond-forming mechanisms: chain-growth or step-growth polymerizations. In chain-growth polymerization, single-monomer molecules react successively and rapidly with the growing polymer chain, and large macromolecules are formed almost immediately. High molecular weight molecules are obtained even at very low monomer conversions.

In step-growth polymerization, the molecules build up slowly. Polymerization takes place between two monomer molecules, a monomer and polymer segment, or two polymer segments. Usually, step-growth polymerization is relatively slow and must be continued to a very high conversion to attain molecular weights sufficiently high for industrial applications. The classifications differ in the rate of growth of the individual molecules, not necessarily the rate of polymerization of the monomers.

Polymer Properties. Important factors determining the properties of polymers are the composition and structure of the macromolecules, the molecular weight (degree of polymerization), and the molecular weight distribution. The general term structure includes end groups, branching, steric order, cross-

Table 1. Characteristics of Polymerization Processes

Process	Example
Bulk polymerization	
homogeneous, monomer solvent for polymer	methyl methacrylate
homogeneous, solvent additive and monomer soluble in polymer melt	nylon-6,6
Solution polymerization	
homogeneous, monomer and polymer soluble in diluent	styrene in ethylbenzene
heterogeneous, monomers and polymer soluble and catalyst insoluble in diluent	copolymerization of ethylene and propylene in a hydrocarbon solvent
heterogeneous, polymer A dissolved in monomer B, phase separation in polymer	high impact polystyrene with polybutadiene rubber
Suspension or precipitation polymerization	
heterogeneous, polymer insoluble in monomer	vinyl chloride, no diluent
heterogeneous, monomer miscible with diluent, polymer insoluble	acrylonitrile in water
heterogeneous, monomer and polymer insoluble in dispersant	vinyl chloride in water
Emulsion polymerization	
heterogeneous, monomer and polymer dispersed in a diluent (eg, water) with a surfactant, colloidal dispersion	butadiene and styrene comonomers
Interfacial polymerization	
heterogeneous, monomers A and B dissolved in immiscible solvents	isophthaloyl dichloride in methylene chloride and *m*-phenylenediamine in water
Solid-state polymerization	
condensation of polymer molecules via reaction of bifunctional end groups	nylon-6,6

linking, and in the case of copolymers, the sequence of monomer units. Thus the quality of the polymer is determined not only by the nature of the monomer building blocks, but also by the kinetics of the polymerization. Monomer concentration, temperature, pressure, and extent of conversion are among the variables affecting the polymerization kinetics, which are also affected by such physical processes as the transport of mass and energy and the momentum of mixing and flow. Detailed information on the effects of transport processes (mass, momentum, and energy) on the chemical reaction path have been presented.

Methods of Polymerization. Four general methods of polymerization are used commercially: bulk, solution, suspension, and emulsion. Interfacial and solid-state polymerizations have also had limited use in commercial processing. Polymerizations are also identified as homogeneous or heterogeneous, depending on the solubility of the reactants and product.

Table 1 summarizes these polymerization methods and their characteristics. A list of polymerization methods used to produce a number of industrial polymers is given in Table 2.

Processes. Polymerizations are carried out in batch, semibatch (or semicontinuous), and continuous systems.

Engineering Aspects of Polymerization Systems

The reaction of monomers to form useful polymers is similar to other chemical reactions. However, the polymerization kinetics and size of the macromolecules result in unique processing problems, which become increasingly important in scale-up to industrial quantities. To produce polymers on a large scale it has been necessary to develop new technologies and design suitable equipment. Overall U.S. production of polymers in 1986 reached 23.7×10^6 t. Production quantities of some of the commodity polymers for 1986 are included in Table 2.

The design of the industrial reactor and polymerization system aims at maximizing the rate of output (quantity of polymer per unit time), without sacrificing safety or the most relevant polymer properties, and concurrently minimizing costs.

Improvements in polymerization and polymer quality have been attained from extensive studies and experience, and the use of mathematical models for polymerization. The quality of

Table 2. Polymerization Methods in Commercial Use[a]

	Polymerization method					1986
Polymer	Bulk	Solution	Suspension	Emulsion	Interfacial	Production, 10^3 t
Chain-growth polymerization						
ABS[b] polymers		O		C		794
acrylic polymers	C	C	C	C		281[c]
butyl rubber[d]		C				
EPDM[e]		C				231
acetal resins	C		O			
polyethylene						
low density[f]	C	O	C	O		4032
high density[g]		C	C			3253
polyisoprene		C				
polypropylene	C	C	C			2635
polystyrene	C	O	C	O		2014
poly(vinyl acetate)	O	O	O	C		
poly(vinyl chloride)	O	O	C	O		3293[h]
fluoroelastomers				C		
Step-growth polymerization						
aliphatic polyamides						1143[i]
nylon-6,6	C[j]	O				
nylon-6	C	C				
aromatic polyamides		C				
phenolic resins		C				1234
polyester	C					576[k]
						1497[i]
polycarbonate		C			O	192
polyimide		C				
polyurethane	C	C				
amino resins		C				690

[a] C identifies a method of polymerization commonly used commercially; O identifies a method of polymerization occasionally used commercially.
[b] ABS = poly(acrylonitrile-*co*-butadiene-*co*-styrene); includes elastomers and thermoplastic polymers and co- and tripolymers, excludes high styrene latex.
[c] Fiber production; includes modacrylic.
[d] Butyl rubber = poly(isobutylene-*co*-isoprene).
[e] EPDM = poly(ethylene-*co*-propylene-*co*-diene) elastomers.
[f] Density ≤ 0.904 g/cm^3.
[g] Density > 0.904 g/cm^3.
[h] Includes copolymers.
[i] Fiber production.
[j] Reaction started in aqueous solution, and the water is removed during polymerization by evaporation.
[k] Thermosetting polymers; unsaturated polyesters.

the mathematical models continues to improve with improved understanding of the process, improvement in measurements, and the rapidly expanding use of computer-simulated studies of the polymerization systems. In recent years computer programs have been developed to help select approximate operating conditions for some polymerization processes (see POLYMER MODELING; PROCESSING, MODELING).

The center of a polymerization system is the reactor (see REACTOR TECHNOLOGY). Important features of a reactor are its geometry, agitation, heat-transfer system, and instrumentation.

Reactor Types

Numerous reactors have been developed for commercial polymerizations; the reactor is usually some variation of a stirred tank reactor, a series of stirred tank reactors, a linear-flow tubular reactor, or a static mold reactor. A hybrid reactor may be constructed by combining features of two different reactors or units of two or more reactors. Reactors are conventionally described by the simple idealized models shown in Figure 1.

Selection of the reactor for a polymerization depends primarily on the polymerization kinetics, satisfactory transport of mass and energy, requirements for the polymer properties, and the state of the polymer product as marketed, ie, solid, latex, or solution. As new technologies evolve, the reactors and ancillary equipment are modified to increase output and improve polymer uniformity. The general type of reactors that are useful for different methods of polymerization are summarized in Table 3.

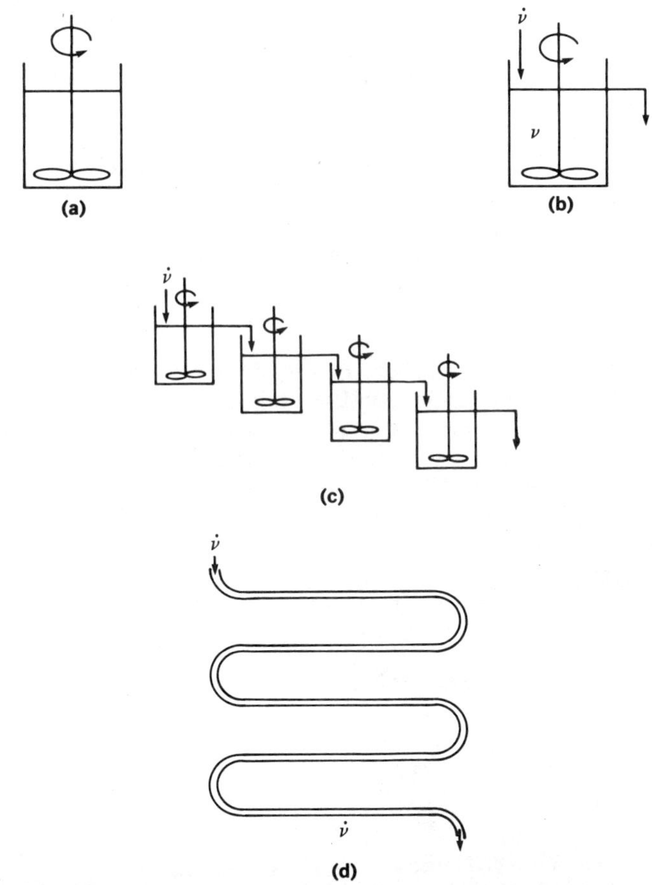

Figure 1. Model reactors: (**a**) stirred tank reactor; (**b**) continuous-flow stirred tank reactor; (**c**) cascade of continuous-flow stirred tank reactors; (**d**) continuous linear-flow reactor. v = volume; $\dot{v}$ = volumetric flow rate.

Reactor Sizing and Scale-up

Design of an industrial polymerization facility aims at maximizing the rate of output (quantity of polymer per unit time),

Table 3. Reactors Used in Different Polymerization Processes

	Stirred tank Reactors					Static reactor[c]	
			Continuous		Linear-		
Polymerization reaction	Batch	Semibatch	Single tank	Multiple tank[a]	flow reactor[b]	Cast	RIM[d]
Chain-growth polymerization							
free-radical							
bulk	x				x	x	
solution	x	x	x	x	x		
suspension							
precipitation	x		x				
bead	x						
emulsion	x	x	x	x			
ionic							
bulk	x						
solution	x	x		x			
precipitation	x		x	x	x		
Step-growth polymerization							
bulk	x			x	x	x	x
solution	x			x	x		
interfacial	x						
solid-phase	x				x		

[a] Cascade reactor.
[b] Continuous; also called a plug-flow reactor.
[c] Batch.
[d] Reaction injection molding.

without sacrificing safety or the most relevant polymer properties, and concurrently minimizing costs.

A preliminary design of a new polymerization process and an economic evaluation are usually begun as soon as successful laboratory work identifies a product with market potential. Although the data may be relatively meager, considerable information about the process can be generated. Areas for additional research or need for more careful calculations may be identified. Most importantly, a preliminary design is useful in eliminating projects having unfavorable economics.

If the preliminary design of the process appears favorable, the assumptions of the design are usually tested in a pilot plant to obtain data that may be used to provide a more precise design and an improved economic evaluation. Construction and operation of a pilot plant is expensive. However, the pilot plant need not include every major step of the commercial process.

Computer models have been developed for a number of polymerization processes. Such models take into account numerous variables and allow selection of approximate operating conditions for initial testing.

To maximize output and minimize costs, the current trends are not only to increase the size of the reactor, but also to reduce diluent:monomer ratios, improve initiators or catalysts, convert more polymerizations to continuous processes, and, where feasible, use an in-line monomer-to-product process.

Instrumentation

Process control is crucial for satisfactory operation of commercial polymerization systems. An increase in control problems has resulted from several factors. The increase in polymer production by the use of larger reactors or reduction in diluent complicates the transfer of mass and heat. Specifications for polymer properties and consequently polymerization conditions have become more critical. Concurrently, government regulations on environmental emissions and safety have tightened. These process control problems require improvement in the performance of process controllers.

Advanced control systems are coming into use as more capacity is added to distributed controllers. Various control loops, which may be combined to provide a system to control the chemical reactor, have been described. The hardware is available and can be programmed by the user. The software is evolving rapidly (see PROCESSING, COMPUTER-CONTROLLED).

The use of computers provides memory capacity for data and recipe storage, high speed solution of complex mathematical models, and optimization of the process operation to improve product yield or quality. Optimizing the process operation may include control loops to program the feed, temperature, and pressure. Control loops are also used to provide recordkeeping.

Reactor Cleaning

Polymerization reactors tend to foul due to the buildup of adherent polymer on heat-transfer surfaces. In solution polymerization, this deposit may consist of polymer gels that are insoluble in the solvent used. In emulsion polymerization, it may be a coagulum formed because of excessive agitator shear or it may be caused by the impingement of sticky semipolymerized droplets against the reactor wall. Whatever the mechanism, most polymer systems must be designed for easy cleaning.

In some batch reactors, spray rinse valves are often installed in the top reactor head to allow washing of the walls of the vessel while emptying the reactor, without opening the reactor manhole. In some modern reactors the hydraulic reactor cleaner is installed inside the reactor. At the end of the cleaning cycle, additives are sometimes incorporated in the pressurized water to condition the reactor wall for subsequent polymerization. If the polymerization is water-sensitive, the reactor must be dried thoroughly after cleaning with the water jet.

Solvents are sometimes employed for cleaning polymerization reactors with heavy wall deposits. Solvent cleaning can be automated. The hot solvent is sprayed onto the reactor surface to swell or dissolve the polymer deposit.

Cleaning of polymerization reactors has been accomplished for many years by manual scraping and chipping. Even with special scraping tools to minimize damage to the reactor surfaces, this type of cleaning tends to cause scratches that foul quickly, resulting in more frequent cleaning schedules.

Safety

AIChE has published information on chemical-plant safety research. The information is available as a computer program and in a series of manuals and reports.

Storage tanks for polymerization reactants and solvents are located a safe distance from production facilities and other buildings. Reactants and solvents are usually pumped to the polymerization facilities. Storage, transfer, and processing of polymerization reactants must be carefully planned because many reactants are potentially hazardous. Some reactants are toxic and highly flammable, and form explosive mixtures with air.

Primary safety considerations in the polymerization system are the prevention of fire or explosion, the proper relief of reactor overpressure caused by uncontrolled polymerization or other accidents, and containment and disposal of vented materials. Sources of safety problems include equipment failure, loss of utilities, human error, and instrument failure.

BIBLIOGRAPHY

J. A. Beisenberger and D. H. Sebastian, *Principles of Polymerization Engineering,* Wiley-Interscience, John Wiley & Sons, Inc., New York, 1983.

H. Gerrens, *Chemtech* **380** (June 1982); **434** (July 1982).

J. P. Clark, *Chemtech* **664** (Nov. 1975); **23** (Jan. 1976); 235 (Apr. 1976).

B. G. Liptak, *Chem. Eng.* **69** (May 26, 1986).

VIRGINIA MENIKHEIM
Consultant

POLYMERIZATION PROCEDURES, LABORATORY

This article is concerned with the procedures and techniques employed to carry out polymerizations in the laboratory, the equipment required, and the manipulation of monomers and the resulting polymers (see also ANIONIC POLYMERIZATION; CATIONIC POLYMERIZATION; EMULSION POLYMERIZATION; RADICAL POLYMERIZATION; INITIATION; INTERFACIAL POLYMERIZATION; SUSPENSION POLYMERIZATION). Textbooks and other references provide excellent background material.

Polymerization Methods

A number of experimental methods have been developed for preparing polymers. Each has been worked out to solve the problems attendant to monomer types, reaction mechanisms, and polymer requirements. Thus although bulk polymerization, direct conversion of monomer into polymer, can be used for many materials, high heat of reaction, high melt viscosity, and the final physical form of the polymer desired have forced investigators to develop other methods. Distinctions among various types of polymerizations are sometimes blurred, and frequently a particular polymerization may have the characteristics of two categories. Methods include bulk polymerization, solution polymerization, suspension polymerization, emulsion polymerization, and interfacial polymerization.

The Course of the Polymerization. The course of a polymerization can be followed in various ways. A change in the density of the reaction medium is easily determined with a dilatometer and provides a quick measure of the reaction rate (see DILATOMETRY). Changes in the viscosity can be followed with a viscometer; this method is used for the alkoxide-initiated polymerization of 1,2-epoxides. The solids content is determined by polymer precipitation from an aliquot of the reaction medium. Variations in refractive index are used to follow the air-catalyzed oligomerization of allylic monomers. Titration is employed wherever applicable, for example, to determine the acid number in the formation of polyesters from anhydrides or diacids and glycols. The oxirane number for epoxide content in polyether formation is another example. Instrumental methods, such as ir or nmr spectroscopy, can be used to follow reduction in percent unsaturation, which is a direct measure of monomer uptake in addition polymerizations. For condensation polymers, the rate of by-product formation is a direct measure of the degree of polymerization.

Reagents and Their Purification

Monomers and Diluents. The purity of monomers and diluents is of utmost importance. The most common methods of purification are distillation, recrystallization, adsorption columns (see CHROMATOGRAPHY), sparging with an inert gas, and chemical treatment. The pure materials are stored in an inert atmosphere and sealed. Common analytical techniques include high pressure liquid chromalography (hplc), gas chromatography, ir spectroscopy, mass spectroscopy, nmr, and combinations thereof.

Many diluents can be purchased in sufficiently high purity so that further purification is unnecessary, eg, hplc-grade solvents. In some cases, it may be necessary to remove preservatives; these are indicated on the label.

Initiators. Unlike catalysts, initiators are consumed in the course of the reaction. Molecular fragments of the initiator attach to the polymer chain ends.

The most important free-radical initiators are the organic peroxides and hydroperoxides (see PEROXY COMPOUNDS).

Most organic peroxides and azonitriles are not soluble in water and thus not suitable for emulsion polymerization. The most common water-soluble initiators for emulsion polymerization are potassium and ammonium peroxydisulfates.

A redox system is often used for low temperature initiation of vinyl polymerization, in which a reducing agent reacts with the peroxide or other oxidizing agent.

Catalysts. The catalyst is chosen according to the monomer and, in some cases, according to the polymer desired.

Molecular-weight Regulators. The molecular weight is controlled by regulators. For polyolefins, such as polyethylene or polypropylene, produced by transition-metal catalysis, hydrogen is commonly used. For free-radical polymerization, aliphatic mercaptans are suitable. The molecular weight of condensation polymers, such as nylons or polyesters, can be regulated by using an excess of one of the monomers. A second method is the use of monofunctional species as chain terminators.

Polymerization Equipment

Equipment includes glassware, vacuum systems, and pressure reactors.

Purification and Characterization

Polymer Isolation from Solution. The products from a solution polymerization are usually isolated by precipitation from a nonsolvent, which removes the polymerization diluent, unreacted monomer, and initiator residues.

Stabilization. Polymers must be stabilized by antioxidants and photochemical stabilizers.

Molecular Weight and Distribution. The molecular weight and molecular weight distribution are among the most important parameters of a polymer (see MOLECULAR WEIGHT DETERMINATION). Absolute values of number-average and weight-average molecular weights are obtained by osmometry and light scattering. For the average laboratory, however, solution viscosity remains the most rapid and convenient method of estimating molecular weight.

Thermal Properties. A rough estimate of the softening and melting points of a polymer may be made by dragging a sample along a thermal gradient bar or watching its behavior on a hot-stage microscope. However, these tests are usually too crude to be of value. With the differential scanning calorimeter (dsc) and similar instruments, thermal measurements on polymers can be made quickly and accurately.

Sample Fabrication. On a laboratory scale, fabrication is usually limited to pressing or casting films, simple fiber spinning and compression molding. Small injection-molding machines can be used for kilogram quantities.

Mechanical Properties. Mechanical and physical properties are determined by several methods. When sufficient quantities are available, the polymer is injection-molded into tensile bars or other shapes, depending on the test.

Informative data are obtained from a stress–strain curve in tension. These data can be determined by an Instron.

Toughness is measured by impact testing. Failure occurs by brittle or ductile failure. Abrasion resistance and the hardness of polymers are related to penetration, scratching, and marring of the surface.

When only small samples are available, testing is carried out on films or fibers by dynamic mechanical analysis with the help of a Rheovibron.

Rheology. The rheological behavior of a polymer greatly affects fabrication. Rheological measurements are important for the characterization of thermoplastic melts, solutions, and latices.

Melt viscosities are commonly measured with a capillary or a cone-and-plate rheometer. The viscosity parameters for solutions, suspensions, and emulsions are usually measured with a coaxial cylinder viscometer.

BIBLIOGRAPHY

D. Braun, H. Cherdron, and W. Kern (translated by K. J. Ivin), *Practical Macromolecular Organic Chemistry,* Harwood Academic Publishers GmbH, New York, 1984.

F. W. Billmeyer, *Textbook of Polymer Science,* 3rd ed., John Wiley & Sons, Inc., New York, 1984.

W. R. Sorenson and W. R. Campbell, *Preparative Methods of Polymer Chemistry,* Wiley-Interscience, New York, 1968.

J. A. Moore, ed., *Macromolecular Syntheses,* Vols. 1–9, John Wiley & Sons, Inc., New York, 1977–1985.

MORRIS R. ORT
OLIVER D. DEEX
Monsanto Chemical Company

POLYMERIZATION THERMODYNAMICS

No chemical reaction can proceed to a substantial extent at a given temperature and pressure unless two conditions are fulfilled: the equilibrium between reactants and products must favor the products, and there must be a mechanism that allows the reaction to occur. A reaction involving only small molecules, such as the synthesis of ammonia from nitrogen and hydrogen, can proceed to some extent even when the thermodynamic change is somewhat unfavorable, that is, when the equilibrium position lies on the side of the reactants; but for the formation of high polymer from a monomer, many hundreds of molecules must come together to form one large molecule. An unfavorable equilibrium position for the propagation reaction $P_n + M \rightleftharpoons P_{n+1}$ is then multiplied many times, which means that if the free energy change for the propagation reaction is only slightly positive, the formation of high polymer will be scarcely detectable. For this reason addition polymerizations exhibit discontinuities of a kind more usually associated with physical aggregation processes. Thus for a given solvent and monomer concentration $[M]$, there exists a sharp temperature, defined as the ceiling temperature T_c, above which the formation of long-chain polymer does not occur. Experimentally, such temperatures can be almost as well defined as melting points. A physical analogy is the condensation of a vapor, which occurs only when the temperature falls below the dew point appropriate to the particular concentration of vapor. Conversely at a given temperature the aggregation process, whether polymerization of monomer or condensation of vapor, occurs only if the concentration of the unaggregated form exceeds a critical value: the equilibrium monomer concentration $[M]_e$ in the case of polymerization, or the saturation vapor pressure in the case of condensation of vapor.

Any addition polymerization to form long-chain polymer occurs only if $[M] > [M]_e$ at a given temperature or if $T < T_c$ at a given monomer concentration. These are two ways of expressing the same thermodynamic truth. Two other types of aggregation process where such effects may be observed are the aggregation of soap molecules to form micelles, which occurs only above the critical micelle concentration, and the aggregation of protein molecules, such as bacterial flagellin, to form flagellar filaments. Thus *Salmonella,* strain SJ814, has a critical concentration of 0.36 *M* at 38.6°C.

All of the above examples are exothermic and exoentropic aggregation processes. There are examples of endothermic and endoentropic aggregation processes that exhibit the reverse effect, a floor temperature (ie, ΔG becomes negative when the temperature is raised above this value). The spontaneous conversion of sulfur to a highly viscous liquid when heated to 159°C is the classic example of this case.

When cyclic compounds are polymerized by ring opening, it is frequently found that, in addition to linear high polymer, a series of cyclic oligomers having the same chemical composition as the monomer is formed. If isolated and if their concentration exceeds the appropriate critical value, the individual cyclic oligomers will undergo polymerization to yield the same linear high polymer and series of cyclic oligomers as before. Equilibrium between the polymer and cyclic oligomers is often readily established (see Ring–Chain Equilibria below and MACROCYCLIC POLYMERS).

Methods for Determining Ceiling Temperatures and Equilibrium Concentrations

The ceiling temperature can be found by varying the temperature while keeping the reactant composition constant. The onset of polymerization may be observed directly if the polymer is insoluble, as in the case of poly(isobutene sulfone), or from dilatometric measurements by gradually lowering the temperature in the presence of an initiator and observing the temperature at which there is a sharp break in the curve of volume against temperature. Alternatively, the yield or the rate or degree of polymerization may be measured at several temperatures below the ceiling temperature T_c, and the results extrapolated to zero yield, zero rate R_p, or zero degree of polymerization. This method has the advantage that the chain length may remain quite high even to within a few degrees of the ceiling temperature. The extrapolated value for T_c then corresponds to the limiting value for long chains.

Another way to determine ceiling temperature is to hold the temperature constant, allow the polymerization to proceed to equilibrium, and then determine the composition of the equilibrium mixture. Alternatively, the initial rate or degree of polymerization may be measured at a series of initial monomer concentrations, and the results extrapolated to zero rate or degree of polymerization. The only difference in these two approaches lies in the choice of independent variable: reactant composition or temperature.

Physical Factors Affecting ΔG

Since $\Delta G = 0$ at T_c, it follows that

$$T_c = \Delta H / \Delta S$$

where ΔH and ΔS are the appropriate heat and entropy changes, respectively, for the polymerization process under the prevailing conditions. Any factor that affects ΔG will also affect T_c through either ΔH or ΔS, or both. The main factors affecting ΔG for a given monomer are the physical state of the monomer, temperature, and hydrostatic pressure. Minor factors are the nature of the solvent and the presence of other solutes, particularly the polymer, or in the case where the polymer is insoluble, its degree of crystallinity. Any given situation may ultimately be analyzed in terms of a thermodynamic cycle containing the step

$$M \text{ (liquid)} \longrightarrow \frac{1}{n} M_n \text{ (amorphous)} \quad \Delta H^0_{ss},\ \Delta S^0_{ss},\ \Delta G^0_{ss}$$

When a solvent is used and the polymer is also soluble, it is more convenient to use the values for the reaction in solution:

$$M \text{ (solution)} \longrightarrow \frac{1}{n} M_n \text{ (solution)} \quad \Delta H^0_{ss},\ \Delta S^0_{ss},\ \Delta G^0_{ss}$$

with appropriately defined standard states. For polymerization in solution, $[M]_e$ generally increases with increasing temperature, corresponding to an exothermic propagation process. Conversely, lowering the monomer concentration makes ΔG less negative until the equilibrium concentration is reached, where ΔG is zero.

Expressing the propagation reaction as an equilibrium:

$$P_n^* + M \underset{k_d}{\overset{k_p}{\rightleftharpoons}} P_{n+1}^* \qquad K = k_p/k_d$$

and remembering that when n is large,

$$[P_n^*] \simeq [P_{n+1}^*],$$

$$K = 1/[M]_e$$

$$\Delta G_{ss}^0 = -RT \ln K = RT \ln [M]_e$$

Also, since at the ceiling temperature $\Delta G = \Delta H_{ss} - T_c \Delta S_{ss} = 0$,

$$T_c = \Delta H_{ss}/\Delta S_{ss} = \Delta H_{ss}^0/(\Delta S_{ss}^0 + R \ln [M]_e)$$

The standard state is taken as a concentration of 1 M. The activity coefficients of the monomer and polymer and their variation with temperature or medium are thus absorbed in the values of ΔH_{ss}^0 and ΔS_{ss}^0.

A thermodynamic analysis of such a system that takes into account the interactions between all the components of the system leads to

$$\Delta G_{lc}/RT = 1 + \ln \phi_1 + \chi_{12}(\phi_2 - \phi_1) + (\chi_{13} - \chi_{32} V_1/V_3)\phi_3 \qquad (1)$$

where χ_{12}, χ_{13}, χ_{32} are the three interaction parameters (1 = monomer, 2 = polymer, 3 = solvent); ϕ_1, ϕ_2, and ϕ_3 are the equilibrium volume fractions; and V_1 and V_3 are the molar volumes. The magnitude of the last term is determined mainly by the difference in monomer–solvent and polymer–solvent interaction parameters. The former generally shows the stronger dependence on solvent and hence is mainly responsible for the variation of equilibrium monomer concentration with solvent. Using a simplified form of equation 1, the solvent effect for styrene can be explained quantitatively. When the variation of ϕ_1 with ϕ_2 is not too large, it can be shown from equation 1 that ϕ_1 will be a linear function of ϕ_2, as found for α-methylstyrene in tetrahydrofuran (THF) and dioxane.

For systems without a solvent but in which the polymer is soluble in the monomer, a more exact thermodynamic treatment can be used, leading to

$$\Delta G_{lc}/RT = 1 + \ln \phi_1 + \chi_{12}(\phi_2 - \phi_1) \qquad (2)$$

where the symbols are as defined in equation 2.

The effect of pressure on melting points is generally much larger than its effect on ceiling temperatures, so that in principle it would be possible by the application of pressure to go from an unpolymerizable liquid to a polymerizable solid.

Since most addition polymerizations are accompanied by a decrease in both volume and enthalpy, the most favorable thermodynamic conditions for polymerization are high pressure and low temperature. Usually, one or the other of these variables has been applied to encourage difficult monomers to polymerize, but seldom have both been applied together. It is clearly a mistake to raise both the pressure and temperature since the two effects on ΔG would then be working in opposition.

The Effect of Chemical Structure on ΔG

A change in the chemical structure of the monomer can influence the magnitude and sign of the free energy of polymerization in a number of ways. The most important factors are strain in the monomer, especially cyclic monomers where ring size is critical, configuration in the monomer, and strain in the polymer arising from interaction of substituents in adjacent units, especially in vinyl polymers. Less important factors are geometrical isomerism (cis–trans) and configurational isomerism in the polymer (tacticity).

Monomers Polymerizing Through a Double Bond. ***Unsubstituted Monomers.*** There is a marked decrease in the polymerizability of the unsubstituted monomers along the series ethylene, formaldehyde, thioformaldehyde.

Substituted Ethylenes. Most monosubstituted ethylenes can be readily polymerized, and the back reaction is unimportant, unless the reaction is carried out at exceptionally low concentration and high temperature, for example, 10^{-3} M and 120°C for styrene.

Substituted Carbonyls. The polymerizability of aldehydes is generally thermodynamically less favorable than that of the corresponding monosubstituted ethylenes. Primary aliphatic aldehydes can only be polymerized at low temperatures or high pressures.

Although substitution by methyl groups has an adverse effect on polymerizability, substitution by halogen atoms has a beneficial effect. The series of monomers from chloral to pivalic aldehyde, shown in Table 1, is noteworthy: here the entropy effect outweighs the enthalpy effect.

Thiocarbonyls. Unlike acetone, thiocetone can be polymerized to a high molecular weight product: T_c is reported to be about 100°C. However, neither aliphatic thioaldehydes nor thioformaldehyde can be polymerized; thus the trend is opposite to that observed in the carbonyl series.

Isocyanates. Certain isocyanates can be polymerized in the presence of sodium cyanide or other catalysts, such as $Ni/(pyr)_2(C_2H_5)_2$, to give 1-nylons (pyr = pyridine). The nonpolymerizability of many other isocyanates, particularly those substituted close to the nitrogen atom, is attributable to a lowering of the ceiling temperature.

Table 1. Thermodynamic Data for the Polymerization of a Series of Aldehydes in THF

Monomer	$-\Delta H_{ss}^0$, kJ/mol[a]	$-\Delta S_{ss}^0$, J/(K·mol)[a]	T_c, °C, for $[M] = 1\ M$
CCl_3CHO	14.6	51.9	11
CH_3CCl_2CHO	17.1	69.1	−24
$(CH_3)_2CClCHO$	19.7	90.4	−54
$(CH_3)_3CCHO$			<−78

[a] To convert J to cal, divide by 4.184.

1:1 Copolymerization. A number of copolymerizations are known in which 1:1 copolymers are formed regardless of the composition of the feed. Thermodynamically, such systems may be treated as homopolymerizations, and the effect of structure on T_c is therefore directly comparable with such effects in homopolymerization. The formation of 1:1 poly(olefin sulfone)s from sulfur dioxide and an ethylene derivative falls into this category.

Monomers Polymerizing by Ring Opening. ***Monocyclic Monomers.*** The operation of both thermodynamic and kinetic restrictions on ring-opening polymerizations is well illustrated by the cycloalkanes and cycloalkenes (Table 2).

Monomers Containing Fused Rings. Any bicyclo compound may be regarded as derived from a disubstituted monocyclic compound by joining the two substituents together. The act of closing the second ring may thus introduce additional strain energy into the molecule, thereby making its polymerizability by ring opening more favorable than for the related disubstituted monocyclic compound.

A molecule containing two fused rings has three bridges joining the two bridgeheads so that three possible modes of addition polymerization can be written, with up to three different possible decreases in free energy, corresponding to the opening of the three bridges. In practice the bridges only open when there is a suitable functional group present, such as a C=C double bond or a heteroatom.

Fused-ring Systems Containing Two Potentially Reactive Nonequivalent Bridges. All of these compounds polymerize.

Isomerism in the Polymer. Various types of structural isomerism can be encountered in ring-opened polymers. With particular monomers and catalysts it is sometimes possible to direct the polymerization selectively so that the polymer may be all cis or all trans, all syndiotactic or all head–tail. However, in no case is there any indication that such effects are thermodynamic rather than kinetic in origin.

Ring–Chain Equilibria

Experimental Observations. When tetraoxane, the cyclic tetramer of formaldehyde, is cationically polymerized in 1,2-dichloroethane at 50°C, the concentration of tetraoxane falls to an equilibrium value of less than 0.008 M, but at the same time an equilibrium concentration of 0.2 M trioxane is built up. At concentrations of tetraoxane below 0.008 M no linear polymer is formed, but cyclic trimer is rapidly produced. Such behavior is typical of a whole range of cyclic compounds.

Cyclic oligomeric species have likewise been identified in the products of ring-opening polymerization of cyclic ethers. Under certain conditions the macrocyclic ethers are the predominant products of polymerization of ethylene oxide and may account for as much as 35 wt % of the polymerization products of oxetane. Cyclic polyethers also yield cyclic oligomers in the products of polymerization.

Theory of Ring–Chain Equilibria. The basic postulates of the Jacobson-Stockmayer theory are first, that the distribution of ring sizes is determined only by entropy considerations, ie, all rings are strainless; second, that the end-to-end distances of linear chains obey Gaussian statistics; third, that the random flight model is assumed, and no allowance is made for the excluded volume effect; fourth, that the probability of ring formation is governed by the frequency with which the ends of a given chain coincide; and fifth, that the probability of bond formation is independent of whether the ends form part of the same chain or are ends of separate chains.

The theory predicts that no linear polymer should be formed below a critical concentration of starting monomer, expressed as moles of repeat units per liter (this is essentially the same prediction as that for a simple polymerization, but generalized to cover a series of cyclic monomeric species); the concentration of rings $c\text{-}M_x$, in equilibrium with linear polymer M_y, should be independent of dilution (in fact, the polymer concentration can affect $[M]_e$ because of polymer–monomer interaction); the equilibrium constant K_x, for the equilibrium $M_y \rightleftharpoons M_{y-x} + c\text{-}M_x$, should be proportional to $x^{-2.5}$ and independent of temperature (since it has been assumed that this equilibrium is thermoneutral). In contrast to the thermodynamic distribution of linear polymers this means that the weight distribution of macrocyclic constituents at equilibrium is predicted to be a monotonically decreasing function of ring size.

Thermodynamics of Copolymerization

Two new variables, namely, monomer and polymer composition, are involved in a consideration of the thermodynamics of copolymerization. Such systems are of particular interest when one or more of the comonomers is polymerizing reversibly. Most models postulated to describe equilibrium copolymerization are based on a kinetic or thermodynamic consideration of reversibility in one or more of the propagation steps. Some experi-

Table 2. Polymerizability of Some Cyclic Hydrocarbons

Cyclic compound	ΔG_{lc}[a]	Mechanism available	Polymerizes
cyclobutane	large negative	no	no
cyclopentane	small negative	no	no
cyclohexane	small positive	no	no
cycloheptane	small negative	no	no
cyclobutene	large negative	yes	yes[b]
cyclopentene	small negative	yes	yes[c]
cyclohexene	small positive	yes	yes[d]
cycloheptene	small negative	yes	yes[c]

[a] Free energy of polymerization of liquid monomer to condensed amorphous polymer.
[b] To high conversion.
[c] To limited conversion.
[d] Polymerizes in solution in toluene at low temperatures.

mental results can be explained with the diad model of binary copolymerization (Steps 1–4):

$$\text{Step 1: } M_1^* + M_1 \longrightarrow M_1^*$$

$$\text{Step 2: } M_2^* + M_1 \longrightarrow M_1^*$$

$$\text{Step 3: } M_1^* + M_2 \longrightarrow M_2^*$$

$$\text{Step 4: } M_2^* + M_2 \longrightarrow M_2^*$$

others require more elaborate schemes involving the effect of the terminal monomer sequence of the active species, the effect of i on the rate constants for specific propagation–depropagation steps, Step 5, for example:

$$\text{Step 5: } (M_2)_i^* + M_2 \rightleftharpoons (M_2)_{i+1}^*$$

Such schemes have been considered, and the experimental results for various copolymerizations have been shown to be adequately described by these models.

Recently, there has been criticism of methods based only on the propagation steps on the grounds that they deal only with the dynamic process of copolymer formation and not with the complete equilibrium state eventually established between the copolymers and the comonomers. A vector analysis of the ultimate dynamic equilibrium state of a living copolymer system has been carried out.

BIBLIOGRAPHY

F. S. Dainton and K. J. Ivin, *Q. Rev. Chem. Soc.* **12,** 61 (1958).

J. Brandrup and E. H. Immergut, eds., *Polymer Handbook,* 3rd ed., Wiley-Interscience, New York, 1989, p. II–295.

H. Sawada, *Thermodynamics of Polymerization,* Marcel Dekker, Inc., New York, 1976.

K. J. Ivin and T. Saegusa, *Ring Opening Polymerization,* Elsevier Applied Science, Publishers, Ltd., London, 1984.

KENNETH J. IVIN
The Queen's University of Belfast

W. K. BUSFIELD
Griffith University, Brisbane

POLYMER MODELING

Molecular modeling utilizes computer-assisted techniques for the study of molecules, including polymeric systems or fragments of such systems. A wide range of techniques is employed, such as molecular mechanics, quantum mechanics, molecular dynamics, intermolecular modeling, and molecular-shape analysis.

Molecular modeling is currently extensively used in the pharmaceutical industry. However, the basic types of modeling for drug design originated mostly in early work on polymers, particularly polypeptides and simple organic homopolymers. Early treatments reflected the computational capabilities at that time and were confined to simple conformational-analysis techniques. The growth of molecular modeling largely follows the development of computers. More aggressive applications of these techniques to polymer research were used as early as 1973.

Theory and Techniques

Evaluation of Energy. A variety of ways can be used to evaluate the energy of a molecule. Each has inherent assumptions, limitations, and different degrees of sophistication. In generic terms, these can be considered *ab initio,* semiempirical, or empirical methodologies.

Searching Conformational Space. The goal when modeling small molecules is to find the lowest energy conformation as well as all possible minima and their relative energies. This gives the most probable conformation as well as information on the flexibility of the molecule and is easy to visualize for small molecules with a few bonds. When looking at a single rotating bond, as in ethane, the relationship of energy to bond rotation is obvious. For more complex molecules, and as the full force field is applied, the calculation becomes extremely complex, depending on the position of each atom in space and its relation to each of the other atoms. The energy can then be thought of as a multidimensional hyperspace with each dimension related to each of the atomic and rotational variables to be studied. The analysis of this hyperspace comprises the science of molecular modeling. Techniques include combinatorial searching, minimization, random searching, and combined search strategies.

Molecular Dynamics. Although molecular modeling as described previously is the evaluation of the potential energy of a molecular system, it is also possible to consider the kinetic-energy contribution, as derived directly from Newton's equations of motion

$$\text{force} = \text{mass} \times \text{acceleration}$$

Since the force is the negative derivative of the potential energy, and the acceleration is simply the second derivative with respect to time of the atomic positions, the result is

$$-\delta V/\delta r = m_i\, \delta^2 r_i/\delta t^2$$

These equations of motion can be integrated with respect to time to compute the trajectories of each atom in the system, the r_i for each atom as a function of time.

Molecular Modeling of Polymers

When modeling polymers, as opposed to small organic molecules, it is important to realize that in terms of the potential-energy functions, polymer modeling is like small-molecule modeling. The main difference is that there is a much larger number of atoms and many additional variables to treat. This makes some of the techniques used for small molecules less tractable. The specific goals of the modeling effort are important to consider from the beginning. Various types of models can be built, depending on the goal. These include models targeted at understanding flexibility, hydrodynamic properties, interaction with small molecules, cross-linking, crystal packing, or interaction with a surface.

Changing the reaction conditions slightly can give rise to various isomerisms that must be taken into account. Each explicit possible configuration and isomer must be treated as a specific model.

Polymer tacticity must also be considered. Depending on the monomers and the reaction conditions, each monomer added can have pseudo-asymmetric centers. If these are all the same, the side chains are all on the same side of the polymer backbone, and the polymer is isotactic (Fig. 1**a**). If the side chains alternate sides, the polymer is syndiotactic (Fig. 1**b**). When placement is random, the polymer is called atactic.

Polymer Construction. There are too many variables to treat true, full-sized, polymer chains directly by most modeling techniques. The best way is to begin with model compounds and build larger and larger models (monomers, dimers, trimers, etc) and apply the experience to the polymer systems. To build realistic models as few simplifications as possible must be used, and information on the isomerization and tacticity is important. Each of these must be treated as a separate parameter in building models. Thus a polymer suspected of being atactic must be modeled as isotactic, syndiotactic, and a mixture thereof.

Applications

Outlook. In many ways, polymer modeling is at a new beginning, waiting for application. Molecular modeling began with polymers but has been extended to drug design. Homopolymer modeling was fairly simple and straightforward. To go beyond requires sophisticated computer hardware and software, much of which is slowly becoming available.

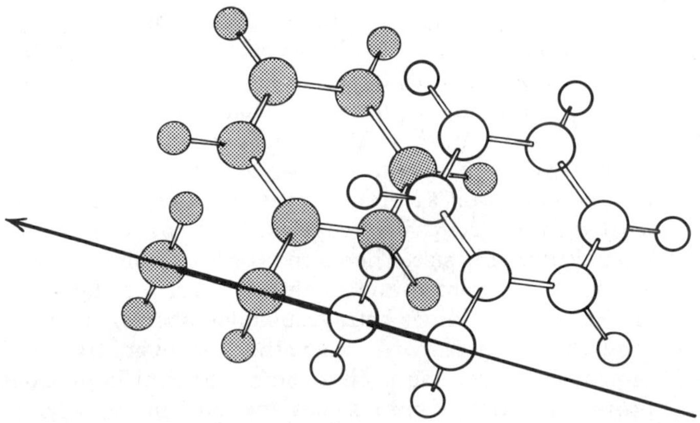

(a)

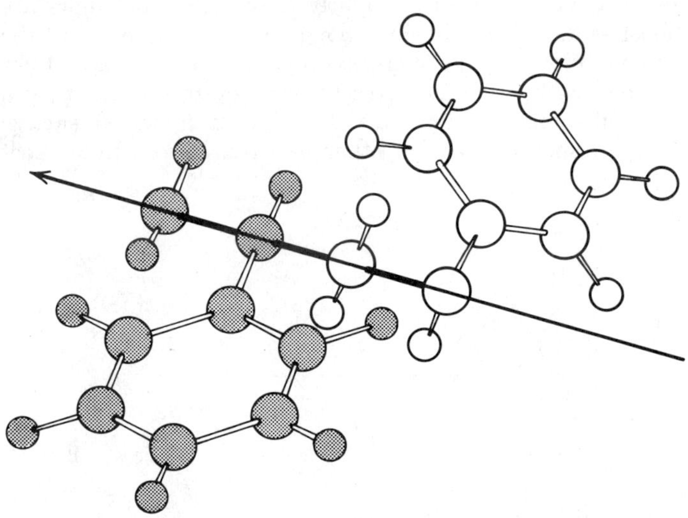

(b)

Figure 1. Styrene dimer: (**a**) isotactic; (**b**) syndiotactic.

Hypothesis testing of molecular speculation is the most significant application of polymer modeling today and will allow the techniques described here to make important contributions to polymer science.

BIBLIOGRAPHY

P. J. Flory, *Statistical Mechanics of Chain Molecules,* Wiley-Interscience, New York, 1969.

A. J. Hopfinger, *Conformational Properties of Macromolecules,* Academic Press, Inc., New York, 1973.

T. Clark, *A Handbook of Computational Chemistry,* John Wiley & Sons, Inc., New York, 1985.

RUDOLPH POTENZONE, JR.
Polygen Corporation

POLYMERS AS CHEMICAL REAGENTS

Ever since solid-phase peptide synthesis was reported in 1963 there has been considerable interest in the application of polymer-supported reactants in organic chemistry.

Various types of reaction systems can be envisaged. Thus either a catalyst, a reagent, or a substrate can be bound to the polymer. Polymer-supported catalysts are a particularly attractive type of polymer-supported reactant, because a relatively small amount of polymer can be used to chemically transform a relatively large amount of substrate (see CATALYSTS, POLYMER-SUPPORTED).

The major attraction of polymer-supported reactions is that after the reaction has been carried out, the polymer-supported species can be separated without difficulty from the low molecular weight species in solution. The polymer, in effect, serves as a handle to allow certain species to be separated efficiently from others. Separation is achieved most readily if the polymer is cross-linked, since it is then insoluble in all solvents. Ideally, the polymer should be in the form of beads ca 50–100 μm in diameter. These are large enough to be filtered off from the soluble species using common laboratory filters, but small enough to minimize diffusion problems. Linear polymers have been used occasionally, but separation is less straightforward. Reactions with linear polymers may be carried out by dissolving the polymer in the reaction solvent. When reaction is complete, the polymer is precipitated out by adding a solution of the products to a nonsolvent. When precipitation is not feasible, the soluble polymer may be separated from low molecular weight species by membrane filtration, but this can be tedious and time-consuming.

The easy separation of polymer-supported species from low molecular weight species can have several practical advantages when polymer-supported reagents or substrates are used. (*1*) Product isolation is simplified, making it possible to avoid exposing the reaction product to water, to avoid a chromatographic separation, or to allow the rapid isolation of an unstable species. (*2*) If the use of excess reagent results in a greater reaction yield, it can be used in excess without causing separation problems. (*3*) When a polymer-supported reagent is used, the spent reagent is recovered easily and possibly can be regenerated, making it economically feasible to prepare complex and sophisticated supported reagents. (*4*) If the reaction proceeds virtually to completion so that filtration and washing of the polymer are all that is required to isolate the product, automation of the process is relatively easy, an obvious industrial attraction. (*5*) If a reaction proceeds particularly rapidly, it possibly can be carried out in one pass of the soluble reactant down a column of the bound reactant.

Another attractive feature that does not concern easy separation, but rather environmental problems, is that since cross-linked polymers are involatile and insoluble, they are odorless and nontoxic. Thus a superior method for carrying out reactions with poisonous reagents, such as arsenic or selenium compounds, is offered. As a consequence of all of these advantages, polymer-supported reagents and substrates have found many applications.

Preparation of Polymers with Reactive Functional Groups

There are two main approaches to the synthesis of polymers with reactive functional groups: direct polymerization of the functional monomers with, if necessary, other monomers and/or a cross-linking agent, or chemical modifications of a preformed polymer.

Monitoring of Reactions on Polymers

With cross-linked, and therefore insoluble polymers, it is difficult to monitor their reactions, and consequently, many techniques have been used. They all have their limitations, and, in general, it is wise to use a combination of techniques to monitor a particular reaction. The following are the most widely used: elemental analysis, gravimetric analysis, titration of reactive groups, cleavage of substrate from polymer, ir and Raman spectroscopy, and nmr. The first four are essentially quantitative, the others are qualitative or only semiquantitative.

Reactivity of Polymeric Reagents

Although the intrinsic reactivity of a functional group bound to a polymer is generally the same as that of a simple low molecular weight model analogue, a complication with polymer-supported reactions is in maintaining access to the bound functional groups. Access must be maintained throughout the reaction. If the polymer is linear, soluble in the reaction medium, and remains soluble throughout the reaction, access usually is no problem. If, however, the linear polymer is insoluble or becomes insoluble during the reaction, or is a cross-linked polymer and is, therefore, always insoluble, there are potential difficulties. Under these circumstances, the polymer is, in effect, a separate phase. This is depicted schematically in Figure 1, where species A is polymer-bound and species B is in solution.

With polymer beads or particles of the sizes commonly used for polymer-supported reagents, <0.1% of the polymer-supported functional groups are on the surface of the beads or particles. Therefore, to react with A, species B must diffuse into the insoluble polymer. If the reaction of A with B is relatively fast compared with the rate of diffusion of B into the polymer, and this is often the case, the rate of diffusion limits the rate of the overall reaction. If, on the other hand, diffusion is relatively rapid compared with the rate of the reaction, the distribution of B between the solution and the polymer becomes particularly important; B may be concentrated in the polymer or the reverse. This effect may lead to an increase or decrease in the overall reaction rate and can, when more than one type of species is in solution, lead to selectivity. The choice of reaction solvent obviously dramatically affects the situation.

Intrapolymeric Reactions: Site Isolation

The ease with which two reactive groups bound to the same polymer can react together has long been a topic of interest. In many cases research workers have sought to use polymers to prevent or minimize reactions between bound species, to achieve high dilution conditions or site isolation. The success rate has varied.

The situation is complex and it is convenient to begin to discuss this topic by considering linear polymers bearing groups that are intrinsically capable of reacting together. If a linear polymer is in solution, has a flexible backbone, and is in the form of a random coil, there is, in most cases, no reason for bound reactive groups not to be able to react together completely. One important exception is the special case where reaction between the groups involves cyclization to give a product with a particular ring size. Then a small fraction of the groups (ca 15%) becomes isolated and unable to react for statistical reasons because by chance both groups neighboring a particular group have reacted with their other neighbors first. Another important exception is that conformational limitations may prevent reaction of groups close together on the polymer chain.

Most applications of polymer-supported reagents use cross-linked polymers. Whereas lightly cross-linked polymers are almost as flexible as the corresponding linear polymers, as the degree of cross-linking increases a point may be reached, for example, with some 20% cross-linked polystyrenes, where a small but significant fraction of the polymer-supported reactive groups cannot reach each other and therefore cannot react

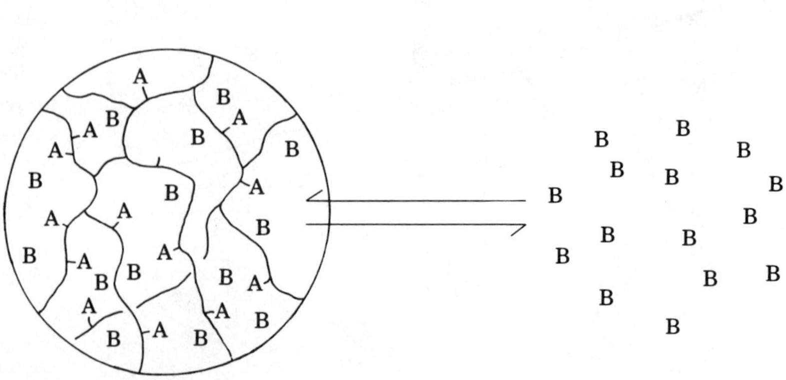

Figure 1. An insoluble polymer represents a separate phase. A is polymer-bound; B is soluble.

together; hence some permanent site isolation is achieved. Low degrees of substitution also assist site isolation. Site isolation is of most practical value with polymer-supported catalysts, since it is acceptable to have low degrees of functionalization and also a situation where some functional groups are isolated and some are not.

Another more common type of site isolation is possible, based on the fact that although polymer-supported reactive groups may be able to react together, in general, they do so more slowly than analogous reactive molecules in solution because the polymer reduces the mobility of the supported reactive groups. The reduced mobility of the polymer chain may allow a reaction between a supported reactive group A and either a low molecular weight species B in solution (Fig. 2**a**) or another group C within the same pendent group as itself (Fig. 2**b**) to compete effectively with the intrapolymeric reaction (reaction 1). Thus, if in this scheme the low molecular weight analogues of reaction 1 would normally proceed at a higher or a similar rate to reaction 2, but as a consequence of using the polymer, reaction 1 slowed down to the point where reaction 2 is actually the faster, then kinetic site isolation is achieved and the polymer-supported reaction system follows a different course from the analogous low molecular weight reaction system.

Reactions Using Polymer-supported Reagents

Polymer-supported Ionic Reagents. These reagents can be prepared easily from commercial ion-exchange resins (see ION-EXCHANGE POLYMERS). In principle the active species bound could be a cation or an anion, but very few examples of cations have been reported. Two examples are the use of Ti^{3+} and Cr^{2+} as reducing agents.

Many reactions using anions bound to commercial strong-base, anion-exchange resins have been reported. The resins usually contain residues (**1**) and are generally supplied in chloride form.

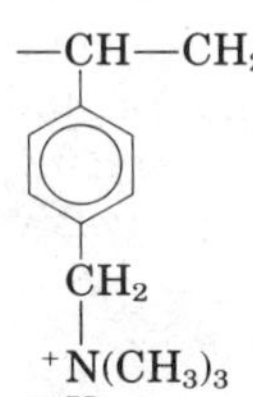

(**1**)

These reagents are used in oxidation, reduction, halogenation, and nucleophilic displacement.

Polymer-supported Reagents with Covalently Bound Reactive Moieties. These reagents are less readily available and more expensive than the ionic reagents. They are used in oxidation (*N*-chloronylon-6,6), reduction (polyvinylpyridine, polymer-supported amino alcohols, or polymer-supported thioethers, each with borane), Wittig reactions (phosphine (**2**)) (Fig. 3), other phosphine reactions, and peptide synthesis.

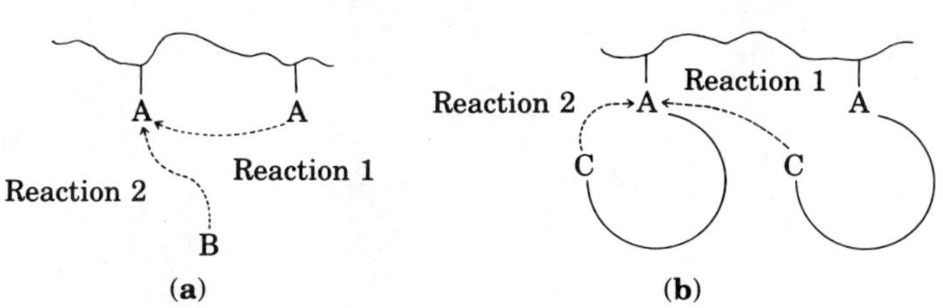

Figure 2. Reactions along a polymer chain (**a**) Chain–chain (A–A); chain–soluble group (A–B). (**b**) Chain–chain (A–C); cyclization of a pendent group.

Reactions with Polymer-supported Substrates

This topic has been reviewed on several occasions. A major problem encountered in such reactions is that any unreacted starting material or by-product cannot be separated from the desired product. Hence, unless the reactions proceed in high yields, each giving just one product, reactions sequences must be short. The classical example of where these difficulties have been largely overcome is in solid-phase peptide synthesis. In other areas, however, much shorter reaction sequences are acceptable, although for most practical purposes the loadings of the substrates on the polymers need to be much higher than in peptide work. The two main attractions of polymer-supported substrates are the ease of product isolation and/or the benefits accruing from site isolation. In all systems it is necessary to have good procedures for attaching substrates to polymers and for detaching them. The requirements are essentially the same as in protecting group technology (see POLYPEPTIDE SYNTHESIS, SOLID-PHASE METHOD).

Two Polymer-supported Reactants: The Three-phase Test

If two polymer-supported reactants on separate cross-linked polymers are mixed in the same reaction vessel, the reactive groups do not come into contact. This has been exploited in both synthetic and mechanistic studies.

So far the most important application of two-polymer systems is the three-phase test for reactive intermediates. In such a test a precursor for a suspected intermediate is bound to one cross-linked polymer (donor polymer), and a trap for the intermediate is bound to a second cross-linked polymer (acceptor polymer). The two polymers are placed in the same reaction vessel and the donor polymer activated to release the intermediate. If the product of the intermediate and the trapping species is found on the acceptor polymer, a free intermediate

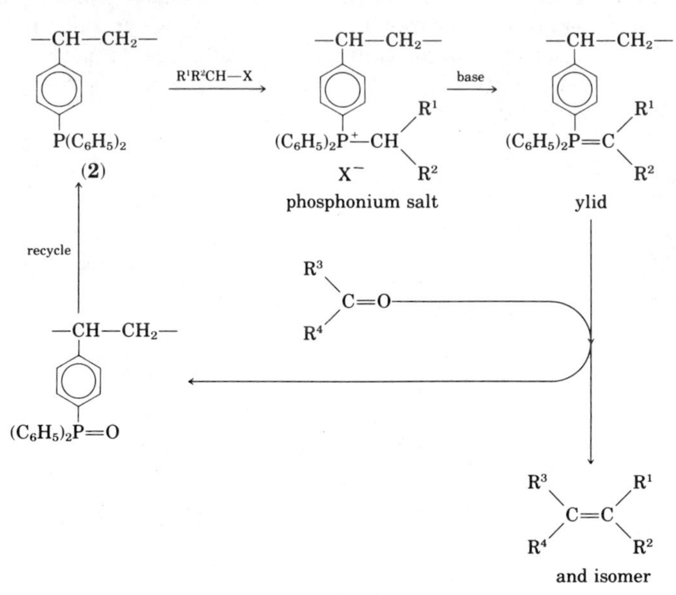

Figure 3. Polymer-supported Wittig reaction.

must have been involved. Moreover, by having the two polymers in separate but connected vessels and arranging a flow system from the donor to the acceptor polymer, it is possible, by varying the time taken in travel from the donor to the acceptor polymer, to estimate the lifetime of the intermediate. The procedure establishes that there is a free intermediate, but does not prove its structure.

BIBLIOGRAPHY

A. Akelah and D. C. Sherrington, *Polymer* **24,** 1369 (1983).

P. Hodge and D. C. Sherrington, eds., *Polymer-supported Reactions in Organic Synthesis,* John Wiley & Sons, Inc., Chichester, UK, 1980.

W. T. Ford, ed., *Polymeric Reagents and Catalysts, ACS Symposium Series, No. 308,* American Chemical Society, Washington, D.C., 1986.

D. C. Sherrington and P. Hodge, eds., *Syntheses and Separations Using Functional Polymers,* John Wiley & Sons, Inc., Chichester, UK, 1988.

PHILIP HODGE
University of Lancaster

POLYMERS FROM RENEWABLE SOURCES

Renewable resources are materials that are produced in the plant and animal kingdoms. The term polymers from renewable resources refers to natural products that are polymeric as grown or can be polymerized. These products may be subjected to purification, further reactions, or a variety of processing steps before final utilization.

In addition to the traditional commercially important materials mentioned above, a number of other natural products are under investigation and have high potential, such as spiderwebs, wild plant oils, rubber from non-Hevea sources, and polymers produced by biochemical engineering processes.

Functional, reactive sites on a natural product, within limits, undergo reactions similar to those found in synthetic molecules. Furthermore, the macromolecular structure produced naturally may exhibit properties unlike those of synthetic materials. A better understanding of the structure–property relationships of natural products facilitates the modification of these materials, as well as new ones.

Major classes of renewable natural products are given below:

Alkaloids
pyrrolidine, pyridine, pyrrolizidine, tropane, quinolizidine, isoquinoline, piperidine, indole, quinoline, quinazoline, acridone, steroids, and terpenes.

Carbohydrates
simple (glucose, sucrose, fructose, lactose, and galactose); complex (starch, cellulose, glycogen, and gums).

Lipids and resins
linseed, cottonseed, castor, tung, soybean, oiticica, perilla, menhaden, sardine, corn, safflower, vernonia, fossil resins (amber, kauri, congo oleoresins), damar, and ester gum; simple (glycerol and cholesterol) esters; phosphoglycerides; sphingolipids (mucolipids, sulfatide, sphingomyelin, and cerebroside); complex lipoproteins, proteolipids, and phosphatidopeptides.

Heme, bile, and chlorophylls

Phenolic plant products, lignins
phenols, resorcinols, anthraquinones, naphthoquinones, hydrangenol, stilbenes, and coumarins.

Polyisoprenes
natural rubber, isoprene, gutta percha, and balata.

Proteins
enzymes (lysozyme, trypsin, and chymotrypsin); transport and storage (hemoglobin, myoglobin, and zein); antibodies; structural (elastin, actin, keratin, myosin, collagen, and fibroin); amino acids.

Purines, pyrimidines, nucleotides, nucleic acids

Steroids
cholesterol, adrenocortical, bile acids, ergosterol, desmosterol tannins, wattle, quebracho.

Polysaccharides

Polysaccharides and their sources are shown in Table 1. Purity, molelcular weight, amount and location of branching and crosslinks vary according to species and location. However, chemical reactivity and physical properties are largely independent of the source. Uses include food (starch, fructose, mannose, and sucrose), building material (cellulose), and clothing (cellulose).

The most important renewable polymers are the polysaccharides (qv), especially cellulose and starch, but also amylopectin, glycogen, chitin, carrageenan, pectic acid, dextran, and agar. Applications of starch derivatives are given in Table 2.

Rubber and Gutta Percha

Natural rubber consists of poly(*cis*-1,4-isoprene).

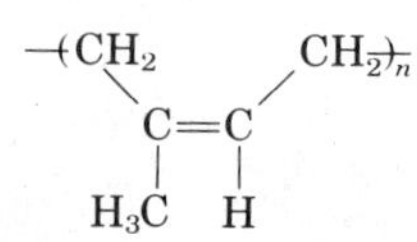

Table 1. Naturally-occurring Polysaccharides

Polysaccharide	Source	Monomeric sugar unit(s)
amylopectin, amylose	corn, potatoes	D-glucose
cellulose	plants	D-glucose
chitin	animals	2-acetamido glucose
glycogen	animals, muscles	D-glucose
gum tragacanth	plant resin	L-arabinose and D-galactose
inulin	artichokes	D-fructose
laminaran	seaweeds	D-glucose
mannan[a]	yeast	D-mannose
xylan[a]	plants	D-xylose
pustulan	lichens	D-glucose

[a] Hemicelluloses.

Table 2. Applications of Starch Derivatives

Starch or derivative	Applications
glycol glucosides	urethane foams, surfactants, alkyd resins
combination with poly(ethylene-*co*-acrylic acid) and polyethylene	biodegradable materials as mulch film
carboxymethylstarch	water-soluble laundry-detergent bags
starch encasement of rubber droplets, powered rubber in latex	assist in rubber manufacture
insoluble starch xanthates	removal of heavy metal contaminants from industrial process water
graft starch products[a]	absorption of body fluids; thickening agents; fire control; removal of water in fuel–alcohol mixtures

[a] Including Super Slurper.

Table 3. Sources of Rubber

Common name	Botanical name	Region
cis-1,4-Polyisoprene		
natural rubber	*Hevea brasiliensis*	Brazil, Malaysia
guayule	*Parthenium argentatum*	Mexico, Texas
dandelion	*Taraxacum*	USSR
trans-1,4-Polyisoprene		
gutta percha	*Palaquim gutta*	Malaysia
	Dichopsis gutta	
balata	*Bolle tree*	Brazil

The cis isomer is usually an amorphous solid with a T_g of $-72°C$; it melts at 27°C if crystallized. The trans isomer, known as gutta percha or balata, is a highly crystalline polymer; sources are given in Table 3. The most important sources are the plantations in Malaysia and Indonesia.

Vulcanization. The term vulcanization refers to cross-linking, also called tanning, drying, or curing. Vulcanization converts the material from a high viscosity liquid to an amorphous solid, suppressing creep and flow, and enhancing elasticity.

Chemical Modification. Rubber–plastic compositions, such as Hevea plus, are prepared by graft copolymerization of natural rubber, starting with the latex.

Chemical modifications have resulted in commercial products, including chlorinated, hydrochlorinated, and cyclized rubber.

Other Natural cis-Polyisoprenes. The Guayule grows in Mexico and the southwestern United States as a desert shrub. Guayule rubber has been grown continuously at low levels since before 1900.

Terpenes. A broad class of naturally occurring hydrocarbons, terpenes have the general formula $(C_5H_8)_n$. Both *iso*-rubber and *trans*-polyisoprenes are terpenes. Some of the terpenes are important as essential oils; many have been or could be polymerized (see HYDROCARBON RESINS).

Proteins

Proteins, which are built up of 23 α-amino acids, are essential to life. To a synthetic polymer chemist proteins appear to be copolymers of nylon-2 derivatives, since each of the 23 amino acids are, in fact, of the general structure

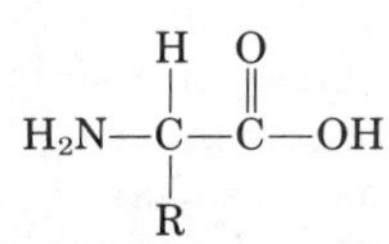

In the simplest amino acid, glycine, R = H and the polypeptide solely composed of this amino acid is nylon-2. In fact, proteins are not random copolymers, but consist of a regular array of amino acids with all chains having the same molecular weight. Except for glycine, the amino acids are optically active, but only the L-isomers occur in nature. Structural proteins are shown in Table 4.

Table 4. Structural Proteins

Polymer	Morphology	Occurrence
collagen	triple helix	skin, tendon, cartilage, bone, teeth
keratin	three α-helices wound around each other in a superhelix (wool)	hair, skin, fur, wool, horn, feathers
fibroin	antiparallel, pleated sheets	silk, spiderwebs
elastin	covalently cross-linked elastomer	arterial walls, ligaments
actin and myosin	myofibrils	striated muscles

Table 5. Polymeric Products Based on Fats and Oils

Product	Uses
acetals of aldehyde esters	plasticizers
fatty aldehydes	feedstock
cyclic fatty acids from alkali treatment of linseed oil and fats or soybean oil soapstock	lubricants in alkyd resins
polycarboxylic acids	lubricant, plasticizer
epoxidized soybean oil polymerized *in situ*	plasticizer–stabilizer for vinyl plastics
products from aldehydic acids and formaldehyde	coatings, plasticizers, lubricants
interpenetrating polymer networks based on castor oil	tough plastics and reinforced elastomers
free fatty acids plus multifunctional acids and alcohols, polymerized	alkyd resins
cross-linked linseed and tung oils	house paint

Natural Oils

Drying Oils. The triglyceride oils constitute the main portion of animal fats, butter, vegetable oils, and margarine, as well as oils from oil-bearing seeds. Although some triglyceride oils such as tallow, are saturated and cannot be polymerized easily, most oils contain at least one double bond per acid residue.

Linseed oil, tung oil, and other highly unsaturated oils are used in various applications (Table 5).

Factices. Like natural rubber, unsaturated oils can be vulcanized with sulfur. Vulcanized, unsaturated triglyceride oils are known as factices.

Functional-group Oils. A few oils contain reactive groups in addition to unsaturation. Castor oil is the most important example in this category.

Potential Applications. Functional-group oils can be polymerized to form soft networks. If another monomer is grafted and polymerized, a series of interpenetrating polymer networks (IPNs) can be formed. If the second monomer forms a glassy polymer, these materials are impact-resistant plastics or reinforced elastomers. Mid-range compositions are tough leathery products.

Japanese Lacquer. Japanese lacquer is a naturally occurring, phenolic coating material originating from the lacquer tree *Rhus vernicifera.* It is known to have been used in continental China for 5000–6000 years. Japanese lacquer confers a beautiful finish.

Lignin

Wood (qv) is made up of cellulose (qv) and lignin (qv). The cellulose component is used for paper (qv), rayon (qv), cellophane (qv), plastics, and a host of other products, but the lignin component finds little application. It is dried and burned as a fuel to provide energy for pulp mills.

Natural Polymers Through Biotechnology

Poly(β-hydroxybutyrate). Poly(β-hydroxybutyrate) (PHB) derives from bacteria. It occurs in soil bacteria, estuarine microflora, blue-green algae, and microbially treated sewage; a good laboratory source is *Alcaligenes eutrophus.*

Although the homopolymer is a highly crystalline thermoplastic melting at 180°C, it tends to be too stiff and brittle for commercial applications. However, it was discovered that copolymers based on 3-hydroxybutyrate and 3-hydroxyvaleric acid can be produced by bacteria. At low comonomer content, the polymers resemble unplasticized poly(vinyl chloride), whereas in midrange they behave as does polypropylene. At higher concentration with reduced crystallinity, biopol copolymers are soft and tough as is polyethylene.

Lactic Acid. Lactic acid is a key biomass intermediate obtained from acetaldehyde or fermentation of hexoses or hexose polymers such as starch or cellulose. Lactic acid forms a lactide through internal esterification, and homopolymers and copolymers with other hydroxyl acids.

Dextran. Dextran consists mostly of α-(1→6)-linked D-glucopyranose residues. These polymers are extracellular polysaccharides synthesized by bacteria grown in sugars such as sucrose. Dextran (qv) was the first microbial polysaccharide used on an industrial scale. It is employed as a pharmaceutical additive and coating, in cosmetics, as food extenders, in oil-well drilling muds, and in synthetic resins.

Biomedical Applications

Rejection of foreign materials for transplants and other *in vivo* applications must be kept to a minimum. Biomedical applications of naturally occurring materials are widespread. Most are aimed at controlled drug delivery and a few are produced as potential biomaterials, such as artificial skin. Macromolecular agents include DNA, RNA, proteins, starch, dextran, cellulose, and chitin (qv).

BIBLIOGRAPHY

T. P. Nevell and S. H. Zeronian, eds., *Cellulose Chemistry and Its Applications,* Ellis Horwood, Chichester, UK, 1985.

E. A. MacGregor and C. T. Greenwood, *Polymers in Nature,* John Wiley & Sons, Inc., New York, 1980.

C. E. Carraher, Jr. and L. H. Sperling, eds., *Polymer Applications of Renewable-Resource Materials,* Plenum Publishing Corp., New York, 1983.

C. E. Carraher, Jr. and L. H. Sperling, eds., *Renewable Resource Materials: New Polymer Sources,* Plenum Publishing Corp., New York, 1986.

LESLIE H. SPERLING
Lehigh University

CHARLES E. CARRAHER
Florida Atlantic University

POLYMER STANDARDS

A variety of polymer samples with stated values of chemical or physical properties and available to the world at large are widely employed in polymer science and technology. Such polymer standards, or standard polymers, are used in three ways. First, they are used to calibrate instruments and techniques that determine physicochemical properties of polymers in a relative sense. Chromatographic methods of molecular weight determination are well-known examples of relative measurements that require calibration. Second, they are used to check the accuracy of instruments and techniques that in principle are absolute but in practice are sufficiently complex to require verification. Finally, such polymers are often used as well-characterized starting materials for research.

The number of polymer standards available has increased greatly since the last edition of this Encyclopedia was published and is still increasing. Nearly all the available standards are linear or essentially linear homopolymers, although a few branched polymers and a few copolymers are available. Almost without exception, the primary certified or otherwise stated properties of these materials are molecular weights of various kinds. A few polymers of specified tacticity are also available and are tabulated separately. The rapid increase in samples of specified molecular weights is a result of the widespread use of gel-permeation chromatography (gpc), also known as size-exclusion chromatography (sec), with its attendant need for calibrants of known molecular weights (see CHROMATOGRAPHY).

A variety of techniques are used to obtain molecular weights for standard polymers. Absolute techniques employed include end-group analysis and membrane osmometry for number-average molecular weight M_n, and light scattering and sedi-

Table 1. Standard Samples of Polybutadiene

Range in molecular weight	No. of samples	Range of M_w/M_n
2×10^2–5×10^2	1	1.07–1.17
5×10^2–1×10^3	8	1.07–1.2
1×10^3–2×10^3	6	1.06–1.2
2×10^3–5×10^3	8	1.08–1.2
5×10^3–1×10^4	3	1.07–1.3
1×10^4–2×10^4	0	
2×10^4–5×10^4	5	1.04–1.11
5×10^4–1×10^5	0	
1×10^5–2×10^5	3	1.26–1.87
2×10^5–5×10^5	8	1.32–4.5

Table 2. Standard Samples of Polydimethylsiloxane

Range in molecular weight	No. of samples	Range of M_w/M_n
1×10^3–2×10^3	1	1.6
2×10^3–5×10^3	2	1.06–1.1
5×10^3–1×10^4	2	1.2–1.4
1×10^4–2×10^4	2	1.4–1.7
2×10^4–5×10^4	2	1.7–1.8
5×10^4–1×10^5	4	2.2–2.5
1×10^5–2×10^5	5	2.4–3.5
2×10^5–5×10^5	2	3.5
5×10^5–1×10^6	3	3.2–4.6
1×10^6–2×10^6	1	5.6

Table 3. Standard Samples of Linear Polyethylene[a,b]

Range in molecular weight	No. of samples	Range of M_w/M_n
5×10^2–1×10^3	4	1.11–1.15
1×10^3–2×10^3	2	1.20
2×10^3–5×10^3	2	1.14
5×10^3–1×10^4	0	
1×10^4–2×10^4	5	1.3–2.0
2×10^4–5×10^4	13	1.1–6.0
5×10^4–1×10^5	3	2.9
1×10^5–2×10^5	2	1.2

[a] For polyethylene, branched, see Table 11.
[b] Samples of hydrogenated polybutadiene are not included in this table; see polybutadiene, hydrogenated, in Table 11.

Table 4. Standard Samples of Poly(ethylene glycol) and Poly(ethylene oxide)[a]

Range in molecular weight	No. of samples	Range of M_w/M_n
Poly(ethylene glycol)		
5×10^1–1×10^2	1	1.0
1×10^2–2×10^2	9	1.0–1.11
2×10^2–5×10^2	13	1.07–1.20
5×10^2–1×10^3	9	1.07–1.10
1×10^3–2×10^3	11	1.05–1.10
2×10^3–5×10^3	11	1.03–1.10
5×10^3–1×10^4	6	1.03–1.10
1×10^4–2×10^4	12	1.04–1.34
2×10^4–5×10^4	3	1.06–1.13
Poly(ethylene oxide)		
1×10^4–2×10^4	3	1.10
2×10^4–5×10^4	9	1.03–1.14
5×10^4–1×10^5	6	1.02–1.10
1×10^5–2×10^5	6	1.03–1.10
2×10^5–5×10^5	6	1.04–1.10
5×10^5–1×10^6	9	1.04–1.10
1×10^6–2×10^6	3	1.1–1.12

[a] Although poly(ethylene oxide) and poly(ethylene glycol) have the same basic repeating unit, (—CH_2CH_2O—), the name poly(ethylene glycol) implies that there are hydroxyl groups on both ends of the chain (which is important for some applications of these materials), whereas poly(ethylene oxide) implies nothing about the nature of the end groups. Therefore, these materials are tabulated by the names used by their suppliers.

Table 5. Standard Samples of Polyisoprene

Range in molecular weight	No. of samples	Range of M_w/M_n
5×10^2–1×10^3	4	1.0–1.2
1×10^3–2×10^3	4	1.05–1.20
2×10^3–5×10^3	4	1.05–1.10
5×10^3–1×10^4	4	1.04–1.20
1×10^4–2×10^4	11	1.04–1.20
2×10^4–5×10^4	6	1.04–1.34
5×10^4–1×10^5	5	1.04–1.16
1×10^5–2×10^5	3	1.04–1.10
2×10^5–5×10^5	5	1.05–1.10
5×10^5–1×10^6	2	1.05–1.10
1×10^6–2×10^6	2	1.08
2×10^6–5×10^6	2	1.15

Table 6. Standard Samples of Poly(methyl methacrylate)[a]

Range in molecular weight	No. of samples	Range of M_w/M_n
2×10^3–5×10^3	2	1.15
5×10^3–1×10^4	2	1.15
1×10^4–2×10^4	4	1.10–1.80
2×10^4–5×10^4	12	1.07–2.40
5×10^4–1×10^5	14	1.08–2.0
1×10^5–2×10^5	12	1.04–2.44
2×10^5–5×10^5	16	1.09–4.10
5×10^5–1×10^6	12	1.10–3.30
1×10^6–2×10^6	4	1.3–4.0
Deuterated poly(methyl methacrylate)		
1×10^4–2×10^4	1	1.1
2×10^4–5×10^4	1	1.1
5×10^4–1×10^5	2	1.1
1×10^5–2×10^5	2	1.1
2×10^5–5×10^5	1	1.1

[a] For stereoregular poly(methyl methacrylate), see Table 12.

mentation equilibrium ultracentrifugation for weight-average molecular weight M_w. Nonabsolute methods, which require calibration with samples of known molecular weight, include vapor-phase osmometry for M_n; limiting viscosity number (LVN) for viscosity-average molecular weight (usually close to M_w); and gpc for M_n, M_w, peak molecular weight and, at least qualitatively, distribution in molecular weight (see MOLECULAR WEIGHT DETERMINATION).

The Encyclopedia tabulates standard polymers available as of mid-1987, with ranges in molecular weight and in polydispersity M_w/M_n, a measure of the breadth of the distribution in molecular weight. Tables are given for standard samples of polybutadiene (Table 1), polydimethylsiloxane (Table 2), linear polyethylene (Table 3), poly(ethylene glycol) and poly(ethylene oxide) (Table 4), polyisoprene (Table 5), poly(methyl methacrylate) (Table 6), poly(α-methylstyrene) (Table 7), polystyrene (Table 8), poly(sodium styrenesulfonate) (Table 9), vinyl polymers (Table 10), miscellaneous polymers (Table 11), and stereoregular polymers (Table 12).

Table 7. Standard Samples of Poly(α-methylstyrene)

Range in molecular weight	No. of samples	Range of M_w/M_n
1×10^3–2×10^3	1	1.15
2×10^3–5×10^3	1	1.04
5×10^3–1×10^4	5	1.04–1.20
1×10^4–2×10^4	2	1.04–1.10
2×10^4–5×10^4	5	1.04–1.15
5×10^4–1×10^5	5	1.05–1.10
1×10^5–2×10^5	5	1.05–1.10
2×10^5–5×10^5	4	1.05–1.10
5×10^5–1×10^6	4	1.05–1.10
1×10^6–2×10^6	1	1.10

Table 8. Standard Samples of Polystyrene[a,b]

Range in molecular weight	No. of samples	Range of M_w/M_n
1×10^2–2×10^2	1	1.0
2×10^2–5×10^2	3	1.1–1.15
5×10^2–1×10^3	17	1.03–1.30
1×10^3–2×10^3	8	1.10–1.30
2×10^3–5×10^3	20	1.04–1.30
5×10^3–1×10^4	22	1.04–1.19
1×10^4–2×10^4	19	1.02–1.15
2×10^4–5×10^4	25	1.01–1.40
5×10^4–1×10^5	21	1.03–1.16
1×10^5–2×10^5	21	1.02–1.18
2×10^5–5×10^5	29	1.04–3.80
5×10^5–1×10^6	20	1.01–1.12
1×10^6–2×10^6	20	1.05–1.85
2×10^6–5×10^6	17	1.04–1.20
5×10^6–1×10^7	9	1.14–1.20
1×10^7–2×10^7	5	1.2–1.3
2×10^7–5×10^7	4	1.2–1.3
Deuterated polystyrenes		
5×10^3–1×10^4	2	1.05
1×10^4–2×10^4	2	1.05–1.10
2×10^4–5×10^4	1	1.05
5×10^4–1×10^5	2	1.05–1.10
1×10^5–2×10^5	2	1.05–1.10
2×10^5–5×10^5	2	1.10
5×10^5–1×10^6	1	1.10
1×10^6–2×10^6	1	1.30
Polystyrene oligomer mixture 13 peaks in distribution in molecular weight, from 208 to 1458		

[a] For polystyrene star molecules, see Table 11.
[b] For stereoregular polystyrene, see Table 12.

Table 9. Standard Samples of Poly(sodium styrenesulfonate)

Range in molecular weight	No. of samples	M_w/M_n
1×10^3–2×10^3	4	1.25
2×10^3–5×10^3	5	1.10
5×10^3–1×10^4	5	1.10
1×10^4–2×10^4	5	1.10
2×10^4–5×10^4	5	1.10
5×10^4–1×10^5	10	1.10
1×10^5–2×10^5	5	1.10
2×10^5–5×10^5	6	1.10
5×10^5–1×10^6	5	1.10
1×10^6–2×10^6	6	1.10

Table 10. Standard Samples of Vinyl Polymers

Range in molecular weight	No. of samples	Range of M_w/M_n
Poly(vinyl acetate)		
1×10^5–2×10^5	9	2.4–9.1
2×10^5–5×10^5	3	2.6–3.2
Poly(vinyl chloride)		
2×10^4–5×10^4	1	1.8
5×10^4–1×10^5	6	1.8–2.5
1×10^5–2×10^5	6	2.1–2.4

Table 11. Standard Samples of Miscellaneous Polymers

Polymer	Range in molecular weight	No. of samples	Range of M_w/M_n (if given)
polyhexamethyleneadipamide (nylon-6,6)	2×10^4–1×10^5	8	2.2–2.8
phenoxy resin	5×10^4–1×10^5	1	3.8
polyacrylamide	2×10^4–1×10^7	19	1.5–2.6
poly(acrylic acid)	2×10^3–5×10^6	16	22
poly(sodium acrylate)	1×10^3–2×10^6	23	1.3–2.4
polyacrylonitrile	5×10^4–5×10^5	3	3.8–5.3
polybutadiene, hydrogenated	5×10^2–5×10^5	10	1.16–2.66
poly(butyl acrylate)	5×10^4–1×10^5	1	3.0
poly(butyl methacrylate)	2×10^5–5×10^5	1	4.4
polycaprolactone	2×10^4–5×10^4	2	3.1–3.2
polycarbonate	2×10^4–1×10^5	9	2.5–4.5
poly(2,6-dimethyl-*p*-phenylene oxide)	5×10^4–5×10^5	2	2.5–7.6
polyepichlorohydrin	1×10^3–5×10^3	2	1.40–1.58
poly(ethyl acrylate)	5×10^4–1×10^5	3	2.5
polyethylene, branched	5×10^4–5×10^5	5	4–16
poly(ethylene terephthalate)	2×10^4–5×10^4	1	1.9
poly(2-ethylhexyl acrylate)	1×10^5–2×10^5	3	2.6
poly(ethyl methacrylate)	2×10^5–5×10^5	3	2.7
poly(isobutyl acrylate)	1×10^5–2×10^5	1	3.7
polyisobutylene	5×10^2–2×10^6	7	1.03–3.5
poly(isobutyl methacrylate)	2×10^5–5×10^5	2	2.1
poly(lauryl methacrylate)	1×10^5–5×10^5	3	1.4–3.2
poly(methyl acrylate)	2×10^4–5×10^4	3	2.9
poly(*n*-butyl acrylate)	5×10^4–2×10^5	2	3.0–3.6
poly(*n*-butyl methacrylate)	2×10^5–5×10^5	1	4.4
poly(octadecyl acrylate)	2×10^4–5×10^4	1	1.8
poly(octadecyl methacrylate)	2×10^5–5×10^5	2	2.2–3.3
polypropylene	2×10^4–5×10^5	14	2.1–8.0
poly(propylene glycol)	5×10^1–1×10^4	10	1.0–1.2
polysaccharide	5×10^3–1×10^6	16	1.06–1.14
polystyrene, star-branched	5×10^3–2×10^5	4	
polysulfone resin	5×10^4–1×10^5	2	3.3
polytetrahydrofuran (poly(butylene oxide))	1×10^3–1×10^6	27	1.03–1.20
poly(vinyl alcohol)	5×10^3–5×10^5	7	1.7–2.2
poly(vinyl butyral)	1×10^5–2×10^5	2	6.1
poly(*N*-vinylcarbazole)	5×10^5–1×10^6	3	3.6
poly(vinyl fluoride)	1×10^5–2×10^5	2	3.2
poly(vinyl formal)	2×10^4–5×10^4	2	2.8
poly(vinyl methyl ether)	5×10^4–1×10^5	2	2.1
poly(2-vinylpyridine)	2×10^3–2×10^6	20	1.05–2.05
poly(vinyl stearate)	5×10^5–1×10^6	1	9.4

Table 12. Standard Samples of Stereoregular Polymers

Polymer
poly(1-butene), isotactic
poly(methyl methacrylate), isotactic
poly(methyl methacrylate), syndiotactic
polypropylene, isotactic
polystyrene, isotactic
poly(2-vinylpyridine), isotactic

BIBLIOGRAPHY

P. J. Flory, *Principles of Polymer Chemistry,* Cornell University Press, Ithaca, N.Y., 1953.

H. Morawetz, *Macromolecules in Solution,* Wiley-Interscience, New York, 1965.

H. Yamakawa, *Modern Theory of Polymer Solutions,* Harper & Row, Publishers, Inc., New York, 1971.

LESLIE E. SMITH
PETER H. VERDIER
National Bureau of Standards

POLYNUCLEOTIDES

Role of Nucleic Acids in Biology

Polynucleotides is the generic term for the nucleic acids and their synthetic derivatives. This term is frequently used for the synthetic polynucleotides to distinguish them from the naturally occurring DNA and RNA. All these compounds have the same basic structure, consisting of very long chains, with the repeating units being the nucleotides (Fig. 1).

Nucleotides are the 3′ or 5′ esters of the nucleosides, which in turn consist of a pentose sugar (ribose or 2′-deoxyribose, respectively) and a heterocyclic base (Fig. 1). The bases in nucleic acids are mainly adenine (A), guanine (G), cytosine (C), and thymine (T) (in DNA) or uracil (U) (in RNA) (Fig. 2).

The fact that the nucleic acids and polynucleotides always occur as the 3′–5′ phosphodiesters gives these chains directionality (or polarity), which is of primary importance for the structure and function of nucleic acids. The central role played by DNA as the carrier and transmitter of hereditary traits has been amply detailed in numerous books and review articles.

Structure: DNA as Double Helix

Each of the two strands of the DNA double helix consists of a sugar–phosphodiester backbone, to which the bases are attached. The backbone has polarity since the connection of the phosphate is to the 3′ group of the sugar in one direction and to the 5′ groups of the sugar in the other. Since the two strands are running in opposite directions, they are of opposite polarity, or antiparallel. The bases are complementary to each other. Adenine is always paired with thymine, and guanine with cytosine, taking into account the equivalence found by Chargaff. The opposite polarity of the two chains has an important structural consequence: since the bases are paired not across the helix, but astride, the double helix has two grooves of unequal depth and height, the large groove being about twice as high as the small one. The opposite polarity of the two strands of the double helix also causes the symmetry pseudodyad axis to be perpendicular to the helix axis: the helix, when turned upside down, shows the two strands again in the same direction. The base pairs are essentially perpendicular to the helix axis. The bases themselves are twisted in a slight propellerlike way.

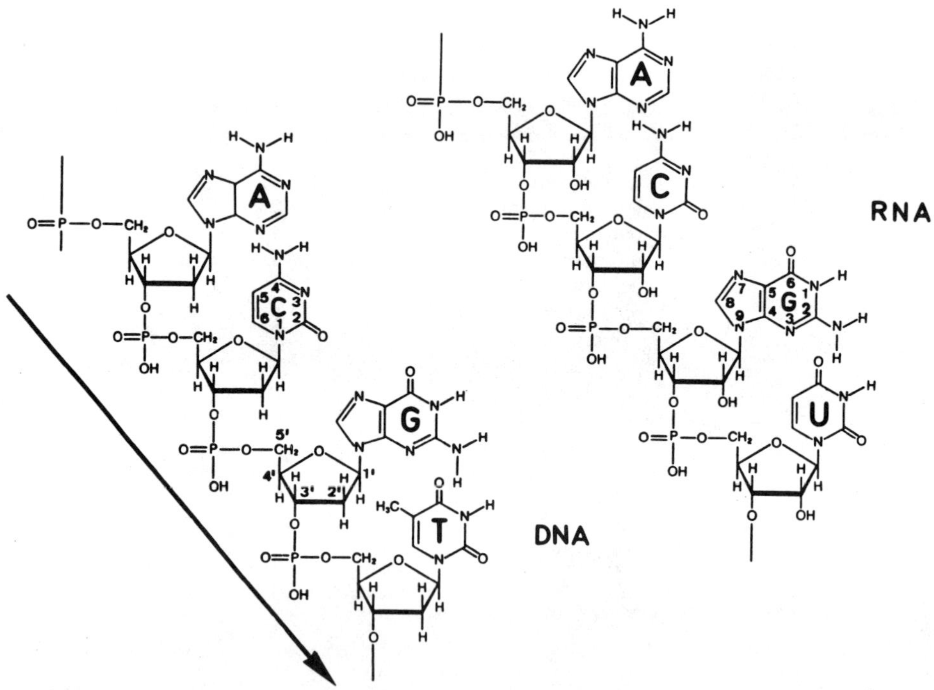

Figure 1. DNA and RNA chains. The direction of the nucleic acids is always read in the 5′–3′ direction. Numbering of bases and sugars is included.

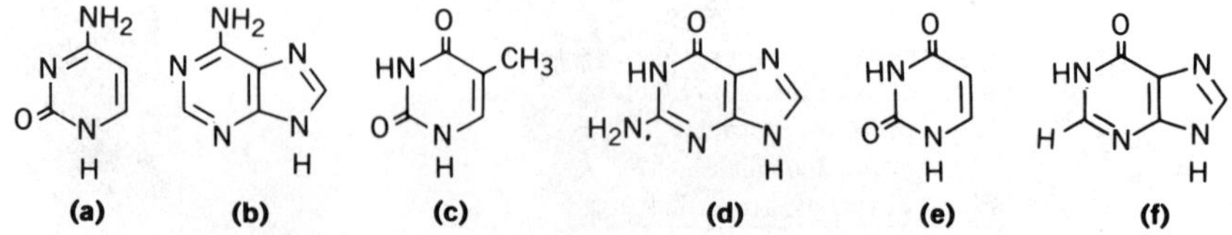

Figure 2. Natural and synthetic bases, nucleosides and nucleotides (in the syn form). (**a**) Cytosine; (**b**) adenine; (**c**) thymine; (**d**) guanine; (**e**) uracil; (**f**) hypoxanthine.

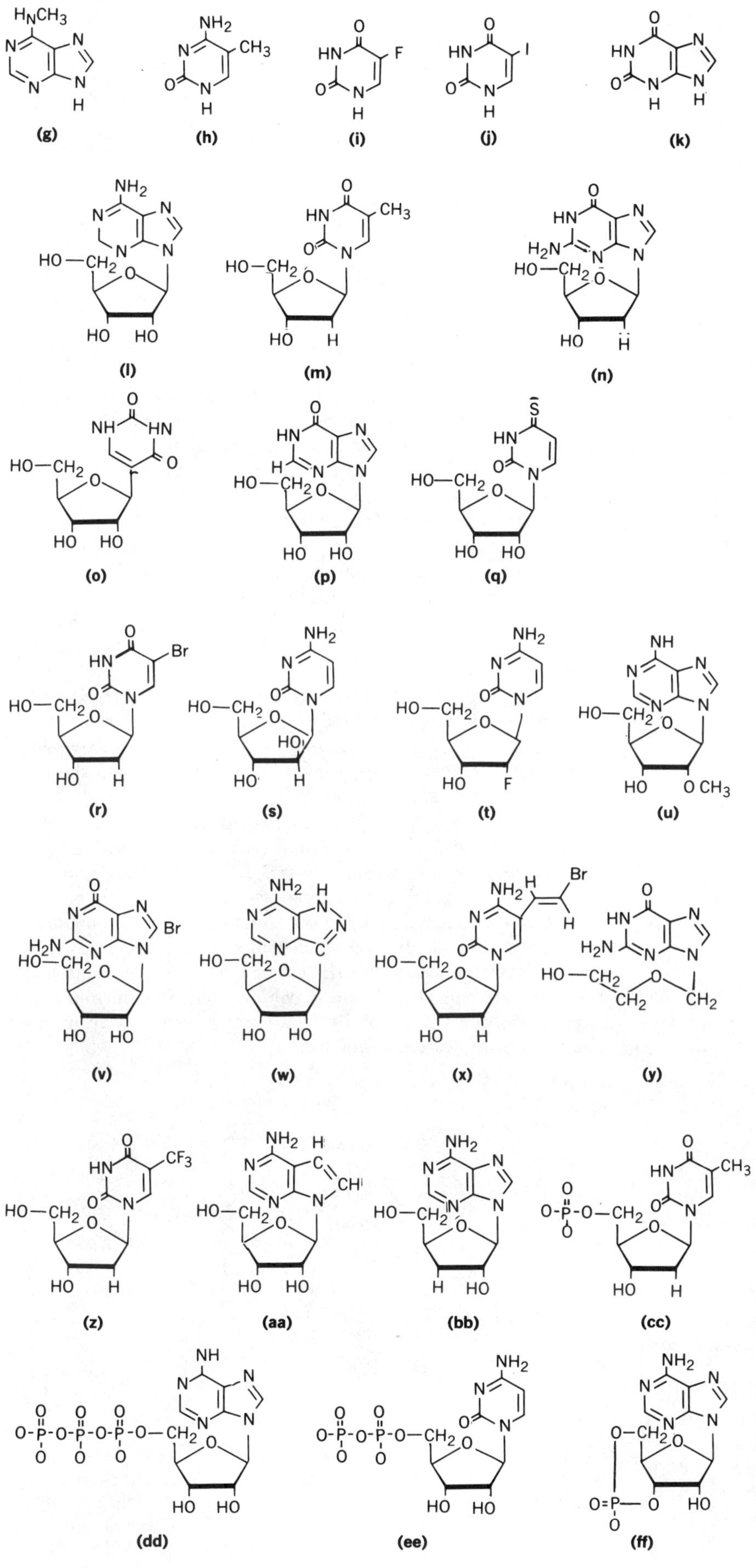

Figure 2 continued. (**g**) N^6-methyladenine; (**h**) 5-methylcytosine; (**i**) 5-fluorouracil; (**j**) 5-iodouracil; (**k**) xanthine; (**l**) adenosine; (**m**) thymidine; (**n**) deoxyguanosine; (**o**) pseudouridine; (**p**) inosine; (**q**) 4-thiouridine; (**r**) 5-bromodeoxyuridine; (**s**) arabinosyl-cytosine (ara-C); (**t**) 2′-deoxy-2′-fluorocytidine; (**u**) 2′-*O*-methyladenosine; (**v**) 8-bromoguanosine; (**w**) formycin A; (**x**) (E)-5-(2-bromovinyl)deoxyuridine; (**y**) acyclovir (9-(2-hydroxyethoxymethyl)guanine); (**z**) 5-trifluoromethyldeoxyuridine (trifluoridine); (**aa**) tubercidine; (**bb**) cordycepine (3′-deoxyadenosine); (**cc**) thymidine-5′-phosphate (thymidylic acid, TMP); (**dd**) adenosine triphosphate (ATP); (**ee**) cytidine diphosphate (CDP); (**ff**) adenosine-3′,5′-cyclic phosphate (cyclic AMP, cAMP).

After the Watson-Crick proposal of the DNA double helix, considerable effort was made to investigate the various aspects of this model. The finding that DNA could assume two forms gave rise to questions of the reasons why and the circumstances under which such a change would take place and of its structural implications. The questions of why and in which respect RNA was different from DNA were raised in this context. It also was of interest to know whether modifications of the basic nucleic acid chain could be effectuated and tolerated. Such modifications could be important to understanding the forces determining nucleic acid structure. In view of the central importance of nucleic acids in the cellular machinery, interference with the structure would be expected to have grave repercussions on their function, could yield valuable clues on mutations, and could open ways to pharmacological applications and the study of cancer. For this reason the natural and modified nucleosides and nucleotides, and then homopolynucleotides and polynucleotides with defined repeating sequences, have been studied. Modifications of virtually every part of nucleic acids (Fig. 2), the phosphate group, the sugar moiety, and the bases, have been introduced and their effects studied. Interactions with drugs, metal ions, metal complexes, and proteins or peptides have given valuable information on their mode of action.

Techniques

Experimental techniques include x-ray fiber diffraction, uv absorbance and optical activity, ultracentrifugation and hydrodynamic studies, gel electrophoresis, nmr, and ir and Raman spectroscopy.

Monomers

Studies on nucleosides and nucleotides have afforded valuable information on the chemical and physical properties of the bases, sugars, and phosphate esters, in both solution and solid state. These studies have permitted some general rules about the chemistry, conformation, ionization properties, and electronic structure of the building blocks of nucleic acids.

Chemical Reactions. Polynucleotides are stable for short times against acid at low temperature; RNA and DNA are precipitated by acid. Prolonged exposure to acid hydrolyzes the glycosidic linkage of DNA.

Polynucleotides can be distinguished by their stability against alkali. Polydeoxynucleotides (DNAs) are denatured above pH 11, whereas the polyribonucleotides (RNAs) are hydrolyzed to their nucleoside 3′-(2′-)phosphates.

The action of many ribonucleases (RNAase T_1 RNAaseA) splits in analogous fashion, yielding termini with Guo or Urd 2′-(3′-)phosphates, respectively.

The reactivity of the building blocks of nucleic acids, the bases, sugars, and phosphates, toward specific reagents has been extensively studied. In one of the oldest analytical methods for nucleic acids, phosphate is determined by the color reaction with molybdate after ashing the sample.

The Bases. Pyrimidine and purine bases have been investigated individually or as their glycosides (Fig. 2). The heterocyclic aromatic rings of purines and pyrimidines are approximately planar. The amino groups in cytosine and the purines are integrated into the resonance system. Bond distances of the amine groups indicate a certain degree of double-bond character. The exocyclic C═O bond, on the other hand, shows normal double-bond character. Protonation and deprotonation take place between pH 3.5 and 10.5, indicating amide-like hydrogens. Protonation always takes place on the ring nitrogens and not on the amino groups; pK values are influenced by chemical substituents on the bases.

Tautomers have been the subject of a longstanding debate. Tautomerism may be responsible for mutations in DNA, as the cause of cancer, and other phenomena.

Nucleosides and Nucleotides. The linkage of the base to the sugar changes their behavior considerably; the solubility in water increases, that in aprotic solvents decreases, and the pK changes considerably. In the nucleotides the water solubility increases further, and the phosphate group itself is charged. In strong alkaline solution the secondary hydroxyl groups of the riboses are deprotonated. Phosphate groups have two values, one around pH = 1.5, the other around pH = 6. The absorption spectra of the bases are modified considerably when glycosylated. The introduction of an asymmetric center ($C_1{}'$ of the pentose sugar) confers optical activity to the nucleosides and nucleotides.

Conformation of Nucleosides and Nucleotides. There are six bonds (corresponding to the angles α to ζ in Fig. 3) that determine the conformational freedom of a nucleotide along the nucleic acid chain, which can be rotated, plus the glycosidic linkage (torsion angle χ, Fig. 3). The highly repetitive chain cannot have unlimited degrees of freedom, if a regular structure is to be formed.

X-ray structure determinations on hundreds of nucleosides and nucleotides have yielded a fairly coherent picture of the conformation of these compounds. These studies have made it clear that the bases are essentially planar, whereas the five-membered sugar ring is puckered.

The puckers of nearly all nucleosides and nucleotides in crystals and solution have been found in conformations clustered in the two canonical regions S, 2′-endo(3′-exo), and N,3′-endo(2′-exo). Notable exceptions are the various 2′-deoxy-2′-fluoronucleosides, which have the unusual 4′-exo-O′-endo form. Numerous nmr studies on nucleosides and nucleotides in solution have concluded that the sugar moiety is in a rapid equilibrium between the N and S conformers. Furthermore, the substitution on the sugar by polar groups strongly influences the conformational equilibrium.

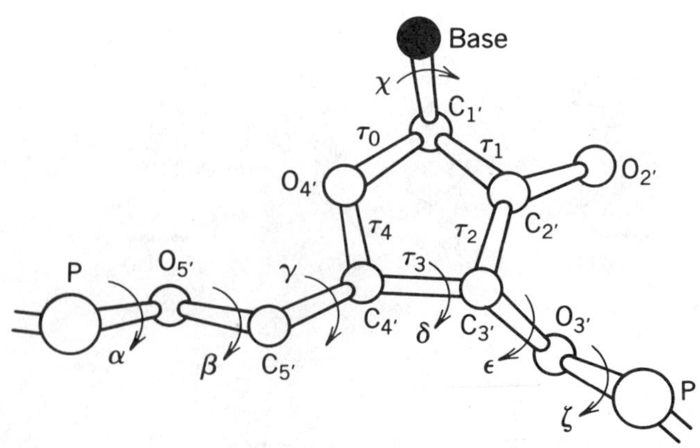

Figure 3. Six conformational bonds and nomenclature of bond angles along the phosphate–sugar–phosphate chain (α to ζ). The seventh conformational angle is the glycosidic torsion angle χ. The dihedral angles of the pentose are designated τ_0 to τ_4. The conformational angle δ is related to the sugar torsion angles by the relation $\delta = 120° - \tau_3$.

Interactions Between Bases or Nucleosides. In nucleic acids two kinds of interactions between bases occur: horizontal, in the plane of the bases, and vertical, perpendicular to the bases. The first kind is exemplified by the hydrogen-bonding base pairing of A–T and G–C bases, the second by the stacking of the bases in the helix. In monomers these phenomena can also be observed. In nonpolar solvents, bases associate to form hydrogen-bonded complexes, whereas in water and other polar solvents stacking dominates, forming a hydrophobic stack or core, while the amino, amido, and keto sites of the bases, as well as the OH groups of the sugars, are solvated.

Polynucleotide Complexes

Polynucleotide complexes include polymers of natural nucleotides, homopolynucleotides, (polyuridylic acid, polycytidylic acid, polyadenylic acid, polyinosinic acid, and polyguanylic acid), and polynucleotide duplexes (poly(A)·poly(U) and related deoxyduplexes, and poly(G)·poly(C) and related duplexes).

Sequence Determines the Structure of Polynucleotide Duplexes. Polydeoxynucleotides with short repeating sequences were first made using DNA polymerase. This enzyme synthesizes poly-(dA–T) or poly(dI–dC) from the corresponding nucleoside triphosphates without a template. Normally, however, DNA polymerase mimics the replication process and copies a given template. These templates can be short oligonucleotides (usually chemically synthesized), which the enzyme copies. By a "slipping mechanism" rather large polymers (up to mol wt = 10^3–10^6) with the basic repeating unit are formed. Polydeoxynucleotides containing two, three, or four repeating nucleotides have thus been synthesized and studied.

It became apparent that duplexes with the same base composition but different sequence had different physical properties, such as T_m values, buoyant densities, or cd spectra. Also included are polynucleotides with modified bases, polynucleotides with modified sugars, and polynucleotides with modified phosphates (polythiophosphonucleotides and polymethylphosphononucleotides).

Polynucleotide Polymorphism. Early x-ray fiber diffraction studies greatly contributed to understanding DNA structure. Although fiber diffraction patterns cannot give results with the same precision as single-crystal data, they have clearly established the variety of DNA helices (Fig. 4). A dozen structures with different helix symmetries and conformations have been observed for DNA and its synthetic analogues. The salient features of these observations are shown in Table 1, together with more recent data.

These forms can be classified into three groups: the A family is characterized by the 3′-endo (N) puckering of the sugar, whereas those of the B family have a 2′-endo (S) conformation. Both these families are right-handed helices. The third family is the left-handed Z-DNA.

Z-DNA. When increasing the salt concentration in a solution of poly(dG–dC) to 4 *M*, a dramatic inversion of the cd spectrum was observed. This transition was cooperative with slow kinetics and an activation energy on the order of 100 kJ/mol (24 kcal/mol). The low salt form was called R, and the high salt form L; a right–left-hand transition in the sense of the helix was proposed. Other agents were also found to convert poly(dG–dC) from the R to the L form, including alcohols at 60–65%, di- and trivalent cations, as well as amine complexes of cobalt, platinum(II), zinc, and copper. It took several years before a small fragment $(dC–dG)_3$ was synthesized and studied by x-ray crystallography; it was shown to form a left-handed helix, called Z-DNA.

In addition to the left-handedness of the helix, Z-DNA exhibits several striking features. The repeating unit of the double helix is not a mononucletide pair, but a pair of dinucleotides, the dyad is between the bases, not in the plane of the bases. The orientation of the bases around the glycosidic linkage is anti for the deoxycytidine residue, but syn (ie, the base is above the sugar) for the deoxyguanosine residue. The sugar puckering is alternatingly 2′-endo in deoxycytidine, but 3′-endo for deoxyguanosine. These changes also affect the ir and Raman spectra of poly(dG–dC). The sugar–phosphate backbone of the double helix is not continuous as in the right-handed A and B forms of DNA, but follows a zigzag pattern to accommodate the unusual glycosidic linkages and puckers with the two sugars facing each other. The dCpdG sequence is thus not equivalent to dGpdC. This nonequivalence of the phosphates gives rise to two distinct phosphorus resonances in Z-DNA. Although in A- and B-DNA two distinct grooves are observed, only one can be distinguished in Z-DNA, which is much deeper. The second is a broad outside crest, caused by the flip over of the base pairs outside the helix.

Supercoiled DNA. Synthetic polynucleotides are linear by definition. The situation is quite different in natural DNAs, which are circular and supercoiled with very few exceptions.

Supercoiling results when the two strands of the double helix of a closed circular DNA are underwound (negative su-

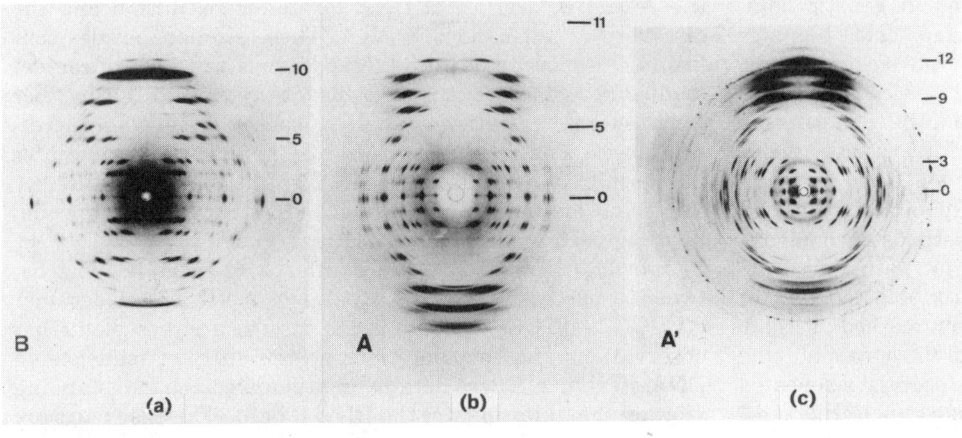

Figure 4. X-ray fiber diffraction patterns of (**a**) B-DNA (10-fold helix); (**b**) A-DNA (11-fold helix); and (**c**) A′-RNA (poly(I)·poly(C), 12-fold helix). Courtesy of Prof. S. Arnott.

Table 1. Helical Forms of DNA and RNA in Fibers

Polynucleotide	Counterion	Relative humidity, %	Form[a]	Helix		Turn angle	Tilt γ, °
				Symmetry	Rise, nm		
native DNA	Na	75	A	11/1	0.256	32.7	20.3
	Na	92	B	10/1	0.338	36.0	−5.9/−2.1
	Li	66	B	10/1			
	Li	57–66	C	28/3	0.332	38.6	−8.0/−1.0
	Li	44	C	28/3			
poly(dA)·poly(dT)	Na	70	B′	10/1	0.329	36.0	7.9/1.0
	Na	92	αB′	10/1			
		71	H(A/B)	10/1	0.323		26.5/18.0
poly(dG)·poly(dC)	Na	75	A	11/1	0.256	32.7	
	Na	92	B	10/1			
poly(dA–dT)	Na	75	D	8/1	0.304	45.0	−16.4
	Na	<98	A	11/1			
	Li	66	B	10/1			
	Li	56	W	8/2	0.61/2	43.5/46.5	15.0/16.2
	Li	56	P	24/4	1.85/4	36.5–54.4	18.0–23.3
poly(dG–dC)	Na	<92	A	11/1			
	Li	81	B	10/1			
	Li	56	W	10/2	0.672/2		
	Na	43	Z	6/5	0.363	−30.0	
poly(dA–dC)·poly(dG–dT)	Na	66	A	11/1			
	Na	66–92	B	10/1			
	Na	66	Z	6/5			
poly(dA–dG)·poly(dC–dT)	Na	66	C″	9/2	0.323	48.0	
	Na	95	B	10/1			
poly(A)·poly(U)	Na	<92	A	11/1	0.256	32.7	16.0
poly(I)·poly(C)	Na	<92	A′	12/1	0.300	30.0	10.0
poly(rI)·poly(dC)	Na	75	A′	12/1			
poly(dA)·poly(rU)			H(B/A)	11/1	0.306	32.7	13.0
poly(dI)·poly(rC)			H(B/A)	10/1	0.306	32.7	15.0
DNA–RNA hybrid	Na	33–92	A	11/1			
poly(U)·poly(A)·poly(U)	Na	72	A	11/1	0.304	32.7	11.7/13.6
	Na	92	A′	12/1	0.304	30.0	10.2/12.0
poly(dT)·poly(dA)·poly(dT)	Na		A		0.326		7.2/9.2

[a] H: heteromerous helices; (A/B) indicates that the first strand is in the A conformation (3′-endo) and the second in the B conformation (2′-endo). W: wrinkled helices; dinucleotide repeat. Z: left-handed helices; dinucleotide repeat. P: pleiomorphic helix; hexanucleotide repeat.

percoiling) or overwound (positive supercoiling). In both cases an unfavorable free energy change relative to the relaxed closed circular DNA exists.

Oligonucleotides

Chemical Synthesis of Oligonucleotides. The chemical synthesis of oligonucleotides has been dramatically improved in the last decade, primarily because of the large demand for oligonucleotides for biological applications in genetic engineering. The improvement of syntheses and yields has also permitted the study of oligonucleotides by physical methods, principally x-ray crystallography and nmr.

Oligonucleotides can be synthesized chemically in solution or on solid support. The latter method has increased the yields and speed of the coupling reactions.

The two principal methods are the phosphotriester method and the phosphite triester method. Both methods have advantages and inconveniences and follow similar paths: first, attachment of the 3′ terminal nucleotide to the solid support (or protection of the 3′OH group in the solution method); second, coupling with the following nucleotide; and third, capping and removal of the unreacted nucleotide. This process is repeated with the coupling and capping steps of the next nucleotide until the entire chain has been built up. Finally, there is deprotection in several steps.

Single-crystal X-ray Studies of Oligonucleotides. *Molecular Conformation of A-, B-, and Z-DNA.* The development of synthetic methods for the preparation of oligonucleotides has permitted for the first time the synthesis of sizeable amounts (tens of milligrams) of DNA. Although the first synthetic DNA pApTpApT was not a regular double helix, x-ray analysis of single crystals has since permitted the characterization of both the A and B families of DNA to atomic resolution and the discovery of left-handed Z-DNA. These studies, besides confirming most of the results obtained from x-ray fiber diffraction studies, revealed structural details totally inaccessible by fiber work, such as variations in helix-turn angles between successive bases. The resolution of tightly bound water molecules has permitted investigation of DNA hydration. These data have suggested mechanisms by which individual base pairs could be recognized by enzymes or drugs.

Structural Details. The large differences in helix twist between successive base pairs in the dodecamer d(CGCGAATTCGCG) have given rise to a series of further observations and considerations on the effect of sequence on DNA structure. These details have changed the longstanding view on the uniformity of the DNA B helix. The base pairs are

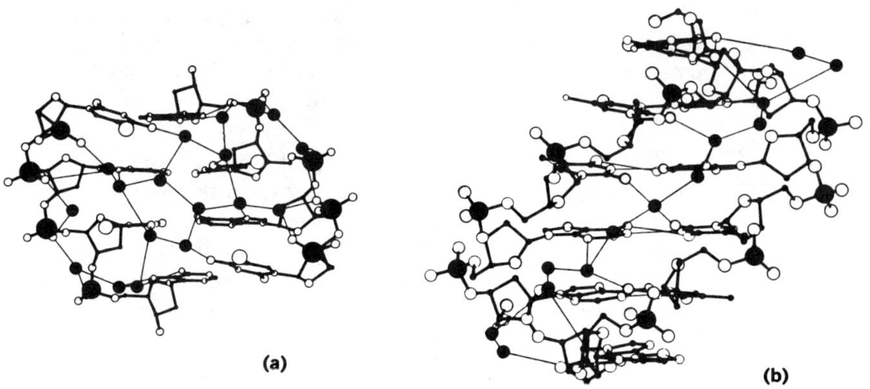

Figure 5. (a) Hydration network across the large groove in A-DNA. Courtesy of Adenine Press, Guilderland, New York. (b) Hydration spine in the minor groove of B-DNA. Courtesy of Academic Press, London. ●, phosphorus atoms; •, water molecules.

not flat, but have a "propeller twist" between the two bases of 10–20°. This propeller twist enables a more efficient overlap of neighboring bases in a given strand. In homopolymer duplexes the propeller twist is a strong stabilizing principle in nucleic acid structure, even with the bases of the opposite strand twisted in the opposite sense.

Hydration of DNA. The existence of two forms of DNA (Fig. 5) as a function of relative humidity was revealed early by x-ray fiber diffraction. The stabilizing role of hydration in DNA structure was revealed by the discovery of a waterspine in the small groove of A·T regions of DNA.

Nuclear Magnetic Resonance Studies. Clearly, the best technique to study the dynamics of nucleic acids is high resolution nmr. The advent of new nmr technologies combined with the availability of sizeable amounts of oligonucleotides has led to innumerable nmr studies on the structure and dynamics of nucleic acids.

The most interesting and successful methods are the use of nuclear Overhauser effect (NOE) measurements, the development of two-dimensional nmr methods, and the use of H_2O as solvent to study the dynamics of exchangeable protons. The first is by far the most interesting method, both in the standard one-dimensional as well as in the two-dimensional technique (NOESY). The sensitivity of NOE measurements is due to their $1/(r_{ij})^6$ relationship. Any measureable NOE indicates a distance between the irradiated and the enhanced proton of less than 0.4 nm.

The investigation of oligonucleotides by nmr with standard base pairings have all turned out to be in the standard B-DNA form. Also studied are double helices that contain mismatches or insertions, both of which can result from replication errors. Mismatches occur transiently when a wrong base is incorporated in the first round of replication. Genetic experiments have demonstrated that certain mismatches are repaired, whereas others are not. The mismatch is thus transformed into a permanent mutation, which becomes a hereditary trait of the genome. The understanding of these mechanisms, why certain mismatches are more resistant to repair than others, is of primary importance for the comprehension of mutation mechanisms and the causes of cancer. Much effort is thus invested into the study of mismatches from the biological, structural, and dynamic standpoints.

Thermodynamics and Kinetics of Helix Formation

For a long time it was believed that the hydrogen bonds between the bases constitute the essence of the stabilizing forces. Several important studies in the 1960s demonstrated that two other contributions were of considerably greater importance: (a) the charge repulsion between the phosphate ions and the afferent electrostatic free energy change, exemplified by the strong ionic strength dependence of the T_m values and (b) the stacking forces between successive bases, which are present even in the absence of a double helix, exemplified by the persistence of hypochromicity, even above the T_m. These stacking forces were studied by theoretical calculations and experimentally demonstrated and quantified by the chain-length dependence of cd in homo- and heterooligonucleotides.

Kinetics of Helix Formation. The formation of a double helix from the separated complementary strands is a second-order reaction. The beginning duplex requires the formation of a stable nucleus, which is the rate-limiting step. The further growth of the chain is cooperative and thus rather rapid. If the base pair sequence is not a repeating sequence as in regular homopolynucleotide duplexes, the stability constant s is not the same in each step; the nucleation of the first base pair becomes dominant and thus very slow. False pairings redissociate until the exact base pairing is obtained. This had been observed in the renaturation of biologically active DNA from many sources.

Polynucleotide–Ligand Interactions

Interactions with Cations. The polyanionic character of polynucleotides favors the electrostatic binding of small cations to the negatively charged phosphates. These cations are usually hydrated. The importance of electrostatic effects on the stability of helical polynucleotides has long been recognized. Studies of DNA and two- or three-stranded polynucleotide complexes have shown that the transition temperature T_m is a linear function of the logarithm of the activity (concentration) of monovalent cations.

Interaction with Small Organic Compounds. Considerable progress has been made on the interaction of small organic compounds, both charged and uncharged, on nucleic acids. Many compounds studied are either antibiotics (or otherwise pharmacologically significant) or carcinogenic (or otherwise harmful) drugs. Many of them bind with remarkable efficiency and high specificity to DNA, but frequently not to RNA. They have thus also become sensitive sensors for the structure and conformation of polynucleotides. Most of the drugs studied can be classified into three groups: intercalators, groove binders, and covalent linkers.

Nucleic Acid–Protein Interactions. Gene expression is usually controlled by specific proteins, which recognize a nucleic acid base sequence or the modification of the structure of this target sequence. These promotor sequences are turned off by the binding of a specific repressor protein.

The problem of protein–DNA recognition has kinetic and thermodynamic aspects. The problem of how a protein "finds" its (frequently unique) target sequence out of millions of base pairs of a DNA sequence in a biologically acceptable time lag has been the subject of much debate.

Restriction Enzymes and Genetic Engineering

Nucleic acids are cleaved by enzymes, either endo- or exonucleases. Exonucleases attack nucleic acid chains from a given end, eg, 3′-exonucleases from the 3′ end, 5′-exonucleases from the 5′ end, and never from both. Most exonucleases are processive enzymes, ie, once a nucleic acid chain is attacked by them, the enzyme–nucleic acid complex does not dissociate until the entire chain is digested. Thus they do not follow the classical Michaelis-Menten kinetics of enzymes.

Endonucleases attack within a chain and frequently at specific sites. Some nucleases split only double-stranded DNA; others attack only single-stranded DNA. Most RNAases specifically attack single-stranded RNA. A notable exception is, for instance, micrococcal nuclease, which splits both single- and double-stranded DNA at all sites, yielding 3′-phosphates.

A particular class of endonucleases comprises the so-called restriction enzymes, which were named after the genetic phenomenon of restriction that was first observed in bacteria. Foreign DNAs (eg, from bacterial viruses) are destroyed under certain physiological conditions when entering the bacterial cell. It soon became apparent that restriction was a defense mechanism of the bacterium against invading genetic material. The question of how the bacterium avoided digestion of its own DNA arose. It was found that the state and degree of chemical modification of the host DNA (mainly base methylation) distinguished it from foreign DNA.

Since the discovery of the first restriction on nuclease from *Escherichia coli,* Eco RI, several hundred restriction enzymes (restrictases) have been isolated, all from bacteria. No analogous enzymes have yet been found in eukaryotes. Most of these enzymes have a very similar mechanism of action: a given palindromic DNA sequence, four to 10 bases long, is the specific recognition sequence, and cleavage takes place generally in a staggered fashion at one and only one specific sequence, yielding protruding ends on both strands.

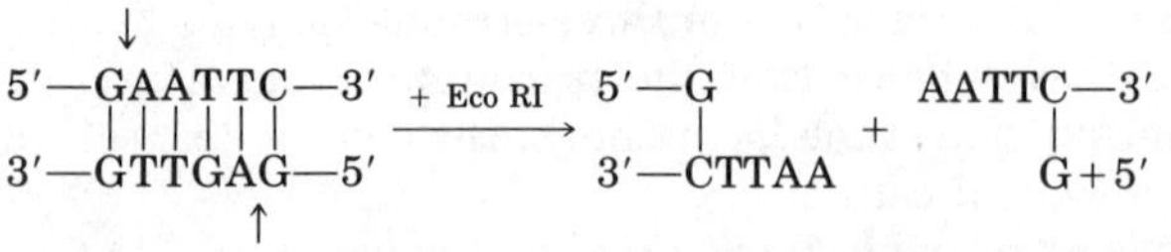

This opens the possibility of inserting a piece of foreign DNA, cleaved by the same enzyme, into the original host DNA. After reassociation, phosphodiester bonds are reestablished by ligases. This methodology is now routinely used to insert genes of proteins into bacterial vectors. This has become known as genetic engineering, which has proven useful in the industrial production of pharmacologically interesting proteins and has also found applications in agriculture. Nitrogen-fixating enzymes have been cloned in bacteria that normally do not metabolize molecular nitrogen. This clearly has enormous growth potential. The scientific usage of restriction enzymes has permitted the elucidation of numerous biological and genetic mechanisms.

The use of synthetic DNA fragments will permit dealing with protein engineering and prying open the astronomically complex problem of nucleic acid–protein interactions. The understanding of how proteins bind to DNA will give perceptions of how they regulate the expression of the genomic message and, eventually, insight into what life is all about.

BIBLIOGRAPHY

J. D. Watson, *Molecular Biology of the Gene,* 3rd ed., W. A. Benjamin, Menlo Park, Calif., 1976.

B. Alberts, D. Bray, J. Lewis, M. Raff, K. Roberts, and J. D. Watson, *The Molecular Biology of the Cell,* Garland Publishing, New York, 1983.

I. Rosenfeld, E. Ziff, and B. van Loon, *DNA for Beginners,* Writers and Readers, London, and W. W. Norton & Co., New York, 1983.

W. Guschlbauer and W. Saenger, eds., *DNA Ligand Interactions: From Drugs to Proteins,* NATO-ASI Series, Vol. 137, Plenum Press, New York, 1987, 286 pp.

R. E. Dickerson, *Sci. Amer.* **249**(6), 3 (1983).

WILHELM GUSCHLBAUER
Centre d'Etudes Nucleaires de Saclay, Gif-sur-Yvette, France

POLYPEPTIDES

Present knowledge of the spectroscopic characteristics of ordered structures of natural polypeptides and proteins derives from structural and spectroscopic studies of synthetic polypeptides. These macromolecules can be synthesized by various methods and, under proper conditions, undergo conformational changes from ordered to disordered forms. These changes are useful for the elucidation of the compositional factors that determine protein structure (see HELIX–COIL TRANSITIONS).

Synthetic Procedures

Using appropriate synthetic procedures, polypeptide chains of various composition, such as homopolypeptides, random or block copolypeptides, sequential copolypeptides, and sequences reproducing naturally occurring bioactive peptides and polypeptide chains may be obtained (Table 1). It is also possible to synthesize branched copolypeptides containing side-chain sequences from the sequence of the peptide backbone.

Polymers and Copolymers of α-Amino Acids. Polymers and copolymers of α-amino acids are usually synthesized by polymerization of α-amino acid N-carboxyanhydrides (NCAs):

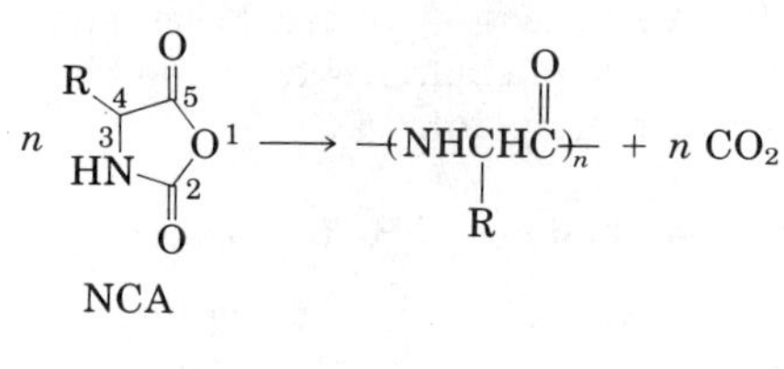

NCA

Table 1. Polypeptide Chains of Various Compositions

Composition	Description
–Glu–Glu–Glu–Glu–Glu–Glu–	poly(L-glutamic acid)
–Ala–Ala–Glu–Ala–Glu–Glu–Glu–Ala–	random copolymer of alanine and glutamic acid
–(Ala–Ala–Ala–Ala)$_n$–(Glu–Glu–Glu)$_m$–	block copolymer of alanine and glutamic acid
–(Ala–Gly–Ala–Gly–Ser–Gly)$_n$–	sequential copolymer reproducing the sequence of silk fibroin
p-Glu–His–Trp–Ser–Tyr–Gly–Leu–Arg–Pro–Gly–Gly–NH_2	sequence of a naturally occurring peptide hormone: LH–RF

Polymerization

Initiation by Primary Amines. In the presence of primary amines, NCAs polymerize according to the so-called normal mechanism.

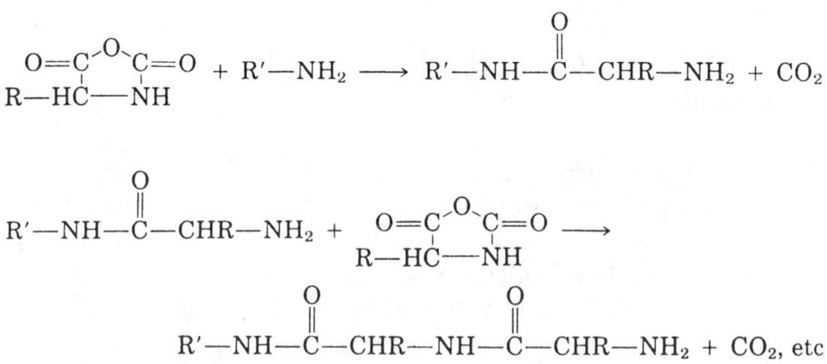

The reaction is a multistep nucleophilic addition of terminal amino group to carbonyl-5 of the NCA ring, followed by ring opening and the evolution of carbon dioxide.

Initiation by Strong Base or Tertiary Amine. High molecular weight polypeptides are prepared from NCAs with the assistance of strong bases or aprotic initiators. In the presence of carbanions, alkoxides, alkaline metal hydroxides, and aprotic bases (such as tertiary amines), the polymerization of NCAs takes a course different from that of polymerization initiated by primary amines. The reaction is much faster and the degree of polymerization (DP) is much higher than the monomer–initiator molar ratio. Under appropriate conditions, strong bases and tertiary amines initiate the polymerization by extracting the proton from the nitrogen of the NCA ring to produce an active monomer.

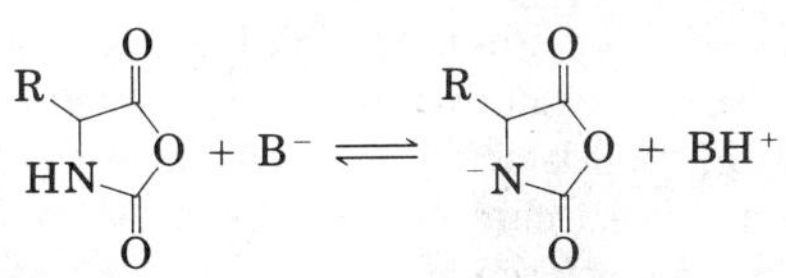

NCA   NCA$^-$ active monomer

Random Copolymerization. Copolymers of α-amino acids are usually obtained via NCA copolymerization with any of the initiators described above. In most cases a random sequence distribution has been assumed in copolypeptides prepared by NCA copolymerization. However, at present, the copolymerization mechanism is not completely understood in terms of copolymerization reactivity ratios of the various monomers, and a complete reactivity scale of NCA monomers has not been determined.

Peptides with Predetermined Primary Sequence. Many natural peptides have been isolated and characterized. However, the synthesis of optically pure homogeneous peptides of intermediate chain length (5–25 residues) in good yields remains a laborious task, requiring detailed knowledge of the specific characteristics of each amino acid and a strategy for use of protected derivatives. These syntheses employ the selective protection of end groups, a coupling step resulting in peptide bond formation, and deprotection steps to permit chain elongation. Ideally, each step in the synthesis proceeds in high yield without side reactions. Most critically, no racemization of the α-carbon chiral centers can be tolerated from cleavage or coupling reactions.

Peptides can be synthesized by two radically different strategies. In one procedure the peptide chain is extended by sequential addition of an *N*-blocked amino acid to the free amino terminus of the growing chain. In the second method small peptide fragments are coupled to prepared larger fragments of the desired peptide. A third route is solid phase peptide synthesis in which amino acid or peptide derivatives are attached to a solid resin support. Repetitive addition of the units and cleavage from the resin leads to a sequential peptide (see POLYPEPTIDE SYNTHESIS, SOLID-PHASE METHOD).

Sequential Polypeptides. Preparing polypeptides with a controlled amino acid sequence by the first two methods above requires the stepwise synthesis of oligopeptides, which can be condensed to a high molecular weight macromolecule. Oligopeptide is prepared by stepwise addition or fragment condensation techniques. The carboxyl terminus of this monomer is activated and, after cleavage of the *N*-terminal protecting group, polycondensation is carried out in dimethyl sulfoxide (DMSO) or *N,N*-dimethylformamide (DMF). With these procedures, a number of biologically significant model compounds have been prepared.

Sequence-controlled polymeric species have also been prepared by template polymerization (qv). These reactions are also used for the preparation of sequential copolypeptide models of protein structures, eg, (Ala–Pro–Gly)$_n$, (D-Ala–Gly–Pro)$_n$, (Ala–Ala–Gly)$_n$, (Ala–Gly–Gly)$_n$, (Gly–Ala–Hyp)$_n$, (Gly–Pro–Leu)$_n$, (Gly–Pro–Hyp)$_n$, (Gly–Pro–Gly)$_n$, and (Gly–Pro–Pro)$_n$ for collagen.

Conformation of Polypeptide Chains

The Peptide Bond and Its Structural Implications. Polypeptide chains fold into different ordered conformations, usually called secondary structures, which are largely determined by the particular residues that make up the polypeptide. The characteristics of the peptide group, obtained from a large body of crystallographic data on amides, are fundamental to polypeptides and proteins (Fig. 1). The C′–N distance is lower by ca 10% than the normal value observed in amines, which is one of the most important aspects of this structure. This is a consequence of the partial double-bond character of the amide linkage caused by resonance between two limiting structures. Thus free rotation around this bond is not allowed with the consequent planarity of the peptide bond, and the possibility of cis and trans configurations.

Secondary Structures of Polypeptide Chains. The particular geometry of the peptide bond has important structural implications. In particular, the planarity forces six atoms into a plane: those of the peptide group and the two adjacent C^{α}-carbons. Thus the structure of each amino acid residue is completely defined if two torsional angles φ and ψ are specified. According to the IUPAC convention, for an all-trans peptide chain, both φ and ψ angles are assumed to be 180°. Thus for any polypeptide chain constituted by n residues, the geometry and, therefore, the conformation (secondary structure) of the peptide backbone are completely defined by a set of n pairs of torsional angles. By far, the most famous secondary structure is the right-handed α-helix (see PROTEINS).

Conformational Analysis. In principle, a polypeptide chain can assume a large number of secondary structures, each one characterized by a pair of dihedral angles. However, only a limited number of secondary structures is actually found in natural and synthetic sequences, ie, in α-helical and β-pleated sheet structures. The reasons for such a limited conformational preference can be derived by taking steric and energetic factors into account. A conformational analysis that considers only steric factors was developed years ago. It is necessary to calculate the conformational energy as a function of φ and ψ dihedral angles.

Experimental Methods. In principle, any physicochemical measurement sensitive to conformation can be used to study peptide structures. X-ray diffraction (qv) is important for studies of polypeptides and proteins in the solid state. Recent work has demonstrated that absorption spectroscopy, optical rotatory dispersion (ord), circular dichroism (cd), and ir, Raman, and nmr spectroscopy are all sensitive to the secondary structures of polypeptides. Most importantly, these techniques complement each other and allow determination of the conformational properties of polypeptides in solution.

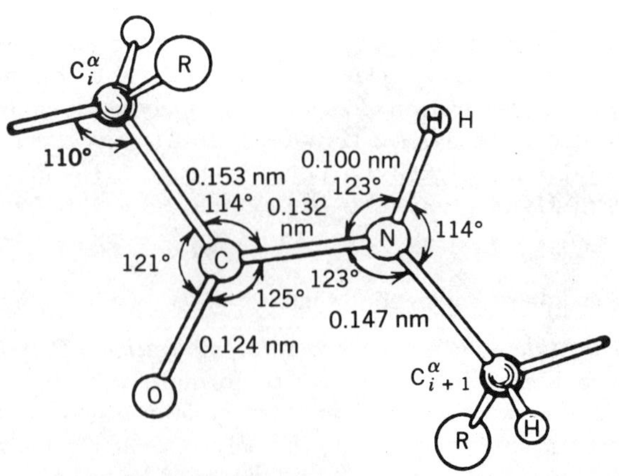

Figure 1. The peptide group indicating angles and distances for the trans configuration.

BIBLIOGRAPHY

E. Katchalski and M. Sela, *Adv. Protein Chem.* **13,** 249 (1958).

M. Szwarc, *Adv. Polym. Sci.* **4,** 1 (1965); M. Goodman and E. Peggion, *Pure Appl. Chem.* **53,** 669 (1981).

C. R. Cantor and R. P. Schimmel, *Biophys. Chem.,* Vol. 2, W. H. Freeman Publishers, San Francisco, Calif., 1980, Chapts. 7–9 and references therein.

R. W. Woody, *J. Polym. Sci. Macromol. Rev.* **12,** 181 (1977).

J. A. Schellman and C. Schellman in H. Neurath, ed., *The Proteins,* 2nd ed., Vol. 2, Academic Press, Inc., Orlando, Fla., 1967, p. 1.

EVARISTO PEGGION
University of Padua, Italy
MURRAY GOODMAN
University of California at San Diego

POLYPEPTIDE SYNTHESIS, SOLID-PHASE METHOD

Polypeptides comprising 3–50 amino acid residues play key structural and functional roles in biochemistry, pharmacology, and neurobiology. The amino acid building blocks can be among the 20 genetically encoded L-residues or unusual ones, and the sequences can be linear, cyclic, or branched. A rapid, efficient, and reliable method for the chemical synthesis of peptides is of utmost importance. Synthesis can confirm structures of naturally occurring substances and make them available in greater quantities for further investigation, can delineate antigenic determinants to allow preparations of artificial vaccines, and can lead to potent new drugs by judicious chemical substitutions that change functional groups and/or conformations of the parent peptide. These methods can occasionally be extended for the study of proteins, which may be defined as linear polypeptide chains of 50 or more residues.

The twentieth century has witnessed the development of an extraordinary array of techniques for the chemical assembly of amino acids to form peptides. They can be subdivided into solution and solid-phase methods.

Solid-phase synthesis is based on the covalent attachment (anchoring) of the growing peptide chain to an insoluble polymeric support or resin carrier, so that unreacted soluble reagents can be removed by filtration and washing without manipulative losses. Subsequently, the insoluble peptide resin is extended by a series of addition cycles, which are required to proceed with high yields and fidelities. Excess soluble reagents drive the reactions to completion. Because of the speed and simplicity of the repetitive steps, the solid-phase procedure is amenable to automation. After the chain has been established, it is necessary to release (cleave) the crude peptide from the support under conditions that do not affect sensitive residues in the sequence. Purification and characterization follow to ensure and verify that the desired structure has been obtained.

Strategy and Tactics

A general scheme for stepwise solid-phase synthesis of linear polypeptides is shown in Figure 1. An insoluble support is functionalized with group X. Building blocks are the amino acids with temporary protection of the α-amino groups and permanent protection of the side chains. The C-terminal amino

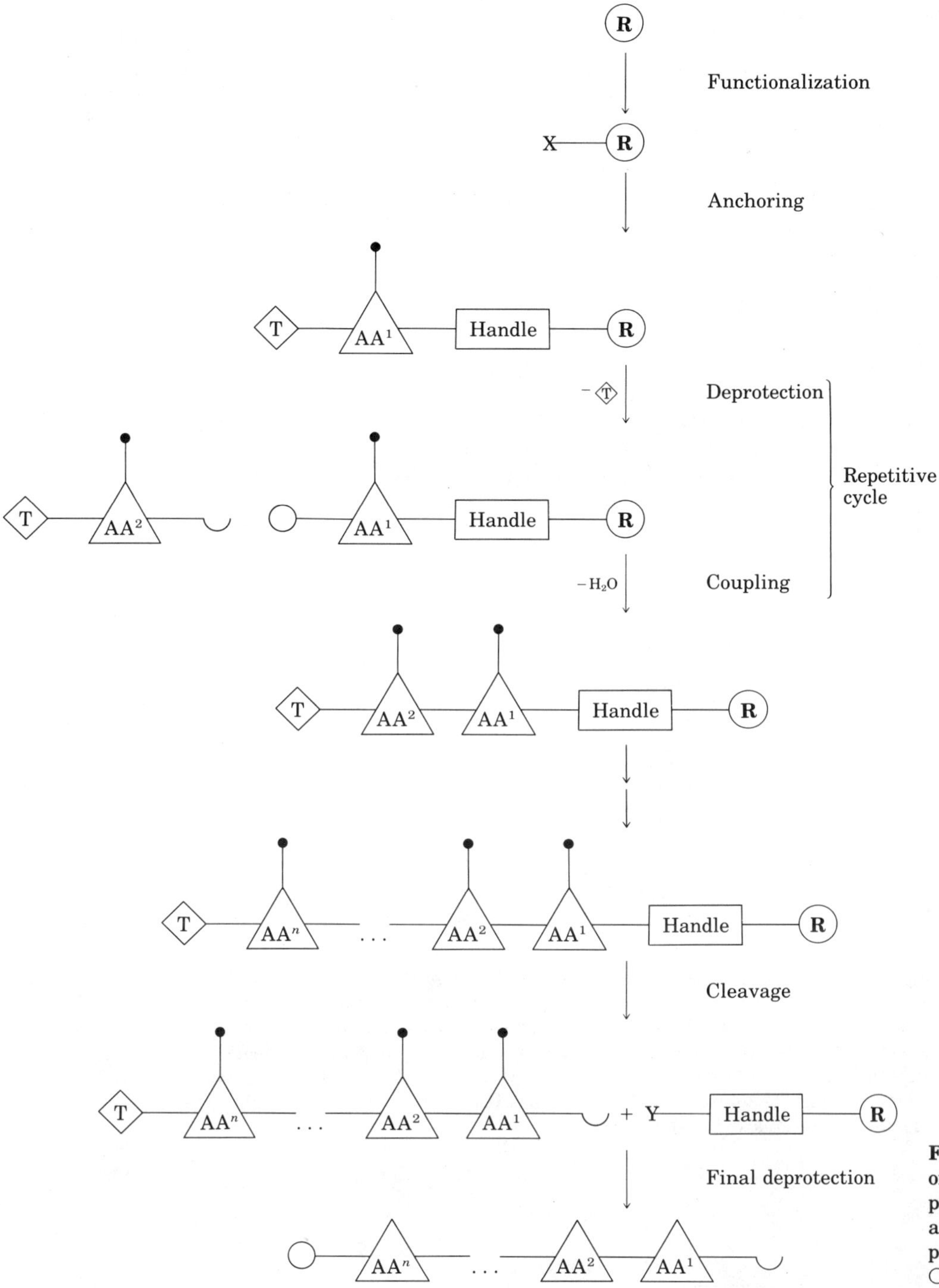

Figure 1. Stepwise solid-phase synthesis of linear polypeptides. (R), insoluble polymeric support; AA^1, AA^2, . . . , AA^n, amino acids; T, temporary protection; ●, permanent protection; ⌣, free carboxyl; ○, free amino group.

acid AA^1 is covalently attached to the polymeric support by a bifunctional spacer (handle); one end of the handle reacts with the polymer bound functional group X and the other with the free α-carboxyl of the protected amino acid derivative. After selective removal of the temporary protecting group to release the free α-amino group, the second partially protected amino acid AA^2 is introduced and a coupling step carried out in which the elements of water are removed to form the peptide bond. This deprotection/coupling cycle is repeated until all the amino acids required are introduced. The peptide is cleaved from the polymeric support and the protecting groups are removed; these two steps are often combined in a single procedure.

Resins with good swelling characteristics are best. After reaction and batchwise washing with gentle agitation, the resins are filtered either under suction or positive nitrogen pressure. In the continuous-flow mode, reagents and solvents are pumped through columns packed with resins.

Regardless of the structure and nature of the polymeric support, it must contain functional groups onto which the first amino acid can be anchored. Among more than 50 functionalization methods, Friedel-Crafts reactions introducing chloromethyl, aminomethyl, and benzhydrylamino groups are the most widely applied.

The next stage is the systematic elaboration of the growing

peptide chain. This involves repeated deprotection/coupling cycles, which are carried out in a single reaction vessel at ambient temperature. A temporary protecting group, such as N^{α}-*t*-butyloxycarbonyl (Boc), N^{α}-9-fluorenylmethyloxycarbonyl (Fmoc), or N^{α}-dithiasuccinoyl (Dts) is quantitatively removed by acidolysis, treatment with base, or thiolysis to liberate the N^{α}-amine of the peptide resin.

Completion of the assembly of the polymer-bound protected polypeptide chain is followed by cleavage of the anchoring linkage. Depending on the chemistry of the original handle and on the cleavage reagent selected, the product from this step can be a C-terminal peptide acid, amide, ester, or hydrazide.

The choice and optimization of the protection-scheme chemistry is perhaps the key factor influencing the successful outcome of any synthetic effort. At least two levels of protecting group stability are required. The permanent groups used to prevent branching or other problems on the side chains must withstand repeated quantitative removal of the temporary N^{α}-amino protecting group. On the other hand, structures of the permanent groups must be such that conditions can be found to remove them with minimal side reactions affecting the desired product. The necessary stability is often achieved by kinetic "fine-tuning," which relies on quantitative rate differences whenever the same chemical mechanism (usually acidolysis) serves to remove both classes of protecting groups. An often limiting consequence of schemes based on graduated lability is that they force adoption of relatively severe final deprotection conditions.

In the alternative orthogonal protection schemes, classes of groups are removed by different chemical mechanisms, and therefore can be removed in any order and in the presence of the other classes. Orthogonal schemes offer the possibility of substantially milder overall conditions. They also provide a method for the preparation of partially protected peptide segments.

Several protection schemes are currently used. Figure 2 shows an improved version of the standard Merrifield system. Additional themes of the standard system are shown in Figures 3 and 4: Figure 3 shows a mild orthogonal alternative; a scheme based on the acid-stable Dts group appears in Figure 4.

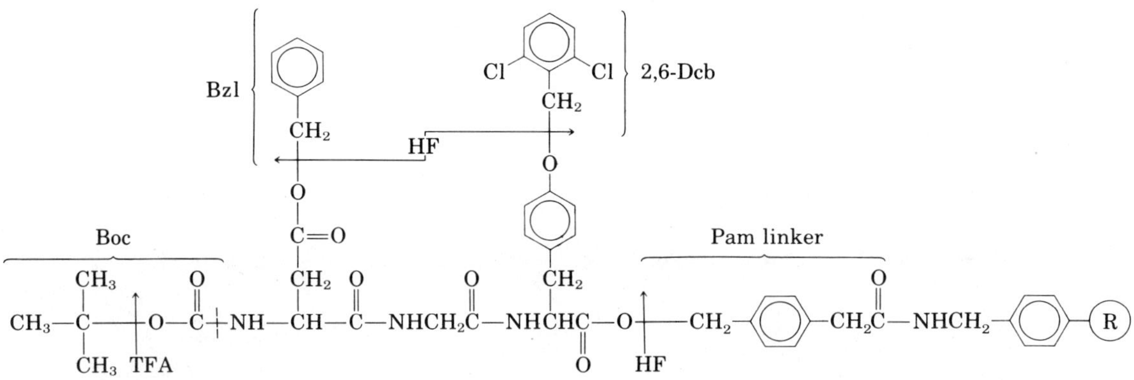

Figure 2. Commonly used protection scheme based on graduated acidolysis. Temporary α-amino protection is provided by the *tert*-butyloxycarbonyl (Boc) group, removed at each step by the moderately strong acid trifluoroacetic acid (TFA). Permanent benzyl-based side-chain protecting groups and the phenylacetamidomethyl (Pam) ester handle linkage are cleaved simultaneously at the end of the synthesis by HF or other strong acids to yield the free peptide acid. Some workers prefer 2-bromo-benzyloxycarbonyl (BrZ) over 2,6-dichlorobenzyl (Dcb) for the protection of tyrosine, since the latter may be marginally too stable to HF.

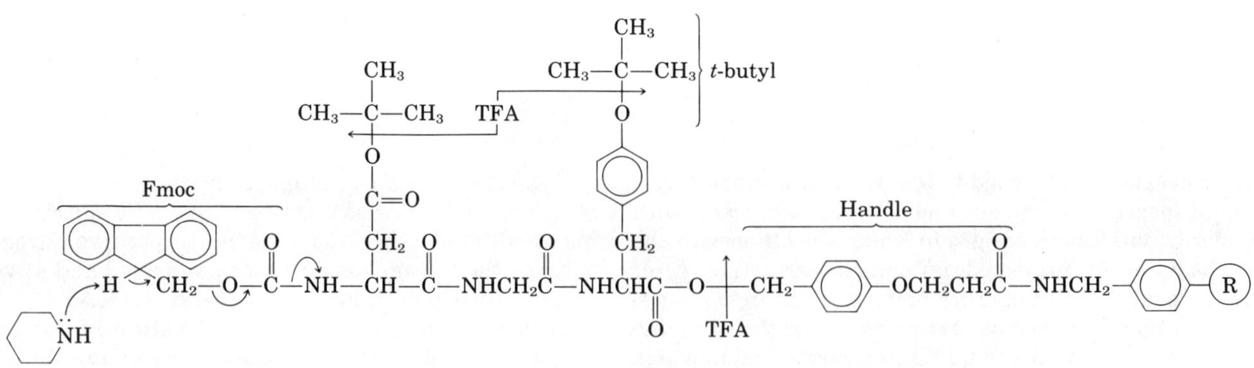

Figure 3. A mild two-dimensional orthogonal protection scheme for solid-phase synthesis. Temporary α-amino protection is provided by the 9-fluorenylmethyloxycarbonyl (Fmoc) group, removed by the indicated base-catalyzed elimination mechanism. Permanent *tert*-butyl-based side-chain protecting groups and the *p*-alkoxy-benzyl ester handle linkage are both cleaved by treatment with TFA to yield the free peptide acid at the end of the synthesis.

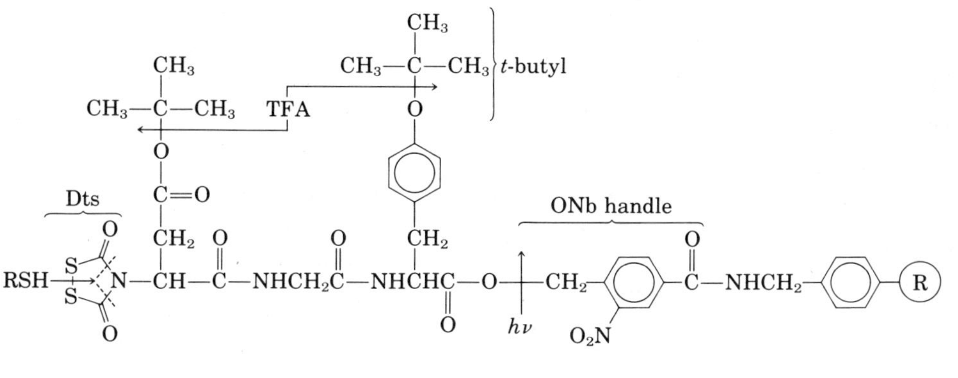

Figure 4. A mild three-dimensional orthogonal protection scheme for solid-phase synthesis. Temporary α-amino protection is provided by the dithiasuccinoyl (Dts) protecting group, which is removed by thiolysis. Permanent *tert*-butyl-based groups are removed as before by TFA, whereas the *o*-nitrobenzyl ester anchoring linkage can be cleaved by photolysis at 350 nm. Since any one of the indicated cleavages can be performed in any order without affecting the other classes of protecting groups, it is possible to use this scheme to obtain a full range of partially or fully deblocked peptide acids.

Procedures

The standard Merrifield system as it had evolved in its first decade involved perhaps one principal and no more than two auxiliary choices of apparatus (manual or automated), resin, protection scheme, anchoring regimen, coupling conditions, monitoring, and cleavage procedure. This original methodology is unquestionably successful and remains viable in many laboratories, but its repeated use revealed problems requiring the development of a new generation of improvements. There are now a staggering array of choices, each dogmatically championed by its innovator. All of these can be used successfully within reasonable limitations, and one approach should not be promoted in the absence of carefully controlled comparative studies.

Side Reactions

There are a number of stages in the solid-phase synthesis of polypeptides where problems may arise. In some cases, pinpointing the nature and cause of a side reaction has led to alternative synthetic and/or purification approaches that have served to mitigate or eliminate its occurrence.

Cumulative Errors. These can arise at each deprotection/coupling cycle and represent the summation of deviations from 100% yields per step. Some hypothetical calculations may be instructive: an overall repetitive yield of 99% per step would lead to 90% of a correct 10-mer, 78% of a correct 25-mer, and 37% of a correct 100-mer. If the repetitive yield should drop to 95% per step, the corresponding values would be 60, 28, and 0.6%. Although per step yields in the 95–99% range are regarded as outstanding in synthetic chemistry, the corresponding values are unacceptable in peptide chemistry. Also, the incorrect products often differ only slightly from the desired one, making the isolation of a pure product from a microheterogeneous mixture difficult or impossible.

Residue or Sequence-specific Errors. The introduction of certain amino acids entails risks. Even when a residue has been safely incorporated into the growing resin-bound polypeptide chain, it may still undergo irreversible structural modification or rearrangement during subsequent synthetic steps. The vulnerability to damage is particularly pronounced at the final deprotection/cleavage step, which often takes place under severe conditions. Consequently, the extent of residue-specific errors depends on the protection scheme chosen. Some of these are quite subtle in that they do not result in any changes in the amino acid composition or sequence. Some of the more urgent side reactions include: diketopiperazine formation during attempted incorporation of the third amino acid; dehydration at asparagine and glutamine; pyroglutamyl formation at N-terminal glutamine, or occasionally N-terminal protected glutamate; succinimide formation and $\alpha \rightarrow \beta$ rearrangement, particularly at aspartylglycine sequences; acylium ion formation and anisylation at glutamic acid; N → O acyl shift at serine and threonine; intramolecular C-alkylation at tyrosine; ornithine formation from arginine; and problems at methionine and cysteine.

Applications

The first biologically active peptide prepared by solid-phase synthesis was the smooth muscle hypotensive agent bradykinin. In the next 20 years, thousands of naturally occurring peptides or their analogues have been made by the original procedure or its variations. With current methods, any sequence of 40 or less residues, as well as occasional longer sequences, can be reliably assembled. For some time, it was feasible to issue reasonably complete tabulations. Several recent as well as older highlights include conformationally restricted enkephalin analogues, oxytocin and vasopressin, deamino-oxytocin carba analogues, [β,β-trithiobis(alanine)^{1,6}]-oxytocin, LH- and FSH-releasing hormones; gramicidin S, conotoxin GI, somatostatin, gastrin, apamin, alamethicin, thymosin, glucagon, β-endorphin, adrenocorticotropin (ACTH), synthetic vaccines for foot and mouth disease, staphylococcal nuclease fragments, trypsin inhibitor, parathyroid hormone, β-lipotropin, ribonuclease A, interleukin-3, and growth hormone.

A diametrically opposite concern is the very large scale production of peptides for commercial purposes. Although little of this work is published, several companies, such as Armour, Smith-Kline-Beckman, Monsanto, and Parke-Davis are known to have programs providing peptides such as ACTH, atriopeptin, calcitonin, and parathyroid hormone in annual amounts up to 200 g by solid-phase synthesis.

BIBLIOGRAPHY

G. Barany, N. Kneib-Cordonier, and D. G. Mullen, *Int. J. Peptide Protein Res.* **30,** 705–739 (1987).

R. B. Merrifield, *J. Am. Chem. Soc.* **85,** 2149 (1963).

R. B. Merrifield, *Science* **232,** 341 (1986).

G. Barany and R. B. Merrifield in E. Gross and J. Meienhofer, eds.,

The Peptides: Analysis, Synthesis, Biology, Vol. 2, Academic Press, Inc., New York, 1979, pp. 1–284.

GEORGE BARANY
NANCY KNEIB-CORDONIER
DANIEL G. MULLEN
The University of Minnesota

POLY(PERFLUOROBUTYL ACRYLATE).

See ACRYLIC ELASTOMERS.

POLYPEROXIDES.

See PEROXY COMPOUNDS.

POLY(PHENYLENE ESTERS)S.

See POLYESTERS.

POLY(PHENYLENE ETHER)

Poly(phenylene ether) alloys have developed into important engineering thermoplastics in the short period of time since the discovery of the oxidative coupling of phenols. The largest commercial usage of poly(phenylene ether) is in Noryl (General Electric Co.) engineering resins, which are alloys of poly(2,6-dimethyl-1,4-phenylene ether) and polystyrene. The trademark of the General Electric Co. for poly(2,6-dimethyl-1,4-phenylene ether) is PPO. Since poly(2,6-dimethyl-1,4-phenylene ether) and polystyrene are thermodynamically compatible over the complete composition range, a family of Noryl resins can be produced that offers a combination of physical properties and processibility.

Noryl resins include both filled and flame-retarded grades. They are processed by injection molding, foam molding, extrusion, thermoforming, or blow molding and are widely employed in automotive, electrical, construction, and liquid-handling applications, as well as appliances and business machines.

Few data are available on other poly(phenylene ether)s. Poly(2,6-diphenyl-1,4-phenylene ether) was commercially produced for a short time. Copolymers of 2,6-dimethylphenol and 2,6-diphenylphenol and the homopolymer of 2-methyl-6-phenylphenol have been prepared and properties reported.

Monomer

Properties. Although many phenol monomers can be coupled oxidatively to form poly(phenylene ether), the only commercially important phenol is 2,6-dimethylphenol (mol wt 122.2) (Table 1).

Table 1. Physical Properties of 2,6-Dimethylphenol

Property	Value
boiling point at 101.3 kPa[a], °C	201
freezing point, °C	45
flash point, °C	73
liquid density, g/cm^3	
at 38°C	1.00
at 149°C	0.91
viscosity, mPa·s (= cP)	
at 38°C	5.5

[a] To convert kPa to mm Hg, multiply by 7.5.

Manufacture. Because of the high separation and purification costs, coal tar is not an economical source of high purity 2,6-dimethylphenol. Poly(2,6-dimethyl-1,4-phenylene ether) is usually made by the vapor-phase alkylation of phenol by methanol.

Properties of Poly(2,6-dimethyl-1,4-phenylene ether)

Thermal Properties. Poly(2,6-dimethyl-1,4-phenylene ether) has a glass-transition temperature T_g of 205–210°C. Unlike polystyrene, it is ductile below its T_g and has high impact strength even down to −200°C. Thermal properties are given in Table 2.

Table 2. Thermal Properties of Poly(2,6-dimethyl-1,4-phenylene ether)

Property	Value
glass-transition temperature T_g, °C	205
crystal melting temperature T_m, °C	267
thermal conductivity, W/(m·K)	0.192
coefficient of thermal expansion per K	5.2×10^{-5}

Solubility. Poly(2,6-dimethyl-1,4-phenylene ether) is soluble in toluene and benzene and halogenated hydrocarbons.

Molecular Weight. Poly(2,6-dimethyl-1,4-phenylene ether), which is suitable for injection molding or extrusion, has a weight-average molecular weight $\bar{M}_w$ of about 40,000 and a number-average molecular weight $\bar{M}_n$ of about 18,000. The viscosity of higher molecular weight polymers makes processing difficult with little gain in material properties.

Chemical Properties. Poly(2,6-dimethyl-1,4-phenylene ether) is a linear amorphous thermoplastic with a free phenolic hydroxyl on the head group of each polymer chain. The hydroxyl group undergoes the usual reactions, such as ester formation with anhydrides or acid chlorides.

Physical Properties. Poly(2,6-dimethyl-1,4-phenylene ether) is a hard, ductile material with a high heat-deflection temperature under load (DTUL, 179°C). Physical properties are given in Table 3.

Table 3. Physical Properties of Poly(2,6-dimethyl-1,4-phenylene ether)[a]

Property	Value	ASTM method
density, g/m^3		
at 23°C	1.06	
HDT at 0.46 MPa, °C	179	D648
tensile strength, MPa[b]	80	D638
elongation at break, %	20–40	D638
tensile modulus, MPa[b]	2690	D638
flexural strength, MPa[b]	114	D790
flexural modulus, MPa[b]	2590	D790
compressive stress at 10% deformation, MPa[b]	114	D695
shear strength, MPa[b]	76	D732

[a] At 23°C

[b] To convert MPa to psi, multiply by 145.

Compatibility. Poly(2,6-dimethyl-1,4-phenylene ether) is completely miscible with polystyrene in all proportions.

Properties of Blends. Almost all of the poly(2,6-dimethyl-1,4-phenylene ether) production is used in blends with other polymers. The neat compound is not suitable for injection molding because of its high melt viscosity. Blending with high impact polystyrene (HIPS) improves processibility and impact properties. Different products can be made by changing the ratio of poly(2,6-dimethyl-1,4-phenylene ether) to HIPS.

Polymerization

Polymerization is frequently based on the oxidative coupling of 2,6-disubstituted phenols with hydrogen atoms in the 4 position. Some phenols with substituents in the 4 position can also be polymerized, such as phenols with a 4-methyl group or 4-aryloxy groups.

Another method is based on a free-radical-initiated displacement process in which a halo group is removed from the 4 position of a 2,4,6-trisubstituted phenol salt. Oxidizing agents are only used in catalytic amounts as initiators.

A third process, also a displacement polymerization of 4-halophenols, is based on the Ullmann ether synthesis instead of a radical-anion chain process. It does not require a 2,6-substituted phenol, but high temperatures are necessary; stoichiometry must be carefully controlled and the system must be free of moisture and oxygen.

A large number of poly(phenylene ether)s have been prepared by the oxidative coupling polymerization of 2,6-disubstituted phenols.

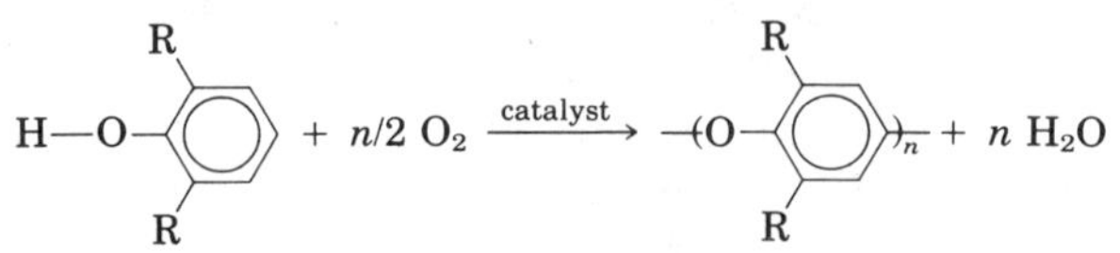

(1)

Mechanisms. A variety of mechanisms have been proposed for oxidative-coupling polymerization, including ionic intermediates, but most often free-radical processes. First, aryloxy radicals are formed by oxidation of the phenol with the oxidized form of the copper-amine complex, or with other catalytic or oxidizing agents. The aryloxy radicals form cyclohexadienones, which undergo enolization and redistribution.

A second process occurring concurrently with the dissociation–redistribution process is an intramolecular rearrangement in which cyclohexadiene groups appear to move along a polymer chain. The rearrangement may involve two consecutive sigmatropic rearrangements with an intermediate biscyclohexadienone. When the cyclohexadienone reaches a terminal position, it is converted to a phenol by enolization.

This process, which predominates at low temperatures, leads to migration of the internal ketal structures toward the end of the chain without requiring the dissociation to separate aryloxy radicals, which normally occurs in the redistribution process. At higher temperatures, the quinone ketal may dissociate before reaching the end of the chain. Then both mechanisms become important.

A number of the reactions of poly(phenylene ether)s proceed via the steps and intermediates of the redistribution mechanism. The type of product generated from a poly(phenylene ether) depends on the catalyst and reagents.

The mechanism of the halogen-displacement polymerization appears to differ significantly from the oxidative coupling polymerization, yet there are a number of common features. The reaction proceeds by a combination of the redistribution steps described previously and a sequence in which a phenoxide ion couples with a phenoxy radical and then expels a bromide ion. The resultant phenoxy radical can couple with another phenoxide or it can redistribute with other aryloxy radicals.

Kinetics. Molecular weight buildup in the oxidative coupling polymerization follows a step-growth pattern. Michaelis-Menten kinetics provides a good model for the reaction rate.

Catalyst Systems and Reaction Variables. The oxidizing agent for the polymerization of phenols is usually molecular oxygen and a transition-metal catalyst such as a copper or manganese compound combined with ligands. Stoichiometric amounts of metal oxides and certain other oxidizing agents are also employed. The type of oxidizing system, its composition (eg, metal-to-ligand ratio), concentration, and other reaction variables can have a large effect on reaction rate, selectivity, and product quality.

Polymer Structure. In a typical poly(2,6-dimethyl-1,4-phenylene ether), the 2,6-dimethyl-1,4-phenylene ether unit is the principal repeat unit. Other units occur in small amounts.

Copolymers. Copolymers are prepared from mixtures of phenols. The phenylene ether repeating units are usually arranged randomly in the backbone chain, although block-type copolymers have also been found. Typical examples include polymers from the following phenol pairs:

2,6-dimethylphenol–2,6-diphenylphenol
2,6-dimethylphenol–2-methyl-6-phenylphenol
2,6-diphenylphenol–2-methyl-6-phenylphenol
2,6-dimethylphenol–2-methyl-6-methylthiomethylphenol
2,6-dimethylphenol–2-allyl-6-methylphenol

Laboratory Procedure. For a typical laboratory-scale oxidative-coupling polymerization, 2,6-dimethylphenol (DMP) is purified by recrystallization from hexane until at least 99.5% purity is achieved. Reagent-grade solvents and catalyst components are used.

Industrial Production. Commercial poly(2,6-dimethyl-1,4-phenylene ether) is prepared with a catalyst system based on a hydrolytically stable, copper–diamine catalyst. The polymerization rates can be high, even at low catalyst concentration. To avoid an explosive mixture of oxygen and toluene vapor in the reactors, nitrogen is swept through the headspace at all times and oxygen is introduced at a rate sufficiently low to prevent buildup in the headspace. The polymerization is terminated by introducing nitrogen instead of oxygen and by removing the copper with an aqueous chelating solution. The aqueous phase is separated by centrifugation and the polymer is precipitated continuously with methanol. The polymer is washed with methanol and dried to produce a fine, relatively dense powder that contains only a trace of volatiles.

Processing of Blends

Melt Rheology. Poly(2,6-dimethyl-1,4-phenylene ether) has a high melt viscosity below 300°C, but like other amorphous thermoplastics exhibits shear thinning at high shear rates. It may be processed by injection molding, foam molding, engineered blow molding, and extrusion and thermoforming.

Finishing. Noryl resin parts are finished by common techniques; they can be painted, hot-stamped, and sputtered or plated. Articles can be bonded by adhesives or ultrasonic welding.

Health and Safety

Phenols are toxic materials and must be handled with caution. Before handling phenols, a detailed Material Safety Data Sheet should be consulted.

No known acute or chronic health hazards are associated with exposure to Noryl resins. Extruder vents, extruder dies, molds, and areas where gases from melted polymer may be produced should be adequately vented.

BIBLIOGRAPHY

A. S. Hay, *Polym. Eng. Sci.* **16**(1), 1 (1976).

H. L. Finkbeiner, A. S. Hay, and D. M. White in C. E. Schildknecht and I. S. Skeist, eds., *Polymerization Processes, High Polymers,* Vol. 29, John Wiley & Sons, Inc., New York, 1977.

W. J. MacKnight, F. E. Karasz, and J. R. Friend in D. R. Paul and S. Newman, eds., *Polymer Blends I,* Academic Press, Inc., Orlando, Fla., 1978.

J. R. Fried, G. A. Hanna, and H. Kalkanoglu in K. Solc, ed., *Polymer Compatibility and Incompatibility,* Vol. 2, Harwood Academic Publishers, New York, 1982.

G. E. Feth in J. M. Margolis, ed., *Engineering Thermoplastics,* Marcel Dekker, Inc., New York, 1985.

DAVID AYCOCK
VISVALDIS ABOLINS
DWAIN M. WHITE
General Electric Company

POLY(PHENYLENE OXIDE).

See POLY(PHENYLENE ETHER).

POLYPHENYLENES.

See OXIDATIVE POLYMERIZATION.

POLY(PHENYLENE SULFIDE)S.

See POLY(ARYLENE SULFIDE)S.

POLYPHOSPHATES.

See PHOSPHORUS-CONTAINING POLYMERS.

POLYPHOSPHAZENES

Although some ambiguity exists in the literature, the term polyphosphazene is generally reserved for high molecular weight linear or macrocyclic polymers that have repeating structures of the type shown in (**1**) where the side group R is an alkoxy, aryloxy, amino, alkyl, aryl, or an inorganic or organometallic unit.

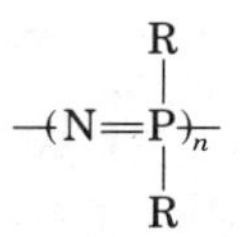

(**1**)

The weight-average molecular weights of these polymers are usually very high and correspond to 15,000 or more repeating units per chain.

Synthesis

Three different synthesis routes to polyphosphazene have been developed. The first is a macromolecular substitution route, (Fig. 1), that yields a broader range of polymers than the other two. The second and third methods (polymerization of organo-substituted trimers and direct synthesis of polyphosphazenes) allow the preparation of polymers that are inaccessible via the first route. A great many different poly(organophosphazene)s have been synthesized (Table 1).

Polyphosphazene Elastomers

Many single-substituent polyphosphazenes are microcrystalline, film-forming materials. However, if microcrystallinity is absent and if the side groups are sufficiently flexible to permit backbone mobility at room temperature or below, then the polymers exist as elastomers. Two molecular characteristics favor the absence of crystallinity: side-group flexibility and the presence of two or more different side groups dispersed randomly along the chain. In some cases, elasticity is retained at low temperatures and T_g values as low as −60 to −100°C are common. The elastomers may be divided into three classes; nonfluorinated elastomers, fluoroalkoxyphosphazenes, and mixed-substituent aryloxyphosphazene polymers.

Crystalline Polyphosphazenes

Polymers such as $[NP(OCH_2CF_3)_2]_n$ or $[NP(OC_6H_5)_2]_n$ are microcrystalline (Table 1). The microcrystallinity is associated with the molecular symmetry of these single-substituent polymers and their ability to assume a preferred (cis–trans planar) conformation in the microcrystalline domains.

Biocompatible or Bioactive Polyphosphazenes

Biocompatible Polymers. The ease with which the bulk and surface properties of polyphosphazenes can be modified by changes in the side groups has generated some interest in the development of these polymers as inert, biocompatible mate-

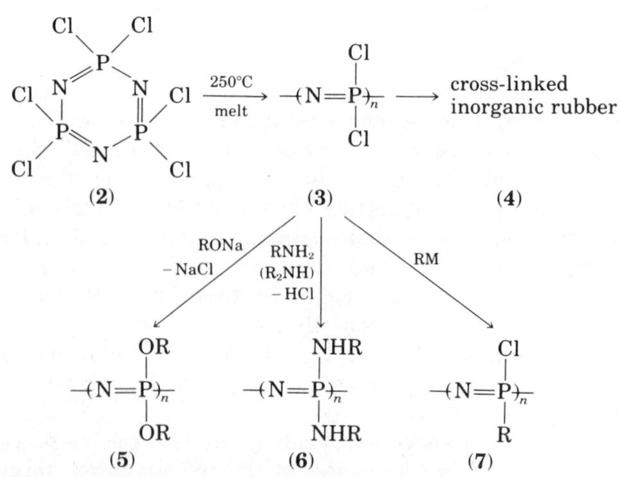

Figure 1. Macromolecular substitution route to poly(organophosphazene)s.

Table 1. Polyphosphazene Properties

Formula	T_g, °C	T_m, °C	Properties
$[NP(OC_3H_7)_2]_n$	−100		elastomer
$(NPF_2)_n$	−96	−68, −40	hydrolytically unstable elastomer
$[NP(OC_2H_5)_2]_n$	−84		elastomer
$[NP(OCH_2CH_2OCH_2CH_2OCH_3)_2]_n$	−84		water-soluble elastomer
$[NP(OCH_3)_2]_n$	−76		elastomer
$(NPCl_2)_n$	−66	−7.2 (+39)[a]	hydrolytically unstable elastomer
$[NP(OCH_2CF_3)_2]_n$	−66	+242	microcrystalline film-forming polymer
$[NP(OCH_2CF_3)(OCH_2(CF_2)_xCF_2H)]_n$	−60[b]		elastomer
$[NP(OC_9H_{19})_2]_n$	−56		elastomer
$(NPBr_2)_n$	−15		hydrolytically unstable, leathery material
$[NP(OC_6H_5)(OC_6H_4C_2H_5)]_n$	−10[c]	[d]	elastomer
$[NP(OC_6H_5)_2]_n$	−8	+390	microcrystalline, film-forming polymer
$[NP(OC_6H_4CH_3)(OC_6H_4CHO)]_n$	+11		thermoplastic
$[NP(NHCH_3)_2]_n$	+14		water-soluble glass and film former
$[NP(NHC_6H_5)_2]_n$	+91		glassy thermoplastic
$[NP(OC_6H_4C_6H_5\text{-}p)_2]_n$	+93		microcrystalline thermoplastic

[a] For the stretched polymer.
[b] Value varies with values of x and ratio of side groups.
[c] Varies with ratio of side groups.
[d] Complex melting phenomena.

rials for use in cardiovascular and other biomedical devices (see BIOMATERIALS).

Biodegradable Polyphosphazenes. Certain side-group structures impart hydrolytic sensitivity to polyphosphazenes, examples being those with amino acid ester or imidazolyl side groups. They are being investigated as an erodable matrix for the controlled release of pharmacological agents (see BIODEGRADABLE POLYMERS; CONTROLLED RELEASE).

Bioactive Polyphosphazenes. Biologically active side groups can be attached as cosubstituent groups by the same reactions as those illustrated in Figure 1. Steroids, local anesthetics such as procaine, catecholamines, and heparin have been incorporated in this way.

Transition-metal Derivatives

Polymers that bear transition metals in the side groups are of interest as macromolecular catalysts, as prospective electroactive materials, and possibly as chemotherapeutic agents. Three approaches have been explored in some detail: coordinative binding of the transition metal to the polymer through a skeletal nitrogen atom or a pendent phosphine unit; linkage of the transition metal directly to the phosphorus atoms of the skeleton; and linkage of metallocenyl units to the polyphosphazene backbone.

BIBLIOGRAPHY

H. R. Allcock, *Chem. Eng. News* **63,** 22 (1985).

D. P. Tate, *J. Polym. Sci. Polym. Symp.* **48,** 33 (1974).

R. E. Singler, G. L. Hagnauer, and R. W. Sicka, *ACS Symp. Ser.* **260,** 143 (1984).

P. Wisian-Neilson, R. R. Ford, R. H. Neilson, and A. K. Roy, *Macromolecules* **19,** 2089 (1986).

HARRY R. ALLCOCK
Pennsylvania State University

POLYPROPYLENE.

See PROPYLENE POLYMERS.

POLY(PROPYLENE OXIDE).

See 1,2-EPOXIDE POLYMERS.

POLYPROPYLENEIMINE.

See ALKYLENEIMINE POLYMERS.

POLYPYRAZOLES.

See HEAT-RESISTANT POLYMERS.

POLYPYRROLES

Monomers

Pyrrole is a constituent of coal tar and bone oil:

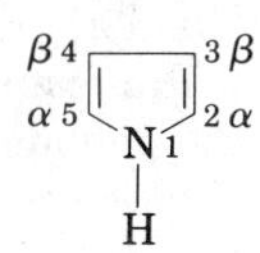

and isolatable by fractional distillation. Thermal decomposition of ammonium muccate with glycerol or mineral oil also yields pyrrole.

Physical Properties. Pyrrole is a colorless flammable liquid when freshly distilled, but darkens upon exposure to air. By excluding oxygen with an inert gas, the rate of decomposition in air is decreased. Some of the physical properties of pyrrole are listed in Table 1.

Polymerization

Over 70 years ago it was reported that pyrrole can be polymerized by hydrogen peroxide in acetic acid, yielding an insol-

Table 1. Physical Properties of Pyrrole

Property	Value
formula weight	67.09
boiling point, °C	130
vapor pressure at 13°C, kPa[a]	1.33
flash point, closed cup, °C	35
vapor density, air = 1	2.31
specific gravity d_4^{20}	0.969
critical temperature, °C	366
refractive index n_D^{20}	1.5082
melting point, °C	−24
dipole moment, C·m[b]	5.14×10^{-30}
resonance energy, kJ/mol[c]	100
dielectric constant at 20°C	8.00
dissociation constant[d],	−0.27
heat of atomization, kJ/mol[c]	4315
heat of combustion, kJ/mol[c]	2352

[a] To convert kPa to mm Hg, multiply by 7.5.
[b] To convert C·m to debyes, divide by 3.336×10^{-30}.
[c] To convert kJ to kcal, divide by 4.184.
[d] −3.8 in aqueous solution at 25°C.

uble brittle polymer. This material, designated as pyrrole black, has a varying elemental analysis, depending on the conditions of its formation $(C_{4.0-4.4}H_{3.0-4.4}N_{1.0}O_{1.1-1.5})_x$.

Pyrrole blacks have also been obtained by the reaction of pyrrole with benzoquinones. The reaction of pyrrole with diazonium salts is reported to yield polymeric oxygen containing products of unknown structure. Polymerization of pyrrole also occurs in the presence of sulfuric acid.

Electrically Conductive Polymers of Pyrrole

In the past decade significant research has been performed on electrochemically synthesized pyrrole (co)polymers because in the oxidized form polypyrrole is an intrinsically conductive organic polymer. A large number of conjugated polymers have been considered as candidates in this area of research. Pyrrole polymers have been in the forefront because of their overall stability in their doped (conductive) state.

Mechanism of Formation. The mechanism for the formation of the electrodeposited polymer has been studied. Film formation occurs only when the monomer is oxidized above its oxidation potential, as demonstrated by the lack of polymer formation when pyrrole is held at −0.2 V vs SCE. The degree of oxidation of the polymer film is a thermodynamic state and depends on the nature of the polymer and anion.

Electrochemical Growth Conditions. *Electrolyte Concentration.* The electrochemical growth conditions have an effect on the conductivity of the resulting film. A strong dependence of the conductivity on the electrolyte concentration has been reported. As expected, the rate of growth of polymer depends directly on the current density; the film thickness has a linear dependence on the reaction time. The current density and the electrical conductivity of the product increase with the electrolyte concentration.

Temperature. The temperature of the polymerization also plays a role in this process. Films prepared at a lower temperature have improved mechanical properties compared to the films prepared at the higher temperature.

pH. Tenside (surfactant) anions were employed as counterions for pyrrole polymerization. A strong influence on the conductivity of the resulting polymer was observed by modification of the pH of the aqueous polymerization solution. The optimum pH is between 2 and 5.5; at higher and lower pHs, the conductivity of the product drops.

Structural Studies. The dimer is monoclinic, whereas the trimer shows orthorhombic structure. Both structures are linear and planar with an alternating arrangement of the pyrrole rings. A linear chain structure for polypyrrole linked only through the α,α'-positions of the heterocyclic rings has been proposed. The unit translation sector along the chain is 0.735 nm. Electron diffraction studies were also performed on poly(β,β'-dimethylpyrrole) and were used to evaluate the structure of polypyrrole. These analyses suggest that the polypyrrole chains lie with the pyrrole rings coplanar to the film surface. Diffraction patterns from cross sections of neutral polypyrrole (obtained by the reduction of conducting polypyrrole) suggest that the structure of the polymer consists of chains stacked in an ABABAB sequence with the alternate planes aligned so that the "A" pyrrole ring is over the "B" location (Fig. 1). The volume of the unit cell is 3.66 nm^3. The spacing between the planes of the chains is 0.341 nm, the value determined by the π,π interaction of the pyrrole rings.

Spectroscopic Studies. Polypyrroles have been studied by a variety of spectroscopic techniques. ^{13}C nmr of various forms of polypyrrole confirm that the pyrrole rings are basically intact and that the bonding of these rings emanates from the α,α'-positions.

Improved Processibility

A major effort in recent research has been directed toward the preparation of processible polymers that maintain their high conductivity. For example, pyrrole has been electrochemically polymerized in the presence of latices with anionic surface characteristics. The resulting polymer compositions which precipitate onto the anode can be removed, ground into small pieces, and then dispersed in a variety of common solvents and cast into free-standing conductive polymer films.

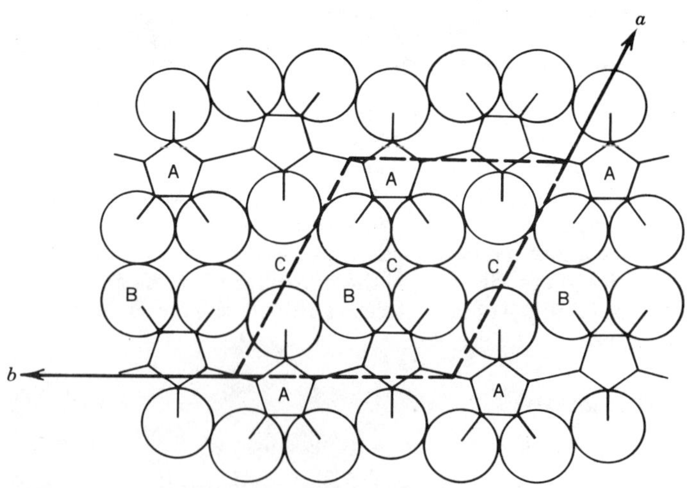

Figure 1. Proposed arrangement of the chains in the first layer of neutral polypyrrole. The circles represent the van der Waals diameters of the hydrogen atoms. The locations marked B represent the positions of the pyrrole rings (A) in the second layer of the ABAB stacking. Positions marked C are probable locations for anion intercalation. The a and b axes of the proposed monoclinic unit cell are indicated. The c axis is normal to the plane of the figure.

Uses of Conductive Polypyrrole

Numerous uses of conductive polymers have been proposed. The focus has been to combine their electrical properties with the processibility of polymers. One of the potential uses of these materials is in the preparation of lightweight secondary batteries. Another area where this product might be useful is for electromagnetic interference (EMI) shielding; hybrid compositions of metal particles with a conductive polymer binding may prove effective in obtaining higher attenuations. Coatings of conductive polymers could find uses in the area of electrostatic discharge.

BIBLIOGRAPHY

E. M. Genies, G. Bidan, and A. F. Diaz, *J. Electroanal. Chem.* **149,** 101 (1980).

W. Wernet, M. Monkenbusch, and G. Wegner, *Mol. Cryst. Liq. Cryst.* **118,** 193 (1985).

R. M. Geiss, G. B. Street, W. Volksen, and J. Economy, *IBM J. Res. Dev.* **27,** 321 (1983).

STANLEY JASNE
CIBA-GEIGY Corporation

POLYQUINOLINES

Quinoline polymers were developed during the 1970s in response to increasing demand for thermally stable materials. Their value as heat-resistant polymers was recognized early in part due to decomposition temperatures that, for some samples, exceeded 600°C. Further evidence of their stability was manifested by weight losses in nitrogen of only 15–30% at 800°C and useful lifetimes in air at 300°C. The source of this stability is the quinoline ring. Pyrolysis studies of model compounds for heat-resistant polymers have shown that quinoline is one of the most thermally stable structures known. Quinoline was the only compound examined that remained wholly intact when pyrolyzed at 870°C.

Polyquinolines are characterized by repeating quinoline units that display catenation patterns of 2,6-, 2,4-, or 3,6-quinoline substitution. The quinoline ring can be connected with a variety of linkages that control chain stiffness, which in turn influence solubility, thermal transitions, thermal stability, and thermomechanical properties. Most polyquinolines have a single phenyl group substituent on the quinoline ring; an additional phenyl group is optional.

Polyquinolines are made by a step-growth polymerization of *o*-aminophenyl(aryl) ketone monomers and ketone monomers with alpha hydrogens (mostly acetophenone derivatives). Both AA–BB and AB polyquinolines are known as well as a number of copolymers. The stoichiometry of the polymerization can be readily controlled in order to make oligomers of any desired molecular weight. These oligomers have been derivatized with reactive groups and used as cross-linkable resins. Polyquinolines have also been used as conductive and photoresponsive materials, as polymer supports for transition-metal catalyzed reactions, and as resins for fiber, film, and composites.

The method most often used for the preparation of polyquinolines involves the Friedlander reaction, which can be illustrated by the condensation of acetophenone with *o*-aminobenzophenone. Yields in excess of 98% for 2,4-diphenylquinoline suggested the viability of this approach for making high molecular weight polyquinolines. Various catalysts and reaction conditions were studied until polyquinolines with high degrees of polymerization were consistently obtained. In 1976 a patent claiming composition of matter as well as methods for making polyquinolines was granted and polyquinolines have been reviewed.

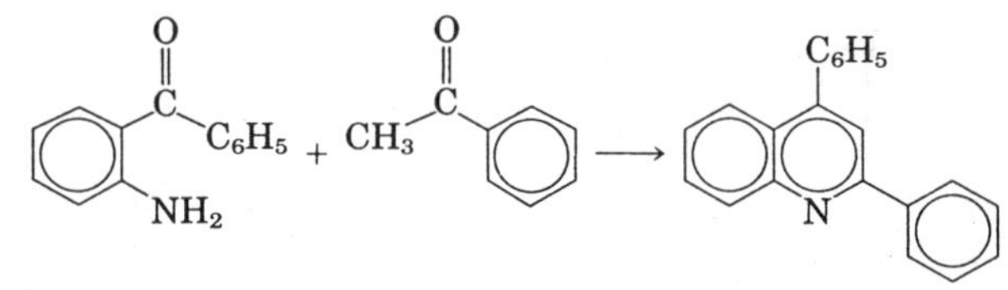

Polymer Synthesis

Polyquinolines are most often prepared by solution polymerization in a mixture of *m*-cresol–di-*m*-cresyl phosphate at 135–140°C for a period of 24–28 h.

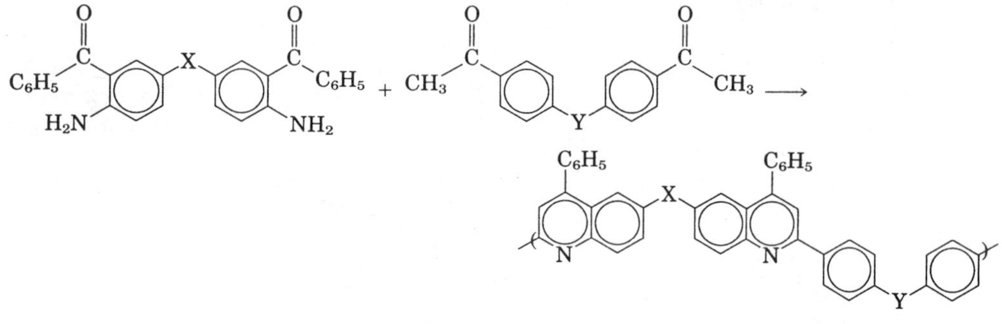

Although it functions as cosolvent for the polymerization, the main role of di-*m*-cresyl phosphate is to catalyze the Friedlander reaction. Polymers are isolated by coagulating the solution in an ethanol–triethylamine mixture. Alternatively, the polymer solution can be used for the direct fabrication of fiber and film.

Polymer Properties

Thermal Transition Temperatures. Backbone stiffness is strongly influenced by catenation pattern, phenyl substitution pattern, and the type of linkage that connects quinoline units in the polymer chain. Polyquinolines can vary in degree of stiffness from flexible to semirigid and rigid. The wide range of thermal transition temperatures displayed by polyquinolines is caused by these differences in backbone rigidity.

Thermal Stability. Thermogravimetric analysis (tga) has shown that polyquinolines have excellent stability in air and nitrogen. The onset of decomposition (T_d) in air for most polyquinolines is well above 510°C. Some are stable to temperatures as high as 605°C. Weight loss at 800°C ranges from 15–30% when polyquinolines are heated in nitrogen.

Isothermal aging results are better indicators of a polymer's stability. These types of study have shown that rigid polyquinolines have the highest stability. Isothermal aging tests have shown that polyquinolines can have a life of 10 years at a service temperature of 200°C.

Thermomechanical Properties. Because of their excellent thermal stability, the use of polyquinolines is limited only by their T_g, if amorphous, or their T_m, if crystalline. Dynamic thermomechanical analysis of polyquinolines has shown that when an amorphous polymer traverses its T_g, a sudden loss in mechanical properties occurs. The amplitude of the loss depends on the amount of order in the polymer sample. In contrast, highly crystalline, rigid polyquinolines retain most of their mechanical properties above T_g.

Solution Properties. Regardless of catenation, all flexible polyquinolines are readily soluble in chlorinated hydrocarbons such as methylene chloride and chloroform. They are totally insoluble in hydrocarbons and alcohols. As chain rigidity increases, polyquinolines become more difficult to solubilize. Only strong acids such as sulfuric and trifluoromethanesulfonic acid can dissolve rigid polyquinolines.

Morphology. Most polyquinolines, as prepared, are largely amorphous and have less than 20% crystalline content. Rigid polyquinolines, however, can be annealed above the T_g to high degrees of crystallinity.

BIBLIOGRAPHY

W. H. Beever and J. K. Stille, *J. Polym. Sci. Polym. Symp.* **65,** 41 (1978).

J. K. Stille, *Macromolecules* **14,** 870 (1981).

J. K. Stille, *Contem. Top. Polym. Sci.* **5,** 209 (1984).

W. Wrasidlo and J. K. Stille, *Macromolecules* **9,** 505 (1976).

DIRK M. SUTHERLIN
Consultant

POLYQUINOXALINES

Polyquinoxalines are a family of polymers that contain quinoxaline moieties in the main chain. The polymers may be unsubstituted, have no pendent groups on the quinoxaline ring, or be substituted with pendent groups on the quinoxaline ring. The properties of the polyquinoxalines can be altered by changing the chemical structure of the polymer backbone and also the chemical nature of the pendent group on the quinoxaline ring. This unique feature provides versatility in designing polyquinoxalines with an attractive combination of properties for use in specific applications. Variations of quinoxaline polymers include copolymers and oligomers end-capped with reactive groups such as vinyl, ethynyl, and phenylethynyl.

A unique feature of linear polyquinoxalines is their excellent solubility in the fully cyclized form. Unlike most other aromatic heterocyclic polymers, high molecular weight polyquinoxalines form stable solutions that can be used in a variety of applications to form films, ultrafiltration membranes, bioreactor cells, protective coatings, and to impregnate reinforcements. Polyquinoxalines are an important family of polymers that offer high chemical and thermal stability coupled with high mechanical properties. They are potentially useful as functional resins (eg, films and coatings) and structural resins (eg, adhesives and composite matrices) for many applications that demand stability in harsh environments.

More polyquinoxalines have been prepared from 3,3′,4,4′-tetraaminobiphenyl than all of the other tetraamines combined, undoubtedly because of availability and cost. 3,3′-4,4′-tetraaminobiphenyl became commercially available because it is used in the synthesis of polybenzimidazoles produced in various forms by Celanese Corp. Aromatic tetraamines are suspected carcinogens and should be handled accordingly.

Aromatic bis(α-dicarbonyl) monomers (bisglyoxals) have been used in the synthesis of unsubstituted polyquinoxalines. Because the anhydrous form is unstable, the bisglyoxals are isolated in the form of the dihydrate. Different melting points, presumably due to polymorphism, have been reported for two of the bisglyoxals.

Bis(phenyl-α-diketone)s have been employed to prepare polyphenylquinoxalines. Depending on the recrystallization solvent and cooling rate, many of the bis(phenyl-α-diketone)s display kinetically favored or thermodynamically favored crystalline forms.

Several bis(phenyl-α-diketone)s containing imide, amide, and ester groups between the α-diketones have been employed in the synthesis of polyphenylquinoxalines.

Polymer Synthesis

The unsubstituted polyquinoxalines (PQs) are usually prepared by the solution polymerization of aromatic bis(*o*-diamine)s such as 3,3′,4,4′-tetraaminobiphenyl and aromatic bis(glyoxal hydrate)s such as 4,4′-oxybis(phenylglyoxal hydrate) (eq. 1):

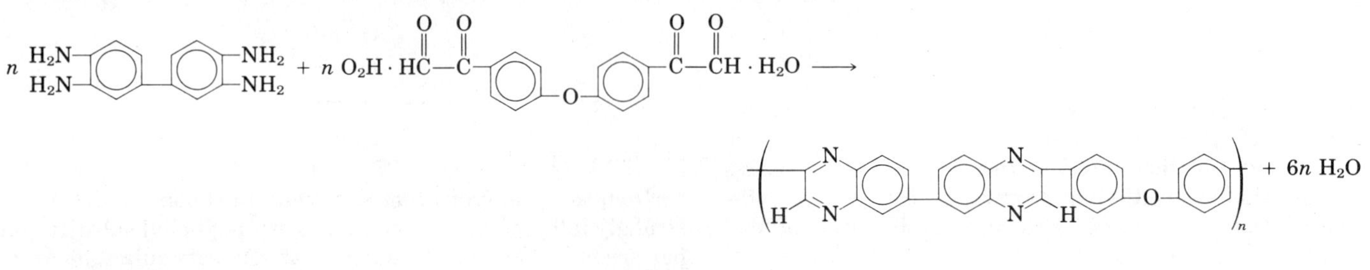

(1)

Polyphenylquinoxalines. Polypenylquinoxalines (PPQs) are prepared by the solution polymerization of aromatic bis(*o*-diamine)s such as 3,3′,4,4′-tetraaminobiphenyl and aromatic bis(phenyl-α-diketone)s such as 4,4′-oxydibenzil (eq. 2):

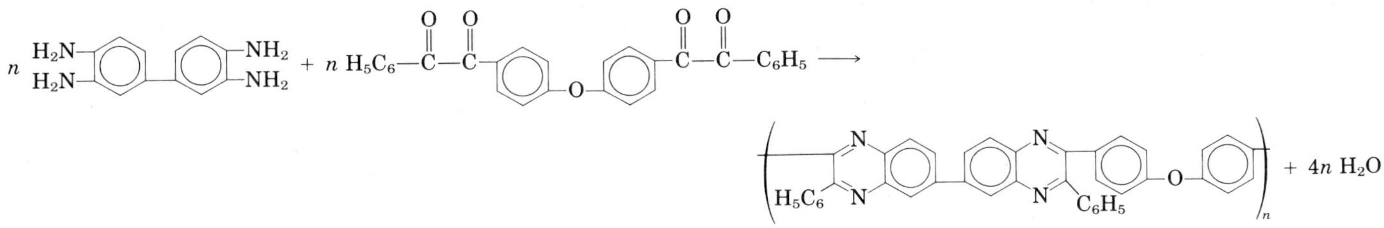

(2)

Reactive Quinoxaline Oligomers. Quinoxaline oligomers terminated with reactive groups are generally prepared by adjusting the stoichiometry of the aromatic bis(*o*-diamine) and the bis(α-dicarbonyl) monomer to obtain the molecular weight desired.

Polymer Properties

Glass-transition Temperatures. Although the very rigid PQ (**1**) and PPQ have been reported to display some crystalline character, the vast majority of quinoxaline polymers show no crystallinity in wide-angle x-ray diffraction (waxs) studies.

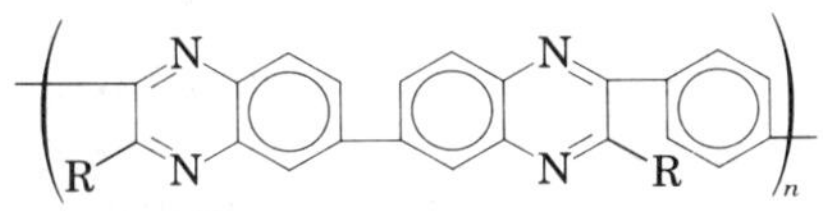

where (**1**) R = H and (**2**) R = C_6H_5

Their amorphous nature is due to configurational disorder (isomers) in the backbone of the polymer chains. As in other amorphous polymers, the glass-transition temperatures (T_gs) of PQs are controlled primarily by the chemical structure, molecular weight, and thermal history.

Thermal Stability. The thermal stability of quinoxaline polymers has been determined by thermogravimetric analysis (tga), isothermogravimetric analysis (itga), and the retention of mechanical properties at elevated temperatures after exposure for various times at elevated temperatures. Tga of analogous PQs and PPQs seems to indicate similar thermal stabilities, but itga shows the PPQs to be significantly more stable than PQs.

Tga and itga serve only as screening tests for polymers and can often be misleading in the prediction of the performance of a material in an applied form. The true test is the mechanical performance of a polymer in the intended form of use (eg, film, composite matrix, adhesive, molding, etc) under the intended

Table 1. Thin Film Properties of 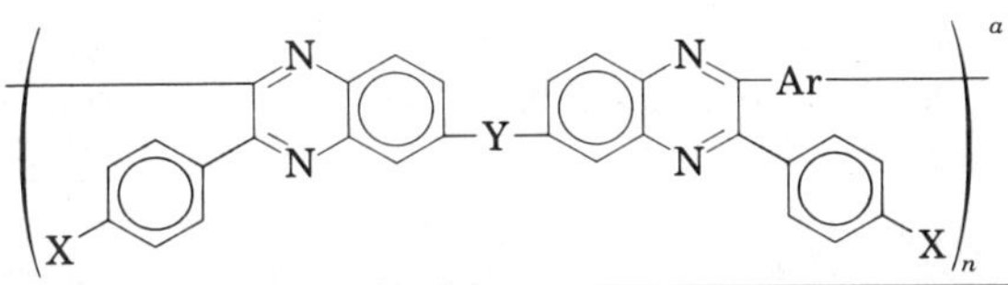 [a]

Ar	Y[b]	T_g, °C[c]	Test temp, °C	Tensile strength[d], MPa	Tensile modulus[d], GPa	Break elongation, %
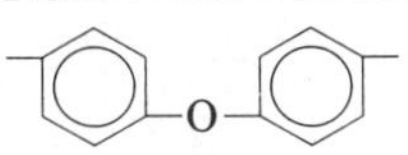		290	25	117.2	2.76	23
			177	65.5	2.28	51
			232	35.8	1.67	83
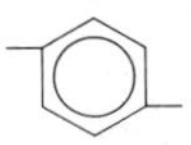		357	25	127.5	2.83	7.1
			232	87.6	1.70	12.9
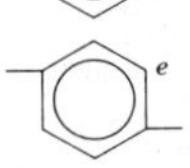 [e]		394	25	122.7	3.10	4.8
			232	100.0	2.34	8.2
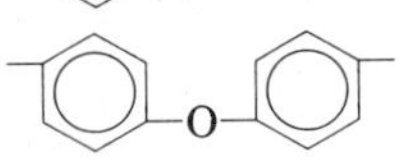	O	261	25	111.0	2.44	7.1
			200	50.3	1.64	12.3
			232	29.3	1.37	24.0
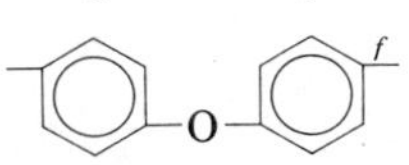 [f]	O	277	25	111.0	2.57	6.2
			200	56.5	1.78	10.7
			232	48.0	1.76	14.5

[a] X = H unless otherwise stated.
[b] When no Y is specified, the phenyl rings are directly joined.
[c] Determined by dsc at a heating rate of 20°C/min.
[d] ASTM D882; to convert MPa to psi, multiply by 145; to convert GPa to psi, multiply by 145 × 10^3.
[e] X = 90% H, 10% $C{\equiv}CC_6H_5$.
[f] X = 95% H, 5% $C{\equiv}CC_6H_5$.

Table 2. Laminate Properties

Test condition	Flexural strength[a], GPa[b]	Flexural modulus[a], GPa[b]
PQ glass[c] laminate[d]		
25°C	0.687	25.1
371°C after 1 h at 371°C	0.425	16.8
PQ boron unidirectional laminate[e,f]		
25°C	1.74	291.6
316°C after 10 min at 316°C	1.52	166.8

[a] ASTM D790.
[b] To convert GPa to psi, multiply by 145,000.
[c] 1581 Style AF-994 glass fabric with an HTS finish.
[d] Press cured for 10 min at 232°C and for 4 h at 407°C under 1.4 MPa (200 psi), followed by postcure unrestrained of 4 h each at 205 and 232°C, 8 h at 288°C, 4 h each at 316 and 344°C, and 8 h at 371°C under a nitrogen atmosphere. Resin content, 33.1%; density, 1.71 g/cm³; and calculated void content, 8.6%.
[e] Tungsten core; filaments, ~0.01-cm thick; tensile strength, ~0.345 GPa[c], and tensile modulus, ~386.1 GPa[c].
[f] Press cured by 5-min contact at 382–394°C, then 4 h at 394°C under 3.4 MPa (490 psi); postcured same as reported in footnote *d*.

use conditions. Polymers in the form of neat resins often exhibit different thermooxidative behavior from those in other forms (eg, adhesives or composites).

Chemical Stability. PPQs exhibit excellent stability toward strong bases and acids.

Mechanical Properties. ***Films.*** Films of linear PQs and PPQs can be prepared by melt pressing at temperatures above the T_g or more preferably by solution casting. Solutions at concentrations of 10–20% solids in solvents, such as chloroform, *sym*-tetrachloroethane, *m*-cresol, and mixtures of *m*-cresol with xylene or toluene, can be readily used to prepare films. The solution is cast onto an appropriate substrate (eg, plate glass) and the solvent removed by careful heating to avoid bubbles. Little work has been done with PQ films, whereas the film properties of many structurally different PPQs have been determined (Table 1). The film properties were obtained without optimization or orientation.

Fracture Energy. The fracture toughness (K_{IC}, stress intensity factor, or G_{IC}, critical strain energy release rate) for PQs has not been reported. PPQs are very tough materials as evidenced by the room temperature G_{IC} of 4000 J/m² (22.9 in lb/in²) as determined by ASTM E399 on compact tension specimens of an all-aromatic PPQ shown in equation 1.

Adhesives. Linear PQs and PPQs are excellent high temperature adhesives for joining plastic, ceramic, composite, and metal to themselves and to each other. They are generally processed at temperatures above their T_g under low to moderate pressure. Adhesive work with PPQs has been more extensive than with PQs.

Composites. PQs and PPQs have undergone preliminary evaluation as composite matrices using glass, different types of carbon–graphite, and boron filament reinforcements. When comparing the properties of different laminates, factors such as type of fiber, surface treatment, sizing, prepreg quality, process conditions, etc are important.

Glass, boron, and carbon–graphite (Modmor II) filament-reinforced laminate properties of the PQ of equation 1 are presented in Tables 2 and 3.

Ladder Polymers

Several different ladder-type polymers containing quinoxaline rings have been reported. Ladder- and step-ladder-type poly-

Table 3. PQ Modmor II[a] Unidirectional Laminate[b] Properties

Test conditions	Flexural strength[c], GPa[d]	Flexural modulus[c], GPa[d]	Short-beam shear strength[e], MPa[f]
25°C	0.842	106.9	88.2
315°C after 1 h at 315°C, air	0.815	104.8	59.3
315°C after 200 h at 315°C, air	0.650	100.6	48.3
371°C after 1 h at 371°C, air	0.747	104.1	61.4
371°C after 50 h at 371°C, air	0.713	100.0	60.7

[a] Tensile strength, ~2.41 GPa; tensile modulus, ~262.0 GPa; and density, 1.71 g/cm³. Resin content, 34.3%; density, 1.50 g/cm³; and calculated void content, 4.3%. Press cured at 232°C for 10 min; 316°C for 2 h with intermittent bumping; and 395°C for 1 h under 1.7 MPa; postcured through 4 h at 395°C in a nitrogen atmosphere unrestrained.
[b] Tensile strength, 0.887 GPa; tensile modulus, 131.7 GPa. Compression strength, 0.512 GPa; compression modulus, 102.0 GPa.
[c] ASTM D790.
[d] To convert GPa to psi, multiply by 145,000.
[e] ASTM D2344.
[f] To convert MPa to psi, multiply by 145.

mers, in general, are difficult to synthesize, have poor solubilities and high T_gs, and are exceedingly difficult or virtually impossible to fabricate into shapes. The thermal stability as measured by tga or itga can be misleading since degradation can occur without a weight loss.

BIBLIOGRAPHY

G. P. deGaudemaris and B. J. Sillion, *J. Polym. Sci. Part B* **2,** 203 (1964).

P. M. Hergenrother and H. H. Levine, *J. Polym. Sci. Part A-1* **5,** 1453 (1967).

P. M. Hergenrother, *Macromolecules* **14,** 891 (1981).

P. M. Hergenrother, *Macromolecules* **7,** 575 (1974).

PAUL HERGENROTHER
Langley Research Center

POLYRADICALS.

See RADICAL POLYMERIZATION.

POLYSACCHARIDES

Polysaccharides are naturally occurring carbohydrate polymers in which monosaccharide residues are linked directly through glycosidic linkages. They are found in the plant, animal, and microbial kingdoms, where they may serve for energy storage, as structural materials, or confer specific biological properties such as immunological reactions.

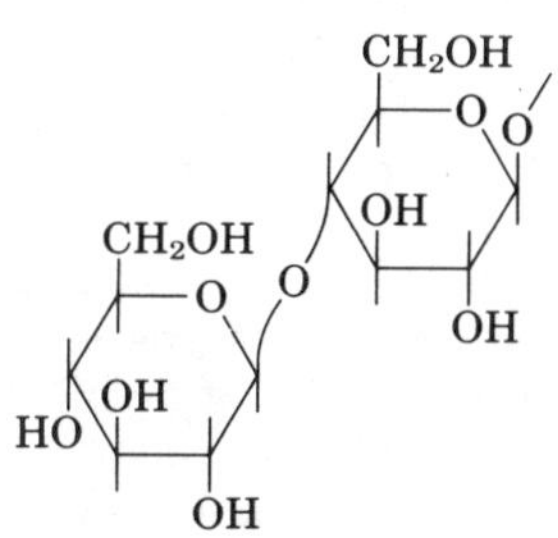

(1)

In a few instances (eg, teichoic acids) (**2**), the carbohydrate units are not linked directly to each other but through a polyol and a phosphate group.

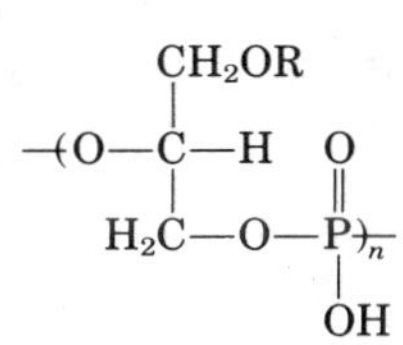

(2)

where R = glycosyl

Technically, these materials are not polysaccharides, but they are often, for convenience, included under this heading, whereas polynucleotides, which represent an analogous situation, are not; it is largely a matter of usage and convention.

Carbohydrate Nomenclature

Detailed rules of carbohydrate nomenclature approved by the International Union of Pure and Applied Chemistry and the International Union of Biochemistry have been published. A brief summary of those aspects pertinent to polysaccharides is presented here.

Monosaccharides. Monosaccharides are polyhydroxy carbonyl compounds and are classified as tetroses, pentoses, hexoses, etc, according to the number of carbon atoms in the molecule with a prefix to indicate the nature of the carbonyl group; thus aldohexose or ketohexose (hexulose).

Glycans. The generic name for polysaccharides is glycans, with the characteristic suffix "an." Strictly, the names of all polysaccharides should end thus (eg, D-glucan, D-mannan, etc) and some trivial names have been modified in accordance with this system (eg, laminaran rather than laminarin). As in other areas of organic chemistry, certain historic names such as cellulose, amylose, and amylopectin have been retained.

Structure

Since polysaccharides are biopolymers and thus produced by living systems, many parameters may affect the details of a polysaccharide structure. The molecular weight (M_w) of a polymer may increase with the age of the organism.

Young plant tissue is often rich in pectic substances, and changes in the nature of the polysaccharides in a bamboo-like reed *Arundo donax* have been studied as a function of evolution and the length of individual shoots. Likewise, it must not be assumed that the leaves and stems of plants have similar compositions.

In some plants there may be diurnal variation in polysaccharide content, but seasonal change is more common.

When comparing data on polysaccharides isolated from human (and other animal) sources, it is particularly important to know whether the source was an infant, child, or adult.

Biopolymers may be linear as in Figure 1**a**, or certain units may constitute branch points to give a polymer with a long side chain as in (**b**), or many short lateral chains (**c**) that may or may not be arranged in a stereoregular manner. The side chains may themselves be branched, thus giving rise to a tree-like structure (**d**). Whereas commonly there is only one lateral chain on any monosaccharide residue, several cases are known with two such substituents, giving a cross-like pattern (**e**).

Structural Determination

Since about 1970 there have been major advances in the methodology of polysaccharide structure determination, and structures published before that date, although correct in broad outline, may suffer from varying degrees of inaccuracy with respect to detail. Many investigations of bacterial polysaccharides have substantiated the concept of a repeating unit associated with a regular structure. By contrast, the majority of polysaccharides of plant origin lack any well defined regularity.

The determination of the structure of a polysaccharide must be carried out on pure material, isolated from its natural source with a minimum of degradation. General methods for determining polysaccharide structures have been reviewed and include extraction, purification, composition via hydrolysis, nmr, methylation, periodate oxidation, base-catalyzed β-elimination, partial hydrolysis, and enzymatic degradations.

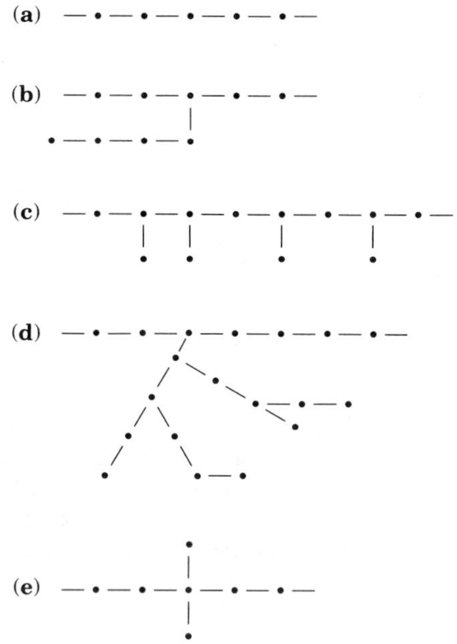

Figure 1. Structural patterns of polymer chains.

Plant Polysaccharides

This group includes the two major industrial polysaccharides, starch and cellulose, together with hemicelluloses, pectins, and plant exudates (see GUMS, INDUSTRIAL).

Starch. This is the principal food-reserve polysaccharide in plants and serves as the main source of carbohydrate in the diet of man and animals. Commercial sources include cereal grains such as corn (maize) and wheat, in addition to potatoes.

Starch is a mixture of two polysaccharides, amylose and amylopectin. Both are D-glucans, wherein the sugar residues are joined by α-(1→4) linkages, but the latter also contains some α-(1→6) bonds and, in this regard, bears a superficial resemblance to glycogen.

Most uses of starch capitalize on the high viscosity of its solution and gelling characteristics. These properties may be altered by physical or chemical modifications. The former include conditions of drying or, by more extended heating, degradation to dextrins. Chemically, starch may be converted to ester or ether derivatives, commonly acetate or hydroxyethyl or hydroxypropyl, to amino derivatives, and to inorganic esters such as phosphates and sulfates.

Starch is a major source of corn syrup and corn sugar (dextrose or D-glucose).

Cellulose. Cellulose is a (1→4)-β-D-glucan commonly obtained from cotton or wood, where it amounts to less than 50% (see CELLULOSE, STRUCTURE AND PROPERTIES; CELLULOSE, BIOSYNTHESIS; CELLULOSE ETHERS; CELLULOSE ESTERS).

Hemicelluloses. Associated with cellulose in many plants is a group of structurally similar polysaccharides; those from wood, grasses, and cereals have been the most studied.

The composition of the hemicellulose mixture depends on the source, and for deciduous trees, the main component is the glucomannan type with some glucuronoxylan. Coniferous trees yield less of the glucomannan, and the xylan has both arabinofuranosyl and acidic side chains.

The hemicelluloses of annual plants, such as grasses and cereal crops, are predominantly of the xylan type with different degrees of substitution.

The pulping of wood to produce cellulose pulp is carried out with one of two main objectives in view. The pulp can be used for making paper, either fine paper or cardboard, or for dissolving pulp.

Arabinogalactans. The wood of most coniferous trees contains a water-soluble L-arabino-D-galactan generally present only in small amounts, but in the genus Larix (larch) it may amount to 25% of the wood. Despite its occurrence in wood, it bears little structural similarity to the hemicelluloses, but more resembles the plant exudates.

Pectins. Pectic substances comprise a group of plant polysaccharides in which D-galacturonic acid is the principal component. The term pectin is used for the water-soluble, gel-forming polysaccharides of this group; pectinic acids have a portion of the uronic acid groups esterified, usually as methyl esters, and pectic acids are acidic materials devoid of ester groups.

Pectic substances are found in the primary cell wall of land plants, and commercial sources include citrus rinds (30% content), sugar beet pulp (25%), and apples (15%). Pectins are widely used for gelling fruit and juices, as in the manufacture of jams and jellies.

The basic structure of a pectic acid is a polymer of (1→4)-linked α-D-galacturonic acid residues, but a range of structures is possible, depending on the frequency with which rhamnose units are inserted in the chain and the degree to which both residues are substituted by chains of neutral sugars.

Plant Exudates. A gum is any material that swells or dissolves in water and exhibits gelling characteristics, acts as an emulsifier, or possesses adhesive properties. This wide-ranging term covers not only plant gums but also materials such as alginates, cellulose derivatives (eg, carboxymethylcellulose) (see CELLULOSE ETHERS), and modified starches. It is convenient to distinguish between plant exudates and gums obtained from seeds or bark since the chemical structures of the two groups are different.

From the point of view of chemical structure, plant exudates may be divided into three main groups and one minor one. Group A has a main chain →3)-β-D-Gal-(1 with side chains consisting of β-D-GlcA-(1→6)-β-D-Gal-(1, with some α-L-Ara*f* and/or α-L-Ara*p* (eg, gum arabic). Group B has a main chain →4)-β-D-GlcA-(1→2)-D-Man-(1 with side chains containing L-Ara*p* and/or D-Gal and/or D-GlcA (eg, gum Ghatti). Group C has a main chain →4)-α-D-GalA-(1, or →4)-α-D-GalA-(1→2)-α-L-Rha with side chains of D-glucuronic acid or its 4-methyl ether and/or D-Gal (eg, gum tragacanth).

Seed and Bark Mucilages. Polysaccharides from these sources may be divided into those that are neutral and those that are acidic. Important members of the former group are D-galacto- and D-gluco-D-mannans, of which guaran or guar gum, a galactomannan, is commercially important.

Algal Polysaccharides

Seaweeds represent a source of many different polysaccharides, several of which are of commercial importance, eg, alginic acid, a polyuronide, and agar and carrageenan, sulfated galactans. There are three main families of seaweeds: the Phaeophyceae (brown), Rhodophyceae (red), and Chlorophyceae (green), each with its own characteristic mix of polysaccharides.

Polyuronides. Alginate is used largely in the food industry, where it was introduced in 1929 as a stabilizer for ice cream. Because of its water-holding, gelling, and emulsifying properties, it is also used in paper coating, the manufacture of explosives, and latex emulsions.

Alginic acid is found in all brown seaweeds at depths not exceeding 40 m, the depth to which sunlight can penetrate. The principal source is the giant kelp *Macrocystis pyrifera* found along the coasts of North and South America, New Zealand, Australia, and Africa.

Sulfated Galactans. There are three main types, known as agar, porphyran, and carrageenan, which represent specific examples of a range of alternating (1→3)- and (1→4)-linked galactose units that differ in fine detail.

Agar. Agar is obtained from *Gelidium* and *Gracilaria* species with Japan as the main producer; others include Korea, the Soviet Union, Spain, and the United States. Agar is a mixture of agarose and agaropectin with the content of the former varying from 28% (*Acanthopeltis japonica*) to over 80% (*Gelidium subcostatum*); the average appears to be 55–65%.

Agar is composed of D-galactose and 3,6-anhydro-L-galactose units.

Porphyran. Porphyran is another sulfated galactan resembling agarose in that 3,6-anhydro-L-galactose and 6-*O*-methyl-D-galactose are present, and also carrageenan because of the content of galactose-6-sulfate which, in this instance, is of the L-configuration.

Carrageenan. Carrageenan differs from agar in that 3,6-anhydro-α-D-galactose replaces the L enantiomer and there is a higher (ca 24%) sulfate content. Because of its solubility in the cold, carrageenan is used as a stabilizer or protective colloid, such as in chocolate milk drinks.

Bacterial Polysaccharides

The bacterial kingdom is a source of many types of polysaccharides, several of which have attained commercial importance, and the majority are of medical interest because of their behavior as antigens and their use in the preparation of vaccines. Bacterial polysaccharides often contain rare and unusual sugars.

Bacteria are classified as Gram-positive or Gram-negative, and the nature of the carbohydrate polymers produced by the two groups is different. Gram-positive bacteria, eg, *Streptococcus pneumoniae,* are characterized by a thick cell wall consisting of a peptidoglycan, where polysaccharide chains are cross-linked by short peptide chains, forming a two-dimensional network. Their cell walls and membranes may also contain teichoic acids.

Gram-negative bacteria often contain two distinct types of carbohydrate polymer. Lipopolysaccharides are present in the cell wall and represent the somatic or *O*-antigen and, in many cases, there is also present a capsular polysaccharide corresponding to the *K*-antigen. Some strains of bacteria produce so much capsular polysaccharide that the excess, over the amount required to form a capsule, is elaborated into the growth medium as a slime. The composition of the capsular polysaccharide and the slime has been shown to be the same.

Fungi and Lichens

Fungi produce a wide variety of polysaccharides, often as mixtures, which makes the isolation of single, pure polymers difficult. Discussion of the different groups of polysaccharides is organized as in a review of this group.

Glucans. *α-D-Glucans.* Amylose and glycogen have been detected in many fungal preparations although they are not important sources of these compounds.

β-D-Glucans. Polysaccharides with a single linkage include cellulose, which is widely distributed, and pachyman, a tree-rot fungus that is predominantly a 3-linked polymer but also contains a few 1→6 linkages. Most of the β-D-glucans have well-defined mixed links, eg, 1→3 and 1→6 or 1→3 and 1→4.

Mannans. *Linear Mannopyranans.* When the growth medium is devoid of phosphate, species of the yeast *Hansenula* produce linear mannans with 1→2 and 1→6 linkages (**3**), and species of *Rhodotorula* form a linear mannan (**4**) with alternating 1→3 and 1→4 linkages.

-α-D-Man-(1→2)-α-D-Man-(1→2)-α-D-Man-(1→6)-
(**3**)

-α-D-Man-(1→3)-α-D-Man-(1→4)-
(**4**)

Branched Mannopyranans. A large number of yeasts produce branched mannans with 6-linked α-D-mannopyranosyl units in the main chain substituted at O-2 with other mannopyranosyl residues. The side chains may be single or several units in length and two distinct series may be recognized: those in which the side chains are joined by α-linkages and those with β-linkages.

Phosphonomannans. Many yeasts produce phosphate-containing mannans with ratios of mannose to phosphate typically in the range of 10 to 20:1.

Heteroglycans with Mannan Main Chains. Heteromannans are known where the side chains contain rhamnose, glucose, galactose, xylose, arabinose, 2-acetamido-2-deoxyglucose, glucuronic acid, and fucose; they have been reviewed.

Mammalian Polysaccharides

Two homopolymers are found in animals, glycogen and chitin, but the majority of mammalian polysaccharides are heteroglycans (glycosaminoglycans) based on a disaccharide repeating unit.

Glycogen. This D-glucan has both α-(1→4) and α-(1→6) linkages. The structure is similar to that of amylopectin, but the average chain length (CL) is 10–12 α-D-glucopyranosyl units instead of 20–25 for the starch polysaccharide. This greater degree of branching gives glycogen a higher degree of solubility in cold water than amylopectin. The highest concentration of glycogen is found in muscle, but the greatest amounts are to be found in liver, the tissue from which it is normally isolated.

Chitin. This homopolymer has units of 2-acetamido-2-deoxy-D-glucose (*N*-acetylglucosamine) 1→4-linked in the β-configuration; it is therefore an amino sugar analogue of cellulose. The native material usually exhibits different degrees of N-acetylation and the term chitosan is used for a polymer that has been N-deacetylated. Chitin is widely distributed and is found in bacteria, fungi, and crustaceans, but only crustaceans are an important commercial source.

Proteoglycans

Proteoglycans are carbohydrate–protein polymers in which many polysaccharide chains are covalently linked to a protein core;

the percentage of protein is low, as reflected in the name. The name proteoglycan is now preferred to the historical term mucopolysaccharide. Many different patterns of attachment are possible (Fig. 2). Proteoglycans are widely distributed in mammals, especially in connective tissue, cartilage, tendon, and in sinovial fluid. Heparin is a powerful anticoagulant. The carbohydrate components are referred to as glycosaminoglycans.

Glycoconjugates

Glycoconjugates, eg, glycoproteins and glycolipids, are biopolymers about which much has become known in the last few years. The construction of their names is meant to indicate that the carbohydrate is the minor component.

Glycoproteins. As the name implies, the protein portion constitutes the major part of the molecule, but the realization that the carbohydrate moiety is responsible for many of the specific biological, physiological, and enzymatic activities has resulted in detailed studies of these polymers. The bonds between the carbohydrate and protein segments are of the O- and N-glycoside types as in proteoglycans, and many glycoproteins have in common a sequence of five sugars referred to as the "inv" (invariant) fraction. Each specific glycoprotein is distinguished by modifications to this basic structure by the incorporation of "var" (variable) fractions which may take the form of extensions, not necessarily equally, on the two mannose units and/or by the introduction of an additional branch. These patterns are referred to as di- and tri-antennary structures.

Glycolipids. In these glycoconjugates, a glycosyl unit is covalently linked to a lipid. In comparison to glycoproteins, these compounds are of low molecular weight, nonpolymeric.

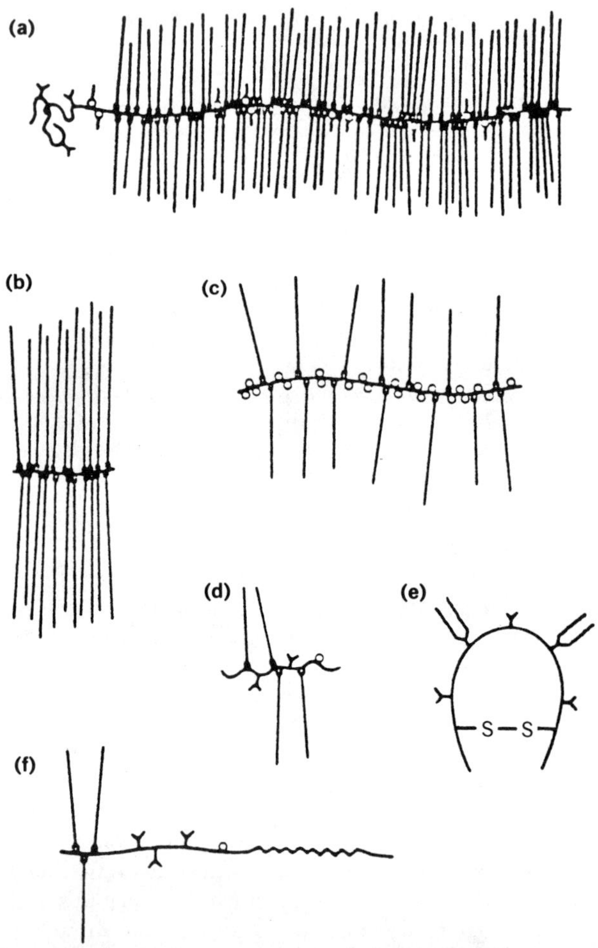

Figure 2. Different designs in proteoglycan structure. The proteoglycans hitherto encountered vary in size of the protein core, number and type of carbohydrate prosthetic groups, and locations of these groups along the polypeptide: (**a**) cartilage proteoglycan; (**b**) heparin proteoglycan; (**c**) a proteodermatan sulfate that also contains mucinlike oligosaccharides; (**d**) a common chondroitin sulfate- or dermatan sulfate-bearing proteoglycan; (**e**) tentative model for corneal proteokeratan sulfate; (**f**) tentative model for cell-surface proteoheparan sulfate; Y, N-linked, mannose-rich oligosaccharides; ○, O-linked, mucin-type oligosaccharides; □, GlcA–Gal_2–Xyl–Ser linkage region; —, glycosaminoglycuronan chain; ∿, keratan sulfate chain; ∿, hydrophobic peptide portion. Courtesy of Academic Press, Inc.

BIBLIOGRAPHY

G. O. Aspinall, ed., *The Polysaccharides,* Vols. 1–3, Academic Press, Inc., New York, 1982, 1983, and 1985.

J. F. Kennedy and C. A. White in E. Haslam, ed., *Comprehensive Organic Chemistry,* Vol. 5, Pergamon Press, Oxford, UK, 1979.

Advances in Carbohydrate Chemistry and Biochemistry, Academic Press, Inc., New York.

Methods in Carbohydrate Chemistry, Academic Press, Inc., New York.

G. G. S. Dutton
University of British Columbia

POLYSACCHARIDES, SYNTHETIC

Polysaccharide synthesis can be considered formally as the elimination of one molecule of water between two sugar units repeated many times over, as in condensation polymerization. The reaction is acid-catalyzed with the elimination of a hydroxyl group from the acetal center, C-1, of one sugar unit and a hydrogen from one of the other hydroxyl groups of another sugar unit. The products from free sugars are multibranched polymers with a variety of linkages and are of relatively little current interest.

Methods for the synthesis of polysaccharides of *controlled* structure fall into two categories: methods of sequential synthesis and methods of polymerization. In recent years, sequential syntheses of oligosaccharides up to hexa- or heptasaccharides have been accomplished. Polymerization methods have also been highly developed and are of two types: condensation and ring-opening polymerization.

Condensation. Condensation polymerization requires repetitive elimination of a small molecule by reaction of a leaving group on C-1 with a free alcoholic hydroxyl on another sugar molecule. All other hydroxyl groups must be blocked with a stable substituent during polymerization and there must be complete steric control of bond formation at the acetal center. This method of polymerization has the advantage that copolymers of regular repeating sequence can be prepared by coupling dimers, trimers, or tetramers, but the molecular weight of the products tends to be limited, even at high conversion.

Ring-opening Reactions. To obtain polysaccharides of controlled structure, ring-opening polymerizations are also carried out on sugar derivatives that have blocking substituents on the hydroxyl groups and with steric control at the acetal carbon. These reactions have the advantage that they can proceed by a chain-propagation mechanism that produces relatively high molecular weight products throughout the entire polymerization. However, when mixtures of monomers are copolymerized, copolymers of random, not regular, distribution are formed. Two classes of compounds primarily have been used, internal esters and anhydro sugar derivatives.

However the polymer may be formed, the final step in the synthesis is the removal of the blocking substituent groups to obtain the free polysaccharide. Some polysaccharides not found in nature, as well as some more regular in structure than corresponding natural products, have been prepared.

Characterization, Properties, and Uses. Polysaccharides prepared from orthoesters and cyanoethylidene derivatives have generally been characterized by classical physical methods: optical rotation for stereoregularity, methylation and hydrolysis for the identification of linkage position, and end-group analysis for molecular size. In more recent work, gel chromatography and nuclear magnetic resonance, especially carbon-13, have been the most important methods for proof of molecular size and structure. In general, the structures of stereoregular polysaccharides prepared from anhydrosugar derivatives have been established similarly. In a number of cases enzymic analysis has confirmed the identity and high stereoregularity indicated by ^{13}C nmr. In general, the polysaccharides formed have been linear or with a simple comb-like distribution of branches.

Synthetic polysaccharides have proven useful in the correlation of structure and physical properties of polysaccharides, and in basic research on the relationship between structure and function in physiology, biochemistry, enzymology, allergy, and immunology. Some possible biomedical applications are under investigation.

BIBLIOGRAPHY

I. J. Goldstein and T. L. Hullar, *Adv. Carbohydr. Chem.* **21**, 431 (1966).

H. Paulsen, *Angew. Chem. Int. Ed. Engl.* **21**, 155 (1982).

A. F. Bochkov and G. E. Zaikov, *Chemistry of the O-Glycosidic Bond*, Pergamon Press, Inc., Elmsford, New York, 1979.

C. Schuerch, *Adv. Carbohydr. Chem. Biochem.* **39**, 157 (1981).

C. SCHUERCH
State University of New York, Syracuse
College of Environmental Science and Forestry

POLYSILANES AND POLYCARBOSILANES

Polysilanes have backbones that consist entirely of a framework of catenated Si—Si bonds. These silicon-based polymers are chemically different from polysiloxanes (silicones), which contain both silicon and oxygen atoms alternating in the polymer chain, or their carbon-based analogues, the polyalkylenes. The long chains of catenated Si atoms in the polymer backbone of polysilanes are photochemically active and display a significant degree of σ-electron delocalization.

Polycarbosilanes contain both silicon and carbon atoms in the polymer backbone. The electron delocalization along the backbone is usually much less apparent, as it is disrupted by the inclusion of carbon atoms into the polymer chain. Polycarbosilanes are usually much less photoactive than polysilanes. Electronic and chemical properties of these two classes of polymers are different.

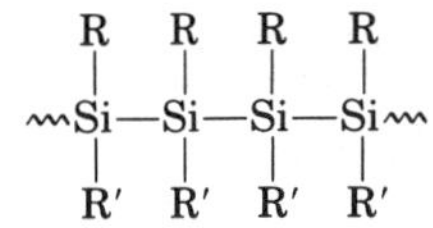

polysilane

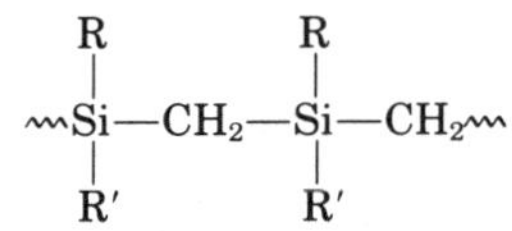

polycarbosilane

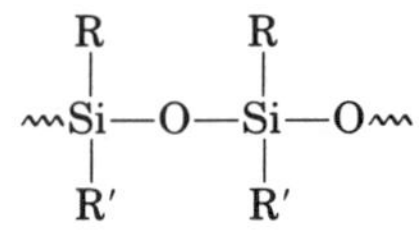

polyorganosiloxane

POLYSILANES

Synthesis

The synthesis of formable, soluble, linear, high molecular weight polysilanes is based on the condensation of organosilicon dihalide monomers with sodium metal:

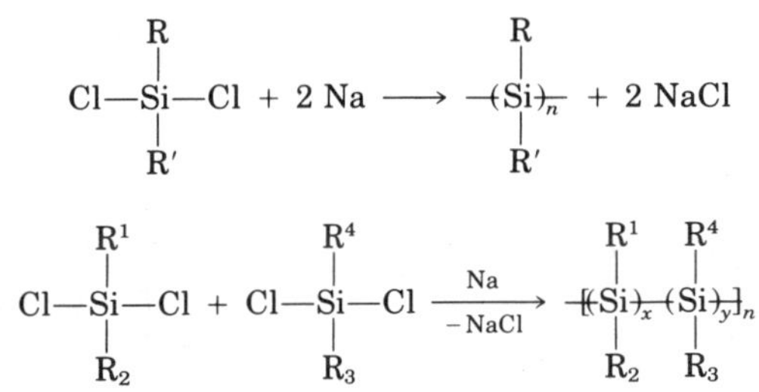

Copolymers are prepared similarly from a mixture of organosilicon dihalide monomers. Copolymers prepared in such a manner may be random or blocklike in composition, depending on the reactivities of the two monomers.

Characterization. Molecular weights have been determined by a variety of methods, especially gel-permeation chromatography (gpc) and light scattering.

Properties

Physical Properties. The solubility properties and melting points of polysilanes depend on the substituents bonded to the polymer backbone and the molecular weight.

The glass-transition temperature T_g of soluble polysilanes ranges from about -50 to $>120°C$, depending on the structure of the organic substituents attached to the Si atoms in the chain.

Electronic Properties. The interesting characteristics of polysilanes are due to a unique electronic structure. Unlike paraffins that absorb in the vacuum uv region, long-chain Si catenates display a characteristic intense low energy uv absorption between 290 and about 400 nm, the result of low ionization energies and readily accessible excited states.

The location of the low energy absorption maxima λ_{max} shows a regular bathochromic shift with increasing chain length n, which levels off and becomes constant as $n \rightarrow \infty$, similar to the behavior of conjugated polyenes, such as polyacetylene. The location of λ_{max} in polysilanes at room temperature strongly depends on the nature of the substituent attached to the backbone. This property is most apparent when the electronic spectra of aryl and alkyl polysilanes of high molecular weight polymers (where λ_{max} is constant with chain length) are compared.

Poly(di-*n*-alkylsilylenes) with longer alkyl groups also show an unusual, reversible discontinuous thermochromic behavior in solution with the formation of a new intense band of longer wavelength (ca 354 nm) upon cooling below a certain critical temperature. In solid films, unusual, reversible discontinuous thermochromic behavior is also observed, with cooling leading to the formation of a new long-wavelength band at approximately 373 nm.

Chemical Properties. *Thermolysis.* Polysilanes are stable in air and inert to most common solvents. Although they are stable at room temperature, heating above 220°C for extended periods can cause thermolysis.

Oxidation. Polysilanes are slowly oxidized by air at elevated temperatures or under mild solution conditions by peroxides, such as *m*-chloroperbenzoic acid (MCPBA):

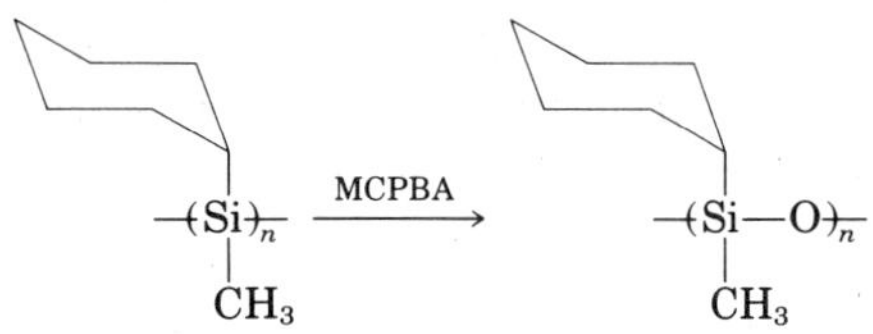

Conductivity. Although polysilanes are normally insulators, addition of oxidizing agents, such as AsF_5 and SbF_5, can alter the electronic properties to form semiconducting materials.

Chemical Modification. Most functional organic groups cannot survive the vigorous synthesis conditions of polysilanes, restricting the variety of possible reactive Si-bound substituents and limiting the options of reactions to chemically modify polysilanes.

Cross-linking. The presence of the unsaturated phenyl and cyclohexenyl substituents enhances the photochemical cross-linking yield, especially under argon with irradiation wavelengths >300 nm.

Photochemistry

Many compounds containing three or more catenated Si—Si bonds are photoactive to uv radiation. Photolysis of organosilanes has given synthetic routes to a variety of reactive compounds, including silyl radicals, silylenes, silenes $R_2Si{=}CR_2$, and disilenes $R_2Si{=}SiR_2$. Polysilane polymers are also photoactive and might be used in applications where photosensitivity is required.

The mechanism for photochemical scission of alkyl polysilanes in solution has been studied. The most likely mechanisms for photoscission are the photoreactions A, B, and C:

Reaction A:

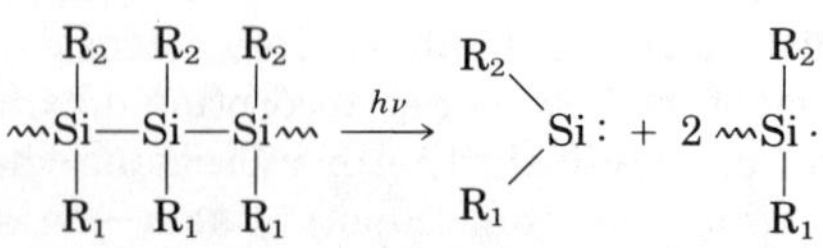

Reaction B:

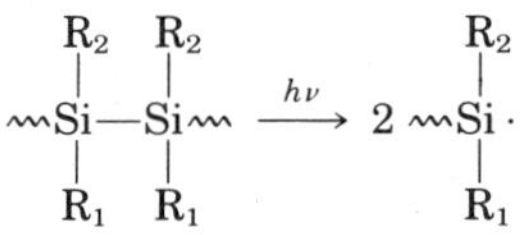

Reaction C:

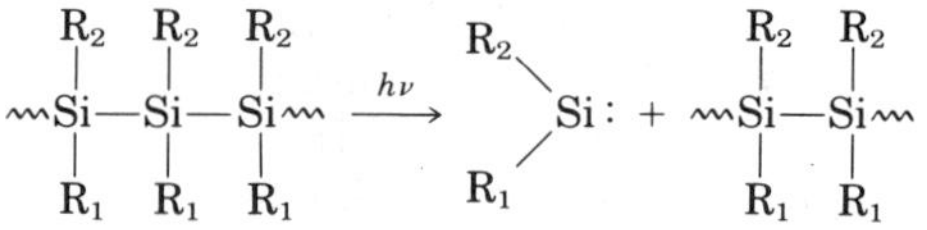

Reaction A leads to both highly reactive silyl radicals and silylenes, whereas B yields only silyl radicals and C only silylenes.

As a result of photochemical scission of the Si—Si bonds in the polysilane chain, the long-wavelength absorption is reduced with increasing radiation dose. The λ_{max} and the absorptivity per Si—Si bond decrease as the chain (and thus the σ-conjugation) is shortened. This behavior allows light to photochemically penetrate through a thick film, enhancing the usefulness of polysilanes as photoresists, and enables these polymers to be used as contrast-enhancing materials.

POLYCARBOSILANES

Synthesis

Polycarbosilanes can be synthetized by Wurtz coupling, hydrosilation, and ring-opening polymerizations.

Branched Polymers. Branched carbosilanes often give higher ceramic yields than linear polymers. They can be prepared via the Wurtz coupling reaction of a mixture of dihalide monomer and a small amount of trihalide comonomer with an alkali metal.

Properties

Thermal Stability. Polydimethylsilmethylene shows remarkable thermal stability in an inert atmosphere, with a weight loss of 5% at 600°C as measured by tga at 10°C/min for a fractionated high molecular weight sample. However, in air an exothermic reaction occurs at 220°C, with weight loss beginning at ca 325°C.

Physical Properties. At room temperature, most high molecular weight polydimethylsilalkylenes are waxy gums or thick viscous oils, with glass-transition temperatures below 20°C.

Photochemistry

In contrast to polysilanes, linear saturated polycarbosilanes absorb only in vacuum uv (λ <210 nm). Although the ground states of polysilalkylenes are apparently destabilized compared to polyalkylenes, a large energy gap separates the ground and excited states in the polymer backbone. The inclusion of carbon atoms into the polymer chain disrupts σ-electron delocalization, obscuring delocalization more than in polysilanes.

Applications

Polycarbosilanes are used in silicon carbide ceramics, photoresist applications, and as photoinitiators.

BIBLIOGRAPHY

R. West, "Organopolysilanes" in G. Wilkinson, F. G. A. Stone, and E. W. Abel, eds., *Comprehensive Organometallic Chemistry,* Vol. 9, Pergamon Press, Oxford, UK, 1983, pp. 365–397.

P. Trefonas III, J. R. Damewood, Jr., R. West, and R. D. Miller, *Organomet.* **4,** 1318 (1985).

P. Trefonas III, R. West, and R. D. Miller, *J. Am. Chem. Soc.* **107,** 2737 (1985).

R. D. Miller, D. Hofer, G. N. Fickes, C. G. Willson, E. Marinero, P. Trefones III, and R. West, *Proceedings of the SPE Regional Technical Conference, Photopolymers Principles-Processes and Materials,* Ellenville, N.Y., Oct. 28–30, 1985.

PETER TREFONAS
Monsanto Electronic Materials Company

POLYSILICATES. See SILICATES.

POLYSILOXANES. See SILICONES.

POLYSTYRENE. See STYRENE POLYMERS.

POLY(STYRENE-*co*-BUTADIENE). See BUTADIENE POLYMERS.

POLY(STYRENE-*co*-DIVINYLBENZENE). See STYRENE POLYMERS.

POLY(STYRENE-*co*-MALEIC ANHYDRIDE). See MALEIC AND FUMARIC POLYMERS.

POLYSULFIDES

Polysulfide polymers, first marketed in 1929, have arrived at a mature position in specialty applications. Although the original polysulfide polymers were solid rubbery materials, today the predominant product, discovered some 20 years later, is the mercaptan-terminated liquid polymer (LP). It can be transformed *in situ* from a liquid state into a solid elastomer, even at low temperatures, which makes its use convenient for adhesives, coatings, and sealants. Its success is due to good moisture, solvent, and ozone resistance, and good weathering and low temperature properties. These characteristics depend on the presence of polysulfide linkages. However, these linkages also contribute to poor high temperature stability and high compression set because of stress relaxation.

Monomers

Polysulfide polymers are usually prepared from bis(2-chloroethyl)formal (BCEF) and sodium polysulfide. Since the former is difunctional, a controlled amount of 1,2,3-trichloropropane is added to promote branching, which leads to higher functionality. Dihalides formerly employed in this reaction include bis(2-chloroethyl) ether, ethylene dichloride, and propylene dichloride; the last two are still used in water dispersions.

Polymerization

The polysulfide polymerization is most conveniently carried out in aqueous suspension and may be classified as a condensation reaction. The BCEF is the dihalide monomer most often used with 0–2 mol % of 1,2,3-trichloropropane (TCP) cross-linking agent.

The particle size of the suspension is critical for the optimization of the polymerization rate and the subsequent settling rate in the purification.

Clarified Polysulfide Liquid Polymers. A more uniform, higher quality product has recently been introduced with three grades of LP (LP-2C, LP-12C, and LP-32C), which may eventually replace the unclarified types.

Properties

The higher the initial molecular weight of the polymer, the higher the tensile strength. This is thought to be due to structure defects. If the initial molecular weight is high, the number of defects is low, and the cured network approaches an idealized one. At a lower initial molecular weight, the number of these defects is increased, resulting in inferior properties.

Solvent Resistance. The polysulfide polymers exhibit superior solvent resistance compared to natural or conventional synthetic rubbers. This is due to a backbone structure with short (usually C_2) sequences interspersed with formal and S_x, in contrast to the longer hydrocarbon structures found in traditional rubbers.

Low Temperature Properties. The glass-transition temperature (T_g) is a function of both the structure of the organic halide residue and the rank of the sulfur in the polysulfide sequence. The dependence on cross-link density is low at the TCP concentration usually encountered.

High Temperature Properties. The thermal and thermal oxidative stabilities of the cured polysulfide networks are mainly a function of the polymer backbone, although the curing agent also plays a role. Most sulfide polymers are based on BCEF with a formal linkage that weakens at high temperature. The formal group is highly susceptible to acid-catalyzed hydrolytic attack, which ruptures the chain, yielding formaldehyde and hydroxyl-terminated polysulfide species. The latter are readily discerned in the ir spectrum and are often seen if the polymer is "burnt" during drying.

Odor. Oxidative cure may lead to volatilization and odor formation by cyclodepolymerization.

Specifications and Quality Control

The three types of polysulfide rubbers produced are for different applications, requiring different specifications and different test methods. The LP specifications are more complex than those for the solid elastomers, because they originate as liquids, which must be polymerized in order to achieve rubbery properties.

Health and Safety Factors

Polysulfide polymers are regarded as safe. However, they give off odors of the type usually associated with sulfur compounds. In high concentrations, they may cause nasal irritation and should be used with adequate ventilation.

Acute oral toxicity tests of polysulfide liquid polymers have indicated an LD_{50} above 3.5 g/kg. Gloves should be worn when handling polysulfides and direct contact with the skin should be avoided.

Applications

The polysulfide liquid polymers are by far the most important polysulfide polymers in terms of volume. At present, the largest single use is in the insulating glass industry, which has grown rapidly in recent years because of the emphasis on energy conservation. In the construction industry, they are used to fill irregular joints and to follow joint movement without separation. In addition, polysulfides have the ability to stress relax, which allows stress relief before joint failure.

In the aircraft industry, the excellent fuel resistance and ease of application have made LPs the sealant material of choice.

The uses for LP include dental molding compounds, cold molding, potting, formed-in-place gaskets, concrete coatings, and bondings. Polysulfides are also used in crude rubbers and water dispersions.

BIBLIOGRAPHY

U.S. Pat. 2,466,693 (Apr. 12, 1949), J. C. Patrick and H. R. Ferguson (to Thiokol Chemical Corp.).

M. B. Berenbaum, "Polymers Containing Sulfur, Polysulfide" in A. Standen, ed., *Kirk-Othmer Encyclopedia of Chemical Technology,* 2nd ed., Vol. 16, John Wiley & Sons, Inc., New York, 1968, pp. 253–272.

S. M. Ellerstein and E. R. Bertozzi, "Polymers Containing Sulfur, Polysulfides" in M. Grayson, ed., *Kirk-Othmer Encyclopedia of Chemical Technology,* 3rd ed., Vol. 18, John Wiley & Sons, Inc., New York, 1982, pp. 814–831.

Morton Thiokol LP Liquid Polysulfide Polymer Bulletin, LP-2/85, Moss Point, Miss., Feb. 1985.

S. Ellerstein
Morton Thiokol Inc.

POLYSULFONES

Polysulfones are clear, rigid, tough thermoplastics with glass-transition temperatures of 180–250°C. Chain rigidity is derived from the relatively inflexible and immobile phenyl sulfone groups, and toughness from the connecting ether oxygens. These groups also impart the thermal stability and chemical inertness that characterize these resins. Those properties, along with the high glass-transition temperatures, make possible continuous use in the 150–200°C range.

Good thermal oxidative stability allows the usual range of thermoplastic processing operations in spite of the high melt temperatures encountered. Applications particularly suited for these polysulfones include molded articles and extrusions for which hydrolytic stability and good electrical properties are required.

Properties

Physical and Mechanical Properties. The physical properties of four commercial polysulfones are given in Table 1; they are transparent and light yellow to amber in color.

Impact resistance and ductility are retained to temperatures below 0°C. However, like most thermoplastics, polysulfones exhibit notch sensitivity as can be seen by comparing the values for notched and unnotched Izod impact strength given in Table 2. In design, care should be taken to avoid sharp radii and abrupt changes in cross section. Polysulfones can be considered nonlinear and elastic to a good approximation over a wide range of temperature and strain rates.

Electrical Properties. Polysulfones have low dielectric constants and loss tangents even in the GHz (microwave) range. Resistivity and dielectric strength are also high.

Flammability. Polysulfones exhibit excellent resistance to burning and in most applications require no flame-retardant additives. Polyphenylsulfone and polyethersulfone are UL 94 V-0; they show particularly low smoke emission by the National Bureau of Standards test.

Solvent Resistance. Polysulfones and poly(aryl ether)s, in general, have excellent hydrolytic stability compared to polycarbonates, polyesters, and polyetherimides, and may undergo many hours' exposure to hot water and steam. Within a broad range, pH has little effect on the rate of hydrolysis and, consequently, polysulfones are highly resistant to aqueous mineral acid, alkali, and salt solutions.

As with all amorphous glassy polymers, polysulfones may exhibit stress cracking and crazing in organic solvents.

The solubility of polysulfones in common solvents ranks according to bisphenol A polysulfone > polyarylethersulfone > polyethersulfone > polyphenylsulfone.

Polymerization

A disodium salt of bisphenol A is prepared *in situ* by reaction of bisphenol A with 2 mol equivalent of aqueous sodium or potassium hydroxide. The resultant salt is then allowed to react with 4,4′-dichlorodiphenylsulfone according to the following reaction scheme which takes place at 160–250°C:

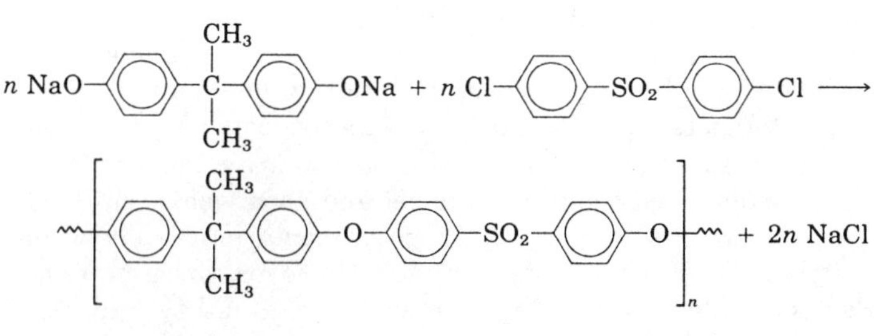

A dipolar aprotic solvent, such as dimethyl sulfoxide or 1-methyl-2-pyrrolidinone, is required for this polymerization. Besides providing the necessary solubility, the aprotic solvent is apparently essential for rapid polymerization rates.

Other polysulfones are prepared by substituting alternate aromatic dihalides and/or bisphenols for dichlorodiphenylsulfone and bisphenol A.

Table 1. Physical Properties of Commercial Polysulfones

Property	Test method	Udel P1700	Radel A300	Radel R5000	Victrex PES 4100G
color		light yellow	amber	amber	amber
clarity		transparent	transparent	transparent	transparent
refractive index		1.63			1.65
density[a], g/cm³	ASTM D1505	1.24	1.37	1.29	1.37
glass transition, °C		185	217	220	220
water absorption, %					
in 24 h	ASTM D570	0.22			0.61
at equilibrium	ASTM D570	0.62	1.85	1.1	2.1
melt flow, g/10 min[b]		8 at 343°C	30 at 380°C	10 at 400°C	25 at 380°C
mold shrinkage, cm/cm	ASTM D955	0.005	0.006	0.006	0.006
Rockwell hardness	ASTM D785	M69			M88

[a] Increases with increasing sulfone content.
[b] Similar to melt-index test (ASTM D1238) for lower melting thermoplastics. Number indicates rate of flow from 2.10-mm dia orifice under 298 kPa (43 psi).

Processing

Rheology. In contrast to many other polymers, the melt viscosities of polysulfones are relatively insensitive to shear. In this respect, polysulfones are similar to bisphenol A polycarbonate. Although low shear sensitivity confers high melt viscosity at processing conditions, it improves important properties. The low degree of molecular orientation during flow results in molded parts with uniform physical properties which vary little with direction of flow. Polysulfones can be processed by injection molding, blow molding, extrusion, thermoforming, annealing, joining, electroplating, and vacuum metallizing.

Machinability. Polysulfones can be machined without difficulty using ordinary metal-working tools. The high softening point of polysulfones allows deep, smooth cuts to be made without chipping or shattering (see also MACHINING).

Economic Aspects and Grades

All polysulfones are available from the manufacturer in various molecular weight grades as natural or pigmented resins, and in glass- and mineral-reinforced grades. Udel bisphenol A polysulfone is also available in a medical grade and in polymer blends under the trade name Mindel.

Uses

Polysulfones are employed in applications requiring prolonged exposure to steam and hot water. High HDTs combined with

Table 2. Mechanical Properties of Commercial Polysulfones

Property	Test method	Udel P1700	Radel A400	Radel R5000	Victrex PES 4100G
tensile strength (yield), MN/m²[a]	ASTM D638	70.3	82.7	71.7	84.0
tensile modulus, GN/m²[b]	ASTM D638	2.48	2.66	2.14	2.4
elongation at break, %	ASTM D638	75	40	60	40
flexural strength (yield), MN/m²[a]	ASTM D790	106	111	85.5	129
flexural modulus, GN/m²[b]	ASTM D790	2.69	2.75	2.28	2.6
compressive strength (yield), MN/m²[a]	ASTM D695	96			108
compressive modulus, GN/m²[b]	ASTM D695	2.58			2.1
shear strength (yield), MN/m²[a]	ASTM D732	41.4			50
tensile impact, kJ/m²[c]	ASTM D1822	420	343	340	337
Poisson ratio at 0.5% strain		0.37			0.4
notched Izod, J/m[d]	ASTM D256				
6.2 mm[e]		64			84
3.1 mm[e]		69	85	640	83
3.1 mm[e] at −40°C		64		425	
unnotched Izod, 3.1 mm[e]	ASTM D256	NB[f]	NB[f]	NB[f]	NB[f]

[a] To convert MN/m² to psi, multiply by 145.
[b] To convert GN/m² to psi, multiply by 145,000.
[c] To convert kJ/m² to ft·lbf/in.², divide by 2.10.
[d] To convert J/m to ft·lbf/in., divide by 53.38.
[e] Thickness.
[f] No break.

excellent hydrolytic stability and the ability to retain mechanical properties in hot, wet environments are the keys to their use in medical and food service applications which require repeated cleaning with hot water or sterilization. Microwave cookware is a significant market for polysulfones. Electrical and electronic applications constitute a growing market, particularly injection-molded printed circuit boards and connectors.

BIBLIOGRAPHY

L. M. Robeson and S. T. Crisafulli, *J. Appl. Polym. Sci.* **28,** 2925 (1983).

R. N. Johnson, A. G. Farnham, R. A. Clendinning, W. F. Hale, and C. N. Merriam, *J. Polym. Sci. Part A-1* **5,** 2375 (1967).

J. B. Rose, *Polymer* **15,** 456 (1974).

T. E. Attwood, D. A. Barr, G. G. Feasey, V. J. Leslie, A. B. Newton, and J. B. Rose, *Polymer* **18,** 354 (1977).

J. E. HARRIS
R. N. JOHNSON
Union Carbide Corporation

POLYTETRAFLOUOROETHYLENE.

See TETRAFLUOROETHYLENE POLYMERS.

POLY(TETRAMETHYLENE OXIDE).

See TETRAHYDROFURAN POLYMERS.

POLYTHIAZOLES.

See POLYBENZOTHIAZOLES AND POLYOXADIAZOLES; POLYBITHIAZOLES.

POLYTRIAZOLES.

See POLYAMINOTHIAZOLES.

POLY(TRIFLUORONITROSOMETHANE).

See NITROSO POLYMERS.

POLY(TRIFLUOROPROPYLMETHYLSILOXANE).

See SILICONES.

POLYUREAS

Polyureas may be classified as heterochain macromolecular compounds, which contain urea groups in their structure. Urea was first synthesized by Wöhler in 1829. Structurally, urea may be considered as the diamide of the carbonic acid.

Linear polyureas are thermoplastic polycondensation products with aliphatic or aromatic structures.

$$\text{-}\!\!\left(\text{NH}-\overset{\overset{\displaystyle\text{O}}{\|}}{\text{C}}-\text{NH}-\text{R}\right)_{\!n}\!\!\text{-}$$

where R = alkyl, aryl

Polyureas or copolyureas containing aliphatic structures exhibit a difference of 50–100°C between melting points and the beginning of decomposition; they are used for castings. Polyureas containing aromatic structures melt near their decomposition temperatures. They are soluble in some organic solvents and can be used for preparation of lacquers, varnishes, and coatings. The relationship between the softening points and structure is shown in Table 1. Melting points also depend on the molecular weight.

Polyureas were first produced on a commercial scale at I.G. Farben, employing the reaction between diisocyanates and diamines. Mitsui Toatsu Chemical, Ltd. produces synthetic fibers from nonamethylenediamine and urea. Today, polyureas and copolyureas (especially polyurethane polyureas) have many practical applications as foams, elastomers, adhesives, fibers, and others.

Table 1. Alternating Copolyureas from Diamines and Diisocyanates

Diamine	Diisocyanate	Inherent viscosity[a], dL/g	Softening temp, °C
piperazine	1,4-cyclohexylene	0.60	324
piperazine	4-methyl-1,3-phenylene	0.51	295
2,5-dimethylpiperazine	1,4-cyclohexylene	0.48	390
2,5-dimethylpiperazine	1,4-phenylene	0.56	376
2,5-dimethylpiperazine	bis(4-isocyanatocyclohexyl)-methane	1.00	291
2,5-dimethylpiperazine	1,2-bis(4-isocyanatophenyl)-ethane	0.33	362(dec)
2,5-dimethylpiperazine	bis(4-isocyanatophenyl)-methane	1.30	350(dec)
bis(4-aminocyclohexyl)-methane	bis(4-isocyanatophenyl)-methane	0.24	359(dec)
1,4-phenylenediamine	bis(4-isocyanatophenyl)-methane	0.36	400(dec)
bis(4-aminophenyl)ether	4-methyl-1,3-phenylene	0.40	320
1,3-phenylenediamine	hexamethylene	0.69[b]	255–260
hexamethylenediamine	4-methyl-1,3-phenylene	0.47	225–230

[a] Except where otherwise stated.
[b] Reduced viscosity.

The worldwide consumption was below 4.8×10^6 t in 1984 (in Europe 2×10^6 t) and is projected to be about 5.9×10^6 t in 1995; more than 75% of consumption is for foams. The number of products and producers continues to increase and new processing technology for thermosets, such as reaction injection molding (RIM), liquid injection molding (LIM), and others, are being developed.

Synthesis

The synthesis of polyureas is based on the reactions between polyamines (especially diamines) with aliphatic, aromatic, or heterocyclic structures and carbon dioxide, carbon oxysulfide, carbonic esters, phosgene, urea, urethane, and isocyanates. Preparations include polyaddition of diamines and diisocyanates, diamines and carbon dioxide or its derivatives, and diisocyanates and water.

Polyureas as Starting Materials for Other Polymers

Polyhydrouracils. Polyhydrouracils may be obtained by the cyclization of the 2-carbomethoxyethyl-substituted polyureas with polyphosphoric acid (PPA).

Polyhydantions. Aromatic polyhydantions of high molecular weight have been prepared by a polyaddition–cyclocondensation reaction of diethyl *m*-phenylene-bisiminoacetate and an aromatic diisocyanate in *m*-cresol at reflux.

Polyquinazolinediones. Aromatic diamino dicarboxylic acids and diisocyanate give polyureas. For example, polyureas are obtained from 3,3′-benzidinedicarboxylic acid and diisocyanate with a triethylamine catalyst in *N*-methylpyrrolidinone. These polyureas are cyclized by heating at 220°C to polyquinazolinediones.

Homopolyureas

Homopolyureas, $-\!\!(\text{HN—R—NHCO})_{\overline{n}}$, have a polar structure and hydrogen bonds, which result in high melting points and high crystallinity (Table 2).

Copolyureas

Copolyureas have many practical applications. They have the formula $-\!\!(\text{NH—R—NHCONH—R'—NHCO})\!\!-$ where R and R′ can be aliphatic, aromatic, or heterocyclic, and contain phosphorus, sulfur, silicon, or halogen (simple copolyureas). Polyureas may contain other functional groups such as urethane, amide, imide, and carbonate. The functional groups can be in alternating or in random distribution along the polymer chain, depending on the method of preparation.

Simple Copolyureas. Simple copolyureas are prepared by the reaction of diisocyanates with diamines in solution or by interfacial polycondensation. They include copolyureas from diamines and long-chain diisocyanates, copolyureas containing phosphorus, halogens, silicon, and heterocyclic moieties, copolyureas with acid or ester groups, copolyureas containing crown ethers in the main chain, and copolyureas with cyano groups.

Copolyureas Containing Other Functional Groups. ***Polyurethaneureas.*** Polyurethaneureas are the most important class of copolyureas. They are prepared from macrodiols (polyester diols, polyether diols, etc) and diisocyanates (in excess), resulting in prepolymers with NCO end groups which react with diamines (chain extenders). At the end of the reaction a chain terminator (ROH, RNH_2) is added.

Polyamideureas. Polyamideureas with aromatic structure are prepared from 4-isocyanatobenzoyl chloride and diamines by an interfacial polyaddition–polycondensation. These polymers exhibit excellent thermal stability and good solubility in *m*-cresol and amide solvents.

Polyimideureas. Polyimideureas are made by a polyaddition reaction between oligomers and diisocyanates.

Polyureacarbonates. Polyureacarbonates are prepared from bis(4-aminophenyl)carbonate and dissocyanates. They have properties similar to those of aromatic polyureas, but with higher solubility, forming a viscous fluid on heating.

Polyureasulfonamides. Polyureasulfonamides are prepared from 4-isocyanatobenzenesulfonyl chloride (obtained from sulfanilic acid and phosgene) and a diamine by interfacial polymerization in aqueous methylene chloride, using NaOH or Na_2CO_3 as acid acceptor. These copolymers, with inherent viscosities >0.30, produce tough films from DMF solutions.

Structure

The structures of the polyureas and the copolyureas influence chemical and mechanical properties. Polyurethaneureas consist of an alternating sequence of two chemically different hard and soft segments. The hard segments are often composed of an aromatic diisocyanate that reacts with a diamine chain extender, whereas the soft segments consist of a low molecular weight hydroxyl-terminated polyether or polyester. During solidification from solution or melt, the hard and soft segments undergo phase separation from which elastomeric properties derive. The hard domains serve as cross-links and filler particles in the soft matrix. Both the soft and the hard domains in segmented polyurethaneureas (SPUU) can be amorphous or partially crystalline, depending on the particular system.

The properties of SPUU depend largely on the extent of the microphase segregation which is due to thermodynamic incompatibility.

Table 2. Homopolyureas

Structure of main chain	Mp, °C	Method of preparation
aliphatic	180–230	diamines and urea
aromatic	260–320	diamines and carbon dioxide
aliphatic–aromatic	200–240	aromatic–aliphatic dihalide and alkali cyanate
unsaturated	240–260	diamines and urea
naphthalene	350–380	diamines and urea
heterocyclic	330–350	diamines and diisocyanates

Table 3. Mechanical Properties of Polyurea and Copolyurea Films

	Copolyurea[a]	Polynonamethyleneurea[b]	Polyurethaneurea[c]	Polyureas[d]
tensile strength, MPa[e]	44.6	58.8	39.7	22.8–28.3
elongation at break, %	630	215–238	850	280–360
tear strength, MPa[e]	8.7			16.3–17.1
notched impact strength, MPa[e]		0.80		
modulus at 100%, MPa[e]	6.3		6.1	

[a] From poly[tetramethylene glycol bis(4-aminobenzoate)] and diphenylmethane-4,4′-diisocyanate.
[b] From nonamethylenediamine and urea.
[c] From poly(ethylene adipate), TDI, and butane-1,4-diamine.
[d] Average values for polyureas obtained from amine-terminated polyethers and polyisocyanates.
[e] To convert MPa to psi, multiply by 145.

Properties

Physical Properties. *Melt Temperature.* The polar structure of polyureas results in melting points higher than those of the corresponding polyimides because of greater molecular cohesion.

Solubility. Polyureas, having polar structures, are difficult to dissolve in organic solvents. Copolyureas, particularly aliphatic, are usually more soluble than polyureas.

Water Absorption and Permeability. The water absorption of polyureas is low and depends on the structure. The water permeability of polyurea films is higher than that of polyamides.

Chemical Properties. *Thermal Stability.* Polyureas have a higher thermal stability than polyurethanes, but lower than polyamides. The thermal stability of the urea group depends on the phase at the temperature of heating.

Hydrolytic Stability. Polyureas have excellent chemical stability to hydrolysis under acidic and alkaline conditions.

Biodegradation. Poly(urethaneurea acylsemicarbazide)s and other copolyureas containing amino acids or peptides in their chain are degraded by living organisms.

Mechanical Properties. The mechanical properties of polyureas and copolyureas are shown in Table 3.

Applications

The urea linkages in polyurethaneureas or polyureas extended with diamines cause an increase in phase separation compared to polyurethanes extended with diols. Higher segregation improves mechanical properties such as elongation at break, hysteresis, and stress relaxation. Polyureas are used in foams, (eg, polyurethaneurea foams, flexible foams, rigid foams, polyurea foams), moldings, fibers, films and membranes, coatings, lubricants, biomedical applications, and other applications. High modulus polyurethaneurea and polyurea RIM (reaction injection molding) elastomers have many practical applications; the most important are in the automotive industry.

BIBLIOGRAPHY

P. W. Morgan, *Condensation Polymers by Interfacial and Solution Methods,* Interscience Publishers, New York, 1965, pp. 209–259, 493.

G. S. Paik Sung and co-workers, *Macromolecules* **12,** 538 (1979); *Polymer Alloys,* Plenum Press, New York, 1977, p. 261.

Polyplastie Plastici Rinforzati **343,** 47 (May 1986).

C. I. CHIRIAC
Petru Poni Institute of Macromolecular Chemistry
Jassy, Romania

POLYURETHANES

The polymers known as polyurethanes include materials that incorporate the carbamate functional group as well as other functional groups, such as ester, ether, amide, and urea. These polymers were discovered by Otto Bayer and co-workers in 1937. They are usually produced by the reaction of a polyfunctional isocyanate, most often with hydroxyl compounds. Since the functionality of the hydroxyl-containing reactant or the isocyanate can be varied, a wide variety of branched or cross-linked polymers can be formed. Hydroxyl-containing components cover a wide range of molecular weights and types, including polyester and polyether polyols. The polyfunctional isocyanates can be aromatic, aliphatic, cycloaliphatic, or polycyclic in structure and can be used directly as produced or modified. This flexibility in the selection of reactants leads to the wide range of physical properties that allows polyurethanes to play an important role in the world market for quality products from synthetic polymers.

The recent growth in the commercial importance of polyurethane products formed by reaction injection molding (RIM) has provided the impetus for basic studies of relationships between molecular structure and bulk properties.

The ability to select RIM formulations for applications covering a range of required physical properties is facilitated by the understanding of the segmented structures of such polyurethanes and how the chemical nature and length of these segments affect the morphology.

In brief, polyurethane elastomers may be considered to be block copolymers consisting of alternating polyurethane and polyol segments. The polyurethane and polyol segments are designated hard and soft segments, respectively, since they are below and above their softening temperatures at the normal use temperatures. The elastomeric properties of these block copolymers are generally attributed to the phase separation of the polyurethane and polyol segments, whereby the hard polyurethane domains serve as cross-links and filler particles in the matrix of the polyol soft segment. Studies by both physical and chemical methods have shown that bulk properties depend on the nature and extent of phase separation.

Raw Materials

Isocyanates. Organic isocyanates, the A side of polyurethane reactant systems, contain the functional group —NCO. For general references, *Chemical Abstracts* classifies isocyanates as esters of isocyanic acid, HNCO. Specific compounds are indexed using the term isocyanato, analogous to halogen-substituted compounds.

Preparation and Manufacture. Polyurethane isocyanates are usually produced commercially by the phosgenation of amines.

Polyols. By convention, the B side of a liquid, two-component polyurethane system consists of a compound or mixture of compounds containing functional groups that react with isocyanates. The functional groups usually contain active hydrogen atoms as defined by the Zerewittioff reaction. In most compounds, a hydrogen is connected to oxygen, nitrogen, or sulfur. Most often, the B reactant is a polyfunctional alcohol or polyol whose reaction product with the A-side isocyanate yields the carbamate or urethane linkage. The second most frequent B-side components are polyfunctional primary or secondary amines. When the B side consists of a blend of polyamine and polyol, it is still referred to as a polyol or polyol blend. Although the reaction product of an amine with an isocyanate yields the urea linkage, the polymer produced from a mixed polyol–polyamine blend may be referred to as a polyurethane.

Polyesters. Polyester polyols are prepared by condensation or step-growth polymerization. The formation of polyester chains is a random process and leads to a broad distribution of molecular weights. Polyester polyols used for polyurethanes are low molecular weight (<3500) hydroxyl-terminated polymers, produced by using an excess of hydroxyl equivalents.

Polyethers. Polyether polyols are prepared by an anionic, ring-opening, addition polymerization of epoxides.

Filled polyethers are important ingredients of some special flexible and semiflexible foams. These polyols contain finely divided solids in the form of a dispersed phase in the base polyol. Such polyols give polyurethanes with exceptional properties. For example, filled polyols significantly increase the resilience and load-bearing properties of flexible foams.

Recent advances in RIM technology include the development of new polyurea systems for the production of automobile body panels. These catalyst-free systems are based on amine-terminated polyethers.

Polyurethane Foams

The polyurethanes, originally intended as fibers competitive with nylon, found greatest commercial success as foams. In 1985, U.S. sales of flexible and rigid polyurethane foams were estimated at 630×10^3 and 328×10^3 metric tons, respectively. Use of both types and numerous specialized foams has grown steadily since the original technology was developed in the FRG over 30 years ago. Acceptance can be attributed to excellent properties at reasonable cost and simple, easily controlled production processes. Foam products are made from liquid materials and are available in a wide range of types and forms designed for specific applications. Processes have been adapted to in-plant production of large continuous blocks and molded articles, as well as to on-site spraying and foam-in-place operations. No other foam polymer system provides the flexibility in control of properties and process (see also CELLULAR MATERIALS).

Reactants. Polyurethane foams, or urethane foams as they are often called, are prepared, except for a few specialties, from aromatic isocyanates and mainly aliphatic polyols. Toluene diisocyanate (TDI), usually the commercial blend of 80% 2,4- and 20% 2,6-isomers, is chosen for most flexible foams. Rigid foams can be made from TDI prepolymers but are most often based on a polymeric diphenylmethane 4,4′-diisocyanate (polymeric MDI) with a number-average functionality of 2.5-3. The polyols are usually polyethers prepared by adding 1,2-propylene oxide and ethylene oxide to polyalcohols or polyamines.

Flexible urethane foams are formed by a process of simultaneous polymerization and expansion. The gas for expansion is primarily carbon dioxide produced by the reaction of isocyanate and water. Rigid foams are formed by a similar process but depend on vaporization of a low boiling solvent for expansion. The most common blowing agent is trichlorofluoromethane, refrigerant 11, but its damaging effect on the ozone layer means that it must be replaced.

Much of the commercial acceptance of urethane foams can be attributed to catalysts that precisely control the polymerization and blowing processes. This control is essential for flexible urethane foams. Imbalance between polymerization and foam expansion can lead to foam collapse, serious imperfections, and cells that open prematurely or not at all.

The inherent tendency of polyurethane foams to burn is markedly reduced by flame retardants or combustion modifiers. These include either organic halogen compounds containing chlorine or bromine, or organic phosphates and phosphonates. Nonreactive compounds such as trichloroethyl phosphate are well-known flame retardants and also act as plasticizers. Hydroxy-functional compounds, such as *O*,*O*-diethyl-*N*,*N*-bis(2-hydroxyethyl)aminomethyl phosphate or halogen-containing polyesters, can be built into the polymer. Inorganic compounds, including ammonium polyphosphate and even red phosphorus, are effective combustion inhibitors.

Although most urethane foams are made from raw materials without modification, solid fillers are added for special applications. Finely ground $CaCO_3$ or baryte increases hardness and resilience. Glass fibers are added to reinforce high density, specialized systems. Soluble or liquid extenders have been studied as a means of cost reduction. Most were natural by-products of agricultural or wood-processing industries; none of these are widely used, however.

Chemistry. The combination of chemical and physical changes in foam formation is very complex and, even today, is not completely understood.

The mechanism of formation of water-blown, flexible polyurethane foams has been thoroughly investigated during the past 20 years. Although some points are still debated, it is generally agreed that the reaction of isocyanate groups with water is more rapid than that with hydroxyl groups. The expansion and initial stabilization of the foam depend on the water reaction and its formation of carbon dioxide and aromatic ureas. Some authors report separation of a partly crystalline polyurea microphase which tends to promote cell opening near the end of foam rise.

The chemistry of solvent-blown rigid foam formation is simpler and depends on the isocyanate–hydroxyl reaction. The polyurethane network is strengthened, however, by isocyanurate, allophanate, and carbodiimide structures.

Physics. The physics of foam formation is closely tied to the chemistry but is in many ways more descriptive. On a macroscopic scale, mixed starting materials react, expand, and flow. The polymer increases in molecular weight, grows in viscosity, and ultimately forms a cross-linked gel which cures to the final product. Microscopically, cells nucleate, grow in size, coalesce, stabilize, and in the case of flexible foams, open. The exothermic chemical reactions produce a marked increase in temperature, which affects both chemical and physical processes. Foam formulations are often characterized by laboratory measurements of the rate of expansion and rate of temperature increase.

Processing. Foam machines appear to be complex but actually are based on a few simple principles. Components common to all foam machines include at least two feed tanks, metering pumps, temperature controls, and mixers.

Forming Devices. The forming section is the most important part of a foam machine. The rate of delivery of mixed materials must be sufficiently high to allow foam to fill the desired space before viscosity becomes too high or the material gels. The heat generated by the foaming reactions must be controlled. The thermal conductivity of foamed materials is low and, depending on product density and dimensions, the temperature can be too low for proper cure or so high that serious degradation occurs. A closed forming device must withstand the significant but not excessive pressure which builds up during foam formation.

In most polyurethane foam manufacture, one-shot, one-step techniques are used. The reactants are combined simultaneously, and no chemical reaction takes place before mixing.

Continuous Slab or Block. Mixed materials are spread on a downward-moving conveyor. Liquid flows and is carried downward until foaming begins. Expansion balances the flow and conveyor travel, forming a stationary foam front that rises until expansion is complete and a continuous block is formed. Finished blocks are cut into appropriate lengths and set aside for ca 24 h to cure and cool before cutting into desired shapes.

Molding. Accurately metered amounts of mixed foam formulations are poured into molds which are then closed. Foam expansion fills the mold, forming, for example, a contoured automobile cushion. Thin sections and slowly reactive formulations may require postcure.

Foam-in-place. Expanding foam tends to fill space. Rigid urethane foam, in particular, is often poured into existing cavities, providing both thermal insulation and physical strength. Most refrigerators and freezers are insulated by in-place applied rigid urethane foam.

Spraying. Rigid urethane foam formulations can be sprayed by special two-stream spray guns in a process similar to paint spraying. The technique is useful for uneven surfaces.

Properties. *Flexible Foams.* The properties of flexible polyurethane foam vary widely with formulation and density. Standard polyether foams produced in slab form have densities of 16–24 kg/m^3; molded foams from similar formulations have higher densities, 32–40 kg/m^3. The physical properties of slab products are shown in Table 1.

Rigid Foams. The isocyanate-based rigid polyurethane and polyisocyanurate foams typically have low densities of 16–48 kg/m^3 and are closed cellular; chlorofluoroalkanes, usually $CFCl_3$, are used as blowing agents.

The important properties of a polyol blend are functionality, molecular weight, and viscosity. Properties of rigid polyurethane foams are shown in Table 2.

Applications. *Flexible Foams.* These low density, resilient foams serve mainly as protective and cushioning materials. The largest single U.S. market, over 40% of the total, is upholstered furniture, where urethane foam is used alone or in combination with other materials for cushions, backs, and arms.

Rigid Foams. Most applications of rigid urethane foams depend on their superiority as thermal insulating materials. Recent concerns about energy consumption have increased interest in thermal insulation, and urethane foams, because of their combination of properties and production methods, have received particular attention.

Construction, particularly commercial roofing and residential sheathing, continues as the largest U.S. market area, consuming over 60% of the total production.

Elastomers

A major breakthrough occurred when Bayer and associates at I.G. Farbenindustrie prepared a material with rubberlike elasticity from naphthalene 1,5-diisocyanate and a linear, hydroxyl-terminated polyester (mol wt 2000) of adipic acid and ethylene glycol. A new synthetic rubber had been discovered. The addition of low molecular weight diols as chain extenders provided a wide range of hardnesses and moduli of elasticity. The properties of these new polyurethane elastomers exceeded those of vulcanized natural and synthetic rubbers, therefore initiating the development of the endless variety of solid polyurethane elastomers and polyurethane foams offered today. Polyurethane elastomers based on naphthalene 1,5-diisocyanate are still among the top performers. The components are sold by Bayer AG under the trade name Vulkollan. Most current polyurethane elastomers, however, are based on MDI or TDI.

Cast Elastomers. The ability to prepare the new polyurethane elastomers from liquid starting materials differentiates them from other natural or synthetic rubbers, which require milling or mastication followed by shaping and compression molding. Polyurethane elastomers are formed by a liquid casting process which led to the designation cast elastomers. Today, complete

Table 1. Physical Properties of Slab Foam

Property	Low density, combustion-modified	Medium density, firm
density, kg/m^3	16.8	22.2
IFD[a], 10-cm thickness, kPa[b]		
at 25% compression at rest	1.9	5.4
at 65% compression at rest	3.7	9.8
tear strength, N/m[c]	322	473
tensile strength, kPa[b]	55.1	134.4
elongation, %	180	275
90% original compression set, %	8.8	4.1

[a] Indentation-force deflection.
[b] To convert kPa to psi, multiply by 0.145.
[c] To convert N/m to ppi, divide by 175.

Table 2. Physical Properties of a Rigid Polyurethane Foam Molding System[a]

Property	ASTM test method	Value
density, kg/m^3	D1622	
core		28.9
overall		33.4
compression strength, MPa[b]	D1621	
at 7.4% yield		0.13
at 5%		0.12
at 10%		0.14
cellularity, % closed cells	D2856-70	90
K factor[c], W/(m·K)[d]		0.017
dimensional stability, % volume change	D2126-66	
at 70°C, 100% rh		
1 d		5.0
7 d		6.2
14 d		8.3
28 d		8.6
at 100°C		
1 d		−0.1
7 d		0.4
14 d		1.8
28 d		0.3
at −30°C		
1 d		0.1
7 d		0.4
14 d		1.8
28 d		0.3

[a] ASTM standard test methods.
[b] To convert MPa to psi, multiply by 145.
[c] Thermal conductivity (ASTM C518).
[d] To convert W/(m·K) to (Btu·in.)/(h·ft²·°F), divide by 0.144.

product lines of MDI and TDI prepolymers based on linear polyesters as well as polyethers are available, enabling producers to obtain any desired elastomer hardness.

Thermoplastic Elastomers. Polyurethanes based on MDI, linear hydroxyl-terminated polyesters, and glycol chain extenders, with a 1:1 ratio of NCO and OH groups (not chemically cross-linked by an excess of isocyanate) have excellent elastomer properties and can be processed as thermoplastics.

Today, TPUs are well-established as synthetic rubbers as well as engineering plastics for a multitude of applications and as compounds for polymer alloys (see ELASTOMERS, THERMOPLASTIC).

Millable Gums. Following the sequence from cast elastomers, cross-linked with excess diisocyanate, to the TPUs, with practically stoichiometric amounts of NCO and OH groups, millable gums are obtained with an even larger excess of hydroxy-functional components and with no or only small amounts of chain extender. Their molecular weight is lower than that of TPUs. The ideal cross-linking agent for millable polyurethane gums is the dimer of 2,4-TDI. Millable polyurethane gums are used in synthetic rubber applications, eg, for machine parts, where only polyurethane chemistry fulfills the needs.

Molecular Models. The outstanding properties of polyurethane elastomers based on naphthalene 1,5-diisocyanate can be attributed to chemical cross-linking and to the bulky diisocyanate molecules that separate the mobile chains of the polyester.

The original model of the widely meshed network for cross-linked polyurethane elastomers was replaced when it was shown that the elastomeric properties derive from the segmented polymer structure. Today, it is generally accepted that the excellent properties of polyurethane elastomers result from the segmented block copolymer structure. The forces holding the hard segments together derive from primary chemical bonds or hydrogen bonds, or can be, like van der Waals forces, of physical nature. All provide the elastomeric character. The proof for this elastomer model originated from studies of thermoplastic polyurethane elastomers.

Molecular Structure. The viscoelastic behavior of linear segmented urethane elastomers was investigated with the help of modulus–temperature experiments. Unusual properties were discovered, such as an extended high modulus plateau region between two thermal transitions. By selectively solvating the aromatic urethane and the polyester segments with different swelling agents, it was demonstrated that an association of the hard segments in the solid state (hard-segment domain) was a prerequisite for the occurrence of a higher temperature transition. This domain formation and phase segregation are responsible for the properties of these elastomers.

Composition. The main components of polyurethane elastomers include polyester and polyether diols, diisocyanates, and chain extenders. The linear, aliphatic polyesters required to form the soft segments of these elastomers are often formed by the reaction of adipic acid with simple glycols; polymers of ϵ-caprolactone can also be used.

Physical Properties. The segmented molecular chains and two-phase domain morphology of polyurethane elastomers provide the characteristic physical properties of these materials (Tables 3 and 4). Specific properties vary widely with composition, but urethane elastomers generally have excellent tear, abrasion, impact, and wear resistance; resistance to hydrocarbon and aromatic oils is also good.

Table 3. Physical Properties of MDI–Polyester Cast Polyurethane Elastomers, Chain-extended with 1,4-Butanediol

	Shore hardness	
Property	93–95A	52–55D
tensile strength, MPa[a]	51.7	35.2
tensile modulus, MPa[a]		
at 100%	10.3	16.2
at 200%	17.2	20.6
at 300%	26.2	26.5
elongation, %	3.1	3.1
die C tear, kN/m[b]	123	116
split tear, kN/m[b]	70	105
compression set[c], %	32	28
resilience, %	18	43
demold time, min	30	10

[a] To convert MPa to psi, multiply by 145.
[b] To convert kN/m to ppi, divide by 0.175.
[c] Method B, ASTM D395.

Applications. The applications of polyurethane elastomers are both numerous and varied. Most depend on the excellent toughness and wear resistance with a broad temperature range for use.

Cast-elastomer applications

forklift tires and wheels	elevator wheels
roller-skate wheels	brake diaphragms
special running shoes	snowplow blades
oil and mud hydrocyclones	drilling-pipe thread protectors
grain buckets	grain and coal chutes
classifier shoes	helicopter-blade sleeves
hydraulic seals	bumper pads
wheel chocks	die-cut pads (paper box industry)
bowling-ball cover stock	large rollers for steel and paper mills
shaft couplers	copier rolls
sheet goods	encapsulated gate valves
rod stock	encapsulated transponders (cattle tags)
mining screens	encapsulated concrete-mixer blades
conveyor belts	drive belts
skate-board wheels	dead-blow hammers
coated conveyor belts	sprockets
gears	O-rings
pipeline pigs	
boat fenders	
pump liners	
crane-shock absorbers	
sound-dampening pads	
potting compounds	

Thermoplastic polyurethane applications

adhesives	magnetic-tape coatings
automobile fascia	cattle-ear tags
golf-ball cover stock	fluorescent golf balls
roller-skate wheels	skate-board wheels
top lifts for shoes	mine screens
elevator wheels	conveyor-drive belts
ski boots	hockey skate boots
gears	sprockets
dust covers	automotive grease seals
hose jackets	hose lines
airline arm rests	fuel-line tubing
water-line tubing	grommets
films for diapers	films for laminations
automotive parts	small rollers
football cleats	striker sleeves
industrial wheels	

Table 4. Physical Properties of MDI–Polyester Thermoplastic Polyurethane Elastomers

		Shore hardness[a]	
Property	ASTM method	91A	55D
tensile strength, MPa[b]	D412	41.4	41.4
modulus, MPa[b]	D412		
100%		8.3	11.0
elongation at break, %	D412	540	550
tear strength, die C, kN/m[c]	D624	131.3	140.1
flexural modulus, MPa[b]	D790		
at 23°C		37.9	124.1
at 70°C			51.7
compressive load, MPa[b]	D575		
at 2% deflection		0.7	1.0
at 10% deflection		2.7	4.8
at 20% deflection		5.5	8.6
compression set, %	D395 Method B		
22 h at 23°C		16	20
postcured[d]			
22 h at 23°C		12	15
Taber abrasion[e], mg loss	C501	40	100

[a] ASTM D2240.
[b] To convert MPa to psi, multiply by 145.
[c] To convert kN/m to ppi, divide by 0.175.
[d] For 16 h at 104°C.
[e] H-18 wheel, 100-g load, 1000 cycles.

Reaction Injection Molding

Reaction injection molding, or RIM, is a process in which a mixture of two or more reactive materials is injected into a mold where the materials react to form a polymer. Commercial application of RIM originated in 1964 at Bayer AG, Leverkusen, FRG. Rigid, structural foam systems contain a polyether polyol blend, surfactants, tertiary amine or organotin catalysts, flame retardants, a blowing agent, and polymeric MDI. Such formulations give cross-linked, network structures. Elastomeric systems consist of polyether diols or triols (4000–6000 mol wt), chain extenders (ethylene glycol, 1,4-butanediol, or diamines), tertiary amine or organotin catalysts, and a modified MDI-type polyisocyanate. This combination produces the unique segmented morphological structures that provide tough resilient materials.

Physical Structure and Properties. The RIM process produces the characteristic sandwich structure of a solid skin and a lower density microcellular core in a single operation. The skin is formed by the blowing agent condensing on the cooler mold surface under the increasing pressure of the expanding foam.

Accordingly, skin formation and thickness are affected by factors that influence the condensation of the blowing agent at the mold interface, such as shot weight (increasing shot weight produces thicker skins); formulation (high blowing-agent concentration increases skin thickness); mold temperature

Table 5. Physical Properties of Structural Foam Materials

Properties	Polyurethane structural foam	Polystyrene	ABS	Modified PPO
specific gravity	0.85	0.85	0.89	0.85
thickness, mm	63	63	63	63
flexural modulus, GPa[a]	1.65	1.89	1.65	1.79
flexural strength, MPa[b]	65.5	39.3	41.3	46.8
tensile strength, MPa[b]	39.9	15.8	20.6	23.4
elongation, %	10	20		16
Gardner impact, N·m[c]	8.3	3.2	5.8	6.2
falling ball, N·m[d]				24.4
dart impact, N·m[d]			13.5	
heat-distortion temperature, °C at 455 kPa[e]	100	83	80	96

[a] To convert GPa to psi, multiply by 145,000.
[b] To convert MPa to psi, multiply by 145.
[c] To convert N·m to in.·lbf, divide by 0.113.
[d] To convert N·m to ft·lbf, divide by 1.35.
[e] To convert kPa to psi, multiply by 0.145.

(40–80°C; lower temperatures give thicker skins); and mold thermal conductivity (a highly heat-conductive mold is necessary for controlling skin thickness).

Polyurethane structural foams utilize the smooth, hard skins as a source of impact strength, heat-distortion temperature, and combustion resistance (Table 5).

The development of systems based on polymeric isocyanates, polyether polyols, and diethyltoluenediamine and other amine chain extenders led to the use of the RIM process by the automotive industry. The amine-extended systems exhibit the much faster reactivities needed to achieve a 30–60-s demold time without tearing, distortion, or delamination. These materials can be used with soap-based, water-washable mold releases. Physical properties such as flexural modulus and impact strength show flatter temperature-response curves over a broader range of temperatures (−30–65°C) than glycol-extended systems and provide improved resistance to deformation on cars in the summer and better resistance to breaking or cracking in the winter. Properties of amine-extended materials are given in Table 6.

The stiffness required for automotive body panels is obtained by reinforcement with inorganic fillers (eg, glass fibers, glass flakes, Wollastonite, calcium carbonate, mica).

Internal Mold Release. In 1984, RIM polyurethane systems with an internal mold release (IMR) were introduced, greatly reducing the need to apply an external mold-release agent from every shot to every 80th shot. Processibility, paintability, and physical properties are maintained. An IMR system offers reduced cycle times, increased productivity (40–60%, depending on part design), less frequent mold cleaning, and lower scrap rate (see RELEASE AGENTS).

Polyrurea RIM Systems. In automotive body panels, polyurea RIM materials offer greater impact resistance at equivalent glass-flake reinforcement; improvement in thermal stability for less distortion during paint cure cycles; and lower water absorption for better environmental resistance. These polyureas are prepared by the reaction of an MDI-based isocyanate with a blend of an amine-terminated polyether (replacing the standard hydroxyl-terminated polyether) and an aromatic diamine chain extender.

Processing. Rigid structural and elastomeric RIM materials are processed by similar methods and equipment.

Equipment. The metering units are either "pump" equipment or "lance cylinder" units. The first RIM machines were equipped with piston or axial metering pumps. Sophisticated versions of these units are still widely used. The RIM process

Table 6. Properties of Typical Elastomeric RIM Systems[a]

Property	Density, kg/m³ 1040	998.4	1040	998.4
hardness, D scale Durometer	50	50	60	65
tear strength, die C, kN/m[b]	86	91	105	
ultimate tensile strength, MPa[c]	24.1	23.4	26.2	24.1
elongation at break, %	340	260	280	110
flexural modulus, MPa[c] at 25°C	172	248	358	517
flexural modulus ratio (−30°C/65°C)[d]	3.55	3.2	3.43	3.1
heat sag[e], 1 h at 121°C, mm	10.2	5	5.1	˙3

[a] All samples are postcured for 1 h at 121°C; sample thickness = 31 mm.
[b] To convert kN/m to ppi, divide by 0.175.
[c] To convert MPa to psi, multiply by 145.
[d] Ratio of modulus at −30°C to modulus at 65°C.
[e] 10-cm overhang.

requires that components A and B (isocyanate and polyol, respectively) be maintained in separate, temperature-controlled tanks. Feed pumps transfer the material from the tanks to metering pumps. The speed of the metering pump, along with its adjustable vernier, controls the amount of each component sent to the mixhead. Machine throughput capacity is determined by the size of the metering pumps and the delivery ratio of the system being processed.

Finishing. Many RIM parts can be trimmed and put into service; however, where appearance is a consideration, painting is recommended.

Applications. *Structural.* The first commercial applications of the RIM process appeared in Europe during the early 1970s in furniture and skis using rigid structural foam, followed by electronic, telecommunication, agricultural, sporting goods, construction, medical, and material-handling markets.

Elastomeric. The largest use for elastomeric RIM is found in the automotive industry in the form of bumper and fascia parts.

Foamed Elastomeric. Foamed elastomeric parts are also used in a broad range of applications. In automobiles, they are found in instrument panels, arm rests, steering wheels, and rear deck air domes. In the furniture industry, RIM is used to encapsulate metal to produce arm rests, chair bases, and desk supports; beer kegs are encapsulated for damage resistance.

Coatings

Polyurethane coatings are preferred for many applications because of a unique combination of performance and application properties. These products provide excellent abrasion resistance, flexibility, hardness, and chemical and solvent resistance; light stability and weatherability are excellent. Most polyurethane coatings are referred to as reactive coatings because lower molecular weight oligomers react to form a polymeric network on the substrate; energy requirements for cure are low. Raw materials include polyisocyanates and coreactants, eg polyester and polyether polyol.

Two-component Coatings. The two-component (or two-package) systems are prepared by mixing a polyisocyanate with a polyol before application to the substrate. Upon mixing, the isocyanate and polyol react immediately, resulting in continual viscosity buildup and eventual gelation. Thus two-component systems have a limited pot life.

One-component Polyurethane Coatings. *Moisture Cure.* Isocyanate-terminated prepolymers that dry, harden, and cure at ambient temperature by reaction with ambient moisture are classified as one-component or one-package systems.

Air Drying. Another type of one-component coating system consists of solvent-borne, urethane-modified oils and alkyds. These systems air dry at ambient temperatures to give coatings. They are often used as interior or exterior paints and varnishes.

Blocked Isocyanates. A third type of one-component coating system contains a "blocked" or "capped" isocyanate. A blocked polyisocyanate is stable at RT, but dissociates to regenerate the isocyanate at elevated temperatures. Storage-stable, one-component coatings that are heat curable can be prepared by mixing blocked polyisocyanates and polyols. These convenient one-package systems are well-suited for applications that demand the performance properties of polyurethane coatings and can be cured at high temperatures. Blocked isocyanates have been used for many years in wire coatings and coil coatings.

Powder Coatings. Powder coatings are solvent-free coating systems that are applied to a substrate as a dry powder. The solid particles adhere to the substrate upon melting or by electrostatic attraction. Upon heating, the solid particles fuse and, depending on the resin, undergo further curing.

Fully Reacted Polyurethane Coatings. *Solvent-borne Systems.* Solvent-borne, fully reacted polyurethane coatings are solutions of high molecular weight polyurethanes that form films solely by loss of solvent. With fast evaporating solvents, films dry in a few minutes and exhibit good mechanical properties. However, because of inherently poor solvent resistance, these coatings are limited to special applications.

Aqueous Dispersions. Waterborne coatings eliminate solvent emissions and reduce costs. Coatings formed from linear thermoplastic polyurethane dispersed in water cure at ambient temperature by evaporation of water. The dispersed particles coalesce to form a film.

Aqueous polyurethane dispersions have found applications in areas previously dominated by solvent-borne polyurethane coatings. They are used as coatings for wood; glass fibers; flexible substrates such as textiles, vinyl, leather, rubber, paper, and foil; finishing of business-machine housings and plastic parts; and as metal primers.

Other Applications

Other applications include adhesives, sealants, binders, fibers, and mine reinforcements.

Health and Safety

Raw Materials. *Isocyanates.* The primary route of exposure to most isocyanates is through inhalation. Isocyanate vapors or aerosol can irritate the mucous membranes in the respiratory tract (nose, throat, lungs), causing nasal irritation, sore throat, coughing, chest discomfort, shortness of breath, and reduced lung function (breathing obstruction).

Because of these adverse health effects, the American Conference of Governmental Industrial Hygienists (ACGIH) has set the following threshold limit values (TLV) for most diisocyanates: 0.005 ppm, time-weighted average over 8 h; and 0.02 ppm, short-term exposure limit. The OSHA permissible exposure limit (PEL) for both TDI and MDI is 0.02 ppm, expressed as a ceiling value.

Upon skin contact, isocyanates produce local discoloration, in some cases reddening, swelling, rash, scaling, or blistering, and occasionally skin sensitization.

Liquid, aerosols, or vapors can cause severe eye irritation, pain, tearing, reddening, and swelling; the cornea can be damaged. Prolonged vapor contact may cause conjunctivitis.

Although not a common route of exposure, ingestion can result in irritation and corrosive action in the mouth, stomach, and digestive tract. Symptoms can include sore throat, abdominal pain, nausea, vomiting, and diarrhea.

Polyols. Most polyols are very low in acute oral toxicity and are considered only mildly irritating to eyes and skin. However, they should be handled with appropriate caution and in accordance with the manufacturer's recommendations.

Catalysts. Tertiary amine and alkyltin compounds are used to catalyze the polyurethane reaction. Some tertiary amines are corrosive to the skin and eyes and can cause burns. Some amines cause blurred vision and an effect called "halo vision," where the vapors cause a temporary swelling of the cornea.

Inhalation of tertiary amines may cause irritation of the upper respiratory tract and may lead to sensitization. Skin sensitization has also been reported. Many amines have flash points from −6 to 46°C and are classified as flammable or combustible liquids.

Alkyltin (or organotin) salts may penetrate the skin and mucous membranes and can cause local inflammation as well as skin sensitization. Skin and eye contact must be avoided.

Blowing Agents and Solvents. Both chlorofluorocarbons and methylene chloride are used as blowing agents. Inhalation of high concentrations of chlorofluorocarbons can be dangerous and may cause anesthesia and unconsciousness. (The TLV for trichlorofluoromethane is 1000 ppm.) Severe acute exposures have caused cardiac arrhythmias, including ventricular fibrillation. In addition, when confined and subject to high temperatures, chlorofluorocarbons may present an explosion hazard.

Chlorofluorocarbons, including those used in polyurethane foams, may lead to reduction of the ozone layer surrounding the earth. Regulatory action has been taken in the United States and other countries.

Methylene chloride is used as a blowing agent and solvent. Upon skin contact, it can cause irritation, drying, and weakening of the skin. Inhalation can cause central nervous system depression, resulting in impaired judgment and reduced motor ability. Inhalation of high concentrations causes cancer in experimental animals. It is therefore extremely important to follow the manufacturer's recommendation for the safe use and handling of methylene chloride.

Dust. Polyurethane chemicals and foams can present a hazard if improperly used and handled. Dust produced during the cutting and sawing of rigid polyurethane foam can irritate the eyes and the mucous membranes of the nose and throat. In addition, finely divided airborne polyurethane can cause a dust explosion. For this reason, the accumulation of dust should be controlled and good housekeeping practiced.

Combustibility. Although the products of combustion vary, virtually all polyurethane foams and materials burn, often generating a variety of toxic, potentially lethal gases and dense smoke.

BIBLIOGRAPHY

J. H. Saunders and K. C. Frisch, *Polyurethanes: Chemistry and Technology,* Vol. XVI of *High Polymers,* John Wiley & Sons, Inc., New York, 1962, Pt. I.

G. Oertel, *Polyurethane Handbook: Chemistry—Raw Material—Processing-Application—Properties,* Hanser Publishers, New York, 1985.

G. Woods, *The ICI Polyurethanes Book,* John Wiley & Sons, Inc., New York, 1987.

J. K. Backus
C. D. Blue
P. M. Boyd
F. J. Cama
J. H. Chapman
J. L. Eakin
S. J. Harasin
E. R. McAfee
C. G. McCarty
N. H. Nodelman
J. N. Rieck
H. G. Schmelzer
E. P. Squiller
Mobay Corporation, a Bayer USA Company

POLYURETHANE TECHNOLOGY. See Polyurethanes.

POLY(VINYL ACETATE). See Vinyl ester polymers.

POLY(VINYL ACETATE-*co*-VINYL CHLORIDE). See Vinyl chloride polymers.

POLY(VINYL ALCOHOL). See Vinyl alcohol polymers.

POLY(VINYL ALCOHOL-*co*-VINYL ACETATE). See Vinyl alcohol polymers; Vinyl ester polymers.

POLY(VINYL ALKYL ETHER-*co*-MALEIC ANHYDRIDE). See Maleic and fumaric polymers; Vinyl ether polymers.

POLY(VINYL ALKYL ETHER)S. See Vinyl ether polymers.

POLY(VINYL AMIDE)S. See *N*-Vinyl amide polymers.

POLY(VINYLAMINE)S. See Polyamine and polyquaternary ammonium salts.

POLY(VINYL BUTYRAL). See Vinyl acetal polymers.

POLY(VINYLCARBAZOLE). See Vinylcarbazole polymers.

POLY(VINYL CHLORIDE). See Vinyl chloride polymers.

POLY(VINYL CHLORIDE-*co*-ACRYLONITRILE). See Acrylic fibers; Vinyl chloride polymers.

POLY(VINYL CHLORIDE-*co*-PROPYLENE). See Vinyl chloride polymers.

POLY(VINYL CHLORIDE-*co*-VINYL ACETATE). See Vinyl chloride polymers.

POLY(VINYL ESTER)S. See Vinyl ester polymers.

POLY(VINYL ETHER)S. See Vinyl ether polymers.

POLY(VINYL FLUORIDE). See Vinyl fluoride polymers.

POLY(VINYL FORMAL). See VINYL ACETAL POLYMERS.

POLY(VINYLIDENE CHLORIDE). See VINYLIDENE CHLORIDE POLYMERS.

POLY(VINYLIDENE FLUORIDE). See VINYLIDENE FLUORIDE POLYMERS.

POLY(VINYL ISOCYANATE). See ISOCYANATE-DERIVED POLYMERS.

POLY(VINYL KETONE)S. See VINYL KETONE POLYMERS.

POLY(VINYL PROPIONATE). See VINYL ESTER POLYMERS.

POLY(VINYLPYRIDINE). See VINYLPYRIDINE POLYMERS.

POLY(*N*-VINYLPYRROLIDINONE). See *N*-VINYL AMIDE POLYMERS.

POLY(VINYLTOLUENE). See STYRENE POLYMERS.

POPCORN POLYMER. See PROLIFEROUS POLYMERIZATION.

POROMERIC MATERIALS. See LEATHERLIKE MATERIALS.

POROSITY. See TRANSPORT PROPERTIES.

POSITRON ANNIHILATION

The annihilation of positrons in condensed matter provides a unique way of obtaining information about the internal structure of a material. This information is transmitted from the material by γ-rays emitted when the positron is annihilated. It may be obtained by measuring three fundamentally different quantities: the angular correlation between these γ-rays, the energy distribution of the γ-rays, or the lifetime distribution of the positrons.

The angular correlation of annihilation radiation gives independent data that can be combined with lifetime results to deduce features of the distribution of free-volume sizes in polymers. Positron lifetimes are also sensitive to other properties of molecular solids, such as cross-link densities in epoxy resins. The positron method has not yet been extensively applied, but results obtained so far show that it is capable of obtaining new, useful information.

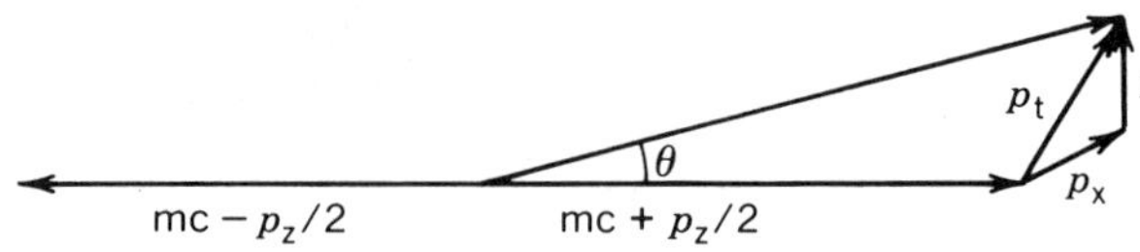

Figure 1. Relation between momentum components and angular correlation of annihilation radiation. Magnitudes of x, y, and z components of resultant momentum p are exaggerated for clarity.

Fundamentals

Energy and Momentum. Most annihilations of positron–electron pairs in a solid or liquid occur from a singlet (spins antiparallel) state with the emission of two photons (γ-rays). These photons carry the total energy and momentum of the annihilated pair. Because the rest energy of an electron or positron is 0.511 MeV, the total energy of the two photons is at least 1.022 MeV.

Angular Correlation of Annihilation Radiation (ACAR). By measuring the angle between the directions of the two photons, two components of the net photon momentum can be found; the angle usually differs from 180° by a very small angle θ (Fig. 1).

Doppler Broadening. The momentum component p_z (Fig. 1) causes a Doppler shift in the energy of both photons. The photon energies have a spectrum of values related to the magnitude and direction of the momenta of the annihilating pairs.

Positron Lifetimes. In a homogeneous environment, each positron has the same annihilation probability per unit time as every other positron, and the lifetime distribution of the positrons is a pure exponential like that describing the decay of a radioactive nuclide. However, the lifetime distribution of positrons in matter seldom follows such a simple curve. When the material is inhomogeneous, different positrons may have different environments.

As a result, the positron lifetime distribution is a sum of several exponentially decaying components, each with its own characteristic value of τ.

Experimental Methods

Most positron studies of condensed matter have been made with positron-emitting isotopes, usually ^{22}Na in the form of ^{22}NaCl. Sodium-22 emits positrons with kinetic energies of up to 540 keV; positrons from other radionuclides have similar energies. In a polymer, these positrons can penetrate to a depth on the order of 1 mm before reaching thermal energy. Thermalization is very rapid, requiring less than 100 ps; thus very few positrons are annihilated at high energies. However, because the emitted positrons have a spectrum of energies, they penetrate to various depths in the sample. Monoenergetic beams of positrons can now be produced for use when positrons are required at a precise depth in the sample.

BIBLIOGRAPHY

J. R. Stevens, *Methods Exp. Phys.* **16A,** 371 (1980).

B. D. Malhotra and R. A. Pethrick, *Macromolecules* **16,** 1175 (1983).

A. P. Mills, Jr. in P. G. Coleman, S. C. Sharma, and L. M. Diana, eds., *Positron Annihilation, Proceedings of the Sixth International Conference on Positron Annihilation,* Elsevier Science Publishers BV, Amsterdam, the Netherlands, 1982, pp. 121–137.

JOHN D. MCGERVEY
Case Western Reserve University

POSTFORMING. See THERMOFORMING.

POTTING. See EMBEDDING.

POWDER COATINGS. See COATINGS.

POWDER TECHNOLOGY. See COATING METHODS.

PRECERAMIC POLYMERS

The usually extremely high softening temperature of ceramics precludes their being shaped into fibers, films, or other complex shapes through melt processing common to organic polymers. The use of preceramic polymers (ceramic precursors), which are processed as ordinary polymers to the desired shape and then pyrolyzed to ceramics, can overcome the high temperature shaping problem.

Ceramic Technology

The special requirements of preceramic polymers are based on the requirements for the specific ceramic application. Heat engines and similar applications require structural ceramics stable in combustion atmospheres above 1200°C. Examination of the thermodynamic stabilities of these ceramic systems is helpful. Currently, most ceramics derived from preceramic polymers are silicon-based. Ceramic alloys with varying Si:C:N:O ratios can be prepared by processing preceramic polymers. A calculation of the vapor pressure of all species found in Si—C—N—O systems showed that N_2, CO, and SiO are the most prevalent species at 1200°C with vapor pressures of 3000, 1000, and 1 Pa (22.5, 7.5, and 0.0075 mm Hg), respectively. Therefore, for highest thermal stability, oxygen and nitrogen content should be minimized.

The brittle nature of ceramics can be overcome by forming a composite of ceramic fibers in a ceramic matrix. Ceramic fibers are made from preceramic polymers. The toughness of ceramic composites is controlled by the interaction of the fibers with the matrix. The composite derives its toughness from an energy-absorbing mechanism involving the separation (pull out) of fibers from the matrix. Therefore, it is important in the design of ceramic-fiber precursors to consider the effects of the ceramic on the matrix–fiber interface.

The extremely high softening point precludes the molding of complex shapes and fibers from the melt. Sintering aids used to densify SiC and Si_3N_4 powders usually compromise properties such as strength at high temperatures and oxidation resistance because of the formation of low melting glassy phases at grain boundaries.

Because of these difficulties, attention is focused on new chemical routes employing organometallic and inorganic polymers as precursors. Controlled pyrolysis of these polymers has revealed several promising directions such as formation of fibers, ultrafine powders, thin films and layer structures, and high strength, lightweight monolithics and composites.

The route illustrated in Figure 1 is very similar to the controlled pyrolysis of organic polymers such as polyacrylonitrile to produce carbon fibers (qv).

Organometallic preceramic polymers offer several ways to overcome the fundamental problems of shaping ceramics: tra-

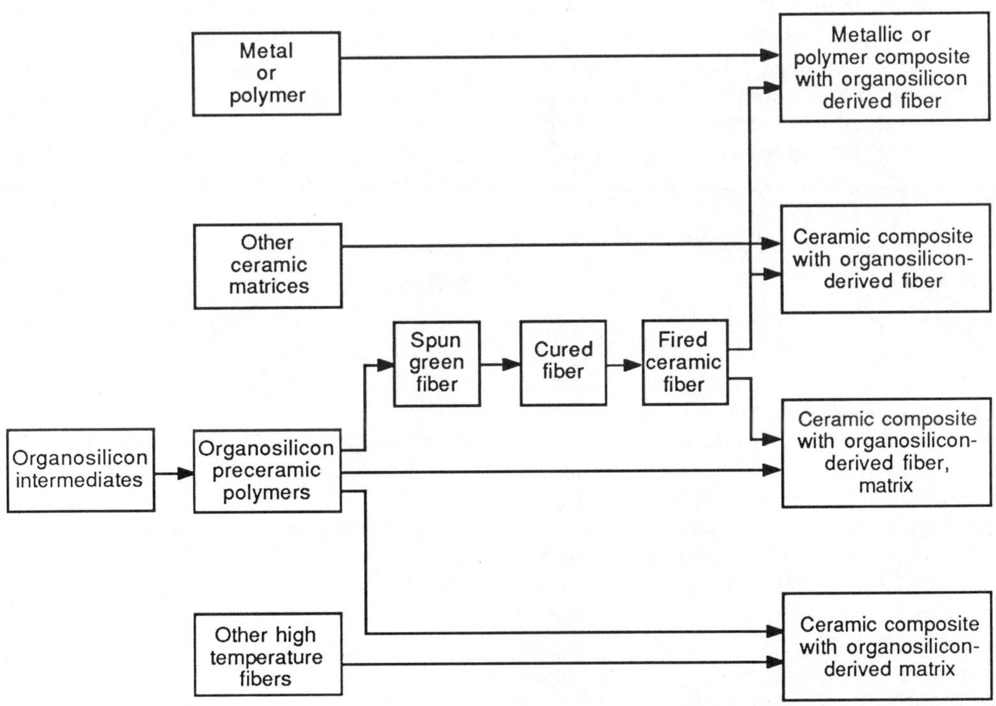

Figure 1. Organosilicon-derived ceramic fiber and composites.

ditional low temperature processing techniques are used to form complex shapes such as fibers and yarns; fabricated shapes are converted to ceramic forms at lower temperatures than those of traditional ceramic processing.

Economical production of silicon-based ceramics via a polymer pyrolysis route requires low cost raw materials and processing. High char (ceramic) yield is needed for low shrinkage upon pyrolysis and for minimum cost.

In general, branched-ring (polycyclic or cage) polymers with low molecular weights ($\bar{M}_n$ 1000–2000) provide the desired tractability and favorable pyrolysis with high char yield. Most effective silicon-based preceramic polymers developed thus far are in this category. For high quality ceramic fibers with high mechanical strength, high polymer purity and lack of spinning-induced imperfections are essential.

Organosilicon Technology

Most polymeric ceramic precursors are organosilicon-based materials.

Organic groups are attached to silicon by a Si—C bond by reaction of an organometallic compound, such as Grignard reagent or organolithium compound, with chlorosilanes. Organic groups can also be attached to silicon by hydrosilylation.

$$\equiv SiH + \gt C{=}C\lt \longrightarrow \equiv Si{-}\overset{|}{\underset{|}{C}}{-}\overset{|}{\underset{|}{C}}{-}H$$

This reaction can be initiated by free radicals or platinum catalysts. Chlorosilanes undergo many nucleophilic displacement reactions, such as hydrolysis, ammonolysis, and alcoholysis to form SiOH, $SiNH_2$, and SiOR structures, respectively.

Silicon–Carbon Ceramic Precursors

Polycarbosilanes. Although polycarbosilanes have been defined as polymers containing the Si—C—Si skeleton, polymers with $Si{-}(C)_x{-}Si$, $x > 1$, are included here.

Thermolysis of Polydimethylsilanes. The conversion of polysilane polymers to silicon carbide is based on permethylpolysilanes, derived from readily available $(CH_3)_2SiCl_2$; this process was commercialized in Japan by the Nippon Carbon Co., Ltd. to produce high strength silicon carbide fiber (see POLYSILANES AND POLYCARBOSILANES). Thermal decomposition of linear permethylpolysilanes, or the cyclic $[(CH_3)_2Si]_6$, in an autoclave at 450–470°C yields a polycarbosilane, PCS.

The structure of PCS polymer has been studied in detail by nmr (^1H, ^{13}C, ^{29}Si), ir, and uv spectra, and intrinsic viscosity measurements. These studies support a structure containing both linear and ring segments; there are 3–4 linkages between chains in the polymer unit of 10 silicon atoms.

Polycarbosilanes can be melt-spun into continuous fibers at 250–350°C. The fiber is cross-linked by oxidation of Si—H structures by heating (110–190°C) in air and pyrolyzed to 1100–1200°C under nitrogen or vacuum to black Si—C—O fibers.

The cross-linking of the uncured fibers introduces oxygen, which lowers the mechanical properties. It is an essential step, however, to make the fibers infusible, thereby retaining their shape during pyrolysis:

$$PCS(SiH) \xrightarrow{O_2} PCS(Si\text{-}OH) \xrightarrow{-H_2O} PCS(SiOSi)PCS$$

Polycarbosilanes prepared at higher temperatures with higher concentration of the polyborodiphenylsiloxane catalyst incorporate less oxygen during the cross-linking step.

The basic technology of making Si—C—O fibers from polycarbosilanes was commercialized by Nippon Carbon Co. (Japan) which supplies Si—C—O ceramic fibers under the trademark Nicalon. Nicalon fiber (10–20-μm dia, round cross section) has excellent tensile strength of 2500–3200 MPa (360×10^3–470×10^3 psi) and elastic or Young's modulus of 180–200 GPa (26×10^6–29×10^6 psi). It resists acid and alkali attack and oxidation. Yield and chemical nature of the ceramic produced from Nicalon precursor depend on the pressure and the atmosphere of the pyrolysis. The mechanical properties of Si—C—O fibers remain essentially unchanged upon neutron irradiation.

Polycarbosilanes are also used to synthesize sinterable, utrafine β-SiC powders, flaky β-SiC, and dense sintered SiC bodies. These polymers serve as sintering aids and binders for SiC and Si_3N_4 powders to produce sintered moldings with high strength and oxidation resistance.

Chemical modification of polycarbosilanes by metal alkoxides or siloxides results in metallopolycarbosilanes which, upon pyrolysis, give mixed carbides of the type M_xC_y—SiC and, as expected, contain a high concentration of oxygen. Reaction of PCS ($\bar{M}_n$ ~950) with $Ti(On\text{-}C_4H_9)_4$ gives a cross-linked polytitanocarbosilane with the following structural units:

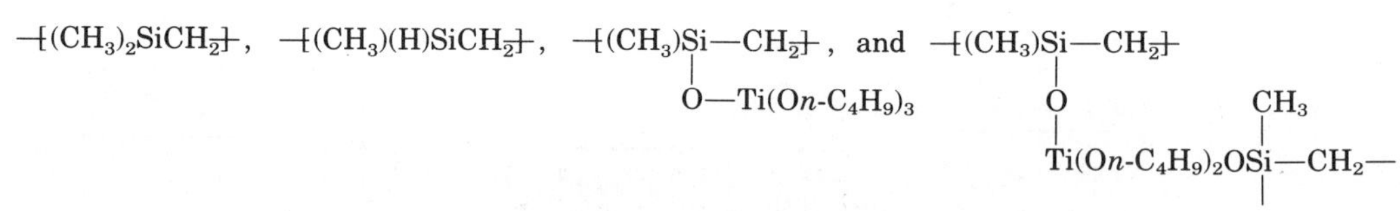

The mechanical properties of cured polytitanocarbosilane fibers improve with increasing pyrolysis temperature.

Ube Industries Ltd. (Japan) developed these fibers and has several patents claiming novel polycarbosilanes containing titanium, zirconium, or vanadium.

Thermolysis of Silane Monomers. A 1974 patent by Bayer AG describes the preparation of a polycarbosilane resin by passing an organosilicon compound, such as $(CH_3)_4Si$ or an alkyl–aryl chlorosilane, through a hot tube at ~700°C. The resin is processed into fibers, filaments, films, coatings, foams, and other forms, and pyrolyzed in an inert atmosphere between 800 and 2000°C to SiC-type ceramics. A linear high molecular weight organic polymer, such as polystyrene or poly(ethylene oxide), is used as a spinning aid at 1–2% to shape these brittle resins into fibers.

Potassium Dechlorination of Chlorosilanes. Linear polycarbosilanes yield little or no ceramic char when pyrolyzed at 1200°C; branching is critical for char yield.

$$-\!\!\!\{CH_2CH[Si(CH_3)_3]\}_m[Si(CH_3)_2]_n\!\!\!- \xrightarrow{1200°C} \text{trace SiC}$$

$$-\!\!\!\{CH_2CH[Si(CH_3)_3]\}_m[(CH_3)\overset{|}{Si}]_n\!\!\!- \xrightarrow{1200°C} \beta\text{-SiC } (\sim 41\%)$$

Several branched polycarbosilanes, as copolymers or terpolymers, have been synthesized by a one-step process of reaction of chloromethyl- or vinylsilanes with other methylchlorosilanes and potassium. Disilylation of the vinyl group is responsible for introducing branching, which converts $CH_2{=}CHSiCH_3Cl_2$ to a tetrafunctional reactant.

The SiC ceramics obtained from the potassium-derived branched polycarbosilanes are thermally unstable in air when crushed, but stable as monoliths. Tractable branched polycarbosilanes can be shaped as fibers and films and are useful as binders for ceramic powders.

Organopolysilanes. Current widespread interest in organopolysilanes is due to their use as ceramic precursors and their potential applications as photoinitiators and photoresist materials.

Sodium Dechlorination of Diorganodichlorosilanes. Organopolysilanes are prepared by alkali metal dechlorination of R_2SiCl_2:

$$(CH_3)_2SiCl_2 \xrightarrow[\text{toluene}]{\text{Na}} -\!\!\!\left[(CH_3)_2Si\right]\!\!\!-_n + 2\ NaCl$$

Long-chain polydimethylsilanes are insoluble, infusible, crystalline white solids. They must be first converted to polycarbosilanes which are precursors to Nicalon fibers.

Tractability is increased by introducing other groups besides methyl. About 50 new polysilane homopolymers and copolymers have been prepared with different groups on silicon including cycloalkane and unsaturated alkyl side groups.

Polysilastyrene is a colorless resinous material, soluble in benzene, toluene, ether, THF, and acetone, but insoluble in hexane and alcohols. With strong uv absorption near 330 nm, polysilastyrene is photoreactive and undergoes chain scission when its solution or solid samples are exposed in air to uv.

With recognition of the importance of cross-linking for ceramic fiber production and other applications, several new polysilanes with different functionalities have been synthesized. These functionalities undergo catalytic, thermal, oxidative, or photochemical cross-linking.

Methylphenylpolysilanes are used to strengthen Si_3N_4 ceramics. Branched polymethylsilanes, useful as binders of inorganic heat-resistant materials, are prepared by treating methylchlorodisilanes with a methyl Grignard to increase the methyl content and reduce the chlorine content; subsequent heating with an alkali metal gives polysilanes.

Redistribution of Methylchlorodisilanes. Polymethylchlorosilanes, MPS-Cl-T, are synthesized from a mixture of methylchlorodisilane (MCDS) (bp = 145–152°C) from the direct process. The MCDS typically consists of 55% $(CH_3Cl_2Si)_2$, 35% $(CH_3)_2ClSiSiCH_3Cl_2$, and 10% $[(CH_3)_2ClSi]_2$.

The structure of MPS-Cl-250 exhibits 2.65 Si—Si bonds per Si atom, indicating a high degree of branching. Based on chlorine determination, gc analysis of the distillate, mass-balance considerations, and molecular weight measurements (1000–1300 g/mol) on methyl derivatives, the polymer is believed to have a polycyclic cage structure containing about 20 silicon atoms:

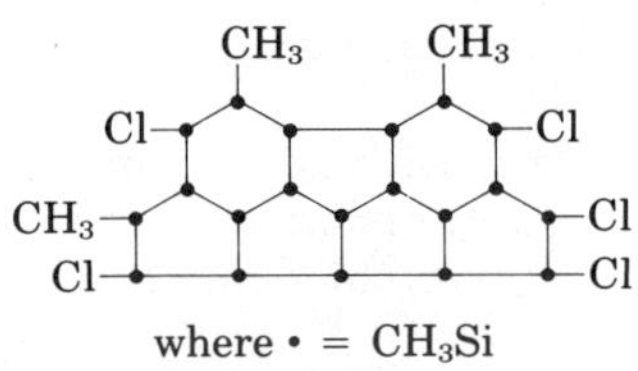

where • = CH_3Si

Handling. Great care must be taken in the handling, storage, and disposal of polymethylchlorosilanes, their derivatives, and their intermediates. They can be spontaneously ignited, particularly when in high surface area forms, such as fine fibers or powders. Inert conditions (<10 ppm water) are recommended for preserving the chemical identity and for safe handling of these materials.

Silicon–Nitrogen Ceramic Precursors

The preparation of stable precursors for silicon nitride is difficult. The average atom coordination in Si_3N_4 is 3.4; the coordination must be closer to 2.0 for a polymer to have the tractability required for processing the precursor polymers into complex ceramic shapes, including filaments. Monofunctional organic substituents, such as methyl groups, employed to lower the functionality, usually give carbon-containing ceramic residues upon pyrolysis. This leaves only hydrogen, halogen, or pseudohalogens as choices for substituents. Nitrogen, hydrogen, and halogen substituents undergo rapid redistribution reactions on silicon. For these reasons, only moderately stable silicon nitride precursors have been described.

Synthesis of these procedures includes ammonolysis of silicon tetrachloride and ammonolysis of dichlorosilane.

Silicon–Carbon–Nitrogen Ceramic Precursors

Silsesquiazanes and Triaminosilanes. ***Ammonolysis of RSiCl₃.*** Treatment of methyltrichlorosilane with ammonia in methylene chloride gives methylsilsesquiazanes.

$$CH_3SiCl_3 + 6\ NH_3 \xrightarrow{CH_2Cl_2} CH_3Si(NH)_{3/2} + 3\ NH_4Cl$$

The ammonium chloride is filtered and leaves a methylene chloride solution of methylsilsesquiazanes, which are presumably polycyclic moieties terminated with $SiNH_2$.

Aminolysis of CH_3SiCl_3 and $SiCl_4$. The earliest preceramic polymer research is described in a patent, which discloses treatment of methyltrichlorosilane with methylamine to form methyltris(methylamino)silane. Thermal rearrangement in a column heated to 520°C gave a polycarbosilazane.

$$CH_3SiCl_3 + 6\ CH_3NH_2 \longrightarrow CH_3Si(NHCH_3)_3 + 3\ CH_3NH_3Cl$$

$$CH_3Si(NHCH_3)_3 \xrightarrow{520°C} \text{polycarbosilazane}$$

Silazanolysis of CH_3SiCl_3 and $HSiCl_3$. Methylsilsesquiazane has a tendency to gel during ammonolysis of methyltrichlorosilane. This can be partly alleviated by reaction of SiCl (obtained from CH_3SiCl_3) with $[(CH_3)_3Si]_2NH$.

$$CH_3SiCl_3 + 3/2\ [(CH_3)_3Si]_2NH \longrightarrow CH_3Si(NH)_{3/2} + 3\ (CH_3)_3SiCl$$

Little or no NH_4Cl has to be removed, but some $(CH_3)_3SiNH$ groups are included in the polymer. Hexamethyldisilazane also reacts with mixtures of silanes such as CH_3SiCl_3 and $(CH_3)_2SiCl_2$.

Polysilazanes. ***Ammonolysis of CH_3HSiCl_2.*** A preceramic polymer can be prepared by ammonolysis of methyldichlorosilane, followed by polymerization of the resulting cyclic catalyzed by potassium hydride.

$$CH_3HSiCl_2 \xrightarrow{NH_3} (CH_3HSiNH)_{3-4} + NH_4Cl$$
$$\downarrow {-H_2} \; KH$$
$$(CH_3HSiNH)_a\,(CH_3SiN)_b$$

Aminolysis of H_2SiCl_2. Methylamine and dichlorosilane give *N*-methylpolysilazanes.

$$H_2SiCl_2 + CH_3NH_2 \longrightarrow CH_3NH{+}SiH_2N(CH_3){+}_nH + CH_3NH_3Cl$$

About 70% of the polymer consists of linear silazanes with CH_3NH endblocks and a number-average molecular weight M_n of 570.

Silazanolysis of Methylchlorodisilanes. Heating methylchlorodisilanes with hexamethyldisilazane gives methylpolydisilylazanes through Si—Cl/Si—N redistribution:

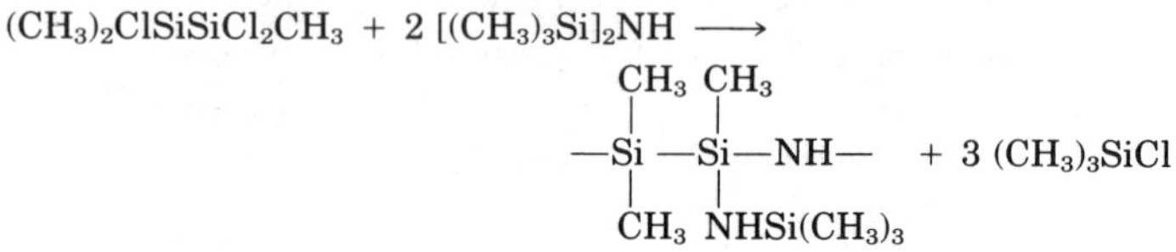

Chemical Vapor Pyrolysis. Silicon nitride precursors can be prepared from $[(CH_3)_2SiNH]_n$ and its isomer $(CH_3HSiNCH_3)_n$. Polymers are prepared from these cyclic compounds by catalytic reequilibration at 225–300°C. Complete conversion to ceramics occurs after heating to 700°C.

Plasma Pyrolysis. A glow-discharge polymerization has been described of hexamethyldisilazane in a mixture of gases consisting of ammonia, argon, nitrogen, and hydrogen.

Silicon–Carbon–Oxygen Ceramic Systems

Hydrolysis of Trialkoxysilanes. Sol–gel routes to ceramics are currently the subject of extensive research. This technology involves the hydrolysis of metal alkoxides (usually with the formation of inorganic aquasols), followed by gelation and ultimately thermolysis of the gels. Films, fiber, and monolithic articles have been prepared by this route. The advantage to ceramic technology is the lower temperature required to convert the precursor to dense ceramics. This allows applications and compositions that do not tolerate higher temperatures, such as ceramic fibers or electronic devices.

Hydrolysis of $RSi(OR')_3$ with Inorganic Sols. Methylsilsesquioxanes containing silica or alumina were prepared by hydrolysis of methyltrimethoxysilane in alumina or silica aquasols.

Other Precursors

The search for suitable low temperature polymer synthesis and fabrication routes to high technology ceramics of boron, aluminum, and a few transition metals is showing some promise.

Boron-containing systems include borosiloxane polymers, carborane–siloxane polymers, and boron–nitrogen polymers.

Aluminum-containing systems include organometallic aluminum precursors.

Systems Containing Transition Metals. Low temperature pyrolysis of organometallic complexes could provide an inexpensive fabrication method for stoichiometric transition-metal ceramics for electronic, structural, or refractory applications.

Though there are significant challenges to this technology, opportunities to use polymer processing for ceramics are only beginning to be tapped. Japanese researchers described the use of preceramic polymers to produce sinterable powders. Preceramic polymers may be used for ceramic coatings or as binders. Ceramic foams may supply high temperature insulating materials. Polymer forms such as fibers, foams, films, binders, and adhesives are well known to the polymer chemist, but are only now beginning to be utilized by ceramists through preceramic polymer technology.

BIBLIOGRAPHY

K. J. Wynne and R. W. Rice, *Ann. Rev. Mater. Sci.* **14,** 297 (1984).

S. Yajima, J. Hayashi, M. Omori, and K. Okamura, *Nature* **261,** 683 (1976).

R. H. Baney, *Chem. Tech.* **17,** 738 (1988).

RONALD BANEY
GRISH CHANDRA
Dow Corning Corporation

PRECIPITATION POLYMERIZATION.

See ACRYLONITRILE POLYMERS; RADICAL POLYMERIZATION; SOLUTION POLYMERIZATION.

PREFORMS.

See REINFORCED PLASTICS.

PREIMPREGNATED MATERIALS.

See REINFORCED PLASTICS.

PREPEGS.

See REINFORCED PLASTICS.

PREPOLYMER

A prepolymer is a partially polymerized substance, or one polymerized to a low degree of polymerization, for subsequent conversion to a high polymer (see also POLYMER).

PRESERVATIVES.

See BIOCIDES.

PRESSURE BAG MOLDING.

See REINFORCED PLASTICS.

PRESSURE-SENSITIVE ADHESIVES AND PRODUCTS

Pressure-sensitive adhesives (PSAs) are used mainly for adhesive tapes and labels. Lesser but nonetheless important applications are for medical products, protective masking sheets, and specialty products.

The term pressure-sensitive designates adhesives "which in dry form are aggressively and permanently tacky at room temperature and firmly adhere to a variety of dissimilar surfaces upon mere contact without the need of more than finger or hand pressure". Unlike early materials such as pitches and gums, and unlike self-bonding cements or even sticky substances like fly paper, a PSA requires no activation by solvents or heat and exerts a strong adhesive force toward a variety of surfaces, including paper, glass, wood, metals, plastics, skin, and others. Moreover, according to the Pressure Sensitive Tape Council (PSTC) "they have a sufficiently cohesive holding and elastic nature so that, despite their aggressive tackiness, they can be handled with the fingers and removed from smooth surfaces without leaving a residue".

Construction and Components

Pressure-sensitive products consist of a flexible backing coated with an adhesive; many require a release liner or coating for ease of delivery and application. By definition, the release surface confers low adhesion to the PSA. A typical construction of a PSA tape is shown in Figure 1. Label construction is similar, but always includes a release liner. The distinction between labels and decals is not clear, although the latter connotes display or advertising.

In the manufacture of tapes, the adhesive is coated on a continuously moving web. Frequently, the adhesive is coated on a release liner, paper, or film with an adherent coating, which is then laminated, using pressure applied by roller nips, to a backing to which it has high adhesion. For the manufacture of labels and many medical products, the release liner is never removed but becomes an integral part of the final package, whether in roll or sheet form. The adhesive usually is applied directly to the backing of tapes.

Adhesives are supplied to the coating operations as solutions in organic solvents, aqueous emulsions, or 100% solids. An emerging technology, not yet of industrial importance, is radiation polymerization of adhesive components on a moving web. Radiation cross-linking of preformed polymers is of minor importance.

Backings. The choice of backing material is governed by the PSA application.

The most widely used backing in both tapes and labels is paper. It constitutes perhaps 80% of the backing market for labels. Second in volume is oriented polypropylene, used mainly for carton and box sealing. Other widely used backings are glass filament-reinforced films and papers, polyethylene-coated cloth, acetate film (which is now the principal backing for the ubiquitous cellophane tape), poly(ethylene terephthalate), poly(vinyl chloride), and aluminum foil. Specialty backings in flexible-film form include foams, fabric-reinforced epoxies, nylons, polyurethanes, and polyethylene.

Release Coatings. Release coatings are of great importance for pressure-sensitive products. Tapes in roll form usually carry release coatings on their reverse sides or a release-coated interliner. The release treatment on self-wound tapes may use silicones, alkyds, stearyl derivatives of vinyl polymers, stearate chromic chloride, or agents like stearamides. All labels and some specialty products (ostomy devices, medical devices, etc), and double-faced tapes require a release-coated sheet to permit delivery. Release liners, which are based on paper or films, are almost always silicone-coated.

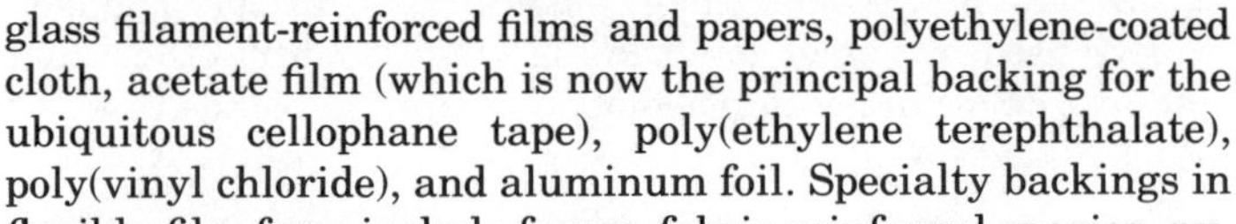

Figure 1. Laminate construction of pressure-sensitive tape.

Tackifiers. All rubber-based adhesives require resin tackifiers as a main component. These materials impart tack or stickiness. The tackifier content of rubber-based adhesives is usually 60–110 pph elastomer. Acrylic PSAs utilize tackifiers in smaller amounts.

By far the largest volume tackifiers are the hydrocarbon resins obtained by oligomerization of the C_5 and C_9 fractions of petroleum. The C_5 fraction tackifiers, including hydrogenated fractions, are most compatible with PSA base polymers.

Other tackifiers of significance are rosin esters, coumarone–indene resins and copolymers of α-methylstyrene and vinyltoluene, which resemble the C_9 hydrocarbon tackifiers (see HYDROCARBON RESINS).

Raw Materials

Natural Rubber. Natural rubber (NR), derived from the latex of the *Hevea brasiliensis* tree, is all *cis*-1,4-polyisoprene (see RUBBER, NATURAL). Blending NR with tackifiers produces high quality PSAs, the properties of which are determined by the tackifier. A mixture of tackifiers often gives better results than a single resin. Fillers, such as clays, modify properties and reduce costs. Antioxidants are included to protect the unsaturated backbone polymer from degradation. Plasticizers, such as petroleum oils or low molecular weight polyisobutylene, are added to many formulations.

Block Copolymers. Triblock copolymers of the form A–B–A, where A is polystyrene (S) and B is a polydiene, have recently surpassed NR as the most widely used hydrocarbon rubber in PSAs. These polymers are trademarked as Kraton by Shell, the only domestic producer. S–I–S triblock copolymers behave as thermoplastic elastomers with a low melt viscosity; they are well suited for hot-melt adhesives. These characteristics permitted development of 100%-solids PSA formulations. At present, hot-melt PSAs, which make up about 20% of the market, are almost entirely based on these block copolymers.

A significant portion of block copolymer is used in solution adhesives, primarily for paper-backed tapes.

Typical styrene–isoprene–styrene block copolymer PSA formulations are given in Table 1. (Note that end-block tackifiers are sometimes omitted.)

Styrene–Butadiene Rubber. The styrene–butadiene random copolymer used for adhesives differs from the typical elastomer SBR. Because of the higher polymerization temperature (50–60°C), the PSA has a broader molecular weight distribution and often a gel fraction, both of which are desirable.

Solution adhesives using gum SBR are more widely used than SBR latex.

Polyisobutylene. The polymerization of isobutylene yields polyisobutylene (PIB) in a broad range of molecular weights. The vulcanizable elastomer polymerized with 0.8–2% isoprene is called butyl rubber. In many applications, PIB and butyl rubber are interchangeable (see ISOBUTYLENE POLYMERS).

Butyl rubber is used in pipe-wrap adhesives, which are calendered onto PVC or polyethylene backings. A wide variety of tackifiers are used with butyl. The principal application of PIB adhesives, usually as solutions in petroleum fractions (naph-

Table 1. PSAs Based on Styrenic Block Copolymers

Applications and ingredients	Parts by weight
Hot-melt formulations	
S–I–S Kraton 1107	100
hydrocarbon resin, 95°C[a]	100
coumarone–indene resin, 155°C[a]	60
naphthenic oil	40
antioxidant	5
S–I–S block copolymer	100
hydrocarbon resin, 95°C[a]	140
naphthenic oil	40
antioxidant	2
Solution formulation	
S–I–S block copolymer	100
polyterpene, 105°C[a]	100
α-methylstyrene polymer	20
antioxidant	1
hexane–toluene, 70:30	200

[a] Softening or melting point.

Table 2. Typical Acrylic PSA Formulations

Applications and ingredients	Parts by weight
General purpose	
2-ethylhexyl acrylate	97
acrylic acid	3
2-ethylhexyl acrylate	68
vinyl acetate	30
acrylamide	2
Cross-linkable	
2-ethylhexyl acrylate	89.5
dimethylaminoethyl methacrylate	7.0
acrylic acid	3.0
glycidyl methacrylate	0.5
2-ethylhexyl acrylate	78
vinyl acetate	20
maleic anhydride	2
3-methacryloxypropyltrimethoxysilane	0.02
Skin adhesive	
2-ethylhexyl acrylate	50
vinyl acetate	30
N-t-butylacrylamide	20
Conformable, high tack	
isooctyl acrylate	96
acrylic acid	4
hydrogenated rosin ester, 85°C[a]	50
Water soluble	
butyl acrylate	65
hydroxyethyl methacrylate	15
acrylic acid, neutralized with aminopropanol	20

[a] Softening point.

thas), is for removable labels where low peel adhesion is required; tackifiers are C_5 hydrocarbon resins and polyterpenes.

Reclaim Rubber. Reclaim rubber is obtained in various ways from used rubber, mostly tires, primarily by digestion at elevated temperatures. For PSAs it is normally blended with virgin elastomers. Reclaim rubber, particularly butyl, is used in pipe-wrap tapes, and duct and friction tapes (see RUBBER RECLAIMING).

Ethylene–Vinyl Acetate (EVA) Copolymers. Highly tackified EVA-based adhesives blended with petroleum wax are used extensively in the hot-melt adhesive industry, but very little in the PSA market because of poor high temperature shear and low tack.

Acrylics. Currently, acrylics (Table 2) account for about one-quarter of the domestic PSA market (see ACRYLIC AND METHACRYLIC ESTER POLYMERS).

Butyl and 2-ethylhexyl esters of acrylic acid are the dominant raw materials. The 3M Company, the largest tape producer in the United States, manufactures the ester of isooctyl alcohol (from the oxo process) for captive consumption. Other acrylate esters give rise to the required low T_g homopolymers, but they are more expensive.

Acrylics are used for a multitude of tapes, especially transparent, strapping, transfer, medical, and metal-foil tapes. They are also widely used for paper and film-backed labels and decals.

Vinyl Acetate Copolymers. Poly(vinyl acetate) has a glass-transition temperature that is too high for PSA use, but it can be copolymerized to yield products competitive with acrylics. These copolymers are available only as emulsions and made by Air Products; they have T_gs close to −30°C, are uv-resistant, and can be cross-linked readily.

The principal use, although still small, is in permanent labels.

Silicone Polymers. Silicone polymers are used in sealants as well as PSAs. The materials for their preparation are supplied by General Electric and Dow Corning. The high price of the adhesives, about $22/kg, limits their use to applications in which their remarkable use temperature range, from −100 to 250°C, is important. Silicone PSA adhesives are derived from an MQ resin and a gum.

Solution adhesives are used primarily in tapes. Polyester film is the preferred backing, although polytetrafluoroethylene and polyimide films are also used.

Poly(vinyl alkyl ether)s. The vinyl alkyl ether polymers, although little used in the United States, are unusual in that poly(vinyl ethyl ether) has a water permeability similar to that of skin and hence finds use in medical applications. The most widely used monomer for PSAs is vinyl isobutyl ether; high and low molecular weight homopolymers are blended (see VINYL ETHER POLYMERS).

Physical Properties

Unlike structural adhesives, which change from a liquid to a solid state by loss of solvent or because of chemical reaction, PSAs do not undergo a phase change from the initial stage of adhesion upon wetting the surface to the final rupture of the adhesive bond.

A balance of cohesive strength and viscoelastic properties is required, allowing the PSA to spread over a surface with application of mininum pressure and yet be removable from that surface without leaving an adhesive residue. The characteristics of these adhesives are based on three parameters: tack, or the ability to wet surfaces instantaneously; adhesion, or the ability to remain permanently attached to surfaces in the absence of excessive forces; and cohesive strength, sometimes called shear resistance, or the ability to resist flow or creep under an applied load.

Test Methods

Pressure-sensitive adhesives are characterized in terms of their ability to wet out on surfaces, resistance to peel from an attached surface, and response to shear forces. Accordingly, test procedures are designed to measure tack, peel adhesion, and shear resistance.

Not all PSAs require high values in all three parameters. For example, low peel adhesion may be necessary for permanent labels and low tack for repositionable tapes. Nevertheless, PSAs are described in terms of values for these parameters. In the United States, standardized methods have been designed by the Pressure Sensitive Tape Council (PSTC) and the American Society for Testing and Materials (ASTM). Similar standardized procedures have been developed in other countries.

Economic Aspects

In 1985 the estimated volume of pressure-sensitive articles in the United States was over 2.5×10^9, based on current prices. Tapes account for about 60%, labels and decals for about 30%. In the same year, over $2.5 \times 10^9\,m^2$ of tapes, about half of which were single-faced, and over one billion square meters of labels and decals were produced in the United States. Furthermore, ca 7×10^6 kg of release coatings were consumed by the pressure-sensitive tape industry.

BIBLIOGRAPHY

D. Satas, ed., *Handbook of Pressure-Sensitive Adhesive Technology*, Van Nostrand Reinhold, New York, 1982.

M. A. Krenceski, J. F. Johnson, and S. C. Temin, *J. Macromol. Sci. Rev. Macromol. Chem. Physics* **C26(1)**, 143 (1986).

C. A. Dahlquist in R. L. Patrick, ed., *Treatise on Adhesion and Adhesives,* Vol. 2, Marcel Dekker, New York, 1969, Chapt. 5.

M. Toyama and T. Ito, *Polym. Plast. Tech. Eng.* **2(2)**, 161 (1973).

SAMUEL C. TEMIN
Consultant

PRICE-ALFREY EQUATION. See COPOLYMERIZATION.

PRINTING. See DECORATING; PRINTING-INK VEHICLES.

PRINTING-INK VEHICLES

Printing Processes and Equipment

A prerequisite for understanding printing ink and the polymers used in ink construction is a knowledge of the basic printing processes, which can be classified into six basic systems (Fig. 1): letterpress and flexography (typographical, relief, and aniline); gravure (intaglio); lithography (chemical); screen (stencil); electrostatic; and jet printing (nonimpact).

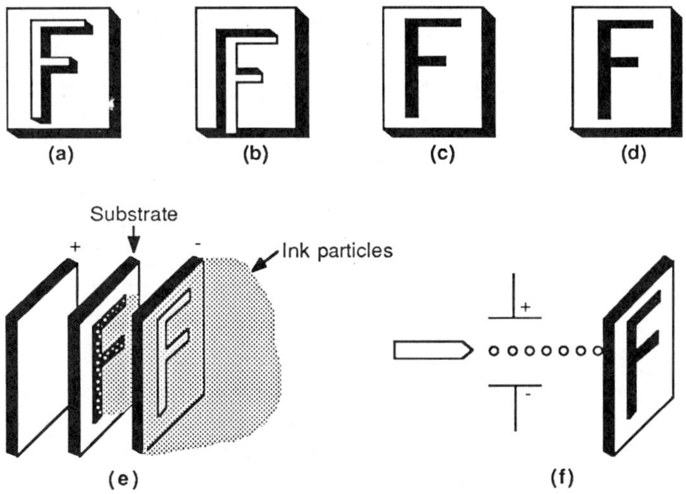

Figure 1. The principles of the printing processes: (**a**) letter-press (relief); (**b**) gravure (intaglio); (**c**) offset lithography (planographic); (**d**) screen (stencil); (**e**) electrostatic (reprography); (**f**) jet printing (ink spray).

Ink Drying

The various methods by which inks are dried in the printing processes described above are shown in Table 1. The different vehicles used in these drying methods are given in Table 2.

Vehicles and Formulations

Printing-ink vehicles are chosen according to cost and availability; pigment-dispersing efficiency; end-product requirements (gloss, chemical resistance, etc); adaptation to printing processes; adaptation to substrate (coated or uncoated paper plastic, metal, etc); press speed; drying system (Table 2); and environmental concerns. The following characteristics are inherent in their construction and are essentially unchanged by the printing process: The monomer or building block should be a bulky molecule or of relatively high molecular weight. The finished resin should be of low molecular weight (1000–3000) and soluble at high solids concentrations to give low viscosity solutions. The resin should be a balanced system of hydrophilic and hydrophobic sections which impart surfactant-like properties. The functions of the vehicle are dispersing the pigment, carrying it to the substrate (printing), binding to the substrate, and providing for finished print properties.

Letterpress and Lithographic Inks. ***Vehicles.*** Among the drying oils, linseed oil is used in the largest volume, whereas soybean, safflower, and China wood oil are for special applications. Drying oils are suitable binders for printing ink since they are available at 100% solids and convert to tough, yet flexible, films. Their pigment-dispersing properties are excellent although they dry slowly. Since slow-drying oil film is not adaptable to high speed printing equipment, their use in the printing-ink market has declined with the shift away from letterpress printing.

Oleoresinous Systems. Drying oils can be upgraded by the addition of modified rosins to improve gloss, drying speed, water resistance, chemical resistance, and rub resistance. Rosin esters modified with phenol–formaldehyde resins are widely employed in conjunction with linseed and China wood oil to produce gloss oleoresinous varnishes.

Table 1. Methods of Drying Inks Used in Printing Processes

Ink-drying method	Letter-press	Flexo-graphic	Litho-graphic	Gravure	Silk screen	Electro-static
oxidation	X		X		X	
evaporation—low temp (70–80°C)		X		X	X	
heat set—high temp (120–130°C)	X		X			
moisture set	X					
polymerization			X		X	
absorption	X		X			
fusion	X		X			X
filtration	X		X			

Gloss varnishes are also made from maleic-modified rosin esters and drying oils. The Diels-Alder reaction of 5–20% maleic anhydride (or fumaric acid) with rosin is well known and is of commercial importance.

Hydrocarbon Resins. Hydrocarbon resins (qv), derived from dicyclopentadiene-rich refinery streams, treated with drying oils give dark, inexpensive vehicles which are useful in heat-set inks as gloss varnishes.

Hydrocarbon resins are also used in varnishes for heat-set and sheet-fed processes. These products are hard, brittle materials with number average molecular weights between 400 and 1500. They are formed by polymerization of unsaturated by-product feedstreams of ethylene and propylene manufacture derived from the cracking of naphtha and gas oil fractions.

Alkyd Resins. Alkyd resins, particularly those with more than 65% oil content, are used extensively in letterpress and offset printing inks. They are normally used at 75–100% solids and are superior to bodied linseed oil or oleoresinous varnishes. In letterpress and offset printing, alkyd resins improve rub resistance without decreasing pigment dispersibility and other essential properties. Alkyd resins usually contain 5–20% isophthalic or phthalic constituents and are used principally in air-drying compositions for offset and letterpress printing. Isophthalic alkyd resins (qv) are more resistant to water and harder drying, ie, give tougher films.

Resin–Solvent Systems. In rotary heat-set web offset and letterpress printing, the ink is dried by solvent evaporation on heating. Heat is provided by a flame, a steam-heated drying drum, or heated air. Vehicular systems for these processes consist of high melting modified rosin esters or hydrocarbon resins dissolved in high boiling petroleum solvents.

Radiation-curable Resins. Radiation curing, both uv and electron beam, affords high quality prints and coatings on packaging, magazine covers, and other applications. These essentially 100% solids systems provide a gloss and print quality difficult to attain with other systems. In addition, the systems satisfy the most stringent antipollution regulations. The vehicles for these inks and coatings are usually acrylate or methacrylate functional oligomers, polyfunctional acrylate monomers, photoinitiators, stabilizers, pigments, and waxes. The oligomers most commonly used are the acrylate and methacrylate derivatives of saturated polyesters, aromatic, aliphatic, and cycloaliphatic epoxy resins, and aromatic and aliphatic polyether polyols. Film properties such as toughness, adhesion, gloss, and flexibility depend on the oligomer.

Flexographic and Gravure Inks. ***Vehicles.*** Flexographic and gravure vehicles have many similarities and can be discussed as a single class. Both employ volatile solvents and are dried by forced-air, solvent evaporation, and absorption. Solvent release of the vehicular system is important since both processes utilize high speed roll-to-roll printing. The polymers employed are of higher molecular weight than those in letterpress and offset systems. Vehicles for flexography must be soluble or dispersible in lower alcohols or water; only limited quantities of hydrocarbon solvents and esters can be used because of the sensitivity of the rubber plate.

Nitrocellulose. Nitrocellulose is widely used in both flexo-

Table 2. Classification of Ink Vehicles by Drying Method

Drying method	Ink vehicles
oxidation	linseed, tung and hydrocarbon oils, oleoresinous products with maleic, phenolic, and modified phenolic resins, rosin esters, alkyd resins with 10–20% isophthalate and *o*-phthalate, and soya or linseed oil
heat set	rosin soaps, rosin esters, modified phenolic resins, maleic resins, natural resins, hydrocarbon resins
moisture set	maleic and fumaric modified rosins, shellac, alkyds
polymerization	special acrylics, alkyds and polyester resins, alkylated urea and melamine–formaldehyde resins, oleoresinous varnishes, multifunctional acrylates, epoxies, and epoxy acrylates
evaporation	shellac, rosin soaps and esters, chlorinated rubber, cyclized rubber, gilsonite, polyamides, polyamide–imides, acrylic resins, vinyl resins, cellulose derivatives, ketone resins, polystyrenes, styrene–maleics, poly(vinyl butyral), polyurethanes
absorption	petroleum oils, nondrying vegetable oils, water dispersions
fusion	plastisols, poly(vinyl chloride) and polystyrene
filtration	cyclized rubber, gelled oils

graphic and gravure inks; it wets pigments effectively and dissolves easily in alcohol and ester combinations. The grades with lower nitrogen content (10.5–11.2%) are more soluble in alcohols and are used in flexographic inks. The grades with medium nitrogen content (11.2–11.8%) have good solubilities in ester–alcohol combinations and can be used in both flexographic and gravure ink. The high nitrogen content grades (11.8–12.2%) are soluble in ester–hydrocarbon mixtures and are usually restricted to gravure. A wide range of viscosities is available in each class. The most widely used viscosities for printing ink are ½ s, ¼ s (falling-ball method), 30–35 mPa·s (= cP) and 18–25 mPa·s (= cP). Because of its high melting point, nitrocellulose can be employed in heat-resistant compositions, which are of particular importance in flexible packaging materials. Nitrocellulose has a sufficiently high molecular weight to accept liquid plasticizers and produce a highly elastic, tack-free film. It is used extensively in flexographic inks for paper, cellulosics, and other plastic films and in gravure inks for paper, board, and flexible-film packaging. Other resins, eg, maleic-modified rosin esters, polyamides, and acrylic resins, are frequently blended in for special applications.

Plasticized nitrocellulose-based inks have excellent rub resistance when dry, even in the presence of high humidity. In addition, scratch and grease resistance are good, and printability is excellent (see CELLULOSE ESTERS, INORGANIC). Materials used in these inks include ethylcellulose and ethylhydroxyethylcellulose (EHEC), maleic-modified rosin esters, rosin soaps, shellac, polyamides, acrylic and methacrylic ester resins, chlorinated rubber, hydrocarbon resins, and modifying resins (eg, polyurethanes, polyamide-imides, vinyl resins, ketone resins, and polystyrene and copolymers.)

Screen Printing Inks. ***Vehicles.*** Screen printing utilizes a large variety of polymers, depending on the end use. In the decoration of plastic containers, for example, adhesion to the substrate is of utmost importance. Similarly, outdoor plastic signs require inks of maximum durability. The screen-printing process provides thicker ink films than letterpress or offset; hence, thorough drying is essential.

Oleoresinous esters and alkyd resins are widely used in screen inks and are very similar to paint vehicles of the same class; the solvent is a mineral spirit similar to that used for paints.

For the screen printing of polyethylene bottles, long oil alkyds have been replaced by epoxy–ester oil resins which afford superior chemical resistance after drying in the presence of suitable catalysts. Recent developments include two-part polyisocyanate inks and one-part systems containing blocked acid catalysts.

The printing of textiles constitutes the largest single use of screen inks with most of the work done on the finished products, eg, clothing, hats, shirts, etc. In addition, toweling and sheeting fabrics are screen printed, producing a deep, rich array of colors. Since flexibility and washability are required, the vehicles are cross-linkable acrylic copolymers in aqueous dispersions, and heat-curable plastisol systems based on PVC homopolymer dispersed in liquid plasticizers.

Screen inks are also used in the manufacture of printed circuit boards, requiring primary image inks to prepare circuitry; solder masks to protect the copper and tin–lead circuitry; and legend inks to indicate component types.

BIBLIOGRAPHY

D. E. Bisset and co-workers, *The Printing Ink Manual,* 3rd ed., Northwood Books, London, 1979.

Printing Ink Handbook, 4th ed., National Association of Printing Ink Manufacturers, 1980.

J. W. Vanderhoff, "Chemistry and Technology of Printing Inks" in J. K. Craver and R. W. Tess, eds., *Applied Polymer Science,* American Chemical Society, 1975, pp. 493–511.

S. G. Wentenk and S. D. Koch, eds., *Practical Guide to UV Curing in Screen Printing for Printed Circuits and the Graphic Arts,* Technology Marketing, Norwalk, Conn., 1981.

THEODORE SULZBERG
ALBERT KVEGLIS
Sun Chemical Corporation

PROCESSING, COMPUTER CONTROL

Many devices are available for process control, such as sensors, actuators, or computers, which can be used by the plastics industry. These devices can be connected with the automation apparatus and integrated into a procedure. Automation improves process efficiency and product quality, and reduces costs.

Early analogue systems have been replaced by digital automatic controls. The applications range from front-end computers which serve a single control circuit to process guidance systems which coordinate a complete operation. Digital systems offer the advantage of flexible parameterization and connection of devices over the exchange of digital data.

Computer-controlled Processes

Connections Between the Process and the Computer. The process and its control system can only function with the help of appropriate instrumentation for the exchange of data. The information gained must be recorded by sensors and translated into a form that the computer can process. Connections must be made for the reverse flow of information in order for the process to receive instruction from the computer. Numerous sensors are used to record the process conditions; process variables are given in Table 1.

Table 1. Process Variables and Instrumentation

Process variables	Instrumentation
temperature	thermoelements resistance elements, PT 100 radiation recorder
pressure	strain gauges resistance elements
rotational speed	a-c/d-c-generators incremental counters
weight	strain gauges Piezo elements
velocity	a-c/d-c-generators incremental counters correlators
torque	strain gauges scales
time	impulse counters
logic state (on/off)	switches

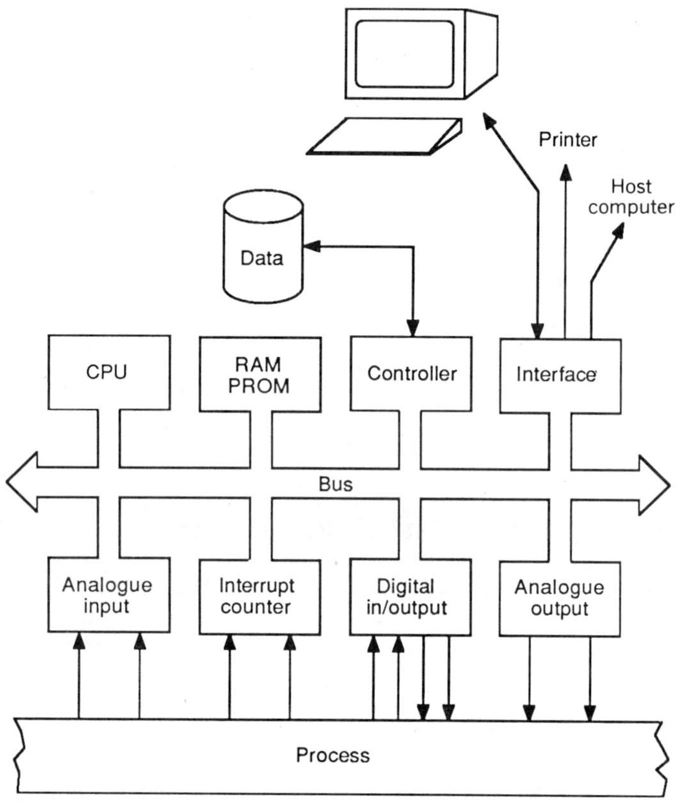

Figure 1. Flow sheet of a process computer.

Computer Hardware. The flow sheet of a process computer is shown in Figure 1.

Computer Software. The software establishes the predetermined function procedure. There are differences between the operating system software that controls the start-up phase, administers all periphery units, and guides the user programs. All periphery-specific activities are planned in the user programs.

Rules for Digital Process Data Processing

Digital Signal Processing. The digital working method of a computer requires a basic knowledge of the theory of digital signal processing. Both the transfer from analogue to digital signals and their return conversion, as well as the algorithmic processing of the digital information, takes place according to certain mathematical rules.

As illustrated in Figure 2, the analogue entry signal, which moves over the measuring amplifier to the analogue entry card, passes through a low-pass filter and a sample-hold circuit before it is translated into data by the analogue-digital converter. The design of these three transfer sections is crucial for the presentation precision and the data rate.

Discrete Signals and Processes. The algorithmic treatment of signals and the description of digital systems, such as process, filters, or controllers, require a brief introduction to the mathematical principles of digital signal processing.

A sequence is produced from a continuous signal through the sampling, which, over the Dirac function

$$\delta(t - kT) = \begin{matrix} 1 \text{ for } t = kT \\ 0 \text{ for } t \neq kT \end{matrix} \tag{1}$$

can be expressed by

$$f^*(t) = f(t) \sum_{k=0}^{\infty} \delta(t - kT)$$
$$= \sum_{k=0}^{\infty} f(kT)\, \delta(t - kT) \tag{2}$$

In a similar way, this expression can be translated to the continuous signals by the Laplace transformation in the frequency range

$$F^*(s) = L[f^*(t)] = \sum_{k=0}^{\infty} f(kT) \exp(-kTs) \tag{3}$$

Control Engineering

Models of Real Processes. A model-type description of the real process relationships is required for the understanding of process control of optimal process operation. The most common form of process presentation is a linear, time-invariable model (LTI), as described in equation 4. This allows a sufficiently accurate mathematical formulation of the physical process and a numerical solution of the control problem. The greatest difficulty with modeling a technical process is with the degree of generality and the clear reduction of the actual (usually complex) system.

$$G(z) = \frac{y(z)}{u(z)} = \frac{b_0 + b_1 z^{-1} + b_2 z^{-2} + \cdots + b_n z^{-n}}{1 + a_1 z^{-1} + a_2 z^{-2} + \cdots + a_n z^{-n}} \tag{4}$$

The description of a process as a LTI system is usually only valid for small signal changes, which are related to a fixed operational point. In the technical process, operational point alterations, which encompass the complete operational field, must also be controlled. For this reason, the nonlinear behavior is frequently considered; equation 4, enlarged by such a func-

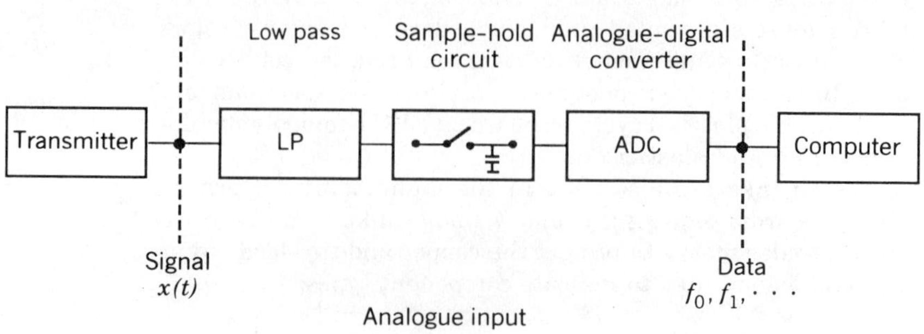

Figure 2. Signal-to-data conversion of an analogue input.

tion, would read as follows:

$$G_p = FG_p(z) \quad (5)$$

After the model structure is established, the model parameters are indentified, that is, they are estimated in line with the smallest square-root error according to a regression process.

Controlled Processes. The simplest form of process control comprises a control unit. The manipulated value is given over the required value of the control, which is necessary in order to achieve a desired real value.

Controller Structures. The most important control feature is that information is available in the form of a measuring signal and compared with the required value. A controller that is preswitched to the process evaluates the control deviations, which are corrected as needed through a changed required value of the working point of the process. The compensation of disturbance influences that affect the process is a distinct advantage.

Controller Optimization. After the design of a control and the selection of controller types, the control parameters are developed. This can only take place with reference to the process characteristics. The transfer behavior of a circuit must be examined and described. However, only in few cases can the process action be described from physical statements with sufficient accuracy. Therefore, experimental methods are used, which produce easily adjusted controllers. Various experiments have led to similar parameter values.

Adaptive Controllers. The transition from the analogue to the digital technique has increased the flexibility of controller design. Adaptive controllers have been developed, which allow optimization of control parameters independently of the process. The adaptation consists of the identification of the process parameters and their adjustment to better known processes.

Process Control for Typical Problems in Plastics Processing

Computer-controlled process automation can be applied to any process engineering problem, including typical tasks in plastics processing, eg, injection molding machine, pressure control, dependence, dead-time control on extrusion lines, and adaptive control for haul-off control.

Process Control Using a Process Model. Automation can be installed if the process is easy to survey and can be described by experimental data. Clear physical relationships can seldom be set up in plastics processing because the flow processes are coupled with thermal and mechanical models. A valid description is usually only achieved by experiments. The important process parameters are changed one after another to the limits of the working point. The results are summarized by mathematical formulas in the form of linear, logarithmic, and power functions. If a physical formula can be assumed, eg, an Arrhenius function, it can be included. Since these experiments are mainly related to stationary working points, the function F is described by equation 5. The dynamic behavior is described separately if applicable. For reliable automation the process should be invariant or known in its invariance described by the model.

BIBLIOGRAPHY

A. W. M. van den Enden and L. A. M. Verhoeckz, *Philips Technol. Rev.* **42(4)**, 110 (Dec. 1985).

N. Ahmad and T. Natarajin, *Discrete-time Signals and Systems,* Reston Publishing Co., Reston, Va., 1983.

H. Unbehauen, *Regelungstechnik I,* Vieweg Verlag, Braunschweig, Wiesbaden, FRG, 1985.

G. BOLDER
Institut fur Kunststoffverarbeitung of the RWTH Aachen

PROCESSING, MODELING

A model is a quantitative abstraction of a physical process, in which the description of the process is represented by the solution to a set of mathematical equations. The model equations represent the behavior of the real process to the extent that the equations embody an accurate description of the underlying physical and physicochemical phenomena. The mathematical formulation enables the model to be used for a variety of purposes, including design, control, and exploration of operating strategies; the effect of changes in process variables can be inferred from the model response without extensive experimentation. Mathematical models are widely used for these purposes in the chemical and petrochemical industries, but less so in polymer processing. Computer-aided design and computer-aided manufacturing (CAD/CAM) are now possible in some polymer processing operations.

The essential elements of any model of a physical process are threefold: the geometry, the relevant laws of physical conservation (momentum, mass, and energy), and the specific constitutive relations. In polymer processing applications, the last of these could include the stress-deformation equations for the melt or solution (the material rheology) and the kinetic equations for phase change.

Rheology

Models of the mechanics of polymer processing operations can be categorized by the rheological description used. Low molecular weight liquids are Newtonian, which means that the stress at any point in a deformation field is a linear function of the instantaneous rate of strain; the stress in an incompressible liquid is characterized by a single material parameter, the viscosity (η), which can be temperature- and pressure-dependent but is independent of the deformation rate:

$$\sigma = -pI + \tau \quad (1)$$

$$\tau = \eta[\nabla v + (\nabla v)\dagger] \quad (2)$$

Here, σ is the total stress, p the isotropic pressure, I the identity (unit) tensor or matrix, and τ the extra stress (ie, the stress in excess of the isotropic pressure). ∇ is the gradient differential operator, and v the velocity; $\dagger$ denotes the transpose of a tensor or matrix.

In contrast, the stress in a macromolecular liquid depends in a nonlinear manner on the entire history of deformation. The nonlinearity is manifested by phenomena such as the strain-rate dependence of the viscosity. The dependence on history is demonstrated by the transient buildup and relaxation of stresses following changes in flow variables, and by the existence of the dynamic storage modulus in oscillatory shearing. There have been many routes to the development of appropriate stress-constitutive equations for use in processing flows of polymeric liquids, including general continuum mechanics and molecular and quasimolecular theories. No constitutive theory developed to date is adequate in all processing flows, but some stress equations have been useful in particular classes of flows.

Most modeling that has attempted to account for the strain-history dependence of macromolecular liquids has utilized some version of a transient-network equation, in which the equation for the extra stress τ is of the general form

$$A(\tau/G) \cdot \frac{\tau}{G} + \lambda \frac{\mathcal{D}(\tau/G)}{\mathcal{D}t} = \lambda[\nabla v + (\nabla v)^{\dagger}] \quad (3)$$

$A(\tau/G)$ is a nonlinear function of the invariants of the extra stress; λ is the time constant for stress relaxation, and G the shear modulus, which is a material property; the viscosity, η, is equal to the product λG. $\mathcal{D}/\mathcal{D}t$ denotes a nonlinear time derivative which accounts for the invariance of the physical properties under changes of the frame of reference of the observer; $\mathcal{D}/\mathcal{D}t$ contains terms of the form $\tau \cdot \nabla v$, but the precise form depends on the particular theory.

If the spectrum of relaxation times characterizing the linear viscoelastic response is to be included, the stress is comprised of a sum of terms, where each term in the sum satisfies an equation of the form of equation 3.

Classification

Classifications of viscoelastic flow are useful for analysis, mainly in determining conditions under which a viscoelastic constitutive equation that accounts for fluid memory can be replaced by a much simpler purely viscous equation.

The broadest classification that has proved to be useful in modeling is based on the Deborah number, which is loosely defined as the ratio of the fluid time scale to that of the process. Small Deborah numbers correspond to situations in which stress transients occur rapidly compared with the total processing time, and hence the "memory" of the fluid is short and can be neglected. Large Deborah number flows take place over a time scale in which stress transients cannot be neglected, and a viscoelastic description like that in equation 3 is essential.

The concept of strong flows has been introduced as one attempt to quantify the Deborah number concept unambiguously. This classification is based on the notion that shearing flows are an ineffective (weak) means of stretching polymer molecules, whereas extensional flows can be effective (strong) in elongating molecules to nearly maximum extensibility.

Flows must also be classified on the basis of whether the melt is isotropic or anisotropic at rest, and whether the material is homogeneous or heterogeneous.

Numerical Methods

The equations describing polymer processing operations are usually coupled, nonlinear partial-differential equations in which two or three spatial directions and perhaps time appear as independent variables. Fully three-dimensional problems can rarely be solved, and substantial simplification is usually required even for two-dimensional problems because of limitations of computer speed and memory. In some situations the geometry provides simplifications that lead to closed-form analytical solutions, but these are rare and numerical methods are usually required in order to obtain process information. The three following broad classes of numerical methods are in common use: finite difference methods, finite element methods, and boundary element methods.

Continuous Low Deborah Number Processes

The most successful process modeling has been carried out on continuous, low Deborah number processes. These include the metering region in melt extrusion, calendering, and many coating operations. The fluid deformation in these flow fields is closely approximated by combined drag and pressure-driven flow between nearly parallel surfaces, and the lubrication approximation, first developed by Reynolds for the analysis of lubrication flows, is applicable as a first approximation.

Periodic Low Deborah Number Processes

Mold filling, which is a repetitive transient process, can often be treated as a low Deborah number process when the fill time is long relative to a characteristic time for stress relaxation, so that only purely viscous rheology need be considered. The fluids are usually of very high viscosity, and the characteristic gap size small; inertial effects are thus usually negligible compared with viscous stresses. The transient terms in the momentum balance are inertial, so the momentum equation becomes quasistatic if inertial effects can be neglected. Mold filling under these conditions is simulated by solving the steady-state momentum equation for an inelastic liquid, with the front movement calculated at each instant in time from the mass balance. The numerical methods used for continuous processes can thus be applied, with (conceptually) minor modification to follow the moving front. Special techniques must be implemented to follow temperature transients.

Free-surface Processes

Spinline models are based on thin-filament equations, in which the essentially elongational character of the deformation is exploited and the conservation and constitutive equations are averaged over the diameter.

The blown-film process is kinematically similar to fiber spinning, and thin-sheet equations analogous to the thin-filament equations have been employed in process modeling.

The modeling of polyester fiber spinning from the melt is well developed for normal processing conditions, where crystallinity is low; commercial computer codes are available for the prediction of temperature and diameter profiles along the spinline and of the radial birefringence distribution (through a stress-optical coefficient) at takeup. Crystallization kinetics have been included in some models described in the literature, so that it is possible in principle to simulate the spinning of highly crystalline polymers, but the absence of stress-dependent crystallization rates makes such calculations less useful. Predictive modeling of the high speed spinning of polyesters, where rapid stress-induced crystallization is known to occur, has not been reported.

Numerics

The numerical realization of a model requires the availability of algorithms that can solve the model equations accurately. Modeling of the detailed flow and stress fields in confined geometries has been severely limited by the fact that iterative numerical algorithms for solving the momentum and continuity equations together with the nonlinear constitutive equations typically fail to converge to a solution at a critical value of the recoverable shear.

This convergence failure appears to be a numerical artifact of the techniques in use, which propagate the large stresses generated near corners and lips into the bulk of the flow and degenerate the solution quality.

Boundaries

The tradition in fluid mechanics, dating to the mid-nineteenth century, is that a fluid adheres to a solid boundary during flow, and that the fluid at the boundary therefore takes on the velocity of the solid. This "no-slip" condition has been used routinely in the modeling of polymer flow, despite occasional hints that there might be "slip" at the wall. It has now been shown conclusively that the properties of a polymer extrudate can be controlled by changing the materials of construction of the metal die-face, thus establishing that the boundary is not a passive element in the flow field. The experimental evidence indicates that the no-slip condition is valid up to a critical stress that is of order 0.2 MPa. Since the shear modulus of most polymers at processing conditions is of the same order, this indicates that no-slip can be used with confidence under conditions where $\|\tau\|/\overline{G} < 1$. The appropriate boundary condition at higher stresses is not known, although there is limited experimental evidence for a polymer slip velocity at the wall that is proportional to the stress in excess of the critical stress.

Instabilities

Dynamic models are routinely used in engineering applications to study conditions under which stable operation cannot be maintained. Attempts have been made to use flow models to study polymer processing instabilities. The most common technique has been based on linear stability theory, in which the dynamical response to infinitesimal disturbances can be obtained by solution of a set of approximate linear equations; if infinitesimal disturbances die out, the steady mode of operation is stable. Linear stability theory establishes absolute instability, but only relative stability; a process that is unstable to infinitesimal disturbances is unstable to all disturbances, but a process that is stable to infinitesimal disturbances might still be unstable to some finite disturbances.

Dynamic modeling has been reasonably successful in describing the onset of a flow instability known as draw resonance that occurs in fiber spinning and extrusion coating.

Coprocessing

Modeling of coprocessing, in which two or more immiscible liquids flow together in a mold or a die, is focused on the prediction of interface location. In principle, the modification of existing finite element codes is straightforward, but there are theoretical issues associated with the treatment of the pressure at the interface.

Short-fiber Composites

Attempts have been made to model the flow and fiber orientation in molding of short-fiber composites. In nearly all these studies it was assumed that the flow field is the same as the one that would exist in the absence of the fibers; equations were then used for the mechanics of noninteracting fibers to determine the orientation distribution. This "weakly coupled" approximation is grossly in error for flow through a contraction, where the presence of less than 0.1 volume percent of fibers causes a considerable change in the flow. The approximation is probably quite good in a purely extensional flow, where there is a strong tendency toward alignment of fibers with the streamlines, and by implication in other flows without vorticity; the latter could arise in some molding operations because of the formation of a lubricating layer along the wall. The possible range of applicability of the weakly coupled approximation does not seem to have been explored in the published literature.

BIBLIOGRAPHY

C. L. Tucker, ed., *Fundamentals of Computer Modeling for Polymer Processing,* Carl Hanser Verlag, Munich, 1989.

J. R. A. Pearson and S. Richardson, eds., *Computational Analysis of Polymer Processing,* Elsevier Applied Science Publ., London, 1983.

J. R. A. Pearson, *Mechanics of Polymer Processing,* Elsevier Applied Science Publ., London, 1985.

M. M. Denn, "Issues in Viscoelastic Fluid Mechanics," *Ann. Rev. Fluid Mech.* **22,** 13 (1990).

MORTON M. DENN
Lawrence Berkeley Laboratory and University of California at Berkeley

PROCESSING AIDS

During the conversion of raw synthetic polymers into desired objects, various melt-processing procedures are involved, such as extrusion, injection molding, calendering, blow molding, and thermoforming. The inherent viscoelastic properties of each polymer type lead to certain undesirable processing defects, so additives are used to ease these processing-related problems. Heat and light stabilizers, antioxidants, and lubricants are well-defined, well-established additives (qv). Processing aids are another important plastic additive product. Processing aids can be made of small molecules, oligomers, or high molecular weight polymers; this article covers only polymeric processing aids because of the many advantages that polymeric processing aids exhibit over nonpolymeric materials.

Functions of Processing Aids

Many plastic materials have limited applications due to either their undesirable physical properties or their poor processing properties. These limitations have been greatly reduced through the use of polymer blends and alloys. By mixing two or more polymers, the overall balance of the physical and processing properties can be optimized. The functions of processing aids are based on the idea of polymer blends, that is, changing the processing behavior of one polymer by adding another polymer. However, there is a major difference that distinguishes processing aids blended with polymers from standard polymer blends. This difference is that processing aids are used in such small quantities that their effect on the final physical properties of the major polymer or polymer mixture is minimized. Therefore, the function of processing aids is to improve the

Table 1. Processing Aids Classified by Function

Type	Function	Processing application
fusion promotion	promote the fusion process of PVC, chlorinated PVC, vinyl chloride copolymers, PVC blends	all applications
melt rheology modification	increase melt elasticity, improve melt strength, control die swell	blow molding, calendering, thermoforming, and foam extrusion
	reduce melt viscosity and increase melt flow	injection molding and extrusion
	reduce melt fracture and improve surface quality	extrusion, calendering, blown film, and blow molding
lubrication	prevent the polymer melt from sticking to hot metal surfaces and assist mold release	calendering, extrusion, and injection molding
dispersion promotion	aid dispersion of fillers, cross-linked impact modifiers, and pigments in the matrix polymer	all applications

Table 2. Commercial Processing Aids

Trademark	Company	Type of product
Amoflo	Amoco (U.S.)	poly(α-methyl styrene) processing aid
Arvis	ARCO (U.S.)	polypropylene-based lubricating processing aid
Bärorapid	München Otto Bärlocher GmbH (FRG)	acrylic processing aid
Constab	Constab Polymer (Italy)	processing aid for polyolefins
Degalan	Degussa (FRG)	acrylic processing aid
Diakon	ICI (UK)	acrylic processing aid
Dynamar	3M (U.S.)	fluoroelastomer processing aid for LLDPE
Epolene	Eastman Kodak (U.S.)	processing aid for LLDPE
Evathlene	Dainippon (Japan)	ethylene–vinyl acetate (EVA) copolymer processing aid
Kane Ace	Kanegafuchi (Japan)	acrylic processing aid, lubricating processing aid
Metablen	Mitsubishi Rayon (Japan)	acrylic processing aid, lubricating processing aid
Modarez	Protex S.A. (France)	acrylic processing aid
Paraloid	Rohm and Haas (U.S.)	acrylic processing aid, lubricating processing aid
Ucarsil	Union Carbide (U.S.)	processing aid for LLDPE
Umakryl	Synthesia Semtin (Czechoslovakia)	acrylic–styrene copolymer processing aid
Vestiform	Huls AG (FRG)	acrylic processing aid, lubricating processing aid
Viton	DuPont (U.S.)	fluoroelastomer processing aid for LLDPE

processability of plastic materials without modifying end-use properties.

Processing behavior involves the melting process of the polymer and the melt rheology of the polymer. It also includes melt mixing with other ingredients, such as fillers, pigments, or impact modifiers. The functions of processing aids can be classified into four general types: promotion of fusion, modification of melt rheology, lubrication, or promotion of dispersion (Table 1). Each processing aid may act by more than one function. Multiple functionality is particularly true for processing aids used in poly(vinyl chloride) (PVC).

Commercial Sources of Processing Aids

Table 2 lists major processing-aid suppliers. Many of them offer products worldwide and many offer products with different molecular weights. Processing aids are usually sold in powder form. Some companies also sell packages of processing aids and impact modifiers for specific applications. FDA-approved processing aids are available for food-contact plastics.

BIBLIOGRAPHY

R. H. Burgess, ed., *Manufacture and Processing of PVC,* Macmillan, New York, 1982, pp. 1–61, 151–213.

K. D. Bohme in R. Gachter and H. Muller, eds., *Plastic Additives Handbook,* 2nd ed., Carl Hanser Verlag, Munich, 1983, pp. 341–360.

Additives For Plastics, 1st ed., The International Plastics Section Inc., D.A.T.A. Inc., San Diego, Calif., 1987, pp. 226–231.

HSING-YEH PARKER
JUDITH L. ALLISON
Rohm and Haas Company

PROCESSING PLASTICS.

See PLASTICS, PROCESSING.

PROCESSING RUBBER.

See RUBBER CHEMICALS.

PROCESSING TEXTILES.

See TEXTILE PROCESSING.

PRODUCT DESIGN

Designing is defined as a process through which a designer creates a form or device that satisfies all its functional requirements while in its intended environment. A well-designed product has good sales appeal and market acceptance; performs functionally and reliably; and can be manufactured, marketed, and sold at a profit. The designer is the person who is capable of creating such a product.

The best designers in the plastics industry work as members of a team, which exploits new developments in materials and processing. The designer should be available for consultation throughout the entire project, to the day the parts are assembled, packed, and shipped. A good designer should be able to recommend a number of reliable sources for mold building, processing, and material supplies.

Guidelines and Parameters

This phase covers the gathering and sorting of information, which enable the designer to become familiar with the scope of the project, its application, and the market. Considerations include application and environment, styling, volume and cost guidelines, part life expectancy, standards, compatibility, and timetables.

Preliminary Design Considerations

In this phase the guidelines established in the preliminary investigation are used to lay out the overall project, develop the design, select the manufacturing process and materials, and evaluate the estimated tooling and part costs. The experienced designer attempts to accomplish this task with readily available plastic materials and molding processes, although this is not always possible and the project may require a modification to the process or a new material. During this phase, enough information should become available to allow an informed decision to be made regarding the feasibility of continuing the project or bringing it to a close before entering the phase where most of the actual costs will be incurred.

Design of Details

During this phase, expertise of the designer with the molding process, materials, and innovation is tested. Cooperation with the production team is essential. Experienced molders and mold builders usually help identify potential processing problems in part design. Considerations include moldability, assembly details, structural details, part weight, cycle time, and obtaining balance between appearance and functional detail.

BIBLIOGRAPHY

R. Beck, *Plastic Product Design,* 2nd ed., Van Nostrand Reinhold Co., New York, 1980.

W. Chow, *Cost Reduction in Product Design,* Van Nostrand Reinhold Co., New York, 1978.

Plastics Design Forum.

Design News.

CHARLES M. CAMPBELL
Tradewinds Technologies Inc.

PROLIFEROUS POLYMERIZATION

Proliferous polymerization is an abnormal reaction encountered in free-radical polymerization systems. A proliferous polymer has a popcorn- or cauliflowerlike configuration, a voluminous structure due to many included voids, and a white opaque appearance that distinguishes it from a glassy, transparent polymer obtained by a normal polymerization reaction. Proliferous polymers are insoluble in organic solvents and exhibit very low swellability in solvents. A proliferous polymerization produces a popcorn polymer under certain limited conditions. Formation is based on cross-linking between polymers and growth of branched chains on the network of cross-links. The resulting proliferous polymer is further characterized by its ability to initiate proliferous polymerization of certain vinyl monomers.

Popcorn Polymers

Proliferous polymers can be classified in terms of their monomers into three groups: compounds having two or more double bonds, such as butadiene and isoprene; monovinyl compounds with components acting as cross-linking agents, such as the styrene–divinylbenzene system; and monovinyl compounds of certain classes, such as alkyl acrylates (Fig. 1).

Mechanism of Polymerization

Proliferous polymerization forming popcorn polymers can be spontaneous polymerization from monomers or seed polymerization in which popcorn polymers are used as seeds.

Influence of Solvent

The inhibition of the proliferous polymerization of butadiene by hydrogen sulfide indicated that a chain-transfer agent strongly suppresses the proliferation of a popcorn polymer. Nitrogen dioxide effectively inhibits the popcorn polymerization of a butadiene–styrene system. In a seed polymerization, addition of a solvent having a large chain-transfer constant usually supresses or inhibits the proliferation of seed popcorn polymer.

Decomposition. When styrene popcorn polymer is treated at room temperature with a carbon disulfide solution of aluminum bromide, the solid popcorn polymer is dissolved to form a blackish brown solution. A polymer soluble in benzene can be obtained from such solution. This means that styrene popcorn polymer can be decomposed with the aluminum bromide–carbon disulfide system. Decomposition of popcorn polymer by means of chemical processes offers a useful way to study the structure of popcorn polymer.

Reaction Rate and Esr Spectra

The heterogeneous nature of proliferous polymerization makes it hard to analyze the reaction kinetically. However, the time

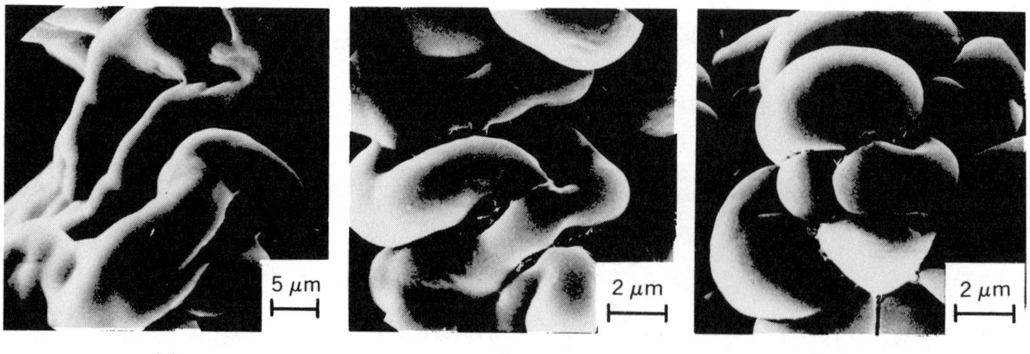

Figure 1. Scanning electron micrographs of surface views of butyl methacrylate–divinylbenzene popcorn polymer system. Divinylbenzene content in popcorn polymer: (**a**) 0.1 wt %; (**b**) 0.5 wt %; (**c**) 1.0 wt %.

at which the mass of popcorn polymer becomes doubled (t_2) has been estimated. The t_2 value of butadiene popcorn polymer is about 140 h. Carothers reported that the t_2 value of chloroprene popcorn polymer at a temperature of 291 K is 15 h.

Proliferous polymerization is established as a radical polymerization reaction because a high radical density of a popcorn polymer is observed from esr during the proliferation process. An esr signal is observed not from the monomer phase, but from solid popcorn polymer during proliferation.

Application

Styrene popcorn polymer is widely used as a material for chemical modification. In order to use by-products from the synthetic rubber industry, attempts have been made to improve the popcorn polymer of the styrene–butadiene system for inclusion in ion-exchange resins. Popcorn polymers also have been tried as support material in organic synthesis reactions.

BIBLIOGRAPHY

J. W. Breitenbach, *Br. Polym. J.* **6,** 119 (1974).

G. H. Miller, R. R. Eliason, and G. O. Pritchard, *J. Polym. Sci. Part C* **4,** 1109 (1963).

M. Tatsumi and S. Yamamoto, *J. Chem. Soc. Jpn.,* 1386 (1982).

T. G. Fox and S. Gratch, *Ann. N. Y. Acad. Sci.* **57,** 367 (1953).

S. Yamamoto
M. Tatsumi
Kansai University

PROPAGATION. See Kinetics of polymerization.

PROPIOLOACTONE POLYMERS. See Polyesters.

PROPIONATE PLASTICS. See Cellulose esters, organic.

PROPYLENE–ETHYLENE COPOLYMERS. See Ethylene–propylene elastomers.

PROPYLENE POLYMERS

The stereospecific polymerization of propylene by Natta in 1954 was a scientific breakthrough with immediate industrial significance. By obtaining the alignment of propylene molecules, a crystalline polymer, isotactic polypropylene, was synthesized at high yields in a manner that could be exploited commercially. This achievement, which spurred interest in organometallic chemistry, as well as in polymerization, was recognized by the award of the 1963 Nobel prize in chemistry to Natta and Ziegler, who discovered the basic catalyst system for polymerizing ethylene.

Interest in isotactic polypropylene was so high that only three years elapsed between the laboratory discovery and the first commercial production. The excellent combination of thermal and mechanical properties was quickly exploited in a variety of applications, such as injection molding and film and fiber production. This versatility, combined with favorable economics, encouraged a rapid expansion in the use of this material in the 1960s and the early 1970s. At present, polypropylene is one of the most important commercial thermoplastics; its consumption is still increasing more rapidly than the total for all thermoplastics. This situation is likely to continue into the future for the following reasons:

1. The relatively low cost of the product is due to low monomer cost and efficient polymerization technology, compared with other thermoplastics. Competition in licensing technology and production methods promotes further cost reduction by technical improvements.

2. The polymer can be modified for a variety of applications. Through copolymerization, orientation, and other techniques the physical properties of the product can be varied to meet a wide range of thermal and mechanical requirements.

3. The ease in processing these polymers allows their economic use in most commercial fabrication techniques. Modifications of the melt rheology increase this versatility.

The versatility of this product, coupled with its low cost, encourages its introduction into newly developing applications. As a result, new applications of polypropylene will use forms that are significantly different from those available today.

Monomer

The physical properties of propylene are given in Table 1.

Table 1. Physical Properties of Propylene

Property	Value
molecular weight	42.078
boiling point at 101.3 kPa[a], °C	− 47.7
melting point, °C	−185.3
critical temperature, °C	92
critical pressure, MPa[b]	4.65
critical density, g/mL	0.233
critical compressibility	0.275
dipole moment, 10^{-30} C·m[c]	1.3
refractive index, n_D	1.3567
explosion limit, % by volume in air	
lower	2.4
upper	11.1
autoignition temperature, °C	224
solubility in water (at 20°C, 101.3 kPa[a]), mL gas/100 mL	44.6

[a] To convert kPa to mm Hg, multiply by 7.5.
[b] To convert MPa to atm, multiply by 9.87.
[c] To convert C·m to D, divide by 3.336×10^{-30}.

Preparation and Manufacture. Propylene can be produced by a variety of laboratory methods. The most common and practical system is the dehydration of *n*-propanol and 2-propanol with sulfuric acid and an aluminum sulfate catalyst.

Propylene is a by-product of ethylene produced by steam cracking or of gasoline produced by catalytic cracking. Worldwide, 80% of the propylene production is by steam cracking; however, in North America, only 30% is obtained by this method. The main reasons for this difference are the large U.S. gasoline market and the higher molecular weight feedstocks in Europe and Japan in the steam cracking process.

Purity. Refinery, chemical, and polymerization grades are available, containing 50, 93, and 99.8%, respectively. Production capacities, production, and consumption are given in Tables 2 and 3.

Table 2. Capacities, Production, and Consumption of Propylene, 10^3 t

	Actual		Projected	
	1983	1984	1990	1995
capacities				
steam cracker	22,352	22,194	23,321	24,920
catalytic cracker	6,170	6,332	7,057	7,407
propane dehydration			535	535
Total	*28,522*	*28,526*	*30,913*	*32,862*
production				
steam cracker	15,576	16,906	18,400	21,124
catalytic cracker	5,016	5,568	6,149	6,702
propane dehydration			297	410
Total	*20,592*	*22,474*	*24,846*	*28,236*
consumption				
United States	6,410	6,805	7,605	8,420
Canada	398	419	442	467
Western Europe	6,524	7,136	7,353	7,694
Japan	2,638	2,925	2,985	3,035
Latin America	702	732	1,066	1,426
Eastern Europe	1,911	2,051	2,730	3,423
Africa	16	16	103	324
Middle East	13	17	184	270
Asia and Pacific	1,195	1,379	2,150	3,213

Health and Safety. Propylene is a colorless gas with anesthetic properties at high concentration (60% by volume or more) in air. The gas has no known toxic effect, but is only an asphyxiant. Evaporation can cause skin burns from freezing. Propylene should be used under ventilation because of its explosive properties when mixed with air.

Polymer Properties

Molecular Structure. The molecular structure is determined mainly by the size and structural order of the macromolecule, on which its physicomechanical properties depend.

Macromolecular Structure. Information on the macromolecular structure can be obtained from ir spectra, which, for isotactic polypropylene, are affected by vibrations of the regular spiral chain and not by the polymer crystallinity. Typical bands are detectable at 1167, 997, and 841 cm^{-1}. High resolution proton nmr and particularly ^{13}C nmr provide detailed information on configuration distributions along the chain.

The main crystalline form of the isotactic polypropylene is the α-form; other forms, such as β, γ, and smectic, exist also.

The properties of isotactic, syndiotactic, and atactic polypropylene are given in Table 4. Syndiotactic polypropylene is of a more scientific than practical interest. Furthermore, it is difficult to synthesize (particularly because of the required temperature of −78°C).

Table 3. Worldwide Production of Chemicals from Propylene, 10^3 t

	Actual		Projected	
Product	1983	1984	1990	1995
polypropylene	6,836	7,761	9,534	11,700
acrylonitrile	3,622	3,837	4,058	4,536
cumene	1,818	1,956	2,044	2,263
2-propanol	1,144	1,163	1,291	1,252
oxo alcohols	2,151	2,337	2,661	2,986
propylene oxide	1,822	1,881	2,210	2,490
oligomers	774	788	708	703
others	1,640	1,757	2,112	2,342
Total	*19,807*	*21,480*	*24,618*	*28,272*

Table 4. Properties of Isotactic, Syndiotactic, and Atactic Polypropylene

Property	Isotactic	Syndiotactic	Atactic
density, g/cm^3	0.92–0.94	0.89–0.91	0.85–0.90
melting point, °C	165	135	
solubility in hydrocarbons at 20°C	none	medium	high
yield strength	high	medium	very low

The degree of crystallinity is determined by a number of analytical procedures based on different criteria. The most widely used are fractional solubilization, density determination, x-ray diffraction, and thermal analysis.

Copolymer Structure. The molecular structure of polypropylene can be modified by the introduction of comonomers. This allows the production of polymer grades with a wide range of physical properties. Table 5 shows the propylene–ethylene combinations most frequently used in practice. For every grade of copolymer several variations are possible with a different ratio of components and different molecular weight and molecular weight distribution.

Physical Structure of Polymer Particles. Morphological control of the polymer particles, in particular of shape, size, size distribution, and compactness, influences the polymerization process. In addition, the commercial use of the polymer in the form produced in the reactor is of interest. The pelletization equipment currently found at the end of polymerization plants may be eliminated. However, for this purpose it is necessary for the particles to exhibit characteristics similar to those of the pellets.

Because of the mimicking phenomenon, the morphological characteristics are strictly dependent on the catalyst. That is, the catalyst reproduces its shape in the polymer on an obviously greater scale according to its activity, the quantity of polymer produced by a unit of catalyst. The essential requirement for mimicking is that the catalyst particles not be broken apart by mechanical stress.

The mimicking phenomenon is governed by the following equation:

$$R_t = \sqrt[3]{\frac{3}{4\pi\delta N}\int_0^t R_p dt}$$

where

R_t = total polymerization rate (grams of polymer per liter of reagent phase per time unit),
N = number of polymer granules per liter of reagent phase,
δ = polymer density (g/cm^3) referred to polymerization conditions,
R_p = propagation rate, and
t = time.

Chemical Properties. Polypropylene is highly resistant to chemicals. It dissolves in high boiling aliphatic and aromatic hydrocarbons at high temperatures. It is, however, attacked by oxidizing agents, such as concentrated sulfuric acids and fuming nitric acid.

Thermodynamic Properties. The thermodynamic melting point for pure crystalline polypropylene obtained by extrapolation of isothermally crystallized polymer is 187.7°C. This temperature is about 23–28°C higher than that determined on commercial samples under normal analysis conditions.

Reported values of the glass-transition temperature (T_g) vary widely because of the use of different measurement techniques and the influence of tacticity and molecular weight. For isotactic polypropylene, T_g values from −13 to 0°C are reported, and for the atactic from −18 to −5°C.

Physical Properties and Commercial Specifications. Hundreds of polypropylene grades are sold in the United States, including all the copolymers, fillers, and additive packages. The basic types are homopolymer, random copolymer, impact (heterophasic) copolymers, and filled polymers. They are usually tested by ASTM methods; however, various users have developed their own specifications. A number of grades are approved for food contact, as specified in the Code of Federal Regulations, Title 21, Section 177.1520. Typical properties of general-purpose homopolymers are shown in Table 6. Many of these polymers also carry a UL94 flammability class HB rating.

The random copolymers are primarily used in applications requiring lower melting temperatures than homopolymers or where exceptional clarity is necessary. These and most other commercially available products are random copolymers with ethylene; some terpolymers contain butene.

Polymerization

Catalysts. Catalysts include $TiCl_3$-based catalyst, and $MgCl_2$-supported catalyst.

Catalyst Evaluation. To evaluate a catalyst system, the following characteristics must be known: yield or productivity, that is, the amount of polymer produced per unit of total catalyst or single component; the polymerization kinetics, or yield development with time; stereospecificity; sensitivity to the molecular weight regulator; ability to copolymerize; and the polymer morphology.

Table 5. Propylene–Ethylene Copolymerization

Monomer sequences[a]	Products
PPPPPPPPPPPPPPP	homopolymer polypropylene
PPPEPPPPPPEPPPP	polypropylene randomly modified with ethylene
PPPPPPPP + EPEPEP PPPPP + EPEP + EEEE	impact-resistant polypropylene
EPEPPEEPEPEPEPE	elastomer
EEEEPEPEPEEEEEP	thermoplastic elastomer
EEEEEPEEEEEEPEE	polyethylene randomly modified with propylene
EEEEEEEEEEEEEEE	homopolymer polyethylene

[a] P = propylene, E = ethylene.

Table 6. Properties of Homopolymers

	Value				
Property	Extrusion, sheet, profiles		General purpose, injection molding		Injection molding, thin complex parts
melt flow, g/10 min	0.4	0.8	4	12	35
density, g/cm^3	0.902	0.903	0.903	0.903	0.902
tensile strength[a], MPa[b]	37	35	35	34.8	33.2
elongation[a], %	13.5	13	12	11	12
flexural modulus 1% secant, MPa[b]	1655	1666	1724	1793	1470
Rockwell hardness, R scale	95	95	99	100	98
deflection temperature at 455 kPa[c], °C	90	95	100	102.2	90
notched Izod impact at 23°C, J/m[d]	161	133	42.7	37.3	32

[a] At yield.
[b] To convert MPa to psi, multiply by 145.
[c] To convert kPa to psi, multiply by 0.145.
[d] To convert J/m to ft·lbf/in., divide by 53.38.

Reaction Mechanisms. The global polymerization rate, R_p in Ziegler-Natta catalysis, is generally represented by

$$R_p = k_p[C^*][M]$$

where k_p is the propagation constant, $[C^*]$ the concentration of active sites, and $[M]$ the concentration of monomer.

The polymerization rate changes with time. A period of increasing rate is usually followed by a decline and, eventually, by a stationary state. The initial increase may be due to the progressive activation of new active centers. The deactivation, particularly evident in supported catalysts, is attributed to variations in both number and chemical nature of the centers.

Kinetics are affected by catalyst system and polymerization conditions. Effects are due to the chemical and physical structure of the catalyst as well as to the nature of the activator. Important parameters include the ratio between catalyst and activator and their concentrations, hydrogen concentration, temperature, stirring rate, and type and amount of Lewis base. The effects vary with the polymerization medium, ie, a diluent or the monomer in liquid or gas phase.

Stereochemistry of Monomer Insertion. The propylene molecule may be inserted in the chain by various procedures, giving rise to complex phenomena of structural and steric isomerism.

A first possibility, common to all olefins, is encountered at the opening of the double bond and involves the spatial location of valences. The opening is of the cis type; that is, the valence releases in the same semispace. Another possibility, principally for α-olefins, is given by the monomer disposition in the metal–carbon bond. Such disposition is called primary if the CH_2 group of the monomer is located near the metal and secondary if the CH group occupies this location.

primary $\quad Mt{-}C_2H_5 + CH_2{=}CH(CH_3) \longrightarrow Mt{-}CH_2{-}\dot{C}H(CH_3){-}C_2H_5$

secondary $\quad Mt{-}C_2H_5 + CH(CH_3){=}CH_2 \longrightarrow Mt{-}\dot{C}H(CH_3){-}CH_2{-}C_2H_5$

where Mt = metal. To construct regular structural macromolecules, the insertion must occur in the same way.

It has been proposed that the stereochemistry of the isotactic insertion is governed by the asymmetric structure of the active site or the asymmetric carbon atom of the last chain-inserted monomer unit. It was proved experimentally that the regulating factor is the active center because the steric order is transferred also through ethylene units and because an occasional error is not perpetuated in the chains.

Copolymerization. The composition of the feed mixture of two monomers and the composition of the product copolymers are correlated by the following equation:

$$\frac{m_1}{m_2} = \frac{M_1\,(r_1M_1 + M_2)}{M_2\,(r_2M_2 + M_1)}$$

where m_1 and m_2 are the molar fractions of the two monomers in the polymer, M_1 and M_2 the feed molar fractions, and r_1 and r_2 are the reactivity ratios of the monomers defined by

$$r_1 = \frac{k_{11}}{k_{12}};\ r_2 = \frac{k_{22}}{k_{21}}$$

where k_{ij} are the rate constants of the four probable addition reactions of two monomers. The copolymer composition is, therefore, determined from the composition of the feed mixture and the reactivity ratios.

Processes. The first industrial plant for the production of polypropylene went on stream in Italy (Montecatini) in 1957 and was followed, some months later, by a Hercules plant in the United States and those of several other companies. A hydrocarbon diluent was used in which isotactic polypropylene remained in suspension, whereas the atactic dissolved. The isotactic fraction was separated from the atactic fraction and the diluent, freed from catalyst residues, and dried. The process fluids were purified before being reused. The plants differed in some operating conditions, such as temperature, concentration of reagents, diluent, deashing, and reactors. However, they may be represented by the Montedison process of Figure 1 where the polymerization section is small compared with the other sections.

In the 1960s, two processes were commercialized in which the monomer is either in the gaseous or the liquid state.

The Novolen gas-phase process was developed by BASF.

The range of products produced by this process was limited by the low stereospecificity and activity of available catalysts. The limited stereospecificity gave polymers with 85–90% iso-

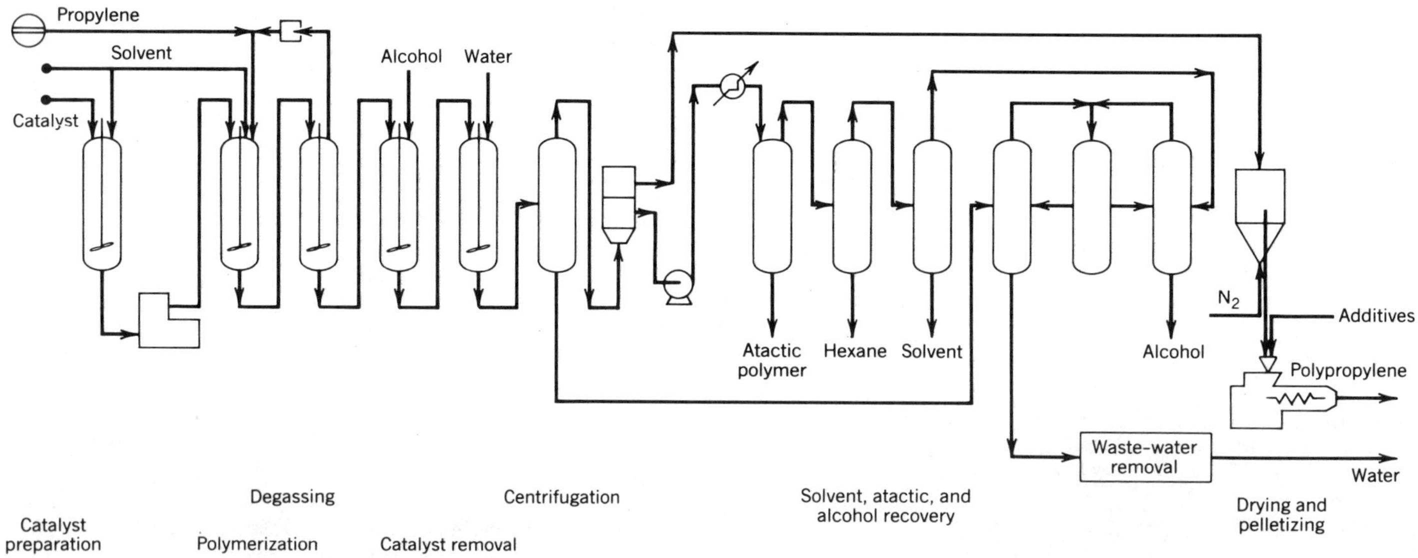

Figure 1. Montedison conventional slurry, low-yield polypropylene process. Courtesy of Gulf Publishing Co.

tacticity, from which the atactic faction could not be separated. For this reason the products were restricted to a few limited markets.

Around the same time, Phillips Petroleum and Rexall Drug and Chemical developed processes for polymerization in liquid propylene. These liquid propylene processes suffered from poor catalyst performance in terms of stereospecificity and yield, and postpolymerization treatments were necessary to reduce the catalyst residue.

Recent Processes. The energy crisis of the 1970s resulted in a dramatic increase in energy and monomer costs. The costs of the monomer rose from about 10% of the selling price of the polymer to 50% in a short time. In addition, the many steps of extraction, separation, and distillation required a large amount of energy. Another factor was the market demand for higher quality products and a wider range of properties, particularly of copolymers. Environmental regulations increased costs even further. Improvement of catalyst performance became critical.

In 1975 Montedison and Mitsui Petrochemical introduced third-generation catalysts, which gave high enough yields to eliminate deashing. However, stereospecificity was insufficient to produce atactic-free polymer. Nevertheless, catalyst performance justified the design of a simplified process in hydrocarbon diluent, free of deashing sections. This process could be used in existing plants, allowing considerable savings.

These third-generation catalysts can also be used in liquid monomer processes, in the El Paso process, where polymerization is carried out at 60–80°C and 2.5–3.5 MPa. The unconverted monomer is removed from the polymer suspension and recycled.

The critical development was the discovery of the superactive third-generation catalysts, which gave optimal yield and stereospecificity, together with morphological control of the polymer in terms of shape, size, and particle size distribution. In both gas-phase and liquid monomer processes, deashing and atactic polymer removal were eliminated. The simplified process train consists primarily of the polymerization section for both gas-phase and liquid monomer processes.

The Spheripol process, licensed by Himont, Inc., takes advantage of the capabilities of the superactive third-generation catalysts.

Polymerization (homo- and random) takes place in one or more loop-tubular reactors in liquid propylene. At the loop outlet of the polymeric slurry any unconverted monomer is eliminated by flashing from the polymer and recycled by reducing the temperature without recompression. The polymer reaches a fluid-bed gas-phase reactor, where the rubbery phase for impact-resistant copolymers is produced. A steam treatment completes monomer removal and terminates catalyst activity.

Because the process is so simple, operating and maintenance costs are very low.

Union Carbide, by exploiting the know-how previously achieved in ethylene gas-phase polymerization, developed a gas-phase process using Shell Chemical's highly active catalyst systems. This process, called Unipol PP, is extremely simple, economical, and versatile. The process for the production of homopolymer and random copolymer is shown in Figure 2.

The more recent combined bulk and gas-phase processes or fully gas-phase processes are difficult to evaluate. However, because of outstanding catalysts, they are vastly superior to earlier processes.

Processing

Propylene polymers are processed by injection molding, extrusion, thermoforming, and blow molding.

Fibers. Fibers are processed by melt spinning, slit and split films, and melt blowing.

Films. The first commercial polypropylene films were produced by extrusion casting. Polymer is extruded through a slit or a tubular die and quenched by cooling on chill rolls or in a water bath. The lack of orientation caused by rapid quenching is responsible for many of the properties desired in cast polypropylene film. Cast films can be heat sealed over a wide range

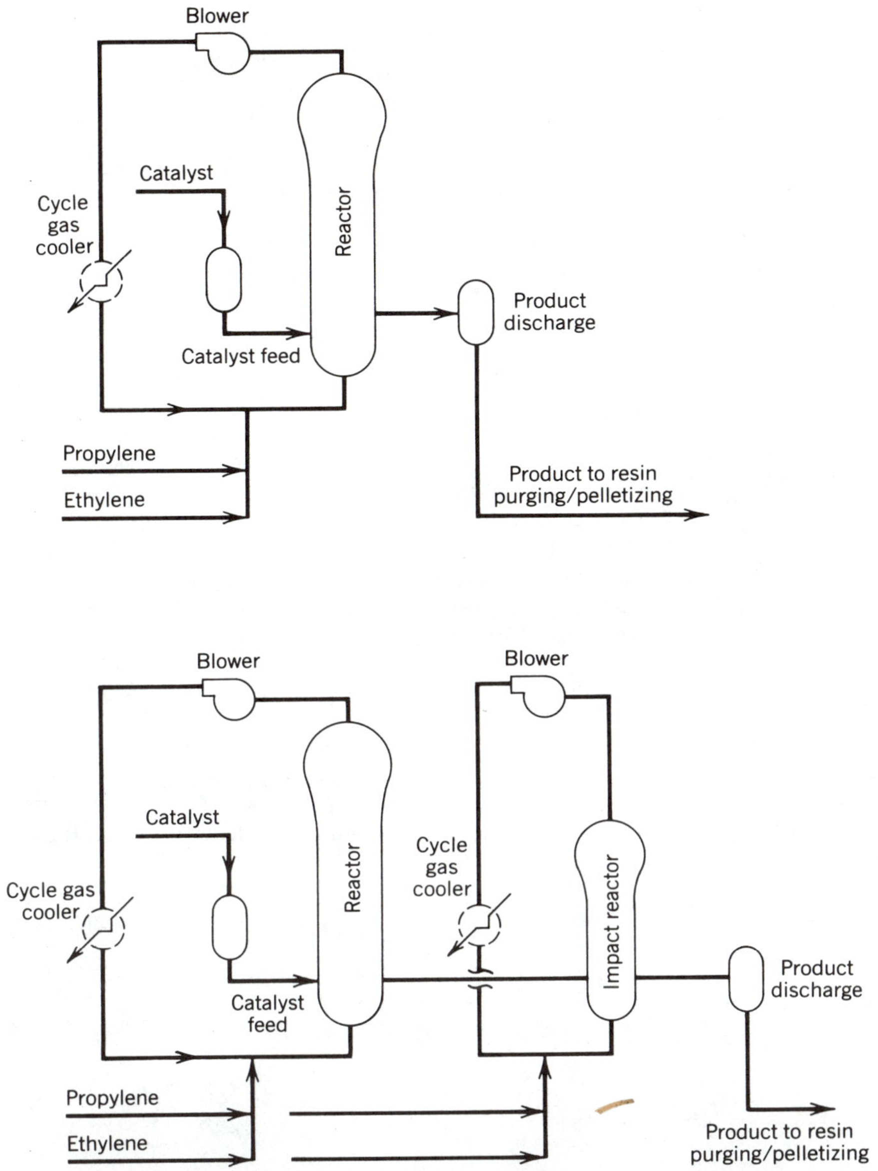

Figure 2. Unipol PP process. Courtesy of European Chemical News.

of temperatures and do not shrink at steam autoclave temperatures.

Orientation improves the strength of polypropylene films; the stiffness of oriented polypropylene (OPP) film is three times cast film. In the tenter frame process a cast film is oriented by extension through a series of rollers in the longitudinal or machine direction and by stretching with a series of clips or tenter frames in the transverse direction. The tenter process is commonly used for thicker oriented films, but can be used for a wide range of film gauges. In the bubble process a tube is extruded, quenched, and oriented by inflation with air. As the tube is radially oriented, extension in the machine direction by the nip rollers provides axial orientation. The tube is then flattened and slit into flat film. This process is best suited for the production of thin films.

Stabilization. Stabilizers minimize the loss of molecular weight and physical properties caused by oxidation, radiation, and thermal degradation. Stabilizers are added in the polymerization plant, usually in the drying section, to protect against degradation during drying and storage. Typically, a small quantity of hindered phenol is used for this purpose. The bulk of the stabilizers, designed to protect the polymer during processing and application, are incorporated during pelletization or in fabrication.

Economic Aspects

Polypropylene has evolved from a new, high growth product in the 1960s to a mature, high volume product in the 1980s. Consumption has increased at an annual rate of 8% between 1974 and 1985, which is higher than the growth rate of the U.S. economy, but not as high as the 50% annual growth in the 1960s. Polypropylene is now used so widely that the overall growth rate depends on the economy.

As shown in Table 6 the annual growth rate of polypropylene consumption is higher than that of the plastics industry as a whole, and sales are over 14% of those of all thermoplastics.

Table 6. U.S. Polypropylene Consumption

Consumption	1974	1975	1976	1985
sales and internal use, 10^3 t				
polypropylene	998	863	1,150	2,393
total thermoplastics	9,743	7,854	10,001	16,364
polypropylene, % of total consumption	10.24	10.99	11.50	14.62

The introduction of high impact copolymers and controlled rheology resins contributes to the continued growth of polypropylene sales. Prices for polypropylene, propylene monomer, and crude oil are shown in Table 7. The distribution of polypropylene consumption is shown in Table 8.

Exports became increasingly important to resin producers after the oil price rise in 1979. The Japanese use much less polypropylene fiber than Western European countries. Injection-molded parts for the automotive and electronic industries and packaging film are the principal markets in Japan. The pattern of polypropylene consumption in Europe is similar to that in the United States.

The principal U.S. producers of polypropylene in order of capacity are Himont, Amoco, Shell, Exxon, Aristech, Fina, El Paso, Phillips, Northern Petrochemical, Soltex, and Eastman. Himont Italia and Himont Belgium together have the largest production in Western Europe, followed by ICI, Hoechst, and a number of others. Japanese manufacturers include Mitsubishi Petrochemical, Chisso, Sumitomo, and Mitsui Petrochemical.

Table 7. Price Development of Polypropylene, Monomer, and Crude Oil

	1979[a]	1981[b]	1983[a]	1986[b]
polypropylene[c], ¢/kg	77	88	83.6	79.2
propylene, ¢/kg	33	51.7	42.9	36.3
crude oil[d], $\$/m^3$	150.9	213.2	183.0	118.9

[a] Fourth quarter.
[b] First quarter.
[c] Market price for general-purpose homopolymer.
[d] Saudi bench-mark crude.

Table 8. Distribution of U.S. Polypropylene Sales and Internal Use, 10^3 t

Application	1974	1976	1984	1985
injection molding	345	418	585	578
appliances	34	39	55	52
consumer products	51	72	132	141
rigid packaging	45	76	164	175
transportation	128	156	143	154
other	87	75	91	56
blow molding	16	17	48	47
film	64	79	182	176
sheet, >254 μm			23	20
fiber and filaments	303	317	607	650
extruded products[a]	69	57	51	40
resellers and compounders			261	388
all other uses[b]	92	102	73	96
total domestic consumption	889	989	1830	1996
exports	109	161	355	397
Total consumption	*998*	*1150*	*2185*	*2393*

[a] Includes wire and cable, pipe and conduit, extrusion coating, and sheet before 1977.
[b] Includes resellers and compounders before 1983.

BIBLIOGRAPHY

M. Ahmed, *Polypropylene Fibers—Science and Technology,* Elsevier, Amsterdam, The Netherlands, 1982.

R. B. Seymour and T. Cheng, ed., *History of Polyolefins,* D. Reidel, Dordrecht, The Netherlands, 1986.

P. C. Barbe, G. Cecchin, and L. Noristi, "The Catalytic System Ti-Complex/$MgCl_2$," *Advances in Polymer Science,* Vol. 81, Springer-Verlag Berlin Heidelberg, 1987, pp. 1–81.

RICHARD B. LIEBERMAN
Himont Research & Development Center

PIER CAMILLO BARBE
Himont Italia
Centro Recerche Guilo Natta, Italy

PROPYLENEIMINE POLYMERS.

See ALKYLENEIMINE POLYMERS.

PROTECTIVE COLLOIDS.

See COLLOIDS.

PROTEINS

Proteins are essential for life. All characteristics commonly attributed to living things, such as reproduction, growth, movement, metabolism, and sensory perception, are entirely dependent on the functions of proteins. Indeed, many proteins have some of the attributes of life and in the laboratory are commonly said to have "died" when they have lost their biological function. Each protein generally has one specified biological function.

Virtually all the chemical reactions that occur in living organisms require a protein catalyst. The entire subject of biochemistry is primarily a description of these protein-catalyzed chemical reactions. In the absence of such catalysts most such reactions occur at negligible rates that must be increased at least a millionfold to make them occur at the physiologically required rates. Those proteins that catalyze chemical reactions are known as enzymes. They catalyze reactions in an extremely specific manner, each enzyme usually catalyzing only a single reaction with only one set of reactants.

Some proteins transport and store small molecules. For example, the red-colored protein of blood, hemoglobin, transports oxygen in erythrocytes, whereas the related protein, myoglobin, stores oxygen in muscle. Another class of proteins, antibodies, confers immune protection on organisms by combining with foreign substances, such as viruses, bacteria, and toxins. Movement of organisms is usually produced by a complex assembly of proteins in muscle. Another method of integrating complex organisms is hormonal regulation, and many hormones, such as insulin, corticotropin, and various growth factors, are proteins. Proteins also regulate growth and development at all biological levels, from the control of gene expression to the interactions between cells and the organism in large organisms. At the other extreme some proteins serve only structural roles; the keratins of hair, nails, and skin are examples (see KERATIN). The high tensile strength of skin and bones is due to the presence of collagen, another fibrous protein. Although all proteins can serve a nutritional role after suitable cooking and denaturation, some appear to have this as their prime biological role, examples being casein of milk and ovalbumin of eggs.

In spite of their multitude of functions, all proteins are basically similar: they are linear polymers of 20 different amino acids. They differ from most other polymers in having a strictly defined length and sequence of amino acid monomers, which they acquire from their unique mode of biosynthesis (see PROTEINS, BIOSYNTHESIS). They differ from each other in that each protein has a unique amino acid sequence. Proteins are generally of biological origin, although it is possible to synthesize them chemically (see POLYPEPTIDES, SYNTHESIS). Proteins are distinguished from polypeptides by their unique, well-defined, and largely unrepetitive sequence of amino acid residues. Proteins usually consist of at least 50 amino acid residues; shorter chains are generally known as peptides. The other distinguishing feature is that proteins generally have well-defined three-dimensional conformations, whereas peptides do not. Nevertheless, the distinction between a peptide and a protein need not be clear cut, as in the case of some small hormones that have considerable conformational flexibility.

This article describes the common structural, physical, and chemical properties of proteins to provide a general understanding. Nevertheless, each protein is unique, so there are exceptions to every statement; indeed, it is often the exceptions that provide proteins with their unique biological properties.

Occurrence

Proteins are ubiquitous in all living organisms, but occur in widely varying amounts. Animal muscle tissues generally contain protein to the extent of 20–30% of their wet weight. At the other extreme the protein content of most plant materials is roughly 1%.

Uses

The potent biological and catalytic properties of proteins give them great potential for pharmaceutical and industrial uses. This potential, however, has not been fully realized thus far because of the technical problems in handling these complex and labile molecules. Nevertheless, such uses for proteins are expanding greatly. The proteins of most obvious use to industry are undoubtedly the enzymes, and their potent and specific catalytic powers are being used increasingly (see ENZYMES and ENZYMES, IMMOBILIZED). Proteases, which hydrolyze proteins and polypeptides, represent about 60% of the total enzyme market.

The biological properties of many proteins make them of great interest pharmaceutically, but the complexity of biological systems is usually the factor that most limits their use.

Of course, proteins have always been used nutritionally, but collectively, as a class, without regard for their individual identities, since they serve in this way merely as sources of the amino acids to which they are degraded. The primary concern in this case is their content of the various amino acids that are essential in the diet. One other protein that has been used for a long time is silk, but it is used almost directly as produced by the silkworm.

Covalent Structure

Amino Acid Monomers. Proteins are synthesized biologically as linear polymers of 20 amino acids. Nineteen of these amino acids have the basic structure

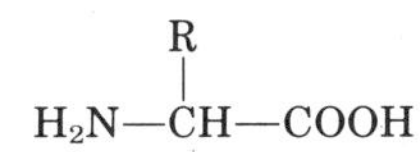

and differ only in the chemical structure of the side chain R. The exception, proline, is similar, but has the side chain cyclized and bonded to the nitrogen atom to give the amino acid

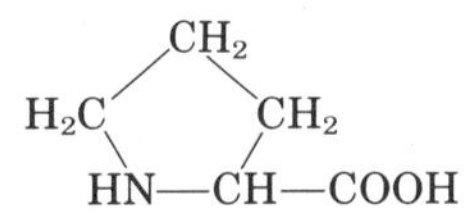

Except in glycine, where the side chain is simply a hydrogen atom, the α-carbon atom is asymmetric and always of the L configuration:

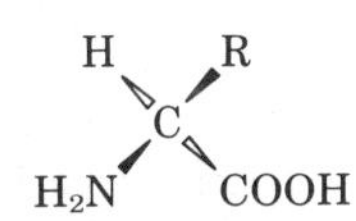

The structures of the side chains of the 20 amino acids are illustrated in Figure 1. Also given are the three- and one-letter abbreviations commonly used.

The amino acids are assembled into proteins by linking them together by peptide bonds:

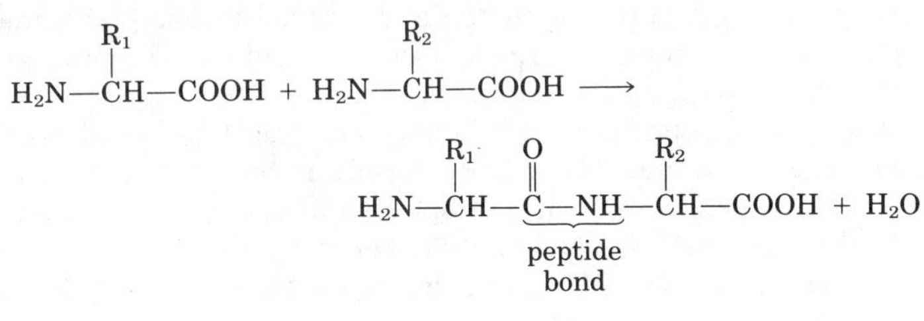

All proteins have this simple structure and differ only in the number and sequence of the 20 amino acids.

Some properties of the 20 amino acid residues are summarized in Table 1. Chemical properties have been described, and the chemistry of the amino acids comprehensively reviewed.

Posttranslational Modifications. Although proteins are normally synthesized as linear polymers with just the 20 amino acids of Figure 1, many alterations of the covalent structure may occur subsequently. Residues are usually not added, but

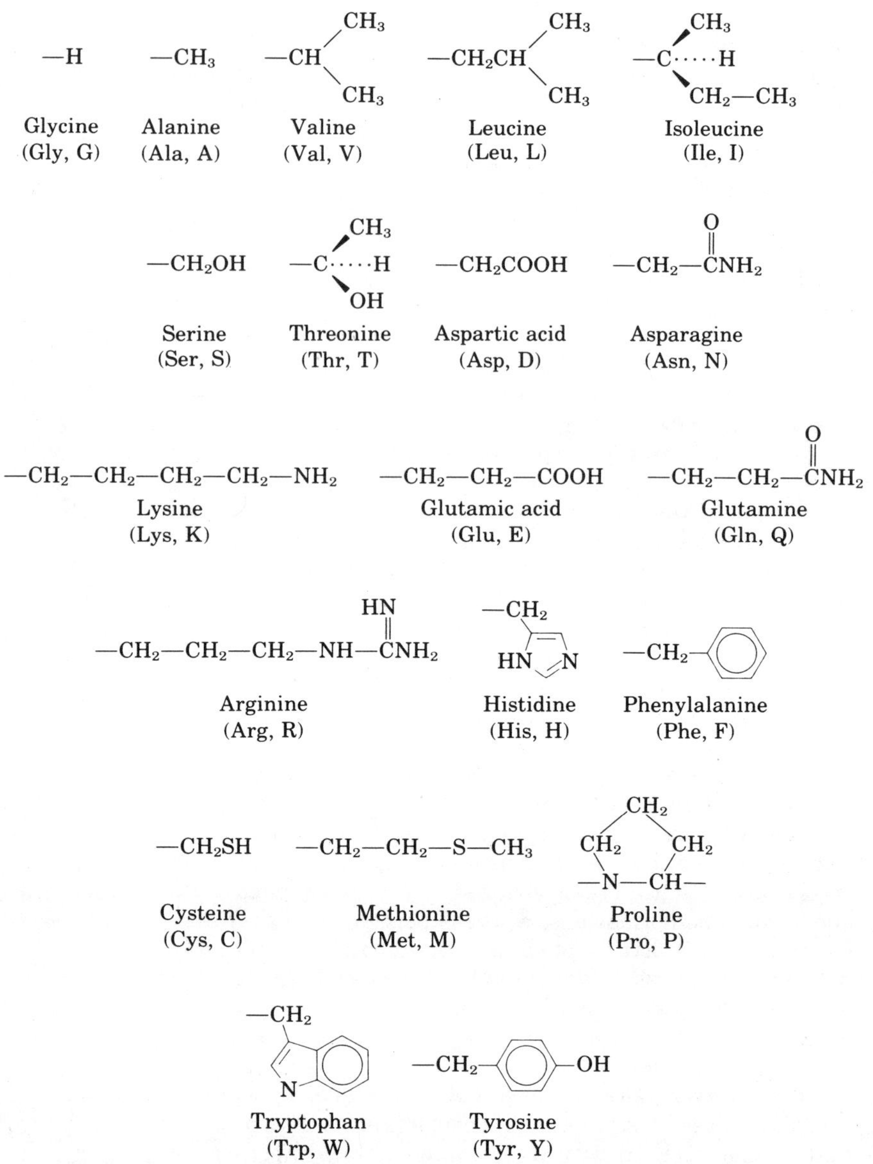

Figure 1. Side chains of the 20 amino acids used to synthesize proteins. The α-carbon and N atoms of the backbone are also included in the unique case of Pro. Ile and Thr have asymmetric centers in their side chains, but only one isomer is used biologically.

they may be removed from either or both ends. The polypeptide chain may be cleaved in other places. In concanavalin A the original single polypeptide chain is cleaved in one place and the original termini are joined in a new peptide bond to produce the single polypeptide chain with a permuted sequence. In other proteins the amino terminus can be acetylated, or blocked in other ways, and the carboxyl terminus can be amidated.

Some modifications occur nonenzymatically and nonspecifically, but most are made only to certain individual residues of specific proteins by specific enzymes; the vast majority of residues are not altered.

Length of Polypeptide Chain. Proteins vary tremendously in length, from about 50 amino acid residues up to at least 4500; therefore, the most pertinent property of a new protein to be determined initially is its size. Proteins often exist in large aggregates, with multiple copies of the same chain or as a complex with other materials, so it is normally necessary to dissociate such complexes into the individual chains.

The mass of an individual polypeptide chain may be determined by the usual hydrodynamic methods, such as ultracentrifugation, or by gel filtration, if it is kept unfolded in a denaturing solvent. Guanidinium chloride or thiocyanate at concentrations of 6 *M* are the most commonly used solvents.

The easiest and most widely used method for obtaining approximate molecular weights is an electrophoretic one in which the usual dependence of electrophoretic mobility on both conformation and net charge is minimized by adding the detergent sodium dodecyl sulfate (SDS).

Amino Acid Composition. Determining the amino acid composition of a protein is now a standard laboratory procedure. The polypeptide chain is generally hydrolyzed to the constituent amino acids by 6 *M* HCl at 110°C for 24–72 h. With this procedure most of the amino acids are recovered quantitatively, but tryptophan is generally destroyed, so that it must be detected by other techniques. Alkaline or enzymatic hydrolysis can give good yields of tryptophan, or it can be measured by spectral or chemical techniques.

Primary Structure. The amino acid sequence identifies a protein, distinguishing it from all other proteins, and determines

Table 1. Properties of Isolated Amino Acid Residues

Residue	One-letter abbreviation	Residue[a] mass, daltons	Accessible surface area[b], nm^2	van der Waals volume, cm^3/mol	Partial molar volume, cm^3/mol	Partial specific volume, cm^3/g	Occurrence in proteins, %
Ala	A	71.09	1.15	40.6	52.0	0.732	9.0
Arg	R	156.19	2.25	89.2	118.9	0.756	4.7
Asn	N	114.11	1.60	57.5	69.6	0.610	4.4
Asp	D	115.10	1.50	54.8	65.4	0.573	5.5
Cys	C	103.15	1.35	51.9	65.0	0.630	2.1
Gln	Q	128.14	1.80	68.5	85.5	0.667	3.9
Glu	E	129.12	1.90	65.7	77.5	0.605	6.2
Gly	G	57.06	0.75	28.8	34.8	0.610	7.5
His	H	137.15	1.95	71.3	90.4	0.659	2.1
Ile	I	113.17	1.75	74.4	99.1	0.876	4.6
Leu	L	113.17	1.70	74.4	99.1	0.876	7.5
Lys	K	128.18	2.00	81.6	100.1	0.775	7.0
Met	M	131.20	1.85	74.6	96.9	0.739	1.7
Phe	F	147.18	2.10	81.3	112.8	0.766	3.5
Pro	P	97.12	1.45	54.1	72.6	0.748	4.6
Ser	S	87.09	1.15	44.2	51.9	0.596	7.1
Thr	T	101.11	1.40	56.0	68.4	0.676	6.0
Trp	W	186.22	2.55	98.1	135.5	0.728	1.1
Tyr	Y	163.18	2.30	85.2	114.7	0.703	3.5
Val	V	99.14	1.55	63.4	82.4	0.831	6.9
weighted average[c]		108.78	1.58	61.1	71.5	0.703	

[a] Molecular weight of nonionized amino acid minus that of water.
[b] Surface area accessible to water of residue X in a tripeptide Gly–X–Gly.
[c] Weighted by occurrence in proteins to give value for average residue in globular proteins.

all its chemical and biological properties. Appropriately, the sequence is also known as the primary structure, since it is the most basic level of structure and also defines the higher levels of structure (secondary, tertiary, and quaternary). The following procedures make it possible to determine the amino acid sequence of a protein: identifying the amino-terminal and carboxyl-terminal residues, using the Edman degradation and mass spectrometry to determine the sequence of a protein after fragmentation into smaller peptides, and identifying disulfide bonds, when present. At the present time, protein sequences are most readily determined from gene sequences.

Nature of Amino Acid Sequences. ***Known Protein Sequences.*** The known sequences of proteins have been compiled in several books, but are now deposited in computer databases. As of the middle of 1989 the sequences of about 6000 proteins were known, comprising nearly two million amino acid residues. Many of the protein sequences were determined only from their gene sequences.

Most proteins have very similar amino acid compositions; the frequencies with which the amino acids occur are tabulated in Table 1. Substantial deviation from this average composition is rare for globular proteins that normally are dissolved in aqueous solution. For such proteins it is possible to describe an average amino acid residue; it has a mass of 109 daltons, a van der Waals volume of 61.1 cm^3/mol, and a surface accessible to water of 1.58 nm^2 (Table 1).

Proteins that are normally embedded in membranes have somewhat higher levels of hydrophobic amino acids (Leu, Ile, Val, Met, Phe, Tyr) and slightly lower levels of those normally ionized (Lys, Arg, Glu, Asp, His).

The most atypical compositions are usually those of fibrous proteins, such as collagen, that have regular conformations and repetitive amino acid sequences.

Proteins Related Through Evolution. Proteins with very similar amino acid sequences are most likely to be so similar because they arose from a common evolutionary ancestor; they usually have similar, if not identical, biological functions. Conversely, proteins with the same role from closely related species invariably have very similar amino acid sequences. For example, the *c*-type cytochromes, the hemoglobin α, β, and γ chains, and the fibrinopeptides from chimpanzees are identical to those of humans. Their myoglobin and hemoglobin δ chains have only one amino acid difference each. These two species are believed to have diverged evolutionarily some 4–8 million years ago. The sequences of proteins, along with those of nucleic acids, provide the most convincing evidence for the validity of Darwinian evolution.

Genetic Manipulation of Protein Structure. Mutation of the genetic information is the most specific way of altering the amino acid sequence of a protein. With the recent revolution in manipulation and sequencing of nucleic acids, it is now possible to produce any desired alteration of the protein amino acid sequence, from insertion, deletion, or rearrangement of blocks of residues to changing one amino acid residue to any other. The results of this new technology are just now becoming available.

Three-dimensional Structures

Proteins of biological origin differ dramatically from most other polymers, and from synthetic polypeptides of random or repetitive sequence, in that they usually have unique folded conformations. Structural proteins tend to adopt relatively simple, regular structures in which most of the residues have the same conformation. This leads to extended, linear structures, the basis for which can often be found in regularities in the amino acid sequence. Globular proteins have complicated, irregular three-dimensional structures that usually give them compact, roughly spherical overall shapes. Their well-defined folded conformations can be determined by x-ray or neutron diffraction analysis.

Protein crystals are unlike those of small molecules in that the crystal lattice interactions are relatively few and have only a small effect on the conformation of the protein. In general, crystalline proteins are not very different from when they are in solution. Their crystal structures give detailed information not only about their three-dimensional structures, but also about their flexibilities and interactions with the solvent.

Polypeptide Conformation. The folded conformations of proteins are possible because the polypeptide backbone is sufficiently flexible, simply because of rotations about single covalent bonds, for it to turn back on itself to give a globular, close-packed structure.

A portion of the backbone of a polypeptide chain is shown in Figure 2.

Levels of Protein Structure. It is customary to discuss protein structure in terms of four levels. The primary structure is the amino acid sequence. Secondary structure is any regular local structure of a linear segment of a polypeptide chain, such as a helix or an extended strand. Tertiary structure is the overall topology of the folded polypeptide chain. Quaternary structure is the specific aggregation of the folded units, either of the same or different polypeptide chains. The crystal structure of a protein gives information about all four levels of protein structure, although independent knowledge of the primary structure is generally required, and interactions between different protein molecules must be measured quantitatively in solution. All the levels of protein structure are related, but the division into the four levels is generally worthwhile.

Although protein structures are extremely diverse in their molecular architecture, they have some common properties. These general properties present a consensus picture of a protein structure, to which there are many exceptions and of which there are many variations. The common properties illustrate the general rules of protein architecture, but each protein is unique and generally attains its functional properties by incorporating specific exceptions to these generalities. This section briefly describes the common properties, whereas the remainder of this article is more concerned with the exceptions.

Tertiary Structure. The overall folded structures of small proteins are generally roughly spherical in overall shape, but with a very irregular surface, and are remarkably compact. When a protein consists of more than about 200 residues, the structure usually appears to consist of an assembly of two or more roughly spherical tertiary structural units, generally referred to as domains. The individual domains are often associated, but they interact less extensively than do portions within the domains.

Secondary Structure. The extended segments of the polypeptide chain often have rather regular conformations like those observed in model polypeptides. Most prominent are the right-handed α-helices (Fig. 3) and the extended β-strands,

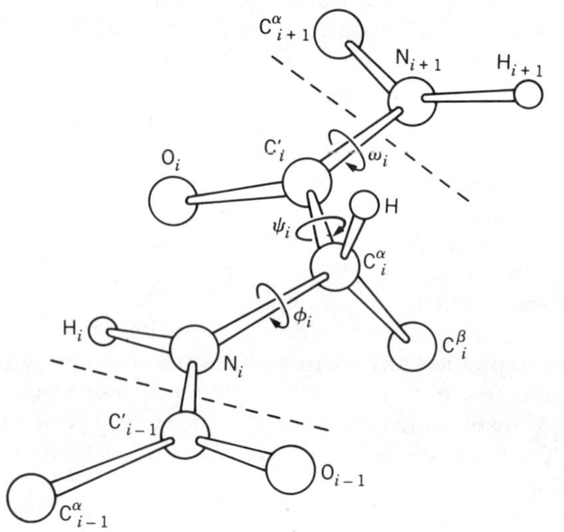

Figure 2. The torsional angles of the polypeptide backbone that determine the backbone conformation of a protein. The limits of one residue, number i of the chain, are indicated by the dashed lines. The usual notations for atoms and torsion angles are indicated; the polypeptide chain is shown in the fully extended conformation, where $\phi = \psi = \omega = \pm 180°$.

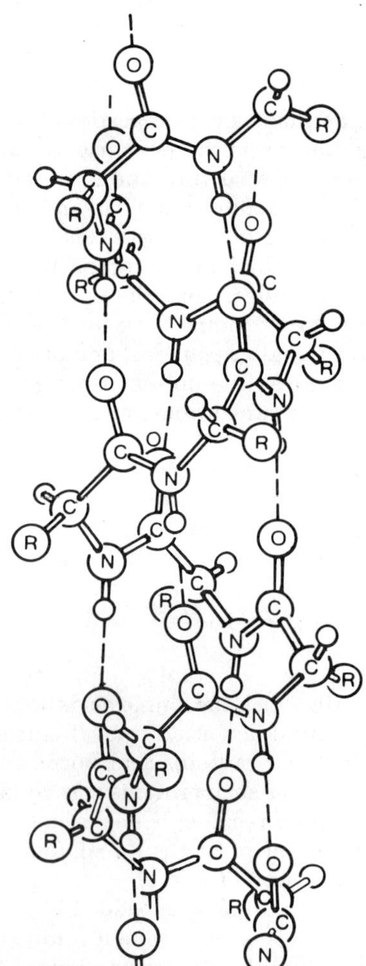

Figure 3. The structure of a right-handed α-helix.

which generally associate side by side into parallel or antiparallel β-sheets (Fig. 4). Approximately 31% of the residues in known proteins occur in α-helices; 28% occur in β-structures.

Regular secondary structure is a useful focal point in attempts to comprehend a complex protein structure, so many studies of it have been made, and most schematic drawings of proteins emphasize it, often in an idealized and attractive manner.

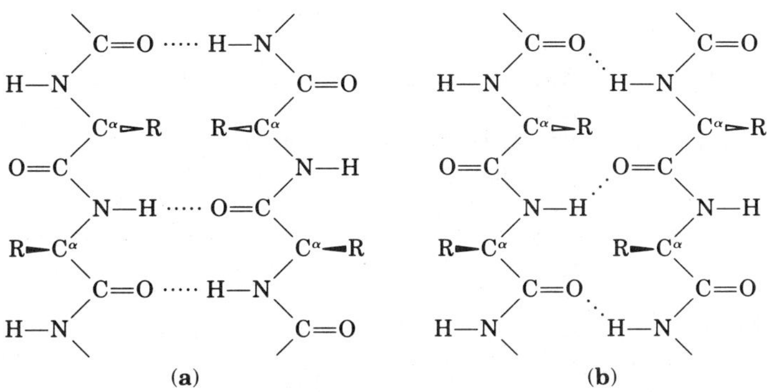

Figure 4. The hydrogen-bonding patterns of (**a**) antiparallel and (**b**) parallel β-sheets. Those found in proteins are usually not planar, but have right-handed twists of varying extents. In three dimensions the amino acid side chains R of each strand alternately project above and below the plane of the sheet.

Reverse Turns. Proteins have roughly spherical structures, in spite of being composed of relatively straight segments of polypeptide chain, because the polypeptide chain makes relatively sharp bends at the surface, thereby reversing the direction of the polypeptide chain. These reverse turns give proteins their globularity, and a substantial fraction of residues of every protein are involved in them. Reverse turns can consist of large loops or tight turns with a minimum of residues; the latter are also known as hairpin bends, β-bends, and β-turns because they often connect antiparallel β-strands. The polypeptide chain is generally rather extended, so sharp chain reversals require somewhat special geometry; not surprisingly, residues of the special amino acids Gly and Pro are often involved.

Supersecondary Structures. Certain assemblies of a number of secondary structure elements, including the connecting segments of polypeptide chain, have been observed a sufficient number of times that they are becoming recognized as a further level of structure, termed supersecondary structure. These structures are at a higher level than secondary structure, but do not constitute entire structural domains.

Some general rules have been observed. Segments of secondary structure that are adjacent in the sequence are also often adjacent in the three-dimensional structure, and the connections between secondary structures do not cross each other or form knots. The most commonly observed units are α-α, β-β, and β-α-β units.

Protein architecture appears to consist primarily of segments of secondary structure packing together, so current efforts are directed at understanding the basis of the interactions between helices, between β-sheets, and between helices and sheets. Because of their basic architecture, the interactions between them are governed primarily by the amino acid side chains on their surfaces.

Interiors and Exteriors. Detailed analyses of the very complex surfaces of folded proteins and of the internal packing of atoms have generally used the procedure described in Figure 5.

Quaternary Structure. Many proteins are observed in solution to exist as aggregates of two or more polypeptide chains, either identical or different. Different polypeptide chains are designated by letters; for example, hemoglobin is normally $\alpha_2\beta_2$, aspartate transcarbamylase is r_6c_6, and many proteins are dimers, trimers, tetramers, or even higher order aggregates of identical polypeptide chains. In many instances, such quaternary structure is vital for the structure and function of the protein. Because it is invariably observed in the crystal structure, the detailed nature of these interactions is known.

Each subunit is generally folded into an apparently independent globular conformation, which then interacts with other

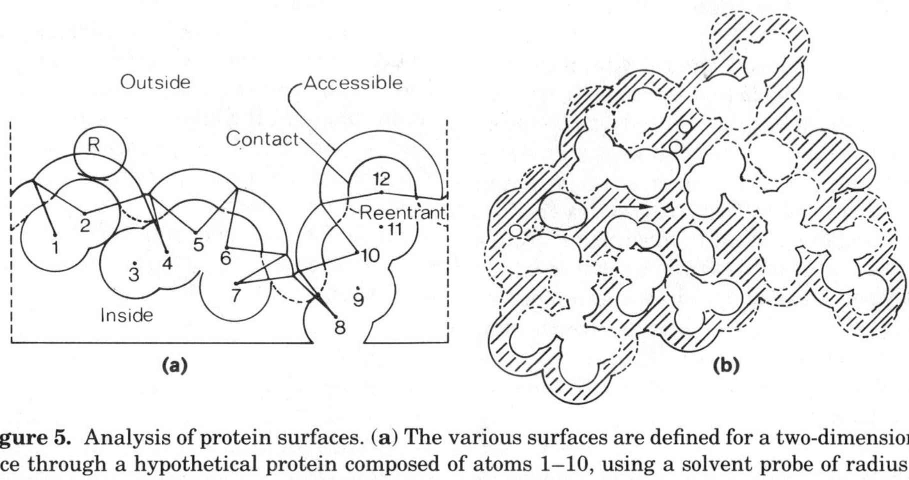

Figure 5. Analysis of protein surfaces. (**a**) The various surfaces are defined for a two-dimensional slice through a hypothetical protein composed of atoms 1–10, using a solvent probe of radius R. Courtesy of Annual Reviews, Inc. (**b**) The van der Waals and accessible surfaces of a section through ribonuclease S. The shaded region is enclosed by the accessible surface; the van der Waals volumes of the atoms are in white. The solid lines enclose carbon and sulfur atoms, the dashed lines nitrogen and oxygen atoms. In places, the accessible surface is determined by atoms above or below the section shown. The arrow indicates a cavity inside the protein large enough to accommodate a probe molecule of radius 0.14 nm, such as a water molecule. Courtesy of Academic Press Inc. (London) Ltd.

subunits; the greatest exception is the dimeric *trp* repressor, where the two subunits are intimately intertwined. The interfaces between different subunits are generally very similar to the interiors of the individual molecules; they are closely packed and involve both hydrophobic and polar interactions. Consequently, the interacting surfaces of the individual molecules are highly complementary, both in shape and in pairing of polar groups.

Two fundamental types of interactions between identical subunits are possible, designated isologous and heterologous. Isologous association involves the same surfaces on both monomers, which associate to produce a dimer with a twofold rotation axis of symmetry. The two monomers are equivalent. This requires that the two halves of the interface of the monomer be complementary about a "mirror plane" that coincides with the symmetry axis in the dimer.

Heterologous association involves association of two different sites that are complementary and largely nonoverlapping. In such a dimer, the two monomers are not equivalent, and each has one binding site unpaired, which tends to bind to another monomer, unless sterically blocked.

Flexibility. The protein models resulting from crystallographic structure determination are invariably static, but this should not be taken to imply that the molecule is equally as static. It is clear that proteins in a crystal lattice and in solution have varying degrees of flexibility.

Solvent Structure. The structure of the aqueous solvent surrounding protein molecules is an important aspect of protein structure. Because about half of a protein crystal is solvent, some aspects of the solvent structure can be inferred from protein crystal structures. However, the electron density maps are averaged over both the extensive time required to collect the crystallographic data and the many molecules of the crystal lattice. Consequently, the bulk solvent gives only a uniform, average electron density throughout the nonprotein areas of the crystal. Of course, the electron density map does not identify the molecules, and crystal solvents usually contain high concentrations of salts or other agents to induce crystallization, but most such fixed molecules appear to be water molecules, and this is generally assumed in the absence of indications to the contrary.

The degree of order of the solvent is generally proportional to its proximity to the protein surface and to its participation in hydrogen bonding. The most highly ordered water molecules, with B values as low as $0.13\ \mathrm{nm}^2$, are those extensively hydrogen bonded in crevices on the protein surface or bridging different molecules in the crystal lattice. Almost invariably, they are anchored by hydrogen bonding to fixed polar groups of the protein surface, and one or more firmly anchored water molecules can apparently serve to fix adjacent water molecules in a hydrogen-bonded lattice. Beyond this there is a continuum of degree of order of solvent molecules, with B values increasing to the point where it is no longer deemed worthwhile to include them in the structure.

The solvent around protein molecules has also been examined with other physical methods, with results broadly in line with the crystallographic observations. Three types of water molecules have been distinguished: the first is like bulk water, with a rotational relaxation time of only 10^{-11} s, and constitutes the largest fraction. The second type is bound water, with a longer relaxation time of about 10^{-9} s and an altered freezing point, presumably produced by interactions with the protein surface. The amount of this water type is approximately 0.3 g per gram of protein, or roughly 2 water molecules per average amino acid residue, which is somewhat greater than the number observed crystallographically. The third type of water molecule comprises those bound very tightly, not rotating significantly with respect to the protein. This type represents only a small fraction of the total solvent and presumably corresponds to those water molecules observed crystallographically to be most firmly fixed.

Proteins with Similar Folded Conformations. With the wide diversity in amino acid sequences observed in proteins, it might be expected that there would be a corresponding diversity in folded conformations. In contrast, many similarities between the folded conformations of proteins have been observed, and their tertiary structures appear to have been much more conserved during evolution than have their primary structures. The degree of this evolutionary conservation is a matter of intense debate and speculation, as similarities in tertiary structures are being found in cases where there was no evolutionary or functional reason to expect them. These findings raise fundamental questions about the evolution and the physical basis of protein structure that are still far from being answered.

Proteins in Aqueous Solution

Proteins occur naturally in a variety of environments, such as the solid state of keratins and other structural proteins (see KERATIN; COLLAGEN; SILK), large supramolecular structures like viruses, the interiors or surfaces of membranes, or simply aqueous solutions. The majority of proteins, certainly of those studied extensively, normally exist in the soluble state in aqueous solution, and this is the form in which they normally are studied. It is now evident that the structures of proteins in solution are basically the same as those determined in the crystal, except that the conformations of flexible groups on the surface may be frozen in specific conformations by interactions in the crystal lattice. The same basic conformation of the core of the structure is observed even when a protein is crystallized in very different ways, and radically different folded conformations have never been demonstrated for a single protein. The folded conformation of each structural domain appears to be unique, and the changes in folded conformations that have been demonstrated unequivocally are either small local changes of a few groups or large rearrangements of mostly unaltered subunits or domains that correspond to quaternary structure changes. The other conformational change that has been observed experimentally is complete unfolding to a random coil-like state.

Solubility. The solubilities of proteins in aqueous solutions vary enormously, depending primarily on the protein, but also on the conditions. Some proteins are immensely soluble in water and can compose up to 35% of the volume of a saturated solution. Others, especially structural proteins, are essentially insoluble and exist normally as solids, aggregated into complexes of varying sizes and specificities. Many other proteins are relatively insoluble in water, but are sequestered into membranes. However, most proteins can be solubilized in aqueous solutions by adding detergents or denaturants, such as urea or guanidinium salts. On the other hand, even highly soluble proteins can be precipitated from aqueous solutions by adding salts such as ammonium sulfate.

Physical and Chemical Properties. ***Hydrodynamic Properties.*** The volume occupied by a protein in aqueous solution is usually close to that calculated from the volumes in aqueous solution of the individual amino acid residues that make up the protein. The partial specific volumes of proteins calculated from their amino acid composition are generally within 2% of the values measured experimentally and are commonly used in hydrodynamic measurements. The agreement between the computed and measured volumes indicates that the packing within protein interiors and at the surface, including the hydration solvent, is as efficient, on average, as the packing of liquid water molecules around the isolated amino acid in solution; it is consistent with the close packing and absence of cavities observed crystallographically.

Because of their compact conformations, folded proteins are usually resistant to hydrolysis by proteases. Only large flexible loops on the surface and flexible joins between domains tend to be cleaved readily. The cleaved products often remain associated and folded, depending upon the degree of association between them.

Spectral Properties. The spectral properties of folded proteins are usually very different from those of random polypeptides because of the variety of environments for the chromophores in the folded structure. The absorbance of Phe, Tyr, and Trp residues buried in the interior is usually red shifted and insensitive to small changes in the solvent. The fluorescence of these residues can be either increased or decreased, often drastically, depending on their immediate environment. Circular dichroism and optical rotatory dispersion spectra of proteins can vary widely, especially in the aromatic wavelength region; in the far ultraviolet, the polypeptide backbone predominates, so such spectra often can be used to estimate the fraction of residues that are in the various types of secondary structure. Laser Raman spectra also are capable of measuring secondary structure contents of proteins.

The nmr spectra of proteins are also affected by their three-dimensional structures, and the high resolution of this technique has made it the most extensively studied at the present time, especially proton nmr. In particular, it can be used to elucidate the three-dimensional structures of small proteins in solution.

Chemical Reactivities. The unique environments of reactive groups in folded proteins can have substantial effects on their chemical properties. Their reactions with chemical reagents can be entirely blocked or greatly enhanced, and they can react readily with reagents with which normal groups show little or no reactivity. In addition to environmental effects, it appears that such cases of hyperreactivity usually result from high local effective concentrations of the reagent because it binds to the protein; the reactions then become essentially intramolecular. Unexpected intramolecular reactions in proteins can also result from groups being held in suitable proximity in the folded protein conformation.

Flexibility. Folded proteins have an immense range of different types of flexibility. Groups on the surface are often as flexible as those in an unfolded polypeptide or in any other polymer, but within the protein interior the close packing requires that many atom movements be considerably restricted and occur only in concert with adjacent parts of the structure. The measured adiabatic compressibilities of proteins are low and usually in the range 1 to 10×10^{-7} cm²/N, which probably indicates only small fluctuations in volume.

The most convincing evidence of protein flexibility is the ability of all labile hydrogen atoms (those bonded to peptide, amide, hydroxyl, and amino groups) to exchange with hydrogen isotopes of the solvent. Even the most buried groups exchange at a nonzero rate, although it might be only 10^{-8} of the rate exhibited by a solvent-accessible group.

Unfolding and Refolding. ***Equilibrium Transitions.*** The folded states of proteins are only marginally stable, even under optimal conditions. Consequently, proteins may be denatured, ie, unfolded, merely by altering the environment, by raising or lowering the temperature, varying the pH, increasing the pressure, or adding to the solvent a variety of chemicals, known as denaturants.

Unfolding can be followed experimentally by many techniques, and usually all detect the same single transition with small single-domain proteins. This is consistent with there being a simple two-state equilibrium between the fully folded (N) and unfolded (U) states, with partially unfolded states not populated substantially.

With some small proteins under certain conditions, a third, intermediate conformational state is most stable. It is nearly as compact as the native state and appears to have similar amounts of secondary structure, but does not have the amino acid side chains in asymmetric environments, undergoes rapid hydrogen exchange, and unfolds noncooperatively. It has been named the "molten globule" state, but its three-dimensional structural properties are otherwise unknown.

Kinetic Pathways of Unfolding and Refolding. In all cases studied thus far, protein folding is a self-assembly process that occurs spontaneously under the appropriate conditions. Proteins that cannot refold are assumed to have undergone covalent modification or to be irreversibly aggregated, although there are only a few well-documented examples. The single-chain precursor, proinsulin, was postulated because mature two-chain insulin does not refold and reassemble efficiently.

The kinetics of unfolding are usually simple and two-state, with a single rate-limiting step apparent; partial unfolding does not seem to occur to any significant extent. In contrast, the kinetics of refolding are usually complex. Many such kinetic complexities are due to two or more populations of unfolded protein molecules that differ in their rates of refolding. The simplest kinetic scheme with fast (U_F) and slow (U_S) refolding species is

$$U_S \underset{\text{slow}}{\overset{\text{slow}}{\rightleftharpoons}} U_F \xrightarrow{\text{fast}} N$$

If the initially folded protein N is rapidly unfolded and then immediately refolded, all the molecules refold rapidly because only U_F is formed. However, if refolding is delayed sufficiently for $U_F \rightleftharpoons U_S$, a fraction of the molecules refold slowly. Consequently, the interconversion of U_F and U_S is slow in both directions.

Stability of the Folded State. The net stability of the folded state depends on the difference in free energies of the folded and unfolded states, including the interactions of both with the solvent. The optimal net difference of 30–70 kJ/mol (7.2–17 kcal/mol) is a relatively small difference between much larger sums that include many contributions to the free energy of each state. The principal contributions to the favorable free energy of the folded state are believed to be the many noncovalent interactions, especially hydrogen bonds and van der Waals interactions, that occur simultaneously in folded structures. The unfolded state differs primarily in having a much greater conformational entropy and many more interactions with the solvent, which are unfavorable in the case of nonpolar groups.

Prediction of Protein Conformation. Since the specific folded conformation of a protein is believed to be dictated solely by its amino acid sequence, it should be possible to predict the folded conformation from just the sequence. However, this has not yet been possible, primarily because of the extremely large number of conformations theoretically possible and uncertainty about how and why folding occurs. Also, the large variation in amino acid sequence possible with essentially the same folded conformation indicates that there is substantial redundancy in the rules relating primary and tertiary structures.

The one instance in which a tertiary structure can be predicted with reasonable certainty is when the amino acid sequence is homologous to that of a protein of known structure, for the two will almost certainly have essentially the same backbone conformation.

Membrane Proteins

Membrane Properties. Membranes are vital for living organisms, providing an impermeable barrier between each cell and its environment. They also divide many cell interiors into various compartments, such as the nucleus, endoplasmic reticulum, Golgi apparatus, lysosomes, secretory vesicles, etc. Proteins compose between 25 and 75%, but typically closer to 50%, of the mass of various membranes and mediate the various functions of the membrane, such as transport of specific molecules into and out of the cell, catalysis of chemical reactions, receiving and transducing chemical signals from the cell environment, or maintaining the membrane structure.

Membranes are lipid bilayers, composed of two-dimensional aggregates of lipid molecules, which are amphipathic in having nonpolar hydrocarbon "tails" and polar "head" groups. The bilayers are arranged so that the two nonpolar surfaces of both layers are in contact in the middle and the polar surfaces are in contact with water.

Insertion of Proteins into Membranes. Integral membrane proteins are found arranged in membranes with a variety of topologies. They can span the membrane once or several times, with the amino and carboxyl termini on either side of the membrane (Fig. 6). Polypeptide loops exposed on the extracellular side of the membrane often have oligosaccharides attached. How the proteins attain these topologies is currently being studied intensively.

Structures of Membrane Proteins. Membrane proteins do not readily form three-dimensional crystals, but can form two-dimensional arrays within lipid bilayers. Nevertheless, the crystal structure of one extremely important membrane protein, the photosynthetic reaction center, has been determined. The membrane proteins of known structure consist primarily of nonpolar helices within the hydrophobic membrane interior.

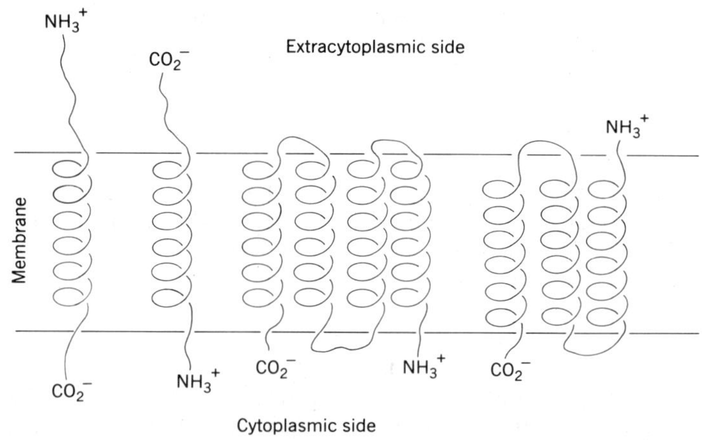

Figure 6. Observed topologies of proteins in membranes. The polypeptide chain traversing the membrane is depicted as a helix.

Identifying Amino Acid Sequences Likely to Traverse Membranes. Even though membrane proteins are not markedly different from water-soluble proteins in amino acid composition, considerable success has been achieved simply by looking for hydrophobic stretches of amino acid residues, like those observed in membrane proteins. Success has been possible especially in those cases where the segments are very nonpolar, with no hydrophilic residues.

Dynamic Behavior in Membranes. Membrane proteins generally diffuse rapidly within the two-dimensional plane of the membrane, with diffusion coefficients of about 10^{-10} cm²/s, unless interacting with other molecules inside or outside the membrane. On the other hand, they usually retain their vertical orientation within the membrane and do not flip between the two surfaces. The lipids move even more rapidly, with diffusion coefficients of 10^{-8} cm²/s, but only very infrequently move from one side of the bilayer to the other. Proteins within a membrane induce disorder into the lipid bilayer and restrict the diffusion of neighboring lipids. However, the restricted lipids exchange rapidly with others of the membrane, so the interactions with the protein are weak and nonspecific.

Functions of Membrane Proteins. Relatively little is known at the molecular level about how membrane proteins function. Those involved in transport of molecules across the membrane, either passively or actively against a concentration gradient, form a continuous protein pathway through the membrane, presumably incorporating an appropriate channel. In some cases the channel must be formed by a single polypeptide chain; in others it is probably between multiple subunits making up the active structure. Even with passive transport, these channels are usually open only under specific circumstances, and how this is controlled is extremely important, but not clear.

Interactions with Other Molecules

The biological functions of proteins invariably depend on their direct, physical interaction with other molecules of widely varying natures: from electrons, to ions, to sugar molecules, to polymers such as polysaccharides and nucleic acids, to other proteins, to membranes and other surfaces. Virtually every substance found within a cell is bound either by the enzyme that produced it or by the receptor protein on the cell surface that enabled it to be taken up into the cell. Living organisms are coordinated by interactions between cells and molecules, virtually all involving at least one protein.

Binding Ligands. The term "ligand" is used here to designate any molecule with which a protein interacts directly and specifically. Interactions with water or membrane lipid molecules of the solvent are not included, nor is protonation of ionic groups on proteins.

The binding of a ligand to a protein is characterized by the site on the protein to which it is bound, by its affinity for the site, ie, the value of its equilibrium association or dissociation constant, and by the rates of its binding and its dissociation.

Structures of Protein Complexes. ***Experimental Determination.*** The structures of many complexes of proteins with many ligands have been determined by x-ray crystallography; often, the structure of the unliganded form is also available for comparison. Chemical methods also can be used to identify the parts of a protein to which a ligand binds.

Binding Sites. The binding sites for ligands tend to be on the surface of the protein, unless the ligand is very small or is held very tightly and an integral part of the protein structure. Very small ligands, such as electrons or O_2 molecules, tend to be bound by special prosthetic groups in proteins, probably because a simple protein alone could not provide a suitably specific binding site.

In general, ligands are held by multiple noncovalent interactions within specific binding sites. Small molecules are usually bound in slight depressions on the surface, and polymeric ligands in grooves across the surface. Large ligands, such as other proteins, are usually bound by multiple interactions over one side of the protein, generating a large interface between the two.

Most proteins have one binding site for each ligand on one polypeptide chain, but there are exceptions.

The interactions between proteins and ligands are extremely diverse, but are based on steric and physical complementarity between the two. The interface between them is closely packed, polar groups are paired in hydrogen bonds, and electrostatic charges are generally neutralized or dissipated. The interface between two interacting protein molecules is basically the same as the interior of either.

Binding sites are usually, but not always, preexistent in the folded protein, but binding always involves some, usually small, conformational adjustments of both protein and ligand; the adjustments may be important for the kinetics of association and dissociation.

Relationship Between Protein Conformation and Binding. Binding sites compose relatively little of the protein structure and are usually on or near the surface, so relatively small localized alterations of the protein can produce large changes in affinities for ligands. Consequently, homologous proteins can bind very different ligands. Conversely, the same ligand can be bound by different proteins with different conformations.

Antigenicity of Proteins. Proteins usually make excellent immunogens and antigens, readily eliciting an immune response under the appropriate circumstances. The antigenicity of proteins is important medically since proteins and peptides can be used as vaccines, but their immunogenicity also can limit the pharmaceutical use of proteins. Antibodies are also useful tools for studying proteins because they can be extremely specific and sensitive. This is also an interesting example of protein–protein recognition, since antibodies are also proteins.

Cooperativity in Ligand Binding: Allostery. ***Interactions Between Binding Sites.*** The binding of one ligand to a protein can alter its affinity for another ligand molecule. The simplest case is where two ligands occupy the same or overlapping binding sites and consequently compete directly for binding. However, much more interesting and important are interactions between different binding sites, a phenomenon known as allostery. The interacting ligands can be identical or unrelated, known, respectively, as homotropic and heterotropic interactions. The binding of one ligand can either increase or decrease the affinity of the protein for the second ligand, respectively, positive and negative cooperativity. The classic example is hemoglobin, which displays both homotropic and heterotropic interactions between its ligands O_2, organic phosphates, H^+ and Cl^- ions, and CO_2.

Ligand Binding and Protein Flexibility. Since proteins are inherently flexible, it might be thought likely that binding of a ligand anywhere on a protein would invariably affect all other binding sites present. Only small structural perturbations would be required because ligand affinities in allosteric proteins are observed usually to vary only about 100-fold, although there are some larger effects of 10^4- to 10^5-fold. Allosteric effects are not the rule, however, and many proteins bind two or more ligands independently. The reason for this appears to be that ligand binding to a monomeric, single-domain protein usually produces minimal perturbations of its folded conformation detectable by crystallography. Any small changes are usually localized near the binding site.

The Concerted Allosteric Model. Proteins demonstrating interactions between nonoverlapping binding sites are invariably oligomeric, consisting of at least several identical polypeptide chains and often also of two different polypeptides, or have multiple domains. Upon binding an appropriate ligand they demonstrate very large conformational changes sufficient to disrupt a crystal lattice; in contrast, nonallosteric proteins generally bind ligands within an unaltered crystal lattice, unless the lattice contacts are directly perturbed by the ligand. The large conformational changes in allosteric proteins involve rearrangements of relatively unaltered subunits or domains, primarily quaternary structure changes.

These properties of allosteric proteins led to proposal of the concerted model for allosteric interactions that envisages the binding of ligand at one site to have no direct effect on the affinity of the other sites, but only to alter the conformational equilibrium between two alternative quaternary conformations, T (tense) and R (relaxed), of the protein.

The Sequential Model. The other extreme allosteric model envisages direct effects of ligand binding at one site on all other binding sites. In this case virtually any type of allosteric interaction would be possible, depending only on what conformational changes are induced upon ligand binding. Each binding site on a protein could be affected independently, its affinity for its particular ligand being either increased, decreased, or unaffected, and binding different ligands at a site could have different effects.

At the present time it is impossible to conclude that any allosteric protein obeys exactly one model and not the other. The best-understood proteins, such as hemoglobin, can be explained to a large extent by the concerted model, but it is likely that some direct effects between binding sites also occur, as in the sequential model.

Negative Cooperativity. Homotropic allosteric interactions in which the binding of one molecule of ligand decreases the affinity of other sites for the same ligand are unlikely with the concerted model; binding of ligand can pull the T↔R transition only toward the high affinity state and can give only positive cooperativity. Nevertheless, negative cooperativity is observed with some proteins; in the extreme cases of half-of-the-sites reactivity, only half the expected sites on an oligomeric protein bind ligand or react with a modifying reagent. Such findings are usually taken as evidence against the concerted model.

Catalysis. The interactions of protein with other molecules are very often followed by covalent changes in these molecules. Such a protein is then classified as an enzyme (E), and these ligands are designated as either its substrate (S) or product (P), depending on the direction in which the chemical reaction is considered:

$$E + S \rightleftharpoons E{\cdot}S \rightleftharpoons E{\cdot}P \rightleftharpoons E + P$$

Enzymes are only catalysts and do not alter the overall equilibrium constant between the substrates and products, but they do increase the rate of the reaction in both directions, usually by many orders of magnitude. They accomplish this by bringing substrates into correct proximity within the unique environment of the active site, with functional groups at high effective concentrations, where a simultaneous reaction between one or more substrate molecules and numerous catalytic groups of the enzyme becomes essentially unimolecular.

Degradation

Chemical and Physical Destruction. Proteins have only limited life spans. They are subject to a wide variety of chemical modifications. Even under optimal conditions their sulfur atoms become oxidized, Gln and Asn residues become deamidated, peptide bonds can be hydrolyzed, their α-carbon atoms racemize to the D-isomer, and the asymmetric side chains of Thr and Ile residues epimerize. The folded conformation of a protein can produce high effective concentrations between groups and, consequently, unexpected reactions between them. In the presence of other materials they can be subject to a wide variety of covalent modifications. These changes can have varying effects on the folded conformations and biological activities of proteins, but most of them are deleterious. Such changes are particularly important in that they impose limitations on the industrial and pharmaceutical use of proteins, and are important for the food industry since cooking at high temperatures accelerates most of the chemical reactions involving proteins.

Detection and Isolation

Measurement of Amino Acids, Peptides, and Proteins. A number of different procedures are available for determining the total protein content of samples. The most convenient and accurate uses their ultraviolet absorbance, either around 280 nm, where the aromatic side chains absorb, or at lower wavelengths such as 210 nm, where the peptide bond also absorbs. Proteins vary widely in their spectral properties; for quantitative results, the molar absorbance value of each protein must be determined. The most widely used reagents that detect most amino acids, peptides, and proteins are ninhydrin, fluorescamine, dyes such as Coomassie brilliant blue, and silver, which produces a photographic type of reaction with proteins.

Detection of Specific Proteins. Specific proteins are usually detected initially by their biological activities. This is often an enzymatic activity, but many different assays are possible.

Proteins can sometimes be identified by their physical properties. The size and hydrodynamic properties of a protein determine its sedimentation, gel filtration, and electrophoretic behavior.

Immunochemical techniques are very specific, and antibodies can be produced that recognize only individual proteins. They can be used quantitatively to measure the quantities of a protein, as with radioimmunoassay, or qualitatively to detect only specific proteins, as with "Western blots."

Purification. A pure protein is usually required in order to determine its covalent structure and physical and biological properties. A huge variety of techniques are available to isolate a protein from a biological source and to purify it to homogeneity, eg, extraction, size fractionation, reversible precipitation, chromatography, and affinity chromatography.

Toxicity. Being biological products with biological functions, proteins have widely varying biological effects and toxicities. At one extreme many of the most potent toxins found in nature are proteins, and they can be lethal to humans. At the other extreme, many proteins are harmless. The most universal problem with proteins is their immunogenicity. They can elicit an immunological response in humans, and many allergies are caused by proteins. This is least likely to be a problem when the protein is from a human source.

BIBLIOGRAPHY

H. Neurath and R. L. Hill, eds., *The Proteins,* 3rd ed., Vols. I–V, Academic Press, New York, 1975–1982.

P. D. Boyer, ed., *The Enzymes,* 3rd ed., Vols. I–XV, Academic Press, New York, 1970–1983.

T. E. Creighton, *Proteins: Structures and Molecular Properties,* W. H. Freeman, New York, 1983.

THOMAS E. CREIGHTON
Medical Research Council
Laboratory of Molecular Biology

PROTEINS, BIOSYNTHESIS

Components and Overall Process

The synthesis of a protein is a complex sequence of highly coordinated events. In all cells, the general scheme of information transfer is probably very similar. The genetically stable component controlling the genotype of each cell is deoxyribonucleic acid (DNA) (see POLYNUCLEOTIDES). The information contained in the DNA of the chromosome, coded in a precise sequence of its four chemical constituents, or nucleotides, is transmitted to the intermediate messenger RNA (mRNA) molecule (transcription). Decoding of the mRNA into proteins occurs on particulate structures, the ribosomes, two of which are shown in Figure 1. The code for each amino acid of a protein is contained in a specific sequence of three nucleotides (triplet) in the mRNA molecule. The specific triplet code for an amino acid is known as a codon. Each triplet is recognized, not by the amino acid itself, but by a specific adapter molecule called transfer RNA, or tRNA, to which the amino acid is linked via the process of amino acid activation. The decoding process involves the recognition of specific mRNA codons by complementary triplets on the decoding region (anticodon) of the tRNA.

In many organisms, the DNA is complexed with histones and constitutes a large fraction of the chromosomes. The DNA serves as its own template for replication of identical molecules, which are passed on from generation to generation as the hereditary material.

The ultimate expression of the genetic information in an organism, its phenotype, is in large part determined by the sum total of the protein constituents of the cell (see PROTEINS). The information flow from DNA to proteins occurs via an intermediate messenger ribonucleic acid (mRNA). The use of cell-free systems of protein synthesis together with genetic studies has resulted in remarkable advances in our knowledge of the mechanism of protein biosynthesis and the genetic code. Cell-free systems dependent on the presence of natural mRNA have been used for translation of mRNA into proteins with high fidelity. They have also been used for translating synthetic polyribonucleotides. This has made possible the deciphering of the genetic code, the assignment of nucleotide triplets (codons) to all the amino acids, and has illuminated other features of messenger translation such as the direction of reading and the mechanisms of polypeptide chain initiation, chain growth (referred to as elongation in the biochemical literature), and chain termination.

In the biosynthesis of a protein a large number of cellular constituents interact (Table 1). The events in which they participate are as follows:

1. acylation of tRNA, amino acid activation
   - formylation of fMet–tRNA
   - deformylation
2. initiation
   - binding of initiator tRNA to ribosome
   - binding of mRNA to ribosome
3. elongation
   - translation, codon–anticodon recognition
   - transpeptidation
   - translocation
4. termination
   - binding of release factors
   - hydrolysis of peptidyl–tRNA
   - dissociation of ribosomal subunits
5. co- and posttranslational modification of proteins
6. secretion of proteins (if appropriate)

Figure 2 shows the universal genetic code for prokaryotes and eukaryotes.

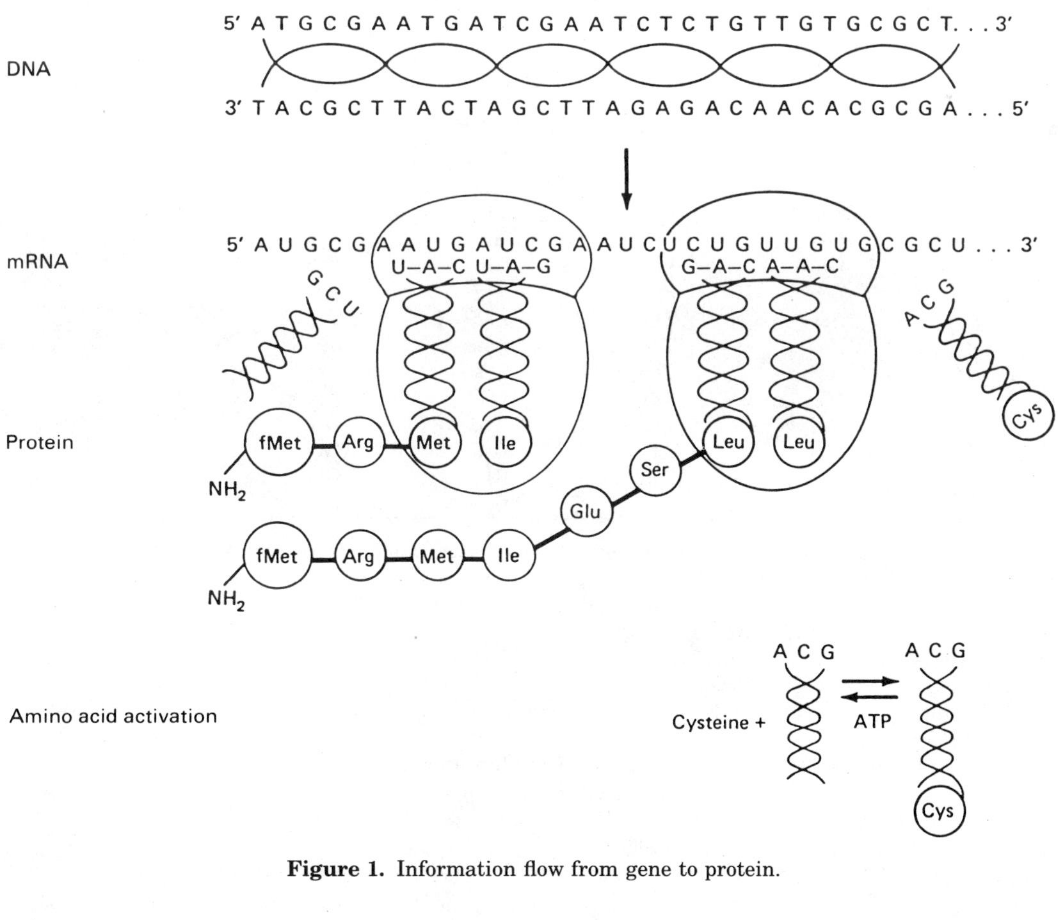

Figure 1. Information flow from gene to protein.

Table 1. The Macromolecular Elements of Protein Synthesis in Prokaryotes and Eukaryotes

Element	Prokaryotes	Eukaryotes
ribosomes, subunits	$30s + 50s \rightleftharpoons 70s^a$	$40s + 60s \rightleftharpoons 80s^a$
ribosomal ribonucleic acid, rRNA	16*s*, 23*s*, 5*s*	18*s*, 28*s*, 5*s*
ribosomal proteins, numbers	30*s*:21	40*s*:c.31
	50*s*:32	60*s*:c.49
initiation factors	IF-1; IF-2; IF-3	eIF-1; eIF-2; eIF-3; eIF-4A,B,C; eIF-5; eIF-6
initiator tRNA	*N*-formyl-Met-$tRNA^{Met}_{init}$	Met-$tRNA^{Met}_{init}$
elongation factors	EF-Tu; EF-Ts; EF-G	eEF-1α; eEF-1β; eEF-2
tRNA molecules	aminoacyl-tRNA	
codons	messenger RNA molecules	
release (termination) factors	RF-1; RF-2	RF

[a] *s* designates sedimentation coefficient in Svedberg units (10^{-13} s).

Cellular Organization of Protein Synthesis

Both prokaryotic and eukaryotic organisms have mechanisms to distinguish between ribosomes making cytoplasmic proteins and those making proteins destined for secretion, the cellular membrane and, in the case of eukaryotes, the internal membrane compartments of the cell (endoplasmic reticulum, Golgi apparatus, lysosomes). The result of this selection is the existence of two classes of ribosomes: free ribosomes and membrane-bound ribosomes. Those that are free in the cytoplasm are involved in the synthesis of cytoplasmic proteins, whereas those that are membrane bound are involved in the synthesis of secreted and membrane proteins.

BIBLIOGRAPHY

K. Moldave, *Ann. Rev. Biochem.* **54,** 1109 (1985).

J. A. Lake, *Ann. Rev. Biochem.* **54,** 507 (1985).

M. Kozak, *Microbiol. Rev.* **47,** 1 (1983).

P. Walter, R. Gilmore, and G. Blobel, *Cell* **38,** 5 (1984).

Michael Green
Richard A. Mazzarella
Saint Louis University School of Medicine

First position 5′	Second position U	C	A	G	Third position 3′
U	Phe	Ser	Tyr	Cys	U
	Phe	Ser	Tyr	Cys	C
	Leu	Ser	Stop	Stop[1]	A
	Leu	Ser	Stop	Trp	G
C	Leu[2]	Pro	His	Arg	U
	Leu[2]	Pro	His	Arg	C
	Leu[2]	Pro	Gln	Arg	A
	Leu[2]	Pro	Gln	Arg	G
A	Ile[3,4]	Thr	Asn	Ser	U
	Ile[3]	Thr	Asn	Ser	C
	Ile[5]	Thr	Lys	Arg[6,7]	A
	Met	Thr	Lys	Arg[6]	G
G	Val	Ala	Asp	Gly	U
	Val	Ala	Asp	Gly	C
	Val	Ala	Glu	Gly	A
	Val[8]	Ala	Glu	Gly	G

Figure 2. The genetic code. Shown is the universal genetic code for prokaryotes and eukaryotes where each nucleotide triplet or codon is assigned to an amino acid or to its function in protein synthesis. The codons in which the mitochondrial genetic code differs from the universal code are indicated by the footnote at the codon assignment.

[1]In all mitochondria, UGA encodes Trp instead of termination.

[2]In mitochondria from *Saccharomyces cerevisiae,* the CUN triplets encode Thr instead of Leu.

[3]In mammalian mitochondria, AUU and AUC can function as initiator codons and thus code for *N*-formylmethionine.

[4]In *Drosophila* mitochondria, AUU can function as an initiator codon.

[5]Mammalian, *S. Cerevisiae,* and *Drosophila* mitochondria use AUA to code for Met rather than Ile.

[6]Mammalian mitochondria use AGA and AGG as stop codons rather than as Arg codons.

[7]*Drosophila* mitochondria use AGA to code for Ser.

[8]In prokaryotes, GUG can function as an initiator codon and thus code for *N*-formylmethionine.

PULLULAN

The name pullulan was first proposed in 1959 to designate the water-soluble extracellular polysaccharide synthesized by *Aureobasidium pullulans*. A common and widespread fungus, *A. pullulans,* better known as *Pullularia pullulans,* is present in sewage and untreated water.

Pullulan is currently accepted as being a water-soluble neutral glucan which consists of linear chains of D-glucopyranosyl units that alternate regularly between one (1→6)-α-D and two (1→4)-α-D linkages, or a linear polymer of maltotriosyl units connected by (1→6)-α-D linkages (Fig. 1).

Properties

Physical Properties. Freeze-dried pullulan in the form of white strands or ground into powder is tasteless, odorless, and nonhygroscopic under normal atmospheric conditions. It is insoluble in methanol or acetone, but dissolves readily in water to form a transparent, colorless, viscous, adhesive solution. Some properties can be improved, and new ones can be incorporated. Partial esterification or etherification of the polymer chain can reduce solubility in water; complete esterification or etherification confers insolubility; hydrogenation increases heat resistance.

Powdered pullulan, mixed with a small amount of water, can be formed into compression moldings, membranes, films, or fibers. As a molding plastic, it resembles polystyrene in transparency, gloss, hardness, strength, and toughness, but is far more elastic. Pullulan solutions at concentrations of 5–10% applied to a smooth surface and heat-dried under pressure, form a film as thin as 0.01 mm. This film is colorless, transparent, tasteless, odorless, tenacious, resistant to oil and grease, unaffected by thermal variation, impermeable to oxygen, nontoxic, and edible. Water-resistant or water-insoluble films are prepared by partial or complete esterification or etherification of pullulan.

The size of pullulan molecules is influenced by the fermentation conditions and differs according to the strain of fungus. The molecular weight increases gradually during microorganism cultivation, reaches a maximum, and decreases.

The rheological properties of pullulan (produced by *A. pullulans* 2552 in a sucrose medium) solutions of different concentrations (0–2%) have been studied at various stages of fermentation; the apparent viscosity always increases nonlinearly with increasing concentration.

Chemical Properties. Periodate oxidation of pullulan yields formic acid as a by-product. Complete acid hydrolysis (for example, by 1 *N* HCl at 100°C for 120 min) gives only glucose as a by-product. Partial acid hydrolysis yields various mono- and oligosaccharides, depending on the conditions and procedures of hydrolysis. Heat of combustion is approximately 16.7 kJ/g (4 kcal/g).

Biochemical Properties. Pullulan is enzymatically hydrolyzed by both (1→6)-α-D- and (1→4)-α-D-pullulanases.

Preparation

Laboratory Scale. Pullulan has been prepared in the laboratory mainly through fermentation of *A. pullulans* cultures grown in a carbohydrate medium.

Pilot-plant Scale and Industrial Manufacture. Following a pilot-plant trial test (one metric ton per month), pullulan is now manufactured by Hayashibara Biochemical Laboratories, Inc. in Japan. Acid or enzymatic hydrolyzed starch (of dextrose equivalent 20–70) is the carbon source. Other ingredients for the culture medium are K_2HPO_4, NaCl, $MgSO_4 \cdot 7H_2O$, and $FeSO_4 \cdot 7H_2O$. The initial pH is set between 5.0 and 7.5. The culture is allowed to ferment for 3–8 days at 27–30°C, heated to deactivate the enzyme, and centrifuged to remove the cells from the fermentation broth. Pullulan is precipitated from the supernatant by methanol at 1:1 vol/vol.

The wholesale price for industrial and food-grade pullulan of medium molecular weight (average mol wt = 200,000) is $10/kg for one-ton orders (FOB Okayama, Japan). Reagent-grade pullulan is available at $1280/kg.

Applications

At the present time, no commercial and industrial uses of pullulan are reported. However, a large number of patents indicate a variety of potential applications, such as in food, in industrial applications, and in medicine.

BIBLIOGRAPHY

H. Bender, J. Lehmann, and K. Wallenfels, *Biochim. Biophys. Acta* **36,** 309 (1959).

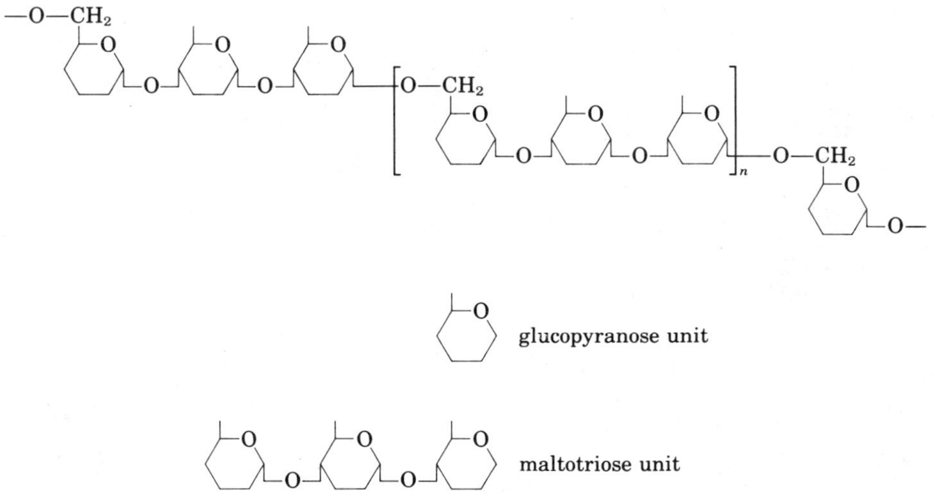

Figure 1. Chemical structure of pullulan.

S. Yuen, *Process Biochem.*, **7,** 22 (Nov. 1974).
A. LeDuy, A. A. Marsan, and B. Coupal, *Biotechnol. Bioeng.* **16,** 61 (1974).
J. E. Zajic and A. LeDuy, *Appl. Microbiol.* **25,** 628 (1973).

ANH LEDUY
LIONEL CHOPLIN
Laval University
JAMES E. ZAJIC
University of Texas
JOHN H. T. LUONG
Biotechnology Research Institute
National Research Council of Canada

PULTRUSION

Pultrusion is a method for producing continuous lengths of fiber-reinforced laminates. They are currently competing with aluminum extrusions and wood and steel parts in a variety of applications.

Material Reinforcements

Most pultruded articles are E-glass fiber-reinforced, in the form of continuous-roving or continuous-strand mat (CSM) and thermosetting polyester resin (Fig. 1). These materials produce acceptable properties economically.

Figure 1. From left to right: continuous-filament reinforced mat (CFRM); roving; continuous-strand glass mat (CSM).

Resins

In recent years, several resins have been manufactured specifically for pultrusion. Polyester resin in various forms accounts for about 90% of pultrusion processes (see POLYESTERS).

Laminate Construction and Properties

The properties obtained in a pultruded article depend on the materials. Fiber loadings of 20–80% affect the properties. Longitudinal and tranverse properties are important. Continuous rovings provide longitudinal strength, whereas mats or resins provide the transverse strength.

For this reason most profiles are manufactured with continuous-strand mats or other forms of transverse reinforcement. These materials are sandwiched between layers of roving. This ply configuration produces a laminate in which the physical properties (Table 1) are more evenly distributed.

Processing and the Equipment

Several companies offer a complete pultrusion line (Fig. 2). However, frequently the equipment is built by the user.

Table 1. Typical Properties of Pultruded Laminate

	35–45% glass 1		45–55% glass 2	
Property[a]	Longitudinal	Transverse	Longitudinal	Transverse
tensile strength, MPa[b]	207	45	310	55
tensile modulus, GPa[c]	15.8	5.5	17.2	6.9
flexural strength, MPa[b]	207	69	310	103
flexural modulus, GPa[c]	8.9	4.1	12.4	5.5
Izod impact, J/m[d]	1068	214	1495	214
compressive strength, MPa[b]	138	69	207	103
compressive modulus, GPa[c]	15.8	5.5	17.2	6.9
shear strength, MPa[b]	31	31	59	59
dielectric strength, V/m	200		200	
specific gravity	1.61–1.75		1.68–1.88	
coefficient of thermal expansion	5.2×10^{-6}		5.2×10^{-6}	
water absorption, % max	0.7		0.5	

[a] Coupon values.
[b] To convert MPa to psi, multiply by 145.
[c] To convert GPa to psi, multiply by 145,000.
[d] To convert J/m to ft·lb/in., divide by 53.38.

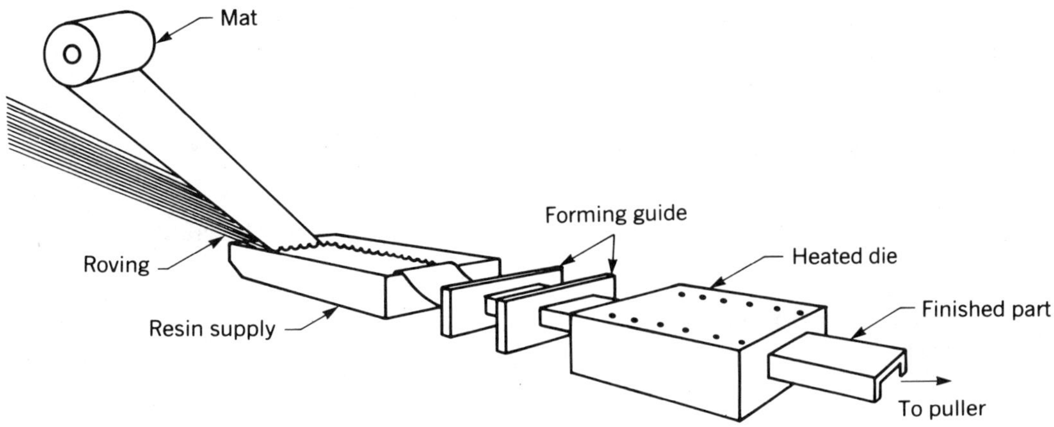

Figure 2. Pultrusion line.

Applications

Pultrusions are used in electrical insulation, for corrosion resistance, low thermal conductivity, and part consolidation.

BIBLIOGRAPHY

U.S. Pat. 2,571,717 (Oct. 16, 1951), A. M. Howald and L. S. Meyer.

L. H. Miner, "Aramid" in *Modern Plastics Encyclopedia 1986–7,* Vol. **3,** No. 10, McGraw-Hill, New York, Oct. 1986.

T. S. McQuarrie, "New Generation Resins for Pultrusions," *SPI RP/CI 33rd Meeting,* Feb. 7–10, 1978, Sect. 8-E.

R. W. Meyer, *Handbook of Pultrusion Technology,* Chapman and Hall, New York, Chapt. 2.

DAVID EVANS
Creative Pultrusion, Inc.

PYROLYSIS. See DEGRADATION.

PYRRONES. See LADDER AND SPIRO POLYMERS.

Q

Q-e SCHEME. See Copolymerization.

QUALITY CONTROL. See Technical services.

QUATERNARY AMMONIUM POLYMERS. See Polyamines and polyquaternary ammonium salts.

QUICK-MOLD CHANGE. See Injection molding.

QUINAZOLINE POLYMERS. See Polybenzoxazinones.

R

RADIATION-RESISTANT POLYMERS

Ionizing radiation can profoundly alter the molecular structure and macroscopic properties of polymeric materials. In numerous applications (eg, space vehicles and nuclear power plants), polymeric materials are specifically needed in environments of ionizing radiation because of various polymeric properties such as elasticity, light weight, formability, and others. Such cases require polymeric materials that are as radiation resistant as possible.

Polymeric materials exhibit a wide range of radiation stabilities. Radiation resistance is strongly influenced by the basic macromolecular structure, the presence of certain types of additives, and the particular environment exposure conditions.

The Interaction of Radiation with Polymers

The deposition of radiation energy in an organic material triggers a complex set of events, the predominant features of which are outlined below. High energy photons, such as γ or x rays, are deposited in a material by three physical mechanisms: the photoelectric effect, by which a photon is absorbed and an orbital electron is ejected, having the energy of the incident radiation minus the binding energy; Compton scattering, by which an electron is ejected, while the photon (of reduced energy) is scattered; and pair production (important above 1.022 MeV), by which a positron–electron pair is produced, and where subsequent annihilation of the positron generates γ-rays of 0.511 MeV that may further interact by the first two mechanisms (see POSITRON ANNIHILATION). The energy of the incident radiation and the atomic composition of the absorbing medium determine which of the above mechanisms predominate. In most cases, Compton scattering is the principal mechanism for organic polymers. Regardless of which mechanism applies, when high energy radiation impinges on a polymeric material, the primary event in radiation damage is the same: namely, the ejection of a high energy electron.

$$\mathrm{R} \rightsquigarrow \mathrm{R}^+ + e^-$$

This primary electron can then cause ionization of further molecules, producing still more electrons. As a result, a localized chain reaction of ionization events is set in motion.

$$e^- + \mathrm{R} \longrightarrow \mathrm{R}^+ + 2e^-$$

Coulombic attraction causes the positively charged ions produced to rapidly disappear as electrons recombine with them. This results in the production of highly excited electronic states.

$$\mathrm{R}^+ + e^- \longrightarrow \mathrm{R}^*$$

In addition, a portion of the interactions of the radiation with the material can result in energy transfer that is insufficient to cause ionization, but that results directly in an electronically excited state of the molecule.

$$\mathrm{R} \longrightarrow \mathrm{R}^*$$

The excited-state molecules may undergo radiationless decay to the ground state. At the same time, heat is produced, which can result in a significant rise in the temperature of the sample under conditions of high radiation flux (ie, at high dose rate). Depending on the particular molecular structures involved, the excited states may also undergo radiative decay (ie, phosphorescence or fluorescence), thereby emitting light. Molecules that behave in this way with high quantum efficiency provide the mechanism by which scintillometry (radiation counting) functions and by which luminescent paints (as on clocks or watches) function. Excited states can also decay by means of chemical reaction via heterolytic bond cleavage, leading to ions, or by homolytic bond cleavage, leading to the formation of free radicals.

$$\mathrm{RX}^* \longrightarrow \mathrm{R}\cdot + \mathrm{X}\cdot$$

Although ions and excited-state molecules can give rise directly to chemical reactions, evidence indicates that in most cases, for organic materials, the predominant intermediate in radiation chemistry is the free radical. The unit used widely to express radiation chemical yields for production of radicals, various gaseous products, cross-links, etc, is the G value. It is defined as the number of molecular changes of a given type resulting per 100 eV of absorbed energy. The main units used to describe absorbed radiation are interrelated by the following equation

$$1\ \mathrm{Gy} = 100\ \mathrm{rad} = 1\ \mathrm{J/kg} = 6.24 \times 10^{15}\ \mathrm{eV/g} = 10^4\ \mathrm{erg/g}$$

Free-radical Reactions in Irradiated Polymers. Homolytic bond cleavage from excited states in irradiated polymers can lead to a pair of free radicals via bond scission, involving the main-chain or side-chain substituents.

The fate of long-lived radicals in irradiated polymers has been extensively studied by electron spin resonance and uv spectroscopy, especially in the case of allyl or polyene radicals. Radicals in both crystalline and amorphous regions of polymers can be identified. Radicals formed in crystalline regions have much lower mobility and longer lifetimes; such radicals can migrate slowly to the amorphous region, where they subsequently react. Radical lifetimes are strongly temperature dependent and can be hours or even weeks at room temperature. Specific radicals can be identified, and less stable radicals can be observed to convert to more stable radicals with time.

Radiation-induced Degradation in the Absence of Oxygen

Scission and Cross-linking. The most important chemical reactions leading to changes in mechanical properties of polymers undergoing radiation degradation are those involving chain scission and cross-linking.

Polymers subjected to radiation degradation in the absence of oxygen can be grouped into two categories, in terms of the predominance of scission vs cross-linking (Table 1). A very useful rule of thumb, for predicting whether cross-linking or scission prevails, holds for a large number of polyolefins: Polymers with a high concentration of quaternary carbon atoms

Table 1. Classification of Polymers According to Their Predominant Degradation Mode When Irradiated in an Inert Atmosphere

Chain-scission polymers	Cross-linking polymers
polyisobutylene	polyethylene
poly(α-methylstyrene)	polypropylene
poly(vinylidene chloride)	polystyrene
poly(vinyl fluoride)	poly(vinyl chloride)
polychlorotrifluoroethylene	poly(vinyl alcohol)
polytetrafluoroethylene	poly(vinyl acetate)
polyacrylonitrile	poly(vinyl methyl ether)
polyvinylformal	polybutadiene
polyvinylbutyral	polychloroprene
poly(methyl methacrylate)	poly(styrene-*co*-acrylonitrile)
polymethacrylamide	poly(styrene-*co*-butadiene)
polymethacrylonitrile	poly(butadiene-*co*-acrylonitrile)
polyoxymethylene	natural rubber
poly(propylene sulfide)	chlorinated polyethylene
poly(ethylene sulfide)	chlorosulfonated polyethylene
cellulose	polyamides
polyalanine	polyesters
polylysine	polyurethanes
DNA	polysulfones
	polyacrylates
	polyacrylamides
	polydimethylsiloxane
	polymethylphenylsiloxane
	phenol–formaldehyde resins
	urea–formaldehyde resins
	melamine–formaldehyde resins

along the chain (polyisobutylene, poly(methyl methacrylate), poly(α-methylstyrene)) undergo primarily scission; polymers lacking this structural feature (polyethylene, polystyrene, natural rubber) undergo primarily cross-linking.

The presence of an aromatic ring in the polymer exerts a strongly stabilizing influence on the yield of radiation-induced cross-linking or scission, and polymers bearing aromatic functionality are particularly useful in applications where radiation resistance is required.

In general, the modulus of flexible polymers undergoing predominantly cross-linking increases whereas the elongation at break decreases. The material becomes harder, stiffer, and eventually brittle.

For flexible polymeric materials undergoing predominantly scission, the modulus, ultimate elongation, and tensile strength typically decrease. The material becomes progressively softer and weaker as a function of absorbed dose and may ultimately be converted into a viscous liquid.

Unsaturation and Color Changes. The formation of unsaturated sites is another important structural change induced by irradiation. Unsaturation is primarily due to loss of adjacent substituents on the chain or side group (for instance, adjacent H atoms in polyethylene) or to radical–radical termination reactions that occur by disproportionation.

Polymers usually undergo color changes upon irradiation, because of the formation of conjugated, unsaturated chromophores or trapped radicals. Poly(vinyl chloride) darkens (brown, red, or black) at doses of 50–150 kGy (5–15 Mrad) and many polyolefins tend to yellow at doses on the order of 100 kGy (10 Mrad). Polystyrene and particularly polysiloxanes are resistant to color formation.

Gaseous Products. The formation of gaseous products accompanies the radiation degradation of polymers. For polyolefins, H_2 typically predominates, together with the lower alkanes and alkenes (CH_4, C_2H_6, C_2H_4, C_3H_8, and others). For other macromolecules, the gaseous products reflect both the atomic composition and the molecular structure of the polymers. Gaseous irradiation products are produced in substantial quantities, which must be considered for irradiations conducted in sealed containers. Gaseous products may lead to bubble formation in glassy materials.

Effects of Temperature, Mechanical Stress, and Dose Rate. Differences in specimen temperature can affect the radiation-degradation chemistry of polymers, primarily because of differences in radical mobility, with sharp changes often occurring near physical transition temperatures. Mechanical stress and high energy irradiation may be interactive in their effect on polymers. Creep and stress relaxation of samples held under a load are both enhanced by irradiation. Reduction in physical properties can proceed at an enhanced rate under the combined influence of applied stress and radiation. Possible factors in radiation-stress effects include a weakening of certain chemical bonds in macromolecules placed under stress and plasticization of the polymer by gaseous radiochemical products.

Under an inert atmosphere, radiation effects on polymers depend on the integrated absorbed dose and not on the dose rate. Consequently, straightforward short-term accelerated aging experiments can be readily employed, using high dose rates, to predict aging behaviors and lifetimes for polymer materials in long-term inert-atmospheric applications involving low radiation dose rates.

Methods for Studying Radiation Degradation. As stated earlier, radiation-induced structural change in polymers is due to cross-linking or scission. The study of these molecular changes is important because they strongly influence macroscopic properties. For cross-linked materials, changes in cross-link density are reflected by the extent to which the material is swelled by soaking in a compatible solvent; this is determined by comparing the weight of a solvent-swelled material with its dry weight. If cross-linking predominates, the cross-link density increases, leading to a decrease in the observed extent of swelling; if scission predominates, the opposite occurs.

For soluble materials below the gel dose (or for the soluble fraction above the gel dose), the effects of cross-linking or scission can be seen by following the molecular weight distribution as a function of irradiation. This is conveniently done by gel-permeation chromatography or by solution viscosity measurements. Mechanical property changes, such as elongation and tensile strength, are commonly monitored for flexible materials; bend strength is a useful parameter for hard materials.

Improvement of Radiation Resistance. The radiation resistance of most polymer types can be significantly improved by stabilizing additives, sometimes referred to as antirads. There are two principal mechanisms of radiation stabilization. These are based on energy scavenging involving the initially formed excited states, and on radical scavenging.

Irradiation of Polymer Solutions. For polymers in solution, several special effects may arise. First of all, the radiation-induced chemistry of the polymer involves the radicals formed initially in the macromolecule (direct action) as well as the radicals formed initially in the solvent (indirect action). Different organic solvents may enhance or retard the degradation rate. In aqueous solutions, highly reactive radicals (such as ·OH) often sharply increase degradation rates. The cross-linking yield for dissolved polymers depends strongly on concentration and may drop to very low values for dilute solutions. The radiation resistance of polymer solutions is often enhanced by dissolving stabilizers in the mixture.

Radiation-induced Conductivity. Under normal conditions, dielectric materials exhibit a low level leakage current, which is proportional to the applied electric field. When placed in an environment of ionizing radiation, the electrical conductivity increases significantly, approaching a new steady-state value. The magnitude of the radiation-induced response is a material-specific property, which is a function of the dose rate and is termed the radiation-induced conductivity. When irradiation is discontinued, the conductivity decays with a $1/t$ dependence to its original value. The decay rate is also material dependent, and may range from a fraction of a second up to months. Polymers can be made more resistant to induced conductivity by dosing with low concentrations of electron donor or acceptor molecules, which serve as deep, extrinisic trap sites.

Radiation Resistance of Specific Polymers. The selection of polymers for use in radiation environments usually involves other criteria besides radiation resistance. For instance, property requirements such as elasticity, chemical resistance, coefficient of friction, coefficient of thermal expansion, impact resistance, electrical properties, moldability, cost, and other factors may strongly limit the choice of materials. Frequently a compromise is necessary between radiation sensitivity and one or more other factors such as those listed above.

Polystyrene, because of its high aromatic content, is the most radiation-resistant of the widely used, inexpensive thermoplastic molding materials. Other polymers containing aromatic rings, such as aromatic polyamides, poly(phenylene sulfide), polyetheretherketone, and aromatic polysulfones, are extremely radiation resistant. Many highly crosslinked thermoset resins exhibit little change at high doses, including phenolics, epoxies, and urethanes. Aromatic polyamides and poly(ethylene terephthalate) are among the more radiation-resistant fibers. Among elastomers, a number of the polyurethane rubbers exhibit excellent radiation resistance.

Radiation Degradation in Air

Degradation can be dramatically affected by ionizing radiation in air, under conditions where extensive oxidation can take place deep within the material. Under such conditions, oxygen becomes heavily involved in the degradation chemistry. The resulting material property changes may be quite different from those obtained in an oxygen-free atmosphere. For many polymers, radiation in the presence of oxygen strongly favors chain scission. Frequently, materials that cross-link when irradiated under inert conditions undergo predominantly scission under oxidizing conditions. For many materials, the presence of oxygen strongly enhances the extent of degradation; this can apply when oxygen is present during or after irradiation. Thus, in many cases, extensive damage may occur at much lower absorbed doses than those required to cause significant changes under inert conditions. Furthermore, the influence of time and sample geometry can become of paramount importance, resulting in large dose-rate effects, postirradiation effects, and environmental synergisms of radiation and temperature. These effects greatly increase the difficulty of predicting long-term radiation-degradation behavior under a particular set of environmental conditions.

Radiation-oxidation Mechanisms. Oxygen reacts with the free radicals generated on irradiation. The degradative reactions resemble those that occur in other environments (elevated temperature, uv light, mechanical stress, chemical initiation involving radicals, etc).

Because of its high diffusional mobility and its high reactivity toward radicals, oxygen serves to trap radicals in an efficient manner, greatly impeding the radical reaction pathways that occur in the absence of oxygen, and channeling the course of the reactions into predominantly oxidative routes. Chemical reaction products include oxidized structures within the polymer, such as ketone, carboxylic acid, and alcohol functionalities, as well as peroxidic species. Gaseous products include those arising under inert-atmosphere irradiation, together with significant amounts of CO_2, CO, and H_2O.

Heterogeneous Oxidation Effects. Heterogeneous oxidation occurs whenever oxygen diffusion into a sample undergoing irradiation becomes a rate-limiting step. Frequently, polymers placed into a radiation environment have previously equilibrated with the surrounding atmospheric gases (usually air) and contain a quantity of dissolved oxygen. Under irradiation, the dissolved oxygen may be quickly consumed by reaction with the free radicals generated. As irradiation continues, subsequent oxidation may occur only in the surface regions of the sample.

Heterogeneous oxidation appears frequently. Many applications involve material thicknesses and dose rates that result in heterogeneous oxidation. In addition, for applications that may involve dose rates so low as to give homogeneous oxidation, accelerated radiation-aging experiments, performed to predict degradation behaviors, which employ short time periods and elevated dose rates frequently give rise to heterogeneous oxidation.

Methods for Studying Radiation-induced Oxidative Degradation. The methods described above for studying radiation degradation in an inert atmosphere also apply to degradation in the presence of air. Additional techniques are useful for monitoring oxidation, including ir spectroscopy and chemiluminescence. Special techniques have been developed for quantitatively profiling heterogeneous oxidation. Qualitative identification of heterogeneous oxidation can often be achieved by polishing the cross-sectional surface of a degraded sample; areas with different extents of oxidation are apparent on optical examination.

Dose-rate Effects. When polymeric materials are irradiated in the presence of oxygen, the rate of degradation and the nature of degradation can be strongly influenced by the dose rate. Lower dose rates often increase oxidation and chain-scission yields, and result in more extensive material degradation per equivalent absorbed dose. Dose-rate effects arise (*1*) due to oxygen diffusion effects, or (*2*) whenever one or more steps in the oxidation chemistry become rate-limiting on a time scale comparable to oxygen diffusion, peroxide breakdown, and radical migration.

Postirradation Effects. Degradation of irradiated materials often continues, in storage or in use, long after the material has been removed from the radiation environment. This effect is caused by continuing free-radical-mediated oxidation. Significant postirradation oxidation effects may take place over weeks or years, and may result in degradation that is much more extensive than that which was the immediate effect of the irradiation.

Temperature Effects. In the presence of oxygen, temperature often influences radiation degradation. Strong temperature effects can arise caused by changes in the rate of key processes that affect the oxidation, such as peroxide decomposition, radical migration, oxygen diffusion, diffusive loss of stabilizer additives, and others.

Stabilizers for Radiation-oxidative Degradation. The resistance of most polymers to radiation oxidation is significantly improved by stabilizers. The stabilization effect can be larger than that achieved by inert-atmosphere irradiation. Stabilizers are usually antioxidants, particularly radical scavengers, that interrupt the radical-mediated oxidation chain reaction.

Radiation Resistance of Specific Polymers. Many materials exhibit much less resistance to radiation under oxidizing conditions, and it would be convenient to have tables of radiation resistance in air available. However, this becomes extremely complicated because of parameters that play no role (or have a comparatively smaller effect) under inert atmosphere, but that dramatically affect the radiation resistance of polymers in air. These parameters include dose rate, material thickness and geometry, formulation (particularly stabilizers), temperature, and postirradiation time. For most polymers, data referring to these variables are scant. Figure 1 illustrates the complexity that emerges when radiation resistance is evaluated under different conditions.

Accelerated Radiation-degradation Tests. Because of the limited data available and the strong influence of many factors specific to material type, sample geometry, and environmental variables, prediction of the radiation resistance of a specific material in air under particular environmental conditions usually requires a laboratory aging test. Most time scales of practical importance span years and require accelerated aging techniques. Accelerated tests are run under short-term, high stress-

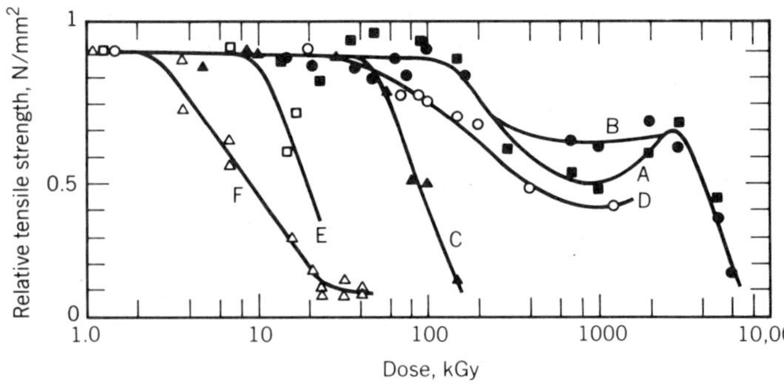

Figure 1. Changes in relative tensile strength of a polypropylene material as a function of irradiation. A, 0.3-mm sample thickness, irradiated at 5×10^6 Gy/h under vacuum; B, 0.3-mm sample thickness, irradiated at 5×10^6 Gy/h under vacuum, followed by heating 1 h at 80°C; C, 0.3-mm sample thickness, irradiated in air at 10^4 Gy/h; D, 1.0-mm sample thickness, irradiated in air at 10^6 Gy/h and left standing two months before testing; E, 1.0-mm sample thickness, irradiated in air at 4 Gy/h; F, 0.4-mm diameter wire, irradiated in air at 4 Gy/h (1 Gy = 100 rad). To convert N/mm² to psi, multiply by 145.

rate conditions that are higher than in the application environment. However, precisely those parameters that must be altered to achieve an accelerated test are also the parameters that can dramatically affect degradation rates and behaviors, and therefore the design and interpretation of accelerated aging experiments is complicated. A number of methodologies have been developed for this purpose.

BIBLIOGRAPHY

A. Charlesby, *Atomic Radiation and Polymers,* Pergamon Press, Elmsford, N.Y., 1960.

A. Chapiro, *Radiation Chemistry of Polymeric Systems,* Wiley-Interscience, New York, 1962.

M. Dole, ed., *The Radiation Chemistry of Macromolecules,* Academic Press, New York, Vol. 1, 1972, and Vol. 2, 1973.

F. A. Makhlis, *Radiation Physics and Chemistry of Polymers,* John Wiley & Sons, Inc., New York, 1975.

ROGER CLOUGH
Sandia National Labs

RADIATION-SENSITIVE POLYMERS. See RADIATION-RESISTANT POLYMERS.

RADICAL-ION. See ANIONIC POLYMERIZATION.

RADICAL POLYMERIZATION

Free radical polymerization has received more intensive study than any other chemical chain reaction; countless investigations have contributed to industry and to the chemistry of free radicals. Advances in both fields continue to be made, with a particular emphasis on the growth of special polymers. These require more refined synthetic routes and are often necessarily free-radical on account of the reactive natures of "active" components. Nevertheless, free-radical polymerization still accounts for a large proportion of mass-produced polymers: in 1985 over 46% (by weight) of the total U.S. production of plastics involved this type of polymerization. Currently, optimizing established processes is a primary and continuing endeavor.

Monomers

Monomers in free-radical polymerization are commonly monosubstituted or unsymmetrically (1,1-) disubstituted ethylenes, $CH_2{=}CHX$ or $CH_2{=}CXY$. The essential polymerization step is repeated free-radical addition to monomer double bonds, forming chains of carbon atoms constructed of units $-\!(CH_2{-}CHX)\!-$ or $-\!(CH_2{-}CXY)\!-$ linked in predominantly head-to-tail fashion (the substituted carbon atom being designated the head).

The substituents X,Y in an olefinic monomer form side-chains attached to the primary chains of the polymer. There are few restrictions on the nature of the substituents. These groups determine the reactivities of the propagating radicals and monomers and also, to a large extent, the properties of the final polymers.

Dienes constitute an important class of vinyl monomers. Those having isolated double bonds are typified by *p*-divinylbenzene, ethylene glycol dimethacrylate (**1**), and *N,N'*-methylenebisacrylamide (**2**). Either one or both double bonds may participate in polymerization; when both react, cross-linked structures are generated which confer enhanced mechanical properties and increased solvent resistance on the products.

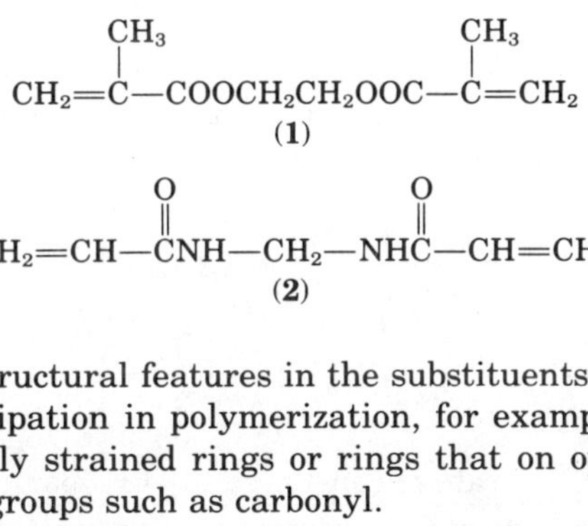

(**1**)

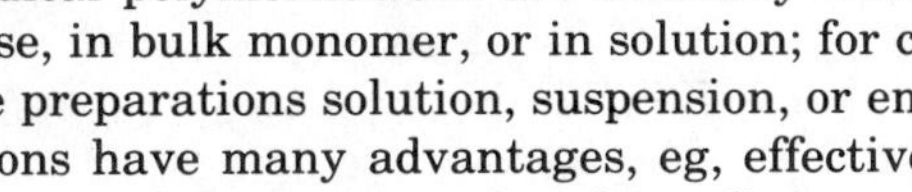

(**2**)

Special structural features in the substituents may induce major participation in polymerization, for example the presence of highly strained rings or rings that on opening form very stable groups such as carbonyl.

Free-radical polymerizations are normally effected in the liquid phase, in bulk monomer, or in solution; for commercial large-scale preparations solution, suspension, or emulsion polymerizations have many advantages, eg, effective heat dissipation, and are frequently employed (see SUSPENSION POLYMERIZATION; EMULSION POLYMERIZATION).

As a result of its great versatility and relative simplicity, free-radical polymerization is employed in the synthesis of countless homo- and copolymers with properties distributed over an almost unlimited range.

Component Reactions

A typical free-radical polymerization comprises four elementary steps: initiation, propagation, termination, and chain transfer.

In initiation, a free radical $\dot{R}_0$ derived from an initiator, the "primary radical," adds to a monomer molecule to form an active center, a free radical that subsequently propagates (eq. 1):

$$\dot{R}_0 + CH_2{=}CXY \longrightarrow R_0{-}CH_2{-}\dot{C}XY \qquad (1)$$

Other initiating reactions, such as addition to the head of the molecule

$$\dot{R}_0 + CH_2{=}CXY \longrightarrow R_0{-}CXY{-}\dot{C}H_2 \qquad (1a)$$

or hydrogen abstraction, are possible but less common; their importance depends on the structures of the radical and the monomer.

Propagating radicals are denoted here by $\dot{R}_r$, in which the subscript indicates the number of monomer units which have added to the initial radical $\dot{R}_0$. Thus if M represents a monomer molecule, equations 1 and 1a may be summarized as

$$\dot{R}_0 + M \longrightarrow \dot{R}_1 \qquad (2)$$

The propagation or growth reaction consists of the (rapid) addition of monomer molecules to the radical species. Usually it occurs in head-to-tail fashion, as indicated in equations 3:

$$\begin{aligned} &R_0{-}CH_2{-}\dot{C}XY + CH_2{=}CXY \longrightarrow \\ &\qquad R_0{-}CH_2{-}CXY{-}CH_2{-}\dot{C}XY \\ &R_0{-}\!(CH_2{-}CXY)\!_{r-1}CH_2{-}\dot{C}XY + CH_2{=}CXY \longrightarrow \\ &\qquad R_0{-}\!(CH_2{-}CXY)\!_r CH_2{-}\dot{C}XY \end{aligned} \qquad (3)$$

However, as with initiation, alternatives are possible and head-to-head, tail-to-head, and tail-to-tail modes occur, usually to minor extents (see HEAD-TO-HEAD POLYMERS).

In termination, growth of polymer chains is brought to an end by the destruction of propagating radicals. Normally, in the absence of retarding species that destroy growing radicals, chain termination occurs by bimolecular interaction of radicals. These processes, analogous to those taking place with simple alkyl radicals, consist of radical combination (eq. 4a) or disproportionation (eq. 4b):

$$\begin{aligned} &R_{r-1}{-}CH_2{-}\dot{C}XY + R_{s-1}{-}CH_2{-}\dot{C}XY \longrightarrow \\ &\qquad R_{r-1}{-}CH_2{-}CXY{-}CXY{-}CH_2{-}R_{s-1} \end{aligned} \qquad (4a)$$

or

$$\longrightarrow R_{r-1}{-}CH_2{-}CHXY + R_{s-1}{-}CH{=}CXY \qquad (4b)$$

In these and many subsequent equations, the number of monomer units in R (or $\dot{R}$) is denoted by a subscript, eg, *r*. Disproportionation leads to two polymer molecules, one saturated and one containing a terminal double bond; the hydrogen atom transferred need not come from a methylene group as in equation 4b but may originate in an X or Y substituent. Many monomers show both types of termination; thus in methyl methacrylate, disproportionation is approximately twice as rapid as combination at room temperature and increases in importance as the temperature is raised. On the other hand, styrene and acrylonitrile radicals terminate preponderantly by combination.

Atom (often hydrogen) abstraction from saturated molecules is a well-known reaction of free radicals and, as would be expected, is important in free-radical polymerization. It leads to the chain-transfer process, first recognized by Flory in 1937, which brings about the cessation of growth of a propagating radical and at the same time produces a new small radical which may propagate. In general, the effect of chain transfer on the overall polymerization is determined by the reactivity of the transfer radical formed.

Chain transfer to polymer normally gives rise to branched or cross-linked molecules. Termination by combination or disproportionation yields cross-linked products or graft copolymers, respectively.

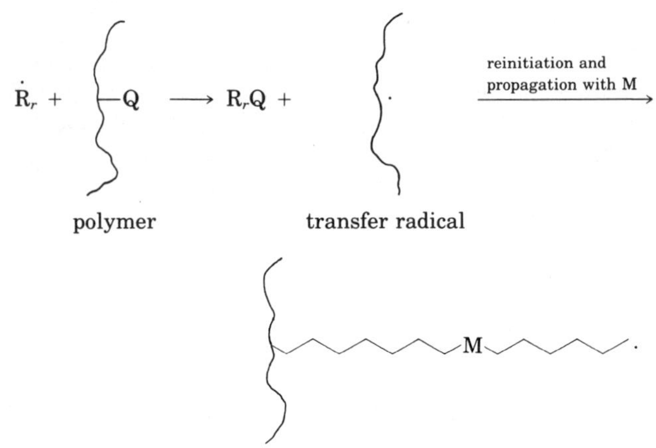

(5)

Ideal Polymerization Mechanism

The four component reactions discussed are set out formally in the following:

Transfer to monomer or substrate

$$\dot{R}_r + M \xrightarrow{k_{fMr}} P_r + \dot{M} \qquad (6a)$$

$$\dot{R}_r + S \xrightarrow{k_{fSr}} P_r + \dot{S} \qquad (6b)$$

Reinitiation by transfer radicals

$$\dot{M} + M \xrightarrow{k_{pM}} \dot{R}_2 \qquad (6c)$$

$$\dot{S} + M \xrightarrow{k_{pS}} \dot{R}_1 \qquad (6d)$$

Termination

$$\dot{R}_r + \dot{R}_s \xrightarrow{k_{t_c rs}} P_{r+s} \qquad (6e)$$

$$\dot{R}_r + \dot{R}_s \xrightarrow{k_{t_d rs}} P_r + P_s \qquad (6f)$$

Initiation

$$I \xrightarrow{fk_d} 2\,\dot{R}_0 \qquad (6g)$$

$$\dot{R}_0 + M \xrightarrow{k_i} \dot{R}_1 \qquad (6h)$$

Propagation

$$\dot{R}_1 + M \xrightarrow{k_{p1}} \dot{R}_2 \qquad (6i)$$

$$\dot{R}_r + M \xrightarrow{k_{pr}} \dot{R}_{r+1} \qquad (6j)$$

In these equations I, M, and S represent initiator, monomer, and solvent (or other substrate), respectively, and P_r stands for a dead polymer molecule containing r monomer units. Subscripts attached to rate coefficients refer to the type of reaction (i,p,f,t_c,t_d), the molecule reacting with the radical (M,S), and the size of the participating radical. f(eq. 6a) is an efficiency factor defined more precisely in the upcoming section on Idealized Kinetics.

The Stationary State Hypothesis. To derive kinetic expressions from a scheme such as equation 6, it is customary to use the stationary state hypothesis. A stationary state is most usefully defined as a phase of reaction in which the rate of change of radical intermediates is much smaller than their rate of formation or rate of destruction. Let $[\dot{R}]$ be the total propagating radical concentration, so that

$$[\dot{R}] = \sum_r [\dot{R}_r] \qquad (7)$$

If $\mathcal{J}$ is the rate of radical formation and R_d the total rate of radical destruction in the termination reactions (eqs. 6i, 6j), then in the stationary state

$$\frac{d[\dot{R}]}{dt} = \mathcal{J} - R_d \ll \mathcal{J}, R_d \qquad (8)$$

Under these conditions, to a good approximation

$$\frac{d[\dot{R}]}{dt} = \mathcal{J} - R_d = 0 \qquad (9)$$

Equation 9 may be regarded as the formal definition of the stationary state.

Idealized Kinetics. Development of kinetic expressions from the ideal scheme (eqs. 6a–j) is enormously simplified by making certain assumptions. Chief among these are the following: (*1*) The velocity coefficients are independent of radical size. (*2*) A stationary state is established for all radical intermediates. (*3*) The mean kinetic chain length is large. Some important results are given below.

$$R_p \equiv -\frac{d[M]}{dt} = k_p[\dot{R}][M] \qquad (10a)$$

$$= k_p[M]\left(\frac{\mathcal{J}}{k_{t_c} + k_{t_d}}\right)^{1/2} \qquad (10b)$$

$$\overline{P_n} = \frac{\text{rate of consumption of monomer}}{\text{rate of formation of polymer molecules}}$$

$$= \frac{k_p[M][\dot{R}]}{(k_{fM}[M] + k_{fS}[S])[\dot{R}] + (\tfrac{1}{2}k_{t_c} + k_{t_d})[\dot{R}]^2} \qquad (11)$$

$$= \frac{k_p[M]}{k_{fM}[M] + k_{fS}[S] + \dfrac{(\tfrac{1}{2}k_{t_c} + k_{t_d})\mathcal{J}^{1/2}}{(k_{t_c} + k_{t_d})^{1/2}}}$$

$$\frac{1}{\overline{P_n}} = \frac{1}{\overline{P_n^0}} + \frac{k_{fS}[S]}{k_p[M]} \qquad (11a)$$

In these equations (which give instantaneous values), R_p is the rate of polymerization, $\mathcal{J}$ is the mean kinetic chain length, and $\overline{P_n}$ the number-average degree of polymerization.

Temperature and Pressure Effects. The influence of temperature on the quantities considered is determined by the activation energies of the appropriate component reactions. Thus the overall activation energies associated with the rate of polymerization and the mean kinetic chain length, E_0 and E_ν, respectively, are

$$\begin{aligned} E_0 &= \tfrac{1}{2}E_i + E_p - \tfrac{1}{2}E_t \\ E_\nu &= -\tfrac{1}{2}E_i + E_p - \tfrac{1}{2}E_t \end{aligned} \qquad (12)$$

where E_i, E_p, and E_t refer to initiation, propagation, and termination, respectively. E_p and E_t are normally small (~30, ~12 kJ/mol or ~7, ~3 kcal/mol, respectively), so that for thermally initiated reactions, E_i is often the largest (~120 kJ/mol or ~30 kcal/mol). For polymerizations initiated photochemically or by high-energy radiation E_i is close to zero and E_0 is, therefore, much smaller.

Relaxation of Simplifying Approximations. The validity of the assumption that the mean kinetic chain length is large depends on negligible consumption of monomer by reactions other than propagation. When the average kinetic chain length is small, this does not hold and allowance must be made for reaction of monomer in initiation and transfer processes.

A general treatment of free-radical polymerization with no assumptions about the form of the dependence of radical reactivity on chain length is impracticable. The validity of the assumption that the velocity coefficients are independent of radical size was questioned at an early stage, since it seems intuitively unlikely to be true. The propagation reaction would not be expected to be very sensitive to viscosity changes in the normal range encountered up to medium degrees of conversion, since it involves one small and rapidly diffusing species (the monomer).

A different situation prevails for termination, which is much more rapid and involves two macromolecules and which, therefore, might be anticipated to be markedly sensitive to viscosity changes. It is currently supposed that termination is diffusion-controlled; however, viscosity changes arising from polymers in solution have relatively little effect on the rate coefficient, mainly because they influence the macroscopic viscosity of the system, whereas the rates of segmental rearrangement of the interacting chains appear to play a more important role in termination, except in highly viscous systems. The assumption of a constant rate coefficient for termination (in a given solvent) is a reasonable approximation for small to moderate conversions, provided the mean radical size is not too small.

Relaxation of the assumption that a stationary state is established for all radical intermediates does not introduce any new matters of chemical principle, although it may cause greatly increased mathematical complexities. Nonstationary states are most commonly encountered when there is a rapid change in the rate of initiation, as in photochemically initiated reactions when the light is turned on or off or in thermal reactions with a rapidly decaying initiator.

Pre- and Aftereffects. The pre-effect and the aftereffect are the best-known examples of nonstationary processes in free-radical polymerization and may be treated rather simply. For the pre-effect in a system with photochemical initiation, assume a stationary state has been set up with a (low) rate of initiation $\mathcal{I}_1$ and at time $t = 0$ the light intensity is increased instantaneously to give a new rate of initiation $\mathcal{I}_1 + \mathcal{I}_2$. In the nonstationary state that ensues the radical concentration builds up, relatively quickly at first, but subsequently more slowly until a second stationary state is reached. If the values of the radical concentration in the initial and final stationary states are $[\dot{R}]_1$ and $[\dot{R}]_2$, respectively,

$$[\dot{R}]_1 = \left(\frac{\mathcal{I}_1}{k_t}\right)^{1/2} \qquad [\dot{R}]_2 = \left(\frac{\mathcal{I}_1 + \mathcal{I}_2}{k_t}\right)^{1/2} \qquad (13)$$

For $t \geq 0$ the radical concentration follows equation 14:

$$\frac{d[\dot{R}]}{dt} = \mathcal{I}_1 + \mathcal{I}_2 - k_t[\dot{R}]^2 \qquad (14)$$

integration of which with the boundary condition $[\dot{R}] = [\dot{R}]_1$ at $t = 0$ yields

$$[\dot{R}] = \theta\left(\frac{\mathcal{I}_1}{k_t}\right)^{1/2} \left\{\frac{(\theta + 1)\exp\,[2\theta\mathcal{I}_1^{1/2}k_t^{1/2}t] - \theta + 1}{(\theta + 1)\exp\,[2\theta\mathcal{I}_1^{1/2}k_t^{1/2}t] + \theta - 1}\right\} \qquad (15)$$

In this equation θ is given by

$$\theta = \left(\frac{\mathcal{I}_1 + \mathcal{I}_2}{\mathcal{I}_1}\right)^{1/2} \qquad (16)$$

θ is therefore equal to the ratio of the final to the initial radical concentration or the final to the initial rate of polymerization. The total pre-effect ΔM_{pre} is then

$$\Delta M_{pre} = \frac{k_p[M]}{k_t} \ln \frac{2\theta}{\theta + 1} \qquad (17)$$

The aftereffect may be treated in a completely analogous way by considering the changes in a system, initially in a stationary state with rate of initiation $\mathcal{I}_1 + \mathcal{I}_2$ brought about by reducing the rate of initiation instantaneously at $t = 0$ to $\mathcal{I}_1$. The relation determining the radical concentration at $t \geq 0$ corresponding to equation 14 is

$$\frac{d[\dot{R}]}{dt} = \mathcal{I}_1 - k_t[\dot{R}]^2] \qquad (18)$$

Hence,

$$[\dot{R}] = \left(\frac{\mathcal{I}_1}{k_t}\right)^{1/2} \left\{\frac{(\theta + 1)\exp\,[2\mathcal{I}_1^{1/2}k_t^{1/2}t] + \theta - 1}{(\theta + 1)\exp\,[2\mathcal{I}_1^{1/2}k_t^{1/2}t] - \theta + 1}\right\} \qquad (19)$$

The total aftereffect M_{aft} is then

$$\begin{aligned} \Delta M_{aft} &= k_p[M] \int_0^\infty ([\dot{R}] - [\dot{R}]_2)dt \\ &= \frac{k_p[M]}{k_t} \ln \frac{\theta + 1}{2} \end{aligned} \qquad (20)$$

Other Nonstationary Systems. The behavior of polymerizing systems in which the initiators have relatively short half-lives has been discussed by several authors. Tobolsky remarked that "when fast initiators are used in styrene at moderate temperature the polymerization stops at quite low conversions." This phenomenon was termed dead-end polymerization and attributed to early depletion of the initiator at a stage when appreciable monomer remains unreacted.

Relationships have been derived for polymerization with first- or second-order initiator decay, notably moments of the molar mass distribution, using stationary state treatment. By considering general solutions of equation 21, results also have been developed for first-order initiator decay in which there were no restrictions on the half-lives of the initiator or the kinetic chains:

$$\frac{d[\dot{R}]}{dt} = nk_d[I]_0e^{-k_dt} - k_t[\dot{R}]^2 \qquad (21)$$

Equation 21 was transformed into the modified Bessel equation (eq. 22):

$$\frac{d^2y}{du^2} + \frac{1}{u}\frac{dy}{du} - y = 0 \qquad (22)$$

where

$$[\dot{R}] = \frac{1}{k_t y}\frac{dy}{du}$$

$$u = 2\left(\frac{n[I]_0k_t}{k_d}\right)^{1/2} e^{-k_dt/2} \qquad (23)$$

Hence,

$$[\dot{R}] = -\frac{uk_d}{2k_t}\frac{I_1(u) + \lambda K_1(u)}{I_0(u) + \lambda K_0(u)} \qquad (24)$$

where $I_1(u)$ and $K_1(u)$ are the modified Bessel functions of order one.

Inhibition and Retardation

Kinetics of Ideal Inhibition. Inhibitors and retarders are substances which when added to polymerization reactions in very low concentrations produce large decreases in rate. Mechanisms responsible for this behavior are deactivation of initiating centers or reduction in their rate of generation (inhibitors) and the interruption of propagating chains (retarders). The kinetic behavior of ideal inhibitors and retarders is of interest in that it provides an important method of determining rates of initiation $\mathcal{I}_0$ in polymerization reactions.

The relevant reactions are

$$I \xrightarrow{fk_d} 2\,\dot{R}_0 \qquad (6a)$$

$$\dot{R}_0 + M \xrightarrow{k_i} \dot{R}_1 \qquad (6b)$$

$$\dot{R}_0 + Z \xrightarrow{k_z} \text{inactive products} \qquad (25)$$

where Z is the inhibitor and equation 25 is the inhibition process.

The distinction between inhibitors and retarders, although clear in classical chain theory, is generally rather blurred in polymerizations. A single substance may function either as inhibitor or retarder or both, depending on its concentration, so that distinction becomes a matter of degree. In terms of gross behavior, addition of inhibitor effectively brings a polymerization to a stop, whereas addition of a retarder merely reduces the rate.

Ideal inhibitors and retarders destroy active radicals without producing any species capable of reinitiation or further reaction with radicals. A stable free radical may act in this way. In the ideal kinetics of these systems, therefore, the nature of the bimolecular termination between propagating radicals remains unchanged, although it may be drastically reduced or even eliminated, but there are no interactions between growing chains and species formed in the inhibition or retardation processes.

Kinetics of Ideal Retardation. The reaction scheme for ideal retardation is set out below.

$$I \xrightarrow{fk_d} 2\,\dot{R}_0 \qquad (6a)$$

$$\dot{R}_0 + M \xrightarrow{k_i} \dot{R}_1 \qquad (6b)$$

$$\dot{R}_r + M \xrightarrow{k_p} \dot{R}_{r+1} \qquad (6d)$$

$$\dot{R}_r + Z \xrightarrow{k_z} P_r + \text{inert products} \qquad (26)$$

$$\dot{R}_r + \dot{R}_s \xrightarrow{k_t} \text{polymer} \qquad (6i, 6j)$$

Equation 26 represents the retardation reaction and the inert products include a dead polymer P_r. In general, the two termination processes (eq. 26 and eqs. 6i, 6j) occur simultaneously, but if retardation is sufficiently strong, equation 26 will predominate.

Determination of Rates of Initiation

Rates of initiation may be determined by use of inhibitors and retarders, from the mean kinetic chain length, and by use of initiators giving known rates of initiation.

Nonideal Free-radical Polymerization

Departures from the classical relations of free-radical polymerization hitherto discussed may be of two types. First, systems may deviate from thermodynamic ideality, and secondly, additional reactions may participate. A single system may show nonclassical behavior of both types.

Medium Effects. Polymerizations, like other reactions in solution, are subject to medium effects. The assumption often made that solvents which do not participate chemically are merely inert diluents is equivalent to supposing that the polymerization obeys gas-phase kinetics. In fact, the component reactions in polymerization are subject to medium effects similar to those encountered with low-molecular-weight reactions. The rate equations for reactions in solution contain activity coefficients, and unless these are maintained effectively constant, relations involving concentrations can be obscured. Changes in activity coefficients arising from marked differences in properties (eg, polarities) of monomers and solvents, associated with significant deviations from Raoult's law, or from large concentration changes may lead to deviations from simple kinetic relations.

Diffusion Control. The sensitivity of the conformations of macromolecules to the environment introduces important medium influences into polymerizations. Chain termination, which is diffusion-controlled, is particularly sensitive to medium changes. Diffusion control is manifest in two respects, involving diffusion of the whole molecule and segmental diffusion, respectively.

Low polymer concentrations may exert a rather subtle effect. The dimensions of the growing chains may change with increasing (dead) polymer concentration, so that the diffusion coefficient and hence k_t increase. Thus, the rate of polymerization may fall over the first few percent conversion.

The nature of the solvent also plays an important role in determining polymer coil dimensions and so has a significant influence on k_t. A decrease in the solvent quality for the polymer brings about an increase in k_t and hence decreases the rate of polymerization. Radical size also influences k_t.

Ionization and Enolization. Solvent changes may induce structural modifications in monomers and radicals, for example ionization and enolization, which may greatly influence rates of polymerization. In copolymerization, similar effects may produce changes in the values of reactivity ratios.

Radical Complexation. There has been a considerable discussion of the extent to which polymer radicals can be assumed to be free. It has been pointed out that the electron affinities of typical propagating radicals are similar to those of tetracyanoethylene and chloranil, compounds which form charge-transfer complexes with many of the common solvents used in free-radical polymerization. An early theory assumed complex formation between radicals and monomer or solvent, propagation being confined to monomer-complexed species.

$$\begin{array}{ccccc} \dot{R}_rM & \underset{K_M}{\overset{M}{\rightleftharpoons}} & \dot{R}_r & \underset{K_S}{\overset{S}{\rightleftharpoons}} & \dot{R}_rS \\ \downarrow k' & & & & \\ \dot{R}_{r+1} & & & & \end{array} \tag{27}$$

From equation 27 it follows that

$$-\frac{d[M]}{dt} = k'[\dot{R}M] = k_p[\dot{R}]_t \frac{K_M[M][M]_0}{K_S[S] + K_M[M]} \tag{28}$$

where $[\dot{R}M]$ is the total concentration of radical–monomer complex, $[\dot{R}]_t$ the total radical concentration, K_M, K_S the equilibrium constants, and k' a rate coefficient. Small solvent effects on k_p have been deduced from rotating-sector measurements in several systems.

Complex formation between propagating radicals and Lewis acids leads to major changes in reactivity, which are the basis of an important technique for synthesizing alternating copolymers (see COPOLYMERS, ALTERNATING).

Polymerizations with Slow Reinitiation. Complicated situations arise when a retardation product is able to reinitiate only slowly. In an early suggestion, Flory proposed the mechanism set out below. Here slow reinitiation by retarder radicals $\dot{Z}$ is represented by equation 30 and bimolecular termination between retarder radicals by equation 31. These two processes distinguish this mechanism from others previously discussed.

$$I \xrightarrow{fk_d} 2\,\dot{R}_0 \tag{6a}$$

$$\dot{R}_0 + M \xrightarrow{k_i} \dot{R}_1 \tag{6b}$$

$$\dot{R}_r + M \xrightarrow{k_p} \dot{R}_{r+1} \tag{6d}$$

$$\dot{R}_r + Z \xrightarrow{k_z} \dot{Z}_r \tag{29}$$

$$\dot{Z}_r + M \xrightarrow{k_{pz}} \dot{R}_{r+1} \tag{30}$$

$$\dot{Z}_r + \dot{Z}_s \xrightarrow{k_{tzz}} \text{products} \tag{31}$$

Equation 29 has been written as an addition reaction, but the kinetic results are also valid if it is a transfer.

A more complete analysis has been carried out employing the Flory scheme with addition of the two termination reactions

$$\dot{R}_r + \dot{Z}_s \xrightarrow{k_{trz}} \text{inert products} \tag{32}$$

$$\dot{R}_r + \dot{R}_s \xrightarrow{k_t} \text{polymer} \tag{33}$$

Thus all three possible bimolecular termination steps between propagating and retarder radicals are included. The relation (eq. 34) was derived by use of stationary state equations:

$$\frac{\mu^2[Z]}{1-\mu^2}\left[1 + \left\{1 + \frac{1-\mu^2}{\varphi^2\mu^2}\right\}^{1/2}\right] = \frac{k_t R_p}{k_p k_z}\left\{1 + \frac{(1-\mu^2)}{\varphi^2\mu^2}\right\}^{1/2} + \frac{k_{pz}k_t[M]}{k_{trz}k_z} \tag{34}$$

In equation 34 $\mu = R_p/R_p^0$ (ie, the ratio of the rates in the presence and absence of the retarder Z, respectively) and $\varphi = k_{trz}/(k_t k_{tzz})^{1/2}$, a parameter familiar in copolymerization kinetics.

The above treatment has some disadvantages in practice that have limited its general use. Among these is the necessity for a knowledge of R_p^0, the rate of the unretarded polymerization, which cannot always be obtained. To overcome those difficulties, an analysis for a similar scheme was carried out, but without the $\dot{Z} + \dot{Z}$ termination. The complete scheme is given below (eqs. 35). The nomenclature is slightly different from that used originally to make it more compatible with the preceding discussion, but S, representing a retarding solvent, replaces Z, used above for a retarder.

$$I \longrightarrow \dot{R}_0 \tag{35a}$$

$$\dot{R}_r + M \xrightarrow{k_i} \dot{R}_1 \tag{35b}$$

$$\dot{R}_r + M \xrightarrow{k_p} \dot{R}_{r+1} \tag{35c}$$

$$\dot{R}_r + S \xrightarrow{k_z} (P_r +)\dot{S} \tag{35d}$$

$$\dot{S} + M \xrightarrow{k_{zm}} \dot{R}_1 \tag{35e}$$

$$\dot{R}_r + \dot{R}_s \xrightarrow{k_t} \text{polymer} \tag{35f}$$

$$\dot{R}_r + \dot{S} \xrightarrow{k_{trs}} P_r \tag{35g}$$

$$\dot{S} + \dot{S} \xrightarrow{k_{tss}} \text{inactive products} \tag{35h}$$

The parentheses apply to degradative addition in which no nonradical polymer is formed. The stationary state treatment with omission of equation 35h leads to the relation (eq. 36)

$$\frac{b}{a^3}R_p^3 + \left(1 + 2abc\frac{[S]}{[M]}\right)\frac{R_p^2}{a^2} - \frac{b\mathcal{I}}{a}R_p - \mathcal{I} = 0 \tag{36}$$

In this equation

$$a = k_p[M]/k_t^{1/2}; \qquad b = k_{trs}/(k_{zm}k_t^{1/2}[M]); \qquad c = k_z/k_p \tag{37}$$

The dimensionless quantity

$$abc = \frac{k_{trs}k_z}{k_{zm}k_t} \equiv \mathcal{R} \tag{38}$$

which determines the extent of retardation and enters into

many kinetic relationships, has been called the intrinsic retardation of the system.

Degradative Reactions Involving Monomer. The formal kinetics of a retarded polymerization do not depend on the nature of the retardation step, both chain transfer and addition giving the same relationships, but when approximations have to be made to obtain tractable expressions, significant differences may appear.

The polymerization of 1-vinylimidazole (VIM) in ethanol solution has been studied. This polymerization is markedly nonclassical; for a given initiator concentration, the order of the rate of polymerization in monomer concentration decreases with increasing $[M]$, tending towards zero at high $[M]$. Such behavior is characteristic of retardation by monomer. Propagating radicals attack the 2-position of the monomer; introduction of a methyl substituent in this position completely removes the retardation, the rate of polymerization of 2-methyl-1-vinylimidazole in ethanol at 70°C being directly proportional to the monomer concentration. Evidence from polymer molar masses suggest that the retardation process consists of addition of propagating chains to the 2-position (rather than hydrogen abstraction) with formation of a relatively unreactive radical $\dot{Z}$:

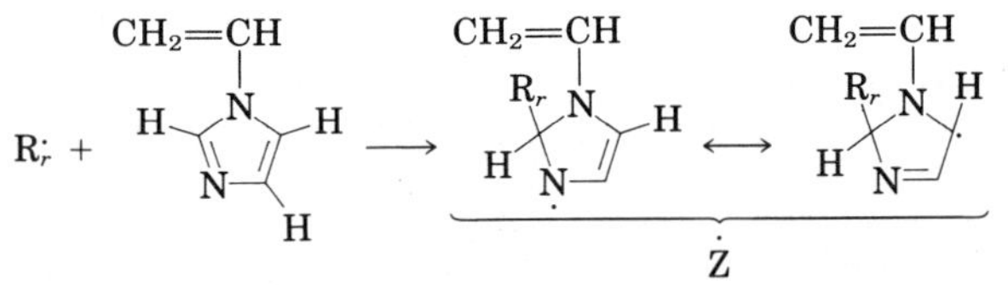

(39)

This type of process has been termed degradative addition.

The Kinetics of Polymerization with Degradative Addition. Complete kinetic analysis of a retarded polymerization including three possible termination reactions suffers from some practical disadvantages. Omission of the interactions between retarder radicals ($\dot{Z}$ or $\dot{S}$) produces considerable simplification, but it becomes necessary to examine *a posteriori* to what degree results obtained justify the approximation. It has been shown that the ratio $[\dot{S}]/[\dot{R}]$ (or $[\dot{Z}]/[\dot{R}]$) is inversely proportional to $\lambda = k_{trs}/k_t$. Two cases of interest are distinguished.

1. If $\lambda \sim 1$, the concentration of $\dot{S}$ is not negligible and may exceed that of $\dot{R}$. Neglect of the $\dot{S} + \dot{S}$ termination can therefore only be justified in terms of a sufficiently small k_{tss}.
2. If $\lambda \gg 1$, $[\dot{S}]$ is relatively small, and the $\dot{S} + \dot{S}$ termination is negligible unless k_{tss} is unexpectedly large.

The simplest assumption to make is that in degradative addition polymerizations $k_t = k_{trs} = k_{tss} = k_t'$ with all three reactions being diffusion-controlled. Since these involve polymer chains of identical type (with different terminal groups), the assumed equality is all the more plausible.

Primary Radical Termination. In addition to geminate recombination, a primary radical derived from an initiator is confronted by several possibilities. It may interact with a monomer to form a propagating radical (eq. 2) or it may enter into primary radical termination by interacting with a growing chain or with a second primary radical. Often primary radical termination is negligible because the radicals are rapidly scavenged by the monomer, but in some circumstances it becomes significant and leads to deviations from ideal kinetic behavior. Deviations are favored by very high rates of initiation and/or low monomer concentrations, since these conditions correspond to a relatively high concentration of primary radicals. Similarly, when the monomer or the primary radical is relatively unreactive, primary termination may become significant.

The qualitative effects of primary termination on polymerization kinetics are easily predictable. The true rate of initiation of propagating chains is no longer proportional to the initiator concentration, but falls progressively below this with increasing concentration so that the initiator exponent falls below 0.5. Further, the rate of initiation becomes dependent on monomer concentration, and the rate of polymerization becomes proportional to a power of [monomer] exceeding one. In the limiting (but practically very improbable) case when all propagating chains end by primary radical termination, the initiator and monomer exponents become zero and two, respectively.

Chemistry of the Component Reactions

Initiation. Polymerization of a reactive vinyl monomer is initiated by free radicals having sufficiently high activity. These may be generated from initiators in diverse ways, among which thermal or photochemical intramolecular bond cleavage, redox reactions, and photochemical hydrogen abstraction are the commonest, but other processes such as use of γ-radiation or electron beams find applications. Initiators belonging to the first group are discussed here, principally (see also ELECTROCHEMICAL INITIATION; PHOTOPOLYMERIZATION).

Thermal Initiators: Intramolecular Bond Cleavage. *Nonaqueous Solutions.* Azo initiators have a long history in polymer technology (see AZO COMPOUNDS). Since nitrogen is evolved on decomposition they have been used in Germany as blowing agents (qv) in the preparation of lightweight plastics and rubbers.

Most of the compounds are represented by the formula

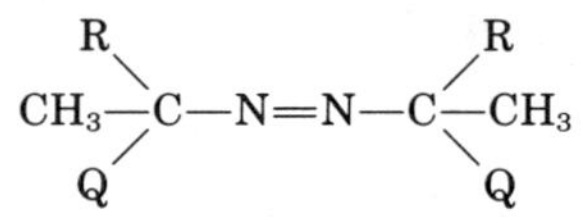

where R = alkyl and Q is a simple carboxylic acid residue or a derivative (nitrile, ester). Azobisisobutyronitrile (AIBN) (R = CH_3, Q = CN) is very widely used as an initiator; 4,4′-azo-4-cyanopentanoic acid (R = CH_3, Q = $(CH_2)_2COOH$) has rather similar initiating properties and, unlike AIBN, is soluble in water.

Thermal decomposition of AIBN and its analogues is generally considered to produce cyanoisopropyl radicals (or analogues) according to

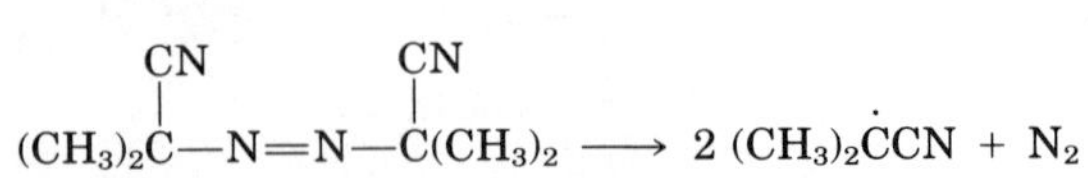

The transient existence of the $(CH_3)_2CN_2\cdot$ radical has been suggested on the basis of esr observations, so that cleavage of

the two C—N bonds may occur consecutively rather than simultaneously.

Peroxy compound chemistry, including decomposition, has been reviewed (see PEROXY COMPOUNDS).

Benzoyl peroxide is a well-established initiator of polymerization, and its reactions have been extensively studied. As with other peroxides, the primary step in the thermal decomposition is scission of the —O—O— bond (eq. 40a) to give two acyloxy radicals. A number of secondary processes may follow; in addition to reacting with monomer, benzoyloxy radicals may recombine (eq. 40b) or undergo β-scission to phenyl radicals and carbon dioxide (eq. 40c). Further recombination reactions giving biphenyl (eq. 40d) and phenyl benzoate (eq. 40e) may then occur.

$$C_6H_5\overset{\displaystyle O}{\overset{\|}{C}}—O—O—\overset{\displaystyle O}{\overset{\|}{C}}C_6H_5 \underset{b}{\overset{a}{\rightleftharpoons}} 2\ C_6H_5CO\dot{O} \qquad (40a)$$

$$C_6H_5CO\dot{O} \xrightarrow{c} C_6H_5^{\cdot} + CO_2 \qquad (40b)$$

$$2\ C_6H_5^{\cdot} \xrightarrow{d} \qquad (40c)$$

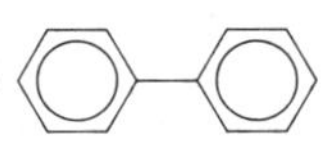

$$C_6H_5CO\dot{O} + C_6H_5^{\cdot} \xrightarrow{e} C_6H_5COOC_6H_5 \qquad (40d)$$

Induced decomposition is an important part of the reaction with many peroxides. This is the interaction with any radical $\dot{R}$ present in the system:

$$\dot{R} + C_6H_5\overset{\displaystyle O}{\overset{\|}{C}}—O—O—\overset{\displaystyle O}{\overset{\|}{C}}C_6H_5 \longrightarrow C_6H_5COOR + C_6H_5CO\dot{O} \qquad (41)$$

$\dot{R}$ may be phenyl, but since benzoyloxy radicals are active in hydrogen abstraction, many different radicals may arise through interaction with the solvent.

Peresters, which are structurally intermediate between diacyl and dialkyl peroxides, have also received attention as initiators.

Aqueous Solutions. Potassium persulfate is a commonly used initiator in aqueous solution: it has long been employed as an initiator in emulsion polymerization, eg, in the original GR-S (styrene–butadiene synthetic rubber) recipe. ("Activated" recipes, in which ferrous salts are added to form redox systems, are also used.) The following reaction scheme (eqs. 42) for the thermal decomposition of persulfate in neutral and alkaline solutions, has been proposed.

$$^-O_3SO—OSO_3^- \longrightarrow 2\ SO_4^{\overline{\cdot}} \qquad (42a)$$

$$SO_4^{\overline{\cdot}} + H_2O \longrightarrow HSO_4^- + \dot{O}H \qquad (42b)$$

$$2\ \dot{O}H \longrightarrow H_2O + \tfrac{1}{2}O_2 \qquad (42c)$$

Redox Initiators. The discovery in 1940 that addition of small quantities of reducing agents to peroxides greatly accelerates radical generation opened the important field of redox initiation. Redox initiators, systems based on mixtures of oxidizing and reducing agents, have acquired great importance, particularly in industry, eg, in low-temperature emulsion polymerization. They initiate through the occurrence of one-electron transfer steps that form free-radical intermediates. Free-radical polymerization has been used in the elucidation of many redox mechanisms, as for example in the much-studied system ferrous ion plus hydrogen peroxide (Fenton's reagent), since it provides a demonstration of the intervention of free radicals and allows their rate of formation to be measured. Many redox initiators are known and numerous "recipes" are current in emulsion polymerization technology.

Redox initiators may be classified according to their solubilities (in water or organic liquids) or their mode of radical generation.

Aqueous Redox Initiators Forming Single Radicals. The powerfully oxidizing properties of mixtures of hydrogen peroxide and ferrous salts, discovered by Fenton in 1894, have been attributed to the participation of $\dot{O}H$ and $H\dot{O}_2$ radicals. Later work has largely confirmed that view and scheme 43 for the principal component reactions is now accepted.

$$Fe^{2+} + H_2O_2 \longrightarrow Fe^{3+} + OH^- + \cdot OH \quad (\text{or } FeOH^{2+} + \cdot OH) \qquad (43a)$$

$$Fe^{2+} + \cdot OH \longrightarrow Fe^{3+} + {}^-OH \qquad (43b)$$

$$\cdot OH + H_2O_2 \longrightarrow H_2O + HO_2^{\cdot} \qquad (43c)$$

$$Fe^{2+} + HO_2^{\cdot} \longrightarrow Fe^{3+} + HO_2^- \qquad (43d)$$

$$Fe^{3+} + HO_2^{\cdot} \longrightarrow Fe^{2+} + H^+ + O_2 \qquad (43e)$$

Organic peroxides or persalts such as potassium persulfate can enter into similar reactions, which are essentially one-electron transfers with concomitant cleavage of the —O—O— bond.

Other transition metal ions such as Ti^{3+} can enter into similar reactions.

Aqueous Redox Initiators Forming Pairs of Radicals. Systems in which two relatively stable salts form a redox pair are commonly used in industrial emulsion polymerizations. Typical oxidizing agents are potassium persulfate, potassium ferricyanide, ceric sulfate, potassium permanganate, *t*-butyl hydroperoxide, cumene hydroperoxide, pinane hydroperoxide, and diisopropylbenzene hydroperoxide. Reducing agents include sodium hyposulfite, sodium metabisulfite, sodium sulfide, sodium thiosulfate, and hydrazine hydrate. No transition metal derivatives are included in these examples. Both oxidizing and reducing components form free radicals, which, in principle, may initiate polymerization, although the behavior in any given system depends on the radical and monomer reactivities.

Typical redox processes in this class, the interactions of cumene hydroperoxide with thiosulfates and of persulfates with thiosulfates and metabisulfites, are shown in equations 44.

$$(CH_3)_2\overset{\displaystyle C_6H_5}{\overset{|}{C}}OOH + S_2O_3^{2-} \longrightarrow (CH_3)_2\overset{\displaystyle C_6H_5}{\overset{|}{C}}O\cdot + S_2O_3^{\overline{\cdot}} + {}^-OH \qquad (44a)$$

$$S_2O_8^{2-} + S_2O_3^{2-} \longrightarrow SO_4^{\overline{\cdot}} + SO_4^{2-} + S_2O_3^{\overline{\cdot}} \qquad (44b)$$

$$S_2O_8^{2-} + S_2O_5^{2-} \longrightarrow SO_4^{\overline{\cdot}} + SO_4^{2-} + S_2O_5^{\overline{\cdot}} \qquad (44c)$$

Nonaqueous Redox Initiators. Organic peroxides may react in nonaqueous solution by redox processes. It has long been known that benzoyl peroxide can enter into relatively rapid reactions with primary, secondary, and tertiary amines.

The most familiar systems include diacyl peroxides and tertiary amines, of which benzoyl peroxide and dimethylaniline are typical. The reactants form a complex which cleaves into radicals:

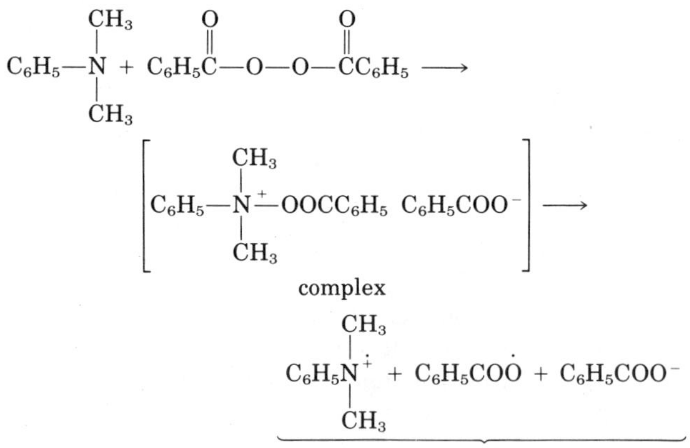

(45)

(3)

In the absence of monomer further reactions occur:

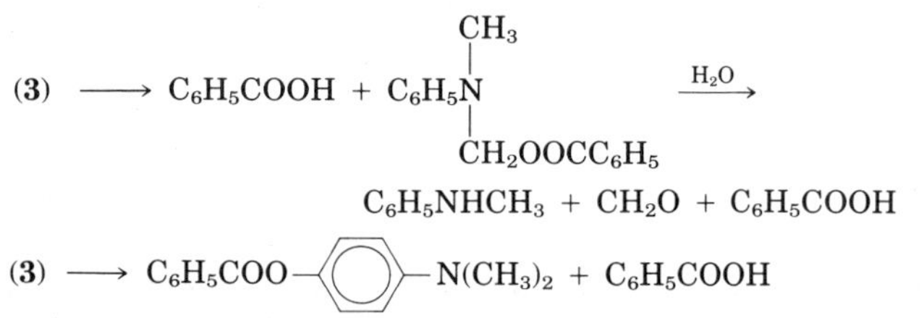

(46)

Decomposition is accelerated by electron-donating substituents in the dimethylaniline and retarded by electron withdrawal. Electron-withdrawing groups in the peroxide accelerate the reaction. These observations all support the occurrence of complex formation.

Initiators Based on Transition Metal Derivatives. Transition metal derivatives with the metal atoms in a low (preferably zeroth) oxidation state may function as components of initiating systems in two distinct ways. Both require the presence of a coinitiator; in type 1 the coinitiator is an organic halide (see below) and in type 2 it is an olefin carrying electron-attracting groups or acetylene or an acetylene derivative. Both types of systems are active as thermal and photoinitiators. Systems of type 2 have so far been used mainly as photoinitiators and will be discussed later.

Transition Metal Derivative plus Organic Halide Initiators. In 1959 the use of metal carbonyls in the presence of high concentration of carbon tetrachloride in the synthesis of ethylene telomers was reported. It was found that although the carbonyls alone are not initiators, they and many other transition metal derivatives become active free-radical sources in the presence of low concentrations of organic halides. An extensive study has been made of these initiators.

Nonaqueous Redox Initiators. Intramolecular Processes. Metal chelates have attracted considerable attention as initiators and appear to be able to initiate different types of polymerizations according to the conditions. Cationic or coordinated cationic polymerization initiated by acetylacetonates (particularly $MoO_2(acac)_2$) has been reported; $Mn(II)(acac)_2$ and $Co(II)(acac)_2$ bring about the polymerization of chloral at room temperatures, probably by a coordinated anionic mechanism. Some chelates produce radicals on heating. Further, several chelates initiate polymerization of styrene and other vinyl monomers on heating. The following order of rates of polymerization has been given.

$$V(III) < Cr(III) = Al(III) < Fe(III) < Co(III) < Ce(IV) < Mn(III)$$

In all the chelate reactions discussed, the initiating radicals are derived from the ligands, so that the processes can be used for incorporating such groups into the terminal positions in polymer chains.

Photoinitiators. Polymerization may be initiated by high- or low-energy radiation. γ-Rays from ^{60}Co or high-energy electrons may initiate ionic polymerization in special circumstances, notably at low temperatures with ultrapure monomers in media of high dielectric constant. High-energy irradiation commonly initiates free-radical polymerization and is much employed for the synthesis of graft copolymers (see GRAFT COPOLYMERS).

Initiation by low-energy radiation, visible or ultraviolet light, will be discussed here. Apart from its value in mechanistic studies, photoinitiation is the basis of important commercial processes including microlithography and photocuring. These processes rely on the cross-linking of multifunctional monomers by free-radical or cationic reactions (see PHOTOCHEMISTRY; PHOTOPOLYMERIZATION; LITHOGRAPHIC RESISTS). Note that photoinitiated polymerization is to be distinguished from photopolymerization, in which each propagation step is effected by the radiation.

Photoinitiators for free-radical polymerization fall into two classes: those which on irradiation undergo intramolecular bond cleavage leading to radical generation and those which when photoexcited abstract hydrogen atoms from H-donors and so form radicals. Among the former are azo compounds (eg, azobisisobutyronitrile), peroxides (eg, benzoyl peroxide), and aliphatic carbonyl compounds (ketones and diketones). The latter group includes aromatic ketones such as benzophenone and derivatives.

Photosensitized Initiation: Polymeric Photosensitizers and Photoinitiators. In a photosensitized reaction, light is absorbed by the photosensitizer and the excitation energy is transferred to an acceptor molecule which subsequently reacts. The photosensitizer is not chemically changed in the process (although, of course, accompanying reactions may result in its decomposition). Photosensitization by excited triplet species, such as the (n,π^*) and (π,π^*) triplets of aromatic ketones and hydrocarbons, respectively, is particularly well documented (see PHOTOCHEMISTRY). These processes are, of course, quite distinct from the hydrogen abstraction reactions discussed in the preceding section, in which the molecule photoexcited by light absorption reacts chemically with another species.

Polymeric photosensitizers and photoinitiators have been described. Many of these are polymers based on benzophenone, eg, poly(*p*-divinylbenzophenone) (DVBP). Such rigid polymers are reported to be effective sensitizers since hydrogen abstraction from the backbone by excited benzophenone is less likely.

Photoinitiators have been developed in which different types of photoinitiating functionality are combined in one molecule, for example, peresters of the type

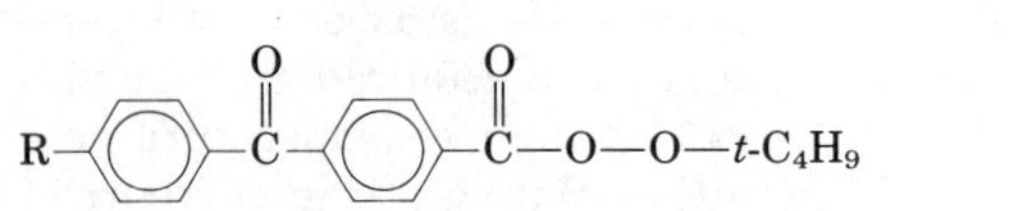

(4)

where R = CH_3, CH_3O, $BrCH_2$, t-C_4H_9O—O—C(=O), and also the peroxide

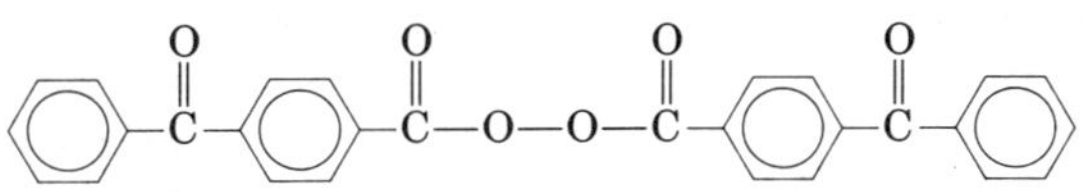

(**5**)

In other systems, the benzophenone chromophore is replaced by (π,π^*) absorbers such as fluorenone, anthracene, and pyrene. When irradiated by light of λ = 366 nm, benzophenone derivatives (**4**) (R = H) and (**5**) initiate free-radical polymerization with effectively classical kinetic features ($R_p \alpha [M],[I]^{1/2}$) at low monomer and initiator concentrations. Polymers prepared from these initiators carry benzophenone end groups.

Redox-type Photoinitiators. Aqueous systems based on metal ion reactions and photoinitiation by dyes and by transition metal derivatives are now discussed.

From a study of the photodecomposition of water in the hydration shells of metal cations Mt^{n+} and anions A^{n-} it was concluded that hydrogen atoms are generated and can initiate polymerization (eq. 47).

$$\begin{aligned} Mt^{n+}\ H_2O + h\nu &\longrightarrow Mt^{(n+1)+} + OH^- + \dot{H} \\ A^{n-}\ H_2O + h\nu &\longrightarrow A^{(n-1)-} + OH^- + \dot{H} \\ A^{n-}\ H_3O^+ + h\nu &\longrightarrow A^{(n-1)-} + H_2O + \dot{H} \end{aligned} \tag{47}$$

In some systems (eg, with cuprous ions) complex formation between monomer and ions interfered with initiation.

Many ion pairs have been shown to photoinitiate, including $Fe^{3+}OH^-$, $Fe^{3+}Cl^-$, $Fe^{3+}F^-$, $Fe^{3+}N_3^-$, $Fe^{3+}C_2O_4^{2-}$, Fe^{3+}H citrate^{2-}, $Pb^{2+}Cl^-$, $Ce^{4+}OH^-$, $Co^{3+}(H_2O)_6(NO_3)_3$; and electron transfer may involve the ligand (eq. 48).

$$Fe^{3+}X^- + h\nu \longrightarrow Fe^{2+}\dot{X} \longrightarrow Fe^{2+} + \dot{X} \tag{48}$$

Dye-sensitized photoinitiation is of practical interest in rapid imaging processes for displays and holography. One familiar system employs aqueous solutions of barium acrylate or acrylamide with a photoredox pair composed of a phenothiazine dye and benzene sulfinate ions.

Photoinitiation by Transition Metal Derivatives. Many transition metal derivatives of the types discussed (eg, carbonyls) photoinitiate free-radical polymerization in the presence of organic halides. Early observations have been described and possible reaction mechanisms outlined. It seems likely that all these reactions are basically similar to the type 1 thermal initiation already described.

Manganese and rhenium carbonyls are probably among the most useful photoinitiators of this group. The long-wave limits of absorption are approximately 460 and 380 nm, respectively, and photoinitiation occurs up to these wavelengths. Manganese carbonyl is therefore active in visible light. Quantum yields are of the order unity.

Recognition of type 2 photoinitiation started with the observation that pure liquid tetrafluoroethylene at −93°C containing a low concentration of $Mn_2(CO)_{10}$ or $Re_2(CO)_{10}$ polymerizes rapidly when irradiated. It was concluded that the reaction is a free-radical polymerization, with C_2F_4 acting as coinitiator to form initiating radicals of type $(CO)_5MnCF_2\dot{C}F_2$.

Efficiency of Initiation. The initiator efficiency f was introduced to accommodate the experimental observation that the rate of initiation is often less than anticipated from a knowledge of the rate of dissociation of the initiator $k_d[I]$. Here k_d refers only to the primary dissociation and not to the total decomposition, which may include induced decomposition (as, for example, with diacyl peroxides). If each molecule dissociates into n primary radicals, the efficiency of initiation is defined by equation 49.

$$f = \frac{\mathcal{I}}{nk_d[I]} \tag{49}$$

Values of f less than unity may arise from partial decomposition into noninitiating products or from participation of the primary radicals in "wastage" (ie, noninitiating) processes. Primary radical termination and reaction of primary radicals with inhibitors are wastage reactions; these have already been considered and are not included in the present discussion.

Many common initiators have efficiencies in the range 0.3–0.8; azobisisobutyronitrile, for example, has $f \sim 0.6$ over a wide range of monomer concentrations. It is generally agreed that these low efficiencies are often the result of cage phenomena. A pair of primary radicals formed in a solvent cage is confronted by several possibilities: (*1*) mutual interaction and destruction, (*2*) reaction of one (or both) radicals with a species (eg, monomer or solvent) in the walls of the cage, (*3*) diffusion out of the cage into the body of the solution. Of these, (*1*) does not lead to chain initiation; the process is conveniently described as geminate interaction, signifying that the partners originate from a single molecule. If the interaction (*1*) is a simple recombination, the parent molecule is reformed and the process, which is obviously without overall chemical effect, does not contribute to radical wastage and reduce f. On the other hand, if (*1*) yields a noninitiating species or an initiator of lower activity, it represents a wasteful process which decreases the efficiency of initiation. Clearly, if radical generation is accompanied by the simultaneous elimination of a stable small molecule such as N_2 or CO_2, geminate recombination cannot reform the parent molecule.

Reactions Between Primary Radicals and Monomers. The relative rates and regioselectivities of addition and transfer reactions of free radicals are matters of central importance in free-radical polymerization. This section deals with addition and hydrogen-abstraction reactions between primary radicals and monomers.

It was long assumed that primary radicals react with monomers almost exclusively by addition, normally to the tail-end of the double bond. Developments in spin-trapping and esr techniques have greatly facilitated experimental investigations, and it turns out that the above hypothesis is broadly applicable, although important deviations have come to light.

More recent and detailed work reveals some complications that can be encountered, especially with oxygen-centered radicals. Practical consequences arising from the introduction of defect groups into polymers through the initiation process have been discussed.

Initiation by t-butoxy radicals generated by thermal decomposition of di-t-butyl peroxalate has received considerable attention. Decomposition of di-t-butyl peroxalate (0.02 mol/dm^3) at 60°C in methyl methacrylate in the presence of the spin-trap 1,1,3,3-tetramethylisoindoline-2-oxyl (0.044 mol/dm³) yields six products which may be translated into the reaction pathways summarized in scheme 50. The expected tail addition to monomer forming radical (**7**) accounts for only 63% of the total. Hydrogen abstraction from both methyl groups occurs, that from the allylic position giving (**8**) (29%) and that from the ester methyl forming (**6**) (4%). This result is of practical importance since the structure and properties of the final polymer can be influenced by the nature of the initiating radicals. Thus (**8**) may be incorporated as an unsaturated terminal group by initiation (a fairly probable process) or by combination with a growing chain; radical (**6**) has been shown to initiate and would also form unsaturated end groups. It is well known that such groups have a deleterious influence on polymer stability; they also increase the possibility of branching during polymerization since the polymer molecule behaves as a macromer. Hydrogen abstraction from the ester moiety increases in importance with the size of the alkyl group.

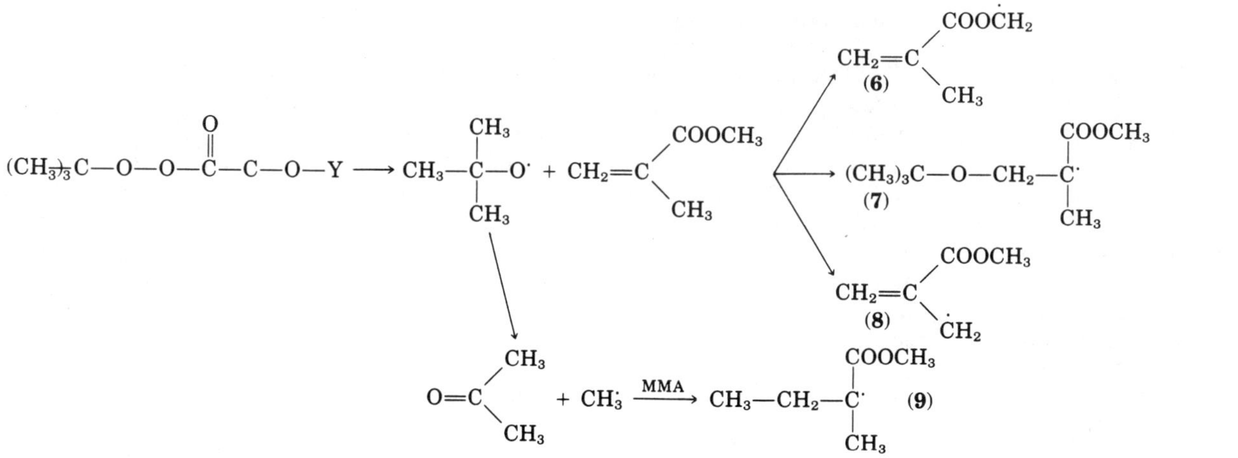
(50)

Solvent effects on reactions of *t*-butoxy radicals with methyl methacrylate have been investigated, and the importance of hydrogen abstraction relative to addition increases with increasing solvent polarity. A greater degree of charge separation may exist in the transition state for hydrogen abstraction or the relative rates may be influenced by interaction of the solvation shells of the reactants. The latter would lead to the observed result if addition were more sterically demanding. The solvent effect clearly has implications in commercial polymer preparation.

Rates of addition of radicals to double bonds are dependent on several parameters, among which polar and steric factors, stabilization of the radical products, and strengths of bonds being formed and broken all play a part. These are discussed more fully later under Propagation.

In an interesting study of the regioselectivity of free-radical processes, reactions of the three electrophilic radicals benzoyloxy, phenyl, and *t*-butoxy with the three α–β unsaturated esters (monomers) methyl acrylate, methyl methacrylate, and methyl crotonate were examined. For the three monomers, the ratio of hydrogen abstraction to double bond addition increased in the series benzoyloxy < phenyl < *t*-butoxy, and the ratio of head-to-tail addition decreased in the same series. Relative rates of head and tail addition were determined by competition experiments, the initiator being decomposed in a mixture of methyl acrylate, methyl methacrylate, and methyl crotonate. Results are presented in Table 1 together with data for cyclohexyl radicals.

The overall rates of addition of radicals to various monomers, collected from various sources, presented in Table 2, illustrate the influence of polar factors.

Spontaneous Thermal Polymerization. A few very highly purified monomers, of which styrene is the best-known example, undergo genuine thermal polymerization in the absence of added initiators. These reactions are uncommon and normally very slow, so that they are highly sensitive to traces of inhibitors or initiators and their existence is not easy to establish. A summary has been given. The mechanism of initiation in the thermal polymerization of styrene has been the subject of much debate. Mayo observed that the rate of initiation is third-order in styrene and that low molecular weight products are also formed. He suggested that radical generation involves formation of a Diels-Alder dimer, which subsequently reacts with styrene.

Strong confirmatory evidence of the correctness of Mayo's proposal has recently been reported, based on the analysis of ¹H nmr spectra of styrene oligomers composed of 6–14 units, prepared by thermal polymerization in the presence of ferric chloride as ideal retarder (see OLIGOMERS).

Table 1. Relative Rate Coefficients for Tail and Head Addition of Radicals (k_T, k_H, respectively) to Acrylate Monomers at 60°C

	$C_6H_5COO\cdot$		$C_6H_5\cdot$		$(CH_3)_3C\dot{O}$		$C_6H_{11}\cdot$	
Monomer	k_T	k_H	k_T	k_H	k_T	k_H	k_T	k_H
methyl acrylate	1.0	0.2	1.0	0.03	1.0	0.02	1.0	0.002
methyl methacrylate	4.5	0.35	1.6	≤0.01	2.9	0.0	0.71	≤0.001
methyl crotonate	1.3	1.6	0.12	0.07	0.3	≤0.03	0.011	0.001

Table 2. Reactivities of Monomers Toward Radicals

Radical	Temperature, °C	Relative reactivities		
		St	MMA	MAN
cyanoisopropyl	60	1	0.56	0.34
polyMAN·	60	1	0.25	0.16
benzoyloxy	60	1	0.12	
benzoylthiyl	20	1	0.12	
t-butoxy	20	1	0.06	0.03
methyl	65	1	1.82	2.68
phenyl	60	1	1.78	2.46

Propagation. The propagation or growth reaction is the basic polymer-forming process. Considerations of polymer structures and stabilities and reaction mechanisms are all facets of chain propagation. Three aspects are discussed here: propagating radical–monomer reactions, the stereochemistry of propagation and polymer microstructures, and the thermodynamics of polymerization and depropagation.

Rates and Regioselectivity of Propagating Radical–Monomer Reactions. The characteristics of chain propagation are essentially similar to those of primary radical additions to monomer. Structure features determining rates and regioselectivity of addition are of prime concern.

The alternatives of head and tail addition to vinyl monomers lead to the four possibilities in propagation shown in equations 51–54.

$$\sim\!\!\sim CH_2{-}\dot{C}XY + CH_2{=}CXY \longrightarrow \sim\!\!\sim CH_2{-}CXY{-}CH_2{-}\dot{C}XY \quad (51)$$

$$\sim\!\!\sim CH_2{-}\dot{C}XY + CXY{=}CH_2 \longrightarrow \sim\!\!\sim CH_2{-}CXY{-}CXY{-}\dot{C}H_2 \quad (52)$$

$$\sim\!\!\sim CXY{-}\dot{C}H_2 + CXY{=}CH_2 \longrightarrow \sim\!\!\sim CXY{-}CH_2{-}CXY{-}\dot{C}H_2 \quad (53)$$

$$\sim\!\!\sim CXY{-}\dot{C}H_2 + CH_2{=}CXY \longrightarrow \sim\!\!\sim CXY{-}CH_2{-}CH_2{-}\dot{C}XY \quad (54)$$

Exclusive propagation by reaction 51 or 53 would lead to polymers of regular head-to-tail structure, whereas reactions 52 and 54 would produce head-to-head and tail-to-tail units. The detailed structure of the polymer chain is determined by the relative rates of the four reactions.

There is compelling experimental evidence from electron spin resonance studies that equation 53 is the major propagation reaction.

Polymers with exclusive head-to-tail addition have been termed "isoregic"; those with half of the units reversed, "syndioregic"; and those with a random arrangement of reversed units, "aregic."

Equally important with regioselectivity are the rates of addition of propagating radicals to monomer. Both are dependent on the same structural features in the radical and monomer and they cannot profitably be discussed separately. From information provided by small-radical studies, important general features may be identified: (*1*) stabilization of radical reactants and products by electron delocalization, (*2*) steric effects, (*3*) polar effects, and (*4*) the strength of the bond being formed. Combinations of these are usually required to rationalize the experimental data and quantitative treatment has so far not been satisfactory. The strength of the new bond (*4*) is a function of (*1*), (*2*), and (*3*). Specific influences, such as the presence of halogen substituents (which lead to weaker bonds), also operate, but they are less common in polymerization since addition to the substituted carbon atoms is exceptional. Its occasional occurrence may be practically important in introducing weak links into the chains.

Free-radical additions are normally facile and exothermic so that according to the Hammond postulate, the transition states are likely to be "early," i.e. to involve little electron and atom redistribution compared to slower processes. The transition states therefore tend to resemble the reactants rather than the products with respect to bond lengths and angles. This is consistent with the small value of the secondary kinetic isotope effects observed when the olefin is substituted with deuterium at the reaction center. Steric and polar interactions, (*2*) and (*3*), respectively, between radical and olefin would therefore be expected to be decisive and the influence of stability of the addend radical of less importance.

Wrong-way Additions. Although the interplay of factors (*1*)–(*4*) that determines the nature of the propagation reaction results in predominantly head-to-tail addition, "wrong-way" additions in polymer chains are well known. In 1948 Flory and Leutner demonstrated chemically the existence of 1,2-glycol units in poly(vinyl alcohol) prepared by hydrolysis of poly(vinyl acetate) and estimated that the polymer contained about 1% of these units. Thus in the polymerization of vinyl acetate approximately 1% of the additions follow equation 52 to give head-to-head groupings. More recently nmr studies have produced ample evidence for the existence of head-to-head units in polymers. Poly(vinyl fluoride), poly(vinylidene fluoride), and their copolymers have received considerable attention; all have been found to contain such units, and in poly(vinylidene fluoride) it has been estimated that 3.5–6% of the monomer units in the chain are reversed (depending on synthesis conditions) (see HEAD-TO-HEAD POLYMERS; VINYL FLUORIDE POLYMERS; VINYLIDENE FLUORIDE POLYMERS).

Lewis Acid Complexes. Complex formation between reacting species and Lewis acids, generally through interaction with side chains carrying groups such as ester or nitrile, can lead to large changes in the reactivities of radicals and monomers which are the basis of an important general method for synthesizing strongly alternating copolymers.

Stereochemistry of Propagation: Polymer Microstructures. The substituted carbon atom of a monomer unit in a polymer chain is a chiral center (unless it is situated at the precise midpoint of a chain) and therefore has a *d* or *l* configuration, depending on the stereochemistry of addition. The assembly of configurations determines the chain microstructure (see MICROSTRUCTURE).

The fine structure of a polymer chain provides a time sequence of the propagation steps by which the chain was formed, and a main concern is to relate microstructure to the stereochemistry of propagation. The sequence of propagation reactions is generally considered in terms of Bernoulli trial or Markov statistics.

Thermodynamics of Polymerization. The Depropagation Reaction. Enthalpy and entropy changes in the propagation reaction are, for long chains, effectively those of the overall polymerization (ΔH_p, ΔS_p respectively for the prevailing experimental conditions). Both are negative, so that the free-energy change,

$$\Delta G_p = \Delta H_p - T\Delta S_p \tag{55}$$

which is negative for a polymerizing system, becomes less negative as the temperature is raised. When a specific temperature T_c is reached such that

$$T_c - \Delta H_p/\Delta S_p \tag{56}$$

ΔG_p changes from negative to positive, being 0 at T_c, so that the system is in equilibrium. Thus no polymerization can occur at temperatures higher than T_c, which is called a ceiling temperature (see POLYMERIZATION THERMODYNAMICS). The existence of such a temperature is a direct consequence of the fact that polymerization is an aggregation process.

Ceiling temperature phenomena may readily be visualized kinetically. It is necessary to take into account the depropagation reaction (eq. 57):

$$\dot{R}_{r+1} \xrightarrow{k_{dp}} \dot{R}_r + M \tag{57}$$

The energy of activation of reaction 57 (E_{dp}) is generally much higher than that of propagation E_p, the difference being equal to the enthalpy change on polymerization (for long chains)

$$E_p - E_{dp} = \Delta H_p \tag{58}$$

which is a relatively large quantity. Consequently, with increasing temperature k_{dp} increases rapidly relative to k_p, and although depropagation is normally negligible, it becomes significant at elevated temperatures.

Early experimental evidence pointed to the existence of a ceiling temperature when, during studies of the copolymerizations of low alkenes with sulfur dioxide, copolymerization only occurred below a fairly well-defined temperature characteristic of the particular olefin. Results were independent of the initiating system employed, indicating that the ceiling temperature phenomenon is a genuine characteristic of the system not caused by impurities.

With many monomers the equilibrium monomer concentrations at ambient temperatures are too small to measure directly since the ceiling temperatures are relatively high; they may be obtained by extrapolation. The ceiling temperature is a function of monomer concentration and is independent of mechanism. It is given by

$$T_c = \frac{\Delta H_p}{\Delta S_p^0 + R \ln [M]} \tag{59}$$

Chain Transfer. Chain transfer is an atom abstraction process that may involve any species present in the polymerization, including monomer and initiator. The transfer constant $C = k_f/k_p$, as defined, is a measure of the relative reactivities of the propagating radicals in transfer and in addition to the monomer. Transfer constants play an important part in discussions of the reactivities of radicals and monomers, and since polymer molar masses are dependent on the incidence of chain transfer, knowledge of the transfer activities of the components of a system is essential if polymers of specified mean size are to be prepared. Transfer constants cover an extremely wide range, from low values of a few units of 10^{-4} up to several hundreds or even thousands. Equation 11a is probably the most widely used relation in determining transfer constants.

Catalyzed Chain Transfer to Monomer. A report in 1981 on the catalyzed chain transfer to monomer in the free-radical polymerization of methyl methacrylate stimulated interest in this type of process. The remarkable catalytic activity of Co(II) hematoporphyrin IX tetramethyl ether (Co-Por) at concentrations in the range 10^{-4}–10^{-3} M was demonstrated in the polymerization of MMA at 60°C initiated by azobisisobutyronitrile; it was found that the degree of polymerization (at high conversion) fell from 4,000 in the absence of Co-Por to approximately 8 when the cobalt catalyst was present.

The effect was shown to be genuinely catalytic; the concentration of Co-Por estimated spectrophotometrically did not change during reaction, and effectively all the material could be extracted unchanged at the end of the process. Co-Por is not an initiator and it was reported to have little influence on the rate of reaction. Since the only potential transfer agent present in sufficient concentration in these experiments was monomer, the results must be attributable to chain transfer to monomer catalyzed by Co-Por. The rate of polymerization was proportional to $[AIBN]^{1/2}$ throughout, indicating the absence of retardation, and the gel effect, normally so prominent in the polymerization of MMA at high conversion, was eliminated by the presence of Co-Por at a concentration 6.6×10^{-4}. The latter effect arises from the reduction in polymer molar mass.

Nmr studies of the polymers revealed that the transfer process is hydrogen atom transfer from propagating radical to monomer:

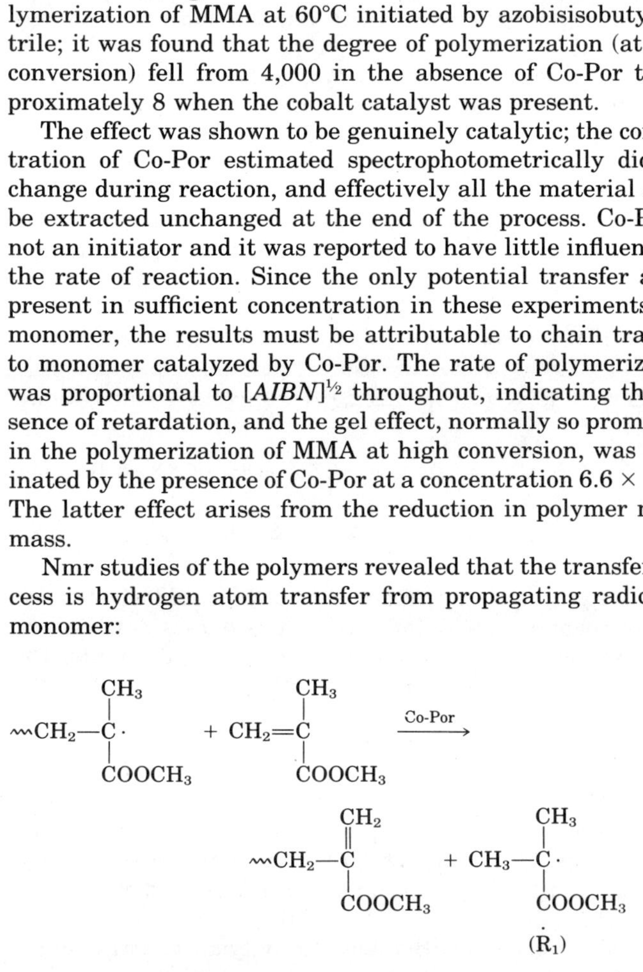

(60)

The transfer constant $C_{por} = 2.4 \times 10^3$ in the range 40–70°C. Reactions of this type have potential uses in synthesis, eg, macromer preparation.

Reactivities of Radicals in Transfer Processes. Systematic consideration of atom abstraction reactions started with the formulation in 1938 of the Evans-Polanyi relation (eq. 61), which relates the activation energy E_A to the total enthalpy change ΔH^0. This relation is still the most general basis for discussion of transfer reactions.

$$E_A = \alpha\Delta H^0 + C \tag{61}$$

It was derived from consideration of the potential energy curves of closely similar reactions; provided these curves do not change shape basically from one reaction to another, equation 61 follows from geometric arguments, α and C being constants. The parameter α is less than unity and should be larger for series of reactions with greater activation energies.

Relations of the Evans-Polanyi type are obscured if the potential energy curves for a series of reactions are dissimilar in shape. This may be brought about by (irregular) polarity changes from one reaction to another or by steric or electron delocalization changes.

Polar factors are important in transfer as in propagation processes. An example in small-radical chemistry, where the effect is well documented, is provided by hydrogen abstraction from methane and hydrogen chloride by methyl and trifluoromethyl radicals. The reactions are almost thermoneutral, and the activation energies are shown in Table 3. The low activation energies of the reactions with HCl are attributable to polar factors in the transition state:

$$\dot{C}F_3 + \overset{\delta+}{H}—\overset{\delta-}{Cl} \longrightarrow [\overset{\leftrightarrow}{CF_3 \cdots \overset{\delta+}{H} \cdots \overset{\delta-}{Cl}}]^{\ddagger} \longrightarrow CF_3H + \dot{C}l$$

$$\dot{C}H_3 + \overset{\delta+}{H}—\overset{\delta-}{Cl} \longrightarrow [\overset{\leftrightarrow}{CH_3 \cdots \overset{\delta+}{H} \cdots \overset{\delta-}{Cl}}]^{\ddagger} \longrightarrow CH_4 + \dot{C}l$$

Steric effects may be of three types: (*1*) steric hindrance, impeding the approach of the radical to the reaction site; (*2*) steric inhibition of resonance; and (*3*) steric compression. The latter seems to be of rather general importance; release of steric compression on radical formation accounts in part for the progressive decrease in C—H bond strength from primary to secondary and tertiary. The high transfer activities of carbon tetrabromide and mercaptans must be partly attributable to the relative weakness of the C—Br and S—H bonds undergoing scission in the compounds, but release of steric compressions in CBr_4 on radical formation is also likely to contribute. In the copolymerization of styrene and methyl methacrylate the penultimate unit effect in chain transfer with CBr_4 probably arises partly from steric compression in the transition state by close contacts of α-CH_3 and Br and partly from steric hindrance.

Electron delocalization in the addend or attacking radical influences the rate of transfer through its effects on bond dissociation energies. In the addend, delocalization reduces the strength of the bond being broken and so enhances transfer.

Chain transfer to halogen derivatives is frequently encountered in polymerizations. Halogen abstraction has been studied in connection with the radical chain addition of polyhalomethanes to olefins. Abstraction follows the order of strengths of the bonds being broken, except for hydrogen abstraction:

$$I > Br \sim H > Cl > F$$

Abstraction from halides by methyl radicals has been investigated. It appears that $\dot{C}H_3$ experiences a polar repulsion by the filled orbitals of the halogen atom. Since this would not arise between $\dot{C}H_3$ and H—R, the rate of abstraction of hydrogen is greater than expected on the basis of bond dissociation energy alone. Other results with $\dot{C}F_3$ are also consistent with this view.

Quantum mechanical studies of hydrogen abstraction from alkenes have been published.

Empirical Relations for Radical and Monomer Reactivities. Empirical schemes have existed for many years; they do not have a firm theoretical basis, but they have proved of value, both in making predictions and in summarizing data.

The Q-e Scheme. The alternating tendency in radical copolymerization, first demonstrated by Mayo and Walling, is most easily accounted for in terms of differing electrical polarities. Alfrey and Price embodied this idea into the *Q-e* scheme, which was developed to allow semiquantitative prediction of reactivity ratios.

Three aspects of the *Q-e* scheme have attracted criticism: first, the assumption of equal charges for radical and monomer; secondly, the assumption of permanent charge, present in the ground-state reactants, rather than transition state polarity; and thirdly, the observed (effective) independence of the dielectric constant of the medium shown by reactivity ratios. Although these criticisms are mainly valid, they do not reduce the practical value of the *Q-e* scheme or alter the fact that it is in harmony with much experimental data.

Patterns of Free-radical Activity. The "Patterns" scheme was devised as a semiempirical approach in which arbitrary features were reduced to a minimum. As in the *Q-e* treatment, the reaction rate was considered to be determined by the general reactivities of the radical and the substrate together with the appropriate polarities, but both types of property were so defined as to make independent experimental measurements possible. The considerations in "Patterns" apply equally well to propagation and transfer reactions.

Other Models and Treatments. Several models for the theoretical treatment of reactivity ratios have been proposed, including the electronegativity model, the charge-transfer model, and a model based on a modified Hammett equation.

Bimolecular Termination. Polymer radicals, like small radicals in the gas or liquid phases, may interact by combination and/or disproportionation. The two processes give rise to unusual groups in the polymer chains—a head-to-head sequence statistically in the middle of the chain or a terminal double bond and a saturated end group, for combination or disproportionation respectively. Such groups influence the properties of the polymers, sometimes to an unduly large extent. Unsaturated end groups in poly(methyl methacrylate) are preferred sites for the initiation of thermal degradation; they may also participate in polymerization and so influence the molar weight

Table 3. Hydrogen Abstraction Reactions

Reaction	Activation energy, kJ/mol[a]
$CH_4 + \dot{C}H_3 \longrightarrow \dot{C}H_3 + CH_4$	58.6
$CH_4 + \dot{C}F_3 \longrightarrow \dot{C}H_3 + CF_3H$	46.9
$HCl + \dot{C}H_3 \longrightarrow \dot{C}l + CH_4$	10.5
$HCl + \dot{C}F_3 \longrightarrow \dot{C}l + CF_3H$	20.9

[a] To convert kJ/mol to kcal/mol, divide by 4.184.

distribution and contribute to branching and gelation. Head-to-head sequences can also lead to thermal instability by virtue of the steric compression they introduce. Further, the molar mass distribution of the polymer is determined by the mode of termination, in the absence of predominating transfer.

The nature of the termination reaction is therefore not merely a matter of academic interest but, like initiation, has numerous practical implications.

Determination of the Disproportionation:Combination Ratio. A number of techniques for measuring $x = k_{td}/(k_{tc} + k_{td})$ are available, but there is a disappointing lack of firm information for many monomers. Techniques include calculation from DP, measurement of $\bar{P}_w/\bar{P}_n$, the kinetic method, coupling experiments, grafting tests, gelation experiments, and model experiments with small radicals.

Diffusion Control in Bimolecular Termination. It is generally agreed that termination reactions of small radicals in solution are diffusion-controlled.

Diffusion control occurs when the rate of an inherently fast process is limited by the rate of approach of the reacting entities to the proximity required for reaction.

Bimolecular termination reactions between macromolecular radicals may be visualized as follows. First, two macromolecules must come into contact by translational diffusion; secondly, the reactive ends of the chains must diffuse sufficiently closely and in a suitable configuration to allow reaction, which is stage three. If chemical reaction is slow compared to either diffusion process, it is rate-controlling, but if the chemical reaction is relatively fast, the rate of the overall process becomes diffusion-controlled by either of the diffusion steps mentioned. Several lines of experimental evidence point to segmental diffusion as the critical step.

Theoretical calculations have given complicated results, but suggest that the termination coefficient is inversely proportional to solvent viscosity (as found experimentally) and is influenced by the molar masses of the polymer radicals, decreasing somewhat as the latter increase. It is also affected by excluded volume, chain flexibility, monomer size, and the thermodynamic quality of the solvent.

Theoretically derived equations are too complex to use in kinetic schemes. An empirical relation which has received considerable attention is the geometric mean expression

$$k_t \propto (rs)^{-b}$$

where r,s are the sizes of the reacting radicals and b is a small positive quantity of the order 0.1.

Polymerization at High Conversion

Optimization of a commercial process that necessitates achievement of high conversions requires a model that enables the course of reaction to be calculated over the whole range of conversion. The model should contain the minimum number of adjustable parameters, be independent of type and concentrations of initiator, and allow calculation as functions of time of molecular weight distributions as well as conversions.

Development of a suitable model evidently involves estimation of termination coefficients at high polymer concentrations, but account must also be taken of diffusion control of the propagation reaction which sets in at very high viscosities. When polymerization at a temperature below the glass transition temperature of the polymer is taken to high conversion, a stage is reached at which the reaction mixture assumes glassy conditions, with residual monomer acting as plasticizer. The propagation rate then becomes diffusion-controlled and may fall to near-zero values, so that polymerization virtually ceases, even though the conversion is less than 100%. This phenomenon is the glass effect; it must not be confused with dead-end polymerization arising from initiator depletion, discussed earlier.

Rather than attempt a mechanistic model, necessarily very complex, it may be more realistic to adopt a semiempirical approach. The "free-volume" treatment assumes that control by translational diffusion occurs when the diffusion coefficient of a polymer radical D_P becomes less than or equal to a critical value D_{Pcr}. After the start of this diffusion control, k_t is assumed proportional to the diffusion coefficient

$$k_t = k_1 D_P \qquad (D_P < D_{Pcr}) \tag{62}$$

k_1 being a constant. D_P is related to the free-volume fraction V_F by the equation

$$D_P = \frac{\varphi_0 \delta^2}{k_2 M} \exp(-A/V_F) \tag{63}$$

where M is the (homogeneous) polymer molar mass; φ_0 and δ the jump frequency and distance respectively; k_2 and A constants. The free volume is a function *inter alia* of the volume fraction of polymer, and so decreases during polymerization in a manner which may be represented by the relation:

$$V_F = \{0.025 + \alpha_P(T - T_{gP})\}\frac{V_P}{V_T} + \{0.025 + \alpha_M(T - T_{gM})\}\frac{V_M}{V_T} + \{0.025 + \alpha_S(T - T_{gS})\}\frac{V_S}{V_T} \tag{64}$$

in which V, T_g, and α represent volume, glass-transition temperature, and coefficient of thermal expansion respectively; and the subscripts P, M, S denote polymer, monomer, and solvent respectively. V_T is the total volume. The best correlation with experimental data is achieved when M in equation 63 is replaced by the square root of the cumulative weight-average molar mass $\overline{\overline{M}}_w$; hence

$$k_t = k_1 \left(\frac{\varphi_0 \delta^2}{k_2 \overline{\overline{M}}_w^{1/2}}\right) \exp(-A/V_F) \tag{65}$$

in the absence of chain entanglements. When chain entanglements occur, probably soon after k_t becomes controlled by translational diffusion, then the relation

$$k_t = \bar{k}_1 \left(\frac{\varphi_0 \delta^2}{\bar{k}_2 \overline{\overline{M}}_w^{1.75}}\right) \exp(-A/V_F) \tag{66}$$

in which $\bar{k}_1$ and $\bar{k}_2$ are constants is used. If it is assumed that termination becomes translation-diffusion-controlled before occurrence of entanglements, then at the onset of the gel effect

$$\overline{\overline{M}}_w^m \exp(A/V_F) = \frac{k_1 \varphi_0 \delta^2}{k_{t0} k_2} = K_3 \tag{67}$$

K_3 being a (temperature-dependent) constant and k_{t0} the value of k_t not influenced by entanglements. According to equation

65, $m = 0.5$, but in equation 67 this quantity is kept as an adjustable model parameter. The cumulative molecular weight and free volume at this critical point are designated $\overline{\overline{M}}_{wcr1}$ and V_{Fcr1} respectively; these quantities satisfy equation 67. For each weight-average molecular weight, at constant temperature there is one and only one conversion satisfying equation 67. As the polymerization proceeds, from equations 66 and 67

$$\frac{k_t}{k_{t0}} = \left(\frac{\overline{\overline{M}}_{wcr1}}{\overline{\overline{M}}_w}\right)^n \exp\left[-A\left(\frac{1}{V_F} - \frac{1}{V_{Fcr1}}\right)\right] \qquad (68)$$

in which $n = 1.75$ according to equation 66 but is retained as a model parameter.

An expression analogous to equation 68 may be written for the propagation coefficient, when this becomes diffusion-controlled, with the aid of the expression for the diffusion coefficient D_M of a small molecule in a polymer solution (φ_2, δ_2 being jump frequency and distance, respectively, and B a constant):

$$D_M = \left(\frac{\varphi_2 \delta_2^2}{6}\right) \exp\left(-\frac{B}{V_F}\right) \qquad (69)$$

If V_{Fcr2} is the critical volume corresponding to the conversion at which diffusion control sets in,

$$k_p = k_{p0} \exp\left[-B\left(\frac{1}{V_F} - \frac{1}{V_{Fcr2}}\right)\right] \qquad (70)$$

k_{p0} being the propagation coefficient at low conversions.

Insertion of k_t and k_p given by equations 68 and 70 into the classical rate expression gives for the rate of reaction

$$\frac{dx}{dt} = \left(\frac{k_{p0}^2}{k_{t0}}\right)^{1/2} \left(\frac{\overline{\overline{M}}_w}{\overline{\overline{M}}_{wcr1}}\right)^{n/2} \exp\left[-B\left(\frac{1}{V_F} - \frac{1}{V_{Fcr2}}\right)\right]$$

$$\times \exp\left[\frac{A}{2}\left(\frac{1}{V_F} - \frac{1}{V_{Fcr1}}\right)\right]\left[\frac{fk_d[I]_0}{1 - \epsilon x}\right]^{1/2} (1 - x)\exp\left(-\frac{k_d t}{2}\right) \qquad (71)$$

In this equation x is the fractional conversion and ϵ is the volume contraction factor ($= (d_p - d_M)/d_p$, the ds being densities).

Since k_t is a function not only of conversion and temperature but also of molecular weight, the latter must be estimated. Instantaneous values of $\overline{M}_w$ are calculated in the conventional manner and the required cumulative values obtained from the relation

$$\overline{\overline{M}}_w = \frac{1}{x} \int_0^x \overline{M}_w dx \qquad (72)$$

Conversion is calculated as a function of time by simultaneous solution of equations 71 and 72 with the aid of an iterative procedure.

The model has been tested for a variety of polymerizations taken to high conversion, including those of methyl methacrylate and styrene and has given good agreement with experimental data obtained in bulk and solution polymerization over wide ranges of temperature and initiator and solvent concentrations.

Heterogeneous Radical (Precipitation) Polymerization

Radical Occlusion and Living Macroradicals. The discussions on the termination reaction and on polymerizations carried to high conversions have illustrated the importance of physical conditions on the course of polymerization. Heterogeneous radical polymerizations supply further illustrations, sometimes of an extreme kind. Heterogeneous polymerizations may be defined as those polymerizations in which propagating radicals pass from one phase to another. Often heterogeneity results from the separation of solid polymer from monomer initially present in bulk or in solution or in the gas phase.

Heterogeneous polymerizations as a class show characteristic types of behavior, among which are autoacceleration with time, from the onset of heterogeneity to appreciable degrees of conversion; initiator exponents exceeding 0.5 (the classical value); and unexpectedly large photochemical aftereffects ΔM_{aft}, which increase with the duration of irradiation. These features are not due to the initial presence of impurities but are associated with polymer separation during reaction.

The notion of radical occlusion arose from consideration of the behavior of acrylonitrile on polymerization. Radicals are trapped in the polymer during photopolymerization at 25°C; the concentrations of such radicals have been determined by "titration" with DPPH and by esr observations. The active polymer is indefinitely stable in contact with monomer at 25°C, but rapidly becomes deactivated by contact with air or oxygen. Clearly, at 25°C the radicals are protected from contact with monomer by the surrounding polymer chains, the system showing glass effect behavior. At 60°C the polymer swells sufficiently to admit monomer and allow reaction (the so-called "fast reaction"). At the end of the fast reaction at 60°C, no radicals survive. The active polymer is to be regarded as a living radical polymer.

Occlusion and Polymerization Kinetics. The "occlusion" picture of heterogeneous polymerization was built up from consideration of phenomena of the kind described. It is based on the idea that many (or all) polymer radicals precipitate during growth; in the nonsolvent environment, the precipitated chains tend to be tightly coiled and since the radical end has a high probability of being occluded ("shut up") within the coil, it will be shielded from attack and so appear unreactive.

The mean degree of occlusion influences the kinetic coefficients k_p and k_t and is thus responsible for the characteristic kinetic features discussed. The incidence of occlusion is also evident in the effects of low concentrations of swelling agents on the rate of polymerization of acrylonitrile at 25°C and the unusual dependence on temperature of the molecular weight of polyacrylonitrile prepared at constant initiator concentration.

The influence of occlusion on the overall kinetics of polymerization is easy to appreciate in a qualitative fashion. As polymerization proceeds, the number of polymer particles present and their average size increase. Both factors increase the mean degree of occlusion; hence the effective k_t diminishes and the rate of polymerization increases.

Block Copolymers from Macroradicals. Living polyacrylonitrile and polyacrylamide have been prepared and the macroradicals used to initiate block polymerization. Methyl methacrylate and methyl acrylate were more active in these reactions than vinyl acetate, vinylidene chloride, or styrene.

Molar Mass Distributions

As has been discussed, stationary conditions in a free-radical polymerization are normally established in a short time after the start of reaction. Subsequently, in the stationary phase of the reaction, the molar masses of the polymeric products change only slowly, corresponding to the consumption of the reactants. If the concentrations change only slightly or are maintained constant, the molar masses remain effectively constant and the number-average degrees of polymerization $\overline{P}_n$ are given by equation 11. This is the origin of the familiar comment that high polymers are formed (almost) instantaneously in radical polymerizations, but in living ionic systems or condensation polymerizations, the molar masses increase throughout the course of the reaction.

Polymerization is governed by statistical laws, and since a growing radical is confronted by several possibilities, the radicals present and also the dead polymer formed have sizes distributed over a range extending from a single unit upwards. (Note the difference from living systems, in which, ideally, propagation is the only reaction possible.) When all concentrations are effectively constant or maintained constant, the distribution (at low conversion) is given by

$$P_r = \frac{\Delta M}{k_p M}(1-\xi)\xi^r \left\{ k_{fM}M + k_{fS}S + k_{t_d}\left(\frac{\mathcal{I}}{k_t}\right)^{1/2} + \frac{k_{t_c}}{2}\left(\frac{\mathcal{I}}{k_t}\right)^{1/2}(r-1)\frac{1-\xi}{\xi}\right\} \quad (73)$$

where M is the concentration of monomer polymerized and P_r is the concentration of polymer molecules of size r formed. ξ is the probability of propagation:

$$\xi = k_p M/\{(k_p M + k_{fM}M + k_{fS}S + (Jk_r)^{1/2}\} \quad (74)$$

For termination by disproportionation, with or without transfer, the number and weight averages are given by

$$\begin{aligned} \bar{r}_{\text{(disp,trans)}} &= \frac{1}{1-\xi} \\ \bar{r}_{w\,\text{(disp,trans)}} &= \frac{1+\xi}{1-\xi} \sim \frac{2}{1-\xi} \text{ for long chains} \end{aligned} \quad (75)$$

while for termination by exclusive combination without transfer

$$\begin{aligned} \bar{r}_{\text{(comb)}} &= \frac{2}{1-\xi} \\ \bar{r}_{w\,\text{(comb)}} &= \frac{2+\xi}{\xi(1-\xi)} \sim \frac{3}{1-\xi} \text{ for long chains} \end{aligned} \quad (76)$$

In a homogeneous polymer of degree of polymerization r, $\bar{r} = \bar{r}_w = r$; for heterogeneous polymers the magnitude of $\bar{r}_w/\bar{r}$ is often taken as a measure of heterogeneity, or breadth of the distribution. For disproportionation (transfer) and combination polymers of the ratios $\bar{r}_w/\bar{r}$ are respectively 2 and 1.5 (from eqs. 75, 76), so that on this criterion the latter has a narrower distribution.

Determination of Absolute Rate Coefficients

The most direct method is measurement of the propagating radical concentration, from which the absolute propagation coefficient may be evaluated by use of equation 10a if the rate of polymerization is known. Although this possibility has been recognized for many years, the low radical concentrations involved ($\sim 10^{-7}$ molar) have delayed its application; however, sufficiently sensitive techniques for observing esr spectra have been developed.

The second type of method is based on kinetics. Inspection of the equations for stationary state condition shows that all the quantities that may be readily determined—the rate of polymerization, kinetic chain length, and polymer molar mass—involve the propagation and and termination coefficients as $k_p/k_t^{1/2}$, so that the two quantities cannot be separated on the basis of these observations. Measurements of transfer coefficients provide ratios (k_p/k_{fM_P} k_p/k_{fS}) and not absolute values. On the other hand, properties measured under nonstationary conditions are expressible in terms of k_p/k_t, eg, the pre- and aftereffects, so that combination of stationary and nonstationary data allows separate evaluations of the rate coefficients. A number of procedures for making the nonstationary observations have been devised, including direct measurement of radical concentrations, and kinetic methods (eg, measuring, pre- and aftereffects, the viscosity method), and the rotating sector technique. Note that the kinetic problem of separating the rate coefficients is encountered in radical chain reactions in general.

BIBLIOGRAPHY

P. J. Flory, *Principles of Polymer Chemistry,* Cornell University Press, Ithaca, N.Y., 1953.

C. H. Bamford, W. G. Barb, A. D. Jenkins, and P. F. Onyon, *The Kinetics of Vinyl Polymerization by Radical Mechanisms,* Butterworths, London, 1958.

C. H. Bamford and C. F. H. Tipper, eds., *Comprehensive Chemical Kinetics,* Vol. 14A, Elsevier, Amsterdam, 1976.

G. Odian, *Principles of Polymerization,* 2nd ed., John Wiley & Sons, Inc., New York, 1981.

C. H. Bamford
University of Liverpool

RADIOPAQUE POLYMERS

The search for a nondestructive method of polymer evaluation has led to a new area of research, that of radiopaque polymers. Based on casting shadows, radiographic imaging techniques incorporate the principle of radiopacity, which is the physical property of absorbing x rays or reflecting ultrasound waves. Both of these techniques depend on variations in densities between a specimen and its surroundings. Sharp x-ray imaging is obtained from materials of high electron density; effective ultrasound is observed only when significant variations in specific gravity exist between an object and its surroundings.

Conventional polymers cannot be detected by imaging techniques because they mainly contain the elements carbon, hydrogen, oxygen, and nitrogen, which exhibit both low electron density and low specific gravity. The recent research into radiopaque polymers explores methods of increasing the average electron density and the specific gravity of polymers by incorporating heavy elements into these systems.

Physical Principles of X-ray and Ultrasound Imaging

X-ray Imaging. Examination of structures by casting shadows was the underlying principle of all x-ray imaging for approximately 80 years, until in the mid-1970s a new imaging principle was established with the introduction of computed tomography for medical diagnosis. The energy of diagnostic x-ray photons ranges from 10 keV to 100 keV. These photons are produced either by the "bremsstrahlung" process or by characteristic radiation. Characteristic radiation produces a spectrum with several discrete energies. The resulting x-ray photons are absorbed by the specimen through collision with its electrons at a rate proportional to the concentration of electrons in the specimen. The electron concentration is proportional only to the density of the material; consequently, a high atomic number is associated with effective x-ray absorption. Iodine and barium sulfate are, therefore, most frequently used to attain high radiopacity.

Ultrasound Imaging. Such imaging is produced by the reflection of sound waves from interfaces between two media. When an ultrasound wave is incident on an interface, a fraction of the wave is reflected; some of it is absorbed and some is transmitted. In diagnostic ultrasound, only the intensity of the reflected beam is of interest. Ultrasound reflection increases with the difference in densities between the two media. Incorporation of heavy-metal salts into polymers increases a specimen's specific gravity and thus improves the detection of polymers by ultrasound techniques.

Radiopacity Determination

The imaging techniques used with radiopaque polymers are the commonly employed roentgenographic and ultrasound techniques. The various specimens are radiographed either by x ray or ultrasound, and the resulting images, which are obtained on film, are compared with the opacity of an aluminum wedge. A minimum radiopacity equivalent to 2.0 mm of aluminum (as specified by the International Standards Organization, ISO 4049-1978, recently revised) is necessary for detecting the thinnest sections normally encountered in polymeric resins. For complete characterization, radiopaque polymers can be compared to the ANSI/ASTM F640-79 Method C.

Classification of Radiopaque Polymer Systems

Three main prototypes of radiopaque polymer systems have been introduced: radiopaque polymer blends, polymerization products of radiopaque monomers, and radiopaque polymer–salt complexes.

Radiopaque polymer blends are produced by incorporating the radiopacifying agent as a physical mixture with the polymer to form a heterogeneous dispersion. Most of these agents are incompatible with the polymer, and the product possesses reduced mechanical properties. None of the radiopaque additives described so far are chemically incorporated into the resin. Many of them leach out with time; thus their radiopacity is temporary.

Polymerization products of radiopaque monomers are produced by introduction of the radiopacifying element either electrovalently (as a gegenion) or covalently (as a substituent) into the monomer unit prior to polymerization. The disadvantage of the electrovalently bound system is its susceptibility to hydrolysis. This may prevent permanent radiopacity and may lead to a complete loss of radiopacity with time. The disadvantage of covalently bound systems is their relatively high cost. Nonetheless, this area of research may be of future promise.

Radiopaque polymer–salt complexes are produced by the incorporation of small amounts of radiopaque, heavy-metal salts into appropriate polymer ligands via chelation. The resulting systems are homogenous and possess both polymer and ionic character. Permanent radiopacity is ensured by entrapment of the polymer–salt complex in an interpenetrating, cross-linked network. It is anticipated that no undue sacrifice of the polymers' physical properties will be observed, because the amounts of incorporated radiopacifiers are low. Complexation can be applicable to a broad range of polymers, and prefabrication as well as polymerization *in situ* are possible. Polymer–salt complexes are desirable for industrial usage because they offer an advantageous form for easy handling, recovery, and reuse.

Uses

The establishment of a reliable, quick, nondestructive imaging method for detecting changes in polymer systems can benefit numerous industries. Biomedical polymer, defense, and plastic manufacturing are three industries where applications of such a technique are now found.

BIBLIOGRAPHY

P. Sprawls, Jr., *The Physical Principles of Diagnostic Radiology,* University Park Press, Baltimore, 1977.

S. C. Bushong, *Radiopaque Science for Technologists,* 3rd ed., C. V. Mosby Co., St. Louis, 1984, p. 554.

K. W. M. Davy and B. E. Causton, *J. Dent.* **10**(3), 254 (1982).

D. W. Xia, R. Silberman, I. Cabbaso, and J. Smid, *Polym. Prepr. Am Chem. Soc. Div. Polym. Chem.* **26**(1), 72 (1985).

RUTH SILBERMAN-HAZONY
Consultant

RAMAN SPECTROSCOPY

The Raman effect was first observed in 1928 by C. V. Raman and K. S. Krishman, but wide application was delayed until the development of the laser. Since the advent of the laser, progress has been rapid and what was predominately a laboratory curiosity has become a useful analytical tool. It is possible to analyze Raman spectra in terms of the molecular components, or functional groups, composing the molecule if the vibrational normal modes are reasonably well isolated energetically from one another. As in infrared absorption spectroscopy, there are tables of compiled observations relating the vibrational motions of molecular subunits to frequency ranges.

Because Raman spectroscopy is sensitive to molecular motions, any environmental perturbations or changes in molecular structure that modify these motions can be detected. By careful analysis of the resulting spectra, the source of the modification can be identified. Thus normal coordinate analysis is an integral part of the process used to interpret Raman spectra. Such studies have been very useful for polymers, because the morphology of the polymer is often of more interest than the molecular constituents of the polymer.

What is the Raman effect? First, it is important to realize that Raman scattering is a different effect from infrared absorption. The Raman effect is a true scattering effect and as such the light involved interacts with the electronic structure of the molecule that is being modulated by the molecular vibrations, but the light does not undergo any form of energy conversion followed by reconversion and subsequent emission (see SCATTERING). The Raman effect can be considered classically as an inelastic scattering at the beat frequencies; the sum frequency is called the anti-Stokes shift, the difference frequency is called the Stokes shift.

The interaction of light with the electronic structure of the molecule involves the interaction of the oscillating electric field of the light with the charge distribution or multipole structure of the molecule. In the case of the Raman effect the principal source of interaction is with the induced dipole, which is produced by the oscillations of the electric field of the light wave driving the electron cloud of the molecule. The electron cloud of the molecule is also subject to driving forces from the molecular vibrations, which to a first approximation are linear with the nuclear displacements. This gives rise to a beat frequency between the two driving forces, the electric field of the light wave and the vibrational modes of the molecule. To calculate the correct results for Raman intensities it is necessary to take into account some of the quantum mechanical features of the problem because the transitions involved are quantized.

What Information Is Obtained from a Raman Spectrum?

Isotropic Systems. The Raman effect is due to an interaction of light with the molecular vibrational motions. The vibrational motions that give rise to strong scattering intensity have certain well-defined symmetry properties. By analyzing the polarization properties of the scattered radiation it is possible to determine the symmetry of the vibrational motion giving rise to a particular scattering frequency. Isotropic systems produce Raman scattering in which the modes can be separated into two groups with different symmetry properties corresponding to the rotational invariants of an object in free space. In general, as the ordering of the sample is increased, the number of groups of motions with different symmetry properties that can be spectroscopically separated increases also. For samples that can be fully oriented relative to a laboratory coordinate system, such as molecular single crystals, it is often possible to determine the symmetry of the normal modes exactly. Similar information can be obtained for polymers that have been oriented by processing or crystallization conditions.

Ordered Systems. Extensive studies have been made of highly ordered systems, such as molecular crystals. For single crystals with one or more sites a subset of the polarizability tensor components contributes to the Raman scattering. These components are determined by the relationship of the space-fixed axes to that of the crystal axes.

For polymers the situation is frequently somewhere between highly ordered and isotropic. A processed crystalline or semicrystalline polymer is partially aligned relative to a set of material axes (ie, draw direction) because of shear forces encountered in drawing, injecting, or blowing operations. The advantage of ordering a system is that it is possible to extract much more detailed spectroscopic information from the additional polarization experiments available. This leads to detailed considerations of the molecular symmetry and the symmetry of the molecular vibrations.

Tools Used to Analyze Raman Spectra

Molecular Symmetry and Group Theory. Group theory provides a method for correlating the symmetries of the molecular vibrations with the components of the derived polarizability tensor; consequently it is possible to determine for a given Raman experiment the molecular vibrations that should be present in the spectrum. Essentially, a given Raman experiment excites a subset of the components of the derived polarizability tensor. The only molecular vibrations that can contribute to the Raman scattering involve the same molecular coordinates as those of the derived polarizability tensor components excited. Numerous sources cover the many aspects of group theory needed to analyze Raman spectra.

Line Groups. Line groups are, strictly speaking, applicable only to one-dimensional systems. However, many useful descriptions of polymer spectroscopy refer to the group analysis of the polymer chain in terms of a line group because many isolated polymer chains can be thought of as roughly linear structures. In reality, two subgroups of the three-dimensional space group are dealt with: the factor group, which is the point group of the repeat unit, and the translational group. The factor group is isomorphous with the crystallographic point group and is consequently very useful because molecular interactions transverse to the chain axis are usually small and can be neglected, leaving the internal motions of the repeat unit coupled to the one-dimensional set of translations parallel to the chain. Furthermore, many of the internal motions of the repeat unit are independent of the chain and consequently can be treated as isolated molecular motions. Therefore, the normal modes of the polymer chain can be handled as if the repeat unit is an isolated molecule for many modes, with the remainder of the modes (primarily motions of the chain backbone) approximating those of a simple line group.

Intensities. In general, intensities have been difficult to calculate because a detailed knowledge of the elements of the polarizability tensor of the molecule as a function of the nuclear coordinates is needed. This would require detailed electronic structure calculations, which, unfortunately, for most molecules are still limited by computational restraints.

Applications of Raman Scattering to Polymers

Chemical Structure and Interactions. Chemical structure refers to local bonding and functional groups. Frequently the Raman spectrum contains bands that can be associated with bonding and functional groups as if these were independent of the overall structure. Although this is not precisely true, it is often the case that the bulk of the normal mode displacements occur in a limited region of the system because the energy of the given vibration is decoupled from its surroundings by an impedance mismatch.

As a molecular characterization tool, Raman spectroscopy can be used in the identification of bonding and functional groups. This is certainly important for new polymers; when combined with other characterization tools, a complete picture of the polymer repeat unit is revealed.

Interactions between molecular units and their surroundings can be readily detected by perturbations in the Raman spectrum. Hydrogen bonding, crystal field splitting, chain conformation, chain packing, and solvation are all examples of modifications to the surrounding electrical field of the molecular unit that can affect the Raman scattering.

Morphology. In many senses it is in this area that Raman spectroscopy is most useful to the polymer scientist. The Raman effect is sensitive to the molecular environment in many ways. Examples of environmental effects that perturb the Raman scattering and that can be used to study polymer morphology include crystal packing, orientation effects, processing history, doping, and surface interactions. Because the bulk properties of a polymer can be drastically modified by variations in morphology, knowledge of and control over the morphology are also technologically important.

Raman spectroscopy has some distinct advantages over other techniques, especially in comparison to infrared methods when morphology is the subject of interest. Raman spectra are frequently obtained with no modification to the sample, which is critical to the preservation of the morphology. Unusual shapes and sizes pose little or no problem because only a surface is needed for sampling.

Crystalline Structure. Crystalline polymers have a number of morphological features that can be studied by Raman spectroscopy. The thickness of the lamellar structure, actually the length of the chain stem that can be at an angle to the surface normal, can be determined by measuring the longitudinal acoustic mode (LAM), which is the accordionlike stretching motion of the chain stem. Likewise, an estimate of the bulk modulus can be obtained from LAM measurements. Oriented polymer samples give Raman spectra that can be analyzed to determine structural details. Examples of such local structure determinations include the separation of planar zigzag from helical chain conformations, molecular packings of the chains, and the number of chains per unit cell. Figure 1 shows spectral changes as a function of orientation; the observed symmetries and mode intensities allow the determination that the sample of ethylene–tetrafluoroethylene copolymer has planar zigzag chains rather than helical chains.

Amorphous Structure. A feature analogous to LAM in the crystalline regions has been identified in the amorphous regions of polymers. It has been named the disordered longitudinal acoustic mode (DLAM) and seems to obtain its Raman intensity from a statistical distribution of all-trans chain segments in the amorphous region.

Processing Effects and Orientation. A typical goal of polymer processing is strength enhancement. One means by which this occurs is through molecular orientation of the chains that can be induced by various techniques of flowing or elongating the polymer. Raman spectroscopy is an excellent tool for studying such orientation effects.

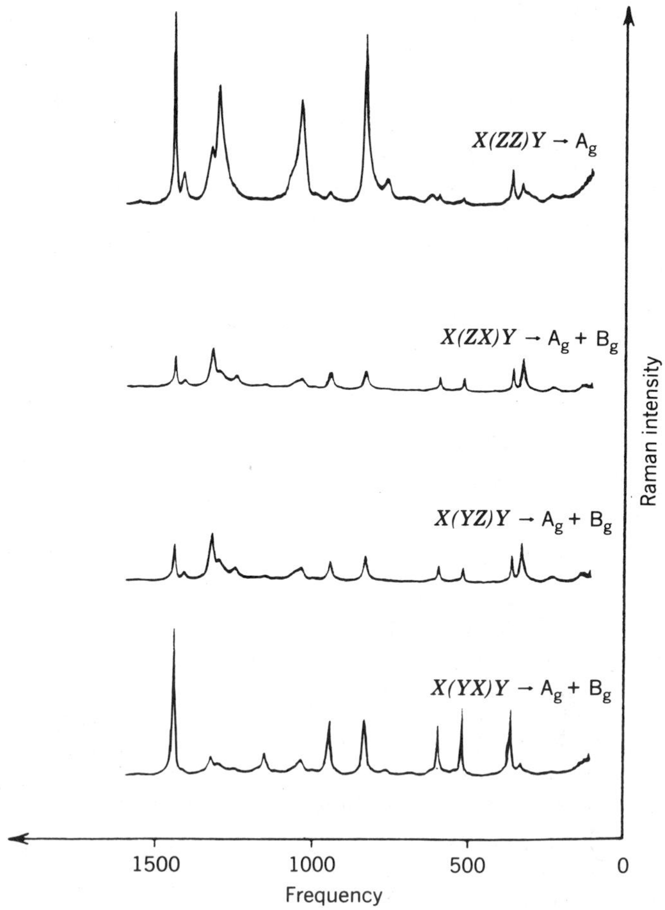

Figure 1. Polarized Raman spectra of ethylene–tetrafluoroethylene (E–TFE) alternating copolymer (Tefzel) filament in a 90° scattering geometry. Symmetry types contributing to each experiment are indicated based on a planar zigzag polymer backbone and give the most consistent interpretation of the data.

Conjugated and Optically Nonlinear Polymers. Conjugated polymers, which have continuous delocalization along the chain axis, have many unusual properties. Because of their potential technological uses, these properties have attracted a great deal of interest. Properties such as electrical conduction that approaches that of metals and nonlinear optical responses exceeding most inorganics are the main focuses of research (see ELECTRICALLY CONDUCTIVE POLYMERS). Because many conducting and nonlinear optical polymers have extended electronic conjugation, which gives rise to visible absorption, resonance Raman spectroscopy has been used as a probe of the electronic structure. General characterization of conjugated polymers by Raman spectroscopy is also useful. However, because of sample absorption and degradation in many conjugated polymers, this has been applied only to a limited extent.

Polymers on Surfaces. Surface coatings of polymers, ranging from a few 10ths of a nanometer to several micrometers in thickness, are of interest both scientifically and technologically. Molecular interactions at interfaces, surface wetting, surface-induced orientation, and surface chemistry are open areas for research and are critical industrially in adhesives, coatings, surface modification, and catalysis. Raman techniques are useful in the study of thin films because of the ability of the laser to probe the surface nondestructively. With care the Raman effect can be measured without disturbing the film or interface.

New Developments

New methods include Fourier Transform Raman spectroscopy, array detectors, waveguide Raman sampling, and micro Raman. In the area of interpretation of Raman spectra normal coordinate analysis is a mature field, but still underdeveloped. Calculations are only adequate for a few classes of molecules, such as hydrocarbons, some conjugated structures, and several functional groups.

Raman spectroscopy has many unique sampling techniques that provide vibrational information that would be difficult to obtain in any other manner. However, as with any technique, it must be combined with other analyses to obtain the full picture. In the past there has been a tendency to overlook Raman because of the difficulties in obtaining spectra and interpreting them. Current technologies are solving many of the earlier problems, and it may be possible in the future to obtain Raman spectra as quickly and easily as infrared spectra.

BIBLIOGRAPHY

E. B. Wilson, Jr., J. C. Decius, and P. C. Cross, *Molecular Vibrations: The Theory of Infrared and Raman Vibrational Spectra,* McGraw-Hill Inc., New York, 1955.

D. A. Long, *Raman Spectroscopy,* McGraw-Hill Inc., New York, 1977.

J. C. Decius and R. M. Hexter, *Molecular Vibrations in Crystals,* McGraw-Hill Inc., New York, 1977.

F. A. Cotton, *Chemical Applications of Group Theory,* 2nd ed., Wiley-Interscience, New York, 1971.

NICHOLAS E. SCHLOTTER
Bell Communications Research

RATES OF POLYMERIZATION. See KINETICS OF POLYMERIZATION.

RAYON

By far the main production of rayon today is by the viscose process; therefore, a large part of this review is dedicated to viscose technology. However, a significant effort is being invested in the search for more efficient routes to make rayons. These involve the new *N*-methylmorpholine *N*-oxide solvent system for cellulose being developed by Courtaulds Co. in the UK and the spinning of cellulose carbamate by Neste Oy in Finland.

Viscose Rayon

Most of the world's rayon is made by the viscose process. The term rayon designates a range of products that have widely varying properties, for the most part owing to the versatility of the viscose process. This versatility, which is the result of the many process steps and spinning changes that can be made, is a mixed blessing, because each stage of processing and spinning requires close attention to guarantee the desired product properties. The viscose process is a demanding process that requires continuous, year-round operation to prevent gelling of the system and to yield high quality products.

Process. The stages in the preparation of a satisfactory cellulose xanthate solution are given in Figure 1.

Spinning. Once the viscose is ripened to the proper xanthate level and the 2- and 3-xanthate wedges are removed so that proper rates of coagulation can be obtained, the viscose is ready for spinning. Chief goals in spinning are control of coagulation versus regeneration rates and the maximum size of the differences in these rates to obtain maximum responses to stretching. Stretching a soft gel does not give permanent alignment of molecules; when the gel hardens too much, further stretching is virtually impossible. In the spinning process, viscose is extruded into a bath containing both salt and acid. The salt usually is sodium sulfate at sufficiently high concentrations to ensure that the sulfate solution behaves as a dehydrating system that removes water from the entering 7–10% cellulose xanthate solution. Sodium sulfate has an RT solubility of 280 g/L; normally, it is used at 120–200 g/L in the coagulation baths. If greater dehydration action is desired, ammonium sulfate is used as the salt in the coagulation bath; its solubility is 440 g/L at RT, and it can be used at 230–350 g/L as a coagulant. Sulfuric acid is present to regenerate cellulose from the coagulated xanthate. The concentration of sulfuric acid relative to the concentration of salt is critical in controlling the extent of water removal from the extruded viscose (by the salt) before the system is rigidly fixed by the acid removing the solubilizing xanthate groups. The control of densification of the extruded viscose by removing water with salt and stretching the densified mass versus the rate of acid hardening is the gist of the spinning process. The denser the viscose, the higher the resulting wet modulus. A high wet modulus (HWM) fiber with minimal stretch has been made in this manner. The higher the density, the more effective the stretching in aligning cellulose molecules; the final dry and wet tenacities of the product will also be higher. In practice, the interplay of simultaneous reactions and other processes leads to significant interactions among the various factors governing spinning.

Spinning Additives. Rayon manufacturers became acutely aware of the need for controlling the complex interactions of coagulation, regeneration, and stretching when they first tried to prepare HWM fibers to improve rayon's competitive position

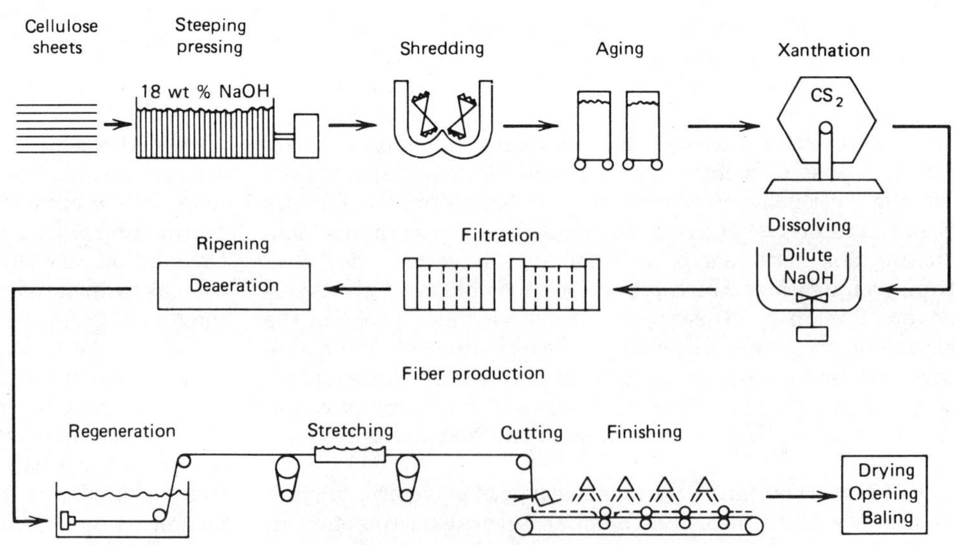

Figure 1. Viscose process.

relative to cotton. The simple approaches of using only Na_2SO_4 and H_2SO_4 were altered by the use of additives in both the viscose dope and the spin bath.

The best systems make use of some type of amine, eg, dimethylamine, and a polyether glycol of ca 1500 mol wt added to the viscose as well as zinc in the coagulation bath. Proper combinations of these ingredients can increase regeneration times as much as 400–500% over unmodified systems.

Stretching. An advantage of retarding regeneration is to make use of the densification of the extruded dope to obtain better molecular alignment on stretching. A fixed amount of stretching can occur; this is balanced between what is called jet stretch and godet stretch. Jet stretch is the ratio of the linear speed at the first godet roll to the average linear flow rate out of the spinnerette holes. Godet stretch is the stretching between the first and last godet rolls. The higher the godet stretch, the less residual elongation is left in the final fiber. Therefore, the manufacturer balances fiber strength and residual elongation.

Spinnerette Design. There are probably as many individual spinnerette designs as there are fiber producers; yet in all cases, certain basic design parameters are dictated by the viscosity-shear nature of the viscose dope. Each spinnerette hole has a defined length to diameter (L/D) ratio and a defined conical entrance angle.

Finishing. After the fibers are spun, cut, and washed, finishes are applied to give the proper frictional performance in subsequent textile processing. Slipperiness presents problems in spinning where physical properties of yarn are closely related to surface characteristics and friction, whereas stickiness results in improper fiber movement during processes such as carding and can lead to nep formation and fiber damage. Over the years, a wide variety of multicomponent finishes have been proposed; one of the original and still one of the best is based on sulfated castor oil (see TEXTILE RESINS).

Modified Rayons. Control of the regeneration conditions coupled with specific additives gives fiber manufacturers the opportunity to produce a host of modified fibers (see FIBERS, MANUFACTURE). An excellent review has been given. Modifications include high wet modulus, crimped fibers, hollow fibers, and flame-resistant fibers. Certain rayon fibers are used to make carbon fibers and are the sole acceptable material for making rocket motors for NASA's space program.

Test Methods. Test methods for viscose include plugging index, xanthate sulfur, total sulfur, salt index, particle count by Coulter, HIAC, and laser scanners, while fiber tests include cuen (copper ethylenediamine) DP, neutralization point, gel swell, rewet swell, and solubility in 6.5% caustic.

The American Enka Co. has issued reports about the Newcell process, which is based on spinning solutions of up to 20 wt % cellulose dissolved in hot aqueous *N*-methylmorpholine *N*-oxide. This system is amenable to closed-loop recovery recycle and relatively rapid spinning speeds. Although most previously investigated solvent-spun rayon systems were based on the use of unstable, easily regenerated cellulose derivatives, the Newcell process involves no discernible derivative formation but rather uses a solution of cellulose; this leads to important economic advantages for the system.

Courtaulds has also spent considerable effort on making a "reconstituted cellulose" fiber from *N*-methylmorpholine *N*-oxide (NMMO) and has constructed a pilot plant at Durbin, South Africa, and a new multimillion-dollar commercial plant in the UK. The fiber has high wet modulus and exhibits wet strengths that are 90% of the conditioned values. This new fiber, which is "reconstituted" rather than "regenerated" cellulose has been given the generic name of Genesis rather than rayon. Courtaulds is calling their product ten-cell and it represents the greatest advance in cellulose fiber technology in the last 50 years. Furthermore, the total process is completely closed loop with all components being recycled and only water

Table 1. Physical Properties of Commercial Rayons

		Viscose				
			High tenacity			
Property	Cuprammonium	Regular	LWM[a]	IWM[a]	HWM	Polynosic, unmodified
stress–strain behavior tenacity, mN/tex[b]						
conditioned	150–200	106–238	265–503	441–574	397–883	300–485
wet	84–120	62–159	132–380	291–353	309–706	238–353
standard loop	88–230	88–132	203–318	88–221	62–132	62–106
standard knot	62–150	62–124	194–309	141–282	106–247	106–256
elongation at break, %						
conditioned[c]	7–23	15–30	9–26	14–18	5–10	6.5–10
wet	16–43	20–40	14–34	17–22	6–11	7–12
specific gravity	1.52–1.54	1.50–1.53	1.52	1.53	1.53	1.51
moisture regain[d]	12.5	13	13	11–13	11–12.5	11.5–12.5
birefringence, by refractive index	0.026	0.018	0.036	0.039–0.044	0.046–0.057	0.040–0.045
water retention, %	100	90–100	65–80	65–75	60–70	55–70
solubility in 6.5% NaOH at 20°C, %	20–30	20–30	30–40	15–20	1.0–5.0	4–6
fiber DP[e]	450–550	330–350	300–500	350–600	500–800	550–650
wet modulus, mN/tex[b]	9–18	16–18	18–26	44–53	71–177	71–177

[a] Low and intermediate wet modulus.
[b] To convert N/tex to g/den, multiply by 11.33.
[c] Conditioned at 21°C and 65% rh.
[d] After centrifuging at 1000 *g* (ASTM D2402-65T); commercial standard is 11%.
[e] Tenacity at 5% stretch.

being flashed off to reactivate the NMMO cellulose solvent. Spinning speeds are about 350 m/min, three times faster than the viscose process and high wet modulus fibers are produced directly without the need for zinc or other additives normally employed in the viscose process.

Another recent entry is the process being promoted by Neste Oy in Finland, which uses ammonia and urea to make a stable cellulose carbamate that can be dissolved in dilute caustic and spun like regular viscose. Success with this system will depend not only on the recycle of components but also on the complete removal of the carbamate groups, because any slight residual functionality is bound to lower the wet strength of the fibers significantly.

Cuprammonium Rayon

Essentially all of the rayon made today is produced by the viscose process; nevertheless, there are some uses where the cuprammonium process has retained an advantage because of specific performance factors (Table 1). Fibers from cuprammonium rayon are significantly more supple than viscose fibers and are used where a very soft hand is desired. The use of films and hollow fibers from the cuprammonium process for making artificial kidneys is critically important in this branch of medicine (see MEDICAL APPLICATIONS).

Ecological and Pollution Considerations

In most processes developed many years ago, little consideration was given to pollution and ecological impact; the older rayon-manufacturing processes are not exceptions. The cuprammonium and the viscose processes consume huge amounts of water for each kilogram of product; ca 420–750 L (110–200 gal) of water per kilogram of rayon is needed directly for processing and 8–10 times that amount of water must be handled to provide the plant supplementary-service facilities. When water effluents did not require special treatments, overall process economics were favorable. However, treating ca 835 L (220 gal) of water per kilogram of fiber imposes on such processes new penalties that cannot be overlooked. Similar considerations exist for air handling in present-day plants.

Energy Requirements for Rayon vs Synthetics

Because rayon is made from trees, no petroleum is used to make the original polymer; also a significant part of the energy needed for separation and purification of cellulose is derived from pulping by-products as energy sources. These factors give rayon a favorable position relative to synthetic fibers with regard to total energy needed for fiber production.

BIBLIOGRAPHY

A. F. Turbak, *Amer. Chem. Soc. Symp. Ser.* **58**, (1977).

A. F. Turbak, R. B. Hammer, R. E. Davies, and H. L. Hergert, *Chemtech* **10**, 51 (Jan. 1980).

ALBIN TURBAK
Georgia Institute of Technology

REACTION FOAM MOLDING.

See CELLULAR MATERIALS; STRUCTURAL FOAMS.

REACTION INJECTION MOLDING

Reaction injection molding (RIM) is a polymer process operation whereby reactive liquid components are mixed by impingement, injected into a mold, and polymerized therein to form a plastic article. It is a recent development in plastics technology, which was commercialized in the late 1960s and early 1970s; acceptance in the United States lagged behind Europe by several years. The greatest utilization has been in the automotive industry, where it is used for bumpers and fascia. The process has shown rapid growth, and predictions indicate continued growth as it expands into new applications such as business machine housings, home construction, and structural parts.

Polyurethane is the principal material used, with more than 95% of the total processed; RIM processing of polyesters, epoxies, nylons, and dicyclopentadienes has been introduced. Of the polyurethane market, approximately 85% of the material is elastomeric; the remainder is rigid. The rigid materials are usually structural foams, a composite morphology of a solid skin encasing a foamed core. Polyurethanes are suitable for physical blowing agents such as halocarbons. Foaming is an integral part of the RIM process, even for ostensibly solid articles, because it compensates for the shrinkage that occurs during polymerization. For this reason most elastomeric articles are also foamed.

The RIM equipment consists of metering and mixing units combined with a mold-and-clamp facility. The unique feature of RIM equipment is the mixhead, where the two or more components are dynamically mixed. Unlike injection molding, the press does not have to be close to the material source; the components can be transferred safely across a factory floor. A metering unit can accommodate as many as five mixheads or molding stations because the elapsed time for the metering shot is only a small fraction of the overall cycle time. Capital expenditures are therefore minimized.

In comparison to other polymer process methods RIM offers several advantages that have sustained its growth. It is well suited to the manufacture of large articles, which is impractical or uneconomical with conventional injection or compression processes. Start-up time for RIM is rapid and tooling inexpensive. The low filling pressures allow the use of aluminum tooling, which can be machined faster and is less expensive than steel tools used for injection molding. For limited production runs or prototyping, metal-filled epoxy or castable metal tooling can be used to provide capital and item savings. Furthermore, RIM is one of the least energy intensive manufacturing processes.

Polymerization of a monomer mixture in the mold allows for the custom formulation of material properties and kinetics to suit a particular application. For the reinforcement of RIM materials, milled glass and short glass fibers are added to one or both components in a process known as reinforced RIM (RRIM).

The North American market for reaction injection molding in 1986 was 72.5×10^3 metric tons. This is an approximate 12% annual increase since 1981, a rate that exceeds the overall growth of the plastics industry. In North America more than 500 RIM machines are spread over 128 plant locations. The U.S. auto industry is by far the largest user of reaction injection molding, consuming nearly 75% of the material processed.

The Process

The RIM process consists of the impingement mixing of two or more reactive components, the injection of this reactive mixture into a mold where polymerization proceeds under static conditions, and the ejection of the article after the required extent of conversion has been attained. A schematic representation of the RIM process is given in Figure 1.

Mixing. Mixing theory and experimentation have been thoroughly covered in the literature; some is relevant to RIM. Complete mixing of two components cannot be achieved in the short time between fluid impingement and polymerization to a hard plastic. Parameters such as the scale and intensity of mixing have therefore been employed to quantify the extent of mixing. The scale is a characteristic size of the segregated regions and the intensity is the concentration difference between these regions. In RIM, the scale, or striation thickness, would have to reach a value where diffusion could complete the mixing over the time for polymerization. The following expression has been derived for the striation thickness as a function of RIM process parameters:

$$S_f = \left(\frac{ndD^3}{1 + 1/r}\right)^{1/4} \frac{1}{Re^{3/4}} \tag{1}$$

Cavity Filling. The flow of the reactive fluid in the mold cavity is important from both scientific and industrial perspectives. Gelation times below 5 s are beginning to appear, yet fill times for large molds can be 10 s or longer. In some RIM applications the flow is interrupted by advanced polymerization. The solution of reactive flow problems is also important in many technologies that concern the flow of polymerizing fluids, such as the encapsulation of microelectronic devices.

The material viscosity increases as the molecular weight increases with advancing degree of conversion. Viscosity during cure, which has been termed chemorheology, is required to analyze reactive fluid-flow phenomena. It can be treated by deriving relations for the degree of conversion during cure, the molecular weight as a function of the degree of conversion, and the viscosity as a function of the molecular weight. The degree of conversion is obtained from an expression for the chemical kinetics of the polymerization reaction. An nth-order kinetic rate expression has often been found suitable for engineering analysis.

$$C_o \frac{dX}{dt} = \frac{-dC}{dt} = K_o \exp\left(\frac{E_a}{RT}\right) C^n \tag{2}$$

The weight-average molecular weight can be determined from the degree of conversion X. For a stoichiometric ratio of a trifunctional component of molecular weight M_a and difunctional component of molecular weight M_b, this expression is

$$\bar{M}_w = \frac{2/3(1 + X^2)M_a + (1 + 2X^2)M_b + 4XM_aM_b}{(2/3 M_a + M_b)(1 - 2X^2)} \tag{3}$$

The viscosity dependence on the molecular weight is often represented as

$$\begin{aligned} \eta &= K_\eta \bar{M}_w^{1.0} \qquad M_w < M_{w_c} \\ \eta &= K_\eta \bar{M}_w^{3.4} \qquad M_w > M_{w_c} \end{aligned} \tag{4}$$

where M_{w_c} is a critical molecular weight often associated with the onset of chain entanglements. Most reactive polymers used in RIM would flow into the mold cavity under conditions of $M_w < M_{w_c}$.

The dependence of viscosity on temperature must be accounted for as well. This dependence is often included through a flow-activation energy.

Polymerization in the Mold. In the RIM process, the static polymerization of the reactive liquid in the mold cavity governs productivity, morphology, and properties. The mold functions as a batch reactor for conducting a chemical reaction. The process is therefore suitable for chemical reaction engineering. Polymerization reactions of the polyurethanes, epoxies, and nylons used in the RIM process are strongly affected by temperature. These reactions have a negative activation energy, which means that the reaction rate increases with increasing temperature. Many reactions are also strongly exothermic,

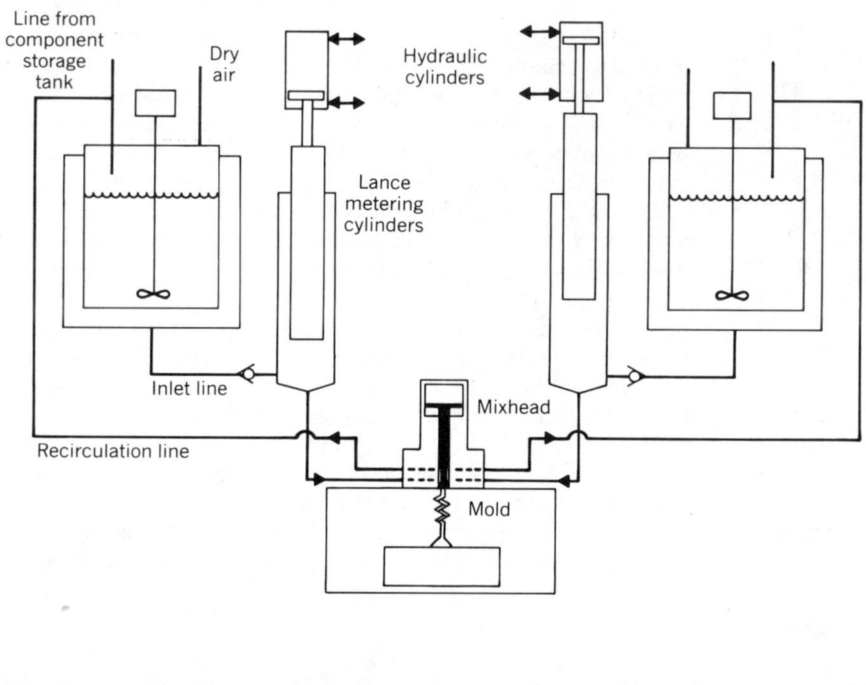

Figure 1. The reaction injection molding process.

evolving heat during the chemical transition to polymer. The materials themselves, however, provide effective thermal insulation, and heat generated from the chemical reaction may not be adequately dissipated. The temperature increase caused by the exotherm is an important component of the RIM process because it accelerates the reaction in the mold. The fluid flows into the cavity at a relatively low temperature and with little chemical conversion, but the reaction goes to completion in an economically feasible time because of the exotherm.

Processibility Analysis. Of a large number of reactive liquid monomer systems that may possibly be used in RIM, only a few have been found suitable. Several criteria determine processibility in RIM, and satisfying all of them is more difficult than it appears; these include effective mixing on impingement, filling a large cavity without flow interruption caused by advanced polymerization or excessive defects caused by turbulence, and polymerization to near complete chemical conversion in an economically acceptable time interval without thermal degradation caused by excessive temperature. The utility of a material can also be limited by the appearance of the molded article, the ability to be foamed, and the development of the required morphology and mechanical properties.

Materials and Chemistry

Polyurethane is the most important material for the RIM process, accounting for approximately 95% of the material molded in 1986. Polyurethanes are formed by the step polymerization of an isocyanate and a polyhydroxyl prepolymer (polyol) (see also ISOCYANATE-DERIVED POLYMERS). In reaction molding, a second short-chain, or monomeric, hydroxyl is added.

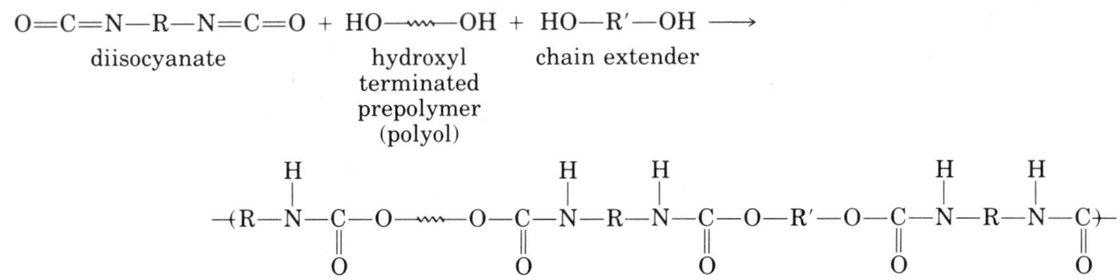

polyurethane

The ratio of hard to soft segments, as well as the specific domain structure or morphology that develops, determines the mechanical properties of the plastic material. The process and material parameters that influence phase separation therefore have a profound effect on the plastic material obtained. The relative rates of mixing at the mixhead and polymerization in the mold correlate with the extent of phase separation.

Health and Safety Factors

The health and safety considerations of reaction injection molding are more serious than in most other polymer process methods. The liquid reagents used in RIM are more noxious and have higher vapor pressures than other plastic molding systems. Reactivity at room temperature adds further to handling problems. Isocyanate is the most hazardous material used in reaction molding, and the dangers of exposure to this chemical are well known. Isocyanates are irritants that can cause severe dermatitis and bronchial spasm and, in some individuals, sensitization. Injury to the mucous membranes of the nose, throat, and lungs is possible.

The polyol component of reaction injection molding is of less concern, but its hazards should not be overlooked. It contains amines that can be irritating to the eyes and skin. Other safety considerations of the RIM process include the high pressure hydraulic and component lines.

BIBLIOGRAPHY

L. J. Lee, *Rubber Chem. Technol.* **153**(3), 542–599 (1980).

W. E. Becker, ed., *Reaction Injection Molding,* Van Nostrand Reinhold Co., Inc., New York, 1979.

L. T. Manzione, *Polym. News* **9,** 38–46 (1983).

L. M. Alberino, *Polym. News* **11,** 135–142 (1985).

L. T. MANZIONE
AT&T Bell Laboratories

REACTIONS OF POLYMERS

Polymers become involved as reactants in chemical reactions for a number of reasons and under a wide variety of circumstances. The final form of a macromolecular species may be obtained by a reaction of a polymer or prepolymer. For example, telechelic polyesters or polyethers can be chain extended with isocyanates to form polyesterurethane or polyetherurethane elastomers (see POLYURETHANES). Fluid phenol–formaldehyde novolac and resole resins can be cured or crosslinked to form rigid thermoset materials (see PHENOLIC RESINS); glassy polystyrene can be grafted with an elastomeric polydiene to increase its impact strength. In a closely related process, polystyrene prepared by a living carbanion technique can be used to initiate polymerization of a diene monomer to form a diblock or triblock copolymer with useful interfacial properties (see BLOCK COPOLYMERS). To some extent the reactions of polymer reactants described might be regarded as part of the inherent overall polymerization process.

There are, however, situations where the reaction of the polymer reactant is a distinctly separate process. For example, a polymer designed and synthesized for one application may prove to be a convenient raw material for chemical modification to produce another polymer. Poly(vinyl acetate) is a useful starting material for poly(vinyl alcohol), which itself is a valuable precursor for poly(vinyl formal) (see VINYL ALCOHOL POLYMERS; VINYL ESTER POLYMERS; VINYL ACETAL POLYMERS).

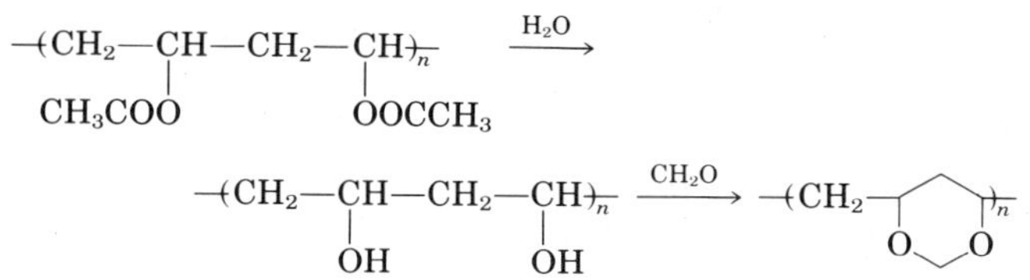
(1)

Polymers may also be chemically modified to introduce reactive or catalytic groups to enable them to function as polymeric reagents and catalysts. Important examples are sulfonic acid derivatives of cross-linked polystyrene, which are used widely as industrial acid catalysts (see CATALYSTS, POLYMER-SUPPORTED); and

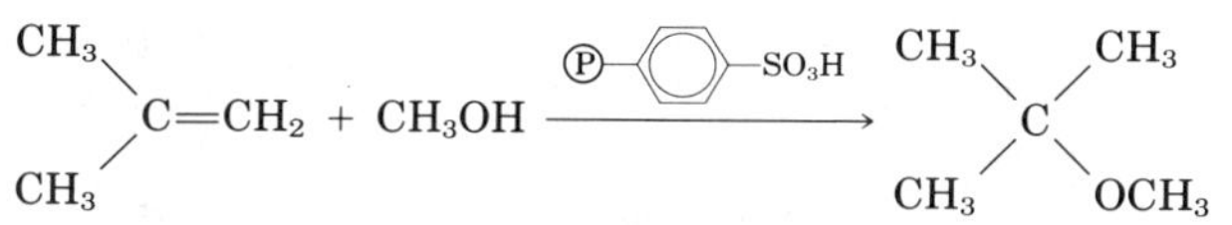
(2)

chloromethylated polystyrene, which is used as a support for Merrifield polypeptide synthesis (qv). Other polymers may be similarly modified with ligands to function as affinity supports in the isolation and purification of enzymes and species involved in immuno interactions. Indeed, such modified polymers can be used directly in immunodiagnostic systems. Alternatively, the ligands introduced may be designed specifically to chelate metal ions for application in water treatment, hydrometallurgy, or analytical chemistry (see CHROMATOGRAPHY, AFFINITY; ENZYMES, IMMOBILIZED).

Considerable ingenuity has been shown in designing polymers that have convenient solubility properties enabling their fabrication but that subsequently undergo facile chemical reaction to generate the final, essentially intractable products. Examples are poly(amic acid)s as precursors to polyimides and unconjugated polyenes, which undergo retro Diels-Alder reactions to form polyacetylenes (see ACETYLENIC POLYMERS). The synthesis of polyenaminonitriles uses a similar philosophy.

There are a wide variety of situations in which the surfaces of polymers are selectively chemically modified to enhance adhesion to the surface (eg, in the case of printing on plastics), or, conversely, to inhibit unwanted adhesion to the surface, such as in release coatings (see ADHESION AND BONDING; RELEASE AGENTS; SURFACE MODIFICATION). An area of increasing importance in this context is the interaction of living cells with polymer surfaces. In some instances a strong and compatible interaction is required (eg, bone adhesives), whereas in others, the primary objective is to eliminate interaction and cell accumulation (eg, contact lenses and arterial replacements) (see BIOMATERIALS).

Finally, there is a wide variety of applications of radiation-sensitive polymers that undergo reactions stimulated by electromagnetic radiation (see PHOTOCONDUCTIVE POLYMERS). These range from spurious ultraviolet environmental degradations (see WEATHERING), through low resolution lithographic processes to high resolution electron-beam microlithography, which is currently being employed to produce the most modern microchips (see LITHOGRAPHIC RESISTS).

Although these situations may appear to have little in common, all the examples mentioned involve at least one reaction where a polymer is a reactant. Many of the areas mentioned are discussed in detail elsewhere in the *Encyclopedia* as the cross-references indicate. This article addresses in a broad sense the science and general principles involved in reactions of polymers.

Synthetic Routes to Functional Polymers

Polymers containing functional groups can be prepared either by chemical modification of a precursor polymer or by a (co)polymerization process that introduces the required functional group during (co)polymer formation itself. For example, polystyrene derivatized with sulfonic acid groups can be prepared via sulfonation of polystyrene or via polymerization of 4-vinylbenzenesulfonic acid.

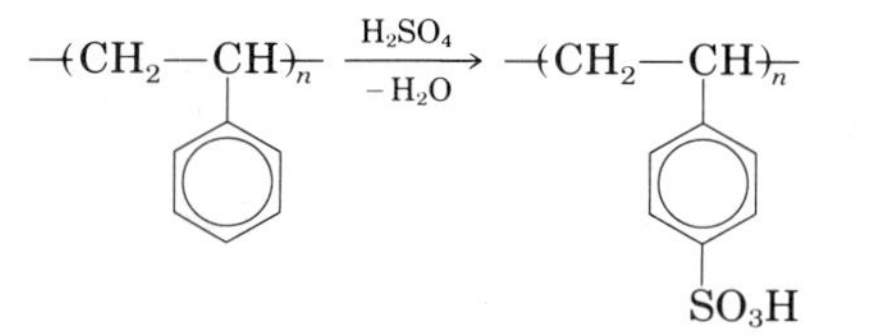
(3)

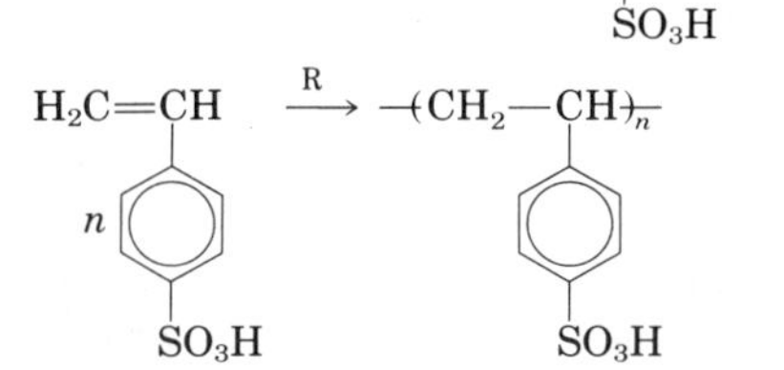
(4)

The degree of sulfonation of the polymer may be controlled by the conditions and time of the reaction; subtle control of sulfonation is difficult. In the polymerization route, different degrees of functionalization can be achieved more readily by copolymerization of the functional monomer with a nonfunctional comonomer, eg, styrene in this example. If the reactivity ratios of the chosen comonomers are reasonably similar, a random distribution of functional groups will result.

Much less work has been carried out on the introduction of specific functional groups into condensation polymers. Most derivatization that has been done has been by chemical modification of existing polymers, eg, the controlled hydrolysis of nylons to improve the binding of dye molecules or affinity chromatography ligands.

Reactivity

Model Reactions and Theoretical Studies. In the case of chemical reactions involving a functional macromolecule as one reactive species, a number of reaction types can be envisaged. The most common are shown schematically in Figure 1.

Functional Groups on Linear Polymers. ***Compatibility Factors.*** In the absence of intermolecular cross-links, random macromolecular coils can migrate apart in a suitable medium, when the overall polymer concentration is sufficiently low. The question arises: Can such coils always be readily penetrated by a potential reactant? Some physicochemical evidence (eg, fluorescence quenching measurements) suggests that they can, even when the coils are present in a theta solvent, a solvent whose free energy of mixing with the polymer is zero at a given

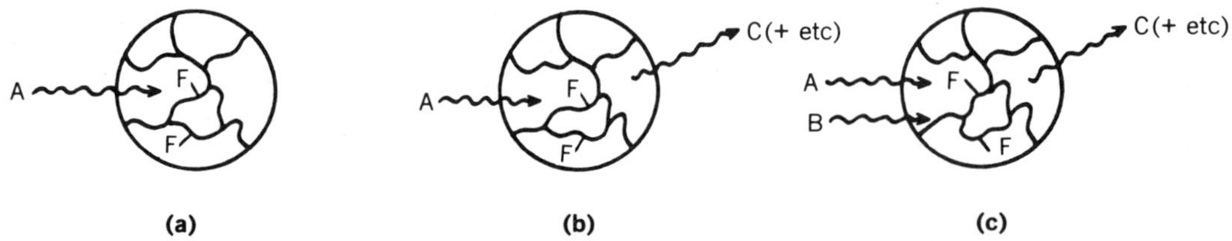

Figure 1. Model reactions involving polymer-bound functional groups F. (**a**) Unimolecular reaction of A with bound functional group; product remains attached to resin. (**b**) Unimolecular reaction of A to produce C; catalyzed or mediated by the bound functional group. (**c**) Biomolecular reaction of A with B to produce C; catalyzed or mediated by the bound functional group.

temperature (see SOLUBILITY OF POLYMERS; SOLUTION PROPERTIES). However, experience with chemical reactions of polymer-bound groups suggests that there are occasions when reagents do appear to have difficulty in penetrating individual polymer coils, as evidenced by rates of reaction or degrees of chemical conversion. Thus even linear chloromethylated polystyrene reacts sluggishly with amino acids, whereas if the dimethylsulfonium salt is formed first by reaction with dimethyl sulfide, the reaction proceeds rapidly and in good yield.

Freely Penetrable Polymer Coils. Although compatibility factors can create problems, more often than not, small reactant molecules can penetrate very freely into polymer coils expanded in a good solvent. The diffusion coefficient of a small molecule such as a solvent in a polymer rises dramatically even with relatively small mass fractions of solvent (~0.1), and indeed becomes indistinguishable from the self-diffusion value in pure solvent when the mass fraction is ~0.5. In dilute polymer solutions, therefore, solvent molecules can move with collisional frequencies unimpaired by the presence of dissolved polymer coils. In the case of freely penetrable polymer coils, the question of reactivity is therefore perhaps the simplest and the one that is most often desirable.

Electrostatic Effects. The electrostatic effect involving ionic reactants and linear polyelectrolytes where the bound group F is also charged represents perhaps the most obvious example where the distribution coefficient of the reactant between the pseudophases deviates substantially from unity. From a simplistic point of view it might be anticipated that ionic reactants of opposite charge to the bound ion will be relatively concentrated in the region of the polymer coil, whereas ionic reactants of the same charge will tend to be repelled into the intervening solvent. This picture predicts that the rates of bimolecular reactions between two ions of the same charge and both opposite in sign to the bound ion would be relatively enhanced owing to this local concentrating effect. On the other hand, reactions involving two ions of opposite charge would be adversely influenced, because the polyion would act to keep these apart. Such effects have been clearly demonstrated in a variety of systems. Similar effects are also seen in micelle-catalyzed reactions. In the case of dipolar species rather than fully charged ones, such effects are likely to be substantially reduced.

Changes in Activation Parameters. In addition to the physical factors that can influence the overall reactivity of a bound functional group, the inherent chemical reactivity of such a group might also be changed simply by its immobilization on a polymeric backbone.

Neighboring Group Effects. Probably the single most important chemical factor giving rise to the modified reactivity of a functional group bound to a polymer, and indeed to low yields and conversions, and a multiplicity of side reactions, is the involvement of the neighboring groups on a macromolecule. In terms of reactivity, the simplest example is the ionization of carboxylic acid groups, eg, in poly(acrylic acid). Clearly the ease of ionization of carboxylic acid groups will decrease as the degree of ionization of neighboring carboxylic acid groups increases, because the polymeric backbone becomes progressively more negatively charged.

Indeed, it is most likely that the term neighboring should be interpreted very loosely. It may refer to a group on an adjacent polymer segment; it may also be applicable to a group in close proximity to the bound group of interest as a result of the conformation adopted by the macromolecular chain. In this case, even a relatively lightly loaded polymer backbone may be subject to these effects. In a few important cases, neighboring groups have been shown to cooperate to produce a pronounced increase in the reactivity of bound functional groups. Perhaps the most important example of this effect is in the dehydration of *t*-butyl alcohol catalyzed by sulfonic acid polymers.

A somewhat different category of neighboring group effect may also give rise to complicating side reactions involving some bound functional groups, eg, additional cross-linking reactions.

Functional Groups on Cross-linked Polymers. So far, functional groups bound to soluble polymers that, depending on the overall concentration of polymer in the bulk solution, may or may not interpenetrate one another have been considered. In the case of cross-linked resins, large collections of macromolecules are permanently interpenetrated and locked together indefinitely by discrete covalent cross-links. It is important to realize that all the factors already discussed in the case of linear polymers can, and very often do, arise with resin-bound species. In some cases the effects are obvious, whereas in others they are difficult to demonstrate or at least to quantify because of the insoluble nature of the resin. Furthermore, because of their cross-linked nature the effects may be masked or swamped by other more dramatic changes. For example, the common occurrence of side reactions resulting from intramolecular reaction of more than one bound functional group, described above, can go completely undetected in the case of resins, where no obvious change in the physical properties may arise with the additional cross-links that are formed. Where such side reactions are suspected, careful chemical and physical analysis can expose them, though precise quantification can remain elusive. This situation is particularly relevant when attempting to exploit the potential of the "site isolation" of groups.

Pseudohomogeneous Systems. These systems represent perhaps the simplest situation that can arise with cross-linked polymers and the one that is most often desirable. In this case the polymeric network plays no role other than to limit the volume in which the bound functional group is found and can move (albeit to a limited extent). The molecular picture is very close to the case of linear polymers with interpenetrating coils, although there is a subtle difference (Fig. 2). Whereas in the latter case the coils can exist throughout the whole volume of the reaction medium, in the former the group of coils forming one resin bead may be well separated from the groups constituting other resin particles. From a thermodynamic point of view, this arrangement represents an intriguing difference. With overlapping coils of linear polymer filling the whole solvent volume, there is no net driving force for further coil expansion (polymer dilution). However, in the case of resin particles not filling or packing into the whole volume of the solvent medium, a concentration driving force remains for further coil dilution, opposed by the mechanical force exerted by the permanent cross-links. Thus it is possible for considerable osmotic forces to be generated within resins, and these may well influence the reactions of bound groups.

Nevertheless, a very close analogy does exist between the two systems, and in the case of the cross-linked polymer the solvent within the polymer matrix may be considered as a "homogeneous" phase within which reaction occurs and wherein reactants and products are in rapid thermodynamic distribution equilibrium with the external solution. The latter acts essentially as an inert reservoir and functions only to supply reactant and store any free product or by-products that are formed. Thus for a bimolecular reaction between polymer resin-bound functional group Ⓟ–F and a free reactant A, the rate-controlling process is the same as in the corresponding homogeneous model reaction involving free F and A, the bimolecular collision itself.

Diffusional Effects. Often the objective of using a polymer or resin-bound species as a reagent or catalyst is to combine the technical convenience of a supported or heterogeneous system with the activity of a homogeneous one, but this is realized on relatively few occasions. More commonly, the reactivity of species bound to a cross-linked polymer is altered substantially relative to the same species in homogeneous solution, and indeed relative to the same species attached to a soluble linear polymer. The change in reactivity is often downward, although there are a number of notable exceptions. It is not unusual to find that experimental kinetic data vary with the particle size of the resin, with its porosity and cross-link ratio, and with the size of the reactant molecules. Under these circumstances, it is usual to suggest that the polymer matrix provides a diffusional barrier to the reactants, generally in some ill-defined manner.

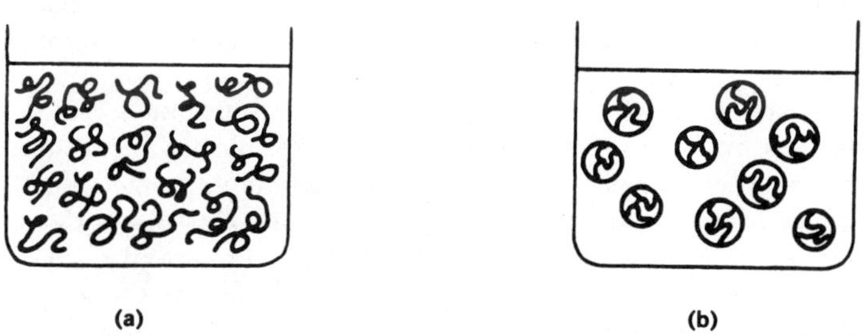

Figure 2. (**a**) Distribution of soluble macromolecular species throughout an entire reaction solution volume. (**b**) Restrictions of macromolecular polymer coils with the particle volume of cross-linked resins.

Heterogeneous Models. Under some circumstances reactions involving functional groups on polymers behave essentially as though the system was truly heterogeneous, with the bound groups residing on a well-defined surface and the reactant molecules becoming adsorbed before reaction. Superficially, such a description fits gas-phase reactions involving groups bound to macroporous polymers most adequately.

Site–Site Interaction and Site Isolation. When two functional groups, E and F, are attached to an isolated linear macromolecular chain, their interactions are limited statistically. This structure is true for like groups as well as unlike ones, providing the accessibility of all the groups is the same and remains constant. Flory has shown that the maximum possible number of interactions (in the case of chemical reaction, for example) is 86.5% for a regular alternating structure

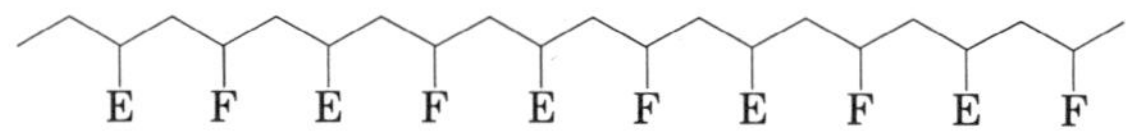

and 81.6% for a statistically random array.

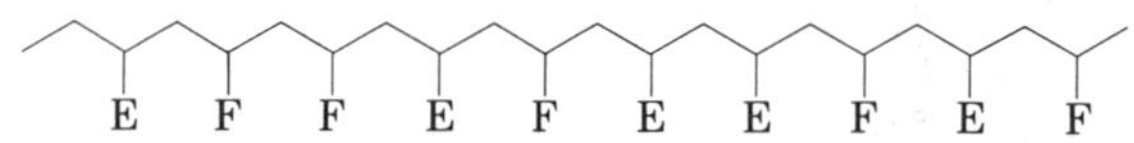

This analysis applies for reactions of E and F that are quantitative. Interaction of 100% is possible only for a regular vicinal structure.

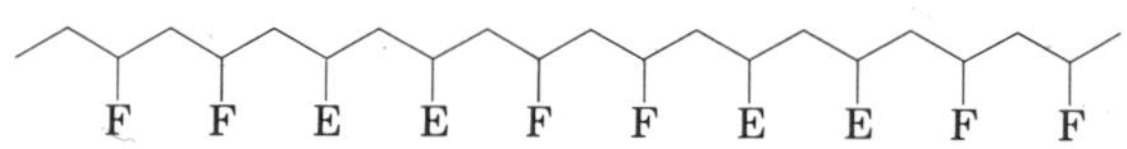

In the case of the collection of interpenetrating polymer chains permanently cross-linked and typical of resins, there has been considerable experimentation and debate concerning the possibility of achieving site isolation of a bound functional group. The situation has been reviewed. In principle it should be possible to introduce bound groups onto a permanent three-dimensional polymer network such that each group experiences a state analogous to that of infinite dilution in a homogeneous solution. Such a situation could be important for producing species that under normal homogeneous conditions would be destroyed by rapid self-condensation reactions, eg, coordinatively unsaturated transition metal complex catalysts. It also has considerable potential for allowing the attachment of symmetrical polyfunctional compounds by one group only, leaving the others free to be chemically transformed. Furthermore, such site isolation of groups might inhibit bimolecular reactions of these and favor some unimolecular process not normally observed in solution. It may also prevent unwanted site reactions such as some of those described earlier.

Synthetic Aspects

Many reactions of polymers are aimed simply at converting one polymer structure into another using a polymer analogous reaction. A number of reviews concentrate mainly on chemical modification of polymers. In this context many of the quantitative correlations described in the foregoing may not seem relevant. However, a major objective is always to achieve a high conversion and selective modification in the reactions undertaken. Thus most of the concepts presented remain of qualitative importance and provide a logical basis on which to select particular polymer types and to optimize reaction conditions. For many organic chemists, macromolecules have long been regarded as a group of molecules quite distinct from low molecular weight species. However, there has been a growing realization that organic macromolecules are really no different from small molecules in terms of chemical manipulation. They are subject to all the conventional rules of organic functional group chemistry, and with care and adaptation much of the methodology of organic synthesis can be applied to macromolecular reactions.

To achieve chemical reaction between two organic groups, they must undergo a collisional interaction. This interaction is most likely if both groups are mobile and within the same phase. With one functional group attached to a linear polymer, reaction is likely to be optimized if the polymer is dissolved in a solvent in which it forms a true molecular solution. When one group is attached to a cross-linked polymer, optimum conditions are most likely to be those where the polymer is highly swollen in the reaction mixture, but this goal is not always attainable. Highly problematic reactions are those where the small reagent is not soluble or is only partially soluble, eg, hydrogenations and hydroformylations catalyzed by supported metal species, reductions using metal hydrides, and reactions using metals themselves. In these cases, sufficient intimate molecular contact may be impossible and other strategies must be adopted. Other troublesome reactions are those in which a solid is formed as a coproduct, eg, metal oxides in many metal hydride reductions. These solids can become trapped within porous cross-linked polymers and interfere with subsequent chemistry. Despite these situations, the literature on reactions on polymers is now vast, and an enormous variety of novel macromolecular species have been generated from a relatively small number of precursor macromolecules.

Styrenic Polymers. Linear polystyrene and cross-linked styrene–divinylbenzene resins have been subject to more chemical transformations than any other group of polymers because they form the basis of water-treatment resins and have attracted wide interest as heterogeneous supports for reagents and catalysts (see CATALYSTS, POLYMER SUPPORTED). Furthermore, they have been used increasingly in the synthesis of a growing variety of selective separation media. The options for initial chemical modification rely largely on the electron-rich nature of the aromatic pendent group and its susceptibility to electrophilic substitution. Some of these, eg, the chloromethylation and sulfonation reactions, are the basis of major industrial processes (see STYRENE POLYMERS).

Polyvinylpyridines. Polymers and copolymers based on both the 2-vinyl and 4-vinylpyridine monomers have been investigated; the latter has found more application. In general, chemical modification has centered on the basic nitrogen heteroatom of the pyridine ring rather than on the aromatic group itself (see VINYLPYRIDENE POLYMERS). This behavior differentiates these macromolecules from the polymers in the previous section.

Polydienes. Polymers from butadiene, isoprene, and chloroprene are elastomers (see BUTADIENE POLYMERS; CHLOROPRENE POLYMERS; ISOPRENE POLYMERS). For appreciable elastomeric properties the polymers must contain significant numbers of *cis*-1,4-linked residues giving rise to carbon–carbon double bonds in the backbone. Inevitably, some 1,2 linkages arise with their accompanying pendent vinyl groups. Chemical reactions of polydienes are generally directed toward tuning the physical properties of the material to meet the needs of a particular application. Usually modifications are based on the unsaturated linkages in the polymer, although modification of too many of these destroys the inherent elastomeric properties. The unsaturated linkage is the site for initial atmospheric oxidation of polydienes and effectively provides a foothold for degradation processes. Destruction of some of the unsaturation therefore leads to some improvement in oxidation stability. The major strategies in modifying polydienes have been reviewed. Natural rubber itself is chlorinated for some applications, and the chemical reactions of this modified material have also been reviewed (see RUBBER, NATURAL: RUBBER DERIVATIVES).

Polyolefins. Polyolefins such as polyethylene, polypropylene, and polyisobutylene and their copolymers are useful macromolecules because of their relative chemical inertness arising from the absence of reactive functional groups. Unfortunately, however, when chemical modification is required the necessary reaction conditions are often extreme.

Poly(vinyl chloride) and Other Halogen-containing Polymers. The presence of the alkyl chloride moiety in poly(vinyl chloride) makes it potentially very reactive. In practice, in the solid state most of these groups are not readily accessible and the polymer is relatively inert. However, the potential for chemical reaction exists and a symposium was dedicated to this topic (see VINYL CHLORIDE POLYMERS).

Acrylic Polymers. The polymeric alkyl esters of acrylic and methacrylic acid represent an important group of industrial polymers (see ACRYLIC AND METHACRYLIC ESTER POLYMERS). Hydrolysis of these esters generates more hydrophilic species than the starting polymer with varying amounts of carboxylic acid or carboxylate salt functions. Methyl trifluoromethacrylate yields alternating copolymers with α-olefins, and hydrolysis of these forms hydrophilic fluoropolymers. Controlled hydrolysis of poly(styrene-*co*-*n*-butyl methacrylate) simultaneously impregnated with carbon black provides a convenient method of controlling the degree of dispersion of the carbon black, and hence the overall triboelectrical properties of these composites used in xerography. Similar shifts in these properties can be achieved by amination of the ester groups with diamines and subsequent acylation or dye coupling of the free amine. The hydrolysis of ester groups in a terpolymer of ethyl acrylate, acrylonitrile, and divinylbenzene has also been reported.

Other Vinyl Polymers. Vinyl acetate is a cheap monomer manufactured on a vast scale via Wacker oxidation of ethylene. As a result, there is a continuing growth in polymers and copolymers made from this monomer. Hydrolysis, or transesterification with methanol, of the homopolymer yields poly(vinyl alcohol), typically available in a number of hydrolysis grades and molecular weights (see VINYL ALCOHOL POLYMERS). Saponification with alkali tends to produce a very "blocky" structure, whereas with transesterification this tendency is somewhat reduced. Acid catalyzed hydrolysis gives the closest approach to random copolymers. Quantification of the reactivity differences between isolated and vicinal vinyl acetate residues in high pressure vinyl acetate–ethylene copolymers has recently been reported.

Polycondensation Polymers. Chemical reactions of polycondensation species have not been as widely studied as those of vinyl and olefin polymers. Generally, the backbones of the latter contain no hetero atoms or functional groups, which facilitates reaction on the pendent groups. In some instances extremely reactive reagents and vigorous conditions can be employed without attack on the backbone. Paradoxically, many polycondensation polymers are considered to be more stable than vinyl addition polymers despite the functionality of the backbone of the former group. Part of the reason for this is the highly crystalline nature of many of these polymers, or at least the high intermolecular forces generated between individual polymer chains, arising from their backbone dipoles. In practice, if condensation polymer molecules become isolated, eg, in homogeneous solution, considerable care must be exercised not to induce chain degradation reactions, eg, hydrolysis. Some polycondensation species, however, such as aromatic amides, ether sulfones, ether ketones, imides, and benzimidazoles have very high intrinsic chemical stability even as isolated molecules.

Polymers Derived from Cyclic Monomers. The ring-opening polymerization of ethyleneimine is complicated. As a result of transfer reactions in which the active center attacks amino groups on its own and other chains, the final product has a highly branched structure containing primary and tertiary amino groups as well as the expected secondary functions (see ALKYLENEIMENE POLYMERS). Despite this, the water-soluble polyamine has proved a useful starting point to develop other species. For example, the polymer has been modified to carry biologically active groups such as nucleoside residues and nucleic acid base derivatives.

Polymers Leading to Conducting Species. The structural and physical bases of conducting polymers, including dopant aspects, have been reviewed. A basic structural requirement for significant electrical conductivity in a polymer is a delocalized conjugated π-electron system. Polypyrrole and polyaniline represent such species and have been made by direct electrochemical reactions involving pyrrole and aniline. Apparently, no controlled synthesis has been reported involving a nonconducting precursor polymer, but this is an active research area.

BIBLIOGRAPHY

G. G. Cameron in A. D. Jenkins and J. F. Kennedy, eds., *Specialist Periodical Report—Macromolecular Chemistry,* Vols. 1 and 2, Royal Society of Chemistry, London, 1979, Chapt. 16, and 1981, Chapt. 15.

C. E. Carraher, Jr. and M. Tsuda, eds., *Modification of Polymers, ACS Symp. Ser. No. 121,* American Chemical Society, Washington, D.C., 1980.

D. C. Sherrington and P. Hodge, eds., *Syntheses and Separations Using Functional Polymers,* John Wiley & Sons, Ltd., Chichester, UK, 1988.

W. T. Ford, ed., *Polymeric Reagents and Catalysts, ACS Symp. Ser. No. 308,* American Chemical Society, Washington, D.C., 1985.

DAVID C. SHERRINGTON
University of Strathclyde

REACTIVE EXTRUSION

Reactive extrusion, also termed reactive processing or reactive compounding, refers to the performance of chemical reactions during extrusion processing of polymers. In this case an extrusion device is used as a chemical reactor instead of only as a processing aid. The chemical reaction may take place in the polymer melt phase or, less commonly, in the liquid phase, as when bulk polymerization of monomers is performed in an extruder, or in the solid phase when the polymer is conveyed through the extruder in a solvent slurry.

The Extruder as a Chemical Reactor

An extruder may be considered to be a horizontal reactor with one or two internal screws for conveying reactant polymer or monomer in the form of a solid or slurry, melt, or liquid. The most common reactants are polymer or prepolymer melts and gaseous, liquid, or molten low molecular weight compounds. A particular advantage of the extruder as a chemical reactor in this context is the absence of solvent as the reaction medium. No solvent stripping or recovery process is required, and product contamination by solvent or solvent impurities is avoided.

Because of their versatility, most extruder reactors are twin-screw extruders that possess a segmented barrel, each segment of which can be individually cooled or heated externally. In addition to external heating, a molten material may be shear heated by the resistance of viscous material to the conveying motion of the screw; these processes provide energy for chemical reaction. Extruder screws often have specialized sections or configurations, eg, high shear mixing sections. Twin-screw extruder screws may be equipped with interchangeable screw elements that provide different degrees of mixing and surface area exposure by varying the depth between screw flights, the individual flight thicknesses, and direction and degree of flight pitch. Kneading blocks may be included as screw elements to provide intensive mixing. By varying the external heating, the screw element configuration, and the clearance between screw and barrel wall in individual barrel segments, the total energy and the degree of mixing of material in each barrel segment may be varied. In this way an extruder may be transformed into a chemical reactor with controlled reaction zones made up of individual barrel segments. In each of these segments sequential chemical processes can take place.

In a typical reactive extrusion process, the reactants are fed into the extruder feed throat where the material is usually heated to initiate reaction or increase the reaction rate. The reactant mixture is conveyed through sequential barrel segments where degree of mixing and specific energy input bring the reaction to the desired degree of completion within the limits of residence time in the extruder. At this stage the reaction may be quenched by cooling or addition of a catalyst quencher where applicable, and volatile by-products or excess reactants may be removed. Molten polymer is forced from the extruder through a die with one or more openings. The geometry of the die openings is one factor determining the pressure against which the extruder has to pump by the conveying motion of the internal screw. Polymer melt issuing from the die is usually rapidly cooled by contact with a fluid medium such as water. Cooling and solidification quench any chemical reaction still occuring that may not have been specially quenched in the extruder.

An advantage of an extrusion device as a reactor is the combination of several chemical process operations into one piece of equipment with accompanying high space–time yields of product. An extruder reactor is ideally suited for continuous production of material after equilibrium is established in the extruder barrel for the desired chemical processes.

Table 1. Types of Chemical Reaction Performed by Reactive Extrusion

Type	Description
bulk polymerization	preparation of high molecular weight polymer from monomer or low molecular weight prepolymer or from mixtures of monomers or monomer and prepolymer
graft reaction	formation of graft copolymer from reaction of polymer and monomer
interchain copolymer formation	reaction of two or more polymers to form random or block copolymer through ionic or covalent bonds
coupling or branching reaction	reaction of homopolymer and polyfunctional coupling or branching agent to build molecular weight by chain extension or branching, or reaction of homopolymer with condensing agent to build molecular weight by chain extension
controlled molecular weight degradation	molecular weight degradation of high molecular weight polymer
functionalization or functional group modification	introduction of functional groups into polymer backbone, end group, or side chain, or modification of existing functional groups

Reaction Types

The types of reactions that have been performed by reactive extrusion may be conveniently divided into six categories as shown in Table 1.

BIBLIOGRAPHY

W. A. Mack and R. Herter, *Chem. Eng. Prog.* **72,** 64 (1976).

L. Wielgolinski and J. Nangeroni, *Adv. Polym. Technol.* **3,** 99 (1983).

Z. N. Frund, Jr., *Plastics Compounding,* 24 (Sept./Oct. 1986).

M. Ratzsch, *Makromol. Chem. Macromol. Symp.* **12,** 165 (1987).

S. Bruce Brown
Charles M. Orlando
General Electric Research and Development Center

REACTOR DESIGN AND SELECTION.

See Reactor technology.

REACTOR TECHNOLOGY

Reactor technology comprises the underlying engineering principles of chemical reaction engineering and the practices used in their application. It constitutes the evolutionary basis for selecting and designing reactors, their configurations, and their operating characteristics. In addition to stoichiometry and kinetics, reactor technology includes requirements for introducing and removing reactants and products, efficiently supplying and withdrawing heat, accommodating phase changes and material transfers, assuring efficient contact among reactants, and providing for catalyst replenishment or regeneration. These issues are taken into account when reaction rates and bench-scale data are translated into designs of pilot plants, larger sized units, and commercial plants.

Reactors are employed in polymer manufacture and utilization, from the preparation of feeds and the polymerization reactions through product finishing and upgrading to their ultimate disposal as waste. The basic principles and underlying reactor technology are the same, whether dealing with initiation and propagation of polymer reactions or the production of their feeds.

Polymerization Classifications

The type of reactor employed is often determined by the polymerization classification, which includes bulk, solution, emulsion, and suspension polymerizations. If gas-continuous rather than liquid-continuous media are used, the classification is gas-phase polymerization in place of bulk polymerization. For monomers that are homogeneous with their polymers or suspended in solvents during reaction, the polymerization is classified as bulk (gas-phase) or solution polymerization, respectively. However, if the polymerization occurs in a separate phase, eg, an oil phase after diffusion of the monomer from a water phase, emulsion or suspension polymerization is the designated classification, as appropriate.

A commercial reactor favors one or possibly two classifications, for instance, bulk polymerization is used more frequently for step polymerization (normally a condensation reaction) than for chain (addition) polymerization because its high reaction rate, large activation energy, and viscous product make reactor control difficult.

Distinctions among classifications are not necessarily clear nor is usage consistent; eg, some heterogenous polymerizations are further classified as precipitation polymerizations when the final products are formed as powders or granules, eg, the polymerization of acrylonitrile in water or the bulk polymerization of vinyl chloride.

Reactor Types and Characteristics

Many reactor configurations and designs have evolved. The specific characteristics, including size, depend on the particular application as laboratory reactor, pilot plant, or commercial unit. Selection is based on the physical properties of the feed and products (vapor, liquid, solid, or combinations), the characteristics of the chemical reactions (reactant concentrations, reaction paths and rates, operating conditions, and heat addition or removal), the nature of the catalyst (activity, life, and physical form), and the requirements for intimate contacting of reactants (flow characteristics, transport phenomena, and mixing and separation mechanisms). These factors are interdependent and must be considered together. The requirements for reactant contact and product removal are the paramount focus of reactor technology; the other factors are usually determined by the selection of the reacting system and desired conversion and product selectivity.

Reactors are classified into three basic types: well-stirred batch reactor, continuous-flow stirred-tank reactor, and tubular reactor (Fig. 1). A reactor may be represented by, or modeled after, one type or a combination of the three types. The semibatch reactor is considered here as a subcategory of the well-stirred batch reactor. The acceptability of a model depends on its accuracy in depicting reactions and thermal and transport processes.

Multiphase Reactors. The presence of more than one phase, whether fixed or flowing, complicates analyses of reactor performance and increases the number of possible reactor configurations. Gases, liquids, and solids each flow in a characteristic fashion, dispersed in other phases or separately. Flow patterns in these reactors are complex, and the phases rarely exhibit plug-flow behavior. Some important reactor configurations are shown in Figures 2 and 3.

Reactors as Process Systems

The interacting effects of mixing, heat and mass transfer, and kinetics are essential in establishing reactors as process systems. The degree of backmixing directly affects overall reactor size and catalyst requirements for reactions with greater than zero-order dependencies on limiting reactants. Backmixing does not necessarily imply reduced conversions. Complete backmixing reduces catalyst deactivation by diluting catalyst poisons. Because total conversion is a function of the kinetic behavior of the entire system, the overall effects of backmixing must be considered.

The full range of flows, from plug to well mixed, can be obtained by operating several reactor vessels in sequence or by recycling. Thus the backmixing character of individual reactors is only consequential to their specific design. The degree of backmixing determines criteria for system selection, but not necessarily for selecting the reactor type. The impact of heat generation and removal on kinetics similarly affects the criteria for selecting reactor systems. Problems of maximizing outputs from exothermic or endothermic reactions occur with reversible reactions (ammonia synthesis or water gas shift reactions), temperature-selective reactions (reforming or hydroformylation), or rapidly deactivating side reactions (catalytic cracking of petroleum or polymerizations).

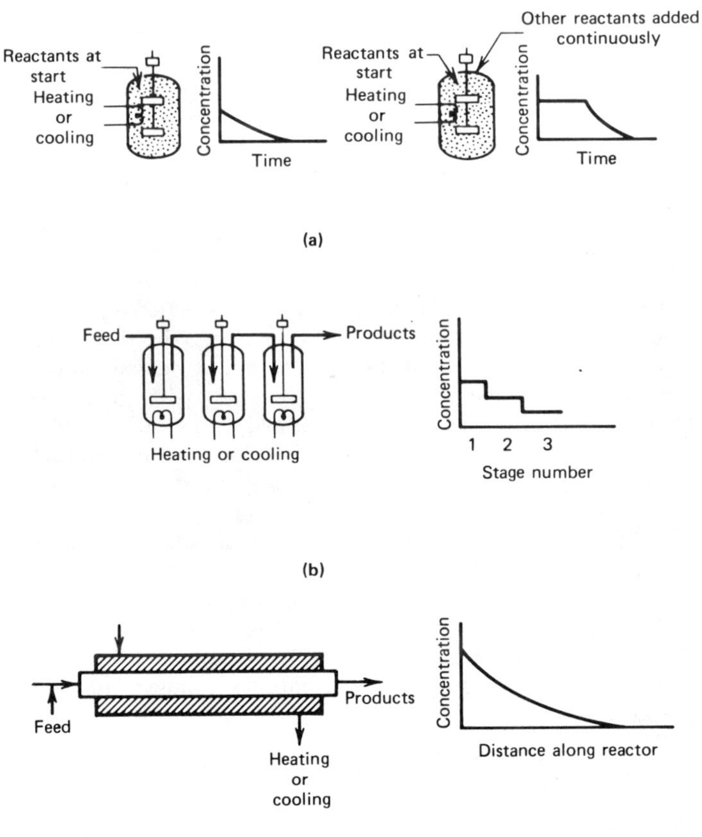

Figure 1. Reactor types: (**a**) batch and semibatch; (**b**) continuous-flow stirred-tank; (**c**) tubular.

Process Improvements and New Technologies

Process improvements and the introduction of new technologies markedly affect reactor selection and usage. Other technologies may remain competitive and not necessarily be displaced. On the other hand, new configurations may develop as a consequence of new technologies. The manufacture of polypropylene, the fastest growing plastic, provides examples of both older technologies remaining competitive and new technologies emerging (see PROPYLENE POLYMERS).

Scale-up

In scale-up the objective is to select designs and operating conditions to achieve comparable yields, product distributions, and time effects in different-size units. Where uncertainties exist or large quantities of products are needed for market evaluations, demonstration units intermediate in size between pilot and commercial plants are useful. In the design and scale-up of reactors, the key factors are the nature of the reaction sites, the specific reaction rates, and the mass- and heat-transport rates to and from the reaction sites.

Minimum Pilot Plant Size. Minimum sizes of laboratory and pilot units are often determined by operability factors not directly involving the reactor. These include feed- and product-transfer line diameters, inventory control in feed and product separation systems, and preheat and temperature maintenance requirements. Most of these extraneous factors favor large units. Large commercial plants can be operated with high service factors for years, whereas it is not unusual for pilot units to operate at sustained conditions for only days or even hours.

Impact of Fluid Dynamics on Scale-up. Matching of fluid-flow characteristics within the reactor might determine the operative scale-up criteria. Ideally, the smaller reactor should act as a volume segment of the larger one. Flow distributions should not be markedly influenced by the wall, turbulent eddies should be approximately the same size, and bubbles in bubble columns and fluid beds should be the same size. Under these conditions, the residence-time distributions of feeds, intermediates, and products should be comparable. In catalytic processes, catalyst particles should remain in the reactor for the same length of time.

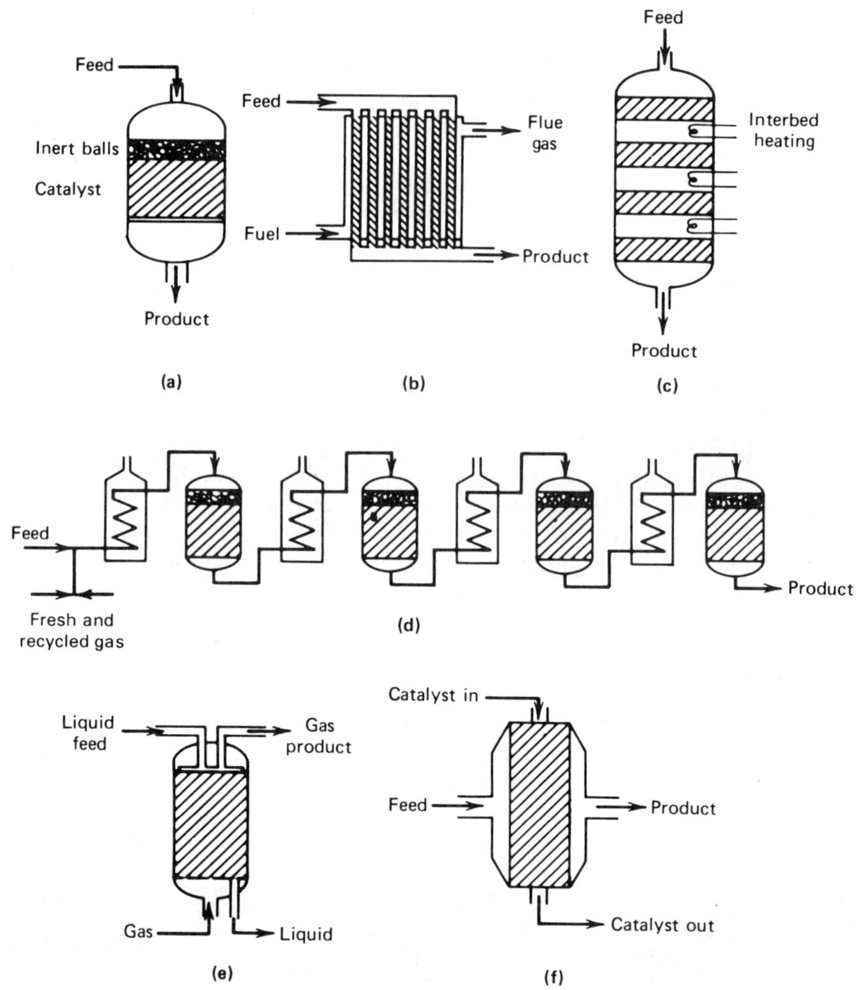

Figure 2. Multiple fixed-bed configurations: (**a**) adiabatic fixed-bed reactor, (**b**) tubular fixed beds, (**c**) adiabatic fixed beds in series, (**d**) trickle beds or pulsed columns in series, (**e**) countercurrent packed-bed absorber, (**f**) moving radial or panel fixed-bed reactor.

Application of Fluid Dynamics to Fluid-bed Scale-up. Scaling up fluid-bed processes is difficult because fluid-bed behavior is as much a function of bed dimensions as of physical properties of the gas and solids. Thus wall effects cannot be neglected and various mechanisms that limit contact must be taken into account. Scaling is possible if the effects of the important variables that influence the reactions are known and a means for predicting conditions can be specified.

Modeling Transient Processes. Reactor technology can also be applied to time-dependent or transient chemical processes, for example, in the recovery of heavy crude oils by water flooding or underground burning, ie, *in situ* combustion in massive unconsolidated sands, and in similar fashion to laminate processing of thermosetting resins.

Experimental Reactors. Data for the chemical kinetics that are operative in a process and the associated physical properties are best obtained with an experimental reactor that is different from the one to be used on a large scale or in process development. The reactor should be similar to one of the basic types, have known flow patterns, operate isothermally, and provide data over wide ranges of variables. The range of the conditions to be studied must be sufficiently large for adequately defining the effects of variables, but the rate equations need not accurately reflect true reaction mechanisms. However, true rate equations can be used to understand the process and extrapolate results to conditions that have not been studied. Recommended experimental reactor types are shown in Table 1.

Reaction and Laboratory Models. Independent measurements of interfacial areas are difficult to obtain in liquid–gas, liquid–liquid, and liquid–solid–gas systems. Correlations developed from studies of nonreacting systems may be satisfactory. Comparisons of reaction rates in reactors of known small interfacial areas, such as falling-film reactors, with the reaction rates in reactors of large but undefined areas can provide an effective measure of such surface areas. Another method is substitution of a model reaction whose kinetics are well established and where the physical and chemical properties of reactants are similar and limiting mechanisms are comparable. The main advantage of employing a model reaction is the use of easily processed reactants, less severe operating conditions, and simpler equipment. Another technique employed for reactions occurring at near ambient conditions utilizes a laboratory-scale reactor and reactants with known interfacial properties.

Complex Flow Behavior

The concepts of well mixed and plug flow become inadequate when flow patterns deviate significantly from ideal behavior. Adsorption and subsequent desorption from within catalyst pellets in a trickle bed, solids recirculating downward to com-

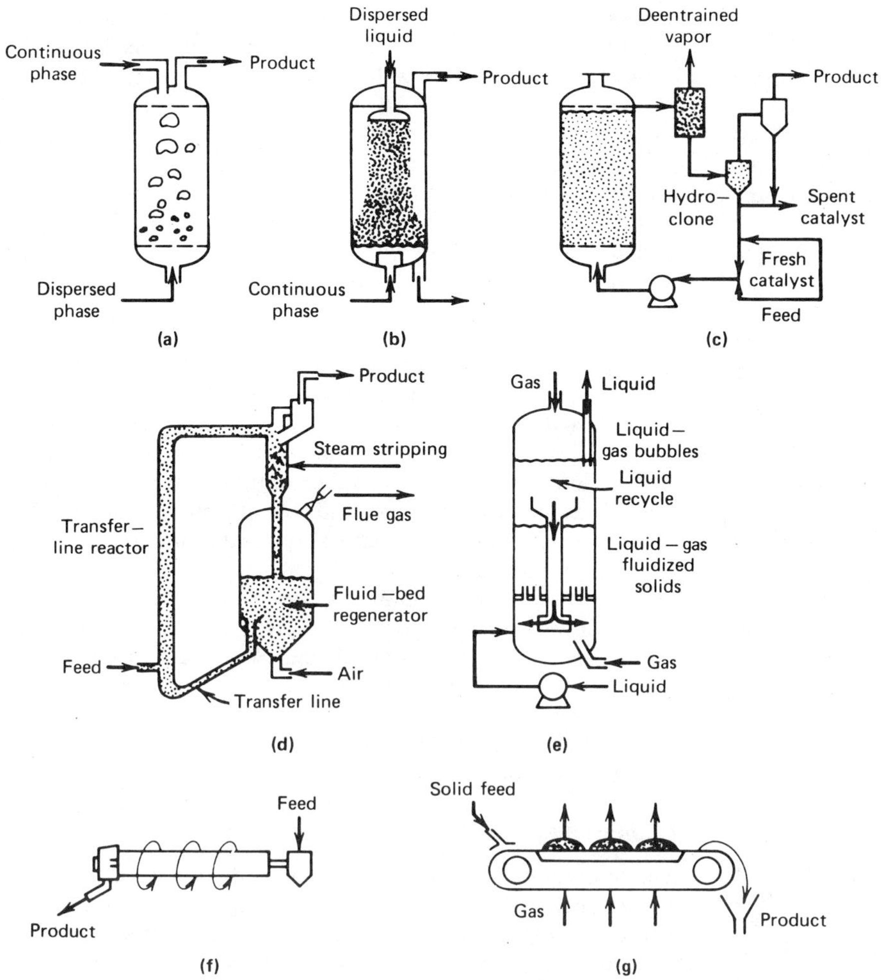

Figure 3. Multiphase fluid and fluid–solids reactors: (**a**) bubble column, (**b**) spray column, (**c**) slurry reactor and auxiliaries, (**d**) fluidization system, (**e**) gas–liquid fluidized system, (**f**) rotary kiln, (**g**) traveling grate or belt dryer.

pensate for those entrained into rising bubbles in a fluid bed, and liquid held in the wake of a solid slurry exemplify backmixing or by-passing in mixed-phase tubular reactors. Practical considerations, such as rapid mixing requirements, heating and reaction quenching demands, variations in flows through inlet distributors, the presence of internal baffles or structural obstructions, and inadequacies in locations of tank inlets and outlets, can effect substantial deviations.

Residence-time Distributions. A range of residence times results from departures from ideal behavior. Residence times can be coupled with reaction kinetics and used in assessing reactor performance. The techniques for measuring residence times are used to define residence-time distributions (RTDs) of flowing and confined circulating elements. The RTD in a flow reactor can be determined by analyzing the effluent pattern resulting from a step function or pulsed change in an inlet flow

Table 1. Experimental Reactors

Reaction mixture	Reactor	Remarks
gas	tubular	integral or differential
liquid	batch	slow reactions
	continuous-flow stirred-tank	moderately fast reactions
	tubular	fast reactions
solid	batch fluidized bed	inert gas or liquid for fluidization
gas–liquid	gas–liquid contactor[a]	slow reactions
	laminar film or jet	fast reactions
liquid–liquid	batch	good agitation
	continuous-flow stirred-tank	no settling of dispersed phase
	film	fast reactions
gas–solid	tubular	integral or differential with fixed bed
liquid–solid	batch	good dispersion of solids
	tubular	fixed or fluidized bed

[a] Interfacial area must be known.

property, such as the concentration of one component of an added tracer. Periodic functions, eg, sine waves, can also be used to measure time-averaged amplitude ratios and phase shifts. Thus they provide inherently higher accuracy but require complex experimental technique. Comparable methods are applicable to analyze nonuniform batch reactors, but these require some means for introducing the tracer transient and for sampling or monitoring the resultant response.

The concept of residence-time distribution is not limited to fluids. Varying catalyst deactivations in a reactor as results from the loss of catalyst surface area by coking or other aging mechanisms can be evaluated with particle population-balance models. Other applications include monitoring particle breakup and agglomeration, drop coalescence, other mixing phenomena, and biochemical and crystal growth.

Cold-flow Models. Use of tracers to obtain RTD characteristics followed by inferential analysis of flow patterns establishes conceptual but not actual detailed flow characteristics of a system. This limitation can be removed for reactions that are zero- or first-order or that can be so treated, eg, pseudofirst-order reactions, with the aid of cold-flow models. These are replicas of reactors or portions of reactors that operate at ambient conditions and are designed for detailed observation of flow patterns. The models are usually large, and little scaling is required. Similarly, fluids with known physical properties that are as similar as possible to those of the real system are used to minimize fluid dynamic scaling. Often the model is in part constructed of glass or plastic to permit the photographing of phenomena. Cold-flow models offer many advantages, including the potential for greater understanding, faster data collection because steady operation and accurate measurements are more easily obtained, and a greater range of geometries and flow conditions that can be studied. The development times should be shorter and research and development costs reduced as a result of adding this method.

Mechanical Fluid Mixing. Mechanical mixing of fluids is employed in emulsion and suspension polymerizations for blending, contacting, or dispersing both miscible and immiscible liquids; for suspending solids and catalysts to promote reaction; for particle dissolution, reduction, or growth; and for enhancing heat transfer. Both axial- and radial-flow mixing impellers are used. Axial-flow impellers have blades or turbines, inclined less than 90° from the plane of rotation; radial-flow impellers use turbines or paddles parallel to the axis. Mixing tanks are often equipped with two or four vertical strips on their periphery, about 8–10% of the tank diameter in width. These so-called full baffles suppress deep vortex formation and oscillating surface flow surges, but more power is used. Mounting the impeller off center serves the same purpose as these baffles.

Reactor Stability and Control. Polymerizations are exothermic reactions, and therefore temperature uniformity and thermal stability are critical to reactor performance and safe operation. The larger the reactor, the more difficult it is to control these variables. Volumetric heat generation increases with temperature as S- or multiple S-shaped curves, while surface heat removal increases linearly. The shapes of these heat-generation curves and the slopes of the heat-removal lines depend on reaction kinetics, activation energies, reactant concentrations, flow rates, and the initial temperatures of reactants and coolants. The intersections of the heat-generation curves and heat-removal lines represent possible steady-state operations called stationary states. Multiple stationary states are possible. The stationary states at the higher temperatures are usually to be avoided because product degradation or explosive reactions may result from the consequent excessive reactor temperatures. Thus control is introduced to establish the desired steady-state operation and obtain desired conversions and products at targeted rates. Control is also needed to provide safe start-up and shutdown. The control methods can affect overall performance by their way of adjusting temperature and concentration variations and upsets, and by the closeness to which critical variables are operated near their limits.

Computer Simulation. Computer simulation (numerical experimentation) reduces dependence on idealized models. Modern quantitative representations of laminar and turbulent flows are combined with new finite difference algorithms and other mathematical maneuvers in order to solve coupled nonlinear differential equations; advanced computer graphics is used to visualize the results. The reduced cost of computation and the increased cost of laboratory experimentation make the former increasingly useful.

Application of mathematical formulations and computational techniques to reactor problems is recent and few validated examples have been published. Some data are available on the performance characteristics of industrial furnaces and gas turbines; systems operating with turbulent diffusion flames have been studied for simple two-dimensional geometries and selected conditions.

Commercial Reactor Designs and Uses

Most reactors have evolved from concentrated efforts focused on one type. A few processes have emerged from parallel developments using markedly different reactor types. In most cases, the reactor selected for laboratory study has become the type used commercially. Further development usually favors extending this technology. Important industrial petrochemical processes and their reactors are described in several texts. Commercial reactor designs include batch reactors, continuous-flow stirred-tank reactors, thermal tubular reactors, bubble columns, spray columns, tubular fixed-bed reactors, fixed-bed reactors, and fluid-bed reactors.

Thermoplastic molds exemplify modern use of reactor technology. In molding thermoplastics, large clamping forces of up to 5000 metric tons are needed to keep molds together, while highly viscous polymers are forced into their cavities. Heat transfer is critical; if the molds are too cold, polymers solidify before filling is completed; if they are too hot, the time required for cooling delays production. Reaction injection molding (RIM) circumvents these problems and provides the technology for fabricating large articles, such as polyurethane fascia panels and automobile fenders and bumpers.

In RIM reactants, catalysts, cross-linking agents, and chain extenders are rapidly mixed and injected into the molds for polymerization. If fibers are included to improve flexibility of the polymerized products and reduce thermal expansion, such batch processes are termed reinforced reaction injection molding (RRIM).

BIBLIOGRAPHY

G. F. Froment and K. B. Bischoff, *Chemical Reactor Analysis and Design,* John Wiley & Sons, Inc., New York, 1979.

O. Levenspiel, *Chemical Reaction Engineering,* 2nd ed., John Wiley & Sons, Inc., New York, 1972.

J. L. Throne, *Plastics Process Engineering,* Marcel Dekker, Inc., New York, 1979.

G. Odian, *Principles of Polymerization,* 2nd ed., John Wiley & Sons, Inc., New York, 1981.

BARRY L. TARMY
Exxon Research and Engineering Company

RECOVERY AND ISOLATION TECHNIQUES

Many processes for the manufacture of synthetic chemicals require a final step of isolation and recovery of that chemical from the matrix of formation. Isolation and recovery is complex for polymers, because of their high molecular weight and the inherent ability of large molecules to change shape. These fundamental characteristics result in physical properties such as high viscosity, low solubility, stickiness, agglomeration, and low (or no) volatility. For these reasons, specialized recovery processes have been developed allowing these materials to be handled and converted into salable form.

The methods for recovery of polymers from the polymerization medium have undergone remarkable changes in recent years, resulting in process simplification and, to a lesser extent, in the improvement of product quality. These simplifications save energy and lower capital investment.

Traditional Recovery Processes

The first step in traditional recovery is the precipitation of the polymer in a separate operation downstream of the reactor by mixing the reactor effluent with a large volume of nonsolvent diluent; chemical coagulation of the polymer with an inorganic agent, for example, acid, brine, or base; or steam stripping (vaporization) of the reaction medium, leaving the polymer dispersed as agglomerated particles in a dilute water slurry. Which of these is used depends on the chemistry of the polymerization process and on the rheology and agglomeration properties of the polymer. Because polymers encompass a broad range of chemistry and properties, a large number of precipitation processes have been developed.

Newer Developments

More recent methods eliminate the common feature of the older methods, namely, precipitation of the polymer outside the reactor to form a new two-phase system that must be filtered and worked up in several steps.

The new processes offer the economic benefits of lower capital investment and manufacturing costs. In addition, elimination of the two-phase separations reduces adverse effects on the environment; consumption of water is also reduced. Most traditional separation equipment, for example, filters, slurry tanks, and drying extruders, are open. In contrast, most newer processes employ closed systems, which are much less likely to discharge emissions into the atmosphere. New processes include bulk polymerization (qv), and direct devolatilization (qv).

Outlook. Considering the advances made in recent years in applying direct devolatilization to different polymers, and considering the substantial increases in cost of energy over the past 10–15 years and forecasts for the future of energy prices, direct devolatilization must be regarded as the successor to steam stripping processes, wherever the technology for specific polymers can be developed. Its chief competition is bulk polymerization, which goes a step further in simplifying the recycling of the reaction medium. For certain polymers the two new technologies can be combined.

BIBLIOGRAPHY

Proctor and Schwartz, Inc., *Bulletin 562,* 1976.

U.S. Pat. 4,508,592 (Apr. 2, 1985), R. C. Kowalski.

U.S. Pat. 4,003,712 (Jan. 18, 1977), A. R. Miller.

R. J. Nichols, R. K. Senn, and F. Kheradi, *Adv. Polym. Tech.* **3**(1) (1983).

RONALD KOWALSKI
Exxon Chemical Company

RECREATIONAL SURFACES

Recreational surfaces are synthetic, durable areas of consistent properties designed for various recreational activities, including football, soccer, field hockey, cricket, baseball, tennis, track, jumping, golf, wrestling, and others. Recreational surfaces can include indoor–outdoor carpets, patio surfaces, and similar materials designed for low maintenance in light recreational service. The characteristics of the playing surface may be selected to match natural surfaces under ideal conditions or may have special features. In all cases, the intent is to provide durability for the intended uses.

Types

Recreational surfaces must provide certain performance characteristics at acceptable cost, with a reasonable lifetime, and with acceptable appearance. For classification, arbitrary but useful distinctions, depending on the primary function of the surface, may be made: a covering intended primarily to provide an attractive surface for private leisure activities, eg, patio surfaces; a surface designed for service in a specific sport; or a grasslike surface designed for a broad range of heavy-duty recreational activities, including professional athletics, eg, artificial turf for outdoor sports.

Performance Characteristics

User-related Properties. The most important element in the player's contact with the surface is traction. Shoe traction for light-duty consumer purposes requires reasonable footing and reasonable surface uniformity. The frictional characteristics are obviously of much greater importance in surfaces designed for athletic use. They can be significantly affected by pile density and height, and other aspects of fabric construction that determine the coefficient of static friction between the playing surface and the shoe.

Typical static friction coefficients are given in Table 1. These data demonstrate that the absolute traction values for synthetic surfaces are satisfactory in comparison with natural turf, provided that shoes with the appropriate surfaces are employed.

Table 1. Typical Traction Characteristics of Surfaces[a]

Surface	Static friction coefficient[b] Dry	Wet	Directionality index[c]
recreational surface			
tufted polypropylene	1.7–2.0	1.8–2.1	0.1–0.2
knitted nylon-6,6	1.8–2.0	1.6–1.8	0.10–0.25
tufted nylon-6	1.9–2.1		0.05–0.15
woven nylon-6	1.9–2.1		0.05–0.15
indoor–outdoor carpeting			
tufted polypropylene	0.4–1.5	0.4–1.5	
natural grass[d]	1.0–2.2	0.7–1.4	not applicable

[a] Ranges measured with appropriate sports shoes for the indicated surfaces.
[b] Defined as the average value measured in the four principal directions parallel to the fabric surface: two across the pile, one with, and one against the pile.
[c] Defined, from the same data as (*b*), as the average absolute deviation of the four traction values from the mean.
[d] The range is determined by the type and condition of grass.

Game-related Properties. For some activities, such as running and wrestling, the only consideration is the direct impact by the player. For others, eg, baseball or soccer, the system must also provide acceptable ball-to-surface contact properties. Important ball-response properties on the artificial surface are coefficients of restitution and friction, because these directly determine the angle, speed, and spin of the ball.

Artificial surfaces can be designed to match certain desirable game-response parameters of natural grass surface and provide these properties consistently. Response is confined to a narrower range and is less affected by weather. As a general rule, however, artificial turf surfaces tend to be somewhat livelier in ball response, velocity, and distance of roll, with coefficients of friction lower than those for natural grass.

Shock Absorption. Artificial playing surfaces for moderate to heavy use must provide shock absorbency for player comfort and safety. This is achieved by incorporation of a resilient layer. For heavy use, a resilient underpad is installed, which provides a distinct layer between playing surface and substrate.

The shock-absorbing characteristics of underpad materials or resilient surfaces are functions of material selection, physical composition, thickness, and temperature. The sensitivity of performance to physical characteristics of the shock-absorbing medium is illustrated in Table 2.

Durability. Grasslike surfaces intended for heavy-duty athletic use should have a service life of at least five years. Lifetime is more or less proportional to the amount of face ribbon available for wear, but pile density and height also have an effect. In addition, different materials respond differently to abrasive wear. These effects cannot be measured except in simulated and controlled laboratory experiments, which do not necessarily reflect field conditions.

Materials and Components

A recreational surface system includes the top surface material directly available for use and observation; backing materials that serve to hold together or reinforce the system; fabric-backing finish; a shock-absorbing underpad system, if any; and adhesives or other joining materials.

Surface Materials. Pile materials used in grasslike surfaces are selected from fiber-forming synthetic polymers, such as polyolefins, polyamides, and polyesters (see FIBERS, ELASTOMERIC). These polymers exhibit good mechanical properties for the necessary functions. For the smooth recreational surfaces of running tracks and tennis courts, polyurethanes, polyolefins, and other flexible, durable elastomers or composites are employed.

Backing Materials. Any fiber-forming polymer of reasonable strength may be used in backing materials, including polyamides, polyesters, and polypropylenes.

Backing Finish. Backing materials are usually applied as a coating, which is subsequently heat cured. For tufting, preferred choices are poly(vinyl acetate), poly(vinyl chloride),

Table 2. Properties of Typical Underpad Materials

Property	Foam, % closed cells 75–85	90–95	Poured elastomer
thickness, cm	1.6	1.6	1.0
tensile strength, kPa[a]	620	655	2700
density, kg/m^3	96	256	1300–1400
g_{max} at °C			
21	85[b]	125[b]	105[c]
−12	105[b]	150[b]	
49	120[b]	150[b]	

[a] To convert kPa to psi, multiply by 0.145.
[b] 60-cm drop height of 9-kg flathead missile (ASTM F355).
[c] 22.8-cm drop height of hemispherical body.

polyurethane resins, and latex formulations. For knitted fabrics, poly(vinylidene chloride) and acrylics or polystyrene–rubber latices are used.

Underpads. Shock-absorbing underpad material is usually made of foamed elastomer, which provides good energy absorption at reasonable cost (see CELLULAR MATERIALS). The foamed materials may be poly(vinyl chloride), polyethylene, polyurethanes, or combinations of these and other materials.

Adhesives and Joining Materials. Grasslike surfaces are employed over substantial areas, and lengths of rolls must be joined, glued, or sewn together. A variety of adhesives, ranging from low cost poly(vinyl acetate) materials to cross-linked epoxy cements, are available.

Fabrication and Installation

Recreational surfaces are made by tufting, knitting, and weaving processes.

Finishing. In each of the processes used, the artificial turf fabric is subjected to a finishing operation in which an adhesive, usually poly(styrene–butadiene), polyurethane, or poly (vinylidene chloride), is applied to the back side, bonding the components and stabilizing the material.

Underlay. An installed artificial turf system may or may not include components between the fabric and the subbase. Such components are not required for light-duty applications, but are essential in attaining the shock-absorbing properties required by heavy-duty surfaces.

Installation. Grasslike surfaces are usually glued down or otherwise affixed to a subbase. For light-duty applications, it may suffice to tack the edges to the perimeter of the area to be covered. For heavy-duty systems, a solid subbase of asphalt or other material is first installed. The asphalt may be permeable, allowing vertical water drainage. The shock-absorbing underpad is glued to this surface; the turf is then glued on top of the pad. Other constructions with partial or complete gluing are also possible. However, it is essential to anchor the perimeter securely.

For artificial surfaces in the athletic category, eg, running tracks, the installation techniques are different. A poured-in-place or interlocking tile technique may be employed; the latter is used for tennis courts. Adequate provision for weathering and water drainage is essential. In general, the resilient surfaces are installed over a hard base that contains the necessary curbs to provide the finished level.

BIBLIOGRAPHY

J. H. Saunders and K. C. Frisch, *Polyurethanes, Chemistry, and Technology,* Vol. 1, John Wiley & Sons, Inc., New York, 1963; Vol. 2, 1964.

T. A. Orofino, *Polym. News* **10,** 294 (1985).

T. A. OROFINO
W. F. HAMNER
Astroturf Industries

REDOX POLYMERS. See ELECTRON TRANSFER POLYMERS.

REFRACTION. See OPTICAL PROPERTIES; REFRACTOMETRY.

REFRACTOMETRY

One particularly useful and simple technique for quantifying the structural state of a polymer sample is polarized refractometry using an Abbe refractometer. This technique is important because it is nondestructive, inexpensive, simple to use, easy to understand, and generates considerable information about the structural state of the polymer.

A partial list of the structural information derivable from the refractive index(es) measured with an Abbe refractometer using polarized monochromatic light include three principal refractive indexes (N_X, N_Y, and N_Z), axiality (isotropic, uniaxial, multiaxial), average refractive index, density, crystallinity, molecular weight, glass-transition temperature, copolymer composition, and three-dimensional molecular orientation. In this article, the theory and practice of polarized refractometry are presented along with applications to real polymer systems.

Isotactic polypropylene is used throughout as the model polymer. However, the techniques described are applicable to any polymer system that is translucent to visible light. Polyethylene, poly(ethylene terephthalate) (PET), and hydroxypropylcellulose are three polymers that have been studied using polarized refractometry.

Review of the Polarized Refractive Index Method

The theory of polarized refractometry was first developed in the 1930s by Okajima.

Figure 1 presents the typical geometry of a polymer sample studied in an Abbe refractometer. N_Z, N_Y, and N_X are the three principal refractive indexes which lie either along or perpendicular to the optical symmetry axis (defined here as the Z axis) of the sample. For uniaxially drawn samples, both the optical symmetry axis and the principal refractive index N_Z lie along the stretch or machine direction of the sample. N_Y is the principal refractive index transverse to the optical symmetry axis in the plane of the sample, and N_X is the refractive index through the plane of the sample. For isotropic samples, all three principal refractive indexes are equal, ie, $N_X = N_Y = N_Z$. For uniaxial samples, two refractive indexes are the same and different from the third, eg, $N_X = N_Y \neq N_Z$. For biaxial samples, all three principal refractive indexes are dif-

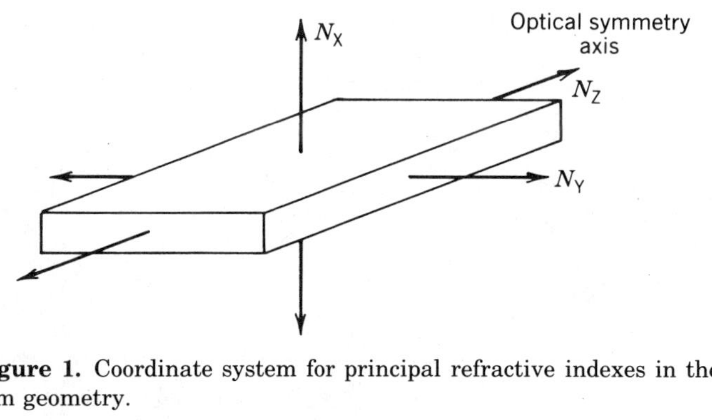

Figure 1. Coordinate system for principal refractive indexes in the film geometry.

ferent, and the direction of the optical symmetry axis in the Z-Y plane must be determined.

Measuring Refractive Index in the Abbe Refractometer

In the Abbe Refractometer, the sample (liquid or solid) is placed between the top and bottom prism, and then the top prism is placed on top of the sample (see Fig. 2). Light is introduced to the sample through either the top or bottom prism and collected by a mirror which directs the light up the telescope to the eyepiece.

The normal light entry point is the top prism, but the bottom prism light entrance is useful for those samples that are too opaque to transmit light through the sample. The light is then reflected off the bottom surface to yield the sample's refractive index.

For polymer samples (films, sheets, or moldings), regardless of whether the light is introduced through the top or bottom prism, the measured refractive index is that of the bottom surface. Polarized refractometry is a surface characterization technique. The heterogeneity of the sample can therefore be studied by comparing the surface information yielded by refractive index to similar information for the sample's bulk using such bulk characterization techniques as x-ray diffraction, polarized microscopy, polarized transmission infrared spectroscopy, etc.

Since many samples exhibit some dispersion with wavelength, ie, the refractive index changes as a function of wavelength, it is preferable to use a monochromatic light source for the refractive index measurement. Monochromatic light of known wavelength both prevents any dispersion effects and allows corroboration of experimentally determined refractive indexes by different laboratories using the same wavelength of light. The sodium D line (589 nm wavelength) is a common optical standard; therefore, the recommended monochromatic light source is a low pressure sodium vapor lamp.

Multiple Refractive Index Layers. Figure 3 shows a rather common experimental setup for measuring the refractive index

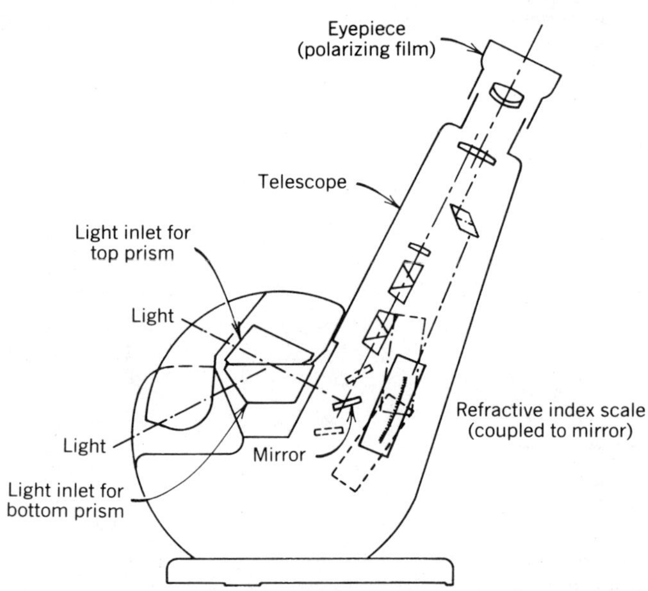

Figure 2. Schematic of the Abbe refractometer.

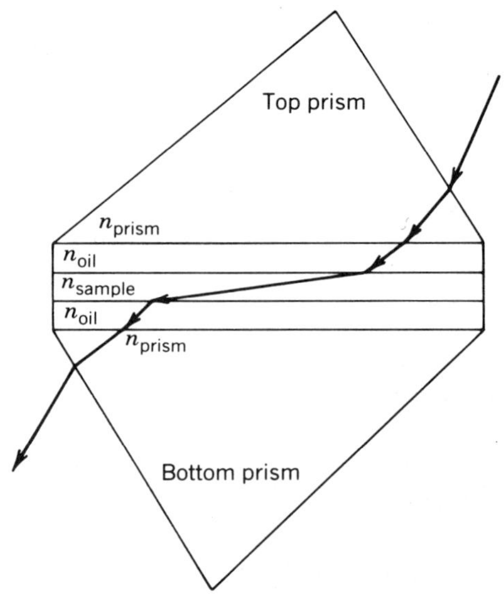

Figure 3. Light propagation through the contact oil and sample in the Abbe refractometer.

of a solid sample in an Abbe refractometer. The sample is coated with a contact oil on both sides to provide a continuous path for the light by excluding air between the prisms and the sample. Without a contact oil, it is often impossible to transmit light through the sample due to air scattering and/or sample surface roughness. An analysis of Figure 3 according to Snell's law yields two practical considerations.

$$(n_1 \sin \theta_1 = n_2 \sin \theta_2)$$

First, the contact oil must have a refractive index greater than the sample's in order for the sample's critical angle interface to be seen. If $n_{oil} < n_{sample}$, then the oil refractive index will be measured instead of the sample's index. Therefore, the contact oil's refractive index must be intermediate between the prism's and the sample's.

Secondly, since $n_{oil} > n_{sample}$, Snell's law would predict that only the light/dark interface corresponding to the sample's refractive index should be seen since it is the lowest one. However, in reality, the interface due to the oil is visible below that of the sample. Therefore, before starting measurement, it is important that the oil's refractive index be measured and recorded such that it will not be confused later with the refractive index of the sample.

Measuring Optically Anisotropic Samples. For isotropic samples, there is only one refractive index to be measured. For anisotropic samples, however, the measured refractive index is dependent upon the direction of the electric vector, **E**. In Figure 1 N_Z is the refractive index along the optical symmetry axis (primary orientation direction) in the sample plane. N_Y is the refractive index in the plane and transverse to the optical symmetry axis direction, and N_X is normal to the sample plane.

Since the sample must always lie flat on the bottom refractometer prism, rotation of the sample on the prism about the X direction should always yield the same N_X value. However, the exact direction of the optical symmetry axis must be

Table 1. Structural Information Derivable from the Principal Refractive Indexes, N_X, N_Y, and N_Z

Information	Definition	Application
optical symmetry axis	β direction	primary orientation direction
axiality	isotropic, uniaxial, biaxial	describes results of processing
average refractive index	$n = (N_X + N_Y + N_Z)/3$	yields density
crystallinity	V_c	volume fraction of crystal phase—derived from average refractive index, n
three birefringences	$N_Z - N_Y$, $N_Z - N_X$, $N_Y - N_X$	3-D surface orientation descriptor—used to generate Hermans' orientation functions in 3-D
fractional orientation	$f(N_Z)$, $f(N_Y)$, $f(N_X)$	3-D proportional orientation descriptor—independent of V_c

found in order to determine the two in-the-plane principal refractive indexes, N_Z and N_Y.

Structural Information Derivable from Polarized Refractometry

Polarized refractive index measurements can yield the structural information given in Table 1 and discussed below.

Optical Symmetry Axis Direction. The optical symmetry axis directions of nine points in the surface of an injection-molded plaque from a fan-gated mold were determined by polarized refractometry. It was apparent from the direction of the optical symmetry axis arrows that, under the process conditions, the flow was not parallel, but rather spurting through the gate and trough like a fountain. Further, the greater elongation of the refractive index ellipses in the middle of the plaque, as compared with the entrance and end sections, show that the degree of polymer orientation toward the individual polymer flow directions was also highly variable across the plaque. The simple visual map provided by refractive index measurements explained the lack of correlation of the impact results with process conditions and indicated that the fan gate and entrance trough needed redesigning to obtain uniformity across the plaque surface.

Axiality and the Three Principal Refractive Indexes. N_X, N_Y, and N_Z are indicators of the degree of orientation of the polymer chains in the three orthogonal directions of a fabricated sample (X, Y, and Z) and provide a quick determination of whether a fabrication process has given the desired axiality (or type of orientation), ie, isotropic, uniaxial, or biaxial. Knowledge of the sample's axiality is important because of its impact on other physical properties, eg, tensile modulus. If a film sample is uniaxial, its modulus is higher in the fabrication draw direction than in the transverse direction. If a film is equally biaxially oriented, or balanced, in the plane, then its modulus is equal in all directions in the film plane. Unbalanced stretching produces greater modulus in one direction than in the other. Polarized refractometry provides a quick and accurate method for the determination of the sample axiality.

Average Refractive Index and Crystallinity. The average refractive index, n, is defined as:

$$n = \tfrac{1}{3}(N_X + N_Y + N_Z)$$

For semicrystalline polymers, the average refractive index of an optically anisotropic polymer sample is the same refractive index as that which an optically isotropic polymer sample of the same volume fraction crystallinity, V_c, would have. V_c is linearly related to the average refractive index, n, by

$$V_c = \frac{n - n_{nc}}{n_c - n_{nc}}$$

where n_c and n_{nc} are the average refractive indexes of the crystalline and noncrystalline phases. These are determined by separate experiments and are material constants which must be determined separately for each different polymer.

Three-dimensional Molecular Orientation. In addition to the three-dimensional orientation information given directly by the three birefringences shown in Table 1, the three birefringences can also be used to generate Hermans' orientation functions in three dimensions (see ORIENTATION). Such information is extremely useful in coupling the three-dimensional surface orientation results from polarized refractometry with similar orientation results obtained for the bulk phase using pole-figure x-ray diffraction or polarized infrared spectroscopy.

Samuels' Fractional Orientation Parameter. A three-dimensional fractional orientation parameter, $f(N_i)$ (i = X, Y, or Z), which is a measure of the degree of optical anisotropy in the sample has been developed. $f(N_i)$ is an average of the anisotropy of both the crystalline and noncrystalline phases; there are no separate crystalline and noncrystalline phase parameter. The usefulness of $f(N_i)$ is its ability to normalize out the effect of crystallinity and allow polymer samples made by different processes to be compared purely on the basis of the degree and type of their optical anisotropy. Figure 4 is an example. The crystallinity difference between the two samples, 55% and 31%, is large and the large differences in the two series' principal refractive index values masks the fact that the anisotropy of both series may be very similar. The fractional orientations plotted in Figure **4b**, however, demonstrate that, with the crystallinity effects removed, the equivalent degrees of optical anisotropy are obvious.

In addition to injection moldings, fractional orientation has been used to compare polypropylene samples produced by such diverse processes as hydrostatically extruded sheets and blown bottle sections. It has also been shown to yield the same fractional surface orientation information as is obtained from polarized ATR (attenuated total reflection spectroscopy) infrared measurements.

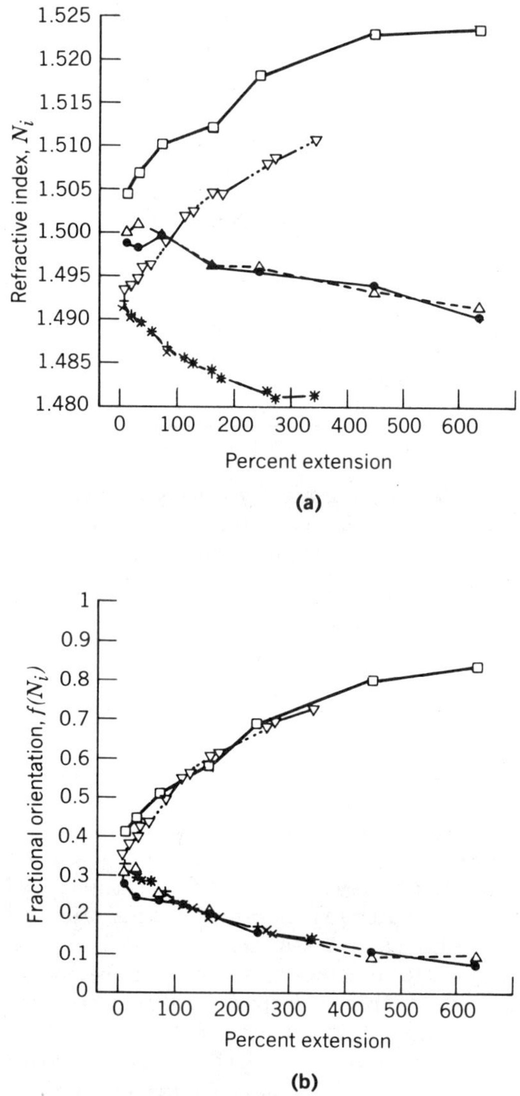

Figure 4. Comparison of uniaxially oriented PP films from different manufacturing processes: (**a**) refractive index vs percent extension of the samples. □ = N_Z; △ = N_Y; ● = N_X; ▽ = N_Z; + = N_Y; × = N_X. (**b**) Samuels fractional orientation parameter vs percent extension of the samples. □ = $f(N_Z)$; △ = $f(N_Y)$; ● = $f(N_X)$; ▽ = $f(N_Z)$; + = $f(N_Y)$; × = $f(N_X)$.

BIBLIOGRAPHY

F. D. Bloss, *An Introduction to the Methods of Optical Crystallography*, Holt, Rinehart, and Winston, New York, 1961.

R. J. Samuels, *Structured Polymer Properties*, John Wiley & Sons, New York, 1974.

R. E. Pepper and R. J. Samuels, *Society of Plastics Engineers Annual Technical Conference (ANTEC)* **34**, 884 (1988).

RANDY E. PEPPER
ROBERT J. SAMUELS
Georgia Institute of Technology

REGENERATION TEMPLATES. See SKIN, REGENERATION TEMPLATES.

REGULATORY CONSIDERATIONS

More than 30 federal departments and agencies commissioned by the U.S. Congress currently have significant regulatory responsibility for interpreting and enforcing public law. These regulatory agencies create the rules and standards that govern our businesses and personal lives, either directly or because they are superimposed on state and municipal laws and ordinances.

The laws that have had the greatest impact on the chemical and plastics industries are those related to health, safety, and the environment. For the most part, these laws have evolved only over the past few decades in the wake of a new environmental consciousness that stirred Congress to correct past industrial abuses and forestall future ones. One consequence of the new laws is a dramatic increase in the cost of doing business.

The health, safety, and environmental laws of most concern to plastics processors are the

Federal Food, Drug, and Cosmetics Act
Toxic Substance Control Act (TSCA)
Resource Conservation and Recovery Act (RCRA)
Comprehensive Environmental Response, Compensation and Liability Act (CERLA or "Superfund")
Clean Air Act
Clean Water Act
Safe Drinking Water Act
Occupational Safety and Health Act
Consumer Product Safety Act

Plastics Packaging

Approximately 10 percent of the nation's food supply is now packaged in plastic material, and there is every reason to believe that both the volume and market share will increase. The plastics food-packaging market represents a significant factor in resin consumption (see PACKAGING MATERIALS).

Substances intended to come into contact with food are regulated by the FDA, the U.S. Department of Agriculture (USDA) and, in the special case of alcoholic beverages, the Bureau of Alcohol, Tobacco, and Firearms (BATF), which is part of the Department of the Treasury. Only the FDA has a published codification relating to packaging materials, Title 21 of the CFR, and the USDA and BATF usually defer to this code.

The Federal Food, Drug, and Cosmetics Act. Under the provisions of the 1938 Federal Food, Drug, and Cosmetics Act, no premarket clearance was required for any type of packaging material. The FDA's authority to regulate packaging derived solely from its authority to remove adulterated food from the marketplace. Thus if a packaging material rendered a food unfit for human consumption, the FDA could move against the food thus packaged by seizure and prohibition.

The 1958 Food Additives Amendment addressed the need for new regulatory authority to control the intentional and direct addition of various substances to food, and ordered the FDA to regulate packaging materials and food-processing equipment in precisely the same way as substances deliverately added to food, to regulate indirect additives. This is the current law and its rules are embodied in Title 21 of the CFR.

Under present law, any substance that comes into contact with food, with certain exceptions, is considered unsafe unless it is used in conformity with a regulation issued by the FDA. Many packaging applications, primarily plastics, provide for safe use of a wide variety of materials (Table 1).

In petitioning for the use of a packaging material, if the producer can trace the composition of matter and the customer's application through the CFR, it is sufficient to draft a letter stating that "this material is considered safe for use according to the following provisions" with citations to the appropriate sections and provisions in the CFR. If no such provisions exist, the producer must petition the FDA to amend the regulations to allow for safe usage. This is a task that usually falls to the plastic-additive supplier or base-polymer producer who has the most to gain by the potential new commerce. The Food-Additive Petition (FAP) becomes a matter of public record and is necessary to market a new food-contact substance for a new application. The FAP must include all the necessary information to allow the FDA to make a determination on safety under the intended conditions of the product's use. A short description of the FAP is published in the FR, but much more information (a summary of extraction and toxicological data) is usually available on request under the Freedom of Information Act. The guidelines for the safety data required in a FAP will change as progress is made in toxicological understanding and analytical methodology; it is always advisable to contact the FDA before initiating any expensive testing program.

An FAP, and particularly one leading to a successful amendment, has some commercial value. It shows the customers that a producer is aware and active, and that the specific product is or will soon be government-approved, with both its efficacy and safety attested to by outside authorities. A deficient FAP, however, may result in publication of a withdrawal, which, even without prejudice toward future refiling, often leads the uninitiated to conclude that the product is unsafe. The quality of the petition therefore often determines whether the commercial impact of the product will be positive or negative. There are many deficiencies in the present regulatory system for indirect food additives, and changes have been called for by the Society of the Plastics Industry, among others.

The FDA and the Bureau of Alcohol, Tobacco, and Firearms. Liquor packaging presents an additional problem because containers for distilled spirits must be approved not only by the FDA but also by the Bureau of Alcohol, Tobacco, and Firearms (BATF), which is part of the Department of the Treasury and is charged with protecting tax revenues from alcohol. In addition to safety, the BATF considers taste and packaging aesthetics.

Drug and Biologics Packaging. The packaging of drugs and biological materials such as blood, plasma, and intravenous solutions for humans requires an entirely different process than those described above. Although the packaging materials are usually selected from those included in Title 21, the listing in the CFR is often not necessary or even desirable.

Guidelines for drug and biologics packaging have recently been published by the FDA. Proposed packaging is usually petitioned as part of the New Drug Application (NDA), along with the drug or biologic to be considered. The salient testing is described in the *United States Pharmacopoeia* (USP) and involves animal implantation–extraction with progressively aggressive solutions, followed by injection of these extracts into the test rats or rabbits. The least stringent is USP class I, and the most, USP Class VI testing.

Industrial Packaging. Industrial containers, especially for the shipment of hazardous goods, fall under the aegis of the Department of Transportation (DOT). Although the Interstate Commerce Commission (ICC) relinquished its regulatory authority on interstate shipment of hazardous wastes to the EPA early in 1982, the DOT still maintains control over shipments of regular or nonwaste materials, hazardous or not. This authority was conferred by the Hazardous Substances Act, which predates the Toxic Substance Control Act and even the existence of the EPA.

Because of the highly technical nature of packaging standards and specifications, the Research and Special Programs Administration must work closely with the technical experts of the container manufacturers. Cooperation seems to be key, and the DOT appears to be much less adversarial than some other regulatory agencies.

Recent regulations of interest to plastics processors include the conversion of individual exemptions to rules of general applicability and the extension of standards for polyethylene containers to allow 0.21-m^3 (55-gal) drums for shipment of hazardous materials, up from the previous limitation to 0.11-m^3 (30-gal) drums. Finally, rules clarifying procedures, exemptions, and enforcement standards have been promulgated.

The Environmental Protection Agency

Perhaps no other federal agency issues as many new regulations or interacts with as many other regulatory bodies as the EPA. While the FDA essentially has only one law to administer, the EPA has a spate of acts, all affecting environmental and ecological issues and touching on all other health and safety legislation. All major environmental laws and regulations that deal with air and water quality, hazardous wastes, and toxic substances are the province of the EPA. EPA regulations have had their heaviest impact on the chemicals and plastics industries. The largest incremental costs, companies' expenditures that, in their judgment would not have been made

Table 1. Regulations for Indirect Food Additives

Title 21, CFR Section	Materials
175.105	adhesives
175.125	pressure-sensitive adhesives
176	paper and paperboard components
177 subpart B	substances for use as basic components of single and repeated use food-contact surfaces
178	adjuvants, production aids, and sanitizers
178.2010	antioxidants and/or stabilizers for polymers
178.2650	octyltin stabilizers in vinyl chloride plastics
178.3740	plasticizers in ploymeric substances

without regulation, are credited to compliance with EPA regulations and amounted to over $3 billion through 1977, mostly in capital projects; this figure does not include the huge tax burden of staffing and running an enormous agency such as the EPA.

The Toxic Substance Control Act. The TSCA, sometimes referred to as TOSCA was signed into law in October 1976 and became effective on January 1, 1977. In cooperation with other federal agencies, TSCA enabled the EPA to carry out programs to obtain more and better information about health and ecological effects, levels of exposure, distribution and use, potential by-products, and disposal of potentially hazardous chemicals; to help prevent hazards and harm to health and the environment through the screening of new chemicals prior to marketing; to balance the costs, risks, and benefits posed by hazardous chemicals; and to coordinate the fragmented efforts at federal and state levels to control such hazards. To accomplish these objectives, the various provisions of TSCA delineate four major functions: information gathering, regulation, premanufacturing requirements, and cooperation of federal agencies through dissemination of data derived from the act, research, data systems, etc.

The Resource Conservation and Recovery Act. The Resource Conservation and Recovery Act (RCRA), signed into law in 1976, regulates the management of hazardous waste. Many of its provisions were not implemented until May 19, 1980, however, and only within the last few years have the final rules been promulgated. The seventh and last set of regulations for the original RCRA, containing standards for permits for land disposal of hazardous waste, were promulgated in July 1982. Tougher than expected, these rules are expected to cost industry up to $500 million more each year than the $700 million now spent. On the other hand, among RCRA's many provisions is the rule that requires that synthetic liners (eg, plastic sheets and films) be used to underlay all new land waste-disposal facilities, and this may prove a boon to the plastics industry.

Current regulations require generators of hazardous waste to obtain an EPA identification number from the administrator before treating, storing, transporting, or disposing of the hazardous waste, to maintain manifests and comply with pretreatment requirements (packaging, marking, labeling, and placarding), and to submit annual reports. Individuals, including generators of waste, who store, treat, and dispose of hazardous wastes, must obtain an EPA permit and comply with the appropriate RCRA requirements for transport, storage, treatment, and disposal.

Generators, including small businesses, are afforded some relief from RCRA compliance through an exclusion that allows all generators to accumulate wastes on site for up to 90 days without a permit, provided that the wastes are placed in clearly labeled and dated containers or tanks that meet RCRA standards. In addition, generators must comply with some of the requirements for owners and operators of hazardous waste facilities, contingency plans, emergency procedures, etc. Other provisions exclude, with certain exceptions, small businesses and other generators who produce less than 1000 kg of hazardous waste a month.

The list of hazardous wastes published in the CFR now includes more than 500 substances. These wastes are listed as hazardous because of ignitability, corrosivity, reactivity, and EP toxicity, or because they meet other criteria for listing. There are also provisions for exclusion or removal from the list in specific cases for individual facilities if the petitioner can show that a listed waste at their facilities lacks all of the relevant criteria that led the EPA to list the waste in the first place. Under another exclusion, wastes that are used, reused, recycled, or reclaimed (or that are stored before such activities) are excluded.

Superfund. The real name of the regulation known as Superfund is the Comprehensive Environmental Response, Compensation, and Liability Act of 1980 (CERCLA). Whereas RCRA is designed to prevent future environmental contamination, Superfund provides a means to clean up existing hazardous waste dumps and to rectify future spills of hazardous substances. The major provision of the law called for the establishment of a five-year, $1.6 billion fund for this purpose, and most of the money came from a tax on about 42 chemicals designated as hazardous. This taxing authority expired on September 30, 1985.

The Clean Air Act. The Clean Air Act gives the EPA authority to regulate in two main areas: automobiles and new industrial plants (where the agency pushes for accelerated development of antipollution technology). Plastics processors are affected by both. The EPA does not intend to monitor emissions themselves, but to effect implementation by states through approval of State Implementation Plans (SIP) for compliance with the National Ambient Air Quality Standards. This is accomplished through the issuance of Control Technique Guidelines (CTG) prepared to assist the states in determining reasonably available control technology (RACT) for control of volatile organic compound (VOC) emissions from organic chemical and polymer producers. CTGs listed in the FR are of interest to manufacturers of HDPE, polypropylene, and polystyrene resins; and synthetic fiber producers. Other final rules have been published on vinyl chloride emissions and test methods for measuring these.

The Clean Water Act. Under the Clean Water Act, which is a 1976 court order modified in 1979, and subsequent Congressional reauthorizations and extensions, the EPA is required to adhere to a strict schedule in promulgating regulations for a vast list of priority pollutants. The administration of this act has been quite painful to the EPA and controversial to other parties.

Yielding to judicial pressure, the EPA has been issuing rules and proposals for clean water standards on an industry-by-industry basis. Final rules and standards have been promulgated for inorganic chemicals manufacturing (involving 144 plants), and for gases and pigments from metal-bearing ores. The EPA has estimated compliance costs for this at a modest $29 million in capital expenditures for additional treatment equipment. The final rule is based on eight treatment options: (*1*) equalization, alkaline precipitation, and settling or clarifying to remove toxic and other heavy metals and fluorides; (*2*) dual media filtration to remove additional precipitated toxic metals; (*3*) sulfide precipitation to knock down additional mercury, lead, copper, silver, cadmium, and arsenic; (*4*) use of coagulants to improve the settling or filtering of the metal precipitates; (*5*) sulfur dioxide, ferrous iron, or sulfide reduction to reduce hexavalent chromium in preparation for alkaline precipitation; (*6*) air oxidation to reduce chemical oxygen demand (COD), to oxidize ferrous iron to ferric iron to make it able to settle, and to coprecipitate additional metals; (*7*) alkaline chlorination, which raises pH to 10–11 and adds chlorine to convert toxic cyanide to the less toxic cyanate; and (*8*) dechlorination, which uses thermal, chemical, or catalytic decomposition to destroy free chlorine.

Because the standards and procedures of the Clean Water Act are still in a state of evolution and have a direct effect on plant design consideration, the FR remains the most valuable source of current information, along with news media coverage of congressional developments.

The Occupational Safety and Health Administration

The Occupational Safety and Health Act was enacted for the purpose of ensuring safe and healthful working conditions for every working man and woman in the nation by authorizing enforcement of standards developed under the act; by assisting and encouraging the states in their efforts to ensure safe and healthful working conditions; by providing research, information, education, and training; and for other related purposes. The Occupational Safety and Health Administration (OSHA) is part of the Department of Labor, and its regulations and standards are found in Title 29 of the CFR. With a strong congressional mandate, OSHA's actions seemed excessive in its early days, when it drafted rules for everything in the workplace from the placement of safety showers and eye-wash basins to the conformation of toilet seats. Many of the unnecessary standards have since been rescinded. But OSHA is still a hotly contested battleground between the opposing forces of industry and labor.

Despite the long list of toxic substances to which workers may be exposed, permissible exposure levels (PELs) and time-weighted averages (TWAs) for such exposure have been developed for only very few, such as vinyl chloride and ethylene oxide. OSHA's principal involvement in such rule development and guidance is usually taken from the recommendations of the National Institute for Occupational Safety and Health (NIOSH), a division of the Centers for Disease Control, which in turn is part of the Public Health Service, Department of Health and Human Services.

OSHA's Cancer Policy. Cancer in the workplace has been and is a serious concern. Estimates of the entire nation's rate of cancer due to occupational exposure vary from 5 to 40%. On the basis of toxicological data, OSHA establishes standards for PELs to minimize risk, within reason. The standards are expressed as TWAs over eight-hour periods, usually in weight per unit volume of air. OSHA sometimes permits short exposures, usually for 15 minutes and slightly above the PEL. It may also mandate initial monitoring when exposure is suspected; if the amounts found exceed a stipulated action level (some fraction of the PEL), continuous monitoring and record keeping are required. This applies not only to suspected carcinogens, but to all hazardous chemicals. It is estimated that about 60% of all occupational illnesses, excluding cancer, can be traced to chemical exposure.

The Hazard Communication Rule. The hazard communication standard, also known as the workers' right to know, became effective on November 25, 1985. The rule is predicated on the belief that more thorough employee communication is needed and problems of interstate conflicts and irregular enforcement can be solved only by a uniform federal rule.

OSHA's hazard communication rule requires chemical manufacturers, chemical importers, and all employers with workplaces in the manufacturing division (Standard Industrial Classification (SIC) codes 20 through 39) to assess the hazards of the chemicals and to provide information to employees concerning hazardous chemicals. The means of these hazard communication programs include labels, material safety data sheets (MSDS), training, and access to written records. In addition, distributors of hazardous chemicals are required to ensure that containers they distribute are properly labeled, and that an MSDS is provided to their customers in the manufacturing division SIC codes. The customers, as employers with chemicals in their workplaces, must establish programs to use the MSDSs to educate their workers and train them in the safe handling of the chemicals.

Occupational Exposure to Lead. OSHA defined the current standard for occupational exposure to lead in 1978, and for 46 specified industries and occupations set a PEL of 50 μg of lead per square meter as an eight-hour TWA. Other OSHA requirements cover environmental and biological monitoring, medical surveillance, medical removal protection, employee education and training, and hygiene facilities. Its compliance hierarchy requires implementation of engineering and workplace practice controls to achieve the PEL without the use of respirators.

Occupational Exposure to Noise. In its efforts to require employers to establish a continuing, effective hearing-conservation program, OSHA set forth specific requirements for noise monitoring, and audiometric testing, hearing protectors, training, warning signs, and record keeping to cover all workers exposed to noise at or above an eight-hour TWA of 85 dB.

The Consumer Product Safety Commission

The Consumer Product Safety Act was signed into law on October 27, 1972. Its most important provision was the creation of the Consumer Product Safety Commission (CPSC), a five-member regulatory body appointed by the president and confirmed by the Senate. The commission must be bipartisan, containing no more than three members of the same political party, and it operates independently of any executive department. Congress emphasizes this independence by including clauses in the law that direct the commission to submit both budget requests and legislative proposals concurrently to the president and to the Congress.

The products subject to the consumer product act are those offered for sale for the personal use or enjoyment of residential, recreational, or institutional consumers. The act does not cover motor vehicles, tobacco, pesticides, aircraft, boats, drugs, foods, cosmetics, and radiation products, which were already subject to federal regulations. The act does preempt some laws previously administered by other agencies, however, including parts of the Federal Hazardous Substance Act, the Toy Safety and Children Protection Act, the Flammable Fabrics Act, the Refrigerator Safety Act, and the Poison Prevention Packaging Act.

BIBLIOGRAPHY

Federal Food, Drug, and Cosmetics Act of 1938, 52 Stat. §1040 (1938); 21 USC §301 *et seq.*; Food Additives Amendment of 1958, Pub. L. 85-929, 72 Stat. §1784 (1958).

Toxic Substance Control Act, Pub. L. 94-569, 90 Stat. §20003 (1976); 15 USC §2601 *et seq.*

Resource Conservation and Recovery Act of 1976, Pub. L. 94-580, 90 Stat. §2795 (1976); 42 USC §6901 *et seq.*

Comprehensive Environmental Response, Compensation, and Liability Act of 1980, Pub. L. 96-510, 94 Stat. §2767 (1980); 42 USC §9601 *et seq.*

Occupational Safety and Health Act of 1970, Pub. L. 91-596, 84 Stat. §1590 (1970); 29 USC §651 *et seq.*

LEWIS B. WEISFELD
Consultant

REINFORCED PLASTICS

Thermoplastics

Reinforced thermoplastics (RTPs) are produced by compounding a resin with one or more reinforcements. The resins include most commercial thermoplastics. Reinforcements are inert materials that may be in the form of fibers, spheres, powders, or platelets. The choice is usually based on the mechanical or physical property enhancement required. Tensile strength and flexural modulus are improved by the addition of glass or carbon fibers, wollastonite, or mica platelets. Flatness is improved by particulate fillers such as calcium carbonate, talc, or glass spheres. For conductivity, carbon powder, carbon fibers, aluminum flakes, stainless steel fibers, or metal-coated fillers may be added. Properties of various RTPs are given in Table 1.

Resins

Virtually all thermoplastic resins are compounded with some type of reinforcement. Different resins react to reinforcement in different ways. Polyamide is one of the most versatile reinforced resins. Commonly called nylon, these products are usually crystalline in nature, although a few amorphous grades have been developed (see POLYAMIDES). All types of nylon have been reinforced, although nylon-6 and nylon-6,6 are the most commonly used.

The advantages of nylon-based RTPs include very high tensile, flexural, and impact properties, and exceptionally high deflection temperature. The disadvantages include relatively high moisture absorption and a propensity to warp.

Polycarbonate, being an amorphous material, offers less flexibility to the compounder (see POLYCARBONATES). Fiber-reinforced products are produced up to 40 wt %. Other than conductive carbon black powders, nonfibrous reinforcements are not generally used. Polycarbonate RTPs offer good tensile and impact properties. Disadvantages include poor heat and chemical resistance. Polycarbonate is frequently combined with PTFE and silicone lubricants for improved friction and wear.

Polyolefins, particularly polypropylene, are the largest volume resins used in RTPs (see PROPYLENE POLYMERS). They are used in conjunction with most reinforcements. Styrenic resins are filled with fiber reinforcements up to 40 wt %. Some nonfibrous fillers, particularly talc and mica, offer high flexural modulus and are occasionally used.

Acetals are normally chosen for their chemical resistance or surface lubricity (see ACETAL RESINS). Reinforcements (up to 40 wt %) provide a harder surface and better wear resistance but do not improve surface lubricity. Polyurethanes are used in limited quantities with reinforcement loadings of up to 40 wt %. These products offer excellent chemical, impact, and wear resistance, but are somewhat expensive.

Table 1. Properties of Reinforced Thermoplastic Materials

Resin[a]	Specific gravity	Tensile strength, MPa[b]	Tensile modulus, GPa[c]	Elongation, %	Flexural strength, MPa[b]	Flexural modulus, GPa[c]	Izod impact notched, J/m[d]	Deflection temperature under load, °C
nylon-6,6								
unreinforced	1.14	83	2.9	60	119	2.8	53	90
30% glass fibers	1.39	172	9.0	4	248	9.0	107	252
30% carbon fibers	1.28	227	20.7	3	324	20.7	85	263
40% mineral filler	1.50	92	5.5	3	155	7.2	48	249
40% glass–mineral	1.49	124	7.6	3	207	9.7	64	246
polypropylene								
unreinforced	0.9	34	1.4	11		1.6	51	53
30% glass fibers	1.13	52	5.5	2.5	65	4.1	64	137
30% glass fibers chemically coupled	1.13	83	5.9	2.3	110	5.5	107	151
40% mica	1.23	31	4.8	4	48	4.1	37	96
40% talc	1.25	29	3.1	4	48	3.1	27	76
polycarbonate								
unreinforced	1.20	65	2.4	7	93	2.3	801[e]	132
30% glass fibers	1.43	131	9.0	2.5	138	7.6	160	143
30% carbon fibers	1.33	152	17.2	1.8	220	15.2	107	149
5% stainless steel	1.27	68	3.1	5	110	3.1	69	146
polyesters								
30% glass fibers (PBT)	1.51	121	6.9	4	200	8.7	96	206
30% glass fibers (PET)	1.56	158	8.7	3	234	9.1	107	224

[a] 6.4-mm thickness, except where otherwise stated.
[b] To convert MPa to psi, multiply by 145.
[c] To convert GPa to psi, multiply by 145,000.
[d] To convert J/m to ft·lbf/in., divide by 53.38.
[e] 3.2-mm thickness.

Polyesters can be reinforced with up to 55 wt % of fibrous or nonfibrous fillers or a combination of the two. These products compete with nylon compounds. They offer extremely high flexural modulus, good dimensional stability, and chemical resistance. However, they are sensitive to moisture during processing.

Sulfone polymers have been developed only recently. These expensive products are usually reinforced up to 40 wt % fibers; heat and chemical resistance are excellent. Because these materials are inherently flame retardant, no additives are required to meet UL 94V-0. Poly(phenylene sulfide) is commonly filled up to 65 wt %. Both fibrous and nonfibrous reinforcements as well as various lubricants are used. However, impact resistance and elongation are poor (see POLY(ARYLENE SULFIDE)S).

Reinforcements and Fillers

Commercial RTP compounds are available in several forms: pellets for injection molding or extrusion, unidirectional tape for filament winding and similar applications, and sheet for stamping or compression molding. Reinforcements improve or modify mechanical properties, whereas fillers merely reduce cost. As resins, additives, and compounding technology are improved, this difference becomes less distinct.

Fibrous Reinforcements. Several different types of fibrous reinforcement are commercially available for compounding RTPs. Although each offers certain special properties and characteristics, they all increase tensile strength and modulus, flexural strength and modulus, deflection temperature under load, and warpage. Properties that usually decrease in value include mold shrinkage, elongation, moisture absorption, and, in some cases, Izod impact resistance (see also FIBERS, ENGINEERING). Fibrous reinforcements include glass fibers, carbon fibers, aramid fibers, thermoplastic fibers, and combinations of carbon–glass, aramid–glass, and aramid–carbon–glass.

Nonfibrous Reinforcements. Nonfibrous reinforcements are employed as reinforcements and fillers. They result in increased tensile strength and deflection temperature, but decrease impact resistance. Nonfibrous reinforcements are preferred when a molding with exceptional flatness is required.

Several different materials are commonly used as nonfibrous reinforcements in thermoplastics, including mica, glass beads, and mineral fibers such as wollastonite. Talc, calcium carbonate, and kaolin are considered fillers. Unlike fibrous reinforcements, nonfibrous reinforcements can be processed by different technologies, such as profile extrusion, sheet extrusion, rotational molding, blow molding, and vacuum forming.

Conductive Fillers. Conductive fillers are combined with plastic resins to form compounds capable of conducting electrical current. Conductive materials are used for shielding of electromagnetic interference or radio-frequency interference (EMI/RFI) and for prevention of electrostatic discharge (ESD). The principal difference between these applications is the degree of conductivity required. In addition, conductive materials are occasionally used to actually conduct electronic signals in applications such as computer keyboards and EKG electrodes. The electrical properties required for these applications vary widely.

Some conductive fillers have other advantages. For instance, carbon fibers, which are valued as reinforcements, are also conductive. Aluminum flakes may be used for either electrical or thermal conductivity.

Conductive fillers tend to be more expensive than other reinforcements or other conductive materials such as metals. However, the economics of plastics production apply to conductive materials as well as to other plastics. Consequently, plastic articles made of conductive materials can often compete with metal articles or plastic articles with a conductive coating despite high costs (see CONDUCTIVE PLASTICS). Conductive fillers include carbon fibers, nickel-coated carbon fibers, stainless steel fibers, nickel-coated mica, carbon powders, and aluminum flakes.

Manufacturing and Processing

Compound Manufacture. Commercial RTP compounds are available as pellets for injection molding, extrusion, and similar applications and as sheet for thermoforming or stamping.

Thermoset prepreg technology has been applied to the production of RTP molding compounds. The pellets produced by this provide a high "fiber wet-out" in the resin. Both this and the earlier pultrusion method produce compounds with superior mechanical properties, particularly impact resistance, compared to those of chopped fiber products.

Economic Aspects, Applications, and Outlook

Government regulatory agencies, economic pressures, technological advances in related fields, and changing consumer requirements are some of the influences that have shaped the RTP market. Table 2 shows the U.S. sales of RTP compounds by resin type for 1985 and 1986.

Electrical and Electronics. Perhaps the most dynamic of all RTP markets is the electrical–electronics market (Table 3).

Automotive. In 1986, the U.S. automotive OEM (original equipment manufacturers) market consumed approximately

Table 2. U.S. Thermoplastic Resin and RTP Compound Sales,[a] 10^3 Metric Tons

	Thermoplastic resins			RTPs		
Material	1985	1986	% Growth	1985	1986	% Growth
nylon	183	193	5.5	61	65	5.2
polycarbonate	144	157	8.8	18	20	10
polyesters[b]	50	56	12.6	47	52	9.6
polypropylene	601	630	4.8	60	63	4.5
styrenic resins	898	982	9.3	21	22	2.1
others[c]	373	393	5.2	14	15	6.7
totals	*2249*	*2410*	*7.2*	*222*	*235*	*6.2*

[a] Injection-molding grades only.
[b] Engineering grades only.
[c] PPO, POM, PVC, sulfone polymers, etc.

Table 3. RTP Growth Applications by Resin Type in the Electrical–Electronics Market

	Resin						
Application	PA	PC	PET–PBT	PP	PS	PSO	Other
business machines							
keyboard frames		X	X		X		
keys, contactors	X	X	X	X	X		
paper tractors	X	X					X
brackets	X	X	X				
gears	X	X					X
bushings, bearings	X	X					X
housings		X		X	X		X
communications hardware							
telephone bezels		X			X		
fiber terminations	X		X	X		X	
microphone housing	X	X			X		
interconnection devices							
connectors	X	X	X			X	
molded circuit boards						X	

123,000 metric tons of reinforced and specialty thermoplastic compounds. This figure does not include mineral-filled polypropylenes. The automotive market for RTPs can be divided into several categories (Table 4).

Durable Consumer Products. In 1986, the U.S. durable consumer products market, which includes appliances, power tools, furniture, cameras, sporting goods, and toys, consumed approximately 180,000 metric tons of reinforced thermoplastic compounds (excluding mineral-filled polyolefins).

Large appliances consumed about 50% of the compounds used, a large portion of which was represented by mineral-filled olefins. Applications include injection molded or thermoformed cabinet and door liners, tubs, and agitators. Injection-molded cabinets are expected to be available in the early 1990s.

Military Applications. Injection-molded RTPs were first used in military applications such as land mine housings. These products, generally glass-reinforced styrenic compounds, are still produced in quantity around the world. Most of the future growth in this market is expected to be in high performance products. Electronic housings, shoulder- and hand-weapon stocks, and personal protective gear, such as flak jackets, are in production or under development.

BIBLIOGRAPHY

The Performance Polymer People, Wilson-Fiberfil International Evansville, Ind.

L. H. Miner, *Modern Plastics Encyclopedia 1986,* Vol. 63, No. 10A, McGraw-Hill, Inc., New York, pp. 132–142.

J. G. Poesch, *Modern Plastics Encyclopedia 1985,* Vol. 62, No. 10A, McGraw-Hill, Inc., New York, pp. 126–127.

Mod. Plast. **64,** 41 (1987).

Steve Gerbig
Akzo Engineering Plastics

Table 4. RTP Growth Applications by Resin Type in the Automotive Market

	Resin						
Application	PA	PC	PET–PBT	PP	PS	PSO	Other
under the hood							
radiator tanks	X		X				
fan shrouds			X	X			
fans	X		X	X			
fuel injectors	X		X			X	
carburetor bodies	X		X				
brake reservoirs	X			X			
valve covers	X		X			X	
exterior body parts							
mirror housings	X		X				
license holders	X						
bumper supports			X				
bumpers			X	X			X
light bezels	X	X	X	X		X	
interior parts							
instrument panels		X		X	X		
radio bezels		X		X	X		
temperature control panel		X		X	X		
inner door panels				X			

Table 1. Fibers and Fillers Used with Thermoset Plastics

Thermosets	Alumina	Calcium carbonate	Carbon black	Clay	Cotton flock	Glass bubbles	Glass fibers	Graphite	Mica	Quartz	Talc	Wollastonite silicate	Wood flour
alkyds		•		•			•		•		•	•	•
diallyl *o*-phthalate		•		•			•				•		
epoxy	•	•	•	•		•	•			•		•	
phenolic	•	•	•	•	•	•	•	•	•		•	•	•
polyester		•		•		•	•					•	•
melamine					•		•		•		•	•	
urea				•	•								•
silicone	•			•			•			•			
urethane	•	•	•	•		•	•			•		•	

Thermosets

Reinforced thermoset plastics (RTS) are plastic resins such as polyester, phenolic, silicone, and epoxy reinforced by fibrous glass, graphite, carbon, or aramid, as well as paper, cotton, flakes, beads, or powders. Most RTS are fibrous, glass-reinforced polyesters. The reinforcement may be in the form of chopped fibers, porous mat, woven fabric, or continuous fibers (see also COMPOSITES; REINFORCEMENT).

The combination of plastics and reinforcements results in versatile materials with unusual characteristics. The reinforcement adds strength and toughness to inherent weather resistance, moldability, and colorability. Thus RTS are used because of increased tensile, flexural, torsional, and impact strengths; increased modulus of elasticity; increased creep resistance; reduced coefficient of thermal expansion; increased thermal conductivity; and, in many cases, lower costs.

A variety of choices is available as to the kind, form, amount of reinforcement and plastics to use (Tables 1 and 2), and the fabrication process. Technical knowledge and experience, as well as adequate equipment and financial backing, are required to develop the best end product (see also COMPOSITES, FABRICATION).

Resins

Thermoset plastics pass once through a soft plastic state and harden irreversibly. Because the first observed irreversible reactions required heat, such plastics were called thermosetting, although the reaction may occur at room temperature. Two

Table 2. Properties of Glass Fiber-reinforced Thermoset (RTS) Polyesters

Polyester types	Characteristics	Applications
general purpose	rigid moldings	trays, boats, tanks, boxes, luggage, seating
flexible resins and semirigid resins	tough, good impact resistance, high flexural strength, low flexural modulus	vibration damping: machine covers and guards, safety helmets, electronic part encapsulation, gel coats, patching compounds, auto bodies, boats
light-stable and weather-resistant	resistant to weather and ultraviolet degradation	structural panels, sky-lighting, glazing
chemical-resistant	highest chemical resistance of polyester group, excellent acid resistance, fair in alkalies	corrosion-resistant pipe, tanks, ducts, fume stacks
flame-resistant	self-extinguishing, rigid	building panels (interior), electrical components, fuel tanks
high heat distortion	service up to 260°C, rigid	aircraft parts
hot strength	fast rate of cure (hot), moldings easily removed from die	containers, trays, housings
low exotherm	void-free thick laminates, low heat generated during cure	encapsulating electronic components, electrical parts, switchgear
extended pot life	void-free and uniform, long flow time in mold before gel	large complex moldings
air dry	cures tack-free at room temperature	pools, boats, tanks
thixotropic	resists flow or drainage when applied to vertical surfaces	boats, pools, tank linings

Table 3. Properties and Manufacturing Processes of Reinforced Thermoset Plastics

Thermosets	Properties	Processes
polyesters	simplest, most versatile, economical, and most widely used family of resins; good electrical properties, good chemical resistance, especially to acids	compression molding, filament winding, hand lay-up, mat molding, pressure bag molding, continuous pultrusion, injection molding, spray-up, centrifugal casting, cold molding, encapsulation
epoxies	excellent mechanical properties, dimensional stability, chemical resistance (especially to alkalies), low water absorption, self-extinguishing (when halogenated), low shrinkage, good abrasion resistance, excellent adhesion properties	compression molding, filament winding, hand lay-up, continuous pultrusion, encapsulation, centrifugal casting
phenolic resins	good acid resistance, good electrical properties (except arc resistance), high heat resistance	compression molding, continuous lamination
silicones	highest heat resistance, low water absorption, excellent dielectric properties, high arc resistance	compression molding, injection molding, encapsulation
melamines	good heat resistance, high impact strength	compression molding
diallyl *o*-phthalate	good electrical insulation, low water absorption	compression molding

similar mechanisms account for the reaction cross-linking and interlinking. Properties and manufacturing processes of reinforced thermoset plastics are given in Table 3.

Properties

By far the biggest market share is in thermoplastics. Most thermosets can be characterized as engineering plastics, competing with engineering thermoplastics. Thermoset consumption, however, is much lower than that of thermoplastics, which are easier to process and which fulfill most performance requirements at low processing costs.

Certain phenolic materials and easy flowing, granular thermoset materials based on urea, melamine, polyester, and epoxide resins provide the basis of numerous technical applications because of their high thermal and chemical resistance, stiffness, surface, hardness, dimensional stability, and low flammability.

An increase in thermoset consumption is expected in sheet molding compounds (SMC) and in bulk-molding compounds (BMC), which are known in Europe as dough-molding compounds (DMC). These basic glass fiber-reinforced polyester materials are used to meet a broad range of properties.

Reinforcements

In general, plastics exhibit good strength properties, but low stiffness as measured by elastic moduli. In structural components, they are frequently combined with other material, often as the matrix or binder constituents of composites. The reinforcements usually are fibrous. Fillers provide strength and lower costs and improve performance. The principal reinforcements are glass fibers (95% of consumption). Higher performing fibers are carbon, graphite, aramid, and boron (see FIBERS, ENGINEERING). Whiskers have a place in high performances RTS.

Construction. High performance glass fibers are used in different constructions, such as continuous strands (rovings, yarns), mats, chopped strands, milled fibers, and fabrics, including tapes and woven rovings.

High Performance Fibers. Composites for aircraft and spacecraft applications call for higher elastic moduli and strengths, frequently at lower weights than can be provided by glass or other common reinforcements. For such applications, high performance fibers are employed, usually at much higher costs in materials and fabrication.

High performance fibers include carbon (amorphous), graphite (crystalline), aramid (organic), boron, beryllium, silica, and silicon carbide, as well as hard-drawn carbon steel, stainless steel, and chromium. High strengths can be achieved by metals, but at an increase in weight. Boron, graphite, and aramid are used most often.

Manufacturing Processes

The selection of the manufacturing process (Table 3) is influenced by the quantity, size, thickness, type of RTS material, design, and cost. Practically all plastic processing techniques can be used for RTS.

Thermosets react in a certain way to heat, pressure, and time cycles. Thermoset plastics are engineering plastics with characteristics between those of noniron metals and ceramics. Molding processes are more complicated than those for thermoplastic materials because of the chemical reactions generating the material during processing. Moldability, heat balance, temperature in the RTS, pressure in the mold, and the state and direction of the molding process must be considered. During the heating of the material, two opposing processes take place: viscosity reduction, by raising the temperature of the melt; and viscosity increase by cross-linking that results in hardening.

BIBLIOGRAPHY

G. Lubin, *Handbook of Fiberglass and Advanced Composites,* Van Nostrand Reinhold Co., New York, 1982.

D. V. Rosato and D. V. Rosato, *Injection Molding Handbook,* Van Nostrand Reinhold Co., Inc., New York, 1986.

D. V. Rosato and D. V. Rosato, *Blow Molding Handbook,* Hanser, 1989.

D. V. Rosato and D. V. Rosato, *Plastics Processing Data Handbook,* Van Nostrand Reinhold Co., Inc., New York, 1989.

D. V. ROSATO
Polysor/Nova Inc. and Rhode Island School of Design

REINFORCEMENT

Reinforcements are discrete inclusions used to improve the structural characteristics of a material. They can be continuous in form (fiber, filament, or tape) or discontinuous in form (whisker, flake, or particulate). The host material, often called the matrix, acts as a binder. Reinforcements permanently combined with a material on a macroscopic scale are referred to as composites. The three basic types of material currently being manufactured into composites are polymers, metal, and ceramics. These materials can be used as reinforcements or matrices.

Reinforcements reduce costs; improve formability and electrical properties; increase the ratios of strength to density and stiffness to density; increase resistance to corrosion, fatigue, creep, and stress rupture; and reduce the coefficient of thermal expansion. Reinforcements allow the material to be tailored to the design, instead of the design being tailored to the material. This often results in better designs that are less complicated to build.

Fillers, a type of nonstructural reinforcement, change the properties of the host material. They serve many purposes, ranging from inexpensive extenders to supports that increase overall strength. The functions of reinforcing fillers are shown below.

Polymer matrix

Extenders: Clay, sand

Rheology control: Mica, asbestos, silica gel

Color: Titanium dioxide, carbon, pigments

Flame and heat resistance: Minerals

Heat distortion resistance: Particulate minerals, beads

Toughness: Fibers, carbon black, disperse rubber phase

Metallic–Inorganic matrix

Hardness, wear resistance: Metal, interstitial powders

Creep resistance: Oxide dispersions

Although reinforcements vary greatly in material type and form, this article focuses on structural composites made with continuous fiber reinforcements. The attractive characteristics of a fibrous composite are usually due to the behavior of its fibers. Although a small whisker or fiber has extremely high in-plane tensile properties, these properties are quickly lost with increasing fiber size. To avoid this problem, the fibers are bound together by a matrix that transfers the load through shear forces. Generally, good interfacial bonding is required for this to work.

Fiber-reinforced Composites

An advanced composite typically refers to a polymeric matrix reinforced with high strength, high modulus glass, boron, carbon, or aramid fibers. These fibers are usually arranged in layers to form an extremely rigid, strong structure. The specific properties required by the structural engineering of advanced composites include strength, stiffness, low density, and useful temperature range. Other important factors are form, fiber diameter and surface texture, and cost. Diameter affects the flexibility and the strain-to-failure, whereas surface texture affects the fiber–matrix interface and determines the ability of the matrix to transfer load between the fibers. The form of the reinforcement structure (for example, fiber or fabric) influences the fabrication of the article.

Metallic composites are based on the characteristics of metals shown in Table 1. Metal can be used both as a fiber and as a matrix material. Fiber and matrix are more compatible in a metallic composite, and they can also operate at higher temperatures than polymeric composites. In some cases, however, galvanic corrosion is a problem limiting the use of some metals.

Plain ceramics are highly temperature resistant but too brittle for most structural applications. The use of reinforcements provides the necessary increased toughness, by acting as barriers to crack propagation, while retaining the desirable characteristics of the ceramic matrix. Ceramic fiber-reinforced

Table 1. Characteristics of Metals Useful in Composites

Metal or alloy	Characteristics
carbon and alloy steels	high strength, high modulus, low cost, ferromagnetic properties
stainless steels	good corrosion resistance
nickel and cobalt base superalloys	strength at elevated temperatures, oxidation resistance
refractory metals and alloys (titanium, tantalum, niobium)	high melting points, good corrosion resistance
copper and copper alloys	high electrical and thermal conductivity
aluminum and aluminum alloys	low density, high electrical and thermal conductivity, low cost, anodization produces extensive color range
precious metals (platinum, silver, gold)	high electrical and thermal conductivity, resistance

metals and metal-impregnated ceramics are used for high temperature, high strength materials capable of withstanding thermal shock and mechanical impact. Most ceramic fibers are oxidation-resistant. They do, however, have creep and grain-growth problems at high temperature and are brittle. Research is now being conducted on single-crystal whisker materials such as silicon carbide, silicon nitride, and aluminum oxide. Other research areas include continuous polycrystalline filaments of alumina, zirconia, and other oxides. The glassy aluminosilicate and silica fibers have been the most extensively developed.

Manufacturing Designs

The use of fibers as reinforcement can result in composites of many forms containing aligned fibers, fabrics, and random mats (Fig. 1). The most common form for structural application, however, are aligned fibers.

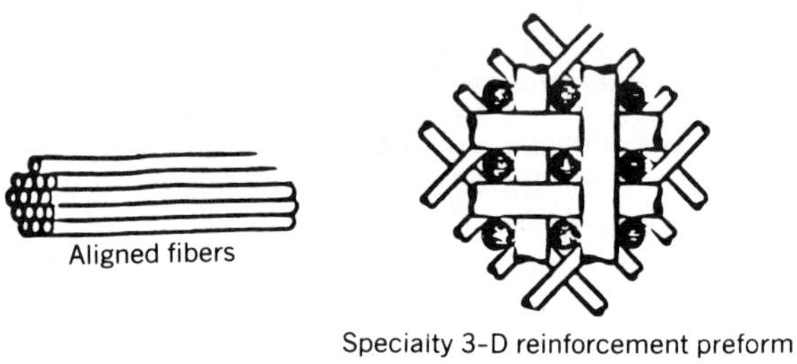

Figure 1. Fiber reinforcement forms.

Economic Aspects

Advanced composites have developed into a multibillion-dollar industry. The combined sales of Kevlar 49, glass, and carbon fibers have grown from \$46 $\times 10^6$ in 1973 to \$295 $\times 10^6$ in 1985.

BIBLIOGRAPHY

H. S. Katz and J. V. Milewski, eds., *Handbook of Fillers and Reinforcements for Plastics,* Van Nostrand Reinhold Co., Inc., New York, 1978, pp. 178–179, 474, 604.

M. M. Schwartz, *Composite Materials Handbook,* McGraw-Hill Inc., New York, 1984.

J. Delmonte, *Technology of Carbon and Graphite Fiber Composites,* Van Nostrand Reinhold Co., Inc., New York, 1981.

G. Lubin, ed., *Handbook of Composites,* Van Nostrand Reinhold Co., Inc., New York, 1982.

JOCELYN M. PATTERSON
STEPHEN TSAI
USAF Materials Laboratory

RELEASE AGENTS

Release agents are substances that control or eliminate the adhesion between two surfaces. They are known by a variety of terms indicative of their function, including abherents, antiblocking agents, antistick agents, external or surface lubricants, parting agents, and slip aids.

Manufacturers and processors use release agents in a wide variety of general handling operations, such as calendering, casting, embossing, extrusion, forming, laminating, machining, molding, packaging and labeling, transference of materials, and protection of processing equipment. Some specific, important polymer-processing applications include pressure-sensitive tape or label release, rubber processing, application of hot stamping foils, and transfer printing of fabrics (see PRESSURE SENSITIVE ADHESIVES).

Polymer scientists and engineers are familiar with release agents not only because manufacturers of polymeric materials and products frequently confront release problems, but also because an efficient, profitable solution to a tacky problem is often a polymer. Silicones, fluoropolymers, and polyolefins are common examples of releasing polymers. Polymers most frequently needing release agents include polyurethanes, polyesters, polyolefins, polycarbonates, and epoxy resins. The importance of these polymers is well illustrated in the extensive bibliography of mold-release agents for rubbers and plastics from the Rubber and Plastics Research Association. Some polymers, particularly silicones, are also used as a self-releasing mold-making material.

Product Types

Release agents come in a wide variety of forms and modes of application. Both neat liquids and solvent solutions, as well as solids such as powders and flakes, compounded mixtures and pastes, emulsions, dispersions, preformed films, *in situ* film formers, and integral migratory additives are available. Many formulated products serve more than one processing function and contain other additives, such as antioxidants. Combined lubricant-stabilizer packages are often used with poly(vinyl chloride). The external treatment types can be applied by any of the standard coating methods, including brushing, dipping, dusting, spraying, electrostatic coating, and plasma arc coating.

Properties

Despite the diversity of products, release agents have some features in common, notably inertness to at least one of the surfaces in question at the temperature of the release process and low surface tension. This latter property is important in obtaining good wetting of a mold. It is also a reflection of the low intermolecular forces desirable in effective release compositions. Not surprisingly, these properties of inertness and weak cohesion are the opposite of those of good adhesives.

The release agent should be chosen with subsequent events in mind, as well as for ease of release and stability. In addition to ease of release, other selection criteria include prevention of buildup; mold cleanability; compatibility with secondary operations such as painting, plating, and ultrasonic welding; mold compatibility; type and time available for application; health and safety requirements; and cost.

Classification of Release Agents

Because of their physical and chemical diversity and the many industries in which they are used, release agents can be classified in several ways. The first classification is by use, which might include subdivisions such as casting, molding, transfer, rubber processing, and packaging. Other classifications are by physical form, with types such as fluids, waxes, greases, emulsions, dry films, and solid powders; by chemical type (Table 1); and by their mode of application and reactivity.

Table 1. Chemical Classification of Release Agents

Chemical class	Chemical subdivision	Specific examples
long-chain alkyl derivatives	fatty ester synthetic waxes	diethylene glycol monostearate, hydrogenated castor oil
	fatty acids	stearic acid, oleic acid
	fatty acid metal salts	zinc stearate, calcium stearate
	fatty amides	ethylenebis(stearamide), oleyl palmitamide
	fatty amines and alcohols	
natural products	petroleum waxes	microcrystalline wax, paraffin wax
	vegetable waxes	carnauba wax
	animal waxes	spermaceti wax
	cellulose derivatives	cellophane, cellulose acetate
	polysaccharides	sodium alginate
	other	flour
synthetic polymers	silicones	polydimethylsiloxane, polyalkylmethylsiloxane
	fluorocarbons	polytetrafluoroethylene
	other fluoropolymers	poly(fluoroacrylates), poly(fluoroethers)
	polyolefins	polyethylene, polypropylene
	other	poly(vinyl alcohol)
fluorinated compounds	fluorinated fatty acids and alcohols	perfluorolauric acid
inorganic materials	silicates	talc
	clays	kaolin, mica
	other	silica, graphite

Mechanism of Release

Release agents should exhibit several of the following features: (*1*) a barrier to mechanical interlocking, (*2*) prevention of interdiffusion, (*3*) poor adsorption and lack of reaction with at least one material at the interface, (*4*) low surface tension and thermodynamic work of separation, (*5*) limited or no electrostatic attraction, (*6*) nonsetting, and (*7*) provision of a weak boundary layer, all the inversion of good adhesive characteristics.

Finely divided solids such as talc provide excellent barriers to mechanical interlocking and interdiffusion. Because compatibility of different polymers is the exception rather than the rule, preformed sheets of adhesive materials usually prevent interdiffusion. Poor adsorption and chemical inertness are associated with low intermolecular forces, which directly results in low surface tension and thermodynamic work of separation.

Health and Safety Factors

Most general purpose release agents have been developed for this market by virtue of their low toxicity and chemical inertness and do not generally present health and safety problems. Some of the solvent dispersions require appropriate care in handling volatile solvents and some of the solids, particularly finely divided hydrophobic solids, can present inhalation problems. The reactive type of release coating with monomers, prepolymers, and catalysts often presents specific handling difficulties. The potential user is advised first to consult the manufacturer's safety data sheets. The principal health and safety concerns involve contact with foodstuffs and drugs.

BIBLIOGRAPHY

S. H. Collins, *Plast. Compounding* **6,** 38 (1983).

"Mold Release Agents for Rubbers and Plastics" in *Citations from the Rubber and Plastics Research Association Data Base 1973–1983,* NTIS Report No. PB84-854850, Springfield, Va., Dec. 1983.

S. L. Clark in G. Lubin, ed., *Handbook of Composites,* Van Nostrand Reinhold Co., Inc., New York, 1982.

R. D. Cowell and B. R. Bluestein in M. Grayson, ed., *Kirk-Othmer Encyclopedia of Chemical Technology,* 3rd ed., Vol. 1, Wiley-Interscience, New York, 1978, pp. 1–9.

MICHAEL J. OWEN
Dow Corning Corporation

REPTATION.

See MACROMOLECULAR DYNAMICS.

RESINOGRAPHY

According to the American Society for Testing and Materials (ASTM), resinography is the science of morphology, structure, and related characteristics as correlated with composition or condition and properties or behavior of resins, polymers, plastics, and their products. Hence resinography, like metallography, belongs to the field of materials science. It applies to the hierarchical structure of polymeric materials on four levels of organization: (*1*) macromolecular domains, (*2*) physical phases, (*3*) surfaces or interfaces, and (*4*) unit materials. In production, quality control, troubleshooting, competition, forensics, research and development, resinography requires a basic understanding and practical applications of chemistry, physics, optics, and electronic theory.

Macromolecular Domains

The structure–property relationships of polymers are based on the chemical structure and physical morphology of the macromolecules and their fundamental domains. Polymeric domains have been studied since the 1940s by electron microscopy.

Polymers possess the characteristic molecular structure of long, entangled chains. As visualized by the modern electron microscope, the fibrous macromolecular domains have been termed fibrils. These fibrils exist in various sizes, ranging from those visible with the scanning electron microscope (sem) to those requiring a transmission electron microscope (tem) at high voltage. Electron diffraction is also used. Fibrils and the spaces between them have also been visualized by means of high resolution scanning electron imaging (sei). High resolution electron microscopy has been applied to radiation-sensitive specimens.

Phases

A phase is a visibly separate portion of a material system, even though it may not be physically separated. Light microscopy can be of great value by providing sufficient resolution and contrast between phases; the process is termed intangible separation.

A gel is a structure intermediate between a solid and a liquid. It consists of long macromolecular chains, cross-linked to create a tangled network immersed in a liquid. The liquid prevents the polymer network from collapsing into a compact mass; the network prevents the liquid from spilling out. Gels vary from viscous fluids to rigid solids. Typically they are soft and resilient. A balance of forces maintains this state, and even a slight disturbance can cause the gel to collapse. For example, the plasma protein fibrinogen is converted into fibrin to form a blood clot, which in time breaks down. Both processes can be understood in terms of the structure of the giant protein molecules.

Surfaces

Surfaces and interfaces are the keys to performance quality. The units of organizational level 3 include films, fibers in yarns, layers in laminates, and coatings. Interfaces include foams, liquid emulsions or suspensions, putties formulated from polymers, and composites.

Even surfaces that seem to be simple may affect properties. For example, a nylon film slips easily when folded in such a way that one surface rubs against itself. However, folded the other way, resistance to slipping is strong. Examination by electron microscope showed this surface to be rough because the crystallites of the polymeric spherulites (shown in Fig. 1) were free to grow up into the air as the molten nylon crystallized on the cooling drum. The other side of the foil was smooth because the drum's surface was highly polished to prevent sticking. Although the macromolecular domains and the crystalline characteristics (also shown in Fig. 1) may or may not be different, the important difference here is in the surfaces.

An uneven surface is sometimes desired to scatter the transmitted light, for example, through a red or amber cover of an automobile light or directional signal, where a zigzag pattern is molded on the light cover. Surfaces are also important in producing plastic or rubbery foams or sponges. Interfaces are the key to colloids and the paper industry. A film, no matter how thin, has two surfaces, which can have the same or different functions. For example, a pressure-sensitive adhesive tape may be coated on one or both sides.

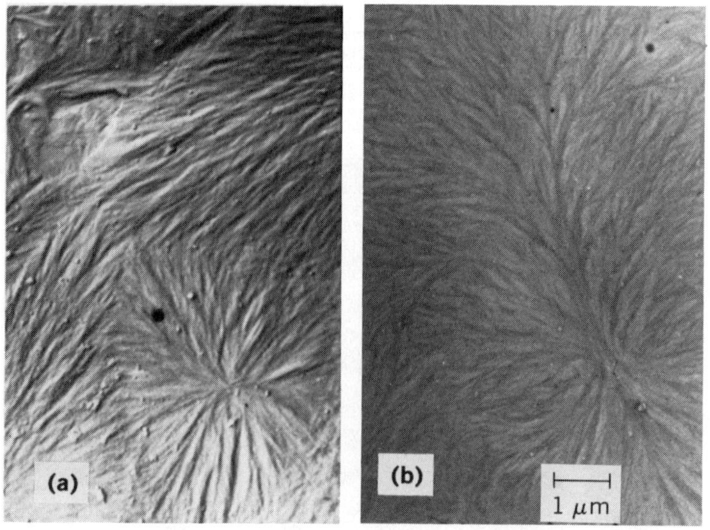

Figure 1. Electron micrograph of nylon film showing difference in texture between the two sides; positive silica replica, no shadowing metal. (**a**) Rough side, cast in air. (**b**) Smooth side, cast against a smooth solid.

Unit Materials

Unit materials may have one, two, or three principal directions. The main unidimensional materials are the staple fibers, continuous filaments, yarns, cords, and ropes. Films, foils, paper, cloth (woven, nonwoven, knitted), carpeting, and laminated sheets are two-dimensional. Among three-dimensional units are moldable pellets, moldings, and assemblies by means of adhesives or fasteners.

BIBLIOGRAPHY

Compilation of ASTM Standard Definitions, ASTM, Philadelphia, Pa., 1986 or later.

T. G. Rochow and E. G. Rochow, *Resinography, an Introduction to the Definition, Identification and Recognition of Resins, Polymers, Plastics, and Fibers,* Plenum Press, New York, 1976.

T. G. Rochow and E. G. Rochow, *An Introduction to Microscopy by Means of Light, Electrons, X-rays, or Ultrasound,* Plenum Press, New York, 1978.

T. G. Rochow, *Light Microscopical Resinography,* Vol. 47, Microscope Publications, Chicago, Ill., 1983.

T. G. Rochow
North Carolina State University at Raleigh

RESINS, NATURAL

The dictionary definition of a natural resin is "any of various solid or semi-solid amorphous, fusible, flammable natural organic substances that are usually transparent or translucent and yellowish to brown, are formed especially in plant secretions, are soluble in organic solvents but not in water, are electrical nonconductors, and are used chiefly in varnishes, printing inks, plastics, and sizes and in medicine." The term natural resin includes tree and plant exudates of recent origin on a geological time scale, fossil resins, mined resins such as asphaltite, and shellac. As articles of commerce, they have been altered from their original state by methods of isolation or processing. Natural resins also include materials that have been chemically modified to increase their industrial utility.

The resins arise from condensation of naturally occurring chemical compounds such as the terpinoids and flavonoids contained in trees. They are complex mixtures that are isolated together because they have similar physical or chemical properties. Although the resins are condensed structures, in most cases they are not intractably cross-linked. The molecular weights of those that flow are low compared with those of synthetic polymers and probably do not exceed 10,000. Being mixtures, the resins do not exhibit simple melting or boiling points, although they do exhibit a predominant glass-transition temperature (T_g), the temperature below which a material exhibits the characteristics of a brittle glass. Above the T_g, the resin may have the properties of a rubber or oil. Depending on its T_g, a natural resin may be like brittle glass at room temperature, or it may be flexible and extensible.

The T_g of natural resins ranges from approximately 0 to 100°C, which is high for low molecular weight condensation polymers. The high T_g is caused by the rigid, cyclic structure of the components. Chemical modification increases the molecular weight and broadens the range of glass-transition temperatures. The resins usually exhibit a sharp drop in viscosity as temperatures are raised above their T_g. This is a consequence of the low molecular weight, which does not allow a rubbery plateau region above the glass-transition temperature. The rapid drop in viscosity is a desirable property for many of the applications.

Rosin and Modified Rosins

Rosin and its derivatives are economically the most important natural resins. Approximately 900×10^3 metric tons of these materials are produced annually and sold throughout the world today. About one-third is produced in the United States, another third in the People's Republic of China, and most of the remainder in Portugal, Mexico, Finland, India, Spain, Greece, and Brazil. Rosin is isolated from pine trees, principally from longleaf *Pinus palustris, Pinus ellioti,* and loblolly pine *Pinus taeda*. Rosin and associated products obtained from pine trees have been used for hundreds of years to caulk the bottoms of wooden sailing vessels and to lubricate the lines. These materials are known as naval stores because of this practice.

Production. Rosins are known as gum, wood, or tall oil rosin, based on the method of isolation and the source. In the gum rosin process, pine trees are wounded to stimulate the flow of gum. V-shaped slashes are cut through the bark and the exudate is collected in a bucket below the lowest slash. In the wood rosin process, rosin is isolated from aged pine stumps that have been left in fields cleared for farming or lumbering operations. Tall oil rosin is a by-product of paper manufacturing.

The three types of rosin perform differently because of different distributions of resin acids, as well as different types and quantities of impurities. These differences are reflected in end use performance.

Properties. Rosin is a brittle, friable solid with a T_g of ca 30°C and a ring-and-ball softening point of ca 75°C. Although the average molecular weight is only about 300, its rigid molecular structure results in a higher T_g than would be expected for an amorphous polymer of the same molecular weight. Heating above softening temperature lowers viscosity sharply as a consequence of the low molecular weight.

Composition. Rosin is primarily a complex mixture of monocarboxylic acids of alkylated hydrophenanthrene nuclei. These constituents, known as resin acids, represent about 90% of rosin.

Modified Rosins. Natural rosins are often modified to improve stability to long-term aging and oxidative degradation. Pale color and color stability are also important. Stabilization methods involve the double bonds in the resin acid molecule that can form a conjugated system highly susceptible to degradation. The double bonds can be stabilized by hydrogenation, disproportionation, or polymerization.

Derivatives. Most rosin products are sold as derivatives, which extends their utility. The carboxyl group or the double bond is usually involved in the reactions. The most important derivatives of the carboxyl group are formed by esterification with monohydric or polyhydric alcohols. Typical alcohols used include methyl alcohol, ethylene glycol, triethylene glycol, glycerol, and pentaerythritol. These rosin esters have a wide range of softening points and compatibilities.

Uses. The largest single commercial use of rosin is in sizing paper to improve its water resistance. Paper size consists of rosin or modified rosins that have been partially saponified with sodium carbonate or sodium hydroxide.

Modified rosins, metal resinates, and resin acid esters find wide application in printing ink formulations. The resins provide pigment wetting, resistance to penetration, adhesion to substrates, hardness, and gloss.

Rosin ester resins are used extensively in pressure-sensitive adhesives as tackifiers. The adhesive is formulated by blending the resin with the polymer in solution or as aqueous emulsions. Rosin, modified rosins, and derivatives are also used in hot-melt adhesives, which function by wetting the substrates as a low viscosity melt and forming a bond when the viscosity increases on cooling. The adhesives are based primarily on ethylene–vinyl acetate copolymers.

Rosin ester resins are used as modifiers in the formulation of chewing gum. The soap of modified rosin has a long history as an emulsifier for the polymerization of styrene–butadiene rubber.

Traditional Natural Resins

There is no simple classification system for these complex natural resins. They can be separated by age into fossil and recent resins. Fossil resins are those that can be found buried underground. Recent resins are those collected as exudates from trees.

Commercial Resins. Commercial resins include manila copal, pontianak, dammar resins, East India resins, gum elemi, sandarac, and mastic.

Less Important Resins. Less important resins include kauri, Congo, accroides, and amber.

Natural Resins in Medicines, Flavors, and Fragrances

Some natural resins have been used for many centuries in medical applications, flavors, and fragrances, possibly before recorded history. With recent advances in pharmacology, the medicinal applications of these resins have mostly disappeared. The only remaining application of importance is the use of guaiac-impregnated paper to detect hemoglobin in stool. This widely used test is helpful in the early diagnosis of colorectal cancer. Resins have also been used externally for topical protective application in skin ulcer therapy and internally as an expectorant.

Mined Resins

Asphalitites and Utah coal resin require mining operations. Both resins have found applications in coatings and inks.

Asphaltites. Asphaltites, of which Gilsonite is the most widely known, were formed in the same manner as oil and coal below the surface of the earth from vegetation that had been subjected to high temperatures and pressures on a geological time scale (see also ASPHALTIC MATERIALS).

Utah Coal Resin. This resin was laid down with coal in veins in the Salina and Huntington region of Utah.

Shellac

Unlike other natural resins, shellac does not originate from plants, but is produced from the secretion of an insect. It is a purified form of lac and is the most widely known lac product. Therefore, the designation shellac has been accepted as the common generic term. There are estimates that as many as 1.5 million insects are required to manufacture enough lac to produce 450 g of shellac.

Cultivation. Lac is produced by the larvae of *Kerria lacca (Laccifer lacca),* an insect of the order Hemiptera. The small insects, 0.5–0.6 mm in length, attach themselves to twigs of the host tree and suck juices from the tender bark. The larva then secretes a coating composed of lac over its body to protect it from weather and enemies. The secretions of neighboring larvae coalesce into one continuous coating over the twig. This coating, known as sticklac, is the raw material from which other grades of lac are refined. It is harvested by cutting the twigs from the trees or by twisting the branches so that the resin falls off.

Production. Raw lac, or sticklac, is collected by scraping or rolling the lac-coated cut twigs. The lac is then sifted in basket trays and washed with water or dilute soda to remove trash. It is kneaded in the water to remove water-soluble carbohydrates and the dye that gives lac its red color. Also removed, by skimming, are the woody material and insect bodies that form a scum on top of the water. At this stage, the dried lac, known as seed lac, is processed further into shellac or converted into bleached lac.

Composition. Shellac is primarily a mixture of aliphatic polyhydroxy acids in the form of lactones and esters.

Uses. Shellac still has significant sales to the protective coatings industry, even though synthetic resins have taken a large share of the market. Shellac is also used as a protective coating for pharmaceuticals, candy, citrus fruits, and apples. Shellac is used in electrical applications. It is also used as a stiffener for felt hat bodies, primarily for recreational hats. It is also used to stiffen playing cards, providing "snap."

BIBLIOGRAPHY

H. J. Enos, Jr., G. C. Harris, and G. W. Hedrich in A. Standen, ed., *Kirk-Othmer Encyclopedia of Chemical Technology,* 2nd ed., Vol. 17, John Wiley & Sons, Inc., New York, 1968, pp. 475–508.

C. L. Mantell, C. W. Kopf, J. L. Curtis, and E. M. Rogers, *The Technology of Natural Resins,* John Wiley & Sons, Inc., New York, 1942.

R. R. Meyers and J. S. Long, eds., *Treatise on Coatings, Vol. 1; Film-Forming Composition,* Part II, Marcel Dekker, Inc., New York, 1968.

J. J. Mattiello, ed., *Protective and Decorative Coatings, Vol. 1: Raw Materials for Varnishes and Vehicles,* John Wiley & Sons, Inc., New York, 1941.

JAY B. CLASS
Hercules Research Center

RETARDATION.

See RADICAL POLYMERIZATION.

RETICULATED FOAMS.

See CELLULAR MATERIALS.

REVERSIBLE POLYMERIZATION.

See DEPOLYMERIZATION; POLYMERIZATION, THERMODYNAMICS.

RHEOLOGICAL MEASUREMENTS

Rheology is the science of the deformation and flow of matter. It is concerned with the response of materials to mechanical force. That response may be irreversible flow, reversible elastic deformation, or a combination of both. An understanding of rheology and the ability to measure rheological properties is necessary before viscous behavior can be controlled. Such control is essential for the manufacture and handling of numerous polymeric materials and products, eg, rubber, plastics, paints, inks, and others. For example, in the coatings industry, rheology is crucial from the time that materials enter the paint plant to the time that the finished paint is applied.

Deformation is the relative displacement of points of a body. It can be divided into two types: flow and elasticity. Flow is irreversible deformation: when the stress is removed, the material does not revert to its original configuration. This means that work is converted to heat. Elasticity is reversible deformation; the deformed body recovers its original shape, and the applied work is largely recoverable. Viscoelastic materials show both flow and elasticity. A good example is Silly Putty, which bounces like a rubber ball when dropped, but slowly flows when allowed to stand. Viscoelastic materials provide special challenges in terms of modeling behavior and devising measurement techniques.

This article is concerned with rheological measurements and the principles on which they are based. The flow properties of a liquid are defined by its resistance to flow, ie, viscosity, and may be measured by determining the rate of flow through a capillary, the resistance to flow when the fluid is sheared between two surfaces, or the rate of motion of a bubble or ball moving through the fluid. The mechanical properties of an elastic solid may be studied by applying a stress and measuring the deformation or strain. Many solids, such as polymers, undergo flow in addition to recoverable elastic deformation. In addition, a number of liquids show elastic as well as flow behavior. These materials are viscoelastic, and additional techniques beyond those indicated for solids and liquids are needed for complete characterization (see VISCOELASTICITY). Examples of such methods are the measurement of response to sinusoidal oscillatory motion; the measurement of flow with time after application of stress, ie, creep; and the measurement of the rate and degree of recovery after removal of stress, ie, stress relaxation.

In addition to viscometers, microscopes are often used for defining and solving flow problems. This is particularly true for formulated materials, such as paints and inks, where the use of a microscope allows the investigator to see the physical structure of the material and its effect on flow.

Viscosity

A liquid is a material that continues to deform as long as it is subjected to a tensile or a shear stress. The latter is a force applied tangentially to the material. In a liquid, shear stress produces a sliding of one infinitesimal layer over another, resulting in the sort of flow illustrated in Figure 1.

For a liquid under shear the rate of deformation or shear rate is proportional to the shearing stress. The original exposition of this relationship is Newton's law, which states that the ratio of the stress to the shear rate is a constant. That constant is viscosity. Under Newton's law, viscosity is independent of shear rate. This is true for ideal or Newtonian liquids, but the viscosities of many liquids, particularly a number of those of interest to industry, are not independent of shear rate. These non-Newtonian liquids may be classified according to their viscosity behavior as a function of shear rate. Some exhibit shear thinning, whereas others give shear thickening. Some liquids at rest appear to behave like solids until the shear stress exceeds a certain value, called the yield stress, after which they flow readily.

Some commonly observed types of flow behavior are shown in Figure 2, in which the shear stress is plotted vs shear rate. These plots are called flow curves and are frequently used to express the rheological behavior of liquids. Newtonian flow is shown by a straight line, and shear thinning and thickening by curves. Yield stresses are shown by intercepts on the stress axis. It should be pointed out that the existence of yield stresses is controversial; they may be artifacts due to high Newtonian viscosity at very low shear rates. However, in many dispersed systems, particularly where severe flocculation occurs, this viscosity is so high that the material does not flow for years. Therefore, in practice, there are true yield stresses. This parameter can be quite useful in characterizing materials.

Viscosity is equal to the slope of the flow curve ($\eta = d\tau/d\dot{\gamma}$). The quantity $\tau/\dot{\gamma}$ is the viscosity η for a Newtonian liquid and the apparent viscosity η_a for a non-Newtonian liquid. The kinematic viscosity is the viscosity coefficient divided by the density, $\nu = \eta/\rho$. The fluidity is the reciprocal of the viscosity, $\theta = 1/\eta$. The SI units for viscosity are pascal seconds (Pa·s and mPa·s). They are replacing poise (P, dyn·s/cm^2) and centipoise (cP) (1 mPa·s = 1 cP). The shear stress units are pascals (1 Pa = 1 dyn/cm^2) and Newtons per square meter (1 N/m^2 = 1 Pa). Shear rate is measured per second (s^{-1}) in both systems. The SI units for kinematic viscosity are square centimeters per second (cm^2/s) (Stokes (St)) and square millimeters per second (mm^2/s) (centistokes (cSt)).

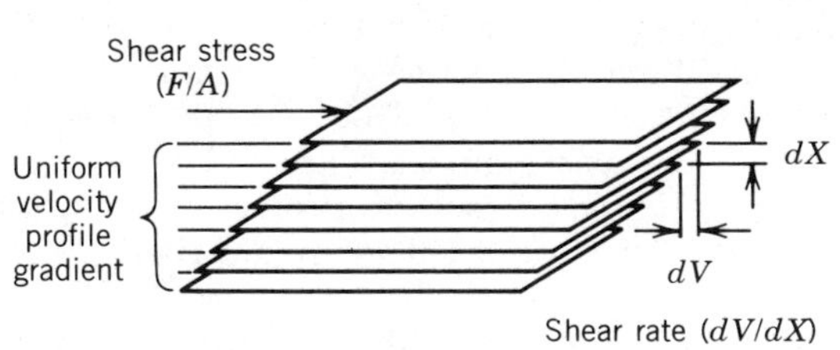

Figure 1. Laminar flow in sample shear. $F/A = \eta\ dV/dX$; η = coefficient of viscosity or Newtonian viscosity.

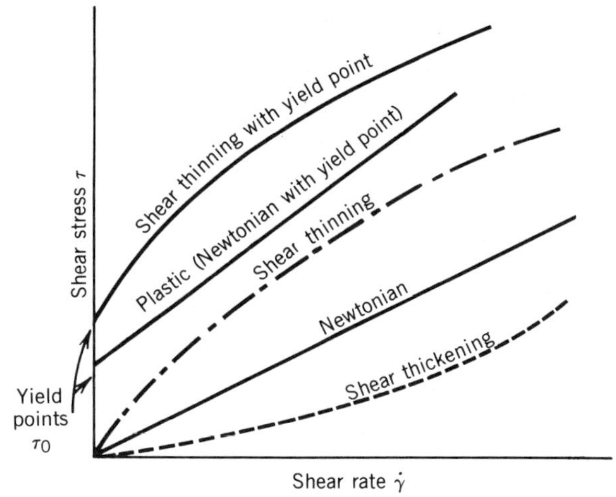

Figure 2. Flow curves (shear stress vs shear rate) for different types of flow behavior.

Many flow models have been proposed and are useful for the treatment of experimental data or for describing flow behavior (Table 1). Probably, no given model fits the rheological behavior of a material over an extended shear rate range. Nevertheless, these models are useful for summarizing rheological data and are frequently encountered in articles relating to the rheology of liquids.

Thixotropy and Other Time Effects. In addition to the nonideal behavior described, many fluids exhibit time-dependent effects. Some fluids increase in viscosity (rheopexy) or decrease in viscosity (thixotropy) with time when sheared at a constant shear rate. These effects can occur in fluids with or without yield values. Rheopexy is a comparatively rare phenomenon, but thixotropic fluids are fairly common. Examples of thixotropic materials are starch pastes, gelatin, mayonnaise, drilling muds, and latex paints. Viscosity–time measurements during or after shearing can be used to show time-dependent effects. Causes of time-dependent behavior include irreversible changes (cross-linking, coagulation, degradation, and mechanical instability) and reversible changes involving the breaking and reforming of colloidal aggregations and networks.

Time-dependent effects are measured by determining the decay of shear stress as a function of time at one or more constant shear rates. A rotational viscometer, which is connected to a recorder, is used. Another method for estimating

Table 1. Flow Equations for Flow Models

Flow model	Flow equation
Newtonian	$\tau = \eta\dot{\gamma}$
plastic (Bingham) body	$\tau - \tau_0 = \eta\dot{\gamma}$
power law	$\tau = k\lvert\dot{\gamma}\rvert^n$
power law with yield value	$\tau - \tau_0 = k\lvert\dot{\gamma}\rvert^n$
casson fluid	$\tau^{1/2} - \tau_0^{1/2} = \eta_\infty^{1/2}\ \dot{\gamma}^{1/2}$
Williamson	$\eta = \eta_\infty + \dfrac{(\eta_0 - \eta_\infty)}{1 + \dfrac{\lvert\tau\rvert}{\tau_m}}$
cross	$\eta = \eta_\infty + \dfrac{(\eta_0 - \eta_\infty)}{1 + \alpha\dot{\gamma}^n}$

thixotropy involves the hysteresis of the thixotropic loop. The area of the thixotropic loop is calculated or measured, which works well with printing inks.

Effect of Temperature. In addition to often being dependent on parameters such as shear stress, shear rate, and time, viscosity is highly sensitive to changes in temperature. Most materials decrease in viscosity as temperature increases. The dependence is logarithmic and can be substantial (up to 10% change/°C). This has important implications for polymer processing and handling and for viscosity measurement.

A common expression relating viscosity to temperature is the Andrade equation:

$$\eta = A \cdot e^{B/T}$$

or

$$\eta = A \cdot 10^{B/T}$$

where A and B are constants characteristic of the polymer or other material and T is the absolute temperature. Estimation of the viscosity of a polymer at a given temperature requires a knowledge of the viscosity at two other temperatures. This knowledge allows calculation of the constants A and B and subsequent determination of viscosities at other, intermediate temperatures.

Dilute Polymer Solutions. The measurement of dilute solution viscosities of polymers is widely used for polymer characterization. Low concentrations reduce intermolecular interactions and allow measurement of polymer–solvent interactions. These measurements are usually made in capillary viscometers, some of which have provisions for direct dilution of the polymer solution. The key viscosity parameter for polymer characterization is the limiting viscosity number or intrinsic viscosity $[\eta]$. It is calculated by extrapolation of the viscosity number (reduced viscosity) or the logarithmic viscosity number (inherent viscosity) to zero concentration.

Concentrated Polymer Solutions. Knowledge of the viscosity behavior of concentrated solutions is important to the manufacture and application of caulks, adhesives, inks, paints, and varnishes. Viscosity behavior can be investigated by a variety of methods, including the use of simple capillary viscometers, extrusion rheometers, and rotational viscometers. Unlike dilute solutions, concentrated polymer solutions show a great deal of interaction between the macromolecules. The degree of interaction is governed by the concentration, the characteristics of the chains, and the nature of the solvent. With concentrated solutions (and polymer melts), there is a critical molecular weight M_c above which there is chain entanglement. The dependence of viscosity on molecular weight changes from a value of one below M_c to 3.4–3.5 above M_c.

Melt Viscosity. The study of the viscosity of polymer melts is important for the manufacturer who must supply suitable materials and for the fabrication engineer who must select polymers and fabrication methods. Thus melt viscosity as a function of temperature, pressure, rate of flow, and polymer molecular weight and structure is of considerable practical importance. Polymer melts exhibit elastic as well as viscous properties. This is evident in the swell of the polymer melt on emergence from an extrusion die, behavior that results from the recovery of stored elastic energy.

A number of experimental methods have been applied to measure the melt viscosity of polymers, but capillary extrusion techniques probably are generally preferred. Rotational methods are also used, and some permit the measurement of normal stress effects resulting from elasticity as well as of viscosity. Oscillatory shear measurements are useful for measuring the elasticity of polymer melts. Recently, controlled stress methods have been applied.

Dispersed Systems. Many fluids of commercial and biological importance are dispersed systems, such as solids suspended in liquids (dispersions) and liquid–liquid suspensions (emulsions). Examples of the former include inks, paints, pigment slurries, and concrete; examples of the latter include mayonnaise and butter.

Dispersion of a solid or liquid in a liquid affects the viscosity. In many cases Newtonian flow behavior is transformed into non-Newtonian flow behavior. Shear thinning results from the ability of the solid particles or liquid droplets to come together to form network structures when at rest or under low shear. With increasing shear the interlinked structure gradually breaks down, and the resistance to flow decreases. The viscosity of a dispersed system depends on hydrodynamic interactions between particles or droplets and the liquid, particle–particle interactions (bumping), and interparticle attractions that promote the formation of aggregates, flocs, and networks.

Dispersions have been studied in greater detail than emulsions, probably because of the greater complexity of the latter. Emulsions tend to be unstable, and frequently, droplets begin to coalesce soon after the emulsion forms. Thus the emulsion is continually changing. In addition, droplets may change shape under shear or when packed tightly together. This makes it difficult to apply theories developed for solid spheres or other well-defined geometries of shape.

Viscosity–Concentration Relationship for Dilute Dispersions. The viscosities of dilute dispersions have received considerable theoretical and experimental treatment, partly because of the similarity between polymer solutions and small particle dispersions at low concentration. Nondeformable spherical particles are usually assumed in the cases of molecules and particles. The key viscosity quantity for dispersions is the relative viscosity or viscosity ratio η_{rel},

$$\eta_{rel} = \frac{\eta}{\eta_0} = \frac{\text{dispersion viscosity}}{\text{viscosity of liquid}}$$

because the effect of the dispersed solid, rather than the dispersing medium, is usually more significant. However, the latter should not be ignored. Many industrial problems involving unacceptably high viscosities in dispersed systems are solved by substituting solvents of lower viscosity.

Other Factors Affecting the Viscosity of Dispersions. Factors other than concentration affect the viscosity of dispersions. A dispersion of nonspherical particles tends to be more viscous than predicted, if the Brownian motion is great enough to maintain a random orientation of the particles. However, at low temperatures or high solvent viscosities, the Brownian motion is small and the particle alignment in flow results in unexpectedly lower viscosities. This is a form of shear thinning.

Emulsions. Because emulsions are different from dispersions, different viscosity–concentration relationships must be used. In an emulsion the droplets are not rigid, and viscosity can vary over a wide range.

Extensional Viscosity. In addition to the shear viscosity η, two other rheological constants can be defined for fluids. They are the bulk viscosity K and the extensional or elongational viscosity η_e. The bulk viscosity relates the hydrostatic pressure to the rate of deformation of volume, whereas the extensional viscosity relates the tensile stress to the rate of extensional deformation of the fluid. Because of the difficulty in measuring these viscosities, little has been published about them. However, extensional viscosity is important in a number of industrial processes and problems. Shear properties alone are insufficient for the characterization of many fluids, particularly polymer melts.

Extensional flows occur when fluid deformation is the result of a stretching motion. The extensional viscosity is related to the stress required for the stretching. This stress is necessary to increase the normalized distance between two material entities in the same plane when the separation is s and the relative velocity is ds/dt. The deformation rate is the extensional strain rate, which is given by

$$\dot{\epsilon} = \frac{1}{s}\frac{ds}{dt}$$

Unlike shear viscosity, extensional viscosity has no meaning unless the type of deformation is specified. The three types of extensional viscosity identified and measured are uniaxial or simple, biaxial, and pure shear.

Elasticity and Viscoelasticity. Elastic deformation is a function of stress and is expressed in terms of relative displacement or strain. Strain may be expressed in terms of relative change in volume, length, or other measurement, depending on the nature of the stress. An ideal elastic body is a material that deforms reversibly and for which the strain is proportional to the stress (Hooke's law) with immediate recovery to the original volume and shape when the stress is released. The relationship between the stress σ and strain ϵ may be written as

$$\sigma = K\epsilon$$

where K is a proportionality constant called the modulus of elasticity. Polymers and many of their solutions are both viscous and elastic, and both types of deformation must be taken into account to explain their behavior (see ELASTICITY; VISCOELASTICITY).

Mechanical Models. Because the complex rheological behavior of viscoelastic bodies is difficult to visualize, mechanical models are often used. In these models the viscous response to applied stress is assumed to be that of a Newtonian fluid and is represented by a dashpot (a piston operating in a cylinder of Newtonian fluid). The elastic response is idealized as an ideal elastic (Hookean) solid and is represented by a spring. The dashpot represents the dissipation of energy in the form of heat, whereas the spring represents a system that stores energy. Stress–strain diagrams for the two elements are given in Figure 3. With the dashpot the stress is relieved by viscous flow and is independent of strain. The result is a plot with constant value for stress. With the spring there is a direct dependence of stress on strain, and the ratio of the two is the modulus E (or G).

The behavior of materials may be approximated by combinations of springs and dashpots. These may be in series, so that the strains (deformations) are additive, or they may be combined in parallel, in which case the stresses are additive. Usually, the inertial effect of the mass is neglected. It is also

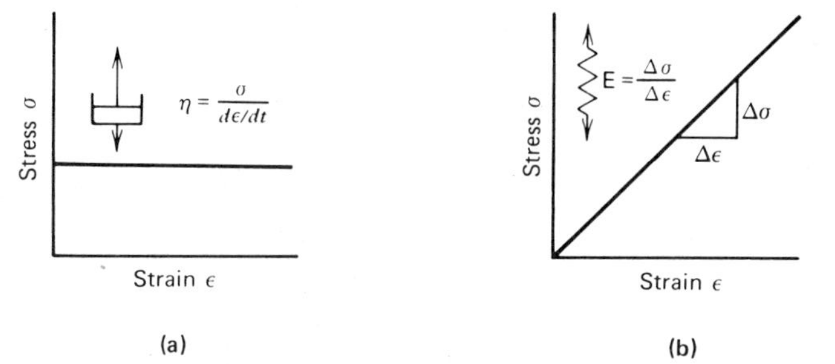

Figure 3. Stress–strain diagrams. (**a**) Dashpot of viscosity η; the stress is independent of strain, but dependent on the speed of testing. (**b**) Spring of modulus E; the slope is the modulus, which is independent of the speed of testing.

the practice to fit experimental data with models that are the sum of many simple viscoelastic elements, each with ideal coefficients independent of stress or strain.

Dynamic Behavior. Knowledge of how mechanical models would behave if they underwent stress–strain and time-dependent measurements is important for the characterization of polymers, particularly in determining their response under conditions of processing or use. For this reason, stress–strain, creep, and stress–relaxation measurements are commonly made to define material properties. The dynamic response of viscoelastic materials to cyclic stresses or strains is also important, partly because cyclic motion occurs in many processing operations and applications, and partly because so much rheological information can be gained from dynamic measurements. By subjecting a specimen to an oscillatory stress and determining the response, both the elastic and viscous or damping characteristics can be obtained. Elastic materials store energy; that is, they convert mechanical work into potential energy, which is recoverable. Liquids do not store energy when stressed, but dissipate it as heat as they flow. This dissipation results in highly damped motion. Viscoelastic materials exhibit both elastic and damping behavior. The latter causes the deformation to be out of phase with the stress applied in the dynamic measurement.

Normal Stress (Weissenberg Effect). Many viscoelastic fluids flow in a direction normal to the direction of shear stress in steady-state shear. Manifestations of the effect include flour dough climbing up a beater, polymer solutions climbing up the inner cylinder in a concentric-cylinder viscometer, and paints forcing apart the cone and plate of a cone–plate viscometer. The normal stress effect has been put to practical use in certain screwless extruders designed in a cone–plate or plate–plate configuration, where the polymer enters at the periphery and exits at the axis.

The two normal stress functions $N_1(\dot{\gamma})$ and $N_2(\dot{\gamma})$ are referred to as the first and second normal stress differences, respectively. The first normal stress difference is a useful quantity, as it often gives a good quantitative measure of viscoelasticity. It can be determined from the normal force, which is measurable with several commercial rotational viscometers. In highly elastic liquids it is common for $N_1(\dot{\gamma})$ to be considerably larger than the shear stress.

Viscometers

To solve a flow problem or characterize a given fluid, an instrument must be carefully selected. Many commercial vis-

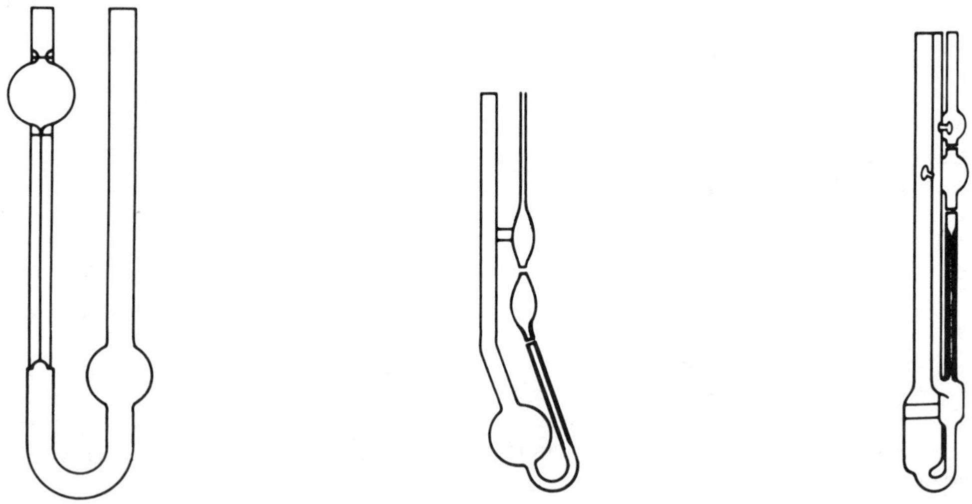

(a) (b) (c)

Figure 4. (**a**) Ostwald glass capillary viscometer; (**b**) Cannon-Fenske viscometer; (**c**) Ubbelohde viscometer.

cometers are available with a variety of geometries for wide viscosity ranges and shear rates. Rarely is it necessary to construct an instrument. However, in choosing a commercial viscometer a number of criteria must be considered. Of great importance is the nature of the material to be tested, its viscosity, its elasticity, the temperature dependence of its viscosity, and other variables. The degree of accuracy and precision required and whether the measurements are for quality control or research must be considered. The viscometer must be matched to the materials and processes of interest; otherwise, the results may be misleading.

Viscometers are available in three types: capillary, rotational, and moving body. Other types are used for special applications. In most cases there is a choice of several different instruments. The choice depends on the particular requirements of the investigator and the price range.

Capillary viscometers include glass-capillary viscometers (Fig. 4), orifice, and piston-cylinder (extrusion) viscometers. Rotational viscometers (Fig. 5) include the coaxial (concentric-cylinder) viscometer, the parallel-plate viscometer, the cone-and-plate viscometer, the dynamic viscometer, and the controlled-stress viscometer. Moving-body viscometers include falling-ball viscometers, the Irvine-Park falling-needle viscometer (FNV), the falling-rod viscometer, and the sliding-plate rheometer.

Other Measurement Techniques

Extensional Viscosity. All three types of extensional viscosity can be measured: uniaxial, biaxial, and pure shear. Only a few commercial instruments are available, however, and most measurements are made with improvised equipment.

Viscoelastic Measurement. A number of methods measure the various quantities that describe viscoelastic behavior. Some require expensive commercial rheometers, some depend on custom-made research instruments, and some require only simple devices. Even qualitative observations can be useful in the case of polymer melts, paints, and resins, where elasticity may in-

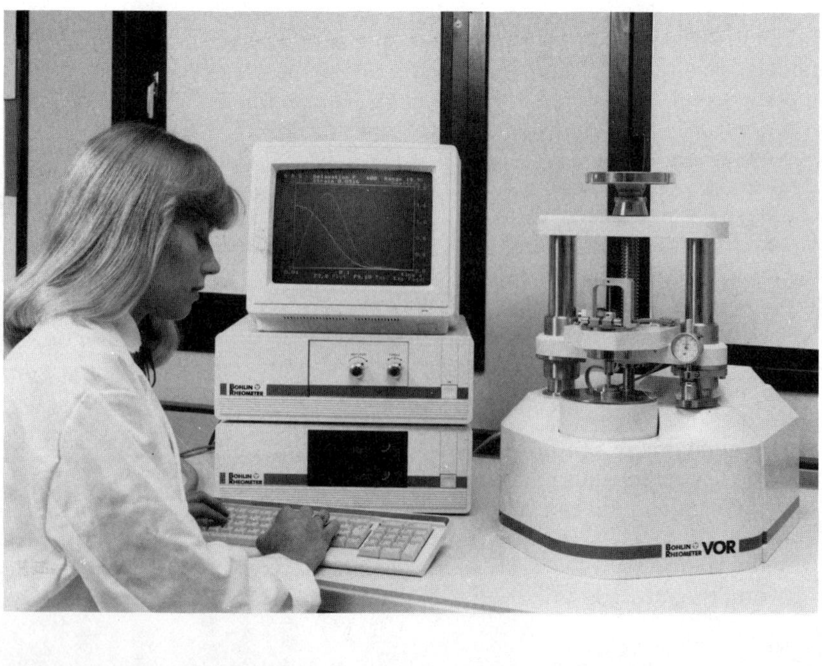

Figure 5. Bohlin VOR viscometer. Courtesy of Pen Kem, Inc.

dicate an inferior batch or unusable formulation. For example, the extrusion swell of a material from a syringe can be observed with a microscope. The Weissenberg effect is seen in the separation of a cone and plate during viscosity measurements or the climbing of a resin up the stirrer shaft during polymerization or mixing.

Dynamic Measurements. Dynamic methods are required for investigating the response of a material to rapid processes, studying fluids, or examining a solid as it passes through a transition region (T_g, melting, etc). Such techniques impart cylic motion to a specimen and measure the resultant response. Dynamic techniques are used to determine storage and loss moduli G' and G'' and the loss tangent tan δ. Some instruments are sensitive enough for the study of liquids and can be used to measure the dynamic viscosity η'. Measurements are made as a function of temperature, time, or frequency, and results can be used to determine transitions and chemical reactions as well as the properties noted above. Dynamic mechanical techniques for solids can be grouped into three main areas: free vibration, resonance-forced vibrations, and nonresonance-forced vibrations.

Fluids. The fluid must be highly viscoelastic for data to register, and absolute modulus values are difficult to determine because of the presence of the support. Different instrumentation is needed for the determination of meaningful modulus values over wide viscosity and elasticity ranges.

Viscoelastic measurements of dilute solutions are useful because they allow studies of the conformational dynamics of long-chain macromolecules in solution without interference from intermolecular interactions. Such measurements require special instrumentation and techniques. An example is the Birnboim-Schrag multiple-lumped resonator (MLR), which has undergone considerable development and modification since it was first described. Viscoelastic fluids that are more concentrated are characterized with devices that are similar to the rotational viscometers described previously.

Time–Temperature Superposition, Master Curves. Because the modulus of a viscoelastic material varies with time and temperature, measurements must be made over wide ranges of these variables for full characterization. This is often impossible, particularly for very long and very short times. Even if it were possible, the amount of experimental work involved would be prohibitive. Therefore, techniques have been developed to determine modulus values and curves at times or temperatures not attainable experimentally. A series of stress-relaxation, indentation, or modulus curves measured at different temperatures can be shifted on a log time axis to give a single modulus–time master curve that covers a very wide time range. A creep-compliance–time master curve can be constructed in the same fashion, and a similar curve can be constructed with the families of modulus curves at different frequencies generated by dynamic mechanical instruments with multiplexing capabilities. These curves can be shifted to a single modulus–frequency master curve because time and temperature have equivalent effects on modulus and other viscoelastic quantities.

BIBLIOGRAPHY

J. R. Van Wazer, J. W. Lyons, K. Y. Kim, and R. E. Colwell, *Viscosity and Flow Measurement, A Laboratory Handbook of Rheology*, Wiley-Interscience, New York, 1963.

J. M. Dealy, *Rheometers for Molten Plastics*, Society of Plastics Engineers and Van Nostrand Reinhold Co., Inc., New York, 1982.

R. W. Whorlow, *Rheological Techniques*, Halsted Press, a division of John Wiley & Sons, Inc., New York, 1980.

L. E. Nielson, *Mechanical Properties of Polymers and Composites*, Vols. 1 and 2, Marcel Dekker, Inc., New York, 1974.

CLIFFORD K. SCHOFF
PPG Industries, Inc.

RHEOOPTICAL PROPERTIES

Rheooptics implies the utility of electromagnetic radiation for probing the structure of a solid or liquid and, in particular, understanding the structural-deformation or molecular-flow behavior of that system. As originally used, the term relates to utilizing electromagnetic radiation for a better understanding of the orientation and general deformation characteristics of polymeric materials, although it certainly could apply to other materials as well.

The term optical does not imply that only the visible part of the electromagnetic spectrum is used; indeed, many other regions of that spectrum are employed in regard to the techniques applied to obtain rheooptical information. The principal methods employed to obtain this information are birefringence, linear dichroism, polarized fluorescence scattering (pfs), polarized Raman scattering (prs), small-angle light scattering (sals), wide- and small-angle x-ray scattering (waxs and saxs), and small-angle neutron scattering (sans). The last technique is different from the others because it concerns the scattering of neutrons (see NEUTRON SCATTERING). However, because the general relationships concerned with scattering phenomena apply to these particles as a result of their deBroglie wavelength, and in view of the importance of sans, it seems appropriate to include some information regarding this method under the subject of rheooptics.

The principal reason for studying the rheooptical properties of polymers is to obtain information concerning their structure and state of molecular orientation. The importance of this orientation is, of course, correlated with the mechanical properties of polymeric materials and often with other properties as well, permeability, dielectric behavior, etc.

Orientation

Figure 1 shows a partially oriented polymer chain with its end-to-end separation vector denoted by **R**. This chain might be in a solid that has undergone uniaxial orientation along the reference axis z, eg, along the machine direction (MD). The state of orientation for this chain may not directly match those of other chains in the system for a variety of reasons, such as differences in entanglement density or cross-linking considerations, molecular weight, relaxation behavior, etc. Furthermore, this chain may possibly be of different chemistry than many of those in the surrounding matrix, as might be the case in a miscible or an immiscible polymer blend. The relevant question is how to define the state of orientation for a given chain or a portion of that chain, as well as for the complete system. It might be desirable to specify the orientation of each backbone bond in this chain relative to the z axis (as for uniaxial orientation, where z would generally represent the principal deformation axis), but this is not an easy task. Statistical aspects might be considered as well as the generation of mo-

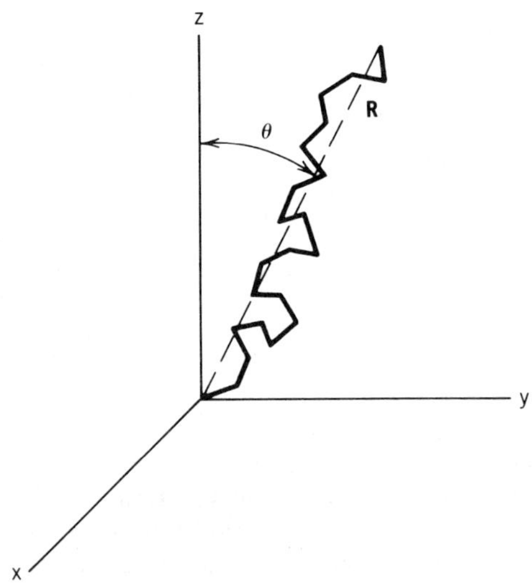

Figure 1. Oriented chain showing preferential orientation along the reference axis z. The end-to-end separation vector of the chain **R** (dashed line) is shown at an angle θ to the reference axis.

ments or average quantities associated with the orientation distribution. The rheooptical methods utilized to obtain orientation information typically provide only average quantities, although in principle the waxs technique can provide the complete orientation distribution for the crystalline phase.

In Figure 1 the angle θ relates the direction of **R**, the end-to-end chain distance, to the reference axis z. Determining this angle or its average for the entire system could be useful information with respect to the state of orientation. An orientation function for uniaxial systems can be defined as

$$f = \frac{3\langle\cos^2\theta\rangle - 1}{2} \qquad (1)$$

where the quantity $\langle\cos^2\theta\rangle$ represents the average value of $\cos^2\theta$. In particular, when **R** is aligned parallel to z, $\theta = 0°$ for all chains, and $f = +1$, whereas for the case of perpendicular orientation, $\theta = 90°$, $f = -\frac{1}{2}$. When the system is randomly oriented, the average value for $(\cos^2\theta) = \frac{1}{3}$, and hence $f = 0$.

Other factors being equal, differences in the value of f are likely to be strongly correlated with mechanical properties such as modulus, strain to break, energy to break, etc. Therefore, much effort has been put forth to develop the rheooptical methods to obtain and quantify information regarding the state of orientation.

Many polymeric materials utilized today have undergone biaxial orientation, simultaneous or sequential, either of which may be balanced or unbalanced in terms of orientation along the two principal orientation axes. Description of the orientation state of such systems is more complicated, and different treatments have been used. A useful mode of describing such systems is provided by the biaxial orientation functions. The relationships of these two functions are expressed by

$$f_z^B = 2\langle\cos_z^2\theta\rangle + \langle\cos_y^2\theta\rangle - 1 \qquad (2)$$

$$f_y^B = 2\langle\cos_y^2\theta\rangle + \langle\cos_z^2\theta\rangle - 1 \qquad (3)$$

The values of θ_z and θ_y relate to the angles between the principal axis of the specific unit (such as **R**) and that of the z and y axes of the sample, respectively. These axes usually relate to the machine direction and the transverse direction, as in drawn film. Only second moments, $\langle\cos^2\theta\rangle$ values, are again indicated in equations 2 and 3 as in the orientation function of equation 1. Hence, only average quantities are obtained from the above two relationships, and no information concerning the orientation distribution itself is provided.

Equilateral Orientation Triangle. A single value of an orientation function f could in principle be specified with respect to the three orthogonal axes of a cartesian coordinate system. The orientation data obtained with respect to these three axes for a given structural unit are conveniently represented by an equilateral triangular plot. Although this method has been utilized primarily for presenting the orientation response of crystals as obtained by x-ray data, it can also be employed for presenting any second-moment data obtained from other rheooptical techniques, eg, polarized Raman scattering, linear dichroism, etc.

Rheooptical Techniques and Their Applications

The rheooptical behavior of polymers can be investigated by many different methods, and it is therefore important to recognize what information is being measured versus what information is sought before applying any of these. This is particularly true for the case of multicomponent or multiphase systems. Although many techniques are intended to supply information on the state of orientation, microscopy techniques also provide some degree of information by direct observation regarding the general deformation behavior of systems. This information can certainly be of utility, for example, in the examination of an injection-molded material containing chopped glass fibers. Because of the flow-induced orientation of the glass filaments, the final material will likely display mechanical anistoropy, and the polymeric matrix may be oriented as well if thermal quenching is sufficiently fast, as is often the case in molding processes. However, when the matrix is fully relaxed, the mechanical anisotropy is likely to persist because of the orientation of the fiber component. Hence, methods applied to determine the molecular orientation of the matrix material may provide little direct explanation of the observed mechanical anisotropy, whereas microscopy methods may provide more information because the orientation of the chopped fibers is likely to be observed.

Rheooptical methods provide information on a dimensional level that is often correlated with the wavelength that is involved with the particular method. Figure 2 illustrates some rheooptical techniques (as well as other important methods) that can be categorized in regard to the dimensional scale over which information is obtained.

Other techniques, however, such as birefringence, linear dichroism, polarized Raman scattering, and polarized fluorescence scattering provide structural information that is not dependent on the size of the structure itself, but rather provide information on the local orientation on the scale of the bond lengths. Birefringence is also known as double refraction and clearly relates to the optical anisotropy of a system (see BIREFRINGENCE; REFRACTOMETRY). With the definition above it should be clear that three values of birefringence exist, but only two are independent, as inferred from equations 4–7. More directly, if

$$\Delta_{ij} = n_i - n_j \qquad (4)$$

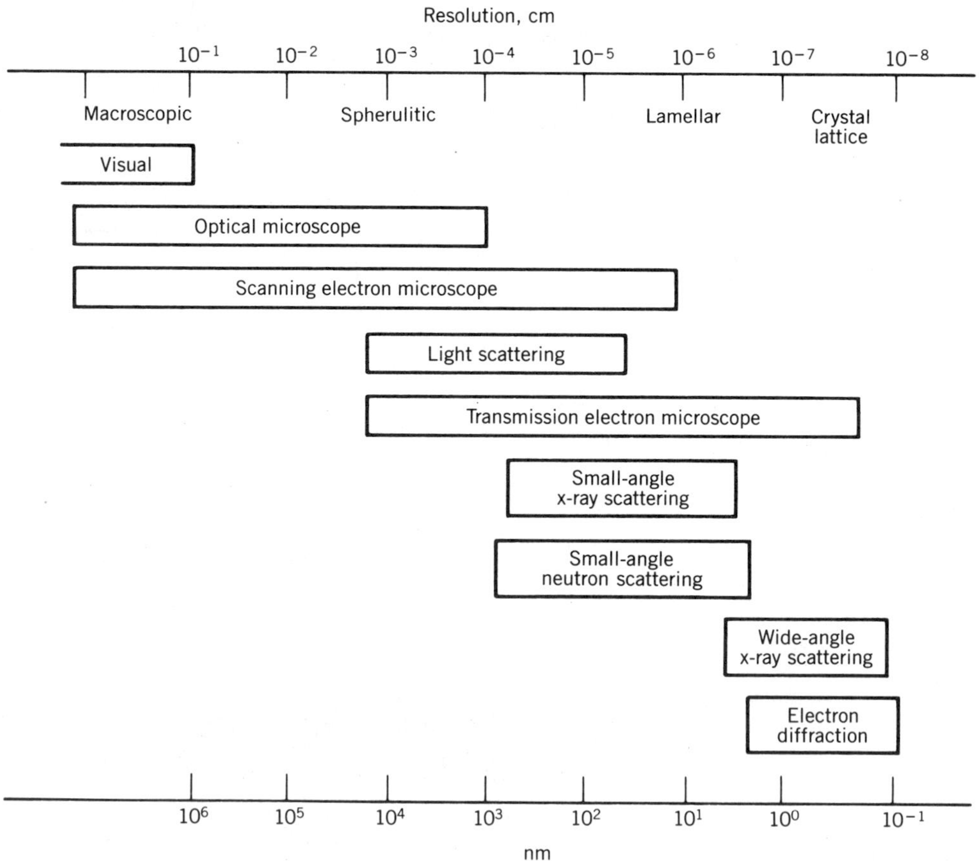

Figure 2. Resolving power of structural characterization techniques, including scales with respect to the types of structure found in polymeric materials.

then
$$\Delta_{xy} = n_x - n_y \tag{5}$$

$$\Delta_{xz} = n_x - n_z \tag{6}$$

and hence
$$\Delta_{yz} = n_y - n_z = \Delta_{xz} - \Delta_{xy} \tag{7}$$

In the case of uniaxial symmetry, as for drawn fibers, only one value of the birefringence has to be specified because any two refractive indexes perpendicular to the fiber axis would be identical. However, although the two values of refractive index orthogonal to a given draw direction may be equal, this does not indicate whether there is a gradient in orientation through that given direction. Skin–core effects may also be present in fiber spinning, for example, caused by variation in thermal quenching due to poor thermal conductivity. Such effects can also be due to shear variation across a fiber or film in processing.

The birefringence technique is a common means of determining the average molecular orientation of a given system. It is important to note that this technique supplies only a second-moment average $\langle \cos^2 \theta \rangle$ response, which takes account of all components within the system. Finally, there are sources of birefringence that are not the result of molecular orientation; their recognition is also important.

Instrumentation and Methodology. The measurement of birefringence is often simple but subject to certain prerequisites. The first and foremost is the ability of the sample to transmit visible radiation. The measurement of the transmitted intensity is the important feature in this technique, whereas in many rheooptical methods it may be the scattered radiation. Hence, samples containing carbon black, opaque fillers, and other structural aspects that reduce optical transmission limit the measurement of birefringence.

If uniaxial symmetry is involved, only one birefringence value is needed. However, if a system has biaxial character, particularly unbalanced biaxial orientation, two values are necessary for further specification of the state of orientation. The absolute values of the refractive indexes permit a more complete specification of the state of orientation.

Shown in Figure 3 is an apparatus utilized to determine birefringence. A source of radiation, such as a mercury arc or similar intense source of visible light, is collimated to direct a beam of radiation on the sample. The radiation is usually polarized at 45° to the reference axis, which is typically the principal deformation direction. A second polarizer (analyzer) is placed on the opposite side of the sample, followed by a photomultiplier or photodiode detector.

Applications: Homogeneous Single-phase Systems. The orientation and general birefringence behavior of oriented rubbery networks (single-component single-phase systems) have been described.

Single-component Multiphase Materials. The principal systems under this topic relate to semicrystalline polymers where both an amorphous and an ordered or crystalline region are present. In such materials the orientability and time-dependent behavior, in orientation as well as during relaxation, are often different for the crystalline and amorphous regions. Although the birefringence behavior of such multiphase single-component systems has been investigated, measurement of the single parameter of birefringence does not separate the orientation behavior of the amorphous phase from that of the

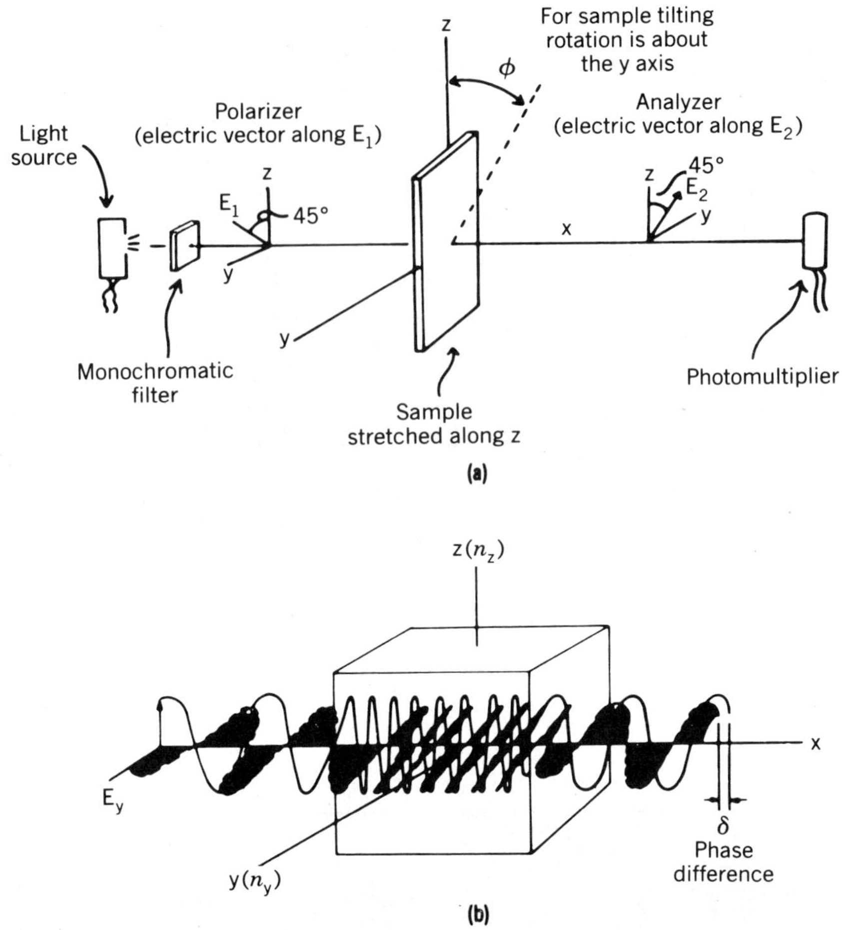

Figure 3. Common optical laboratory device for determining birefringence; E_1 and E_2 represent the optical axes of linear polarizers. The angle θ is usually zero, but for out-of-plane birefringence values, nonzero θ values are utilized. When θ is zero, the birefringence Δ_{zy} is determined.

crystalline phase. To distinguish the orientation response of each phase requires utilization of another technique such as x-ray diffraction or linear dichroism. When these techniques are used together, the orientation state of each phase can be better quantified.

Multicomponent Systems. This category includes materials that display single-phase or multiphase behavior. Miscibility does not imply that the state of orientation of one component is the same as that of any other. Indeed numerous measurements have been made employing other methods to demonstrate that differences in orientation occur between miscible components thereby often limiting the interpretation of observed birefringence values on these systems.

Linear Dichroism. In contrast to birefringence, which provides a system average of the state of orientation, the technique of linear dichroism often provides a quantitative measure of the orientation of individual components or phases. The method is based on the principle of selective absorption of polarized radiation. Measurements are usually made in the infrared (ir) region, but may also utilize the visible or ultraviolet (uv); the choice depends on the nature of the chromophoric groups in the polymer. Because most polymeric materials absorb at frequencies in the ir region because of changes in the vibrational states associated with permanent dipoles, this region is most widely utilized. Commercial instrumentation for ir dichroism measurements is also readily available. Fourier transform infrared spectroscopy (ftir) has particularly increased the accuracy and speed of these measurements.

Instrumentation and Methodology. If the bulk orientation behavior is of interest, it is necessary that the material not be so highly absorbent that the transmitted intensity is too weak to allow measurements of the difference in absorption of polarized radiation. Some progress has been made in utilizing dichroism techniques for determining changes in orientation in the surface regions of a material through the use of attenuated total reflectance (atr) measurements. The determination of orientation by the atr method is not as precise as that for bulk orientation but good correlations have been made. This arises from the fact that, in the methodology applied to surface orientation measurements, potential depolarization effects of the impinging beam are more complex. Similarly, there are greater difficulties associated with determining the details of orientation in biaxially oriented thin films. Although the absorbance characteristics may be favorable for an impinging beam perpendicular to the surface of such a film, a single-direction measurement of dichroism is insufficient to fully characterize the degree of orientation. To employ a sample-tilting procedure in conjunction with extrapolation techniques is not simple but can be applied. It may lead to errors as a result of surface scattering, depolarization, and refraction effects, which can alter the difference in path length for the two states of polarization utilized in the dichroism measurements. Regarding ir dichroism measurements, the principal instrument employed for the collection of data is the Fourier transform infrared spectrophotometer (ftir).

Applications. *Single-component Single-phase Systems.* Atactic polystyrene has been investigated in terms of relaxation behavior and general orientation response at different temperatures and with different molecular weights.

Single-component Multiphase Materials. The principal systems under this category are semicrystalline homopolymers. Numerous dichroism studies demonstrate that orientation during deformation and the subsequent relaxation clearly indicate that the amorphous phase orientation behavior differs from that of the crystalline regions. Many of these studies have been carried out, however, in conjunction with other rheooptical techniques, usually birefringence or waxs.

Multicomponent Single-phase Materials. Both miscible polymer blends and polymeric materials that contain a low molecular weight component, such as a plasticizer or solvent, can be considered to be homogeneous. When a geometrically anisotropic small molecule is contained within a polymeric matrix, the low molecular weight species may show a preferential orientation after this matrix is deformed. As the linearity of this species and its molecular weight increase, the degree of orientation often increases. Hence, dichroism can be utilized to investigate such phenomena.

Multicomponent Multiphase Systems. Block and segmented polymers such as the styrene–butadiene materials and segmented urethane systems have been studied. These materials are well suited to dichroism studies because the chemistry of a given segment or block is different from that of other block components, and hence their different chromophores may be used accordingly to separate their respective orientation responses.

Polarized Fluorescence Scattering and Polarized Raman Scattering. Polarized fluorescence scattering (pfs) and polarized Raman scattering (prs) share certain similarities. These two methods, particularly the former, have had more limited applications in rheooptical studies. The latter method, however, is likely to increase in utility as the techniques and instrumentation for Raman scattering are improved and the knowledge of the origins of Raman lines improves. The prs technique was of little importance until the advent of laser technology, which provided the high power monochromatic and highly polarized sources necessary to provide a sufficient scattering intensity, which is usually on the order of 10^{-8} of that of the incident beam (see RAMAN SPECTROSCOPY).

For Raman or fluorescence scattering to occur, the molecular or atomic system must absorb radiation and rise to a higher energy level. Return to a lower energy level is accompanied by an emission of radiation rather than the transition route common in the ir region, where the higher energy vibration states are dissipated through collisions, leading to a heating of the sample. Hence, prs and pfs each represent a type of emission spectroscopy. This common feature correlates the theory of these two techniques. In particular, the presence of both absorption and emission axes may provide additional information regarding orientation. Both prs and pfs have been applied in conjunction with several other rheooptical techniques to determine both the second moment of the orientation distribution and the values of the fourth moments for single-phase and multiphase systems.

Elastic Scattering. Several scattering methods are of great utility in studying the rheooptical behavior of polymers (see LIGHT SCATTERING; NEUTRON SCATTERING). Even though the origin and characteristics of elastically scattered radiation have different features, depending on whether light, x rays, or neutrons compose the incident beam, the overall amplitude of the coherent angular-dependent scattered beam can be shown to be of the form

$$A_s \sim \sum_i^N \rho_i e^{-ik(\mathbf{r}_i \cdot \mathbf{s})} \tag{8}$$

A_s is the resulting amplitude of the elastically scattered beam arising from the scattering volume composed of N scatterers (elastic or coherent scattering refers to scattered intensity of the same frequency and wavelength as that of the incident intensity); ρ_i is the scattering power of the ith scatterer; k is the wave vector and is $2\pi/\lambda$, where λ is the wavelength in the system (the values of λ differ from that of the incident if the refractive index of the scattering body is higher than unity, as is the case for the optical frequency range); and $\mathbf{r}$ is a vector taken from an arbitrarily established origin to the scattering point in question. Its magnitude is always assumed to be very small compared with the distance between the scattering object and the point of observation. $\mathbf{s} = \mathbf{s}_0 - \mathbf{s}'$ ($\mathbf{s}_0$ and $\mathbf{s}'$ are unit vectors taken along the incident and scattering directions, respectively).

As shown in Figure 4, $\mathbf{s}$ always bisects the angle between $\mathbf{s}_0$ and $\mathbf{s}'$ and possesses a magnitude $|\mathbf{s}| = 2 \sin (\theta/2)$, where θ is the radical scattering angle. Hence, the parameter $\mathbf{s}$ carries information on what scattering angle is under consideration. Sometimes the unit vectors of $\mathbf{s}_0$ and $\mathbf{s}'$ are expressed as equal to the reciprocal of the wavelength, $|\mathbf{s}_0| = 1/\lambda$. In this case $|\mathbf{s}| = (2/\lambda) \sin (\theta/2)$. This means that $\mathbf{s}$ has dimensions of reciprocal length and carries information on the wavelength used as the structural probe as well as on the scattering angle. Because all structural details of the scattering system are contained in the right-hand term of equation 8, this is denoted as the structure factor and is often given the symbol $\mathbf{F}$. From measurements of the angular-dependent scattered intensity $I(\mathbf{s})$, information can be obtained on the structural details of the system.

In analyzing the scattering arising from a given system, its structural features can sometimes be quite precisely determined, as in crystal structure analysis by waxs. In nonperiodic systems, such as glasses or amorphous systems, only limited information can be obtained without additional assumptions or initial knowledge of the molecular texture. Where some

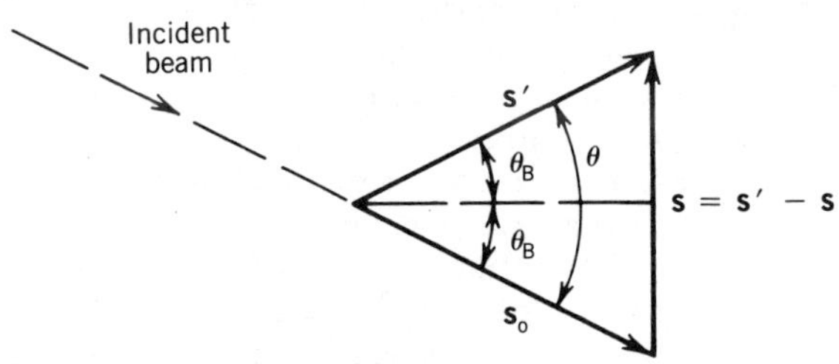

Figure 4. Vectorial plot illustrating the relationship between unit vectors along the incident beam ($\mathbf{s}_0$) and scattered beam ($\mathbf{s}'$). The result of the difference $\mathbf{s}' - \mathbf{s}_0$ produces the vector denoted as $\mathbf{s}$ as shown.

structural information is available, the expected scattering can be calculated in terms of known parameters, such as refractive indexes, electron densities, or scattering lengths. With these assumed parameters the scattering from the model can be compared with that observed. This often permits support of a given structural model.

A statistical treatment is useful for the analysis of scattering arising from a material where little structural information is available. This is especially so when the scattering profile is of monotonic form without shoulders or peaks. Under these circumstances certain statistical or average quantities can be extracted from the scattering profile that may be of utility in characterizing structural features, for example, the variance $\langle \eta^2 \rangle$ in electron density (or refraction index) and the correlation function $\gamma(r)$; the latter provides information about the homogeneity of the system.

In the statistical method these two principal parameters, $\langle \eta^2 \rangle$ and $\gamma(r)$, can be utilized to help describe (or predict) the scattering behavior of a system. This approach has been used to characterize polymer solids. Sometimes other parameters can be used, such as average phase or particle size, smoothness of the interface, interphase thickness, radius of gyration of the scattering particles, and multiple radii of gyration if the azimuthal character of the scattering profile is analyzed. As demonstrated by sans, the scattering function for single oriented chains can be extracted from the scattering profile, thereby providing the major and minor radii of gyration of the oriented chains as determined from the scattered intensity measured parallel and perpendicular to the orientation direction.

Wide-angle X-ray Scattering. The principal application of wide-angle x-ray scattering (waxs) in the characterization of polymeric materials is the determination of crystallinity and information relating to crystallite size and perfection (see CRYSTALLINITY DETERMINATION; X-RAY DIFFRACTION). In rheooptical studies waxs is used to determine the crystal orientation, frequently a single moment of the orientation distribution for a given crystallographic axis. The complete specification of the entire crystal orientation distribution may be obtained and is often presented by so-called pole figures. The technique is based on the use of x rays of small wavelength (0.1–0.3 nm) to probe crystal structure and its degree of ordering. This is the principal method for determining structural features on the order of 0.1–2 nm within a range of easily accessible scattering angles.

Small-angle X-ray Scattering. The development of high intensity x-ray sources and position-sensitive detectors has greatly increased the utility of small-angle x-ray scattering (saxs) for structural analysis of polymeric materials and solutions. Whereas waxs provides information about structures in the range 0.1–2.0 nm, for larger-scale structures, smaller scattering angles are now required.

The measurement and interpretation of saxs data are more often complicated than those of waxs data from crystalline materials. This follows from the Babinet principle of reciprocity, which states that it is impossible to distinguish by angular scattering between a system of particles in space and a corresponding system in which the particles and the space have been exchanged to give a microporous solid. However, with some additional knowledge of the system by other techniques, saxs can often provide much insight into structural details.

The information that can be gained via saxs depends on the system in question, but care must be taken in data acquisition and treatment. In this article emphasis is placed on the determination of morphological behavior, but various data can be obtained including

- Direct particle size
- Macromolecular structure of the particle
- Particle–particle correlation functions
- Matrix–particle correlation function
- Latice-distortion parameters
- Macrolatices
- Particle-size distribution
- Orientation parameters between particles
- Void structure, void–void correlations, and void–particle correlations
- Heterogeneity or structural correlation distances
- Mean electron density fluctuations
- Measurement of interfacial zones and thickness
- Degree of mixing (in blends)
- Dimensionality of small objects (geometric fractal analysis)
- Specific surface
- The onset of phase separation, as in spinodal decomposition or micellar development

Regarding detection of scattered intensity, film techniques and direct intensity measurements can be made with the saxs methodology, although the latter are more commonly used. Most measurements utilize a one- or two-dimensional position-sensitive detector. The ability to discern a given dimensional scale depends on the instrumental resolution, which varies for different devices. Both pinhole and slit illumination procedures are used. With sufficient incident intensity, pinhole illumination is preferred for slit geometries, producing so-called smearing or smeared intensities, which often must be corrected for, particularly for oriented systems.

Applications. *Single-component Amorphous Systems.* The homogeneity of the system can be investigated by determining the correlation function $\gamma(r)$, which provides an index of electron density variations, with separation by a distance r, but averaged over the entire system (pair correlation average). The measured quantity $\gamma(r)$ and its distance dependence express the homogeneity of the structure. The faster $\gamma(r)$ decreased from unity toward zero, the less homogenous the system; this function may show damping oscillations in some materials, particularly in periodic structures that are not found in single-component amorphous systems, eg, block polymers that display microphase morphologies.

Single-component Multiphase System. The principal materials in this category are semicrystalline polymers. Stacked lamellar textures can be viewed as periodic fluctuations in electron density caused by the differences in the densities of the amorphous and crystalline phases. Structural characterization of these systems can be obtained by saxs.

The saxs method can also be used to follow the crystallization process, the development of lamellar textures during crystallization. Such measurements were difficult several years ago when point-by-point intensity measurements with low intensity sources were required. More recently, synchrotrons or rotating anode sources in conjunction with position-sensitive detectors permit easy measurement in kinetic studies and the deformation of semicrystalline polymers can be followed by saxs (including the rearrangement during orientation of the lamellar texture) by utilizing a two-dimensional detector in conjunction with high intensity sources. Such studies provide insight into the temperature, time, and stress–strain response of the crystalline lamellar units, as well as their relaxation characteristics, because the time-dependent scattering intensity can be measured following deformation to a given elongation or stress.

Multicomponent Single-phase Systems. In the absence of true miscibility and where there are composition gradients on the order of 10 nm, the saxs method may be used to detect the occurrence of such fluctuations if the two components of the blend possess different electron densities.

Multicomponent Multiphase Materials. Block and segmented polymers display microphase separation if the difference in solubility between the blocks or segments is sufficient. These materials are particularly suitable for saxs analysis if the interdomain spacing is periodic, as in anionic vinyl–diene block polymers, eg, styrene–butadiene.

Small-angle Light Scattering. Solution light scattering is a standard method for the characterization of the weight-average molecular weight of polymers as well as other parameters of solution behavior. Small-angle laser light scattering (sals) is also important in rheooptical structural and orientational studies of polymer solids. Investigations have focused on semicrystalline solids as well as polymer glasses and rubbery systems. This method was greatly improved by the continuous-wave visible laser; by the late-1960s it was widely used to characterize polymer solids. Basically, the coherent monochromatic high intensity polarized laser greatly enhanced the signal-to-noise ratio compared with older sources, such as the mercury arc. More recent developments in detector technology led to quantitative two-dimensional detector arrays allowing direct, rapid measurements, eg, by the use of television monitoring (Vidicon) systems. However, the simple so-called photographic sals method established in the 1960s is still used today. Such a device is illustrated in Figure 5, where the detector is an open-back Polaroid camera that provides a two-dimensional record of the scattering pattern, often in a fraction of a second, with the help of a medium intensity laser. Dynamic sals as well as relaxation studies can be carried out following the rapid deformation of superstructure in a polymer film. By rotation of the polarizers ahead of and behind the sample, both the isotropic and anisotropic fluctuations (orientation correlations) can be studied as long as they occur over a dimensional scale comparable with that of the wavelength or probe.

Small-angle Neutron Scattering. Small-angle neutron scattering (sans) results from the collision of high speed neutrons with nuclei in the sample investigated. Although both incoherent (inelastic) and coherent (elastic) scattering occur, the coherently scattered neutrons have primary utility in characterizing polymer structure. Polarization effects arise as a result of oriented nuclear spins, but these have been of little use in rheooptical studies to date.

Because of the low neutron scattering power of most polymeric materials and often lower energy (flux) beams relative to x ray or light scattering, thicker samples are necessary and larger incident beams are employed. Several neutron sources are available in the United States and other countries; beam time can generally be acquired for an appropriate project.

As in waxs, incoherent scattering must be corrected for, particularly when absolute intensities are important. In contrast to x-ray scattering, which depends on electron density (which increases with atomic weight), the neutron scattering power of atoms shows no correlation with atomic weight. Isotopes of the same atom show different scattering as a rule and this feature is greatly used in designing experiments.

In contrast to solution light scattering, which cannot be carried out with accuracy except in the limit of dilute solutions, neutron scattering can provide the single-chain scattering function even in concentrated materials; ie, $c > c^*$. This allows specific types of investigation in the semidilute and concentrated regions that may not be possible with light scattering or x-ray scattering. Similar bulk experiments can be carried out with blends of two polymers, where the neutron-scattering power of one species differs from that of the other. Single chain dimensions in semicrystalline polymers have also been studied with sans.

BIBLIOGRAPHY

R. Samuels, *Structured Polymer Properties,* John Wiley & Sons, Inc., New York, 1974.

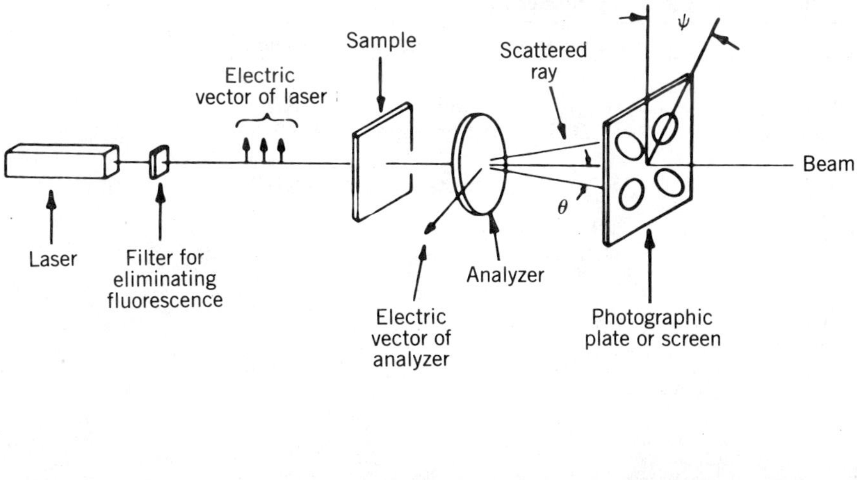

Figure 5. Sals laser apparatus for photographic light scattering measurements. The angle θ represents the radical scattering angle, and the angle ψ the azimuthal angle. The four-leaf clover pattern is typical of undeformed spherulitic textures.

I. M. Ward, *Adv. Polym. Sci.* **66,** 82 (1985).

G. L. Wilkes, *J. Macromol. Sci. Rev. Macromol. Chem.* **C10**(2), 149 (1974).

J. F. Rabek, *Experimental Methods in Polymer Chemistry,* Wiley-Interscience, New York, 1980.

GARTH L. WILKES
Virginia Polytechnic Institute

RIBONUCLEASE. See ENZYMES.

RIBONUCLEIC ACID. See POLYNUCLEOTIDES.

RIM. See REACTION INJECTION MOLDING.

RING-OPENING POLYMERIZATION

Ring-opening polymerization is expressed by the following general equation in which a cyclic compound (monomer) is opened to produce the linear polymer.

$$\widehat{C_n\ X} \longrightarrow \text{-}(C_n\text{—}X)_p\text{-}$$

The functional groups (X) include —CH═CH— and those that consist of one or two hetero atoms such as oxygen, nitrogen, sulfur, phosphorus, silicon, etc. The resulting linear polymers possess the functional groups (X) in the main chains.

The history of ring-opening polymerization of cyclic monomers is not as long as that of addition polymerization of vinyl monomers or condensation polymerization. Ring-opening polymerizations can proceed through a wide variety of mechanisms, depending on the monomer and catalyst. The many kinds of polymers produced have found important industrial applications.

Cyclic Monomers

Table 1 shows the various kinds of cyclic monomers that have been studied for ring-opening polymerization. Cyclic monomers are classified by their ring size and nature of the functional group. These monomers include cyclic olefins where the functional group is the double bond and heterocyclic compounds that may be subdivided according to whether the hetero atom of the functional group is associated with a carbonyl group. The polymerizability of these cyclic monomers is usually controlled by ring size, reactivity characteristics of the functional groups (X), and substituents on the ring. Bicyclic compounds are also used as the reactive monomers for ring-opening polymerization.

Polymerization Mechanisms and Catalysts

Ring-opening polymerizations of heterocyclic monomers are classified into the following seven patterns depending on the nature of the propagating species and the monomers.

polymerization via electrophilic propagating species

$$\sim\!P_E^+ + M_N \longrightarrow \sim\!P_E^+ \quad (1)$$

$$\sim\!P_E + M_N \longrightarrow \sim\!P_E \quad (2)$$

$$\sim\!P_E + M_N^- \longrightarrow \sim\!P_E^- \xrightarrow[HM_N \;\curvearrowright\; M_N^-]{} \sim\!P_EH \quad (3)$$

polymerization via nucleophilic propagating species

$$\sim\!P_N^- + M_E \longrightarrow \sim\!P_N^- \quad (4)$$

$$\sim\!P_N + M_E \longrightarrow \sim\!P_N \quad (5)$$

$$\sim\!P_N + M_E^+ \longrightarrow \underset{M_E^+}{\sim\!P_N} \quad (6)$$

polymerization via zwitterion intermediates

$${}^+M_AM_B\text{-}(M_AM_B)_n\text{-}M_AM_B^- + {}^+M_AM_B^- \longrightarrow {}^+M_AM_B\text{-}(M_AM_B)_{n+1}\text{-}M_AM_B^- \quad (7)$$

Equation 1 represents the conventional cationic polymerization, in which the propagating species bearing a cationic charge ($\sim\!P_E^+$) reacts with a monomer (M_N) of nucleophilic reactivity. In equations 2 and 3, the propagating chain ends ($\sim\!P_E$) are dipoles without electric charge that react respectively with a neutral monomer (M_N) or with an anionically charged monomer (M_N^-). Equation 4 designates anionic polymerization in which an anionically charged center at the propagating chain end ($\sim\!P_N^-$) reacts with a monomer (M_E) of electrophilic reactivity. In equations 5 and 6, the reaction site at the growing chain end ($\sim\!P_N$) is a covalent group having nucleophilic reactivity that reacts either with a neutral monomer (M_E) of electrophilic reactivity or with a monomer (M_E^+) bearing a cationic charge. Equation 7 shows the polymerization via zwitterion

Table 1. Functional Groups (*X*) in the Typical Cyclic Monomers

Name	Structure	Name	Structure	Name	Structure
olefin	—CH═CH—				
ether	—O—	sulfide	—S—	amine	—NH—
formal	—OCH$_2$O—	monothioformal	—OCH$_2$S—	thioformal	—SCH$_2$S—
disulfide	—SS—				
lactone	—OC(═O)—	thiollactone	—SC(═O)—	lactam	—NHC(═O)—
carbonate	—OC(═O)O—	urea	—NHC(═O)NH—	urethane	—NHC(═O)O—
anhydride	—C(═O)OC(═O)—	imide	—C(═O)NHC(═O)—		

intermediates, in which the propagating chain and the monomer bear both cationic and anionic charges. The polymer chain grows by its cation–anion reaction with the monomer. However, the assignment of species, growing chain ends, or monomer, which carry either cationic or anionic charge, is a difficult problem. Among the above seven patterns, cationic and anionic polymerizations are most common in ring-opening polymerizations. The cationic process (eq. 1) has been applied to the polymerization of cyclic ethers, sulfides, disulfides, amines, lactones, lactams, formals, and siloxanes.

Kinetics

In several cationic ring-opening polymerizations, it has been established that the propagation is an S_N2 process between the electrophilic propagating end and the monomer:

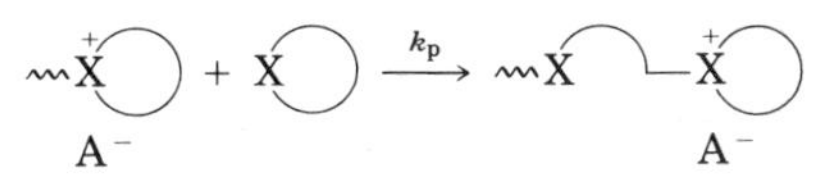

The rate of this cationic ring-opening polymerization is expressed by the second-order rate law:

$$R_p = k_p[P^+][M]$$

where k_p is the rate constant and $[P^+]$ and $[M]$ are the concentrations of propagating onium species and monomer, respectively. Two methods for determination of $[P^+]$ have been developed using end-capping techniques. By using this value, the k_p value can be determined from the rate law. The k_p values are higher for monomers with higher ring strains. In general the order of k_p values for various monomers is cyclic ether > sulfide > amine > imino ether.

Cyclic Ethers

Ring-opening polymerizations of three-, four-, five-, and seven- and higher membered cyclic ethers have been demonstrated. Among them, the polymerization of epoxides (alkylene oxides of three-membered rings) is the oldest example and the most important one in industry.

Cationic ring-opening polymerization of epoxides generally produces low molecular weight polymers. However, this reaction provides the fundamental aspects of cationic polymerizations of cyclic ethers.

Although the cationic polymerization of epoxides is subject to undesirable side reactions due to the high activity of the three-membered cyclic oxonium propagating species, a clean reaction of epoxide polymerization has been demonstrated. In the presence of alcohol and protonic acid and low concentration of an epoxide monomer, the polymerization of epoxide produces only linear polyether having hydroxy end groups at both ends:

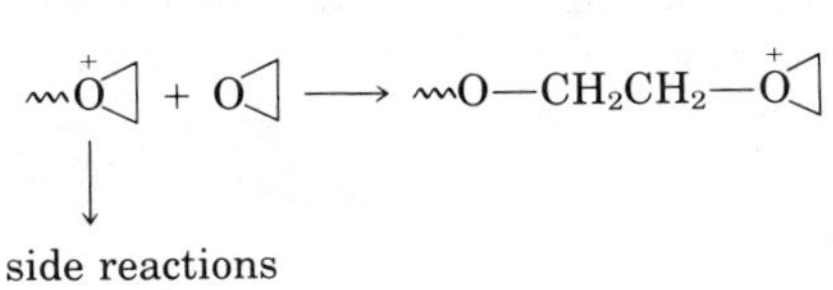

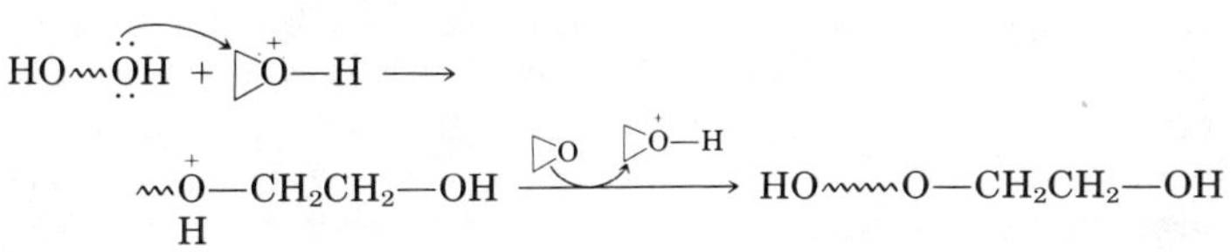

The polymerization producing a clean linear polymer proceeds by the reaction between a covalent (dipole) hydroxy group at the propagating end and a cationically charged monomer. It is appropriate to call it "polymerization via covalent nucleophilic species."

A new concept termed immortal polymerization has been developed in ring-opening polymerization of cyclic monomers. An aluminum–porphyrin complex was used as a catalyst. In this reaction, alcohols or acids act as a chain-transfer agent and the exchange of alcohols and aluminum alkoxide at the propagating end proceeds as shown below, where TPP designates the tetraphenyl–porphyrin ring.

$$(\text{TPP})\text{AlO}\sim\sim + \text{ROH} \rightleftharpoons (\text{TPP})\text{AlOR} + \sim\sim\text{OH}$$

In the usual polymerizations, this reaction kills the propagating species as a termination reaction. However, in the case of aluminum porphyrin the reaction is fast and reversible, and the alcohol-terminated polymer can repolymerize the monomers. This immortal polymerization has two significant characteristics: the resulting polymer has a narrow molecular weight distribution and the number of macromolecules is equal to the combined number of initiator and chain-transfer agent molecules.

Cyclic Sulfides

Among cyclic sulfides for ring-opening polymerization, three-membered ring alkylene sulfides are the most popular. Ethylene sulfide and propylene sulfide are known to undergo anionic, cationic, and coordinate nucleophilic polymerization. Ammonia, amines, trialkylphosphines, metal alkoxides, hydroxides, amides, thiolates, alkyllithium, $(C_6H_5)_3CNa$, sodium naphthalene, and tetrabutylammonium fluorenyl are used as catalysts for the anionic polymerization of alkylene sulfides.

Cyclic Amines

Cyclic amines have been polymerized only by means of cationic initiators. Among various cyclic amines, the ring-opening polymerizations of ethyleneimine (aziridine) and its derivatives have been well studied (see ALKYLENEIMINE POLYMERS). The exact description of ring-opening polymerization is complicated because of the many side reactions.

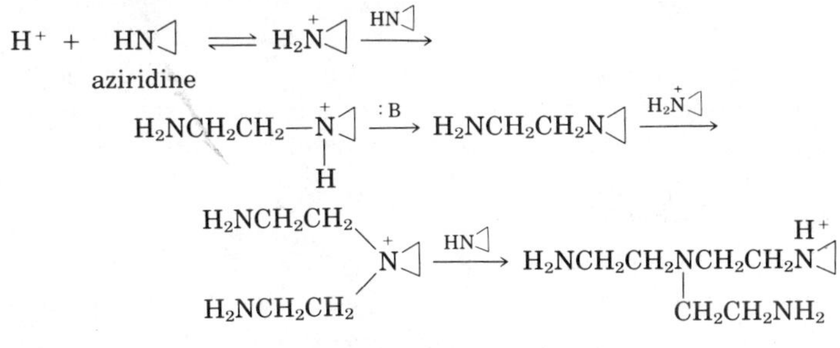

Thus the resulting polyethyleneimine has branched structures.

Polymerization of four-membered cyclic amines, the azetidines, is similar to that of three-membered ones. The final polymer is highly branched and contains primary, secondary, and tertiary nitrogens. Bicyclic amines with three- to six-membered rings, which contain a nitrogen atom at the bridgehead position, can be polymerized by using quaternary ammonium salts, alkyl halides, or BF_3.

Lactones and Lactams

Except for the five-membered ring, cyclic esters (lactones) undergo polymerization through anionic, cationic, and coordinate nucleophilic propagating species. Among various lactams, the polymerization of ϵ-caprolactam producing nylon-6 has been well studied. This is an important industrial reaction. ϵ-Caprolactam is polymerized in the presence of water, and the polymerization consists of the following three equilibrium reactions:

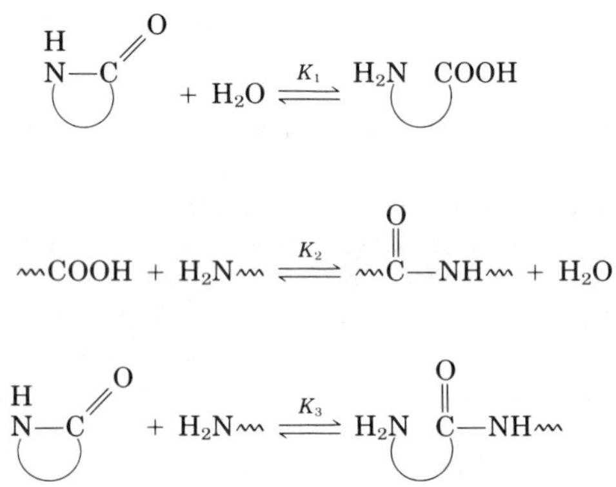

Nonhydrolytic anionic polymerizations of lactams are induced by binary catalyst systems consisting of *N*-acyllactam and a strong base such as an alkali metal, a metal hydride, a hydroxide, or an alkoxide, which abstracts the amide hydrogen in the monomer.

Other Cyclic Monomers

The following cyclic formals have been polymerized by a cationic mechanism.

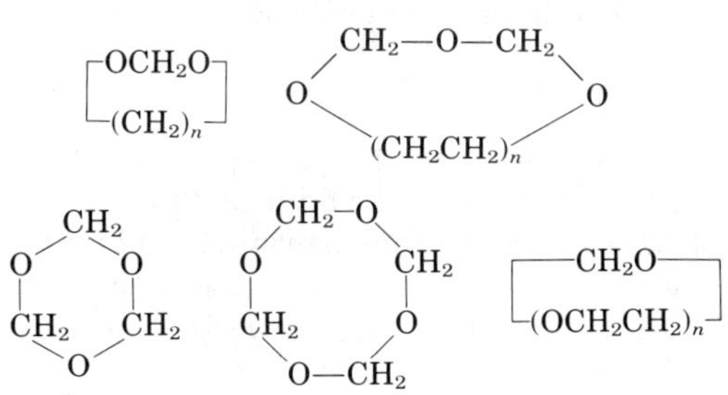

trioxane

Cationic ring-opening polymerization of trioxane, especially copolymerization with ethylene oxide or 1,3-dioxolane is industrially important for the production of acetal resins.

Copolymerization

Block Copolymerization. Some ring-opening polymerizations proceed by a living mechanism, which enables the preparation of block copolymers.

No-catalyst Copolymerization. A new type of copolymerization that does not need any catalyst has been explored. In this copolymerization, nucleophilic monomer M_N reacts with electrophilic monomer M_E to generate a zwitterion $^{+}M_N{-}M_E^{-}$. The propagation of this copolymerization is as follows:

$$^{+}M_N{-}M_E^{-} \longrightarrow {}^{+}M_N{-}M_EM_N{-}M_E^{-} \xrightarrow{n(\ M_N{-}M_E\ )} {}^{+}M_N{-}(M_EM_N)_{n+1}{-}M_E^{-}$$

This copolymerization gives rise to the production of alternating copolymers (see COPOLYMERS, ALTERNATING).

BIBLIOGRAPHY

K. C. Frisch and S. L. Reegen, *Ring-opening Polymerization,* Marcel Dekker, Inc., New York, 1969.

K. J. Ivin and T. Saegusa, *Ring-opening Polymerization,* Elsevier Applied Science Publishers Ltd., Barking, UK, 1984.

T. Saegusa, *Makromol. Chem. Macromol. Symp.* **13/14,** 111 (1988).

T. Saegusa, *Makromol. Chem. (Suppl.)* **4,** 73 (1981).

YOSHIKI CHUJO
TAKEO SAEGUSA
Kyoto University

RNA. See POLYNUCLEOTIDES.

ROBOTICS

Robotics has been defined as "the extension of progarammable computers which allows computers to do physical work as well as process data." Instrument systems using robotics and programmable computers are already being used to improve productivity in scientific laboratories, which can resemble a production environment. Analysts need to identify samples and subsequently weigh, dilute, concentrate, extract, filter, evaporate, manipulate, or analyze them.

Robotic systems for the laboratory must be flexible and user-programmable. Many laboratory robots incorporate tactile sensing and other verification techniques. However, vision capabilities are virtually nonexistent. Automation needs vary from laboratory to laboratory. Quality control laboratories, for example, might emphasize user simplicity and high throughput whereas research laboratories might stress flexibility.

General System Description

In general, a laboratory robot consists of the manipulator, controller, and sensing systems that allow the robot to react to unexpected events. Robotic arms are engineered to service an area that can be described as cartesian, cyclindrical, spherical, or revolute.

The performance of the robot is directly linked to the type of drive mechanism used. Table 1 shows typical characteristics of drive mechanisms used for robotic systems.

A robot controller can be a personal computer, a dedicated microprocessor, or a programmable controller. Although these three types of controller have different capabilities for storing and processing data, they all operate robotic devices in essentially the same manner. However, the language used to drive the system varies depending on the controller.

Flexible Automation

Laboratory robotics provides flexible automation able to meet the changing needs typical of quality control and research laboratories. Flexible automation is programmed by individual users to perform multiple procedures. It can be quickly reprogrammed to accommodate new or revised procedures.

Table 1. Typical Characteristics of Drive Mechanisms Used for Robotic Systems

Drive mechanism	Speed	Load	Travel	Positioning	System support
electric servo	moderate	moderate	unlimited	multi	minimal
hydraulic servo	high	high	unlimited	multi	large
stepper servo	low	low	unlimited[a]	multi	minimal
pneumatic actuator	high	low	to stops	to stops	moderate

[a] Rotary or piston travel

In Figure 1 sample quantity refers to the numbers of analyses using similar procedures. Complexity increases with the required precision, number of steps, and critical timing requirements of the analytical procedure.

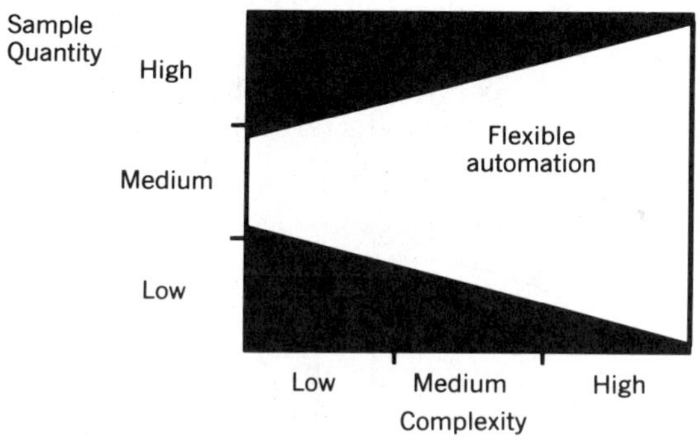

Figure 1. Flexible automation.

Laboratory Unit Operations

The challenge in automating any procedure or operation is to identify the repetitive steps and to create a device that can accomplish the same operation. The idea of automating entire laboratory procedures that appear unique to the compound and matrix being analyzed seems like an impossible task. However, most laboratory procedures consist of sequences of common steps or building blocks called laboratory unit operations (LUOs). Thus laboratory procedures can be systematically specified in terms of the required LUOs, their sequence in the procedure, and what happens at each LUO.

Methods Management

Robotics, automated analytical instruments, and laboratory computers are now routinely used to automate each of the individual components of an analytical method. Complete automation of a method requires total integration of laboratory robotics, analytical instruments, and computers. By integrating these components, samples may be prepared and analyzed automatically. The resulting data may then be used to control the remainder of the procedure automatically. The goal of this type of integrated methods management system is to provide timely quality decisions based on valid data. The ability to interface with a wide variety of standard laboratory instruments and computers greatly increases the flexibility of a laboratory robot.

Precision and Accuracy

In addition to improved productivity, the quality of the data is a key criterion on which many laboratory robotics applications are judged. Automated methods are generally validated by verifying that the individual workstations are functioning properly and then comparing the results of the totally automated procedure to those obtained manually.

Numerous comparisons of manual vs robotic results are available in the literature. In one example, a laboratory robot prepares polyester polymer samples (pellets, powders, films, fibers, and liquids) for glycol determination by gas chromatography. The automated method reduces the variability of results by approximately 50%. This improved precision is due to the elimination of the variability associated with multiple analysts in the manual procedure. An associated improvement in product quality is assumed.

Applications

Laboratory robots perform automated procedures for a broad range of applications in scientific laboratories for almost every industry. Table 2 summarizes a cross section of published articles for laboratory robotics procedures in polymer science.

Table 2. Laboratory Robotics Applications for Polymer Analysis

Assay procedure	Sample type
pH, Brookfield viscosity, percent solids	viscous, adhesive polymer emulsions
gc determination of glycols	polyester polymers
tensile testing	rigid polymers
spectrophotometric measurement of trace levels of polyacrylamide	water solution
gc headspace analysis of residual monomers	plastics
thermal analysis (dsc)	polymers
hplc and gc	polymers
film thickness, gloss, and MEK rub	automotive coatings
immersion testing	composite materials
particle size distribution (screen test)	polymer resins
flexural testing	plastics

BIBLIOGRAPHY

F. H. Zenie in G. L. Hawk and J. R. Strimaitis, eds., *Advances in Laboratory Automation—Robotics 1984,* Zymark Corp., Hopkinton, Mass., 1984.

R. Dessy, *Anal. Chem.* **55,** 1100A (1983).

G. D. Owens and R. J. Eckstein, *Anal. Chem.* **54,** 2347 (1982).

M. Markelov, M. Antloga, and S. A. Schmidt in J. R. Strimaitis and G. L. Hawk, eds., *Advances in Laboratory Automation—Robotics 1985,* Zymark Corp., Hopkinton, Mass., 1985.

JANET STRIMAITIS
Zymark Corporation

ROSIN AND ROSIN DERIVATIVES. See RESIN.

ROTATIONAL CASTING. See CASTING; ROTATIONAL MOLDING.

ROTATIONAL MOLDING

Rotational molding, also referred to as rotomolding or rotational casting, is a processing method that dates back to the early 1900s. Today the rotational molding industry is one of the fastest growing segments of the plastics industry, with an annual growth of 15%. This is largely due to the efforts of the Association of Rotational Molders (ARM), an international organization, which includes molders, resin and equipment suppliers, professional consultants, and design firms. ARM holds meetings twice a year to update its members on new developments in the industry.

Initially only vinyl plastisols were used for rotational molding: however, in the early 1960s, polyethylenes, the first polyolefin powders, were employed. This created the opportunity for molders to enter markets where vinyl products could not compete. In the early 1970s cross-linkable and modified polyethylenes opened up more new areas, especially in the agriculture-tank market. In the mid-1970s, linear low density polyethylene, a major development in resins, was developed. In the 1980s nylon, polypropylene, and polycarbonate were introduced for rotational molding applications. Most resin is used as a ground powder, ranging in particle size from 74 to 2000 μm (200–7 mesh). Rotational molding provides a more uniform wall thickness, for both single- and double-wall construction, than other methods. Thick corners impart strength. Inserts, ribs, and undercuts are easily included.

The range of designs is limitless. The ARM organization has published a design manual for rotational molding. This was the first publication to address the design guidelines required in the rotational molding industry.

Molds and tooling cost are lower than those of other processing methods, because channels for cooling water and resistance to a clamping force are not required. Different articles and colors may be molded on the same machine and in the same cycle. Quick mold changes are possible when several short production runs are required. Large, hollow products are conveniently made by rotomolding. The largest article molded to date is an 85-m^3 tank. Trimming can be eliminated because very little flash is produced. Rotomolded parts are comparatively stress free. Corner sections are thicker than with other processes, which increases strength. Undercuts, intricate contours, molded-in inserts, and double-wall construction are routinely included; wall thicknesses are uniform.

Process

In most rotomolding operations the molds or cavities are filled with a certain amount of powder or liquid (charging the molds); the mold halves are tied together by bolts or clamps, placed in a heated oven, and rotated biaxially. During heating the resin melts, fuses, and densifies into the shape of the mold. The molds are moved into a cooling chamber, where they are slowly cooled by air and water. After removal from the cooling chamber, the molds are opened and the finished articles are released.

Equipment. The equipment used in rotational molding is simple; many variations are available. The most common type is the so-called carousel type (Fig. 1). The shuttle-type machine (Fig. 2) is used for large products, such as tanks.

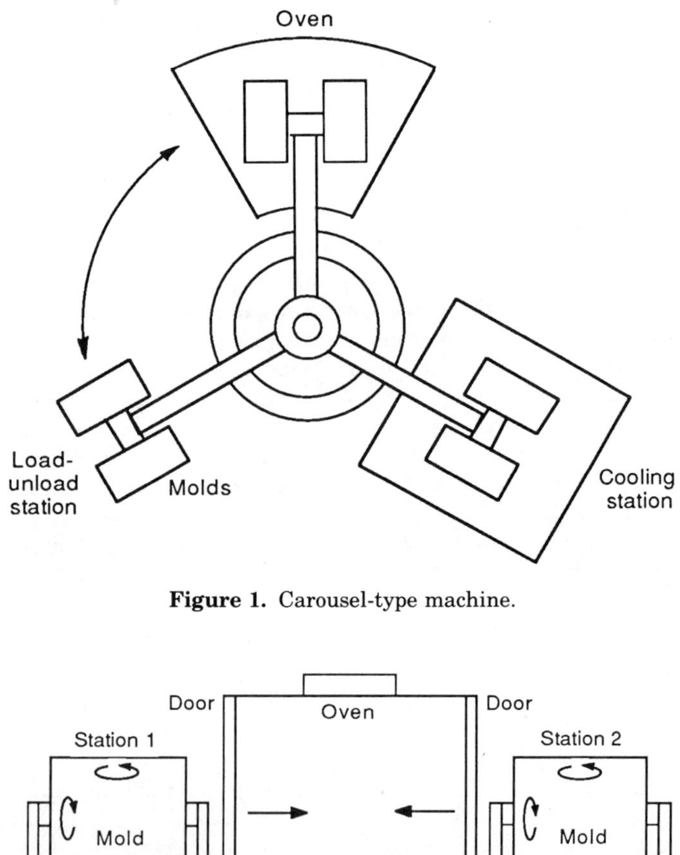

Figure 1. Carousel-type machine.

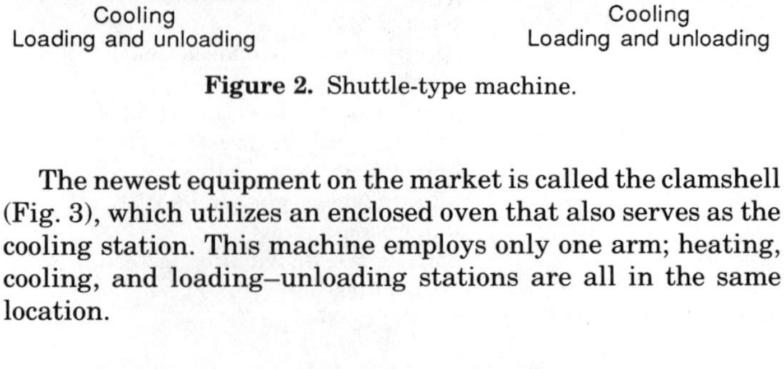

Figure 2. Shuttle-type machine.

The newest equipment on the market is called the clamshell (Fig. 3), which utilizes an enclosed oven that also serves as the cooling station. This machine employs only one arm; heating, cooling, and loading–unloading stations are all in the same location.

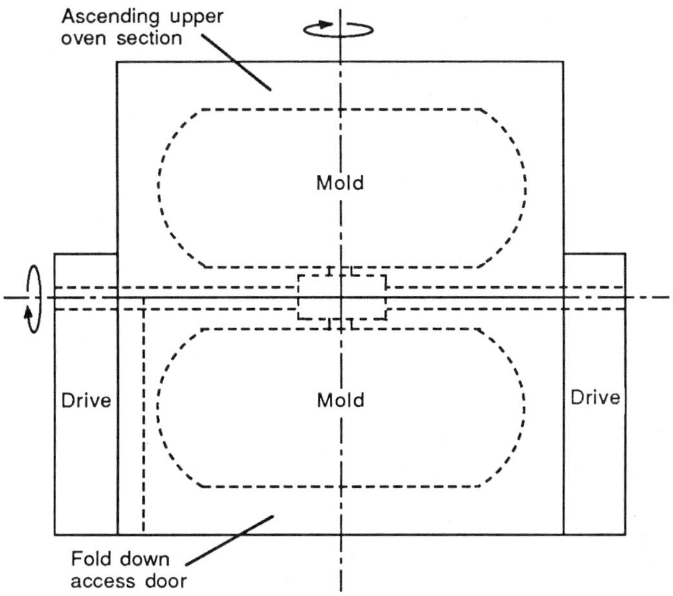

Figure 3. Clamshell-type machine.

Machine maintenance is of extreme importance. A weekly inspection, as well as a good inventory of spare parts, provides a molder with fast and efficient repair in case of emergency.

Molds. The molds used in rotational molding are among the easiest and least expensive to fabricate; two-piece molds are standard, but three-piece molds are sometimes required to remove the finished products. Molds can be as simple as a sphere, or complex, with undercuts, ribs, and tapers. Design considerations include heat transfer, mounting, parting lines, clamping mechanisms, venting, and material stability. Cast aluminum is the most widely used mold material.

Clamping Systems. The most common clamping system for small to medium parts is the C-vise clamp, which is available in hardware stores.

Venting System. Most rotational molds require a venting system to remove the gas that develops in the heating cycle.

Mold Release. Because most rotational molds are designed with little or no draft angle, they must be treated with a release agent.

Secondary Operations. Secondary operations may be required, such as automatic or manual drilling. Additional fittings can be added by spin welding.

BIBLIOGRAPHY

Engineering Design Handbook, Rotational Molding of Plastic Powders, U.S. Army Material Command, Alexandria, Va., 1975.

G. Beall, *The Engineers' Guide to Designing Rotationally Molded Plastic Parts,* Association of Rotational Molders, Chicago, Ill., 1982.

P. T. Dodge, *Materials for Rotational Molding,* Plastics Design Forum, Denver, Colo., 1984.

Association of Rotational Molders Library, Chicago, Ill. The library offers many articles on design, equipment, grinding, mold releases, and molds as well as resins.

PHILIP T. DODGE
Quantum Chemical Company USI Division

RUBBER, CHLORINATED. See RUBBER DERIVATIVES.

RUBBER, CYCLIZED. See RUBBER DERIVATIVES.

RUBBER, DEPOLYMERIZED. See RUBBER DERIVATIVES.

RUBBER, HARD

Hard rubber, rubber-based material that is rigid at room temperature in its finished state, is the oldest synthetic rigid polymer; it was first made in the 1940s by Thomas Hancock in the UK and Charles Goodyear in the United States.

For about a century, this material consisted essentially of rubber hardened by hot vulcanization (qv) with a much higher content of sulfur than that used in normal (soft) rubber products. This type of hard rubber (ebonite) is still important. Its characteristics include high resistance to chemicals, good electrical insulating properties, and the ability to be machined easily. In a later type of hard rubber (pseudoebonite or hard resin rubber), an appreciable contribution to the hardness is made by a resin; some types are free of sulfur.

A true ebonite has a Young's modulus of at least 500 MPa (42,500 psi), corresponding to 100 (absolutely hard) on the international rubber hardness degrees (IRHD) and Shore A durometer scales, or to about 75 Shore D durometer.

Raw Materials

Unlike most polymeric products, hard rubbers are not made directly from a monomer. They can be made from low molecular weight precursors (liquid rubber), but natural or synthetic rubber of molecular weight of 10^5–10^6 is normally used. There are no recently published statistics on hard rubber production or prices.

Ebonite. Ebonite is formed by combining the rubber with a large proportion of sulfur; cross-linking reactions are involved. Ebonites are made from highly unsaturated rubbers, eg, diene polymers or copolymers, containing approximately one double bond per monomer unit.

Hard Resin Rubbers. For these the rubber need not be highly unsaturated; hence butyl rubber, neoprene, and chlorosulfonated polyethylene can be used. The other main organic constituent is a thermoplastic resin, eg, styrene–butadiene or cyclized natural rubber, or a thermosetting resin.

Vulcanization

To convert an elastomer into ebonite, its glass-transition temperature must be raised above 20°C or in practice above the operating temperature of the product in order to remain rigid in use. This is achieved by reaction with a large amount of sulfur. At elevated temperatures, eg, above 140°C, sulfur forms cross-links with highly unsaturated diene rubbers (Fig. 1).

Compounding

Ebonite. Additives used include vulcanization accelerators, fillers (eg, whiting (chalk), clay, baryte, or silica), softeners, (oils, pitches, and waxes), and pigments.

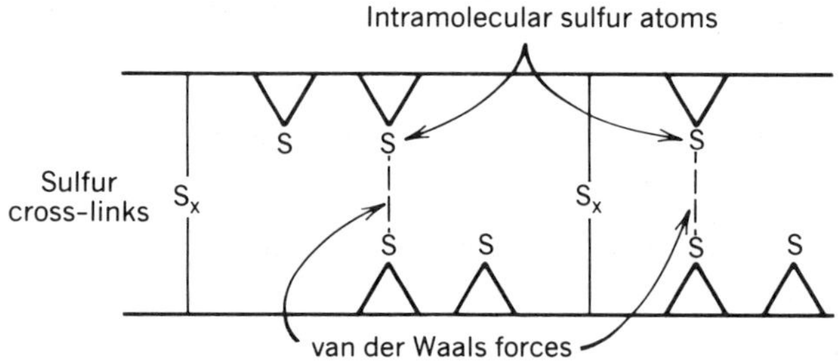

Figure 1. Molecular structure of ebonite. Courtesy of Maclaren and Sons Ltd.

Hard Resin Rubbers. These contain 50–100 phr of a thermoplastic or thermosetting resin. The vulcanizing agent is normally sulfur; metal oxides are used for neoprene and chlorosulfonated polyethylene.

Fillers include carbon blacks, clays, silica, aluminum silicate, and calcium silicate. A greater variety of color is obtainable than with ebonite.

Fabrication

Ebonite articles are fashioned from sheets, rods, or tubes with common machining methods. Hard resin rubbers are likewise easily machined.

Properties

The properties of hard rubber vary widely, depending on the raw rubber, sulfur content, ingredients, and vulcanizing conditions. Table 1 gives properties of unloaded ebonites, those without mineral fillers. Some properties of ebonites containing mineral fillers can be calculated from the properties of the unloaded ebonite and fillers. Properties of hard resin rubbers can be calculated in the same way.

Table 1. General Properties of Unloaded Ebonites[a]

Property	Value
density, g/mL	1.08–1.20[b]
specific heat, kJ/(kg·K)	1.38–1.42
thermal conductivity, W/(m·K) $\times 10^{-2}$	2.9–3.8
coefficient of expansion, vol $\times 10^5$, per °C	
at room temperature	20–24
above T_g	60
refractive index	1.6–1.66
permeability[c] to gases[d], $\frac{\text{mol}}{\text{m·s·Pa}} \times 10^{-17}$	
hydrogen	50
helium	160
nitrogen	<1
permeability[c] to water vapor[e], $\frac{\text{mol}}{\text{m·s·Pa}} \times 10^{-16}$	2–10

[a] Based on natural rubber, except where otherwise stated.
[b] Up to 1.25 for acrylonitrile–butadiene rubber ebonites; density increases with sulfur content.
[c] Approximation.
[d] At 67°C.
[e] At 20–25°C.

Optical Properties. Ebonite has an intense red-brown color by transmitted light. Light transmission of natural rubber ebonite (55-μm thickness) is given below:

wavelength, nm:	<518	545	559	566	571	581.5	613	700
transmission, %:	<0.01	0.4	0.9	2.0	2.9	5	10	25–30

Heat-deformation and Yield Temperature. *Ebonite.* The deformation of ebonite, under stress conditions, increases with temperature. Heat resistance is assessed by the yield temperature or heat-deformation (distortion) temperature at which deformation reaches an arbitrarily chosen value. A better measure is the intercept yield temperature, which, with stressing periods of about 30 min, approximately coincides with the glass-transition temperature as determined from density changes. The intercept yield temperature increases with vulcanization, but is not defined by sulfur content.

Hard Resin Rubbers. In the cantilever heat-deformation plastic-yield test of British Standard BS903:Part D2 at 100°C, hard resin rubbers give deflections (of the end of the cantilever) of 0.7–10 mm. No corresponding data have been recorded for ebonite.

Mechanical Properties. *Ebonite.* The strength properties of ebonites are good, though as usual for materials in the glassy state, impact strength is lowered by notching and all strength values are reduced by the incorporation of mineral fillers. The tensile stress–strain curve is approximately linear up to a stress below the breaking stress, and deformation within the linear range is reversible.

Other properties include the Poisson's ratio: unloaded ebonite, 0.39 (static) and 0.46 (dynamic at 10 MHz); loaded ebonites, 0.2–0.3 (static)).

Ebonite is stable under load at room temperature.

Hard Resin Rubber. These generally have lower tensile strengths than true ebonite, 4–22 MPa (580–3190 psi). However, elongation at break can range from a few percent, typical of ebonite, to almost 500%, even for materials within the hardness range of 98–100 IRHD; much of this elongation is irreversible. Flexural strength is likewise low, eg, 8–16 MPa (1160–2320 psi). On the other hand, impact strength of notched specimens can be as high as ca 25 kJ/m^2 (12 ft·lbf/in^2), compared with 1–7 kJ/m^2 (0.5–3.3 ft·lbf/in^2) for ebonite. Young's modulus of flexure has been reported from 315 to 900 MPa (45,000–130,000 psi).

Electrical Properties. Ebonites made from hydrocarbon rubbers (natural or synthetic polyisoprene, polybutadiene, and styrene–butadiene copolymers) are excellent electrical insulators. However, they suffer surface deterioration during exposure to light and air.

Hard resin rubbers can be compounded to possess good electrical properties.

Chemical Resistance. *Ebonite.* The resistance of ebonite to liquids is a function of composition and vulcanization. Polar elastomers (eg, acrylonitrile–butadiene) and a higher state of vulcanization improve resistance.

Hard Resin Rubbers. The resistance of hard resin rubbers is influenced by the nature of the rubber, amounts of resin and fillers, and degree of vulcanization. Although generally not as resistant as true ebonite, they can be used for chemical plants and battery boxes involving exposure to corrosive chemicals, gasoline, and oils.

Aging and Deterioration. Under normal atmospheric conditions ebonite is highly stable, with no change in physical properties over many years. However, prolonged exposure to elevated temperatures (eg, 70°C) raises the intercept yield temperature and reduces impact strength. Hard resin rubbers likewise have good aging resistance; incorporation of a modified phenolic resin enhances resistance to high temperature aging.

Uses

The principal uses of hard rubbers are those based on chemical resistance, eg, battery boxes and parts; linings and coverings for chemical plants; chemically resistant pipes, valves, pumps, and fittings; tanks, eg, for acid pickling of metals; and corrosion-resistant coverings for rollers. The machinability and acid resistance of ebonite have been exploited in formers for rubber latex dipping. Solvent resistance has led to wide use as the anchoring medium for bristles in paintbrushes.

Many uses depend on mechanical properties such as strength and toughness. Ebonite is used for combs, surgical appliances, pipe stems, and water-meter parts. Hard resin rubbers are used for automobile mudguards and corner panels, dumper bodies, and hospital uses.

BIBLIOGRAPHY

P. Schidrowitz and T. R. Dawson, eds., *History of the Rubber Industry,* Institution of the Rubber Industry, London, 1952.

J. R. Scott, *Ebonite: Its Nature, Properties and Compounding,* Maclaren, London, 1958.

W. J. S. Naunton, ed., *The Applied Science of Rubber,* Edward Arnold, London, 1961.

J. R. Pyne
Consultant

RUBBER, NATURAL

Natural rubber (NR) (*cis*-1,4-polyisoprene) occurs in over 200 species of plants. The *Hevea brasiliensis* tree accounts for over 99% of the world's natural rubber production, which, in 1988, amounted to over 4.9×10^6 t.

Agriculture

Producing *Hevea* trees grow to heights of 15–20 m and require a well-distributed annual rainfall of at least 180–250 cm. They grow well where precipitation is 500 or even 750 cm, but excessive rainfall interferes with tapping. The *Hevea* tree thrives at altitudes up to 300 m and is adaptable to most soils, provided drainage is adequate. It is well documented that there is an inverse correlation between yield per tree and planting density. It has also been proved that yield per hectare (10,000 m^2) increases with planting density up to a point. The current practice on rubber estates calls for planting 400–500 trees/ha. Smallholders, whose average holding in Malaysia is 2.5 ha, may plant as many as 740 trees/ha in order to maximize family income over the short term.

Rubber trees can be propagated by sexual methods, normal or controlled crossing followed by the planting of seeds derived from the crosses, or by vegetative methods, eg, bud grafting. In normal practice, the trees are propagated by grafting buds from a mother tree onto the stem of a seedling. A tree derived either directly or indirectly by this method from a single mother tree is called a clone.

The *Hevea* tree, although robust, is subject to diseases that affect the leaves, bark, and roots. Leaf diseases include *Oidium heveae* and *Colletotrichum gloeosportioides.* Both diseases cause defoliation of the new leaves that form in March and April after the annual wintering. The most devastating leaf disease is the South American Leaf Blight caused by the fungus *Microcyclus ulei.* The disease, which is mainly confined to Central and South America, has prevented significant expansion of rubber growing in this area. Red and White Root Diseases are caused by the attack of *Rigidoporus lignosus* and *Ganoderma pseudoferrum,* respectively. Both diseases prevent the transfer of nutrients from the soil and, if left untreated, eventually kill the tree.

Exploitation

Natural rubber occurs as particles dispersed in an aqueous serum, ie, latex, which is contained in bundles of tubular vessels in the tissue of the tree (Fig. 1). In the trunk, these vertical bundles are inclined from right to left at an angle of ca 5°. Latex is harvested from the tree by a process called tapping, which can be described as controlled wounding of the tree. A specially designed tapping knife is used to remove shavings of bark from the surface of a groove made into the tree to a depth ca 1 mm from the cambium. The groove is made downwards from left to right at an angle of ca 30° to the horizontal across half the tree. The latex flows down the groove and into a collection cup.

Natural rubber arrives at the processing factory as field latex or field coagula. Field coagula generated on a rubber estate consist exclusively of cup lump and tree lace. Cup lump is formed by the natural (bacterial) coagulation of latex which drips into the cup after the morning collection. Tree lace is the thin skin of rubber which forms on the tapping cut, sealing the latex vessels and stopping the flow of latex. Field latex and field coagula are the base raw materials for all types and grades of natural rubber.

Latex Composition

The latex of the *Hevea brasiliensis* tree has been described as a cytoplasmic system containing rubber and nonrubber particles dispersed in an aqueous serum phase.

Freshly tapped *Hevea* latex has a pH of 6.5–7.0, a density of 0.98 g/cm^3, and a surface free energy of 4.0–4.5 μJ/cm^2 (0.96–1.1 μcal/cm^2). The total solids of fresh field latex vary typically from 30 to 40%, depending on the clone, weather, stimulation, tapping frequency, and other factors. The dry rubber content is ca 3 wt % less than the total solids. The nonrubber portion is made up of a large number of substances with proteins and the cyclic polyol quebrachitol predominating. The solids, both rubber and nonrubber, in freshly tapped latex are distributed through three phases, which can be separated by ultracentrifugation. They are the top or rubber phase, the middle or serum phase, and the bottom or lutoid phase.

The serum phase, sometimes referred to as the C-serum or

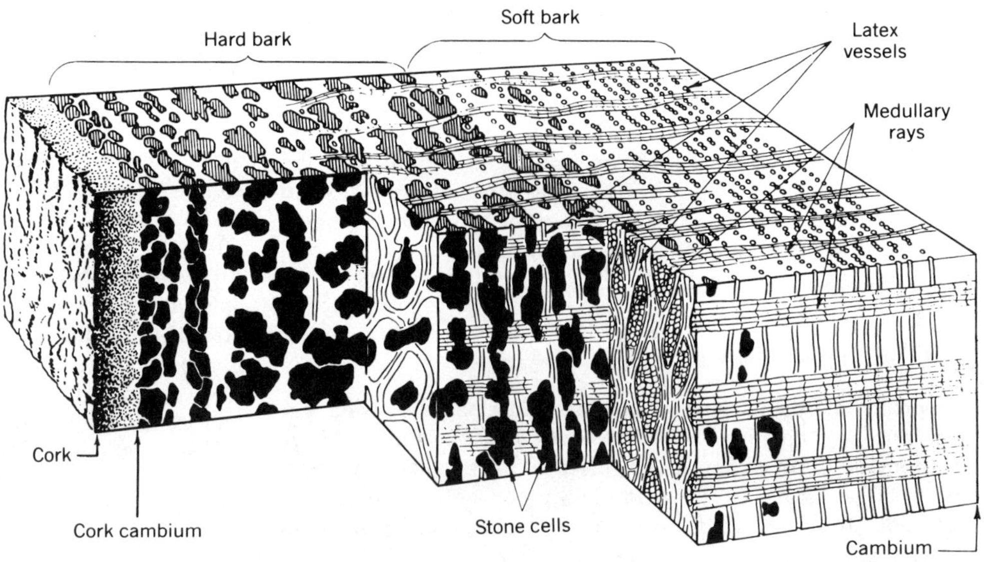

Figure 1. The bark of *Hevea brasiliensis*. Hard bark consists of stone cells, parenchyma, and disorganized sieve tubes and latex vessels. Soft bark consists mainly of vertical rows of sieve tubes. The sieve tubes that make up the bulk of the soft bark and the parenchyma in both soft and hard barks are not shown. The black areas are clusters of stone cells.

aqueous phase, is a dilute aqueous solution with a density slightly over 1.0 g/cm^3. The serum contains many different classes of compounds, including carbohydrates, proteins, amino acids, enzymes, and nitrogenous bases. The bottom fraction or lutoid phase consists mainly of lutoid particles. Lutoid particles contain 20 wt % water-soluble protein, of which ca 70 wt % is hevein (mol wt ca 10,000). The yellow color of the lutoid phase is caused by Frey Wyssling particles, which are spherical, nonrubber particles, whose color is due to the presence of carotenoid pigments. The rubber phase typically contains 96 wt % rubber hydrocarbon, 1 wt % protein, and 3 wt % lipids along with trace amounts of magnesium, potassium, and copper.

The molecular weight and molecular weight distributions of natural rubber can be determined by gel-permeation chromatography (gpc) on fresh samples of raw latex where the gel or insoluble portion is small. The molecular weight distribution (M_w/M_n) is extremely wide for natural rubber, ca 9. Molecular weights range from 3×10^4 to an estimated upper limit of 10^7. The molecular weight distribution curves for natural rubber are bimodal in nature. The bimodal distribution is significant and influences strength and processibility. The high molecular weight fraction imparts strength and other important physical properties; the low molecular weight fraction contributes to processibility.

Types and Grades

Visually Graded Rubber. Visual inspection is the oldest method of grading rubber. There are 8 types: ribbed smoked sheets (RSS), white and pale crepes, estate brown crepes, compo crepes, thin blanket crepes (remills), thick blanket crepes (ambers), flat bark crepes, and pure smoked blanket crepes. The type of rubber is defined by the raw material and the method of production, whereas the grade refers to quality subdivisions within a type. At the factory, each sheet or blanket is visually inspected and graded. The grading is confirmed again by the consumer, usually at the port of entry.

Only two of the eight types of visually graded rubber, ribbed smoked sheet and pale crepes, are of commercial significance.

Technically Specified Rubber. Technically specified rubber (TSR) was introduced by the Malaysians in 1965 under the Standard Malaysian Rubber (SMR) scheme, which is representative of all TSR schemes (Table 1). Other natural rubber-producing countries soon followed with their own version: Indonesia (SIR), Singapore (SSR), and Thailand (TTR). The introduction of TSR brought innovations in processing, packaging, and quality control to the natural rubber industry. In 1987, TSR accounted for ca 45% of the world's natural rubber production. In 1987 Malaysia was the largest producer of TSR, producing 913,000 metric tons, with Indonesia second at 889,000 t.

All versions of TSR are analyzed with the same set of tests to determine quality, but small differences exist in the specification limits and the permissible raw materials.

Production

Visually Graded Rubber. *Ribbed Smoked Sheets.* Ribbed smoked sheets (RSS) are made from intentionally coagulated field latex. The latex is strained through a 250-μm (60-mesh) screen into factory bulking tanks where the dry rubber content (DRC) is determined and the latex diluted with clean water to ca 15% solids. The latex is strained again as it is transferred into coagulation tanks ca 3 m long, 1 m wide, and 40 cm deep. The sides are slotted every 4 cm for aluminum partitions. The pH is adjusted to 4.8 with a 5 wt % formic acid solution. After the acid is stirred into the latex and the surface foam is skimmed, aluminum partitions are inserted into the tank before the latex coagulates. When the coagulum is firm enough, it is flooded with water to prevent oxidation. The rubber is usually allowed to coagulate overnight. The gelatinous slabs are transferred to a sheeting battery along water-filled troughs. The sheeting battery is made up of a series of contra-rotating rollers that rotate at the same speed and compress the coagulum into a sheet 2–3 mm thick. The coagulum is kept immersed in water between each set of rollers to prevent oxidation and to wash the serum (nonrubbers) from the surface of the sheet. The final set of rollers is embossed with the characteristic ribbed pattern, which is impressed in the soft sheet. The sheets are dried for 4–7 d in large smokehouses which may be heated by oil or wood, but enough wood is always burned to smoke the rubber, since smoke acts as a fungicide that prevents mold growth.

Table 1. Standard Malaysian Rubber Specifications

	Latex grades				Field grades		
Grade	SMR CV[a]	SMR L[a]	SMR WF[a]	SMR 5	SMR 10	SMR 20	SMR GP[b]
dirt retained in 44-μm aperture, max wt %	0.03	0.03	0.03	0.05	0.10	0.20	0.10
ash content, max wt %	0.50	0.50	0.50	0.60	0.75	1.00	0.75
volatile matter, max wt %	0.80	0.80	0.80	0.80	0.80	0.80	0.80
Wallace plasticity, min, P_0		30	30	30	30	30	
Plasticity Retention Index (PRI), min %	60	60	60	60	50	40	50
nitrogen content, max wt %	0.60	0.60	0.60	0.60	0.60	0.60	0.60
color limit (Lovibond scale)		6.0					
Mooney viscosity at 100°C, ML (1 + 4)	55–65[c]						58–72

[a] Cure information is provided in the form of a rheograph.
[b] Produced from a combination of latex and field coagulum.
[c] CV 50 and CV 70 are available with viscosity limits of 45–55 and 65–75, respectively.

The smokehouses are divided into sections with temperatures ranging from 40–50°C. The rubber remains 1–2 d in each section before being moved to the next hotter section. Temperatures higher than 55°C cause the finished sheet to blister and have a tacky surface.

Pale Crepes. Pale crepes are produced from freshly coagulated field latex that is carefully selected and preserved. Field latex is selected from clones that produce rubber with a low carotene content. Sodium bisulfite is added, usually in the factory bulking tanks, to prevent enzyme darkening. The latex is strained several times during the processing to remove impurities. The field latex is diluted to 20 wt % DRC with clean water and transferred to the coagulation tanks where formic acid is added to reduce the pH to ca 4.5. When the coagulum is firm enough, it is flooded with water to prevent oxidation.

The coagulum is ready for milling after 3–5 h. The slabs are processed through a series of mills under a water spray which prevents warming and darkening, while it washes away serum and impurities. The coagulum slabs are converted to thick blankets on a series of macerator mills. The rolls on a typical macerator mill have a diamond groove pattern. The crepes are passed through a finisher mill ca 10 times with smooth rolls and a roll speed ratio from 1.6–1.9 to 1. The finished crepes are 1–3 mm thick. The smooth crepes are hung in drying sheds for 6–7 d at 31–34°C.

Technically Specified Rubber. Technically Specified Rubber (TSR) is produced from intentionally coagulated field latex (latex grades) or estate and smallholder coagula (field grades) by means of a crumbling process. The coagulum is converted to blanket form and washed. The blanket is reduced to a crumb which is placed in open metal baskets that travel through an air-circulating drier oven at 110–125°C. The warm, dry crumb is compressed into bales in a hydraulic press and the bales are wrapped in strippable (0.1-mm) or regular (0.03-mm), low melting (108°C max) polyethylene film. The bales measure ca 67 × 33 × 20 cm and weigh between 33 and 35 kg; 30–36 bales are packaged in a polyethylene-lined wooden pallet crate. Shrinkwrapping of the unit with a heavy heat-shrinkable polyethylene film is another method of packaging. This reduces the possibility of wood contamination and eliminates the costly disposal of the pallet crates.

A typical TSR factory has a combination of rubber-processing equipment to convert the latex or field raw material to crumb for drying. This equipment may include crushers, prebreakers, crepers, granulators, hammermills and creper hammermills, and extruders.

This equipment can be assembled in many configurations to produce technically specified natural rubber. A typical TSR L process might involve passing a thick slab of latex coagulum through a crusher, two crepers, and a creper hammermill. The rubber is transported through the process via water or conveyor belts.

Latex Concentrate

Preservation. The preservation of field latex used in the production of latex concentrate may begin with the addition of an ammonia solution to the cup prior to tapping. If the latex is held overnight before centrifuging, it may be treated with aqueous formaldehyde (ca 0.1%) in the factory bulking tanks. Field latex containing significant amounts of magnesium is diluted to ca 30% solids and 0.05 wt % aqueous diammonium phosphate (DAP) is added. The DAP complexes with the magnesium and settles as a sludge.

Centrifuged Latex. The most common method of concentrating field latex is centrifugation with de Laval and Westphalia centrifuges.

Creamed Latex. Ammoniated latex creams slowly on its own, but the rate is not sufficient to obtain a product economically with the required solids content. Creaming agents such as ammonium alginate accelerate the rate of creaming.

Field latex for the creaming process is prepared with ca 1.25 wt % ammonia. This is sufficient for the long-term preservation that is necessary since creaming takes several weeks.

Type. ASTM D1076 sets forth the minimum technical requirements for three types of latex concentrate. Type 1 high ammonia latex represents ca 80% of the market for centrifuged products (Table 2).

Processes and Products. Natural latex is processed to a variety of products (Table 3). In 1988, U.S. latex shipments increased by more than 13% to over 77,000 t (dry). A significant factor in this increase was the demand for rubber gloves for dentists, due to concern over the possible spread of the AIDS virus. The dipped goods segment has increased significantly over the last 5 yr.

Chemistry

Commercial grades of *Hevea* natural rubber contain 93–95 wt % *cis*-1-4-polyisoprene. The nonrubber portion consists of moisture (0.30–1.0 wt %), acetone extract (1.5–4.5 wt %), protein (2.0–3.0 wt %), and ash (0.2–0.5 wt %). Natural rubber is soluble in most aliphatic, aromatic, and chlorinated solvents, but

Table 2. Requirements for Type 1 Concentrated Latex[a]

total solids, min, %	61.5
dry rubber content, min, %	60.0
total solids content minus dry rubber content, max, %	2.0
total alkalinity[b], %	0.60 min
sludge content, max, % of wet mass	0.10
coagulum content, max, % of wet mass	0.050
KOH number, max	0.80
mechanical stability, min, s	650
copper content, max, % of total solids	0.0008
manganese content, max, % of total solids	0.0008
color on visual inspection	no pronounced blue or gray
odor after neutralization	no putrefractive odor

[a] ASTM D1076.
[b] Calculated as ammonia, expressed as a percentage of the latex.

Table 3. U.S. Natural Latex Market

	Percent
dipped goods[a]	67
adhesives	12
carpet industry	5
foam products	7
thread	9

[a] Dentists', surgeons', and household gloves, balloons, condoms, and tubing.

its high molecular weight makes it difficult to dissolve. It is common practice to masticate natural rubber before dissolution. The double bonds in natural rubber undergo the usual chemical reactions such as hydrogenation, addition, substitution, and epoxidation. Fully chlorinated natural rubber, 65 wt % chlorine, is inert and is used in chemically resistant paints and adhesives. The reaction of natural rubber with peracids yields a new polymer, epoxidized natural rubber, which is now available from Malaysia in development quantities at 25 and 50 mol % epoxidation. Epoxidized natural rubber retains good strength properties since it can still crystallize under strain. Oil resistance increases and air permeability decreases with the degree of epoxidation. At 50 mol % epoxidation, the rubber has properties comparable to butyl rubber and a medium acrylonitrile nitrile rubber. The most significant chemical reaction of natural rubber is vulcanization. Unvulcanized natural rubber is soft, easily deformed, tacky when hot, and hard and rigid when cold.

Oxidation. The oxidation of a natural rubber compound is a complex process involving many reactions which are influenced by processing conditions, metal catalysts, temperature, and formulation. Oxidation proceeds through chain scission of the polymer backbone.

An antioxidant provides protection to a rubber vulcanizate by two different mechanisms.

Ozone Attack. In the presence of ozone, the surface of natural rubber vulcanizates develops cracks. The concentration of ozone in the atmosphere varies with geographical area and meterological conditions; concentrations of 0–6 pphm are commonly found. Ozone concentrations are high in areas around electric motors.

Ozone attack on natural rubber can be prevented through the use of wax and chemical antiozonants (qv).

Physical Properties

The physical properties of unvulcanized natural rubber vary slightly, depending on the clonal mix, nonrubber content, and degree of crystallinity and storage hardening. Average physical properties of raw rubber are shown in Table 4. Most applications for natural rubber require that the rubber be vulcanized. The physical properties of a vulcanizate are different from those of the raw rubber. Table 5 gives typical values for the physical constants of a gum and a 50-phr carbon black-filled vulcanizate.

Table 4. Physical Properties of Unvulcanized Natural Rubber

Property	Value
specific gravity	
at 0°C	0.950
at 20°C	0.934
refractive index at 20°C	
RSS	1.5195
pale crepe	1.5218
coefficient of cubical expansion, °C^{-1}	0.00062
heat of combustion, J/g[a]	44.129
specific heat, J/(g·K)[a]	0.502
thermal conductivity, W/(m·K)	0.13
dielectric constant	0.14–0.20
volume resistivity, Ω·cm	10
dielectric strength, V/mm	3.937
cohesive energy density, J/cm^3 [a]	266.5
glass-transition temperature, °C	−72

[a] To convert J to cal, divide by 4.184.

Processing

The manufacture of rubber products from dry natural rubber can be divided into four stages: mastication, incorporation, shaping, and vulcanization.

Applications

Over 60% of the natural rubber consumed in the United States is used in tires.

Aircraft tire treads are also commonly made of all natural rubber, as are automobile engine mountings and suspension systems. Adhesives for surgical, masking, and duct tape as well as packaging adhesives are made from natural rubber. Other products of commercial significance include rubber printing rolls, baby bottle nipples, rubber bands and thread, tank linings, and tennis balls, as well as a wide variety of molded and extruded specialties.

Economic Aspects

The large and sometimes rapid price fluctuations in natural rubber prices from 1940 through 1980 led to the signing of the

Table 5. Physical Properties of Vulcanized Rubber

Property	Gum vulcanizate	Black-filled vulcanizate
density, g/mL	0.95	1.12
Young's modulus, MPa[a]	2	6
bulk modulus, MPa[a]	2000	2200
Poisson's ratio	0.4998	0.4995
sound transmission velocity, m/s	54	37
glass-transition temperature, °C	−70	−70
specific heat, J/(g·K)[b]	1.83	1.50
thermal conductivity, W/(m·K)	0.15	0.28
coefficient of cubical expansion, 10^{-5}/°C	67	56
volume resistivity, ohm·m	10^{14}	10
dielectric constant	3	15
power constant	0.002	0.1

[a] To convert MPa to psi, multiply by 145.
[b] To convert J to cal, divide by 4.185.

International Rubber Agreement (INRA) in 1980. This agreement, developed under the auspices of the United Nations Conference on Trade and Development (UNCTAD), was a cooperative effort of producing and consuming countries whose objective is the stabilization of the natural rubber price and the continuation of adequate natural rubber supplies. The accord had 33 signatories who accounted for over 80% of natural rubber production and 90% of consumption.

Natural rubber producers and consumers adopted a new agreement, INRA II, on March 20, 1987 at the fourth session of negotiations held in Geneva under the auspices of UNCTAD. This agreement replaced the original INRA pact which expired on October 22, 1987. It uses the buffer stock as the sole instrument for market intervention, and specifically excludes export quotas and production control.

BIBLIOGRAPHY

A. D. Roberts, ed., *Natural Rubber Science & Technology,* Oxford University Press, 1988.

C. Barlow, *The Natural Rubber Industry,* Oxford University Press, Kuala Lumpur, 1978.

The Natural Rubber Formulary and Property Index, Malaysian Rubber Producers' Assoc., Brickendonbury, Hertford, UK, 1984.

David R. St. Cyr
The Goodyear Tire and Rubber Company

RUBBER, SYNTHETIC. See Elastomers, synthetic.

RUBBER CHEMICALS

Rubber processing chemicals comprise a group of more than 200 organic materials that are compounded into rubber in varying amounts and combinations to produce vulcanized rubber products. Mixtures of rubbers are often used to achieve desired performance properties. These include natural rubber (NR), synthetic polyisoprene (IR), styrene–butadiene rubber (SBR), butadiene rubber (BR), ethylene–propylene–diene terpolymer (EPDM), halobutyl rubber (XIIR), and butyl rubber (IIR). The materials employed in the fabrication of a commercial tire are given below in approximate decreasing order of amounts used.

- Carbon black
- Sulfur
- Titanium dioxide
- Zinc oxide
- Stearic acid
- Waxes, paraffinic and microcrystalline
- Resins
- Antiozonants
- Antioxidants
- Primary accelerator
- Secondary accelerator
- Retarder
- Bonding agents
- Fabrics: rayon, nylon, polyester, aramid, brass-coated steel, reinforcing fiber glass
- Bead wire: brass-coated steel

Despite the small amounts used, rubber chemicals provide the greatest benefit in terms of gains in physical properties. Among these are state of cure, rebound, aging resistance, and hysteresis. Vulcanization accelerators speed the sulfur cross-linking reaction; activators enhance the accelerator system and are sometimes called secondary accelerators; retarders delay vulcanization and impart scorch protection or process safety; organic sulfur donors and peroxide vulcanization agents promote vulcanization; peptizers reduce viscosity and promote processibility; adhesive systems and fabrics provide reinforcement; antioxidants, antiozonants and waxes protect the rubber during processing and use.

Thus rubber chemicals are used in rubber processing and fabrication; regulation of the rate and state of cure; and for the improvement of vulcanizate properties, product quality, and stability in service (see also Antioxidants; Antiozonants; Calendering; Cross-linking; Curing; Degradation; Vulcanization).

Additional organic materials used in synthetic rubber production but not discussed here include catalysts such as peroxides and metal systems that initiate polymerization; mercaptan chain-transfer agents that regulate molecular weight and polymerization of monomers; fatty acid soaps, surfactant activators, and catalyst neutralizers; plasticizers and polymer extenders; and antioxidant systems, which are added to raw polymers to stabilize them during isolation, drying, and storage (see also ACRYLONITRILE POLYMERS; BUTADIENE POLYMERS: CHAIN-REACTION POLYMERIZATION; CHAIN TRANSFER; COPOLYMERIZATION; EMULSION POLYMERIZATION; ISOBUTYLENE POLYMERS; ISOPRENE POLYMERS; LATICES).

Accelerated Sulfur Vulcanization

Sulfur vulcanization of a rubber polymer is a chemical process by which individual polymer chains are united into a three-dimensional network by connecting sulfur atoms. Other vulcanization systems are also possible where the cross-linking entity is a carbon–carbon bond, a divalent organic radical, or a polyvalent metal. Before vulcanization, rubber can be manipulated and formed into a variety of molded shapes. If left to relax, the rubber frequently loses its predetermined shape and slowly returns to a nondescript form. This process is sometimes called cold flow. After vulcanization, the rubber retains its predetermined shape because the cross-linking sulfur atoms essentially lock or fix the network structure in place (Fig. 1). A vulcanized rubber retracts forcibly to its molded shape after being deformed by some type of mechanical force. Therefore, this process increases the elasticity of the rubber while decreasing its plasticity.

The sulfur vulcanization of rubber is a dynamic process in which the desired state of cure is obtained by heating the product to be vulcanized for a specific length of time. A typical cure system consists of an MBT accelerator, sulfur, and a solubilized zinc salt. These ingredients are trapped in the rubber matrix after the vulcanization. Further changes in the state of cure are possible by reheating this vulcanized system. As is obvious from Figure 2, these changes may depart from the original desired cure state and the properties of the vulcanizate may therefore suffer. This process is called reversion. The compounder must recognize this dynamic nature of the cure system and optimize the choice of ingredients to produce a stable vulcanizate.

Complex chemical transformations and equilibria occur before cross-linking is initiated. During this induction period, very little change occurs in polymer elasticity. In actual production, most of this scorch period is used as process safety time during which compounding, extruding, and mixing as well as fabrication of the product are accomplished. The control of scorch, cure rate, cure state, and reversion depends to a large extent on the accelerators.

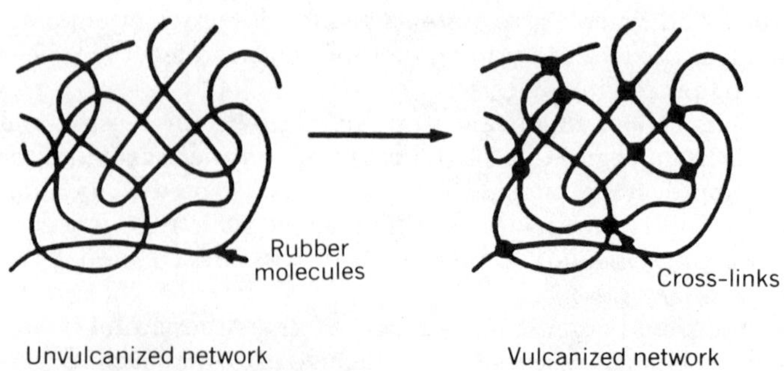

Figure 1. Network formation by vulcanization.

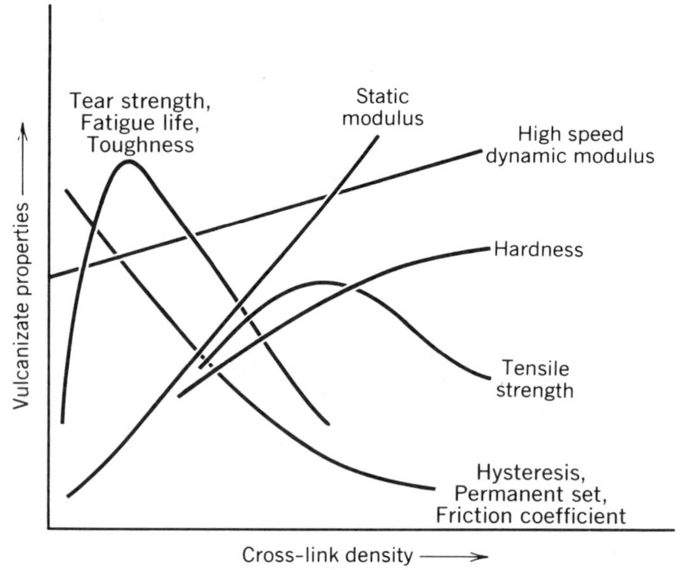

Figure 2. The effects of vulcanization. Courtesy of Academic Press, Inc.

Modern delayed-action accelerators are usually based on mercaptobenzothiazole. These accelerators permit the production of low cost vulcanizates because they allow the practical use of sulfur. The vulcanizates exhibit uniform and desirable wear and aging properties. Furthermore, different cross-sectional areas in a fabricated product can be vulcanized to different states of cure, depending on the sulfur and accelerator concentrations; in addition, the scorch period can be controlled. Typically, a rubber product such as a tire is composed of several distinct regions, for example, tread, carcass, sidewall, apex, and bead areas. These regions are composed of different mixtures of rubbers. Thickness must also be taken into account because the heat-transfer and transport properties are different and depend on the cross section of the product. To compensate for differences in heating time and rubber type, the compounder uses different MBT-based accelerators in different parts of the tire. This allows the inside and outside of a vulcanizate to cure at relatively the same rate.

Accelerators

Rubber accelerators can be categorized into five principal groups: mercaptobenzothiazoles and derivatives, dithiocarbamates and derivatives, sulfur donors, guanidines, and aldehyde–amine reaction products. Thioureas, xanthates, and mercaptothiazolines are less important, and are used in specialty rubbers. Estimated total U.S. accelerator production for 1988 was over 36×10^3 metric tons. This is considerably less than the all-time high of over 60×10^3 t in 1979. This drop is due to the energy crisis, smaller automobiles and tires, and the change in tire construction from bias-ply to radial. The latter last longer, are of thinner gauge, and are used all year round.

2-Mercaptobenzothiazole and Derivatives. The most important commercial cure accelerators are the sulfenamide derivatives of 2-mercaptobenzothiazole. As delayed-action accelerators, these derivatives provide excellent scorch protection. They are active in a broad spectrum of rubbers and rubber blends and permit safe processing at high temperatures. Different derivatives can be used in different components. Yet because the vulcanization process depends on the liberation of MBT, the ultimate cure in the finished product is uniform.

Mercaptobenzothiazole (**1**) is prepared by an autoclave process from aniline, carbon disulfide, and sulfur at 250°C. The product is separated from tars by dissolving it in base, extracting the tars, and precipitating the MBT by acidification.

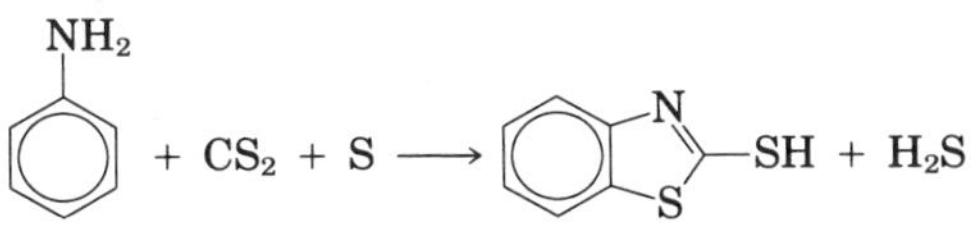

(**1**)

Dithiocarbamates and Derivatives. The reaction of carbon disulfide and an amine produces unstable dithiocarbamic acids, which can be stabilized as amine or metal salts. The most important of these are the zinc dialkyldithiocarbamate salts. They provide a fast cure rate and short scorch period, and are therefore called ultraaccelerators. Several heavy-metal derivatives are also available.

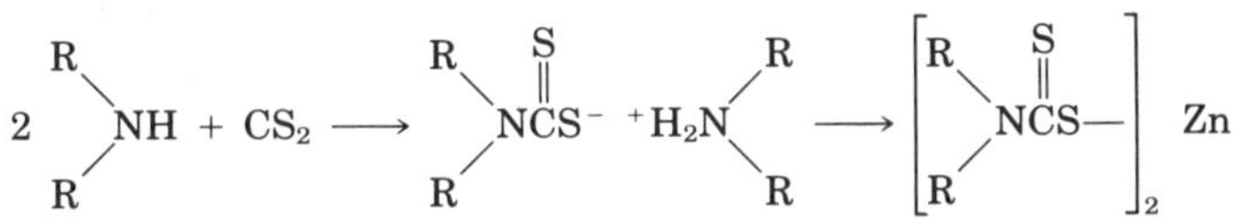

Sulfur Donors. Efficient vulcanization (EV) can be achieved by using high accelerator-to-sulfur ratios or ultrasystems such as the thiuram disulfides. The thiuram disulfides are easily prepared by the oxidation of the dithiocarbamic acids above. Sulfur-donating derivatives include tetrabutylthiuram disulfide, tetraethylthiuram disulfide, dipentamethylenethiuram disulfide, and dipentamethylenethiuram tetrasulfide. Another sulfur-donor system is 2-(4-morpholinyldithio)benzothiazole, which is prepared from MBT, morpholine, sulfur, and bleach. It has much better scorch properties than the thiuram disulfides and can be used with or without sulfur.

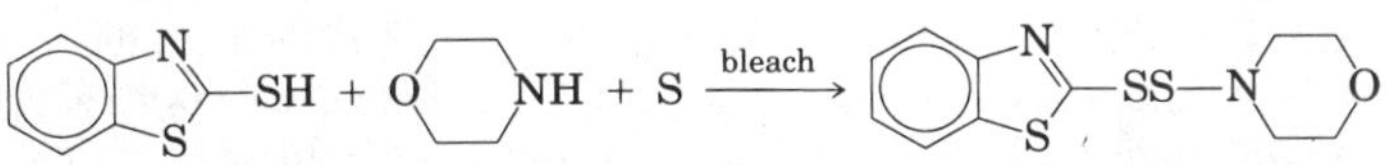

Guanidines. The reaction of aniline and cyanogen chloride gives 1,3-diphenylguanidine (DPG). With *o*-toluidine, the corresponding di(*o*-tolyl) derivative (DOTG) is formed.

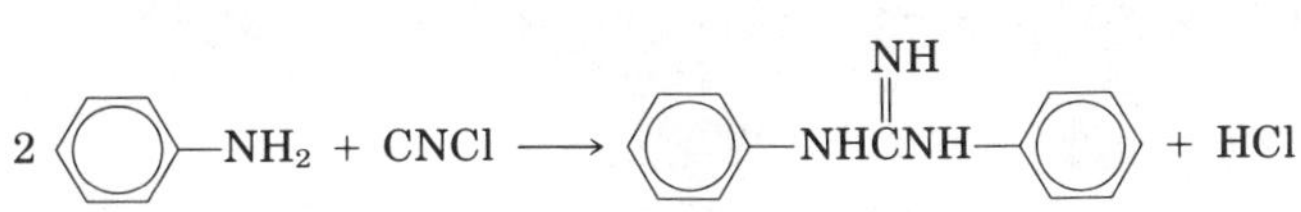

DPG

These basic materials are seldom used as primary accelerators because of their long cure times. When used alone, they produce polysulfidic cross-links and their vulcanizates age poorly. They are frequently used as secondary accelerators or "kickers" for thiazole accelerator systems.

Aldehyde–Amine Reaction Products. Various aldehyde–amine condensation products find use as secondary accelerators. Reaction of aniline, *p*-toluidine, or *n*-butylamine with butyraldehyde, acetaldehyde, formaldehyde, or 2-ethylhexenal in the presence of an acid catalyst leads to the formation of Schiff's base, which is unstable and undergoes self-polymerization.

Cure Systems

Peroxide Cures. Organic peroxides can be used as cross-linking agents forming carbon–carbon cross-links. These vulcanizates are even more stable than those formed in an EV sulfur system. Peroxide vulcanizates can be made transparent or white and will resist discoloration from exposure to ultraviolet light. In black stock, dicumyl peroxide gives vulcanizates with good low temperature flexibility, compression set, and aging. This is true for all rubber polymers except butyl rubber, which undergoes chain scission in the presence of peroxide initiators.

The mechanism for peroxide cure is straightforward. Decomposition of the peroxide produces alkoxy radicals, which abstract hydrogen atoms from the polymer. The polymeric alkyl radicals degenerate by coupling and cross-links form. Therefore, peroxides give uniform cures and better physical properties.

Other Cure Systems. Other chemical cross-linking systems include phenolic resins, quinone derivatives, maleimides, and nitroso derivatives.

Retarders

Retarders inhibit premature vulcanization during processing. Thus they extend the scorch period during which compounds are mixed, milled, calendered, and built into "green" products. Protection continues until the products are ready for heat treatment and vulcanization. Conventional retarders include benzoic acid, salicylic acid, phthalic anhydride, and nitrosodiphenylamine (NDPA). More recent additives include *N*-(cyclohexylthio)phthalimide (CTP), sulfonamide derivative, Vulkalent E/C (Mobay/Bayer), *N*,*N*′,*N*″-tris(isopropylthio)-*N*,*N*′,*N*″-triphenylphosphoric triamide (**2**), and *N*,*N*′,*N*″-hexaisopropylthiomelamine (HTM) (**3**).

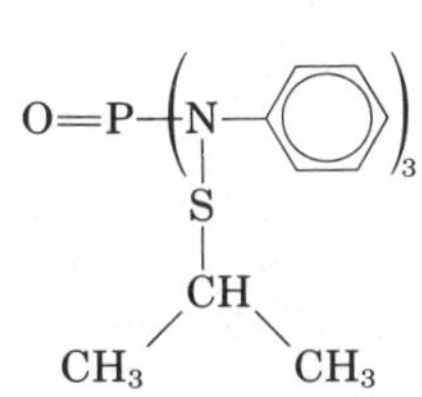

(**2**)

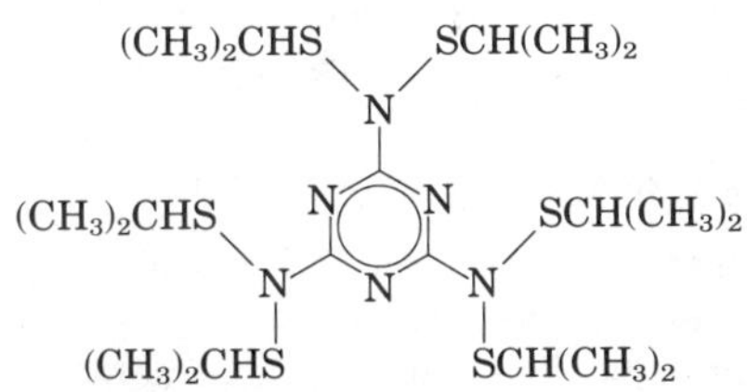

(**3**)

The exact mechanism of acid retarders is uncertain.

Prevulcanization Inhibitors. All prevulcanization inhibitors have a sulfenamide or disguised sulfenamide N-S-type linkage in their substructure. The first such material reported was *N*-(trichloromethylthio)phthalimide, a good retarder for natural and synthetic rubber. Subsequently *N*-(cyclohexylthio)phthalimide (CTP) was introduced as an efficient commercial retarder. Today CTP is the most widely used prevulcanization inhibitor in the United States. Its synthesis from phthalimide and cyclohexylsulfenyl chloride is given below.

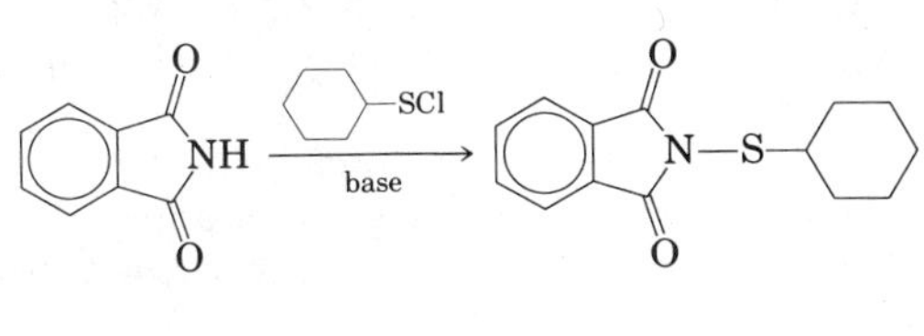

The prevulcanization inhibition effect of CTP may occur by the competitive reaction of CTP for MBT in the initiation step of vulcanization.

Peptizers

The mechanical processing of rubber often requires viscosity modification to improve workability, reduce the mixing temperature, improve dispersion and incorporation of fillers, increase polymer tack, and improve lubricity for extrusions and molding. On mastication peptizers rapidly reduce molecular weight and Mooney viscosity by chain cleavage. This treatment does not adversely affect the vulcanization rate or the physical properties of the cured rubber. Several aromatic disulfides and mercaptans are peptizers; small amounts of MBT have also been used. Peptization can occur under high shear mixing or milling, resulting in chain rupture due to homolytic carbon–carbon bond cleavage and the production of polymeric radicals. Other processing aids include softeners, extenders, and plasticizers, as well as oils, tars, resins, pitches, and synthetic organic esters.

Bonding Agents

Tires, belts, and hoses usually contain fabric reinforcement. The composition of this fabric can be as diverse as organic textiles, glass fiber, or steel wire. These reinforcement materials must adhere firmly to the rubber after being subjected to repeated and varying strains. Organic textiles are treated with adhesive, which is usually proprietary information of the tire manufacturer. The treatment consists of a single or double dip, followed by stretching to remove twisting, and drying.

Bonding agents, such as resorcinol, hexamethylenetetramine, and hexamethoxymethylmelamine are mixed directly into the rubber recipe. Strong adhession results because the resorcinol–formaldehyde resin that forms *in situ* provides strong hydrogen bonding with the fabric and undergoes partial cure to the rubber. For steel cord reinforcement, the adhesive preparation of choice is brass plating.

Antioxidants and Antiozonants

Elastomers undergo aging and deterioration that are characteristic to their structure and the environment in which they are used. The common detrimental environmental factors include oxygen, ozone, heat, light, mechanical stress, fatigue, trace metal contamination, and oxidizing and reactive chemicals. Oxidation can occur during polymer preparation, storage, and processing. Some oxidative degradation is beneficial because it permits controlled breakdown of polymer during mixing and mastication. Oxidation of the finished product should be prevented, however, because it affects physical properties, such as modulus, tensile strength, elongation, compression set, elasticity, and hardness.

Antioxidant Synergism. When a mixture of antioxidants provides more protection than the sum of the individual components, the effect is said to be synergistic. Typically, this involves antioxidants that function by different protective mechanisms, for example, a hydroperoxide decomposer and a radical scavenger.

Ozonation. The problem of ozone cracking of tires was not recognized until after the introduction of amine antioxidants and accelerators in the 1930s. This was because tires wore out faster than they were cracked by ozone. Today, service expectancy is over 80,000 km. Early attempts to protect against ozone involved waxes as nonreactive barrier films. Waxes are classified into paraffins, microcrystalline (amorphous), and intermediate types. The paraffins are pure linear C_nH_{2n+2} hydrocarbons (n = 18–50). They are obtained from petroleum and are separated as specific melting point cuts, eg, 54/56°C, 56/58°C, 58/60°C, etc. The use of wax is a mechanical solution that works well under static, but not under dynamic, conditions. Waxes are used frequently in conjunction with a chemical antiozonant system (see ANTIOZONANTS).

BIBLIOGRAPHY

A. Y. Coran in E. R. Eirich, ed., *Science and Technology of Rubber,* Academic Press, Inc., Orlando, Fla., 1978, Chapt. 7.

R. Taylor and P. N. Son in M. Grayson, ed., *Kirk-Othmer Encyclopedia of Chemical Technology,* Vol. 20, 3rd ed., John Wiley & Sons, Inc., New York, 1982, pp. 337–364.

R. R. Barnhart in M. Grayson, ed., *Kirk-Othmer Encyclopedia of Chemical Technology,* Vol. 20, 3rd ed., John Wiley & Sons, Inc., New York, 1982, pp. 365–468.

JOSEPH A. KUCZKOWSKI
The Goodyear Tire & Rubber Company

RUBBER DERIVATIVES

The natural rubber hydrocarbon is a 1,4-polyisoprene with an almost 100% cis content (**1**) or more skeletally, (**2**):

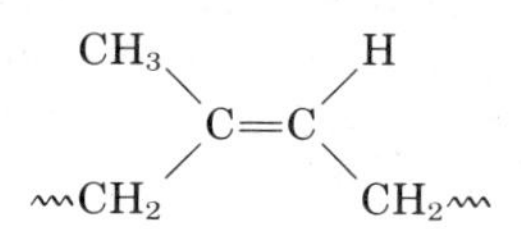

(**1**)

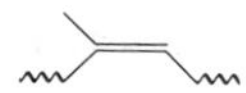

(**2**)

In its chemical reactions, it behaves as a simple trialkylethylene. However, the reactions are influenced by two factors compared to the reactions of simple olefins. The high molecular weight and the presence of nonrubber substances may result in side reactions, such as cross-linking, degradation, and cis–trans isomerization of natural rubber during chemical reactions, and finally in a reduction in activity. The nonrubber substances also prevent the occurrence of certain reactions that can be carried out with synthetic *cis*-1,4-polyisoprene, eg, metalation and anionic graft polymerization.

Bond Rearrangement Reactions

Isomerized Rubber. The 1,4-polyisoprenes occur as isomers with different cis–trans ratios, ranging from the 100% cis structure of natural rubber to the 100% trans structure of gutta percha (**3**).

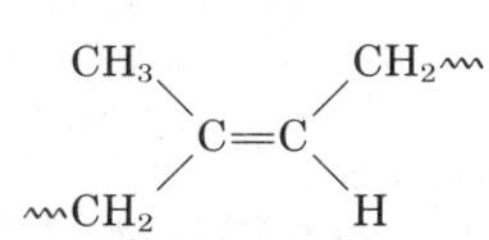

(**3**)

The first successful cis–trans isomerization of natural rubber was reported in 1959. The transformation can be carried out with the help of catalysts on natural rubber in the form of a solid, solution, or latex. Isomerization takes place when thin films, sheets, or crumbs are heated with sulfur dioxide above 100°C. In commercial practice, butadiene sulfone is used, which produces sulfur dioxide *in situ*. Natural rubber is mixed with about 2% butadiene sulfone in an internal mixer or on a two-roll mill and heated at 160°C. This method is also used in the cis–trans isomerization during vulcanization.

Commercially, this material has not been successful because of the availability of low-cost oil-extended styrene–butadiene elastomers that do not crystallize.

Cyclized Rubber. Cyclized rubber, the first chemically modified derivative of natural rubber, is the hard, resinous product obtained by treating the rubber with strong acids or Lewis acids. Sulfuric acid, titanium chloride, ferric chloride, stannic chloride, *p*-toluenesulfonic acid and its chloride, and boron trifluoride have been employed. Cyclizations are carried out on solid rubber, solutions, or latex, depending on the catalyst.

Cyclized and partially cyclized rubbers have been used for shoe soles, hard moldings, heavy-duty industrial rollers, adhesives, bonding agents, reinforcing resins, corrosion-resistant surface coatings, and printing inks. However, in most of these uses, they have been replaced by block copolymers, eg, high styrene–butadiene copolymer resins and cyclized synthetic *cis*-1,4-polyisoprene.

Attachment of Pendent Groups

Hydrogenated Rubber. Complete hydrogenation of natural rubber would give an alternating copolymer of ethylene and propylene.

$$(\mathbf{1}) + H_2 \longrightarrow \sim CH_2{-}\underset{}{\overset{CH_3}{\overset{|}{C}}}H{-}CH_2{-}CH_2\sim$$

Hydrogenated rubber is more crystalline and has a slightly higher glass-transition temperature than does natural rubber. It can be vulcanized with the conventional systems. The vulcanizates show good ozone resistance at a high degree of hydrogenation. Hydrogenated rubber is not produced commercially.

Halogenated Rubber. Natural rubber reacts readily with fluorine, chlorine, and bromine. Chlorination, in particular, has been extensively studied.

Hydrohalogenated Rubber. Hydrogen chloride adds readily to natural rubber in the form of a thin film, in solution or latex, to give the derivative rubber hydrochloride. Both the solution and latex processes have been used in the manufacture of rubber hydrochloride.

Rubber hydrochloride is a highly crystalline ($T_m \sim 115$°C), tough, semielastic, film-forming material (sp gr 1.16). Its solubility is similar to that of chlorinated rubber. It loses hydrogen chloride on heating to about 100°C or on treatment with bases. It was widely used as packaging film, but has been replaced by polyethylene and similar materials. It has been used in compositions for bonding metal to rubber and as a component of laminates with paper, cellulose films, or metal foils.

The addition of hydrogen bromide proceeds like the addition of hydrogen chloride. Rubber hydrobromide $(C_5H_9Br)_n$ has been obtained as a powder; it is unstable. Rubber hydrofluoride is made by adding a dioxane solution of hydrogen fluoride to a xylene solution of rubber at −15°C.

Alkylhalogenated Rubber. Polyhalogen derivatives of alkanes containing at least one bromine atom (eg, CBr_4 and $CBrCl_3$) add to rubber in the presence of a peroxide. The products are elastic and vulcanizable and show some flame resistance. The best combination of physical properties and flame resistance is exhibited by a product containing 15–25% by weight of the halogen compound. The latex derivative is suitable for the production of rubber-bonded hair pads and flame-resistant latex foam for spreading on carpets or fabric.

Epoxidized Rubber. Natural rubber and other unsaturated polymers can be epoxidized in solution or the latex stage by peracids. The reagent may be conveniently prepared *in situ*.

Epoxidized natural rubber (ENR) is prepared from latex with performic acid generated *in situ*. Commercial production of ENR began at the end of 1988; three grades have been available in small tonnage quantities for four years: ENR-10, ENR-25, and ENR-50, corresponding to epoxidation of 10, 25, and 50 mole %, respectively (ENR-10 has 10 moles of epoxide groups per 100 moles of isoprene groups).

Epoxidized natural rubber is an elastic solid with a higher density (sp gr 0.94–1.03) than natural rubber. Epoxidation increases the T_g linearly by about 1°C for every mole % of epoxidation.

Potential applications for ENR are in engineering components, the inner liner of tubeless tires, oil seals, and tire treads. Epoxidized natural rubber has good adhesive properties and can be used to bond poly(vinyl chloride) (PVC) to noncompatible materials such as natural rubber or styrene–butadiene rubber (SBR) and to bond rubber to metals. However, these applications are only in the exploratory stage and commercialization will depend on the availability of the rubber at competitive prices.

Maleic Derivatives. Maleic anhydride reacts with natural rubber under different conditions to give different derivatives. The reactions are of interest because they impart polarity to the nonpolar natural rubber. Moreover, the anhydride groups introduced can serve for further modification. In the presence of free-radical initiators, reaction occurs in solution at 70–80°C to give pendent succinic anhydride groups without loss of unsaturation; the reaction is usually less than 25%. Gelation is a serious problem, but it can be reduced by adding diethylamine.

The products from the above reaction range from elastic rubbers to hard resins, depending on the amount of maleic anhydride involved. They can be vulcanized by metal oxides, diols, diamines, and diisocyanates and show good flex cracking resistance, aging resistance, and solvent resistance.

"Ene" Derivatives. Workers at the Malaysian Rubber Producers' Research Association have prepared several derivatives of natural rubber using the ene thermal addition reaction. The general ene reaction with natural rubber can be represented as follows:

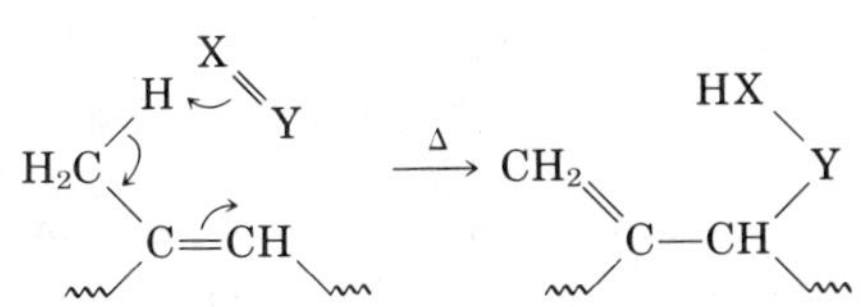

(X=Y can be —N=O, —N=N—, >C=O, >C=C<, >C=S)

These reactions proceed in a predictable manner and do not

involve ionic or free-radical catalysts that can be poisoned by impurities.

Derivatives of Thiols and Sulfur Compounds. Certain thiols and thiol acids add readily to natural rubber by a free-radical mechanism.

$$\text{initiator} + \text{RSH} \longrightarrow \text{RS}\cdot$$

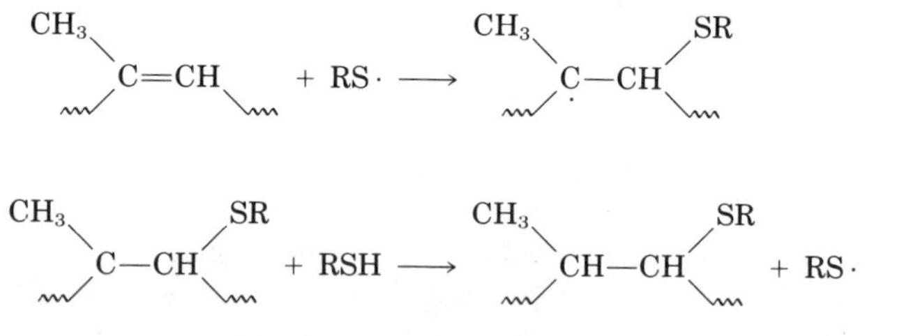

The reaction can take place in solution or in latex. Cis–trans isomerization takes place during the thiol addition. Thiols of the general structure $HS{+}CH_2{+}_nX$ have been used to prepare natural rubber derivatives with reactive functional groups and special properties. Thiols and certain disulfides are used as chemical peptizers in the rubber industry.

Carbene and Nitrene Derivatives. Carbenes (>C:) and nitrenes (—N:) react easily with natural rubber. A dichlorocarbene derivative of natural rubber (**4**) has been prepared by the reaction of a strong aqueous solution of sodium hydroxide with a chloroform solution of natural rubber in the presence of decyltrimethylammonium bromide as phase-transfer catalyst. The derivative shows poor thermal stability and processing behavior. However, with a chlorine content above 25%, it shows flame-retardant properties.

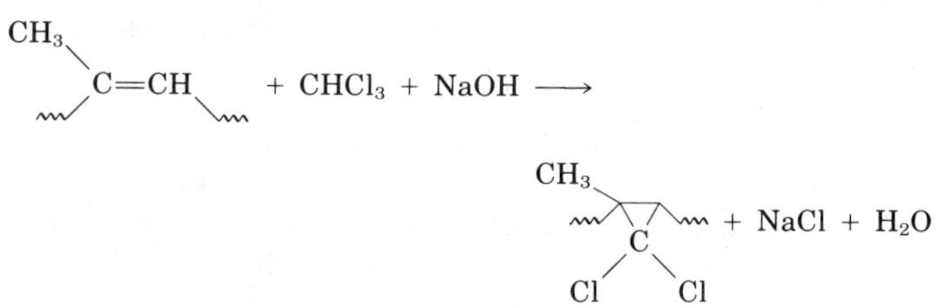

(**4**)

The carbene derived from the photolytic or thermal decomposition of 3,5-di-*tert*-butylbenzene-1,4-diazooxide has been used to introduce a rubber-bound hindered-phenol antioxidant.

Derivatives of Aldehydes. Natural rubber reacts with some aliphatic aldehydes in the presence of acid catalysts. Derivatives with formaldehyde, glyoxal, chloral, bromal, and fluoral have been studied.

Surface Modification

Surface chlorination of natural rubber crumbs with sodium hypochlorite and dilute acid and surface bromination with bromine water are used in the production of free-flowing rubber. Surface chlorination is a preparation for painting with flexible urethane paints. Chlorination initiated by γ-irradiation decreases the coefficient of friction and improves resistance to accelerated aging. Surface bromination with bromine water imparts hardness to certain rubber articles. Bromination with an aqueous solution containing 3–7% of bromine salts and a catalyst reduces the friction, resulting in a smooth surface. Bonding to metal and other rubbers and plastics is improved. Treatment of the vulcanized rubber surface with trichlorofluoromethane at −20°C improves thermal resistance. Surface friction is reduced by fluorination with antimony pentafluoride vapor.

Graft Copolymers

Graft copolymers of natural rubber and vinyl polymers are prepared by polymerization of the monomers initiated by free radicals in the presence of natural rubber, or by attaching polystyrene to the rubber backbone by chemical reactions. Commercial graft copolymers of natural rubber with poly(methyl methacrylate) (PMMA) include Heveaplus MG 30 and MG 49, containing 30 and 49%, respectively, of PMMA.

MG copolymers are used in the manufacture of impact-resistant articles such as cutting boards and rigid moldings, eg, electric plugs, roller skates, and castor wheels. In solution or latex form the copolymer is used as an adhesive or bonding agent to bond rubber to PVC, leather, textiles, and metals.

The graft copolymer with polystyrene has been used in the production of impact-resistant articles. Other graft copolymers show improved oil resistance.

Other Modifications

Liquid or Depolymerized Natural Rubber. Liquid natural rubber is here defined as depolymerized natural rubber that has a sufficiently low viscosity to be able to flow at room temperature. Liquid or depolymerized natural rubber is prepared by the oxidative scission of natural rubber under controlled conditions.

Liquid or depolymerized natural rubber is soluble in most organic solvents except alcohols and the lower ketones. It is used as a rubber softener (for improved processibility), a lubricating oil additive, a binder in the manufacture of grinding wheels and bowling balls, a cement for brush bristles, in corrosion resistant paints, a sealant for pipe works, an insulating material for low power electrical components, and in the manufacture of printing rollers and battery boxes.

Superior Processing Rubbers. Superior processing natural rubber or SP rubber contains a mixture of vulcanized and unvulcanized rubber coprecipitated at the latex stage. Superior processing rubbers retain the normal physical properties of natural rubber but show improved processing characteristics as displayed by reduced die swell and mill shrinkage, better shape retention, and better surface finish. They can be used in the production of extruded articles, calendered sheets, and adhesives.

BIBLIOGRAPHY

D. Barnard, K. Dawes, and P. G. Mente, *Proc. Int. Rubber Conf., Kuala Lumpur, 1975* **4**, 215–234 (1976).

M. Porter, *Plast. Rubber Mater. Appl.* **3**, 32–36 (1978).

D. Barnard, *Kautsch. Gummi Kunstst.* **35**, 747–752 (1982); *Plast. Rubber Singapore* **1**, 23–29 (1982).

C. S. L. Baker and D. Barnard, *Polym. Prepr. Am. Chem. Soc. Div. Polym. Chem.* **26**(2), 29–30 (1985).

J. A. Brydon, *Rubber Chemistry,* Applied Science, London, 1978, pp. 161–193.

A. SUBRAMANIAM
The Rubber Research Institute of Malaysia

RUBBER RECLAIMING

In the mid-1800s Charles Goodyear developed a technique for vulcanizing rubber. He blended natural rubber with sulfur and placed it on his wood stove, where the rubber cured into a water-impermeable sheet. Because of the shortage of natural rubber at that time, a rubber-reclaiming process, based on steam pressure to devulcanize used rubber, was developed in 1858. The cured ground rubber was subjected to steam pressure for 48 h. Today, high cost silicone rubber polymers are reclaimed in much the same way, although most synthetic polymers require more complicated techniques. With today's technological advances, recycled rubber will be used more and more in the the United States, but with regulations for handling and disposing of worn tires. Crude oil costs have dropped from about $189/m^3 ($14/bbl) over the last few years, making it seemingly unprofitable to recycle rubber, especially tires. However, tires discarded in landfills tend to float on top; mosquito breeding and illegal tire disposal are causing problems that could be alleviated by recycling. Approximately 70% of the scrap rubber, primarily as tires, is discarded in landfills. Private landfills may charge up to $3 per tire, and disposal costs at municipal landfills are ca $0.30–0.60 per tire. These costs encourage illegal disposal of tires. As the tire piles grow, rubber recovery should become more economical. Companies entering into multimillion-dollar energy recovery projects will be encouraged by an adequate supply of old tires.

Scrap Rubber as Fuel Source

The use of scrap rubber for fuel offers the best alternative for reusing rubber, as fuel costs increase and tire disposal problems become more serious. Tires in the form of 2.50-cm chips are most economical because shredding is not expensive. Tires contain more than 90% organic materials and have a heat value of ca 32.6 MJ/kg (ca 14,000 Btu/lb), compared with coal values of 18.6–27.9 MJ/kg (ca 8,000–12,000 Btu/lb). A cyclonic, rotary-hearth boiler fired with whole tires was operated by the Goodyear Tire and Rubber Co. from 1975 to 1977. It was designed to burn 1400 kg/h and generated 11,300 kg of steam per hour.

Pyrolysis

Scrap-tire pyrolysis has been the subject of several studies by rubber, oil, and carbon black industries. The Tosco-II process pyrolysis study sought to develop equipment to maximize carbon black production and quality.

The Tosco-II project was completed, but additional projects have been discouraged, because the U.S. carbon black industry already has excess capacity. Furthermore, the types of blacks and residue constituents, eg, ZnO, and other inorganic material and glass fiber from old tires, are unpredictable. However, carbon must be carefully chosen to achieve specific properties in compounded rubbers. Thus the residue mixture is useful only as a low grade filler and cannot be used as a carbon black source.

American Ligurian, Inc. (Stamford, Conn.) is marketing a pyrolysis process, which was developed in Italy, that generates steam for hot water, air heating, dryers, kilns, and similar installations. A modular plant produces 8000 kg/h of steam from 1 t/h of tires. The pyrolysis process produces 0.9 t of fuel oil, 270 t of steel, and 54 t of ash annually. Gas emissions meet the strictest environmental standards.

Depolymerized Scrap Rubber

In the depolymerized scrap rubber (DSR) experimental process, ground scrap-rubber tires produce a carbon black dispersion in oil. Depolymerized scrap rubber has a heat value of 40 MJ/kg (17,200 Btu/lb) and is blended with No. 2 fuel oil as fuel extender.

Asphalt–Rubber

In 1986, an estimated 25,200 t of rubber were used in asphalt. Asphalt–rubber is mixed and applied to roadways by several techniques. In one method, rubber and asphalt are mixed at ca 175–220°C for 1–2 h. The hot mixture is applied to the roadway and covered with a layer of stone chips to form a chip seal. The rubber crumb usually consists of scrap tires that are ground into particles less than 2 mm in diameter. Asphalt–rubber is also used for water proofing membranes, crack-and-joint sealers, hot-mix binders, and roofing materials. The rubber improves asphalt ductility and increases the temperature at which the asphalt softens. The aggregate adhesive bond is stronger, and the asphalt lasts longer. Production of asphalt–rubber has increased from 405 t in 1970 to 27,000 t in 1980. Typically about 2 t of rubber is used for 1 km of roadway. If it is assumed that asphalt–rubber contains ca 25% rubber and 75% asphalt, the potential demand for scrap rubber is ca 40,500 t/yr, or ca 2% of the amount available.

Cryogenic Pulverizing and Mechanical Tire Shredding

Tires must be pulverized or shredded before they can be reclaimed by devulcanization or used in asphalt and other recycling processes. The tires are mechanically ground, sometimes using cryogenic freezing or solvent-swelling techniques to enhance grinding efficiency. In one process, a polar solvent is used to swell the rubber, followed by shearing to reduce particle size. Ground tire-crumb rubber is commonly referred to as rubber reclaim, even though the rubber has not been devulcanized.

Reclaiming

Originally only solid rubber scrap was reclaimed, but with the advent of fiber-reinforced pneumatic tires, methods for removing the fiber had to be developed. Acids and alkalis were used to decompose the fiber to cellulose. The alkali digester process, developed in 1899, is still used today. Fiber glass reinforcement has to be removed mechanically before the rubber can be reclaimed.

Today's scrap tires contain a mixture of synthetic and natural rubbers. They must be ground to a finer particle size for

Table 1. Properties of Reclaimed Rubbers

Polymer	Specific gravity	Mooney viscosity	Tensile strength, MPa[a]	Elongation, %	Ash, %	Acetone extract, %
NR[b]						
neutral	1.19	45	9.0	500	29	9
black	1.19	25	8.6	400	13	11
SBR, whole tire[b]	1.18	50	6.9	280	7	22
IIR, tubes[c]	1.17	65	9.0	540	7	11

[a] To convert MPa to psi, multiply by 145.

[b] Rubber Reclaimers Association cure test recipe for mixed elastomer and NR reclaim. Cured 20 min at 141°C.

[c] Rubber Reclaimers Association cure test recipe for IIR reclaim. Cured 30 min at 160°C.

better oil absorption, and stronger chemical reclaiming agents must be used. Alkylphenol sulfides, aromatic amines, chlorinated mercaptans, and other strong plasticizing agents are absorbed evenly into the rubber particles, resulting in bond cleavage and a uniform smooth reclaim.

The properties of devulcanized reclaimed rubber are predictable, although less so than vulcanized rubber. However, the chemical reactions taking place during reclaiming cannot be accurately studied because of the complex nature of the devulcanized polymers. Although the reclaiming reactions cannot be proved by analysis, it is suspected that depolymerization occurs via oxidation of the rubber bonds; heat, pressure, chemical agents, and mechanical (milling) shearing contribute.

Natural rubber (NR) scrap is reclaimed with solvent naphthas, terpenes, dipentenes, and resins that swell, tackify, and promote bond cleavage (see RUBBER, NATURAL). Tires, mainly synthetic styrene–butadiene rubber (SBR), are less susceptible to oxidation (see BUTADIENE POLYMERS). Aryl disulfides, phenyl disulfides, high molecular weight mercaptans, and other sulfur-containing chemicals are used to swell and lubricate the bonds and devulcanize the rubber by chemical oxidation and bond cleavage. Reclaimed rubber can be used in adhesives, in tire formulations, or with virgin rubber. Compounding is based on rubber hydrocarbon content (RHC).

Tires, natural rubber tubes, and butyl tubes are the main sources of scrap and reclaim (see ISOBUTYLENE POLYMERS; ISOPRENE POLYMERS). Specialty reclaim materials are made from scrap silicone, chloroprene (CR), nitrile–butadiene (NBR), and ethylene–propylene–diene terpolymer (EPDM) rubber scraps (see CHLOROPRENE POLYMERS: ETHYLENE–PROPYLENE ELASTOMERS). Tires, hoses, belts, molded and extruded goods, and asphalt products consume ca 80% of the reclaimed rubber manufactured. Reclaimed rubber is used in the tire carcass and sidewall compounds; reclaimed natural rubber in cements, dispersions, and pressure-sensitive tape; and butyl reclaim in tubes and inner liners for tires. Typical properties are shown in Table 1.

Tire Retreading and Other Uses

Retreading has been the most cost-effective alternative to recycling rubber. However, worn retreaded tires usually are discarded in a landfill. Currently, ca 10% of discarded automobile tires and 55–70% of truck and bus tires are retreaded. Scrap whole tires also have been used to form artificial fishing reefs, oyster beds, and as a floating breakwater.

BIBLIOGRAPHY

M. B. Sikora, *Tire Recovery and Disposal, A National Problem With New Solutions,* Resource Recovery Report, Washington, D.C., June 1986.

J. P. Paul, *CHEMTECH* **9,** 104 (1979).

F. T. Ryan, *Tire Manufacturers' Perspectives on the Legislation Options for Scrap Tire Disposal, paper presented at Disposal Techniques with Energy Recovery for Scrapped Vehicle Tires Workshop,* Feb. 12, 1987, Denver, Color.

Industrial Recovered Materials Utilization Targets for the Rubber Industry, report prepared for U.S. DOE, Assistant Secretary for Conservation and Solar Energy, Office of Industrial Programs, by Hittman Associates, Inc., 1980.

JOHN PAUL
PEI Associates, Inc.

RUNNERLESS MOLDING. See INJECTION MOLDING.

S

SAFETY

Safety in chemical plants and other materials processing industries can be divided into two distinct fields: industrial safety and process safety. Industrial safety includes protection of employees and other personnel from the hazards of being "struck against, struck by, caught in, falling, slipping, overexertion, contact with temperature extremes, contact with electric current, and inhalation, absorption, or ingestion." Fire protection also could be included. These hazards usually are confined within the site boundary and do not affect the public.

Most important in the development of safe work practices and a safe work environment are the development of a safety organization and its continuing activities. It should include the highest levels of management to demonstrate that safety has a priority equal to that of production, quality, efficiency, cost control, and employee relations. Only in this way can responsibility and authority be accepted or shared throughout the organization to develop and maintain attitudes essential to safe work practices and conditions.

Other important aspects are inspections of equipment, audits of work practices, guards on moving equipment, training, accident investigations, record keeping, personal hygiene, personal protective equipment, fire prevention, noise control, and control of thermal radiation, ionizing radiation, and chemical exposures. Process safety includes the safe transportation, handling, and storage of chemicals, the control of reactors and other processing steps, the protection of employees and the public from fires, explosions, and releases of toxic materials, and the protection of property from the effects of such incidents.

Chemical Hazards

Manufacturers and users of chemicals are exposed to toxicity and reactivity hazards inherent in the production of goods, intermediates, and construction materials. Consideration of toxicity, flammability, chemical instability, reactivity and reaction energy, operating conditions, and corrodibility of construction materials aids in analyzing the hazard potentials of chemical operations.

Toxic Materials. Materials encountered in industry come in contact with individuals by ingestion, inhalation, skin absorption, and subcutaneous injection. Unless employees are careful about washing before eating, chemical substances on their hands may be ingested. With most substances, a single occurrence may not have serious effects, but the continuous daily ingestion of small amounts may result in harmful accumulation. Inhalation of gases, vapors, and dusts may also have serious, acute, and cumulative toxicologic effects. Toxicity data are available for a wide variety of substances (see also TOXICOLOGY).

Strong acids and strong alkalies can severely burn the skin. Other chemicals may produce rashes that may cause discomfort and infection. Repeated exposure to solvents removes natural oils from the skin, which then tends to crack and become irritated and possibly infected. Absorption through the skin is possible with materials that are appreciably soluble in either water or oil. Other materials can be absorbed if first dissolved in good solvents. Exposure may also occur accidentally through cuts or scratches.

The eyes are particularly susceptible to liquids, gases, and some solids. The conjunctival membrane surrounding the eye is easily irritated; alkaline materials, eg, hydroxides and amines, rapidly destroy the eye tissues, frequently causing partial or complete loss of vision.

Flammability

Engineering and operational controls are usually effective in fire prevention. In modern continuous processing, flammable materials are confined in closed systems, reducing the likelihood of their mixing with air or coming in contact with ignition sources. However, in batch processing or under emergency conditions, flammable materials may be released and ignited, resulting in vapor cloud explosions. The pertinent hazardous properties of flammable materials have been compiled by the National Fire Protection Association (NFPA).

Reactions. Certain reactions are difficult to control, particularly if there is a failure of instrumentation, cooling, or agitation. The conditions of chemical reactions are often extreme, eg, temperatures well above 500°C or approaching absolute zero. Changes in metal properties at these temperatures can be dramatic. Ordinarily, steel loses much of its tensile strength with increasing temperature and becomes brittle at low temperatures. Stainless steels or other alloys may be used at elevated pressures, and nonferrous materials are needed at cryogenic temperatures. Pressures range from high vacuum to hundreds of megapascals (thousands of atmospheres) in laboratory and commercial reaction. Choices of metals and design are commonly guided by standards. Unusual stresses, eg, temperature cycling, vibrations, load bearing, wear, and earthquakes, must also be considered.

Corrosion. Corrosion and erosion attack chemical processing equipment. Special alloys resistant to stress corrosion may be needed, for example, to prevent corrosion by chloride in stainless steel vessels. Hydrogen can induce similar stress-corrosion cracking and blistering. Lining vessels with glass preserves the purity and color of the product and usually prevents corrosion. Chemical vapors or fumes, eg, hydrogen chloride, corrode piping, valves, and other equipment. Periodic inspection and testing of all vessels and piping are essential preventive maintenance.

Process Design and Operation

Plant Site and Layout. The choice of a location for a chemical plant depends on a number of factors and their effects on plant personnel and the surrounding community. These factors include the hazards associated with flammability and toxicity of materials and reaction energy. Consideration should be given to the possible effect on the community of the worst possible accident, based on prevailing winds and other atmospheric conditions. An overcrowded plant can lead to damage or shutdown of adjacent units and may impede the movement of vehicles and materials in an emergency. Water supply for cooling and fire fighting is a vital necessity.

Equipment and Control System Design. Successful operation of a chemical process depends on the care used in selecting and arranging the process equipment. Flow sheets with material balances aid in equipment design, particularly for piping, pumps, heat exchangers, and the control systems. Models can be of great benefit in the design of spacing, access, piping, exits, and walkways.

Failure of utilities, such as electric power, instrument air, process and cooling water, refrigeration and heating media, and inert gases, must be considered; control devices should include fail-safe design, for example, cooling water valves should fail open, whereas steam valves should fail closed. Providing appropriate backup sensors and controls for continuous processes and sequence-control systems for batch operations can reduce the possibility of improper operation.

Personnel Selection and Training. Trained and knowledgeable personnel are essential for the operation of industrial and chemical equipment. Reliability of operations requires thorough training in normal and emergency procedures and the absence of drug abuse. Medical screening avoids damaging exposures to susceptible individuals; eg, people with respiratory ailments should not be exposed to potentially corrosive or toxic atmospheres.

Process Start-up, Operation, and Shutdown. Written start-up procedures should be available, allowing the operating personnel to study them and correct any flaws before start-up. Following start-up and check-out of processing procedures and equipment, operations are usually routine and uneventful. Standard operating procedures, specified limits for operating parameters (temperature, pressure, flows, levels, etc), log sheets, and checklists are employed. Written procedures for normal and emergency shutdowns should be prepared, rehearsed, and kept up to date.

Maintenance. Maintenance in the chemical industry is unique because of the potentially hazardous nature of the materials, processes, and equipment involved. Maintenance procedures must be carefully developed, including confined-space entry, line-breaking, locking and tagging, and control of ignition sources.

Medical Programs. Immediate treatment is the essence of first aid; this includes artificial respiration and cardiac resuscitation. First aid, as taught by the American Red Cross or the U.S. Bureau of Mines, and rescue techniques should be an integral part of chemical-plant training programs. Many large chemical plants have a full-time physician in attendance during working hours and on call at other times. Smaller plants have part-time physicians and take injured employees to a nearby hospital or clinic. However, when part-time physicians or outside medical services are used, there is little opportunity for medical personnel to become familiar with plant operations or to assist in improving the health aspects of plant work. Therefore, it is essential that chemical hazards manuals that explain symptoms and methods of treatment are available for emergencies.

Disaster Planning. Plant managers must formulate a plan of action in case of disaster. All critical personnel must be familiar with this plan. Fire, explosion, and toxic-release drills should be held to ensure that all personnel can be accounted for and are instructed in emergency procedures.

Accident Investigation. Study of accidents and injuries can lead to correction of unsafe conditions and prevention of recurrence. Accident investigation involves examination of site, equipment, process conditions, and records.

System Safety Analysis. Safety assessments of entire processes begin with quantification of flammability and explosion hazards by measurements of vapor pressure, flash points, and flammability limits. Process safety data are now rather easily obtained for reaction rates and exothermic reactions. Also available are the temperature limits for unstable chemicals (beyond which explosive decomposition can occur); rates of gas or vapor generation (for proper design of emergency pressure-relief devices); recommended limits for exposures to toxic materials, radiation, noise, and heat; and strengths and corrosion rates of materials of construction.

System safety analysis was developed for potential process malfunctions, and can be applied to the chemical industry. The four principal methods are failure mode and effect, fault tree, human-error prediction, and hazard and operability studies.

Polymerization

Reactor Design. Several aspects of polymerization reactor design are important to process safety, including controls, cooling devices, pressure relief, and fire and explosion prevention (see POLYMERIZATION PROCEDURES, INDUSTRIAL; REACTOR TECHNOLOGY). Polymerization reactors often contain significant amounts of unreacted monomer; thus there is potential for runaway reactions. These uncontrolled reactions cannot be stopped by merely turning off the feeds.

Except for single-phase reactors, which are completely filled with gas or liquid, there is a vapor space above the liquid. Precautions should be taken to ensure that the mixture of gases and vapors in this space is not explosive, particularly during preparation for start-up. Otherwise, potential ignition sources, such as worn agitator bearings, pyrophoric catalyst residues, or condensed and reacting monomer (polymer "popcorn"), may cause combustion or explosion. The precautions may include sweeping any air (or other oxidant) out of the vapor space with an inert gas such as nitrogen.

Proper selection and close control of temperature and pressure in polymerization reactions are essential to safe operation. Testing of reactants, fail-safe design of controls, alarm and interlock-shutdown systems, overpressure protection, and reaction inhibitors may be required.

Reactor Operation. Start-up sequences and precautions are specific to each type of reactor or auxiliary operation. In many reaction systems, interlocks are needed to ensure that all ingredients are present in the correct ratios before start-up. To initiate the start-up sequence, it may be necessary to bypass these interlocks; a timer can be used to limit the bypass operation.

Control systems are usually designed to cope with upsets in coolant flow (or heat input), changes in reactant ratios, and power or instrument-air failure. Shutdown caused by power or instrument failure, by upsets that trigger interlock shutdown, or by agitation failure should be considered as serious incidents; restart should not be attempted without careful examination of the causes and effects of the shutdown.

The hazards attending shutdown may be as serious as those of start-up. Unreacted monomers need to be removed and safely separated from catalysts or inhibitor and stored. They are drained or pumped from the reactor or flushed with an inert material.

If polymerization reactors cannot be cleaned with a solvent, acids, caustic solutions, or oxidizers are used. These methods may present hazards from corrosive or flammable liquids.

Overpressure Relief. Relief valves are used to provide protection of reactors against overpressure; if an explosion is likely, a rupture disk of adequate area may provide protection. Relief valves vent the vapors or gases causing the overpressure, but reseat when the pressure drops to the normal range and prevent loss of reactor materials.

Barricades. Barricades are usually provided for research or pilot-plant reactors. Their strength is evaluated in terms of TNT equivalent, because the rupture of process vessels (and vapor/air explosions) can also be assessed in terms of kilograms of TNT. Barricades improve the efficiency of exhaust ventilation, which may be needed to remove accidental releases of materials and to maintain a safe atmosphere for periods of barricade entry.

Water Deluges. Cooling of process vessels and supporting structures is usually provided when flammable liquids are handled. Similar protection for flammable gases is frequently not provided, but careful attention is given to flange orientations, to ensure that jets of burning gas would not be directed toward the supporting structure, cable trays, and feed piping. Water deluges may also be provided for external cooling of runaway reactions.

Polymer Processing. The final product of polymerization processes is usually in the form of pellets, powder, or flakes, although some products are handled molten or as slurries or solutions. The polymer is fabricated into industrial or consumer products by mixing, blending, compounding, and molding in a wide variety of equipment (see PLASTIC PROCESSING). These operations may be accompanied by hazards of moving equipment, hot surfaces, electrical shock, and fire and explosion.

Equipment Guarding. Operating personnel should be protected by suitable guards from rotating and reciprocating equipment, hot surfaces such as heater bands, and electrical connections. Emergency stop devices and "dead man" jog switches are provided where there is exposure to these hazards.

Fire and Explosion Protection. Protection alternatives include reducing the oxygen content of the atmosphere with nitrogen or other inert gas, providing explosion vents of adequate size, installing explosion and fire suppression systems, and installing explosion-proof electrical equipment.

Protection from Toxic Materials. To ensure minimal exposure of operating personnel to toxic materials, efficient ventilation should be provided. In some cases, special precautions are needed to avoid contamination of food and clothing with polymer dusts.

Process Safety Legislation

In the United States, the Occupational Safety and Health Administration (OSHA) has limited its activity to enforcement of codes or consensus standards concerning toxicity and flammability; however, there has been a recent trend toward applying the "general duty" clause more broadly to management of hazardous chemical processes. The responsibility of chemical process managers for public safety has also increased in the past few years, particularly in New Jersey and California. Regulatory legislation was expanded by the Health, Education, and Welfare Department (now Health and Human Services) and the EPA to include air and water quality, waste disposal, noise, manufacture and use of toxic materials, and assignment of responsibility to manufacturers for product safety. Several organizations, eg, the Chemical Manufacturers Association (CMA), have been established to assist manufacturers in coping with these regulations. The Chemical Industries Institute of Toxicology assists the industry in investigating work environment toxicity hazards.

BIBLIOGRAPHY

Registry of Toxic Effects of Chemical Substances, National Institute for Occupational Safety and Health, Washington, D.C., 1983.

Fire Protection Guide on Hazardous Materials, 9th ed., National Fire Protection Association, Boston, Mass., 1986.

W. J. Fenlon, *Plant/Operations Progress* **3,** 197 (1984).

RICHARD W. PRUGH
Hazard Reduction Engineering, Inc.

SANDWICH STRUCTURES. See HONEYCOMB; LAMINATES.

SARAN. See VINYLIDENE CHLORIDE POLYMERS.

SCANNING ELECTRON MICROSCOPY. See ELECTRON MICROSCOPY.

SCATTERING

A scattering experiment in which the sample under investigation (the target) is bombarded by a defined beam of particles yields a wealth of information not only about the molecular structure of the target, but also about its dynamics as influenced by the forces and interactions between atoms, molecules, and nuclei. Usually, the momentum and energy of the incident beam of particles, and sometimes the polarization of the magnetic moment or electrified vector, are known. The particles are scattered by target atoms, which may be in the gaseous, liquid, crystalline, or amorphous state; the scattered particles are subsequently detected by devices giving the intensity of the scattered beam as a function of the scattering angle or the energy of scattered particles. In experiments involving polarized beams the degree of polarization before and after the scattering process must be determined. The intensity of the incident beam is usually a compromise between obtaining statistically good data and using a low enough number density of particles in the beam for them to be treated as noninteracting.

The equations governing the scattering process are of general application:

$$
\begin{aligned}
E - E_0 &= \hbar\omega \qquad \text{for inelastic scattering} \\
E - E_0 &= 0 \qquad \text{for elastic scattering} \qquad (1) \\
\hbar\mathbf{k} - \hbar\mathbf{k}_0 &= \hbar\mathbf{Q}
\end{aligned}
$$

Here, $\hbar\omega$ and $\hbar\mathbf{Q}$ are the momentum and energy transfers, respectively; the initial energy and wave vector are denoted by a subscript 0; and the unsubscripted quantities refer to the scattered beam. $\hbar = h/2\pi$, where h is the Planck constant. If the particles employed possess a rest mass m, their velocities before and after scattering are denoted by $\mathbf{v}_0$ and $\mathbf{v}$, in which case $\hbar\mathbf{k} = m\mathbf{v}$. With elastic scattering both energy and momentum are conserved, whereby $|\mathbf{k}| = |\mathbf{k}_0|$. With inelastic scattering momentum and energy transfer occurs between the target and the beam. Depending on the nature of the particles used, equation 1, although generally valid, is subjected to different boundary conditions and yields different time–space domains. Most scattering studies on polymer systems have been conducted using x rays, neutrons, or light. The domains of energy compatibility of different types of radiation can be seen in Figure 1. Clearly, x-ray and visible light energy ranges are limited, whereas neutrons obtained from a pulsed source span a vast energy range.

General Theory of Scattering

According to quantum mechanics, a particle of momentum $\mathbf{p}$ can be represented by a wave of the de Broglie wavelength $\lambda = h/[\mathbf{p}]$. Scattering theory uses the concept of wave–particle duality extensively. The scattering process itself is theoretically described in terms of waves or fields that influence the target, whereas in an experimental situation both the incident and scattered beams are considered to be composed of particles. A crucial connection between these two descriptions is provided by the differential scattering cross section $d\sigma/d\Omega$, defined as

$$d\sigma/d\Omega = I(\theta,\phi)/I_0 \qquad (2)$$

and is obtained directly from the measurement. Here, $I(\theta,\phi)d\Omega$ represents the number of particles scattered into a narrow cone of solid angle $d\Omega$ having an axis directed toward the detector, and I_0 is the number of incident particles. The scattered intensity $I(\theta,\phi)$ is given in terms of spherical polar coordinates (r,θ,ϕ) with the z axis directed along $\mathbf{k}_0$, the wave vector of the incident beam. The distance from the target to the detector is large compared with the dimensions of both, so that $d\Omega$ is well defined.

By relaxing the spatial constraint of a narrow cone $d\Omega$ and counting the total number of particles scattered in all directions, the total scattering cross section σ is obtained:

$$\sigma = \int_{\text{all directions}} (d\sigma/d\Omega)\, d\Omega \qquad (3)$$

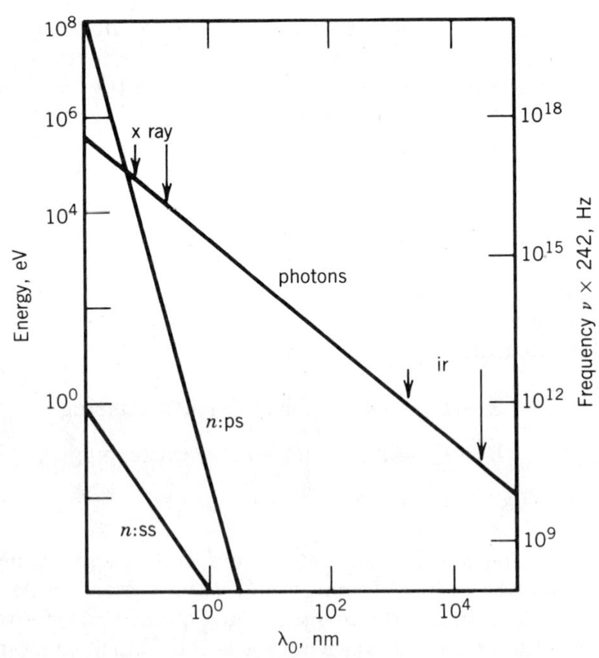

Figure 1. The energy-wavelength dependence for different types of radiation, photons, ir, and neutrons. For neutrons (n), ps denotes the pulsed source and ss the steady source.

Elastic and inelastic scattering are conveniently treated separately (cf eq. 1). For fuller theoretical treatments the reader is referred to standard textbooks on quantum mechanics, general treatises on scattering, and monographs and reviews on neutron, x-ray, and light scattering. Several books describing the applications of these techniques to systems of interest for polymer scientists have appeared.

Scattering Laws and Correlation Functions. The introduction has been quite general; in principle it is valid for both photons and particles. However, in the following the discussion is biased toward neutron scattering, which is unique among diffraction and spectroscopic techniques. For neutrons the relationship between the energy and wave number is such that atomic or molecular dimensions correspond to wavelengths associated with thermal energies. Since the scattered intensity depends on the wave vector $\mathbf{k}$ and the frequency of the wave while passing through the scatterer, thermal neutrons can show appreciable shifts of both wavelength and energy after scattering from the sample. Therefore, in principle both the spatial atoms and molecules and the energy levels in the scatterer can be studied in the same experiment. Because they possess zero charge, neutrons interact weakly with matter and penetrate it easily. The spin-½ of the neutron confers a magnetic moment, which then interacts with the spins of unpaired electrons and permits investigation of magnetic phenomena. Although such investigations are seldom of value in polymer science, there is one application of importance in which spin polarization analysis allows one to analyze the scattered intensity from polymers and extract unambiguous structural information free from the problem of the subtraction of an incoherent background, as shown below.

From equation 2, elastic scattering can be sufficiently analyzed in terms of $d\sigma/d\Omega$, which is directly related to the number of counts measured in the scattered beam, whereas investigation of inelastic processes must also involve energy analysis. In the latter case the partial differential scattering cross section $d^2\sigma/d\Omega\, dE_f$ is related to the scattering function (law)

$$d^2\sigma/d\Omega\, dE_f = N/\hbar|\mathbf{k}|/|\mathbf{k}_0|[\sigma_{\text{coh}}S_{\text{coh}}(\mathbf{Q},\omega) + \sigma_{\text{inc}}S_{\text{inc}}(\mathbf{Q},\omega)] \qquad (4)$$

written for N independent scatters. The subscripts coh and inc denote the coherent and incoherent parts of the scattering, respectively. $S(\mathbf{Q},\omega)$ is often termed the dynamic structure factor.

Although there are several equivalent ways of expressing $S(\mathbf{Q},\omega)$, the physical meaning of the two terms constituting the scattering law (eq. 4) is most easily understood from a description in terms of the Van Hove correlation function $G(\mathbf{r},t)$. $G(\mathbf{r},t)$ is in general a complex function, but it reduces to a real positive function within certain ranges of $\mathbf{r}$ and t, where quantum mechanical effects are insignificant. Physically, $G(\mathbf{r},t)$ is a time-dependent pair-correlation function and is often displayed as the sum of distinct and self-correlation functions $G_d(\mathbf{r},t)$ and $G_s(\mathbf{r},t)$, respectively. The interpretation of these correlation functions is clear in the classical limit, where for all equivalent nuclei $G(\mathbf{r},t)$ is the probability that, given a particle at the origin ($\mathbf{r} = 0$) at time $t = 0$, any particle is found at $\mathbf{r}$ at a later time t. Correspondingly, $G_s(\mathbf{r},t)$ yields the probability that the same particle is found at $\mathbf{r}$ at time t. Provided atomic diffusion can be neglected, $G_s(\mathbf{r},t)$ is applicable directly to both crystalline and amorphous solids.

The importance of these correlation functions $G(\mathbf{r},t)$ stems from the fact that, in accord with the general predictions of the Wiener-Khinchin theorem, any spectrum represented as spectral density in the frequency domain may be expressed as the Fourier transform of the correlation function describing the decay in time of the quantity linking the target to the incident radiation. The intermediate scattering law $S(\mathbf{Q},t)$ derived from the $\mathbf{r} \rightarrow Q$ Fourier transform serves as such a coupling function for neutrons:

$$S_{\mathrm{coh}}(\mathbf{Q},t) = \frac{1}{N} \sum_{j,j'} \langle e^{-i\mathbf{Q}[\mathbf{R}_j(0)-\mathbf{R}_{j'}(t)]} \rangle$$
$$S_{\mathrm{inc}}(\mathbf{Q},t) = \frac{1}{N} \sum_{j} \langle e^{-i\mathbf{Q}[\mathbf{R}_j(0)-\mathbf{R}_j(t)]} \rangle \qquad (5)$$

These quantities are measured directly in dynamic light-scattering experiments and in neutron spin-echo work, two techniques that have a great deal in common. If, however, a full four-dimensional Fourier transform of $G(\mathbf{r},t)$ in space and time is carried out, this yields the coherent scattering function $S_{\mathrm{coh}}(\mathbf{Q},\omega)$:

$$S_{\mathrm{coh}}(\mathbf{Q},\omega) = \frac{1}{2\pi} \int_{-\infty}^{\infty} dt \int_{\mathbf{r}} G(\mathbf{r},t) e^{i(\mathbf{Qr}-\omega t)}\, d\mathbf{r} \qquad (6)$$

where $\mathbf{r}$ is a vector describing the position of a scattering nucleus.

Geometrical Properties. ***Bragg Scattering.*** In regular crystals where crystal planes are separated by a distance d, a special vector termed the reciprocal lattice vector $\mathbf{G}$ can be defined. It is perpendicular to the set of crystal planes, and its magnitude is

$$|\mathbf{G}| = n2\pi/d \qquad (7)$$

where n is an integer. When the scattering occurs at a special angle θ such that the relationship (Bragg's law)

$$n\lambda_0 = 2d \sin \theta/2 \qquad (8)$$

holds, the pattern obtained can be analyzed in terms of the Miller indexes (hkl) of the reflection.

Refractive Index. The refractive index is defined as a ratio of the moduli of the scattered and incident wave vectors:

$$\tilde{n} = |\mathbf{k}|/|\mathbf{k}_0| \qquad (9)$$

Dynamics of the System. ***Inelasticity.*** The analysis of laws governing the dynamics of polymer chains is based on several assumptions: one of the most crucial is that of the dynamical uncoupling of different types of motion that a chain may undergo. This results in a separation of translations, vibrations, and rotations and is usually well fulfilled because of the very different time scales over which they occur. The total scattering law is then a product of the respective translational, vibrational, and rotational laws, and an experimental separation is obtained by choosing the proper energy and time domains.

Large Scale: Main-chain Motion. The intermediate scattering structure factor $S_{\mathrm{coh}}(\mathbf{Q},t)$ represents the average decay in time of collective particle fluctuations of scattering vector $\mathbf{Q}$ (as compared with the same number of independently scattering particles). The intermediate scattering function $S_{\mathrm{inc}}(\mathbf{Q},t)$ represents the dynamics of a single particle (often a whole polymer chain in solution). Fick's law of diffusion in a liquid for which the Green-Kubo relation can be written requires that the single particle fluctuations decay exponentially and that $S_{\mathrm{inc}}(\mathbf{Q},t)$ take a simple form:

$$S_{\mathrm{inc}}(\mathbf{Q},t) = \exp(-D\mathbf{Q}^2 t) \qquad (10)$$

where D is a diffusion coefficient. A Fourier transform of equation 10 gives a Lorentzian-shaped scattering function (spectral density)

$$S_{\mathrm{inc}}(\mathbf{Q},\omega) = 1/\pi \cdot D\mathbf{Q}^2/[\omega^2 + (D\mathbf{Q}^2)^2]$$

Since a polymer chain moving in a melt or solution has no fixed center of mass, liquid-state theories can be applied. There is no ambiguity in experimental separation of the distances involved, because the Q values used imply the scale. Fickian diffusion of the center of mass is dominant for Q values less than $1/R_g$, where R_g is the radius of gyration of the polymer chain. This diffusion is conventionally observed in polymer solutions by photon correlation spectroscopy (pcs) and for solutions and melts by the neutron spin-echo technique. For x rays, inelastic scattering does not contribute any information on the time scale considered here because there is a great energy transfer mismatch between the beam and the sample.

Local Scale: Side-chain Motion. If there is no translation and a particle is confined to a limited region in space, then for very long time the self-correlation function $G_s(\mathbf{r},t) \neq 0$ and $G_s^{\infty}(\mathbf{r})$ is a constant independent of $\mathbf{r}$. The intermediate scattering function (structure factor) $S_{\mathrm{inc}}(\mathbf{Q},t)$ can then be divided into a time-independent part $S_{\mathrm{inc}}(\mathbf{Q},\infty)$ and a time-dependent part $S_{\mathrm{inc}}(\mathbf{Q},t)$, when the scattering law $S_{\mathrm{inc}}(\mathbf{Q},\omega)$ becomes a sum of a δ-function (Fourier transform of a constant) and a quasielastic scattering law:

$$S_{\mathrm{inc}}^{\mathrm{tot}}(\mathbf{Q},\omega) = A_0(\mathbf{Q})\, \delta(\omega) + S_{\mathrm{inc}}^{\mathrm{qel}}(\mathbf{Q},\omega) \qquad (11)$$

with $A_0(\mathbf{Q}) \equiv S_{\mathrm{inc}}(\mathbf{Q},\infty)$. In neutron scattering, $A_0(\mathbf{Q})$ is termed the elastic incoherent structure factor (EISF). It is a direct measure of the time-averaged spatial distribution of the moving particle and so encompasses the geometry of the motion. On the other hand it is the dynamical behavior of the particle that determines the shape and width of the quasielastic component indicated in the second term. It is a unique feature of neutron scattering that information about the geometry and the time domain of the motion is delivered in the same experiment. The spectrum obtained consists of an elastic peak, the intensity of which is a direct measure of the EISF plus a quasielastic broadening.

Molecular Scale: Vibrations. The thermal motion of atoms smears out the Bragg peaks. In terms of the scattering cross sections, this motion is described by the exponential term in the expression for the incoherent elastic cross section $(d\sigma/d\Omega)_{\mathrm{inc\ el}}$:

$$(d\sigma/d\Omega)_{\mathrm{inc\ el}} = \sigma_{\mathrm{inc}}/4\pi \cdot N e^{-2W} \qquad (12)$$

where exp $(-2W)$ is a Fourier transform of a self-correlation function in the limit of infinitely long time given by

$$e^{-2W} = \int G_s(\mathbf{r},\infty)e^{i\mathbf{Qr}}\,d\mathbf{r} \qquad (13)$$

This exponential term is termed the Debye-Waller factor and is usually given in terms of the mean square displacement $\langle u^{-2}\rangle$ of a nucleus from equilibrium.

Small-angle Light Scattering from Solutions. In considering a dilute solution containing n solute molecules per unit volume illuminated by a linearly polarized beam of light, the intensity of light scattered at an angle θ is calculated in exactly the same way as for an ideal gas, provided the presence of the solvent is ignored.

For convenience the molecules are initially considered as isotropic particles that are much smaller in size than the wavelength of the incident beam. This (Rayleigh) condition ensures that the transient electromagnetic field of an incident wave is uniform throughout the particle. Exposure of the particle to a weak electromagnetic field varying sinusoidally with time induces it to become dipolar with a moment μ given by a relationship

$$\mu = \alpha E_0 \cos 2\pi(\nu t - x/\lambda_0) \qquad (14)$$

where E_0 is the amplitude of the field and α is the particle polarizability. $\nu = \omega/2\pi$ and λ_0 are the respective frequency and wavelength of the incident wave, and x is a distance from the origin to the point along a propagation line. The induced dipole oscillates synchronously and in the same direction as the incident electric field and acts as a source of electromagnetic radiation, which contributes to the scattering and is manifested as such. In the case of an incident light beam that does not transfer energy to the sample by inducing an internal transition and is thus said to be elastically scattered, the emitted waves are of the same frequency and wavelength, although of considerably lower intensity than the incident beam.

If an ideal solution is assumed, the presence of the solvent must be acknowledged and incorporated into the model. The assumption that the molecules behave as point sources is, however, retained. For a dilute solution the excess polarizability α', the difference in polarizability between the solute and solvent, can be described either in terms of a difference in refractive index $\tilde{n}$ or in terms of the dielectric constant D. Since $D = \tilde{n}^2$ in the nondispersive region of the visible light, it then follows that

$$\alpha' = \frac{\tilde{n} - 1}{\tilde{n}^2 - \tilde{n}_0^2} \qquad (15)$$

where $\tilde{n}$ and $\tilde{n}_0$ are the respective refractive indexes of the solution and solvent.

Light Scattering from Bulk Polymers. When a light beam impinges on a crystal, the photons may scatter elastically from the sample, or some photons may be absorbed, in which case an excitation, termed a phonon (or quantum of lattice vibration), is created with conservation of energy and (pseudo-) momentum. In the former case there is reflection and, in the latter, absorption of radiation. This process of absorption does not involve electronic excitations and is similar to the inelastic scattering of neutrons considered above. Light scattering by acoustic modes, the passage of which involves density changes, is termed Brillouin scattering, whereas scattering by optic modes in which adjacent subunits move in opposition is referred to as Raman scattering. The conservation laws are the appropriate modifications of equation 1:

$$\begin{aligned} E - E_0 &= \pm\hbar\omega_j(\mathbf{q}) \\ \hbar\mathbf{Q} &= \hbar(\mathbf{G} \pm \mathbf{q}) \end{aligned} \qquad (16)$$

where $\omega_j(\mathbf{q})$ is the j branch of the frequency of the phonon wave with a corresponding wave vector $\mathbf{q}$. Since the wavelength of the visible light is large compared with the unit cell dimension q_c this equation has a solution only for $q \ll 2\pi/a_c$ and $\mathbf{G} = 0$, where $\mathbf{G}$ is a reciprocal lattice vector, namely, at the center of the Brillouin zone.

Instrumentation

Sources of Radiation. Neutrons are produced either in moderated nuclear fission reactors (the High Flux Beam Reactor at Brookhaven National Laboratory (Upton, N.Y.) or the High Flux Reactor at the Institut Laue-Langevin (ILL) (Grenoble, France) or in spallation sources (Los Alamos, N.M.), KENS (Japan), ISIS (UK). In the laboratory, x rays are produced by the bombardment of a target anode with high energy electrons.

In recent years lasers have increasingly replaced mercury lamps ($\lambda_0 = 436$ nm) in light-scattering instruments. Laser light sources produce very intense, stable, narrow beams typically 0.5–1.5 nm in diameter of either linearly or randomly polarized light.

Monochromation. The nature of the anode in laboratory x-ray generators determines the wavelength of the emitted x rays. A beam that contains both K_α and K_β emissions is normally passed through a thin metal foil chosen to absorb the latter component.

In contrast, monochromation of polychromatic x-ray beams produced by synchrotrons is normally achieved by a combination of mirrors and crystals, both of which are bent. This combination acts not only as a monochromator but also as a device for demagnifying the beam before it enters the collimator. This is achieved by positioning the reflecting planes of the crystal at right angles to that of the mirror. In this way a beam with a cross section of 6×30 mm can be reduced to 0.8×1.0 mm. The crystal monochromates the beam by reflecting x rays of a given wavelength into the collimator when positioned at an appropriate angle to the incident beam in accordance with Bragg's law (Fig. 2).

Collimation. The simplest type of collimation system for either x rays or neutrons consists of a pair of pinholes or slits. The resolution of such a system depends on the width of the slits or the diameter of the pinholes and on the sample–detector distance, the maximum resolution being obtained when the slits or pinholes are small and there is a long sample–detector distance. However, such simple collimators give rise to parasitic scattering around the primary beam, as shown in Figure 3.

Detectors. Most small-angle instruments now use solid-state or gas-filled position-sensitive detectors (psds). In the silicone solid-state detector, absorption of the x rays takes place in a reverse-biased silicon diode. Such a detector is energy, and

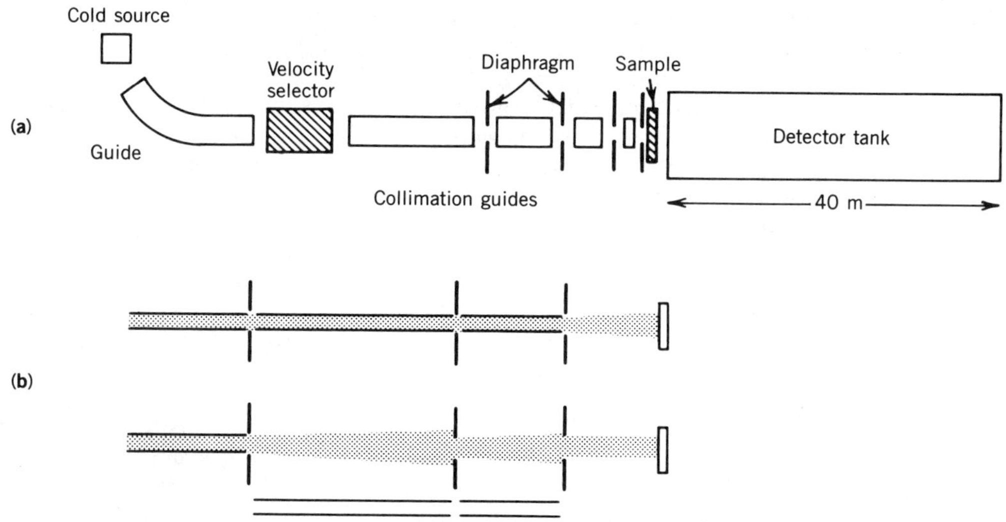

Figure 2. (**a**) A schematic representation of the small-angle diffractometer D11 at the ILL. (**b**) The use of movable neutron guides in the collimator allows adjustment for optimum resolution.

therefore wavelength, discriminating when used in conjunction with a pulse height discriminator.

Diffractometers (Elastic Scattering). ***Small-angle Scattering Instruments.*** Figure 2 shows the arrangements of the small-angle camera and diffuse scattering apparatus D11 at the ILL.

Spectrometers (Inelastic Scattering). The aim of a diffraction experiment is to obtain structural information contained in the interference pattern and given by the scattering function $S_{coh}(\mathbf{Q},\omega)$. In this analysis the $S_{inc}(\mathbf{Q},\omega)$ term is usually neglected under various assumptions. However, it is precisely the latter that yields the desired information on the motion of the nuclei and transitions among excited states. There are essentially two methods of energy analysis of the scattered neutrons: time of flight (tof) and the use of a triple-axis arrangement.

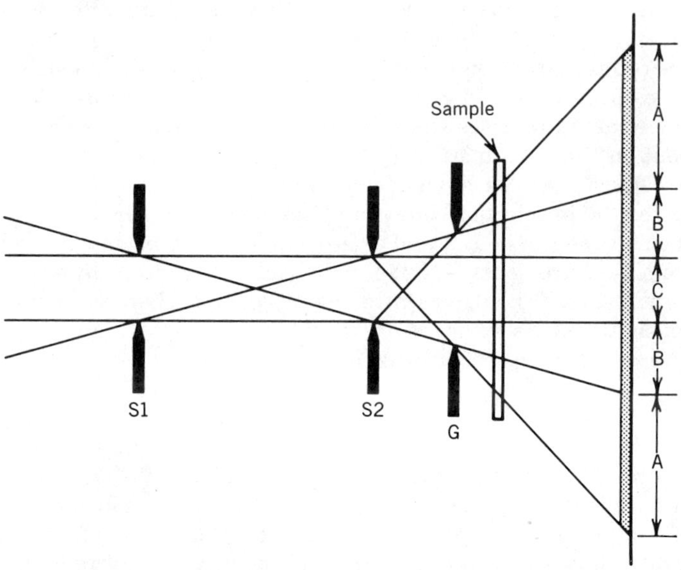

Figure 3. A slit collimation system for a small-angle camera. Slits S1 and S2 define the primary beam C, but give rise to secondary (parasitic) scattering in the small-angle region, which can be reduced by a third or guard slit G. A and B represent weak and strong zones of parasitic scattering.

Light Scattering. To date, most of the data on the dynamics of polymer chains in solution have been collected using light-scattering techniques. Figure 4 shows a typical experimental arrangement.

In a light-scattering experiment, essentially two quantities can be measured: the intensity of the scattered beam as measured by the time-averaged photomultiplier output and the spectral distribution of the scattered light. Three different methods are used to measure the spectral distribution of the scattered light. They are termed filter on the one hand, and homodyne and heterodyne beating on the other. Filter methods are used to study processes occuring on a time scale of $<10^{-6}$ s (relatively fast molecular dynamics, which cause a large Doppler shift in the frequency of the scattered light), whereas the optical mixing or beating methods are normally used for investigation of processes on a time scale $>10^{-6}$ s.

Experimental Requirements

Some care must be exercised in the preparation of samples for use in scattering experiments. For light-scattering measurements the samples must be cleaned of dust by filtering or centrifugation. The need to have clean samples stems from the low contrast value of $d\tilde{n}/dc$, which can result in a low signal-to-noise ratio when coupled with the fact that dust particles are excellent scatterers. However, it has been shown that dust

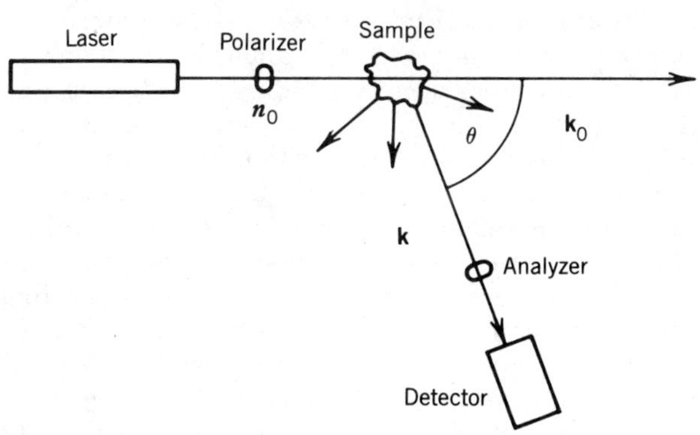

Figure 4. Geometry of the light-scattering experiment. The incident beam wave number $|\mathbf{k}_0| = (2\pi/\lambda_0)\tilde{n}_0$ and the scattered wave number $|\mathbf{k}| = (2\pi/\lambda)\tilde{n}$, where $\tilde{n}$ is the refractive index of the scattering medium.

particles can sometimes be made invisible by adjusting the solvent refractive index. In concentrated solutions the removal of contaminants and even bubbles can be difficult, as the mixtures tend to be viscous or even gel-like. The presence of voids in samples gives rise to intense low-angle scattering, particularly as the scattered intensity is proportional to R^6 (where R is the void radius). The presence of voids in a sample in addition to the excess scattering also makes accurate determination of the sample thickness difficult. Other considerations include sample geometry, contrast, sample–detector distance, and resolution.

Data Treatment

The approximations made in data treatment can generally be divided into two classes. The first concerns the particulars of the system. An example of this is the determination of the amount of scattering due to polymer in a solvent or a blend. The second involves the technique used and the experimental arrangement itself.

Detector Response. In small-angle scattering the detector response is determined by exposing the detector to scattering, which is essentially independent of Q. For neutrons either vanadium or water samples are used, since they are almost purely incoherent scatterers, and for x rays the detector is exposed to a β-source.

Examples

A few selected examples follow, showing the possibilities available with scattering techniques. The division into elastic and inelastic scattering is often artificial, as is the division into coherent and incoherent scattering. Rather, a part of the total scattering is chosen for an investigation and subsequent analysis, neglecting the unwanted part under different assumptions. For instance, in a typical neutron scattering analysis, the total scattered intensity is always measured. Only by using the spin polarization option can the scattering be separated into the coherent and incoherent parts.

Elastic Scattering. Regular Bragg peaks in a diffraction pattern obtained from a solid sample occur only for highly ordered crystalline substances. When a wide scattering angle is used, information on the molecular and atomic scale can be extracted. However, among polymers only polydiacetylenes can be obtained as perfect good single crystals, although a few other polymers, like high density, unbranched polyethylene are sufficiently crystalline to show a distinct x-ray pattern. Wide-angle x-ray patterns obtained from amorphous polymers frequently show broad peaks corresponding to distances greater than those expected from van der Waals radii, but the assignment of those halos to a short- or medium-range order has long been a subject of controversy. Regularities in x-ray scattering patterns from amorphous polymers, particularly n-alkyl acrylate and methacrylate polymers have been analyzed. A distinct shift toward lower angles of maxima corresponding to larger than van der Waals distances with the increasing number of pendent groups, clearly indicates some local order.

Although local order in polymers is of great importance since it ultimately determines the macroscopic structural and dynamical properties, there has not hitherto been sufficiently clear evidence for such order as has now been obtained from spin polarization analysis. This was one of the reasons why polymer scientists have mainly followed a universal approach in which only the whole chain-length scale is important. Such a universal scale lends itself naturally to small-angle light and neutron scattering experiments, where for technical reasons polymer solutions were preferred for investigation.

Small-angle elastic scattering measurements can be used to obtain information on the conformation of the scatterers, ie, their shape, dimensions, and molecular weight; the size and shape of domains in phase-separated or ordered, ie, micellar or liquid crystalline systems; and the thermodynamic parameters that describe the free energy of the sample.

Conformation. Information on the conformation of the scatterers relies on the ability to resolve the scattering from individual chains. In dilute solution where the scatterers are widely dispersed, the scattering functions for polymer chains or particles can be determined in principle by small-angle light, x-ray, or neutron scattering. However, in practice the shape of scatterers with a radius of gyration less than ca 20 nm cannot be determined by light scattering, as measurements cannot be made beyond the Guinier range (where $QR_g \leq 1$). There are no such limitations on Q for either x rays or neutrons, where the choice of radiation often depends on the contrast factor. The intensity of the scattering is proportional to the molecular weight of the scatterer; therefore, a good signal-to-noise ratio for low molecular weight polymers can only be achieved if there is sufficient contrast between the solvent and solute. For neutrons this is generally easy to achieve since either deuterated solute or solvent can be used.

The conformation of a polymer chain is in general determined by two independent factors: the local chain stiffness due to steric hindrances between adjacent segments and the excluded volume effect, which influences segments that are far apart. In a saxs study of poly(sodium acrylates) in dilute aqueous salt solutions, the effect of electrostatic repulsion of charged groups on local chain stiffness has been assessed by determining the persistence length from a Kratky plot as a function of both the ionic strength of the solution and the charge density of the polyelectrolyte. It was concluded that chain stiffness is largely unaffected by electrostatic repulsions, and therefore, the conformation is primarily determined by the excluded volume effect, confirming an earlier light scattering study. However, there are cases where the chain conformation is dependent on local chain stiffness.

Phase Size. The Flory-Huggins parameter χ is used to describe the miscibility behavior of polymers in a solvent where the solvent can be a liquid or a polymer. χ can be determined from measurements of the second virial coefficient A_2 by either light, x-ray, or neutron scattering in systems where the solute concentration is below c^*, the onset of the semidilute regime. The relationship between A_2 and χ is given by

$$A_2 = v_1^2/v_2 M_w^2 (0.5 - N_2\chi) \tag{17}$$

where v_1 and v_2 are the respective molar volumes of the solute and solvent, N_2 is the number of segments in the solvent, and M_w is the molecular weight of the solute. Measurements of the light and x-ray scattering from more concentrated systems are difficult primarily because of a lack of contrast.

Inelastic Scattering. ***Side-chain Motion.*** Because of the molecular sizes and energy transfers involved, it is mainly neutron scattering that can be used for studies of side-chain motion

(Fig. 1). In dynamic investigations there are two principal areas of interest: ins, which involves rather large energy transfers, and quasielastic (nearly elastic) scattering. The former gives information on torsional and vibrational motion, the latter on the geometry and rates of rotational motion. Methyl and phenyl groups are particularly well suited for the ins studies on account of the large incoherent scattering cross section of hydrogen and its large amplitude of motion. Moreover, some torsional bands that can be detected by ins are often weak or forbidden in Raman or ir scattering.

Main-chain Motion. Light scattering studies of polymer dynamics in solution, gels, and colloids have been extensively discussed.

Phonon Modes. Coherent ins provides information about phonon modes in crystalline substances. The samples used are usually deuterated in order to reduce incoherent scattering.

Backbone Motion in Solutions and Melts. These are perhaps the most investigated states of polymer matter to date. Because of the Q and energy resolution requirements, only two neutron scattering spectrometers are suitable for such studies: the neutron spin-echo and the high resolution back-scattering spectrometer (IN10).

BIBLIOGRAPHY

S. W. Lovesey, *Theory of Neutron Scattering from Condensed Matter,* Clarendon Press, Oxford, UK, 1984.

A. Guinier and G. Fournet, *Small Angle Scattering of X-rays,* John Wiley & Sons, Inc., New York, 1955.

W. Hayes and R. Loudon, *Scattering of Light by Crystals,* John Wiley & Sons, Inc., New York, 1978.

J. S. Higgins and A. Maconnachie in W. Forsman, ed., *Polymers in Solution, Theoretical Considerations and Newer Methods of Characterisation,* Plenum Press, New York, 1986.

Barbara Gabrys
Brunel University

Paul E. Tomlins
National Physical Laboratory

SCHIFF BASES. See Azo polymers.

SCREEN PRINTING. See Decorating; Fine arts; Printing-ink vehicles.

SCREW DESIGN

Single Screw

The extruder screw is the single most important mechanical element of a screw extruder. Its design directly determines the conveying, heating, melting, mixing, devolatilization, and pressure development of the polymeric material. Thus, a proper screw design is of critical importance to the proper functioning of the extruder. Selection of an appropriate screw design for a certain application is essential, and recognition of whether an extrusion problem can be solved by changing operating conditions or whether it requires a new screw design is necessary.

Mechanical Considerations

A screw must have sufficient mechanical strength to withstand the stresses imposed by the conveying process in the extruder. Even the best screw design for functional performance is useless if it cannot resist failure or wear for a substantial period of time, preferably at least one year.

Torsional Strength. The screw must have sufficient torsional strength to transmit the torque required to turn the screw. The highest torque is transmitted in the feed section of the screw, at the drive end of the screw. Unfortunately, it is in the feed section that the torsional strength of the screw is usually the lowest because the root diameter of the screw there is usually less than in any other screw section. As a result, the feed section is the most critical region as regards torsional strength.

Strength of the Screw Flight. One of the causes of stresses in the screw flight is the pressure differential between the leading and trailing side of the flight (Fig. 1). In addition, there is a loading from the shear stress acting on the tip of the flight.

The ratio of flight height to flight width can be thought of as a slenderness ratio. In general, the ratio of the allowable stress to the pressure difference $\tau_a/\Delta P$ is larger than 25. Thus, a slenderness ratio of 2, a common value used in screw design, yields more than adequate strength if the shear stress at the flight tip is negligible.

Functional Considerations

In a plasticating extruder, a number of different functions are performed by the screw: solids conveying, plasticating, melt conveying, mixing, and in vented extruders, devolatilization. The design of the screw directly affects how well these functions are performed.

Solids Conveying. Solids conveying in the screw channel occurs by frictional drag, based on solid-to-solid friction of the polymer particles against the metal surfaces of the screw and barrel. If the polymer particles are assumed to move in plug

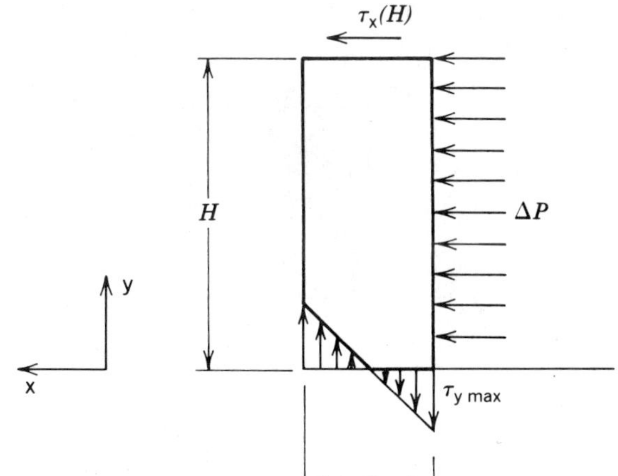

Figure 1. Pressure differential between the leading and trailing edge of the screw flight.

flow and if gravitational and contrifugal forces are negligible, the solids conveying rate can be expressed as:

$$M_s = \rho HWp\pi DN \frac{\sin\theta}{\sin(\theta + \phi)} \tag{1}$$

where ρ is the bulk density of the polymer particles at the appropriate pressure, H the channel depth, p the number of parallel flights, W the channel width, D the screw diameter, N screw speed, θ the solids conveying angle, and ϕ the flight helix angle. The solids conveying angle θ is determined by the direction of the vector representing the relative velocity between the solid bed and the extruder barrel (Fig. 2). It can be calculated from:

$$\theta = \arcsin\left[\frac{(1 + f_s^2 - k^2)^{1/2} - f_s k}{1 + f_s^2}\right] - \phi \tag{2}$$

where:

$$\frac{H}{f_b z}\ln\frac{P}{P_0} + \frac{f_s}{f_b}\left(1 + \frac{2H}{W}\right)$$

where f_s and f_b are the coefficients of friction on the screw and barrel, and z is the down-channel coordinate. With these equations the effect of several screw design variables on solids conveying performance can be evaluated.

Plasticating. The plasticating or melting process in a single screw extruder is based on the mechanism of drag-induced melt removal. Melting takes place because of viscous heat generation in the melt film and the heat conducted from external heaters around the extruder barrel (and sometimes because of the heaters mounted inside the extruder screw). The effects of helix angle, multiple flights, and flight width on melting performance can be deduced from Figure 3.

Melt Conveying. The melt conveying rate in the melting section can be expressed as:

$$\dot{V} = \frac{1}{2}pWHv_{bz} - \frac{pWH^3 g_z}{12\mu} \tag{3}$$

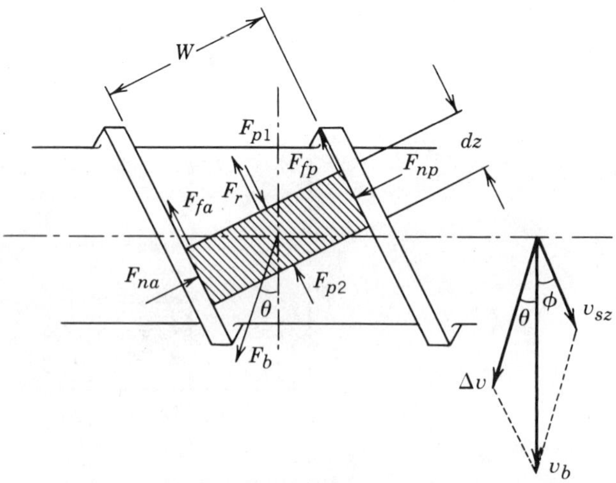

Figure 2. Force and angle diagrams in solids conveying.

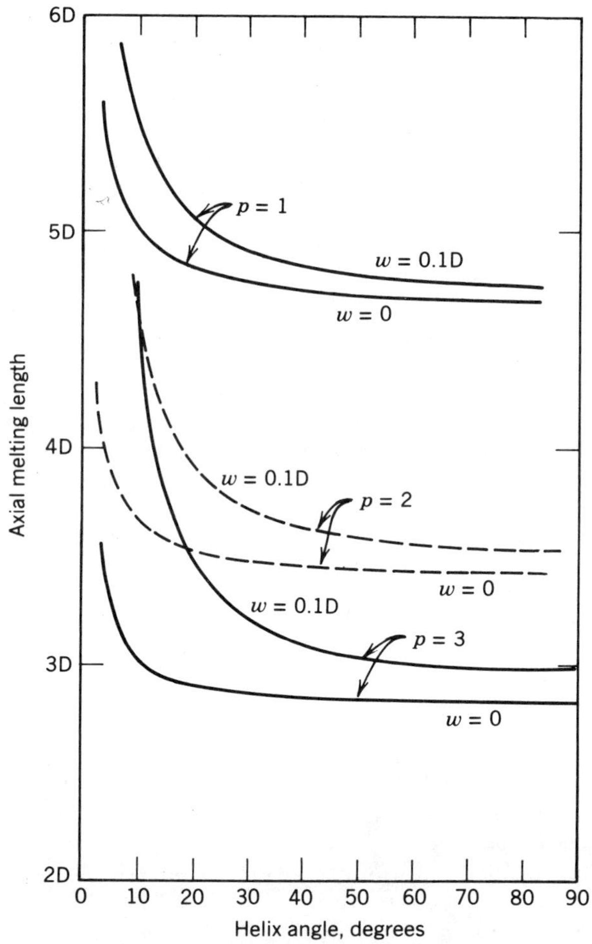

Figure 3. The axial melting length as a function of helix angle. p is the number of flights, and w indicates flight width.

where g_z is the down-channel pressure gradient; the first term on the right-hand side of the equation is the drag-flow rate and the second term is the pressure-flow rate. In equation 3 the polymer melt is assumed to be a Newtonian fluid of constant viscosity and the leakage through the flight clearance is assumed to be negligible. For a power law fluid, the approximate output can be written as:

$$\dot{V} = \frac{4 + n}{10}pWHv_{bz} - \frac{pWH^3 g_z}{4(1 + 2n)\mu} \tag{4}$$

where

$$\mu = m\left(\frac{v_{bz}}{H}\right)^{n-1}$$

Both n and m are constants in the power law equation; n is the power law index and m is the consistency index. The shear stress τ as a function of shear rate as expressed by the power law equation is

$$\tau = m\dot{\gamma}^n \tag{5}$$

The optimum channel depth for melt conveying can be formed

by setting

$$\frac{\partial \dot{V}}{\partial H} = 0 \tag{6}$$

This results in the following expression for the optimum channel depth:

$$H^* = \left[\frac{(4 + n)(1 + 2n)2\mu v_{bz}}{15g_z}\right]^{1/2} \tag{7}$$

By the same method, the optimum helix angle can be determined to be

$$\phi^* = \frac{1}{2}\arctan\left[\frac{(4 + n)(1 + 2n)2\mu v_b}{5H^2 g_a}\right] \tag{8}$$

where g_a is the axial pressure gradient. Optimum values for the channel depth and helix angle can be obtained from equations 7 and 8.

Devolatilization. Devolatilizating extruder screws, or vented extruders, are used to extract volatiles from the polymer in a continuous fashion. A typical two-stage devolatilizing screw is shown in Figure 4. The screw consists of at least five distinct geometrical sections. The first three sections, feed, compression, and metering, are the same as on a conventional screw. After the metering section, there is a rapid decompression followed by the extraction section which, in turn, is followed by a rapid compression and a pump section. Three important functional requirements for efficient degassing are zero pressure in the polymer under the vent port, completely molten polymer under the vent port, and frequent surface renewal of the polymer melt in the extraction zone of the extruder.

Mixing. The mixing capacity of standard extruder screws is relatively limited and many modifications have been made to improve mixing. The number of mixing elements that have been used on extruder screws is nearly infinite. This discussion is limited to the more common types of mixing elements. Before selecting a mixing element, it is important to determine whether distributive or dispersive mixing is required. Distributive mixing is mixing of fluids that do not exhibit a yield point. If one of the components of the mixture is a solid exhibiting yield stress, the mixing is referred to as dispersive mixing (see also MIXING).

Dispersive Mixing Elements. Dispersive mixing elements are used when agglomerates or gels need to be broken down. This is particularly important in small or thin gauge extrusion, eg, fiber spinning, and thin film extrusion.

The most common dispersive mixing section is the fluted or splined mixing section, in which one or more barrier flights are placed along the screw such that the material must flow over the barrier flight(s). In the barrier clearance the material is subjected to a high shear rate; the corresponding shear stress should be large enough to break down the particles in the polymer melt. A well-known fluted mixing section is the Union Carbide (UC) mixing section, often referred to as the Maddock mixing section (Fig. 5**a**). Another fluted mixing section is the Egan mixing section (Fig. 5**b**). The blister ring is another dispersive mixing section (Fig. 5**c**).

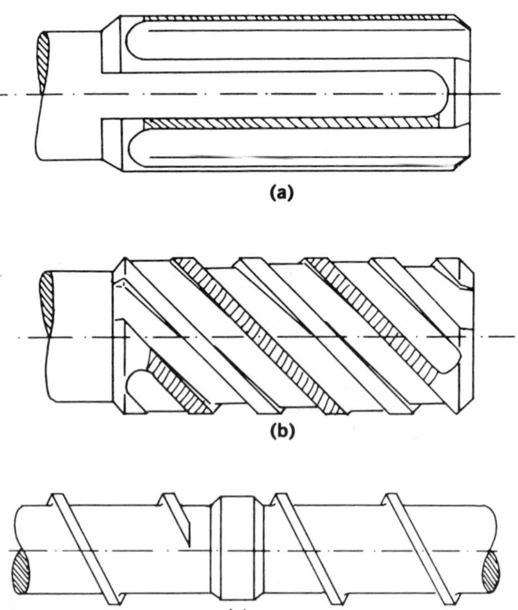

Figure 5. Dispersive mixing sections. (**a**) Union Carbide mixing section; (**b**) Egan mixing section; (**c**) blister ring.

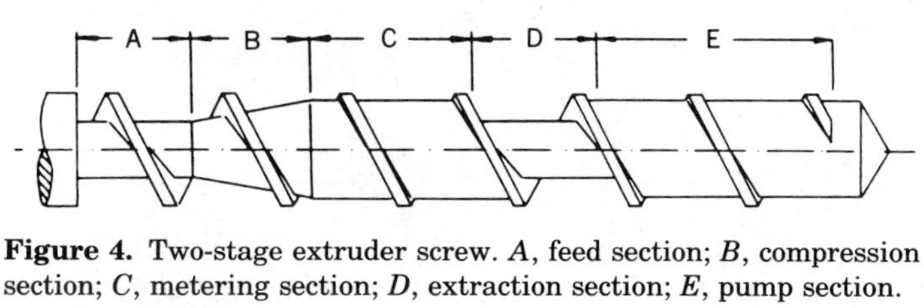

Figure 4. Two-stage extruder screw. *A*, feed section; *B*, compression section; *C*, metering section; *D*, extraction section; *E*, pump section.

Distributive Mixing Elements. Distributive mixing is needed when different polymers with similar flow properties are blended together. Distributive mixing is easier to achieve than dispersive mixing. Disruption of the velocity profiles in the screw channel causes distributive mixing. Dispersive mixing is more difficult to achieve because a certain minimum stress is required to break down agglomerates. Thus, dispersive mixing sections also perform as distributive mixing elements; however, the opposite usually is not true.

One of the simplest distributive mixing elements is the pin mixing section shown in Figure 6**a**; many different patterns have been used to place the pins (eg, Fig. 6**b**, **c**).

Energy Consumption. An additional important performance requirement of the screw is that it does not require substantially more energy consumption than that needed to extrude the polymer and to raise the polymer stock temperature to the desired processing temperature. Control of energy consumption through screw design is important in keeping down operating cost and obtaining high product quality.

Commercial Extruder Screws

The Standard Extruder Screw. Typical features of a standard screw (Fig. 7) are

flighted length $20D$ to $30D$
length of feed section $4D$ to $8D$

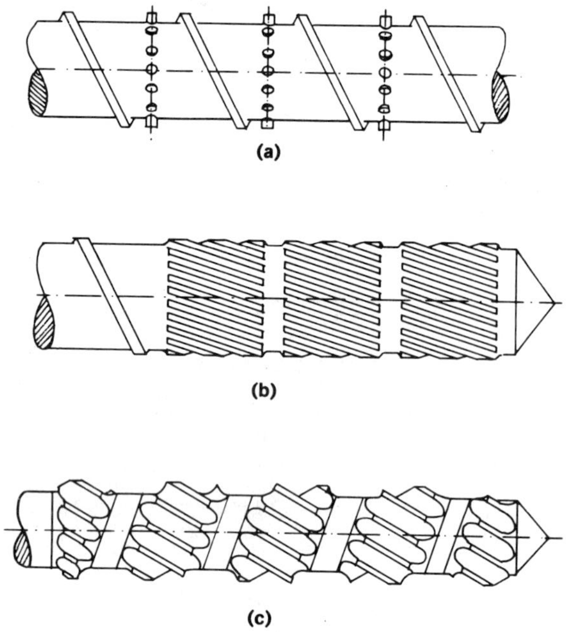

Figure 6. Distributive mixing elements. (**a**) Pin mixing section; (**b**) Dulmage mixing section; (**c**) Saxton mixing section.

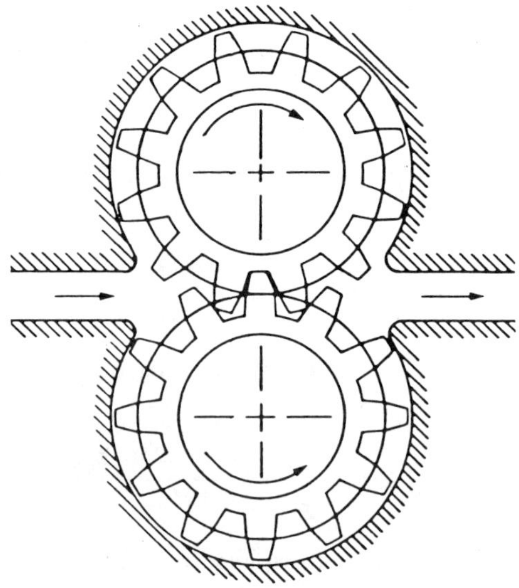

Figure 8. Schematic of a gear pump.

length of metering section $6D$ to $10D$
number of parallel flights 1
flight pitch $1D$ (helix angle 17.66°)
flight width $0.1D$
channel depth in feed section $0.10D$ to $0.15D$
channel depth ratio 2:1 to 4:1
radial flight clearance $0.001D$

Based on these characteristics, a standard screw can be designed in seconds. However, significant improvements in performance can often be made by functional optimization, as discussed earlier. For instance, decreasing the flight width and increasing the helix angle may yield substantial reduction in power consumption. Consequently, a screw should not be designed by using a list of general features. Rather it should be done by analyzing the details of the specific application and performing a functional analysis to obtain the best possible design. Types of screws include the barrier flight extruder screw, the Maillefer screw, the Barr screw, the Dray and Lawrence screw, the Kim screw, and the Ingen Housz screw.

Gear Pump Assisted Extrusion

Gear pumps are used in some single-screw or twin-screw extrusion operations at the end of a plasticating extruder. Strictly speaking, the gear pump (Fig. 8) is a closely intermeshing counter-rotating twin-screw extruder. However, since gear pumps are solely used to generate pressure, they are usually not referred to as an extruder even though the gear pump is an extrusion device. In addition to its good pressure-generating capability the gear pump maintains a relatively constant outlet pressure even if the inlet pressure fluctuates considerably. Some fluctuation in the outlet pressure results from the intermeshing of the gear teeth. This fluctuation can be reduced by a helical rather than an axial orientation of the gear teeth.

Gear pumps can be used advantageously on extruders with poor pressure-generating capability (eg, corotating twin screws, multistage vented extruders), or when output stability must be better than 1% in close tolerance extrusion (eg, fiber spinning, cable extrusion, medical tubing, coextrusion). Gear pumps can cause problems when the polymer contains abrasive components. Because of the small clearances, the gear pump is very susceptible to wear. Also, the polymer's susceptibility to degradation presents problems because gear pumps are not self-cleaning and can cause exposure of the polymer to high temperatures.

Scale-up of Extruder Screws

Scale-up from small laboratory size extruders to large production size extruders is a procedure of great practical importance. To help evaluate the effect of a certain scale-up theory on extruder performance, Table 1 shows various scale-up theories (designated by Roman numerals) with corresponding functional performance characteristics.

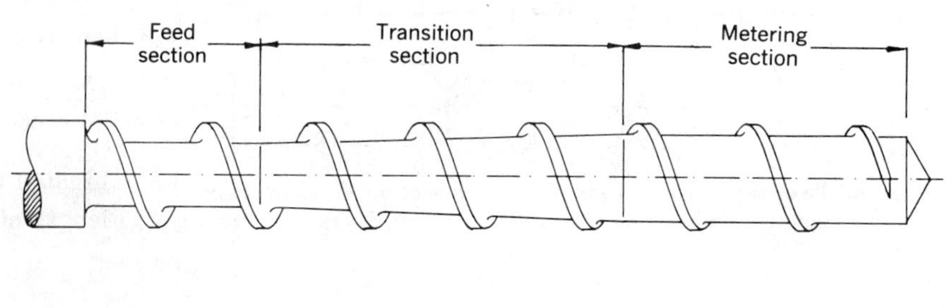

Figure 7. Standard screw geometry.

Table 1. Scale-up Theory Parameters

	Basic relationships	Theory I	Theory II	Theory III	Theory IV	Theory V	Theory VI	Theory VII	Theory VIII	Theory IX
channel depth	h	1	0.5	$\frac{1+n}{1+3n}$	0.3	α	0.7	0.7	$\frac{1+n}{1+2n}$	$\frac{1}{2n}$
axial length	l	1	1	1	1	α	1	1	1	$\frac{1+n}{2n}$
screw speed	ν	0	−0.5	$-\frac{2+2n}{1+3n}$	−0.3	$\alpha - 1$	−1.4	−0.6	$-\frac{1+n}{1+2n}$	−1
shear rate	$1 + \nu - h$	0	0	$-\frac{2}{1+3n}$	0.4	0	1.1	−0.3	$-\frac{1}{1+2n}$	$-\frac{1}{2n}$
pumping rate	$h + 2 + \nu$	3	2	$\frac{1+5n}{1+3n}$	2	$2\alpha + 1$	1.3	2.1	2	$\frac{1+2n}{2n}$
melting rate at low Brinkman number	$1 + \frac{1}{2}\nu + l_t$	2	1.75	$\frac{1+5n}{1+3n}$	1.85	$1\frac{1}{2}\alpha + \frac{1}{2}$	1.3	1.7	$\frac{3+7n}{2+4n}$	$\frac{1+2n}{2n}$
melting rate at high Brinkman number	$2 + \nu + \frac{1}{2}n\nu + l_t$	3	$2.5 - \frac{n}{4}$	$\frac{-n^2+6n+1}{1+3n}$	$2.7 - 0.15n$	$1 - \frac{n}{2} + \alpha\left(2 + \frac{n}{2}\right)$	$1.6 - 0.7n$	$2.4 - 0.3n$	$\frac{-n^2+9n+6}{2+4n}$	$\frac{-n^2+3n+1}{2n}$
solids conveying rate	$h + 2 + \nu$	3	2	$\frac{1+5n}{1+3n}$	2	$2\alpha + 1$	1.6	2.1	2	$\frac{1+2n}{2n}$
residence time	$l - 1 - \nu$	0	0.5	$\frac{2+2n}{1+3n}$	0.3	0	1.4	0.6	$\frac{1+n}{1+2n}$	$\frac{1+n}{2n}$
shear strain	$l - h$	0	−0.5	$\frac{2n}{1+3n}$	0.7	0	0.3	0.3	$\frac{n}{1+2n}$	$\frac{1}{2}$
power consumption	$l + n + 2 + n\nu + \nu - nh$	3	2.5	$\frac{1+5n}{1+3n}$	$2.7 + 0.4n$	$2\alpha + 1$	$1.6 - 1.1n$	$2.4 - 0.3n$	2	$\frac{1+2n}{2n}$
spec. energy consumption	$l - h + n + n\nu - nh$	0	0.5	0	$0.7 + 0.4n$	0	$0.3 - 1.1n$	$0.3 - 0.3n$	0	0
area/throughput	$l - h - 1 - \nu$	−1	0	$\frac{1+n}{1+3n}$	0	0	0.7	−0.1	0	$\frac{1}{2}$

BIBLIOGRAPHY

C. Rauwendaal, *Polymer Extrusion,* Carl Hanser Verlag, Munich, 1986 (in U.S.: Oxford University Press, New York).

Z. Tadmor and I. Klein, *Engineering Principles of Plasticating Extrusion,* Van Nostrand, New York, 1970.

F. Hensen, W. Knappe, and H. Potente, eds., *Kunststoff-Extrusiontechnik II, Extrusionanlagen,* Carl Hanser Verlag, Munich, 1986.

CHRIS J. RAUWENDAAL
Raychem Corporation

Multiscrew

A screw profile is designed to accomplish several unit operations along the extruder length: feeding, melting, mixing, venting, die pressurization, etc. The type of screw element used to perform each unit operation is based on a fundamental understanding of the working principles of the different types of screws.

Both co- and counter-rotating intermeshing machines have certain general screw geometries in common. Conveying screws are used to transport material, characterized by pitch and conveying direction (forward or reverse). Screw pitch is used to control the residence time distribution and degree-of-fill. Conveying screws can have single, double, or triple flights.

Nonintermeshing counter-rotating extruders vary the screw channel depth to effect changes in volume rather than by varying screw pitch or the number of screw flights. Thus, residence time and degree-of-fill are controlled by selection of screw channel depth.

Reverse pitch screw elements are used to create back pressure. Degree-of-fill increases to 100% upstream of the reverse pitch screw; the preceding forward conveying screws must overcome this resistance. The resulting pressure drop is used for creating a melt seal. Reverse pitch screws have the significant impact of increasing residence time distribution and shear input.

Kneading elements are used for melting, mixing, dispersing, and homogenizing. The kneading elements are comprised of individual kneading discs of various widths, which are offset from each other. The kneading discs can be staggered for forward conveying, reverse conveying, or zero conveying and may have one, two, or three lobes. The level of shear input and type of mixing action are controlled by selecting a disc width, offset angle, number of kneading discs, and direction of conveying.

Various other types of screw geometries utilize interrupted or cut screw flights. The working principle is the same as for kneading elements; conveying efficiency is reduced to the benefit of backflow for mixing. The degree to which the screw flight is interrupted directly influences the degree of mixing.

BIBLIOGRAPHY

C. Rauwendaal, *Polymer Extrusion,* Macmillan, New York, 1986, pp. 458–504.

F. Martelli, *Twin-screw Extruders: A Basic Understanding,* Van Nostrand Reinhold Co., Inc., New York, 1983.

L. P. B. M. Janssen, *Twin-screw Extrusion,* Elsevier, Amsterdam, The Netherlands, 1978.

ADAM DREIBLATT
Werner Pfleiderer Corporation

SEALANTS

Sealants are used to isolate an environment, serving as a barrier to the passage of gases, liquids, and solid particles. Sealants also serve to maintain a pressure differential (pressurized aircraft cabins); attenuate mechanical shock, vibration, and sound; protect electronic subcomponents mechanically, electrically, and thermally; and, with specially formulated conductive sealants, provide electrical continuity.

The largest use of sealants (74%) is in commercial and home construction and repair. Transportation (automobile, truck, train, aircraft, marine) is another field using a large volume of sealants. Sales in 1985 were approximately $600 × 10^6$, representing nearly 225×10^3 t of one- and two-component sealants; distribution by market segment is given in Table 1.

Table 1. U. S. Consumption of Sealants, 1985

Market segment	Wt %	Percentage of sales
construction	42	39
transportation	15	17
manufacturing	11	11
consumer retail	32	33

Market growth is projected at approximately $20 × 10^6/yr$ through 1995. The use of polyurethanes for building and construction is expected to exceed that of silicones in tonnage by the early 1990s; ca 5% of construction sealants are polysulfides. The principal suppliers are given in Table 2.

Performance and Applications

Exterior commercial construction sealants must accommodate large changes in joint width (gap) from temperature cycling and wind loads and must resist rain, heat, ultraviolet light,

Table 2. Suppliers of Sealants

Type	Manufacturers
silicones	Dow Corning, General Electric, Mobay, Tremco
polysulfides	Chemseal, Morton Thiokol, Pecora, Products Research & Chemical Corp.
acrylics	Beecham Home Improvement Products Inc. (DAP), 3M, Tremco
polyurethanes	Essex Chemical Corp., Mameco, Products Research & Chemical Corp., Sika, Woodmont, Bostik
tapes, putties, sealants	Products Research & Chemical Corp., 3M, Tremco
butyls (hot melt)	Fuller

oxygen, and ozone. Building materials (stone, marble, aluminum, steel, glass, plastics) differ widely in thermal coefficients of expansion. As sections are fitted together, joint design becomes very important. The gap width between sections is based on material type and length of section; the larger the section, the wider the gap.

Only high performance sealants (silicones, polyurethanes, polysulfides) can be used in commercial building and construction for exterior sealing. Silicones and polyurethanes are first and second in demand. High performance sealants can accommodate 25% + joint movement in compression or tension with good recovery (above 80%). Medium performance (accommodation of 10–25% movement) and low performance sealants are used only in the interior of commercial buildings and in residential (home) construction and repair. The latter are brittle and exhibit large shrinkage on cure or significant creep. Recently, high performance sealants have been used for home and consumer applications, because of price reductions and the introduction of siliconized acrylics.

Polymer Systems and Sealant Formulations

Sealant properties must be matched with expected requirements to select the optimal polymer system and particular sealant for the job. Defining chemical and mechanical property requirements in terms of numerical values is often difficult, even for the experienced materials engineer; however, numerous government and industry specifications relate sealant properties to application and manufacturers supply information. Homeowners can draw from a variety of do-it-yourself manuals for guidance.

Polymer systems include silicones (Table 3), polysulfides (Tables 4 and 5), polyurethanes (Table 6), acrylics (Table 7), butyls (Table 8), chlorosulfonated polyethylene (Hypalon) (Table 9), styrene triblock copolymers (Kraton), anaerobic sealants, bitumens, and mastics.

Preparation of Sealant Formulations, Mixing, and Application

The ingredients of sealant formulations are mixed in ribbon blenders, paste mixers, and occasionally in double-sigma-blade mixers.

Mixing of Two-component Kits. The two-component sealants consist of a base compound and a catalyst, which are matched chemically and must be mixed in the prescribed ratio to produce the desired reaction rate and mechanical/chemical properties.

Surface Preparation and Application of Sealants. For good adhesion, the surface must be clean. Concrete is sometimes waterblasted or sandblasted, cleaned, and dried. Fluorinated plastic surfaces are flame-etched. Composite surfaces are scuff-sanded and wiped with solvent. Metal surfaces are cleaned and rinsed with organic solvents and wiped dry. Primers and adhesion promoters are also available for metal, glass, concrete, plastic, or wood surfaces. Some sealants, such as those used for automobile bodies, are designed to adhere in the presence of oil and other contaminants. Commercial construction involves laying large beads of sealant with a caulking gun in joints 0.635–3.81 cm wide on the face of buildings.

Health and Safety

Only broad statements can be made here as to precautions. Adequate ventilation is always necessary; dust masks, rubber gloves, protective boots, and safety glasses are necessary for manufacturing and cleaning operations.

Table 3. Silicone Sealant Formulation

Ingredients	Wt %	Effect
polydimethylsiloxane diol as binder	80–85	phenyl substitution lowers brittle point; trifluoropropyl or cyanoethyl substitution gives fuel resistance
cross-linker	5–7	strengthens, stiffens, prevents cold flow; polyfunctional organosilane, generally acetoxy or 2-ethylhexanoic substituents, produces acid by-products; amino-substituted alkoxy silanes produce alcohol by-products
catalyst		
tin carboxylates or titanate esters	0.05–0.10	accelerates cure
filler		
fused silica, carbon black, glass microballoons	6–10	extenders, reinforcing fillers, flow control
plasticizer		
nonreactive silicone fluid	0–10	improves uncured flow, reduces modulus
other additives		flame retardants, fungicides, adhesion promoters, pigments, thickeners

Table 4. Applications of Polysulfide Sealants

Application	Polysulfides, % of sealants used	Binders, %	System
aerospace	90	60	two-component
glass doors and windows	55	20–40	two-component
construction	5	25	one- and two-component

Table 5. Polysulfide Formulation

Ingredient[a]	Wt %	Function
base compound		
blend of LPs	15–60	binder
calcium carbonate[b]	30–60	filler
titanium dioxide	<10	pigment
silanes, titanates	<10	adhesion promoters
phenyl-2-naphthylamine	<10	antioxidant
methyl tuads, *p*-quinonedioxime (GMF), diphenylguanidine (DPG)	<10	accelerator
Cabosil, Ircogel	<10	thixotropes
stearic acid, stearates	<10	retarders
toluene, methyl ethyl ketone	<10	solvents
catalyst		
manganese dioxide	50–60	catalyst
dibutyl phthalate, HB-40	40–50	plasticizer

[a] The weight ratio of base compound to formulated catalyst is on the order of 100:10.
[b] Also glass bubbles, carbon spheres, vermiculite, metal powders.

Table 6. One-component Formulation Polyurethane Sealant

Ingredient	Wt %
diisocyanate prepolymer[a]	50–60
carbon black or other filler	40–50
silica, Santocel Monsanto	2–4
toluene	1–2
special-purpose additives[b]	

[a] Diphenylmethane diisocyanate of equivalent wt 1272.
[b] Uv absorbers, antioxidants, flame retardants, fillers, pigments.

Table 7. Formulation of One-component Latex Acrylic Sealant

Ingredient	Wt %
latex acrylic polymer 50% solids	30–40
calcium carbonate filler, or aluminum silicate, talc	30–45
titanium dioxide pigment	1.0–1.5
surfactant[a]	1.0–1.5
ethylene glycol freezing-point depressant	20–25
plasticizer	7.–8.
thixotropes	1–2
mineral spirits	0.2
biocidal preservatives	<1.0
hindered-phenol antioxidants	<1.0
substituted benzotriazoles uv absorbers	<1.0
mildewcides	<1.0

[a] To promote mixing and wetting.

Table 8. Formulation of Butyl Sealants

Ingredient	Wt %
butyl, isobutylene, or polybutene binder	40–45
clays, silicas, calcium carbonate, carbon black filler	30–40
titanium dioxide or zinc oxide pigment	2–4
rosin–pentaerythritol ester tackifier	0.5–1.0
fiber thickener	10
cyclohexane solvent	10

Table 9. Formulation of Chlorosulfonated Polyethylene Sealants

Ingredient	Wt %
chlorosulfonated polyethylene	15–20
chlorinated paraffins	15–20
fillers	15–25
pigments	10–15
oxide catalysts[a]	6–8
dibutyl phthalate plasticizer	15–20
isopropyl alcohol solvent	4–5

[a] Usually lead oxide.

BIBLIOGRAPHY

R. Meyer, *SAMPE J.* **18,** 6 (1982); **12,** 27 (1983).

A. Damusis, *Sealants,* Reinhold Publishing Co., New York, 1967.

Air Force sponsored handbook prepared by Rockwell Int., Los Angeles, Calif., *Aircraft Integral Fuel Tank Design Handbook, AFWAL-TR-87-3078,* Wright Patterson Air Force Base, Ohio. Sealant fillets, topcoats, faying surface applications discussed in detail.

Aerospace Recommended Practice (AIR 4069), *Preferred Practice for Sealing Integral Fuel Tanks,* Society of Automotive Engineers, Aerospace Division, Warrendale, Pa., 1988.

Ralph E. Meyer
Products Research & Chemical Corporation

SEALING. See Adhesion and bonding; Embedding.

SEAWEED COLLOIDS. See Polysaccharides.

SECOND-ORDER TRANSITION. See Glass transition; Transitions and relaxations.

SEDIMENTATION. See Flocculation; Ultracentrifugation.

SEGMENTED POLYURETHANES.

See FIBERS, ELASTOMERIC.

SELF-EXTINGUISHING PROPERTIES.

See FLAMMABILITY.

SELF-REINFORCING PLASTICS.

See REINFORCEMENT.

SEMICONDUCTIVE POLYMERS.

See ELECTRICALLY CONDUCTIVE POLYMERS.

SEMIPERMEABLE MEMBRANES.

See MEMBRANES.

SEQUENCE COPOLYMERS.

See BLOCK COPOLYMERS; GRAFT COPOLYMERS.

SHEETING

Sheeting or sheet is defined as shaped plastic having a large width and length in relation to thickness, with a thickness greater than 250 μm. Plastic films are differentiated from sheet only in thickness (less than 250 μm) (see also FILMS). Sheeting is formed from synthetic polymers, including polyolefins, poly(vinyl chloride) (PVC), polyesters, cellulosic esters, polystyrene, poly(methacrylate), and polycarbonate resins. Cross-linked or thermoset resins, including reinforced composite materials, can be formed into sheet structures; foamed plastic sheeting is also produced (see also POLYURETHANE; STRUCTURAL FOAMS). Laminar sheet materials are important, with many new market opportunities.

Sheeting is used in packaging, construction, electronics, photography, electrical insulation, glazing, identification cards, and printing stock. The market for plastic sheeting in packaging alone is over 415 $\times$ 10^6 metric tons per year.

Manufacturing Techniques

Sheeting is manufactured by solvent casting, calendering (qv), and melt-extrusion (qv) and casting (qv) processes. Acrylic sheeting and glass-reinforced unsaturated polyester sheeting are produced by *in situ* polymerization batch or continuous processes (Fig. 1).

Sheet products may be thermoformed, coated with functional coatings, laminated to other sheets, printed, embossed, polished, or metallized. They can be made to be weather resistant, transparent, pigmented or dyed, heat stabilized, or bondable.

Foaming. Foamed sheet is an important product for thermal insulation, packaging, cushioning, and buoyant articles. It is produced by extrusion under carefully controlled conditions. The foams are formed by the action of gas-producing blowing agents (qv) that are mixed into the polymer melt. Temperature and pressure control during extrusion is critical to prevent premature release of the gas and control the density and surface finish of the sheet. Foamed sheet is mostly produced from polystyrene or poly(vinyl chloride); polyethylene, polypropylene, and ABS are also used (see also CELLULAR MATERIALS).

Multilayer Sheet. Multilayer sheets can be produced in continuous or press laminations, or by coextrusion. Continuous laminations are possible in-line with the sheet-forming process or in a separate off-line process using heat or adhesives to bond the separate layers. Press laminations are used for individual decorative sheeting where heat, pressure, and time are required for cross-linking the various layers (see FILMS, MULTILAYER; LAMINATES).

Properties and Test Methods

The properties of the more important classes of sheet are given in Tables 1–4 (see also ACRYLIC AND METHACRYLIC ESTER POLYMERS; ACRYLONITRILE–BUTADIENE–STYRENE POLYMERS; CELLULOSE ESTERS, ORGANIC; CHLOROTRIFLUOROETHYLENE POLYMERS, THERMOPLASTIC; POLYSULFONES; PROPYLENE POLYMERS; STYRENE POLYMERS; TETRAFLUOROETHYLENE POLYMERS; VINYL CHLORIDE POLYMERS).

Physical properties of plastic sheet vary considerably, depending on method and temperature of testing, thickness of the sheet, and treatment. Some physical, electrical, and transport properties vary with thickness. Prototypes should be prepared for actual performance testing.

Properties of commercial sheet may be modified by fillers, reinforcements, impact modifiers, nucleants, plasticizers, colorants, and other additives. For example, PVC sheet may range from flexible to leatherlike to rigid, depending on the additives.

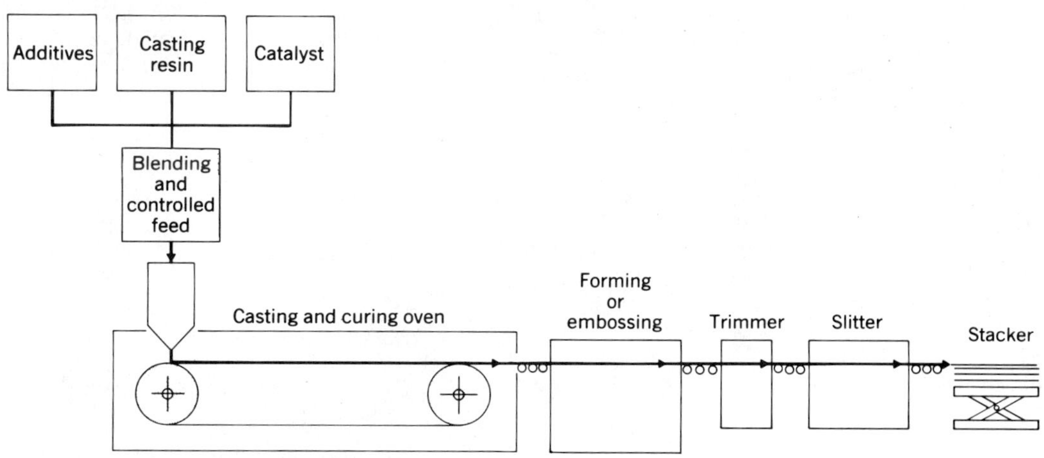

Figure 1. *In situ* polymerization continuous-casting process.

Table 1. Typical Properties of Plastic Sheet[a]

	Acrylonitrile–butadiene–styrene[b]	Acrylic	Cellulose acetate[b]	Cellulose triacetate	Cellulose acetate–butyrate	Poly-sulfone[b]
Characteristics						
thickness range, mm	0.25–0.75	0.75–50	0.25–0.75	0.50	0.5–6.3	0.75
maximum width, m	2.66	2.84	1.52	1.19	1.82	1.32
specific gravity	1.040	1.18–1.19	1.28–1.31	1.29	1.19–1.23	1.24–1.37
Mechanical properties						
tensile strength, MPa[c]						
at yield	40–45	37–86				
at break	30–60	40–96	33–113	60–110	30	
tensile modulus, GPa[d]	2–3	2–4			35–60	50–80
elongation, %	10–50	4–10	15–55	10–50		
flexural modulus, GPa[d]	2.3	3–4			50–100	20–150
Izod, notched, J/m[e]	250–380	20–100	200		1.4	
hardness, Rockwell	R190–105	M90–100	R80		220 R88	
Thermal properties						
thermal conductivity, W/(m·K)	0.24	0.20			0.25	
burning rate, cm/s	0.05–0.12	0.25	0.5–5.5	0.5–1.0	0.5–3.0	
Electrical properties						
dielectric constant						
1 kHz	2.5–3.0	3.5–4.0	3.6	3.2–4.5	2.9	3.07–3.5
dissipation factor						
1 kHz		0.040	0.013	0.016	0.013	0.008–0.0035
dielectric strength, MV/m		16–20			14–15	
volume resistivity, ohm-cm	10^{16}	10^{15}–10^{15}	10^{13}–10^{15}	10^{14}–10^{15}	10^{11}–10^{18}	

[a] Standard temperature and test speed; usually at a thickness of 25 μm.
[b] FDA grade available.
[c] To convert MPa to psi, multiply by 145.
[d] To convert GPa to psi, multiply by 145,000.
[e] To convert J/m to ft·lbf/in., divide by 53.38.

Table 2. Properties of Fluorinated Plastic Sheet[a,b]

	Polytetra-fluoro-ethylene	Poly-vinylidene fluoride	Ionomer	Polychloro-trifluoro-ethylenc	Fluorinated ethylene–propylene copolymer
Characteristics					
thickness range, mm	0.75–50	0.75	1	1.5–25	0.75
maximum width, m	1.22	1.27	1.83	1.22	1.22
specific gravity	2.1–2.2	1.76	0.94	2.08–2.15	2.15
Mechanical properties					
tensile strength, MPa[c]					
at yield					
at break	10–28	40	35	35–70	12–20
tensile modulus, GPa[d]					
elongation, %	100–400	25–500	250–450	50–175	300–350
flexural modulus, GPa[d]	0.55–1.0			1.12	0.66
Izod, notched, J/m[e]	160–320	120–180		390	
hardness, Rockwell	R62			R80	
Thermal properties					
thermal conductivity, W/(m·K)	0.24–1.2	0.24			0.20
linear thermal expansion $\times 10^{-5}$	8–14	14			0.4
change in linear dimension		0.2		+2 to −2	<1
Electrical properties					
dielectric constant					
1 kHz	2.0–2.1	8.4	2.4	2.5–2.7	2.0–2.5

Table 2. ***(Continued)***

	Polytetra-fluoro-ethylene	Poly-vinylidene fluoride	Ionomer	Polychloro-trifluoro-ethylene	Fluorinated ethylene–propylene copolymer
Electrical properties					
dielectric constant					
1 MHz	2.0–2.1		2.4	2.3–2.4	2.0–2.05
1 GHz	2.0–2.1	6.43	2.4	2.3	2.05
dissipation factor					
1 kHz	<0.0001	0.019	0.002	−0.022–0.024	<0.0002
1 MHz	<0.002		0.007	0.009–0.027	0.0003
1 GHz	<0.0005	0.159	0.007	0.004	0.0015
dielectric strength, MV/m	16–20	32		>20	19
volume resistivity, ohm-cm	>10^{18}	10^{14}	10^{16}	10^{18}	10^{18}

[a] Standard temperature and test speed; usually at a thickness of 25 μm.
[b] FDA grade available.
[c] To convert MPa to psi, multiply by 145.
[d] To convert GPa to psi, multiply by 145,000.
[e] To convert J/m to ft·lbf/in., divide by 53.38.

Table 3. Properties of Olefin and Vinyl Sheet[a,b]

	Low density poly-ethylene	High density poly-ethylene	Poly-propylene	Poly-styrene	Rigid poly-(vinyl chloride)	Flexible poly(vinyl chloride)	Poly(vinyl chloride-vinyl acetate) copolymer
Characteristics							
thickness range, mm		3–50	1.8–5	0.25–8	0.75	1	0.75
maximum width, m		2.43	1.52	1.93	2.13	2.03	1.83
specific gravity	0.91–0.92	0.926–0.94	0.905	1.05–1.06	1.2–1.5	1.2–1.4	1.3–1.4
Mechanical properties							
tensile strength, MPa[c]							
at yield							
at break	27.5	14–28	30–70	20–50	50–70	10–70	38–55
tensile modulus, GPa[d]							
elongation, %	100–700	10–650	550–1000	35–40	25–120	100–500	3–100
flexural modulus, GPa[d]	0.17	1.52	1.72	2.2	2.6		2.4–3
Izod, notched, J/m[e]				78–80	850		30–800
hardness, Rockwell		R50	R92	M80–95	R107–120		R118
Thermal properties							
thermal conductivity, W/(m·K)		0.52	0.17	0.15	0.1		
burning rate, cm/s					0.5–4.3	0.5–4.3	
Electrical properties							
dielectric constant							
1 kHz	2.2	2.3	2.2	2.4–2.7	3.0–3.3	4–8	3.0–3.3
1 MHz	2.2	2.3	2.2	2.4–2.7	2.8–3.1	3.3–4.5	2.8–3.1
1 GHz	2.2	2.3	2.2	2.4–2.7	2.8		2.8
dissipation factor							
1 kHz	0.0003	0.0005	0.0003	0.0005	0.009–0.017	0.07–0.16	0.009–0.0017
1 MHz	0.0003	0.0005	0.0003	0.0005	0.006–0.017	0.04–0.19	0.006–0.0019
1 GHz	0.0003	0.0005	0.0003	0.0005	0.019		0.019
dielectric strength, MV/m	24	28	26		16		
volume resistivity, ohm-cm	10^{18}	10^{16}	10^{18}	10^{7}–10^{16}	10^{16}	10^{11}–10^{14}	10^{16}

[a] Standard temperature and test speed; usually at a thickness of 25 μm.
[b] FDA grade available.
[c] To convert MPa to psi, multiply by 145.
[d] To convert GPa to psi, multiply by 145,000.
[e] To convert J/m to ft·lbf/in., divide by 53.38.

Table 4. Properties of Polyamide and Polyester Sheet[a]

	Nylon-6[b]	Nylon-6,6	Poly-carbonate[b]	Poly-(ethylene terephthalate) oriented[b]	Polyester[b]
Characteristics					
thickness range, mm	1.5–75	1.5–75	0.75–32	0.36	0.75–50
maximum width, m	0.61	0.61	2.43	3.04	1.82
specific gravity	1.12	1.13	1.20	1.40	1.85
Mechanical properties					
tensile strength, MPa[c]					
at yield	110	110			
at break	110	110	65	240	−200
tensile modulus, GPa[d]					
elongation, %	25	35	−110	50–150	
flexural modulus, GPa[d]	1.7	3.2	2.4		2–14
Izod, notched, J/m[e]	133	69	850		300–1000
hardness, Rockwell	R107	R118	R120		R90–115
Thermal properties					
thermal conductivity, W/(m·K)	0.24		0.19		0.17
linear thermal expansion $\times 10^{-5}$			6.75		3–10
change in linear dimension at 100°C for 30 min, %				1.27	
Electrical properties					
dielectric constant					
1 kHz	3.7	4.0	2.99	3.2	
1 MHz	3.0	3.9	2.93	3.0	
1 GHz		3.6	2.89	2.8	
dissipation factor					
1 kHz	0.016	0.01	0.0015	0.005	
1 MHz	0.036	0.02	0.010	0.016	
1 GHz	0	0.04	0.012	0.003–0.008	
dielectric strength, MV/m	40	40	15		14–16
volume resistivity, ohm-cm	10^{15}	10^{14}	10^{16}	10^{18}	

[a] Standard temperature and test speed; usually at a thickness of 25 μm.
[b] FDA grades available.
[c] To convert MPa to psi, multiply by 145.
[d] To convert GPa to psi, multiply by 145,000.
[e] To convert J/m to ft·lbf/in., divide by 53.38.

Foamed sheet properties depend on the foam density that is controlled by process conditions as well as polymer properties and the formulation. Solid-skin layers of foamed sheet strongly influence rigidity and permeability properties. The properties of composite or laminated sheet depend on the thickness of each layer and the bonding between them.

Economic Aspects and Outlook

The plastic sheet industry has grown steadily since the 1960s. Cellulose esters have lost market share, while ABS polymers, polycarbonate, thermoplastic polyesters, polystyrene, and PVC consumption for sheet applications has increased. Foamed sheet has increased in volume as a result of many new applications.

Many of the new high performance engineering resins should find markets as sheet, including the liquid crystal polyesters, polyether ketones, and polyarylates. On a cost-performance basis, polymer sheet will make significant inroads into markets of the traditional metallic materials.

BIBLIOGRAPHY

Modern Plastics Encyclopedia, Vol. 62, McGraw-Hill, Inc., New York, 1985.

Film, Sheet, Laminates—Desk Top Data Bank, International Plastics Selector, San Diego, Calif., 1979.

Mod. Plast. **64,** 55 (1987).

K. J. MACKENZIE
Hoechst-Celanese Corporation

SHEET MOLDING COMPOUNDS

Sheet molding compound (SMC) is a polymeric composite material. The term SMC was adopted by the Society of the Plastics Industry to denote a resin-impregnated chopped-strand fiber glass in dry sheet form (see also REINFORCED PLASTICS). During the last decade, the annual consumption of SMC in the automotive industry alone (including trucks) grew from 50 × 10^3 t in 1975 to more than 120 × 10^3 t in 1985. Other applications range from aircraft parts to appliances, business equipment, and electrical components.

Chemistry

The chemistry of SMC is complex because of the nature of the composite material. In general, SMC is made up of a styrenated unsaturated polyester resin, a low profile thermoplastic polymer, an inert filler, a fibrous reinforcement, a polymerization initiator, metal oxide maturation agent(s), and lubricant(s). It can be formulated in an almost infinite number of ways to produce materials with a wide range of properties for a variety of applications. A typical formulation is shown in Table 1.

Table 1. SMC Formulation

Material	Wt %
styrenated unsaturated polyester	24
thermoplastic polymers	3
filler, $CaCO_3$	41
glass fibers, 12–50 mm	30
polymerization initiator	0.3
maturation agent, MgO	0.7
lubricant, zinc stearate	1

Manufacture

Manufacture of SMC is a multistage process (Fig. 1).

Properties

Like other thermoset polymeric materials, SMC is inert compared with most thermoplastic polymers. Resistance to chemicals, including solvents, is excellent, as are thermal properties, with a heat deflection temperature above 200°C. The coefficient of thermal expansion is reported to be 1.4×10^{-5} mm/mm°C, which indicates superior dimensional stability.

Physical and mechanical properties depend on formulation (especially glass fiber content), fabrication, and other variables (Table 2).

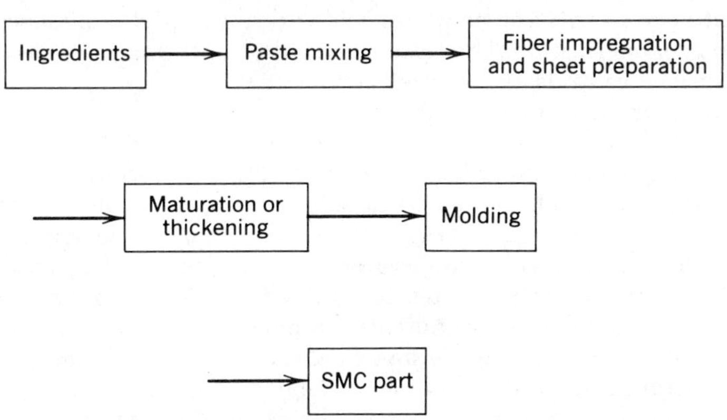

Figure 1. Flow diagram of SMC manufacturing process.

Table 2. Physical and Mechanical Properties of SMC

Properties	Value
specific gravity	1.6–2.6
hardness, Barcol	50–70
dielectric strength, V/mm	9.7–12.7
tensile strength, MPa[a]	55.16–172.37
tensile modulus, MPa[a]	9653–17,237
flexural strength, MPa[a]	68.95–248.22
flexural modulus, MPa[a]	6895–15,169
compressive strength, MPa[a]	103.42–206.85
Izod impact, notched, J/m[b]	373.6–1174.3

[a] To convert MPa to psi, multiply by 145.
[b] To convert J/m to ft·lbf/in., divide by 53.38.

BIBLIOGRAPHY

K. Sawallisch, *Polym. Plast. Technol. Eng.* **23,** 1 (1984).

Modern Plastics Encyclopedia **62,** 422 (1985–1986).

R. Burns, *Polyester Molding Compounds,* Marcel Dekker, Inc., New York, 1982, pp. 181–227.

R. W. Meyer, *Handbook of Polyester Molding Compounds and Molding Technology,* Chapman and Hall, New York, 1987.

T. J. Mao
General Motors Research Laboratories

SHELF AGING.
See Antioxidants; Antiozonants.

SHELLAC.
See Resins, natural.

SHOCK RESISTANCE.
See Impact resistance.

SHRINKPROOFING.
See Textile resins.

SILICA FIBERS.
See Fibers, engineering.

SILICATES

The commercial soluble silicates have the general formula $M_2O \cdot mSiO_2 \cdot nH_2O$, where M is an alkali metal and m and n are the number of moles of SiO_2 and H_2O, respectively, per mole of M_2O. The composition of commercial alkali silicates is typically described by the weight ratio of SiO_2 to M_2O; sodium silicates are the most common. Potassium and lithium silicates are manufactured to a limited extent for special applications. These materials are usually manufactured as glasses that dissolve in water to form viscous, alkaline solutions. The ratio of SiO_2 to M_2O in commercial sodium silicate products typically varies from 0.5 to 4.0. The most common form of soluble silicate, sometimes called waterglass, has a ratio of 3.2.

These glasses, which consist of the salt of a strong base and a weak acid, give highly alkaline solutions that make them an economical builder for synthetic detergents. The glassy nature of highly concentrated sodium silicate solutions led to the development of various adhesive and binder applications. A third major use is as a raw material for the manufacture of synthetic pigments and fillers; silica gels and sols; and clays and zeolites. Soluble silicates are also used in cements, coatings, bleaching, water treatment, soil stabilization, and ore beneficiation. These applications are based on the ability of silicates to form gels or react with multivalent metal ions in solutions or at or near oxide surfaces. Sodium silicates and their derivatives constitute a versatile, stable commodity and are ranked among the top 50 commodity chemicals.

The importance of the different uses of silicates has shifted since World War II, when adhesives and soap comprised the principal markets. Currently, soluble silicates are used primarily as sources of reactive silica (45%), in detergency (26%), paper and board adhesives (5%), and other applications (24%). As the structure and chemistry of solutions containing polymeric silicate species are elucidated, new and growing markets emerge, eg, enhanced oil recovery and peroxide pulp bleaching. Their versatility ensures continued market growth for silicates.

Structure

The structures of silicate glasses and solutions can be investigated with the help of chemical derivatization and gas chromatography, and laser Raman and ^{23}Si Fourier transform nuclear magnetic resonance (ft-nmr) spectroscopy.

Silicate Glasses. Synthetic silicates and silica are composed of oligomers of the orthosilicate ion, SiO_4^{4-}. Orthosilicate monomers have a tetrahedral structure. Silicate tetrahedra can be used to construct more complex structures according to Pauling's rules, ie, Si—O—Si bonds are permitted only at polygon corners. The physical and chemical properties of silicate glasses depend on the composition of the material, ion size, and cation coordination number.

Silicates in Solutions. The distribution of silicate species in aqueous sodium silicate solutions has long been of interest because of the wide variations in properties that these solutions exhibit with different moduli.

Research using light-scattering techniques showed that stable silicate solutions did not contain large particles; however, aggregation of particles was detected in highly concentrated, high ratio silicate solutions. Polymers in solutions of a variety of alkali silicates were studied by preparing, separating, and identifying trimethylsilyl derivatives using gas chromatography. This work gave strong evidence for the presence of a variety of multimeric silicate structures, even in highly alkaline, dilute silicate solutions. The results of investigations based on trimethylsilyl (TMS) derivatization and other indirect methods, eg, gel chromatography, indicated that the solutions contained a complex mixture of silicate anions in dynamic equilibrium.

Crystalline Soluble Silicates. The most common crystalline soluble silicates belong to the metasilicate family, $Na_2O{\cdot}SiO_2{\cdot}nH_2O$. Anhydrous sodium metasilicate, Na_2SiO_3, contains SiO_2 chains, whereas the hydrates $Na_2H_2SiO_4{\cdot}xH_2O$, where $x = n - 1$, contain only monomeric silicate. The structures of the series sodium metasilicate pentahydrate ($x = 4$), sodium metasilicate hexahydrate ($x = 5$), sodium metasilicate octahydrate ($x = 7$), and sodium metasilicate nonahydrate ($x = 8$) vary primarily by the order and coordination of the hydrated sodium ion. However, some differences in the symmetry and interatomic bonding of the $Si(OH)_2O_2^{2-}$ groups are observed. Only the anhydrous and tetrahydrate forms are of commercial importance.

Dissolution

The dissolution of soluble silicates is of considerable commercial importance. Its rate depends on the glass ratio, solids concentration, temperature, pressure, and glass particle size. Dissolution of sodium silicate glass proceeds through a two-step mechanism that involves ion exchange and network breakdown.

Ion exchange

$$\equiv\text{Si—ONa} + H_2O \rightleftharpoons \equiv\text{Si—OH} + Na^+ + OH^-$$

Network breakdown

$$\equiv\text{Si—O—Si}\equiv + OH^- \rightleftharpoons \equiv\text{Si—O}^- + \text{HO—Si}\equiv$$

Thus, silica is removed from the glass following leaching of the alkali cations. The rate of SiO_2 extraction increases as the pH of the solution rises above 9. Extraction of alkali occurs more readily at pH values below the pK_a of the glass surface silanol functionalities. If sufficient hydroxide activity is not generated in the ion exchange step, the silica network does not decompose and glass leaching results.

Polymerization in Solution

Polymerization and depolymerization of silicate anions are of great interest in gel and catalyst manufacture, detergency, and other applications where ion complexation and surface chemistry are of importance.

The complex silanol condensation process may be represented empirically by

$$\equiv\text{Si—OH} \overset{K_a}{\rightleftharpoons} \equiv\text{Si—O}^- + H^+$$

$$\equiv\text{Si—OH} + \equiv\text{Si—O}^- \underset{k_{-2}}{\overset{k_2}{\rightleftharpoons}} \equiv\text{Si—O—Si}\equiv + OH^-$$

Condensation occurs most readily at a pH value equal to the pK_a of the participating silanol group. This representation becomes less valid at pH values above 10, where the rate constant of the depolymerization reaction (k_{-2}) becomes significant and at very low pH values where acids exert a catalytic influence on polymerization. The state of ionization of the silica particle surface controls the rate of polymerization following homogeneous nucleation.

Extensive research on sol–gel processing of silicic acid esters, eg, tetraethoxysilane $Si(OC_2H_5)_4$, TEOS, in alcohol–water mixtures has elucidated silica polymerization in nonaqueous solvents. The relative rates of hydrolysis and condensation depend on the substrate, water, and catalyst (acid or base) concentrations; these rates determine the polymer structure.

Silicate solutions of equivalent composition may exhibit different physical properties and chemical reactivities because of differences in the distributions of polymer silicate species. This effect is observed in commercial alkali silicate solutions with compositions that lie in the metastable region near the solubility limit of amorphous silica.

Chemical Activity

Silica Polymer–Metal Ion Interactions in Solution. The reaction of metal ions with polymeric silicate species in solution may be viewed as an ion-exchange process. Silicate species acting as ligands would exhibit a range of reactivities toward cations in solution due to the wide range of silanol pK_a values. In general, metal ion activities at a given pH value are 1 to 2 decades lower in the presence of multimeric silicate species when compared to results predicted by hydrolysis reactions.

Effect on Oxide-Water Interfaces. The adsorption of ions at clay mineral and rock surfaces is an important step in natural and industrial process. Silicates are adsorbed on oxides to a far greater extent than would be predicted from their concentrations. This adsorption maximum at a given pH is independent of ionic strength, and maximum adsorption occurs at a pH near the pK_a values of the adsorbing species (between 6 and 10).

Characteristics

The characteristics of soluble silicates relevant to various uses include the pH behavior of solutions, the rate of water loss from films, and dried film strength.

Manufacture and Processing

Soluble silicate glasses are usually manufactured in oil- or gas-fired, open-hearth regenerative furnaces. Recent continuous-flow glass melters are equipped with high intensity gas burners or plasmas. These technologies offer significant advantages over conventional batch-melting processes. Glass composition and shutdown and start-up procedures can be changed rapidly because no large molten glass reservoir is maintained. In conventional processes, the glass is made by the reaction of sand and sodium carbonate (soda ash) at a temperature sufficiently high to provide a reasonable rate of quartz dissolution in the molten batch and manageable melt viscosity.

Commercial Products

The average composition and pertinent properties of commercial soluble silicates are given in Table 1. The largest volume of these materials is sold as liquids, which are differentiated by the SiO_2/Na_2O ratio and specific gravity. The powdered forms are useful as ingredients in dry-blended products; to control silicate reactivity; for convenience in handling and storage; or to increase silicate solution concentrations above the ranges available in liquid form.

Handling and Storage

Liquid silicates are available in drums or bulk quantities. Bulk shipments are transported by truck, rail, or tanker; large volumes are transferred by gravity or pumping.

Regulatory Status

Additives to Food and Potable Water. Sodium silicate is "generally recognized as safe" (GRAS) by the FDA in fabrics and when exposed to food. It is also recognized as a secondary direct food additive when used in boiler water for food-contact steam.

Table 1. Typical Properties of Commercial Sodium and Potassium Silicates

Commercial silicates	$SiO_2:M_2O^a$ Wt ratio	$SiO_2:M_2O^a$ Mole ratio	M_2O^a, wt %	SiO_2, wt %	Baumé[b,c]	d^{20}_{20}, g/cm³	Viscosity[b] Pa·s[d]	pH
anhydrous glasses								
sodium silicates	3.22	3.33	23.5	75.7				
	2.00	2.06	33.0	66.0				
potassium silicates	2.50	3.92	28.3	70.7				
hydrated amorphous powders								
sodium silicates	3.22	3.33	19.2	61.8				
	2.00	2.06	27.0	64.0				
solutions								
sodium silicates	1.60	1.65	19.7	31.5	58.5	1.68	7.00	12.8
	2.00	2.06	18.0	36.0	59.3	1.69	70.00	12.2
	2.50	2.58	10.6	26.5	42.0	1.41	0.06	11.7
	2.88	2.97	11.0	31.7	47.0	1.49	0.96	11.5
	3.22	3.32	8.9	28.7	41.0	1.39	0.18	11.3
	3.75	3.86	6.8	25.3	35.0	1.32	0.22	10.8
potassium silicates	2.50	3.93	8.3	20.8	29.8	1.259	0.04	11.30
	2.20	3.45	9.05	19.9	30.0	1.261	0.01	11.55
	2.10	3.30	12.5	26.3	40.0	1.381	1.05	11.70
	1.80	2.83	16.4	29.5	47.7	1.490	1.30	12.15
crystalline solids								
sodium orthosilicate		0.50	60.8	28.8				
anhydrous sodium metasilicate		1.00	51.0	47.1				
sodium metasilicate pentahydrate		1.00	29.3	26.4				
sodium sesquisilicate		0.67	36.7	24.1				

[a] M = Na or K.
[b] At 20°C.
[c] To convert Be° to sp gr, divide 145 by (145 − Be°).
[d] To convert Pa·s to P, multiply by 10.

Table 2. 1986 Prices of Silicate Products

Product	$/t Bulk	Bags
sodium silicate glass (3.22 ratio)	346	611
sodium silicate glass (2.00 ratio)	447	487
anhydrous sodium metasilicate	557	600
sodium metasilicate pentahydrate	378	417
sodium orthosilicate (technical grade, anhydrous)		760

In addition, it has unpublished GRAS status as a corrosion inhibitor in potable water. The safety of sodium silicate in these and other food-related applications has been reviewed recently.

Transportation and Disposal. Only highly alkaline forms of soluble silicates are regulated by the Department of Transportation (DOT) as hazardous materials for transportation.

Production and Economic Aspects

Worldwide silicate production is estimated to be approximately 4.0×10^6 t/y; the 1986 prices for typical products are given in Table 2. The price for a liquid product, normalized to a 100% solids basis, is usually close to the price of the corresponding glass.

The main producers of soluble silicates in the United States are the PQ Corporation, Occidental Chemical Co., Power Silicates, PPG Industries, W. R. Grace, and Chemical Products.

Health, Safety, and Environmental Aspects

The primary hazard of commercial soluble silicates is their moderate-to-strong alkalinity. Contact-exposure effects can range from irritation to corrosion, depending on the concentration of the silicate solution, the silica-to-alkali ratio, the sensitivity of the tissue exposed, and the duration of exposure.

Derivatives

The largest use of soluble silicates is as an active source of SiO_2 for the manufacture of silica-based products, such as gels, sols, precipitates, zeolites, clays, and metal compounds.

BIBLIOGRAPHY

J. S. Falcone, ed., *Soluble Silicates, ACS Symposium Series 194*, American Chemical Society, Washington, D.C., 1982.

R. K. Iler, *The Chemistry of Silica*, John Wiley & Sons, Inc., New York, 1979.

F. Liebau, *Structural Chemistry of Silicates*, Springer-Verlag, Berlin, 1985.

J. G. Vail, *Soluble Silicates, ACS Monograph No. 116*, Van Nostrand Reinhold Co., Inc., New York, 1952, 2 Vols.

JAMES S. FALCONE, JR.
SCOTT D. BOYCE
The PQ Corporation

SILICONES

The term silicone denotes a synthetic polymer $(R_nSiO_{(4-n/2)})_m$, where n = 1–3 and $m \geq 2$. A silicone contains a repeating silicon–oxygen backbone and has organic groups R attached to a significant proportion of the silicon atoms by silicon–carbon bonds. In commercial silicones most R groups are methyl; longer alkyl, fluoroalkyl, phenyl, vinyl, and a few other groups are substituted for specific purposes. Some of the R groups can also be hydrogen, chlorine, alkoxy, acyloxy, or alkylamino, etc. These polymers can be combined with fillers, additives, and solvents to result in products loosely classed as silicones.

Silicones have an unusual array of properties. Chief among these are thermal and oxidative stability and physical properties little affected by temperature. Other important characteristics include a high degree of chemical inertness, resistance to weathering, good dielectric strength, and low surface tension. As the general formula implies, the molecular structure can vary considerably to include linear, branched, and cross-linked structures. These structural forms and the R groups provide many combinations of useful properties that lead to a wide range of commercially important applications. Silicones include fluids, resins, and elastomers. Many derived products, eg, emulsions, greases, adhesives, sealants, coatings, and chemical specialties, have been developed for a large variety of uses.

Silicones are unique among the important commercial polymers in both basic chemistry and variety of industrial applications. They were the first useful polymers based on organometallic chemistry.

Nomenclature

Polymer nomenclature is inherently complex and difficult to use; that of silicones is simplified by the use of the letters M, D, T, and Q to represent monofunctional, difunctional, trifunctional, and quadrifunctional monomer units, respectively. Primes, eg, D′, are used to indicate substituents other than methyl; symbols are shown in Table 1.

The meaning of the primes must be specifically indicated in the text. The MDTQ nomenclature can be used to identify silicones (polysiloxanes) with as little ambiguity as the more conventional systems. Proper nomenclature follows IUPAC rules, but in the written or spoken language of the industry,

Table 1. Formulas and Symbols for Silicones

Formula	Functionality	Symbol
$(CH_3)_3SiO_{0.5}$	mono	M
$(CH_3)_2SiO$	di	D
$(CH_3)SiO_{1.5}$	tri	T
$(CH_3)(C_6H_5)SiO$	di	D′
$(C_6H_5)_2SiO$	di	D′
$(CH_3)(H)SiO$	di	D′
SiO_2	quadri	Q

Table 2. Identification of Siloxanes

Chemical name	Structural formula	MDT formula	Common name
hexamethyldisiloxane	$(CH_3)_3SiOSi(CH_3)_3$	MM	mono
octamethyltrisiloxane	$(CH_3)_3SiOSi(CH_3)_2OSi(CH_3)_3$	MDM	linear trimer
decamethyltetrasiloxane	$(CH_3)_3SiO[Si(CH_3)_2O]_2Si(CH_3)_3$	MD_2M	linear tetramer
octamethylcyclotetrasiloxane	$(CH_3)_2Si—O—Si(CH_3)_2$ O O $(CH_3)_2Si—O—Si(CH_3)_2$	D_4	cyclic tetramer
octaphenylcyclotetrasiloxane	$(C_6H_5)_2Si—O—Si(C_6H_5)_2$ O O $(C_6H_5)_2Si—O—Si(C_6H_5)_2$	D'_4	cyclic phenyltetramer
2,4,6,8-tetramethyl-2,4,6,8-tetraphenylcyclotetrasiloxane	$(CH_3)(C_6H_5)Si—O—Si(CH_3)(C_6H_5)$ O O $(CH_3)(C_6H_5)Si—O—Si(CH_3)(C_6H_5)$	D'_4	cyclic methylphenyltetramer
methyltris(trimethylsiloxy)silane	$[(CH_3)_3SiO]_3SiCH_3$	M_3T	
1,1,1,3,5,5,5-heptamethyltrisiloxane	$(CH_3)_3SiOSi(H)(CH_3)OSi(CH_3)_3$	MD'M	
1,1,3,5,5-pentamethyl-1,3,5-triphenyltrisiloxane	$(CH_3)_2(C_6H_5)SiOSi(CH_3)(C_6H_5)$ $OSi(C_6H_5)(CH_3)_2$	M'D'M'	

it is much more common, for example, to see D_4, methyltetramer, or simply tetramer than the proper octamethylcyclotetrasiloxane. In another example, $(CH_3)_3SiOSi(CH_3)_2OSi(C_6H_5)$-$[OSi(CH_3)_3]_2$ is designated $MDT'M_2$, where T′ is $C_6H_5SiO_{1.5}$. Table 2 shows the various ways of identifying siloxanes.

Structure and Properties

Silicone properties can be interpreted in terms of structural bond concepts. The silicon–oxygen chain that constitutes the backbone of these polymers is predominantly responsible for their uniqueness. A segment of methylsiloxane chain is shown below:

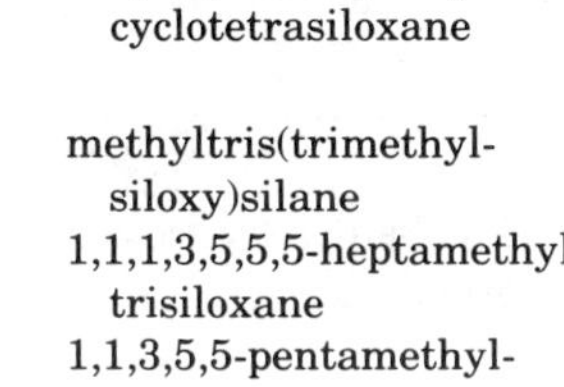

Commonly accepted bond angles are C—Si—C = 112°, Si—O—Si = 143°, and O—Si—O = 110°; bond lengths are Si—C = 0.188 nm and Si—O = 0.163 nm. The siloxane chain flexes, and rotation is fairly free about the Si—O axis, especially with small substituents, eg, methyl, on the silicon atoms. Rotation is also free about the Si—C axis in methylsilicon compounds. As a result of the freedom of motion, the intermolecular distances between methylsiloxane chains are greater than between hydrocarbons, and intermolecular forces are smaller. The small rotational barriers contribute to properties such as low modulus, low glass-transition temperature, and high permeability. The relationships between ring and chain forms in siloxane systems can be understood by the high flexibility of the chains.

Differences between silicon and carbon chemistry can be explained by electronegativity and orbital bonding effects. Silicon is more electropositive than carbon. Bonds between silicon and Cl, N, O, and S are more ionic and have higher energies than those to carbon. Analogously, bonds between silicon and C and H have lower energies than those to carbon.

Unlike carbon, silicon has available *d* orbitals. Electrons in these *d* orbitals can participate in bonding with π electrons from other atoms. The resulting $d\pi$–$p\pi$ involvement is often consistent with partial double-bond character.

The influence of the orbital effect is usually stronger than that of the electronegativity. Accordingly, the oxygen atom in ethers is more basic than the oxygen in siloxanes. Similarly, the rate of electrophilic, eg, dichlorocarbene, addition to $CH_3(CH_2)_3CH{=}CH_2$ is about twice that for $(CH_3)_3SiCH{=}CH_2$. In the absence of orbital effects, electronegativity effects explain behavior, eg, the greater basic character of nitrogen in $(CH_3)_3SiCH_2NH_2$ over nitrogen in $CH_3(CH_2)_2NH_2$.

The Si—O bond is ca 50% ionic, with silicon being the positive member. It has a high heat of formation of 452 kJ/mol (108 kcal/mol) and is resistant to homolytic cleavage. However, it is susceptible to heterolytic cleavage, ie, to attack by acids and bases. In this respect the —Si—O—Si—O— backbone of silicones is very different from the —C—C—C—C— backbone of hydrocarbon polymers.

The Si—Cl bond is also important in silicone chemistry. Silicones are usually made from organochlorosilanes. The first step is a hydrolysis reaction. Such reactions occur very fast compared with hydrolyses of corresponding alkyl halides:

$$(CH_3)_3SiCl + H_2O \xrightarrow[\text{solvent}]{\text{very fast}} (CH_3)_3SiOH + HCl$$

$$(CH_3)_3CCl + H_2O \xrightarrow[\text{solvent}]{\text{slow}} (CH_3)_3COH + HCl$$

Here the hydrogen in SiOH is more acidic than in COH. A hydroxyl group attached to silicon is generally referred to as a silanol. Most silanols are unstable and readily condense (with the elimination of water) to form siloxane linkages, ie, Si—O—Si.

Process

Silica, a material of mineral origin, is the natural source of silicon. It is usually reduced in an electric furnace process to silicon metal.

$$SiO_2 + 2\,C \xrightarrow[\text{furnace}]{\text{electric}} Si + 2\,CO$$

Silicon is treated with an organic halide to form organohalosilanes. A direct process provided the impetus for the commercial manufacture and marketing of silicones.

$$Si + 2\,RCl \xrightarrow{\text{catalyst}} R_2SiCl_2$$

where RCl is usually methyl chloride.

Hydrolysis of organochlorosilanes gives siloxane structures, which are the basis of silicone products:

$$2\,R_2SiCl_2 + 4\,H_2O \longrightarrow \underset{\text{unstable}}{2\,[R_2Si(OH)_2]} + 4\,HCl$$

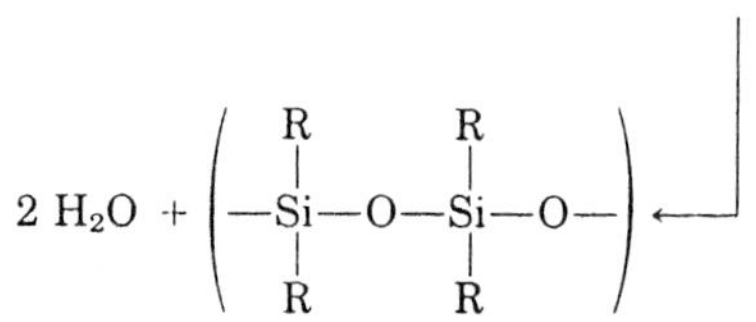

The main by-product of silicone manufacture is HCl. Silicone and HCl are made in almost equal amounts. HCl can be recycled by reaction with methyl alcohol to produce methyl chloride, however, large amounts often remain as waste to be treated and disposed of in an environmentally responsible manner.

Methylchlorosilanes are the starting materials for methylsilicones. They are made industrially by the copper-catalyzed exothermic reaction of methyl chloride with silicon at ca 300°C.

$$x(CH_3Cl + Si) \xrightarrow[300°C]{Cu} (CH_3)_2SiCl_2,\ CH_3SiCl_3,\ (CH_3)_3SiCl$$

To minimize high temperature side reactions the heat generated is removed with the help of stirred or fluidized beds. The reaction produces chlorosilanes, chlorodisilanes, hydrocarbons, and complex silicon compounds (Table 3).

The highest demand is for dimethyldichlorosilane, which is a raw material for silicone fluids, elastomers, and resins; operating conditions are chosen to optimize its formation. Of these conditions the catalyst is most important, followed by particle size distribution, the condition of the particles in the bed, and methyl chloride quality.

The production of the individual compounds in this complex reaction mixture cannot be balanced with demand. Several interconversion processes can be used to achieve a balance.

Redistribution

$$(CH_3)SiCl_3 + (CH_3)_3SiCl \underset{150°C}{\overset{\text{catalyst}}{\rightleftharpoons}} 2\,(CH_3)_2SiCl_2$$

Cleavage

$$(CH_3)_2Si_2Cl_4 + HCl \xrightarrow{\text{catalyst}} CH_3SiCl_3 + CH_3SiHCl_2$$

Other Chlorosilanes and Derivatives. Except for methyl, ethyl, and phenyl systems, the direct reaction is inefficient for making organochlorosilanes because of extensive degradations. Trichlorosilane is made by the reaction of elemental silicon with anhydrous hydrogen chloride at elevated temperature:

$$Si + 3\,HCl \longrightarrow HSiCl_3 + H_2$$

Condensation reactions with hydrogen functional chlorosilanes give phenylchlorosilanes.

$$CH_3(H)SiCl_2 + C_6H_6 \xrightarrow[300°C]{BCl_3} (CH_3)(C_6H_5)SiCl_2 + H_2$$

$$HSiCl_3 + C_6H_5Cl \xrightarrow{500°C} C_6H_5SiCl_3 + HCl$$

Phenyl groups in phenylchlorosilanes can be chlorinated.

Vinylchlorosilanes are made similarly from vinyl chloride in a hot tube without a catalyst,

$$(CH_3)_2(H)SiCl + CH_2{=}CHCl \xrightarrow{500°C} HCl + (CH_3)_2(CH_2{=}CH)SiCl$$

or by the addition of Si—H compounds to acetylene,

$$(CH_3)(H)SiCl_2 + C_2H_2 \xrightarrow{Pt} (CH_3)(CH_2{=}CH)SiCl_2$$

This type of addition reaction to carbon–carbon multiple bonds is termed hydrosilation.

Hydrosilation reactions are widely used for the preparation of silanes, including many important organofunctional species. The reaction can be initiated by heat, light, radiation, and peroxides, but catalysis by bases and especially noble metals is more common.

Catalysis by platinum (homogeneous and heterogeneous forms) is usually preferred; chloroplatinic acid is frequently employed. Catalysis by platinum is not always effective.

$$(CH_3)SiH(Cl)_2 + CH_2{=}CHCF_3 \xrightarrow{Pt} (CH_3)(CF_3CH_2CH_2)SiCl_2$$

$$(CH_3)SiH(Cl)_2 + CH_2{=}CHCN \xrightarrow{\text{base}} (CH_3)(CNCH_2CH_2)SiCl_2$$

Table 3. Properties of Methylchlorosilanes

Compound	Boiling point, °C	Density d^{20}, g/cm^3	Refractive index, n_D^{20}	Assay, %
$(CH_3)SiCl_3$	66.4	1.273	1.4088	95–98
$(CH_3)_2SiCl_2$	70.0	1.067	1.4023	99–99.4
$(CH_3)_3SiCl$	57.9	0.854	1.3893	90–98
$(CH_3)SiH(Cl)_2$	41.0	1.110	1.3982	95–97
$(CH_3)_2SiH(Cl)$	35.0	0.854	1.3820	

The vinyl compounds are vehicles for incorporating organofunctionality into siloxane compositions. Competitive reactions reduce hydrosilation yields. The platinum-catalyzed addition of silicon hydrides to 3-chloropropene, eg, gives both 3-chloropropyl and propyl derivatives.

The hydrolyzable chlorine groups in chlorosilanes can be substituted to give derivatives that are similarly hydrolyzable, but do not yield HCl as a by-product. This reduces corrosion and aids the preparation of reactive silicone compositions and of siloxanes that are unstable in the presence of strong acids. Alkoxylation, acyloxylation, and amination are employed for commercial products. Similar reactions lead to substitution by amides, ketoximes, and other active hydrogen compounds. Continuous processes give silanes (and some siloxanes) containing Si—O—C linkages in good yields. Properties of industrially important silanes are shown in Table 4. Almost any functional group can be introduced into the side chain of an alkyl- or aryl-substituted silane.

Chlorosilane Reactivity. Chlorosilanes are reactive toward polar reagents. The first step leads to the replacement of one or more halogens, usually by an attacking nucleophile.

$$R\overset{|}{\underset{|}{Si}}Cl + HOH \longrightarrow R\overset{|}{\underset{|}{Si}}OH + HCl$$

Such reactions provide a number of intermediates that can serve as starting materials for polymer synthesis, eg, silanols, alkoxy- and acyloxysilanes. The most reactive organochlorosilanes are the methylchlorosilanes, in which low steric bulk facilitates reactivity, especially toward nucleophilic substitution. Substitution of more electronegative groups for methyl (such as $ClCH_2$) increases the reactivity even further.

The higher alkylchlorosilanes are far less reactive, mainly because of steric inhibition and the effect of enhanced electron release. In branched chain alkylsilanes, eg, *t*-butylchlorosilanes, greater steric inhibition and electron release reduce the reactivity toward nucleophilic substitution and in condensation reactions. The phenylchlorosilanes are much less reactive than methylchlorosilanes. The bulkiness of the phenyl group more than offsets its electron-withdrawing power.

Table 4. Properties of Silanes

Compound	Boiling point, °C	Density,[a] g/cm³	Refractive index,[a] n_D
$HSiCl_3$	32	1.3298	1.3983
$(C_2H_5)SiCl_3$	100	1.2342	1.4257[b]
$(C_2H_5)_2SiCl_2$	129	1.0472	1.4291
$(C_2H_5)_3SiCl$	146	0.8977[b]	1.4299
$(C_2H_5)SiH(Cl)_2$	74.5	1.0926[b]	1.4148[b]
$(C_6H_5)SiCl_3$	201.5	1.3185	1.5245
$(C_6H_5)_2SiCl_2$	305	1.218	1.5765[c]
$(C_6H_5)SiH(Cl)_2$	184	1.2115	1.5257[b]
$(CH_3)(C_6H_5)SiCl_2$	205	1.174	1.5180[b]
$(CH_3)_2(C_6H_5)SiCl$	193.5	1.032[c]	1.5082
$(CH_3)(C_6H_5)_2SiCl$	93	1.085[c]	1.5742[b]
$(C_2H_3)SiCl_3$	92	1.265	1.4330
$(CH_3)(C_2H_3)SiCl_2$	93	1.085	1.4200
$(CH_3)_2(C_2H_3)SiCl$	83	0.884	1.4141
$(CH_3)(CF_3CH_2CH_2)SiCl_2$	122	1.211[b]	1.3817
$(CNCH_2CH_2)SiCl_3$	224	0.9699[b]	1.4103[b]
$(CH_3)(CNCH_2CH_2)SiCl_2$	215	1.187	1.4564[b]
$(CH_3)_2Si(OCH_3)_2$	80.5	0.8646[b]	1.3708[b]
$(CH_3)Si(OCH_3)_3$	103.5	0.955	1.3687
$(CH_3)_3Si(OCH_3)$	56.5	0.7537	1.3678[b]
$(CH_3)Si(O\overset{\overset{O}{\|}}{C}CH_3)_3$	95	1.1677	1.407[b]
$(CNCH_2CH_2)Si(OC_2H_5)_3$	224	0.978	1.4160
$(NH_2CH_2CH_2CH_2)Si(OC_2H_5)_3$	217	0.943	1.4190
$[(CH_3)_3Si]_2NH$	125.5	0.774	1.4078

[a] At 25°C unless otherwise noted.
[b] At 20°C.
[c] At 26°C.

Hydrolysis of Reactive Organosilicon Compounds. The silicone oligomers and polymers are formed by reaction between an organohalogenosilane or other reactive organosilane and water.

$$n\ X{-}\overset{R}{\underset{R'}{Si}}{-}X + n\ H_2O \longrightarrow {-}\!\!\left(\overset{R}{\underset{R'}{Si}}{-}O\right)_{\!n}\!\! + 2n\ HX$$

where X may be a halogen or any other hydrolyzable group and R and R′ may be alkyl or aryl groups or other hydrolyzable groups. In the latter case further reactions leading to cross-linking can occur. When X is a halogen, this reaction is normally heterogeneous, highly exothermic, and difficult to control. Mechanistically, it probably never conforms to this simplified equation. The process can be modified and controlled.

Heat of Reaction and Free Energy. For the following equilibrium, $2\ R_3SiCl + H_2O \rightleftharpoons (R_3Si)_2O + 2\ HCl$, the heat of reaction is approximately -100 kJ/mol (-24 kcal/mol) when $R = CH_3$. The free energy of the hydrolysis reaction is highly sensitive to the nature of the organic substituents, varying from -65.7 kJ/mol (-15.7 kcal/mol) for $(C_2H_5)_3SiCl$ to -81.2 kJ/mol (-19.4 kcal/mol) for $ClCH_2Si(CH_3)_2Cl$ at 25°C.

Silsesquioxanes. The oligomeric organosilsesquioxanes $(RSiO_{1.5})_n$ can be prepared by base-catalyzed siloxane rearrangements of hydrolysates of organotrichlorosilanes or their derivatives. The known crystalline oligomeric silsesquioxanes may be visualized as composed of fused tetracyclosiloxane rings as shown below.

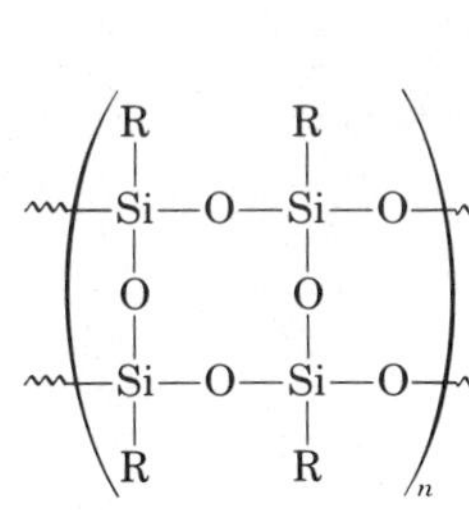

When $n = 2$, tetragonal symmetry is indicated. When $n = 3$ or 4, hexagonal or octagonal symmetry, respectively, is expected. Crystallographic relations are less clear-cut in the known decamers and dodecamers.

Silanols. Silanol groups are present in varying concentrations at various stages of the condensation of silicones, particularly in arylsilicone resins and alkylarylsilicone coreactants, and less in methylsilicone resins. The silanol function is used to control physical and chemical properties and the cure rates of silicone resins and other silicone compositions.

The physical properties of silanols are similar to those of analogous carbinols. They are, eg, associated in the liquid state, presumably through hydrogen bonding. Ir studies indicate that silanols are much more acidic than related carbinols, but only slightly less basic. Silanols include silanemonols, silanediols, polysiloxanediols, silanetriols, silanetetrol, disiloxanetetrols, and cyclotetrasiloxanetetrols.

Functionality. The role of functionality in silicone chemistry can be illustrated by the hydrolysis of methylchlorosilanes. Monofunctional reactants can be condensed only to disiloxanes.

$$(CH_3)_3SiCl \xrightarrow{H_2O} (CH_3)_3SiOH \xrightarrow{\text{condensation}} (CH_3)_3SiOSi(CH_3)_3$$

Difunctional reactants are converted through chlorosilanes or silanediols to linear polymers with reactive end groups or to monocyclic compounds.

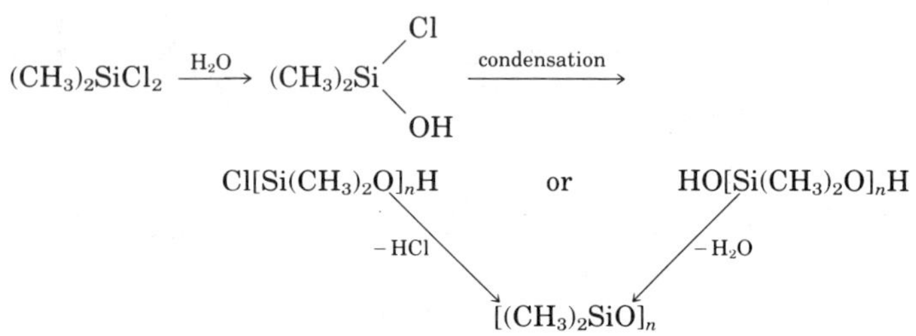

Trifunctional reactants can form dichlorosilanols, chlorosilanediols, or silanetriols, and these, upon subsequent condensation, give highly cross-linked gels (exceptions noted earlier); many reaction paths are available to trifunctional silanes.

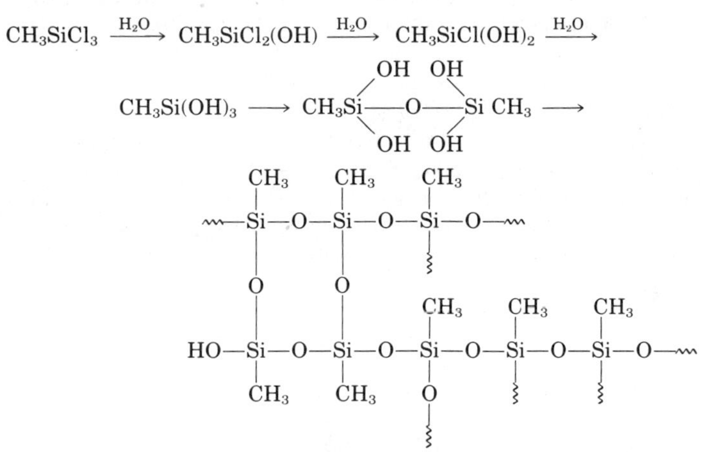

Tetrafunctional reactants give highly complex, usually highly hydrated, gel structures.

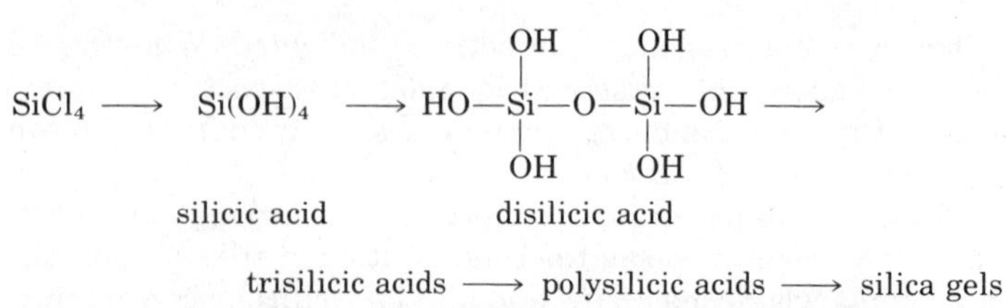

The characteristics and composition of end products can be controlled and modified in various ways. Chain lengths can be shortened and average molecular weights controlled by addition of small amounts of monofunctional reagents to difunctional reactants. Cross-linking can be controlled by addition of proper amounts of tri- or tetrafunctional reactants to difunctional reactants. Gelation can be retarded or prevented by adding appropriate amounts of monofunctional reactants to a system with an average functionality well above 2.0. Solubility of a polymerizing system with an average reactant functionality above 2.0 can be maintained by adjusting concentration and solvent polarity; cure can be affected by physical changes that allow further chemical reactions, usually rearrangement and cross-linking.

The hydrolysis of difunctional reactants gives a mixture of siloxanes of moderate average molecular weight with the properties of a fluid. The mixture contains both linear and cyclic oligomers. Higher average molecular weight materials, ie, more viscous fluids and gums, are prepared by polymerizing purified lower cyclic compounds (mainly the $n = 4$ oligomer), which can be distilled from the mixture. When equilibrium is reached, there is usually a Gaussian distribution of molecular weights among the chain molecules and a decreasing distribution of cyclic oligomers with increasing molecular weight.

Organofunctionality in silicon compounds and siloxane systems makes it possible to design formulations that contain reactive groups, which later undergo further reactions effecting cross-linking and curing. Common examples include hydrosilation,

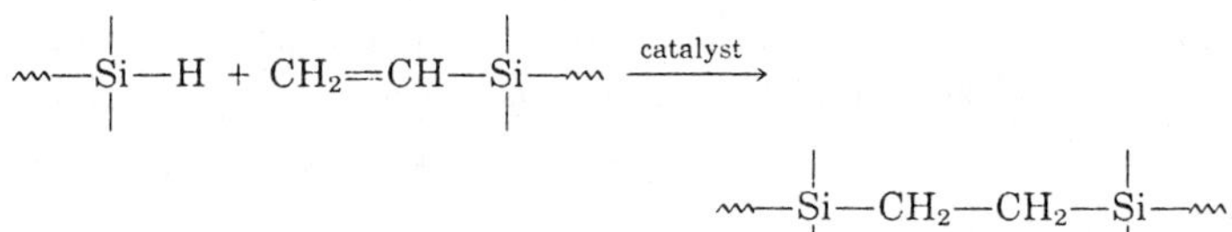

and silanol condensation,

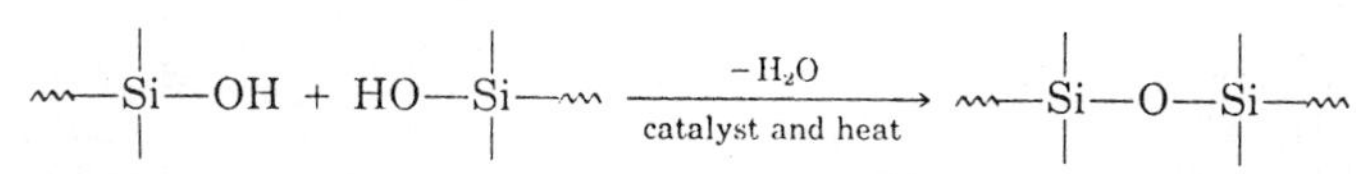

Silicone Polymers

The three commercially important classes of silicone polymers include silicone homopolymers, silicone random copolymers, and silicone–organic (block) copolymers. Polydimethylsiloxanes constitute by far the largest volume of homopolymers produced today.

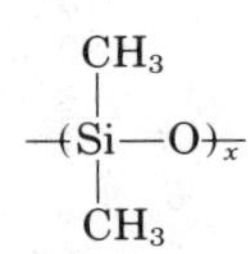

PDMS is usually the principal component of the random copolymers and the principal siloxane component of most silicone–organic copolymers. The production of commercial silicone homopolymers and random copolymers and the synthesis of siloxane intermediates for silicone–organic copolymers usually involves acid- or base-catalyzed equilibration of linear or cyclic siloxane oligomeric intermediates. These intermediates are formed during hydrolysis of difunctional silane reactants. The lower molecular weight cyclic compounds can be stripped from the hydrolysate. Cyclic compounds are also obtained by depolymerization of linear polymers, which readily crack when heated with caustic.

$$HO[(CH_3)_2SiO]_mH \xrightarrow[200°C]{KOH} [(CH_3)_2SiO]_3 + [(CH_3)_2SiO]_4 + [(CH_3)_2SiO]_5 + \cdots + H_2O$$

Cyclic tetramer is the main component obtained from base-catalyzed thermal cracking; the reaction is driven by removal of the product. Ring-opening polymerization of purified cyclic compounds leads to higher molecular weight materials.

$$H_2O + [(CH_3)_2SiO]_4 \xrightarrow{base} HO[(CH_3)_2SiO]_xH$$

The highest molecular weight silicone polymers (gums) are made by using strong bases such as KOH as catalysts, usually in batch processes. Large volume grades are also produced in continuous reactors. Molecular weight is controlled by chain-terminating groups.

Hydrolysis and Equilibration. The first step in the conversion of chlorosilanes to useful polymer products is hydrolysis. The chlorosilanes are hydrolyzed to silanols, which condense rapidly to siloxanes:

$$\equiv SiCl + H_2O \longrightarrow \equiv SiOH + HCl$$

$$\equiv SiOH + HOSi\equiv \longrightarrow \equiv SiOSi\equiv + H_2O$$

$$\equiv SiOH + ClSi\equiv \longrightarrow \equiv SiOSi\equiv + HCl$$

When dimethyldichlorosilane is hydrolyzed with excess water, the hydrogen chloride evolved is partially or completely dissolved in the aqueous phase, and a light, colorless siloxane oil separates as a second phase. Both linear and cyclic polydimethylsiloxanes are formed.

$$(CH_3)_2SiCl_2 + H_2O \longrightarrow HO[(CH_3)_2SiO]_mH + HCl \text{ (linear)}$$

$$(CH_3)_2SiCl_2 + H_2O \longrightarrow [(CH_3)_2SiO]_n + HCl \text{ (cyclic)}$$

Approximately half of this oil consists of a mixture of cyclic species of empirical composition $[(CH_3)_2SiO]_n$, mainly tetramer ($n = 4$).

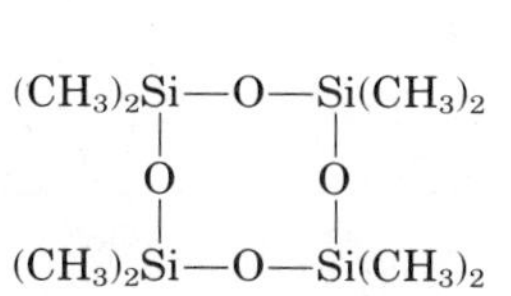

cyclic tetramer

Base-catalyzed Polymerization. Strong inorganic alkalies and quaternary ammonium hydroxides and phosphonium hydroxides polymerize octamethylcyclotetrasiloxane to a high molecular weight polymer. The polymer thus formed has a number-average molecular weight of about 10^6.

The effectiveness of alkali metal hydroxides as catalysts increases in the order $Li^+ < Na^+ < K^+ < Rb^+ < Cs^+$, which is also the order in which the degree of ionization of their salts or silanolates would be expected to increase. Silanolates are good catalysts for the polymerization of cyclic polysiloxanes. The observed order of reactivity for siloxane oligomers with base is $D_3 > D_4 > D_5 > MD_2M > MDM > M_2$.

Acid-catalyzed Polymerization. With protonic acids, the idea that protonation of the oxygen in a siloxane bond initiates polymerization is consistent with the observed basicity of oxygen in siloxanes and the order of reactivity found for siloxane oligomers with sulfuric acid: $D_3 > M_2 > MDM > MD_2M > D_4$.

Brønsted and Lewis acids are also effective catalysts. There is doubt whether Lewis acids (including hydrogen chloride) are catalysts when completely pure, dry, and not exposed to light. Cocatalysts such as water, carboxylic acids, or hydrogen chloride must be present in small amounts. The effective catalysts are presumably complexes such as $HAlCl_4$, $HFeCl_4$, $HFeCl_3OH$, or $HSnCl_4OH$.

Because of the reactivity of the Si—H bonds with strong bases, polymethylhydrogensiloxane homopolymers and random copolymers can only be prepared by acid equilibration at low temperatures.

Copolymerization. Copolymerization of siloxanes extends the possibility of modification to provide specific properties for particular applications. In simple cases different cyclic siloxanes and catalysts are mixed, and random copolymerization occurs. Base-catalyzed copolymerization is preferred for higher molecular weight products. It depends on inherent copolymerization ratios; the various cyclic siloxanes do not react at the same rates because of electronic and steric effects. Typically, copolymers of dimethylsiloxane containing diphenyl-, methylphenyl- and methylvinylsiloxane units are prepared in this manner. Final equilibrated products have a random distribution of siloxane units.

Another technique is to polymerize a single cyclic siloxane that contains the desired substituents, such as a mixed methylphenyl, eg,

$$[(CH_3)_2SiO]_2[(C_6H_5)_2SiO]_2$$

a mixed ethylmethyl, or a mixed methylvinyl cyclic trimer or tetramer.

A third method of copolymerization gives block copolymers, such as silicone blocks of different compositions or silicone heteropolymers; specific catalysts are required. An example is the formation of block copolymers by the reaction of linear polydimethylsiloxanes with silanol end groups with *p*-bis(dimethylhydroxysilyl)benzene (a disilphenylenediol), using quaternary ammonium salt catalysts such as the *n*-hexylamine, or the tetramethylguanidine salts of di-2-ethylhexanoic acid.

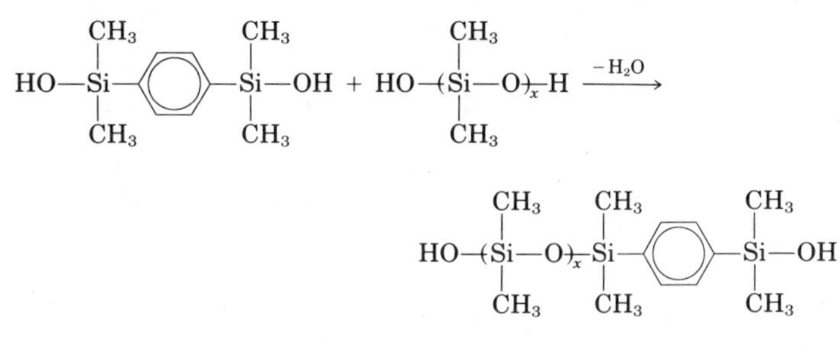

Table 5. Properties of MD_xM and D_n

MDT formula	Melting point, °C	Boiling point, °C	Density d^{20}, g/cm^3	Refractive index, n_D^{20}	Viscosity (η) at 25°C, mm^2/s	Flash point, °C
MM	−67	99.5	0.7636	1.3774	0.65	−9
MDM	−80	153	0.8200	1.3840	1.04	37
MD_2M	−76	194	0.8536	1.3895	1.53	70
MD_3M	−80	229	0.8755	1.3925	2.06	94
MD_4M	−59	245	0.8910	1.3948	2.63	118
MD_5M	−78	270	0.9012	1.3965	3.24	133
MD_6M	−63	290	0.9099	1.3970	3.88	144
MD_7M		307.5	0.9180	1.3980	4.58	159
D_3	64.5	134	1.12[a]			
D_4	17.5	175.8	0.9561	1.3968	2.30	69
D_5	−44	210	0.9593	1.3982	3.87	
D_6	−3	245	0.9672	1.4015	6.62	
D_7	−32	154[b]	0.9730	1.4040	9.57	
D_8	31.5	290	1.1770[b]	1.4060	13.23	

[a] Crystals.
[b] At 2.7 kPa (20 mm Hg).

Silicone-Organic Copolymers. Silicone-organic block copolymers have been made with a variety of organic components. Both copolymerization and grafting prepolymerized segments together are employed. Block copolymers of silicones and organic polymers can provide physical strength as well as the durability and surface properties of silicones. Some compositions are compatible with organic plastics and are used as additives to improve processibility. Others have unique properties for special applications. Silicone–polyethers are used as surfactants with polyurethane foam, eg.

Properties. Linear polydimethylsiloxanes have been extensively studied. Because of weak intermolecular forces, the polymers have low melting points and second-order transition temperatures. For dimethylsilicone gums (rubber polymers) these are −50 and −85°C, respectively. They do not crystallize under ordinary conditions. The low intermolecular forces result in low boiling points for the lower molecular weight oligomers, low activation energy for viscous flow, high compressibility, small change of viscosity with temperature, and for resins and unreinforced elastomers, poor physical properties despite high molecular weights. Structural interpretation of many properties includes the high flexibility of siloxane chains; factors influencing the spatial extension of polysiloxane molecules are now better understood.

Volatile oligomers in the products of equilibration of MM and D_4, eg, MD_xM and D_n, can be separated and purified by fractional distillation. Their properties show the effects of molecular size and structure in these systems (Table 5).

The physical properties of several oligomers with various substituents are shown in Table 6. Many such oligomeric compounds have been prepared and described, and several are used as intermediates in the preparation of polymers.

In higher linear dimethylsilicone polymers, the number-average molecular weight is related to the bulk viscosity:

$$\log\,(\text{viscosity in mm}^2\text{/s at }25°\text{C}) = 1.00 + 0.0123\bar{M}_n^{0.5}$$

$$\text{where } \bar{M}_n > 2500$$

The intrinsic viscosity [η] in dL/g, determined by the extrapolation of the viscosity to zero concentration, has the following values:

$$[\eta] = 2 \times 10^{-4}\bar{M}_n^{0.66}\ \text{(toluene)}$$

and

$$[\eta] = 8 \times 10^{-4}\bar{M}_n^{0.5}\ \text{(methyl ethyl ketone)}$$

Branched polymers, made by introducing T or Q units, have lower bulk or intrinsic viscosities than linear polymers of the same average molecular weight. Even slight branching causes a noticeable decrease in the bulk viscosity.

Table 6. Properties of Siloxane Oligomers

Compound	Boiling point, °C$_{kPa}$[a]	Density d^{20}, g/cm^3	Refractive index, n_D^{20}	Melting point, °C
$[(C_6H_5)_2SiO]_4$	$335_{0.13}$	1.23[b]		200
$[(CH_3)(C_6H_5)SiO]_4$	$237_{0.13-0.67}$	1.183	1.5461	99
$(CH_3)_3SiOSi(C_6H_5)_2OSi(CH_3)_3$	$172_{2.4}$	0.984	1.4927	
$[(CF_3CH_2CH_2)(CH_3)SiO]_4$	$134_{0.4}$	1.255	1.3724	
$[(CH_2{=}CH)(CH_3)SiO]_4$	$111_{1.3}$	0.9875	1.4342	−43.5
$(CH_3)_3Si[OSiH(CH_3)]_2OSi(CH_3)_3$	177	0.8559	1.3854	
$[(CH_3)(H)SiO]_4$	134	0.9912	1.3870	−69

[a] To convert kPa to mm Hg, multiply by 7.5.
[b] Crystals.

The unique flexibility of the siloxane chain and the fact that there is essentially no energy barrier for rotation around the siloxane bond result in low glass-transition temperatures, small changes in physical properties (except viscosity) with molecular weight, and small changes (compared with other polymer systems) with temperature.

The effect of temperature on the viscosity of polydimethylsiloxanes is much smaller than for other organic polymers. Silicone fluids flow almost as readily at low temperatures as at high. They can be uniquely suited to applications that require uniform characteristics over wide temperature ranges.

The surface properties of polydimethylsiloxanes are the basis for their applications as antifoam agents, foam stabilizers, internal lubricants and external release or slip agents. The foam-stabilizing properties of the silicone–polyether copolymers further extend the use of these materials. The low surface tensions and mild impact of molecular weight are characteristic of the silicones. The interfacial tension against water is ca 42 mN/m (= dyn/cm); such fluids repel water when spread on glass.

Polydimethylsiloxanes are insoluble in water, about 30 ppb for the trimethylsiloxy-terminated polymer. Solubility in water increases when polar groups, including silanols, are incorporated into the polymer.

Silicones have high gas solubility compared with other liquids and polymers. At 25°C and 101 kPa (1 atm), the air, nitrogen, and carbon dioxide solubilities are 0.17, 0.17, and 1.0 cm^3/cm^3, respectively.

Silicones have excellent electrical properties and are widely used as fluids, resins, elastomers, and dielectric compounds in electrical and electronic insulation applications. These properties are affected by moisture and, in the case of elastomers and dielectric compounds, by compounding ingredients and cure conditions.

Chemical Properties. The silicon–carbon bond and the siloxane bond have substantial thermal stability. The siloxane bond is less thermally stable than the silicon–carbon bond; however, the silicon–carbon bond is vulnerable to oxidation near 200°C. In the absence of catalysts the hydrolytic depolymerization of silicones is very slow. They are attacked by steam and depolymerize when heated with water in a closed vessel.

Polysiloxane Cures. Curing (cross-linking) processes for polysiloxanes of commercial importance include condensation and silane-olefin addition (hydrosilation) reactions. Peroxides are widely used to cross-link heat-cured rubber products. Peroxide fragment radicals abstract hydrogen from silicone methyl groups, which leads to coupling of the chains. Radiation cure with uv light is receiving a great deal of attention and some commercial products cured with this technology are available.

Silicone Fluids

A large number of silicone fluids have been synthesized and studied. Many are manufactured commercially, including dimethyl, methylalkyl, and dimethyl–diphenyl copolymers and silicone–polyether copolymers. Reactive fluids with silanol, alkoxy, or vinyl end groups share many properties with nonreactive fluids terminated with trimethylsiloxy groups. However, the reactive fluids are used in products where they are cured by appropriate methods.

Dimethylsilicone fluids and many reactive fluids are made by catalyzed equilibration of dimethylsilicone stock (hydrolysate or distilled cyclic compounds) with a source of chain terminator. With trimethylsiloxy, $(CH_3)_3SiO_{0.5}$, as chain terminator, this reaction produces mixtures of MD_xM and D_n polymers. The ratio of M to D in the charge controls the average molecular weight and the viscosity of the product. For example, a 50-mm^2/s fluid has an average molecular weight of ca 3,000, a 350-mm^2/s fluid ca 15,000, and a 1,000-mm^2/s fluid ca 25,000.

In most instances the equilibrated fluid is devolatilized by heat and vacuum after catalyst deactivation. Fluids can be blended to fluids of intermediate viscosities. Since the properties of a polymer depend on molecular weight distribution and average molecular weight, blending can affect physical properties appreciably. Blends of fluids of widely different viscosities are less Newtonian in behavior than those with normal molecular weight distribution.

Properties. Properties of commercial silicone fluids are given in Table 7.

Table 7. Properties of Silicone Fluids

$(CH_3)_2SiO$, mol %	Copolymer silicone	Viscosity at 25°C, mm^2/s	d^{25}, g/cm^3	n_D^{25}	Pour point, °C	Flash point, °C	Surface tension at 25°C, mN/m(= dyn/cm)
100	none	10	0.940	1.399	−73	210	20.0
100	none	100	0.968	1.4030	−55	302	20.9
100	none	1,000	0.974	1.4035	−50	315	21.1
100	none	10,000	0.975	1.4035	−47	315	21.3
100	none	100,000	0.978	1.4035	−40	315	21.3
50	$CH_3(C_6H_5)SiO$	125	1.07	1.495	−45	302	24.7
91.2	$CH_3(C_6H_5)SiO$	50	0.99	1.425	−73	282	25.0
91.2	$CH_3(C_6H_5)SiO$	100	0.99	1.425	−73	293	24.1
95.6	$(C_6H_5)_2SiO$	100	1.00	1.421	−73	302	24.0
>90	tetrachlorophenyl-siloxane	70	1.045	1.428	−73	288	21
0	$CH_3(H)SiO$	25	0.98	1.397			20
0–10[a]	$CF_3CH_2CH_2(CH_3)SiO$	300	1.25		−48		26
92	$(CH_3)SiO_{1.5}$	50	0.972	1.403	−84	315	21.0

[a] Approximate value.

Derived Products. Several types of greases, including lubricating, insulating, and water-repellent greases, are made from mixtures of silicone fluids and fillers. Silicone emulsions in water are vehicles for applying small amounts of silicone to textiles, paper, and other surfaces. As antifoaming agents they disperse rapidly in aqueous systems.

Silicone Resins

Silicone resins are highly cross-linked siloxane systems. The cross-linking components are introduced as trifunctional or tetrafunctional silanes in the first stage of manufacture. For example, a solution of CH_3SiCl_3, $(CH_3)_2SiCl_2$, $C_6H_5SiCl_3$, and $(C_6H_5)_2SiCl_2$ or $CH_3(C_6H_5)SiCl_2$ in toluene is hydrolyzed to form a complex copolymer mixture, which remains in solution in toluene. The aqueous hydrochloric acid is separated, and the resin solution is washed and heated in the presence of a mild condensation catalyst to adapt (body) the resin to the proper viscosity and cure time. It is finally adjusted to specifications by distilling or adding solvents. The properties of the resins depend on the choice of chlorosilanes, the degree of cure, and the processing conditions.

Properties. The chlorosilanes for a particular resin formulation determine its characteristics. Monomethyl-, dimethyl-, monophenyl-, diphenyl-, methylphenyl-, monovinyl-, and methylvinylchlorosilanes, together with silicon tetrachloride, have been most widely used. Prediction of specific resin properties as a function of composition frequently fails because processing and cure influence the final molecular configuration and related characteristics. However, some generalizations can be made: trifunctional siloxy units produce harder, less flexible resins, which are frequently immiscible with organic polymers; difunctional siloxy units increase softness and flexibility; and phenylsiloxanes are more miscible with organic polymers than methylsiloxanes and produce resins that are less brittle and have superior thermal resistance. Alkyl groups other than methyl also increase the compatibility with other organic materials. The effects of silanes on the properties of a film are shown in Table 8. Properties vary considerably; some resins are soft and flexible and others are hard and glassy.

Silicone Elastomers

Cross-linking silicone polymers of appropriate molecular weight provides elastomeric properties. Fillers increase strength through reinforcement, and extending fillers and additives, eg, antioxidants, adhesion promoters, and pigments, provide specific properties.

Many curing (cross-linking) systems have been developed commercially for silicone elastomers. Different silicone elastomers are conveniently distinguished by their cure system chemistries and can be categorized by the temperature conditions needed for proper cure. Most compositions are based on polydimethylsiloxanes:

$$RO\!-\!\!\left(\begin{matrix} CH_3 \\ | \\ Si \\ | \\ CH_3 \end{matrix}\!-\!O\right)_{\!x}\!\!-\!R$$

R is determined by the cure system chemistry. It can be hydrogen, an organic radical, or a silyl radical. The silyl radicals can contain single or multiple reactive groups like vinyl or alkoxy. Small amounts of reactive functionality are sometimes present in the chain in (copolymerized) units such as $(CH_2CH)(CH_3)SiO$. The value of x varies mainly with the type of product. For room temperature vulcanizing (RTV) products, x is in the 200–1,500 range; for heat-cured products x is ca 3,000–11,000.

Fillers. The reinforcing fillers for silicone elastomers are finely divided silicas made by fume or wet processes. Particle sizes $<0.05\ \mu m$ give the best reinforcement.

Nonreinforcing fillers, such as iron oxide or titanium dioxide, may stabilize or color the product or lower the cost per unit volume. Carbon black or carbon fibers provide electric conductivity.

Silicone elastomers differ in several important ways from most organic elastomers. The most striking difference is the degree to which the strength of the final compound depends on the reinforcement conferred by the incorporation of fillers. Typical unfilled silicone gums, when cross-linked, are weak and soft, with tensile strengths on the order of 0.34 MPa (50 psi). Compounding with suitably reinforcing fillers may increase the tensile strength as much as fiftyfold. The selection of the filler is therefore extremely important for applications where strength is required. Furthermore, because of the excellent thermal and electrical properties of the polymers, performance may in some cases be limited by the stability of the fillers rather than the polymer.

RTV Rubbers. Room temperature vulcanizing (RTV) silicone elastomers are supplied as uncured rubbers with liquid or pastelike consistencies. They are based on polymers of intermediate molecular weights and viscosities, eg, 100–1,000,000 mm^2/s at 25°C. Curing is based on chemical reactions that increase molecular weights and provide cross-linking; catalysts ensure cure control. The RTV silicone rubbers are available in two modifications. The cure reactions of one-component products are triggered by exposure to atmospheric moisture. Those of two-component products are triggered by mixing the two components, one of which consists of or contains the catalyst; the two components are supplied separately.

Table 8. Effect of Silanes on the Properties of Silicone Resin Films

Property	CH_3SiCl_3	$C_6H_5SiCl_3$	$(CH_3)_2SiCl_2$	$(C_6H_5)_2SiCl_2$	$CH_3(C_6H_5)SiCl_2$
hardness	increase	increase	decrease	decrease	decrease
brittleness	increase	great increase	decrease	decrease	decrease
stiffness	increase	increase	decrease	decrease	decrease
toughness	increase	increase	decrease	decrease	decrease
cure speed	much faster	some increase	slower	much slower	slower
tack	decrease	some decrease	increase	increase	increase

Fluids with silanol end groups are employed for most one-component products. The most widely used products are cured by reactions involving acetoxysilanes. Tin soaps are commonly used as catalysts.

$$HO[Si(CH_3)_2O]_nH + 2\ CH_3Si(CH_3CO_2)_3 \longrightarrow$$

$$(CH_3CO_2)_2(CH_3)SiO[Si(CH_3)_2O]_nSi(CH_3)(CH_3CO_2)_2 + 2\ CH_3COOH$$

$$CH_3CO_2Si{\equiv} + H_2O \longrightarrow HOSi{\equiv} + CH_3COOH$$

$$CH_3CO_2Si{\equiv} + HOSi{\equiv} \longrightarrow {\equiv}SiOSi{\equiv} + CH_3COOH$$

Products based on alkoxy-, amino-, ketoxy- and other silanes are also available. The one-component RTV silicone rubbers are mostly used in adhesive and sealant applications. Other uses include formed-in-place gasketing, protective coatings, and encapsulation; bonding properties are important.

The two-component RTV silicone rubbers do not require atmospheric moisture to trigger cure. Several different curing systems are employed, with different advantages. For example, the silanol-terminated silicone polymer is treated with alkoxy-functional silicon curing agents:

$$3\ HO[Si(CH_3)_2O]_nH + CH_3Si(OC_2H_5)_3 \xrightarrow{\text{catalyst}}$$

$$CH_3Si[O[(CH_3)_2SiO]_nH]_3 + 3\ C_2H_5OH$$

The catalyst is usually a tin salt. The product can be designed in such a way that the polymers and curing agent are in one package and the catalyst alone or the curing agent and the catalyst in the other. Cure is triggered by mixing the contents of the two packages. Fillers and additives are incorporated in the formulations according to the desired product properties.

Hydrosilation curing gives two-component RTV rubbers without liberating a by-product. This cure system is based on the addition of silicon hydride to olefin. In practice, vinyl-functional silicone polymers are used with hydride-functional siloxanes as curing agents.

$${\equiv}SiCH{=}CH_2 + HSi{\equiv} \longrightarrow {\equiv}SiCH_2CH_2Si{\equiv}$$

The reaction proceeds at room temperature in the presence of a catalyst, eg, chloroplatinic acid or another solubilized platinum compound. The two-component RTV silicone rubbers are extensively used in encapsulating and molding applications, as well as for bonding, sealing, protective coatings, and electrical insulation.

Typical properties of cured RTV silicone rubber formulations are shown in Table 9.

Heat-cured Rubbers. Unlike RTV compositions, most heat-curing silicone rubbers are based on high molecular weight polymer gums. Gums, fillers, and additives are mixed in dough mixers or Banbury mills. Catalysts (usually peroxides) are added on water-cooled rubber mills, which can be used for the complete process in small-scale operations.

Silicone rubbers are commercially available as gums, filler-reinforced gums, dispersions, and uncatalyzed and catalyzed compounds; the last are ready for use. Dispersions or pastes are stirred with solvents such as xylene. The following types of gums are commercially available: general purpose (methyl and vinyl), high and low temperature (phenyl, methyl, and vinyl), low compression set (methyl and vinyl), low shrink (devolatilized), and solvent resistant (fluorosilicone); product properties are shown in Table 10.

Properties. The properties of cured silicone elastomers are temperature dependent; eg, Young's modulus decreases from ca 10,000 to 200 MPa (145×10^4 to 2.9×10^4 psi) between −50 and 25°C and remains fairly constant to 260°C. Resistivity decreases; electric strength does not change much; the dielectric constant increases at 60 Hz current and decreases at 10^4 Hz current and above; and the power factor increases considerably. Tensile strength decreases from ca 6.9 MPa (1000 psi) at 0°C to 2.1 MPa (300 psi) at 300°C. The thermal conductivity of silicone rubber is usually ca 1.5–4 W/(m·K) and increases with increasing filler content.

Silicone rubber (gum) films are permeable to gases and hydrocarbons. Solvents diffuse into silicone rubber and swell, soften, and weaken it. Silicone elastomers appear completely hydrophobic to liquid water.

Heat alone causes depolymerization and volatilization; the rate increases appreciably above 315°C, even in neutral compounds. Silicone rubber can burn, but the residue is siliceous rather than carbonaceous; it has some structural integrity and is a nonconductor of electricity. Electron or gamma radiation damages cured silicone rubber. Elongation is reduced, but tensile strength is not greatly affected.

General purpose silicone rubber performs over wide temperature ranges. Estimated service life is from 5 to 10 years at 150°C and 40 years at 90°C. Heat aging in air at 125°C causes some elongation loss but does not affect tensile strength appreciably. Properties are maintained at low temperatures as well. Physical properties of some formulations do not deteriorate at −100°C.

Table 9. Cure Properties of Typical RTV Silicone Rubbers

	One component		Two components	
Property	General purpose	Construction sealant	Adhesive sealant	Molding compound
hardness, Shore A, durometer	30	22	50	60
tensile strength, MPa[a]	2.4	1.0	3.4	5.5
elongation, %	400	850	200	220
tear strength, J/cm^2 [b]	0.80	0.35	0.52	1.75

[a] To convert MPa to psi, multiply by 145.
[b] To convert J/cm^2 to lbf/in., multiply by 57.1.

Table 10. Properties of Silicone Rubber Classes

Class	Hardness, durometer	Tensile strength, MPa[a]	Elongation, %	Compression set at 150°C for 22 h, %	Useful temperature range, °C Minimum	Maximum	Tear strength, J/cm[b]
general purpose	40–80	4.8–.70	100–400	15–50	−60	260	0.9
low compression set	50–80	4.8–7.0	80–400	10–15	−60	260	0.9
extreme low temperature	25–80	5.5–10.3	150–600	20–50	−100	260	3.1
extreme high temperature	40–80	4.8–7.6	200–500	10–40	−60	315	
wire and cable	50–80	4.1–10.3	100–500	20–50	−100	260	
solvent resistant	50–60	5.8–7.0	170–225	20–30	−68	232	1.3
high-strength flame retardant	40–50	9.6–11.0	500–700				2.8–2.8

[a] To convert MPa to psi, multiply by 145.
[b] To convert J/cm^2 to lbf/in., multiply by 57.1.

Silicone polymers are electrical insulators. The dielectric constant is ca 2.7 at 25°C. Ozone and corona resistance is excellent, approaching that of mica in some cases. The overall weatherability of silicone elastomers is exceptional. Fully cured rubber has no odor or taste. It is not toxic and can be formulated for physiological inertness.

Economic Aspects

The four principal U.S. manufacturers of silicones include Dow Corning, General Electric, Union Carbide, and Wacker Silicones (formerly Stauffer-Wacker and SWS Silicones); large producers are also found in the UK, France, the FRG, the GDR, Japan, and the USSR, and small manufacturers in Belgium, Italy, Czechoslovakia, and elsewhere.

Worldwide production of silicones was estimated at 4.5×10^5 annually in 1986, with the United States accounting for approximately half of the total. Average U.S. prices of $3–6/kg for fluids and resins and $6–9/kg for elastomers imply that total worldwide sales are now over three billion dollars annually.

Silicone products include fluids (50–55 wt %), resins (5–10 wt %), and elastomers (40–45 wt %). Other products, eg, emulsions, greases, pressure-sensitive adhesives, and dispersions, are derived from these three basic types. Organosilanes, of growing importance, are listed as fluids, chemicals, or silanes by most producers.

Environmental Aspects

Silicone polymers are generally considered nonhazardous. Many of the raw materials and intermediates, however, are hazardous substances with the potential of an adverse impact on the environment, if not properly handled.

Silicone manufacture produces a collection of wastes, including residues and by-products, for disposal; related waste water streams require treatment prior to discharge. Most solid and liquid wastes and process vent gases can be incinerated. Comprehensive waste water treatment involves a combination of steps that include neutralization, precipitation, clarification, and removal of organic materials (eg, air stripping with incineration, biological treatment).

Health and Safety Factors

Polydimethylsiloxanes generally are not linked to harmful physiological effects. The harmless character of the methyl polymers is shared by alkyl-, fluoroalkyl-, and phenyl-substituted polymers. Certain low molecular weight oligomers from this group are known to be biologically active. The compound *cis*-2,6-diphenylhexamethylcyclotetrasiloxane, a notable example, is an active estrogen and has been used as an antifertility agent and for prostate cancer treatment. Methylphenyl and methyltrifluoropropyl trimers and tetramers are biologically active. Formulated silicones may contain ingredients that can be irritating or toxic.

Applications

Silicones are widely used in industry and consumer products. Applications are based on properties and cost effectiveness. Savings in assembly, installation, and labor costs, combined with longer service life, offset any price premium initially charged for silicones. Some applications require unique properties of silicones not offered by other materials. Salient properties of dimethylsilicones include thermal and oxidative stability; little dependence of physical properties on temperature; resistance to weathering, ozone, and radiation; low surface tension; high surface activity; good spreading power; and chemical and biological inertness.

The range of silicone applications is extremely wide, as shown below:

Fluids
plastic additives
hydraulic fluids
vibration damping
release agents
antifoamers
dielectric media
greases
coagulants
particle and fiber treatments
cosmetic and health product additives
heat-transfer media
polishes
water repellents
surfactants
lubricants

Resins
varnishes
paints

molding compounds
protective coatings
encapsulants
junction coatings
electrical insulation
pressure-sensitive adhesives
laminates
release coatings
adhesives

RTV rubbers
sealants
adhesives
conformal coatings
gaskets
foams
molding parts
encapsulants
electrical insulation
glazing
medical implants
surgical aids
mold making

Heat-cured rubbers
tubing and hoses
belting
wire-cable insulations
surgical aids
fuel-resistant rubber parts
penetration seals
molded parts
embossing–calendering rollers
autoignition cable and spark-plug boots
extruding
medical implants
laminates
electrically conducting rubber
fabric coating
foams

BIBLIOGRAPHY

F. O. Stark, J. R. Falender, and A. P. Wright in G. Wilkinson, F. G. A. Stone, and E. W. Abel, eds., *Comprehensive Organometallic Chemistry,* Vol. 2, Pergamon Press, New York, 1982, pp. 305–363.

R. J. H. Voorhoeve, *Organohalosilanes, Precursors to Silicones,* Elsevier Science Publishing Co., Inc., New York, 1967.

H. A. Liebhafsky, *Silicones under the Monogram,* John Wiley & Sons, Inc., New York, 1978.

W. Noll, *Chemistry and Technology of Silicones,* Academic Press, Inc., New York, 1968.

BRUCE HARDMAN
ARNOLD TORKELSON
General Electric Company

SILICON-NITROGEN POLYMERS.

See PRECERAMIC POLYMERS.

SILICON POLYMERS.

See POLYSILANES AND POLYCARBOSILANES; PRECERAMIC POLYMERS; SILICATES; SILICONES.

SILK

Silk is the solidified viscous fluid excreted from special glands or orifices by many types of insects and spiders. It is a protein; the amino acids glycine and alanine are the main components. The only significant source for textile usage is the silk-moth caterpillar, commonly referred to as the silkworm. Several varieties are known; the most valuable is the caterpillar of the moth *Bombyx mori,* which was domesticated centuries ago. A somewhat different product known as wild silk, sometimes called tussah, is produced by species of moths that are not domesticated. The growth of domesticated silkworms is known as sericulture.

Silk is available for textiles as a continuous filament and as staple yarn. Although not truly continuous, a filament may be 300–1200 m long. The strand produced by the silkworm consists of two filaments, called fibroin, encased in a protein gum called sericin. The silkworm uses this product to form the cocoon in which it encases itself for metamorphosis. If the moth is allowed to emerge alive, the filaments are ruptured. These and other damaged filaments, called silk waste, are used in producing staple yarns, which are formed by twisting short lengths of filament together.

Physical Properties

In Table 1 the physical properties of silk are compared with those of cotton and polyester.

Chemical Properties

Raw silk consists primarily of the proteins sericin and fibroin. The bulk is fibroin or fiber, which is coated with 15–25% sericin or silk gum. The principal amino acid constituents are glycine, alanine, serine, and tyrosine. The fiber is composed of long-chain amino acid units joined by peptide links with hydrogen bonding between parallel chains.

Although not significantly affected by dilute acids, silk is readily hydrolyzed by concentrated sulfuric acid. It is highly sensitive to concentrated alkaline solutions and calcium thiocyanate but withstands treatment with hot dilute (1%) sodium hydroxide solutions.

Processing

Processing steps include reeling, spinning, preparation (scouring and bleaching), weighing, dyeing, and finishing.

Fiber Identification

Fibers and fiber blends may be identified by specific mechanical, chemical, or microscopical tests. All protein fibers upon burning impart the odor of burning hair or feathers. Silk, unlike other protein fibers, is soluble in 59.5 and 70% sulfuric acid. As the only protein fiber without surface scales, silk can be identified microscopically (see IDENTIFICATION OF FIBERS).

Table 1. Physical Properties of Selected Textile Fibers

Property	Cotton	Silk	Polyester
abrasion resistance	good	fair	good
moisture regain, % at 65% rh	8.5	11	1
drapability	fair	excellent	fair
elasticity at 21°C, 65% rh			
elongation at break, %	3–10	20	19–23[a]
recovery, %	75 45		97 80
from % strain	2 5	2[b]	2 8
environmental resistance			
mildew resistance	poor	good	excellent
renovation[c]	W, DC	W, DC	W, DC
safe iron limit, °C	204	148	163
sunlight resistance	fair	poor	good
hand	good	excellent	fair
pilling resistance	good	good	bad
resiliency	poor	fair	excellent
specific gravity	1.54	1.30	1.22–1.38
static resistance	good	fair	bad
strength, N/tex[d]	0.26–0.44 good	0.34–0.39 good	0.22–0.83 excellent
strength loss when wet, %[e]	10	15	0
thermoplasticity	no	no	yes

[a] Regular.
[b] Poor if stretched beyond.
[c] W = wash, DC = dry clean.
[d] To convert N/tex to gf/den, multiply by 11.33.
[e] Approximate values.

Economic Aspects and Uses

World production of silk increased steadily through the 1970s and the early 1980s, reaching pre-World War II levels in 1985. In the United States, silk imports increased rapidly from 1900 to 1930, but declined appreciably during the 1930s and drastically after 1940, mainly because of the development of synthetic fibers. Currently, only a modest amount of silk is imported into the United States. The limited use of silk in the United States is due to its very high price and the competition of synthetic fibers at far lower prices.

BIBLIOGRAPHY

W. F. Leggett, *The Story of Silk,* Lifetime Editions, New York, 1949.

M. S. Otterburn in R. S. Asquith, ed., *Chemistry of Natural Protein Fibers,* Plenum Press, Inc., New York, 1977.

Text. Organon **48,** 107 (1977); **51,** 105 (1980); **57,** 101 (1986).

C. D. Livengood in M. Grayson, ed., *Kirk-Othmer Encyclopedia of Chemical Technology,* 3rd ed., Vol. 20, John Wiley & Sons, New York, 1982, pp. 973–981.

CHARLES D. LIVENGOOD
North Carolina State University

SILK SCREEN PRINTING.

See DECORATING; FINE ARTS; PRINTING-INK VEHICLES.

SILOXANES.

See SILICONES.

SISAL.

See FIBERS, VEGETABLE.

SIZE ENLARGEMENT.

See PELLETIZING.

SIZE EXCLUSION CHROMATOGRAPHY.

See CHROMATOGRAPHY.

SIZING.

See TEXTILE RESINS.

SKIN, REGENERATION TEMPLATES

Regeneration templates are polymers of natural origin that induce synthesis of physiological tissue when brought in close contact with the surface of a surgical lesion (wound) in living, adult, mammalian tissue. The unusual biological activity exhibited by these polymers has encouraged investigators to question certain premises of developmental biology and has led to the introduction of novel devices for tissue and organ repair in medicine.

The polymers that have shown this unusual biological activity are copolymers formed by grafting chains from chondroitin 6-sulfate, a glycosaminoglycan (GAG), on the fibrous protein collagen. Following nomenclature suggested by IUPAC, these polymers are referred to as collagen-*graft*-chondroitin 6-sulfate copolymers, or generically, as collagen-*graft*-glycosaminoglycan copolymers, or briefly, as CG copolymers.

Synthesis

Polymers of natural origin that function as regeneration templates have so far been synthesized as graft copolymers of collagen and chondroitin 6-sulfate, and the subsequent discussion is accordingly limited to these macromolecular species. There is currently no evidence that such a chemical composition is a necessary prerequisite for template activity. Future studies may possibly point to other chemical and physical structures that are also active as templates.

A template measuring $\sim$400 cm^2 in area with a thickness of $\sim$1 mm and a pore volume fraction of $\sim$0.99 can be prepared as follows. Collagen (moisture content $\sim$10 wt %) is pulverized in a mill and 1.65 g are dispersed in 600 mL of 0.05 M acetic acid (pH ca 3.2) by stirring in a refrigerated homogenizer for 60 min. During this period, collagen fibers swell in the acid solvent along the direction lateral to the fiber axis and, in the process, the lateral packing order is almost completely erased whereas the triple helical structure (configuration of individual collagen molecules) remains intact. At the end of the swelling step 0.13 g GAG in 120 mL of 0.05 M acetic acid is metered into the collagen suspension and a white collagen–GAG coprecipitate forms. The coprecipitate is an ionic complex that can be resuspended readily and the macromolecular components can be separated reversibly, either by increasing the ionic strength of the aqueous component or by increasing the acidic pH to neutral.

The composition of collagen–GAG coprecipitates varies with the amount of GAG added to the collagen dispersion and with the pH (Fig. 1). The process of coprecipitation can be modeled roughly as a Langmuir adsorption, which reaches an asymptotic limit when a sufficient amount of GAG adsorbate has saturated the surface of the collagen adsorbent.

Collagen–GAG coprecipitates are converted into highly porous foams by freeze-drying (lyophilization).

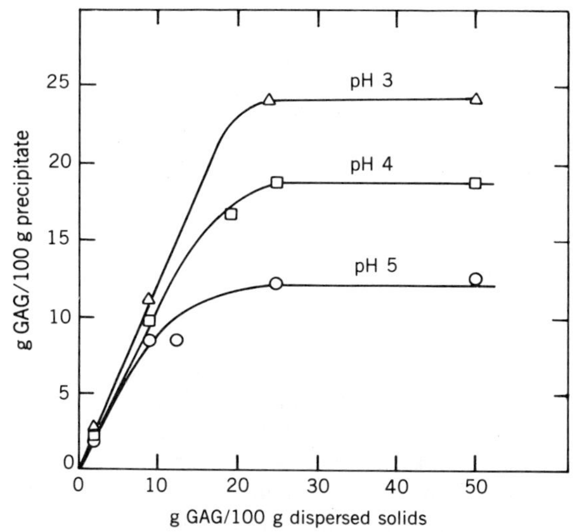

Figure 1. The composition of collagen–GAG coprecipitates, plotted on the ordinate, depends on the amount of GAG added to the bovine hide collagen dispersion (abscissa). The GAG is chondroitin 6-sulfate. Adsorption isotherms were obtained at 4°C at the pH levels indicated. No chondroitin 6-sulfate coprecipitated with collagen at pH 7.

Grafting of GAG on collagen chains prevents elution of GAG, which occurs when the uncross-linked coprecipitate is exposed to neutral pH following implantation in animals or humans. Grafting is accomplished either by exposure of the dry foam to 105°C under 6.7 Pa (50 mtorr) or by exposure to aqueous glutaraldehyde. Both processes are currently used in the preparation of medical devices.

Characterization Methods

The physiochemical characteristics of insoluble collagen-*graft*-glycosaminoglycan copolymers can be determined by procedures that have been described in detail. A GAG assay is based on the well-known Elson-Morgan hexosamine procedure. Assays for measuring the content of triple helical (nongelatinized) collagen are based on infrared spectroscopy of thin, nonporous films prepared by casting.

The average pore diameter $\bar{d}$ of CG copolymers is determined by viewing specimens in a scanning electron microscope and applying stereological methods of analysis. The cross-link density is determined by tensile testing of a gelatinized specimen whereas the degradation rate is measured by incubation in a standardized bath of collagenase.

Fabrication of Devices for Implantation

Devices for grafting of full-thickness skin wounds are two-layered membranes. These have been referred to as Stage 1 membranes when unseeded with cells prior to grafting and Stage 2 membranes when seeded with autologous cells.

Stage 1 membranes are prepared by knife coating a porous collagen-*graft*-glycosaminoglycan sheet with a thin layer of a viscous, moisture-curing polydimethylsiloxane prepolymer (Silastic Medical A, Dow Corning, Midland, Mich.) (Fig. 2). The silicone layer controls the loss of moisture from the wound, contributes to prevention of bacterial contamination of the wound, and provides the relatively weak, highly porous copolymer sheet with sufficient mechanical strength to make it capable of being handled surgically and sutured without rupture.

These devices have been used very successfully in treatment of patients with massive burns ("artificial skin," "artificial dermis"). Stage 2 membranes are Stage 1 membranes that have been inoculated with autologous, uncultured, dermal cell suspensions.

Biological Specificity

Certain collagen-*graft*-glycosaminoglycans can induce synthesis of dermis within a wound bed in guinea pigs in which the epidermis and the complete dermis have been removed by surgical excision. These copolymers are also capable of delaying significantly or arresting wound contraction. Studies of the kinetics of skin wound contraction have been used to screen copolymers that possess biological activity from those that do not. CG copolymers appear to fall into three classes. Inactive copolymers delay the kinetics of wound contraction insignificantly. Active polymers that are not seeded with a minimum number of cells induce a significant delay in onset of wound contraction, thereby causing a substantial increase in wound half-life; however, after the wound perimeter begins to contract, it eventually closes up, yielding a linear scar. Active, seeded polymers show the most complex behavior. Following a significant delay in onset of contraction, these grafts induce arrest and reversal of wound contraction (expansion of wound perimeter) while nearly intact skin is being synthesized in the wound bed.

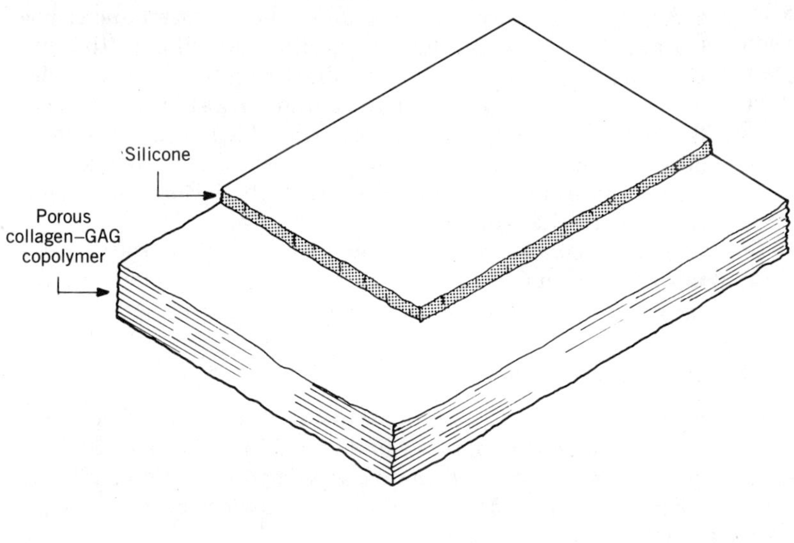

Figure 2. Schematic of a bilayer membrane used to graft full-thickness skin wounds in animals and severely burned humans (Stage 1 artificial skin). The elastomeric silicone layer controls the moisture flux, bacterial contamination, and mechanical behavior of the device. The collagen–glycosaminoglycan (GAG) layer interacts biologically with the wound bed, induces regeneration of skin if designed as a template, and undergoes enzymatic degradation while being replaced with new tissue. Courtesy of AAAS.

Active copolymers are characterized by narrowly defined properties. The average pore diameter of copolymers active in skin regeneration lies between 20 and 120 μm whereas the degradation rate in collagenase is low enough to ensure that insoluble fragments of copolymer persist as implants. These fragments apparently induce cell–matrix and cell–cell interactions to occur, which facilitate regeneration rather than conventional wound healing to occur.

Evidence that the newly synthesized tissue is quite distinct from scar tissue, although not identical to intact skin, has been presented. Ultrastructural studies of intact skin, newly synthesized skin, and scar show that the cellularity, constituent vasculature, and neural structures allow for an easy separation of the newly formed dermis from scar. The new skin does not contain residual fragments of CG copolymer.

Devices, in the form of porous, cylindrical bridges, which induce regeneration of peripheral nerve in animals have been also studied. The CG copolymers, which are best suited for this purpose, degrade more rapidly in collagenase and have smaller pores than the copolymers that induce skin regeneration. It is intriguing to speculate that different copolymers are best suited for each tissue type.

BIBLIOGRAPHY

I. V. Yannas, J. F. Burke, C. Huang, and P. L. Gordon, *Polym. Prepr. Am. Chem. Soc. Div. Polym. Chem.* **16,** 209 (1975).

I. V. Yannas, J. F. Burke, P. L. Gordon, C. Huang, and R. H. Rubenstein, *J. Biomed. Mater. Res.* **14,** 65 (1980).

I. V. Yannas, J. F. Burke, D. P. Orgill, and E. M. Skrabut, *Science* **215,** 174 (1982).

I. V. Yannas, E. Lee, D. P. Orgill, E. M. Skrabut, and G. F. Murphy, *Proc. Natl. Acad. Sci. USA,* **86,** 933 (1989).

IOANNIS V. YANNAS
Massachusetts Institute of Technology

SKIN, SYNTHETIC

Human skin is a stratified bilayered (dermis and epidermis) organ consisting of several cell types intimately colinked and interdigitated, bearing a variety of adnexal structures and performing a variety of functions. No artifice of man can yet duplicate this organ. The closest approach has been the extension of autologous skin-derived cells (fibroblasts and keratinocytes) to sheets in tissue-culture systems. Tissue-cultured cells have been used as skin substitutes. Synthetic (or artificial) skin, therefore, is a misnomer referring to a temporary skin substitute, a nonliving manufactured membrane, cover, or biosynthetic dressing with one or more of the physical attributes and functions of skin, ready for emergency use on open wounds whose edges cannot be directly apposed, and capable of remaining in place for an extended (indefinite) period of time with no harm to the recipient.

Wound Healing

Reduced to a quintessential core, the physician's role in wound care is to debride, decontaminate, and decide whether or not a wound cover is needed and, if it is needed, what the ideal choice is. Ideally, these processes present a smooth overlapping continuum from cleanup to cover-up.

Desirable Wound Cover Attributes

Currently available wound covers have improved patient care significantly. The properties that serve to produce an ideal skin substitute or an optimally useful general dressing for discard and replacement as needed during wound care have been enumerated. The dressing must be rapidly and uniformly adherent and conform to wound bed topography and contour so as to prevent air or fluid pocket formation. Good adherence reduces pain, facilitates decontamination, prevents peripheral channeling into the wound by bacteria, and promotes bonding to tissues. The dressing must be an absolute barrier to bacterial ingress, resist bacterial degradation, and prevent egress of wound organisms to the dressing's surface. It must be permeable to water vapor to the extent that a moist exudate under the dressing is maintained without pooling, but excess fluid absorption and evaporation and evaporation leading to desiccation of the wound bed are prevented. Restricting the rate to that of normal water transmission through skin may be insufficient to prevent pooling over a wounded area. The dressing must be compatible with body tissues (eg, permit ingrowth of vascular tissue), be nontoxic, nonantigenic, and nonallergenic.

The dressing should be a durable, stress-resistant (flexible and pliable) elastic material, easy to apply and remove without trauma during dressing changes, nonadherent over healed areas in the wound bed, or able to be sloughed off with the scab. It would be advantageous for the material to be antiseptic in nature and not only stop bacterial growth but also reduce bacterial density in the wound, or at least be permeable to applied antibacterials. It would also be an advantage for the material to be hemostatic, transparent (in order to indicate the presence of infection), able to provide thermal insulation, accelerate healing, minimize scar formation, and finally, in some cases, be incorporated by the body (biodegradable) as it restores normal function to the skin.

Materials

At present there is a limited but growing number of commercially produced or experimentally available skin-substitute wound covers that may be classified as synthetics, organics, or composites of synthetic and organic origin. These covers include: **synthetics** such as Hydron (methacrylate-PEG mixture), Epigard (polypropylene film over polyurethane foam), Opsite, Tegaderm, Bioclusive (polyurethane films with adhesive that sticks to normal skin), Vigilon (poly(ethylene oxide)-water gel backed by polyethylene film); **organics** such as collagen (cross-linked) sponge, Poly-Nag (chitin or chitosan powders or mats), Stra Cor II (chitosan-keratinate), blended chitosan-gelatin film; and **composites** such as Biobrane (nylon-collagen peptides-silicone membrane), Stage I, MIT (polysiloxane cover bonded to cross-linked collagen-chondroitin 6-sulfate), Duoderm (adhesive polyurethane-backed gelatin-pectin-carboxymethylcellulose-polyisobutylene mixture).

Stra Cor Membranes

Hydrogels refer to polymers that can swell from 30 to >95% in water but do not dissolve in it. An ability to swell with body fluids without chemical change gives hydrogels (qv) an innate advantage over other materials in biocompatibility.

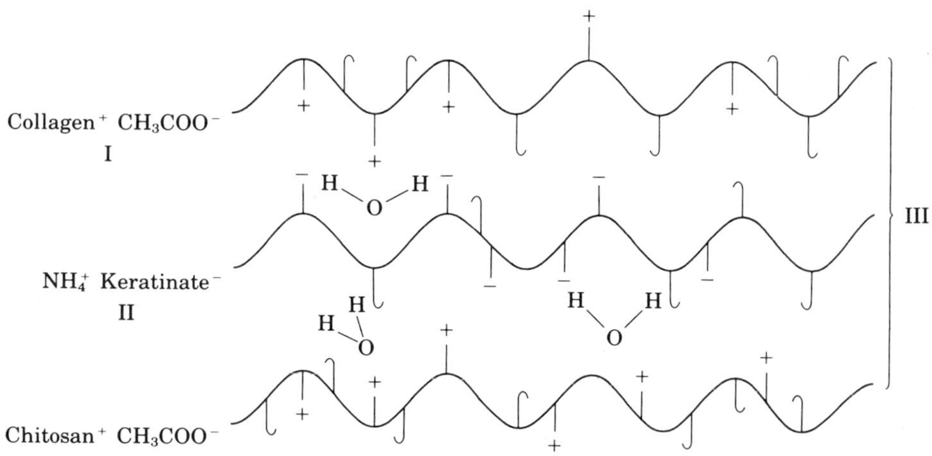

Figure 1. Stra Cor membranes are polyelectrolyte complexes that can bind water.

Stra Cor membranes are a series of three semiocclusive biodegradable hydrogels formed by the spontaneous coprecipitation of at least two oppositely charged polyelectrolytes; ie, solubilized polymers containing charged backbone or side groups. They are derived from a group of three biopolymers: keratin (qv), collagen (qv), and chitin (qv). The membranes are formed as follows: Stra Cor I from ammonium keratinate (α-keratose) and collagen acetate, Stra Cor II from ammonium keratinate and chitosan acetate, and Stra Cor III from collagen acetate and chitosan acetate colinked through ammonium keratinate (Fig. 1) (see also POLYELECTROLYTE COMPLEXES). These membranes are truly formulated bioplastics.

BIBLIOGRAPHY

H. I. Maibach and D. T. Rovee, eds., *Epidermal Wound Healing*, Year Book Medical Pub., Inc., Chicago, Ill., 1972.

T. K. Hunt in W. Van Winkle and J. E. Dunphy, eds., *Fundamentals of Wound Management in Surgery, Wound Healing: Normal Repair*, Chirurgecom, Inc., South Plainfield, N.J., 1976, pp. 54–62.

R. H. Demling, *N. Engl. J. Med.* **313,** 1389 (1985).

S. de Riel in D. L. Wise, ed., *Burn Wound Coverings*, Vol. I, CRC Press, Inc., Boca Raton, Fla., 1984, pp. 1–37.

ABE WIDRA
University of Illinois College of Medicine, Chicago

SLIP AGENTS.
See RELEASE AGENTS.

SMC.
See SHEET MOLDING COMPOUND.

SODIUM CARBOXYMETHYLCELLULOSE.
See CELLULOSE ETHERS.

SOFTENERS.
See PLASTICIZERS; TEXTILE RESINS.

SOIL STABILIZATION.
See FIBERS, ENGINEERING; MINING APPLICATIONS.

SOL.
See COLLOIDS.

SOLID-PHASE FORMING.
See COLD FORMING.

SOLID-STATE EXTRUSION

Solid-state extrusion (SSE) has been investigated extensively for the deformation and evaluation of uniaxial molecular chain orientation and extension for a wide range of semicrystalline and amorphous thermoplastics. High density polyethylene (HDPE) has been most widely studied, in order to understand the morphological transformations. It can now be drawn into fibers with some of the highest specific tensile moduli and strengths (see HIGH MODULUS POLYMERS).

A two-step drawing process was developed for the preparation of polyoxyethylene, polypropylene, and polyethylene fibers. The polymer is first drawn up to the natural draw ratio at a fast rate and subsequently slowly superdrawn at a temperature that depends on the crystalline dispersion temperature; ordered structures can be produced.

Extrusion

A highly oriented extrudate can be obtained by extruding through a capillary rheometer with a conical die at temperatures close to the melting point. Initial work led to the development of transparent and fibrous linear polyethylene extrudates. These were obtained by extruding HDPE from the molten state (132–136°C) above a critical shear rate in a capillary rheometer and through a conical die.

This procedure was subsequently modified by processing HDPE exclusively in the solid state where the polymer is semicrystalline before extrusion through the die. This modification produced continuous transparent fibers with moduli in the range of 30 to 70 GPa (4.35×10^6–10×10^6 psi) for linear polyethylene.

Deformation Mechanism

To obtain a highly oriented, chain-extended conformation, the isotropic and spherulitic structure must be examined during deformation. A spherulite may be envisaged as an isolated structure of fibrils arranged radially about its center. The fibrils consist of stacks of folded-chain lamellae with the molecular chains oriented perpendicular to the radius. Stacks of lamellae and lamellae of the same stack are separated by the amorphous component of the semicrystalline polymer. The lamellae reorient on extrusion in the stress direction. At yield, when necking occurs, chain-folded blocks tear away from the lamellae and, through partial unfolding, convert to microfibrils. The microfibrils are thus made of folded crystalline blocks (10.0–30.0 nm in dia) with their chain axes oriented in the extrusion direction. The amorphous regions of partly unfolded chains generate the mechanical connection between the chain-folded blocks. The microfibrils form bundles known as fibrils. These fibrils deform further by a sliding displacement over each other and are responsible for the observed macroscopic deformation.

Molecular weight significantly affects the solid-state extrusion process and the mechanical properties of the extrudates. Extrusions of bulk-crystallized polyethylenes of high molecular weight proceed at slower rates and lower maximum extrusion draw ratios than lower molecular weight grades. The deformation resistance imposed by the higher degree of entanglement between molecular chains tends to inhibit high elongation.

Extrusion Draw Ratio

The extrusion draw ratio (EDR) describes the extent of the solid-state deformation in the solid-state extrusion process. It is most important for the attainment of a high tensile modulus. The EDR is defined as the ratio of die entrance to exit cross-sectional areas. Assuming volume conservation and cylindrical symmetry,

$$\text{EDR} = \left(\frac{R_\alpha}{r}\right)^2$$

where R_α and r are the radii of the entrance and exit cross-sectional areas, respectively. This ratio determines the maximum deformation that can be produced through a particular die.

Die Geometry

In conventional extrusion, the die geometry determines the shape of the extrudate. Each piece may be designed to have two or three deforming zones. The principal characteristics of a die are the nominal EDR; the capillary length; and the half-die angle, determined by the inclination of the deforming zone from the axis or plane of symmetry along the die (in wedge-shaped dies there may be more than one half-die angle, depending on the number of deforming zones). The optimal die geometry for the most efficient deformation depends on the polymer and the extrusion conditions.

Extrusion Variables

Extrusion variables include pressure, external loading, split billet, and push–pull.

Powder Extrusion. Powder processing is used extensively for metals and ceramics, but so far has been applied only rarely to polymers, although granular polytetrafluoroethylene has been molded and extruded.

Solid-state Coextrusion

In solid-state coextrusion a polymer that is difficult to process is extruded within a polymer that is easy to process (sandwich arrangement). The process is a form of substrate drawing and has been used for the extrusion of films (split-billet coextrusion) and fibers (concentric-billet coextrusion).

Reversible Plasticization

Polymers of strong lateral interactions, eg, nylons, are difficult to draw, often requiring a plasticizer. However, plasticizers affect mechanical performance and should be removed after the polymer is processed. Volatile plasticizers are easily removed. Reversible plasticization has been demonstrated for the solid-state deformation of aliphatic polyamides with iodine or with anhydrous ammonia, which are absorbed selectively.

Properties

Thermoplastics that have been processed by solid-state extrusion include polyethylenes, polypropylene, poly(1-butene), poly(4-methyl-1-pentene), polystyrene, poly(vinylidene fluoride), polyamides, polyetheretherketone, poly(ethylene terephthalate), poly(ethylene oxide), poly(methyl methacrylate), and poly(*p*-phenylene sulfide). The tensile properties achieved are equal to or exceed those attained by other techniques for uniaxial deformation; examples are given in Table 1. Highly drawn thermoplastics can result in ratios of tensile to shear moduli approaching 100, and compressive buckling by a shear mechanism can thus occur at high EDR. For PE, compressive buckling limits the development of morphologies in a single-step of EDR >40. Thus PE and PP are best extruded to moderate EDRs of ≥6, followed by tensile elongation in ambient atmosphere at controlled rates and temperatures.

Table 1. Experimental C-Axis Young's Moduli[a]

Polymer	Conformation	X ray, nm	Neutron scattering, nm	Raman, nm	Solid-state extrusion product: Young's modulus, GPa[b]	Solid-state extrusion product: Theoretical tensile strength GPa[b]
polyethylene	planar zigzag	24.0	32.9	35.6	220	6.4
isotactic polypropylene	3/1 helix	3.5		8.8	34	2.4
poly(ethylene oxide)	7/2 helix	1.0		2.2		3.5

[a] Measured at room temperature.
[b] To convert GPa to psi, multiply by 145,000.

Applications

Commercial applications of solid-state extrusion are being developed. The polyethylene fibers made by a two-stage process in Japan have been cited to exhibit the highest specific strength of any fiber.

Other applications are being considered, particularly for polyolefins. High modulus ribbons and tapes have been prepared for use as strapping and electrical insulation. Tubular forms, produced by planar deformation, may be used for pipes.

BIBLIOGRAPHY

H. Mark, *Trans. Faraday Soc.* **32,** 144 (1936).

I. Sakurada and K. Kagi, *J. Polym. Sci. Part C* **31,** 57 (1970).

G. Capaccio and I. M. Ward, *Nature (London) Phys. Sci.* **243,** 143 (1973).

J. H. Southern and R. S. Porter, *Chem. Eng. News* **45** (Sept. 22, 1969); *J. Macromol. Sci. Phys.* **B4,** 541 (1970); *J. Appl. Polym. Sci.* **14,** 2305 (1970).

A. E. Zachariades and R. S. Porter, eds., *High Modulus Polymers: Approaches to Design and Development,* Marcel Dekker, Inc., New York, 1988.

JOHN H. SOUTHERN
Monsanto Textiles Company

NUMA J. CAPIATI
PIDCOP

TETSUO KANAMOTO
Science University of Tokyo

ROGER S. PORTER
University of Massachusetts

ANAGNOSTIS E. ZACHARIADES
IBM Corporation

SOLID-STATE POLYMERIZATION

For propagation in the solid state, multiple conditions must be satisfied. The diffusional freedom of the reactants in the solid state is restricted or nonexistent. Consequently, a monomer that is highly reactive in the fluid state may be unreactive in the solid state. Conversely, a reaction that proceeds efficiently in the solid state by virtue of crystalline symmetry and arrangements may not occur in the fluid phase.

Polymerization in the solid state produces macroscopic polymer single crystals and in some cases crystals of high optical quality comparable to that of commercial optical materials. Because of this unique capability the solid-state method has been investigated from various research viewpoints. The complex and distinct reaction mechanism in the solid state has been an active field of research in chemistry as well as solid-state physics. Since the polymer single crystals are naturally anisotropic, with covalent bonding in one direction and van der Waals interaction in the others, these systems are considered as ideally one-dimensional. Consequently, these single crystals are used as test materials for theoretical predictions in low dimensional physics. Conjugated polymers, which have interested physical chemists and physicists over the years, are now available as macroscopic single crystals by solid-state synthesis. Considerable attention has been focused recently on these conjugated polymeric single crystals from the standpoints of linear and nonlinear optics. Thus, solid-state polymerization is today an intensely active interdisciplinary field of science.

Topotaxy and Topochemistry

A reaction that leads to a product that is crystallographically related to the precursors is called topotactic and the process topotaxy. Reactions in a single crystal are usually lattice-controlled. Depending on the lattice spacings and the overall packing of the monomers in the unit cell, a reaction may or may not proceed. If it proceeds, its propagation direction is controlled by the crystal structure and symmetry of the monomer. In an unfavorable geometric arrangement, a reaction may proceed with equal probability in arbitrary directions and thus lead to amorphous products. In more specific cases, the propagation may be favored in a specific direction leading to a crystal-to-crystal transformation.

Reactivity and Crystal-to-crystal Transformation. The following criteria were derived for a crystal-to-crystal transformation by solid-state polymerization: (*1*) in the monomer crystal, the distance between the reactant unsaturated bonds of the neighboring molecules should not exceed 0.4 nm, and (*2*) the mutual orientation of the unsaturated bonds should be close to parallel and preferably coplanar.

Diffusion-assisted Reactions

Most of the materials that were studied in early solid-state polymerizations were made by diffusion-assisted reactions, where the propagation is facilitated by the translational diffusion of the molecules. These reactions are usually not topotactic and give amorphous products. Most of these reactions involved addition at the carbon–carbon double bonds to form saturated polymer backbones. Some thermally initiated polycondensation reactions are also included within this class. Many monomers that can be polymerized in the fluid state were studied (Table 1).

Substrate-assisted Reactions

The reactions in substrate-assisted polymerization are topotactic, but may not be topochemical.

Inclusion Polymerization. In this method, monomers are included inside the channel-like cavities of specific host crystals before polymerization (see POLYMERIZATION IN CLATHRATES).

Polymerization of Langmuir-Blodgett Films. Certain amphiphilic molecules can be organized as mono- and multilayer films following the well-known Langmuir-Blodgett (L-B) method. By introducing polymerizable groups within such molecules, polymerization in the solid state can be achieved subsequent to formation of the L-B films, giving ultrathin oriented polymeric films.

Simultaneous Polymerization and Crystallization: Addition via Ring Opening. Polymerization reactions can proceed simultaneously with the crystallization of the monomers. The driving force associated with the crystallization process promotes the covalent linking between the monomers and crystallization becomes an elementary step of polymerization. Although diffusion-assisted, the reaction is crystallographically controlled, and the polymer is obtained as a single crystal of well-defined orientation.

Table 1. Diolefinic Monomers Undergoing Topochemical Reactions

Monomer	Reactivity[a]	Product[b]
2,5-distyrylpyrazine	P	C
p-phenylenediacrylic acid, dimethyl ester	P	C
m-phenylenediacrylic acid, dimethyl ester	O	A
4-(β-carboxyvinyl)-α-cyanocinnamic acid, dimethyl ester	O	A

[a] P, high reactivity resulting in high polymer; O, only oligomers are obtained.
[b] C, crystalline; A, amorphous.

Diffusionless Reactions

The reactions of this class proceed without appreciable translational diffusion of the monomers. The criteria for crystal-to-crystal transformation are exactly satisfied, and these reactions are truly topochemical in nature. The crystal symmetries and the structures of the polymers are predictable from the crystal structures of the corresponding monomers. In diffusionless polymerization specific monomers, by virtue of their spatial proximity and mutual orientation in the crystalline state, react with each other without interruption and produce fully extended polymer chains along well-defined crystallographic directions. Such a reaction is promoted by simple rotational and torsional movements of the monomer molecules without involving any significant movement of the corresponding centers of mass. The benefits of solid-state polymerization are best realized in this class of reactions.

Topochemical Polymerization

Diolefinic Monomers. The topochemical polymerization of diolefinic monomers proceeds via a [2+2] photocycloaddition of the double bonds of nearest-neighbor molecules resulting in cyclobutane linkages, so-called four-center-type polymerization; it is initiated by high energy photoradiation.

In studies of topochemical reactions of various diolefinic compounds, significant conversions were achieved for only a few monomers (Table 1). These monomers all possessed centrosymmetric unit cells. Topochemical reactions of specific unsymmetrical diolefinic monomers give optically active polymer crystals.

Diacetylenes. In the topochemical polymerization of diacetylenes, the reaction proceeds without any appreciable change in the original shape and size of the monomer crystal. Therefore, the resulting polydiacetylene crystals can be macroscopic in size (>1 cm in certain directions), comparable to single crystals of small organic or inorganic molecules.

Application and Outlook

The solid-state method of polymerization has led to many new concepts, novel materials, and promises for applications in ultrafast optical signal processing.

BIBLIOGRAPHY

G. Wegner, *Faraday Discuss. Chem. Soc.* **68,** 494 (1979).

D. J. Sandman, ed., *Crystallographically Ordered Polymers,* American Chemical Society, Washington, D.C., 1987.

C. H. Bamford and G. C. Eastmond, *Quart. Rev. Chem. Soc.* **23,** 271 (1969).

M. Thakur, B. Verbeek, G. C. Chi, and K. J. O'Brien, *MRS Proc.* **109,** 41 (1988).

MRINAL THAKUR
AT&T Bell Laboratories

SOLUBILITY OF POLYMERS

A polymer solution refers to a uniform molecular dispersion of a macromolecular solute in a solvent usually of much lower molecular weight. The solubility is the measure of the extent to which the two pure components can be mixed homogeneously.

The process of dissolving a pure polymer begins with solvent molecules permeating bulk polymer. The polymer near the surface swells to accommodate incoming molecules of solvent whereas individual long-chain molecules may be freed from near the surface and diffuse into the solvent phase. If the solubility limit is not exceeded, the system eventually becomes homogeneous. The extent to which the solution process can occur depends especially on the temperature, the chemical nature of the solute and solvent, the molecular weight of the polymer, and on the degree of crystallinity for semicrystalline polymers. Highly cross-linked polymers can imbibe solvent and swell, forming a saturated polymer gel. Solvent added beyond this point remains as a second phase.

Generally, polymers are less soluble than are their low molecular weight analogs; the solubility of most polymers decreases as their molecular weight is increased. Temperature-vs-weight fraction phase diagrams showing the solubility of amorphous polymers are skewed; the region of limited miscibility typically occurs at low concentrations of polymer rather than near the middle of the phase diagram. Consequently, mixing polymer with "poor" solvent sometimes leads to a swollen polymer phase in equilibrium with almost pure solvent. As with mixtures of low molecular weight liquids, the solubility of polymers in most cases increases with increasing temperature.

The saying "Like dissolves like" ("*Similia similibus solvuntur*") applies to polymer solutions as well as to mixtures of small molecules. Also, solution is favored when specific interactions such as hydrogen bonding are formed between solvent and polymer.

Thermodynamics of Mixing

Some relationships from the thermodynamics of mixing are important for understanding polymer solubility. The free energy change ΔG_{mix} for mixing two pure substances at temperature T is given by the familiar equation

$$\Delta G_{mix} = \Delta H_{mix} - T\,\Delta S_{mix} \tag{1}$$

The heat of mixing ΔH_{mix} is a positive quantity for most solutions, ie, the formation of polymer solutions is usually an endothermic process. The entropy of mixing ΔS_{mix} is positive owing to the more random nature of solutions compared to that of the unmixed components. Both ΔH_{mix} and ΔS_{mix} vary with temperature and concentration. A necessary condition for the formation of a thermodynamically stable solution is that $\Delta G_{mix} \leq 0$.

Flory-Huggins Model

Mathematical models of polymer solutions have increased our understanding of how various factors affect polymer solubility. The Flory-Huggins theory for noncrystalline polymers is valuable in this capacity. An algebraic expression for ΔG_{mix} derived from the model can be combined with general thermodynamic relationships for calculating the phase behavior of polymer solutions.

The Flory-Huggins model uses combinatorial analysis to estimate the increase in configurations available to the system when a flexible polymer in a disordered state is mixed with solvent. The result leads to a predicted increase in entropy of

$$\Delta S_{mix} = -R(n_1 \ln \phi_1 + n_2 \ln \phi_2) \quad (2)$$

where n_1 and n_2 are the numbers of moles of solvent and polymer, respectively; ϕ_1 and ϕ_2 are their volume fractions; and R is the gas constant. In almost all polymer solutions this increase in entropy is the principal driving force for the mixing process.

The heat of mixing in the Flory-Huggins theory is the van Laar heat:

$$\Delta H_{mix} = z\ \Delta\epsilon\ n_1\phi_2 \quad (3)$$

In this equation z is the coordination number for the solvent molecule, ie, the average number of solvent molecules and solvent-sized segments of polymer in contact with one molecule of solvent. The energy $\Delta\epsilon$ is the difference between the intermolecular contact energy between a segment of the polymer and a solvent molecule and the average interaction for a pair of polymer segments and a pair of solvent molecules. Usually $\Delta\epsilon$ is positive corresponding to endothermic mixing; however, it is negative in the unusual instances of exothermic mixing.

The ratio $z\Delta\epsilon/RT$ is generally written as the unitless Flory-Huggins interaction parameter χ:

$$\chi = z\ \Delta\epsilon/RT \quad (4)$$

The Flory-Huggins equation for the free energy of mixing becomes

$$\Delta G_{mix} = RT(n_1 \ln \phi_1 + n_2 \ln \phi_2 + \chi n_1\phi_2) \quad (5)$$

The smaller that χ is, the more stable is the solution relative to the pure components and the more likely that the system is miscible over a wide (or the entire) range of concentrations. Values of χ are typically in the range 0.2–0.9. For most systems χ decreases with increasing temperature and increases with increasing concentration of polymer.

The Flory-Huggins theory is applicable to amorphous polymers. Crystalline polymers require an additional energy, namely, the heat of fusion, in order to mix with solvent and therefore tend to have lower solubilities than their amorphous counterparts. For systems in which there are strong specific interactions between solvent and solute, such as hydrogen bonding, mixing can be exothermic and/or can occur with an entropy change that differs markedly from that predicted by Flory and Huggins.

Solubility Parameter Method

The solubility parameter method can be used to predict solubilities of amorphous polymers and sometimes, with less justification, crystalline polymers. It affords an approximate means of estimating whether ΔH_{mix} in equation 1 is large or small for a particular polymer–solvent pair. This approach utilizes the energy per unit volume required to separate the molecules of a pure liquid, ie, the cohesive energy density E^{coh}. Its square root is known as the solubility parameter δ:

$$\delta = (E^{coh}/V)^{1/2} \quad (6)$$

For volatile liquids E^{coh} is often taken as the molar energy of vaporization. Values of the E^{coh}/V include 170 J/cm^3 for neopentane, which is near the low end of the range of cohesive energy densities for organic liquids, to 1100 J/cm^3 for glycerol near the high end. Water with its extensive hydrogen bonding has an extraordinarily high cohesive energy density of 2300 J/cm^3.

The heat of mixing can be expressed in terms of the solubility parameters for polymer and solvent:

$$\Delta H_{mix} = n_1 V_1^0 \phi_2 (\delta_1 - \delta_2)^2 \quad (7)$$

This is an important result in the solubility-parameter approach. It shows that when the values of the solubility parameters (widely available in standard references) of amorphous polymer and solvent approach one another, ΔH_{mix} should decrease toward zero. Since the free energy of the mixture decreases with decreasing ΔH_{mix}, systems that mix with small heats are expected to have high or complete miscibility. Miscibility is predicted when δ_1 and δ_2 are within 2 or 3 J$^{1/2}$/cm$^{3/2}$ of one another. Since the right-hand side of equation 7 cannot be negative, the solubility parameter method does not comprehend exothermic mixing.

Other Aspects of Polymer Solubility

Figure 1 shows the solubility as a function of temperature of four different molecular weight fractions of polystyrene in cyclohexane. It illustrates the common type of partial miscibility found in binary systems. At temperatures and volume fractions corresponding to the regions under the curves, the system exists in two phases. The highest temperature at which two phases can coexist is referred to as the upper critical solution temperature (UCST). Systems with UCST's are likely to have positive heats of mixing and intermediate-to-large entropies of mixing. In some systems the form of the phase diagram is inverted from that shown in Figure 1; ie, the solubility curves are concave upward. In these instances, the *lowest* temperature at which the system can form two liquid phases is called the lower critical solution temperature (LCST). LCST's are found in systems for which the heat of mixing is negative (exothermic mixing) and the entropy of mixing is unusually small.

Although the solvents in solutions described up to this point have been low molecular weight liquids, they need not be. Homogenous mixtures of two polymers, ie, polymer blends, can be prepared in some instances. In most blends $T\Delta S_{mix}^{comb}$ is negligible compared to ΔH_{mix}, a consequence of the small number of molecules involved in the mixing. Since positive heats are more common than negative, mixing is impossible for a majority of polymer–polymer pairs. Those pairs that are miscible usually have complementary constituent groups that form favorable intermolecular interactions on mixing.

Some polymers that are not soluble in pure liquids can be dissolved in a multicomponent solvent mixture. Crystalline

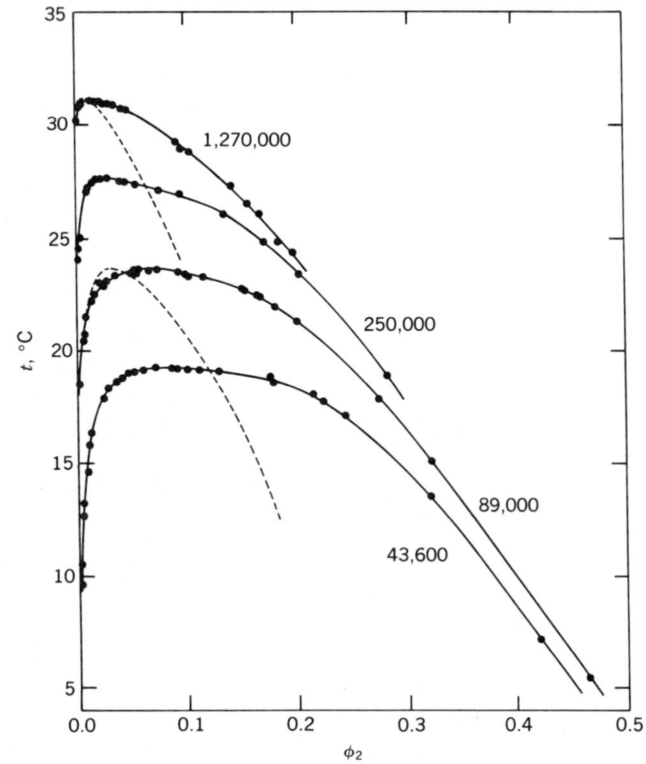

Figure 1. Phase diagrams for four fractions of polystyrene mixed with cyclohexane plotted against the volume fraction of polystyrene. The molecular weight of each fraction is given. The dashed lines show the predictions of the Flory-Huggins theory for two of the fractions. Courtesy of American Chemical Society.

polymers with backbones comprising rigid segments such as *p*-phenylene or fused aromatic rings are typically difficult to dissolve in organic solvents but are sometimes soluble in solvents containing salts or in strong acids.

BIBLIOGRAPHY

O. Fuchs, *Löslichkeitstabellen von Makromolekulen,* Vol. 2 of *Lösungsmittel und Weichmachungsmittel,* 8th ed., Wissenschaftliche Verlagsgesellschaft, Stuttgart, FRG, 1980.

J. S. Rowlinson and F. L. Swinton, *Liquids and Liquid Mixtures,* 3rd ed., Butterworth Scientific Publications, London, 1982.

P. J. Flory, *Principles of Polymer Chemistry,* Cornell University Press, Ithaca, N.Y., 1953, pp. 495–548.

A. F. M. Barton, *Handbook of Solubility Parameters and Other Cohesion Parameters,* CRC Press, Inc., Boca Raton, Fla., 1983.

ROBERT A. ORWOLL
College of William and Mary

SOLUTION POLYMERIZATION

Solution polymerization is the earliest method of addition polymerization and is still widely used today. Many solution polymers of commercial interest are prepared by free-radical addition-polymerization. These processes may be run in any solvent (water or organic solvent) in which both the monomers and resultant polymers are soluble. The advantages of free-radical solution polymerization over free-radical emulsion or suspension processes include solubility of monomer and polymer, as well as control of polymer molecular weight and molecular weight distribution (see also EMULSION POLYMERIZATION; SUSPENSION POLYMERIZATION).

Free-radical solution polymerization provides a route to industrial polymers. Molecular weights are between 500 and 500,000. These polymers are used in adhesives, coatings, water treatment, pigment dispersants, detergent additives, and other applications. Producers of water-soluble polymers include Alco, Allied Colloids, BASF, Diamond Shamrock, GAF, Nalco, and Rohm & Haas. Polymers produced by nonaqueous processes are manufactured by DuPont, ICI, SC Johnson, and Rohm & Haas. Estimated volumes are given in Table 1.

Solvents and Process Conditions

Solution polymers prepared in aqueous solution find utility in detergents, dispersants, thickeners, and leather and water treatments. Those in nonaqueous solutions are commonly used in adhesives and coatings. Utility and performance are dramatically affected by copolymer composition and molecular weight. The unusual properties of solution polymers are due to their molecular weights and molecular weight distributions. Composition and molecular weight are manipulated by synthesis parameters.

Polymers that are soluble in aqueous solution include poly(acrylic acid), poly(methacrylic acid), poly(maleic acid), polyacrylamide, polymethacrylamide, poly(vinyl alcohol), poly(*N*-vinylpyrrolidinone), and copolymers of these. The acrylic, methacrylic, or maleic acid products may be isolated in the acid form or as salts by neutralization with base, depending on end use.

Table 1. Solution Polymerization, 1986 Production, 10^3 Metric Tons

Polymer	United States	Total
Nonaqueous process		
high density polyethylene	440	499
linear low density polyethylene	313	400
low density polyethylene	925	1152
acrylic resins	68	113
vinyl chloride–vinyl acetate copolymers	36	59
fluoropolymers	14	25
Aqueous process		
poly(acrylic acids)[a]	21	

[a] Includes poly(methacrylic acid) and copolymers.

Table 2. Monomers Suitable for Free-radical Solution Polymerization

Monomer	Manufacturer	Solvent: Organic	Solvent: Water
acrylic acid	Rohm & Haas	+	+
	BASF		
	Hoechst-Celanese		
methacrylic acid	Rohm & Haas	+	+
methyl methacrylate	Rohm & Haas	+	−
methyl acrylate	Rohm & Haas	+	+
ethyl acrylate	Rohm & Haas	+	−
butyl acrylate	Rohm & Haas	+	−
maleic acid	Monsanto	+	+
itaconic acid	Pfizer	+	+
acrylonitrile	American Cyanamid	+	+
styrene	Dow	+	−
	USX Corporation		
vinyl acetate	Hoechst-Celanese	+	+
methyl vinyl ether	GAF	+	+
butadiene	Exxon	+	−
α-methylstyrene	USX Corporation	+	−
vinyl pyrrolidinone	GAF	+	+
vinyl chloride	Dow	+	−
vinylidene chloride	Dow	+	−

Solutions of polymers not soluble in water can be prepared in aliphatic and aromatic hydrocarbons, alcohols, ketones, esters, and ethers, depending on end use. Solvent choice is also influenced by cost, flash point (fire hazard), toxicity, solvency, and chain-transfer efficiency.

The technique used for preparation of commercial products depends on the solubility of monomers, initiators, chain-transfer agents, and the resultant polymer in the solvent. Some solvents are reactive and serve to lower the polymer molecular weight, which is often desirable. Solvents also serve as diluents, provide viscosity control, and dissipate the heat of polymerization. In a reflux process, heat is removed by cooling as well as evaporation.

Solution polymerizations are performed in stainless steel or glass-lined reactors, depending on corrosion behavior. These reactors are usually able to withstand pressure. Commercial processes are typically run in batches, although numerous continuous processes are patented. The important parameters in reactor selection and design are temperature control, mixing, and scale-up, thermal conductivity, heat of polymerization, viscosity limits, fluid behavior (Newtonian vs non-Newtonian), and shear rate (see POLYMERIZATION PROCEDURES, INDUSTRIAL; REACTOR TECHNOLOGY).

Monomers

Monomers that are suitable for solution polymerization are given in Table 2. All are soluble in organic solvents, but few are water-soluble. This limits the number of compositional variations that can be made solely by solution polymerization in water as a solvent.

Polymerization

Free-radical solution polymerization takes place in four steps: radical generation by initiator decomposition; initiation; propagation; termination by disproportionation and combination (see RADICAL POLYMERIZATION).

Initiators. A wide variety of radical sources, called initiators, are commercially available (Table 3). Homolytic cleavage by radiation, or thermal or chemical reaction, produces radicals capable of initiating polymerization. Solubility limits the solvents in which the initiators can be used. As with monomers, only a few are truly water-soluble.

Copolymerization

By appropriate selection of monomers, the properties of copolymers can be tailored to suit a wide range of applications. The choice of monomers also controls the solubility of copolymers in various solvents, including water. The reactivity of monomers is dependent on resonance stability, polarity, and steric effects.

Copolymers have several chain sequences:

Random copolymers, —ABAABAAAABBAB—

Alternating copolymers, —ABABABAB—

Table 3. Free-radical Initiators

Initiator	Manufacturer	Water solubility	Temperature range, °C
potassium persulfate	FMC	+	40–80
benzoyl peroxide	Lucidol	−	40–90
cumene hydroperoxide	Lucidol	−	50–100
di-*t*-butyl hydroperoxide	Lucidol	−	80–150
t-butyl peroctoate	Lucidol	−	60–120
2,2′-azobisisobutyronitrile	DuPont	−	20–100
t-butyl hydroperoxide	Lucidol	+	100–200

Block copolymers, —AAABBBAAABBB— (see COPOLYMERIZATION; COPOLYMERS, ALTERNATING; BLOCK COPOLYMERS).

Composition. The copolymerization of two monomers may produce polymer chains of a composition different from the monomer mixture. At one extreme, two homopolymers are formed; at the other, an alternating polymer is produced. Frequently, the result is somewhere in between. The composition varies throughout the polymerization, and different copolymer compositions are formed.

BIBLIOGRAPHY

F. W. Billmeyer, Jr., *Textbook of Polymer Science,* 3rd ed., Interscience Publishers, New York, 1984.

P. J. Flory, *Principles of Polymer Chemistry,* Cornell University Press, Ithaca, N.Y.

B. Vollmeit, *Polymer Chemistry,* transl. by E. H. Immergut, Springer-Verlag, New York.

G. Odian, *Principles of Polymerization,* McGraw-Hill Inc., New York.

GRAHAM SWIFT
KATHLEEN A. HUGHES
Rohm & Haas

SOLUTION PROPERTIES

The main types of thermodynamic theories of polymer solutions are: Flory-Huggins theory, equation of state theory, dilute-solution theory, scaling theory, and renormalization group theory.

Flory-Huggins theory deals successfully with the large negative deviations from the ideal solution laws that polymer solutions frequently exhibit. In this theory such deviations are attributed to the large entropy increase that occurs when chain molecules dissolve in a low molecular weight solvent. The reason for the abnormally large increase in entropy occurring when polymer chains mix with solvent molecules is to be found in the requirement that all of the units of a flexible chain must necessarily be connected by chain bonds. Every chain configuration requires a contiguous series of spatial locations that are occupied neither by segments of other chains nor by solvent molecules. For a large collection of chains in an undiluted sample this is a restriction that seriously lowers the total number of configurations that would obtain in the absence of this requirement. For a given collection of chains there is a larger number of configurations accessible if some solvent molecules are present. The presence of the latter, which are not constrained to occupy adjacent spatial locations, allows the collection of chains to assume all of those configurations that were accessible in the undiluted sample, plus those new configurations that can occur in the mixed state. This effect, which has to do entirely with the configurational entropy, becomes more important as the chain length increases.

The successes of the Flory-Huggins theory are these: (a) it accounts in a rational way for the large negative deviations from Raoult's law that are usually observed in polymer solutions; (b) it predicts the general characteristics observed in systems exhibiting upper critical temperatures in liquid–liquid phase separation; (c) it provides a framework for interpretation of fractionation by liquid–liquid phase separation; and (d) it provides a simple set of equations that can be applied to other special situations encountered in both experimental and theoretical applications. The deficiencies are these: (a) it fails to account for the observed concentration dependence of the reduced residual chemical potential; (b) it predicts entropies of dilution that are too positive; (c) it is unable to account for nonzero volume changes of mixing; and (d) it is unable to predict the occurrence of lower and upper critical solution phenomena in the same polymer–solvent system.

The equation of state theory removes some of the deficiencies of the Flory-Huggins theory by relating the properties of the solutions to the experimentally known properties of the pure unmixed liquids. In this way the persistence of the structural characteristics of the constituents is partially accounted for. The terminology employed (equation of state theory) is related to the way in which these properties are extracted from equations of state for the constituents and also to the way in which equations for the mixture are written in a form similar to that for the pure constituents.

By incorporating the individual characteristics of the pure liquids into its equations, equation of state theory satisfactorily accounts for all the phenomena for which the Flory-Huggins theory fails in addition to those for which it is successful. The two main disadvantages of the equation of state theory appear to be these: (a) a large amount of precise data on the pure liquids is required in order to apply the equations, (b) the equations are much more cumbersome to use in theoretical applications than are those of the Flory-Huggins theory. Neither theory can be expected to apply in the dilute-solution range because they both make assumptions about the distribution of chain segments, which are appropriate for concentrated solutions but not for dilute solutions.

Dilute-solution theory is designed to treat the situation that prevails at high dilutions, where overlapping of polymer chains is rare and the distribution of segments throughout the solution is not uniform. To deal with this, a completely different approach is required, which focuses attention on the interactions of small numbers of polymer molecules with each other, since simultaneous interactions of many chain molecules in dilute solutions are extremely infrequent. The mathematical techniques and the physical perspectives of this type of polymer solution theory are reminiscent of theories of dilute imperfect gases and the second virial coefficient, and its relation to molecular parameters is one of the primary objects of concern. In its most popular version the equations for the solution properties can be expressed in terms of two parameters, one of which is the size of a chain molecule and the other of which is related to the strength of polymer–solvent interactions. Although this theory is an improvement over the Flory-Huggins theory and equation of state theory in the dilute concentration range, careful experimental studies indicate that dilute-solution theory needs further refinement.

Scaling and renormalization group theory encompass both the dilute and the concentrated regimes of the concentration range; they also are the most successful theories developed so far that give a reasonably satisfactory account of the intermediate concentration range between these extremes, ie, "semidilute" solutions, where partial overlap of chains is important. Whereas the other theories of solutions considered above seek equations that relate experimental properties such as osmotic pressure (and other colligative properties), light-scattering intensities, and phase equilibrium relationships to molecular quantities such as molecular weight, radius of gyration, and molecular interaction parameters, scaling theory deals with remarkably simple power law relationships between measured properties. The popularity of scaling theory is due to the fact that by starting with simple physical pictures and using plausible arguments that should hold under certain limiting conditions it has derived a number of relationships that give good agreement with experiment.

Despite its successes, scaling theory suffers several defects. Renormalization group theory (RGT) can be viewed as an attempt to remove these deficiencies. This is a technique that involves replacing a very detailed description of a system that requires knowledge of a large number of molecular parameters with an equivalent but "coarser" description that contains fewer parameters. The replacement of a real polymer chain with a succession of Kuhn equivalent segments is an example of this first step. RGT goes further; this coarser description is in turn replaced (renormalized) with yet another equivalent coarser description with even fewer parameters, etc, until the description of the system can, with confidence, be identified with a known correct phenomenological description. This, together with the transformations (group operations) that connect the successive coarser descriptions, produces equations that show under what conditions scaling laws should hold.

BIBLIOGRAPHY

P. J. Flory, *Principles of Polymer Chemistry,* Cornell University Press, Ithaca, N.Y., 1953, Chap. 12.

P. J. Flory, R. A. Orwell, and A. Vrij, *J. Am. Chem. Soc.* **86,** 3507 (1964).

H. Yamakawa, *Modern Theory of Polymer Solutions,* Harper and Row, New York, 1971, Chapt. 4.

K. H. Freed, *Renormalization Group Theory of Macromolecules,* Wiley-Interscience, New York, 1987.

DEWEY K. CARPENTER
Louisiana State University at Baton Rouge

SOLUTION PROPERTIES, POLYDISPERSITY EFFECTS

Synthetic polymers always exhibit polydispersity; they consist of a large number of similar chemical species with different molar mass distribution density function that can be obtained experimentally by fractionation or gel-permeation chromatography.

Polydispersity has a strong influence on some thermodynamic properties, especially the liquid–liquid equilibrium of polymer solutions and mixtures. This effect is often important for practical problems. In the high pressure synthesis of low density polyethylene (LDPE), eg, the polyethylene is obtained as solute in supercritical ethylene. For its precipitation the pressure is lowered, resulting in separation into a polymer-rich and a polymer-lean phase. The molar mass distributions of the polyethylene in these phases depend on the precipitation conditions and influence the quality of the final product. Another example is polymer fractionation, where the distribution of the precipitated polymer depends strongly on the composition of the mixed solvent applied and on the temperature. In mixtures of two polymers (polymer blends), it is important whether these two polymers form a thermodynamically stable mixture or tend to phase separation (polymer–polymer compatibility). Again, this depends on the molar mass distributions of the applied polymers (see COMPATIBILITY; MISCIBILITY).

In addition to the polydispersity caused by molar mass, polydispersity may be due to branching (numbers, position, and length of side chains) or to tacticity. These sources of polydispersity usually exert a minor influence on thermodynamic properties and are not discussed here. Copolymers exhibit additional polydispersity owing to the two kinds of monomer units; this subject is also not discussed here.

Pseudocomponent Method and Continuous Thermodynamics

The method for calculating solution properties, such as the liquid–liquid equilibrium, is provided by chemical thermodynamics. For a system formed by a solvent 1 and a polydisperse polymer 2 the condition for equilibrium between phase ′ and phase ″ is as follows:

$$\mu_1' = \mu_1'' \qquad (1)$$
$$\mu_{2,i}' = \mu_{2,i}''; \qquad i = 1, \ldots, N$$

where μ, the chemical potential, is equal to the partial molar Gibbs free energy and i indicates the different polymer species whose number is indicated by N.

Solutions of polydisperse polymers, however, present a special problem: because of the very large number N of the occurring species, a mathematical solution of equation 1 is too complicated. The same difficulty arises for all polydisperse system, such as polymer mixtures, petroleum fractions, shale oils, and others. These problems can be solved by the pseudocomponent method or by continuous thermodynamics.

Because of the very large number N of occurring species, characterization of a polymer by fractionation or gel-permeation chromatography does not lead directly to the amounts of the individual polymer species but to a continuous molar mass distribution density function. The amounts of the individual species can be approximately obtained from this curve by dividing the area below it into N bars, the breadth of which equals the molar mass of a monomer unit.

In the pseudocomponent method, the area below the continuous distribution density function is divided into a smaller number of broader bars, as shown in Figure 1. Hence, a pseudocomponent is an imagined combination of a large number of neighboring species. The thermodynamic calculations are based on these arbitrarily chosen pseudocomponents. To obtain precise results, a large number of pseudocomponents have to be chosen, leading to considerable numerical complexity.

To overcome these disadvantages, continuous thermodynamics was established in 1980; preliminary attempts were made earlier. In this method, the continuous distribution density function is used directly without separation into pseudocomponents for thermodynamic calculations. This is possible by converting chemical thermodynamics into a suitable form.

Cloud-point Curve and Shadow Curve by Continuous Thermodynamics

To treat the liquid–liquid equilibrium by continuous thermodynamics, the phase-equilibrium condition (eq. 2) is com-

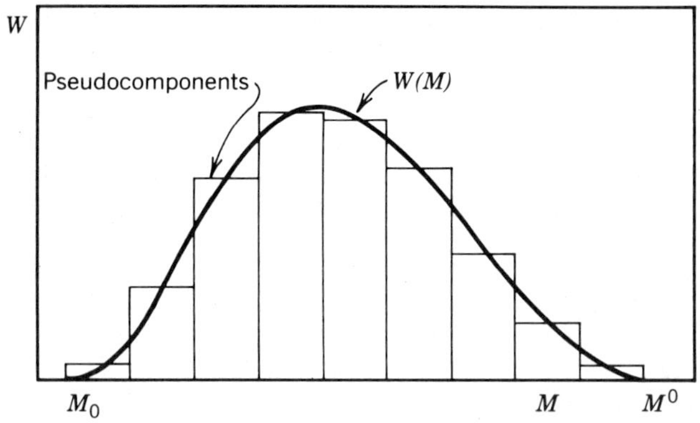

Figure 1. Pseudocomponent method. The molar mass distribution represented by the distribution density function $W(M)$ is divided into an arbitrarily chosen number of pseudocomponents. M_0 and M^0 are the lower and upper limits, respectively, of the molar mass M of the polymer species.

bined with explicit expressions of the chemical potentials. For the Flory-Huggins theory, these expressions are provided by equations 3 and 4. In the case of the cloud-point curve and the shadow curve, the composition of the coexisting phase ′ (ϕ_2' and $W'(M)$) is specified because this phase is identical with the initial homogeneous system. Hence, the unknowns are the composition of the coexisting phase ″ (ϕ_2'' and $W''(M)$) and the equilibrium temperature because $\chi = \chi(T)$.

$$\mu_1' = \mu_1'' \qquad \mu_2'(M) = \mu_2''(M); \qquad M_0 \le M \le M^0 \tag{2}$$

$$\mu_1 = \mu_1^0 + RT\left[\ln(1-\phi_2) + 1 - \frac{1}{\bar{x}_n} + \chi\phi_2^2\right] \tag{3}$$

$$\mu_2(M) = \mu_2^0(M) + RT\left[\ln \phi_2 W(M) + 1 - \frac{x_2(M)}{\bar{x}_n} + x_2(M)\chi(1-\phi_2)^2\right] \tag{4}$$

Combining equations 2, 3, and 4, the chemical reference potentials cancel because they are identical for two liquid phases. Hence, forming the exponential gives

$$1 - \phi_2'' = (1 - \phi_2')\exp \rho_1 \tag{5}$$

$$W''(M) = \frac{\phi_2'}{\phi_2''} W'(M) \exp\left(\frac{M}{M_S}\rho_2\right) \tag{6}$$

For the Flory-Huggins theory ρ_1 and ρ_2 are given by

$$\rho_1 = \left(1 - \frac{M_S}{\bar{M}_{2,n}'}\right)\phi_2' + \chi\phi_2'^2 - \left[\left(1 - \frac{M_S}{\bar{M}_{2,n}''}\right)\phi_2'' + \chi\phi_2''^2\right] \tag{7}$$

$$\rho_2 = \left(1 - \frac{M_S}{\bar{M}_{2,n}'}\right)\phi_2' + \chi(1-\phi_2')^2 - \left[\left(1 - \frac{M_S}{\bar{M}_{2,n}''}\right)\phi_2'' + \chi(1-\phi_2'')^2\right] \tag{8}$$

More sophisticated theories (see SOLUTION PROPERTIES) lead to relations for ρ_1 and ρ_2 similar in structure but more complicated.

From a mathematical point of view, equations 5 and 6 are quite different. Equation 5 results from the phase-equilibrium condition for solvent 1 and therefore is an equation between numbers. Equation 6, however, results from the phase-equilibrium condition for polymer 2 and therefore applies to all M values within the occurring molar mass interval. This is called a function equation, an equation between two functions of M. In the cloud-point curve and the shadow curve, the function equation 6 interrelates the unknown distribution density function $W''(M)$ for the polymer in the newly formed phase and the known distribution density function $W'(M)$ of the original polymer.

Applications

The method of continuous thermodynamics may easily be applied to other problems. To calculate coexistence curves, the mass balance must be accounted for as well as the phase-equilibrium condition. To treat phase-separation behavior in mixtures of two polydisperse polymers, polymer–polymer compatibility, component 1 is also considered as a polydisperse ensemble. Solutions and mixtures of random copolymers can be treated in this way by applying divariant distribution density functions, accounting additionally for differences in the chemical composition of the species. Furthermore, spinodals and critical solution points for polydisperse polymer systems may be calculated using stability considerations on the basis of continuous thermodynamics.

BIBLIOGRAPHY

H. Kehlen and M. T. Rätzsch, *Proc. 6th Int. Conf. Thermodyn. Merseburg,* 1980.

H. Kehlen and M. T. Rätzsch, *Z. Phys. Chem. (Liepzig)* **264,** 1153 (1983).

M. T. Rätzsch and H. Kehlen, *J. Macromol. Sci. Chem.* **22,** 323 (1985).

M. T. Rätzsch and H. Kehlen, *Progr. Polym. Sci.* **74,** 1 (1989).

MARGIT T. RÄTZSCH
HORST KEHLEN
Carl Schorlemmer Technical
University at Merseburg

SOLUTION RHEOLOGY

Discussion of rheological properties of polymer solutions can be organized into different concentration regimes: isolated chains; dilute and semidilute solution; and concentrated solutions. The first two refer to situations where the chains are, respectively, noninteracting and weakly interacting through direct interactions (hard-sphere repulsions, van der Waals forces, electrostatic or electrodynamic forces, hydrogen bonding) as well as indirect (hydrodynamic) interactions. The latter two regimes refer to higher concentration domains where topological interactions of polymer molecules (chain entanglements, restricted rotational and translational degrees of freedom) become important. The differentiation between semidilute and concentrated regions is defined by the appearance of phenomena related to new states of existence of the polymeric solute (eg, glass-transition behavior, liquid crystal transitions, gelation). In evaluating the success of theoretical models for rheological properties, a more definitive test is obtained if the comparison is extended to more than one viscoelastic function and to other transport properties, eg, the diffusion and sedimentation coefficients of the polymer in solution.

Mechanical Response Behavior of Polymer Solutions

Rheological characterization of polymer solutions involves analysis of the flow behavior under a well-specified stress field (Fig. 1). For steady-shear flow, the components of the shear-stress tensor σ_{ij} are solely a function of the shear rate $\dot{\gamma} = d\gamma/dt$, with γ the shear strain:

$$\sigma_{12} = \eta(\dot{\gamma})\dot{\gamma} \tag{1}$$

$$\sigma_{11} - \sigma_{22} = \psi_1(\dot{\gamma})\dot{\gamma}^2 \tag{2}$$

$$\sigma_{22} - \sigma_{33} = \psi_2(\dot{\gamma})\dot{\gamma}^2 \tag{3}$$

where $\eta(\dot{\gamma})$ is the shear-rate dependent viscosity coefficient, and $\psi_1(\dot{\gamma})$ and $\psi_2(\dot{\gamma})$ are the first and second normal stress coefficients, respectively. In the limit of zero shear rate, these material functions become constants, designated $\eta(0)$, $\psi_1(0)$, and $\psi_2(0)$.

The viscoelastic behavior of liquids is described as linear or nonlinear, depending on the relationship between the stress and the imposed deformation history. For small deformation, the material parameters are linear functions of the imposed strains. Properties such as the shear-rate dependence of viscosity and of normal stress involve large deformations and hence are nonlinear properties.

The viscoelastic properties of polymeric fluids are determined by the distribution of relaxation times that characterize the configurational dynamics of the polymer chains. If $H(\tau)d\ln\tau$ is defined as the fraction of the initial modulus contributed by processes with relaxation times in the range $\ln\tau < \ln t + d\ln\tau$, and assuming these processes relax exponentially, then

$$G(t) = \int_{-\infty}^{+\infty} H(\tau)e^{-t/\tau}\, d\ln\tau \tag{4}$$

The decay of $G(t)$ in dilute polymer solutions is determined by the fast processes involving configurational rearrangements of isolated polymer chains. As polymer concentration increases, intermolecular interactions result in a gradual increase in the relaxation times for chain dynamics. When the concentration increases substantially beyond the value where the chains overlap, a new group of slow relaxation mechanisms may appear, which are interpreted to be due to molecular dynamic phenomena associated with the entanglement network (terminal relaxations). The $G(t)$ decays initially from G^0 to a plateau value G_N^0, called the plateau modulus, and then decays further to zero at longer times. Combinations of η_0, G_N^0, and J_e^0 can be used to define average configurational relaxation times in the terminal region:

$$\tau_n = \frac{\int_{-\infty}^{+\infty} \tau H(\tau)\, d\ln\tau}{\int_{-\infty}^{+\infty} H(\tau)\, d\ln\tau} = \frac{\eta_0}{G_N^0} \tag{5}$$

$$\tau_w = \frac{\int_{-\infty}^{+\infty} \tau^2 H(\tau)\, d\ln\tau}{\int_{-\infty}^{+\infty} \tau H(\tau)\, d\ln\tau} = \eta_0 J_e^0 \tag{6}$$

The mean times τ_n and τ_w correspond, respectively, to the first and second moments of $H(\tau)$, and thus the ratio τ_w/τ_n is a measure of the width of the terminal relaxation distribution.

Viscosity

The viscosity coefficient relates the applied stress (force per unit area) in a flow experiment to the rate of deformation of the fluid:

$$\sigma_{12} = \eta\dot{\gamma} \tag{7}$$

where σ_{12} is in mN/m (= dyn/cm), η in Pa·s (= 10P), and $\dot{\gamma}$ in s^{-1}. Two well-defined geometries can be distinguished: shear viscosity, where the fluid is subjected to laminar-shearing flow, and extensional or elongational viscosity, where the fluid experiences uniaxial extension in the direction of the applied force. Most experiments on polymer solutions concern the measurement of shear viscosity, which can be obtained by a variety of techniques (Fig. 2).

Behavior of Isolated Molecules, Intrinsic Viscosity. When a macromolecular solute is dispersed in a fluid medium, the fluid viscosity increases because of the increased resistance to flow. Einstein derived an equation that correctly describes this viscosity increment when the solute can be modeled as impermeable, nondeformable, and noninteracting spheres of radius R:

$$\eta = \eta_s(1 + 2.5\phi) \tag{8}$$

where η_s is the viscosity of pure solvent and ϕ the volume fraction of spheres; ϕ can be expressed in terms of the solute mass concentration, c(g/cm³), as

$$\phi = \frac{N_A V_h c}{M} \tag{9}$$

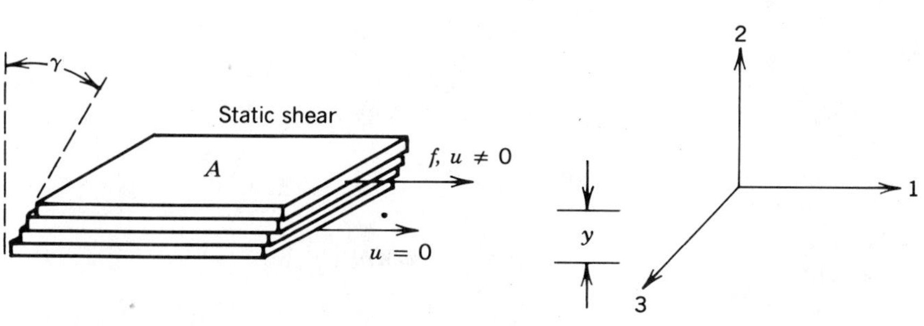

Figure 1. Shear deformation in laminar, unidirectional flow. Axes shown define the components of the stress tensor σ_{ij}. γ = shear strain, dimensionless; $\dot{\gamma}$ = rate of shear = du/dy = u(velocity)/y(thickness); σ = shear stress = f(force)/A(area); η = shear viscosity = $\sigma/\dot{\gamma}$.

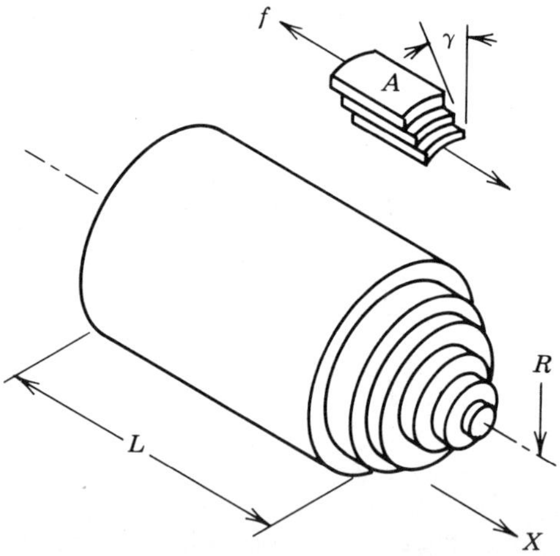

Figure 2. Laminar flow in capillary geometry. f = force, A = area, L = length, R = radius, γ = shear strain, X = direction of flow.

where $V_h = (4/3)\pi R^3$ is the molecular volume of the solute, M its molecular weight, and N_A Avogadro's number. Equations 8 and 9 indicate that the quantity

$$\frac{\eta - \eta}{\eta_s c} = \frac{2.5 N_A V_h}{M} \tag{10}$$

is uniquely related to the particle structure.

The above discussion provides a rationale for using a quantity, named the intrinsic viscosity

$$[\eta] = \lim_{c \to 0} \left(\frac{\eta - \eta_s}{\eta_s c} \right) \tag{11}$$

to relate the viscosity of dilute polymer solutions to the structure of the macromolecular solute. Experimentally, $[\eta]$ may be determined by Huggins or Kraemer plots, utilizing, respectively, the equations

$$\frac{\eta_{sp}}{c} = [\eta] + k_1[\eta]^2 c + \cdots \tag{12}$$

and

$$\frac{\ln \eta_r}{c} = [\eta] - k_1'[\eta]^2 c + \cdots \tag{13}$$

where the specific viscosity $\eta_{sp} = \eta_r - 1$ and the relative viscosity $\eta_r = \eta/\eta_s$. The coefficients k_1 and k_1' are a measure of the direct and indirect (hydrodynamic) interactions between pairs of macromolecules.

Dilute, Semidilute, and Concentrated Solutions. The concentration dependence of the zero-shear viscosity of dilute polymer solutions may be expressed in terms of an expanded form

$$\eta_r = 1 + [\eta]c + k_1[\eta]^2 c^2 + k_2[\eta]^3 c^3 + \cdots \tag{14}$$

where each term of order c^n reflects interparticle hydrodynamic and thermodynamic interactions of order n. In dilute solutions, these interactions are essentially analogous to those between hard spheres of radius equal to the hydrodynamic radius of the coil, $R_h \sim \sqrt[3]{[\eta]M}$. The coefficients k_1 and k_2 are only slightly dependent on solvent or temperature variation. Experimentally, values of k_1 are in the range of 0.3–0.5. The precise details of the theoretical interpretation of k_1 and k_2 remain to be established. Equation 14 indicates that the solution viscosity is a monotonically increasing function of polymer concentration. Furthermore, factors such as chain flexibility, chain expansion, and solvent draining affecting the magnitude of $[\eta]$ have a much stronger influence on the solution viscosity at high concentration. Thus solvent quality affects the concentration dependence of η_r mainly via the changes in $[\eta]$ accompanying the expansion or contraction of the coil.

An extension of equation 14, sometimes utilized in the discussion of more concentrated solutions, is the Martin equation:

$$\eta_r = 1 + c[\eta] \exp (K_M k_1 c[\eta]) \tag{15}$$

It provides a simple empirical representation of viscosity data at higher concentrations.

Theoretical efforts have been made to relate polymer solution viscosity to molecular properties. A corresponding states theory has been extended from dilute to concentrated solutions by incorporating the concept of free volume. The expression derived for the scaled viscosity, $\tilde{\eta} = \eta_{sp}/c[\eta]$ is:

$$\ln \tilde{\eta} = D_0 + D_1 Y\phi$$

where

$$Y\phi = [1 + D_2^+ \phi]$$

and

$$D_0 = \frac{-6(D_0^1 - 2)^2}{[(D_0^1)^2 - 12D_0^1 + 12]}, \qquad D_1 = -D_0,$$

$$D_2 = \frac{D_2^1(D_0^1 - 2)}{(2D_0\phi^*)} = \frac{D_2^+}{\phi^*} \tag{16}$$

where ϕ^* is the scaling volume fraction such that $\phi = \phi/\phi^*$.

An alternative method represents an extension of earlier experimental and theoretical studies of the viscosity of polymer liquids. The suggested relation is

$$\eta = \frac{1}{6} N_A x \left(\frac{x}{x_c} \right)^{a-1} \zeta \tag{17a}$$

with

$$a = \begin{cases} 3.4, & x > x_c \\ 1.0, & x < x_c \end{cases} \tag{17b}$$

and

$$x = g \left(\frac{R_g^2}{M} \right) Z\phi\rho_p \tag{17c}$$

or, for linear chains ($g = 1$):

$$x = \alpha_\phi^2 \phi M \rho_p \left(\frac{C_n}{6M_L} \right) \tag{17d}$$

where ρ_p is the density of the (undiluted) polymer, $C_n = 6R_{g,0}^2/nl^2$ is the characteristic ratio of the unperturbed chain with n bonds of length l, M_L is defined as M/nl, ζ is the friction factor per chain atom, the parameter $g = 1$ for linear chains, and $g = R_{g,0,br}/R_{g,0}$, for branched chains, x_c is the constant

$$x_c = M_e\rho_p\left(\frac{C_n}{6M_L}\right) \tag{17e}$$

and $\alpha_\phi = R_g/R_{g,0}$ is the (apparent) expansion factor for the chain dimensions in a solution with concentration $c = \phi\rho_p$.

A new molecular dynamic description of the transport properties of entangled polymer solutions utilizes the concept of reptation or snakelike motion of the chains. Detailed predictions of a variety of experimentally accessible parameters related to the linear and nonlinear viscoelastic properties of polymer solutions have been developed. The reptation model postulates that at high concentrations the polymer matrix provides a random meshlike structure. An individual chain is free to move but cannot cross the surrounding mesh lines. When the lifetime of the mesh constraints is sufficiently long, the chain acts as if it diffuses laterally along a tube of length $L = N\xi$, where the primitive path step length ξ is of the order of the mesh size, and $N = M/M_e$. The predicted Newtonian viscosity is

$$\eta = \eta_s(M/M_e)^3 \tag{18}$$

Experimental results and related theoretical modeling have been reported on solution viscosities for macromolecular architectures other than linear flexible coils. At low concentrations, branched chains are expected to exhibit lower viscosities than linear chains of similar molecular weight. However, when concentration and/or molecular weight increase to the extent that the chains between branch points become longer than the entanglement molecular weight M_e, the viscosity should increase over that of linear chains.

In solutions of rigid linear chains at high concentration c^{**}, a transition to a lyotropic liquid crystal phase is often observed. Above the entanglement concentration c_e, the viscosity increases approximately as C^3M^9 until $c \geq c^{**}$. At this point, a sudden and dramatic reduction takes place to values comparable to the solvent viscosity. Viscometric data for semidilute solutions of stiff-chain polymers are in qualitative agreement with theoretical predictions that assume rodlike behavior.

Linear Viscoelasticity of Polymer Solutions

Dilute Solutions. The viscoelastic behavior of dilute polymer solutions can be elucidated by instruments sensitive enough to measure the storage and loss shear moduli, $G'(\omega)$ and $G''(\omega)$. Extensive data have been reported on the intrinsic dynamic moduli,

$$[G']_\omega = \lim_{c\to 0}\frac{G'(\omega)}{c} \tag{19a}$$

$$[G'']_\omega = \lim_{c\to 0}\frac{G''(\omega) - \eta_s\omega}{c} \tag{19b}$$

for polymers in good and poor solvents. The observed moduli usually accurately fit the Zimm extension of the Rouse bead-spring model for configurational dynamics of Gaussian chains using the hydrodynamic interaction variable $h^* = h/\sqrt{N}$, where N is the number of Gaussian subchains (springs) and

$$h = \frac{n\zeta_0}{(12\pi^3)^{1/2}\eta_s R_g} \tag{20}$$

is an adjustable parameter. A further modification has been developed for self-avoiding chains in which the excluded volume exponent appears. In equation 20, n is the number of monomers in the chain and ζ_0 is the monomeric frictional coefficient.

Semidilute and Concentrated Solutions. As concentration and viscosity increase, the solution may exhibit a glass transition in the viscoelastic functions that is reflected by the appearance of a constant modulus G_g at short times (high frequencies). Provided the polymer molecular weight exceeds a certain value, M_e, the solution also exhibits rigidity at intermediate time (frequency) scales because of formation of an entanglement network. This occurs when the concentration rises above a value c_e, related to M_e by equation 21.

$$\log M_e = \log M_e^0 - \nu \log c \tag{21}$$

The entanglement rigidity manifests itself in the appearance of a constant value of the storage modulus, ie, the plateau modulus G_N^0, when the solution is deformed on a time scale that is short compared with the relaxation time(s) characteristic of the entanglement network. The corresponding behavior of the loss modulus $G''(\omega)$ exhibits two relaxation peaks associated with the dissipation of energy via molecular motions associated with the glass transition and viscous flow (terminal relaxation spectrum), respectively.

Only a few studies have been reported of rigid rodlike or branched polymers in semidilute or concentrated solutions. The concentration of rodlike polymers cannot be increased very much before spontaneous separation into a concentrated liquid crystal in a dilute isotropic phase takes place. However, because of the large hydrodynamic volume of elongated polymers the effect of intermolecular interactions is seen at very low concentrations. Thus, above a critical concentration c_e, which is much smaller than that for a flexible chain of comparable molecular weight, features appear that are caused by entanglement coupling and are very similar to those observed in concentrated solutions of flexible polymers.

The manifestation of chain entanglements in solutions of branched polymers is delayed to higher concentrations than in solutions of linear flexible chains of similar molecular weights. Again, theoretical discussions of the viscoelasticity of such systems have been published. The concentration and molecular weight dependence of the shear moduli solutions of branched polymers at $cM > (cM)_e$ is much stronger than for linear chains. Thus, the storage shear modulus of concentrated solutions of branched molecules is smaller than that of linear chains of the same molecular weight until the molecular weight of the arm becomes extremely high ($M_a > M_e$), when the situation becomes reversed.

Nonlinear Viscoelasticity

Steady-state Experiments: Non-Newtonian Viscosity and Steady-state Recoverable Compliance. As is well known, the steady-state shear viscosity of polymer solutions usually decreases

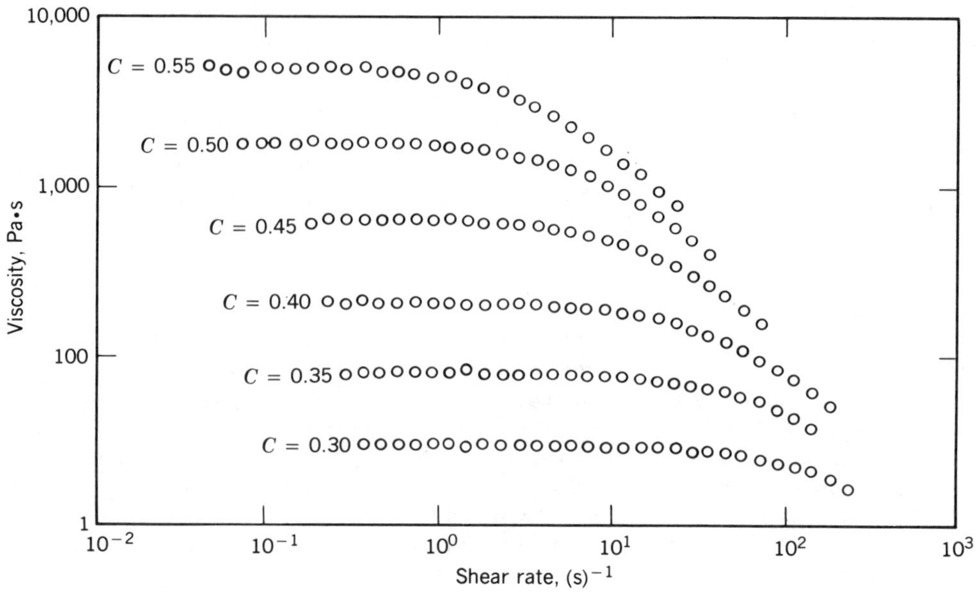

Figure 3. Double logarithmic plots of non-Newtonian viscosity against shear rate for different concentrations (g/mL) of polystyrene (M_w = 411,000) in n-butylbenzene at 30°C (1 Pa·s = 10 P).

with shear rate $\dot{\gamma}$ when $\dot{\gamma}$ is larger than some value that is strongly dependent on polymer concentration and molecular weight. Typical concentration-dependent behavior is shown in Figure 3, which exhibits data for polystyrene in n-butylbenzene. As concentration increases, the Newtonian viscosity η_0 increases and shear-thinning behavior begins at lower shear rates $\dot{\gamma}$. Similar effects are produced by an increase in molecular weight.

A few polymer–solvent systems exhibit shear-thickening behavior. Generally, such behavior is interpreted theoretically by a model in which transient intramolecular associations are replaced by intermolecular associations as the polymer chains are deformed under the action of shear forces.

Nonsteady-state Experiments. Nonlinear viscoelasticity can be conveniently studied by two kinds of stress-relaxation experiments. The first involves observation of stress relaxation after the imposition of a large sudden shear strain:

$$G(t,\gamma) = \frac{\sigma(t)}{\gamma} \tag{22}$$

The function $G(t,\gamma)$ decreases dramatically with increasing γ.

An alternative approach to investigating the transient viscometric functions involves analysis of stress growth and relaxation following initiation and cessation, respectively, of a constant shear rate. These properties can be described by the time-dependent viscosity and normal stress functions:

$$\eta^+(\dot{\gamma},t) = \frac{\sigma^+(\dot{\gamma},t)}{\dot{\gamma}} \tag{23a}$$

$$\eta^-(\dot{\gamma},t) = \frac{\sigma^-(\dot{\gamma},t)}{\dot{\gamma}} \tag{23b}$$

$$\psi_1^+(\dot{\gamma},t) = \frac{\sigma_{11}^+(\dot{\gamma},t) - \sigma_{22}^+(\dot{\gamma},t)}{\dot{\gamma}^2} \tag{23c}$$

and

$$\psi_1^-(\dot{\gamma},t) = \frac{\sigma_{11}^-(\dot{\gamma},t) - \sigma_{22}^-(\dot{\gamma},t)}{\dot{\gamma}^2} \tag{23d}$$

Elongational Viscosity

Elongational viscosity is also referred to as extensional or tensile viscosity of polymer solutions. Elongational flow implies shear-free flow where the stress tensor contains only diagonal components, $\sigma_{ii} \neq 0$, $\sigma_{ij} = 0$, and for the uniaxial case, $\sigma_{11} = \overline{\eta}\dot{\epsilon}_{11}$, where $\overline{\eta}$ is the elongational viscosity, and σ_{11} and ϵ_{11} refer to the tensile stress and tensile strain rate along axis 1 in Figure 1. Theoretically, continuum mechanics requires that, for Newtonian fluids, the Trouton result holds, $\overline{\eta} = 3\eta$. It has also been demonstrated that, for the linear viscoelastic case, the time-dependent elongational viscosity functions characterizing stress growth and stress relaxation, $\overline{\eta}^+(t)$ and $\overline{\eta}^-(t)$, are related to those for shear viscosity by $\overline{\eta}^+(t) = 3\eta^+(t)$ and $\overline{\eta}^-(t) = 3\eta^-(t)$. For nonlinear effects, and dilute polymer solutions, all continuum and molecular models predict that, in contrast to the predominantly shear-thinning behavior of η, $\overline{\eta}$ is an increasing function of strain rate. Thus the ratio $\overline{\eta}/\eta$ increases above 3 at large strains. The strain rate dependence of $\overline{\eta}$ is predicted to be a universal increasing function of $\dot{\epsilon}\tau_c$ where τ_c is the configurational relaxation time of the polymer chains. As polymer concentration increases, the magnitude of $\overline{\eta}$ is anticipated to increase in a fashion at least qualitatively similar to that of η.

BIBLIOGRAPHY

J. D. Ferry, *Viscoelastic Properties of Polymers,* 3rd ed., John Wiley & Sons, Inc., New York, 1980, Chapt. 1.

R. B. Bird, R. C. Armstrong, and O. Hassager, *Dynamics of Polymeric Liquids,* Vols. I and II, John Wiley & Sons, Inc., New York, 1977.

C. J. S. Petrie, *Elongational Flows,* Pitman Publishers Ltd., London, 1979.

ALEX M. JAMIESON
Case Western Reserve University

SOLVENT CASTING. See FILMS, MANUFACTURE.

SOLVENT RESISTANCE. See CHEMICALLY RESISTANT POLYMERS.

SOLVENTS FOR POLYMERS. See SOLUBILITY OF POLYMERS.

SORPTION. See TRANSPORT PROPERTIES.

SOUND ABSORPTION. See ACOUSTICAL PROPERTIES.

SPACE ENVIRONMENT PROCESSING

One of the most exciting goals of materials processing in space is the production of high value materials that cannot be made on Earth.

Space processing provides several benefits. First, there is the lack of gravity-driven sedimentation. Secondly, convective stirring may occur in the earth's gravity if fluids of different concentrations are mixed. Convective stirring also occurs when a chemical reaction or crystallization process occurs. The process lowers the local concentration of the material and sets off the instability. Such convective forces may be disastrous to a process that needs a calm environment. Also, in the absence of gravity-driven convection, diffusion of molecules occurs without disturbing the local environment.

The number of investigations on polymers has been limited because low gravity experiments are much more expensive than comparable experiments in a laboratory. However, it is not necessary to go into orbit to carry out a low gravity experiment. Alternate low gravity research opportunities are available, eg, drop towers, sounding rockets, and parabolic airplane flights. These alternate opportunities give only short periods of microgravity, making them unsuitable for many kinds of experiments, eg, crystal growth. Drop tower and parabolic flight are relatively inexpensive and have been useful in solidification and demixing experiments.

Monodisperse Latex Particles. Monodisperse polystyrene latices are used as calibration standards by microscopists. For this application samples of latex particles of various sizes are needed. It is important, of course, to have particles with a low standard deviation in the particle diameter. There being no sieving techniques for particles in the nanometer and micrometer range, production of these calibration standards has relied on direct production of the latices through seeded emulsion polymerization techniques. When space processing was first considered for monodisperse latex particles, many experts believed that an upper limit to particle diameter had been reached using conventional seeded emulsion polymerization technology because of gravity-induced problems. Space processing was attractive because, in the absence of gravity, the latex particles should show no tendency to cream or settle. Thus the polymerization reactor must be stirred only to ensure good mixing and heat transfer.

As of late 1989, the latex reactor had been successfully operated on four shuttle flights. In a fifth flight an equipment failure led to incomplete polymerization. The first experiments produced particles in the 2–5 μm range. By the third flight particles in the 10–18 μm range were being produced from seeds grown during earlier flights. The most recent flight produced particles with diameters of up to 30 μm. Space-manufactured latices and their ground-based counterparts are shown in Figure 1.

Polymer Solidification. There have been preliminary indications from parabolic airplane (KC-135) flight experiments that solidification of a polymer may be affected by space processing. Polyethylene samples processed in a microgravity appear to have different morphological patterns than comparison samples processed at 1 *g*. However, significant differences were not confirmed using x-ray, ftir, and dsc analysis. The preliminary results need to be verified by longer duration experiments where equilibrium heat transfer can occur, allowing for slow solidification.

Multiphase Composites and Polymeric Blends. Dissimilar polymers undergo phase separation when mixed and allowed to stand. Many such blends have been made in order to eval-

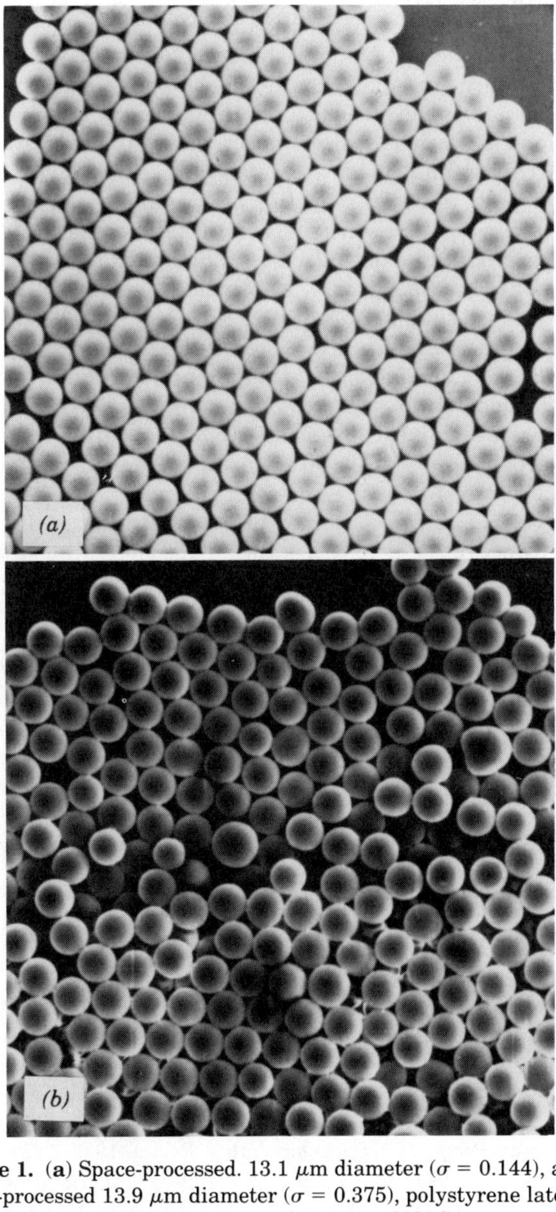

Figure 1. (**a**) Space-processed. 13.1 μm diameter (σ = 0.144), and (**b**) Earth-processed 13.9 μm diameter (σ = 0.375), polystyrene latex particles at a magnification of 500 ×. Courtesy of NASA.

uate properties of blends. There are obvious property differences expected of blends that are multiphase and those that are phase homogeneous. Space processing offers a way to obtain phase homogeneous blends.

Attention has also been directed toward identifying the effects of space processing on composites. Some investigators are interested in determining if processing in a microgravity environment can lead to the formation of better composites. Others are concerned about the space process because, in the course of the installation of the U.S. space station, components employing composites may be fabricated in space.

Polymeric Foams. Like composites, certain insulating and structural foams may be processed during space station deployment. It is therefore necessary to know the effect of gravity and vacuum on the properties of foams manufactured in space. A recent experiment on a sounding rocket flight proved that processing at low gravity leads to more spheroidal cell shape.

Protein Crystal Growth. The main reason that space processing seems attractive for protein crystals is that it is difficult in a normal gravity environment to grow protein crystals to a size and homogeneity needed for x-ray crystallographic studies. Protein crystals, being stabilized by weak crystalline interactions, are extremely fragile. Thus it is expected that crystal formation is very sensitive to fluctuations in the growth environment, such as sedimentation and convection. Also, under a gravitational field, a crystal's own mass may produce enough sag to allow introduction of crystal defects.

Protein crystal growth experiments flown on the space shuttle have produced exciting results. Several large crystals of high quality have been obtained in space when efforts on Earth have produced smaller or poorer quality crystals. The prospects for space processing are therefore bright.

Purification of Biological Materials by Partitioning Between Two Polymer Solutions. Space processing has been considered because phase partitioning is affected by gravity in two ways. First, separation of cells often involves particles (eg, megakaryocytes or cells cultured on carrier beads) which may settle to the bottom of the container before phase demixing occurs and thus prevent efficient separation. Secondly, the efficiency of the separation is apparently affected by a randomizing effect owing to rapid fluid movement during phase emulsion demixing at unit gravity. It is not certain if movement is the result of gravity-induced convection, but this is a possibility. Low gravity conditions would reduce the effect of these factors.

To date no space experiment to examine the efficiency of cell separation by phase partitioning has been conducted. However, extensive ground experiments have been done and several low gravity experiments, including one shuttle experiment, have been aimed at understanding the demixing process.

BIBLIOGRAPHY

R. J. Naumann and H. W. Herring, *Materials Processing in Space: Early Experiments,* NASA Sp-443, National Aeronautics and Space Administration, Washington, D.C., 1980.

B. Feuerbacher, H. Hamacher, and R. J. Naumann, eds., *Materials Science in Space,* Springer-Verlag, Berlin, Berlin, 1986.

H. U. Walter, ed., *Fluid Sciences and Materials Sciences in Space—A European Perspective,* Springer-Verlag, Berlin, 1987.

C. E. Bugg, *J. Cryst. Growth* **76,** 535 (1986); L. J. DeLucas, F. L. Suddath, R. Snyder, R. Naumann, M. B. Broom, M. Pusey, V. Yost, B. Herren, D. Carter, B. Nelson, E. J. Meehan, A. McPherson, and C. E. Bugg, *J. Cryst. Growth* **76,** 681 (1986).

SAMUEL P. MCMANUS
University of Alabama in Huntsville

SPANDEX FIBERS.
See FIBERS, ELASTOMERIC.

SPHEROPROTEINS.
See PROTEINS.

SPHERULITES.
See MORPHOLOGY.

SPINNING.
See FIBERS, MANUFACTURE.

SPLIT FILM.
See OLEFIN FIBERS.

SPRAY COATING.
See COATING METHODS.

SPRAYED FABRIC.
See NONWOVEN FABRICS.

SPRAYED FIBERS.
See OLEFIN FIBERS.

SPRAY LAY-UP.
See REINFORCED PLASTICS.

SPRAY MOLDING.
See REINFORCED PLASTICS.

SPUNBONDED FABRICS.
See NONWOVEN FABRICS.

SPUTTERING.
See METALLIZING.

STABILIZATION

All polymers, natural and synthetic, degrade under normal use conditions with progressive loss in aesthetic appearance, mechanical strength, and dielectric integrity. Degradation, the result of irreversible chemical reactions or physical changes, ultimately leads to failure. Polymers vary considerably in the rate at which they degrade, but all known polymers will fail eventually when exposed under normal conditions of use. For many applications, degradation must be inhibited by stabilizers to assure the required life expectancy. Mechanisms by which a few of the common polymers degrade have been established, and in these instances, stabilization can be approached on a scientific basis rather than a trial and error method.

The mechanism responsible for the thermal oxidation of hydrocarbon polymers has been established. Its origins are in the observation by Hoffman in 1861 that the deterioration of natural rubber is accompanied by reaction with oxygen. A series of publications led to acceptance of the following kinetic scheme for the oxidative degradation of hydrocarbon polymers:

initiation

$$RH \longrightarrow R^{\cdot}$$

propagation

$$R^{\cdot} + O_2 \longrightarrow ROO^{\cdot}$$

$$ROO^{\cdot} + RH \longrightarrow ROOH + R^{\cdot}$$

oxidative chain branching

$$ROOH \longrightarrow RO^{\cdot} + HO^{\cdot}$$

$$RO^{\cdot} + RH \longrightarrow ROH + R^{\cdot}$$

$$HO^{\cdot} + RH \longrightarrow HOH + R^{\cdot}$$

In this mechanism, initiation is depicted as the removal of hydrogen from a hydrocarbon molecule (RH). Although this reaction may be responsible for initiation of oxidative degradation in pure, low molecular weight hydrocarbons, initiation in polymers is considerably more complex. It is now generally believed that impurities such as traces of hydroperoxides are responsible for initiation in commercial polymers.

Oxidative degradation will terminate eventually through combination or disproportionation of propagating radicals. However, critical properties of the polymer must be expected to degenerate long before autotermination takes place. Stabilization, an induced termination reaction, is the obvious route to restricting the extent of degradation and thus extending the useful life of a polymer.

Stabilization may be required in all stages of a polymer's life cycle. Polymers that are particularly sensitive to oxidative degradation often require protection during processing and fabrication into final products. Polymers are fabricated at elevated temperatures and oxygen is always present. Stabilizers have been developed to protect the polymer under the extreme conditions encountered during processing. To be effective, these short-term stabilizers must be sufficiently mobile within the molten polymer to reach sites of incipient degradation before extensive reaction can occur. Therefore, the short-term stabilizers are low molecular weight compounds that are usually lost through evaporation or extraction during long-term exposure under normal use conditions. Long-term stabilizers must be incorporated into the polymer to protect it from degradation during extended exposure. These stabilizers are of higher molecular weight so that they are not as readily lost by evaporation or extraction. Stabilizers have been incorporated into the polymer structure or chemically bound to polymer molecules to achieve maximum retention.

There are two general approaches to the stabilization of polymers: structural modification and the use of additives. In the first case, the chemical structure of the polymer is modified to eliminate or reduce trace impurities or weak links. Bonding of a stabilizer moiety to polymer molecules should also be classified as stabilization through modification of polymer structure, but this approach has some limitations. The second general approach to stabilization is the most commonly used and involves additives blended into the polymer mass. A wide variety of additives has been developed to stabilize polymers against the several modes of degradation. Thus, for example, there are additives to protect against thermal oxidation, ultraviolet degradation, burning, and the attack of ozone (see also ANTIOXIDANTS; ANTIOZONANTS; FLAMMABILITY).

Structure Modification

It has long been suspected that commercial polymers degrade rapidly because of the presence of "weak links" or structural defects that are more reactive than the normal segments of the molecules. Oxidative degradation might be expected to be initiated at these sensitive sites.

Surface Modification. Polymers that are susceptible to oxidative degradation can be stabilized to a limited extent by restricting the penetration of oxygen into the polymer bulk. Radiation can lead to a denser cross-linked surface layer, which could restrict permeation to oxygen. A densified surface has also been produced in polyethylene by molding this polymer against a noble metal surface. The transcrystalline surface layer that is formed is less permeable than the normal crystalline structure but rapidly reverts to the normal structure.

Thermal Oxidation

All known polymers are susceptible to thermal oxidative degradation, although there is considerable variation in their degree of stability. Stabilization through incorporation of additives into the polymer matrix is the most common method employed to protect against thermal oxidation. Structure modification, usually applied in protection against nonoxidative degradation, can also be effective in protecting against thermal oxidation, but this approach has only limited application.

Additives

Chain-breaking Antioxidants. Both phenols and aromatic amines, classified as labile-hydrogen donors, inhibit the thermal oxidation of natural rubber. Chain-breaking antioxidants perform a sacrificial role. Once the labile hydrogens have been removed, their primary mechanism for stabilization is ended. Secondary products, however, may enter into the reaction, either providing additional stabilization or perhaps serving as catalysts for degradation. As an example, the short-term antioxidant 4-methyl-2,6-di-*tert*-butylphenol (**1**) can terminate one oxidative chain by providing its labile hydrogen to a propagating radical and a second, by addition at the para position:

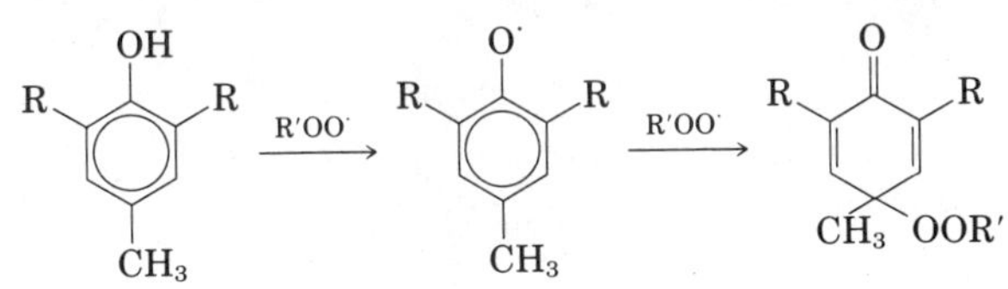

(**1**)

where R = $-C(CH_3)_3$

Preventive Antioxidants. These antioxidants (Table 1) are effective in the initiation step. Advances in the development of these stabilizers arose from attempts to stabilize the synthetic hydrocarbon polymers, polyethylene and polypropylene. The basic reaction for stabilization with preventive antioxidants is

$$ROOH \xrightarrow[\text{antioxidants}]{\text{preventive}} \text{inert (nonradical) products}$$

Hydroperoxides are represented as the species responsible for initiation, but these antioxidants may also deactivate other groups or impurities that would otherwise initiate degradation.

Antioxidant Combinations. Combinations of two or more antioxidants are sometimes used to protect polymers that are particularly sensitive to thermal oxidation, eg, polypropylene and poly(1-butene). Since there are several types of antioxidants available, combinations are expected to give different effects. The effect may be additive, antagonistic, or synergistic.

Ultraviolet stabilizers, metal deactivators, and hydroperoxide decomposers can function individually or cooperatively to reduce the rate of initiation. Each of these preventive antioxidants is capable of forming synergistic combinations with chain-breaking antioxidants.

Nonmigrating Antioxidants. Loss of antioxidants by evaporation, washing away during outdoor weathering, or extraction by various solvents that may come in contact with stabilized polymers are major factors contributing to failure. This loss is accentuated by the poor miscibility of antioxidants in polymers, which can lead to migration of the antioxidant to the polymer surface, a phenomenon known as blooming. The chemical bonding of antioxidants to polymer molecules is a promising approach to improved retention.

Bound Antioxidants. A variety of polymerizable, monomeric antioxidants has been synthesized. Many of these stabilizers have been used to protect natural rubber and synthetic elastomers against thermal oxidation.

Carbon Black as a Thermal Antioxidant. Carbon black has been used extensively to protect polymers against ultraviolet-induced degradation. Though this physical effect is its most familiar role in stabilization, the structure of carbon black below suggests that it might also have important chemical activity.

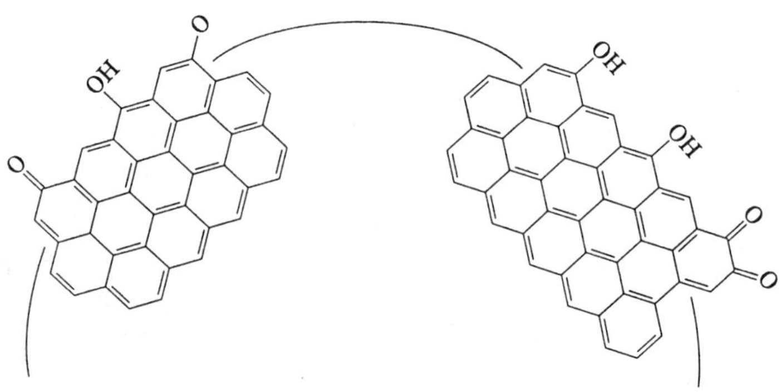

Particles of carbon black are comprised of many condensed aromatic rings. Depending on the method by which it is made, there may be 1% or more of chemically combined oxygen on

Table 1. Typical Preventive Antioxidants

Name	Structure
Hydroperoxide decomposers	
dilauryl-β,β-thiodipropionate	$S(CH_2CH_2\overset{\overset{O}{\|}}{C}OC_{12}H_{25})_2$
β-naphthyl disulfide	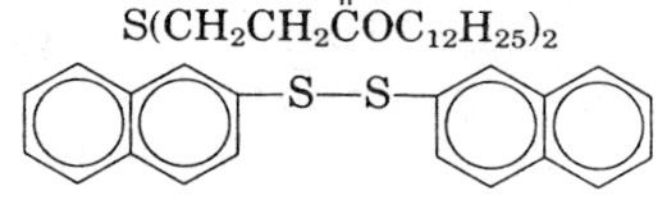
thiol-β-naphthol	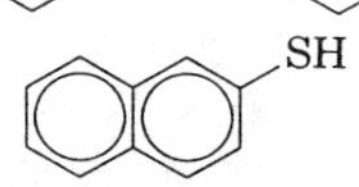
2-mercaptobenzothiazole	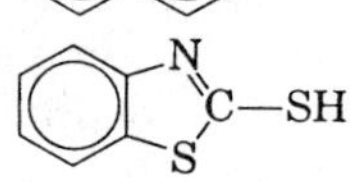
benzothiazyl disulfide	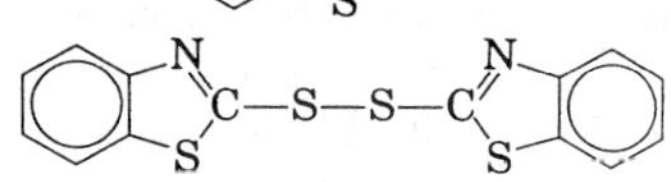
phenothiazine	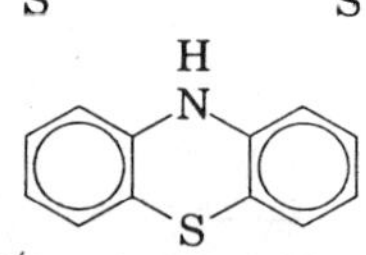
tris(*p*-nonylphenyl)phosphite	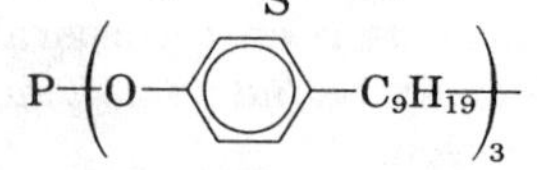
zinc dimethyldithiocarbamate	$(CH_3)_2N{-}\overset{\overset{S}{\|}}{C}{-}S{-}Zn{-}S{-}\overset{\overset{S}{\|}}{C}{-}N(CH_3)_2$
Metal deactivators	
oxamide	$H_2N{-}\overset{\overset{O}{\|}}{C}{-}\overset{\overset{O}{\|}}{C}{-}NH_2$
oxanilide	$C_6H_5NH{-}\overset{\overset{O}{\|}}{C}{-}\overset{\overset{O}{\|}}{C}{-}NHC_6H_5$
N,N′-dibenzaloxalyl dihydrazide	$C_6H_5{-}CH{=}N{-}NH{-}\overset{\overset{O}{\|}}{C}{-}\overset{\overset{O}{\|}}{C}{-}NH{-}N{=}CHC_6H_5$
benzotriazole	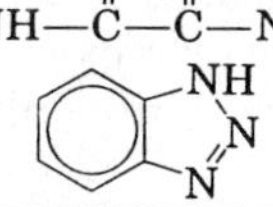

the surface of the particles. Phenolic, carbonyl, carboxyl, and other oxygen-containing groups have been identified on carbon black surfaces.

Since there is strong evidence for the presence of phenolic groups on the surface, carbon black can be considered a giant phenol. As such, it might be expected to function as a labile-hydrogen donor, capable of breaking oxidative chains.

Stabilization of Halogenated Polymers. Poly(vinyl chloride) (PVC) must be stabilized in all of its applications. Mechanisms used to stabilize PVC are representative of those employed to protect other halogenated polymers. Stabilization of PVC by structural modification has been discussed earlier. As with the hydrocarbon polymers, however, the most general approach to PVC stabilization is through the use of additives. Organic salts of barium, cadmium, and tin have come into general use. These may be salts of maleic, phthalic, or other organic acids.

Evidence has been presented for attachment of organic ligands to PVC molecules through reaction with organometallic stabilizers. Apparently, these stabilizers reduce color formation by attacking double bond sequences in the polyene product.

Oxygen is another catalyst for dehydrochlorination. Liberation of hydrogen chloride proceeds at a much faster rate in oxygen than in an inert atmosphere. Reaction with oxygen reduces the length of conjugated sequences, and also leads to formation of hydroperoxides. These intermediates may then undergo typical decomposition reactions and peroxide decomposers can be used to inhibit these secondary reactions of degradation.

There are important differences in the stabilization of rigid and flexible PVC formulations. Flexible PVC formulations contain large amounts of plasticizers. Conventional antioxidants, such as bisphenol A, must be used in flexible formulations to inhibit oxidation of the plasticizers. Oxidation or loss of plasticizers reduces flexibility.

The most frequently used ultraviolet absorbers are derivatives of *o*-hydroxybenzophenone (**2**), *o*-hydroxyphenyl salicylates (**3**) or 2-(*o*-hydroxyphenyl)benzotriazoles (**4**).

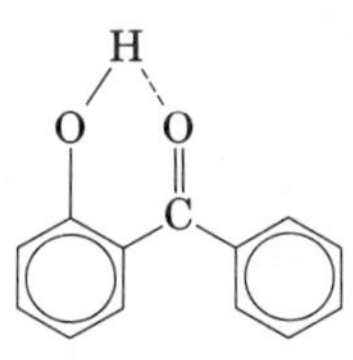

(**2**)

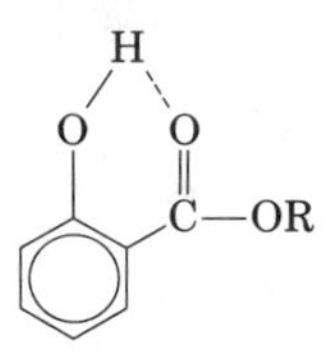

(**3**)

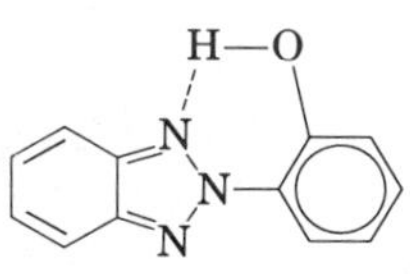

(**4**)

High Energy Radiation. Reactions that take place when polymers are exposed to high energy radiation are usually more destructive and much more complex than those occurring during exposure to uv radiation. Sources of high energy radiation can originate in the natural environment or they may be artificial, eg, x rays, gamma rays, and radiation emitted from nuclear reactors.

Radical scavengers, including labile-hydrogen-donating antioxidants, can provide limited protection against high energy radiation. Energy dissipation within aromatic structures as a method of stabilization has been demonstrated in the radiation resistance of isobutylene–styrene.

Ozone Attack

Despite the relatively few polymers in which degradation by ozone is encountered, stabilization against ozone attack is of considerable importance. It is essential for the protection of natural rubber, which remains the ideal elastomer based on elastomeric properties. Natural rubber and synthetic elastomers containing ethylenic unsaturation are rapidly degraded by ozone when under stress.

Polymers can be protected against degradation by ozone, either by physical or chemical processes. Blending of waxes with elastomers is the most common approach to stabilization by physical processes.

Additives that protect elastomers against degradation by ozone through chemical processes are called antiozonants. Structures of some of the more common antiozonants include the following:

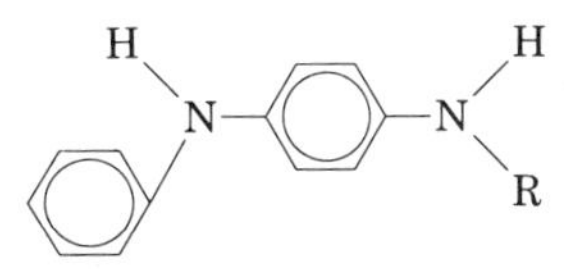

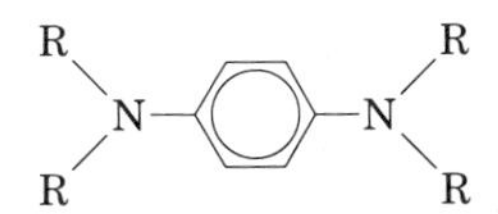

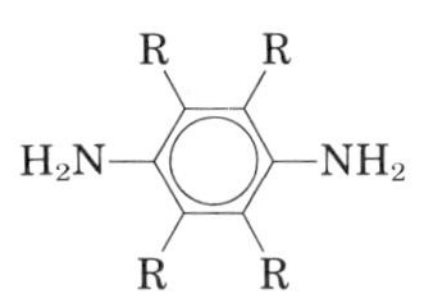

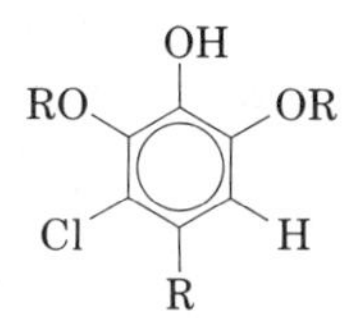

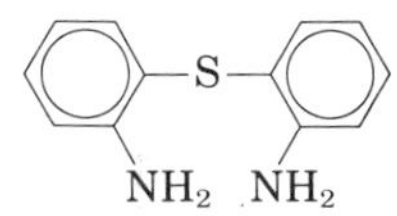

The R groups in these structures are mostly alkyl but, in the substituted *p*-phenylenediamines, a combination of alkyl and aryl substituents may be present.

Antiozonants are often used in combination with waxes. Since less ozone reaches the elastomer when it is protected by a wax film, antiozonants are more effective. Structure modification is also a route to stabilization against ozone attack.

Burning

All polymers burn in oxygen when exposed to sufficient heat. As in thermal oxidation, polymers vary over a wide range in their resistance to burning, fluorinated polymers being the least combustible. Because of their expanding use in construction, there is considerable concern over the formation of toxic products during combustion of certain polymers. Stabilization by structure modification or with flame retardants are promising approaches, but the adverse effect that such modifications may have on other critical properties must be taken into account.

Maximum stability of flexible PVC against burning can involve a combination of approaches. When the maximum effect is reached by reducing plasticizer content to the lowest level acceptable, additives can then be incorporated to raise the level of flame retardance still further. These additives could be inert fillers, those capable of evolving diluents into the burning zone, or true flame retardants such as antimony oxide. When all of these methods for stabilizing flexible PVC are used together, however, resistance to burning is still not comparable to that obtained with partially fluorinated copolymers of ethylene and propylene.

Chemical Reagents

Oxygen in its various forms is the most common chemical reagent responsible for polymer degradation, and stabilization against degradation by oxygen has been discussed in previous sections. Water is the second most important chemical reagent involved in polymer degradation and is a significant contributing factor to adverse outdoor weathering. Polymers that are formed by condensation reactions are most vulnerable to degradation by water. Primary bonds are ruptured in this class of polymers. Addition polymers may also be degraded by water, but only through cleavage of secondary bonds.

Urethane polymers containing ether groups are much less vulnerable to hydrolysis than those based on ester groups. Thus, polyurethanes with predominantly ether bonds are preferred for applications where high humidity might be encountered, despite the superior initial properties of ester-based polyurethane. Nylons that contain long hydrocarbon segments in their structure are more resistant to hydrolysis than those with shorter hydrocarbon segments. The hydrolyzable amide bonds occur along the backbone chain of nylon-11 less frequently than in nylon-6.

Biodegradation

Stabilization of polymers against attack by biological agents has not been widely studied despite the damage caused by various organisms. Although it is believed that only the naturally occurring polymers are susceptible to attack by microorganisms, certain of the synthetic polymers also undergo biodegradation (see BIODEGRADABLE POLYMERS; BIOCIDES).

Since very little is known about the chemical mechanisms by which enzymes degrade polymers, stabilization against biodegradation has been based on trial-and-error methods. Structure and formulation modification can be used within the limitations previously discussed for this approach to stabilization. Recognition of the role played by enzymes in the biodegradation of polymers has led to the use of toxins designed to poison these catalysts or at least to reduce their activity.

In addition to degradation by microorganisms, many polymers are attacked by macroorganisms. The use of repellents is the only method available for protection against macroorganisms. Metal shields offer only limited protection, since many rodents will even gnaw their way through metals.

BIBLIOGRAPHY

W. L. Hawkins, ed., *Polymer Stabilization,* Wiley-Interscience, New York, 1972.

N. Grassie, ed., *Developments in Polymer Degradation,* Vol. 3, Applied Polymer Science Publishers, London, 1981.

C. C. Davis and J. T. Blake, eds., *The Chemistry and Technology of Rubber,* Van Nostrand Reinhold Co., Inc., New York, 1937.

G. Scott, *Atmospheric Oxidation and Antioxidants,* Elsevier Science Publishing Co. Inc., Amsterdam, The Netherlands, 1965.

W. LINCOLN HAWKINS
Consultant

STAMPING. See COLD FORMING.

STATISTICAL MECHANICS

Statistical mechanics is designed to describe systems that have many degrees of freedom and a wide range of possible states. An exact classical or quantum mechanical description of the full system is usually impossible, but a great deal can be understood about the average properties of these systems by using the concepts and methods of statistical mechanics. Polymer systems are naturally adapted to be studied by statistical mechanics and may be used for readily illustrating its general principles.

Polymers are found in solutions, melts, blends (polymer liquids), and crystalline and amorphous (glassy) solids. Some features of statistical mechanics can be illustrated by considering the idealized limit of a single polymer in solution. Polymers with carbon backbones generally have freedom of rotation about the carbon–carbon single bonds. This leads to an approximation of the chain conformations in which the bond lengths and bond angles are fixed, and the freedom of rotation permits conformations about the single bonds to lie close to the potential minima of the corresponding trans and gauche configurations, as depicted in Figure 1. Since there are two gauche configurations, this common rotational isomeric model contains three allowed configurations about each bond. Although the probabilities of occurrence of trans and gauche bonds are not identical and neighboring conformations are interdependent, a single polymer that, eg, in this model, has 10^3 units would produce 3^{10^3} different possible conformations of the chain. Because the substitution of a gauche bond for a trans bond substantially changes the spatial configuration of the whole chain, these 3^{10^3} conformations have widely differing spatial monomer distributions.

To deal with the enormous amount of information that systems such as the rotational isomeric model example contain, statistical mechanics introduces the two key concepts of ergodicity and ensemble averaging.

Ergodicity and Ensembles

Experimental measurements on a single polymer chain generally involve time scales that are very long compared to the time scales of individual or correlated trans → gauche transitions. This means that the polymer undergoes an enormous number of transitions between the allowed 3^{10^3} conformations during the course of the measurement.

The ergodic hypothesis assumes that the time average, im-

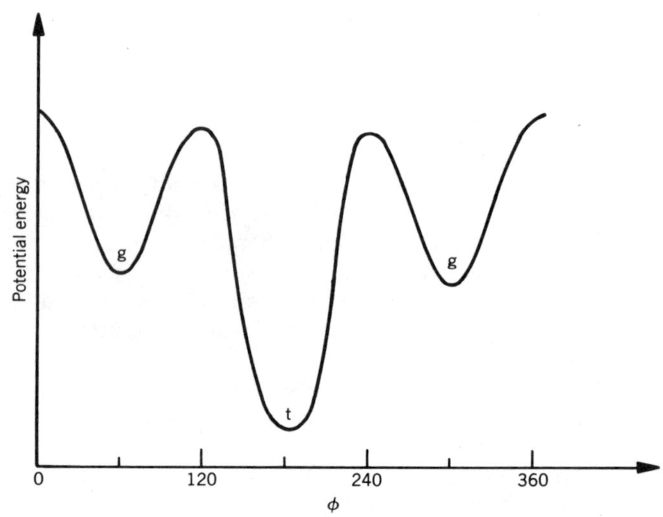

Figure 1. Schematic representation of potential energy for rotation about backbone single bonds. The minima correspond to the trans (t) and gauche (g) states.

plicit in the long time-scale measurements for a single polymer chain, is equal to the ensemble average over a set of identical polymers prepared under identical conditions. Ensemble averaging assigns a weight to each of the configurations that is proportional to the probability of occurrence of that configuration. This weight in equilibrium systems is proportional to the Boltzmann factor

$$\exp(-\beta V)$$

where $\beta = (k_B T)^{-1}$ with k_B Boltzmann's constant, T the absolute temperature, and V the potential energy for the configuration.

This introductory example is a useful prelude to a discussion of the general principles that guide equilibrium statistical mechanics. The principles are fairly straightforward in nonconstrained systems where individual configurations are assigned Boltzmann weight factors, but the presence of a large number of degrees of freedom leads to such enormous complexity in the evaluation of many measurable properties of the system that it is almost always necessary to introduce simplifying models in order to perform any useful calculations.

Polymer Models

The examples chosen here are from the equilibrium theory of polymer solutions, melts, and blends, and the theory of polymer dynamics in dilute solutions. They are from the author's research interests; a more extensive set of references may be found in the sources cited. The important areas of polymer glasses, rubbers, and crystalline polymers are not included.

Equilibrium Statistical Mechanics of Dilute Polymer Solutions

The conformational statistics of polymers in solution may, in principle, be determined from the probability distribution function $P(\{\mathbf{R}_i\})$, where $P(\{\mathbf{R}_i\})d\{\mathbf{R}_i\}$ is the probability that the positions of the monomer units lie between $\mathbf{R}_i$ and $\mathbf{R}_i + d\mathbf{R}_i$. Knowledge of $P(\{\mathbf{R}_i\})$ permits the evaluation of polymer properties such as the mean square radius of gyration $\langle S^2 \rangle$, the light-scattering intensity, the second virial coefficient, and the average dipole moment of the polymer. As an intermediate step in determining $P(\{\mathbf{R}_i\})$ from $V(\{\mathbf{R}_i\},\{\mathbf{q}_\alpha\})$, the Boltzmann weight $\exp(-\beta V)$ is first integrated over the solvent degrees of freedom $\{\mathbf{q}_\alpha\}$:

$$\begin{aligned} G(\{\mathbf{R}_i\}) &= \prod_\alpha \int d\mathbf{q}_\alpha \exp[-\beta V(\{\mathbf{R}_i\},\{\mathbf{q}_\alpha\})] \\ &\equiv \exp[-\beta U(\{\mathbf{R}_i\})] \end{aligned} \tag{1}$$

where $U(\{\mathbf{R}_i\})$ is called the potential of mean force. The polymer distribution function is then

$$P(\{\mathbf{R}_i\}) = Q^{-1}G(\{\mathbf{R}_i\}) \tag{2a}$$

and the partition function is calculated from

$$Q = \int d\{\mathbf{R}_i\}G(\{\mathbf{R}_i\}) \tag{2b}$$

All the information on the equilibrium average configurations of the polymer is contained in $P(\{\mathbf{R}_i\})$, but evaluating the properties of interest is, of course, the main problem. Even if the true $V(\{\mathbf{R}_i\},\{\mathbf{q}_\alpha\})$ were known, it would be impossible to evaluate the integral in equation 1, so models are usually introduced for $U(\{\mathbf{R}_i\})$. In addition, for high polymerization indexes n (for high molecular weights), $P(\{\mathbf{R}_i\})$ contains more information than is needed for most practical applications. For instance, the equilibrium averages of interest can be derived from reduced distribution functions. Particularly useful is the internal vector distribution $P(\mathbf{R}_i,\mathbf{R}_j)$, giving the probability density that the ith and jth monomers are at the positions $\mathbf{R}_i$ and $\mathbf{R}_j$, respectively. In the absence of external fields or confining walls, $P(\mathbf{R}_i,\mathbf{R}_j)$ depends only on the modulus of the intermonomer separation vector $|\mathbf{R}_i - \mathbf{R}_j|$. The calculation of $P(\mathbf{R}_i,\mathbf{R}_j)$ proceeds directly from $P(\{\mathbf{R}_i\}$ as

$$\begin{aligned} P(\mathbf{R}_i,\mathbf{R}_j) &= Q^{-1}\left[\prod_{k(\neq i,j)=1}^{n} \int d\mathbf{R}_k\right] \exp[-\beta U(\{\mathbf{R}_k\})] \\ &= \left[\prod_{k(\neq i,j)=1}^{n} \int d\mathbf{R}_k\right] P(\{\mathbf{R}_k\}) \end{aligned} \tag{3}$$

and an observable quantity such as $\langle S^2 \rangle$ can be evaluated from $P(\mathbf{R}_i,\mathbf{R}_j)$ through

$$\langle S^2 \rangle = (2n^2)^{-1} \sum_{i,j=1}^{n} \int d\mathbf{R}_i \int d\mathbf{R}_j \, |\mathbf{R}_i - \mathbf{R}_j|^2 P(\mathbf{R}_i,\mathbf{R}_j) \tag{4}$$

whereas the theoretically interesting mean-square end-to-end vector is

$$\langle |\mathbf{R}|^2 \rangle = \int d\mathbf{R}_1 \int d\mathbf{R}_n \, |\mathbf{R}_1 - \mathbf{R}_n|^2 P(\mathbf{R}_1,\mathbf{R}_n) \tag{5}$$

The evaluation of the integral in equation 3 is not easy for general models of the potential of mean force $U(\{\mathbf{R}_i\})$.

Models. Polymer models vary greatly in faithfulness to the structure and microscopic interactions between the monomers. All models formally separate $U(\{\mathbf{R}_i\})$ into one portion describing chain connectivity and another involving the nonbonded interactions between monomers. The latter interactions are usually modeled in terms of pairwise additive potentials, although nonpairwise additive terms cannot always be safely ignored.

The simplest models treat only chain connectivity. They, as well as most other models, describe the polymer in terms of monomer units, where $\mathbf{R}_i$ designates the position of the unit center of mass. Such a description is meaningful, provided the model is used to treat polymers on length scales that exceed a few monomer units; otherwise, the atomic structure of individual monomers also must be included. Perhaps the simplest possible model of chainlike polymers is the freely hinged chain model. As depicted in Figure 2, this model introduces structureless monomer units that are connected by bonds of fixed length l. Thus,

$$|\mathbf{R}_i - \mathbf{R}_{i+1}| \equiv l$$

but bond angles and rotations about individual single bonds are completely random. Such a model represents the configurations of the freely hinged chain as equivalent to the paths executed by a random walker with steps of equal length but arbitrary direction.

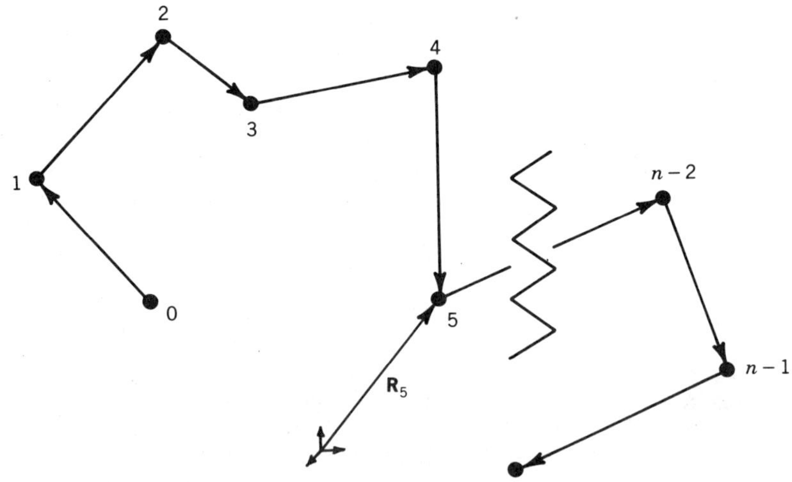

Figure 2. Random-walk model of a polymer with bonds of fixed length.

A generalization of this model is the freely rotating chain model, which includes the fixed bond angles of real molecules. A further generalization introduces potentials describing the hindered rotations about the backbone bonds of the chain. The common rotational isomeric model contains the additional realistic feature of the interdependence of neighboring rotational angles; it assigns different conditional probabilities to the successive pairs of configurations that can be generated from among the nine possibilities in polymers with gauche and trans conformations.

Excluded Volume in Polymers. Kuhn's early work on random-walk models and excluded volume constraints attempted to explain how the Mark-Houwink exponent a in the dependence

$$[\eta] = KM^a$$

of the instrinsic viscosity on the molecular weight M varies with polymer, solvent, and temperature. Since that time, considerable theoretical effort has been devoted to developing statistical mechanical treatments of the influence of excluded volume on polymer properties in dilute and concentrated solutions. A lattice model is but one example of these efforts; others are provided by continuum theories that introduce models for the potential of mean force $U(\{\mathbf{R}_i\})$ in equation 1.

Scaling and Renormalization Group Theories of Excluded Volume in Dilute and Semidilute Solutions

Polymer scientists are interested in scaling limits, where polymer properties display simple and universal power law dependences, and in the crossover regions in between. A description of crossover requires a theory more general than scaling theory; one such theory uses renormalization group methods. All these crossover regions correspond mathematically to the "critical domain" of critical phenomena as long as the molecular weight of the polymers is high enough. Consequently, the analogy between polymers and critical phenomena is useful for most of the range of polymer molecular weights and temperatures up through semidilute concentrations. (The term semidilute refers to concentrations where the chains are mutually interpenetrating.) The growing number of renormalization group calculations for polymers provides quantitative descriptions of both the scaling and crossover regions.

It is impossible in this brief account to describe the scaling and renormalization group theories in any mathematical detail, but their physical content can be summarized. Scaling theories are more heuristic. Given a model for the geometrical structure and interactions of polymers, the large-length-scale property P, as calculated from the model, would depend, in principle, on the microscopic parameters $\{\alpha\}$ of the model as well as on macroscopic variables, such as the chain concentration c, the dimensions D confining the system, and the thermal energy k_BT. This dependence may be expressed as

$$P = P(\{\alpha\}, c, D, k_BT, \ldots) \qquad (6)$$

Because scaling only treats the asymptotic limits in which the radius of gyration $R_G \to \infty$, it is convenient to scale all lengths by R_G. Similarly, all energies are scaled by k_BT. If P depends on the dimensionless quantity α/R_G with positive powers, then as $R_G \to \infty$, the dependence of P on this dimensionless parameter must disappear. On the other hand, a dependence of P on negative powers of α/R_G implies that it must remain as a relevant parameter for describing P in the scaling limits.

After this simple dimensional analysis, the fundamental assumption of scaling theories is introduced, that all dependences on microscopic parameters $\{\alpha\}$ disappear in the limit of $R_G \to \infty$ except for the unknown dependence of R_G on these $\{\alpha\}$.

Renormalization group theory can be expressed in two apparently dissimilar but equivalent formulations. The first was originally introduced in particle physics. This treatment is more mathematical in formulation and focuses on the removal of singularities that arise in perturbation expansions. The second method was developed for critical phenomena. It employs the coarse-graining concept reminiscent of the effective-segment construction of Kuhn. This method presents a physical picture in which the large-length-scale properties of systems with long-range correlations are treated as unchanged by a coarse graining of computed properties with a given model. This coarse graining is an averaging over length scales small compared with the correlation length.

Although both formulations of the renormalization group have been applied to treat polymer properties, the first is more powerful and convenient for treating crossovers and calculating polymer-distribution functions. The second method provides a more physical interpretation of the renormalization process.

Polymer Dynamics in Solution

The viscosity and diffusion coefficients are important large-length-scale properties of polymer solutions; they have been the subject of extensive theoretical investigations. In fact, attempts to explain the molecular weight dependence of the intrinsic viscosity,

$$[\eta] \equiv \lim_{c \to 0} \frac{(\eta_s - \eta_0)}{c\eta_0} \tag{7}$$

where η_s is the solution viscosity, η_0 the solvent viscosity, and c the polymer mass concentration, have been at the center of the development of modern models of polymer structure and dynamics. An analysis of experimental data for $[\eta]$ of many low molecular weight polymers led to the Staudinger law $[\eta] \propto M$. Staudinger believed that this law applied to all polymers because of their postulated rodlike configurations in solution. However, Kuhn and other workers obtained different results by introducing random-flight models for the polymer configurations, elementary concepts of excluded volume, and simple models of the viscosity.

The solvent plays a passive role in the statistical mechanics of equilibrium polymer properties through its implicit presence in the potential of mean force $U(\{\mathbf{R}_i\})$ in equation 1, a presence that is felt in the excluded volume models through a solvent dependence of the z parameter that describes the strength of excluded volume. The solvent must, however, play an active role in polymer dynamics because the motion of a single monomer or group of monomers in a polymer requires that solvent molecules take their place. This type of concerted motion is obviously very complicated at the molecular level, but such a detailed description is unnecessary for evaluating a large-length-scale polymer property like $[\eta]$. Again, it should be permissible to introduce a course-grained model, where averages are applied over distances large compared with monomer or solvent dimensions, but small compared to the polymer (small compared to R_G). The solvent is described on such a coarse-grained length scale by continuum hydrodynamics. Models include the Kirkwood-Riseman model, and the Rouse-Zimm model.

Polymers in Concentrated Systems

Concentrated polymer solutions, melts, and blends are examples of polymer systems in which the volume fraction of polymers approaches unity. Consequently, the nonzero volume of monomers introduces important excluded volume constraints that, for instance, limit the maximum density of the system and that must be incorporated in any statistical mechanical model. The simple model, which is so useful for describing polymer-excluded volume in dilute and semidilute solutions, is inapplicable to dense polymer systems because it does not explicitly contain the hard-corelike interaction between monomers. Consequently, other models are required for concentrated systems.

Treating a single polymer with excluded volume would, at first sight, suggest a highly complicated situation for dense polymer systems where more polymers exclude each other and where the hard-core interactions must introduce additional complexities. The actual situation in the dense polymer systems is qualitatively different, owing to the extended nature of the polymer chains. Polymers at semidilute and higher concentrations are interpenetrating; a single polymer chain thus experiences excluded volume interactions with a large number of other polymers. Under such circumstances, it is permissible to consider average properties of these strongly interacting systems because the fluctuations about the average are small compared to the averages. In concentrated systems, a single chain is embedded in the surroundings of its neighboring polymers, and the overall polymer volume fraction is fairly uniform throughout the system. Because a pair of strongly interacting polymer segments in concentrated systems cannot discern whether they belong to different polymer chains or to distant portions of the same polymer chain, Flory predicted that the average dimensions of polymers in the melt must be governed by Gaussian statistics, despite the obvious fact that much more polymer volume is excluded in the melt than in dilute solutions; this prediction was subsequently verified.

An important concept is the small role of fluctuations in many properties of concentrated polymer systems and the consequent good approximation provided by considerations of average characteristics of the interactions, polymer densities, and other variables. These features permit the use of mean-field theories in statistical mechanics, which focus on the replacement of intricate microscopic interactions by average interactions. However, mean-field theories encompass a wide range, from crude qualitative to detailed models, with significant predictive powers.

BIBLIOGRAPHY

H. Yamakawa, *Modern Theory of Polymer Solutions,* Harper & Row, New York, 1971 and references therein.

P. J. Flory, *Statistical Mechanics of Chain Molecules,* Wiley-Interscience, New York, 1969 and references therein.

K. F. Freed, *Renormalization Group Theory of Macromolecules,* John Wiley & Sons, Inc., New York, 1987.

M. Doi and S. F. Edwards, *The Theory of Polymer Dynamics,* Clarendon Press, Oxford, UK, 1986 and references therein.

KARL F. FREED
University of Chicago

STATISTICAL THERMODYNAMICS

Objective and Mode of Thinking

Statistical thermodynamics is a discipline that tries to compute macroscopic properties of materials from more basic structures of matter. These properties are not necessarily static properties as in conventional statistical mechanics. The problems in statistical thermodynamics fall into two categories. The first category involves the study of the structure of phenomenological frameworks and the interrelations among observable macroscopic quantities. The second category involves the calculation of the actual values of phenomenological parameters appearing in the phenomenology, such as viscosity or phase transition temperatures, from more microscopic parameters.

The fundamental wisdom of statistical thermodynamics can be summarized as follows. Understanding general relations requires only a model specified by fairly broad and abstract conditions. Realistically detailed models are not needed to understand general properties of a class of materials. Understanding more specific relations, on the other hand, requires microscopically detailed models.

Equilibrium Problems

The main theoretical framework of the static part of statistical thermodynamics is the statistical mechanics of Gibbs and Boltzmann. Gibbs statistical mechanics has not been convincingly derived from underlying mechanics.

The universal behavior of macroscopic objects can be described phenomenologically with the aid of thermodynamics or, more precisely, thermostatics, with five laws (axioms): the existence of thermal equilibrium and temperature (the zeroth law); the conservation of energy (the first law); the variational principle for entropy (the second law); the inaccessibility of the absolute zero temperature (the third law); and the clear distinction between extensive and intensive quantities (sometimes called the fourth law). The equilibrium states of macroscopic objects comprise a universality class characterized by equilibrium thermodynamics. Thermodynamics cannot give any equation of state of matter, nor explain, eg, why the fourth law is valid for macroscopic systems.

All macroscopic observables are obtainable from the distribution function of microscopic states of elements (in the classical case, the density distribution function on the phase space in which each microscopically distinct state corresponds to a distinct point). Strictly speaking, this is the most fundamental and the most important assumption in statistical thermodynamics. We can compute Helmholtz free energy from the canonical partition function. This is one of the key relations in equilibrium statistical thermodynamics.

Nonequilibrium Problems

Most likely there is no universality class that can be called the macroscopic nonequilibrium class. However, a well-defined class can be recognized that may be called the linear nonequilibrium class. Its phenomenological framework is the so-called linear nonequilibrium thermodynamics, which contains several fundamental laws: the second law in terms of entropy produced inside the system; phenomenological linear force-flux relations; and the Onsager reciprocity stating the symmetry relations among kinetic coefficients.

Since there is no reliable general phenomenological framework for nonlinear nonequilibrium phenomena, a statistical thermodynamical framework cannot be constructed according to any guiding principle. Thus the basis of nonlinear nonequilibrium statistical studies is less solid than that of equilibrium statistical thermodynamics. However, the study of dynamical critical phenomena reveals a class that may be called the nonlinear Langevin class.

There are three levels of description of nonequilibrium systems. At the microscopic level a dynamical (mechanical) description is used, and at the macroscopic level, the hydrodynamic description. The latter description contains no stochastic element, which is why it is called hydrodynamic. At the intermediate level the kinetic description is used, where only a few variables (fluid velocity, order parameter, etc, are needed) to describe the system. In nonequilibrium statistical physics the dynamics of the system can be separated into the systematic part in terms of more or less phenomenological variables and the stochastic noises excluded from the systematic part. The projection operator method is the fundamental tool used to accomplish this separation. Thus, in principle, the kinetic level description may be derived from mechanics. Since this description is coarse grained compared to the mechanical description, a stochastic description is obtained often in terms of nonlinear Langevin equations (differential equations with stochastic elements, ie, noise, are called Langevin equations). Whether the resultant Langevin equation is useful or not depends critically on the choice of variables. If the separated noise is almost white, ie, the correlation time is negligibly small, the resultant kinetic level description is useful.

BIBLIOGRAPHY

R. Kubo, H. Ichimura, T. Usui, and N. Hashitsume, *Statistical Mechanics,* North Holland, Amsterdam, 1965.

H. B. Callen, *Thermodynamics,* Interscience Publishers, New York, 1960.

R. Kubo, H. Ichimura, T. Usui, and N. Hashitsume, *Thermodynamics,* North Holland, Amsterdam, The Netherlands, 1968.

R. Kubo, M. Toda, and N. Hashitsume, *Statistical Physics II Nonequilibrium Statistical Mechanics,* Springer-Verlag, New York, 1985.

Yoshitsugu Oono
University of Illinois

STEP-REACTION POLYMERIZATION

Step-reaction polymerization is one of the two principal classifications of the mechanism of the growth reaction in polymerization. The other class is chain-reaction polymerization. Equivalent terms are either step polymerization or step-growth polymerization for the former and either chain polymerization or chain-growth polymerization for the latter.

In step-reaction polymerization the growth reaction proceeds by combinations of monomer, oligomer, or polymer molecules with other monomer, oligomer, or polymer molecules. The consumption of reactive groups is a prerequisite for polymer formation in step-reaction polymerization, whereas in chain-reaction polymerization the reactive groups are continuously regenerated during the growth process. In step-reaction polymerization, average molecular weight increases continuously with monomer consumption.

Each polymer chain grows at a relatively slow rate over a much longer period of time in step-reaction polymerization compared to a chain-reaction polymerization, where there is rapid growth of high molecular weight polymer. There is generally only one type of reaction in step-reaction polymerization, and the same basic mechanism is involved in the reactions of either two monomer units with each other, or a monomer with the end group on a polymer chain, and/or two polymer-chain end groups with each other.

In this manner, all polymer chains start growing at approximately the same time and continue to grow at the same relatively slow steady rate, one step at a time, as indicated in Figure 1 by a schematic representation of the reactions occurring in a small region of the polymerization medium.

The random or haphazard growth of the polymer permits only a very gradual increase in molecular size. For attainment

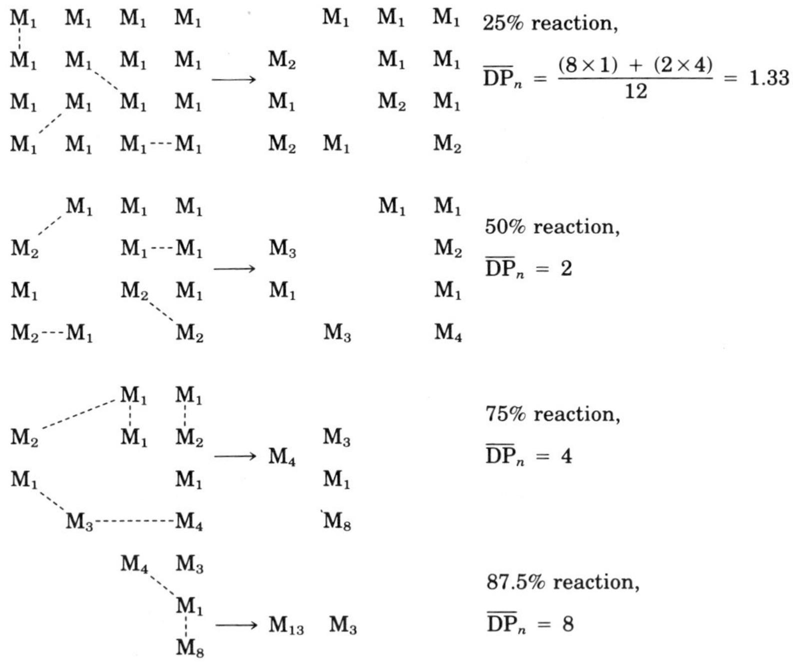

Figure 1. Representation of step-reaction polymerization. The notation $\overline{DP}_n$ signifies the number-average degree of polymerization of all molecules in the products at each point in the reaction, as indicated by the percent conversion. The subscript to the symbol M indicates the degree of polymerization of that particular molecule; eg, M_1 is a monomer unit and M_4 is a tetramer.

of quantitative conversions and, therefore, high molecular weight polymers, the products of each of the localized areas must combine. For the overall polymerization reaction, the relationship between $\overline{DP}_n$ and reaction conversion p for the preparation of linear polymers is

$$\overline{DP}_n = 1/(1 - p)$$

This simple equation dictates the rigorous requirements that must be met by the basic reaction of the step-growth polymerization in order to attain a reasonable degree of polymerization. According to this limiting equation, a step-growth polymerization reaction that has gone to 95% completion forms a polymer with only 20 repeating units in the average chain. The linear polymer that shows interesting physical and mechanical properties at a $\overline{DP}_n$ of 20 is rare; a $\overline{DP}_n$ of 50 is a minimum for all but the most polar of repeating units. Nevertheless, the achievement of a $\overline{DP}_n$ of 50 requires a reaction conversion of 98%. Very few reactions can be forced to and beyond 98%, and yet this requirement is the first characteristic that should be considered when a step-growth polymerization is proposed. Other requirements must also be met within very narrow limits, particularly an exact equivalence of functional groups, an absence of side reactions, high monomer purity, reasonably high reaction rates, and little or no tendency toward ring formation by the bifunctional monomers involved.

There are two notable exceptions to this relationship between degree of polymerization and conversion; one is for step-growth polymerization reactions carried out by interfacial polycondensation and the other is for polymerization reactions involving monomers whose average functionality is greater than two. In both cases, very high molecular weight polymers can be obtained at relatively low conversions.

BIBLIOGRAPHY

R. W. Lenz, *Organic Chemistry of Synthetic High Polymers,* Wiley-Interscience, New York, 1967.

D. H. Solomon, ed., *Step-Growth Polymerizations,* Marcel Dekker, Inc., New York, 1972.

A. Schoberl, *Makromol. Chem.* **37,** 64 (1967).

R. W. Lenz, "Polymerization Mechanisms and Processes" in M. Grayson, ed., *Encyclopedia of Chemical Technology,* 3rd ed., Vol. 18, John Wiley & Sons, Inc., New York, 1982, pp. 722–727.

STEREOREGULAR LINEAR POLYMERS

A stereoregular polymer, according to the International Union of Pure and Applied Chemistry (IUPAC) definition, is a polymer "the macromolecules of which can be described in terms of only one species of stereorepeating unit, in a single sequential arrangement." A stereorepeating unit is a configurational unit having a defined configuration at all sites of isomerism in the main chain of a polymer molecule. A complete nomenclature and definitions concerning the stereoregular polymers are reported in the literature.

Stereoregular polymers include both natural (*Hevea* rubber, cellulose, fibroin of silk, etc) and synthetic (isotactic polypropene, *cis*-1,4-polybutadiene, etc) polymers. Only the latter will be considered in this article.

In a synthetic stereoregular polymer some "errors" are normally present in the polymer molecule. The errors can be both of a chemical nature (head-to-head linking, isomerized monomer units) and of a steric nature at the sites of stereoisomerism. Each synthetic stereoregular polymer is therefore characterized by some degree of stereoregularity. The nomenclature recommended applies to the predominant structural features of the real polymer molecule. Terms such as "highly isotactic," "high cis," and "almost completely syndiotactic," although imprecise, are currently in use.

The degree of stereoregularity of a given polymer may be determined by extraction with solvents; measurement of crystallinity by the x-ray diffraction technique; measurement of the absorbance bands related to stereoregularity or crystallinity; and H- and ^{13}C-nmr.

At present, for the determination of the degree of stereoregularity, the ^{13}C-nmr techniques are the most widely used (see NUCLEAR MAGNETIC RESONANCE). The information given is limited to sequences of a few monomeric units, up to nonads in the case of polypropylene (see also MICROSTRUCTURE).

In particular, nmr techniques allow the determination of the content of pairs of monomeric units (diads) and of three adjacent units (triads) arranged in one of the following ways:

isotactic (meso) diads (*m*)
syndiotactic (racemic) diads (*r*)
isotactic triads (*mm* or *i*)
syndiotactic triads (*rr* or *s*)
heterotactic triads (*mr* or *h*)

The degree of stereoregularity (eg, isotactic or syndiotactic) of a given polymer is the percent amount of diads, triads, etc, of the same type present in the polymer.

Some synthetic polymers, which contain two or more sites of stereoisomerism in each monomeric unit, show regularity at only one site. Although they cannot rigorously be regarded as stereoregular, they have been included in this article.

Several classes of monomers have been polymerized to stereoregular polymers: unsaturated compounds, ie, 1-alkenes, conjugated dialkenes, aldehydes, and vinyl monomers containing heteroatoms; and saturated cyclic monomers such as epoxides, episulfides, β-lactones, *N*-carboxyanhydrides, cyclic acetals, and formals. Stereoregular copolymers have also been prepared.

Catalysts or initiators used in these polymerizations depend on the nature of the monomer. Most of the hydrocarbon monomers are polymerized with Ziegler-Natta catalysts (qv), which are prepared from a transition-metal compound and an organometallic compound of a metal belonging to one of the first three groups of the periodic system. Some of these catalysts act in the solid phase (such as $TiCl_3$–$Al(C_2H_5)_3$ used in the polymerization of 1-alkenes), and others in solution (such as $CoCl_2$–$Al(C_2H_5)_2Cl$ used in the polymerization of butadiene to *cis*-1,4-polybutadiene).

Several monomers containing heteroatoms (eg, vinylpyridine, acrylates, aldehydes, etc) arc polymerized in a stereospecific way by organometallic compounds, such as LiC_4H_9, Grignard compounds, etc, acting through an anionic mechanism, or by cationic-type catalysts, such as $AlBr_2C_2H_5$, or complexes, such as $BF_3{\cdot}O(C_2H_5)_2$, used in the polymerization of vinyl ethers. In several cases these catalysts are soluble.

At least partially stereoregular polymers of some monomers can be obtained by free-radical polymerization or by radical polymerization of monomers included in clathrates (see POLYMERIZATION IN CLATHRATES).

The factors that determine the formation of stereoregular polymers vary with the type of catalyst or initiator used and in several cases have not been fully clarified.

Stereoisomerism in Polymers

Unsaturated Monomers. ***Vinyl Monomers.*** A monomer of the $CH_2{=}CHR$ type, by opening its double bond, can yield two enantiomeric monomer units (Fig. 1). The head-to-tail arrangement of the units, all of which have the same steric configuration (I or II), produces an isotactic polymer (Fig. 2**a**). The regular head-to-tail arrangement of monomer units having alternately each of the two enantiomeric configurations I and II yields a syndiotactic polymer (Fig. 2**b**). The irregular arrangement of units of the two types (I and II) yields an atactic polymer (Fig. 2**c**).

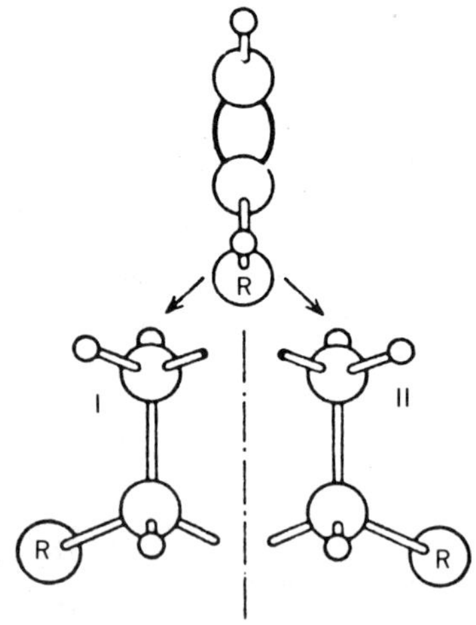

Figure 1. Enantiomeric units derived from a vinyl monomer $CH_2{=}CHR$.

Polymers containing two distinct series of tertiary carbon atoms alternating with each other are called hemitactic: in one series the arrangement of the substituents is exactly defined, whereas in the other series their arrangement is random. In terms of microtacticity a hemiisotactic polymer consists of a statistical succession of the *mm* and *rr* triads, whereas a hemisyndiotactic polymer consists of statistical successions of nonoverlapping *mr* and *rm* triads.

The chain conformation of most of the stereoregular vinyl polymers in the crystalline state is a helix.

Aldehydes. The same stereoisomerism phenomena exhibited by polymers of vinyl monomers are also found in polymers of

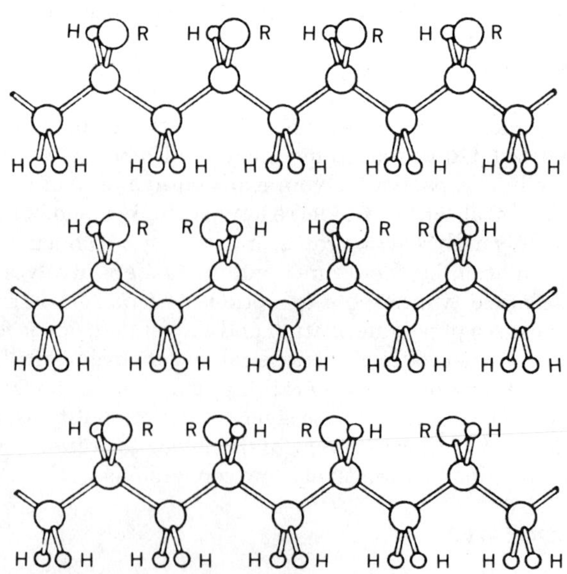

Figure 2. Configuration of polymer chains of $CH_2{=}CHR$ monomers (in a flat zigzag form). (**a**) Isotactic, (**b**) syndiotactic, (**c**) atactic.

aldehydes (except for formaldehyde); in fact, by opening the C=O bond, two enantiomeric units may be obtained.

1,2-Disubstituted Ethylene Monomers. Opening the double bond of a monomer of the CHR=CHR′ type can yield four types of monomeric units, eg, two pairs of enantiomers (Fig. 3). The head-to-tail arrangement of units of the same type yields diisotactic polymers. These may be threo-diisotactic, a succession of units having configuration I or II (Fig. 3), or erythro-diisotactic, a succession of units having configuration III or IV. Chains of these polymers in the crystalline state assume a helical conformation. The regularly alternating head-to-tail arrangement of a pair of enantiomeric units (eg, units having configurations I, II, I, II, etc, or III, IV, III, IV, etc) gives rise to disyndiotactic polymers.

Conjugated Dialkenes. Isomerism phenomena in conjugated dialkene polymers are more complex than those observed in monomers having only one double bond. Three types of monomer units may be obtained from butadiene, the simplest conjugated dialkene (Fig. 4). Therefore, four stereoregular polymers of butadiene can exist: isotactic 1,2; syndiotactic 1,2; cis-1,4; and trans-1,4. All four types of polymer have been prepared.

The stereoisomerism phenomena of isoprene and other substituted butadiene polymers are even more complex.

Saturated Cyclic Monomers with Heteroatoms. ***Monocyclic Monomers.*** Higher 1,2-epoxides and higher 1,2-episulfides that contain at least one asymmetric carbon atom, eg, propylene oxide, give rise to stereoregular polymers. Other monomers belonging to this class are β-lactones, *N*- and *S*-carboxyanhydrides, and some substituted cyclic formals, which give monomeric units containing an asymmetric carbon atom; consequently, iso-, syndio-, or heterotactic-type polymers can be obtained.

Bicyclic Monomers. These monomers, eg, some acetalic derivatives, give monomeric units containing asymmetric carbon atoms and also residual rings. Starting from some of these monomers, polysaccharides exhibiting the typical stereoisomerism phenomena for these substances can be obtained (see POLYSACCHARIDES).

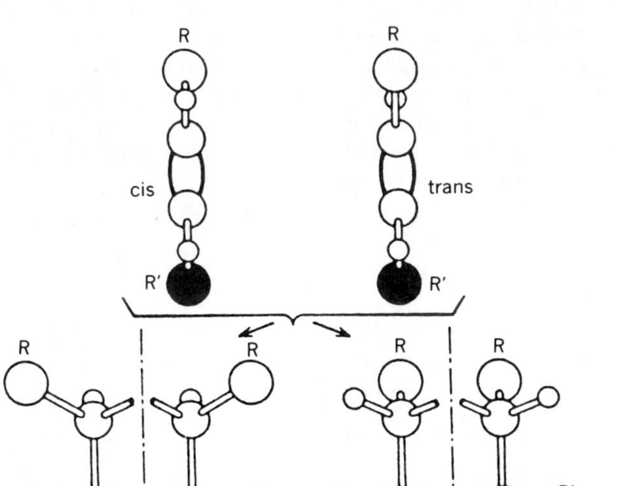

Figure 3. Monomer units derived from 1,2-disubstituted ethylene monomer CHR=CHR′.

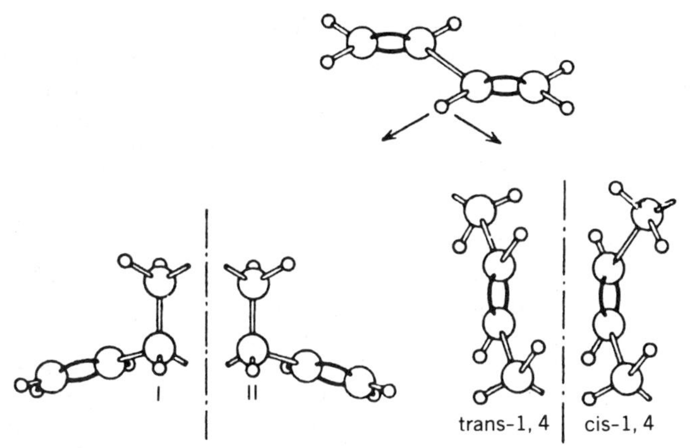

Figure 4. Monomer units derived from butadiene: the pair of enantiomeric units resulting from 1,2 openings and the cis and trans isomeric units resulting from 1,4-opening.

POLYMERS PREPARED WITH ZIEGLER-NATTA CATALYSTS AND OTHER IONIC-TYPE CATALYSTS

Polymers from Unsaturated Hydrocarbon Monomers

Isotactic Polymers of Aliphatic 1-Alkenes. The first isotactic polymers of 1-alkenes were obtained by Natta, Pino, and Mazzanti in 1954. As a general rule, all 1-alkenes either linear or branched, aliphatic or alicyclic, may be polymerized to isotactic polymers. However, the reactivity of the double bond decreases drastically as the size of the pendent group (R in CH_2=CH—R) increases and as the branching in R moves close to the double bond; it is completely inhibited when the carbon atom in the 3-position is fully substituted. Some of these polymers and their melting temperatures are shown in Table 1. All these polymers have a helical chain conformation in the crystalline state. Isotactic polypropylene is the most important among them from a practical standpoint and is also the most extensively studied (see PROPYLENE POLYMERS).

Catalytic Systems. *Unsupported Heterogeneous Catalysts.* The most important heterogeneous catalytic systems originally used

Table 1. Isotactic Polymers of 1-Alkenes

Monomer	Mp[a], °C
(ethylene)	(138)
propylene	177
1-butene	136
1-pentene	80
1-hexene	−55
1-heptene	40
1-octadecene	71
3-methyl-1-butene	ca 300
4-methyl-1-pentene	ca 235
4,4-dimethyl-1-pentene	>320
vinylcyclopropane	228–230
vinylcyclohexane	ca 370
3-methyl-1-vinylcyclohexane	276–355
4-methyl-1-vinylcyclohexane	[b]
2-methyl-1-vinylcyclohexane	[c]

[a] Melting point of the polymer obtained by conventional methods, by subjecting the sample to one slow annealing.
[b] Slightly crystalline polymer.
[c] Insoluble, probably cross-linked.

for the polymerization of aliphatic and alicyclic 1-alkenes to isotactic polymers contain a halide of a transition metal of the fourth or fifth group of the periodic table (eg, titanium or vanadium) and an organometallic compound of a metal belonging to one of the first three groups of the periodic table (eg, lithium, beryllium, or aluminum). The first commercial system was α-$TiCl_3$ + $Al(C_2H_5)_3$. Properties of several of these systems have been reviewed.

High Activity, Supported Catalysts. Numerous improvements of the traditional Ziegler-Natta catalysts were insufficient to simplify polypropylene production in the most common commercial processes based on δ-$TiCl_3$ and $Al(C_2H_5)_2Cl$. This goal was achieved when high activity, supported catalysts were discovered. These catalysts consist mainly of titanium halides, alkylaluminums, and Lewis bases supported on activated $MgCl_2$.

Soluble Catalysts. Isotactic polymers of propylene and 4-methyl-1-pentene were obtained for the first time with soluble catalysts by means of Ti and Zr benzyl and allyl derivatives. High activity homogeneous catalysts are based on oligomeric methylaluminoxane and Ti, Zr, or Hf metallocene derivatives.

Polymerization Mechanism. Polymer chain growth is believed to take place by the repeated insertion of the alkenic double bond into a previously formed transition metal–carbon bond (see INSERTION POLYMERIZATION). The simplest and most widely accepted mechanistic scheme for propylene polymerization with the $TiCl_3$-based Ziegler-Natta catalysts implies the reversible addition of an alkene molecule to a coordinatively unsaturated octahedral transition-metal complex with the formation of a coordinatively saturated π complex. Cis migration of the alkyl group (previously attached to the transition-metal atom) to the π-alkene ligand, with π–σ rearrangement, gives a new coordinatively unsaturated complex that is able to add another alkene ligand. Through a series of similar steps, a high molecular weight isotactic polymer chain is produced. There is some disagreement as to whether the σ-bonded ligand migrates to the site of the π-bonded alkene or whether the reaction takes place at the coordinate site of the σ-bonded ligand.

The stereochemistry of polymerization can be controlled by the asymmetry of the catalytic center or by the asymmetry of the methine carbon of the last inserted monomer unit.

Stereochemical control by the intrinsic chirality of the catalytic centers seems to occur in highly isospecific alkene polymerization initiated either by heterogeneous catalysts or by homogeneous catalysts based on chiral *rac*-ethylene-bridged, bis-indenyl (or tetrahydroindenyl) Zr(Ti) dichloride and oligomeric methylaluminoxane. This assumption is supported by experimental evidence.

Stereoselectivity and Stereoelectivity-Optically Active Polymers. The polymerization of racemic 1-alkenes containing asymmetric carbon atoms by means of unsupported and supported heterogeneous Ziegler-Natta catalysts and homogeneous isospecific catalysts based on 1,2-ethylene-bis-tetrahydroindenylzirconium dichloride and polymeric methylaluminoxane has been thoroughly investigated (see OPTICALLY ACTIVE POLYMERS). All isospecific catalysts studied are highly stereoselective; that is, in the presence of racemic 1-alkenes, polymers are obtained that can be separated into fractions having optical activity of opposite sign. The separation has been achieved by elution on an optically active polymeric crystalline support. Each macromolecule consists of units derived largely from the same enantiomer.

The use of organometallic compounds containing optically active groups as cocatalysts or the presence in the catalyst system of an optically active Lewis base allows the stereoelective polymerization of racemic 1-alkenes. In this process, an optically active polymer is obtained, and the optically active unreacted monomer is recovered, thus suggesting that the monomer antipode is preferentially polymerized. Only racemic alkenes with an asymmetric carbon atom in the α- or β-position to the double bond have been stereoelectively polymerized.

In general, stereoelectivity is lower than stereoselectivity, the enantiomeric prevalence in the polymer being less than 20%. Another approach to stereoelective polymerization is the copolymerization of an optically active 1-alkene (R_1^*) with another racemic 1-alkene (R_2 + S_2).

Syndiotactic Polypropylene. Syndiotactic polypropylene is the only crystalline syndiotactic poly(1-alkene) known at present (see PROPYLENE POLYMERS). Syndiotactic polypropylene has a binary helix conformation in the solid state. A crystalline modification in which the macromolecules have a planar zigzag conformation is also known.

Catalytic systems for the synthesis of syndiotactic polypropylene are prepared and utilized at low temperature and are based on soluble vanadium compounds, eg, VCl_4; $VOCl_3$; $V(acac)_3$, where acac = 2,4-pentanedionate (acetylacetonate); $V(mbd)_3$, where mbd = 1-methyl-1,3-butanedionate; and aluminum dialkyl monochlorides. These catalysts are homogeneous.

The polymerization rate, stereoregularity, and molecular weight of the polymer are markedly influenced by several factors: the polymerization temperature at which the catalyst was prepared, the nature of the vanadium and aluminum compounds, the ratio between the components of the catalyst system, the nature of the polymerization solvent (aliphatic or aromatic), the nature and amount of Lewis bases added, and the polymerization time.

Syndiotacticity is believed to be due to both the cis addition and the secondary insertion of monomer into the double bond. In this case the stereospecificity probably is controlled by the chirality of the last carbon atom of the growing chain, whose configuration influences directly the chirality of the catalytic intermediate complex, which in turn determines the configuration of the entering monomer unit.

The above conclusions explain the microstructure of ethylene–propylene copolymers obtained with syndiospecific catalysts. These products contain both meso (*m*) and racemic (*r*) structures when the sequence of propylene units is interrupted by isolated ethylene units.

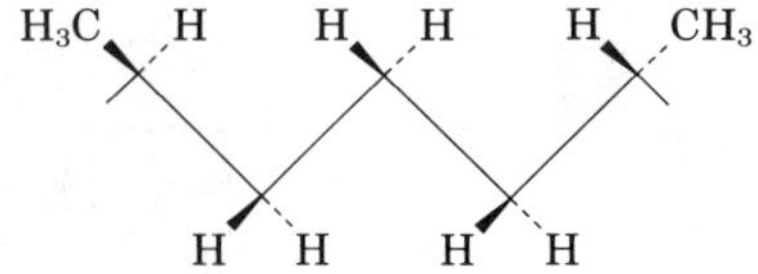

r

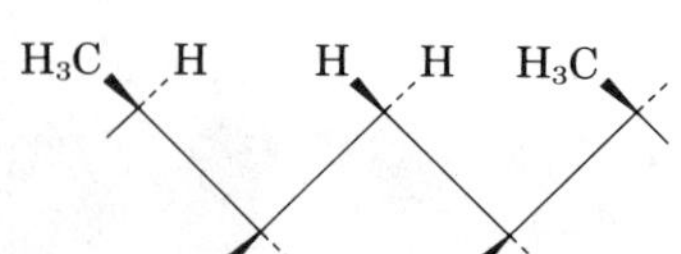

m

The existence of a steric interaction between the entering monomer molecule and the last added monomer unit is confirmed by the fact that the propagation rate for secondary insertion of propylene in syndiospecific polymerization is lower than that for primary insertion in nonstereospecific polymerization.

In a model proposed for the active center, the V^{3+} atom is pentacoordinated. Its ligands include three chlorine atoms (two of which are bridge-bonded to the aluminum atom), the asymmetric carbon atom of the last monomer unit of the growing chain, and the alkene molecule. Prior to alkene coordination and after its insertion, the vanadium atom is tetracoordinated.

In an alternative model two chlorine atoms are substituted by a bidentate dionate, and a chlorine atom is bridge-bonded to the Al in dimeric AlR_2Cl.

Hemiisotactic Polypropylene. Hemiisotactic polypropylene has been prepared by hydrogenation of isotactic *trans*-1,4-poly(2-methylpentadiene) with *p*-toluenesulfonyl hydrazide. The hemiisotactic structure of the polymer has been demonstrated by examination of the methyl region in the ^{13}C-nmr spectrum.

Isotactic Polymers of Vinylaromatic Monomers. Vinylaromatic monomers can be polymerized to isotactic polymers by the same Ziegler-Natta catalyst systems that are active in the isospecific polymerization of 1-alkenes. Partially isotactic polystrenes can also be obtained with homogeneous catalytic systems consisting of C_4H_9Li solution in toluene used at $-40°C$, with heterogeneous catalysts based on $C_5H_{11}Na$–NaCl–hexane (used at $-40°C$), or with allylsodium–sodium isopropoxide (1:1)-NaCl–hexane (Alfin catalyst, used at $-20°C$). No exact parallel exists between the catalyst requirements for stereospecific polymerization of vinylaromatic monomers and for that of 1-alkenes. For example, propylene polymerization with the systems based on $TiCl_4$ and $Al(C_2H_5)_3$ is much less stereospecific than polymerization with the systems based on violet $TiCl_3$. Such a difference is not observed in the polymerization of styrene; in both cases highly isotactic polystyrene can be obtained. This indicates that the stereospecific polymerization is more facile with vinylaromatic monomers than with 1-alkenes.

In these polymers high degrees of stereoregularity do not always correspond to high degrees of crystallinity. The chain conformation of crystalline isotactic polymers of vinylaromatic monomers is the same as that reported for other polymers of vinyl monomers (see also STEREOREGULAR STYRENE POLYMERS).

All the results obtained suggest that a given catalyst system has essentially the same stereospecificity for all the substituted styrene monomers examined. Some polymers obtained from these monomers are listed in Table 2; the melting temperatures of the crystallizable products are also reported. The molecular weights of these polymers are usually very high, even on the order of several million.

The kinetic characteristics of the polymerization processes of vinylaromatic monomers with Ziegler-Natta catalysts are similar to those of the isospecific polymerization of 1-alkenes. The chemical mechanism of the isotactic polymerization of vinylaromatic monomers also is of the same type as that of 1-alkenes.

Syndiotactic Styrene Polymers. Syndiotactic polystyrene has been obtained in the presence of a catalyst containing a titanium compound and an organoaluminum compound used at 50–60°C. This new polymer, with an average number molecular weight of 82,000, is insoluble in methyl ethyl ketone (MEK) and melts at 270°C. A planar zigzag conformation has been suggested for this polymer.

Polymethylstyrenes with a predominantly syndiotactic structure (the highest value of the racemic diad amount is 81%) have been obtained with n-C_4H_9Li in toluene at -25 to $+30°C$. The degree of syndiotacticity increases in the order *p*-, *m*-, and *o*-derivatives. With the same catalyst system, syndiotactic polystyrene can be obtained with racemic diad content not higher than 70%.

n-C_4H_9Li in tetrahydrofuran (THF) at $-75°C$ also polymerizes *n*-propyl *p*-vinylbenzene sulfonate. The polymer obtained can be hydrolyzed to poly(styrene sulfonic acid), having a molecular weight of approximately 4000–5000. The structure proposed for the polymer is syndiotactic.

α-Methylstyrene and some of its derivatives yield syndiotactic polymers. The polymerization may be initiated by catalysts of the cationic type, eg, $BF_3 \cdot O(C_2H_5)_2$, $TiCl_4$, $AlCl_3$, $SnCl_4$, and triphenylmethyl salts of various metals, as well as by anionic initiators, eg, n-C_4H_9Li.

The $Al(C_2H_5)_3$–H_2O–CH_3OCH_2Cl system at $-78°C$ in CH_2Cl_2 is also active in polymerization of α-methylstyrene. For this system too a cationic-type polymerization mechanism has been proposed. Ziegler-Natta catalyst systems (eg, $Al(C_2H_5)_3$–$TiCl_4$) give poorly stereoregular poly(α-methylsytrene)s. With the Friedel-Crafts catalysts the syndiotacticity of poly(α-methylstyrene) increases with decreasing polymerization temperature. The degree of stereoregularity of poly(α-methylstyrene) is also dependent on the type of catalyst (eg, $BF_3 \cdot O(C_2H_5)_2$, $TiCl_4$, or $AlCl_3$) and the polarity of the solvent.

In general, the polymers prepared with cationic catalysts are more stereoregular than those obtained with anionic ones, but in both cases the steric hindrance at the propagating chain end must be the main factor controlling the stereospecific polymerization.

Isotactic Polymers of 1-Alkenes Containing Aromatic End Groups. Ziegler-Natta systems based on titanium or vanadium halides and organometallic compounds of aluminum or lithium are also active in the polymerization of 1-alkenes containing an aromatic end group. Crystalline or crystallizable polymers of allylbenzene, 4-phenyl-1-butene, and 5-phenyl-1-pentene have been prepared. Other monomers have been polymerized: 3-phenyl-1-butene; *o*-, *m*-, and *p*-allyltoluene; 2-allyl-*p*-xylene; 4-allyl-*o*-xylene; 5-allyl-*m*-xylene; 4-(*o*-tolyl)-1-butene; 4-(*p*-tolyl)butene; 9-allylanthracene; 4-phenyl-1-hexene; and 5-phenyl-1-heptene.

These polymers are usually isotactic, but in some cases syndiotactic products can be obtained.

Cycloalkene Polymers. Several cycloalkenes can be polymerized to linear polymers using catalysts based on transition metals. Some of these polymers have a stereoregular structure. The polymerization of cycloalkenes having a highly strained ring, such as cyclobutene and norbornene, is in some aspects different from that of higher homologues, such as cyclopentene, cyclooctene, etc.

Strained-ring Cycloalkenes. *Monomers Containing a Cyclobutene Ring.* Cyclobutene and its derivatives can give three types of monomeric units (**1**)–(**3**):

Table 2. Isotactic Vinylaromatic Polymers

Monomer		Physical state of the polymer	Polymer mp[a], °C
Compound	Formula		
styrene	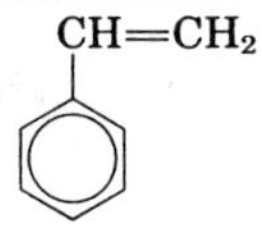	crystalline	240
o-methylstyrene	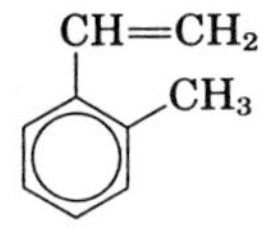	crystalline	350
m-methylstyrene	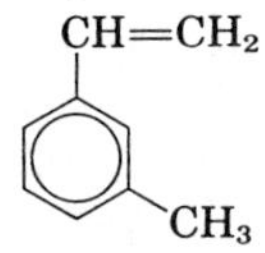	crystalline	215
p-methylstyrene	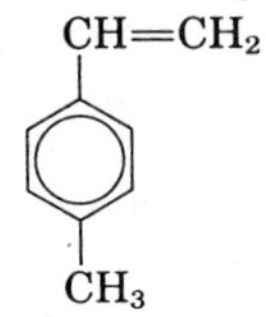	amorphous	
2,4-dimethylstyrene	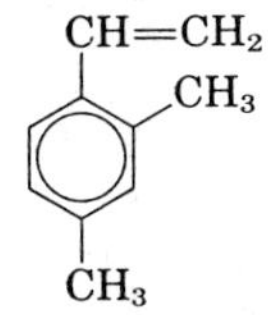	crystalline	310
2,5-dimethylstyrene	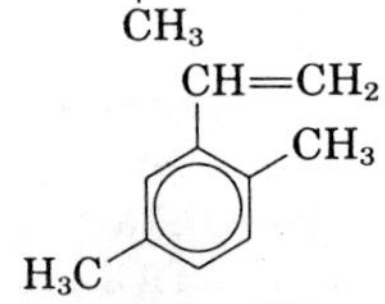	crystalline	330
3,4-dimethylstyrene	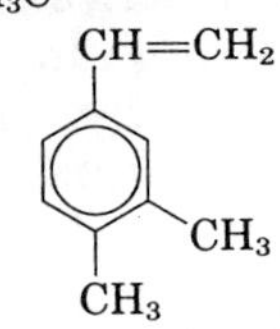	crystalline	240
3,5-dimethylstyrene	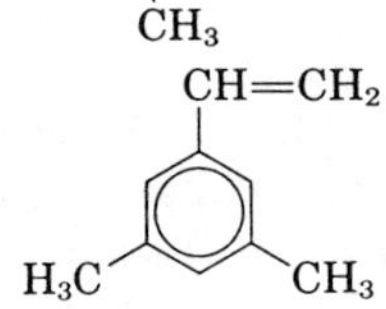	crystalline	290
p-ethylstyrene	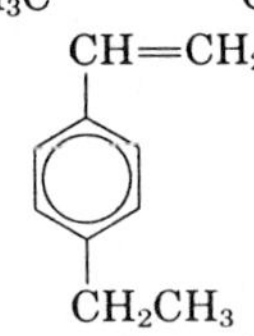	amorphous	
p-isopropylstyrene	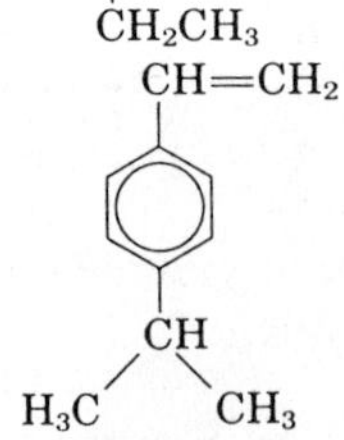	amorphous	
p-*sec*-butylstryene[b]	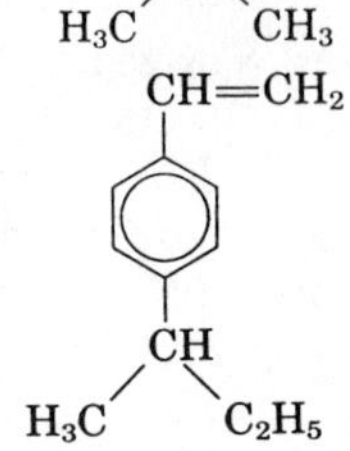	amorphous	

Table 2. (***Continued***)

Monomer		Physical state of the polymer	Polymer mp[a], °C
Compound	Formula		
p-cyclohexylstyrene		amorphous	
o-fluorostyrene	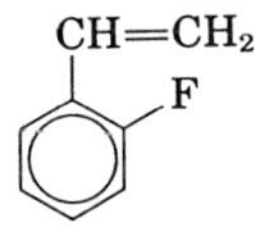	crystalline	270
m-fluorostyrene	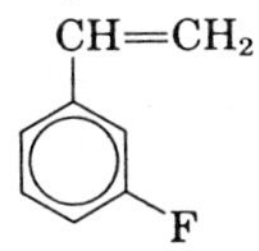	amorphous	
p-fluorostyrene	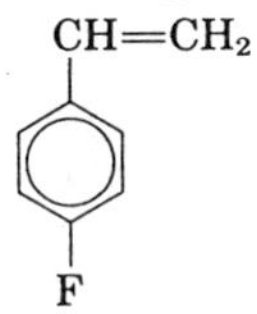	crystalline	265
2-methyl-4-fluorostyrene	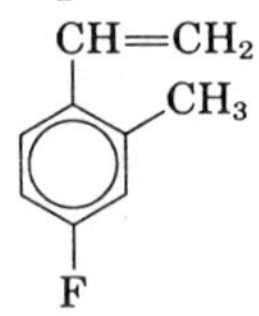	crystalline	360
m-chlorostyrene	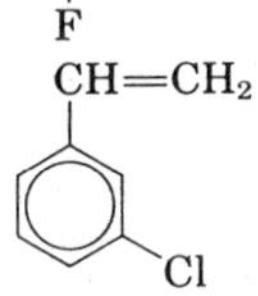	amorphous	
p-chlorostyrene	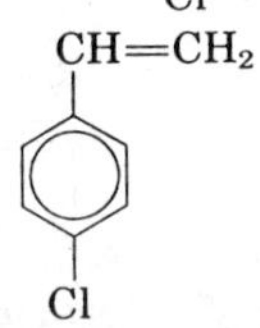	amorphous	
p-bromostyrene	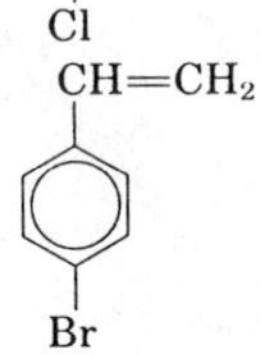	amorphous	
1-vinylnaphthalene	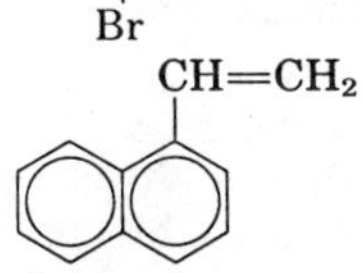	crystalline	360
2-vinylnaphthalene	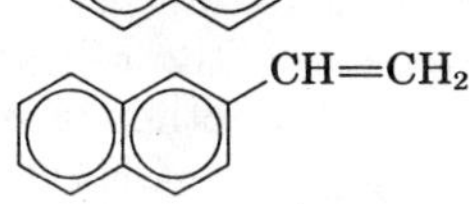	amorphous	
1-vinyl-4-chloronaphthalene	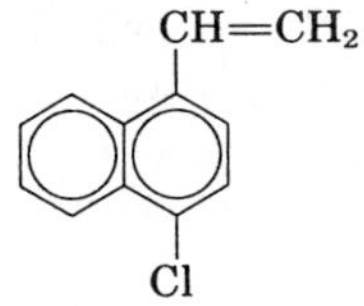	amorphous	
6-vinyl-1,2,3,4-tetrahydronaphthalene	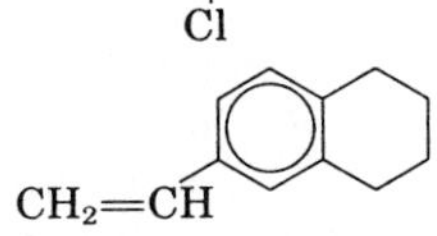	amorphous	

Table 2. (*Continued*)

Monomer		Physical state of the polymer	Polymer mp[a], °C
Compound	Formula		
4-vinylbiphenyl	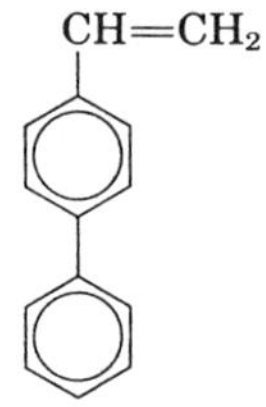	amorphous	
9-vinylphenanthrene	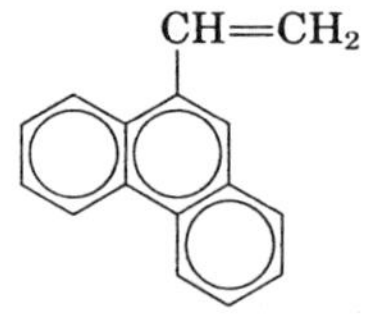	amorphous	

[a] Determined by polarizing microscope; heating rate of 1–2°C/min.
[b] Optically active polymer.

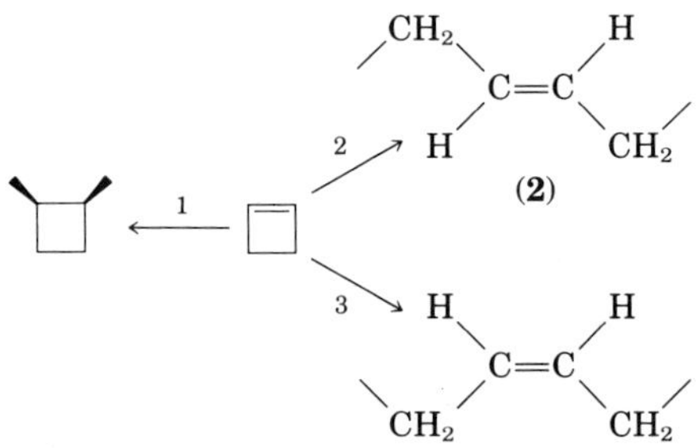

(1) (3)

Scheme 1 leads to two different types of stereoregular polymers (polycyclobutenamer), which were tentatively assigned an erythro-diisotactic (**4**) and an erythro-disyndiotactic structure (**5**):

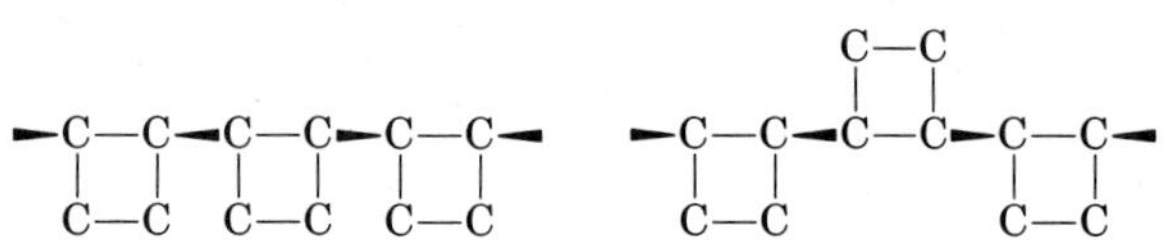

(**4**) (**5**)

Polymerization by ring opening (schemes 2 and 3) yields polymers with the systematic name poly(1-butenylene) that have the structure of a 1,4-polybutadiene. Depending on the catalyst used, cis or trans polymers are obtained as well as polymers of a mixed structure. A stereoregular polymer consisting essentially of cis units is obtained with the complex $C_6H_5(CH_3O)C{=}W(CO)_5$ in toluene at 50°C.

Norbornene and Its Derivatives. Like cyclobutene, norbornene can give three types of monomeric units:

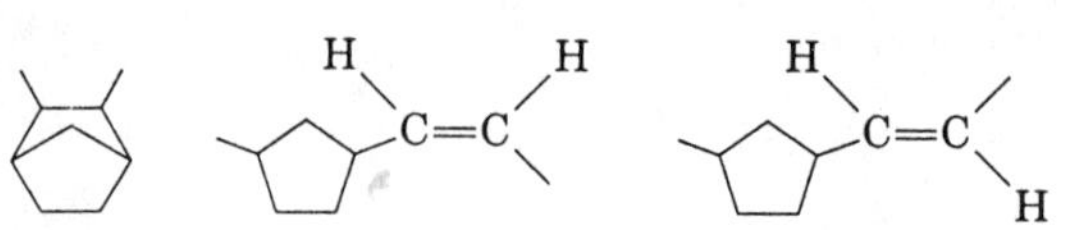

(**6**) (**7**) (**8**)

Polymers consisting of (**6**) units have been obtained with various catalysts, eg, $Al(i\text{-}C_4H_9)_3$–$TiCl_4$ (Al:Ti = 0.5:1), $Al(C_2H_5)_3$–$TiCl_4$ (Al:Ti = 1:1), (η^3 allyl)NiX (X = Br, I), $PdCl_2$, and bisallylic Pd complexes. Stereoregular diisotactic or disyndiotactic structures can be predicted for polymers made up of (**6**) units.

Several catalysts can bring about the polymerization of norbornene by ring opening, eg, $LiAlR_4$–$TiCl_4$, $Al(C_2H_5)_3$–$TiCl_4$–donor, alkene complexes of iridium, chlorides of Ir, Ru, or Os in H_2O or C_2H_5OH, allyl derivatives of Ru, and complexes of $RuCl_2$ with phosphines.

Cyclopentene and Higher Homologues. Of the monomers of this class, only cyclopentene has been polymerized by addition across the double bond. This monomer at very high pressure (~6,500 MPa or 64,000 atm) is converted to a high molecular weight, semicrystalline, saturated polymer. However, copolymers containing saturated cycloolefin units have been obtained not only from cyclopentene, but also from higher homologues. Using soluble vanadium catalysts, stereoregular crystalline alternating ethylene–cyclopentene and ethylene–cycloheptene copolymers with an erythro-diisotactic structure have been prepared.

Polymerization of this class of monomers by ring opening has a more general significance:

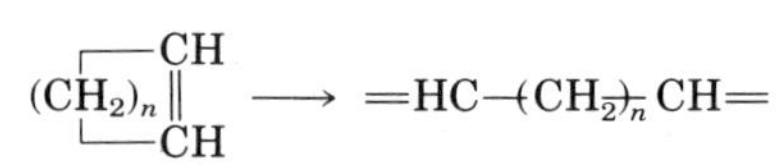 (1)

Ring opening occurs by cleavage of the carbon–carbon double bond (see also METATHESIS POLYMERIZATION). Therefore, the repeating unit may be represented as in equation 1, but other modes of representation are commonly used, eg, $-CH{=}CH-(CH_2)_n-$. Polymers of this type are commonly named polyalkenamers (qv), the systematic name being poly(1-alkenylene)s.

Catalysts. The catalysts used for the polymerization of nonstrained cycloalkenes can be grouped into three classes: (*1*) catalysts derived from transition-metal compounds and organometallic compounds of the 1a–4a groups; (*2*) catalysts based on carbenes or carbynes; (*3*) catalysts based on oxides of transition metals.

Properties of Polyalkenamers. All high *trans*-polyalkenamers are crystalline. They can be grouped into two classes: those with an odd number and those with an even number of CH_2

Table 3. Crystalline Structures and Melting Points of Homologous Series of *trans*-Polyalkenamers[a]

trans-Polyalkenamer	Crystalline modification	Mp, °C
polypentenamer	I, orthorhombic	23
polyhexenamer	III, triclinic	61
	IV, monoclinic	41
polyheptenamer[b]	I, orthorhombic	51
polyoctenamer	III, triclinic	67
	IV, monoclinic	62
polydecenamer	III, triclinic	86
	IV, monoclinic	67
polydodecenamer	IV, monoclinic	80
	III, triclinic	

[a] Trans content 85–94%.
[b] Obtained by alternating butadiene–ethylene copolymerization.

groups between two double bonds. Two members of the odd series have been examined in detail: *trans*-polypentenamer and *trans*-polyheptenamer.

The following polymers of the even series were examined: *trans*-polyoctenamer, *trans*-polydecenamer, *trans*-polydodecenamer. The polymers of this series show polymorphism. Table 3 reports some physical data of *trans*-polyalkenamers.

Polymers of Conjugated Dialkenes. Conjugated dialkenes may polymerize stereospecifically to polymers of various types (see also INSERTION POLYMERIZATION). Most of the stereoregular dialkene polymers are obtained with systems based on transition metals. This section deals with the stereoregular homopolymers of butadiene, isoprene, pentadiene, and other dialkenes. The copolymerization of these monomers is described in other articles (see BUTADIENE POLYMERS; ISOPRENE POLYMERS; PENTADIENE POLYMERS).

Butadiene Polymers. Four highly stereoregular polymers have been obtained from butadiene: trans-1,4; cis-1,4 (Tables 4 and 5); 1,2 isotactic; and 1,2 syndiotactic.

1,2-Polybutadienes. Both isotactic and syndiotactic 1,2-polybutadienes have been prepared. The melting points of the isotactic and syndiotactic polymers are 128 and 210–215°C, respectively.

Catalyst systems that are active in the polymerization of butadiene to stereoregular 1,2 polymers are listed in Tables 6 and 7.

Isoprene Polymers. Isoprene has been polymerized to crystalline trans-1,4 and cis-1,4 polymers and to amorphous polymers consisting predominantly of 3,4 units (Table 8). 1,2 Polymers have not been prepared with Ziegler-Natta catalysts or other ionic initiators.

1,3-Pentadiene Polymers. 1,3-Pentadiene can give different types of stereoregular polymers. The following structures have been prepared: trans-1,4 isotactic, cis-1,4 isotactic, cis-1,4 syndiotactic, and 1,2 syndiotactic. The most significant catalysts for obtaining these polymers are given in Table 9.

Table 4. Polymerization of 1,3-Butadiene to Trans-1,4 Polymers

Catalyst system	Diluent	Polymerization temperature	Structure of crude polymer, %[a]		
			Trans-1,4	Cis-1,4	1,2
VCl_3 plus AlR_3 (or AlR_2Cl) (heterogeneous)	aliphatic hydrocarbon	room temperature	≥99		≤1[b]
$VOCl_3$ or VCl_4 plus AlR_3 or AlR_2Cl (heterogeneous)	aliphatic hydrocarbon	room temperature	ca 95	1	4[b]
$TiCl_3$ (α, γ, or δ) plus AlR_3 or AlR_2Cl (heterogeneous)	aliphatic hydrocarbon	room temperature	70–80	12–22	ca 8[b]
$VCl_3 \cdot 3THF$ or $V(acac)_3$ plus AlR_2Cl (homogeneous)	aromatic hydrocarbon	≤0	>99		<1
$[(\eta^3\text{-}C_4H_7)NiI]_2$ (homogeneous)	hydrocarbon, ethers, C_2H_5OH	room temperature	>99		<1[c]
$RhNO_3$ (soluble)	water	0–50	≥99		≤1[d]

[a] By ir analysis.
[b] Percentage of amorphous, diethyl ether-extractable fraction: VCl_3 catalyst, ca 2%; $VOCl_3$ or VCl_4 catalyst, 12–15%; $TiCl_3$ catalyst, 30–50%.
[c] $\overline{M}_n$: ca 2500.
[d] $[\eta]$: 0.2–0.4 dL/g.

Table 5. Polymerization of 1,3-Butadiene to Cis-1,4 Polymers

Catalyst system	Structure of crude polymer, %		
	Cis-1,4	Trans-1,4	1,2
AlR_3–TiI_4	92	3	5
AlR_3–$TiI_4 \cdot O(i\text{-}C_3H_7)_2$	92–94	2–3	5
AlR_3–$TiCl_4$–TiI_4	93–94	2–3	4
alanes–AlI_3–$TiCl_4$	93–95	2–3	ca 4
$Al(C_2H_5)_2Cl$–H_2O–$Co(acac)_2$ (or $CoCl_2 \cdot 2Py$[a], $Co(OCOR)_2$)	98–99	ca 1	ca 1
AlR_3–$Ni(OCOR)_2 \cdot BF_3 \cdot O(C_2H_5)_2$	97–98	1–1.5	1–1.5
$Al(C_2H_5)_2Cl$–$Nd(OCOR)_3$–$Al(i\text{-}C_4H_9)_3$	99		1
AlR_3–$NdCl_3$–donor	99		1

[a] Py = pyridine.

Table 6. Al–Cr Catalysts for 1,2-Polymerization of Butadiene[a]

			Crystalline polymer[b]	
Cr compound	Al:Cr molar ratio	Aging of catalyst, min	%	Structure
$Cr(CNC_6H_5)_6$	2.5	2–3	10	syndiotactic[c]
	5	2–3	14	syndiotactic + isotactic
	5	480	46	isotactic[d]
$Cr(acac)_3$	3	2–3	15	syndiotactic[c]
	10	2–3	2	isotactic[d]
$Cr(CO)_5$·pyridine	3	2–3	31	isotactic + syndiotactic
	5	2–3	15	isotactic[d]

[a] Syndiotactic, isotactic, or atactic polymers in the presence of catalyst systems prepared from AlR_3 and of a Cr compound (T = 15°C); solvent is toluene.
[b] Ir analysis of the amorphous fraction: 70–78% 1,2; 20–25% cis-1,4; 5–8% trans-1,4.
[c] Ir analysis of the crystalline fraction: 90–95% 1,2; 4–6% cis-1,4; 1–3% trans-1,4.
[d] Ir analysis of the crystalline fraction: 97–99% 1,2; 0.3% cis-1,4; traces of trans-1,4.

Table 7. Catalyst for the Preparation of 1,2 Syndiotactic Polybutadiene

Catalyst system			
Type	Al:Mt molar ratio	Aging time, min	Syndiotactic polymer content, %
$AlH_3{\cdot}N(CH_3)_3–Ti(N(C_2H_5)_2)_3$	10	2–3	ca 58
$Al(C_2H_5)_3–V(acac)_3$	6	2–3	25
	6	180	40
	10	2–3	41
	10	180	60
$AlR_3–MoO_2(acac)_2$	4.5	2–3	ca 75
$AlR_3–Co(acac)_3–CS_2$	50	2–3	ca 99
$(\eta\text{-}C_8H_{13})Co(C_4H_6)–CS_2$ (in chlorinated solvents)			100

Table 8. Polymerization of Isoprene in Hydrocarbon Diluents

Catalyst system[a]	Type of polymer
AlR_3 (or AlR_2Cl) plus VCl_3	≥99% trans-1,4
AlR_3 plus supported VCl_3	≥98% trans-1,4
$AlR_3–VCl_3–Ti(OR)_4$[b]	≥99% trans-1,4
Li, LiR[b]	92–94% cis-1,4
$AlR_3–TiCl_4$[c]	96–97% cis-1,4
AlR_3 (or AlR_2Cl)–β-$TiCl_3$	96–97% cis-1,4
$-\!\!(\underset{H}{Al}-\underset{R}{N})_x\!\!-$ plus $TiCl_4$	97% cis-1,4
AlR_2Cl (or $AlRCl_2$) plus $Ti(OR)_4$	96–97% cis-1,4
$Al(i\text{-}C_4H_9)_3–NdCl_3{\cdot}D$ (D = donor)	96–97% cis-1,4

[a] Heterogeneous unless otherwise noted.
[b] Homogeneous.
[c] Al:Ti molar ratio = 1.

Polymers of Other 1,3-Dialkenes. *Monomers of Formula* $CH_2{=}CH{-}CH{=}CHR$. Stereoregular crystalline polymers with trans-1,4 isotactic structure have been obtained from several monomers of formula $CH_2{=}CH{-}CH{=}CHR$ ($R = C_2H_5$, $n\text{-}C_3H_7$, $i\text{-}C_3H_7$, $n\text{-}C_4H_3$, $i\text{-}C_4H_9$) using the $Al(C_2H_5)_3–VCl_3$ system. The polymers are always obtained as mixtures with amorphous polymers and can be separated by extraction with diethyl ether.

Monomers of Formula $CH_2{=}CR{-}CH{=}CH_2$. Several stereoregular polymers have been obtained from the monomers $CH_2{=}CR{-}CH{=}CH_2$ ($R = C_2H_5$, $n\text{-}C_3H_6$, $t\text{-}C_4H_9$, C_6H_5). Cis-1,4 polymers have been obtained from 2-ethyl-, 2-butyl-, and 2-phenylbutadiene. Catalysts based on $TiCl_4$ and AlR_3 or aluminum hydrides were used. The polymers of 2-ethyl- and 2-propylbutadiene are rubbery products; the polymers from 2-*t*-butylbutadiene are highly crystalline, with mp 103–106°C.

2,3-Dimethylbutadiene. Two stereoregular polymers, with a cis-1,4 and a trans-1,4 structure, have been obtained from this monomer. The trans polymer (trans content >95%) has been obtained with $AlR_3–TiCl_4$ at an Al:Ti molar ratio of approximately 0.5. The melting point is ca 270°C.

2-Methyl-1,3-pentadiene. A trans-1,4 polymer and a cis-1,4 polymer have been obtained from this monomer. The trans polymer (mp 182–192°C) was produced with the $Ti(O\text{-}n\text{-}C_4H_9)_4–VCl_3–Al(i\text{-}C_4H_9)_3$ system (Ti:V:Al = 1:2:20), at room temperature and the cis polymer with two different systems based on Ti and Nd.

4-Methyl-1,3-pentadiene. A stereoregular polymer with a 1,2 isotactic structure has been obtained from this monomer using different catalyst systems: $Al(C_2H_5)_3$–α-$TiCl_3$, $Al(C_2H_5)_3–TiCl_4$ (Al:Ti = 1.2), and $Al(C_2H_5)_3–VCl_3$.

2,4-Hexadiene. A stereoregular polymer with a trans-1,4 threo-diisotactic structure has been obtained from this monomer. Several catalyst systems have been used: $Al(C_2H_5)_2Cl–Co(acac)_2$, $Al(i\text{-}C_4H_9)_3–TiCl_4$, $Al(C_2H_5)_2Cl–Ti(acac)_3$, $Al(C_2H_5)_2Cl–Nd(OCOR)_3–Al(i\text{-}C_4H_9)_3$.

1,1,4,4-Tetramethylbutadiene. A stereoregular polymer with a trans-1,4 structure has been obtained from this monomer using BF_3 at −70°C. The polymer is crystalline and has mp 260–265°C (with decomposition), $[\eta]$ = 0.8–1.5 dL/g.

Table 9. Polymerization of 1,3-Pentadiene to Stereoregular Polymers

Catalyst system	Polymerization temperature, °C	Crystalline polymer Type	Mp, °C
AlR_3–VCl_3 in heptane	0–70	trans-1,4 isotactic[a]	ca 96
AlR_3–$Ti(OR)_4$ in heptane (Al:Ti ca 7)	−15 to +20	cis-1,4 isotactic[b]	ca 45
$Al(C_2H_5)_2Cl$–$Nd(OCOR)_3$–$Al(i\text{-}C_4H_9)_3$ in heptane	−50 to +50	cis-1,4 isotactic[c]	ca 45
aluminoxanes–$Co(acac)_3$ (in heptane or *cis*-butene)	−10 to +20	cis-1,4 syndiotactic[d]	ca 53
$Al(C_2H_5)_3$–Ni octanoate–BF_3·phenolate (10:1:15) in heptane	50	cis-1,4 syndiotactic	ca 51
$Al(C_2H_5)_2Cl$–$Co(acac)_3$ in heptane	18	1,2 syndiotactic[e]	
$Al(C_2H_5)_3$–Fe octanoate–butyl thiocyanate (3:1:11)	50	1,2 syndiotactic	

[a] Fraction insoluble in $(C_2H_5)_2O$, 80–90% of the total crude polymer.
[b] Fraction insoluble in MEK (ca 35% of the total), containing 85–90% cis units.
[c] The cis content is ca 85% for the polymer obtained at 20°C and ca 95% for that obtained at −18°C.
[d] The crude polymer has 89–93% cis-1,4 units.
[e] The crude polymer has about 90% 1,2 units.

Polymers of Nonconjugated Dienes: Cyclopolymerization. Nonconjugated α,ω-dialkenes are polymerized by Ziegler-Natta catalysts. The polymerization generally leads to the formation of polymers containing cyclic repeating units. For example, 1,6-heptadiene, 1,5-hexadiene, and 2,5-dimethyl-1,5-hexadiene can be polymerized with the $Al(i\text{-}C_4H_9)_3$–$TiCl_4$ system in heptane at 0–60°C to yield benzene-soluble polymers, the chains of which contain cyclic rings formed by cyclopolymerization (qv).

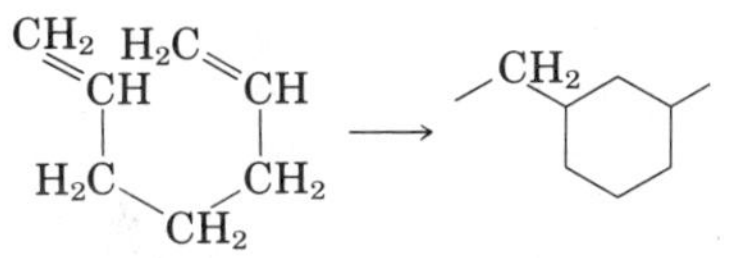

(2)

Polymers of Acetylenic Monomers. Acetylene was first polymerized to a crystalline polymer by Natta and collaborators using the $Al(C_2H_5)_3$–$Ti(OC_4H_9)_4$ catalyst at an Al:Ti molar ratio of ca 4. The polymer obtained at temperatures higher than 150°C has an all trans structure with conjugated double bonds, whereas that obtained at −78°C is about 88% cis. It has been demonstrated that monomer insertion occurs by cis addition: hence, a cis polymer is initially formed. The increasing trans content with increasing polymerization temperature has been attributed to thermal isomerization of the chain segments before crystallization.

Because of the interest in polyacetylene as an electroconducting polymer, much work has been done on its preparation and properties. The simplest way to obtain the polymer is by polymerization of acetylene. Several Ziegler-Natta catalysts have been used as well as other systems, such as NaBH in combination with $CoCl_2$ or $NiCl_2$ (see ACETYLENIC POLYMERS).

Other Homopolymers of Hydrocarbon Monomers. Several other hydrocarbon monomers have been homopolymerized in the presence of Ziegler-Natta catalysts, generally based on alkylaluminum compounds and titanium or vanadium compounds or other ionic catalysts. Polymerization of various conjugated or nonconjugated polyalkenes has been reviewed.

Stereoregular Copolymers of Hydrocarbon Monomers. Stereoregular linear copolymers may be classified into the following groups: stereoregular random copolymers, in which the monomer units are similar, both in chemical nature and in shape and dimension, so that stereoisomerism phenomena among the monomer units are allowed; stereoregular alternating copolymers, in which the monomer units follow one another along the polymer chains in a regularly alternating manner; and block copolymers, whose chains consist of relatively long stereoregular sequences of one or the other type of monomer unit.

Polymers with a Carbon–Carbon Main Chain from Unsaturated Monomers Containing Heteroatoms

Monomers containing heteroatoms, such as O, N, or halogen, can interact with at least one of the components of the Ziegler-Natta catalyst systems (eg, alkylaluminum compounds) to prevent the formation of active catalyst complexes and, therefore, polymerization. However, some 1-alkenes substituted with heteroatoms, which do not give rise to considerable interactions with any of the components of the Ziegler-Natta catalyst systems, may be polymerized in the presence of such systems. For example, this is the case with some monomers containing silicon or tin atoms, but usually is not the case with 1-alkenes containing O, N, or halogen atoms. However, this problem can be overcome in one of the following ways: by the use of a monomer in which the heteroatom (O, N) is shielded by substituents, preventing its interaction with the catalyst components; by preliminary complexation of the monomer with the organometallic compound used in the preparation of the catalyst; by the use of monomers in which the oxygen or the nitrogen atom is bound to the Si atom, which reduces its electron-donor character; in the case of the polymerization of some ω-halogenated 1-alkenes, by preliminary complexation of the organometallic compounds with Lewis bases; or by operating in the presence of an excess of an olefin.

For the stereospecific polymerization of other vinyl monomers containing heteroatoms (eg, vinyl ethers, vinylpyridine, acrylates, etc), it is better to use catalysts of the ionic type, based on a single compound or a complex, such as organometallic, Grignard compounds or BF_3 etherates, acting in a homogeneous phase. These compounds interact with the monomer without losing their activity.

Isotactic Polymers of $CH_2{=}CH(CH_2)_n$ X Monomers Containing Heteroatoms. The crystalline polymers of the monomers considered in this section are isotactic. The polymerization conditions are largely the same as those adopted for the polymerization of 1-alkenes to isotactic polymers. The polymerization mechanism may be considered anionic-coordinated, similarly to that reported for aliphatic 1-alkenes.

Polymers of 1-Alkenes Substituted with a Silane Group. The polymerization of allyl- and vinylsilane and other 1-alkenes substituted with a silane group in the presence of Ziegler-Natta catalysts based on AlR_3 and $TiCl_4$, violet $TiCl_3$, or VCl_3 gives isotactic polymers.

Polymers of 1-Alkenes Substituted with Tertiary Amino Groups. Several monomers of the type (**9**) have been polymerized in the presence of systems based on δ-$TiCl_3$ and on an alkylaluminum compound. In these monomers the nitrogen atom is shielded by groups sufficiently bulky to hinder its interaction with the catalyst.

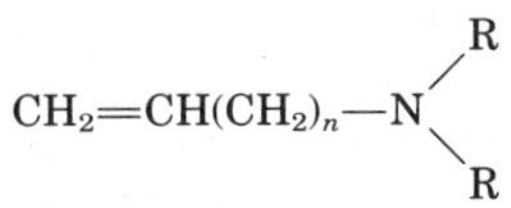

(**9**)

Polymers of 1-Alkenes Substituted with $OSi(CH_3)_3$ or $(CH_3)_2CHNSi(CH_3)_3$ Groups. The complexing power of the heteratom in these monomers is decreased because of both the steric hindrance caused by the $Si(CH_3)_3$ group and the presence of d–π bonds between the silicon atom and the heteroatom.

Polymers of ω-Halogenated 1-Alkenes and of Other Halogenated Monomers. The polymerization of halogenated monomers in the presence of Ziegler-Natta catalyst systems encounters some difficulties because of the reactions of these monomers with the organometallic compounds. Halogenated styrenes represent an exception because their carbon–halogen bond is stabilized by resonance.

Polymers of Vinyl and Alkenyl Ethers. Vinyl ethers and other alkenyl ethers can be polymerized stereospecifically to isotactic and, in some cases, syndiotactic polymers. Linear as well as cyclic β-alkyl- and β-halogen-substituted vinyl ethers and trimethylvinyloxysilanes (TVOSs) have been polymerized stereoregularly. First, partially crystalline poly(vinyl ether)s, poly(isobutyl vinyl ether), and poly(methyl vinyl ether) were obtained at low temperature (−25 to −80°C) in homogeneous systems in the presence of $BF_3{\cdot}O(C_2H_5)_2$ in a nonpolar solvent (eg, propane). These polymers are isotactic.

The approximate melting temperatures of some isotactic poly(vinyl ether)s are 130–144°C for poly(methyl vinyl ether); 98°C for poly(isopropyl vinyl ether); 117°C or higher for poly(isobutyl vinyl ether); 160°C for poly(t-butyl vinyl ether); and 155°C for poly(neopentyl vinyl ether).

Catalytic Systems and Polymerization Conditions. The degree of stereoregularity of poly(vinyl ether)s greatly depends on the nature of the catalyst, monomer, and solvent, on the monomer concentration, and on the polymerization temperature. A comparison of the stereospecificities of various catalyst systems can be made more rigorously by considering their behavior in the polymerization of a single monomer, eg, isobutyl vinyl ether (Table 10).

Monomer concentration affects poly(vinyl ether) stereoregularity, in particular in the polymerization of isobutyl, isopropyl, and ethyl vinyl ethers carried out at −78°C in the presence of $BF_3{\cdot}O(C_2H_5)_2$. In such cases significant increases in the degree of isotacticity are obtained by decreasing the monomer concentration.

Table 10. Isospecific Polymerization of Isobutyl Vinyl Ether with Friedel-Crafts, Aluminum Alkyl, and Modified Friedel-Crafts Catalysts in the Homogeneous Phase at −78°C in Toluene

Catalyst	Conversion, %	Acetone-insoluble fraction, %
$Al(C_2H_5)_3$	0	
$AlBr(C_2H_5)_2$	98	96[a]
$AlBr_2C_2H_5$	98	97[a]
$AlBr_3$	70	10[b]

[a] High crystallinity.
[b] Very low crystallinity.

Modified Friedel-Crafts catalysts are highly isospecific in the polymerization of n-butyl vinyl ether when used at −78°C in toluene, but give amorphous polymers in the presence of aliphatic solvents, eg, heptane or propylene. The influence of the nature of the solvent is further evidenced by the following. (*1*) The AlR_3–H_2O system, with or without cocatalysts, is isospecific in the polymerization of isobutyl vinyl ether carried out at −78°C in toluene, but gives amorphous polymers in CH_2Cl_2. (*2*) The polymerization of t-butyl vinyl ether in the presence of $BF_3{\cdot}O(C_2H_5)_2$ at −78°C gives partially isotactic polymers in toluene and at least partially syndiotactic polymers in polar solvents, eg, methylene chloride, nitroethane, or nitrobenzene; the same behavior is shown by $[(CH_3)_3SiOCH{=}CH_2]$, ie, trimethylvinyloxysilane.

Diisotactic and Optically Active Polymers: Mechanistic Aspects. The stereospecific polymerization of β-substituted vinyl ethers, R—CH=CH—OR′, is of particular interest because it yields diisotactic polymers. For example, threo-diisotactic polymers are obtained from *trans*-1-methyl-2-alkoxyethylenes, *trans*-1-chloro-2-alkoxyethylene, and 1,2-dimethoxyethylene, and erythro-diisotactic polymers are obtained from *cis*-1-chloro-2-alkoxyethylenes.

Diisotactic polymers have also been prepared from cyclic unsaturated ethers, such as 2,3-dihydropyran, and probably also from 2,3-dihydrofuran, 5-methyl-2,3-dihydrofuran, and 2,2-dimethyl-2,3-dihydrofuran. Benzofuran polymerized with $AlCl_2C_2H_5$ at −100 and −80°C gives amorphous polymers. However, by complexing alkylaluminum halide with an optically active Lewis base, an optically active polymer has been obtained. Of the possible chiral erythro- and threo-diisotactic forms, the threo appears more probable. The polymer has a rigid backbone and is usually amorphous, but can be converted into a crystalline material by annealing. An analogous structure was observed in the polymers obtained from α- and β-naphthofuran in the presence of the $AlCl_3$–phenylalanine complex.

Optically active poly(vinyl ether)s with different structures have been obtained by different polymerization methods. The polymerization of racemic 1-methylpropyl vinyl ether by the stereospecific catalyst system $Al(i\text{-}C_3H_7)_3$–H_2SO_4–$Al(i\text{-}C_4H_9)_3$–tetrahydrofuran complex is stereoselective. Moreover, the copolymerization of racemic 1-methylpropyl vinyl ether with several optically active vinyl ethers with the same system gives polymer mixtures including copolymers of the optically active comonomers (with the enantiomer of the racemic monomer having the same configuration) and homopolymers of another enantiomer of the racemic monomer. These facts are analogous to those observed in the copolymerization of racemic and optically active 1-alkenes with Ziegler-Natta catalysts.

Methoxystyrene Polymers. The *o*- and *p*-methoxystyrenes have been polymerized at −70°C in toluene and in toluene–propylene mixtures in the presence of modified Friedel-Crafts catalysts (eg, $AlCl_2C_2H_5$, $AlCl(C_2H_5)_2$, $TiCl_2(O\text{-}n\text{-}C_4H_9)_2$). The polymerization mechanism is coordination cationic, and the meta isomer does not polymerize in the presence of the same catalysts. This may be explained by the electronic nature of aromatic compounds with functional groups in the meta position. Coordination of the ethereal oxygen of the meta-substituted monomer with the catalyst induces polarization of the vinyl double bond opposite to that occurring when the substituent is in the ortho or para position. This prevents cationic polymerization. The polymers of *o*- and *p*-methoxystyrene are not crystalline; however, a crystalline poly(*o*-methoxyvinylcyclohexane) has been obtained by hydrogenation of the phenyl group of poly(*o*-methoxystyrene). This shows that, at least in the case of the polymerization of *o*-methoxystyrene, the polymerization is stereospecific.

***N*-Vinylcarbazole and Other *N*-Vinylamine Polymers.** *N*-vinylamines are in many respects similar to vinyl ethers and, by analogy, can be polymerized in the presence of cationic catalysts. Crystalline polymers of *N*-vinyldiphenylamine have been obtained by polymerization with the following systems: BF_3, $BF_3{\cdot}O(C_2H_5)_2$, $Al(C_2H_5)Cl_2$, $FeCl_3$, $SnCl_4$, $TiCl(O\text{-}i\text{-}C_3H_7)_3$, $TiCl_2(OCOCH_3)_2$, or (in toluene or, in some cases, anhydrous diethyl ether). Unlike the vinyl ethers, the stereoregularity of *N*-vinyldiphenylamine is practically independent of the polymerization temperature in the range between −80 and +15°C. Crystalline poly(*N*-vinyldiphenylamine) melts at 315–320°C; it is soluble at room temperature in chloroform and methylene chloride, but insoluble in ether, dimethylformamide, acetone, aliphatic hydrocarbons, and alcohols, even at their boiling temperatures.

The polymerization of *N*-vinylcarbazole can be initiated by several metal salts and by organometallic catalysts, some of which are typical cationic catalysts, eg, $BF_3{\cdot}O(C_2H_5)_2$, $AlCl_3$, $AlC_2H_5Cl_2$, $Al(C_2H_5)_2Cl$, $Ti(n\text{-}OC_3H_7)_2Cl_2$, $Ti(CH_3COO)_2Cl_2$, $Fe(CH_3COO)Cl_2$, $Sn(CH_3COO)_2Cl_2$, $TiCl_4$–LiC_4H_9, $TiCl_4$–AlR_3, $TiCl_3$–AlR_2Cl, $Zn(C_2H_5)_2$–$CoCl_2$–quinoline or –pyridine complex, LiCl, LiBr, and LiI. With several of the systems the polymerization can be carried out at different temperatures (from −78°C to 70°C).

The stereoregularity of the polymers depends first on the nature of the catalyst system and the solvent. In particular, the stereoregularity of the poly(*N*-vinylcarbazole) obtained with $BF_3{\cdot}O(C_2H_5)_2$ is greatly affected by the polarity of the solvent used. The polymers prepared in toluene and dichloromethane appear to have a prevailingly isotactic configuration, whereas those obtained in nitrobenzene seem to be syndiotactic (see also VINYLCARBAZOLE POLYMERS).

Vinyl- and α-Alkenylpyridine Polymers. A crystallizable isotactic polymer of 2-vinylpyridine was first prepared by polymerization in the presence of anionic catalysts in aromatic hydrocarbons at 45°C. This polymer has a melting point in the 190–212°C range. It is insoluble in acetone and aliphatic hydrocarbons, but is soluble in polar solvents such as methanol and chloroform. The best catalysts for the preparation of isotactic 2-vinylpyridine polymers include organometallic or metal amide compounds of groups IIA and IIIA of the periodic table: $Be(C_2H_5)_2$, $Mg(C_2H_5)_2$, $Mg(C_6H_5)_2$, $MgBrC_6H_5$, $MgClC(CH_3)_3$, $LiAlH_4$, $LiN(C_6H_5)_2$, $Be[N(CH_3)_2]_2$, MgCl (or Br) $N(C_2H_5)_2$, $MgBrN(C_6H_5)CH_3$, and $Al[N(CH_3)_2]_3$. The polymerizations are done at a temperature of 45°C in toluene.

Although the stereospecific catalysts have low solubility in the solvent usually employed (toluene), the polymerization reaction occurs in solution because the catalyst is brought into solution through formation of a complex with the monomer.

A mechanism has been suggested for these polymerizations. Further information on the polymerization mechanism has been obtained through kinetic and viscosimetric data and from spectroscopic studies (^1H- and ^{13}C-nmr or uv–visible) (see also ANIONIC POLYMERIZATION; VINYLPYRIDINE POLYMERS).

3- and 4-Vinylpyridines also polymerize in the presence of the catalytic systems listed for 2-vinylpyridine, but the polymers obtained are amorphous.

Polymers of *trans*-2-propenylpyridine can also be obtained in the presence of the same catalysts. 3-Methyl-2-vinylpyridine gives stereoregular oligomers and polymers by anionic polymerization with Li compounds.

Polymers of Acrylic Esters. Methacrylates and other α-substituted acrylates and some acrylates give isotactic or syndiotactic polymers with several ionic-type catalysts (Tables 11–13) and, in some cases, also by radical polymerization (see also

Table 11. Stereospecific Polymerization of Alkyl Acrylates to Isotactic Polymers

Monomer	Catalyst system	Polymerization temperature, °C
t-butyl acrylate	$n\text{-}C_4H_9Li$ in hexane	30–50
i-propyl acrylate[a]	C_6H_5MgBr in toluene	−65
i-propyl acrylate *t*-butyl acrylate	LiN–carbazyl in toluene	−10 to +30
i-propyl acrylate *t*-butyl acrylate	$(C_2H_5)_2NMgBr$ in toluene	−78
methyl α,β-d_2-acrylate	$LiAlH_4$ in toluene	−78

[a] Same conditions have been applied to *t*-butyl, *sec*-butyl, and cyclohexyl acrylate.

Table 12. Stereospecific Polymerization of Methyl Methacrylate

Catalyst	Solvent	Polymerization temperature, °C	Polymer structure
9-fluorenyllithium	toluene	−78	isotactic
	1,2-dimethoxyethane, NH_3, or $C_2H_5NH_2$	−78 to +60	syndiotactic
$LiAlH_4$	aromatic hydrocarbon or NR_3	room	isotactic
	$(C_2H_5)_2O$ or toluene	−70	isotactic
	THF	−70	syndiotactic
N-carbazyl lithium	toluene	0–30	isotactic
butyllithium	toluene	−78 to 0	isotactic[a]
	THF or pyridine	−78	syndiotactic
sodium	aliphatic hydrocarbon or NR_3	room	syndiotactic
$C_5H_{11}Na$	toluene	−70 to 0	isotactic
	pyridine	−60 to 0	syndiotactic
C_6H_5MgBr	toluene	30	isotactic
(t-C_4H_9)MgBr	toluene	−78	isotactic
$(C_2H_5)_2NMg$	toluene	−78 to +30	isotactic
C_2H_5MgBr benzalacetophenone	heptane	30	isotactic
dipiperidylmagnesium	toluene	−78	syndiotactic
ethyl piperidylmagnesium	toluene	−78	syndiotactic
ethylpentamethylene-iminomagnesium	toluene	−78	syndiotactic
bis(pentamethyleneimino)-magnesium	toluene	−78	syndiotactic
bis(1-pyrrolyl)-magnesium	toluene	−78	isotactic
ethyl 6-pyrrolyl magnesium	toluene	−78	isotactic
ethyl vinylene-iminomagnesium	toluene	−78	isotactic
bis(divinyleneimino)-magnesium	toluene	−78	isotactic
C_4H_9MgCl	toluene[b]	−50	syndiotactic
	toluene[c]	−50	isotactic
	toluene	−78	syndiotactic
C_4H_3MgCl[d]	toluene + $(C_2H_5)_2O$	−78	isotactic + syndiotactic
$VOCl_3$–AlR_3[e]	*n*-hexane	40	syndiotactic
$TiCl_4$–AlR_3[e]	*n*-hexane	40	syndiotactic
$Cr(allyl)_3$ or $Cr(methallyl)_3$	toluene	0	

[a] Depending on monomer concentration.
[b] Preformed isotactic poly(methyl methacrylate) present.
[c] Preformed syndiotactic poly(methyl methacrylate) present.
[d] Containing 0.1 mol $(C_2H_5)_2O$/mol C_4H_9MgCl.
[e] Heterogeneous.

ANIONIC POLYMERIZATION; ACRYLIC AND METHACRYLIC ESTER POLYMERS).

Polyacrylates. Acrylates with branched alkyl groups may be polymerized to isotactic polymers in homogeneous systems in the presence of anionic-type catalysts in hydrocarbon solvents. Isotactic poly(*i*-propyl acrylate), poly(*sec*-butyl acrylate), poly(*t*-butyl acrylate), and poly(cyclohexyl acrylate) have been prepared with these systems (Table 11). In the presence of the same systems, linear acrylates yield amorphous polymers, but an isotactic poly(methyl α,β-d_2-acrylate) was prepared by polymerization at −78°C in the presence of $LiAlH_4$ in toluene.

Mechanistic Aspects. Several studies of the polymerization mechanism of acrylates and α-substituted acrylates in the presence of organometallic compounds have been reported. The polymerization process is complicated by side reactions taking place between the initiator and the polar substituent of the monomer; therefore, the fraction of the initiator producing growing chains is quite low.

Polymers of α,β-Unsaturated Lactones and α,β-Unsaturated Cyclic Ketones. α,β-Unsaturated lactones contain the group (**10**), similar to that of the esters of acrylic and methacrylic acid.

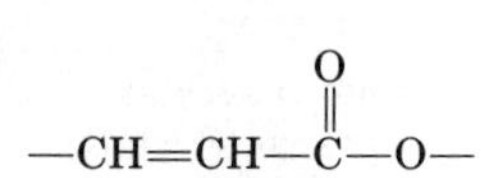

(**10**)

These monomers can be polymerized in the presence of anionic catalysts consisting of organometallic and amide compounds of lithium and sodium, which are active in the polymerization of acrylates.

Table 13. Stereospecific Polymerization of Other Acrylic Esters

Monomer	Catalyst, solvent	Polymerization temperature, °C	Polymer structure
diphenylmethyl methacrylate	n-C_4H_9Li, toluene	−95 to +30	isotactic
	n-C_4H_9Li, THF	−95 to −30	syndiotactic
triphenylmethyl methacrylate	n-C_4H_9Li, toluene	−78 to 0	isotactic
	n-C_4H_9Li, THF	−78	isotactic
α-methylbenzyl methacrylate	cyclohexylmagnesium chloride	−78	isotactic
benzoethyl methacrylate	$(n$-$C_4H_9)_2CuLi$, toluene	−10	isotactic
diphenyl-2-pyridylmethyl methacrylate	n-C_4H_9Li, toluene	−78	isotactic
β-(*N*-carbazyl)ethyl methacrylate	C_6H_5MgBr, toluene	0 to 20	crystalline
trimethylsilyl methacrylate	n-C_4H_9Li, toluene	−70	isotactic
	n-C_4H_9Li, THF	−70	syndiotactic
methyl α-ethylacrylate	n-C_4H_9Li, toluene	0	isotactic
methyl α-*n*-propylacrylate	n-C_4H_9Li, toluene	0	isotactic
methyl α-phenylacrylate	n-C_4H_9Li, toluene	−78	isotactic
allyl acrylate	n-C_4H_9Li or C_6H_5MgBr, toluene	−50	isotactic
ethyl α-chloroacrylate	C_6H_5MgBr, n-C_4H_9Li, or fluorenyllithium, THF	78	syndiotactic[a]
ethyl and methyl α-chloroacrylate	C_2H_5MgCl or C_6H_5MgBr benzalacetophenone, *n*-heptane	30	isotactic
alkyl α-bromoacrylates[b]	CH_3MgCl benzalacetophenone, heptane–ether	35	isotactic
alkyl thioacrylates[c]	n-C_4H_9Li, C_6H_5MgBr, or other organometallic compounds, toluene	−78	isotactic
ethyl thioacrylate	n-C_4H_9Li or C_6H_5MgBr, toluene	−78	amorphous
2-methoxyethyl methacrylate	t-C_4H_9Li, benzene	20	isotactic[d] syndiotactic[e]

[a] Low molecular weight polymers.
[b] Methyl, ethyl, *i*-propyl, *n*-propyl, *n*-butyl, and *n*-pentyl.
[c] *n*-Propyl, *i*-propyl, *n*-butyl, *i*-butyl, *sec*-butyl, and *t*-butyl.
[d] At the beginning of polymerization.
[e] At high conversions.

α-Alkenyl Ketone Polymers. Crystalline isotactic polymers of vinyl ketones (**11**)

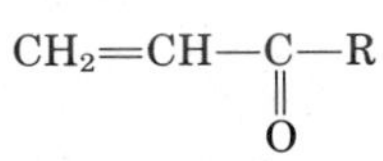

(**11**)

(with R = CH_3, C_2H_5, i-C_3H_7, *sec*-C_4H_9, t-C_4H_9, or C_6H_5) have been obtained by polymerization with anionic catalysts such as Li, $LiAlH_4$, alkyllithium, or alkylzinc compounds in hydrocarbon solvents, eg, toluene, at room temperature or below. These polymers are insoluble in acetone. The mechanisms of these polymerization reactions have some common features with the mechanism of anionic polymerization (qv) of other monomers.

α-Alkylacrylonitrile and Crotonitrile Polymers. The best catalytic systems for the stereospecific polymerization of methacrylonitrile and other α-alkylacrylonitriles, corresponding to the general formula (**12**),

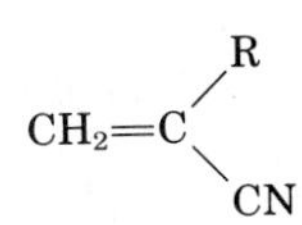

(**12**)

where R = CH_3, C_2H_5, n-C_3H_7, or i-C_3H_7, are based on magnesium or beryllium alkyls, alkali metal or magnesium alkylaluminum compounds, alkali metal aluminum alkoxides, and some other compounds (see METHACRYLONITRILE POLY-

MERS). Catalysts for the stereospecific polymerization of α-alkylacrylonitriles include

$Mg(C_2H_5)_2$	$LiAl(C_2H_5)_3C_3H_7$
$Mg[C(CH_3)_3]_2$	$C_2H_5MgAl(C_2H_5)_4$
$Mg(C_6H_5)_2$	$NaAl(C_2H_5)_3N(C_6H_5)_2$
$Mg(C_6H_5)Br$	$LiZn(C_2H_5)_2N(C_6H_5)_2$
$Be(C_2H_5)_2$	$LiAl(C_2H_5)_3N(C_6H_5)_2$
$NaAl(C_2H_5)_4$	$C_2H_5Mg[Al(C_2H_5)_3N(C_6H_5)_2]$
$Mg[Al(C_2H_5)_4]_2$	$LiOAl(C_2H_5)_2$
$LiZn(C_2H_5)_2C_3H_7$	

Substitution of an alkyl by a diphenylamino group yields more stereospecific but less active systems.

The stereoregularity and molecular weight of polymethacrylonitrile prepared with these catalytic systems increase with increasing polymerization temperature.

The stereoregular polymers of α-alkylacrylonitrile are insoluble in the usual solvents, eg, acetone, cyclohexanone, nitromethane, and dimethyl sulfoxide, but are soluble in trifluoroacetic and dichloroacetic acids and in inorganic acids, such as sulfuric and nitric acids. These polymers are isotactic; their crystallinity increases in the following order; polymethacrylonitrile (mp ca 250°C) < poly(α-ethylacrylonitrile) (mp ca 200°C) < poly(α-*n*-propylacrylonitrile) (mp ca 210°C) < poly(α-*i*-propylacrylonitrile) (mp ca 310°C).

The mechanism of these polymerizations is assumed to be anionic-coordinated.

Acrylamide Polymers. Crystalline polymers of *N*-isopropylacrylamide have been obtained by anionic polymerization with alkylmetal–transition-metal halide catalysts in such solvents as hydrocarbons, chlorinated hydrocarbons, and ethers.

Polycarboalkoxybutadienes and Poly(1-cyanobutadiene). Crystalline polymers have been obtained from some esters of sorbic acids, $CH_3—CH{=}CH—CH{=}CH—COOR$, and of β-styrylacrylic acid using catalysts of the anionic type. The structure of poly(methyl sorbate) has been examined in some detail, by both x ray and nmr. The trans-1,4 erythro-diisotactic structure was assigned to the polymers obtained by butyllithium in hexane at −50°C, whereas the polymers obtained with butyllithium–tetramethyl urea in toluene at −50°C have the trans-1,4 threo-diisotactic structure.

Polyisocyanides. Poly(iminomethylene)s or polyisocyanides (qv), or more specifically, poly(carbonimidoyl)s (**13**),

(**13**)

can be obtained by polymerization of isocyanides: $R—N{\equiv}C{:}$. These polymers have a rigid helical structure with four monomer units per helix turn. The polymers do not show any stereoisomerism; the carbon atoms of the main chain are not asymmetric. However, the polymerization process is stereoselective with respect to the screw sense of the polymeric helix (atropisomerism) as demonstrated by investigations carried out with polymers obtained from chiral monomers and by the resolution of polymers of achiral monomers.

Various catalyst systems are suitable for polymerization of isocyanides, including coordination complexes, organometallic complexes, Lewis acids, and protonic acids.

Homopolymers of Other Monomers. Other than those mentioned above, few stereoregular polymers obtained from unsaturated monomers containing heteroatoms are known. Stereoregular, but amorphous, polymers seem to have been obtained from 2-isoprenylfuran and 2-isopropenyl-5-methylfuran by polymerization in the presence of Ziegler-Natta catalyst systems such as $VOCl_3$–$Al(i\text{-}C_4H_9)_3$ or $(C_2H_5)_2AlO\text{-}t\text{-}C_4H_9$. Poly(vinyl acetate) with some degree of isotacticity has been obtained by polymerization at −12 to +20°C in the presence of C_{4-6} alkylmagnesium halides. Polymers of bis(β-chloroethyl) vinylphosphonate obtained with Ziegler-Natta systems probably also have a regular structure.

Copolymers of Vinyl-type Monomers Containing Heteroatoms. These are random, alternating, and block copolymers in which the monomer units deriving from at least one comonomer show a certain degree of stereoregularity. The following description concerns stereoregular random or alternating copolymers. Various block copolymers obtained by different methods are described in several reviews (see also BLOCK COPOLYMERS; ANIONIC POLYMERIZATION; and CATIONIC POLYMERIZATION).

Random coisotactic copolymers have been obtained by copolymerization of *p*-methoxystyrene with (−)(*S*)-4-methyl-1-hexene in benzene at 25°C in the presence of δ-$TiCl_3$-$Al(i\text{-}C_4H_9)_3$ as a catalyst. Copolymers of this type were also obtained by copolymerization of *p*-iodostyrene with 1-alkenes.

The alternating copolymerization of several monomers is treated in some reviews (see also COPOLYMERS, ALTERNATING). *o*- And *p*-methoxystyrenes give alternating copolymers with 1,1-diphenylethylene when copolymerized with LiC_4H_9 or other anionic initiators in THF. The copolymers with *p*-methoxystyrene are crystalline. In aromatic hydrocarbons the copolymerization with *o*-methoxystyrene gives a polymer having a higher *o*-methoxystyrene content, whereas an almost alternating copolymer is obtained with *p*-methoxystyrene.

Polymers from Unsaturated Monomers, Containing Heteroatoms in the Main Chain

Stereoregular polymers containing heteroatoms in the main chain have been obtained by polymerization of various types of unsaturated monomers, such as aldehydes, ketones, ketenes, or isocyanates. The polymerization mechanism may be anionic or cationic.

The best catalysts for these monomers, as well as for the vinyl monomers containing heteroatoms, are usually based on a single compound or a complex, acting in a homogeneous phase.

Aldehyde and Ketone Polymers. The first important investigations of the polymerization of formaldehydes were carried out by Staudinger. This monomer, however, because of the symmetry of the molecule, does not produce stereoregular polymers. The first studies of the polymerization of higher aldehydes were carried out in 1929 and 1936. The polymers obtained were amorphous. More recently, stereoregular polymers of aldehydes have been prepared by polymerization in the presence of ionic catalysts. Polymerization of aldehydes has been reviewed (see also ALDEHYDE POLYMERS).

The macromolecular chains of aldehyde polymers are thermodynamically stable with respect to the monomer only at relatively low temperature (the ceiling temperature is 137°C for polyformaldehyde and −40°C for polyacetaldehyde); this is the reason why polymerization of aldehydes, in particular higher aldehydes, is possible only at low temperatures. Both anionic and cationic catalysts are used in the stereospecific polymerization of higher aldehydes. Anionic catalysts are highly stereospecific for the polymerization of acetaldehyde and higher aldehydes. The cationic catalysts are stereospecific only for higher aldehydes. The stereoregular aldehyde polymers obtained so far are isotactic.

Acetone polymers with a polyketalic chain have been obtained by polymerization of liquid acetone in the presence of metallic magnesium on a cold surface. The polymer obtained decomposes rapidly even at room temperature.

Ketene Polymers and Copolymers. Ketenes can be polymerized by opening either the C=O or the C=C bond.

Ketene Polymers. Crystalline polymers of ketene have been obtained both by γ-ray polymerization of the monomer in the molten state or solution and by polymerization at −40°C in polar media in the presence of solid potassium hydroxide. The polymers obtained under these conditions have a polyester structure.

Dimethylketene Polymers. The polymerization of dimethylketene is much more complex than that of ketene. It yields crystalline polymers having polyketonic, polyketalic, and polyester structures.

Copolymers. Crystalline copolymers of dimethylketene have been obtained with various carbonyl compounds such as acetone, benzaldehyde, and furfural. Copolymers with formates have also been prepared. These polymers have the polyester structure.

Thioaldehyde and Thioacetone Polymers. In contrast to aldehydes, higher thioaldehydes polymerize with difficulty, especially because of their marked tendency to trimerize. Polymers of thioacetone have been obtained by spontaneous polymerization of the purified monomer.

Isocyanate Polymers. Isocyanates, R—N=C=O, contain two adjacent double bonds. As in the case of symmetric ketenes, the monomer units derived from the opening of one of the two double bonds of the isocyanate molecule do not contain asymmetric atoms. A mere regularity of chemical structure may be sufficient to yield crystalline polymers.

High polymers of alkyl-, aryl-, and substituted aryl isocyanates have been prepared by polymerization at low temperature (−60 to −78°C) in the presence of sodium cyanide, alkyllithium and sodium compounds, and other anionic initiators. The polymerization is carried out in the presence of various types of solvents, such as toluene, carbon disulfide, THF, methylene chloride, or dimethylformamide. Phenyl isocyanate can also be polymerized in the presence of cationic catalysts, eg, $AlBr_3$.

Polymers from Saturated Heterocyclic Monomers

1,2-Epoxide Polymers. The stereospecific polymerization of higher 1,2-epoxides is treated in several reviews (see also 1,2-EPOXIDE POLYMERS). These monomers allow formation of isotactic optically active or optically inactive polymers. Amorphous disyndiotactic polymers have been prepared from *cis*-2-butene epoxide. Crystalline poly(*t*-butylethylene oxide) prepared in the presence of potassium *t*-butoxide was assumed to have a structure consisting of alternating isotactic and syndiotactic diads ···*ddllddllddll*··· (isosyn sequences).

Chemical Nature of Active Species. The importance of the catalyst structure is evident from the observation that amorphous $Zn(OCH_3)_2$ prepared from $Zn(C_2H_5)_2$ and CH_3OH is very active in the polymerization of epoxides, whereas crystalline $Zn(OCH_3)_2$ used under the same conditions is practically inactive. The true active species may be soluble complexes consisting of mixtures of $C_2H_5ZnOCH_3$ and $Zn(OCH_3)_2$, with excess methanol as a solvating medium, but not the dimethoxide itself, which is insoluble in organic solvents.

Mechanistic Aspects. Mechanistic studies with several catalytic systems and with several types of monomers demonstrate that the polymerization is either stereoelective (the optically active catalyst polymerizes preferentially one of the two optical isomers of the racemic monomer) or stereoselective (the polymerization of a racemic monomer gives a racemic mixture of optically active macromolecules).

1,2-Episulfide Polymers. Polymerization of higher 1,2-episulfides presents the same stereoisomerism phenomena as the polymerization of 1,2-epoxides. In this case too, isotactic polymers with or without optical activity have been prepared.

Polyesters from β-Lactones. β-Lactones (**14**), substituted in the α- or β-position, or in both positions, can be polymerized by several cationic and anionic initiators.

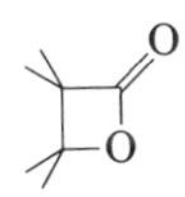

(**14**)

The monomers and the repeating units of the resulting polyesters contain asymmetric carbon atoms; stereoregular and optically active polymers can be obtained by using appropriate catalytic systems.

Polymers from Cyclic *N*- or *S*-Carboxyanhydrides. Polypeptides (qv) are synthesized by polymerization of *N*-carboxy-α-aminocarboxylic acid anhydrides (NCA) (4-substituted oxazolidin-2,5-diones) (**15**), the corresponding *N*-thiocarboxyanhydrides (substituted 1,3-thiazolidin-2,5-diones) (**16**), or the heterocyclic 5-oxo-2-thiooxazolidine (**17**) derivatives by concurrent elimination of CO_2 or COS:

(**15**) (**16**) (**17**)

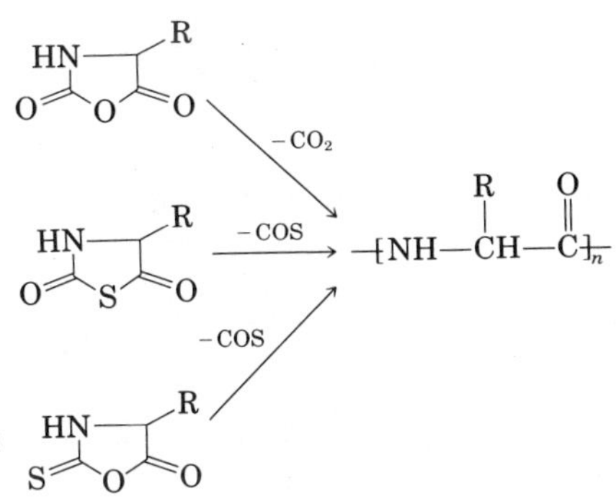

(3)

Table 14. Stereoregular Polymers Obtained by Free-radical Polymerization

Monomer	Polymerization conditions[a]	Polymer structure
methyl methacrylate	at −115 to +90°C with several initiators	syndiotactic
	at 40°C and 400–750 MPa	partially isotactic
	at 25–35°C in the presence of preformed syndiotactic polymer in polar solvent with (*p*-*t*-butylcyclohexyl)dipercarbonate	isotactic
	at 25–35°C in the presence of preformed isotactic polymer in polar or apolar solvents with (*p*-*t*-butylcyclohexyl)dipercarbonate	syndiotactic
ethyl, *i*-propyl, and *t*-butyl methacrylate	at 60–70°C with BPO	syndiotactic
n-butyl, *d*-bornyl, *dl*-isobornyl, and 2,4,6-triphenylbenzyl methacrylate	at 70–80°C with AIBN in toluene	syndiotactic
triphenylmethyl methacrylate	at 30–60°C with AIBN	slightly isotactic
diphenylmethyl methacrylate	at 30–60°C with AIBN	syndiotactic
menthyl methacrylate	at −78°C, γ-rays	slightly syndiotactic
β-phenylethyl methacrylate	at −47°C, γ-rays	syndiotactic
hydroxyethyl methacrylate	at −40°C in methanol, uv photolysis with AMIB	highly syndiotactic
diphenyl-2-pyridylmethyl methacrylate	at 60°C with AIBN in toluene	highly isotactic; low molecular weight
benzyl methacrylate	at 60°C with BPO in bulk	moderately syndiotactic
p-phenylbenzyl, α-methylbenzyl, α,α-dimethylbenzyl, and 1,1-diphenylethyl methacrylate	at 60–85°C with AIBN in toluene	moderately syndiotactic
α-phenoxy methacrylate	at 27°C, uv radiation with benzoin in *n*-hexane	syndiotactic
isopropyl acrylate	at −105 and −78°C, uv radiation with benzoin in toluene	moderately syndiotactic
trimethylsilyl acrylate	at −100°C, uv radiation with benzoin in toluene	moderately syndiotactic
alkyl α-bromoacrylates[b]	at −40°C, photopolymerization	syndiotactic
methyl, ethyl, and *i*-propyl α-chloroacrylate	at 60–70°C with BPO	moderately syndiotactic
methacrylic acid	at −78°C to +60°C in polar solvents with several initiators	syndiotactic
acrylic acid	at −78°C, photopolymerization with benzoin in C_2H_5OH	syndiotactic
	at 20–76°C, irradiation of the liquid monomer	syndiotactic
acrylonitrile	at −196 to +65°C with several initiators	moderately syndiotactic
methacrylonitrile	at 0–80°C with several initiators	moderately syndiotactic

Table 14. (*Continued*)

Monomer	Polymerization conditions[a]	Polymer structure
styrene	at −78 to +80°C with several initiators	moderately syndiotactic
	at −10 to +60°C with solid nickel peroxide	partially isotactic
o-methylstyrene	at 80°C with AIBN in toluene	highly syndiotactic
o-methoxystyrene	at 80°C with AIBN in toluene	highly syndiotactic
α-methylstyrene	at 100°C and 600 MPa with AIBN	moderately syndiotactic
N-vinyl-2-pyrrolidinone	at 17°C in water, photochemical with AIBN	moderately syndiotactic
4-vinylpyridine	at 60°C with AIBN in CH_3OH	moderately syndiotactic
2-vinylpyridine	at 60°C with AIBN in CH_3OH	moderately syndiotactic
vinyl chloride	at −78 to +50°C with several initiators	moderately syndiotactic
vinyl fluoride	at 0–80°C with Ziegler-Natta catalysts	slightly syndiotactic
chlorotrifluoroethylene	at −10 to +100°C with several initiators	slightly syndiotactic
vinyl esters	at −50 to +80°C with several initiators	slightly syndiotactic
vinyl trifluoroacetate	at −40 to +60°C in aldehyde with AIBN	slightly syndiotactic
propylene	at 100–200°C and 700 MPa with or without radical initiators	slightly isotactic

[a] BPO = benzoyl peroxide; AIBN = 2,2′-azobisisobutyronitrile; AMIB = Azobis(methyl isobutyrate).

[b] Methyl, ethyl, *n*-propyl, isopropyl, *n*-butyl, and *n*-pentyl derivatives.

Poly(β-amides) are prepared analogously from the corresponding six-membered cyclic anhydrides and thioanhydrides (substituted perhydro-1,3-oxazine (or thiazine-2,6-diones). An optically active polythiolactide has been obtained by polymerization of thiolactic acid *S*-carboxyanhydride (**18**) initiated by optically active amines.

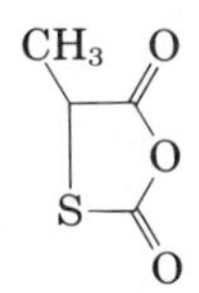

(**18**)

The reported methods of preparing poly(α- and β-amino acid)s present the advantage of preserving the chirality of the asymmetric carbon atoms of the optically active monomer.

The possibility of achieving stereoselective and stereoelective polymerization of racemic monomer mixtures has attracted significant interest.

Polysaccharides and Other Polymers from Cyclic Acetals and Formals. The polymerization of the dioxacyclo and bicyclo compounds can, in theory, give different types of stereoisomeric polymers because of the presence of asymmetric carbon atoms and the formation of oxycyclohexane and/or oxycyclopentane rings having equatorial and/or axially positioned acetal protons.

Condensation Polymers

Syndiotactic all-*ortho* phenol–acetaldehyde oligomers were separated from low molecular weight polymers prepared in the presence of phenoxymagnesium bromide precursors in benzene. Optically active condensation polymers have been obtained by condensation of *cis*-*d*-camphoryl dichloride with 1,4-butanediol, bisphenol A, and piperazine hexahydrate.

FREE-RADICAL OR PHOTO- OR RADIATION-INDUCED POLYMERIZATION

The stereochemical features of free-radical polymerizations have been reviewed. The relationships between stereoregularity and crystallinity in polymers obtained by free-radical or photo- or radiation-induced polymerizations have also been reported. Polymerizations of vinyl monomers, conjugated diolefins, isocyanates, and isocyanides are briefly considered in this section (see also RADICAL POLYMERIZATION).

Polymers of Vinyl Monomers. Most vinyl monomers polymerize with a predominantly head-to-tail arrangement of the monomer units; however, in only a few cases do free-radical or photo- or radiation-induced polymerizations yield polymers with some degree of stereoregularity. In radical polymerization the factors controlling the steric configuration of the monomer units entering into the chain (steric and/or dipole interactions) are less important than in the processes initiated by ionic-type catalysts.

Some examples of polymers with a certain degree of stereoregularity obtained by free-radical or photo- or radiation-induced polymerizations are given in Table 14; others are reported in the literature.

Most polymerizations follow the first-order Markov mechanism. With increasing temperature, more heterotactic than isotactic triads form at the syndiotactic radical end group, but the addition of a monomer at an isotactic radical end group favors isotactic triads.

The degree of syndiotacticity increases with decreasing polymerization temperature, and in some cases, eg, for vinyl chloride, predominantly syndiotactic polymers are obtained at a sufficiently low temperature (eg, −78°C). However, poly(vinyl chloride) with a significant level of syndiotacticity has also been obtained by polymerization at temperatures above room temperature by operating under particular conditions, eg, by polymerization in the presence of aldehydes.

Polymers of Conjugated Dialkenes. Polymers of various conjugated dialkenes obtained by radical polymerization contain a fairly high amount of trans-1,4 units, which generally increases with decreasing polymerization temperature. According to some reports, the monomer maintains its steric configuration during polymerization. Consequently, the cis : trans ratio of the units of polybutadiene or polyisoprene is equal to the relative proportions of the monomer configurations at the polymerization temperature (see BUTADIENE POLYMERS).

Polymers of Isocyanates and Isocyanides. Isocyanates can be polymerized by radiation. Some initiators of the polymerization of isocyanides, based on protonic acids (eg, H_2SO_4), promote radical processes if activated by O_2 or benzoyl peroxide.

Alternating Copolymers. Some acrylic monomers (acrylates, alkyl acrylates, acrylonitrile, methacrylonitrile) give complexes with Lewis acids, such as $RAlCl_2$, R_2AlCl, $AlCl_3$, BF_3, $SnCl_4$. These complexes, in the presence of other monomers, eg, styrene, diolefins, and olefins, give alternating copolymers (see COPOLYMERS, ALTERNATING). These copolymers are generally amorphous, but can exhibit a certain degree of stereoregularity, both in the case of vinyl-type monomers (coisotacticity or cosyndiotacticity) and with regard to the diene units (which are often present in a single form, cis-1,4 or trans-1,4). These polymerizations proceed by a radical mechanism.

Polymers Prepared by Matrix Control

Stereoregular polymers can be obtained through the control of polymerization by rigid matrices; this polymerization can also be of the radical type. Some important results have been obtained in this field, especially in polymerization of monomers in canal complexes (see POLYMERIZATION IN CLATHRATES). Other techniques, such as polymerization of monomers in the crystalline state or in the presence of polymer surfaces, give interesting results, particularly in the case of oxetanes and diketene. Polymerization of acrylonitrile oriented on polycaprolactam fibers and preparation of isotactic polystyrene in the presence of binary potassium–graphite inclusion compounds have been described.

BIBLIOGRAPHY

G. Natta, *Science* **147,** 261 (1965).

IUPAC Commission on Macromolecular Nomenclature, "Stereochemical Definitions and Notations Relating to Polymers," *Pure Appl. Chem.* **53,** 733 (1981).

I. Pasquon and U. Giannini, "Catalytic Olefin Polymerization," in J. R. Anderson and M. Boudart, ed., *Catalysis, Science and Technology,* Vol. 6, p. 65, Springer-Verlag, Berlin, 1984.

L. Porri and A. Giarrusso, "Conjugated Diene Polymerization," in G. Allen, *Comprehensive Polymer Science,* Vol. 4, Part 2, p. 53, Pergamon Press, Oxford, 1989.

STEREOREGULAR STYRENE POLYMERS

Styrene is an exceptionally versatile monomer. It can be polymerized by all known mechanisms, including radical, anionic, and with transition metal-containing catalysts. Whereas radical, cationic, and anionic polystyrenes (PS), as a rule, lack steric regularity and are completely amorphous, styrene polymerization with various Ziegler-Natta catalysts produces both highly isotactic, partially crystalline products and highly syndiotactic crystalline polymers. Many physical properties of isotactic PS and syndiotactic PS differ significantly from the properties of amorphous PS (see STYRENE POLYMERS). A large number of ring-substituted styrenes can also be polymerized with Ziegler-Natta catalysts to isotactic and syndiotactic crystalline products. Crystalline polystyrenes have not found industrial application and are not produced on a commercial scale.

Isotactic PS was obtained for the first time by Natta and his co-workers in 1955, simultaneously with isotactic polypropylene and poly(1-butene). Because PS was the first polymer available in both the isotactic and atactic forms, it provided a vivid example of the remarkable differences existing between the physical properties of crystalline and amorphous polymers. Crystalline isotactic PS has a density of 1.08 g/cm³, a melting temperature >210°C, and is insoluble in ketones and esters. The density of amorphous atactic PS is 1.04–1.06 g/cm³, T_g = 70–100°C, and it is soluble in ketones and esters. Syndiotactic PS was prepared for the first time in 1986.

Structure and Properties

Molecular and Crystalline Structure. PS produced with heterogeneous Ziegler-Natta catalysts is a regular head-to-tail polymer with a repeating monomer unit $—CH_2—\underset{|}{\overset{C_6H_5}{CH}}—$. A significant fraction of the polymer, ca 70–80%, has a nearly perfect isotactic structure; the meso-triad content is higher than 95% (see STEREOREGULAR LINEAR POLYMERS).

The highest melting point T_m reported for isotactic PS is 240–242°C, although usually the polymer melts in the 210–220°C range. Crystallization of isotactic PS, both from the melt and from solutions, proceeds very slowly.

Syndiotactic PS can be produced with some homogeneous Ziegler-Natta catalysts. It has a melting point of 266°C and rapidly crystallizes from the melt.

Solutions and Gels. Crystalline isotactic PS is virtually insoluble in any known solvent at ambient temperature. In this respect it strongly differs from sterically irregular PS, which is easily soluble in many organic solvents. At increased temperatures, isotactic PS dissolves in naphthalenic and aromatic hydrocarbons, their halogen- and nitro-derivatives, and esters (benzene, toluene, tetralin, xylenes, chlorobenzenes, nitrobenzene, phthalates), but remains insoluble in alkanes, ethers, and ketones. Isotactic and atactic PS have very similar dependencies of the intrinsic viscosities of their dilute solutions, $[\eta]$, on molecular weight.

Concentrated (3–10%) solutions of isotactic PS in various naphthalenic and aromatic solvents, if cooled rapidly to 15–20°C, form stable transparent gels. The gels exhibit several features typical for crystals, including a specific x-ray diffraction pattern and sharp melting points.

Processing and Uses

Isotactic and syndiotactic PS and some of its crystalline ring-substituted analogues are potentially valuable materials because of their high heat resistance and good dielectric properties. Low crystallinity (40–60%), the low rate of crystallization, and high brittleness represent the principal obstacles for commercial application of isotactic PS as a general-purpose plastic.

BIBLIOGRAPHY

G. Natta, P. Pino, P. Corradini, F. Danusso, E. Mantica, G. Mazzanti, and G. Moraglio, *J. Am. Chem. Soc.* **77,** 1708 (1955).

N. Ishihara, M. Kuramoto, and M. Uoi, *Macromolecules* **21,** 3356 (1988).

A. M. Usmani and I. O. Salyer, *J. Elastomers Plast.* **12,** 90 (1980).

Y. V. Kissin, *Isospecific Polymerization of Olefins with Heterogeneous Ziegler-Natta Catalysts,* Springer-Verlag, New York, 1985.

Y. V. KISSIN
Mobil Chemical Company

STRESS-INDUCED REACTIONS.

See MECHANOCHEMICAL REACTIONS.

STRETCH BLOW MOLDING.

See BLOW MOLDING.

STRUCTURAL FOAMS

Structural foams (SF), also called integral skin foams, are characterized as polymer structures with nearly uniform-density foam cores and integral near-solid skins. The Structural Foam Division of the Society of the Plastics Industry, Inc. (SPI) defines structural foam as a plastic product with integral skins, a cellular core, and enough strength-to-weight ratio to be classed as structural. When these foams are used in load-bearing applications, the foam bulk density is typically 50–90% of the unfoamed bulk density. The ratio of foamed bulk density to unfoamed bulk density is called reduced density, $\phi = \rho_f/\rho_0$. The mechanical properties of structural foam are related to reduced density.

Most structural foam products (ca 85%) are made of thermoplastics. Polyurethane (PUR) is the primary thermoset. Unfilled and unreinforced polymers represent about 70% of SF products. The principal method of conversion (75%) is modified injection molding. Extrusion and reaction injection molding (RIM) account for about 10% each. Small amounts are made by blow molding and rotational molding. Foams are produced in a single-step process by volatilization or decomposition of a measured amount of low molecular weight material.

Structure

Typical structural-foam cells are nearly spherical and about 100 μm in dimension near the center of the cross section and elongated, elliptical, or teardrop in shape. They are smaller near the skin region (Fig. 1) (see also CELLULAR MATERIALS).

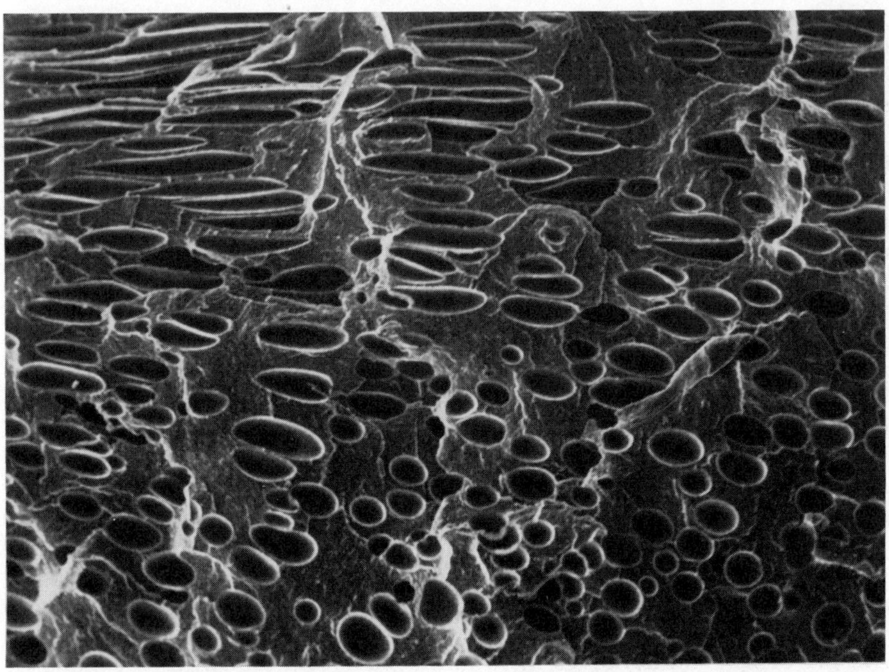

Figure 1. Cross section of polystyrene structural foam showing near-spherical cells close to the centerline and elongated, teardrop cells adjacent to the skin-core transition zone.

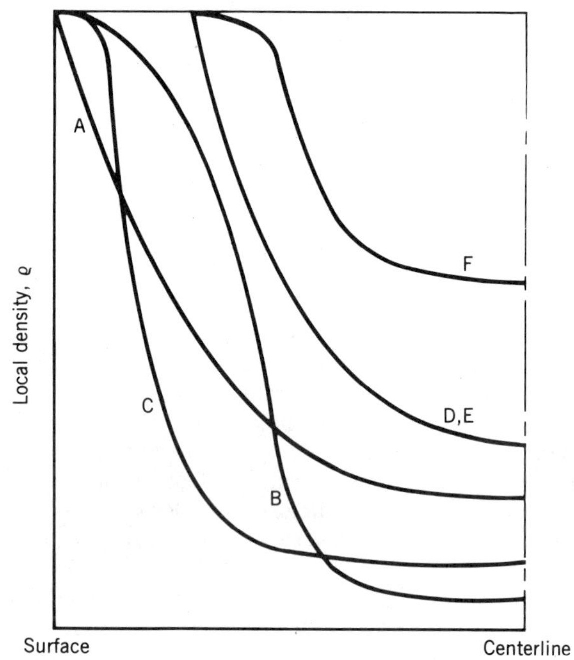

Figure 2. Cross-sectional density gradient for A, crystalline thermoplastic; B, amorphous thermoplastic; C, RIM-type thermoset; D, thermoplastic, moving mold; E, thermoplastic, gas counterpressure; F, thermoplastic, co-injection (two-component).

The near-solid skin is typically 0.25–1.0-mm thick and may contain microbubbles about 10 μm in dimension and surface irregularity or roughness (splay). Amorphous polymers yield broader transition zones than crystalline polymers (Fig. 2).

Processing

Structural-foam processing requires dissolution of a carefully metered amount of gas-producing chemical (foaming or blowing agent) into a softened or molten polymer at processing conditions that are similar to those of conventional injection molding or extrusion. Depending on the process conditions, the blowing agent can be a pure, thermally unstable chemical such as azodicarbonamide (AZ), a volatile liquid such as a chlorofluorocarbon, a low boiling hydrocarbon such as butane, or a gas such as nitrogen. Pressure is applied to the polymer melt to keep the foaming gas or liquid dissolved during resin conditioning and transfer from the extruder or molding machine to the die or mold cavity. Foaming occurs by a sudden drop in pressure. For structural foaming, the amount of foaming agent and the extent of foam expansion are restricted. About 80% of thermoplastic and thermoset RIM structural foam is made by low pressure foam molding.

Other Techniques. In low pressure SF injection molding, the molding-cavity volume is always larger than the volume of resin in an unfoamed state. The low cavity pressure allows microbubbles to nucleate and grow. Foam expansion occurs during filling and growing bubbles are carried to the mold surface, resulting in surface irregularities and imperfections called splay. Many variants of the basic process have been developed to improve surface appearance. Patented technologies include cyclic heating and cooling of the mold surface direct injection of blowing agent gas into the melt as it is being injected into the mold, degassing the foamable resin surface prior to injection, extracting a "packing core" from the mold to allow controlled expansion, and overpacking followed by egression of a small amount of polymer from a fixed-mold configuration.

Molds. Low pressure SF articles can be molded in jig or cast aluminum (type 6061-T651 or A356) and Kirksite molds. Prototypes can be fabricated on metallized epoxy molds. Articles with complex geometries or those requiring special sliders, core pulls, or ejectors should be fabricated on steel molds. Higher pressure foams require conventional injection-molding tool steels (P20, H13). Tool steels are also recommended for SF extrusion dies.

Finishing. For articles where appearance is important, such as cabinets, the surface requires finishing. The low pressure SF surface is first coarsely sanded and primed with a high solids, solvent-based urethane. The finish coat is usually a medium solids, water-based (2000 mPa·s (= cP)) urethane paint, which is oven cured for 30 min at 50–60°C.

Foaming Process

Cellular thermoplastic polymers are formed by dissolution of dissolved gas (or vaporization of dissolved liquid) from polymer melt as the result of a sudden drop in applied pressure. Cells are created by nucleation of microbubbles (microvoids) in the polymer melt as the pressure drops. Bubbles form on nucleating surfaces of inorganic materials such as talc or titanium dioxide that have been added to promote bubble formation. The decomposition of many chemical foaming agents creates "hot spots," which promote early bubble formation. In thermoset-reaction molding, nucleation occurs on inorganic sites and in hot regions that are the result of the polymerization reaction.

Cell formation is the result of phase separation between the dissolved gas and the polymer melt. The initiation of the bubble depends on the differential pressure between the gas within the microbubble and the surrounding resin as well as on the surface temperature. Bubble growth, on the other hand, depends on rheological and mass-transfer phenomena in the resin region around the bubble.

In the thermoplastic-injection process, the foamable resin is rapidly transferred from a high pressure accumulator to a low pressure mold. During transport, resin pressure drops and bubbles are initiated and grow. Ideally, the resin is injected in the mold cavity, which is then sealed to allow isobaric resin foaming to fill the mold cavity. In practice, the ideal is rarely achieved. High speed resin transport is required for uniform cell structure, density, and density distribution throughout the final product. Surface appearance and impact resistance are improved.

Expansion. In low pressure thermoplastic SF injection molding, the time to transfer foamable resin from the accumulator to the mold cavity is on the order of 0.2–5 s. The time to fill the cavity by gas expansion is on the order of 10–30 s. High density skin forms partly because the foam nearest the wall cools rapidly and partly because pressure builds rapidly as the flow front passes.

Surface roughness or splay, characteristic of slow, low pressure thermoplastic SF injection molding, is probably due to bubble rupture at the wave front. In most newer SF techniques formation of these larger bubbles is inhibited, providing an improved surface appearance.

Cycle Time. Cooling accounts for about 90% or more of the total cycle time of low pressure SF injection-molded products. Although the thermal diffusivity of any foamed resin is equal to or higher than that for the unfoamed resin, cooling times at the same wall thickness are usually somewhat longer. As the foam cools, internal gas pressure decreases, and the amount of resin with temperature below the heat deflection temperatures increases. At some point, the demolded article can withstand the internal gas pressure without later distortion. This is the minimum cycle time for that wall thickness and foaming agent concentration (Fig. 3). Cycle time increases with increased blowing agent concentration, melt temperature, wall thickness, and mold temperature.

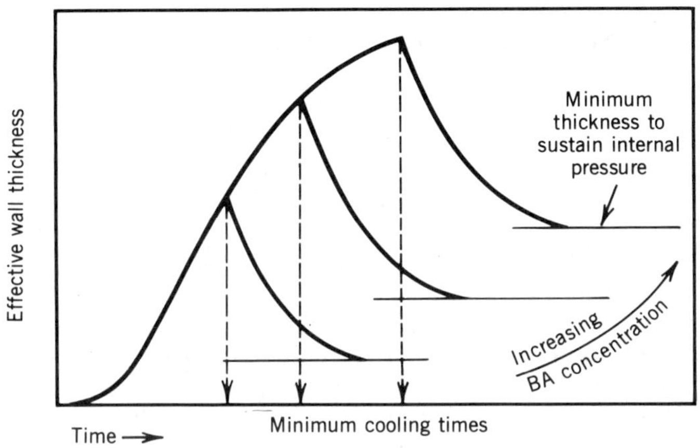

Figure 3. The development of time-dependent wall thickness for minimum cycle time needed to sustain internal gas pressure. The effect of increasing blowing agent (BA) concentration is shown.

Mechanical Properties

Although high density foams are composite structures, they are usually treated as continuous when analyzing response to applied load. Many structural models have been proposed. Models based on a spherical cell in a polymer matrix cube and on series-parallel resistance to applied load have been used to predict Young's modulus of uniform-density cellular plastics. The simplest model relates specific mechanical property to reduced density in a power-law fashion:

$$X = \phi^n$$

where X is the specific mechanical property (the property of the foam divided by the property of the unfoamed resin) and ϕ the reduced density. For uniform-density foams, short-term properties such as tensile, flexural, and compressive strengths and moduli show a square-law dependence on reduced density ($n = 2$). This simple empirical model yields results nearly identical to more complex models. Because it ignores density gradient from skin to core, it provides the lower expected limits on these properties.

Because most structural foams are used in load-bearing applications, flexural properties are important. The stiffness S of a foam beam is obtained from

$$S = E_{\text{eff}}I_{\text{eff}} = \int_{A'} E_y y^2 \, dA'$$

where E_{eff} is an effective foam flexural modulus, I_{eff} an effective moment of inertia, E_y the local modulus at position y across the beam, and A' the beam cross section.

Simple continuum models cannot be used to predict impact and flexural fatigue strengths of structural foams. Stress concentration around individual cells results in crack initiation. Thick skins are desirable for high impact strength.

Applications and Economic Aspects

The structural-foam market can be divided into twelve principal segments: appliances, building and construction, business machines, consumer electronics, furniture, maintenance and waste containers, materials handling, medical and analytical, recreational, tools, transportation, utility enclosure and conduit, miscellaneous. The principal markets are business machines, building and construction, and utility enclosures and conduits.

Structural foam technology competes well with injection molding, profile extrusion, thermoset sheet molding, RIM (unfoamed), and sheet metal and die casting. The economic advantages include the production of large heavy-walled, lightweight articles with low molded-in stresses and low tooling costs. Disadvantages include high finishing costs, long cycle times, and special processing equipment. Most structural-foam processes are competitive where appearance is important.

BIBLIOGRAPHY

F. A. Shutov, *Integral/Structural Polymer Foams: Technology, Properties and Applications,* Springer-Verlag, New York, 1986.

B. C. Wendle, *Structural Foam: A Purchasing and Design Guide,* Marcel Dekker, Inc., New York, 1985.

S. Semerdjiev, *Introduction to Structural Foam,* Society of Plastics Engineers, Brookfield Center, Conn., 1982.

N. C. Hilyard, ed., *Mechanics of Cellular Plastics,* Applied Science Publishers, London, 1982.

James L. Throne
Sherwood Technologies, Inc.

F. A. Shutov
Mendeleev Institute of Chemistry and Technology

STRUCTURAL RELAXATION. See Aging, physical.

STYRENE–ACRYLONITRILE COPOLYMERS. See Acrylonitrile polymers; Styrene polymers.

STYRENE–BUTADIENE–ACRYLONITRILE COPOLYMERS. See Acrylonitrile–butadiene–styrene polymers.

STYRENE–BUTADIENE COPOLYMERS. See Butadiene polymers.

STYRENE POLYMERS

Introduction

The year 1988 was the 50th anniversary of the successful commercialization of polystyrene in the United States. The gigantic business that has developed for polystyrene, rubber-modified polystyrene (HIPS), acrylonitrile–styrene copolymers (SAN), and rubber-modified acrylonitrile–styrene copolymers (ABS) originated with the simple addition polymerization of styrene:

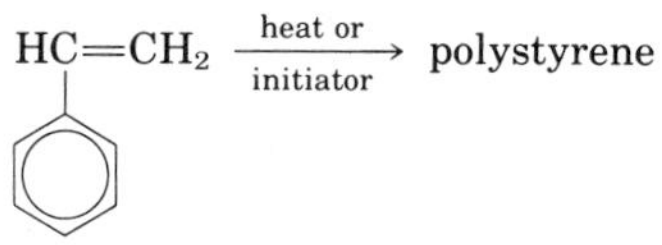

Discovery of the monomer is credited to Newman who, prior to 1786, isolated styrene by steam distillation from liquidambar, a solid resin obtained from a family of trees or shrubs native to the Far East and California. Liquidambar has a vanillalike odor and was used in medicine and perfumery. It contains cinnamic acid, which is easily decarboxylated to yield styrene. The first polymerization of styrene is credited to E. Simon in 1839. He gave the name of styrol to the solid compound obtained after steam distillation from a resinous gum.

Commercialization of Styrene and Polystyrene

In the intervening years between discovery and commercialization, styrene was encountered as an impurity in several commercial chemical processes and was prepared in small quantities by a number of interesting routes. It was not until the cracking of ethylbenzene and the development of successful polymerization inhibitors, however, that styrene (and consequently polystyrene) became a commercial reality.

Globally, today's modern polystyrene plants use a more sophisticated process, producing 6,653,000 t in 1985. The average growth has been 4.5% per year over the last 15 years. Sophisticated computer control systems and rapid, precise analytical techniques supply the product quality information needed for control of the process.

Most general purpose polystyrene is now produced by solution polymerization in a continuous process consisting of one or more vessels (usually in series) followed by very complete volatile removal at high temperatures and high vacuum. The molten polymer is then cooled and pelletized. Suspension polymerization is now used primarily for products for which a small spherical form is desirable such as foam-in-place beads or ion-exchange resins. Rubber-modified polystyrene is manufactured in facilities that are similar to those for the unmodified product except that multiple agitated reactors in series are generally used to give the desired rubber particle size, shape, and occlusion density. Emulsion polymerization remains the principal mode for production of ABS resins; lesser quantities are produced by the solution process.

Characteristics and Uses of Polystyrene

Styrene polymers have some unique properties that make them useful in a wide range of products. The single most important characteristic of general purpose polystyrene is that it is a glasslike solid below 100°C. Above this temperature, commonly called the glass-transition temperature, the polymer chain (on a molecular level) has rotational freedom, which allows large-chain-segment mobility. The polymer is thus fluid enough to be easily shaped into useful forms. Below the glass-transition temperature, polystyrene possesses considerable mechanical strength, allowing it to be used in load-bearing tasks in thousands of applications.

EUGENE R. MOORE
The Dow Chemical Company

Monomers

Styrene ($C_6H_5CH{=}CH_2$) has grown to be a world-scale commodity chemical over the last 50 years. A plant of any size can be purchased on a turnkey basis from engineering contractors, putting the manufacture of styrene within the reach of any company with the available capital. The Badger Co., Inc. has built the largest number of plants based on technology from Cosden Chemical Co., Union Carbide Corp., and Mobil Chemical Co., Lummus Crest, Inc. has built many plants based on Monsanto Chemical Co. technology. This helped make styrene the global 10 million t/yr business that it was in 1986.

Properties of Styrene

Styrene is very reactive and readily undergoes homopolymerization and copolymerization ($Q = 1.0$, $e = -0.8$). Strong acids and peroxides can initiate polymerization. Although styrene does not react with bases, bases react with the *t*-butylcatechol (TBC) inhibitor, allowing the styrene to polymerize. Styrene oxidizes to form various compounds, including benzaldehyde, formaldehyde, formic acid, and polystyrene peroxide (1:1 styrene:oxygen copolymer). Styrene also reacts with chlorine and other halogens. Table 1 lists the physical properties of styrene monomer.

Manufacture

All current styrene manufacture starts with ethylbenzene, followed by conversion to a crude styrene, which then requires finishing to produce the pure product.

Dehydrogenation of Ethylbenzene. Direct dehydrogenation of ethylbenzene to styrene accounts for most of the commercial production of the monomer.

Oxidative Processes. Styrene can also be manufactured, along with a coproduct, by oxidation of ethylbenzene.

Finishing. The distillation of the crude styrene product from the dehydrogenation reactors is also an evolving technology. Time and temperature are both minimized to maintain the styrene as a monomer and prevent premature polymerization. At least three steps are involved. Benzene and toluene are removed, and then ethylbenzene is separated out for recycling

Table 1. Physical Properties of Styrene Monomer

Property	Value
molecular weight	104.153
boiling point, °C	145.15
freezing point, °C	−30.6
critical density D_c, g/mL	0.297
critical pressure P_c, MPa[a]	3.83
critical temperature T_c, °C	362.1
critical volume V_c, mL/g	3.37
flammable limits, vol % in air	1.1–6.1
flash point (tag closed cup), °C	31.1
autoignition, °C	490
heat of combustion[b] ΔH_c, kJ/g[c]	−40.93
heat of formation ΔH_f, kJ/g[c] at 25°C	
gas	1.415
liquid	0.9928
heat of fusion ΔH_m, kJ/g[c]	−0.1056
heat of polymerization ΔH_p, kJ/g[c] at 25°C	−0.6702
heat of vaporization ΔH_v, kJ/g[c]	
25°C	0.4217
typical volume shrinkage on polymerization, %	17.0
vapor pressure, kPa, Antoine equation[d]	$\log_{10}P = [6.08201 - 1445.58/(209.43 + T\,°C)$[e]
refractive index, n_D[f]	1.54969
specific heat (liquid), J/(g·K)[g]	
0°C	1.6367
40°C	1.7489
80°C	1.8779

[a] To convert MPa to psi, multiply by 145.
[b] At constant pressure and 25°C. All reactants and products are gases.
[c] To convert kJ/g to Btu/lb, multiply by 430.4.
[d] $\text{Log}_{10}P$ mm Hg = $[6.95711 - 1445.58/(209.43 + T\,°C)]$.
[e] To convert kPa to psi, multiply by 0.14503.
[f] $n_D = 1.55830 - 0.000574T\,°C$.
[g] To convert J/(g·K) to Btu/(lb·°F), multiply by 0.2389.

to the reactors. Finally, styrene is distilled away from the tars and polymers. The distillation is performed under a vacuum to keep the temperature below 110°C at the bottom of the columns.

Economic Aspects

Table 2 lists the principal U.S. manufacturers and capacities. The price of styrene has fluctuated widely in the last few years. About 60–65% of the cost of styrene is in its raw material, benzene. The benzene cost is tied to the price of crude oil, which has been quite volatile. Styrene manufacturing capacity has been abundant for some time, but recent plant closings, along with a steady demand, pushed the price of monomer to close to $1/kg in mid-1987.

Virtually all styrene produced is used in the manufacture of plastics, synthetic rubbers, and latices.

Table 2. Principal U.S. Styrene Manufacturers, 1986

Company		Capacity, 10^3 t/yr
Cos-Mar, Inc.	Carville, La.	680
Sterling Chemical Co.[a]	Texas City, Tex.	680
The Dow Chemical Company	Freeport, Tex.	590
Lyondell Petrochemical Co.[b]	Channelview, Tex.	454
Hoechst Celanese[c]	Bayport, Tex.	408
Amoco Corp.	Texas City, Tex.	358
Chevron Corp.	St. James, La.	272
El Paso Products Co.	Odessa, Tex.	145
Atlantic Richfield Co.	Beaver Valley, Pa.	100

[a] Formerly owned by Monsanto Chemical Co.
[b] A subsidiary of Atlantic Richfield Co.
[c] Formerly American Hoechst Corp. All production is tolled for Huntsman Chemical Ltd.

Health and Safety

Styrene monomer is regarded as a relatively safe organic chemical: its single-dose oral toxicity is low. Prolonged exposure to the skin may cause irritation, or even a burn, but styrene is unlikely to be absorbed through the skin in harmful amounts. Styrene can cause moderate irritation to the eyes, with possible corneal injury.

Styrene has a flash point of 31°C and a flammable range of 1.1–6.1 vol % in air. It undergoes exothermic polymerization quite readily, so uncontrolled polymerization can give a vapor cloud with explosive potential or pressure buildup in a closed container.

Styrene Derivatives

Vinyltoluene (Methylstyrene). Vinyltoluene (VT, MS) ($CH_3C_6H_4CH{=}CH_2$) is a specialty monomer with properties similar to those of styrene. The vinyl and methyl side chains may be ortho, meta, or para. Commercially, it is available in two different isomer mixes. VT proved to be more difficult to make than styrene; thus it has not gained a large market share. However, it has an enduring specialty position because of its unique properties.

Properties. Table 3 lists the physical properties of VT and *p*-MS.

Manufacture. The three isomers of vinyltoluene are formed by dehydrogenation of the corresponding isomers of ethyltoluene, so the mix of this feedstock determines the vinyltoluene mixture.

Applications. Vinyltoluene is currently available for about $1.35/kg and is generally used in copolymers and as a specialty monomer in paint, varnish, and unsaturated polyester formulations. Its lower vapor pressure is of particular value in the latter applications.

Divinylbenzene. Divinylbenzene (DVB) ($C_6H_4(CH{=}CH_2)_2$) is a cross-linking agent that may improve polymer properties. It is manufactured by dehydrogenation of a mixture of diethylbenzene isomers. The commercial monomer is mainly a mixture of *m*- and *p*-DVB diluted with ethylvinylbenzenes. Table 4 lists the physical properties of the 55% commercial mixture.

Health and Safety. Divinylbenzene is similar to styrene monomer in its toxicological properties, although the ACGIH TLV for DVB is 10 ppm.

Applications. DVB is used in ion-exchange resins.

Table 3. Physical Properties of Vinyltoluene (VT) and *p*-Methylstyrene (*p*-MS)

Property	VT	*p*-MS
molecular weight	118.18	118.18
boiling point, °C	167.7	170.0
freezing point, °C	−77	−34.
flash point (tag closed cup), °C	52.8	
heat of combustion[a] ΔH_c, kJ/g[b]	−41.17	
heat of formation ΔH_f, kJ/g[b] gas at 25°C	0.9743	
heat of polymerization ΔH_p, kJ/g[b] at 25°C	−0.592	−0.592
heat of vaporization ΔH_v, kJ/g[b] at 25°C	0.4261	0.4109

[a] At constant pressure and 25°C. All reactants and products are gases.
[b] To convert kJ/g to Btu/lb, multiply by 430.4.

Table 4. Physical Properties of Divinylbenzene[a]

Property	Value
molecular weight	130.191
boiling point, °C	195
freezing point, °C	−45
flash point (tag closed cup), °C	65.6
autoignition, °C	505

[a] All properties, except molecular weight, apply to DVB 55 (mixture of 55% DVB and 45% ethylvinylbenzene).

α-Methylstyrene. α-Methylstyrene ($C_6H_5C(CH_3){=}CH_2$) is a monomer with a much slower polymerization rate than that of styrene. It was once commercially produced by the dehydrogenation of isopropylbenzene (cumene). However, it is now produced by cumene oxidation, since it occurs as a by-product in the manufacture of phenol and acetone by this process. Table 5 lists the physical properties of α-methylstyrene.

Table 5. Physical Properties of α-Methylstyrene

Property	Value
molecular weight	118.18
boiling point, °C	165
freezing point, °C	−23.2
flash point (Cleveland open cup), °C	57.8

Health and Safety. α-Methylstyrene is similar to styrene in its toxicological properties. The ACGIH TLV is 50 ppm.

Applications. α-Methylstyrene homopolymers are used as plasticizers in paints, waxes, and adhesives. The monomer is also used to form a copolymer with methyl methacrylate, which has a high heat-distortion temperature.

Chlorostyrene. The ortho and para isomers of chlorostyrene ($ClC_6H_4CH{=}CH_2$) are reactive monomers produced in research quantities by several specialty companies. *p*-Chlorostyrene has a boiling point of 192°C and a flash point of 60°C. It is combustible. Chlorostyrene is more reactive than styrene and is easily polymerized under thermal conditions.

Vinylbenzyl Chloride (Chloromethylstyrene). Vinylbenzyl chloride ($CH_2{=}CHC_6H_4CH_2Cl$) is a useful monomer because it contains a polymerizable double bond and a reactive benzylic chloride group. The commercial monomer is a mixture of 70% meta and 30% para isomers. Vinylbenzyl chloride has a high boiling point (98.9°C at 1.33 kPa) and low flammability.

Vinylbenzyl chloride is made by The Dow Chemical Company in the United States and Siemi in Japan and is available for about $60/kg. Vinylbenzyl chloride is used primarily as a comonomer in a variety of specialty plastics. The monomer is a lacrimator.

BIBLIOGRAPHY

R. H. Boundy and R. F. Boyer, eds., *Styrene, Its Polymers, Copolymers, and Derivatives,* Reinhold Publishing Corp., New York, 1952; reprinted by Hafner Publishing Corp., Darien, Conn., 1970, p. 31.

J. Santodonato and co-workers, Monograph on Human Exposure to Chemicals in the Workplace; Styrene, PB86-155132, Syracuse Research Corp., Syracuse, N.Y., July 1985.

K. E. Coulter, H. Kehde, and B. F. Hiscock, "Styrene and Related Monomers," in E. C. Leonard, ed., *Vinyl Monomers,* Wiley-Interscience, New York, 1969.

P. J. Lewis, C. Hagopian, and P. Koch in M. Grayson, ed., *Kirk-Othmer Encyclopedia of Chemical Technology,* Vol. 21, 3rd ed., John Wiley & Sons, Inc., New York, 1983, pp. 793, 795.

DENIS H. JAMES
JOHN B. GARDNER
ERNEST C. MUELLER
The Dow Chemical Company

Polymerization

Styrene, along with most of its substituted derivatives, can act as either an electron-donating or an electron-withdrawing center. Therefore, in addition to free-radical polymerization, these monomers can be polymerized by all other primary progagation mechanisms: anionic, cationic, and coordination (Ziegler-Natta).

Free-radical Polymerization

Styrene monomer can generate enough free radicals during heating that high conversion and high molecular weight polymer can be prepared without chemical initiators. A Diels-Alder reaction between two monomer molecules provides free radicals for the polymerization reaction:

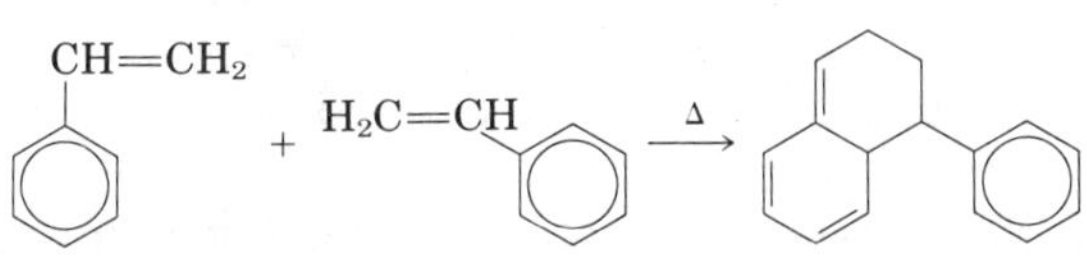

This dimerization produces two stereoisomers of 1-phenyltetralin; axial 1-phenyltetralin and equatorial 1-phenyltetralin:

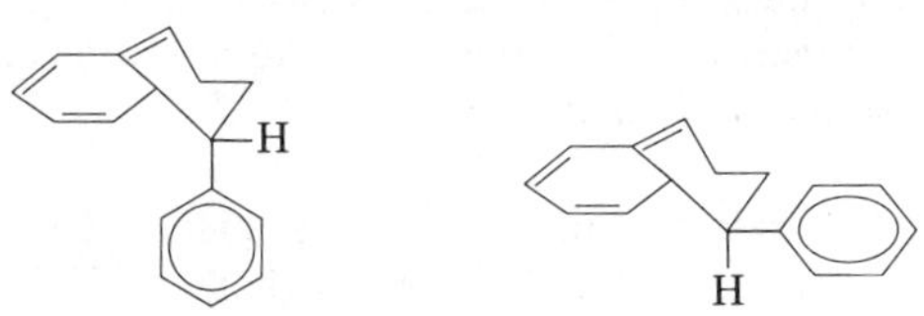

Only the axial isomer is able to react with another styrene monomer to produce the two monoradicals that are responsible for the thermal initiation of chain propagation:

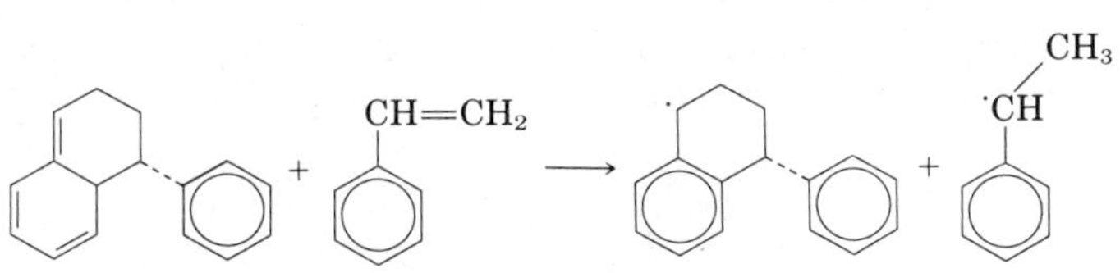

The propagation mechanism for the chain growth of polystyrene proceeds predominantly by head-to-tail addition:

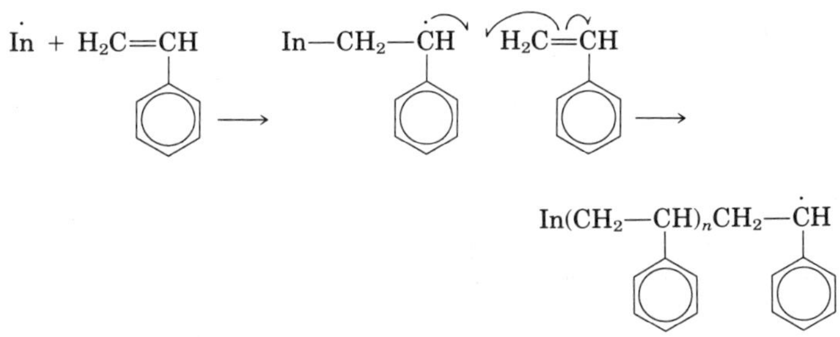

The addition occurs in this manner because of the greater stability of the benzylic radical over the methylene radical. The benzylic radical is stabilized by the resonance structures possible with the phenyl ring:

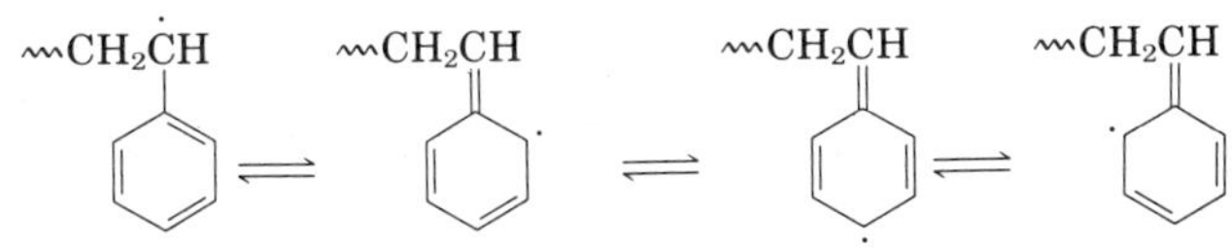

A small number of head-to-head or tail-to-tail linkages may occur in a given polymer chain.

Initiators. Molecular weight, molecular weight distribution, and conversion rate at a given temperature can be manipulated by using initiators. The addition of a free-radical initiator at various stages of conversion can also be used to produce novel bimodal molecular weight distributions, increase conversion, and reduce residuals in suspension polymerization. Suspension and emulsion systems invariably need initiators.

Over the years, hundreds of different initiators have been used for polystyrene. Commercially, the most important initiator types for styrenics have been peroxide and azo compounds (see AZO COMPOUNDS; PEROXY COMPOUNDS).

Inhibition and Retardation. Because styrene monomer undergoes thermal self-initiation, it requires the use of an inhibitor during manufacturing steps such as distillation and during shipping and storage to avoid polymer formation.

There are many materials that inhibit the polymerization of styrene monomer. Most of these are highly efficient radical scavengers that react with free radicals to form stable moieties incapable of further reaction with styrene. The majority are quinones, hindered phenols, or amines and require traces of oxygen to function.

If the inhibitors are not removed before polymerization, they create an induction period in batch reactions during which little or no polymer is formed until all of the inhibitor has been consumed. In a continuous system, the effect is similar to batch under plug-flow conditions, but acts more like retardation in a well-mixed reactor. Retardation is a more serious problem than inhibition in styrene polymerization because conversion rate is lost.

Chain Transfer. As with all radical polymerizations, some chain-transfer reactions occur in the polymerization of styrene monomer (see CHAIN TRANSFER). Chain transfer is a reaction of the free radical of the propagating polymer P· with another compound HA such that an inactive polymer PH is formed along with a new initiator fragment A·:

$$P\cdot + HA \longrightarrow PH + A\cdot$$

In styrene polymerization, the chain-transfer agent can be solvent, monomer, initiator, polymer, or an added chemical agent. The chain-transfer constant C_Q is the ratio of k_{tr}/k_p, where k_{tr} and k_p are the rate constants for chain transfer and propagation, respectively. As C_Q increases, the chain-transfer agent becomes more effective.

The most important aspect of chain-transfer reactions for the commercial polymerization of styrene is the molecular weight and molecular weight distribution control available through the proper use of added agents. Mercaptans are by far the most widely used chain-transfer agents for polystyrene. α-Methylstyrene dimer may also be used in some commercial systems.

Termination Reactions. In the free-radical polymerization of styrene at temperatures up to 80°C, termination of the growing chain occurs almost exclusively by combination, according to chemical evidence.

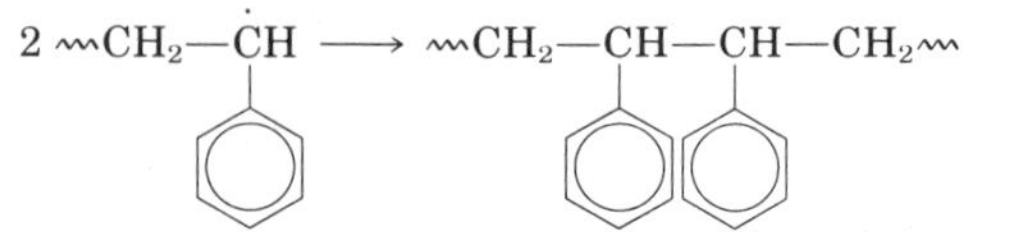

In addition, at temperatures above 80°C, kinetic modeling has shown that molecular weight data are best fit by using combination over disproportionation as the principal mode of termination, although this has not been proven to occur exclusively. There has been no good evidence to support a significant amount of the disproportionation reaction for polystyrene termination.

Ionic Polymerization

The phenyl group of the styrene monomer is uniquely able to act as an electron-donating or an electron-withdrawing center:

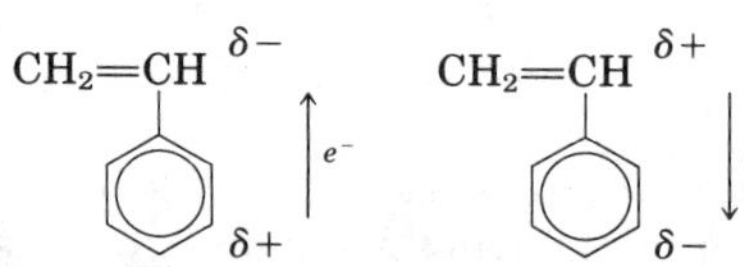

This electronic effect allows the growing end of the polymer chain to be either a carbocation or carbanion with an associated counterion. It also allows styrene to be polymerized by coordination (Ziegler-Natta) catalysis.

Anionic Polymerization. ***Initiation.*** The initiation of anionic polymerization of styrene involves the reaction of the highly purified monomer with an extremely strong base. Several types of initiator are known; however, organolithium compounds are the most studied and best understood.

Propagation. The rate of polymerization for this type of system is expressed as

$$R_p = k_p[M^-][M]$$

where $[M^-]$ = concentration of polystyryllithium. Because there are no chain-transfer reactions present, $[M^-]$ is given by the concentration of initiator. The rate constant k_p is affected by several factors, particularly the solvent.

The rate constant of propagation (k_p) for anionic polymerization is closer to the k_p for free-radical polymerizations than the overall rate of polymerization would indicate. The much larger rates observed in anionic polymerizations are due to a larger concentration of active species. In typical free-radical systems, $[M\cdot]$ is only about 10^{-8}–10^{-7} $\bar{M}$, whereas for propagating anions, $[M^-]$ may be as much as 10^{-3} to 10^{-2} $\bar{M}$. This means that the rate of polymerization R_p is 10^4–10^6 times faster in most anionic polymerizations than in radical polymerizations.

Dianionic Initiators. Difunctional initiators are also used in anionic polymerization of styrenics. One of the earliest systems involves the radical anions produced by reaction of an alkali metal with naphthalene at −78–0°C. The initiator is generated in ether solvents by direct reaction of sodium or potassium metal with naphthalene. The metal napthalenide is formed. When this initiator is used with styrenics, the radical anions of styrene are formed, which quickly combine to form a dianion. The dianion is then free to propagate in both directions at the same rate. The molecular weight reflects the dianionic nature of these systems and is given by the equation:

$$M_w = 2 \text{ monomer (g)/initiator (mol)}$$

Block Copolymers. Several commercial uses for anionic polymerization of styrene are tied directly to the living nature of the polystyryl anion. This property allows for the unique control of block copolymer structures such as polystyrene–isoprene rubbers. Diblock, triblock, and other multiblock copolymers are the subject of several reviews and have been used in the commercial manufacture of such products as Kraton rubber (Shell Oil Co.) (see ELASTOMERS, THERMOPLASTIC).

Chain-end Functionalization. The other use of living anionic polystyrene involves chain-end functionalization. In principle, the polystyryl anion end groups can be converted to a wide variety of functional, or reactive, end groups. The number of end-group possibilities is limited only by the imagination of the researcher.

Uses of the reactive anion end group fall into three categories dependent on the nature of the terminating agent: (a) coupling reactions for chain extension for controlled polystyrene structures, eg, star-branched polymers (see DENDRITIC POLYMERS); (b) coupling or linking reactions with other polymers or oligomers, eg, graft copolymers; and (c) reaction in further polymerizations as another monomer or as an initiator, eg, macromonomers (see MACROMERS).

Cationic Polymerization. The active species in styrene polymerization may be a carbocation (carbonium ion) associated with an anionic counterion. The initiation of cationic styrene polymerization is carried out in the presence of strong acids, traditional protonic acids such as perchloric, hydrochloric, or sulfuric acids, or strong Lewis acids such as BF_3, $AlCl_3$, PF_5, or $TiCl_4$. Additionally, alumina, silica, and molecular sieves will initiate cationic polymerization.

The protonic acids initiate by a straightforward mechanism of protonation:

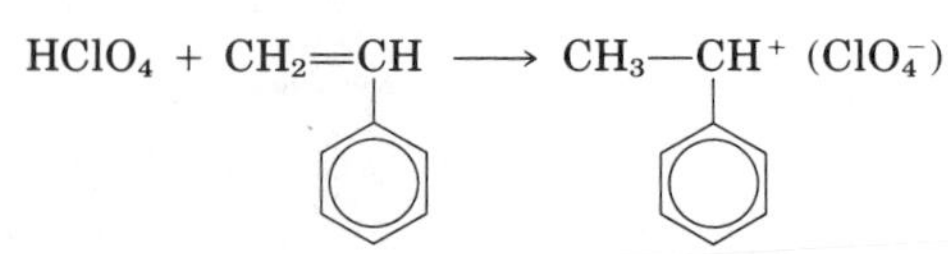

These polymerizations can be very rapid and release a large amount of heat very quickly. Molecular weight control is a significant problem, because low temperatures are necessary to obtain high molecular weight. The polarity of the solvent plays a role in solvating the ion pair and, in general, the greater the dielectric constant of the solvent, the greater the rate of polymerization and the higher the molecular weight of the polymer.

Chain Transfer. Chain-transfer reactions are common and difficult to control in cationic polymerization of styrene, making molecular weight difficult to control. Although transfer to polymer and solvent are known to occur, chain transfer to monomer is the most important mechanism in cationic styrene polymerization.

Living Systems. A living carbocationic polymerization for styrene has been described. The chemistry involves the reaction of an organic acetate with BCl_3 and styrene at about −30°C in chloromethane.

Coordination Polymerization

Styrene monomer can also be polymerized to stereoregular structures through the proper use of coordination catalysts. Until recently, only isotactic polystyrene had been described (see INSERTION POLYMERIZATION; STEREOREGULAR LINEAR POLYMERS). More recently the preparation of syndiotactic polystyrene has also been discovered. Syndiotactic polystyrene is prepared with homogeneous coordination catalysts (see STEREOREGULAR STYRENE POLYMERS). Highly isotactic polystyrene is prepared using heterogeneous catalysts obtained from the reaction of aluminum alkyls with titanium chlorides.

The syndiotactic polymer has a melting point of about 270°C with a glass-transition temperature for the amorphous phase at about 100°C. Isotactic polystyrene melts at about 220–230°C. The crystallization rate for syndiotactic polystyrene also appears to be much greater than that of the isotactic material, which may give it more commercial potential.

Copolymerization

The variety of copolymers that can be prepared with styrene have greatly expanded the use of the monomer. Dramatic improvements or modifications of physical properties can be achieved by choosing the right comonomer. Table 1 lists reactivity ratios for typical styrene comonomers. A more extensive list has been published.

The reactivity ratios define the composition of the copolymer that is instantaneously produced from a monomer mixture. This is true in either a batch reactor or a continuous stirred-tank reactor. However, in the batch reactor, the copolymer composition drifts from that produced by the initial monomer composition to that produced by the monomer composition at the end of the reaction. In the ideal continuous stirred-tank reactor, all copolymer produced is of the same composition, which is determined by the ratio of the unreacted monomers. Homogeneity in a continuous reactor system is controlled by the intensity of mixing and the number and location of addition points for the more reactive monomer (see COPOLYMERIZATION).

Copolymers of styrene include styrene-*co*-acrylonitrile (SAN), styrene-*co*-methyl methacrylate (SMMA), styrene-*co*-butadiene (SB), acrylonitrile–butadiene–styrene (ABS), styrene-*co*-maleic anhydride (SMA), styrene-*co*-divinylbenzene, and styrene-*co*-α-methylstyrene (SAMS).

Rubber-modified Polystyrene. Rubber is incorporated with polystyrene in commercial high impact polystyrene (HIPS). In order to provide a rubber with better compatibility with the continuous polystyrene phase, it is advantageous to graft the rubber with polystyrene. The most common commercial rubbers are pure polybutadiene or copolymers with styrene; *cis*-polybutadiene is particularly popular. Ethylene–propylene–diene terpolymer (EPDM rubber) is used to prepare impact-resistant polymers that are resistant to outdoor aging (see ETHYLENE–PROPYLENE ELASTOMERS).

In mass-produced rubber-modified polystyrene, the rubber is dissolved in the styrene. As the polymerization of the styrene proceeds, two phases are formed: a rubber-rich phase and a polystyrene-rich phase. Grafting occurs during the polymerization of styrene when some of the free radicals react with the rubber (see GRAFT COPOLYMERS). Grafting is important because the graft copolymer concentrates at the interface and strongly affects particle size, morphology, and toughness.

Commercial Processes

As with most polymeric materials, process and product considerations for polystyrene are intimately related. The choice and evolution of processes and range of possible operating conditions all weigh heavily on the quality, cost, and range of products that can be produced. Process evolution is closely tied to the capabilities of producing a wide range of products with useful combinations of properties at minimum cost.

Although polystyrene can be produced by free-radical, anionic, cationic, and Ziegler mechanisms, only the free-radical process is of major commercial interest. Anionic polymerization of styrene is limited to the production of narrow molecular weight distribution polystyrenes in laboratory quantities. These polymers are used for instrument calibration and theoretical studies of properties as functions of molecular weight. The anionic polymerization process is used commercially only for block copolymers of styrene with butadiene. Cationic polymerization is only used to produce low molecular weight polystyrenes for coatings and glues. All commercial molding and extrusion resins have weight-average molecular weights between 100,000 and 400,000 and distribution indexes (M_w/M_n) between 2 and 4. This range of molecular weights can be produced

Table 1. Styrene Comonomer Reactivity Ratios[a]

Monomer 2	r_1 [b]	r_2 [c]	Temperature, °C
acrylonitrile	0.4	0.04	60
butadiene	0.5	1.40	50
maleic anhydride	0.014	0.0	80
methyl methacrylate	0.52	0.46	60
m-divinylbenzene	0.65	0.60	60

[a] Styrene is monomer 1.
[b] $r_1 = k_{11}/k_{12}$.
[c] $r_2 = k_{22}/k_{21}$.

by free-radical polymerization at economic production rates. An overview of processes for producing polystyrene is given in Table 2.

Continuous Solution Reactors. The three main characteristics that distinguish the various solution-process reactors are the mode of heat transfer, the degree of axial (end-to-end) mixing, and the degree of radial (side-to-side) mixing. Types of reactors include tower reactors, stirred-tank reactors (including recirculated loop), stratifier reactors, and axially segregated agitated reactors.

Rubber-modified Polystyrene Polymerizers. The current process for rubber-modified polystyrene (HIPS) evolved from early work, demonstrating that if rubber was dissolved in the feed and the polymerization was carried out in agitated tower-type reactors, the final polystyrene product contained small rubber domains that contributed significant toughness to the product. Both stratifier and backmixed reactors have been used to produce rubber-modified styrenics. Some rubber-modified polystyrene is still produced by the mass-suspension process.

Devolatilization. Devolatilization is the process of removing volatiles from the reactor effluent to put the polymer into usable form. Flash tank evaporation, wiped-film evaporators, and devolatilizing extruders have all been used commercially to devolatilize styrene plastics.

Mixing. Other than polymerization and devolatilization, the most important unit operation in polystyrene processes is mixing. Mixing is used to blend partial polymers of different compositions, molecular weights, or morphologies; control morphology of two-phase systems; blend in devolatilization aids; or blend in additives and plasticizers either before or after devolatilization. Mixers can be divided into two categories, rotational or static. Rotational mixers generally consist of a set of moving elements interacting with static elements on walls at fairly close tolerances. Power requirements for the rotational units may be a factor of 10 higher than for static mixers, but residence times can be short and are less than a minute in many cases. Static mixers are mixers with internal flow paths that promote mixing but have no moving parts. They have come into much wider use in the last 10 years and at least 20 different geometries are sold by equipment vendors.

Process Control. The early polystyrene processes required little process control because they were run at rates that were inherently stable. As control systems moved from jacket temperature control to cascade control units controlling jacket temperature by monitoring polymer temperature, production rates

Table 2. Polymerization Techniques for Producing Polystyrene

Method	Polymerizing system	Advantages	Disadvantages	Cost factors
bulk (batch)	pure styrene monomer	simplicity; high clarity	poor heat control	noncommercial
solution (batch)	styrene diluted with solvent	much better temperature control	solvent reduces molecular weight and rate	noncommercial
continuous bulk–solution	styrene monomer with or without solvent fed continuously to reactor	more uniform product; high clarity; excellent color; low volatile levels	mechanical problem in moving highly viscous finished polymer	expensive equipment
suspension	water used as a carrier with stabilizing agents to keep material in suspension, and one or more catalysts to speed polymerization	excellent heat of polymerization removal; simple kettles; very high conversions to polymer	possible contamination from water and stabilizing agent; batch operation	catalysts; stabilizers; washing and drying of polymer; may require further devolatilization or reextrusion; waste-water treatment
emulsion	water used as carrier with emulsifying agent to give extremely small particles	rapid reaction and excellent heat of polymerization removal; some copolymerizations not possible by other techniques proceed readily in emulsion; simple kettles	contamination of polymer with water and emulsifying agents; color and clarity of polymer deficient; broad molecular weight distribution; usually a batch operation	catalysts; emulsifiers and stabilizers; coagulants; washing and drying of polymer; pelleting of dried polymer; waste-water treatment
solution, ionic	styrene and other monomers diluted with solvent	can produce copolymers that cannot be produced readily by other polymerization method, eg, styrene–butadiene block copolymer; rapid polymerization rate	difficulty in maintaining good color; solvent and catalyst residue removal; high sensitivity to impurities	expensive equipment; recovery of solvent and recycle; catalyst; purification

could move into the region of higher temperature differences that is "inherently" unstable. Experience with active control systems was a major contributor to the reduction in average residence time by a factor of four to six since the solution process was developed. Today, most manufacturing plants use computer control.

BIBLIOGRAPHY

G. Odian, *Principles of Polymerization,* 2nd ed., John Wiley & Sons, Inc., New York, 1981, pp. 274, 382.

M. Morton, *Anionic Polymerization Principles: Principles and Practice,* Academic Press, New York, 1977.

R. H. M. Simon and D. C. Chappelear in J. N. Henderson and T. C. Bouton, eds., *Polymerization Reactors and Processes, ACS Symposium Series,* Vol. 104 American Chemical Society, Washington, D.C., 1979, pp. 71–112.

BERNARD J. MEISTER
MICHAEL T. MALANGA
The Dow Chemical Company

Properties

GENERAL-PURPOSE POLYSTYRENE

Although general-purpose polystyrene is often called crystal polystyrene, this designation refers only to the clarity of the resin and not to its molecular order. The commercial success of polystyrene is largely due to transparency, lack of color, ease of fabrication, thermal stability, low specific gravity, relatively high modulus, excellent electrical properties, and low cost. Typical applications for general-purpose polystyrene resins include packaging products, disposable medical ware, toys, tumblers, cutlery, tape reels, storm windows, and consumer electronics. Foam applications include egg cartons, meat-packaging trays, building insulation, "clamshells" for fast-food packaging, and expanded polystyrene cushioning materials for packaging.

Currently, all commercially produced polystyrene is atactic (nonstereospecific) and is produced by free-radical polymerization. Much of polystyrene's advantage over other thermoplastics is due to its amorphous nature that leads to both clarity and fabrication ease.

Commercial Grades

There are three common commercial grades of general-purpose polystyrene: easy flow, medium flow, and high heat. The choice of resin grade depends mostly on the fabrication method, but also on the end-use application. Easy flow and medium flow polystyrenes are used primarily for injection molding; high heat polystyrene is used mostly for extrusion applications. Table 1 lists typical properties of specific resins of each resin type. Most commercial polystyrene producers have resins in each of these grades.

Specialty Grades

Besides the common grades of general-purpose polystyrene, there are several specialty grades that are produced by only one or two manufacturers. These have some unique properties and are usually more expensive than the larger volume grades (Table 2). They include fast-cycle resins, food-contact-grade resins, low molecular weight resins, and poly(*p*-methylstyrene).

Polystyrene is a nonpolar polymer and is soluble in, and subject to attack by, many nonpolar organic solvents. The solubility of polystyrene in organic solvents is largely a function of the difference in solubility parameters of the polystyrene and the solvent (see SOLUBILITY OF POLYMERS).

Effect of Molecular Weight Distribution on Properties

The shape of the molecular weight distribution curve and the magnitude of the average molecular weight have a large influence on the thermal, mechanical, and rheological properties of polystyrene. Even if the average molecular weight, the

Table 1. Properties of Typical Commercial Grades of Polystyrene

Property	ASTM test method	High heat[a]	Medium flow[b]	Easy flow[c]
melt flow rate, g/10 min (condition G)	D1238	1.6	7.5	16
Vicat softening temperature, °C	D1525	108	102	88
deflection temperature under load,[d] °C	D648	103	84	77
tensile strength at break, MPa[e]	D638	56.6 (44.1)[f]	44.8	35.9
flex strength at break, MPa[e]	D638	82.8	82.8	
tensile elongation at break, %	D638	2.4 (1.4)[f]	2.0	1.6
tensile modulus, MPa[e]	D638	3,340 (3,240)[f]	2,450	3,100
flexural modulus, MPa[e]	D790	3,155	3,170	
Izod impact strength notch, J/m[g]	D256	24 (14)[f]	16	19
hardness, Rockwell, M scale	D785	76	75	72
molecular weight				
$\overline{M}_w$	[h]	300,000	225,000	218,000
$\overline{M}_n$	[h]	130,000	92,000	74,000

[a] Data provided by Dow Chemical Co. (Styron 685D); injection-molded specimens.
[b] Data provided by Mobil Chemical Co. (PS 2124); injection-molded specimens.
[c] Data provided by Polysar, Inc. (555 Resin); injection-molded specimens.
[d] Injection-molded, annealed specimens; load = 1.82 MPa (264 psi).
[e] To convert MPa to psi, multiply by 145.
[f] Data in parentheses are from compression-molded specimens.
[g] To convert J/m to ft-lb/in., divide by 53.38.
[h] Determined by size-exclusion chromatography at Dow Chemical Co.

Table 2. Properties of Several Specialty Grades of Polystyrene

Property	ASTM test method	Fast cycle[a]	Low residual[b]	Increased heat[c]
melt-flow rate, g/10 min (condition G)	D1238	9.0	1.6	4.5
Vicat softening temperature, °C	D1525	106	108	115
deflection temperature under load,[d] °C	D648	100	103	97
tensile strength at break, MPa[e]	D638	40.0	44.1	47.5
flex strength at break, MPa[e]	D638	51.0	82.8	79.3
tensile elongation at break, %	D638	1.3	1.4	
tensile modulus, MPa[e]	D638	3,170	3,240	2,410
flexural modulus, MPa[e]	D790	3,100	3,155	2,900
Izod impact strength (notched), J/m[f]	D256		14	16
hardness, Rockwell, M scale	D785	74	76	81
molecular weight				
$\bar{M}_w$	[g]	195,000	300,000	
$\bar{M}_n$	[g]	95,000	130,000	

[a] Data provided by Dow Chemical Co. (Styron 666 APR); compression-molded specimens.
[b] Data provided by Dow Chemical Co. (Styron LR–175); compression-molded specimens.
[c] Poly(*p*-methylstyrene) data provided by Mobil Chemical Co. (PMX H50); injection-molded specimens.
[d] Injection-molded, annealed specimens; load = 1.82 MPa (264 psi).
[e] To convert MPa to psi, multiply by 145.
[f] To convert J/m to ft-lb/in., divide by 53.38.
[g] Determined by size-exclusion chromatography.

stereoregularity, and the amounts and types of additive are kept constant, the inherent properties of general purpose polystyrene can be altered by simply changing the shape of the molecular weight distribution curve. The development of anionic polymerization and gel-permeation chromatography have aided the study of the effect of molecular weight distribution on polystyrene properties.

Despite these new developments, the effect that molecular weight distribution has on polymer properties is often complicated and has been difficult to predict quantitatively. Simple relationships between molecular weight averages and properties often break down when the shape of the distribution changes.

For compression-molded or solvent-cast specimens, the mechanical strength of nearly monodisperse polystyrenes asymptotically approaches a limiting value at or above 160,000 weight average molecular weight (M_w). Some researchers have claimed that mechanical properties of low molecular weight polystyrenes are essentially zero below ca 80,000 M_w. However, others claim that low molecular weight polystyrene has a small, but measurable, mechanical strength. Figure 1 illustrates the changes in tensile strength of monodisperse polystyrenes with changes in molecular weight.

Although mechanical properties such as hardness and modulus of elasticity seem to be independent of molecular weight, flexural and impact strength of monodisperse polystyrenes show S-shaped relationships with molecular weight that are similar to those for tensile strength. Molecular weight distribution may also indirectly affect the mechanical properties of formed articles by influencing the orientation of the polystyrene. There are several reviews on mechanical and thermal properties of polymers.

BIBLIOGRAPHY

J. R. Martin, J. F. Johnson, and A. R. Cooper, *J. Macromol. Sci. Rev. Macromol. Chem.* **8**, 57, 59 (1972).

R. W. Nunes, J. R. Martin, and J. F. Johnson, *Polym. Eng. Sci.* **22**, 205 (1982).

Jerry L. Hahnfeld
Brian D. Dalke
The Dow Chemical Company

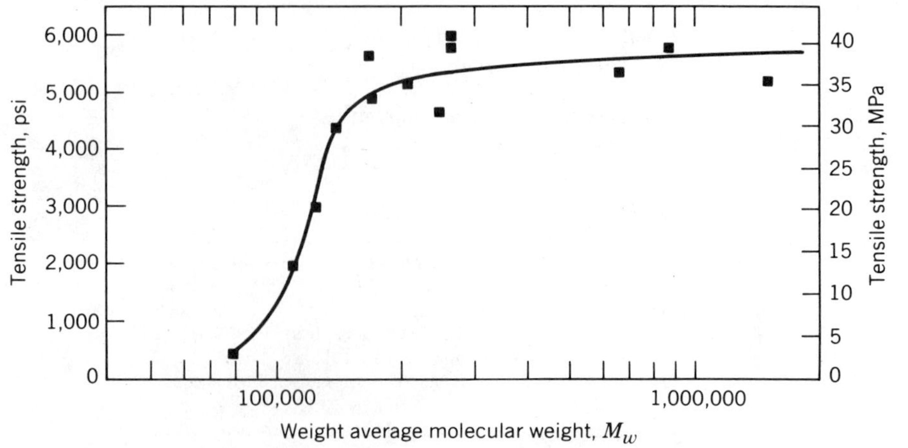

Figure 1. Tensile strength of monodisperse ($\bar{M}_w/\bar{M}_n$ = 1) polystyrene as a function of molecular weight.

COPOLYMERS

One attraction of styrene as a monomer stems from the wide range of properties attainable by copolymerization. The homopolymer can be improved through copolymerization and still maintain excellent economics. The result is materials as diverse as automobile instrument panels and ion-exchange beads.

Styrene–Acrylonitrile Copolymers (SAN)

Acrylonitrile-containing resins have gained an important place in the plastics industry. In 1985, the United States produced about 1.1 billion kg of acrylonitrile, 25% of which was used in styrene and rubber-modified styrene copolymers.

Rubber-modified SAN Copolymers (ABS)

General-purpose SAN copolymers have improved properties when compared with polystyrene, but still fail in a brittle fashion when subjected to sudden, high speed impact (see ACRYLONITRILE POLYMERS). Rubber reinforcement of SAN increases its ability to withstand high speed impact; these materials are called acrylonitrile–butadiene–styrene (ABS) polymers (qv). ABS has the processibility of styrene-based resins and excellent toughness.

Poly(styrene-*co*-methyl methacrylate)

The properties of styrene–methyl methacrylate (SMMA) copolymers generally fall between those of the individual homopolymers. Thus the weatherability and solvent resistance of this copolymer are superior to those of polystyrene homopolymer. Figure 1 shows that SMMA copolymers do exhibit a substantial weathering advantage when compared with polystyrene. Table 1 lists some physical properties of commercial grades of SMMA resins.

Poly(styrene-*co*-divinylbenzene)

Divinylbenzene is a diolefinic monomer capable of addition polymerization involving both of its double bonds. In reactions with styrene, at concentrations above about 0.06%, a cross-linked structure is obtained that is no longer thermoplastic,

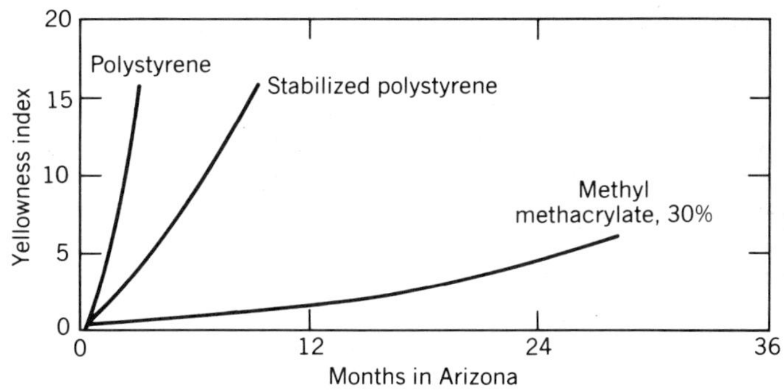

Figure 1. Outdoor weatherability of unmodified general-purpose polystyrene, light-stabilized polystyrene, and a 30% methyl methacrylate copolymer (with stabilizers). A higher yellowness index indicates more yellow resin.

but has particular uses. One of the most important uses of styrene–divinylbenzene copolymer is in ion-exchange beads (see ION-EXCHANGE POLYMERS).

Poly(styrene-*co*-butadiene)

Random Copolymers. The copolymerization of butadiene with styrene results in a product that has a lower T_g than polystyrene and can be vulcanized (cross-linked).

Styrene–butadiene (SB) copolymers can be tailored to fit end-use applications by careful attention to the monomer ratio (including additional monomers), average molecular weight, molecular weight distribution, linearity of molecular structure, extent of gel, cis–trans vinyl group ratios, and distribution of styrene and butadiene units (block vs random copolymers) (see also BUTADIENE POLYMERS). Random SB copolymers are made on a commercial scale by either emulsion or, more recently, solution polymerization methods.

Block Copolymers. Block SB copolymers, made by anionic solution polymerization, have found commercial application because of their thermoplastic and elastomeric property combinations. Morphology studies have revealed that the reversible vulcanization exhibited by some of these elastomers is caused by the phase separation of the styrene blocks into distinct and discrete domains (see ELASTOMERS, THERMOPLASTIC).

Table 1. Physical Properties of Commercial SMMA Resins

	SMMA			
	Extrusion[a]	Injection molding[b]	TABS[c]	TIPS[d]
tensile strength, MPa[e]	68.2	57.2	50.5	40.8
tensile elongation, %	5.0	2.0	30.0	60.0
tensile modulus, MPa[e]	3300	3500	2600	2500
flexural strength, MPa[e]	116	103	85	67
flexural modulus, MPa[e]	3300	3500	3000	2200
Izod impact (notched), J/m[f]	20	20	107	91
specific gravity	1.13	1.09	1.10	1.11
Rockwell hardnes, M scale	80	64	110[g]	

[a] P-205 UVA (Richardson Polymer Corp.) extrusion grade acrylic polymer.
[b] NAS (Richardson Polymer Corp.) injection-molding grade acrylic polymer.
[c] Magnum CLR-95 (The Dow Chemical Co.) transparent ABS, injection-molding grade.
[d] Experimental resin EX-836 (The Dow Chemical Co.) transparent impact polystyrene, extrusion grade.
[e] To convert MPa to psi, multiply by 145.
[f] To convert J/m to ft-lb/in., divide by 53.38.
[g] Rockwell R scale.

Poly(styrene-*co*-maleic anhydride) (SMA)

One commercially significant approach to raising the heat resistance of styrenics is to copolymerize styrene monomer with maleic anhydride. Along with increased temperature resistance, maleic anhydride imparts moderate increases in strength and stability with increased density, large decreases in coefficient of expansion, and increases in solvent resistance. This is accompanied by higher viscosity (see MALEIC AND FUMARIC POLYMERS).

Reactive Polystyrenes

An interesting approach to promoting adhesion in polystyrene blends has recently been commercialized as reactive polystyrene (or RPS) (see REACTIVE EXTRUSION). A polystyrene copolymer containing 2-isopropenyl-2-oxazoline was found to react with acid or anhydride polymers during melt processing. Specifically, blends of RPS with maleic anhydride-modified polyethylenes and RPS–poly(phenylene ether)–ethylene–acrylic acid blends show better toughness when compared with blends with general-purpose polystyrene.

Ionomers

The unique properties of ionomers are due to ionic interactions between the polymer molecules containing anionic groups and added metal cationic species. Polymers containing low concentrations of ionic groups exhibit properties that are radically different from the non-ion-containing homopolymers or copolymers. Generally, the copolymerization of ionic comonomers with styrene increases the glass-transition temperature.

Poly(styrene-*co*-alkylstyrene)

Alkylstyrenes that have short-chain alkyl groups bonded to either the carbon of the double bond or the benzene ring itself increase the softening point of styrene copolymers and decrease solvent resistance. Importantly, styrene–alkylstyrene copolymers maintain many of the styrene polymer processing advantages.

p-Methylstyrene, although susceptible to gel formation, has gained attention as an interesting comonomer in styrene polymerizations. Although the flexural strength is reduced (to about 75%) from that of general-purpose polystyrene of the same melt flow rate, the heat-distortion temperature is 8–10°C higher. At high shear the *p*-methylstyrene resins have lower melt viscosities.

α-Methylstyrene adds heat resistance to styrene copolymers. It is inexpensive, but changes the polymerization kinetics considerably. Most commercial polymers are grafted emulsion terpolymers with SAN and butadiene. Random styrene–α-methylstyrene copolymers can be made at excellent rates by using a continuously stirred tank reactor and butyllithium anionic initiator. These copolymers can be blended with emulsion polystyrene-grafted polybutadiene to form a series of high heat (127°C Vicat), impact-resistant resins. *t*-Butylstyrene also increases the heat-distortion temperatures of styrene resins.

POLYMER BLENDS

Polymer blends (qv) can be divided into two distinct classes. Most of the combinations of generically different polymers result in compositions in which the polymer molecules phase-separate into distinct domains containing nearly exclusively molecules of only one component. These types of blends are termed immiscible (or incompatible). Miscible, or compatible, polymers appear to mix at the molecular level (see COMPATIBILITY; MISCIBILITY). Styrenic polymer blends have played a historically significant role and continue to add to a growing roster of commercially significant blends.

Blends of Polystyrene with Poly(phenylene ether)s

Blends of polystyrene and poly(phenylene ether) (PPE), commonly called poly(phenylene oxide), are an example of a family of commercially successful, compatible polymer blends. Most of these blends are with HIPS and contain an immiscible rubber phase, which increases toughness. Typical properties of

Table 2. Properties of Some Heat-resistant Polystyrene–PPE Blends

	ASTM test	Medium impact[a]	High impact[b]
Thermal properties			
heat-deflection temperature (1.82 MPa[c]), °C	D648	113	113
coefficient of thermal expansion, m/m/°C	D696	7.7×10^{-5}	7.2×10^{-5}
Mechanical properties			
Izod impact strength (notched), J/m[d]	D256		
23°C		267	350
−40°C		133	190
tensile strength at yield (23°C), MPa[c]	D638	48	45
elongation at break (23°C), %	D638	60	
flexural strength (23°C), MPa[c]	D790	66	65
flexural modulus (23°C), MPa[c]	D790	2300	2200
General properties			
specific gravity	D792	1.06	1.06
mold shrinkage, 10^{-3} m/m	D955	5.7	6.7
water absorption (24 h, 23°C), %	D670	0.07	

[a] Noryl 884 (General Electric Co.).
[b] Prevex W40 (Borg-Warner Chemicals, Inc.).
[c] To convert MPa to psi, multiply by 145.
[d] To convert J/m to ft-lb/in., divide by 53.38.

commercially available polystyrene–PPE blends are shown in Table 2 (see also POLY(PHENYLENE ETHERS).

Blends of Polystyrene with Elastomers

One of the most successful blends is that of polystyrene and SB block rubbers. The combination usually results in an economical material with modest impact resistance, but low haze. Impact strength reaches a maximum at about 40 wt % styrene in the copolymer elastomer.

Polystyrene–SB blends can be fabricated by several methods: injection molding, blow molding, blown film, and sheet extrusion. The morphology and properties are greatly dependent on fabrication conditions, with higher shear fields yielding smaller particles (Table 3).

Table 3. Properties of Polystyrene Blends[a] with Polystyrene–SB Block Copolymers

	Stereon 874[b]	KR05[c]
polystyrene, %	50	40
dart impact (method B), g	315	350
MIT flex (0.046-cm specimen), folding endurance, cycles to failure	250	500
elastic modulus, MPa[d]	104	106
tensile at yield, MPa[d]	36	36
elongation (1.3 cm/min), %	<10	<15

[a] Extruded sheet.
[b] Firestone Synthetic Rubber and Latex Co. (multiblock, 70% styrene).
[c] Phillips Petroleum Co. (radial block).
[d] To convert MPa to psi, multiply by 145.

Blends of SAN and ABS with PC

Commercial products based on PC and ABS blends were introduced several years ago and have gained popularity among resin fabricators. The principal macromolecular components are immiscible with one another. Despite the immiscibility, an excellent balance of properties occurs in the region of 35–65 vol % PC.

Another study of the effect of processing conditions on the morphology of the fabricated part showed that the final blend morphology is determined primarily by the last fabrication step. The best morphology is a colamellar state achieved with 35–65 vol % PC. The toughness of PC–ABS blends is optimized when a colamellar state is achieved (see also POLYCARBONATES).

Poly(styrene-*co*-maleic anhydride) Blends

The terpolymers of styrene–maleic anhydride–acrylonitrile, styrene–maleic anhydride–methyl methacrylate, styrene–maleic anhydride–ethyl acrylate, and styrene–maleic anhydride–isobutylene have been blended with ABS resins and their properties determined. Blending maleic anhydride terpolymers with ABS allows the potential of increased solvent resistance, toughening, modified processibility, and improved heat-distortion properties.

Blends of Polystyrene with Polyethylene

The crystalline nature of polyolefins is responsible for a very large change in viscosity at the melting point. This forces the fabricator to take special measures when attempting to thermoform these resins. The potential fabrication benefits from blending with polystyrene have encouraged many studies in both industrial and academic laboratories.

BIBLIOGRAPHY

G. E. Molau and H. Keskkula, *J. Appl. Polym. Sci.* **7**, 35 (1968).

L. Holliday, ed., *Ionic Polymers,* Halsted Press, a division of John Wiley & Sons, Inc., New York, 1975.

D. R. Paul and S. Newman, eds., *Polymer Blends,* Vol. 1, Academic Press, New York, 1978.

J. D. Keitz, J. W. Barlow, and D. R. Paul, *J. Appl. Polym. Sci.* **29**, 3131 (1984). An excellent PC–ABS study.

THOMAS D. TRAUGOTT
The Dow Chemical Company

HIGH IMPACT POLYSTYRENE

Polystyrene and styrene copolymers are noted for their high tensile strength and optical clarity, but they are brittle under many normal use conditions. Rubber modification opens up many markets unavailable to unmodified polystyrene.

Synthesis

Commercial high impact polystyrene (HIPS) products are made by polymerization of an unsaturated rubber dissolved in styrene in a solution or mass-suspension process. The rubber, generally polybutadiene, is dispersed throughout the polystyrene matrix in the form of discrete particles.

The rubber-in-styrene polymerization usually results in a superior product when compared with blended products; that is, higher modulus and equivalent impact strength are achieved with less rubber. The property improvement is due to a chemical interaction (grafting) between the growing polystyrene chains and the rubber, chemical cross-linking of the rubber, and occlusion of the continuous-phase polymer inside the rubber particles, which increases the effective volume of the rubber phase. Rubber reinforcement produces other physical property improvements such as increased elongation, ductility, and environmental stress crack resistance (ESCR). These improvements in the composite polymer are accompanied by loss of clarity and a large decrease in tensile strength and modulus from the unmodified polymer.

Morphology

Many electron microscopy techniques have been developed for evaluating morphology. The differences in structure among the HIPS resins are primarily a result of differences in production processes, types of rubber, and rubber concentrations. Most commercial HIPS resins have particle sizes ranging from about 1 to 10 μm.

Mechanism of Rubber Reinforcement

The role of rubber in reinforcing glassy polystyrene and the mechanism of rubber particle formation have been described in detail. For HIPS the multiple craze theory seems to best explain the elongation, impact, and tensile properties of the resin, including stress whitening and the decrease in density that occurs without lateral contraction. This theory is based

Table 1. Properties of Conventional and Specialty Commercial Grades of HIPS[a]

Properties[b]	ASTM test method	Extrusion	Injection molding	High heat	High melt flow	High gloss	ESCR[c]	IR[d]
Styron resins[e]		484	479	492U	410	8028	469R	6061
melt flow rate[f], g/10 min	D1238	3.0	7.5	3.0	13.5	3.5	2.5	5.0
Vicat softening point, °C	D1525	100	94	104	94	103	101	102
deflection temperature under load at 1.8 MPa[g], °C	D648	92	82	96	81	88		
tensile rupture, MPa[g]	D638	16.6	15.9	20.7	17.9	23.0	17.2	15.2
tensile yield, MPa[g]	D638	17.9	18.6	22.1	22.8	25.5	13.8	20.0
tensile modulus, MPa[g]	D638	1655	1793	1793	2207	2241		1862
elongation, ·	D638	40	40	40	20	20	30	25
flexural modulus, MPa[g]	D790	1931	1793	1793	2276	2138		
flexural strength, MPa[g]	D790	38.6	32.4	41.7	42.8	51.7		
notched Izod at 23°C, J/m[h]	D256	80.1	85.5	85.5	37.4	122.8	74.7	74.7
Gardner impact,[i] cm·kg		138	121	173	29	173		34.6
Rockwell hardness	D785							
L scale		73	65	66	88	85		
M scale		29	25	40	65	45		
specific gravity	D792	1.05	1.05	1.05	1.05	1.05	1.05	1.16
other companies with product[j]		ACFHMP	ACFHMP	CFHMP	ACFHMP	M		H

[a] Data supplied by The Dow Chemical Company.
[b] All test specimens were prepared by compression molding.
[c] Improved ESCR resin.
[d] IR = ignition-resistant grade. Additives do not make these polymers noncombustible, but rather increase their resistance to ignition and reduce the rate of burning on minor fire sources. The organic resin material is flammable and will burn under proper conditions of heat and oxygen supply.
[e] The Dow Chemical Company trademark.
[f] Test condition G; 200°C/5 kg.
[g] To convert MPa to psi, multiply by 145.
[h] To convert J/m to ft · lb/in, divide by 53.38.
[i] 5 cm diamter × 0.32 cm thickness.
[j] A, Amoco Chemicals Corp.; C, Chevron Corp. (Gulf); H, Huntsman Chemical Corp.; F, American Petrofina, Inc.; M, Mobil Oil Corp.; P, Polysar Ltd.

on the premise that rubber particles both initiate and control craze growth.

Rubber-toughened styrenics other than HIPS exhibit similar, but not identical, behavior. Most rubber-reinforced styrene–acrylonitrile copolymer (ABS) resins, eg, deform by a more complicated mechanism. Both stress whitening (crazing) and necking (shear yielding) are observed. The growth of crazes is controlled by both rubber particles and the developing shear band, with the rubber particles playing the primary role.

These mechanisms explain the failure of fabricated articles subjected to instantaneous large deformations. Molded parts also fail because of repeated cyclic loads. This type of failure has been described as fatigue. Much recent research has focused on developing an understanding of the fatigue failures of polymers (see FATIGUE).

Properties of HIPS

Standard HIPS materials are used in both injection molding and extrusion processes. Generally, resins with low melt viscosity (melt flow rate (MFR) greater than 3) are used for injection-molding applications; resins with high melt viscosities (MFRs less than 3) are more suitable for extrusion applications. Typical end uses are toys, appliance parts, packaging, and furniture. These applications require property improvements over polystyrene, such as increased impact strength and ductility. A balance of flow and heat resistance is needed, but optical clarity is not required. Table 1 lists typical properties for extrusion, injection-molding, high heat, high melt flow, high gloss, ESCR, and ignition-resistant HIPS resins. The specific physical properties in this table are representative of the grades of resins available from various manufacturers.

BIBLIOGRAPHY

C. B. Bucknall, *Toughened Plastics,* Applied Science Publishers Ltd., London, 1977, pp. 182–212.

J. A. Sauer and C. C. Chen in H. H. Kausch, ed., *Advances in Polymer Science; Crazing and Fatigue in One and Two Phase Glossy Polymers,* Springer-Verlag, New York, 1983, p. 169.

E. J. Kramer in K. Solc, ed., *Mechanisms of Toughening in Polymer Mixtures, Polymer Compatibility and Incompatibility Principles and Practices,* Harwood Academic Publishers, New York, 1983, p. 251.

R. P. Levek, *Additives for Plastics,* Vol. 1, Academic Press, New York, 1978.

MARK E. SODERQUIST
ROBERT P. DION
The Dow Chemical Company

ENVIRONMENTAL STRESS CRACKING

The premature fracture of polymers at stresses considerably below the nominal yield stress, while in contact with aggressive environments, is a well-established observation and a practical engineering problem (see CRAZING). This process is a result of enhanced craze growth and breakdown under load and under the influence of an agent that is often only moderately interactive with the polymer. Crazes are cracklike features with microfibrils that bridge the top and bottom surfaces of the interfaces between the fibrillar matter and the bulk polymer. The degree to which the aggressive agent promotes the formation, drawing, and breakdown of the craze fibrils largely determines the extent to which the critical stress for environmental crazing is decreased below the nominal yield stress of the polymer. This fracture process is called environmental stress cracking.

There are two types of environmental stress cracking, characterized by either a liquid or a gas as an aggressive agent. Liquid-induced environmental stress cracking is given much greater emphasis in the literature.

Liquid Agents

An aggressive liquid agent acts on the polymer at a flaw located at or near the polymer surface and causes localized plasticization and suppression of the glass-transition temperature T_g under an applied stress. The critical stress σ_c at which this process can lead to the onset of craze and crack formation is considerably less than the yield stress and is a function, to a first approximation, of the difference between the solubility parameter of the polymer δ_p and of the "solvent" δ_s as shown in Figure 1. Decreasing the interfacial surface tension can also decrease critical stress. The surface tension of polystyrene against air is 33 to 36 mN/m (= dyn/cm). The effects of surface tension, solvent polarity, and molecular size and shape of the solvent, as well as δ_s must be considered in an overall assessment of the crazing and cracking tendency of a given aggressive liquid.

Once the deformation zone at the forming environmental craze tip is plasticized, the craze begins to advance, forming new fibrils by the Taylor meniscus instability process. At this point, craze-growth kinetics are governed by liquid transport through the craze-fibril structure, where capillary pressure is the driving force.

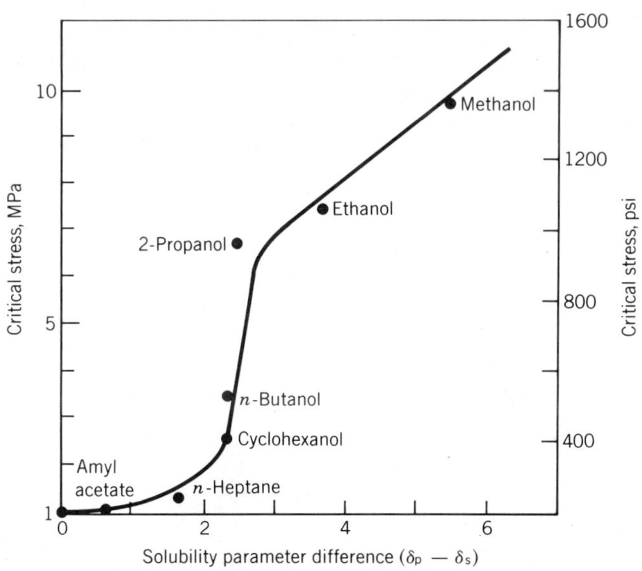

Figure 1. Critical stress needed to cause crazing of polystyrene as a function of the absolute value of the difference in solubility parameters, $(\delta_p - \delta_s)$, between polymer and solvent; δ_p for polystyrene = 9.1 $(cal/cm^3)^{1/2}$ [18.6 $MPa^{1/2}$]. The point at the origin corresponds to decalin and ethyl acetate. To convert MPa to psi, multiply by 145.

Molecular variables that can affect environmental stress cracking in nonrubber-modified styrenics are molecular weight, molecular weight distribution, and composition, such as acrylonitrile content in a styrene–acrylonitrile copolymer (SAN). Although uniaxial orientation enhances craze and environmental stress-cracking propagation in the direction parallel to the orientation, biaxial orientation suppresses environmental stress cracking (see also CRAZING and ORIENTATION).

A principal means of improving the environmental stress-cracking resistance of styrenic polymers is by rubber modification to make high impact polystyrene (HIPS). The polymer parameters used to optimize environmental stress-cracking resistance are molecular weight, rubber particle size and morphology, rubber level, and mineral oil (plasticizer) level.

Environmental stress-cracking resistance is enhanced in HIPS by varying the polymer parameters that govern the incubation time for craze nucleation (t_i), as well as craze microstructure and stability. The t_i is influenced by the compressive prestress in the polystyrene matrix imposed by the rubber particles. Craze microstructure and stability are controlled primarily by molecular weight, rubber particle size f_r, and mineral oil content. In commercial practice, it is assumed that the rubber or gel particles have sufficient cross-linking in the rubber to endow the particles with the required mechanical tenacity, and sufficient chemical graft to couple the particles to the matrix effectively.

Gaseous Agents

Near their boiling points, nitrogen, oxygen, helium, argon, and carbon dioxide have been found to be strong crazing agents for a fairly wide range of amorphous and semicrystalline polymers. It has been proposed that diffusion of the gas at the craze tip controls the craze growth velocity in contact with a gaseous agent. Unlike the case with liquid agents, the craze fibril structure poses very little resistance to gas flow.

Oxidation-induced Embrittlement

The ultraviolet degradation of polystyrene and rubber-modified styrenics involves the photooxidation of the rubber phase and chain scission of the polystyrene. This general condition leads to weaker crazes that are poorly stabilized, resulting in enhanced stress cracking and embrittlement.

BIBLIOGRAPHY

E. J. Kramer in E. H. Andrews, ed., *Environmental Cracking of Polymers, Developments in Polymer Fracture,* Applied Science Publishers, Barking, UK, 1979.

E. J. Kramer in H. H. Kausch, ed., *Crazing in Polymers, Advances in Polymer Science,* Springer-Verlag, Berlin, 1983, pp. 1–56.

ROBERT A. BUBECK
The Dow Chemical Company

THERMAL PROPERTIES

Because the glass-transition temperature (T_g) determines the temperature range in which a polymer can be used, it is the most important thermal property for any amorphous polymer such as polystyrene (PS).

From a molecular standpoint, the glass transition has been viewed as the temperature above which large-scale chain segments (20–50 monomeric units in a polymer chain) develop mobility that permits conformational rearrangements of the chain backbones. Because additional molecular motions are prevalent above T_g, a discontinuity in the thermal coefficient of expansion and specific heat is found at T_g.

Values from 365 to 385 K have been cited for the glass-transition temperature of polystyrene. From studies of the effects of such factors as molecular weight and diluents, the glass-transition temperature of high molecular weight polystyrene has been reported as 373 K.

Factors Affecting the T_g of Polystyrene. ***Molecular Weight.*** The glass-transition temperature of polystyrene is a function of molecular weight. This dependence of T_g (determined volumetrically) can be expressed

$$T_g\,(\mathrm{K}) = 373 - \frac{1.9 \times 10^5}{M_n}$$

where M_n is the number-average molecular weight.

Diluents. Materials such as mineral oil, paraffin oil, etc, may be purposely added to lower the glass-transition temperature and processing temperature of styrenic polymers. Diluents can also be reactants, solvents, or by-products formed during the polymer synthesis (eg, styrene monomer, dimer, trimer, ethylbenzene) or additives such as stabilizers, ignition-resistant additives, or antioxidants. Diluents affect the free volume of the polymer. As free volume increases, the glass-transition temperature tends to decrease (Table 1).

Cross-linking. The effect of cross-linking on the glass-transition temperature of a polymer has been examined. Chemical cross-linking agents can affect T_g by changing the backbone composition and by forming links between the molecules. Cross-link formation inhibits molecular motion and, therefore, always causes a net increase in T_g. Compositional changes can cause an increase or decrease in T_g for the resultant copolymer, so the net effect of chemical cross-linking can be either an increase or decrease in T_g.

Scan Rate. Determining T_g by the volumetric method is accurate, but tedious, because it is a very slow process. Faster methods for determining the glass-transition temperature are differential scanning calorimetry (dsc), where smaller samples are used, and dynamic mechanical spectroscopy (dms), where excellent automation has been developed. However, these methods create other difficulties in that they are not equilibrium methods and the temperature scan rate or the frequency of the experiment can affect the measured T_g value.

Substitution. Table 2 lists glass-transition temperatures for polystyrene and various substituted derivatives.

Copolymers. Random copolymers show glass-transition temperatures that are related to the weight fraction and the T_g value of the homopolymers of the monomers used to make the copolymer. Block copolymers behave more like blends, and separate T_g values for each component can be observed.

Blends. The two most common blends of polystyrene are with polybutadiene to produce high impact polystyrene (HIPS) and with poly(phenylene ether). These two blends represent extreme cases in that polystyrene is immiscible with polybutadiene and totally miscible with poly(phenylene ether).

Specific Heat of Polystyrene

Specific heat values for polystyrene have been studied by several authors and an excellent review of the data has been published. Dsc has been used to examine the effects of sample size, scan rate, and molecular weight on specific heat values. Molecular weight becomes a significant factor only for molecular weight values below 5000; thermal history, sample size, and scan rate have the greatest effect on the specific heat determinations for higher molecular weights.

Density of Polystyrene Melts

Density measurements have been done on polystyrene above T_g (M_n = 51,000; $M_w/M_n < 1.06$). The density at 135°C is given as 1.0054 ± 0.0001 g/cm³. Volume changes in the range of 100 to 322°C were made at small (approximately 4 degree) intervals and the data were found to fit a second order polynomial:

$$\rho = 1.0865 - 6.19 \times 10^{-4}\,T + 1.36 \times 10^{-7}\,T^2$$

where ρ is in g/cm³ and T is in °C.

Table 1. Effect of Diluents (1 wt %) on the Glass-transition Temperature of Polystyrene

Diluent	Decrease in T_g, K
ethylbenzene	6.0
methyl acetate	5.8
ethyl acetate	5.8
carbon disulfide	5.6
diethylbenzene	5.4
n-butyl acetate	4.5
amyl butyrate	4.5
benzene	4.2
toluene	4.2
methylene chloride	3.8
chloroform	3.4
nitrobenzene	3.4
carbon tetrachloride	2.8
phenyl salicylate	2.8
tricresyl phosphate	2.8
naphthyl salicylate	2.2
paraffin oil	2.0
benzyl benzoate	1.9

Table 2. Effect of Ring and Chain Substituents on Glass-transition Temperature of Polystyrene Derivatives

Polymer	T_g, K
polystyrene	373
poly(α-methylstyrene)	441
poly(2-methylstyrene)	409
poly(3-methylstyrene)	370
poly(4-methylstyrene)	366–374
poly(2-chlorostyrene)	392
poly(3-chlorostyrene)	363
poly(4-chlorostyrene)	383
poly(2,4-dichlorostyrene)	406
poly(2,5-dichlorostyrene)	379
poly(2,6-dichlorostyrene)	440

Table 3. Thermal Conductivity and Thermal Diffusivity for Polystyrene and Polystyrene Foams

	Polystyrene		Foamed polystyrene[a]		Foamed polystyrene[b]	
T, K	Thermal conductivity, W/(m·K)[c]	Thermal diffusivity, $\times 10^{-8}$ m^2/s	Thermal conductivity, W/(m·K)[c]	Thermal diffusivity, $\times 10^{-8}$ m^2/s	Thermal conductivity, W/(m·K)[c]	Thermal diffusivity, $\times 10^{-8}$ m^2/s
100	0.123	24.52	0.011	36.7	0.014	203
200	0.141	16.42	0.020	40.0	0.025	285
300	0.154	12.35	0.032	44.2	0.042	359
400	0.160	8.28				
500	0.164	8.02				

[a] Polystyrene, foam voids filled with air, density at 300 K = 60 kg/m³ (3.7 lb/ft³).
[b] Polystyrene, foam voids filled with air, density at 300 K = 10 kg/m³ (0.6 lb/ft³).
[c] To convert W/(m·K) to Btu·inch/(h·ft²·°F), multiply by 6.940.

Thermal Expansion of Glassy Polystyrene

The volumetric coefficient of thermal expansion α at constant pressure is defined as

$$\alpha = \frac{1}{V}\left(\frac{\partial V}{\partial T}\right)_P$$

A value for α of 2.085×10^{-4}/°C below T_g has been reported. The compressibility for the polystyrene glass at T_g is given as 3.31×10^{-10}/Pa (2.28×10^{-6}/psi).

Thermal Conductivity and Thermal Diffusivity

Much of the thermal conductivity and thermal diffusivity data in the literature has been reviewed. The values for polystyrene are presented in Table 3. Thermal conductivity increases until the glass-transition temperature is reached, and then decreases slightly.

Thermal diffusivities can be calculated according to the equation

$$TD = \frac{\kappa}{(C_p \cdot d)}$$

where TD is the thermal diffusivity, κ is the thermal conductivity, C_p is the specific heat, and d is the density of the polymer. Calculated thermal diffusivity data have been compared to experimental data and found to be in reasonably good agreement. Thermal conductivity and thermal diffusivity are important factors in the use of foamed polystyrene for insulation.

Crystalline Polystyrene

Although crystallization is not expected and has never been detected in atactic polystyrene, both isotactic and syndiotactic polystyrene can be crystallized. No thermal property data for crystalline syndiotactic polystyrene have been reported. Properties for crystalline isotactic polystyrene and several poly(methyl-substituted styrenes) have been reviewed.

BIBLIOGRAPHY

T. G. Fox, Jr., and P. J. Flory, *J. Appl. Phys.* **21**, 581 (1950).

I. Abu-Isa and M. Dole, *J. Phys. Chem. Part 2* **69**, 2668 (1965).

C. Y. Ho, P. D. Desai, K. Y. Wu, T. N. Havill, and T. Y. Lee, *Proc. Symp. Thermophys. Prop.* **7**, 198, 217 (1977).

J. Brandrup and E. H. Immergut, eds., *Polymer Handbook*, 3rd ed., John Wiley & Sons, Inc., New York, 1989.

ANDREW J. PASZTOR, JR.
The Dow Chemical Company

RHEOLOGY

Polystyrene is probably the most studied of all polymers because it has been the best characterized, most available linear amorphous polymer; has low intermolecular attractions; and is soluble in a wide range of solvents. In developing knowledge and theories about linear polymers, polystyrene is often the test case.

The processor of polystyrene is primarily interested in melt rheology in the temperature range of 180 to 250°C and in materials of commercial molecular weights and molecular weight distributions. The manufacturer is probably more interested in solution rheology in the 120 to 180°C range because viscosity is a primary determinant of equipment sizing and capabilities for the solution polymerization process. A comprehensive overview of the field of rheology and its applications can be obtained from several books. Polystyrene data are used in these references to illustrate many of the concepts of rheological behavior.

State of the Molecules

Although polystyrene molecules are generally linear, the entropy of long-chain molecules is maximized when the chain occupies a spherical random coil shape, whether in solution or in a melt. Coil dimensions, as defined by radius of gyration R_G, can be measured by neutron-scattering data on deuterated polystyrene molecules in a polystyrene melt or solution.

Linear Viscoelastic Behavior

In this article, a linear viscoelastic measurement is defined as any test that does not change the structure of the material. By this definition, any deformation that does not distort the isotropic random coil configuration of the material, such as zero-shear viscosity and steady-state compliance, in addition to the usual oscillatory and stress-relaxation tests, are considered linear tests. The response of polystyrene to these linear viscoelastic tests is considered to be typical of linear amorphous polymers. Two primary distinguishing features are molecular weight and molecular weight distribution.

Nonlinear Viscoelastic Properties

Both molten polystyrene and concentrated polystyrene solutions disentangle under conditions of large strain. In steady-flow situations, when the disentanglement time forced by the velocity gradient is slower than the reentanglement by molecular diffusion, random coil configurations can be maintained, the state of entanglement remains constant and the measured viscosity remains at the zero-shear value. As the shear rate is increased, the rate of disentanglement is increased and reptation back to random entangled coils can no longer be maintained. The state of entanglement decreases and the viscosity decreases. If the relaxation time spectrum, $H(\tau)$, is measured while the material is being sheared, the longest time elements are found to no longer contribute to the response. The shape of the resulting viscosity/shear-rate curve depends on the shape of the relaxation time spectrum, which, for all polystyrenes with $M > M_c$, depends only on the molecular weight distribution. If the molecular weight distribution is narrow, the relaxation spectrum is narrow and the viscosity/shear-rate curve is steeper than if the molecular weight distribution is broad. Therefore, if data are shifted for temperature, pressure, molecular weight, and concentration, all polystyrene data for the same shape molecular weight distribution can be placed on the same plot. Conversely, all polystyrene data can be read from such a plot.

A plot from such data makes rheological calculations with polystyrene seem extraordinarily simple. Unfortunately, that is not the case because polymer processing is not usually done under steady-shear, isothermal conditions. The reason for the focus on constitutive equations in the literature is because the application of steady-shear data is so difficult.

Most polymer processing operations are highly transient, if not in a space–time sense like injection molding, then in a material-history sense. As elements of fluid pass through an extruder, they are continually subjected to shear fields of varying intensity. The state of entanglement of the fluid element depends on the history as well as the instantaneous state. This is apparent from the high stress levels obtained when high shear rates are initiated on a polystyrene melt or solution starting from rest. Stresses starting from rest can reach values 10 times higher than the steady-state value, because the stresses build up before the state of entanglement is reduced to the steady-state value. This causes what is known as entrance effects in capillary rheometer measurements and exists to an even greater extent in high speed processing equipment.

The most widely available rheological parameter is the melt flow rate, obtained by a constant stress test that is performed by forcing polymer melt through a short capillary ($L{:}D = 3.8$) at a controlled temperature and determining the grams of effluent collected over 10 min. Because shear rate in the capillary is generally in the range of 1 to 10/s for melt flow rate values in the normal range of 2 to 20, this test is in the non-Newtonian region and includes entrance effects. Therefore, it is molecular weight distribution sensitive and does not correlate directly with weight-average molecular weight.

Because molecular weight distributions do not vary a great deal within commercial polystyrenes, it has been possible to develop an extremely useful master plot based on melt flow rate. These plots are shown in Figure 1 for polystyrene and in Figure 2 for styrene–acrylonitrile resins.

Concentrated Solutions

The effect of polystyrene concentration on the rheology of polystyrene solutions is relatively complex, as shown by the plot in Figure 3.

In the concentrated entangled regime, the relaxation time and zero-shear viscosity do not have the same concentration dependence. Most data indicate that

$$\frac{\eta_0}{\tau} \propto c^2 \qquad (1)$$

Equation 1 suggests that the solution modulus is proportional to the square of concentration, which is consistent with the kinetic theory of entanglement.

The viscosity/shear-rate dependence for concentrated polystyrene solutions shows identical behavior to melts, with the shape of the shifted curves depending only on molecular weight

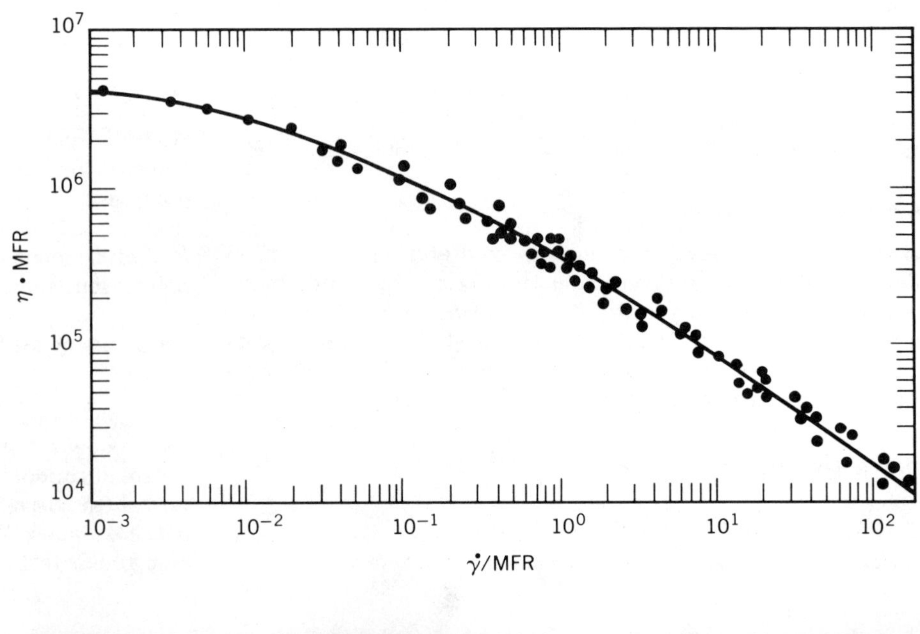

Figure 1. Conversion of melt flow rate MFR (g/10 min) (5 kg weight of condition "G" except at any temperature) to apparent melt viscosity η (dPa·s = P) for commercial polystyrene resins. Shear rate $\dot{\gamma}$ in s^{-1}. Courtesy of *Rheologica Acta*.

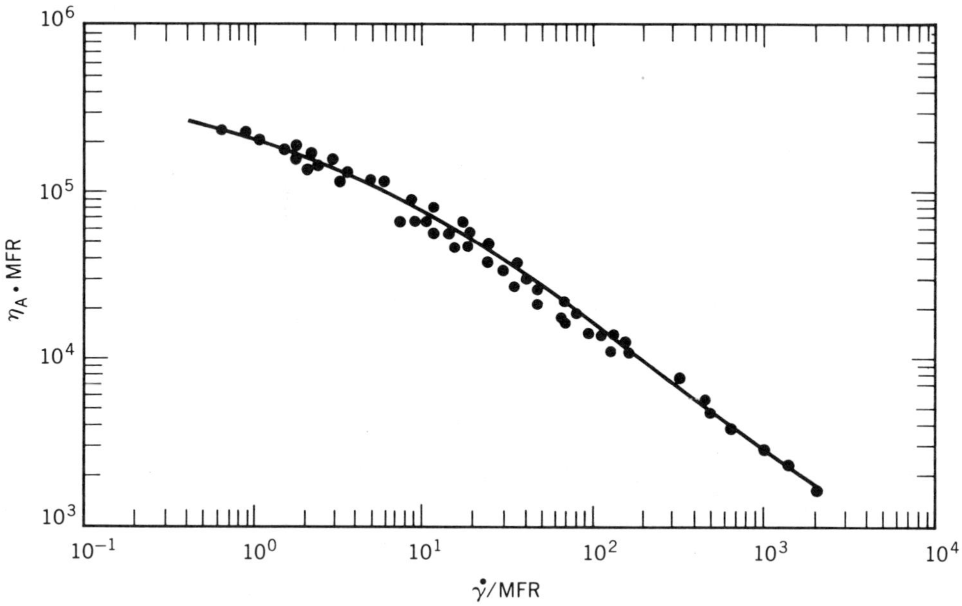

Figure 2. Conversion of melt flow rate MFR (g/10 min) (3.8 kg weight of condition "I" except at any temperature) to apparent melt viscosity η (dPa·s = P) for commercial 25 wt % acrylonitrile poly(styrene-*co*-acrylonitrile). Shear rate $\dot{\gamma}$ in s^{-1}. Courtesy of *Rheologica Acta*.

distributions. The effect of temperature on viscosity in concentrated solutions is much less than the effect in melts because of the large effect of solvents on the glass-transition temperature. If comparison is made at the same number of degrees above the glass transition of the solution, then the shift factors are similar to those in the melt. The interaction between concentration and temperature dependence is apparent in most solution-viscosity data.

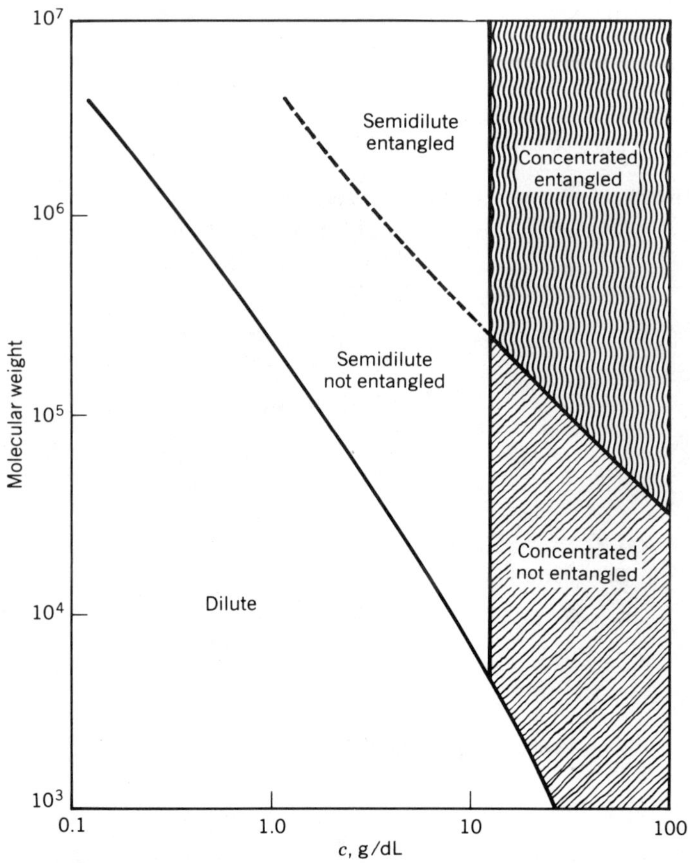

Figure 3. Concentration–molecular-weight diagram of viscoelastic regimes for polystyrene in a good solvent. Courtesy of *Polymer*.

Rubber-modified polystyrene (HIPS)

The introduction of rubber particles into polystyrene brings about only minor changes in the melt rheological behavior. Because the rubber particles in high impact polystyrene (HIPS) are generally cross-linked, they undergo only minor deformation during shear flow and the majority of the velocity gradient must be incurred by the matrix (polystyrene phase). The effect of the rubber particles on the viscosity is usually treated by analogy to suspension rheology. Melt viscosity η and melt elasticity β for several HIPS materials have been compared to measurements made on the matrix phase.

Copolymers

Copolymerized monomers can have a significant effect on the rheology of polystyrene. The main effects are caused by changes in the glass-transition temperature T_g, changes in the molecular weight between entanglements M_e, and changes in the monomeric friction coefficient ξ_0. An increase in the glass-transition temperature introduces a decrease in the rate of all molecular motions; this raises the relaxation time τ and viscosity η. The molecular weight between entanglements and the monomeric friction coefficient have a significant effect on the viscosity of a copolymer. The following correlation has been suggested:

$$\eta_0 = K\xi_0 M(M/M_e)^{2.5} \tag{2}$$

The general dependence of copolymer rheology on molecular weight, molecular weight distribution, and concentration are the same as for polystyrene. The time–temperature shift factors for many of them are identical to those for polystyrene.

Polystyrene Ionomers

Although no polystyrene ionomers are produced commercially, they have attracted much research interest. They are of particular interest to the rheologist because they are so unusual.

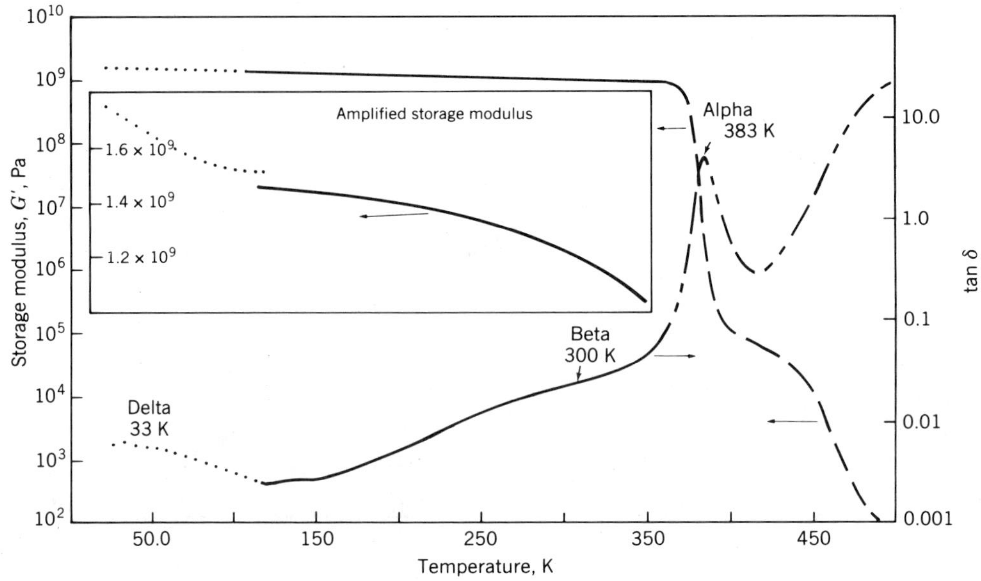

Figure 1. Dynamic mechanical properties for polystyrene as a function of temperature. Data below 120 K were measured at 1.7 Hz. Data above 120 K were measured at 0.16 Hz. Data above 400 K were obtained using parallel plate geometry. The inset shows G' plotted on a linear scale (1.0 Pa = 10 dyn/cm^2 = 0.000145 psi).

Although all polystyrenes have the same temperature function, ionomers often do not. Often, time–temperature shifting of data breaks down. Melt viscosity increases rapidly with ion content, as does dynamic viscosity.

BIBLIOGRAPHY

F. H. Cogswell, *Polymer Melt Rheology,* George Godwin Ltd., London, 1981. An introduction to rheology.

C. D. Han, *Rheology in Polymer Processing,* Academic Press, New York, 1976. An introduction to applying rheological data.

J. D. Ferry, *Viscoelastic Properties of Polymers,* 3rd. ed., Wiley-Interscience, New York, 1980. The prime reference for linear viscoelasticity.

W. W. Graessley, *Adv. Polym. Sci.* **16,** 14, 55, 64 (1974).

Bernard J. Meister
The Dow Chemical Company

DYNAMIC MECHANICAL PROPERTIES

The dynamic mechanical properties of polymers provide the means toward a fundamental understanding of the relationships between molecular structure and macroscopic mechanical properties. Knowledge of the dynamic mechanical properties for a particular polymer can also help determine suitable applications. In analytical applications, dynamic mechanical properties are used to investigate the structure and morphology of a polymeric system. This article discusses the dynamic mechanical properties of polystyrene and high impact polystyrene (HIPS) (see also Dynamic mechanical properties and Viscoelasticity).

Polystyrene

Although polystyrene has two semicrystalline forms, isotactic and syndiotactic, neither of these are yet of commercial im-

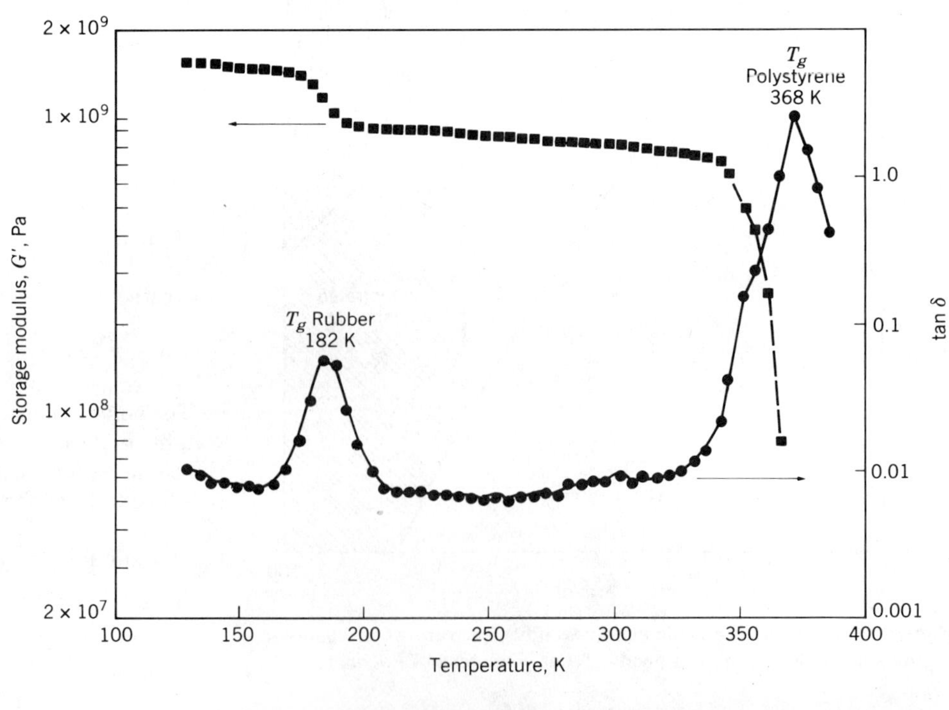

Figure 2. Dynamic mechanical properties as a function of temperature for a typical HIPS high impact polystyrene; frequency = 0.16 Hz. ● tan δ; ■ G'. To convert Pa to dyn/cm^2 multiply by 10. To convert Pa to psi, divide by 6896.

portance; most of this discussion, therefore, deals with atactic polystyrene. Figure 1 shows a composite of data for atactic polystyrene. The data from about 20 to 100 K were obtained with an inverted torsion pendulum at a frequency of 1.7 Hz. The data above 120 K are for an anionic polystyrene with M_w = 210,000 and M_w/M_n = 1.06. Measurements were made using a forced oscillation type instrument (Rheometrics RDS II) at a frequency of 0.16 Hz. A rectangular bar sample was used below 393 K, and parallel plate geometry was used at higher temperatures.

High Impact Polystyrene (HIPS)

Various rubbers are often added to improve the impact strength of polystyrene. The morphology of the system can be quite complex but generally consists of dispersed rubber domains that may contain occlusions of polystyrene. At temperatures between the glass-transition temperatures of the rubber and polystyrene, HIPS can be treated as a composite material consisting of a hard continuous matrix with dispersed soft domains. It is therefore possible to examine the structure of the rubber phase with dynamic mechanical techniques. Data for a typical HIPS material are shown in Figure 2.

BIBLIOGRAPHY

K. M. Sinnot, *SPE Trans.* **2,** 65 (1962).

J. M. Crissman, A. E. Woodward, and J. A. Sauer, *J. Polym. Sci.* **A3**(7), 2693 (1965).

C. G. Seefried, Jr., and J. V. Koleske, *J. Polym. Sci. Polym. Phys. Ed.* **14,** 663 (1976).

N. Bach Van and C. Noel, *J. Polym. Sci. Polym. Chem. Ed.* **14,** 1627 (1976).

Charles A. Berglund
The Dow Chemical Company

OPTICAL PROPERTIES

The optical properties of a material are governed by the way in which light interacts with the material. The significant interactions for polymeric materials are refraction, reflection, absorption, and scattering (qv). More commonly these optical properties of polymers are observed in terms of color, haze, gloss, transparency (percent transmission), and refractive index (see Optical properties).

Properties at Visible Wavelengths

Refractive Index. Atactic polystyrene is well known for its transparency to visible light (400–700 nm). The transmission of light through a transparent material is characterized by the refractive index, n. Refractive index is the ratio of the speed of light in a vacuum to the speed of light in the material. The high polarizability of the phenyl ring at visible wavelengths gives styrene polymers a relatively high refractive index when compared with other polymers. Phenyl ring substitution can increase or decrease refractive index. Refractive index increases with wavelength (optical dispersion) and decreases with temperature.

Color. Color is due to selective transmission or reflection of visible light. Polystyrene is normally colorless.

Many styrene copolymers, such as styrene-*co*-acrylonitrile, have some inherent yellowness. Pigments or dyes are often added to polymers before forming to produce brightly colored parts that require no painting (see Colorants).

Haze. Haze can be the result of crystallites in isotactic polystyrene, rubber particles in HIPS, or the separate phases in any incompatible polymer blend. Because haziness arises from variation in the index of refraction of a material, it is possible to produce a transparent HIPS by matching the refractive index of the rubber with that of the rigid phase.

Gloss. Rubber-modified materials can have high or low gloss depending on the size and integrity of the rubber particles near the surface.

Birefringence. Birefringence (qv) can be induced in polymers by stressing the polymer above or below the glass-transition temperature. Most melt processed polymers display birefringence due to orientation of polymer chains. Melt-drawn polystyrenes typically display values of birefringence in the range of −0.001 to −0.02; highly oriented fibers display values of birefringence up to −0.03. The maximum birefringence for

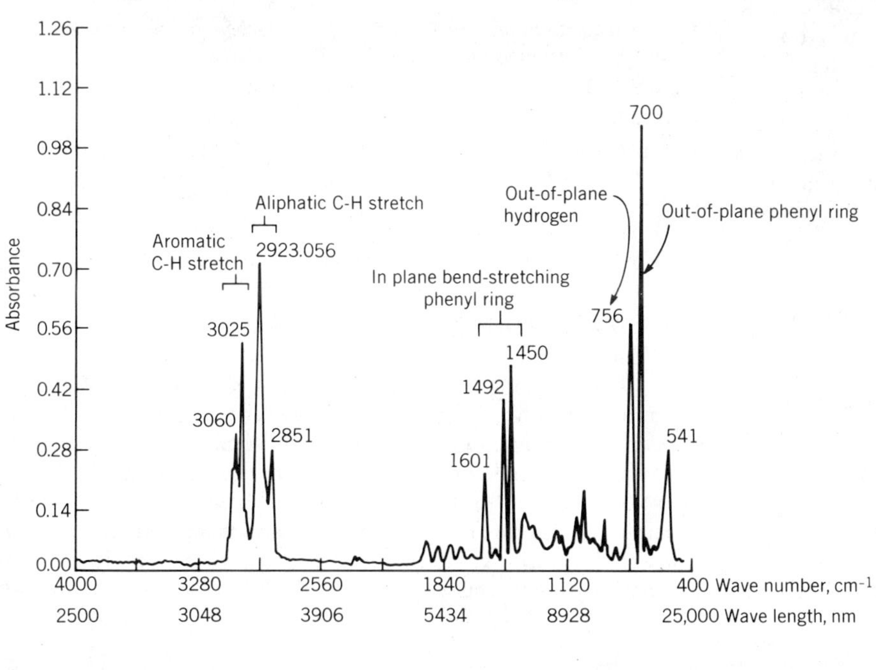

Figure 1. Infrared absorption curve for atactic polystyrene, with major vibrational modes. Courtesy of R. Peters, The Dow Chemical Company.

amorphous polystyrene calculated from experimental data is -0.1 ± 0.05; this is called the intrinsic birefringence ($\Delta°$).

Properties at Infrared Wavelengths

The optical properties of a polymer at infrared wavelengths are dominated by vibrational mode absorption. Figure 1 is the infrared absorption curve for polystyrene with the identification of the specific vibrational mode for several of the major peaks.

Polarized infrared spectroscopy has been used to study the molecular orientation of polystyrene, particularly infrared dichroism, which measures the difference in absorption of polarized light in the machine direction vs the cross direction.

BIBLIOGRAPHY

J. Brandrup and E. H. Immergut, ed., *Polymer Handbook,* 3rd ed., John Wiley & Sons, Inc., New York, 1989, pp. VI 451–VI 461.

A. I. Isayev, *Injection and Compression Molding Fundamentals,* Marcel Dekker, Inc., New York, 1987, p. 227.

H. Janeschitz-Kriegl, *Polymer Melt Rheology and Flow Birefringence,* Springer-Verlag, New York, 1983, p. 60.

P. C. Painter, M. M. Coleman, and J. L. Koenig, *The Theory of Vibrational Spectroscopy and its Application to Polymeric Materials,* John Wiley & Sons, Inc., New York, 1982.

MITCHELL G. DIBBS
The Dow Chemical Company

ORIENTATION AND PHYSICAL PROPERTIES

When amorphous polystyrene is subjected to extensional forces above its glass-transition temperature, its polymer chains align in the direction of the extensional forces. The degree of alignment is a function of the applied stress. If the sample is cooled quickly below its glass-transition temperature before the stress significantly relaxes, the orientation of the polymer chains is frozen in. This results in the sample having quite different properties in the orientation direction compared to the cross-orientation direction.

Commercially this has great consequences; if the orientation is not properly controlled, a fabricated part may have a relatively low strength in some directions. On the other hand, if the orientation is preferentially directed, parts with great strength and toughness in a desired direction can be made. Processes that give this type of enhancement include foam production, blow molding, melt spinning, film blowing, injection molding, and sheet tentering (see FILMS, MANUFACTURE).

Measurement of Orientation

There are a number of methods for measuring the orientation in amorphous materials such as polystyrene; the most often used optical methods are birefringence and infrared dichroism. When the material is opaque or very thick, optical methods cannot be used; rather, orientation must be measured by stress release, contraction, or calorimetry.

Uniaxial Orientation

In uniaxial orientation of polystyrene under normal tensile testing rates, the onset of brittle to ductile behavior begins in the draw direction at a birefringence of -0.002. At this point, crazing is suppressed and shear yielding is the predominant mode of deformation. In contrast to the other mechanical properties, the yield strength and the modulus are slowly rising functions of the birefringence.

Although it has been reported that there is no one-to-one relationship between the mechanical properties and the birefringence, data obtained from melt-oriented material, which is drawn at reasonably high rates and at temperatures just above the glass-transition temperature, seem to show a fairly good relationship between the birefringence and the mechanical properties. In addition to the mechanical properties, the thermal expansion and the thermal conductivity are also affected by the orientation.

Biaxial Orientation

In the case of biaxial orientation there are three principal orthogonal axes. It is customary to label the two draw directions in the plane of the film or sheet as 1 and 2, and the direction through the thickness is labeled 3. The three values of birefringence involved, Δn_{13}, Δn_{23}, Δn_{12}, can each be related to an orientation function by the equation $F = \Delta n/\Delta°$:

$$F_1^B = \frac{\Delta n_{13}}{\Delta°} \qquad F_2^B = \frac{\Delta n_{23}}{\Delta°}$$

The value of Δn_{12} represents the uniformity of orientation in the plane of the film. In the very special case of equal orientation in directions 1 and 2, $\Delta n_{12} = 0$ and $\Delta n_{13} = \Delta n_{23}$. This equal planar orientation is often sought in commercial applications.

In equal biaxial orientation of polystyrene, the brittle to ductile behavior is reported to occur at F_1^B or F_2^B of 0.04 at room temperature. Impact strength is shown to increase with birefringence but not nearly as dramatically as with uniaxial orientation. Because of the suppression of crazing, the environmental stress crack resistance is greatly enhanced for biaxially drawn sheet.

BIBLIOGRAPHY

I. M. Ward, ed., *Structure and Properties of Oriented Polymers,* Applied Science Publishers Ltd., London, 1975, p. 57.

K. Oda, J. L. White, and E. S. Clark, *Polym. Eng. Sci.* **18**(1), 53 (1978).

J. L. White and J. E. Spruiell, *Polym. Eng. Sci.* **21**(13), 859 (1981).

K. Matsumoto, J. F. Fellers, and J. L. White, *J. Appl. Polym. Sci. Part 1* **26,** 85 (1981).

JERRY T. SEITZ
The Dow Chemical Company

DIELECTRIC PROPERTIES

The dielectric constant is determined by the additive effects of electronic and dipolar polarizabilities; the latter is due to the presence of permanent electric dipoles. Since polystyrene is essentially nonpolar (with an effective dipole moment of only $6.7–10 \times 10^{-31}$ C·m (0.2 to 0.3 D) per repeating unit), its dielectric constant at most frequencies and temperatures is very close to the classic relationship of $\epsilon_\infty = n^2$, where n is the optical refractive index and ϵ_∞ is the dielectric constant at optical frequencies. In fact, the dielectric constant of atactic polystyrene of about 2.55 is similar to $\epsilon_{\infty} = 2.5$ to 2.6, based on its refractive indexes in the 436 to 769 nm wavelength range (hydrogen A through G line).

Changes in the dielectric constant of polystyrene with temperature are described to a good approximation by changes in density and neglecting dipolar orientation effects. Because of the low polarity of polystyrene, dielectric losses are generally so low that impurities are often as important as the polymer in determining the magnitude of the losses.

Although the electrical properties of polystyrene do not account for a large volume of sales, the dielectric properties are important in special electrical applications, such as the manufacture of capacitors. With the trend toward miniaturization of electrical circuitry, there is a need for components that will withstand higher temperatures. A number of styrene copolymers are available that have higher use temperatures but still retain most of the excellent dielectric properties of the homopolymer (Fig. 1).

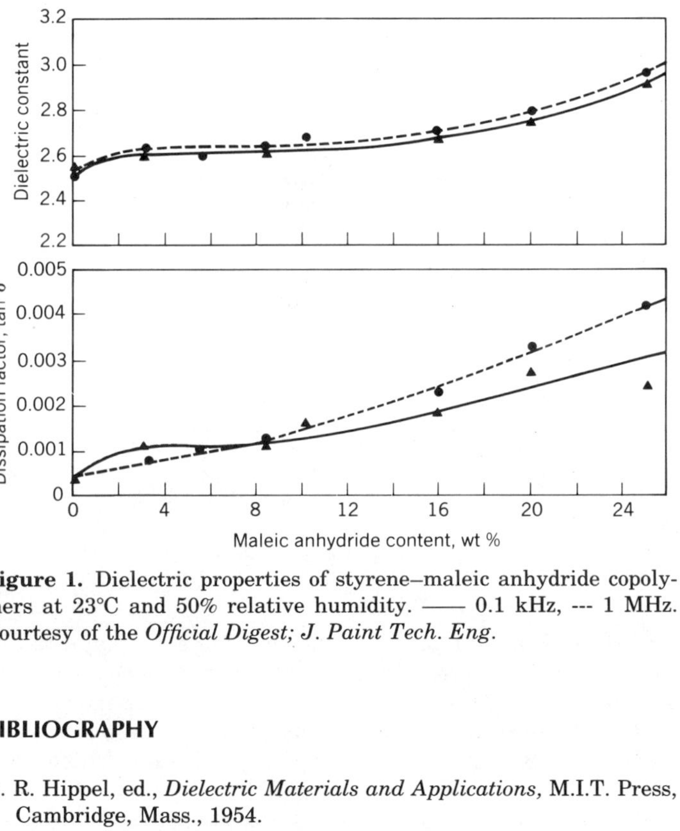

Figure 1. Dielectric properties of styrene–maleic anhydride copolymers at 23°C and 50% relative humidity. —— 0.1 kHz, --- 1 MHz. Courtesy of the *Official Digest; J. Paint Tech. Eng.*

BIBLIOGRAPHY

A. R. Hippel, ed., *Dielectric Materials and Applications,* M.I.T. Press, Cambridge, Mass., 1954.

M. E. Lyssy, M. P. Kubisiak, and P. D. Aldrich, *Proc. Int. Elec. Pack. Conf.,* San Diego, Calif., 1986, pp. 101, 102.

PETER ALDRICH
The Dow Chemical Company

DIFFUSION

This article discusses principally the diffusion of small molecules at low concentrations in a polystyrene host and associated phenomena. Mass transport in polystyrene is an active research area.

Theory

Diffusion (qv) describes the movement of one material within the matrix of another material. Permeation describes the movement of the material to and from the interfaces of the matrix, as well as within the matrix. For gas permeation through a polymer film, a permeant molecule collides with the "upstream" surface of the film and adsorbs. Then the molecule dissolves into the film. The molecule then moves randomly from one solution site to another. The statistical process of this moving is "diffusion." For an individual molecule, the movement is equally probable in all directions. However, whenever a concentration gradient exists in the polymer, there is net movement from the area of high concentration to the area of low concentration. When the molecule reaches the "downstream" surface, it volatilizes. For most cases, the permeation process is controlled by the solution and diffusion components. The equilibria between the surfaces of the polymer and the adjacent phases control the number of molecules in the film for diffusion. Diffusion controls the speed of the motion.

In the simplest case, where the diffusion coefficient is not influenced by concentration, the permeability (P), diffusivity (or diffusion coefficient) (D), and solubility coefficient (S) are related by the following equation:

$$P = DS$$

Four variables govern diffusion in a polymer host: temperature, concentration of penetrant (which is dependent on solubility), molecular composition and geometry of the penetrant, and structure of the polymer, including molecular and morphological features.

Diffusion Variables and Data

General Data. Tables 1 and 2 contain permeability and vapor-transmission rate data for some styrene-based resins with data for some other polymers given for comparison. The vapor-transmission rates in Table 2 were determined over neat solvents and may not apply to dilute or mixed systems. The reported failures were due to stress cracking.

Prediction of Permeability. Three simple methods are available for predicting the permeability of simple molecules in polystyrene. One proposal is an empirical equation that calculates a "permachor" value after values have been assigned to all the chemical functions of the polymer. This method assigns a high permachor value to polystyrene and therefore a high relative permeability. The method is related to the cohesive energy density and the free volume of the polymer. The model has been modified to liquid permeation with some success.

A less empirical model based solely on the specific free volume of the polymer has been proposed. Once the densities and the intrinsic volumes of the polymer components are known, the specific free volume can be calculated. This method assigns a high specific free volume to polystyrene and, therefore, a high permeability. This model is for simple gases such as oxygen, nitrogen, and carbon dioxide.

The diffusion coefficients of many small molecules in polystyrene have been reported. Elongated molecules have larger diffusion coefficients than spherical molecules with the same mean diameter. The activation energy for diffusion as a function of the penetrant size has also been reported. With these

Table 1. Permeability and Water Vapor Transmission Rate Data for Selected Polymers at Room Temperature

Polymer	Gas permeability, $\frac{nmol^a}{m \cdot s \cdot GPa}$			WVTR[b], $\frac{nmol}{m \cdot s}$ at 90% rh
	O_2	N_2	CO_2	
polystyrene (various types)	600–800	40–50	2000–3000	0.5–2.5[c]
styrene–acrylonitrile				
low VCN[d]	60–200	20	800	1.2–3.5[c]
medium VCN[d]	80–140			
styrene–maleic anhydride (25 wt %)	200–220		160–240	
ABS				
low VCN	400–520	50–70	1800–2400	1.2–4.0[c]
medium VCN	240–280	20–30	800–1200	
low density polyethylene	500–700	200–400	2000–4000	0.25–0.38[e]
vinylidene chloride resin[f]	0.3	0.06	0.50	0.025[e]
polypropylene	300	60	1200	0.08–0.18
poly(ethylene terephthalate)	10–18	2–4	30–50	0.45–0.8

[a] To convert nmol/m·s·GPa to cc·mil/100 in.^{2}·d·atm, divide by 2.
[b] WVTR = water vapor transmission rate. To convert nmol/m·s to g·mil/100 in.^{2}·d, multiply by 3.95.
[c] 24–38°C.
[d] VCN represents acrylonitrile.
[e] 38°C.
[f] Saran resin (The Dow Chemical Co.).

Table 2. Organic Vapor Transmission Rates[a] for Selected Polymers

Vapor	Polystyrene	Styrene–acrylonitrile copolymers	Low density polyethylene
methanol	0.14–0.84	[b]	0.84–1.12
ethanol	0.1[b]	[b]	0.2–0.4
n-heptane	[b]	5.2–52	16–26
ethyl acetate	[b]	[b]	1.5–15
formaldehyde	0.6–0.8	0.8–1.5	0.3–0.8
tetrachloroethylene	[b]	[b]	12–18
acetone	[b]	[b]	1–4
benzene	70[c]		35[c]

[a] At 23°C unless specified, over pure liquid, nmol/m·s. To convert to g·mil/100 in.^{2}·d, divide by 4.5 and multiply by the molecular weight of the penetrants.
[b] Sample failed because of stress cracking.
[c] At 35°C.

correlations, D in polystyrene at any temperature below T_g may be estimated.

Polymer Molecular Weight. There is a dependence of permeation on molecular weight for carbon dioxide in polystyrene. As M_n decreases from 8.4×10^6 to 1.0×10^4 daltons, the permeability increases nearly 40%. The diffusion coefficient does not vary with molecular weight. Therefore, the solubility coefficient is responsible for the enhanced permeability of carbon dioxide in the low molecular weight samples. For narrow molecular weight distribution polystyrene, the diffusion coefficient of oxygen in polystyrene at 25°C depends on molecular weight below 2×10^4.

In contrast, after orientation, high molecular weight polystyrene films absorb alkanes more rapidly than low molecular weight films. This effect disappears after annealing. This observation has been explained as a free volume relaxation phenomenon and not as a direct result of molecular weight. There is a slight dependence of the diffusion coefficient of organic vapors in polystyrene on M_n.

Orientation. Diffusion in some polymers is unaffected by orientation; in others, increases or decreases are observed. The circumstances of the orientation (elongation) are important.

Heat Treatment. Heat treatment causes only minor changes in the permeation of oxygen or nitrogen in unoriented polystyrene films.

Crystallinity. Crystallinity reduces the permeability in polymers.

Blends. For miscible blends, a plot of log P (or log D) versus the volume composition should yield a nearly straight line. When strong interactions between the polymers exist, the curve is concave upward. When weak interactions exist, the curve is concave downward. For two-phase blends, an S-shaped curve is usually expected.

Copolymers. Copolymers represent an interesting "blend" of polymers at the molecular level. Permeabilities in copolymers of styrene and methacrylonitrile, both random and block, have been studied in detail. For oxygen and carbon dioxide, the random copolymers behave like weakly interacting miscible blends of polystyrene and polymethacrylonitrile. The block copolymers behave like two-phase blends of the parent homopolymers, showing a typical S-shaped curve on a log (permeability) versus composition graph.

Radiation. Radiation can cause changes in a polymer that can, in turn, cause changes in the permeability (see RADIATION-RESISTANT POLYMERS). The permeation of polystyrene to deuterium decreased by about 10% after β-radiation. Uv radiation has been found to increase the permeability of polystyrene films to air.

Applications

Diffusion processes are important in several commercial applications of polystyrene, including barrier and migration properties of food packaging and foamed products.

BIBLIOGRAPHY

H. B. Hopfenberg and H. L. Frisch, *J. Polym. Sci.* **7,** 405, 406 (1969).

A. R. Berens and H. B. Hopfenberg, *J. Membr. Sci.* **10,** 283, 292, 294, 296 (1982).

J. Crank and G. S. Park, eds., *Diffusion in Polymers,* Academic Press, Inc., Orlando, New York, 1968, p. 42.

PHILLIP T. DELASSUS
The Dow Chemical Company

DEGRADATION

Polystyrene is easy to fabricate thermally into a wide variety of shapes without significant decomposition. The inherent stability of polystyrene has played a significant role in its remarkable commercial success. However, conditions do exist where polystyrene will degrade thermally in the absence of oxygen or in the presence of oxygen by thermal-oxidative degradation, by photodegradation, or as a result of ionizing radiation.

Thermal Degradation

The thermal degradation of polystyrene proceeds by a free-radical chain mechanism. The steps involve initiation of free radicals, propagation (during which intermolecular and intramolecular chain transfers are very active), and a first-order termination. In the absence of oxygen, polystyrene usually begins to exhibit significant weight loss at about 300°C, but molecular weight reductions by random chain scissions are observed at much lower temperatures. Rates of degradation depend on the technique by which the polymer was synthesized. Anionically initiated polystyrene is initially appreciably more stable than that produced thermally or from free-radical initiators, such as benzoyl peroxide or azobisisobutyronitrile (AIBN). After about 10% weight loss, however, all polystyrenes degrade at about the same rate.

Rubber Modification. Polybutadiene and polyisoprene rubbers are often added to polystyrene to provide impact strength. These rubbers and blends of the rubbers with polystyrene have been thermally degraded in the absence of oxygen, and the reaction products have been examined. Although the presence of rubber accelerates property loss on thermal aging, thermal degradation of the polystyrene phase in the blends, as shown by total weight loss and appearance of volatile products, is dramatically slowed by the presence of the rubber phase. However, the same degradation products are observed whether the polymers are heated as blends or individually.

Thermal-oxidative Degradation

Oxygen plays a very important role in the degradation of polystyrene. The activation energies for thermal degradation of polystyrene above 350°C in the absence and presence of oxygen have been thoroughly studied using thermogravimetry. In vacuum or flowing nitrogen, the activation energy is 188 kJ/mol (44.9 kcal/mol), but with an excess of oxygen, it decreases to 90 kJ/mol (21.5 kcal/mol).

The postulated reaction steps for the free-radical degradation of polystyrene in the presence of oxygen are very similar to those for the photodegradation of polystyrene. First, a polystyryl radical is generated thermally. The polystyryl radical reacts 10^6 times faster with an oxygen molecule than with a polystyrene molecule. The resulting peroxy radical rapidly abstracts a tertiary hydrogen, forming polystyrene hydroperoxide. This propagation step is repeated as long as oxygen remains. Thermal decomposition of the hydroperoxide can produce a variety of oxygenated species. Benzaldehyde, benzoic acid, acetophenone, phenol, benzyl alcohol, cinnamaldehyde, styrene, styrene dimer, and styrene trimer are thermal-oxidative products of polystyrene.

Antioxidants. Hindered phenolic species, such as 2,6-di-*t*-butylcresol, retard oxidative degradation by quickly scavenging peroxy radicals as they form. The reactions are as follows.

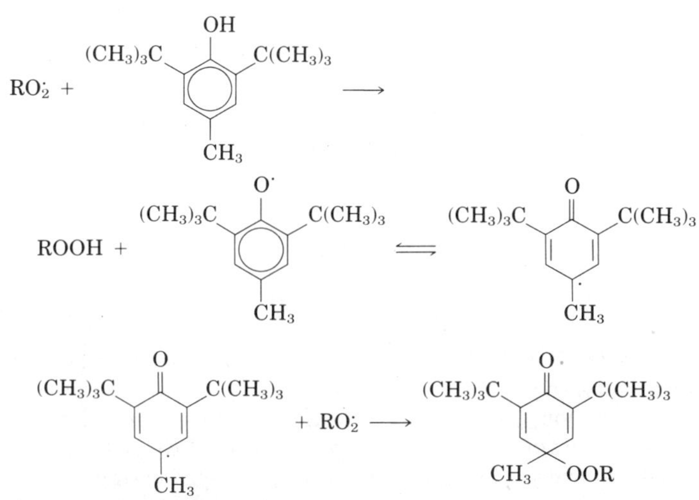

One technique to reduce oxidation during molding or extrusion is to remove absorbed oxygen from the resin before melting it.

Photodegradation

The most important variable in polystyrene photodegradation is the wavelength of the incident radiation. The ultraviolet (uv) absorption spectrum of polystyrene (Fig. 1) shows that the polymer absorbs strongly at wavelengths less than 290 nm. Terrestrial sunlight does not contain radiation at wavelengths absorbed by polystyrene, so short-wavelength conditions and outdoor weathering are considered separately.

Short-wavelength Conditions. The absorption of short-wavelength radiation by polystyrene is entirely due to the excitation

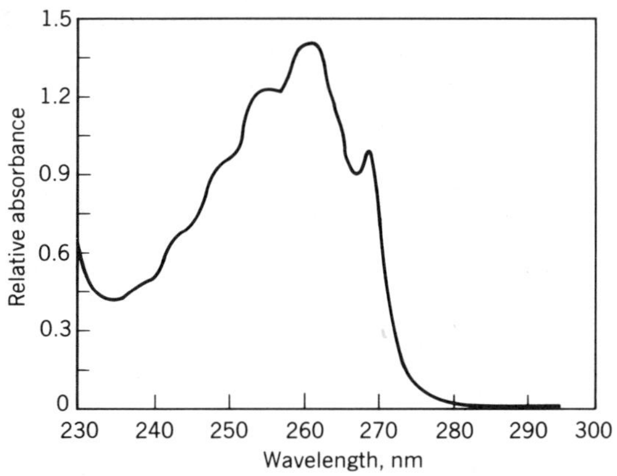

Figure 1. Ultraviolet absorption spectrum of polystyrene, 1.115 mg/mL THF.

of the phenyl rings. Deactivation of the excited phenyl groups occurs preferentially by nondestructive photophysical processes. However, scission of bonds adjacent to the phenyl ring may also occur. The most important of these reactions is the scission of the tertiary C—H bond to generate a polystyryl radical.

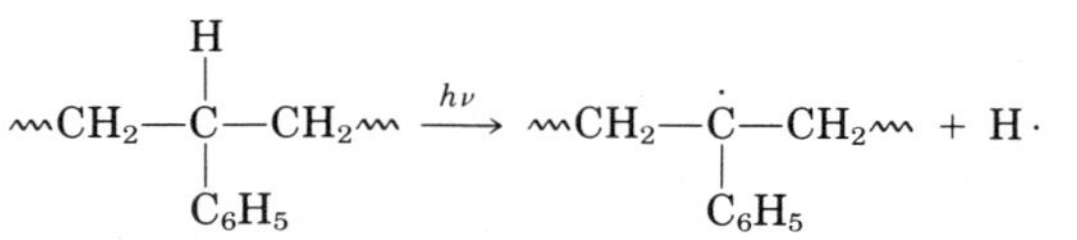

Short-wavelength irradiation causes yellowing of polystyrene under vacuum, in an inert atmosphere, or in the presence of oxygen. There is strong experimental evidence that yellowing is primarily due to the formation of conjugated polyene structures of the type

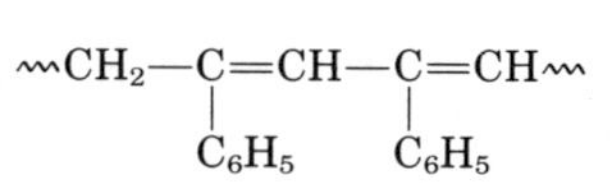

Outdoor Weathering Conditions. The predominant reactions in polystyrene exposed to outdoor weathering conditions are chain scission, yellowing, and formation of oxygenated structures. Initiation is due to the presence of chromophoric impurities formed during polymerization and processing.

HIPS and ABS both contain polybutadiene rubber, which oxidizes rapidly and leads to poor weatherability. To overcome this deficiency of HIPS and ABS, polymers based on a styrene–acrylonitrile matrix in combination with a more stable rubber phase have been developed. Commercially available materials of this type include acrylonitrile–(ethylene–propylene–diene rubber)–styrene (AES) and acrylonitrile–styrene–acrylate rubber (ASA). The weatherability of all styrenic polymers is vastly improved by the use of appropriate stabilizers.

Degradation by Ionizing Radiation

Polystyrene is highly resistant to degradation by ionizing radiation, particularly under vacuum or in inert atmospheres. For comparisons of the radiation resistance of various polymers, see DEGRADATION; CROSS-LINKING BY RADIATION; and RADIATION-RESISTANT POLYMERS. This remarkable stability is due to the phenyl groups, which confer radiation protection in polystyrene (and similar aromatic polymers) by dissipating excitation energy without decomposing.

BIBLIOGRAPHY

A. Guyot, *Polym. Deg. Stab.* **15,** 219 (1986).

M. Guaita, *Br. Polym. J.* **18**(4), 226 (1986).

Y. Kamiya and E. Niki in H. H. G. Jellinek, ed., *Aspects of Degradation and Stabilization of Polymers,* Elsevier Science Publishing Co., Inc., New York, 1978, p. 79.

A. Davis and D. Sims, *Weathering of Polymers,* Applied Science Publishers, Ltd., London, 1983, pp. 208–215.

NORBERT L. MAECKER
DAVID N. ARMENTROUT
The Dow Chemical Company

Processing

POLYSTYRENE FOAMS

The properties of polystyrene that make it useful for many applications as a solid polymer also make it very desirable as a foam. Polystyrene has high modulus, good water resistance, low moisture transmission, ease of fabrication, and low cost. Closed-cell foams have excellent thermal-insulating capabil-

Table 1. Polystyrene Foam Uses in the United States, 1986

Type	Density range,[a] kg/m³	1986 usage in U.S., 10^3 t	Applications
foam sheet	32–160	235	food packaging
molded expandable	16–160	223	packaging, buoyancy, insulation boards
extruded board	27–66	58	insulation boards, decorative, buoyancy
expanded loose	8–30	18	loose-fill packaging
injection molded (structural foam)	> 300	5	cabinets, pallets, boxes

[a] To convert kg/m³ to lb/ft³, divide by 16.

ity, low weight, and good cushioning characteristics. The combination of these properties provides a wide spectrum of products. A number of processes have been developed over the last 40 years to prepare polystyrene foams for a multitude of applications that consumed 494,000 t of polystyrene in the United States in 1985 and 539,000 in 1986. Table 1 shows the types of foam products and their volumes.

BIBLIOGRAPHY

Mod. Plast. **64**(1), 59 (1987).

C. Benning, *Plastic Foams,* Vol. 1, John Wiley & Sons, Inc., New York, 1969, pp. 90–100.

K. C. Frisch and J. A. Saunders, eds., *Polystyrene and Related Foams, Part II, Plastic Foams,* Marcel Dekker Inc., New York, 1973.

RICHARD E. SKOCHDOPOLE
GARY C. WELSH
The Dow Chemical Company

EXTRUSION AND THERMOFORMING

Extrusion

Fifty-five percent of all styrene polymers and copolymers are processed by extrusion and mainly by single-screw extruders. Because the resins are easy to plasticate and have good melt strength, they are readily fabricated into sheet, film, rod, pipe, profiles, tubing, and monofilament. More complicated shapes and geometries can be made by postextrusion treatments such as thermoforming or embossing. The process is very economical because single-screw extrusion is a relatively low cost technique and trim and scrap can be efficiently ground and reextruded, with little change in appearance or properties.

Coextrusion of styrenic polymers is a rapidly advancing science. The current number of barrier sheet coextrusion lines for commercial production of food packaging is reported to be between 15 and 30 worldwide; this number is expected to double within the next year. Although commercial coextruded products may have as many as nine layers, the largest volume of extruded styrenic polymers is used in simple two- and three-layer structures. The applications for these relatively simple composites include dairy containers, two-color drink cups and dinnerware, and sheet for outdoor use. An example of a more complex coextrusion structure for barrier packaging consists of polystyrene–poly(vinylidene chloride)–polypropylene–high density polyethylene and appropriate adhesive layers. Each layer carries out a specific function in the composite (see BARRIER POLYMERS; FILMS, MULTILAYER).

Screw plasticating is also used for injection molding, most frequently accomplished by a reciprocating screw. Extrusion is also used to develop the parison for blow molding. Projected bosses, recesses, and undercuts, which are easily made by injection and blow molding, are not possible with extrusion.

Thermoforming

Styrenic polymers are thermoformed by heating sheet above the glass-transition temperature and then shaping in a male or female mold using vacuum, air pressure, or mechanical force (see THERMOFORMING). The polymer is then cooled in the mold. Most thermoformed products require trimming after forming. This can be accomplished in the mold or outside the mold in a punch press, using robotic positioning of lasers or water jet systems, as well as manually operated saws, routers, drills, and other trimming devices. It is relatively easy to carry out these trim operations on impact-modified styrenics (HIPS and ABS), but general-purpose styrenics (GPPS and SAN) tend to craze and crack because of their brittleness. This problem can be overcome if the sheet is biaxially oriented to improve its ductility.

Thermoforming is most often carried out in a separate operation from extrusion of the sheet. Pressure forming is now being practiced more widely in thermoforming. The quality of thermoformed parts depends on sheet quality, tooling, and forming temperature.

The plastics thermoforming industry is dominated by styrenics, including oriented polystyrene sheet (OPS), impact-resistant polystyrene, and acrylonitrile–butadiene–styrene (ABS). It is estimated that 75% of the approximately 7×10^5 t of plastics thermoformed annually in the United States are styrene-based products (Table 1). Thermoforming is often the least expensive process for fabricating plastic items because of its ability to easily make thinner container walls (which decreases raw material costs), the relatively low cost of forming equip-

Table 1. Thermoforming Plastics Consumption, 10^3 t

Material	1984	1989 (estimated)	Average compounded annual growth, %
high acrylonitrile copolymers	8.6	8.6	0
cellulosics	3.2	2.3	−6.5
high density polyethylene	4.5	5.0	2.0
K-Resin[a]	18.2	26.4	7.7
polyesters	7.3	26.4	29.0
polypropylene	5.9	16.8	24.0
polystyrene	485.0	597.0	4.2
poly(vinyl chloride)	46.0	61.4	6.0
XT-Polymer[b]	6.4	5.0	−4.5
Total	*585.0*	*749.0*	

[a] Trademark of Phillips Petroleum Co. for 70 styrene–30 butadiene block copolymers.

[b] Trademark of American Cyanamid for transparent polymers containing styrene, methyl methacrylate, and butadiene.

ment, and the high output rates achieved. Major areas of use are packaging, appliances, building and construction, disposables, and housewares.

Oriented Polystyrene Sheet

Where a clear formed product is desired, many fabricators use oriented polystyrene sheet (OPS). Because of the low elongation of general-purpose polystyrene, it is necessary to biaxially orient the sheet to obtain the required physical properties for satisfactory part trimming.

OPS is commercially available in 0.13–0.76 mm (5–30 mil) gauge. The tentering-frame process for manufacturing OPS is the most common process used today.

OPS is usually made of a high heat distortion, unplasticized, high molecular weight, general-purpose polystyrene resin. For most food packaging applications, the resin is low in residual ethylbenzene and styrene monomer so that no detectable taste or odor is transferred to the foodstuff. Scrap resin can be added back to the process as either ground sheet or repelletized material.

OPS is a particularly desirable packaging material because of its high flexural modulus, excellent clarity, low taste and odor transfer, and relatively low cost. A summary of OPS properties is shown in Table 2.

The OPS market has realized significant growth since its introduction in the late 1950s. In 1986, the U.S. volume of OPS produced was 73,000 t. Most of this volume is used in bakery goods trays, making OPS the largest volume thermoplastic material used in this packaging application. OPS used in food packaging applications requires that all resins and additives be approved for food contact.

Table 2. Typical Properties of Biaxially Oriented Polystyrene (OPS)

Property	Value
tensile strength, MPa[a]	62–83
tensile modulus, MPa[a]	3100
tensile elongation, %	5–60
light transmission, %	92
specific gravity	1.05
material yield, m^2/N at 0.025 mm	3.8[b]

[a] To convert MPa to psi, multiply by 145.
[b] 26,200 in^2/lb at 1 mil.

BIBLIOGRAPHY

Mod. Plast., 98 (Jan. 1988).

Z. Tadmor and I. Klein, *Engineering Principles of Plasticating Extrusion,* Van Nostrand Reinhold Co., Inc., New York, 1970.

N. P. Cherimisinoff, *Polymer Mixing and Extrusion Technology,* Marcel Dekker, Inc., New York, 1987, p. 304.

J. L. Throne, *Thermoforming,* Hanser Publishers, New York, 1986, p. 21.

Richard E. Skochdopole
Theodore C. Wallace
Gary C. Welsh
The Dow Chemical Company

INJECTION MOLDING

Injection molding is an important process for converting about 32% of all plastic materials into useful plastic products. An

Figure 1. Office supplies from injection-molded polystyrene.

estimated 70,000 injection-molding machines are reportedly operational in the United States to service this billion-dollar business. The injection-molding use for polystyrene in the United States was 114,000 t in 1986. Injection-molded parts provide unique properties in a wide range of home and commercial service such as housewares, electronics, medical ware, office supplies (Fig. 1), toys (Fig. 2), and home appliances.

Figure 2. Toys from injection-molded polystyrene.

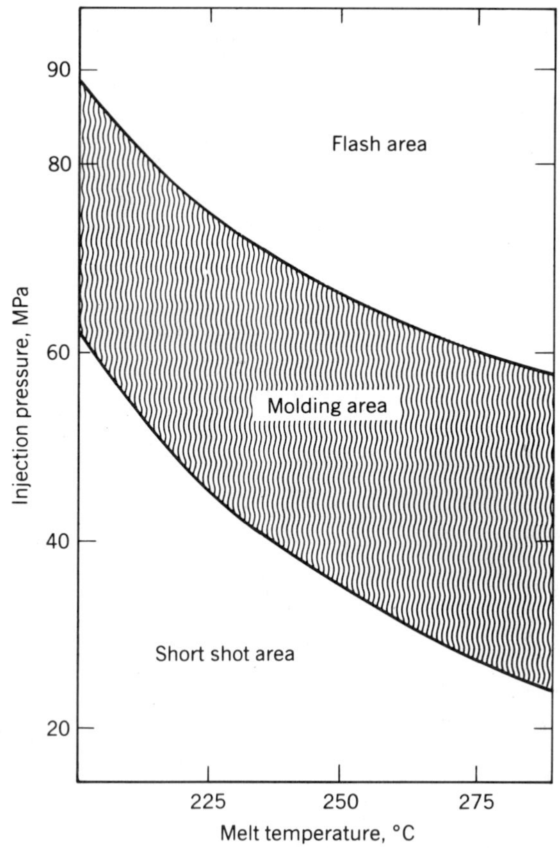

Figure 3. Two-dimensional molding-area diagram that plots injection pressure (ram pressure) vs melt (resin) temperature. Within this area, all parts meet performance requirements. However, at the edge of the diagram, tolerance for machine and plastic variations is narrow, so rejects increase. To convert MPa to psi, multiply by 145.

Molding Process

During the injection-molding process plastic granules are melted and injected into a mold containing a cavity in the shape of the part desired; the molded plastic is cooled to retain the part shape and the part is ejected from the mold. While the molded part is cooling, the next shot of polymer is being melted, so that another injection can take place as soon as the cooled part is ejected and the mold is closed. This allows for high speed production of identical parts.

Resin Selection and Quality Control

The commonly published melt flow rate (MFR) value for each resin product is used as an indication of the ease of resin flow and is generally accepted for approximating the performance of polystyrene in an injection-molding machine.

Another tool for resin selection and quality control is the molding-area diagram (Fig. 3), which is drawn from the results of a series of molding tests conducted with a mold of specified difficulty.

The spiral mold provides another evaluation of moldability. The distance that the resin flows at specified melt temperatures and injection pressures is compared to the values obtained for another grade of resin when using the same spiral mold and molding machine.

Now that greater uniformity of temperature, pressure, and cycle time is possible with modern control systems, the current emphasis centers on minimizing cycle times for cost reduction.

Molded Properties

Orientation. Orientation of molecules during injection of the molten polymer into the mold cavity has been a matter of considerable study. Generally, strength is developed in the part by this orientation, particularly in the direction of resin flow from the gate or entrance to the mold. The amount of orientation in each area or zone in a part is a function of the amount of resin flow or elongation while cooling through a critical temperature range.

Part Defects. Molding problems and part defects come in many forms, too great in number to be adequately discussed here. However, there are good sources of information from books, resin suppliers, mold components suppliers, and professional and trade organizations.

BIBLIOGRAPHY

D. V. Rosato and D. V. Rosato, *Injection Molding Handbook,* Van Nostrand Reinhold, Inc., New York, 1986, p. 314.

I. I. Rubin, *Injection Molding,* John Wiley & Sons, Inc., New York, 1972, pp. 3–8.

FRANK L. BURKETT
The Dow Chemical Company

Economic Aspects

The rapid growth of styrene polymers during the 1940s, 1950s, and 1960s led to a significant decline in selling prices as volumes increased. This article describes developments that have taken place since 1970.

Styrene Monomer

Table 1 shows U.S. and world capacities for styrene monomer. U.S. capacity has increased only slightly since 1975, but total world capacity has grown at an annual rate of 4.2%. A breakdown of production of the monomer and its derivatives for each global region is shown in Table 2.

Table 1. U.S. and World Plant Capacities (10^3 t) for Styrene Monomer

Year	United States	World
1975	3,406	8,892
1976	3,682	9,061
1977	4,162	9,942
1978	3,890	11,000
1979	3,927	11,292
1980	4,000	11,635
1981	4,000	11,710
1982	3,346	11,053
1983	3,705	11,522
1984	3,719	12,394
1985	3,782	12,622
1986	3,798	13,955

Styrene Polymers

Global polystyrene capacities and domestic consumption levels for two recent years appear in Table 3. The table includes all grades of polystyrene; general purpose, impact, and expandable polystyrene (EPS) beads. Europe, the Pacific, and the United States together account for over 90% of both world capacity and consumption. The Pacific region has shown the most rapid rate of growth in consumption during the 1980s. The Dow Chemical Co. and BASF rank as first and second in global polystyrene capacity.

Polystyrene. Table 4 indicates annual polystyrene price trends since 1971. Large price increases took place in 1974 and 1979 because of significant upward shifts in crude oil and benzene prices. Lower prices for these feedstocks since 1981 and intense competition caused resin prices to fall through 1985. However, higher monomer prices, stronger market demand, and almost complete use of effective capacity resulted in about 13¢/kg increase in solid resin prices in the last half of 1986. Additional price increases were posted in 1987.

Table 5 shows polystyrene application patterns for the three largest regions. In each case, packaging is the most important market segment.

Table 3. Polystyrene Global Production, 10^3 t[a]

Region	Nameplate capacity[b] 1985[c]	Nameplate capacity[b] 1986[d]	Domestic consumption 1985[c]	Domestic consumption 1986[d]
Canada	178	163	144	149
Europe[e]	3408	3414	2725	2797
Latin America	472	475	347	381
Pacific	1910	2010	1599	1684
United States	2458	2433	1838	1982
Total	*8426*	*8495*	*6653*	*6923*

[a] All grades including general purpose, impact, and expandable polystyrene.
[b] Announced production capacity.
[c] Actual.
[d] Estimated.
[e] Including Mid-East and Africa.

ABS. The Pacific region not only consumes the most ABS, but it is also the area with the highest growth rate. Large volumes of resin are used in automobiles, radio and television cabinets, and other goods that are exported from this region. The U.S. and European regions have similar capacity and consumption.

As in the case of polystyrene, ABS prices tend to correlate

Table 4. U.S. Polystyrene Price Trends, ¢/kg

Year	SPI average[a]	ITC average[b,c,d]	Chem Data[d,e]
1971	32.1[c]		
1972	29.5[c]		
1973	35.0[c]		
1974	63.1[c]		
1975	65.8		
1976	65.6	62.7	
1977	63.1	60.5	
1978	65.3	62.7	56.3
1979	88.9	86.9	84.5
1980	106.0	110.0	93.3
1981	106.5	102.3	92.3
1982	99.9	91.3	85.8
1983	98.8	85.8	83.6
1984	99.7	88.0	84.5
1985	82.9	77.2	75.5

[a] SPI average price equals gross dollar revenues divided by total domestic merchant volume of all grades.
[b] U.S. International Trade Commission (ITC) average equals unit value of all sales of solid polystyrene, including exports.
[c] EPS not included.
[d] Not available before 1978.
[e] Chem Data price is estimated average delivered bulk contract price for large volume purchasers of general-purpose polystyrene.

Table 2. Global Production (10^3 t) of Styrene Monomer and Styrene Derivatives, 1984[a]

Region	Styrene monomer	ABS	Polystyrene	SBR	S/B copolymers[b]
Canada	406	39	145	105	
Europe[c]	2986	453	2018	1132	
Latin America	317	50	269	348	
Pacific	1835	561	1417	788	
United States	3495	512	1742	1388	338
Total	*9039*	*1615*	*5591*	*3761*	

[a] Data on East Bloc region are incomplete.
[b] Data outside the U.S. are not available.
[c] Including Mid-East and Africa.

Table 5. Regional End-use Patterns for Polystyrene, % of Region's Total Domestic Consumption

Market segment	Japan 1984, %	United States 1984, %	Western Europe 1982–1984, %
General purpose and high impact			
packaging	37	37	44–45
serviceware and flatware		8	
furniture and housewares		9	16–19
electrical–electronic	34	7	8–9
toys	5	8	5–6
appliances	6	4	8–9
construction		3	
other	18	24	12–19
Total	*100*	*100*	*100*
Expandable polystyrene bead			
packaging	67	53	32–36
insulation	33	40	62–66
other		7	2
Total	*100*	*100*	*100*

with feedstock costs and the relative balance between market demand and resin supplies. Prices moved sharply higher in 1974 after monomer prices rose rapidly. Average ABS prices increased at about 8% per year between 1976 and 1985. However, after adjusting for inflation, the change in real dollar terms was less than 1% per year. During the first quarter of 1987 prices moved higher, reflecting rising feedstock costs and fairly strong market demand.

SAN. SAN (styrene-*co*-acrylonitrile) is a relatively small volume specialty polymer. It is used primarily in injection-molding applications that require the combination of very good chemical resistance and clarity, good tensile strength, and high modulus. Table 6 shows regional capacity and consumption data for recent years. The Pacific region is the largest consumer of SAN. In the United States, SAN has had a negative growth rate since 1981.

Table 6. Global SAN Data, 10^3 t

Region	Nameplate capacity 1985[a]	Nameplate capacity 1986[b]	Domestic consumption 1985[a]	Domestic consumption 1986[b]
Canada	5	5	7	7
Europe[c]	80	80	76	77
Latin America	11	11	10	10
Pacific	131	136	98	101
United States	93	96	38	32
Total	*320*	*328*	*229*	*227*

[a] Actual.
[b] Estimated.
[c] Including Mid-East and Africa.

BIBLIOGRAPHY

Digest of Polymer Developments, Styrenics & Acrylics, Series III, Number 41, Springborn Labs., Inc., Enfield, Conn., Dec. 1986, pp. 2, 48, 77.

WILLIAM PEARSON
RICHARD H. SIEMER
The Dow Chemical Company

Analytical Methods

Physical Property Testing

Selection of a plastic for a specific application requires an understanding of how well its property profile fits with existing or proposed part design. Good physical property testing is a vehicle through which successful decisions can be made on performance and long-term quality control. The importance of proper polymer testing along with an understanding of how to interpret the resulting data cannot be overemphasized. This article covers only the more common tests useful with polystyrene resins. However, there are several in-depth reviews of physical property testing techniques for polymers. Manufacturers that fabricate general-purpose polystyrene (GPPS) or high impact polystyrene (HIPS) are usually interested in melt-flow rate, tensile properties, notched Izod impact, Vicat softening temperature, and practical heat distortion characteristics. These tests are easy to conduct with the proper equipment. However, care must be exercised to ensure reproducibility of results as well as accuracy.

Impact Testing. Impact testing (ASTM D256–84, ASTM D3029–84, and ASTM D3763–84) includes notched Izod, dart drop, and high speed instrumented puncture impact.

Tensile Properties. Tensile properties (ASTM D638–84) are determined by pulling a sample at a constant rate.

Flexural Properties. It has been found that measurement precision for flexural strength and tensile strength are similar (ASTM D790). The greater spread in flexural strength in the strength range of commercial interest makes it a more powerful tool to differentiate between stronger and weaker materials.

Thermal Properties. Heat distortion tests (ASTM D1525–82 and ASTM D648–82) are used to predict the temperature at which a fabricated part will distort.

Rheology. Rheology test techniques (ASTM D1238–85) include melt flow rate and capillary rheometry.

Complementary Analytical Techniques

There are many analytical techniques that can be used to characterize a polymer for low molecular weight species, macromolecular structure, or compositional analysis. Periodic reviews of the latest technologies and advancements in these areas appear in most journals dealing with analytical or polymer chemistry. Although different techniques may be used to obtain the same data about a polymer, the initial state of the sample (latex, granules, film, foam) and sample preparation may limit the usefulness of a particular method.

Thermal analysis techniques, such as differential thermal analysis (dta), differential scanning calorimetry (dsc), or thermal gravimetric analysis (tga) are useful for measuring heats of fusion, crystallization rates, glass-transition temperatures, other transitions, and thermal stability. Several authors have described specific techniques with examples of the application of thermal methods to polymer systems for differentiating polymer performance on a molecular level.

BIBLIOGRAPHY

R. Evans in M. D. Baijal, ed., *Plastics Polymer Science and Technology,* John Wiley & Sons, Inc., New York, 1982, p. 779.

V. Shah, *Handbook of Plastics Testing Technology,* John Wiley & Sons, Inc., New York, 1984.

E. Galli, *Plastics Compounding,* 18–28 (Sept.–Oct. 1982) (good impact testing review).

C. VanHall, ed., *Modern Methods of Analytical Chemistry,* The Dow Chemical Co., Midland, Mich., 1986, pp. 89–102.

ROBERT L. WINTERMYER
WALTER C. BUZANOWSKI
The Dow Chemical Company

Health and Safety

Polystyrene is considered physiologically inert. Solids or dusts may cause eye irritation or corneal injury due to mechanical action. Skin absorption and vapor inhalation are unlikely because of the physical properties of polystyrene. Dusts may cause irritation to the upper respiratory tract. Implant studies with polystyrene have produced tumors consistent with carcinogenesis caused by physical rather than chemical factors.

Polystyrene and rubber-modified polystyrene may be safely used as components of articles and packaging materials in contact with food and are regulated for this use by the Food and Drug Administration in 21 CFR 177.1640. Other styrene copolymers are regulated for food contact applications by other regulations. Under fire conditions, the hazardous decomposition products of polystyrene are smoke and noxious gases, primarily carbon monoxide and hydrocarbons.

BIBLIOGRAPHY

Material Safety Data Sheet for Polystyrene, The Dow Chemical Company, 1986.

Patty's Industrial Hygiene and Toxicology, 3rd rev. ed., Vol. 2C, John Wiley & Sons, Inc., New York, 1982, pp. 4337, 4339.

Code of Federal Regulations, Food and Drugs, Vol. 21, Part 170 to 199, April 1987, p. 259.

M. LOUISE RAINEY
The Dow Chemical Company

Uses

Polystyrene has more diverse uses than any other polymer because of its excellent physical properties, ease of fabrication with a variety of equipment, and low cost. The principal applications for polystyrene include packaging, appliances, construction, automotive parts, toys, housewares, luggage, medical ware, disposable food service, and electronic equipment. Shape, property requirements, aesthetics, and cost must all be considered before making a material choice and deciding on the fabrication method. Table 1 relates fabrication methods to end uses.

Polystyrene use has grown rapidly in many market segments. However, the most significant growth has been in the packaging industry, primarily because of the low cost of polystyrene and its ease of fabrication at high speeds.

In the automotive industry, growth (primarily ABS and SAN) has been spurred by the need to produce lighter, more energy-efficient vehicles. The cost to produce many of these parts and the labor requirements for assembly are considerably less than for some of the parts they are replacing.

Growth in the electronics market arises from the many new devices being produced, such as computers, videocassette recorders, and new telecommunications equipment. Medical applications have expanded because of the increased diagnostic testing being done for all patients. Polystyrene is a good candidate for these applications because of its clarity, inertness, and ease of fabrication into complex parts.

PHILLIP A. WAGNER
JACK H. BENSON
The Dow Chemical Company

Table 1. Fabrication Methods and Uses for Polystyrene

Fabrication method	Use
injection molding	furniture, toys, radio and television cabinets, automotive parts, medical ware, housewares, bottle caps, containers
blow molding	bottles, containers, furniture, automotive parts
extrusion	film (including multilayer and oriented), profiles, light diffusers, wall covering
extrusion and thermoforming	refrigerator and freezer parts, luggage, food containers both solid and formed, disposable cups and dinnerware, large automotive parts

SULFIDES, POLYMERS.
See POLYSULFIDE POLYMERS; POLY(ARYLENE SULFIDES); SULFUR-CONTAINING POLYMERS.

SULFOCHLORINATED POLYETHYLENE.
See ETHYLENE POLYMERS.

SULFONAMIDE–FORMALDEHYDE RESINS.
See AMINO RESINS.

SULFONE POLYMERS.
See OLEFIN–SULFUR DIOXIDE COPOLYMERS; POLYSULFONES.

SULFUR-CONTAINING POLYMERS

Polymeric Elemental Sulfur

The thermal "gelling" of elemental sulfur was first described a century ago. This phenomenon was attributed later to the equilibrium polymerization of cyclooctasulfur into a high polymer:

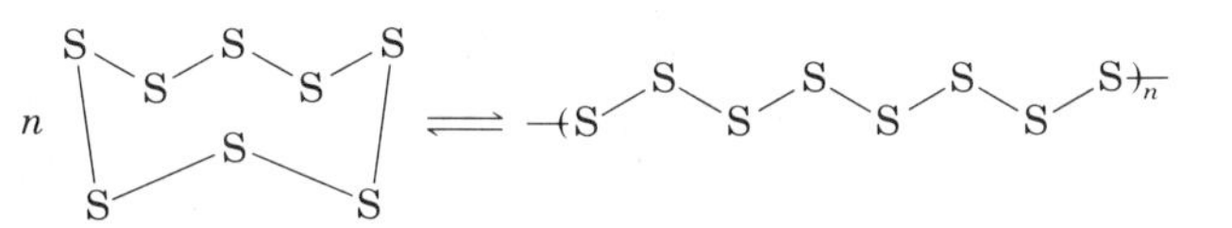

The polymerization of elemental sulfur and properties of the monomeric species and of the polymer have been extensively reviewed.

Monomers and Their Preparation. Under normal conditions, the most stable form of sulfur is the crystalline orthorhombic α-S_8, which is transformed on heating above 368.5 K to the monoclinic β-S_8, melting at 392.2 K. On further heating, the liquid sulfur polymerizes at a temperature depending on the heating rate. Extrapolation to zero rate results in the onset of polymerization at 432 K.

According to more recent findings, other isomers of elemental sulfur, namely cyclic S_n with $n \neq 8$, are also able to polymerize. The only commercial monomeric sulfur is S_8.

Polymerization. The ability of any given S_n to polymerize can be judged from the ΔG_p of the equilibrium

$$\cdots\text{—}S_x\text{—}\cdots + S_n \rightleftharpoons \cdots\text{—}S_{x+n}\text{—}\cdots$$

$$\Delta G_{p(n)} = \Delta H_{p(n)} - T\,\Delta S_{p(n)}$$

The polymerization of S_8 is endothermic and endotropic ($\Delta H^0_{p(8)} > 0$, $\Delta S^0_{p(8)} > 0$, see Table 1). Thus, $\Delta G_{p(8)}$ is negative only above a certain temperature T_f, called the floor temperature, given by the equation

$$T_f = \frac{\Delta H^0_{p(8)}}{(\Delta S^0_{p(8)} + R\ln[S_8]_0)}$$

The polymerization is driven by the positive change of entropy, due to the higher number of conformational degrees of freedom in polyS compared to the more rigid monomer S_n; this effect outweighs the decrease of the translational entropy.

At 432 K and normal pressure, almost all physical properties (viscosity, specific heat, electric conductivity, density, magnetic susceptibility) characterizing liquid sulfur exhibit a discontinuity. PolyS content rises with increasing temperature, reaching the maximum value of 40 wt % at about 513 K.

Almost all theories to date describe the polymerization of elemental sulfur in terms of the equilibria between S_n rings and polyS diradicals. The complete polymerization scheme should include formation of free radicals by ring opening and by polyS chain splitting, displacement of sulfur–sulfur bonds by radicals in cyclic S_n molecules (propagation) and in polyS chains, ring interconversion, and propagation by ring expansion.

Polymeric Sulfur. *Preparation and Properties.* Both experimental data and thermodynamic considerations indicate that polyS is stable at normal pressure only above T_f = 432 K. On the other hand, it is possible to obtain almost pure (up to ca 99 wt %) polymer below T_f by temporary freezing of the equilibrium attained at higher temperatures or by kinetic enhancement of polyS content.

Depolymerization. Although polyS is relatively stable below its T_g, it gradually depolymerizes:

$$\text{polyS} \longrightarrow S_6 + S_7 + S_8 + S_9 + \cdots$$

with the rate increasing with temperature.

PolyS is used mainly in rubber vulcanization. Other applications include removal of Hg from waste waters and in coatings of controlled-release fertilizers.

Polysulfides

During the last several decades the synthesis of polymers with high sulfur content was investigated with the aim of improving chemical stability and viscoelastic properties over those of polymeric sulfur. Polymers that are structurally closest to polyS

Table 1. Standard Thermodynamic Parameters and Transition Temperatures in the Polymerization of Cyclic Sulfur Molecules, S_n

	n							
	5	6	7	8	9	10	11	12
$\Delta H^0_p(n)$, kJ/mol[a]	−34.9	−8.2	−4.6	19.0	−8.0	−6.0	−7.0	17.0
$\Delta S^0_p(n)$[b], J/(mol·K)[a]	−27.2	−2.6	20.7	44.7	69.0	93.0	117.0	141.0
T_c[c], K[d]	1283	2928						
T_f, K[d,e]			[e]	430	[e]	[e]	[e]	120

[a] To convert J to cal, divide by 4.184.
[b] Standard state at weight fraction of S_n equal to 1.
[c] T_c = ceiling temperature of S_n polymerization.
[d] Evaluated for bulk polymerization.
[e] No thermodynamic limitations of the polymerization; calculated T_f is below 0 K.

are polysulfides (qv) (**1**), which contain various units R (most often alkylene, oxaalkylene, or arylene groups):

$$\text{-(}R\text{—}S_x\text{)}_n\text{-} \qquad x = 1, 2, 3 \ldots$$
(**1**)

The first example of a polysulfide synthesis was the polycondensation of 1,2-dichloroethane with sodium tetrasulfide, followed by polysulfides based on bis(2-chloroethyl) ether and bis(2-chloroethyl) formal:

$$n\,(Cl\text{—}CH_2CH_2OCH_2)_2O + n\,Na_2S_x \longrightarrow \text{-(}CH_2CH_2OCH_2OCH_2OCH_2CH_2\text{—}S_x\text{)}_n\text{-} + 2n\,NaCl$$
(**2**)

Polyaddition. Remarkable progress has been made recently on the synthesis of polysulfide polymers by other routes, such as ring-opening polymerization of cyclic polysulfides (eg, eq. 1) as well as addition and condensation copolymerization of elemental sulfur.

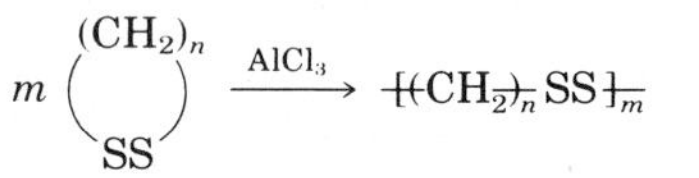
(1)

Copolymers of S_8 with Inorganic Comonomers. The copolymerization of S_8 with selenium (Se_8) is an example of an equilibrium copolymerization:

$$S_8 + Se_8 \rightleftharpoons \text{-(}S_xSe_y\text{)}_n\text{-} + S_{8-z}Se_z$$
(**3**) (**4**)

Calorimetric and differential thermal analysis (dta) studies show that the T_f in selenium–sulfur copolymerization is between that of S_8 and that of Se_8; it decreases with increasing concentration of Se_8. In addition to the linear copolymer (**3**), cyclic sulfur–selenium compounds (**4**) are formed. Copolymerization of S_8 with arsenic or phosphorus has also been reported.

Copolymers of S_8 with Unsaturated Comonomers. The vulcanization (qv) of rubber can be considered as copolymerization of sulfur with an olefin. In addition, S_8 can be copolymerized with monomers having carbon–carbon multiple bonds; poly(alkylene polysulfide)s are the products:

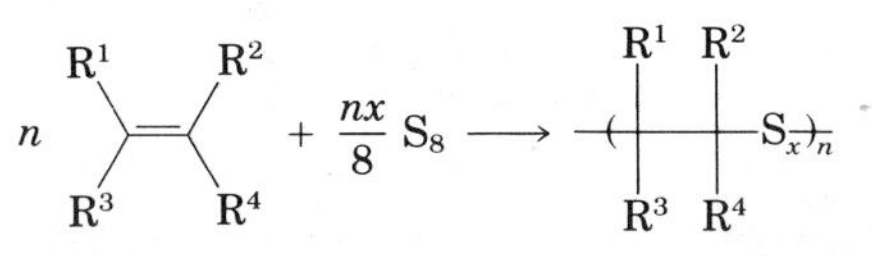

The sulfur rank (average x value) and molecular weights of the polysulfide products depend strongly on temperature, feed composition, and mode of initiation.

Potential applications of the copolymers of S_8 with unsaturated monomers have recently been covered by numerous patents. Examples include acid- and salt-resistant concrete, asphalt replacements, and impregnation of fire-resistant materials.

Copolymers of S_8 with Cyclic Mono- and Polysulfides. The anionic copolymerization of S_8 with thiiranes and 3,4,5-trithiolanes gives high molecular weight poly(alkylene polysulfide)s:

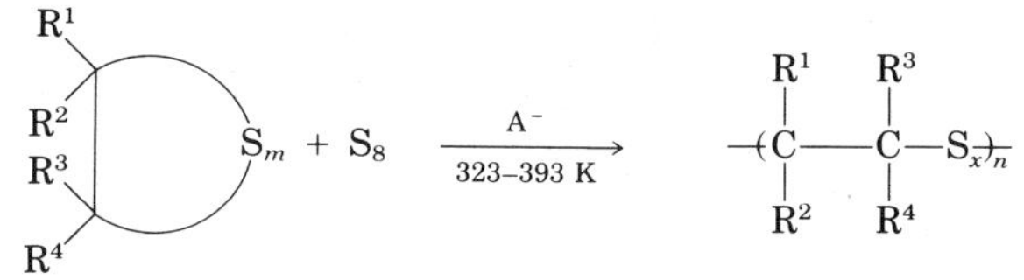

$m = 1, 3$

where A^- = anionic initiator; average $x \le 9$; $\bar{M}_n(\text{gpc}) \le 5 \times 10^4$.

The sulfur in the resulting linear polysulfides ((**5**)–(**10**)) is chemically bonded, as shown by ^1H- and ^{13}C-nmr and Raman spectra of the polymerization mixtures and isolated products.

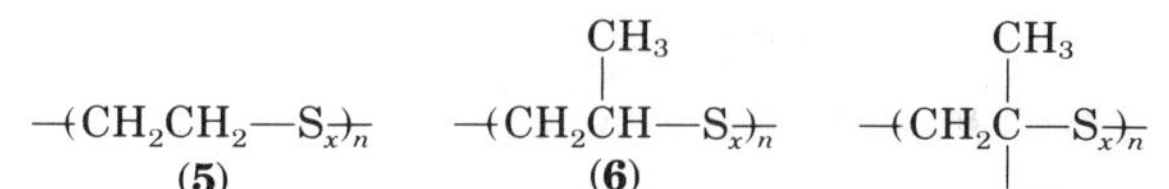
(**5**) (**6**) (**7**)

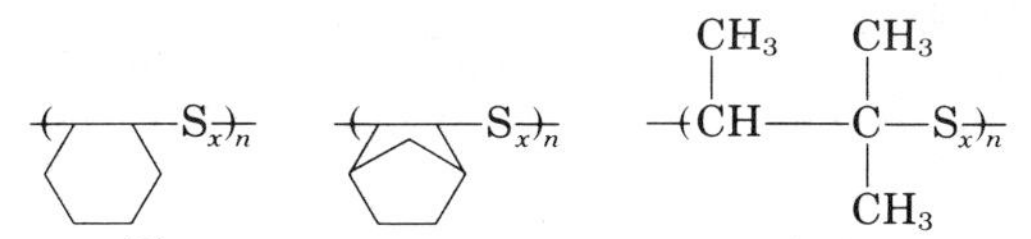
(**8**) (**9**) (**10**)

These analyses, particularly the Raman spectra, bear on the question of whether the cyclic sulfides' reaction products are true copolymers or mixtures.

Copolymerizations carried out in dilute solutions of comonomers at high S_8–thiirane molar ratios yield, in addition to the linear polysulfide, low molecular weight cyclic oligomers.

Polycondensation. ***Reductive and Oxidative Polycondensations.*** The most widely used synthetic route for the preparation of polysulfide polymers is based on the reaction between a dihalide and sodium polysulfide. Other methods, such as oxidative coupling of dithiols, are used only rarely. Oxidation of alkanedithiols by iodine or oxygen gives oligomeric products. Oxidizing agents such as bis(oxycarbonyl)disulfide, dithiobis(carbonyl chloride), thiuram disulfides, dixanthogens, and biscarbamoyl disulfide give poly(alkylene disulfide)s with η_{inh} up to 0.6 dL/g (in tetrachloroethane, 0.5%, 303 K). However, the polycondensation of 1,3-propanedithiol and 1,4-butane yields cyclic disulfides as side products. Oxidation of dithiols with hot DMSO may give higher molecular weight poly(disulfide)s.

Condensation Copolymerization of S_8. Condensation copolymerization of S_8 is also used for the synthesis of polysulfide polymers. These copolymers, however, have ill-defined structures and low molecular weights; hydrogen sulfide is frequently evolved. Condensation of S_8 with dicarbanionic oligomers of vinyl monomers give the best results.

Polythiols and Polydisulfides

Polythiols cannot be prepared by the polymerization of thiol-containing monomers because of the high reactivity of the sulfhydryl group toward free radicals and electrophiles and nucleophiles. Protecting groups are required (acetate, alkoxymethyl, O-alkylcarbonate), which are removed after polymerization. In an alternative method, a thiol group is introduced into a polymer.

The high reactivity of the thiol group creates various applications of polythiols, eg, as redox resins, ion-exchange resins, heavy-metal ion absorbers, antiradiation agents, unblocking reagents, hydrolysis catalysts, and substrates in graft copolymer synthesis.

The thiol group participates in the redox equilibrium with the disulfide moiety, eg

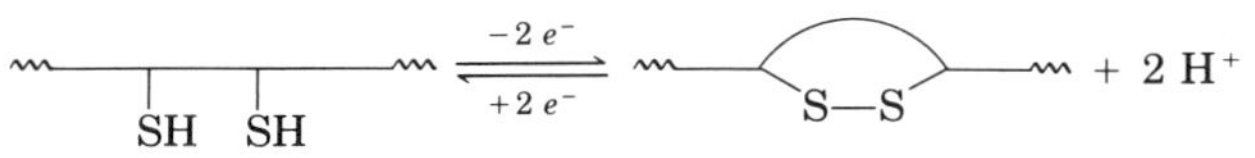

$$\text{(P)—SH} + \text{HS—R} \underset{+2\,e^-}{\overset{-2\,e^-}{\rightleftharpoons}} \text{(P)—S—S—R} + 2\,H^+$$

Thiol and Disulfide Groups in Biopolymers. The presence of sulfur, particularly in the form of disulfide bridges, plays an important role in biopolymers. These linkages are easily formed by oxidation of thiol functions. Thus, a tertiary structure can be built at the end of a sequence of reactions by building the structural features responsible for the primary and secondary structures and using thiol groups to link units in order to get the required three-dimensional structure.

Peptide Synthesis on Solid Polymeric Supports. The synthesis of peptides, an important process occurring in nature, can be replicated in the laboratory. Lipoic acid (**11**) has been used for this purpose; it is an important coenzyme in the acyl transfer and redox reactions in living systems.

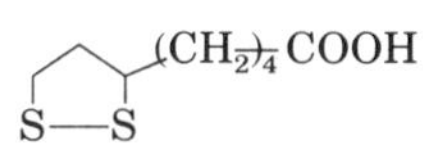

(**11**)

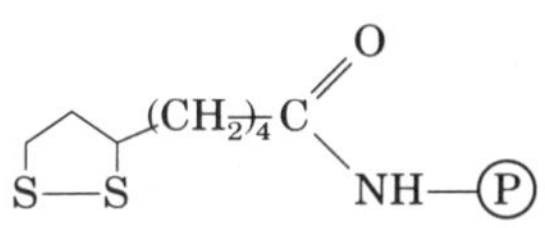

(**12**)

Thus (**11**) and its derivatives, such as lipoamides (**12**), have been immobilized on polymeric supports and tested as polymeric reagents.

Biological Systems. The structure of the naturally occurring biological systems is usually complex. Therefore, studies are often conducted on simpler model systems. The influence of the backbone flexibility of the polymeric matrix on the redox interconversions between polymeric thiols and disulfides has been investigated. The models allow the redox reactions to take place on the macromolecular moiety and to simulate the function of enzymes. These studies were conducted with polymers derived from (**13**) and (**14**).

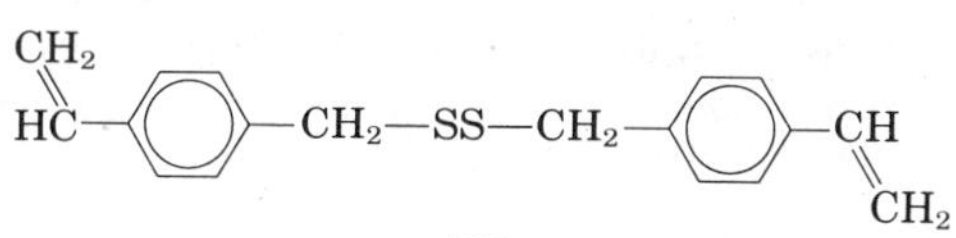

(**13**)

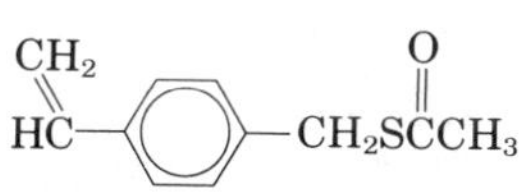

(**14**)

Biomedical Applications. Disulfide spacers have been used to attach bioactive compounds to polymers, particularly to polypeptides.

Polymeric Catalysts and Ion Sorbers. Polymeric thiols and disulfides have also been used in other chemical processes. Thus lipoamide, incorporated into various hydrophilic polymers, eg, polyallylamine, polyvinylaniline, polyethyleneimine, and chitosan, is an effective catalyst in the presence of ferrous ions for the reduction of *O*-benzylhydroxylamine to NH_3 and benzyl alcohol or of diphenylacetylene to *cis*-stilbene as well as of nitrobenzenes to anilines.

Specific reactivity of thiols is also used for reactions with ions, eg, heavy metal ion sorption. Thus, cyanuric chloride in polycondensation with polyamines and thiourea forms polymeric isothiouronium salts, which after hydrolysis give polyamines containing thiol groups. The resulting polymers are insoluble in common organic solvents and adsorb ions of cadmium, copper, zinc, nickel, cobalt, mercury, and lead.

Polymonosulfides

The mechanisms of polyreactions giving polymers with monosulfide linkage in the main chain or in the pendent group are well understood. Polysulfides with controlled and well-defined structures are easily prepared. Nevertheless, the only commercial polymonosulfide is, eg, Ryton, poly(phenylene sulfide), produced by Phillips Petroleum Co. The other polysulfides, based on thiirane or 2-methylthiirane, were evaluated for industrial applications, but have not yet been commercialized because of a lack of a convenient monomer synthesis and a persistent odor of the polymers. Thiiranes cannot be produced with satisfactory yields directly from the respective olefin and elemental sulfur. Moreover, polymeric thiiranes are not sufficiently stable toward oxygen and heat.

Polyaddition. ***Ring-opening Polymerization.*** Among the cyclic monosulfides, only three- and four-membered monomers, thiiranes and thietanes, respectively, are able to polymerize. Polymerization proceeds via both anionic and cationic mechanisms and is practically irreversible. Only tetrafluorothiirane polymerizes radically.

Copolymerization of Cyclic Disulfides. Cyclic disulfides, in contrast to polysulfides or elemental sulfur, are able to copolymerize radically with vinyl monomers because of the high reactivity of a thiyl radical toward a carbon–carbon multiple bond. The resulting copolymers contain monosulfide linkages. Nevertheless, this method has been used only rarely for the preparation of polymonosulfides. Cyclic disulfides have been copolymerized with styrene, vinyl acetate, methyl methacrylate, acrylonitrile, and others by initiation with AIBN at 333 K or thermally at 403 K; up to 20 mol % of sulfide moieties is introduced.

Polymonosulfides from Thiols and Olefins. Addition of dithiols to diolefins or polyaddition of unsaturated thiols is a versatile method of building macromolecules with the desired repeating unit.

Other Polyaddition Methods. For some polyadditions leading to polymonosulfides, only certain monomers are used; for example, diolefins with structures restricting cyclization, such as butadiene and dicyclopentadiene, form polymers with sulfur dichloride:

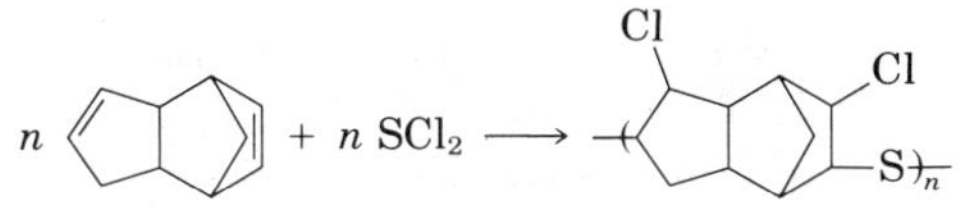

Polymonosulfides by Polycondensation. ***Aliphatic Polymonosulfides.*** Poly(alkylene monosulfide)s, in contrast to polysulfide polymers, are rarely prepared by polycondensation. Reactions of polymethylene dihalides with polymethylene dithiols or their salts or with sodium monosulfide give linear polymers of the general formula (**15**) with low molecular weights ($\bar{M}_n \sim 10^3$).

$$-\!\!\!+\!(CH_2)_m S (CH_2)_n S \!+\!\!\!-_p$$

(**15**)

$m = 2, 4, 6, \quad n = 1, 2, 3, 5, 6, 10$

High molecular weight polymonosulfides are prepared with the help of phase-transfer catalysis (PTC).

Mixed Aliphatic-aromatic Polymonosulfides. The PTC method of polycondensation produces polymonosulfides, with fairly high molecular weights in almost quantitative yields.

Poly(phenylene sulfide)s. Poly(phenylene sulfide)s (PPS) exhibit exceptional chemical stability and a high melting point. This allows the conversion of linear polymers into useful products by melt processing. The presence of S atoms between the aromatic rings provides flexibility. Applications of PPS include molding resins, fibers, and matrices for thermoplastic composites.

The polycondensation of aromatic dihalides with alkali metal sulfides in highly solvating media (eg, NMP) leads to poly(arylene sulfide)s (qv) and is the route to poly(1,4-phenylene sulfide), marketed under the trademark Ryton:

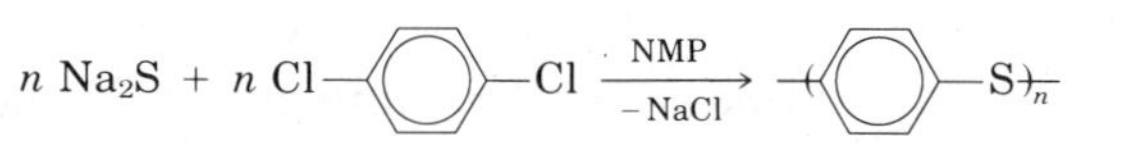

(**16**)

Poly(phenylene sulfide) is a thermoplastic polymer, melting at about 558 K and showing no appreciable thermal decomposition below 773 K.

Indications are that PPS (doped with AsF_5, I_2, or SO_3 and less reactive dopants) could provide electrically conductive materials. Commercial PPS contains up to a few percent of disulfide linkages. (see also ENGINEERING PLASTICS; POLY(ARYLENE SULFIDES)).

Polythiophenes. Thiophene is a potential structural unit of thermostable polymers. Thiophene polymers are expected to be better electrical conductors than the phenylene analogues because of more delocalized π-orbitals and lower oxidation potentials (see ELECTRICALLY CONDUCTIVE POLYMERS).

Electrochemical polycondensation is the most commonly used method of making highly conducting, homogeneous, flexible polythiophene films. This process is usually carried out in a one-compartment cell with an indium–tin oxide conducting glass anode and a platinum cathode.

Polytetrathiafulvalenes. Tetrathiafulvalenes (TTF) form highly conducting anisotropic charge-transfer (CT) complexes. To obtain compounds with three-dimensional conductivity, the polymeric forms of TTF were prepared. Copolymerization of 2,6(7)-bis(*p*-hydroxyphenyl)tetrathiafulvalene with sebacyl or terephthaloyl chlorides and diisocyanates or disulfonyl chlorides gave yields above 90%. All TTF polymers are soluble in CF_3COOH or in dipolar aprotic solvents (HMPA, DMSO, DMF); thermal stability is up to about 540 K. Polyamides with tetrathiafulvalene linkages have been described.

Polymonosulfides with the Sulfide Function in the Pendent Group. Polymers with pendent groups containing sulfide functions are prepared by polymerization of monomers with sulfide substituents or introduction of a sulfide group into a polymer. Polymerization of alkyl or aromatic vinyl sulfides has been initiated with both cationic and free-radical initiators. However, only the latter give high molecular weight polysulfides. Results of copolymerization of methyl vinyl sulfide, unsubstituted and substituted vinyl phenyl sulfides, and divinyl sulfides with styrene, methyl methacrylate, and vinyl acetate indicate stabilization of the free radical by the sulfur atom in the macroradicals formed after the addition of the sulfide monomers.

The net result of this stabilization is a decrease in propagation rate constants, compared with vinyl monomers containing a carbon atom in the α position. On the other hand, vinyl sulfides polymerize faster than their oxygen analogues. The introduction of sulfide groups into polymers usually improves properties such as thermostability and solubility.

Polythioacetals

Polythioacetals have received less attention than polyacetals, which are large-scale industrial thermoplastics (see ACETAL RESINS). This is partly due to the instability of thiopolymers in the molten state. In addition, some monomers and degradation products have an offensive odor. The synthesis and properties of polythioacetals have been reviewed.

Linear polymeric thioacetals are prepared by polymerization of thioaldehydes or thioketones via the carbon–sulfur double bond; ring-opening polymerization of cyclic trimers of thioaldehydes or thioketones; and copolycondensation of aldehydes or ketones with α,ω-dithiols, hydrogen sulfide, or their salts.

Sulfonium Polymers

The chemistry of polyelectrolytes (qv), to which tertiary sulfonium polymeric salts belong, has been studied exhaustively during recent years. Sulfonium polyelectrolytes are less stable than others, such as ammonium derivatives, and are less readily available. Nevertheless, some potential applications of polysulfonium salts have been investigated. They can be used as ion-exchange resins, polymer supports in peptide synthesis, polymeric reagents, and conducting materials. The simplest method of preparing sulfonium salts is by alkylation of the sulfur atoms in polysulfides.

Polythioesters

Studies on polythioesters were attempted with the aim of obtaining polymers with better high melting thermal and chemical resistance than that of their oxygen counterparts. Polythioesters are prepared by the condensation of the diacid chlorides with dithiols in the presence of base, polyaddition of bis(thiocarboxylic S-acid)s with compounds containing two double bonds, condensation of phenyl esters of aromatic diacids with dithiols, and ring-opening polymerization of thiolactones. Other routes are given in the example of polythioglycolate (**17**), which is prepared from thioglycolic acid (**18**); the methyl ester of thioglycolic acid (**19**); the cyclic dimeric anhydride of thioglycolic acid (**20**); 2,5-dioxo-1,3-oxathiolane (**21**); or *S*-trimethylsilylthioglycolic acid chloride (**22**):

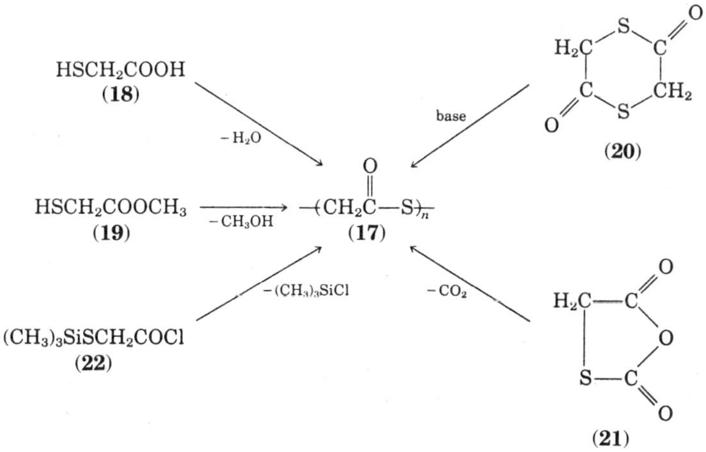

Polythiocarbonates

Among the polythioesters, polythiocarbonates exhibit specific structural features and are prepared by special methods. Depending on the number of sulfur atoms and on their arrangement, the following thiocarbonate units can be considered: oxycarbonylthio (**23**), oxythiocarbonyloxy (**24**), thiocarbonylthio (**25**), oxythiocarbonylthio (**26**), and thiothiocarbonylthio (**27**).

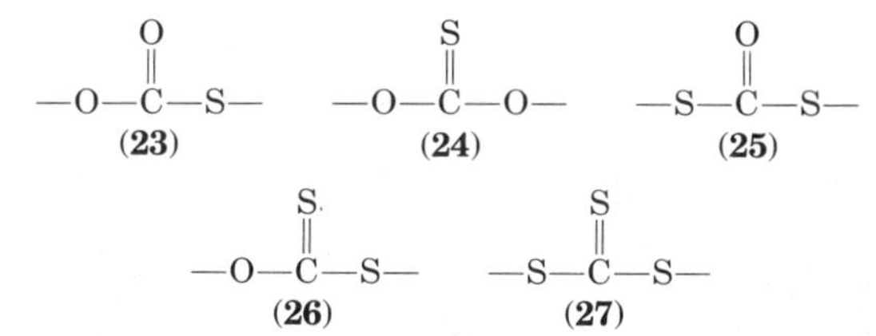

Many kinds of polythiocarbonates (**23**)–(**27**) have been described in the literature.

Polyazothiones

Polythioamides are structurally the simplest polymers containing the azothione moiety. These polymers became interesting after the discovery of polyamide fibers. The patent assigned to DuPont in 1939 disclosed the synthesis of polythioamides by the reaction of dinitriles with diamines in the presence of H_2S above 523 K. Later publications describe the synthesis of aliphatic bisthioesters and their use in the preparation of polythioamides by condensation with diamines at ambient temperatures. The products were low molecular weight oligomers. More recently, high molecular weight polythioamides have been prepared from bisdithioesters and diamines:

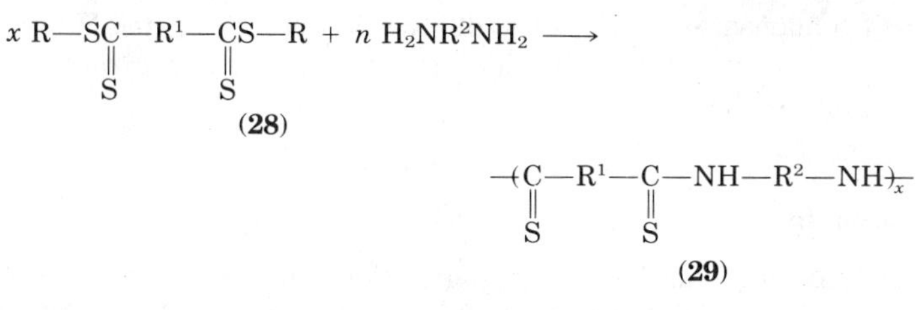

Other polyazothiones are often used as reactive polymers and low molecular weight [2*H*]-1,2,4-triazoline-3-thione may have application in photographic systems. Some polymers with dithiocarbamate functions are used to bind metal ions.

Thiazo Polymers

Poly(sulfur nitride). Polymeric sulfur nitride (polythiazyl) exhibits unusual properties, especially an electrical conductivity comparable with that of metals (600 S/cm at ambient temperatures). The conductivity of macroscopically oriented crystals of $-\!(SN)_n\!-$ is anisotropic and increases with decreasing temperature up to the superconducting transition near 0.3 K. Poly(sulfur nitride) is usually prepared from tetrasulfur tetranitride (cyclotetrathiazene) by a method discovered in 1910:

$$ {}^{n}\!/_{4}\, S_4N_4 \xrightarrow{Ag,\ 573\ K} {}^{n}\!/_{2}\, S_2N_2 \xrightarrow{273\ K} -\!(SN)_n\!- $$

Poly(1,3-thiazole)s. After the synthesis of the first all-aromatic-chain polymers, like polyimide or polybenzimidazole, other related systems were investigated. Inclusion of an S atom in the ring was believed to improve heat and oxidation stability. Current research focuses on aromatic poly(1,3-thiazole)s, which belong to an important class of high performance polymers (see also POLYBENZOBISTHIAZOLES).

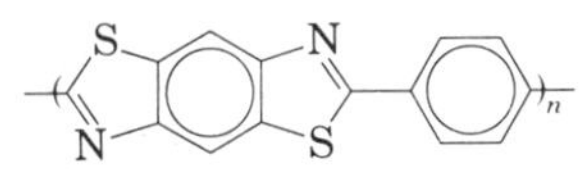

(**30**)

Poly(1,3-thiazolium) Salts. Polymeric polymethine dyes bearing benzothiazole units (**31**)–(**34**) exhibit electrical photoconductivity.

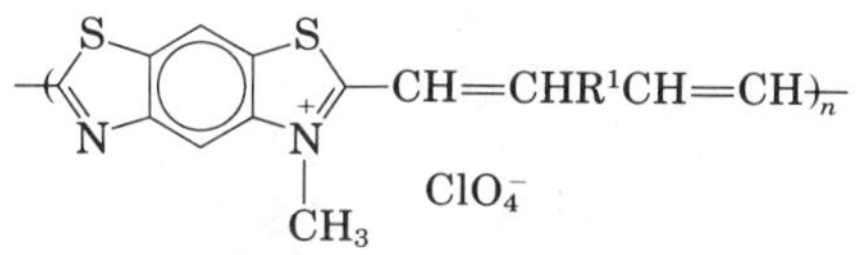

(**31**)

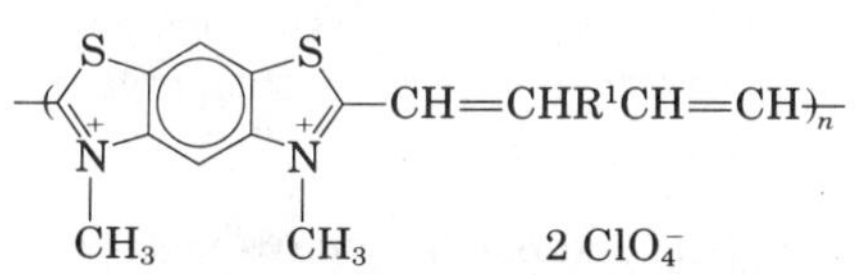

(**32**)

R^1 = 1,4-phenylene, 1,4-(2,5-methoxyphenylene), 2,5-thienylene

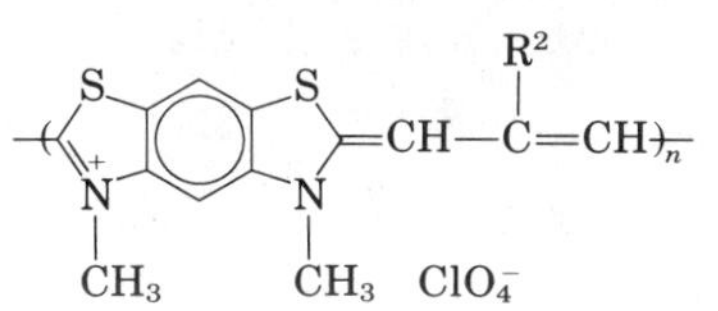

(**33**)

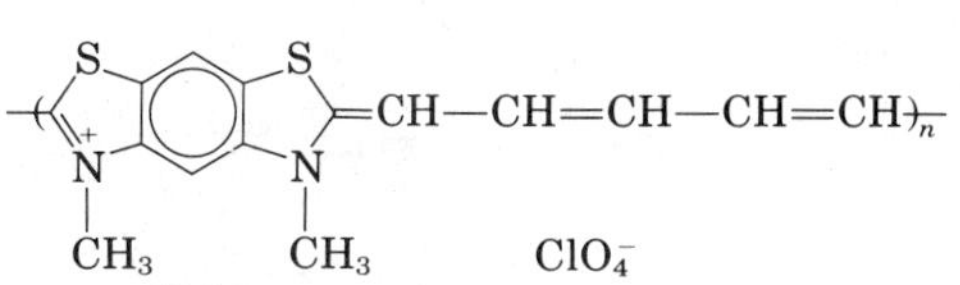

(**34**)

R^2 = H, CH_3, C_2H_5, C_6H_5

Poly(thiazolium) salts (**31**) and (**32**) are synthesized via Knoevenagel condensation of 2,3,6-trimethylthiazolo[4,5-*f*]-benzothiazolium perchlorate and from 2,3,5,6-tetramethylthiazolo[4,5-*f*]benzothiazolium diperchlorate with dialdehydes (OHC—R^1—CHO) at 423 K in the presence of benzoic anhydride. Polymeric cyanines with trimethine (**33**) or pentamethine (**34**) units are prepared by condensation of the thiazolium diperchlorate derivatives with the respective ortho esters of 1,1,3,3-tetramethoxypropane in pyridine–triethylamine at 423 K. The structures of the resulting polymers were confirmed by spectroscopic studies of (**31**)–(**34**) and of model compounds.

Polymer-bound thiazolium ions have been used as models of coenzymes for decarboxylation and acyloin-type condensation enzymes.

Polyphenothiazines. Polymers with phenothiazine electron-acceptor groups might be applied as redox reagents, charge-transfer agents, antioxidants, electroconductors, and pharmacologically active macromolecular compounds. Poly(3,7-*N*-alkylphenothiazyl methylene) (**35**) is prepared by the condensation of *N*-alkylphenothiazine with paraformaldehyde in dioxane in the presence of H_2SO_4 at 363 K. It is readily oxidized by photoirradiation in $CHCl_3$ to give colored products ($\lambda_{max} \sim$ 500 nm). However, the low viscosity $[\eta] \approx 0.1$ dL/g and mp (for $n = 0$, $T_m = 303$ K) suggest that polyphenothiazines (**35**) are low molecular weight oligomers.

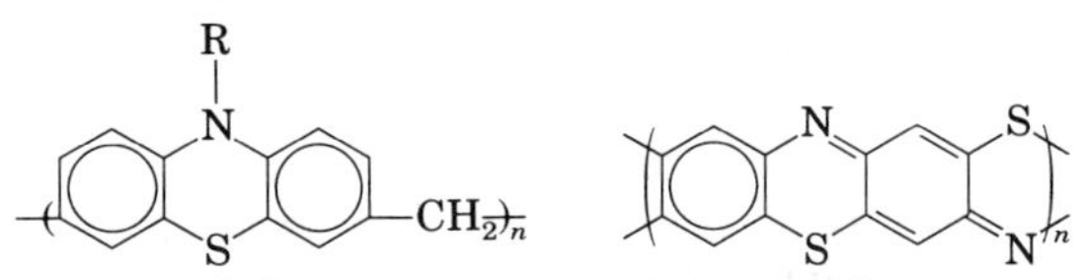

(**35**) (**36**)

$R = CH_3(CH_2)_n$
$n = 0, 1, 2, 3, 5$

The electroconductive ($\delta = 10^{-5}$–10^{-6} S/cm) phenothiazine ladder polymer (**36**) was prepared by polycondensation of 2,5-diamino-1,4-benzenedithiol with 2,5-dichloro-*p*-benzoquinone in poly(phosphoric acid) or DMA at 523 K.

Radical homo- and copolymerizations of various acrylic monomers with an attached *N*-phenothiazinyl moiety have been described.

Polymers Containing Sulfur–Phosphorus Linkages

In one of the first attempts to prepare polythioesters of phosphoric acid, anionic ring-opening polymerization of cyclic alkylene or arylenethiono esters (**37**) was used. This process led to linear, stereoregular, high molecular weight polymers (**38**):

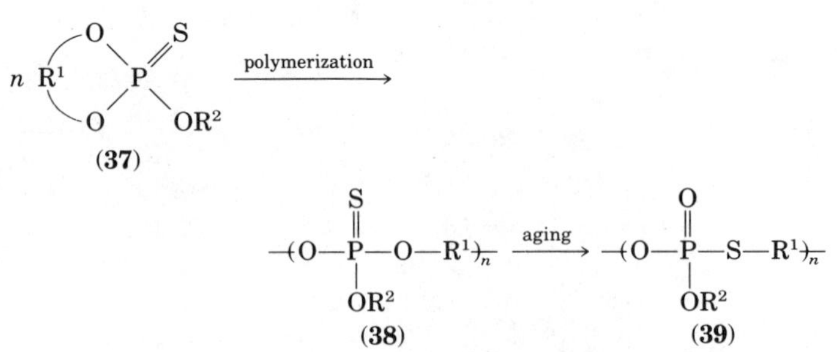

$R^1 = -(CH_2)_n-$ (n = 1, 2, 3), 1,2-cyclohexylene, *o*-phenylene

$R^2 = CH_3, C_2H_5, CH_2CH_2Cl, C_6H_5$

Polymerization was initiated by an R_2Mg—NH_3 system, by R_2Mg alone (where R = ethyl, *n*-octyl), or butyllithium. 2-Phenoxy-2-thiono-1,3,2-dioxaphospholane ($R_1 = -(CH_2)_2-$, $R_2 = C_6H_5$) gave (**38**) with $\eta_{inh} = 2.2$ dL/g (DMF). Ring-opening of (**37**) proceeded exclusively by P—O bond cleavage. Structures (**38**) are based on the ^{1}H- and ^{31}P-nmr spectra. However, for polymers derived from 2-methoxy-2-thiono-1,3,2-dioxaphospholane ($R_1 = -(CH_2)_2-$, $R_2 = -CH_3$), thiono–thiolo rearrangement was observed, resulting in 40 mol % of structures (**39**).

More recently, copolymerization involving zwitterions has been described as a promising method for the preparation of polythioesters of phosphorous acid.

Polymers bearing a P—S moiety in the pendent group have been reported. Polythiophosphonate (**40**) was prepared by the anionic polymerization of 2-methyl-1,3-butadienethiophosphonic acid dimethyl ester.

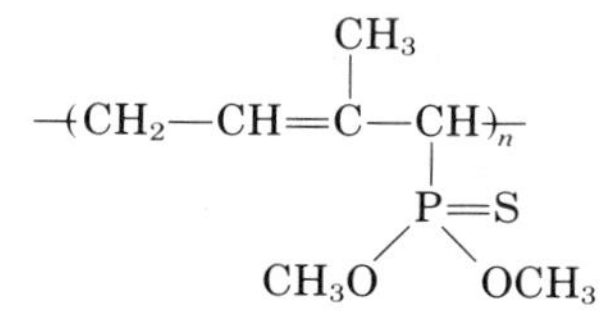

(**40**)

Polysulfoxides

Few polymers containing a sulfoxide group have been reported; they are used as functional polymers. Polysulfoxides are catalysts for reactions conducted under phase-transfer conditions. Poly(sulfinyl acrylate)s exhibit some biological activity. Membranes bearing sulfoxide moieties are permeable, particularly to polar compounds.

The first reports on the synthesis of polysulfoxides showed that monomers with a sulfoxide function attached directly to the vinyl group (**41**) do not easily homopolymerize. Polymerization of monomers containing a sulfoxide group linked to the unsaturated carbon–carbon bond via aromatic or aliphatic units (**42**) was more successful.

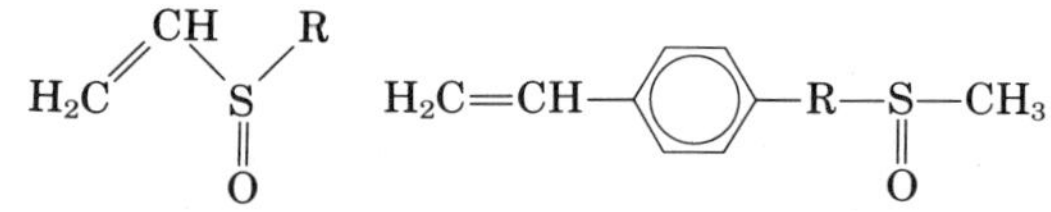

(**41**) (**42**)

Polysulfoxides are also prepared by a Michael-type addition of dithiols to diolefins with a double bond activated by an adjacent sulfoxide group and by polycondensation, including the reaction of phosgene with bis(4-hydroxyphenyl) sulfoxide or self-condensation of benzenesulfinic acid in concentrated H_2SO_4. Oxidation of polymonosulfides is another synthetic route to polysulfoxides.

Polysulfones

Both aliphatic and aromatic polysulfones (qv) have important applications. Poly(arylene sulfone)s belong to the class of high performance engineering thermoplastics that retain useful mechanical properties up to about 473 K. Their thermal and oxidative stability is due to the presence of the sulfonyl group. Poly(olefin sulfone)s, in contrast to their aromatic counterparts, are thermally unstable, giving on irradiation highly volatile components, an olefin and SO_2. This property is utilized in the application of aliphatic polysulfones as resist polymers in microelectronic devices. Poly(aliphatic sulfone)s are made by copolymerization of sulfur dioxide with unsaturated monomers (see also OLEFIN–SULFUR DIOXIDE COPOLYMERS).

Copolymers of Sulfur Dioxide with Unsaturated Monomers. A number of aliphatic polysulfones were prepared by means of radical copolymerization of SO_2 with alkenes, dienes, and alkynes. These studies led to the establishment of the ceiling temperature of polymerization (T_c). Above T_c poly(olefin sulfone)s are thermodynamically unstable. Ceiling temperatures are between 298 and 373 K.

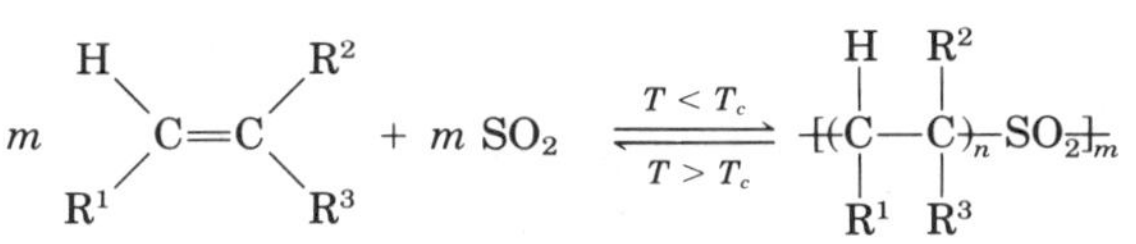

(43)

Poly(arylene sulfone)s. Poly(arylene sulfone)s are important high performance thermoplastics. They are prepared by polycondensation, either by a Friedel-Crafts reaction of sulfonyl halides with arylene compounds or by nucleophilic substitution of chloroarylenes with phenoxide derivatives.

Poly(sulfonic acid)s and Their Derivatives

Polymers substituted by sulfo groups have unusual properties, allowing a wide range of applications. Water-soluble polymers are used as thickeners, flocculants, emulsifiers, tanning agents, and wetting or cleaning agents. Introduction of the sulfo group into fibers improves dyeability and imparts antistatic properties. Copolymers of unsaturated sulfonates with water-insoluble vinyl comonomers give stable synthetic latices. Insoluble formulations are used as ion-exchange resins and find application as membrane materials, such as Nafion, a sulfonated perfluoropolymer produced by Du Pont. Hypalon, another Du Pont product, is a highly stable chlorosulfonated polyethylene elastomer that retains useful mechanical properties at elevated temperatures. Sulfopolymers are employed in biomedical systems, including biomembranes, "photozymes" mimicking natural photosynthesis systems, and blood-compatible materials. The poly(3-alkylsulfonate thiophene)s are the first water-soluble conductive polymers.

Sulfo groups are introduced into the polymers by homo- or copolymerization of monomers containing sulfo groups or by direct sulfonation of the polymer. The choice of method depends on the particular application.

The sulfo group can be in the form of the free acid, a salt, or an ester. The presence of even small amounts of ionic groups in hydrocarbon polymers has a strong effect on physical properties (see ETHYLENESULFONIC ACID POLYMERS).

Polysulfonates. Polysulfonates of the general structure (**45**) exhibit excellent thermostability. (R^1 = phenylene, biphenylene, oxadiphenylene, methane, etc; R^2 = phenylene, diphenylene-2-propylidiene, etc.) These polymers are usually prepared by interfacial copolycondensation of disulfonyl dichlorides with diphenols or diphenoxides. Some thermoplastics have low crystallinity and melt below 523 K and are thermally stable up to 573 K (under N_2); molecular weights reach 5×10^4.

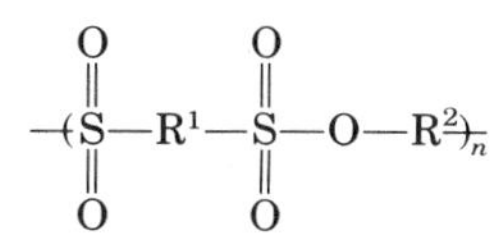

(45)

Polysulfonamides. Polymers containing a sulfonamide function in the backbone or pendent groups are easily processed. They are usually lower melting and more readily soluble in organic solvents than the analogous amide or sulfone polymers.

Initially, high molecular weight polysulfonamides were prepared by the interfacial polycondensation of aromatic disulfonyl dichlorides with aliphatic amines. Later, wholly aromatic polysulfonamides with high molecular weights were obtained by solution polycondensation of di-1-benzotriazolyl disulfonates or aromatic disulfonyl dichlorides with aromatic or aliphatic amines.

Polysulfates

Sulfated polysaccharides are the most widely studied polysulfates. Some, such as heparin, carrageenan, and agarose sulfate have naturally bound sulfate functions. Their analogues, such as derivatives of dextran, chitosan, or lignin have been prepared by sulfation of the parent polymer (see POLYSACCHARIDES).

Organometallic Polymers Containing Sulfur

Various metals belonging to the Ib, IIb, IVa, VIIIa, and VIII groups of the periodic table form covalent and coordination bonds with divalent sulfur. This ability is used in the separation of metal ions by means of their fixation on a sulfur-containing polymer. Both mono- and polydentate ligands bearing sulfur atoms give polymers with metallic salts or complexes. The presence of readily polarizable sulfur atoms confers electroconductivity. Other polymers belonging to this class exhibit catalytic activity due to their polynuclear structure and can be used as models of metalloenzymes. Insolubility in common inorganic and organic solvents is a common property of polymers containing sulfur–metal linkages in the main chain and precludes their precise characterization (microstructure, molecular weights, etc).

BIBLIOGRAPHY

E. J. Goethals, *Top. Sulf. Chem.* **3,** 1 (1977).

V. A. Sergeev and V. I. Nedel'kin, *Sulfur Containing Polymers* in V. V. Korshak, ed., *Progress in the Area of Synthesis of Elementoorganic Polymers* (in Russ.), Nauka, Moscow, 1980, 1988.

ANDRZEJ DUDA
STANISŁAW PENCZEK
Polish Academy of Sciences

SULFUR DIOXIDE COPOLYMERS.

See OLEFIN–SULFUR DIOXIDE COPOLYMERS.

SUPERCRITICAL FLUIDS

Supercritical-fluid processing of polymers and monomers has received increasing attention during the 1980s. Supercritical fluids exhibit a unique combination of solvent and transport properties and have been used to purify reactive monomers; fractionate ultralow vapor-pressure oils and polymers; extract residual solvents, monomers, and catalysts from solid and molten polymers; and recrystallize polymers from solution. Many of these processing operations are combined when a supercritical fluid is used as the reaction medium for polymerization. For example, by proper choice of operating conditions a monomer can be polymerized to a controlled molecular weight, residual monomer can be separated from the polymer, and the polymer can be precipitated to form particles of a controlled size distribution.

Properties and Processing Characteristics

A gas or liquid at ambient conditions becomes a supercritical fluid (SCF) when it is compressed and heated to conditions above its critical point. As the critical point of a substance is approached, its isothermal compressibility tends to infinity, and, thus, its molar volume or density changes dramatically. In the critical region a substance that is a gas at normal conditions exhibits a liquidlike density and a much increased solvent capacity that is pressure-dependent. The variable solvent capacity of a supercritical fluid is the basis on which separation processes can be devised.

The critical temperatures of gases and liquids can differ by hundreds of degrees, and this difference suggests the use of specific supercritical fluids in specific applications. For example, because the critical temperatures of carbon dioxide, ethane, and ethylene are near ambient, they are attractive solvents for processing heat-sensitive flavors, pharmaceuticals, labile lipids, and reactive monomers. Substances that are less temperature-sensitive, such as most industrial chemicals and polymers, are readily treated with the C_3 and C_4 hydrocarbons with critical temperatures in the range of 100–150°C; the C_3 and C_4 hydrocarbons are generally better solvents for polymers than the C_2 hydrocarbons. Still higher molecular weight hydrocarbons, such as hexane and benzene, with high critical temperatures of 250–300°C, are used to process nonvolatile substances such as coal and high molecular weight petroleum fractions. Supercritical water, at the end of the temperature spectrum of common solvents, has a critical temperature of 374°C and is being tested for hazardous-waste detoxification and hydrocarbon reforming.

Applications to Monomer and Polymer Processing

Extraction of Monomers from High Molecular Weight Oils and Polymers. The use of supercritical fluids for extracting monomers and low molecular weight oligomers from polymers provides technical, economic, and scale-up advantages, especially in the processing of certain silicone polymers and other ultralow vapor-pressure oils.

Fractionation of Polymers. Supercritical fluids are being used to fractionate polymers. The separation into narrow fractions of a high molecular weight, ultralow vapor pressure perfluoroalkylpolyether oil, that is virtually impossible to fractionate even by molecular distillation, is an especially dramatic demonstration of the applicability of supercritical fluids to the processing of refractory polymers. Figure 1 shows a schematic diagram of a continuous supercritical fluid fractionation process for separating polymers composed of a wide range of molecular weight oligomers into fractions with narrow molecular weight distribution.

Purification of Reactive Monomers. Acrylate and methacrylate monomers present an especially intractable purification problem, which can, however, be solved by supercritical fluid extraction. The monomers can be processed with supercritical fluids at moderate temperatures using ethylene, carbon diox-

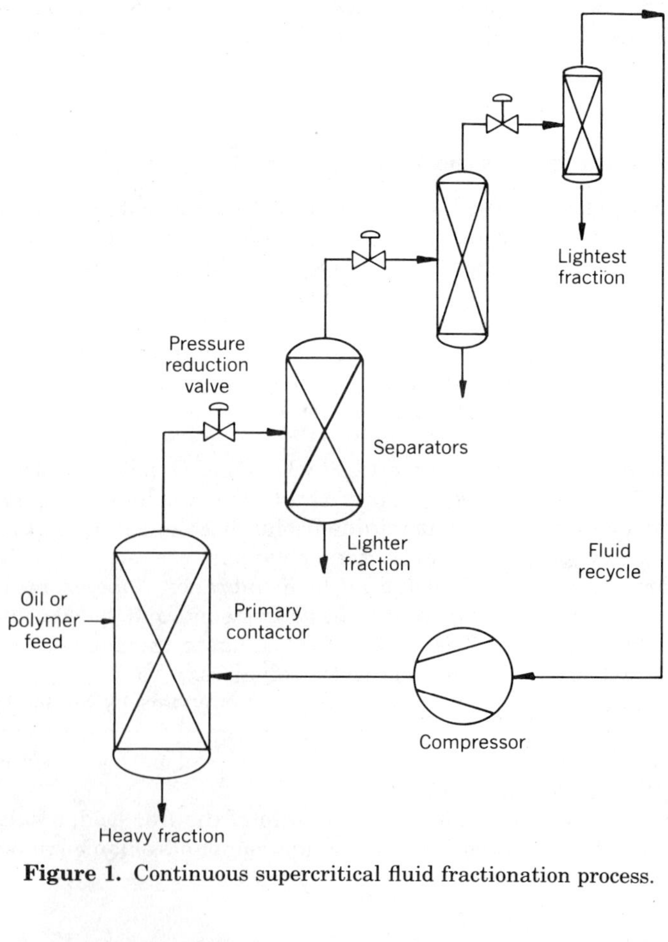

Figure 1. Continuous supercritical fluid fractionation process.

ide, and ethane, the three "ambient-temperature" gases mentioned earlier; all are effective solvents for purifying reactive monomers.

Polymerization at Supercritical Conditions. The earliest and perhaps still the most dramatic example of polymerization in a supercritical fluid is the high pressure, low density polyethylene process now almost 50 years old (Fig. 2). More recently the free-radical polymerization of styrene in supercritical ethane has been examined.

Precipitation of Polymers from Solution by Supercritical Antisolvents. Solutions of polymers and organic solvents can be separated with the help of a supercritical fluid, which acts as an antisolvent. The supercritical fluid preferentially dissolves in the organic liquid and therefore causes it to dilate. In the expanded state the organic liquid is no longer a solvent for the polymer, and the solution separates into two phases. In essence, this phase-splitting technique is an adaptation of the well-known, lower critical solution temperature (LCST) phase-splitting technique.

Absorption of Gases in Polymers. At sufficiently high pressure, the absorption of certain gases can result in substantial swelling of the polymer, which can be beneficial or deleterious. In the case of elastomeric materials used for seals and O-rings in supercritical-fluid processes, the absorption of a gas at high pressure and the subsequent release of the dissolved gas at ambient pressure (when an extraction is completed, for example) causes many polymeric sealing materials to degrade. When the system pressure is reduced, adsorbed gases nucleate into bubbles and cause blistering of the seals. In addition, the supercritical fluid can extract the plasticizing agents in the seals and thus cause embrittlement.

Schroeder and Arndt reported on the solubility of compressed gases in polymers, and described the ability of high pressure carbon dioxide to act as a plasticizer. They also recognized that sufficiently high temperatures are needed to reduce the interactions between polymer segments in the polymer-rich phase before any reasonable amounts of polymer can be dissolved in the supercritical fluid. The dissolution of gases in polymers can also be beneficial for gas separation by membranes. The amount of gas sorbed into the polymer must be known for the design of polymeric membranes used for separating certain gas mixtures.

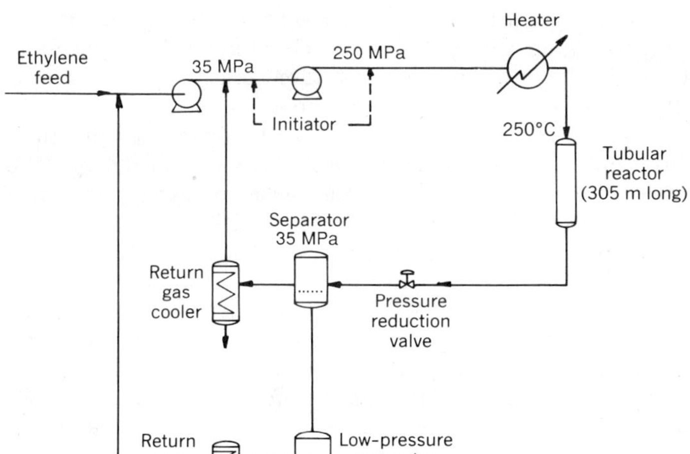

Figure 2. High pressure low density polyethylene polymerization process. To convert MPa to bar, multiply by 10.

Solubility of Polymers in Supercritical Fluids. In the 1980s substantial experimental work has been in progress on the measurement of polymer solubility in supercritical carbon dioxide. The goal is the improvement of the properties of carbon dioxide for enhanced oil recovery applications. Carbon dioxide exhibits some advantageous economic and technical characteristics for extracting oil that cannot be readily recovered from reservoirs by normal pumping operations. It can modify the hydrodynamic characteristics of the oil so that it can be forced to the surface. However, the low, gaslike viscosity of carbon dioxide causes it to bypass much of the oil through microcracks in the substrata, resulting in efficient utilization. Hydrodynamic calculations show that if the viscosity of the carbon dioxide could be increased, bypassing would be reduced. Therefore, increasing the viscosity of carbon dioxide by dissolution of a polymer is receiving considerable attention.

Other Applications

Some applications of supercritical fluids in the processing of polymers are not typical; they do not entail purification or fractionation of polymer products in the traditional sense. They include particle formation, impregnation of polymers into microporous solids, and extraction of oligomers from microlithographic films.

BIBLIOGRAPHY

M. A. McHugh and V. J. Krukonis, *Supercritical Fluid Extraction: Principles and Practice,* Butterworths, Boston, 1986.

J. M. L. Penninger, M. Radosz, M. A. McHugh, and V. J. Krukonis, eds., *Supercritical Fluid Technology,* Elsevier, New York, 1985.

MARK A. MCHUGH
The Johns Hopkins University
VAL J. KRUKONIS
Phasex Corporation

SURFACE ANALYSIS

In 1971 the first x-ray photoelectron spectrum from a polymer was obtained. X-ray photoelectron spectroscopy (xps, also commonly referred to as esca) is ideally suited to polymer surface analysis and has subsequently been widely used. The technique is not without limitations, however, and more recently, the considerable potential of secondary ion mass spectrometry (sims) has been demonstrated. The first sims studies of polymers were carried out as recently as 1980.

This article discusses the physical bases of these two techniques and their application in the field of polymer surface characterization.

X-ray Photoelectron Spectroscopy

Physical Basis. In xps the sample, inside a high vacuum system (pressure $< 10^{-5}$ Pa $\approx 10^{-7}$ mm Hg), is irradiated with soft x rays, usually MgK_α (1253.6 eV) or AlK_α (1486.6 eV). The primary event is photoemission of a core (atomic) electron, but relaxation processes lead also to emission of Auger electrons, as shown in Figure 1. The emitted electrons are collected by an electrostatic energy analyzer and detected as a function of kinetic energy E_K. The spectrum is a plot of binding energy vs electron intensity. The intense, narrow peaks are due to photoelectrons emitted from core orbitals (eg, carbon 1*s*). The binding energies E_B of these electrons, obtained from the Einstein relation

$$E_B = h\nu - E_K - \phi \qquad (1)$$

where $h\nu$ is the x-ray photon energy and ϕ the sample work function are highly characteristic and allow the identification of all elements except hydrogen. The peak intensities are proportional to the number of atoms sampled, and with the aid of appropriate sensitivity factors, atomic compositions can be calculated, with detection limits of typically 0.2 atom %. The much weaker series of peaks at very low binding energy is due to photoelectron emission from the valence band (molecular orbitals). The other, broader peaks are due to Auger electrons, for which the kinetic energy is given by

$$E_K(ABC) = E_A - E_B - E_C - \phi \qquad (2)$$

where *A*, *B*, and *C* are the photoionized level, the level of the electron that fills this vacancy, and the level of the emitted Auger electron, respectively (Fig. 1). These kinetic energy values are also element characteristic, although not all elements produce strong Auger signals under typical xps conditions, and are obviously independent of $h\nu$.

The information from xps is surface specific because the electrons that give rise to the useful peaks in the spectrum have emerged from the material elastically, ie, without loss of energy.

Instrumental Aspects. The photon sources commonly used are $MgK_{\alpha_{1,2}}$ and $AlK_{\alpha_{1,2}}$. The doublet nature of the x-ray line is not resolved, and the single line widths are 0.7 and 0.85 eV, respectively. Thus MgK_α has an advantage in terms of spectral resolution. However, using a quartz crystal monochromator, AlK_α can be monochromated to give a significant decrease in line width (eg, to ~0.4 eV). Some experiments have also been carried out with the higher energy $TiK_{\alpha_{1,2}}$ (4.5 keV). For any given core level the sampling depth is significantly greater than for the lower energy photons because of the increased kinetic energy of the photoelectrons. However, the $TiK_{\alpha_{1,2}}$ line is a broad resolved doublet (causing the photoelectron spectra to be complex), and photoemission cross sections for core levels of interest in polymer studies (C 1*s*, O 1*s*, N 1*s*) are much reduced.

All insulating surfaces charge up slightly during xps. This is not normally a problem since the charge buildup is small (a few electron volts) and reaches an equilibrium rapidly. For various reasons the use of a monochromated AlK_α source exacerbates the effect greatly, and an external source of charge-compensating electrons has to be provided, usually from an electron "flood gun."

The most useful instrumental facility is the ability to vary the electron takeoff angle so that changes in composition within the available sampling depth can be studied. This experiment is appropriate only on samples with reasonably flat surfaces.

Core Levels. The binding energies, BEs, of the core-level peaks are highly element characteristic, and all elements except hydrogen can be detected. Relative sensitivity factors are such that, except for Li and Be, which have low photoemission cross sections, the most intense core levels show a sensitivity range of roughly one order of magnitude. Peak overlap problems are relatively rare, particularly in polymer surface studies. When photoelectron peaks and Auger peaks overlap (eg, between C 1*s* and Na KLL and Na 1*s* and Cl LMM using MgK_α) use of an alternative photon source (in this case AlK_α) moves the photoelectron peaks on the kinetic energy scale. Relative sensitivity factors for the appropriate instrument can either be generated *in situ* by studying homogeneous standards, taken from published tables, or calculated. Peak areas can then be transformed into surface atomic composition data.

Any given core level also shows relatively small variations in BE, depending on the chemical environment, referred to as chemical shifts. A large body of literature on this subject has been created through the study of pure polymers and model small molecules coupled with theoretical calculations. From these data some simple summaries can be made.

C 1s Binding Energies. Carbon of any hybridization bound to itself and/or hydrogen only gives C 1*s* = 285.0 eV. This is often used as a BE reference. Halogens induce shifts to a higher BE, which can be broken down into a primary substituent effect on the carbon atom directly attached and a secondary substituent effect on the neighboring carbon atom(s). Oxygen induces shifts to higher BE by 1.5 eV per C—O bond. Nitrogen functionalities have a primary substituent effect, which is markedly dependent on the nature of the substituent.

O 1s Binding Energies. Oxygen 1*s* BEs from most functionalities fall within a narrow range of ~2 eV around 533 eV. The extremes are seen in carboxyl and carbonate groups, in which the singly bound oxygen has the higher BE.

N 1s Binding Energies. Many common nitrogen functionalities give N 1*s* BEs in the narrow region of 399–401 eV.

Other Core Levels. The only other atoms with a variable oxidation state commonly encountered in polymers are sulfur and silicon. The primary effect of sulfur on the C 1*s* BE is very small (~0.4 eV measured in a polysulfone). However, S 2*p* BEs cover a reasonable range: R—S—R (~164 eV), R—SO_2—R (~167.5 eV), and R—SO_3H (~169 eV).

Halide ions can be distinguished from covalently bound halogen. In the case of fluorine the F 1*s* BE for F^- is ~4 eV lower than for C—F (typically 689 eV), whereas the Cl 2*p* BE for Cl^- is ~2 eV lower than C—Cl (typically 201 eV).

Small charging effects move all peaks by the same amount.

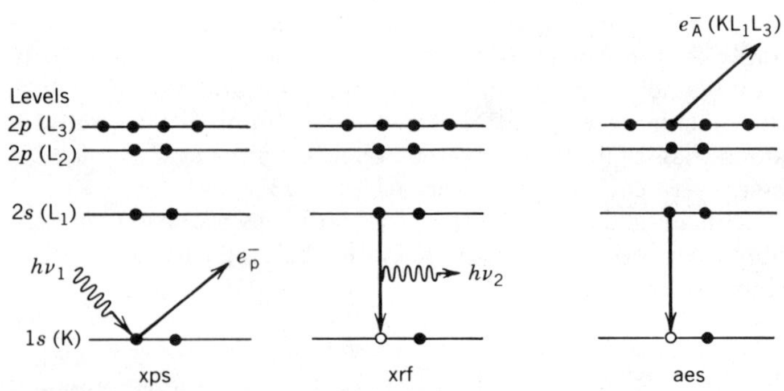

Figure 1. Schematic of the processes involved in x-ray photoelectron emission and subsequent relaxation by x-ray fluorescence and Auger electron emission. Courtesy of Orbital Press.

Absolute BEs are established by referencing to a peak of known energy, usually to C 1s (CH_2) = 285.0 eV. This structure is present in many polymers, but a hydrocarbon peak is frequently seen from adventitious contamination on surfaces and may even appear *in situ* if the sample is left in the xps instrument long enough, particularly with the x-ray source operating.

Strengths and Limitations of Xps. The principal strengths of xps are its ability to obtain elemental and chemical information for materials with $Z > 2$; ease of sample handling, permitting almost any type of vacuum-compatible polymeric material to be investigated irrespective of shape, surface morphology, etc; minor problems with sample charging; negligible radiation damage in most cases; and ease of data quantification. The limitations are its very poor spatial resolution (~200 μm minimum diameter examined, in most instruments several square millimeters area) and relatively poor degree of molecular specificity.

Secondary Ion Mass Spectrometry

Physical Basis. In sims the sample is bombarded by a primary ion beam. About 5% of the material sputtered from the surface consists of positively and negatively charged atoms (elemental ions) or molecular fragments (cluster ions), and these are mass analyzed to produce positive and negative secondary ion mass spectra. In the case of polymer surface characterization, it is essential that the experiment be carried out in the static mode; ie, the accumulated ion dose during spectral acquisition be sufficiently low for the surface to be essentially unperturbed during the measurement. This threshold dose is a function of polymer structure and bombardment conditions, but is as low as 10^{13} ions/cm^2 for many polymers.

Instrumental Aspects. Vacuum requirements for sims are rather more stringent than for xps because of the greater need to avoid surface contamination *in situ*. This is due to the greater surface sensitivity of sims and the greater confusion of the sims spectrum by fragment ions from contamination. True UHV operating pressures of $<10^{-7}$ Pa ($<10^{-9}$ mm Hg) are therefore required.

Ion sources can be operated in two ways. The aim is to maintain static conditions and the appropriate (very low) ion current density can be obtained either by using a stationary defocused beam or by using a focused beam that is scanned raster fashion (video rate) over a large area. The advantage of the latter approach is that it also allows, in principle, microanalysis and imaging to be carried out.

Quadrupole mass spectrometers, employing a variety of energy filter designs, have been used almost exclusively to date for static sims experiments. These give adequate mass resolution, have a mass range of several hundred daltons (up to 1000), and are compact. However, the collection efficiency is relatively low (~0.1% of emitted secondary ions detected), and spectrometer transmission falls with increasing mass ($T \propto 1/M - 1/M^2$). Interest is growing in the use of time-of-flight (tof) mass spectrometers (see MASS SPECTROMETRY).

Charge neutralization is effected by means of a relatively low energy electron flood.

Polymer Fingerprint Spectra. As a general rule it is found that sims spectra from pure polymers contain peaks due to multiple repeat units (main-chain scission), intact side chains, and fragments of these. Fragmentation pathways have much in common with electron impact (ei) mass spectrometry, and in several cases a rather close resemblance between sims and pyrolysis–ei mass spectra has been observed. Different classes of polymers are therefore readily distinguishable, and individual members of a class can usually be identified by careful spectral interpretation. The information content of the combined positive and negative ion spectra is so high that fingerprinting techniques can be used. The strong influence of the repeat unit on the spectra clearly means that copolymer identification is also possible.

Quantitative Aspects. Quantification of sims intensity data is difficult in general because of the strong matrix effects on secondary ion yields. The case of organic surface analysis might be expected to offer even more difficulty. However, it has been found that for closely related structures relative peak intensity data can be used quantitatively.

Strengths and Limitations of Sims. The principal advantages of sims are its high absolute sensitivity and low information depth; molecular specificity; and imaging potential. The main disadvantages are the requirement of charge compensation and the difficult optimization of secondary ion yields and nonstraightforward quantification of data.

Applications of Xps and Sims

In polymer technology a large number of material properties involve control of surface composition. These include optical properties (eg, haze, gloss, stains), adhesive properties (eg, wettability, printability, bondability, heat sealability, and releaseability), electrical properties (eg, static chargeability), general processing and machine handling properties (eg, friction and "blocking" of film in reels or stacks), biocompatibility, resistance to weathering, and so on. Specific contaminants can also lead to surface crazing and cracking of many plastics.

Additives in plastics and polymeric materials include agents to prevent oxidation (thermal and photochemical), neutralize acidity, promote fire retardancy, aid processing, lower surface friction, increase surface conductivity, and prevent blocking. Other additives originating in polymerization processes, eg, stabilizer and emulsifier molecules, may be present. The presence on the material surface of these additives may or may not be intended, but their identification is necessary in order to establish structure–property relationships and in problem solving.

Surface properties of polymeric materials can also be modified after fabrication, thereby optimizing surface and bulk properties of an article semiindependently. Modification methods include chemical (wet) treatment, etching, electrical discharge treatment ("corona," glow discharge), flame treatment, and thin film deposition by a variety of methods. Establishing the changes taking place during these modification processes, eg, depth of change and alteration of chemical functionality, is clearly crucial to understanding optimum behavioral properties.

It is in such areas that xps and sims have made a dramatic impact in the last 10 years or so. Since xps has been used for the longest period of time, the literature is most extensive and has been reviewed. Most of the sims literature has already been covered in this article, but its most fruitful area of application to date has been the specific analysis of segregated small molecular additives and very thin films. Although developments in technology will enhance performance, the relative strengths and weaknesses of xps and sims are reasonably

Table 1. Comparison of Xps and Sims of Polymer Surfaces

Advantages	Disadvantages
xps	
quantitative elemental analysis	chemical shift dynamic range very low
few charging problems	sampling depth kinetic energy dependent
little radiation damage	poor spatial resolution[a]
some structural information	hydrogen not detected
sims	
very high molecular structure information content	not all elements detected as such
sampling depth of 1 nm	quantification difficult
imaging and microanalysis possible	charge neutralization essential
hydrogen detected	ion beam damage limitations
isotope discrimination	

[a] Major improvements from new designs of instrumentation (1988).

well established. As Table 1 shows, the strengths of the two techniques are highly complementary, and consequently, polymer surface analysis is best pursued by a combination approach. This is most effective when both techniques are combined into one UHV system.

BIBLIOGRAPHY

D. Briggs and M. P. Seah, eds., *Practical Surface Analysis by Auger and X-ray Photoelectron Spectroscopy,* John Wiley & Sons, Inc., Chichester, UK, 1983.

D. Briggs, *Polymer* **25**, 1379 (1984).

D. Briggs, *Brit. Polym. J.* **21**, 3 (1989).

DAVID BRIGGS
ICI

SURFACE COATINGS. See COATINGS.

SURFACE MODIFICATION

The surface properties of solid organic polymers have a strong impact on many if not most of their applications. All objects have surfaces through which they communicate with their surroundings. The properties and structure of these surfaces are therefore of obvious importance. The chemical structure and thermodynamic state of polymer surfaces are important factors that determine many of their practical characteristics. Examples of properties affected by polymer surface structure include adhesion, wettability, friction, coatability, permeability, dyeability, gloss, corrosion, surface electrostatic charging, cellular recognition, and biocompatibility. Interfacial characteristics of polymer systems control domain size and stability of polymer–polymer dispersions, adhesive strength of laminates and composite materials, cohesive strength of polymer blends, and mechanical properties of adhesive joints (see INTERFACIAL PROPERTIES).

A surface is defined as that part of an object that affects or can be affected by its environment. Polymer surfaces are unique among solid surfaces in that chains near the surface can be mobile and resemble in many aspects a viscous fluid. The degree of this mobility depends on the environment of the sample as well as the morphology, molecular weight, and glass-transition temperature T_g of the polymer.

This article discusses synthetic organic polymers and reviews the three most common surface-modification techniques: plasma modification, surface graft polymerization, and chemical reaction. Other methods include ion-beam modification, adsorption and polymerization of amphiphiles, *in situ* deposition of metal oxides, reaction with metal vapors, entrapment of terminally functionalized oligomers, photoisomerization, and flame treatments. Surface properties can be controlled by adjusting the concentration of one component in a two-component blend or the orientation of polymer molecules near surfaces.

Plasma Modification

The principal advantage of polymer surface modification with the help of plasmas is experimental simplicity. Inexpensive equipment can be used to generate a variety of plasmas and scale-up for industrial applications is straightforward. The disadvantage, from an academic viewpoint, is that chemically well-defined surfaces cannot be designed and prepared. The technique is severe and plasma species (electrons, ions, and metastables) and uv radiation induce cascades of reactions involving homolytic bond cleavage, fragmentation, and cross-linking in the polymer surface. Inert gases (helium and argon), reactive gases (O_2, N_2, and F_2), and polymerizable gases (tetrafluoroethylene, methane, and acetylene) can be used to generate plasmas.

A wide range of modified surfaces has been prepared with this technology. Inert gas plasmas (helium, neon, argon, and krypton) have been used to modify surfaces of polyethylene, poly(ethylene terephthalate), polytetrafluoroethylene, poly(ethylene-*co*-tetrafluoroethylene), poly(tetrafluoroethylene-*co*-hexafluoropropylene), Kevlar, poly(*p*-xylylene), polysulfone, silicone rubber, polycarbonate, and polystyrene. Nitrogen plasmas have been used to modify polyethylene, Kevlar, poly(ethylene terephthalate), poly(*p*-xylylene), polysulfone, and silicone rubber. Oxygen and hydrogen plasmas are reported to modify a range of polymers. Methane, ethylene, and acetylene plasmas have been used to modify the surfaces of polyethylene, polytetrafluoroethylene, poly(vinyl chloride), and poly(vinyl fluoride). Organosilane plasmas are used to modify a variety of polymers.

Surface Graft Polymerization

Numerous references describe the modification of polymer surfaces by graft polymerization of a monomer or monomers from active sites generated on solid polymer surfaces. The most common technique is γ-radiation, which generates surface radicals, but thermal, mechanical, photochemical, plasma, and wet chemical methods are also used to introduce sites for initiation. Monomers can be present in the gas phase (sublimed solids), in solution, or as neat liquids. Reactive species can be generated prior to monomer exposure (preirradiation grafting), during contact with monomer, or after the polymer surface has been saturated with monomer and isolated (postirradiation grafting). Most reported work had as an objective the modification of a particular surface property. Interesting questions such as controlling the diffuseness of the polymer–polymer interface with temperature or solvent selection have, for the most part, been ignored.

Chemical Modification

In many reports of polymer surface modification standard organic reactions are employed in which reagents in solution are allowed to react with solid organic polymers at the solution–polymer interface. This field of science must still be considered in its infancy. The criteria that should be considered in carrying out a heterogeneous organic reaction at an organic solid–solution interface have not generally been included and, indeed, these criteria are not known. Most of the reports involve conditions under which the solution interacts strongly with the polymer surface chains. The interface between the solid and the solution is diffuse and the reaction can be viewed as a solution reaction. If organic chemistry is to be used to rationally control polymer surface properties, modification reactions must change only the chemical structure of the surface. In this regard, reactions at sharp (not diffuse), well-defined solid–solution interfaces are of interest.

Among the chemical modifications reported are oxidations, especially of polyethylene, reductions of fluoropolymers, sulfonations, dehydrohalogenations, and hydrolyses. Ozonations have been carried out, polyesters have been modified with amines, alcohol-containing polymers have been modified with diisocyanates, unsaturated polymers have been epoxidized and have reacted via ene reactions, and aromatic polymers have been electrophilically substituted. Polyethylene has been surface-modified by reaction with carbenes, nitrenes, and sulfur atoms. A variety of reactions have been reported as spectroscopic labeling techniques.

BIBLIOGRAPHY

D. T. Clark and W. J. Feast, eds., *Polymer Surfaces,* John Wiley & Sons, Inc., New York, 1978.

William J. Ward
Thomas J. McCarthy
University of Massachusetts

SURFACE PROPERTIES. See Adhesion and bonding; Interfacial properties.

SURFACTANTS. See Antifoaming agents.

SUSPENSION POLYMERIZATION

The term suspension polymerization refers to a system in which monomer(s) are suspended as the discontinuous phase of droplets in a continuous phase and polymerized. The reactor product is a slurry of suspended polymer particles. The monomers suitable for suspension polymerization usually can be polymerized by free-radical mechanisms. The continuous phase is usually water, as most monomers are relatively insoluble in water. The terms pearl polymerization and bead polymerization describe the smooth, spherical particles obtained from some suspension polymerizations.

Suspension particle morphology (the particle size, shape, and internal structure) affects how the resin performs in many mixing operations and applications and affects its economic value. Suspension particles are 50–200 μm in diameter, with the exact size depending on the monomer type, surfactants, and agitation in the reactor. The range can be extended to 400 μm for some products. These particles are much larger than those formed in emulsion polymerization (20–1000 nm) and usually move as a free-flowing powder when dry and are relatively easy to recover from water slurries by centrifugation. They can be conveyed by and recovered from air streams. The small average particle size produced by suspension processes aids in mixing other components with the resin such as in compounding operations. In many cases, particle morphology makes important contributions to the value of suspension polymer products.

Microsuspension processes can be used to make particles in the range of 20–30 μm, but these constitute a small percentage of the commercial market. These materials [poly(vinyl chloride)] require much higher surfactant levels and more agitation than conventional suspension resins. The particles are nonporous and are blended with other resins for plastisol applications.

A number of important polymer products are made by suspension polymerization, including styrenic ion-exchange resins, poly(vinyl chloride) for extrusion and injection-molding applications, and extrusion grade poly(vinylidene chloride-*co*-vinyl chloride).

Several characteristics of the suspension polymerization method are common to most systems. The weight ratio of the continuous water phase to the discontinuous monomer phase varies from 1:1 to 4:1 in most commercial processes. The bulk viscosity of the slurry is near that of water during most of the polymerization. The low bulk fluid viscosity allows good mixing of the reactor contents at modest energy inputs and can improve heat transfer in the polymerization reactor. Water is a good medium for removing heat from polymerizing droplets because it has both a high heat capacity and a high thermal conductivity. However, suspension polymers must be separated and dried from the water phase. Suspension droplets are not thermodynamically stable, and their coalescence is controlled by balancing the agitation system and the suspending agents.

A number of commodity polymers are made by suspension polymerization in water. The potential engineering advantages over bulk, solution, and emulsion polymerizations are

1. Low cost of conversion with flexibility to vary the particle properties.
2. Excellent heat transfer.

3. Particle-size control.
4. Low levels of additives in the polymer.

Most suspension processes are batch, but research is continuing on continuous methods. Control of coalescence and prevention of polymer buildup on reactor surfaces are major engineering challenges.

Process Description and Theory

The steps of a typical polymerization process are preparation of the suspending agents, buffers, and initiators; formation of the monomer suspension in the water continuous phase; initiation and polymerization; monomer recovery, dewatering, and powder storage. However, the details vary among the polymers produced and among different commercial manufacturers. Suspension droplets are thermodynamically unstable, unlike emulsion systems in which monomer and polymer particles can remain dispersed for long periods of time. A suspension of monomer droplets in the size range 50–200 μm coalesces rapidly in the absence of agitation to form a distinct monomer phase. The particle-size distribution achieved is often considered to represent a dynamic equilibrium between particle dispersion by shear or turbulence forces in the reaction vessel and coalescence by impingement or wetting of reactor surfaces. Droplets that wet reactor surfaces can polymerize and result in polymer buildup or "adhesions."

Each particle can be considered a mini-"bulk" reactor for modeling purposes. Except near the completion of the polymerization, most of the unreacted monomer is in the droplets or polymerizing particles because it is sparingly soluble in the water phase. Initiators soluble in the monomer phase are used to begin the polymerization. Polymerizations usually are conducted above room temperature but below the boiling point of the continuous phase (100°C for water). Reactor pressures for homopolymerizations are typically several atmospheres at the beginning of the process (owing to the vapor pressure of the monomer) and decrease to atmospheric pressure or less at the end. Some copolymer processes actually increase in pressure, eg, copolymers of vinyl chloride and olefins. Polymerization is often preceded by a lag time during which oxygen and other free-radical traps are consumed. At high conversions, the polymerization rate may be lowered because of the slow diffusion of monomer to reacting chain ends. Many free-radical polymerizations exhibit a sharp pressure drop that occurs when the last liquid monomer is consumed and only monomer soluble in the water and polymer phases remain. When the reactor pressure is reduced to atmospheric pressure, rates are often too low to be economical and the reactor is discharged.

At the end of the reactor cycle, the polymer phase has solid properties and may trap unreacted radicals. These usually need to be destroyed for safe storage and transport of the polymer, by raising the temperature, by adding a free-radical trap, or by adding a terminating agent. Residual monomer is recovered and recycled after free radicals have been destroyed. The suspended solid polymer can be dewatered by centrifugation and drying. The dried solid polymer particles can be conveyed by mechanical or pneumatic methods and stored in silos (see also POLYMERIZATION PROCEDURES, INDUSTRIAL).

Droplet Formation. A principal advantage of suspension polymerization over emulsion polymerization is that fewer additives are required to make the polymer. A key problem is the formation of a "uniform" suspension of monomer droplets in the aqueous phase and the reduction of coalescence of the droplets during polymerization. In practice, monodisperse distributions, narrow distributions, broad distributions, and bimodal distributions of particles have been reported.

The monomer must be relatively insoluble in water for the droplets to form. A typical water-insoluble organic monomer has a lower surface tension than water. When such a monomer is mixed continuously as a dispersed phase in a continuous phase of water with no surfactants present, an unstable dispersion forms due to the continuous breakup and coalescence of monomer droplets. If the agitation stops, the monomer–water system will separate into its two phases. Particle coalescence is not prevented but is controlled by a combination of surfactant and agitation-system design. Much research was done during the 1950s and 1960s on the relation of agitation to particle size.

Initiators and Initiation. The usual range of reaction temperatures for suspension polymerizations has been from 40 to 90°C, although the suspension copolymerization of styrene–acrylonitrile (SAN) has been carried out from 60 to 150°C. Many commercial free-radical initiator systems have been tailored to function between 40 and 90°C in a controlled manner. Temperatures below 0°C are not feasible when water is the continuous phase. Temperatures above 100°C require that the system be under pressure, which usually is the case. The temperature chosen for a specific polymerization depends on the initiator system and the mechanisms and kinetics of nonpropagation reactions such as branching and termination.

Most of the monomers used in suspension polymerizations can decompose thermally to initiate free-radical reactions, but the rates are usually low compared to standard free-radical initiators.

Initiators are added in the range of 0.1 wt % of the monomer or less. They are often charged to the reactor before the suspension is brought to the reaction temperature. The initiators usually are dissolved in hexane or other organic solvents for ease of handling and storage. This practice leaves low amounts of the carrier solvent in the polymer. Some systems for forming water emulsions of polymerization initiators do not require solvents.

Polymerization Rates and Kinetics. The polymerization affects the performance properties of the suspension product, including molecular weight distribution, side reactions, residual monomer, and, in some cases, particle morphology. Models of the polymerization rates are used to predict molecular weight distributions and conversion as a function of time for specific reactor conditions. In addition, rate models can be used to assist in control of reactors. The polymerization kinetics of a suspension process are similar to those for the bulk polymerization of the monomers. The monomer droplets are large enough so that more than one growing chain exists at a time (in contrast to the normal assumption of one or zero growing chains per droplet in emulsion systems). The monomer in the droplet usually is the only participant in the polymerization; very little monomer and initiator are soluble in the water phase.

Systems in which the polymers are soluble in the monomers such as styrene–polystyrene can be described as a single phase in the monomer droplet. These systems often produce spherical particles, or beads. Systems in which the monomer swells the polymer, such as vinyl chloride–poly(vinyl chloride), may contain both monomer and precipitated polymer phases in the polymerizing droplet. These two phases require different kinetic models. Suspension poly(vinyl chloride) particles are composed of smaller "primary" particles that have been related to the precipitated polymer. Systems in which the monomer is relatively insoluble in the polymer, such as vinylidene chloride in poly(vinylidene chloride), are difficult to model because polymerization and crystallization seem to be coupled. A number of polymerization rate equations that may be applied to suspension polymerizations have been reviewed.

Copolymerization. Copolymers can be made by suspension methods for monomers with good solubility in their polymers (15% ethyl acrylate–85% methyl methacrylate and styrene–divinylbenzene) and those with little solubility in their polymers (the Saran family, including vinylidene chloride–vinyl chloride). One difficulty in most copolymerizations is controlling the composition drift caused by the different reactivities of a free-radical chain end with the two different monomers. The usual approach to modeling these types of polymerizations has been discussed. The average monomer content in the copolymer needs to be fairly uniform in order to get the desired properties in the final product. This usually can be done by feeding the monomer that is being depleted more rapidly. For systems in which the monomer is soluble in the polymer, fresh liquid feed of one monomer can be distributed uniformly through the reacting phase (the monomer/polymer droplet). For systems in which the polymer precipitates or crystallizes quickly, monomer distribution in the reacting phase can be fairly uniform. For the case in which polymer is swollen by the monomer, two different monomer compositions could exist in the polymerizing macroparticle: one in the polymer-rich phase and one in the monomer-rich phase. If the compositions were sufficiently different, different polymer compositions could be formed by polymerizations in the two phases. Several methods have been developed to produce homogeneous copolymers.

Monomer Removal. The polymerization rate slows as monomer is nearly depleted from the reactor. At some point it is not economical to continue the polymerization, even though not all monomer has reacted. The conversion at which polymerizations are stopped depends on the system, but the reactor may contain unreacted monomer between 1 and 10% of the original monomer charged. A practical stopping point is when the reactor pressure falls below atmospheric so that organic vapor losses are minimized. Monomer is partitioned between the polymer and water phases and usually is recovered from the system for recycle.

The amount of unreacted monomer recovered depends on economic and health factors. Monomers may be removed to the parts per million level as part of the polymer production process. This is the typical residual level for vinyl chloride in suspension poly(vinyl chloride) used for pipe applications. Monomer can be stripped from the suspension particles by a variety of methods, including pulling a vacuum over the suspension slurry and/or boiling the water. Either method reduces the partial pressure of monomer outside the resin and provides a driving force for monomer removal from the polymer.

Product Handling. A principal advantage of the suspension polymerization process is the ease of handling the particles produced. This particle size range is simple to remove from liquid streams by centrifugation. It can be conveyed by air streams and can be separated from them by bag filters and cyclones. It can be stored in tank cars and silos. The particle size is small enough to be convenient for mixing with other compounding ingredients and gives few dusting problems.

Health and Safety

Health and safety issues have had significant effects on industrial practices. Major changes in the poly(vinyl chloride) business occurred in the mid-1970s as a result of health issues. This section concentrates on health and safety issues in the production of poly(vinyl chloride). Other polymer systems have similar problems.

Health issues associated with a number of vinyl monomers have led to process and processing changes. In the mid-1960s, exposure to vinyl chloride monomer by workers who cleaned reactors by hand was linked to a bone deficiency (acro-osteolysis). The degeneration of the bones in the fingertips could be reversed by reassigning the workers to other duties. By 1974, a doctor for the BF Goodrich Company had established a linkage between exposure to vinyl chloride monomer and angiosarcoma of the liver. Such a linkage had been reported earlier, but the findings were not well disseminated in the chemical industry or to regulatory agencies. These findings led to the need to remove monomer from polymers so that the polymers contain very low levels of monomer.

A number of safety issues are associated with batch process equipment containing organic monomers under pressure. Automatic valves need to be installed with safety interlocks to prevent opening or closing them at the wrong times. Some methods for stopping or slowing the reaction must be available in case of equipment failure.

BIBLIOGRAPHY

J. A. Davidson and K. L. Gardner, "Vinyl Polymers, Poly(vinyl Chloride)," in M. Grayson, ed., *Kirk-Othmer Encyclopedia of Chemical Technology,* 3rd ed., Vol. 23, Wiley-Interscience, New York, 1983, p. 886.

G. C. Portingell, *Processing Properties,* in G. Butters, ed., *Particulate Nature of PVC,* Applied Science Pub. Ltd., London, 1982, Chapt. 4.

F. M. Peng, "Acrylonitrile Polymers: survey and styrene–acrylonitrile copolymers," in M. Grayson, ed., *Kirk-Othmer Encyclopedia of Chemical Technology,* 3rd ed., Vol. 1, Wiley-Interscience, New York, 1978, p. 433.

D. S. Gibbs and R. A. Wessling, "Vinylidene Chloride and PVDC," in M. Grayson, ed., *Kirk-Othmer Encyclopedia of Chemical Technology,* 3rd ed., Vol. 23, Wiley-Interscience, New York, 1983, p. 766.

ERIC A. GRULKE
Michigan State University

SUTURES

Surgical sutures are sterile filaments used to hold tissues together until they heal adequately for self-support or to join tissues with implanted prosthetic devices. They are normally attached to needles for coaptation of the edges of wounds or surgical incisions. As ligatures, they are generally used without a needle, to tie off ends of severed tubular structures such as blood vessels and ducts to prevent bleeding or other fluid leakage.

Sutures are characterized according to type of material,

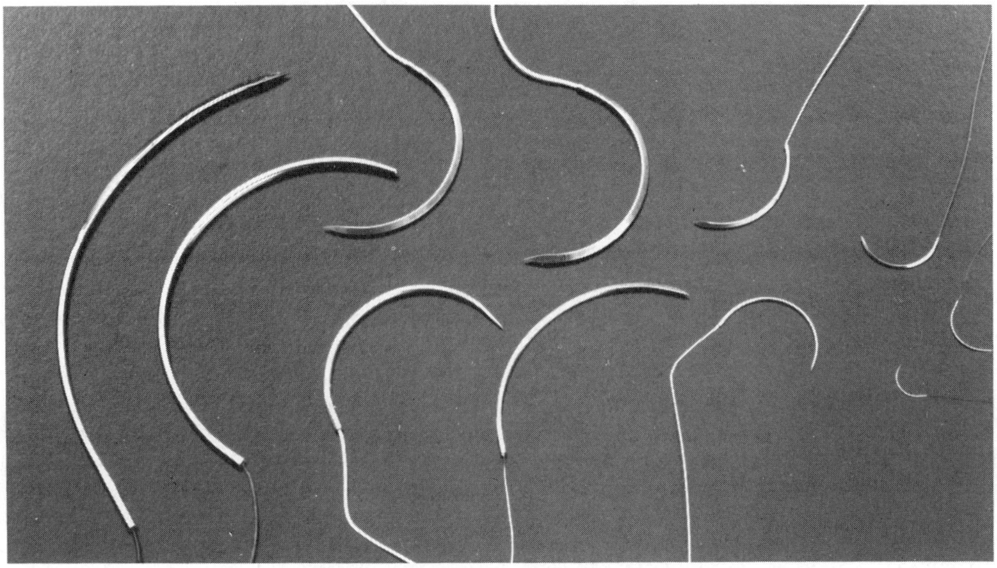

Figure 1. Surgical needles for sutures.

physical form, biodegradability, size, surgical use, and other criteria. The materials utilized are natural products, such as surgical gut or silk, many common synthetic fibers, and some fibers from synthetic polymers which have been synthesized specifically for use as sutures (see FIBERS). Sutures may be fabricated as monofilaments or multifilaments; the latter are usually braided, but sometimes twisted or spun, and may be coated with waxes, silicones, fluorocarbons, or other polymers to reduce capillarity and improve handling or functional properties.

Materials

Components. *Needles.* Surgical needles may be considered as integral parts of sutures. They are generally made from heat-treatable (400 series) stainless steel or carbon steel.

The traditional eyed needle, threaded in the hospital, has been increasingly replaced by the eyeless types.

Filament. The threads are either absorbable or nonabsorbable. The suture may be single monofilament or braid made out of multifilament or may be twisted. Commercial suture materials are described below. Trade names are indicated by initial capital letters throughout this article.

Absorbable

natural, degraded by enzymolysis
- catgut
- reconstituted collagen

synthetic, degraded by hydrolysis
- Dexon, poly(glycolic acid)
- Vicryl, glycolide–lactide copolymer
- Maxon, copolymer of glycolide and trimethylene carbonate
- PDS, polydioxanone

Nonabsorbable, not degraded over a long period of time

natural
- cotton
- linen
- silk

synthetic
- polypropylene
  - Prolene
- Surgilene
- Deklene

polyester
- poly(ethylene terephthalate), Ti.Cron, Mersilene, Ethiflex, Ethibond, Polydek, Tevdek, Mirafil
- poly(butylene terephthalate), Miralene
- copolymer of tetramethylene ether terephthalate and tetramethylene terephthalate, Novafil

polyamide
- nylon-6,6, Surgilon, Dermalon
- nylon-6, Nurolon, Ethilon (both nylon-6 and nylon-6,6)

polytetrafluoroethylene
- Gore Tex

metals
- 316 stainless steel
  - Flexon
  - surgical stainless steel

Colorants. Although some surgical procedures require sutures that are not colored, dyed or pigmented sutures are frequently used to enhance their visibility while surgery is in progress.

Coatings. Braided sutures are often coated to improve handling properties, facilitate knot tying, and reduce tissue drag.

Manufacturing Process

Natural material, such as surgical gut, is made from natural raw materials, followed by purification, shaping to monofilament form, and chemical treatment. The manufacture of synthetic sutures starts with polymers, which are extruded via a fiber spinning process, followed by various treatments.

The finished product is sterilized by the manufacturer, which eliminates a procedure that had mostly been done in the hospital. In the United States, steam and dry-heat sterilization have been largely replaced by ethylene oxide sterilization or ionizing radiation. The choice depends on economics and the nature of the sutures.

Packaging Materials. Sutures are often packaged in dual packets. The operating room nurse opens the outer envelope and aseptically transfers the sealed inner envelope onto the sterile field where it is opened. Thus, product sterility is maintained throughout.

In addition to sterility assurance, synthetic absorbable sutures need to be protected from moisture. Therefore the material for the inner envelope is made of water-vapor impermeable metallic foils (eg, aluminum foil).

Standards

Sutures are subjected to many controls for quality assurance. Inspections and testing are rigorous for raw materials, in-process materials, and final products.

Regulatory Requirements

Until 1970, only two suture materials, Dexon poly(glycolic acid) and Prolene polypropylene sutures, had been subjected to a complete premarket clearance procedure as new drugs for the FDA and had received approval. In 1971, the FDA classified the older suture materials as drugs and specified abbreviated new drug application requirements for manufacture and sale [Fed. Register 36 (218), 21612 (1971)]. The manufacture and sale of sutures have since been subject to regulatory control by the FDA, as have pharmaceutical and biological agents. Since the enactment of the Medical Device Amendments to the Food and Drug Act in 1976, sutures have been considered Class-III, "premarket approval" devices and are now under the control of the FDA Center for Devices and Radiological Health. The introduction of a new polymeric suture material, considered a Class-III product, entails a demonstration of safety and surgical efficacy through controlled laboratory, animal, and clinical studies.

Mechanical Properties

The mechanical properties of commercially available sutures are shown in Tables 1 and 2.

Suture–Tissue Interactions

As with all surgical implants, some tissue response to sutures is inevitable. Trauma caused by the passage of needle and thread can be minimized, but not eliminated, by using materials of optimum design and size, proper knot tension, and other aspects of meticulous technique. The initial tissue reaction, of 5–7 days duration, is essentially the same for all sutures. Beyond this point, the tissues respond to the foreign body by absorbing or encapsulating it in an envelope of fibrous connective tissue; the difference in the degree, intensity, and type of cellular reaction depends on the nature of the material used. Surgical gut and its related collagens (qv) evoke the strongest reaction, followed by silk (qv), cotton (qv), and other natural fibers. The synthetic fibers currently used, including polyolefins, polyesters (qv), and polyamides (qv), as well as the newer synthetic absorbable sutures composed of homo- and copolymers of glycolic acid, result in relatively bland reactions, as do metallic wires.

Suture Selection

Selection of the best suture material for a given procedure is difficult, if not impossible, and is a subject of lively controversy among surgeons. In some cases, any of a variety of materials might perform equally well, and the choice would depend not only on the purpose and particular technique employed, but also on the training and experience of the user. A material that one surgeon can handle well (tie secure knots with ease) may feel uncomfortable in the hands of another surgeon.

Commercial Products and Outlook

The search for new biomaterials is progressing at a fast pace, as judged from the number of patents that have been issued recently. Suture materials based on monomers such as oxalate, succinate, and other copolymers having both amide and ester linkages in the polymer chain have been patented for suture use. Attempts to develop radiation-sterilizable, synthetic absorbable suture materials are continuing.

It appears that sutures have reached a mature market. The U.S. growth rate in dollar sales was about 5.7% for 1982–1986 (IMS-America Hospital Supply Index), and the outlook for higher

Table 1. Mechanical Properties of Absorbable Sutures[a,b]

Class	Trade name	Type[c]	Supplier	Break strength, MPa[d]: Straight pull	Break strength, MPa[d]: Knot pull	Elongation to break, %	Young's modulus, GPa[e]
Polyesters							
poly(*p*-dioxanone)	PDS	M	Ethicon	450–560	240–340	30–38	1.2–1.7
poly(glycolic acid)	Dexon "S"	B	American	760–920	310–590	18–25	7–14
	Dexon Plus	B	Cyanamid				
poly(glycolide-*co*-lactide)	Vicryl	B	Ethicon	570–910	300–400	18–25	7–14
poly(glycolide-*co*-trimethylene carbonate)	Maxon	M	American Cyanamid	540–610	280–480	26–38	3.0–3.4
Proteins		Tw	Davis & Geck	310–380	110–210	15–35	2.4
catgut							
regenerated collagen			Ethicon				
			Deknatel				

[a] Reproduced from A. F. von Recum, ed., *Handbook of Biomaterials*. Courtesy of Macmillan and the author.
[b] These properties are typical for sizes 0 through 3-0, but differ for finer or larger sizes.
[c] Types: B = braid, M = monofilament, Tw = twisted monofilament.
[d] To convert MPa to psi, multiply by 145.
[e] To convert GPa to psi, multiply by 145,000.

Table 2. Mechanical Properties of Nonabsorbable Sutures[a,b]

Class	Trade name (commercial name)	Type[c]	Supplier	Break strength, MPa[d] Straight pull	Knot pull	Elongation to break, %	Young's modulus, GPa[e]
Cellulosics							
cotton		Tw	Davis & Geck Ethicon	280–390	160–320	3–6	5.6–10.9
linen		Tw	Ethicon	similar to cotton			
Metals							
stainless steel, 316L S.S.	Flexon	Tw	American Cyanamid	540–780	420–710	29–65	200
	(surgical stainless steel)	M,Tw	Ethicon				
Polyamides							
nylon-6,6	Surgilon	B	American Cyanamid	460–710	300–330	17–65	1.8–4.5
	Dermalon	M					
nylon-6	Nurolon	B	Ethicon				
	Ethilon	M					
	Supramid	S/C	S. Jackson				
Polyesters							
poly(butylene terephthalate)	Miralene	M	B. Braun	490–550	280–400	19–22	3.6–3.7
poly(ethylene) terephthalate[f]	Dacron	B	American Cyanamid	510–1060	300–390	8–42	1.2–6.5
	Ti.Cron	B					
	Mirafil	M	B. Braun				
poly(tetramethylene ether) terephthalate-*co*-tetramethylene terephthalate	Novafil	M	American Cyanamid	480–550	290–370	29–38	1.9–2.1
Polyolefins							
polypropylene	Surgilene	M	American Cyanamid	410–460	280–320	24–62	2.2–6.9
	Prolene	M	Ethicon				
Proteins	(silk)	B	Davis & Geck	370–570	240–290	9–31	8.4–12.9
silk	(surgical silk)	B	Ethicon				
	Dermal	Tw					
	(virgin silk)	Tw					

[a] Reproduced from A. F. von Recum, ed., *Handbook of Biomaterials*. Courtesy of Macmillan and the author.
[b] These properties are typical for sizes 0 through 3-0, but differ for finer or larger sizes.
[c] Types: B = braid, M = monofilament, Tw = twisted monofilament.
[d] To convert MPa to psi, multiply by 145.
[e] To convert GPa to psi, multiply by 145,000.
[f] Other braided types include Ethicon's Ethibond, Mersilene, and Ethiflex, also Deknotal's Polydek and Tevdek.

sales growth in the near future is not promising. The reasons for this trend are cost containment in the health care industry, reduction in the number of elective surgeries, and other means of wound management, such as staplers. The current worldwide wound-closure market is about $\$10^9$ (U.S. dollar sales); sutures represent about 78%, mechanical wound closures the rest.

BIBLIOGRAPHY

The United States Pharmacopia, XX(USP XX-NFXV), the U.S. Pharmacopeial Convention, Inc., Rockville, Md., 1980.

A. F. von Recum, ed., *Handbook of Biomaterials,* Macmillan, New York, 1986, pp. 88–89.

A. R. Katz, D. P. Mukherjee, A. L. Kaganov, and S. Gordon, *Surg. Gynecol. Obstet.* **161,** 213–222 (1985).

F. W. Billmeyer, Jr., *Textbook of Polymer Science,* 3rd ed., John Wiley & Sons, Inc., N.Y., 1984, pp. 502–503.

D. P. MUKHERJEE
Dow Corning Wright

SYNDIOTACTIC POLYMERS. See INSERTION POLYMERIZATION; MICROSTRUCTURE; STEREOREGULAR POLYMERS.

SYNTACTIC FOAMS. See CELLULAR MATERIALS.

SYNTHESIS, EQUILIBRIUM CONTROL. See POLYMERIZATION, THERMODYNAMICS.

T

TACTICITY. See MICROSTRUCTURE; STEREOREGULAR POLYMERS.

TANNING. See LEATHER.

TAPE. See FILMS; SHEETING.

TECHNICAL SERVICE

In the process of marketing plastic materials (molding compounds and formed products such as film, sheet, and rod) technical service is frequently the component that allows a plastics supplier to retain an advantage over a competitor. Effective and efficiently delivered technical service may outweigh a minor price advantage or slightly better performance characteristic of a competitor's product. Although the nature of technical services supplied to a customer varies from one segment of the plastics industry to another, their common purpose is to meet customer requirements related to the plastics or plastic products being marketed.

Functions of the Technical Service Organization

The functions of the technical service organization are numerous and include training of personnel, testing of materials, observation of industry trends, and assistance to customers by providing training, processing technology, and problem solving. Information about products, processing technology, and applications may be provided to the general public by presenting technical papers at meetings and addressing students and interested groups.

Contact with a customer or potential customer is usually made by a salesperson or a market development engineer (also referred to as an application development engineer). At this stage the need for a technical service engineer may be identified. The salesperson usually introduces the technical service engineer to the customer.

Organizational Structures

All technical service engineers or personnel work with customers on technically related issues. Some technical service organizations are a part of the sales and marketing department, and others are a unit of the research and development (R&D) department. Other technical service organizations consist of personnel from two or more groups (research and development, marketing, and manufacturing), representing a loose confederation of human resources. The traditional business organization, on the other hand, has a director, managers, supervisors, and individual contributors. Both types can be effective if their efforts are focused on the customer and on finishing programs in due time, with the requisite pool of technical talents to handle the programs in a competent manner. The traditional business organization does not contain the potential for jurisdictional disputes between groups. In any case both functional and support groups must be available.

Other Considerations

Requirements for Technical Service Engineers. Most technical service engineers working in the U.S. plastics industry have at least one degree in a technical discipline, such as chemistry, material science, polymer technology, chemical engineering, mechanical engineering, or electrical engineering; many have advanced degrees. Training is important because of the growing complexity of the plastics industry.

Backup Groups. The services of analytical, physical testing, and polymer characterization laboratories are frequently needed. Code and agency specialists and quality-assurance personnel assist with customer needs and requests. A comprehensive list of competent outside testing laboratories or consultants is valuable.

BIBLIOGRAPHY

V. Shah, *Handbook of Plastics Testing Technology,* John Wiley & Sons, Inc., New York, 1984.

W. E. Deming, *Quality, Productivity, and Competitive Position,* Massachusetts Institute of Technology, Cambridge, Mass., 1982.

FRANCIS B. MCANDREW
Hoechst Celanese Corporation

TELECHELIC POLYMERS

Telechelic polymers can be used for chain extension by means of bifunctional linking reagents, network formation by means of multifunctional linking agents, and the synthesis of block copolymers by the reaction of telechelics with different end groups (see also BLOCK COPOLYMERS; NETWORKS).

Industrial interest in telechelics was stimulated by the development of thermoplastic elastomers (TPE), which consist of ABA block and multiblock copolymers. Liquid telechelic polymers are the basis for reaction injection molding (qv) (RIM). Liquid telechelics that can be used for network formation offer processing advantages and may result in materials with improved properties.

A polymer can be considered to be telechelic if it contains end groups that react selectively to give a bond with another molecule. Depending on the functionality, which must be distinguished from the functionality of the end group itself, telechelics can be classified as mono-, di-, tri-, and polytelechelics. The functionality is defined as

$$f = \frac{\text{number of functional groups}}{\text{number of polymer chains}}$$

The functionality of the end group itself is important. When such groups are bifunctional (eg, vinyl groups) and participate in polymerization reactions, yielding graft copolymers or networks, such telechelic polymers are called macromolecular monomers, macromonomers, or macromers (qv).

Radical Polymerization

Telechelics can be synthesized by radical polymerization (qv) in two ways. End groups can be controlled with large amounts of initiators (dead-end polymerization) (see Table 1), or polymerization can be conducted in the presence of suitable transfer agents (telomerization).

The first method has serious limitations because well-defined end groups can be observed only if not more than one type of primary radical is formed that does not cause hydrogen abstraction. Termination occurs only by combination, involving two macroradicals or a macroradical and a primary radical.

In telomerization the amount of nonfunctional end groups (formed by the initiator) can be reduced to trace quantities.

Anionic Polymerization

Anionic polymerization and, in particular, living polymerization can be used for the synthesis of telechelics.

Living polymerization is characterized by the absence of termination and transfer, by a narrow dispersity that can be described by a Poisson distribution, and by a linear increase of molecular weight with conversion. The propagating species are carbanions or other negatively charged species. Bivalent growing chains allow the generation of two terminal end groups and the facile synthesis of telechelic polymeric materials, depending on the monomer, initiator, solvent, and termination agent.

Carbocationic Techniques

The preparation of telechelics by carbocationic polymerization of vinyl monomers has received little attention because of the high reactivity and low selectivity of the growing species. The synthesis of useful telechelics requires the control of molecular weight and end groups; this was not possible, until recently, by carbocationic polymerization (see LIVING POLYMER SYSTEMS).

Macromonomers

Macromolecular monomers, called macromonomers or macromers (qv), can be defined as oligomers or polymers with a polymerizable end group. Such groups may be vinyl, acrylic, or heterocyclic (ring-opening polymerization), or dicarboxylic, or dihydroxylic (step-growth polymerization). The increasing interest in these materials stems from the growing need for well-defined graft copolymers for which macromonomers often are an ideal starting material. The length and the number of branches of the graft copolymers can be controlled by the molar mass and the feed ratio of macromonomers to comonomers.

Anionic Polymerization. Anionic methods for the synthesis of macromonomers can be divided into initiation and end capping (Table 2).

Table 1. Initiators and Monomers Used for the Synthesis of Telechelics

Initiator	Functional end group	Monomer[a]
$H_2O_2—Fe^{2+}$	—OH	MMA, MA
H_2O_2	—OH	Bu
$H_2O_2—BF_3$	—OH	Bu
$(NC—C(CH_3)_2—N)_2$	—CN	S E, Bu, IS
$(HOOC—C(CH_3)_2—N)_2$	—COOR	S, Bu
$(RO—C(=O)—O)_2$	—OC—OR	E, Bu, S MMA
$[(CH_3)_3Si—O—C(C_6H_5)_2]_2$	$—O—Si(CH_3)_3$	MMA
$[HO(CH_2)_3C(CH_3)_2—N]_2$	—OH	S, S—AN
$[HOOC(CH_2)_2C(CH_3)_2—N]_2$	—COOH	VAC, S—AN MMA, MA AN
$(ClCH_2—C_6H_4—C(=O)—O)_2$	$—CH_2Cl$	S

[a] AN, acrylonitrile; Bu, 1,3-butadiene; E, ethylene; IS, isoprene; MA, methyl acrylate; MMA, methyl methacrylate; S, styrene; VAC, vinyl acetate.

Table 2. Macromonomers by Anionic Polymerization

Initiator	End capping
Styrene	
$CH_2{=}CHLi$	$CH_2{=}CH{-}CH_2Br$, $CH_2{=}C(CH_3){-}COCl$
$CH_2{=}CHCH_2Li$	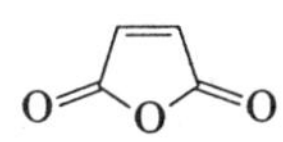 $ClCH_2CH_2{-}O{-}CH{=}CH_2$
	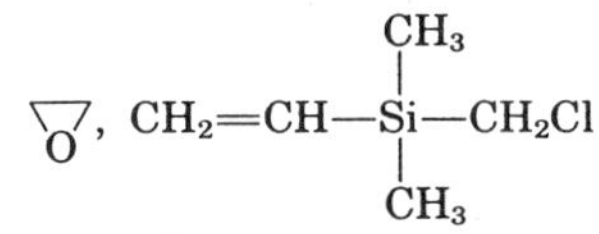
	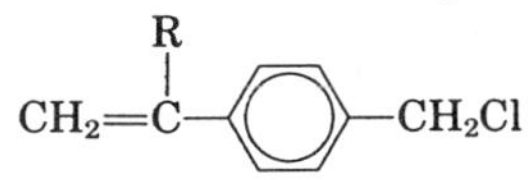
	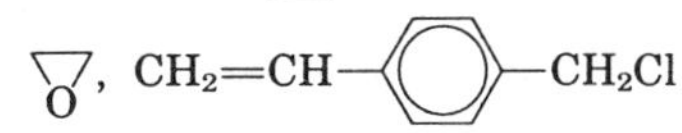
Methyl methacrylate	
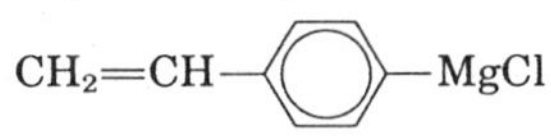	$CH_2{=}CH{-}CH_2Br$
	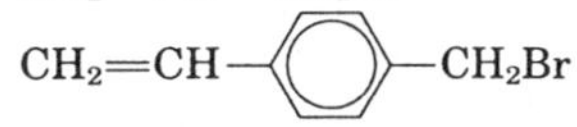
	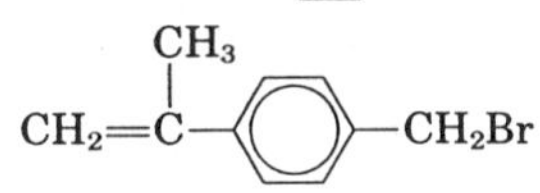
Ethylene oxide	
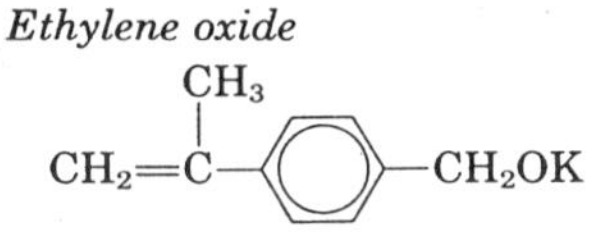	$CH_2{=}C(CH_3){-}COCl$
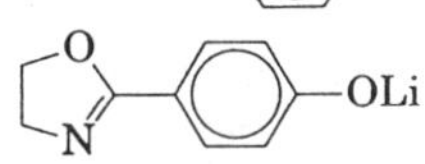	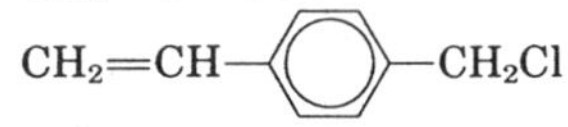
Vinylpyridine	
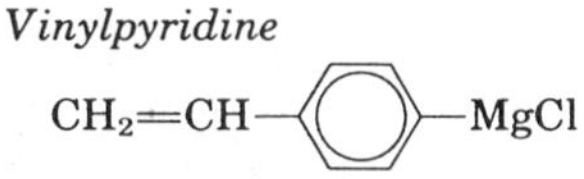	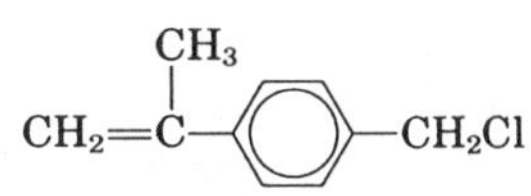
	$CH_2{=}C(CH_3){-}COCl$
	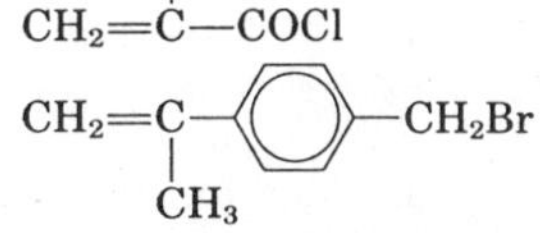
Hexamethylcyclo-trisiloxane	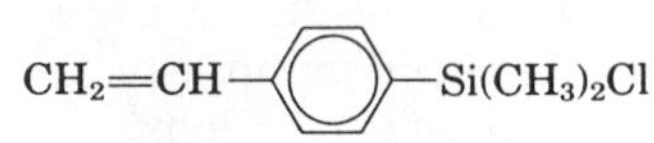
	$CH_2{=}CH{-}Si(CH_3)_2Cl$
	$CH_3{=}C(CH_3){-}COO(CH_2)_3{-}Si(CH_3)_2Cl$

Although the initiation method is widely used, side reactions of the unsaturated head groups often lead to low functionality.

Polar monomers, such as MMA, rarely give high yields of macromonomers by straightforward anionic polymerization. However, group-transfer polymerization (qv), which is a special case of anionic polymerization, gives macromonomers with perfect functionality.

Cationic Polymerization. Both initiator and end-capping methods can be used to obtain macromonomers by cationic polymerization. Macromonomers can be made this way from polytetrahydrofuran (PTHF) (Table 3). A typical reaction scheme is given below:

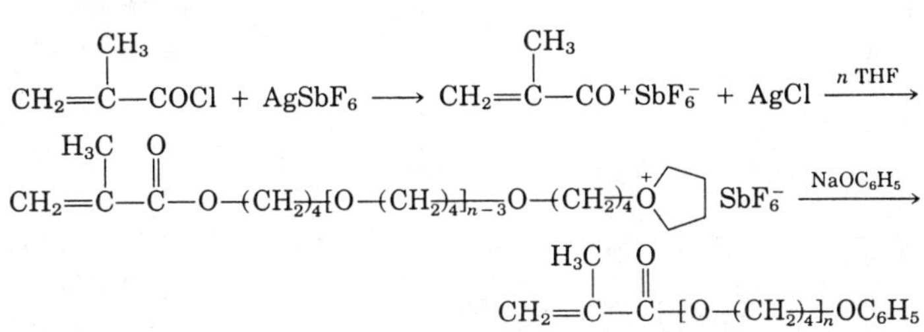

Table 3. Macromonomers from PTHF

Initiator	End capping

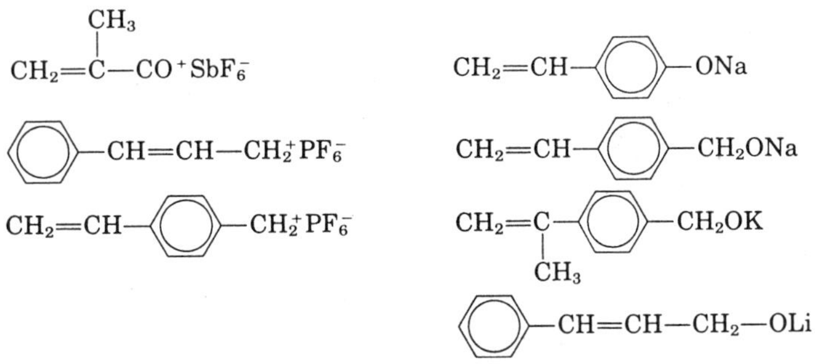

Radical Polymerization. Macromonomers can also be synthesized by radical polymerization:

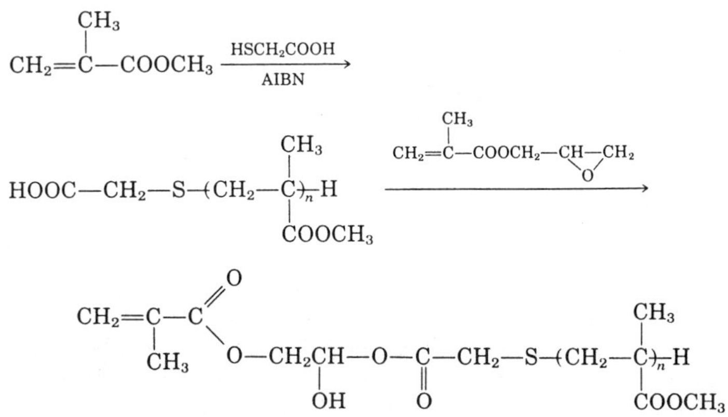

With thioglycolic acid as a chain-transfer agent, macromonomers are prepared from dimethylaminoethyl methacrylate and other acrylate and methacrylate derivatives. Iodoacetic acid is used as the transfer agent for polystyrene macromonomers.

Ring-opening Polymerization

The synthesis of telechelics by ring opening has attracted great interest because of the commercial potential of the resulting compounds, such as polyether polyols. The end groups are introduced by initiation, end capping, or transfer reactions.

Chain Scission

Telechelics can also be prepared by the cleavage of high molecular weight moieties (Table 4) by oxidation, reduction, rearrangements, and irradiation.

Step-growth Polymerization

Step-growth polymerization is a simple method of preparing reactive oligomers (qv), many of which cannot be prepared by the methods described previously. Increasing commercialization of the thermoplastic elastomers (TPE), based on tri- or multiblock copolymers, has led to a growing interest in telechelics of polymers with high glass-transition temperatures, which may be used as rigid blocks (see ELASTOMERS, THERMOPLASTIC). Polymers with high glass-transition temperatures are frequently prepared by step-growth polymerization.

Almost all functional groups can, in principle, be utilized for step-growth polymerization. However, side reactions such as cyclization and decarboxylation limit the molecular weight, functionality, and yield.

AR^1A and BR^2B are allowed to react ($x < y$),

$$x\,AR^1A + y\,BR^2B \longrightarrow B{-}R^2{-}(BAR^1ABR^2)_n{-}B$$

and the molecular weight of the products is given by

$$\overline{DP}_n = \frac{(1 + r)}{(1 + r - 2rp)}$$

where $\overline{DP}_n$ = number-average degree of polymerization, $r = (AR^1A):(BR^2B)$, and p = conversion of AR^1A.

At 100% conversion of the minor reactant, the molecular weight of the product is determined by r, and the end groups are those of the major component.

This method can be applied for the preparation of α,ω-dicarboxyoligoesters, α,ω-dicarboxyoligoamides, and phenoplasts and aminoplasts having reactive end groups. It has also been applied to the preparation of α,ω-diphenol oligocarbonates and epoxy resins.

In an alternative method of achieving nonstoichiometry, AR^3X or BR^4X (where X represents a functional group which does not react with A or B) is added to the reaction system. At complete conversion, X forms the end groups of the polymer.

$$n\,AR^1A + n\,BR^2B + m\,BR^4X \longrightarrow XR^4B{-}(AR^1ABR^2B)_n{-}R^4X$$

Table 4. Telechelics Prepared by Chain Scission

Structure	End group
$-(CH_2-CHCl)_n-$	COOH
	COCl
	OH
$-(CH_2-CH{=}CH-CH_2)_n-$	OH
$-[CH_2-C(CH_3)_2]_n-$	COOH
	$COCH_3$
	CH_2OH
	$CHOHCH_3$
$-(CH_2-CH_2)_n-$	COOH
$-[CH_2-CH(C_6H_5)]_n-$	OH
$-(CH_2-CH_2-O-CH_2CH_2S_x)_n-$	SH
$-[(CH_2CH_2)_n(CH_2-CH(CH_3))_p]_q-$	OH

The effect of nonstoichiometry on the number-average degree of polymerization is given by

$$\overline{\mathrm{DP}}_n = \frac{(1 + q)}{(1 - 2pq + q)}$$

where p = conversion of the minor component and q = [A]/[B], $q < 1$.

The latter method is also useful for preparing monofunctional telechelics. In this case a monofunctional reactant is used. For example, R^4B gives

$$n\ \mathrm{AR^1A} + n\ \mathrm{BR^2B} + \mathrm{R^4B} \longrightarrow \mathrm{R^4{-}\!\!(BAR^1ABR^2)_{\overline{n}}B}$$

BIBLIOGRAPHY

E. J. Goethals, ed., *Telechelic Polymers,* CRC Press, Boca Raton, Fla., 1989.

C. M. Starks, *Free Radical Telomerization,* Academic Press, New York, 1974.

V. Percec, C. Pugh, O. Nuyken, and S. D. Pask in G. Allen and J. C. Bevington, eds., *Comprehensive Polymer Science,* Vol. 6, Pergamon Press, Oxford, 1989.

OSKAR NUYKEN
Universität Bayreuth

STEPHEN PASK
Bayer AG, Dormagen

TELOMERIZATION

The terms telomerization and telomer were defined many years ago to describe an innovative method for the free-radical polymerization of ethylene. Since DuPont released these patents in the late 1940s, the telomerization reaction has expanded from the original free-radical mechanisms for which the early definitions were designed to include other polymerization mechanisms. This article reviews the literature in the areas of free-radical, anionic, cationic, and transition metal catalyzed telomerization and the applications of telomers.

The following excerpt from one of the original DuPont patents explains this remarkable reaction.

> Telomerization is defined as the process of reacting, under polymerization conditions, a molecule YZ which is called a telogen with more than one unit of a polymerizable compound having ethylenic unsaturation called a taxogen to form products called telomers having formula $Y(A)_nZ$ wherein $(A)_n$ is a divalent radical formed by chemical union, with the formation of new carbon bonds, of n molecules of the taxogen, the unit A being called a taxomon, n being any integer greater than one, and Y and Z being fragments of the telogen attached to the terminal taxomons.

The original definition conceived by Hanford and Joyce was intended for free-radical polymerizations, especially of ethylene.

Free-radical Telomerization

Free-radical telomerization has been reviewed through 1973 with a very good account of basic ideas involved in the synthesis of telomers and the kinetics of telomerization reactions.

Synthesis. The interesting feature of telomerization of many different monomers is the variety of telomer products that can be produced from the same monomer by simply altering the mode of initiation or the telogens being used. The radically initiated telomerization of ethylene with CCl_4 (eq. 1) proceeds via the following mechanism.

$$\underset{\text{taxogen}}{n\ \mathrm{CH_2{=}CH_2}} + \underset{\text{telogen}}{\mathrm{CCl_4}} \xrightarrow{\text{initiator}} \underset{\text{telomer}}{\mathrm{CCl_3{-}\!\!(CH_2CH_2)_{\overline{n}}Cl}} \qquad (1)$$

$$\mathrm{R\cdot} + \mathrm{CCl_4} \longrightarrow \mathrm{RCl} + \mathrm{\cdot CCl_3} \qquad (2)$$

$$\mathrm{\cdot CCl_3} + \mathrm{C_2H_4} \longrightarrow \mathrm{CCl_3CH_2CH_2\cdot} \qquad (3)$$

$$\mathrm{CCl_3CH_2CH_2\cdot} + n\ \mathrm{C_2H_4} \longrightarrow \mathrm{CCl_3(CH_2CH_2)^{\cdot}_{n+1}} \qquad (4)$$

$$\mathrm{CCl_3(CH_2CH_2)^{\cdot}_{n+1}} + \mathrm{CCl_4} \longrightarrow \mathrm{CCl_3(CH_2CH_2)_{n+1}Cl} + \mathrm{\cdot CCl_3} \qquad (5)$$

$$2\ \mathrm{\cdot CCl_3} \longrightarrow \mathrm{Cl_3CCCl_3} \qquad (6)$$

Small quantities of hexachloroethane have been found, but no evidence of the products that would result from a different initiation–chain-transfer mechanism, ie, $Cl(CH_2CH_2)_nCl$ or $Cl_3C(CH_2CH_2)_nCCl_3$, have been found.

Initiation of telomerization by oxidation–reduction (redox) mechanisms has also been widely studied.

Another method employed for the radical telomerization of vinyl monomers is the use of radiation, in particular, γ-radiation and uv radiation.

The telomerization reaction is greatly dependent on the nature of the telogens used. Most telogens are polyhalogenated methanes, eg, CCl_4 and $CHCl_3$. Numerous other telogens can be used, such as alcohols to prepare telomeric alcohols, carboxylic acids to prepare telomeric acids, and even telomers themselves can be used as telogens (eq. 7):

$$\mathrm{CCl_3CH_2\underset{\displaystyle COOCH_3}{\underset{|}{C}}HCl} + \mathrm{CH_2{=}\underset{\displaystyle CH_2Cl}{\underset{|}{C}}H} \xrightarrow{\text{redox}} \mathrm{\overset{\displaystyle Cl}{\overset{|}{\underset{\displaystyle COOCH_3}{\underset{|}{C}}}}HCH_2CCl_2CH_2\underset{\displaystyle CH_2Cl}{\underset{|}{C}}HCl} \qquad (7)$$

Kinetics. The mechanistic steps for a typical telomerization reaction are outlined below; I = initiator, YZ = telogen, and M = monomer.

Initiation

$$\mathrm{I} \longrightarrow 2\ \mathrm{R\cdot}$$

$$\mathrm{R\cdot} + \mathrm{YZ} \longrightarrow \mathrm{Y\cdot} + \mathrm{RZ}$$

Propagation

$$\mathrm{Y\cdot} + \mathrm{M} \longrightarrow \mathrm{YM_1\cdot} \xrightarrow{\mathrm{YZ}} \mathrm{YM_1Z} + \mathrm{Y\cdot}$$

$$\mathrm{YM_1\cdot} + \mathrm{M} \longrightarrow \mathrm{YM_2\cdot} \xrightarrow{\mathrm{YZ}} \mathrm{YM_2Z} + \mathrm{Y\cdot}$$

$$\mathrm{YM}_n\cdot + \mathrm{M} \longrightarrow \mathrm{YM}_{n+1}\cdot \ldots$$

$$\mathrm{YM_1\cdot} + n\ \mathrm{M} \longrightarrow \mathrm{Y{-}\!\!(M)_{\overline{n+1}}\cdot}$$

Termination

$$2\ \mathrm{R\cdot} \longrightarrow \text{nonradical products}$$

All steps are similar to a normal free-radical polymerization reaction, except that the first few addition and chain-transfer steps are more important.

Anionic Telomerization

Anionic polymerization is one of the best methods for obtaining polymer products of known architecture and molecular weight. The degree of polymerization can be shown as the initial molar ratio of monomer to initiator because in the simplest case one molecule of initiator initiates one polymer chain. In order to obtain polymers of very low molecular weight, the ratio of initiator to monomer must be increased dramatically. The majority of initiators used in anionic polymerization are expensive organolithium compounds. Thus, using anionic polymerization to prepare low molecular weight polymers is economically unfeasible because of the amount of initiator required. If the desired polymer product can be polydisperse, chain-transfer mechanisms can be investigated, and this is where anionic telomerization reactions come into play.

Cationic Telomerization

In the vast majority of studies of a cationic polymerization mechanism in the preparation of telomers, Lewis acids have been used as the initiators. Various monomers have been used in these telomerizations, including isoprene, 2-methyl-2-butene, styrene, and isobutene. The field of cationic telomerization has been reviewed; information on butadiene, isoprene, and styrene monomers and on various initiators, telogens, and solvents is included.

Transition Metal Catalyzed Telomerization

The use of transition metal catalyzed telomerization is relatively new compared with free-radical telomerizations. Telomerization by homogeneous transition metal catalysis has been defined as "the oligomerization of dienes under the concomitant addition of a nucleophilic reagent." If the same terminology is applied for transition metal catalyzed telomerizations as was originally developed, the diene is the taxogen, or monomer, and the nucleophilic reagent is the telogen. In the following reaction butadiene is telomerized by some catalyst in the presence of a telogen, HY.

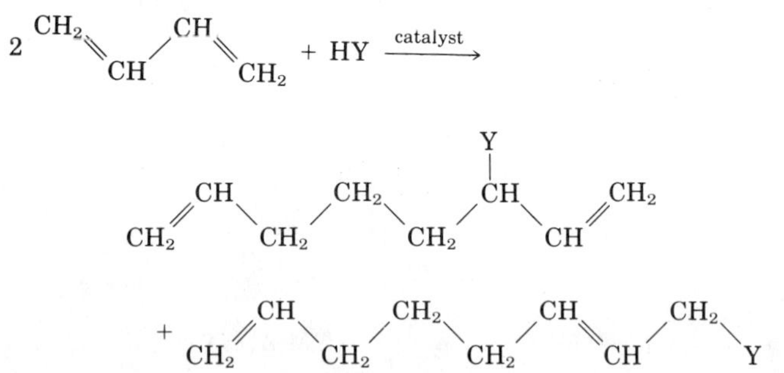

(8)

The telogen is most often a nucleophile such as water, an amine, an alcohol, or an acid. The catalyst can be various organometallic compounds of the transition metals, where the metal is often nickel or, more frequently, palladium.

Transition metal catalyzed telomerization reactions of dienes, specifically homogeneous reactions, have been reviewed.

Mechanisms. As is the case for many organometallic reactions, the mechanism of telomerization is not entirely agreed on. There appear to be three major mechanisms for palladium-catalyzed telomerization reactions and only one for nickel-catalyzed reactions. The nickel-catalyzed telomerization mechanism is more well known because of the numerous studies done on nickel-catalyzed oligomerization reactions. The three mechanisms suggested for palladium catalysis involve two monometallic pathways and one bimetallic pathway.

Synthesis. The use of transition metal catalyzed telomerization reactions seems to be almost entirely confined to 1,3-dienes, especially isoprene and butadiene. There are numerous examples of telomerization reactions of these 1,3-dienes using various nucleophilic telogens, almost all of which are catalyzed by either nickel or palladium complexes.

Applications. Telomers, both modified and unmodified, have found many applications in industry. These applications range from using modified telomers as monomers in various polymerization reactions to using unmodified telomers as lubricants.

Telomers are frequently used as additives in various systems to impart special properties to that system. One example is the use of telomeric materials as high molecular weight plasticizers (qv) to improve the processibility of various polymers. Better properties have also been obtained by adding telomers to various polymer formulations.

Uses of Telomers

Not only are telomers useful for a host of applications, but the products from telomer modification reactions are also very useful. Two modification reactions performed on telomers are end group modifications and pendent group modifications.

Fluorotelomers are used as coatings for glass to improve oil and water repellency and for printing plates as high temperature protective layers. Some tetrafluoroethylene telomers have even been used as substrate coatings where, after application, paint will not adhere to the surface.

The coatings industry has also discovered many uses for telomeric materials, eg, the preparation of photocross-linkable telomers. Telomers have also been used to carry photosensitive groups into coating formulations containing cross-linkable monomers.

BIBLIOGRAPHY

U.S. Pat. 2,457,229 (June 13, 1946), W. E. Hanford and J. R. Roland (to DuPont).

C. M. Starks, *Free Radical Telomerization,* Academic Press, Inc., Orlando, New York, 1974, p. 1.

R. Milner, R. N. Young, and A. R. Luxton, *Polymer* **26,** 1265 (1985).

A. Behr, *Aspects Homogen. Catal.* **5,** 3 (1984).

BERNARD GORDON III
JAMES E. LOFTUS
Pennsylvania State University

TEMPLATE POLYMERIZATION

Polymerizations in which polymer chains are able to propagate along template macromolecules during most of their lifetime are called template (matrix, replica) polymerizations. The peculiar mode of propagation is made feasible through specific cooperative interactions between the growing chain and the template. The template is usually a linear macromolecule that is added as a preformed polymer to the polymerization system. The polymer produced is often named the daughter polymer, and the template occasionally the parent polymer. Unfortunately, the term template polymerization is also applied to other systems, such as polymerization in the presence of a low molecular weight removable template to produce polymers having well-defined cavities (see POLYMERIZATION IN CLATHRATES).

Although the concept of template polymerization originates from natural processes like the self-replication of DNA and the biosynthesis of proteins, only nonbiological systems are discussed here (see PROTEINS, BIOSYNTHESIS).

Template polymerization is intimately connected with the ability of two different polymers to form stoichiometric interpolymer complexes or polycomplexes. Based on this relationship, a template system can be composed as follows. Assuming poly(A) can form a polycomplex with poly(B) through interaction between A and B units, such a complex is also obtained by polymerizing monomer A in the presence of poly(B). Sometimes, the reverse or mirror system B–poly(A) is also feasible. Another way to arrive at a template poly(T) may involve complexation of a low molecular weight substance T with monomer M followed by propagation of the latter.

In contrast to conventional polymerization, a template may influence the reaction kinetics (polymerization rate, reaction order), the average molecular weight and molecular weight distribution, the microstructure and stereoregularity, and the reactivity ratios in copolymerization.

Mechanism

In certain template polymerizations the monomer molecules may be preadsorbed onto a template, ie, the constant K_M for the (Langmuir-type) adsorption equilibrium M + —T— $\rightleftharpoons$ M(—T—) (—T— is a template unit) is infinite or zero. If $K_M = \infty$, adjacent molecules are linked to one another according to a step-growth mechanism, or adjacent molecules are added to the active chain ends according to a chain-growth mechanism (type I or zip mechanism). If $K_M = 0$, template polymerization occurs only when the growing oligomer in the bulk solution is able to complex with the template. Since complexation is a cooperative process, the oligomer should attain a critical chain length before its attachment to the template. Template propagation proceeds in this case by picking up monomer from the surrounding solution (type II or pick-up mechanism) (Fig. 1). The reactions invariably end with the *in situ* production of a polycomplex. An exception is a variant of type I, in which the growing chain cannot complex with the template, but detaches itself as it is formed. This variant has been suggested as possibly operating in the template polymerization of phenylalanine *N*-carboxyanhydride (NCA).

The strength of template interaction with a monomer and its resulting polymer determines the mechanism of the process. Strong ionic interactions may favor type I, whereas the weaker hydrogen bond tends to favor type II.

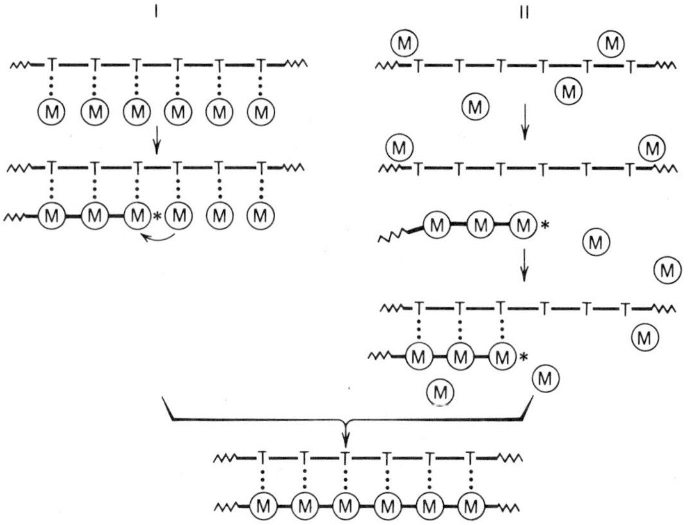

Figure 1. Idealized template polymerization mechanisms of type I and type II. —T— = template unit; M = monomer; —M— = monomeric unit; * indicates the active (propagating) chain end.

Chain-growth Reaction

Addition Polymerization. Template polymerizations are widely used for the addition polymerization of vinyl monomers. In most cases the polymerizations proceed according to a radical mechanism, usually by a thermal initiator. An initiating group can be covalently attached to the template macromolecule at one of its ends to start initiation of a monomer array from one end of the template. In evaluating kinetic results it is assumed that the initiator decomposition rate and activation parameters are unaffected by the template. Chain transfer may occur with the template, resulting in grafting with daughter chains, especially at high temperatures.

Systems of acid monomer and polybase or basic monomer and polyacid usually show a type-I behavior. Systems in which hydrogen bonding is dominant are very sensitive to reaction conditions. This may be seen by comparing the polymerization of methacrylic acid in the presence of poly(*N*-vinyl-5-methyloxazolidone), poly(*N*-vinylpyrrolidinone), poly(4-vinylpyridine), poly(2-vinylpyridine), poly(ethylene oxide), poly(vinyl alcohol), and isotactic poly(methyl methacrylate) (PMMA). Depending on the type of template, solvent, and temperature, the mechanism may be type I or II.

In template polymerizations the template is commonly added as a prepolymer to the reaction mixture, but in some instances a polymer formed *in situ* may function as a template for its own monomer. The unusual kinetics observed in the radiation-initiated polymerization of methacrylic acid and acrylonitrile was ascribed to a template effect. It has been suggested that in the early stages of the reaction the oligomers already form associates with the monomer that then propagates according to a ziplike mechanism (type I).

The presence of a template lends, in fact, a new dimension to a polymerizing system because the template can be varied with regard to its structure. Apart from its chain length and stereochemistry, it can be modified through copolymerization or cross-linking.

Ring-opening Polymerization. Template polymerizations by ring opening are so far limited to those involving NCAs of α-amino acids. In general, any α-amino acid NCA with an unsubstituted NH group, $R^1R^2\underline{CC(O)O(O)CNH}$, in which R^1 and R^2 = alkyl, aryl, or H as monomer may be used in combination with a template possessing hydrogen-accepting groups such as polypeptide or poly(2-vinylpyridine).

Step-growth Reaction

Step-growth polymerization of a template (template polycondensation) was first attempted in the so-called template-directed synthesis of oligonucleotides. The principle is based on the specific association between nucleic acid bases, eg, thymine and adenine. It was utilized in the synthesis of polyamides.

A more complicated system is the template polycondensation of an aromatic dicarboxylic acid like terephthalic acid (TPA) and a diamine with the aid of triphenylphosphite in the presence of, eg, poly(4-vinylpyridine) in *N*-methylpyrrolidinone as solvent containing LiCl.

Template polycondensation of urea and formaldehyde along poly(acrylic acid) in water is based on the polycomplex formation of urea–formaldehyde (UF) oligomers with hydrogen-accepting polymers.

Stereochemical Effects

The tacticity of the template can affect the polymerization rate, such as in the polymerization of *N*-vinylpyrrolidinone where syndiotactic poly(methacrylic acid) gives the highest rates. Other examples include the systems methacrylic acid–isotactic poly(2-vinylpyridine) and methacryloyloxy (MAO)–adenine with the template poly(MAO–uracil). The rate enhancement increases in the order atactic < syndiotactic < isotactic at 20°C, but changes to syndiotactic < isotactic < atactic above 40°C. Optimal stereochemical fitting of the daughter and parent polymer may account for these phenomena.

Only two systems exhibit a significant stereochemical effect. In a type-I system the presence of an ionene template increases the heterotactic and isotactic triad contents at the expense of syndiotactic triads in the poly(vinylsulfonic acid) produced. In the type-II template system of MMA–isotactic PMMA, the daughter PMMA is highly syndiotactic, especially at low conversions.

In other template systems the stereochemical effect is moderate to slight or absent. It is therefore impossible at this time to make general statements concerning the factors underlying stereochemical control by a template.

Applications

In situ formed polycomplexes have not found any application, although various possibilities are emerging. The construction of several membranes has been described recently.

In imaging, a thin layer consisting of monomer (acrylic acid), a basic template such as poly(dimethyldiallylammonium chloride), and a photoinitiator (eg, a benzoin ether) is applied to a transparent (polyester) carrier. On uv irradiation through a screen, a polyelectrolyte complex (qv) forms by template polymerization. The unexposed areas are removed by washing, and after immersion in a dye bath (eg, methylene blue) the image becomes visible. The maximum rate of the template reaction is at a template-to-monomer ratio of 1:1; the process therefore conforms to type I.

Other possibilities include fiber spinning from solutions of a complex produced by polymerizing methacrylic acid in the presence of an ionene, modification of polymer surfaces by application of a very thin film, and the creation of interfaces within solid microparticles by template polymerization.

BIBLIOGRAPHY

C. H. Bamford in R. N. Haward, ed., *Developments in Polymerization,* Vol. 2, Applied Science Publishers, London, 1979, p. 215.

V. A. Kabanov, *Makromol. Chem. Suppl.* **3,** 41 (1979).

Y. Y. Tan and G. Challa, *Makromol. Chem. Makromol. Symp.* **10/11,** 215 (1987).

Y. Y. Tan, "Template Polymerization" in G. Allen and J. C. Bevington, eds., *Comprehensive Polymer Science,* Vol. 3, Part 1, Pergamon Press, Oxford, 1989, p. 245.

Y. Yong Tan
Ger Challa
Groningen University

TENSILE PROPERTIES.

See Fracture; Mechanical properties; Test methods.

TERMINATION.

See Kinetics of polymerization.

TERPENE RESINS.

See Hydrocarbon resins.

TERPOLYMERIZATION.

See Copolymerization.

TERTIARY STRUCTURE.

See Proteins.

TEST METHODS

The purpose of testing a sample of a plastic, elastomer, composite, or raw material is to determine its properties. The choice of sampling procedure and test methodology is generally related to the following purposes:

- Quality assurance of the manufacturing system.
- Quality control of the production line, including inspection of raw material, in-process manufacturing, and the final product.
- New product research or development by the plastic manufacturer.
- Engineering assessment or vendor selection.
- Verification of conformity of product to buyer specifications or military specifications.
- Development of sales specifications.
- Comparison of product or performance with competitive products.
- Follow-up of complaints or performance failures.
- Product liability assessment before product release or after an accident.
- Resolution of buyer–seller or plaintiff–defendant disagreement.
- Compliance with federal, state, and city laws and codes, guidelines, or industry standards.

The test methods fall under the following general categories: physical (mechanical, electrical, optical, thermal, appearance, permeability, blocking); chemical (analytical, photochemical, radiochemical, resistance to chemical attack, flammability (qv), toxic gas evolution); biological (microbiological, toxicological, ecotoxicological); and performance (fitness for purpose, effects of use and foreseeable abuse, compatibility with end-use environment) (see also MECHANICAL PROPERTIES; ELECTRICAL PROPERTIES; OPTICAL PROPERTIES; THERMAL PROPERTIES; BARRIER POLYMERS; CHEMICALLY RESISTANT POLYMERS; DEGRADATION; RADIATION-RESISTANT POLYMERS; BIODEGRADABLE POLYMERS; WEATHERING).

The available test methods range from rudimentary physical tests such as determination of size, weight, color, and density to analyses of complex problems dealing with thorough on-line and off-line testing of the product being manufactured (see also CHARACTERIZATION OF POLYMERS; IMPACT RESISTANCE; and NONDESTRUCTIVE TESTING).

A sampling and testing strategy is chosen after consideration of the reason for the testing. The test method is chosen according to the parameters to be tested and the number and types of samples. For a large number of samples, automation reduces costs. The following steps are basic to testing:

Review of the purpose of testing.
Selection of the test method.
Selection of the samples.
Implementation of the test procedure.
Production of the test report.
Review of the results in light of the objectives.

BIBLIOGRAPHY

Annual Book of ASTM Standards, American Society for Testing and Materials, Philadelphia, Pa., updated regularly.

ANSI methods, American National Standards Institute, New York, updated regularly.

J. M. Juran, ed., *Quality Control Handbook,* 3rd ed., McGraw-Hill Inc., New York, 1974, sect. 24.

Title 16, Code of Federal Regulations: Commercial Practices, Part 1000–END, Title 21, Code of Federal Regulations: Food and Drugs, Parts 0–1299. Office of the Federal Register, National Archives & Records Service, General Services Administration, U.S. Government Printing Office, Washington, D.C., updated regularly.

G. NEIL SPOKES
Chemetrics, Inc.

TETRAFLUOROETHYLENE POLYMERS

POLYTETRAFLUOROETHYLENE, HOMOPOLYMER OF TETRAFLUOROETHYLENE

Polytetrafluoroethylene (PTFE) has a unique position in the plastics industry due to its chemical inertness, heat resistance, excellent electrical insulation properties, and the low coefficient of friction in a wide temperature range. Polymerization of tetrafluoroethylene (TFE) monomer gives this perfluorinated straight-chain high polymer with the formula $-(CF_2-CF_2)_n-$. The white to translucent solid polymer has an extremely high molecular weight, in the 10^6–10^7 range, and consequently has a viscosity of about 10 GPa·s (10^{10}–10^{12} P) at 380°C. It is a highly crystalline polymer and has a crystalline melting point. Its high thermal stability is due to the strong carbon–fluorine bond and characterizes PTFE as a very useful high temperature polymer.

Because of its chemical inertness and high molecular weight, PTFE does not flow and cannot be fabricated by conventional techniques. Therefore, an extensive processing technology had to be developed.

The suspension-polymerized PTFE polymer (referred to as granular PTFE) is usually fabricated by modified powder metallurgy techniques. Emulsion-polymerized PTFE behaves entirely differently from granular PTFE. Coagulated dispersions are processed by a cold extrusion process (like lead). Stabilized PTFE dispersions, made by emulsion polymerization, are usually processed according to latex processing techniques.

Manufacturers of PTFE include Daikin Kogyo (Polyflon), DuPont (Teflon), Hoechst (Hostaflon), ICI (Fluon), Ausimont (Algoflon and Halon), and the USSR (Fluoroplast). The People's Republic of China also manufactures some PTFE products.

Monomer

Preparation. The manufacture of tetrafluoroethylene involves the following steps:

$$CaF_2 + H_2SO_4 \longrightarrow CaSO_4 + 2\ HF$$

$$CH_4 + 3\ Cl_2 \longrightarrow CHCl_3 + 3\ HCl$$

$$CHCl_3 + 2\ HF \longrightarrow CHClF_2 + 2\ HCl$$

$$2\ CHClF_2 \longrightarrow CF_2{=}CF_2 + 2\ HCl$$

Properties and Reactions. Tetrafluoroethylene (mol wt 100.02) is a colorless, tasteless, odorless, nontoxic gas; physical properties are given in Table 1. It is stored as a liquid (vapor

Table 1. Physical Properties of Tetrafluoroethylene

Property	Value
boiling point at 101.3 kPa[a], °C	−76.3
freezing point, °C	−142.5
liquid density at t°C, g/mL $-100 < t < -40$	$= 1.202–0.0041\ t$
vapor pressure at T K, kPa[b] $196.85 < T < 273.15$	$\log_{10} P_{kPa} = 6.4593–875.14/T$
critical temperature, °C	33.3
critical pressure, MPa[c]	39.2
critical density, g/mL	0.58
dielectric constant at 28°C at 101.3 kPa[a]	1.0017
thermal conductivity at 30°C, mW/(m·K)	15.5

[a] To convert kPa to atm, multiply by 0.01.
[b] To convert kPa to psi, multiply by 0.145.
[c] To convert MPa to atm, divide by 0.101.

pressure at −20°C is 1 MPa or 9.9 atm). It is usually polymerized above its critical temperature of 33.3°C and below its critical pressure of 3.94 MPa (38.9 atm). The polymerization reaction is highly exothermic.

Tetrafluoroethylene undergoes additional reactions typical of an olefin.

Uses. Besides polymerizing TFE to various types of high PTFE homopolymer, TFE is utilized as a comonomer in a large number of other polymers.

Manufacture of PTFE

Engineering problems involved in the production of TFE seem simple compared to those associated with polymerization and processing of PTFE resins. The monomer must be polymerized to an extremely high molecular weight in order to achieve the desired properties. The low molecular weight polymer does not have the strength needed in end-use applications.

Polytetrafluoroethylene is manufactured and sold in three forms: granular, fine powder, and aqueous dispersion; each requires a different fabrication technique. Granular resins are manufactured in a wide variety of grades to obtain a different balance between powder flows and end-use properties. Fine powders that are made by coagulating aqueous dispersions also are available in various grades. Differences in fine-powder grades correspond to their usefulness in specific applications and to the ease of fabrication. Aqueous dispersions are sold in latex form and are available in different grades. A variety of formulation techniques are used to tailor these dispersions for specific applications.

Properties

The two types of valence force binding the PTFE structure are the carbon–carbon bonds, which form the backbone of the polymer chain, and the carbon–fluorine bonds. These chemical bonds are extremely strong and are the key contributors in imparting an outstanding combination of properties to PTFE. The fluorine atoms form a protective sheath over the chain of carbon atoms. If the atoms attached to the carbon-chain backbone were smaller or larger than fluorine, the sheath would not form a regular, uniform cover. This sheath shields the carbon chain from attack and confers chemical inertness and stability. It also reduces the surface energy resulting in a low coefficient of friction and nonstick properties.

Transitions. Effects of structural changes on properties, eg, specific heat, specific volume, and dynamic mechanical and electrical properties, are observed at various temperatures. Transitions observed by various investigators, their interpretation, and modes of identification are shown in Table 2.

Mechanical Properties. The mechanical properties of PTFE depend on processing variables, eg, preforming pressure, sintering temperature and time, cooling rate, void content, and crystallinity. Properties, such as the coefficient of friction, flexibility at low temperatures, and stability at high temperatures, are relatively independent of fabrication. Molding and sintering conditions affect flex life, permeability, stiffness, resiliency, and impact strength. The physical properties of PTFE have been reviewed and compiled in several publications (Table 3).

Filled Resins. Filled compositions meet the requirements of an increased variety of mechanical, electrical, and chemical applications. Physical properties of filled granular compounds are shown in Table 4.

Chemical Properties. Vacuum thermal degradation of PTFE results in monomer formation. The degradation is a first-order reaction.

Radiation Effects. Polytetrafluoroethylene is attacked by radiation. In the absence of oxygen, stable secondary radicals are produced. An increase in stiffness in material irradiated in vacuum indicates cross-linking. Degradation is due to random scission of the chain; the relative stability of the radicals in vacuum protects the materials from rapid deterioration. Reactions take place in air or oxygen and accelerated scission and rapid degradation occur.

Absorption and Permeation. Polytetrafluoroethylene is chemically inert to industrial chemicals and solvents even at elevated temperatures and pressures. Gases and vapors diffuse through PTFE more slowly than through most other polymers. The higher the crystallinity, and the less space between polymer molecules, the slower the permeation.

Table 2. Transitions in Polytetrafluoroethylene

Temperature, °C	Region affected	Technique
1st order		
19	crystalline, angular displacement causing disorder	thermal methods, x ray, nmr
30	crystalline, crystal disordering	thermal methods, x ray, nmr
90 (80 to 110)	crystalline	stress relaxation, Young's modulus, dynamic methods
2nd order		
−90 (−110 to −73)	amorphous, onset of rotational motion around C—C bond	thermal methods, dynamic methods
−30 (−40 to −15)	amorphous	stress relaxation, thermal expansion, dynamic methods
130 (120 to 140)	amorphous	stress relaxation, Young's modulus, dynamic methods

Table 3. Typical Mechanical Properties of Molded and Sintered PTFE Resins

Property	Granular resin	Fine powder	ASTM method
tensile strength at 23°C, MPa[a]	7–28	17.5–24.5	D638-61T
elongation at 23°C, %	100–200	300–600	D628-61T
flexural strength at 23°C, MPa[a]	does not break		D790-61
flexural modulus at 23°C, MPa[a]	350–630	280–630	D747-61T
impact strength, J/m[b], at 21°C	106.7		D256-56
24°C	160		
77°C	>320		
hardness durometer, D	50–65	50–65	D1706-59T
compression stress, MPa[a]			
at 1% deformation at 23°C	4.2		D695-52T
at 1% offset at 23°C	7.0		D695-52T

[a] To convert MPa to psi, multiply by 145.
[b] To convert J/m to ft·lbf/in., divide by 53.38.

Table 4. Properties of Filled PTFE Compounds

Property	Unfilled	Glass fiber, wt % 15	Glass fiber, wt % 25	Graphite, 15 wt %	Bronze, 60 wt %
specific gravity	2.18	2.21	2.24	2.16	3.74
tensile strength, MPa[a]	28	25	17.5	21	14
elongation, %	350	300	250	250	150
stress at 10% elongation, MPa[a]	11	8.5	8.5	11	14
thermal conductivity, mW/(m·K)	0.244	0.37	0.45	0.45	0.46
creep modulus, kN/m[b]	2	2.21	2.1	3.4	6.2
hardness, Shore durometer, D	51	54	57	61	70

[a] To convert MPa to psi, multiply by 145.
[b] To convert kN/m to lbf/in., divide by 0.175.

Electrical Properties. Polytetrafluoroethylene is an excellent electrical insulator because of its mechanical strength and chemical and thermal stability as well as excellent electrical properties (Table 5).

Fabrication

Granular Resins. These resins are sold in different forms; an optimum balance between handleability and product properties is desired. A free-flowing resin is used in small and automatic moldings. A finely divided resin is more difficult to handle but it distributes evenly in large moldings and has superior properties in sintered articles; it is used for large billet- and sheet-molding operations. A presintered resin with low crystallinity and superior handleability is highly suitable for ram extrusion.

Virgin PTFE melts at about 342°C; viscosity, even at 380°C, is 10 GPa·s (10^{11} P). This eliminates processing by normal thermoplastic techniques, and other fabrication techniques had to be developed; the dry powder is compressed into a handleable form by heating above the melting point. This coalesces the particles into a strong homogeneous structure; cooling at a controlled rate achieves the desired degree of crystallinity.

Fine-powder Resins. Fine-powder resins are extremely sensitive to shear. They must be handled gently to avoid shear, which prevents processing. However, fine powder is suitable for the manufacture of tubing and wire insulation for which compression molding is not suitable. A paste-extrusion process may be applied to the fabrication of tubes with diameters from fractions of a millimeter to about a meter, walls from thicknesses of 100 to 400 μm, thin rods with up to 50-mm diameters, and cable sheathing. Calendering unsintered extruded solid rods produces thread-sealant tape and gaskets.

Dispersion Resins. Polytetrafluoroethylene dispersions in aqueous media contain 30–60 wt % polymer particles and some

Table 5. Electrical Properties of Polytetrafluoroethylene

Property	Granular	Fine powder	ASTM method
dielectric strength, short time, 2-mm thickness, V/mm	23,600	23,600	D149-55T
surface arc-resistance, s	>300	>300	D495-55T
volume resistivity, Ω·cm	$>10^{18}$	$>10^{18}$	D257-57T
surface resistivity at 100% rh, Ω/sq	$>10^{16}$		D257-57T
dielectric constant, at 60 to 2×10^9 Hz	2.1	2.1	D150-59T
dissipation factor, at 60 to 2×10^9 Hz	0.0003		D150-59T

surfactant. The type of surfactant and the particle characteristics depend on the application. These dispersions are applied to various substrates by spraying, flow coating, dipping, coagulation, or electrodeposition. Aqueous dispersions are also used for spinning PTFE fibers.

Effects of Fabrication on Physical Properties of Molded Parts. The physical properties are affected by molecular weight, void content, and crystallinity. Molecular weight can be reduced by degradation but not increased during processing. These factors can be controlled during molding by the choice of resin and fabricating conditions. Void distribution (or size and orientation) also affects properties; however, it is not easily measured.

Testing and Standards

A description of PTFE resins and their classification is given in ASTM D1457–83.

Economic Aspects

Polytetrafluoroethylene homopolymers are more expensive than other thermoplastics because of high monomer refining costs. For extremely high molecular weight, ingredients and manufacturing process must be free of impurities, which increases costs. In the United States, the polymer sold between 8.8 and 22 $/kg in 1988, depending on the resin type.

Health and Safety

Exposure to PTFE can arise from ingestion, skin contact, or inhalation. The polymer has no irritating effect to the skin, and test animals fed with the sintered polymer have not shown adverse reactions.

Applications

Consumption of PTFE increases continuously as new applications are being developed. Electrical applications consume half of the PTFE produced; mechanical and chemical applications share equally the other half.

Micropowders. The PTFE micropowders, also called waxes, are tetrafluoroethylene homopolymers with molecular weights significantly lower than that of normal PTFE.

These micropowders are commonly used as additives in a wide variety of applications, where, for example, chemical resistance, weathering resistance, and a wide temperature application range may be required. The low friction properties of micropowders are an advantage in many applications.

BIBLIOGRAPHY

S. V. Gangal and R. L. Johnson, "Fluorine Compounds, Organic," in M. Grayson, ed., *Kirk-Othmer Encyclopedia of Chemical Technology,* Vol. 11, Wiley-Interscience, New York, 1980, pp. 1–49.

C. A. Sperati in J. Brandrup and E. H. Immergut, eds., *Polymer Handbook,* 3rd ed., John Wiley & Sons, Inc., New York, 1989, pp. V35–V44.

Mechanical Design Data, Teflon Fluorocarbon Resins, bulletin, E. I. du Pont de Nemours & Co., Inc., Wilmington, Del., 1964.

Teflon Occupational Health Bull. **17**(2) (1962) (published by Information Service Division, Dept. of National Health and Welfare, Ottawa, Can.).

SUBHASH V. GANGAL
E. I. du Pont de Nemours & Co., Inc.

TETRAFLUOROETHYLENE–HEXAFLUOROPROPYLENE COPOLYMERS

Perfluorinated ethylene–propylene (FEP) resin is a copolymer of tetrafluoroethylene (TFE) and hexafluoropropylene (HFP). Its branched structure contains units of $—CF_2—CF_2—$ and units of $—CF_2—CF(CF_3)—$. It retains most of the desirable characteristics of polytetrafluoroethylene (PTFE) but with a melt viscosity low enough for conventional melt processing. The introduction of hexafluoropropylene lowers the melting point of PTFE from 325°C to about 260°C.

As a true thermoplastic it can be melt processed by extrusion and compression, injection, and blow molding. Films can be heat bonded and sealed, vacuum formed, and laminated to various substrates. Chemical inertness and corrosion resistance make FEP highly suitable for chemical services; its dielectric and insulating properties favor it for electrical and electronic service; and its low frictional properties, mechanical toughness, thermal stability, and nonstick quality make it highly suitable for bearings and seals, high temperature components, and nonadhesive surfaces.

Mechanical properties are retained up to 200°C, even in continuous service, which is better than with most plastics. At high temperatures, these copolymers react with fluorine, fluorinating agents, and molten alkali metals. They are commercially available under the DuPont trademark Teflon FEP fluorocarbon resin. A similar product is manufactured by Daikin Kogyo of Japan and sold under the trademark Neoflon. The People's Republic of China also manufactures some FEP products.

Monomers

Preparation. The preparation, properties, and uses of tetrafluoroethylene have been described previously. Hexafluoropropylene (HFP) was initially prepared by pyrolysis of PTFE and by fluorination of 1,2,3-trichloropropane followed by dehalogenation. A number of other routes are described in the patent literature.

Properties and Reactions. The properties of HFP are shown in Table 1.

Copolymers

Hexafluoropropylene and tetrafluoroethylene are copolymerized with trichloroacetyl peroxide as the catalyst, at low temperature. Newer catalytic methods, including irradiation, achieve copolymerization at different temperatures. Aqueous and nonaqueous dispersion polymerizations appear to be the most convenient routes to commercial production. The polym-

Table 1. Properties of Hexafluoropropylene

Property	Value
molecular weight	150.021
boiling point at 101 kPa, °C	−29.4
freezing point, °C	−156.2
critical temperature, °C	85
critical pressure, kPa[a]	3254
critical density, g/cm³	0.60
vapor pressure at K, kPa[a]	
243.75 < T < 358.15	log P(kPa) = 6.6938 − 1139.156/T

[a] To convert kPa to mm Hg, multiply by 7.5.

erization conditions are similar to those of TFE homopolymer dispersion polymerization. The copolymer of HFP–TFE is a random copolymer, ie, HFP units add to the growing chains at random intervals. The optimal composition of the copolymer requires that the mechanical properties are retained in the usable range and that the melt viscosity (MV) is low enough for easy melt processing.

The extrusion grade is suitable for tubing, wire coating, and cable jacketing. It is less suitable for injection molding than the low MV resin because of its relatively high melt viscosity. The high MV resin is used as liners for process equipment. Product properties are given in Table 2.

Properties. The crystallinity of FEP polymer is significantly lower than that of PTFE (70 vs 98%). The structure resembles that of PTFE, except for a random replacement of a fluorine atom by a perfluoromethyl group (CF_3). The crystallinity after processing depends on the rate of cooling the molten polymer. The presence of HFP in the polymer chain tends to distort the highly crystallized structure of the PTFE chain and results in a higher amorphous fraction.

Transitions and Relaxations. Only one first-order transition is observed, the melting point. The relaxation temperature appears to increase with increasing HFP content. Relaxation involves 5–13 of the chain carbon atoms. Besides α and γ relaxations, one other dielectric relaxation was observed below −150°C, which did not vary in temperature or in magnitude with comonomer content or copolymer density.

Thermal Stability. The polymer is thermally stable and can be processed at ca 270°C.

Radiation Effects. The primary effect of radiation is the degradation of large molecules to small molecules. Molecular weight reduction can be minimized by excluding oxygen. If FEP is lightly irradiated at elevated temperatures in the absence of oxygen, cross-linking offsets molecular breakdown.

Mechanical Properties. Most of the important properties of FEP are similar to those of PTFE; the main difference is the lower continuous service temperature of 204°C of FEP compared to that of 260°C of PTFE. The flexibility at low temperatures and the low coefficients of friction and stability at high temperatures are relatively independent of fabrication conditions. Unlike PTFE, FEP resins do not exhibit a marked change in volume at room temperature, because they do not have a first-order transition at 19°C. They are useful above −267°C and are highly flexible above −79°C.

Optical Properties. Teflon FEP fluorocarbon film transmits more ultraviolet, visible light, and infrared radiation than ordinary window glass.

Economic Aspects. Because of the high cost of hexafluoropropylene, FEP is more expensive than PTFE. In the United States in 1987, FEP sold at prices up to $26.4/kg, depending on the type and quantity. Most grades are marketed in a colorless, translucent, extruded pellet form. In the last 10 years FEP sales have increased rapidly, but as other polymers are used for plenum wire coating, the growth rate is expected to slow down.

Health and Safety. The safety precautions required in handling TFE–HFP copolymers are the same as those applied to handle PTFE.

Table 2. Properties of Teflon FEP Fluorocarbon Resin[a]

Mechanical property	ASTM method	Teflon 100	Teflon 160
melt flow number, g/10 min	D2116	7.0	1.5
specific gravity	D792	2.13–2.17	2.13–2.17
tensile strength at 23°C, MPa	D1708	23	31
elongation, 23°C, %	D1708	325	305
compressive strength, MPa	D695	21	23
flexural strength at 23°C, MPa	D790	18	18
flexural modulus at 23°C, MPa	D790	620	586
hardness durometer, Shore D	D2240	56	57
coefficient of friction, metal/film	D1894	0.27	0.235
deformation under load at 23°C, 6.9 MPa, 23 h, %	D621	0.5	0.5
water absorption, 24 h, %	D570	0.004	0.004
linear coefficient of expansion per °C × 10^{-5}	E831		
0–100°C		13.5	7.6
100–150°C		20.8	11.5
150–200°C		26.6	14.2

[a] Compression-molded specimens; property data on extruded wire specimens are similar.

BIBLIOGRAPHY

Teflon–FEP Fluorocarbon Resin, Techniques for Injection Molding, information bulletin 95d, E. I. du Pont de Nemours & Co., Inc., Wilmington, Del., 1969.

Safe Handling Guide, Teflon Fluorocarbon Resins, DuPont Materials for Wire and Cable, bulletin E-85433, E. I. du Pont de Nemours & Co., Inc., Wilmington, Del., 1986.

Teflon Fluorocarbon Resins, Mechanical Design Data, 2nd ed., E. I. du Pont de Nemours & Co., Inc., Wilmington, Del., 1965.

SUBHASH V. GANGAL
E. I. du Pont de Nemours & Co., Inc.

TETRAFLUOROETHYLENE–PERFLUOROVINYL ETHER COPOLYMERS

Perfluoroalkoxy (PFA) fluorocarbon resins are designed to meet industry's needs in chemical, electrical, and mechanical applications. These melt processible copolymers contain a fluorocarbon backbone in the main chain and randomly distributed perfluorinated ether side chains:

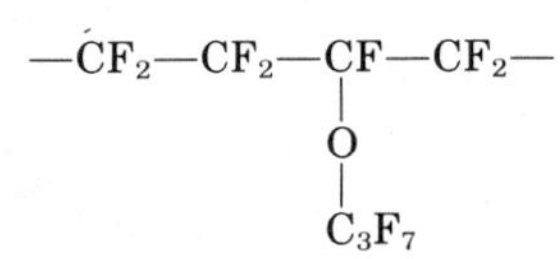

A combination of excellent chemical and mechanical properties at elevated temperatures results in reliable, high performance service to the chemical processing and related industries.

Chemical inertness, heat resistance, toughness and flexibility, stress-crack resistance, excellent flex life, antistick characteristics, little moisture absorption, nonflammability, and exceptional dielectric properties are among the characteristics of these resins.

Tetrafluoroethylene and perfluorovinyl ether are copolymerized in aqueous or nonaqueous media. The polymer is separated and converted into various forms, such as extruded cubes, powders, beads, or dispersions. This family of products is manufactured by Du Pont, Daikin, and Hoechst and sold under the trade names of Teflon PFA, Neoflon AP, and Hostaflon TFA, respectively.

Monomers

Preparation. The preparation of tetrafluoroethylene has been described previously. Perfluorovinyl ethers are prepared by the following steps:

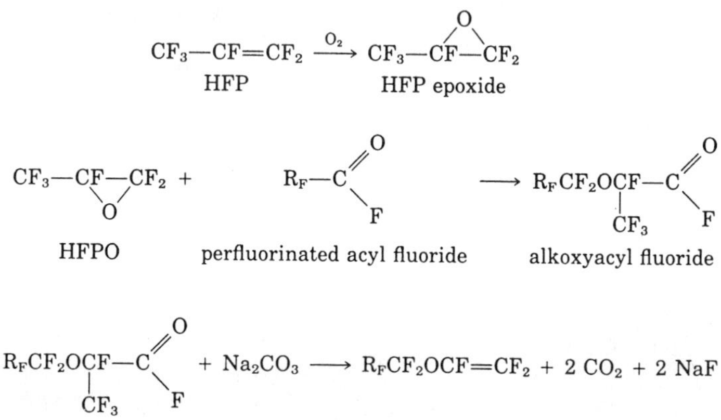

Properties. Properties of perfluoropropyl vinyl ether (PPVE), a colorless, odorless liquid (mol wt 266), are shown in Table 1. Perfluoropropyl vinyl ether is an extremely flammable liquid and burns with a colorless flame.

Copolymers

Properties. The melting point of commercial Teflon PFA is 305°C, ie, between those of PTFE and FEP. Second-order transitions are at −100, −30, and 90, as determined by a torsion pendulum. The crystallinity of the virgin resin is 65–75%. Specific gravity and crystallinity increase as the cooling rate is reduced. An ice-quenched sample with 48% crystallinity has a specific gravity of 2.123, whereas the press-cooled sample has a crystallinity of 58% and a specific gravity of 2.157.

Table 1. Properties of Perfluoropropyl Vinyl Ether, $F_3C—CF_2—CF_2—O—CF{=}CF_2$

Property	Value
critical temperature, K	423.58
critical pressure, MPa[a]	1.9
critical volume, $cm^3/(g{\cdot}mol)$	435
surface tension, mN/m (= dyn/cm)	9.9
boiling point, °C	36
specific gravity at 23°C,	1.53
vapor density at 75°C, g/cm^3	0.2
vapor pressure at 25°C, kPa[b]	70.3
solubility in water	0
odor	none
color	colorless
flash point, °C	−20
flammable limits in air[c], % by vol.	1

[a] To convert MPa to atm, divide by 0.1013.
[b] To convert kPa to psi (psia), multiply by 0.145.
[c] Extremely flammable.

Mechanical Properties. At 20–25°C the mechanical properties of PFA, FEP, and PTFE are very similar; differences between PFA and FEP become significant as the temperature is increased. The latter should not be used above 200°C, whereas PFA can be used up to 260°C. Tests at liquid nitrogen temperature indicate that PFA performs well in cryogenic applications.

Fabrication. Teflon PFA resins are fabricated by the conventional melt-processing techniques used for thermoplastics (Table 2).

Health and Safety

Safe practices employed for handling PTFE and FEP resins are adequate for Teflon PFA; adequate ventilation is required for processing above 330–355°C.

Applications and Economic Aspects

The perfluorovinyl ether comonomer used for PFA is very expensive, and so is PFA. Most PFA grades are sold as extruded, translucent cubes in various colors at 39.6–48.4 $/kg. Some PFA types are also marketed in nonextruded forms.

BIBLIOGRAPHY

U.S. Pat. 3,536,733 (Oct. 27, 1970), D. P. Carlson (to E. I. du Pont de Nemours & Co., Inc.).

Table 2. Teflon PFA Products and Their Applications

Product	Form	Melt flow[a], g/10 min	Application
PFA 310	extruded cubes	6	wire coating
PFA 340	extruded cubes	14	injection molding
PFA 350	extruded cubes	2	transfer molding
TE-9724	powder	14	spray coating
TE-9725	powder	2	spray coating
TE-9738	beads	6	rotomolding
PFA 335	dispersion	2	coating and cocoagulation

[a] At 372°C and 5-kg load.

Handbook of Properties for Teflon PFA, E. I. du Pont de Nemours & Co., Sales Brochure, E46679, Oct. 1987.

Handling and Use of Teflon Fluorocarbon Resins at High Temperatures, Bulletin, E. I. du Pont de Nemours & Co., Inc., Wilmington, Del., 1961.

SUBHASH V. GANGAL
E. I. du Pont de Nemours & Co., Inc.

TETRAFLUOROETHYLENE–ETHYLENE COPOLYMERS

Modified ethylene–tetrafluoroethylene copolymer has good tensile strength, moderate stiffness, high flex life, and outstanding impact strength, abrasion resistance, and cut-through resistance. Electrical properties include low dielectric constant, high dielectric strength, excellent resistivity, and low dissipation factor. Thermal and cryogenic performance and chemical resistance are good. These properties combined with elasticity make this material an ideal candidate for heat-shrinkable film and tubing. This family of copolymers can be processed by conventional methods such as melt extrusion, injection molding, transfer molding, and rotational molding.

Ethylene and tetrafluoroethylene are copolymerized in aqueous, nonaqueous, or mixed medium with free-radical initiators. The polymer is isolated and converted into extruded cubes, powders, and beads, or a dispersion. This family of products is manufactured by DuPont, Hoechst, Daikin, Asahi Glass, and Ausimont and sold under the trade names of Tefzel, Hostaflon ET, Neoflon EP, Aflon COP, and Halon ET, respectively.

Monomers

Preparation. The preparation of tetrafluoroethylene has been described previously. Tetrafluoroethylene of purity suitable for granular or dispersion polymerizations is acceptable for copolymerization with ethylene. Polymerization-grade ethylene is suitable for copolymerization with tetrafluoroethylene (see ETHYLENE POLYMERS). Modifying termonomers are incorporated by free-radical polymerization.

Manufacture

Tetrafluoroethylene–ethylene copolymers have tensile strengths two to three times as high as the tensile strength of polytetrafluoroethylene or of the ethylene homopolymer. Because these copolymers are highly crystalline and fragile at a high temperature, they are modified with a third monomer, usually a vinyl monomer free of telegenic activity. The termonomer

Table 1. Typical Properties of Tefzel[a]

	ASTM method	Tefzel 200, 280	Tefzel[b] HT-2004
ultimate tensile strength, MPa[c]	D638	44.8	82.7
ultimate elongation	D887-64T	200[d]	8
compressive strength, MPa[c]	D695	48.9	68.9
shear strength, MPa[c]		41.3	44.8
heat deflection temp, °C	D648		
at 0.45 MPa		104	265
at 1.8 MPa		74	210
low temp embrittlement	D746	below −100°C	
tensile modulus, MPa[c]	D638	827	8270
flexural modulus, MPa[c]	D790	96.5	6550
impact strength notched Izod	D256		
at −54°C, J/m[e]		>1067	373
at 23°C		no break	485
deformation under load, 13.7 MPa at 50°C, %	D621	4.11	0.68
coefficient of linear expansion per °C × 10^{-5}	D696-70		
20–30°C		9	3
50–90°C		9.3	1.7
140–180°C		14	3.2
specific gravity	D792	1.70	1.86
refractive index, n_D		1.4028	
melting point, dta peak, °C		270	270
water absorption at saturation, %	D570	0.029	0.022
hardness			
Rockwell	D785	R50	R74
Durometer D		D75	
coefficient of friction[f]			
dynamic, 689 kPa (at >3 m/min)		0.4	0.3
static, 689 kPa			0.3

[a] At 23°C and 50% rh, unless otherwise specified.
[b] Reinforced with 25 wt % glass fiber.
[c] To convert MPa to psi, multiply by 145.
[d] Elongations between 100 and 300% are achieved with varying methods of sample fabrication.
[e] To convert J/m to ftlbf/in., divide by 53.38.
[f] Mating material AISI 1018 Steel, Rc20, 16AA.

provides the copolymer with side chains of at least two carbon atoms, such as perfluoroalkylvinyl or vinylidene compounds, perfluoroalkyl ethylenes, and perfluoroalkoxy vinyl compound. For high tensile properties and cut-through resistance, a molar ratio of ethylene and tetrafluoroethylene between 60:40 and 40:60 is required.

Copolymerization is effected by suspension or emulsion techniques under such conditions that tetrafluoroethylene, but not ethylene, may homopolymerize.

Properties

The equimolar copolymer of ethylene and tetrafluoroethylene is isomeric with poly(vinylidene fluoride) but has a higher melting point and a lower dielectric loss. A copolymer with the degree of alternation of about 0.88 was used to study the structure. Its unit cell was determined by x-ray diffraction. Despite irregularities in the chain structure and low crystallinity, a unit cell and structure were derived that gave a calculated crystalline density of 1.9 g/cm^3. The unit cell is believed to be orthorhombic or monoclinic (a = 0.96 nm, b = 0.925 nm, c = 0.50 nm; γ = 96°).

Transitions. Samples containing 50 mol % tetrafluoroethylene with ca 92% alternation were quenched in ice water or cooled slowly from the melt to minimize or maximize crystallinity, respectively. Internal motions were studied by dynamic mechanical and dielectric measurements, and by nuclear magnetic resonance. The dynamic mechanical behavior showed that the α relaxation occurs at 110°C in the quenched sample; in the slowly cooled sample it is shifted to 135°C. The β relaxation appears near −25°C. The γ relaxation at −120°C in the quenched sample is reduced in peak height in the slowly cooled sample and shifted to a slightly higher temperature.

Physical and Mechanical Properties. Modified ethylene–tetrafluoroethylene copolymer has a good combination of mechanical properties, including excellent cut-through and abrasion resistance, high flex life, and exceptional impact strength (Table 1).

Fabrication

Modified ethylene–tetrafluoroethylene copolymers are commercially available in a variety of physical forms (see Table 2) and can be fabricated by conventional thermoplastic techniques. Like other thermoplastics, they exhibit melt fracture above certain critical flow rates.

Health and Safety

Large quantities of Tefzel have been processed and used in many demanding service applications. No cases of permanent injury have been attributed to these resins, and only limited instances of temporary irritation to the upper respiratory tract.

Bulk quantities of Tefzel fluoropolymer resins should be stored away from flammable materials. In the event of fire, personnel entering the area should have full protection, including acid-resistant clothing and self-contained breathing apparatus with a fullface piece operated in the pressure-demand or other positive-pressure mode. All types of chemical extinguishers may be used to fight fire involving Tefzel resins. Large quantities of water may be used to cool and extinguish the fire.

Economic Aspects

Commercial ETFE resins are marketed in melt-extruded cubes that are sold in 20-kg bags or 150-kg drums. In the United States, the 1987 price was 26.4–33.0 $/kg, depending on volume and grade; color concentrates are also available. Some ETFE grades are available in powder form.

BIBLIOGRAPHY

Tefzel Fluoropolymer, Design Handbook, E. I. du Pont de Nemours & Co., Inc., Wilmington, Del., 1973.

Tefzel Fluoropolymers, Safe Handling Guide, Bulletin E-85785, E. I. du Pont de Nemours & Co., Inc., Wilmington, Del., May 1986.

Modified ETFE-Fluorocarbon Molding and Extrusion Materials, ASTM D3159-83, American Society for Testing and Materials, Philadelphia, 1987.

Subhash V. Gangal
E. I. du Pont de Nemours & Co., Inc.

PERFLUORINATED IONOMERS

The field of aqueous electrochemistry has been revolutionized by the introduction of perfluorinated ion-exchange membranes. In addition to the dramatic improvement of the commercial process for the production of chlorine and caustic, new opportunities have been opened for the reclamation of heavy metals from plant effluents, the regeneration of process streams in the plating and metal industry, and in hydrometallurgy. New types of batteries and fuel cells have been made possible, and research in neurochemistry has been stimulated by the availability of selective membrane electrodes. The resins containing sulfonic acid groups have been used as powerful catalysts.

Table 2. Forms of Modified ETFE Resins

Tefzel grade	Form	Melt flow,[a] g/10 min	Application
210	extruded cubes	45	injection molding, thin coating
200	extruded cubes	8	general purpose, insulation, tubing, fasteners
280	extruded cubes	3	chemical resistance, jacketing, heavy-wall logging cables
HT-2000	compacted powder	8	compounded products
HT-2010	compacted powder	3	compounded products, coating lining
HT-2010	compacted powder	45	coating

[a] At 297°C and 45 N (5 kgf) load.

Synthesis and Composition

Perfluorosulfonate Ionomers. These resins were first developed in the early 1960s by DuPont and are based on a copolymer of tetrafluoroethylene (TFE) and a perfluorinated vinyl ether (bp = 132°C) containing a terminal sulfonyl fluoride group.

$$CF_2{=}CFOCF_2CF(CF_3)OCF_2CF_2SO_2F$$

The resulting precursor polymer is readily melt-fabricable. After fabrication into the desired physical shape, the pendent sulfonyl fluoride groups are converted to sulfonate groups by reaction with a solution of sodium or potassium hydroxide. Further conversion to sulfonic acid groups or other ionic forms is possible by ion exchange. These products are available from DuPont in sheet form as Nafion perfluorinated membranes. More recently, Dow Chemical Company has developed a perfluorosulfonate ionomer of similar composition.

In attempting to synthesize perfluorosulfonate ionomers, special safety aspects must be considered. The hazards of TFE and sulfur trioxide are well known; in addition, the reaction of the two can be explosive. Many reaction by-products are highly toxic; their formation must be considered whenever fluoride ion catalyzed additions of hexafluoropropylene oxide are conducted.

Perfluorocarboxylate Ionomers. In the mid-1970s, perfluorocarboxylate ionomers were developed to meet the needs of the chloralkali industry for improved rejection of hydroxide ions.

In contrast to polystyrene-based ion-exchange resins, which use cross-linking to control properties, perfluorinated ionomers are not covalently cross-linked. As a result, perfluorinated ionomers, in the nonionic precursor form (—SO_2F or —COOR), are readily melt-fabricable. A limited amount of fabricability is even retained in the ionic form (particularly H^+ or organic cation form), permitting heat sealing.

Solubility. At temperatures above 230°C, perfluorosulfonate polymers form liquid compositions in polar solvents such as alcohol/water mixtures at autogeneous pressure. It is believed that these compositions contain micelles consisting of fluorocarbon cores surrounded by a layer of ionic groups. These liquid compositions can be used to coat electrodes or other substrates.

Properties. Some representative properties of unreinforced film (Nafion 117) and reinforced membrane (Nafion 417) are shown in Table 1.

Health and Safety

The toxicological properties of Nafion are similar to those of other perfluorinated polymers.

Table 1. Properties of Nafion

Property	N-117	N-417
resistance in 0.6 *N* KCl, Ω · cm²	1.9	4.1
thickness, mm	0.19	0.43
linear expansion on wetting, %		
MD[a]	16	9
TD[a]		12
thermal conductivity, W/(m·K)	0.18	
tensile strength, N/cm[b]		
50% rh	44	105
wet	26	105

[a] MD = machine direction; TD = transverse direction.
[b] To convert N/cm to ppi, multiply by 57.

Applications

For most applications as a separator in electrolytic cells, perfluorinated ionomers are supplied in sheet form, containing a reinforcing fabric made from a perfluorinated polymer. The price range is $520–800/m², depending on type and volume. Perfluorosulfonate ionomers in bead form are priced at $525–725/kg, depending on volume.

Chloralkali Electrolysis. By far the most important application for perfluorinated ionomers is as a membrane separator in chloralkali cells.

Other Electrochemical Applications. A separator between the anode and cathode compartments of an electrochemical cell prevents interference between the two electrode reactions and mixing of the reactants and products of one side with those of the other side. Perfluorinated ionomer membranes are ideally suited as separators because of their exceptional chemical stability, their good mechanical properties, and nonporous structure. Sulfonate copolymers are generally preferred because they can be used throughout the entire pH range.

Nonelectrolytic Applications. A gas dryer has been developed based on the fact that water can diffuse readily through perfluorinated sulfonate ionomers (Perma Pure Products, Inc., Toms River, N.J.). Extensive literature references are available on the use of perfluorinated sulfonate ionomers as strong acid catalysts.

BIBLIOGRAPHY

R. L. Dotson and K. E. Woodard in A. Eisenberg and H. L. Yeager, eds., *Perfluorinated Ionomer Membranes, ACS Symposium Series 180,* 1982.

W. G. Grot, "Nafion as a Separator in Electrolytic Cells," *paper presented at the Meeting of the Electrochemical Society,* Boston, Mass., May 7, 1986.

G. A. Olah, *Synthesis,* 513 (July 1986).

J. Electrochem. Soc., abstracts 899, 1308, 1414, and 1685 (Aug. 1987).

WALTHER GROT
DuPont Experimental Station

TETRAHYDROFURAN POLYMERS

Tetrahydrofuran (THF, tetramethylene oxide, diethylene oxide, 1,4-epoxybutane, tetrahydrofurane, oxolane) is a cyclic ether containing four methylene groups connected by an oxygen bridge. It has a slightly puckered conformation with some internal strain resulting from repulsion of the eclipsed hydrogen atoms. Normally the monomer is stable but under appropriate conditions it can be polymerized to form linear polytetrahydrofuran (PTHF, poly(tetramethylene oxide), PTMO, poly(tetramethylene ether) glycol, PTMEG, polytetrahydrofurane, poly(oxy-1,4-butanediyl)).

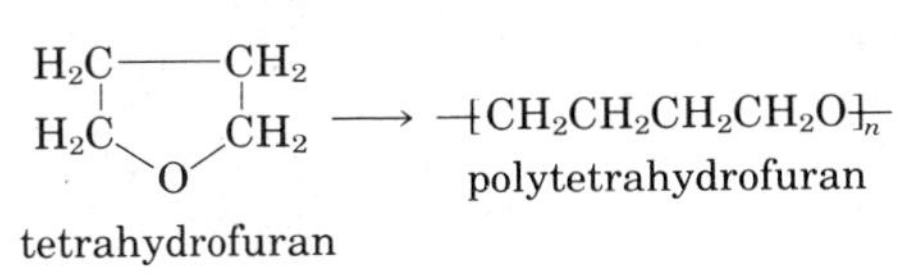

(1)

Tetrahydrofuran Monomer

Physical Properties. THF is an industrial solvent with the properties shown in Table 1. It is miscible with water and

Table 1. Physical Properties of Tetrahydrofuran

Property	Value
molecular weight	72.108
boiling point at 101.3 kPa[a], °C	66
freezing point, °C	−108.5
vapor pressure, kPa[a]	
0°C	7.52
50°C	61.2
liquid density, g/mL	
20°C	0.888
vapor density, relative (air = 1)	2.49
evaporation rate, relative (*n*-butyl acetate = 1)	8.0
viscosity, 20°C, mPa·s (= cP)	0.48
surface tension in air, 25°C, mN/m (= dyn/cm)	26.4
refractive index, n_D^{20}	1.4073
heat of combustion ($-\Delta_c^\circ$), at 25°C, kJ/g[b]	34.72
specific heat, liquid, 20°C, kJ/(kg·K)[c]	1.97
coefficient of thermal expansion 10–20°C, av/°C	0.00126
flash point (TCC), °C	−14.4
autoignition temperature, °C	321

[a] To convert kPa to mmHg, multiply by 7.50.
[b] To convert kJ/g to kcal/mol, multiply by 17.24.
[c] To convert kJ/(kg·K) to cal/(g·°C), divide by 4.184.

common organic solvents (esters; ketones; alcohols; diethyl ether; and aliphatic, aromatic, and chlorinated hydrocarbons). Because THF is a flammable liquid, a fire hazard exists and reasonable precautions and care in handling are necessary. THF can be stored and handled in ordinary steel tanks and piping. However, because THF is an excellent solvent, welded piping is better than screwed fittings and pump and valve packings must be tight and nonsoluble.

Chemical Properties. THF is a saturated five-membered cyclic ether, inert to chemicals such as lithium, sodium, or potassium, dilute mineral acids, Grignard reagents, oxidizing agents, and metal hydrides. Because it also has high solvent power for organomagnesium chlorides, alkyl and aryl alkali metal compounds, lithium, aluminum and boron hydrides, and diborane, it is used as a reaction solvent for Grignard reactions, reactions of alkali metals, anionic polymerization, metal hydride reactions, and hydroborations.

Purification. Because of the ease with which THF forms peroxides, it is normally shipped and stored in the presence of a stabilizer such as 2,6-di-*t*-butyl-4-methylphenol (butylated hydroxytoluene, BHT). In addition, THF readily absorbs water and oxygen. Because these impurities hinder polymerization, it is essential to purify the THF before it can be used as a polymerization monomer.

Manufacture. Tetrahydrofuran is manufactured predominantly by the acetylene–formaldehyde route: acetylene reacts with two moles of formaldehyde to give 1,4-butynediol, which is hydrogenated to 1,4-butanediol. THF is formed by subsequent cyclodehydration. DuPont, BASF, and GAF currently use this technology.

QO Chemicals, a division of Great Lakes Chemicals, formerly The Quaker Oats Co., still manufactures some THF from furfural, obtained from organic raw materials containing pentosanes, such as bagasse, corn cobs, oat hulls, etc. In the only other route used commercially, butadiene is converted catalytically to 1,4-diacetoxy-2-butene in the presence of acetic acid and oxygen. After hydrogenation and hydrolysis, the resulting 1,4-butanediol is cyclized to THF. Currently only Mitsubishi uses this method.

Health and Safety Factors. THF exhibits relatively low toxicity. Prolonged exposure to THF vapor may produce narcosis, nausea, dizziness, and headache. However, these symptoms disappear after moving to a fresh-air atmosphere.

Liquid THF defats the skin, which may lead to irritation. Hands can be protected by wearing solvent-resistant gloves made of poly(vinyl alcohol). THF may cause eye burns. Goggles will protect the eyes. If THF gets into the eyes, they should be flushed with water immediately. THF should not be taken internally. If THF is swallowed, vomiting should not be induced because of the danger of aspiration of the solvent into the lungs, where it can produce severe lung irritation.

Polymer Physical Properties

Bulk Properties. THF polymers are low melting semicrystalline materials which have the chemical formula $X(OCH_2CH_2CH_2CH_2)_nY$. X-ray studies have shown, and ir and Raman spectroscopic data have confirmed, that the PTHF molecule has a planar zigzag conformation. Crystal structure parameters are given in Table 2. Introduction of copolymeric units lowers the melting temperature and reduces the crystallinity.

The polymers can be prepared with any desired molecular weight ranging from oligomers to polymers with molecular weights in the millions. As the molecular weight increases the

Table 2. Typical Properties of Polytetrahydrofuran[a]

Property	Value
melting temperature T_m, °C	43
	58–60[b]
glass-transition temperature T_g, °C	−86
density	
amorphous at 25°C, g/cm³	0.975
crystalline at 25°C, g/cm³	1.07–1.08
tensile strength, MPa[c]	
high mol wt	29.0
cured	16.8–38.3[d]
elongation, %	
high mol wt	820
cured	400–740[d]
modulus of elasticity, MPa[c]	97.0
Shore A hardness	95
thermal expansion coefficient $\alpha = (1/V)(\delta V/\delta T)_p$, K^{-1}	$4–7 \times 10^{-4}$
coefficient of expansion (dV_s/dT), cm³/(g·K)	7.3×10^{-4}
refractive index at 20°C	1.48
dielectric constant (k_c) at 25°C	5.0
solubility parameter δ_p, $(J/cm^3)^{1/2}$	17.3–17.6
unit cell (oriented)	
monoclinic	$C_{2/c}–C_{2n}^6$
	$C_{2/c}$
a,nm	0.548–0.561
b,nm	0.873–0.892
c,nm	1.297–1.225
β, degrees	134.2–134.5

[a] For PTHF elastomers.
[b] The melting temperature most often reported for PTHF is about 43°C, although these higher values have been observed for elastomers, especially after annealing.
[c] To convert MPa to psi, multiply by 145.
[d] Varies with curing system.

polymers change from sticky, viscous oils at room temperature, first to waxes that flow easily on melting, and then to tough materials that retain their shape on melting. Table 2 lists typical physical properties of PTHF. PTHF glycols, the only PTHFs of commerce, are waxy solids when crystallized and colorless viscous fluids when melted. Like PTHF, the PTHF glycols normally contain butylated hydroxytoluene to prevent peroxide formation.

Solution Properties. The ether oxygen atoms of PTHF greatly increase the solubility of PTHF in polar solvents compared to hydrocarbon polymers. Solubility varies with molecular weight but there are solvents that will dissolve PTHFs of all molecular weights. Lower molecular weight PTHFs dissolve more readily in more solvents and form less viscous solutions. Solvents for nearly all molecular weights include benzene, toluene, chlorobenzene, nitrobenzene, methylene chloride, chloroform, carbon tetrachloride, nitromethane, THF, ethyl acetate, and other esters, and liquid sulfur dioxide. Ethyl ether, alcohols, acetone, and water are solvents for lower molecular weight PTHF, although PTHF of even low molecular weight precipitates in an ice–water slurry. Nonsolvents are aliphatic hydrocarbons, including pentane, petroleum ether, and hexane. The glycols are soluble in aromatic and chlorinated hydrocarbons, alcohols, esters, ketones, and nitroparaffins.

Chemical Properties

The backbone of PTHF consists of a series of linear aliphatic ether sequences. Like monomeric ethers it is subject to oxidation, forms hydroperoxides, and subsequently undergoes thermal degradation. The addition of common antioxidants, such as amines or pyrocatechol, inhibits these reactions and thereby imparts adequate stability to the polymer for commercial application. PTHF can be prepared with a wide variety of end groups depending on the choice of initiator and termination reagent. The nature of the end groups determines what, if any, further reactions can occur.

Polymerization

The THF molecule is a slightly puckered five-membered ring with a small amount of internal strain due to repulsion of eclipsed hydrogens. The ring contains an oxygen atom with two unshared pairs of electrons, and THF is, therefore, a nucleophilic monomer with little steric interference toward potential electron acceptors. The free energy of polymerization is only about −4.2 kJ/mol (−1 kcal/mol) at ambient temperatures. Consequently, a polymerization characterized by only modest rates and heats of polymerization is not surprising. Simply substituted tetrahydrofurans, in general, do not polymerize, although oligomers have been reported for monomethyl derivatives. This was predicted from thermodynamic considerations. Many bridged (bicyclic) derivatives polymerize readily. Purity of reagents, dryness of apparatus, choice of initiator, atmosphere, monomer concentration, and solvent greatly influence the course of the polymerizations.

Cationic ring-opening polymerization is the only mechanism available to tetrahydrofuran. The propagating species is a tertiary oxonium ion that is associated with a negatively charged counterion in order to achieve electrical neutrality.

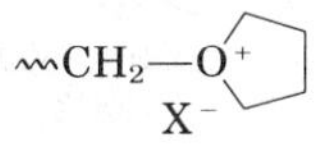

For continuing polymerization to occur the ion pair must be reasonably stable. In order to achieve reproducible and predictable results in the cationic polymerization of THF it is necessary to use pure, dry reagents, dry apparatus, and preferably high vacuum techniques. Careful work in an inert atmosphere, such as dry nitrogen, is satisfactory for some purposes.

The polymerization of THF is an equilibrium polymerization (see CATIONIC POLYMERIZATION; DEPOLYMERIZATION). It fits the equation that relates the enthalpy of polymerization, ΔH_p, and entropy of polymerization at 1 M, ΔS_p°, to the equilibrium monomer concentration $[M]_e$ as a function of the absolute temperature T

$$\ln\,[M]_e = \frac{\Delta H_p}{RT} - \frac{\Delta S_p^\circ}{R} \tag{2}$$

where R is the gas constant. Thus to obtain significant conversions to polymer, it is essential both to choose the polymerization temperature carefully and to limit the amount of solvent. The polymerization of bulk THF has a relatively low ceiling temperature, T_c, of 83 ± 2°C. At T_c or above no polymer is formed. At temperatures below T_c, there is an equilibrium monomer concentration below which no further polymerization occurs.

Initiation. The basic requirement for polymerization is that a THF tertiary oxonium ion must be formed by some mechanism. If a suitable counterion is present, polymerization follows. The requisite tertiary oxonium ion can be formed in any of several ways, eg, direct alkylation or acylation of the oxygen atom of THF by exchange or addition.

Propagation. The tertiary THF oxonium ion undergoes propagation by an S_N2 mechanism as a result of a bimolecular collision with THF monomer.

Termination. THF can be polymerized in the virtual absence of termination and transfer reactions. Under these conditions a living polymerization results and the number average molecular weight of the polymer produced can be calculated from the number of active sites introduced and the amount of polymer obtained (or produced at equilibrium). Any strong nucleophile can be used for chain termination.

Chain Transfer. A number of materials act as true transfer agents in THF polymerization. Notable examples are dialkyl ethers and orthoformates.

Kinetics. The details of the kinetics of the polymerization of THF have been reviewed. The principal conclusions are as follows:

Macroions are the principal propagating species in all systems.

With "stable" complex anions, such as PF_6^-, SbF_6^-, and AsF_6^-, the polymerization is living under normal polymerization conditions. When initiation is fast, kinetics of polymerizations in bulk can be closely approximated by the expression

$$k_p t = \left[\frac{1}{[I]_0}\right]\left[\ln\frac{[M]_0 - [M]_e}{[M]_t - [M]_e}\right] \tag{3}$$

k_p is the specific rate constant of propagation, t is time, I_0 is the initiator concentration at $t = 0$, and $[M]_0$, $[M]_e$, and $[M]_t$ are the monomer concentrations at $t = 0$, equilibrium, and t, respectively.

PTHF does not behave ideally in solution, and the equilibrium monomer concentration varies with both solvent and temperature, as stated above. Kinetics of THF polymerizations fit equation 3 provided the equilibrium monomer concentration is determined for the conditions used.

When initiation is not fast, the number of growing centers as a function of time must be determined in a separate step before the kinetic analysis can be carried out. Several different methods are available.

Copolymerization

Copolymers provide a way of reducing the melting temperature of THF polymers to room temperature and below. The crystallization tendency of THF segments in products is thereby overcome and properties are maintained. Copolymers also provide a way of introducing unsaturated units for vulcanization purposes or introducing active groups such as $—CH_2N_2$, $—NO_2$, and $—CH_2OCH_2C(NO_2)_2CH_3$ along the backbone, which convert polymeric binders into energetic polymers that can be used for the preparation of propellants and other explosives.

Many copolymerization studies have been made. A detailed discussion and critique of the results has been published and the breadth of the comonomers studied has been summarized. Among the comonomers used are oxiranes, oxetanes, 1,3-dioxolane, substituted tetrahydrofurans, 1,4-dioxane, ϵ-caprolactone, and cyclopentadiene. Random copolymerization with cyclic sulfides and most vinyl compounds does not occur. Some monomers that will not homopolymerize do copolymerize. In addition to chemical functionality, relative basicity, ring strain, and ring size seem to be important factors in determining the reactivity of different monomers in copolymerizations.

Block, Graft, and Star Copolymers. A host of copolymers of this type have been prepared in the last decade or so. They include block copolymers from ϵ-caprolactam and PTHF glycol, and block copolymers from PTHF and other cationically polymerizable heterocycles.

"One-," "two-," "three-," and "four-arm" stars have been prepared with PTHF arms. Graft copolymers with PTHF branches have been prepared from a variety of hydrocarbon backbones. Graft copolymers with poly(vinyl chloride) branches have been prepared from PTHF backbones. The properties determined for some of these copolymers have proven to be very attractive, but so far none has achieved commercial importance.

Commercial Processes

Of the many possible polymerization initiators for THF, only a very small number are actually used on a commercial scale. Such initiators have to meet many criteria, eg, formation of strictly bifunctional polymers, easily adjustable molecular weights of product, no side products or at least only low concentration of easily removeable side products, and low cost.

The most widely used polymerization system for THF is still based on fluorosulfonic acid, HSO_3F, a catalyst originally described by H. Meerwein. In this reaction, the primary polymerization products are poly(tetramethylene ether) chains with sulfate ester groups. In a subsequent step, the sulfates are hydrolyzed and the acid is removed by water extraction. After neutralization and drying, the product is filtered and loaded into drums or tank trucks under nitrogen.

Storage, Handling, and Transfer

PTMEG is susceptible to both thermal and oxidative degradation. It usually contains antioxidants such as BHT (2,6-di-*t*-butyl-4-hydroxy toluene) at concentrations between 0.05 and 0.1% to prevent oxidation under normal storage and handling conditions. Thermal decomposition in the absence of air starts at 210–220°C with formation of highly flammable THF. In the presence of acidic impurities the decomposition temperature can be significantly reduced, and contact with acids should therefore be avoided. Storage temperatures must be controlled, and only low pressure steam should be used as a heat source to avoid overheating and formation of THF.

Oxidative degradation occurs when PTMEG is heated in contact with air. Under these conditions decomposition can start at temperatures as low as 100°C, with evolution of THF, and formation of aldehydes and ketones. Degradation is particularly pronounced when PTMEG is in contact with high surface area materials, eg, pipe insulation. The heat of oxidative degradation is normally sufficient to sustain the degradation reaction, and contaminated fibrous insulation should immediately be cooled and discarded.

PTMEG is hygroscopic and should be stored in completely enclosed tanks under a dry nitrogen blanket. Mild steel tanks, tanks lined with phenolic resin, or stainless steel (type 304) tanks are usually satisfactory.

Economic Aspects

The bulk of commercially produced and used PTMEG is 1000 and 2000 molecular weight grade, with 650 and 2900 produced only on a small scale. Other molecular weight grades are sometimes manufactured as specialty products. Current prices range from $3.52/kg to $4.18/kg. DuPont is the world capacity leader, with the largest PTMEG plants both in the United States as well as in Europe. DuPont's capacity in 1986 was about 48,000 t, with about 30,000 tons in the United States and about 18,000 t in Europe. DuPont has a large internal demand for PTMEG; the rest is sold on the merchant market under the trade name Terathane.

Analytical and Test Methods

A number of test methods have been developed specifically for the analysis of polyether glycols, although in many cases general analytical procedures are applicable. One of the most important tests is the determination of the hydroxyl number, ie, the number of milligrams of KOH (formula weight: 56.1) equivalent to the hydroxyl content of 1 g of the diol sample (ASTM D4274). Other important tests are for water content (ASTM D4672), and for acid and alkalinity number (ASTM 4662), because they all influence the reaction with isocyanates.

Health and Safety Factors

PTMEG polyether glycols show very low oral toxicity in animal tests. Vapor pressures are very low, and no adverse effects of inhalation have been observed. The polymers are mild skin and eye irritants, and contact with skin, eyes, and clothing should be avoided. Tests with bacterial or mammalian cell cultures demonstrated no mutagenic activity.

Applications

Poly(tetramethylene ether) diols are mainly used as chain segments in polyurethanes and polyesters. Formation of polyurethanes involves reaction with diisocyanates, such as methylene diphenylene-4,4′-diisocyanate (MDI); formation of polyesters

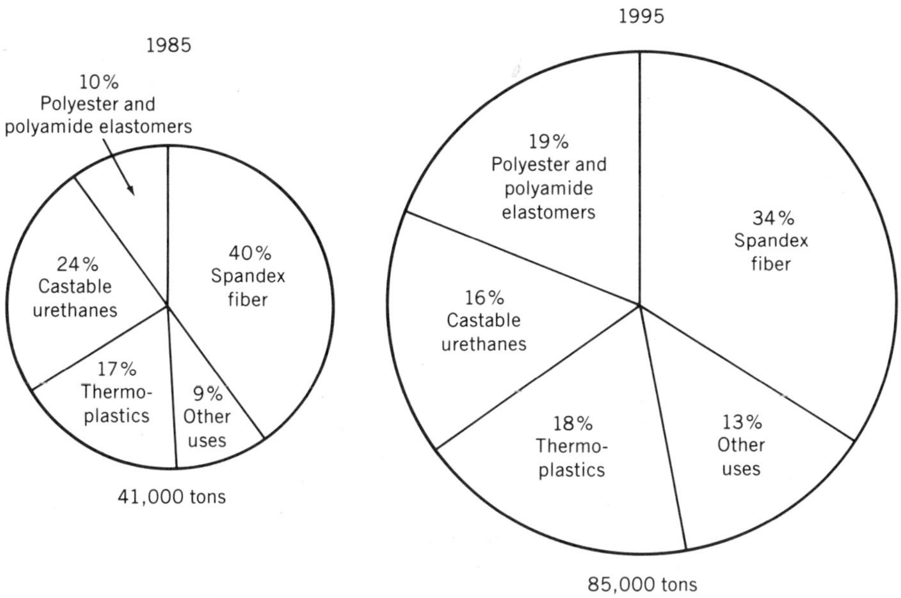

Figure 1. Markets for PTMEG.

involves reaction with diacids or diesters. The diols become "soft segments" in the resulting elastomeric materials.

The largest end-use areas are in spandex fibers for apparel, which in 1985 accounted for about 40% of the PTMEG market, followed by castable urethanes at about 24%, and thermoplastics at 17% (Fig. 1).

BIBLIOGRAPHY

P. Dreyfuss, *Poly(tetrahydrofuran)*, Gordon & Breach, New York, 1982.

S. Penczek, P. Kubisa, and K. Matyjaszewski, *Adv. Polym. Sci.* **68/69**, 1 (1985).

S. Penczek, P. Kubisa, and K. Matyjaszewski, *Adv. Polym. Sci.* **37,** 1 (1980).

P. Dreyfuss and M. P. Dreyfuss, *Adv. Polym. Sci.* **4,** 528 (1967).

P. Dreyfuss
M. P. Dreyfuss
Michigan Molecular Institute
Midland, Michigan

G. Pruckmayr
E. I. du Pont de Nemours & Co., Inc.
Wilmington, Delaware

TEXTILE RESINS

The finishing of textiles includes any process that improves the properties of a knitted, woven, or bonded fabric. The principal classes of textile fibers include cellulosics (rayon (qv), cotton (qv)), synthetics, proteins (wool, silk (qv)), glasses, and blends. Because of the great differences in the chemical and physical nature of these fibers, finishing processes and resins vary widely. Numerous mechanical, heat-related, and chemical treatments have been developed ever since the first attempts to convert fibers into yarns and fabric.

Thus, there are at least 10 commercially important textile fibers sold on the world market. Finishing processes are available to enhance crease retention, ease of care, dimensional stability, abrasion resistance, flame retardance, smolder resistance, odor control, soil release, control of moisture, static charges, stretch, water repellency, waterproofing, and resistance to weathering (including sunlight and microorganisms). A number of processes are designed to enhance the hand (or feel) and appearance of fabrics. Annual sales for the U.S. textiles market are estimated at more than $\$50 \times 10^9$ with about $\$330 \times 10^6$ spent on textile chemicals in the United States.

Some of the more important mechanical finishing processes are compressive shrinkage (Sanforized process) and calendered finishes, which include schreinerizing (produces a silklike appearance), chintz finishing (a glazed fabric with bright prints), and embossing. These finishes are semidurable but are made more durable, especially if the base fabric is cotton or rayon, by including a treatment with an *N*-methylol resin.

Mechanical finishing processes employ many types of rollers. Although trends in fashion have reduced the demand for calendered fabrics, these processes are still of considerable interest. Permanent effects with fabrics of thermoplastic material are usually obtainable, if the heat and pressure applied soften the material. Calendering operations improve the surface appearance of polyester and nylon fabrics and their blends. Processes dealing with carpets, nonwovens, and chemical modifications or additions that occur before the fiber is formed are not included here.

Treatments with Chemicals or Resins

Resin treatments are divided into topical and chemical treatments. Most chemical treatments of synthetic fibers are topical because of poor penetration and the inert nature of the substrate.

Textile Finishing. The conventional finishing process for cellulosic fabrics involves three steps, commonly referred to as pad–dry–cure. The fabric is first immersed in an aqueous finishing formulation and then squeezed to remove excess formulation and ensure even distribution of the reagents. The increase in weight (wet pickup) at this point is carefully controlled, but is usually 60–100%, depending on the process. Newer technology with smaller increases offers some advantages.

A finishing pad bath usually contains a cross-linking agent, a curing catalyst, and modifiers for "hand." For conventional durable-press processes of cellulosics and cellulosic blends (60–100% wet pickup), the bath contains about 5–9% cross-linking agent, sufficient acidic catalyst to affect a cure (0.4–3.5%), 1–3% of an emulsified polymer, which acts as fiber lubricant or stiffening agent, and a wetting agent (0.1%). A topical treatment for a synthetic fabric generally contains 1–3% emulsified polymers.

Textile finishing frequently refers primarily to chemical finishing (also called resin finishing). The chemicals used in the early finishing processes were usually aminoplasts capable of homopolymerization and of cross-linking. These chemicals were capable of improving the appearance and the dimensional stability of the fabric, and also promoted pigment retention. After some controversy, it was decided that cross-linking of cellulose, and not polymer formation within or about the cellulose molecule, was primarily responsible for the desired smooth drying properties.

Cross-linking of Cellulosic Fabrics. Reaction with any bifunctional compound, except those that cannot penetrate the fiber or where the reactive groups are widely separated, improves smooth drying of cotton cellulose fabrics. Fabric finished in this manner is said to possess a memory.

$$2\ \text{CellOH} + y\text{—}X\text{—}y \rightleftharpoons \text{CellO}X\text{OCell} + 2\ \text{Hy} \qquad (1)$$

Cross-linking cellulose may reduce breaking and tearing strength as well as abrasion resistance. In addition, moisture retention or regain is reduced. Moisture regain properties can be retained by using a special cross-linking agent or a swelling agent during cross-linking. The addition of poly(ethylene glycol) improves antistatic performance and thermal storage and release properties.

Cross-linking of cellulosic fibers also reduces accessibility. Therefore, fabrics are usually dyed before cross-linking by a heat-cure because the dye cannot penetrate the cross-linked fiber.

The loss of strength and abrasion resistance caused by cross-linking can be minimized, but with a reduction in smooth drying performance. This problem can be solved by blending with polyester.

Melamines and Other Amino-s-triazines. Commercial melamine finishes are usually mixtures. Specially prepared non-methylated dry methylolmelamines have been commercially available, but have not been widely used in finishing. Melamine resins have been used to shrink-proof wool, rot-proof cellulosics, as a binder for pigments, and for numerous other applications. Guanamines have been suggested for many of these uses. Their chemistry is similar, but hypochlorite yellowing is a problem (see AMINO RESINS).

Alkyleneureas, Triazones, and Urons. Many cyclic ureas are available and a few were the first commercial *N*-methylol products offered as cross-linking agents to eliminate the hypochlorite yellowing. They are still used to a limited extent worldwide. A slight excess of formaldehyde gives the fully methylolated compound. Systematic names for the cyclic ureas of the 1,3-bishydroxymethyl compounds and their common names are as follows: 2-imidazolidinone or ethyleneurea, tetrahydro-2-(1*H*)-pyrimidinone or propyleneurea, tetrahydro-1,3,5-triazin-2(1*H*)-one or triazone, and tetrahydro-4*H*-1,3,5-oxadiazin-4-one or uron. Mixtures of these cyclic ureas with urea–formaldehyde or melamine–formaldehyde are used in a number of resin formulations.

Of the alkyleneureas, DMEU (1,3-dimethylolethyleneurea) was widely used in the United States instead of urea– and melamine–formaldehyde; DMPU (1,3-dimethylolpropyleneurea) was used more in Europe. Most of these materials have been replaced with systems that give a finish with better resistance to acid hydrolysis.

The main problem with the triazones and some of the other agents is the low resistance to acid hydrolysis. Low acid resistance of a finish indicates high formaldehyde release. Triazones are no longer used in the United States, but may be used again in the future.

The use of urons in resin finishing is of interest because they provide an excellent cross-linking system based on inexpensive starting materials, ie, urea, formaldehyde, and methanol. Crude reaction mixtures and a fairly pure *N,N'*-bismethoxymethyluron were developed for commercial use. The crude product, which contains large quantities of methylated methylolureas, imparts smooth drying properties to cotton fabrics, but the finished fabric loses strength when bleached with hypochlorite and ironed. The pure product, which is readily obtained by vacuum distillation, is an effective cross-linking agent for cotton, giving a finished fabric resistant to chlorine through repeated washes.

Carbamates and DMDHEU. A glyoxal–formaldehyde–urea adduct called DMDHEU or 1,3-dimethylol-4,5-dihydroxyethyleneurea, is the principal cross-linking agent used in the United States, because of its low formaldehyde release and superior hydrolysis resistance. Another finish, based on *N,N*-dimethylol carbamates, also affords durable low formaldehyde treatment but is used less than DMDHEU.

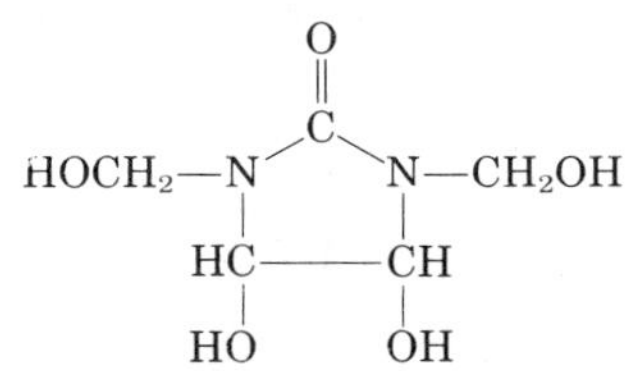

DMDHEU

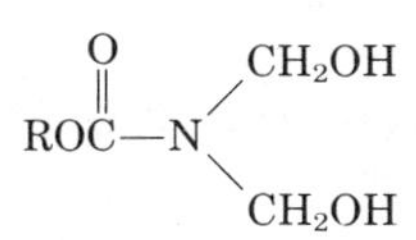

N,N-dimethylol carbamates

Unlike triazone and uron cross-linking agents, pure DMDHEU can be prepared commercially from glyoxal and urea–formaldehyde and no purification steps are necessary. However, commercial glyoxal is a mixture of products and its shelf life is limited.

Because protonation of pendent methylols or ring hydroxyls in DMDHEU leads to the formation of resonance-stabilized immonium–carbonium ions, the ring hydroxyls alone can be used to cross-link cellulose. 1,3-Dimethyl-4,5-dihydroxyethyleneurea (DMeDHEU) is used worldwide as a nonformaldehyde cross-linking agent for cellulose.

The chemistry of carbamate and other monoamide finishes differs significantly from finishing cellulosics with DMDHEU. In the presence of excess formaldehyde, the amide–aldehyde reaction approaches completion. However, in practice a finishing bath cannot tolerate a large excess of formaldehyde, and the *N*-methylol and *N,N*-dimethylol amides and unreacted formaldehyde remain. Little free amide remains, but it is expected to form by demethylolation during drying or curing.

Methyl, ethyl, isopropyl, isobutyl, hydroxyethyl, and methoxyethyl carbamates are used commercially; mixtures of alkyl carbamates are sold in large quantities. The type of carbamate in the resin formulation is frequently not specified. Little ethyl carbamate is used because of its toxicity.

The main advantage of carbamates is their low price. These agents cost less than DMDHEU. For sheeting fabrics, carbamates are preferred. They may be used in a two-stage curing process or a delayed cure. Carbamates reduce the lightfastness of some dyed fabrics; manufacturers and formulators should be consulted in this respect.

Carbamate finishing agents do not form a single, simple cross-link, as shown in eq. 1. Many different types of reaction can occur. Wrinkle recovery and the durable-press ratings are not as high as those obtained with the urea-based agents. In addition, the possibility of formaldehyde cross-links and the loss of strength has, for the most part, limited carbamates to cotton–polyester blends.

Other N-Methylol Agents. Cost is an important factor in the choice of the finishing system. The use of different dyes or finishing of white fabrics may require a change in a formulation. Other changes may be required because of toxicity.

Dimethylolpropyleneurea (DMPU) is used only to a limited extent in the United States. Other propyleneureas, such as the 5-hydroxy- and the 4-methoxy-5,5-dimethyl derivatives, have been developed, providing a wide selection of chemical reactivity. The characteristics of internally buffered *N*-methylol reagents and their reactions with cotton have been reviewed.

In addition to DMeDHEU, a number of other glyoxal-based cross-linking systems with little or no formaldehyde have been evaluated, as well as linear and cyclic compounds containing $=N-CH(OH)-CH(OH)-N=$ groups.

Chemistry of N-Methylol Cross-linking Systems. The urea–formaldehyde reactant system is usually a mixture of methylolated ureas with a ratio of formaldehyde to urea of 1.3–1.8. This ratio provides a stable pad bath and the reactant system is capable of cross-linking cellulose and polymerization. Pure urea–formaldehyde adducts have a low solubility in water. Catalysis is usually provided by metal salts or amine hydrochlorides.

Chloroamide formation can be reduced or eliminated by complete methylolation, which limits the number of NH groups in the finished fabric. The amide–aldehyde reaction is reversible and can approach completion in some cases in the presence of excess formaldehyde.

Glyoxal is also used to produce substituted *N*-methylol reactants capable of reaction with cellulose. The addition of amides to formaldehyde and glyoxal has been studied for imides, carboxylic acid amides, carbamates, and ureas. Substitution of all amido hydrogens available is limited by steric effects or by low amide acidity. Generally, all amido hydrogens are substituted if cyclization occurs during addition to the aldehyde or if a cyclic amide with nitrogen in the ring, rather than an *N*-alkylamide, acts as nucleophile.

A fully methylolated cross-linking agent does not ensure freedom of damage by hypochlorite bleaching. The curing step does not proceed to completion, even for the most effective cross-linking agent, and some NH groups (formed by loss of formaldehyde) and some $N-CH_2OH$ are found in the finish. Furthermore, hydrolysis of a link to cellulose, resulting in the easily hydrolyzable $N-CH_2OH$ group, occurs readily in some finishes.

$$-\overset{\overset{\displaystyle O}{\|}}{C}-\underset{|}{N}-CH_2OCell + H_2O \overset{H^+}{\rightleftharpoons} -\overset{\overset{\displaystyle O}{\|}}{C}-\underset{|}{N}-CH_2OH + CellOH \qquad (2)$$

Formation and Hydrolysis of Finishes. Formation and acid hydrolysis reactions pass through the same intermediate immonium–carbonium ion. Therefore, a highly reactive agent forms an easily hydrolyzable product.

Hydrolysis resistance of finishes is important during storage of cured fabrics and during the final pressing and creasing of the garment. Cotton and rayon fabrics regain moisture on storage; the amount depends on the blend and the finishing process. Because cured, flat fabrics or finished garments are rarely washed prior to sale to the consumer, residues from the acid curing catalyst remain on the fabric until it is laundered the first time.

Errors made in processing are often corrected by removal of the finish by acid hydrolysis or stripping under conditions that do not cause strength reduction.

Leaving Group Effect. The deactivating effect of the methoxy group is weaker than that of the hydroxyl group. This provides a way to circumvent the direct relationship between the reactivity of *N*-methylol agents and hydrolysis rates of the cellulose derivatives. The factors involved are illustrated by the following equations:

$$\begin{array}{c} R(CO)NR'CH_2OA + H^+ \rightleftharpoons R(CO)NR'CH_2\overset{H}{\underset{+}{O}}A \\ \Updownarrow \\ [R(CO)\overset{+}{N}R'{=}CH_2 \longleftrightarrow R(CO)NR'\overset{+}{C}H_2]^+ + HOA \\ \text{where A = alkyl or acyl} \end{array} \qquad (3)$$

Ion stability is largely determined by the electronic and steric effects produced by R and R′. Activation by the leaving group depends on its basicity.

Formaldehyde Analysis and Source of Formaldehyde in Finished Fabrics. Formaldehyde can be present in the air, in cross-linking reagents and the pad baths, and in finished fabrics and garments. Studies indicate the allowable maximum concentration of formaldehyde in the atmosphere of working places should be specified. Concentrations in air above 1 ppm are unusual in work areas. The free formaldehyde in finishing formulations has been reduced and conditions in finishing areas have been greatly improved.

Finishing Techniques

Mathematical treatments and computer programs assist the formulator and the finisher to obtain fabrics with the desired physical and chemical properties.

Finishes requiring chemical modification of the fiber are applied by an aqueous pad–dry–cure process. These procedures result in an acceptable, economical cellulose-containing fabric that can be tumble or air dried. If energy costs are excessive, wet-cure processes are available.

Similar treatments have been applied to chemically modified celluloses. Pretreatments of cotton cellulose, such as etherification of the hydroxyl groups to methyl, hydroxyethyl, cyanoethyl, and carboxymethyl, affect cross-linking with DMEU. Chemically modified celluloses of these types are of interest for a number of practical uses, but they have not been used commercially as a substrate for finishing.

Synthetics, Wool, and Special Fabrics

Synthetics and Blends. The finishing of synthetic fibers is limited primarily to heat-setting, soil release, hand or comfort effects, and antistatics. Strength, abrasion resistance, chemical inertness, and thermoplasticity are increased.

Obtaining Properties of Synthetic Fibers in Cotton. Some of the finishing treatments described above are intended to make polyester more like cotton or other natural fibers. On the other hand, cotton and cotton–polyester blends have been chemically modified to accept disperse dyestuffs and allow transfer printing. Successful treatments include the use of glycols with cross-linking agents, use of melamine–formaldehyde resins, and acetylation. Wrinkle resistance is improved by treatment with polyacrylates, polyurethanes, ethylene–vinyl acetate copolymers, and others.

Outdoor Fabrics. Water repellency has been reviewed. Some fabrics, such as polyalkylenes and similar synthetics, are sufficiently hydrophobic to be water repellent without addition of a special finish. Similarly, tightly woven fabrics of synthetic or natural fibers retaining natural or processing oils are often water repellent.

Outdoor fabrics must withstand the effect of mildew, rot, sunlight, airborne acids, abrasion, and wetting and drying. Cotton is widely used in weather-resistance applications, such as awnings, tarpaulins, and tents. The lighter, stronger, microbial-resistant synthetic fabrics have made large inroads into markets held by cotton even though the service life of outdoor cottons has been significantly increased since 1960.

Methylolmelamines in outdoor fabrics protect the cotton against actinic degradation. Flame-retardance finishes may provide protection during weathering. Experimental and commercial finishes have been compared.

Flame Retardants. Flame retardance can be conferred in many different ways. Most textiles are organic polymers, which ignite and burn under many conditions. Resistance to smoldering appears to be independent of flame retardance. Textile resins are useful components of smolder-resistant finishes.

Finishing of Wool. Wool competes for markets primarily with noncellulosic, synthetic fibers where warmth and wrinkle recovery are desired. The disadvantages of wool relate to fuzzing, shrinkage, and difficulty in creasing. Chlorination, modification of the disulfide cross-links, and polymer treatments appear to be chemical reactions of choice.

Delayed Cure, Permanent Press, and Formaldehyde Release. High performance permanent-press garments with durable properties do not normally require touch-up ironing. As a result, the resistance to chlorine damage does not have to be as high as in the earlier garments requiring some touch-up. However, the release of formaldehyde has to be kept to a minimum.

Curing Catalysts. Finishing formulations need acids to produce a durable-press finish. Choice of the specific acid catalyst is determined by effluents, dyes, formaldehyde release, discoloration, chlorine retention, odor formation, and storage conditions. Chemical suppliers often specify a certain catalyst for their durable-press formulations. Metal salts, such as zinc nitrate and magnesium chloride, are the most widely used.

Hand and Comfort. Appearance and comfort are frequently the main factors in commercial success. Fabrics that irritate the skin are unacceptable. Thermal and moisture transport are simple to evaluate; stiffness is also of major concern.

A variety of chemical products and fabrics claim to have antibacterial finishes and prevent odors and the spread of infections; one such finish is based on an organosilicon quaternary ammonium chloride compound.

Dyeing and Reactive Grafting. Garments made of noncross-linked fabrics can be dyed. However, the consumer prefers garments that are smooth drying and shrink resistant and yet retain the advantage of dyeability. For this purpose, textile finishes are required that permit fabrics to be dyed subsequent to cross-linking.

BIBLIOGRAPHY

H. Tovey, *Text. Res. J.* **31,** 185 (1961).

S. L. Vail in T. P. Newell and S. H. Zeronian, eds., *Cellulose Chemistry and Its Applications,* Ellis Horwood Ltd., Chichester, UK, 1985, pp. 384, 393.

H. Mark, S. M. Atlas, and E. Cernia, eds., *Man-Made Fibers,* Vol. 3, Wiley-Interscience, New York, 1968.

W. A. Reeves, G. L. Drake, and R. M. Perkins, *Fire Resistant Textiles Handbook,* Technomic, Westport, Conn., 1974.

Sidney L. Vail
Robert J. Harper, Jr.
Agricultural Research Service, U.S.D.A.
New Orleans, Louisiana

THERMAL ANALYSIS

Thermal analysis is defined by the International Confederation for Thermal Analysis (ICTA) as including all techniques in which a physical property of a substance and/or of its reaction products is measured as a function of temperature while the substance is subjected to a controlled temperature program. This definition appears to encompass much of scientific experimentation in which properties are measured with respect to time and temperature, as it even includes sets of isothermal experiments at differing temperatures. However, in actuality, those who consider themselves engaged in thermal analysis restrict their work to a much more limited area. First, thermal analysis is conducted only on condensed phase systems, ie, solids, glasses, liquids, and solutions. Second, only a few techniques are employed in almost all of the experiments reported in literature as thermal analysis. These are thermogravimetry (tg), which gives gain or loss of weight (mass change); differential scanning calorimetry (dsc) and differential thermal analysis (dta), which give change in specific-heat capacity (rate of enthalpy change); thermomechanical analysis (tma) and thermodilatometry, which give penetration or expansion (dimension change); dynamic mechanical analysis (dma) and dielectric thermal analysis (deta), which give loss moduli; evolved gas detection (egd) and evolved gas analysis (ega); and various multiple techniques. The last type may be performed on the same specimen (eg, simultaneous thermogravimetry and differential scanning calorimetry), the same sample in different instruments connected through an interface [eg, thermogravimetry and Fourier transform infrared spectroscopy (ftir)], or on the specimen plus its gaseous products (eg, pyrolysis followed by mass spectrometric and gas chromatographic analyses). The most common thermal analysis techniques are tg and dsc, although the others mentioned above are gaining in popularity. (Acronyms used in thermal analysis are usually written with capital letters. The use of lowercase acronyms herein conforms to the style of this publication.)

Most of the basic techniques, methods, and analyses of data used in thermal analysis are equally applicable to all materials, inorganic, organic, or polymeric. However, there are a number of important differences in thermal behavior between polymers and other substrates. These differences influence both the design of thermal analysis procedures and the interpretation of experimental results.

Most inorganic systems have their atoms held in rigid matrices. Phase changes are usually sharp, and when these substances react chemically at high temperatures, it is often the result of the greater freedom of movement of the reacting species that is allowed by the onset of such cooperative phenomena. Low molecular weight decomposition products are formed at temperatures near equilibrium temperatures so that chemical reactions are reversible. (Physical and geometrical changes are usually not reversible, which adds a further complication.) These factors lead to systems that are very dependent on partial pressures of reactant and product gases, geometry, rates of material and thermal diffusion, etc.

On the other hand, polymeric materials are, by definition, made up of long chains of atoms usually connected by covalent linkages and are partially or completely noncrystalline. Some, and occasionally all, of the complications described above for inorganic materials may be present during their reactions but, in general, these factors have a lesser and often insignificant effect on their reactions and transformations. Polymeric materials usually do not degrade at a measurable rate until they are heated to about 100°C above their equilibrium or ceiling temperature, so that degradation reactions are usually irreversible. Polymers are often in the liquid state at reaction temperature, and even below their melting temperatures there can be considerable segmental motion. Therefore, conditions can usually be found where the reactions are slow enough so that thermal and material diffusion are not rate limiting. Because of the complex molecular structure of polymers, physical changes take place much more sluggishly and over broader temperature ranges than is the case for lower molecular weight substances. Glass transitions and even melting transitions extend over broad temperature ranges and exhibit fine structure (often due to variations in configuration and conformation and distributions of molecular weight within a single specimen). Furthermore, many of the changes that affect their properties take place in the condensed phase without significant mass change so that other thermoanalysis techniques besides thermogravimetry are employed in their study.

Thermogravimetry

The official acronym for this technique is tg. However, many polymer scientists prefer to use the older term and acronym, thermogravimetric analysis (tga), especially in oral presentation to avoid confusion with glass-transition temperature.

Thermogravimetry, by definition, can be used to measure any reaction involving mass change. Probably the simplest of these reactions is sorption–desorption, and tg has been employed to measure rates of sorption; amounts of volatile contaminants, monomer, plasticizers, additives, etc; solubilities of gases and vapors in polymers; and diffusion and permeation constants of gases and vapors through polymer matrices. A method has been developed quite recently to measure the glass-transition temperature from the temperature at which the rate of escape of an inert, nonplasticizing vapor from a polymer rapidly accelerates. The traditional role for tg investigation of polymers is the study of their thermal degradation in vacuum and inert and active (eg, oxygen) atmospheres. Simple tg analysis in this area has been superseded by or combined with techniques that measure change in functional groups and volatile products. For example, tg is run in tandem with ftir or other spectroscopic techniques, or the volatile products from tg are identified and measured by mass spectroscopy and gas chromatography.

However, thermogravimetry is still widely used both in studies of degradation mechanisms and for methods for service lifetime prediction and thermal breakdown measurement. It is also employed in many analytical test methods such as those for carbon content of rubbers, compositional analysis of polymer composites, and others.

Apparatus. Thermobalances in almost all modern commercially produced tg systems are null-type instruments that permit continuous weighing. Null detectors are electrooptical systems (light beam–shutter–photocell). Horizontal torque and top-loading beam balances are used in some instruments, but the most common type is the vertical bottom-loading beam balance. In these instruments, the specimen container is suspended from the beam and may be heated by a microfurnace inside the glass envelope or by an exterior furnace. Most temperature controllers now are solid state, feedback, proportional types.

Problems in Thermogravimetric Measurement. Besides problems in temperature calibration and measurement, polymeric materials pose other problems due to their physical and thermal characteristics. Many of these are a result of their low thermal conductivity.

Kinetics. The determination of kinetics parameters and their extrapolation beyond the experimental range have been reviewed recently.

Reactions taking place in condensed media are complicated by many factors, both from the nature of the materials and the effects of procedural variables on reactions in the condensed phase. For the theoretical aspects of kinetics, eg, molecular modeling and determining mechanism, it is necessary to consider the effects of each of the procedural factors, such as specimen weight and dimensions, heating rate, purge gas composition and flow rate, and others. Thermal analysis systems may also be used to reveal new aspects of the reaction kinetics, for example, cycling and jumping temperature, pressure, and purge gas composition, relaxation techniques, and controlled transformation-rate thermal analysis. In the last technique, a physical or chemical property is controlled so that it follows a predetermined program by the appropriate heating of the sample.

In more practical applications, such as failure and service life prediction, oxidative stability, thermal breakdown, quality assurance and control, and the optimization of conditions for industrial synthesis and fabrication, less rigorous and exhaustive studies are adequate. Even methods that produce theoretically incorrect kinetics parameters may be sufficient if great care is taken not to allow any of the procedural parameters (sample weight, purge gas conditions, etc) or temperature program to change from experiment to experiment.

Differential Scanning Calorimetry

Differential scanning calorimetry (dsc) is the dominant technique for the thermal analytical investigation of polymeric materials. It measures the difference in energy inputs into a substance and a reference material as they are subjected to a controlled temperature program. Practically all physical and chemical processes involve changes in enthalpy or specific heat, and the applicability of dsc to condensed-phase systems is almost universal. Its measurement process is quantitative and the change of enthalpy is usually a linear function of the reaction coordinate.

Differential scanning calorimetry (dsc) has almost completely superseded differential thermal analysis (dta), except for some very high temperature or pressure applications.

Differential scanning calorimetry has, to a large extent, superseded classic calorimetric techniques for the determination of heat capacity, heats of transition, heats of reaction, and temperatures of transition (eg, glass-transition temperatures) for polymeric materials. Its great advantage, of course, is speed. Furthermore, its accuracy of measurement of thermal properties of very small specimens is sufficient even for most fundamental studies. Thermodynamic constants, as well as the effects of preparation, treatments (annealing), molecular weight distribution, and compositions on these constants can be determined. Fundamental studies include kinetics measurements in which dsc is used to obtain cure rates of resins and monomers; rates of thermal, oxidative, and radiative degradation; and rates of physical and chemical changes occurring in polymeric systems.

Differential scanning calorimetry is used extensively in industrial polymer laboratories for characterizing materials and products, developing quality assurance methods, monitoring quality control of production, troubleshooting, and other testing applications.

Apparatus. The three main types of scanning enthalpic instrument are shown in Figure 1.

Thermal Mechanical Analysis

Thermal mechanical analysis (tma) is a technique in which the deformation of a substance is measured under nonoscillatory (static) load as the substance is subjected to a controlled temperature program. When used at negligible load to measure expansion or contraction, it is a programmed temperature form of dilatometry (qv) that measures volume expansion. Thermomechanical analyzers are used in compression (or penetration), extension, and flexure. Measurement under a dynamic or oscillating load during programmed temperature change is called dynamic mechanical analysis (dma).

The compression mode is used to measure glass transitions (change in expansion) and softening points (penetration). Films, fibers, and sheet materials may be examined under tension in the extension mode, measuring the change in strain at constant stress. When used in this mode at high sensitivity, dimensional changes due to mechanical shrinkage and evaporation of water and other volatiles may be observed.

Flexure is not as widely utilized, but results from this mode can be related to such parameters as the Vicat softening point and the elastic modulus of the specimen.

Apparatus. Many different thermomechanical analyzers are marketed by instrument manufacturers. A loaded probe rests on the specimen in the penetration mode. In the extension mode, the specimen is clamped at two points, the lower being fixed to the push rod and the upper to the rigid support in such a way that tension is applied to the specimen when the push rod is loaded. In the flexure mode, a strip specimen is supported on two knife edges and the loaded probe is placed on the strip midway between the supports. Probes may be flat ended (for

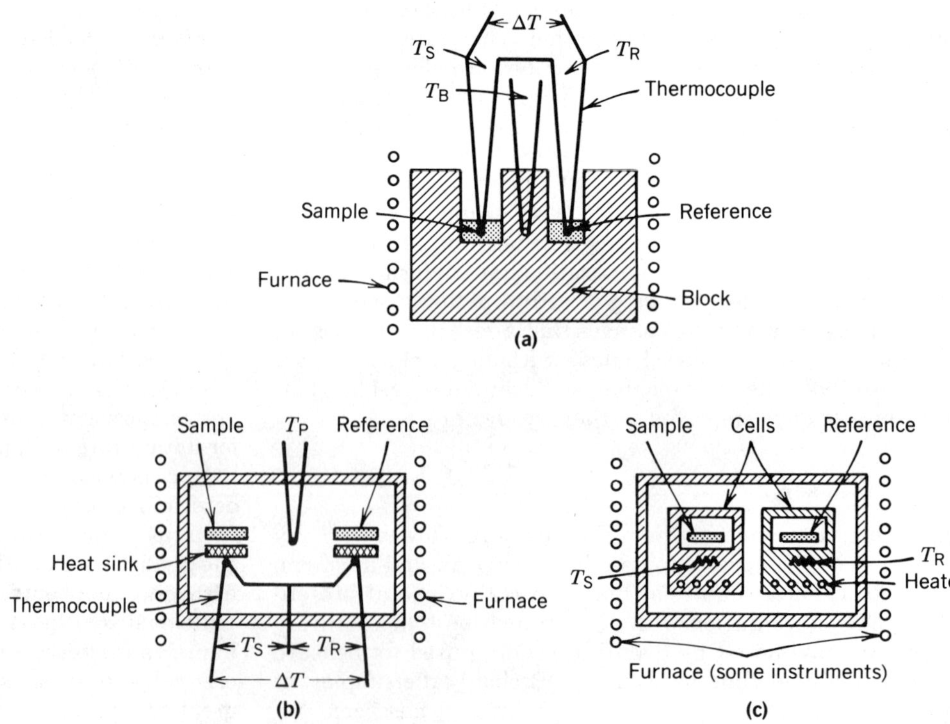

Figure 1. Scanning enthalpic instruments. (**a**) Classic dta; (**b**) heat flux dsc; (**c**) power-compensated dsc. T_B = block temperature; T_P = programmed temperature; T_S = sample temperature; T_R = reference temperature; $\Delta T = T_S - T_R$.

compression), tipped, tapered, or pointed (for penetration or flexure), or fitted with hooks or clips (for extension). The various instruments accept specimens in different forms, eg, rods, plugs, films, powders, or fibers.

Dynamic Mechanical Analysis and Dielectric Thermal Analysis

Dynamic mechanical analysis (dma) is the term most often applied to the technique in which the dynamic moduli and damping of a substance are measured under oscillatory load as a function of temperature and frequency as the substance is subjected to a controlled temperature program. Dielectric thermal analysis (deta) (dea is also used as an acronym), which measures dielectric properties of a material as a function of temperature and frequency, is a parallel technique in which electrical rather than mechanical stress is applied to the specimen. Both techniques are forms of relaxation spectroscopy in which onset of molecular motions and the resulting physical transitions are detected. The deta technique is limited to dipole activity or charge displacement (see DIELECTRIC RELAXATION).

In dma, an oscillatory stress is applied to the specimen in the bending or tensile mode of deformation and the lag of the resulting oscillatory strain is measured. Modern dma instruments measure changes in viscoelastic properties of materials as a function of temperature, time, or frequency of oscillation at constant oscillatory strain. In the cases of the free torsional pendulum, torsional braid analysis (tba), and some older dma apparatuses, the system is allowed to oscillate at its natural resonance frequency.

In deta, electrical rather than mechanical stress is applied to the specimen. The electrical stress is measured using bridge techniques to obtain the dielectric constant e', the dielectric loss e'', and the loss factor $\tan \delta = (e''/e')$.

Apparatus. In the commercial dma apparatuses oscillatory strains are applied to the specimen in the bending, tensile, or shear mode as a function of time or temperature and frequency. The manner by which these strains are applied differs in the various apparatuses, eg, a free or driven torsion pendulum, an electromechanical vibrator, an oscillating cone and plate, and others. The magnitude and resulting phase angle of the stress is measured by transducers and other techniques. Glass ribbons and stainless steel mesh are also used as substrates for dma measurements.

In the deta apparatus, an oscillating electrical field is applied to a specimen between two electrodes. Commercial instruments cover a frequency range from 10 Hz to 100 kHz; some research instruments cover wider ranges. Commercial instruments cover the temperature range of −150 to 500°C and are operated both isothermally and at programmed heating rates. The method is usually applied to thin sheets, films, and fibers. These (and dma) systems also have been adapted to the measurement of liquid specimens for cure studies.

Applications. Differential mechanical analysis and dielectric thermal analysis instruments are used by scientists and engineers to investigate rheological properties, glass and other transitions, and other mechanical and dielectric phenomena in polymeric materials, and to monitor the cure of resins.

BIBLIOGRAPHY

J. Sestak in G. Svehla, ed., *Thermal Analysis Part D, Thermophysical Properties of Solids, Their Measurement and Theoretical Analysis, Wilson and Wilson's Comprehensive Analytical Chemistry,* Elsevier, Amsterdam, The Netherlands, 1984.

E. Turi, ed., *Thermal Characterization of Polymeric Materials,* Academic Press, New York, 1981.

Journal of Thermal Analysis, published six times per year by John Wiley & Sons, Inc., Chichester, Budapest.

Thermochimica Acta, published semimonthly by Elsevier, Amsterdam, The Netherlands.

JOSEPH H. FLYNN
Sci. Thermal Research & Data Analysis

THERMAL PROPERTIES

Thermal properties of polymeric materials are always important and often critical in both processing stages and product uses. Most polymeric substances are processed and fabricated at elevated temperatures, often as a melt, necessitating the transfer of substantial quantities of heat first to the material to reach the desired temperatures and then to return the formed or cured product to a solidified state at room temperature. Because of the low thermal conductivities and thermal diffusivities of polymers in general, this process can be both slow and expensive, especially when large masses are involved. The heaters and heat-exchange equipment must, therefore, be carefully designed to ensure efficient and effective heat transfer, particularly if phase changes take place.

After formation or curing processes, most polymer products are cooled from relatively high temperatures, accompanied by shrinkage and dimensional changes. Clearly, these effects must be carefully considered during the design stage of processing equipment and can become even more critical with multiphase products, such as filled polymers or composite systems. In addition, accurate thermal property data are required for modeling processing steps such as extrusion or molding.

Other more subtle heating and cooling effects may also play an important role. For example, when a mass of crystallizable polymer is cooled from its melt, the inner regions of the mass tend to cool more slowly than the other surfaces because of a slow overall heat-transfer process. This leads to a more highly crystalline, and hence brittle, inner core. This cooling-induced morphology may be beneficial or detrimental, depending on the use of the product.

Thus, an intimate relationship exists between thermal properties and energy or dimensional considerations on the one hand and equipment design and fabrication techniques on the other. Ultimately, process optimization rests on these relationships.

Definitions

The general term thermal properties includes a wide range of properties and phenomena. In this article, however, the discussions are confined to four polymer properties: thermal conductivity, thermal diffusivity, specific heat capacity, and linear thermal expansivity. These properties are sometimes referred to as thermophysical properties.

Thermal Conductivity. The Fourier law of heat conduction is applicable to steady-state heat transfer in polymers. For unidirectional, rectilinear heat flow, it leads to the well-known relationship between heat flux and temperature gradient,

$$q = -k\,\frac{dT}{dx} \tag{1a}$$

or, in integrated form,

$$q = -k\,\frac{\Delta T}{\Delta x} \tag{1b}$$

where q is the heat flux (W/m^2) and dT/dx and $\Delta T/\Delta x$ are the temperature gradient and temperature drop per unit length (K/m), respectively. Equation 1a may be considered to constitute the definition of the thermal conductivity, k, expressed in SI units as W/(m·K), in English units as Btu·in./(ft^2·h·°F), and in cgs units as cal/(cm·s·°C). The right side of equation 1b is often expressed in the form $kA\Delta T/\Delta x$, where A is the cross-sectional area (m^2). Then the left side is not strictly a flux but a heat flow or rate (W or J/s). The dimensions and numerical values of k, however, remain unchanged.

In general, the thermal conductivity values of polymers are low, on the order of 0.1–0.5 W/(m·K). Among polymeric substances, crystalline polymers tend to have higher thermal conductivities than amorphous polymers.

Thermal Diffusivity. In much of the literature dealing with the thermal properties and characteristics of polymers, thermal diffusivity is either neglected or treated as an easily derivable quantity or mathematical parameter. This is unfortunate because thermal diffusivity is an important material property, determining temperature distributions in nonsteady-state heat conduction. Like thermal conductivities, thermal diffusivities of polymers are low. Therefore, most processing and fabricating techniques involve nonsteady-state heat transfer where temperatures are a function of both position and time.

In terms of temperature, T, the equation governing nonsteady-state heat transfer in one dimension may be written

$$\rho C_p \frac{\partial T}{\partial t} = \frac{\partial}{\partial x}\left(k\,\frac{\partial T}{\partial x}\right) \tag{2}$$

where ρ is the density (kg/m^3), C_p the specific heat capacity at constant pressure (J/(kg·K)), and t the time (s). Performing the indicated differentiation of the right side yields

$$\rho C_p \frac{\partial T}{\partial t} = k\,\frac{\partial^2 T}{\partial x^2} + \frac{\partial k}{\partial T}\left(\frac{\partial T}{\partial x}\right)^2 \tag{3}$$

For polymeric systems the second term of the right side of equation 3 is generally small or negligible, yielding

$$\rho C_p \frac{\partial T}{\partial t} = k\,\frac{\partial^2 T}{\partial x^2} \tag{4}$$

or

$$\frac{\partial T}{\partial t} = \alpha\,\frac{\partial^2 T}{\partial x^2} \tag{5}$$

where $\alpha = k/\rho C_p$ is the thermal diffusivity, expressed in SI units as m^2/s, in English units as ft^2/s, and in cgs units as cm^2/s.

Equation 5 may be considered to provide the formal definition of thermal diffusivity.

Specific Heat Capacity. Specific heat capacity is, by definition, the quantity of heat required to raise the temperature of a unit mass of material by one degree. As such, this quantity may be considered a measure of the thermal energy storage capacity of a substance. Specific heat capacity is expressed in SI units as J/(kg·K), in English units as Btu/(lb·°F), and in cgs units as cal/(g·°C). Polymers have relatively large specific heat capacity values, typically on the order of 750–2500 J/(kg·K).

Linear Thermal Expansivity. For linear thermal expansivity, a variety of other terms are in common use, including thermal expansion, coefficient of expansion, and coefficient of thermal expansion. Frequently it is difficult to know whether linear or volume expansivity is the subject, even when units are given. To further complicate the picture, the temperature coefficient of specific volume is often referred to as thermal expansion or thermal expansion coëfficient, although the units usually imply this distinction.

If a bar of material of initial length l_0 undergoes a change in dimension to a new length l, as a result of a temperature change ΔT, the linear thermal expansivity, denoted a_L, is defined as

$$a_L \equiv \frac{l - l_0}{l_0 \Delta T} \tag{6}$$

where a_L is assumed to be constant over the temperature range ΔT. Thermal expansivity is often denoted by the symbol α, but so, unfortunately, is thermal diffusivity. To avoid confusion the symbol a_L is used for linear thermal expansivity in this article. Linear thermal expansivity is expressed in SI units as m/(m·K), in English units as ft/(ft·°F), and in cgs units as cm/(cm·°C). Values of thermal expansivity of polymers are relatively large, with a_L typically on the order of 25–250 $\times$ 10^{-6} m/(m·K).

Experimental Techniques

A large number of experimental techniques and test methods have been developed for determining thermal properties of materials in general, and of polymers in particular, especially thermal conductivity and thermal diffusivity.

Among the thermal properties discussed here, thermal conductivity and linear thermal expansivity are far more frequently reported than either specific heat capacity or thermal diffusivity. However, the two latter properties have great significance, and systematic studies deserve attention.

Thermal Conductivity. Techniques used to determine thermal conductivity include flat-plate methods (the guarded hot-plate method, the biguarded hot-plate method, and the guarded heat-flow-meter method), the radial-heat-flow method, and transient-heat-flow methods (the line-source (or hot-wire) method, and the plane-source method).

Thermal Diffusivity. The relative advantages and disadvantages of thermal diffusivity vs thermal conductivity measurements are summarized in Table 1. Both thermal diffusivity and thermal conductivity should be measured directly. Techniques for measuring thermal diffusivity include the pulse method, the monotonic heating regime method, and the temperature-wave method.

Table 1. Thermal Diffusivity and Thermal Conductivity Measurements

	Thermal diffusivity	Thermal conductivity
experimental design	relatively simple[a]	relatively complex to very complex
mathematical treatment	relatively complex to very sophisticated	straightforward[a]
experimental time requirements	relatively fast to very rapid[a]	relatively slow to very slow
accuracy of results	less accurate	more accurate[a,b]
specimen size	small[a]	large

[a] Advantages, other factors being equal.
[b] In principle.

Specific Heat Capacity. The advent of the differential scanning calorimeter has revolutionized the area of heat-capacity measurement in general, and certainly in the field of polymer thermal properties, where it has become an area of specialization in itself. The technique is both reasonably rapid and accurate and is suitable for careful scientific work and routine testing. Adiabatic calorimetry and drop calorimetry are not widely used today for polymer measurements.

A fourth technique, differential thermal analysis, measures a complex thermal property function. In general it is not possible to deduce heat capacity from the experimental data. Nevertheless, differential thermal analysis has been, and probably continues to be, widely used, especially for detecting transitional phenomena (see also THERMAL ANALYSIS).

Linear Thermal Expansivity. The linear thermal expansivity of polymeric materials, either homopolymers, copolymers, or composites, is usually measured by thermodilatometry using a vitreous silica dilatometer or a device referred to as a thermomechanical analyzer (tma). Other methods, such as those involving optical microscopes or interferometers, afford high precision, which, however, is rarely needed for polymers with relatively high thermal expansivities. These methods are sometimes used for soft or low viscosity, liquidlike polymers. Pycnometry also represents a very accurate method of measuring volume thermal expansivity of both solid and molten polymers.

BIBLIOGRAPHY

J. Brandrup and E. H. Immergut, eds., *Polymer Handbook,* 3rd ed., Wiley-Interscience, New York, 1989.

R. P. Tye, ed., *Thermal Conductivity,* Vols. 1 and 2, Academic Press, New York, 1969.

K. D. Maglic, A. Cezairliyan, and V. E. Peletsky, eds., *Compendium of Thermophysical Property Measurement Methods,* Vol. 1, Plenum Press, New York, 1984.

Y. S. Touloukian and coworkers, eds., *Thermophysical Properties of Matter,* Vols. 2 and 5, 1970, Vol. 10, 1973, Vol. 13, 1977, Plenum Press, New York.

EDWARD V. THOMPSON
University of Maine

THERMAL STRESSES

Polymer processing operations are usually carried out under highly nonisothermal conditions in which polymeric melts pass through a complicated temperature history with solidification. Depending on the process, cooling rates can be as high as several thousand degrees per minute. During the initial stage of cooling, the surface layer cools so fast that the temperature difference between the surface layer and core attains a maximal value. Then the core cools faster than the surface layer until uniform temperature is attained. During the initial stage of cooling, the contraction of the surface layer is greater than that of the core. This introduces tensile stresses in the surface and compressive stresses in the core region. During a later stage, the core contracts more than the surface, leading to tensile stresses in the core and compressive stresses in the surface. These stresses, arising during rapid inhomogeneous cooling of the melt through the glass-transition temperature or melting point, are called thermal or cooling stresses (stresses due to a change in temperature). Therefore, in this case, thermal stresses are introduced in a material that was originally stress-free and externally unconstrained. Thermal stresses also arise in externally constrained polymeric products subjected to a change in temperature (cooling or heating) even without the occurrence of a solidification or phase transition. Thermal stresses arising in polymeric products are developed from a melt that is not originally stress-free. They occur in the presence of unrelaxed flow stresses. Accordingly, development of residual stresses in polymeric products during processing is due to the coupled effect of thermal stresses arising from a temperature gradient and an external constraint at the boundary and flow stresses. At the present time, the precise nature of the coupling is not well understood. However, it is well known that the presence of residual stresses greatly affects the mechanical properties and performance characteristics of polymeric products. Thus, understanding the factors governing the residual-stress development in polymeric products is of great importance to the design of polymeric articles. It is also important to be able to measure and predict residual-stress development during polymer processing.

Experimental Techniques

Various techniques for residual-stress measurement have been described in the literature. Most of those adopted to study polymeric products were taken from a similar area of development in metals. They can be classified into intrusive and nonintrusive techniques. The former involve the removal of internal constraints by machining or sectioning the polymeric article.

Intrusive techniques include the hole-drilling method, stress relaxation, and the layer-removal technique. Nonintrusive techniques include the photoelastic method, x-ray diffraction, and infrared spectroscopy.

Residual Stresses in Shaped Articles

The thermal stresses in polymeric articles are reduced during storage, even at temperatures below the glass-transition temperature (Fig. 1). This is due to the viscoelastic nature of polymers leading to stress relaxation, which proceeds quickly during the first several weeks following processing, after which it slows down. In contrast to the thermal stresses, the residual stresses due to flow do not relax during storage of the article at a temperature below the glass-transition temperature. The residual flow stresses can only relax when the ambient temperature is above the glass-transition temperature.

The residual stresses are greatly affected by the nature of the process and processing conditions. Various processing operations have been studied, including cold rolling, extrusion, and compression and injection molding. Cold rolling has been performed at room temperature by pulling a strip of PVC, ABS, or PC under tension through a rolling mill or by a conventional rolling mill. Toughness is enhanced by the presence of residual stresses, which are dependent on the thickness reduction. The magnitude and gradient of residual stress decrease as the thickness reduction increases.

Residual-stress Theories

Theories for the residual flow stresses started to evolve in the 1980s. These stresses appear as a consequence of the development of the high normal and shear stresses during shaping operations and their incomplete relaxation after completion of the process. Available theories for the frozen-in flow stresses have mainly been developed for the injection-molding process.

Theoretical analyses of residual stresses arising in curing systems have been carried out in several investigations. The treatments dealing with evaluation of thermal stresses during quenching differ from those concerned with residual stresses arising during curing. In particular, during curing the stresses arise because the material hardens while chemical reaction with heat generation takes place, which in turn promotes further hardening. Thus, a constitutive equation that describes the hardening process is required together with a suitable curing kinetic equation. In this case, the modulus of the hardening medium is dependent on the state of cure and temperature.

Theories developed for injection molding were extended to extrusion and blow molding. Predictions following from these theories were verified by comparison with various experimental measurements of birefringence in molded products. In most studies, verification is made for injection-molded PS strips. Relations between various components of stress and birefringence were based on the stress-optical rules.

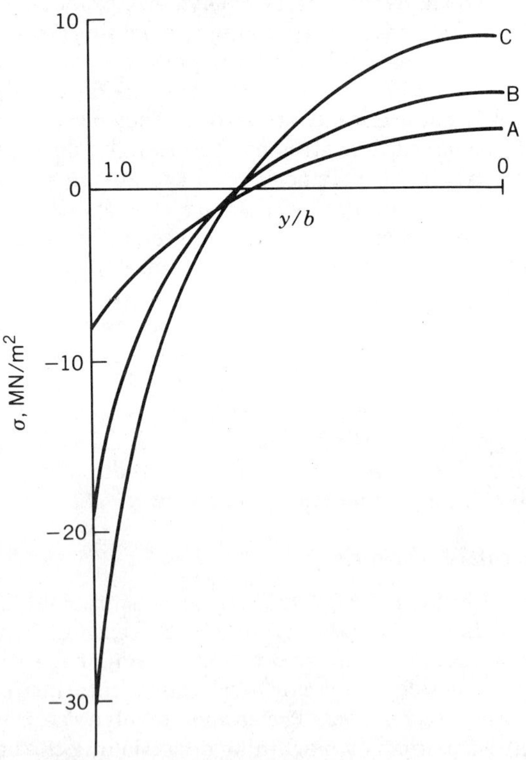

Figure 1. Experimental thermal-stress profile (curve A) and predictions based on theories of Indenbom (curve B) and Lee, Rogers, and Woo (curve C) for 0.26-cm-thick PS quenched from 150 to 23°C.

BIBLIOGRAPHY

A. I. Isayev, ed., *Injection and Compression Molding Fundamentals*, Marcel Dekker, New York, 1987, chapt. 3, p. 227.

H. Janeschitz-Kriegl, *Polymer Melt Rheology and Flow Birefringence*, Springer, Berlin, 1983.

J. R. White, *Polymer Testing* **4**, 165 (1984).

H. J. Kang and J. L. White, *Int. Polym. Process.* **1**, 12 (1986).

A. I. Isayev
University of Akron

THERMODYNAMIC PROPERTIES

The thermodynamic properties of macromolecules cover a broad field of materials science. As with any material, macromolecules are well described if two sets of information are available: structure and energetics. Each of the two information sets can be separated into a microscopic and a macroscopic part. The microscopic description deals with matter on an atomic scale and with sizes in the nanometer range; the macroscopic description, on a phase or larger scale, with sizes in the micrometer range. The thermodynamic properties cover the macroscopic description of energetics. For full understanding, however, the connection to the microscopic description is needed, which is achieved through statistical thermodynamics (qv). Another link to the microscopic description is offered by the kinetics that can be observed macroscopically, but for its interpretation and understanding, a microscopic-level model must be available.

Equilibrium thermodynamic data are presented here to provide a base on which to judge the material on hand. It must always, however, be kept in mind that the actual sample analyzed is usually not in equilibrium, ie, it is not 100% crystalline, does not have perfect crystals that melt at the equilibrium line, does not phase separate or mix where the thermodynamics are favorable. Similarly, many data must still be questioned as to their closeness to equilibrium or to the perfection of their extrapolation to equilibrium.

Basic to the thermodynamic description is heat capacity. A close link between microscopic and macroscopic description is possible for this fundamental property. The derived thermodynamic functions include enthalpy, entropy, and free enthalpy (Gibbs function), as well as PVT information. The transition parameters of pure components are revealed by first-order and glass transitions. The very new field of mesophase transitions was reviewed recently and the effect of conformational disorder is described. The broad field of multicomponents is particularly troubled by nonequilibrium behavior. Polymerization thermodynamics (qv) relies on the properties of the monomers and does not have problems with nonequilibrium.

The instrumentation for thermal properties (qv) is covered mainly by calorimetry, thermogravimetry, and dilatometry (qv). Calorimetry is separated into adiabatic calorimetry, covering the temperature range from 10 to 400 K, differential scanning calorimetry, covering the temperature range from 200 to 1000 K, and, close to isothermal combustion, reaction and dissolution calorimetry. Much information on thermal analysis is collected in the proceedings of the International Confederation for Thermal Analysis (ICTA) and the North American Thermal Analysis Society (NATAS). The broad range of literature is abstracted selectively. The thermal characterization of polymeric materials has been described in an extensive treatise (see also THERMAL ANALYSIS).

Heat Capacities of Solids and Liquids

Heat capacity is of importance not only for the evaluation of thermodynamic properties, such as enthalpy, entropy, and free enthalpy (Gibbs function), but also for the precise separation of latent heats of transition from motional energies and for the determination of glass transitions. Experimentally, in the low temperature region (0–100 K), heat capacity measurements are dominated by adiabatic calorimetry. Above that region, differential scanning calorimetry (dsc) has gained increasing prominence. Typical precision of measurements is now ±3% in the temperature range above 10 K for both solid and liquid heat capacities.

The theory of the heat capacity of solids is well established. The heat capacity is largely of vibrational origin and well described by the harmonic oscillator approximation. Separating the vibrations into skeletal and group vibrations is useful. The former are of relatively low frequency, strongly coupled, and represent the intra- and intermolecular vibrations of the chain molecules. The skeletal vibrations dominate the heat capacity in the low temperature region (<100 K). All efforts to calculate the intermolecular, skeletal vibrations from crystal-structure and force-constant data were unsuccessful in the past. Heat capacity measurements still provide the best means for evaluation of the low frequency, integral vibrational spectrum. The intramolecular skeletal vibrations are somewhat higher in frequency and are accessible in their upper frequency range through isolated chain normal-mode calculations. Such calculations also permit the evaluation of the numerous group vibrations. After an approximate vibrational spectrum has been established, the heat capacity at constant volume can be calculated up to, and even above, equilibrium melting temperatures of the polymeric materials. Figure 1 shows a typical example of the cumulative frequency spectrum and Figure 2 illustrates the heat capacity contributions of the skeletal and group vibrations for polyoxymethylene (POM).

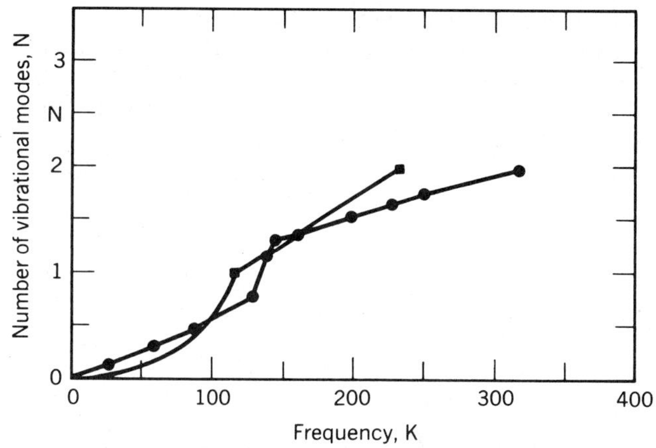

Figure 1. Cumulative vibrational spectrum of polyoxymethylene. ●, Experimental crystalline POM; ■, calculated by Tarasov equation. 1 Hz = 4.799×10^{-11} K.

The Einstein function inverts frequency to heat capacity. Integrating over all frequencies leads to the heat capacity:

$$\frac{C_v}{NR} = \int_0^{\theta_{\max}} \frac{(\theta/T)^2 \exp(\theta/T)}{[\exp(\theta/T) - 1]^2}\, d\theta \tag{1}$$

where T is the temperature and θ is the frequency, both expressed in Kelvin ($\theta = h\nu/k$, with ν representing the frequency in Hz; h and k are Planck's and Boltzmann's constants, respectively; 1 Hz is 4.799×10^{-11} K; 1 cm^{-1} is 1.4388 K).

The integration of equation 1 is evaluated in steps. The lowest vibrational frequencies (skeletal) usually follow a quadratic function up to a frequency limit called θ_D or θ_3. This is the well-known Debye approximation of low temperature heat capacities:

$$D_3\left(\frac{\theta_D}{T}\right) = \frac{C_v}{NR} = \left(\frac{12T^3}{\theta_3^3}\right) \int_0^{\theta_3/T} \frac{x^3 dx}{\exp(x - 1)} - \frac{3\theta_D/T}{\exp(\theta_D/T) - 1} \tag{2}$$

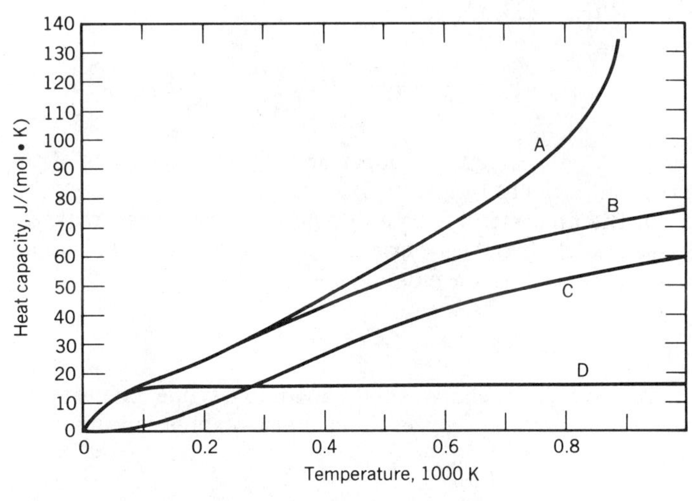

Figure 2. Heat capacity contributions of the skeletal and group vibrations in the solid state of polyoxymethylene, calculated by advanced thermal analysis (ATHAS). A, C_p; B, C_v; C, group; D, skeletal. To convert J to cal, divide by 4.184.

For two- or one-dimensional structures, such as are found in crystals of layer and chain molecules, the frequency distribution changes to linear and constant functions, respectively. The corresponding integrals are called the two- and one-dimensional Debye functions:

$$D_2\left(\frac{\theta_2}{T}\right) = \frac{C_v}{NR}$$
$$= \left(\frac{6T^2}{\theta_2^2}\right)\int_0^{\theta_2/T} \frac{x^2\,dx}{\exp(x) - 1} - \frac{2\theta_2/T}{\exp(\theta_2/T) - 1} \tag{3}$$

$$D_1\left(\frac{\theta_1}{T}\right) = \frac{C_v}{NR}$$
$$= \left(\frac{2T}{\theta_1}\right)\int_0^{\theta_1/T} \frac{x\,dx}{\exp(x) - 1} - \frac{\theta_1/T}{\exp(\theta_1/T) - 1} \tag{4}$$

The N in equations 1–4 refers to the appropriate number of vibrators and x stands for $h\nu/kT = \theta/T$. The sum of all N is three times the number of atoms in a repeating unit, the number of degrees of freedom. Tarasov proposed a special combination of equations 2 and 4 for the skeletal vibrations of a linear chain:

$$T\left(\frac{\theta_3}{T}, \frac{\theta_1}{T}\right) = \frac{C_v}{NR}$$
$$= D_1\left(\frac{\theta_1}{T}\right) - \left(\frac{\theta_3}{\theta_1}\right)\left[D_1\left(\frac{\theta_3}{T}\right) - D_3\left(\frac{\theta_3}{T}\right)\right] \tag{5}$$

The Tarasov equation holds for most linear macromolecules. Its limits are reached when phenylene groups are included in the backbone chain as in poly(oxy-1,4-phenylene) or poly(ethylene terephthalate), or alternating heavy and light mass backbone units occur, as in poly(vinylidene fluoride) or poly(vinylidene chloride)s.

For group vibrations, which are usually of much narrower distribution, it is sufficient to use single-frequency Einstein terms or to average over a frequency range that leads to a box-distribution function:

$$B\left(\frac{\theta_U}{T}, \frac{\theta_L}{T}\right) = \frac{C_v}{NR}$$
$$= \frac{\theta_U}{(\theta_U - \theta_L)}\left[D_1\left(\frac{\theta_U}{T}\right) - \frac{\theta_L}{\theta_U} D_1\left(\frac{\theta_L}{T}\right)\right] \tag{6}$$

where θ_U and θ_L are the upper and lower frequencies of the group vibrational range.

In the final step of computation, C_v is connected with the measured C_p; both the thermal expansivity α and the isothermal compressibility κ must be known:

$$C_p - C_v = \alpha^2 vT/\kappa \tag{7}$$

Since α and κ are usually, at best, available at one temperature, the semiempirical Nernst-Lindemann expression must be utilized:

$$C_p - C_v = C_p^2(T/T_m^0)A_0 \tag{8}$$

where T_m^0 is the equilibrium melting temperature and A_0 is a constant which for many polymers has a value close to 5.11×10^{-3} (K·mol)/J (per mole of heavy atoms) [2.14×10^{-2} (K·mol/cal)].

In the temperature region from 100 K to the glass or melting transition, the intramolecular skeletal vibrations of most polymers can be evaluated after computation of the group vibrations. The intermolecular vibrations are commonly already excited at these temperatures, and thus contribute a constant amount to the heat capacity NR.

Between the glass and melting transitions of semicrystalline macromolecules, special problems occur. These macromolecules have a microphase-separated structure with most molecules bridging both phases in a nonequilibrium state. This special influence of phase boundaries on molecular mobility gives rise to positive or negative C_p deviations from the value computed under the assumption of a macroscopic two-phase system (crystallinity model). These deviations permit the analysis of such structure-sensitive properties as rigid amorphous behavior, premelting, crystal perfection, reorganization, and recrystallization, and cold crystallization.

The heat capacity of liquid macromolecules can be measured above the glass-transition or the melting-transition temperature. The group vibrations change little on fusion. The changes of the skeletal vibrations in frequency due to the volume expansion also have little influence on heat capacity, since, once excited, the heat capacity contribution of a vibrational degree of freedom remains unchanged on lowering of the frequency.

Other Thermodynamic Functions

Enthalpy, Entropy, and Free Enthalpy. By knowing solid and liquid heat capacities of linear macromolecules, other thermodynamic functions can be obtained through the following relationships:

$$H_T - H_0 = \int_0^T C_p\,dT \tag{9}$$

$$S_T - S_0 = \int_0^T \frac{C_p}{T}\,dT \tag{10}$$

$$G_T = H_T - TS_T$$

Pressure, Volume, and Temperature. The pressure–volume–temperature (PVT) diagrams of linear macromolecules in different states provide another set of important thermodynamic functions (Fig. 3). Using high pressure volume dilatometry, over 10 macromolecules have been analyzed in their liquid states. All data can be fitted to the empirical Tait equation:

$$v(p, T) = v_0(T)\{1 - 0.0894 \ln[1 + p/B(T)]\} \tag{11}$$

where $v_0(T)$ is the zero-pressure volume and $B(T)$ is the Tait parameter. Both constants are exponential functions of temperature.

Transition Properties

First-order Transitions. The principal transitions in macromolecules are those concerned with enthalpy (latent heat) and entropy changes. They are called first-order transitions according to a classification requiring the first derivative of the change in free enthalpy with respect to temperature not to be zero. For a single-component system (pure compound), ther-

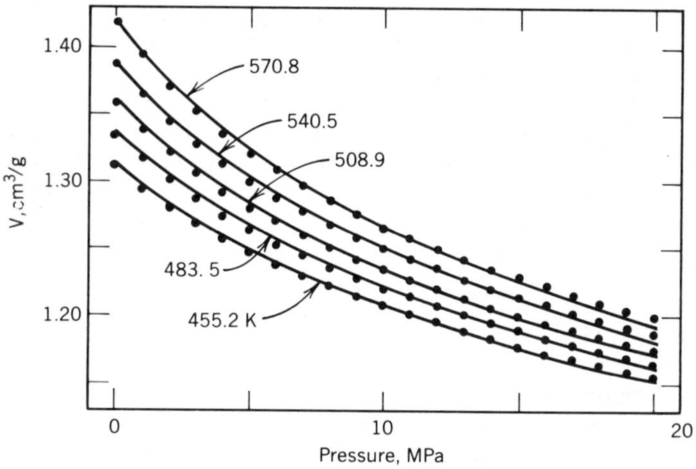

Figure 3. Experimental (circles) and theoretical (curves) isotherms of polypropylene. Calculations based on Simha-Somcynsky theory show good agreement. $V^* = 1.1954$ cm^3/g; $P^* = 530$ kPa (5417 kg/cm^2); $T^* = 11{,}155$ K.

modynamic equilibrium states are always at the lowest Gibbs energy. A crystal must melt at a sharp melting temperature T_m^0. The transition shows a change in slope of free enthalpy in going from one phase to the other. The relationship of the melting temperature to the changes in enthalpy and entropy is

$$T_m^0 = \frac{\Delta H_f^0}{\Delta S_f^0} \tag{12}$$

where ΔH_f^0 and ΔS_f^0 are the heat and entropy of fusion, respectively.

Glass Transitions. The glass-transition temperature is the main characteristic temperature of the amorphous solid and liquid states. A liquid becomes a solid on cooling through the glass-transition temperature if crystallization does not intervene. The microscopic process involved is the freezing of large-scale, molecular motion without change in structure. Since the heat capacity of the glass is always lower than that of the liquid at the same temperature, and since there is no latent heat involved in stopping molecular motion, the glass transition takes superficially the appearance of a thermodynamic second-order transition. The freezing of molecular motion is, however, time-dependent, and therefore the glass transition must be called an irreversible process. The glass–liquid transition occurs at a recognizable transition temperature because of a rather large temperature dependence of the relaxation time for large-scale molecular motion (micro-Brownian motion). Since the glass transition is usually a temperature region, a full characterization of the glass-transition region must quote five temperatures, as shown in Figure 4. The glass-transition temperature T_g is commonly the temperature of half-freezing, as it can be determined by heat capacity, expansion coefficient, or compressibility, as examples.

Multicomponent Systems. For the description of multicomponent systems, the number of moles N_i or concentrations of each of the components is added to temperature and pressure as variables of state. For each extensive function of state, such as V, U (internal energy), S, G, H, C_p, and C_v, for example, for each component A, B, etc, derivatives can be developed:

$$\left(\frac{\partial X}{\partial N_A}\right)_{T,p,N_B} = X_A, \qquad \left(\frac{\partial X}{\partial N_B}\right)_{T,p,N_A} = X_B \tag{13}$$

The quantities X_A, X_B, etc, are called the partial molar quantities. In general, they are not additive, but the following relationship can be derived between different components (shown for two components, but easily generalized for multicomponent systems):

$$dX = X_A\,dN_A + X_B\,dN_B \tag{14}$$

Integration of equation 14 without changing the composition gives

$$X = N_A X_A + N_B X_B \tag{15}$$

The total differential of equation 15 must be

$$dX = X_A\,dN_A + N_A\,dX_A + X_B\,dN_B + N_B\,dX_B \tag{16}$$

Comparison of the coefficients of equations 16 and 14 reveals

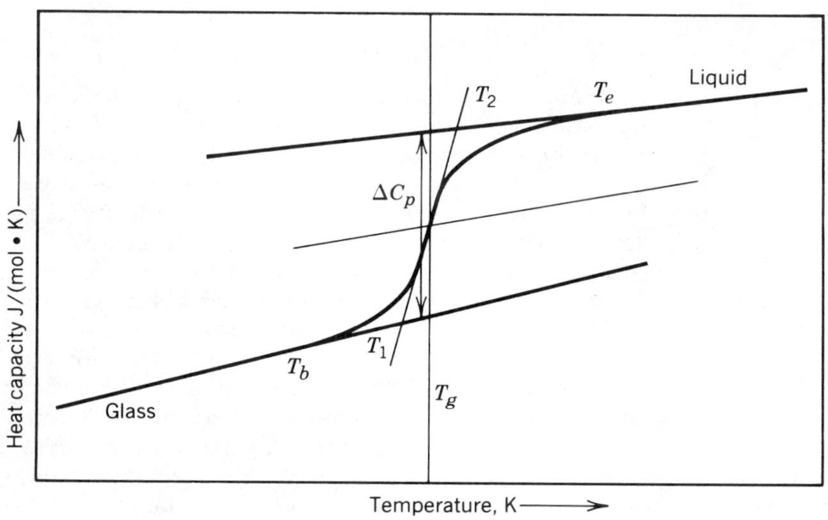

Figure 4. Heat capacities in the glass-transition region. The first perceptible beginning of the glass transition, T_b, is judged by the first increase in C_p from that of the solid state; T_1 and T_2 are the extrapolated beginning and end of the glass transition. The difference between T_1 and T_2 is indicative of the broadness of the main portion of the glass transition. The glass-transition temperature T_g is chosen at half-devitrification when judged by the C_p increase. The temperature T_e, the end of the glass transition, is reached when C_p reaches the liquid C_p.

the important equations

$$N_A\, dX_A + N_B\, dX_B = 0 \qquad \text{or} \qquad dX_A = -\left(\frac{N_B}{N_A}\right) dX_B \tag{17}$$

which permit the calculation of the change in partial molar quantities of one component from the other.

Whether a single phase or a multiphase is formed in a multicomponent system depends on compatibilities among the components. Generally speaking, three kinds of compatibilities are usually involved: compatibility of small molecules with macromolecules, compatibility within the macromolecules on a repeating unit scale, and compatibility among macromolecules as a whole.

Thermodynamics of Polymerization

In a polymerization, two types of behavior should be studied: thermodynamics and kinetics. A reaction with unfavorable thermodynamic criteria should not occur. If a reaction scheme fails kinetically, the reaction is still possible via another mechanism, provided the thermodynamics is favorable. A typical example is the preparation of polypropylene. Although the polymerization of propylene is possible thermodynamically, it was not achieved until the work of Ziegler and Natta who discovered the catalyzed mechanism with favorable kinetics. If kinetics is favorable, thermodynamics must also be favorable. Thus, much effort has been devoted to kinetics, pioneered by Flory.

A chemical or physical reaction can only occur if there is a decrease in the free energy, $\Delta G < 0$. The free energy is composed of two parts: the enthalpic ΔH and the entropic $T\Delta S$:

$$\Delta G = \Delta H - T\Delta S \tag{18}$$

In most polymerization reactions, the entropy decreases because the polymer lacks translational and rotational motion. It has been argued that the loss of the rotational entropy is balanced by the gain of internal rotational and vibrational entropy in the polymer, and that the total loss is, therefore, equal to the translational entropy of the monomer.

Since the entropic contribution to ΔG in equation 18 is thus positive, the enthalpy of polymerization must be negative and more than compensate for the entropic term for the reaction to occur. Indeed, polymerizations are typically exothermic reactions.

From equation 18, it follows that above a certain temperature T_c, ΔG becomes greater than zero and polymerization is no longer possible. Instead, the equilibrium of the polymerization reaction: $M \rightleftharpoons (1/n)[M_n]$ is shifted toward the monomer and a depolymerization reaction occurs. The temperature at which ΔG is zero is called the ceiling temperature, and is a characteristic of a polymerization reaction. It depends on the pressure of the reactant and obeys a Clausius-Clapeyron-type relation:

$$\ln T_c = \ln T_c^0 + p\Delta V/\Delta H \tag{19}$$

where T_c^0 is the ceiling temperature at $p = 0$ and ΔV is the difference in molar volume between polymer and monomer.

In equation 19, ΔH is negative since the concept of ceiling temperature is valid for exothermic reactions. Polymerizations are typically accompanied by a decrease in the system volume; thus ΔV is also negative. Therefore, an increase in pressure raises the ceiling temperature. It follows also that higher pressures increase the yield of the reaction.

In some special cases, however, the polymerization is an endothermic reaction (positive ΔH) compensated by an increase in entropy. Such a case is the polymerization of elemental, eight-membered, cyclic sulfur (S_8). It follows from equation 18 that, in such a polymerization, a minimum temperature exists below which polymerization is not possible. Therefore, an inverse ceiling temperature exists, called the floor temperature. This situation can also occur in ring-opening polymerizations, where prior predictions of the signs in ΔS and ΔH are not possible.

The qualitative discussion can be quantified for the prediction of the extent of a polymerization reaction. The change in Gibbs free energy is related to the equilibrium constant of the reaction (K_e) through the equation

$$\Delta G = -RT \ln K_e \tag{20}$$

There is, however, a difference between step (condensation) and chain (addition) polymerization.

BIBLIOGRAPHY

B. Wunderlich, *Macromolecular Physics,* Vols. 1–3, Academic Press, New York, 1973, 1976, 1980.

B. Wunderlich and J. Grebowicz, *Adv. Polym. Sci.* **60/61,** 1 (1984); see also other articles in *Adv. Polym. Sci.* **59–61** (1984) on liquid crystals.

B. Wunderlich, M. Moeller, J. Grebowicz, and H. Baur, *Adv. Polym. Sci.* **87,** 1 (1988).

B. Wunderlich, *Thermal Analysis,* Academic Press, New York, 1990.

BERNHARD WUNDERLICH
Oak Ridge National Laboratory and
The University of Tennessee in Knoxville

STEPHEN Z. D. CHENG
The University of Akron

KYRIAKOS LOUFAKIS
Institut fur Makromolekulare Chemie
University of Freiburg

THERMOFORMING

Thermoforming is the process of manufacturing products from thermoplastic sheet heated to its softening point, formed by low pressure into a mold, cooled, die cut from the sheet, stacked, and packaged. All thermoplastic resins that can be processed into sheet can be thermoformed, provided that the heated area to be formed does not exceed the hot strength capabilities of the material to support itself (see also SHEETING).

Specific variations of basic thermoforming include processing sheet cut-to-size, or from rolls, or directly from an extrusion line. Sheet is heated to forming temperature in radiation or convection ovens, by contact with a heated platen, or by utilizing the heat of extrusion. Forming is accomplished by vacuum between sheet and mold, by adding external air pressure, or by matched male and female molds. Part separation from the web (trimming) may be done with steel rule dies in the mold or at a downstream station in the forming machine, with close tolerance punches and dies in an external trim press, or with band saws and/or routers.

The decision to manufacture a product by thermoforming depends on part design; the selected thermoplastic; the cost of material, tooling, and fabrication; and the anticipated sales volume. Products formed by thermoforming may be characterized by shape as containers: very shallow as lids, or deep as boat hulls, but with the greatest area at the container mouth, and the wall and bottom thickness generally thinner than the starting sheet. The ratio of mouth area to depth, the sidewall draft angle, and the presence and amount of reverse taper or undercuts are important considerations. All of these considerations determine the specific thermoforming variation as well as the general process.

Thermoforming has many advantages over other methods of thermoplastic fabrication:

1. Parts with a large surface area can be formed with relatively low mold and equipment cost, because of the low pressures required.

2. Very thin walled parts can be readily formed, which is not feasible by any other method.

3. High volume thin-walled products, such as drinking cups, can be produced at the lowest cost per capital investment, at production rates of 50,000 to over 200,000 units per hour.

4. Low volume heavy-gauge products, such as computer housings, are competing favorably with injection molding in price through lower tooling cost and in product detail with 690 kPa (100 psi) forming pressures.

A wide variety of consumer and commercial products are thermoformed, ranging from the 28 mL condiment container to the bed liner for the half-ton pickup truck. The total consumption of thermoplastics processed by thermoforming in 1989 is estimated at 1042×10^3 t. Polystyrene continues to dominate the thermoforming markets in the form of high impact styrene, OPS (oriented polystyrene), and polystyrene foam at 780×10^3 t or 75% of the total. Similarly, packaging remains the largest user of thermoformed products, around 75% of the total, ca 780×10^3 t.

Advances in thermoforming technology in the 1980s in the various facets of the industry may be summarized as follows:

Materials

Multilayer extruded sheet provides synergism of the physical and chemical-resistant properties of olefins and styrenics. It includes barrier layers of EVOH (ethylene–vinyl alcohol copolymers) or Saran in a strongly growing market of aseptically packaged food products with a long shelf life at room temperature. Regrind is buried in the structure.

CPET (crystallized polyethylene terephthalate) is a breakthrough in dual-ovenable thermoformed trays for packaging frozen food products from the processor to the dinner table. These trays can be heated in microwave as well as conventional ovens. A market estimate of 2×10^9 CPET units in 1992 must be modified downward, in favor of polypropylene, as microwave ovens dominate reconstitution use and do not require trays of higher heat resistant characteristics.

Equipment

The most pronounced improvements in thermoforming machinery relate to computer control. The processing functions of oven-heat profile, sequence of the forming cycle, and integration with the extrusion process, coupled with data management, job file, manufacturing reports, and data storage, are leading to CIM (computer-integrated manufacturing).

Production rates of integrated in-line extrusion thermoforming have achieved 1135 kg/h of trimmed product because of improved sheet temperature control, faster forming and cooling, trim-press speed increases, and mechanized packaging.

Rotary thermoformers have increased in size of forming area to 300 by 840 cm with a large number of machines in the 243 by 350 cm size for forming olefin sheet of 6 mm thickness. The largest operating rotary thermoformer in the United States is the Brown Model R–244TS at the Penda Corp., capable of forming sheet sizes up to 190 cm wide, 342 cm long, and up to 1.25 cm thick. Pickup truck cargo bed liners, encapsulated pallets, and large double-walled parts are produced automatically.

Pressure forming at 690 kPa (100 psi) affords detail not previously obtainable to compete with injection molding.

Tooling

The precision and reproducibility of thermoformed tooling has dramatically improved owing to the use of computer-assisted design (CAD) and computer-assisted manufacturing (CAM). Mold base and cavity dimensions and trim tool punch and die clearances, formerly 10% of the precision of injection molds, are narrowing the tolerance gap by holding to nanometers rather than micrometers.

Techniques

The central element of any thermoforming system is the tooling, comprising the mold and trimming means. The design of the product will determine the thermoforming technique to be used. Cost and volume will influence the size of the forming machinery, the number of mold cavities, and the rate of production.

The most elementary thermoforming techniques include basic vacuum forming, drape forming, pressure forming, and matched-mold forming. These terms describe the single operation of clamping the heated sheet against the male or female mold and removing air between sheet and mold surface by vacuum, external air pressure, or both. All thermoforming molds contain fine holes for air evacuation. A secondary operation within the forming cycle prestretches the hot sheet in order to control thickness in the final product. Prestretching may be accomplished with a plug or ring in plug-assist or ring-assist forming, or by air in pressure-bubble plug-assist forming, vacuum snapback forming, pressure-bubble vacuum snapback, and air slip forming. Another popular technique is trapped-sheet contact heat pressure forming.

There are several specialized techniques that have achieved some commercial success, including solid-phase pressure forming (limited to polypropylene), scrapless pressure forming (specific to square blanks of multilayer sheet), and cuspation forming (prestretching hot sheet on cam activated blades).

Figure 1. Roll-fed continuous in-line thermoformer and trim press for high volume disposable HIPS containers. Courtesy of Brown Machine.

Materials

The selection of a thermoplastic material for a thermoformed product requires, first, that it be extrudable into sheet form. Then its cost, availability, and manufacturing continuity are considered. The performance requirements of the product dictate the selection of the lowest cost polymer that offers adequate formability; tensile, elongation, and impact strength; chemical and low or high temperature resistance and other properties that may apply; clarity; dielectric strength; moisture, vapor, and oxygen permeability; and recyclability. If the properties of expensive engineering polymers are required, it may be possible through multilayer extrusion or lamination or by alloying to achieve such properties at lower cost with a structure composed mainly of commodity resins. Other alternatives to lower the product material cost include density reduction by introducing foaming agents, and incorporating inexpensive fillers to extend the primary resin.

Thermoplastic resins of large volume usage in thermoforming include high impact and high heat polystyrene, high density polyethylene, polypropylene, poly(vinyl chloride), ABS (acrylonitrile–butadiene–styrene), CPET (crystallized poly(ethylene terephthalate)), and poly(methyl methacryalate). Other polymers of lesser usage are transparent styrene–butadiene block copolymers, acrylic multipolymer, polycarbonate, cellulosics, and ethylene–propylene thermoplastic vulcanizates (TPE). Coextruded structures of up to seven layers include barriers of Eval, Saran, or nylon, glued to polyolefins, and/or styreneics for functional properties and decorative esthetics at reasonable cost.

Equipment

All thermoforming systems include a means of receiving sheet cut to size, from rolls, or directly from an extrusion line. Generally, sheet thickness over 1.5 mm is cut to size, and thinner sheet is supplied on rolls. In line extrusion, forming, obviously, is not limited by gauge. The thermoformer contains the mold, may or may not include the trimming means, and provides the pin chain or gripping system that indexes the sheet through the heating, forming, and in certain cases, the trimming operations (Figure 1).

The most common source of heat is the radiant heat tunnel with several index stops, utilizing calrod, ceramic, panel, and quartz elements, sometimes combined. Newer forming machines provide better sheet temperature uniformity with more zones individually controlled by microprocessor. Convection oven heating is available, but rare. Contact heat against the forming station platen is used in shallow draw trim-in-place machines, but with increased cycle time. One system for high volume packaging applications maintains the heat of extrusion into the forming station.

The forming station is actuated by pneumatics with toggle clamps, by hydraulics in conjunction with pneumatics, or by mechanical cam actuation. The mechanical platen drive leads the technology, but is the most expensive, and is best suited to high speed, large volume applications.

Product may be separated from the web at the forming station, or in the forming machine an index downstream, with steel rule dies. In higher speed, high-volume applications, the formed sheet is fed into a separate trim press using precision punches and dies trimming up to four formed rows at a time at typical speeds of 110 strokes per minute. The horizontal press is preferred, allowing easier handling of the nested stacks of parts. Heavy gauge parts may be trimmed with profile cutters, routers, or band saws, and by punch press. In certain in-line forming operations, decorating, labeling, notching, slotting, and reforming can be done prior to final trimming.

Tooling

Molds may be single or multiple cavity according to the size and volume of the product. For prototyping and short runs it is feasible to use wood, plaster, epoxy, and unsaturated polyester. Mold material for production runs is aluminum, cast and machined, or machined from denser rolled plate. Mold cooling is generally by temperature controlled circulating water through deep-hole-drilled mold platens, through tubing welded to the mold shell, or by actually flooding around insert cavities.

In multiple cavity tooling, clamping each cavity individually provides better product weight uniformity, and good thickness distribution at the lip for rim rolling or heat sealing. Many thermoformed articles have undercuts for stacking, lid seating, or other purposes. At the conclusion of the forming cycle, a stripper plate or knockout assembly is required to free the parts from the mold. Usually pneumatically activated, the stripper assembly must be cooled, as it is equally important to the optimum cycle to set the web as well as the article. Undercuts are best used with female molds to take advantage of plastic shrinkage on cooling. With long production runs, and in particular with ABS, it may be desirable to harden undercut surfaces by aluminum oxide impregnation to reduce wear.

BIBLIOGRAPHY

D. Irwin in *Modern Plastics Encyclopedia,* McGraw-Hill, New York, 1989, pp. 301–304.

R. Whiteside in *Modern Plastics Encyclopedia,* McGraw-Hill, New York, 1990, pp. 318–322.

J. L. Thorne, *Thermoforming,* Hanser Verlag, Munich, 1987.

S. Senzer, *Thermoforming of Packaging and Disposables,* Sabre Associates, Inc., Norwalk, Conn., 1985.

JOSEPH N. MCDONALD
Joseph McDonald Associates, Inc.

THERMOGRAVIMETRIC ANALYSIS. See DEGRADATION; THERMAL ANALYSIS.

THERMOMECHANICAL PROPERTIES. See MECHANICAL PROPERTIES; THERMAL PROPERTIES.

THERMOPLASTIC POLYMER

A thermoplastic polymer is one that is capable of being repeatedly softened by heating and hardened by cooling through a characteristic temperature range, and that in the softened state can be shaped by flow into articles by molding or extrusion. Thermoplastic applies to those materials whose change upon heating is substantially physical (see also THERMOSETTING POLYMER).

THERMOSETTING POLYMER

A thermosetting polymer is one that is capable of being changed into a substantially infusible or insoluble product when cured by heat or other means. The cured polymer may be termed thermoset (see also CURING; THERMOPLASTIC POLYMER).

THETA PARAMETER. See SOLUTION PROPERTIES.

THIAZOLE POLYMERS. See POLYBITHIAZOLES; POLYBENZOTHIAZOLES AND POLYBENZOXAZOLES.

THIN-LAYER CHROMATOGRAPHY. See CHROMATOGRAPHY.

THIOKOL POLYMERS. See POLYSULFIDE POLYMERS.

THIOUREA RESINS. See AMINO RESINS.

THIXOTROPY. See COLLOIDS; RHEOLOGICAL MEASUREMENTS; SOLUTION RHEOLOGY.

THREE-DIMENSIONAL NETWORK. See CROSS-LINKING; NETWORKS.

TILE. See FLOORING MATERIALS.

TIRE CORD. See TIRES.

TIRES

The pneumatic tire is a toroidal, high performance polymeric composite, which exhibits characteristics of a flexible-membrane pressure container providing load carrying capability, and cushioning and road-handling abilities. It transmits driving and braking torque, cornering forces, and performs all functions essential to locomotion. These features can be attributed to its deformability and damping characteristics. The pneumatic tire is a complex system of interacting components, each with specific properties for maximum effectiveness. The performance of a tire depends on the properties, interactions of the components, and the service conditions.

Tire Profile

The tire profile or aspect ratio is the ratio of the cross-section height from tread to bead to the cross-section width from sidewall to sidewall. The aspect ratio is expressed as a percent and is defined by equation 1.

$$\text{Aspect ratio} = \frac{\text{Tire section height}}{\text{Tire section width}} \times 100 \qquad (1)$$

A low aspect ratio normally provides greater stability, handling, traction, and high speed capabilities. A racing tire, for example, can have an aspect ratio as low as 30; that is, its height is 30% of its width.

Tire Types

A tire is made from rubber, fabric, chemicals, and metal (steel). The outer surface in contact with the road is the tread, which is covered on each side by a sidewall. On the inside are the rubber-coated cords, which give the tire its dimensional integrity. Tubeless tires have a thin layer of impermeable rubber, called the inner liner, which covers the plies. Figure 1 is a schematic of the Goodyear Eagle GT+4 all-season high performance radial passenger tire. The semirigid base of the tire in contact with the rim is the bead, consisting of high tensile steel wire (commonly brass-coated) encased in a rubber compound. The casing is the skeletal structure of the tire which contains all components except the tread. The basic strength of a tire is derived from its skeletal structure, that is, its layer of rubberized cord reinforcement. The cord reinforcement provides stability and resistance to bruises, fatigue, and heat. The layers or plies of rubber-coated reinforcing cords are assembled

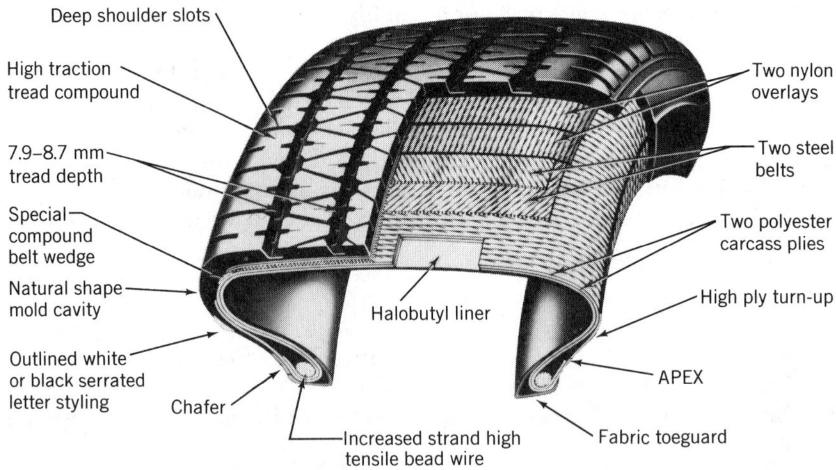

Figure 1. Cross section of Goodyear's all-season high performance passenger tire, the Eagle GT+4.

in three basic constructions, resulting in three types of tires: bias ply, bias belted, and radial ply.

In radial tires, the most common type of passenger tires, the plies of the reinforcing cords in the casing run in a radial direction (from bead to bead), with a 90 degree bias or crown angle in relation to the axis of rotation of the tire. Above the plies (under the tread) is an inextensible belt package comprised of layers of rubber-coated cords nearly parallel to the circumferential direction of the tire. The belt cords are laid at various angles (10–30 degrees) and serve to constrain the carcass plies. The belt rigidity is essential to the functioning of the tire.

Components

A tire is a composite of components, each serving a specific and unique function, yet all functioning synergistically to produce the desired performance. It consists of the tread area, which includes the tread, the base, and the cushion; and the casing, which includes the carcass plies, belts, sidewall, bead, inner liner, apex, chafer, toeguard, and other constituents. These components or parts are developed, designed, and fabricated to meet the mobility requirements of vehicles such as load carrying, vehicle control, handling, ride, traction, durability, etc. Tire components are made from natural or synthetic rubbers; chemicals such as antioxidants, antiozonants, and processing aids; accelerators; sulfur and other cross-linking materials; fillers such as carbon black and silica; and cords (textile, fiber glass, and steel wire usually brass coated) (see also CARBON; RUBBER CHEMICALS). These materials are selected for their physicochemical properties and interactions with other constituent materials to provide a broad range of mechanical properties. Figure 1 shows the components of the Goodyear Eagle GT+4 all-season high performance radial passenger tire.

Manufacture

Rubber Processing. Tire manufacturing begins with preparation of specific rubber compounds for individual components. The process begins with the testing of the elastomers, fillers, and chemical additives. Ingredients are weighed and blended in an internal mixer such as a Banbury mixer in order to produce a compound with sufficient homogeneity for easy handling in subsequent operations. Sequence time, energy of mixing, and the stock temperature are important variables since physical properties of the compound and the final tire performance depend upon exact processing conditions as well as upon the type and ratios of ingredients that comprise each compound formula. The compound is stretched into sheets in preparation for additional mixing in the Banbury mixer or downstream processing such as extruding. The sheets are formed into components, and masticated on a mill where mechanical and thermal energy reduces plasticity (hot feed system). The compound is taken off the mill as a continuous ribbon directly to the calender, extruder, or bead insulator. Alternatively, the sheet can be fed directly to a cold feed extruder. All materials are cut to the required sizes at the desired angles before processing by the tire building machine.

Fabric Preparation. Initial manufacture consists of the spinning and drawing of yarn and application of the finish (see FIBERS, MANUFACTURE). The yarn is twisted into plies, cabled into cords, woven into a fabric, and treated with an adhesive dip (see FABRICS, COATED; ADHESION AND BONDING). The fabric is coated with rubber in a four-roll calender to form a composite for components, such as plies, breakers, chippers, and belts.

Wire Preparation. The steel wires for tires are manufactured by the cold-drawing process to the typical specification shown in Table 1. Steel rod is brass plated to enhance cord-to-rubber adhesion, drawn into filaments, and twisted into strands that are combined into cords. The wire cords are wound on spools and rubber coated with a calender.

Tire Building. The green (uncured) tire is built from the inside out on a machine equipped with a steel or inflated rubber drum. Precut calendered inner liner, then plies, are wrapped tightly around the drum, beads are affixed near each end, and the area is constructed (apex, wedge, chafer), and sidewall strips wrapped. The building of radial tires requires partial shaping to closer resemble ultimate dimensions before the belts

Table 1. Wire Specification

carbon, %	0.7	chromium, max %	0.05
manganese, %	0.5	copper, max %	0.02
silicon, %	0.3	sulfur, max %	0.03
		phosphorus, max %	0.03

and tread are wrapped. The building drum is then collapsed, permitting the green tire to be removed.

Large tires can be assembled by sliding bands or preformed sleeves of coated fabric over the drum.

Curing. In the curing process, specified temperatures and pressures are applied over a period of time to effect vulcanization. The rubber becomes cross-linked in a series of reactions with sulfur, accelerators, and activators. The process is usually combined with shaping as a preliminary step, in which the green tire is compressed and forced against the mold where it is to be cured. The cured tire has the final form and physical properties required for service (see RUBBER CHEMICALS; VULCANIZATION).

Testing

The importance of safe operation of automobiles, trucks, and other vehicles, and the safety of their occupants require that tires perform reliably over their entire life span and endure the hazards of rough roads and treacherous terrains. As such, tire tests are an essential part of tire research, development, and manufacturing. Even after a particular tire construction has been on the market for some time, laboratory and field tests are performed regularly, because of design modifications resulting from new materials, vehicle designs, and performance requirements. Initially new designs and materials are screened in the laboratory where field-type failures are simulated. Road and field testing is expensive but necessary for assessing safe performance under service conditions. Road tests are conducted on public highways and proving grounds. The latter are preferred for specific studies under controlled variables such as speed, road texture, and curves. Tire tests can be divided into different categories, such as endurance, treadwear, handling, ride and harshness, and rolling resistance. Laboratory tests are further tailored for testing a specific tire property under controlled conditions. The U.S. Department of Transportation (DOT) specifies tests which tires are required to pass before they are released for sale. Included are plunger energy, high speed, and endurance. Tires must also be rated for treadwear, traction, and temperature. The automobile manufacturers also require that tires be tested for mechanical characteristics such as vertical stiffness, lateral stiffness, steering effort, cornering power, and self-aligning torque as well as handling characteristics such as stability, passing maneuvers, evasive maneuvers, and ultimate stability. These tests are conducted over dry, wet, icy, and snowy pavements. Traction, braking, and skid tests are also conducted on these surfaces. Tire manufacturers are required to submit the test data along with the tires to the automobile manufacturers who may or may not conduct their own tests before certifying a tire for use as original equipment on the vehicle.

Economic Aspects

The tire industry has gone through radical changes since the turn of the century. The cost per kilometer for a passenger tire has been substantially reduced. For example, in 1900 a passenger tire cost $100, required an innertube worth $20 to be fully operational, and lasted only 800 km. A radial tire produced in 1987 can cost less than $70 and last more than 80,000 km. The evolution of the pneumatic tire has followed the diversification in vehicles ranging from passenger automobiles to trucks, airplanes, farm tractors, and mining and earthmover vehicles. Improvements in tire properties have paralleled the advances made in vehicle design and performance.

It is expected that, before 1995, the growth rate will fall to 1.5%/yr, based on reduced motor vehicle demands in heavily industrialized countries such as the United States, Canada, Japan, and Western Europe. This low growth rate reflects a low growth rate of the global economy, increased penetration of longer-life radial tires, lower average distances driven because of high oil prices, and growth of the retreading industry.

BIBLIOGRAPHY

F. J. Kovac, *Tire Technology,* The Goodyear Tire & Rubber Co., Akron, Ohio, 1978.

S. K. Clarke, *Mechanics of Pneumatic Tire,* U.S. Government Printing Office, Washington, D.C., 1981.

Motor Vehicle Safety Standard No. 109, New Pneumatic Tires—Passenger Cars, U.S. Dept. of Transportation, Federal Highway Administration, National Highway Safety Bureau, effective January 1, 1968.

W. H. WADDELL
R. S. BHAKUNI
S. K. MOWDOOD
I. S. RAI
D. L. KNIGHT
The Goodyear Tire & Rubber Company

TISSUE REGENERATION.

See SKIN, REGENERATION TEMPLATES.

TOLYLENE DIISOCYANATE.

See POLYURETHANES.

TOOLING WITH PLASTICS

The term plastic tooling covers a variety of applications. It includes the construction of master models, patterns, trim tools, drill jigs, dies, prototypes, and checking and assembly fixtures and anything else required to produce a final article. Liquid resins are used that harden into approximately the desired shape. Hand finishing is limited to repair and minor modifications. Because tooling resins shrink slightly on curing, care must be taken to reproduce shapes within close tolerances. The shrinkage of plastics is uniform in all directions and the dimensions on full cure can be accurately predicted.

All types of tooling are derived from the master model, which is a critical element and the basis for all reference measurements. The use of plastics has progressed from a "black art" to a modern technique based on sound chemical and engineering principles. Plastics have advantages and limitations (Table 1) and their selection by the tool engineer should be made on a cost performance basis.

Materials

For most applications, thermosetting resins are used to construct plastic tooling. These materials offer the advantages of strength, rigidity, dimensional stability, and higher operating

Table 1. Plastic and Metal Tooling

Plastic	Metal
Advantages	
lower cost	longer tool life
lower weight	established performance history
similar coefficient of thermal expansion to composite parts	separate master may not be required
easy to repair and modify	
equipment for composite-part fabrication can be used for tool fabrication	
Disadvantages	
separate master required for tool fabrication	higher cost
limited performance history in some applications	higher weight
limited tool life in some applications	thermal expansion difference between tool and composite part
	repaired and modified usually by outside personnel
	specialized equipment required for tool fabrication
	tool quality and delivery are controllable by outside personnel

temperatures than thermoplastics. They are generally supplied as liquid two-component systems except for modeling blocks and prepregs. Plastic tools can be constructed from various chemical classes of plastics. For rigid tool construction the most common types are epoxy and unsaturated polyester resins. Flexible tooling applications may utilize modified versions of these or use silicone or polyurethane elastomers. For high-use-temperature systems, bismaleimides and cyanate esters are currently being investigated for possible tooling applications. Although in many cases the plastic is used in conjunction with wood, plaster, or metal, the finished product is usually classed as a plastic tool. Materials include phenolic resins, unsaturated polyester resins, epoxy resins, polyurethanes, and silicones.

Functional Classification of Materials

Classifications of materials include laminating compounds (the wet lay-up systems, and the prepreg laminating systems), surface coats (gel coats), casting systems, paste systems, and modeling blocks.

Tool Fabrication

Plastic tools are fabricated by five methods: lamination, surface casting, mass casting, construction from paste materials, and numerical or computer-controlled machining. The method is selected according to the criteria shown in Table 2.

Applications

Plastics are used in tooling to best advantage when complex contours and fine detail are encountered. Their ability to be cured in the desired shape accounts for significant savings compared to the costs of machining metal. When a multiple number of tools is required, a simple procedure can be employed to build plastic replicas. It is often less expensive to build several plastic tools than to invest in one metal tool. Savings can also be made in lead time, because plastic tools can be constructed quickly and are easy to repair or modify. They are lighter than metals, which is advantageous when tools are moved or heated. They are used in foundry applications, fixtures, dies, molds, and prototypes.

BIBLIOGRAPHY

H. Lee and K. Neville, *Handbook of Epoxy Resins,* McGraw-Hill Inc., New York, 1967, chapt. 18.

Silicone Moldmaking Materials from Dow Corning, Dow Corning Corp., Midland, Mich., 1984.

Epoxy Plastic Tooling Manual, CIBA-GEIGY Formulated Systems Group, Lansing, Mich., 1982.

J. J. Morena, *Advanced Composite Moldmaking,* Van Nostrand Reinhold, New York, 1988.

George Sollner
CIBA-GEIGY Corporation

Table 2. Desired Characteristics of Plastics for Tools[a]

	Lamination	Metal-core surface casting	Mass casting	Paste construction	Machined blocks
dimensional stability	1	2–3	4	2–3	2
shrinkage during cure	1	2–3	4	2–3	1
least weight	1	4	3	2	2
labor cost	4	3	1	2	1
material cost, wt basis	4	3	1	2	3
strength	1	2	3	4	3
toughness	1	2	4	3	3

[a] 1 = most desirable, 2 = satisfactory, 3 = fair, 4 = least desirable.

TORSIONAL BRAID ANALYSIS.

See CHARACTERIZATION OF POLYMERS; DYNAMIC MECHANICAL PROPERTIES; TEST METHODS; VISCOELASTICY.

TOXICITY.

See SAFETY; TOXICOLOGY.

TOXICOLOGY

Definition and Concept

The precise definition of toxicology varies among different authors and teachers, depending on their background, specialization, and breadth of interest and knowledge. However, central to any meaningful definition is the concept that toxicology is concerned with assessing the potential of chemicals, or mixtures of them, to produce harmful effects in living organisms. The following definition covers the various activities in the discipline:

> Toxicology is a study of the interactions between chemicals and biological systems in order to quantitatively determine the potential for the chemical(s) studied to produce injury that results in adverse effects in living organisms and to investigate the nature, incidence, mechanism of production, factors influencing, and reversibility of such adverse effects.

In the context of this definition, adverse effects are taken to mean those that are detrimental to the survival or normal functioning of the individual. By this definition toxicity is the potential for a chemical to produce adverse effects. This definition highlights the following important aspects of toxicity:

1. Materials, or their conversion products, must come into close structural or functional relationship with the tissue(s) or organ(s) they may affect. In this manner they can physically or chemically interact with particular biological components in order to effect the toxic response.

2. If possible, there should be measurement of the observed toxic effect or of a correlated result of the effect in order to quantitatively relate the observation made to the degree of exposure to the chemical (exposure dose).

3. Laboratory studies to determine the toxicity of a material include observations on exposed whole animals (*in vivo* procedures) and studies on isolated tissues, organs, or cells, homogenates of tissues, chemical extracts of tissues, and cultured lower organisms (*in vitro* procedures). The primary aim of such studies is to determine the potential for harmful effects on the intact living organism, usually with an ultimate goal of assessing the significance of the findings with respect to humans.

4. Any series of toxicological studies on a material should ideally attempt to define the following characteristics of a toxic response: the nature of the harmful effect(s), ie, the basic structural or biochemical injury produced; the incidence and severity of the effects produced as a function of the exposure dose, ie, the dose–response relationships; the mechanism(s) by which the toxic effect(s) are produced, ie, the fundamental chemical and biological interactions and the consequent biochemical and biophysical aberrations that are responsible for the initiation and maintenance of the toxic response; any factors that modify the toxic response in addition to dose, such as route of exposure, species, sex, method of presentation (vapor vs particle), or the presence of other chemicals; the detection of particular toxic effects by the development of methodologies and approaches for the recognition and quantification of specific adverse effects produced; and whether the toxic effect is reversible spontaneously after cessation of exposure (healing) or induced by antidotal or other procedures (treatment).

Toxicity describes the nature of the adverse effects produced and the conditions necessary for their induction; it is the potential of a material to produce injury to biological systems. Toxicity is to be clearly differentiated from hazard, which is the likelihood that a particular material exhibits its known toxicity under specific conditions of use. There is also a need to differentiate between toxicological and pharmacological effects, as the latter may occur in the context of certain occupational exposures. Pharmacological effects imply an interaction between the active chemical (usually a drug) and receptor molecules on or in cellular structures, the result of which is either to simulate normal physiological or biochemical effects (an agonist action) or to block these effects (an antagonist action) with the prime practical objective usually being a therapeutic one. The principal determinants of a pharmacological action are that it is reversible because of the dissociation of the receptor–effector molecule complex, and that at therapeutic doses it should occur without any toxic effects.

The current markedly expanded practice of toxicology has led to its subdivision into various specialist areas. According to professional involvement, toxicology can be classified into the following subspecialities: clinical toxicology, veterinary toxicology, forensic toxicology, occupational toxicology, aquatic toxicology, pharmacological toxicology, toxinology, regulatory toxicology, and environmental toxicology. Toxicology may be also subdivided into specialties according to tissue or functional concerns.

Classification of Toxic Effects

Figure 1 gives the basis for the general classification of toxic effects according to the degree of exposure and site affected. To be able to cause a toxic effect, a substance must come into contact with an exposed body surface. This may be the skin, eye, or the lining membranes of the respiratory or alimentary tracts. These are referred to as the routes of exposure. In some experimental and therapeutic situations there may be special routes of exposure, including injection under the skin (subcutaneous), into the venous system (intravenous), into muscles (intramuscular), or into the abdominal cavity (intraperitoneal). Toxic effects may be produced where a substance initially comes into contact with the body; these are referred to as local effects. However, material may be absorbed into the circulatory system from the exposure site and thus disseminated to various body organs and tissues. As a result, toxic injury may be produced in tissues and organs remote from the site of primary contamination; these are referred to as systemic effects. They may be produced by the absorbed parent material or by products of conversion usually formed after exposure. Systemic toxicity may be restricted to one organ or tissue or may be present in multiple tissues and organs. Many materials may produce both local and systemic toxicity.

Because the nature of a toxic effect and the probability of its developing is often related to the number and frequency of exposures, as well as the magnitude and route of exposure,

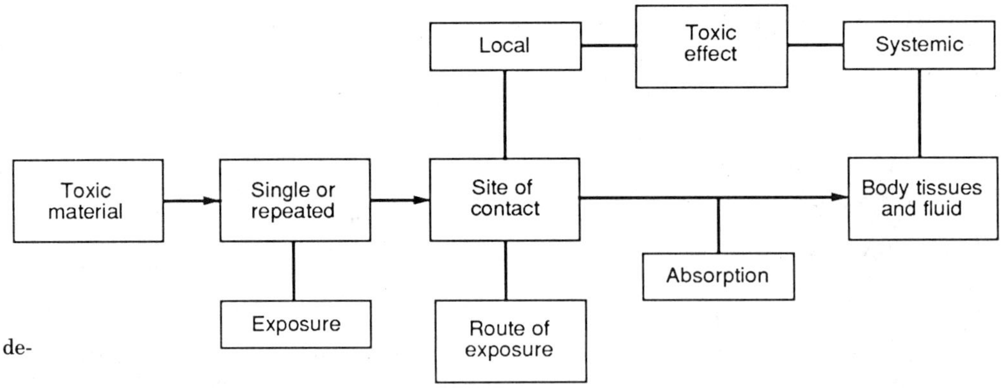

Figure 1. Basis for the classification and description of toxic effects.

toxic effects may also be classified according to the number of exposures that elicit a toxic response.

An overview of the classification of toxic effects is given in Table 1; examples of different types of toxicity classified according to the above definitions are given in Table 2. The same material may produce different toxicities depending on the precise conditions of exposure. Therefore, when describing the toxicity of a material, it is necessary to note whether the effects are local, systemic, or mixed; the nature of the injury; the organs and tissues affected; the conditions of exposure (including route, number, and magnitude of exposure); and the species investigated.

Nature, Expression, and Mechanisms of Toxicity

The response of biological systems and processes to injurious chemicals may take many forms depending, among a number of other factors, on the physical and chemical properties of the toxicant and the conditions of exposure. The more important and frequently seen types of morphological and biochemical injuries include inflammation, degeneration, necrosis, enzyme inhibition, biochemical uncoupling, lethal synthesis, lipid peroxidation, covalent binding, receptor interactions, immune-mediated hypersensitivity reactions, immunosuppression, neoplasia, genotoxicity, developmental and reproductive toxicity, and pharmacological effects. They may take the form of tissue pathology, aberrations of growth processes, altered or aberrant biochemical pathways, or extreme physiological responses. Various mechanisms may produce a toxic effect, eg, enzyme inhibition, uncoupling, lethal synthesis, lipid peroxidation, and receptor interaction.

Table 1. Classification of Toxic Effects

Basis	Description	Comments
location	local	site of primary contamination
	systemic	single or multiorgan
	mixed	local and systemic
exposure	acute	single exposure
	short-term	few days to few weeks
	subchronic	10–15% of life-span
	chronic	life-span
onset	immediate	during or immediately after exposure
	latent	delay between exposure and toxic effects
duration	transient	reversible
	persistent	permanent

Dose–Response Relationships and Their Toxicological Significance

Toxicity describes the nature of biological injury produced and the conditions necessary for its induction. Dose is an important aspect; that is, the amount of material to which the organism is exposed. Dose is usually expressed as exposure dose, the total amount of material to which the organism is exposed. Although exposure dose may be highly relevant and correlated with local toxicity at the site of exposure because of the effects of various exogenous and endogenous effects on absorption, it may not necessarily be a good index of the dose of material (or reactive metabolite) received at a specific systemic target organ. Clearly, if such information exists, it may be scientifically more appropriate to relate systemic toxicity to absorbed and target-organ dose and, in turn, relate these to the environmental exposure dose. In this manner the most reasonable assessment of hazard and the relationship between environmental exposure dose and probability of tissue injury can be made. However, because of time and economic constraints, most information on dose in toxicology studies relates to the exposure dose.

Dose–response considerations are important. Because of the influence of biological variability on toxicity, a clear positive dose–response relationship is good evidence for a causal relationship between exposure to a given material and the response (injury) obtained. A plateau to such a relationship may be imposed by the limitation of toxic species generation following saturation of metabolic pathways. The quantitative dose–response information on a given chemical provides valuable information on the average and range of sensitivity of the population to the chemical, the possible existence of a higher than normal susceptible population, threshold doses, and no-effect doses. By appropriate and careful examination of data, collected as far as possible under similar conditions, it is possible to compare and contrast quantitatively the toxicity of a given material between species, or of different materials to the same species.

Acute lethal toxicity studies are conducted by giving different doses of the chemical under controlled conditions. Lethal toxicity may be compared by using particular levels of mortality; the most frequently used numerical comparison is mortality at the 50% level, as the largest proportion of deaths is distributed (usually) equally around this point. This numerical

Table 2. Toxicity Classification According to Exposure Time Scale and Site Affected

Exposure	Site	Example	Chemical
acute	local	tissue corrosion	sulfuric acid
		lung damage	hydrogen chloride
	systemic	hemolysis	arsine
		kidney injury	ethylene glycol
	mixed	lung injury	nitrogen oxides
		methemoglobinemia	
		skin injury	hydrogen cyanide
		cytotoxic hypoxia	
short-term	local	sensitization	ethylenediamine
		ulceration	chrome salts
	systemic	peripheral neuropathy	ethylene oxide
		ototoxicity	toluene
	mixed	respiratory irritation	pyridine
		kidney injury	
		skin irritation	boric acid
		neurobehavioral	
chronic	local	bronchitis	sulfur dioxide
		laryngeal carcinoma	nitrogen mustard
	systemic	liver angiosarcoma	vinyl chloride
		leukemia	benzene
	mixed	emphysema	cadmium
		kidney injury	
		pneumonitis	manganese
		central neurotoxic	
latent	local	pulmonary edema	phosgene
	systemic	peripheral neuropathy	organophosphates
		lung fibrosis	paraquat

reference point is referred to as the (median) lethal dose$_{50}$ (LD_{50}) or concentration$_{50}$ (LC_{50}).

Factors Influencing Toxicity

During the design, conduct, and interpretation of toxicology studies, it is important to be aware of the fact that a large number of factors and variables may influence the nature, severity, probability of induction, and incidence of toxic injury. Factors to consider include species and strain, age, nutritional status, time of dosing, environmental physical factors, and dosing characteristics.

Absorbed Chemicals and Systemic Toxicity

The induction and severity of systemic toxicity may result from a complex interrelationship between the absorbed parent material and tissue-conversion products, their distribution and concentration in body tissues and fluids, and their excretion from the body. The pathway followed by a chemical after its absorption is shown in Figure 2.

Routes of Exposure

In order to induce a toxic effect, local or systemic, the causative material must first come into contact with an exposed body surface, presenting the routes of exposure. Under normal circumstances, depending on the nature of the material, the practical routes of exposure are by swallowing, inhalation, and skin and eye contact. In addition, for investigational and therapeutic purposes, it may be necessary to consider intramuscular, intravenous, and subcutaneous injections as routes of administration.

Both the metabolism of a material and its potential to cause toxic injury may vary with the route of exposure, although the magnitude of the dose and duration of dosing may influence this relationship. For example, materials that are metabolically activated by the liver are likely to exhibit greater toxicity when given perorally than when absorbed across the lung or skin. This is mainly due to the amounts distributed to the liver by the different routes of exposure.

Exposure Conditions and Absorbed Dose. With respect to environmental exposure conditions, the probability of adverse effects occurring depends on many factors, especially magnitude, duration, and number of exposures. These determine the amount of material to which an organism is exposed (the environmental exposure dose), and hence to the amount of material available for absorption (the absorbed dose). The absorbed dose determines the amount of material available for distribution for toxic metabolite formation, and thus the likelihood of inducing a toxic effect. Opposing absorption and metabolite accumulation is elimination. Thus, for a given environmental exposure situation, the probability of inducing toxicity depends on a dynamic equilibrium between the rate of absorption and the rate of excretion of absorbed parent material and its metabolites.

The principal factors that determine the likelihood and severity of toxic effects being produced are the rates at which the causative substances (parent molecule, toxic metabolite, or both) reach the tissues and the absolute amounts of materials to which the tissues are exposed. These determine the dose of material received at the target tissues.

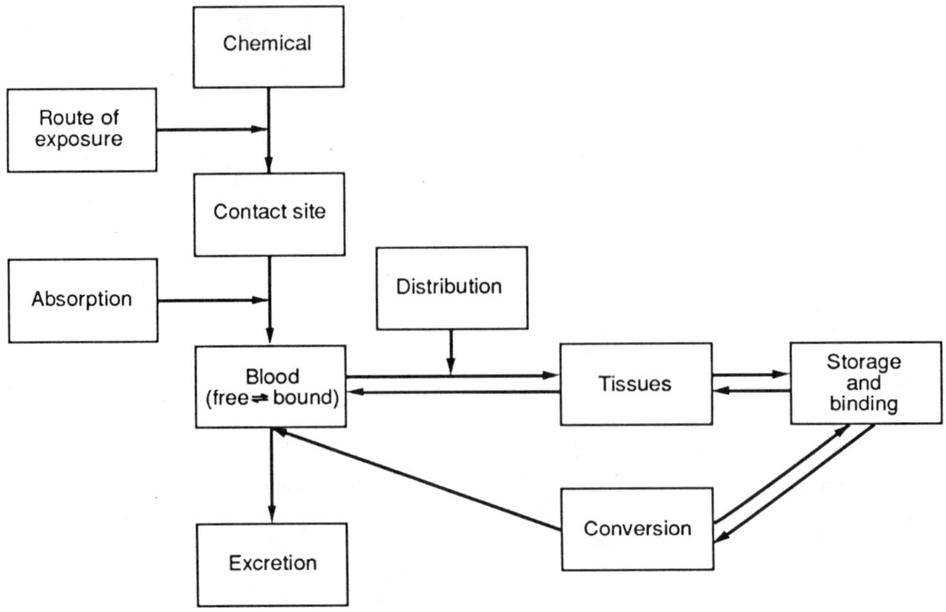

Figure 2. Schematic representation of the possible fate of a chemical absorbed from the primary contact site.

Mixed Exposures

Although most toxicology studies are carried out on single chemicals, or a material in an inert vehicle, in most occupational situations exposure to several chemicals is involved. Such exposures may be coincidental, prior, or successive, affording a potential for complex biological interactions. Examples are enhancement of 2,5-hexanedione-induced peripheral neurotoxicity by acetone and the influence of toluene on blood ethanol concentrations because of the reciprocal competitive inhibition in their metabolisms. The evaluation of hazard from multiple chemical exposures can be a much more demanding procedure than that from exposure to single chemicals and is currently a much less satisfactory procedure.

Testing

The goal of toxicology studies should be to allow a quantitative determination of the potential for a chemical to produce local and systemic adverse effects and permit the definition of factors that may influence the nature, severity, and possible reversibility of these effects. A toxicology testing program should allow a determination of the nature of the adverse effects; relationship of the adverse effects to in-use situations; dose–response relationships (average, range, hypersensitive groups, no-effects, and minimum effects doses); modifying factors; effects of gross overexposure and repeated overexposure; recognition of adverse effects; definition of allowable and nonallowable exposures; definition of monitoring requirements; guidance on personal and collective protection as well as first-aid and medical management; and definition of "at-risk" individuals (sex, preexisting disease, genetic susceptibility, etc).

The cost of toxicology studies varies from a few thousand dollars for simple acute toxicity tests to perhaps $\$10^6$ or more for an extensively monitored chronic inhalation study.

Testing procedures are either general or specific. In general toxicological studies animals are exposed to a test material under appropriate conditions and examined for all types of toxic effects that the monitoring procedures allow. In specific toxicological studies the exposed animals are monitored specifically for a defined toxic end point or effect.

Guidelines. Many guidelines need to be followed. The number of animals should be sufficient to allow a quantitative determination of the average response and the range of responses, including the possible presence of hypersensitive individuals. The choice of species should, where possible, allow hazard evaluation and risk assessment from animal models most closely approximating humans. Objective procedures should allow valid statistical comparison to be made between treated and control groups.

To assess the functional status of the living animal and detect injury in dead animals various observations are made, eg, body weight, food and water consumption, hematology, chemical pathology, urinalysis, pathology, and organ weight determinations. Frequently employed specific toxicology studies include those for primary irritation, sensory irritation, immune-mediated hypersensitivity, neurological and behavioral toxicology, teratology, reproductive toxicity tests, metabolism and pharmacokinetics, genotoxicity, combustion toxicology, and antidotal procedures.

Review and Interpretation of Toxicology Studies

The review and interpretation of toxicology studies is a professional matter, requiring experience in both the laboratory conduct of such studies and the practice of applied toxicology. Although all studies should be reviewed on a case-by-case basis, some general considerations should be kept in mind. It should be established that the laboratory reporting the study has the necessary scientific credibility, capabilities, experience, and expertise in the area being investigated. The objectives of the study should be precisely stated, and it should be confirmed that the overall design of the study satisfies the objectives. The work should be presented in a clear and unambiguous way, with all necessary detail to allow the reviewer to undertake his or her own assessment and conclusions.

It should be confirmed that adequate quality control procedures and good laboratory practices have been followed.

The study should be designed to allow a determination of the significance of results and permit hazard assessment processes. And it should be possible to determine whether any injury produced is a direct result of toxicity or a secondary effect.

Extreme caution is needed to determine if numerical data have been correctly assessed.

The above considerations indicate a need for careful design of toxicological studies, taking into account all factors inherent in the defined objectives of the study. The relevance and credibility of a study can be no better than its study design and conduct allow.

Polymers

Although polymers are generally regarded as being biologically inert or of low toxicity because of their high molecular weight and nonabsorbable characteristics, several aspects of polymer production and use are of toxicological relevance and concern, including the high toxicity of many monomers, catalysts, and additives used in the synthesis of polymers; the presence of extractable residual monomer in the final polymer product; the toxicological features of low molecular weight oligomers and additives (eg, stabilizers, catalysts); toxicity by metabolism of the polymer reaching the systemic circulation from unusual routes of exposure; toxicity by specific route of exposure; solid-state carcinogenesis and foreign body reactions by chronic exposure; and toxic products from the combustion of polymers.

BIBLIOGRAPHY

B. Ballantyne, ed., *Perspectives in Basic and Applied Toxicology,* John Wright & Sons Ltd., London, 1988.

E. Hodgson and P. E. Levi, eds., *A Textbook of Modern Toxicology,* Elsevier, New York, 1987.

United States Environmental Protection Agency, *EPA Toxicology Handbook,* Government Institutes, Inc., Maryland, 1988.

P. L. Williams and S. L. Burson, eds., *Industrial Toxicology,* Van Nostrand Co., Inc., New York, 1985.

BRYAN BALLANTYNE
Union Carbide Corporation

TOYS

Plastics have been used in toys from the beginning of the plastics industry. In response to an identified market demand for a less expensive material to substitute for ivory balls in a somewhat sophisticated toy, the game of billiards, John Wesley Hyatt discovered cellulose nitrate in 1868. This ushered in the beginning of the plastics industry (see CELLULOSE ESTERS, INORGANIC).

The acceptable cost for volume production of realistic looking, and to an ever increasing degree, functional, participatory, and interactive products was the driving force that pulled plastics into the toy industry. Easy moldability of plastics into complex shapes contributed substantially to both cost and realism factors. The wide range of color variability available with plastics contributed similarly to realism. The improved safety and hygiene offered by plastics over competitive materials became positive factors assisting in market acceptance of the use of plastics in the toy industry.

In addition to the wide use of plastics in conventional toys (dolls, model vehicles, games, musical instruments, etc) an important effect of the advent of plastics has been to facilitate the use of toys as a part of the educational process.

Several of the basic plastic processing techniques have themselves been simulated into toy replicas of industrial equipment. Examples include toys featuring such functions as the molding of articles of a liquid poly(vinyl chloride) (PVC) plastisol or a doughlike plastic material, thermoforming of plastic sheet, and the extrusion of simple shapes.

Materials

The principal plastics used in toys are the commodity resins: high density polyethylene (see ETHYLENE POLYMERS), polystyrene (see STYRENE POLYMERS), polypropylene (see PROPYLENE POLYMERS), and poly(vinyl chloride) (PVC) (see VINYL CHLORIDE POLYMERS).

Although used in lower volume than the commodity resins, specialty plastics such as nylon, acrylonitrile–butadiene–styrene (ABS), and acetal resins have found acceptance in toys for such mechanical working parts as gears, cams, bushings, drives, clutches, and levers. Acrylonitrile, methyl methacrylate, and polycarbonate compounds are used for varying degrees of transparency and mechanical strength. Polyurethane foams are used for lightweight throwing balls, as body board and surfboard filler, and as sound- and vibration-damping material for mechanical drive mechanisms. Expanded polystyrene moldings have been used for toy furniture blocks, to provide simulated bones in animal skeletons, and for packaging. Extruded fibers of various fineness of nylon, poly(vinylidene chloride) (PVDC), and polypropylene are used as simulated hair and bristles and for fabrics. Extruded poly(methyl methacrylate) rods are used to transmit colored patterns.

Toys that use properties peculiar to plastics include the small figures of animals molded of cross-linked polyethylene. After molding, the figures are pressed into small squares under high pressure. When the squares are heated in a toy oven, a child can watch the "plastic memory" of the parts turn the squares back into the shape of the original figure.

Processing

A wide variety of processing techniques is used in the toy industry. Injection molding, blow molding, and rotational molding—all of which operate directly from compounded polymers and have the capability of producing substantially finished products as well as parts for subsequent assembly—are the most prominently used. Various other processes are used to produce special effects, sometimes in combination with one or more other major processes.

BIBLIOGRAPHY

"A Quick Test for Toy Safety," *Consumer Reports,* 653 (Nov. 1987).

"Outlook on Toy Packaging," *Playthings,* 58 (Sept. 1987).

"High-Tech Toys Need Constant Technological Innovations to Deliver Big Numbers," *Toy and Hobby World,* 26 (Feb. 1987).

"Toys—A Triumph of Design," *Mod. Plast.,* 42 (Dec. 1987).

GEORGE E. PICKERING
Arthur D. Little, Inc.

TRANSESTERIFICATION. See POLYESTERS.

TRANSFER. See CHAIN TRANSFER.

TRANSFER MOLDING. See COMPRESSION AND TRANSFER MOLDING.

TRANSITIONS AND RELAXATIONS

TRANSITIONS AND RELAXATIONS

Transition regions are narrow intervals of temperature and frequency over which polymer properties change by large amounts, the most prominent examples of which are the glass transition (qv) and the crystal melting transition (see MORPHOLOGY; CRYSTALLIZATION KINETICS, and CRYSTALLINE POLYMERS). The term transition is employed in these cases to signify the dramatic alteration of macroscopic material properties as the temperature interval is crossed, and to emphasize the similarity of the transformation to true thermodynamic phase transitions in small-molecule systems. Although the midpoints of the temperature regions associated with the glass–liquid (T_g) and crystal–liquid (T_m) transitions are moderately sensitive to measurement frequency, the relationship between transition temperature and heating or cooling rate is essentially logarithmic and normal laboratory variations in these rates cause transition temperature shifts of only a few degrees. Figure 1 shows the temperature variation of Young's modulus E_1 of atactic polystyrene of molecular weight $M_n = 217{,}000$ through its glass transition region. As the polymer is viscoelastic, its tensile modulus E is time-dependent (see MECHANICAL PROPERTIES; VISCOELASTICITY) and the three curves in Figure 1 display the temperature dependence of E measured at 10, 100, and 1000 s after application of the load.

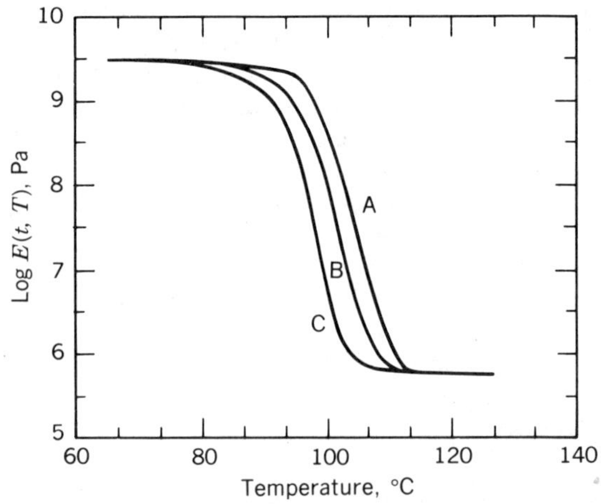

Figure 1. Tensile modulus of atactic polystyrene at different times (A, 10 s; B, 100 s; C, 1000 s) as a function of temperature near the glass transition. To convert Pa to dyn/cm^2, multiply by 10.

If the glass transition temperature T_g is defined as the point of maximum slope on the modulus–temperature plot, Figure 1 indicates that a hundredfold increase in measurement rate shifts T_g by about 6°C.

Although mechanical response is the most sensitive indicator of thermal transitions, the glass transition in uncrosslinked amorphous polymers is detected in nearly all measurements of system properties.

Thermodynamic Changes at T_g

The onset of solidlike rigidity in an amorphous polymer at T_g is accompanied by sharp reductions in heat capacity C_P, thermal expansion α_P, and compressibility κ_T. Taken together with the change in shear modulus, these phenomena provide clear evidence for the loss of molecular configurational freedom which takes place near the transition, although the precise nature of the freezing mechanism and exact specification of the degrees of freedom affected are not known either from direct observation or via computer–mathematical modeling (for realistic molecular models). This dearth of microscopic information is consistent, however, with the relatively undeveloped state of knowledge of equilibrium first-order processes such as melting and fusion of small organic compounds. Experimental and computational difficulties arise in studying phase transformations in organic materials because of the smallness of the cohesive energies (which makes them difficult to measure calorimetrically or spectroscopically), the importance of surface effects, and the (computational) need to study large samples with coexisting phases.

An early, useful approach to the thermodynamics of the glass transition is the order parameter model, which assumes that a configurational equilibrium process occurring in the high temperature (liquid) region freezes out at T_g, thereby leading to reduced thermal expansion, compressibility, and heat capacity. From the dramatic rise in viscosity which is always seen, it is clear that translational freedom of individual polymer segments disappears or is sharply reduced near T_g. Figure 2 illustrates a simple (free) energy diagram which can form the basis for analyzing segment motions and equilibria in a supercooled polymer melt, and may therefore be used to calculate a positional order parameter and to estimate thermodynamic discontinuities at T_g.

Overall, the order parameter model provides both a simple physical interpretation of thermodynamic changes at T_g and a semiquantitative estimate of their magnitude. It does not, however, explain why segmental motion freezes in, and in the absence of knowledge of the two-state parameters $\delta\epsilon$ and δs, it does not lead to predictions of T_g, and therefore is unable to explain how T_g will vary with molecular weight, composition, and chemical structure. The free-volume theory and the Gibbs-DiMarzio configurational entropy theory are the two most important attempts to explain why molecular motions eventually stop in a supercooled liquid, and hence why the glass transition takes place.

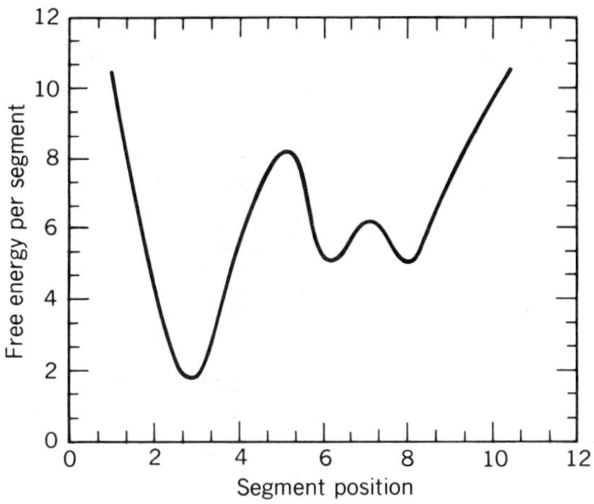

Figure 2. Schematic free-energy surface for segment configurational displacements.

Free-volume Theories of T_g

Suggestions that volume might be a better variable than temperature in terms of which to discuss liquid flow date at least to 1913 when Batchinski proposed that the viscosity η of a liquid and its volume V were related by

$$\eta = \frac{\eta_0}{V - b} \tag{1}$$

where b is approximately the van der Waals excluded volume for the liquid in question. Equation 1 implies a direct (linear) relationship between the mobility of the liquid molecule and the free or excess volume over that of the hard-core diameter. In a study of the influence of pressure on liquid viscosities, it was found that for nonpolar liquids, temperature coefficients of viscosity at constant volumes are small. This suggests that the temperature variation of viscosity at constant pressure is chiefly due to the thermal expansion, and that the direct effect of temperature is less important. Careful studies of n-alkanes revealed that their viscosities are nearly exponential rather than linear functions of the "free volume" as equation 1 suggests. Therefore, equation 2 was proposed instead:

$$\eta = Ae^{BV_0/V_f} = Ae^{B/f} \tag{2}$$

as a general relation where V_0 and V_f are the "occupied" and "free" volumes of the liquid respectively, $f \equiv V_f/V_0$ is the fractional free volume, and A and B are empirical constants. The van der Waals b for occupied volume is not used, but if V_0 is defined as the volume of the liquid extrapolated to zero degrees Kelvin, and $V_f = V - V_0$, equation 2 gives excellent fits for the viscosities of the n-alkanes. Equation 2 forms the basis for the Williams-Landel-Ferry equation (WLF), discovered a few years later and found to describe the temperature dependence of relaxation times in polymers and other glass-forming liquids above T_g. The ratio of a mechanical or dielectric relaxation time, τ_m or τ_d, at a temperature T, to its value at an arbitrary reference temperature T_0 can be represented by a simple empirical, nearly universal function. Writing the ratio as a_T,

$$a_T \equiv \frac{[\tau_m]_T}{[\tau_m]_{T_0}} = \frac{[\tau_d]_T}{[\tau_d]_{T_0}} \tag{3}$$

results can be represented by

$$\log_{10} a_T = \frac{-C_1^0(T - T_0)}{(C_2^0 + T - T_0)} \tag{4}$$

with $C_1^0 \approx 17.44$ and $C_2^0 \approx 51.6$ K providing a good representation of relaxation data as a function of temperature for a selection of more than 20 polymer systems.

A significant conclusion of the WLF analysis is that the free-volume fraction at T_g is nearly a constant, $f_g = 0.025$. The resulting isofree-volume picture of the glass transition is of both theoretical and practical importance. Theoretically it suggests an equilibrium criterion for the glass transition and (in principle) a methodology for calculating T_g and its thermodynamics from a specific model. This approach requires a microscopic definition of free volume, and statistical mechanical models have been developed. In practice, it provides a clear and simple way to rationalize and predict influences of molecular weight, diluents, and copolymer composition on the glass temperature T_g.

If the free-volume fraction f is constant at T_g (approximately 2.5% of the total volume), it follows that the glass transition temperature for a polymer can be changed by altering its free-volume fraction at a fixed temperature. One way to accomplish this is to "dissolve" or mix with the polymer on a molecular level another molecular unit which has a larger free-volume fraction at the same temperature. Chain ends have larger free volumes since they are only attached by one covalent bond to the rest of the polymer molecule. If the extra free volume of a chain end is designated θ, then the extra free volume per cm^3 due to chain ends in a polymer sample with bulk density ρ and molecular weight M is

$$\left(\frac{2\rho N_A}{M}\right)\theta \tag{5}$$

where N_A is Avogadro's number. A polymer of finite molecular weight M_n must then be cooled to a temperature T_g below that of an infinite molecular weight polymer T_g^∞ so that the extra free volume is just compensated by the extra cooling due to thermal contraction:

$$\left(\frac{2\rho N_A}{M}\right)\theta = \alpha_l(T_g - T_g^\infty) \tag{6}$$

where α_l is the thermal expansivity of the melt. Thus

$$T_g = T_g^\infty - \frac{K}{M} \tag{7}$$

where $K = 2\rho N_A\theta/\alpha_f$, which states that a straight line should result if T_g values of different molecular weights M are plotted versus M_n^{-1}. Equation 7 is known as the Flory-Fox equation, and Figure 3 shows measured glass temperatures for several different styrene fractions studied by Flory and Fox, along with a least-squares straight line fit of the data.

In spite of its success, the free-volume theory has a number of shortcomings (as do all current theories of the glass transition) which are difficult to remedy or ignore, and this has led to alternative approaches to explain and predict transition behavior. Problems for the free-volume theory include the fol-

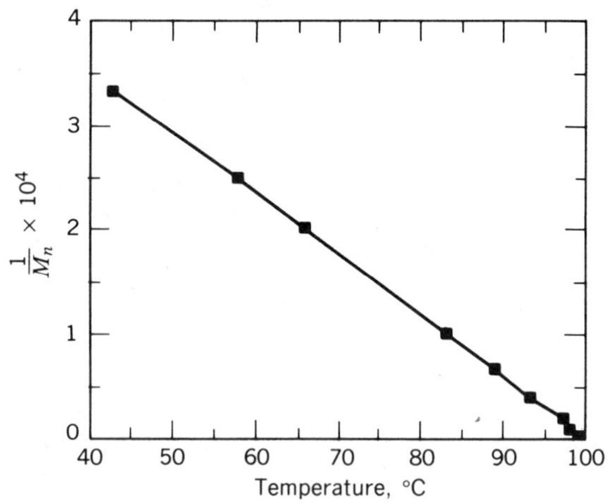

Figure 3. Glass-transition temperatures of polystyrene fractions plotted vs $1/M_n$.

lowing: (*1*) the definition of free volume is not unique—at least four different choices have some degree of acceptance; (*2*) it attributes molecular "freezing-in" at T_g to intermolecular packing difficulties, and appears to ignore the role of internal degrees of freedom in the glass-forming tendency; (*3*) it emphasizes kinetics, thereby downplaying the role of thermodynamics and the Kauzmann paradox; (*4*) it predicts that the viscosity of a supercooled liquid is independent of temperature at constant volume, which is not true.

The most notable alternative to the free-volume theory is the Gibbs-DiMarzio (GD) configurational entropy model. These authors presented theoretical arguments suggesting that even though the laboratory transition is kinetically controlled, it has as an underlying thermodynamic basis, a genuine second-order transition in the Ehrenfest sense. The latter refers to an order–disorder transition at which the first derivatives of the Gibbs free energy G, such as volume V and enthalpy H, are continuous:

$$V = \left(\frac{\partial G}{\partial P}\right)_T \tag{8}$$

$$H = G - T\left(\frac{\partial G}{\partial T}\right)_P \tag{9}$$

while the second derivatives, such as heat capacity and compressibility, are discontinuous:

$$C_P = \left(\frac{\partial H}{\partial T}\right)_P = -T\left(\frac{\partial^2 G}{\partial T^2}\right)_P \tag{10}$$

$$\kappa_T = -\frac{1}{V}\left(\frac{\partial V}{\partial P}\right)_P = \left(\frac{\partial^2 G}{\partial P^2}\right)_T \tag{11}$$

Although glass-forming substances show nearly discontinuous changes in their second derivative properties at T_g, it is technically incorrect to identify T_g as a second-order transition, since all measured properties are history-dependent and therefore not thermodynamic quantities. However, Gibbs and DiMarzio reasoned that if experiments could be carried out infinitely slowly, T_g would reach T_2, the "true" second-order phase transition temperature, and the discontinuities in C_P, κ_T, and α_P would also reach limiting equilibrium values. The origin of this theoretical second-order transition, according to GD, is collapse of the amorphous liquid structure to a hypothetical glass phase with zero (or relatively small) entropy. In spite of its original approximate character, the entropy approach has much to recommend it, and recent studies of relaxation and recovery kinetics near T_g point rather clearly toward the entropy as a critical parameter.

Enthalpy Relaxation and Volume Recovery

Both the free-volume and configurational entropy theories described above are idealized (or rather incomplete) since they treat an equilibrium (supercooled) polymer liquid or a frozen glass state, and explicitly avoid discussion of the complicated nonlinear, nonexponential behavior found in real systems near T_g. From a practical standpoint, knowledge of the approach of a polymer to equilibrium is needed to understand molding, extrusion, and thermoforming processes so as to be able to predict influences of process variables such as cycle time and mold pressure on product properties and quality. Manufacturing operations often take advantage of polymer properties near or above the T_g to fabricate parts in desired sizes and shapes easily, and sometimes modify postprocessing conditions to allow for the thermal and mechanical relaxation phenomena which accompany return of the polymer to room temperature. The nonlinear, nonexponential relaxation behavior near T_g results in final properties of the polymer varying from point to point in the sample, depending on part thickness, thermal history, solvent, and stress history. Formerly, it was customary to treat kinetic aspects of the glass transition as nonessential complications and to adopt simplified models appropriate for limiting (glass or liquid) behavior which, naturally, is never observed for real polymers in actual experiments. The last 20 years have seen considerable progress toward realistic modeling of glass transition kinetics. While borrowing from the classical models, these phenomenological approaches have also helped to clarify and refine the concepts of free-volume and configurational entropy, and have focused attention on the nature of the relaxation spectrum. When a liquid in equilibrium at a temperature T_1 is suddenly cooled to a temperature T_2, its structure has no time to adjust and its properties (such as volume V, enthalpy H, or index of refraction n) exhibit instantaneous, solidlike changes characteristic of the glassy state as illustrated in Figure 4. For example, the instantaneous change in the enthalpy H may be written

$$H_{U,T_2} - H_{R,T_2} = C_{U,P}(T_1 - T_2) \tag{12}$$

where $C_{U,P}$ is the unrelaxed (glassy) heat capacity at pressure P (ie, the rate of change of H with respect to temperature in the glass). $H_{R,T}$ and $H_{U,T}$ are the relaxed (equilibrium) and unrelaxed (glassy) values of the enthalpy H at temperature T. The instantaneous (glasslike) change in H is followed by a slower, liquidlike structural relaxation to a new equilibrium value, H_{R,T_2} appropriate to the liquid at temperature T_2 (Fig. 5). The total change in H is characteristic of the liquid slope:

$$H_{R,T_1} - H_{R,T_2} = C_{R,P}(T_1 - T_2) \tag{13}$$

where $C_{R,P}$ is the relaxed liquid heat capacity, which includes the glasslike (vibrational) changes as well as the structural

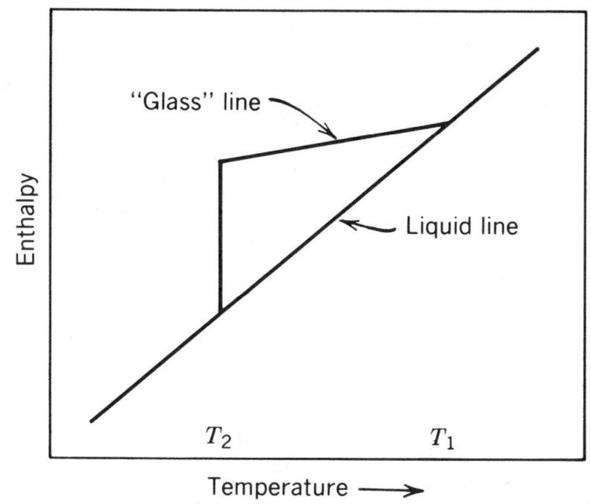

Figure 4. If a liquid in equilibrium at temperature T_1 is suddenly cooled to T_2, its enthalpy H must follow the "glass" curve since the structure has no time to adjust.

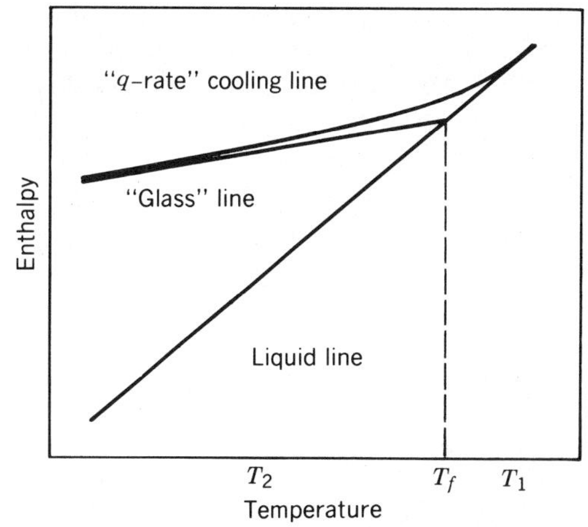

Figure 6. Evolution of the enthalpy during cooling history of Fig. 5.

relaxation contribution. The purely structural contribution is therefore

$$(C_{R,P} - C_{U,P})(T_1 - T_2) = \Delta C_P(T_1 - T_2) \tag{14}$$

If temperature equilibration is rapid compared with the structural changes, the time variation of H displayed in Figure 5**b** represents the isothermal structural kinetics. Instead of rapid temperature jumps, it is usually convenient to cool the liquid at a constant rate $q = dT/dt$ through the glass transition as shown in Figure 6 (again the enthalpy H is selected as the property to study).

During the early phase of cooling, the temperature is sufficiently high that the structure and enthalpy follow the liquid line. At lower temperatures the structural relaxation falls behind the temperature and the change of H with temperature is slower than if it followed the liquid line. Finally, far below the transition region, little structural relaxation occurs during the time of the experiment and the H–T curve follows the glassy slope.

To date, the most successful approach to modeling this behavior employs both the fictive temperature, T_c and the relaxation-time distribution method.

Secondary Relaxations

The thermodynamic changes which accompany relaxations other than T_g are smaller, and much less is known about them. Most experiments are relaxational, employing dielectrics or dynamic mechanical spectroscopy. Much attention has been directed toward low temperature relaxations of engineering polymers such as polycarbonate and polysulfone in the belief that these local motions are related to the superior mechanical properties of the glass. Selective deuteration, magic angle spinning, and multiple-pulse techniques have been employed to unravel the mechanism in polycarbonate.

BIBLIOGRAPHY

C. A. Angell and M. Goldstein, *Ann. N.Y. Acad. Sci.* **484** (1986).

G. W. Scherer, *Relaxation in Glass and Composites,* John Wiley & Sons, Inc., New York, 1986.

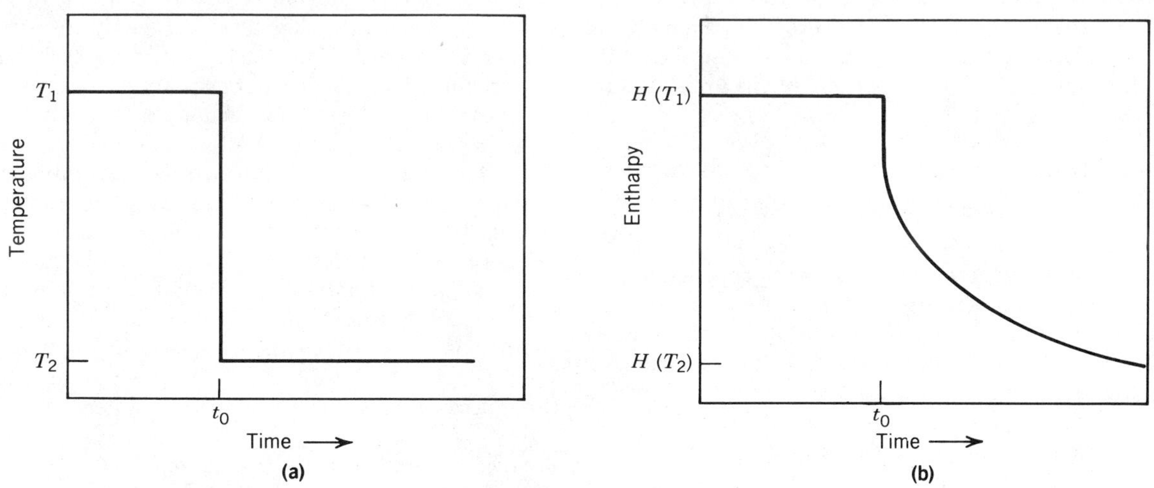

Figure 5. (**a**) Thermal history corresponding to Fig. 4. t_0 is the time of temperature jump; (**b**) change in enthalpy with temperature jump at t_0.

G. B. McKenna in C. Booth and C. Price, eds., *Comprehensive Polymer Science, Vol. 2: Properties,* Pergamon Press, Oxford, England, 1988.

N. G. McCrum, B. E. Read, and G. Williams, *Anelastic and Dielectric Effects in Polymeric Solids,* John Wiley & Sons, Inc., New York, 1967.

JOHN T. BENDLER
General Electric Company

T_{ll} AND RELATED LIQUID-STATE TRANSITIONS AND RELAXATIONS

This article is concerned with three liquid-state events occurring above T_g in amorphous polymers and copolymers: the intermolecular T_{ll} transition, the intramolecular $T_{l\rho}$ and $T_{l\phi}$ transitions, and the global T_f relaxation where the subscript f indicates a flow temperature. The primary emphasis is on T_{ll} and T_f. Crystalline polymers appear to exhibit a local motion associated with T_{ll}, but not motion of the entire macromolecule below T_m.

T_{ll} is now considered to arise from intermolecular segment–segment interactions, producing a three-dimensional, mobile physical network that "melts" at T_{ll}. This network is both thermal and shear reversible. T_{ll} increases with $\bar{M}_n$ and then levels off just as T_g does; it increases with frequency and with hydrostatic pressure; T_{ll} is decreased by diluents and suppressed by cross-links. T_{ll}, like T_g, represents an iso-free-volume transition.

$T_{l\rho} > T_{ll}$ is intramolecular in origin, independent of $\bar{M}_n$ and cross-links. It involves rotational motions about the chain axis of short segments, likely monomer units, similar to the T_β motion. T_f, on the contrary, involves melt flow and increases with molecular weight without limit. It represents an isoviscous state and hence is classified as a relaxation.

T_{ll} can be distinguished from other liquid-state events, including artifacts, by its tendency to occur at $(1.20 \pm 0.05)T_g$ (K), regardless of the structural type or nonflow method used to determine it. T_{ll} is of scientific interest not only because it appears to be a characteristic feature of amorphous polymers and copolymers, but also because of its implications for structure in atactic polymers; it is also of practical interest because its effects must be considered in many melt fabrication processes. Crystalline polymers appear to have an intermolecular transition T_u occurring near $1.2T_m$ (K) where T_m is the melting point; $T_u > T_{ll}$. Double glass transitions have been found in poly(dialkyl itaconate)s, but the upper one seems unrelated to T_{ll}.

Phenomenological Aspects

A long series of experimental programs by a number of authors, going back to 1949, have revealed the detailed nature of the polymeric liquid-state behavior with regard to dependence on molecular weight, frequency, hydrostatic pressure, diluent content, melt flow, and other variables. Many of these phenomena have contributed to an understanding of the nature of T_{ll}, $T_{l\rho}$, $T_{l\phi}$, and T_f. Figure 1 is a schematic plot showing T_g, T_{ll}, $T_{l\rho}$, $T_{l\phi}$, and T_f as a function of $\bar{M}_n$ and $\bar{M}_w$. The T_g locus is described by the classical Fox and Flory equation:

$$T_g(\text{K}) = T_g(\infty) - K_g\bar{M}_n^{-1} \tag{1}$$

where $T_g(\infty)$ is the asymptotic value of T_g at high $\bar{M}_n$ and K_g is a constant that, almost without regard to structure, increases exponentially with T_g. T_{ll} follows a similar pattern:

$$T_{ll}(\text{K}) = T_{ll}(\infty) - K_{ll}\bar{M}_n^{-1} \tag{2}$$

The structural dependence of K_{ll} is not yet known. Preliminary evidence indicates that $K_{ll} \cong K_g$; K_{ll} is expected to be slightly greater than K_g. T_{ll} and T_g for equations 1 and 2 should be determined by quasiequilibrium nonflow methods such as V_{sp}–T, C_p–T, and others where V_{sp} is specific volume and C_p specific heat.

The gradual evolution of the T_{ll} concept over the past 25 years has been attended by controversy, as summarized in *Polymer Yearbook,* Vol. 2. This controversy has raised three significant issues. (*1*) Is the normally weak event noted above T_g an artifact of the test method and hence not real? (*2*) If real, is it a relaxation, ie, a rate effect, and hence not a transition? (*3*) If it is a transition, what are its origin and nature? Brief answers are as follows. (*1*) If seen by different instruments, especially by different authors, and always at about the same temperature with respect to T_g, it is likely that it is real. (*2*) If real, and if it follows ABD of Figure 1 as a function of log molecular weight, it is a relaxation. (*3*) If it follows path ABC, it is likely that it is a transition. In view of the linear $T_{ll} \sim T_g$ relationship (eq. 3), T_{ll} is closely related to T_g. Several theoretical approaches are now available.

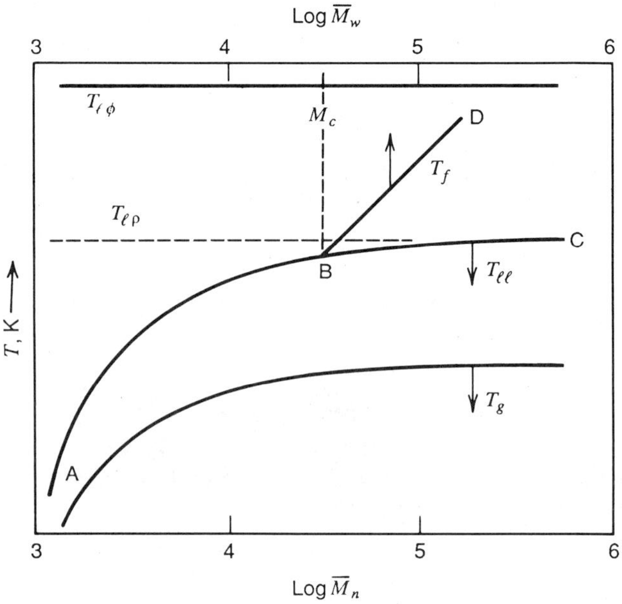

Figure 1. Schematic loci for T_g, T_{ll}, $T_{l\rho}$, and $T_{l\phi}$ as a function of $\bar{M}_n$ and for T_f and entanglement molecular weight M_c as a function of $\bar{M}_w$. T_g, T_{ll}, and T_f are intermolecular in origin; $T_{l\rho}$ and $T_{l\phi}$ are intramolecular. T_{ll}, $T_{l\rho}$, and $T_{l\phi}$ appear in a single V_{sp} vs T plot at $\bar{M}_n = 52{,}000$ and in a single log η_0 vs T^{-1} plot at $\bar{M}_n = 16{,}500$. $T_{l\phi}$ for a-PS is found at about 460–480 K, being higher by dsc than by log η_0. $T_{l\phi}$ in other polymers is 30–50 K above T_{ll}. $T_{l\phi}$ was originally designated T_{ll} (prime). Later, it was designated $T_{l\rho}$. $T_{l\phi}$ likely arises from motion of $>$CHR about the chain axis.

Frequency Dependence of T_β, T_g, and T_{ll}

Figure 2 is a schematic relaxation map plotting log f_{max} (Hz) against 1000/T (K) showing the well-known intersection of T_β and T_g at high frequency. The frequency at the maximum in

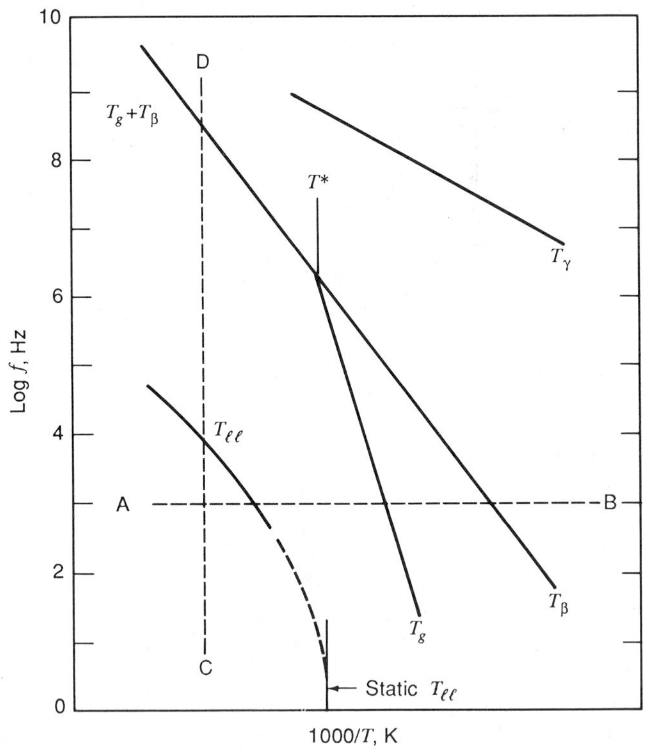

Figure 2. Schematic plot showing frequency dependence of T_γ, T_β, T_g, and T_{ll}. T^* is the Lobanov-Frenkel intersection for T_β and T_g at log $f \sim 6 \pm 1$ with $T^* \equiv T_{ll}$ (static). T_{ll} (static) is from C_p vs T or V_{sp} vs T data. T_{ll} is determined by dynamic mechanical loss. T_β and T_g loci are based on combined dielectric, dynamic mechanical, ultrasonic, and hypersonic loss and on spectroscopy. T_γ represents side-group motion and should not be used to locate T^* in the absence of T_β.

dielectric or mechanical loss is f_{max}. Such plots have been related to T_{ll} by showing empirically that the intersection temperature T^* is numerically equal to T_{ll} static. This implies a basic contradiction, namely, T^* ($f = 10^6$ Hz) $\equiv T_{ll}$ (static), which ignores the frequency dependence of T_{ll}. This dilemma was resolved by proposing that the catastrophic disappearance of a physical network at T_{ll} caused the change in behavior at T^*, whose characteristic frequency happened to be $\sim 10^6$ Hz. T^* is linear when plotted against T_g (static) with α = 76 K and $\beta = 1$ in equation 3. Figure 2 implies that log f_{max} at T^* is 6. Actually, log f_{max} can be as high as 7 for very flexible chains and as low as 5 for stiff chains such as PIB. In general, log f at T^* is 6 ± 1.

$$T_{ll}(\infty)(\mathrm{K}) = \alpha + \beta T_g(\infty)(\mathrm{K}) \quad (3)$$

$$\frac{T_{ll}(\infty)(\mathrm{K})}{T_g(\infty)(\mathrm{K})} = 1.20 \pm 0.05 \quad (4)$$

Some common polymers do not exhibit a T_β process, which may be a result of structure, unfavorable thermal history, suppression of T_β by diluent, or lack of suitable data. Those polymers lacking T_β still show the sharp slope change at T^*, with the new slope about where a T_β slope would be expected.

Zero Shear Melt Viscosity η_0 vs T

The first tentative identification of T_{ll} with the intersection of two Arrhenius lines in a plot of log η_0 vs T^{-1} was made for an a-PS fraction. This was later extended to additional PS data of $\bar{M}_w/\bar{M}_n \sim 2$, with η_0 estimated by a long extrapolation. However, the crucial tests came with two sets of high precision data based on anionically prepared PSs of $\bar{M}_n$ from 600 to 390,000 with $\bar{M}_w/\bar{M}_n$ ca 1.06, determined under almost zero shear conditions. Values of T_{ll}, $T_{l\phi}$, and the associated enthalpies of activation, ΔH_a, were tabulated. A new source of quasiequilibrium T_{ll} values for PS is now available. Avoidance of η_0 values

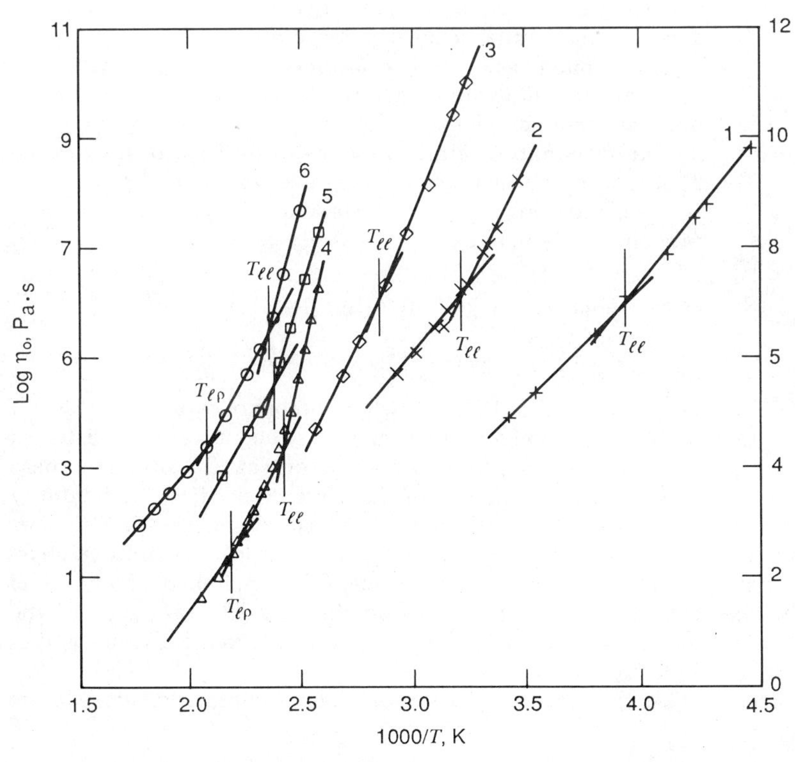

Figure 3. Plots of log η_0 vs 1000/T in kelvin, based on literature data where η_0 is the zero shear melt viscosity. Polymers are 1, PIB, $\bar{M}_n$ = 4900; 2, a-PP, $\bar{M}_n$ = 60,000; 3, poly(n-butyl methacrylate) (P(n-BMA)) $\bar{M}_\nu$ = 1.11 × 10^5; 4, a-PS, $\bar{M}_n$ = 16,200, $\bar{M}_w/\bar{M}_n$ = 1.06; 5, a-PS, $\bar{M}_n$ = 1.25 × 10^5, $\bar{M}_w/\bar{M}_n$ = 1.06, displaced to the right 0.1 unit; 6, a-PS, $\bar{M}_n$ = 1.83 × 10^5, $\bar{M}_w$ = 3.86 × 10^5. Low temperature slope changes occur at T_{ll}, and high temperature ones at $T_{l\phi}$. Log η_0 at T_{ll} is not constant.

taken too close to T_g is advised because of the marked curvature of log η_0 vs T^{-1} plots near T_g.

This technique was subsequently extended to other polymers reported in the literature: *cis–trans*-vinyl polybutadiene (PBD), PIB, a-polypropylene, poly(*n*-butyl methacrylate), a broad molecular weight distribution a-PS, and both a polyarylate and a polysulfone. Moreover, an analysis of η_0 data on 14 polymers showed a common change in behavior at $1.2T_g$, ie, at T_{ll}. Again, it seemed that log η_0 vs T^{-1} could be generalized as a source of T_{ll}. It had been shown earlier for poly(ethylene glycol phthalate) of unstated T_g that η_0 above T_g followed Vogel behavior at lower temperatures and Arrhenius behavior at higher temperatures. This difference was attributed to the loss of structure at T_f. Slope changes have been located in the various sets of log η_0 vs T^{-1} data cited above.

Only recently have the structural implications of log η_0 vs T^{-1} data in terms of the Frenkel theory of T_{ll} been emphasized. Figure 3 is a composite log η_0 vs T^{-1} plot for six polymers. All show a distinct slope change at a temperature near $1.2T_g$. A second slope change is found in several cases for $T_{l\phi}$. There is no discontinuity in η_0 at T_{ll}, but there is a discontinuity in its first derivative, $d(\log \eta_0)/d(T^{-1})$, ie, in ΔH_a. This is typical of a second-order transition, which T_g was once called. T_g was first (1930–1950) designated in the literature as a second-order transition because of its discontinuity in dV_{sp}/dT. There is some evidence that T_{ll} is a third-order transition. It is currently believed that equilibrium data show T_{ll} to be like T_g, whereas dynamic data, as with dsc, may reveal it to be third-order. The greater the difference ΔH_a (below T_{ll}) $-$ ΔH_a (above T_{ll}), the higher the T_g. The melt viscosity at T_{ll} is not constant, but is polymer specific. Being above or below M_c, as with a-PSs of $\bar{M}_n$ = 16,000 and 125,000, has no significant effect on ΔH_a. Values of $\bar{M}_w/\bar{M}_n$ for a-PS, as in curves 5 and 6 of Figure 3, do not influence T_{ll}, although η_0 at T_{ll} differs 10-fold.

There is a final key point about η_0 data: if the behavior patterns are indicative of the structure present below T_{ll} and absent above it, such structure need not disappear at finite rates of shear. Various sets of published η vs T data as a function of shear rate, molecular weight, and temperature have been analyzed by a mathematical technique called a "double-shift procedure." One shift is the common correction for temperature used with rheological data, as in the WLF procedure. The new shift is a correction for shear. This latter shift shows clear evidence for T_{ll} at finite rates of shear, whereas such evidence tends to be weak or absent in the WLF plots.

Activation enthalpies ΔH_a can be estimated from the slopes of the T_β and T_g loci or, alternatively, from the loci below and above T^*.

Polymer–Diluent Systems

A brief review of this topic is recapitulated here by the general schematic diagram shown in Figure 4.

Effect of Hydrostatic Pressure on T_{ll}

Studies of published PVT data by several objective statistical methods have permitted calculation of T_{ll} at elevated pressure. The pressure coefficient $d(T_{ll})/dP$ tends to be about twice that for dT_g/dP. A plot of $d(\ln V_{sp})/DP = \kappa$ for a-PS, where κ is the isothermal compressibility, shows evidence for $T_{l\phi}$ and T_{ll} by distinct slope changes at $P > 0$. T_{ll} from low pressure isobaric data has been found. Hydrostatic pressure presumably forces closer intermolecular segmental contacts, thus raising T_{ll}; diluents, by contrast, tend to break up such contacts, thus lowering T_{ll}.

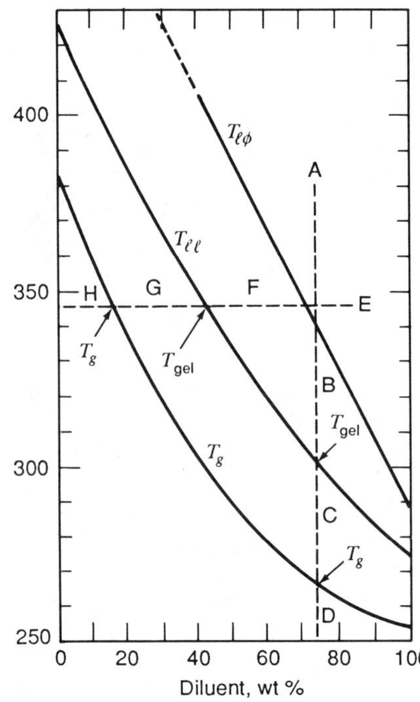

Figure 4. Schematic plot of T_g, T_{ll}, and $T_{l\phi}$ as a function of diluent content based on actual data for a-PS of $\bar{M}_n$ = 37,000. Pure diluent of molecular weight 447 exhibits all three transitions. Vertical line ABCD represents cooling or heating for 75 wt % diluent in a-PS with T_{gel} as indicated. The horizontal line EFGH represents isothermal evaporation of a dilute PS solution.

Intensity of T_{ll} vs Structure

Stereoregularity of the polymer backbone, short of that producing crystallinity, leads to a more intense T_{ll}.

T_{ll} of Copolymers. It was assumed that random vinyl copolymers might exhibit very weak T_{ll}. Thus far there have been minimal problems in locating T_{ll} by dsc and dynamic mechanical analysis (dma). Moreover, the T_{ll} locus mimics the T_g locus, in one instance a sharp maximum. The $T_{ll} \sim 1.2T_g$ relation still holds. It is not clear at the present time why random copolymers exhibit a readily detectable T_{ll}.

Experimental Techniques for T_{ll} and T_f

T_{ll} can be detected by most, if not all, of the methods traditionally used for T_g, keeping in mind several provisos: T_{ll} is generally weaker than T_g, which probably explains its lack of early recognition and the controversy about its very existence; T_{ll} is subject to artifacts not present at T_g, although these artifacts can be circumvented once recognized; the high fluidity of polymers at T_{ll} compared to T_g requires mechanical support or other means to reduce flow, such as mild cross-linking, physical entanglements, fillers, encapsulation, bimodal blends, and others; any method involving shear at $M > M_c$, such as the torsion pendulum, increases T_{ll} over its zero shear value, giving T_f rather than T_{ll}.

At the same time some methods are uniquely suited to the

liquid state and may locate T_{ll}, but not T_g: zero shear melt viscosity η_0; ^{13}C nmr; and the T^* method of Figure 2. It is beyond the scope of this article to discuss detailed methodology, except in a few cases where the technique is both relatively new and ideally suited to T_{ll}. A general discussion of methods for T_{ll} has appeared. Various techniques that have been employed with PIB, and especially the several types of dynamic mechanical loss results, have been discussed.

Practical Aspects of T_{ll}

Since $1.2T_g$ is in the midrange for melt fabrication of thermoplastics, it is natural to ask if and how T_{ll} influences melt forming. This topic has been covered previously with regard to extrusion and injection molding. The empirical answer for fabrication, given before T_{ll} was recognized, advised fabrication temperatures well above T_{ll}. Extended studies include the disappearance of orientation and birefringence and new results on melt elasticity as it affects extrudate swell, commonly called die swell, and bottle blowing.

Newer data at high strain reveal two relaxation phenomena, the one of lower temperature associated with T_{ll} and the one of higher temperature with relaxation of the entanglement network and possibly with $T_{l\phi}$. A better understanding of melt fabrication is the immediate result. Better insight into rubber elasticity occurring at much lower temperatures in conventional elastomers should also follow.

Retraction of Orientation. Uniaxial orientation induced in a-PMMA by hot stretching and quench cooling relaxes in two stages, part at T_g and the remainder in the T_{ll}–$T_{l\phi}$ region. Loss of birefringence induced in a-PS during solvent drying on glass disappears at T_{ll}.

T_{ll} effects may be manifested to varying degrees in all common melt fabrication methods except vacuum forming, which occurs near T_g.

Molecular Aspects of T_{ll}

There are four proposed molecular interpretations for the origin of T_{ll}: the energetic mechanism, the phase dualism mechanism, the energetic kinetic network (EKNET) mechanism, and a nonequilibrium theory of continuum treatment.

BIBLIOGRAPHY

R. F. Boyer in R. A. Pethrick, ed., *Polymer Yearbook,* Vol. 2, Harwood Academic Publishers, New York, 1985.

J. K. Gillham and R. F. Boyer, *J. Macromol. Sci. Phys.* **B13,** 497 (1977).

R. F. Boyer, *J. Macromol. Sci. Phys.* **B18,** 461 (1980).

R. F. Boyer and J. B. Enns, *J. Appl. Polym. Sci.* **32,** 4075 (1986).

RAYMOND F. BOYER
Michigan Molecular Institute

TRANSITION TEMPERATURES.

See GLASS TRANSITION; TRANSITIONS AND RELAXATIONS.

TRANSMITTANCE OF LIGHT.

See OPTICAL PROPERTIES.

TRANSPORT PROPERTIES

Transport properties determine a polymer's ability to move through some medium or to have some penetrant medium move between its constitutent segments. This definition encompasses processes with diverse driving forces such as concentration and pressure gradients and even electrical or temperature gradients capable of motivating one component relative to another (see also BARRIER POLYMERS; DIFFUSION; MEMBRANES). This article emphasizes polymer transport properties under conditions of low to intermediate penetrant concentration where extraordinary differences can exist between the diffusivities of penetrants having relatively small differences in molecular size or shape.

Relationships between molecular events controlling markedly different processes such as a polymer's response to impact energy and the ease of movement of small molecular penetrants within the material have become clearer in recent years. Noting the relationship between mechanical and transport properties provides a consistent qualitative or semiquantitative molecular picture of factors affecting both amorphous and semicrystalline polymers. The discussion of transport properties will occur within a framework defined by both phenomenological and spectroscopic parameters for polymers from a variety of sources. The principles and insights that emerge from such considerations are intuitively appealing and help guide material scientists in the molecular design of advanced polymers for diverse applications including gas separation membranes, barrier polymers, and interfacial polymer–polymer adhesives (see also INTERFACIAL PROPERTIES).

The movement of a penetrant molecule, or a segment of a large penetrant molecule, requires a minimum characteristic packet of unoccupied volume. Except for very small gas molecules, similar types of polymer motion and locally available volume packets are likely to be involved in both the transport process and in the mechanical response to stress. This concept was the original basis for various formulations of the so-called free-volume theories used to explain many aspects of mechanical and transport properties. One of the several versions of these free-volume theories of diffusion has been reviewed (see DIFFUSION).

A molecular scale view of the process is attractive as efforts gain sophistication in an attempt to tailor barrier and membrane properties. This approach, coupled with spectroscopic information, has indeed been useful in material selection, but as yet has not been very useful in terms of actual detailed *a priori* predictions of diffusion coefficients. In this regard, quantitative free-volume discussion and a more molecularly based but nonpredictive treatment are complementary and both will be considered.

Terminology

Reference Frames and Fluxes. In speaking of a diffusional flux, it is necessary to specify a reference frame from which the diffusion process is to be observed. Generally, a reference velocity such as the mass, molar, or volume average bulk velocity, v, v^*, or $v^\blacklozenge$, respectively, in the system is selected, and movement of the component of interest relative to this reference velocity is defined to be true diffusion. For practical reasons, a fixed reference frame is generally also considered to relate the mathematical treatment of this molecular scale transport process to an actual physical system with well-defined dimensions. In this case, the total flux relative to the fixed reference frame is partitioned into two parts, bulk flow and true molecular diffusion. This partitioning is necessary, because even in the absence of an externally imposed bulk flow, interdiffusion of molecules with respect to each other produces an effective bulk flow relative to fixed coordinates if the molecules have different masses.

The definitions of the mass, molar, and volume average bulk velocities are given in Table 1 along with selected mass and molar flux expressions related to each of the specified reference velocities for a binary system of components A and B. The so-called mutual diffusion coefficient, D_{AB} is the diffusivity of component A and B in the mixture as defined below. The mutual diffusion coefficients appearing in Table 1 are identical in all of the expressions and $D_{AB} = D_{BA}$.

Mutual Diffusion Coefficients. A better appreciation of the separation of bulk and true diffusive fluxes and the significance of the mutual diffusion coefficient D_{AP} is useful prior to considering factors that determine this coefficient. Consider, for example, the transient problem of a block of polymer (subscript P) placed in contact with an external solvent (penetrant, subscript A) phase. When a small penetrant moves into a polymer under transient conditions, most of the polymer movement, measured relative to a fixed coordinate, is thought to be due to bulk flow arising from the outward swelling of polymer segments into the region that was occupied previously by the external solvent phase. Due to its intrinsically high mobility, the penetrant (solvent) tends to interpose itself between the polymer segments and invade the domain initially occupied solely by the polymer, thereby acting as the prime mover for the mixing process.

The initial invasion of the solvent effectively induces a local swelling stress. The polymer sample responds to this swelling stress and moves in a direction opposite to the invading solvent flux. This backward bulk motion thereby carries entrained solvent molecules against the direction of the simple concentration-driven diffusive flux. Unlike the steady-state membrane case, it is clearly not acceptable to assume the mass flux n of polymer is zero in this situation, and some relationship between n_A and n_P must be defined. This issue becomes important for cases involving the interaction of strong swelling solvents with glassy polymers in transient sorption processes. Also in this case, a very strong concentration dependence of D_{AP} may exist, in some cases approaching a step function.

Self- vs Mutual Diffusion Coefficients. The so-called self-diffusion coefficient D_A^* is measured by observing the rate of diffusion of a small amount of radioactively tagged component A in a system composed of a uniform chemical composition of untagged components A and B. Due to the essentially identical physical natures of the tagged and untagged penetrant, observing the process of interdiffusional exchange of the tagged and untagged molecules allows measurement of the true mobility of the tagged molecules with respect to the stationary solution of known concentration. By varying the concentration of A and B in the presence of a small amount of tagged component A, the concentration dependence of the diffusion coefficient of the tagged molecule can be obtained, uncomplicated by bulk-flow considerations, namely

$$D_A^* = RT\mathcal{M}_A = \frac{RT}{\zeta_A} \tag{1}$$

where $\mathcal{M}_A$ is the so-called mobility of A, ζ_A is the so-called friction coefficient, the product of the effective viscosity of the medium and the effective diameter of the penetrant.

The self-diffusion coefficient of a polymer in the presence of a low molecular weight solvent is many orders of magnitude lower than that of the solvent. Therefore, when $D_P^* = D_B^*$ is neglected, D_{AP} can be expressed in terms of the mass fraction of component A as follows:

$$D_{AP} = D_A^* \, \omega_P \left(\frac{\partial \ln a_A}{\partial \ln \omega_A}\right) = D_A^* \, (1 - \omega_A) \left(\frac{\partial \ln a_A}{\partial \ln \omega_A}\right) \tag{2}$$

Amorphous Rubbery Polymers

Gases and Low Activity Vapors. At low concentration ($\omega_A \to 0$) and in systems where Henry's law applies ($\partial \ln a_A/\partial \ln \omega_A \to 1$), D_{AB} approaches the infinite-dilution self-diffusion coefficient $(D_A^*)^0$. Therefore, the mutual diffusion coefficient is a good estimator of the infinite-dilution self-diffusion coefficient for most gases in polymers at atmospheric pressure and below. Even at relatively high pressures with condensable gases and low activity vapors such as CO_2, SO_2, and propane, deviations from Henry's law can generally be described in terms of the Flory-Huggins isotherm or a similar simple expression. Therefore, self-diffusion coefficients and thus the mobility of the penetrant can be calculated using permeation data in conjunction with equilibrium sorption data to determine ($\partial \ln a_A/\partial \ln \omega_A$) at the appropriate ω_A value.

Concerted movements of several adjacent segments of chain comprising the sorbed cage of a simple gaseous penetrant are believed to provide the source of the penetrant-scale hole needed for diffusion. Lennard-Jones collision diameters (σ), van der Waals' volumes (b), and actual measured values of the partial molar volumes of various gases in polydimethylsiloxane (PDMS) are shown in Table 2. These partial molar volumes in PDMS are close to the van der Waals volumes, very close to infinite-dilution partial molar volumes in several low molecular weight

Table 1. Average Velocities and Forms of Fick's First Law for Binary Diffusion[a]

Average velocity	Definition of average velocity	Form of Fick's first law[b]
mass average	$v = \omega_A v_A + \omega_B v_B$	$j_i = -\rho D_{AB} \nabla \omega_i = \rho_i (v_i - v)$
molar average	$v^* = x_A v_A + x_B v_B$	$J_i^* = -M_A C D_{AB} \nabla x_A = C_i (v_i - v^*)$
volume average	$v^\blacklozenge = \rho_A v_A V_A / M_A + \rho_B v_B V_B / M_B$	$j_i^\blacklozenge = -D_{AB} \nabla \rho_i = \rho_i (v_i - v^\blacklozenge)$

[a] ω_i and ρ_i refer to the mass fraction and mass density of component i, respectively; ρ refers to the total solution mass density, and V_i and M_i are the molar volume and molecular weight of component i, respectively. The velocity of component i, v_i, represents the average velocity of component i relative to fixed coordinates due to both bulk and true diffusive movement.

[b] j_i and $j_i^\blacklozenge$ are the mass fluxes of the component i relative to the mass average and volume average velocities, respectively, and J_i^* is the molar flux relative to the molar average velocity.

Table 2. Various Molecular Volumes for Gaseous Penetrants

Volume measurement	He	N_2	O_2	CH_4	C_2H_4	CO_2
van der Waals volume b, nm^3	0.0396	0.0642	0.0529	0.0709	0.0856	0.0711
Lennard-Jones diameter σ, nm	0.258	0.368	0.343	0.382	0.422	0.400
Lennard-Jones occupied volume, nm^3	0.00899	0.0261	0.0211	0.0292	0.0393	0.0335
partial molar volume[a] in silicone rubber (PDMS), nm^3		0.0549		0.0792	0.0940	0.0767
free volume in sorbed state[b], nm^3		0.0288		0.0500	0.0547	0.0432
required free volume for diffusional jump in PVC–PVA copolymer, nm^3	0.075		0.299			0.690

[a] Experimental.
[b] Partial molar volume minus occupied volume.

liquids, and are measures of the volume of the cage in which the sorbed penetrant resides at equilibrium. The occupied volume is calculated by viewing the molecule acting as a freely rotating sphere of diameter equivalent to the Lennard-Jones diameter σ. Based on these data, it is clear that only a fraction of the actual equilibrium cage volume is occupied and the remainder is in a sense free volume available to be shared with the neighboring polymer segments to facilitate the mutual diffusion process.

Although the data in Table 2 are for two different polymers, for an order of magnitude analysis they can be taken as representative of movement in typical rubbery polymers. The required free volumes for diffusional jumps are significantly greater than the penetrant volumes. Therefore, most of the free volume involved in a given jump must be supplied by a momentary fluctuation in the rubbery polymer segmental position due to a local thermal fluctuation, rather than being locally present in the equilibrium sorbed cage. This fact can be treated in several ways, eg, via Activated State Theory or Free Volume Theory.

Membrane and Barrier Implications. In the case of typical rubbers, crankshaft and other related rotational motions of the repeat units are so large that little difference in diffusion coefficients exists for simple gaseous penetrants, the sizes of which differ by less than a few hundredths of a nm. Of course, since permeation is a solution-diffusion process, membranes or selective barriers do not operate strictly on the basis of size selection. Under a fixed partial pressure driving force for component i, the flux through a material of given thickness ℓ is determined by the permeability:

$$N_i = \Delta p_i P_i / \ell \tag{3}$$

The permeability can be expressed as the product of a diffusivity and solubility coefficient:

$$P_i = \bar{D}_i \bar{S}_i \tag{4}$$

The diffusivity and solubility coefficients in the above expression are effective averages applying across a polymer film between its upstream and downstream faces. The presence of other copermeating components can often be neglected for gases and low activity vapors permeating through rubbers. The average mutual diffusion coefficient D_i is given below for penetrant A in the polymer, component P:

$$\bar{D}_i = \frac{\int_{\omega_{A1}}^{\omega_{A2}} D_{AP}\, d\omega_A}{\omega_{A2} - \omega_{A1}} \tag{5}$$

When the presence of other components cannot be neglected, the system can still be treated approximately by considering the membrane in the presence of the additional component to be a new effective medium; however, this is at best a rough approximation since it neglects the bulk-flow contribution and thermodynamic influence of additional components on the flux of component A. Such issues are beyond the scope of this discussion, but have been treated.

Plasticization and Hydrostatic Effects

New complexity arises in the basic transport properties of a polymer at higher penetrant activities, since the local concentration of penetrant within the polymer can become quite high. At higher pressures (approaching the vapor pressure) these high concentrations can occur in cases where favorable polymer–penetrant interactions promote solubility. This is common for organic vapors and even for small polar or quadrupolar molecules such as CO_2, H_2S, and SO_2. Transport plasticization is defined as a significant increase in the diffusivity of a penetrant due to facilitation of local polymer segmental motion caused by another penetrant molecule in its neighborhood. This definition applies both to the increase in diffusivity of a penetrant caused by the presence of its own kind and that due to increases caused by a different component. Even for pure component penetrants, a detailed fundamental analysis of this phenomenon on a molecular basis has not been achieved; however, various free-volume analyses are available.

Plasticization is typically an unfavorable phenomenon in separation applications because permselectivity is reduced as the diffusivity of a slower penetrant is increased. In the case of gases, plasticizing phenomena are suppressed due to hydrostatic compression in some polymers despite rather high sorption levels. Analysis in terms of the free-volume approach requires consideration of the compressibility of the rubbery matrix. In fact, exposure to the hydrostatic pressure of low solubility gases such as helium or nitrogen may actually compress out free volume and thereby reduce the ability of polymer segments to open gaps for movement of other segments or sorbed penetrants.

The local solubility and diffusivity must be considered to see when true transport plasticization occurs. For this it is more useful to consider the local mutual diffusion coefficient $D_{AB}(C)$ rather than the average value. The local mutual diffusion coefficient is a measure of the ability of a penetrant to move through the membrane at a point where the local conditions are well defined in terms of the local penetrant concentration C_A, volume fraction ϕ_A, weight fraction ω_A, or penetrant fugacity, whichever is most convenient.

Penetrant Clustering. Clustering may reduce the effective mobility of a penetrant such as water or methanol that can self associate. The activated state representation of diffusion is typically used to discuss the suppressed diffusion coefficient for a clustered penetrant. If the diffusion process involves primarily jumps by monomeric, unassociated penetrants, energy must be available both to make a gap to allow the penetrant to jump into and to cause the dissociation of one of the penetrants in a local region adjacent to the gap. Therefore, higher activation energies for diffusion and, hence, lower diffusion coefficients at activities above which clustering occurs are expected if this visualization is correct. Work to document this point of view has been done for silicone rubbers, poly(alkyl methacrylates), and hydrocarbon rubbers.

Intermediate Size Penetrants. These penetrants are defined as molecules with sizes approaching a polymer segment, but much smaller than an entire polymer chain. The infinite-dilution mutual diffusion coefficients for increasingly large, but not macromolecular, penetrants approach the diffusion coefficient for the segment of a polymer diffusing within itself. Movement of the entire polymer molecule itself is clearly a slower and more complex concerted process than the local movement of its segments; however, the segmental mobility and the self-diffusion coefficient of the entire molecule can be related.

For intermediate-size penetrants such as typical organic solvents, the self-diffusion coefficient of the polymer is much smaller than that for the penetrant. Consistent with this expectation, equation 2 has been found to be appropriate for mass fractions of toluene as high as 85%. Therefore, for low and intermediate penetrant concentration, equation 2 should still apply as it did in the case of gases and other small penetrants.

Semicrystalline and Cross-linked Rubbery Polymers

Gases and Low Activity Vapors. Regions of crystallinity or points of chemical cross-linking have somewhat similar effects on the transport properties of a rubber causing restriction of swelling and suppression of long-range chain segmental motion. Semicrystalline polymers, however, are more complex due to tortuosity caused by the presence of the typically impermeable crystalline regions. Similarly, adding fillers such as glass, talc, or mica particles to an amorphous matrix prior to cross-linking can make the transport properties of the resultant material take on a semicrystallinelike character if the interfacial adhesion between the filler and matrix phase is good. Promotion of adhesion can be achieved by using interfacial wetting agents such as silanes and titanates that bond to the inorganic surface and can either chemically bond or physically interact with the elastomer matrix (see COUPLING AGENTS). In the absence of such adhesion promoters, the barrier properties may be degraded due to interfacial defects not found in semicrystalline materials under most conditions.

The effects of fillers (qv) or crystallites can be influenced considerably not only by the volume fraction of this phase, which affects the solubility of a component directly, but also the area-to-thickness aspect ratio and orientation of the filler with respect to the direction of permeation. The orientation factor affects the apparent diffusivity by determining how the crystallites or fillers behave as tortuous impediments to penetrants.

Studies of gas sorption and transport strongly support the notion of the impenetrability of crystalline domains by even tiny gas molecules. As shown in Figure 1, the solubility coefficient ($\overline{S}_i = C_i/p_i$) of both gases and the low activity vapors are essentially proportional to the volume fraction of amorphous material Φ_a and the intrinsic solubility coefficient for the totally amorphous material $\overline{S}_a$, ie,

$$\frac{\overline{S}}{\overline{S}_a} = \Phi_a \qquad (6)$$

The effects of crystallinity on the diffusion coefficient can be more complex than this simple volumetric exclusion, since the crystallites may act not only as tortuous barriers, but also as effective restrictors of chain motion analogous to chemical cross-linking. This twofold effect has been treated in terms of a tortuosity factor τ, and a chain immobilization factor β^*, both of which increase with increasing crystalline fraction. The parameters D and D_a are the diffusion coefficients in the actual sample and in a totally amorphous sample, respectively:

$$D = \frac{D_a}{\tau\beta^*} \qquad (7)$$

High Activity Vapors and Liquids. In considering the transport behavior of high activity vapors and liquids in semicrystalline polymers at low solubility (eg, water in polyethylene), small degrees of transport plasticization or clustering may need to be considered to generalize the $\overline{D}$ and $\overline{S}$ to include depen-

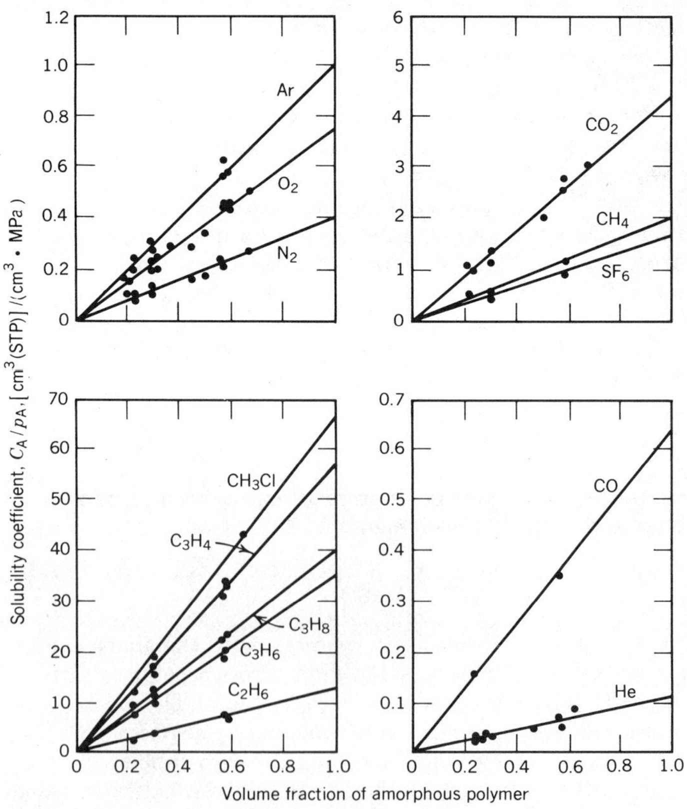

Figure 1. Linear relationship between the infinite dilution solubility coefficient and the amorphous fraction in polyethylene at 25°C. To convert [cm³(STP)]/(cm³·MPa) to [cm³(STP)]/cm³·atm), multiply by 0.1013.

dence on activity. Combination of these more general relations for $\bar{D}$ and $\bar{S}$ into the permeability relationship of equation 4 should approximately describe the steady-state permeability of the amorphous fraction of the material. Generally, the permeability of the crystalline fraction is assumed to be zero. Tortuosity and chain-restriction effects should be of decreasing importance as the extent of swelling increases. At high degrees of swelling (above 70% solvent fraction), frame of reference corrections must be incorporated in formulating steady-state permeability expressions.

For high solubility penetrants, great complexity is often observed in sorption–desorption kinetics or transient permeation due to time-dependent rearrangements of crystalline regions in response to swelling-induced stresses. These are so-called non-Fickian diffusional behaviors. Protracted approaches to equilibrium in the surface concentration may make the approach to sorption equilibrium or steady-state permeation behavior much slower than expected on the basis of the effective diffusion coefficient calculated from certain procedures.

Film formation protocols may cause significant nonuniformity in structure between a thin surface layer and the bulk of the film, especially when casting on a chilled roll. This effect can cause apparent thickness-dependent behavior of permeability and diffusivity parameters. Effectively, the film behaves as a laminate with different transport properties that can be treated according to a simple laminate model involving series resistances.

Large (Polymeric) Penetrants. When the penetrant considered is either a linear or a branched molecule of a length approaching that of the polymer matrix in which it diffuses, a considerably different situation exists compared to previous cases. Several problems of commercial importance are concerned, including crack healing, adhesion building between two polymeric surfaces placed in contact, and the creep compliance and zero shear viscosity limit behavior of macromolecular samples. In a pioneering study of the self-diffusion coefficient of one polymeric chain moving among others it was found that D_A^* is proportional to $M^{-5/3}$, where M is the penetrant molecular weight (see MACROMOLECULAR DYNAMICS).

Interdiffusion of Polymers in Miscible Polymer Blends

An important issue related to an aspect of polymer–polymer diffusion arises from consideration of the free energy of mixing relationship for a binary polymer blend of monodisperse components 1 and 2:

$$\Delta G_{\text{mix}}/kT = N_1^{-1}\phi_1 \ln \phi_1 + N_2^{-1}\phi_2 \ln \phi_2 + 2\,\phi_1\phi_2\chi \quad (8)$$

where χ is the Flory-Huggins interaction parameter due to the segment–segment interactions from the enthalpy of mixing and nonideal entropy of mixing, and k is Boltzmann's constant. The first two terms in the above equation contain factors of N_1^{-1} and N_2^{-1} where N_i refers to the number of monomer segments per chain of each component. The combinatorial entropy contribution represented by these two terms is very small and the thermodynamic behavior is often dominated by the term containing χ. An elegant series of experiments confirmed that this factor can dominate the behavior of polymer–polymer interdiffusion. The relationship for the mutual diffusion coefficient in such systems using the volume fractions ϕ_i and the numbers of monomer segments per chain of each component N_i (defined in such a way that the molar volume V_0 of each monomer segment of the two types is equal) is

$$D_{12} = \phi_1^2\phi_2^2\,[N_2\phi_2^{-1}D_2^* + N_1\phi_1^{-1}D_1^*] \times [N_1^{-1}\phi_1^{-1} + N_2^{-1}\phi_2^{-1} \ln \phi_2 - 2\chi] \quad (9)$$

For the limit of large values of N_1 and N_2, the expression simplifies to

$$D_{12} = -2\chi\phi_1\phi_2\,[\phi_1 N_2 D_2^* + \phi_2 N_1 D_1^*] \quad (10)$$

For cases where $\chi < 0$, typical of miscible polymer pairs, the diffusion coefficient has the expected positive sign, and is more strongly influenced by relatively small nonideal solution effects as compared to low molecular weight systems.

Glassy vs Rubbery Materials

Glassy materials are differentiated from rubbery materials in discussions of physical properties because their responses in most experiments are quite different. Essentially all of the differences arise from the characteristic scales of the micromotions that occur at the segmental level in rubbery and glassy polymers. The high molecular weight liquidlike nature of rubbers provides them with the ability to adjust their segmental configurations rapidly over significant distances by large-scale motions.

The frequency and amplitude of the torsional motions increase as the glass transition is approached. Close to T_g the frequency increases dramatically, and gives rise to true rotational motions typical of rubbery materials. In glasses, the intramolecular backbone motions occurring over time scales of seconds or microseconds are much less extensive than in rubbers and are believed to be primarily torsional oscillations.

Measurements of transport properties provide a tool to characterize the properties of glasses that is complementary to many spectroscopic methods. Polymer chain motions probed by methods such as dynamic mechanical and dielectric spectroscopy are important in the diffusion of penetrants. The real and loss moduli change orders of magnitude within a short temperature interval near the T_g for typical glasses (Fig. 2). The real modulus G' characterizes the elastic nature of a material and measures the amount of deformation induced by a given stress over a given time characterized by the frequency of the sinusoidal perturbing stress. The loss modulus G'', on the other hand, characterizes the viscous or dissipative nature of a material and measures the amount of stress required to cause a given rate of deformation of the material (see DYNAMIC MECHANICAL PROPERTIES). The ratio G''/G' of these material parameters, called tan δ, characterizes the ratio of the input energy dissipated to that stored during a given cycle of the microdeformational elongation that is generally less than $\pm 0.1\%$ of the length. The higher the testing rate, the higher the temperature needed to achieve the resonance condition, which corresponds to the point at which the input frequency matches the frequency of chain oscillations and at which the maximum in tan δ is observed. Practical testing rates from 0.1 Hz to 1000 Hz can be considered using this technique, so a measure of the activation energy of the rotational process in the range of T_g can be made. In this case, the apparent activation energy changes dramatically near T_g. A free-volume description of the mobilization of chain segments is useful for describing the strongly temperature-dependent activation energies in the range of T_g.

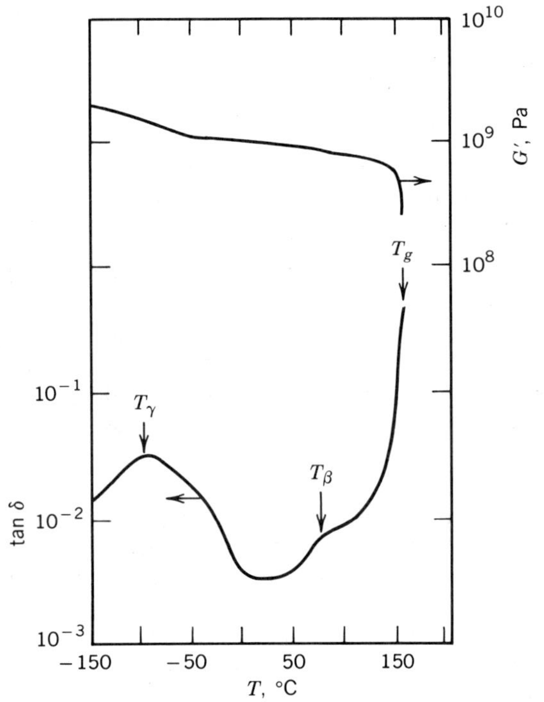

Figure 2. Dynamic mechanical spectra for bisphenol-A polycarbonate at 1 Hz. To convert Pa to psi, multiply by 1.45×10^{-4}.

Transport properties change less dramatically at the glass-transition temperature of a polymer relative to the mechanical properties. Nevertheless, the transport properties of glasses depend on the chemical structure and even prior history of the sample over time scales that are experimentally accessible. This history dependence is often referred to as the nonequilibrium nature of glasses. Commonly available glasses consist of simple, relatively compact, repeat units that produce low penetrant diffusion coefficients. Glasses consisting of more complex repeat units can have higher diffusion coefficients that can approach those of rubbery materials, for small and intermediate size penetrants.

The distribution of subtle differences in packing in glasses is expected to produce a corresponding distribution of local torsional oscillatory motions even if the influence of polymer–penetrant interaction is neglected. The average of this distribution is, of course, what is sensed by most probes of the material, including diffusion experiments. A corresponding cumulative distribution of average gap sizes forms spontaneously and randomly throughout the glass in response to the concerted torsional motion of repeat units. For penetrants smaller than the limiting moving segment size, the tendency of the penetrant mobility to decrease with increasing penetrant size should be clear. This visualization is in effect the basis for free-volume theories discussed in the context of rubbery materials.

If the available gap is larger than the critical dimension of a penetrant, a diffusional step can be taken. If the available gap is too small, a jump is precluded. The fraction of diffusional jumps available to penetrant A vs B that are actually selective is small in rubbers, unless the size difference between penetrants is very large. For glassy materials, on the other hand, the range of motions, and hence the size of transiently opening gaps, are much more narrowly distributed, thereby providing a much more selective environment for penetrant diffusion.

The higher diffusivity selectivities often make glasses more favorable than rubbers for small molecule permselective membranes. The lower diffusivities also favor glasses over rubbers for barrier applications. On the other hand, rubbers are used in some membranes when a very large solubility selectivity is available, such as in separating organic vapors from air. Complex non-Fickian and plasticization phenomena can occur in applications where a significant amount of penetrant is present in a glass. The diffusion coefficients of penetrants in glasses may increase by several orders of magnitude if the weight fraction of the penetrant reduces the T_g to the use temperature.

Nonequilibrium Nature of Glassy Polymers

Plasticization phenomena can be observed in some polymers exposed to high pressures of condensable gases such as CO_2. Moreover, a pressure dependence of permeability occurs for glassy polymers whereas rubbery materials tend to have essentially pressure independent permeabilities. The decrease in permeability of component A with increasing partial pressure of either component A or B prior to the onset of plasticization shown in Figure 3 is characteristic of glassy polymers.

The pressure at which plasticizing effects begin to overcome permeability depression depends primarily on the ratio of the partial pressure of the penetrant in the feed stream to the corresponding vapor pressure of the penetrant at the system temperature. As the pressure approaches the vapor pressure, solubility tends to increase rapidly, thereby inducing plasticization or clustering responses such as those discussed for rubbery polymers.

Dual-mode Sorption and Transport Models. The concept of unrelaxed volume $V_g - V_\ell$, where V_g is the specific volume of the glass and V_ℓ is the extrapolated specific volume of a hypothetical rubber, is illustrated in Figure 4. The extremely long relaxation times for segmental motion in the glassy state lead to trapping of nonequilibrium chain conformations in glasses, thereby permitting miniscule gaps to exist between chain segments. This results in the additional unrelaxed volume, which causes the specific volume of a glass to be higher than the extrapolation for a hypothetical rubber. This unrelaxed volume is part of the physical basis for the dual-mode sorption model.

Isotherms for gas sorption in glassy polymers are concave to the pressure axis at low pressures with a tendency to approach a straight line at high pressure. The dual-mode sorption

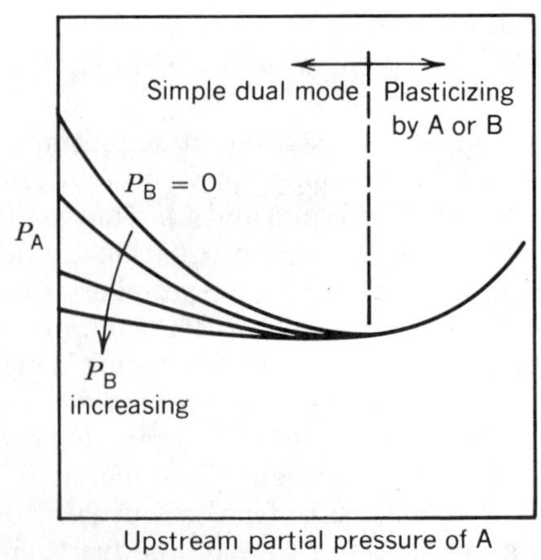

Figure 3. Permeability of a glassy polymer to penetrant A (P_A) in the presence of varying partial pressures of penetrant B.

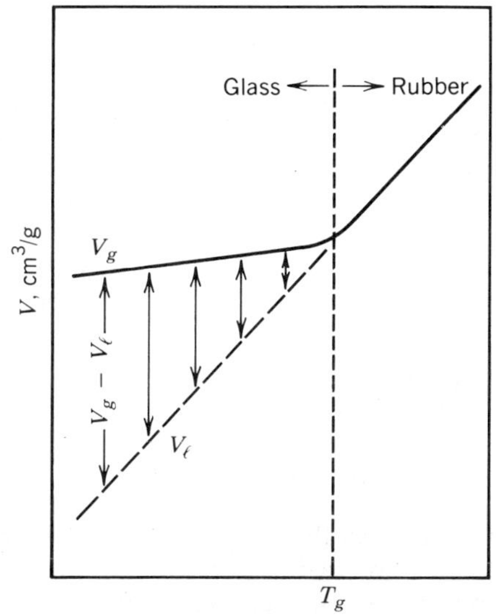

Figure 4. Schematic representation of the unrelaxed volume, $V_g - V_l$, in a glassy polymer. V_g = actual glass specific volume; V_l = equilibrium volume of densified glass. The unrelaxed volume disappears at the glass-transition temperature.

model considers the glassy solid to consist primarily of an equilibrium densified matrix with a small volume fraction of uniformly distributed molecular-scale gaps, or holes, throughout the matrix. Gaps that are smaller in size than a penetrant molecule may be locally redistributed during the sorption process in the initial exposure of the polymer to a penetrant. This model has proved to be very useful for interpreting a wide spectrum of phenomena including pure component and multicomponent permeation and sorption and penetrant-induced dilation in gas–glassy polymer systems. The development of the dual-mode model has been reviewed.

The dual-mode model has also been applied with some success to the sorption and transport behavior of small and intermediate size organic molecules in glasses. Above some size, however, the penetrant may be too large to be accommodated solely within individual holes, and two truly distinct environments may not actually exist for these penetrants. Even in this case, however, the concept of unrelaxed volume remains valid, and observations of sorption and dilation behavior that apply to gases also apply qualitatively to larger penetrants up to the point of extreme plasticization.

Simple Sorption Competition Transport Models. As a direct extension of the dual-mode sorption model, it is convenient to treat pure gas permeability as the sum of two terms, which are characteristic of the two sorbed penetrant populations. Petropoulos first developed such an expression in terms of the gradient of the thermodynamic activity of the sorbed gas. For the case where there is a vacuum downstream and a pressure p_{A_2} of component A upstream the following expression results:

$$P_A = k_{D_A} D_{TD_A} + \frac{C'_{H_A} D_{TH_A}}{p_{A_2}} \ln\,(1 + b_A p_{A_2}) \tag{11}$$

where D_{TD_A} and D_{TH_A} are diffusion coefficients characteristic of the Henry's law (dissolved), and the Langmuir (hole-filling) populations of sorbed gas molecules. Extension of the Petropoulos model to include gas mixtures involves numerical integration of the flux equations for components A and B. These equations can be solved using a Runge-Kutta or similar numerical integration technique to give the steady-state concentration profile across the polymer membrane corresponding to any specified upstream and downstream partial pressures of each component. The steady-state permeability can then be calculated by a trial-and-error numerical method.

Combined Direct Sorption and Diffusion Competition Model. To account for the effect of nearby gas molecules on the ability of a given gas molecule to find sites into which it might jump, a model that adopts a somewhat more complex description of the diffusion process in glassy materials has been proposed. A gas molecule is capable of executing four distinct types of diffusional jumps, depending on which of the two distinct sorption environments it is diffusing from and diffusing into. Denoting the Henry's law and Langmuir environments as D and H, respectively, the four possible jumps can be depicted as follows:

$D \longrightarrow D$ (dissolved to dissolved)	$D \longrightarrow H$ (dissolved to hole)
$H \longrightarrow D$ (hole to dissolved)	$H \longrightarrow H$ (hole to hole)

Each jump is described by a distinct primitive diffusion coefficient D_{DD}, D_{DH}, D_{HD}, or D_{HH} that is related to the average mobility of a penetrant in each of the two environments D_{D_A} and D_{H_A} as shown below:

$$D_{D_A} = D_{DD_A} + D_{DH_A}\,(1 - \theta_A) \tag{12}$$

$$D_{H_A} = D_{HD_A} + D_{HH_A} \tag{13}$$

where $\theta_A = C_{H_A}/C'_{H_A}C'$ represents the fraction of Langmuir sorption sites occupied by sorbed gas A molecules. Consequently, the model predicts that D_{D_A} decreases as the concentration of gas in the polymer increases. In effect, this model predicts a reduction in the number of diffusional pathways available to a penetrant molecule. When combined with the dual-mode sorption model for describing the fraction of filled holes, the flux equations can be integrated to give the resulting competitive sorption–diffusion model for the permeability of a pure gas A:

$$P_A = k_{D_A} D_{DD_A} + \frac{C'_{H_A} b_A (D_{HH_A} + D_{HD_A}) - k_{D_A} D_{DH_A}}{1 + b_A p_{A_2}} + 2k_{D_A} D_{DH_A} \frac{\ln(1 + b_A p_{A_2})}{b_A p_{A_2}} \tag{14}$$

Using nonlinear regression techniques to fit the pure gas permeability vs pressure data for gas A, values for D_{DD_A}, D_{DH_A}, and $(D_{HH_A} + D_{HD_A})$ can be determined.

The extension of the competitive sorption–diffusion model to mixed gases is considerably more difficult than the extension of either the Petropoulos or the simple competitive sorption model. To include gas mixtures, the model must be modified to accurately describe the fraction of filled Langmuir sorption sites when more than one gas species is present in the polymer. To account for the presence of a second gas, the flux expressions can be rewritten using the following definition for the fraction of filled holes:

$$\theta_{\text{total}} = \frac{C_{H_A}}{C'_{H_A}} + \frac{C_{H_B}}{C'_{H_B}} \tag{15}$$

where θ_{total} represents the fraction of the total unrelaxed volume occupied by either penetrant A or B, and therefore is not available for allowing a penetrant to jump into it.

This model results in a system of nonlinear differential equations that must be solved numerically. Figure 5(**a**)–(**c**) shows pure gas permeability data for CO_2 in poly(phenylene oxide) along with the theoretical fits for the Petropoulos, competitive sorption, and competitive sorption–diffusion models, respectively. The competitive sorption–diffusion model provides a slightly more accurate fit of the permeability data than either of the simpler models, especially in the low pressure region. This is expected because of the additional model parameter.

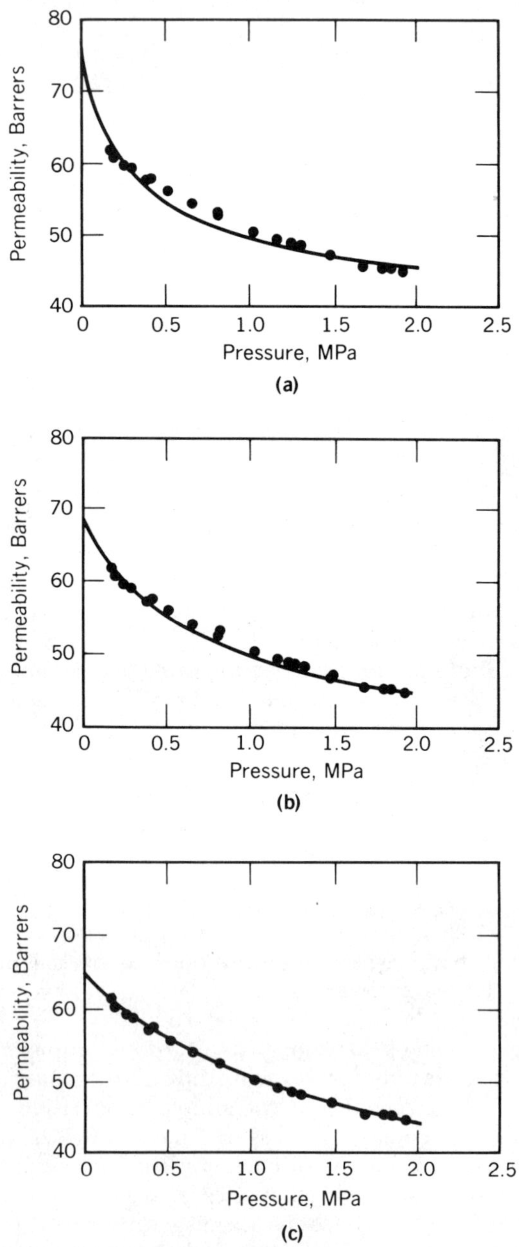

Figure 5. Permeability of pure CO_2 in poly(phenylene oxide). The solid lines represent (**a**) the simple competitive sorption model, (**b**) the Petropoulos model, and (**c**) the competitive sorption–diffusion model. To convert MPa to atm, multiply by 10. 1 Barrer = 10^{-10}cm^3(STP)cm/cm^2·s·cmHg = 0.335 mmol/m·s·TPa.

History Effects in Glassy Polymers

Penetrant-induced Effects. History-dependent properties of glassy polymers are well documented in terms of thermal and mechanical properties. Studies have related excess enthalpy changes measured by differential scanning calorimetry to changes in small molecule sorption. Many studies have reported penetrant-induced history-dependent (or conditioning) effects for gas and vapor sorption and transport properties in glassy polymers. The ability to display long-lived changes in sorption, diffusion, and permeation properties distinguishes glassy materials from rubbery ones and is believed to arise from the long relaxation times in glasses that reflect their inhibited segmental motions.

Orientation-induced Effects. Orientation and combined heat and orientation processing affect the transport properties of glassy polymers. Especially when crystallites are present, the effects can become surprisingly large. As noted, for rubbery semicrystalline materials, the obvious improvements in barrier properties associated with organization of lamellar crystalline domains with their platelets perpendicular to the direction of penetrant flow can produce significant reductions in permeability. Simple crystallinity appears to act principally as tortuous blocking elements to increase the path length of penetrants in their random walks across the film.

Relationship Between Transport and Other Physical Properties

The ability of polymers to deform under stress is fundamentally related to their ability to rotate around constituent bonds comprising the chain backbone. In the solid state or in concentrated solutions, intersegmental attractions and impedances to rotations can influence this rotational ability as profoundly as do intrasegmental barriers to motion. Amorphous materials maintain high elastic moduli below the glass-transition temperature, because rotational motions of the chain backbone are suppressed. In many situations, ductility and impact strength are equally important and somewhat at odds with achieving a high modulus. Large-scale rotations around chain bonds are not necessary to achieve adequate impact resistance; however, motions involving only side chains do not impart impact resistance. Apparently, energy transmission down the backbone through intrasegmental motion is more effective at maintaining the local internal energy below a critical level that causes macroscopic failure. The dissipation mechanism is also intermolecular in its nature, since even small-scale motions in the glass are believed to be coupled to motions of neighboring segments due to efficient packing in the solid state.

This coupling promotes energy transmission and dissipation, since it is necessary to generate a gap of sufficient size in the matrix adjacent to a moving segment to permit the motion to occur. Parallels are thought to exist between these mechanically active motions and transport properties.

A penetrant molecule requires a minimum characteristic packet of unoccupied volume to move in the same way that a mechanically active segmental motion requires the localization of unoccupied volume to permit movement of its center of mass. These ideas can be used to understand transport properties of materials for gas separation applications. Typically, it is difficult to achieve simultaneous increases in permselectivity and permeability in materials used in polymeric membranes. However, systematic variations of the intrasegmental mobility and the chain packing have achieved a simultaneous improvement in a series of polyimides. Significant increases in diffusivities and diffusivity selectivities have resulted from variations in the chemical structure of the polymers. Two principles for material selection for membranes have emerged from these studies: (*1*) substitutions on the polymer backbone that simultaneously inhibit the chain packing and the rotational mobility about flexible linkages lead to an increase in both permeability and permselectivity; (*2*) substitutions that reduce the concentrations of the more mobile linkages in the polymer backbone lead to increased permselectivity without a large reduction in permeability if the moiety added does not significantly affect intersegmental packing.

Non-Fickian Transport Behavior

The topic of non-Fickian transport encompasses all cases in which the simple form of the flux equation with a concentration-dependent diffusion coefficient and constant system boundary conditions, is insufficient to describe the observed time-dependent behavior of the system. This definition includes cases with position-dependent diffusion and solubility coefficients. For example, a film conditioned with a swelling agent (eg, CO_2) used to induce different properties at the upstream and downstream faces fall in this category. Another variation of this problem includes cases in which quenching or orientation produces transverse swelling stresses during the sorption kinetics process. Such position dependencies, superimposed on concentration-dependent transport coefficients, considerably complicate the response characteristics of a system. Nevertheless, mathematical analysis of the problem is possible by generalizing procedures used for the case of concentration-dependent, time-invariant coefficients if the position-dependent profile of the transport properties can be defined.

Unlike the situation with Fickian systems, when dealing with non-Fickian systems data derived from the kinetic responses of a system cannot be readily extended to other configurations with significantly different boundary conditions or sample dimensions. By far the most difficult problems encountered in the domain of non-Fickian processes involve time-dependent changes in transport and sorption properties, ie, additional time dependence, beyond that caused by simple time variation in the local concentration. The additional time-dependent problem is the most difficult to treat because it generally involves introducing additional physical expressions. Transverse swelling stresses typify such cases. In this example, the additional physics are accommodated by a linear viscoelastic approximation to account for the time-dependent relaxation of stresses induced in the medium by swelling.

The frontiers in the field of transport properties of polymers involve a better integration of the useful free-volume theory with the ideas of activation energies and jump lengths involved in the actual primitive diffusion process. Encouraging advances have been made in the study of non-Fickian phenomena, but this field needs much more work because of the extremely close coupling of large-scale segmental motions responsible for time-dependent mechanical properties with the subtle micromotions involving molecular diffusion. Ultimately, the goal for both Fickian and non-Fickian processes is to relate their phenomenological parameters to the detailed molecular motions of the polymer matrix in the presence of the penetrant, whether it be a small molecule or another macromolecule. The challenge will be met by a complementary use of classical diffusion methods and modern spectroscopy.

BIBLIOGRAPHY

H. J. Bixler and O. J. Sweeting in O. J. Sweeting, ed., *The Science and Technology of Polymer Films,* Vol. 2, John Wiley & Sons, Inc., New York, 1971.

W. J. Koros, G. K. Fleming, S. M. Jordan, T. H. Kim, and H. H. Hoehn, *Prog. Polym. Sci.* **13,** 339 (1988).

J. Crank and G. S. Park, *Diffusion in Polymers,* Academic Press, London, 1968.

H. B. Hopfenberg and V. Stannett in R. N. Haward, ed., *The Physics of the Glassy State,* Applied Science Publishers Ltd., London, UK, 1973, chapt. 9.

William J. Koros
Mark W. Hellums
The University of Texas at Austin

TRIAZOLE POLYMERS. See Heat-resistant polymers; Polyaminotriazoles.

TRIMETHYLENE OXIDE POLYMERS. See Oxetane polymers.

TRIOXANE POLYMERS. See Acetal resins.

TROMMSDORFF-NORRISH EFFECT. See Gel effect; Radical polymerization.

TUBING

Plastic tubings were introduced in the late 1930s as replacements for rubber tubing and hose. Today, cured elastomer tubings of natural or synthetic rubber, silicone rubber, and fluoroelastomer materials are used, but the growth of the tubing market has been principally in thermoplastic materials that can be extruded into long lengths without the need for postcuring. Because virtually all thermoplastics can be extruded, the choice depends on the application. Unlike pipe (qv), which is used mostly for plumbing, tubing is used in many different applications. As a result, no general specifications and classification standards for tubing have been promulgated, except for some government specifications such as the Federal Specification L-T-790B: Tube, Nonmetallic, Poly(vinyl chloride), Flexible (for Laboratory and Medical Use), and the United States Military Specification MIL-T-36647: Tube, Intravenous, Plastic, Radiopaque, Disposable. However, these are written for specific application areas and specific polymers. The product engineer must therefore evaluate the strengths and limitations of the individual polymers and determine their suitability.

Tubings are defined as long, hollow, flexible cylinders used to transport fluids or solids. Pipe is rigid and larger in size. Tubing is flexible with outside diameters up to 150 mm. A flexible material can be stressed or bent without breaking. The extent of flexibility is important in selecting a tubing material. Associated with flexibility is kinking. Kinks are caused by reducing the bend radius to a point where the tubing wall collapses.

Properties

Requirements of the end use application should be used to specify tubing properties. Properties to consider include flexibility (bend radius), hardness, chemical resistance, environmental stress-crack resistance, permeability and chemical absorption, burst pressure and working pressure, recommended use temperatures, flammability, electrical conductivity, peristaltic pumpability, surface wettability, abrasion resistance, dielectric strength, fatigue resistance, and appearance.

Applications

Applications include tubing for food-processing plants and beverage dispensers, plumbing, gas transmission lines, hydraulic lines, automotive applications, medical applications (catheters, infusion and implant tubing, dialysis), insulation, fuel and vacuum lines, and hoses.

ROBERT A. CHARTON
Norton Company

TUNG OIL. See DRYING OILS.

TURBIDIMETRIC TITRATION. See FRACTIONATION.

U

ULTIMATE PROPERTIES, UNIAXIAL SYSTEMS

Liquid crystalline solutions, liquid crystalline melts, and solutions of flexible high molecular weight polymers can be converted into products of exceptionally high strength and modulus. This remarkable technological achievement in fiber technology renewed interest in the theoretical limits of strength of uniaxial polymers. These limits are necessary to assess the potential for improvements in fiber strength beyond the levels currently achieved. It has been known since the early work of H. Mark that the maximum theoretical tensile strength of uniaxially oriented polymers is almost two orders of magnitude higher than that of the corresponding fibers. Although a great deal has been written about the maximum achievable strength, the question of how this limit in strength should be estimated is still not settled.

At present, one of the most relevant questions of fiber science is whether the strength of current high performance fibers is sufficiently close to the theoretical limits to consider the current processes as the last step in ultrastrong fiber technology, or whether the gap between current strength and theoretical potential is so large that new processes should be developed allowing another quantum jump in strength.

Because of the simplicity of its structure and the remarkable strength achieved in industrial and academic laboratories, uniaxial polyethylene is the logical choice for theoretical studies of strength. The results depend mostly on the methods of calculation. On the basis of published data it is impossible to set the limits in fiber strength with sufficient confidence and assess on an absolute scale the merits of current fiber processes.

The theoretical limits in modulus are calculated from the stretching, bending, and torsional force constants. The low temperature, high rate of deformation modulus of ultrastrong PE fibers of the Spectra type is so close to the theoretical limits that extensive research to further increase the modulus is not warranted. The measured crystalline modulus is also within the limits predicted by theories, indicating that the structure of a large volume fraction (~75%) of the fiber is nearly perfect. This conclusion is confirmed by the crystalline orientation function, crystalline density, and the very high melting point of the fiber.

Theoretical Strength of Fibers

The calculations of the theoretical limits of fiber strength are carried out here for a perfect uniaxial structure with randomly placed chain ends (Fig. 1). The calculations of theoretical strength of fibers involve two steps: calculation of the strength of an individual molecule under stress and of an ensemble of perfectly aligned extended chain molecules. The strength of an individual molecule is a problem of quantum mechanics. It has been studied by several authors, and the results have been reviewed and expanded.

Calculation of the Strength of Uniaxial Polyethylene

The strength of an ensemble of perfectly aligned molecules is calculated assuming that the crack is circular and that it propagates perpendicularly across the specimen. At each molecule the crack follows the path of minimum energy, ie, chain pullout vs chain breakage. The free energy Δf_r associated with such a crack is

$$\Delta f_r = 2\pi\rho^2\rho - \frac{8(1-\mu^2)r^3\sigma^2}{3E} \qquad (1)$$

where E is Young's modulus, r the radius of the crack, ρ the specific surface energy, σ the applied stress, and μ Poisson's ratio.

Assuming that the growth of cracks is a thermally activated process, the time to rupture equals the time to form an unstable crack, and the free-energy change associated with the formation of cracks can be expressed by equation 1, the following expression for the time to break t_B of a specimen under constant stress is obtained:

$$t_B = \left(\frac{h}{kTZV}\right) \exp\left\{\left(\frac{1}{kT}\right)\left[\Delta f_r^* + \Delta F^* - \frac{(v\sigma^2)}{2E}\right]\right\} \qquad (2)$$

where k and h are Boltzmann and Planck constants, respectively, T is the absolute temperature, Z the concentration of crack nucleation sites, V the volume of specimen, ΔF^* the free energy of activation associated with the process of crack growth, v the effective volume of the vacancy created by the breakage

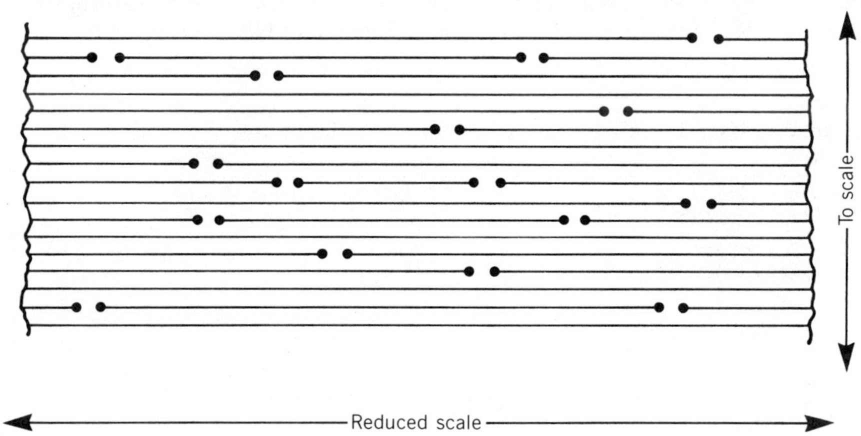

Figure 1. Perfect uniaxial fiber structure assumed in the calculation of theoretical strength. The chain ends are randomly placed.

of a chain segment, and

$$\Delta f_r^* = \frac{\pi^3 \rho^3 E^2}{6(1-\mu^2)^2 \sigma^4} \quad (3)$$

In a tensile test performed at a constant rate of stress increase, $\sigma(T) = \dot{\sigma} \cdot t$ where $\dot{\sigma}$ is the rate of stressing and

$$\frac{VZkT}{h} \exp\left(-\frac{\Delta F^*}{kT}\right) \int_0^{t^*} \exp\left\{\frac{1}{kT}\left[\frac{v\sigma^2 t^2}{2E} - \frac{\pi^3 \rho^3 E^2}{6(1-\mu^2)^2 \sigma^4 t^4}\right]\right\} dt = 1 \quad (4)$$

The upper limit of the integral in equation 4, t^*, is the time to failure in a tensile test. The breaking stress $\dot{\sigma}^*$ can be calculated from

$$\sigma^* = \dot{\sigma} t^* \quad (5)$$

The molecular weight dependence (MWD) of strength is derived from the effect of molecular weight on the fracture surface energy ρ. The results can be plotted in the dimensionless form as reduced fracture energy (ρ_r) vs reduced chain length (L_r). The strength of uniaxial polyethylene is calculated by means of equation 4 using the values of parameters shown in Table 1.

Using activation energies and moduli, these calculations yield the values of strength and elongation at break given in Table 2.

With regard to strength at infinite molecular weight, the difference between the findings of this author and the results of others discussed below depend primarily on the choice of the activation energy of the C—C bond under stress.

Irreconcilable differences exist, however, in the effect of molecular weight on strength. These results are shown in the plots of Figure 2, corresponding to moduli and activation energies for C—C bonds obtained by various authors. The main finding is that $M_n = 60{,}000$ is sufficient to achieve a strength exceeding 95% of the theoretical potential at infinite molecular weight.

Effect of Strain Rate. The effect of strain rate on the strength of viscoelastic fibers is well known. A strong strain-rate dependence of melt-spun and drawn polyethylene fibers has been recently demonstrated.

Figure 3 shows the strain-rate dependence of the stress–strain curves for low molecular weight monodisperse polyethylene ($M = 2.2 \times 10^4$). According to this model the strain rate has no effect on the modulus. The maximum strength and the mode of failure, by contrast, display a marked dependence on the strain rate ϵ.

Effect of Temperature. The effect of temperature predicted by theory on tensile properties for high molecular weight polyethylene ($M_n = 3.3 \times 10^5$) is shown in Figure 4. The temperature does not affect the modulus but has a very large effect on strength and elongation at break.

Table 2. Fracture Surface Energy ρ, Activation Energy ΔF^*, Strength σ^*, and Elongation at Break ϵ_B for Uniaxial Extended Chain PE

σ, J/m^2 [a]	ΔF^*, J/bond [b]	σ^*, GPa [c]	ϵ_B, %
1730	3.39×10^{-19}	72	24
519	1.02×10^{-19}	26	20
1313	2.57×10^{-19}	56	19
1153	2.26×10^{-19}	50	17

[a] To convert J/m^2 to ftlbf/in^2, divide by 2100.
[b] To convert J to erg, multiply by 10^7.
[c] To convert GPa to psi, multiply by 145,000.

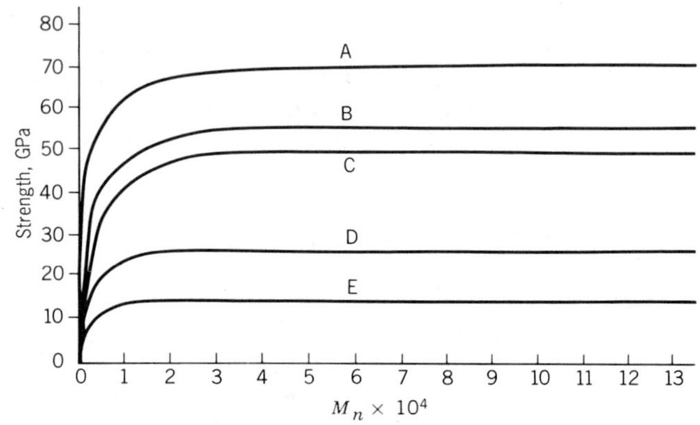

Figure 2. Random placement of chain ends in a PE extended chain model. Dependence of strength on molecular weight, modulus, and activation energies taken from the studies of various authors. To convert GPa to psi, multiply by 145,000.

Interpretation

Comparison of theoretical calculations and experimental data shows that the theory does not accurately predict the magnitude of the strain rate and temperature dependence of fiber properties. This casts doubt also on the predicted molecular weight dependence (the power law, $M^{0.4}$) and the strength at infinite molecular weight of 24 GPa (3480×10^3 psi).

Table 1. Numerical Values of Parameters Used in Calculations of Theoretical Maximum Strength of Uniaxial Extended Chain PE

Temperature T, K	100
Poisson's ratio μ	0.4
Volume of specimen V, cm^3	3×10^{-5}
Concentration of nucleation sites Z, N/cm^3	10^{19}
Volume of an elementary vacancy v, cm^3	2×10^{-23}
Average distance between polymer chains l, cm	4×10^{-8}
Stress concentration factor Q	1
Stress rate $\dot{\sigma}$, Pa/s [a]	4×10^{-5}

[a] To convert Pa/s to dyn/cm^2·s, multiply by 10.

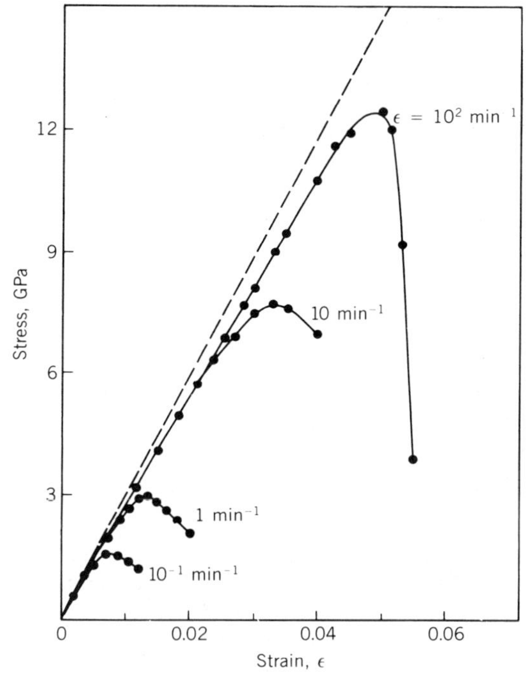

Figure 3. Strain–stress dependence. Courtesy of Marcel Dekker, Inc. To convert GPa to psi, multiply by 145,000.

It must be recognized that for the molecular weight dependence of strength, agreement with the experimental data is a weak supporting argument. Experiments concerned with the effect of molecular weight on fiber strength are very difficult (or virtually impossible) to conduct in such a way that fiber morphology, nonuniformity, flaws, and other characteristics

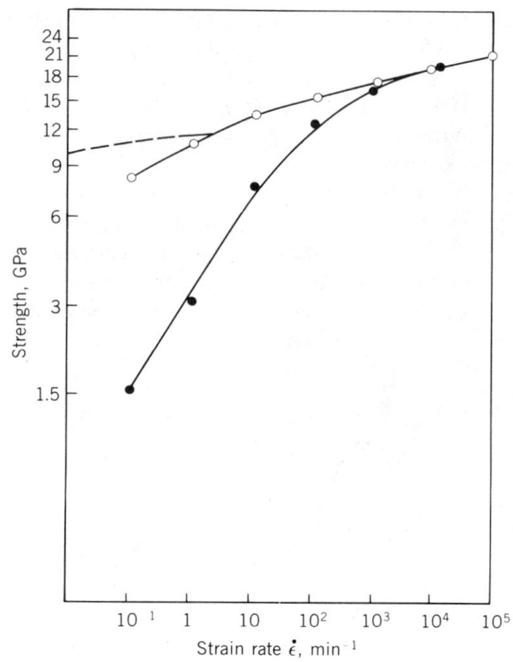

Figure 4. Stress–strain curves calculated for polyethylene with $M = 3.3 \times 10^5$ at different temperatures; strain rate ϵ = 1/min and ϵ = 0.02 s. The dashed line indicates a slope equal to the theoretical modulus (= 300 GPa). To convert GPa to psi, multiply by 145,000. Courtesy of Marcel Dekker, Inc.

remain the same as the molecular weight is changed. Thus, in addition to changes in molecular weight, these strength data reflect many other factors. Therefore the molecular weight dependence is not a good choice for verifying the theories. No such problems are encountered with the temperature and strain-rate dependence of strength where the experiments are conducted with the same type of fiber.

BIBLIOGRAPHY

U.S. Pat. 4,344,908 (Aug. 17, 1982), P. Smith and P. J. Lemstra (to Stamicarbon B. V.).

U.S. Pat. 4,413,110 (Nov. 1, 1983), S. Kavesh and D. C. Prevorsek (to Allied Corp.).

D. C. Prevorsek, *J. Polym. Sci. Part A* **1,** 993 (1963); *J. Polym. Sci. Polym. Symp.,* **32,** 343 (1971).

Y. Termonia and P. Smith in A. E. Zachariades and R. S. Porter, eds., *High Modulus Polymers,* Marcel Dekker, Inc., New York, Chapt. 11.

D. C. PREVORSEK
Allied Corporation

ULTRACENTRIFUGATION

Macromolecules of biological origin, especially the proteins (qv), are often monodisperse in size and molecular weight (see MONODISPERSE POLYMERS). Measurement of a given average property of a protein solution thus serves to specify that property for each molecule in the solution. In contrast, most synthetic high polymers are polydisperse. It is therefore important to determine distributions of properties of these solutes as well as their average properties. The measurement of molecular weight distributions has been the major objective of much of the ultracentrifugal analysis of synthetic high polymers (see also MOLECULAR WEIGHT DETERMINATION).

A number of difficulties are associated with investigations of synthetic polymers. Solutions of these molecules in many solvents are highly nonideal. Often, conditions (θ conditions) can be found in which the effects of nonideality are minimized. High temperatures or difficult solvents, however, are frequently required to obtain such θ conditions, and these requirements bring technical difficulties with them. Another problem results from the relatively high compressibilities of the solutes or solvents employed. Most theoretical treatments of sedimentation behavior assume incompressible systems. The investigator must be careful that the effects of differential compressibility of solvent and solute are taken into account in the interpretation of results or else are minimized by thoughtful experimental design (see also SOLUTION PROPERTIES).

Equipment

The analytical ultracentrifuge is a device that permits observation of the concentration distribution within a solution while it is being subjected to a centrifugal field. The solution is contained within a cell that is fitted into a rotor, and the rotor is then driven at speeds up to 60,000 or 68,000 rpm by a geared electric motor.

Optical Systems. Light sources are arranged so that the beam passes through a collimating lens in the bottom of the vacuum chamber and strikes the rotor, which occludes the light beam except when the cell is in the optical path. The beam that passes through a condensing lens in the top of the chamber is bent through 90° by a mirror above the drive, passes through the remainder of the optical system inside a horizontal tube, and strikes either a photographic plate or the photomultiplier of the absorption scanner at the upper right-hand corner of the instrument.

The desired measurement in any analytical centrifuge experiment is that of the solute concentration (or its concentration gradient) as a function of the radial position. Three optical systems are in common use for observations of solute redistribution. The first two employ the refractive-index increment as a measure of solute concentration; the third utilizes light absorption. The schlieren optical system is the most commonly employed. This system displays the refractive-index gradients as a function of radial position and thus, since refractive index is closely proportional to concentration, provides values of the first derivative of the concentration with radius throughout the solutions. The second optical system is a variant of the classical Rayleigh interferometer, and is more sensitive and precise than the schlieren system.

The third optical system is a light-absorption scanning device that produces a record of optical transmission or absorbance as a function of radius. The refractive-index-based and light-absorption-based optical systems provide complementary information, and each is useful for particular applications.

General Techniques

Despite the great variety of techniques employed with the ultracentrifuge, there are certain common operations that can be discussed in a general way.

In sedimentation equilibrium experiments, the time required to reach equilibrium varies from a few hours to a few days. It depends on the length of the solution column, the viscosity of the solution, and the diffusion coefficient of the solute. Photographs should be taken at several times during the run. These records should then be compared to determine when equilibrium has been attained.

Data consisting of concentration (zero arbitrary) as a function of radius, or of the first derivative of concentration as a function of radius, or of true concentration as a function of radius are then read from the photographic plate or scan. If weight-average molecular weights are to be obtained from refractometric records, and unless a high-speed technique is used, it is necessary to find the initial concentration of the solution by a separate synthetic boundary experiment or with a differential refractometer.

Sedimentation velocity experiments are usually performed at relatively high rotor speeds for, at most, a few hours. Photographs or scans are taken automatically at known time intervals to follow the progress of the boundary or boundaries that are formed. The experimental records obtained are a series of pictures of the concentration or its gradient as a function of radius at a sequence of times. These records can be treated in a number of ways.

It is now common practice in many laboratories to make use of the digital computer to perform much of the labor of data reduction. A number of appropriate programs have been written, and reference to them will be found in much of the literature.

Sedimentation Equilibrium

Molecular Weight Averages. The determination of molecular weights and their distributions by sedimentation equilibrium is based soundly on classical thermodynamics. The relationship between the information from sedimentation analysis and that obtained from colligative properties or light scattering has been reviewed. There has been much progress in overcoming interpretational and experimental difficulties associated with sedimentation equilibrium, and the breadth of chemical systems for which this approach is useful makes it the method of choice for determining molecular weight averages.

Molecular Weight Distribution. The determination of the molecular weight distribution of a heterogeneous polymer is a formidable task at present. Several procedures have been adopted to characterize such distributions from data obtained at sedimentation equilibrium (see also FRACTIONATION).

If one has *a priori* knowledge of the form of a distribution, one can estimate the width of the distribution and the location of its maximum from a knowledge of several molecular weight averages. One such treatment was carried out in which $\bar{M}_n$, $\bar{M}_w$, $\bar{M}_z$, and $\bar{M}_{z+1}$ were used to fit a very flexible assumed form of distribution based on Laguerre polynomials (see MOLECULAR WEIGHT DETERMINATION).

Some effort has been directed at finding experimental procedures that will allow molecular weight distributions of arbitrary shape to be determined. A method using Laplace transforms has yielded moderate success. Unfortunately, because of the extreme complexity of the mathematics involved the method

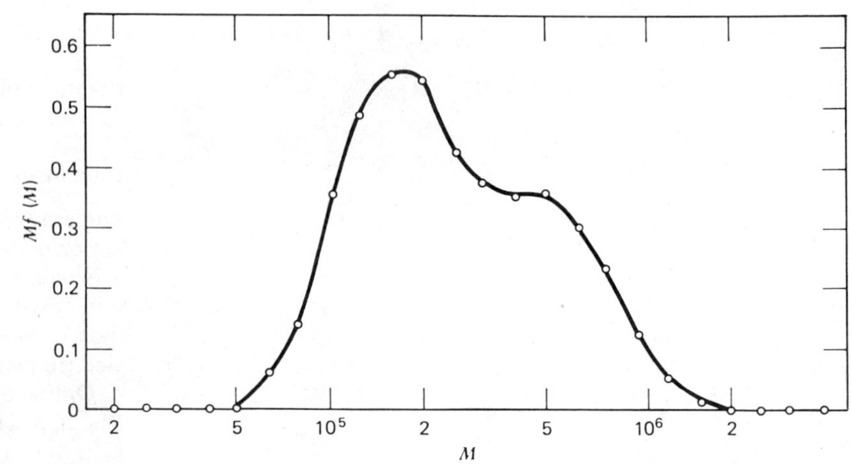

Figure 1. Molecular weight distribution of a polystyrene sample, obtained by the method of Scholte. Measurements performed in cyclohexane at 35°C. The distribution was calculated from 44 values of $U(\lambda, \xi)$.

has not been widely accepted. A different approach has been taken, in which the rotor speed is varied, and data from several different radii in the solution column are used at each of the speeds.

Results for a sample of polystyrene in cyclohexane at 35°C are shown in Figure 1. The values of $\bar{M}_n$, $\bar{M}_w$, $\bar{M}_z$, and $\bar{M}_{z+1}$ calculated from the distribution agree well with values obtained from ordinary sedimentation methods. At present, this scheme for obtaining molecular weight distributions is restricted to ideal systems. Its extension to nonideal solutions appears to offer considerable computational difficulty, but is not to be considered impossible. An approximate correction for nonideality is not sufficient to make Scholte's method suitable for determining molecular weight distributions of dextran samples exhibiting a B_{LS} of 1 mol-L/g^2. Moreover, Scholte's method is better suited for examining samples with rather broad molecular weight distributions.

Sedimentation Velocity

Velocity sedimentation yields results less rigorously interpretable in principle than those of equilibrium sedimentation. In practice, however, it has been easier to use and has therefore been more widely employed. The primary measurement made by this technique is of the distribution of sedimentation coefficients $g(s)$. This distribution, once determined, can be transformed by a suitable relation to a distribution of molecular weights $f(M)$. The procedures commonly in use seem to involve a number of uncertainties as well as much data manipulation. Large corrections must be introduced for the effects of concentration, pressure, and diffusion.

Sedimentation in a Density Gradient

If a mixed solvent composed of materials of different densities is subjected to prolonged sedimentation, an equilibrium density gradient is established by redistribution of the components within it. If a macromolecular solute is also present, this solute will redistribute to form a band in the gradient that is centered about the effective buoyant density of the macromolecule in the solvent. In the ideal case, this band is Gaussian in shape and its width is inversely proportional to the square root of the molecular weight of the macromolecule. The basic theory and methods for this technique have been worked out. The technique has been much used in biochemical investigations, where aqueous solutions of various salts are used to form the density gradients.

BIBLIOGRAPHY

J. W. Williams, K. E. Van Holde, R. L. Baldwin, and H. Fujita, *Chem. Rev.* **58,** 715 (1958).

H. Fujita, *Mathematical Theory of Sedimentation Analysis,* Academic Press, Inc., New York, 1962.

J. W. Williams, ed., *Ultracentrifugal Analysis in Theory and Experiment,* Academic Press, Inc., New York, 1963.

D. A. Yphantis, *Biochem.* **3,** 297 (1964).

THOMAS M. LAUE
University of New Hampshire

ULTRAFILTERS. See MEMBRANES.

ULTRAFILTRATION

Ultrafiltration is a pressure-driven filtration separation occurring on a molecular scale (see also MEMBRANES). Typically, a process fluid containing dissolved and/or suspended material contacts one side of a porous membrane. A pressure gradient is applied across the membrane. The liquid, including small dissolved molecules, is forced through the pores. Large dissolved molecules, colloids, and suspended solids are retained by sieving.

Ultrafiltration operates on species having molecular dimensions ranging from ca 2 to 20 nm. Microfiltration is a filtration process operating on species whose size ranges from ca 20 to 200 nm. Transport through ultrafiltration and microfiltration membranes is best described by pore-flow models. Reverse osmosis or hyperfiltration separates species smaller than ca 2 nm. That process, dominated by interactions between the membrane material, solute, and solvent, is best described by solution–diffusion mechanisms.

The process fluid retained by the membrane is called concentrate or retentate. That portion passing through the membrane is called filtrate, ultrafiltrate, or permeate. The rate at which material permeates the membrane is called flux. The ultrafiltration objective may be to recover or concentrate materials in the retentate (eg, in cell harvesting, pigment recovery, protein recovery from whey, etc) or to produce a purified permeate (eg, in waste treatment, sterile water production, enzyme purification, etc).

Retained materials accumulate at the membrane surface, forming a gel-polarization layer. When the hydraulic permeability of this layer is low, it can reduce the membrane flux. The gel-polarization thickness can be controlled by flowing the process fluid tangentially across the membrane.

Diafiltration is a specific ultrafiltration process in which the retentate is further purified or the permeable solids are extracted further by the addition of water to the retentate. It is analogous to conventional filter cake washing. Membrane reactors combine chemical or biological processes with ultrafiltration. Systems have been engineered for reactions both upstream and within the membrane. Electroultrafiltration is an emerging process that combines electric fields with ultrafiltration to improve flux or separation specificity.

In spite of significant equipment and process performance improvements, the industrial ultrafiltration market has not grown as predicted. The 1988 global annual industrial market was estimated at ca $80 million. The laboratory market is an order of magnitude larger. The average membrane life for industrial applications ranges from one to five years, depending on the severity of service. Approximately half of the annual industrial sales are replacement membranes and parts for the installed capacity base.

Media

Ultrafiltration membranes can be integrally skinned, homogeneous, or nonintegrally skinned (composite and dynamic).

Process

Pore-flow models most accurately describe ultrafiltration processes. Other transport mechanisms, which may occur simultaneously, although generally at a much lower rate, include dialysis, osmosis, reverse osmosis (osmosis retarded by a pressure gradient), electrodialysis, electroosmosis, piezodialysis,

Donnan effects, Knudsen flow, thermal effects, chemical reactions, and active transport.

For pure water transport, Darcy's law states that the flow rate is directly proportional to the pressure gradient:

$$J = \frac{V}{At} = \frac{K_m \, \Delta P}{\mu} \tag{1}$$

where J is permeate flux in units of volume V per membrane area A at time t. K_m is the membrane hydraulic permeability, μ is the fluid viscosity, and ΔP is the pressure drop across the membrane.

Membrane hydraulic permeability is a function of the number of pores, their size, their tortuosity, their length, and any substructural resistance. Most polymeric, ultrafiltration membranes yield (compact) or creep under pressure. K_m is therefore a function of the pressure history. Compaction can result from either a dynamic pressure drop (flow through a membrane) or a static pressure drop (eg, across a fouling film). Initial compaction occurs rapidly during start-up, whereas long-term compaction accumulates slowly over the membrane's operating life. Membrane swelling agents, particularly hypochlorite, can sometimes reverse compaction. Polymeric membranes are generally limited to operating pressures of less than 400 kPa (<4 atm).

Fouling

If the gel-polarization layer is not in hydrodynamic equilibrium with the fluid bulk, the membrane is fouled. Fouling is caused by either adsorption of material on the membrane or consolidation (fusing, polymerization, etc) of the gel layer. Fouled systems respond differently than nonfouled systems. Flux and K become independent of the hydrodynamic conditions, but vary as a function of total permeate production. Theoretical permeate flux (TPF) decreases with time. A sensitive test for predicting fouling or process instability is to measure the TPF change after subjecting the system to process extremes (eg, high pressure with no flow).

Fouling is controlled by selecting appropriate membrane media that do not adsorb species in the process stream, modifying the membrane surface, pretreating the process fluid, and adjusting the operating conditions.

Cleaning

Fouling films are removed from membrane surfaces by chemical and mechanical methods. A cleaning regimen must dissolve and remove the fouling film and be compatible with the materials used to construct the membrane module and equipment. The equipment manufacturer recommends cleaning methods and chemicals for specific applications. A system is clean when its original water flux is restored.

Membrane Equipment

Commercial industrial membrane equipment first became available in the late 1960s. Five different configurations are available today. No configuration is superior in all respects to any other. Equipment that can use flat sheet membrane generally offers a broader selection of membrane type than configurations with cast or extruded membranes. Wide channel spacing is more tolerant of suspended solids, but costs more to build and operate.

Systems include the parallel-leaf cartridge, the plate-and-

Figure 1. A production-scale membrane bioreactor system, capable of producing 50–100 × 10^3 kg per year of chiral intermediates. Courtesy of Sepracor Inc.

frame, the spiral wound cartridge, the supported tube, and self-supporting tubes.

Systemization

Systems comprise assemblies of the aforementioned modules in combinations of series and parallel flow paths. The maximum allowable pressure for specific modules and the cross-flow pressure drop for the specific geometry limit the number of modules in a series flow path. Systems are engineered with pumps, valves, heat exchangers, pretreatment systems, instrumentation, and controls to conform to the system type, size, operation mode, and industry standards for specific applications. The International Dairy Federation has drafted 3-A Sanitary Standards for dairy membrane systems.

Since theoretical models cannot predict performance from physical data, plant design parameters are obtained from laboratory testing, pilot plant data, or, in the case of established applications, the performance of operating plants.

Flux response to concentration, cross-flow velocity, pressure, and temperature should be determined for allowable plant excursions. Fouling must be quantified, and cleaning procedures proven. The final design flux should reflect long-range variables such as feed composition changes, long-term compaction, membrane fouling, etc.

Because of the differences in ultrafiltration equipment, the performance of one type of device cannot be used to predict the performance of different devices. Comparisons should only be made on an economic basis when the performance of each is known. Figure 1 is an example of an ultrafiltration system.

Membrane Reactors

There are two basic types of membrane reactors: those that contain the reaction in the retentate and those that convert the permeate.

Applications

The best application knowledge resides with the equipment suppliers and the end users who develop and practice the processes. The principal industrial applications are in electrocoat paint, food processing, waste recovery, and pharmaceutical and biotechnology applications.

BIBLIOGRAPHY

K. K. Sirkar and D. R. Lloyd, *AIChE Symp. Ser.* **84**(261), 1 (1988).

A. R. Cooper, ed., *Ultrafiltration Membranes and Applications*, Proceedings of the 178th National ACS Meeting, Washington, D.C., 1979, Plenum Press, New York, 1980.

E. Drioli and M. Nakagaki, eds., *Membranes and Membrane Processes*, Plenum Press, New York, 1986.

PETER R. KLINKOWSKI
Innotech Corporation

ULTRAVIOLET-RADIATION ABSORBERS.
See STABILIZATION.

ULTRASONIC WELDING.
See BONDING in ADHESION AND BONDING.

UNIDIRECTIONAL TAPE.
See COMPOSITES, FABRICATION.

UNZIPPING.
See DEGRADATION; DEPOLYMERIZATION.

UREA–FORMALDEHYDE RESINS.
See AMINO RESINS.

UREAS, POLYMERIC.
See POLYUREAS.

URETHANE POLYMERS.
See POLYURETHANES.

V

VACUUM BAG MOLDING. See REINFORCED PLASTICS.

See REINFORCED PLASTICS.

VACUUM FORMING. See THERMOFORMING.

VACUUM METALLIZING. See METALLIZING.

VAPOR BARRIERS. See BARRIER POLYMERS.

VARNISH. See COATINGS; DRYING OILS.

VEGETABLE FIBERS. See FIBERS, VEGETABLE.

VESICLES

Vesicles are self-supported closed bilayer assemblies of several thousand lipid molecules (amphiphiles) that enclose an aqueous interior volume. The lipid bilayer is a two-dimensional fluid composed of lipids with their hydrophilic head groups exposed to the aqueous solution and their hydrophobic tails aggregated to exclude water. The bilayer structure is highly ordered yet dynamic because of the rapid lateral motion of the lipids within the plane of each half of the bilayer. The term liposomes is frequently interchanged with the term vesicles and is usually reserved for vesicles of glycerophospholipids or other natural lipids.

Supramolecular Assemblies: Micelles, Monolayers, and Membranes

In addition to their usefulness as model membranes, vesicles and BLMs (bilayer lipid membranes) represent two classes of supramolecular assemblies, which also include micelles, monomolecular layers, multilayers, extended bilayers, and nonspherical microstructures (tubules) (Fig. 1) (see also MEMBRANES).

Polymerization of Vesicle Assemblies

The polymerization of lipid assemblies was first demonstrated in monolayers, vesicles, and extended bilayers and subsequently in cast multilayers, BLMs, and tubules. The successful polymerization of monolayers and vesicles contrasts with the inherent difficulties in polymerizing single-chain amphiphiles in micelles. The significant difference between the two systems is the higher rate of exchange for monomers from micellar than from vesicular assemblies. In micelles the exchange rate is competitive with the rate of polymerization, whereas in bilayer and monolayer structures the exchange rate is too low to compete significantly with the polymerization reaction.

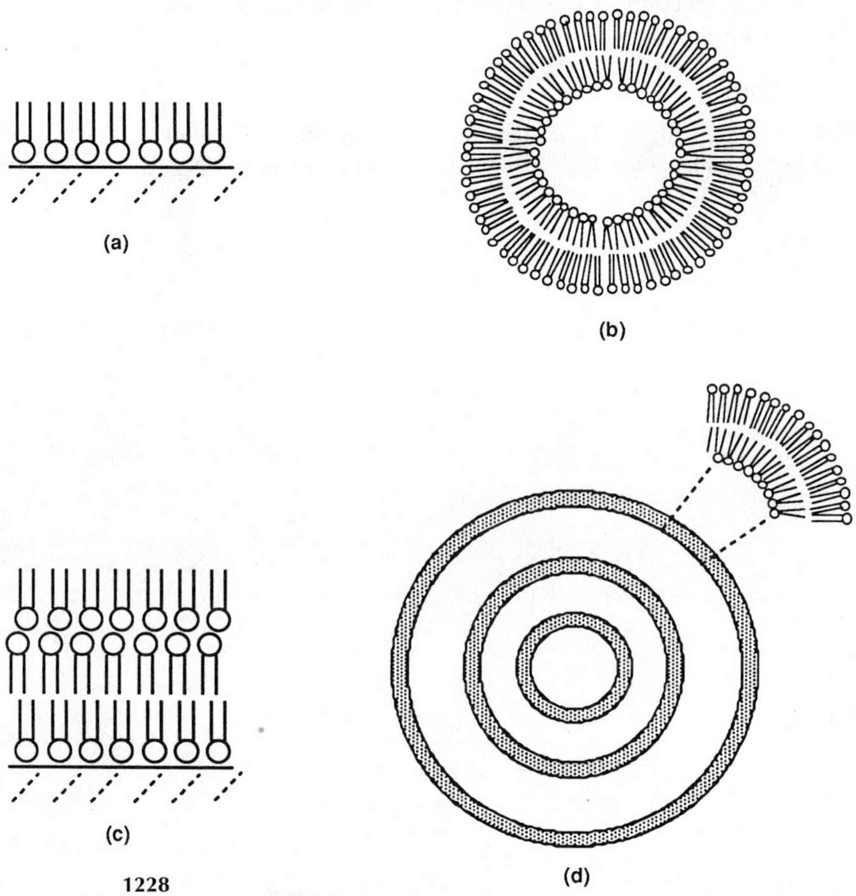

Figure 1. Supramolecular assemblies. (**a**) Monolayer; (**b**) unilamellar vesicle; (**c**) multilayer; (**d**) multilayer vesicle.

Strategies. The formation of polymerized supramolecular assemblies may be achieved by a variety of experimental strategies developed in the 1980s.

An important method for polymerizing vesicles and other supramolecular assemblies is the preparation of polymerizable lipid monomers, the formation of the lipid assembly from the monomer, and the subsequent chain polymerization of the monomers in the assembly. A variation of this method is to incorporate reactive moieties that allow step polymerization of the lipid assembly.

In another method the amphiphile is polymerized in isotropic solution, and the supramolecular assembly is formed from the prepolymerized amphiphiles. This technique has been extensively studied in monolayers; the final assembly is usually termed a polymeric membrane.

In a third method the monomers or polymers are associated with preformed assemblies. Polymerizable species may be electrostatically associated with the amphiphile either before or after assembly formation (polymerizable gegenions) and polymerized in the assembly. Alternatively, polymers and oligomers may be adsorbed or anchored onto the surface of membrane assemblies.

Polymerizable Vesicles from Polymerizable Amphiphiles. Polymerizable amphiphiles have been prepared by the introduction of the reactive group into different regions of the lipid molecule. The polymerizable group may be at the chain terminus in the middle of the bilayer (type A), near the middle of the hydrophobic chains (type B), attached to the hydrophilic head group (type C), electrostatically associated with a charged lipid (type D), or in the center of a bolaamphiphile (type E). Methods A, B, and E have no direct influence on the membrane–water interface. The mobility of the lipid chains is significantly decreased by polymerization in these systems. In contrast, methods C and D alter the membrane–water interface, but have less effect on the hydrophobic interior of the membrane. These examples with only one reactive group per lipid form linear polymer chains in supramolecular assemblies. The presence of a second polymerizable group per molecule allows cross-linking of the polymer chains.

A variety of lipid classes is used to form polymerizable lipids including phospholipids, dialkyldimethylammonium salts, dialkyl zwitterionic taurines, dialkyl phosphates, glycerol ethers, glutamates, and bolaamphiphiles. Although the hydrophobic lipid chains (tails) are usually hydrocarbon, fluorocarbon-substituted lipids, which display enhanced chain hydrophobicity, are known as well.

Characterization of the Polymerized Vesicle. Polymerization of lipid vesicles by the above methods alters the assembly characteristics. Preliminary characterization of the resultant vesicles usually includes determination of their size and phase behavior.

It is necessary to demonstrate that the vesicle size and shape are not significantly changed by the polymerization reaction. Optical microscopy, electron microscopy, and light scattering are commonly used for these purposes.

Polymerization of vesicles composed of type A or B lipids has a strong effect on the cooperative lipid chain motions and resultant phase behavior of the assembly, which is determined by dsc. The polymerization of vesicles from type C lipids has less effect on the chain motions, especially if the polymerizable group is separated from the head group by a spacer linkage.

Permeability. The permeability of membranes to water-soluble markers is proportional both to the partition coefficient of the compound from the aqueous phase into the lipid bilayer and to the rate of diffusion of the compound across the bilayer membrane.

The formation of sizable linear polymer chains at the surface of the vesicle bilayer or the membrane interior of the bilayer moderately reduces the vesicle permeability. Significant reductions (greater than one order of magnitude) in permeability require cross-linked polymer networks in or at the surface of the vesicle.

Chemical Stability. The decrease in vesicle permeability in cross-linked vesicles is paralleled by increases in membrane stability to detergents and organic solvents. The polymerization of monosubstituted lipids in vesicles leads to moderate changes in stability, whereas more significant changes are found only after polymerization of bis-substituted lipids.

Polymer Characterization. The degree of polymerization (DP) is known for only a few vesicle and bilayer systems. The DP of linear polymers obtained from the thermal polymerization of some monomethacryloyl lipids in vesicles has been estimated. The isolated linear polymers have a number-average molecular weight consistent with a DP of about 500. These data were independently confirmed for a similar methacryloyl lipid. To date, these are the largest linear polymers reported in vesicles or extended bilayers. The size of the vesicles and length of the polymer chains indicate that 20–100 polymer chains are formed in each vesicle. Therefore, even the formation of the highest molecular weight linear polymers yields a vesicle composed of several polymers. In contrast to the methacryloyl lipids, photopolymerization of vinylbenzoylammonium halide lipids in vesicles produces polymers of chain lengths from 10 to 20, which would be at least 500 polymer chains per vesicle. Thiosubstituted PCs in vesicles yield short polymers with DPs of 17–25.

The rate of polymerization of lipids in assemblies has been reported only for vesicles of styrene-containing lipids. In contrast to the rate of polymerization of isotropic ethanol, which is dependent on monomer concentration, the rates in vesicles are higher and linearly dependent only on the light intensity.

Prepolymerized Amphiphiles. The preparation of polymeric membranes is based on the use of prepolymerized amphiphiles. Since polymers are three-dimensional molecules, they are more difficult to accommodate in two-dimensional assemblies than the individual monomers. The organizational problem may be alleviated by the utilization of polymers with a low DP or circumvented by the incorporation of flexible spacers into the monomer and polymer. The spacers function to decouple the motions of the alkyl chains that form the assembly from the polymer chain. The spacer concept is used extensively in the synthesis of liquid crystalline polymers.

There are three possible locations for spacer groups in amphiphiles:

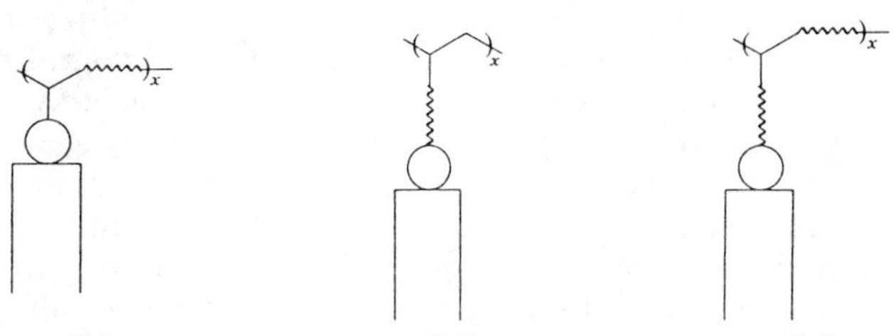

C-1 C-2 C-3

These are all derived from type C amphiphiles where the reactive entity is in the head group. The spacer is readily incorporated into the main chain by copolymerization of the polymerizable lipid with hydrophilic comonomers (type C-1). To date, this has been accomplished by solution radical chain polymerization. The lipid may be prepared with a hydrophilic spacer group as part of the head group. This side-chain spacer separates the reactive moiety from the bilayer-forming chains (type C-2). The combination of both main-chain and side-chain spacers is made possible by the copolymerization of the monomer from C-2 with hydrophilic comonomers (type C-3). Most of the research reported on prepolymerized amphiphiles emphasizes monolayer and multilayer assemblies.

Surface-associated Polymers. Bilayer membrane surfaces are hydrophilic by the nature of the head groups designed to interface with water. Head groups may be ionic, zwitterionic, or dipolar, eg, phosphatidic acid (PA), phosphatidyl choline, or phosphatidylinositol, respectively. Natural and model membranes are frequently composed of representatives of each type. Specific and general associations of enzymes, cofactors, etc, with the surfaces of biological membranes are well known. During the past few years, polymer chemists have used analogous methods to associate polymers with the surfaces of bilayer vesicles. It is convenient to divide these methods into three categories:

1. Counterions (gegenions) to ionic lipids may be exchanged for a polymerizable counterion. Subsequent matrix polymerization at the membrane surface yields a surface-associated polyelectrolyte.
2. Some water-soluble polymers adsorb to neutral or charged membrane surfaces by hydrophilic or ionic interactions.
3. Water-soluble polymers that bear periodic hydrophobic anchor groups can be bound to neutral or charged membrane surfaces by insertion of the anchor groups into the membrane.

Destabilization of Polymerized Membranes

The interaction of polymers with vesicles may serve to destabilize as well as stabilize bilayer membranes. Different chemistries can be designed into the lipid assembly to increase the lability to various agents, such as light, enzymes, reducing agents, etc. Each of these allows the encapsulation of water-soluble substances, eg, dyes, drugs, nucleic acids, and enzymes, in vesicles and their release upon subsequent application of the desired stimulus or reagent. Several methods rely on the phase separation of lipids as described in the literature.

Polymerized Model Membranes

The reconstitution of membrane proteins into the lipid bilayer of liposomes is frequently used to prepare model membranes. Proteins are conveniently classified as water-soluble, membrane-surface-associated, and membrane-penetrating proteins. The latter two are termed peripheral and integral proteins, respectively. Integral proteins are held in the bilayer by hydrophobic interactions with lipids and frequently span the membrane thickness. The bilayer is a two-dimensional solvent for these proteins. Reconstitution allows the examination of specific surface and membrane interactions that are difficult to isolate in the more complex natural membrane and are sometimes not observed in detergent preparations of membrane proteins.

The availability of polymerizable lipids makes it possible to reconstitute membrane proteins in polymerized bilayer membranes. The enhanced chemical stability of the membranes from some polymerizable lipids opens the way to the preparation of robust model membranes, and initial reports indicate significant success for the incorporation of membrane proteins, eg, ATP synthetase, bacteriorhodopsin, and rhodopsin, into polymerized membranes.

Outlook

Several methods are now available for the modification of vesicle properties by formation of or interaction with polymers. Many of these enhance the chemical and colloidal stability of vesicles and thereby expand the range of conditions for their use. Others perturb or disrupt the vesicle membranes in response to various stimuli in a manner that seeks to mimic biological membranes. The initial understanding of the nature of the polymer chemistry that has occurred with these systems has been of great benefit in designing new materials. A more complete comprehension of the nature of two-dimensional polymerizations will undoubtedly aid the future development of the field.

This is a relatively new field, which has expanded rapidly. Already there are strong parallels between the chemistry of vesicle assemblies and that of other supramolecular assemblies, eg, liquid crystals, monolayers and Langmuir-Blodgett (LB) multilayers, BLMs, tubules, etc. Potential applications of these polymerized vesicles and other assemblies from these amphiphiles are varied and include biocompatible surfaces; catalysis; drug delivery and surface targeting; magnetic particles; reactivity control; reagent release by light, acid, or reducing agent; separation of photogenerated charge carriers; and surface modification.

BIBLIOGRAPHY

H. Ringsdorf, B. Schlarb, and J. Venzmer, "Molecular Architecture and Function of Polymeric Oriented Systems: Models for the Study of Organization, Surface Recognition, and Dynamics of Biomembranes," *Angew. Chem. Int. Ed. Engl.* **27,** 113 (1988).

J. H. Fendler and P. Tundo, "Polymerized Surfactant Aggregates: Characterization and Utilization," *Acc. Chem. Res.* **17,** 3 (1984).

D. F. O'Brien, R. T. Klingbiel, D. P. Specht, and P. N. Tyminski, "Preparation and Characterization of Polymerized Liposomes," *Ann. N.Y. Acad. Sci.* **446,** 282 (1985).

S. L. Regen, "Polymerized Liposomes" in M. J. Ostro, ed., *Liposomes: From Biophysics to Therapeutics,* Marcel Dekker, Inc., New York, 1987, p. 73.

D. F. O'Brien
V. Ramaswami
University of Arizona

VINAL FIBERS.

See Vinyl alcohol polymers.

VINYL ACETAL POLYMERS

The poly(vinyl acetal)s are a family of polymers with the general structure shown in Figure 1. They are prepared by the

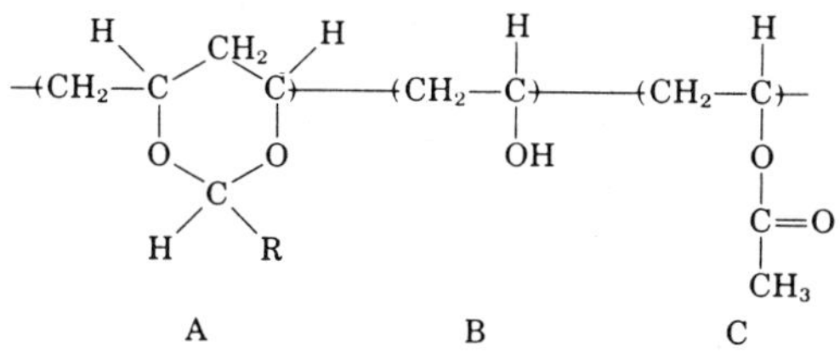

Figure 1. Generalized structure for the poly(vinyl acetals). For commercial grades: A, poly(vinyl acetal), is 50–75 mol % or 75–90 wt %; B, poly(vinyl alcohol), is 12–50 mol % or 6–25 wt %; C, poly(vinyl acetate), is 0–15 mol % or 0–13 wt %.

acid-catalyzed acetalization of poly(vinyl alcohol). The predominantly 1,3-glycol structure of poly(vinyl alcohol) strongly favors the formation of the cyclic 1,3-dioxane ring.

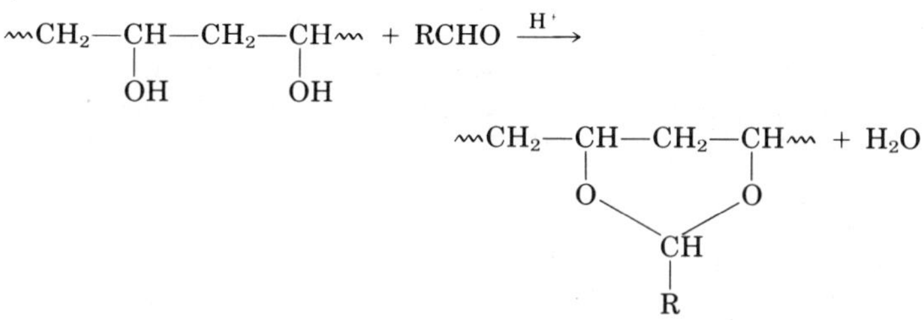

Poly(vinyl acetal)s are characterized by toughness, chemical resistance, and strong adhesion to a wide variety of surfaces. As a component in coatings, they improve adhesion and flexibility. Their pigment-binding properties have led to their widespread use as a component in inks, electrographic toners, magnetic tapes, and as a fugitive binder in the preparation of high quality ceramics for the electronics industry.

Stereochemistry of the Acetal Ring

The relationship between the stereostructure of the starting poly(vinyl alcohol) and the course of the formalization reaction indicate that cis-formalization at isotactic portions of the poly(vinyl alcohol) chain is faster than trans-formalization at syndiotactic parts of the chain. Using poly(vinyl alcohol) derived from three sources, varying ratios of high isotactic (I-type), high atactic (A-type), and high syndiotactic (S-type) polymer were obtained (Tables 1 and 2).

Table 2. Configuration of Poly(vinyl formal) Obtained by Formalization of Poly(vinyl alcohol) of Different Tacticity

Poly(vinyl alcohol), tacticity	Degree of formalization, mol %: Total	*cis*-Formal	*trans*-Formal
I	84	70	14
A	87	59	28
S	87	50	37

Manufacture

Although there are a number of potential methods for the manufacture of poly(vinyl acetal)s, in practice, only two basic methods with variants are used commercially today. The solution method involves the formation of the acetal as a solution in organic solvent followed by precipitation and purification of the polymer. In the aqueous process, the polymer precipitates as formed by the reaction of the aldehyde with an aqueous solution of poly(vinyl alcohol).

Properties

The properties of the poly(vinyl acetal)s are a function of molecular weight and the final chemical composition. Molecular weight is determined largely by the molecular weight of the poly(vinyl alcohol), which is a function of the conditions of polymerization and hydrolysis of the starting poly(vinyl acetate). Minor differences in molecular weight, molecular weight distribution, and polymer branching can be introduced during the acetalization reaction. The chemical properties are a function of the chain length of the aldehyde component and the relative concentrations of the three components, poly(vinyl acetal), poly(vinyl alcohol), and poly(vinyl acetate).

Poly(vinyl formal)s. Formvar is the Monsanto trademark for poly(vinyl formal). Formvar grades 5-, 6-, 7-, and 15-95E have the same chemical composition but differ in molecular weight and viscosity. This is an outgrowth of the principal applications of these materials and the processing problems inherent in producing resins of significantly different chemical composition. As the percentage of poly(vinyl acetate) increases, the resin becomes softer, and heat-distortion temperature and tensile and impact strengths are reduced. However, solubility range increases and the resin becomes difficult to process and dry. The water sensitivity of the resin increases with increasing poly(vinyl alcohol) content. Resistance to acids and bases, regardless of strength, is excellent. Resistance to most common organic solvents is good to excellent, except for some highly polar compounds, most notably phenolics, lower aliphatic halogenated solvents, and certain cyclic ethers (eg, tetrahydrofuran).

Formvar resins are insoluble in most common organic solvents, but soluble in acetic acid, aniline, cresylic acid, dimethyl sulfoxide, *N,N*-dimethylacetamide, *N,N*-dimethylformamide, furfural, methyl benzoate, methyl butynol, methylene chloride, methyl pentynol, *N*-methyl-2-pyrrolidinone, phenol, tetrachloroethane, tetrahydrofuran, and toluene–ethanol.

Table 1. Polymerization Conditions Giving Poly(vinyl alcohol) of Different Stereoregularities

Monomer	Medium	Initiator	Temperature, °C	Poly(vinyl alcohol): Type	DP	Tacticity: I	A	S
vinyl *tert*-butyl ether	toluene	$BF_3{\cdot}O(C_2H_5)_2$	−78	I	1000	56	32	12
vinyl acetate	methanol	AIBN[a]	60	A	1700	23	47	30
vinyl formate	methyl formate	$B(C_2H_5)_3O_2$	−78	S	970	17	46	37

[a] Azobisisobutyronitrile.

Poly(vinyl formal) resins are at least partially compatible with a variety of important polymers. This property allows blending with other resins to enhance the properties of the final formulation. Poly(vinyl formal)s have adequate thermal stability for many applications.

Butvar resins (Monsanto's poly(vinyl butyral)) fall into two distinct groups: those with a high (ca 19 wt %) residual poly(vinyl alcohol) content and those with a low (ca 12 wt %) poly(vinyl alcohol) content. The latter are soluble in a wider range of solvents than are the former. Tensile strength increases with poly(vinyl alcohol) content at a given molecular weight. Also, at a given molecular weight viscosity increases with increasing hydroxyl content. For a given hydroxyl content, viscosity is proportional to molecular weight. For a given grade, viscosity increases with increasing molecular weight of the solvent alcohol. Solvent blends are commonly used, the exact composition dictated by the end use of the solution and the required drying properties.

Resistance of both Butvar and Formvar resins to acids and bases is excellent; resistance to nonpolar organic solvents is generally good, but resistance to polar organic solvent is fair to poor, depending on solvent and resin type. Solubility in organic solvents increases with increasing acetylation and decreasing molecular weight. Chlorinated organic solvents swell or dissolve the resins depending on resin grade and the specific solvent.

Chemical Properties. The chemical reactions of poly(vinyl butyral) and poly(vinyl formal) are the characteristic reactions of the functional groups: the cyclic acetal, alcohol, and acetate. The most reactive moiety is the hydroxyl group that undergoes reactions typical of a hindered secondary alcohol. Reaction via the hydroxyl group with difunctional reactants provides a convenient method of cross-linking.

Plasticizers

A wide variety of plasticizers is suitable for poly(vinyl butyral)s and formals. Suggested for poly(vinyl butyral) are dibutyl and butyl benzyl phthalates (Santicizer 160, Monsanto Chemical Co.), phosphates, eg, tricresyl phosphate and Santicizer 141 (2-ethylhexyl diphenyl phosphate), butyl ricinoleate, castor oil, dibutyl sebacate, tributyl citrate, dihexyl adipate (Santicizer 367), and various polyester-, rosin-, and linseed-oil-based materials. Plasticizers for poly(vinyl formal) include diethyl, diphenyl, and dicyclohexyl phthalates, butyl benzyl phthalate (Santicizer 160), phosphates such as tributyl, tricresyl, and triphenyl phosphates as well as Santicizer 141. Other useful plasticizers include polyesters, chlorinated naphthalene derivatives, furfuryl oleates, sulfonamides, epoxidized soybean oil, and adipate diesters. By judicious choice of plasticizer type and concentration, the mechanical properties, adhesion, and chemical resistance of the polymers can be tailored for many applications.

Analytical Methods

Structural Studies. The infrared spectrum of poly(vinyl formal) has been studied and band assignments have been made. Structure and degradation mechanisms have been elucidated with the help of thermooxidation; nmr and infrared studies have compared the polymer to model compounds. The kinetics of formation of the formal have been studied.

Analysis. Poly(vinyl acetal)s can be analyzed by the hydrolysis of the acetal moiety in strong aqueous acid, followed by the determination of the aldehyde via the 2,4-dinitrophenylhydrazone.

Health and Safety

Butvar resins are regulated by the Food and Drug Administration under the Code of Federal Regulations as indirect food additives. Subject to the appropriate regulations, they are useful in a number of packaging applications for both fatty and aqueous foods. Because of frequent changes in these regulations, up-to-date information is best obtained by contacting the manufacturer (Monsanto Co.). Both Butvar and Formvar resins have flash points above 370°C. The lower explosive limit (lel) for Butvar dust in air is 20 g/m^3.

Economic Aspects

Poly(vinyl formal). The 1988 estimated worldwide production of poly(vinyl formal) was ca 3500–4000 t. Monsanto Co. is the only U.S. producer.

Poly(vinyl butyral). The 1988 worldwide production of poly(vinyl butyral) resin was estimated to be ca 45,000–50,000 t; DuPont, Monsanto, and Union Carbide are the U.S. suppliers.

Poly(vinyl butyral) Dispersions. In addition to the normal dry powder forms of poly(vinyl butyral), these resins are also available in the form of an aqueous dispersion from Monsanto Co.

Applications

The principal applications of the poly(vinyl acetal)s depend on their unique combination of properties: adhesion to a wide variety of surfaces, chemical and solvent resistance (especially when cross-linked), heat stability, film clarity, and physical toughness. These properties, alone or in combination, make these polymers useful in certain coating and adhesive formulations as the principal or as a minor constituent. Their compatibility with other resins, due largely to their unique chemical compositions, permits blending with other resins, resulting in properties not achievable with either component alone.

Poly(vinyl butyral). The principal single use of this resin is in the lamination of safety glass, mostly for automotive windshields, but its use in architectural applications is growing rapidly. Because of their excellent adhesion to a wide variety of surfaces, specialty grades of poly(vinyl butyral) find broad application as pigment binders, coatings, and adhesives.

Poly(vinyl formal). The principal use for poly(vinyl formal) is as a main component of oil-resistant magnet wire enamels, where it provides the required abrasion and thermal resistance and toughness.

BIBLIOGRAPHY

A. F. Fitzhugh, E. Lavin, and G. O. Morrison, *J. Electrochem. Soc.* **100,** 351 (1953).

Butvar, Poly(vinyl Butyral), Formvar, Poly(vinyl Formal), Tech. Bulls. Nos. 8084 and 8052, Monsanto Chemical Co., St. Louis, Mo., 1989.

P. M. Cotts and A. C. Ouano in P. Dubin, ed., *Microdomains in Polymer Solutions,* Plenum Press, New York, 1985, pp. 101–119.

P. H. Farmer and B. A. Jemmott in I. Skeist, ed., *Handbook of Adhesives,* 3rd Ed., Van Nostrand Reinhold, New York, 1990, pp. 423–436.

THOMAS P. BLOMSTROM
Monsanto Chemical Company

VINYL ACETATE–ETHYLENE COPOLYMERS.

See ETHYLENE POLYMERS.

VINYL ACETATE POLYMERS.

See VINYL ESTER POLYMERS.

VINYL ACETATE–VINYL CHLORIDE COPOLYMERS.

See VINYL CHLORIDE POLYMERS.

VINYL ALCOHOL POLYMERS

Poly(vinyl alcohol) (PVA), a polyhydroxy polymer, is the largest volume, synthetic, water-soluble resin produced in the world. It is commercially produced by the hydrolysis of poly(vinyl acetate), as the vinyl alcohol monomer does not exist in the free state, although traces have been detected.

The discovery of PVA dates back to 1924 when Herrman and Haehnel added alkali to a clear alcoholic solution of poly(vinyl acetate) and obtained the ivory-colored poly(vinyl alcohol). Staudinger studied the reversible transformation between alcohol and poly(vinyl acetate) via esterification and saponification. The first scientific reports on poly(vinyl alcohol) were published in 1927.

The excellent chemical resistance and physical properties of PVA resins have led to their broad industrial use. The polymer is an excellent adhesive and possesses solvent, oil, and grease resistance matched by few other polymers. Poly(vinyl alcohol) films have high tensile strength and abrasion resistance; oxygen-barrier properties under dry conditions are superior to those of any known polymer. Because of low surface tension, emulsification and protective colloid properties are excellent.

The main uses of PVA are in fibers, adhesives, emulsion polymerization, production of poly(vinyl butyral), and textile and paper sizing. Significant volumes are also used in joint cements for building construction; water-soluble films for hospital laundry bags; cold-water-soluble packaging for pesticides, herbicides, and fertilizers; emulsifiers in cosmetics; temporary protective films; soil binding to control erosion; and photoprinting plates.

Physical Properties

The physical properties of poly(vinyl alcohol) depend to a greater extent on the method of preparation than is the case with other polymers. The final properties are affected by the polymerization conditions of the parent poly(vinyl acetate) as well as by the hydrolysis conditions, drying, and grinding. It is difficult to assign specific physical properties to solid poly(vinyl alcohol) as this designation refers to an array of products, including copolymers of vinyl acetate–vinyl alcohol.

The effect of hydrolysis and molecular weight is illustrated in Figure 1.

Table 1 gives representative properties.

Crystallization and Melting Point. The degree of crystallization has a pronounced effect on solubility, water sensitivity, tensile strength, oxygen-barrier properties, and thermoplastic properties. The melting point of the crystal depends on perfectness and size. Various experimental values of the melting point of poly(vinyl alcohol) have been reported, ranging between 220 and 267°C for fully hydrolyzed PVA. Exact determination of the crystalline melting point using a normal dta technique is difficult as decomposition takes place above 130°C.

Glass-transition Temperatures. The glass-transition temperature of fully hydrolyzed PVA has been determined to be 85°C for high molecular weight material. The glass-transition temperature for 87–89% hydrolyzed PVA is 58°C; it is dependent on the degree of polymerization

$$T_g = 58 - (2.0 \times 10^{-3}/\text{DP})°\text{C}$$

Solubility. Poly(vinyl alcohol) is soluble in highly polar and hydrophilic solvents, such as water, DMSO, acetamide, glycols, and DMF. The solvent of choice is water.

Solution Viscosity. The viscosities of PVA solutions are mainly dependent on molecular weight, concentration, hydrolysis, and temperature. Materials with a high degree of hydrolysis increase in viscosity on standing and may even gel.

Mechanical Properties. The tensile strength of unplasticized PVA depends on degree of hydrolysis, molecular weight, and relative humidity. Heat treatment increases the tensile strength; plasticizers reduce tensile strength disproportionately, due to increased water sensitivity.

Solvent Resistance. Poly(vinyl alcohol) is virtually unaffected by hydrocarbons, chlorinated hydrocarbons, carboxylic acid esters, greases, and animal or vegetable oils. Resistance to organic solvents increases with increasing hydrolysis.

Gas-barrier Properties. The oxygen-barrier properties of PVA

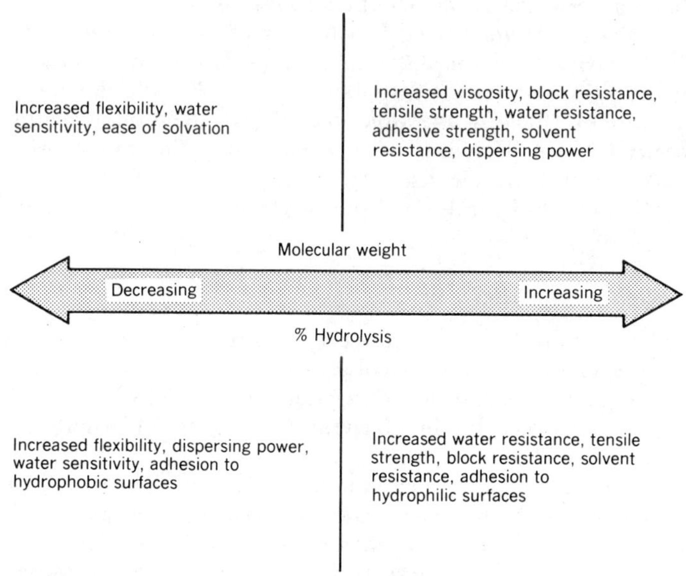

Figure 1. Effect of molecular weight and hydrolysis on the properties of poly(vinyl alcohol).

Table 1. Physical Properties of Poly(vinyl alcohol)

Property	Value	Remarks
appearance	white to ivory white granular powder	
specific gravity	1.27–1.31	increases with degree of crystallinity
tensile strength, MPa[a]		
98–99% hydrolyzed	67–110	increases with degree of crystallinity (heat treatment), decrease in humidity, and increase in molecular weight
87–89% hydrolyzed	24–79	increases with molecular weight and decrease in humidity
elongation, %	0–300	increases with humidity
thermal coefficient of expansion $\times 10^{-5}$, per °C	7–12	
specific heat, J/(g·K)[b]	1.67	
thermal conductivity, W/(m·K)	0.2	
glass transition, °C		
98–99% hydrolyzed	85	
87–89% hydrolyzed	58	
melting point[c], °C		
98–99% hydrolyzed	230	
87–89% hydrolyzed	180	
electrical resistivity, Ω·cm	$(3.1–3.8) \times 10^7$	
thermal stability		gradual discoloration above 100°C, darkens rapidly above 150°C; rapid decomposition above 200°C
refractive index, n_D	1.55	
degree of crystallinity	0–0.54	increases with heat treatment and degree of hydrolysis
storage stability of solid		indefinite when protected from moisture
flammability		burns similarly to paper
stability in sunlight		excellent

[a] To convert MPa to psi, multiply by 145.
[b] To convert J to cal, divide by 4.184.
[c] Unplasticized.

at low humidity are the best of any synthetic resin. However, barrier performance deteriorates above 60% rh.

Surface Tension. The surface tension of aqueous solutions of PVA varies with concentration, temperature, the degree of hydrolysis, and acetate distribution on the PVA backbone.

Intrinsic Viscosity and Molecular Weights. The intrinsic viscosity of PVA decreases with temperature. The relationship between the intrinsic viscosity and molecular weight changes with degree of hydrolysis of the polymer.

Chemical Properties

Poly(vinyl alcohol) undergoes chemical reactions in a manner similar to other secondary polyhydric alcohols. Of the greatest commercial importance are reactions with aldehydes to form acetals, such as poly(vinyl butyral) and poly(vinyl formal) (see VINYL ACETAL POLYMERS).

Esterification. Boric acid and borax form cyclic esters with poly(vinyl alcohols). The reaction is markedly sensitive to pH, and an insoluble gel is formed above pH 4.5–5.0. Poly(vinyl nitrate) has been prepared and studied for use in explosives and rocket fuel. Poly(vinyl alcohol) and sulfur trioxide give poly(vinyl sulfate). Poly(vinyl sulfonate)s have been prepared from poly(vinyl alcohol) and alkanesulfonyl chloride. In the presence of urea, poly(vinyl alcohol) and phosphorus pentoxide or phosphoric acid give poly(vinyl phosphate)s.

An unlimited number of organic esters can be prepared by reactions of poly(vinyl alcohol) employing standard synthesis. Chloroformate esters react with poly(vinyl alcohol) to yield poly(vinyl carbonate). Poly(acrylic acid) or poly(methacrylic acid) reacts with poly(vinyl alcohol) to form insoluble gels which, depending on the degree of cross-linking, can be used as superabsorbents. Urea and poly(vinyl alcohol) yield a polymeric carbamate ester.

Etherification. Ethers of poly(vinyl alcohol) are easily formed. Insoluble internal ethers are formed by elimination of water, a reaction catalyzed by mineral acids and alkali.

Acetalization. Poly(vinyl alcohol) and aldehydes give compounds of industrial importance.

Intramolecular acetalization

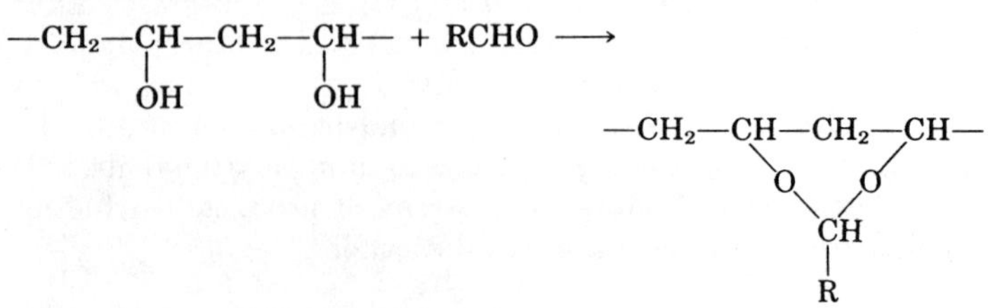

Intermolecular acetalization

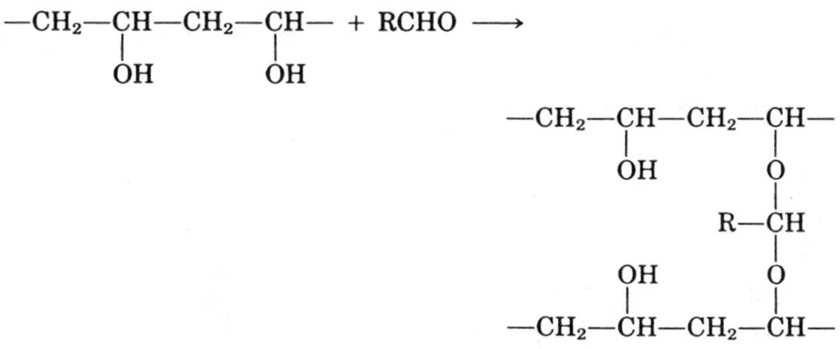

Thermal Decomposition. The thermal decomposition of PVA in the absence of oxygen occurs in two stages: The first stage begins at about 200°C and is mainly dehydration, accompanied by the formation of volatile products. The residue is predominantly macromolecules of polyene structure. Further heating to 400–500°C yields carbon and hydrocarbons. The published data on the decomposition products show some disagreement, which could be explained by differences in the manufacturing conditions of the PVA.

Biodegradation. Poly(vinyl alcohol) is one of the few completely biodegradable synthetic polymers. A strain of *Pseudomonas* 0–3, similar to the strain of *Pseudomonas boreopolis,* in combination with *Pseudomonas genous* (both isolated from soil) are needed in the biodegradation of PVA.

Cross-linking. Poly(vinyl alcohol) can be readily cross-linked to improve water resistance. Glyoxal, urea–formaldehydes, and melamine–formaldehyde are the most common cross-linking agents used for this reaction. Trimethylolmelamine is preferred if low temperatures are required. However, as a rule, all multifunctional compounds that react with hydroxyl groups can be used.

Strongly chelating metal salts of copper and nickel, eg, cupric ammonium complexes, chromium complexes, and organic titanates and dichromates, can be effective insolubilizers for PVA.

Manufacture

Production of Poly(vinyl acetate). Both batch and continuous processes are used for the polymerization of vinyl acetate. A batch or semibatch process is suitable if a range of poly(vinyl alcohol)s is desired on a small scale. The continuous process is more economical for large-scale production (Fig. 2).

Production of Poly(vinyl alcohol). Poly(vinyl acetate) is converted into the corresponding poly(vinyl alcohol) by direct hydrolysis or catalyzed alcoholysis. The reaction can be catalyzed by strong acids such as sulfuric or hydrochloric acid or strong bases such as sodium hydroxide, sodium alcoholates, or potassium hydroxide. On an industrial scale alkaline alcoholysis is generally used.

The base-catalyzed alcoholysis of poly(vinyl acetate) in an alcohol solution is an extremely fast reaction. Insolubility of the poly(vinyl alcohol) in the resulting mixture of alcohol and alkyl acetate leads to a dramatic increase in viscosity followed by the precipitation of a swollen PVA gel, which can be either broken up during the hydrolysis process or ground before washing. The degree of hydrolysis is controlled by residence time, catalyst concentration, temperature, and additives. Reduction of available alcohol for the transesterification by substituting alcohol with the generated ester limits the hydrolysis and yields a blocky product. Other methods claim the addition of both organic and inorganic acids to neutralize the catalyst. Both continuous and batch processes are in commercial use, with the continuous process dominating.

Solvent Recovery. With methanol as the solvent, the alcoholysis process yields methyl acetate and sodium acetate as the principal by-products. The methyl acetate can be used as a solvent or hydrolyzed, using a cation-exchange column or strong acid, into acetic acid and methanol. The methyl acetate is separated from the methanol by extractive distillation and the addition of water before the hydrolysis step. The resulting dilute acetic acid solution is concentrated by extraction or azeotropic distillation. Sodium acetate is converted into sodium sulfate and acetic acid by adding sulfuric acid.

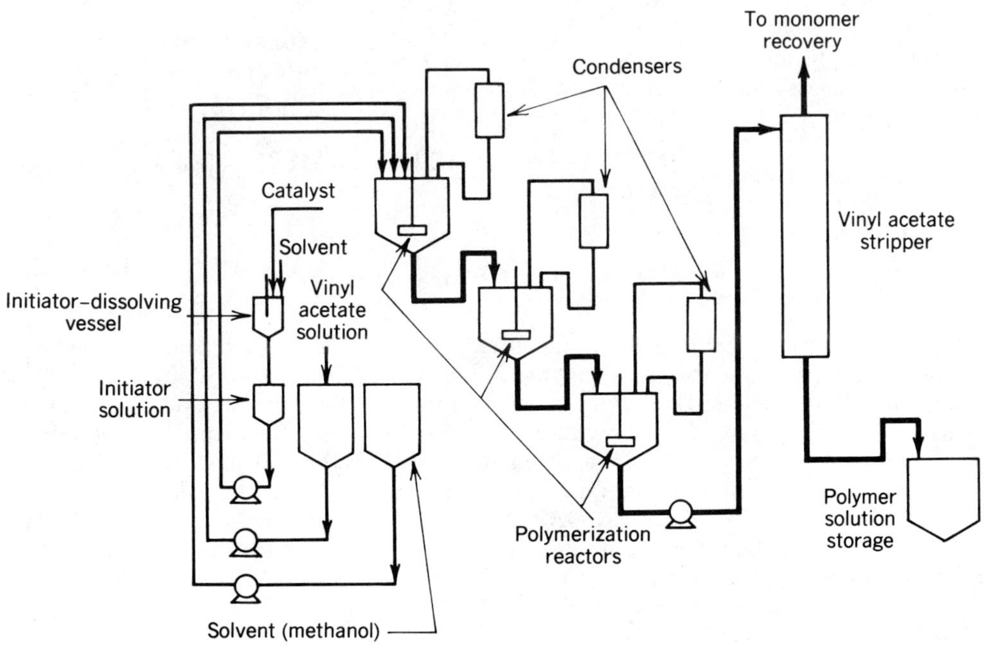

Figure 2. Typical polymerization flow diagram for the continuous polymerization of vinyl acetate.

Copolymers

Numerous vinyl alcohol copolymers have been prepared. Copolymers with ethylene and methacrylates have found commercial utility. Ethylene–vinyl alcohol (EVOH) copolymers containing 20–30 mol % ethylene are used as an oxygen barrier in food packaging. A five-layer coextruded structure with polyolefins is, however, needed to protect the EVOH from moisture. Vinyl alcohol–methyl methacrylate copolymers are used as alkaline-resistant textile-sizing agents. Copolymers with poly(alkyleneoxy acrylates) can be thermoplastically processed by blow molding (qv), injection molding (qv), or extrusion (qv).

Processing

Poly(vinyl alcohol) is not considered a thermoplastic copolymer as the degradation temperature is below the melting point. Thus industrial applications of poly(vinyl alcohol) are based on and limited by the use of water solutions.

Solution Preparation and Handling. Poly(vinyl alcohol) should be completely dispersed in water at room temperature or lower before heating. Good agitation is important to prevent lumping during the addition of solid PVA to the water.

The slurry temperature must be increased after dispersing the PVA. In order to solubilize the PVA, direct stream injection, jet cooking, or heating with jacket and coil are commonly employed. Aqueous solutions of PVA are stable on storage but must be protected from bacterial growth.

Extrusion. Several patents claim the preparation of extrudable PVA, although no such compound is commercially available. The main obstacles are thermal stability during extrusion and gels in the final product.

Economic Aspects

The capacity of PVA in 1987 in non-Communist countries was greater than 360,000 t. This is essentially unchanged from 1980 despite capacity increases by Air Products & Chemicals, Inc. and Chang-Chun as Monsanto withdrew from the PVA market in 1985.

Approximately two-thirds of capacity is in Japan and Taiwan, with the remainder in the United States and Europe.

Poly(vinyl alcohol) is an innocuous material with unlimited storage life. It is most commonly supplied in 20 kg, 22.7 kg, and 25 kg bags equipped with a moisture barrier to prevent caking.

The PVA price has historically reflected the cost of ethylene, acetic acid, and energy. It has risen from $0.77/kg in 1970 to $2.20/kg in 1980 and $2.75/kg in 1988 for a medium mol wt fully hydrolyzed grade.

Specifications and Standards

The important commercial grades of PVA are distinguished by the degree of hydrolysis and molecular weight. The resins are most often categorized by degree of hydrolysis, ie, mole percent of alcohol groups in the resin.

Analytical and Test Methods

The important test methods relate to the degree of hydrolysis, pH, viscosity, ash, and volatiles.

Health and Safety Factors

Poly(vinyl alcohol) is a nonhazardous material according to the American Standard for Precautionary Labeling of Hazardous Industrial Chemicals (ANSI 2129.1–1976).

Uses

The main applications for PVA are in textile sizing, adhesives, polymerization stabilizers, and paper coating (Table 2). The largest application is textile sizing, which accounts for nearly half the U.S. consumption of PVA, excluding poly(vinyl butyral) production.

Table 2. Distribution of PVA Applications

Application	Percent
textile sizes	30
adhesives	25
fibers	15
paper sizes	15
poly(vinyl butyral)	10
other	5

Other uses include water-soluble films, nonwoven fabric binders, thickeners for latex coatings, binders for phosphorescent pigments in television picture tubes, slow-release binders for fertilizers, photoprinting plates, and sponges for cosmetic and health care applications.

BIBLIOGRAPHY

J. G. Pritchard, *Poly(vinyl alcohol) Basic Properties and Uses,* Gordon & Breach, New York, 1985.

I. Sakurada, *Poly(vinyl alcohol) Fibers,* Marcel Dekker, Inc., New York, 1958.

C. A. Finch, ed., *Polyvinyl Alcohol,* John Wiley & Sons, Inc., London, 1973.

F. Lennart Marten
Air Products and Chemicals

VINYL ALKYL ETHER POLYMERS.

See Vinyl ether polymers.

N-VINYL AMIDE POLYMERS

Polyvinylamides and polyvinylimides are highly polar, amphoteric polymers, which have been known since the 1930s. The polymers derived from cyclic structures are more important. Poly(*N*-vinyl-2-pyrrolidinone), also called polyvinylpyrrolidinone, povidone, or PVP, is the only one available in commercial quantities. It is manufactured and sold in the United States by GAF Chemicals Corp. under the trade names Plasdone (soluble) and Polyclar and Polyplasdone (insoluble). It is manufactured in the FRG by BASF as Luviscoll, Kollidon, and Igecoll.

Polyvinylpyrrolidinone exhibits a unique combination of properties, including solubility in water and in organic solvents, very low toxicity, high complexing ability, good film-forming characteristics, and the ability to adhere to a number of substrates. As a result, PVP finds use in a wide variety of industries. Its film-forming and adhesive properties are utilized in hair sprays, adhesives, and photographic and lithographic coatings. As a protective colloid, it is used in drug and detergent formulations, emulsions and suspensions, and in pigment and dyestuff dispersions. Its complexing ability is utilized in the textile industry for improving dyeability and dye stripping, in the pharmaceutical industry and agriculture for drug dissolution and sustained release of drugs and chemicals, and in the food industry for the clarification of beer and wine. It was used as a blood plasma extender during World War II.

Physical Properties. Vinylpyrrolidinone is a clear, colorless liquid; the commercial monomer sometimes has a light straw color. It is miscible with water and organic solvents in all proportions. In water–organic systems, vinylpyrrolidinone prefers the organic phase. The physical properties of the monomer are given in Table 1.

Chemical Properties. The double bond of the vinylpyrrolidinone is reactive. In the presence of dilute aqueous acid, it hydrolyzes to pyrrolidinone and acetaldehyde; treatment with dry HCl leads to the formation of a dimer. With fluoroacetic acid catalyst at 85–90°C the yield of dimer is 90%.

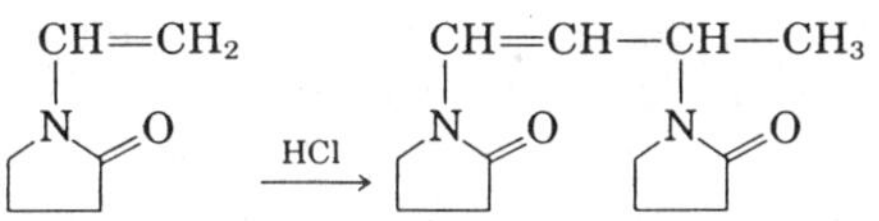

Amides react with the double bond of vinylpyrrolidinone to form amidoethylpyrrolidinones. 2-Pyrrolidinone and vinylpyrrolidinone react in aqueous mineral acid. The 2-pyrrolidinone is formed by partial hydrolysis of vinylpyrrolidinone.

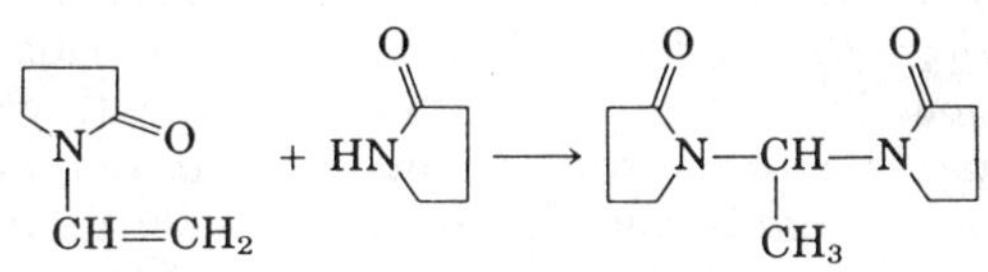

The reaction probably proceeds via the acid-catalyzed addition of water to the double bond with the formation of an unstable hemiacetal, which decomposes to 2-pyrrolidinone and acetaldehyde.

The reaction with alcohols requires an acid catalyst. Phenols ($R = C_6H_5$), however, because of their acid character, add to the double bond in an exothermic reaction without any catalyst. The reactions follow Markovnikov's law.

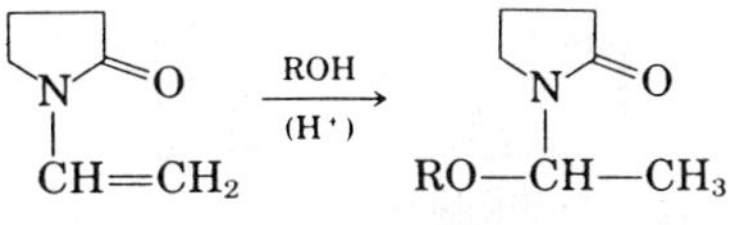

The lactam ring is stable under neutral or acidic conditions, but the ring is split with strong alkalies at elevated temperatures to yield the corresponding linear amino acid salts.

Polymerization. *Bulk Polymerization.* In the absence of an initiator, pure vinylpyrrolidinone is fairly stable in the dark at room temperature. However, at elevated temperatures, as well as during prolonged exposure to light, polymerization takes place. In air, pure vinylpyrrolidinone may form a peroxy compound, which acts as a polymerization catalyst.

Polymerization in bulk can be carried out with azo initiators; kinetic data are reported. The rate of polymerization is proportional to the square root of the initiator concentration and to the temperature of the reaction.

Solution Polymerization. *N*-Vinyl-2-pyrrolidinone (VP) and other vinylamides show high electron density on the β-carbon, whereas the nitrogen atom is attached to a strongly electron-withdrawing group. Thus the e values of these monomers are usually negative. They polymerize readily and with high yields.

Polymerization of *N*-vinylpyrrolidinone may be carried out by a free-radical or ionic mechanism. The latter gives only low molecular weight products and is of no practical importance.

In solution polymerization, the heat of the reaction can be controlled. The product has a much narrower molecular weight distribution and is free of color. The moderately viscous solution is easy to handle.

Storage and Shipping. Vinylpyrrolidinone should be stored in a dry, cool place, and never be kept above 37°C for prolonged periods.

Vinylpyrrolidinone is stored in stainless steel or carbon steel tanks that may be lined with aluminum or phenolic resins.

Toxicity. Under good industrial or laboratory handling, vinylpyrrolidinone is not considered hazardous to health. Normal hygienic practice is recommended, including prompt removal from the skin and avoidance of inhalation, swallowing, or other unnecessary contact. Vinylpyrrolidinone is a severe eye irritant. Inhalation studies indicate that absorption may cause liver damage in some animals.

Table 1. Properties of Vinylpyrrolidinone

Property	Value
molecular weight	111
assay, % minimum	98.5
moisture content, % maximum	0.2
appearance	clear liquid
color, APHA	100 maximum
vapor pressure, Pa[a]	
17°C	6
24°C	13
45°C	60
54°C	133
64°C	260
77°C	666
boiling point at 5.8 kPa[a], °C	193
freezing point, °C	13.5
flash point, open cup, °C	98.4
fire point, °C	100.5
viscosity at 25°C, mPa·s (= cP)	2.07
specific gravity, 24/4°C	1.04
refractive index n^{25}	1.511

[a] To convert Pa to mm Hg, multiply by 0.0075.

Poly(*N*-vinyl-2-pyrrolidinone)

Physical Properties. Poly(*N*-vinyl-2-pyrrolidinone) is a white (or slightly yellow) hygroscopic powder, forming hard clear films. Physical properties are determined on films or powder.

The polymer strongly interacts through dipole–dipole attraction; the melt viscosity of the polymer is therefore too high for typical thermoplastic-forming operations.

Glass-transition Temperature. The glass-transition temperature of linear PVP increases with increasing molecular weight and is calculated according to the following equation:

$$T_g\ (°C) = 175 - \frac{9685}{K^2}$$

where K is the Fikentscher K value of the polymer. The glass-transition temperatures of various commercial grades are given in Table 2.

Hygroscopicity. The amount of water absorbed is nearly independent of the molecular weight of PVP. The equilibrium water content varies with the ambient relative humidity and is approximately one-third of the latter value. Approximately 0.5 mol of water is associated with one monomer unit, which corresponds to the values given for the hydration of proteins.

Molecular Weight. Polyvinylpyrrolidinone can be prepared in a variety of molecular weights, depending on the conditions of polymerization. The hydrogen peroxide–ammonia process gives a molecular weight range of 2500 to about 1.1×10^6. Molecular weight is expressed by the so-called K-value, based on kinematic viscosity measurements and calculated according to the Fikentscher equation.

$$\frac{\log \eta_{rel}}{c} = \frac{75\,K_0^2}{1 + 1.75\,K_0 c} + K_0$$

where c is the concentration in g/100 mL solution, η_{rel} is the viscosity of the solution compared to that of the solvent, and $K = 1000\,K_0$.

Rheology. The rheology of PVP is investigated in aqueous media, because melts are difficult to handle. Viscosities have been determined in a variety of solvents, but because of the polar nature of the polymer, measurements are most conveniently made in water or methanol. Application of the Mark-Houwink equation gives the following results:

$$[\eta]_{CH_3OH} = 2.3 \times 10^{-2} \times \bar{M}_w^{0.65}$$

$$[\eta]_{H_2O} = 5.65 \times 10^{-2} \times \bar{M}_w^{0.55}$$

These results indicate that methanol is a better solvent for PVP than water. The effect of temperature on the viscosity depends on the polymer concentration.

Solubility. Polyvinylpyrrolidinone consists of hydrophobic methylene groups and a strongly hydrophilic imide group. As a consequence of the hydrophobic–hydrophilic balance, the polymer is soluble in a variety of organic solvents as well as in water. The heat of solution is −4.81 kJ/mol (−1.15 kcal/mol).

Compatibility. Polyvinylpyrrolidinone is compatible in solutions and in films cast from the solutions with a wide variety of synthetic resins. It is also compatible with inorganic salt solutions, as well as with other chemicals.

Chemical Properties. In accordance with the characteristics of vinylpyrrolidinone, PVP is chemically inert. The dry polymer can be stored under normal conditions without decomposition, degradation, or structural change. Heat sensitivity is low, and it is stable at 130°C for short intervals. However, at 150°C in air, solubility is reduced and color increased; after an extended period of time, the polymer becomes insoluble. Aqueous solutions are stable for a long time if protected from molding. Preservatives such as benzoic acid, sorbic acid, esters of *p*-hydroxybenzoic acid, hexachlorophene, and others can be used.

Complexation. The individual unit in the PVP chain, consisting of the polar imide group, four nonpolar methylene groups, and a methine group, has an amphiphilic character. The dipole moment of this unit must be close to that of its small-molecule analogue, the *N*-methyl-2-pyrrolidinone: $\mu = 13.5 \times 10^{-30}$ C·m (4.07 D).

Because of its unique structure, the exposed cyclic imide group and the balance of hydrophilic–hydrophobic segments, PVP forms complexes with a variety of compounds. The hydrophobic character is attributed to the methylene groups in the ring and the linear aliphatic backbone, whereas the hydrophilicity is the result of the polar lactam moiety of the pyrrolidinone ring. The complexes are usually formed through hydrogen bonding to which the pyrrolidinone structure is highly susceptible. However, in certain cases other forces, such as hydrophobic attraction, van der Waals forces, dipole forces, Debye forces, dispersion forces, ion–dipole forces, and charge-transfer interactions may also be responsible for the strong complexing ability of PVP.

Cross-linking. Lightly cross-linked PVP can be prepared by treatment of linear PVP with persulfates, hydrazine, and hydroperoxide; with diolefins in the presence of peroxides; or with γ-irradiation. The products gel when treated with water. Copolymerization with multifunctional, unsaturated compounds gives densely cross-linked polymers swellable in water and organic solvents. Heavily cross-linked popcorn polymers (see PROLIFEROUS POLYMERIZATION) can be made by treating vinylpyrrolidinone above 100°C with an alkali metal hydroxide and a small amount of water.

Specifications and Analysis. The specifications for technical- and pharmaceutical-grade PVP are given in Tables 3 and 4.

Table 2. Glass-transition Temperatures of PVP

Sample	Measured K value[a]	T_g, °C
Plasdone K-15	14.0	126
PVP K-15	14.9	130
Plasdone K-25	22.5	160
PVP K-30	27.5	163
Plasdone K-29/32	28.7	164
PVP K-60	55.5	170
PVP K-90	89.6	174

[a] Fikentscher value.

Table 3. Specifications of Technical-grade PVP[a]

Grade	Form	K range[b]	Water, % maximum
PVP K-15	powder	12–18	5
PVP K-30	powder	26–35	5
PVP K-60	aqueous solution	50–62	55
PVP K-90	aqueous solution	80–100	80
PVP K-90	powder	80–100	5
polyclar AT	powder	cross-linked	5

[a] Ash, % maximum = 0.02; monomer residue, % maximum = 1.0.
[b] Fikentscher value.

Table 4. Specifications of Pharmaceutical-grade PVP

Assay	Value
K value[a]	
10–15	85–115% of stated value
16–90	90–107% of stated value
moisture, % maximum	5
pH[b]	3.0–7.0
residue on ignition, % maximum	0.02
aldehydes, %[c] maximum	0.02
N-vinyl-2-pyrrolidinone, % maximum	0.20
lead, ppm, maximum	10
arsenic, ppm, maximum	1
nitrogen, %	11.5–12.8
hydrazine	<1 ppm

[a] Fikentscher value.
[b] Of a 5% solution in distilled water.
[c] Calculated as acetaldehyde.

Polyvinylpyrrolidinone–Iodine Complex

The most widely used complex of vinylpyrrolidinone is that with iodine. It is prepared by thoroughly mixing dry elemental iodine with PVP powder or by reaction of the ingredients in homogeneous solution. When PVP and iodine react, the iodine becomes partially linked to the polymer; some of it reacts with water and is reduced to iodide. Iodine undergoes a number of reactions in aqueous solution, including hydrolysis, dissociation of hypoiodous acid formed by hydrolysis, and perhalide complex formation. Consequently, in aqueous solution, iodine is present in seven different forms: I_2, HOI, OI^-, H_2O^+I, I_3^-, IO_3^-, and I^-. Only the hydrated molecular iodine (I_2) and hypoiodous acid (HOI) are significant and bactericidal. However, the ionization constant of hypoiodous acid (5×10^{-13}) shows that HOI is almost completely nonionized in the biological pH range. In the completed reaction, most of the iodine is attached to the polymer in the form of triiodide, which is in equilibrium with iodide and iodine. The PVP–triiodide complex is water-soluble and, although not itself bactericidal in solution, rapidly releases iodine, acting like an iodine reservoir.

$$I_3^- \rightleftharpoons I^- + I_2$$

In the equilibrium state, the rate of dissociation and association is constant. Because the amount of free iodine in the PVP–iodine complex is low, the toxicity and irritating effect of the product is lower than that of iodine tinctures. Although this iodophor is active toward microorganisms in the same way as iodine, the iodine is kept in the form of nontoxic, nonirritating complex form until needed. The PVP–iodine complex, listed in the *U.S. Pharmacopeia* as povidone–iodine, is a free-flowing, brown powder that is soluble in water and lower alcohols.

The clinical administration of PVP–iodine can result in systemic absorption of iodine. The amount absorbed depends on the concentration of iodine and the frequency and method of application.

Graft Copolymers of Vinylpyrrolidinone

Ganexes. Because of the activating effect of the carbonyl group, the pyrrolidinone ring is susceptible to radical-producing agents (eg, organic peroxides). Such attack can remove a proton and create a grafting site. In this way lipophilic radicals can be attached to the hydrophilic polymer. By varying the molecular weight of the lipophilic group and its ratio to the vinylpyrrolidinone, a variety of polymers can be made. These polymers have different surface activities and different solubilities (Table 5).

Table 5. Physical Properties of Ganex Polymers[a]

Property	V-220	V-216	V-516	V-816	P-804	P-904
physical form at 25°C	waxy solid	straw-colored viscous liquid	off-white solid	white granular solid	off-white granular solid	white granular solid
relative viscosity[b]	1.007	1.009	1.008	1.003	1.005	1.008
density, g/ml, 25°C	0.95	0.90	0.98	1.00	1.09	1.09
average mol wt[c]	8,600	7,300	9,500	20,000	19,000	16,000
softening point, °C	30		30	120	121	157
melting range, °C[d]	32–36		83–90	146–152	131–140	186–195

[a] With 100% activity.
[b] 0.1% in solvent.
[c] High speed membrane osmometry.
[d] Mannheim block.

Toxicity. The acute oral toxicity determined on rats is given below.

Ganex	V-216	64 mL/kg	V-816	5.76 g/kg
	V-220	17 g/kg	P-804	16.0 g/kg
	V-516	12.8 g/kg	P-904	16.0 g/kg

Neither intact nor abraded skin showed any skin irritation. Similarly, no corneal, iridial, or conjunctival effect was detected in eye-irritation studies. Besides aliphatic hydrocarbons, allyl alcohol, allylamine, and other monomers can be grafted on PVP to make functional Ganexes.

Polectrons. Polectrons are emulsion polymers prepared by grafting hydrophobic monomers (M_2) onto polyvinylpyrrolidinone (M_1). The polymer is activated by treatment with ammonium persulfate at elevated temperature (100°C), cooling to the temperature of the reaction (65–80°C), and gradually adding the hydrophobic monomer. The product is a latex, which consists of graft copolymer and some homopolymers of vinylpyrrolidinone and the hydrophobic monomer. These polectrons are latices of moderate viscosity with 40% solid content. They have good mechanical, freeze–thaw, and chemical stabilities. Properties are given in Table 6. These products are fluid, milky white emulsions, with a particle size of $<0.5\ \mu m$ (95% min).

Only Polectron 430 is available in commercial quantities. The latex is used in household products, eg, as opacifier in liquid detergents, in acid bowl cleaners, and in floor polishes.

Toxicity. Animal tests show an LD_{50} for Polectron 430 above 4 g/kg. Dermal irritation studies on humans indicate no primary irritation or sensitization. Primary ocular irritation testing showed no effects.

Applications of Polyvinylpyrrolidinone

Because of its unique physical and chemical properties, particularly its good solubility in water and organic solvents, its outstanding chemical stability, its strong complexing ability toward both hydrophobic and hydrophilic substances, PVP is one of the most widely used specialty polymers. It has found application in the textile, paper, adhesive, membrane, suspending aid, plastic, cosmetic, detergent, ceramic, electrochemical, and photochemical industries, and in water treatment, metallurgy, and radiation curing. Of particular importance is the use of PVP in the pharmaceutical industry and medicine.

Copolymerization

Because of the electron distribution around the reactive vinyl imide grouping, vinyllactams and particularly vinylpyrrolidinone copolymerize readily with a wide variety of monomers. The Q and e values are reported to be $Q = 0.093$ and $e = -1.17$.

Copolymers are usually made in solution with a free-radical initiator. The comonomer is added gradually to the vinylpyrrolidinone, in order to overcome the great difference in copolymerization parameters and facilitate the formation of a more uniform structure. The more important comonomers are as follows: vinyl acetate, ethyl acrylate, methyl acrylate, methyl methacrylate, dimethylaminoethyl methacrylate, acrylamide, methacrylamide, acrylonitrile, ethylene, styrene, maleic anhydride, acrylic acid, sodium vinylsulfonate, vinyl chloride, vinylpyridine, trimethyl (siloxy) vinylsilane, vinyl propionate, vinyl caprolactam, and methyl vinyl ketone.

Vinylpyrrolidinone is a weak electron donor and copolymerizes readily with acceptor monomers as well as with stronger donors. The copolymerization may be carried out in solution, emulsion, or suspension. The free-radical initiators for PVP are usually also suitable for copolymerizations. *Chemical Abstracts* lists over 3000 different copolymer systems containing vinylpyrrolidinone as main ingredient. In spite of this great number, only a few are produced commercially. The purpose of these copolymers is to modify or improve the inherent properties of PVP. All VP copolymers are more or less hygroscopic, because of the highly hydrophilic character of the imide group. The degree of hygroscopicity depends on the nature of the comonomer. For instance, a copolymer with vinyl acetate is more hygroscopic than a copolymer with styrene with the same comonomer ratio.

Various initiators are used for solution polymerization including peroxy types and azo compounds; transition metal catalysts are also employed. These copolymers can also be prepared in suspension or emulsion, and by γ-radiation or photopolymerization. They can also be prepared as block and graft copolymers.

Vinylamides

Polyvinylcaprolactam. Vinylcaprolactam and its polymer are made in pilot-plant quantities. The monomer, bp 110.8–111.2°C at 458 Pa (34.5 mm Hg), $n_D^{20} = 1.5057$, is prepared from caprolactam by vinylation with acetylene; transvinylation with vinyl acetate or any vinyl ester and a heavy metal catalyst; and vinylation with acetaldehyde or acetal and an aluminum oxide catalyst. These copolymers are recommended for medical, cosmetic, and textile applications.

Other Vinylamide Monomers. Vinylamides are liquids or low melting solids, soluble in water and organic solvents; properties are given in Table 7.

Table 6. Properties of Polectron Vinylpyrrolidinone Graft Copolymers

	M_2 [a]		
Property	130, ethyl acrylate	430, styrene	825, vinyl acetate
primary emulsifier	anionic	anionic	none
viscosity range[b] at 25°C, mPa·s (= cP)	200–500	300–1500	approximately 5000
pH	6–8	2–4	3–4
appearance of films	soft, hazy, continuous, flexible	hard, brittle, cloudy	milky white, glossy, translucent

[a] The first digit designates the monomer; the last two digits indicate the percentage PVP.
[b] Based on Brookfield Instrument, No. 3 spindle of 30 rpm.

Table 7. Physical Properties of Vinylamides

N-Vinyl, derivative of	Boiling point, °C at Pa[a]	n_D^{20}
linear		
acetamide	107–109	
methylacetamide	70_{332}	1.4806
ethylacetamide	69_{239}	1.4750
phenylacetamide	101_{40}	
methylpropionamide	69_{173}	1.4737
ethylpropionamide	74_{160}	1.4737
methylisobutyramide	75_{252}	1.4739
methylbenzylamide	$137–139_{186}$	1.5670
cyclic		
2-piperidinone	$109–111_{160}$	
5-methyl-2-pyrrolidinone	$90–93_{66}$	1.4996
3,3,5-trimethyl-2-pyrrolidinone	$50–52_{6.6}$	1.4837
5-methyl-2-pyrrolidinone	86_{133}	1.4989

[a] To convert Pa to mm Hg, multiply by 0.0075.

BIBLIOGRAPHY

F. Haaf, A. Sanner, and F. Straub, *Polym. J.* **17**(1), 143 (1985).
Vinylpyrrolidinone, Bull. 7543-037, GAF Corp., New York.
PVP. Polyvinylpyrrolidone, Bull. 7543-113, GAF Corp., New York, 1964.
Kollidon Grades, Bull. MEF 129e, BASF, Bedford, Mass., 1986.

E. S. Barabas
GAF Chemicals Corporation

VINYLAMINE POLYMERS. See Polyamines and polyquaternary ammonium salts.

VINYLBENZENE POLYMERS. See Styrene polymers.

VINYL BUTYRAL POLYMERS. See Vinyl acetal polymers.

VINYLCARBAZOLE POLYMERS

Poly(*N*-vinylcarbazole) is a specialty rather than a commodity material. In the late 1980s, no commercial applications of the polymer were known.

Monomer

The synthesis, purification, and properties of *N*-vinylcarbazole and 3-vinylcarbazole, and a number of their derivatives, are well documented. Monomers with vinyl groups in the 2- and the 4- positions are more difficult to synthesize; preparative routes have been described.

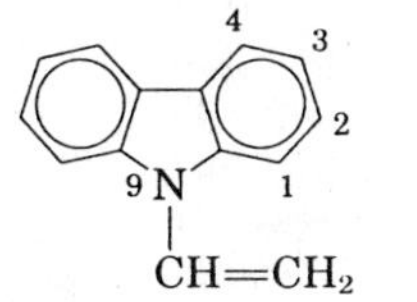

Polymerization

The monomer *N*-vinylcarbazole is readily polymerized by free-radical and cationic methods. This results from its ability to stabilize electron-deficient centers by resonance involving the nonbonding electron pair on the N atom of the carbazole ring. Attempts to polymerize the monomer anionically were unsuccessful. Polymerization by Ziegler-Natta type catalysts has been questioned. The extremely facile cationic polymerization can dominate the process, and a coordinated Ziegler-Natta mechanism is not believed to be involved. The related *N*-alkyl-2-, -3-, and -4-vinylcarbazole monomers can be polymerized by radical, cationic, and anionic methods.

Charge-transfer Polymerization. Polymerization reactions involving electron donor–acceptor interactions (charge-transfer interactions) have been extensively studied since it was reported in 1963 that *N*-vinylcarbazole could be polymerized spontaneously in the presence of electron-acceptor molecules.

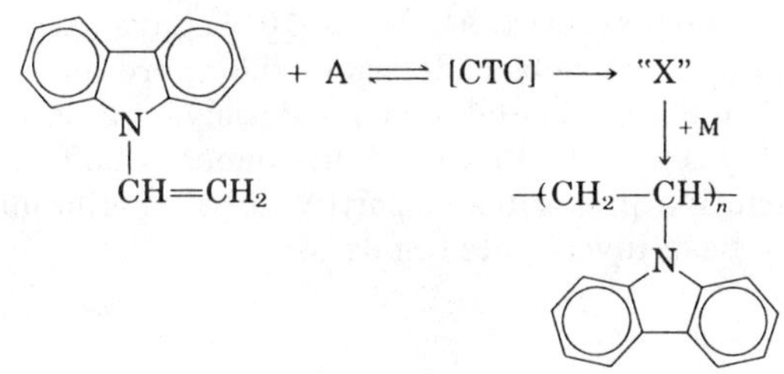

The mechanism proposed originally involved the formation of an initiating species from the charge-transfer complex (CTC) (qv) by complete (monomer radical cation) or partial (mesomeric polarization) electron transfer. Both thermally and photochemically induced processes have been used. Attempts to establish a mechanistic framework for the wide range of charge-transfer (CT) polymerizations, which have been described, have been complicated by unreliable and incomplete data. Because *N*-vinylcarbazole forms colored charge transfer complexes even with weak electron-acceptor molecules, the observation of color was too frequently taken as evidence for participation of the complex in the polymerization reaction. Impurity effects and alternative nonpolymer-producing reaction pathways were often completely ignored.

Electrochemical Polymerization. The electrochemical polymerization of *N*-vinylcarbazole in acetone with $ZnBr_2$ as electrolyte with Pt electrodes gives cyclodimer or polymer in high yields, depending on the current density. The initiating species is believed to be the monomer radical cation formed from the monomer–$ZnBr_2$ complex at the electrode. Polymer is produced by cationic propagation.

Solid-state Polymerization. *N*-Vinylcarbazole has been polymerized in the crystalline state using high energy radiation, cationic catalysts, halogen vapors, and redox initiators. Free-radical and cationic polymerization processes have been proposed, based on kinetic and molecular weight analyses.

Polymerization in the solid state initiated by surface reactions with species such as BF_3 and Cl_2 are thought to occur by an electron transfer to form the monomer radical cation. Propagation proceeds from the surface into the crystal bulk. Esr evidence for formation of the *N*-vinylcarbazole radical cation has been presented in the γ-ray-initiated reaction.

Copolymerization. Most of the early work on the copolymerization of *N*-vinylcarbazole was directed toward improving the poor mechanical properties of the polymer while maintaining its desirable photo- and electrical properties. Q values in the range 0.28 to 0.43 (0.25 theoretical) and e values from -1.0 to -1.6 (-1.4 theoretical) have been reported.

Solution and Solid-state Properties

Solution Properties. The solution properties of poly(*N*-vinylcarbazole) reflect the bulky nature of the carbazole group and the restricted rotation around the C—N bond that controls internal chain freedom. The intrinsic viscosity–molecular weight (Mark-Houwink) relation

$$[\eta] = KM^a$$

has been determined in a number of solvents (Table 1). For these particular solvents the a values suggest a random-coil structure. Polymers with $\bar{M}_w$ below 10^5 behave as rigid rods. The θ conditions have been determined for pure solvents, such as toluene at 38 ± 2°C, and solvent–nonsolvent mixtures, such as chlorobenzene with 14.1 vol % methanol at 25°C.

The solution properites of poly(*N*-vinyl-3,6-dibromocarbazole) have been investigated in detail.

Table 2. Physical Constants for Poly(*N*-vinylcarbazole)

Property	Value
tensile modulus, N/mm^2 [a]	$2.5–4.2 \times 10^3$
tensile strength, N/mm^2 [a]	14
oriented	140
flexural strength, N/mm^2 [a]	35–55
compressive strength, N/mm^2 [a]	30–35
shear strength, N/mm^2 [a]	20–30
elongation, %	
amorphous	<<1
oriented	1
hardness, N/mm^2 [b]	137.2
refractive index, n_D^{20}	1.69
density, g/cm^3	
amorphous	1.184
oriented	1.191
glass-transition temperature, °C	
amorphous	227
syndiotactic	276
isotactic	126
heat capacity, kJ/(kg·K)	0.18
linear coefficient of thermal expansion per °C between 20 and 100°C	5×10^{-5}
thermal conductivity, W/(cm·K)	
at 20°C	1.26×10^{-3}
at 170°C	1.68×10^{-3}
specific heat, J/(g·°C) [c]	1.26

[a] To convert N/mm^2 to psi, multiply by 145.
[b] To convert N/mm^2 to kgf/mm^2, divide by 9.8.
[c] To convert J to cal, divide by 4.184.

Mechanical and Rheological Properties. Poly(*N*-vinylcarbazole) is a glassy polymer with good thermal stability. It is extremely brittle, and many attempts have been made to improve the mechanical and processing characteristics by plasticization, orientation, blending, and copolymerization. The physical constants are given in Table 2. Inconsistencies with other published data are most likely due to differences in purity, tacticity, molecular weight, and reliability of characterization.

Rheological studies tend to be complicated by degradation effects at the temperature required to achieve a melt (>200°C). The zero shear viscosity at 285°C for poly(*N*-vinylcarbazole)

Table 1. Mark-Houwink Constants for Poly(*N*-vinylcarbazole)

Solvent	$K \times 10^4$, cm^3/g	a	Temperature, °C
benzene	3.05	0.58	25
	3.35	0.58	25
cyclohexanone	2.00	0.61	25
THF	1.44	0.65	25
chloroform	1.36	0.67	25
tetrachloroethane	1.29	0.68	25
toluene	7.62	0.50	37
bromobenzene	0.52	0.76	25
chlorobenzene	0.59	0.74	25
nitrobenzene	0.93	0.69	25
1,3-dichlorobenzene	0.57	0.75	25
1,2-dichloroethane	1.10	0.68	25
1,4-dioxane	1.91	0.63	25
	1.77	0.63	45

samples with molecular weights $\bar{M}_w$ in the 10^6–10^5 range, have been measured as 2×10^6 and 1.5×10^4 Pa·s (2×10^7 and 1.5×10^5 P), respectively. Dynamic viscoelastic measurements in the shear mode on a series of polymer fractions covering a molecular weight range $\bar{M}_w$ from 10^3 to 10^6, gave a value for $\bar{M}_c$, the average molecular weight between entanglements, of 3.5×10^4, a value similar to that reported for polystyrene.

Dielectric Properties. The dielectric constant (ϵ') and the loss factor (ϵ'') for poly(*N*-vinylcarbazole) are remarkably constant over a wide temperature and frequency range, and the resistivity is high, $>10^{14}$ Ω·cm. These properties, as well as others, shown in Table 3, make the polymer suitable for many electrical applications.

Glass-transition Temperature. Glass-transition temperatures for poly(*N*-vinylcarbazole) ranging from 150 to 250°C have been reported. The T_g is sensitive to impurities, such as residual monomer or solvent. Molecular weight dependence is described by the Fox-Flory relationship, $T_g = T_{g\infty} - A/M_n$, where $A = 2.3 \times 10^5$ and the limiting value of the T_g at infinite molecular weight is $T_{g\infty} = 227$°C. In addition to the molecular weight dependence, the T_g is also strongly dependent on chain tacticity. This provides an explanation for the multiple T_g values (up to 3) seen in many polymer samples prepared cationically.

Orientation. Melt extrusion of high molecular weight poly(*N*-vinylcarbazole) introduces a high degree of structural order. X-ray diffraction patterns are similar to those reported for bulk and solution crystallized samples. Density, crystallinity, and average orientation function, determined from ir dichroic spectral analysis, increase significantly following extrusion. The densities of extruded materials are 1.191 g/cm³ compared to values around 1.184 g/cm³ for amorphous polymer. Thermal annealing of the extruded samples reduces the density.

Crystallinity. The molecular structure and morphology of poly(*N*-vinylcarbazole) is dominated by the bulky pendent carbazole units and their limited rotation around the C—N bond attaching them to the backbone. X-ray diffraction patterns are composed of a sharp crystalline peak superposed over a broad amorphous halo for thermally treated and oriented samples. A pseudohexagonal packing of rigid, rodlike chains with a lateral interchain spacing of 1.07 nm was proposed. Examination of models suggested a stereoblock structure consisting of isotactic 3/1 and syndiotactic 2/1 helical sequences. Analysis of small single crystals grown from solution confirmed the rodlike packing in a hexagonal array, but specified $a_1 = a_2 = a_3 = 1.23$ nm. In addition, some evidence supports an isotactic 3/1 helical polymer structure with a 0.74 nm periodicity. On examination of thermally crystallized bulk samples, and on the basis of x-ray and density data, the repeat distance was revised to 0.647 nm. This value is more consistent with numbers reported for other isotactic vinyl polymers such as polystyrene.

Table 3. Electrical Properties of Poly(*N*-vinylcarbazole)

Property	Value
dielectric strength, at 25–150°C, mV/cm	1.1–0.86
2.5-mm-thick sample	0.4
0.225-mm-thick sample	1.2
resistivity, Ω·cm	10^{16}–10^{17} [a]
25–150°C	$(0.05–8) \times 10^{15}$
loss factor	
10^3–3×10^8 Hz	$<10^{-3}$
10^4 Hz[a]	$(2–6) \times 10^{-4}$
10^3 Hz, 200°C	50×10^{-4}
permittivity at 20°C, 50 Hz–1 MHz	3
arc resistance,[b] 3.84-mm thick sample, s	20
dielectric constant, 10^4 Hz	3[a]
breakdown field strength, kV/mm	50[a]

[a] Deutsche Industrie Normen, DIN 57303.
[b] ASTM D495-42.

Conformational Studies

A conformational analysis of poly(*N*-vinylcarbazole) was based on the skeletal bond rotations ϕ_i and ϕ_{i+1} for meso and racemic dyad units. The methodology was similar to that for other vinyl polymers with planar pendent side groups. Interactions between the bulky carbazole groups makes the g conformations very high in energy and they can therefore be excluded. The energy calculations indicate a very high population of tt conformation in both meso and racemic dyads. In the meso dyad the perfectly staggered tt state is of high energy, but reaches a minimum energy state by a slight synchronized rotation of about 20°. The minimum for the gt state in the meso dyad is 9.6 kJ/mol (2.3 kcal/mol) higher than the tt minimum. For the racemic dyads the conclusions are similar, with the tg state being 6.7 kJ/mol (1.6 kcal/mol) higher than the tt minimum. Both syndiotactic and isotactic dyads have a high population of tt conformation.

The high tt population has provided a rationalization for the unique fluorescence behavior and for the unusually high value of the characteristic ratio, determined from solution property measurements, of poly(*N*-vinylcarbazole). In addition, this model provides strong support for the polymer crystal structure based on a threefold helical conformation with a repeat distance of 0.647 nm.

Degradation

Poly(*N*-vinylcarbazole) exhibits unusually high thermal stability for a vinyl polymer. The T_g is a function of molecular weight and tacticity, but a value of 220°C is commonly cited for high molecular weight polymers prepared by free-radical procedures. Minimal degradation occurs below 300°C for reasonable treatment times. The degradation mechanism is dependent on the environment. Poly(*N*-vinylcarbazole) is also degraded by mechanical energy.

Spectroscopic Studies

Absorption Spectra. The absorption spectrum of a thin poly(*N*-vinylcarbazole) film (40 nm) and poly(*N*-vinylcarbazole) in solution are very similar and match very closely with a solution spectrum of *N*-isopropylcarbazole. These findings indicate very weak ground-state mixing of the carbazole chromophores in the polymer, with spectra characteristic of isolated, noninteracting carbazole units.

Emission Spectra. The unusual photophysical properties of poly(*N*-vinylcarbazole) are the subject of considerable debate. The polymer is unique among carbazole-containing polymers and other vinyl aromatic polymers in that it does not exhibit any locally excited-state-emission characteristic of the carbazole chromophore, even in solution. The fluorescence spectrum is composed of two structureless bands centered around 370 and 420 nm, assigned to excimer states. These bands are formed between adjacent carbazole units only when the carbazole is attached directly to the chain backbone at the *N*-position.

Energy Transfer. The earliest photophysical studies of poly(*N*-vinylcarbazole) established that electronic excitation energy migration takes place in the solid state. The recent use of nano- and picosecond laser spectroscopic techniques allowed the detection and monitoring of energy transfer in the polymer molecules in solution.

Nmr Spectroscopy. The microstructure of poly(*N*-vinylcarbazole) prepared by a variety of synthetic methods has been thoroughly studied by nmr spectroscopic methods. The proton assignments have been made: aromatic (carbazole) $\delta = 8\text{–}6, 5$; methine $\delta = 3.5\text{–}2.5$; and methylene $\delta = 2\text{–}1$ ppm. The large and unusual upfield shift of one of the aromatic protons, $\delta = 5$, is attributed to the restricted internal rotation of the bulky carbazole group. It can be shown that the methine and methylene proton patterns are dependent on the method of polymerization, and hence the tacticity, of the polymer. Analysis of the diastereomeric dimeric model compounds, meso- and racemic-2,4-di-*N*-carbazolylpentane (equivalent to the isotactic and syndiotactic polymer sequences, respectively), provided the appropriate stereoassignments for the polymer. Optimal proton spectral resolution for polymer samples is obtained using 1,2-dibromoethane-d_4 as solvent at 120°C.

Charge-transfer Complexes

Poly(*N*-vinylcarbazole) forms colored solutions and films when combined with electron-acceptor molecules. The absorption spectra of these polymeric charge-transfer complexes (qv) are similar to those of acceptor–carbazole molecules, eg, *N*-ethylcarbazole combinations. The charge-transfer (CT) absorption bands of carbazole-containing molecules, including polymers, are broad and usually asymmetric.

Electrical Properties

Conductivity. Poly(*N*-vinylcarbazole) is an insulator that can be readily converted into a p-type semiconductor.

Photoconductivity. Poly(*N*-vinylcarbazole) was the first known organic polymeric photoconductor. It is an insulator in the dark and becomes conductive on exposure to uv radiation. The photoconductive response can readily be extended into the commercially important visible and near ir region of the spectrum by incorporating sensitizing dyes or electron acceptors into the polymer. The photoconductive process can be described in terms of the generation of free-charge carriers (electron–hole) by the absorption of radiation and the subsequent transport of these carriers through the material to the electrodes.

Applications

Poly(*N*-vinylcarbazole) has been widely investigated as a photoconductor. The complex with TNF was used in the Copier 1 series of electrophotographic IBM machines. Several other applications for the polymer have been proposed. The facile polymerization of the monomer has made it the basis for a number of nonsilver image-recording technologies.

Health and Safety

There are no reports on the toxicology of vinyl carbazole-containing polymers, which appear to be nontoxic. However, the polymers may contain residual carbazole monomer, and extreme caution should be exercised in their handling. *N*-Vinylcarbazole has been reported to exhibit mutagenic and allergenic activity.

BIBLIOGRAPHY

J. M. Pearson and M. Stolka, *Poly(N-vinylcarbazole),* Polymer Monographs, Vol. 6, Gordon & Breach, New York, 1981.

Y. Shirota and H. Mikawa, *J. Macromol. Sci. Rev. Macromol. Chem.* **16,** 129 (1978).

J. M. Pearson, S. R. Turner, and A. Ledwith in R. Foster, ed., *Molecular Association,* Vol. 2, Academic Press, London, 1979, p. 79.

J. Mort and G. Pfister, eds., *Electronic Properties of Polymers,* Wiley-Interscience, New York, 1982.

J. M. PEARSON
Eastman Kodak Company

VINYL CHLORIDE–ACRYLONITRILE COPOLYMER FIBERS. See ACRYLIC FIBERS.

VINYL CHLORIDE POLYMERS

INTRODUCTION

Poly(vinyl chloride) (PVC) is one of the world's major bulk polymers; 16.4 million t were produced worldwide in 1987. This might appear surprising because the polymer suffers from a number of disadvantages: the raw polymer starts to decompose at about 100°C, well below its melting point of 150–200°C, and it degrades in the presence of light. It has poor mechanical properties and a relatively low softening point, between 70 and 80°C.

The polymer owes its popularity to its versatility. The polymer itself is chemically inert and nonflammable, burning only in the presence of a source of ignition. It is compatible with many additives, including plasticizers, heat stabilizers, lubricants, fillers, and other polymers. These additives enable the polymer to be processed easily by a wide range of techniques without the fear of degradation to produce articles with a variety of mechanical properties. The articles may be rigid, such as house sidings, rain gutters, and window frames, or flexible, as in cable insulation, medical tubing, and plastic gloves. PVC is transparent, and can be used to make clear bottles and sheet. Finally, it is one of the least expensive plastic resins.

Approximately 75% of the world's PVC is produced by the suspension polymerization (S-PVC) process in which the polymerization is carried out inside vinyl chloride monomer (VCM) droplets dispersed in water. The product is in the form of porous 100–150-μm-diameter grains.

Attempts to reduce the cost of the suspension PVC process, by eliminating the need for water, led to development of a bulk or mass polymerization process which accounts for approximately 10% of the world's production of PVC. One of the earliest methods of polymerizing VCM was by the emulsion route. It accounts for the remaining 15% of PVC produced. The polymerization product is in the form of an aqueous latex dispersion of PVC particles 0.1–2.0 μm in dia. These are spray-dried to form larger agglomerates, which are milled before being dispersed in solvents and plasticizers to give plastisols, used in a variety of spreading and dipping applications. The development of the seeded emulsion and microsuspension processes has provided methods of producing latices of different sizes and size distributions, to give plastisols of different viscosities.

P. V. SMALLWOOD
ICI Chemicals & Polymers Limited

VINYL CHLORIDE MONOMER

The increasing volume of sales of poly(vinyl chloride) (PVC) has led to the search for more productive and cost-effective manufacturing methods for the monomer, vinyl chloride. The major cost of manufacturing the polymer is the cost of the monomer feedstock, so vinyl chloride has become one of the world's most important commodity chemicals.

Combining the individual processes of direct chlorination of ethylene to EDC, thermal pyrolysis of EDC, and oxychlorination of the co-produced HCl to more EDC provided the so-called "balanced" process of making VCM with no net consumption or production of HCl:

Direct chlorination $C_2H_4 + Cl_2 \longrightarrow C_2H_4Cl_2$

EDC pyrolysis $C_2H_4Cl_2 \longrightarrow C_2H_3Cl + HCl$

Oxychlorination $C_2H_4 + 2\ HCl + \frac{1}{2}\ O_2 \longrightarrow C_2H_4Cl_2 + H_2O$

Overall reaction $2\ C_2H_4 + Cl_2 + \frac{1}{2}\ O_2 \longrightarrow 2\ C_2H_3Cl + H_2O$

Most of the world's production is now based on this process, which was first put in place in the early 1960s. Since then, the process has been refined and the scale of operation has greatly increased, but no fundamentally new processes have achieved commercial viability.

Physical Properties

The physical properties of vinyl chloride are listed in Table 1. VCM is a colorless gas at ambient temperatures under at-

Table 1. Physical Properties[a] of Vinyl Chloride

Property	Value	Value at −40°C	Value at −20°C	Value at 0°C
molecular weight	62.500			
melting point, °C	−153.8			
normal boiling point, °C	−13.4			
critical pressure, kPa[a]	5600			
critical temperature, °C	156.6			
critical volume, m³/mol	169×10^{-6}			
refractive index at 15°C	1.398			
dielectric constant at 15°C	6.26			
latent heat of fusion, kJ/kg[b]	76.0			
standard enthalpy of formation, kJ/mol[b]	35.991			
standard Gibbs free energy, kJ/mol[b]	51.517			
flash point (open cup), °C	−77.75			
autoignition temperature, °C	472			
lower flammable limit in air, % v/v	4.0			
upper flammable limit in air, % v/v	22.0			
solubility of VCM in water at 20°C, mg/kg	1100			
solubility of water in VCM at −15°C, mg/kg	300			
Liquid properties				
latent heat of vaporization, kJ/kg[b]		374	362	349
liquid density, kg/m³		1014.4	983.4	947.1
viscosity, mPa·s (=cP)		0.35	0.29	0.23
thermal conductivity, W/(m·K)		0.14	0.136	0.129
surface tension, mN/m (=dyn/cm)		26	23	20
specific heat, kJ/(kg·K)[b]		1.361	1.375	1.394
liquid saturated vapor pressure, kPa				
$P° = A + B/(T + C)$, for T in K				
A	21.331			
B	−2366.4			
C	−18.455			
Gas properties at 25°C, 101.3 kPa[a]				
viscosity, mPa·s (=cP)	0.011			
thermal conductivity, W/(m·K)	0.010			
specific heat, kJ/(kg·K)[b]	0.872			

[a] To convert kPa to atm, divide by 101.3.
[b] To convert kJ to kcal, divide by 4.184.

mospheric pressure and has a slightly sweet odor. It is flammable and forms explosive mixtures with air. It is soluble in most common organic liquids, eg, hydrocarbons, oils, alcohol, and other chlorinated hydrocarbons, and is only sparingly soluble in water.

Chemical Properties

The vinylic double bond of the vinyl chloride molecule enables the polymerization reactions to occur. Also, addition reactions with chlorine or HCl are possible, yielding 1,1,2-trichloroethane and 1,1-dichloroethane, respectively. These addition reactions are normally ionic in path and involve sparging the reacting gases into a volume of the liquid product containing dissolved ferric chloride as catalyst. Substitution of the chlorine atom is more difficult to achieve.

If oxygen and water are excluded, pure vinyl chloride is very stable and noncorrosive to mild steel, which is the usual construction material for storage and transportation equipment containing vinyl chloride.

Manufacturing Processes

All current production plants for vinyl chloride depend on the use of a C_2 hydrocarbon feedstock. The choices are acetylene, ethylene, and ethane. Manufacture from acetylene is a relatively simple single-stage process, but the cost of acetylene is high. By contrast, ethane is by far the least expensive C_2 hydrocarbon, but it cannot at present be converted to vinyl chloride with high selectivity. Therefore, well over 90% of the world's vinyl chloride is produced by the balanced ethylene route.

Chlorine, the other principal feedstock, is derived almost exclusively from brine electrolysis. In areas of high electrical power costs, chlorine is an important cost element in vinyl chloride production; this accounts for interest in developing a route that does not involve electrolysis.

Environmental Protection

Current legislation in the United States, Japan, and most European countries requires the incineration of all continuous or semicontinuous vents containing chlorinated hydrocarbons in significant concentration. The remaining emissions are from loading and transportation operations, equipment decommissioning before maintenance activity, fugitive emissions, and infrequent discharges from pressure-relief valves. If, as is common practice, EDC is stored in large atmospheric pressure stock tanks, there will be emissions through the vents as the tanks breathe.

Aqueous process streams originate from the oxychlorination and (low temperature) direct chlorination processes. These are typically neutralized and, together with contaminated surface water, are steam-stripped to remove organics. If a fluidized bed oxychlorination process is used, it may be necessary to precipitate copper salts; the most modern facilities may have a biological oxidation unit to remove trace organics and to achieve a suitably low biological oxygen demand (BOD) of the final waste water.

BIBLIOGRAPHY

J. A. Cowfer and A. J. Magistro in M. Grayson, ed., *Encyclopedia of Chemical Technology,* Vol. 23, 3rd ed., John Wiley & Sons, Inc., New York, 1983, p. 865.

Process Economics Program, Rep. Ser. 5, Supplements 5A-5C, Vinyl Chloride, Stanford Research Institute, Menlo Park, Calif., Dec. 1982.

C. G. Bertram, *Environ. Sci. Technol.* **11,** 864 (1977).

Ullmann's Encyclopedia of Industrial Chemistry, 5th ed., Vol. A6, VCH Verlagsgesellschaft mbH, New York, 1986, pp. 263–270, 285–294.

M. W. Newman
European Vinyls Corporation

SUSPENSION AND MASS POLYMERIZATION

Most of the world's homopolymers and copolymers of vinyl chloride are produced by the suspension polymerization route. Like the suspension process, the mass polymerization of vinyl chloride involves the use of monomer-soluble free-radical initiators. The products of the two processes are similar and are used in similar applications. Chlorinated PVC is produced from these resins and thus will also be considered here.

The object of the suspension and mass polymerization processes is to produce resin in a safe and efficient way, that can be handled and processed easily, to give finished articles with the required properties. This means that the polymer must consistently meet the required specifications. The following properties must be optimized: (*1*) the molecular weight determines the ease of processing the polymer, and properties of the finished article; (*2*) the chemical composition is of relevance to copolymers and determines the flow properties of the polymer melt; (*3*) the grain size and size spread determine the ease of powder handling and processing; (*4*) the grain porosity must be optimized to facilitate the removal of unreacted VCM and the uptake of plasticizers and processing aids; (*5*) the density determines the ease and speed of processing; (*6*) purity—the resin should be free of impurities, unreacted monomer, and dispersion faults; (*7*) good color and thermal stability are needed to enable polymer processing without degradation and to optimize the appearance of the finished article; and (*8*) good electrical resistivity is required for cable insulation. Thus the physical properties of the grains of PVC are almost as important as the chemical properties.

Suspension Polymerization of Vinyl Chloride

Vinyl chloride, which boils at −13.4°C, is polymerized in a batch process by dispersing the liquid monomer, under pressure, in water, in a stirred jacketed reactor (or autoclave) (Fig. 1). One or more protective colloids (or suspending agents) are added to control the size and morphology of the grains. The other essential ingredient is the monomer-soluble free-radical initiator. Other additives, such as buffers, chain transfer or extending agents, and comonomers can be added if required. A guide to the quantities of each reagent used is given in Table 1, but the exact amounts vary from grade to grade and from plant to plant. Reactors are normally run at 85–95% capacity.

Table 1. Typical Suspension Polymerization Recipe

Ingredient	VCM, wt %
water	100–130
VCM	100
protective colloid(s)	0.05–0.15
initiator	0.03–0.07

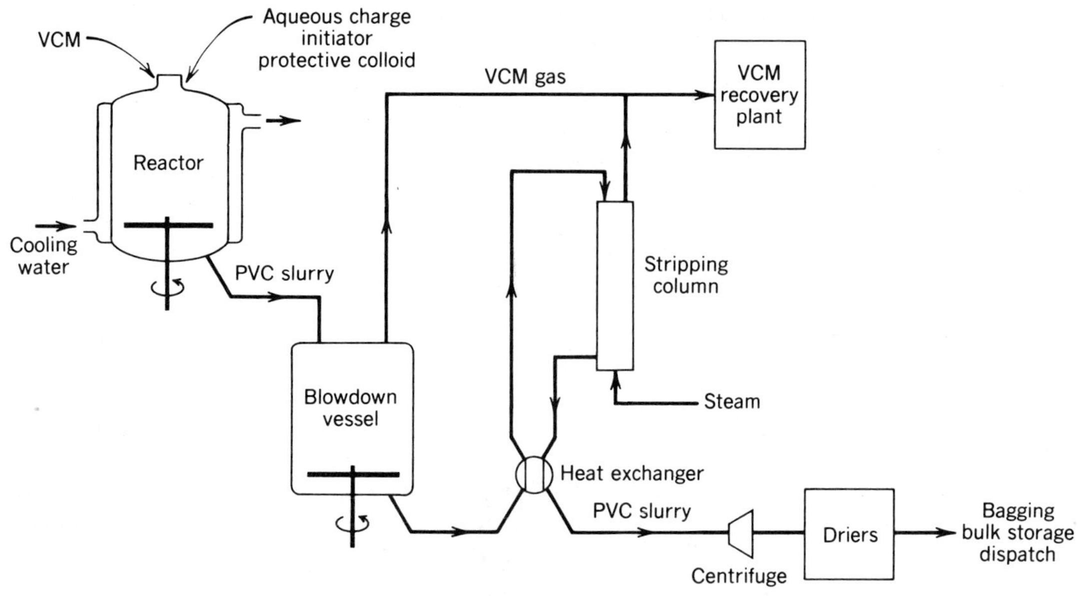

Figure 1. Suspension PVC plant.

Grains of suspension PVC are about 100–180 μm in dia and their structure has been shown in scanning electron micrographs. Each grain consists of an agglomerate of a number of ~40-μm-dia monomer droplets. The section of the grain shows that each agglomerate is in turn composed of a large number of ~3–5-μm PVC particles, aggregated into a porous network. The grain is surrounded by a wall or skin approximately 2 μm thick.

The properties of the PVC grains are determined at the polymerization stage. The particle size, porosity, and apparent density of the grains are primarily determined by the type and concentration of the protective colloid system used and by the agitation regime present in the reactor.

The molecular weight of PVC is determined by the polymerization temperature. Increasing the temperature reduces the molecular weight of the PVC, but unfortunately also decreases its porosity.

The color and thermal stability of the PVC is maintained by controlling polymerization conditions to ensure defects are not introduced into the growing polymer chains or that the polymer is not overheated during stripping and drying. This is not always easy to arrange and, as a compromise, antioxidants can be added to minimize degradation.

Contamination of the product by foreign substances is eliminated by good housekeeping and plant management.

It is important to choose additives carefully. Many, particularly ionic impurities, reduce the electrical resistance of PVC, restricting its use in cable insulation. Others can adversely affect the color or heat stability of the product. As PVC is often used in food contact applications, all additives must be approved by the appropriate government agencies (eg, FDA and EEC).

Although all suspension PVC processes are basically similar, each producer has developed a reactor design and combination of protective colloids and agitation system to suit customer requirements. These details are all closely guarded secrets.

Additives. ***Protective Colloids.*** Protective colloids used in the suspension polymerization process are of two types: the primaries and secondaries. The main function of the primaries is to control particle size, although they also influence porosity. They can be used singly or in combination with other primaries and their use in polymerization is essential. The demand for polymers with reduced residual VCM, and thus even higher porosities, has led to the use of secondary stabilizers, added to increase porosity.

There are reports in the older patent literature of solids and water-soluble polymers being used to stabilize the VCM droplets, but the substances in most common use today are the cellulose ethers and partially hydrolyzed poly(vinyl acetate)s (PVA). Their effect on PVC product properties is determined by their molecular weight and degree of substitution or hydrolysis.

Two types of secondary stabilizers are added to suspension polymerizations to enhance the effect of the primary by giving further increase in porosity. The first type is the low molecular weight surfactants. The second type is low hydrolysis PVAs which are much more effective at increasing porosity than many of the surfactants.

The use of both types of secondary stabilizers often affects the particle size of the PVC; thus the concentration of the primary protective colloid must be adjusted to compensate.

Initiators. Free radical initiators used in vinyl chloride suspension and mass polymerizations are azo, or more commonly, peroxide initiators (see AZO COMPOUNDS; PEROXY COMPOUNDS). The most important property of the initiator is its half-life ($t_{1/2}$) which ideally should be 2–3 h at the required polymerization temperature. To optimize the reaction time and fully utilize the cooling capacity of the reactor, mixtures of initiators can be used.

Mechanism of Grain Formation. Polymerization occurs inside the VCM monomer droplets. PVC is insoluble in its monomer (although it is swollen by it) and precipitates at very low conversions. These PVC particles grow and aggregate to give 3–5-μm particles visible inside the final porous grains. The size of the initial VCM droplets is determined by the type and concentration of the protective colloid and the reactor agitation. The final PVC grains are agglomerates of these smaller monomer droplets.

Kinetics of Suspension and Mass Polymerization. A typical reaction profile for the suspension polymerization of vinyl chloride is shown in Figure 2. The reactor is charged with the appropriate ingredients and is then heated by passing hot water through the reactor jacket. Increasing the temperature of the reactor increases the vapor pressure of the VCM, and thus the pressure in the reactor rises. At the required polymerization temperature the heating is turned off and colder water is passed through the jacket to remove the heat of polymerization and maintain constant reactor temperature. As the conversion increases so does the rate of polymerization, as is shown by the increased demand for cooling water. In the example in Figure 2, the flow of cooling water is constant and the increased cooling is achieved by reducing the cooling water temperature. The maximum polymerization rate occurs near the pressure drop. Kinetic models of the polymerization must account for these changes.

Classical homogeneous free-radical polymerizations can be described by the following equations.

Initiation occurs as a result of the thermal decomposition of the initiator to produce free radicals:

$$I \xrightarrow{k_d} 2I\cdot \quad (1)$$

where $[I]$ and $[I\cdot]$ are the concentrations of the initiator and initiator radicals, respectively, and k_d is the rate constant of initiator decomposition.

The rate of formation of radicals is given by

$$R_i = 2k_d\,[I] \quad (2)$$

or more usefully,

$$R_i = 2k_d\,f[I_0]\,\exp\,(-k_dt) \quad (3)$$

where $[I_0]$ is the initial concentration of initiator, t is the time, and f is the initiator efficiency factor, introduced because not

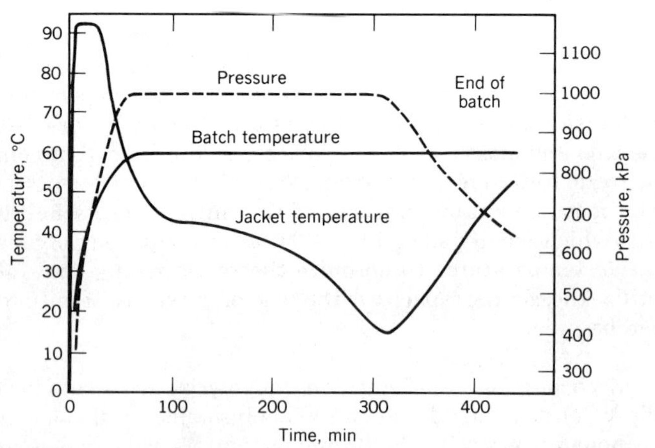

Figure 2. Reaction profile for a typical suspension polymerization. To convert kPa to bar, divide by 100. Courtesy of Elsevier Applied Science Publishers.

all of the initiator radicals react to produce a growing chain. The constant k_d is related to the half-life of the initiator $t_{1/2}$ by:

$$k_d = \frac{\log_e 2}{t_{1/2}} \quad (4)$$

Propagation is a step process involving the addition of an initiator radical I·, or more generally a polymer radical $R_n^{\cdot}$, to a molecule of monomer (M):

$$R_n^{\cdot} + M \xrightarrow{k_p} R_{n+1}^{\cdot} \quad (5)$$

The rate of propagation is given by

$$R_p = k_p\,[R_n^{\cdot}]\,[M] \quad (6)$$

Transfer involves the transfer of the radical from a growing chain to another species (eg, a monomer molecule). In VCM polymerizations, transfer to the monomer is particularly important and can be represented as in equation 7, although its mechanism is now believed to involve intramolecular rearrangement following the relatively uncommon head-to-head addition of monomer units:

$$R_n^{\cdot} + M \xrightarrow{k_{tr}} P_n + R^{\cdot} \quad (7)$$

where P_n is the dead polymer chain and k_{tr} is the transfer rate constant.

The rate of transfer is given by

$$R_{tr} = k_{tr}\,[R_n^{\cdot}]\,[M] \quad (8)$$

The rate of transfer divided by the rate of propagation, the chain transfer coefficient is given by

$$C_{tr} = k_{tr}/k_p \quad (9)$$

Termination can occur as a result of disproportionation or radical combination. The molecular weight distribution of PVC and the frequency of groups at the ends of polymer chains suggest that disproportionation is the most common mechanism:

$$R_nCH_2CHCl\cdot + R_mCH_2CHCl\cdot \longrightarrow R_nCH{=}CHCl + R_mCH_2CH_2Cl$$

or kinetically

$$R_n^{\cdot} + R_m^{\cdot} \xrightarrow{k_t} P_n + P_m \quad (10)$$

Thus the rate of termination will be

$$R_t = 2k_t\,[R_n^{\cdot}]\,[R_m^{\cdot}] = 2k_t\,[R\cdot]^2 \quad (11)$$

Polymerization Rate. The overall rate of polymerization can be obtained by applying the classical stationary state hypothesis, which states that the concentration of radicals remains constant and thus their rate of formation must be the same as their rate of loss. Thus from equations 2 and 11,

$$R_i = 2k_d\,[I] = 2k_t\,[R\cdot]^2 \quad (12)$$

Therefore,

$$[R\cdot] = \sqrt{\frac{k_d\,[I]}{k_t}} \tag{13}$$

Substituting into equation 6 and setting the rate of polymerization equal to the rate of loss of monomer,

$$\frac{-d[M]}{dt} = k_p\,[M]\left\{\frac{k_d\,[I]}{k_t}\right\}^{1/2} \tag{14}$$

This equation only predicts the rate of polymerization up to low degrees of conversion; thus the simple model needs improvement. The most obvious change is that polymerization occurs in two phases, the monomer-rich phase and the polymer-rich gel phase. Termination occurs primarily by the diffusion together and collision of two bulky radicals. This is considerably slower in the viscous polymer phase (ie, k_t is reduced) and as a result the rate of polymerization in this phase is much faster. Thus as the amount of polymer increases, the rate of polymerization accelerates (Fig. 2). Further refinements have involved considering the distribution and transport of initiator and polymer radicals between the two phases.

Process Engineering. The main requirements of the engineering needed to achieve the commercial production of PVC are that the polymerization be done as economically, efficiently, and safely as possible. Over the past 30 years the size of installed reactors has increased from 10 m^3 or less up to 100 m^3 and even 200 m^3. Environmental concerns have resulted in modern reactors being run automatically for many batches without being opened, ie, closed-lid operation. A simplified flow diagram of a suspension PVC plant can be seen in Figure 1.

Reactor Charging. The automatic metering of reactants demands that ingredients be in a liquid form. Many plants use demineralized water, which gives more consistent granulation, to make up the protective colloid and buffer solutions, and to charge the reactor. The water can be deaerated before use to reduce its oxygen content.

Reactor Configuration, Heating, Cooling, and Agitation. As a polymerization reactor becomes larger, its surface/volume ratio decreases and thus it becomes increasingly difficult to transfer heat through the reactor wall. The conventional way of heating is to pass hot water through the reactor jacket. Ways of reducing reactor heat up time include charging the reactor with warm water (eg, at 40–50°C), the addition of very hot (≥100°C) water, or by the direct injection of steam into the reactor's contents.

Once "on temperature," the heat of polymerization, 1540 kJ/kg (368 kcal/kg), must be removed to maintain constant temperature. The polymerization rate increases around pressure drop. Thus the cooling capacity of the reactor must be sufficient to remove the heat generated at pressure drops. Ways of increasing the heat removed from the reactor include the use of water cooled baffles, reflux condensers, and refrigerated cooling water.

Because reactor agitation has such an important effect on product properties, PVC manufacturers give few details of the systems they use. Older, smaller reactors often had top-entry stirrers with a variety of paddle shapes. The higher power inputs and longer stirrer shafts required by larger reactors have resulted in the installation of bottom entry stirrers.

Reactor Fouling. The formation of buildup, or deposits of PVC, on the internal surfaces of reactors has always been a problem in the manufacture of PVC because of reduced heat transfer, large alteration of the agitation regime present in the reactor, and broken-off fragments blocking transfer systems and contaminating the product. Up until the early 1970s these deposits were removed manually and PVC workers were exposed to high levels of VCM, which caused a number of them to contract angiosarcoma. Early attempts to reduce buildup formation involved the use of glass-coated reactors, reactor internals with smooth surfaces, and particular types of stainless steel linings. Buildup can also be reduced by the correct choice of initiator, buffer, and aqueous phase pH. Further impetus for the elimination of buildup came from the introduction of new, larger, computer-controlled plants, designed to operate for many batches without opening the reactor. Probably the most effective way of reducing reactor fouling is by the use of chemicals (build-up suppressants) to coat the reactor wall or to add to the reactor charge water.

Stripping and Drying. In order to reduce VCM emissions to the atmosphere during the drying and processing of PVC, unreacted monomer is removed from the polymer at the slurry stage and is recovered and returned to the polymerization cycle. Thus the VCM must be removed from the PVC grain into the aqueous phase and then into the gas phase. The transfer of VCM out of the PVC grain is a diffusion-controlled process and the mechanism of diffusion depends on whether the monomer-swollen PVC is above or below its T_g. Under the stripping conditions practiced industrially at temperatures above 70–80°C, PVC is above its T_g. In this case diffusion corresponds to a simple Fickian relationship.

The most common method of removing unreacted VCM is by sparging the polymer slurry with steam. The steam is separated from the effluent gas by condensation before the VCM is liquified and used in later polymerizations.

Steam stripping can be used batchwise or continuously. Batch stripping demands heating up large volumes of slurry with steam sparging and evacuating to remove the steam/VCM mixture. It can be carried out in the reactor or in a separate stripping vessel. However, most modern plants are fitted with continuous stripping columns.

Relatively coarse grains of suspension PVC are separated from the bulk of the water by centrifugation. This dewatering process is unable to remove the water trapped in the pores of the grains or on their surface and thus the damp polymer must be heated to remove the residual moisture. A number of types of driers are available: *flash driers, rotary driers,* and *fluid bed driers.*

It is possible to use one or more types of driers in series, although flash driers cannot be used on their own. Overdrying can lead to the buildup of static electricity, which adversely affects the flow of the PVC granules.

Suspension Copolymerization of Vinyl Chloride

Vinyl chloride can copolymerize with a wide variety of monomers, but very few of these copolymers are of commercial significance. Most have been developed to improve the melt flow characteristics of unplasticized PVC. The comonomer units in the polymer chains prevent the vinyl chloride repeating units from crystallizing into high melting flow units. However, with improvements in the manufacture, properties, and processing of unplasticized homopolymers, as well as with changes in the marketplace, the proportion of copolymers made commercially has decreased. The most important are vinyl acetate copolymers with applications in records, flooring, and packaging. They are made by a suspension process similar to that used for homopolymers. Some suspension propylene copolymers have been manufactured. Small quantities of other copolymers are available but are often made by an emulsion or solution route for special applications such as surface coatings.

In addition to random copolymers, block and graft copolymers of vinyl chloride have also been prepared.

Bulk or Mass Polymerization of Vinyl Chloride

Approximately 10% of the world's PVC is produced by the mass or bulk polymerization process. All operators of the bulk process use technology licensed by ATO of France, one of whose predecessors, Pechiney St Gobain, developed the process. The history of the process has been reviewed. As the name implies, the polymerization is carried out in VCM in the absence of water. It takes place in two stages: the first takes the product to 7–12% conversion and the second to completion at 70–90% conversion. The layout of a typical mass plant is shown in Figure 3. The mechanism and kinetics of mass polymerizations are, except for the absence of water, very similar to suspension polymerizations and have been described earlier.

Control of Product Properties. As with all types of vinyl chloride polymerizations, the molecular weight (K value) is determined by the polymerization temperature. As most of the polymerization occurs in the second stage, it is the temperature of this part of the process that determines the K value of the product.

The particle size of the final product is mainly determined by the tip speed of the agitator, and by the conversion in the first stage of the process. Additives can also be used to increase or decrease particle size.

Although the first stage does affect the product porosity, this property is in practice controlled in the second stage. As in suspension polymerizations, the higher the polymerization temperature the lower the porosity. The product's porosity can also be altered to a certain extent by degassing the product during the second stage polymerization, and by decreasing the conversion.

A Comparison of the Suspension and Mass Polymerization Processes

In principle, resins from the two processes are similar and can be used in the same applications, but in practice there are differences. Mass resins are especially suited for injection molding and give transparent articles of excellent clarity, probably because they do not contain the protective colloids and other additives used in the suspension process. It is not possible to mix suspension and mass resins because static effects lead to poor powder flow, so converters tend to use one or the other type of resin in their process. As they are more generally available, most PVC processors prefer suspension resins.

Chlorinated PVC

PVC (56.7% chlorine) reacts with chlorine radicals, produced from chlorine gas, to give chlorinated PVC (CPVC), a polymer with a higher T_g and heat distortion temperature. Two types

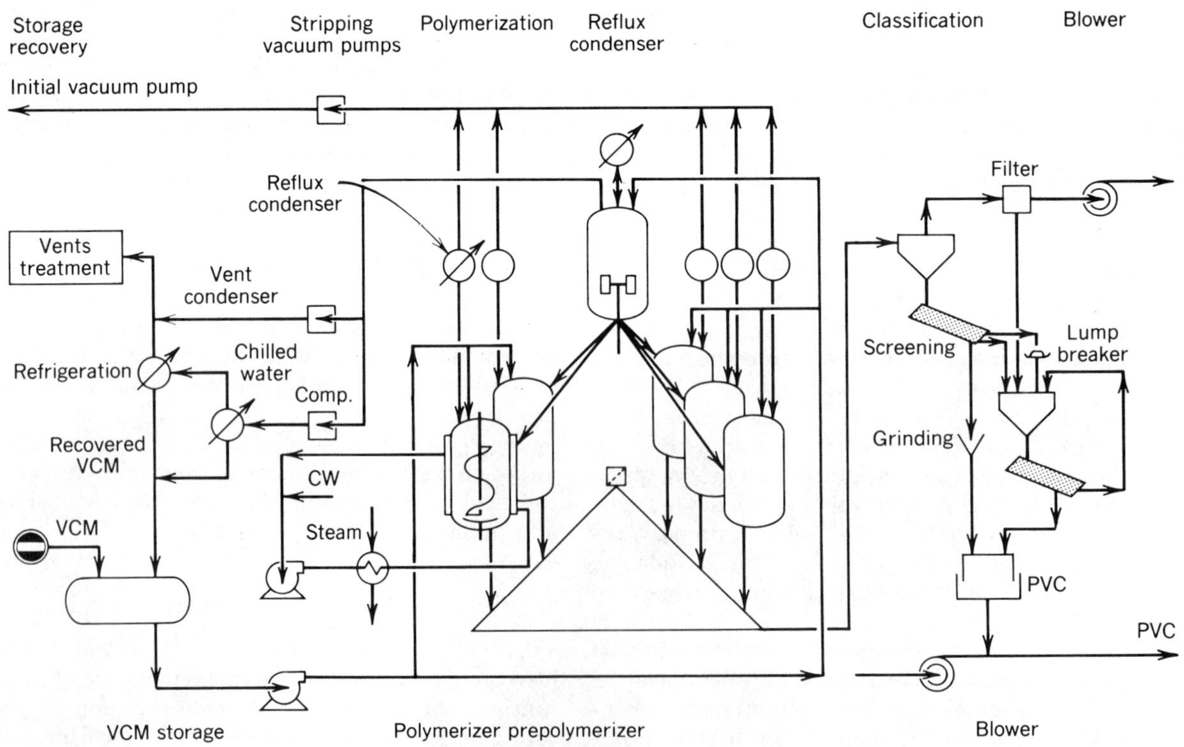

Figure 3. Layout of a mass PVC plant.

of polymer are produced commercially, those containing 63–64% and 67–68% chlorine.

The most common commercial process in use today involves dispersing the PVC particles in water and swelling with a suitable swelling agent (eg, a chlorinated hydrocarbon). The reactor is degassed to remove O_2, is heated to the required temperature (eg, 50°C), chlorine is added, and the system is irradiated with uv light to produce the chlorine radicals. Further chlorine is added to maintain the pressure. The HCl produced dissolves in the water and is removed later by washing the product with an inorganic base. The product, which is preferentially chlorinated on its surface, is then dried.

A more recent process dispenses with the use of a swelling agent.

BIBLIOGRAPHY

R. H. Burgess, ed., *Manufacturing and Processing of PVC,* Applied Science Publishers, London, 1982.

E. D. Owen, ed., *Degradation and Stabilization of PVC,* Elsevier Applied Science Publishers, Ltd., Barking, UK, 1984.

P. V. Smallwood
ICI Chemicals and Polymers

STRUCTURE AND PROPERTIES

The polymerization of vinyl chloride can result in the formation of molecules with a number of isomeric forms and structural defects. The structure $CH_2{=}CHCl$ can add onto the polymer chain in either of two ways: head-to-head or head-to-tail; polymer characterization studies show that PVC grows mainly by head-to-tail addition. Besides addition of new monomer units, side reactions can occur that introduce defects into the polymer chain. Similar defects can occur by the incorporation of impurities. These factors are of vital importance to the users of poly(vinyl chloride) (PVC), because they affect the color and thermal stability of the polymer, its crystallinity, processing behavior, and the mechanical properties of the finished article. Studies of defects also give an insight into the nature of the side reactions occurring during polymerization.

Tacticity

Because there is free rotation around the carbon–carbon bond of each terminal monomer unit on a growing PVC chain, PVC segments can adopt three configurations (see Microstructure). The polymer segment is syndiotactic when adjacent chlorine atoms occur on opposite sides of the chain, isotactic when they occur on the same side, and atactic when the chain contains both types in a random fashion.

Defect Structure of PVC

Typical PVC molecules are built up of approximately 1000 vinyl chloride monomer units, joined in a predominantly head-to-tail fashion. Other structural arrangements are considered defects. The types and concentrations of these defects have been summarized and the effect of preparing PVC samples under different conditions on the defect structure has been reported. Irregular structures determined in ordinary PVC include $\sim CH_2{-}CHCl{-}CH_2Cl$, 0.8–0.9/molecule; $\sim CH_2{-}CH{=}CH{-}CH_2Cl$, approximately 0.7/molecule; initiator residues, approximately 0.2/molecule; total unsaturation, 1/molecule; internal double bonds, 0.1–0.2/1000 VC; polyene sequences, $<5 \times 10^{-4}$/1000 VC; chloromethyl branches, 4/1000 VC; 2-chloroethyl branches, <0.5/1000 VC; 2,4-dichlorobutyl branches, 1/1000 VC; long branches, <0.5/1000 VC; tertiary chlorine, 0.5–1.5/1000 VC; and head-to-head groups, <0.2/1000 VC.

Crystallinity

Crystallinity in commercial poly(vinyl chloride) is generally associated with syndiotactic sequences. Syndiotacticity of PVC increases with decreasing polymerization temperature. At the usual polymerization temperature of 55°C it is 0.55. The sequences are rather short, putting restrictions on the size and perfection of the crystallites. However, there are strong attractive forces between chains associated with the C—Cl dipoles that provide for stable structures.

Melting Behavior. The theoretical melting point of a perfect crystallite of poly(vinyl chloride) has been estimated by various workers; values of 260–310°C are indicated. In practical materials the small size and concentration of imperfections in the ordered regions both reduce the melting point and broaden the melting range.

Measurement of Crystallinity. X-ray scattering, the classical technique used to assess the crystallinity of materials, presents some difficulties with poly(vinyl chloride). Most estimates for commercial polymers are of 7–20% crystallinity, increasing for materials polymerized under conditions intended to generate increased syndiotacticity up to about 65%. The unit cell dimensions for syndiotactic poly(vinyl chloride) have been established on the basis of x-ray analysis, which suggest an orthorhombic unit cell with *a*-, *b*-, and *c*-dimensions of 1.06, 0.54, and 0.51 nm, respectively, in size. Single crystals, grown from low molecular weight material, have been observed.

Characteristics of the Amorphous State

Most of the material is in the amorphous state, and has a dominating influence on, eg, the deformation resistance and the diffusion characteristics. The amount of free volume in the glassy state of the amorphous material is influenced by the thermal history near the glass transition temperature. The influence of this on, eg, the ability to remove monomer from the polymer has been studied. The glass transition temperature of poly(vinyl chloride) depends on the molecular structure. The effects of molecular weight and tacticity have been separated; reductions in either lower T_g. The exact value of T_g

Table 1. The Syndiotacticity of PVC as a Function of Polymerization Temperature[a]

Polymerization temperature, °C	Syndiotacticity
55	0.55
25	0.57
0	0.60
−30	0.64
−50	0.66
−76	0.68

[a] From ^{13}C-nmr data.

depends on the method used to measure it. Generally, values of 70–80°C are obtained. The addition of plasticizers lowers the T_g of the mixture to below room temperature.

Chemical and Solution Properties

Unplasticized poly(vinyl chloride) has good resistance to a wide range of chemicals: inorganic solutions, detergents, and aliphatic hydrocarbons including oils and waxes. Water absorption by poly(vinyl chloride) is very low and many properties are insensitive to humidity. However, this statement is no longer applicable if the poly(vinyl chloride) composition contains materials, such as fillers, that themselves absorb water. Unplasticized poly(vinyl chloride) is attacked by aromatic and chlorinated hydrocarbons, organic materials derived from them, and highly polar organic materials such as ketones, esters, cyclic ethers, nitro compounds, and halogens. Strong oxidizing agents also attack poly(vinyl chloride). In stressed applications, it is necessary to consider the combined effects of stress and environment (environmental stress cracking).

A commercially important method of chemically modifying poly(vinyl chloride) is postchlorination to form chlorinated poly(vinyl chloride) (CPVC). The chlorine content is increased from 56.7% to typically between 63 and 68%.

Solubility. It is difficult to prepare a true solution from poly(vinyl chloride). Even the best solvents, such as THF, cyclohexanone, and EDC, require heating and agitation. The solubility parameter of poly(vinyl chloride) has been measured as being approximately 9.5–9.7 (see SOLUBILITY OF POLYMERS).

Plasticizers. Conventional plasticizers (qv) such as dioctyl phthalate (DOP) swell rigid poly(vinyl chloride) grains, particularly when the temperature is in the region of the glass-transition temperature of the resin. At higher temperatures the degree of swelling increases until a transition is observed between the swollen gel and an apparent solution.

Polymer–Polymer Miscibility. Complete miscibility in polymers is quite rare; however, poly(vinyl chloride) is miscible with a number of other polymers and some of these mixtures have wide commercial application (see COMPATIBILITY; MISCIBILITY).

Permeability. Permeability is defined as the product of diffusivity D and solubility S. As with many other properties, the permeability of poly(vinyl chloride) is altered by the addition of plasticizers and by copolymerization, so that a range of properties needs to be considered.

Thermal Stability of Unstabilized PVC

Heating unstabilized PVC above 100°C, or exposing it to uv light or gamma rays, causes the polymer to degrade.

The thermal stability of PVC is considerably less than that of its lower molecular weight analogues. Thus attempts to explain the instability of PVC have involved examining irregular structures in the polymer that could act as sites to initiate the dehydrochlorination.

Extrapolation indicates that even in the absence of labile chlorine atoms, PVC is unstable, suggesting that other factors are operative. Random dehydrochlorination from ordinary monomer units has been suggested as being important and could perhaps explain the observation that dehydrochlorination is catalyzed by HCl. In summary, the contribution of different PVC defect structures to dehydrochlorination is as follows: tertiary chlorine, 70–80%; allylic chlorine, 10–15%; vinyl chloride monomer unit, 10–15%.

Physical Properties

Physical properties are given in Table 2.

Table 2. Typical Properties of Vinyl Chloride Homopolymers

Property	Value
Thermal	
thermal conductivity, W/(m·K)	0.16
thermal diffusivity, m^2/s	1.2×10^{-7}
coefficient of thermal expansion	$6\text{–}7 \times 10^{-5}$
glass-transition temperature, °C	75–85
crystalline melting point, °C	120–210
Mechanical	
tensile modulus, GPa[a]	2.8
shear modulus, GPa[a]	1.0
Poisson's ratio	0.4
tensile yield strength, MPa[b]	55
Shore D hardness	65–85
Electrical	
volume resistivity, Ω·cm	10^{-16}
dielectric strength, V/cm	$1\text{–}5 \times 10^5$
Other	
density, kg/m^3	1390–1430
refractive index	1.54
water absorption, %	<0.5

[a] To convert GPa to psi, multiply by 145,000.
[b] To convert MPa to psi, multiply by 145.

BIBLIOGRAPHY

L. I. Nass and C. A. Heiberger, eds., *Encyclopedia of PVC,* Vol. 1, 2nd ed., Marcel Dekker, Inc., New York, 1986.

R. Gachter and H. Muller, *Plastic Additives Handbook,* Hanser, Munich, 1984.

R. C. STEPHENSON
European Vinyls Corporation
P. V. SMALLWOOD
ICI Chemicals and Polymers Ltd.

PROCESSING AND APPLICATIONS OF SUSPENSION AND MASS POLY(VINYL CHLORIDE)

Poly(vinyl chloride) resins are utilized in a large number of applications. Although the basic properties of the material are unattractive, the use of additives to facilitate processing and modify the properties has generated a large international industry. The success of PVC rests on its good economics, and on the vast range of possible product properties: from rigid, engineering thermoplastics to rubbery materials, when plasticized.

Poly(vinyl chloride) polymers are notable for the wide range of applications for which they are suited, and for the associated diversity of processing techniques. Even excluding the techniques used to process plastisols, PVC compositions are converted into products using all the standard techniques developed by the plastics processing industry, plus a few unique to PVC.

Properties of PVC Compositions

Rheology. The rheology of PVC is more complicated than that of most other thermoplastics because of the hierarchical particulate structure present in the original grains of suspension or bulk PVC. If a material melts during processing, the original structure is lost. However, the limited heat stability of PVC precludes total melting, and the structures laid down during polymerization persist.

The terms gelation and fusion are widely used in the industry to describe the conversion of the grains to a uniform mass.

The persistence of the particulate structures influences both the rheology of the material during processing and the ability of the material to develop optimum physical properties. The wide range of crystalline melting points results in a progressive modification of the material state under normal processing conditions. The phenomena associated with the rheology of PVC are based on this feature.

Processing

A wide variety of poly(vinyl chloride) resins are commercially available. A choice must be made to meet the combinations of

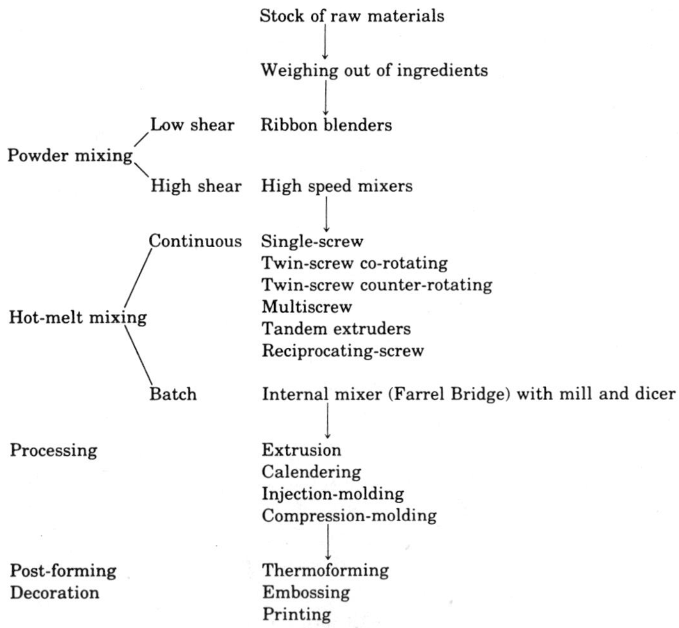

Figure 1. Summary of processing alternatives with suspension and mass poly(vinyl chloride).

Table 1. Application Areas for Suspension and Mass Homopolymers

	Application	
Processing technique	Rigid	Plasticized
injection molding		
	pipe fittings	footwear
	electrical and equipment housing	novelties
	bottles: injection stretch blow	components
calendered sheet		
	flooring	flooring
	packaging sheet	sheet and foil
	film and foil	swimming pool liners
	decorative laminates	waterproof membranes
	thermoforming sheet	toys
	rigid sheet	
extrusion		
	bottles: blow molding	hoses, tubes, film
	OPVC bottles	cables
		belting
	pipe: pressure	waterstops and seals
	water	trim
	irrigation	
	drain/vent/waste	
	conduit	
	sewer/drain	
	profiles: GP profiles	
	window profile	
	composite windows	
	siding/cladding	
	drawer components	
	curtain rails	
	sheet and foils	
	foam sheet, pipe, and profile	

processing methods and end uses. The most important property influencing processing and the mechanical properties is the molecular weight, characterized by *K*-value; resins are available with *K*-values from below 50 to above 90.

Low *K*-value materials are easily processed but have poor mechanical properties, whereas high *K*-value materials are difficult to process; few rigid applications are served by resins with *K*-values >70, since the difficulties of obtaining full gelation outweigh the potential improvement in some mechanical properties. The processing of plasticized formulations is less restricted and higher *K*-values may be used. The efficiency of the polymerization process rapidly declines, however, and resins with *K*-values >80 are regarded as specialty resins.

The apparent density of polymers depends on the porosity, and grain size and shape distribution. The density of most resins is between 450 and 600 kg/m^3, although specialty resins such as filler or blending resins for plastisols reach 700 kg/m^3; the density of high *K*-value resins can be as low as 400 kg/m^3. Commercial mass polymerized resins are slightly denser than an equivalent *K*-value suspension resin. Most resins have mean grain sizes of 100–200 μm.

A number of copolymer products are also available that are intended to provide a different balance of properties than homopolymer PVC. Most important are the vinyl chloride–vinyl acetate copolymers that have found applications in flooring, thermoformed sheets, and records. Copolymers with improved solution properties are also manufactured.

Mixing Techniques. Suspension and mass polymerized poly(vinyl chloride) resins must be mixed with additives during processing. These additives facilitate the subsequent processing (eg, heat stabilizers or lubricants), or are intended to develop desirable properties in the final product (eg, plasticizers or impact modifiers). No fundamental distinction between mixing, compounding, and processing can be drawn. Some conversion processes require several pieces of equipment, whereas the same processes may be carried out on a single unit of alternative design (Fig. 1).

Applications

The main application areas for suspension and mass PVC homopolymers are given in Table 1. Copolymers are used in film, foil, and records. Solution copolymers may be found in inks, adhesives, varnish, fibers, sealants, and cast film. CPVC is used in pipes, valves, tanks, and high specification applications.

Additives

Typical compositions based on vinyl chloride polymers contain a number of additives to provide the desired service performance, processing characteristics, and cost. The choice of additives is governed by their function and cost; other factors such as government regulations, quality, and availability of supply may also need to be considered.

The principal additive systems are shown in Table 2. Many additives may be regarded as multifunctional, for example, a stearate-coated calcium carbonate filler affects rheology, stability, and mechanical properties of the product; however, the primary function is usually obvious.

BIBLIOGRAPHY

R. H. Burgess, ed., *Manufacturing and Processing of PVC,* Applied Science Publishers, London, 1982.

G. Butters, ed., *Particulate Nature of PVC,* Applied Science Publishers, London, 1982.

W. V. Titow, ed., *PVC Technology,* 4th ed., Elsevier Applied Science Publishers, Ltd., Barking, UK, 1984.

I. Luis Gomez, *Engineering with Rigid PVC,* Marcel Dekker, Inc., New York, 1984.

R. C. Stephenson
European Vinyls Corporation

EMULSION POLYMERIZATION AND PLASTISOLS

The commercial scale emulsion polymerization of vinyl chloride normally involves the polymerization of monomer in an aqueous medium containing surfactant and water-soluble free-radical initiators, producing colloidal dispersions of 0.1–3.0 μm spherical particles called latices.

Commercial PVC latices are also produced by a nonemulsion polymerization route using monomer-soluble initiators. This process is usually referred to as microsuspension or sometimes as miniemulsion.

Most PVC homopolymer latices are spray-dried and milled to form fine powders which are used in plastisols or pastes for plasticized products such as floor and wall coverings, artificial leather or leathercloth, soft molded products, and dipped items such as protective gloves. Some is also used for blending with suspension PVC for extrusion applications to assist extrusion

Table 2. Main Additive Classes Used in Suspension PVC Compositions[a]

Unplasticized (UPVC)		
Processing enhancers	Property modifiers	Plasticized (PPVC)
lubricants	impact modifiers	primary and secondary plasticizers
(internal, external)	fillers	flame retardants
process aids	miscible blends	pigments
	gases (foam)	gases (foam)
	smoke retardants	fungicides
	pigments	
	ultraviolet stabilizers	
	antistatic agents	
	reinforcements	

[a] In both UPVC and PPVC, the polymer is combined with a heat stabilizer.

or improve the surface finish of the product. Improvements to suspension polymers over the last few years to increase extrusion rate and other properties have, however, reduced the need for extrusion emulsion grades. Poly(vinyl chloride) emulsion polymers have only limited application in nonplasticized forms but are often used in the manufacture of battery separator plates where the dry polymer is fused into porous sintered plaques. Similarly, little if any homopolymer latex is sold directly in latex form. Latex applications usually take advantage of the relatively high glass-transition temperature (84°C) of PVC and the controlled colloidal nature of the particles to act as slip and antiblocking additives in coatings and inks. A variety of copolymers of vinyl chloride are produced by emulsion, but again these tend to be for specialized applications at low tonnages. In the case of the plastisol grades, vinyl acetate is probably the only comonomer in use and these copolymers constitute only a very small proportion of the plastisol grades offered. Although the vinyl acetate reduces the gelation temperature of the pastes, it causes marked increases in viscosity during the processing lifetime of the plastisol. A number of vinyl chloride–acrylate and vinyl chloride–ethylene or propylene copolymer latices are produced for specialty coatings and adhesives (see COATINGS; ADHESIVE COMPOSITIONS). Similarly, vinyl chloride is used as the minor comonomer in other copolymer latices and products, eg, vinylidene chloride–vinyl chloride copolymers for packaging films, moisture barrier coatings, and paints (see PACKAGING MATERIALS; BARRIER POLYMERS; VINYLIDENE CHLORIDE POLYMERS).

Emulsion Polymerization Processes

There are several polymerization processes for commercial production of poly(vinyl chloride) latices. Not all of these involve true emulsion polymerization, as colloidal PVC particles can also be produced by a special variant of the suspension process. In the normal emulsion polymerization process the latex particle formation, and hence the particle size and distribution, is controlled by a surfactant or emulsifier. A more complex variant of this process is the seeded emulsion polymerization in which a controlled amount of well characterized seed latex is added to the reactor at the start of polymerization. The subsequent latex particle size distribution is a function of the seed size and amount, together with the amount and rate of addition of the surfactant. A third process is the microsuspension polymerization of vinyl chloride using monomer soluble initiators. In this case, polymerization takes place within the monomer droplets in a manner similar to the normal suspension process, but a coarse emulsion of the monomer containing the initiator and surfactant is broken down into fine droplets by passing through a mechanical homogenizer. This produces latices with broad particle size distributions in the range 0.2–3.0 μm. A similar result can be achieved using miniemulsion techniques where fine emulsions are produced under normal low shear agitation conditions. Here the droplet breakdown is achieved mainly by chemical means, using a combination of a synthetic surfactant such as sodium lauryl sulfate and a long chain fatty alcohol. This combination produces a fine emulsion with only moderate agitation which can then be polymerized as above using a monomer-soluble initiator. Polymerization of these fine emulsions can also be achieved by using conventional water-soluble initiators because of the high surface area of the fine droplet emulsion. This takes place by the "stinging" or initiation of the droplets from the water phase by the water-soluble initiator or waterborne oligomeric radicals, resulting in the polymerization of the droplets, as well as normal aqueous phase initiation to produce a wide particle size distribution.

The following polymerization techniques are used for these processes: in batch polymerizations all of the monomer is charged initially; in semibatch or semicontinuous polymerization the monomer is added in stages or continuously until a high reactor charge is achieved and then allowed to finish; in continuous polymerization monomer and other components are added continuously and latex is withdrawn from the reactor at a rate such that the volume of the reactor is replaced several times during the course of many hours or days.

Each of these processes and operating procedures has certain advantages and disadvantages. All are in current use in order to achieve the wide range of rheological properties required of PVC plastisols. The normal emulsion polymerization process in the presence of surfactants is the simplest and easiest, but tends to produce small latex particle sizes which, after drying and milling, produce plastisols with high viscosities at low shear rates and pseudoplastic rheology. This is an advantage in certain dipping and screen-printing applications but at the lower plasticizer concentrations required for other applications the resulting pastes are too difficult to handle. Apart from size adjustments in the 0.05–0.35 μm range it is difficult to produce larger particle sizes by this technique. Larger, 0.5–3 μm, particles and low viscosity at low shear rates can readily be obtained from polymers produced by the microsuspension process but these plastisols are dilatant unless blended with a smaller particle size fraction. Some of this dilatancy can be overcome in the normal formulation process when mixed with inorganic pigments and filler particles. The resulting plastisols have certain advantages because of the broad particle size distribution of the polymer and the lack of polar initiator end groups. This can lead to improved plasticizer "dry-up" and better gelation characteristics during the early stages of gelation and good release from heated gelling drums compared to seeded products. Seeded polymers, however, are capable of producing plastisols with Newtonian rheology and low viscosity even at low plasticizer concentrations, but the polymerization process is more complex on the commercial scale and requires careful control of a wide range of variables.

Modern plants are highly automated with computer control of valves and pumps based on feedback from meters, gauges, and other probes. This includes not only the primary polymerization process but also the ancillary operations such as the preparation and storage of additive solutions, monomer recovery, venting, rundown, cleaning, etc. Although the manual addition of small quantities was quicker than the current computerized systems, the automatic control has many advantages including increased safety, better reproducibility, and faster operation with reduced manpower. For safety reasons, it is now common practice to have an automatic system for the injection of inhibitor if temperature or pressure exceeds certain preset values. In the unlikely event of inhibitor failure, pressure is relieved first through a reseatable relief valve and then via a bursting disk. Emissions to the atmosphere have therefore been virtually eliminated, unlike 20 years ago when relief to the atmosphere was often the first step.

Table 1 shows typical stages and cycle times for computerized batch-operated reactors of 30–50 m^3 volume, producing typically 40–45% solids latices with initial filling ratios of 80%. In Table 1, polymerization temperatures of 50–60°C are assumed, with unchilled cooling water of 15–20°C, and a modest use of condensers.

Polymerization Kinetics. The classical Smith-Ewart Case 2 kinetics describes the emulsion polymerization of water-insoluble monomers, such as styrene, but vinyl chloride does not follow this kinetic model because of the increased water solubility of the monomer. Case 2 assumes that the radicals are immobilized within the polymer particles and that mutual termination occurs if more than one radical enters a particle at any one time. Case 2 therefore assumes an average of 0.5 radicals per particle or droplet but other workers found average radical concentrations in active particles of only 10^{-1} to 5×10^{-4} for poly(vinyl chloride). Although Case 2 is most widely described, Smith-Ewart theory also considers the transfer of radicals to and from the aqueous phase and the polymerization site in relation to their rate of mutual termination in these centers. Case 1 describes the situation where the rate at which radicals are transferred out of the particles is higher than that at which they are gained from the water phase (ie, number of radicals per particle $\ll 0.5$). The Case 1 model describes vinyl chloride emulsion polymerization. Theories involving the desorption (and reabsorption) of oligomeric PVC radicals have been proposed, and modified rate equations have been developed. These show that although the rate is not as directly linked to the number of particles as in the case of polystyrene (Case 2), the rate expression for vinyl chloride emulsion polymerization depends on the number of particles to the one-sixth power.

The third Smith-Ewart model (Case 3), where the average number of radicals per particle is greater than one, is used to describe the microsuspension polymerization system. For microsuspension polymerization, the initiator is monomer-soluble; it is dissolved in the small droplets produced by mechanical homogenization. Radicals are therefore generated in pairs within these small droplets by the thermal decomposition of the initiator. Because Smith-Ewart predicts mutual termination of radicals in normal (0.1–0.3 μm) latex particles, it is necessary to consider how this is avoided. In the case of normal emulsion polymerization, a mechanism of rapid desorption has been proposed as described above, but this is not thought to occur in the microsuspension process as this would lead to aqueous phase initiation and the formation of buildup. Therefore, the extent to which the propagating centers are compartmentalized must be considered. A subdivision factor has been proposed in relating particle size and polymerization rate for systems where radicals are acquired either singly or in pairs. This is shown in Figure 1 where the subdivision factor z and parameter a are defined as follows:

$$z = \frac{n}{a/4}$$

where n = average number of radicals per particle; k_i = initiation rate; k_t = termination rate; and

$$a = 4(k_i/2k_t)^{1/2}\, N_A V$$

where N_A = Avogadro's number; V = volume of particle or droplet.

Initiators. Slightly water-soluble organic peroxides, such as methyl ethyl ketone peroxide and 1-hydroperoxy-1′-hydroxydicyclohexyl peroxide, may be used as emulsion polymerization initiators for vinyl chloride but inorganic peroxydisulfate (or persulfate) salts are most widely used in the form of activated (or redox) systems. For microsuspension polymerization the monomer-soluble lauroyl peroxide is mainly used.

Emulsifiers. A wide range of emulsifiers (also called surfactants) are used in the emulsion polymerization of vinyl chloride. Those used during the polymerization are usually anionic in type, but nonionic surfactants are sometimes added later to the latex. Typical emulsifiers used in poly(vinyl chloride) emulsion polymer production are sodium alkylbenzenesulfonates, sodium alkyl sulfates, sodium alkyl sulfonates, sodium dialkyl sulfosuccinates, fatty acid soaps, sodium alkyl ethoxysulfates, alkyl phenol ethoxylates, and fatty acid ethoxylates, ie, esters with polyethylene glycols.

The surfactants used during the polymerization are chosen not merely to initiate particle formation and control the latex particle-size distribution, but for a wide range of other reasons, some of which involve the latex; others are concerned with the effect that the surfactant has on the properties of the final product, including latex particle size and distribution, latex mechanical stability, reactor wall buildup and "pebble," powder flow, plastisol formation and mixing rate, plastisol viscos-

Table 1. Polymerization Stages

Stage	Operation	Time, min
1	reactor closing, prebatch safety and services checks, including additive availability	10–15
2	initial charge	15–30
3	degassing to remove oxygen	15–25
4	monomer charge	10–15
5	heating	15–25
6	injection of initiator and other additives	10–15
7	polymerization	400–500
8	venting	20–30
9	latex rundown	20–30
10	evacuation and sparging to remove vinyl chloride from reactor before opening	20–40
11	cleaning	45–60

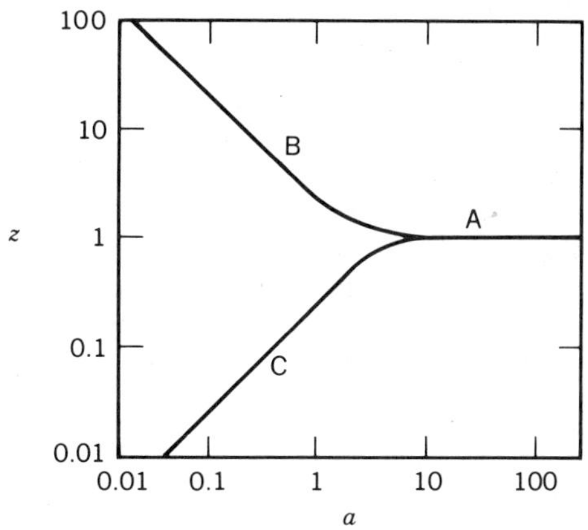

Figure 1. Subdivision factor z as a function of parameter a. Radicals are generated in, or enter, the particles singly (curve AB) or in pairs (curve AC).

ity, plastisol aeration characteristics, heat and color stability, foam-cell structure, water absorption and whitening, leathercloth "fogging," coat adhesion, and gloss and surface finish.

Emulsifier selection is based on a balance of cost, polymerization suitability, and the required end product properties.

Plastisols

Poly(vinyl chloride) plastisols and organosols are dispersions of very fine PVC particles in liquid organic media. In the case of plastisols, high boiling liquids are employed, which attack and plasticize the particles and remain in the final product; these liquids are referred to as plasticizers (qv). The continuous phase in the organosols are lower boiling liquids, which usually do not attack or plasticize the particles and are lost during gelation and fusion. Examples of nonsolvent dispersants include white spirit and aromatic hydrocarbons.

Formulation. Plastisol formulations (Table 2) vary tremendously, depending on the viscosity requirements and the application. Phthalate esters, such as dioctyl phthalate (DOP), are the most widely used plasticizers (qv).

Mixing. Originally a wide range of mixers were used for PVC plastisol preparation, but over the years these have been reduced to three types: low-speed planetary mixers, high speed mixers or dissolvers, and horizontal turbomixers.

Viscosity and Rheology. The viscosity of PVC pastes is rarely, if ever, truly Newtonian and displays instability with time due to interactions between the plasticizer and the polymer. It is therefore important to assess the viscosity of plastisols over a wide range of shear rates and to carry out these measurements at carefully controlled periods of time after mixing in order to obtain meaningful comparisons.

The viscosity of plastisols is affected by most aspects of the formulation, in particular by the type and amount of dispersion resin; plasticizer; surfactants, dispersants, and viscosity depressants; filler, pigment, and blending resin; mixing procedure; and temperature.

The dispersion resin controls the overall rheology of the paste, whereas the other ingredients influence the magnitude of the viscosity. High polymer and filler loadings increase viscosities and exaggerate inherent rheological trends in the system. Similarly, reducing the amount of plasticizer or changing to a more active type produces a higher viscosity and accelerates viscosity aging because of the increased volume fraction of the particles and the increased solvation of the polymer.

Processing and Applications. *Gelation and Fusion.* The gelation and fusion characteristics of a plastisol are often critical in end-product applications and can affect the cell structure in foam coats, the "bubble through" resistance of wear layers for sealing foam coats, the residence time required to achieve the correct wall thickness in slush-molded products, and many other effects. These characteristics may be examined with a torque rheometer in order to study the changes in complex viscosity and viscoelastic properties over the temperature range of 23–230°C. Figure 2 shows a typical curve of complex viscosity against temperature. The viscosity drops initially between 25 and 50°C as the sample remains as a dispersion and thins with temperature. Above 50°C the viscosity increases rapidly, particularly between 70 and 90°C. This corresponds to the normally observed "drying-up point" where coatings

Table 2. Plastisol Ingredients

Additive	phr[a]
dispersion (paste) polymer	60–100
extender (filler) polymer	40–0
plasticizer	40–110
heat and light stabilizers	1–5
inorganic filler	0–100
pigment	1–5
volatile diluent	0–10
other additives	0–5

[a] Parts per hundred resin.

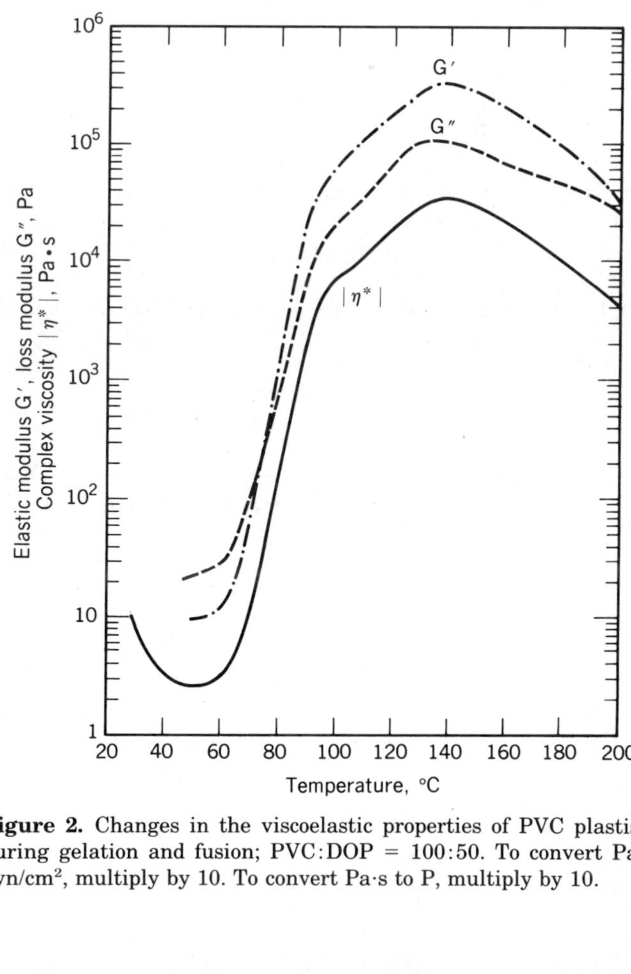

Figure 2. Changes in the viscoelastic properties of PVC plastisols during gelation and fusion; PVC:DOP = 100:50. To convert Pa to dyn/cm^2, multiply by 10. To convert Pa·s to P, multiply by 10.

become a dry, semigelled mass, the plasticizer has diffused into the outer shell of the particles, and the film has lost its tackiness. This process continues during the latter stage of gelation and the polymer chains become progressively more entangled as interparticle boundaries disappear. The viscoelastic properties reach a maximum between 110 and 160°C, where the gelation stage is regarded as complete. The rate of increase in the viscoelastic properties in the range of 70–90°C and the maximum value attained is influenced by the polymer molecular weight and type, ie, homopolymer or copolymer, microsuspension or seeded. This is not always readily apparent from the viscosity curve alone because of the rapid increase over several orders of magnitude. A more detailed study of the elastic modulus (G'), the loss modulus (G''), and tan γ reveals, as expected, that copolymers and low molecular weight products gel more rapidly than their higher molecular weight or homopolymer counterparts. This information is important in terms of the processing operation, eg, the sticking to heated rolls, and of the end-product properties.

After reaching a maximum the melt viscosity begins to decline as small crystallites melt, particle boundaries disappear at 160°C, and full tensile properties are developed at 190–200°C when fusion is complete.

Plastisols are used in coatings to produce vinyl floor and wall coverings, leathercloth, conveyor belting, and tarpaulins. Other applications include dipping to produce industrial and surgical gloves, molding or casting to produce hollow articles such as balls and dolls, and foaming to produce a variety of foams. Foams are produced using a wide range of techniques including mechanical whipping, chemical foaming using the thermal decomposition of azodicarbonamide, and the direct injection of gases. See also BLOWING AGENTS; CELLULAR MATERIALS.

BIBLIOGRAPHY

D. C. Blackley, *Emulsion Polymerization: Theory and Practice,* Applied Science Publishers Ltd., London, 1975.

R. H. Burgess ed., *Manufacture and Processing of PVC,* Applied Science Publishers Ltd., London, 1982.

L. I. Nass and C. A. Heiberger, eds., *Encyclopedia of PVC,* 2nd ed., Vol. 1, Marcel Dekker, Inc., New York, 1986, p. 149.

H. A. Sarvetnick, *Plastisols and Organosols,* Van Nostrand Reinhold, New York, 1972.

M. J. BUNTEN
European Vinyl Corporation (UK) Ltd.

ECONOMIC ASPECTS

Worldwide poly(vinyl chloride) (PVC) consumption has seen a steady growth since World War II, with two major exceptions. The first significant downturn occurred in the mid-1970s as a result of the earlier dramatic rise in oil prices and an economic recession. Consumption rose again until the early 1980s when a further drop in PVC consumption occurred as a result of a second economic downturn. The situation was made worse by the large number of new PVC plants brought on-line at the same time. During 1983 and 1984, the market recovered and since then demand has continued to rise dramatically (Fig. 1). The discovery in the late 1960s that exposure to high concentrations of vinyl chloride monomer (VCM) was harmful resulted in major changes in production methods and product quality, but did not greatly affect output, because the industry was able to adapt.

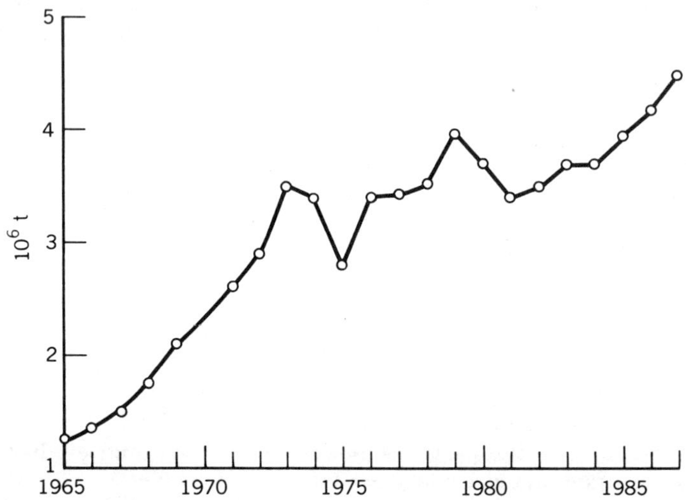

Figure 1. PVC consumption in Western Europe.

World production statistics are given in Table 1, along with a breakdown of resin type for the United States and Western Europe.

The principal applications for PVC in the United States, Europe, and Japan are given in Tables 2 and 3.

P. V. SMALLWOOD
ICI Chemicals and Polymers Ltd.

Table 1. World PVC Production, 10^3 t

Region	Resin type	1970	1975	1980	1985	1987
World		*6,100*	*8,200*	*11,500*	*14,300*	*16,400*
United States	suspension[a]	1,010	1,300	2,090	2,670	3,275
	copolymer[b]	235	180	200	210	170
	paste	165	175	190	190	195
	Total	*1,410*	*1,655*	*2,480*	*3,070*	*3,640*
Western Europe	suspension[a]			3,330	3,700	4,020
	copolymer[b]			130	125	150
	paste			420	495	550
	Total	*2,555*	*3,100*	*3,880*	*4,320*	*4,720*
Japan	*Total*		*1,105*	*1,415*	*1,550*	*1,660*

[a] Suspension and mass resin.
[b] Vinyl acetate copolymers.

Table 2. PVC Consumption by Application, 1986

Application[a]	United States[b], %	Western Europe[c], %	Japan[d], %
rigid			
pipes and fittings	47.2	27.0	28
film and sheet	3.1	11.4	15
bottles	2.6	8.8	
records	1.0	1.0	
profiles	10.0	14	
injection molding	3.4		
others	5.1	1.8	10
Total	*72.4*	*64*	*53*
flexible			
film and sheet	5.2	8.4	14
synthetic leather	2.5		
flooring	2.5	5.6	
cables	6.4	8.8	14
tubes and profiles		3.5	
coating		4.2	
supported sheet			5
others	11.0	5.9	
wire cable and others			14
Total	*27.6*	*36.4*	*47*

[a] The definition of each application varies between regions, which must be considered when comparing data.
[b] Total consumption: 3.2×10^6 t.
[c] Total consumption: 4.3×10^6 t.
[d] Total consumption: 1.4×10^6 t.

Table 3. PVC Consumption by Business Sector, 1986

Business sector[a]	United States, %	Western Europe, %
building and construction	65.6	50.0
packaging	8.8	18.0
electrical	7.4	10.0
home furnishings	4.8	3.0
consumer goods	3.5	
transportation	2.1	3.0
leisure		4.0
clothing and footware		2.0
domestic applications		1.0
other	7.8	9.0

[a] The definition of each business sector can vary between the regions, which must be considered when comparing data.

TEST METHODS AND SPECIFICATIONS

The quality of a poly(vinyl chloride) (PVC) resin is important to both the producer and user. Over the years many test methods have been developed to allow the determination of PVC properties and its behavior in a variety of applications. These tests have been developed by individual PVC producers and users to suit their own requirements. In order to make these tests universally applicable, national and international standards organizations have defined techniques for measuring resin properties and for assigning them to particular categories. In the United States, the American Society for Testing and Materials (ASTM) is the relevant standards authority, and, because of the country's economic importance; it has become the accepted standard for many other countries. The ASTM, along with other national standards authorities, is a member of the International Organization for Standardization (ISO), which has defined tests and standards that are accepted throughout the world.

Various schemes exist for classifying PVC resins: ASTM D1755 defines a system for classifying general-purpose (mass and suspension) and dispersion (paste) resins, and ASTM D2474 defines a scheme for classifying some vinyl chloride copolymers.

Molecular Weight

The molecular weight of PVC is determined by the usual methods. Osmometry (qv) is used to determine the number-average, and light-scattering, the weight-average molecular weight. Gel-permeation chromatography (gpc) gives a complete molecular weight distribution, although the method is not absolute. Because the molecular weight distribution of commercial PVC is not sensitive to manufacturing conditions, it is not measured routinely.

The most common method of measuring molecular weight of PVC is by the determination of its solution viscosity. The results are expressed in different ways. In the ASTM test method D1243 the inherent viscosity is quoted for a 0.2% solution of

PVC in cyclohexanone at 30°C. In Japan, the specific viscosity (JIS K6721) is used. In Europe the K value, a number calculated from the relative viscosity, is widely used.

Particle Size

The techniques used to measure particle size depend on the particle size of the resin. Mass and suspension PVC resins with particles 50–200 μm in diameter can be measured using sieving techniques as described in ASTM D1921 and D1705 and ISO 4610 and 1624. Sieving is time consuming, and more modern automatic techniques are available, such as laser diffraction techniques. A widely used method is based on the resistive pulse technique or Coulter counter principle. It gives a good measure of mean size and size spread, but sieving gives a more accurate measure of the fine- and coarse-particle content of a sample.

The determination of the particle size of paste resins is more difficult, because these particles are smaller. The size of spray-dried and milled agglomerates can be determined by wet sieving down to a particle diameter of about 30 μm. Coulter counters can measure particles down to about 5 μm diameter and laser diffraction particles of 1 μm diameter.

Apparent Density

The bulk or apparent density of mass or suspension resins is the weight of a given volume of resin in the unconsolidated state. The measurement is carried out by allowing the powder to flow through a funnel into a metal cup of known weight and a certain volume (eg, 100 mL). Excess powder is scraped off the cup with a straightedge and the cup is reweighed. Details of the test method are given in ASTM D1895-69 and ISO 60-1977.

Grain Porosity

The standard method of measuring the porosity of a mass or suspension resin is by mercury intrusion porosimetry, using an instrument such as the Carlo Erba mercury intrusion porosimeter. A more rapid method, often used for quality control testing, is the cold plasticizer absorption (CPA) method. It involves determining the weight of plasticizer that fills the grain pores. Numerically the results are similar to those given by mercury porosimetry.

Torque Rheometer Tests for Fusion and Dry-up Times

The processing characteristics of poly(vinyl chloride) resins depend on several factors that cannot be completely defined. Therefore, the industry has developed a number of tests that simulate some aspects of processing on a small scale with the aim of assessing the resin. The most common type of test is based on the use of a torque rheometer, which provides a method of mixing the material under a predetermined set of conditions, while measuring the torque needed. The torque values may be related to the material's viscosity, and transitions in torque values are used to identify fusion, gelation, powder mixing characteristics, and dynamic heat stability.

Dispersion Testing

To determine the dispersion faults in a polymer sample, the resin is mixed into a typical formulation and processed, for example, by milling or extrusion. The processed polymer is examined to determine the number of faults or fisheyes present.

Contamination

Methods of identifying contamination depend on the type, but a commonly used method is described in ISO 1265. A resin specimen is flattened under a glass plate with a grid of 30 mm × 30 mm squares etched onto its surface. The number of contaminating particles in a given area can then be counted.

Color and Heat Stability

The heat stability is often determined by the dehydrochlorination test in which the rate of HCl evolution from a sample of heated PVC (eg, at 170°C) is measured. However, it is often more appropriate to formulate and process the PVC. The color of the processed polymer can be measured before it is given further heat treatment and its color redetermined.

The Determination of Vinyl Chloride Monomer

The techniques for measuring residual vinyl chloride monomer (VCM) in PVC are relatively simple, although they are continually being improved to measure smaller and smaller amounts. The PVC sample (raw resin or a processed sample) is dissolved in a solvent in a sealed vial and allowed to equilibrate under standard conditions of temperature and pressure. The VCM gas in the head space is determined by gas chromatography and the result is used to calculate the VCM content of the original sample. The accuracy of the method can be improved by sparging the VCM from the solution and trapping the gas in cold ethanol as in ASTM D4443-84.

Viscosity of PVC Plastisols

The plastisol viscosity of paste resins can be used to predict their behavior in coating and spreading operations. Testing at very low (0–10 s^{-1}) shear rates shows how a resin might behave during dip coating or low speed spreading. Testing at higher shear rates (100–10,000 s^{-1}) shows how the resin might behave in high speed coating or spraying applications. Viscosities at low shear rates are determined using a Brookfield viscometer. At intermediate shear rates a coaxial instrument such as a Haake Rotovisco can be used. At higher shear rates a Severs viscometer is used.

BIBLIOGRAPHY

L. I. Nass and C. A. Heiberger, eds., *Encyclopedia of PVC,* Vol. 1, 2nd ed., Marcel Dekker, Inc., New York, 1986.

G. Butters, ed., *Particulate Nature of PVC,* Applied Science Publishers, Barking, UK, 1982.

P. V. SMALLWOOD
ICI Chemicals and Polymers Ltd.

HEALTH, SAFETY, AND ENVIRONMENTAL ASPECTS

Vinyl Chloride Monomer

Vinyl chloride is a gas at room temperature with a boiling point of −13.4°C. Industrially, it is usually handled as a liquid under pressure; thus the normal precautions need to be taken when handling a material of this type. However, industrial users have long been aware of other problems associated with the monomer (VCM). The gas is explosive, forming an explosive mixture at 3.6–25% v/v VCM/air; 12% oxygen is required for ignition. It also has a narcotic effect at concentrations of 8–12% and can cause death at higher concentrations.

However in the mid 1960s evidence became available that VCM might cause problems below its narcotic limit. In December 1973 the first correlation between exposure to VCM and cancer in humans was made when doctors discovered that three PVC workers had died from the very rare angiosarcoma of the liver (ASL).

The knowledge that VCM is a human carcinogen has led to the introduction of stringent controls by governmental and regulatory authorities to limit the exposure of the work force and the general public to the monomer.

Atmospheric Pollution. In the United States OSHA sets the permitted concentration of VCM in the atmosphere in a VCM or PVC plant. Before 1973 the threshold limit value (TLV) was 500 ppm, but this was reduced to 50 ppm after the first evidence of carcinogenicity in VCM was obtained. This figure has been progressively reduced; the present regulations require that an employee's exposure to VCM not exceed an 8-h TWA of 1 ppm VCM in any 8-h shift of a 40-h working week. During any shift, an employee's exposure must not exceed 5 ppm over a period of 15 min or less.

Vinyl Chloride in PVC Articles. The most stringent regulations apply to PVC intended for use in food-packaging applications. In 1975 the FDA proposed to ban the use of rigid and semirigid but not flexible PVC in food-contact applications, although it gave a "prior sanction" for the use of rigid PVC in certain applications. In 1986 the FDA suggested withdrawing the proposal to ban food-contact use of PVC, provided the residual VCM in plasticized PVC articles was ≤ 5 ppb and in rigid PVC articles ≤ 10 ppb. To meet these limits, the residual monomer in the unprocessed PVC should be at similar levels, because at these low levels little VSM is lost during processing.

Poly(vinyl chloride) in Fires

Like all carbon-containing materials PVC is combustible, but because of its high chlorine content it is inherently fire resistant. Thus, although it burns when touched by a strong flame, it stops burning when the flame is removed and thus the fire does not spread. Additives such as plasticizers may reduce this flame retardancy, although this can be compensated for by the addition of flame retardants.

The two gases that contribute to the toxicity of PVC in fires are carbon monoxide and hydrogen chloride. Carbon dioxide and smaller quantities of other gases are also produced. No evidence has been found for the production of phosgene. Chlorine and VCM are not formed.

Thus, although use of PVC is criticized along with many other plastics because of perceived dangers in fires, PVC is in fact less of a hazard than many natural materials and other plastics.

PVC Dust

Dust can present an explosion hazard. However, high energies are required to ignite PVC [1500–10,000 J (350–2300 cal) for 20–100-μm size particles]; thus it does not present an explosion hazard.

Disposal of PVC Waste

Concern has been expressed in some parts of the United States and Europe about the environmental effects of the disposal of PVC in municipal waste. This concern has led to restrictions in the use of some PVC articles, especially PVC packaging, in some countries. The major concern is the toxic effects of gases emitted during the incineration of municipal waste.

One problem is the emission of HCl and the fear it will contribute to acid rain. However, the contribution of HCl derived from PVC to acid rain is significantly less than 1%. Furthermore, the gas can and should be easily removed from incinerator effluent gas streams by scrubbers. The other fear is the emission of dioxins from incinerators burning PVC. The feed to incinerators contains organic material and chlorides from a variety of sources and they have been shown to emit very small quantities of dioxins whether or not PVC is present. The amount of dioxins emitted can be significantly reduced by operating the incinerator at the correct temperature.

Recycling of PVC

Because PVC materials are thermoplastic, they can in principle be recycled. A number of schemes have been initiated to collect plastics waste and to reprocess it into useful articles. The most advantageous situation occurs when a source of a single material type can be identified, for example, scrap from plastics converters or particular types of packaging, eg, bottles, agricultural film, and window profiles. Such materials can often be simply reformulated and converted into high quality products based on reduced cost feedstock.

The situation is more complicated when the waste contains many different materials. If an initial sorting is not possible, separation by washing, grinding, flotation, etc, may be attempted or processes that can handle mixed material feedstocks must be developed. Potential applications include fencing, posts, etc.

Providing the essential steps to protect the environment are taken, the burning of plastics is an attractive use of plastics waste. The energy stored in plastics can be recovered by using them as fuel in incinerators (see DISPOSAL AND REUSE OF PLASTICS).

Other Problems

Other environmental problems associated with PVC often involve additives used in processing. Some heat stabilizers are toxic (eg, lead and cadmium salts), and are being replaced by nontoxic alternatives, eg, calcium and zinc.

Under certain circumstances low molecular weight plasticizers can diffuse out of plasticized PVC articles. This is of particular concern with food packaging, and although there is no evidence these plasticizers are toxic, systems using polymeric plasticizers that have a lower rate of diffusion have been developed as a precaution.

BIBLIOGRAPHY

R. D. Jones, D. M. Smith, and P. G. Thomas, *Scand. J. Work Environ. Health* **14**, 153 (1988).

M. Hirschler, *Fire Prevention* **204,** 19 (1987).
New York State Energy Research and Development Authority, *Plast. Technol.* **33,** 120 (1987).

P. V. SMALLWOOD
ICI Chemicals and Polymers Ltd.

VINYLCYCLOALKANE POLYMERS

There was very little interest in the oligomeric polyvinylcycloalkanes until the discovery of the Ziegler-Natta coordination catalysts, which could effect the stereoregular polymerization of α-olefins. High molecular weight polymers have now been prepared from all of the common vinylcycloalkanes by employing a variety of transition metal catalysts. Vinylcyclohexane yields a high melting, crystalline polymer that may be useful as an insulator. However, none of the vinylcycloalkane polymers are sold commercially.

Monomers

Physical Properties. Vinylcycloalkanes exhibit physical properties typical of hydrocarbons, ie, they are colorless, flammable liquids with low densities. Purification is accomplished by fractional distillation. The infrared spectra of vinylcyclopropane, vinylcyclopentane, and vinylcyclohexane, along with several related cycloalkane derivatives, have been summarized. Proton nmr studies have been employed to determine the conformation of the vinyl substituent relative to the ring; the *s*-trans conformer is the most stable.

Chemical Properties. Vinylcycloalkanes undergo the normal electrophilic reactions of olefins. The most significant reaction of vinylcycloalkanes derived from stable rings is the isomerization of the double bond to ethylidene and 1,3- or 1,4-ethylcycloalkenes.

Monomer Synthesis. Most of the common techniques for synthesizing vinyl compounds have been applied to the preparation of vinylcycloalkanes. Pyrolysis techniques are favored because isomerization of the product mixture can be minimized. Dehydrohalogenations in the presence of strong bases yield ethylidene derivatives as the major products. The synthesis of vinylcyclopropane requires special techniques, because of the reactivity of the cyclopropyl substituent.

Polymerization of Vinylcycloalkanes

Ziegler-Natta Catalysts. Vinylcycloalkanes undergo 1,2 addition polymerization to form isotactic polymers. The olefins also rearrange to various internal olefins on the surface of heterogeneous Ziegler-Natta type catalysts. The polymers are mixtures of low molecular weight amorphous fractions and high molecular weight crystalline fractions. For a given catalyst system, the yield and molecular weight of the crystalline fraction varies, depending on the size of the cycloalkyl substituent (Table 1).

Kinetics. The rate of vinylcyclohexane polymerization in the presence of the $TiCl_3$–$Al(C_2H_5)_3$ catalyst has been studied in the temperature range of 30–70°C, in cyclohexane or *n*-heptane. The polymerization was first order in monomer for monomer concentrations greater than 0.8 *M* but in dilute solutions the reaction appeared to be second order. The initial rate of polymerization exhibited a linear dependence on the $TiCl_3$ concentration; a slight maximum was observed in the rate at Al/Ti = 0.28. The intrinsic viscosity of the polymer increased with reaction time and monomer concentration and was essentially independent of the $Al(C_2H_5)_3$ concentration. The Arrhenius activation energy was calculated to be 21.8 kJ/mol (5.2 kcal/mol).

Cationic Polymerization. The polymerization of vinylcycloalkanes can be initiated by Friedel-Crafts catalysts such as $AlBr_3$ or $AlCl_3$. The molecular weight of the polymers increases when the polymerization temperature is lowered to −100°C, but no polymers with molecular weights greater than 10,500 have been isolated. The vinylcycloalkanes can be considered analogues of 3-methyl-1-butene, which isomerizes before propagation to yield 1,3-enchainment (see ISOMERIZATION POLYMERIZATION). Indeed, the polymerizations of vinylcyclohexane and vinylcyclopropane are accompanied by hydride transfer reactions, which lead to rearranged polymers with cycloalkyl groups as integral parts of the chain.

The polymerization of vinylcyclopropane may proceed via a nonclassical bicyclobutonium ion, which leads to the incorporation of either olefin groups (**1**) or cyclobutane units (**2**) into

Table 1. Synthesis of Polyvinylcycloalkanes with Heterogeneous Catalysts

Monomer	Catalyst type[a]	Temperature, °C	Time, d	Yield, %	Crystallization, %[b]	$[\eta]$, dL/g[c]	Mp, °C
vinylcyclopropane	A	68	2	35.4	63		>260 dec
vinylcyclobutane	B	75	7	24.5	66		228
vinylcyclopentane	B	68	5	13.7	87.5	0.56	>260 dec
	C	40	0.5		100		292
vinylcyclohexane	A	68	1.5	46.5	93.5		>300
	C	40	0.5		93.2	5.02[d]	383
	D	50	2	15.1	80.1	1.9	385
4-methylvinylcyclohexane	B	80	1.5	30.0		0.45	225–250
3-methylvinylcyclohexane	B	70	1.5	42.5	30.0	0.74	276–355
vinylcycloheptane	B	72	7	9.8	58.8	0.39	>300

[a] Solvent, heptane; A, freshly ground $TiCl_3$, $Al(i\text{-}C_4H_9)_3$ (mole ratio Ti/Al = ¼); B, $TiCl_4$, $Al(i\text{-}C_4H_9)_3$ (Ti/Al = ½); C, "*AA*" $TiCl_3$, $Al(C_2H_5)_2Cl$ (Ti/Al = ¼); D, $TiCl_4$ $Al(C_2H_5)_3$ (Ti/Al = ⅓).
[b] Insoluble fraction remaining after continuous extraction with ether for at least 24 h.
[c] Measured in benzene unless otherwise noted.
[d] Toluene.

the polymer backbone. Vinylcyclopropane yields a copolymer containing 60% of unrearranged repeat units, 10% of structure (**1**), and 40% of structure (**2**).

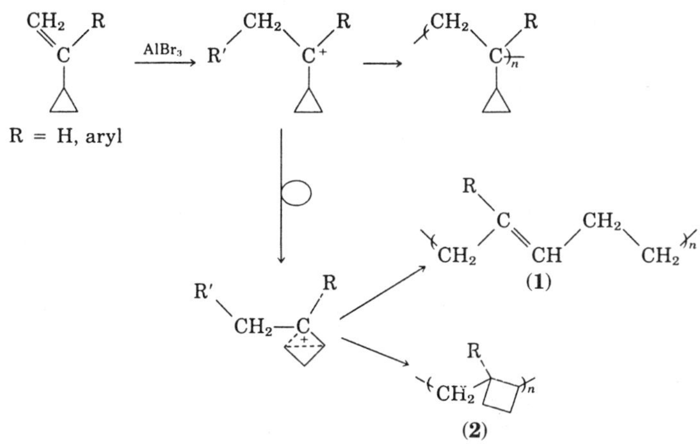

Vinylcyclopropane Derivatives. There are a few exceptions to the general rule that vinylcycloalkanes do not polymerize in the presence of free-radical initiators. The polymerization of vinylcyclopropane derivatives with polar substituents in the ring can be initiated by benzoyl peroxide, AIBN, uv-irradiation, or cumene hydroperoxide in an aqueous emulsion. Two geminal substituents promote cyclopropane ring opening and high molecular weight polymers can be obtained in high conversions.

Copolymerization. Detailed investigation of the copolymerization ability of vinylcycloalkanes has been focused on Ziegler-Natta systems. The vinylcycloalkanes copolymerize with 1-butene, 3-methyl-1-butene, and 4-methyl-1-pentene to yield crystalline copolymers. The most effective catalyst composition includes $Al(C_2H_5)_2Cl$ and $TiCl_3$(3:1). Since the vinylcycloalkanes are much less reactive in copolymerization than the listed comonomers, incremental addition of the comonomers is required to obtain copolymers with uniform sequence distributions. Copolymers containing at least 66% vinylcyclohexane have exhibited melting points comparable with nylon or polyester fibers. Thus, the excellent mechanical properties of the copolymers are achieved without sacrificing the high melting characteristics of the corresponding homopolymers.

Ziegler-Natta systems have been employed in the preparation of copolymers of vinylcyclohexane with ethylene, propylene, and styrene. The ethylene–vinylcyclohexane copolymers were useful in reinforcing paraffin coatings on paper, cardboard, fabrics, or foods.

Free-radical initiators effect the copolymerization of vinylcyclohexane with acrylonitrile and vinyl chloride.

Physical Properties of Polymers

The most outstanding property of isotactic polyvinylcycloalkanes is their high crystallinity. Polymers derived from nonbranched acyclic α-olefins tend to be rubbery noncrystalline materials if the monomer contains more than five carbon atoms. Vinylcycloalkanes fall into a general class of branched α-olefins, which exhibit a high propensity to crystallize even when incorporated into copolymers.

The dynamic mechanical properties of polyvinylcyclohexane at acoustic frequencies have been determined in conjunction with a study of polymers containing cycloalkyl groups. All polymers except polyvinylcyclohexane exhibited a γ-relaxation connected with the internal motion of the cycloalkyl substituent. The activation parameters of this relaxation depended exclusively on the nature of the cycloalkyl ring. However, polyvinylcyclohexane did exhibit a secondary relaxation effect (δ-process) at lower temperatures.

The dissipation factor, tan δ, and specific volume electrical resistance have been determined for polyvinylcyclohexane in the temperature range from −180°C to +160°C and from 50 to 10^6 Hz. The dielectric properties of polyvinylcyclohexane were superior to those of polystyrene or polytetrafluoroethylene; the dielectric permeability coefficient was −0.2 in the glass phase and increased to −0.6 in the more elastomeric phase (at ca 100°C). The dielectric coefficient combined with the high melting point of the polymer make it ideal for insulator applications.

Polyvinylcycloalkanes are generally soluble in aromatic solvents, decalin, tetralin, and chlorinated ethylenes at high temperatures. The intrinsic viscosity–molecular weight relationship for atactic polyvinylcyclohexane, which was prepared by hydrogenation of polystyrene, has been determined in tetrahydrofuran. Assuming that $\alpha = 0.5$ in theta solvents, the value for K_θ was calculated to be 0.0788 mL/g.

Chemical Properties of Polyvinylcycloalkanes

The alkanes are practically inert to chemical modification unless the cycloalkyl substituent is a strained ring.

Polyvinylcyclopropane can be modified by treatment with hydrogen bromide; addition across the cyclopropyl ring yielding poly(3-bromo-1-pentene) occurs to only 85% of the theoretical ring openings and the crystallinity of the polymer is destroyed.

Safety Precautions

The normal precautions for handling toxic, flammable hydrocarbons should be observed. The LD_{50} value of vinylcyclohexane vapor for mice is 13.7 mg/L. Daily 4-h exposures to concentrations averaging 0.42 mg/L produce pathological effects. The maximum vapor concentration recommended is 0.01 mg/L.

BIBLIOGRAPHY

C. G. Overberger, A. E. Borchert, and A. Katchman, *J. Polym. Sci.* **44,** 491 (1960).

V. I. Kleiner, B. A. Krentsel, and L. L. Stotskaya, *Eur. Polym. J.* **7,** 1677 (1971).

G. B. Kharas, Y. V. Kissin, V. I. Kleiner, B. A. Krentsel, L. L. Stotskaya, and R. Z. Zakharayan, *Eur. Polym. J.* **9,** 315 (1973).

William H. Daly
Louisiana State University

VINYL DISPERSIONS.
See Vinyl chloride polymers.

VINYLENE CARBONATE POLYMERS.
See Polycarbonates.

VINYL ESTER POLYMERS

VINYL ACETATE POLYMERS

Vinyl acetate (VA), $CH_2{=}CH\overset{\overset{\displaystyle O}{\|}}{O}CCH_3$, is a colorless, flammable liquid with an initially pleasant odor that quickly becomes sharp and irritating. Its chief use is as a monomer for making poly(vinyl acetate) (PVAC) and vinyl acetate copolymers, which are widely used in water-based paints, adhesives, paper coatings, or nonwoven binders, and applications not requiring service at extreme temperatures. Poly(vinyl acetate) is the precursor for poly(vinyl alcohol) (PVA) and poly(vinyl acetal) resins (see VINYL ALCOHOL POLYMERS; VINYL ACETAL POLYMERS). Vinyl acetate is also copolymerized as the minor constituent with vinyl chloride or ethylene to form commercial polymers and with acrylonitrile to form acrylic fibers (qv).

Physical Properties. Physical properties of vinyl acetate are shown in Table 1. The values of vapor pressure, heat of vaporization, vapor heat capacity, liquid heat capacity, liquid density, vapor viscosity, liquid viscosity, surface tension, vapor thermal conductivity, and liquid thermal conductivity over temperature ranges have been calculated and graphed.

Vinyl acetate is completely miscible with organic liquids but not with water.

Table 1. Physical Properties of Vinyl Acetate

Property	Value
boiling point, °C	72.7
melting point, °C	−100, −93
specific gravity, 20/20	0.9338
refractive index, n_D^{20}	1.3952
viscosity at 20°C, mPa·s (=cP)	0.42
flash point, tag open cup, °C	0.5–0.9
heat of vaporization at 72°C, J/g[a]	379.1
heat of combustion, kJ/g[a]	24.06
heat of polymerization, kJ/mol[a]	89.12
critical temperature, °C	252[c]
critical pressure, kPa[b]	4200
critical density, g/cm³	0.324[c]

[a] To convert J to cal, divide by 4.184.
[b] To convert kPa to psi, multiply by 0.145.
[c] Estimated.

Polymerization can be initiated by organic and inorganic peroxides; azo compounds (qv), redox systems, many of which include organometallic components; light; and high energy radiation. Polymerization is inhibited or strongly retarded by aromatic hydroxyl, nitro, or amino compounds and by oxygen, quinone, crotonaldehyde, copper salts, sulfur, conjugated polyolefins, and enynes.

Vinyl acetate is polymerized in bulk, suspension, solution, and emulsion. It copolymerizes readily with only a few monomers. Reactivity ratios for common comonomers are given in Table 2. The Q (monomer reactivity factor) and e (electronic factor) values are 0.026 and −0.22, respectively.

Table 2. Reactivity Ratios (r_1,r_2) in Vinyl Acetate (M_1) Copolymerizations

Comonomer (M_2)	r_1	r_2	Temperature, °C
acrylic acid	0.1	2	70
acrylonitrile	0.07	6	70
chloroprene	0.01	50	65
diallyl phthalate	0.72	2.0	
diethyl fumarate	0.01	0.44	60
ethylene	1	1	90–150
ethyl vinyl ether	3	0	60
isobutyl methacrylate	0.025	30	60
isopropenyl acetate	1	1	75
maleic anhydride	0.07	0.01	
styrene	0.01	55	60
vinyl chloride	0.6	1.4	40
N-vinylpyrrolidinone	0.2	3.3	50

Chemical Properties. The most important chemical reaction of vinyl acetate is free-radical polymerization.

$$n\ CH_2{=}CHO\overset{\overset{\displaystyle O}{\|}}{C}CH_3 \longrightarrow {-}\!\!\left(CH_2{-}\underset{\underset{\displaystyle O{-}\underset{\underset{\displaystyle O}{\|}}{C}{-}CH_3}{|}}{CH}\right)\!\!_n{-}$$

The scientific, as distinguished from the technological, aspects of the subject have been reviewed. Other chemical reactions of vinyl acetate are those common to an ester and a compound containing a double bond.

Manufacture. The manufacturing process most widely used for vinyl acetate is the oxidative addition of acetic acid to ethylene in the presence of a palladium catalyst; both liquid-phase and vapor-phase processes are used.

$$CH_3\overset{\overset{\displaystyle O}{\|}}{C}OH + H_2C{=}CH_2 + \tfrac{1}{2}\,O_2 \longrightarrow CH_3\overset{\overset{\displaystyle O}{\|}}{C}OCH{=}CH_2 + H_2O$$

Economic Aspects. The 1988 U.S. production of vinyl acetate by the acetic acid–ethylene route is shown in Table 3.

Capacity in Japan in 1988 was ca 524,000 t (90% ethylene-based). Western European overall 1988 capacity was ca 510,000 t (60% ethylene-based). In 1987, 1.1×10^6 t of vinyl acetate was produced in the United States.

Specifications and Standards. Vinyl acetate monomer is supplied in two grades, which differ only in the amount of inhibitor but otherwise have identical specifications. A low hydroquinone grade (3–7 ppm) is preferred if it is expected to be used within two months of delivery. For storage up to four months before polymerization, a grade containing 12–17 ppm hydroquinone is used.

Storage and Shipment. Vinyl acetate is commonly stored in carbon steel tanks. In the wake of recent findings about vinyl chloride, vinyl acetate as well as acrylic esters and other monomers have been under study as carcinogens via animal testing in several laboratories. No definitive conclusions have been reached as yet.

Uses. The only application for vinyl acetate is the production of poly(vinyl acetate).

Health and Safety Factors, Toxicology. Vinyl acetate is only moderately toxic if ingested or absorbed through the skin. It is not a severe irritant on contact with skin or eyes; however,

Table 3. U.S. Production of Vinyl Acetate Monomer by the Acetic Acid–Ethylene Process

Company	Location	Process	Production 10^3 metric tons
Hoechst-Celanese	Bay City, Texas, and Clear Lake, Texas	modified Bayer	486
Union Carbide	Texas City, Texas	modified Bayer	249
E.I. du Pont de Nemours & Co., Inc.	LaPorte, Texas	modified Bayer	238
Quantum Chemical	Strange, Texas	National Distillers	272
Total			*1245*

prolonged or repeated contact with the skin should be avoided. The vapor is not especially irritating to the nose and throat, but good industrial hygiene practices should be followed and adequate ventilation provided. Repeated or prolonged exposure should be avoided.

The flammable limits of vinyl acetate vapors in air are 2.6–13.4 vol %. At normal storage temperatures, an explosive vapor–air mixture can exist in the vapor space of a storage tank, unless nitrogen is used to exclude air. All tanks and lines should be electrically bonded and grounded, and explosion-proof equipment (UL class 1, group D) should be used.

Vinyl acetate requires a red precautionary label for shipment. It should not be stored without an inhibitor or transferred between vessels by means of air pressure. It is shipped in tank cars, tank trucks, and drums.

Analytical and Test Methods. Gas chromatography is an excellent method for simultaneous determination of vinyl acetate and its volatile impurities.

Vinyl Acetate Polymers

Properties. Poly(vinyl acetate)s vary with increasing molecular weight from viscous liquids and low melting solids to hard, brittle polymeric materials. They are neutral, water-white to straw-colored, tasteless, odorless, nontoxic products. The resins have no sharply defined melting points but become softer with increasing temperature. They are soluble in organic solvents, eg, esters, ketones, aromatics, halogenated hydrocarbons, carboxylic acids, etc, but insoluble in the lower alcohols (excluding methanol), glycols, water, and nonpolar liquids such as ether, carbon disulfide, aliphatic hydrocarbons, oils, and fats. Poly(vinyl acetate)s dissolve in alcohols, eg, ethyl, propyl, and butyl, containing 5–10 wt % water. Butyl alcohol or xylene, both of which only swell the polymer at normal temperatures, dissolve it when heated. Physical properties are shown in Table 4.

The chemical properties of PVAC are those of an aliphatic ester. Thus, acidic or basic hydrolysis produces PVA and acetic acid or the acetate of the basic cation. Industrially, poly(vinyl alcohol) is produced by a base-catalyzed ester interchange with methanol, where methyl acetate forms in addition to the polymeric product. The chemical properties of PVAC can be modified by copolymerization. A comonomer containing a carboxylic or sulfuric acid group increases the solubility of the copolymer in dilute aqueous alkali or ammonia. These copolymers adhere better to metals than homopolymers or neutral copolymers because of the interaction between the acid groups and the metal surface.

Many properties of PVAC and various copolymer emulsions are determined by the characteristics of the aqueous phase and its contents.

Table 4. Physical Properties of Poly(vinyl Acetate)

Property	Value
absorption of water at 20°C for 24–144 h, %	3–6
coefficient of thermal expansion, K^{-1}	
cubic	6.7×10^{-4}
linear, below T_g	7×10^{-5}
above T_g	22×10^{-5}
cohesive energy density, $(MJ/m^3)^{1/2}$ [a]	18.6–19.09
compressibility, $cm^3/(g \cdot kPa)$ [b]	17.8×10^{-6}
decomposition temperature, °C	150
density, g/cm^3	
at 20°C	1.191
at 200°C	1.05
dielectric constant at 2 MHz	
at 50°C	3.5
at 150°C	8.3
dielectric dissipation factor at 2 MHz, tan δ	
at 50°C	150
at 120°C	260
dielectric strength, V/L	
at 30°C	0.394
at 60°C	0.307
dynamic mechanical loss peak at 100 Hz, °C	70
elongation at break at 20°C, 0% rh, %	10–20
glass-transition temperature, T_g, °C	28–31
pressure dependence, °C/100 MPa[c]	0.22
hardness at 20°C, Shore units	80–85
heat capacity at 30°C, J/g[a]	1.465
heat distortion point, °C	50
heat of polymerization, kJ/mol[a]	87.5
refraction index, n_D	
at 20.7°C	1.4669
at 142°C	1.4317
interfacial tension at 20°C, mN/m[d]	
with polyethylene	14.5
with polydimethylsiloxane	8.4
with polyisobutylene	9.9
with polystyrene	4.2
modulus of elasticity, GPa[e]	1.275–2.256
notched impact strength J/m[f]	102.4
softening temperature, °C	35–50
surface resistance, Ω/cm	5×10^{11}
surface tension, mN/m[d]	
at 20°C	36.5
at 140°C	28.6
at 180°C	35.9
tensile strength, MPa[c]	29.4–49.0
thermal conductivity, mW/(m·K)	159
Young's modulus, MPa[c]	600

[a] To convert J to cal, divided by 4.184.
[b] To convert kPa to atm, divide by 101.3.
[c] To convert MPa to psi, multiply by 145.
[d] mN/m = dyn/cm.
[e] To convert GPa to psi, multiply by 145,000.
[f] To convert J/m to lbf/in, divide by 53.38.

By the judicious selection of monomer concentration, protective colloid, emulsifying and wetting agents, method of polymerization and posttreatment, the properties can be varied to suit the intended application. Such properties include average particle size and particle-size distribution, molecular weight, pH, emulsion viscosity, particle charge, adhesion, speed of tack, solvent tolerance, film characteristics, water resistance, and stability to storage, freezing, dilution, mechanical action, and compounding. Various resins, plasticizers, thickening agents, solvents, pigments, extenders, and some dyes may be added, but caution is necessary and it is advisable to follow closely manufacturer's directions. Plasticizers are often added to increase flexibility of the dried film, but they are usually not incorporated before the final formulation is made, since their presence may adversely affect stability, particularly in cold weather.

Poly(vinyl acetate) emulsion films adhere well to most surfaces and have good binding capacity for pigments and fillers. Plasticized films are strong and flexible at ordinary temperatures. The films are unaffected by light, oxygen, moderate amounts of chlorine, and dilute solutions of acids, alkalies, and salts. The films are also inert to oils, fats, waxes, and greases, unless these are mostly aromatic. Acetone, alcohol, ethyl acetate, benzene, and toluene dissolve or at least swell the films. Permeability to water vapor allows the film to be laid down on a damp surface with trapped moisture gradually passing through the film without causing lifting or blistering. The permeability to saturated water vapor at 40°C is 2.1 g/(h·m^2) for a film 0.025 mm thick.

Polymerization Processes. Vinyl acetate is polymerized industrially in bulk, solution, suspension, and emulsion. Most of the material identified as poly(vinyl acetate) or copolymers that contain predominantly vinyl acetate are made by emulsion techniques. The patent literature abounds in descriptions of this technology, and recipes and procedures are readily available in the brochures of monomer-producing companies as well as in the literature (see EMULSION POLYMERIZATION).

The initiators or catalysts used in vinyl acetate polymerizations are the familiar free-radical generating types, eg, hydrogen peroxide, peroxysulfates, and various redox combinations.

Buffers are frequently added to emulsion recipes and serve two main purposes. The decomposition rates of some initiators are affected by pH and can be stabilized by the addition of buffer, since catalyst decomposition frequently causes changes in pH. The hydrolysis rate of vinyl acetate and of other monomers is also pH-sensitive, and it is desirable to minimize monomer hydrolysis since it produces acetic acid which can affect the catalyst. Monomer hydrolysis also produces acetaldehyde, which may lower the polymer molecular weight.

A chain-transfer agent is often added to vinyl acetate polymerizations, whether emulsion, suspension, solution, or bulk, to control the molecular weight. Aldehydes, thiols, carbon tetrachloride, and others have been used.

A polymerization may be carried out by charging all ingredients to the reactor, heating to reflux, and stirring until the reaction is over; however, this simple procedure is seldom followed. Typically, only a part of the monomer and catalyst is initially charged; the remainder is added during the course of the reaction. Better control of the polymerization rate can be maintained in this fashion, which is particularly important for industrial-scale operations where large quantities of material are involved and heat-dissipation capacity may be limited.

Mechanistic Aspects. At the molecular level, vinyl acetate polymerizations are free-radical reactions characterized, in particular, by a high degree of chain transfer (qv). The high reactivity of the PVAC growing chain radical is attributed to its low degree of resonance stabilization. The high reactivity of the vinyl acetate radical also contributes to the high rate constant for propagation in vinyl acetate polymerization, compared with styrene, acrylates, and methacrylates. Chain transfer to monomer and other small molecules results in lower molecular weight products. When polymerization occurs in the absence of monomer and other transfer agents, such as solvents, chain transfer becomes more important. As a result, branched polymer chains tend to form toward the end of batch-suspension or emulsion polymerizations. In suspension and emulsion processes where monomer is fed continuously, the products tend to be more branched than in polymerizations conducted in the presence of an abundant supply of monomer. Chain transfer also occurs with emulsifying agents, leading to their incorporation into the polymer chain.

Grafting During Polymerization. The degree of grafting of PVAC or PVA during emulsion polymerization strongly affects latex properties, eg, viscosity, rheology, and solubility. The composition and structure of PVA significantly affect vinyl acetate emulsion polymerization. Thus, the presence of acetaldehyde condensates as a PVA impurity strongly retards polymerization.

Copolymerization. Block copolymers of vinyl acetate with methyl methacrylate, acrylic acid, acrylonitrile, and vinylpyrrolidinone have been prepared by copolymerization in viscous, poor solvents for the vinyl acetate macroradical. The copolymerization of vinyl acetate with methyl methacrylate is enhanced in acetonitrile and acetone solvents and decreased by propanol. Copolymers of vinyl acetate containing cyclic functional groups in the polymer chain have been prepared by copolymerization of vinyl acetate with *N,N*-diallylcyanamide and *N,N*-diallylamine.

Alternating equimolar copolymers of vinyl acetate and ethylene and alternating copolymers of vinyl acetate and acrylonitrile have been reported. Vinyl acetate and vinylidene cyanide form highly alternating copolymers.

Polymer Blends. It has long been common practice to modify latex film properties by blending one polymer latex with another, ie, a "soft" polymer is made slightly less soft and less blocking by a "hard" latex is observed.

Poly(ethyl methacrylate) (PEMA) yields truly compatible blends in solvents with PVAC up to about 20% PEMA concentration. The mechanical properties of these blends are optimal at 20% PEMA in PVAC, with a ratio of molecular weights, PEMA/PVAC, below 4.9/1. The optimal mechanical properties (tensile strength, initial modulus, % elongation at break) were higher than for either component alone, owing to the presence of favorable and strong polymer chain interactions.

Blends of PVAC with cholesterol and certain cholesterol derivatives are compatible; cholesterol at 3–5 wt % has a distinct antiplasticization effect. The T_g as determined by dielectric studies is increased to ca 60°C. The normal segmental motion of PVAC is hindered by cholesterol and cholesterol acetate, oleate, nonanoate, and benzoate; a molecular complex was proposed to explain the observed data.

Fibers formed from a phase-separated blend of acrylonitrile–vinyl acetate copolymer (I) and acrylonitrile–methyl acrylate–Na methylpropenyl sulfonate copolymer (II) can be produced by solvent spinning. A vinyl acetate content of I above 12.5 wt % gives phase-separated fiber blends which are porous and therefore retain water. Fiber blends of II with pure PVAC were completely incompatible.

Poly(ethylene oxide) forms compatible blends with both syndiotactic and atactic poly(methyl methacrylate) and poly(vinyl acetate), which are homogeneous one-phase systems. The T_g of the blends and the PEO crystallizability depend on composition. Dielectric relaxation studies of both blends and core–shell composites of PVAC and poly(butyl acrylate) (PBA) structures indicated a mutual antiplasticizing effect of the polymer chains upon each other. The β-relaxation temperature of each is shifted to higher temperatures by ~10–15 K; no mechanism was proposed to explain this effect.

In addition, copolymers were unexpectedly found in the core–shell mixture identical to those of a core–shell composite sample made by a three-step synthesis. In each case it was proposed that the morphology of the particles was that of an inner PBA core, an intermediate shell of PVA–PBA copolymer, and an outer shell of PVAC.

Economic Aspects. The prices of PVAC polymers depend on the form in which they are sold (resin or emulsion, homopolymer or copolymer), as well as on the specific product. The 1987 emulsion prices were $1.00–2.00/kg dry resin. Prices of copolymer emulsions tend to be higher than those of the homopolymer by 20¢/kg. Large shipments of bead resins and other dry forms are priced at $2.00–3.00/kg.

Emulsions continue to dominate the adhesives and paint markets. The principal suppliers, among numerous others, of poly(vinyl acetate)s and vinyl acetate copolymers are shown in Table 5. Many other producers consume these polymers internally.

Specifications and Standards. Typical specifications of commercial PVAC emulsions are given in Table 6.

Many ASTM tests are used to evaluate materials containing PVAC, such as paints and lacquers. Poly(vinyl acetate) and its copolymers are nontoxic and are approved by the FDA for food-packaging applications under regulations 21 CFR 75.105, 176.170, and 176.180.

Table 6. Emulsion Specifications of Poly(vinyl Acetate)

Property	Range
solids, wt %	48–55
viscosity, mPa·s (= cP)	200–4500
pH	4–6
residual monomer, % max	0.5
particle size, μm	0.1–3.0
particle charge	neutral or negative
density at 25°C, g/cm³	0.92
stability to borax	stable or unstable
mechanical stability	good or excellent

Storage and Shipping. Emulsions are shipped in 19-L pails, 209-L drums, and 11.4–18.9-m³ railroad tank cars. Large shipments are thermally insulated. Precautions should be taken to avoid the formation of foam during unloading. Latices corrode ordinary steel and must be stored in tanks lined with stainless steel, glass, plastic, or coated surfaces such as baked phenolic, epoxy phenolic, or PVC. Storage temperatures should be about 20–25°C; excessive exposure to extreme temperatures may cause phase separation. Containers should be tightly closed to prevent solvent evaporation resulting in skin formation.

BIBLIOGRAPHY

M. K. Lindemann in G. E. Ham, ed., *Vinyl Polymerization,* Vol. 1, Marcel Dekker, Inc., New York, 1967, Part 1, chapt. 4.

M. El-Aasser and J. W. Vanderhoff, eds., *Emulsion Polymerization of Vinyl Acetate,* Applied Science Publishers, Inc., Englewood, N.J., 1981.

WILEY A. DANIELS
Air Products and Chemicals

Table 5. Principal Poly(vinyl Acetate) Producers

Country	Company	Trade names
United States	Air Products	Vinac, Flexbond, Airflex
	Rohm & Haas	
	E.I. du Pont de Nemours & Co., Inc.	Elvacet, Elvax
	W. R. Grace (Dewey and Almy)	Darex, Daratak, Evenflex
	Monsanto	Gelva
	National Starch and Chemical	Resyn
	Reichhold Chemical	Wallpol
	Union Carbide Corp.	Bakelite, Ucar, Co-Mer
	Union Oil of California	
Japan	Kobunshi Chem	
	Nippon Carbide	
	Nippon Gosei	
	Sekisui	
	Showa Denko	
	Sumitomo	Sumikaflex
United Kingdom	Imperial Chemicals Industries, Ltd.	
	Revertex	Emultex
Federal Republic of Germany	Farbwerke Hoechst	Mowilith
	Wacker-Chemie	Vinnapas
Switzerland	Ebnoether	
Canada	Shawinigan	Gelva
France	Rhône-Poulenc	Rhodopas
Italy	Monte-Edison	Vinavil

OTHER VINYL ESTER POLYMERS

Economically, vinyl esters other than vinyl acetate are not important in the United States. Small amounts of vinyl Versatate and vinyl laurate are imported into the United States for special applications, at prices of approximately $2.90/kg and $6.60/kg, respectively. Other vinyl esters, such as vinyl stearate and vinyl myristate, are specialty chemicals which can be bought through organic chemicals catalogs. The world production of vinyl esters other than vinyl acetate was estimated to be <23,000 t in 1987.

Preparation of Higher Vinyl Esters

Industrially, vinyl esters have been prepared by the reaction of acetylene with carboxylic acids, although transvinylation has also been used on a large scale, especially when purification by distillation was difficult. In the laboratory, it is more convenient to prepare vinyl esters by transvinylation than by the reaction of acetylene and carboxylic acids. Other methods have been proposed that are useful in special cases.

Only vinyl propionate and the vinyl esters of Versatic acids (C_9 to C_{19} trialkylacetic acids) are prepared industrially directly from acetylene and the corresponding acid. Vinyl stearate, vinyl myristate, and vinyl laurate are prepared by the ester interchange or transvinylation method.

Purification. The lower boiling vinyl esters can be purified by distillation, usually under reduced pressure. To prevent polymerization, about 200-ppm copper resinate is added as an inhibitor. Vinyl pelargonate, laurate, palmitate, and stearate have also been purified by preparing urea complexes, which form in good yields. These complexes can be purified by recrystallization.

Properties

Physical Properties. A general compilation of boiling points of many vinyl esters and divinyl esters is given in Tables 1, 2, and 3.

Infrared Spectra. Vinyl esters show the characteristic vinyl absorptions at 1650, 945, and 878 cm^{-1}. These bands are not present in the poly(vinyl esters). The very strong carbonyl absorption is often shifted from 1750 cm^{-1} for vinyl esters to 1730 cm^{-1} for poly(vinyl esters).

Nuclear Magnetic Resonance Spectra. The nuclear magnetic resonance spectra of various esters have been published. Nuclear magnetic resonance spectra of poly(vinyl ester)s have also been determined, mainly in connection with investigations of the stereoconfiguration of poly(vinyl alcohol)s derived from vinyl esters.

Chemical Properties. Most of the chemical reactions of vinyl acetate also occur with other vinyl esters. The rates of hydrolysis, however, are quite different from those of vinyl acetate. Vinyl formate hydrolyzes very readily in water whereas vinyl stearate can only be hydrolyzed with great difficulty. Difficult hydrolysis is also true for the vinyl esters of branched carboxylic acids such as vinyl pivalate and the vinyl Versatates. The halogen-substituted vinyl esters, such as vinyl monofluoroacetate, hydrolyze very readily. The vinyl esters of hydroxy acids, eg, vinyl glycolate, hydrolyze on coming in contact with atmospheric moisture. Poly(methyl methacrylate) is more resistant to hydrolysis than poly(vinyl pivalate). On the other hand, vinyl pivalate is more resistant than methyl methacrylate.

Toxicity. Vinyl esters are more irritating to the mucus membranes than the corresponding ethyl esters. The toxicity of vinyl acetate is relatively low; the LD_{50} for rats is about 2.9 g/kg. Vinyl propionate likewise has a LD_{50} of 3 g/kg for rats.

Polymerization

The mechanism of polymerization and copolymerization of the higher vinyl esters is quite similar to those of vinyl acetate, which have been reviewed. Some vinyl ester monomers such as vinyl propionate and vinyl benzoate and other vinyl esters of aromatic carboxylic acids exhibit a somewhat different polymerization behavior, which will be described.

Vinyl esters of aromatic carboxylic acids have been prepared in great numbers. In general, the rate of polymerization of

Table 1. Physical Properties of Vinyl Esters of Normal Saturated Aliphatic Acids

	Mp of ester, °C	Bp of ester,[a] °C	Density,[b] g/mL	Refractive index[c]
formic		46.6	0.9592[d]	1.3855(21)
propionic		94.9	0.9203[d]	1.4003(20)
butyric		116.7	0.9012	1.4102(20)
valeric		134		
caproic		166	0.8837	1.4159(30)
enanthic		36–37(0.93)		
caprylic		90–91(2)	0.8719	1.4256(30)
pelargonic		65(0.2)	0.8689	1.4291(30)
capric		148(6.7)	0.8670	1.4320(30)
undecanoic		125(1.3)	0.8799	1.4442(30)
lauric		142(1.3)	0.8639	1.4387(25)
myristic		150(0.4)	0.8617	1.4410(25)
palmitic	26.7–27.1	165(0.27)	0.8602	1.4444(25)
stearic	35–36	187–188(0.57)		

[a] At 101.3 kPa (= 1 atm) unless otherwise noted in kPa in parentheses.
[b] At 30°C unless otherwise noted.
[c] At °C in parentheses.
[d] 20°C.

Table 2. Boiling Points and Melting Points of Vinyl Esters of Aromatic Acids

	Mp of ester, °C	Bp of ester,[a] °C	Pressure of bp determination, Pa[b]
benzoic		80.5; 203[a]	1200
o-chlorobenzoic	−21	115	1600
m-chlorobenzoic		86–88	3
p-chlorobenzoic	49.5	122	2700
2,3,4,5-tetrachlorobenzoic	48–49	125–134	13
m-nitrobenzoic	54.5	117–119	270
p-nitrobenzoic	74.5–75.5	102–104	67
p-cyanobenzoic	44–45		
trimellitic anhydride	127–128		
o-toluic		92.5	1200
m-toluic		102	1470
p-toluic		103	1530
m-anisic		104–106	4
p-anisic	59	135	1470
salicylic	32–33	97–98	133
p-acetylbenzoic	42.5–43	110–111	12
phenylacetic		88–90	533
p-dimethylaminobenzoic	68.5–69.5	130–140	13
o-acetylsalicylic		109–110	3
3,4,5-trimethyoxybenzoic	82.5–83.5	142	93
p-methoxycinnamic	65.5–66	150–155	13
1-naphthoic		126–128	7
2-naphthoic	44–45	144	27
p-phenylbenzoic	51	166	40
2-naphthylacetic	57.5–58	152	53

[a] At 101.3 kPa (= 760 mm Hg) unless noted in the next column.
[b] To convert Pa to mm Hg, divide by 133.

these esters is low and the resulting molecular weights are low. The polymers are hard and brittle substances.

The polymerization of vinyl benzoate has also been studied extensively. Its rate of polymerization is proportional to the square root of the initiator concentration and to the reciprocal of the degree of polymerization. The rate of polymerization is much lower than that of vinyl acetate and the molecular weight of the resulting poly(vinyl benzoate) are quite low. This has been explained by postulating a "complexed radical" mechanism. The complexed radical, being quite stable, participates rarely in the propagation step, but readily in termination steps. The degree of polymerization of poly(vinyl propionate) is much

Table 3. Boiling and Melting Points of Esters of Dicarboxylic Acids

Acid	Monovinyl ester mp, °C	Divinyl ester bp, °C (kPa[a])	Vinyl alkyl ester: Alkyl	Vinyl alkyl ester: Bp, °C (kPa[a])
oxalic			ethyl	60(0.7)
succinic		88(1.3)	methyl	86–88(1.3)
			allyl	110(1.3)
fumaric	97–98.5		methyl, mp 32°	
			ethyl	95(1.7)
			allyl	100–101(1.2)
maleic	bp 80–82(0.13 kPa)		methyl	82.5(1.3)
glutaric		86–91(0.4)		
adipic	38; bp 127(0.2 kPa)	105–106(0.5)	allyl	135–136.5(1.2)
pimelic		132–135(0.4)		
sebacic		162(1.3)	allyl	143–144(0.27)
			methyl	103–106(0.2)
phthalic			ethyl	173(0.4)
			octyl	138–142(0.001)
			2-ethylhexyl	144(0.01)
isophthalic	146–147	mp 59–60° 120–125(0.01)		
terephthalic	194–195	mp 83–84° 110–120(0.07)	methyl, mp 61.5–62°	153(2) 100(0.005)

[a] To convert kPa to mm Hg, multiply by 7.5.

lower than those of other poly(vinyl esters) prepared under similar conditions. Also, the rate of polymerization of vinyl propionate is much lower than that of vinyl acetate.

Copolymerization of Vinyl Esters

The copolymerization behavior of vinyl esters of other fatty acids is very similar to that of vinyl acetate, ie, they enter copolymers at a slower rate than do most monomers. Vinyl esters of aromatic acids, such as vinyl benzoate, have a somewhat different tendency in copolymerization. Vinyl benzoate is more conjugated than is vinyl acetate despite the separation of the phenyl from the vinyl group by the ester linkage. This results in a larger Q value.

Polymer Properties and Uses

Poly(vinyl esters) are thermoplastic resins that range in properties from hard to soft to waxy, depending on the nature of the ester group. The glass-transition temperatures of the straight-chain fatty acid vinyl esters go through a minimum at about 12 carbons (lauric acid) and then rise again because side-chain crystallinity becomes an important factor.

The dilute-solution properties of many poly(vinyl esters) have been studied, mainly to obtain molecular weight–intrinsic viscosity relationships.

Homopolymers of vinyl esters other than vinyl acetate are not commercially available. Poly(vinyl ester)s have more recently been used in the preparation of magnetic tapes functioning as both magnetic oxide binders and adhesives for polyester bases, anti-graffiti coatings, crack-resistant coatings for concrete structures, lipstick bases, and as compatibilizing agents for mineral fillers yielding improved mechanical properties of various filled plastics.

The most recent uses reported for poly(vinyl ester)s reflect an increasing breadth of applications areas: as comonomers for improved stability of liposome structures, polymeric binders for sustained release medication formulations, binders for photoelectric phonetic imaging films, photoelectric imaging formulations, jet printing inks with good water resistance, and dye barriers and adhesion-promoting subbing layers in thermal dye transfer printing processes.

Long chain poly(vinyl ester)s such as poly(vinyl stearate) have been claimed as waterproofing agents for skin moisturizing and sunscreen formulations and as thermoplastic bandages for casts, splints, etc, which are easily molded in place with only brief warming to 60 or 70°C to give wound coverings with a soft feel that are nonirritating to the skin.

WILEY A. DANIELS
Air Products and Chemicals

VINYL ETHER POLYMERS

Polymerization of vinyl ethers introduces polar alkoxy groups along the polymer chain. The most important commercial materials are vinyl ether–maleic anhydride copolymers.

Monomers

Vinyl ether geometrical isomers have been prepared by the Horner-Wittig reaction. Treatment of $CH_3OCHRP(O)(C_6H_5)_2$ (R = H, CH_3) with $LiN(CH_2CH_3)_2$ yields $CH_3OCRLiP(O)(C_6H_5)_2$, which reacts with aldehydes and ketones to afford separable diastereomeric adducts. Addition of a base yields the geometrical isomers.

Fluorovinyl ethers $RCF_2OCF{=}CF_2$ are receiving increasing attention from a commercial viewpoint. One synthesis involves dehalogenation of $RCF_2O(CF_2)_2I$ in the presence of a selective metallic or organometallic catalyst (eg, magnesium, copper, zinc, and tin) or organomagnesium or organoaluminum compounds in 45% yield.

Manufacture. New synthetic routes include vinylation of alcohols or vinyl exchange with vinyl acetate.

Polymerization

Catalysts and Initiators. Cationic initiators, eg, Lewis acids, are the preferred catalysts for preparing alkyl vinyl ether homopolymers. Conventional Friedel-Crafts catalysts, such as BF_3, $AlCl_3$, or $SnCl_4$, in bulk or in dry solvents, are widely used and detailed kinetic characterizations have been done.

Molecular sieves, V_2O_5, and carbon black are heterogeneous catalysts for the atactic homopolymerization of vinyl ether monomers at room temperature. Coinitiators such as benzaldehyde increase the polymerization rate of V_2O_5-initiated isobutyl vinyl ether, whereas triethylaluminum in place of benzaldehyde induces stereoregular polymerization. The cationic polymerization of isobutyl vinyl ether by carbon black in CCl_4 at 20–50°C is inhibited by pyridine and DMF.

Stereoregular Homopolymerization. The significant factor in the steric control of the cationic chain growth of alkyl vinyl ethers is the degree of association of the growing cation and the counteranion. The tightness of the growing ion pair and the steric limitations of the monomer approach by side-chain substituents control the front or back-side attack of the monomer. Highly polar solvents that stabilize separated ion pairs favor the frontal attack leading to syndiotactic polymer. Nonpolar solvents tend to associate the propagating cation–counteranion pair, thereby facilitating the back-side approach, leading to isotactic polymer. When the ion pair is crowded by a bulky substituent (*t*-butyl), the coulombic attraction between the components is reduced and frontal attack leading to syndiotactic placement is favored.

Oxychlorides of phosphorus, sulfur, and chromium attack vinyl ethers when used alone, but in the presence of triethylaluminum they initiate the stereoregular polymerization of isobutyl vinyl ethers. Addition of oxychloride to mixtures of isobutyl vinyl ethers and triethylaluminum increases the stereoregularity but reduces conversion.

An example of a heterogeneous stereoregulating catalyst is the H_2SO_4–Al salt system.

Living Polymerization. The first truly living cationic polymerization of vinyl compounds was achieved with isobutyl vinyl ether and $HI–I_2$. Conditions for living polymerization of alkyl vinyl ethers are shown in Table 1. Primary and secondary alkyl vinyl ethers are polymerized by specific initiating systems into well-defined monodisperse living polymers with controlled molecular weights (see LIVING POLYMER SYSTEMS).

Quasi-living Cationic Polymerization. Allyl cations, presumably arising during reaction of active allylic chlorine compounds, $(CH_3)_2C{=}CHCH_2Cl$, $CH_2{=}CHCH(Cl)CH_3$, with $AgPF_6$, induce the quasi-living polymerization of isobutyl vinyl ether.

Likewise, the *p*-dicumyl chloride–$AgSbF_6$ system promotes

Table 1. Conditions for Living Cationic Polymerization of Vinyl Ethers[a]

Initiating systems	Solvents	Temperature, °C
$HI-I_2$	n-C_6H_{14}, toluene, CH_2Cl_2	−5 to −40
$HI-ZnX_2$[b]	toluene, CH_2Cl_2	+25 to −40
$HI-SnCl_2$	toluene, CH_2Cl_2	0 to −40
$C_2H_5AlCl_2$–ester[c,d]	n-C_6H_{14}	+25 to −40
$C_2H_5AlCl_2$–ether[d,e]	n-C_6H_{14}	0
$C_6H_5CHO-ZnBr_2$[f]	CH_2Cl_2	+30, −60

[a] Alkyl and functionalized.
[b] X = Cl, Br, I.
[c] Ethyl acetate, ethyl benzoate, etc, in ether[d].
[d] Dioxane, tetrahydrofuran, and diethyl ether.
[e] Protogen = water or protonic acids such as acetic acid.
[f] For trialkyl silylvinyl ethers.

the quasi-living polymerization of isobutyl vinyl ether or methyl vinyl ether.

Radiation-induced Polymerization. Cobalt-60 γ-radiation is used to induce the cationic homopolymerization and copolymerization of alkyl vinyl ethers.

Cyclohomopolymerization of Divinyl Ethers. Divinyl ether and the divinyl ethers of ethylene glycol, diethylene glycol, and 1,4-butanediol are polymerized by free-radical or cationic initiators to give soluble polymers with highly cyclized low molecular weight structures. With $BF_3\cdot O(C_2H_5)_2$ catalyst, the ethylene glycol–divinyl ether copolymer is insoluble. Structurally, cation-initiated cyclopolymers of divinyl ethers of aliphatic diols differ from those obtained by free-radical initiation (see CYCLOPOLYMERIZATION).

Homopolymer Properties. Poly(alkyl vinyl ethers) may be formed as viscous, sticky liquids or rubbery or brittle solids; those with long chains are waxy. Glass-transition temperatures and crystalline melting points are shown in Table 2. Table 3 shows some useful properties of commercial poly(methyl vinyl ether).

Table 2. Glass-transition Temperatures and Melting Points of Poly(vinyl Ether)s

Alkyl group	Amorphous, T_g, °C	Crystalline, mp, °C
methyl	−34	144
ethyl	−42	
isopropyl	−3	191
n-butyl	−55	
isobutyl	−19	170
2-ethylhexyl	−66	
n-pentyl	−66	
n-hexyl	−77	
n-octyl	−80	
t-butyl		238

Uses

Poly(alkyl vinyl ethers) find wide applications as adhesives, surface coatings, lubricants, greases, elastomers, molding compounds, fibers, and films as well as in chemical processing.

Copolymerization

In the copolymerization of alkyl vinyl ethers both ionic and free-radical initiation have been used, although formation of copolymers is difficult with vinyl ethers. A homopolymer frequently forms first; at higher temperature and increased initiator concentration, the comonomer enters the polymer. Alternating copolymers of alkyl vinyl ethers and acrylates have been prepared by cationic initiation with careful choice of conditions. Block copolymers have been prepared with p-methoxystyrene with I_2 in CCl_4 at −15°C, resulting in a living cation that initiates isobutyl vinyl ether polymerization. Block copolymers of the AB and ABA type have been prepared by the living polymerizations of vinyl ethers, and p-methoxystyrene in a few cases, by $HI-I_2$ and related systems. A copolymer of methyl vinyl ether, cetyl vinyl ether, and methyl vinyl ether (60:10:6) has been prepared with a molecular weight distribution (MWD) of 1.3–3.00. Most of these polymers have a narrow MWD (1.1–1.3) and moderate molecular weights. Block polymers such as

$$-(CH_2-\underset{OR_1}{\underset{|}{CH}})_m-(CH_2-\underset{OCH_2CH_2OH}{\underset{|}{CH}})_n-$$

are amphiphilic in nature, dissolve in water between 15 and 90°C, and exhibit surface activity, reducing the surface tension of water from 72.8 to 30 mN/m (= dyn/cm) or below (0.1 wt %, 25°C).

Divinyl Ethers and Maleic Anhydrides. Radical copolymerizations of divinyl ether or substituted divinyl ethers with maleic anhydrides yield cyclized, alternating copolymers of divinyl ether–maleic anhydride in a 1:2 ratio.

Photoconductive Vinyl Ether Polymers. A few potentially photoconductive vinyl ether homopolymers and copolymers have been prepared, eg, 2-(9-carbazolyl)ethyl vinyl ether; homopolymerization is initiated with $BF_3\cdot O(C_2H_5)_2$ and $C_2H_5AlCl_2$. A photosemiconductive copolymer of 4,5,6,7-tetrahydro-1-vinylindole and ethylene glycol glycidyl vinyl ether, sensitized with

Table 3. Properties of Commercial Poly(methyl Vinyl Ether)s

Property	Gantrez M-154	Gantrez M-574	Gantrez M-555	Gantrez M-550
soluble in	water	toluene	toluene	toluene
solids, %	50	70	50	50
specific viscosity[a]	0.47	0.47	0.77	
viscosity, Pa·s[b]	ca 40	ca 30	ca 15	
bulk density, g/cm^3	1.03	0.96	0.94	0.94
flash point[c], °C	none	22	22	22

[a] 1 g in 100 mL benzene.
[b] To convert Pa·s to P, multiply by 10.
[c] Tagliabue open cup.

2,4,7-trinitrofluorene, has been described (see also PHOTOCONDUCTIVE POLYMERS; VINYLCARBAZOLE POLYMERS).

Optically Active Vinyl Ether Copolymers. The radical copolymerization of 1-menthyl vinyl ether with vinylene carbonate or indene initiated by AIBN yields optically active products by asymmetric induction. Copolymerization of isobutyl vinyl ether with maleic anhydride using AIBN gives an alternating, optically active copolymer. The specific rotation of the copolymer decreases with increasing reaction temperature.

Copolymer Properties and Uses

Applications of commercial vinyl ether copolymers are shown in Table 4.

Health and Safety Factors

Poly(methly vinyl ether) in powder form or aqueous solutions is not a primary irritant or a sensitizer. White rats and guinea pigs tolerate 90–100 cm^3/kg of a 25% aqueous solution in oral toxicity tests. Poly(methyl vinyl ether-*co*-maleic anhydride) and its monoalkyl ester are neither primary irritants nor primary sensitizers. Oral toxicity values for white rats are 29 g/kg and 25 g/kg body weight, respectively, for the two polymers tested.

BIBLIOGRAPHY

M. Grayson, ed., "Vinyl Ether Monomers and Polymers," *Encyclopedia of Chemical Technology,* 3rd ed., Vol. 23, John Wiley & Sons, Inc., New York, 1983, p. 937.

M. Biswas and N. C. Maity, *J. Macromol. Sci. Chem.* **A15,** 1153 (1981).

T. Higashimura and M. Sawamoto, *Makromol. Chem. Suppl.* **12**, 153 (1985); *Adv. Polym. Sci.* **64**, 49 (1984).

J. P. Kennedy and S. S. Plamthottam, *J. Makromol. Sci. Chem.* **A14**, 729 (1980).

MUKUL BISWAS
AMITAVA MAZUMDAR
PRADYOT MITRA
Indian Institute of Technology, Kharagpur

Table 4. Vinyl Ether Copolymer Applications

Copolymer	Applications
poly(ethyl vinyl ether-*co*-maleic anhydride)	soil redeposition prevention agent in household detergents
poly(hexafluoropropene-*co*-butyl vinyl ether-*co*-trimethoxysilane)	coating material with excellent weather and solvent resistance
poly(vinyl methyl ether-*co*-maleic anhydride)	in lens arrays for optical devices, ink-jet recording films
poly(butyl vinyl ether-*co*-styrene)	viscosity modifier in lubricant oils
poly(tetrafluoroethylene-*co*-perfluoroalkyl vinyl ether)	lubricant for bearings of porous sintered metals
poly(trifluorochloroethylene-*co*-ethyl vinyl ether-*co*-cyclohexyl vinyl ether-*co*-hydroxybutyl vinyl ether-*co*-succinic anhydride)	automobile top coat with good weather resistance and recoat adhesion
poly(tetrafluoroethylene-*co*-olefin-*co*-perfluoroalkyl vinyl ether)	elastomer for long-term service in severe environments
poly[$CF_2{=}CFO(CF_2)_3SC_2H_5$-*co*-$CF_2{=}CF_2$]	cation exchange membranes for the electrolysis of aqueous sodium chloride
poly(methyl vinyl ether-*co*-maleic anhydride) (PVM–MA)	photographic film
poly(isobutyl vinyl ether-*co*-monoethyl maleate)	aerosol hair-spray resin
poly(octadecyl vinyl ether-*co*-maleic anhydride) (POVE–MA)	release coat for pressure-sensitive adhesive tape backings
poly(isobutyl vinyl ether-*co*-vinyl chloride)	electrostatic image-developing powders, toners, binders; adhesive tape for roofing membranes
poly(cetyl vinyl ether-*co*-octadecyl vinyl ether-*co*-*N*-vinylpyrrolidinone)	protective colloid in suspension polymerization of vinyl chloride
poly(methyl vinyl ether-*co*-monoethyl maleate)	copolymer coating on polyester film
poly(hexadecyl vinyl ether-*co*-octadecyl vinyl ether-*co*-styrene-*co*-maleic anhydride)	thickening agent for heavy-duty cleaners and ammonia-based fertilizers
poly(2-chloroethyl vinyl ether-*co*-*N*-vinylpyrrolidinone)	photochemical cross-linking of polymers to immobilize enzymes
poly(methyl vinyl ether-*co*-monobutyl maleate)	thermographic copying material

VINYL FLUORIDE POLYMERS

The vinyl fluoride polymers include homopolymers and copolymers. The base unit of the homopolymers is shown by structure (**1**). Both head-to-head and head-to-tail repeating units are possible.

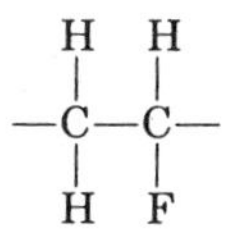

(**1**)

Polymerization of vinyl fluoride usually requires high pressure processes similar to those used in the preparation of polyolefins such as linear low density polyethylene (LLDPE). Ziegler-Natta type catalysts allow the preparation of vinyl fluoride polymers at lower pressures and temperatures. DuPont first commercialized a poly(vinyl fluoride)-based film in 1961 under the trade name Tedlar.

Vinyl fluoride homopolymers and copolymers have excellent resistance to sunlight degradation, chemical attack, water absorption, and solvents, and a high solar energy transmittance rate. These properties have spurred the use of the poly(vinyl fluoride) (PVF) films in outdoor and indoor functional and decorative applications. These films have found use where exceptional outdoor longevity, stain resistance, adherence, and release propensity are needed. Poly(vinyl fluoride) has a greater tendency to crystallize than poly(vinyl chloride). It is stable at high temperatures, which is important in many of its applications.

Monomer

Properties. Vinyl fluoride (VF), $CH_2{=}CHF$, is a colorless inert gas with an etherlike odor. It is insoluble in water at atmospheric pressure but slightly soluble in acetone and ethanol. Properties are given in Table 1.

Flammability. Vinyl fluoride has been classified as a flammable gas by the Interstate Commerce Commission.

Preparation

The commercial synthesis process of vinyl fluoride has not been described in the literature for proprietary reasons. Addition of hydrogen fluoride to acetylene and fluorination of vinyl chloride are the most likely industrial routes to VF production. Commercial vinyl fluoride is usually stabilized with terpenes such as *d*-limonene to inhibit polymerization.

Table 1. Physical Properties of Vinyl Fluoride

Property	Value
molecular weight	46.046
boiling point, °C	−72.0
freezing point, °C	−160.5
critical temperature, °C	54.7
critical pressure, MPa[a]	5.1
critical density, g/cm^3	0.320
vapor pressure at 21°C, MPa[a]	2.5
liquid density at 21°C, g/cm^3	0.636

[a] To convert MPa to atm, multiply by 10.

Polymer

Properties. The physical, chemical, and electrical properties of poly(vinyl fluoride) film are shown in Table 2.

Poly(vinyl fluoride) (PVF) is a semicrystalline polymer with a planar zigzag chain configuration. The degree of crystallinity depends on the method of polymerization and the thermal history, with reported values ranging from 20 to 60%. Wide-line nmr and x-ray diffraction studies show two monomer units in the unit cell with cell dimensions $a = 0.857$ nm, $b = 0.495$ nm, and $c = 0.252$ nm; similarity to the phase I crystal form of poly(vinylidene fluoride) shows that the unit cell is orthorhombic.

Polymerization

Vinyl fluoride is more difficult to polymerize than other vinyl halides because of the electronegativity of the fluorine atom. Low boiling point (−72°C) and high critical temperature (54.7°C) necessitate high pressure polymerization similar to that required for ethylene.

The propagation of the poly(vinyl fluoride) chain appears to proceed via a free-radical mechanism in both head-to-tail and head-to-head configurations. Typically, commercial PVF contains 10–12% of inverted (head-to-head and tail-to-tail) units.

Suspension Polymerization. Vinyl fluoride may be polymerized by suspension polymerization. Typically, liquid VF is suspended in water with the help of a dispersion stabilizer. Polymerization is initiated by an organic peroxide such as diisopropyl peroxydicarbonate below the critical temperature of vinyl fluoride. The polymerization reaction can also be initiated by uv light and ionizing radiation. Vinyl fluoride dispersions, in most cases, are stabilized by water-soluble polymers such as cellulose derivatives (cellulose ester and sodium carboxymethylcellulose) and poly(vinyl alcohol). Inorganic salts such as magnesium carbonate, barium sulfate, and alkylsulfoacids are also used.

Bulk Polymerization. In the bulk polymerization of VF, a glass ampule is filled with vinyl fluoride containing 6.5×10^{-2} mol/L of di-*tert*-butyl peroxide. The ampule is irradiated by uv light from a mercury lamp below 25°C. A highly porous polymer at conversions of over 90% is obtained; this product is insoluble in VF.

Emulsion Polymerization. Vinyl fluoride can be polymerized by emulsion polymerization at greatly reduced pressures and lower temeratures compared to those employed in suspension polymerization. Polymerization in an aqueous emulsion facilitates the process control and removal of the reaction heat, and increases the molecular weight of the product; high reaction rates and high yields are possible.

Radiation-induced Polymerization. Exposure to uv radiation in the absence of free radicals does not polymerize vinyl fluoride, but decomposes it into acetylene and hydrogen fluoride. Radiation can be used in the presence of initiators, which decompose into free radicals.

Graft Polymerization. The most effective method for graft polymerization of VF is radiation polymerization. Vinyl fluoride has been grafted to low density polyethylene, polyisobutylene, and polyamides.

Continuous Polymerization. A process for continuous polymerization of vinyl fluoride in aqueous medium has been described. A mixture of VF, water, and a water-soluble catalyst is stirred at 50–250°C and 15–100 MPa (150–1000 atm). A small amount of a monoolefin (three to five carbon atoms) is continuously introduced into the reactor to inhibit the bulk

Table 2. Properties of Poly(vinyl fluoride) Film

Property	Value	ASTM test method[a]
bursting strength, kPa[b]	200–450[c]	D774
coefficient of friction with metal	0.18–0.21	D1894-78
density, g/cm^3	1.38–1.72	weighed samples
impact strength, kJ/m[d]	43–90	D3420-80
refractive index, n_D^{30}	1.46	D542; Abbe refractometer
tear strength, kJ/m[d]		
propagated	6–22	D1922-67
initial	129–196	D1004-66
tensile modulus, MPa[e]	44–110	D882
ultimate elongation, %	115–250	D882
ultimate yield strength, MPa[e]	33–41	D882
linear coefficient of expansion, cm/(cm·°C)	0.00005	air oven, 30 min
useful temperature range, °C		
continuous use	−70 to +107	
short cycle (1–2 h)	175	
zero strength	260–300	hot bar
thermal conductivity, 1°C/cm, W/(m·K)		
at −30°C	0.14	
at 60°C	0.17	
self-ignition temperature, °C	390	D1929
solar energy transmittance, 359–2500 nm, %	90	E427-71
chemical resistance after 1 yr immersion at 25°C in acids (10%), bases (10%), or solvents	no visible effect	
chemical resistance after 2 h immersion at boiling temperature in acids (10%), bases (10%), or solvents	no visible effect	
electrical[f]		
corona endurance, h at 60 Hz, 40 V/μm, h	2.5–6.0	D2275
dielectric constant		D150-81
1 MHz at 23°C	6.2–7.7	
dielectric strength		
short-term ac, kV/μm	0.08–0.13	
dissipation factor		D150-81
1 MHz at 23°C	0.17–0.28	
1 MHz at 100°C	0.09–0.21	
volume resistivity, GΩm		D150-81
23°C	2,000–700	
surface resistivity, GΩm		D257
23°C	60,000–20,000	

[a] Unless otherwise stated.
[b] To convert kPa to psi, multiply by 0.145.
[c] Range dependent on composition and tensile modification.
[d] To convert kJ/m to ft·lbf/in., divide by 0.534 (see ASTM D256).
[e] To convert MPa to psi, multiply by 145.
[f] When range of electrical properties is given, the first value refers to 54.8 μm transparent film, and the second value to 54.8 μm white pigmented film.

polymerization of VF to low molecular weight polymer. The water-soluble catalyst generates free radicals, which initiate the polymerization.

Effect of Polymerization Variables. The polymerization variables include temperature, pressure, medium, impurities, telogen, and catalyst (initiator).

Temperature. The polymerization temperature has a significant effect on the molecular weight of poly(vinyl fluoride). Molecular weight generally decreases as the polymerization temperature is raised.

Pressure. Polymerization of vinyl fluoride requires high pressures. The rate of polymerization and the intrinsic viscosity of the polymer increase when pressure is increased.

Medium. Water is the best medium for the polymerization of VF; organic solvents have also been used.

Impurities. Oxygen, acetylene, and 1,1-difluoroethane are

among the impurities most likely found in vinyl fluoride. Typically, impurities affect the polymerization of vinyl fluoride adversely and act as inhibitors.

Telogen. The melt viscosity of poly(vinyl fluoride) can be best controlled by a chain-terminating compound (telogen), such as isopropyl alcohol or 1,3-dioxolane; toughness is not affected.

Catalysts and Initiators. The type of free-radical initiator employed can have a profound effect on PVF properties, such as thermal stability and wettability.

Ziegler-Natta Catalysts. Ziegler-Natta type catalysts are used to polymerize vinyl fluoride. These catalysts are a mixture of transition metal halides (titanium halides) and trialkylaluminum (triethylaluminum). Ziegler-Natta catalysts allow preparation of poly(vinyl fluoride) at lower temperatures and pressures. In contrast to polyolefins, no increase in the stereoregularity of vinyl fluoride polymers using these catalysts was observed.

Copolymerization

Homopolymers of fluorinated organic monomers have poor solubility, high cure temperatures, and high melt viscosity. To improve these properties, vinyl fluoride has been copolymerized with a number of other monomers under conditions similar to those employed for homopolymerization. Vinyl chloride, vinylidene carbonate, acrylonitrile, vinylidene fluoride, vinyl acetate, vinyl formate, hexafluorocyclobutene, ethylene, chlorotrifluoroethylene, and hexafluoropropene are among the large number of monomers that have been copolymerized with vinyl fluoride. Acrylic monomers include ethyl acrylate, acrylic acid and perfluoromethacryloyl fluoride. Free-radical initiators can be used in the copolymerization. Ziegler-Natta systems and alkylboron catalysts activated by O_2 have been used with several monomers over the entire composition range. The reaction proceeds by a free-radical mechanism, much the same as the homopolymerization reaction.

Vinyl fluoride copolymers exhibit good properties. A copolymer of VF–C_2F_3Cl, soluble in acetone, forms a transparent, thermally stable film. A copolymer of VF with two other monomers containing etheric and hydroxylic, carboxylic, or ethoxylic functional groups exhibits chemical resistance, weatherability, and resistance to stain and heat. However, copolymers of vinyl fluoride are not commercially available.

Fabrication and Processing

Poly(vinyl fluoride) is fabricated into self-supporting films and coatings. Moldings of PVF have been reported but are not significant.

A structurally modified PVF has been extruded. Thin films are manufactured by extrusion of a dispersion of poly(vinyl fluoride) in a latent solvent. Pigments, stabilizers, flame retardants, plasticizers, deglossing agents, and other additives are incorporated in the dispersion prior to extrusion. The solvent is removed by evaporation. The extruded film can be biaxially oriented to varying degrees. Solvent removal is usually completed after orientation has taken place. Physical properties of the film depend on the extent of its orientation.

Poly(vinyl fluoride) homopolymers and copolymers can be applied to substrates as a dispersion or by powder-coating methods.

DuPont supplies Tedlar poly(vinyl fluoride) film and a dispersion-grade resin. Tedlar is available as types 1, 2, 3, 4, and 5, which differ in the degree of orientation; orientation decreases with increasing numbers. Tensile strength, impact resistance, and resistance to fatigue increase with orientation. Lower orientation results in higher tear strength, greater formability (extensibility at elevated temperatures), and less shrinkage. Tedlar is produced in multiple variations of gloss, color, transparency, uv absorption, flame retardancy, fluorescence, thermal reflectivity, etc.

Economic Aspects

Poly(vinyl fluoride) is available from DuPont both as transparent and pigmented films under the trademark Tedlar PVF film; several tensile modifications and thicknesses are available with either adhering or nonadhering surfaces. The 1989 prices ranged from $24 to 62/kg ($13 to 24/kg in 1979); prices for specially tailored films were significantly higher.

Health and Safety

Vinyl fluoride is flammable in air between the limits of 2.6 and 22% by volume. The Department of Transportation has classified the inhibited monomer as a flammable gas.

An extensive toxicity testing program of vinyl fluoride, including inhalation teratology and carcinogenicity studies, is currently in progress.

BIBLIOGRAPHY

G. H. Kalb and co-workers, *J. Appl. Polym. Sci.* **4**, 55 (1960).

K. U. Usmanov and co-workers, *Russ. Chem. Rev.* **46**(5), 462–478 (1977); trans. of *Usp. Khim.* **46**, 878–906 (1977).

DONALD BRASURE
SINA EBNESAJJAD
E. I. du Pont de Nemours & Co., Inc.

VINYL FORMAL POLYMERS.

See VINYL ACETAL POLYMERS.

VINYL FORMATE POLYMERS.

See VINYL ESTER POLYMERS.

VINYLIDENE CHLORIDE POLYMERS

The commercialization of vinylidene chloride (VDC) polymers under the trademark Saran began in 1939. In the United States, Saran is a generic term for polymers with high vinylidene chloride content. In other countries, Saran is still a trademark of The Dow Chemical Co. Of the many copolymers that have been prepared, only three types are commercially important: vinylidene chloride–vinyl chloride copolymers, vinylidene chloride–alkyl acrylate and methacrylate copolymers, and vinylidene chloride–acrylonitrile copolymers.

The most valuable property of poly(vinylidene chloride) (PVDC) and its copolymers is low permeability to a wide range of gases and vapors. The most serious deficiency is thermal instability at melt-processing temperatures.

Monomer

Properties. Pure vinylidene chloride (1,1-dichloroethylene) is a colorless, mobile liquid with a characteristic sweet odor. Its properties are summarized in Table 1. Vinylidene chloride is soluble in most polar and nonpolar organic solvents. Its solubility in water (0.25 wt %) is nearly independent of temperature at 16–90°C.

Manufacture. Vinylidene chloride monomer can be conveniently prepared in the laboratory by the reaction of 1,1,2-trichloroethane with aqueous alkali.

Vinylidene chloride is prepared commercially by the dehydrochlorination of 1,1,2-trichloroethane with lime or caustic in slight excess (2–10%). A continuous liquid-phase reaction at 98–99°C yields about 90% VDC. Caustic gives better results than lime. Vinylidene chloride is purified by washing with water, drying, and fractional distillation. It forms an azeotrope with 6 wt % methanol. Purification can be by distillation of the azeotrope followed by extraction of the methanol with water; an inhibitor is usually added at this point. Commercial grades contain 200 ppm of the monomethyl ether of hydroquinone (MEHQ). Many other inhibitors for the polymerization of vinylidene chloride have been described in patents, but MEHQ is the one most often used. The inhibitor can be removed by distillation or by washing with 25 wt % aqueous caustic under an inert atmosphere at low temperatures.

Health and Safety Factors. Vinylidene chloride is highly volatile and, when free of decomposition products, has a mild, sweet odor. Its warning properties are ordinarily inadequate to prevent excessive exposure. Inhalation of vapor presents a hazard, which is readily controlled by observance of precautions commonly taken in the chemical industry.

A single, prolonged exposure and repeated short-term exposures can be dangerous, even when the concentrations of the

Table 1. Properties of Vinylidene Chloride Monomer[a]

Property	Value
molecular weight	96.944
odor	pleasant, sweet
appearance	clear, liquid
color (APHA)	0–10
solubility of monomer in water, at 25°C, wt %	0.25
solubility of water in monomer, at 25°C, wt %	0.035
normal boiling point, °C	31.56
freezing point, °C	−122.56
flash point (tag closed-cup), °C	−28
flash point (tag open-cup), °C	−16
flammable limits in air (ambient conditions), vol %	6.5–15.5
autoignition temperature, °C	513[b]
latent heat of vaporization, ΔH_v°, kJ/mol[b]	
at 25°C	26.48 ± 0.08
at normal boiling point	26.14 ± 0.08
latent heat of fusion (at freezing point), ΔH_m°, J/mol[b]	6514 ± 8
heat of polymerization (at 25°C), ΔH_p°, kJ/mol[b]	−75.3 ± 3.8
heat of combustion, liquid monomer (at 25°C), ΔH_c°, kJ/mol[b]	1095.9
heat of formation	
liquid monomer (at 25°C), ΔH_f°, kJ/mol[b]	−25.1 ± 1.3
gaseous monomer (at 25°C), ΔH_f°, kJ/mol[b]	1.26 ± 1.26
heat capacity	
liquid monomer (at 25°C), C_p°, J/(mol·K)[b]	111.27
gaseous monomer (at 25°C), C_p°, J/(mol·K)[b]	67.03
critical temperature, T_c, °C	220.8
critical pressure, P_c, MPa[c]	5.21
critical volume, V_c, cm^3/mol	218
liquid density, g/cm^3	
−20°C	1.2852
0°C	1.2499
20°C	1.2137
index of refraction, n_D	
10°C	1.43062
15°C	1.42777
20°C	1.42468
absolute viscosity, mPa·s (= cP)	
−20°C	0.4478
0°C	0.3939
20°C	0.3302
vapor pressure, T, °C (P measured from 6.7–104.7 kPa)[d]	$\log P_{kPa} = 6.1070 - 1104.29/(T + 237.697)$

[a] Inhibited with methyl ether of hydroquinone.
[b] To convert J to cal, divide by 4.184.
[c] To convert MPa to atm, divide by 0.101.
[d] To convert kPa to mmHg, multiply by 7.5 (add 0.875 to the constant to convert $\log_{kPa}$ to $\log_{mmHg}$).

vapor are too low to cause an anesthetic effect. They may produce organic injury to the kidneys and liver. For repeated exposures, the vapor concentration of vinylidene chloride should be much lower. The American Conference of Governmental Industrial Hygienists (ACGIH) threshold limit value (TLV) of 5 ppm and short-term exposure level (STEL) of 20 ppm have been established to provide an adequate margin of safety. Investigations of vinylidene chloride exposure have been reviewed.

The liquid is irritating to the skin after only a few minutes of contact. The inhibitor MEHQ may be partly responsible for this irritation. Inhibited vinylidene chloride is moderately irritating to the eyes. Contact causes pain and conjunctival irritation and, possibly, some transient corneal injury and iritis; permanent damage is not likely.

Peroxide Formation. In the presence of air or oxygen, uninhibited vinylidene chloride forms a violently explosive complex peroxide at temperatures as low as −40°C. Copper, aluminum, and their alloys should not be used in handling vinylidene chloride. Copper can react with acetylenic impurities forming copper acetylides, and aluminum can react with the vinylidene chloride forming aluminum chloroalkyls. Both these compounds are extremely reactive and potentially hazardous.

Polymerization

Vinylidene chloride polymerizes by both ionic and free-radical reactions. Processes based on the latter are far more common. Vinylidene chloride is of average reactivity when compared with other unsaturated monomers. The chlorine substituents stabilize radicals in the intermediate state of an addition reaction. Because they are also strongly electron-withdrawing, they polarize the double bond, making it susceptible to anionic attack. For the same reason, a carbonium ion intermediate is not favored.

Homopolymerization. The free-radical polymerization of VDC has been carried out by solution, slurry, suspension, and emulsion methods.

Copolymerization. The importance of VDC as a monomer results from its ability to copolymerize with other vinyl monomers. Its Q value equals 0.22 and its e value equals 0.36. It most easily copolymerizes with acrylates, but it also reacts, more slowly, with other monomers, eg, styrene, that form highly resonance-stabilized radicals. Reactivity ratios (r_1 and r_2) with various monomers are listed in Table 2. Many other copolymers have been prepared from monomers for which the reactivity ratios are not known. The commercially important copolymers include those with vinyl chloride (VC), acrylonitrile (AN), or various alkyl acrylates, but many commercial polymers contain three or more components, VDC being the principal component. Usually one component is introduced to improve the processability or solubility of the polymer; the others are added to modify specific-use properties. Most of these compositions have been described in the patent literature, and a list of various combinations have been compiled. A typical terpolymer might contain 90 wt % VDC with the remainder made up of acrylonitrile and an acrylate or methacrylate monomer.

Table 2. Reactivity of Vinylidene Chloride (r_1) with Important Monomers (r_2)

Monomer	r_1	r_2
styrene	0.14	2.0
vinyl chloride	3.2	0.3
acrylonitrile	0.37	0.91
methyl acrylate	1.0	1.0
methyl methacrylate	0.24	2.53
vinyl acetate	6	0.1

Polymer Structure and Properties

Chain Structure. The chemical composition of poly(vinylidene chloride) has been confirmed by various techniques, including elemental analysis; x-ray diffraction analysis; ir, Raman, and nmr spectroscopy; and degradation studies. The polymer chain is made up of vinylidene chloride units added head-to-tail:

$$—CH_2CCl_2—CH_2CCl_2—CH_2CCl_2—$$

Because the repeat unit is symmetrical, no possibility exists for stereoisomerism. Variations in structure can occur only by head-to-head addition, branching, or degradation reactions that do not cause chain scission, including such reactions as thermal dehydrochlorination, which creates double bonds in the structure to give, for example, $—CH_2CCl_2CH{=}CClCH_2CCl_2—$ and a variety of ill-defined oxidation and hydrolysis reactions that generate carbonyl groups.

Crystal Structure. The crystal structure of PVDC is fairly well established. Several unit cells have been proposed. The unit cell contains four monomer units with two monomer units per repeat distance. The calculated density, 1.96 g/cm^3, is higher than the experimental values, which are 1.80–1.94 g/cm^3 at 25°C, depending on the sample.

The properties of PVDC (Table 3) are usually modified by copolymerization. Copolymers of high VDC content have lower melting temperatures than PVDC. Copolymers containing more than about 15 mol % acrylate or methacrylate are amorphous. Substantially more acrylonitrile (25%) or vinyl chloride (45%) is required to destroy crystallinity completely.

Morphology and Transitions. The highly crystalline particles of PVDC precipitating during polymerization are aggregates of thin lamellar crystals. The substructures are 5–10 nm thick and 100 or more times larger in other dimensions. In some respects, they resemble the lamellar crystals grown from dilute solution. The single crystals are better characterized than the as-polymerized particles. They are highly branched, with branching angles of 65–70°; the angle appears to be associated with a twin plane in the crystal.

Melting temperatures of as-polymerized powders are high, ie, 198–205°C as measured by differential thermal analysis (dta) or hot-stage microscopy.

Solubility and Solution Properties. Poly(vinylidene chloride), like many high melting polymers, does not dissolve in most common solvents at ambient temperatures. Copolymers, particularly those of low crystallinity, are much more soluble. However, one of the outstanding characteristics of vinylidene chloride polymers is resistance to a wide range of solvents and chemical reagents. The insolubility of PVDC results less from its polarity than its high melting temperature. It dissolves readily in a wide variety of solvents above 130°C.

The polarity of the polymer is important only in mixtures with specific polar aprotic solvents; many solvents of this general class solvate PVDC strongly enough to depress the melting temperature by more than 100°C. A list of good solvents is given in Table 4. The relative solvent activity is characterized by the temperature at which a 1 wt % mixture of polymer in solvent becomes homogeneous when heated rapidly.

Table 3. Properties of Poly(vinylidene Chloride)

Property	Best value	Reported values
T_m, °C	202	198–205
T_g, °C	−17	−19 to −11
transition between T_m and T_g, °C	80	
density at 25°C, g/cm^3		
amorphous	1.775	1.67–1.775
unit cell	1.96	1.949–1.96
crystalline		1.80–1.97
refractive index (crystalline), n_D	1.63	
heat of fusion (ΔH_m), J/mol[a]	6275	4600–7950

[a] To convert J to cal, divide by 4.184.

Table 4. Solvents for Poly(vinylidene Chloride)

Solvents	T^a, °C
nonpolar	
1,3-dibromopropane	126
bromobenzene	129
1-chloronaphthalene	134
2-methylnaphthalene	134
o-dichlorobenzene	135
polar aprotic	
hexamethylphosphoramide	−7.2
tetramethylene sulfoxide	28
N-acetylpiperidine	34
N-methylpyrrolidinone	42
N-formylhexamethyleneimine	44
trimethylene sulfide	74
N-*n*-butylpyrrolidinone	75
diisopropyl sulfoxide	79
N-formylpiperidine	80
N-acetylpyrrolidinone	86
tetrahydrothiophene	87
N,N-dimethylacetamide	87
cyclooctanone	90
cycloheptanone	96
di-*n*-butyl sulfoxide	98

[a] Temperature at which a 1 wt % mixture of polymer in solvent becomes homogeneous.

Copolymers with enough vinylidene chloride content to be quite crystalline behave much like PVDC. They are more soluble, however, because of their lower melting temperatures.

Mechanical Properties. Because of the difficulty of fabricating PVDC into suitable test specimens, very few direct measurements of its mechanical properties have been made. In many cases, however, the properties of copolymers have been studied as functions of composition, and the properties of PVDC can be estimated by extrapolation. Some characteristic properties of high-VDC-content, unplasticized copolymers are listed in Table 5. The performance of a given specimen is very sensitive to morphology, including the amount and kind of crystallinity, as well as orientation.

Barrier Properties. Vinylidene chloride polymers are more impermeable to a wider variety of gases and liquids than other polymers. This is a consequence of the combination of high density and high crystallinity in the polymer. An increase in either tends to reduce permeability (see BARRIER POLYMERS). A more subtle factor may be the symmetry of the polymer structure.

The inherent barrier in VDC copolymers can best be exploited by using films containing little or no plasticizers and as much VDC as possible. However, the permeability of even completely amorphous copolymers, eg, 60-VDC–40-AN or 50-VDC–50-VC, is low compared to that of other polymers. The primary reason is that diffusion coefficients of molecules in VDC copolymers are very low. This factor, together with the low solubility of many gases in VDC copolymers and the high crystallinity, results in very low permeability. Permeability is affected by the kind and amounts of comonomer as well as crystallinity.

Degradation Chemistry

Vinylidene chloride polymers are highly resistant to oxidation, permeation of small molecules, and biodegradation, which makes them extremely durable under most use conditions. However, these materials are thermally unstable and, when heated above about 120°C, undergo degradative dehydrochlorination. For this reason the superior characteristics of the homopolymer cannot be exploited. It degrades with rapid evolution of hydrogen chloride within a few degrees of its melting temperature (200°C). As a consequence, the copolymers of VDC with vinyl chloride, alkyl acrylates or methacrylates, acrylonitrile or methacrylonitrile rather than the homopolymer have come to commercial prominence. Such copolymers have often served as substrates for a study of the degradation reaction. The thermal degradation of vinylidene chloride copolymers occurs in two distinct steps. The first involves degradative dehydrochlorination via a chain process to generate poly(choroacetylene) sequences. Subsequent Diels-Alder-type condensation between conjugated sequences affords a highly cross-linked network,

Table 5. Mechanical Properties of High Vinylidene Chloride (VDC) Copolymers

Property	Range
tensile strength, MPa[a]	
unoriented	34.5–69.0
oriented	207–414
elongation, %	
unoriented	10–20
oriented	15–40
softening range (heat distortion), °C	100–150
flow temperature, °C	>185
brittle temperature, °C	−10 to 10
impact strength, J/m[b]	26.7–53.4

[a] To convert MPa to psi, multiply by 145.
[b] To convert J/m to ft·lbf/in., divide by 53.38 (see ASTM D256).

which on further dehydrochlorination leads to the formation of a large-surface-area, highly absorptive carbon.

To some extent, the stability of VDC polymers is dependent on the nature of the comonomer present. Copolymers with acrylates degrade slowly. Apparent degradation propagation rates decrease somewhat as the acrylate content of the copolymer increases. Copolymers with acrylonitrile or methacrylate undergo degradation more readily. In addition, the degradation is more complex than that observed for acrylate copolymers. The degradation of VDC polymers in nonpolar solvents is comparable to degradation in the solid state. However, these polymers are unstable in many polar solvents.

Polymers of high VDC content are reactive toward strong bases to yield elimination products and toward nucleophiles to yield substitution products. Agents capable of functioning as both a base and a nucleophile react with these polymers to generate a mixture of products. Amines can also swell the polymer, leading to very rapid reactions.

Lewis acids, particularly transition metal salts, strongly promote the thermal degradation of VDC polymer. The rate of initiation of degradation is greatly enhanced in the presence of metal ions.

Stabilization. The stabilization of VDC polymers toward degradation is a highly developed art and is responsible for the widespread commercial use of these materials. Although the mode of action is often not understood, some general principles of effective stabilization have been established. The ideal stabilizer system should absorb or combine with evolved hydrogen chloride irreversibly under conditions of use, but not strip hydrogen chloride from the polymer chain; act as a selective uv absorber; contain a reactive dienophilic moiety capable of preventing discoloration by reacting with and disrupting the color-producing conjugated polymer sequences; possess nucleophilicity sufficient for reaction with allylic dichloromethylene units; possess antioxidant activity so as to prevent the formation of carbonyl groups and other chlorine-labilizing structures; be able to scavenge chlorine atoms and other free radicals efficiently; and chelate metals, eg, iron, to prevent chlorine coordination and the formation of metal chlorides.

Acid acceptors are of two general types: alkaline-earth and heavy-metal oxides or salts of weak acids, such as barium, cadmium, or lead fatty-acid salts; and epoxy compounds, such as epoxidized soybean oil or glycidyl ethers and esters.

Antioxidants are generally of two types: those that react with a free radical to stop a radical chain, that is, to scavenge chlorine atoms or peroxy radicals; and those that reduce hydroperoxides to alcohols (see ANTIOXIDANTS; ANTIOZONANTS).

Applications

Vinylidene chloride polymers are used in molding resins, extrusion resins, multilayer film, rigid barrier containers, lacquer resins, and vinylidene chloride copolymer latex and foams. Vinylidene chloride copolymers offer significant ignition resistance.

BIBLIOGRAPHY

R. A. Wessling, *Polyvinylidene Chloride,* Gordon & Breach, New York, 1977.

P. T. DeLassus, *J. Vinyl Technol.* **1**, 14 (1979).

P. T DeLassus, J. C. Tou, M. A. Babinec, D. C. Rulf, B. K. Karp, and B. A. Howell in J. H. Hotchkiss, ed., *Food and Packaging Interactions,* ACS Symposium Series No. 365, American Chemical Society, Washington, D.C., 1987, chapt. 2.

R. A. WESSLING
D. S. GIBBS
P. T. DELASSUS
The Dow Chemical Company
B. A. HOWELL
Central Michigan University

VINYLIDENE FLUORIDE POLYMERS

Poly(vinylidene fluoride) (PVDF) is the addition polymer of 1,1-difluoroethene, $CH_2{=}CF_2$, known commonly as vinylidene fluoride and abbreviated VDF or VF_2.

Vinylidene fluoride is polymerized readily by free-radical initiators to form a high molecular weight, partially crystalline polymer that contains 59.4 wt % fluorine and 3 wt % hydrogen. The spatially symmetrical disposition of the hydrogen and fluorine atoms along the polymer chain gives rise to unique polarity influences that affect solubility, dielectric properties, and crystal morphology; the dielectric constant is unsually high. The polymer has application in many areas of the chemical-processes industry, in a variety of electrical and electronic devices, as a weather-resistant binder for exterior architectural finishes, and in many specialized uses requiring high performance thermoplastic properties and fluoroplastic attributes.

Poly(vinylidene fluoride) is readily melt-processed by standard methods of molding or extrusion; coatings or porous membranes are prepared from solutions using polar solvents. Stability of poly(vinylidene fluoride) to harsh chemical, thermal, ultraviolet, weathering, and oxidizing or high energy radiation environments approaches the essentially inert attribute of fully fluorinated polymers. The mechanical properties are those typifying a tough engineering thermoplastic that may be used in structural designs for elevated temperature environments. Poly(vinylidene fluoride) has useful mechanical integrity from −60 to 150°C and a high heat-deflection temperature. It is prepared by methods that assure exceedingly low contami-

Table 1. Properties of Vinylidene Fluoride

Property	Value
molecular weight	64.04
melting point, °C	−144
boiling point, °C	−84
critical constants	
T_c, °C	30.1
P_c, Pa[a]	0.443
D_c, kg/m^3	417
density, liquid, at 23.6°C, kg/m^3	617
explosive limits, vol % in air	5.8–20.3
heat of formation, $H_f^\circ(g)$, kJ/mol[b]	−335
specific heat, $C_p(g)$ at 25°C, J/(mol·K)[b]	60
Polymerization	
activation energy, E_a, kJ/mol[b]	161
radiation-chem. yield, mol/100 eV	4.3×10^4

[a] To convert Pa to mm Hg, multiply by 7.5.
[b] To convert J to cal, divide by 4.184.

nation, and does not require additives for stabilization, thereby qualifying for applications, such as ultrapure water systems, where high purity is demanded for materials of construction.

A search of information data bases covering 1982–1986 yielded more than 900 U.S. patents pertaining to vinylidene fluoride or poly(vinylidene fluoride); the search of the general literature for the same period yielded almost 2000 references. This interest reflects both the growing commercial importance of vinylidene fluoride and copolymers of vinylidene fluoride. Whereas Pennwalt Corporation (Philadelphia) was the only commercial producer of poly(vinylidene fluoride) two decades ago, in 1988 two Japanese and two European manufacturers were also producing PVDF. The estimated combined annual production capacity is approximately 9000 metric tons of vinylidene-fluoride-based resin.

Monomer

Vinylidene fluoride is a colorless, nearly odorless gas boiling at −82°C. It can be prepared by dehydrohalogenation of 1-chloro-1,1-difluoroethane, 1-bromo-1,1-difluoroethane, or 1,1,1-trifluoroethane, or by the dehalogenation of 1,2-dichloro-1,1-difluoroethane.

$$CH_3{-}CClF_2 \longrightarrow CH_2{=}CF_2 + HCl$$

$$CH_3{-}CBrF_2 \longrightarrow CH_2{=}CF_2 + HBr$$

Table 2. Properties of Poly(vinylidene Fluoride)

Properties	Value or description
clarity	transparent to translucent
melting point, crystalline, °C	155–192
specific gravity	1.75–1.80
refractive index, n_D^{25}	1.42
mold shrinkage, average, %	2–3
color possibilities	unlimited
machining qualities	excellent
flammability	self-extinguishing, nondripping
tensile strength, MPa[a]	
at 25°C	42–58.5
at 100°C	34.5
elongation, %	
at 25°C	50–300
at 100°C	200–500
yield point, MPa[a]	
at 25°C	38–52
at 100°C	17
creep, at 13.79 MPa[a] and 25°C for 10,000 h, %	2–4
compressive strength, at 25°C, MPa[a]	55–90
modulus of elasticity, at 25°C, GPa[b]	
in tension	1.0–2.3
in flexure	1.1–2.5
in compression	1.0–2.3
Izod impact, at 25°C, J/m[c]	
notched	75–235
unnotched	700–2300
durometer hardness, Shore, D scale	77–80
heat-distortion temperature, °C	
at 0.455 MPa[a]	140–168
at 1.82 MPa[a]	80–128
abrasion resistance, Tabor CS-17, 0.5 kg load, mg/1000 cycles	17.6
coefficient of sliding friction to steel	0.14–0.17
thermal coefficient of linear expansion, per °C	$0.7–1.5 \times 10^{-4}$
thermal conductivity, at 25–160°C, W/(m·K)	0.17–0.19
specific heat, J/(kg·K)[d]	1255–1425
thermal degradation temperature, °C	390
low-temperature embrittlement, °C	−60
water absorption, %	0.04
moisture vapor permeability, for 1 mm thickness, g/(24 h)(m²)	2.5×10^{-2}
radiation resistance (^{60}Co), MGy[e,f]	10–12

[a] To convert MPa to psi, multiply by 145.
[b] To convert GPa to psi, multiply by 145,000.
[c] To convert J/m to ftlbf/in., divide by 53.38.
[d] To convert J to cal, divide by 4.184.
[e] Retains tensile strength of about 85% of its original value.
[f] To convert Gy to rad, multiply by 100.

$$CH_3{-}CF_3 \longrightarrow CH_2{=}CF_2 + HF$$

$$CH_2Cl{-}CClF_2 \longrightarrow CH_2{=}CF_2 + Cl_2$$

The preferred monomer precursor, 1-chloro-1,1-difluoroethane (refrigerant 142b), is prepared from acetylene, vinylidene chloride, and 1,1,1-trichloroethane:

$$CH{\equiv}CH \xrightarrow{+2\,HF} CH_3{-}CHF_2 \xrightarrow{+Cl_2} CH_3{-}CClF_2 + HCl$$

$$CH_2{=}CCl_2 \xrightarrow{+2\,HF} CH_3{-}CClF_2 + HCl$$

$$CH_3{-}CCl_3 \xrightarrow{+2\,HF} CH_3{-}CClF_2 + 2\,HCl$$

Vinylidene fluoride is stored and shipped without polymerization inhibitor. If desired, terpene or quinone inhibitors can be added. The properties of the monomer are given in Table 1.

Polymer Properties

Typical physical properties of poly(vinylidene fluoride) are shown in Table 2.

The electrical properties shown in Table 3 are typical values obtained on homopolymers without additives or special treatments.

Fabrication and Processing

Equipment with materials of construction similar to those used for processing PVC and polyolefins is adequate.

Poly(vinylidene fluoride) resins are usually sold as powder or pellets; latices are also commercially available. A range of melt viscosities fulfills fabrication equipment requirements. High flow grades are used for injection molding complex parts. Higher viscosity grades offer the melt strength necessary to form various profiles in extrusion processes, such as rods, film, tubing, pipe, wire insulation, and monofilament.

Special carbon-filled grades reduce mold shrinkage and improve thermal conductivity. Carbon-fiber reinforced compositions have been prepared. Foamable compositions reduce density and dielectric loss factor.

Health and Safety

Based on toxicity studies (including acute oral, systemic, subchronic, subacute contact, implantation, and tissue culture tests), poly(vinylidene fluoride) is nontoxic. Poly(vinylidene fluoride) may be safely used in articles intended for repeated contact with food.

Poly(vinylidene fluoride) is not hazardous under typical processing and use conditions. Mechanical malfunctions or human error may lead to thermal decomposition with evolution of toxic hydrogen fluoride.

Some additives, such as mica, glass fibers, and titanium dioxide, catalyze thermal decomposition of PVDF and should be used with caution. Fabricators should consult with the resin producer about safe processing procedures.

Applications

The combination of PVDF properties has led to many diverse industrial applications that require excellent mechanical properties and resistance to severe environmental stresses. Most important are long-lasting architectural metal finishes; wire and cable insulation for the electrical and electronics industry; fluid-handling systems for the chemical processing industry; and rapidly growing specialized uses.

BIBLIOGRAPHY

J. E. Dohany in R. B. Seymour and G. S. Krischenbaum, eds., *High Performance Polymers: Their Origin and Development,* Elsevier Science Publishing Co., Inc., New York, 1986, p. 287.

A. J. Lovinger in G. C. Bassett, ed., *Developments in Crystalline Polymers,* Vol. 1, Elsevier Applied Science Publishers, Ltd., Barking, UK, 1982, pp. 195–273.

J. E. Dohany and L. E. Robb in M. Grayson, ed., *Kirk-Othmer Encyclopedia of Chemical Technology,* 3rd ed., Vol. 2, Wiley-Interscience, New York, 1980, p. 64.

A. J Lovinger, *Science* **220**, 1115 (1983).

JULIUS E. DOHANY
JAMES S. HUMPHREY
Pennwalt Corporation

VINYLIMIDAZOLE POLYMERS.

See POLYIMIDAZOLES.

VINYL ISOCYANATE POLYMERS.

See ISOCYANATE-DERIVED POLYMERS.

VINYL KETAL POLYMERS.

See VINYL ACETAL POLYMERS.

VINYL KETONE POLYMERS

The polymerization of α,β-unsaturated ketones can be initiated by free-radical, cationic, and anionic catalysts. Both vinyl and isopropenyl ketones are reactive monomers, polymerizing spontaneously on exposure to heat or sunlight. The investigation of vinyl ketone polymerization has been instrumental in the development of fundamental polymer chemistry. It has been shown that poly(methyl isopropenyl ketone) is a high molecular weight polymer by converting the polyketone to a

Table 3. Electrical Properties of Poly(vinylidene Fluoride) Homopolymer

Property	60 Hz	10^3 Hz	10^6 Hz	10^9 Hz
dielectric constant at 25°C	9–10	8–9	8–9	3–4
dissipation factor	0.03–0.05	0.005–0.02	0.03–0.05	0.09–0.11
volume resistivity, ohm·m				2×10^{12}
dielectric strength, short time, V/2.54×10^{-5} m				
0.003175 m thickness				260
0.000203 m thickness				1300

series of polymeric derivatives that display common physical properties and that vinyl monomers undergo head-to-tail addition by cyclizing poly(vinyl methyl ketone) to form enchained six-membered rings. The clear, tough polymers formed from methyl isopropenyl ketone exhibit an elasticity similar to ivory; thus the first attempt at commercialization involved the manufacture of billiard balls. This application, as well as most subsequent uses, was thwarted by the poor thermal and photochemical stability of vinyl ketone polymers. Currently, applications that exploit their inherent instability, ie, polymeric photosensitizers and photoresists, are revitalizing interest in polyketones.

Monomers

Physical Properties. Vinyl ketones are colorless, flammable liquids with strong irritating odors. With the exception of methyl vinyl ketone, the monomers exhibit limited (<5%) solubility in water and are miscible with most organic solvents. Vinyl methyl ketone (VMK) is miscible with water in all proportions, but the monomer can be isolated by saturation of the solution with salt.

The physical constants of several polymerizable α,β-unsaturated ketones are summarized in Table 1.

The conformation, energies, and electronic absorption spectra of vinyl alkyl ketones have been calculated and correlated with experimental data.

Chemical Properties. Vinyl ketones serve as important chemical intermediates in the synthesis of steriods, vitamin A, bicyclic derivatives, and heterocyclic compounds. In addition to the normal reactions of the vinyl and the carbonyl groups, α,β-unsaturated ketones readily undergo 1,4-additions. Their activity in Diels-Alder reactions is unusual in that they behave both as dienes and dienophiles.

Monomer Synthesis

Most of the syntheses are carried out at low pH to minimize the base-catalyzed condensation of the vinyl ketones. Alternatively, vinyl ketones can be generated *in situ* from β-(diethylaminoethyl)ketones if the reaction must be run in basic media. One useful technique for preparing both vinyl and isopropenyl derivatives involves a base-catalyzed condensation of formaldehyde with methyl or ethyl ketones, respectively. Preparation of vinyl ketones via the Mannich reaction is the most general approach to the synthesis of α,β-unsaturated ketones. A variant of the Mannich synthesis using *N,N*-dimethylmethyleneimmonium chloride as an intermediate has proved more amenable to large-scale isopropenyl ketone preparations.

Polymerization of α,β-Unsaturated Ketones

Free Radical. Vinyl ketones are among the most reactive monomers and tend to polymerize spontaneously when uninhibited. Among the common radical initiators, azobisisobutyronitrile (AIBN) is considered the best radical source for producing polymers with good color stability; all other initiators tend to leave acid residues or to degrade the polymer during the polymerization. Although bulk polymerization is feasible, better results are obtained in solvents such as benzene, cyclohexane, or petroleum ether that dissolve the monomer but not the polymer. Precipitation polymerization produces higher rates of reaction as well as higher molecular weights than homogeneous solutions under comparable conditions. However, solution polymerization in toluene or acetone yields polymer with satisfactory properties for most studies.

Ionic Polymerization. Ionic initiators can also effect the polymerization of vinyl ketones. Only one cationic system has been reported; vinyl methyl ketone yields a solid polymer when

Table 1. Physical Properties of Polymerizable α,β-Unsaturated Ketones, $R^1{-}C({=}O){-}R^2$

R^2	Bp, °C[a]	d_4^{20}	n_D^{20}	Yield[b], %
R^1 = vinyl				
methyl	81.4	0.8636	1.4086	87
	32 (16)			
ethyl	102.2 (99)	0.8468	1.4192	
	38 (8)			
isopropyl	49 (10)	0.847	1.4303	83.4
tert-butyl	59–60 (14)		1.4222[c]	61.7
phenyl	108–112 (1.7–1.8)	1.060	1.5522	82.2
	58–60 (0.03)		1.5580	
vinyl	41 (8.3)	0.8839	1.4440	54.6
isopropenyl	59 (13)	0.8906	1.4540	64.3
R^1 = isopropenyl				
methyl	97–98	0.8551	1.4161	83
ethyl	119–120	0.8473	1.4361	70
isopropyl	64–65 (13)	0.8579	1.4410	38
tert-butyl	64–65 (8.8)	0.9357	1.4211	46
phenyl	60–61 (0.3)	1.0071	1.5379	78
benzyl	94–95 (0.3)	1.0060	1.4840	42
R^1 = 1-fluorovinyl				
methyl	71.3			58[d]

[a] At 101.3 kPa (760 mm Hg) unless otherwise noted in parentheses in kPa. To convert kPa to mm Hg, multiply by 7.5.
[b] Maximum yield obtained from the corresponding methyl or ethyl ketone via a Mannich reaction, unless otherwise noted.
[c] At 15°C.
[d] Phosphoric acid dehydration of 1-fluoro-2-hydroxyethyl methyl ketone.

boron trifluoride etherate is added to a mixture of monomer and CO_2 in petroleum ether. Oligomers can be produced from VMK using group-transfer techniques. However, the primary focus of ionic-initiator studies has been on anionic systems. Stereoregular polymers are formed in the presence of lithium alkyls, and Grignard or zinc alkyl/alkali metal complexes.

Kinetics of Polymerization. The uncatalyzed thermal polymerization of isopropenyl methyl ketone (IMK) has been studied in bulk and in either aromatic or aliphatic hydrocarbons. No induction period is detected nor are the initial rates influenced by the nature of the solvent. Moreover, the polymerization in cyclohexane is autocatalytic; this is consistent with the gel effect observed in precipitation polymerization. In a study of both bulk and solution processes, it has been shown that polymerization rates are one-half order in initiator and first order in monomer. The energy of activation has been calculated to be 21 kJ/mol (5 kcal/mol), a value that compares favorably with the value of 20 kJ/mol (4.8 kcal/mol) reported for ^{60}Co-initiated polymerization of VMK.

Copolymerization. The copolymerization of α,β-unsaturated ketones has been explored in efforts to improve the poor chemical and thermal stability of homopolymers. Copolymers containing various ketonic monomers have been studied as models for the photodegradation of polymers in the environment as well as materials for photoresists. Even ketonic monomers that do not homopolymerize can be incorporated into copolymers.

Anionic copolymerization of vinyl methyl ketone with styrene yields copolymers containing a minimum of 98% vinyl methyl ketone for all initial monomer compositions. β-Substituted-α,β-unsaturated ketones can also be copolymerized. Benzalacetophenone copolymerizes with butadiene, styrene, and methyl methacrylate when emulsion techniques are used. The copolymerizability of 4-cyclopentene-1,3-dione, a monomer similar in structure to maleic anhydride, has been studied; acrylonitrile and methyl methacrylate yield normal monomer feed to copolymer composition relationships, but anomalous results are obtained for the styrene copolymers.

Polymer Properties

All vinyl ketone polymers prepared with free-radical initiators are amorphous materials with low softening points and poor thermal or chemical stability. Amorphous poly(vinyl methyl ketone) that softens between 40 and 60°C is soluble in the monomer, acetone, methyl ethyl ketone, tetrahydrofuran, dimethylformamide, acetic acid, ethyl acetate, dioxane, pyridine, and chloroform, and is insoluble in aromatic and aliphatic hydrocarbons, water, ethyl ether, and carbon tetrachloride. The density of the polymer is 1.12 g/mL, corresponding to a 25% contraction during polymerization. Colorless samples have a refractive index of 1.5 and exhibit a strong ultraviolet absorption around 290 nm.

The molecular weights of vinyl ketone polymers are relatively low due to the lability of the protons α to the carbonyl group. The chain-transfer constant to methyl isopropenyl ketone (C_M) is 4.0×10^{-4} at 80°C. Chain-transfer constants to initiator and several solvents also have been reported. The molecular weight–viscosity relationships used to estimate the degree of polymerization have been summarized.

Degradation Reactions

Thermal and Chemical Processes. Both vinyl and isopropenyl alkyl ketones lose water at temperatures above 250°C to yield glassy, red, noncross-linked products, probably via intramolecular aldol reactions. Chlorinated solvents such as tetrachloroethane or chloroform also catalyze the aldol reaction of poly(methyl isopropenyl ketone). In contrast to the thermal degradation, amine- or alkali-catalyzed degradation occurs by a chain process.

Photochemical Processes. Irradiation of dioxane solutions or thin films of poly(vinyl methyl ketone) with monochromatic light at 313 nm produces an initial rapid reduction in molecular weight, followed by a slower process that is accompanied by the evolution of acetaldehyde, carbon monoxide, and methane. These results can be interpreted by assuming a concomitant occurrence of Norrish Type I (α-scission) and Type II (γ-H abstraction and scission) ketone cleavages (see PHOTOCHEMISTRY).

Chemical Transformations. The carbonyl functional group is one of the most reactive groups that can be incorporated into a polymer substrate. Oximes of vinyl ketones have been prepared in 70–90% yields. However, the reaction of poly(vinyl methyl ketone) with hydrogen cyanide, phenylhydrazine, semicarbazide, mercaptans, or amines is complicated by the simultaneous cyclization of adjacent methyl ketone groups. Although the expected derivative is formed to some extent, the major reaction is usually cyclization. A unique cyclization reaction is observed when poly(phenyl vinyl ketone) is treated with trityl perchlorate in refluxing nitromethane–acetic acid mixtures; a polymer composed of up to 50% 3,5-disubstituted 2,6-diphenylpyrylium perchlorate units is isolated.

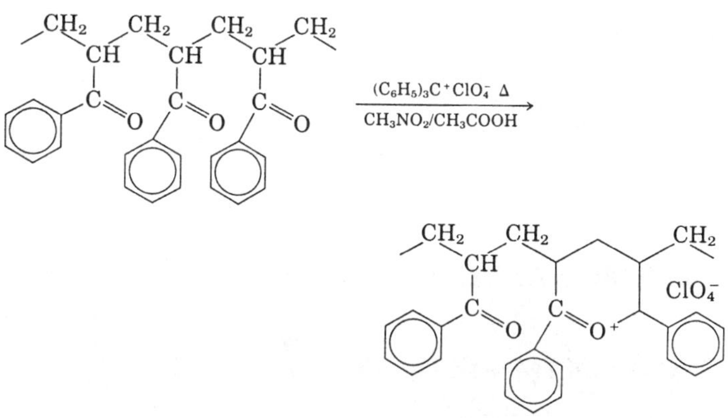

Derivatives of poly(vinyl ketones) also have been used as precursors for the preparation of polypyridines.

Polyketones can be reduced to secondary alcohols by catalytic hydrogenation over platinum oxide or reaction with lithium aluminum hydride or potassium borohydride. The products are poly(ketoalcohols), since a maximum conversion of only 50% is attained.

A promising application of polyketones is their use as polymeric photosensitizers. Poly(vinyl phenyl ketone) (PVPK) dissolved in deaerated benzene with an excess of d,l-phenylethanol undergoes nearly complete photoreduction to a polypinacol on irradiation with a strong ultraviolet source.

Poly(vinyl acetophenone) as well as copolymers of vinyl acetophenone and styrene have been prepared. Treatment of polymers containing acetophenone substituents with benzaldehyde in the presence of concentrated H_2SO_4 introduces benzalacetophenone units that dimerize on exposure to light. Photosensitive cross-linking agents of this type are very useful in the preparation of new photoresist resins for use in photocopiers or printing plates.

Analytical Techniques

α,β-Unsaturated ketones can be analyzed by polarographic reduction, bromate bromide titration, and quantitative oxime formation. The ketones, as well as their intermediate Mannich bases, can also be analyzed by gas chromatography.

Health and Safety

Vinyl ketones are generally toxic, lacrimatory compounds. Dermatitis results from exposure to vinyl methyl ketone vapor or liquid. Methyl isopropenyl ketone is less irritating, but prolonged exposure to its vapor should be avoided. The physiological properties of the higher homologues have not been reported. Normal precautions for handling toxic, flammable liquids should be observed.

BIBLIOGRAPHY

T. Tsuruta, R. Fujio, and J. Furukawa, *Makromol. Chem.* **80**, 172 (1964).

A. Allio and P. Pino, *Helv. Chim. Acta* **57**, 616 (1974).

T. Otsu and H. Tanaka, *J. Polym. Sci. Polym. Chem. Ed.* **13**, 2605 (1975).

WILLIAM H. DALY
Louisiana State University

VINYL PLASTICS

In common usage, the term vinyl plastics (or "vinyls") refers especially to those plastics derived from vinyl chloride and vinyl acetate (see VINYL CHLORIDE POLYMERS; VINYL ESTER POLYMERS).

VINYL PLASTISOLS. See VINYL CHLORIDE POLYMERS.

VINYL POLYMERIZATION

The term vinyl polymerization is used to denote addition polymerization of vinyl compounds, $CH_2{=}CHX$, where X is an alkyl group, aryl group, OR, OCOR, Cl, COOR, CN, etc (see also ADDITION POLYMERIZATION).

VINYL PROPIONATE POLYMERS. See VINYL ESTER POLYMERS.

VINYLPYRIDINE POLYMERS

Vinylpyridines are an important class of polymers exhibiting interesting properties due to the presence of the nitrogen atom in the pyridine ring. The weakly basic nitrogen atom makes possible a variety of reactions on vinylpyridine polymers, eg, reaction with acids, quaternization, and complexation of metals. The heteroatom is also responsible for unique characteristics in the polymerization mechanism. Vinylpyridine polymers are particularly important in applications as polyelectrolytes (qv), polymeric reagents, and in electrical applications.

A large number of vinylpyridine monomers have been prepared and polymerized. The two most common ones are 2-vinylpyridine (**1**) and 4-vinylpyridine (**2**).

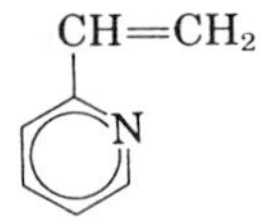
(1)

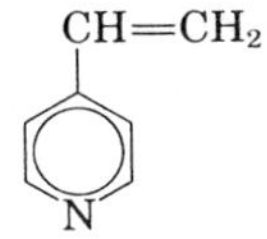
(2)

Monomer

Vinylpyridine monomers are generally prepared by dehydrogenation of the corresponding alkylpyridine or by the dehydration of hydroxyalkylpyridine according to equation 1, where X is either —OH or hydrogen.

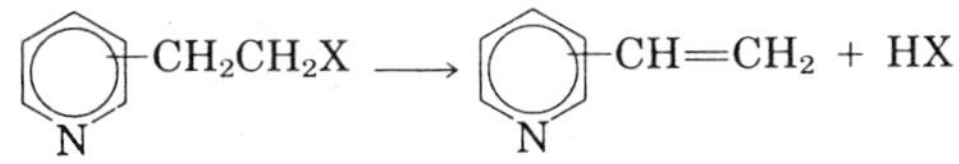
(1)

Toxicity of Monomers. The toxicity of 4-vinylpyridine and 4-β-hydroxylethylpyridine is approximately twofold higher than 2-vinylpyridine and 2-β-hydroxylethylpyridine. All compounds show narcotic effects in rats and mice after administration via stomach or inhalation. These compounds also cause irritation of the skin and mucous membrane. The monomer 4-vinylpyridine is classified as highly toxic material and its safe exposure level in the working zone is 0.5 mg/m^3; 2- and 4-β-hydroxylethylpyridine are moderately toxic with a safe exposure level of 10 mg/m^3.

Polymerization

The electronic structure of vinylpyridine monomers is similar to that of styrene and thus the radical polymerizabilities of the two are comparable. However, the presence of the heteroatom on the pyridyl ring results in interesting mechanistic implications during anionic or coordination polymerization.

Stereoregular Polymers

The first stereoregular poly(2-vinylpyridine) was prepared with organometallic initiators, eg, $Mg(C_2H_5)_2$. The overall stereochemistry of polyvinylpyridines is a function of the type of initiator, solvent, temperature, etc. In the ^{13}C-nmr spectrum of poly(2-vinylpyridine) obtained in CD_3OD, the C_2 and C_6 ring carbons clearly indicate splitting due to stereochemistry. For the C_2 carbon, the isotactic peak is at the lowest field and the syndiotactic at higher field. Stereochemical resolution is also observed in the methine and methylene regions. Poly(2-vinylpyridine) obtained by radical polymerization is atactic, whereas when Grignard reagents are used as catalysts, predominantly isotactic polymers are obtained. With poly(4-vinylpyridine) only atactic polymers are obtained with both radical and Grignard initiators. The location of the nitrogen atom is responsible for the difference in the stereoregularity of the two polymers; intramolecular coordination favors the formation of isotactic polymers. Some new organometallic initiators such as $Cp^*Hf(C_6H_{10})CH_3$ and $Cp^*Hf[(C_6H_{10})(CH_3)_2]_2Mg$, where $Cp^* = \eta^5\text{-}C_5(CH_3)_5$, at 25°C in toluene give highly (>97%) isotactic poly(2-vinylpyridine) in good yields. Predominantly syndiotactic poly(2-vinylpyridine) has been obtained using $t\text{-}C_4H_9Li/Al(C_2H_5)_3$ as the initiator in toluene at −78°C.

Block Copolymers

Block and graft copolymers of vinylpyridine monomers are important as emulsifying agents, thermoplastics, and membranes. Block polymers are most conveniently prepared by the sequential addition of monomers to anionic initiators.

The interesting properties of block copolymers result from the presence of two phases. Differential thermal analysis of poly(styrene-*b*-2-vinylpyridine) block copolymers show two transition temperatures, indicating the formation of two phases.

Vinylpyridines in Polymer-supported Catalysis

An important application of vinylpyridine polymers is as polymer supports in reaction catalysis. Vinylpyridine polymers have been used as supports in oxidation and reducing reactions.

Metal Adsorption and Flocculation

The presence of the weakly basic nitrogen in the ring makes vinylpyridine polymers of use in flocculation and adsorption of metals. High molecular weight poly(2- and 4-vinylpyridine)s in small amounts (3 parts/10^3 parts solids) are used to destabilize clay dispersions and clarify SiO_2 dispersions. The flocculation efficiency increases with the molecular weight of the polymer.

Electrical Properties of Vinylpyridine Polymers

Doping of poly(2-vinylpyridine)s with iodine or organic electron acceptors (eg, TCNQ) results in the formation of semiconductors with conductivities of 10^{-4} and 10^{-2} S/cm ($ohm^{-1}cm^{-1}$). Block copolymers of 2-vinylpyridine and styrene doped with I_2 have been studied as potential semiconductors; the conductivity values are quite low due to the presence of the nonconducting polystyrene block. Polyvinylpyridines have been employed as electron beam resists, an attractive feature being the outstanding plasma etch resistance due to the aromatic nature of the polymer.

BIBLIOGRAPHY

I. M. Khan, T. E. Hogen-Esch, *Macromolecules* **20,** 2335 (1987).

K. Matsuzaki, T. Matsubara, and T. Kanai, *J. Polym. Sci. Polym. Chem. Ed.* **15,** 1573 (1977).

J. M. J. Frechet, *Polym. Prepr. Am. Chem. Soc. Div. Polym. Chem.* **24**(2), 340 (1983).

M. L. Hallensleben, *Eur. Polym. J.* **7,** 1197 (1971).

ISHRAT M. KHAN
Atlanta University

VINYLPYRIDINIUM SALTS

The preparation and polymerization of 2-vinylpyridinium, 4-vinylpyridinium, and 2-methyl-5-vinylpyridinium salts have generated considerable academic and industrial interest. The polymers of these isomeric vinylpyridinium salts have been employed commercially as flocculating agents in water treatment, in photography, in ion-exchange resins, and in electron beam resists. *N*-Vinylpyridinium salts are readily made and polymerized by added radical initiators. These polymers are not considered in this article; rather, the spontaneous ionic polymerization of the salts named above is covered. There is controversy concerning the mechanism of the polymerizations.

Monomer Preparation

Many stable vinylpyridinium salts have been prepared through alkylation, protonation, or acylation reactions of 2-vinylpyridine, 4-vinylpyridine, and 2-methyl-5-vinylpyridine. For the synthesis of 2- and 4-vinylpyridinium salts in organic or aqueous media, facile quaternization or protonation rates that give counterions of low nucleophilicity and precipitation of the monomer are required, thus circumventing spontaneous polymerization.

Pyridylethylation Initiation and Zwitterionic Propagation

Originally it was reported that stable 4-vinylpyridinium salts could not be prepared from alkylation or protonation (with either monomeric or polymeric acids) since facile spontaneous polymerization would occur, giving the corresponding poly(4-vinylpyridinium salt). Initiation was thought to result from a nucleophilic attack of a counterion on the β-position of the vinylpyridinium's activated double bond, resulting in the formation of an active zwitterion existing as a pyridone methide through resonance stabilization, as in equation 1.

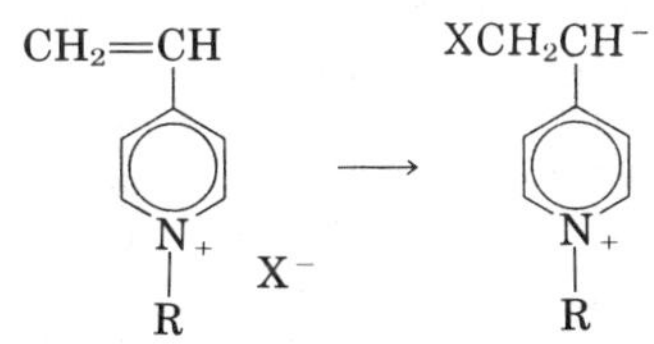

(1)

Poly(4-vinylpyridinium salts) could then be formed from the zwitterion adding to the polarized double bond of a 4-vinylpyridinium ion, thereby regenerating the active zwitterion end group that can further propagate in the same fashion.

Contrary to the previously reported results, stable monomeric salts were prepared from 4-vinylpyridine with several alkylating agents and monomeric acids.

The propagation that forms the poly(4-vinylpyridinium salt) occurs by the specific addition of the zwitterion to a 4-vinylpyridinium ion maintaining the zwitterionic end group.

Further investigation has revealed that 2-vinylpyridine is also capable of forming stable monomeric salts and undergoes pyridylethylation reactions when quaternized, thus yielding the corresponding polyelectrolyte. However, when alkylated, 2-methyl-5-vinylpyridine forms stable monomeric salts that cannot undergo pyridylethylation initiation, since stable pyridone methides cannot be formed to give a resonance stabilized propagating anion.

Organization vs Kinetic Considerations

Data suggest that the spontaneous polymerization of vinylpyridinium salts in water (in which no added nucleophile is present) depends on monomer concentration above 1 *M*, and that no polymerization is observed below this 1 *M* concentration.

However, an experiment conducted at 25°C for 10 days has shown that polymers can in fact be obtained for 1,2-dimethyl-

5-vinylpyridinium methyl sulfate (DMVPMS) at concentrations as low as 0.75 *M*. Also, the yield decreases considerably as the monomer concentration decreases. This is a consequence of reaction rate rather than a lack of a critical monomer concentration. The kinetic order of 2.7 with respect to monomer concentration explains clearly why such a drastic change occurs at monomer concentrations below 1.0 *M*.

However, because of the current inconsistencies in the literature regarding the spontaneous polymerization of the salt DMVPMS, in which pyridylethylation reactions leading to zwitterionic propagation are excluded, the initiation of this mechanism is as yet unknown; the propagation is most likely by radical means.

Copolymerization

Attempts have been made to copolymerize other monomers in the spontaneous polymerization process to determine the mechanistic pathway of the quaternization of 4-vinylpyridine. During the reaction between ethyl bromide and 4-vinylpyridine in acrylonitrile and stryene media, the polymers produced contain no nitrile or styrene units. Thus, under the reaction conditions employed, styrene and acrylonitrile behave as solvents and are not copolymerized with the 4-vinylpyridinium salt. This supports the fact that the usual anionic and cationic mechanisms are not operative.

During studies of the spontaneous polymerization mechanism, a new class of zwitterionic monomeric salts in which both cationic and anionic components of the salt are individually polymerizable has been prepared. The ion pair-comonomer 4-vinylpyridinium *p*-styrenesulfonate has been isolated from the protonation of 4-vinylpyridine with styrene sulfonic acid in an inert organic solvent. The isolation is difficult because this compound polymerizes readily when solvated, particularly by water, or when heated. The resulting polymer is equimolar in both components.

Similarly, α-methylstyrene can be copolymerized with 4-vinylpyridinium chloride producing alternating copolymer. In each instance an equimolar copolymer results, suggesting that a charge-transfer initiation may be operative in the mechanism of this alternating copolymerization.

Recently, an anionic–zwitterionic copolymer of equimolar composition has been prepared as a result of the spontaneous nature of the polymerization of 1(3-sulfopropyl)-2-vinylpyridinium betaine with potassium *p*-styrenesulfonate.

BIBLIOGRAPHY

J. C. Salamone, B. Snider, and W. L. Fitch, *Macromolecules* **3**, 707 (1970).

J. C. Salamone, B. Snider, and W. L. Fitch, *J. Polym. Sci. Part A* **9**, 1493 (1971).

J. C. Salamone, M. K. Raheja, Q. Anwaruddin, and A. C. Watterson, *J. Polym. Sci. Polym. Lett. Ed.* **23**, 655 (1985).

V. A. Kabanov, K. V. Aliev, T. I. Patrikeeva, O. V. Kargina, and V. A. Kargin, *J. Polym. Sci. Part C* **16**, 1079 (1967).

JOSEPH C. SALAMONE
W. C. RICE
University of Lowell

N-VINYLPYRROLIDINONE POLYMERS. See *N*-VINYLAMIDE POLYMERS.

VINYL STEARATE POLYMERS. See VINYL ESTER POLYMERS.

VINYL SULFONE POLYMERS. See SULFUR-CONTAINING POLYMERS.

VINYLSULFONIC ACID POLYMERS. See ETHYLENESULFONIC ACID POLYMERS.

VINYL TILE. See FLOORING MATERIALS.

VINYLTOLUENE POLYMERS. See STYRENE POLYMERS.

VINYON FIBERS. See VINYL CHLORIDE POLYMERS.

VIRUSES. See POLYNUCLEOTIDES.

VISCOELASTICITY

The classical theory of elasticity deals with mechanical properties of elastic solids, for which, in accordance with Hooke's law, stress is always directly proportional to strain in small deformations but independent of the rate of strain or the strain history. The classical theory of hydrodynamics deals with properties of viscous liquids, for which, in accordance with Newton's law, the stress is always directly proportional to rate of strain but independent of the strain itself. These categories are idealizations, however; although the behavior of many solids approaches Hooke's law for infinitesimal strains, and that of many liquids approaches Newton's law for infinitesimal rates of strain, under other conditions deviations are observed. Two types of deviations may be distinguished.

First, when finite strains are imposed on solids (especially those soft enough to be deformed substantially without breaking) the stress–strain relations are much more complicated (non-Hookean); similarly, in steady flow with finite strain rates, many fluids (especially polymeric solutions and melts) exhibit marked deviations from Newton's law (non-Newtonian flow). The dividing line between infinitesimal and finite depends, of course, on the degree of precision under consideration, and varies greatly from one material to another.

Second, even if both strain and rate of strain are infinitesimal, a system may exhibit behavior that combines liquidlike and solidlike characteristics. For example, a body that is not quite solid does not maintain a constant deformation under constant stress but continues to slowly deform with time, or creeps. When such a body is constrained at constant deformation, the stress required to hold it diminishes gradually or relaxes. On the other hand, a body that is not quite liquid may, while flowing under constant stress, store some of the energy input instead of dissipating it all as heat; it may recover part of its deformation when the stress is removed (elastic recoil). When such bodies are subjected to sinusoidally oscillating stress, the strain is neither exactly in phase with the stress (as it would be for a perfectly elastic solid) nor 90° out of phase (as it would be for a perfectly viscous liquid) but is somewhere in between. Some of the energy input is stored and recovered in each cycle, and some is dissipated as heat. Materials whose behavior exhibits such characteristics are called viscoelastic. If both strain and rate of strain are infinitesimal, the behavior is linear viscoelastic; then, in a given experiment the ratio of stress to strain is a function of time (or frequency) alone, and not of stress magnitude.

The relations between stress, strain, and their time dependencies are in general described by a constitutive equation or rheological equation of state. If strains or rates of strain are finite, the constitutive equation may be quite complicated. If they are infinitesimal, however, corresponding to linear viscoelastic behavior, the constitutive equation is relatively simple; most of the material in this article falls under its jurisdiction.

Measurements of viscoelastic properties of polymers impart information about the nature and rates of the configurational rearrangements, and the interaction of the macromolecules in both their short-range and long-range interrelations. From the point of view of the physical chemist and the physicist, this provides a field of inquiry with unique features of interest. Investigation of viscoelastic properties of polymers has also been greatly stimulated by the practical importance of the mechanical behavior in the processing and utilization of rubbers, plastics, and fibers. As a result, a high proportion of all studies on viscoelasticity in the past four decades has been devoted to the viscoelasticity of polymers (see also AGING, PHYSICAL; FRACTURE; FATIGUE; MECHANICAL PROPERTIES; POLYMER CREEP; RHEOLOGICAL MEASUREMENTS).

Strain, Stress, and Linear Constitutive Equations for Simple Shear

The principal purpose of this article is to relate the viscoelasticity of polymers to their molecular structure and modes of molecular motion, and to describe the dependence of viscoelastic properties on molecular weight, molecular weight distribution, temperature, chemical structure, and other variables. Emphasis is given to the descriptive theories of viscoelasticity. For treatments of the corresponding molecular theories, numerous references are available.

Equations of Change. Experimental measurements of mechanical properties are usually made by observing external forces and changes in the external dimensions of bodies with simple shape, eg, a cube, disk, rod, or fiber. The connections between forces and deformations depend not only on the stress–strain relations (the constitutive equation), but also on a number of other relations, eg, the equation of continuity, expressing the conservation of mass,

$$\frac{\partial \rho}{\partial t} = -\sum_{i=1}^{3} \frac{\partial(\rho v_i)}{\partial x_i} \tag{1}$$

and the equation of motion, expressing the conservation of momentum

$$\rho\left\{\frac{\partial v_j}{\partial t} + \sum_{i=1}^{3} v_i \frac{\partial v_j}{\partial x_i}\right\} = \sum_{i=1}^{3} \frac{\partial \sigma_{ij}}{\partial x_i} + \rho g_j \tag{2}$$

where ρ is the density, t is the time, g_j is the component of the gravitational acceleration in the j direction, and σ_{ij} is the appropriate component of the stress tensor; x_i represents the three Cartesian coordinates in a space-fixed coordinate system; and v_i represents the components of the velocity of the material particles in the respective directions at these coordinates. A third condition concerns the symmetry of the stress tensor, expressing the conservation of angular momentum:

$$\sigma_{ij} = \sigma_{ji} \tag{3}$$

A fourth restriction, conservation of energy, can be omitted for a process at constant temperature, since the discussion in this article is limited to isothermal conditions.

However, many experiments are designed so that both sides of equation 1 are zero, and that in equation 2 the inertial and gravitational forces represented by the left side and by the last term of the right side are negligible. In this case, the internal states of stress and strain can be calculated from observable quantities by the constitutive equation and the simple conditions $\Sigma\, \partial\sigma_{ij}/\partial x_i = 0$. In the most simple cases in infinitesimal deformations, when the stress field is factorable in the following form,

$$\sigma_{ij}(x_1, x_2, x_3, t) = \sigma_{ij}^{0}(x_1, x_2, x_3)\cdot f(t) \tag{4}$$

the appropriate relations for viscoelastic materials involve the same geometrical form factors as in the classical theory of elasticity.

Infinitesimal Strain Tensor. In a viscoelastic body, like in a perfectly elastic one, the state of deformation at a given point is specified by a strain tensor that represents the relative changes in dimensions and angles of a small cubical element cut out at that position. Similarly, the state of stress is specified by a stress tensor that represents the forces acting on different faces of the cubical element from different directions.

For an infinitesimal deformation, the components of the strain tensor in Cartesian coordinates are

$$\gamma_{ij} = \begin{bmatrix} 2\dfrac{\partial u_1}{\partial x_1} & \dfrac{\partial u_2}{\partial x_1} + \dfrac{\partial u_1}{\partial x_2} & \dfrac{\partial u_3}{\partial x_1} + \dfrac{\partial u_1}{\partial x_3} \\ \dfrac{\partial u_1}{\partial x_2} + \dfrac{\partial u_2}{\partial x_1} & 2\dfrac{\partial u_2}{\partial x_2} & \dfrac{\partial u_3}{\partial x_2} + \dfrac{\partial u_2}{\partial x_3} \\ \dfrac{\partial u_1}{\partial x_3} + \dfrac{\partial u_3}{\partial x_1} & \dfrac{\partial u_2}{\partial x_3} + \dfrac{\partial u_3}{\partial x_2} & 2\dfrac{\partial u_3}{\partial x_3} \end{bmatrix} \tag{5}$$

where x_i and u_i are the coordinates of the point where the strain is specified and its displacement in the strained state, respectively, ie, $u_i = x_i - x_i^0$, where the superscript zero refers to the position of the particle in the unstrained state. The rate of strain tensor $\dot{\gamma}_{ij}$ is formulated similarly, with u_i replaced by v_i, the velocity of displacement.

Stress Tensor. Associated with a strain and its time dependence is a state of stress:

$$\sigma_{ij} = \begin{pmatrix} \sigma_{11} & \sigma_{12} & \sigma_{13} \\ \sigma_{21} & \sigma_{22} & \sigma_{23} \\ \sigma_{31} & \sigma_{32} & \sigma_{33} \end{pmatrix} \tag{6}$$

where σ_{ij} is the component, parallel to the j direction, of the force per unit area acting on the face of a cubical element perpendicular to the i direction. The normal stresses are taken as positive for tension and negative for compression. This convention, however, is not universal; the opposite signs are used in some treatments. The components of the stress tensor are illustrated graphically in Figure 1.

Constitutive Equation for Linear Viscoelasticity in Simple Shear. If the deformation is uniform (homogeneous), stress and strain

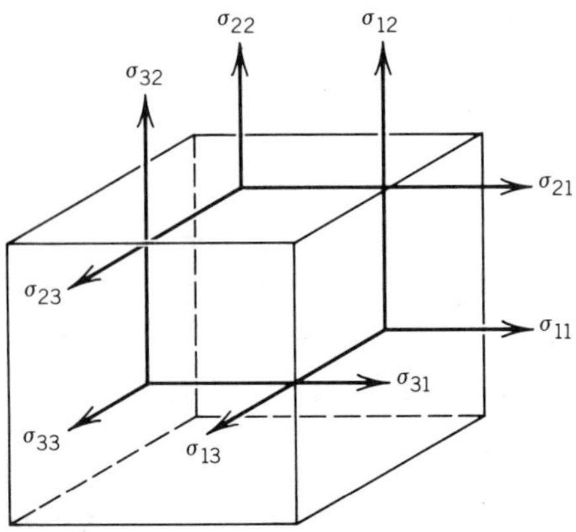

Figure 1. The components of the stress tensor.

components do not vary with position (are independent of x_i). If the movements are sufficiently slow to allow for the neglect of the effects of inertia and if gravitational effects do not play a role, a uniform stress system represents a solution of the equations of motion (2). For several specific types of uniform deformation, strain and stress tensors assume a relatively simple form. A form that corresponds to a commonly used experimental geometry is simple shear, where two opposite faces of the cubical element are displaced by sliding; stress and strains are then

$$\gamma_{ij} = \begin{pmatrix} 0 & \gamma_{12} & 0 \\ \gamma_{21} & 0 & 0 \\ 0 & 0 & 0 \end{pmatrix} \tag{7}$$

where $\gamma_{21} = \gamma_{12} = \partial u_1/\partial x_2 = \tan \alpha$, and

$$\sigma_{ij} = \begin{pmatrix} 0 & \sigma_{12} & 0 \\ \sigma_{21} & 0 & 0 \\ 0 & 0 & 0 \end{pmatrix} \tag{8}$$

in which the two nonzero components can be identified in Figure 1. Only simple shear is discussed in this article. To avoid awkward notation, the indexes are dropped and the abbreviations $\gamma_{12} = \gamma$ and $\sigma_{12} = \sigma_{21} = \sigma$ are used. The strain γ and the stress σ are functions of time and are connected by a constitutive equation that takes a very simple form for linear viscoelasticity and for simple shear.

This constitutive equation is based on the principle of superposition, stating that the effects of sequential changes of time-dependent strains are additive with respect to the resulting time-dependent stresses. This principle was stated as early as 1874 by Boltzmann; in mathematical form it reads

$$\sigma(t) = \int_{-\infty}^{t} G(t - t')\dot{\gamma}(t')\, dt' \tag{9}$$

where $\dot{\gamma} = d\gamma/dt$ is the shear rate, the function $G(t)$ is called the relaxation modulus for reasons that will be apparent later, and the integration is carried out over all past times t' up to the current time t. In equation 9, the prescribed strain history is assumed to be a piece-by-piece differentiable function of time without discontinuities. Therefore, equation 9 cannot be used directly for step-strain functions. However, partial integration of equation 9 gives the alternative formulation

$$\sigma(t) = G_0\gamma(t) + \int_{-\infty}^{t} \dot{G}(t - t')\gamma(t')\, dt \tag{10}$$

In this form, the superposition principle may be used for all strain histories that are piece-by-piece differentiable and do not contain more than a finite number of discontinuities of finite jumping heights; $\dot{G}(t - t')$ is an abbreviation for the time derivative of the relaxation modulus taken at elapsed time $t - t'$

$$\dot{G}(t - t') = \left[\frac{dG(y)}{dy}\right]_{y=t-t'} \tag{11}$$

whereas G_0 is an abbreviation for the expression

$$G_0 = \lim_{t\to+0} G(t) \tag{12}$$

An alternative constitutive equation can be written to express the strain in terms of the history of stress:

$$\gamma(t) = \int_{-\infty}^{t} J(t - t')\dot{\sigma}(t')\, dt' \tag{13}$$

or the equivalent:

$$\gamma(t) = J_0\sigma(t) + \int_{-\infty}^{t} \dot{J}(t - t')\sigma(t')\, dt' \tag{14}$$

where $J(t)$ is the creep compliance; $\dot{J}(t - t')$ is the creep rate taken at the elapsed time $t - t'$; and J_0 is the limiting value of $J(t)$ for $t \to 0$.

From the knowledge of the shear–relaxation modulus or the shear–creep compliance of a particular material, its stress–strain relations for any kind of experiment in shear with a prescribed time dependence of stress or strain can be predicted under the validity of the principle of superposition (ie, as long as the motions are sufficiently small and sufficiently slow). Materials obeying the laws of equations 10 and 14 are called linear viscoelastic materials.

Time-dependent Experiments in Shear

Shear–Relaxation Experiment. It is assumed that a shear strain is imposed suddenly at time $t = 0$ and held constant at the value γ_0:

$$\begin{aligned} \gamma(t) &= 0 \qquad \text{for } t < 0 \\ \gamma(t) &= \gamma_0 \qquad \text{for } t > 0 \end{aligned} \tag{15}$$

The stress necessary to comply with this condition is found, from equation 10, to be

$$\begin{aligned} \sigma(t) &= 0 \qquad \text{for } t < 0 \\ \sigma(t) &= \gamma_0 \cdot G(t) \qquad \text{for } t > 0 \end{aligned} \tag{16}$$

The physical meaning of the relaxation modulus $G(t)$ is apparent in terms of this experiment: it is the stress as a function of time, divided by the constant strain γ_0 necessary to create a stepwise increase of the strain from zero to γ_0 at time $t = 0$, and maintain the strain constant thereafter; $G(t)$ is a monotonically nonincreasing function of time.

It is, of course, not possible physically to realize the step-strain function in equation 15 because of the finite value of the inertia of the specimen and the clamp, which must be moved suddenly. Instead of equation 15, a very steep ramp function with curvatures at the beginning and the end results from this experiment. However, the resulting stress does not significantly differ from that in equation 16 for time arguments that are sufficiently longer than the time t_0 necessary to reach the constant strain value γ_0.

Creep Experiment. Alternatively, the shear stress may be applied suddenly at time $t = 0$ and held constant at the value σ_0:

$$\begin{aligned} \sigma(t) &= 0 \qquad \text{for } t < 0 \\ \sigma(t) &= \sigma_0 \qquad \text{for } t > 0 \end{aligned} \tag{17}$$

The resulting strain is found from equation 14 to be

$$\begin{aligned} \gamma(t) &= 0 \qquad \text{for } t < 0 \\ \gamma(t) &= \sigma_0 \cdot J(t) \qquad \text{for } t > 0 \end{aligned} \tag{18}$$

This shows the physical meaning of the creep compliance $J(t)$ that has the dimension of a reciprocal modulus and is a monotonically nondecreasing function of time. It is the strain, divided by the constant stress σ_0, under a stress, increasing stepwise from zero to σ_0 at the time $t = 0$, and remaining constant thereafter.

For a perfectly elastic solid, $J(t)$ and $G(t)$ are both time-independent and $J = 1/G$. However, for a viscoelastic material $J(t) \neq 1/G(t)$, because of the difference between the two experimental time patterns. It can be shown easily that the following integral equation, valid for all $t \geq 0$, relates $G(t)$ and $J(t)$

$$\int_0^t G(t')J(t - t')\,dt' = \int_0^t J(t')G(t - t')\,dt' = t \tag{19}$$

from which an inequality is easily derived:

$$J(t) \cdot G(t) \leq 1 \tag{20}$$

Stress–relaxation and creep experiments do not yield the characteristic functions $G(t)$ and $J(t)$ over the entire time domain $0 \leq t < \infty$ but only over a finite domain of the time scale $a < t < b$ called the experimental window of the measuring technique. The short-time limit of the experimental window is determined by the eigenfrequency of the initial vibrations of the system; it is usually situated between 0.01 and 1 s. The long-time limit is determined by the patience of the experimenter, the stability of the equipment and thermostating system, and often by the thermal stability of the sample. In very favorable cases, a creep measurement yields information from 0.1–10^7 s, thus over eight orders of magnitude of the time scale.

Sometimes, it is possible to continue the creep or stress–relaxation experiment sufficiently long to decide between the two possibilities in the long-time behavior of polymers: viscoelastic behavior without flow, shown by cross-linked polymeric systems; and viscoelastic behavior with flow, shown by uncross-linked polymeric systems.

Materials showing linear viscoelastic behavior without flow are called linear viscoelastic solids and are characterized by a finite value of the long-time limit of the creep compliance

$$\lim_{t=\infty} J(t) = J_e \tag{21}$$

designated as ultimate compliance. This behavior is connected with a positive nonzero value of the long-time limit of the relaxation modulus

$$\lim_{t=\infty} G(t) = G_e \tag{22}$$

designated as ultimate modulus or equilibrium modulus. Ultimate modulus and ultimate compliance are reciprocal: $G_e \cdot J_e = 1$.

Materials showing linear viscoelastic behavior with flow are called linear viscoelastic liquids. For those materials, the creep compliance shows the following asymptotic behavior:

$$J(t) \approx J_e + t/\eta \qquad \text{for } t \to \infty \tag{23}$$

The constants η and J_e are called (Newtonian) viscosity and steady-state compliance, respectively. In this case, the stress–relaxation modulus relaxes to the value zero:

$$G_e = 0 \tag{24}$$

Both cases may be treated simultaneously by using equations 22 and 23 as definitions. Then the class of viscoelastic solids is characterized by $G_e > 0$ and $\eta = \infty$, the class of viscoelastic liquids by $G_e = 0$ and $\eta =$ finite. Therefore, in both cases the following equation holds:

$$G_e/\eta = 0 \tag{25}$$

If J_0 and η are known, the creep compliance may be separated into three components:

$$J(t) = J_0 + \Psi(t) + t/\eta \tag{26}$$

that differ in behavior in a recovery experiment; J_0 is the instantaneous elastic compliance, $\Psi(t)$ the retarded viscoelastic compliance, and t/η the flow term. Comparison with equation 23 shows that $\Psi(t)$ has a finite long-time limit, equal to

$$\lim_{t\to\infty} \Psi(t) = J_e - J_0 \tag{27}$$

and therefore

$$\lim_{t\to\infty} \dot{\Psi}(t) = 0 \tag{28}$$

However, it rarely occurs that J_0 and η at the same temperature are known. In most cases, either the value of J_0 or the viscosity term t/η, or both, fall outside the experimental window of the method. If the creep measurement has been continued long enough to allow description of the compliance by means of equation 23, the viscoelastic liquid is said to be in the steady state of flow. Then the value of the viscosity may be determined from the creep rate. This, however, occurs only in the molten state.

Periodic or Dynamic Experiments. To supplement the above transient (ie, nonperiodic) experiments and provide information corresponding to shorter times, the stress may be varied periodically, usually harmonically with a frequency ν in cycles per second (Hz) or a circular frequency ω $(= 2\pi\nu)$ in radians per second. A periodic experiment at circular frequency ω is qualitatively equivalent to a transient experiment at time $t = 1/\omega$.

A large number of different dynamic experiments are in use. They may be classified according to the value of the ratio of the wavelength of the elastic waves to the dimensions of the specimen. Since the reciprocal of the wavelength and frequency is a crude approximation, this classification is also based on the frequency of the method.

Viscoelastic Recovery. Time profiles of stress and strain in a recovery experiment are shown in Figure 2, where the concepts of creep time t_0, recovery time t', residual deformation $\gamma(t_0,t')$, and recovery $r(t_0,t')$ are illustrated. The recovery experiment is defined by the following stress history:

$$\begin{aligned} \sigma(t) &= 0 \quad && \text{for } t < 0 \\ \sigma(t) &= \sigma_0 && \text{for } 0 < t < t_0 \\ \sigma(t) &= 0 && \text{for } t > t_0 \end{aligned} \tag{29}$$

which, after insertion into equation 14, yields for the strain during recovery

$$\gamma(t) = \sigma_0 \cdot [J(t) - J(t - t_0)] = \sigma_0 \cdot [J(t_0 + t') - J(t')]$$

or, by use of the separation in equation 26 gives

$$\gamma(t_0,t') = \sigma_0 \cdot [\Psi(t_0 + t') - \Psi(t') + t_0/\eta] \tag{30}$$

$$r(t_0,t') = \sigma_0 \cdot [J_0 + \Psi(t_0) + \Psi(t') - \Psi(t_0 + t')] \tag{31}$$

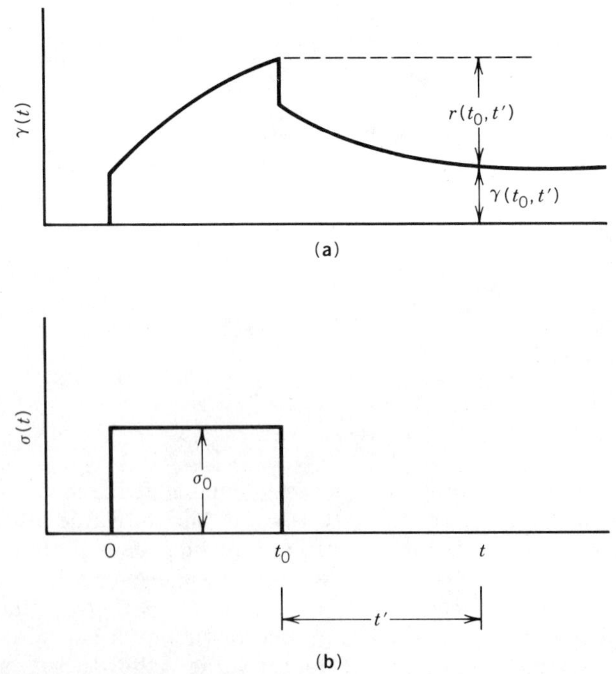

Figure 2. Time profile of (**a**) strain and (**b**) stress in a recovery experiment.

Therefore, the residual deformation is the sum of the flow term $\sigma_0 t_0/\eta$ and the difference of the retarded viscoelastic components of two creep experiments, one starting at time $t = 0$, the other at time $t = t_0$. The construction of the recovery curve is indicated in Figure 3.

Linear Viscoelastic Behavior under Three-dimensional Stress Distributions

Linear Constitutive Equations for Isotropic Viscoelastic Substances. If the superposition principle is generalized to a state of three-dimensional stress in an isotropic viscoelastic material, the following relations hold between the components of the stress and the strain tensor:

$$\begin{aligned} \sigma_{ij}(t) = {} & G_0 \left[\gamma_{ij}(t) - \frac{1}{3}\, \delta_{ij} \sum_k \gamma_{kk}(t) \right] + \frac{1}{2} K_0\, \delta_{ij} \sum_k \gamma_{kk}(t) \\ & + \int_{-\infty}^{t} \dot{G}(t - t') \left[\gamma_{ij}(t') - \frac{1}{3}\, \delta_{ij} \sum_k \gamma_{kk}(t') \right] dt' \\ & + \frac{1}{2}\, \delta_{ij} \int_{-\infty}^{t} \dot{K}(t - t') \sum_k \gamma_{kk}(t')\, dt' \end{aligned} \tag{32}$$

where δ is the Kronecker delta; $G(t)$ is the shear relaxation modulus (eq. 16); $K(t)$ has the physical significance of a volumetric relaxation modulus; and G_0 and K_0 are the corresponding limits of those functions for $t \to +0$. With the help of equation 32, it is possible to express the components of the

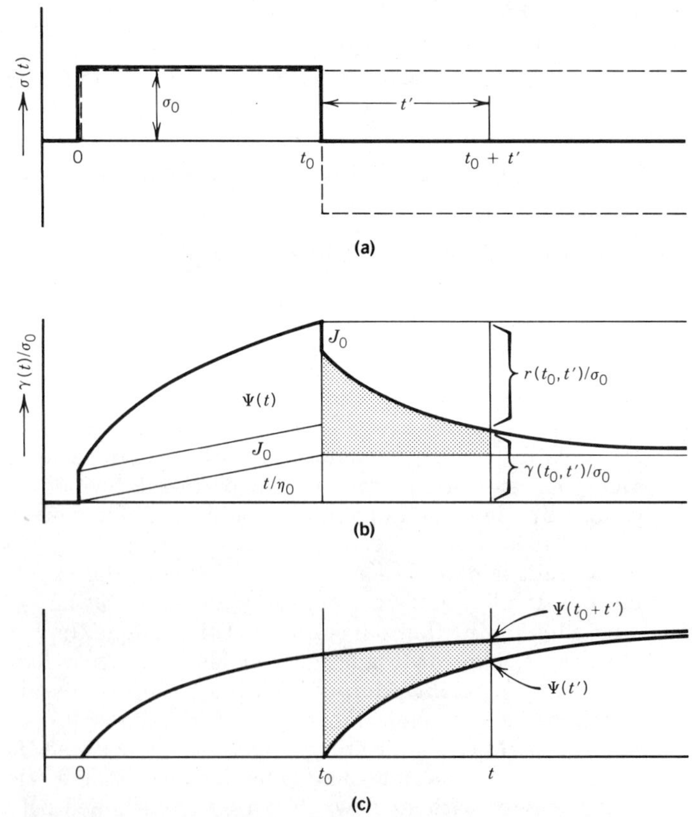

Figure 3. Recovery curves: (**a**) recovery experiment as sum of two creep experiments; (**b**), separation of the deformation in equation 26; (**c**) construction of the retarded viscoelastic component of the recovery.

strain tensor by those of the stress tensor:

$$\gamma_{ij}(t) = J_0 \left[\sigma_{ij}(t) - \frac{1}{3}\,\delta_{ij} \sum_k \sigma_{kk}(t) \right] + \frac{2}{9} B_0 \delta_{ij} \sum_k \sigma_{kk}(t)$$
$$+ \int_{-\infty}^{t} \dot{J}(t - t') \left[\sigma_{ij}(t') - \frac{1}{3}\,\delta_{ij} \sum_k \sigma_{kk}(t') \right] dt'$$
$$+ \frac{2}{9}\,\delta_{ij} \int_{-\infty}^{t} \dot{B}(t - t') \sum_k \sigma_{kk}(t')\, dt' \qquad (33)$$

where $J(t)$ is the shear creep compliance (eq. 18); $B(t)$ has the physical significance of a volumetric creep compliance; and J_0 and B_0 are the limits of these functions for $t \rightarrow +0$.

The requirements for the validity of equations 32 and 33 are the following:

1. The principle of superposition holds in its general form, ie, not only two different shear-stress histories should superimpose undisturbed, but also a shear-stress history and a history of isotropic pressure. This means that, for instance, the shear compliance should not be influenced by a superimposed hydrostatic pressure. When perceptible changes in volume occur, $G(t)$ and $K(t)$ are found to depend strongly on density (or hydrostatic pressure). Hence, the application of the linear equation 32 to viscoelastic behavior must be limited to extremely small volume changes. Under this restriction, equations 32 and 33 form the basis of the technical mechanics of polymers.

2. The deformations are infinitesimal; otherwise the use of the tensor components defined by equation 5 would not be justified. This condition is always fulfilled in the hard region of polymers. In the glass–rubber transition, in the rubbery region, or in the flow region, large deformations are possible without rupture, and this condition might be violated. For these cases, it is possible to generalize the linear equations to finite deformations, at least in the incompressible case. This becomes more complicated mathematically, and leads to the theory of viscoelastic liquids by Lodge.

However, finite deformations are often accompanied by nonlinear phenomena and the breakdown of the principle of superposition. In those cases, it is no longer possible to construct a useful general framework for viscoelastic behavior. Treatments must be limited from the beginning to certain types of material behavior, such as rubber–elastic behavior or nonlinear viscoelastic flow of polymer melts.

3. The material is isotropic. When polymers are processed by extrusion, injection molding, or sheet blowing, the materials are unlikely to behave isotropically. For those cases, it is possible to apply the constitutive equations of linear viscoelastic materials to anisotropic behavior. This does not lead to major mathematical complications. However, the number of independent characteristic functions, which is two for isotropic materials, becomes larger according to the symmetry relations left; accordingly, the difficulty in the experimental determination of the characteristic functions increases considerably.

Bulk Compression or Dilatation. In a cube of length a, under hydrostatic pressure p, with edges parallel to the coordinate axis, the components of the stress tensor are

$$\sigma_{ij} = -p\delta_{ij} \qquad (34)$$

Because of the pressure, all edges of the cube are shortened in the same ratio $1{:}(1 - b)$, and the components of the strain tensor become

$$\gamma_{ij} = -2b\delta_{ij} \qquad (35)$$

The volume of the cube decreases from $V_0 = a^3$ to $V = V_0 \cdot (1 - b)^3 \simeq V_0 \cdot (1 - 3b)$. Hence, the relative decrease in volume is

$$\Delta \equiv \frac{\Delta V}{V_0} = -3b - \frac{1}{2} \cdot (\gamma_{11} + \gamma_{22} + \gamma_{33}) \qquad (36)$$

Insertion of these equations into equations 32 and 33 gives the relations between the (time-dependent) hydrostatic pressure and the (time-dependent) relative decrease in volume:

$$p(t) = K_0\Delta(t) + \int_{-\infty}^{t} \dot{K}(t - t')\Delta(t')\, dt' \qquad (37)$$

$$\Delta(t) = B_0 p(t) + \int_{-\infty}^{t} \dot{B}(t - t')p(t')\, dt' \qquad (38)$$

These equations express the principle of superposition for isotropic compression; they are analogous to equations 10 and 14, respectively. All conclusions that have been drawn for simple shear may be readily derived for this case.

Bulk compression would not be expected to involve changes in long-range configuration or contour shape, and in fact the differences between polymers and simple liquids and solids are not so striking in compression as they are in shear.

Simple Extension. If a specimen of an isotropic material is elongated in one direction by application of a unidirectional tensile stress, the dimensional changes in the two mutually perpendicular directions are equal; this experiment is called simple extension. This state of stress is found in the thin part of a tensile specimen tested under simple tension.

Examples of Viscoelastic Behavior

Linear Viscoelastic Behavior of Amorphous Polymers in Shear. When the creep compliance or the relaxation modulus for a polymer is plotted against time, or any of the dynamic functions is plotted against frequency, the most striking feature is the enormous range of magnitudes the property can assume, changing over many factors of 10. Concomitantly, a still larger range of time or frequency is required on the abscissa scale to encompass these changes. Therefore, an overall survey of the behavior is only obtained when these functions are plotted on a logarithmic scale versus a logarithmic time or frequency axis.

The logarithm of the shear–creep compliance is plotted vs the logarithm of the creep time in Figure 4 for an amorphous polymer. The creep compliance shows a number of dispersion (transition) regions. Each region corresponds to a molecular movement that starts to contribute to the process of deformation at the particular time region. From left to right, the parts of the molecule involved increase in size.

The Temperature Dependence of the Glass–Rubber Transition. Because of the time–temperature shift superposition principle, a change in temperature shifts the position of the creep compliance, when plotted against the logarithm of time along the logarithmic time axis without any change in shape. Although this principle is valid only approximately and only within a

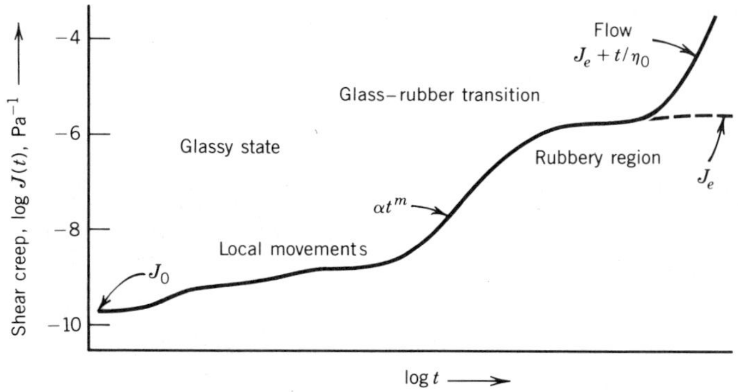

Figure 4. Shear–creep compliance as function of creep time. —, Amorphous uncross-linked polymers; ---, amorphous cross-linked polymers. To convert Pa^{-1} to in.^{2}/lb, multiply by 7000.

section of the time scale where the characteristic times of one single molecular process dominate the behavior, it is of utmost practical importance. This principle reduces the expenditure for the characterization of the shear behavior of the material. Instead of the determination of a function $J(t,T)$ that depends on two independent variables, the time t and the temperature T, it is sufficient to determine two functions, each of which depends on one independent variable only. These are the master creep compliance as a function of the reduced time, and the time–temperature shift function as a function of the temperature.

Linear Viscoelastic Behavior in Compression. Shear and bulk deformations are essentially different in character: one involves a change in shape at constant volume; the other involves a change in volume without any change in shape. Although certain molecular processes such as, for instance, the onset of the internal rotations of the backbone chain may be assumed to contribute to change in shape and volume, their effect on the respective compliances is different. The shear–creep compliance is changed by this process for several orders of magnitude (in the glass–rubber transition); the bulk compliance changes, approximately by a factor of 2.

Fewer experimental data are available on viscoelastic properties in bulk than in shear. Notwithstanding, the following qualitative rules may be stated: in general, $K(t)$, $[K'(\omega)$, $(\tan\delta)_k]$, and $G(t)$, $[G'(\omega)$, $(\tan\delta)_s]$ may be expected to show dispersions in the same regions of time (frequency) and temperature; the same molecular movements may contribute to a change in shape and volume. The dispersion steps in $K(t)$ and $K'(\omega)$ are much less pronounced than the dispersion steps in $G(t)$ and $G'(\omega)$; correspondingly, the maxima in $(\tan\delta)_k$ are much lower than those in $(\tan\delta)_s$ or $(\tan\delta)_e$. Viscous flow does not occur under hydrostatic pressure. Opposite to E and G, $K(t)$ and $K'(\omega)$ do not show the steep fall with temperature in the melt but remain approximately constant.

Viscoelastic Behavior in Large Deformations. *Strain Tensor.* When strains are not infinitesimal, the definitions of equation 5 are inapplicable; there are several alternative ways to define finite strain, eg, the strain tensor by Finger. If

$$x_i = x_i(x'_1, x'_2, x'_3; t, t') \tag{39}$$

are the Cartesian coordinates in a space-fixed coordinate system of the material point at time t, which had been at the coordinates x'_1, x'_2, x'_3 at a time t' in the past, then t' is called the reference time and t the observation time; the coordinates x'_i serve to specify the material point considered, and are called material coordinates or Lagrange coordinates. The x_i are called space coordinates or Euler coordinates. If equation 39 is specified for one fixed value of the reference time t', for all triplets x'_1, x'_2, x'_3, and for all t, the movement of all particles of the body is known.

The components of Finger's strain tensor are defined by

$$B_{ik}(x_1, x_2, x_3; t, t') = \sum_{m=1}^{3} \frac{\partial x_i}{\partial x'_m} \cdot \frac{\partial x_k}{\partial x'_m} \tag{40}$$

where the differentiations are to be performed at constant values of t and t', and where afterward the Euler coordinates are introduced as independent variables, using the equations inverse to equation 39. The three invariants of the Finger tensor are

$$I_1 = B_{11} + B_{22} + B_{33} \tag{41}$$

$$I_2 = \begin{vmatrix} B_{11}B_{12} \\ B_{21}B_{22} \end{vmatrix} + \begin{vmatrix} B_{22}B_{23} \\ B_{32}B_{33} \end{vmatrix} + \begin{vmatrix} B_{33}B_{31} \\ B_{13}B_{11} \end{vmatrix} \tag{42}$$

$$I_3 = \det(B_{ik}) \tag{43}$$

For most viscoelastic substances sufficiently soft to support substantial deformations without breaking, $K \gg G$ and hence volume changes can be neglected. In this case each admitted set of equation 39 must represent an isochorus deformation. The condition of incompressibility expresses itself by

$$I_3 = 1 \qquad \sum_{i=1}^{3} \frac{\partial v_i}{\partial x_i} = 0 \tag{44}$$

The Viscoelastic Liquid Model of Lodge. For infinitesimal deformations, the condition of incompressibility, equation 44, degenerates into

$$\sum_{i=1}^{3} \gamma_{ii} = 0 \tag{45}$$

Therefore, for an incompressible, linear viscoelastic material under infinitesimal deformations, equation 32 becomes simply

$$\sigma_{ik}(t) + p(t)\cdot\delta_{ik} = G_0\cdot\gamma_{ik}(t) + \int_{-\infty}^{t} \dot{G}(t - t')\cdot\gamma_{ik}(t')\,dt' \tag{46}$$

where

$$p(t) = -[\sigma_{11} + \sigma_{22} + \sigma_{33}]/3 \tag{47}$$

is the mean hydrostatic pressure. The value of the hydrostatic pressure remains undetermined, since an arbitrary hydrostatic pressure may always be superimposed on σ_{ik} without effecting a change of the state of strain.

For finite deformations, an equation of the same form as equation 46 may be written in coordinates that deform with the material and are transformed back into space-fixed coordinates. Several procedures give equations that have the appropriate invariance properties and reduce to equation 46 for infinitesimal deformations. A widely used nonlinear extension of equation 46 is that proposed by Lodge, which describes the rheological behavior of an elastic liquid. It can be represented by the constitutive equation

$$\sigma_{ik}(t) + p(t)\cdot\delta_{ik} = \int_{-\infty}^{t} m(t - t')\cdot B_{ik}(x_1, x_2, x_3; t, t')\, dt' \quad (48)$$

Here, $p(t)$ is the undetermined additive hydrostatic component of the stress tensor, but not identical to one third of its negative trace in this case; $B_{ik}(x;t;t')$ is the Finger tensor as defined in equation 40, and $m(t)$ is the memory function that is equal to the negative time derivative of the stress relaxation modulus of the liquid:

$$m(t) = -\dot{G}(t) \quad (49)$$

Since the material is assumed to be a liquid with fading memory, the following conditions are imposed additionally:

$$m(\infty) = 0 \qquad G(\infty) = G_e = 0 \quad (24)$$

Viscoelastic Liquid Model of Wagner. Wagner has generalized Lodge's theory by assuming that the entanglements are disentangled also by the reaction of the stressed molecular strands between them during the stressing period. The memory function in equation 48 depends on the state of strain of the network via both invariants of the Finger tensor:

$$\sigma_{ik}(t) + p(t)\cdot\delta_{ik} = \int_{-\infty}^{t} m(t - t'; I_1, I_2)\cdot B_{ik}(x_1,x_2,x_3; t,t')\, dt' \quad (50)$$

Furthermore, Wagner's theory assumes a factorization of the memory function into a factor depending on the elapsed time $t - t'$ (and the temperature) and a factor depending on the state of strain of the considered network (probably independent of temperature):

$$m(t - t'; I_1, I_2) = m(t - t')\cdot h(I_1, I_2) \quad (51)$$

The factor $h(I_1,I_2)$ has been called the damping function; it varies between unity (in the region of validity of Lodge's theory) at moderate strains, and zero for very large strains. Its meaning is the probability that the network junctions survive the state of strain characterized by the invariants I_1 and I_2.

For simple shear, both invariants are equal:

$$I_1 = I_2 = 3 + [\gamma(t) - \gamma(t')]^2 \quad (52)$$

and the damping function depends on the value of $|\gamma(t) - \gamma(t')|$ only. Equation 50 becomes

$$\sigma_{12}(t) = -\int_{-\infty}^{t} \dot{G}(t - t')\cdot h_s(|\gamma(t) - \gamma(t')|)\cdot[\gamma(t) - \gamma(t')]\, dt' \quad (53)$$

$$\sigma_{11}(t) - \sigma_{22}(t) = -\int_{-\infty}^{t} \dot{G}(t - t')\cdot h_s(|\gamma(t) - \gamma(t')|)\cdot[\gamma(t) - \gamma(t')]^2\, dt' \quad (54)$$

where

$$h_s(|\gamma(t) - \gamma(t')|) = h(I_1, I_2) \quad (55)$$

is the damping function in shear.

BIBLIOGRAPHY

J. D. Ferry, *Viscoelastic Properties of Polymers,* 3rd ed., John Wiley & Sons, Inc., New York, 1980.

J. Janeschitz-Kriegl, *Polymer Melt Rheology and Flow Birefringence,* Springer-Verlag, Berlin, 1983.

A. S. Lodge, *Body Tensor Fields in Continuum Mechanics,* Academic Press, Inc., New York, 1974.

M. H. Wagner, *Rheol. Acta* **15,** 133, 136 (1976); **16,** 43 (1977).

FRIEDRICH R. SCHWARZL
University of Erlangen-Nurnberg

VISCOSE.

See RAYON.

VISCOSITY-AVERAGE MOLECULAR WEIGHT.

See MOLECULAR WEIGHT DISTRIBUTION.

VOLUME RECOVERY.

See AGING, PHYSICAL.

VULCANIZATION

Useful rubber articles, such as tires and mechanical goods, cannot be made without vulcanization. Unvulcanized rubber is generally not very strong, does not maintain its shape after a large deformation, and can be very sticky; it has about the same consistency as chewing gum.

The accelerated sulfur vulcanization of general purpose high diene rubbers [eg, natural rubber (NR), styrene–butadiene rubber (SBR), and butadiene rubber (BR)] by sulfur in the presence of organic accelerators and of other rubbers, which are vulcanized by closely related technology [eg, ethylene–propylene–diene monomer (EPDM) rubber, butyl rubber (IIR), halobutyl rubbers, and nitrile rubber (NBR)] comprises more than 90% of all vulcanizations.

Definition

The term vulcanization is generally applied to rubbery or elastomeric materials. These materials forcibly retract approximately to their original shape after a large mechanically imposed deformation. Vulcanization can be defined as a process that increases the retractive force and reduces the amount of permanent deformation remaining after removal of the deforming force. Thus, vulcanization increases elasticity while it decreases plasticity. It is generally accomplished by the formation of a cross-linked molecular network (Fig. 1).

Effects of Vulcanization on Properties

At the molecular level, vulcanization causes profound chemical changes. The long rubber molecules (molecular weight usually

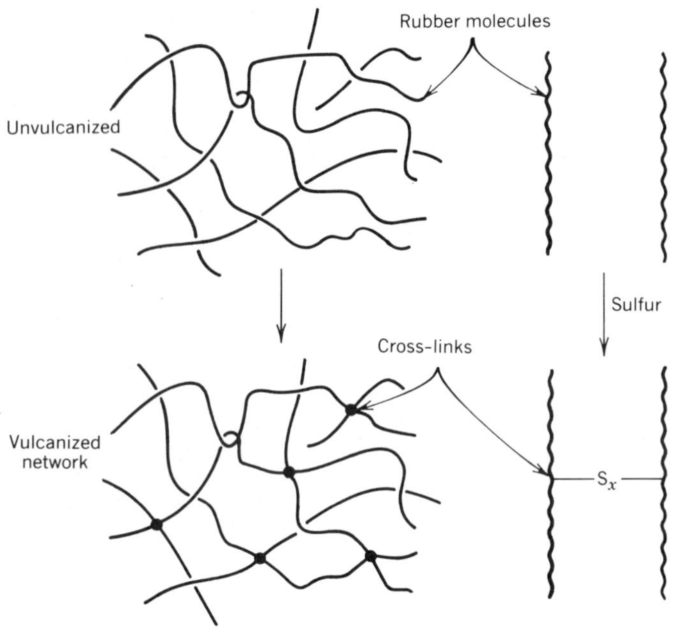

Figure 1. Network formation.

100,000–500,000) become linked in junctures (cross-links) spaced along the polymeric chains, with the average distance between junctures corresponding to a molecular weight between cross-links of about 4,000–10,000. As a result of this network formation, the rubber becomes essentially insoluble in any solvent and cannot be processed by any means that requires it to flow, eg, in a mixer, in an extruder, on a mill, on a calender, or during shaping, forming, or molding. Thus vulcanization can only occur after the rubber article is in its final form.

The main effects of vulcanization on use-related properties are illustrated by the idealization of Figure 2.

The properties shown in Figure 2 are functions not only of cross-link density, but are also affected by the types of cross-links, polymers, fillers, and other additives.

Reversion. Reversion is a term generally applied to the loss of network structures by nonoxidative thermal aging. It is usually associated with isoprene rubbers vulcanized by sulfur. It can be the result of too long a vulcanization time (overcure) or of hot aging of thick sections. It is most severe above about 155°C and occurs in vulcanizates containing a large number of polysulfidic cross-links. Although the mechanism is complex, most of the chemical changes occurring during the reversion of natural rubber are understood.

The term reversion is sometimes applied to other types of nonoxidative degradation, especially with respect to rubbers not based on isoprene. For example, thermal aging of SBR, which can increase cross-link density and cause hardening, has been called reversion because it can be the result of overcure.

The Vulcanization Process

Important characteristics related to the vulcanization process are the time elapsed before cross-linking starts, the rate of cross-link formation, and the extent of cross-linking at the end of the process. There must be sufficient delay or scorch resistance (resistance to premature vulcanization) to permit mixing, shaping, forming, and flowing in the mold. Then the formation of cross-links should be rapid and the extent of cross-linking must be controlled.

Both the rate of vulcanization after the scorch period and the final extent of vulcanization are measured by so-called cure meters. The most widely used are oscillating disk rheometers (Fig. 3) of the type introduced by the Monsanto Company around 1965.

The rheometer chart, or cure curve, gives a complete picture of the overall kinetics of cross-link formation and even cross-link disappearance (reversion) for a given rubber mix. In some cases, instead of reversion, a long plateau or even a marching cure can occur. The cure meter is, therefore, extensively used to control the quality and uniformity of rubber stocks.

Sulfur Vulcanization Without Accelerator

Initially, vulcanization was accomplished by heating elemental sulfur at a concentration of 8 parts per 100 parts of rubber (phr) for 5 h at 140°C. The addition of zinc oxide reduced the time to 3 h. Accelerators in concentration as low as 0.5 phr have since reduced the time to as little as 1–3 min. As a result, elastomer vulcanization by sulfur without accelerator is no longer of commercial significance. An exception is the use of about 30 or more phr of sulfur, with little or no accelerator, to

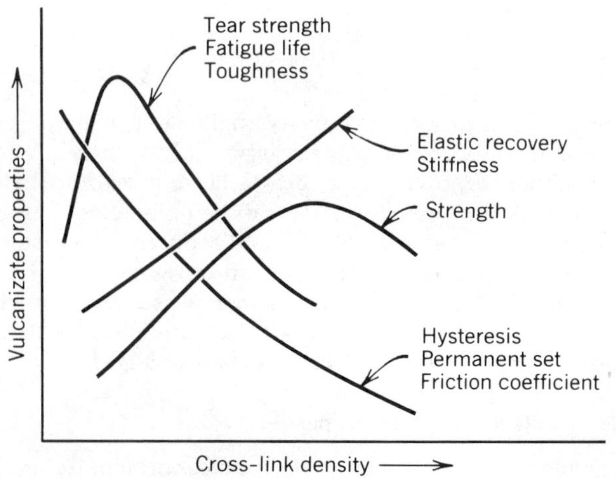

Figure 2. Vulcanizate properties as a function of the extent of vulcanization.

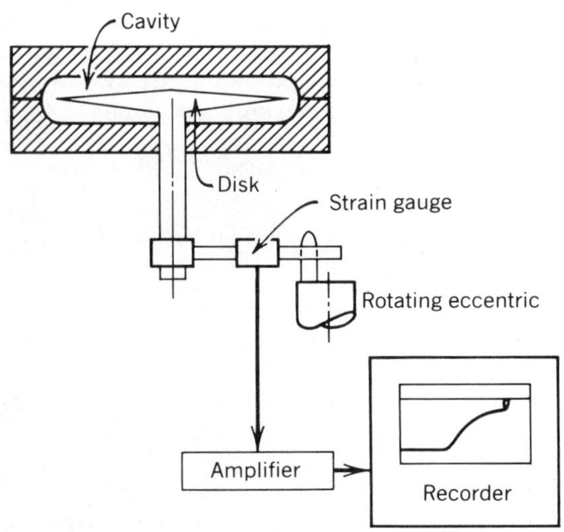

Figure 3. Oscillating disk rheometer.

produce molded products of hard rubber called ebonite (see RUBBER, HARD).

Even though unaccelerated sulfur vulcanization is of no commercial significance, its chemistry has been the object of much research and study, and it has remained controversial. Many slow reactions occur over the long period of vulcanization. Some investigators have felt that the mechanisms involved free radicals; others have promoted mechanisms involving ions.

Accelerated Sulfur Vulcanization

Accelerated sulfur vulcanization is the most widely used method. For many applications, it is the only rapid cross-linking technique that can, in a practical manner, give the delayed action required for processing, shaping, and forming before the formation of the intractable vulcanized network. It is used to vulcanize natural rubber (NR), synthetic isoprene rubber (IR), styrene–butadiene rubber (SBR), nitrile rubber (NBR), butyl rubber (IIR), chlorobutyl rubber (CIIR), bromobutyl rubber (BIIR), and ethylene–propylene–diene monomer (EPDM) rubber. The reactive moiety for all these elastomers can be represented by

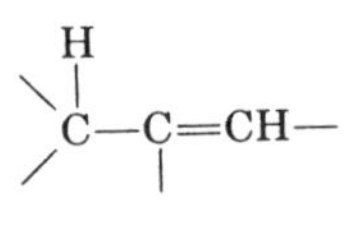

Typically, a recipe for the vulcanization system for one of the above elastomers contains:

	phr
zinc oxide	2–10
fatty acid (stearic)	1–4
sulfur	0.5–4
accelerator	0.5–2

Zinc oxide and the fatty acid are vulcanization activators. The fatty acid with zinc oxide forms a salt that can form complexes with accelerators and reaction products formed between accelerators and sulfur.

Frequently, mixtures of accelerators are used. Different types of accelerators impart vulcanization characteristics that differ with respect to scorch resistance and cross-linking rate. The premature-vulcanization inhibitor (PVI) *N*-cyclohexylthiophthalimide (CTP) is frequently used to independently control scorch resistance with little or no effect on the cross-linking rate.

Chemistry. The general reaction path of accelerated sulfur vulcanization is thought to be as follows. The accelerator reacts with sulfur to give monomeric polysulfides of the structure Ac—S_x—Ac, where Ac is an organic radical derived from the accelerator (eg, benzothiazyl). The monomeric polysulfides interact with the rubber to form polymeric polysulfides, eg, rubber—S_x—Ac. During this reaction, 2-mercaptobenzothiazole (MBT) is formed with a benzothiazole-derivative accelerator if the elastomer is natural rubber. (In SBR the MBT becomes bound to the elastomer molecular chain, probably as the thioether rubber—S—Ac.) With MBT itself as the accelerator in natural rubber, it first disappears and reforms with the formation of rubber—S_x—Ac. Finally, the rubber polysulfides react, either directly or through an intermediate, to give cross-links, rubber—S_x—rubber. There are obvious differences between accelerated and unaccelerated vulcanization. Better cross-linking efficiencies and higher cross-linking rates are obtained with accelerated vulcanization.

Zinc in Benzothiazole-accelerated Vulcanization. An increase in the concentration of fatty acid and hence an increase in the concentration of available Zn^{2+} raises the overall rate in the early reactions during the delay period that lead to the formation of rubber—S_x—Ac. However, the rate of cross-link formation is reduced but the extent of cross-linking is increased.

The lower specific rate of cross-link formation and the increased extent of cross-linking due to the presence of Zn^{2+} in benzothiazole-accelerated vulcanization have been explained as follows:

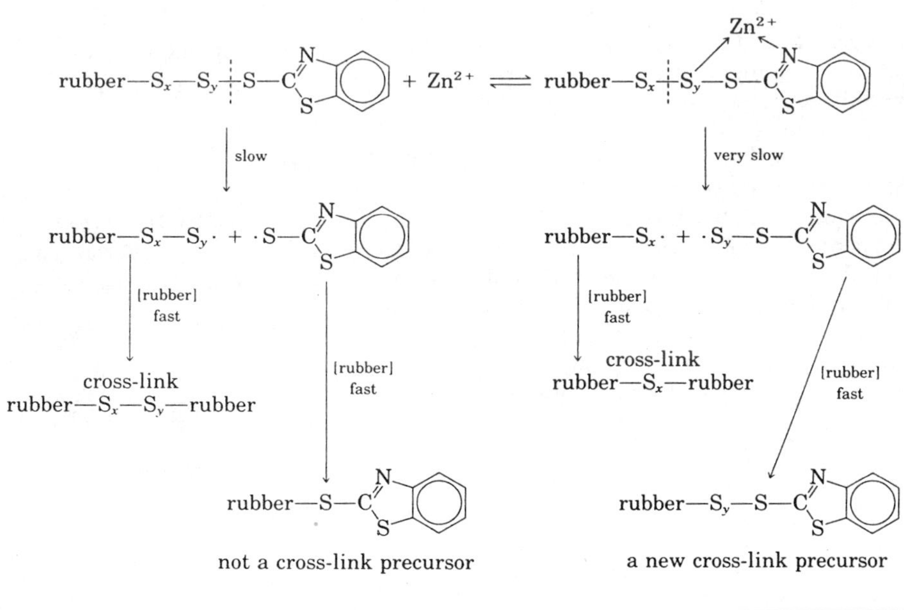

Here zinc chelation changes the position of the S—S bond most likely to break. Because a stronger bond must break, the rate is lower. Although the cross-linking rate is reduced, the extent of cross-link formation is increased because less sulfur is used in each cross-link.

Specific Vulcanization Characteristics. For many years, it was difficult to independently control the two main vulcanization characteristics—scorch resistance (processing safety) and cross-link formation rate. In the case of natural rubber (NR), a fast accelerator system chosen to obtain short curing times in the press would have greatly reduced process safety. A delayed-action acceleration system, however, would have limited the rate of vulcanization in the press. The development of the highly efficient premature-vulcanization inhibitor *N*-cyclohexylthiophthalimide (CTP) resulted in great improvements in scorch resistance with little or no change in cross-linking rate. Thus, the cross-link formation rate and the scorch resistance can be independently controlled.

Effects on Adhesion to Brass-plated Steel. The adhesion between the rubber and brass-plated steel, such as in steel tire cords for belted radial tires, has been the subject of much study and speculation. Brass plating is at present the only important method of obtaining adhesion between natural rubber and the steel of tire cords. Over the years, there has been much speculation about its mechanism, but agreement has been reached on one aspect of the adhesion of natural rubber to brass-plated steel: The actual adhesive between the natural rubber and the brass-plated cord, formed *in situ* during the vulcanization process, is an interfacial layer of sulfides and oxides of copper.

Failure rarely occurs between the rubber and the copper sulfide film. It generally occurs cohesively within the sulfide film or adhesively in a layer below the sulfide film.

Sulfidation of the brass surface is not due to its interaction with elemental sulfur, but is the result of the interaction between the brass surface and accelerator–sulfur reaction products that can be represented by the general structures Ac—S_y—Ac and Ac—S_y—H, where Ac is an accelerator-derived moiety (eg, benzothiazolyl group).

The choice of accelerator also has an effect on the quality of adhesion between cord and rubber. The accelerator should not form a stable copper complex that dissolves in the rubber and would be corrosive to the brass plating. In this respect, benzothiazoles and their sulfenamides are superior to dithiocarbamates; eg, DCBS is an excellent accelerator for rubber-to-brass adhesion.

Effects on Vulcanizate Properties. Increases in sulfur and accelerator concentrations result in higher cross-link densities and, therefore, higher moduli, stiffness, hardness, and similar properties. However, as the ratio of the concentration of accelerator to the concentration of sulfur increases, the proportion of monosulfidic cross-links increases in natural rubber stocks. Greater amounts of accelerator (with respect to sulfur) also increase the amount of pendent groups of the type —S_x—Ac, which are attached to and "dangle" from the rubber molecular chains. Higher ratios of sulfur concentration to accelerator concentration give both more polysulfide cross-links and more sulfur combined with the rubber chains to form sulfur-containing six-membered heterocyclic rings along the rubber molecular chains. In addition, conjugated olefinic double bonds appear along the polymer backbone chain. Such changes in the vulcanizate network structure, no doubt, are responsible for changes that occur in vulcanizate properties as a result of changes made in the curing-system recipe. The best mechanical properties are obtained with rather long sulfur chain cross-links, but with as little rubber main-chain modification (dangling groups, cyclic structures) as possible. Too much combined sulfur in the network gives poor oxidative stability.

Unsaturated Rubbers

The chemistry of accelerated vulcanization of BR, SBR, and EPDM appears to have much in common with the vulcanization of natural rubber. Before the formation of cross-links, the rubber is first sulfurated by accelerator-derived polysulfides (Ac—S_x—Ac) to give macromolecular, polysulfidic intermediates (rubber—S_x—Ac). However, whereas in the case of MBTS- or benzothiazolesulfenamide-accelerated sulfur vulcanization of natural rubber, MBT is given off during the formation of rubber—S_x—BT from the attack of rubber by BT—S_x—BT, in the case of BR and SBR, MBT is not eliminated and remains unextractable, presumably because it becomes bound as the macromolecular thioether rubber—S—BT (where BT is a 2-benzothiazolyl group). As in the case of natural rubber, the average length of a cross-link (its sulfidic rank, the value of x in the cross-link, rubber—S_x—rubber) increases with the ratio of sulfur concentration to accelerator concentration (S/Ac) in the compounded rubber mix. However, in the case of BR or SBR, the cross-link sulfidic rank is not as sensitive to S/Ac as it is in the case of natural rubber. Model compound studies of the vulcanization of EPDM, wherein ethylidenenorbornane was used as a model for EPDM, indicate that the polysulfidic rank of the EPDM cross-links probably responds to changes in S/Ac in a natural rubber-like fashion. A difference here is that evidence for model monosulfidic cross-links was lacking, whereas the model disulfidic cross-links were more apparent than in the case of natural rubber vulcanization.

Other Vulcanization Agents

Other agents include phenolic curatives, benzoquinone derivatives, bismaleimides, metal oxides, and organic peroxides.

BIBLIOGRAPHY

A. Y. Coran, *Chemtech* **23,** 106 (1983).

W. J. van Ooij, *Rubber Chem. Technol.* **52,** 605 (1979); **57,** 421 (1984).

A. D. Roberts, ed., *Natural Rubber Science and Technology,* Oxford University Press, Oxford, 1988.

AUBERT Y. CORAN
Monsanto Chemical Company

W

WALLING-MAYO THEORY. See COPOLYMERIZATION.

WASTE TREATMENT

Industry's residual wastes are found in three physical states: (*1*) liquid, (*2*) solid, and (*3*) gaseous (Table 1). Each of these emanate at different locations in industrial plants and exert separate and many environmental impacts. In order to prevent industrial pollution—the existence of excessive amounts of contaminants—a thorough investigation of the types and amounts of pollutant as well as their specific impact on the environment must be carried out and evaluated carefully and completely. The environment may be the water, land, or air surrounding the industry. A condition of pollution exists when excessive contaminants are discharged by industry into the water, land, or air environment.

Table 1. Sources, Characteristics, and Treatment of Industrial Wastes

Waste-producing industry and materials	Origin	Characteristics	Treatment and disposal methods
agriculture	chemicals, irrigation return flows, crop residues, and liquid and solid animal wastes	highly organic and BOD detergent cleaning solutions	biological oxidation basins, some composting and anaerobic digestion, land application
fertilizer	spills, cooling waters, washing of products, boiler blowdowns	sulfuric, phosphoric, and nitric acids; mineral elements, P, S, N, K, Al, NH_3, NO_3, etc; F; some suspended solids	neutralization, detain for reuse, sedimentation, air stripping of NH_3, lime precipitation
pesticides	washing and purification products such as 2,4-D and DDT	high organic matter, benzene ring structure, toxic to bacteria and fish, acidic	dilution, storage, activated-carbon adsorption, alkaline chlorination
asbestos	cleaning and crushing ore	suspended asbestos and mineral solids	detention in ponds, neutralization and landfilling
acids	dilute wash waters; many varied dilute acids	low pH, low organic content	upflow or straight neutralization, burning when some organic matter is present
candles	wax spills, stearic acid condensates	organic (fatty) acids	anaerobic digestion
cement and concrete blocks	fine and finish grinding of cement, dust-leaching collection, dust control	heated cooling water, suspended solids, some inorganic salts	segregation of dust-contact streams, neutralization and sedimentation
coal processing	cleaning and classification of coal, leaching of sulfur strata with water	high suspended solids, mainly coal; low pH; high H_2SO_4 and $FeSO_4$	settling, froth flotation, drainage control, and sealing of mines
energy	scrubber power-plant wastes, scrubbing of gaseous combustion products by liquid water	particulates, SO_2, impure absorbents or NH_3, NaOH, etc	solids removal usually by settling, pH adjustment, and reuse
nuclear power	processing ores; laundering of contaminated clothes, research-laboratory wastes, processing of fuel, power-plant cooling waters	radioactive elements, can be very acidic and "hot"	concentration and containing, or dilution and dispersion
steam power	cooling water, boiler blowdown, coal drainage	hot, high volume, high inorganic and dissolved solids	cooling by aeration, storage of ashes, neutralization of excess acid wastes
explosives	washing TNT and guncotton for purification, washing and pickling of cartridges	TNT, colored, acidic, odorous, and contains organic acids and alcohols from powder and cotton, metals, acids, oils, and soaps	flotation, chemical precipitation, biological treatment, aeration, chlorination of TNT, neutralization, adsorption
formaldehyde	residues from manufacturing synthetic resins and from dyeing synthetic fibers	normally high BOD and HCHO, toxic to bacteria in high concentrations	trickling filtration, adsorption on activated charcoal
food			
bakeries	washing and greasing of pans	high BOD, grease, floor washings, sugars, flour, detergents	amenable to biological oxidation
beet sugar	transfer, screening, and juicing waters; draining from lime sludge; condensates after evaporator; juice and extracted sugar	high in dissolved and suspended organic matter, containing sugar and protein	reuse of wastes, coagulation, and lagooning
brewed and distilled beverages	steeping and pressing of grain, residue from distillation of alcohol, condensate from stillage evaporation	high in dissolved organic solids, containing nitrogen and fermented starches or their products	recovery, concentration by centrifugation and evaporation, trickling filtration, use in feeds digestion of slops
cane sugar	spillage from extraction, clarification; evaporator entrainment in cooling and condenser waters	variable pH, soluble organic matter with relatively high BOD of carbonaceous nature	neutralization, recirculation, chemical treatment, some selected aerobic oxidation

Table 1. (*Continued*)

Waste-producing industry and materials	Origin	Characteristics	Treatment and disposal methods
canned goods	trimming, culling, juicing, and blanching of fruits and vegetables	high in suspended solids, colloidal and dissolved organic matter	screening, lagooning, soil absorption of spray irrigation
coffee	pulping and fermenting of coffee bean	high BOD and suspended solids	screening, settling, and trickling filtration
cornstarch	evaporator condensate or bottoms when not reused or recovered, syrup from final washes, wastes from bottling process	high BOD and dissolved organic matter; mainly starch and related material	equalization, biological filtration, anaerobic digestion
dairy products	dilutions of whole milk, buttermilk, and whey	high in dissolved organic matter, mainly protein, fat, and lactose	biological treatment aeration, trickling filtration, activated sludge
fish	rejects from centrifuge, pressed fish, evaporator and other wash water wastes	very high BOD, total organic solids, and odor	evaporation of total waste, barge remainder to sea
meat and poultry products	stockyards, slaughtering of animals, rendering of bones and fats, residues in condensates, grease and wash water, picking chickens	high in dissolved and suspended organic matter, blood, other proteins, and fats	screening, settling or flotation, trickling filtration
rice	soaking, cooking, and washing of rice	high BOD, total and suspended solids (mainly starch)	lime coagulation, digestion
soft drinks	bottle washing, syrup storage-tank drains	high pH, suspended solids, and BOD	screening plus discharge to municipal sewer
yeast	residue from yeast filtration	high in solids (mainly organic) and BOD	anaerobic digestion, trickling filtration
fuel oil	spills	high in emulsified and dissolved oils	leak and spill prevention, flotation
glass	many sources	red color, alkaline nonsettleable suspended solids	calcium chloride precipitation
glue	lime and acid washes	high COD, BOD, pH	to aerobic biological pond
hospital	washing, sterilization of facilities, used solutions, spills	bacteria, various chemicals, radioactive materials	discharge to municipal sewers, holding and biological aeration in large facilities
iron-foundry products	sand from hydraulic discharge	high suspended solids, mainly sand; some clay and coal	selective screening, drying of reclaimed sand
laundry	washing of fabrics with soap or detergents	high turbidity, alkalinity, and organic solids	screening, chemical precipitation, flotation and adsorption
detergents	washing and purifying soaps and detergents	high in BOD and soaps	flotation and skimming, precipitation with $CaCl_2$
leather	unhairing, soaking, deliming, and bating of hides	high total solids, hardness, salt, sulfides, chromium, pH precipitated lime, and BOD	equalization, sedimentation, and biological treatment
metal container	cutting and lubricating metals, cleaning can surfaces	metal fines, lubrication oils, variable pH, surfactants, dissolved metals	oil separation, chemical precipitation, collection and reuse, lagoon storage, final carbon absorption
metal-plated products	stripping of oxides, cleaning and plating of metals	acid, metals, toxic, low volume, mainly mineral matter	alkaline chlorination of cyanide, reduction and precipitation of chromium, lime precipitation of other metals
mortuary	body fluids, wash waters, spills	blood salt, formaldehydes, high BOD, infectious diseases	discharge to municipal sewer holding and chlorination
naval stores	washing of stumps, drop solution, solvent recovery, and oil-recovery water	acid, high BOD	by-product recovery, equalization, recirculation, and reuse, trickling filtration
oil fields and refineries	drilling muds, salt, oil, and some natural gas; acid sludges and miscellaneous oils from refining	high dissolved salts from field; high BOD, odor, phenol, and sulfur compounds from refinery	diversion, recovery, injection of salts; acidification and burning of alkaline sludges
paints and inks	solvent-based rejected materials scrubbers for paint vapors, refining or removing inks	contain organic solids from dyes, resins, oils, solvents, etc	settling ponds for detention of paints, lime coagulation of printing inks
petrochemicals	contaminated water from chemical production and transportation of second-generation oil compounds	high COD, TDS[a], metals, COD/BOD ratio, and compounds inhibitory to biological action	recovery and reuse, equalization and neutralization, chemical coagulation, settling or flotation, biological oxidation
pharmaceutical products	mycelium, spent filtrate, and wash waters	high in suspended and dissolved organic matter, including vitamins	evaporation and drying, feeds
phosphate and phosphorus	washing, screening, floating rock, condenser bleed-off from phosphate reduction plant	clays, slimes and tall oils, low pH, high suspended solids, gypsum, phosphorus, silica and fluoride, radioactivity	lagooning, mechanical clarification, coagulation and settling of refined waste

Table 1. (*Continued*)

Waste-producing industry and materials	Origin	Characteristics	Treatment and disposal methods
photographic products	spent solutions of developer and fixer	alkaline, containing various organic and inorganic reducing agents	recovery of silver, discharge of wastes into municipal sewer
plastics and resins	unit operations from polymer preparation and use, spills and equipment wash downs	acids, caustic, dissolved organic matter such as phenols, formaldehyde, etc	discharge to municipal sewer, reuse, controlled discharge
plywood	glue washings	high BOD, pH, phenols, potential toxicity	settling ponds, incineration
pulp and paper	cooking, refining, washing of fibers, screening of paper pulp	high or low pH; color; high suspended, colloidal, and dissolved solids; inorganic fillers	settling, lagooning, biological treatment, aeration, recovery of by-products
rubber	washing of latex, coagulated rubber, exuded impurities from crude rubber	high BOD and odor, high suspended solids, variable pH, high chlorides	aeration, chlorination, sulfonation, biological treatment
steel	coking of coal, scrubbing of blast-furnace flue gases, and pickling of steel	low pH, acids, cyanogen, phenol, ore, coke, limestone, alkalies, oils, mill scale, and fine-suspended solids	neutralization, recovery, and reuse, chemical coagulation
textiles	cooking of fibers, desizing of fabric	highly alkaline, colored, high BOD and temperature, high suspended solids	neutralization, chemical precipitation, biological treatment, aeration or trickling filtration
toxic chemicals	leaks, accidental spills, and refining of chemicals	various toxic dissolved elements and compounds such as Hg and PCBs	retention and reuse, change in production
water production	filter backwash, lime-soda sludge, brine, alum sludge	minerals and suspended solids	direct discharge to streams or indirectly through holding lagoons
wood furniture	wet spray booths and laundries	organics from staining and sealing wood products	evaporation or burning
wood preserving	steam condensates	high in COD, BOD, solids, phenols	chemical coagulation, oxidation pond, and other aerobic biological treatment

[a] TDS = total dissolved solids.

BIBLIOGRAPHY

N. L. Nemerow, *Industrial Water Pollution*, Krieger Publishing Co., Melbourne, Fla., 1987.

N. L. Nemerow, *Industrial Solid Wastes*, Ballinger Publishing Co., Cambridge, Mass., 1983, chapts. 2–7 and p. 94.

Composting—Theory and Practice for City, Industry, and Farm, The J. G. Press, Emmaus, Pa., 1981.

P. A. Vesiland and A. E. Rimer, *Unit Operations in Resource Recovery Engineering*, Prentice-Hall Co., Englewood Cliffs, N.J., 1981.

NELSON L. NEMEROW
University of Miami
Florida International University

WATER-SOLUBLE POLYMERS

Macromolecules exhibiting solubility in aqueous solutions represent a diverse class of polymers ranging from biopolymers that mediate life processes to synthetic resins of immense commercial utility. In this article water-soluble polymers have been grouped into the categories biopolymers, nonionic, ionic, and associative, based on key features. Structure and solution properties are discussed with particular emphasis on industrially significant polymers (see also ACRYLAMIDE POLYMERS; CELLULOSE; GUMS; POLYELECTROLYTES; POLYNUCLEOTIDES; POLYSACCHARIDES; and PROTEINS).

General Considerations

Structure. Solution properties and performance of water-soluble polymers are dictated by structural characteristics of the solvated polymer chain. Primary structure depends directly on the nature of the repeating units (bond lengths, valence bond angles), effective compositions, and locations along the backbone. The polymer structure may derive from a single monomer [ie, poly(ethylene oxide) or polyacrylamide] or from multiple monomers. These units may be placed to yield random, alternating, or graft copolymers (Fig. **1a**). Biopolymers such as proteins and polynucleotides have multiple repeating units specifically ordered by template polymerization. Additionally, soluble polymers may be linear or branched (Fig. **1b**).

Secondary structure in water-soluble polymers is related to configuration, conformation, and intramolecular effects such as hydrogen bonding and ionic interactions. Tertiary structure involves intermolecular and water–polymer interactions; quaternary structure requires multiple chain aggregation or complexation.

A large number of functional groups (Fig. 2) can impart water solubility in copolymers. The degree of solubility is dependent on the number, position, and frequency of these moieties. Hydration relies on interaction at polar (ionic and hydrogen bonding) sites.

Hydrodynamic Volume. Polymers are often described in terms of hydrodynamic volume (HDV) or that volume occupied by the solvated chain. HDV and molecular shape, determined from light scattering measurements, may be used along with chemical microstructure to predict rheological behavior.

Ionic interactions (repulsive or attractive) can also dramatically affect HDV. For charged polymers, ionic effects often dominate behavior, especially in aqueous solutions.

A number of synthetic strategies may be employed to increase macromolecular coil size (HDV) for water-soluble polymers. Monomers with water-soluble moieties may be polym-

ABABABABABABABABABABABABAB
Alternating copolymer

AABABBBABAABAAABBABBBAAB
Random copolymer

AAAAAABBBBBBAAAAAABBBBBB
Block copolymer

AAAAAAAAAAAAA
B B B
B B B
B B B
B B B
B B B
Graft copolymer
(**a**)

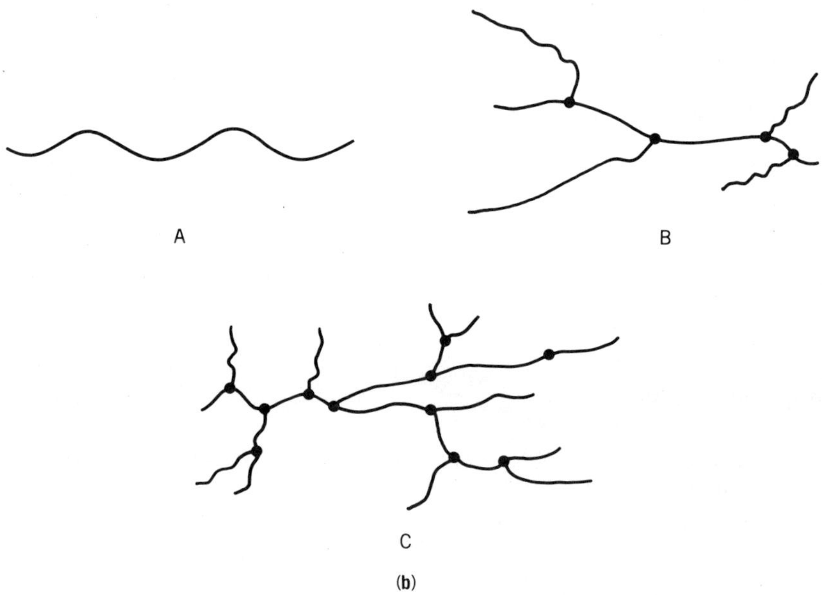

Figure 1. (**a**) Monomer distributions; (**b**) segmental distribution in soluble copolymers: A, linear; B, branched—long branches; C, branched—branches protruding from branches give a dendritic structure.

erized to high molecular weight. Effective bond lengths along the backbone may be increased by introducing chain stiffening elements. The shape of the solvated coil is determined by placement of charged groups, hydrophobic moieties, hydrogen bonds, chiral centers, or restrictive rings along the molecular backbone.

Polymers dissolved in water can have structures ranging from random coils to microheterogeneous polymeric vesicles (qv). Solution behaviors of the various types are quite diverse. Water-soluble random and extended coil polymers are often used for rheology control in various applications. Extended rods such as cellulose derivatives in aqueous solution may exhibit lyotropic liquid crystalline behavior. Amphophilic molecules such as hypercoils, polymeric micelles, and vesicles are used in formulation, drug delivery, and phase-transfer catalysis.

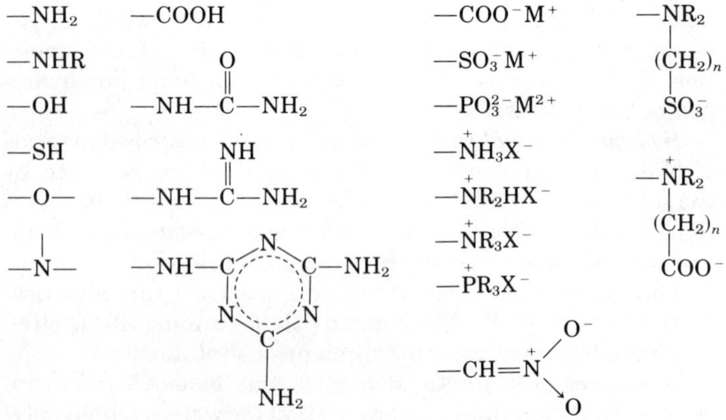

Figure 2. Functional groups imparting water solubility.

Role of Water in Solvation/Phase Behavior. Water plays an extremely important role in determining the properties and ultimate utility of polymers in solution. Solvation of polymer chains may involve simple interaction of ionic, polar, or hydrogen-bonded hydrophilic segments with water or more complex solvation of amphiphilic structures. For amphiphilic polymers, water structure around the hydrophobic portion of the chain is thought to be more ordered than in the bulk, whereas hydrophilic portions disorder water structure.

Some polymers such as poly(acrylic acid) or polyacrylamide precipitate from aqueous solutions when cooled (normal solubility behavior) whereas others such as poly(ethylene oxide), poly(propylene oxide), or poly(methacrylic acid) precipitate when heated (inverse solubility behavior). Solution turbidimetry is often used to obtain plots of phase-separation temperatures termed "cloud point" vs concentration for fixed solvent conditions. Changes in ionic strength, molecular weight, and addition of cosolvents or structure-breakers affect the shapes of phase-behavior curves. The important conclusion of such studies is that the total free energy of the polymer and water must be considered to predict phase behavior.

The structure and dynamics of water surrounding polynucleotides, proteins, polysaccharides, and lipids are also major determinants of biological activity.

Viscosity and Rheology. Viscosity yields important information as to the disposition of the polymer chains in solution and is routinely used to evaluate polymers for particular applications. Dilute solution measurements can yield intrinsic viscosity $[\eta]$, which is a direct indication of the hydrodynamic volume of an isolated polymer chain. This fundamental parameter is related to molecular weight M through the Mark-Houwink-Sakurada (MHS) relationship:

$$[\eta] = KM^a \tag{1}$$

The parameters K and a are characteristics of a polymer chain under specific conditions of solvency and temperature. Values of a can be related to chain extension in dilute solution.

Synthetic Methods. Water-soluble copolymers are prepared by step-growth or chain-growth mechanisms. Linear or branched systems may be formed from single monomers or from a large number of monomers. Distribution of monomers, along the backbone or side chain, can be controlled in a number of ways. In nearly all cases, sequence selection is obtained by carefully controlling monomer reactivity, concentration, addition order, and reaction conditions.

Major commercial synthetic water-soluble polymers are made by chain-growth polymerization of functionalized alkenes, carbonyl monomers, or strained ring compounds as illustrated below.

$$n\,\mathrm{CH_2{=}}\underset{\displaystyle \mathrm{R}}{\underset{|}{\mathrm{C}}}\mathrm{H} \longrightarrow -\!\!(\mathrm{CH_2}-\underset{\displaystyle \mathrm{R}}{\underset{|}{\mathrm{C}}}\mathrm{H})_{\overline{n}} \tag{2}$$

$$n\,\mathrm{R}-\underset{\displaystyle \mathrm{H}}{\underset{|}{\mathrm{C}}}{=}\mathrm{O} \longrightarrow -\!\!(\overset{\displaystyle \mathrm{R}}{\overset{|}{\underset{\displaystyle \mathrm{H}}{\underset{|}{\mathrm{C}}}}}-\mathrm{O})_{\overline{n}} \tag{3}$$

$$n\,\mathrm{R}-\underset{\displaystyle \mathrm{H}}{\underset{|}{\mathrm{C}}}\overset{\mathrm{X}}{\frown}\mathrm{CH_2}\ (\text{cyclic}) \longrightarrow -\!\!(\overset{\displaystyle \mathrm{R}}{\overset{|}{\underset{\displaystyle \mathrm{H}}{\underset{|}{\mathrm{C}}}}}-\mathrm{CH_2}-\mathrm{X})_{\overline{n}} \tag{4}$$

These may be initiated free radically, anionically, or cationically, depending on monomer structure. In some cases coordination catalysts are used. Copolymers or terpolymers with random, alternating, block, or graft sequences can be prepared under appropriate reaction conditions (see ANIONIC POLYMERIZATION; CATIONIC POLYMERIZATION; RADICAL POLYMERIZATION).

Of particular commercial interest is synthesis of water-soluble polymers in solutions, dispersions, suspensions, or emulsions.

Naturally Occurring Polymers

A large number of water-soluble polymers are derived from biological sources. Termed "biopolymers," this class includes polynucleotides, polypeptides, and polysaccharides. Because these polymers perform special biological functions, they have specific microstructures and are often monodisperse. In the following section the general structural features of major biopolymer types will be reviewed as related to water solubility (see also POLYNUCLEOTIDES; POLYSACCHARIDES; PROTEINS).

Polynucleotides. ***Structure.*** Nucleic acids are biopolymers that carry genetic information involved in the processes of replication and protein synthesis. The primary structural units of the nucleic acids, the mononucleotides, are composed of a 5-carbon cyclic sugar with a phosphate ester at the C-5 carbon. A heterocyclic amine base (purine or pyrimidine) is attached at C-1.

RNA and DNA are terpolymers with sugar phosphate backbones (3′ and 5′ position) and pendent bases on the 1′ position.

Deoxyribonucleic acids (DNAs) contain the deoxyribose sugar (no hydroxyl group at the 2′ position on the sugar ring). The usual bases substituted at the 1′ position are adenine (A), guanine (G), thymine (T), and cytosine (C). Ribonucleic acids contain a ribose sugar and the principal bases A, G, C, and uracil (U).

Studies of the structure of DNA by a large number of investigators led to the proposal by Watson and Crick of the classic DNA double helix. Major contributions as to base content and of x-ray structure were made by the groups of Chargaff and Wilkens. This proposal not only allowed explanation of DNA stability but also provided a framework for postulation of template polymerization during the replication and transcription processes. Molecular weights of DNA can be extremely large, depending on sources. Some DNAs are linear; however, a number are cyclic.

Ribonucleic acids are synthesized from the DNA template through transcription. The mononucleotides contain ribose. The phosphodiester linkage is located between 3′ and 5′ positions on adjacent ribose rings with the bases adenine, uracil, guanine, or cytosine attached at C-1. Unlike DNAs, RNAs are single stranded, although twisting may occur in a complementary fashion matching base pairs. RNAs are much lower in molecular weight than DNAs. The three major types—messenger RNA, transfer RNA, and ribosomal RNA—are involved in a multistep process of protein synthesis.

Synthesis. Recently, nucleic acids have been synthesized from mononucleotides, some with altered or substituted bases. Synthetic methods, along with nucleic acid sequencing methods, have allowed rapid advancement in identifying gene sequences responsible for specific protein synthesis. Kornberg and coworkers first prepared synthetic DNAs polymerized from mononucleotides using isolated enzymes, DNA polymerases. Likewise, RNA polymerases have been found for synthesis of oligomeric RNAs. In both cases, no template is involved, so homopolynucleotides are usually prepared. Oligonucleotides can be synthesized in solution or on a solid support, usually polymeric. Two principal methods are the phosphotriester method and the phosphite-triester method.

Polypeptides and Proteins. Most polypeptides and proteins are water-soluble or water-swellable. Enzymes are proteins that catalyze all chemical reactions of biological origin. Enzyme functions include oxygen transport, muscle movement, nerve response, nutrient digestion and storage, hormonal regulation, gene expression, and protein synthesis.

Structure. Despite the large number of functions, all proteins are similar with repeating structures along the backbone chosen from 20 amino acid monomers. These polymers, the structures of which are assembled from a template coded by messenger RNA, are monodisperse. Each protein has a unique sequence and molecular weight.

The 20 amino acids, capable of appearing in various microstructural combinations, sequence lengths, and total molecular lengths, allow assembly of an infinite number of distinct proteins.

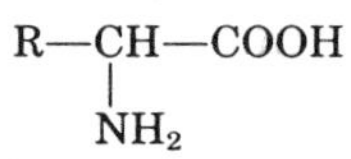

α-amino acid

There are four major types of side chains (R): hydrophobic, polar, acidic, and basic, eg,

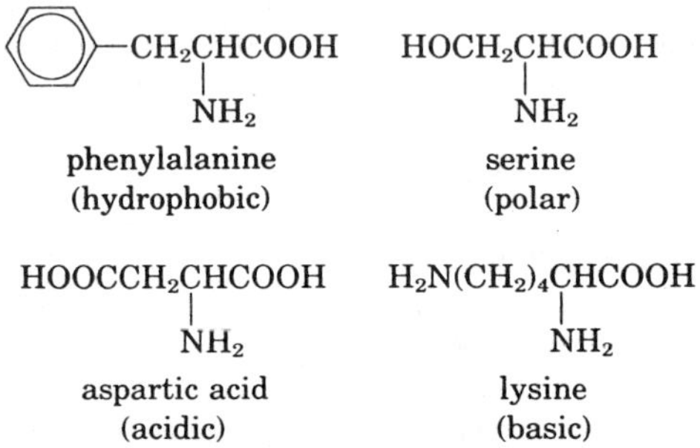

The solubilities of proteins vary considerably based on composition and conditions of ionic strength, pH, and concentration. Those with the highest density of polar groups or electrolyte character are most soluble. Therefore, solubility in water is lowest at the isoelectric point and increases with increasing basicity or acidity.

Protein Synthesis. Protein biosynthesis is complex, involving more than 300 macromolecules. Five stages can be identified, involving (*1*) activation of amino acid monomers and transfer to ribosomes, (*2*) initiation of polymerization, (*3*) propagation, (*4*) termination and release, and (*5*) folding and processing. Ordering of the monomers is dictated by operation of the triplet code in which a sequence of three consecutive nucleotide units on messenger RNA positions a specific amino acid for polymerization (see PROTEINS, BIOSYNTHESIS).

The development of recombinant DNA techniques has led to cloning of a number of genes and their expression in the form of protein products from bacterial or yeast cells. Large numbers of proteins are now being produced with enormous commercial potential. Among the first available genetically engineered proteins are insulin, the pituitary growth hormone, and interferon.

Commercial Use. In addition to the rapid growth of genetically engineered enzymes and hormones for medical and agricultural applications, other water-soluble proteins are isolated from biological sources in a more traditional manner for commercial application. Enzymes are used as detergent additives, for hydrolyzing polysaccharides and proteins, to isomerize various glucose and sucrose precursors, for wine and beer making, for leather tanning, and for mineral recovery. Support-bound enzymes are becoming commercially significant for large-scale substrate conversion (see ENZYMES; ENZYMES, IMMOBILIZED).

Polysaccharides. Water-soluble polysaccharides are a diverse class of biological macromolecules with a wide range of structural and behavioral characteristics. More than 100 sugars and sugar derivatives comprise the monomers available for polysaccharide synthesis. Covalent linkages between repeating units may occur at different ring positions; linear and branched structures with single or multiple monomers may be formed.

Industrial polysaccharides have traditionally been extracted from renewable resources in the plant and animal kingdom. Examples include starch and gums (qv) from plant seeds, pectin from fruits, and algin and carrageenan from algae. Recently, microbial sources have produced commercially useful polysaccharides such as dextran (qv), curdlan, pullulan (qv), xanthan (qv), and pharmacologically active oligosaccharides.

Polysaccharides (qv) are cyclolinear or branched polymers formed by enzyme-directed step-growth condensation reactions of various activated sugar molecules. Some common monomer units are shown in Figure 3; other units include glycerol and other polyhydroxyalcohols, phosphate, sulfate, malonate, and pyruvate.

Aqueous solution properties, including solubility, phase behavior, and viscosity, are highly dependent on the macroscopic nature (linear, branched) of the chain as well as the chemical microstructure (polar characteristics, sequencing) of the repeating units. As with proteins, the presence of acidic or basic functionality causes pH, electrolyte, and temperature-dependent behavior. Degrees of polymerization may vary from 30 to 1×10^5. Of primary importance in the solution behavior of water-soluble polysaccharides are the configurations of asymmetric carbon atoms in the cyclic monomers and the ring conformations of the resulting covalent backbone units.

Water-soluble polysaccharides can be classified according to structure and commercial applications. Classes include storage polysaccharides, pectins, plant gums, seed and bark mucilage, algal polysaccharides, bacterial and fungal polysaccharides, mucopolysaccharides, and synthetically modified polysaccharides.

Nonionic Polymers

A number of commercially important nonionic acrylic, oxide, imine, and vinyl polymers are soluble in water at normal temperatures. Water solubility for these materials is the result of a high number of polar or hydrogen-bonding functional groups per repeating unit.

Polyacrylamide (PAM). ***Synthesis/Manufacture.*** Acrylamide (AM) monomer is polymerized by free-radical initiators, eg, azo compounds, redox catalysts, light, and radiation.

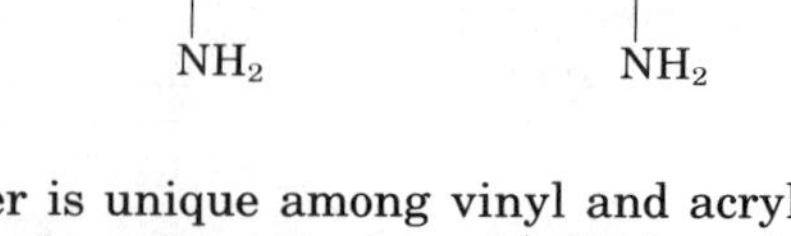

(5)

This monomer is unique among vinyl and acrylic monomers because it can be polymerized to ultrahigh molecular weight (10^6–10^7). This extraordinary feature of acrylamide polymerization is attributed, in part, to the ease of purification of AM monomer and to the unusually high ratio of its propagation to

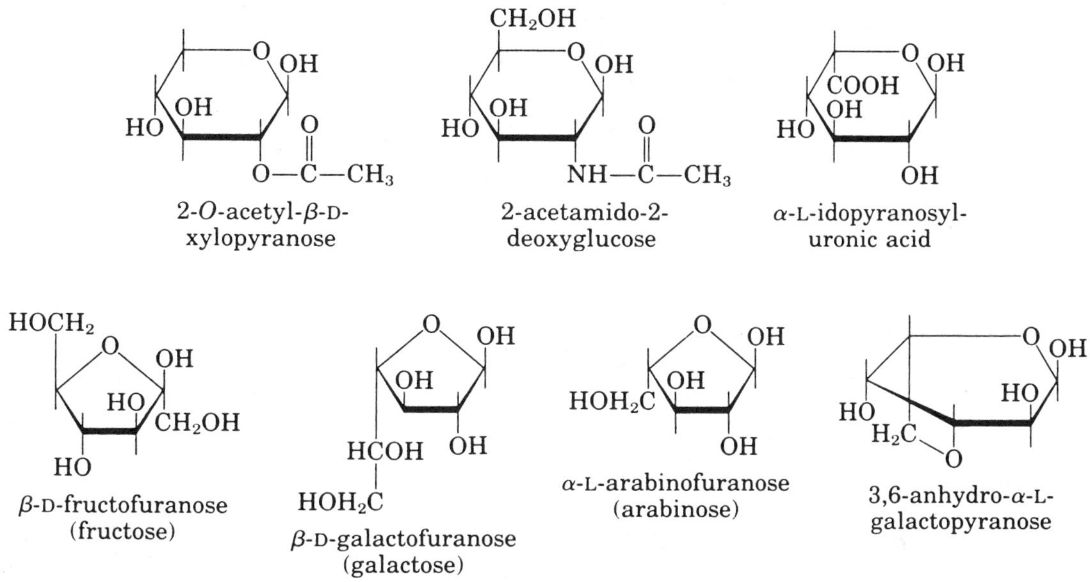

Figure 3. Common structural units in naturally occurring polysaccharides.

termination rate constants (k_p/k_t). In fact, AM has the highest k_p/k_t of any free radically polymerizable monomer. Polyacrylamide can be prepared via solution, inverse emulsion, inverse microemulsion, or precipitation techniques.

Properties. Polyacrylamide (PAM) has reported T_g values of 165 and 188°C. No matter which value is correct, it is clear that PAM remains glassy to relatively high temperatures. Although PAM is slow to dissolve, it is soluble in water in all proportions. However, PAM solution viscosities show a time dependence, attributed to intramolecular conformational changes.

Because PAM can be polymerized to very high molecular weight, it is a highly efficient aqueous viscosifier.

Applications. Polyacrylamides function as flocculating agents, in rheology control, and as adhesives.

Poly(acrylic acid). ***Synthesis/Manufacture.*** Poly(acrylic acid) (PAA) can be prepared by direct free-radical polymerization in aqueous solution or by precipitation polymerization in benzene (eq. 6). Alcohols and mercaptans are commonly used chain-transfer agents for regulating polymer molecular weight. Alternatively, PAA can be prepared by hydrolysis of poly(alkyl acrylates) (eq. 7).

$$\underset{\displaystyle COOH}{CH_2{=}\underset{|}{CH}} \xrightarrow{I^{\cdot}} -\!\!(CH_2-\underset{\underset{\displaystyle COOH}{|}}{CH})\!\!- \qquad (6)$$

$$-\!\!(CH_2-\underset{\underset{\displaystyle COOR}{|}}{CH})\!\!- \xrightarrow[\Delta]{H_2O} -\!\!(CH_2-\underset{\underset{\displaystyle COOH}{|}}{CH})\!\!- \qquad (7)$$

Properties. The T_g of PAA has been variously reported as 75, 106, and 126°C, depending on the mode of measurement. However, the highest value is probably the most accurate. Solid polymers are hard, clear, brittle materials. In aqueous solutions, viscosity increases with increasing molecular weight. PAA undergoes a number of reactions in solution, ie, hydrolysis, esterification, dehydration, and complex formation with polyethers.

Applications. PAA is an excellent thickener for latices. PAA has been used in oil recovery, as a dispersant for inorganic pigments, as a flocculant, and as an adhesive (see ACRYLIC ACID POLYMERS).

Poly(ethylene oxide). ***Synthesis/Manufacture.*** Poly(ethylene oxide) (PEO) is prepared by ring-opening polymerization of ethylene oxide monomers.

$$n\ \overset{O}{CH_2-CH_2} \longrightarrow -\!\!(CH_2-CH_2-O)\!\!-_n$$

Polymers of molecular weight greater than 1×10^5 are prepared by heterogeneous catalysis (eg, alkaline earth carbonates) in low-boiling aliphatic hydrocarbons. Few of the details of the commercial manufacture have been made public.

Properties. Poly(ethylene oxide) is a white free-flowing powder with commercial grades from 100,000 to 5 million molecular weight. It has a T_m of about 65°C and a T_g of −45 to −53°C. Above the melting point it can be processed as a thermoplastic, ie, molded, extruded. However, owing to its high melt viscosity, incorporation of plasticizer is often desired.

Poly(ethylene oxide) resins are completely soluble in water at room temperature, but show a lower critical solution temperature (LCST) near the boiling point of water (see 1,2-EPOXIDE POLYMERS).

Poly(vinyl alcohol). ***Synthesis/Manufacture.*** Poly(vinyl alcohol) (PVA) is manufactured by alcoholysis/hydrolysis of poly(vinyl acetate), which is, in turn, produced by free-radical polymerization of vinyl acetate monomer.

$$\underset{\displaystyle OCOCH_3}{CH_2{=}\underset{|}{CH}} \xrightarrow[\text{initiator}]{\Delta} -\!\!(CH_2-\underset{\underset{\displaystyle OCOCH_3}{|}}{CH})\!\!- \qquad (8)$$

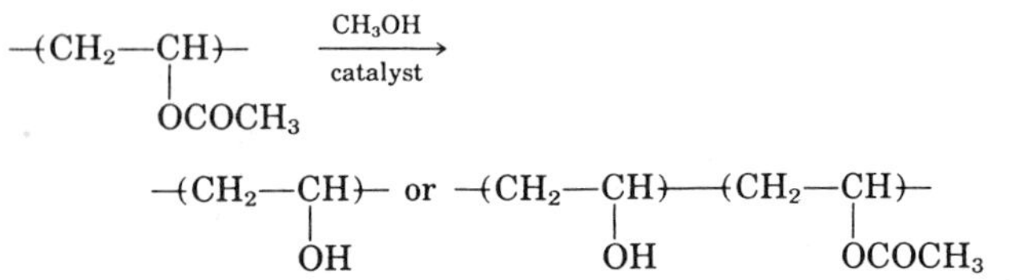

(9)

Typical commercial producers and trade names are Dupont (Elvanol), Air Products (Vinol), Monsanto (Gelvatol), Nippon Gohsei (Gohsenol), Kuraray (Poral), and Hoescht (Mowiol) (see VINYL ESTER POLYMERS).

Properties. Properties of PVA depend on the degree of alcoholysis/hydrolysis and polymer viscosity/molecular weight. Bulk/film properties (eg, tensile strength, tear resistance, elongation, and flexibility) of PVA increase with increasing extent of alcoholysis/hydrolysis and with increasing viscosity/molecular weight. The tensile strength is exceptional compared with other water-soluble polymers.

Water solubility/sensitivity is at a maximum at 88% alcoholysis/hydrolysis.

Other noteworthy properties of PVA are its film-forming ability, its barrier, adhesive, and emulsifier properties, and its grease, oil, and solvent resistance.

Applications. Poly(vinyl alcohol) is used alone or combined with extenders, pigments, etc, in the preparation of high wet-strength adhesives for paper. It is an excellent binder for textiles and sizing agent for paper and can be used to emulsify a wide range of materials including vegetable oils, mineral oils, solvents, plastics, waxes, and resins. Its emulsifying, binding, film-forming, and thickening behaviors are useful in cosmetic formulation. PVA films can also be used for making oxygen tents. However, lack of FDA approval has excluded use for direct-contact food (eg, meat) packaging (see VINYL ALCOHOL POLYMERS).

Poly(*N*-vinylpyrrolidinone) (PNVP). ***Synthesis/Manufacture.*** *N*-Vinylpyrrolidinone (NVP) monomer polymerizes under free radical conditions via bulk, solution, and suspension methods.

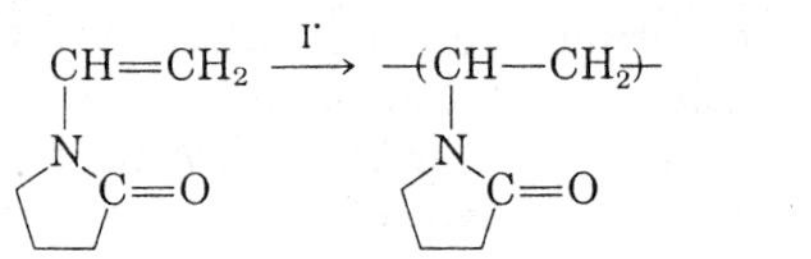

(10)

Properties. The T_g of PNVP is 175°C, but this value is reduced considerably by small amounts of water. Dry-cast films of PNVP are hard and transparent. However, the presence of strong intermolecular dipole–dipole interactions causes it to have a high processing temperature. Consequently, PNVP has never found acceptance for molded parts.

PNVP has interesting solution properties. It is readily soluble in water and forms complexes with a wide variety of substances, eg, iodine, polyacids, phenolics. Solutions of PNVP are stable to electrolytes. Solution viscosities increase with increasing polymer molecular weight. PNVP (bulk or solution) is also characterized by high thermal or thermohydrolytic stability and chemical resistance. PNVP is also compatible with a wide range of hydrophilic and hydrophobic resins and modifiers.

Applications. PNVP has found applications in a wide variety of industries, including medicine, pharmaceuticals, cosmetics, textiles, beverages, adhesives, and paper.

Polyelectrolytes

Polyelectrolytes (qv) are polymers with charged functional groups attached along the chain. These polymers are usually classified as polyanions (negative charges), polycations (positive charges), or polyampholytes (qv) (both positive and negative charges). Associated with the polyion are counterions or gegenions of the opposite charge in sufficient numbers to maintain electroneutrality.

Water-soluble polyelectrolytes exhibit a number of common traits with water-soluble nonionics. However, differences arise from the presence of charges on the macromolecular backbone and more mobile counterions electrostatically bound to an extent determined by pK_a, solvent, and local dielectric effects. Generally, phase behavior and enhanced solubility result from increased segmental hydration and increased free energy of mixing.

The interactions between fixed charges on the polymer chain in dilute solution normally expand (repulsive) or contract (attractive) coil dimensions. Counterion binding also influences hydrodynamic volume and involves specific ion binding as well as atmospheric ion binding. Theoretical discussions of these effects can be found.

Polyelectrolytes with flexible chains and high charge density are more expanded in water than nonionic polymers, especially at low ionic strength. Determination of intrinsic viscosity is difficult in this regime. Electrostatic repulsions not only cause increases in hydrodynamic volume but also increased shear sensitivity or non-Newtonian behavior.

The extent of ionization of polybases or polyacids depends on the relative base or acid strength, degree of solvation, and dielectric constant of the solvent. Poly(acrylic acid), for example, ionizes progressively in aqueous basic solution to yield a copolymer with ionized acrylate units and unionized acrylic acid units along the backbone; neighboring group hydrogen-bonding effects accelerate initial ionization. Eventually, however, ionization becomes more difficult owing to excessive buildup of charge along the backbone.

Polyelectrolytes have been studied extensively because molecular structure can be tailored to allow large conformational changes with pH, temperature, or added electrolytes. Molecular parameters that influence behavior include number, type, and distribution of charged mers on the chain, hydrophobic/hydrophilic balance, distance of charged moiety from the backbone, and counterion type. Solution properties including phase behavior, hydrodynamic volume, and binding can be altered, offering utilization in flocculation, adhesion, stabilization, compatabilization, viscosification, suspension, etc.

Ionic Polyelectrolytes. Anionic polyelectrolytes include poly(acrylic acid), poly(vinylsulfonic acid), poly(methacrylic acid), poly(styrenesulfonic acid), and their salts.

Cationic polymers are a class of polyelectrolytes that derive their unique properties from the density and distribution of positive charges along a macromolecular backbone as well as molecular weight. Chain conformation and solubility depend on the extent of ionization and interaction with water. Cationic functional groups can strongly interact with suspended, negatively charged particles or oil droplets and are useful for many applications, including waste treatment and paper making.

Water-soluble polymers containing cationic charge can be

segregated into three main categories: ammonium (amines and quaternaries), sulfonium, and phosphonium quaternaries.

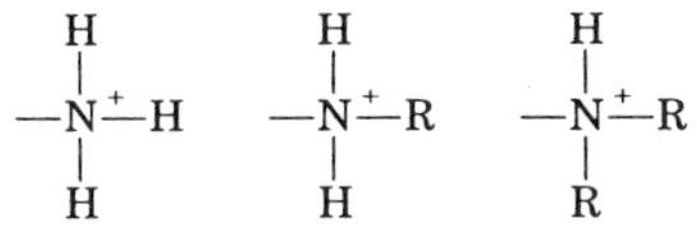

ammonium

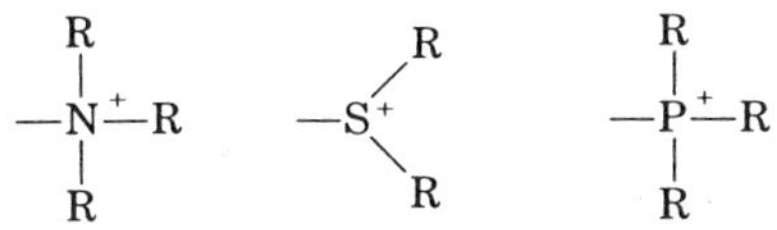

sulfonium phosphonium

The commercially significant ammonium polymers include cationic or quaternary polyacrylamides and ammonium polymers and polyamines and polyimines, including polyvinylammonium polymers (see POLYAMINES AND POLYQUATERNARY AMMONIUM SALTS).

Amphoteric Water-soluble Polymers. Amphoteric water-soluble polymers are polymeric systems containing both anionic and cationic charges. Such materials can be polymeric zwitterions with positive and negative charges on the same pendent group (betaines) or on the same backbone (ampholytes). Alternatively, they can be complexes of positively charged chains and negatively charged chains (interpolymer complexes). The betaines necessarily have an equal balance of anionic and cationic groups. The ampholytes can be charge-balanced or charge-imbalanced. The interpolymer complexes must be charge-imbalanced to maintain solubility. However, all amphoteric water-soluble polymers have a high charge density to ensure water solubility (see POLYAMPHOLYTES).

Synthesis. Polybetaines are prepared by free-radical solution polymerization of carboxy or sulfobetaine monomers. They can be homopolymers or copolymers with other water-soluble monomers (eg, acrylamide, *N*-vinylpyrrolidinone).

Polyampholyte homo- and copolymers are also synthesized via free-radical solution polymerization of anionic and cationic monomers. However, because of the inherent donor–acceptor nature of positively and negatively charged monomers, ampholytic monomer pairs are prone to spontaneous polymerization. Interpolymer complexes of water-soluble anionic and cationic polymer chains are formed by simply mixing the two oppositely charged polymers. Care must be taken to make sure that the stoichiometry is not exact. Otherwise, water-insoluble polysalts form.

Properties. Polybetaines and polyampholytes have unusual phase-behavior and solution properties. Unlike polyelectrolytes, polybetaines and charge-balanced polyampholytes are frequently more soluble and show higher viscosities in salt than in deionized water solution. Hence, these polymers are known as "antipolyelectrolytes."

Applications. Polybetaines, polyampholytes, and interpolymer complexes have found utility as water and brine viscosifiers, brine drag-reduction agents, and oil-field chemicals. All of these applications are the result of the unusual interplay of positive and negative charges on the same group or backbone, between chains, and/or between chains and external electrolytes.

Hydrophobically Associating Polymers

Synthetic polymers containing nonpolar groups that aggregate in polar media were first used as model compounds in studies of the conformation of proteins. These copolymers, prepared from *n*-dodecylation of poly(2-vinylpyridine) or hydrolysis of alternating copolymers of maleic anhydride and alkyl vinyl ethers, possessed surfactantlike properties.

Hydrophobically associating polymers fall into two categories: (*1*) those that form intramolecular associations and (*2*) those that form intermolecular associations. Each type possesses unique solution characteristics based on hydrophobic interactions and associated water ordering as discussed earlier. The parameters determining inter- or intramolecular aggregation include the molar volume of the copolymer, chemical composition and the relative placement of the hydrophobe, hydrophilic segmental length and hydration, and total molecular weight.

These parameters may be varied to optimize a copolymer for a particular application. In general, intramolecular micellelike associations occur for high concentrations of long, pendent hydrophobic groups on flexible hydrophilic chains. Intermolecular interactions require a critical concentration of polymer molecules with relatively small molar concentrations (0.1–2 mol %) of long alkyl chains.

Polymer Synthesis and Structural Characteristics. There are two major synthetic routes to water-soluble, hydrophobically associating polymers: post polymerization modification and copolymerization of water-soluble and hydrophobic monomers.

Polymer Properties. Hydrophobically associating polymers are used mainly in applications that require control of rheological properties of water-based fluids. These properties depend on the solubility characteristics of the polymers. Solutions are actually microheterogeneous with specific polymer–polymer and polymer–solvent interactions. The nature of hydrophobic associations and their effect on the structure and properties of water have been extensively studied for small molecules. However, hydrophobic associations of synthetic, water-soluble polymers have received limited study. Polymer solubility depends on the chain length and concentration of hydrophobic functionality as well as the nature of the water-soluble portion of the polymer molecule.

The introduction of hydrophobic groups into a water-soluble polymer has a different effect on the dilute solution viscometric behavior than on rheological properties above a critical concentration c^* (the overlap concentration). Below the overlap concentration, hydrophobic associations tend to create hypercoiling of the polymer molecules. This results in a decrease in the intrinsic viscosity and an increase in the Huggins' coefficient owing to intramolecular associations.

The most interesting feature of the rheology of hydrophobically associating polymers is the unique response to shear rate and solvent quality (eg, salt content). The viscosity can be independent (Newtonian), decrease (pseudoplastic), or even increase (dilatant) with shear rate depending on the amount (and type) of hydrophobe, polymer concentration, and solvent quality. As the hydrophobe level and polymer concentration increase beyond a critical value, a shear-rate region in which viscosity increases with shear is observed. Below and above this shear rate region, the viscosity behaves normally with shear; the time required to reach a steady state or equilibrium condition is relatively short. However, within the boundaries of the shear regime for shear thickening, polymer relaxation times increase significantly. An explanation of the molecular behavior of the system for shear thickening involves a change in the relative amount of intra- and intermolecular associations with shear. As the polymer chains are extended under a shear field, the number of intermolecular associations increase, resulting in enhanced viscosity. A similar explanation has been used for the shear-thickening behavior of ionically associating polymers in hydrocarbon solvents.

Applications. Many of the studies of hydrophobically modified water-soluble polymers were motivated by the need for improved mobility control agents for enhanced oil recovery (EOR). These studies have demonstrated the enhanced viscosification due to intermolecular hydrophobic associations. In addition, the presence of salt or brine in the reservoir fluid leads to further enhancements in viscosity due to increased associations. The potential of these polymers for EOR is promising, but further work is required before the applicability of the concepts can be demonstrated.

The use of associative thickeners to control the rheology of water-based latex paints is the application that has received the most attention. Other applications for hydrophobically associating polymers have been suggested. Recently, hydrophobically modified systems have been studied for controlled release of bioactive materials, for drag reduction in water-based fluids, as polymeric surfactants, and as phase-transfer agents in catalysis.

BIBLIOGRAPHY

P. Molyneaux, *Water-Soluble Synthetic Polymers, Properties and Behavior,* Vols. 1 & 2, CRC Press, Inc., Boca Raton, Fla., 1984.

A. G. Walton and J. Blackwell, *Biopolymers,* Academic Press, New York, 1973.

C. Tamford, *Physical Chemistry of Macromolecules,* John Wiley & Sons, Inc., New York, 1961.

CHARLES L. MCCORMICK
University of Southern Mississippi
J. BOCK
D. N. SCHULZ
Exxon Research

WAXES

Waxes are usually classified by their source as natural or synthetic waxes. The waxes obtained from natural sources include animal waxes, vegetable waxes, mineral waxes, and petroleum waxes. These waxes are sometimes classified by their chemical composition, with the animal, vegetable, and some mineral waxes (ie, montan) referred to as ester waxes and the petroleum waxes and some mineral waxes (ie, ozokerite and ceresin) referred to as hydrocarbon waxes. Sometimes synthetic waxes are also classified in this manner.

Petroleum waxes are by far the largest market of the naturally occurring waxes.

Petroleum Waxes

Paraffin Wax. Paraffin wax is obtained from the distillation of crude oil. The distillate is solvent extracted to remove oil. The resulting wax is decolorized by hydrogenation and bauxite filtration. The processes of producing paraffin wax have been described. Paraffin waxes consist mainly of straight-chain, saturated hydrocarbons.

Microcrystalline Wax. Microcrystalline waxes are produced by deoiling the petrolatums or greases obtained by dewaxing deasphalted residual lube stocks or by deoiling the deasphalted tank bottoms that settle out during the storage of crude oil. The refining of wax has been described. These waxes are referred to as microcrystalline because the crystals are much smaller than those of paraffin wax. Microcrystalline waxes are composed predominantly of isoparaffinic and naphthenic saturated hydrocarbons along with some *n*-alkanes. The molecular weights range from 450 to 800 (C35 to C60). Microcrystalline waxes are produced in two grades. The lower melting, softer, flexible waxes are used for laminations in the paper industry. The harder, higher melting, less-flexible waxes are used in coatings and polishes.

Typically, the microcrystalline waxes are higher melting and have a higher viscosity than the paraffin waxes. Because of their inertness and excellent vapor-barrier properties, petroleum waxes are used in laminations and coatings for food-packaging applications. Other applications include hot-melt adhesives, printing inks, protective coatings (eg, rustproofing formulations), molded products (candles, crayons, dental wax, investment casting, etc), and antichecking waxes for automobile tires.

Chemically Modified Petroleum Waxes. Paraffin waxes show the same chemical inertness as the *n*-alkanes, which are their principal components. However, paraffin waxes undergo a number of chemical reactions, including formation of adducts, cracking reactions, and free-radical substitution reactions. The *n*-alkane components in paraffin wax react with urea to give crystalline clathrated adducts. Furthermore, *n*-alkanes can be separated from hydrocarbons by molecular sieves, such as Linde's type 5A, Linde Molecular Sieves, Union Carbide Corp. The thermal cracking of paraffin wax is a commercial process for making alpha olefins.

Food-grade Wax. Federal Regulation 21 CFR 172.886 contains a specification that describes food-grade wax as a "mixture of solid hydrocarbons, paraffinic in nature, derived from petroleum and refined to meet specifications." In addition to conventional guidelines, these U.S. Food and Drug Administration specifications and their counterpart in the UK impose stringent controls on the polynuclear aromatic hydrocarbon content of food-grade waxes made for direct use in food or indirect contact in food-packaging materials. Current practice requires that the wax refiner certify the ultraviolet absorbance limits mandated in CFR 172.886.

Animal, Vegetable, and Mineral Waxes

The consumption of these waxes has declined in recent years. However, their dollar value remains significant.

Animal Waxes. Animal waxes may be of insect or mammalian origin. The most important commercial animal waxes are beeswax and wool grease (unrefined lanolin).

Vegetable Waxes. Vegetable waxes are obtained either from leaves and stems or from fruits and seeds. Candelilla and carnauba waxes are obtained from the leaves and stems of the candelilla plant and the carnauba palm. These two waxes are the most important commercial vegetable waxes.

Candelilla wax is composed of hydrocarbons (50%), nonglyceride esters, alcohols, and free acids. Candelilla wax is used as a component of chewing gum base and in the cosmetic industry.

Carnauba wax is the hardest and highest melting of the vegetable waxes. It is composed primarily of nonglyceride esters with small amounts of free acids, resins, and hydrocarbons. Carnauba waxes are used in paper coating, cosmetics, and polishes. It is the largest volume vegetable wax in use today.

Mineral Waxes. Mineral waxes are obtained from fossil remains such as lignite, bitumens, peat, and shale, excluding the waxes derived from petroleum. The waxes, eg, montan wax, can be obtained by direct mining. The most important mineral wax is montan wax. Its main components are nonglyceride esters of carboxylic acids, alcohols, acids, resins, and hydrocarbons. Montan wax is used extensively in polishes and carbon paper.

Synthetic Waxes

Fischer-Tropsch Wax. The Fischer-Tropsch wax is a by-product in the synthesis of liquid fuels, such as gasoline and diesel oils. It is white with a melting point of 95–110°C (ASTM D938) and a needle penetration of 0.5 (ASTM D1321). This wax is composed of *n*-alkanes in the molecular weight range of 600–950. It is used with other waxes and wax–polymer blends in hot-melt coatings and adhesives to increase melting points and impart hardness. In textiles it is used for impregnation and sealing, in candles to increase the melting point and opacity, in carbon paper coating to increase hardness and eliminate adherence, in rubber and plastic as a mold release, and in other lesser applications.

Polyethylene Waxes. Polyethylene waxes are used in the following applications.

Plastic manufacturing, processing aids, mold and extrusion lubricant, color dispersant	25%
Modification of petroleum waxes, hot-melt coatings, waxed paper, cable compounds, candles	21%
Paints, inks, antislip, antimar additive	16%
Textiles, impregnant, sealant, lubricant	13%
Polishes, industrial maintenance	8%
Rubber, processing and mold release	8%
Crayons, cosmetics, chewing gum base, dental wax, and others	9%

Fatty Acids and Amides. Stearic, palmitic, and myristic acids as well as hydrogenated tallow fatty acids are waxy materials that serve as lubricants and mold-release agents in acid form or as metal salts. The fatty acid amides such as stearamide and oleamide are used as lubricant additives; they are good antislip agents and impart antistatic properties.

BIBLIOGRAPHY

A. H. Warth, *The Chemistry and Technology of Waxes,* Van Nostrand Reinhold Publishing Co., Inc., New York, 1956.

H. Bennett, *Industrial Waxes,* Vols. 1 & 2, Chemical Publishing Co., Inc., New York, 1963.

W. A. Gruse and D. R. Stevens, *Chemical Technology of Petroleum,* 3rd ed., McGraw-Hill Inc., New York, 1960.

George V. Dyroff, ed., *Manual on Significance of Tests for Petroleum Products,* 5th ed., ASTM, Philadelphia, PA, 1989.

John J. Kaufman
Alan W. Stratton
Witco Corporation

WEAR RESISTANCE.

See Abrasion and wear.

WEATHERING

The weatherability of a material is its ability to withstand environmental conditions during exposure. Both natural and synthetic polymers in the form of paints, plastics, rubbers, and textiles are subject to deterioration by the elements, but the extent varies with the type of material and its formulation. Weathering lifetimes and susceptibilities to degradation of some of the common types of polymer are shown in Tables 4 and 5 of the article, Degradation (pp. 249 and 250). The useful lifetime of a polymeric material depends on the rate and type of change that occur in the structural components leading to changes in appearance and mechanical or dielectric properties. It also depends on the application.

Metals, ceramics, and glass are being replaced by polymeric materials at a rapid rate. In 1986, 75% of the 20×10^6 t of plastics sold in the United States were designed for long-life applications, whereas 25% were utilized in packaging and other short-term uses. For the former applications, good weathering characteristics are essential to commercial acceptability. However, where durability is not important, polymers are needed that photodegrade quickly after their intended use in order to protect the environment.

Factors Responsible for Weathering

Weather Factors. The components of the weather responsible for the deterioration of polymeric materials are radiation, atmospheric temperature and its cycles, moisture in its various forms, wind, and the normal constituents and contaminants. Although the ultraviolet portion of solar radiation is mainly responsible for initiating weathering effects in most polymers, the visible and near-infrared portions also contribute to the degradative processes. Colored materials are susceptible to visible radiation, and near-infrared radiation can accelerate chemical reactions by raising the temperature. The other factors act synergistically with solar radiation to significantly influence the weathering processes. In condensation-type polymers, moisture alone can be a major cause of degradation by promoting hydrolysis. In PVC formulations, elevated temperatures cause evaporation of the plasticizer that leads to failure of the material by embrittlement. All weather factors, including the quality and quantity of sunlight, vary with geographic location, time of year, time of day, and climatological conditions. Therefore, in order to understand and predict the effect of weather on materials, data are required on each factor that may contribute to degradation.

Material Factors. ***Absorption of Radiation.*** According to the first law of photochemistry (qv), the Grotthus-Draper law, only light absorbed by a material can have any effect on it. Therefore, the absorption properties of a material are critical to the effect of light on it. The more light is absorbed, the higher the potential for degradation if the energy of the light absorbed is sufficient to break some of the chemical bonds.

The absorption properties are determined by the chemical structure of the material. Colorless polymers transmit or reflect all wavelengths in the visible region but absorb ultraviolet radiation. However, the wavelengths of ultraviolet radiation absorbed by polymers vary widely. Most saturated aliphatic-type polymers have absorption characteristics similar to those of poly(vinyl chloride) and polyethylene. They are almost completely transparent to solar ultraviolet radiation. Therefore, the main structural components of these polymers cannot account for their sensitivity to sunlight. The vulnerability of saturated aliphatic polymers to deterioration by outdoor conditions is due mainly to the presence of uv-absorbing impurities, both physical and molecular imperfections.

Although aromatic polymers absorb much more ultraviolet radiation than aliphatic polymers, some of the former cannot absorb solar radiation directly. Thus, uv-absorbing impurities, similar to those present in the aliphatic-type polymers, are largely responsible for the weathering of polystyrene and polycarbonate.

Absorption properties depend on the polymer type and its thickness. With an increase in thickness from 0.1 to 2.5 mm of the aromatic polyester, the long wavelength edge of the absorption band is shifted about 25 nm to longer wavelength. The thin film absorbs only the short wavelengths of terrestrial solar radiation, the wavelengths that are strongly absorbed by the material. It is almost completely transparent to wavelengths above 320 nm. The longer wavelengths are more weakly absorbed and therefore can penetrate more deeply. Thus only the thicker specimen is capable of effectively absorbing the longer wavelengths. The 2.5-mm specimen absorbs wavelengths as long as 370 nm.

Because of the differences in depth of material penetration between short and long wavelengths, the effects of short wavelengths usually predominate in the surface layers of an aromatic polymer, whereas the longer wavelengths interact with the bulk.

Stability to Absorbed Radiation. The stability of a material to the absorbed radiation depends on the energies of the photons absorbed in relation to the bond energies as well as on the electronic structure of the molecule. All wavelengths in the ultraviolet do not have the same effect. Rate, mechanism, and type of degradation often vary with the wavelengths absorbed. There is evidence in many polymers for dual mechanisms of degradation with longer wavelengths being responsible for photooxidation reactions and shorter wavelengths being responsible for photo-Fries rearrangements or direct scission reactions of the main structural components. Photooxidation is usually initiated by the uv-absorbing impurities present as imperfections in the molecular structure of the polymer. Because of the differences in degradation mechanism with wavelength, the type of stabilizers required may depend on the wavelengths that initiate the degradation. Short wavelengths usually cause more damage than long wavelengths. Because of the higher energy of the photons, they are capable of breaking more types of bond.

Weathering Tests

Weathering tests are important in the development and manufacture of materials designed for outdoor exposure. In research and development, the tests are carried out to rank the stabilities of different materials and formulations, determine the effectiveness of stabilizers, and predict the lifetimes of materials. Such tests are also used for quality control and material certification. Outdoor exposure tests are carried out under more severe conditions than those to which the materials are normally exposed. However, because of the need for more rapid evaluations of the resistance of materials to weathering than can be obtained by outdoor exposure tests and the need for controlled conditions, test devices with artificial light sources are generally used to accelerate the degradation for initial evaluations.

Outdoor Exposures. ***Exposure Sites.*** The leading U.S. outdoor exposure sites are south Florida and the Phoenix, Arizona area. The former has a subtropical climate with high temperature, humidity, and solar radiation. The latter has a climate typical of the hot and dry conditions of the desert with high solar radiation. At both of these exposure sites the weather elements responsible for degradation are intensified. Other test sites are in tropical and temperate climates, industrial areas, and salt air locations. The main criterion in selection of the test site is the sensitivity of the material to moisture and the type of climate in which it will be used. Often, materials are tested in a number of different climates.

Types of Exposure. The test can take numerous forms. Some of the variables include tilt angle, backing, protection by a glass shield, temperature, intensification of solar radiation, and specialized environmental conditions. The results may differ significantly. The particular test conditions depend on the application of the material and should simulate the in-service conditions as closely as possible. Because of the variability of outdoor conditions, repeated testing during different seasons and over a period of at least two years is required to obtain meaningful results. Control samples with known stabilities should be exposed at the same time.

Timing of Exposures. Traditionally, outdoor durability tests have been timed in calendar units, ie, weeks, months or years, although it is well known that the quantity and quality of sunlight and other weather factors vary considerably with the season and even from one year to the next. Timing based on the exposure to the total spectrum of irradiance, both direct and diffuse, was introduced in the 1960s using a pyranometer to measure exposure at the tilt angle of the test racks. Although this was an improvement over elapsed-time testing, it does not provide information on the radiation that is mainly responsible for weathering, ie, the ultraviolet component of sunlight. The percentage of ultraviolet varies significantly and changes in the ultraviolet portion are not detected in measurements of the full spectral range of solar radiation.

Various schemes have been devised to correct for this deficiency. However, studies have shown that timing exposures on the basis of the amount of measured incident ultraviolet radiation, particularly in the wavelength band responsible for the degradation, can significantly improve correlations between different types of exposure. It is also important for comparing natural weathering results obtained under different seasonal and site conditions and in predicting weathering based on accelerated tests. Currently, hemispherical ultraviolet irradiance between 295 and 385 nm at the tilt angle of the samples is continuously monitored and integrated at several U.S. exposure sites using the Eppley Model TUVR radiometer.

Accelerated Indoor Weathering Tests. Artificial weathering devices have been used since the early 1920s for rapid testing of materials durability, both for quality control and to screen new materials and formulations. The importance of these devices lies in their ability to accelerate the weathering processes under controlled and reproducible conditions. However, they are useful only if the results correlate with those under natural exposure, ie, if the chemical and physical changes induced in materials duplicate those caused by outdoor exposure and the accelerated tests correctly rank the stabilities of materials. These relationships must hold over the full service life of a material in order to correctly predict its lifetime under use conditions based on an accelerated test.

These requirements can only be realized if the balance of weather factors in nature is correctly simulated and if intensification is not excessive. A five to tenfold acceleration appears to be the maximum that can be imposed without distorting the results. The main causes of degradation, ie, ultraviolet radiation, temperature, and moisture, are included in the basic design of almost all artificial weathering devices.

Evaluations

Stability Rankings and Quality-control Tests. The resistance of new polymeric materials and formulations to weathering is usually tested both outdoors and in artificial weathering devices by exposing them for certain time periods or radiant exposure. This method is also used for quality-control tests of weatherability. Because of seasonal variations in ultraviolet radiation and other weather factors, one or preferably two years of outdoor weathering are an absolute minimum exposure time period. Variations in climatic conditions even at the same exposure site affect not only the lifetime of a paint film, but also the ranking of performance. For optimum reproducibility, tests should be started during the same season of the year. The timing of both natural and artificial weathering tests on the basis of ultraviolet radiant exposure, particularly in the actinic region primarily responsible for weathering, is preferable. Reliable evaluations by accelerated tests require conditions that closely simulate natural exposures in terms of radiation, temperature, and moisture levels.

Stability tests are also carried out by exposing test specimens until a desired change is produced. These are referred to as end-point tests. The relative stabilities are expressed quantitatively in terms of the ratios of the exposure times or radiant exposures required to produce the same amount of change in different materials or formulations. This is often a more sensitive method of ranking. However, it is time consuming and requires frequent measurements to identify the end point. If the latter is passed by, interpolation can only give an estimate because of the nonlinearity of degradation as a function of exposure (see also STABILIZATION).

Durability. The ultimate goal of weatherability tests is to predict long-range performance under normal use conditions based on relatively short-term tests. The prediction of service life by extrapolation of data obtained under accelerated test conditions is not a simple procedure. Ever since accelerated test methods were first developed, attempts have been made to establish a single acceleration factor for extrapolating test results to predict lifetimes under natural weathering conditions for a variety of materials. However, for most polymeric materials the degradation rate is not a simple function of the irradiance level or other stress factors.

Accurate predictions of durabilities, whether they are based on an empirical approach involving the relationship between accelerated and natural tests or on fundamental principles using mathematical modeling, depend on valid accelerated tests. Degradation is the result of complex interactions of the synergistic effects of radiation, temperature, and moisture as well as other components of weather. Therefore, all important weather factors must be closely simulated by the accelerated test.

BIBLIOGRAPHY

A. Davis and D. Sims, *Weathering of Polymers,* Applied Science Publishers Ltd., Barking, UK, 1983.

M. P. Morse in *Permanence of Organic Materials,* Special Technical Publication 781, ASTM, Philadelphia, Pa., 1982, pp. 43–66.

G. A. Zerlaut and M. L. Ellinger, *J. Oil Colour Chem. Assoc.* **64**(10), 387–397 (1981).

N. D. Searle in J. M. Fitzgerald, ed., *Analytical Photochemistry and Photochemical Analysis: Solids, Solutions and Polymers,* Marcel Dekker, Inc., New York, 1971, chapt. 9.

N. D. SEARLE
Consultant, Plastics and Chemicals

WEIGHT-AVERAGE MOLECULAR WEIGHT. See MOLECULAR WEIGHT DETERMINATION.

WELDING. See ADHESION AND BONDING; DIELECTRIC HEATING.

WET LAY-UP. See REINFORCED PLASTICS.

WETTING. See ADHESION AND BONDING; INTERFACIAL PROPERTIES.

WHISKERS. See FIBERS, ENGINEERING.

WIRE AND CABLE COVERINGS

In the design of insulated wires, flexibles, and cables, three types of coverings are of crucial importance: insulation, possibly in conjunction with semiconductive inner or outer layers; internal coverings and fillers (lapped or extruded); and sheaths and external protective coverings. The large number of applications requires a wide range of varied designs, such as heat-resistant wiring cables, highly flexible lift cables, mining cables subject to high mechanical loads, simple installation cables, underground cables, and cables for highest operating voltages and maximum transmission capacities. Recent developments have produced materials that are tailored to these different requirements. These materials have frequently replaced products such as natural rubber and paper insulation. Today paper insulation is still mainly used for extra-high-voltage cables where it is protected by a metal sheath. However, even in this application, it is being replaced by cross-linked polyethylene.

According to their physical properties, the polymers of technical importance can be classified into thermoplastics (plastomers), elastomers, and thermosetting plastics (duromers). Table 1 shows some polymers currently in widespread use in cable engineering. Typical thermosetting plastics in use are epoxy resins (EP) and polyurethane (PU). Increasing use is being made of materials that cannot be classified as thermoplastics, elastomers, or thermosetting plastics. Cross-linked polyolefins (cross-linked polyethylene and ethylene copolymers), for example, behave similarly to elastomers above the crystalline melting point. Thermoplastic elastomers with thermoplastic properties at processing temperatures and elastomeric properties at normal operating temperatures are also gaining in importance.

Chemical Aging

The term aging (qv) describes the way in which the properties of a material change with time. When in use, polymers are subject to chemical changes that adversely affect mechanical and electrical properties. The chemical aging processes accelerate with increasing temperature. Therefore, polymers exposed to elevated temperatures must be protected by stabilizers to give adequate service life for the end products (see DEGRADATION).

Insulating coverings and sheaths exposed to direct sunlight (uv radiation) must be protected by adding light stabilizers such as carbon black, which is used particularly for polyolefins; a 2–3% (by weight) addition of well-dispersed carbon black gives effective protection (see STABILIZATION; WEATHERING).

In the presence of atmospheric oxygen, the chemical aging of many polymers, especially polyolefins, is based on oxidation processes that are triggered or accelerated by heat and light. The antioxidants (qv), used for stabilizing, are added to the polymers in a proportion of 0.1–0.5% (by weight).

Effect of Moisture

Practical experience and long-term tests have shown that water has an adverse effect on polyolefin insulating materials such as PE and XLPE. Using suitable coloring methods, treelike structures can be detected in insulating materials that have been exposed to electrical voltage stress. These structures branch out from microscopically small imperfections in the direction of the electrical field. This phenomenon, known as water treeing (WT), differs from electrical treeing (ET), which is caused, for example, by partial discharge.

The following precautions can be taken for the construction, production, and installation of cables with PE or XLPE insulation:

1. Imperfections in the insulation and at the interfaces of the semiconductive layers are kept to a minimum by optimizing the quality of the insulating and semiconductive layer material and the production process; special attention is given to extruded semiconductive layers.

2. Water and moisture are prevented from penetrating into conductors and screen area during production, storage, transport, installation, and operation as a result of damage to the sheath. Outer sheaths must be strong, eg, PE sheaths. A longitudinal watertight screen area limits water penetration in the event of sheath damage. High voltage cables of at least

Table 1. Polymers Used in Cable Engineering

Thermoplastics[a]		Thermoplastic	Elastomers	
Polymer	Abbreviation	elastomers	Polymer	Abbreviation
poly(vinyl chloride)	PVC	blends of polyolefins and rubber	natural rubber	NR
polyethylene	PE		butyl rubber	IIR
ethylene–vinyl acetate copolymers[b]	EVA	triple-block copolymers[c]	styrene–butadiene rubber	SBR
ethylene–alkyl acrylate copolymers		polyurethanes, polyesters	nitrile–butadiene rubber	NBR
ethylene–ethyl acrylate copolymers	EEA		ethylene–propylene rubber	EPM[d]
			ethylene–propylene terpolymer rubber	EPDM[d]
ethylene–butyl acrylate copolymers	EBA		polychloroprene	CR
			chlorosulfonated polyethylene	CSM
polypropylene	PP			
polyamide	PA		chlorinated polyethylene	CM
ethylene–tetrafluoroethylene copolymers	ETFE		silicone rubber	SR
tetrafluoroethylene–hexafluoropropylene copolymers	FEP		epichlorohydrin rubber	ECO
			ethylene–vinyl acetate copolymers[e]	EVA

[a] Plastomers.
[b] Vinyl acetate <30%.
[c] Styrene–diene–styrene.
[d] The collective term for EPM and EPDM is EPR.
[e] Vinyl acetate >30%.

36/60 kV should always be provided with an aluminum composite layer sheath and a longitudinal watertight screen.

Intensive research is directed toward improving the water-tree-retardant (WTR) compound of XLPE insulation with the assistance of additives (see also ELECTRICAL PROPERTIES).

Electrical Screening of Insulating Material

The partial electrical discharge that occurs at high voltages in cavities or gaps in the insulating material or at the transitions from the conductor to the insulating material and from the insulating material to the metal screen can eventually lead to electrical breakdown. Therefore, in plastic insulations conductive layers must be provided on the inside and outside of the insulation.

Sheaths and Outer Coverings

In some specifications for cable designs, a distinction is made between sheaths and protective coverings. The latter afford protection from corrosion in the form of a metal sheath or light mechanical protection for installation cables. Sheaths, on the other hand, are provided for higher mechanical stresses or, for example in cables with paper insulation, as a protection against the ingress of moisture, solvents, etc. As the properties of plastic coverings and sheaths coincide (apart from their dimensions), the collective term sheath is used here.

Plastic sheaths are made from poly(vinyl chloride), polyethylene, polyamide and polyurethane, and polypropylene. Rubber sheaths are made from polychloroprene, chlorosulfonated polyethylene, chlorinated polyethylene, and nitrile–butadiene rubber.

Sheath Materials for Special Applications

Special sheath materials have been developed for cables and wires with flame-retardant characteristics, such as reduced fire propagation even in bundled configuration, no corrosive components liberated by combustion, low smoke emission, and preservation of insulation. Elastomeric olefinic copolymers such as EVA, EPR, or EEA are used as basic materials.

BIBLIOGRAPHY

M. T. Shaw and S. H. Shaw, *IEEE Trans. Electr. Insul.* **EI-19,** 419 (1984).

E. Peschke and R. Wiedenmann, *Elektrizitaetswirtschaft* **86,** 226 (1987).

H. Stoger, R. Stubbe, and M. Ulrich, *Elektrizitaetswirtschaft* **84,** 792 (1985).

M. Grayson, ed., "Insulation, Electric (Wire and Cable)," *Encyclopedia of Chemical Technology,* 3rd ed., Vol. 13, John Wiley & Sons, Inc., New York, 1981, p. 564.

RUDOLF WIEDENMANN
Siemens AG

WOOD

Structure

The chemical and anatomic structure of wood affects strength properties; appearance; resistance to penetration by water, chemicals, and decay; pulp quality; and chemical reactivity. To use wood effectively requires a knowledge of the amounts and distribution of constitutive substances in the wood cells.

Woods are divided into hardwoods and softwoods. Hardwood trees (angiosperms) generally have broad leaves, are deciduous in the temperate regions of the world, and contain a special cell called a vessel element. Softwood trees (conifers or gymnosperms) are cone-bearing, generally have scalelike or needlelike leaves, are not deciduous, and do not contain vessel elements. The terms hardwood and softwood have no direct application to the hardness or softness of the wood. In fact, hardwood trees such as cottonwood, aspen, and balsa have softer wood than the western white pines and true firs; certain softwoods, such as longleaf pine and Douglas fir, produce wood that is much harder than that of basswood or yellow poplar.

Just under the bark of a tree is a thin layer of cells called the cambium. Here cells divide and eventually differentiate to form bark tissue on the outside of the cambium and wood or xylem tissue on the inside. This newly formed wood contains living cells and conducts sap upward in the tree, and hence is called sapwood. Eventually the inner sapwood cells become inactive and are transformed into heartwood. Often this transformation is accompanied by the formation of extractives that darken the wood, make it less porous, and sometimes provide more resistance to decay.

Because of the extreme structural variations in wood, species can be selected for a specific purpose. Some species, like spruce, combine light weight with relatively high stiffness and bending strength. Very heavy woods, like lignum vitae, are extremely hard and resistant to abrasion. Balsa, a very light wood, has high thermal insulation value; hickory is extremely shock resistant and mahogany has excellent dimensional stability. Moreover, many mechanical properties of wood, such as bending strength, crushing strength, and hardness, depend on its density; the heavier woods are generally stronger. Wood density is determined largely by the relative thickness of the cell wall and by the proportion of thick-walled to thin-walled cells.

Cell Wall. The main building block of wood is the cell, and the cell wall is the main structural component. The cells that make up wood are of various sizes and shapes. However, the cells that give the wood strength are long, pointed cells called fibers or tracheids. These cells vary greatly in length within a tree and among species. Hardwood tracheids average about 1 mm in length; softwood tracheids are 3–8 mm long. Wood cells may be partly filled with deposits such as gums, resins, or other extraneous organic or inorganic substances.

The unique properties of the cell, and hence of wood, arise from the organization of the chemical components that make up the cell wall and the systematic way in which the wall is constructed. The principal compound in mature wood cells is cellulose (qv), a polysaccharide of repeating glucose molecules that may reach 4 μm in length. These cellulose molecules are arranged in an orderly manner into structures about 10–25 nm wide called microfibrils. In certain parts (micelles) the ordered arrangement is crystalline and can be detected in polarized light. The microfibrils wind together to form macrofibrils, which measure about 0.5 mm in width and may reach 4 mm in length. These macrofibrils are as strong as an equivalent thickness of steel. The framework of cellulose macrofibrils is cross-linked with hemicelluloses, pectins, and lignin. Lignin (qv), the second most abundant polymer found in plants, confers rigidity to the cell wall.

Hardwood. Figure 1 illustrates the cellular structure of a yellow poplar, an example of a hardwood.

Softwood. Figure 2 illustrates the cell structure of white pine, an example of a softwood.

Chemical Composition

The principal chemical components of wood are carbohydrates (70–80%), lignin (20–30%), and extraneous materials (5–10%). Because the chemical composition varies between and among species, it is impossible to define absolute compositions. Nevertheless, the following generalizations can be made: ranges of composition values can be defined within species, and the average lignin content of softwood is slightly higher than that of hardwood. The carbohydrate consists primarily of cellulose, a linear polymer of anhydroglucose units linked by β-1→4 glycosidic bonds, and lesser amounts of hemicelluloses, branch-chained polymers of several different compositions that include five or six anhydro-sugars plus uronic anhydride and acetyl substituents. The chain length of wood cellulose polymers is between 7,000 and 10,000 glucose units, whereas the hemicelluloses have chain lengths of 200 or less sugar units. Lignins are complex phenolic polymers of irregular structure. The basic structure unit is a mono- or dimethoxy-substituted propylphenol unit. Molecular weights of isolated lignins have been measured up to 16,000. The extraneous materials are of much lower molecular weight (<1000) and can be removed from the wood with organic or aqueous solvents. They are not an integral part of the wood structure and include a wide variety of organic compounds and a small amount of inorganic salts.

Physical Properties

The physical properties of wood include some that are directly related to water content and movement, such as hygroscopicity; mechanical, electrical, and thermal properties; corrosiveness;

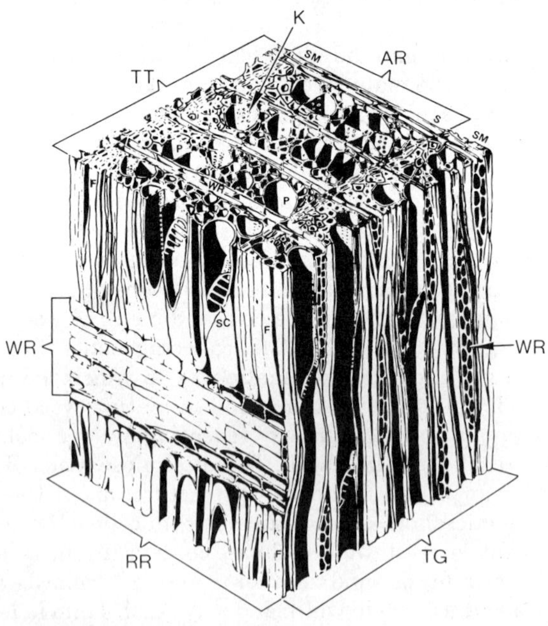

Figure 1. Cellular structure of a hardwood (yellow poplar); enlargement of a block ca 1 mm high. AR, annual rings; F, fiber; K, pits; P, pores; RR, edge grain; S, earlywood; SC, grating separating vessels; SM, latewood; TG, flat grain; TT, end grain; WR, wood rays.

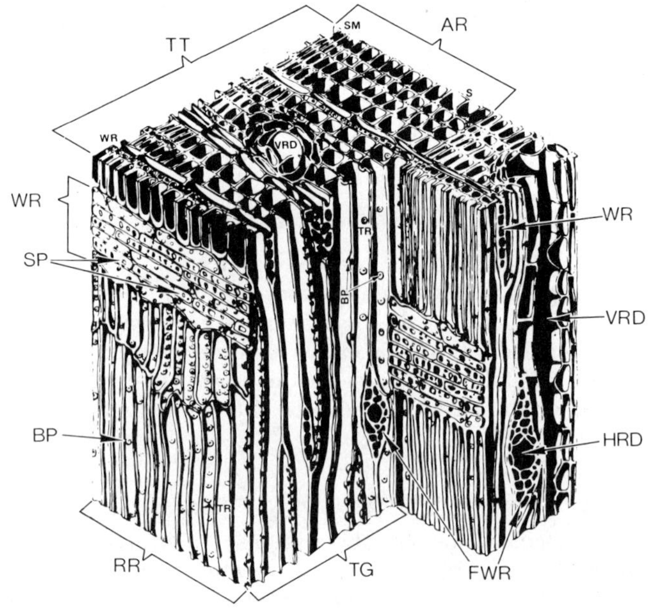

Figure 2. Cellular structure of a softwood (white pine); enlargement of a block ca 1 mm high. AR, annual rings; BP, border pit; FWR, fusiform wood rays; HRD, horizontal resin duct; RR, edge grain; S, earlywood; SM, latewood; SP, simple pits; TG, flat grain; TR, tracheid; TT, end grain; VRD, vertical resin duct; WR, wood rays.

and chemical resistance. Because of the great effects water has on wood, it is dried before use for a number of important reasons. Dry wood is lighter than wet wood and generally stronger, less prone to staining, mildew, or decay, has greater electrical resistance, and provides better thermal insulation. Moreover, the shrinkage that accompanies drying can occur before the wood is put to use. Wood that is dried before assembly has stronger joints made with fasteners, such as nails and screws, than wood dried after assembly. Ideally, temperature and relative humidity should be controlled during drying. If wood dries too rapidly, it is likely to split, check, warp, or honeycomb because of stresses. The two most widely used methods are air drying and kiln drying.

Mechanical Properties. The source of wood strength and stiffness is the fiber. Wood can be viewed as a series of tubular fibers or cells cemented together; each fiber is composed of various quantities of polymers. Together, the tubular structure and polymeric construction are responsible for most physical and chemical wood properties. Wood is an anisotropic material; it has unique and independent mechanical properties in the directions of three mutually perpendicular axes: longitudinal, radial, and tangential. The longitudinal axis is parallel to the fiber or grain direction; the radial axis is normal to the growth rings (perpendicular to the grain in the radial direction); and the tangential axis is perpendicular to the grain but tangent to the growth rings. The strength of wood depends on the direction of loading with respect to these three axes.

Extensive data on basic mechanical properties of different wood species are given in the *Wood Handbook*.

The use of wood as a structural material is influenced by environmental factors such as temperature and moisture content. These factors have a direct effect on mechanical properties and promote the growth of decay fungi that degrade wood.

Mechanical properties may be further altered by chemical treatments used to protect wood from decay and fire.

Moisture Content. Most basic mechanical properties of clear wood increase markedly (Table 1) as moisture content is reduced below the fiber-saturation point (at about 30% moisture content, based on ovendry weight). In structural lumber, the increase in clear-wood mechanical properties caused by drying is partially offset by a weakening as a result of drying checks, particularly in association with knots and cross grain.

Temperature. Clear-wood mechanical properties tend to deteriorate when wood is heated and improve when it is cooled. At temperatures below approximately 100°C, the effects are reversible and the properties return to the values at the original temperature. This response to temperature change is termed an immediate effect. The immediate effect of temperature on clear-wood strength and modulus of elasticity is dependent on moisture content and varies from species to species.

Chemical Treatments. The effect of treatments on strength properties largely depends on the chemical nature and reactivity of the absorbed liquid. In general, neutral, nonswelling liquids have little if any effect on the strength properties of wood. Any liquid that causes wood to swell causes a reduction in strength. This effect may be temporary, and only while the liquid remains in the wood. It may also be permanent, as in the case of chemically reactive liquids (strong acids and bases), and of a magnitude dependent on time and temperature of exposure and concentration of the solution.

Wood preservatives are either oil-borne or waterborne. The former, such as coal tar creosote, creosote–petroleum oil mixtures, and pentachlorophenol dissolved in petroleum, are practically inert to wood and do not chemically affect its strength. However, many chemicals used in waterborne salt preservatives, including those containing chromium, copper, ammonia, zinc, and boron, react with wood and can affect mechanical properties.

Wood is also sometimes impregnated with monomers, such as methyl methacrylate, which are subsequently polymerized. Some mechanical properties of the resulting wood–plastic composite are superior to those of the original wood, because the void spaces in the wood structure are filled with plastic. The polymerization process and the chemical nature and quantity of monomers influence the mechanical properties of the composite product.

Decay. Decay or wood-destroying fungi may seriously reduce strength. Brown-rot fungi primarily metabolize carbohydrates; white-rot fungi metabolize lignin as well. Decay is inhibited by keeping the wood dry, ie, at a moisture content below 20%. In an environment conducive to decay, the wood should be treated with a preservative.

Electrical Properties. Electrical properties include conductivity, dielectric constant, loss tangent, dielectric strength, and piezoelectric activity. These properties are affected by moisture and temperature.

Conductivity. Conductivity depends on moisture content. Ovendry wood has a conductivity on the order of 10^{-16} S/m; at 30% moisture content, it has about 10^{-3} to 10^{-4} S/m.

Dielectric Constant. The dielectric constant of wood varies greatly with moisture content; frequency has a strong influence, temperature has less influence. At frequencies higher than about 1 MHz, the dielectric constant of moist wood appears to be essentially the sum of that of dry wood and water. At lower frequencies, the dielectric constant of moist wood can become very large. The very low frequency dielectric constant of wet wood can be as large as 10^6 at room temperature, whereas typical corresponding values for dry wood are less than 10.

Loss Tangent (Dissipation Factor). Under most conditions, the loss tangent of wood increases as the moisture content increases. However, under some combinations of frequency and temperature, the loss tangent goes through maximum or minimum values as the moisture content is increased.

Dielectric Strength. The dielectric strength of dry wood, with the field perpendicular to the grain, is on the order of 10–20 kV/mm. At 15% moisture content, it is about 1.0–2.0 kV/mm, and at 20% moisture content, about 0.5 kV/mm.

Piezoelectric Activity. The semicrystalline structure of wood causes some piezoelectric activity, most readily observed at a transverse voltage gradient resulting from a longitudinal stress. The effect decreases as the wood moisture content increases and increases with increasing wood density, temperature, and degree of crystallinity. Maximum piezoelectric moduli are of the order of 10^{-13} C/N.

Thermal Properties. ***Specific Heat.*** The specific heat of dry wood from 20 species between 106 and 0°C is 1.32–1.41 kJ/(kg·K)

Table 1. Approximate Average Effects of Moisture Content on Mechanical Properties of Clear Wood at about 20°C

Mechanical property	Relative change in property from 12% moisture content, %: At 6% moisture	At 20% moisture
modulus of elasticity parallel to the grain	+9	−13
modulus of elasticity perpendicular to the grain	+20	−23
shear modulus	+20	−20
bending strength	+30	−25
tensile strength parallel to the grain	+8	−15
compressive strength parallel to the grain	+35	−35
shear strength parallel to the grain	+18	−18
tensile strength perpendicular to the grain	+12	−20
compressive strength perpendicular to the grain at the proportional limit	+30	−30

[0.31–0.33 kcal/(kg·K)]. It is practically independent of species and specific gravity.

Thermal Conductivity. Compared to most other structural materials, the thermal conductivity of wood is low; the organic components have low inherent thermal conductivity across the grain. For woods commonly used in construction, thermal conductivity is about 9.2×10^{-2} to 1.7×10^{-1} W/(m·K).

Coefficient of Thermal Expansion. The thermal expansion of wood is negligible compared to that of moisture-induced swelling and shrinking. The coefficients of thermal expansion of common woods in the longitudinal direction are in the range of 2×10^{-6} to 6×10^{-6}/°C and are practically independent of species and specific gravity.

Thermal Degradation and Fire Resistance. Except for loss of absorbed water and volatile oils, the composition of wood is not significantly affected by temperatures up to about 100°C. As the temperature increases to about 175°C, some of the components of wood begin to break down. Between 100 and 200°C, noncombustible products such as carbon dioxide, traces of inorganic compounds, and water vapor are produced. Above 200°C, the celluloses break down and produce tars and flammable volatiles. Above 450°C, all volatile material is driven off, and in the presence of air the residual char undergoes oxidative reactions. The oxidation of the char is termed afterglow.

Building codes require fire-resistance characteristics to be determined with standard fire test methods, such as those developed by ASTM. Materials used in building construction must be tested for noncombustibility, fire endurance, and flame spread. The results determine where wood products can be used in building construction.

The fire hazard classification (flame-spread, smoke, and heat-release rates) can be reduced by treating the wood material with flame retardants. These treatments involve chemical impregnation or coatings.

Chemical Properties. ***Corrosiveness.*** Wood and metal are compatible in most construction and furniture uses. However, if there is sufficient moisture at the wood–metal interface, some corrosion can be expected with susceptible metals.

Wood that has been treated with certain preservatives such as chromated copper arsenate and certain flame retardants is usually more corrosive to some metals than untreated wood.

Chemical Resistance. Wood is widely used in the chemical industry as a structural material because of its resistance to a variety of chemicals. Different species vary in their resistance, depending on wood structure and cell-wall composition. Wood is more resistant to mild acids than common steel but not as resistant as some of the more expensive acid-resistant alloys. Wooden storage tanks for cold, dilute acid have a relatively long service life, which, however, is reduced as acid concentration and temperature increase.

Impregnation with resin generally improves the resistance of wood to chemicals. The chemical resistance of the resin protects the underlying wood and reduces movement of liquid into the wood. Such wood has high resistance to acids.

Biological Deterioration. Microorganisms (fungi and, to a relatively minor extent, bacteria), insects, and marine borers attack wood.

Wood Pulp

Wood pulp is primarily used for producing paper and paperboard products. Paperboard products include linerboard and corrugating medium for corrugated boxes and boxboard for cartons. Pulp is also used for producing a relatively pure cellulose, which is converted into cellulose derivatives. Pulp is produced from wood by a number of methods that separate the wood fibers or tracheids. Softwood fibers are longer (3–8 mm) than hardwood fibers (1–2 mm) and produce papers with high tear strength. Hardwood fibers produce papers with good surface smoothness for printing. Methods for separating wood fibers range from purely mechanical to primarily chemical. Mechanical processes are indiscriminate in their action and produce pulps with significant proportions of both fiber bundles and fragments. Mild chemical or thermal treatments prior to mechanical pulping aid in separating the fibers more completely with less fiber damage.

Mechanically and chemically produced pulps are primarily distinguished by weight-percent yield and lignin content, which influence fiber strength and papermaking properties.

Bleaching. The objectives of bleaching (qv) vary from temporary brightening for newsprint to providing aging resistance for valuable books. For newsprint, the pulp can be brightened by treatment with a mild oxidative (hydrogen peroxide) or reductive (sodium hydrosulfite) agent. If strength is not of prime importance, for example in pulp used for the manufacture of tissue or toweling, sodium hypochlorite is an economical bleaching agent. It may be used alone or as part of a bleach sequence (eg, following chlorination and alkaline extraction).

Chlorine and hypochlorite bleaching chemicals are very effective but are more environmentally troublesome because of the formation of organochlorine compounds. Chlorine dioxide is not as troublesome because the chlorine does not bind with the organic compounds.

Chemicals from Wood

The main products derived from wood extractives are turpentine and tall oil; other extractive components include tannins, resins, gums, and various oils. The chief products derived from wood hydrolysis are sugars for fermentation and molasses.

Modified Wood

The properties of wood can be modified for purposes other than preservation or protection from fire. Different methods can be used to reduce swelling and shrinking under conditions of fluctuating relative humidity by bulking the cell wall with leachable, nonleachable, or bonded chemicals that leave the wood in a swollen state. Because the wood is essentially at its green volume, little additional swelling can occur when it comes in contact with water.

The extent to which a procedure reduces the tendency of wood to swell on contact with water is usually referred to as antishrink efficiency (ASE); it is defined as

$$1 - \frac{\%\ \text{swelling of treated specimen}}{\%\ \text{swelling of control specimen}} \times 100$$

Composites

Wood properties vary among species, between trees of the same species, and between pieces from the same tree. Therefore, the properties of comminuted and reconstituted wood can be better controlled in processing than those of natural wood. With suitable processing variables, composites can sometimes surpass nature's best effort. The basic element for reconstituted wood

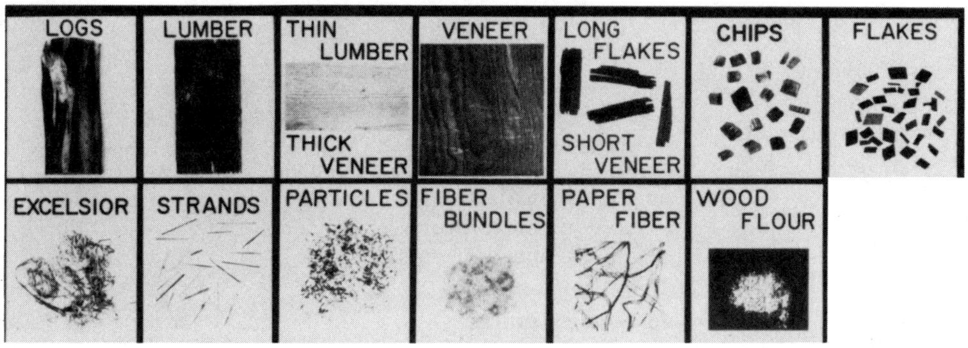

Figure 3. Basic wood elements, from largest to smallest.

can be the individual fiber or larger wood particles composed of many fibers and varying in size and geometry. Changes in properties are partially created by combining, reorganizing, or stratifying these elements. Although the molecular structure of fibers, particles, or flakes is considered fixed, the wood can be modified by chemically changing the wood element itself or by adding chemicals to improve resistance to decay, insect attack, and fire. Such modifications are easily made when wood is in the comminuted form, before it is converted into the final product.

A number of wood components or elements can be used to produce wood-based composite products (Fig. 3). The designation of wood composites varies in different parts of the world, and no standard system has been adopted. The categories of composite materials are given in Table 2 (see also COMPOSITION BOARD).

Fuel Properties

The fuel properties of wood depend on particle size, density, moisture content, and chemical composition. The particle size of wood fuel ranges from sander dust to whole logs and trees. Sander dust completely burns in tenths of a second, whereas large-sized material burns at a surface thickness loss rate of about 10 cm/h. The heating value of softwoods is slightly higher than that of hardwoods on a weight basis (moisture-free), ca 20,900 vs 20,000 kJ/kg (4990 vs 4780 kcal/kg). On a volume basis, however, the heating value of hardwoods is generally higher because many hardwoods have a higher density than softwoods.

Charcoal, made from wood by heating in a reducing atmosphere, has a heating value of over 28,000 kJ/kg (6690 kcal/kg). Charcoal is usually made from wood-processing residue and low quality roundwood.

Economic Aspects

The economic importance of wood in the United States can be evaluated through an assessment of wood products as an industrial raw material, such as that published by the USDA Forest Service in 1980. Approximately one half of all wood-based economic activity is divided into two main categories, primary manufacturing and secondary manufacturing. Primary manufacturing involves the processing of logs and related products into lumber, veneer, plywood, pulp and paper, turpentine, rosin, and other products. Secondary manufacturing involves the processing of lumber, plywood, paper, and other products into finished goods such as furniture, toys, and containers. To arrive at the full value attributed to wood-based activity in the United States, five additional categories must be added: timber management, harvesting, construction, transportation, and marketing.

BIBLIOGRAPHY

Agricultural Handbook 72, U.S. Department of Agriculture, Washington, D.C., 1987 rev.

H. A. Core, W. A. Cote, and A. C. Day, *Wood Structure and Identification,* 2nd ed., Syracuse University Press, Syracuse, N.Y., 1979.

R. R. Rowell, ed., *The Chemistry of Solid Wood,* American Chemical Society, Washington, D.C., 1984.

Table 2. The Family of Composite Materials

Veneer-based materials	*Laminates*
plywood	laminated beams
laminated-veneer lumber (LVL)	overlayed materials
parallel-laminated veneer (PLV)	(panels or shaped materials combined with nonwood
Composition materials	materials such as metal, plastic, and fiber glass)
fiberboard	*Edge-glued material*
insulation board	lumber panels
hardboard	*Structural components*
medium density fiberboard	I-beams
particleboard	T-beam panels
wafer board	stress-skin panels
oriented wafer board (OWB)	
oriented strand board (OSB)	
COM-PLY	
cement board	

J. P. Casey, ed., *Pulp and Paper: Chemistry and Technology*, Vols. 1–4, John Wiley & Sons, New York, 1980.

THEODORE H. WEGNER
United States Department of Agriculture

WOOD, POLYMER-IMPREGNATED

A range of new wood treatments was introduced during the period 1930–1960 (Table 1). Some of the monomers employed are of the condensation type and react with the hydroxyl groups in the wood; other chemicals react with the hydroxyl groups to form cross-links. Another group of compounds simply bulk the wood by replacing the moisture content in the cell walls.

During the early 1960s a new class of chemicals containing one or more double bonds was used to treat wood. The chemicals consisted of vinyltype monomers that could be polymerized into solid polymers by means of free radicals. Vinyl polymerization was an improvement over condensation polymerization because the free-radical catalyst is neither acidic nor basic; no residue, such as water, is left behind that must be removed from the final composite. The acid and base catalysts used with other treatments degrade the cellulose chain and cause brittleness of the composite. Vinyl polymers range from soft rubber to hard, brittle solids, depending on the groups attached to the carbon–carbon backbone.

Examples of vinyl monomers used in wood–polymer composites (WPC) include styrene, *t*-butylstyrene, methyl methacrylate, and other acrylates, vinyl acetate, acrylonitrile, diallyl phthalate, vinyltoluene, and various combinations of unsaturated monomers. Because most vinyl monomers are nonpolar, little if any interaction takes place with the hydroxyl groups attached to the cellulose molecule. In general, vinyl polymers simply bulk the wood structure by filling the capillaries, vessels, and other void spaces in the wood structure. The free radicals used for the polymerization reaction are produced by temperature-sensitive catalysts or cobalt-60 γ radiation; the chemistry is the same.

Impregnation Process

In both processes the impregnation of wood is carried out by first evacuating the air from the wood vessels and cell lumens. Any type of mechanical vacuum pump is adequate that can reduce the pressure in the apparatus to 133 Pa (1 mm Hg) or less. The catalyzed monomer containing cross-linking agents, and occasionally dyes, is introduced into the evacuated chamber through a reservoir at atmospheric pressure. The wood must be weighted down so that it does not float in the monomer solution.

After the wood is covered with the monomer solution, air is admitted at atmospheric pressure, or dry nitrogen in the case of the radiation process. The monomer solution begins to flow immediately into the evacuated wood structure to fill the void spaces. The soaking period, like the evacuation period, depends on the structure of the wood. Maple, birch, and other open-celled woods are filled in about 30 min; other woods require more time. A block of maple wood 7.6 × 7.6 × 17.6 cm in dimensions absorbs about 300 mL of monomer in less than 10 min.

Table 1. Processes for Wood Modification

Process and chemical	Description
acetylation	hydroxyl groups react with acetic anhydride–pyridine catalyst to form esters; capillaries empty; antishrink efficiency (ASE) about 70%
ammonia	evacuated wood exposed to anhydrous ammonia vapor or liquid at 1.03 MPa (150 psi); 12.7-mm stock bends up to 90°
phenol–formaldehyde	Impreg, a noncompressed wood–phenolic-formaldehyde composite; thin veneers soaked, dried, and cured under mild pressure; swells cell wall, capillaries filled; ASE ~75%
	Compreg, a compressed wood–phenolic-formaldehyde composite; thin veneers soaked, dried, and cured under high pressure to collapse cell structure; density 1.3–1.4; ASE ~95%; cutlery handles
cross-linking	wood treated with 2% zinc chloride catalyst exposed to paraformaldehyde, heated to 120°C for 20 min; ASE 85%; drastic loss of toughness and abrasion resistance
cyanoethylation	reaction with acrylonitrile (ACN) with NaOH catalyst at 80°C; fungi resistant, impact strength loss
ethylene oxide	high pressure gas treatment, amine catalyst; ASE increased to 65%
irradiation	exposure to 10^4 Gy (10^6 rad) slightly increases mechanical properties; above this level cellulose is degraded and mechanical properties decrease rapidly; low exposure used to temporarily inhibit fungi growth
ozone	gas-phase treatment degrades cellulose and lignin; pulping action
beta-propiolactone	wood treated with β-propiolactone diluted with acetone and heated; polyester side chains grafted on swollen cell-wall cellulose; carbonyl end groups react with copper or zinc, reducing fungi attack; compression strength increase
heat treatment	Staybwood, a heat-stabilized wood, heated to 150–300°C
	Staypak, a heat-stabilized compressed wood, heated to 320°C, compressed at 2.7–27 MPa (400–4000 psi), cooled, and pressure released; handles and desk legs

Monomers for Wood–Polymer Composites

Many different vinyl monomers have been used to make wood–polymer composites during the past 20 years, but methyl methacrylate (MMA) appears to be the preferred monomer for both the catalyst–heat and radiation processes.

All vinyl monomers contain inhibitors to prevent polymerization during transport and storage. These inhibitors, such as hydroquinone (HQ), hydroquinone monomethyl ether (MEHQ), *t*-butylcatechol (TBC), and 2,4-dimethyl-6-*t*-butylphenol (DMTBP), must be removed before impregnation, or excess catalyst must be added to use up the inhibitor before polymerization will begin.

Physical Properties

Dry wood has unique physical properties: its tensile strength, bending strength, compression strength, impact resistance, and hardness per unit weight are the highest of all construction materials. Hydrogen bonding, the unique helical structure of the cell walls, the combination of the linear cellulose with the three-dimensional lignin molecules, and the low molecular weight extractives contained in the lumens make wood an infinitely variable resource. The unusual features of wood are the basis of the "art" of wood treatment.

Polymer loaded wood shows increased density, compression strength, and tangential hardness.

Applications

Commercial production of wood–polymer composites (WPC) began in the mid-1960s using the radiation process. Parquet flooring was the principal product, where the increased hardness and abrasion resistance offered an advantage in high-traffic commercial installations.

In the late 1980s, the catalyst–heat process was widely used for making wood–polymer composites, but in smaller volume than the radiation process. Because of the simplicity of the catalyst–heat process and the low capital investment, these products are used by small companies that make high cost, low volume articles. For example, billiard cues and experimental bowling alley flooring was produced by the AMF Bowling Division.

Today the physical properties of WPC are better understood and specific commercial products are taking advantage of the desirable appearance, high compression strength, increased hardness and abrasion resistance, and improved dimensional stability.

BIBLIOGRAPHY

J. A. Meyer, *For. Prod. J.* **15**(9), 362 (1965).

J. A. Meyer in I. S. Goldstein, ed., *Wood Technology: Chemical Aspects,* American Chemical Society Symposium Series No. 43, American Chemical Society, Washington, D.C., 1977, pp. 301–325.

J. A. Meyer, *Wood Sci.* **14**(2), 49 (1981).

J. A. Meyer, *For. Prod. J.* **32**(1), 24 (1982).

John A. Meyer
SUNY College of Environmental
Science and Forestry

X

XANTHAN

Traditionally, industrially useful polysaccharides have been derived from botanical sources, but more recently several microbial polysaccharides have achieved commercial status. These include dextran (qv), xanthan gum, and welan gum, the generic name for polysaccharide S-130, which was commercialized in 1985.

Manufacture

An appropriate medium for xanthan production comprises carbohydrate and other nutrients, which include ammonium ion, a phosphate buffer, magnesium ion, and trace elements. A pure *Xanthomonas campestris* culture is grown in a seed vessel which is then used to inoculate a large fermentor. Broth pH decreases during the fermentation, because anionic groups are formed as part of the polysaccharide molecules, and it is necessary to add caustic to maintain the pH within the range of 6.0 to 7.5. A suitable fermentation temperature is about 28°C. Glucose, sucrose, and starch are similar with respect to polysaccharide production efficiency.

At completion, that is, when the carbohydrate source is exhausted, the fermentation broth is heat treated at near-boiling temperature. Xanthan gum is recovered by precipitation with isopropyl alcohol, and the polysaccharide is dried and milled to the desired particle size distribution.

In another commercial process, pasteurization, alcohol precipitation, and drying are omitted, biocide is added, and the liquid broth is shipped as the final product. Alternatively, the broth is ultrafiltered and shipped as a liquid concentrate.

Alternative methods have been suggested for the recovery of xanthan gum.

Structure

Xanthan gum is a high molecular weight heteropolysaccharide comprising three different monosaccharides: mannose, glucose, and glucuronic acid (as the mixed potassium, sodium, and calcium salt).

Numerous studies have indicated a molecular weight of approximately 2 million, but values as high as 13–50 million have been measured. These differences probably reflect different degrees of association of the polymer chains.

An ordered conformation of the xanthan gum molecule was proposed as a result of x-ray-diffraction studies using oriented fibers. A molecular conformation of a right-handed, fivefold helix with a rise per backbone disaccharide residue of 0.94 nm, ie, a fivefold helix with a pitch of 4.7 nm, was postulated. In this conformation, the trisaccharide side chains are believed to align with the backbone and presumably stabilize the overall conformation by various noncovalent interactions. Multistranded structures are consistent with the x-ray data and represent alternative conformations.

Solution studies on the conformation of xanthan gum suggest that the molecule is rodlike with some degree of flexibility and is related to the solid-state conformation.

Earlier studies suggested a double or multistranded assembly, whereas more recent studies argue for a single helical entity in solution.

Pyruvate and Acetate Content. The pyruvate content of xanthan gum varies and is dependent on the strain, fermentation conditions, and recovery conditions employed. Pyruvate content influences the viscosity in salt solutions and the thermal stability of xanthan gum.

Although no variability in the *O*-acetyl content of xanthan gum has been noted, these ester groups can be removed by the use of mild alkali. An increased synergism between deacetylated xanthan gum and galactomannans has been reported.

Properties

The commercial importance of xanthan gum results from several unique rheological properties and its stability and compatibility over a wide range of solution conditions, including ionic strength variations, heat, pH changes, shear, enzymes, and additions of salts, acids, or bases.

Rheology. Xanthan solutions display high viscosity at low concentration, high viscosity at low shear rates, a high degree of pseudoplasticity, and high elastic modulus.

Compatibility. Xanthan gum solutions have excellent compatibility and stability in the presence of many chemicals, eg, acids, bases, salts, oxidizing and reducing agents, thickeners, enzymes, surfactants, preservatives, latex emulsions, and cationic dyes.

Xanthan and Proteins. Interactions between polysaccharides and proteins are critically dependent on hydrocolloid concentrations, ionic strength, pH, other ingredients, and processing conditions. An empirical approach has successfully identified combinations in which the interaction between xanthan gum and proteins has been prevented or, alternatively, used to best advantage.

Xanthan and Starch. Upon storage, many starch-containing foods exhibit a phenomenon known as retrogradation. Retrogradation is a major factor in bread staling and, in frozen foods, can reduce freeze–thaw stability. Xanthan gum has been successfully used in a number of foods to reduce the adverse effects of retrogradation.

Applications

The unique physical and rheological properties of xanthan gum make it one of the most versatile hydrocolloids for use in a host of food, pharmaceutical, and personal care products (Table 1). These properties result in products with longer shelf life, improved flow, consistent viscosity, better texture, and a pleasing appearance. The excellent suspension properties at low concentration and compatibility with salts, acid, and alkali have resulted in the use of xanthan gum in a variety of industrial applications.

Regulatory Status

Food-grade xanthan gum products meet the following requirements: (*1*) U.S. Food and Drug Administration (FDA) food additive order for xanthan gum, 21 CFR 172.695; (*2*) *Food Chemicals Codex* (FCC) requirements and the *National Formulary*;

Table 1. Food and Industrial Applications of Xanthan Gum

Food	
pourable salad dressing (high oil, low oil, no oil)	dry mix products (desserts, gravies, beverages, sauces, dressings)
relishes and sauces	farinaceous foods (cakes)
syrups and toppings	beverages
starch-based products (canned desserts, sauces, puddings and fillings, retort pouches)	dairy products (ice cream, shakes, processed cheese spread, cottage cheese)
	confectionery
Personal care	*Industrial*
lotions	flowable pesticides
creams	liquid feed supplements
toothpaste	cleaners, abrasives, and polishes
	metalworking
Oilfield	ceramics
drilling fluids (muds)	foundry coatings
workover and completion fluids	texturized coatings
stimulation	slurry explosives
hydraulic fluids	dye and pigment suspensions
acidizing	
enhanced oil recovery—polymer flooding	

(*3*) EEC purity criteria for xanthan gum, laid down in EEC Directive 82/504/EEC.

Economics

Although no domestic production figures are available, the current U.S. demand for xanthan gum has been estimated at about 3600 t/yr. The food-grade fraction of the total market is estimated at 60%.

The majority of the product is sold in dry powder form, but broth is also available, generally in a concentrated form, for domestic enhanced oil recovery operations. The largest U.S. producer is Kelco Division of Merck & Co., Inc., with production facilities in San Diego; Okmulgee, Oklahoma; and Liverpool, UK. There is excess production capacity.

The estimated current U.S. market is \$30–40 million with a projected 5% annual growth rate.

BIBLIOGRAPHY

P. Jansson, L. Kenne, and B. Lindberg, *Carbohydr. Res.* **45,** 275 (1976).

R. Moorhouse, M. D. Walkinshaw, and S. Arnott in P. A. Sandford and A. Laskin, eds., *Extracellular Polysaccharides,* ACS Symposium Ser. 45, American Chemical Society, Washington, D.C., 1977, pp. 90–102.

P. A. Sandford, I. W. Cottrell, and D. J. Pettitt, *Pure Appl. Chem.* **56**(7), 879 (1984).

G. R. Sanderson, *Food Technol. Chicago* **35**(7), 50 (1981).

JOHN K. BAIRD
Kelco Division of Merck & Co., Inc.

XEROGRAPHIC MATERIALS

Since the invention of electrophotography by Chester Carlson in 1938, polymers have been used in place of sulfur photoconductors and lycopodium toners. Present-day toners are polymers that may be homopolymers, copolymers, or polymers that are cross-linked to achieve the desired fusing characteristics. In order to fuse plastic toners, fuser rolls and pressure rolls made of elastomers are required. A xerographic process consists of multiple steps; in each step the polymer properties are important.

Elastomers are used in several key areas of xerographic machines: paper-feed devices; paper transporting, blade cleaning, and metering devices; and fusing subsystems. The elastomers are subject to thermal oxidative degradation, release-agent swelling, and cyclic fatigue because of crack propagation. In addition to the mechanical properties of elastomers in fuser rolls, surface energies are important considerations in the fuser life.

Xerographic materials generally contain polymers that are fabricated in a variety of ways so as to optimize their performance in a xerographic machine. The quality of the xerographic image depends on the properties of developers and fuser roll materials. Both toners and carriers undergo collisions in a developer housing before the toner is transferred to the paper for fusing. Fusing models predict that image quality on the paper depends on time-dependent pressure on the particle, particle size, and functional dependence of temperature and deformation rate or viscosity. Toner particles undergo deformation during collisions between the toner and carrier particles in a developer housing and during cold-pressure fixing of toner, between the rolls. Generally, toners are composites of polymers. Their moduli, as well as their strengths, are important parameters predicting deformation and development properties. Collision of carrier beads with each other can result in degradation and loss of polymer coating. Deformation and degradation of the carrier coating depend on the size of the carrier beads, the bead velocities, and the yield strengths of the polymers. Degradation and loss of polymer coating affects the electrical and development properties of the toner.

Viscoelastic properties of toners are dependent on the chemical structure, molecular weight, and molecular weight distribution of the polymer and on the carbon black dispersed in the polymer. Molecular weight and molecular weight distribution affect the viscoelastic moduli, depending on the weight fraction of the molecular size present in the polymer. The effect of carbon black on the viscoelastic properties of the toner is influenced by the toner polymer structure, the type of carbon black, and the loading. At loadings of 10–20% by volume, a polymer–carbon black network is formed. Carbon black thus affects the viscosity of the toner and consequently its fusing behavior.

Toners are processed and manufactured by melt-mixing polymer with carbon black in a Banbury or roll mill mixer or in an extruder. Carbon black aggregates are broken and dispersively mixed with a polymer. After the toner slabs or strands are made, they are ground in a jet mill. The rate of breakage depends on the particle concentration in the mill, average velocity of jet stream, impact strength, and temperature. Spray drying is an alternative process where the polymer is dispersed with the carbon black. The spray process depends on the flow rate of dispersion, the temperature, and the type of solvent.

Process Steps

The basic process steps of transfer xerography have not changed since the original invention proposed by Carlson; the key process steps are shown in Figure 1.

Fusing Subsystems and Materials

The four basic types of fuser systems in common use are hot roller fusing, cold-pressure roll fusing, radiant fusing, and flash fusing. A fifth system, vapor or solvent fusing, is seldom used in commercial copiers or printers. Hot roll fusing is the most frequently chosen process for both low and high volume copiers and printers because of its energy efficiency, cost, and compactness.

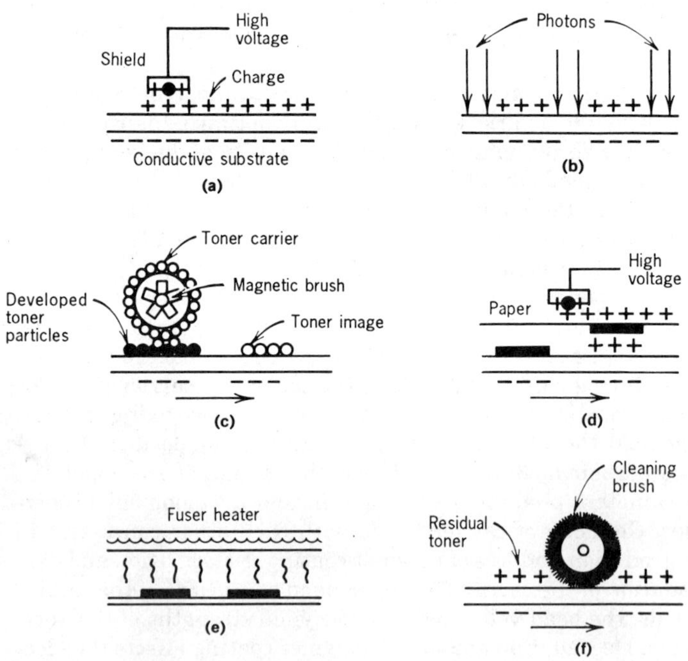

Figure 1. The basic steps in the xerographic process. (**a**) Charging the photoconductor; (**b**) exposing to form the latent electrostatic image; (**c**) developing the latent image into a real image; (**d**) transferring the image to paper; (**e**) fusing the image to paper; (**f**) cleaning residual toner from the photoconductor.

Developer Materials

Xerographic developer materials of two or a single component are designed to satisfy copy quality and developer life requirements for a certain fusing process and process speeds with rigid hardware constraints.

It is frequently necessary to control the toner electrostatic charge by coating the carrier made of ferrite or steel with a thin layer of polymer.

In general, highly halogenated homopolymers and copolymers are used to give carriers a negative charge and copolymers of poly(styrene-*co*-alkyl methacrylates and acrylates) are used to give carriers a positive charge.

Toner particles are generally composed of a thermoplastic containing typically 5–10% by weight of colorant. Black toners often contain well dispersed, finely divided particles of carbon black less than 1 μm in size. Pigments, dyes, or mixtures of pigments and dyes are used in fabricating color toners. Pigment concentration is generally determined by the desired contrast or color density and the toner charging requirements. A thermoplastic is required primarily for its softening characteristics, which help to fix the image to the substrate by using heat, pressure, or a combination of heat and pressure. Although the thermoplastic is chosen primarily for its fusing or softening properties, other characteristics such as charging capabilities, mechanical properties, cost, processing constraints, and systems compatibility must be considered carefully. Polymers frequently used for toner fabrication are the poly(styrene-*co*-alkyl methacrylate)s and acrylates.

Plastics as Xerographic Engineering Materials

Plastic parts, complex in shape and function, are readily produced to accurate dimensions and repeatable tolerances through a variety of production techniques.

Typical polymeric materials and uses are shown in Table 1. The materials are primarily filled thermoplastics; acetal, polycarbonate, ABS, and high impact polystyrene are the most widely used.

Elastomeric Materials

In many uses, some degree of compliance or conformability between two surfaces is required. In such situations, elastomer properties of high compliance (reciprocal of modulus) and reversible and quickly achieved regain become important when the part must perform critical functions. In other situations, high friction coefficient and flex fatigue life are the desirable features. Elastomers are used in paper-feed devices, paper transporting subsystems, blade cleaning and metering devices, and fusing subsystems.

BIBLIOGRAPHY

S. K. Ahuja, *Rheol. Acta.* **19,** 299, 307 (1980).

M. E. Scharfe, D. M. Pai, and R. J. Gruber, *Imaging Processes and Materials,* Neblette's 8th edition, Van Nostrand-Rheinhold, New York, 1989, pp. 135–180.

L. B. Schein, *Electrophotography and Development Physics,* Springer Verlag, Inc., New York, 1988.

Table 1. Plastics as Xerographic Engineering Materials

Plastic	Application
epoxy resin	structural[a]
polyolefin	structural, developer rolls
polyethylene	insulation
acetals	drives, optics cavity, pulleys, rollers
filled acetals (PTFE and mineral)	structural, insulation, drives
nylon	drives
filled nylon	drives, bushings
polyester	structural, fuser covers
polyamideimides	fuser, stripping fingers, optic carriage rolls
polycarbonates	structural, optics, paper trays, housings, high voltage insulators
glass-filled polycarbonate	drives, structural, covers
HIPS[b], ABS	covers, optics cavity, paper trays
ABS, SAN	guides, covers, bearings, mountings
poly(phenylene sulfide)	fuser, optics cavity, bearings, bearing sleeves, thermostat holders
Noryl[c]	
PEEK[d]	fuser stripper fingers
polyurethane	cleaning blades, rolls, friction feeds
fluoroelastomer	fuser rolls, pressure rolls
silicones	fuser rolls, pressure rolls, paper-handling rolls
PTFE	fuser rolls, pressure roll coatings, filler for other plastic parts
PFA[e]	fuser roll/pressure roll coatings
EPDM[f] rubber	paper-handling rolls/belts, feed belts
polyisoprene	paper-handling belts
EPDM/silicones	paper-handling belts/rolls
thermoplastic elastomers	paper-handling rolls
Hypalon[g]	paper-handling belts

[a] Frames, baffles, gates, mountings, guides.
[b] High impact polystyrene.
[c] Poly(2,6-dimethyl-1,4-phenylene ether)–HIPS blends.
[d] Polyetheretherketone.
[e] Perfluorinated alkoxy polymer.
[f] Ethylene–propylene–diene monomer.
[g] Chlorosulfonated polyethylene.

F. R. Eirich, *Science and Technology of Rubber,* Academic Press, New York, 1978.

Robert J. Gruber
S. Ahuja
D. Seanor
Xerox Corporation

X-RAY DIFFRACTION

The method of x-ray diffraction and scattering (qv) is one of the oldest and most widely used techniques available for the study of polymer structures. A beam of x rays incident to a material is partly absorbed and partly scattered, and the rest is transmitted unmodified. The scattering of x rays occurs as a result of interaction with electrons in the material. The x rays scattered from different electrons interfere with each other and produce a diffraction pattern that varies with scattering angle. The variation of the scattered and diffracted intensity with angle provides information on the electron density distribution, and hence the atomic positions, within the material. The words diffraction and scattering are used interchangeably in this article. The word diffraction is generally preferred when the specimen under study has a regularity in its structure so that the detected x rays exhibit well-defined intensity maxima. Other scattering techniques are also employed in the study of polymers, ie, the scattering of light, neutrons, and electrons (see Light scattering; Neutron scattering; Electron diffraction). The basic principles governing the scattering and diffraction of these different types of electromagnetic waves and particles are very similar. The differences in the wavelength and the mode of interaction with matter, however, make one radiation more suitable than another for studying some particular aspects of polymer structure.

X-ray scattering (or diffraction) techniques are usually categorized into wide-angle x-ray scattering (waxs) and small-angle x-ray scattering (saxs). In the former, the desired information on the polymer structure is contained in the intensities at large scattering angles and, in the latter, at small scattering angles. In general terms, waxs is used to obtain structural information on a scale of 1 nm or smaller, and saxs on a scale of 1–1000 nm. The distinction between waxs and saxs is required because the instrumental (especially collimation) requirements are often very different and the method of analyzing the data can also be different, although, of course, the basic principles governing the scattering and diffraction of x rays are the same in both cases.

Principles of X-Ray Scattering and Diffraction

In the spectrum of electromagnetic radiation, x rays lie between the ultraviolet rays and gamma rays. Those x rays used for structure analysis have wavelengths λ in the range of 0.05–0.25 nm. Most work on polymers is done with the Cu Kα emission line, a doublet with an average wavelength equal to 0.154 nm. In view of the wave–particle duality, it is in some cases useful to consider x rays to consist of photons of energy $h\nu$, where h is Plank's constant and the frequency ν is given by c/λ (c = velocity of light). Thus, the Cu Kα line consists of photons of energy 8.04 keV. A high intensity x-ray beam is one with a high flux of photons.

Normally the sample is irradiated with a collimated beam of x rays, and the intensity of the scattered x rays is measured as a function of scattering direction. The scattering angle, that is, the direction of the scattered beam in relation to the incident beam, is customarily denoted by 2θ (Fig. 1). If $\hat{\mathbf{s}}_0$ is defined as the unit vector in the incident beam direction and $\hat{\mathbf{s}}$ as the unit vector in the scattered beam direction, the relation between the incident and scattered beams can be conveniently expressed by means of the scattering vector **s**:

$$\mathbf{s} = (\hat{\mathbf{s}} - \hat{\mathbf{s}}_0)/\lambda \qquad (1)$$

The magnitude of **s** is related to the scattering angle θ by

$$s = |\mathbf{s}| = 2\,(\sin\,\theta)/\lambda \qquad (2)$$

In more recent literature the use of scattering vector **q**, defined by

$$\mathbf{q} = 2\pi\mathbf{s} \qquad (3)$$

is gaining acceptance in place of **s**. Different authors often use different symbols for the scattering vector, eg, **h**, **Q**, **k**, $\boldsymbol{\kappa}$, and **S**, all of which are identical to **q** defined in equation 3.

If an (unpolarized) x-ray beam of intensity I_0 irradiates a free electron, the intensity I scattered at angle 2θ and measured at distance r from the electron is given, according to classic theory, by

$$I = I_0 \frac{K}{r^2}\left(\frac{1 + \cos^2 2\theta}{2}\right) \qquad (4)$$

which is called the Thomson formula. Here, the constant K is defined by

$$K = e^4/m^2c^4 \qquad (5)$$

where e and m are the charge and mass of an electron and c is the velocity of light. The square root of K is often referred to as the classic electron radius and is equal to 2.82×10^{-13} cm. The factor $(1 + \cos^2 2\theta)/2$ arises from the polarization effect and would be somewhat different if a partially polarized incident beam such as that reflected from a monochromator crystal is used.

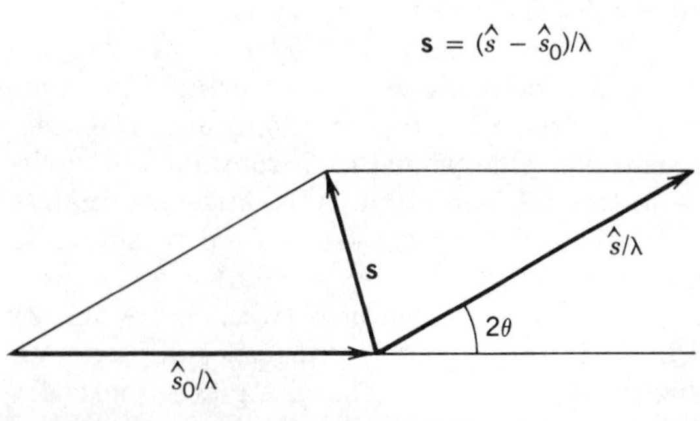

Figure 1. Definition of the scattering vector **s**, where $\hat{\mathbf{s}}_0$ and $\hat{\mathbf{s}}$ are the unit vectors in the direction of the incident and scattered beams, respectively, and λ is the wavelength of the x rays. The length of the vector **s** is related to the scattering angle 2θ by $|\mathbf{s}| = 2(\sin\,\theta)/\lambda$.

Instrumentation

The measurement of x-ray scattering from a sample requires an x-ray source to irradiate the sample and a method of determining the intensities of x rays scattered at different angles. As the x-ray source, a sealed fixed-focus tube generator or a rotating-anode tube generator is generally used in the laboratory. More recently, synchrotron radiation sources in large national laboratories have increasingly become available to the polymer science community. The methods for determining the diffracted intensities can be classified into a film method and a detector method. Although the film method is simpler, and easier and cheaper to implement, the detector method generally yields data of superior quality, especially as a result of recent advances in the electronics technology. A variety of geometric arrangements are possible for the positioning of the x-ray source, the sample, and the detecting device (film or detector). The equipment for this arrangement is called a camera or diffractometer. A large number of such cameras and diffractometers, differing greatly in their sophistication, are available commercially. These cameras are classified according to whether they are used for waxs or saxs, because the beam-collimation needs are fairly distinct between these two cases.

Crystal Structure Determination

The term crystal structure determination denotes the procedure of determining the positions of atoms contained in a crystal from the knowledge of the intensities of the observed Bragg diffraction peaks. This is the technique traditionally developed and practiced for the determination of the crystal structure of low molecular weight materials and lately of large biological molecules. The technique has evolved and has been refined to a high degree of sophistication ever since the possibility was first suggested by Bragg. As applied to polymeric materials, there is, of course, no difference in the principles involved, but in actual practice there are differences and difficulties caused by the characteristics specific to polymers. The main difficulty arises from the fact that polymer crystals are seldom prepared to the high degree of crystalline perfection that is possible with crystals of low molecular weight materials. As a result, the number of Bragg peaks that can be observed with precision is limited. This leads to a reduced resolution of the deduced atomic positions. Yet, an amazing amount of detail emerges frequently even from such a limited number of Bragg peaks observable.

The task of determining the atomic positions of the unit cell content begins with the measurement of the intensities of the Bragg peaks. The integrated intensity or the area of the curve under the peak is determined when the measurement is done by a diffractometer or by a combination of a photographic film and a microdensitometer. The primary data collected are thus the series of intensity values $I(hkl)$ for various h, k, and l. Before processing further, the observed $I(hkl)$ values may have to be corrected for absorption effect, Lorentz-polarization effect, and multiplicity of the crystal planes. The electron density distribution within a unit cell of a crystal is given by

$$\rho(\mathbf{r}) = \frac{1}{V} \sum_h \sum_k \sum_l F_{hkl} \, e^{i2\pi(hx+ky+lz)} \tag{6}$$

where x, y, and z are the components of the position vector $\mathbf{r}$ (in the coordinate frame defined by $\mathbf{a}$, $\mathbf{b}$, and $\mathbf{c}$), and V is the volume of the unit cell. The summation is for all values of h, k, and l from $-\infty$ to ∞; F_{hkl} is the structure factor, whose square is equal to $I(hkl)$. Taking a square-root of $I(hkl)$, however, does not give F_{hkl}, because the latter is, in general, a complex number and the information on only its modulus (absolute value) and not on its phase is contained in $I(hkl)$. Therefore, to use equation 6 in order to obtain the electron density distribution, the phase information must be supplied by some other sources.

Once a reasonable estimation of the phases of the structure factors is obtained, a refinement process can proceed by which the correctness of the estimated phases is ascertained and corrections are made to those in obvious error. In the refinement process, the electron density map is constructed by means of equation 6, where F_{hkl} is given the modulus equal to the square root of the observed intensity $I(hkl)$ and the phase as estimated by some method mentioned above. The peaks in the constructed map are now identified as the location of atomic centers. The knowledge of the chemical structure of the molecule is helpful in assigning the peaks to atomic species C, O, N, and others. Using the atomic coordinates (u_n, v_n, w_n) thus determined, the expected structure factor F_{hkl} is calculated by

$$F_{hkl} = \sum_1^N f_n \, e^{-i2\pi(hu_n+kv_n+lw_n)} \tag{7}$$

where f_n is the atomic scattering factor of the nth atom and N is the number of atoms in the unit cell. The agreement between the observed $I(hkl)$ and the calculated value $|F_{hkl}|^2$ is now examined. If it is not satisfactory, for the next round of refinement some of the phases assigned to the structure factors F_{hkl} have to be modified, eg, by equating them to those of the calculated F_{hkl}. This process can be continued until the agreement between the observed and calculated intensities becomes satisfactory.

Degree of Crystallinity

The degree of crystallinity is one of the most important basic parameters characterizing semicrystalline polymers. A number of methods are available for the determination of the degree of crystallinity (see CRYSTALLINITY DETERMINATION). Of these, the one based on x-ray diffraction offers the most fundamental method, against which the results based on other techniques may be compared. This is so because the basic concept of the crystalline and amorphous phases involves the degree of order of packing of atoms and segments in space, which can be revealed most unambiguously by the study of x-ray diffraction patterns.

The x-ray diffraction pattern obtainable from a semicrystalline polymer consists of a superposition of crystalline and amorphous patterns, as shown in Figure 2. The evaluation of the degree of crystallinity involves the solution of two problems: first, how to separate the contributions to the diffraction pattern by the crystalline and amorphous phases, and second, once the separation is achieved, how to evaluate the relative amounts (volume or mass) of the two phases. If the individual crystalline diffraction peaks are well separated from each other so that they leave gaps between them where no crystalline diffraction makes any contributions, it is easy to draw the amorphous background pattern by connecting these successive gaps with a smooth curve. Very often, however, broadening of the diffraction peaks results in extensive overlap of crystalline peaks, making such a procedure difficult. Some independent information is needed about the shape of the amorphous curve. If the same polymer can be prepared in a totally amorphous state by, eg, rapid quenching, its diffraction pattern can give the most unambiguous "reference" amorphous pattern. If such a quench is not possible, the amorphous diffraction patterns could be determined at several temperatures above the melting point and extrapolated to room temperature.

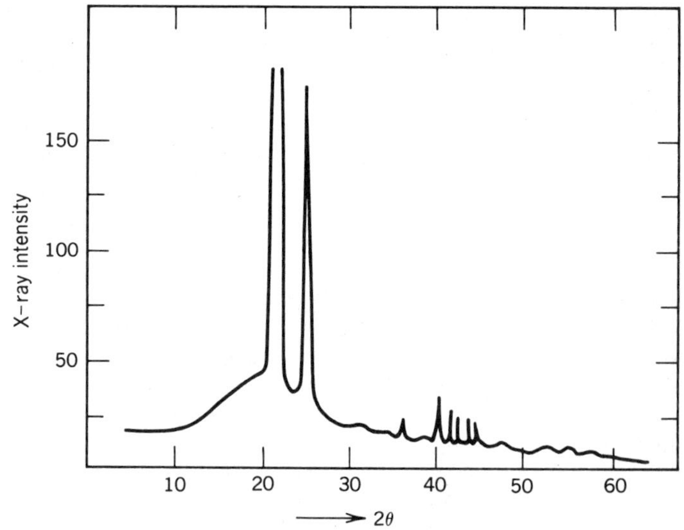

Figure 2. X-ray diffraction pattern of a high density polyethylene, showing the superposition of Bragg diffraction peaks on diffuse amorphous scattering.

After the observed diffraction patterns are separated into crystalline and amorphous contributions, the relative masses of the two phases responsible for these two components must be calculated. This rests, in principle, on the fact that the overall intensity scattered at all angles by each phase is proportional to the amount of material (the number of atoms) contained in the phase. In other words, the weight fraction w_c of the crystalline phase is given by

$$w_c = \frac{\int_0^\infty s^2 I_c(s)\, ds}{\int_0^\infty s^2 I(s)\, ds} \tag{8}$$

where $I(s)$ and $I_c(s)$ are the total and crystalline scattered intensities at scattering angle s. The reason that the factor s^2 arises in the integrals in equation 8 is that the intensity $I(s)$ must be integrated throughout the reciprocal space spanning all values of the scattering vector $\mathbf{s}$; for isotropic samples the volume integral with respect to $d\mathbf{s}$ reduces to a single integral with respect to $4\pi s^2\, ds$. In equation 8, $I_c(s)$ refers to all scattered intensity arising from the crystalline phase. When the crystalline lattice deviates from perfect regularity because of vibrations of atoms and lattice imperfections, some of the scattered energies appear in the diffuse background rather than in the Bragg peaks. Such diffuse scattered intensities are apt to be included in the amorphous components in the scheme of separation of amorphous and crystalline components of diffraction patterns as discussed above. A method of correcting for the effect of such diffuse disorder scattering from the crystalline phase is available.

The scheme to evaluate the degree of crystallinity on the basis of equation 8 is theoretically rigorous but tedious and seldom used. A much simpler way is to compare the height I_{am} of the amorphous halo measured with a semicrystalline sample with the I^0_{am} of a completely amorphous sample of the same thickness:

$$1 - w_c = I_{am}/I^0_{am} \tag{9}$$

The measurements on the semicrystalline and the totally amorphous sample must be made with the same instrument in order to eliminate any effects due to instrumental variations. Otherwise, corrections must be made for the possible differences in the incident beam intensity, air scattering, etc. A number of other approximate but practical methods have been devised and applied by many workers.

Crystallite Orientation

X-ray diffraction offers a direct means of determining the orientation characteristics of semicrystalline polymers. In an unoriented sample, crystallites are oriented in all directions with equal probabilities, and the diffraction pattern obtained with a flat-plate film shows a series of concentric rings, each corresponding to an (*hkl*) Bragg reflection. When the sample is deformed so that the crystallites have a preferred direction of orientation, the circles in the diffraction pattern break up into arcs. The intensity distribution along the length of the arc gives the information on the orientation distribution of the (*hkl*) plane in the sample. Quantitative results on the orientation distribution of a crystallographic plane can be obtained more conveniently by means of a diffractometer. When the detector position (hence the scattering angle 2θ) is fixed, only a particular (*hkl*) plane with its interplanar distance d_{hkl} satisfying the Bragg condition is potentially capable of diffracting the x-ray beam. Among the large number of crystallites in the sample, however, only a small fraction have (*hkl*) planes oriented in the proper direction so as to produce diffraction. The diffracted beam intensity is proportional to the number of crystallites oriented in such a way that the normal to the (*hkl*) plane (or the $\mathbf{r}^*_{hkl}$ vector) bisects the incident and diffracted beam. As the orientational placement of the sample in the diffractometer is altered, the number of crystallites with plane normals in the proper orientation changes and the diffracted intensity changes. The primary data collected for a given (*hkl*) are the diffracted beam intensity as a function of the two angles specifying the geometric orientation of the sample relative to the incident and diffracted beams. These data represent the density distribution of the plane normal $\mathbf{r}^*_{hkl}$ within the sample. A graphic presentation of such data on a stereographic projection is the so-called pole figure. With a single sample, there can be a number of different pole figures, each pertaining to a particular (*hkl*) plane. The set of pole figures thus obtained do not by themselves provide complete information about the orientation distribution of crystallites (or of unit cells) in the sample.

Thus a further mathematical step is needed to combine the information presented by these pole figures of different (*hkl*) planes and to produce information on the orientation distribution of crystallites as a whole. There are two possible methods toward this aim. In the first, a certain average is evaluated from each pole figure. These averages are combined to determine a single (or a set of a small number of) parameters that characterize the orientation of crystallites. In the second, more comprehensive, method the observed pole figures are represented analytically by means of a series of spherical harmonics and attempts are made to derive the complete crystallite orientation distribution function from them.

Amorphous Polymers

The scattering pattern from an amorphous polymer usually consists only of a broad amorphous peak, which, at first glance, appears to contain little useful information in comparison to the diffraction pattern from a crystalline polymer. Detailed analysis, however, provides much useful information on the state of packing of polymer molecules in the bulk amorphous polymers.

Radial Distribution Function Analysis. If an amorphous material (gas or liquid) consists of a single type of atom, the spatial distribution of atoms in it can be described conveniently by means of the radial distribution function $\rho(r)$, which is defined as the number density of atoms to be found at distance r from the center of any one of the atoms. The intensity $I(\mathbf{s})$ of x rays scattered from the material can be represented by

$$I(\mathbf{s}) = Nf^2 \left(1 + \int \rho(\mathbf{r})\, e^{-2\pi i \mathbf{s}\cdot\mathbf{r}}\, d\mathbf{r}\right) \tag{10}$$

where f is the atomic scattering factor and N the number of atoms in the scattering volume, and the integration with respect to $\mathbf{r}$ is throughout the scattering volume. The function $\rho(\mathbf{r})$ approaches the average density ρ_0 of atoms at $r \rightarrow \infty$, and it is more useful to modify equation 10 to

$$I(\mathbf{s}) = Nf^2 \left\{1 + \int [\rho(\mathbf{r}) - \rho_0]\, e^{-2\pi i \mathbf{s}\cdot\mathbf{r}}\, d\mathbf{r}\right\} \tag{11}$$

by recognizing the fact that the scattering from the sample as a whole occurs only at extremely small angles and is usually not observable. If the material is isotropic, as is usually the case, equation 11 can be simplified to

$$I(s) = Nf^2 \left\{1 + \int_0^\infty 4\pi r^2[\rho(r) - \rho_0]\, \frac{\sin 2\pi sr}{2\pi sr}\, dr\right\} \tag{12}$$

Equation 12 can be inverted by using the inversion theorem for Fourier integral

$$4\pi r^2\rho(r) = 4\pi r^2\rho_0 + 8\pi r \int_0^\infty s\, i(s) \sin 2\pi rs\, ds \tag{13}$$

where

$$i(s) = \frac{I(s)}{Nf^2} - 1 \tag{14}$$

Equation 13 thus shows how the radial distribution function $\rho(r)$ can be determined from the experimental intensity $I(s)$.

If the material contains more than one kind of atom, as is always the case with polymers, its amorphous structure can be characterized in principle completely by a set of radial distribution functions $\rho_{nm}(r)$, where $\rho_{nm}(r)$ is the density of atoms of type n to be found at distance r from any one of the atoms of type m.

Density Fluctuation. In liquids, as a result of thermal motions of molecules, the density of atoms at different parts of the material does not remain exactly the same. The magnitude of the density fluctuation in liquids is proportional to the thermal energy kT that induces the thermal motion and is inversely proportional to the bulk modulus that opposes the creation of disparity in the density among neighboring regions. Density fluctuation is present even in glassy materials, not only because of the residual thermal motions of atoms and segments still present, but also because of the density fluctuation frozen in at the time of glass formation. The nonuniformity of the electron density in the material produces x-ray scattering, which is superimposed on the scattering caused by the crystalline order or short-range liquidlike ordering of atoms and molecules. For the purpose of relating to the intensity of x-ray scattering, the density fluctuation can be defined rigorously as follows. A reference volume v of arbitrary size and shape is moved around within the sample. The number N of atoms (or electrons) falling within v fluctuates as a result of the nonuniformity of the density. The density fluctuation $\psi(v)$ is defined by the ratio of the variance $\langle(N - \langle N\rangle)^2\rangle$ to the mean

$$\psi(v) = \frac{\langle(N - \langle N\rangle)^2\rangle}{\langle N\rangle} \tag{15}$$

Defined this way, $\psi(v)$ is only a slowly varying function of v when v is large, and converges to a finite limit when v goes to infinity.

Small-angle Scattering

The technique of small-angle scattering is called for when the structural feature of interest in the sample is of the order of 1 nm or larger. The largest structural features that can be investigated conveniently by this technique are up to about 200 nm, although still larger features could be investigated if extreme care is given to details of the instrumentation. The basic principle of scattering of x rays in the small-angle region is the same as that of wide-angle scattering. The Bragg relation shows that when d is 2–200 nm and $\lambda = 0.154$ nm (Cu $K\alpha$ radiation), the scattering angle θ ranges from 2.2 to 0.022°. Because of the need to determine the intensity of the beam scattered in the direction deviating so little from the transmitted beam, there is a much greater need to collimate the primary beam more narrowly and precisely. As a result, collimating systems, cameras, and detecting systems specifically designed for small-angle scattering measurements have been developed. The methods of analysis of small-angle data differ somewhat according to the types of structures present in the sample even though, as stated earlier, the basic principles of scattering are the same. The samples that can be investigated by means of small-angle scattering can be classified into three categories. Those in the first category have a regularity in their structure, such as a macrolattice, in which the repeat distance of the structure is in the 2–200 nm range. An example is a crystalline polymer in which folded chain lamellae are stacked on top of each other, creating a layered structure with the repeat distance on the order of 10 nm in the direction perpendicular to the fold surface. Another example is offered by block copolymers in which segregated microdomains of regular spherical, cylindrical, or lamellar shape are arranged on a macrolattice. The samples in the second category possess a particulate structure, ie, they contain particles of one type of material suspended (or embedded) in the matrix of another type of material. The size of the particle should preferably be in the 2–200 nm range, but their shape or their positions in the sample do not have to be regular. In addition, the concentration of the particles should be dilute, so that any interaction between the particles can be neglected in the first approximation. An example of such samples would be a dilute solution of a polymer in a solvent. The samples in the third category possess a two-phase structure, consisting of irregular intermixing of two different phases, a phase in this context signifying a region of a fairly uniform electron density. The average dimension of each phase must be again in the 2–200 nm range. Examples would be a semicrystalline polymer consisting of amorphous and crystalline phases, and a porous material in which one of the phases is the empty pores of zero electron density.

BIBLIOGRAPHY

B. D. Cullity, *Elements of X-ray Diffraction,* 2nd ed., Addison-Wesley Publishing Co., Inc., Reading, Mass., 1978.

L. E. Alexander, *X-ray Diffraction Methods in Polymer Science,* Wiley-Interscience, New York, 1969.

M. Kakudo and N. Kasai, *X-ray Diffraction by Polymers,* Elsevier Science Publishing Co., Inc., New York, 1972.

O. Glatter and O. Kratky, *Small Angle X-ray Scattering,* Academic Press, New York, 1982.

RYONG-JOON ROE
University of Cincinnati

X-RAY SCATTERING.

See SCATTERING.

XYLAN.

See POLYSACCHARIDES.

XYLYLENE POLYMERS

In a coating process capable of producing pinhole-free coatings of outstanding conformality and thickness uniformity by the unique chemistry of the xylylene monomer, a substrate is exposed to a controlled atmosphere of pure gaseous monomer, *p*-xylylene (PX) (**2**). The coating process is best described as a vapor deposition polymerization (VDP). The monomer itself is thermally stable but kinetically unstable. Although it is stable as a gas at low pressure, on condensation it spontaneously polymerizes to produce a coating of a high molecular weight, linear poly(*p*-xylylene) (PPX) (**3**). In the commercial Gorham process the extremely reactive PX is conveniently generated by the thermal cleavage of its stable dimer, di-*p*-xylylene (DPX), a [2.2]paracyclophane (**1**). With only one known exception, substituents on the PX nucleus are carried unchanged through to the polymeric repeat unit.

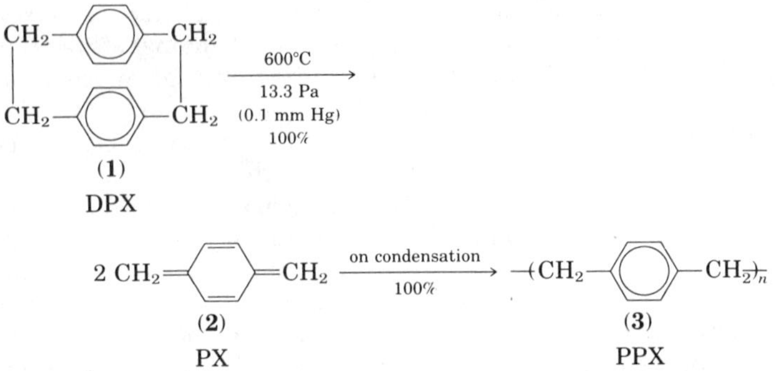

The *p*-xylylene polymers (PPXs) formed by the Gorham process are referred to generically as the parylenes, which is the designation used in this article. The terms Parylene N, Parylene C, or Parylene D refer specifically to coatings produced from the Union Carbide Corporation dimers. The process takes place in two stages that must be physically separate but temporally adjacent. Figure 1 is a schematic of typical parylene deposition equipment and indicates approximate process operating conditions. Current commercial deposition equipment is shown in Figure 2. The chemistry of PX, its polymerization, and its polymers have been the subject of several reviews.

Vapor deposition polymerization processes using means other than the pyrolytic cleavage of DPX to generate the reactive monomer are also known. They include the use of ultraviolet light alone as well as the use of electrically excited plasmas or glow discharges. The products of the last two processes, called plasma polymers, receive considerable scientific attention.

When *p*-xylene is used as the monomer feed in a plasma polymer process, PX may play an important role in the formation of the plasma polymer. The plasma polymer from *p*-xylene closely resembles the Gorham process polymer in the infrared, although its spectrum contains evidence for minor amounts of nonlinear, branched, and cross-linked chains as well. Furthermore, its solubility and low softening temperature suggest a material of very low molecular weight.

The parylene process has certain similarities with vacuum metallizing. The principal distinction is that truly conformal parylene coatings are deposited even on complex, three-dimensional substrates, including on sharp points and into hidden or recessed areas. Vacuum metallizing, on the other hand, is a line-of-sight coating technology. Whatever areas of the substrate cannot be "seen" by the evaporation source are "shadowed" and remain uncoated (see METALLIZING, VACUUM).

Dimer

In contrast to the extreme reactivity of the monomeric PX (**2**) generated from it, the dimer DPX (**1**) feedstock for the parylene process is an exceptionally stable compound. Because of their chemical inertness, dimers in general do not exhibit shelf-life limitations. Although a variety of substituted dimers are known in the literature, at present only three are commercially available: DPXN, DPXC, and DPXD, which give rise to Parylene N, Parylene C, and Parylene D, respectively.

The unsubstituted C_{16} hydrocarbon, [2.2]paracylcophane (**1**), is DPXN. Both DPXC and DPXD are prepared from DPXN by aromatic chlorination and differ only in the extent of chlorination; DPXC has an average of one chlorine atom per aromatic ring and DPXD has an average of two.

Manufacture. *DPXN (di-p-xylylene).* For the commercial production of DPXN two synthetic routes have been used: the direct pyrolysis of *p*-xylene and the 1,6-Hofmann elimination of amine from a quaternary ammonium hydroxide.

Purification. Unsubstituted di-*p*-xylylene (DPXN) is readily purified by recrystallization from xylene. It is a colorless, highly crystalline solid. The principal impurity is polymer, which fortunately is insoluble in the recrystallization solvent and easily removed by hot filtration.

In purifying DPXC and DPXD, care must be taken not to disturb the homologue composition so that product uniformity is maintained.

The principal purification concerns are the removal of insoluble residues and components that contain aliphatic side-chain chlorine.

Properties. The DPXs are all crystalline solids; melting points and densities are given in Table 1; solubility in aromatic hydrocarbons is limited. At 140°C, solubility in xylene is only about 10%. DPXC is more readily soluble in aliphatic solvents, eg, in methylene chloride at 25°C its solubility is 10%. In contrast, the corresponding figure for DPXN is 1.5%.

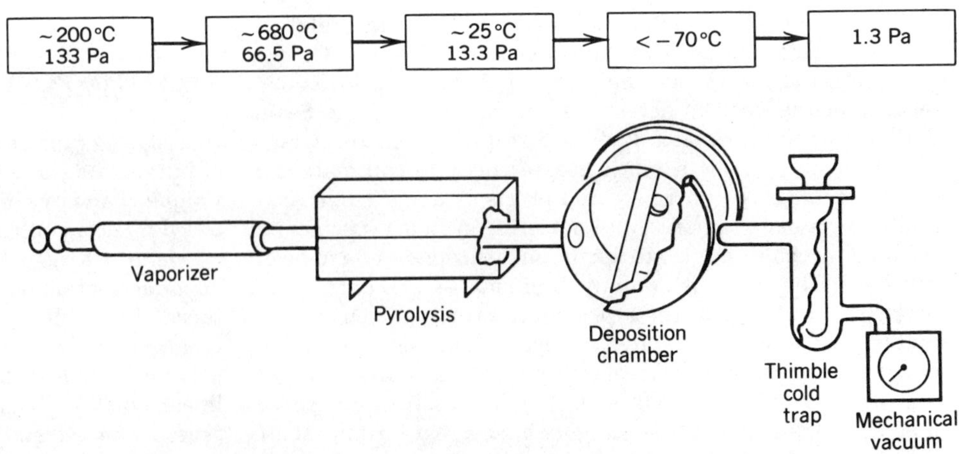

Figure 1. Parylene deposition apparatus. To convert Pa to torr, multiply by 0.0075.

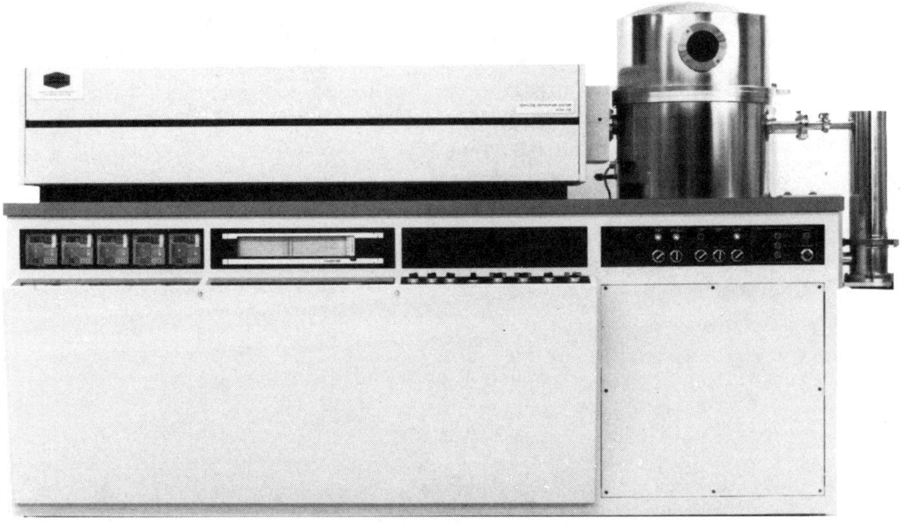

Figure 2. Commercial vapor deposition equipment.

Table 1. Properties of Parylene Dimers

Dimer	Melting point, °C	Density, g/cm^3
DPXN	284 (decomp)	1.22
DPXC	140–160^a	1.30
DPXD	170–195^a	1.45

^{a} Mixture of homologues and their isomers.

Monomer

The eight-carbon monomer, PX, is generated in the first stage of the parylene process by heating gaseous dimer as it passes through a high temperature zone. Its intermediacy in the process was deduced by the earliest investigators. An appreciation of the unusual properties of PX is an important step toward understanding the unique features of the coating process.

Chemical Evidence for PX Monomer. It is important to establish that PX is indeed the pyrolysis product, rather than the molecule formed by breaking only one of the original dibenzyl bonds, such as the dimer diradical (**4**).

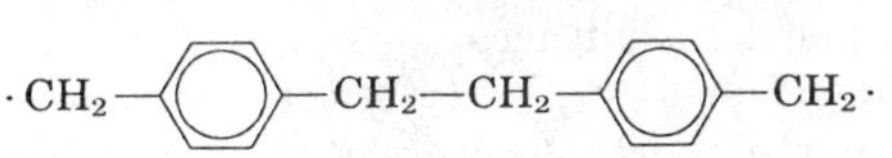

(**4**)

When the pyrolysis gases are quenched with a molar excess of iodine vapor, a yield of greater than 50% *p*-xylylene diiodide is recovered. This observation gives direct chemical support for the belief that pyrolysis results in PX.

Properties. A great deal is known about the structure of the parylene process monomer PX. The eight-carbon framework is planar. The molecule is diamagnetic, ie, all electron spins are paired in the ground state (spectroscopically, a singlet). Although many have ascribed its reactivity to its "biradical nature," the true biradical (triplet) form of the molecule (**5**), an electronically excited state, is substantially more energetic, estimated at ca 50 kJ/mol (12 kcal/mol). Therefore, it cannot constitute any appreciable amount of the monomer at equilibrium, even at pyrolysis temperatures. The PX molecule is rather a conjugated tetraolefin whose particular arrangement gives it extreme reactivity at its end carbons.

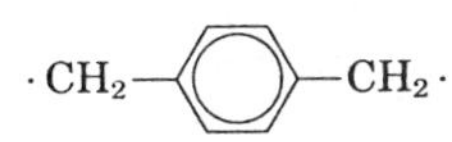

(**5**)

Using a semiempirical molecular orbital technique, which is "tuned" to give good values for heat of formation on experimentally accessible compounds, the heat of formation of *p*-xylylene was recently computed to be 234.8 kJ/mol (56.1 kcal/mol).

Polymer

The linear polymer of PX, poly(*p*-xylylene) (PPX), is formed as a VDP coating in the parylene process. The energetics of the polymerization distinguish it from other polymerizations, and, in fact, enable it to proceed as a vapor deposition polymerization.

Thermodynamic Considerations. *p*-Xylylene polymerization thermodynamics are shown in Figure 3.

Polymerization Mechanism. To understand what happens microscopically during vapor deposition polymerization, the physical processes of condensation and diffusion must be considered as well as the *p*-xylylene polymerization chemistry. This point constitutes an important distinction between VDP and vacuum metallization, in which adsorption is followed by reorganization on the surface of the existing deposited material. In most parylene depositions, a parylene coating forms from gaseous monomer under steady-state conditions.

Gaseous monomer is transported to the location in the coating where it is consumed to produce polymer by an initial condensation followed by diffusion. The net flux of monomer molecules through the growth interface (the outer boundary of the coating, between the gaseous and condensed phases) needed to sustain growth at a given rate can be readily calculated [for Parylene C, 10 μm/h requires $1.55 \times 10^{15}/(\mathrm{cm}^2\cdot\mathrm{s})$]. Comparing a net flux so obtained with the flux of molecules that, according to the kinetic theory of gases, are striking the growth surface [$Z = PN_0/(2\pi MRT)^{0.5}$] for the conditions typical of parylene deposition, a large difference (two or three orders of magnitude) is observed. For Parylene C monomer at a pressure of 1.3 Pa (10 μm Hg) and 25°C, $Z = 6.7 \times 10^{17}/(\mathrm{cm}^2\cdot\mathrm{s})$. For each molecule that eventually enters the coating, some

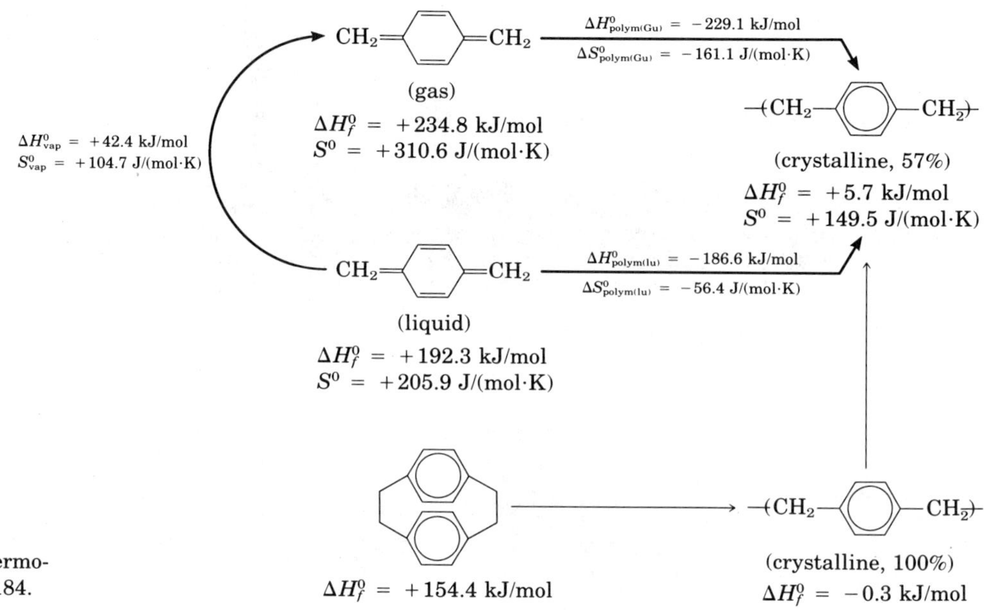

Figure 3. *p*-Xylylene polymerization thermodynamics. To convert J to cal, divide by 4.184.

hundred or thousand molecules strike the growth interface. Those that condense and do not react must, of course, evaporate.

Because most of the condensing molecules evaporate, to a good approximation condensation equilibrium at the growth surface must be assumed. The concentration of monomer dissolved in the coating near the growth interface is, therefore, governed by Henry's law, and monomer concentration in polymer solution increases proportionately to the partial pressure of monomer in the gas phase. Furthermore, as the temperature is lowered, or as higher molecular-weight monomers of lower volatility are selected, monomer concentration at the growth interface increases. In most practical situations, these Henry's law effects dominate in determining growth rates for VDP coatings by regulating monomer concentration within the coating. For each monomer, there exists a threshold condensation temperature T_{tc}, above which the rate of growth of coating is, for all practical purposes, zero.

Once in "solution" in the coating, the monomer moves about in random directions by diffusion until it evaporates or is consumed by chemical reaction. The polymer molecules that have already grown to higher molecular weight cannot relocate appreciably owing to entanglement with their neighbors. The rate of diffusion of monomer through the polymer bulk is adequate for the participation of diffusive transport in the mechanism of VDP (ca 10^{-10} cm²/s at room temperature). This can be confirmed in swelling-rate experiments with solvents having similar physical properties, such as *p*-xylene.

The monomer is consumed by two chemical reactions: initiation, in which new polymer molecules are generated, and propagation, in which existing polymer molecules are extended to higher molecular weight. In steady-state VDP, both reactions proceed continuously inside the polymeric coating, in the reaction zone just behind the growth interface.

During the vapor deposition process, the polymer chain ends remain truly live, ceasing to grow only when they are so far from the growth interface that fresh monomer can no longer reach them. No specific termination chemistry is necessary, although subsequent to the deposition, reaction with atmospheric oxygen, as well as other chemical conversions that alter the nature of the free-radical chain ends, is clearly indicated.

Polymer Properties. The single most important feature of the parylenes, that feature which will most likely decide their use in any specific situation, is the vapor deposition polymerization (VDP) process by which they are applied, providing coatings formed at room temperature of uniform thickness, with thickness control, conformality, and purity. The engineering properties of commercial parylenes are given in Table 2.

Uv and Infrared Spectra; Optical Properties. The parylenes do not absorb visible light, and absorb only at the shorter wavelength, high energy end of the ultraviolet range.

Surface Energy. The surface energies of Parylenes N, C, and D were measured by observing the contact angles for several standard probe liquids. All three have surface energies of approximately 45 mJ/m² (45 dyn/cm); that is, all test liquids having less than 45 mJ/m² surface tension completely wet the as-deposited parylene surfaces.

Crystallinity. The crystallinity of the parylenes determines two of their most important practical characteristics: mechanical strength at elevated temperatures and solvent resistance.

In a manner typical of crystalline polymers, the crystallinity of parylenes is normally limited to small submicrometer domains that are randomly and uniformly dispersed throughout a continuous amorphous phase. Adjacent polymer chains, in order to acquire greater overall system stability, would prefer to be close to one another, but the extent to which they can is limited by the tangles already present. Because a given polymer chain is long enough to participate successively in several crystallites, these crystallites function as cross-links to strengthen the bulk polymer.

Solvent Resistance. At temperatures below the melting of the crystallites, the parylenes resist dissolution. Although the solvents permeate the continuous amorphous phase, they are virtually excluded from the crystalline domains. Consequently, when a parylene film is exposed to a solvent a slight swelling is observed as the solvent invades the amorphous phase. In the

Table 2. Typical Engineering Properties of Commercial Parylenes

Property	Parylene N	Parylene C	Parylene D	ASTM method
General				
density, g/cm^3	1.110	1.289	1.418	D1505
refractive index, n_D^{23}	1.661	1.639	1.669	
Mechanical				
tensile modulus, GPa[a]	2.4	3.2	2.8	D882
tensile strength, MPa[b]	45	70	75	D882
yield strength, MPa[b]	42	55	60	D882
elongation to break, %	30	200	10	D882
yield elongation, %	2.5	2.9	3	D882
Rockwell hardness	R85	R80		D785
coefficient of friction				D1894
static	0.25	0.29	0.35	
dynamic	0.25	0.29	0.31	
Thermal				
melting point, °C	420	290	380	
linear coefficient of expansion at 25°C × 10^{-5}, K^{-1}	6.9	3.5		
heat capacity at 25°C, J/(g·K)[c]	1.3	1.0		
thermal conductivity at 25°C, W/(m·K)	0.12	0.082		
Electrical				
dielectric constant				D150
60 Hz	2.65	3.15	2.84	
1 kHz	2.65	3.10	2.82	
1 MHz	2.65	2.95	2.80	
dissipation factor				D150
60 Hz	0.0002	0.020	0.004	
1 kHz	0.0002	0.019	0.003	
1 MHz	0.0006	0.013	0.002	
dielectric strength at 25 μm, short time, MV/m	275	220	215	D149
dielectric strength at 25 μm, step-by-step, MV/m	235	185		D149
volume resistivity at 23°C, 50% rh, Ω	1.4×10^{17}	8.8×10^{16}	2×10^{16}	D257
surface resistivity at 23°C, 50% rh, Ω	1×10^{13}	1×10^{14}	5×10^{16}	D257
Barrier				
water absorption, %	<0.1	<0.1	<0.1	D570
water vapor transmission at 37°C, ng/(Pa·s·m)[d]	0.0012	0.0004	0.0002	E96 D570
gas permeability at 25°C, amol/(Pa·s·m)[e]				D1434
N_2	15.4	2.0	9.0	
O_2	78.4	14.4	64.0	
CO_2	429.0	15.4	26.0	
H_2S	1590.0	26.0	2.9	
SO_2	3790.0	22.0	9.53	
Cl_2	148.0	0.7	1.1	

[a] To convert GPa to psi, multiply by 145,000.
[b] To convert MPa to psi, multiply by 145.
[c] To convert J to cal, divide by 4.184.
[d] To convert ng/(Pa·s·m) to mol/(100 in.^{2}/d) at 97% rh, multiply by 133.
[e] To convert amol/(Pa·s·m) to (cm^3(STP)·mil)/(100 in.^{2}/d atm), multiply by 0.498.

thin films commonly encountered, equilibrium is reached fairly quickly, within hours. The change in thickness is conveniently and precisely measured by an interference technique.

Circuit Boards. The most important application of parylenes is as a conformal coating for printed wiring assemblies. The parylene VDP process has the ability to produce a continuous, thin, truly conformal coating on geometrically complex, delicate articles. These coatings provide excellent chemical resistance, especially to solvent attack, and resistance to fungal attack. In addition, they exhibit stable dielectric properties over the wide range of temperatures in which military boards are expected to perform, as well as low dielectric constants, which together with thinness minimize the unwanted loading of high frequency tuned circuits. The reliability with which the process operates substantially reduces labor-intensive touch-up and inspection operations required for conventional coatings.

Parylenes are also used in hybrid circuits, semiconductors, capacitors, electrets, the Bathythermograph, miniature electrical components, medical uses, artifact conservation, fusion targets, contamination control, barrier coating, corrosion control, dry lubricants, and pellicles and membranes.

Health and Safety

In a world increasingly conscious of the dangers of contact with chemicals, a process that is conducted within the walls of a vacuum chamber, such as the VDP process for parylene coatings, offers great advantages. Provided the vacuum pump exhaust is appropriately vented and suitable caution is observed in the cleaning out of the cold trap (trace products of the pyrolysis, which may possibly be dangerous, would collect here), the VDP parylene process has an inherently low potential for operator contact with hazardous chemicals.

BIBLIOGRAPHY

U.S. Pat. 3,342,754 (Sept. 19, 1967), W. F. Gorham (to Union Carbide Corp.).

L. A. Errede and M. Szwarc, *Q. Rev. Chem. Soc.* **12,** 301 (1958).

M. Szwarc, *Polym. Eng. Sci.* **16**(7), 473 (1976).

W. F. Beach, *Macromolecules* **11**(1), 72 (1978).

W. F. BEACH
C. LEE
D. R. BASSETT
T. M. AUSTIN
Union Carbide Corporation

R. OLSON
Nova Tran Corporation

Y

YARNS. See FIBERS, MANUFACTURING.

Z

Z-AVERAGE MOLECULAR WEIGHT.

See MOLECULAR WEIGHT DETERMINATION.

ZETA POTENTIAL.

See COLLOIDS.

ZIEGLER-NATTA CATALYSTS

Ziegler-Natta catalysts are named after their originators, Karl Ziegler of the Max Planck Institut für Kohlenforschung in Mülheim and Giulio Natta of the Istituto di Chimica Industriale del Politecnico di Milano. Initially, they consisted essentially of an alkylaluminum compound together with a compound of the titanium group of the periodic table, a typical combination being triethylaluminum and either titanium tetrachloride or titanium trichloride. Subsequently, an enormous variety of mixtures of this general type were prepared and used in polymerization, and these are commonly referred to as catalysts of the Ziegler-Natta type.

The interest in Ziegler-Natta catalysts resides in their remarkable stereospecificity. Their discovery by Ziegler and subsequent utilization by Natta opened entirely new vistas in polymer chemistry and technology. The monumental contributions of Karl Ziegler and Giulio Natta received just recognition when they were jointly awarded the Nobel Prize in chemistry in 1963.

ZWITTERIONIC POLYMERIZATION

Zwitterionic polymerization is defined as an ionic polymerization (or copolymerization) in which both cationic and anionic sites are located at each of the two chain ends of one propagating species (**1**):

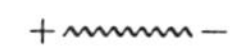

Zwitterionic
propagating species

(1)

It is assumed that the zwitterionic propagating species in the reaction mixture takes an intramolecular cyclic form (**2**) or a form of intermolecular association (**3**).

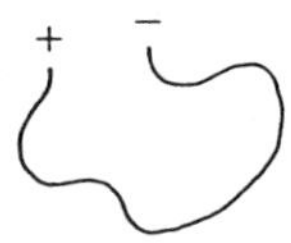

(2)

(3)

There are two possibilities for the propagating reactivities of the cationic and anionic sites. In the first case, propagation takes place at only one site, either cationic or anionic. In the second case, propagation is achieved by the cation–anion reaction between two zwitterions. The types of monomers involved in zwitterionic polymerization include polar vinyl monomers and heterocyclic monomers.

BIBLIOGRAPHY

T. Saegusa, *Angew. Chem.* **89,** 867 (1977); *Angew. Chem. Int. Ed. Engl.* **16,** 826 (1977).

T. Saegusa, *Makromol. Chem. Suppl.* **3,** 157 (1979).

T. Saegusa, *Pure Appl. Chem.* **53,** 691 (1981).

TAKEO SAEGUSA
MASATO SUZUKI
Kyoto University

INDEX